COMPREHENSIVE COORDINATION CHEMISTRY

IN 7 VOLUMES

PERGAMON MAJOR REFERENCE WORKS

Comprehensive Inorganic Chemistry (1973)

Comprehensive Organic Chemistry (1979)

Comprehensive Organometallic Chemistry (1982)

Comprehensive Heterocyclic Chemistry (1984)

International Encyclopedia of Education (1985)

Comprehensive Insect Physiology, Biochemistry & Pharmacology (1985)

Comprehensive Biotechnology (1985)

Physics in Medicine & Biology Encyclopedia (1986)

Encyclopedia of Materials Science & Engineering (1986)

World Encyclopedia of Peace (1986)

Systems & Control Encyclopedia (1987)

Comprehensive Coordination Chemistry (1987)

Comprehensive Polymer Science (1988)

Comprehensive Medicinal Chemistry (1989)

COMPREHENSIVE COORDINATION CHEMISTRY

The Synthesis, Reactions, Properties & Applications of Coordination Compounds

Volume 3

Main Group and Early Transition Elements

EDITOR-IN-CHIEF

SIR GEOFFREY WILKINSON, FRS

Imperial College of Science & Technology, University of London, UK

EXECUTIVE EDITORS

ROBERT D. GILLARD

University College, Cardiff, UK

JON A. McCLEVERTY

University of Birmingham, UK

PERGAMON PRESS

OXFORD · NEW YORK · BEIJING · FRANKFURT
SÃO PAULO · SYDNEY · TOKYO · TORONTO

U.K.	Pergamon Press, Headington Hill Hall, Oxford OX3 0BW, England
U.S.A.	Pergamon Press, Maxwell House, Fairview Park, Elmsford, New York 10523, U.S.A.
PEOPLE'S REPUBLIC OF CHINA	Pergamon Press, Room 4037, Qianmen Hotel, Beijing, People's Republic of China
FEDERAL REPUBLIC OF GERMANY	Pergamon Press, Hammerweg 6, D-6242 Kronberg, Federal Republic of Germany
BRAZIL	Pergamon Editora, Rue Eça de Queiros, 346, CEP 04011, São Paulo, Brazil
AUSTRALIA	Pergamon Press Australia, P.O. Box 544, Potts Point, NSW 2011, Australia
JAPAN	Pergamon Press, 8th Floor, Matsuoka Central Building, 1-7-1 Nishishinjuku, Shinjuku-ku, Tokyo 160, Japan
CANADA	Pergamon Press Canada, Suite No. 271, 253 College Street, Toronto, Ontario M5T 1R5, Canada

First edition 1987

Library of Congress Cataloging-in-Publication Data

Comprehensive coordination chemistry.

Includes bibliographies.
Contents: v. 1. Theory and background – v. 2. Ligands – v. 3
Main group and early transition elements – [etc.]
1. Coordination compounds.
I. Wilkinson, Geoffrey, Sir, 1921–
II. Gillard, Robert D.
III. McCleverty, Jon A.
QD474.C65 1987 541.2′242 86-12319

British Library Cataloguing in Publication Data

Comprehensive coordination chemistry: the synthesis, reactions, properties and applications of coordination compounds.
1. Coordination compounds.
I. Wilkinson, Geoffrey, 1921–
II. Gillard, Robert D.
III. McCleverty, Jon A.
541.2′242 QD474

ISBN 0-08-035946-9 (vol. 3)
ISBN 0-08-026232-5 (set)

Printed in Great Britain by A. Wheaton & Co. Ltd., Exeter

Contents

Preface

Since the appearance of water on the Earth, aqua complex ions of metals must have existed. The subsequent appearance of life depended on, and may even have resulted from, interaction of metal ions with organic molecules. Attempts to use consciously and to understand the metal-binding properties of what are now recognized as electron-donating molecules or anions (ligands) date from the development of analytical procedures for metals by Berzelius and his contemporaries. Typically, by 1897, Ostwald could point out, in his 'Scientific Foundations of Analytical Chemistry', the high stability of cyanomercurate(II) species and that 'notwithstanding the extremely poisonous character of its constituents, it exerts no appreciable poison effect'. By the late 19th century there were numerous examples of the complexing of metal ions, and the synthesis of the great variety of metal complexes that could be isolated and crystallized was being rapidly developed by chemists such as S. M. Jørgensen in Copenhagen. Attempts to understand the 'residual affinity' of metal ions for other molecules and anions culminated in the theories of Alfred Werner, although it is salutary to remember that his views were by no means universally accepted until the mid-1920s. The progress in studies of metal complex chemistry was rapid, perhaps partly because of the utility and economic importance of metal chemistry, but also because of the intrinsic interest of many of the compounds and the intellectual challenge of the structural problems to be solved.

If we define a coordination compound as the product of association of a Brönsted base with a Lewis acid, then there is an infinite variety of complexing systems. In this treatise we have made an arbitrary distinction between coordination compounds and organometallic compounds that have metal–carbon bonds. This division roughly corresponds to the distinction — which most but not all chemists would acknowledge as a real one — between the cobalt(III) ions $[Co(NH_3)_6]^{3+}$ and $[Co(\eta^5\text{-}C_5H_5)_2]^+$. Any species where the number of metal–carbon bonds is at least half the coordination number of the metal is deemed to be 'organometallic' and is outside the scope of our coverage; such compounds have been treated in detail in the companion work, 'Comprehensive Organometallic Chemistry'. It is a measure of the arbitrariness and overlap between the two areas that several chapters in the present work are by authors who also contributed to the organometallic volumes.

We have attempted to give a contemporary overview of the whole field which we hope will provide not only a convenient source of information but also ideas for further advances on the solid research base that has come from so much dedicated effort in laboratories all over the world.

The first volume describes general aspects of the field from history, through nomenclature, to a discussion of the current position of mechanistic and related studies. The binding of ligands according to donor atoms is then considered (Volume 2) and the coordination chemistry of the elements is treated (Volumes 3, 4 and 5) in the common order based on the Periodic Table. The sequence of treatment of complexes of particular ligands for each metal follows the order given in the discussion of parent ligands. Volume 6 considers the applications and importance of coordination chemistry in several areas (from industrial catalysis to photography, from geochemistry to medicine). Volume 7 contains cumulative indexes which will render the mass of information in these volumes even more accessible to users.

The chapters have been written by industrial and academic research workers from many countries, all actively engaged in the relevant areas, and we are exceedingly grateful for the arduous efforts that have made this treatise possible. They have our most sincere thanks and appreciation.

We wish to pay tribute to the memories of Professor Martin Nelson and Dr Tony Stephenson who died after completion of their manuscripts, and we wish to convey our deepest sympathies to their families. We are grateful to their collaborators for finalizing their contributions for publication.

Because of ill health and other factors beyond the editors' control, the manuscripts for the chapters on Phosphorus Ligands and Technetium were not available in time for publication.

However, it is anticipated that the material for these chapters will appear in the journal *Polyhedron* in due course as Polyhedron Reports.

We should also like to acknowledge the way in which the staff at the publisher, particularly Dr Colin Drayton and his dedicated editorial team, have supported the editors and authors in our endeavour to produce a work which correctly portrays the relevance and achievements of modern coordination chemistry.

We hope that users of these volumes will find them as full of novel information and as great a stimulus to new work as we believe them to be.

ROBERT D. GILLARD
Cardiff

JON A. McCLEVERTY
Birmingham

GEOFFREY WILKINSON
London

Contributors to Volume 3

Professor K. W. Bagnall
10 Castle Hill Court, Prestbury, Cheshire SK10 4UT, UK

Dr D. S. Barratt
Johnson Matthey Research Centre, Blount's Court, Sonnig Common, Reading RG4 9NH, UK

Dr F. J. Berry
Department of Chemistry, University of Birmingham, PO Box 363, Birmingham B15 2TT, UK

Dr J. M. Charnock
Department of Chemistry, University of Manchester, Manchester M13 9PL, UK

Dr J. Costa Pessoa
Centro de Quimica Estrutural, Complexo Interdisciplinar, Instituto Superior Tecnico, Av Rovisco Pais, 1096 Lisbon, Portugal

Professor Z. Dori
Department of Chemistry, Technion-Israel Institute of Technology, Haifa 32000, Israel

Dr A. J. Edwards
Department of Chemistry, University of Birmingham, PO Box 363, Birmingham B15 2TT, UK

Professor R. C. Fay
Department of Chemistry, Baker Laboratory, Cornell University, Ithaca, NY 14853, USA

Dr D. E. Fenton
Department of Chemistry, University of Sheffield, Sheffield S3 7HF, UK

Professor C. D. Garner
Department of Chemistry, University of Manchester, Manchester M13 9PL, UK

Dr B. Gyori
Department of Inorganic and Analytical Chemistry, Lajos Kossuth University, H-4010 Debrecen, Hungary

Dr P. G. Harrison
Department of Chemistry, University of Nottingham, University Park, Nottingham NG7 2RD, UK

Dr F. A. Hart
Chemistry Department, Queen Mary College, Mile End Road, London E1 4NS, UK

Professor L. G. Hubert-Pfalzgraf
Laboratoire de Chimie Minérale Moléculaire, Unité Associée au CNRS, Université de Nice, Parc Valrose, 06034 Nice Cedex, France

Dr L. F. Larkworthy
Chemistry Department, University of Surrey, Guildford, Surrey GU2 5XH, UK

Dr G. J. Leigh
AFRC Unit of Nitrogen Fixation, University of Sussex, Falmer, Brighton BN1 9RQ, UK

Professor C. A. McAuliffe
Department of Chemistry, University of Manchester Institute of Science and Technology, PO Box 88, Sackville Street, Manchester M60 1QD, UK

Professor K. B. Nolan
Department of Chemistry, Royal College of Surgeons, St Stephen's Green, Dublin 2, Republic of Ireland

Dr P. O'Brien
Department of Chemistry, Queen Mary College, Mile End Road, London E1 4NS, UK

Professor M. T. Pope
Department of Chemistry, Georgetown University, Washington, DC 20057, USA

Dr M. Postel
Laboratoire de Chimie Minérale Moléculaire, Unité Associée au CNRS, Université de Nice, Parc Valrose, 06034 Nice Cedex, France

Dr R. L. Richards
AFRC Unit of Nitrogen Fixation, University of Sussex, Falmer, Brighton BN1 9RQ, UK

Professor J. G. Riess
Laboratoire de Chimie Minérale Moléculaire, Unité Associée au CNRS, Université de Nice, Parc Valrose, 06034 Nice Cedex, France

Professor R. Schmutzler
Lehrstuhl B für Anorganische Chemie der Technischen Universität, Pockelstrasse 4, Postfach (TU) 3329, 33 Braunschweig, Federal Republic of Germany

Dr E. I. Stiefel
Corporation Research, Room LE242, Exxon Research & Engineering Co., Annandale, NJ 08801, USA

Professor A. G. Sykes
Department of Inorganic Chemistry, University of Newcastle upon Tyne, Newcastle upon Tyne NE1 7RU, UK

Dr M. J. Taylor
Department of Chemistry, University of Auckland, Private Bag, Auckland, New Zealand

Professor D. G. Tuck
Department of Chemistry, University of Windsor, 401 Sunset Avenue, Windsor, Ontario N9B 3P4, Canada

Professor L. F. Vilas Boas
Centro de Quimica Estrutural, Complexo Interdisciplinar, Instituto Superior Tecnico, Av Rovisco Pais, 1096 Lisbon, Portugal

Contents of All Volumes

Volume 1 Theory & Background

Volume 2 Ligands

Volume 3 Main Group & Early Transition Elements

Volume 4 Middle Transition Elements

Volume 5 Late Transition Elements

Volume 6 Applications

Volume 7 Indexes

23

Alkali Metals and Group IIA Metals

DAVID E. FENTON
University of Sheffield, UK

23.1 PREFACE

The chemistry of the alkali and alkaline earth metals has generally been regarded as straightforward and discernible using an ionic model.[1,2] The interaction between metal ions and ligands in coordination complexes can be regarded as ion–dipolar in nature and this leads to the prediction that large cations having low charge (or low ionic potential[3]) will be unlikely to form complexes. As the alkali and alkaline earth metals provide the largest known cations, they were long regarded as being unlikely to form complexes.

In a book review written in 1974,[4] R. J. P. Williams commented that 'the inorganic chemistry of these elements—Na, K, Mg and Ca—is taught no more because there are no spectra to tease the mind' and consequently they remain 'out of sight (spectroscopic), out of mind'. Certainly this bore much truth with regard to coordination chemistry, especially as the watershed established in 1967, when C. J. Pedersen[5] discovered that macrocyclic polyethers can act as efficient ligands for alkali and alkaline earth metals, had barely reached the textbooks.

This discovery, followed by the exploitation of macrobicyclic multidentate ligands (cryptands) as complexing agents,[6] and accompanied by an increasing realization of the need to understand the vital role played by Na, K, Mg and Ca in biological systems, has led to a renaissance of interest in the coordination chemistry of these metals.[7–9]

23.1.1 Historical Perspective

Whilst what has been termed the real scientific advancement in this area took place in 1967,[5] complexes of alkali and alkaline earth metal cations (M^{n+}) with simple monodentate ligands can be traced back, through the 'metal-ammonias', to Faraday.[10] It was not until almost a century later, however, that precise determinations of the stoichiometries of the M^{n+}—NH_3 products were realized and the term coordination was introduced to describe the bonding mode of the ligand.[11–13] The alkali metals themselves were first noted to dissolve in liquid ammonia in 1863[14] and since that time it has been found that the metals also dissolve in amines and ethers.[15]

The interaction of potentially chelating ligands with M^{n+} was first noted by Ettling, in 1840, who reported (salicylaldehydato)sodium(salicylaldehyde), hemihydrate.[16] The monohydrated analogue was later reported by Hantzsch who termed it, and similar compounds, 'acid salts'.[17] Compounds having the general formula ML·*n*HL ($n = 1, 2$) were then extensively studied by Sidgwick and his coworkers,[18,19] who considered the ML·HL complexes to consist of a quadridentate cation coordinated independently by both L and HL, as in (**1**). The ML·2HL series were believed to contain a hexacoordinated cation. Later Pfeiffer suggested that although this bonding mode could operate, it was also possible for L to interact with M^{n+} and HL simultaneously, with the HL having no direct link to M^{n+}.[20] The essential correctness of this proposal has been demonstrated more recently[21] and is discussed later.

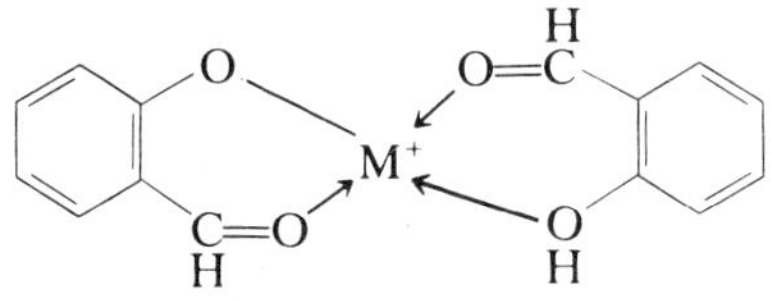

(**1**)

Several isolated reports of stable, solid complexes of M^{n+} with a range of organic solvents (*e.g.* alcohols, ketones, aldehydes, esters, ethers, acetonitrile and pyridine) were made available during this same period. Some were obtained by crystallization of a salt from the relevant solvent, and others more serendipitously through observations made during solubility studies.

The observations made by Sidgwick and Brewer, in 1925, on the solubility charactersitics of alkali metal-β-diketonates are of seminal importance.[19] (1-Phenyl-1,3-butanedionato)sodium was found to be insoluble in toluene, whereas the dihydrated form was toluene soluble. This was explained by proposing the formation of a complex in which the diketonate acts as a chelating anion and the two waters act as monodentate ligands (**2**). This suggestion was vindicated in the 1960s with the solution of the X-ray crystal structure of the corresponding ethylene glycol complex.[22] In this molecule the glycol acts as a bridging, doubly monodentate ligand linking two sodium cations; each sodium is coordinated by one diketonate and two monodentate glycol oxygen atoms.

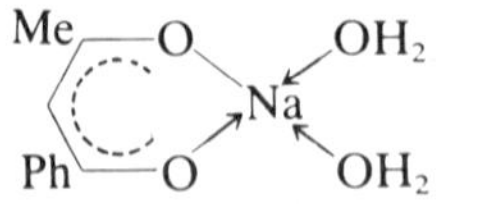

(**2**)

The area then developed in a non-systematic way. Representative examples illustrating the coordination of M^{n+} can be seen in the syntheses of well-defined complexes with the bidentate nitrogen-containing ligands, en and phen;[23] the isolation of metal complexes of sugars;[24] and the preparation of complexes with amino acids, di- and tri-peptides and related small molecules of biological interest.[25,26] Just prior to Pedersen's disclosures concerning the formation of metal complexes involving macrocyclic polyethers,[5] interest in solvent extraction and desalination led to the availability of ligands such as *p,p'*-diamino-2,3-diphenylbutane which readily precipitate

NaCl from water.[27] Moore and Pressman's discovery that the antibiotic valinomycin exhibits alkali metal cation selectivity in rat liver mitochondria then provided a starting point for studies in the area of M^{n+} selectivity by biological and model systems.[28]

Since Pedersen's disclosures, and the subsequent extension into the third dimension through Lehn and his macrobicyclic and macrotricyclic cryptands,[29] there have been a plethora of papers concerning many aspects of M^{n+} complexation, the majority being centred on crown ether and related species. Their literature has been extensively reviewed and attention is drawn to articles by Poonia and Bajaj[30] and by Vögtle and Weber[31] which are comprehensive and provide excellent bibliographies of the area.

There is also an extensive organometallic chemistry of the alkali and alkaline earth metals. Many of these compounds, particularly those of Li, Be and Mg, form coordination complexes with appropriate donor ligands. This topic has been recently surveyed in the companion volume 'Comprehensive Organometallic Chemistry' and so only apposite references will be included here.[32]

23.1.2 General Remarks

The alkali and alkaline earth metals are usually regarded as having 'hard', or class 'a', character and so are capable of binding to 'hard' or class 'a' ligands. The ionic size increases down each group (Table 1), and individually with increasing coordination number. The ionic potential (charge–radius ratio) is diminished in the same direction, and this is reflected in the more covalent character found in compounds of Li, Be and Mg relative to their higher congeners. Li^+ and Be^{2+} have the greatest polarizing power in each series and the highest tendency towards solvation.

Table 1[a] Ionic Radii and Hydration Energies

(a)	Li	Na	K	Rb	Cs
Ionic radius (Å)	0.60	0.95	1.33	1.48	1.69
Hydration energy (kJ mol^{-1})	−498	−393	−310	−284	−251
(b)	Be	Mg	Ca	Sr	Ba
Ionic radius (Å)	0.31	0.65	0.99	1.13	1.35
Hydration energy (kJ mol^{-1})	−2455	−1900	−1565	−1415	−1275

[a] Data compiled from F. A. Cotton and G. Wilkinson, 'Advanced Inorganic Chemistry', 4th edn., Wiley, New York, 1980. Pauling ionic radii are quoted.

There is an increase in coordination number going from $Li^+(4)$ through $Na^+(6)$ to K^+, Rb^+ and $Cs^+(8)$, and from $Be^{2+}(4)$ through $Mg^{2+}(6)$ to Ca^{2+}, Sr^{2+} and $Ba^{2+}(8)$. All of the M^{2+} ions are smaller and considerably less polarizable than the isoelectronic M^+ ions. The oxidation state is +1 for the alkali metals and +2 for the alkaline earth metals. In contrast to the transition metals, there is no preferred coordination geometry for this group of cations except that, in the absence of crystallographically imposed symmetry, a roughly spherical distribution of neighbours is sought.

Beryllium is normally divalent in its compounds and, because of its high ionic potential, has a tendency to form covalent bonds. In free BeX_2 molecules, the Be atom is promoted to a state in which the valence electrons occupy two equivalent *sp* hybrid orbitals and so a linear X—Be—X system is found. However, such a system is coordinatively unsaturated and there is a strong tendency for the Be to attain its maximum coordination of four. This may be done through polymerization, as in solid $BeCl_2$, *via* bridging chloride ligands, or by the Be acting as an acceptor for suitable donor molecules. The concept of coordinative saturation can be applied to the other M^{n+} cations, and attempts to achieve it have led to attempts to deliberately synthesize new compounds.

Multidentate ligands, both acyclic and cyclic, are now abundant and herein the nomenclature used will follow that recently proposed by Vögtle and Weber.[33] Coronands are taken to mean macromonocyclic compounds having any heteroatoms present as potential donors, with crown ethers being reserved for compounds of the same type having only oxygen atoms present as donor species. Cryptands are bi- and poly-cyclic compounds having any heteroatoms present as donors and podands are acyclic coronand and crown ether analogues.

23.2 MONODENTATE LIGANDS

Many salts of the alkali and alkaline earth metals occur as solvated species and are obtained by crystallization of the salt from an appropriate solvent. It is only recently that such compounds have been seriously considered as coordination compounds and factors such as the coordination number of the cation and whether the interaction between Na^+ and H_2O leads to $[Na–H_2O]^+$,[34] a species comparable to the oxonium cation, have been thoroughly investigated. One problem is precise definition, or when does a complex become too complex? Alkali metal compounds range from discrete molecular solvates through various degrees of association to extensive arrays and networks. The Na^+ in NaCl is regarded as six coordinated, having six nearest Cl^- neighbours, but the system is not usually regarded as anything other than an ionic compound. In this review complexes have been generally restricted to discrete molecular entities, and solvated species have been divided into two general categories:[30,35] (i) solvates, or compounds in which the cation is surrounded by the solvent and the anion is separate, and (ii) solvated ion pairs in which both solvent and anion are coordinated to the cation.

23.2.1 Alkali Metal Complexes

A comparison of the commoner alkali metal salts showed that ~76% of the Li salts gave solid hydrates, as did ~74% of the Na salts and ~23% of the K salts; very few hydrates of Rb and Cs are available. The degree of hydration of the cation diminishes down the group, with decreasing ionic potential. The Li salts do not form hydrates with more than four waters present, except for the bromide, iodide and hydroxide at low temperature. The cation is coordinated only to the solvating molecules and so the species is charge separated; the anion, in the case of the protic solvent derivatives, is generally held by the hydrogens of the solvent. An X-ray structure of $Na_2SO_4·10H_2O$ shows the cation to be present in an octahedral environment as $[Na(H_2O)_6]^+$;[36] the K^+ in $CaKAsO_4·8H_2O$ is present as $[K(H_2O)_8]^+$[37] in a square antiprismatic environment.

A related series of complexes is formed with ammonia; some of these complexes have considerable stability such as $LiI·4NH_3$, in which the cation is believed to be tetrahedrally coordinated and charge-separated $[Li(NH_3)_4]^+$. Results of studies on the dissociation properties show that the stability of the ammines falls off in the sequence $Li^+ > Na^+ > K^+$, and with halides of the same metal in the order $I^- > Br^- > Cl^-$.[38] The ammonia may be replaced by primary, secondary and tertiary amines but the products are less stable.

The alkali metals are soluble in liquid ammonia, and certain amines, to give solutions which are blue when dilute. The solutions are paramagnetic and conduct electricity, the carrier being the solvated electron. In dilute solutions the metal is dissociated into metal ions and ammoniated electrons. The metal ions are solvated in the same way that they would be in a solution of a metal salt in ammonia, and so comparison can be made with, for example, $[Na(NH_3)_4]^+I^-$, the IR and Raman spectra of which indicate a tetrahedral coodination sphere for the metal.[39]

Many solid solvates of the alkali metals are known[30] and the monodentate ligands involved include pyridine, MeOH, EtOH and higher alcohols, acetone, dioxane, THF, aryl- and alkyl-phosphine oxides, DMSO, amides (acetamide, NMA, DMA, NMF, DMF, urea, *etc.*) and thiourea. The X-ray structures of several compounds have been solved and representative examples are given in Table 2. Although these fall broadly into the two categories of complex given earlier, one observation is that discrete molecules are not always available. In the complexes LiI·5tppo,[40] $Li[Li\{C(SiMe_3)_3\}_2]·4THF$[41] and LiCl·4NMA[42] the cation is tetrahedrally coordinated as $[Li(solvent)_4]^+$. In the first molecule there is an additional, non-complexed tppo present, and in the last the NMA also hydrogen bonds to the Cl^-. In contrast, in the sodium complexes $NaI·3Me_2CO$[43] and NaI·3DMF,[44] the cation coordination number is six and the geometry is octahedral, and midway between octahedral and trigonal prismatic respectively. The triads of oxygen atoms are shared between pairs of cations giving rise to infinite cationic columns.

All halides of alkali metals, except Li, are practically insoluble in pyridine; this situation is

Table 2 Some Representative Structures of Alkali Metal Solvates

Complex	*Bond lengths* (Å)	*Coordination numbers*	*Comments*	*Ref.*
(i) Solvated species				
$LiI\cdot 5tppo$[a]	$Li—O_{av}$, 1.97	4	Td, one uncomplexed tppo	1
$LiCl\cdot 4nma$[b]	Not given	4	Td	2
$Li[Li\{C(SiMe_3)_3\}_3]\cdot 4THF$	$Li—O_{av}$, 1.97	4	Td, $Li(THF)_4^+$	3
$NaI\cdot 3MeOH$	$Na—O_{av}$, 2.47	6	Oh, shared donor triads	4
$NaI\cdot 3acetone$	$Na—O_{av}$, 2.46	6	Oh, shared donor triads	5
$NaI\cdot 3DMF$	$Na—O_{av}$, 2.40	6	Trigonal prism Oh, shared donor triads	6
$NaClO_4\cdot 3dioxane$	$Na—O_{av}$, 2.43	6	Elongated Oh, bridging dioxane	7
$LiI\cdot 2urea$	Li—O, 1.88, 1.95	4	Td, each urea binds 2 Li^+	8
(ii) Solvated ion pairs				
$LiCl\cdot pyridine$	Li—Cl, 2.38, 2.39 Li—N, 2.01	4	Td: 3 Cl^-, 1 N	9
$LiCl\cdot H_2O\cdot 2pyridine$	Li—O, 1.94 LiCl, 2.33 Li—N, 2.05, 2.06	4	Td: 1 Cl^-, 1 N, 1 O	10
$LiCl\cdot dioxane$	Li—O, 1.88, 1.95 Li—Cl, 2.39, 2.42	4	Distorted Td	11
$NaBr\cdot 2am$[c]	Na—O, 2.33, 2.37	6	Distorted Oh, shared Br^- and O	12
$NaClO_4\cdot 2DMA$	$Na—O_{lig}$, 2.35, 2.39 $Na—O_{ClO_4}$, 2.38	5	Shared DMA and ClO_4^-	13

[a] tppo = triphenylphosphine oxide. [b] NMA = *N*-methylacetamide. [c] am = acetamide.

1. Y. M. G. Yasin, O. J. R. Hodder and H. M. Powell, *J. Chem. Soc., Chem. Commun.*, 1966, 705.
2. D. J. Haas, *Nature (London)*, 1964, **201,** 64.
3. C. Eaborn, P. B. Hitchcock, J. D. Smith and A. C. Sullivan, *J. Chem. Soc., Chem. Commun.*, 1983, 827.
4. P. Piret and C. Mesureur, *J. Chim. Phys. Phys.-Chim. Biol.*, 1965, **62,** 287.
5. P. Piret, Y. Gobillon and M. van Meerssche, *Bull. Soc. Chim. Fr.*, 1963, 105.
6. Y. Gobillon, P. Piret and M. van Meerssche, *Bull. Soc. Chim. Fr.*, 1962, 554.
7. J. C. Barnes and T. J. R. Weakley, *Acta Crystallogr., Sect. B.*, 1978, **34,** 1984.
8. J. Vierbist, R. Meulemans, P. Piret and M. van Meerssche, *Bull. Soc. Chim. Belg.*, 1970, **79,** 391.
9. F. Durant, J. Vierbist and M. van Meerssche, *Bull. Soc. Chim. Belg.*, 1966, **75,** 806.
10. F. Durant, P. Piret and M. van Meerssche, *Acta Crystallogr.*, 1967, **22,** 52.
11. F. Durant, Y. Gobillon, P. Piret and M. van Meerssche, *Bull. Soc. Chim. Belg.*, 1966, **75,** 52.
12. P. Piret, L. Rodrique, Y. Gobillon and M. van Meerssche, *Acta Crystallogr.*, 1966, **20,** 482.
13. J. Bello, D. Haas and H. R. Bello, *Biochemistry*, 1966, **5,** 2539.

generally modified by stepwise addition of water, and the observation served as the basis for an early process for the separation of Li from the other alkali metals.[45] The structure of $LiCl\cdot H_2O\cdot 2py$ has present a four-coordinate cation bound to the three donor atoms and the Cl^- anion and so serves as an example of a solvated ion pair.[46] In NaBr·2acetamide the Na^+ is six-coordinated and this is achieved by the sharing of one Br^- anion and two oxygen atoms between pairs of Na^+ cations to give electrically neutral columns.[47]

For some solvent systems, it has not yet been possible to obtain structural information in the solid state. The ion-pair loosening ability of DMSO has been thoroughly illustrated in studies of M^+–DMSO interactions in solution using several different techniques: NMR;[48,49] solution IR;[50–52] conductometry[53] and calorimetry.[54] In the case of Na^+, a coordination number of six has been confirmed. IR techniques have also been used to monitor alkali metal complexation with the related sulfoxides, diisopropyl and dibutyl sulfoxide.[50] Solid M^+–DMSO species have been recovered from salt solutions in DMSO; the IR spectra suggest that the DMSO molecules present in these complexes are not equivalent but no X-ray structural information is presently available.

Amides and related compounds have been thoroughly investigated as ligands for alkali metals. Studies in solution have been made using various physicochemical techniques; these, and MO calculations, show that the interaction of amides with M^+ is *via* the carbonyl function.[56,57] This leads to a delocalization of the amide N electron pair and so it is possible to follow the decrease in C—O, and the increase in C—N, bond orders using IR spectroscopy. X-Ray crystal structures of amide complexes have shown both types of complexation interaction to be present (Table 2).

The amides have been employed as primitive models for peptides in studies concerning the denaturing of proteins by alkali metal salts; it has been suggested that as alkali metal cations have strong interactions with amides this could indicate that denaturing occurs *via* interactions with peptide units.[58] It is therefore pertinent to comment, at this point, that the crystal structures of NaI(glycine)$_2$·H_2O[59] and LiBr(alanylglycine)·2H_2O[60] have been solved and in each the amino acids function as zwitterions. The M^+ is then coordinated to the carboxylic acid functions and water molecules to give six and four coordination respectively. A survey of the earlier work on metal salt interactions with amino acids has been published.[61]

An interesting structural feature arises in the coordination of Li^+ by urea in the complex LiI·2urea;[62] each urea bridges two cations. This may be compared with LiCl·dioxane[63] in which a related bridging occurs, and contrasted with NaCl·urea·H_2O[64] in which the octahedral metal environment is due to two Cl^- anions, two water oxygen atoms and two urea oxygen atoms.

Thiourea complexes of the type MX(thiourea)$_4$ have been found if the lattice energy of MX is less than 670 kJ mol^{-1}. This leads to complexes of K^+, Rb^+ and Cs^+, especially with Br^- and I^- as anions. The polarizable thiourea separates the anions from the cations and each cation is probably surrounded by eight sulfur atoms at the corner of an approximately regular cube.[65]

The complexation of M^+L^- (M^+ = Li–Rb; L^- = *e.g.* 2,4-dinitrophenolate) by triphenylphosphine oxide (tppo) has been studied conductometrically and shows that M^+L^- association constants decrease down the group.[66] Tri-*n*-octylphosphine oxide (topo) has been used to extract alkali metal cations into organic media; the extractant species are designated as either $[M(topo)_3]^+ClO_4^-$ (M^+ = Li–Cs) or $[M(topo)_4]^+ClO_4^-$ (M^+ = Rb, Cs). The most efficacious transfer occurs for the smaller cations.[67] The study of the synergic effect of various aryl- and alkyl-phosphine oxides on the extraction of alkali metal cations has been reviewed.[68] Using the dibenzoylmethane anion, water-insoluble complexes such as Li(dbm)(topo)$_2$ have been characterized; the use of topo as the extractant in a water–*p*-xylene system gave separation of Li^+ from the other alkali metal cations as this dbm complex.[69]

Solid alkali metal complexes of hexamethylphosphoramide (hmpa) have been isolated and IR studies indicate coordination through the oxygen of the phosphine oxide.[70] Solution studies have also been carried out and show that hmpa coordinates to alkali metal cations in solution.[48] Rather interestingly a solution of sodium metal dissolved in hmpa gives, in the ^{23}Na NMR spectrum, a signal corresponding to Na^-.[71] This provided the first example of the generation of Na^- in a solvent in the absence of an added complexor; here the hmpa can itself coordinate to the Na^+ formed during dissolution.

23.2.2 Alkaline Earth Metal Complexes

The alkaline earth metal cations follow similar trends to the alkali metal cations in their complexation reactions with monodentate ligands. 'Hard', class 'a' donor atoms are preferred, and beryllium and magnesium show a greater tendency to form complexes than do their larger congeners.

The strong hydration of Be^{2+} can be explained by the formation of $[Be(H_2O)_4]^{2+}$ in which coordinate links with water are formed. This contrasts with, for example, Ba^{2+} where the molecular forces would be electrostatic in origin. The presence of the tetrahedral $[Be(H_2O)_4]^{2+}$ unit has been confirmed in the X-ray crystal structures of $BeSO_4$·4H_2O[72] and $Be(NO_3)_2$·4H_2O.[73] Solid hydrates have been obtained for $BeCl_2$ and $BeBr_2$, but not for the fluoride or iodide.[74] The product of direct hydration of $BeCl_2$ has been identified as a cyclic trimer, $\{[Be(OH)(H_2O)_2]Cl\}_3$, by IR and Raman spectroscopic studies.[75] The cationic unit is shown in (**3**). BeF_2·2H_2O has been suggested as the formula for the solution adduct, but attempts to isolate the compound led to HF and basic residues being formed. ^{9}Be NMR studies support the existence of the dihydrate in saturated aqueous solutions. The species are prone to hydrolysis and this increases from the fluoride to the iodide; this hydrolysis has been reviewed by Bell[74] in a survey of the chemistry of beryllium halides, and a review of aqua complexes including Li, Be, Mg and Ca has recently appeared.[76] The predominate species in solution are affirmed as polymeric in nature: $[Be_2OH]^{3+}$ and $[Be_3(OH)_3]^{3+}$.[77]

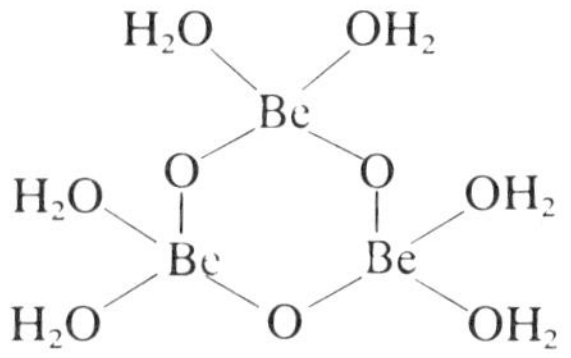

(3)

A recurrent feature in the solid state chemistry of magnesium hydrates is the ready formation of $[Mg(H_2O)_6]^{2+}$. This is evidenced in many X-ray studies, *e.g.* $Mg(NO_3)_2{\cdot}6H_2O$,[78] $MgX{\cdot}6H_2O$ ($X = SO_3^-$[79], $S_2O_3^-$[80]), $Mg_2CdCl_6{\cdot}12H_2O$,[81] $MgCO_3{\cdot}3H_2O$[82] and $MgSO_4{\cdot}7H_2O$.[83] In this latter compound the seventh water molecule is readily lost at 48.3 °C. The tenacity of Mg^{2+} for its solvated water is illustrated in the formation of complexes such as $MgBr_2{\cdot}6H_2O{\cdot}$2caffeine[84] and $MgCl_2{\cdot}6H_2O{\cdot}$12-crown-4[85] in which the $[Mg(H_2O)_6]^{2+}$ unit is present and the second ligands are not coordinated.

About 150 examples of Ca^{2+}–H_2O interactions in crystalline hydrates of Ca^{2+} have been reviewed; higher coordination numbers are generally achieved in these species.[86] $CaX_2{\cdot}6H_2O$ (X = Cl, Br) have present a nine-coordinated cation ligated by both water molecules and halide anions;[87] $CaKAsO_4{\cdot}8H_2O$ has Ca^{2+} eight-coordinated by water molecules.[37] The complexes $CaCr_2O_7{\cdot}7H_2O{\cdot}$2hmt[88] and $CaBr_2{\cdot}10H_2O{\cdot}$2hmt[89] are both solvent-separated systems (hmt = hexamethylenetetramine). The cations have coordination numbers of seven and six respectively and the hmt ligand is too weak a donor to desolvate the cation. $CaCl_2{\cdot}4H_2O$ has present a seven-coordinated cation in which four water molecules and three chloride anions interact with the metal.[90] This involves sharing and the beginnings of the build-up of intricate networks. $BaCl_2{\cdot}2H_2O$ provides an example of such a network, the metal being eight-coordinated by four shared water molecules and four shared chloride anions,[91] as does $Ba(NCS)_2{\cdot}3H_2O$ wherein the cation has tricapped trigonal prismatic coordination sharing water and N-bonded thiocyanate anions to build up continuous chains.[92]

BeF_2 forms the ammine $BeF_2{\cdot}NH_3$ at −78.5 °C but this complex decomposes at higher temperatures; the formation of only a single, unstable ammine is in contrast with the behaviour of the remaining beryllium halides which form a series of ammine complexes increasing in stability with increasing weight of the halogen (Table 3).[1,74,93] The weaker donor power of alkyl- and aryl-amines, together with the introduction of steric constraints on complexation, is seen in the reactions of amines with BeX_2. Only $MeNH_2$ gives a tetramine complex with $BeCl_2$; related primary, secondary and tertiary amines give 1:2 adducts. Benzidine gives a non-chelated, binuclear complex $2BeCl_2{\cdot}$3benzidine. There appears to be no report of characterized amine complexes for $BeBr_2$ and BeI_2.[74]

Table 3 Ammonia Complexes of Beryllium Halides

$BeF_2{\cdot}NH_3$			
$BeCl_2{\cdot}2NH_3$	$BeCl_2{\cdot}4NH_3$	$BeCl_2{\cdot}6NH_3$	$BeCl_2{\cdot}12NH_3$
$BeBr_2{\cdot}4NH_3$	$BeBr_2{\cdot}6NH_3$	$BeBr_2{\cdot}10NH_3$	
$BeI_2{\cdot}1.5NH_3$	$BeI_2{\cdot}4NH_3$	$BeI_2{\cdot}6NH_3$	$BeI_2{\cdot}13NH_3$

(The tetraammines are the most stable species; adducts having more than four NH_3 present are stable only at low temperature. The table is compiled from data given in ref. 74.)

Amine complexes have been used to facilitate the purification of BeH_2[94] and 1:1 species have been found using Me_3N and $PhCH_2NMe_2$.[95] The reaction of $BeCl_2$, NaH and Me_3N gave the complex $ClBeH{\cdot}Me_3N$ and this and the analogous BrBeH complex are dimeric in benzene solution having bridging hydrogen atoms.[96,97] If there is a reactive hydrogen present on the amine used, then an elimination reaction is possible, yielding oligomeric products. BeH_2 and Me_2NH give the trimer $\{(Me)_2N)_2Be\}_3$ in which there are alternatively three- and four-coordinate beryllium atoms present (**4**).[98] The chemistry of alkylamidoberyllium compounds has been reviewed.[99,100]

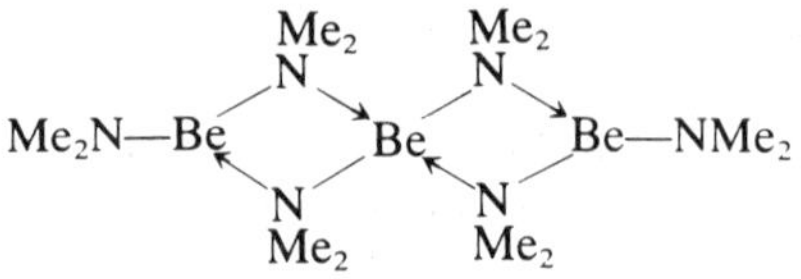

(**4**)

Beryllium acetate has been found to give complexes of the type $Be(OAc)_2 \cdot 2L$ with ammonia and primary amines, and beryllium borohydride forms 1:1 complexes with primary amines.[101] IR studies on the complex $Be(ClO_4)_2 \cdot 2H_2O$ show it to be in the form $[Be(H_2O)_4]^{2+}[Be(ClO_4)_4]^{2-}$; at elevated temperatures $Be_4O(ClO_4)_6$ is formed from the dihydrate.[102]

Nearly all magnesium salts take up NH_3 and some of the resulting ammines are stable to loss of NH_3 on heating, *e.g.* $MgCl_2 \cdot 6NH_3$, and $Mg(ClO_4)_2 \cdot 6NH_3$. Aromatic amines can replace the ammonia in these compounds.[1] The stability of ammine complexes decreases down the group. X-Ray powder studies on $CaCl_2 \cdot 8NH_3$ and $CaCl_2 \cdot 2NH_3$ suggest that in the former the cation is contained in a distorted triangular prism of NH_3 molecules, and in the latter the cation is coordinated by an irregular octahedral array comprising two NH_3 and four, presumably shared, Cl^- anions.[103]

Beryllium halides form 1:2 complexes with monodentate heterocyclic aromatic nitrogen compounds (pyridine, *etc.*) as does $Be(NCS)_2$.[74] The X-ray structure of $MgBr_2 \cdot 6py$ reveals the cation as six-coordinated; there are two non-bonding pyridine molecules present as well as the four which coordinate.[104] Bis-pyridine and bis-imidazole complexes of bis(dinitrophenoxy)magnesium have been proposed as model complexes for the coordination of histamine in Mg-ATPases.[105] The cation is six-coordinated with the two pyridines occupying *cis* octahedral sites. Hydrazine acts as a monodentate ligand in the complex $BeCl_2 \cdot 4N_2H_4$ and the more bulky phenylhydrazine forms a 1:3 complex, together with a very unstable 1:4 species. $BeCl_2$ also forms a 1:4 complex with HCN, but only 1:2 adducts with RCN. With both ligand types, this is believed to be a consequence of increased steric constraint imposed by the substituted ligands.[74] $BeCl_2 \cdot 2L$ (L = MeCN, ClCN) have been synthesized by direct reaction of $BeCl_2$ and the nitrile;[106] the X-ray structure of $BeCl_2 \cdot 2MeCN$ reveals the metals to be in a distorted tetrahedral environment.[107] Ligand exchange occurs with nucleophilic species such as diethyl ether and benzonitrile and the metathetical reaction between $BeCl_2 \cdot 2MeCN$ and KSCN leads to the isolation of $Be(NCS)_2 \cdot 2MeCN$ and of $ClBeNCS \cdot 2MeCN$.[108] A ^9Be NMR study of $BeCl_2$ in MeCN shows the following species to be present in solution: $[BeCl_4]^{2-}$, $[BeCl_3 \cdot H_2O]^-$, $[BeCl_3 \cdot MeCN]^-$, $BeCl_2 \cdot 2MeCN$ and $[Be_2Cl_2 \cdot 4MeCN]^{2+}$.[109] Nitriles also give complexes of the types $MgBr_2 \cdot nMeCN$ ($n = 3, 4$) and $MgI_2 \cdot 6MeCN$. The complex $[Mg(Me(CN)_6]^{2+}[SnCl_4(NO_3)_2]^{2-}$ has been recovered from $Mg(NO_3)_2$–$SnCl_4$–MeCN and its nature established both in solution and in the solid state.[110]

Solid solvates such as $MgX_2 \cdot 6ROH$ ($X = Br^-$, NO_3^-, ClO_4^-; R = Me, Et) and $Ca(NO_3)_2 \cdot EtOH$ are believed to be charge separated in nature as is the more unusual mixed solvate $MgBr_2 \cdot 4THF \cdot 2H_2O$.[111–113] Solvated ion pairs have also been found for Group IIA cations and the structures of $MgBr_2 \cdot 2THF$,[114] $MgBr_2 \cdot 4THF$[115] and $MgBr_2 \cdot 2Et_2O$[116] show coordination numbers of six, six and four respectively for the cation. The sites are composed from anions and donors and so, for example, in $MgBr_2 \cdot 2THF$ the distorted octahedral geometry derives from two Mg—O, two Mg—Br (terminal) and two Mg—Br (bridging) bonds. Solvated organometallic complexes are also known; this aspect has been reviewed but a representative compound is $EtMgBr \cdot 2Et_2O$, which consists of discrete monomers with the bromide anion, ethyl group and two ether oxygen atoms tetrahedrally arranged around the magnesium.[117]

$BeCl_2 \cdot 2Et_2O$ has been extensively studied;[74] it may be prepared by reaction of anhydrous $BeCl_2$ and diethyl ether. A two-phase system is found in which the top layer is a solution of ether in the complex and the bottom layer is composed oppositely. The product is recovered on removal of the solvent. The crystal structure shows it to contain the cation in a distorted tetrahedral environment.[118] $BeBr_2 \cdot 2Et_2O$ is readily available, but although BeI_2 is soluble in ether no solid product has yet been recovered. THF and dioxane complexes of $BeCl_2$ are also known.[74]

It would be inappropriate not to mention the contribution made by the Leiden school to the development of techniques which have made available a wide range of metal complexes containing relatively weak donor molecules.[119] Using strictly anhydrous conditions and by reacting the required anhydrous metal salt with the ligand, species having the general formulae $M^{II}X \cdot 6L$ and $M^{II}X_2 \cdot 6L$ have been prepared from such diverse ligands as ethanoic acid, ethanol, diphenyl sulfoxide, acetonitrile, nitromethane, methyl formate, 2-pyridone, acetaldehyde, THF, *etc.* For the alkaline earths mostly Mg^{2+} and Ca^{2+} have been used, although in some cases Sr^{2+} and Ba^{2+} have also been employed. An interesting synthetic stratagem in the above work has been to react the required anhydrous metal chloride with a Lewis acid such as iron(III) chloride, indium(III) chloride or antimony(V) chloride (usually as its nitromethane adduct) and with the appropriate ligand to give complexes of the type $MX_2 \cdot 6L$ ($X = FeCl_4^-$, $InCl_4^-$ or $SbCl_6^-$).

Beryllium halides have been found to form complexes with several aldehydes, ketones and esters.[74] Antipyrine, which acts as a monodentate carbonyl ligand, dissolves alkaline earth metal salts to give solid products of the type $MX_2 \cdot 6ap$ ($M^{2+} = Mg$, Ca, Sr; $X = ClO_4^-$; $M^{2+} = Mg$, Ca; $X = BF_4^-$). The X-ray structure of $Mg(ClO_4)_2 \cdot 6ap$ confirms the monodentation and shows the magnesium to be octahedrally coordinated.[120]

Magnesium and calcium complexes derived from triphenylphosphine oxide,[121] phenacyldiphenylphosphine oxide[122] and trishydroxymethylphosphine oxide[123] have been characterized. A study of the interaction of trioctylphosphine oxide with alkaline earth salts has shown that the order of extraction into an organic phase is $Be^{2+} > Ca^{2+} > Sr^{2+} \approx Mg^{2+} \approx Ba^{2+}$.[124] The low placement of Mg^{2+} is believed to be due to the exceptionally strong hydration of this cation. The metals are transferred as MX_2, 4 topo ($M^{2+} = Be$–Ba; $X = ClO_4^-$; $M^{2+} = Sr$, Ba; $X = SCN^-$). The extracted species for $Be(NCS)_2$ contains two or three ligands only. The crystal structures of $Mg(ClO_4)_2 \cdot 5Me_3PO \cdot H_2O$ and $Mg(ClO_4)_2 \cdot 5Me_3PO$ show that whereas in the former the cation is octahedrally coordinated by the mixed solvates, in the latter a square pyramidal geometry is found.[125] Exchange studies in d_2-CH_2Cl_2 led to the postulate that such complexes are formed by a dissociation of Me_3PO from a putative $[Mg(Me_3PO)_6]^{2+}$ species.[126]

NMR has been used to study ligand exchange in non-coordinating solvents for a series of $Be(ClO_4)_2 \cdot 4L$ (L = trimethyl phosphate, DMSO, DMA, DMF, NMA, 1,1,3,3-tetramethylurea, dimethyl methylphosphonate and dimethyl phenylphosphonate) complexes. ^1H NMR studies show that the mode of activation for exchange on beryllium varies from dissociative to associative depending upon L and the nature of the non-coordinating solvent.[127,128]

Pyridine *N*-oxide complexes have been prepared for Mg^{2+}, Ca^{2+}, Sr^{2+} and Ba^{2+}, and arsine *N*-oxide complexes for Mg^{2+} and Ca^{2+}.[30] An X-ray structure of $Mg(ClO_4)_2 \cdot 5Me_3AsO$ shows the cation to be in a square pyramidal environment provided by the donor ligands; the charge separation and unusual geometry are believed to be a consequence of the electronic and spatial needs of the ligand.[129]

An interesting method of synthesis has been employed to produce $MgBr_2 \cdot 2HMPA$; the ligand is reacted with EtMgBr in Et_2O to produce unstable crystals of the product.[30] The pathway presumably is by displacement of the Grignard equilibrium. A complex of barium *p*-nitrobenzoate with DMSO has been synthesized and actually separated from DMSO with water present in the system.[130]

Amide complexes of alkaline earth metals have been prepared. The crystal structure of $MgBr_2 \cdot 4urea \cdot 2H_2O$ shows the cation to be octahedrally coordinated to four urea and two water molecules.[131] In $CaBr_2 \cdot 6urea$ the cation is exclusively within the octahedral environment of urea donors, whereas in $Ca(NO_3)_2 \cdot 4urea$[132] and $Ca(NO_3)_2 \cdot urea. 3H_2O$[133] there is a difference in environment. The cation in the former is linked to four urea molecules and two monodentate nitrates; in the latter bidentate nitrates are present and together with a urea and water molecule complete an eight-coordination around the cation. The last complex is not discrete but through sharing of donors and hydrogen bonding builds up a three-dimensional array. A survey of urea–cation interactions is available.[134] The compound formed from the reaction of $Ba(ClO_4)_2$ with DMA has been shown to be the bimetallic complex, bis-μ-(DMA)-bis(aquatris-DMA-barium) tetraperchlorate. Each barium is eight-coordinated by three terminal DMAs, one water molecule, two monodentate perchlorates and two bridging DMA molecules.[135a] Thiourea complexes of the type $CaX_2 \cdot 6TU$ ($X = Cl^-$, I^-) have been reported.[65] The interaction of Ca^{2+} with glycine and glycylglycylglycine parallels that of Na^+ in that the zwitterionic form of the ligand is used. The structure of $CaBr_2 \cdot 3glycine$[135b] shows two independent Ca^{2+} cations each linked to six oxygens from the glycine and a bromide anion. The two coordination polyhedra are then edge-linked to provide a dimer having each Ca^{2+} in a distorted pentagonal bipyramidal geometry. The structure of $CaCl_2(Gly\text{-}Gly\text{-}Gly) \cdot 3H_2O$ has each cation linked to donors from four different peptides in a much more complicated network structure.[135c]

$BeCl_2$ does not dissolve in H_2S; weak 1:2 adducts of H_2S with $BeBr_2$ and BeI_2 have, in contrast, been obtained at low temperatures.[74] Solutions containing adducts of beryllium halides with thioethers have been prepared and $BeCl_2{\cdot}2Me_2S$ has been isolated. The complexes $BeCl_2{\cdot}3R_2S$ (R = Me, Et, Bu^i) have been obtained as needles at low temperatures. Thioamide and thiourea compounds of $BeCl_2$ have been obtained as 1:2 complexes; steric factors are believed to preclude the formation of 1:2 complexes when symmetrically and asymmetrically substituted thioureas are used as donors.[74] Only a few phosphine complexes are known; $Be(BH_4)_2$ forms 1:1 complexes with Ph_3P, Me_3P, Me_2PH and Et_3P, all of which are monomeric in benzene. $Be(BH_4)_2{\cdot}Me_3P$ will add further phosphine but with accompanying decomposition to give $Me_3P{\cdot}BH_3$.[136]

The possibility of forming complex anions is also open to alkaline earth metals. There is an abundant literature concerning fluoroberyllates and this has been reviewed.[74,93] Much work concerns phase distribution in molten mixtures, although fluoroberyllates can be prepared in, and isolated from, aqueous solution. The fluoroberyllates may be compared with silicates: $[BeF_4]^{2-}$ is analogous to orthosilicate $[SiO_4]^{4-}$; $[Be_2F_5]^-$, which has double networks of BeF_4 tetrahedra infinite in two directions, is rather like the Si–O network in dimetasilicates;[137] and $[Be_2F_7]^{3-}$ with two BeF_4 tetrahedra linked at one corner resembles pyrosilicate.[138]

The sequence of halide complexes is shown to be $F^- \gg Cl^- > Br^- > I^-$ which is as expected if the complex formation is due to electrostatic forces. Consequently, the chloroberyllates are formed only under anhydrous conditions. In the alkali metal tetrachloroberyllates there is a distortion of the $[BeCl_4]^{2-}$ tetrahedron on going from the sodium to the lithium salt.[74] The addition of hydrogen halide to anhydrous $BeCl_2$ produces the liquid complexes $HBeCl_2X{\cdot}2Et_2O$ (X = Cl, Br); pyridine can replace the ether molecules to give a complex formulated as $[py\cdots H\cdots py]^+[BeCl_2X]^-$.[139]

A similarity between BeF_4^{2-} compounds and the corresponding sulfates has been established. In general, the solubilities of the salts are similar, and double salts such as $M_2^IBeF_4{\cdot}6H_2O$ are available.[1] The aqueous chemistry of the haloberyllates has been reviewed.[140]

Other anionic complexes of beryllium that have been detected are $(NH_4)_2Be(NO_3)_4$,[141] $(NH_4)_2Be(NCS)_4{\cdot}MeCN$ and $(NH_4)_2Be(NCS)_3{\cdot}MeCN$.[142] ESCA studies on beryllium and magnesium complexes of the type $[M(NCS)_4]^{2-}$ and $[M(CNS)_3L]^-$ (L = DMF, py, MeCN) are consistent with N-bonding thiocyanate anions being present.[142] The structure of $K[Be(NH_2)_3]$ shows the beryllium to be in a trigonal planar unit.[143]

Although various complex magnesium halides have been identified from Raman spectra of mixed salts melts, $(NEt_4)_2MgCl_4$ is the only reported salt containing a complex halide anion. Vibrational spectroscopic studies have shown the anion to be tetrahedral.[144] Studies on Cs_2MgCl_4, Cs_3MgCl_5 and Rb_3MgCl_5 have also shown the presence of discrete $[MgCl_4]^{2-}$ anions, and in melts of high $MgCl_2$ concentration $[Mg_2Cl_7]^{3-}$ has been found in equilibrium with $[MgCl_4]^{2-}$.[145] The crystal structure of tachyhydrate, $Mg(H_2O)_6CaCl_6$, shows it to consist of discrete $[Mg(H_2O)_6]^{2+}$ cations and octahedral $[CaCl_6]^{2-}$ anions.[146] A brief survey of species available from melts involving strontium has been given.[147]

A large number of organometallic derivatives of alkali and alkaline earth cations, particularly those of Li, Be and Mg, form coordination complexes involving monodentate ligands.[32] For Li and Mg, these are primarily oxygen donor ligands such as THF and Et_2O. There is an extensive coordination chemistry of Be compounds of the types BeR_2, $BeRR'$ and RBeX.[32b] The degree of association has been explored for these complexes and steric effects are of paramount importance in influencing the degree of coordination to the beryllium. For Me_2Be a stability sequence of N > O > P, S > As is found and in general there are few complexes of the last three donors. Be—C bonds are polar and so are susceptible to protolysis; it is preferable to use ligands devoid of acidic hydrogens otherwise, even at low temperature, an elimination of hydrocarbon occurs to give Be donor atom bonds. As an example of Re_2Be complexation, R_2Be form 1:2 complexes with pyridine, but only 1:1 complexes with Me_3N. This is seen as a manifestation of steric problems; furthermore if 2-methyl pyridine is used as a ligand only a 1:1 complex results.[32b]

23.3 NEUTRAL ACYCLIC POLYDENTATE LIGANDS

A diverse range of acyclic polydentate ligands has been used to form complexes with alkali and alkaline earth metal cations. For convenience of description these ligands have been subdivided into anionic and neutral species.

23.3.1 Nitrogen Donors

1,2-Diaminoethane (en), which can be regarded as the simplest bidentate ligand of this type, is capable of dissolving various inorganic salts to produce solvates such as LiCl·2en, LiBr·2en, LiI·3en, NaI·3en, $NaClO_4$·3en, $CaCl_2$·6en, $SrCl_2$·6en and $BaCl_2$·6en.[148] $BeCl_2$·en has also been prepared,[149] and in all of these compounds the en may be viewed as paralleling the behaviour of ammonia, or water. The structures of LiCl·2en[150] and LiI·3en[151] reveal the cation to be tetrahedrally and distorted octahedrally coordinated respectively. In the former complex the Li^+ is coordinated by four nitrogens from three en molecules; one en is chelating (*gauche*) and the other two are monodentate (*trans*) and bridging to produce infinite columns. In LiI·3en the ligands are chelating only. This mode of bonding is also seen in $Ba(SbSe)_2$·4en in which the cation is eight-coordinated.[152]

It has been noted that the complexes $M(picrate)_2$·*n*en (M^{2+} = Mg, Ca; $n = 1, 2$) can be prepared in EtOH, and that the IR spectra show bands *ca.* 1900 and 2500 cm^{-1} indicative of $NH_2\cdots$anion interactions. Such an effect is general and has also been noted in the complexes LiX·*n*en ($X = Cl^-$, Br^-; $n = 2$; $X = I^-$; $n = 3$). Conformational studies on these complexes reveal that the ligand is in the *gauche* and *trans* forms in LiX·2en and only in the *gauche* form in LiI·3en; this is as found in the X-ray structures.[30]

Polyethylene polyamines are available as mixtures of isomeric and homologous compounds; by using inorganic salts of alkali and alkaline earth metals individual polyamines have been selectively complexed from several multicomponent samples.[153] Commercial polyamine mixtures containing primary and secondary nitrogens, as well as their *N*-permethylated tertiary amine counterparts can be separated *via* metal complexation, complex isolation and subsequent complex destabilization leading to product recovery. One example of differential selectivity is found with tetramethylcyclohexanediamine where the *trans* isomer complexes with LiBr, but the *cis* isomer prefers $MgCl_2·6H_2O$. Other solid complexes found include NaI·L [L = *iso*-HMTT (**5**), *n*-HMTP (**6**)] and LiCl·L [L = *N,N'*-c-PMPP (**7**), *sym*-c-TMTT (**8**)] which are isolated and used in the separation of permethylated tetramine and pentamine mixtures. With mixtures containing primary and secondary amines, complex formation often varied with reaction time. Commercial trien (**9**) generally contains about 75% of the linear (**9**) and about 8% of the branched tetramine tren (**10**). The remainder is composed of other tri- and tetr-amine isomers. With NaBr the complex isolated after five minutes was essentially pure NaBr·tren; after 75 minutes the complex isolated was mainly NaBr·trien. Similar results were obtained using LiBr. In general in these reactions, complexation was favoured for salts of low lattice energy.

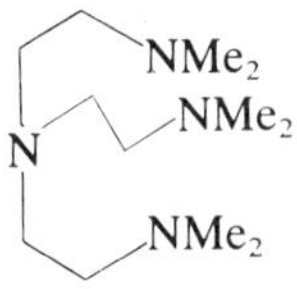

(**5**)

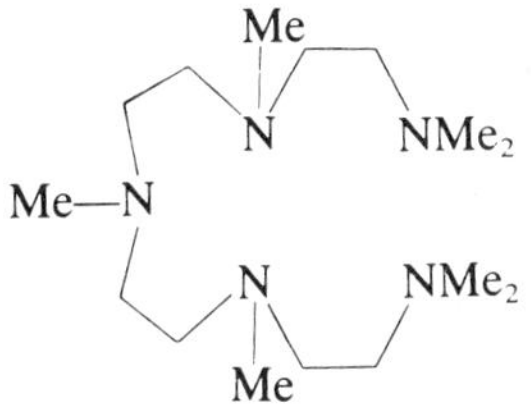

(**6**)

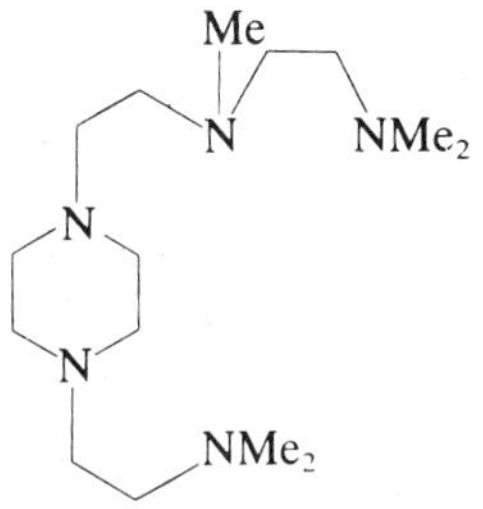

(**7**)

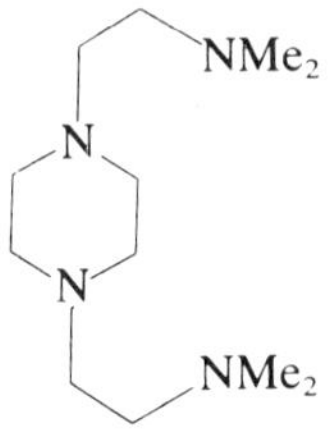

(**8**)

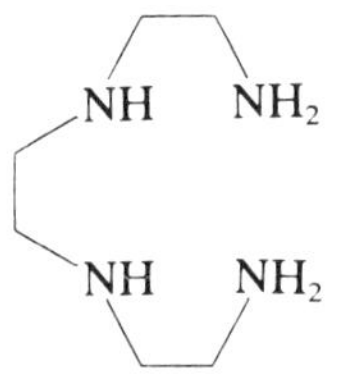

(**9**)

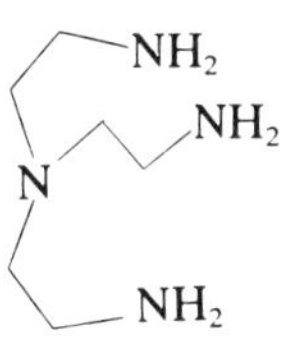

(**10**)

The crystal structure of the complex LiBr·pmdeta [pmdeta = permethylated (**11**)] reveals a dimeric association in the solid state; each Li^+ cation is five-coordinated by the three nitrogen donors and by two bridging bromide anions. The bridging —Li_2Br_2— unit is non-symmetric, and the cation is in a distorted trigonal bipyramidal geometry in which the three nitrogen donors are coordinated facially, with one axial and one equatorial bromide anion. The complex was formed by salt elimination in the reaction of Bu^nBr and Bu^nLi·pmdeta.[154] Solid complexes of alkaline earth metal cations with (**11**) have been characterized as $[M(\mathbf{11})_3]^{2+}X_2$ (M^{2+} = Ca, Sr, Ba; X = ClO_4^-, Br^-, I^-). The cation is proposed as being nine-coordinated.[155]

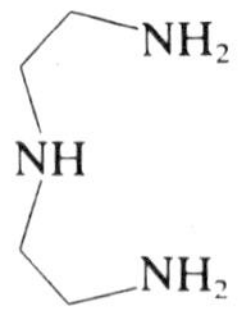

(**11**)

Permethylated diamines and triamines have also been used for the complexation of organometallic derivates of groups IA and IIA.[32,156] This is because of the absence of acidic protons. Well-defined complexes have been isolated in a number of instances and several have been characterized by X-ray structural analysis. *N,N,N′,N′*-Tetramethylethylenediamine (TMEDA) has been used to monomerize alkyllithium clusters, and Bu^nLi·TMEDA is used extensively as a reagent for lithiation in non-polar media. The structure and binding of N-chelated alkali metal complexes have been reviewed.[156] Many of the structures are of contact ions pairs in which the metal is coordinated to the di- or tri-amine and to the carbanion. Consequently the structures have been of great value in giving information on the nature of carbanion–metal interactions. The systems studied are diverse and a representative collection are given in Table 4. In these examples all of the donor nitrogen atoms coordinate the metal; in the cyclopentadienyl complex, $[(Me_3Si)_3C_5H_2]Li$·pmdeta the cyclopentadienyl ring is

Table 4 Polyamine-chelated Alkali Metal Complexes for which X-Ray Structural Data are Available

Triphenylmethyllithium·TMEDA[a]	Chelated TMEDA
Indenyllithium·TMEDA[b]	Chelated TMEDA
Acenaphthyldilithium·2TMEDA[c]	Chelated TMEDA
Naphthalenyldilithium·2TMEDA[d]	Chelated TMEDA
Anthracenyldilithium·2TMEDA[e]	Chelated TMEDA
Bifluorenyldilithium·2TMEDA[f]	Chelated TMEDA
Triphenylmethylsodium·TMEDA[g]	Chelated TMEDA
$[(SiMe_3)_2CH]$lithium·pmdeta[h]	Tridentate pmdeta
{[1.1.0]Bicyclobutanyllithium·TMEDA}$_2$[i]	μ-C, chelated TMEDA
{Phenyllithium·TMEDA}$_2$[j]	μ-C, chelated TMEDA
{Phenylethynyllithium·tmpd}$_2$[k]	μ-C, chelated tmpd
Cyclopentadienylsodium·TMEDA[l]	chelated TMEDA, μ-η^5-Cp
Fluorenylpotassium·TMEDA[m]	Bridging monodentate TMEDA
{(Methyllithium)$_4$·2TMEDA}[n]	C_4Li_4 core, bridging monodentate TMEDA
{(Phenylethynyllithium)$_4$·2tmhd}[o]	C_4Li_4 core
$[(SiMe_3)_3C_5H_2]$Li·pmdeta[p]	Bidentate pmdeta, pentahapto cyclopentadienyl

[a] J. J. Brooks and G. D. Stucky, *J. Am. Chem. Soc.*, 1972, **94**, 7333.
[b] W. E. Rhine and G. D. Stucky, *J. Am. Chem. Soc.*, 1975, **97**, 737.
[c] W. E. Rhine, J. H. Davis and G. D. Stucky, *J. Organomet. Chem*, 1977, **134**, 139.
[d] J. J. Brooks, W. Rhine and G. D. Stucky, *J. Am. Chem. Soc.*, 1972, **94**, 7346.
[e] W. E. Rhine, J. H. Davis and G. D. Stucky, *J. Am. Chem. Soc.*, 1975, **97**, 2079.
[f] M. Walczak and G. D. Stucky, *J. Organomet. Chem.*, 1975, **97**, 313.
[g] H. Koster and E. Weiss, *J. Organomet. Chem.*, 1979, **168**, 273.
[h] M. F. Lappert, L. M. Engelhardt, C. L. Raston and A. H. White, *J. Chem. Soc., Chem. Commun.*, 1982, 1323.
[i] R. Zerger and G. D. Stucky, *J. Chem. Soc., Chem. Commun.*, 1973, 44.
[j] D. Thoennes and E. Weiss, *Chem. Ber.*, 1978, **111**, 3157.
[k] B. Schubert and E. Weiss, *Chem. Ber.*, 1983, **116**, 3212.
[l] T. Aoyagi, H. M. M. Shearer, K. Wade and G. Whitehead, *J. Organomet. Chem.*, 1979, **175**, 21.
[m] R. Zerger, W. E. Rhine and G. D. Stucky, *J. Am. Chem. Soc.*, 1974, **96**, 5441.
[n] H. Koster, D. Thoennes and E. Weiss, *J. Organomet. Chem.*, 1978, **160**, 1.
[o] B. Schubert and E. Weiss, *Angew. Chem., Int. Ed. Engl.*, 1983, **22**, 496.
[p] P. Jutzi, E. Schluter, C. Krüger and S. Pohl, *Angew. Chem., Int. Ed. Engl.*, 1983, **22**, 994.

pentahapto-coordinated to the metal but only one pair of chelating nitrogen donors is involved in binding to the metal, the third nitrogen remaining non-coordinated (Figure 1).[157] A pentahapto cyclopentadienyl ring is also found in the structure of (cyclopentadienyl)-MgBr·teeda (teeda = *N,N,N',N'*-tetraethylethylenediamine) (Figure 1).[158] The area of organoalkaline earth metal complexes has been comprehensively reviewed.[32b,c]

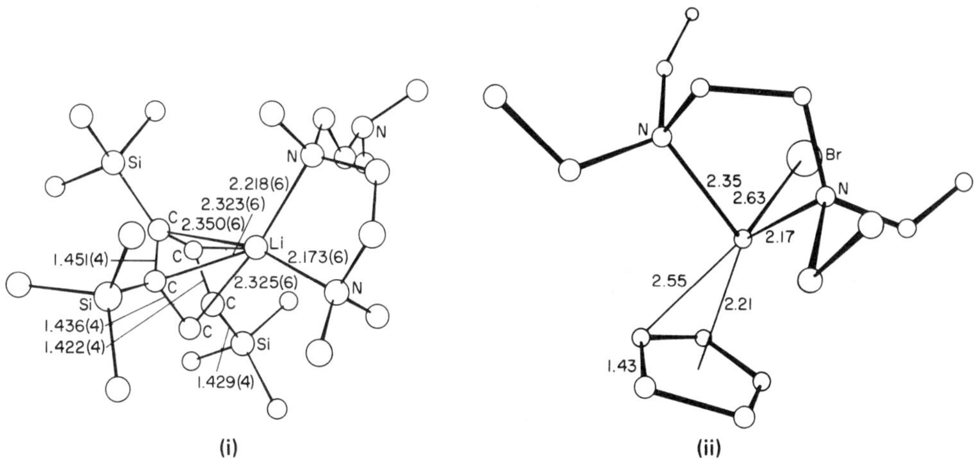

Figure 1 The molecular structures of: (i) $((Me_3Si)_3C_5H_2)Li$·pmdeta and (ii) $(C_5H_5)MgBr$·teeda (reproduced with permission from references 157 and 158)

Early work by Pfeiffer,[23] who isolated crystalline complexes of *o*-phen with the perchlorates of Li^+, Na^+, Ca^{2+}, Sr^{2+} and Ba^{2+}, has been resurrected from obscurity and the results have recently been reconsidered. The range of complexes was extended and the role of the anion and the potential protonation of the anion discussed.[30] *o*-Phenanthroline has also been used to generate M^{2+} complexes and structural studies show the chelating ability of the ligand. Complexes of the type $M(ClO_4)_2$·4phen·$4H_2O$ were characterized and it was suggested that all four phen ligands were associated with the metal. IR and X-ray studies reveal that even for the larger cations the complexes are correctly assigned as $[M(phen)_2(H_2O)_4](ClO_4)_2$·2phen with, in the case of Sr^{2+} and Ba^{2+}, the metal in an eight-coordinated environment and two of the phen molecules non-coordinating.[159] The structures of Na(onp)·2phen and Rb(onp)·2phen (onp = *o*-nitrophenolate) show that in each complex both phen ligands coordinate to the metal. The sodium complex is monomeric, but the rubidium complex dimerizes *via* a nitro group oxygen donor.[160] Bis(2,4-dinitrophenolato)barium·3phen[161] and bis(2,4,6-trinitrophenolato)-barium·2phen·acetone[162] both contain bidentate phen; in the former the Ba^{2+} is nine-coordinated through the three phen molecules, a phenolic oxygen and a nitro oxygen from the anion and a phenolic oxygen from the other; the latter has Ba^{2+} eight-coordinated to the two phen molecules, the acetone oxygen, the phenolic and nitrooxygens from one anion and a phenolic oxygen from the other. The structures of $Ca(NCS)_2$·2bipy·H_2O and *catena*-$Ba(NCS)_2$·2bipy[163] have shown that the Ca^{2+} is seven-coordinated, with N-bonded thiocyanates present, and that the Ba^{2+} is eight-coordinated *via* the six nitrogen donors and two sulfur atoms from adjacent molecules leading to catenation.

In $M(hfacac)_n$ salts, where the coordinating ability of M^{n+} is enhanced by the presence of the fluorinated ligand, addition of further donor atoms leads to coordinative saturation of the central metal. Li(hfacac) forms Li(hfacac)·TMEDA[164] which also exists as the mono- and di-hydrates;[165] Li(hfacac)·phen and Na(hfacac)·phen·H_2O are also available.[165] A comparative study of the role of increasing trifluoromethylation in magnesium β-diketonates shows that whereas $Mg(hfacac)_2$, $Mg(tfacac)_2$ and $Mg(acac)_2$ all form 1:1 complexes with phen and bipy, only the fluorinated compounds yield 1:1 complexes with en, dmen (*sym* and *asym*), TMEDA and dben (dmen = dimethylethenediamine, dben = *N,N'*-dibenzylethylenediamine).[166] $Ca(hfacac)_2$bipy and $Ca(hfacac)_2$phen·H_2O have been isolated[165] and although the complexes have been studied by IR and ^1H NMR, no solid-state structural information is available.

The use of ^7Li, ^{23}Na, ^{39}K and ^{133}Cs NMR allied with ^{13}C NMR has shown that in nitromethane, bipy interacts with M^+ in the sequence $Li^+ > Na^+ > K^+$. Cs^+ does not appear to interact and for Li^+, $[Li(bipy)_2]^+$ was identified.[167]

An interesting area of study was provided by the ligands *rac-p,p'*-diaminodiphenylbutane (DDB)[27] and *p,p'*-methylene-dianiline (MDA).[168] These were the first ligands found to precipitate NaCl from aqueous solution and so were of interest in the potential desalination of brine. IR studies suggested the presence of the then unusual Na–N interaction and this was later confirmed by X-ray studies on the NaCl·3L complexes. The cation was shown to be approximately octahedrally coordinated by six donor nitrogens from separate ligands. The structures therefore consist of columns of cations and anions interspersed with the amino groups to give infinite repetition of the contacts. Each ligand, having two amino functions, can bind to two cations.[169,170]

23.3.2 Oxygen Donors

The interaction of oxygen-containing acyclic ligands with alkali and alkaline earth metal cations has provided a burgeoning area of interest. In historic terms, this was preceded by the advent of crown ethers and the accompanying almost retrospective look at their acyclic precedents. This section is sub-divided into five parts: simple chelates, metal complexes as ligands, podands, polypodands and sugars.

23.3.2.1 Simple chelates

The simplest chelates of this type, ethyleneglycol (eg) and 1,2-dimethoxyethane (glyme-2), give complexes with alkali and alkaline earth metal cations. The complexes MX_2·2eg (M^{2+} = Mg, Ca; X = NO_3^-), MX_2·3eg (M_2^+ = Mg, Ca; X = Cl^-, Br^-), CaX_2·4eg (X = Cl^-, Br^-) and $SrBr_2$·2eg have been isolated and IR studies suggest that the majority are likely to contain monodentate eg.[171,172] The Sr^{2+} complex is, however, believed to have bidentate eg present. $Mg(hfacac)_2$, $Mg(tfacac)_2$ and $Mg(acac)_2$ all form 1:1 complexes with eg; the IR shows the latter to be different and it was suggested that a bridging bis-monodentate eg might be present giving a species related to Na(bzacac)·eg.[166] The X-ray structure of this sodium complex shows the eg to be bridging two sodium cations.[22]

1,2-Dimethoxyethane (glyme-2) also forms complexes and several have been found as a consequence of the use of glyme-2 ('monoglyme') as a solvent medium in many diverse reactions involving M^{n+}.[30] In this context, problems were found in the potential use of glyme-2 with Grignard reagents as precipitation of MgX_2·glyme-2 occurs readily, disturbing the Grignard equilibrium and rendering the system unsuitable.[173] Glymes are weak coordinators and generally require stringent conditions for synthesis. $M(SbCl_6)_2$·glyme-2 (M^{2+} = Mg, Ca) complexes have been prepared under anhydrous conditions and using the glyme as a solvent; the complexes are derived from anhydrous MCl_2 and $SbCl_5$ is added to act as a Lewis acid so generating a bulky anion.[174] Glyme-2 complexes with organometallic species have been proposed and discussed[32] and the chelating potential of the ligand is revealed in the structure of Li(dimesitylborohydride)2glyme-2 wherein the chelated cation also interacts with the hydrogen atoms of the anion (Figure 2).[175] Lithium alkoxide has been found to be dimeric in glyme-2 and is believed to have the formulation (**12**).[176] Beryllium halides have also been shown to give 1:1 complexes with glyme-2, and with the thioether analogue.[74]

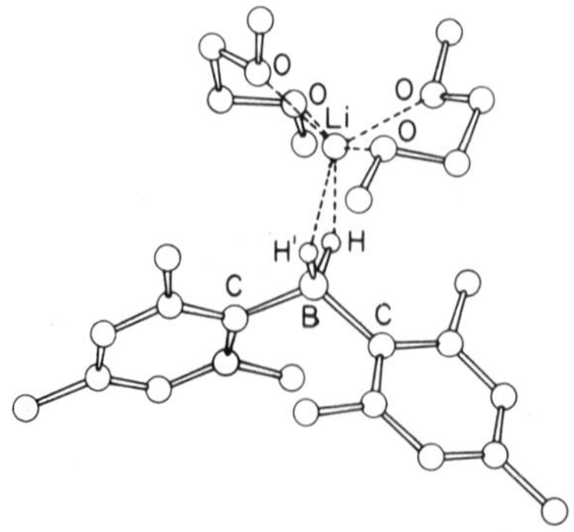

Figure 2 The molecular structure of Li(dimesitylborohydride)·2(glyme-2) (reproduced with permission from reference 175)

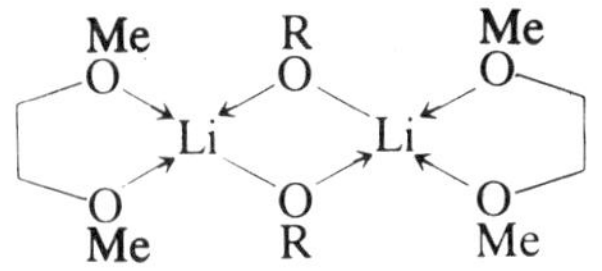

(**12**)

Phosphine oxides, phosphoramides, diacetamide and related species form alkali and alkaline earth metal complexes.[30] The chelation of alkali metal halides with $Ph_2POCH_2POPh_2$, and of NaI with $R_2POCH_2POR_2$ (R = Bu, BuO and *p*-$MeOC_6H_4$) and $Ph_2POCH_2PO(OEt)CH_2POPh_2$ has been followed conductometrically.[177] The structure of NaBr·3($Ph_2POCH_2POPh_2$) shows it to be a charge-separated species with the six-coordinated cation in a chair-shaped trigonal prismatic geometry. Phenacyldiphenylphosphine gives 1:1 complexes with LiBr and LiCl but here the bonding is believed to be monodentate only.[122]

Octamethylpyrophosphoramide (ompa) has been shown to loosen ion pairs in solution[178] and a 1:3 complex has been isolated with $Mg(ClO_4)_2$. The crystal structure shows that the ligand is bidentate and chelates to the cation through the phosphine oxide oxygens.[179]

Diacetamide has been widely used as a ligand and a number of alkali and alkaline earth metal complexes have been characterized.[30] Neither KI·2DA[180] nor NaBr·2DA[181] is charge-separated, the metal interacting with the cation and two chelating DA molecules. DA complexes can be prepared in solution, or by solid reaction; this event suggests caution in using KBr discs for running IR spectra of putative ligands. The salts forming complexes with DA were those with less than 64% ionic character, or differing by two units in electronegativity.[180] With M^{2+} salts, there is a range of stoichiometries in complexes with DA and the coordination number increases from Mg^{2+} to Ba^{2+} (Table 5).

Table 5 Alkaline Earth Cation–Diacetamide Complexes

Complex	*Coordination Number*	*Donor Species*	*Ref.*
$Mg(ClO_4)_2$·4DA·$2H_2O$	6 (octahedral)	2 DA, 2 H_2O (axial)	a
$Ca(ClO_3)_2$·4DA·H_2O	8 (square antiprism)	4 DA	b
$CaBr_2$·4DA	8 (square antiprism)	4 DA	c
$Ca(ClO_4)_2$·5DA	8 (square antiprism)	4 DA	d
$Sr(ClO_4)_2$·4DA·H_2O	9 (monocapped square antiprism)	4 DA, H_2O	e
$Ba(ClO_4)_2$·5DA	10 (symmetrical bicapped square antiprism)	5 DA	f

These complexes are all charge-separated, in contrast to the alkali metal complexes which are ion-paired.

[a] P. S. Gentile, J. G. White, M. P. Dinstein and D. D. Bray, *Inorg. Chim. Acta,* 1977, **21,** 141.
[b] P. S. Gentile and A. P. Ocampo, *Inorg. Chim. Acta,* 1978, **29,** 83.
[c] J. P. Roux and J. C. A. Boeyens, *Acta Crystallogr. Ser. B,* 1970, **26,** 526.
[d] J. P. Roux and G. J. Kruger, *Acta Crystallogr., Sect. B,* 1976, **32,** 1171.
[e] P. S. Gentile, M. P. Dinstein and J. G. White, *Inorg. Chim. Acta,* 1976, **19,** 67.
[f] P. S. Gentile, J. White and S. Haddad, *Inorg. Chim. Acta,* 1975, **13,** 149.

Phenacyl kojate (**13**) was found to form crystalline NaX complexes during its preparation in MeOH.[182] These complexing properties were investigated more fully[183] and the ligand was found to give the complexes MX·PK with CsBr, CsI, RbI, NaNCS, KNCS, RbNCS and CsNCS, and the complexes MX·2PK with NaCl, KCl, RbCl, CsCl, NaBr, KBr, RbBr, NaI and KI. Factors influencing the stoichiometries were discussed and later generalizations suggest that MX·2PK should be charge separated.[21] The structures of KI·2PK,[184] CsSCN·PK[185] and NaI·2PK·$2H_2O$[186] have been solved. The K^+ is eight-coordinated through three tridentate PK molecules and two oxygen atoms from the CH_2OH of adjacent molecules; the Cs^+ is also eight-coordinated due to extensive bridging between the cation, ligand and anion in adjacent molecules; and the Na^+ is six-coordinated through one tridentate and one monodentate PK as well as two water molecules.

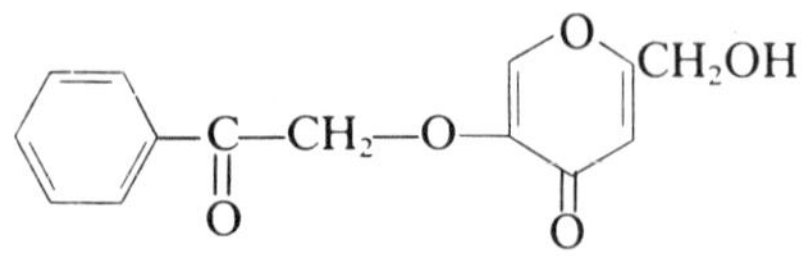

(13)

Dimethyl phthallate acts as a neutral bidentate ligand towards Ca^{2+} and Mg^{2+};[187] the cations are linked to three ligands using both carbonyl groups to give regular octahedral geometry. A large counteranion such as $[InCl_4]^-$ is required. A 1:2 complex is formed between *o,o'*-catecholdiacetic acid and KCl.[188] This has a complicated layer structure in which the cations are enclosed by 10 oxygen atoms, sandwich fashion, in an irregular pentagonal antiprismatic geometry.

23.3.2.2 *Metal complexes as ligands*

Metal complexes have been found to be capable of acting as ligands for alkali and alkaline earth metal cations. In particular, metal complexes of tetradentate Schiff bases (TMSB) such as H_2salen can use the residual lone pairs on the oxygen atoms as donors.[189] Complexes of the type $[(TMSB)_2M^+\cdot x(solvent)]X^-$ and $[(TMSB)_2MX\cdot solvent]$ have been isolated from mixtures of the components. The first group have $[B(C_6H_5)_4]^-$ as the counterion and are charge-separated; X-ray structures of $\{[Co(salen)]_2Na\cdot 2THF\}\cdot BPh_4$,[190] $\{[Ni(salen)_2]Na\cdot 2MeCN\}\cdot BPh_4$[191] and $\{[VO(salen)]_2Na\}\cdot BPh_4$[192] furnish this evidence with the Na^+ being six-coordinated to the TMSB and solvent in the first two instances, and six-coordinated in a cationic polymeric structure in the latter. The second group have ClO_4^- as the counterion and are ion-paired. X-Ray structures again provide the evidence; $\{[Cu(salen)]_2NaClO_4\cdot p\text{-xylene}]$[193] and $\{[Ni(acen)]_2NaClO_4\}$[194] have present Na^+ in a distorted octahedral site generated by the TMSB and a bidentate perchlorate anion. The complex $\{[Co(salen)]_2NaCo(CO)_4\cdot THF\}$ has been synthesized by reaction of NaCo(salen) with CO and shown by X-ray studies to be a complexed ion pair in which the Na^+ is octahedrally coordinated to two TMSB ligands, the THF and to one carbonyl of the $[Co(CO)_4]^-$ anion.[195] In the alkaline earth metal complex $\{[Cu(salen)]\cdot Mg(hfacac)_2\}$ the Mg^{2+} is octahedrally coordinated to the TMSB and the two hfacac anions.[196]

Reduction of Co(salen) with Li or Na in THF leads to the isolation of the bimetallic complexes Co(salen)Na·THF or Co(salen)Li·1.5THF, which are active in CO_2 carriage.[197] The structure of the former consists of infinite chains of the monomer unit $\{[Co(salen)]_2Na_2\cdot 2THF\}$ with distorted octahedra around the Na^+ cations sharing opposite faces and with both THF and Co(salen) acting as bridging ligands. The lithium species has present two different complexes, {[Co(salen)]Li·2THF} and {[Co(salen)]Li·2THF} and $\{[Co(salen)]_2Li_2\cdot 2THF\}$ in 2:1 ratio. In the first the Li^+ is in a distorted tetrahedral environment and in the second one of the THF molecules is replaced by a Co(salen) oxygen atom which then acts in a bridging capacity. This class of complex is active in the bifunctional activation of CO_2, and provides reversible CO_2 carriers. The structure of $\{[Co(pr\text{-}salen)]K\cdot CO_2\cdot THF\}$ has been solved and shows the K^+ coordinated by the TMSB, THF and CO_2 in a dimeric repeat unit (Figure 3). One K^+ is six-coordinated to two CO_2 and two THF molecules and the second K^+ is bound to two bidentate pr-salen units and to two oxygen atoms of two CO_2 molecules. Each CO_2 molecule shares one oxygen between two K^+ ions.[198] The structure of the corresponding Na^+ complex is also available and the general principles of the reaction have been elucidated and discussed.[199] The alkali metal acts as a Lewis acid centre, helping to hold the CO_2 in close proximity to the basic cobalt centre.

$$Co(L)K + CO_2 \xrightleftharpoons{THF} [Co(L)KCO_2THF]$$

L =

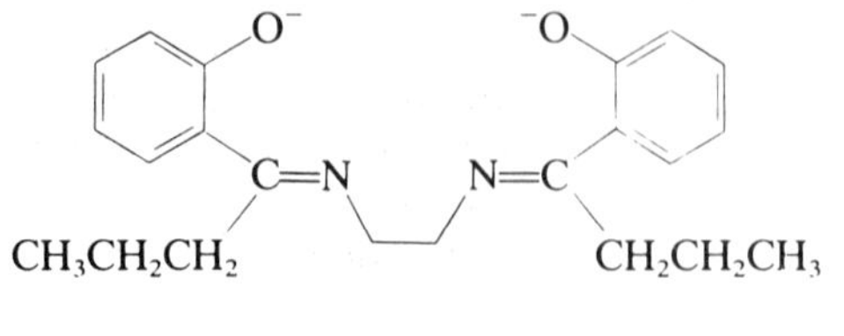

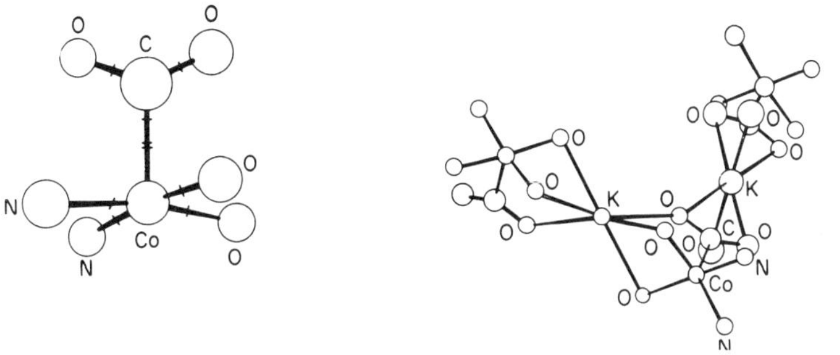

Figure 3 The repetitive unit of polymeric (Co(pr-salen)K·CO_2·THF) and the coordination sphere around the cobalt (reproduced with permission from reference 198)

Alkaline earth metal salts, such as the perchlorates, are also complexed by TMSB,[189] and isolated species such as bis(2-amino-2-hydroxymethyl-1,3-propanediolato(1 −)O,N)-copper(II)·sodium perchlorate·H_2O[200] have also been found.[30]

23.3.2.3 Podands

In a historic context, podands should be preceded by crown ethers and coronands (*vide post*) and much of the work carried out on podands was stimulated by studies on the macrocyclic analogues. Podands have the advantage of relatively facile syntheses, using neither dilution nor template procedures, to provide inexpensive materials of versatile ligand structure. In contrast to macrocycles though, they form complexes of relatively low stability; the hexadentate pentaethyleneglycol dimethyl ether ('pentaglyme'; glyme-6) is found to have a 10^4-fold drop in stability from the cyclic 18-crown-6 in its reaction with K^+, and of 10^3 in its reaction with Na^+.[33] This effect has been termed the 'macrocyclic' effect and was explained originally in terms of entropy factors and later in terms of enthalpy changes coupled with solvation effects.[201] A 'macrobicyclic' effect also operates on going from mono- to bi-cyclic macrocycles.[202]

In keeping with the low stability constants, isolable alkali and alkaline earth metal complexes of acyclic oligoethylene glycols and their ethers, the 'glymes', were unknown until recently although solution studies had given information on probable stoichiometries. Studies on the complexation of difluorenylbarium with glymes[203] of varying chain lengths revealed that glyme-7 (**14**; $n = 5$) and glyme-9 (**14**; $n = 7$) gave 1 : 1 complexes which were mixed tight and loose ion pairs of the type fluorenyl$^-$ Ba^+/glyme/fluorenyl$^-$. Glyme-5 (**14**; $n = 3$) was further able to form a 1 : 2 complex, hinting that glyme-10 (**14**; $n = 8$) should give a 1 : 1 separated ion pair complex. Complexation is generally enhanced on increase of chain length.

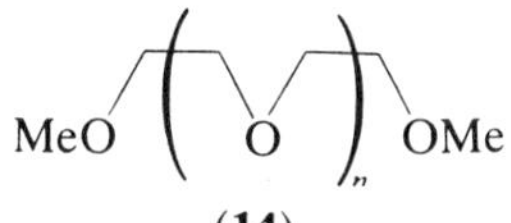

(**14**)

Glyme complexes of M^+ with bulky anions present have been known for some time and tended to arise as a consequence of work in, and recrystallization from, a particular glyme solvent, *e.g.* $[M(diglyme)_2][Ta(CO)_6]$, $[M(diglyme)_2][Mo(CO)_5I]$[204] and $[K(diglyme)][Ce(cot)_2]$.[205] The structure of the latter shows the K^+ coordinated by the diglyme (glyme-3) molecule and by one of the cot rings (Figure 4). The complexes $[M(glyme\text{-}4)_2][SbCl_6]_2$ (M^{2+} = Ca, Sr) have been isolated, and IR studies indicate octacoordination of the metal by the glymes.[206] In $[Mg(glyme\text{-}4)_2][SbCl_6]_2{\cdot}2H_2O$, however, the ligand is proposed to act as a tridentate donor, particularly because of the limited space around the Mg^{2+} cation.

The parent glycols have also been studied, particularly in solution and in connection with their possible role in the template synthesis of crown ethers.[33] Crystalline complexes have been isolated in certain instances and the structures of Ca(picrate)$_2$·glycol-5·H_2O,[207] $Sr(NCS)_2$·glycol-8[208] and $Sr(NCS)_2$·glycol-7[209] have been solved. (Glycol-5 is $HO(CH_2CH_2O)_nH$, $n = 4$, *etc.*) The Sr^{2+} complexes show the metal to be coordinated by the

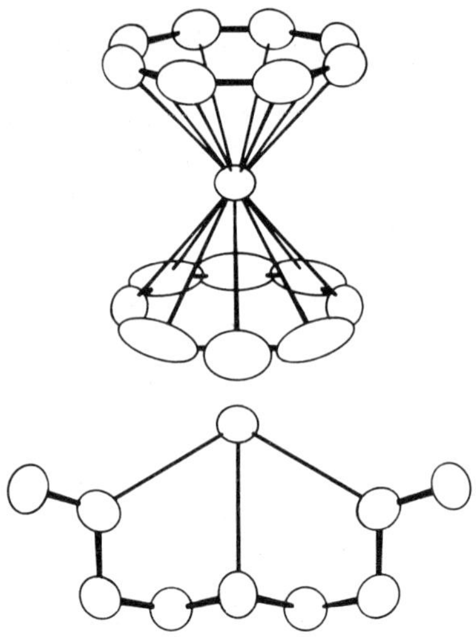

Figure 4 The molecular structure of [K(diglyme)][Ce(cot)$_2$] (reproduced with permission from reference 205)

eight donors of glyme-8, and one NCS^-, and by the seven donors of glyme-7, and two NCS^- respectively to give nine-coordination in each case with the ligand wrapping around the metal in a helical fashion (Figure 5). In the Ca^{2+} complex, the five glycol donor atoms, one phenoxide oxygen, one nitro oxygen and a water molecule give a distorted square antiprismatic cation environment.

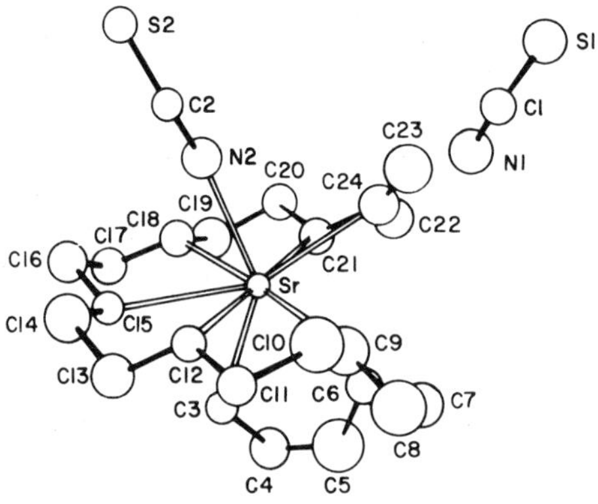

Figure 5 The molecular structure Sr(NCS)$_2$·glycol-7 (reproduced with permission from reference 209)

The attachment of rigid terminal groups such as substituted aromatic rings to the oligo-ethylene glycol units affords neutral ligands which now readily form stable complexes with alkali and alkaline earth metal cations.[33] Many terminal groups have been used, such as 8-quinolyloxy (**15**), methoxyphenoxy (**16**) and *o*-nitrophenoxy groups (**17**). These groups function as anchorage points with locally fixed donor centres on which the cation can take hold. If a polymethylene chain is used, then only a limited ability to form complexes is demonstrated. Symmetrical and non-symmetrical podands of the above type may be synthesized extending the range and versatility of the ligands.

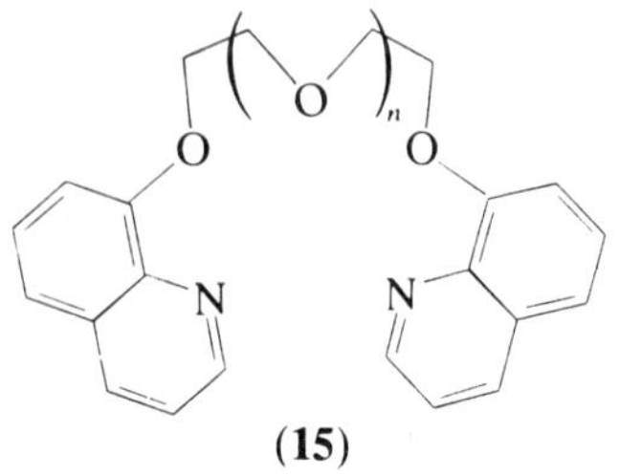

(**15**)
(**a–g**; $n = 0–6$)

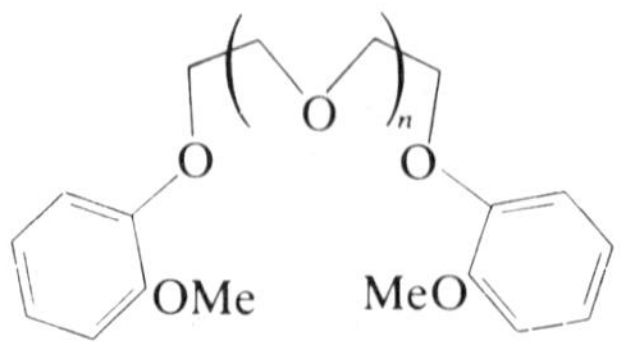

(**16**)
(**a–d**; $n = 0–3$)

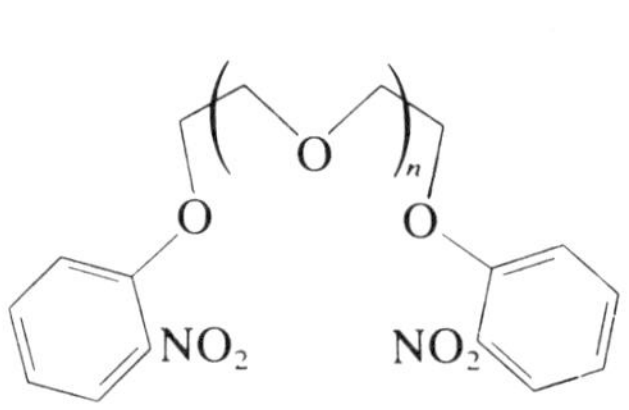

(**17**)
(**a–d**; $n = 0–3$)

The stoichiometries of the complexes formed from podands are usually precise. For example (**15d**) always gives a 1:1 complex with KNCS regardless of the mole ratio of ligand to salt.[210] X-Ray structures are now available for several of these complexes of open-chain ligands. One useful series has been that concerning complexes based on (**15**). Models of the ligand (**15d**) suggested that participation of all seven heteroatom donors in coordination to a central cation would lead to a screw-shaped arrangement of the ligand and this would allow the ligand to wrap around an ion of appropriate radius in a helical fashion, thus providing a plausible model for acyclic ionophores (*vide infra*).[210] The complexation of Rb^+ by the podands (**15b**),[211a] (**15d**)[212] and (**15g**)[213] is illustrated in Figure 6.

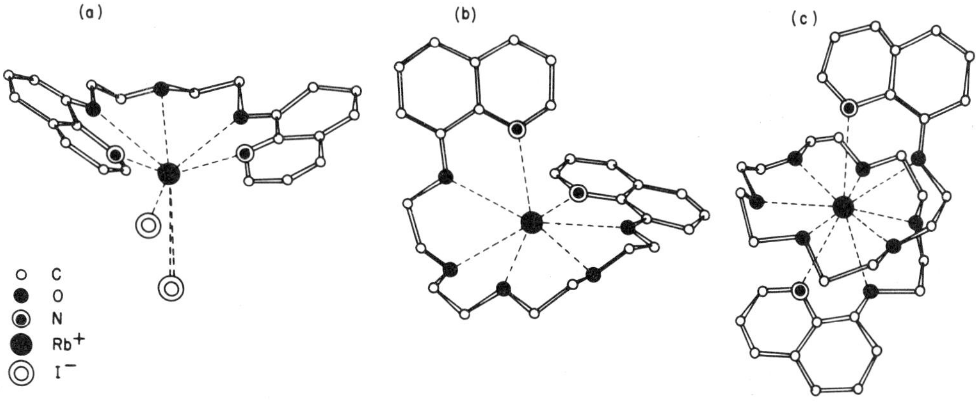

Figure 6 The complexation of Rb^+ by podands; the molecular structures of complexes derived from: (a) (**15b**), (b) (**15d**) and (c) (**15g**) (reproduced with permission from reference 33)

The structure of the complex involving (**15d**) confirms the expected geometry, affording a helical arrangement of the donors around the cation; the complex is also charge-separated. The M—O bond lengths are not all equivalent and the variations are ascribed to the differing donor strengths of aliphatic and aromatic oxygen donors (~2.88–2.99 Å and ~3.07–3.09 Å respectively). The chain is too long to enclose the Rb^+ in a planar donor array, but the major part of the donor array is fixed close to a plane. This helical arrangement is replaced in the shorter (**15b**) ligand by an approximately planar array of donor atoms. These are 'butterfly-folded' and reminiscent of the shape found in the dibenzo-18-crown-6·RbNCS complex (*vide infra*). The ligand is pentadentate and two iodides per cation participate alternately in the complexation. With the decadentate ligand (**15g**), a spherical wrapping of the Rb^+ occurs with the heteroatoms on the surface of a sphere of ~3.0 Å radius. The iodide ions lie in cavities between the 'spheres' and the eight oxygen atoms of the ligand are coiled around the cation in the equatorial plane with the quinolyl moieties coordinating from above and below this plane.

The structure of the related complex (**18**) RbI[211b] shows a significant difference in ligand conformation compared to that in (**15d**) RbI. The heptadentate ligand is arranged as a continuous helix, due presumably to steric hindrance to planar inclusion of the cation by the methyl groups. A continuous helix is also found in the structure of (**16c**) NaSCN,[214] in which the Na^+ is coordinated to all six donors of the ligand and to the thiocyanate anion which lies above the helix. Similar wrap-around structures are found for (**17d**)KSCN,[215] {(**19**)KSCN}$_2$·H_2O[216] and (**20**)RbI.[217]

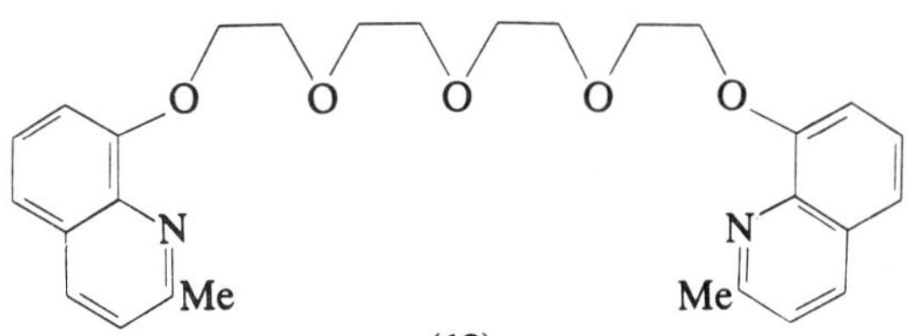

(**18**)

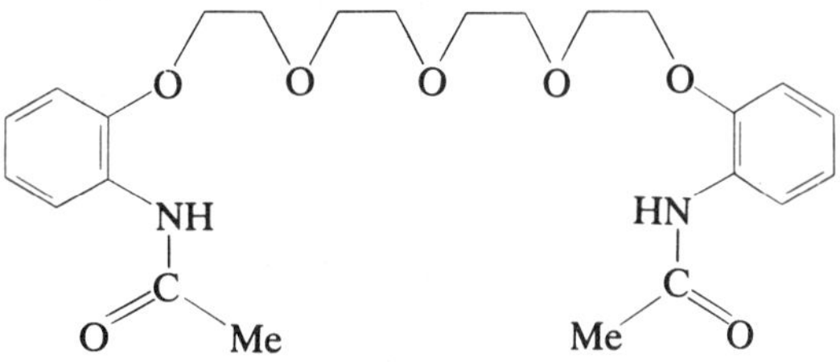

(**19**)

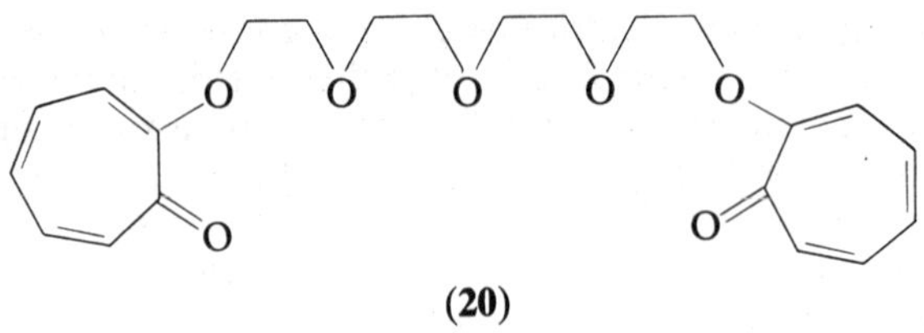

(20)

NMR studies have proved useful in the identification of complexation in solution in these systems.[33] In the ^{1}H NMR of complexes of podands with rigid terminal groups pronounced changes are seen in the splitting patterns of the aromatic region on complexation. The structure of **(15d)**RbI is in full accord with the upfield shift found for the heterocyclic terminal units. This area of study shows much promise for the future with the availability of multinuclear NMR techniques. Studies have also been made on the selectivity between metals, and on the stability of the complexes. Examples of the effects of loss of terminal group, the introduction of rigid ligand spines and the variation of the terminal group are shown in Figure 7.

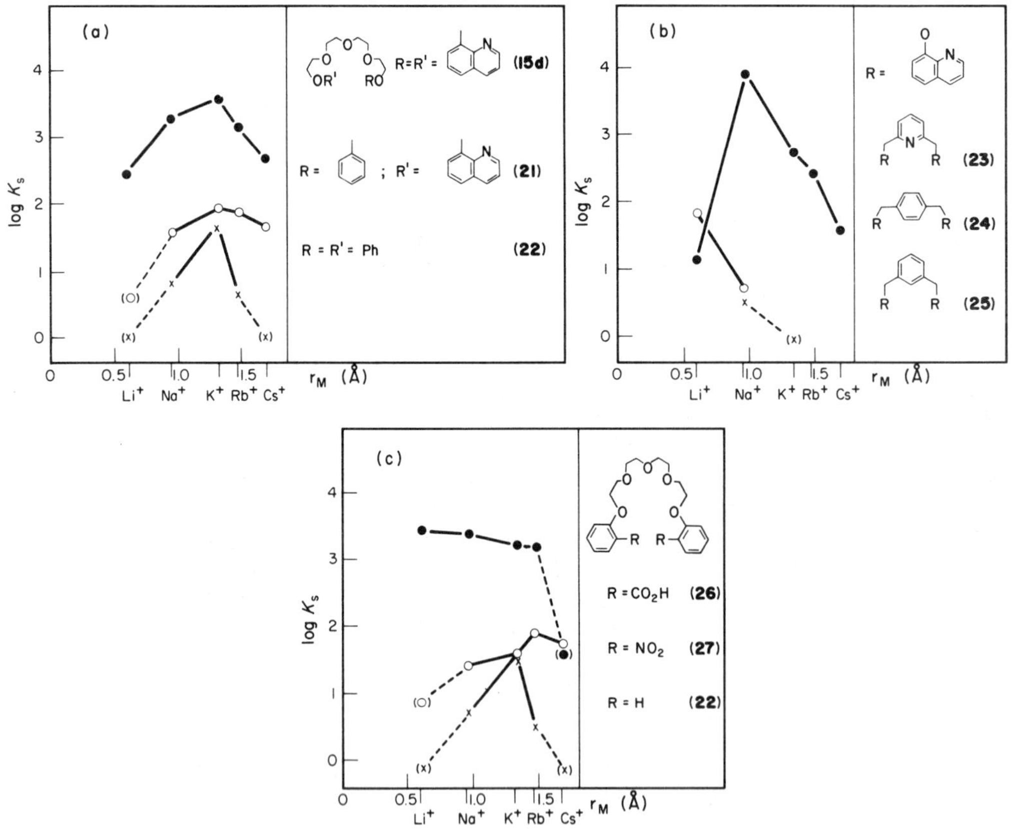

Figure 7 Stability constants (log K_s) of various podand alkali metal complexes as a function of ionic radius: (a) stepwise loss of terminal groups; (b) ligands with stiffened chains; (c) variation of terminal group (reproduced with permission from reference 33)

In the first instance, a relatively small K^+/Na^+ discrimination is maintained (Figure 7a) but stiffening the spine leads to a movement towards the selectivity of individual cations (Figure 7b). Changing the terminal group (Figure 7c) can also introduce a more individual selectivity and also a plateau selectivity.[33] When the quinolyl terminal group is present, then Mg^{2+} complexes are more stable than are the other alkaline earth complexes. This is anticipated from the known propensity of 8-hydroxyquinoline for magnesium.

The process of complexation is believed to occur through the stepwise incorporation of the ligand, *i.e.* stepwise replacement of the existing solvation, or hydration, sheath. The process is likely to be entropy controlled and entropy increases have been noted in thermodynamic studies on these systems.[33]

When the chain length is dramatically increased, as in (**28**), podands are available which could imitate the helical ion channels believed to be available in membranes and so give information on the selective transfer of cations across such barriers. This type of ligand shows pronounced transfer of inorganic salt into organic phases and NMR studies indicate a likely helical coordination of the cation.[33,218] Complexes have been isolated and the structure of (**29**)2KNCS has been solved.[219] In this molecule, the podand wraps, in an S-shaped manner, within the dinuclear complex, and contains a KNCS unit within each loop of the S shape.

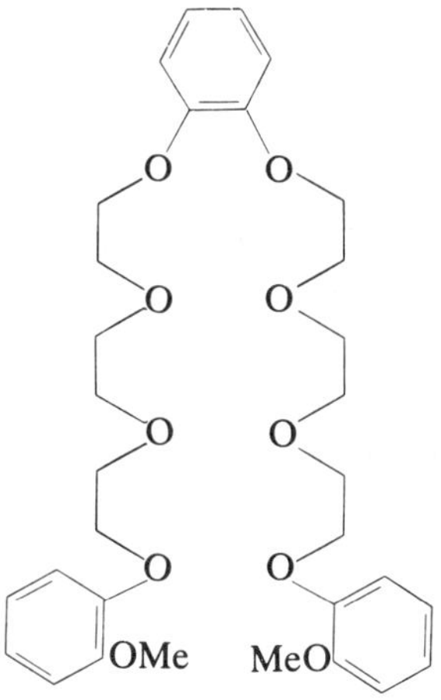

(**28**)

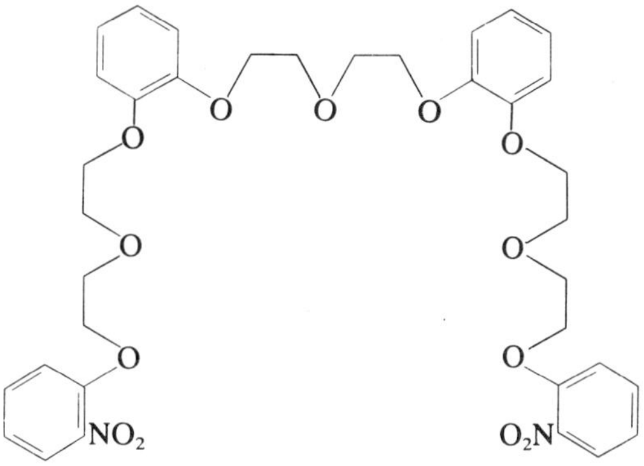

(**29**)

Interactions between terminal groups are of interest, as in the natural ionophores terminal linkage through hydrogen bonding readily occurs to produce a pseudo-macrocyclic structure (*vide infra*). Although the existence of such a link is not yet found in podands, amino-to-carboxy bonding has been found in the $NaClO_4$ complex of (**30**).[220] This ligand is zwitterionic in solution, and in the structure of the complex one of the NH_3^+ hydrogen atoms forms an internal hydrogen bond to a carboxylate oxygen atom. The dicarboxylic acid (**31**) forms a series of crystalline complexes with alkali metal salts in which the ligand acts as a neutral species.[221] However, contrary to expectation the crystal structure of the K(picrate) complex revealed a dimeric unit with no internal, or 'head-to-tail', hydrogen bonding present. Each unit of the dimer is arranged spirally and contains its K^+ in an irregular eight-fold coordination. One carboxyl oxygen from the monomer serves as a bridging unit and binds to the second K^+ (Figure 8).[222]

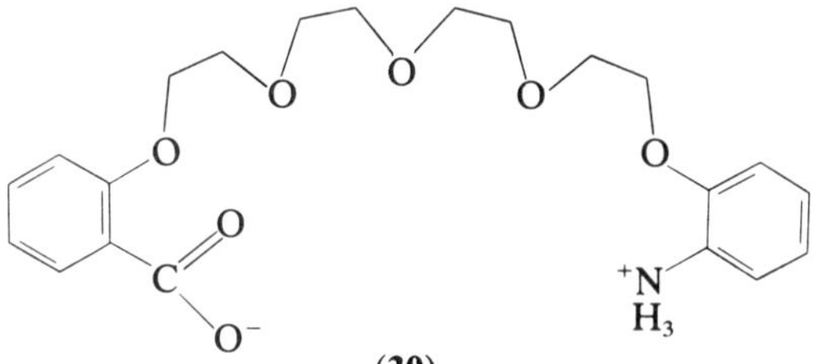

(**30**)

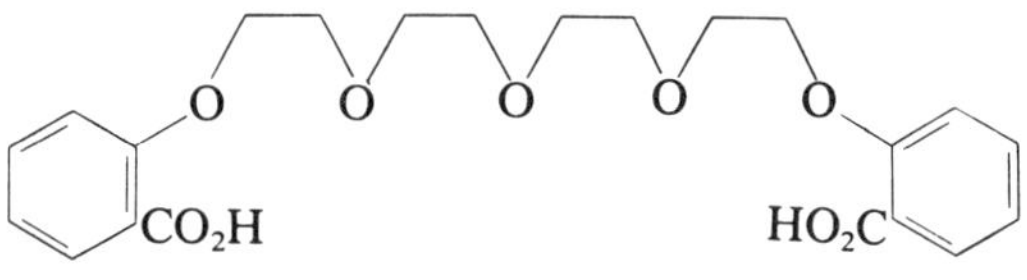

(**31**)

A family of podands containing as terminal groups variously substituted amides, and having a wide range of linking chains have been developed as sensors for particular alkali and alkaline earth cations.[223] Particular emphasis has been directed towards Ca^{2+} and Li^+; for example, a selectivity for Li^+ over other alkali, and alkaline earth cations, in membranes, of the order of 100 and 1000 respectively is induced by *cis*-*N,N,N',N'*-tetraisobutylcyclohexane-1,2-carboxamide.[224] The cation selectivity is dependent mainly upon the nature and number of coordinating sites, the nature of the chain termini and the choice of solvent. In general, divalent cations are preferred over monovalent cations with increasing polarity of both the ligand skeleton and the dielectric constant of the solvent phase.[33,224] Comparative studies have

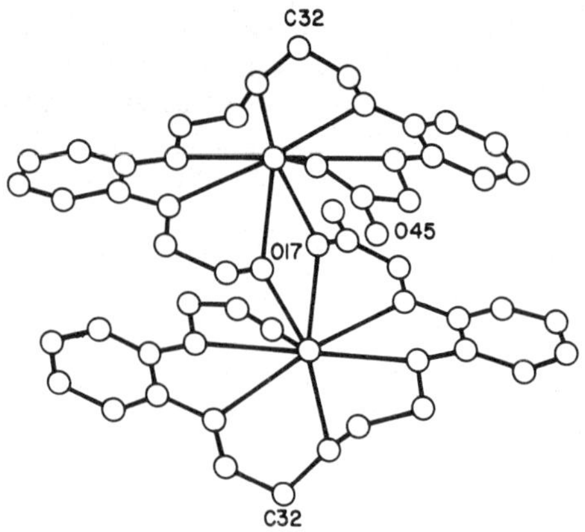

Figure 8 The molecular structure of potassium picrate (**31**) dimer (reproduced with permission from reference 222)

shown (**32a**) to be an Na^+-selective ionophore,[225] (**32b**) a Ca^{2+}-selective ionophore[226] and (**33**) a Ba^{2+}-selective ionophore.[227] The Na^+ ionophore (**35**), in which the tripodal arm helps adaptation of the ligand to the cation size, discriminates against K^+ cations sufficiently strongly to allow its use in intracellular studies under physiological conditions.[228]

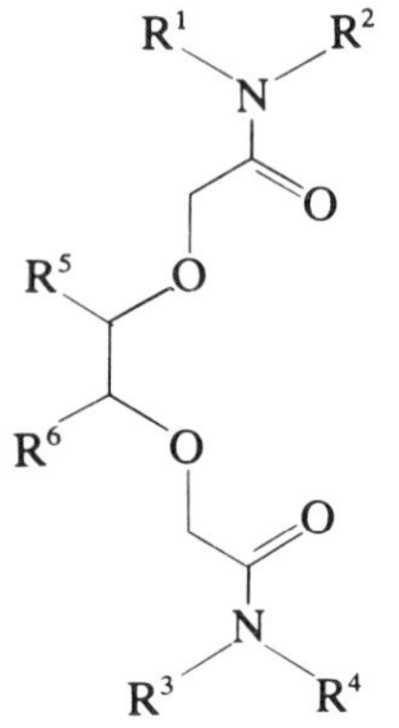

(**32**) **a**; $R^1 = R^2 = R^3 = R^4 = CH_2Ph$, $R^5 = R^6 = H$
b; $R^1 = R^3 = Pr$, $R^2 = R^4 =$ neopentyl, $R^5 = R^6 = H$
c; $R^1 = R^2 = R^3 = R^4 = Ph$, $R^5 = R^6 = H$
d; $R^1 = R^3 = Me$, $R^2 = R^4 = EtO_2C(CH_2)_{11}$, $R^5 = R^6 = Me$
e; $R^1 = R^2 = R^3 = R^4 = Pr$, $R^5 = R^6 = H$

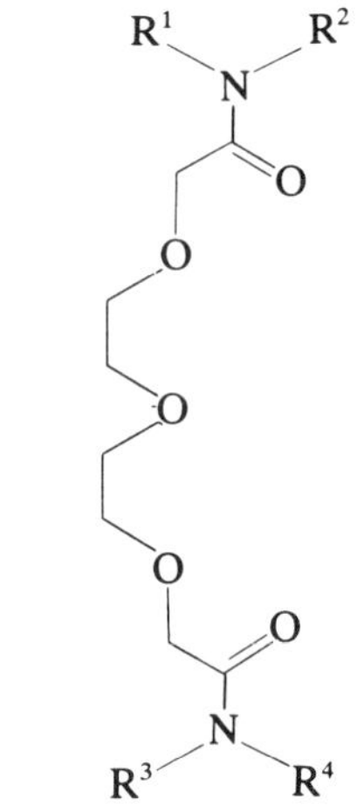

(**33**) $R^1 = R^2 = R^3 = R^4 = Ph$

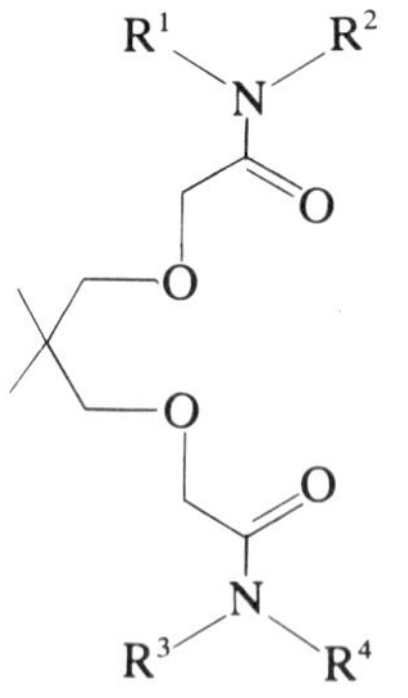

(**34**) $R^1 = R^3 = Me$, $R^2 = R^4 =$ heptyl

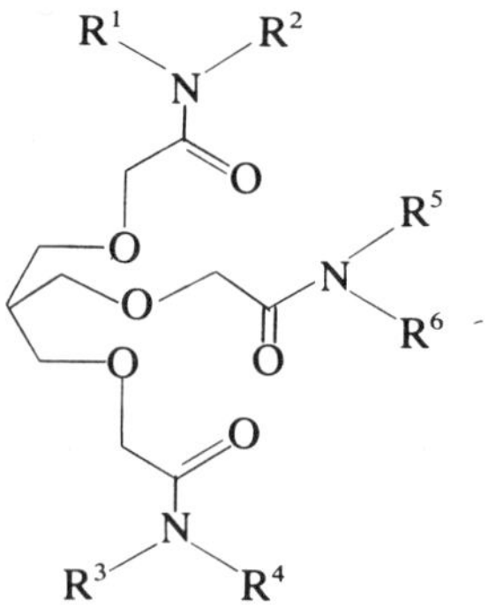

(**35**) $R^1 = R^3 = R^5 = Me$,
$R^2 = R^4 = R^6 =$ heptyl

These comparatively lipophilic ligands have been conceived as ion carriers for systems such as ion-selective electrodes,[229] therefore they do not have high stability constant values, but require fast complexation kinetics in order to achieve rapid equilibration. Nevertheless, it has been possible to recover crystalline species which often have present additional water molecules to help stabilize the crystal lattice. Coordinative participation of the carbonyl oxygen atoms, which was indicated by ^{13}C NMR studies,[230] was confirmed in the solid state by IR studies and by crystal structure analysis of $(\mathbf{32e})_2CaCl_2$.[231] The Ca^{2+} is coordinated to four ether oxygen atoms and four carbonyl oxygen atoms from the pair of symmetrically equivalent ligands and is in a trigonal dodecahedral environment.

23.3.2.4 Polypodands

Polypodands are many-armed neutral ligands. The first described were a series of compounds bearing several donor-containing 'tentacles' attached to a benzene ring *via* sulfur atoms, and reminiscent of octopuses, (**36**).[232] The compounds show phase transfer properties towards cations and (**33a**) transfers alkali and alkaline earth picrates efficiently from aqueous solution into dichloromethane. This phase transfer is inhibited by reducing the number of available donor sites by shortening the tentacles (**36b**) or by reducing the number of tentacles present on the aromatic nucleus.[33] Similar properties are exhibited for hexapodands derived from cyclotriveratrylene (**37**),[233] but if a more rigid central unit is introduced then loss of activity occurs (**38**).[233] Similarly loss of activity is found if the oxygen-free hexasubstituted benzenes are examined.[232]

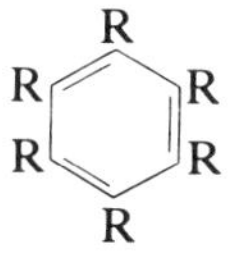

(**36**)

a; $R = Bu^n(OCH_2CH_2)_3SCH_2$
b; $R = Me(OCH_2CH_2)_2SCH_2$

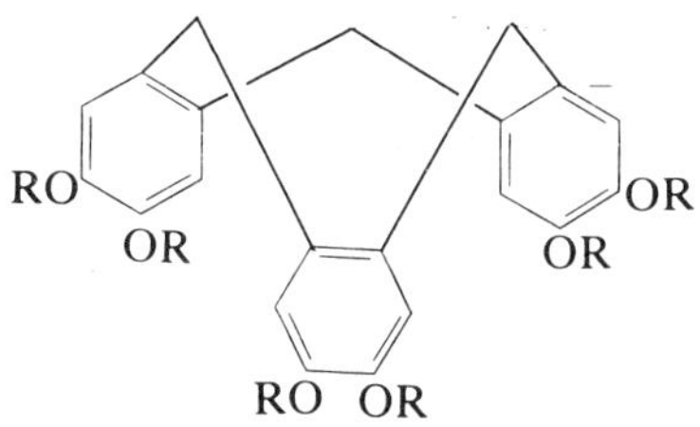

(**37**)

$R = (CH_2CH_2O)_3R'$
$R' = Me, Et, Bu^n$

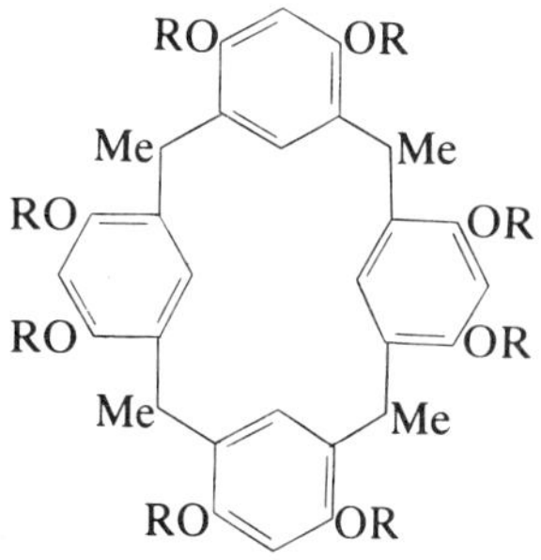

(**38**)

$R = (CH_2CH_2O)_3Me$

Strict stoichiometries have been found for alkali and alkaline earth metal complexes of triethanolamine (te), and 2,2′,2″-trimethoxytriethylamine (tmt). The crystal structure of NaI·tmt consists of discrete molecules in which the Na^+ is pentacoordinated by the heteroatoms of one ligand and by the iodide anion.[234] In the structure of NaI·te each sodium cation is seven-coordinated by the four heteroatoms of one ligand, the iodide anion and two oxygen atoms of a neighbouring ligand.[235] Two OH groups of each molecule bridge two sodium cations and so the structure is catenated. The structure of $Sr(NO_3)_2{\cdot}2te$ has present an eight-coordinated cation, the cubic environment arising from the eight heteroatoms of the ligands,[236] and in $Ba(MeCO_2)_2{\cdot}2te$[237] the cation is bonded to the eight heteroatoms of the two ligands and an acetate oxygen donor. A predominate 1:1 complexation has also been noted in alkali and alkaline earth metal complexes of the related tetra- and penta-podands (**42**, **43**); even addition of excess BaI_2 to (**43**) gave only the 1:1 complex. This type of compound has been viewed as an example of fundamental facile recognition and selectivity processes on a molecular level.[238]

Many polypodands having terminal donor groups are also available, *e.g.* (**39**)–(**41**).[33] These also exhibit cation complexation and phase transfer properties. $KMnO_4$ and aqueous alkali metal picrates are much more readily taken into organic phases in the presence of these podands than with dibenzo-18-crown-6.[239] The KSCN complex of (**39**) exhibits a novel coordination geometry as all 10 donor atoms participate in coordination of the metal cation, and in order to do this the three arms wrap around the cation in a propeller-like fashion.[240]

In terms of selectivity (**40**) selects Na^+ over K^+ and Li^+, but also has a high selectivity for the divalent cations Sr^{2+} and Ba^{2+}. The ligand (**41**) has an even higher complexing capability for the alkaline earth metal cations and a BaI_2 complex is isolable whereas a KNCS complex is not. The ligating differences presumably arise from the inability of the pseudocavity produced on complexation to allow suitable contact with the metal cation by the donor atoms. Ligand–ligand interactions also induce steric constraint to complexing small cations.[239]

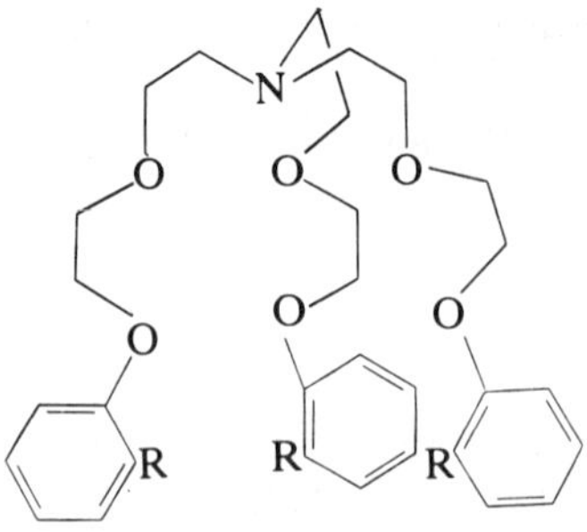

(**39**) R = H, OMe, NO_2

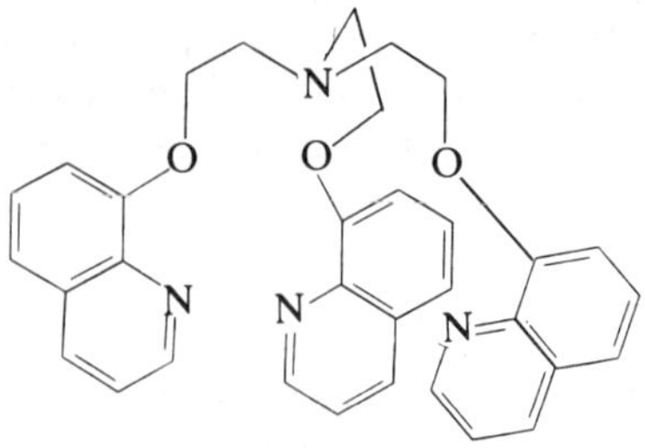

(**40**)

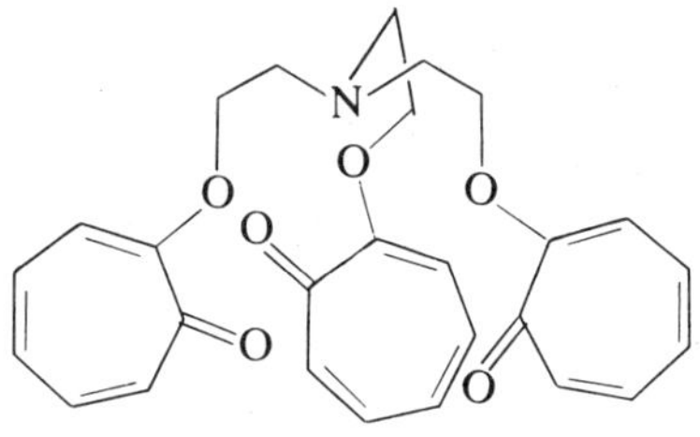

(**41**)

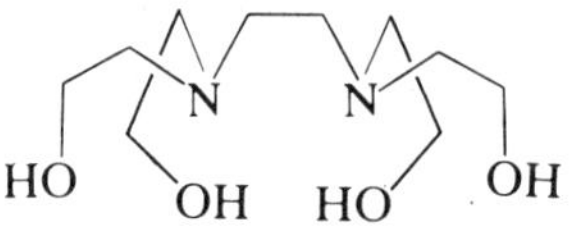

(**42**)

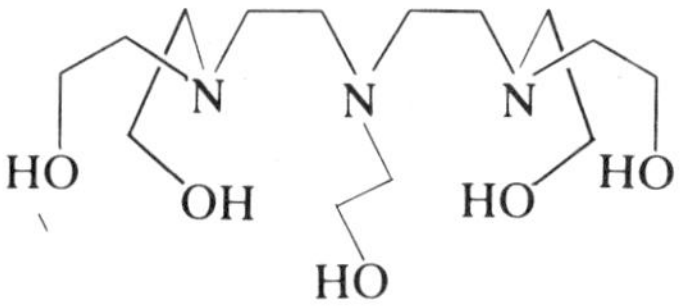

(**43**)

The stability constants of the corresponding tetrapodands (**44, 45**) are generally lower than those for the tripodands, presumably because of more severe steric hindrance to complexation.[241]

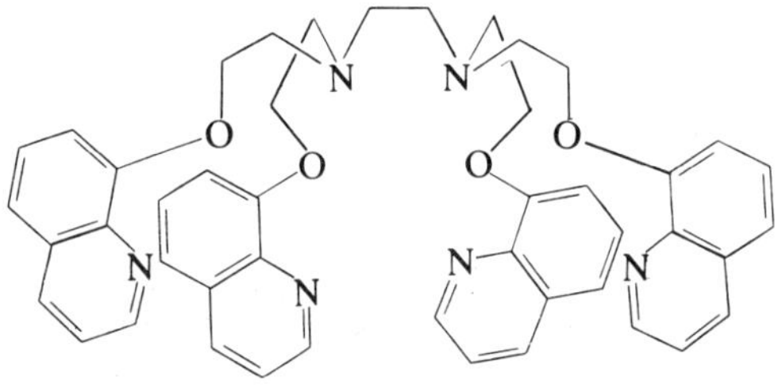

(**44**)

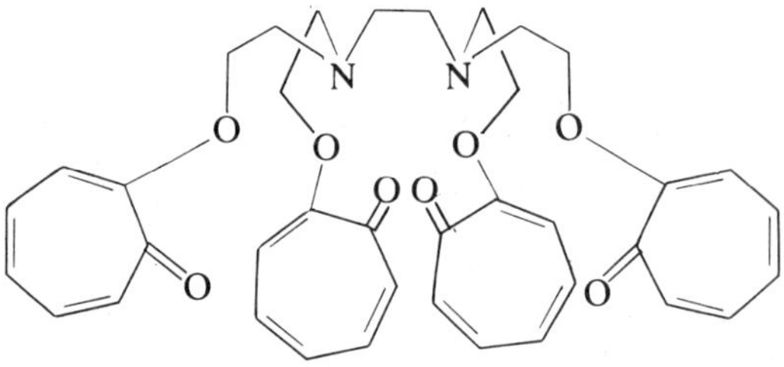

(**45**)

23.3.2.5 Carbohydrates

The complexation of M^{n+} by carbohydrates is now well known.[242] The conditions that influence such complexation have been discussed at length, as have complexation processes in solution with particular reference to NMR studies.[243] Structural aspects of calcium–carbohydrate interactions have also been reviewed.[244]

This area of study has been stimulated by information that sugars and related species can help in the transfer of M^{n+} across membranes.[245] The studies made have generally been concerned with conformational aspects of ligand behaviour under the influence of the cation rather than the nature of the coordination of the metal.[30].

The essential bonding mode of sugars is *via* the hydroxyl groups, and complexation in solution is due basically to the presence of three sequential hydroxyl groups in a *cis–cis* state for a five-membered ring, and an *ax–eq–ax* state for a six-membered ring. These sequences give, for solution complexation, a required equilateral triangle of donor atoms for the cation and favourable entropy factors. In the solid state where lattice packing factors become important, just the pairs of adjacent hydroxyl groups are involved with the cation.[30] If these criteria are not met, as in D-glucose, the sugar is non-complexing. *cis*-Inositol (**46**) with three *ax–eq–ax,* which can act jointly with three axial hydroxyl groups, serves as an example of an alkaline earth metal complexor.

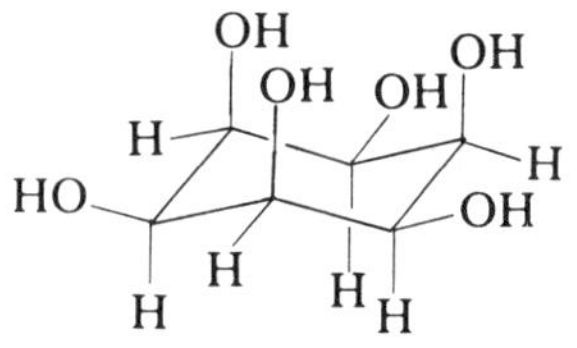

(46)

The first sugar complex structure to be solved was $NaBr\cdot$sucrose$\cdot 2H_2O$, in which the Na^+ is six-coordinated through the water molecules and hydroxyl groups from three different sugar molecules, and is paired with the bromide anion.[246] This extensive sharing is a feature of the solid state chemistry and the X-ray structures of Ca^{2+} carbohydrate complexes reveals the cation to be seven- or eight-coordinated, as in the hydration of Ca^{2+} in the solid state.[244] Representative structures are those of $CaBr_2\cdot$(D-lactose)$\cdot 2H_2O$,[247] $CaBr_2\cdot$(α-D-galactose)$\cdot 3H_2O$[248] and $CaBr_2\cdot$(myoinositol)$\cdot 5H_2O$,[249] in all of which the cation is eight-coordinated in a square antiprismatic geometry. This environment is produced by the water molecules and hyroxyl groups from different sugars. $CaBr_2\cdot$(α-D-fructose)$\cdot 3H_2O$ has a reduced coordination number of seven, but a similar environmental build-up.[250]

Two unusual Ca^{2+} complexes are $CaCl_2\cdot$(β-D-fructose)$_2\cdot 3H_2O$, in which a 1:2 complex is found,[251] and $CaCl_2\cdot$(β-D-mannofuranose)$\cdot 4H_2O$, in which the sugar is triply chelating through sequential hydroxyl groups.[252] Unequivocal coordination of Sr^{2+} in sugar systems is known through structural analysis of $SrCl_2\cdot$(epiinositol)$\cdot 5H_2O$; the cation is nine-coordinated by four water molecules, one bidentate and one tridentate ligand.[253] For Mg^{2+}, $MgCl_2\cdot$(myoinositol)$\cdot 4H_2O$ has been investigated crystallographically and has the cation coordinated to the water molecules and two *cis* hydroxyl groups of the inositol in an octahedral arrangement. In this latter complex the cation is part of an essentially discrete molecule;[254] however, there is an extensive hydrogen bonding network present which leads to an effective extended array.

23.4 ANIONIC ACYCLIC POLYDENTATE LIGANDS

23.4.1 β-Diketonates

The coordination of β-diketonates and related species to alkali and alkaline earth cations has long been recognized. Combes first prepared $Be(acac)_2$ in 1887,[255] and the contribution of Sidgwick and Brewer concerning the nature of dihydrated alkali metal β-diketonates plays a seminal role in the development of this area of coordination chemistry.[19] A review concerning the structure and reactivity of alkali metal enolates has been written,[256] and sections on alkali and alkaline earth β-diketonates can be found within more generalized accounts of β-diketone complexes.[257,258]

2,4-Pentanedione (acetylacetone, acacH) forms strong complexes with alkali metal cations, even in polar media. The stability constants fall in the sequence $Li > Na > K > Rb > Cs$ and this is in keeping with the decreasing ionic potential of the cation.[259] ^1H NMR studies have given similar evidence for the strength of the M^+—$acac^-$ ion pair. Li^+ tends to be completely chelated and Na/Li exchange can be detected when $LiClO_4$ is added to Na(acac). At low concentrations of Li^+ with respect to acacH an anionic species $[Li(acac)_2]^-$ exists in solution; in the presence of excess cation the species $[Li_2(acac)]^+$ is found.[30,98] The $acac^-$ ion can exist in four conformations (**47**); small cations which enforce chelation arrest the anion in the *Z,Z* form whereas larger cations which form weak ion pairs allow *E,Z*, *Z,E* and *E,E* formation. For example, the $acac^-$ in Na(acac) can be converted, in DMF[260] or pyridine,[261] to the *E,E* isomer by addition of 18-crown-6 which complexes the Na^+ and removes the 'ion-pairing' potential of the cation by generating a 'large' cation.

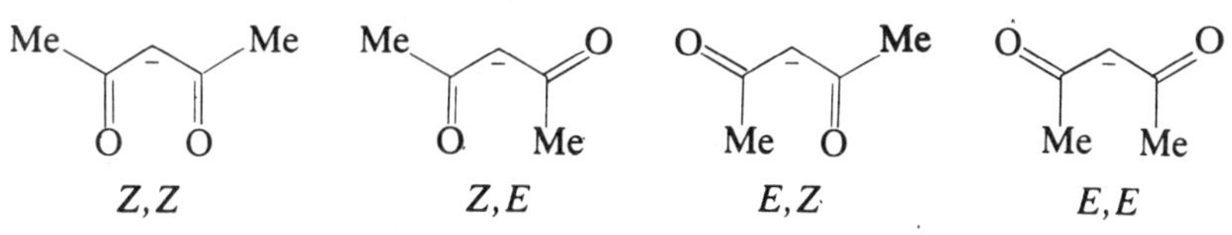

(**47**)

X-Ray crystal structures have been carried out on Li(acac), Na(acac)·H_2O, K(acac)·0.5H_2O and Na(bzac)·eg and show contrasting features. The lithium derivative is composed of chains of $[Li(acac)_2]^-$ anions linked to Li^+ cations; the anions are square planar and the cations tetrahedral in that coordination environments and the cations are bonded to pairs of oxygen atoms from neighbouring anionic units (Figure 9a).[262] The Na(acac)·H_2O structure consists of zigzag chains of Na^+ cations bridged by two water molecules and acac units; the Na—O bonds to the water molecules are slightly longer (2.42 Å) than those to the anionic oxygen atoms (2.31 Å chelated and 2.38 Å bridging) (Figure 9b).[263] The movement towards polymeric lattices is further evidenced in the structure of K(acac)·0.5H_2O, where the cation is surrounded by seven oxygen donors, six from four different $acac^-$ units and one from the water molecule.[264] Although the structure of a dihydrate is not yet available, that of the somewhat analogous Na(bzac)·eg has been determined. The structure is again polymeric with the glycol acting as a bidentate, non-chelating ligand. The cation is five coordinated in a square pyramidal environment consisting of the β-diketonate, two oxygens from bridging glycols and a shared oxygen from a neighbouring chelating anion.[22]

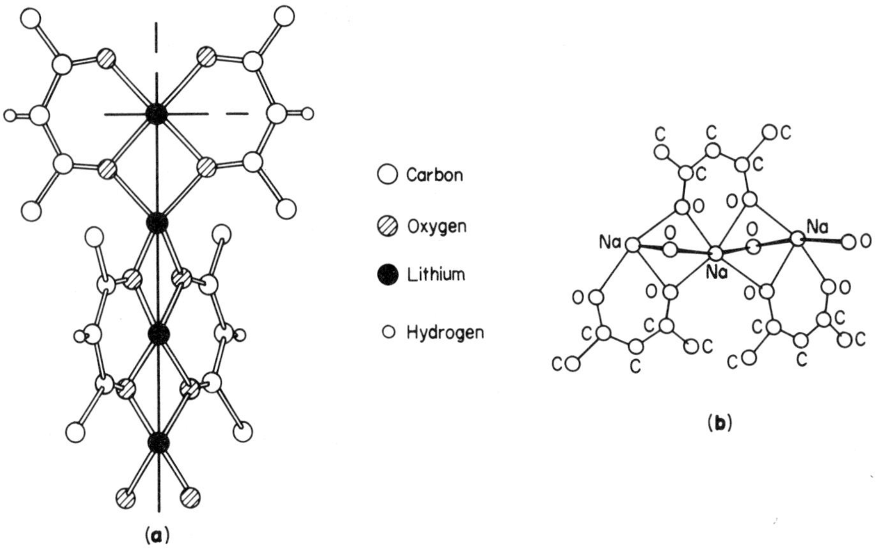

Figure 9 The molecular structures of (a) Li(acac) and (b) Na(acac)·H_2O (reproduced with permission from references 262 and 263)

The alkaline earth metal β-diketonates follow similar stability trends to those of the alkali metals: Be > Mg > Ca > Sr > Ba, with beryllium β-diketonates having extensive covalent character.[259] The crystal structure of $Be(acac)_2$ shows a discrete molecule bearing a tetrahedral beryllium atom,[265] and an electron diffraction study shows retention of this shape in the gas phase.[266] The tetrahedral arrangement of the bonds was first demonstrated in the 1920s.[267] Beryllium complexes such as $Be(bzac)_2$ should exist in optically active forms and resolution into the optical antipodes has been achieved on active quartz, or by coprecipitation.[268,269] Oxy-bridged polymers containing beryllium β-diketonates can be found (**48**), as can coordination polymers derived from bis(1,3-dicarbonyl) compounds (**49**).[93,257] The reaction of N_2O_4 in benzene, or 'acetyl nitrate' generated *in situ,* with $Be(acac)_2$ leads to the introduction of the nitro group in the 3-position and bromination at the same position is obtained using *N*-bromo–succinimide.[270]

Heterobinuclear complexes have been formed between PdCl(acac-C_1), bipy and $Be(acac)_2$.[271] The palladium compound retains an acid character and so replacement of an $acac^-$ from the $Be(acac)_2$ gives a product which can be formulated as (**50**) from MW and ^1H NMR studies.

For the heavier alkaline earth β-diketonates, X-ray studies are available for $Mg(acac)_2·2H_2O$, $Mg(dpp)_2·2DMF$, $Ca(acac)_2·3H_2O$, $Ca(dpp)_2·0.5EtOH$ and $Sr(dpp)_2·0.5$acetone (dpp = 1,3-diphenyl-1,3-propanedione). $Mg(acac)_2·2H_2O$ is *trans* octahedral with the water molecules occupying axial positions; the bond lengths to the axial and equatorial oxygen atoms are different (2.15 and 2.03 Å).[272] In contrast $Mg(dpp)_2·2DMF$ is *cis* octahedral with a similar difference in bond lengths between the $acac^-$ and DMF oxygen to sodium bonds (2.06 and 2.10 Å).[273] $Ca(acac)_2·3H_2O$ is a discrete molecule having *cis* octahedral geometry with the

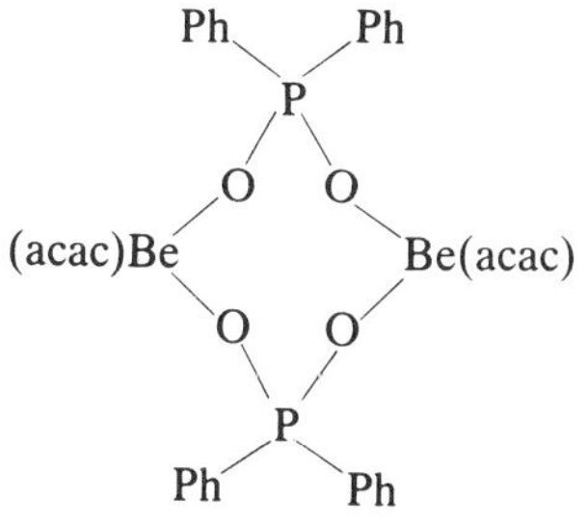

(48)

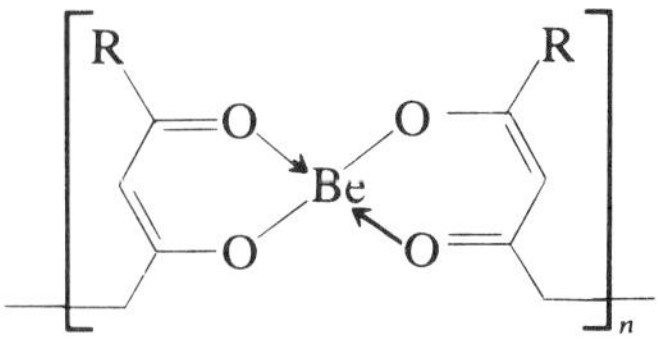

(49)

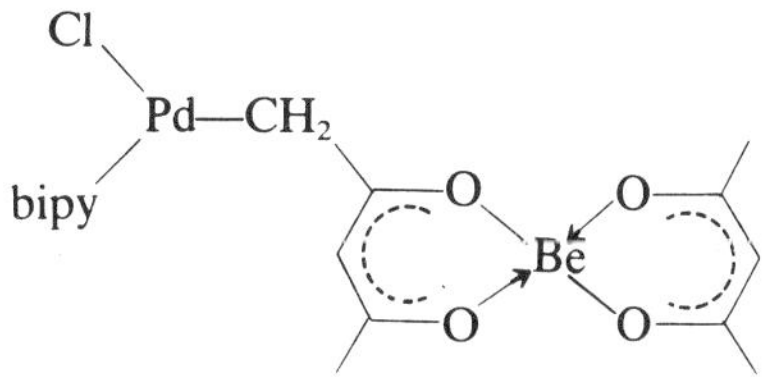

(50)

third water molecule non-coordinating; here the bond lengths become comparable: Ca—O (acac) = 2.32 Å, and Ca—O(H_2O) = 2.36 Å.[274] Ca(dpp)$_2$·0.5EtOH is a tetrameric polymer in which there are two calcium environments, one six- and one seven-coordinate. The former site is built up of dpp oxygen donors and the latter also has present the EtOH. Each Ca^{2+} shares two oxygens with each of two neighbouring Ca^{2+} atoms of the other kind such that four are linked around a centre of symmetry (Figure 10.).[275] Sr(dpp)$_2$·0.5acetone is also a tetrameric polymer having six- and seven-coordinate strontium present.[276] Unusually, perhaps, the acetone is preferred to ethanol which was the bulk medium of recrystallization.

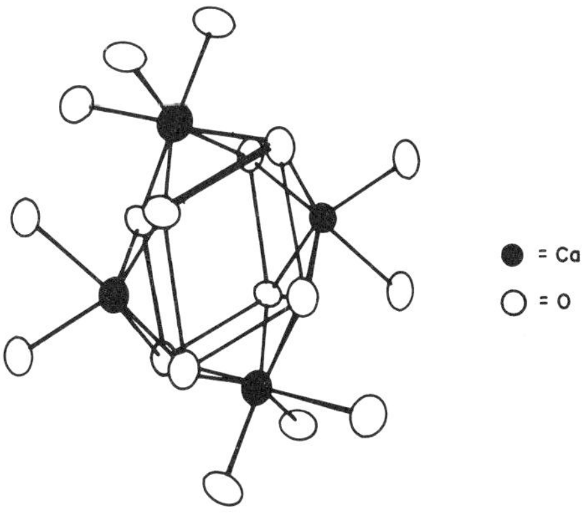

Figure 10 The calcium environment in Ca(ddp)$_2$O·5EtOH (reproduced with permission from reference 275)

^{1}H NMR studies have been carried out on BeL$_2$, MgL$_2$, CaL$_2$, and BaL$_2$ (L = acac, bzac), and on Mg(dpm)$_2$.[277] These show Mg(acac)$_2$, Ca(acac)$_2$ and Ba(acac)$_2$ to be polymeric in solution, but the others are monomeric and tetrahedral. The polymeric nature of M(acac)$_2$ has also been detected in the gas phase by mass spectrometry.[278]

The reaction of neutral donor ligands with metal β-diketonates to effect coordinative saturation has been commented on in Section 23.3.1. If an enolate ion is added to a neutral metal β-diketonate, then an anionic complex results.[165,279,280] Two types of complex have been reported: (i) $(M^+)_m[M^+(hfac)_n^-]$ ($m = 1$ or 2; $n = 2$ or 3) and $M^+[M^{2+}(\beta\text{-diketonate})_3]^-$, and (ii) $TMNDH^+(M^{n+}(hfac)_m^-)$ ($m = 2$, 3 or 4). In the mixed metal complexes, the metal having highest charge density attracts the enolate anion. This is evidenced in the crystal structure of Rb$_2$[Na(hfac)$_3$], which is isomorphous with Cs$_2$[Na(hfac)$_3$].[280] The lattice contains $[Na(hfac)_3]^-$ anions in which the Na^+ is six-coordinated to the enolate anions. The Rb^+ cations interact with both O and F atoms on the top of the trigonal prism, giving a rare example of an alkali cation coordinated by a halogen. Exchange reactions between alkali metal β-diketonates within the ion source of a mass spectrometer also gave species such as $M^1M^2L_2^+$ and $M_2L_2^+$.[281]

For the second group of complexes, the structure of $TMNDH^+[Mg(hfac)_3]^-$ ($TMNDH^+$ = monoprotonated 1,8-bis(dimethylamino)naphthalene) is determined and shows the Mg^{2+} octahedrally coordinated by the anions.[282] It is likely that the structure of NaMg(acac)$_3$ is also of this type.[283]

Neutral diketone complexes are not generally available and so it is of interest to find Mg(ClO_4)$_2$·2acacH·2H_2O. The keto form of the ligand has been shown to predominate in the complex in solution, by use of ^{1}H NMR, and although exchange between free and complexed ligand is fast, the keto–enol tautomerization is slow.[284]

Complexes of related species such as keto acids, keto esters, and acetoacetanilides are known, particularly for beryllium.[93,270,285,286,287] Bromination and nitration studies have been carried out on a wide range of *N*-substituted *N*-acetoacetamides of beryllium and show introduction of bromo and nitro groups at the 3-position.[285,286]

Heterobinuclear complexes containing alkali or alkaline earth metal cations have been derived from mononuclear transition metal complexes of compartmental ligands (**51**).[288] The molar conductivities of the (**51**)–$CuLi_2$ series suggest that the complexes are uniunivalent electrolytes in water and so are present in solution as Li[(**51b**)CuLi]·$n$$H_2O$; the corresponding di-sodium, -potassium or -cesium complexes are unibivalent electrolytes and so likely to be present in solution as M_2[(**51**)Cu]·$n$$H_2O$.

For the alkaline earth metals, only the Mg^{2+} complexes were soluble enough to allow a conductivity determination in methanol. A non-electrolyte was indicated and suggested that the species was dinuclear with Mg^{2+} in the outer cavity (**52**). This is confirmed by the X-ray structure[289] which shows two *trans* water molecules helping to achieve coordinative saturation.

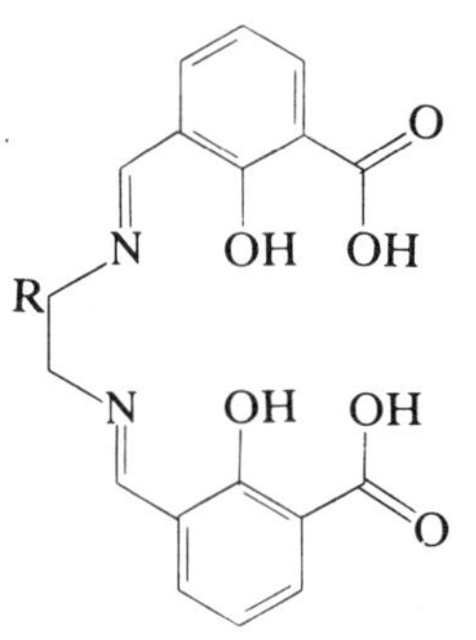

(**51**) **a**; R = H
b; R = Me

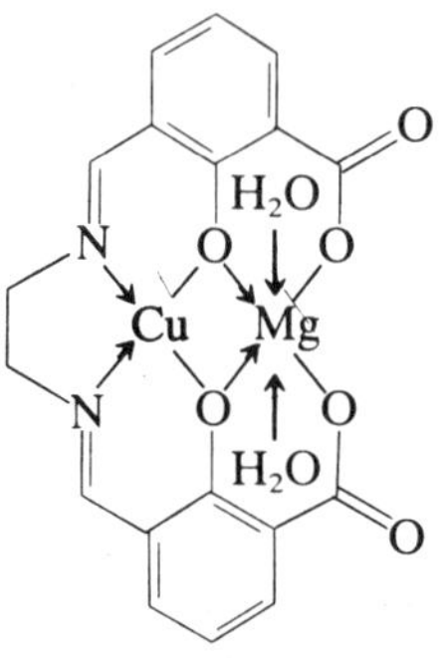

(**52**)

There is a considerable colour change on adding M^+ or M^{2+} to the copper precursors; CuLi and CuMg complexes are red, and as the size of the added metal ion increases, there is a steady reversion to the violet colour of the mononuclear copper(II) complex. This can also be traced by electronic spectroscopy and is viewed as a steady movement away from occupancy of the outer compartment by the second metal.[288]

The purpurate anion (**53**) has a mononegative charge delocalized over more than one donor atom.[30] The kinetics of its complexation with alkali metal cations have been reviewed.[290] The structures of several purpurates have been solved and the ligand consistently acts as a tridentate ligand. However, a variety of bonding modes are exhibited with the cation and polymeric structures resulting.[30]

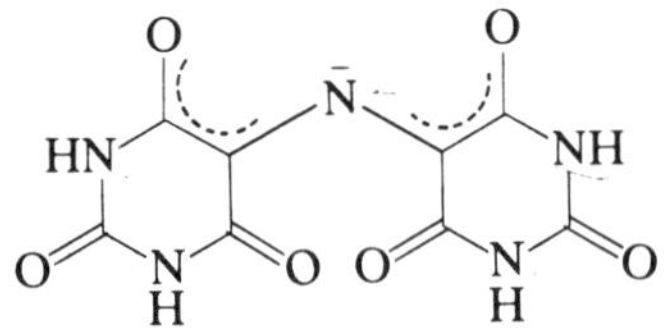

(**53**)

23.4.1.1 *Schiff base complexes*

In contrast to the extensive coverage of transition metal Schiff base complexes, there is a somewhat desultory presentation of alkali and alkaline earth metal complexes of these ligands.

Bis(*N*-alkylsalicylaldiminato)beryllium complexes (**54**) have been prepared and their physico-chemical properties studied. Dipole moment[291,292] and ^{1}H NMR studies[393] are interpreted in terms of a tetrahedral environment for the beryllium. ^{1}H NMR studies on the magnesium and calcium analogues are similarly interpreted. The free energy of enantiomerization has been determined for (**54**; R = isopropyl) using NMR techniques and without needing resolution of the optical antipodes;[294] the inversion has been proposed as proceeding *via* a dimerization step.[295] The fluorescent properties of a range of beryllium chelates have been monitored with a view to developing reagents for the determination of beryllium.[296] The complexes are formulated as having 1:1 stoichiometry and the ligands are of the general type (**55**).

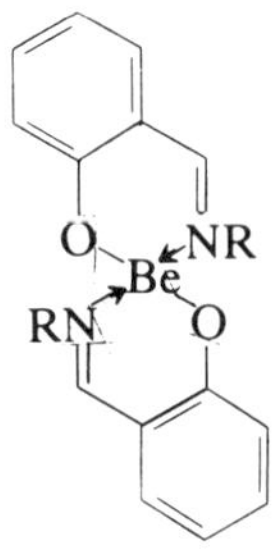

(**54**)

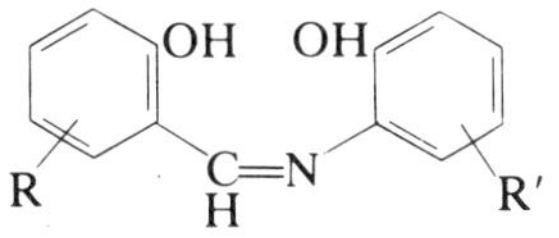

(**55**)

The structure of bis(salicylaldoxime)beryllium has been proposed as being *trans* octahedral by comparison of the space group and unit cell volume with those of related transition metal complexes; it is presumably a dihydrate if it is indeed octahedral in geometry.[297] Stability constants have been reported for a range of beryllium β-ketoamines derived from both salicylaldehyde and acetylacetone precursors. They show strong complexes which are stable to hydrolysis under the conditions used.[298,299]

Salen complexes of beryllium have been reported. The reaction of $BeSO_4$ and H_2salen in alcohol, followed by addition of aqueous alkali to effect precipitation, gave Be(salen)$_2 \cdot H_2O$ in which it was proposed, from IR studies, that each salen was monoanionic and bidentate.[300] Be(salen)$\cdot 2H_2O$ was isolated by addition of salen to an ammoniacal solution of beryllium containing sodium potassium tartrate,[301] and polymeric Be(salen) has been prepared from the reaction of salen and beryllium sulfate (or nitrate) at pH 7.[302] This latter compound has been characterized by IR, ^{1}H NMR and mass spectrometry. A related complex derived from (±)-2,2′-bis(salicylideneamino)-6,6′-dimethyldiphenyl contains beryllium in a tetrahedral environment.[303] Schiff base complexes of Be and Mg, the former four coordinated and the latter six coordinated, have been prepared by the reaction of the metal acetylacetonate with biguanide, and with *N*′-amidinoisourea.[304] The reaction of diacetylacetone with beryllium salts in the presence of ammonia gave the unexpected beryllium complex (**56**) instead of the anticipated diacetylacetonate.[305]

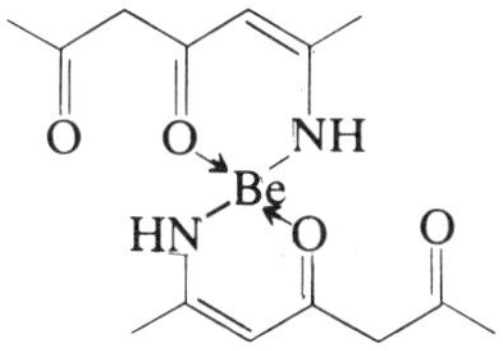

(**56**)

Mg(salen), first reported by Pfeiffer[306] who synthesized it using the classic Schiff procedure of reaction of the metal salicylate with ethylenediamine, has been investigated as a possible reagent for the detection of trace amounts of magnesium. Both fluorimetric and spectrophotometric techniques have been used;[307,308] no structural information is yet available.

Generally, Schiff bases, such as H_2salen, containing phenolic groups can form complexes either by removal of the two phenolic protons to give a neutral metal complex, or by direct ligation as a mono- or non-deprotonated ligand. The latter modes are generally found for A-class metals. In the above discussion fully, and singly, deprotonated H_2salen have been introduced; in the reaction of calcium salts with H_2salen, the non-deprotonated ligand is in evidence. Two types of compound are found: $Ca(H_2L)X_2 \cdot nEt_2O$ ($X = NO_3$, Cl; $n = 0, 2$) and $Ca(H_2L)_2X_2 \cdot nS$ ($X = NO_3$, Cl; $n = 0, 1$; $S = H_2O$, EtOH).[309] The ligands used are H_2salen, H_2salpd, H_2salpn and H_2salhxn (pd = 1,3-propanediamine, pn = 1,2-propanediamine and hxn = 1,6-hexanediamine). The X-ray structure of $Ca(H_2salpd)(NO_3)_2$ has been solved and shows the ligand to be binding to two Ca^{2+} ions *via* two negatively charged oxygen atoms. The azomethine nitrogens are protonated by the proton from the phenolic group and are intramolecularly hydrogen bonded to the oxygen atoms. A bidentate nitrate anion, and a pair of bridging, equivalent nitrate anions complete the eight coordination around the cation.

Several alkaline earth metal complexes of polydentate, acyclic Schiff bases have been prepared and characterized during studies on the generation of tetraimine Schiff base macrocycles. $Ba(ClO_4)_2$ complexes of the ligands **(57)**,[310] **(58)**,[311] **(59)**[312] and **(60)**[313] have been isolated together with $Sr(ClO_4)_2$ complexes of **(58)**, **(59)** and **(60)** and $Mg(ClO_4)_2$ and $Ca(ClO_4)_2$ complexes of **(59)**.

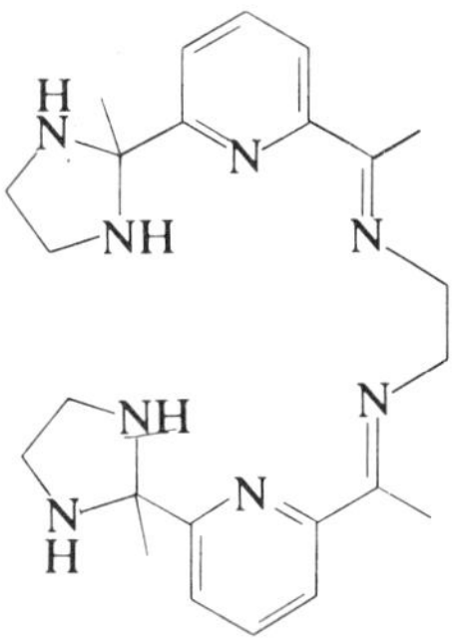

(57)

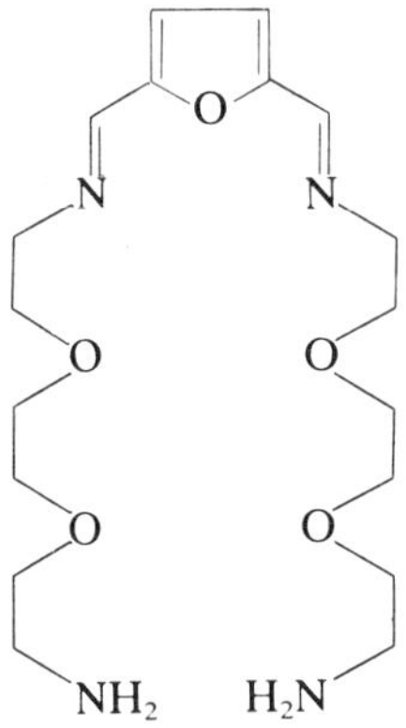

(58)

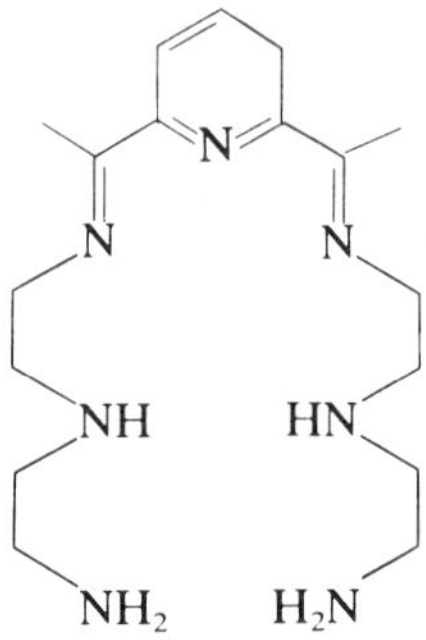

(59)

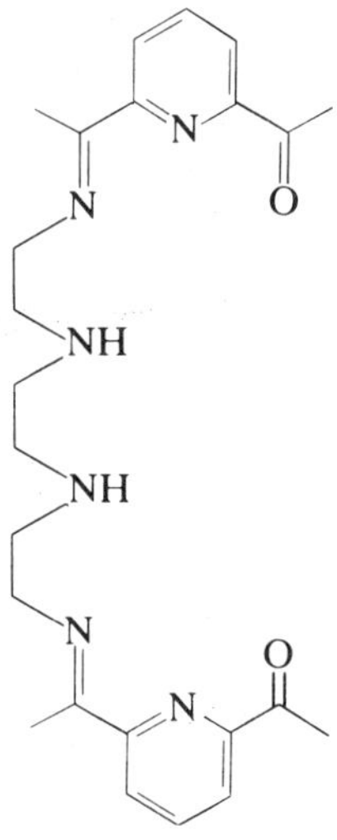

(60)

Such complexes have been postulated as possible intermediates in macrocycle formation and whilst **(58)** and **(59)** can be ring-closed, **(60)** cannot. The crystal structure of **(60)**$Ba(ClO_4)_2$ has been solved and shows the barium to be 10-coordinated by all of the ring donor atoms and two perchlorate anions. The conformation of the ligand is such that two planar keto-pyridinylimine segments are inclined at 80.2° to each other forming a cleft. The failure of the ring to close on further diamine addition is likely to be due to a steric barrier to the entry of the nucleophile into the cleft.[313]

23.4.2 Acid Anions and Acid Salts

In discussing the interaction of chelating anions with alkali metal cations, Truter argues that one possible criterion for designating a compound 'complex', namely that the ligand is held exclusively by one metal, is too restrictive because it excludes simple μ-complexes.[35] The proposal is also made that it is reasonable to state that if all anions contact more than three cations, then the solid does not contain complexed cations, as it begins to move towards an extended lattice.[35] The crystal structure of the mixed ligand species K(*o*-nitrophenolate)(iso-nitrosoacetophenone) is used as an illustration. In this compound the anion is present as the acid anion $[LHL']^-$ and the potassium is coordinated by four donor atoms from this anion **(61)**.[314] This does not provide coordinative saturation at the metal and so in the solid state polymerization brings the coordination number up to seven (if M—O < 3.0 Å) or eight (if M—O include 3.1 Å interactions). Each anion becomes involved in interactions with four cations, some donor atoms being shared.

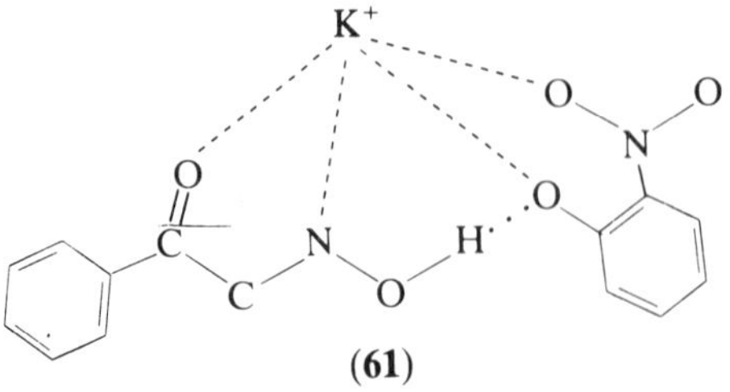

(61)

Compounds similar to **(61)** and containing *o*-nitrophenol, acetylacetone, salicylaldehyde, *etc.* as neutral ligands were described by early workers in the area and considered as complexes of

ML·HL and ML·HL′, where HL and HL′ acted as chelated ligands.[17,18,19] They were referred to as 'acid salts' and their solubility in organic media was taken as a criterion of complexation. In historic terms they play an important role in the development of the area. Generally, these compounds concerned organic anions, and $KIO_3 \cdot HIO_3$ is a unique example of an inorganic 'acid salt'.[315] With the later generation of interest in alkali metal complexation stimulated by a desire to understand membrane transport phenomena and to establish desalinating agents, a large range of 'acid salts' were investigated and the principles of formation have been discussed by Poonia.[21,30] Speakman has reviewed the structures of 'acid salts' of carboxylic acids[316] and uses a delightful quotation from Flecker, 'we have . . . broideries of intricate design', as a preface. This is indeed appropriate as the structures are often infinite, and intricate, networks moving away from discrete molecules, and, although the metal is often chelated, it is considered here that the multiple cation sharing by the anions leads to a complex compound rather than a coordination complex. This is a further extension of the progression seen earlier in β-diketonate chemistry where more intricate polymeric lattices were found on moving to the heavier alkali and alkaline earth metal cations.

23.4.2.1 Basic beryllium carboxylates

An unusual series of beryllium complexes is the basic carboxylates, $Be_4O(RCO_2)_6$, the best known member of which is the acetate, first prepared by Urbain and Lacombe.[317] The principal synthetic route is by reaction of the organic acid or acid anhydride with beryllium oxide or hydroxide;[318] this method does not work for the basic formate. Instead reaction of the organic acid with the basic carbonate will give this compound,[319] as will hcating the normal formate to 250–260 °C.[320] The interaction of an organic acid or acid chloride with another basic carboxylate yields mixed carboxylato species;[93] for example, the range of complexes $Be_4O(OAc)_x(OPr)_{6-x}$ can be so prepared.[321]

The white crystalline solids are readily soluble in organic solvents (even alkanes) but are insoluble in water and the lower alcohols. They are inert to water but are hydrolyzed by dilute acid. In solution they are monomeric and non-ionized. An X-ray crystallographic study of $Be_4O(OAc)_6$ shows the O to be centrally coordinated to four Be atoms in a tetrahedral geometry; each of the Be atoms is tetrahedrally surrounded by four O donors from bridging acetates and the central oxygen (**62**).[322]

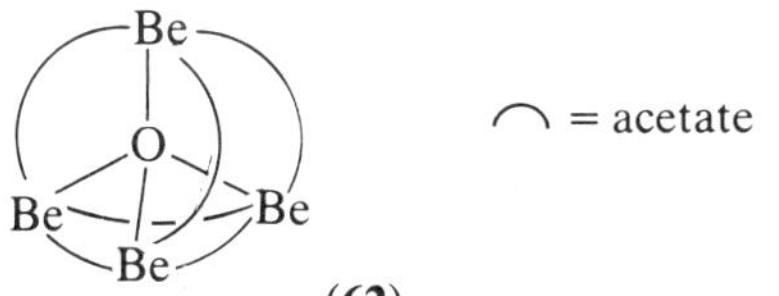

(**62**)

Ammonia and amines form compounds with basic beryllium carboxylates;[323,324] liquid NH_3 gives $Be_4O(OAc)_6 \cdot 12NH_3$ at −33 °C, which then loses ammonia to give $Be_4O(OAc)_6 \cdot nNH_3$ ($n = 3$ or 4). These species lose NH_3 and give the basic acetate at 180 °C. Similar compounds are formed with amines and a study of this reaction suggests that it follows the pathway:[325]

$$Be_4O(OAc)_6 + RNH_2 \rightarrow Be_3O(OAc)_4 \cdot nRNH_2 + Be(OAc)_2RNH_2$$

Thermal decomposition of the tripropylamine or $BuNH_2$ complex at 100 °C gave $Be_3O(OAc)_4$ which was then recrystallized from petroleum ether to give two crystalline forms of a dimeric structure.[325] SO_2 and N_2O_4 have been found to give loose addition products with $Be_4O(OAc)_6$.[326]

The reaction of alcohols with $Be_4O(OAc)_6$ gives a range of novel, but not enirely characterized, products, *e.g.* Be(OR)(OAc), Be(OH)(OAc) and oligomers based on these structures.[93] Be(OEt)(OAc) sublimes at 200 °C to give a reversal of the reaction and formation of $Be_4O(OAc)_5(OEt)$.[328] The related compound $Be_4O(OAc)_5(OH)$ has been prepared from the reaction of $Be_5O_2(OAc)_6$ with water.[327]

NMR studies of the mixed carboxylates containing acetates and trichloroacetates have indicated that ligand scrambling reactions occur in solution; although all of the species were isolated, and isomerization possibilities occur, no discrete isomers were identified.[329] Gas chromatographic studies have shown that the complete series of mixed basic carboxylates generated through ligand distribution in a mixture of the basic acetate and basic propionate can be found and separated clearly.[330] Extraction studies using capric acid have shown that beryllium can be efficiently extracted solely as the basic carboxylate.[331]

An inorganic analogue of these compounds exists in basic beryllium nitrate. This was prepared by heating $Be(NO_3)_2$ to *ca.* 125 °C where a sudden decomposition occurs to give $Be_4O(NO_3)_6$.[332] The structure of this compound is similar to that of the basic acetate;[333] and electron impact studies have confirmed the structural similarity of the two types of basic compound.[334] More complex basic acetates have also been reported. Pyridine-2-carboxylic acid provides a trimeric ring system, the strucure of which has been confirmed by X-ray crystallography (**63**).[335]

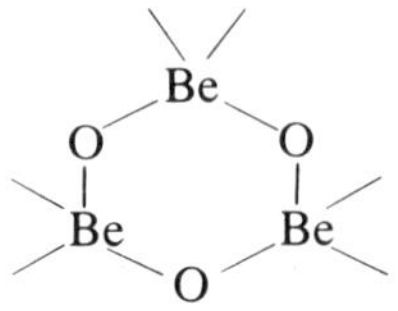

(**63**)

Rings containing beryllium, and derived from alkoxides and amides have been reviewed.[100,336] $Be_6O_2(OAc)_8$ has been shown to be comprised of dimeric $Be_3O(OAc)_4$ units.[325] The X-ray structure shows a polymeric network in which BeO_4 tetrahedra share common edges.[337] This compound forms adducts in liquid NH_3 of the type $Be_6O_2(OAc)_8{\cdot}6NH_3$.[338] Thermolysis of $Be_6O_2(OAc)_8$ yields $Be_4O(OAc)_6$ and $Be_5O_2(OAc)_6$, the X-ray structure of which is illustrated in Figure 11.[339]

Oligomeric beryllium oxycarboxylates containing up to Be_7, Be_9 and Be_{11} fragments have also been obtained by the controlled thermolysis of the products of the reaction of $Be_4O(OAc)_6$ and EtOH.[340]

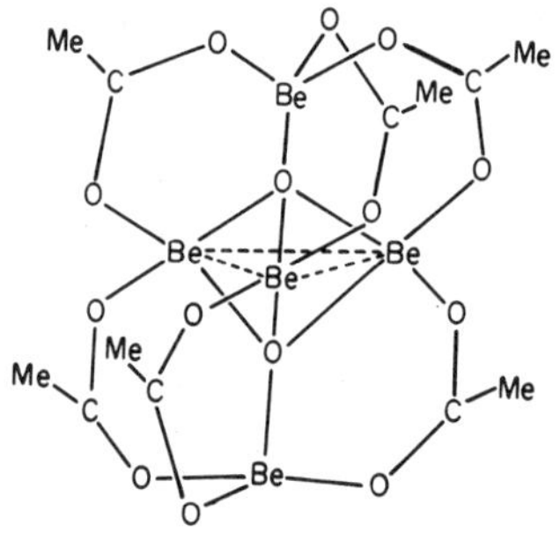

Figure 11 Representation of the molecular structure of $Be_5O_2(OAc)_6$

23.4.3 Polycarboxylic and Phosphonic Acids

An extensive range of functionally substituted polycarboxylic acids have been studied as metal-binding reagents (*e.g.* **64–66**).[341,342,343] Many such organic sequestering agents have been claimed as potential phosphate substitutes in detergent formulations, for ecological reasons; nta, because of its strong capability to sequester Ca^{2+} has seen application in this area.[344] These ligands have also seen application as analytical reagents, and the ability of edta to form water-soluble complexes with alkaline earth metal cations is the basis of the determination of hardness in water caused by these two cations. This was one of the first applications of edta to be described.[345] 1,2-Diaminocyclohexanetetraacetic acid also forms complexes with calcium and although these are stronger than the corresponding edta complexes, they are formed at a lower rate and so are not as useful for analytical purposes.

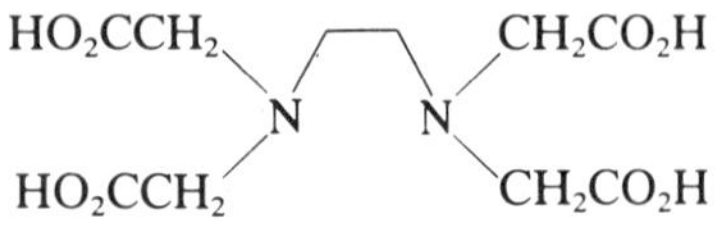

edta
(**64**)

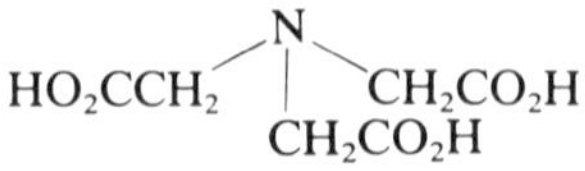

nta
(**65**)

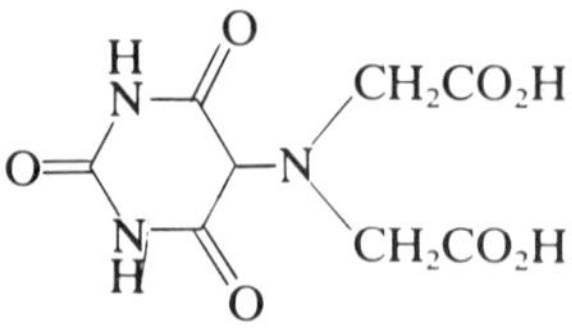

uda
(**66**)

The aminopolycarboxylic acids in particular have been extensively studied in solution;[341,342,346] tabulations of stability constants concerning the interaction with M^+ and M^{2+},[259,341,342] and reviews concerning solid state structures[347,348] and the relationships between solution and solid state structures[349] are available. The stability constant data show that in general $Ca^{2+} > Mg^{2+} \approx Sr^{2+} > Ba^{2+}$ and that group $II^A >$ group I^A. For edta the Mg^{2+} complex is lower in stability than the Ca^{2+} complex owing to the more endothermic nature of the former reaction. A strainless octahedral complex is approached for the Ca^{2+} complex in solution but the presence of a smaller cation results in a strained system being set up. Such strains would be expected to be greater for Be^{2+} and the log K value of 3.8, relative to 8.69 for Mg^{2+} and 10.96 for Ca^{2+}, validates this. In group IA the order of stability is generally $Li^+ > Na^+ > K^+ > Rb^+ > Cs^+$, and so follows the ionic potential decrease.

Beryllium is often left out of stability sequence studies; however, in a comparison of the complexing properties of edta, nta and uda, it was found that the Be^{2+} and Mg^{2+} complexes of uda were more stable than those formed with the higher alkaline earth cations.[350] This is in contrast to the edta and nta sequences. In the analysis of the thermodynamic information, it was found that for uda the enthalpy changes were high and negative, and for alkali metals these increased with increasing ionic radius. The entropy terms were negative for the alkali cations but positive for the alkaline earth cations, decreasing with increasing ionic radius for each series. The enthalpy changes for alkaline earth cations, in edta complexation, were small but the entropy changes were positive, decreasing with increasing ionic radius. The enthalpy terms for Mg^{2+} and Be^{2+} were positive and so were unfavourable. The Be^{2+} was predicted to be tetrahedral in these complexes.[350]

In solution, the potential for binding is six-coordinate for edta and four-coordinate for nta. The total coordination, however, is not well known and water molecules are also involved. $[Ca(edta)]^{2-}$ is often represented as an octahedral complex but it would be anticipated that Ca^{2+} could have a coordination number in excess of six. The denticity of edta has been found to vary from one to six, and the coordination number about the metal from four to nine.[349] A review of the crystal structures of Ca^{2+} complexes of amino acids, peptides and related model systems surveys the coordination patterns found for Ca^{2+}–carboxyl and Ca^{2+}–carbonyl interactions.[351]

In the solid state, the story is quite complicated. The structures of several alkali and alkaline earth metal complexes of acids are known and include those of 2,6-pyridinedicarboxylic acid (H_2dpa), 4*H*-pyran-2,6-dicarboxylate (H_2dpc), oxydiacetic acid (H_2oda), thiodiacetic acid (H_2tda), iminodiacetic acid (H_2ida), ethylenedioxydiacetic acid (H_2edoda), malonic acid, maleic acid, fumaric acid, succinic acid and citric acid, as well as those of edta and nta. These have been reviewed and show a wide variety of structures.[30] These range from $[Mg(H_2O)_6]^{2+}[H_2edta]^{2-}$, which may be crystallized from acid solution,[352] through $Ca(oda)\cdot 6H_2O$, which consists of discrete $Ca(oda)\cdot 5H_2O$ units in which the Ca^{2+} is coordinated by the tridentate ligand and five water molecules and in which there is linking caused by hydrogen bonding involving the sixth non-coordinated water molecule,[353] and $K_2[Be(oxalate)_2]$ and $K_2[Be(malonate)_2]$ which are discrete dianionic beryllium complexes of tetrahedral geometry,[354,355] to the $\{Ca(edoda)\cdot 3H_2O\}_2$ dimer,[356] in which the species are linked together by hydrogen bonding involving the water molecules and eventually to the intricate networks of $Ca_2(edta)\cdot 7H_2O$ and $NaCa(nta)$[357] in which no discrete monomers are detected.

In studies on $Ca(edoda)\cdot 3H_2O$ it was noted that the ethereal oxygen was involved in binding and that there was no effective increase of Ca^{2+} sequestration on going from oda to edoda.[356] The suggestion that the ether oxygen could be involved in complexation was implied from the stability sequence Ca(oda) (log K, 3.4) > Ca(nda) (log K, 2.7) > Ca(glutamate) (log K, 1.1).

In accord with remarks made earlier concerning the nature of complex species containing acid anions (Section 23.4.2) discussion of complexes containing polycarboxylic species has been limited to the general outline above. Recent developments in the area include the derivation of ligands such as (**67**)–(**69**) and their application to complexation. The sequence $Ca^{2+} > Mg^{2+} > Ba^{2+} > Sr^{2+}$ is found for (**67**) and (**68**) with the complexation to (**67**) being stronger, and showing lower complexing powers than edta and nta.[358] For the macrocyclic polycarboxylic acids (**69a**) exhibits the strongest known complexing ability for Ca^{2+}, and the larger (**69b**) shows a noticeable preference for Sr^{2+}.[359]

Polyamino polyalkyl phosphonic acids have been investigated as potential complexors for alkaline earth metal complexes. Ligands investigated include nitrilotrimethylphosphonates (**70**),[360] phosphoryl derivatives of polyamines (**71**)[361] and monoalkylphosphonic acids.[362] They show complexing capability towards Be^{2+}, Mg^{2+}, Ca^{2+}, Sr^{2+} and Ba^{2+}, but are not as efficient

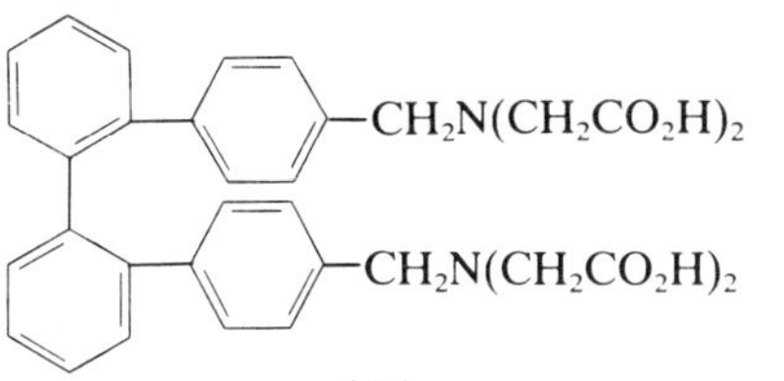

(67)

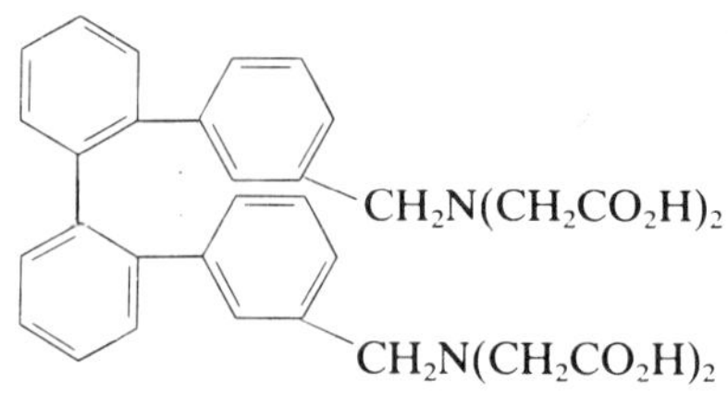

(68)

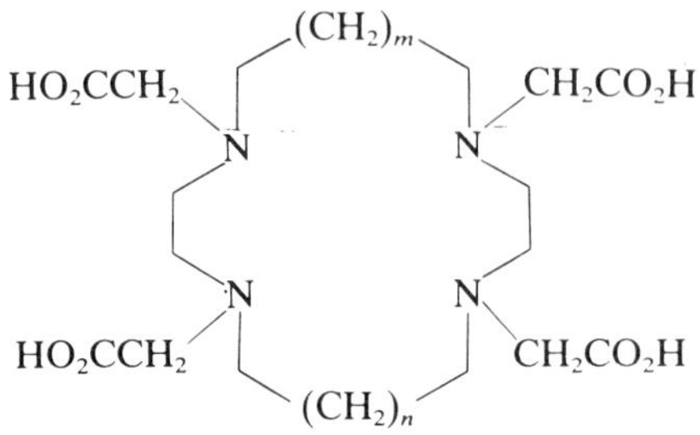

(69) **a**; $n = m = 2$
b; $n = 2,\ m = 3$
c; $n = m = 3$

as the corresponding carboxylic acids. Suggestions have been made concerning the likely nature of species by use of IR techniques but no precise structural statement is available.

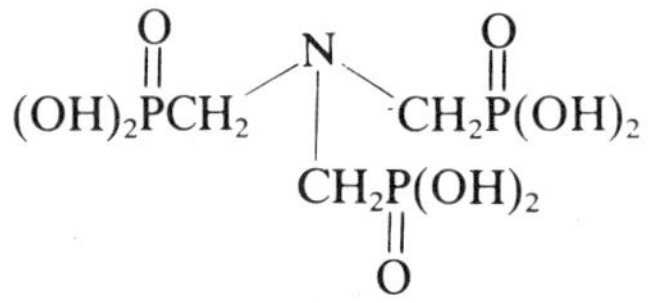

(70)

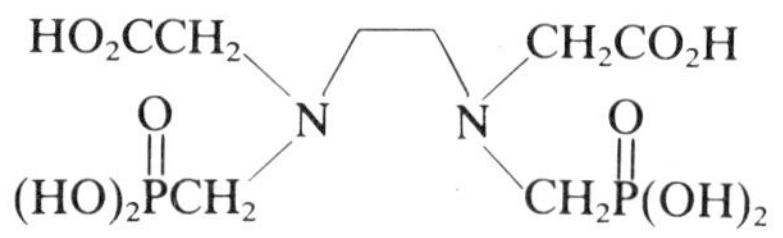

(71)

Alkali and alkaline earth metal complexes of sugar phosphates and related moieties belonging to nucleosides and nucleotides have been studied, and accounts of the nature of the metal–ligand interaction[363,364] and of the stability of the complexes are available.[365,366] Bidentate and tridentate interactions with the cations occur and generally the structures are networks having multiple cation–anion interactions. The structure of $Na_2ATP\cdot3H_2O$ (ATP = adenosine triphosphate) is taken as an example.[367] The structure has two formula units in the asymmetric unit and can be viewed as being dimeric in origin. There are two distinct types of

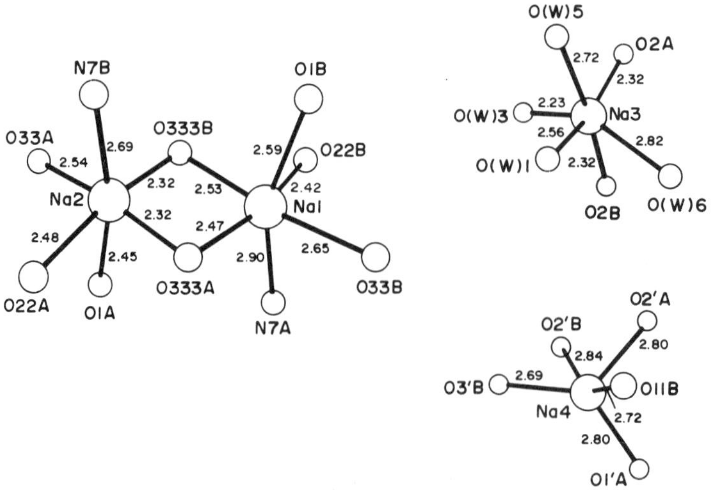

Figure 12 The coordination of the sodium atoms in $Na_2ATP\cdot3H_2O$ (reproduced with permission from reference 367)

sodium coordination; in the first two independent sodium cations Na1 and Na2 are octahedrally coordinated. This is achieved by interaction with two ATP molecules, one of type A and one of type B. For Na1 the coordinating atoms are three phosphate oxygens (O_1, O_{22} and O_{333}) from a molecule B, a phosphate oxygen (O_{33}) from a second molecule of B, the N7 atom and a phosphate oxygen (O_{333}) of molecule A. Na2 is coordinated with the corresponding O_1, O_{22} and O_{333} atoms from molecule A, the O_{33} from a second molecule of A, and the N7 and O_{333} of molecule B. One of the terminal atoms O_{333} in each molecule is coordinated to both sodium cations, and this, together with the N7 coordination, leads to the formation of dimers, which then repeat through the structure. The second type of sodium coordination involving Na3 and Na4 plays a less significant role in the structure of ATP; its main function is to link the dimers and maintain them in the extended lattice. These sodium atoms are six- and five-coordinated respectively; the first to four waters and two O2 from types A and B, the second to a selection of phosphate O donors (Figure 12). The role of the alkali metal cation can be viewed as that of a network assembler.

23.5 MACROCYCLIC LIGANDS

Pedersen's discovery[5] that dibenzo-18-crown-6 (**72**) and a related family of cyclic polyethers could readily complex alkali metal cations provided the stimulus required for a renaissance of interest in alkali metal coordination chemistry. Since that discovery, much effort has been directed towards modifying all possible structural parameters within the ligand system in order to make new, and hopefully improved, ligands. Variable parameters include the number and type of heteroatoms available for coordination, the nature and length of bridging units within the polyether and the general nature of any substituent groups present. This general class of ligands was first termed crown ethers,[5*] and more recently the term coronand has been introduced.[33]

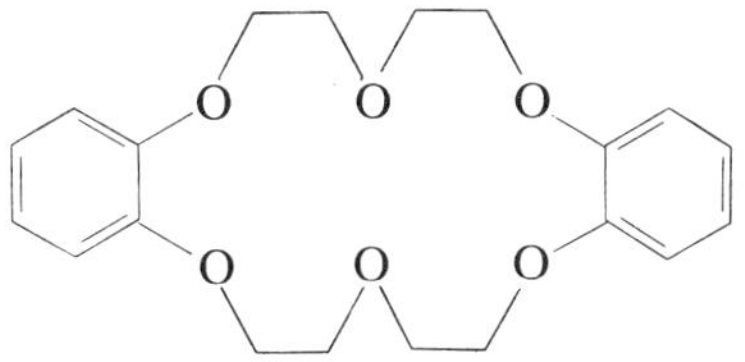

(**72**)

A second important advance was made with the availability of cryptands (**73**), a family of bicyclic polyoxadiamines which have available a three-dimensional cavity for complexation.[6] Again many structural modifications are possible and lead to numerous macropolycyclic ethers.

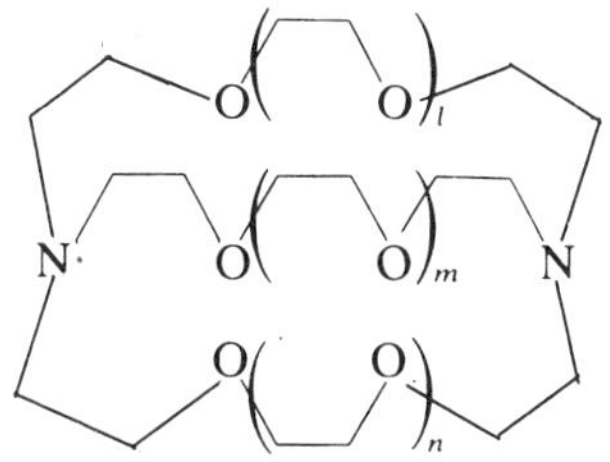

(**73**)

Selected examples of crown ethers and coronands are given in Figure 13. This area of chemistry has been remarkable for its growth, and many excellent reviews covering a wide range of topics are available; that by Vögtle and Weber[31] carries a full bibliography of work carried out until 1979. Subsequent reviews, of interest here, have been concerned with synthesis and general properties,[368,369,370] complexes,[30] structural aspects,[30,371,372] stability and reactivity,[373,374] cryptands,[375] polymeric polyethers,[376,377] NMR studies[140] and applications to analytical chemistry.[378,379,380]

* The IUPAC nomenclature for crown ethers is cumbersome and so was trivialized for common use. Dibenzo-18-crown-6 is named as follows: dibenzo describes the non-ethyleneoxy content; 18, the total number of heteroatoms in the crown ring; and 6, the number of heteroatoms in the ring.

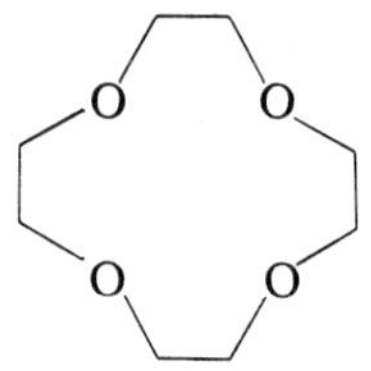

(**74**)

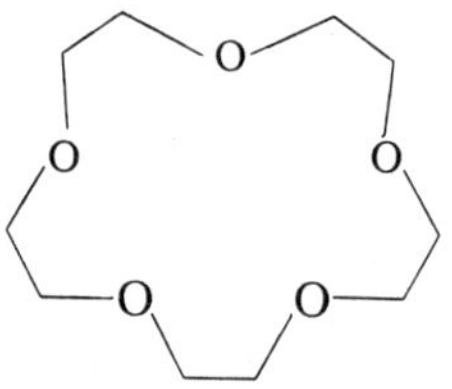

(**75**)

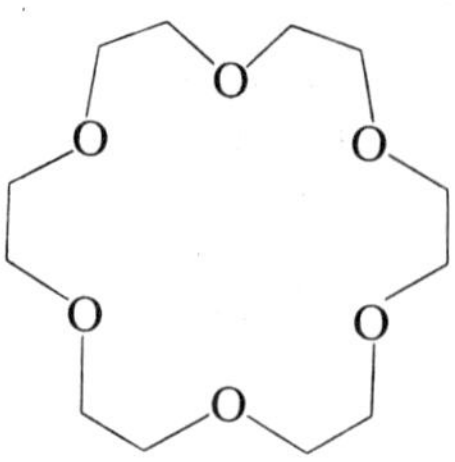

(**76**)

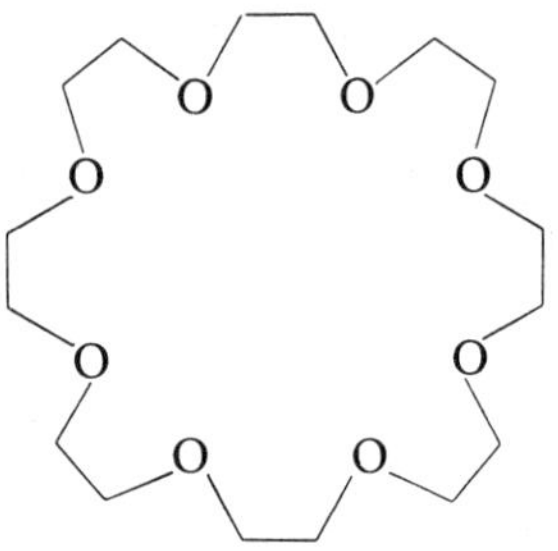

(**77**)

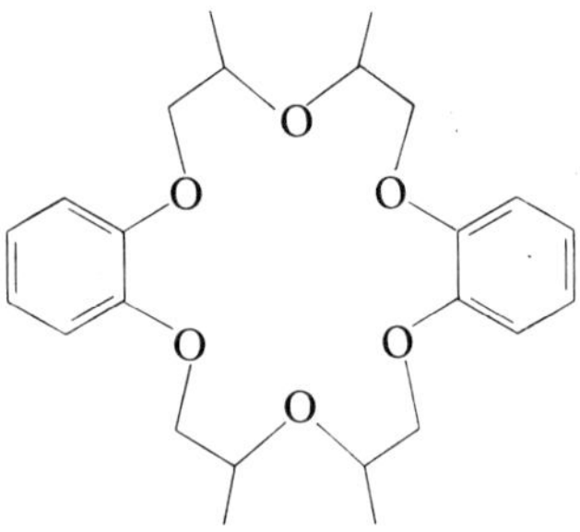

(**78**)

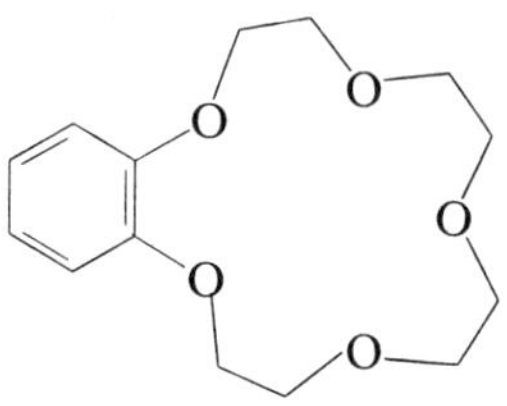

(**79**)

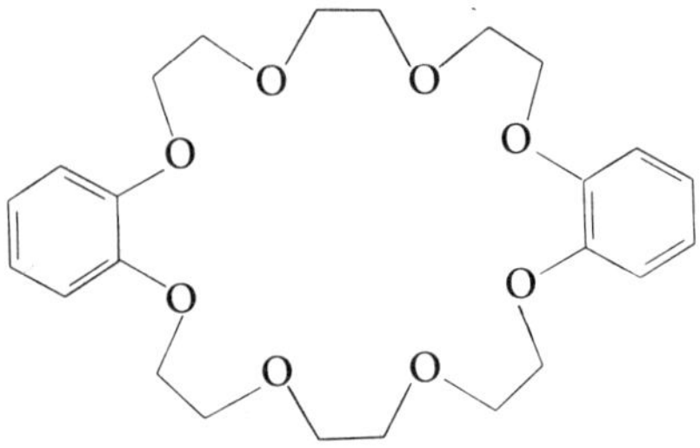

(**80**)

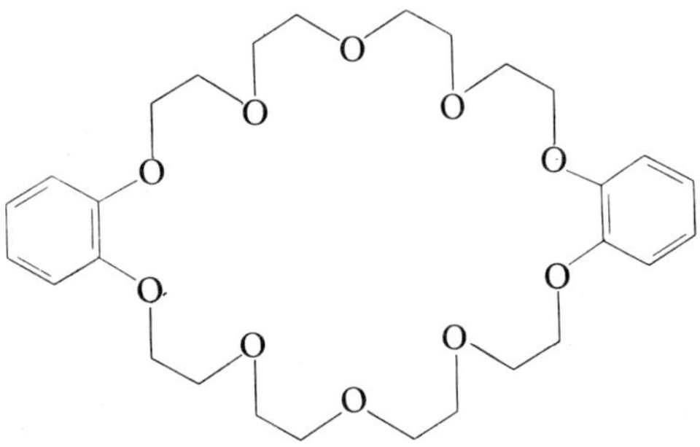

(**81**)

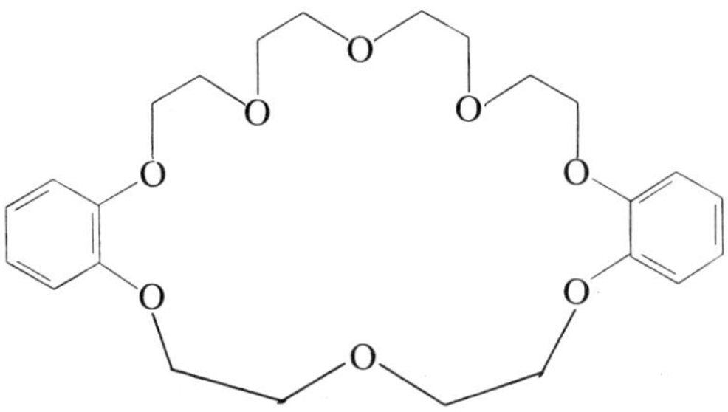

(**82**)

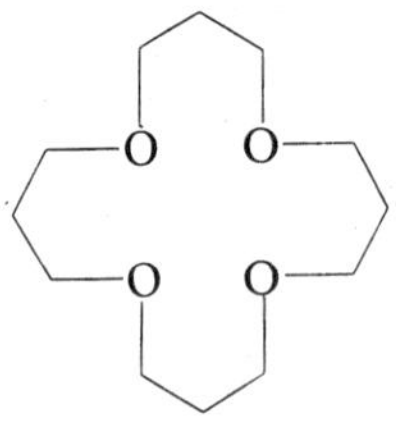

(**83**)

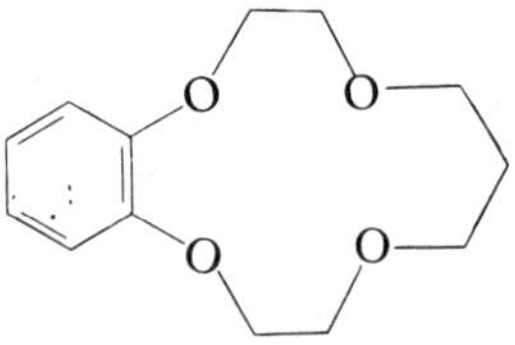

(**84**)

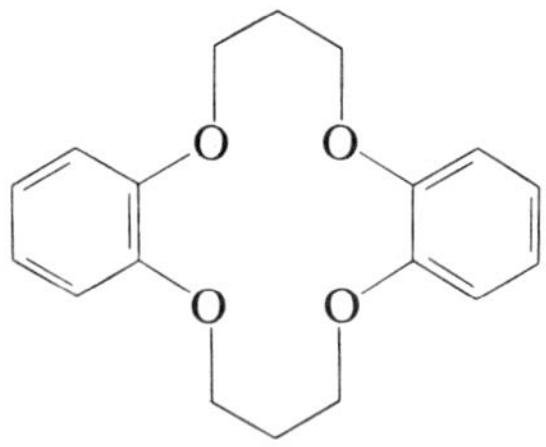

(**85**)

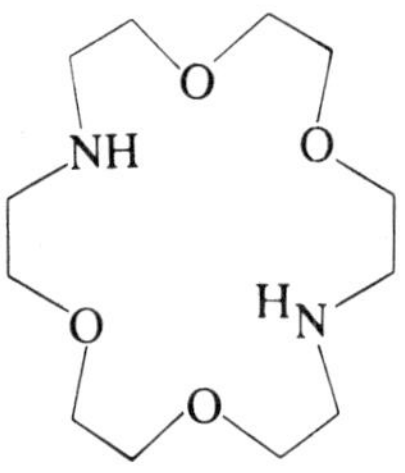

(**87**)

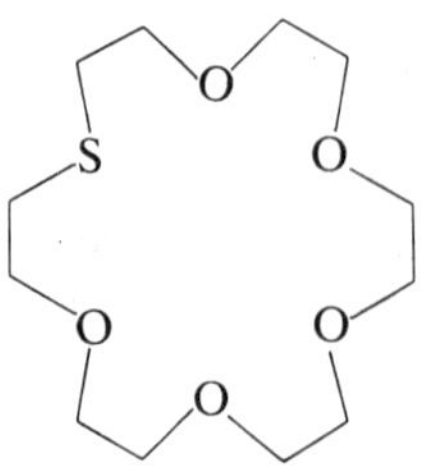

(**88**)

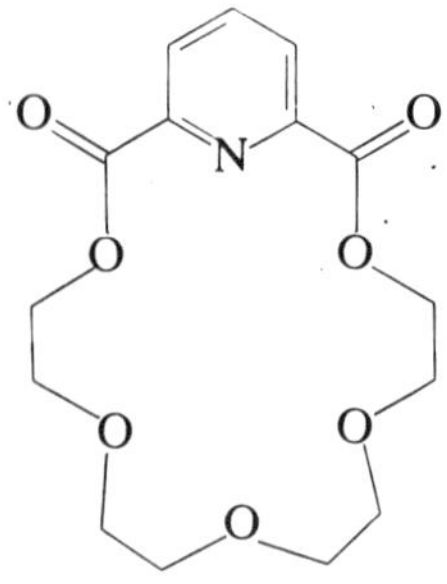

(**109**)

Figure 13 Some crown ethers and coronands

23.5.1 Crown Ethers and Coronands

Crystalline macrocyclic polyether complexes may be prepared by several routes,[5,381] depending upon the solubilities of the precursors and the products. The polyester and the salt may be dissolved in a small amount of solvent and warmed together; the complex can crystallize either on cooling or after evaporation of some of the solvent. It is also possible to warm a suspension of the ligand and the salt to effect solubilization and subsequent crystallization. The reaction can also be carried out by mixing the components in the solid state and heating to melting.

Complex formation is favoured with salts having low lattice energies,[382] and so, although alkali metal fluorides, nitrates and carbonates give complexes with polyethers in alcoholic solution, it is often difficult to isolate species as crystalline solids because the high lattice energies of the alkali salts lead to a reassembling of themselves. Well-defined crystalline crown ether complexes have been found with alkali and alkaline earth metal thiocyanates,[5,383] chlorides,[384] bromides,[5,385] iodides,[5,386,387] polyiodides,[5,387] perchlorates,[388] benzoates,[383a] nitrophenolates,[383a] tosylates,[386b] picrates,[383a,389] tetraphenylborates,[390] nitrites,[5,386a] tetrafluoroborates[301] and molybdates,[392] and representative references have been given. The X-ray crystal structures of many of these complexes have been solved. Schematic representations of several types of crown ether complexes are given in Figure 14.

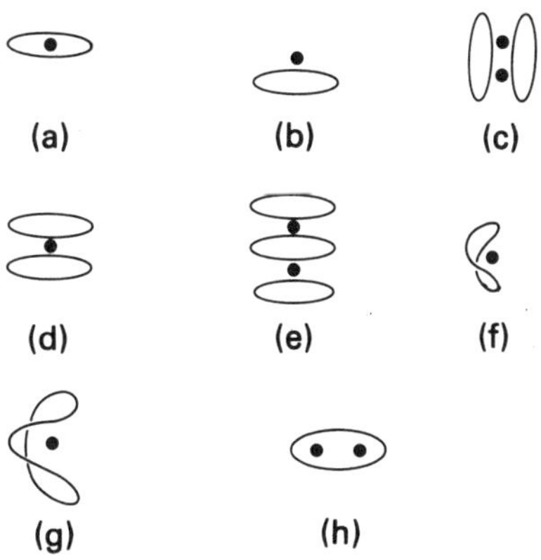

Figure 14 Schematic representations of some bonding modes found in crown ether complexes

For a 1:1 metal–ligand complex, the ideal environment for the cation is at the centre of a torus of donor atoms provided by the cyclic ligand (Figure 14a); this is because the cations, having rare-gas-like electronic structures, have no overriding directional influences on geometry, but, as in the VSEPR approach to covalent interactions, accept the donor atoms at the optimum points of distribution on the surface of a sphere subject to any imposition of steric constraint or uneven distribution of charge from accompanying ligands and anions. This latter effect can cause a shift from the central position (Figure 14b) as can a cation size–cavity size disparity. This occurs when the metal is too large for the cavity and can lead to dimerization (Figure 14c). A further consequence of the cation being too large for the cavity is the formation of 1:2 metal–ligand 'sandwich' complexes (Figure 14d), or 2:3 metal–ligand 'club sandwich' complexes (Figure 14e). If, in contrast, the cation is too small for the cavity then encapsulation of the cation may be effected either in part (Figure 14f) or in full (Figure 14g). Entrapment of two cations within the ligand torus is also possible (Figure 14h). A comparison of structures of free and complexed crown ethers shows that conformational changes occur on complexation.[30,371,372]

By studying a series of complexes, it is possible to observe the differences in structural type that occur with change of cation radius. Table 6 shows the ionic radii for the alkali and alkaline earth metal cations, together with the average ligand cavity radii for simple polyethers.[33] From this information it can be seen that the predicted optimal fit situation for 1:1 complexes would arise for Li^+ and 12-crown-4 (**74**); for Na^+ and 15-crown-5 (**75**); and for K^+ and Ba^{2+} and 18-crown-6 (**76**). For 24-crown-8 (**77**) all of the cations have smaller radii than that of the ligating cavity.

A survey of the structures of complexes derived from 18-crown-6 (**76**) reveals the presence of several of the structural types depicted in Figure 14. The Na^+ cation in NaSCN(**76**)·H_2O is coordinated by all six oxygen atoms of the ligand; five oxygen atoms lie in a plane containing

Table 6 Correlation of Cation and Cavity Radii (Å) of Alkali and Alkaline Earth Metal Ions and some Crown Ethers

Li^+	0.60		
Mg^{2+}	0.65	12-crown-4 (**74**)	0.72
Na^+	0.95	15-crown-5 (**75**)	0.92
Ca^{2+}	0.99		
Sr^{2+}	1.13		
K^+	1.33		
Ba^{2+}	1.35	18-crown-6 (**76**)	1.45
Rb^+	1.48		
Cs^+	1.69	24-crown-8 (**77**)	1.90

the cation, whilst the sixth is folded out of the plane and partially envelops the cation. The coordination environment of the metal is completed by a water molecule, and this type of complexation is a typical example of Figure 14(f) where the cation is smaller than the polyether cavity.[393]

The prediction from Table 6 is that K^+ showed the best 1:1 fit, and this is borne out in the structure of KSCN(**76**) in which the donor oxygen atoms lie in an almost hexagonal plane, coordinating the K^+ at the centre.[394] When the cation is too large for the cavity, as in RbSCN(**76**), or CsSCN(**76**), then the metal lies out of the plane of the ring donor oxygen atoms and a dimeric entity ensues through N-bonded-NCS bridging;[395,396] this process also ensures 'saturation' of each cation from the other 'naked' side of the coordination sphere (Figure 15).

Further, but monomeric, examples of Figure 14(b) have been found in RbNCS(**72**) and K(acetoacetate)(**76**). In the former a 2:3 metal–ligand ratio is indicated by microanalysis[5] and so a 'club sandwich' structure was envisaged. The X-ray structure showed that only a 1:1 complex was present, together with molecules of free crown ether.[397] The Rb^+ is displaced out of the ligand and the seventh coordination site is taken up by the nitrogen-bonded thiocyanate anion, which is approximately perpendicular to the oxygen-atom ring. A similar geometry is found in K(acetoacetate)(**76**) where the anion is O-chelating.[398]

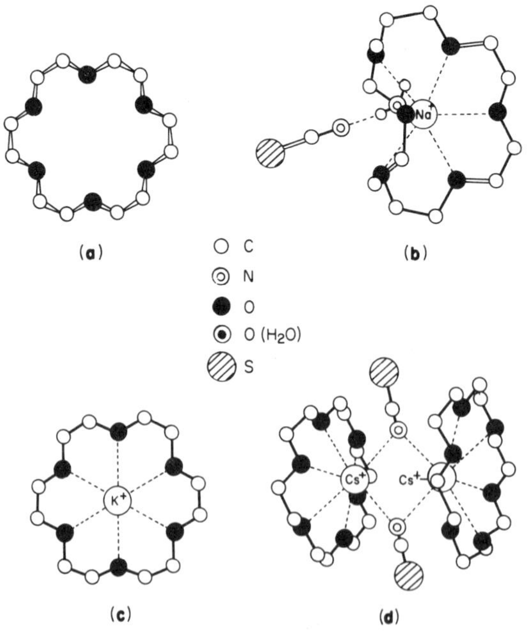

Figure 15 Structures of 18-crown-6 and some 18-crown-6 complexes with different alkali metal salts (reproduced with permission from reference 31)

A dimeric structure has been determined for the Cs(NCS) complex of the *meso* form of tetramethyldibenzo-18-crown-6 (**78**), N-bonded bridging NCS anions being present, and a 1:2 'sandwich' structure with the racemic form.[399] In the dimeric species, the Cs^+ is only octacoordinated. An unusual Cs^+-bearing compound has been isolated with a transition metal cluster as the anion: $[(\mathbf{76})_{14}Cs_9]^{9+}[Rh_{22}(CO)_{33}H_x]^{5-}[Rh_{22}(CO)_{35}H_{x+1}]^{4-}$.[400] In this complex, the cationic moiety has been shown to be comprised of three distinct Cs^+–crown ether units: $Cs^+(\mathbf{76})$, $Cs^+(\mathbf{76})_2$ and $(Cs^+)_2(\mathbf{76})_3$. Of these, the latter two are of interest as they provide a 'sandwich' structure (Figure 14d) and the first genuine example of a 'club sandwich' structure (Figure 14e) respectively. The former is also a new species in that the cation is pulled out of the ligand plane and interacts with two oxygen atoms from C≡O units. An interesting sandwich complex of (**76**) has been found in the reaction of a benzene solution containing excess of (**76**) with potassium phthalocyanine.[401] A discrete trimacrocyclic structure, [(**76**)K—pc—K(**76**)], is found in which the cations are each sandwiched between a terminal crown ether and the central phthalocyanine anion. The benzo-15-crown-5 ligand (**79**) can also be used to illustrate trends in structural type (Figure 16). Na^+ and Ca^{2+} form the 1:1 complexes $NaI(\mathbf{79})\cdot H_2O$ and $Ca(NCS)_2(\mathbf{79})\cdot ROH$ (R = Me, Et); the cations have similar ionic radii. In the Na^+ complex, the cation is coordinated by the five crown oxygen donors and lies out of the plane of the ring in the direction not of the I^- but of the water molecule.[402] In the Ca^{2+} complexes, the Ca^{2+} is displaced further out of the plane of the crown ring and an octacoordination is completed by the two thiocyanates (N-bonded) and the alcohol of solvation.[403] The smaller Mg^{2+} can be found at the centre of the crown ring in $Mg(NCS)_2(\mathbf{79})$, a pentagonal bipyramidal geometry being completed by the two N-bonded thiocyanates occupying the apical positions.[403b]

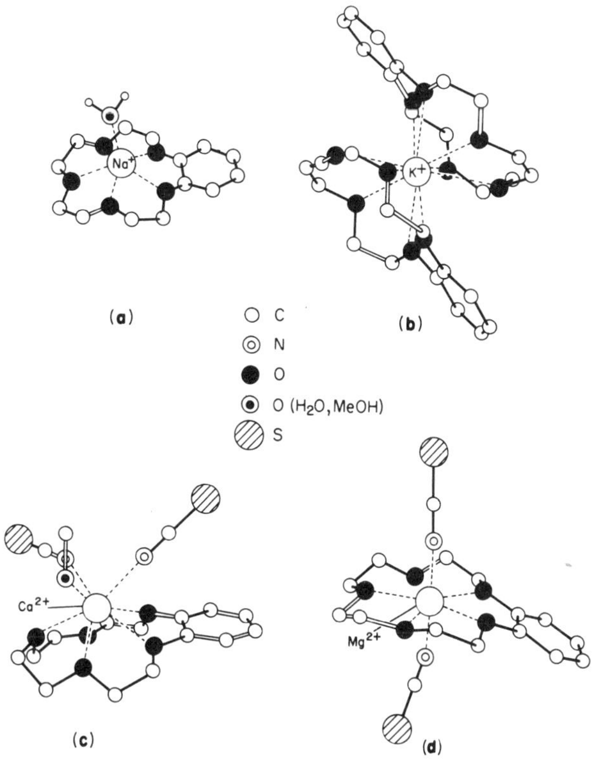

Figure 16 Different types of benzo-15-crown-5 alkali and alkaline earth metal complexes (reproduced with permission from reference 31)

The larger K^+ is found to give a 1:2 metal–ligand complex as in $KI(\mathbf{79})_2$. A sandwich structure is found in which the ligand units are arranged approximately centrosymmetrically relative to one another and the 10 oxygen atoms are at the corners of an irregular pentagonal antiprism.[404] Such a 'sandwich' structure is also found with $BaBr_2(\mathbf{75})_2\cdot 2H_2O$[405] and, rather interestingly, sodium has been found to give 1:2 complexes. The structures of $NaClO_4(\mathbf{79})_2$ and $Na(BPh_4)(\mathbf{79})_2$ both show sandwich arrangements;[406] they differ slightly, and the precise coordination of the metal is uncertain as a range of metal–oxygen distances is found. The preparation of these species depends upon the reaction conditions, requiring concentrated solutions to obtain the 1:2 ratio. If the correct stoichiometric ratio is used, 1:1 solvated complexes are also recovered. A 1:2 complex of Ca^{2+}, $Ca(dinitrobenzoate)_2(\mathbf{79})_2\cdot 3H_2O$, has been found, but in this one of the ethers is not coordinated, rather like the proposed 'club sandwich' of Rb^+—(**72**) mentioned earlier.[407] Sandwich complexes of $Ca(ClO_4)_2$ have been identified; these have varying degrees of solvation and their precise nature is not established.[383b]

The larger crown ethers dibenzo-24-crown-8 (**80**) and dibenzo-30-crown-10 (**81**) provide examples of the encapsulated and bimetallated structural modes (Figure 14g, h). Dibenzo-24-crown-8 takes up two potassium cations in the complex $[KNCS]_2$(**80**); the eight oxygen donor atoms are shared between the cations, and the anions also participate in the bonding as monodentate N-bonded species (Figure 17a). Each K^+ is heptacoordinated and it is proposed that there is a participation of neighbouring benzene rings in the bonding.[408] The disodium *o*-nitrophenolate complex of this crown ether differs in that two ethers of the octadentate ligand are not involved in the bonding; the *o*-nitrophenolate ions serve to bridge the Na^+ cations and complete a coordination of six at each cation.[409] The bis sodium thiocyanate complex of (**81**) also contains two metals within the ligand framework.[410] Alkaline earth metals, in contrast, form, to date, only 1:1 complexes with (**80**) and this may be a reflection of the difficulty of juxtaposing the two higher-charged cations within the polyether framework. $Ba(picrate)_2$(**80**)·$2H_2O$ has the ligand using five of the eight donor oxygen atoms from the ring to coordinate to the cation, together with two water molecules, one of which is simultaneously bound within the polyether torus, two phenolate oxygen atoms and one nitro group oxygen atom giving an overall coordination number of 10.[411] The ligand almost encircles the cation in $Ba(ClO_4)_2$(**80**) using all of the available ring donors to coordinate; the perchlorate anions are also bonded to the cation.[412]

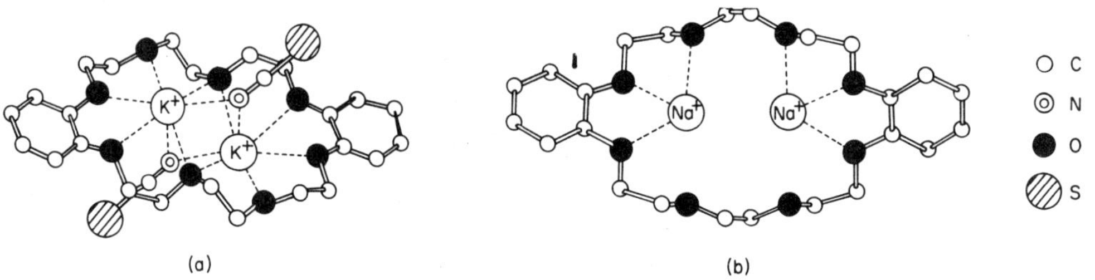

Figure 17 Structures of the Na^+ and K^+ complexes of dibenzo-24-crown-8 (reproduced with permission from reference 31)

Complete encapsulation has been found in the complexes KNCS(**81**)[413] (Figure 18) and RbNCS(**81**).[414] The free ligand[413] shows an open loop and wraps completely around the central metal cation, perhaps not surprisingly as (**81**) is effectively a dimer of (**79**). The central metal atom is wrapped by the ligand in a 'tennis ball seam' configuration and all 10 donor atoms are involved in the coordination sphere (Figure 18). The first example of a metal complex completely encapsulated by a dibenzo-24-crown-8 ligand is the 1:1 $NaClO_4$ complex of the non-symmetric ligand (**82**).

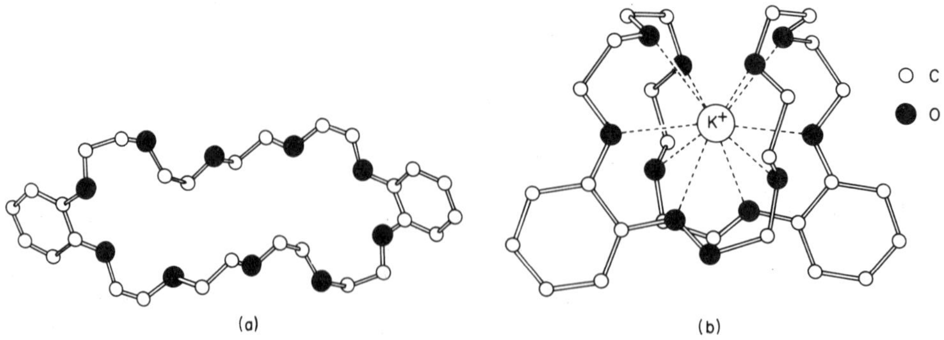

Figure 18 The molecular structures of dibenzo-30-crown-10 and its potassium complex (reproduced with permission from reference 31)

The structures of several Li^+ complexes with crown-4 ethers have been solved. The complexes LiNCS(**83**),[416] LiNCS(**74**),[417] LiNCS(**84**)[418] and LiNCS(**85**)[419] all show Li^+ to be square pyramidally coordinated by the four ring oxygen donors and the nitrogen from the anion. It has been remarked that this geometrical arrangement could be a common feature of Li^+ coordination in crown systems.[418] A dilithium complex of the spiro crown ether (**86**) has been determined. $(LiI)_2$(**86**)·$4H_2O$ has each lithium in a distorted trigonal bipyramidal environment;[420] one water molecule is incorporated in the crown ring, *cf.* that in the barium picrate complex of (**80**), and the cation interacts with three ring oxygens and two water molecules.

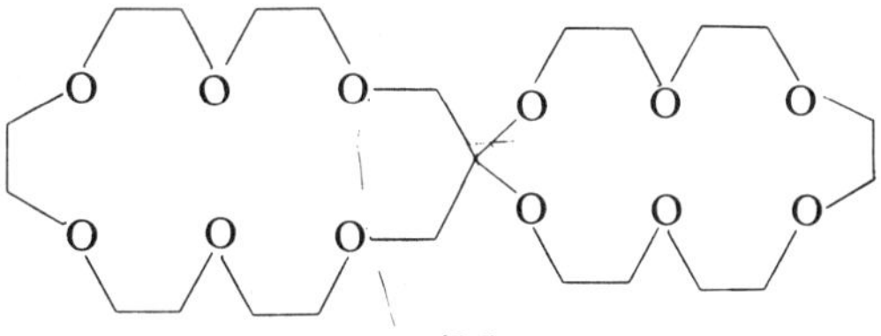

(**86**)

It can be seen from the above discussion that the varying structural modes present in Figure 14 are readily available. For the partially encapsulated complexes there is not only a dependence upon the cation cavity best fit but also on the relative strengths of the competing counter donors, both anions and/or solvent molecules. This is shown in the structures of (**79**)NaI·H_2O, where the water is held by the metal rather than the I^-,[402] and (**72**)NaBr·H_2O, where the sodium is either symmetrically coordinated with two H_2O in axial positions, or non-symmetrically coordinated and lying out of the ligand plane towards the Br^- not the H_2O (Figure 19).[421]

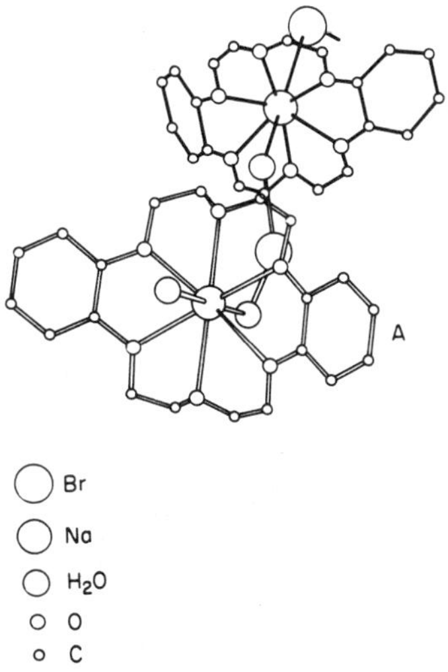

Figure 19 The asymmetric unit in NaBr(**72**)$2H_2O$ showing A, a complex cation Na(**72**)$2H_2O$ and B, a complex ion pair NaBr·(**72**)·H_2O. The fourth water molecule forms a hydrogen-bonded bridge between a coordinated water molecule and the coordinated bromide ion (reproduced with permission from reference 35)

The structures of several related crown ether complexes have also been solved. Ligand (**87**) forms a 1:1 complex with KNCS. In this complex the cation is coordinated to all of the donor atoms of the ligand, with a further weak interaction with the NCS^- anion.[422]

A study of the interaction of ligand (**88**) with alkali metal complexes has been reported.[423] The NaNCS complex contains the ligand bound to the Na^+ by five oxygen donors but the sulfur does not interact at all; the six coordination of the cation is completed by the NCS^- anion.[424] The KNCS complex has the metal at the centre of a hexagon of donors, all of which interact with the metal,[425] and in the RbNCS complex the Rb^+ sits out of the plane towards an N-bonded NCS^- anion and is also coordinated by a sulfur from an adjacent ring to give a dimeric structure.[426]

There has been considerable interest shown in the design and synthesis of functionalized crown ethers, and in whether the appended functional group can then further complex the cation. Examples include the so-called 'lariat' ethers, which may be functionalized on either N or C bridgeheads in the ligand frameworks (**89**).[427]

The interaction of the functional group has been established within the structures of the KI complex of (**90**) and the Na^+ complex of (**91**). In both structures, the metal is held in the ring and the pendant O donor atoms further coordinate to the metal.[428]

(**89**)

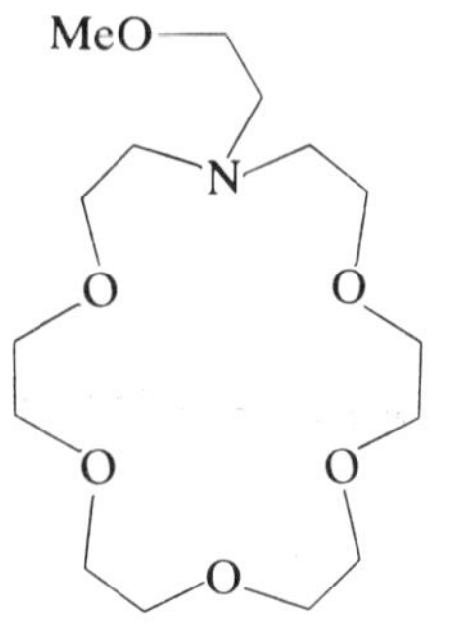

(**90**)

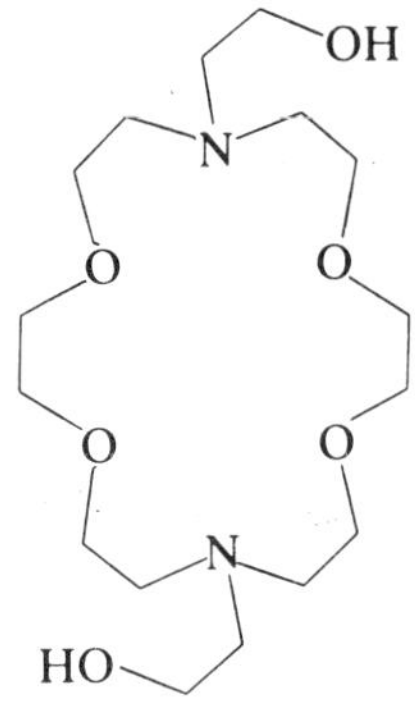

(**91**)

In contrast, in the structures of the RbI complex of *N,N'*-bis(cyanomethyl)-1,10-diaza-4,7,13,16-tetraoxacyclooctadecane, the Rb^+ is complexed by the ring, a water molecule and a further two *inter*molecular interactions with cyanomethyl groups.[429] The structures of bis{4'-OH(**79**)}KNCS[430] and (4'-acetobenzo-18-crown-6)$Sr(ClO_4)_2$[388b] show no interaction of the side chain with the central metal either inter- or intra-molecularly.

In the structure of a 2:1 complex between LiNCS and 1,4,7,10-tetra(2-hydroxyethyl)-1,4,7,10-tetraazacyclodecane the Li^+ is coordinated by four nitrogens from the tetraaza macrocyclic ring and one oxygen from a side arm to give another example of square pyramidal Li^+.[431] The second Li^+ is coordinated by three different side arm oxygens from different ligands and one water molecule. Similar side arm participation is found in the corresponding NaNCS and KNCS complexes, where in the former three side arm oxygens are involved with the Na^+, and in the latter all of the side arm oxygens complex the K^+; in both cases the cation is held by the ring nitrogen donors.[432]

Two interesting examples of alkali metal complexation arise from studies on organometallic systems in which metal acyl binding to the complexed alkali cation is also demonstrated. This type of product has been seen as of value in studies on the kinetics of alkyl migration reactions. A derivative of 1,10-dioxa-4,7-diaza-11-phosphacycloundecane has been shown to give an $Mo(CO)_5L$ complex and this then reacts with PhLi such that the Li^+ is coordinated to two nitrogen atoms and one oxygen atom of the ring and to the PhC=O unit (**92**). The overall structure is dimeric through the acyl binding.[433] The related reaction in which the precursor macrocyclic ring includes an $Mo(CO)_4$ unit gives a product (**93**) wherein the Li^+ is coordinated to the four ring oxygens and to the acyl group oxygen.[434]

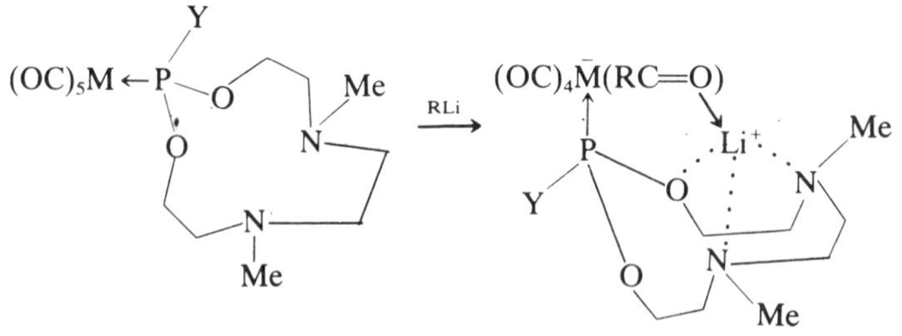

(**92**)

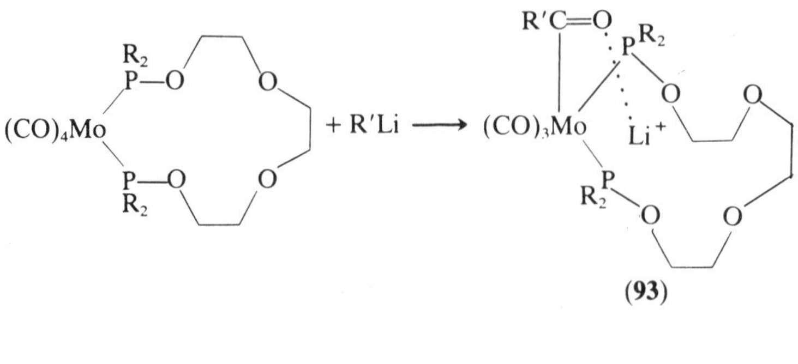

(**93**)

In the derivatives described above, the ligand undergoes conformational change on metal incorporation. Spherands have been described as the first ligand systems fully organized during synthesis rather than during coordination as, unlike the previous examples, the free ligands possess the same conformational organization as do the complexes. The full burden of binding site collection and organization is then transferred from the complexation process to the synthesis of the ligand.[435] This is illustrated in the structures of (**94**), (**94**)LiCl and (**94**)$NaCH_2SO_4$. The Li^+ fills the octahedral cavity of the ligand and causes a reduction in hole diameter from 1.62 to 1.48 Å; the Na^+ sits similarly in its cavity but causes a slight expansion of cavity from 1.62 to 1.75 Å.

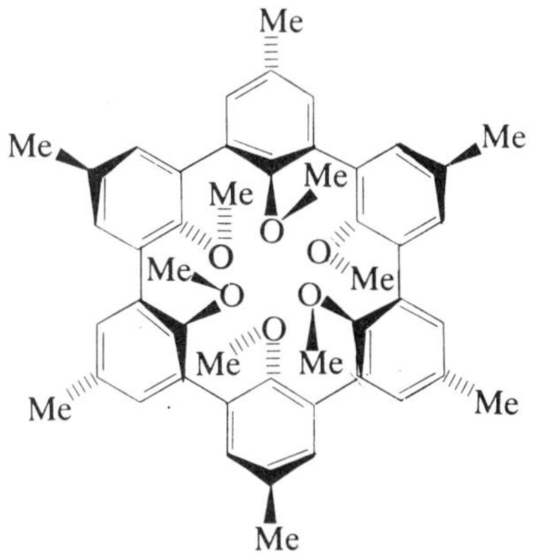

(**94**)

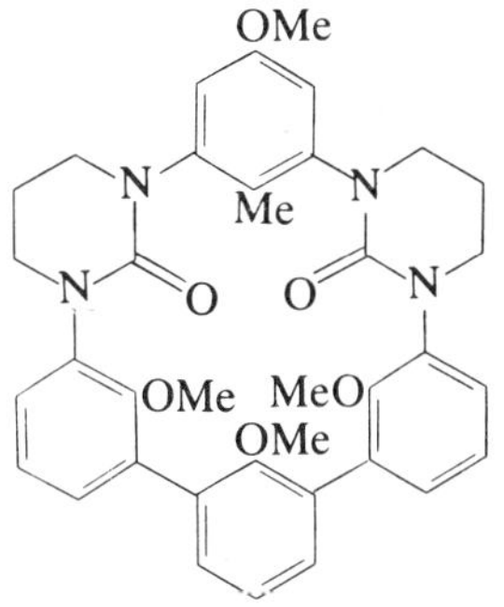

(**95**)

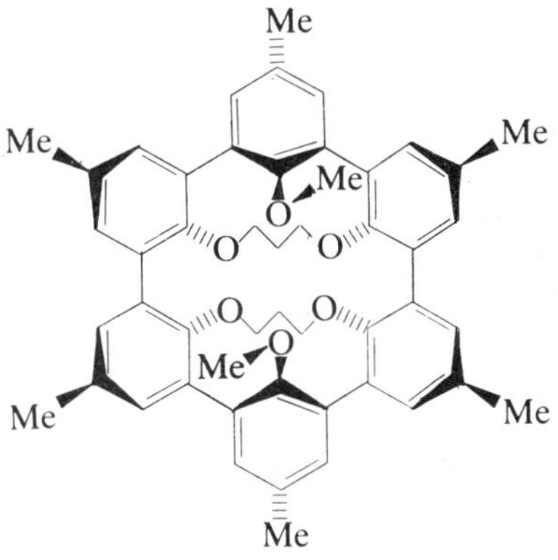

(**96**)

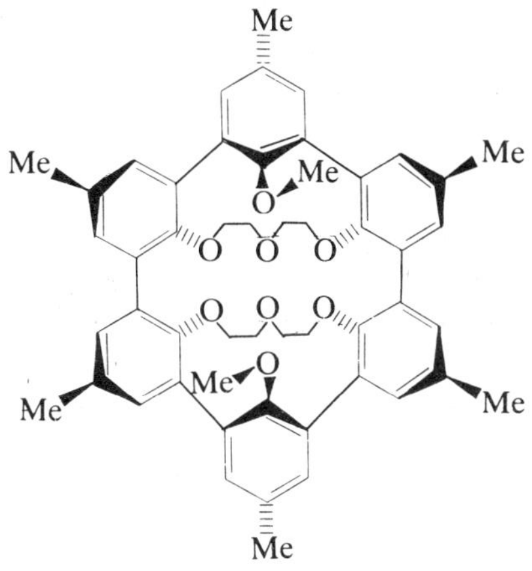

(**97**)

In the NaBr and $CsClO_4$ complexes of the related spherand (**95**) in which *N,N′*-disubstituted cyclic urea moieties have been introduced, the Na^+ forms a nesting complex with five oxygen atoms of the spherand lining a nest capped by a water molecule of solvation. The Cs^+ is too large for the nest and so perches on four *syn* oxygen atoms and is further ligated by a solvent water molecule and the carbonyl from an adjacent spherand to give a dimeric structure, as with cyclic polyethers.[436] The structures of the Li^+ complexes of the augmented spherands (**96**) and (**97**) have been determined.[437] The effect of the bridges is to squeeze one of the methoxy groups of the ligating range (2.48 Å and 3.48 Å respectively). The Li^+ in (**96**)$LiFeCl_4$ is coordinated by the five remaining oxygen atoms to give an Li^+ diameter of 1.27 Å, while in (**97**)LiCl the residual seven oxygen atoms lead to an Li^+ diameter of 1.71 Å and provide a very unusual seven-coordination of the metal. These diameters compare with the value of 1.48 Å given earlier.[436]

Hemispherands have a partial preorganization of the ligand. The X-ray structure of (**98**)$Na(BPh_4)\cdot H_2O$ shows the metal coordinated by the hemispherand donors and by a water molecule.[438] The related complex (**99**)NaNCS·MeOH shows the metal to be coordinated by all six donors of the crown, which is in a boat configuration and the terphenyl is folded away. The metal geometry is a biapical square pyramid.[439]

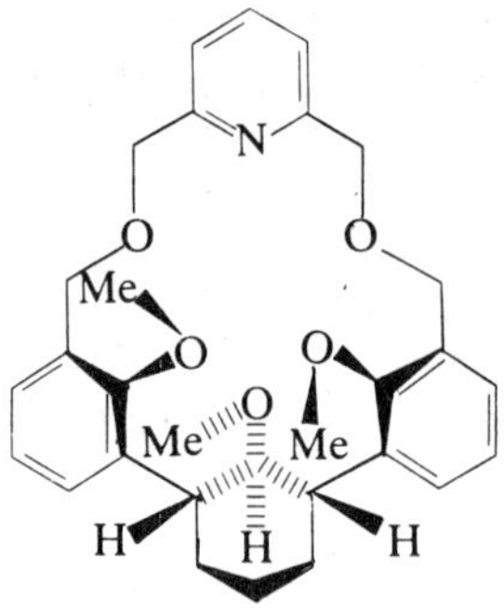

(98)

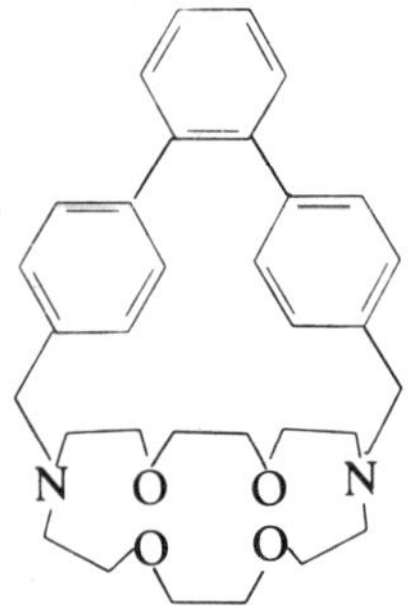

(99)

The crystal structures of alkaline earth metal complexes of several (1 + 1) and (2 + 2) Schiff base macrocycles have been reported. These macrocycles are formed by the metal template-controlled condensation of the required heterocyclic dicarbonyl derivative and α,ω functional diamine in alcoholic solution. (1 + 1) complexes arise from the condensation reaction of one dicarbonyl with one diamine and (2 + 2) complexes from the condensation of two dicarbonyls with two diamines.

The X-ray structures of $[Mg(H_2O)_2(\mathbf{100})]Cl_2$,[440] $Ca(\mathbf{101})(NCS)_2$,[441] $[Sr(\mathbf{101})H_2O](NCS)_2$[441] and $[Sr(\mathbf{102})H_2O](NCS)_2$[442] have been determined. The magnesium is found to lie in the plane of the donor atoms, pentagonally bipyramidal through interaction with two axial water molecules.[440] The calcium is similarly centrally coordinated with N-bonded thiocyanates axially disposed to give a hexagonal bipyramid.[441] The strontium complexes have the metal interacting with all of the ring donors but displaced out of plane towards two N-bonded anions. A water molecule on the opposite side of the ring completes the nine-coordination of the metal.[441,442]

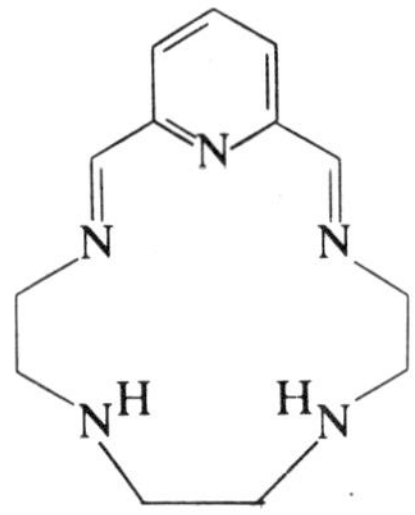

(100)

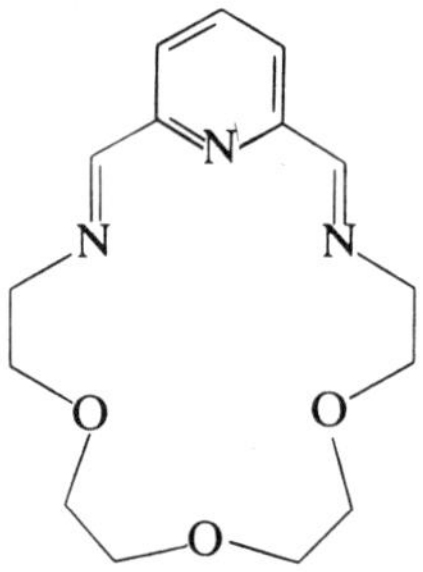

(101)

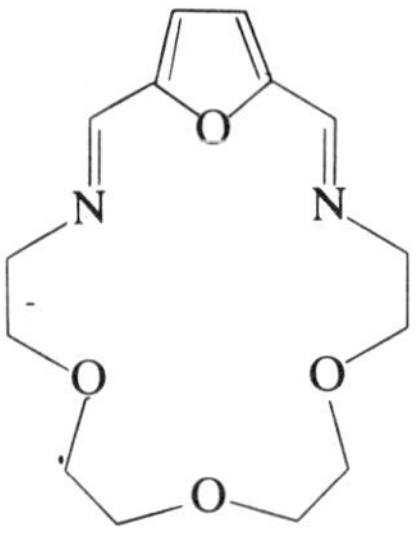

(102)

With (2 + 2) macrocycles several features arise and both 1:1 and 1:2 complexes have been found. The complex $(\mathbf{103})SrCl_2$ has the metal coordinated in a hexagonal bipyramidal geometry comprised of the six ring donor atoms and two apical anions.[443] The barium complexes of (**104**) and (**105**) are, in contrast, sandwich structures: $(\mathbf{104})_2Ba(BPh_4)_2$[444] and $[(\mathbf{105})_2Ba(H_2O)_2][Co(NCS)_4]$.[445] The former has the barium 12-coordinated to the six donors of each ring, whereas in the latter all six donors from one ring are bound and only three from the second ring. The ligand is folded and the complex has fluxional behaviour as determined by NMR studies. The Ba^{2+} is only 11-coordinated, two water molecules occupying the last two sites. Alkaline earth metal complexes of (**106**) have also been reported.[446] One useful role these complexes play is that of precursors for transmetallation reactions leading to dicopper(II) complexes of the macrocycles. These latter complexes serve as speculative models for probing the nature of the bimetallobiosites in copper-containing enzymes and proteins, such as type III oxidases and oxyhaemocyanin.[446]

The barium complex of (**107**) is unusual in that a ring contraction reaction occurs on its formation and the imidazolidine ring presence means that a contraction from a 24-membered macrocycle to an 18-membered macrocycle occurs.[312] The cation in $(\mathbf{107})Ba(ClO_4)_2$ is 10-coordinated to the six in-planar nitrogen donor atoms and to two bidentate perchlorate anions.[312] In the complex $(\mathbf{108})Ba(ClO_4)_2{\cdot}2H_2O$, the metal is almost fully encapsulated by the ligand with the hydroxyl groups participating in the bonding.[447] The cation is 10-coordinated by the eight ring donor atoms and two water molecules with the perchlorate anions solvent

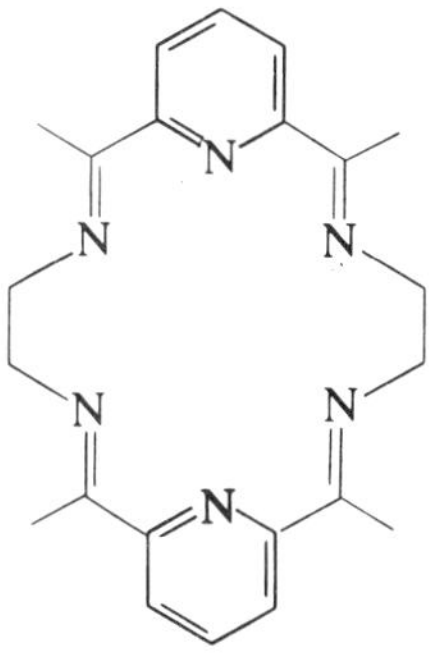

(103)

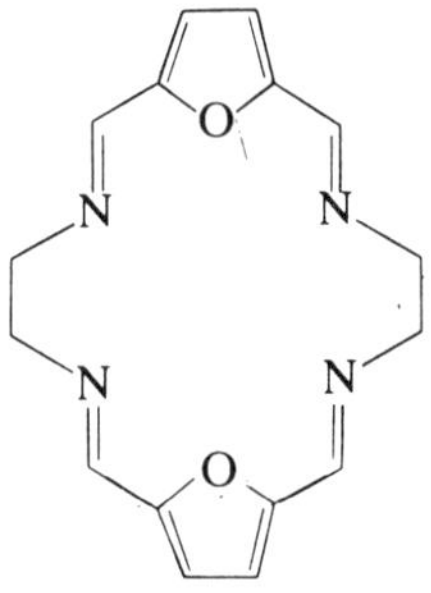

(104)

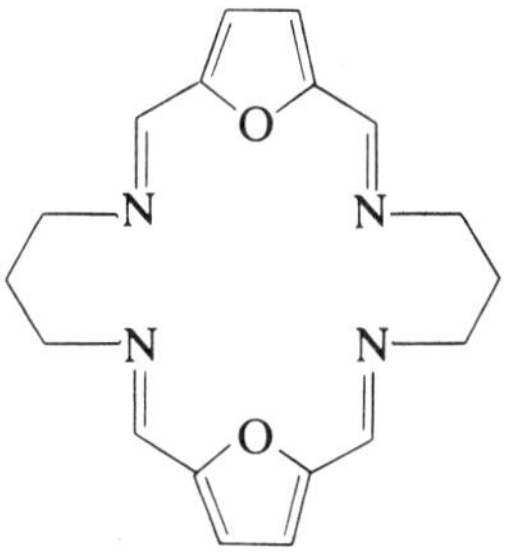

(105)

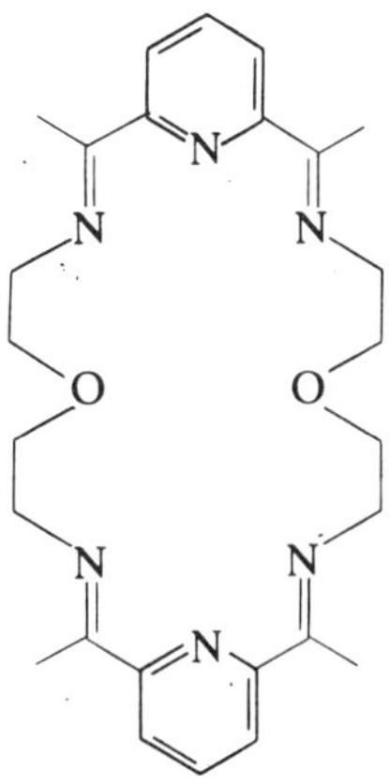

(106)

separated. In the analogous complex having furanyl moieties as head units, instead of pyridinyl units, a similar ligand arrangement is found but the perchlorates are now bound to the cation; this serves as an example of the fine balance that exists between ligand donors, solvent donors and anions in such complexes.

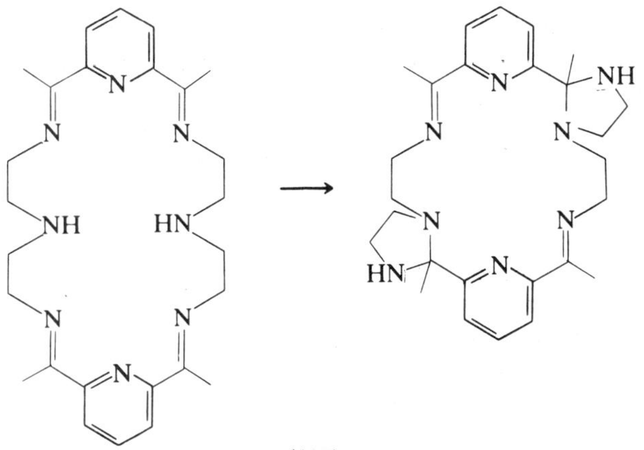

(107)

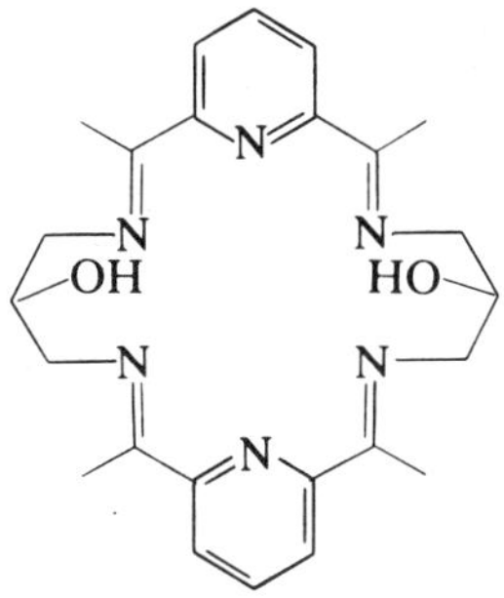

(108)

Coronands containing pyridinyl, thiophenyl and furanyl head units have been prepared,[33] and their complexation studied. The coronand (**109**; Figure 13) gives a 1:1 complex with KNCS in which the metal is coordinated centrally by the six ring donor atoms; the keto groups are not involved in the coordination.[448]

23.5.2 Cryptands

Macropolycyclic ligands containing intramolecular cavities of a three-dimensional nature are referred to as cryptands. The bicyclic cryptands (**73**) exist in three conformations with respect to the terminal nitrogen atoms, *exo-exo, endo-exo* and *endo-endo*;[6] these forms can rapidly interconvert *via* nitrogen inversion but only the *endo-endo* form has been found in the crystal structures of a variety of complexes[372] and for the free ligand ([2.2.2], **73**, $m = n = l = 1$).[449] In their complexes with alkali and alkaline earth cations, the cryptands exhibit an enhanced stability over the crown ethers and coronands due to the macrobicyclic, or cryptate, effect.[33,202]

The configurations of the free ligand, [2.2.2]cryptand, and its RbNCS complex are shown in Figure 20. The rubidium is complexed by four oxygen and two nitrogen donors and is completely trapped in the three-dimensional cavity.[450]

The ligand shape has changed and it has 'inflated' on encapsulating the cation. The structures of a large number of [2.2.2]cryptand alkali and alkaline earth metal complexes have been studied by X-ray crystallography.[372] Using the criterion that the optimum complexation occurs when the cation and cavity radii are similar, it can be predicted that for this ligand potassium is

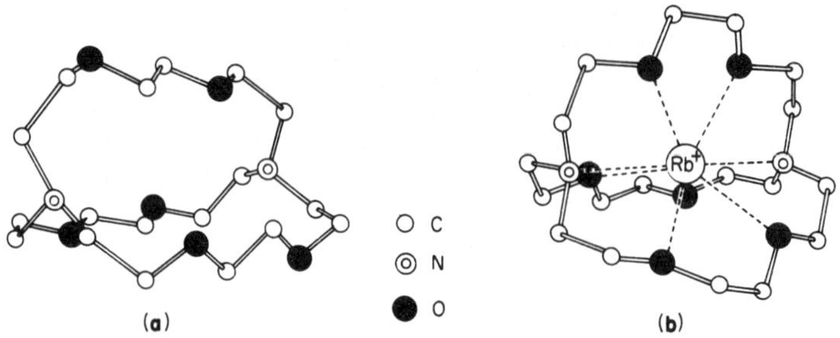

Figure 20 Molecular structures of [2.2.2]cryptand (**73**) and of the rubidium cryptate (reproduced with permission from reference 31)

the best cation (cavity diameter ≈ 2.80 Å, ionic diameter ≈ 2.66 Å). However, the ligand is flexible enough to incorporate the smaller sodium cation by undergoing significant twisting of the ligand.[372] This flexibility is also evidenced with increasing ionic size and coordination number of the embedded cation, as a progressive opening up of the cavity is detected. When this occurs, addition of an anion, or a solvent molecule, is observed. This is a common feature in the alkaline earth metal complexes such as $[Ba(NCS)\cdot H_2O\cdot[2.2.2]]NCS$[451] and $[Ca\cdot H_2O\cdot[2.2.2]]Br_2\cdot 2H_2O$.[452] In the [2.2.2] complex with cesium it was concluded from ^{133}Cs NMR data that in solution the [2.2.2] forms an exclusion complex with the cation. The cation does not occupy the central cavity but is only partly encapsulated by the cryptand.[453]

An exclusion complex of this type has been identified by X-ray crystallography for the KNCS complex of the smaller [2.2.1]cryptand,[454] ([2.2.1]cryptand = **73**, $n = 1$, $m = 1$, $l = 0$). The cation occupies a site in the 18-membered ring provided by the ligand rather than in the central cavity and so resembles the coordination of potassium by 18-crown-6 (Figure 21a). In contrast, the smaller Na^+ cation is of optimal size for the [2.2.1]cryptand and so an inclusion complex occurs with the cation inside the cage (Figure 21b).[454] Binuclear inclusion complexes have not yet been isolated for bicyclic cryptands.

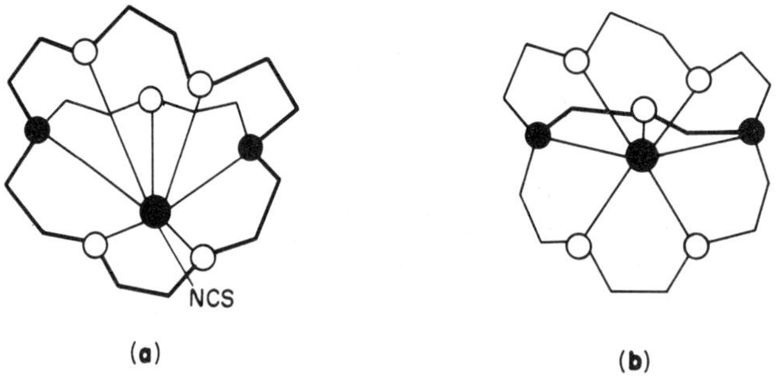

Figure 21 Representatives of the molecular structures of the complexes formed between [2.2.1]cryptand and (a) KSCN and (b) NaSCN

An interesting feature of cryptand complexes has been their use in the stabilization and identification of unusual anions. Structures of interest include those for an alkalide anion, $[2.2.2]Na^+Na^-$,[455] and for polyatomic anions of post-transition metals, $\{[2.2.2]Na^+\}_3Sb_7^{3-}$,[456] $\{[2.2.2]Na^+\}_2Pb_5^{2-}$,[457] $\{[2.2.2]Na^+\}_4Sn_9^{4-}$,[457] $\{[2.2.2]K^+\}_2Sn_4^{2-}\cdot en$,[458] $\{[2.2.2]K^+\}_2Ge_4^{2-}$[458] and $\{[2.2.2]Ba^+\}Se_4^{2-}\cdot en$.[459] The organometallic complex $\{[2.2.2]\}Na^+FeCO_4^{2-}$ was also studied in order to define the stereochemistry of the anion whilst concealing the cation from it.[460] Complexes containing two enclosed cations can be established using macrotricyclic ligands. The structure of (**110**)$(NaI)_2$ shows each sodium to be bound to two nitrogen atoms and five oxygen atoms of each terminal cycle in the ligand, and separated by a distance of ≈6.4 Å (Figure 22).[461]

In the 1:1 $NaBF_4$ complex of (**111**) the cation is completely encapsulated by the tricyclic ligand and bound to all eight donor atoms.[391b] In the structure of the dipotassium complex of

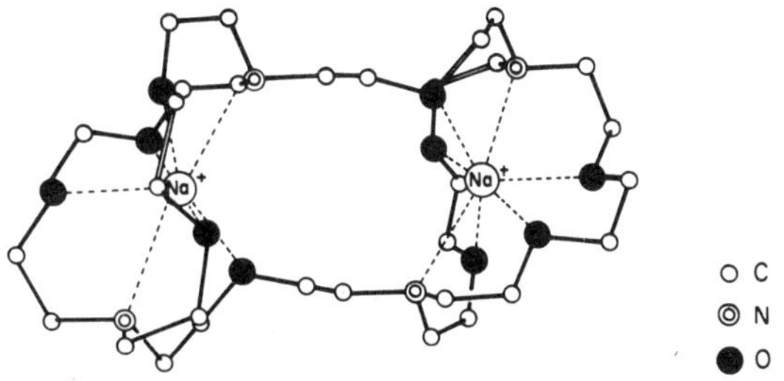

Figure 22 The molecular structure of the binuclear sodium complex of the tricyclic cryptand (**110**) (reproduced with permission from reference 31)

the ligand (**112**),[462] the potassium cations are located in the two '30-crown-10' fragments at a distance of 7.38 Å apart.

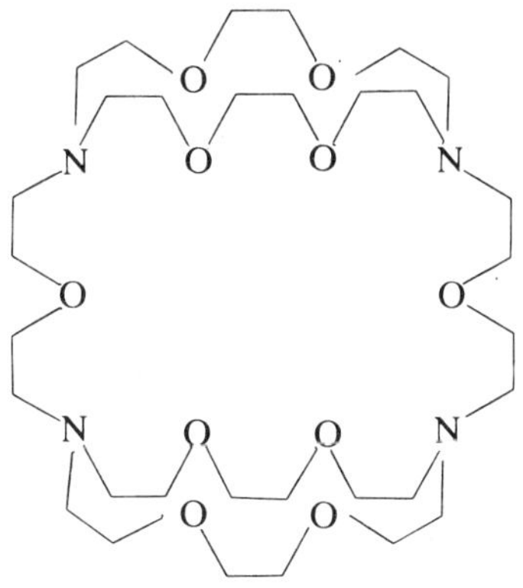

(**110**)

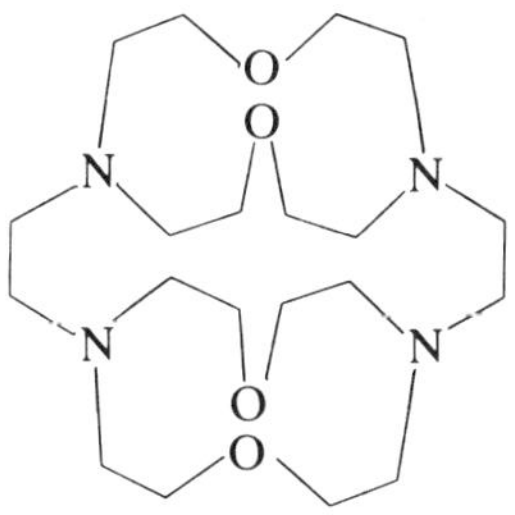

(**111**)

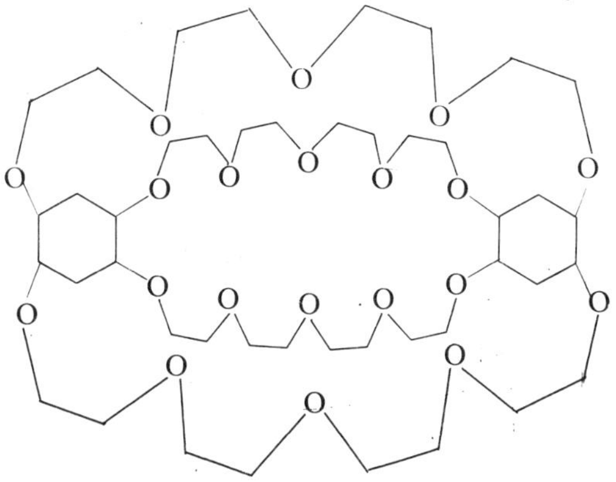

(**112**)

As well as N-terminal, or bridgehead, cryptands, C-bridgehead ligands have also been developed. The X-ray structure of (**113**)$KCl{\cdot}H_2O$[463] shows the potassium coordinated only to the encapsulating ligand. The structure of (**114**)$Mg(ClO_4)_2$ shows the metal to be octacoordinated by the seven donor atoms of the ligand and by a water molecule.[464]

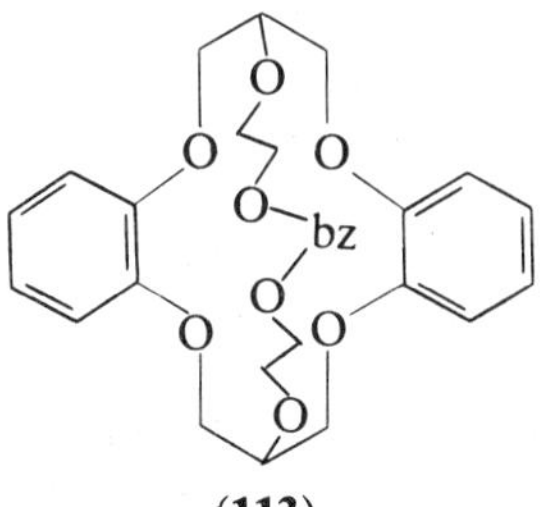

(**113**)

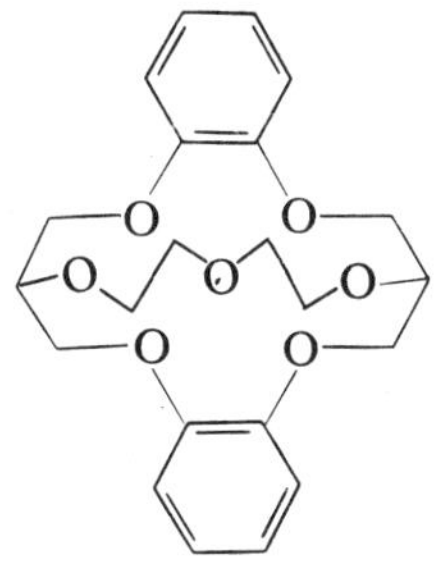

(**114**)

23.5.3 Cation Selectivity

One of the driving forces in the study of alkali metal and alkaline earth metal complexes of polyethers and related species has been the retrieval of information on complex stability and selectivity with its possible bearing on the problems associated with the understanding of biological transfer, with selectivity, of those metals. The above information, derived in the main from studies on the solid state, has helped establish the nature of the complexes and, in parallel with these studies, an abundant literature has appeared on studies in the solution state. Several reviews have discussed the methodology of study, and the techniques used to determine the stability constant, K_s, for the complexes including calorimetry, NMR, pH metric methods, conductimetry, ion selective electrodes, temperature jump techniques, osmometry and solvent extraction.[20,33,465,466,467] Extensive compilation of the retrieved data has followed on Frensdorff's original studies using Pedersen's polyethers,[468] and is available through the review articles. It should be mentioned that the constants obtained are determined using different methods, in different media and so are often difficult to compare with each other. Consequently, the discussion here will centre on factors which influence stabilities and selectivity and numerical data will be given where appropriate.

The enhanced K_s values for the crown ethers relative to their open-chain analogues have been attributed to the so-called 'macrocyclic effect' (Figure 23). The complexation reactions are believed to be almost entirely entropy driven. The entropy of formation depends mostly on the change of the number of degrees of freedom of particles during complex formation and the biggest item represents the translational entropy of the released water molecules. Highly charged cations which are strongly hydrated should give larger values of ΔS° and this is experimentally confirmed on going from K^+ to Ba^{2+} complexes of dicyclohexyl-18-crown-6. The ΔS° value for Ba^{2+} ($-0.8368\,\mathrm{J\,deg^{-1}\,mol^{-1}}$) is more favourable than that for K^+ ($15.899\,\mathrm{J\,deg^{-1}\,mol^{-1}}$). There is also a macrobicyclic, or cryptate, effect detected on moving to the analogous cryptand (Figure 23). This effect is caused by a favourable enthalpic effect attributable to the strong interaction of the cation with the poorly solvated ligand.[29,465]

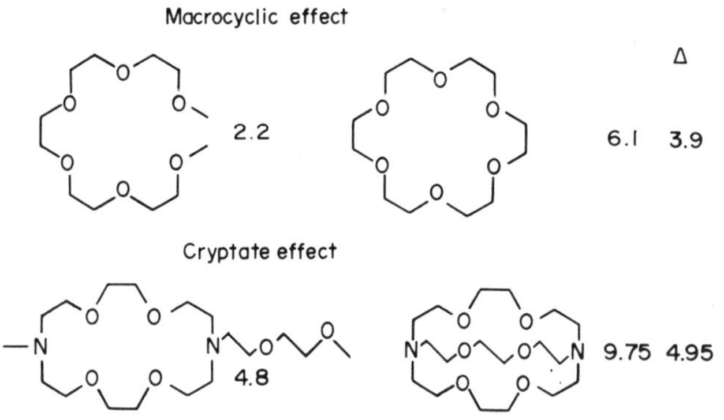

Figure 23 The macrocyclic and macrobicyclic effects (Δ) on complex stability; the values given are the stability constants $\log K_s$ of the K^+ complex in methanol (top) and methanol/water (95.5) (bottom) (reproduced with permission from reference 29)

23.5.3.1 Factors influencing selectivity and stability

(i) The nature of the donor atom

Alkali and alkaline earth metal cations are A-type, 'hard' acceptors and so would be expected to combine most readily with A-type, 'hard' donor atoms. Consequently complexes derived from crown ethers containing solely oxygen donor atoms give high K_s values. The introduction of B-type, 'soft' donors, such as nitrogen, sulfur and phosphorus, gives a destabilizing influence to the complexation process. For example, the K_s values for a series of thia crown analogues having 9-crown-3, 12-crown-4, 15-crown-5, 18-crown-6 and 24-crown-8 skeletons show much reduced K_s values for alkali and alkaline earth cations relative to their oxa crown precursors, but enhanced K_s values for transition metal complexation.[469,470] Substitution of oxygen by an amino (NH) unit in benzo crown ethers has also been seen to lead to a reduction in their complexing ability towards alkali metal picrates and so to affect extraction into an organic phase.[471]

Table 7

(i) A Comparison of log K_s Values for the Complexation of (**73**) and its Aza and Thia Analogues with K^+ and Ag^+

Cation	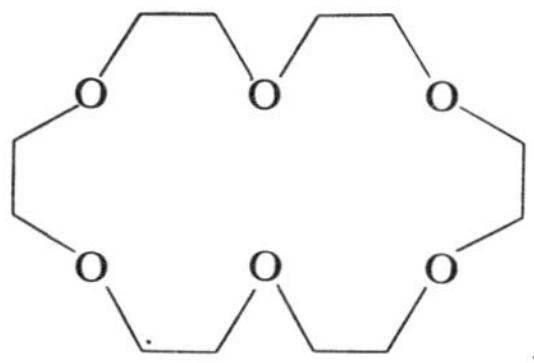	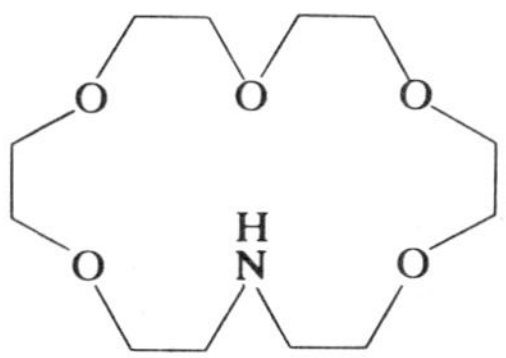	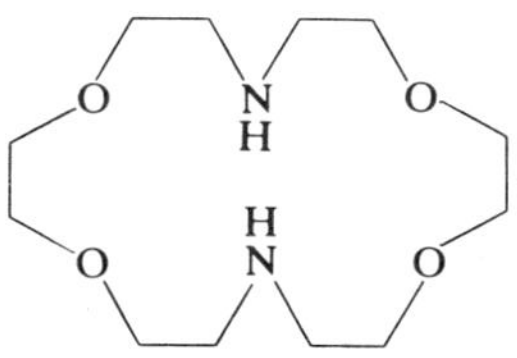	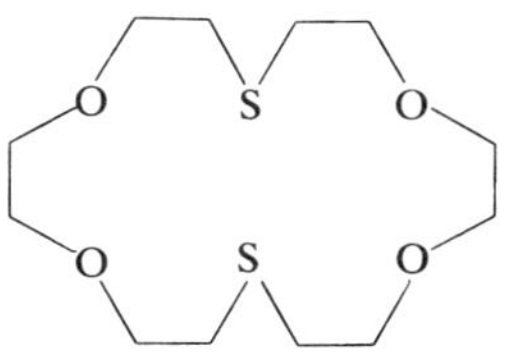	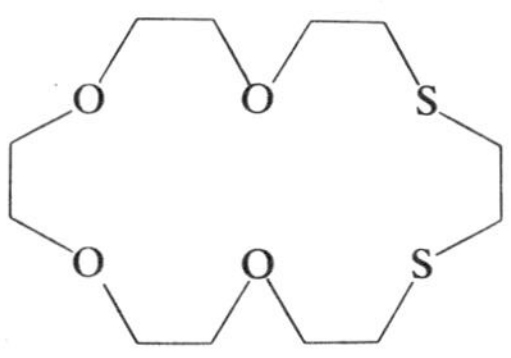
K^+ (in MeOH)	6.10	3.90	2.04	1.15	—
Ag^+ (in H_2O)	1.60	3.30	7.80	4.34	3.0

(ii) A Comparison of log K_s Values for the Complexation of [2.2.2]Cryptand and its Aza Analogues with Alkali and Alkaline Earth Cations and with Ag^+

Cation (in H_2O)	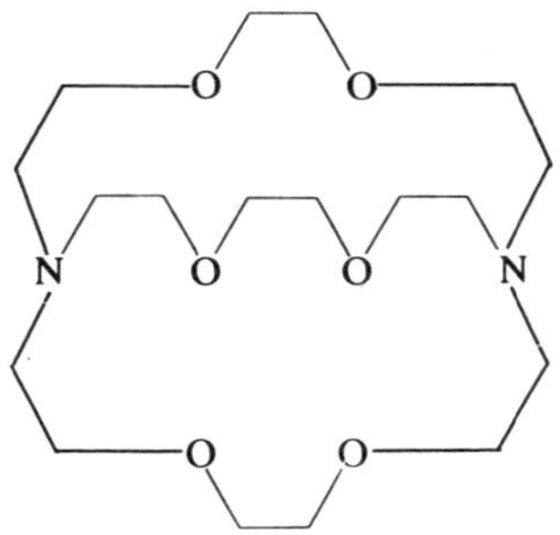	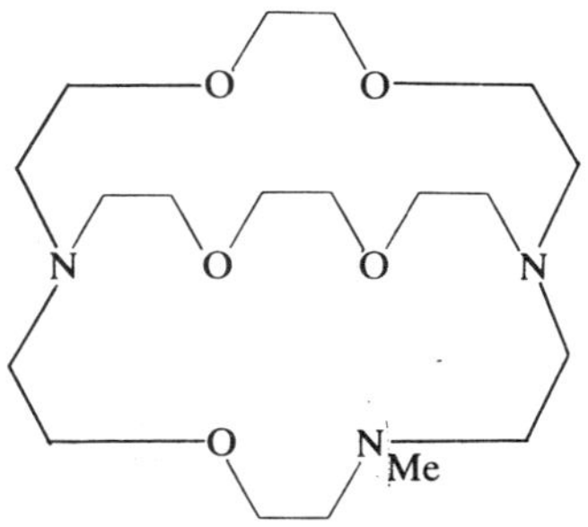	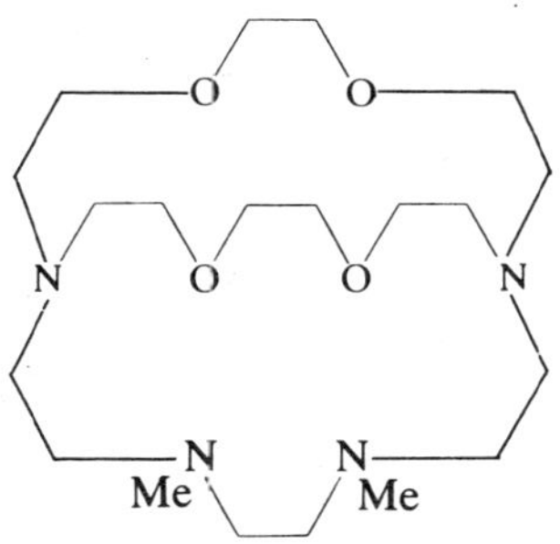	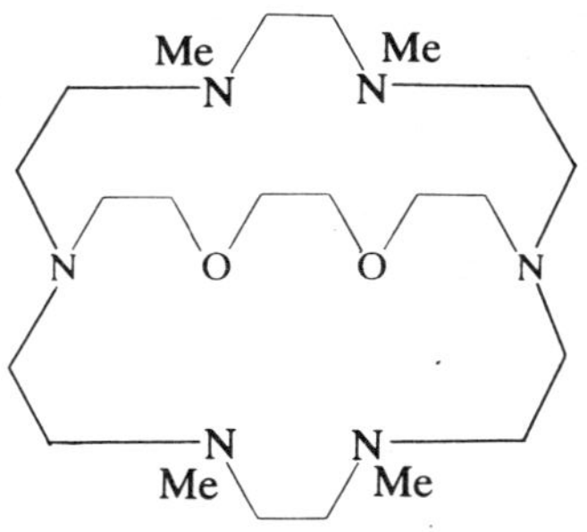
Li^+	<2	1.5	2.4	—
Na^+	3.9	3.0	2.5	—
K^+	5.4	4.2	2.7	1.7
Rb^+	4.3	3.0	2.3	—
Cs^+	<2	<2	<2	—
Mg^{2+}	2	1.9	2.6	—
Ca^{2+}	4.4	4.6	4.3	1.5
Sr^{2+}	8.0	7.4	6.1	1.5
Ba^{2+}	9.5	9.0	6.7	3.7
Ag^+	9.6	10.8	11.5	13.0

Data taken from compilations in ref. 33.

Similar effects are noted with polycyclic systems; the introduction of NR units into cryptands led to reduced K_s values,[472,473] an effect noted particularly for K^+ complexes of methylaza compounds based on [2.2.2]cryptand. Table 7 lists results showing (i) how K_s values for K^+ and Ag^+ complexation vary on introduction of NH and S units into (**76**), and (ii) the variation of K_s on introduction of NMe into [2.2.2]cryptand.

Donors present in appended groups may also cause gradations in stability and selectivity, for example in systems based on (**115**),[386a,474] there is an enhanced Na^+ stability shown, and this is also seen when pyridinyl, furanyl and thiophenyl units are introduced.[475–477] Careful manipulation of the donors, in conjunction with the properties discussed below, can lead to reasonably discrete metal cation specificities.

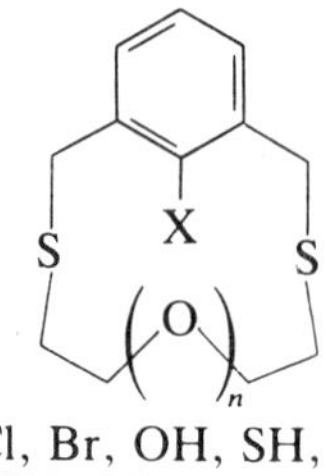

X = H, Cl, Br, OH, SH, NH_2, OMe, CN, CO_2H, CO_2Me

(**115**)

(ii) *The number of, and spatial arrangement of, donor atoms*

The number of donor atoms can be influential in complex formation. Ideally, as the incoming ligand is comparable to an inner solvation sphere, the number of donor atoms should match the preferred coordination number of the cation. An example of this factor comes from a comparison of two similarly sized ligands, [2.2.2]cryptand and [2.2.C_8]cryptand, the latter having one dioxoether chain replaced by a C_8 unit.[478] A reversal of the Ba^{2+}/K^+ selectivity of the order of 10^6 is found, the ratio being 10^4 for [2.2.2]cryptand and $<10^{-2}$ for [2.2.C_8]cryptand; the preferred coordination numbers are six for potassium and eight for barium.

The spatial arrangement for the donor atoms is significant in complexation as every deformation of the donor environment which is not compatible with the required geometry at the metal leads to a reduction in the bonding potential of the ligand and the consequent stability of the complex. For spherical metal cations an optimum donor array should also have a spherical form. The 'soccer ball' molecule (**116**)[479] and the more 'rugby ball'-like cryptands approach this end; the crown ethers, in which the oxygen dipoles are not ideally located in the ring centre, and which have vacant acceptor sites, give somewhat lower K_s values unless they are sufficiently flexible to encapsulate the metal.

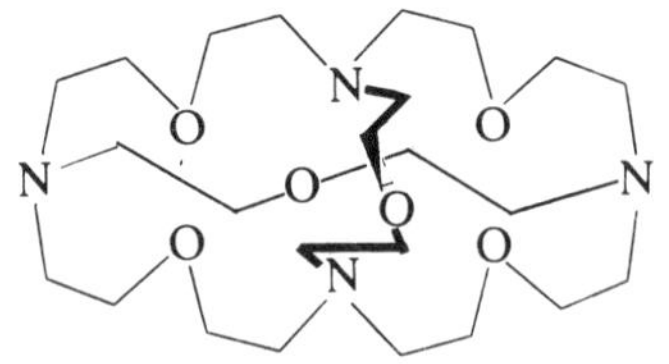

(**116**)

Modification of the torus of crown ethers by replacement of —C_2H_4— units with —C_3H_6— or longer hydrocarbon bridges, or by insertion of aromatic units such as *o*-, *m*- or *p*-xylene, naphthalene and biphenylene,[465–467,480,481] also leads to unfavourable complexation as the chelating ability of the donor atom pairs within the crown ether framework is modified. The ethyleneoxy group plays a crucial role in polyethers and this is believed to be related to the enhanced stability of five-membered chelate rings comprised of the binding sites, one ethylene bridge and the recipient metal over four- and six-membered chelate rings.[202]

(iii) *Cavity shape and size*

As has been indicated in the discussion of solid state systems, the relationship between the cation size and the ligand donor cavity size (the 'best fit' criterion) is rather important, and many correlations have been made between this factor and K_s values.[465]

Figure 24 illustrates the relationship between K_s and the cation radius for dicyclohexyl-18-

crown-6 (isomers A and B) and for benzo-5-crown-5.[465] Figure 25 shows the change in K_s for the reaction of Na^+, K^+ and Cs^+ with several polyethers. It can be seen that for the 18-crown-6 systems, K^+ is preferred, followed by Rb^+, Na^+, Cs^+ and Li^+, and for the 15-crown-5 systems the sequence becomes $Na^+ > K^+ > Rb^+ > Cs^+$. This is a reflection of the diminished ring size and cation compatability. One problem, however, in the latter case, is the necessity to distinguish between the formation of 1:1 and 1:2 complexes in solution.

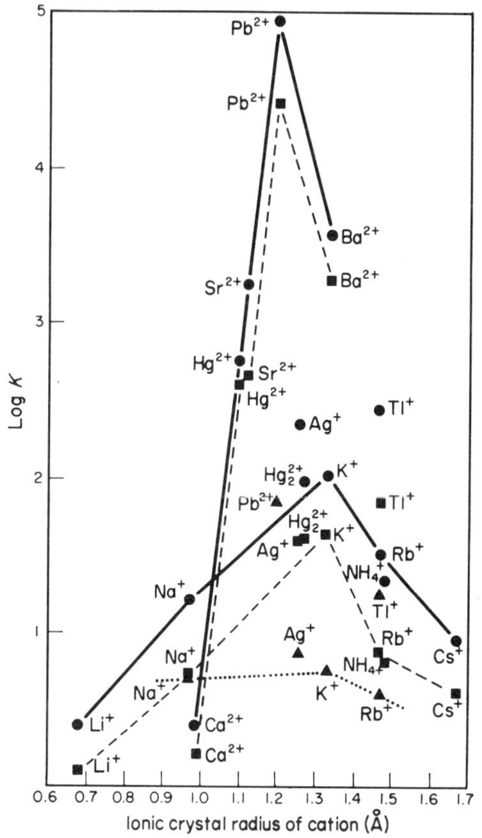

Figure 24 Plot of log *K* *vs.* cation radius for the reaction in aqueous solution, $M^{n+} + L = ML^{n+}$ where L = dicyclohexyl-18-crown-6, isomer A (●), dicyclohexyl-18-crown-6, isomer B (■), or 15-crown-5 (▲). $T = 25\,°C$ (reproduced with permission from reference 465)

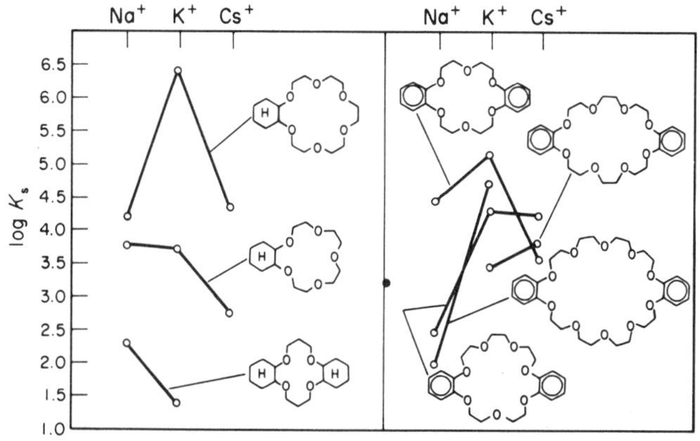

Figure 25 Plots of log K_s (in MeOH at 25 °C) for complex formation between alkali metal cations and several cyclohexano- and dibenzo-crown ethers (reproduced with permission from reference 31)

More recent studies show some variations to the above generalities. ^{13}C and ^{23}Na NMR studies indicate that for Na^+ the complexation sequence is 18-crown-6 > 15-crown-5 > benzo-15-crown-5,[482] hinting at the need for ligand flexibility; conduction data reveal that for 18-crown-6 the stability sequence is $K^+ > Rb^+ > Cs^+ > Na^+$;[483] extraction studies using metal picrates show that for benzo-15-crown-5, $Na^+ > K^+ > Rb^+ > Li^+ > Cs^+$ for 1:1 complexes and that for the corresponding 1:2 complexes $K^+ > Rb^+ > Cs^+$;[484] further extraction work indicated that $K^+ > Rb^+ > Cs^+ > Na^+$ if dicyclohexyl-18-crown-6 was the ligand.[485] In general terms such results follow the sequences predicted from comparison of cation and cavity radii; the larger crown ethers, however, have been found to show discrimination against the smaller cations, but do not show much specific preference between the larger cations.

A more precise cavity control is found with the three-dimensional cryptands and Figure 26 shows the variation of K_s with alkali and alkaline earth metal cations for a range of cryptands.[29] Two types of selectivity are displayed. Peak selectivity occurs when there is an excellent correlation between cavity and cation radius (Table 8) and so a sharp discrimination can occur (Figure 26). Plateau selectivity occurs when there is little discrimination between cations of higher radius but discrimination against cations of small radius which are not well accommodated (Figure 26).

Table 8 Log Stability Constants of Some Cryptate Complexes in Water, Log K_1

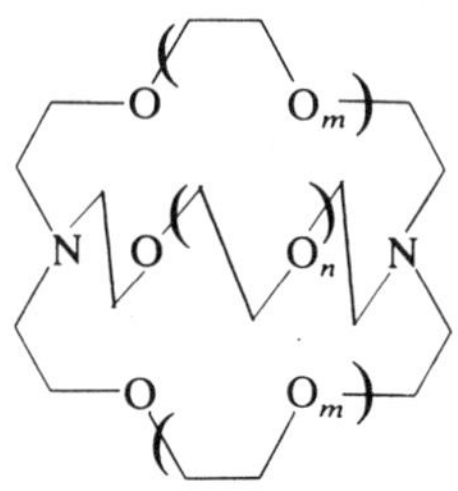

Cation (ionic radius in Å)	*Li$^+$ (0.86)*	*Na$^+$ (1.12)*	*K$^+$ (1.44)*	*Rb$^+$ (1.58)*	*Cs$^+$ (1.84)*
Bicyclic ligand (cavity size in Å)					
[2.1.1], $m = 0$, $n = 1$ (0.8)	4.30	2.80 2.7[a]	2.0	2.0	2.0
[2.2.1], $m = 1$, $n = 0$ (1.15)	2.50	5.40 9.0[a]	3.95	2.55	2.0
[2.2.2], $m = n = 1$ (1.4)	2.0	3.9 9.0[a]	5.40	4.35	2.0
[3.2.2], $m = 1$, $n = 2$ (1.8)	2.0	2.0 4.80[a]	2.2	2.05	2.20
[3.2.2], $m = 2$, $n = 1$ (2.1)	2.0	2.0 2.80[a]	2.0	0.7	2.0
[3.3.3], $m = n = 2$ (2.4)	2.0	2.0	2.0	0.5	2.0

Compiled from J. M. Lehn and J. P. Sauvage, *Chem. Commun.*, 1971, 440.

[a] Recorded in methanol.

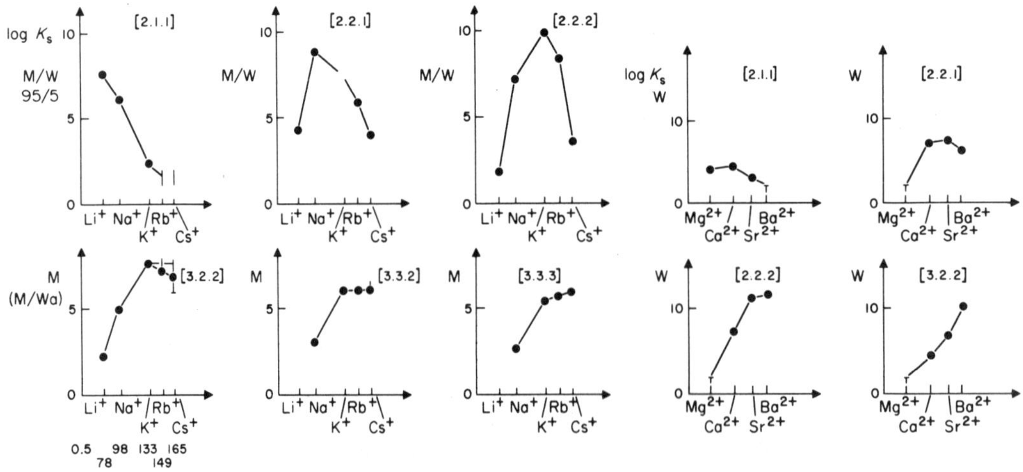

Figure 26 Stability constants (log K_s) of the alkali cryptates (left, in 95:5 methanol/water, M/W, or in pure methanol, M, at 25 °C) and of the alkaline earth cryptates (right, in water at 25 °C) (reproduced with permission from reference 29)

As discussed earlier, the optimum selector for an alkali cation is one in which a spherical cavity is offered for interaction. Ligand **(116)** has a rigid cavity of diameter *ca.* 3.6 Å and so would be predicted as ideal for complexing Cs^+ (ionic radius 1.65 Å). Up to now this ligand is actually the best one for selective complexation of cesium cations.[479]

The barium cation (ionic radius 1.43 Å) has a good compatibility with [2.2.2]cryptands (cavity diameter *ca.* 2.8 Å). This cryptand will solubilize even $BaSO_4$ in aqueous solution,[486] and this phenomenon has been investigated with a view to removal of $BaSO_4$ scale deposited by the use of sea water as an injection fluid in oil-producing formations. The scale deposition can lead to blockage of wells.[487]

(*iv*) *Conformational rigidity and flexibility of the ligand*

These factors are very much linked with cavity size. Ligands with small cavities are often rigid in nature simply because of cavity delineation. Larger ligands become more flexible and capable of undergoing often pronounced conformational changes. This leads to a potential availability of a wide range of cavities of variable dimensions. Rigid skeletons should display higher selectivities for individual cations, *i.e.* peak selectivities, and the more flexible ligands exhibit plateau selectivity. The improved selectivity of cryptands over coronands may be related to their more rigid nature.[33]

The unexpectedly large K_s for dibenzo-30-crown-10 and K^+ is consistent with the crystal structure and due to the wrapping round the metal of the ligand such that all donor sites are involved in complexation.[488]

The addition of rings into the skeleton of the polyether can also lead to a stiffening of the skeleton and produce selectivity shifts.[33]

(*v*) *Lipophilicity*

The lipophilic character of a ligand may be controlleɑ by the nature of the hydrocarbon residues present. Ligands with thick lipophilic shells shield the cations from solvent media and decrease the stability of the complex. This effect is four times more strongly felt by the doubly charged M^{2+} cations and so ligand lipophilicity can influence the selectivity between mono- and di-valent cations; the thicker the organic shell, the smaller the selectivity ratio for M^{2+}/M^+ cations.[478] Ba^{2+} and K^+ are of similar size and so if an appropriate cryptand is used, comparisons can be made. The introduction of benzo rings into [2.2.2]cryptand enhances its lipophilicity (**117, 118**); complexation using (**117**) does not lead to any significant change in Ba^{2+}/K^+ selectivity as the ligand still has an 'open side' to the solvent, but addition of the second benzo ring (**118**) leads to the stabilities becoming similar and a loss of selectivity.[489]

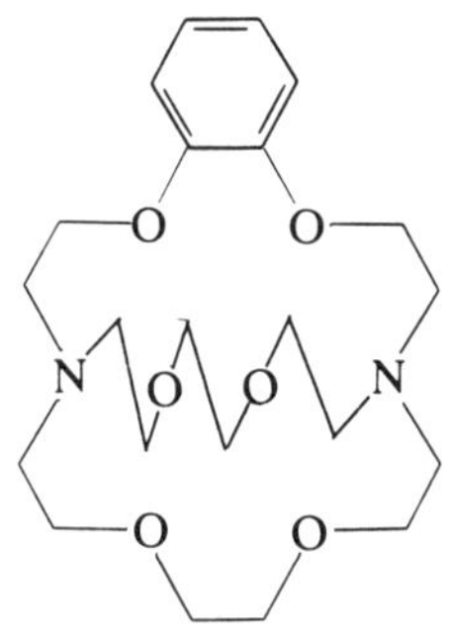

(**117**)

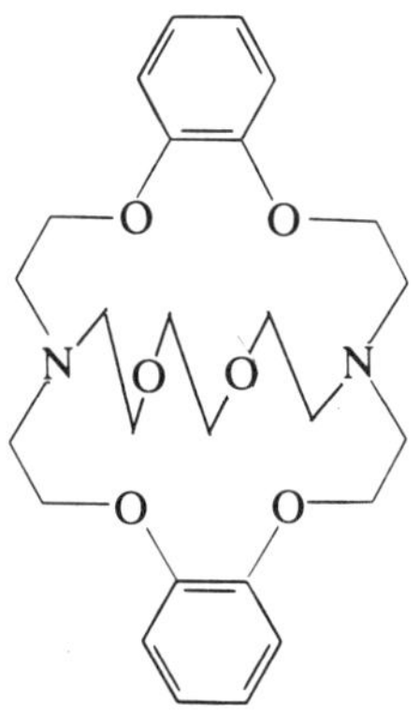

(**118**)

Competition between mono- and di-valent cations has an important role in biological processes. Furthermore, the lipophilicity of a ligand and its complex plays an important role in deciding whether a species is soluble in organic media of low polarity. This has important consequences in areas such as phase-transfer catalysis, the use of crown ethers as anion activators, and in cation transport through lipid membranes. Many crown ethers have now been synthesized with incorporation of long alkyl side chains and enhanced lipophilicity and used successfully in the above areas.

(vi) Electronic influences

The introduction of substituent groups into the benzo rings on dibenzo-18-crown-6 ethers leads to marked effects on their selectivity. A reversal of the usual selectivity sequences occurs when strong, electron-withdrawing substituents are present.[490] Similar effects are found for benzo-15-crown-5, and a good correlation is found with the Hammett functions of the substituents.[491] In contrast for substituted benzo-18-crown-6 systems, no such correlation was found, indicating that caution must be exercised on extrapolating from one crown ether system to another.

23.5.3.2 Comments on cation transfer

The discriminative uptake of alkali metal cations by biological systems, through their membranes, has been an area of much interest. In the membrane, the cations must pass through a lipid bilayer of low dielectric constant and this has led to the proposition that the cation could be selectively transferred *via* a 'carrier' molecule, or through a suitably donor-lined pore.[7–9] As a consequence of their selective properties, the polyethers and cryptands have been investigated as speculative models for the above process and selectivity sequences have been established.

Much of the preceding discussion has concerned the behaviour of alkali metal complexes in the solid state, and in an equilibrium state. Kinetic studies, to understand the dynamic stability of the complexes, have also been made. Such investigations for monocyclic crown ethers are hindered by several factors: the complexes are relatively weak and so must be studied at high cation concentrations; the rate constants are generally very high; and the complexes often do not exhibit properties commensurate with spectroscopic determinations of reaction rate constants.[33,290,492–495] The rates for the complexes studied are practically diffusion-controlled, therefore the complexes require appreciable conformational flexibility. During the complexation of dibenzo-30-crown-10 the cation is simultaneously desolvated and complexed; it has also been proposed that the loss of solvent is a dominant factor and that it could be a stepwise process in which an indeterminate number of water molecules are removed from the hydration sheath of the metal by the sinuous ligand. The energy barrier to decomplexation is believed to be the energy required to produce a conformational rearrangement of the ligand.[488,496]

The more rigid cryptands exchange cations more slowly and a modified stepwise mechanism of incorporation is agreed.[494] The pronounced selectivity of the cryptands for alkali metal cations is reflected in the dissociation steps; the formation rates increase only slightly with increasing cation size (Table 9). For a given metal the formation rates increase with increasing cryptand cavity size and for the [2.2.2]cryptands they are similar to the rates of solvent exchange in the inner sphere of the cations. This suggests that for the larger cations interaction between the cryptand and incoming cation can compensate effectively for the loss of solvation of the cation on complexation. Conformational change can occur during the process of

Table 9 Overall Rates for Complex Formation Between Cryptands and Alkali Metal Cations (MeOH, 25 °C)

Ligand	*Cation*	$\vec{K}$ ($l\,mol^{-1}\,s^{-1}$)	$\overleftarrow{K}$ ($l\,s^{-1}$)
[2.2.1] (**73**; $n = m = 0$, $l = 1$)	Li^+	4.8×10^5	4.4×10^{-3}
	Na^+	3.1×10^6	2.50
[2.2.1] (**73**; $n = 0$, $m = l = 1$)	Li^+	1.8×10^7	7.5×10
	Na^+	1.7×10^8	2.35×10^{-2}
	K^+	3.8×10^8	1.09
	Rb^+	4.1×10^8	7.5×10
	Cs^+	$\sim5\times10^8$	$\sim2.3\times10^4$
[2.2.2] (**73**; $n = m = l = 1$)	Na^+	2.7×10^8	2.87
	K^+	4.7×10^8	1.8×10^{-2}
	Rb^+	7.6×10^8	8.0×10^{-1}
	Cs^+	$\sim9\times10^8$	$\sim4\times10^4$

Data compiled from B. G. Cox, H. Schneider and J. Stroka, *J. Am. Chem. Soc.*, 1978, **100**, 4746.

complexation. In general, the most stable cryptates act as cation receptor complexes which release the metal only slowly, whereas the less stable ones can exchange more rapidly and so act as cation carriers.

Selectivity sequences in solvents such as water, methanol and ethanol do not guarantee a similar behaviour in the lipid membrane. Experiments have been carried out in attempts to investigate the selective transfer of cations across model membranes, and these are exemplified here by reference to an investigation concerning the cryptands [2.2.2], [3.2.2], [3.3.3] and [2.2.C_8]. Two aqueous phases (IN and OUT) were bridged by a chloroform layer into which the carrier can be dissolved. Alkali metal picrate was dissolved in two aqueous layers such that the IN layer was 1000 times more concentrated than the OUT layer. All layers were stirred and the transport monitored *via* increase in picrate in the OUT layer (UV) and increase in potassium in the OUT layer (atomic absorption). The membrane phase was also analyzed at the end of the experiment.[497]

The results (Table 10) show that the cryptands could act to produce carrier-mediated facilitated diffusion and there was no transport in the absence of the carrier. The rate of transport depended upon the cation and carrier, and the transport selectivity differed widely. The rates were not proportional to complex stability. There was an optimal stability of the cryptate complex for efficient transport, $\log K_s \approx 5$, and this value is similar to that for valinomycin (4.9 in methanol). [3.2.2] and [3.3.3] showed the same complexation selectivity for Na^+ and K^+ but opposing transport selectivities. The structural modification from [2.2.2] to [2.2.C_8] led to an enhanced carriage of both Na^+ and K^+ but K^+ was selected over Na^+. The modification changes an ion receptor into an ion carrier, and indicates that median range stability constants are required for transport. Similar, but less decisive, results have been found in experiments using open-chain ligands and crown ethers.[498]

Table 10 Rates and Selectivities of Alkali Metal Cation Transport *via* Cryptate Complexes

Carrier	*Cation*	*log K_s (in MeOH)*	*Cation conc. in membrane* (μ mol l^{-1})	*Carrier saturation* (%)	*Initial transport rate* (μ mol h^{-1})	*Transport selectivity, $K^+:Na^+$*
[2.2.2]	Na^+	7.2[a]	1400	95	0.6	1:20
	K^+	9.7[a]	>1400	>95	0.03	
[3.2.2]	Na^+	5.0	750	60	2.4	1:3.5
	K^+	7.0[a]	1100	75	0.7	
[3.3.3]	Na^+	2.7	110	10	1.5	1:0.55
	K^+	5.4	950	65	2.7	
[2.2.C_8]	Na^+	3.5	140	15	1.6	1:0.45
	K^+	5.2	850	80	3.6	

[a] These values are for methanol containing 5% water; higher values are expected for pure methanol.

Data compiled from M. Kirch and J. M. Lehn, *Angew. Chem., Int. Ed. Engl.*, 1975, **14**, 555.

Molecular channels are also thought to play important roles in ion exchange processes across membranes, and data from physiological and biophysical sources point to the existence of Na^+- and K^+-specific pores. The molecular nature of the ion flow through such channels remains obscure in contrast to the carrier-facilitated processes. The crystal structure of the potassium complex (**119**)1.5KBr·3.5H_2O of the chiral macrocyclic tetracarboxamide (**119**) has recently been solved and shows a solid-state model for a molecular channel in which the macrocyclic units are organized in a polymolecular stack. The K^+ cations are located alternatively inside and on top of successive macrocycles, as in a frozen picture of potassium ion carriage through the channel (Figure 27). The water molecules solvate the cation and amide groups and provide hydrogen-bonded links between the component molecules. The stepwise connection of macrocyclic units through their side chains should allow for the construction of an artificial molecular channel with selective properties dependent upon the choice of basic unit.[499]

Other recent developments include crown ethers which contain potential switch mechanisms for complexation and transport. When photoresponsive chromophores are linked to crown ethers, ion binding can change on photoirradiation. The crown (**120**) is formed as its *trans* isomer with no alkali metal affinity, but photoirradiation gives a *cis* isomer capable of alkali metal complexation;[500] this is an example of a reversible all or none ion-binding capability. A

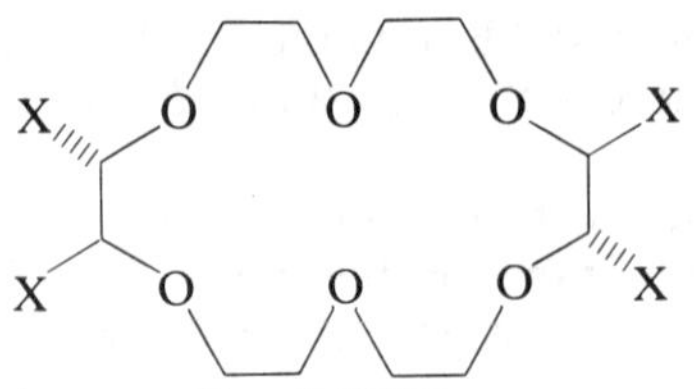

$X = CONMe_2$
(119)

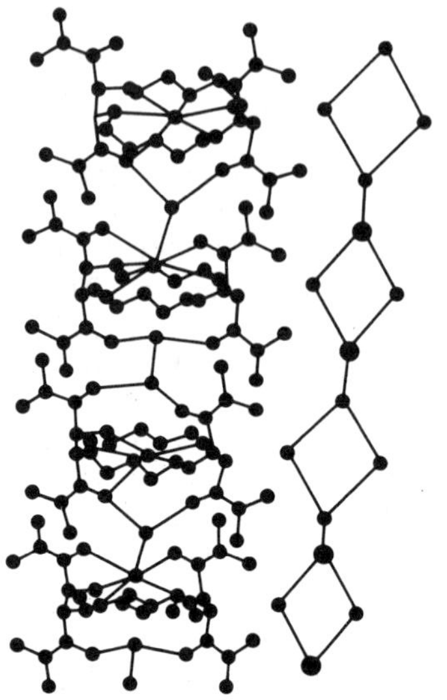

Figure 27 Molecular stacking in the KBr complex of (**119**) (reproduced with permission from reference 499)

redox switch can also be introduced into polyethers (**121**),[501] and solvent extraction studies show the ion affinity of the thiol crown to be significantly higher than the disulfide bis crown. The compounds (**122**) and (**123**) are found to be pre-ionophores, which on oxidation by Fe^{3+} are converted into disulfide-bridged monomacrocycles having M^+ transport capability. These systems have been proposed as models for biological 'ion gates'.[502]

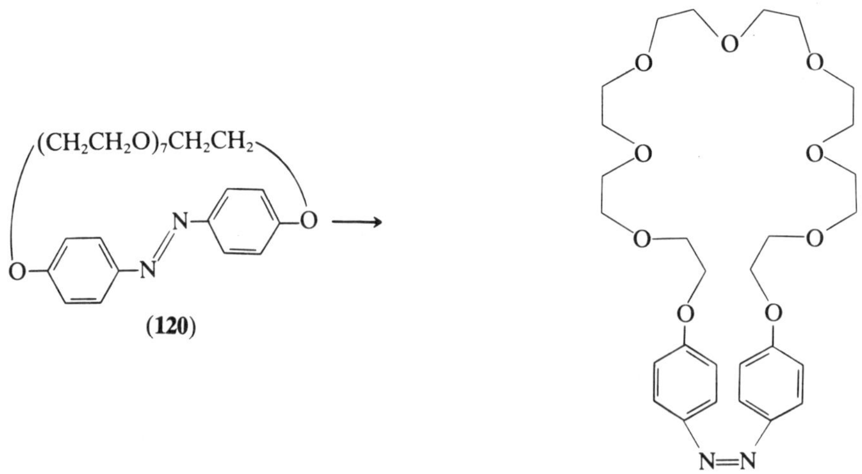

(120)

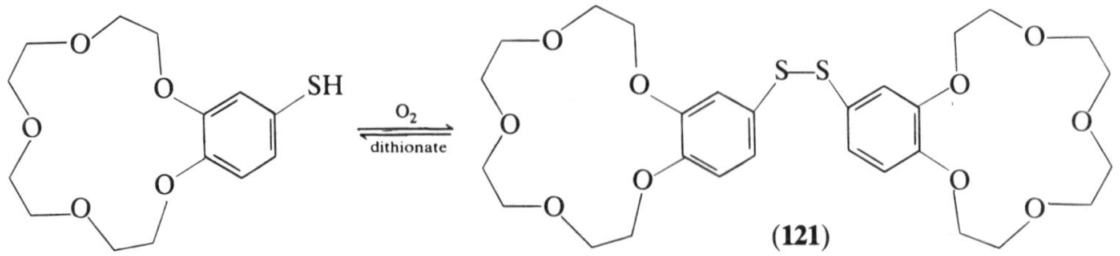

(121)

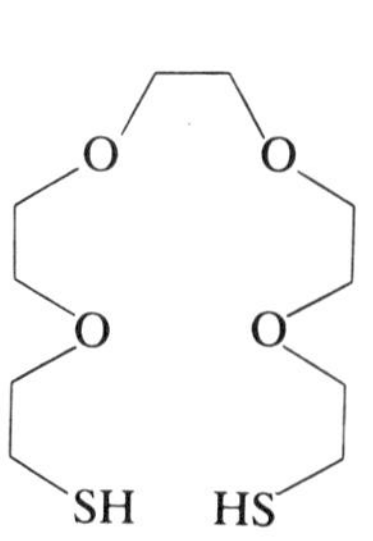

(122)

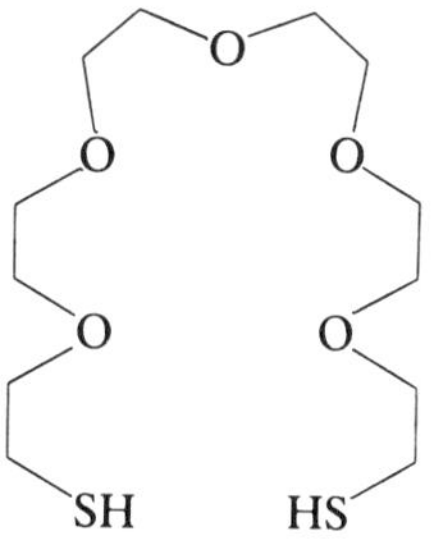

(123)

23.5.4 Quaterenes and Calixarenes

The reaction of furan with acetone leads to the isolation of (**124**); similarly (**125**) can be prepared from pyrrole with acetone, and the mixed species (**126**)–(**128**) are also available.[503] There is a donor cavity of 1.2–1.7 Å radius depending upon the construction and so it was thought that these would be suitable ligating systems for the alkali metals.[504] The fully reduced form of (**124**), (**129**), has been found to form 1:1 complexes with $LiClO_4$ and LiSCN,[505] and lithium and magnesium salts have been found to give enhanced yields of (**124**) leading to an invocation of the metal-template effect in the mechanism of synthesis.[506] There is also an anion effect and a pH control at the same time.[507] Conflicting information was presented on the solution complexation of M^+ by quaterenes. Monitoring of the NMR spectra, of absorption bands in the electronic spectra, and picrate extraction into organic phases indicated that (**124**)–(**129**) did not form any complexes.[504] However, work carried out using higher concentrations of reagents shows that (**129**) was indeed capable of complexing Li^+ and preferred this cation to Na^+ and K^+.[508] The larger quaterenes (**130**) and (**131**) were also investigated and showed K^+ preference in the case of (**130**) and no complexation at all using (**128**).[508]

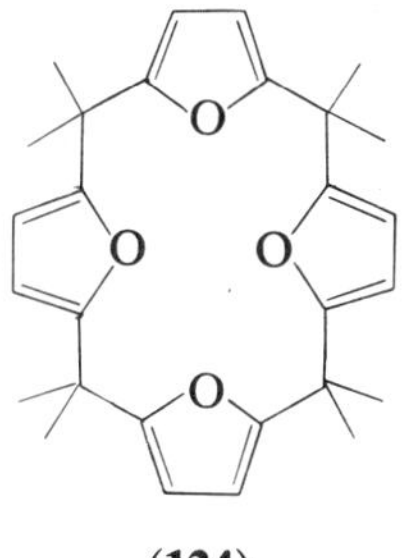
(**124**)

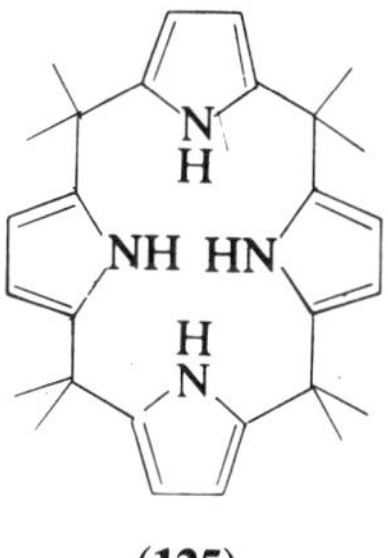
(**125**)

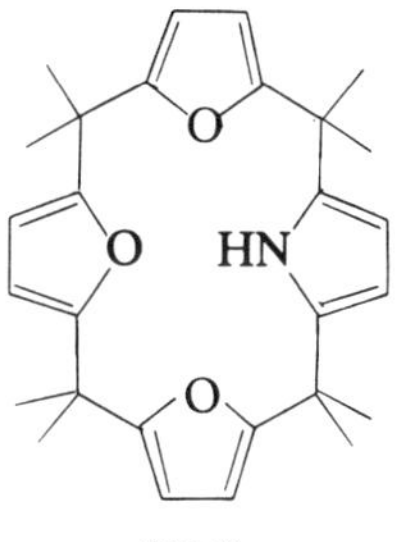
(**126**)

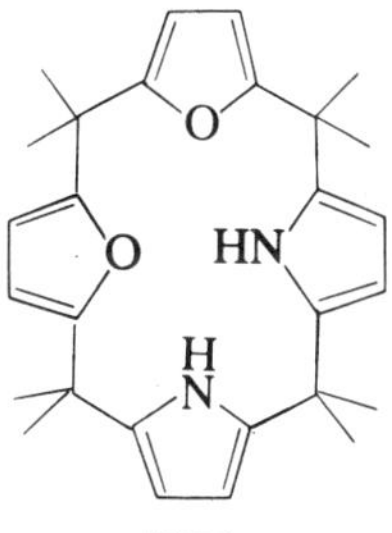
(**127**)

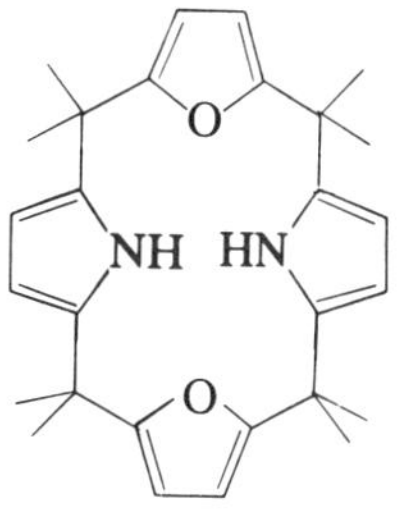
(**128**)

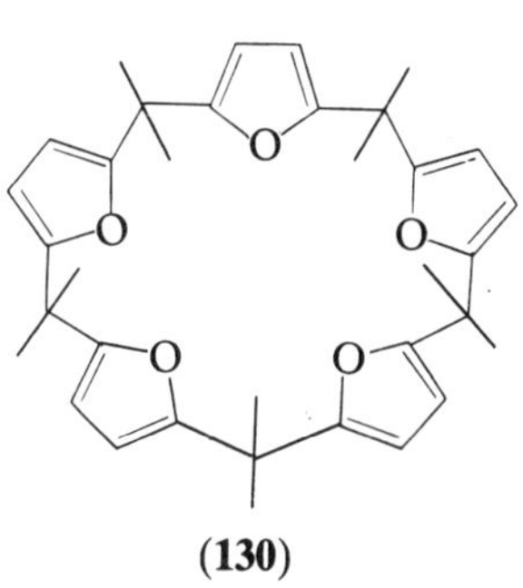
(**130**)

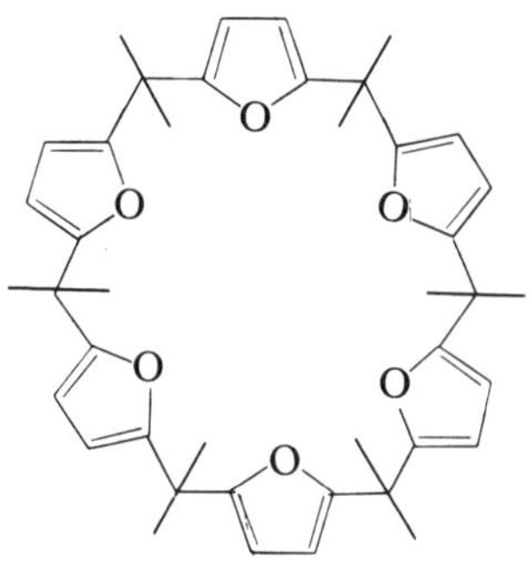
(**131**)

Macrocyclic phenol–formaldehyde condensation products have been termed calixarenes[509] and are capable of providing a cavity for complexation (**132**). Rb^+ was shown to be a good templating device during synthesis and this hinted at complex formation. Transport experiments have shown calixarenes-[4], -[6] and -[8] to be selective for Cs^+; using MNO_3 no transport was detected but with MOH it occurred. 18-Crown-6 behaved in a contrary fashion;

the transport was from an aqueous phase through a chlorocarbon membrane into an aqueous phase. The ligand sequence was found to be calixarene-[4] > calixarene-[6] > calixarene-[8] and cation selectivity was related to the hydration energy of the cation. It was noted that the cation with the lowest hydration energy was that most readily transported.[510]

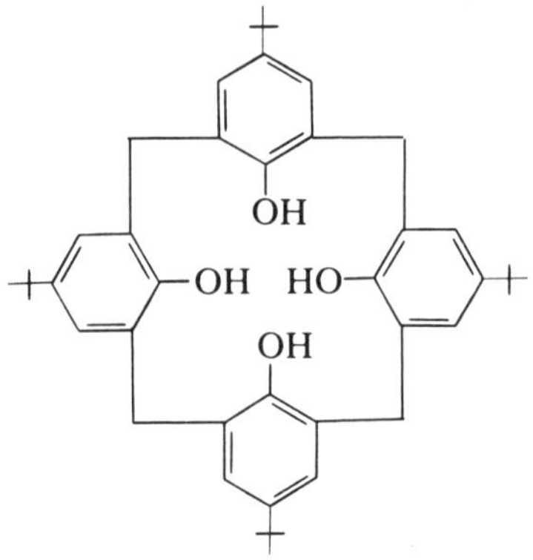

(132) *p-tert*-butylcalix[4]arene

23.5.5 Porphyrins, Phthalocyanines and Chlorophylls

Magnesium complexes of natural origin, the chlorophylls and related pigments, form a vast area of study in both the solid and solution states.[511–513] The chlorophyll ring system is a porphyrin in which a double bond of one of the pyrrole rings has been reduced **(133)**; a fused cyclopentanone ring is also present. Chlorophyll is crucial to the process of photosynthesis; photons of light hitting a molecule of chlorophyll provide the energy for a series of redox reactions to occur. The magnesium can be found either centrally bound as in **(133)**, or peripherally located on the exterior of the macromolecule. The latter interactions influence molecular aggregation and spectral features, whereas the former siting determines the basic chemistry of the pigment.

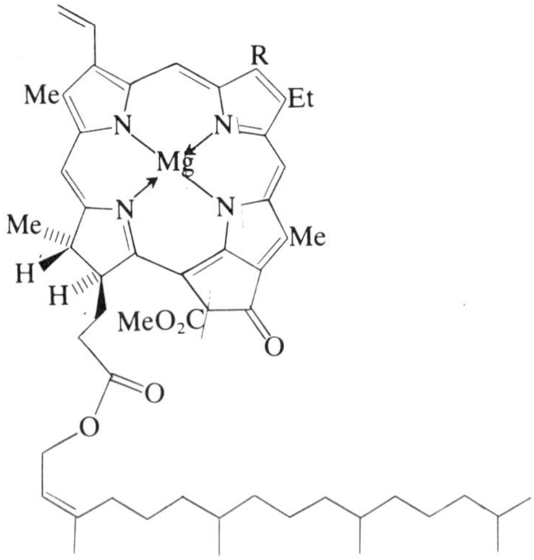

(133) R = Me, chlorophyll *a*
R = CHO, chlorophyll *b*

The molecular structure of ethylchlorophyllide *a* dihydrate,[514] a chlorophyll molecule in which the phytyl side chain has been replaced by an ethyl group, shows one water molecule coordinated to the magnesium and forming a hydrogen bond with the ring V keto group of an adjacent molecule. This in turn has a magnesium-coordinated water molecule hydrogen-bonded to a third molecule and so on to produce a one-dimensional array of overlapping chlorophyll molecules.

Bis-chlorophylls **(134)**[515–517] and 'cyclophane'-chlorophylls **(135)**[518] are reported to exhibit several photochemical properties which mimic the *in vivo* 'special pair' chlorophylls. By constructing these dimeric units it has been possible to account for the redox and spin delocalization properties of the chlorophyll pairs *in vivo* but not yet to produce the unusually red-shifted spectrum. Such synthetic systems are of interest as evidence relating to the molecular organization of chlorophyll in the photoreactive centres of green plants and photosynthetic bacteria suggest that special pairs of chlorophyll molecules are oxidized in the primary light conversion event in photosynthesis.

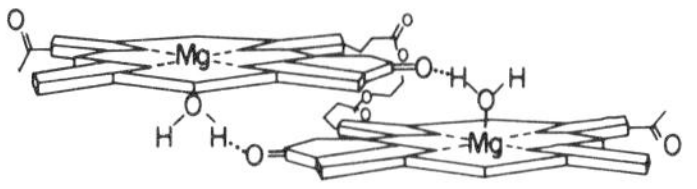

(134)

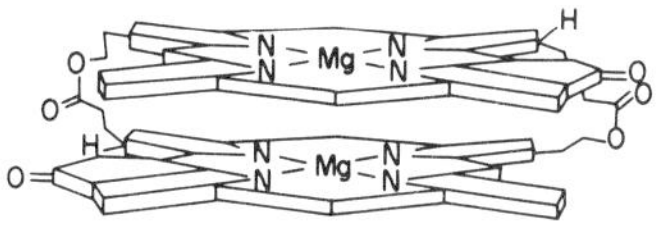

(135)

Studies have also been carried out using porphyrin and phthalocyanine systems as models for the site in chlorophyll. The structure of Mg(tetraphenylporphyrin)·H_2O was solved and showed a disordered system, but it was possible to find the cation in a square pyramidal environment with the water molecule in the apical site.[519] In Mg(phthalocyanine)·H_2O·2py, the magnesium is similarly coordinated, the pyridines not being involved with the metal.[520] The magnesium is displaced from the basal donor plane towards the apical ligand as a common feature. In the latter compound it is a significant displacement of 0.49 Å and is assisted by hydrogen bonding of the water to a proximal pyridine.

The magnesium porphyrin radical, Mg(tetraphenylporphyrin)ClO_4 (**136**), has been used as a model for the structural and stereochemical consequences of loss of an electron in photosynthetic chromophores.[521] The primary photosynthetic reaction in plants and bacteria consists of a transfer of an electron, in the picosecond time domain, from the chlorophyll phototrap to nearby acceptors yielding chlorophyll π-cation radicals. The structure of (**136**) shows a five-coordinated Mg^{2+} cation which is not quite symmetrically sited in the porphyrin ring but has metal–ligand distances similar to those found in the previous structures. The perchlorate anion is tightly bound (Mg—O = 2.01 Å) in a monodentate mode. It was concluded that the porphyrin can act as an electron sink and that no major effects are found in the bond lengths, or on the stereochemistry, of the macrocycle.

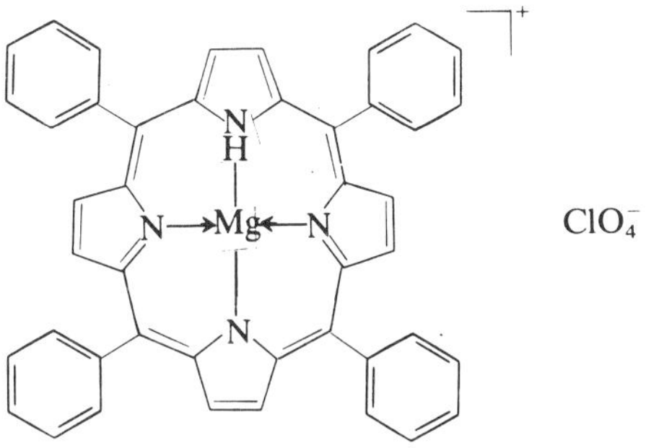

(136)

Phthalocyanines have been found to give other alkali and alkaline earth metal complexes.[522] The reaction of alkali metal amyl oxides (M = Li, Na, K) with phthalonitrile yields M_2(phthalocyanine). In the case of lithium a second, monometallic species believed to be MH(phthalocyanine) is recovered. With the alkaline earth metals Be, Mg, Ca and Ba complexes are available. Be(phthalocyanine) is prepared from the reaction of beryllium and phthalonitrile and gives a dihydrate on standing. The structure of Be(phthalocyanine) shows a unique, square planar environment for the cation.[523] It is likely that in the formation of the dihydrate two bonds in the complex are broken and a tetrahedral (N_2O_2) environment is recovered for the cation. Mg(phthalocyanine) is the only other dihydrated metal derivative of this ligand and it also forms bis adducts with a variety of donor ligands bearing oxygen, nitrogen and sulfur atoms. Mg(tetraphenylporphyrin) behaves in a similar fashion with simple donors; induced Cotton effects in the CD and ORD spectra of Mg(porphyrin) amino acid systems have been interpreted in terms of Mg(porphyrin)·2(amino acid) complexes.

23.6 NATURALLY OCCURRING IONOPHOROUS AGENTS

The 1964 studies, made by Pressman, on the effects of the antibiotic valinomycin on mitochondria showed its ability to induce the selective movement of K^+ into the mitochondria.[28] Valinomycin, and other antibiotics neutral at physiological pH, were considered as discriminatory cation carriers and such species which could carry cations across lipid barriers, such as those found in membranes, as lipid-soluble complexes were collectively termed ionophores, or ionophorous agents.[526] A second group of antibiotics containing carboxylic acid functions was also found to exhibit discriminatory powers, and to reverse the effects of valinomycin.[527] The term ionophore had actually been introduced several years previously in suggested nomenclature for conductance.[528] It was used to designate a species in which ions, and only ions, were present in a crystal lattice; a substance whose molecular crystal lattice produced an electrolyte when dissolved in an appropriate solvent was termed an ionogen. This nomenclature does not appear to have found general use, whereas the more contemporary definition is widely accepted.

Pressman has described his discovery as serendipitous[529] and it is quite remarkable that modern alkali metal coordination chemistry should have as its base the two fortuitous discoveries of Pressman and Pedersen. It is of interest to read the engaging accounts written by these two workers of their early work and to see how work in one area can often lead to major advances in another.[530,531]

There are several excellent reviews concerning ionophorous agents and showing their biological value as well as their structural aspects.[2,226,372,493,529,532–535] Physical studies on these compounds have indicated that the complexation–decomplexation kinetics and diffusion rates of ionophores and their metal complexes across lipid barriers are extremely favourable.[290,493,494] This, along with high cation selectivity, inspired their use as model biological carriers. Representative selectivity sequences for the ionophorous antibiotics, and comparisons with crown ethers are given in Table 11. These sequences correlate well with alkali metal selectivity patterns predicted by Eisenman from studies on ion selective glass electrodes.[536] Using an electromechanical model, it was shown that ion mobility is a function of the ion's mobility in a glass lattice and of its ability to become desolvated and enter the glass. This is not an entirely true analogy as the ionophorous agents are more flexible and mobile than the glass matrix.

Table 11 Representative Ion Selectivities for Ionophorous Agents and Crown Ethers

Valinomycin	Rb > K > Cs > Na > Li	Ba > Ca > Sr > Mg
Enniatin A	K > Rb ≈ Na > Cs ≫ Li	
Enniatin B	Rb > K > Cs > Na ≫ Li	
Beauveracin	Rb ≈ Cs > K ≫ Na > Li	
Nonactin	K ≈ Rb > Cs > Na	
Antamanide	Na > Li > K > Rb > Cs	
Monensin	Na ≫ K > Rb > Li > Cs	
Nigericin	K > Rb > Na > Cs > Li	
Lasolocid	Cs > Rb ≈ K > Na > Li	Ba > Sr > Ca > Mg
A 23187	Li > Na > K	Ca > Mg > Sr > Ba
Dianemycin	Cs > Rb ≈ K > Na > Li	Ba > Ca ≈ Mg
Benzo-15-crown-5	Na > K > Cs > Li	
Dibenzo-18-crown-6	K > Cs > Na > Li	
Dicyclohexyl-18-crown-6	K > Cs > Na > Li	

The ionophorous agents have been divided into three groups: neutral ionophores, carboxylic ionophores and quasi-ionophores.[529] Neutral ionophores include cyclodepsipeptides such as valinomycin (**137**), the enniatins (**138**) and beauveracin (**139**); cyclic peptides such as antamanide (**140**); and the macrotetralide actins (**141**). Carboxylic ionophores are comprised of monovalent polyether antibiotics such as monensin (**142**) and nigericin (**143**); monovalent monoglycoside polyether antibiotics such as dianemycin (**144**); divalent polyethers such as lasolocid (X-537A) (**145**); and divalent pyrrole polyether antibiotics such as the antibiotic A 23187 (**146**). The quasi-ionophores are channel-forming species such as the gramicidins, alamethicin and monazomycin.

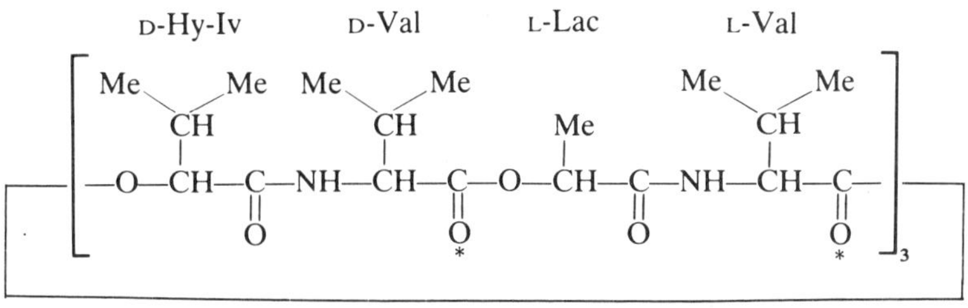

(**137**) Valinomycin

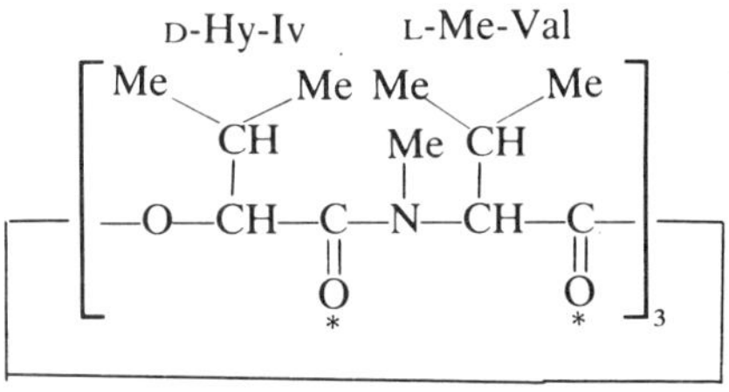

(**138**) Enniatin B

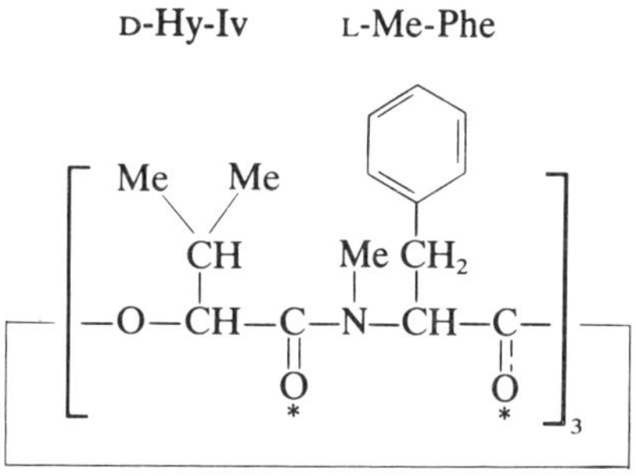

(**139**) Beauvericin

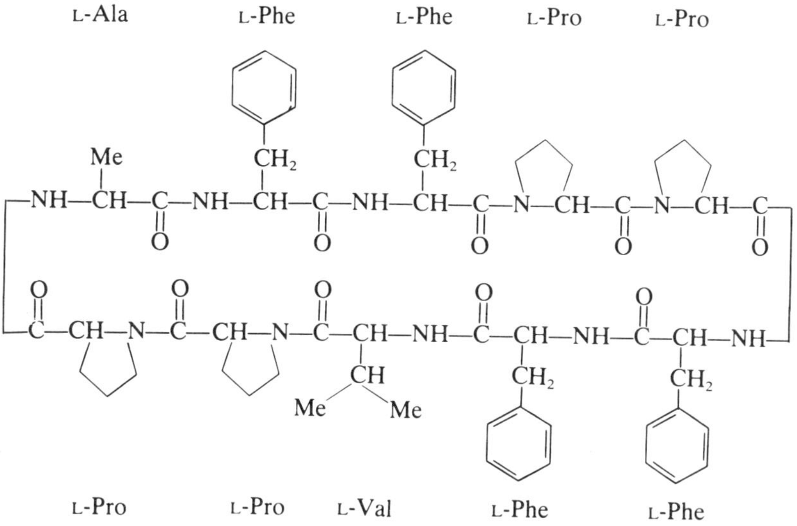

(**140**) Antamanide

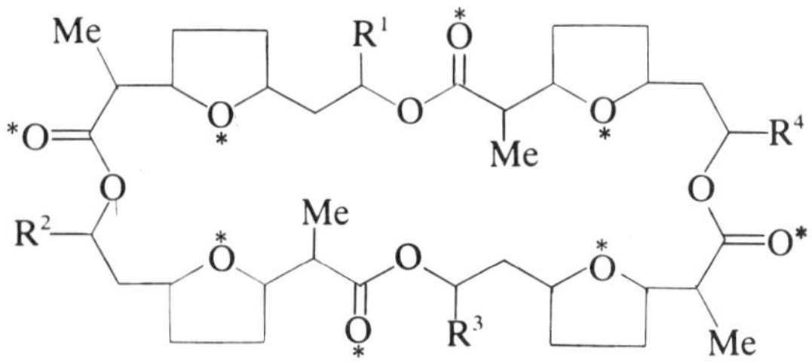

a; $R^1 = R^2 = R^3 = R^4 = Me$ Nonactin
b; $R^1 = R^2 = R^3 = Me$, $R^4 = Et$ Monactin
c; $R^1 = R^3 = Me$, $R^2 = R^4 = Et$ Dinactin
d; $R^1 = Me$, $R^2 = R^3 = R^4 = Et$ Trinactin
e; $R^1 = R^2 = R^3 = R^4 = Et$ Tetranactin

(**141**)

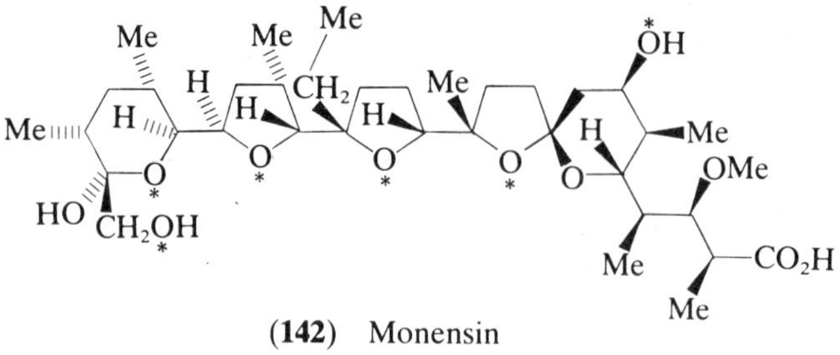

(**142**) Monensin

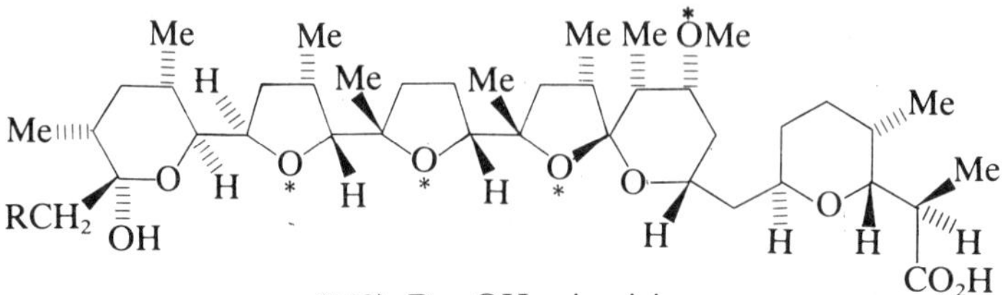

(**143**) R = OH, nigericin
(= Polyetherin A)

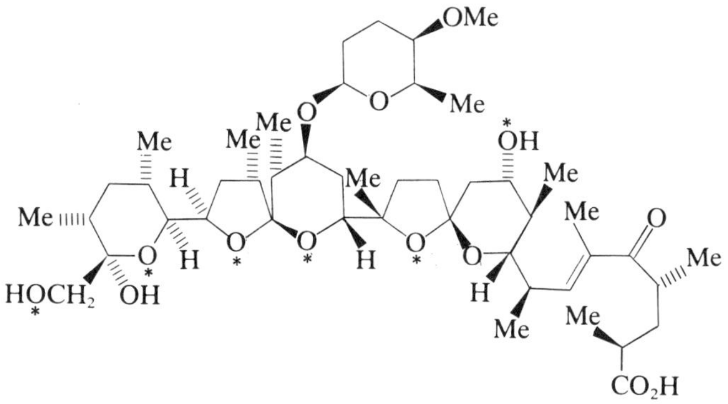

(**144**) Dianemycin

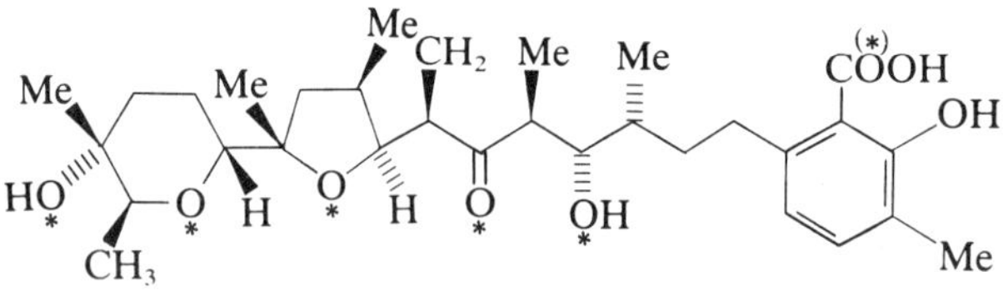

(**145**) X-537 A
(= Lasolacid)

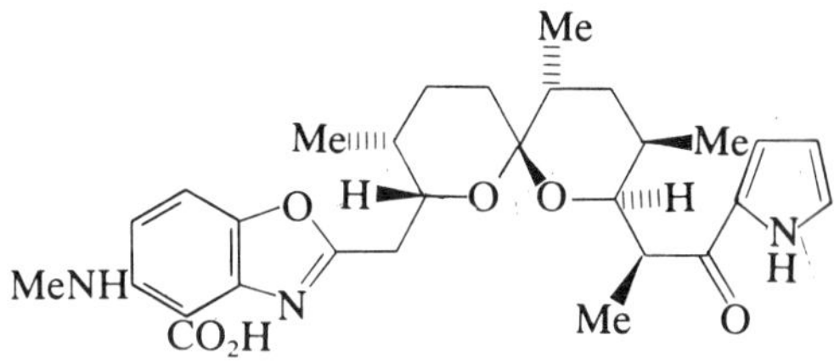

(**146**) Antibiotic A23187

23.6.1 Neutral Ionophores

Valinomycin (**137**) is a macrocyclic dodecadepsipeptide and has shown an unequalled K^+/Na discrimination.[529] As a consequence of this prime ionophorous behaviour many studies have been carried out on both the free ligand and on its metal complexes, and here discussion is restricted to the nature of the metal complexes.

The overall features found in the X-ray structures of (**137**)KX ($X^- = AuCl_4$,[537] I_3/I_5,[538] picrate[539]) are comparable. There is a close to three-fold symmetry of the complex, the cation being located at the centre of the 36-membered ring (Figure 28). The ring is folded around the cation as would three sine waves and these are held in position by six intramolecular hydrogen bonds. The cation is coordinated by six carbonyl oxygens from ester groups in an almost perfectly octahedral arrangement. The cation is thus effectively screened from solvent interaction, although in the structure of (**137**)K(picrate) there is a weak but definite interaction between a picrate oxygen and the cation. All of the lipophilic side chains are orientated on the periphery of the molecule. The structure of (**137**)$RbAuCl_4$ reveals essentially the same features as in the K^+ complex, K^+ and Rb^+ having similar ionic radii.

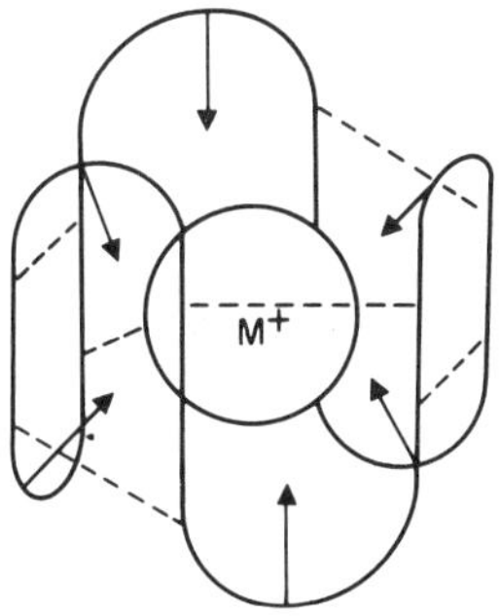

Figure 28 Schematic representation of the valinomycin K^+ complex

The hydrogen bonding in valinomycin leads to a limited flexibility of the ligand in the complex: the free ligand is much more open, and ellipsoidal in nature.[541] Consequently (**137**) is not able to adjust itself to accommodate ions of different radii, and this helps explain its high K^+/Na^+ discrimination. Solution studies on the Na^+ complex indicate an asymmetry of the ester carbonyl stretching frequencies indicating a non-equivalence of these groups in the structure.[542] The smaller size of Na^+ would lead to this observation as the cation would be unable to interact fully with all six ester carbonyls unless there is an accompanying, unfavourable disruption of the hydrogen bonding. Cs^+ would be too large to fit into the cavity and so again would be discriminated against.[372]

The selectivity ratio $K^+:Na^+$ at 25 °C is of the order $10^4:1$ whereas at 0 °C it is reduced to only 2:1.[543] This dramatic decrease has been interpreted as indirect evidence for the possible role of (**137**) as a carrier in membranes as, if it provided a pore mechanism, this would be unimpaired by freezing. In a carrier mechanism which necessarily involves a mobile ligand to effect incorporation and transfer of the metal freezing would cause a loss of mobility and so impair the mechanism. The high K^+ selectivity of (**137**) has led to its employment in ion-selective electrodes.[543]

The enniatins and beauveracin are cyclic hexadepsipeptides and so are 18-membered macrocycles built up of alternating amino acid and carboxylic acid residues. They exhibit lower selectivity than (**137**) and consequently are more broad spectrum ionophores. The structure of (**138**)KI has been solved[544] but it was not possible to conclude, with certainty, whether the metal is entrapped in the central cavity, as in related cyclic polyether structures (Figure 14a), or whether it occupies a site between two adjacent ligand molecules to give an infinite sandwich (**147**) as has been found in the RbNCS complex of a synthetic LDLLDL isomer of (**138**).[545] Adducts of 1:2 'sandwich' and 2:3 'club sandwich' stoichiometry have been proposed for Cs^+ complexes of (**138**),[546a] and the 1:2 K^+ 'sandwich' complex of (**138**) has been proposed as the species which transports K^+ across lipid layers as it shields the metal effectively from solvent interactions.[546b]

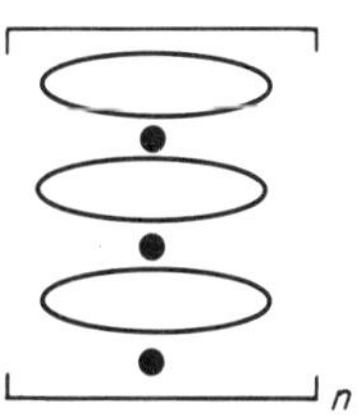

(**147**)

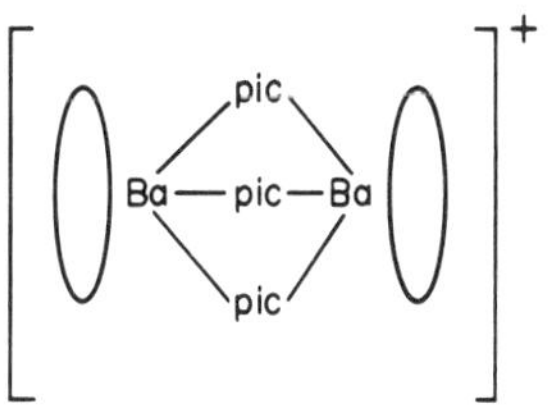

(**148**)

The crystal structure of a barium picrate complex of (**139**) reveals a 2:2 dimeric structure of the form [(**139**)Ba(picrate)$_3$Ba(**139**)]$^+$(picrate)$^-$ (**148**),[547] which is unique as three picrate anions are held in the space between the ligated cations. The cations are not sited within the macrocyclic cavities but are displaced towards the centre of the dimer. This situation is reminiscent of the Cs^+–polyether dimer structure (Figure 15d). The outer surface of the (**139**)Ba(picrate)$_2$ complex is highly lipophilic, and the bulky phenyl residues provide screening for the enclosed anions.

Antamanide is a decapeptide which showed preference for Na^+ and Li^+ in solution and selectivity studies.[548] The crystal structures of (**140**)LiBr·MeCN[549] and (**140**-Phe^4,Val^6)NaBr·EtOH[550] have been solved and show the complexes to be essentially isostructural. The backbone of the antamanide moiety can be described as resembling the periphery of a saddle and the alkali cation is located near the mathematical saddle part. Four carbonyl oxygen atoms coordinate to the metal in an almost coplanar geometry and the complex may be considered as cup-like in this region. A molecule of solvent fills the mouth of the cup and gives the metal a square pyramidal coordination (Figure 29). The cage is stabilized by hydrogen bonding. (**140**) has not yet been observed to act as an ion carrier through membranes.

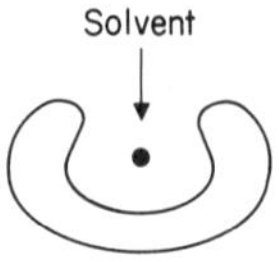

Figure 29 Schematic representation of the antamanide Na^+ complex

The final group of naturally occurring ionophorous agents giving molecular complexes is the macrotetrolide actins (**141**). For this group a more detailed accumulation of structural information allows for a more precise understanding of the selective relationships between the ligands and the cations. The ligands (**141**) are 32-membered cyclic tetralactones, built up of four ω-hydroxycarboxylic acid subunits condensed to each other by esterification. These ionophores exhibit high alkali metal cation selectivity rather like (**137**).[529] Much structural information is available for complexes of nonactin (**141a**) and tetranactin (**141e**).[372] The structures of the free ligands show a large cavity to be present; early consideration of the nature of the alkali metal complexes, based on molecular models, led to the conclusion that the cavity was able to incorporate hydrated cations.[551] However, as in the case of (**137**), the cavity serves as a binding site for the cation and the complexation process is accompanied by much conformational change.

The overall structure found first in the (nonactin)KNCS complex[552] is retained in changing the metal present. The ligand wraps itself around the cation as the seam of a tennis ball (Figure 30) and the cation is thence eight-fold coordinated by the four furanyl and four carbonyl oxygen donor atoms. The periphery of the molecule is lipophilic and the lowered lipophilicity of nonactin relative to tetranactin leads to a diminished transfer of these compounds.

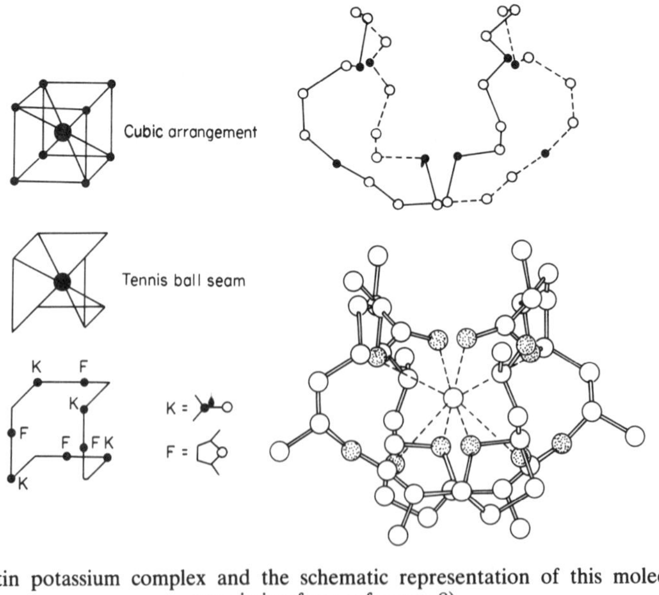

Figure 30 The nonactin potassium complex and the schematic representation of this molecule (reproduced with permission from reference 8)

The different stabilities of the metal nonactin complexes can be rationalized in terms of the interactions within the overall spatial framework. In the K^+ complexes (**141a**)KNCS and (**141e**)KNCS,[553] the distances between the cation and the carbonyl oxygens are slightly less than those to the ether oxygens (Table 12), and the former are close to the sum of the ionic radius of the cation and the van der Waals radius of oxygen. In the (**141a**)–NaNCS complex the carbonyl oxygen atoms approach the cation but the ether oxygens are restricted in their approach by intramolecular steric hindrance and this is seen in the disparate bond distances.[554] The sodium to ether distances are long (2.70–2.94 Å) and in excess of the sum of the relevant radii (2.35 Å), thus indicating a lowered electrostatic interaction and leading to a distorted cubic coordination geometry. The inability of the ligand to readjust to the required optimum cavity for sodium leads to a reduced stability of the sodium complexes. Complexes of the smaller lithium cation have not yet been observed for (**141**). Incorporation of the larger Rb^+ and Cs^+ cations leads to an expansion of the ligand which involves an outward displacement of the carbonyl oxygen atoms.[553,555] In these species, the observation of bond distances lower than the sum of the component radii suggests the presence of steric strain, which would diminish the stability of the complexes. Solution studies reveal, as with valinomycin, almost identical structures in solution compared with those found in the solid state.[372]

Table 12 Structures of Macrotetrolide Actin Complexes

Complex	*Coordination number*	*Bond distances* (Å) *M—O (carb)*	*M—O (ether)*
Nonactin·NaNCS	8	2.40–2.44	2.74–2.79
KNCS	8	2.73–2.81	2.81–2.88
CsNCS	8	3.13–3.18	3.07–3.16
$Ca(ClO_4)_2$	8	2.33–2.36	2.60–2.61
Tetranactin·NaNCS	8	2.43–2.45	2.70–2.94
KNCS	8	2.75–2.81	2.85–2.92
RbNCS	8	2.88–2.93	2.90–2.98
CsNCS	8	3.06–3.16	3.03–3.10

Data compiled from R. Hilgenfeld and W. R. Saenger, *Top. Curr. Chem.*, 1982, **101**, 1.

The structure of (**141a**)$Ca(ClO_4)_2$ has been solved and shows a close similarity to that of the sodium complex. This is not surprising due to the close similarity in ionic radii between sodium and calcium.[556]

Certain common features may be found in the complexation of alkali and alkaline earth metal cations by ionophorous agents and which aid in cation transport.[8]

(1) The cation sits in the ligand cavity at a centre of optimum electron density provided by the donor atoms of the ligand.

(2) A flexible ligand is required to effect the energetically favoured stepwise removal of the cation solvation sheath. Upon complexation the cation must be stripped of its hydration shell, a process requiring the loss of about 322 kJ mol^{-1} for K^+. This loss must be compensated for by interactions between the cation and the donor atoms of the ligand. Complex formation has been shown to be very fast and stepwise incorporation of the cation by the ligand is essential.[290] Models show that the hydrated K^+ cation could be inserted into the ligand cavity (of **137** and **141**) and held there by hydrogen bonding. Through a series of ligand conformational changes, the water molecules could then be removed and the cation held in the cavity by ion–dipolar interactions.

(3) A lipophilic exterior is presented to facilitate solubilization and cation transport.

(4) A 'best-fit' situation is required with regard to the cavity diameter and the diameter of the incoming cation, but ligand–ligand repulsions must be minimized.

(5) The difference between the energy of the ligation and the energy of solvation must be maximized.

On fulfilment of these requirements, the metal ion is efficiently coordinated, and transported, by the ligand.

An extensive range of synthetic analogues of valinomycin and the enniatins has been prepared and investigated; these studies have been fully described in a monograph.[532] Synthetic peptides have also been prepared and their complexing properties investigated; for example $[\text{L-Val-D-Pro-D-Val-L-Pro}]_3$ reacts with potassium picrate to give an isolable 1:1 complex which is expected to have a structure similar to that of the K^+ complex of (**137**).[557] More detailed studies have been made on the complexation properties of the small synthetic peptides $[\text{L-Pro-Gly}]_3$ and $[\text{L-Pro-Gly}]_4$.

The cation-binding properties of [L-Pro-Gly]$_3$ parallel those of the related naturally occurring ionophores. Solution studies have indicated the availability of 1:1 complexes with Li^+, Na^+, Rb^+, Cs^+ and Ca^{2+}, but Mg^{2+} gave three different stoichiometries, 1:2, 1:1 and 2:1.[558] X-Ray structural studies have been carried out on the magnesium and calcium perchlorate complexes.[559] In the former a 1:1 interaction is found between the peptide and the cation, which is in an octahedral environment provided by three glycyl carbonyl oxygens and three water molecules; these water molecules are hydrogen-bonded to an adjacent peptide and so the molecules are arranged in an infinite stack, and the composite is not unlike a channel or pore. In the Ca^{2+} complex a sandwich structure (1:2) is found and the cation is octahedrally coordinated by six glycol carbonyl oxygens, three from each ring.

[L-Pro-Gly]$_4$ also forms complexes with alkali and alkaline earth cations showing a preference, in solution, for the larger cations. Solution studies revealed 1:1 stoichiometries for Li^+, Na^+, K^+, Cs^+, Ca^{2+}, Mg^{2+} and Ba^{2+} in water, and both 1:1 and 1:2 ratios for these cations in acetonitrile. Na^+ also gave a bimetallic 2:1 complex in acetonitrile.[560] The structure of the Rb^+ complex [L-Pro-Gly]$_4$RbNCS·3H_2O was solved and showed a dimeric structure in which Rb^+ cation was octahedrally coordinated by four glycyl carbonyl oxygens from one ligand, one glycyl carbonyl oxygen from the second ligand and a water molecule.[561]

A further sandwich structure was found for [Gly-L-Pro-L-Pro-Gly-L-Pro-L-Pro]$_2$Mg(ClO_4)$_2$·4MeCN. The cation is in a regular octahedral coordination sphere with two glycyl and one prolyl carbonyl oxygen from each peptide providing the donor groups.[562]

A series of synthetic 32-membered ring macrotetrolides related to the actins have been prepared and show cation transfer ability. The K^+/Na^+ selectivity is however inferior to that of the actins and cyclic polyethers.[563]

23.6.2 Carboxylic Ionophores

These are so described because they are linear and contain a terminal carboxy group, one or two hydroxy groups at the other end of the molecule, and several other oxygen donors provided by the constituents of the chain. In contrast to the neutral ionophores which provide cationic complexes, this group of ligands generally provides neutral salts because, at physiological pH, their carboxy group is dissociated. Carboxylic ionophores have been subdivided according to their ability to coordinate to alkaline earth cations and so those ionophores not capable of transporting divalent cations are termed monovalent polyethers and those capable of transporting divalent cations are termed divalent polyethers, and further subdivisions are available according to the nature of the constituents of these compounds.[529,535]

The ion selectivities displayed by these antibiotics are lower than those of the neutral ionophores, and are given in Table 11. Many studies have been made on the properties of these ionophores, particularly with reference to the calcium-transporting abilities of A 23187 (**146**) and lasolocid (**145**). The search for new antibiotics is ongoing and there is constant addition to a list of about 50 distinct polyether antibiotics which have been isolated from various streptomycetes. Representative structures will be discussed here to illustrate the nature of complexation with alkali and alkaline earth metal cations.

Monensin (**142**) provides examples of complexation by both the anionic and neutral ligand. The structure of the Ag^+ salt was first solved and showed that the monensin anion is wrapped around the cation and that it is held in this conformation by two strong head-to-tail hydrogen bonds between the carboxylate group and the hydroxy functions at the opposite terminus (Figure 31a).[564] The cation is in an irregular six-coordination provided by four ether and two hydroxy oxygen donors. The carboxylates do not interact with the cation but are only involved in the hydrogen bonding. This shows that the cation is encapsulated in a pseudo-cavity, and the ligand provides a lipophilic exterior for the carrier. The crystal structures of the anhydrous and monohydrated sodium salts of (**142**) have extremely similar structures,[565] as does that of the NaBr complex of the acid,[566] where the antibiotic acts as a neutral ligand. The structure of the free acid of (**142**) also reveals a pseudo-cyclic conformation held in place by head-to-tail hydrogen bonding, and having a water molecule located in the ligand cavity.[567] This molecule could provide a route into the cavity for an incoming cation by involvement in the solvation shell of the cation prior to ligand stripping and incorporation. Spectroscopic studies provide comparable results for the conformation of the complexes in solution.[372]

The structural features described for monensin have been found in all the monovalent polyether antibiotic complexes studied. In some instances, *e.g.* nigericin[568] and dianemycin,[569]

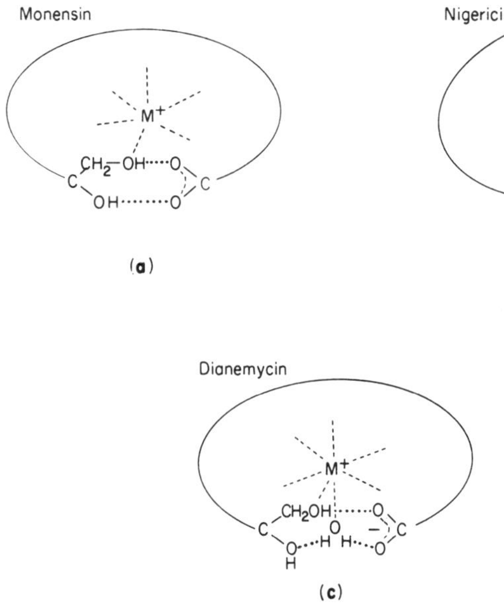

Figure 31 Schematic representations of some metal complexes of acid antibiotics (reproduced with permission from reference 226)

the carboxylic groups are involved in metal complexation as well as in the 'head-to-tail' hydrogen bonding locking scheme (Figure 31b, c). The coordination number of the cation is variable: (monensin)Na^+, 6; (nigericin)Na^+, 5; (nigericin)K^+, 7; (dianemycin)Na^+K^+, 7.[372]

The divalent polyether antibiotics have revealed the existence of dimeric molecules as monomeric complexes in their interaction with alkali and alkaline earth cations. Lasolocid (**145**) forms complexes of the type M^+L^- and these may be monomeric when formed and recovered from methanol, but dimeric when recrystallized from the non-polar carbon tetrachloride. The shortened ligand backbone relative to the monovalent polyethers does not allow complete shielding of the cation from the solvent, and so in non-polar solvents this is overcome by dimerization. If the salicylate and tetrahydropyran moieties present in (**145**) are referred to as the 'head' and the 'tail' of the molecule then two forms of dimerization may be envisaged: 'head-to-tail' and 'head-to-head'.

Discrete dimers of the 'head-to-head' type have been found in the structures of the Ag^+ complex of (**145**)[570] and the Na^+ complex of (**145**)[571] respectively. The complexes were recrystallized from carbon tetrachloride. In both complexes each metal is five-coordinated in the cavity provided by one anion, and there is an additional reaction with the second anion [through an Ag^+–phenyl interaction or an Na^+–carboxylate oxygen atom (Figure 32a)]. When the Na^+ complex was crystallized from a solvent of medium polarity, acetone, the 'head-to-head' dimer was recovered.[571] In contrast, recrystallization from a polar medium, methanol, gave a monomeric complex in which one methanol of solvation was also present.[572] In all of these complexes an intramolecular 'head-to-tail' hydrogen bond was present to hold the ligand in its pseudo-macrocyclic conformation.

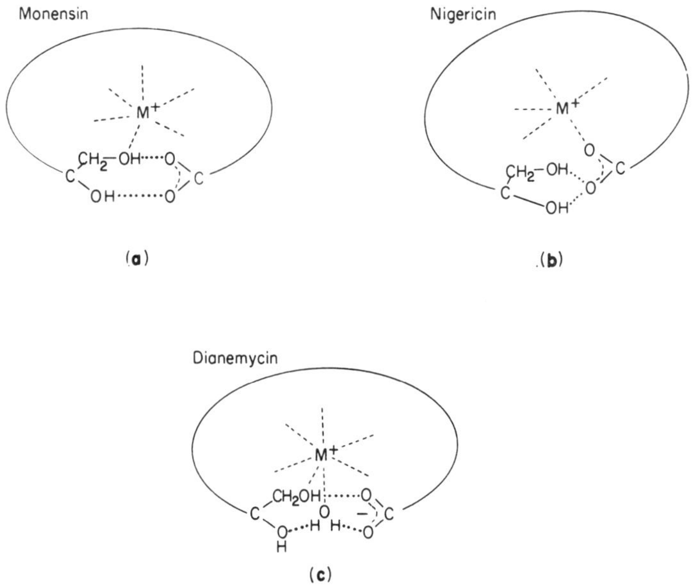

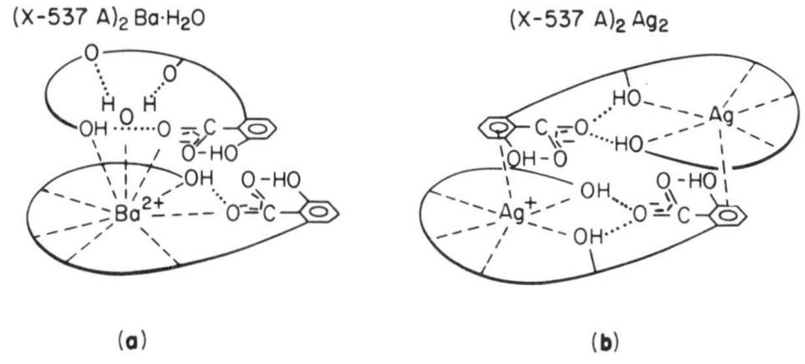

Figure 32 Schematic representations of the structures of the metal complexes of X-537A (reproduced with permission from reference 226)

As a consequence of this structural information, the proposal was made that it would be reasonable to consider that the monomeric forms are involved in cation uptake and release in polar media, such exposure being at the surface of a membrane, but that the actual transfer in the medium of low dielectric constant (the lipid phase) would be the dimeric form.[572]

In the Ba^{2+} complex with (**145**), two anions coordinate to the cation in different ways (Figure 32b). The metal ion sits primarily in a cavity provided by one of the anions and is six-coordinated by two ether, two hydroxy, one keto and one carboxylate oxygen atoms. A nine-fold coordination is completed by further coordination to two oxygen atoms from the second anion and a water molecule.[573] A review of the structures of polyether antibiotic complexes is available and includes a compilation of structural data.[372] The stoichiometries of alkali and alkaline earth complexes of (**145**) in methanol, have been determined potentiometrically and show 1:1 neutral complexes for the alkali metal cations, and high stability 1:1 (charged) and 1:2 (neutral) complexes for the alkaline earth cations.[574]

Recent additions to this array of antibiotics include two unusual ionophores: ionomycin (**149**), which efficiently chelates Ca^{2+} as a dibasic acid, the dianion resulting from ionization of the carboxylic acid and β-diketone moieties,[575] and M-139603 (**150**), which has present an acyltetronic acid moiety.[576] A crystallographic study of the calcium complex of (**149**) revealed an octahedral coordination of the cation provided by a carboxylate oxygen, both β-diketonate oxygens, two hydroxy oxygens and one ether oxygen.[575] In the sodium complex (**150**) the cation is six-coordinated through the chelating anion, two ether and one methoxy oxygen atoms of the ligand, and a solvent water molecule.[576] Solution studies on M-139603 reveal the selectivity sequences $Ca^{2+} > Mg^{2+}$ and $Na^+ > K^+ > Rb^+ \gg Li^+$.[577]

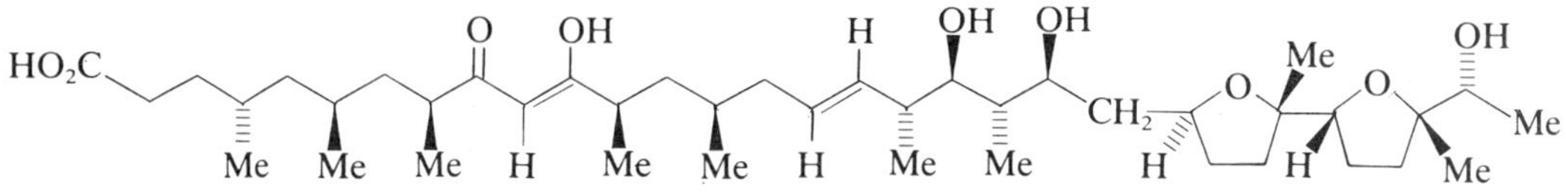

(**149**) Ionomycin

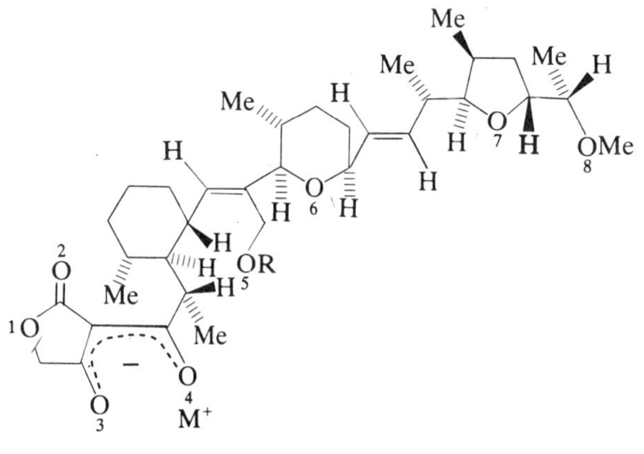

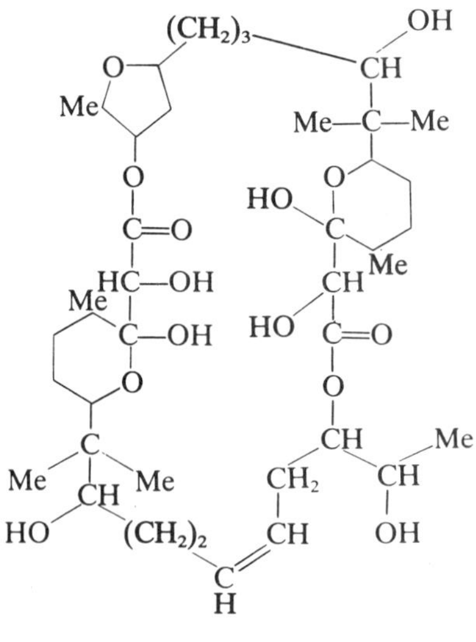

(**150**) M = Na, R = H, M 139603 (**151**)

An interesting antibiotic which does not strictly belong to this group, but which exhibits the features of an anionic ion carrier, is a compound derived from the hydrolysis of boromycin by hydrolytic cleavage of D-valine. Boromycin is itself of interest as it is the first well-defined natural organic compound containing boron. The formula, as determined from the X-ray determination of the structures of the Rb^+ and Cs^+ salts of the des-valine product, is (**151**). This molecule is a Boekesen complex of boric acid with a macrodiolide consisting of two almost identical halves. The overall shape of the anion is roughly spherical with a cleft lined by oxygen atoms and a lipophilic surface. The cations are housed in the cleft in an irregular eight-fold coordination (Figure 33).[578]

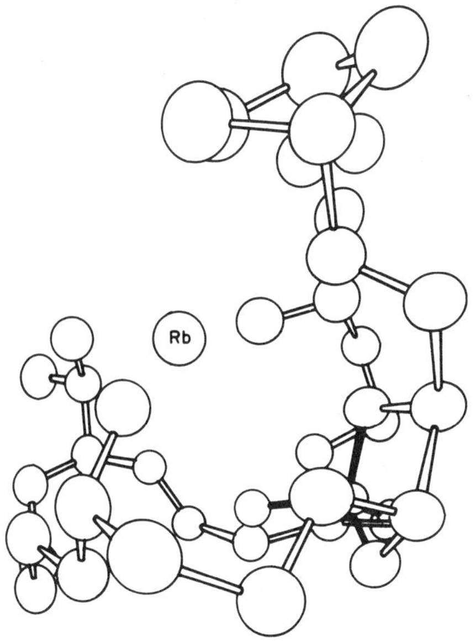

Figure 33 The molecular structure of the Rb^+ salt of desvalinoboromycin (reproduced with permission from reference 578)

23.7 CONCLUDING REMARKS

During the Second World War, Field Marshall Wavell prepared an anthology of verse and used a sentence from the works of Montaigne as his introduction: 'I have gathered a posie of other men's flowers and nothing but the thread that binds them is my own'. This sentence is also appropriately placed here as with any anthology the choice of contents must be both selective, introducing subjectivity, and appropriate. The coordination chemistry of alkali and alkaline earth metal cations covers a very amorphous area and consequently it has been necessary to prune so that the basic principles are revealed with appropriate examples. As is evident from the references, many reviews of the area now exist and these contain comprehensive compilations of data such as full structural information and the statistical analysis of thermodynamic and kinetic data. It was not deemed appropriate, partly from spatial consideration, to reiterate these tabulations in full in this article. An updating of the area can be found in annual reports presented in *Coordination Chemistry Reviews*[579] and by the Royal Society of Chemistry.[580]

23.8 POSTSCRIPT

Recent progress in the chemistry of alkali and alkaline earth metals has centred on the chemistry of oligomeric coordination compounds of lithium. $[Ph_2C{=}NLi(py)]_4$ and $[ClLi(HMPA)]_4$ have been shown to exist as discrete pseudo-cubane structures in which N and Cl atoms respectively bridge the Li_3 triangular faces.[581] An unprecedented pentanuclear Li_5 cluster anion, having present both μ_3 face and μ_2 edge N to Li bonding, has been found in the structure of $[Li(HMPA)_4]^+[Li_5(N{=}CPh_2)_6(HMPA)_3]^-$.[582] The X-ray structure of dibenzylamidolithium itself has been shown to be a trimeric structure having present an N_3Li_3 core; in contrast the adducts $(PhCH_2)_2NLi{\cdot}OEt_2$ and $(PhCH_2)_2NLi{\cdot}HMPA$ are found to be dimeric with N_2Li_2 cores (Figure 34).[583] Dimeric cores have also been revealed in two isomers of $[\{Ph(2\text{-pyridinyl})NLi\}(HMPA)]_2$, which have N_2Li_2 and O_2Li_2 cores respectively; the latter core uses the donor O atom of the HMPA ligand in a novel bridging mode.[584] A further example of this mode of bonding was then detected in the cation $[Li(H_2O)_2(HMPA)_2]_2^{2+}$ isolated as its chloride salt from the reaction of HMPA with deliberately hydrated LiCl.[585]

Related to $[ClLi(HMPA)]_4$ are the cationic species $[Li_4Cl_2(Et_2O)_{10}]^{2+}$ and $[Li_6Br_4(Et_2O)_{10}]^{2+}$ characterized during studies on the preparation of mixed transition metal–alkali metal anionic clusters.[586,587] These solvated salt clusters are viewed as representing early stages of nucleation and growth of the lithium halide as it crystallizes from the solvent.

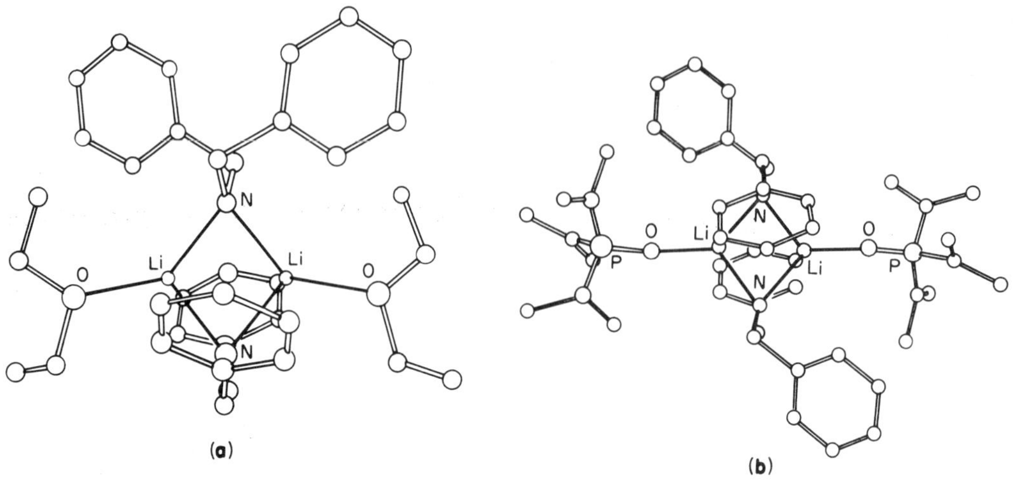

Figure 34 (a) Molecular structure of $[(PhCH_2)_2NLi{\cdot}OEt_2]_2$ and (b) $[(PhCH_2)_2NLi{\cdot}HMPA]_2$ (reproduced with permission from ref. 583)

Evidence has been presented for π bonding around a three-coordinate centre in the beryllium dimer (**152**); shortened bond distances from the terminal trigonal Be atoms to the O and Cl atoms are used as the criterion.[588] $(MO)_2$ cores have been found in the complexes (**153**; M = Mg, Be).[589] Further interesting complexes of Mg have been found in the reactions of transition metal halides with $MgCl_2$; these studies have been used to demonstrate that simple molecular complexes can exist in mixtures claimed to be active in Ziegler–Natta catalysis.[590] $MgCl_2{\cdot}2THF$ and $TiCl_4{\cdot}2THF$ react together to give $[Mg_2Cl_3{\cdot}6THF][TiCl_4{\cdot}THF]$, the cation of which is $(THF)_3Mg(\mu\text{-}Cl)_3(THF)_3^+$.[591] The reaction between $MgCl_2$ and $FeCl_3$ in THF gave an unstable complex $[Mg(THF)_5][FeCl_4]$, which on further reaction with THF was reduced to $FeCl_2(\mu\text{-}Cl_2)Mg(THF)_4$.[592] Mixed ligand complexes of Mg^{2+} and Ca^{2+} involving salicylaldehyde, pentane-2,4-dione and *o*-hydroxyacetophenone have been characterized.[593]

The chemist's ingenuity in devising new coronands and cryptands has been unrelenting and many new systems have been revealed.[594] A representative example is the development of redox-switch crown ethers in which specificity for individual alkali metals accompanies a change in ligand oxidation state; for example (**154**) shows an affinity for Cs^+ only when in the oxidized cyclic form (**154**ox).[595] A further example is the development of an aza macrocycle (**155**) which selectively binds Li^+ and on so doing changes colour from red to colourless.[596] The crystal structure of $[Cs_2(18\text{-crown-}6)][Al_3Me_9SO_4]$ shows a Cs^+—Cs^+ interaction only slightly larger than the sum of the ionic radii of the two cations (3.92 Å *cf.* 3.56 Å) and this is close to that for a full bond between the cations.[597]

In the area of bioinorganic chemistry the constitution of a novel carboxylic acid ionophore, griseochelin, has been elucidated.[598] This ionophore forms stable complexes with alkaline earth metal cations in a 2:1 stoichiometry.

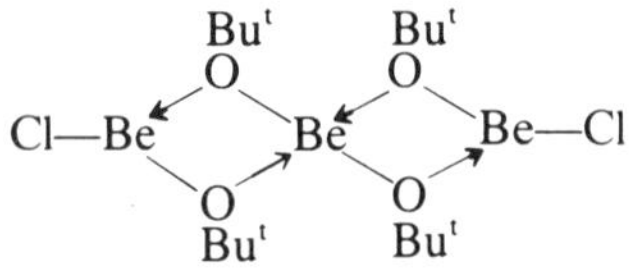

(**152**)

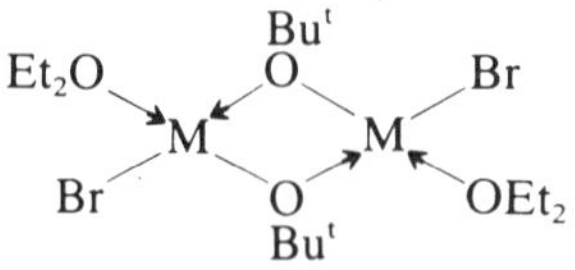

(**153**) M = Mg, Be

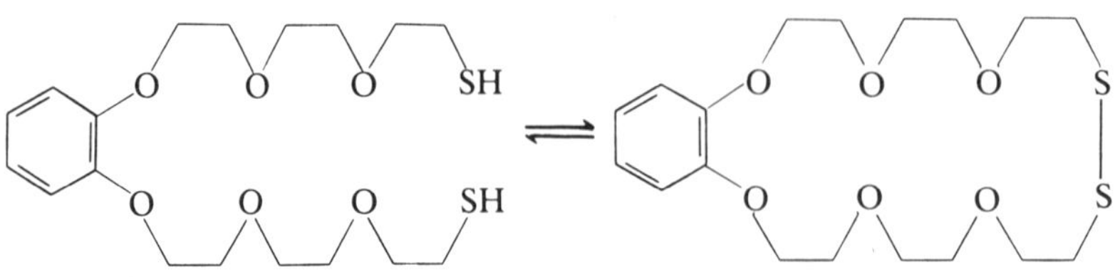

(154 red) **(154 ox)**

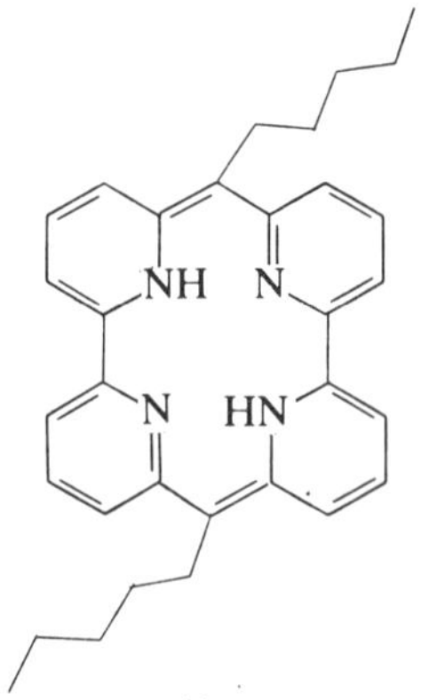

(155)

23.9 REFERENCES

1. N. V. Sidgwick, 'The Chemical Elements and Their Compounds', Oxford University Press, London, 1950, vol. 1.
2. C. S. G. Phillips and R. J. P. Williams, 'Inorganic Chemistry', Clarendon Press, Oxford, 1966, vol. 2.
3. J. E. Huheey, 'Inorganic Chemistry', 2nd edn., Harper & Row, New York, 1978.
4. R. J. P. Williams, *Nature (London)*, 1974, **248**, 303.
5. C. J. Pedersen, *J. Am. Chem. Soc.*, 1967, **89**, 7017.
6. B. Dietrich, J. M. Lehn and J. P. Sauvage, *Tetrahedron Lett.*, 1969, 2885 and 2889.
7. R. J. P. Williams, *Q. Rev., Chem. Soc.*, 1970, **24**, 331.
8. D. E. Fenton, *Chem. Soc. Rev.*, 1977, **6**, 325.
9. P. B. Chock and E. O. Titus, *Prog. Inorg. Chem.*, 1973, **18**, 287.
10. M. Faraday, *Pogg. Annalen*, 1820, **1**, 20.
11. G. F. Huttig, *Z. Anorg. Allg. Chem.*, 1922, **123**, 31.
12. G. F. Huttig and W. Martin, *Z. Anorg. Allg. Chem.*, 1923, **125**, 269.
13. W. Biltz and W. Hansen, *Z. Anorg. Allg. Chem.*, 1923, **127**, 1.
14. W. Weyl, *Ann. Phys.*, 1863, **121**, 601.
15. M. T. Lok, F. J. Tehan and J. L. Dye, *J. Phys. Chem.*, 1972, **76**, 2975 and references therein.
16. H. Ettling, *Liebigs Ann. Chem.*, 1840, **35**, 252.
17. A. Hantzsch, *Chem. Ber.*, 1906, **39**, 3089.
18. N. V. Sidgwick and S. G. P. Plant, *J. Chem. Soc.*, 1925, 209.
19. N. V. Sidgwick and F. M. Brewer, *J. Chem. Soc.*, 1925, 2379.
20. P. Pfeiffer, *Chem. Ber.*, 1914, **47**, 1580.
21. N. S. Poonia, *J. Sci. Ind. Res.*, 1977, **36**, 268.
22. D. Bright, G. H. W. Milburn and M. R. Truter, *J. Chem. Soc. (A)*, 1971, 1582.
23. P. Pfeiffer and W. Christeleit, *Z. Anorg. Allg. Chem.*, 1938, **239**, 133.
24. J. K. Dale, *J. Am. Chem. Soc.*, 1929, **51**, 2788.
25. P. Pfeiffer, 'Organische Molekuleverbindungen', Enke, Stuttgart, 1927.
26. W. K. Anslow and H. King, *Biochem. J.*, 1927, **21**, 1168.
27. N. P. Marullo and R. A. Lloyd, *J. Am. Chem. Soc.*, 1966, **88**, 1076.
28. C. Moore and B. C. Pressman, *Biochem. Biophys. Res. Commun.*, 1964, **15**, 562.
29. J. M. Lehn, *Acc. Chem. Res.*, 1978, **11**, 49.
30. N. S. Poonia and A. Y. Bajaj, *Chem. Rev.*, 1979, **79**, 389.
31. F. Vögtle and E. Weber, in 'The Chemistry of Ethers, Crown Ethers, Hydroxyl Groups and their Sulfur Analogues', Part 1, ed. S. Patai, Wiley, New York, 1980, chap. 2.
32. (a) J. L. Wardell, in 'Comprehensive Organometallic Chemistry', ed. G. Wilkinson, F. G. A. Stone and E. W. Abel, Pergamon, Oxford, 1982, chap. 2; (b) N. A. Bell, chap. 3; (c) W. E. Lindsell, chap. 4.
33. F. Vögtle and E. Weber, *Angew. Chem., Int. Ed. Engl.*, 1979, **18**, 753.
34. J. Celeda, *Collect. Czech. Chem. Commun.*, 1983, **48**, 1680.
35. M. R. Truter, *Struct. Bonding (Berlin)*, 1973, **16**, 1.
36. (a) H. W. Ruben, D. H. Templeton, R. D. Rosenstein and I. Olovsson, *J. Am. Chem. Soc.*, 1961, **83**, 820; (b) H. A. Levy and G. C. Lisensky, *Acta Crystallogr., Sect. B*, 1978, **34**, 3502.
37. B. Dickens and W. E. Brown, *Acta Crystallogr., Sect. B*, 1972, **28**, 3056.
38. A. Regis, J. Limouzie and J. Corset, *J. Chim. Phys. Phys.-Chim. Biol.*, 1972, **69**, 699.
39. (a) W. L. Jolly, *Prog. Inorg. Chem.*, 1959, **1**, 235; (b) J. L. Dye, *Pure Appl. Chem.*, 1977, **49**, 3.
40. Y. M. G. Yasin, O. J. R. Hodder and H. M. Powell, *Chem. Commun.*, 1966, 705.
41. C. Eaborn, P. B. Hitchcock, J. D. Smith and A. C. Sullivan, *J. Chem. Soc., Chem. Commun.*, 1983, 827.
42. D. J. Haas, *Nature (London)*, 1964, **201**, 64.
43. P. Piret, Y. Gobillon and M. van Meerssche, *Bull. Soc. Chim. Fr.*, 1963, 205.
44. Y. Gobillon, P. Piret and M. van Meerssche, *Bull. Soc. Chim. Fr.*, 1962, 554.
45. L. Kahlenburg and F. C. Krauskopf, *J. Am. Chem. Soc.*, 1980, **30**, 1104.
46. F. Durant, P. Piret and M. van Meerssche, *Acta Crystallogr.*, 1967, **22**, 52.
47. P. Piret, L. Rodrique, Y. Gobillon and M. van Meerssche, *Acta Crystallogr.*, 1966, **20**, 482.
48. M. S. Greenberg and A. I. Popov, *Spectrochim. Acta, Part A*, 1966, **31**, 697.
49. B. Maxey and A. I. Popov, *J. Am. Chem. Soc.*, 1968, **90**, 4470.
50. B. Maxey and A. I. Popov, *J. Am. Chem. Soc.*, 1969, **91**, 20.
51. A. I. Popov, *Pure Appl. Chem.*, 1979, **51**, 101.
52. R. C. Paul, P. Singh and S. L. Chadka, *Indian J. Chem.*, 1971, **9**, 1160.
53. J. H. Exner and E. C. Steiner, *J. Am. Chem. Soc.*, 1974, **96**, 1782.
54. A. K. R. Unni, N. Sitaramai and V. C. Menon, *J. Indian Chem. Soc.*, 1972, **54**, 1021.
55. B. Maxey and A. I. Popov, *J. Inorg. Nucl. Chem.*, 1970, **32**, 1029.
56. B. M. Rode, in 'Metal–Ligand Interactions in Organic Chemistry and Biochemistry', ed. B. Pullman and N. Goldblum, Reidel, Dordrecht, 1977, pp. 127–145.
57. C. N. R. Rao, in 'Metal–Ligand Interactions in Organic Chemistry and Biochemistry', ed. B. Pullman and N. Goldblum, Reidel, Dordrecht, 1977, pp. 147–157.
58. J. Bello, D. Haas and H. R. Bello, *Biochemistry*, 1966, **5**, 2539.
59. J. Verbist, J.-P. Putzeys, P. Piret and M. van Meerssche, *Acta Crystallogr., Sect. B*, 1971, **27**, 1190.
60. J. P. Declerq, R. Meulemans, P. Piret and M. van Meerssche, *Acta Crystallogr., Sect. B*, 1971, **27**, 539.
61. J. P. Greenstein and M. Winitz, 'Chemistry of the Amino Acids', Wiley, New York, 1961, vol. 1, chap. 6.
62. J. Vierbist, R. Meulemans, P. Piret and M. van Meerssche, *Bull. Soc. Chim. Belg.*, 1970, **79**, 391.
63. F. Durant, Y. Gobillon, P. Piret and M. van Meerssche, *Bull. Soc. Chim. Belg.*, 1966, **75**, 52.
64. J. H. Palm and C. H. McGillavray, *Acta Crystallogr.*, 1963, **16**, 963.
65. J. C. A. Boeyens and F. H. Herbstein, *Inorg. Chem.*, 1967, **6**, 1408.
66. H. B. Flora, Ph.D. Thesis, Univ. of S. Carolina, 1971; *Diss. Abstr.*, 1971–1972, **32**, 6947B.

67. S. Kusukabe and T. Sekno, *Bull. Chem. Soc. Jpn.*, 1980, **53,** 2081.
68. T. V. Healy, *J. Inorg. Nucl. Chem.*, 1969, **31,** 499.
69. D. A. Lee, W. L. Taylor, W. J. MacDowell and J. S. Drury, *J. Inorg. Nucl. Chem.*, 1968, **30,** 2807.
70. D. C. Luehrs and J. P. Kohur, *J. Inorg. Nucl. Chem.*, 1974, **36,** 1459.
71. P. P. Edwards, S. C. Guy, D. M. Holton and W. McFarlane, *J. Chem. Soc., Chem. Commun.*, 1981, 1185.
72. I. G. Dance and H. C. Freeman, *Acta Crystallogr., Sect. B,* 1969, **25,** 304.
73. V. Dijakovic, A. Edenharter, W. Nowacki and B. Ribar, *Z. Kristallorg., Kristallgeom., Kristallphys., Kristallchem.*, 1976, **114,** 314.
74. N. A. Bell, *Adv. Inorg. Chem. Radiochem.*, 1972, **14,** 255.
75. F. Bertin and J. Deroualt, *Can. J. Chem.*, 1979, **57,** 913.
76. J. P. Hunt and H. L. Friedman, *Prog. Inorg. Chem.*, 1983, **30,** 359.
77. P. L. Brown, J. Ellis and R. N. Sylva, *J. Chem. Soc., Dalton Trans.*, 1983, 2001.
78. A. Braibanti, A. Tiripicchio, A. M. Manotti–Lanfredi and F. Bigoli, *Acta Crystallogr., Sect. B,* 1969, **25,** 354.
79. H. Flack, *Acta Crystallogr., Sect. B,* 1973, **29,** 656.
80. S. Baggio, L. M. Amzel and L. N. Becka, *Acta Crystallogr., Sect. B,* 1969, **25,** 2650.
81. M. Ledesert and J. C. Monier, *Acta Crystallogr., Sect. B,* 1981, **37,** 652.
82. G. W. Stephen and G. H. MacGillavray, *Acta Crystallogr., Sect. B,* 1972, **28,** 1031.
83. G. Ferraris, D. W. Jones and J. Yerkess, *J. Chem. Soc., Dalton Trans.*, 1973, 816.
84. M. B. Cingi, A. M. Manotti–Lanfredi, A. Tiripicchio, G. Bandoli and D. A. Clemente, *Inorg. Chim. Acta,* 1981, **52,** 237.
85. M. A. Neuman, E. C. Steiner, F. P. van Remoortere and F. P. Boer, *Inorg. Chem.*, 1975, **14,** 734.
86. H. Einspahr and C. E. Bugg, *Acta Crystallogr., Sect. B,* 1980, **36,** 264.
87. A. Leclaire and M. M. Borel, *Acta Crystallogr., Sect B,* 1977, **33,** 2938.
88. F. Dahan, *Acta Crystallogr., Sect. B,* 1975, **31,** 423.
89. L. Mazzarella, A. L. Kovacs, P. De Santis and A. M. Liquori, *Acta Crystallogr.*, 1967, **22,** 65.
90. A. Leclaire and M. M. Borel, *Acta Crystallogr., Sect. B,* 1979, **35,** 585.
91. V. M. Padmanabhan, W. R. Busing and H. A. Levy, *Acta Crystallogr., Sect. B,* 1978, **34,** 2290.
92. K. Mereiter and A. Preisinger, *Acta Crystallogr., Sect B,* 1976, **32,** 382.
93. D. A. Everest, 'The Chemistry of Beryllium', Elsevier, New York, 1964.
94. L. H. Shepherd, Jr. and G. Ter Haar, *US Pat.* 3 808 307 (1974) (*Chem. Abstr.*, 1974, **81,** P51 772).
95. G. Ter Haar, *US Pat.*, 3 806 520 (1974) (*Chem. Abstr.*, 1974, **81,** P11 379).
96. J. H. Murib, S. Schott and C. A. Bonecutter, *US Pat.* 3 919 320 (1976) (*Chem. Abstr.*, 1976, **84,** P46 756).
97. L. H. Shepherd, Jr., *US Pat.* 3 483 219 (1969) (*Chem. Abstr.*, 1970, **42,** P80 950).
98. J. L. Atwood and G. D. Stucky, *J. Am. Chem. Soc.*, 1969, **91,** 4426.
99. D. C. Bradley, *Adv. Inorg. Chem. Radiochem.*, 1972, **15,** 259.
100. D. A. Armitage, 'Inorganic Rings and Cages', Edward Arnold, London, 1972.
101. A. I. Grigor'ev and E. G. Pogodilova, *Zh. Strukt. Khim.*, 1969, **10,** 43; A. I. Grigor'ev, E. G. Pogodilova and V. Sinachov, *Russ. J. Inorg. Chem.* (*Engl. Transl.*), 1970, **15,** 902.
102. L. B. Serezhkina, Z. I. Grigorovich, V. E. Serezhkin, N. S. Tamm and A. V. Novoselova, *Dokl. Akad. Nauk SSSR,* 1973, **211,** 123.
103. S. Westman, P.-E. Werner, T. Schuler and W. Raldow, *Acta Chem. Scand., Ser. A,* 1981, **35,** 467.
104. S. Halut-Desportes, *Acta Crystallogr., Sect. B,* 1977, **33,** 599.
105. R. Sarma, F. Ramirez, P. Narayanan, B. McKeever and J. F. Marecek, *J. Am. Chem. Soc.*, 1979, **101,** 5015.
106. J. MacCordick, *C. R. Hebd. Seances Acad. Sci., Ser. C,* 1974, **278,** 1177.
107. G. Chavant, J. C. Daran, Y. Jeannin, G. Kaufmann and J. MacCordick, *Inorg. Chim. Acta,* 1975, **14,** 281.
108. H. Boehland and W. Hanay, *Z. Chem.*, 1974, **14,** 286.
109. F. W. Wehrli and S. L. Wehrli, *J. Magn. Reson.*, 1982, **47,** 151.
110. D. W. Arnos and N. R. Russell, *Spectrochim. Acta, Part A,* 1978, **34,** 397.
111. A. D. van Ingen Schenau, W. L. Groeneveld and J. Reedjik, *Recl. Trav. Chim. Pays-Bas,* 1972, **91,** 88.
112. P. W. N. M. van Leeuwen, *Recl. Trav. Chim. Pays-Bas,* 1967, **86,** 247.
113. P. Piret and C. Mesureur, *J. Chim. Phys. Phys.-Chim. Biol.*, 1965, **62,** 287.
114. S. Halut-Desportes and M. Philoche-Levisalles, *Acta Crystallogr., Sect. B,* 1978, **34,** 432.
115. R. C. van Landschoot, J. A. M. van Hest and J. Reedjik, *J. Inorg. Nucl. Chem.*, 1977, **39,** 1913.
116. Y. Gobillon, P. Piret and M. van Meerssche, *Bull. Soc. Chim. Fr.*, 1962, 551.
117. L. J. Guggenberger and R. E. Rundle, *J. Am. Chem. Soc.*, 1968, **90,** 5375.
118. K. N. Semenenko, E. B. Lobkovska, M. A. Simonov and A. I. Shunakov, *Zh. Strukt. Khim.*, 1976, **17,** 532.
119. This work is presented in a series of papers by W. L. Groeneveld, W. L. Driessen, J. Reedjik *et al.*, in *Recl. Trav. Chim. Pays-Bas,* in the late 1960s and early 1970s.
120. M. Vijayan and M. A. Viswamita, *Acta Crystallogr.*, 1967, **23,** 1000.
121. M. W. G. de Bolster, I. E. Kortram and W. L. Groeneveld, *J. Inorg. Nucl. Chem.*, 1972, **34,** 575.
122. C. N. Lestas and M. R. Truter, *J. Chem. Soc.* (*A*), 1971, 738.
123. C. M. Mikulski, J. S. Skryantz, N. M. Karyannis, L. L. Pytlewski and L. S. Gelfand, *Inorg. Chim. Acta,* 1978, **27,** 69.
124. S. Kusakabe and T. Sekine, *Bull. Chem. Soc. Jpn.*, 1981, **54,** 2930.
125. Y. S. Ng, G. A. Rodley and W. T. Robinson, *Acta Crystallogr., Sect. B,* 1978, **34,** 2835, 2837.
126. O. Pisaniello, S. F. Lincoln and E. H. Williams, *Inorg. Chim. Acta,* 1978, **31,** 237.
127. M. N. Tkaczuk and S. F. Lincoln, *Ber. Bunsenges. Phys. Chem.* 1982, **86,** 147 and references therein.
128. S. F. Lincoln and M. N. Tkaczuk, *Ber. Bunsenges. Phys. Chem.*, 1982, **86,** 221.
129. Y. S. Ng, G. A. Rodley and W. T. Robinson, *Inorg. Chem.*, 1976, **15,** 303.
130. H. Brusser, H. Gillier-Pandraud and A. U. Deschamps, *Bull. Soc. Chim. Fr.*, 1966, 3364.
131. L. Lebioda and K. Lewinski, *Acta Crystallogr., Sect. B,* 1980, **36,** 693.
132. L. Lebioda, *Acta Crystallogr., Sect. B,* 1977, **33,** 1583.
133. L. Lebioda, *Rocz. Chem.*, 1972, **46,** 373.

134. L. Lebioda, *Acta Crystallogr., Sect. B,* 1980, **36,** 271.
135. (a) P. Lemoine and P. Herpin, *Acta Crystallogr., Sect. B,* 1980, **36,** 2608; (b) J. K. Mohana Rao and S. Natajaran, *Acta Crystallogr., Sect. B.,* 1980, **36,** 1058; (c) D. van der Helm and T. V. Willoughby, *Acta Crystallogr. Sect. B.,* 1969, **25,** 2317.
136. L. Banford and G. E. Coates, *J. Chem. Soc.,* 1964, 5591; *J. Chem. Soc. (A),* 1966, 274.
137. Y. Le Fur, and S. Aléonard, *Acta Crystallogr., Sect. B,* 1972, **28,** 2115.
138. J. Vicat, D. Tran Qui and S. Aléonard, *Acta Crystallogr., Sect. B,* 1976, **32,** 1356.
139. J. A. Miliotis, A. G. Galinos and J. M. Tsangaris, *Bull. Soc. Chim. Fr.,* 1961, 1413.
140. J. J. Dechter, *Prog. Inorg. Chem.,* 1982, **29,** 285.
141. L. B. Serezhina, N. S. Tamm and A. I. Grigor'ev, *Russ. J. Inorg. Chem. (Engl. Transl.),* 1975, **20,** 1018.
142. H. Boehland, W. Hanay, A. Meisel and R. Scheibe, *Z. Chem.,* 1981, **21,** 372.
143. M. G. B. Drew, J. E. Goulter, L. Guémas-Brisseau, P. Palvadeau, R. Rouxel and P. Herpin, *Acta Crystallogr., Sect. B,* 1974, **30,** 2579.
144. J. E. D. Davies, *J. Inorg. Nucl. Chem.,* 1974, **36,** 1711.
145. M. H. Brooker and C.-H. Huang, *Can. J. Chem.,* 1980, **58,** 168.
146. (a) A. Leclaire, M. M. Borel and J.-C. Monier, *Acta Crystallogr., Sect. B,* 1980, **36,** 2734; (b) J. R. Clark, H. T. Evans and R. C. Erd, *Acta Crystallogr., Sect. B,* 1980, **36,** 2736.
147. P. Hubberstey, *Coord. Chem. Rev.,* 1982, **40,** 64.
148. H. S. Isbin and K. Kobe, *J. Am. Chem. Soc.,* 1945, **67,** 464.
149. G. E. Coates and S. I. E. Green, *J. Chem. Soc.,* 1962, 3340.
150. F. Durant, P. Piret and M. van Meerssche, *Acta Crystallogr.,* 1967, **23,** 780.
151. H. Gillier-Pandraud and S. Jamet-Delcroix, *Acta Crystallogr., Sect. B,* 1971, **27,** 2476.
152. T. Konig, B. Eisenmann and H. Schaefer, *Z. Anorg. Allg. Chem.,* 1982, **488,** 126.
153. A. W. Langer, *Adv. Chem. Ser.,* 1974, **130,** 113.
154. S. R. Hall, C. L. Raston, B. W. Skelton and A. H. White, *Inorg. Chem.,* 1983, **22,** 4070.
155. P. S. Gentile, J. Carlotto and T. A. Shankoff, *J. Inorg. Nucl. Chem.,* 1967, **29,** 1427.
156. G. D. Stucky, *Adv. Chem. Ser.,* 1974, **130,** 56.
157. P. Jutzi, E. Schleiter, C. Kruger and S. Pohl, *Angew. Chem., Int. Ed. Engl.,* 1983, **22,** 994.
158. C. Johnson and G. D. Stucky, *J. Organomet. Chem.,* 1972, **40,** C11.
159. G. Smith, E. J. O'Reilly, C. H. L. Kennard and A. H. White, *J. Chem. Soc., Dalton Trans.,* 1977, 1184.
160. D. L. Hughes, *J. Chem. Soc., Dalton Trans.,* 1973, 2347.
161. J. A. Kanters, R. Postma, A. J. M. Duisenberg, K. Venkatasubranian and N. S. Poonia, *Acta Crystallogr., Sect. C,* 1983, **39,** 1519.
162. R. Postma, J. A. Kanters, A. J. M. Duisenberg, K. Venkatasubranian and N. S. Poonia, *Acta Crystallogr., Sect. C,* 1983, **39,** 1221.
163. W. H. Watson, D. A. Grossie, F. Vogtle and W. N. Muller, *Acta Crystallogr., Sect. C,* 1983, **39,** 720.
164. K. Shobatake and K. Nakamoto, *J. Chem. Phys.,* 1968, **49,** 4792.
165. D. E. Fenton and R. Newman, *J. Chem. Soc., Dalton Trans.,* 1974, 655.
166. D. E. Fenton, *J. Chem. Soc. (A),* 1971, 3481.
167. E. Schmidt, A. Hourdakis and A. I. Popov, *Inorg. Chim. Acta,* 1981, **52,** 91.
168. T. C. Shields, *Chem. Commun.,* 1968, 832.
169. L. A. Duvall and D. P. Miller, *Inorg. Chem.,* 1974, **13,** 120.
170. (a) J. A. Jarvis and P. G. Owston, *Chem. Commun.,* 1971, 1403; (b) J. W. Swardstrom, L. A. Duvall and D. P. Miller, *Acta Crystallogr., Sect. B,* 1972, **28,** 2510.
171. D. Knetsch and W. L. Groeneveld, *Recl. Trav. Chim. Pays-Bas,* 1973, **92,** 855.
172. D. Knetsch and W. L. Groeneveld, *Inorg. Chim. Acta,* 1973, **7,** 81.
173. G. E. Coates and K. Wade, 'Organometallic Compounds', 3rd edn., ed. G. E. Coates, M. L. H. Green and K. Wade, Methuen, London, 1967, vol. 1, p. 81.
174. M. Dentteijer and W. L. Driessen, *Inorg. Chim. Acta,* 1978, **26,** 227.
175. J. Hooz, S. Akiyama, F. J. Cedar, M. J. Bennett and R. M. Tuggle, *J. Am. Chem. Soc.,* 1974, **96,** 274.
176. (a) L. M. Jackman and B. C. Lange, *Tetrahedron,* 1977, **33,** 2737; (b) L. M. Jackman and N. Szeverenyi, *J. Am. Chem. Soc.,* 1977, **99,** 4954.
177. W. Hewertson, B. T. Kilbourn and R. H. B. Mais, *Chem. Commun.,* 1970, 952.
178. R. H. Cox, H. W. Terry, Jr. and L. W. Harrison, *J. Am. Chem. Soc.,* 1971, **93,** 3297.
179. M. D. Joesten, M. S. Hussain and P. G. Lenhart, *Inorg. Chem.,* 1970, **9,** 151.
180. P. S. Gentile and T. A. Shankoff, *J. Inorg. Nucl. Chem.,* 1965, **27,** 2301.
181. J. P. Roux and J. C. A. Boeyens, *Acta Crystallogr., Sect. B,* 1969, **25,** 1700.
182. C. D. Hurd and R. J. Sims, *J. Am. Chem. Soc.,* 1949, **71,** 2440.
183. D. E. Fenton, *J. Chem. Soc., Dalton Trans.,* 1973, 1380.
184. D. L. Hughes, S. E. V. Phillips and M. R. Truter, *J. Chem. Soc., Dalton Trans.,* 1974, 907.
185. S. E. V. Phillips and M. R. Truter, *J. Chem. Soc., Dalton Trans.,* 1974, 2517.
186. S. E. V. Phillips and M. R. Truter, *J. Chem. Soc., Dalton Trans.,* 1975, 1066.
187. W. L. Driessen and P. L. A. Everstijn, *Z. Naturforsch., Teil B,* 1977, **32,** 1284.
188. E. A. Green, W. L. Duax, G. M. Smith and F. Wudl, *J. Am. Chem. Soc.,* 1975, **97,** 6689.
189. E. Sinn and C. M. Harris, *J. Inorg. Nucl. Chem.,* 1968, **30,** 1805.
190. C. Floriani, F. Calderazzo and L. Randaccio, *J. Chem. Soc., Chem. Commun.,* 1973, 384.
191. N. Bresciani-Pahor, M. Calligaris, P. Delise, G. Nardin, L. Randaccio, E. Zotti, G. Fachinetti and C. Floriani, *J. Chem. Soc., Dalton Trans.,* 1976, 2310.
192. M. Pasquali, F. Marchetti, C. Floriani and M. Cesari, *Inorg. Chem.,* 1980, **19,** 1198.
193. G. H. W. Milburn, M. R. Truter and B. L. Vickery, *J. Chem. Soc., Dalton Trans.,* 1974, 841.
194. L. F. Armstrong, H. C. Lip, L. F. Lindoy, M. McPartlin and P. A. Tasker, *J. Chem. Soc., Dalton Trans.,* 1977, 1771.
195. G. Fachinetti, C. Floriani, P. F. Zanazzi and A. R. Zanzari, *Inorg. Chem.,* 1978, **17,** 3002.

196. D. E. Fenton, N. Bresciani-Pahor, M. Calligaris, G. Nardin and L. Randaccio, *J. Chem. Soc., Chem. Commun.*, 1979, 39.
197. G. Fachinetti, C. Floriani, P. F. Zanazzi and A. R. Zanzari, *Inorg. Chem.*, 1979, **18,** 3469.
198. G. Fachinetti, C. Floriani and P. F. Zanazzi, *J. Am. Chem. Soc.*, 1978, **100,** 7405; S. Gambarotta, F. Arena, C. Floriani and P. F. Zanazzi, *J. Am. Chem. Soc.*, 1982, **104,** 5082.
199. C. Floriani, *Pure Appl. Chem.*, 1982, **54,** 59.
200. G. J. M. Ivarsson, *Acta Crystallogr., Sect. C,* 1984, **40,** 67.
201. D. K. Cabbiness and D. W. Margerum, *J. Am. Chem. Soc.*, 1969, **91,** 6540.
202. J. M. Lehn and J. P. Sauvage, *J. Am. Chem. Soc.*, 1975, **97,** 6700.
203. J. Smid, *Angew. Chem., Int. Ed. Engl.*, 1972, **11,** 112.
204. F. A. Cotton and G. Wilkinson, 'Advanced Inorganic Chemistry', 2nd edn., Wiley, New York, 1966, p. 419.
205. K. O. Hodgson and K. N. Raymond, *Inorg. Chem.*, 1972, **11,** 3030.
206. W. van Gils, A. P. Zuur and W. L. Driessen, *Inorg. Chim. Acta,* 1982, **58,** 187.
207. T. P. Singh, R. Reinhardt and N. S. Poonia, *Inorg. Nucl. Chem. Lett.*, 1980, **16,** 293.
208. H. Ohmoto, Y. Kai, N. Yasuoka, N. Kasai, S. Yanagida and M. Okahara, *Bull. Chem. Soc. Jpn.*, 1979, **52,** 1209.
209. I. Yamaguchi, K. Miki, N. Yasuoka and N. Kasai, *Bull. Chem. Soc. Jpn.*, 1982, **55,** 1372.
210. E. Weber and F. Vögtle, *Tetrahedron Lett.*, 1975, 2415.
211. (a) W. Saenger and B. S. Reddy, *Acta Crystallogr., Sect. B,* 1979, **35,** 56; (b) G. Weber and W. Saenger, *Acta Crystallogr., Sect. B,* 1979, **35,** 1346.
212. W. Saenger and H. Brand, *Acta Crystallogr., Sect. B,* 1979, **35,** 838.
213. G. Weber, W. Saenger, F. Vögtle and H. Sieger, *Angew. Chem., Int. Ed. Engl.*, 1979, **18,** 226.
214. W. Saenger, unpublished results quoted in ref. 33.
215. I. H. Suh, G. Weber and W. Saenger, *Acta Crystallogr., Sect. B,* 1980, **36,** 946.
216. I. H. Suh, G. Weber, M. Kaftory, W. Saenger, H. Sieger and F. Vögtle, *Z. Naturforsch., Teil B,* 1980, **35,** 352.
217. K. K. Chacko and W. Saenger, *Z. Naturforsch., Teil B,* 1980, **35,** 1533.
218. F. Vögtle and U. Heimann, *Chem. Ber.*, 1978, **111,** 2757.
219. G. Weber and W. Saenger, *Acta Crystallogr., Sect. B,* 1980, **36,** 61.
220. K. K. Chacko and W. Saenger, *Z. Naturforsch., Teil B,* 1981, **36,** 102.
221. F. Vögtle and H. Sieger, *Angew. Chem., Int. Ed. Engl.*, 1977, **16,** 396.
222. D. L. Hughes, C. L. Mortimer and M. R. Truter, *Inorg. Chim. Acta,* 1978, **28,** 83.
223. D. Ammann, E. Pretsch and W. Simon, *Helv. Chim. Acta,* 1973, **56,** 1780.
224. D. Erne, D. Ammann, A. F. Zhukov, F. Beton, E. Pretsch and W. Simon, *Helv. Chim. Acta,* 1982, **65,** 538.
225. P. Ammann, E. Pretsch and W. Simon, *Anal. Lett.*, 1974, **7,** 23.
226. W. Simon, W. E. Morf and P. C. Meier, *Struct. Bonding (Berlin)*, 1973, **16,** 113.
227. R. Bissig, E. Pretsch, W. E. Morf and W. Simon, *Helv. Chim. Acta,* 1978, **61,** 1520.
228. M. Guggi, M. Oehme, E. Pretsch and W. Simon, *Helv. Chim. Acta,* 1976, **59,** 2417.
229. W. Simon, E. Pretsch, D. Ammann, W. E. Morf, M. Guggi, R. Bissig and M. Kessler, *Pure Appl. Chem.*, 1975, **44,** 613 and references therein.
230. R. Buchi and E. Pretsch, *Helv. Chim. Acta,* 1977, **60,** 1141.
231. K. Neupert-Laves and M. Dobler, *Helv. Chim. Acta,* 1977, **60,** 1861.
232. F. Vögtle and E. Weber, *Angew. Chem., Int. Ed. Engl.*, 1974, **13,** 814.
233. J. A. Hyatt, *J. Org. Chem.*, 1978, **43,** 1808.
234. J. C. Voegel, J. C. Thierry and R. Weiss, *Acta Crystallogr., Sect. B,* 1974, **30,** 56.
235. J. C. Voegel, J. Fischer and R. Weiss, *Acta Crystallogr., Sect. B,* 1974, **30,** 62.
236. J. C. Voegel, J. Fischer and R. Weiss, *Acta Crystallogr., Sect. B,* 1974, **30,** 66.
237. J. C. Voegel, J. C. Thierry and R. Weiss, *Acta Crystallogr., Sect. B,* 1974, **30,** 70.
238. H. Sieger, W. M. Muller and F. Vögtle, *J. Chem. Res. (S)*, 1978, 398.
239. F. Vögtle, W. Muller, W. Wehner and E. Buhleier, *Angew. Chem., Int. Ed. Engl.*, 1977, **16,** 548.
240. W. Saenger, unpublished results quoted in ref. 33.
241. F. Vögtle, W. M. Muller, E. Buhleier and W. Wehner, *Chem. Ber.*, 1979, **112,** 899.
242. J. A. Rendleman, Jr., *Adv. Carbohydr. Chem. Biochem.*, 1966, **21,** 209.
243. S. J. Angyal, *Pure Appl. Chem.*, 1973, **35,** 131.
244. W. J. Cook and C. E. Bugg, in 'Metal Ligand Interactions in Organic Chemistry and Biochemistry', ed. B. Pullman and N. Goldblum, Reidel, Dordrecht, 1977, part 2, p. 231.
245. C. L. Moore, *Biochem. Biophys. Res. Commun.*, 1971, **42,** 298.
246. W. J. Evans and V. L. Frampton, *Carbohydr. Res.*, 1977, **59,** 571.
247. C. E. Bugg and W. J. Cook, *J. Chem. Soc., Chem. Commun.*, 1972, 727.
248. W. J. Cook and C. E. Bugg, *J. Am. Chem. Soc.*, 1973, **95,** 6442.
249. W. J. Cook and C. E. Bugg, *Acta Crystallogr., Sect. B,* 1973, **29,** 2404.
250. W. J. Cook and C. E. Bugg, *Biochim. Biophys. Acta,* 1975, **389,** 428.
251. D. C. Craig, N. C. Stephenson and J. D. Stevens, *Cryst. Struct. Commun.*, 1974, **3,** 195.
252. D. C. Craig, N. C. Stephenson and J. D. Stevens, *Carbohydr. Res.*, 1972, **22,** 494.
253. R. A. Wood, V. J. James and S. J. Angyal, *Acta Crystallogr., Sect. B,* 1977, **33,** 2248.
254. G. Blank, *Acta Crystallogr., Sect. B,* 1973, **29,** 1677.
255. A. Combes, *C. R. Hebd. Seances Acad. Sci.*, 1887, **105,** 869.
256. L. M. Jackman and B. C. Lange, *Tetrahedron,* 1977, **33,** 2737.
257. J. P. Fackler, Jr., *Prog. Inorg. Chem.*, 1966, **7,** 361.
258. R. C. Mehrotra, R. Bohra and D. P. Gaur, 'Metal β-Diketonates and Allied Derivatives', Academic, New York, 1978.
259. A. Martell and L. G. Sillen, 'Stability Constants', Special Publications 17 and 25, The Chemical Society, London, 1964 and 1971.
260. M. Raban, E. Noe and G. Yamamoto, *J. Am. Chem. Soc.*, 1977, **99,** 6527.
261. E. Noe and M. Raban, *J. Am. Chem. Soc.*, 1974, **96,** 6184.

262. F. A. Schroder and H. P. Weber, *Acta Crystallogr., Sect. B,* 1975, **31,** 1745.
263. J. J. Sahbari and M. M. Olmstead, *Acta Crystallogr., Sect. C,* 1983, **39,** 1037.
264. S. Shibata, S. Onuma, Y. Matsui and S. Motegi, *Bull. Chem. Soc. Jpn.,* 1975, **48,** 2516.
265. J. M. Stewart and B. Morosin, *Acta Crystallogr., Sect. B,* 1975, **31,** 1164.
266. S. Shibata, M. Ota and K. Ijima, *J. Mol. Struct.,* 1980, **67,** 245.
267. W. H. Mills and R. A. Gotts, *J. Chem. Soc.,* 1926, 3121.
268. D. H. Busch and J. C. Bailar, Jr., *J. Am. Chem. Soc.,* 1954, **76,** 5352.
269. E. Ferroni and R. Cini, *J. Am. Chem. Soc.,* 1960, **82,** 2427.
270. D. N. Sen and N. Thankarajan, *Indian J. Chem.,* 1966, **4,** 454.
271. N. Yanase, Y. Nakamura and S. Kawaguchi, *Inorg. Chem.,* 1978, **17,** 2874.
272. B. Morosin, *Acta Crystallogr.,* 1967, **22,** 315.
273. F. J. Hollander, D. H. Templeton and A. Zalkin, *Acta Crystallogr., Sect. B,* 1973, **29,** 1289.
274. J. J. Sahbari and M. M. Olmstead, *Acta. Crystallogr., Sect. C,* 1983, **39,** 208.
275. F. J. Hollander, D. H. Templeton and A. Zalkin, *Acta Crystallogr., Sect. B,* 1973, **29,** 1295.
276. F. J. Hollander, D. H. Templeton and A. Zalkin, *Acta Crystallogr., Sect. B,* 1973, **29,** 1303.
277. A. Recca, F. Bottino, P. Finocchiaro and H. G. Brittain, *J. Inorg. Nucl. Chem.,* 1978, **40,** 1997.
278. G. G. MacDonald and J. S. Shannon, *Aust. J. Chem.,* 1966, **19,** 1545.
279. D. E. Fenton and C. Nave, *Chem. Commun.,* 1971, 662.
280. D. E. Fenton, C. Nave and M. R. Truter, *J. Chem. Soc., Dalton Trans.,* 1973, 2188.
281. J. R. Majer and R. Perry, *Chem. Commun.,* 1969, 271.
282. D. E. Fenton, M. R. Truter and B. L. Vickery, *Chem. Commun.,* 1971, 93; M. R. Truter and B. L. Vickery, *J. Chem. Soc., Dalton Trans.,* 1972, 395.
283. F. P. Dwyer and A. M. Sargeson, *Proc. R. Soc. N.S.W.,* 1956, **90,** 29.
284. P. W. N. M. van Leeuwen and A. P. Praat, *Inorg. Chim. Acta,* 1970, **4,** 101.
285. J. N. Patil and D. N. Sen, *Indian J. Chem.,* 1973, **11,** 782.
286. J. N. Patil and D. N. Sen, *Indian J. Chem.,* 1974, **12,** 189.
287. J. N. Patil and D. N. Sen, *Indian J. Chem.,* 1975, **13,** 291.
288. U. Casellato, P. A. Vigato, D. E. Fenton and M. Vidali, *Chem. Soc. Rev.,* 1979, **8,** 199.
289. J. P. Beale, J. A. Cunningham and D. J. Phillips, *Inorg. Chim. Acta,* 1979, **33,** 113.
290. R. Winkler, *Struct. Bonding (Berlin),* 1972, **10,** 1.
291. R. W. Green, R. J. LeFevre and J. D. Saxby, *Aust. J. Chem.,* 1966, **19,** 2001.
292. R. L. Angel, J. W. Hayes and D. V. Radford, *J. Chem. Soc., Faraday Trans. 2,* 1975, **71,** 81.
293. A. Recca, F. A. Bottino and P. Finocchiaro, *J. Inorg. Nucl. Chem.,* 1980, **42,** 479.
294. F. A. Bottino, A. Recca and P. Finocchiaro, *J. Organomet. Chem.,* 1979, **160,** 373.
295. L. Nivorozhkim, L. E. Konstantinovskii, M. S. Korobov and V. I. Minkin, *Koord. Khim.,* 1979, **5,** 508 (*Chem. Abstr.,* 1979, **91,** 97 344).
296. K. Morisige, *Anal. Chim. Acta,* 1974, **73,** 245 and 1980, **121,** 301.
297. V. C. Syal and D. C. Jain, *Indian J. Chem.,* 1973, **11,** 494.
298. R. W. Green and D. W. Alexander, *Aust. J. Chem.,* 1965, **18,** 651.
299. D. F. Martin, G. A. Janusonis and B. B. Martin, *J. Am. Chem. Soc.,* 1961, **83,** 73.
300. R. W. Green and P. W. Alexander, *Aust. J. Chem.,* 1965, **18,** 1297.
301. K. K. Deshmukh, A. M. Hindekar and D. M. Sen, *J. Indian Chem. Soc.,* 1979, **56,** 411.
302. B. R. Singh and S. Kunar, *Indian J. Chem.,* 1972, **10,** 663.
303. F. Lions and K. V. Martin, *J. Am. Chem. Soc.,* 1957, **79,** 1273.
304. C. R. Saha, *J. Inorg. Nucl. Chem.,* 1980, **42,** 159.
305. F. Sagara, H. Kobayashi and K. Ueno, *Bull. Chem. Soc. Jpn.,* 1973, **46,** 484.
306. P. Pfeiffer, E. Breith, E. Lubbe and T. Tsunaki, *Liebigs Ann. Chem.,* 1933, **503,** 84.
307. C. E. White and F. Cuttita, *Anal. Chem.,* 1959, **31,** 2083.
308. F. Cuttita and C. E. White, *Anal. Chem.,* 1959, **31,** 2087; Y. Takeda, *Top. Curr. Chem.,* 1984, **121,** 1.
309. J. I. Bullock, M. F. C. Ladd, D. C. Povey and H. A. Tajmir-Riahi, *Acta Crystallogr., Sect. B,* 1979, **35,** 2013.
310. S. M. Nelson, F. S. Esho and M. G. B. Drew, *J. Chem. Soc., Dalton Trans.,* 1982, 407.
311. S. M. Nelson, C. V. Knox, M. McCann and M. G. B. Drew, *J. Chem. Soc., Dalton Trans.,* 1981, 1669.
312. M. G. B. Drew, J. Nelson and S. M. Nelson, *J. Chem. Soc., Dalton Trans.,* 1981, 1678.
313. M. G. B. Drew, C. V. Knox and S. M. Nelson, *J. Chem. Soc., Dalton Trans.,* 1980, 943.
314. M. A. Bush and M. R. Truter, *J. Chem. Soc. (A),* 1971, 745.
315. L. Y. Y. Chan and F. W. B. Einstein, *Can. J. Chem.,* 1971, **49,** 468.
316. J. C. Speakman, *Struct. Bonding (Berlin),* 1972, **12,** 141.
317. G. Urbain and H. Lacombe, *C. R. Hebd. Seances Acad. Sci.,* 1901, **133,** 874.
318. T. Moeller, *Inorg. Synth.,* 1950, **3,** 4.
319. H. Lacombe, *C. R. Hebd. Seances Acad. Sci.,* 1902, **134,** 772.
320. A. V. Novoselova, K. N. Semenenko, N. N. Krasovskaya and Yu. P. Simanov, *Vestn. Mosk. Univ. Khim.,* 1956, **11** (3), 87 (*Chem. Abstr.,* 1956, **50,** 15 320a).
321. H. P. Harat, *Z. Anorg. Allg. Chem.,* 1962, **314,** 210.
322. A. Tulinsky, C. R. Worthington and E. Pignataro, *Acta Crystallogr.,* 1959, **12,** 623, 626.
323. A. I. Grigor'ev, A. V. Novoselova and K. N. Semenenko, *Zh. Neorg. Khim.,* 1957, **2,** 1374.
324. A. I. Grigor'ev, Yu. P. Simanov, K. N. Semenenko and N. N. Krasovskaya, *Zh. Neorg. Khim.,* 1956, **1,** 2280.
325. A. I. Grigor'ev and V. A. Sipachev, *Dokl. Akad. Nauk SSSR,* 1970, **192,** 808.
326. A. I. Grigor'ev, V. A. Sipachev and A. V. Novoselova, *Dokl. Akad. Nauk SSSR,* 1969, **185,** 95.
327. A. I. Grigor'ev and L. N. Reshetova, *Russ. J. Inorg. Chem. (Engl. Transl.),* 1974, **19,** 1426.
328. A. I. Grigor'ev, L. N. Reshetova and A. V. Novoselova, *Russ. J. Inorg. Chem. (Engl. Transl.),* 1972, **17,** 933.
329. K. J. Wynne and W. Bander, *Inorg. Chem.,* 1970, **9,** 1985.
330. T. J. Cardwell and M. R. L. Carter, *J. Chromatogr.,* 1977, **140,** 93.
331. N. Kodama, H. Yamada and M. Tanaka, *J. Inorg. Nucl. Chem.,* 1976, **38,** 2063.

332. C. C. Addison and A. Walker, *Proc. Chem. Soc.*, 1961, 242.
333. B. Duffin, M. J. Haley and S. C. Wallwork, personal communication quoted in ref. 334.
334. V. A. Sipachev, N. I. Tuseev and Y. S. Nekrosov, *Polyhedron,* 1982, **1,** 820.
335. R. Fauré, F. Bertin, H. Loiseleure and G. Thomas-David, *Acta Crystallogr., Sect. B,* 1974, **30,** 462.
336. D. C. Bradley, R. C. Mehrotra and D. P. Gaur, 'Metal Alkoxides', Academic, New York, 1978.
337. L. O. Atovmyan, O. N. Krasochka, A. I. Grigor'ev and V. A. Spachev, *Dokl. Akad. Nauk SSSR,* 1975, **225,** 99.
338. A. I. Grigor'ev, L. N. Reshetova and A. V. Novoselova, *Russ. J. Inorg. Chem. (Engl. Transl.),* 1974, **19,** 1093.
339. A. I. Grigor'ev, L. N. Reshetova and A. V. Novoselova, *Dokl. Akad. Nauk SSSR,* 1972, **202,** 85.
340. A. N. Reshetova, A. I. Grigor'ev and V. A. Sipadev, *Zh. Neorg. Khim.*, 1978, **23,** 2626.
341. A. E. Martell and M. Calvin, 'Chemistry of the Metal Chelate Compounds', Prentice-Hall, Englewood Cliffs, New Jersey, 1952.
342. S. Chaberek and A. E. Martell, 'Organic Sequestering Reagents', Wiley, New York, 1959.
343. A. E. Martell, *Pure Appl. Chem.*, 1978, **50,** 813.
344. A. E. Martell, *Pure Appl. Chem.*, 1975, **81,** 44.
345. G. Schwarzenbach, W. Biedermann and F. Bangerter, *Helv. Chim. Acta,* 1946, **29,** 811.
346. D. Midgley, *Chem. Soc. Rev.*, 1975, **4,** 549.
347. N. A. Porai-Koshits, *Zh. Strukt. Khim.*, 1974, **15,** 1117.
348. R. H. Nuttall and D. M. Stalker, *Talanta,* 1977, **24,** 355.
349. A. A. McConnell, R. H. Nuttall and D. M. Stalker, *Talanta,* 1978, **25,** 425.
350. J. J. R. Frausto da Silva and M. C. T. Abreu Vaz, *J. Inorg. Nucl. Chem.*, 1977, **39,** 613.
351. (a) H. Einspahr and C. E. Bugg, 'Calcium-binding Proteins and Calcium Function', ed. R. H. Wasserman, R. A. Corradino, E. Carafoli, R. H. Kretsinger, D. H. MacLennan and F. L. Siegel, North Holland, New York, 1977; (b) C. E. Bugg and H. Einspahr, *Acta Crystallogr., Sect. B,* 1981, **37,** 1044.
352. M. O'D. Julian, V. W. Day and J. L. Hoard, *Inorg. Chem.*, 1973, **12,** 1754.
353. V. A. Uchtman and R. P. Oertel, *J. Am. Chem. Soc.*, 1973, **95,** 1802.
354. M. Jaber, R. Fauré and H. Loiseleur, *Acta Crystallogr., Sect. B,* 1978, **34,** 429.
355. G. Duc, F. Fauré and H. Loiseleur, *Acta Crystallogr., Sect. B,* 1978, **34,** 2115.
356. F. Ancillotti, G. Boschi, G. Perego and A. Zazzetta, *J. Chem. Soc., Dalton Trans.*, 1977, 901.
357. B. L. Barnett and V. A. Uchtman, *Inorg. Chem.*, 1979, **18,** 2674.
358. R. Leppkes, F. Vögtle and F. Luppertz, *Chem. Ber.*, 1982, **115,** 927.
359. H. Stetter and W. Frank, *Angew. Chem., Int. Ed. Engl.*, 1976, **15,** 686.
360. (a) L. V. Nikitina, A. I. Grigor'ev and N. M. Dyatlova, *J. Gen. Chem. USSR (Engl. Transl.),* 1974, **44,** 1641; (b) L. V. Nikitina, N. F. Shugal, A. I. Grigor'ev and N. M. Dyatlova, *J. Gen. Chem. USSR (Engl. Transl.),* 1975, **45,** 546.
361. N. M. Dyatlova, V. V. Medyntsev, T. Ya. Medved and M. I. Kabachnik, *J. Gen. Chem. USSR (Engl. Transl.),* 1968, **38,** 1035.
362. N. M. Dyatlova, V. V. Medyntsev, T. Ya. Medved and M. I. Kabachnik, *J. Gen. Chem. USSR (Engl. Transl.),* 1968, **38,** 1025.
363. R. M. Izatt, J. J. Christensen and J. H. Rytting, *Chem. Rev.*, 1971, **71,** 439.
364. E. A. Merritt and M. Sundaralingan, *Acta Crystallogr., Sect. B,* 1980, **36,** 2576.
365. H. Sigel, *J. Inorg. Nucl. Chem.*, 1977, **39,** 1903.
366. G. L. Eichhorn, in 'Inorganic Biochemistry', ed. G. L. Eichhorn, Elsevier, Amsterdam, 1973, vol. 2, chap. 34.
367. O. Kennard, N. W. Isaacs, W. D. S. Motherwell, J. C. Coppola, D. L. Wampler, A. C. Larson and D. G. Watson, *Proc. R. Soc. London, Ser. A,* 1971, **325,** 401.
368. J. F. Stoddart and D. A. Laidler, in 'The Chemistry of Crown Ethers, Hydroxyl Groups and their Sulfur Analogues', part 1, ed. S. Patai, Wiley, New York, 1980, chap. 1.
369. G. W. Gokel and S. H. Korzeniowski, 'Macrocyclic Polyether Synthesis', Springer-Verlag, Heidelberg, 1982.
370. E. Weber and F. Vögtle, *Top. Curr. Chem.*, 1981, **98,** 1.
371. J. Goldberg, 'The Chemistry of Crown Ethers, Hydroxyl Groups and their Sulfur Analogues', part 1, ed. S. Patai, Wiley, New York, 1980, chap. 4.
372. R. Hilgenfeld and W. R. Saenger, *Top. Curr. Chem.*, 1982, **101,** 1.
373. F. de Jong and D. N. Reinhardt, *Adv. Phys. Org. Chem.*, 1980, **17,** 279.
374. J. M. Lockhart, *Adv. Inorg. Bioinorg. Mech.*, 1983, **1,** 217.
375. J. M. Lehn, *Pure Appl. Chem.*, 1980, **52,** 2303, 2441.
376. J. Smid, *Pure Appl. Chem.*, 1982, **54,** 2129.
377. E. Blasius and K. P. Janzen, *Pure Appl. Chem.*, 1982, **54,** 2115.
378. K. Ueno and M. Takagi, *Stud. Phys. Theor. Chem.*, 1983, **27,** 279.
379. E. Blasius and K. P. Janzen, *Top. Curr. Chem.*, 1981, **98,** 163.
380. K. Tada and H. Hashitami, *Bunseki Kagaku,* 1982, 155 (*Chem. Abstr.*, 1982, **97,** 161 776).
381. C. J. Pedersen, in 'Synthetic Multidentate Macrocyclic Compounds', ed. R. M. Izatt and J. J. Christensen, Academic, New York, 1978, chap. 1.
382. J. M. Lehn, *Struct. Bonding (Berlin),* 1973, **16,** 1.
383. (a) N. S. Poonia and M. R. Truter, *J. Chem. Soc., Dalton Trans.*, 1973, 2062; (b) J. Petranek and O. Ryba, *Collect. Czech. Chem. Commun.*, 1974, **39,** 2033; (c) C. G. Krespan, *J. Org. Chem.*, 1975, **40,** 1205; (d) D. G. Parsons and J. N. Wingfield, *Inorg. Chim. Acta,* 1976, **18,** 263.
384. F. Vögtle and J. P. Dix, *Liebigs Ann. Chem.*, 1977, 1968.
385. F. A. L. Anet, J. Krane, J. Dale, K. Daasvatu and P. O. Kristiansen, *Acta Chem. Scand.*, 1973, **27,** 3395.
386. (a) E. Weber and F. Vögtle, *Chem. Ber.*, 1976, **109,** 1803; (b) J. Dale and P. O. Christiansen, *Acta Chem. Scand.*, 1972, **26,** 1471.
387. C. J. Pedersen, *J. Am. Chem. Soc.*, 1970, **92,** 386.
388. (a) D. G. Parsons and J. N. Wingfield, *Inorg. Chim. Acta,* 1976, **17,** L25; (b) D. E. Fenton, D. Parkin, R. F. Newton, I. W. Nowell and P. E. Walker, *J. Chem. Soc., Dalton Trans.*, 1982, 327.
389. N. S. Poonia, B. P. Yadao, V. W. Bhagwat, V. Naik and H. Manohar, *Inorg. Nucl. Chem. Lett.*, 1977, **13,** 119.

390. D. G. Parsons, M. R. Truter and J. N. Wingfield, *Inorg. Chim. Acta*, 1975, **14**, 45.
391. (a) E. Weber and F. Vögtle, *Liebigs Ann. Chem.*, 1976, 891; (b) P. Groth, *Acta Chem. Scand., Ser. A*, 1981, **35**, 717.
392. O. Nagano, *Acta Crystallogr., Sect. B*, 1979, **35**, 465.
393. M. Dobler, J. D. Dunitz and P. Seiler, *Acta Crystallogr., Sect. B*, 1974, **30**, 2741.
394. P. Seiler, M. Dobler and J. D. Dunitz, *Acta Crystallogr., Sect. B*, 1974, **30**, 2744.
395. M. Dobler and R. P. Phizackerley, *Acta Crystallogr., Sect. B*, 1974, **30**, 2746.
396. M. Dobler and R. P. Phizackerley, *Acta Crystallogr., Sect. B*, 1974, **30**, 2748.
397. D. Bright and M. R. Truter, *J. Chem. Soc. (B)*, 1970, 1544.
398. C. Riche and C. Pascard-Billy, *J. Chem. Soc., Chem. Commun.*, 1977, 183.
399. P. R. Mallinson, *J. Chem. Soc., Perkin Trans. 2*, 1975, 261.
400. J. L. Vidal, R. C. Schoening and J. M. Tronp, *Inorg. Chem.*, 1981, **20**, 227.
401. R. F. Ziolo, W. H. H. Gunther and J. Tronp, *J. Am. Chem. Soc.*, 1981, **103**, 4629.
402. M. A. Bush and M. R. Truter, *J. Chem. Soc., Perkin Trans. 2*, 1972, 341.
403. (a) J. D. Owen and J. N. Wingfield, *J. Chem. Soc., Chem. Commun.*, 1976, 318; (b) J. D. Owen, *J. Chem. Soc., Dalton Trans.*, 1978, 1418.
404. P. R. Mallinson and M. R. Truter, *J. Chem. Soc., Perkin Trans. 2*, 1972, 1818.
405. J. Feneau-Dupont, E. Arte, J. P. Declercq, G. Germain and M. van Meerssche, *Acta Crystallogr., Sect. B*, 1979, **35**, 1217.
406. J. D. Owen, *J. Chem. Soc., Dalton Trans.*, 1980, 1066.
407. P. D. Chadwick and N. S. Poonia, *Acta Crystallogr., Sect B*, 1977, **33**, 197.
408. (a) D. E. Fenton, M. Mercer, N. S. Poonia and M. R. Truter, *J. Chem. Soc., Chem. Commun.*, 1972, 66; (b) M. Mercer and M. R. Truter, *J. Chem. Soc., Dalton Trans.*, 1973, 2469.
409. D. L. Hughes, *J. Chem. Soc., Dalton Trans.*, 1975, 2374.
410. D. L. Hughes, *J. Chem. Soc., Dalton Trans.*, 1979, 1831.
411. D. L. Hughes and J. N. Wingfield, *J. Chem. Soc., Chem. Commun.*, 1977, 804.
412. D. L. Hughes, C. L. Mortimer and M. R. Truter, *Acta. Crystallogr., Sect. B*, 1978, **34**, 800.
413. M. A. Bush and M. R. Truter, *J. Chem. Soc., Perkin Trans. 2*, 1972, 345.
414. J. Hasek, K. Huml and D. Hlavata, *Acta Crystallogr., Sect B*, 1979, **35**, 330.
415. J. D. Owen, *Acta Crystallogr., Sect. C*, 1983, **39**, 861.
416. P. Groth, *Acta Chem. Scand., Ser. A*, 1981, **35**, 460.
417. P. Groth, *Acta Chem. Scand., Ser. A*, 1981, **35**, 463.
418. G. Shoham, W. N. Lipscomb and U. Olsher, *J. Am. Chem. Soc.*, 1983, **105**, 1247.
419. G. Shoham, W. N. Lipscomb and U. Olsher, *J. Chem. Soc., Chem. Commun.*, 1983, 208.
420. M. Czugler and E. Weber, *J. Chem. Soc., Chem. Commun.*, 1981, 472.
421. M. A. Bush and M. R. Truter, *J. Chem. Soc., Perkin Trans. 2*, 1971, 1440.
422. P. Moras, B. Metz, M. Herceg and R. Weiss, *Bull. Soc. Chim. Fr.*, 1972, 551.
423. M. L. Campbell, N. K. Dalley, R. M. Izatt and J. D. Lamb, *Acta Crystallogr., Sect. B*, 1981, **37**, 1664.
424. M. L. Campbell, S. B. Larson and N. K. Dalley, *Acta Crystallogr., Sect. B*, 1981, **37**, 1741.
425. M. L. Campbell, S. B. Larson and N. K. Dalley, *Acta Crystallogr., Sect. B*, 1981, **37**, 1744.
426. M. L. Campbell, N. K. Dalley and S. H. Simonsen, *Acta. Crystallogr., Sect. B*, 1981, **37**, 1747.
427. (a) R. A. Schultz, D. M. Dishong and G. W. Gokel, *J. Am. Chem. Soc.*, 1982, **104**, 625; (b) A. Kaifer, H. D. Durst, L. Echegoyen, D. M. Dishong, R. A. Schultz and G. W. Gokel, *J. Org. Chem.*, 1982, **47**, 3195; (c) D. M. Goli, D. M. Dishong, C. J. Diamond and G. W. Gokel, *Tetrahedron Lett.*, 1982, **23**, 5243.
428. F. R. Fronczek, V. J. Gatto, R. A. Schultz, S. J. Jungk, W. J. Colucci, R. D. Gandow and G. W. Gokel, *J. Am. Chem. Soc.*, 1983, **105**, 6717.
429. G. Weber, W. Saenger, K. Muller, W. Wehner and F. Vögtle, *Inorg. Chim. Acta*, 1983, **77**, L199.
430. E. Weber and M. Czugler, *Inorg. Chim. Acta*, 1982, **61**, 35.
431. P. Groth, *Acta Chem. Scand., Ser. A*, 1983, **37**, 71.
432. P. Groth, *Acta Chem. Scand., Ser. A*, 1983, **37**, 283.
433. J. Powell, K. S. Ng, W. W. Ng and S. C. Nyburg, *J. Organomet. Chem.*, 1983, **243**, C1.
434. J. Powell, M. Gregg, A. Kuskin and P. Meindl, *J. Am. Chem. Soc.*, 1983, **105**, 1064.
435. K. N. Trueblood, C. B. Knobler, E. Maverick, R. C. Helgeson, S. B. Brown and D. J. Cram, *J. Am. Chem. Soc.*, 1981, **103**, 5594.
436. D. J. Cram, I. B. Dicker, C. B. Knobler and K. N. Trueblood, *J. Am. Chem. Soc.*, 1982, **104**, 6828.
437. D. J. Cram, G. M. Lein, T. Kaneda, R. C. Helgeson, C. B. Knobler, E. Maverick and K. N. Trueblood, *J. Am. Chem. Soc.*, 1981, **103**, 6228.
438. D. J. Cram, J. R. Moran, E. F. Maverick and K. N. Trueblood, *J. Chem. Soc., Chem. Commun.*, 1983, 647.
439. G. Weber, *Acta Crystallogr., Sect. B*, 1981, **37**, 1832.
440. M. G. B. Drew, A. Hamid bin Othman, S. G. McFall and S. M. Nelson, *J. Chem. Soc., Chem. Commun.*, 1975, 818.
441. D. E. Fenton, D. H. Cook, I. W. Nowell and P. E. Walker, *J. Chem. Soc., Chem. Commun.*, 1978, 279.
442. D. E. Fenton, D. H. Cook, I. W. Nowell and P. E. Walker, *J. Chem. Soc., Chem. Commun.*, 1977, 623.
443. J. de O. Cabral, M. F. Cabral, W. J. Cummins, M. G. B. Drew, A. Rodgers and S. M. Nelson, *Inorg. Chim. Acta*, 1978, **30**, L313.
444. S. M. Nelson, F. S. Esho and M. G. B. Drew, *J. Chem. Soc., Dalton Trans.*, 1983, 1857.
445. M. G. B. Drew, F. S. Esho and S. M. Nelson, *J. Chem. Soc., Dalton Trans.*, 1983, 1653.
446. S. M. Nelson and C. V. Knox, *J. Chem. Soc., Dalton Trans.*, 1983, 2525.
447. H. Adams, N. A. Bailey, D. E. Fenton, R. J. Good, R. Moody and C. O. Rodriguez de Barbarin, *J. Chem. Soc., Dalton Trans.*, 1987, 207.
448. S. B. Larson and N. K. Dalley, *Acta Crystallogr., Sect. B*, 1980, **36**, 1201.
449. B. Metz, D. Moras and R. Weiss, *J. Chem. Soc., Perkin Trans. 2*, 1976, 423.
450. D. Moras, B. Metz and R. Weiss, *Acta Crystallogr., Sect. B*, 1973, **29**, 388.

451. B. Metz, D. Moras and R. Weiss, *Acta Crystallogr., Sect. B,* 1973, **29,** 1382.
452. B. Metz, D. Moras and R. Weiss, *Acta Crystallogr., Sect. B,* 1973, **29,** 1377.
453. (a) E. Mei, A. I. Popov and J. L. Dye, *J. Am. Chem. Soc.,* 1977, **99,** 6532; (b) E. Meir, L. Lui, J. L. Dye and A. I. Popov, *J. Solution Chem.,* 1977, **6,** 771.
454. F. Mathieu, B. Metz, D. Moras and R. Weiss, *J. Am. Chem. Soc.,* 1978, **100,** 4412.
455. F. J. Tehan, B. L. Barnett and J. L. Dye, *J. Am. Chem. Soc.,* 1974, **96,** 7203.
456. J. D. Corbett, D. G. Adolphson, D. J. Merryman, P. A. Edwards and F. J. Armatis, *J. Am. Chem. Soc.,* 1975, **97,** 6267.
457. J. D. Corbett and P. A. Edwards, *J. Chem. Soc., Chem. Commun.,* 1975, 984.
458. S. C. Critchlow and J. D. Corbett, *J. Chem. Soc., Chem. Commun.,* 1981, 236.
459. T. Konig, B. Eisenman and H. Schaefer, *Z. Anorg. Allg. Chem.,* 1982, **488,** 126.
460. R. G. Teller, R. G. Finke, J. P. Collman, H. B. Chin and R. Bau, *J. Am. Chem. Soc.,* 1977, **99,** 164.
461. J. Fischer, M. Mellinger and R. Weiss, *Inorg. Chim. Acta,* 1977, **21,** 259.
462. J. D. Owen, *Acta Crystallogr., Sect. C,* 1984, **40,** 246.
463. I. R. Hanson, D. G. Parsons and M. R. Truter, *J. Chem. Soc., Chem. Commun.,* 1979, 486.
464. J. D. Owen, *Acta Crystallogr., Sect. C,* 1983, **39,** 579.
465. R. N. Izatt, D. J. Eatough and J. J. Christensen, *Struct. Bonding (Berlin),* 1973, **16,** 162.
466. J. J. Christensen, D. J. Eatough and R. M. Izatt, *Chem. Rev.,* 1974, **74,** 351.
467. J. D. Lamb, R. M. Izatt and J. J. Christensen, *Prog. Macrocycl. Chem.,* 1981, **2,** 41.
468. K. Frensdorff, *J. Am. Chem. Soc.,* 1971, **93,** 600.
469. R. M. Izatt, K. E. Terry, A. G. Avondet, J. S. Bradshaw, N. K. Dalley, T. E. Jensen, J. J. Christensen and B. L. Haymore, *Inorg. Chim. Acta,* 1978, **30,** 1.
470. R. M. Izatt, J. D. Lamb, G. E. Maas, R. E. Asay, J. S. Bradshaw and J. J. Christensen, *J. Am. Chem. Soc.,* 1977, **99,** 2365.
471. J. R. Blackborow, J. C. Lockhart, M. E. Thompson and D. P. Thompson, *J. Chem. Res. (S),* 1978, 53.
472. J. M. Lehn and F. Montavon, *Helv. Chim. Acta,* 1978, **61,** 67.
473. F. Arnaud-Neu, B. Sipas and M. J. Schwing-Weill, *Helv. Chim. Acta,* 1977, **60,** 2633.
474. E. Weber and F. Vögtle, *Angew. Chem., Int. Ed. Engl.,* 1974, **13,** 149.
475. D. J. Cram, R. C. Helgesen, L. R. Sousa, J. M. Timko, M. Newcomb, P. Moreau, F. de Jong, G. W. Gokel, D. H. Hoffman, L. A. Domeier, S. C. Peacock, K. Madan and L. Kaplan, *Pure Appl. Chem.,* 1975, **43,** 327.
476. J. M. Timko, S. S. Moore, D. M. Walba, P. C. Hyberty and D. J. Cram, *J. Am. Chem. Soc.,* 1977, **99,** 4207.
477. (a) J. M. Timko and D. J. Cram, *J. Am. Chem. Soc.,* 1974, **96,** 7159; (b) M. Newcomb, G. W. Gokel and D. J. Cram, *J. Am. Chem. Soc.,* 1974, **96,** 6810.
478. J. M. Lehn, *Struct. Bonding (Berlin),* 1973, **16,** 1.
479. E. Graf and J. M. Lehn, *J. Am. Chem. Soc.,* 1975, **97,** 5022.
480. A. C. Coxon and J. F. Stoddart, *J. Chem. Soc., Perkin Trans. 2,* 1977, 785.
481. F. de Jong, D. N. Reinhardt and R. Huis, *Tetrahedron Lett.,* 1977, 3985.
482. J. D. Lin and A. I. Popov, *J. Am. Chem. Soc.,* 1981, **103,** 3773.
483. Y. Takeda, *Bull. Chem. Soc. Jpn.,* 1981, **54,** 3133.
484. Y. Takeda, Y. Wada and S. Fujiwara, *Bull. Chem. Soc. Jpn.,* 1981, **54,** 3727.
485. A. S. Khan, W. G. Baldwin and A. Chow, *Can. J. Chem.,* 1981, **59,** 1490.
486. B. Dietrich, J. M. Lehn and J. P. Sauvage, *Tetrahedron,* 1973, **29,** 1647.
487. F. de Jong, A. von Zorn, D. N. Reinhoudt, G. J. Torny and H. P. M. Tomassen, *Recl. Trav. Chim. Pays-Bas,* 1983, **102,** 164.
488. P. B. Chock, *Proc. Natl. Acad. Sci. USA,* 1972, **69,** 1939.
489. B. Dietrich, J.-M. Lehn and J. P. Sauvage, *J. Chem. Soc., Chem. Commun.,* 1973, 15.
490. E. Shchori and J. Jagur-Grodzinski, *Isr. J. Chem.,* 1973, **11,** 243.
491. R. Ungaro, B. El Haj and J. Smid, *J. Am. Chem. Soc.,* 1976, **58,** 5198.
492. G. W. Liesegang and E. M. Eyring, in 'Synthetic Multidentate Ligands', ed. R. M. Izatt and J. J. Christensen, Academic Press, New York, 1978, chap. 5.
493. W. Burgermeister and R. Winkler-Ostwatitsch, *Top. Curr. Chem.,* 1977, **69**, 91.
494. M. Eigen and R. Winkler, *Neuro. Sci. Res. Prog. Bull.,* 1971, **9,** 330.
495. J. D. Lamb, R. M. Izatt, J. J. Christensen and D. J. Eatough, 'Coordination Chemistry of Macrocyclic Compounds', ed. G. A. Nelson, Plenum, New York, 1979, chap. 3.
496. G. W. Liesegang, M. M. Farrow, F. A. Vazquez, N. Purdie and E. M. Eyring, *J. Am. Chem. Soc.,* 1977, **99,** 3240.
497. M. Kirch and J.-M. Lehn, *Angew. Chem., Int. Ed. Engl.,* 1975, **14,** 555.
498. N. Yamakazi, S. Nakahama, H. Hirao and S. Negi, *Tetrahedron Lett.,* 1978, 2429.
499. (a) J.-P. Behr, J.-M. Lehn, A.-C. Dock and D. Moras, *Nature (London),* 1982, **295,** 526; (b) A.-C. Dock, D. Moras, J.-P. Behr and J.-M. Lehn, *Acta Crystallogr., Sect. C,* 1983, **39,** 1001.
500. S. Shinkai, T. Minami, Y. Kusaro and O. Manabe, *Tetrahedron Lett.,* 1982, **23,** 2581; S. Shinkai and O. Manabe, *Top. Curr. Chem.,* 1984, **121,** 67.
501. T. Minami, S. Shinkai and O. Manabe, *Tetrahedron Lett.,* 1982, **23,** 5167.
502. M. Raban, J. Greenblatt and F. Kandi, *J. Chem. Soc., Chem. Commun.,* 1983, 1409.
503. (a) R. G. Ackman, W. H. Brown and G. F. Wright, *J. Org. Chem.,* 1955, **20,** 1147; (b) A. Baeyer, *Chem. Ber.,* 1886, **19,** 2184; (c) W. H. Brown and W. N. Freiden, *Can. J. Chem.,* 1958, **36,** 371.
504. A. Corsini and J. M. Parayan, *J. Inorg. Nucl. Chem.,* 1977, **39,** 1449.
505. M. Chastrette and F. Chastrette, *J. Chem. Soc., Chem. Commun.,* 1973, 534.
506. A. J. Rest, S. A. Smith and I. D. Tyler, *Inorg. Chim. Acta,* 1976, **16,** L1.
507. M. de Sousa Healy and A. J. Rest, *J. Chem. Soc., Chem. Commun.,* 1981, 149.
508. Y. Kobuke, K. Hanji, K. Horiguchi, M. Asada, Y. Nakayama and J. Furukawa, *J. Am. Chem. Soc.,* 1976, **98,** 7414.
509. C. D. Gutsche and R. J. Muthukrishnan, *J. Org. Chem.,* 1978, **43,** 4905.

510. R. N. Izatt, J. D. Lamb, R. T. Hawkins, P. R. Brown, S. R. Izatt and J. J. Christensen, *J. Am. Chem. Soc.*, 1983, **105,** 1782.
511. J. J. Katz, in 'Inorganic Biochemistry', ed. G. L. Eichhorn, Elsevier, Amsterdam, 1973, chap. 29.
512. J. H. Fuhrop, *Angew. Chem., Int. Ed. Engl.*, 1976, **15,** 648.
513. C. E. Strouse, *Prog. Inorg. Chem.*, 1976, **21,** 159.
514. H.-C. Chow, R. Serlin and C. E. Strouse, *J. Am. Chem. Soc.*, 1975, **97,** 7230.
515. M. R. Wasielewski, U. H. Smith, B. T. Cope and J. J. Katz, *J. Am. Chem. Soc.*, 1977, **99,** 4172.
516. M. R. Wasielewski, M. H. Studier and J. J. Katz, *Proc. Natl. Acad. Sci. USA*, 1976, **73,** 4382.
517. S. G. Boxer and G. L. Closs, *J. Am. Chem. Soc.*, 1976, **98,** 5406.
518. (a) M. R. Wasielewski, W. A. Svec and B. T. Cope, *J. Am. Chem. Soc.*, 1978, **100,** 1961; (b) R. E. Overfield, A. Scherz, K. J. Kaufmann and M. R. Wasielewski, *J. Am. Chem. Soc.*, 1983, **105,** 4256.
519. R. Timkovich and A. Tulinsky, *J. Am. Chem. Soc.*, 1969, **91,** 4430.
520. M. S. Fischer, D. H. Templeton, A. H. Zalkin and M. Calvin, *J. Am. Chem. Soc.*, 1971, **93,** 2622.
521. K. M. Barkigia, L. D. Spaulding and J. Fajer, *Inorg. Chem.*, 1983, **22,** 349.
522. A. B. P. Lever, *Prog. Inorg. Chem.*, 1965, **7,** 38.
523. R. P. Linstead and J. M. Robertson, *J. Chem. Soc.*, 1936, 1736.
524. R. Y. Lee and P. Hambright, *J. Inorg. Nucl. Chem.*, 1970, **32,** 477.
525. (a) O. C. Choon and G. A. Rodley, *Inorg. Chim. Acta*, 1983, **80,** 177; (b) G. A. Rodley and O. C. Choon, *Inorg. Chim. Acta*, 1983, **80,** 171.
526. B. C. Pressman, E. J. Harris, W. J. Jogger and J. H. Johnson, *Proc. Natl. Acad. Sci. USA*, 1967, **58,** 1949.
527. S. N. Graven, S. Estrada-O. and H. A. Lardy, *Proc. Natl. Acad. Sci. USA*, 1966, **56,** 654.
528. R. M. Fuoss, *J. Chem. Educ.*, 1955, **32,** 527.
529. B. C. Pressman, *Annu. Rev. Biochem.*, 1976, **45,** 4925.
530. B. C. Pressman and D. H. Haynes, in 'The Molecular Basis of Membrane Function', ed. D. C. Tosteson, Prentice-Hall, Englewood Cliffs, New Jersey, 1969, p. 221.
531. C. J. Pedersen, *Aldrichim. Acta*, 1971, **4,** 1.
532. Yu. A. Ovchinnikov, V. T. Ivanov and A. M. Shkrob, 'Membrane-active Complexones', Elsevier, Amsterdam, 1974.
533. M. Dobler, 'Ionophores and their Structures', Wiley, New York, 1982.
534. J. D. Dunitz and M. Dobler, in 'Biological Aspects of Inorganic Chemistry', ed. A. W. Addison, W. R. Cullen, D. Dolphin and B. R. James, Wiley, New York, 1977, p. 133.
535. J. W. Westley, *Adv. Appl. Microbiol.*, 1977, **22,** 177.
536. G. Eisenman, in 'Ion Selective Electrodes', ed. R. A. Durst, National Bureau of Standards, USA, Special Publication 314, Washington DC, 1969.
537. M. Pinkerton, L. K. Steinrauf and P. Dawkins, *Biochem. Biophys. Res. Commun.*, 1969, **35,** 512.
538. K. Neupert-Laves and M. Dobler, *Helv. Chim. Acta*, 1975, **58,** 432.
539. J. A. Hamilton, M. N. Sabesan and L. K. Steinrauf, *J. Am. Chem. Soc.*, 1981, **103,** 5880.
540. L. K. Steinrauf and M. N. Sabesan, in 'Metal–Ligand Interactions in Organic Chemistry and Biochemistry', ed. B. Pullman and N. Goldblum, Reidel, Dordrecht, 1977, part 2, p. 43.
541. (a) W. L. Duax and H. Hauptman, *Acta Crystallogr., Sect. B*, 1972, **28,** 2912; (b) I. L. Karle, *J. Am. Chem. Soc.*, 1975, **97,** 4379; (c) G. D. Smith, W. L. Duax, D. A. Langs, G. T. de Titta, J. W. Edmonds, D. C. Rohrer and C. M. Weeks, *J. Am. Chem. Soc.*, 1975, **97,** 7242.
542. Yu. A. Ovchinnikov and V. T. Ivanov, *Tetrahedron*, 1975, **31,** 2177.
543. E. Eyal and G. A. Rechnitz, *Anal. Chem.*, 1971, **43,** 1090.
544. M. Dobler, J. D. Dunitz and J. D. Krajewski, *J. Mol. Biol.*, 1969, **42,** 603.
545. B. K. Vainshtein *et al.*, 4th European Crystallographic Meeting, Oxford, 1977, quoted in ref. 372.
546. (a) Yu. A. Ovchinnikov, *FEBS Lett.*, 1974, **44,** 1; (b) V. T. Ivanov, A. V. Evstratov, L. V. Sunskaya, E. I. Melnik, T. S. Chumburidze, S. L. Portnova, T. A. Balashova and Yu. A. Ovchinnikov, *FEBS Lett.*, 1973, **36,** 65.
547. B. Braden, J. A. Hamilton, M. N. Sabesan and L. K. Steinrauf, *J. Am. Chem. Soc.*, 1980, **102,** 2704.
548. T. Wieland, H. Faulstich and W. Burgermeister, *Biochem. Biophys. Res. Commun.*, 1972, **47,** 984.
549. I. L. Karle, *J. Am. Chem. Soc.*, 1974, **96,** 4000.
550. I. L. Karle, J. Karle, T. Wieland, H. Faulstich and B. Witkop, *Proc. Natl. Acad. Sci. USA*, 1973, **70,** 1836.
551. P. Mueller and D. O. Rudin, *Biochem. Biophys. Res. Commun.*, 1967, **26,** 398.
552. (a) B. T. Kilbourn, J. D. Dunitz, L. A. R. Pioda and W. Simon, *J. Mol. Biol.*, 1967, **30,** 559; (b) M. Dobler, J. D. Dunitz and B. T. Kilbourn, *Helv. Chim. Acta*, 1969, **52,** 2573.
553. T. Sakamaki, Y. Iitaka and Y. Nawata, *Acta Crystallogr., Sect. B*, 1976, **32,** 768.
554. M. Dobler and R. P. Phizackerley, *Helv. Chim. Acta*, 1974, **57,** 664.
555. T. Sakamaki, Y. Titaha and Y. Nawata, *Acta Crystallogr., Sect. B*, 1977, **33,** 52.
556. C. K. Vishnawath, N. Shamala, K. R. K. Easwaran and M. Vijayan, *Acta Crystallogr., Sect. C*, 1983, **39,** 1640.
557. B. F. Gisin and R. F. Merrifield, *J. Am. Chem. Soc.*, 1972, **94,** 6165.
558. V. Madison, M. Atreyl, C. M. Deber and E. R. Blout, *J. Am. Chem. Soc.*, 1974, **96,** 6725.
559. G. Kartha, K. J. Varughese and S. Aimoto, *Proc. Natl. Acad. Sci. USA*, 1982, **79,** 4519.
560. V. Madison, C. M. Deber and E. R. Blout, *J. Am. Chem. Soc.*, 1977, **99,** 4788.
561. Y. H. Chiu, D. L. Brown and W. N. Lipscomb, *J. Am. Chem. Soc.*, 1977, **99,** 4799.
562. I. L. Karle and J. Karle, *Proc. Natl. Acad. Sci. USA*, 1981, **78,** 681.
563. A. Sanat, M. El Malouli Bibout, M. Chanon and J. Elguero, *Nouv. J. Chim.*, 1982, **6,** 483.
564. M. Pinkerton and L. K. Steinrauf, *J. Mol. Biol.*, 1970, **49,** 533.
565. W. L. Duax, G. D. Smith and P. D. Strong, *J. Am. Chem. Soc.*, 1980, **102,** 6725.
566. D. L. Ward, K.-T. Wei, J. G. Hoogerheide and A. I. Popov, *Acta Crystallogr., Sect. B*, 1978, **34,** 110.
567. W. K. Lutz, F. K. Winkler and J. D. Dunitz, *Helv. Chim. Acta*, 1971, **54,** 1103.
568. Y. Barrans and M. Alleaume, *Acta Crystallogr., Sect. B*, 1980, **36,** 936.
569. E. W. Czerwinski and L. K. Steinrauf, *Biochem. Biophys. Res. Commun.*, 1971, **45,** 1284.
570. C. A. Maier and I. C. Paul, *Chem. Commun.*, 1971, 181.

571. (a) P. G. Schmidt, A. H. Wang and I. C. Paul, *J. Am. Chem. Soc.*, 1974, **96,** 6189; (b) G. D. Smith, W. L. Duax and S. Fortier, *J. Am. Chem. Soc.*, 1978, **100,** 6725.
572. C. C. Chiang and I. C. Paul, *Science,* 1977, **196,** 1441.
573. S. M. Johnson, J. Herrin, S. J. Liu and I. C. Paul, *J. Am. Chem. Soc.*, 1970, **92,** 4428.
574. J. Bolte, C. Demuynck, G. Jeminet, J. Juillard and C. Tissier, *Can. J. Chem.*, 1982, **60,** 981.
575. B. K. Toeplitz, A. I. Cohen, P. T. Funke, W. L. Parker and J. Z. Gougoutas, *J. Am. Chem. Soc.*, 1979, **101,** 3344.
576. D. H. Davies, E. W. Snape, P. J. Suter, T. J. King and C. P. Falshaw, *J. Chem. Soc., Chem. Commun.*, 1981, 1073.
577. J. Grandjean and P. Laszlo, *Tetrahedron Lett.*, 1983, **24,** 3319.
578. W. Marsh, J. D. Dunitz and D. N. J. White, *Helv. Chim. Acta,* 1974, **57,** 10.
579. P. Hubberstey, *Coord. Chem. Rev.*, 1983, **49,** 1.
580. D. E. Fenton, Annual Reports A (1982), Royal Society of Chemistry, London, 1983, p. 3.
581. D. Barr, W. Clegg, R. E. Mulvey and R. Snaith, *J. Chem. Soc., Chem. Commun.*, 1984, 79.
582. D. Barr, W. Clegg, R. E. Mulvey and R. Snaith, *J. Chem. Soc., Chem. Commun.*, 1984, 226.
583. D. Barr, W. Clegg, R. E. Mulvey and R. Snaith, *J. Chem. Soc., Chem. Commun.*, 1984, 285.
584. D. Barr, W. Clegg, R. E. Mulvey and R. Snaith, *J. Chem. Soc., Chem. Commun.*, 1984, 700.
585. D. Barr, W. Clegg, R. E. Mulvey and R. Snaith, *J. Chem. Soc., Chem. Commun.*, 1984, 974.
586. H. Hope, D. Oram and P. Power, *J. Am. Chem. Soc.*, 1984, **106,** 1149.
587. M. Y. Chiang, E. Bohlen and R. Bau, *J. Am. Chem. Soc.*, 1985, **107,** 1679.
588. N. A. Bell, H. M. M. Shearer and J. Twiss, *Acta Crystallogr., Sect. C,* 1984, **40,** 610.
589. (a) N. A. Bell, H. M. M. Shearer and J. Twiss, *Acta Crystallogr., Sect. C,* 1984, **40,** 605; (b) N. A. Bell, P. T. Moseley and H. M. M. Shearer, *Acta Crystallogr., Sect. C,* 1984, **40,** 602.
590. D. Gordon and M. H. G. Wallbridge, *Inorg. Chim. Acta,* 1984, **88,** 15.
591. P. Sobota, J. Utko and T. Lis, *J. Chem. Soc., Dalton Trans.*, 1984, 2077.
592. P. Sobota, T. Pluzinski and T. Lis, *Polyhedron,* 1984, **3,** 45.
593. R. P. Sharma and R. N. Prasad, *Synth. React. Inorg. Met.-Org. Chem.*, 1984, **14,** 501.
594. D. E. Fenton, Annual Reports A (1984), Royal Society of Chemistry, London, 1985, p. 3.
595. S. Shinkai, K. Inuzuka, O. Miyakazi and O. Manabe, *J. Org. Chem.*, 1984, **49,** 3440.
596. S. Ogawa, R. Narushima and Y. Arai, *J. Am. Chem. Soc.*, 1984, **106,** 5760.
597. C. M. Means, N. C. Means, S. G. Bott and J. L. Attwood, *J. Am. Chem. Soc.*, 1984, **106,** 7627.
598. L. Radics, *J. Chem. Soc., Chem. Commun.*, 1984, 599.

24

Boron

JÓZSEF EMRI and BÉLA GYÖRI
Lajos Kossuth University, Debrecen, Hungary

24.1 INTRODUCTION

Tricoordinated boron compounds (boranes) are coordinatively unsaturated and their chemistry is dominated by reactions in which complexes are formed. These complexes are either neutral molecules (borane complexes), anions (borates) or boron cations. Space limitations mean that little or no attention will be paid to complexes containing several boron atoms and to species of the type $L{\cdot}BH_3$, $[BH_4]^-$ and $[L_2BH_2]^+$ (L = neutral ligand), discussed in detail in several books and reviews. Similarly, little attention will be paid to the plethora of metal borates and the cyclic and polymeric amino- and phosphino-boranes.

The whole field of boron chemistry is covered, in a most comprehensive manner, in the 28 volumes on boron of Gmelin's Handbook. Two summarize early results and coordination compounds are dealt with, in part or completely, in recent volumes.[1] Useful chapters, relevant to the coordination chemistry of boron can be found in refs. 2–4. For general information, standard works[5–9] are available while several others contain relevant chapters that will be referred to where appropriate. ^{11}B NMR results are comprehensively surveyed in a recent book,[10] with up to 300 pages of tabulated material, including all the known types of the boron coordination compounds, and serving as a useful guide. In 1964, only one review comprehensively covered the coordination chemistry of boron.[11] Reviews relevant to specialized areas of the field will be referred to under appropriate subheadings.

24.2 NEUTRAL BORANE COMPLEXES

24.2.1 General Considerations and Stability

Since the discovery of $NH_3{\cdot}BF_3$ by Gay-Lussac and Thenard[12] in 1809, many neutral borane complexes have been described, mainly with N, P, As, O and S donor molecules. Their stabilities vary considerably; many are formed at low temperatures only, whereas in other cases the dissociation of the coordinate bond cannot be observed even at the temperature of decomposition. They are of importance in most reactions of boranes, the first step usually being the formation of some kind of borane complex.

The stability of the coordinative bond in the borane complexes is determined by the electrophilicity of the borane, by the nucleophilicity of the ligand and by steric factors. The electrophilicity of the borane, in turn, depends on the electronegativity of the groups bonded to the boron and on the overlap between the unoccupied *p* orbital of the boron and the occupied *p* orbitals of the atoms directly bonded to it ($p\pi$–$p\pi$ interaction). The predominance of the latter is reflected by the fact that the relative acidities of halogenoboranes increase in a sense opposite to that expected from electronegativity considerations and is supported by the very weak Lewis acidity of the alkoxy- and amino-boranes. The importance of steric factors is pointed out by Brown and co-workers.[13–15]

In a number of cases, *e.g.* for CO, PH_3, PF_3 and thioethers, the existence of complexes or their surprising stability was explained in terms of a special π bond.[16] However, experimental data and theoretical considerations seem to contradict the assumption that such kinds of π bonds exist.[25] Borane complexes in which donor atoms of appropriate nucleophilicity are part of the group attached to the boron, *e.g.* $BCl_2(NMe_2)$, $BH_2(PMe_2)$, BBr_2SMe, constitute a special case and exist due to intermolecular coordination, in the form of dimers, trimers or polymers.[17–19]

The amine boranes constitute the largest and most extensively studied group of borane complexes. The B—N coordinate bond is generally formulated[17] as having an N^+—B^- polarity as a result of charge transfer from N to B although MO calculations indicate[2,20] the opposite charge distribution. The strength of the coordinate bond in neutral borane complexes is most adequately characterized by the enthalpy change, $\Delta H_{(g)}$ for the dissociation of the adduct in the gas phase into its components.[11,21] Values of $\Delta H_{(g)}$ and heats of reaction for some borane complexes are collected in Table 1.

Table 1 Complex-forming Reactions Heats of Selected Boranes

Base	ΔH^a	ΔH *values* (kJ mol^{-1}) *for* BH_3^b	BF_3	BCl_3	BBr_3	*Ref.*
MeCN	ΔH_s	−121.2	−111.0	−144.0	−165.0	1–3
C_5H_5N	ΔH_s	—	—	−234.1	−265.9	4
	ΔH_{sol}	−149.3	−137.8	−192.6	−220.7	4–6
Me_3N	ΔH_g	−146.5	−111.4	(−127.7)	—	7
Me_3P	ΔH_s	−334.5	−190.5	−287.2	−512.1	8
Me_3As	ΔH_s	−207.7	−85.4	−193.4	−340.0	8
Me_3Sb	ΔH_s	(−27.6)	(−17.6)	−112.2	−82.9	8

[a] ΔH_g for acid (g) + base (g) → complex (g); ΔH_s for acid (g) + base (g) → complex (s); ΔH_{sol} for acid (g) + base (sol) → complex (sol).
[b] $\Delta H = 74.3$ kJ mol^{-1} for $B_2H_6 \rightarrow BH_3$ from refs. 8 and 9.

1. H. J. Emeleus and K. Wade, *J. Chem. Soc.*, 1960, 2614.
2. A. W. Laubengayer and D. S. Sears, *J. Am. Chem. Soc.*, 1945, **67**, 164.
3. J. M. Miller and M. Onyszchuk, *Can. J. Chem.*, 1965, **43,** 1877.
4. N. N. Greenwood and P. G. Perkins, *J. Chem. Soc.*, 1960, 1141.
5. H. C. Brown and L. Domash, *J. Am. Chem. Soc.*, 1956, **78,** 5384.
6. H. C. Brown and D. Gintis, *J. Am. Chem. Soc.*, 1956, **78,** 5378.
7. R. E. McCoy and S. H. Bauer, *J. Am. Chem. Soc.*, 1956, **78,** 2061.
8. D. C. Mente, J. L. Mills and R. E. Mitchell, *Inorg. Chem.*, 1975, **14,** 123.
9. T. P. Fehlner and G. W. Mappes, *J. Phys. Chem.*, 1969, **73,** 873.

Based on the data in Table 1, on additional semiquantitative data and upon qualitative observations[2,9,11] the following order of relative stabilities could be established for borane complexes:

(a) for $L\cdot BX_3$(X = halogen), $R_3N > R_3P > R_3As > R_3Sb$;
(b) for $L\cdot BF_3$, $R_2O > R_2S > R_2Se$;
(c) for $R_3N\cdot BX_3$, $BBr_3 > BCl_3 > BH_3 > BF_3 > BMe_3 > B(SMe)_3 > B(OMe)_3$;
(d) for $R_3P\cdot BX_3$, $BBr_3 > BH_3 > BCl_3 > BF_3 > BMe_3$.

For borane complexes of intermediate strength the order of relative stabilities can also be estimated from equilibrium data as well, obtained in gas phase or in solution.[11,23] From the temperature dependence of the equilibrium constant for the dissociation in the gas phase the $\Delta H_{(g)}$ values can be determined with great accuracy.[22]

In addition to these limited procedures a number of experimental methods (vibrational spectroscopy, dipole moment measurements, electron diffraction, NMR, *etc.*) have been employed to determine the relative stabilities of these complexes.[11,23] Intense effort has been directed towards establishing some kind of correlation between NMR parameters and stability of the borane complexes. The chemical shifts alone rarely show good correlation. However, complexation shifts (the chemical shift difference between the free and complexed borane or ligand) and various spin–spin coupling constants correlate better with calorimetric data, especially for ligands or boranes belonging to structurally similar series (Table 2).[10,24]

Table 2 Spectroscopic Data for $Me_3N{\cdot}BX_3$ Complexes

Parameter[a] (*unit*)	*Values for* $Me_3N{\cdot}BX_3$ X = H	X = F	X = Cl	X = Br	X = I	*Ref.*
K(B—N) ($kN\ m^{-1}$)	0.259	0.359	0.310	0.398	0.342	1, 2
δ(Me) (p.p.m)	2.583	2.62	2.97	3.13	3.32	3
$\Delta\delta(^{11}B)$ (p.p.m.)	78.1	9.36	36.34	41.96	46.31	4, 5
$J(^{15}N-^{11}B)$ (Hz)	—	−18.70	−16.50	−15.23	−12.09	5
r(B—N) (Å) (microwave)	1.638	1.636	—	—	—	6, 7
(electron diffraction)	1.62	1.664	1.659	—	—	8, 9
(X-ray)	—	1.585	1.610	1.603	1.584	10, 11

[a] K, force constant; δ, chemical shift, downfield from internal TMS in CH_2Cl_2; $\Delta\delta(^{11}B)$, complexation shift measured in CH_2Cl_2; r(B—N), B—N bond distance.

1. J. D. Odom, J. A. Barnes, B. A. Hudgens and J. R. Durig, *J. Phys. Chem.*, 1974, **78**, 1503.
2. P. H. Laswick and R. C. Taylor, *J. Mol. Struct.*, 1976, **34**, 197.
3. W. H. Myers, G. E. Ryschkewitsch, M. A. Mathur and R. W. King, *Inorg. Chem.*, 1975, **14**, 2874.
4. H. Nöth and B. Wrackmeyer, *Chem. Ber.*, 1974, **107**, 3070.
5. J. M. Miller, *Inorg. Chem.*, 1983, **22**, 2384.
6. P. Cassaux, R. L. Kuczkowski, P. S. Bryan and R. C. Taylor, *Inorg. Chem.*, 1975, **14**, 126.
7. P. S. Bryan and R. L. Kuczkowski, *Inorg. Chem.*, 1971, **10**, 200.
8. S. H. Bauer, *J. Am. Chem. Soc.*, 1937, **59**, 1804.
9. M. Hargittai and I. Hargittai, *J. Mol. Struct.*, 1977, **39**, 79.
10. S. Geller and J. L. Hoard, *Acta Crystallogr.*, 1951, **4**, 399.
11. P. H. Clippard, J. C. Hanson and R. C. Taylor, *J. Cryst. Mol. Struct.*, 1971, **1**, 363.

The stability and equilibrium properties of neutral borane complexes are not well known. Instead, the recent chemistry of the borane complexes is predominantly characterized by preparative and structural investigations. This is even more obvious for the charged borane complexes.

24.2.2 Complexes of Hydroboranes

The complexes of borane (BH_3) itself with amines, phosphines, dialkyl sulfides and carbon monoxide, the first representatives prepared only in the 1930s were thoroughly discussed in several books and reviews.[2,9,25–36] Therefore, in this section, only a brief summary will be given outlining the most basic aspects of these compounds, whereas complexes of substituted boranes will be discussed in more detail.

It is generally assumed that the reaction of diborane with neutral Lewis bases (L) results in the formation of $L{\cdot}B_2H_6$ singly bridged complexes ($L{\cdot}BH_2$—H—BH_3) in the first step. These complexes, which could be positively identified in some cases, generally give rise to the formation of neutral ($L{\cdot}BH_3$) or, rarely, ionic ($[L_2BH_2][BH_4]$) end products.[9,25,34] The amine boranes are, in general, prepared through high-pressure hydrogenolysis of amine trialkylboranes, from alkoxyboranes with H_2 and Al in the presence of NR_3,[2,9] by nucleophilic displacement of weak donors,[9,25,37] by displacement of hydride ion from $[BH_4]^-$ with ammonium salts[9,17,38] or using amines and I_2.[39]

The most characteristic reaction of amine boranes is their conversion into aminoboranes and, subsequently, to borazine at higher temperatures.[9,17] For complexes of low stability, the transfer of H from the boron to the donor is a characteristic process,[11,40] as in the utilization of diborane as an electrophilic reducing agent. The neutral complexes of boranes are fairly stable towards hydrolysis. The key step of the hydrolysis was formerly assumed to involve displacement of BH_3 by a proton, whereas in recent studies ionic intermediates, containing five-coordinated boron (R_3N—BH_4^+) are also taken into consideration.[41,42] The hydrolytic stability and good solubility of neutral complexes both contribute to their widespread use as reducing agents.[29,36]

Of the many neutral complexes of borane, $CO\cdot BH_3$[11,31,32] and $Me_2S\cdot BH_3$[35] deserve special attention. The former is of surprisingly high stability; no other borane(3) is able to form such complexes. $Me_2S\cdot BH_3$, being a convenient, stable borane carrier, has an excellent combination of stability and reactivity properties and its use as a borane source[35,36] is on the rise.

The majority of the $L\cdot BH_nX_{3-n}$ (X = halogen, R, OR, SR, CN, CO_2H, *etc.*) borane complexes were described in the last 20–25 years. There exist various possibilities for their preparation, summarized in Table 3.

The preparations of complexes of halogenohydroboranes are conveniently achieved through halogenation of the appropriate BH_3 complexes. For some of the halogenating agents the sequence of reactivity is $SbCl_5 \approx SO_2Cl_2 > SbCl_3 > SOCl_2 > HgCl_2 > HCl > [Me_3NH]Cl$, while for the trimethylamine boranes $Me_3N\cdot BH_3 > Me_3N\cdot BH_2Cl > Me_3N\cdot BHCl_2$ applies. Weaker halogenating agents appear to react *via* a polar hydride transfer mechanism, whereas the strongest ones are likely to halogenate by a free radical chain mechanism. The preparation, properties and applications for hydroboration of the THF and Me_2S complexes of halogenoboranes have been reviewed by Brown.[44] The hydrolytic stability of the amine halogenoboranes is fairly high. The rate of hydrolysis for some Me_3N complexes is as follows: $Me_3N\cdot BH_2Cl < Me_3N\cdot BH_2Br < Me_3N\cdot BH_2I$ and $Me_3N\cdot BH_3 < Me_3N\cdot BHI_2 < Me_3N\cdot BH_2I$. These rates are independent of the hydrogen ion concentration. It has been suggested that the rate-determining step is the cleavage of the boron—halogen bond.[45] Hydrolysis of $Me_3N\cdot BH_2F$ may occur *via* acid-independent and acid-catalyzed pathways. In the latter case the complex becomes protonated on the fluorine followed by a loss of HF in the rate-determining step.[46]

To obtain complexes of monocyanoborane, a solution of $(BH_2CN)_n$ in Me_2S appears to be the most suitable reagent (Table 3). The complexes of monocyanoborane are extremely stable in acidic media in general and its complexes with tertiary amines are particularly stable in basic media as well. The complexes of primary and secondary amines undergo N—H cleavage in basic media to give the $[BH_2(CN)NR_2]^-$ ion which, in turn, rapidly decomposes *via*, most likely, a dissociative mechanism.[47] Amine complexes of $BH(CN)NC_4H_4$ are even more stable in acidic media, as they become protonated at the α carbon atom of the pyrrole.[48] The CN group in the monocyanoboranes can be transformed into CO_2H, CO_2R or CONHR groups (equations 1–3).[49,50]

$$A\cdot BH_2CN \xrightarrow[CH_2Cl_2]{[Et_3O][BF_4]} [A\cdot BH_2CNEt][BF_4] \quad (A = \text{amines})$$

$$[A\cdot BH_2CNEt][BF_4] \xrightarrow{H_2O} A\cdot BH_2CO_2H \qquad (1)$$

$$[A\cdot BH_2CNEt][BF_4] \xrightarrow{ROH} A\cdot BH_2CO_2R \qquad (2)$$

$$[A\cdot BH_2CNEt][BF_4] \xrightarrow{NaOH} A\cdot BH_2CONHEt \qquad (3)$$

The amine carboxyboranes may be considered to be analogues of protonated glycine and *N*-substituted glycines. On the basis of structural[49,51] and equilibrium[52] studies, it appears that these analogues differ in several respects from their glycine counterparts. The amine complexes of cyano- and carboxy-borane, as well as the ester and amide derivatives of the latter, display significant antitumor,[49] antiinflammatory[49,50] and antihyperlipidemic[49,53] activity. Recently, some hydroborane complexes containing a chiral boron were synthesized.[48,54] These may be useful for stereoselective transformations of organic compounds. One diastereoisomer of 1-methylbenzylamine-cyanohydropyrrolylborane could be obtained in pure form.[48]

24.2.3 Complexes of Halogenoboranes

As strong Lewis acids, the halogenoboranes are capable of producing a great number of complexes. The largest subgroup in this class is represented by the complexes of trifluoroborane; their first representative was prepared[12] as early as 1809, whereas stable complexes of BI_3 were synthesized[19,20,55] only in the last decades. Some representative halogenoborane complexes and types of reaction used to synthesize them are summarized in Table 4.

Halogenoboranes are capable of giving weak complexes even with unsaturated hydrocarbons.[55] Similarly, weak van der Waals complexes are produced from BF_3 by 'supersonic expansion' with CO, N_2 and Ar.[56]

Table 3 Neutral Complexes of Hydroboranes and their Preparation

Complex	*Starting material*	*Reagent*	*Ref.*
$R_2O{\cdot}BH_nCl_{3-n}$	$R_2O{\cdot}BH_3$	$R_2O{\cdot}BCl_3$	1
	$M[BH_4]$	$R_2O{\cdot}BCl_3$	1, 2
$Me_2S{\cdot}BH_nX_{3-n}$ (X = Br, I)	$Me_2S{\cdot}BH_3$	X_2 or HX	3
$Me_2S{\cdot}BH_2Cl$	$Me_2S{\cdot}BH_3$	CCl_4	4
$Me_2S{\cdot}BH_2R$	$Me_2S{\cdot}BBr_2R$	$Li[AlH_4]$	5
$Me_2S{\cdot}BHCl(CMe_2CHMe_2)$	$Me_2S{\cdot}BH_2Cl$	$Me_2C{=}CMe_2$	6
$Me_3N{\cdot}BH_nF_{3-n}$	$Me_3N{\cdot}BH_3$	HF	7
$R_3N{\cdot}BH_2X$ (X = Cl, Br, I)	$R_3N{\cdot}BH_3$	$R_3N{\cdot}BX_3$	8, 9
$R_3N{\cdot}BH_2X$ (X = Cl, Br)	$R_3N{\cdot}BH_2SR'$	$R''X$	10
$Me_3N{\cdot}BH_2X$ (X = Cl, Br)	$Me_3N{\cdot}BH_3$	$(CH_2CO)_2NX$	11
	$Me_3N{\cdot}BH_2I$	LiX	12
$Me_3N{\cdot}BH_2I$	$Me_3N{\cdot}BH_3$	HI or I_2	8, 13
$R_2NH{\cdot}BHCl_2$	$BH(Cl)NR_2$	HCl	14
$RNH_2{\cdot}BH_2(OCH_2Ph)$	$RNH_2{\cdot}BH_3$	$PhCOCl \cdots + RNH_2$	15
$Me_3N{\cdot}BH_n(O_3SR)_{3-n}$	$Me_3N{\cdot}BH_3$	RSO_3H	16
$4\text{-Mepy}{\cdot}BH(OCOCF_3)_2$	$Na[BH(OCOCF_3)_3]$	4-Mepy	17
$R_3N{\cdot}BH_2SR'$	$(R'SBH_2)_3$	R_3N	10
$Me_3N{\cdot}BH_2N_3$	$[(Me_3N)_2BH_2]N_3$	130–140 °C	18
$Me_3N{\cdot}BH_2NC$	$Me_3N{\cdot}BH_2I$	$AgCN \cdots + H_2S$	19
$Me_3N{\cdot}BH_2(NCX)$ (X = O, S)	$Me_3N{\cdot}BH_2I$	NaNCO or KSCN	12
$NH_3{\cdot}BH_2(NCS)$	B_2H_6	$[NH_4]SCN$	20
$Me_3E{\cdot}BH_2P(CF_3)_2$ (E=N, P)	$[(CF_3)_2PBH_2]_3$	Me_3E	21
$Me_3N{\cdot}BH_2CN$	$Me_3N{\cdot}BH_3$	AgCN	22
$A{\cdot}BH_2CN$	$Na[BH_3CN]$	[AH]Cl	23, 24
	$(BH_2CN)_n$	A	25
$A{\cdot}BH(CN)_2$	$Li[BH_2(CN)_2]$	$A{\cdot}Br_2 + A$	25
$A{\cdot}BH(CN)Pyl$	$Na[BH(CN)Pyl_2]$	[AH]Cl	26
$(MeO)_3P{\cdot}BH_nBr_{3-n}$	$(MeO)_3P{\cdot}BH_3$	$(CH_2CO)_2NBr$	27
$Me_3P{\cdot}BH_2(NCX)$ (X = S, Se)	$Me_3P{\cdot}BH_2I$	AgXCN	28
$R_3P{\cdot}BH_2CN$	$Na[BH_3CN]$	$Br_2 \cdots + R_3P$	29

A = amine; M = Li, Na; R = alkyl, aryl; py = pyridine; Pyl = pyrrol-1-yl; n = 1, 2.

1. H. C. Brown and P. A. Tierney, *J. Inorg. Nucl. Chem.*, 1959, **9**, 51.
2. H. C. Brown and N. Ravindran, *J. Am. Chem. Soc.*, 1976, **98**, 1785, 1798.
3. K. Kinberger and W. Siebert, *Z. Naturforsch., Teil B*, 1975, **30**, 55.
4. W. E. Paget and K. Smith, *J. Chem. Soc., Chem. Commun.*, 1980, 1169.
5. S. U. Kulkarni, D. Basavaiah, M. Zaidlewicz and H. C. Brown, *Organometallics*, 1982, **1**, 212.
6. H. C. Brown, J. A. Sikorski, S. U. Kulkarni and H. D. Lee, *J. Org. Chem.*, 1980, **45**, 4540.
7. J. M. VanPaasschen and R. A. Geanangel, *J. Am. Chem. Soc.*, 1972, **94**, 2680.
8. H. Nöth and H. Beyer, *Chem. Ber.*, 1960, **93**, 2251.
9. J. N. G. Faulks, N. N. Greenwood and J. H. Morris, *J. Inorg. Nucl. Chem.*, 1967, **29**, 329.
10. B. M. Mikhailov, in 'Progress in Boron Chemistry', ed. R. J. Brotherton and H. Steinberg, Pergamon, Oxford, 1970, vol. 3, p. 313.
11. J. E. Douglass, *J. Org. Chem.*, 1966, **31**, 962.
12. P. J. Bratt, M. P. Brown and K. R. Seddon, *J. Chem. Soc., Dalton Trans.*, 1974, 2161.
13. G. E. Ryschkewitsch and J. W. Wiggins, *Inorg. Synth.*, 1970, **12**, 116.
14. H. Nöth, W. A. Dorochov, P. Fritz and F. Pfab, *Z. Anorg. Allg. Chem.*, 1962, **318**, 293.
15. H. Nöth and H. Beyer, *Chem. Ber.*, 1960, **93**, 1078.
16. G. E. Ryschkewitsch, *Inorg. Nucl. Chem. Lett.*, 1971, **7**, 99.
17. K. Yamada, M. Takeda and T. Iwakuma, *J. Chem. Soc., Perkin Trans. 1*, 1983, 265.
18. N. E. Miller, B. L. Chamberland and E. L. Muetterties, *Inorg. Chem.*, 1964, **3**, 1064.
19. J. L. Vidal and G. E. Ryschkewitsch, *Inorg. Chem.*, 1977, **16**, 1898.
20. V. D. Aftandilian, H. C. Miller and E. L. Muetterties, *J. Am. Chem. Soc.*, 1961, **83**, 2471.
21. A. B. Burg, *Inorg. Chem.*, 1978, **17**, 593.
22. O. T. Beachley, Jr. and B. Washburn, *Inorg. Chem.*, 1975, **14**, 120.
23. P. Wisian-Neilson, M. K. Das and B. F. Spielvogel, *Inorg. Chem.*, 1978, **17**, 2327.
24. B. F. Spielvogel, F. Harchelroad, Jr. and P. Wisian-Neilson, *J. Inorg. Nucl. Chem.*, 1979, **41**, 1223.
25. B. Györi, J. Emri and I. Fehér, *J. Organomet. Chem.*, 1983, **255**, 17.
26. B. Györi and J. Emri, *J. Organomet. Chem.*, 1982, **238**, 159.
27. G. Jugie and J.-P. Laussac, *C. R. Hebd. Seances Acad. Sci., Ser. C*, 1972, **274**, 1668.
28. M. L. Denniston, M. A. Chiusano and D. R. Martin, *J. Inorg. Nucl. Chem.*, 1976, **38**, 979.
29. D. R. Martin, M. A. Chiusano, M. L. Denniston, D. J. Dye, E. D. Martin and B. T. Pennington, *J. Inorg. Nucl. Chem.*, 1978, **40**, 9.

The stoichiometric composition of the halogenoborane complexes is 1:1, as a rule. However, in some cases, 2:1 and 4:1 complexes are also obtained (Table 4). In the latter instances additional ligands are bonded to the 1:1 complex through hydrogen bonds.[2,11,55] With ammonia only the 1:1 complex is stable, whereas $(H_2O)_2{\cdot}BF_3$ is more stable than $H_2O{\cdot}BF_3$. BF_3 produces 2:1 complexes of similar stability with alcohols and carboxylic acids.[2,55,57] At low temperatures some tertiary amines give complexes with halogenoboranes and complexes of

Table 4 Neutral Complexes of Halogenoboranes and their Preparation

By direct synthesis:

Complex	*Ref.*	*Complex*	*Ref.*
$(NH_3)_n{\cdot}BF_3$ ($n = 1, 2, 3, 4$)	1	$RRCO{\cdot}BF_3$ ($R' = H, R$)	9
$PhNH_2{\cdot}BCl_3$	2	$R_2CO{\cdot}BCl_3$ ($R = Me, Et$)	10
$MeCN{\cdot}BX_3$ ($X = Cl$)	3	$Y_3PO{\cdot}BX_3$ ($X = Cl, Br$; $Y = Cl, Br$)	11
$PH_3{\cdot}BX_3$ ($X = Cl, Br$)	4	$Me_2S{\cdot}BX_3$ ($X = F, Cl$)	12
$AsH_3{\cdot}BBr_3$	5	$Me_2E{\cdot}BCl_3$ ($E = Se, Te$)	13
$Me_3E{\cdot}BX_3$ ($E = As, Sb$; $X = Br, I$)	6	$Me_3N{\cdot}BX_n(NCS)_{3-n}$ ($X = Cl, Br, I$)	14
$(H_2O)_n{\cdot}BF_3$ ($n = 1, 2$)	7	$Me_3N{\cdot}BCl_2(NMe_2)$	15
$R_2O{\cdot}BF_3$ ($R = Me, Et, Pr^i$)	8	$Me_3N{\cdot}BCl_n(OMe)_{3-n}$ ($n = 1, 2$)	16

By other methods:

Complex	*Method*	*Ref.*
$L{\cdot}BF_3$ (L = diamines)	$Me_3N{\cdot}BF_3 + L$	17
$Me_2NH{\cdot}BX_3$ ($X = Cl, Br, I$)	$Me_2NH{\cdot}BH_3 + X_2$	18
$Me_3N{\cdot}BF_2Cl$	$(Me_3N)_2{\cdot}B_2F_4 + HCl$	19
$Me_3N{\cdot}BF_nBr_{3-n}$ ($n = 1, 2$)	$Me_3N{\cdot}BF_3 + BBr_3$	20
$Et_3P{\cdot}BBr_nCl_{3-n}$ ($n = 1, 2$)	$Et_3P{\cdot}BH_2Br + (CH_2CO)_2NCl$	21
$Me_2NH{\cdot}BCl_2R$ ($R = Pr, Bu$)	$BCl(NMe_2)R + HCl$	22
$L{\cdot}BCl_2CN$ (L = py, 4-Mepy, Et_3N)	$L{\cdot}BH_2CN + Cl_2$	23

1. H. C. Brown and S. Johnson, *J. Am. Chem. Soc.*, 1954, **76,** 1978.
2. R. G. Jones and C. R. Kinney, *J. Am. Chem. Soc.*, 1939, **61,** 1378.
3. A. W. Laubengayer and D. S. Sears, *J. Am. Chem. Soc.*, 1945, **67,** 164.
4. A. Bensson, *C. R. Hebd. Seances Acad. Sci.*, 1890, **110,** 516; 1891, **113,** 78.
5. A. Stock, *Chem. Ber.*, 1901, **34,** 949.
6. M. L. Denniston and D. R. Martin, *J. Inorg. Nucl. Chem.*, 1974, **36,** 2175.
7. H. Meerwein and W. Pannwitz, *J. Prakt. Chem.*, 1934, **141,** 123; N. N. Greenwood and R. L. Martin, *J. Chem. Soc.*, 1953, 1427.
8. H. C. Brown and R. M. Adams, *J. Am. Chem. Soc.*, 1942, **64,** 2557.
9. H. C. Brown, H. I. Schlesinger and A. B. Burg, *J. Am. Chem. Soc.*, 1939, **61,** 673.
10. E. Wiberg and W. Sütterlin, *Z. Anorg. Allg. Chem.*, 1931, **202,** 22, 31.
11. E. W. Wartenberg and J. Goubeau, *Z. Anorg. Allg. Chem.*, 1964, **329,** 269.
12. H. L. Morris, N. I. Kulevsky, M. Tamres and S. Searles, *Inorg. Chem.*, 1966, **5,** 124.
13. M. E. Peach, *Can. J. Chem.*, 1969, **47,** 1675.
14. H. Mongeot, J. Dazord, H. Atchekzai and J. P. Tuchaques, *Synth. React. Inorg. Metal-Org. Chem.*, 1976, **6,** 191.
15. C. A. Brown and R. C. Osthoff, *J. Am. Chem. Soc.*, 1952, **74,** 2340.
16. E. Wiberg and W. Sütterlin, *Z. Anorg. Allg. Chem.*, 1935, **222,** 92.
17. J. M. VanPaasschen and R. A. Geanangel, *Can. J. Chem.*, 1975, **53,** 723.
18. G. E. Ryschkewitsch and W. H. Myers, *Synth. React. Inorg. Metal-Org. Chem.*, 1975, **5,** 123.
19. B. W. C. Ashcroft and A. K. Holliday, *J. Chem. Soc. (A)*, 1971, 2581.
20. B. Benton-Jones, M. E. A. Davidson, J. S. Hartman, J. J. Klassen and J. M. Miller, *J. Chem. Soc., Dalton Trans.* 1972, 2603.
21. G. Jugie, J.-P. Laussac and J.-P. Laurent, *Bull. Soc. Chim. Fr.*, 1970, 2542.
22. H. Nöth and P. Fritz, *Z. Anorg. Allg. Chem.*, 1963, **324,** 270.
23. D. R. Martin, M. A. Chiusano, M. L. Denniston, D. J. Dye, E. D. Martin and B. T. Pennington, *J. Inorg. Nucl. Chem.*, 1978, **40,** 9.

diethyl ether with three halogenoboranes are also known. One borane is bonded through a strong donor–acceptor bond while the others are held by weak halogen bridge bonds.[8,11,55] Halogen-bridged complexes are involved as intermediates in the formation of mixed halogenoborane complexes[24,34] as well as in the halogen exchange reaction between borane complexes and boranes. Complexes with ethers, amines and phosphines of all the conceivable mixed halogenoboranes are obtained in this reaction, under equilibrium conditions, but only a few have been isolated.[24]

The strength of the boron–halogen bond decreases in the order $F \gg Cl > Br > I$, in the complexes as well as in the boranes themselves.[55] This explains the observation that the BF_3 is capable of forming stable complexes, even with molecules containing active hydrogen, whereas the BCl_3, BBr_3 and BI_3 complexes are generally unstable and protonolysis of the trihalogenoboranes occurs.[15,17,18,55,57] Complexes with oxygen donors devoid of active hydrogen, BF_3 complexes excepted, may undergo decomposition with formation of alkyl- or acyl-halogenides or rearrangement with halogen transfer (equations 4–6).[15,55,57] According to X-ray and NMR studies, the complexes of BF_3 with O donors are covalent as solids, whereas some are ionized as liquids or in solution.[2,55,58] In the light of equilibria (7) and (8) it can be understood that BF_3

alone, or in the presence of cocatalysts such as H_2O, ROH, RCO_2H, *etc.*, catalyzes a great number of organic reactions and its use is widespread in the chemical industry.[55,58]

$$R_2O{\cdot}BX_3 \longrightarrow B(OR)X_2 + RX \quad (4)$$
$$MeCOOR{\cdot}BX_3 \longrightarrow MeCOX + B(OR)X_2 \quad (5)$$
$$Ph_2CO{\cdot}BCl_3 \longrightarrow BCl_2OC(Cl)Ph_2 \quad (6)$$
$$ROH{\cdot}BF_3 \rightleftharpoons H^+ + [BF_3OR]^- \quad (7)$$
$$RCOOMe{\cdot}BF_3 \rightleftharpoons RCO^+ + [BF_3OMe]^- \quad (8)$$
$$2L{\cdot}BX_3 \rightleftharpoons [X_2BL_2]^+ + [BX_4]^- \quad (9)$$
$$L{\cdot}BX_3 \rightleftharpoons [X_2BL]^+ + X^- \quad (10)$$

Equilibria of another type are shown in equations (9) and (10).[24,25] For $R_2NH{\cdot}BX_3$ (X = Cl, Br) complexes[59] and for several BI_3 complexes[24,60] equilibrium (9) prevails in solution. The $[X_2BL]^+$ ion is expected[24,55] to form in reactions of BI_3 complexes containing very weak B—I bonds (equation 10). A similar ion was assumed to have formed as an intermediate during the hydrolysis of $Me_3N{\cdot}BCl_3$.[61] The BF_3 complexes rarely enter into ionic equilibria, although with chelating ligands equilibria similar to (9) were detected.[24]

Owing to the strength of the B—F bond, the BF_3 complexes are of widespread use as model compounds, for investigating Lewis acid–base interactions and the nature of the donor–acceptor bond. BF_3 is frequently employed as a standard Lewis acid, for the quantitative characterization of the Lewis basicity of donor molecules.[62,63] The gas-phase equilibrium constants for some BF_3 complexes are shown in Table 5.

Table 5 Trifluoroborane–Base Equilibria in the Gaseous Phase[a]

Base	*K* (bar^{-1})	*Ref.*	*Base*	*K* (bar^{-1})	*Ref.*
Me_2O	5.80	1	Me_3N	large	5
Et_2O	2.41	1	$MePH_2$	0	6
Pr^i_2O	8.8 (at 50 °C)	2	Me_2PH	0.094	6
$(CH_2)_4O$	68.5	3	Me_3P	14.8	6
$(CH_2)_5O$	29.1	3	$(CH_2)_4PH$	0.38 (at 60 °C)	7
Me_2S	0.17 (at 60 °C)	4	$(CH_2)_5PH$	0.78 (at 60 °C)	7
Et_2S	0.52 (at 60 °C)	4	Me_3As	2.5 (at 5 °C)	8
$(CH_2)_4S$	0.25 (at 60 °C)	5	Me_3Sb	0 (at −78 °C)	8
$(CH_2)_5S$	0.50 (at 60 °C)	4			

[a] *K* refers to equilibria base + $BF_3 \rightleftharpoons$ base·BF_3; temperature, 100 °C unless stated otherwise; vapor pressure method.

1. D. E. McLaughlin and M. Tamres, *J. Am. Chem. Soc.*, 1960, **82,** 5618.
2. H. C. Brown and R. M. Adams, *J. Am. Chem. Soc.*, 1942, **64,** 2557.
3. D. E. McLaughlin, M. Tamres and S. Searles, *J. Am. Chem. Soc.*, 1960, **82,** 5621.
4. H. L. Morris, N. I. Kulevsky, M. Tamres and S. Searles, *Inorg. Chem.*, 1966, **5,** 124.
5. A. B. Burg and A. A. Green, *J. Am. Chem. Soc.*, 1943, **65,** 1838.
6. H. C. Brown, *J. Chem. Soc.*, 1956, 1248.
7. H. L. Morris, M. Tamres and S. Searles, *Inorg. Chem.*, 1966, **5,** 2156.
8. R. H. Harris, Ph. D. Thesis, Purdue University, 1952.

The halogenoboranes, unlike BH_3, form more stable complexes with N and O donors than with P and S donors (see Section 24.2.1 and Table 5). Especially stable complexes are formed with N donors; pyridine–trifluoroborane for instance can be distilled at 300 °C without decomposition.[64]

The strength of the coordinative bond in halogenoborane complexes changes in the order $BBr_3 > BCl_3 > BF_3$ for a given base (see Section 24.2.1). Earlier theories based on electronegativity considerations predicted the opposite order.[65] The correct sequence first indicated by the measurements of Laubengayer and Sears[66] became generally accepted, however, only following comprehensive calorimetric studies by Brown *et al.*[11,67] In the absence of calorimetric and equilibrium data the position of BI_3 in the sequence is uncertain.[2] NMR,[10,24,34,68] IR[55,69,70] and displacement reaction invesgitations[71] make it likely that, of all trihalogenoboranes, BI_3 produces the strongest coordinative bond with N, P and As donor molecules. X-Ray data[72] for $Me_3P{\cdot}BX_3$ (X = Cl, Br, I) are in accord with this assumption, while electron diffraction studies[73] seem not to support it. However, in view of the large atomic radius of iodine it appears likely that BI_3, like BBr_3,[74] displays weaker acceptor ability for bases with larger donor atoms than BCl_3.

24.2.4 Complexes of Boranes with B—O, B—S and B—N Bonds

Owing to the $p_\pi-p_\pi$ interaction between the vacant *p* orbital of boron and the occupied *p* orbitals of the neighbouring atoms, the majority of these boranes are weak Lewis acids.[15,75] The range of possible substituents at the boron is illustrated in Table 6.

Table 6 Neutral Complexes of Boranes with B—O, B—S and B—N bonds

Complex	*Ref.*	*Complex*	*Ref.*
$A\cdot B(OR)_3$ (R = Me, Et, All, Ph)	1, 2	$A\cdot B(SR)_3$ (R = Me, Et)	8, 9
$2NH_3\cdot B(OPh)_3$	1	$Me_3N\cdot B(SCF_3)_3$	10
$NH_3\cdot B(O\text{-}t\text{-}C_4F_9)_3$	3	$py\cdot B(N_3)_3$	11
$A\cdot B(OAc)_3$	1, 4	$py\cdot B(NCO)_3$	11, 12
$py\cdot B(OCOR)_3$ (R = Et, Ph)	4, 5	$L\cdot B(NCS)_3$ (L = A, MeCN)	11, 13
$2A\cdot B_2O(OAc)_4$	4	$Me_3N\cdot B(NCS)_n(SMe)_{3-n}$	14
$A\cdot B(ONO_2)_3$	6	$A\cdot B(NC_4H_4)_3$	15
$L\cdot B(OTeF_5)_3$ (L = py, MeCN)	7	$Me_2NH\cdot BPz_3$	16

A = amine; All = allyl; Pz = pyrazol-l-yl

1. M. F. Lappert, *Chem. Rev.*, 1956, **56,** 959.
2. H. Landesman and R. Williams, *J. Am. Chem. Soc.*, 1961, **83,** 2663.
3. D. E. Young, L. R. Anderson and W. B. Fox, *Inorg. Chem.*, 1971, **10,** 2810.
4. H. U. Kibbel, *Z. Anorg. Allg. Chem.*, 1968, **359,** 246.
5. A. Pelter and T. E. Levitt, *Tetrahedron,* 1970, **26,** 1899.
6. O. P. Shitov, T. N. Golovina and A. V. Kalinin, *Izv. Akad. Nauk SSSR, Ser. Khim.*, 1979, 1883.
7. F. Sladky, H. Kropshofer and O. Leitzke, *J. Chem. Soc., Chem. Commun.*, 1973, 134.
8. B. M. Mikhailov, in 'Progress in Boron Chemistry', ed. R. J. Brotherton and H. Steinberg, Pergamon, Oxford, 1970, vol. 3, p. 313.
9. H. Nöth and U. Schuchardt, *Chem. Ber.*, 1974, **107,** 3104.
10. A. Haas and M. Häberlein, *Chem.-Ztg.*, 1972, **96,** 412.
11. M. F. Lappert and H. Pyszora, *Adv. Inorg. Chem. Radiochem.*, 1966, **9,** 133.
12. M. F. Lappert and H. Pyszora, *J. Chem. Soc.*, (*A*), 1968, 1024.
13. D. B. Sowerby, *J. Am. Chem. Soc.*, 1962, **84,** 1831.
14. H. R. Atchekzai, H. Mongeot, J. Dazord and J.-P. Tuchagues, *Can. J. Chem.*, 1979, **57,** 1122.
15. P. Szarvas, B. Györi and J. Emri, *Acta Chim. Acad. Sci. Hung.*, 1971, **70,** 1.
16. K. Niedenzu, S. S. Seelig and W. Weber, *Z. Anorg. Allg. Chem.*, 1981, **483,** 51.

Of the trialkoxyboranes, only $B(OMe)_3$, $B(OEt)_3$ and $B(OCH_2CH{=}CH_2)_3$ are known to give complexes with strongly basic unhindered amines. The triaryloxyboranes are generally regarded as stronger Lewis acids, for they give, in contrast to the trialkoxyboranes, stable complexes with pyridine. This fact is rationalized[15,76] in terms of resonance interaction between the lone pairs on the oxygen and the π electrons of the aromatic ring. The complexes of the fluoroalkoxyboranes, owing to the inductive effect of the fluoroalkyl group, are more stable than those of their alkyl counterparts; strong complexes are formed even in cases when bulky fluoroalkyl groups are present (Table 6). Despite the bulkiness of the $OTeF_5$ groups, $B(OTeF_5)_3$ forms stable complexes with weak O and N donors (Table 6). In addition to the inductive effect of the TeF_5 groups, an interaction between the oxygen *p* orbitals and the vacant orbitals on the Te is invoked to explain this behavior.

The alkylthioboranes form more stable complexes than the alkoxyboranes do probably due to a B—S π bond weaker than the B—O π bond.[77,78] This argument is analogous to the one generally accepted for explaining the difference in the Lewis acidity of BF_3 and BCl_3.

The cyanato- and, especially, the thiocyanato-boranes form equally stable complexes with amines, nitriles and ethyl acetate (Table 6). The Lewis acidity of these pseudohalogenoboranes and halogenoboranes, with respect to pyridine and ethyl acetate, follows the order Cl > NCS > NCO > F.[79]

The aminoboranes containing NR_2 groups do not form neutral complexes because the B—N bond possesses a higher π order than B—O and B—S bonds. In contrast, the pyrazol-1-yl and pyrrol-1-yl boranes with sp^2 nitrogens bonded to the boron do give stable amine complexes

(Table 6). The majority of the complexes in Table 6 are obtained by the reaction of boranes with ligands. Some other methods of synthesis are illustrated by equations (11)–(14).

$$B_2O(OCOR)_4 + (RCO)_2O + 2py \longrightarrow 2py{\cdot}B(OCOR)_3 \quad (11)$$

$$A{\cdot}BBr_3 + 3N_2O_4 \longrightarrow A{\cdot}B(NO_3)_3 + 3NOBr \quad (12)$$

$$(A = py, 4\text{-}Etpy, Me_3N)$$

$$py{\cdot}BCl_3 + 3LiN_3 \longrightarrow py{\cdot}B(N_3)_3 + 3LiCl \quad (13)$$

$$B(NMe_2)_3 + 3HPz \longrightarrow Me_2NH{\cdot}BPz_3 + 2Me_2NH \quad (14)$$

$$Pz = \text{pyrazol-1-yl}$$

24.2.5 Chelate Complexes

When a group linked to the boron contains donor atoms (usually O, N) within a distance of four to five bonds from the boron, chelate rings are often formed *via* annular coordination. The majority of the known boron chelates contain BPh_2 or dialkylboron groups[80] but a number of chelates containing BX_2 or BXX' groups (X and X′ being mostly F and, more rarely, Cl, H, OH, OR, OAc) are also described, as are bicyclic chelates with B—X groups (Table 7). Many chelates are easily obtained *via* reaction of the borane with the proton complex of the ligand

Table 7 Neutral Chelate Complexes

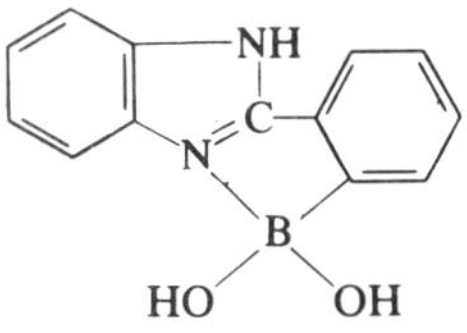

Ref. 1

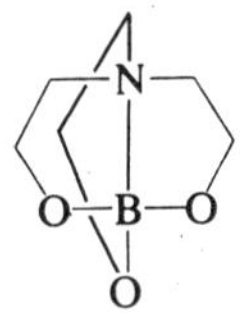

Ref. 2

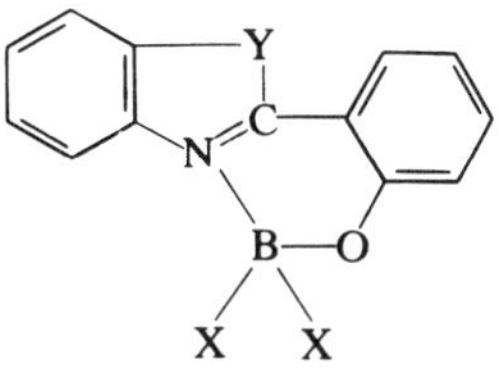

Y = CH_2, NH, O, S
X = F, Cl, Ph

Ref. 3

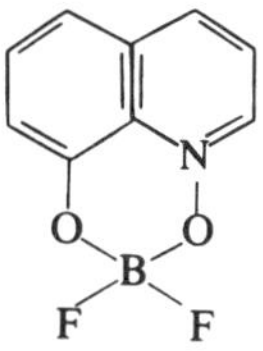

Ref. 3

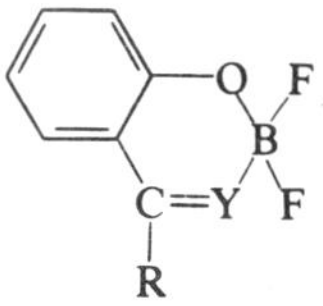

Y = O, NR, N—NHPh

Refs. 3, 4

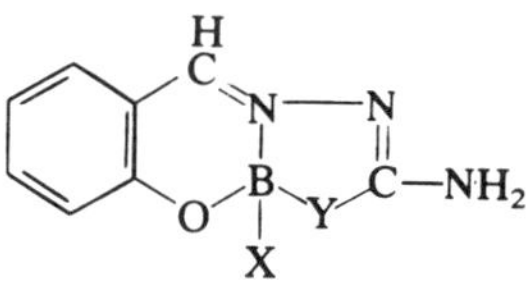

Y = O, S
X = F, OAc

Ref. 5

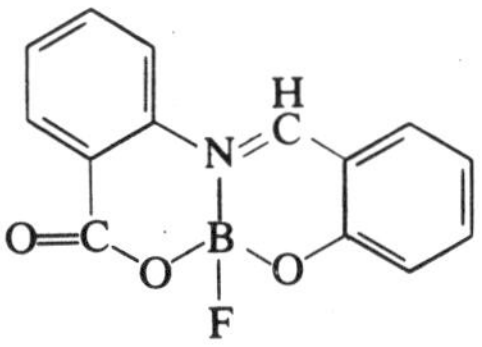

Ref. 3

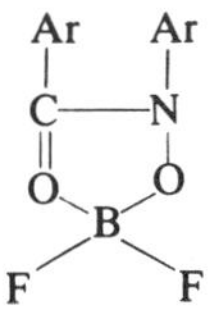

Refs. 3, 6

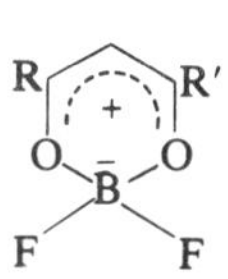

Ref. 7

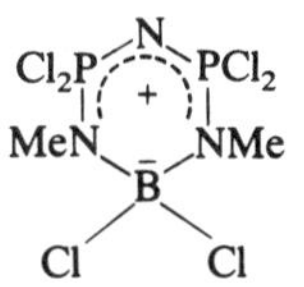

Ref. 8

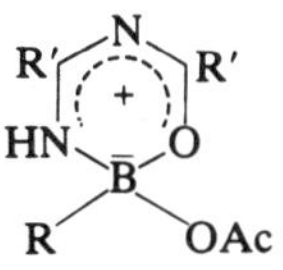

Ref. 9

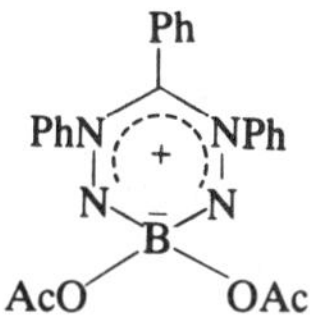

Ref. 10

1. R. L. Letsinger and D. B. MacLean, *J. Am. Chem. Soc.*, 1963, **85**, 2230.
2. H. C. Brown and E. A. Fletcher, *J. Am. Chem. Soc.*, 1951, **73**, 2808.
3. F. Umland and E. Hohaus, 'Untersuchungen über Borhaltige Ringsysteme vom Chelattyp', Westdeutscher Verlag, 1976.
4. S. M. Tripathi and J. P. Tandon, *Inorg. Nucl. Chem. Lett.*, 1978, **14**, 97.
5. P. K. Singh, J. K. Koacher and J. P. Tandon, *J. Inorg. Nucl. Chem.*, 1981, **43**, 1755.
6. D. Mukhopadhyay, B. Sur and R. G. Bhattacharyya, *J. Inorg. Nucl. Chem.*, 1981, **43**, 607.
7. N. M. D. Brown and P. Bladon, *J. Chem. Soc. (A)*, 1969, 526.
8. H. Binder, *Z. Naturforsch., Teil B*, 1971, **26**, 616.
9. V. A. Dorokhov, L. I. Lavrinovich, M. N. Bochkareva, V. S. Bogdanov and B. M. Mikhailov, *Zh. Obshch. Khim.*, 1973, **43**, 1115.
10. B. I. Stepanov and G. V. Avramenko, *Zh. Obshch. Khim.*, 1980, **50**, 358.

(*e.g.* benzazoles, benzimidazoles, salicylaldehyde and derivatives, 8-hydroxyquinoline). Five-membered monocyclic BF_2 chelate rings are generally stable when stabilizing mesomerism in the chelate ring is possible and/or other factors favor chelate formation,[81] as is the case with *N*-aryl-*N*-acylarylhydroxylamine chelates (Scheme 1). As opposed to the metal chelates, the six-membered BF_2 chelates are more stable than their five-membered counterparts in cases when two double bonds or an aromatic system occurs within the chelate ring (Table 7).[81] Seven-membered chelate rings with mesomeric stabilization are also known; they often display considerable stability.[82]

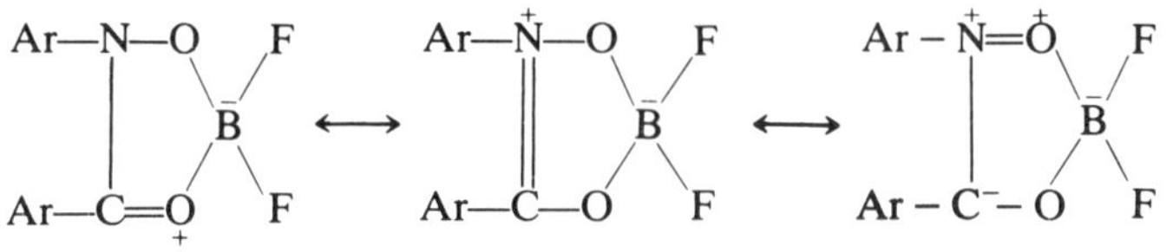

Scheme 1

The BF_2 chelates are hydrolytically stable (the majority of them may be recrystallized from water) whereas the BCl_2 chelates are easily hydrolyzed. The boric acid ester complex triptych boroxazolidines are also stable towards hydrolysis as are other aminoalcohol esters.[83] A great number of the boron chelates are colored; several spectrophotometric methods are based on chelate formation[81,84] for the analytical determination of boric acid, organoboric acids and chelating organic compounds. The boron chelates are remarkable for their pharmacological properties as well.[84] Various aspects of the boron chelates have been reviewed.[81,83–86]

24.3 ANIONIC COMPLEXES

24.3.1 Hydroborates

The preparation and various aspects of the properties of tetrahydroborates have been comprehensively reviewed.[2,7,9,14,87,88] Therefore, only substituted derivatives of the tetrahydroborates will be dealt with in this section. The methods of synthesis of these compounds are summarized in Table 8.

Of the halogenohydroborates $K[BH_3F]$ and salts of $[BH_2X_2]^-$ and $[BHX_3]^-$ ($X = Cl, Br$) were obtained in pure form. During the reaction of $[Bu_4N][BH_4]$ with $[Bu_4N][BCl_4]$ or $[Bu_4N][BBr_4]$, redistribution of the ligands occurs with the formation of mono-, di- and tri-halogenohydroborates depending on the molar ratios used.[89] The bromohydroborates are more stable towards disproportionation than the chloro compounds although the monochloro- and monobromo-borates undergo disproportionation equally at 20 °C. The reaction of $[Bu_4N][BH_4]$ with iodine and chlorinated hydrocarbons has been investigated; $[Bu_4N][BH_2Cl_2]$ could be isolated when using CCl_4.[90]

While the reaction of borane with NaSCN results in the formation of monothiocyanatoborate, that of $Na[BH_4]$ with $(SCN)_2$ gives triisothiocyanatoborate. $Na[BH_3SCN]$, which probably contains some B—NCS isomer, displays a higher hydrolytic stability than $Na[BH_4]$ does.[91] Of the hydropseudohalogenoborates, the most extensive studies were made on the cyanoborates. The large −I effect of the CN group confers an exceptionally high hydrolytic stability on the cyanohydroborates. The hydrolysis and H–D exchange of $[BH_3CN]^-$ were studied in detail.[92,93] The hydrolytic stabilities of $[BH_2(CN)_2]^-$ and $[BH(CN)_3]^-$ are even greater; they do not decompose even in concentrated sulfuric acid. The hydrolytic stabilities of $[BH_2(NC)_2]^-$ and $[BH(NC)_3]^-$, on the other hand, are much lower; they can be thermally isomerized into the corresponding cyanoborates.[94] Of all the substituted hydroborates, $Na[BH_3CN]$ is by far the most important one from the practical point of view; it is extensively employed as a reducing agent in organic syntheses.[95]

The existence of hydrohydroxoborates has been the subject of many investigations in connection with the hydrolysis of the $[BH_4]^-$ ion. $[BH_3OH]^-$, $[BH_2(OH)_2]^-$ and $[BH(OH)_3]^-$ ions (Table 8) were all obtained when $Na[BH_4]$ was subjected to partial hydrolysis in strongly acidic media at −78 °C, followed by neutralization. The hydrolysis kinetics of hydrohydroxoborates were studied.[96] The $[BH_3SH]^-$ ion is more stable than the hydroxoborates; $[Et_4N][BH_3SH]$ can be obtained in solid form (Table 8). The trialkoxyboranes can, in general, be converted into trialkoxohydroborates with the aid of alkali metal hydrides. The rate of

Table 8 Preparation of Hydroborates

Hydroborate	*Reagent*	*Ref.*	*Hydroborate*	*Reagent*	*Ref.*
1. Through substitution of tetrahydroborates					
$M[BHX_3]$[a]	BX_3	1	$Na[BH(NCS)_3]$	$(SCN)_2$	10
$[BH_n(OH)_{4-n}]^-$[b]	HCl · · · + NaOH	2, 3	$Li[BHPyl_3]$	HPyl	11
$Na[BH(OCOCR)_3]$	RCO_2H	4, 5	$K[BH_nPz_{4-n}]$[c]	HPz	12
$Na[BH_2E_3]$	E = S, Te	6, 7	$Na[BH_3CN]$	HCN	13
$[Et_4N][BH_3SH]$	H_2S	8	$Na[BH_n(CN)_{4-n}]$[d]	$Hg(CN)_2$	14
$Li[BH_3N_3]$	HN_3	9	$M[BH_n(SiF_3)_{4-n}]$[e]	SiF_4	15
2. Through reaction of B_2H_6 with anionic Lewis bases					
$K[BH_3F]$	KF	16	$Na[BH_3CN]$	NaCN	17
$Na[BH_3SCN]$	NaSCN	16, 17	$Na[BH_3CNBH_3]$	NaCN	16
$K[BHPyl_3]$	KPyl	18	$K[BH_3GeH_3]$	$KGeH_3$	20
$K[(BH_3)_2PH_2]$	KPH_2	19	$Li[BH_3SnMe_3]$	$LiSnMe_3$	21
3. Through reaction of boranes with alkali metal hydrides					
$Na[BHF_3]$	NaH	22	$Li[BH(SMe)_3]$	LiH	24
$Na[BH(OR)_3]$	NaH	23	$Na[BH(OCOMe)_3]$	NaH	25
4. From neutral borane complexes *via* proton abstraction					
$Na[BH_3NMe_2]$	Na or NaH	16	$[NH_4][(BH_3)_2PH_2]$	NH_3	27
$Li[BH_3PH_2]$	LiBu	26	$K[BH(CN)Pz(Pyl)]$	KH	28

5. Miscellaneous

Hydroborate	*Starting material*	*Reagent*	*Ref.*
$M[BH_nBr_{4-n}]$[f]	$M[BH_4]$	$M[BBr_4]$	29
$Na[BH(OCOH)_3]$	$Na[BH_4]$	CO_2	30
$Ag[BH_n(NC)_{4-n}]$[c]	$Me_2S \cdot BH_nBr_{3-n}$	AgCN	31
$Na[BH_2(3,5\text{-}Me_2Pz)Pz]$	$Na[BH_3(3,5\text{-}Me_2Pz)]$	HPz	32
$Na_2[BH_3P(OMe)_3]_2$	$(MeO)_3P \cdot BH_3$	$NaC_{10}H_8$	33
$K_2[BH_3(CO_2)]$	$CO \cdot BH_3$	KOH	34

Pz = pyrazol-1-yl; Pyl = pyrrol-1-yl. [a] M = $[Bu_4N]$, $[MePPh_3]$; X = Cl, Br. [b] in H_2O, n = 1–3. [c] n = 1, 2. [d] n = 2, 3. [e] M = $[Bu_4N]$; n = 2, 3. [f] M = $[Bu_4N]$; n = 2, 3.

1. M. A. Toft, J. B. Leach, F. L. Himpsl and S. G. Shore, *Inorg. Chem.* 1982, **21**, 1952.
2. F. T. Wang and W. L. Jolly, *Inorg. Chem.*, 1972, **11**, 1933.
3. J. W. Reed and W. L. Jolly, *J. Org. Chem.*, 1977, **42**, 3963.
4. P. Marchini, G. Liso, A. Reho, F. Liberatore and F. M. Moracci, *J. Org. Chem.*, 1975, **40**, 3453.
5. K. Yamada, M. Takeda and T. Iwakuma, *J. Chem. Soc., Perkin Trans. 1*, 1983, 265.
6. J. M. Lalancette, A. Frêche and R. Monteux, *Can. J. Chem.*, 1968, **46**, 2754.
7. J. M. Lalancette and M. Arnac, *Can. J. Chem.*, 1969, **47**, 3695.
8. P. C. Keller, *Inorg. Chem.*, 1969, **8**, 1695.
9. E. Wiberg and H. Michaud, *Z. Naturforsch., Teil B*, 1954, **9**, 499.
10. F. Klanberg, *Z. Anorg. Allg. Chem.*, 1962, **316**, 197.
11. B. Györi, J. Emri and P. Szarvas, *Acta Chim. Acad. Sci. Hung.*, 1975, **86**, 235.
12. S. Trofimenko, *J. Am. Chem. Soc.*, 1967, **89**, 3170.
13. R. C. Wade, E. A. Sullivan, J. R. Berschied, Jr. and K. F. Purcell, *Inorg. Chem.*, 1970, **9**, 2146.
14. J. Emri and B. Györi, *J. Chem. Soc., Chem. Commun.*, 1983, 1303.
15. S. Brownstein, *J. Chem. Soc., Chem. Commun.*, 1980, 149.
16. V. D. Aftandilian, H. C. Miller and E. I. Muetterties, *J. Am. Chem. Soc.*, 1961, **83**, 2471.
17. B. C. Hui, *Inorg. Chem.*, 1980, **19**, 3185.
18. P. Szarvas, J. Emri and B. Györi, *Acta Chim. Acad. Sci. Hung.*, 1970, **64**, 203.
19. N. R. Thompson, *J. Chem. Soc.*, 1965, 6290.
20. D. S. Rustad and W. L. Jolly, *Inorg. Chem.*, 1968, **7**, 213.
21. W. Biffar, H. Nöth, H. Pommerening, R. Schwerthöffer, W. Storch and B. Wrackmeyer, *Chem. Ber.*, 1981, **114**, 49.
22. J. Goubeau and R. Bergmann, *Z. Anorg. Allg. Chem.*, 1950, **263**, 69.
23. H. C. Brown, E. J. Mead and C. J. Shoaf, *J. Am. Chem. Soc.*, 1956, **78**, 3616.
24. T. A. Shchegoleva, E. M. Shashkova and B. M. Mikhailov, *Izv. Akad. Nauk SSSR, Ser. Khim.*, 1963, 494.
25. C. D. Nenitzescu and F. Badea, *Bul. Inst. Politeh. Bucuresti*, 1958, **20**, 93 (*Chem. Abstr.*, 1961, **55**, 2325i).
26. E. Mayer, *Inorg. Chem.*, 1971, **10**, 2259.
27. J. W. Gilje, K. W. Morse and R. W. Parry, *Inorg. Chem.*, 1967, **6**, 1761.
28. B. Györi and J. Emri, *J. Organomet. Chem.*, 1982, **238**, 159.
29. L. A. Gavrilova, L. V. Titov and V. Ya. Rosolovskii, *Zh. Neorg. Khim.*, 1981, **26**, 1769.
30. T. Wartik and R. K. Pearson, *J. Inorg. Nucl. Chem.*, 1958, **7**, 404.
31. B. Györi, J. Emri and I. Fehér, *J. Organomet. Chem.*, 1983, **255**, 17.
32. E. Frauendorfer and G. Agrifoglio, *Inorg. Chem.*, 1982, **21**, 4122.
33. L. A. Peacock and R. A. Geanangel, *Inorg. Chem.*, 1976, **15**, 244.
34. L. J. Malone and R. W. Parry, *Inorg. Chem.*, 1967, **6**, 817.

reaction of NaH with some of the alkoxyboranes decreases sharply in the order OMe > OEt > OPr^i > OBu. The trimethoxo- and triethoxo-borates undergo disproportionation in THF, whereas the two others are stable. The hydride transfer ability of $K[BH(OPr^i)_3]$ is larger than that of $Na[BH_4]$.[97] The preparation of $Li[BH(SMe)_3]$ is analogous to that of the trialkoxo compounds.[98]

The aminoboranes do not form stable complexes with neutral Lewis bases (see Section 24.2.2), but the amidotrihydroborates, $M[BH_3NR_2]$, are relatively stable. Monophosphidoborates are also known but, apparently, hydroborates with more NR_2 or PR_2 groups are not stable. However, di- and tri-amidoborates derived from amines containing sp^2 N (*i.e.* pyrrole, pyrazole, indole, indazole) are well known.

The amidoborates with NR_2 groups are moderately strong bases.[99] Consequently they are transformed, on treatment with B_2H_6, into μ-amido-bisborate(-1), $[R_2N(BH_3)_2]^-$ [commonly known as bis(borate)amide(1-)]. The phosphidotrihydroborates show analogous reactions (Table 8).

The whole series of pyrrol-1-ylborates, $M[BH_n(NC_4H_4)_{4-n}]$, has been synthesized (Tables 9 and 12). The mechanism of their formation and their hydrolysis kinetics have been studied in detail.[100] The hydropyrazol-1-ylborates are very stable compounds and can be prepared in acid form.[101] The di- and tri-pyrazol-1-ylborates are extensively used as complexing ligands (see Chapter 13.6). The photoelectron spectra of the sodium and thallium(I) hydrotris(pyrazol-1-yl)borates and the electronic structure of the anion itself are reported.[102]

Borates containing hydrocarbon groups have recently been comprehensively reviewed.[4,103] Also known are the Si, Ge and Sn analogues of the monoalkylborates (Table 8). The $[BH_3(COY)]^{n-}$ ions may be mentioned as an interesting subclass of hydroborates with B—C bonds. Because BH_3 and O are isoelectronic,[104] they are isoelectronic with the carbonate, O-alkylcarbonate or carbamate ions (for Y = O, OR, and NR_2, respectively). Hydroborates with three or four different substituents have only recently been synthesized (Table 8). Of these, the $[BH_2(CN)NC_4H_4]^-$ ion shows a remarkable hydrolytic stability, giving rise to the formation of the $C_4H_5N{\cdot}BH_2CN$ complex through reversible protonation, in contrast to the $[BH(CN)(NC_4H_4)_2]^-$ ion which has a much lower hydrolytic stability.[105] The cyanohydropyrazolylpyrrolylborate and the corresponding imidazolylborate (Table 8) are the first representatives of hydroborates containing a chiral boron.

24.3.2 Halogenoborates

The most widely known representatives of this group are the borates with four identical ligands. The simplest way of synthesizing them is to add a suitable halide to the corresponding trihalogenoboranes.[2,55] A large cation is necessary when Cl^-, Br^- or I^- ions are used (equation 15), whereas tetrafluoroborates can be obtained using small cations as well and the reactions can often be performed in aqueous media. The tetrafluoroborates of transition metals are stable compounds, the Ag and Cu^{II} salts being soluble in water and organic solvents.[55,57,106] The $[BF_4]^-$ ion is extensively employed for the isolation of cationic transition metal complexes. Several covalent fluorides, such as BrF_3, ClO_2F, O_2F_2, SF_4, NOF and NO_2F, react with BF_3 to give the appropriate BF_4 salts.[55,57,106] In addition to the inorganic tetrafluoroborates, a great number of organic fluoroborates,[57,106] including ammonium, diazonium, oxonium, sulfonium and carbonium tetrafluoroborates, are known and play an important role in synthetic chemistry.[106] The K, Rb and Cs tetrachloroborates can be prepared without solvent or in appropriate organic solvents, whereas the tetrabromoborates and tetraiodoborates can only be isolated using the ions shown in equation (15).[2,53]

$$MX + BX_3 \longrightarrow M[BX_4] \qquad (15)$$
$$M = [R_4N]^+, [pyH]^+, Ph_3C^+, C_7H_7^+; X = F, Cl, Br, I$$

The relative stabilities of tetrahalogenoborates have been studied by ^{11}B NMR. The ^{11}B chemical shift of the $[BF_4]^-$ ion is independent of the F^- concentration, whereas a marked dependence upon halide ion concentration in organic solvents was observed in other cases. These studies established[24,55] the order of stability as $[BF_4]^- > [BCl_4]^- > [BBr_4]^- > [BI_4]^-$. The heats of formation measured in the gas phase also indicate a significantly higher stability for $[BF_4]^-$ than for $[BCl_4]^-$ (402 and 272 kJ mol^{-1} respectively).[107] The thermal stability of the alkylammonium tetrafluoroborates is fairly high, whereas those with other halogens decompose to give borazines (equation 16).[55] In accordance with the stabilities of the complexes, the $[BF_4]^-$ ion is only partially hydrolyzed in aqueous medium (*vide infra*), whereas the other tetrahalogenoborates undergo complete and irreversible hydrolysis.[2,55]

$$3[RNH_3][BX_4] \xrightarrow{\Delta} (RNBX)_3 + 9HX \qquad (16)$$

A great number of mixed tetrahalogenoborates are known. While all of the binary mixed halogenoborates, except the $[BF_nI_{4-n}]^-$ ions, have been detected, of the ternary halogen species only a few, such as $[BF_2ClBr]^-$ and $[BFCl_2Br]^-$, are known.[24,108] The mixed halogenoborates are easily identified[24,108] by ^{11}B NMR using the pairwise additivity rule.[109] The majority of mixed halogenoborates exist only under equilibrium conditions. Of the compounds hitherto isolated, only the $[BF_3Cl]^-$ salts[110] have been well characterized.[24]

Derivatives of tetrahalogenoborates, of $[BF_4]^-$ above all, are known in great number; these are summarized in Table 9. $[BF_n(OH)_{4-n}]^-$ ions are formed, *inter alia,* in the course of reversible reactions that take place in a diluted solution of $[BF_4]^-$ ion. The formation and stability of fluorohydroxyborate species have been the subject of several studies[55,111,112] as was the stepwise replacement of OH^- by F^- in the conversion of $[B(OH)_4]^-$ to $[BF_4]^-$. BF_3 with nitrogen oxides (N_2O_3, N_2O_4, N_2O_5) and HNO_3 gives rise to the formation of various ionic complexes containing monodentate NO_3^-, bidentate NO_2^-, or OH^- ligands (Table 9) that can be conveniently used for nitration or nitrosylation.[113]

Table 9 Isolated Halogenborates

Halogenoborate	*Ref.*	*Halogenborate*	*Ref.*
$K[BF_3OH]$	1	$Li_3[(BF_3)_3N]$	13
$K[BF(OH)_3]$	2	$M[BF_3NH_2]$ (M = Na, K)	14
$NO_2[F_3B(OH)BF_3]$	3	$M[BF_3CF_3]$ (M = K, NH_4)	15
$[R_4N][BF_3ClO_4]$ (R = Et, Bu)	4	$K[BF_nPh_{4-n}]$ (*n* = 2, 3)	16
$[Me_4N][BF_3SO_3Y]$ (Y = F, CF_3)	5	$M[BF_2(CF_3)_2]$ (M = K, Cs)	17
$[Et_4N]_2[(BF_3)_nSO_4]$ (*n* = 2, 3)	6	$[Me_3NH][BF_2(Cl)Et]$	18
$K[BF_3ONH_2]$	7	$[Ph_4As][BX_3GeCl_3]$ (X = F, Cl)	19
$M[(BF_3)_2NO_2]$ (M = NO, NO_2)	8	$[Ph_4As][BX_3SnCl_3]$ (X = F, Cl)	19
$NO_2[BF_3NO_3]$	8, 9	$[R_4N][BCl_3ClO_4]$ (R = Me, Et, Bu)	20
$K_3[(BF_3)_3PO_4]$	10	$[R_4N][BX_3SH]$ (X = Cl, Br)	12
$K[BF_n(OMe)_{4-n}]$ (*n* = 1, 2, 3)	11	$[Me_4N][BCl_3X]$ (X = NCO, N_3)	21
$[Me_4N][BF_3SH]$	12	$[C_7H_7][BI_3Ph]$	22

1. C. A. Wamser, *J. Am. Chem. Soc.*, 1948, **70**, 1209.
2. A. K. Sengupta and S. K. Mukherjee, *J. Indian Chem. Soc.*, 1967, **44**, 658.
3. R. Fourcade and G. Mascherpa, *Bull. Soc. Chim. Fr.*, 1972, 4493.
4. K. V. Titova, E. I. Kolmakova and V. Ya. Rosolovskii, *Izv. Akad. Nauk SSSR, Ser. Khim.*, 1975, 2821.
5. S. Brownstein, J. Bornais and G. Latremouille, *Can. J. Chem.*, 1978, **56**, 1419.
6. V. Gutmann, U. Mayer and R. Krist, *Synth. React. Inorg. Metal-Org. Chem.*, 1974, **4**, 523.
7. I. G. Ryss and S. L. Idel's, *Zh. Neorg. Khim.*, 1965, **10**, 714.
8. R. W. Sprague, A. B. Garrett and H. H. Sisler, *J. Am. Chem. Soc.*, 1960, **82**, 1059.
9. M. Schmeisser, *Angew. Chem.*, 1952, **64**, 616.
10. P. Baumgarten and H. Hennig, *Chem. Ber.*, 1939, **72**, 1743.
11. L. Kolditz and C.-S. Lung, *Z. Anorg. Allg. Chem.*, 1968, **360**, 25.
12. J. D. Cotton and T. C. Waddington, *J. Chem. Soc. (A)*, 1966, 789.
13. R. A. Geanangel, *J. Inorg. Nucl. Chem.*, 1970, **32**, 3697.
14. I. G. Ryss and N. G. Parkomenko, *Zh. Neorg. Khim.*, 1966, **11**, 2052.
15. R. D. Chambers, H. C. Clark and C. J. Willis, *J. Am. Chem. Soc.*, 1960, **82**, 5298.
16. D. Thierig and F. Umland, *Naturwissenschaften*, 1967, **54**, 563.
17. G. Pawelke, F. Heyder and H. Bürger, *J. Organomet. Chem.*, 1979, **178**, 1.
18. F. E. Brinckman and F. G. A. Stone, *Chem. Ind. (London)*, 1959, 254.
19. M. P. Johnson, D. F. Shriver and S. A. Shriver, *J. Am. Chem. Soc.*, 1966, **88**, 1588.
20. K. V. Titova and V. Ya. Rosolovskii, *Izv. Akad. Nauk SSSR, Ser. Khim.*, 1974, 2174.
21. L. Chow, F. Schmock and U. Müller, *Z. Naturforsch., Teil B*, 1978, **33**, 1472.
22. W. Siebert, *Angew. Chem., Int. Ed. Engl.*, 1970, **9**, 734.

BF_3 reacts smoothly, in inert solvents, with alkali metal sulfates and phosphates to give stable 2:1 and 3:1 complexes (Table 9), whereas the intermediate complexes that form with NO_3^-, SO_3^{2-} and CO_3^{2-} decompose to $[BF_4]^-$ and B_2O_3.[55] Brownstein *et al.* have established[114] that BF_3 reacts easily with alkylammonium salts in CH_2Cl_2 or liquid SO_2. $[BF_3A]^-$ complex anions are formed with the salts of strong acids (equation 17) whereas complexes with salts of weak acids easily undergo disproportionation (equation 18) and/or conversion into a 2:1

complex (equation 19). Disproportionation proceeds to a small extent for the CN^- ion, but goes to completion with N_3^- ion.[114] Of the 2:1 complexes, an unequivocal structure has been reported only for the cyanocomplex $[F_3BCNBF_3]^-$.[114]

$$[Me_4N]A + BF_3 \longrightarrow [Me_4N][BF_3A] \quad (17)$$
$$(A = ClO_4^-, NO_3^-, SO_3CF_3^-, SO_3F^-)$$
$$[Me_4N]A + BF_3 \longrightarrow [Me_4N][BF_3A] \longrightarrow [Me_4N][BF_4] + [Me_4N][BF_2A_2] \quad (18)$$
$$A = NO_2^-, OAc^-, CF_3CO_2^-, HCOO^-, SCN^-, N_3^-, CN^-$$
$$[Me_4N]A + 2BF_3 \longrightarrow [Me_4N][A(BF_3)_2] \quad (19)$$
$$(A = F^-, OAc^-, PhS^-, CN^-)$$

While BF_3 forms stable complexes with a number of anionic Lewis bases, BCl_3 and BBr_3, although reacting easily with them, rarely give trihalogenoborates.[24,55] Apparently, the $[BX_3OMe]^-$ ions (X = Br, Cl) disproportionate rapidly to give $[BX_4]^-$ ions.[24,55] With pseudohalogenides the BCl_3 generally gives mixed ions, $[BCl_nY_{4-n}]^-$ (Y = CN, NCO, NCS, N_3)[115] and stable trichloroborates $[Me_4N][BCl_3Y]$ (Y = NCO, N_3) could only recently be isolated in liquid SO_2 (Table 9). The preparation of $[Me_4N][BX_3SH]$ (X = F, Cl, Br) in liquid H_2S and that of $[R_4N][BCl_3ClO_4]$ in $CHCl_3$ has been reported (Table 9).

24.3.3 Borates Containing B—O Bonds

Because of the great volume of literature on boron–oxygen coordination compounds, it has been necessary to be highly selective in the choice of material and to confine our attention predominantly to reviews. Of the metal borates[7,116] only those existing in aqueous media will be briefly mentioned and of the complexes of organoboron oxygen compounds,[75] known in a very large number, the most important types only will be included.

24.3.3.1 Hydroxoborates

The aqueous chemistry of borates is a relatively neglected area of research; the existence or significance of many ionic species postulated previously should now be regarded with suspicion because of incomplete or unreliable evidence. In dilute solutions or at pH > 11 only the monomeric $[B(OH)_4]^-$ exists, while at higher concentrations and at $5 \leqslant pH \leqslant 11$ polyborate ions appear, the most important being $[B_5O_6(OH)_4]^-$ with a spiro structure, the monocyclic $[B_3O_3(OH_4)]^-$ and the bicyclic $[B_4O_5(OH)_4]^{2-}$. Other species such as $[B_2O(OH)_5]^-$, $[B_3O_3(OH)_5]^{2-}$ and $[B_4O_5(OH_3)]^-$ have also been postulated.[117]

Of the experimental techniques employed for studying polyborate equilibria, potentiometry has received the greatest attention although incorrect interpretation of the experimental data often led to the postulation of conflicting species. However, vibrational spectroscopy has proved to be a most useful technqiue for identifying the various species in solution.[117]

A number of peroxo derivatives of hydroxoborates are well known, the most important being a cyclic borate, $Na_2[B_2(O_2)_2(OH)_4]\cdot 6H_2O$, which possesses, according to crystallographic studies,[118] double peroxo bridges. This compound is in widespread use as a bleaching agent.

24.3.3.2 Alkoxo- and aryloxo-borates

Reaction of trialkoxyboranes with metal alcoholates, alcoholysis or hydride transfer reactions of tetrahydroborates with aldehydes or ketones all result in the formation of tetraalkoxoborates. Steric factors play an important role in these reactions. As a consequence, *sec*-alcohols react very slowly and tetra-*t*-alkoxoborates in general cannot be obtained by any of the reactions above. At elevated temperatures the tetraalkoxoborates revert to the trialkoxyborane and metal alkoxide.[75] Thioalcoholysis of tetrahydroborates can also be effected but, in contrast to the situation in alcoholysis, the last hydrogen atom is more difficult to substitute, probably for steric reasons.[119] Tetraalkoxoborates and tetramercaptoborates are readily hydrolyzed by water or moist air.

Complexes with polyhydric alcohols or phenols are much more stable than those with monohydric species, probably due to a more favorable entropy contribution attendant upon chelate formation. Typical structures of polyolborates are exemplified in (**1**)–(**6**). Unsymmetrical substitution in bisdiolborates [structures (**2**) and (**3**)] creates overall molecular asymmetry so that such derivatives could be resolved into optical isomers.[75]

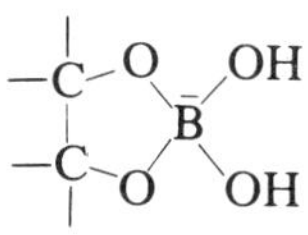

(**1**)

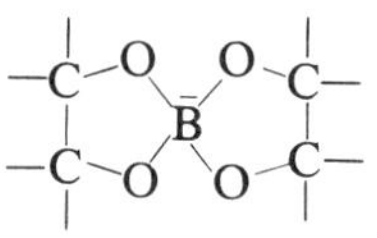

(**2**)

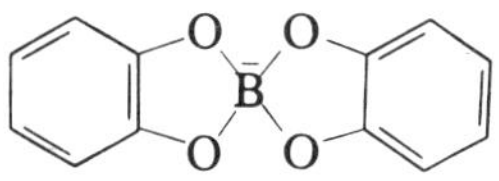

(**3**)

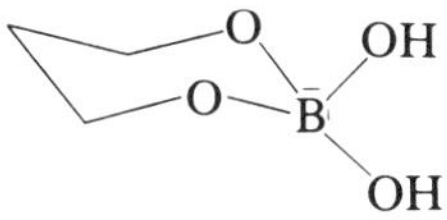

(**4**)

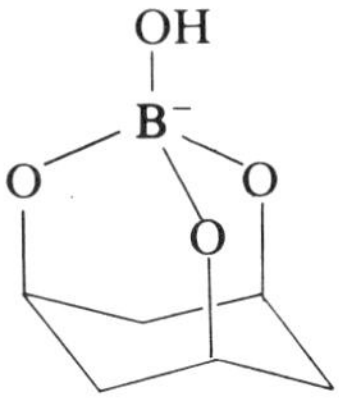

(**5**)

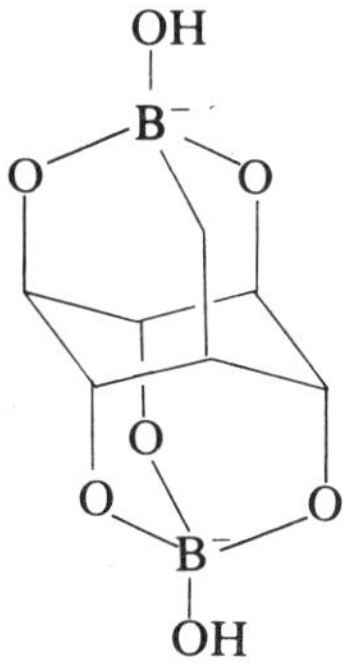

(**6**)

Diols and polyols can participate in equilibria with boric acid in aqueous solution. The stability of polyolborates is determined by the number of OH groups in *cis* positions. Complexes with polyols are more stable than with diols, and 1,2-diol complexes are more stable than their 1,3-diol counterparts (Table 10) since the resulting five-membered chelate ring is unstrained.[75,120] In the case of 1,3,5-triols stable cage-like structures (**5**) and (**6**) are favored. Open-chain or five-membered cyclic polyols form more stable chelate complexes than their six-membered counterparts.[120] Thus, chelates from alditols and ketohexoses are more stable than the corresponding aldose chelates (Table 10). Many polyols allow quantitative titrimetric determination of boric acid. Of these, mannitol remains the most widely used reagent on the basis of availability, cost and ease of handling.[75]

Table 10 Formation Constants of Polyol Borate Complexes[1,2]

Ligand	*log* β_1	*log* β_2	*Ligand*	*log* β_1	*log* β_2
Ethylene glycol	—	−8.53	D-Ribose	—	−1.90
Glycerol	−7.54	−7.17	D-Lyxose	−4.28	—
Pentaerythritol	−6.40	−5.45	D-Xylose	—	−5.09
Catechol	−5.13	−4.84	D-Arabinose	—	−5.82
Pyrogallol	−5.05	−4.40	D-Fructose	−5.68	−4.12
D-Mannitol	−5.99	−3.91	L-Sorbose	—	−3.30
D-Galactitol	—	−3.87	α-D-Glucose	−7.08	−6.47
D-Sorbitol	—	−3.45	α-D-Mannose	—	−4.58
meso-Inositol	−7.46	—	α-D-Galactose	−6.79	−6.51

[1] All data taken from A. E. Martell and R. M. Smith, 'Critical Stability Constants, vol. 3: Other Organic Ligands', Plenum, New York, 1977.
[2] β_1 and β_2 are equilibrium constants for the reactions
$B(OH)_3 + H_2L \rightleftharpoons [B(OH)_2L]^- + H^+$
$B(OH)_3 + 2H_2L \rightleftharpoons [BL_2]^- + H^+$
in 0.1 mol dm^{-3} KCl solutions at 25 °C.

The more stable boron chelates can be isolated even from aqueous solution, whereas those of lower stabilities are only accessible from non-aqueous media. Catechol- and inositol-borates (**3**, **5** and **6**) possesses a well-defined monomeric structure,[75] whereas those obtained from monosaccharides and alditols are polymeric.[121] A crystal structure determination[122] has been carried out for sodium *scyllo*-inositol diborate (**6**).

On the basis of complex-forming properties, boric acid and borates are in widespread use for synthesis, separation, identification and analysis in the chemistry of carbohydrates, nucleosides, nucleotides, polyphenols, flavonoids and other polyols.[84] Differences in the stabilities of the borate complex can be put to good use, especially in ion-exchange chromatographic separation of sugars and derivatives.[123–125] Polyolborates have also aroused some interest for potential biochemical and pharmaceutical applications.[84,126]

The first natural boron compound, boromycin, isolated in 1967, is the boron complex of a macrocyclic polyol betaine incorporating D-valine ($C_{45}H_{74}BNO_{15}$).[84,127] Its structure has been elucidated by X-ray analysis of the Rb and Cs salt of the fragment obtained after hydrolytic cleavage of the D-valine part.[128] The antimicrobial activity[127,129] and biosynthetic pathway[130] of boromycin have been invesigated. Another antibiotic, containing a BO_4 unit, is aplasmomycin ($C_{40}H_{60}BNaO_{14}$) which has a carbon skeleton devoid of D-valine, but otherwise identical to that of boromycin.[131]

24.3.3.3 Carboxylatoborates

Reaction of $B_2O(OAc)_4$ with sodium acetate results in the formation of $Na[B_2O(OAc)_5]$ with O and OAc bridges, whereas acetatoborates, $M[B(OAc)_4]$ (M = K, Rb, Cs, Tl), are obtained when the reaction is conducted in the presence of acetic anhydride. These borates are easily hydrolyzed; in the case of $[B(OAc)_4]^-$ the rate determining step seems to be dissociation of the complex. Binuclear borates $M_2[B_2O(OAc)_6]$ and $M_2[B_2O_2(OAc)_4]$, were detected, *inter alia,* during thermal decomposition of tetraacetatoborates.[132] The crystal structure of $K[B(OAc)_4]$ has been determined.[133]

Boron gives stable complexes with hydroxycarboxylic and dibasic carboxylic acids when the possibility of the formation of five- or six-membered chelate rings exists. Complexes with 1:1 and 1:2 boron to ligand ratio can both be prepared. 1:1 complexes are generally obtained in aqueous media, preferably with acids (lactic, tartaric, salicyclic and oxalic acids) capable of giving five-membered chelate rings.[75] In non-aqueous media five- and six-membered chelates of hydroxycarboxylic[75] and dicarboxylic acids[134] can be synthesized. Complexes with α-hydroxycarboxylic acids, like those with unsymmetrically substituted bisdiols, display optical isomerism, and resolution through brucine and strychnine salts has been reported in some cases.[75] The crystal structure of potassium bis(*para*-aminosalicylato)borate was reported recently.[135]

24.3.3.4 Borates with anions of oxoacids

Summarized in Table 11 are the compounds known to date with the possible ways of synthesizing them. The majority of these borates has been characterized by IR and NMR and, in some cases, X-ray diffraction. The hydrolytic and thermal stabilities of the tetraalkylammonium tetranitratoborates are surprisingly high, whereas the Rb and Cs salts undergo slow decomposition at room temperature.

Table 11 Borates with Oxo-anion Ligands and their Preparation

Borate	*Method*	*Ref.*
$M[B(NO_3)_4]$	$M[BCl_4] + N_2O_4$ (M = $[R_4N]$, Rb, Cs)	1, 2
$M[B(ClO_4)_4]$	$M[BCl_4] + HClO_4$ (M = K, Rb, Cs)	3
	$BCl_3 + MClO_4 + HClO_4$ (M = K, Rb, Cs)	3
$[NH_4][B(SO_4)_2]$	$BN + H_2SO_4$	4
$M[B(SO_4)_2]$	$B(OH)_3 + M_2SO_4 + SO_3$ (M = NH_4 and metal ions)	4
$H[B(HSO_4)_4]$	$BCl_3 + H_2SO_4$	5
$M[B_2O(SO_4)_3]$	$B(OH)_3 + MSO_4 + SO_3$ (M = Ca, Sr, Ba)	4
$M[B(SO_3Cl)_4]$	$M[BCl_4] + SO_3$ (M = Li, Na, K, Rb, Cs)	6
$Cs[B(OTeF_5)_4]$	$B(OTeF_5)_3 + CsOTeF_4$	7

1. C. R. Guibert and M. D. Marshall, *J. Am. Chem. Soc.*, 1966, **88,** 189.
2. K. V. Titova and V. Ya. Rosolovskii, *Zh. Neorg. Khim.*, 1971, **16,** 1450.
3. V. P. Babaeva and V. Ya. Rosolovskii, *Izv. Akad. Nauk SSSR, Ser. Khim.*, 1973, 494.
4. G. Schott and H. U. Kibbel, *Z. Anorg. Allg. Chem.*, 1962, **314,** 104.
5. N. N. Greenwood and A. Thompson, *J. Chem. Soc.*, 1959, 3643.
6. M. Drache, B. Vandorpe and J. Heubel, *Bull. Soc. Chim. Fr.*, 1976, 1749.
7. H. Kropshofer, O. Leitzke, P. Peringer and F. Sladky, *Chem. Ber.*, 1981, **114,** 2644.

The thermochemical properties of tetraperchloratoborates have been studied in detail.[136] The estimated heats of dissociation for $BX_4^-{}_{(g)} \rightleftharpoons BX_{3(g)} + X^-_{(g)}$ decrease in the order F > H > Cl > ClO_4. The sulfatoborates are probably ionic compounds with highly polymeric anions and are sensitive to hydrolysis. The chlorosulfatoborates are stable up to 80–150 °C; their decomposition seems to give, as a first step, $M[BCl_4]$ and SO_3.

24.3.4 Borates with B—N, B—P, B—C and B—Si Bonds

Various kinds of these compounds along with their methods of preparation are summarized in Table 12. The lithium tetraazidoborate is of low hydrolytic stability and detonates easily when it is freed of solvent. The IR spectrum of $Li_3[BN_2]$ is in favor of an $[N{=}B{=}N]^{3-}$ structure isoelectronic and isosteric with the N_3^- and CN_2^{2-} ions. $Li[B(NCS)_4]$ has appreciable stability only in the form of etherate or dioxanate whereas $[(SCN)_3P{=}N{-}P(NCS)_3][B(NCS)_4]$ can be obtained through the reaction of $[Cl_3P{=}N{-}PCl_3][BCl_4]$ with NH_4SCN.[137] The B—NCS bond has been established by IR and ^{11}B NMR studies.

Table 12 Borates with B—N, B—P, B—C and B—Si Bonds and their Preparation

Borate	*Method*	*Ref.*
$Li[B(N_3)_4]$	$Li[BH_4] + HN_3$	1
$Li_3[BN_2]$	$BN + Li_3N$	2
$Li[B(NCS)_4]\cdot 2C_4H_8O_2$	$Li[BH_4] + (SCN)_2$	3
$[Ph_4P][B(NSOF_2)_4]$	$B(NSOF_2)_3 + [Ph_4P]NSOF_2$	4
$K[B(NC_4H_4)_4]$	$K[BF_4] + C_4H_4NMgBr$	5
	$BF_3 + KNC_4H_4$	6
$Na[B(NC_4H_4)_4]$	$Na[BEt_4] + B(NC_4H_4)_3$	7
$K[BL_4]$	$K[BH_4]$ + HL (HL = pirazole, indazole)	8, 9
$Li[B(PEt_2)_4]$	$Et_2O\cdot BCl_3 + LiPEt_2$	10
$M[B(CN)_4]$	BCl_3 + MCN (M = Ag, Cu^I)	11, 12
$Li[B(CN)_3NMe_2]$	$B(CN)_2NMe_2 + LiCN$	11
$Li[B(SiMe_3)_4]$	$B(OR)_3 + LiSiMe_3$	13

1. E. Wiberg and H. Michaud, *Z. Naturforsch., Teil B*, 1954, **9**, 497.
2. J. Goubeau and W. Anselment, *Z. Anorg. Allg. Chem.*, 1961, **310**, 248.
3. F. Klanberg, *Z. Anorg. Allg. Chem.*, 1962, **316**, 197.
4. R. Eisenbarth and W. Sundermeyer, *Angew. Chem.*, 1978, **90**, 226.
5. V. A. Sazonova and V. I. Karpov, *J. Gen. Chem. USSR (Engl. Transl.)*, 1963, **33**, 3241.
6. P. Szarvas, J. Emri and B. Györi, *Acta Chim. Acad. Sci. Hung.*, 1970, **64**, 203.
7. M. A. Grassberger and R. Köster, *Angew. Chem., Int. Ed. Engl.*, 1969, **8**, 275.
8. S. Trofimenko, *J. Am. Chem. Soc.*, 1967, **89**, 3170.
9. K. S. Siddiqi, M. A. Neyazi and S. A. A. Zaidi, *Synth. React. Inorg. Metal-Org. Chem.*, 1981, **11**, 253.
10. G. Fritz and E. Sattler, *Z. Anorg. Allg. Chem.*, 1975, **413**, 193.
11. E. Bessler and J. Goubeau, *Z. Anorg. Allg. Chem.*, 1967, **352**, 67.
12. E. Bessler, *Z. Anorg. Allg. Chem.*, 1977, **430**, 38.
13. W. Biffar and H. Nöth, *Chem. Ber.*, 1982, **115**, 934.

Although $Li[B(NHMe)_4]$ exists,[138] most tetraamidoborates contain ligands derived from five-membered heterocyclics such as pyrrole, indole, imidazole, pyrazole and indazole. The thermal and hydrolytic stabilities of these borates are fairly high. When the heterocycle contains two nitrogens the borates can exist in acidic form and the tetraimidazol-1-ylborate can be used for gravimetric determination of the proton.[139] Another type of borate containing a BN_4 unit is represented by bicyclic derivatives formed with *N,N″,N″*-trimethylbiuret.[140]

Like the cyanohydroborates, the $[B(CN)_3NMe_2]^-$ and $[B(CN)_4]^-$ ions are extremely stable towards hydrolysis. In the series $Li[B(SiMe_3)_nMe_{4-n}]$, the $^1J(^{29}Si-^{11}B)$ coupling constants suggest increasing *s* contribution in the B—Si bond with increasing *n*.

24.4 CATIONIC COMPLEXES

Although some of the cationic chelate complexes have long been known,[141] most of the cationic boron complexes have been synthesized in the last three decades following the structure elucidation[142] of 'diammoniate of diborane'. Most comprehensively studied are the tetracoordinate cations and, among them, the $[H_2B(NR_3)_2]^+$ species have been the subject of several reviews;[3,34,143] thus we shall not discuss them in detail. The number of tricoordinate boron cations known to date (Table 13) is, however, small. The majority of them have not been isolated and the others are not yet well characterized.[144,145] The unusual dicoordinate cationic boron complexes (Table 13) have only recently been obtained either in solution[146] or in solid form.[144]

Table 13 Types of Cationic Boron Complexes

Complex[a]	*Ref.*	*Complex*[a]	*Ref.*
$[X_2B]^+$		$[H_2B(SMe_2)_2]_2[B_{12}H_{12}]$	7
$[(tmp)BOC_6H_2Bu^t_3]^+ + [AlBr_4]^-$	1	$[(Me_3NBH_2)_2SMe][PF_6]$	10
$[(tmp)BNR_2][AlBr_4]$ (R = Me, Et)	2	$[XHB(NMe_3)_2][PF_6]$ (X = F, Cl, Br)	7
$[(Pr^i_2N)_2B]^+ + [AlCl_4]^-$	3	$[XHB(Me_2pip)][PF_6]$ (X = Cl, Br)	11
$[(tmp)BR][AlBr_4]$ (R = Me, Ph)	2	$[ClHB(PMe_3)PPh_3][PF_6]$	6
$[X_2BL]^+$		$[X_2Bdipy][BX_4]$ (X = F, Cl)	12
$[Cl_2B(4\text{-Mepy})]^+ + [Al_2Cl_7]^-$	4	$[F_2B(HMPA)_2]^+ + [BF_4]^-$	13
$[(CH_2(Me)N)_2Bpy]^+ + SO_3CF_3^-$	1	$[Cl_2B(OEt_2)_2][FeCl_4]$	14
$[(CH_2(Me)N)_2BNHMe_2]SO_3CF_3$	1	$[(C_6H_4O_2)BL_2]I$ (L_2 = phen, dipy)	15
$[Me_2N(Me)BNHMe_2]I$	5	$[B(acac)_2]Cl$	16
$[X_2BL_2]^+$		$[(Me_2N)_2Bdipy][BPh_4]$	15
$[H_2B(NMe_3)PMe_2Ph][PF_6]$	6	$[HN(BClNH)_2Bpy_2]Cl$	17
$[H_2BL_2][PF_6]$ (L_2 = dppe, dmpe)	6, 7	$[XBL_3]^{2+}$ *and* $[BL_4]^{3+}$	
$[H_2B(AsMe_3)_2][PF_6]$	7	$[HB(4\text{-Mepy})_2py][PF_6]_2$	18
$[H_2B(OSMe_2)_2]^+ + [BH_4]^-$	8	$[HB(Rpy)_3]Br_2$ (R = H, Me)	19
$[H_2B(THF)_2]^+ + [B_3H_8]^-$	9	$[BL_4]I_3$ (L = Rpy; L_2 = dipy)	20

[a] Ionic form means that the complex exists only in solution; tmp = 2,2′,6,6′-tetramethylpiperidino group, Me_2pip = *N,N′*-dimethylpiperazine.

1. H. Nöth, S. Weber, B. Rasthofer, Ch. Narula and A. Konstantinov, *Pure Appl. Chem.*, 1983, **55**, 1453.
2. H. Nöth, R. Staudigl and H.-U. Wagner, *Inorg. Chem.*, 1982, **21**, 706.
3. J. Higashi, A. D. Eastman and R. W. Parry, *Inorg. Chem.*, 1982, **21**, 716.
4. G. E. Ryschkewitsch and J. W. Wiggins, *J. Am. Chem. Soc.*, 1970, **92**, 1790.
5. H. Nöth and P. Fritz, *Z. Anorg. Allg. Chem.*, 1963, **322**, 297.
6. B. T. Pennington, M. A. Chiusano, D. J. Dye, E. D. Martin and D. R. Martin, *J. Inorg. Nucl. Chem.*, 1978, **40**, 389.
7. N. E. Miller and E. L. Muetterties, *J. Am. Chem. Soc.*, 1964, **86**, 1033.
8. G. E. McAchran and S. G. Shore, *Inorg. Chem.*, 1965, **4**, 125.
9. R. Schaeffer, F. Tebbe and C. Phillips, *Inorg. Chem.*, 1964, **3**, 1475.
10. R. J. Rowatt and N. E. Miller, *J. Am. Chem. Soc.*, 1967, **89**, 5509.
11. G. E. Ryschkewitsch and T. E. Sullivan, *Inorg. Chem.*, 1970, **9**, 899.
12. D. D. Axtell, A. C. Cambell, P. C. Keller and J. V. Rund, *J. Coord. Chem.*, 1976, **5**, 129.
13. J. S. Hartman and P. Stilbs, *J. Chem. Soc., Dalton Trans.*, 1980, 1142.
14. B. M. Mikhailov, V. D. Sheludyakov and T. A. Schegoleva, *Izv. Akad. Nauk SSSR, Ser. Khim.*, 1962, 1698.
15. L. Banford and G. E. Coates, *J. Chem. Soc.*, 1964, 3564.
16. W. Dilley, *Liebigs Ann. Chem.*, 1906, **344**, 300.
17. M. F. Lappert and G. Srivastava, *J. Chem. Soc. (A)*, 1967, 602.
18. M. A. Mathur and G. E. Ryschkewitsch, *Inorg. Chem.*, 1980, **19**, 3054.
19. M. A. Mathur and G. E. Ryschkewitsch, *Inorg. Chem.*, 1980, **19**, 887.
20. R. D. Bohl and G. L. Galloway, *J. Inorg. Nucl. Chem.*, 1971, **33**, 885.

General and specific methods for synthesizing cationic complexes are summarized in equations (20)–(27). Nucleophilic displacement of group X (equations 20 and 21) will be favoured if the B—X bond is weak (*e.g.* B—Br, B—I, B—SR, B—OSO_3CF_3) and an appropriately strong base is used, or if a chelate ring is formed.[143,145] Removal of the X group (equations 22 and 23) can only be effected by the use of strong acceptors (*e.g.* BF_3, BCl_3 $AlCl_3$, $AlBr_3$, SbF_5).[145] Formation of di- or tri-coordinate boron cations is to be expected only when the group bonded to the boron is capable of counterbalancing effectively, by π back-donation, the electron deficiency of the boron (*e.g.* NR_2, OR).[145] Of equal importance is the steric shielding on the boron; bulky groups bonded to the donor atom contribute to the stability (Table 13). Tri- and tetra-coordinate cationic complexes can be obtained through reaction of aminoboranes with the corresponding acids (equations 24 and 25); tricoordinate complexes are formed only when the anion of the acid is of low basicity.[34,145] The methods represented by reactions (26) and (27) are of limited scope and are useful only for synthesizing four-coordinate complexes.[34,143]

$$L{\cdot}BX_3 \xrightarrow{L} [X_2BL_2]X \xrightarrow{L} [XBL_3]X_2 \xrightarrow{L} [BL_4]X_3 \quad (20)$$

$$BX_3 + L \longrightarrow [X_2BL]^+ + X^- \quad (21)$$

$$L{\cdot}BX_2Y + EY_n \longrightarrow [X_2BL]^+ + [EY_{n+1}]^- \quad (22)$$

$$BX_2Y + EY_n \longrightarrow [X{=}B{=}X]^+ + [EY_{n+1}]^- \quad (23)$$

$$BNMe_2\{N(Me)CH_2\}_2 + HSO_3CF_3 \longrightarrow [\{CH_2(Me)N\}_2BNHMe_2]SO_3CF_3 \quad (24)$$

$$(Me_2N)_2BX + 2HY \longrightarrow [XYB(HNMe_2)_2]Y \quad (25)$$
$$X = Cl, H, Me, Et, Ph;\ Y = Cl, Br$$

$$[H_2B(NMe_3)_2]^+ + X_2 \longrightarrow [H_nX_{2-n}B(NMe_3)_2]^+ \quad (26)$$
$$n = 1;\ X = Cl, Br;\ n = 0,\ X_2 = F_2, ICl$$

$$Me_3N{\cdot}BHBr_2 + 3py \longrightarrow [HBpy_3]Br_2 + NMe_3 \quad (27)$$

The stability of boron cations is very much dependent on the coordination number; the di- and tri-coordinate derivatives exist mainly as equilibrium species, their stability being significantly lower than for those with a coordination number of four. They are converted by even weak bases into neutral compounds or cations with higher coordination numbers.[145] The thermal and hydrolytic stabilities of the tetracoordinate complexes decrease[150] in the order bisamine > trisamine > tetrakisamine; for the $[H_2BL_2]^+$ cations the order of stability is $L = R_3N > R_3P > R_3As > R_2O > R_2S$.[143,147] The bisamine cations are hydrolytically the most stable monoboron species, withstanding concentrated acids and bases without decomposition.[34,143] The bisphosphine complexes, although also stable, undergo slow decomposition in base whereas the bisthioether complexes are rapidly hydrolyzed in water.[143,147] The hydrolytic stability of $[XClB(HNMe_2)_2]Cl$ salts decreases in the order $X = H > F > Cl > Ph$.[143]

The four-coordinate amine complexes are very stable to oxidation;[34,143] they do not react with I_2 and undergo only monosubstitution upon treatment with Br_2 (equation 26). Monohalogenation of bisamine complexes of the $[H_2BLL']^+$ type affords cations containing chiral boron;[34,148] the $[XHB(NMe_3)4\text{-}Mepy]^+$ ($X = Cl, Br$) cations were resolved[149] into enantiomers. The crystal structure for $X = Br$ has been reported.[150]

24.5 BORON AS A LIGAND

Various neutral and, especially, anionic boron species are well known as ligands in transition metal complexes. In the majority of cases, however, one or more H, N or other basic atom of the boron compounds is coordinated to the transition metal (*e.g.* Chapters 13.8, 13.6 and 19). In the last 20 years, however, several complexes containing boron–transition metal bonds have been prepared with ligands such as :BX, $\cdot BX_2$ or BX_3 (Table 14).

Compounds $[Fe(BX)(CO)_4]$ are, in all probability, dimers with two Fe—B—Fe bonds. $[Fe(BNR_2)(CO)_4]$ ($R = Me, Et$) could, however, be obtained in monomeric form in low yield. According to IR and ^{11}B NMR studies the $:BNR_2$ ligand is a weak σ donor but seems to be a better π acceptor than CO. When :BX is a bidentate ligand (Table 14) its bonding properties are similar to those of the $\cdot BX_2$ ligands.

Several representatives of boryl–metal compounds containing covalent $M\text{–}BX_2$ bonds are known. ^{11}B NMR and, in the case of carbonyl complexes, IR studies as well indicate back-bonding from the metal atom to the sp^2 boron.[151,152] This may be a reason for the greater stability of boryl carbonylphosphine complexes as compared with boryl carbonyl complexes. The paramagnetic d^7 low-spin complex $[Co(BPh_2)_2\,(diphos)_2]$ behaves somewhat differently.[152] The Co—B bond is extremely sensitive to oxidation, the BPh_2 group can be exchanged for other metal complexes easily (equation 32), and reduction with sodium smoothly yields the corresponding anionic complex (Table 14). In the case of $[Co(BPh_2)_2(diphos)_2]$ the ^{11}B chemical shift values indicate tetravalent boron which means that the boron carries a unit negative charge, while other boryl–metal complexes behave as molecules with typical trivalent

Table 14 Complexes with a Boron—Metal Bond

Type of boron ligand	*Metal complex or complex unit partner*	*Substituents at the boron*	*Metal–boron ratio*	*Ref.*
BX	$[Mn(CO)_5]$	NMe_2	2:1	1
	$[Mn(CO)_4PPh_3]$	Cl, NEt_2	2:1	2
	$[Fe(CO)_4]$	Br, I, NR_2	1:1	3
BX_2	$[TiXCp_2]$, (X = Cl, Br)	Cl, Br, Ph	1:1	4
	$[Mo(CO)_4PPh_3]$	Cl	1:1	5
	$[Mn(CO)_5]$	NMe, Ph	1:1	1, 6
	$[Re(CO)_5]$	NMe_2	1:1	7
	$[Fe(CO)_2Cp]$	Cl, Ph, NMe_2	1:1	5, 6
	$[Co(CO)_4]$	Cl, Ph	1:1	7
	$[Co(diphos)_2]^-$	Ph	1:2	6
	$[Rh(PPh_3)_3]$	Br	1:1	8
	$[IrCl(CO)(PPh_3)_2]$	F	1:2	9
	$[Ni(NO)(PPh_3)_2]$	Cl	1:1	5
	$[PtX(PPh_3)_2]$, (X = Cl, Br)	Br, Ph	1:1	4
	[Mtriars], (M = Cu, Ag)	Br	1:1	8
	$[AuPPh_3]$	Ph	1:1	6
	$[SnMe_{4-n}]$, (n = 1–4)	Ph	1:n	10
BX_3	$[MH_2Cp_2]$, (M = Wo, W)	F, Cl	1:1	11, 12
	$[Mn(CO)_n(PPh_3)_{5-n}]^-$, ($n$ = 4, 5)	H	1:1	13
	$[Re(CO)_5]^-$	H	1:2	13
	$[ReHCp_2]$	F	1:1	12
	$[Co(NO)_3]$	Me	1:1	14
	$[RhX(CO)(PPh_3)_2$, (X = Cl, Br)	Cl, Br	1:1	15
	$[IrCl(CO)(EPh_3)_2]$ (E = P, As)	F	1:2	16
	$[Pt(PPh_3)_3]$	Cl	1:2	17

triars = o-$Me_2AsC_6H_4As(Me)C_6H_4AsMe_2$-$o$

1. H. Nöth and G. Schmid, *Angew. Chem.*, 1963, **75**, 861.
2. H. Nöth and G. Schmid, *Z. Anorg. Allg. Chem.*, 1966, **345**, 69.
3. G. Schmid, W. Petz and H. Nöth, *Inorg. Chim. Acta*, 1970, **4**, 423.
4. G. Schmid, W. Petz, W. Arloth and N. Nöth, *Angew. Chem.*, 1967, **79**, 683.
5. G. Schmid, P. Powell and H. Nöth, *Chem. Ber.*, 1968, **101**, 1205.
6. G. Schmid and H. Nöth, *Chem. Ber.*, 1967, **100**, 2899.
7. H. Nöth and G. Schmid, *Allg. Prakt. Chem.*, 1966, **17**, 610.
8. G. Schmid, *Chem. Ber.*, 1969, **102**, 191.
9. R. N. Scott, D. F. Shriver and L. Vaska, *J. Am. Chem. Soc.*, 1968, **90**, 1079.
10. H. Nöth, H. Schäfer and G. Schmid, *Angew. Chem., Int. Ed. Engl.*, 1969, **8**, 515.
11. D. F. Shriver, *J. Am. Chem. Soc.*, 1963, **85**, 3509.
12. M. P. Johnson and D. F. Shriver, *J. Am. Chem. Soc.*, 1966, **88**, 301.
13. G. W. Parshall, *J. Am. Chem. Soc.*, 1964, **86**, 361.
14. I. H. Sabherwal and A. B. Burg, *Chem. Commun.* 1970, 1001.
15. P. Powell and H. Nöth, *Chem. Commun.*, 1966, 637.
16. R. J. Fitzgerald, N. Y. Sakkab, R. S. Strange and V. P. Narutis, *Inorg. Chem.*, 1973, **12**, 1081.
17. T. R. Durkin and E. P. Schram, *Inorg. Chem.*, 1972, **11**, 1054.

boron.[152] Equations (28)–(32) present some synthetic methods for boryl–metal compounds.[152,153]

$$[PtH(Cl)(PEt_3)_2] + BClPh_2 \longrightarrow [PtCl(BPh_2)(PEt_3)_2] + HCl \quad (28)$$

$$2[CoH(diphos)_2] + 2BX_2Y \longrightarrow [Co(BX_2)_2(diphos)_2] + [CoY_2(diphos)_2] + H_2 \quad (29)$$

$$Na[Mn(CO)_4PPh_3] + BClX_2 \longrightarrow [Mn(BX_2)(CO)_4PPh_3] + NaCl \quad (30)$$

$$[Pt(PPh_3)_4] + BX_2Y \longrightarrow [PtY(BX_2)(PPh_3)_2] + 2PPh_3 \quad (31)$$

$$2[FeCl(CO)_2Cp] + [Co(BPh_2)_2(diphos)_2] \longrightarrow 2[Fe(BPh_2)(CO)_2Cp] + [CoCl_2(diphos)_2] \quad (32)$$

X, Y = halogen

Compounds of transition metal complexes possessing a nonbonding electron pair with boranes (BX_3) can be regarded classically as boron complexes with a transition metal ligand. In a broader sense, however, boranes can be classified as acceptor ligands.[154,155] Thus, coordination of a borane results in a decrease of the electron density on the metal atom. In the case of carbonyl complexes this effect is reflected in the increase, by 20–100 cm^{-1}, of the CO stretching frequency.[154–156] It follows from the foregoing that stable coordination of boranes is to be expected only with transition metal complexes in low oxidation state; this is exemplified by items in Table 14.

Boranes appear to be especially suitable for investigating transition metal basicity[155] but, in spite of this, few relative basicities have been published to date: $[Cr(CO)_3L] < [Mo(CO)_3L] < [W(CO)_3L]$ ($L = Ph_3PC_5H_4$), $[Mn(CO)_5]^- < [Re(CO)_5]^-$, $[Co(CO)_4]^- < [Mn(CO)_5]^-$, $[Mn(CO)_5]^- < [Mn(CO)_4PPh_3]^-$[155,157]. $[Co(NO)_3]$ (Table 14) proved to be an extremely strong base coordinating the weak Lewis acid BMe_3 more strongly than NH_3 or Me_3N does. In the 1:2 complexes, coordination of both boron atoms to the transition metal is certain only in $[Pt(BCl_3)_2(PPh_3)_3]$; a hydrogen bridge structure seems quite plausible in the rhenium complex, $[Re(BH_2—H—BH_3)(CO)_5]^-$, whereas the nature of bonding for the second BF_3 is unknown in the iridium complex (Table 14).

The recently synthesized complexes $K[\mu\text{-}Fe(B_2H_5)(CO)_4]$ and $[\mu\text{-}Fe(B_2H_5)(CO)_2Cp]$ represent the first examples of a substituted diborane (**6**) with a transition metal occupying a bridge site.[158,159] In the case of complexes with $B(NR_2)R_2$ ligands either $M \leftarrow N$ coordination or ethylenic type π coordination may take place. In the second case boron is also involved in the bonding through $M \rightarrow B$ back-coordination.[160] Metal–boron bonds also occur in transition metal clusters known in great number, as well as in π-complexes of aromatic boron compounds such as borazines, borabenzenes, *etc.*; all these have been reviewed elsewhere.[161,162]

24.6 APPLICATIONS

In spite of the vast effort and fascinating results in academic and industrial research in the last 30 years industrial applications of coordination compounds of boron have remained fairly limited. Borax, other polyborates and sodium peroxoborate are the only compounds that have acquired major commercial importance to date.[163–165] Borax, the most important boron mineral, is utilized as a starting material for other boron compounds and plays a significant role in the glass industry. The highly water soluble polyborate mixture of stoichiometric composition 1:4 $Na_2O:B_2O_3$ is extensively employed as a plant nutrient[165] and for timber preservation.[165,166] Sodium peroxoborate serves as a bleaching ingredient in household washing.[164] Fluoroboric acid is a metal cleaning agent as well as a catalyst in organic reactions; alkali fluoroborates are utilized in the electroplating industry.[165]

Sodium tetrahydroborate is widely used as a powerful and selective reducing agent in inorganic and organic systems; for instance in organic reductions, in the purification and stabilization of organic materials, bleaching of wood pulp, separation of metals from very dilute solutions, electrodeless plating, particularly of nickel, to produce sodium dithionite from bisulfite, and in the development process for color reversal photographic films.[165,167,168]

Based on the excellent solubility and reducing properties of amine boranes, they are often used for the stabilization and purification of industrial materials. Further, they are applied as additives in hydrocarbon fuels and in lubricating oils, as polymerization catalysts, as ingredients in photographic processing and in the electroplating industry.[169] High stability towards hydrolysis and oxidation makes cationic boron chelates useful as water-soluble dyes.[170]

Detailed investigations are being conducted in the field of boron–nitrogen polymers with $N \rightarrow B$ coordinative bonds due to their hydrolytic stability (superior to that of the polymers containing tricoordinate boron). Such plastics and elastomers are likely to become candidates for special applications because of the high cross-section of ^{10}B for the (n, α) reaction.[171]

Several complexes have been tested for potential pharmaceutical applications and some of them have been introduced into practice.[84,126] Isolation of antibiotics of boron complex type (Section 24.3.3.2) and results obtained with carboxyborane complexes that are analogues of amino acids (Section 24.2.2) are likely to stimulate further studies. Quite a few boron complexes have found application in ^{10}B neutron capture therapy based on the $^{10}B(n, \alpha)^7Li$ nuclear reaction.[84,126,172]

24.7 REFERENCES

1. Gmelin Handbook Volumes on Boron and its Compounds: (a) Main Volume (1926); (b) Supplement vol. 1 (1954); (c) Compounds 8 (NSSV 33), 'The tetrahydroborate ion and its derivatives' (1976); (d) Compounds 9 (NSSV 34), 'Boron–halogen compounds', Part 1 (1976); (e) Compounds 10 (NSSV 37), 'Boron compounds with coordination number 4' (1976); (f) Compounds 15 (NSSV 46), Amine-boranes' (1977); (g) Compounds 19 (NSSV 53), 'Boron–halogen compounds', Part 2 (1978); (h) Supplement Volumes (1980–82), Springer-Verlag, Berlin.

2. N. N. Greenwood and (in part) B. S. Thomas, in 'Comprehensive Inorganic Chemistry', ed. J. C. Bailar, Jr., H. J. Emeléus, R. Nyholm and A. F. Trotman-Dickenson, Pergamon, Oxford, 1973, vol. 1, p. 665.
3. A. Pelter and K. Smith, in 'Comprehensive Organic Chemistry', ed. D. H. R. Barton and W. D. Ollis, Pergamon, Oxford, 1979, vol. 3, p. 687.
4. J. D. Odom, in 'Comprehensive Organometallic Chemistry', ed. G. Wilkinson, F. G. A. Stone and E. W. Abel, Pergamon, Oxford, 1982, vol. 1, p. 253.
5. A. Stock, 'Hydrides of Boron and Silicon', Cornell University Press, Ithaca, 1933.
6. W. Gerrard, 'The Organic Chemsitry of Boron', Academic, London, 1961.
7. R. M. Adams, 'Boron, Metallo-Boron Compounds and Boranes', Interscience, New York, 1964.
8. 'The Chemistry of Boron and its Compounds', ed. E. L. Muetterties, Wiley, New York, 1967.
9. E. Wiberg and E. Amberger, 'Hydrides of the Elements of Main Groups I–IV', Elsevier, Amsterdam, 1971, p. 81.
10. H. Nöth and B. Wrackmeyer, 'NMR Basic Principles and Progress', Vol. 14: Nuclear Magnetic Resonance Spectroscopy of Boron Compounds', ed. P. Diehl, E. Fluck and R. Kosfeld, Springer-Verlag, Berlin, 1978.
11. T. D. Coyle and F. G. A. Stone, in 'Progress in Boron Chemistry', ed. H. Steinberg and A. L. McCloskey, Pergamon, Oxford, 1964, vol. 1, p. 83.
12. J. L. Gay-Lussac and L. J. Thenard, *Mem. Phys. Chim. Soc. d'Arcueil,* 1809, **2,** 210.
13. H. C. Brown, *Rec. Chem. Prog.*, 1953, **14,** 83.
14. H. C. Brown, 'Boranes in Organic Chemistry', Cornell University Press, New York, 1972.
15. M. F. Lappert, *Chem. Rev.*, 1956, **56,** 959.
16. W. A. G. Graham and F. G. A. Stone, *J. Inorg. Nucl. Chem.*, 1956, **3,** 164.
17. K. Niedenzu and J. W. Dawson, 'Boron–Nitrogen Compounds', Springer-Verlag, Berlin, 1965.
18. G. W. Parshall, in 'The Chemistry of Boron and its Compounds', ed. E. L. Muetterties, Wiley, New York, 1967, p. 617.
19. E. L. Muetterties, in 'The Chemistry of Boron and its Compounds', ed. E. L. Muetterties, Wiley, New York, 1967, p. 647.
20. R. Hoffmann, *Adv. Chem. Ser.*, 1964, **42,** 78.
21. K. F. Purcell and J. C. Kotz, 'Inorganic Chemistry', ed. W. B. Saunders, Philadelphia, 1977, p. 216.
22. H. C. Brown and M. Gerstein, *J. Am. Chem. Soc.*, 1950, **72,** 2923.
23. E. N. Gur'yanova, I. P. Gol'dshtein and I. P. Romm, 'Donor–Acceptor Bond', Wiley, New York, 1975, p. 29.
24. J. S. Hartman and J. M. Miller, *Adv. Inorg. Chem. Radiochem.*, 1978, **21,** 147.
25. L. H. Long, *Adv. Inorg. Chem. Radiochem.*, 1974, **16,** 201.
26. R. G. Pearson, *J. Am. Chem. Soc.*, 1963, **85,** 3533.
27. F. G. A. Stone, *Q. Rev., Chem. Soc.*, 1955, **9,** 174.
28. A. F. Zhigach and L. N. Kochneva, *Usp. Khim.*, 1956, **25,** 1267.
29. C. F. Lane, *Aldrichim. Acta,* 1973, **6,** 51.
30. T. P. Fehlner, in 'Boron Hydride Chemistry', ed. E. L. Muetterties, Academic, New York, 1975, p. 175.
31. H. C. Brown, *Acc. Chem. Res.*, 1969, **2,** 65.
32. T. Onak, 'Organoborane Chemistry', Academic, New York, 1975.
33. M. G. H. Wallbridge and N. Davies, in 'MTP International Review of Science, Inorganic Chemistry Series Two', ed. M. F. Lappert, Butterworths, London, 1975, vol. 1, p. 61.
34. H. Nöth, in 'Progress in Boron Chemistry', ed. R. J. Brotherton and H. Steinberg, Pergamon, Oxford, 1970, vol. 3, p. 211.
35. R. O. Hutchins and F. Cistone, *Org. Prep. Proced. Int.*, 1981, **13,** 225.
36. A. Pelter, *Chem. Ind. (London)*, 1976, 888.
37. C. J. Foret, M. A. Chiusano, J. D. O'Brien and D. R. Martin, *J. Inorg. Nucl. Chem.*, 1980, **42,** 165.
38. G. W. Schaeffer and E. R. Anderson, *J. Am. Chem. Soc.*, 1949, **71,** 2143.
39. K. C. Nainan and G. E. Ryschkewitsch, *Inorg. Synth.*, 1974, **15,** 122.
40. K. R. Sundberg, G. D. Graham and W. N. Lipscomb, *J. Am. Chem. Soc.*, 1979, **101,** 2863.
41. M. M. Kreevoy and J. E. C. Hutchins, *J. Am. Chem. Soc.*, 1972, **94,** 6371.
42. H. C. Kelly and V. B. Marriott, *Inorg. Chem.*, 1979, **18,** 2875.
43. J. W. Wiggins and G. E. Ryschkewitsch, *Inorg. Chim. Acta,* 1970, **4,** 33.
44. H. C. Brown and S. U. Kulkarni, *J. Organomet. Chem.*, 1982, **239,** 23.
45. J. R. Lowe, S. S. Uppal, C. Weidig and H. C. Kelly, *Inorg. Chem.* 1970, **9,** 1423.
46. C. Weidig, J. M. Lakovits and H. C. Kelly, *Inorg. Chem.*, 1976, **15,** 1783.
47. C. Weidig, S. S. Uppal and H. C. Kelly, *Inorg. Chem.*, 1974, **13,** 1763.
48. B. Györi and J. Emri, *J. Organomet. Chem.*, 1982, **238,** 159.
49. B. F. Spielvogel, in 'Boron Chemistry-4', ed. R. W. Parry and G. Kodama, Pergamon, Oxford, 1980, p. 119.
50. I. H. Hall, C. O. Starnes, A. T. McPhail, P. Wisian–Neison, M. K. Das, R. Harchelroad, Jr. and B. F. Spielvogel, *J. Pharm. Sci.*, 1980, **69,** 1025.
51. P. R. Laurence and C. Thomson, *J. Mol. Struct.*, 1982, **88,** 37.
52. K. H. Scheller, R. B. Martin, B. F. Spielvogel and A. T. McPhail, *Inorg. Chim. Acta,* 1982, **57,** 227.
53. I. H. Hall, M. K. Das, F. Harchelroad, Jr., P. Wisian–Neilson, A. T. McPhail and B. F. Spielvogel, *J. Pharm. Sci.*, 1981, **70,** 339.
54. H. C. Brown, J. A. Sikorski, S. U. Kulkarni and H. D. Lee, *J. Org. Chem.*, 1982, **47,** 863.
55. A. G. Massey, *Adv. Inorg. Chem. Radiochem.*, 1967, **10,** 1.
56. K. C. Janda, L. S. Bernstein, J. M. Steed, S. E. Novick and W. Klemperer, *J. Am. Chem. Soc.*, 1978, **100,** 8074.
57. N. N. Greenwood and R. L. Martin, *Q. Rev., Chem. Soc.*, 1954, **8,** 1.
58. A. V. Topchiev, S. V. Zavgorodnii and Ya.M. Paushkin, 'Boron Fluoride and its Compounds as Catalysts in Organic Chemistry', Pergamon, New York, 1959.
59. G. E. Ryschkewitsch and W. H. Myers, *Synth. React. Inorg. Metal-Org. Chem.*, 1975, **5,** 123.
60. E. L. Muetterties, *J. Inorg. Nucl. Chem.*, 1960, **15,** 182.
61. G. S. Heaton and P. N. K. Riley, *J. Chem. Soc. (A)*, 1966, 952.

62. D. P. N. Satchell and R. S. Satchell, *Chem. Rev.*, 1969, **69**, 251.
63. M. Azzaro, J. F. Gal, S. Geribaldi and B. Videau, *J. Chem. Soc., Perkin Trans. 2,* 1983, 57.
64. P. A. van der Meulen and H. A. Heller, *J. Am. Chem. Soc.*, 1932, **54**, 4404.
65. D. R. Martin, *Chem. Rev.*, 1944, **34**, 461; 1948, **42**, 581.
66. A. W. Laubengayer and D. S. Sears, *J. Am. Chem. Soc.*, 1945, **67**, 164.
67. H. C. Brown and R. R. Holmes, *J. Am. Chem. Soc.*, 1956, **78**, 2173.
68. J. M. Miller, *Inorg. Chem.*, 1983, **22**, 2384.
69. J. E. Drake, J. L. Hencher and B. Rapp, *Inorg. Chem.*, 1977, **16**, 2289.
70. J. E. Drake, L. N. Khasrou and A. Majid, *Can. J. Chem.*, 1981, **59**, 2417.
71. M. Schmidt and H. D. Block, *Chem. Ber.*, 1970, **103**, 3705.
72. D. L. Black and R. C. Taylor, *Acta Crystallogr., Sect. B,* 1975, **31**, 1116.
73. K. Iijima, E. Koshimizu and S. Shibata, *Bull. Chem. Soc. Jpn.*, 1982, **55**, 2551.
74. D. C. Mente and J. L. Mills, *Inorg. Chem.*, 1975, **14**, 1862.
75. H. Steinberg and R. J. Brotherton, 'Organoboron Chemistry', Wiley, New York, 1964, vol. 1.
76. J. T. F. Fenwick and J. W. Wilson, *J. Chem. Soc., Dalton Trans.*, 1972, 1324.
77. H. Nöth and U. Schuchardt, *Chem. Ber.*, 1974, **107**, 3104.
78. R. H. Cragg, J. P. N. Husband and P. R. Mitchell, *Org. Magn. Reson.*, 1972, **4**, 469.
79. M. F. Lappert and H. Pyszora, *Adv. Inorg. Chem. Radiochem.*, 1966, **9**, 133.
80. J. H. Morris, 'Comprehensive Organometallic Chemistry', ed. G. Wilkinson, F. G. A. Stone and E. W. Abel, Pergamon, Oxford, 1982, vol. 1, p. 311.
81. F. Umland and E. Hohaus, 'Untersuchungen über Borhaltige Ringsysteme vom Chelattyp', Westdeutscher Verlag, 1976.
82. W. Ried and W. Bodenstedt, *Tetrahedron Lett.*, 1962, 247.
83. H. K. Zimmerman, *Adv. Chem. Ser.*, 1964, **42**, 23.
84. W. Kliegel, 'Bor in Biologie, Medizin und Pharmazie', Springer, Berlin, 1980.
85. W. Kliegel, *Organomet. Chem. Rev. (A)*, 1972, **8**, 153.
86. B. M. Mikhailov, *Pure Appl. Chem.*, 1977, **49**, 749.
87. B. D. James and M. G. H. Wallbridge, *Prog. Inorg. Chem.*, 1970, **11**, 99.
88. K. N. Semenenko, O. V. Kravchenko and V. B. Polyakova, *Usp. Khim.*, 1973, **42**, 3.
89. L. V. Titov, L. A. Gavrilova, K. V. Titova and V. Ya. Rosolovskii, *Izv. Akad. Nauk SSSR, Ser. Khim.*, 1978, 1722.
90. L. A. Gavrilova, L. V. Titov and V. Ya. Rosolovskii, *Zh. Neorg. Khim.*, 1981, **26**, 2070.
91. B. C. Hui, *Inorg. Chem.*, 1980, **19**, 3185.
92. M. M. Kreevoy and J. E. C. Hutchins, *J. Am. Chem. Soc.*, 1969, **91**, 4329.
93. M. M. Kreevoy and E. H. Baughman, *Finn. Chem. Lett.*, 1978, 35.
94. B. Györi, J. Emri and I. Fehér, *J. Organomet. Chem.*, 1983, **255**, 17.
95. R. O. Hutchins and N. R. Natale, *Org. Prep. Proced. Int.*, 1979, **11**, 201.
96. V. S. Khain, F. I. Pekarina and N. N. Antsygina, *Zh. Fiz. Khim.*, 1979, **53**, 1723.
97. C. A. Brown and J. L. Hubbard, *J. Am. Chem. Soc.*, 1979, **101**, 3964.
98. T. A. Shchegoleva, E. M. Shashkova and B. M. Mikhailov, *Izv. Akad. Nauk SSSR, Ser. Khim.*, 1963, 494.
99. J. W. Gilje and R. J. Ronan, *Inorg. Chem.*, 1968, **7**, 1248.
100. J. Emri and B. Györi, *Polyhedron,* 1982, **1**, 673.
101. S. Trofimenko, *Acc. Chem. Res.*, 1971, **4**, 17.
102. G. Bruno, E. Ciliberto, I. Fragala and G. Granozzi, *Inorg. Chim. Acta,* 1981, **48**, 61.
103. R. Köster, in Houben–Weyl, 'Methoden der Organischen Chemie', Thieme-Verlag, Stuttgart, 1983, vol. 13/3b, p. 681.
104. J. C. Carter and R. W. Parry, *J. Am. Chem. Soc.*, 1965, **87**, 2354.
105. J. Emri and B. Györi, *Polyhedron,* 1983, **2**, 1273.
106. D. W. A. Sharp, *Adv. Fluorine Chem.*, 1960, **1**, 68.
107. A. Finch, P. J. Gardner, N. Hill and N. Roberts, *J. Chem. Soc., Dalton Trans.*, 1975, 357.
108. J. S. Hartman and G. J. Schrobilgen, *Inorg. Chem.*, 1972, **11**, 940.
109. T. Vladimiroff and E. R. Malinowski, *J. Chem. Phys.*, 1967, **46**, 1830.
110. T. C. Waddington and F. Klanberg, *J. Chem. Soc.*, 1960, 2332.
111. R. E. Mesmer, K. M. Palen and C. F. Baes, *Inorg. Chem.*, 1973, **12**, 89.
112. V. M. Masalovich, G. A. Moshkareva and G. S. Menshenina, *Zh. Neorg. Khim.*, 1979, **24**, 1494.
113. G. B. Bachman and J. L. Dever, *J. Am. Chem. Soc.*, 1958, **80**, 5871.
114. S. Brownstein and G. Latremouille, *Can. J. Chem.*, 1978, **56**, 2764.
115. H. Landesman and R. E. Williams, *J. Am. Chem. Soc.*, 1961, **83**, 2663.
116. 'Mellor's Comprehensive Treatise of Inorganic and Theoretical Chemistry', ed. R. Thompson, Longmans, Green, New York, 1980, vol. 5, part A.
117. J. B. Farmer, *Adv. Inorg. Chem. Radiochem.*, 1982, **25**, 187.
118. M. A. A. F. de C. T. Carrondo and A. C. Skapski, *Acta Crystallogr., Sect. B,* 1978, **34**, 3551.
119. P. C. Keller, *Inorg. Chem.*, 1969, **8**, 1695.
120. U. Weser, *Struct. Bonding (Berlin)*, 1967, **2**, 160.
121. H. B. Davis and C. J. B. Mott, *J. Chem. Soc., Faraday Trans. 1,* 1980, **76**, 1991.
122. C. T. Grainger, *Acta Crystallogr., Sect. B,* 1981, **37**, 563.
123. H. Bauer and W. Voelter, *Chromatographia,* 1976, **9**, 433.
124. K. Larsson and O. Samuelson, *Carbohydr. Res.*, 1976, **50**, 1.
125. G. Weckbecker and D. O. R. Keppler, *Anal. Biochem.*, 1983, **132**, 405.
126. W. Kliegel, *Pharmazie,* 1972, **27**, 1.
127. R. Hütter, W. Keller-Schierlein, F. Knüsel, V. Prelog, G. C. Rodgers, Jr., P. Suter, G. Vogel, W. Voser and H. Zahner, *Helv. Chim. Acta,* 1967, **50**, 1533.
128. J. D. Dunitz, D. M. Hawley, D. Miklos, D. N. J. White, Yu. Berlin, R. Marusic and V. Prelog, *Helv. Chim. Acta,* 1971, **54**, 1709.

129. W. Pache, 'Antibiotics', ed. J. W. Corcoran and F. E. Hahn, Springer, Berlin, 1975, vol. 3, p. 585.
130. T. S. S. Chen, C. Chang and H. G. Floss, *J. Org. Chem.*, 1981, **46**, 2661.
131. H. Nakamura, Y. Iitaka, T. Kitahara, T. Okazaki and Y. Okami, *J. Antibiot.*, 1977, **30**, 714.
132. H. U. Kibbel, *Wiss. Z. Univ. Rostock, Math.-Naturwiss. Reihe*, 1973, **22**, 385.
133. A. Dal Negro, G. Rossi and A. Perotti, *J. Chem. Soc., Dalton Trans.*, 1975, 1232.
134. E. Bessler and J. Weidlein, *Z. Naturforsch., Teil B*, 1982, **37**, 1020.
135. I. I. Zviedre, V. S. Fundamenskii and G. P. Kolesnikova, *Koord. Khim.*, 1984, **10**, 408.
136. N. V. Krivtsov, K. V. Titova and V. Ya. Rosolovskii, *Zh. Neorg. Khim.*, 1977, **22**, 679.
137. H. Binder, *Z. Anorg. Allg. Chem.*, 1971, **383**, 279.
138. H. Nöth and H. Vahrenkamp, *Chem. Ber.*, 1966, **99**, 1049.
139. S. Chao and C. E. Moore, *Anal. Chim. Acta*, 1978, **100**, 457.
140. J. Bielawski, K. Niedenzu, A. Weber and W. Weber, *Z. Naturforsch., Teil B*, 1981, **36**, 470.
141. W. Dilthey, *Liebigs Ann. Chem.*, 1906, **344**, 300.
142. S. G. Shore and R. W. Parry, *J. Am. Chem. Soc.*, 1955, **77**, 6084; 1958, 80, 8, 12, 15.
143. O. P. Shitov, S. L. Ioffe, V. A. Tartakovskii and S. S. Novikov, *Usp. Khim.*, 1970, **39**, 1913; *Russ. Chem. Rev. (Engl. Transl.)*, 1970, **39**, 906.
144. H. Nöth, R. Staudigl and H.-U. Wagner, *Inorg. Chem.*, 1982, **21**, 706.
145. H. Nöth, S. Weber, B. Rasthofer, C. Narula and A. Konstantinov, *Pure Appl. Chem.*, 1983, **55**, 1453.
146. J. Higashi, A. D. Eastman and R. W. Parry, *Inorg. Chem.*, 1982, **21**, 716.
147. E. L. Muetterties, *Pure Appl. Chem.*, 1965, **10**, 53.
148. N. E. Miller and E. L. Muetterties, *J. Am. Chem. Soc.*, 1964, **86**, 1033.
149. G. E. Ryschkewitsch and J. M. Garrett, *J. Am. Chem. Soc.*, 1968, **90**, 7234.
150. G. Allegra, E. Benedetti, C. Pedone and S. L. Holt, *Inorg. Chem.*, 1971, **10**, 667.
151. H. Nöth and G. Schmid, *Z. Anorg. Allg. Chem.*, 1966, **345**, 69.
152. G. Schmid, *Angew. Chem., Int. Ed. Engl.*, 1970, **9**, 819.
153. C. S. Cundy and H. Nöth, *J. Organomet. Chem.*, 1971, **30**, 135.
154. R. N. Scott, D. F. Shriver and D. D. Lehman, *Inorg. Chim. Acta*, 1970, **4**, 73.
155. D. F. Shriver, *Acc. Chem. Res.*, 1970, **3**, 231.
156. P. Powell and H. Nöth, *Chem. Commun.*, 1966, 637.
157. G. W. Parshall, *J. Am. Chem. Soc.*, 1964, **86**, 361.
158. G. Medford and S. G. Shore, *J. Am. Chem. Soc.*, 1978, **100**, 3953.
159. J. S. Plotkin and S. G. Shore, *J. Organomet. Chem.*, 1979, **182**, C15.
160. G. Schmid, H. Nöth and J. Deberitz, *Angew. Chem.*, 1968, **80**, 282.
161. R. N. Grimes, in 'Comprehensive Organometallic Chemistry', ed. G. Wilkinson, F. G. A. Stone and E. W. Abel, Pergamon, Oxford, 1982, vol. 1, p. 459.
162. G. E. Herberich, in 'Comprehensive Organometallic Chemistry', ed. G. Wilkinson, F. G. A. Stone and E. W. Abel, Pergamon, Oxford, 1982, vol. 1, p.381.
163. R. Thompson, *Chem. Br.*, 1971, **7**, 140.
164. R. Thompson, *Pure Appl. Chem.*, 1974, **39**, 547.
165. R. Thompson, in 'The Modern Inorganic Chemicals Industry', ed. R. Thompson, Chemical Society, London, 1980, p. 302.
166. R. Cockcroft and J. F. Levy, *J. Inst. Wood Sci.*, 1973, **6**, 28.
167. W. Büchner and H. Niederprüm, *Pure Appl. Chem.*, 1977, **49**, 733.
168. Anonymous, Thiokol/Ventron Division brochure, Sodium Borohydride, 1979.
169. W. Frick, *Galvanotechnik*, 1976, **67**, 730.
170. L. P. Hanson, 'Organoboron Technology', Noyes Data Corporation, Park Ridge, New Jersey, 1973, p. 63.
171. I. B. Atkinson and B. R. Currell, *Inorg. Macromol. Rev.*, 1971, **1**, 203.
172. A. H. Soloway, in 'Progress in Boron Chemistry', ed. H. Steinberg and A. L. McCloskey, Pergamon, Oxford, 1964, vol. 1, p. 203.

25.1

Aluminum and Gallium

MICHAEL J. TAYLOR
University of Auckland, New Zealand

25.1.1 ALUMINUM: INTRODUCTION

Coordination chemistry of aluminum prior to 1970 was included in Wade and Banister's account of the element.[1] There are thorough reviews of aluminum chemistry with particular reference to oxygen compounds, halides and their adducts, which cover the period prior to 1972.[2] Very early studies of aluminum coordination compounds can be traced through the formula index of Gmelin's Handbook.[3] A 1963 review treats the adducts formed by $AlCl_3$, $AlBr_3$ and AlI_3 and discusses their chemistry in the context of Friedel–Crafts reactions,[4] and the operation of organoaluminum compounds, including halides R_2AlX, as Lewis acids has also been examined.[5,6] Stability constants have been tabulated for a great many Al^{3+} complexes.[7,8]

More specialized studies have examined quantitative aspects of Lewis acidity of the AlX_3 halides and R_3Al molecules using thermochemical measurements[9] and photoelectron spectroscopy.[10] X-Ray emission spectra have been employed to assess the occurrence of σ and π bonding in some important aluminum compounds including alumina, aluminosilicates, cryolite (Na_3AlF_6) and tris(acetylacetonato)aluminum.[11] Specific reference to these and other fundamental studies is made within the following sections which examine aluminum coordination chemistry ligand by ligand.

In this chapter, XRD indicates X-ray diffraction, and ED indicates electron diffraction.

25.1.2 GROUP IV LIGANDS

25.1.2.1 Cyanides

There is an early record of the preparation of $Al(CN)_3{\cdot}OEt_2$ and of $LiAl(CN)_4$ by the action of HCN on $LiAlH_4$ in ether at low temperature.[12] They are unstable at room temperature and highly sensitive to O_2 and H_2O. Dicyanoaluminum hydride $AlH(CN)_2$, some polymeric cyanides $(R_2AlCN)_x$, and the anion $(Me_3Al(\mu\text{-}CN)AlMe_3)^-$ have been obtained,[1,13] but there is little recent work.

25.1.2.2 Silicon and Heavier Group IV Ligands

The question of the existence of unsolvated silylaluminum compounds $Al(SiR_3)_3$ was considered some years ago.[14] Subsequently, tetrakis(trimethylsilyl)aluminate was obtained as alkali metal salts, $MAl(SiMe_3)_4$ (M = Li, Na or K),[15] by the action of Me_3SiCl in ether or THF on a 5:1 mixture of the alkali metal and aluminum. This complex, isoelectronic with $Si(SiMe_3)_4$, has been characterized by NMR and IR/Raman spectra. Modification of procedure to use THF/benzene as solvent yields the neutral molecule, as the adduct $Al(SiMe_3)_3{\cdot}THF$.[16] Crystal structures of $Al(SiMe_3){\cdot}OEt_2$ and $NaAl(SiMe_3)_4$ show Al to be tetrahedrally coordinated with Al—Si distances in the range 244–249 pm.[17] Attempts to isolate base-free $Al(SiMe_3)_3$ have failed. Several catalytic applications of silylaluminum compounds have been investigated.[18]

Germanium analogues of organoaluminum compounds have been postulated and several compounds of this type have been prepared, *e.g.* $MeAl(GeMe_3)_2$, $Al(GeMe_3)_3{\cdot}THF$ and $Li(H_3AlGePh_3)$. Some analogous group III–tin derivatives, $Li(Me_3MSnMe_3)$ (M = Al, Ga, In or Tl),[19] are mentioned in Section 25.1.4.1.

25.1.3 NITROGEN LIGANDS

25.1.3.1 Ammonia and Amine Ligands

25.1.3.1.1 Ammonia

Early literature records the preparation of solids $AlX_3(NH_3)_n$ with values of n as high as 9 (X = Cl), 14 (X = Br) or 20 (X = I).[4] $AlCl_3{\cdot}NH_3$, obtained from aluminum trichloride using the stoichiometric quantity of ammonia, forms when NH_4AlCl_4 is vaporized.[20] ED measurement reveals a staggered C_{3v} conformation for the H_3NAlCl_3 molecule with the dimensions: Al—N 199.6 pm, Al—Cl 210.0 pm, N—Al—Cl angle 101.2°. The N—Al—Br angle (101.5°) and Al—N distance (199.7 pm) of H_3NAlBr_3 are very similar to those of the chloride. The Al—N distance is shorter than this in $Me_3N{\cdot}AlCl_3$ (194.5 pm), but longer in $Me_3N{\cdot}AlH_3$ (206.3 pm) or $Me_3N{\cdot}AlMe_3$ (209.9 pm); differences here, and in the corresponding gallium compounds, are indicative of trends in the strength of the coordinate nitrogen to metal bond.[21,22] The structure of solid ammonia–aluminum trichloride, H_3NAlCl_3, is the same as that in the gas phase, with Al—N 192.1 pm and Al—Cl (average) 211 pm.[23]

The H_3NAlX_3 molecules have been studied in the gas phase by IR/Raman spectroscopy.[24] NCA yields Al—N force constants of 1.50 N cm^{-1} (X = Cl) and 1.45 N cm^{-1} (X = Br). The analyses of spectra were supported by *ab initio* MO calculations which were also extended to H_3NAlF_3, a molecule which has eluded synthesis. The ammonia adduct of alane (AlH_3) is unknown. A hexaammine $Al(BH_4)_3{\cdot}6NH_3$ is the product of excess ammonia on $KAl(BH_4)_4$.[25] It is believed to contain the cation $Al(NH_3)_6^{3+}$ and it is possible that this species is also present in some of the complexes $AlX_3(NH_3)_n$, noted above, but structural investigation is required.

25.1.3.1.2 Amines

Many complexes of amines with aluminum trihalides, alane and halogenoalanes, AlH_2X and $AlHX_2$, have been described.[26,27] Most are of the form $AlX_3{\cdot}$amine although the 1:2 type is also known, particularly when the ligand is trimethylamine. Direct reaction with the aluminum compound in a suitable solvent under anhydrous conditions is the usual way to prepare amine adducts of the halides AlX_3 (X = Cl, Br or I). Amine adducts of organoaluminum compounds are obtained similarly. Primary, secondary or tertiary amines may be used, but primary and secondary amines will often react further, eliminating HX, or an alkane from $R'_3Al{\cdot}NHR_2$, giving amino $(R'_2AlNR)_x$ or imino $(\text{-}R'AlNR\text{-})_x$ type products (see Section 25.1.3.2).

Amine adducts of alane can be prepared from AlH_3 in ether, the amine displacing the weaker base, Et_2O (equation 1). Alternative methods use $LiAlH_4$, or proceed from $AlCl_3{\cdot}NR_3$ retaining the existing Al—N bond (equation 2). $AlH_3{\cdot}NMe_3$, m.p. 75 °C, subliming readily, slowly decomposes to polymeric $(AlH_3)_x$ (see equation 3). $AlH_3(NMe_3)_2$ is stable in contact with excess trimethylamine. The reaction of solid $AlH_3{\cdot}NMe_3$ with boron trifluoride does not displace alane, but instead proceeds by a halide–hydride exchange mechanism, forming $BH_3 \cdot NMe_3$ and AlF_3.[28]

$$AlH_3 \cdot OEt_2 + R_3N \longrightarrow AlH_3 \cdot NR_3 + Et_2O \quad (1)$$

$$LiAlH_4 + Me_3NHCl \xrightarrow[-60\,°C]{Et_2O} AlH_3 \cdot NMe_3 + LiCl + H_2 \quad (2)$$

$$AlCl_3 \cdot NMe_3 + 3LiH \longrightarrow AlH_3 \cdot NMe_3 + 3LiCl \quad (3)$$

Reactions of bifunctional amines with aluminum compounds have received some attention. Aluminum isopropoxide with ethylenediamine in benzene solution forms a mixture of species, including polymers.[29] Among these, NMR reveals $(Pr^iO)_3AlNC_2H_4NAl(OPr^i)_3$. The triethylenediamine adduct of alane, $AlH_3 \cdot N(C_2H_4)_3N$, can be prepared by the action of the base on aluminum metal under a pressure of H_2. It adopts a polymeric structure in the crystal involving five-coordinate aluminum, with planar AlH_3 units linked by diamine molecules using both donor sites.[30] The Al—N distance of 219.5 pm, similar to that in $AlH_3 \cdot 2NMe_3$,[31] is distinctly greater than in four-coordinate adducts. Tetramethylethylenediamine (TMEDA) forms $AlHBr_2 \cdot$TMEDA which exhibits ν(Al—H) at 1735 cm^{-1}, but in reaction with $AlH_2I \cdot NMe_3$ it gives $(AlH_2TMEDA)I$, [ν(Al—H) near 1890 cm^{-1}], believed to contain a four-coordinate complex cation.[32] A structure with the unusual feature of a single N donor atom coordinated to two Al centres has been proposed on NMR evidence for the aluminoxane adduct $(Et_2Al)_2O \cdot NMe_2Ph$.[33] The TMEDA adduct of $(Me_2Al)_2O$ is conventional, with the donor atoms of $Me_2NC_2H_4NMe_2$ each attached to an Al atom.[34]

25.1.3.2 N-Heterocyclic Ligands

The formation of molecular complexes between aluminum trihalides and pyridine or alkylpyridines has been the subject of systematic studies.[35,36] Calorimetric data yield bond dissociation energies $D(X_3Al—py)$ of 323, 308 and 264 kJ mol^{-1} for X = Cl, Br and I, respectively, and this same order is found for alkylpyridine adducts, although the Al—N bond is weakened in the case of lutidine by the effects of steric hindrance. For gallium halides the values of $D(X_3M—py)$ are smaller: 248, 237 and 195 kJ mol^{-1} for chloride, bromide and iodide adducts, respectively.

Phase studies reveal the existence of AlX_3py_n (n = 1, 2 or 3 when X = Cl or Br; n = 1 or 3 when X = I). For the $AlCl_3$ system, the 1:1 complex is a simple molecular adduct, while the 1:2 and 1:3 complexes have been shown crystallographically to be *trans*-$AlCl_2py_4^+AlCl_4^-$ and *mer*-$AlCl_3py_3$, respectively.[37] In the cationic species, the Al—N bond length is 207.4 pm (average) and Al—Cl is 227.9 pm (average). In $AlCl_3py_3$, Al—N *trans* to Cl, 209.6 pm, is significantly longer than the other two Al—N bonds of 207.2 pm (average). Vibrational spectra of AlX_3py[38] and AlX_3py_2[39] (X = Cl or Br) had earlier been used to distinguish between the molecular nature of the one and the ionic nature of the other. Al—N force constants in the range 0.7–1.9 N cm^{-1}, with those of $AlBr_3$ adducts greater than those of $AlCl_3$ adducts, were obtained for the 1:1 complexes of several pyridines with AlX_3. Whereas the pyridine complexes of $AlCl_3$ or $AlBr_3$ are usually obtained by direct reaction between the two molecules in a suitable solvent, AlI_3py is conveniently prepared from Al + I_2 + pyridine in ether.[40] There is evidence from phase studies of the $M(OPr^i)_3$/pyridine systems that, whereas gallium forms the expected 1:1 complex, the aluminum compound is of the form $2Al(OPr^i)_3 \cdot py$.[41]

The 1,10 phenanthroline and 2,2′-bipyridyl ligands form chelate complexes with $Al(ClO_4)_3$[42] and Me_3Al,[43] as well as with the halides $AlCl_3$, $AlBr_3$ and AlI_3.[44] An apparently zerovalent complex $Al(bipy)_3$ is obtained by the action of the lithium salt of the bipyridyl anion on aluminum in THF medium.[45] More recently, N-heterocyclic bases have been shown to react with $AlCl_3$,[46] AlH_3[47] or Me_3Al[46,48] and alkali metals in THF to yield neutral radicals and radical cations, derived from *e.g.* $AlCl_2^+$. Their existence is confirmed by ESR spectroscopy. The 2,2′,2″-terpyridyl adduct of $AlCl_3$ is a 1:1 molecular complex. It is isomorphous with the compounds formed by this ligand with $GaCl_3$, $InCl_3$ and $TlCl_3$.[49] A 1:2 complex of $AlCl_3$ with morpholine consists of discrete molecules in which the Al atom is five-coordinate.[50] Morpholine is attached through nitrogen: the Al—N bond lengths are 206.4 and 209.3 pm, and Al—Cl averages 218.4 pm. ^{27}Al NMR spectroscopy has been used to investigate the species formed by Al^{3+} in aqueous solutions containing the base imidazolidin-2-one. In this system, complexes with 1:1, 1:2 and 1:3 stoichiometry are observed[51] but have not been isolated.

25.1.3.3 Amido and Imido Ligands

Amidoaluminates $MAl(NH_2)_4$ and $M_2Al(NH_2)_5$ are formed as crystalline solids when aluminum is attacked by alkali metals in liquid ammonia. A better synthetic route employs the reactions of NH_3 with $LiAlH_4$ in THF.[52] Structures based on octahedral $Al(NH_2)_6$ units with bridging NH_2 groups were originally proposed on the basis of IR spectra; however, single crystal X-ray analysis reveals approximately tetrahedral $Al(NH_2)_4^-$ ions in $MAl(NH_2)_4$ (M = Na, K or Cs) and $Ca[Al(NH_2)_4]_2 \cdot NH_3$.[53] The Al—N bond distances of the AlN_4 unit average 185 pm. Precipitates from the reactions of $AlH_3 \cdot OEt_2$ with ammonia at various molar ratios, and the materials to which they give rise through partial loss of H_2 on warming to ambient temperature, were investigated some years ago by Wiberg.[1] The final product obtained by heating to 100–150 °C is aluminum nitride. These systems probably have polymeric structures akin to the alkylamino- and alkylimino-alanes considered below.

Dialkylamido complexes[54] are generally more tractable than those containing NH_2 groups. Lithium tetrakis(dimethylamido)aluminate is obtained by the reaction shown in equation (4).[55]

$$LiAlH_4 + 4Me_2NH \xrightarrow{Et_2O} LiAl(NMe_2)_4 + 4H_2 \qquad (4)$$

$Al(NMe_2)_3$ is dimeric with bridging ligands (see below). However, $Al(NPr^i_2)_3$ contains a single type of ligand, as indicated by the NMR spectrum over a wide temperature range, and appears to be a monomer.[54] Aluminum is three-coordinate in $Al[N(SiMe_3)_2]_3$ where the AlN_3 unit is planar, with short Al—N bonds of 178 pm.[56] The structure of $Cl_2\overline{AlN(Et)C_2H_4N}Me_2$ is monomeric, with four-coordinate aluminum.[57] Whereas the coordinate Me_2N—Al distance is 196 pm (a typical single bond value), the Al—NEt bond length is only 177 pm, showing π interaction between planar nitrogen and the Al atom. The complex $LiAl(N{=}CBu^t_2)_4$ is prepared from $AlCl_3 + 4Bu^t_2C{=}NLi$.[58] Two ligands display near-linear C=N—Al geometry with relatively short Al—N bonds, indicating π bonding from N to Al. The remaining ligands form bridges to the Li^+ ion.

Remaining compounds considered here are alkylaminoalanes, R'_2AlNR_2, $R'Al(NR_2)_2$ and $Al(NR_2)_3$, where R′ can be H, alkyl or a halide ligand and R is usually an alkyl group. The structures are usually di- or tri-meric. Also treated are iminoalanes $(R'AlNR)_n$ (n = 2–35, commonly 4, 6 or 8) which are polymers with structures based on $(AlN)_x$ rings and clusters. Members of both series, where R′ = H, have been reviewed in the context of aluminum hydride derivatives.[26]

N-Alkylaminoalanes are prepared by the reaction of $AlH_3 \cdot NMe_3$ with a secondary amine (equation 5). Reaction of $LiAlH_4$ in ether with the appropriate amine can also be used. The ease of thermal decomposition of the amine adducts $Et_2(X)Al \cdot NHMe_2$ to form $Et(X)AlNMe_2$ by elimination of ethane depends on the group X in the order Et > Cl > Br > I.[59] Direct synthesis of $Al(NMe_2)_3$ from Al + $3Me_2NH$ is possible, and the compounds $HAl(NEt_2)_2$ and H_2AlNEt_2 have been prepared from Al + Et_2NH + H_2 under pressure.[60] Interaction of $Al(NMe_2)_3$ with $AlH_3 \cdot NMe_3$ gives H_2AlNMe_2 quantitatively. Another route to dialkylaminoalanes is from halogenoalanes by reaction with $LiNR_2$, eliminating LiCl. By carrying out the reaction in stages, mixed ligand products such as $HAl(NR_2)NR'_2$ can be obtained.[26] Of the type $R'Al(NR_2)_2$ is $ClAl(N(Me)SiMe_2)_2NMe$, the first cycloalumadisilatriazane, fully characterized by IR/Raman, NMR, MS and XRD methods.[61] The structure is dimeric so that a central Al_2N_2 ring is present and aluminum attains tetracoordination. The mean Al—N distance (194.7 pm) of the central ring is much longer than the Al—N distance (180.3 pm) to nitrogen of the Al—NMe—Si system, no doubt because this latter N atom is three-coordinate and π bonding is involved.

$$AlH_3 \cdot NMe_3 + 3R_2NH \longrightarrow Al(NR_2)_3 + Me_3N + 3H_2 \qquad (5)$$

The NMR spectrum, with two methyl group signals which remain separate up to 150 °C, provided early evidence for the bridged structure of $[Al(NMe_2)_3]_2$.[54] Crystal structure analyses confirm that $[Al(NMe_2)_3]_2$ and $[HAl(NMe_2)_2]_2$ are dimers with four-membered Al_2N_2 rings, whereas $(H_2AlNMe_2)_3$ is a trimer and exists as a six-membered ring of chair conformation.[62] There is a clear difference between exocyclic Al—N bond distances (180.1 to 182.6 pm) and ring Al—N distances (193.7 to 198.0 pm). For the $(H_2AlNMe_2)_3$ trimer there is evidence, from ED in the vapour phase, to suggest transannular interactions between the Al atoms.[63] The molecules $(X_2AlNMe_2)_2$ (X = Cl, Br or I) are dimers of D_{2h} symmetry bridged by the NMe_2

groups, with Al—N bond distances of 190–196 pm.[64] The above compounds react as monomers with pyridine forming 1:1 adducts, *e.g.* $Al(NMe_2)_3 \cdot py$.

Structural data on other alkylaminoalanes are available.[65–67] Of interest is the existence of geometrical isomers where the amido group carries two unlike substituents. For $(Me_2AlNMePh)_2$ variable temperature NMR results yield a value for ΔH of isomerization of 4.5 kJ mol^{-1}.[68] The *cis* isomer with the two phenyl groups on the same side of the Al_2N_2 ring is energetically favoured. $(Et(Br)AlNHBu^t)_2$ exists in solution as a mixture of three isomers, owing to the asymmetry of Al as well as N centres.[69] A single isomer crystallizes, in which both Bu^t and Br pairs are above the plane of the ring.

N-Alkyliminoalanes $(R'AlNR)_n$ are commonly obtained by thermal decomposition of the primary amine adduct of a suitable aluminum compound which proceeds with the elimination of both hydrogens (equation 6). Preparation of some well-defined poly(*N*-alkyliminoalane) oligomers is possible by the direct interaction of Al and RNH_2, with small amounts of $MAlH_4$ as an activator.[70] Alternatively, the reaction of $MAlH_4$ (M = Li or Na) with Bu^tNH_2 has been used. Table 1 gives structural data on various iminoalanes, the structures of which include rings and clusters.[71–79] These clusters have considerable stability; for example, $(HAlNPr^i)_n$ (n = 4 or 6) can be chlorinated by HCl or $HgCl_2$ to replace H by Cl substituents.[80] Although tetracoordination of Al is usual, an instance of higher coordination occurs in the cluster $(HAlNPr^i)_2(H_2AlNHPr^i)_2(HAlNCHMeCH_2NMe_2)$.[77] The molecule is built up from four Al_3N_3 and two Al_2N_2 rings. The side-chain NMe_2 group is linked to an Al atom which is thus five-coordinate.

$$nR'_3Al \cdot NH_2R \xrightarrow{\text{heat}} (R'AlNR)_n + 2nR'H \qquad (6)$$

Table 1 Cluster and Cage Structures with an Al–N (or Ga–N) Framework

Compound	*Structure*	*Al–N bond length* (pm)	*Ref.*
$(HAlNBu^t)_4$	Al_4N_4 cube	—	71
$(PhAlNPh)_4$	Al_4N_4 cube	193	1
$(ClAlNMe_2)_4(NMe_2)$	Al_4N_6 tetrahedron	179, 190–194	72
$(HAlNPr^i)_2(H_2AlNHPr^i)_3$	Al_5N_5 bridged ring	190–198	73
$(HAlNPr^i)_6$	Al_6N_6 prism	190, 196	74, 75
$\{HAlNCH(Me)Ph\}_6$	Al_6N_6 prism	189, 198	76
$\{HAlN(CH_2)_3NMe_2\}_6$	Al_6N_6 cage	185–200	77
$(MeAlNMe)_7$	Al_7N_7 cage	196 (average)	78
$(MeAlNMe)_6(Me_2AlNHMe)_2$	Al_8N_8 bridged prism	191 (average)	79
$(MeGaNMe)_6(Me_2GaNHMe)_2$	Ga_8N_8 bridged prism	197 (average)	79
$(MeAlNMe)_8$	Al_8N_8 cage	—	79

Although less familiar than the poly(*N*-alkylamino)alanes $(R'AlNR)_n$, some lower molecular weight iminoalane derivatives have been prepared. Examples are the dimer $[(H_2Al)_2NC_3H_6NMe_2]_2 \cdot THF$[81] and $(Cl_2AlNBu^n)_3Al \cdot NMe_3$.[82] The latter is believed to exist as a symmetrical cluster, although the nature of bonding to the central Al atom is not definitely established.

25.1.3.4 Azide, Cyanate and Thiocyanate Ligands

Aluminum azide is obtained from reacting $AlCl_3$ with NaN_3 in THF, or from the reaction of $AlH_3 \cdot Et_2O$ with HN_3 in ether at low temperature.[83] When trimethylamine adducts of alane are employed, solids thought to be azido complexes are produced (equation 7).[84] Aluminium (and gallium) compounds $X_2M(N_3)$, where X = Br or I, have been prepared by the reaction of MX_3 with the halogen azide XN_3 in benzene.[85] They are polymeric solids which show $\nu(M—N_3)$ modes near 490 cm^{-1} (M = Al), or 430 cm^{-1} (Ga) in the vibrational spectra.

$$MH_3(NMe_3)_n \xrightarrow[\text{cold THF}]{HN_3} (Me_3NH)_nM(N_3)_{3+n} \text{ (M = Al or Ga; } n = 1 \text{ or } 2) \qquad (7)$$

Organoaluminum azides $R_2Al(N_3)$ and $RAl(N_3)_2$ are known; the former type adopts a cyclic structure, $[Me_2Al(N_3)]_n$ where n = 2 or 3, based on four-coordinate aluminum with bridging azido ligands which exchange on the NMR timescale.[86] Related anions have been characterized, for example in the compounds $Me_4N[Me_2M(N_3)_2]$ (M = Al or Ga) and $Cs[Me_3Al(N_3)]$.[87] The crystal structure of the latter reveals Al—N 197 pm, a typical N donor to aluminum bond length. This may be compared with the bridging N_3 group in $K[(Me_3Al)_2N_3]$ where the Al—N distance is 201.5 pm, and ⟨Al—N—Al is 128 °.[88]

Little is known of aluminum cyanates. There are several routes to thiocyanate complexes of aluminum. Aluminum thiocyanate can be prepared as the ether adduct $Al(NCS)_3{\cdot}Et_2O$. Reaction of thiocyanogen with $LiAlH_4$ in ether yields $LiAl(NCS)_4{\cdot}2Et_2O$, reported to decompose at −50 °C.[89] $K_3Al(NCS)_6$ is prepared from $AlCl_3$ + 6KNCS, using acetonitrile as the solvent.[90] Complexes of Al^{3+} coordinate to thiocyanate through the nitrogen atom. Alkylaluminum complexes, however, display linkage isomerism. In the salt $Me_4N[Me_2Al(NCS)_2]$ the ligand is S-bound, whereas the crystal structure of $Me_2Tl(Me_2Al(NCS)_2)$ shows NCS coordinated through the N atom with r(Al—N) = 208 pm and also reveals a significant thallium–sulfur interaction.[91] A linear bridging thiocyanate ion is found in $K[(Me_3Al)_2(NCS)]$.[92] The Al—N bond length is normal, 195.1 pm; but Al—S is long, 248.9 pm, indicating a weak interaction. The thiocyanate group is N-bonded to aluminum in $Cs(Me_3AlNCS)$, with the Al—N bond length 194.4 pm.[93]

$Me_4N[(Me_3Al)_2NCSe]$ has been obtained, but reports of selenocyanato complexes of Al are scarce.

25.1.3.5 Nitrile Ligands

The early literature has reports of complexes of aluminum chloride with nitriles $AlCl_3{\cdot}nRCN$ (R = Me, Et or Ph; n = 0.5, 1 or 2).[4] Aluminum bromide reportedly forms $AlBr_3{\cdot}2MeCN$ and $AlBr_3{\cdot}nPhCN$ (n = 1, 1.5 or 2). X-Ray study of $AlCl_3{\cdot}2MeCN$ shows it to be $AlCl(MeCN)_5^{2+}(AlCl_4^-)_2{\cdot}MeCN$.[94] Within the cation, four N donors in a plane perpendicular to the Al—Cl bond have an average Al—N distance of 198.6 pm. The unique MeCN *trans* to Cl forms a longer bond, with Al—N = 202.1 pm. The dimensions of a single acetonitrile molecule coordinated to aluminum are revealed in the X-ray structure determination of $Me_3Al{\cdot}MeCN$.[95] Al—N—C—C is almost exactly linear with Al—N = 202 pm and C≡N = 118 pm (compared with 115.7 pm in the free MeCN molecule). The addition compounds formed by ClCN and BrCN with aluminum halides have been studied by IR/Raman spectroscopy which establish the N-coordinated structure XCN—AlX_3 with C_{3v} symmetry.[96] NCA yields Al—N force constants of 2.0 and 2.8 N cm^{-1} for the chloride and bromide systems, respectively. Halogen exchange occurs when the preparation of $ClCN{\cdot}AlBr_3$ (or $BrCN{\cdot}AlCl_3$) is attempted.

Conductimetric measurements of $AlCl_3$ and $AlBr_3$ in MeCN as solvent show that ionic species are formed. IR/Raman and NMR methods have been used extensively to investigate these solutions.[97–101] For the $AlCl_3$–MeCN solutions, ^{27}Al NMR signals are detected for the octahedral species $Al(MeCN)_6^{3+}$, $AlCl(MeCN)_5^{2+}$, *cis*- and *trans*-$AlCl_2(MeCN)_4^+$, as well as the tetrahedral species $AlCl_3(MeCN)$ and $AlCl_4^-$.[100] $AlCl(MeCN)_5^{2+}$ is the major species in terms of concentration. For the $AlBr_3$/MeCN system, it has been deduced from conductivity measurements that $AlBr_2(MeCN)_4^+$ and $AlBr_4^-$ are the most abundant ions.[98]

25.1.4 PHOSPHORUS AND HEAVIER GROUP V LIGANDS

The aluminum halide–phosphine adducts, colorless crystals, $AlX_3{\cdot}PH_3$ (X = Cl, Br or I), date from 1932,[102] but not very many complexes of aluminum with other P, As or Sb Lewis bases have been investigated.[4,103] There are 1:1 complexes of $AlCl_3$ and $AlBr_3$ with Ph_3P, various alkylphosphines and their R_3As and R_3Sb counterparts. A 1:2 complex, $AlCl_3(PMe_3)_2$, has been assigned a five-coordinate D_{3h} structure on the basis of IR/Raman spectra.[104] Aluminum borohydride forms both 1:1 and 1:2 adducts with Me_3P, whereas with Me_3As only $Al(BH_4)_3{\cdot}AsMe_3$ is obtained.[105] The structure of $Me_3Al{\cdot}PMe_3$ has been determined in the vapour phase by ED.[106] The Al—P dative bond length of 253 pm is significantly greater than the Al—P distance of 235.7 pm in aluminum phosphide, in which Al is tetrahedrally coordinated by phosphorus.[107]

$AlH_3{\cdot}PH_3$ is unknown, and some attempts to prepare $AlH_3{\cdot}PMe_3$ were unsuccessful,[108] although the corresponding boron and gallium compounds exist. No reaction occurs in the system Al_2Cl_6/PCl_3; evidently the energy of formation of the Al—P bond is insufficient to offset that needed to release $AlCl_3$ from the dimer. Reaction does, however, occur with PF_3 and NMR data support the molecular structure $F_3P—AlCl_3$.[109] Subsequently, halogen exchange occurs to produce PCl_3 and AlF_3. Phosphorus(V) compounds PCl_5 and $RPCl_4$ react readily with aluminum chloride to give complex ionic structures, *e.g.* $(PCl_4)(AlCl_4)$, $(RPCl_3)(AlCl_4)$ and $(RPCl_3)(Al_2Cl_7)$.[4] Such compounds are liable to be produced from AlX_3 and a phosphorus(III) molecule if halogen, HX, RX or other reagent capable of oxidative addition is present, whether or not the AlX_3–phosphorus donor complex is stable. (Contrast amine adducts, which are cleaved by hydrogen halides HX, but not usually by halogens or RX.) Arsenic(III) halides do not react appreciably with aluminum halides, although if Cl_2 is admitted to the chloride system, crystals of $AlCl_3{\cdot}AsCl_5$ are formed.[110] The likely formulation is $AsCl_4^+AlCl_4^-$. With antimony and bismuth halides, there are reports of $AlBr_3{\cdot}SbBr_3$, $AlI_3{\cdot}SbI_3$, $AlBr_3{\cdot}BiBr_3$ and $AlI_3{\cdot}BiI_3$.[4] Structural investigation of these systems is desirable: it is probable that they are halogen-bridged species, although Al—group V bonds are a possibility.

Adducts of a series of silylphosphine donors have been synthesized.[111] The molecules $Me_2PSiH_nMe_{3-n}$ ($n = 1$ or 2) form adducts with $AlCl_3$ which decompose on heating with loss of the silyl group to give $(Cl_2AlPMe_2)_3$. A structure based on a six-membered Al_3P_3 ring has been proposed, as in the corresponding dimethylamido compounds. Dialkylphosphido compounds of the type $(R_2AlPR'_2)_{2\,or\,3}$ are known in organoaluminum chemistry.[6] Compounds $(R_2MN(PPh_2)_2)_2$, with R = Me or Et and M = Al or Ga, are obtained from $R_3M + (Ph_2P)NH$, with evolution of alkane. Single crystal X-ray diffraction reveals a structure containing AlNPAlNP rings with Al—P = 253.5 pm and Al—N = 189.4 pm (average values).[112] The diphenylphosphido ligand bridges between $Cr(CO)_5$ and the Al centre in $Cr(CO)_5(\mu\text{-}PPh_2)Al(CH_2SiMe_3)_2{\cdot}NMe_3$.[113] The structure, determined by XRD, shows an Al—P bond length of 248.5 pm. The aluminum atom is four-coordinate owing to the presence of the attached molecule of NMe_3 (Al—N bond length = 204.9 pm).

The phosphido complex $LiAl(PH_2)_4$ may be synthesized by reaction of phosphine with $LiAlH_4$.[114] A tetrakis(dimethylphosphido)aluminate $LiAl(PMe_2)_4$ is prepared by metallation of $HPMe_2$ with Bu^nLi in diglyme at −40 °C, followed by reaction with $AlCl_3$.[115] The silyl analogue $LiAl[P(SiH_3)_2]_4$ had been obtained earlier by the action of $P(SiH_3)_3$ on $LiAlH_4$.[116] Finally, mention may be made of a cationic Al—P coordinated species $Al\{P(OMe)_3\}_6^{3+}$ which exists in trimethyl phosphite–water solutions according to NMR evidence.[117]

25.1.5 OXYGEN LIGANDS

25.1.5.1 Oxide, Hydroxide and H_2O as Ligands

Aluminum hydroxide $Al(OH)_3$, frequently encountered as a gelatinous material, is obtained in crystalline form by passing CO_2 into an alkaline aluminate solution at 80 °C. It occurs as gibbsite and in other forms.[1] AlO(OH) is produced as boehmite by adding ammonia to a boiling solution of an aluminum salt, or from gibbsite + water in an autoclave at 300 °C.[118] Alumina, Al_2O_3, is obtained by the dehydration of AlO(OH) or $Al(OH)_3$ in at least six types, of which α-Al_2O_3 (corundum) is the form produced at high temperature and is indefinitely metastable at ordinary temperatures.[1,119] In α-Al_2O_3, the oxide ions form a hexagonal close-packed array in which Al^{3+} ions are distributed symmetrically among octahedral sites.[120] Crystals of gibbsite have a sheet structure based on a double layer of close-packed hydroxyl ions with Al^{3+} in two-thirds of the octahedral interstices.[121] A rarer form of $Al(OH)_3$, nordstrandite, is obtained by controlling the crystallization of the gelatinous hydroxide by adding a chelating agent, ethylene glycol. In crystalline AlO(OH), —Al—O—Al—O— chains are bridged by oxygens of OH groups to a neighbouring chain.

Aluminum forms mixed oxides with many other metals which either replace Al^{3+} in octahedral positions or occupy tetrahedral sites. The spinel structure of $MgAl_2O_4$, an important prototype, has been investigated by photoelectron spectroscopy.[122] Vibrational and solid state NMR methods are also useful for studying aluminate structures.[123–125]

With the alkali metals the simplest mixed oxide type is $NaAlO_2$, prepared by heating Al_2O_3 with the group I oxalate at 1000 °C. The structure of Na_5AlO_4 is orthorhombic: it contains discrete AlO_4 tetrahedra in which r(Al—O) = 176.1–178.9 pm. $Na_7Al_3O_8$, which is triclinic, contains a ring structure derived from six AlO_4 groups linked by oxygen bridges into an infinite chain, whereas $Na_{17}Al_5O_{16}$ has discrete chains of five AlO_4 tetrahedra sharing corners.[126] ^{27}Al NMR spectroscopy applied to polycrystalline aluminates reveals characteristic chemical shifts for AlO_4, AlO_5 and AlO_6 polyhedra,[125] and the technique also enables the formation of Si—O—Al bonds by AlO_4 groups in aluminosilicate solutions to be monitored.[127] The anion in $Cs_2Al_2O(OH)_6$ is composed to two AlO_4 tetrahedra sharing one O atom.[128] The complex anions $[Al_4(OH)_{16}]^{4-}$ and $[AlO(OH)_2^-]_\infty$ (made up of tetrahedral groups sharing corners) have been identified in XRD studies of barium aluminates.[129] The basic sulfate $Al_2(OH)_4(SO_4)\cdot 7H_2O$ (aluminite) contains the cationic complex $[Al_4(OH)_8(OH_2)_6]^{4+}$ built from four edge-sharing AlO_6 octahedra polymerized in chains and interconnected by SO_4^{2-} ions.[130] The calcium aluminates, which are basic components of Portland cement, have been identified by X-ray crystallography.[131] In another area of applied aluminum chemistry, NMR spectra under magic angle spinning conditions have been used to follow hydration and dehydration of Al sites in ZSM-5 zeolites.[132]

The aquated aluminum cation $[Al(OH_2)_6]^{3+}$ is octahedral. Some crystals in which it is encountered are (average Al—O distances in parentheses): $NaAl(SO_4)_2\cdot H_2O$ (187.8 pm); mellite (187.2 pm); $AlCuCl(SO_4)_2\cdot 14H_2O$ (187.8 pm).[133] These crystals have also been investigated by neutron diffraction and ^1H NMR techniques, which confirm the presence of $[Al(OH_2)_6]^{3+}$ and enable H atoms of the coordinated water molecules to be located.[134] The vibrational modes of $[Al(OH_2)_6]^{3+}$ in alum crystals have been assigned:[135] from the Raman data, $\nu_1 = 542$, $\nu_2 = 473$, $\nu_5 = 347\ cm^{-1}$. Assignments from single crystal Raman measurements of $[Al(OH_2)_6]Cl_3$ place $\nu(Al—O)_{sym}$ at $524\ cm^{-1}$.[136] The aluminyl cation AlO^+ isolated from HF solutions as BF_4^- or SbF_6^- salts shows $\nu(Al—O)$ near $650\ cm^{-1}$.[137]

The Raman bands of very concentrated $Al(NO_3)_3$ solutions are consistent with $[Al(OH_2)_6]^{3+}$ and NO_3^- in the form of solvent-separated ion pairs.[138] Hydration-shell and bulk-water protons can be distinguished by ^1H NMR. ^{27}Al NMR signals distinguish $[Al(OH_2)_6]^{3+}$ from other species, such as $[Al(OH_2)_5(SO_4)]^+$ in sulfate solutions, or $[Al_2(OH)_2(OH_2)_8]^{4+}$ and the more highly polymerized complexes which are produced on the ageing of aqueous aluminate solutions.[139–143] According to Raman[144] and NMR data,[141,145] the cation $[Al_{13}O_4(OH)_{25}(OH_2)_{11}]^{6+}$, which forms a sparingly soluble sulfate and a chloride, persists in aqueous solution. With HCl it decomposes into smaller species, but on adding alkali even higher molecular weight particles can be formed. In strongly alkaline solutions of aluminum compounds, however, the predominant species appear to be mononuclear $[Al(OH)_4]^-$ and dinuclear $[Al_2O(OH)_6]^{2-}$.[146–149] The UV spectra of aqueous solutions of $AlCl_3$, $Al(ClO_4)_3$, $Al(NO_3)_3$ and $Al_2(SO_4)_3$ in which the hexaaqua cation predominates, show an absorption band at 240 nm which is thought to arise from charge transfer involving the Al^{III} dication: $[Al^{III}(OH^-)(OH_2)_5]^{2+} \rightarrow [Al^{II}(OH)(OH_2)_5]^{2+}$.[150]

On the theoretical front, attempts have been made using molecular orbital calculations to elucidate the binding energies and stereochemical constraints associated with the structures of hydrates $[Al(OH_2)_n]^{3+}$ ($n = 1$–7),[151] and hydroxo complexes $[Al(OH)_4]^-$, $(Al(OH)_5)^{2-}$ or $(Al_2(OH)_8)^{2-}$.[152] MO calculations have also been performed for the molecules H_3AlOH_2 and $(H_2AlOH)_2$.[153] No evidence was found to support π bonding between O and Al. Small molecules of this kind can sometimes be detected by matrix isolation methods. For example, Al atoms react with H_2O under such conditions to give HAlOH, whereas the heavier group III metals form $M\cdot OH_2$ adducts.[154]

25.1.5.2 Alkoxide Ligands: Aluminoxanes and Related Systems

Aluminum alkoxides,[155] known since 1876, are prepared from the alcohol and aluminum, activated by I_2 or $HgCl_2$. Heating $Al + Pr^iOH + MeOC_2H_4OH$ provides a route to $Al(OPr^i)_3$ which avoids the need to introduce mercury. Reaction of $AlCl_3$ with alcohol, using NH_3 to take up the HCl formed, or with a group I metal alkoxide, can also be used to obtain $Al(OR)_3$. Mixing solutions of $Al(OPr^i)_3$ and $AlCl_3$ yields a crystalline material $AlCl(OPr^i)_2$. Alkoxides of higher alcohols can be made from $Al(OPr^i)_3$ by alcohol displacement. By control of the stoichiometry, mixed alkoxides $Al(OR)_n(OR')_{3-n}$ are obtained. The trimethylsiloxide $Al(OSiMe_3)_3$ is capable of controlled hydrolysis to yield polymeric products.[156]

Aluminum alkoxides are soluble in many organic solvents, and exist in various liquid and solid forms depending on the degree of polymerization. All are hydrolyzed readily, liberating the alcohol and giving white solids in which Al—O—Al bonds are a feature. $Al(OBu^t)_3$ and $Al(OPr^i)(OBu^t)_2$ are dimeric with terminal and bridging ligands which can be distinguished in NMR spectra.[157] Others are tri- or tetra-meric, *e.g.* $(Al(OCH_2Ph)_3)_4$, the structure of which contains an eight-membered Al_4O_4 ring.[1] The alkoxide formed by ethylene glycol appears to contain both four- and six-coordinate Al atoms.[158] The alkoxide oligomers, especially the dimers, are readily cleaved by donor molecules forming adducts, *e.g.* donor·$Al[OCH(CF_3)_2]_3$ (donor = OEt_2, NEt_3 or PEt_3).[159] A crystallographic study of en·$Al(OH)[OCH(CF_3)_2]_2$ shows it to consist of OH-bridged dimers with the AlO_4N_2 coordination sphere.[160] Three compounds of empirical formula acac·$Al(OR)_2$ have recently been studied by XRD.[161] When OR = $OSiPh_3$, the compound is a monomer with four-coordinate Al. With OR = $OSiMe_3$ or OPr^i, the products are a dimer and a trimer, respectively, and involve both four- and six-coordinate Al centres.

Aluminum methoxide $Al(OMe)_3$ is a solid which sublimes at 240 °C in vacuum. Aluminum isopropoxide melts in the range 120–140 °C to a viscous liquid which readily supercools. When first prepared, spectroscopic and X-ray evidence indicates a trimeric structure which slowly transforms to a tetramer in which the central Al is octahedrally coordinated and the three peripheral units are tetrahedral.[162,163] Intramolecular exchange of terminal and bridging groups, which is rapid in the trimeric form, becomes very slow in the tetramer. There is MS and other evidence that the tetramer maintains its identity in the vapour phase.[164] $Al[OCH(CF_3)_2]_3$ is more volatile than $Al[OCH(Me)_2]_3$ and the vapour consists of monomers.[165] Aluminum alkoxides, particularly $Al(OPr^i)_3$, have useful catalytic applications in the synthetic chemistry of aldehydes, ketones and acetals, *e.g.* in the Tishchenko reaction of aldehydes, in Meerwein–Pondorf–Verley reduction and in Oppenauer oxidation. The mechanism is believed to involve hydride transfer between R_2HCO ligands and coordinated $R'_2C{=}O \rightarrow Al$ groups on the same Al atom.[1]

Double alkoxides are formed by $Al(OR)_3$ and alkoxides of main group,[166,167] transition[168] or lanthanide elements.[169] The titration of $Al(OEt)_3$ with NaOEt gives a sharp end point to thymolphthalein forming $NaAl(OEt)_4$. This product is salt-like and may contain the $[Al(OEt)_4]^-$ anion, whereas other double alkoxides are low melting, volatile compounds and commonly have structures in which alkoxide ligands bridge Al to the other metal. Examples are $LiAl(OBu^t)_4$,[170] $Mg\{Al(OPr^i)_4\}_2$ (m.p. 20 °C) and $Mg[Al(OEt)_4]_2$ (m.p. 129 °C)[171] (thought to be dimeric with four-coordinate Mg and Al atoms on NMR evidence).

Alkoxyhydrides, *e.g.* $LiAlH_2(OBu^t)_2$,[170] $MgAlH_2(OPr^i)_3$[172] and $CaAlH_2(OPr^i)_3$[173] have been prepared and shown to have utility as selective reducing agents. Ligand exchange occurs to establish equilibrium between the species $AlH_2(OR)_2^-$, $AlH(OR)_3^-$ and AlH_4^-.

Alkoxyaluminum compounds $ROAlCl_2$ react with $MeAlCl_2$ to produce materials which contain the Al—O—Al linkage. Similarly, when organoaluminum compounds R_3Al (R = Me, Et or Bu^i) are hydrolyzed in ether solution the reaction proceeds in several steps leading to aluminoxane systems. An initial adduct $R_3Al{\cdot}OH_2$ is converted to R_2AlOH which then dimerizes. In systems with the $R_3Al:H_2O$ ratio of 2:1, reaction gives compounds $(R_2Al)_2O$ which themselves tend to form associated structures.[174] The aluminoxane anion $(Me_2AlOAlMe_3)_2^{2-}$ consists of a planar Al_2O_2 ring with Me_3Al molecules coordinatively linked to the O atoms.[175] The structure has a direct analogy in the chloroaluminate dimer $(Cl_2AlOAlCl_2)_2^{2-}$ characterized by XRD as a by-product of reductive Friedel–Crafts synthesis.[176] A second complex $Cl_2AlO(AlCl_3)_2^-$, obtained similarly, contains a quasi-five-coordinate Al atom since a chlorine of each of the $AlCl_3$ groups makes a close approach to this atom. Dimethylaluminum methoxide, $(Me_2AlOMe)_3$, the structure of which has an Al_3O_3 ring with the methoxide group acting as a bridging ligand, is obtained by reaction of Me_3Al with MeOH. The trimeric structure is well established by IR/Raman and gas phase ED methods.[177] The Al—O distance is 185.1 pm. The formally related compound $(Br_2AlOSiMe_3)_2$ is dimeric with Al—O bond lengths close to 180 pm.[178] Various aluminoxane type structures are now known[179] and there are parallels between polymeric alkoxides and aluminoxanes, and the related oligomers with $(AlN)_n$ frameworks which occur when amido and imido ligands are present.

25.1.5.3 β-Ketoenolate and Related Ligands

β-Ketoenolates of aluminum are of long standing, and tris(acetylacetonato)aluminum itself was first prepared in 1887.[1] It is obtained as colourless crystals, m.p. 194 °C, b.p. 315 °C, by the reaction of Hacac with Al amalgam; from $AlCl_3$ + Hacac; or by mixing solutions of $Al_2(SO_4)_3$ and Hacac in aqueous ammonia. Metal atom synthesis can also be employed.[180] Tris(ethylacetoacetonato)aluminum, m.p. 76 °C, is conveniently prepared by heating $Al(OEt)_3$ + 3 mol of ligand.

The structure of $Al(acac)_3$ contains an octahedral AlO_6 core, with Al—O distance of 189.2 pm. Differences of the monoclinic α form and orthohombic γ form (a racemate) lie chiefly in the orientations of molecules within the crystal: bond distances are almost identical.[181] Resolution of the optical isomers of $Al(acac)_3$ has twice been achieved by column chromatography at low temperature.[182,183] The hexafluoro compound, $Al[(OCCF_3)_2CH]_3$, m.p. 74 °C, is more volatile than $Al(acac)_3$. Its structure, determined by vapour phase ED,[184] reveals a slightly distorted AlO_6 unit in which the Al—O distance of 189.3 pm is the same as that in $Al(acac)_3$. Comparative studies have been made of the Raman spectra of $M(acac)_3$ (M = Al, Ga or In)[185] and of other β-diketonates of these metals.[186]

The tris(β-ketoenolato)aluminum compounds are stereochemically non-rigid species and undergo various ligand exchange reactions in solution. A 1976 report[187] summarizes various attempts to establish the stereoisomerization mechanisms and provides access to kinetic and thermodynamic data derived from this work. There is considerable evidence to support an intramolecular twist mechanism, except where fluorocarbon substituents are present. Here, a bond rupture process involving a five-coordinate intermediate is favoured. Later studies[183,188] using variable temperature NMR spectroscopy have estimated the barrier to interconversion between enantiomers to be close to 85 $kJ\,mol^{-1}$ in a number of β-diketonates. When lanthanide reagents are added to a benzene solution of $Al(acac)_3$, the ^{27}Al NMR signals due to free and complexed molecules can be distinguished and paramagnetic shift data can be obtained.[189]

Acetylacetone complexes have been prepared by the action of acac on $MeAlCl_2$, Me_2AlCl and Me_3Al, and on tetraalkylaluminoxanes R_4Al_2O (R = Me, Et or Bu^i).[190] Mixed ligand derivatives *e.g.* $R_2(acac)_2Al_2O$, tend to disproportionate forming $Al(acac)_3$. Metalla-β-diketones, in which the central CH group is replaced by a transition metal moiety, typically $Mn(CO)_4$ can be effectively coordinated to an aluminum or gallium atom.[191] XRD study of the molecule $Al[(OCMe)_2Mn(CO)_4]_3$ reveals that the acyl CO and Al—O distances, and O—Al—O angles, are not significantly different from the values in $Al(acac)_3$. The cationic complex $(acac)_2Al(OH_2)_2^+$ has been isolated as the picrate,[192] but structural data are lacking.

Tris(tropolonato)aluminum precipitates when tropolone and $AlCl_3$ are mixed in methanol. A significant feature of the ligand coordination, in relation to barriers to isomerization, is the twist angle of 48°, compared to 60° in a regular octahedron.[193] The bond length of 188 pm (average) and estimated bond energy of 250 $kJ\,mol^{-1}$ from calorimetric measurements[194] make this a rather strong Al—O bond, and photoelectron spectroscopy indicates that there are both σ and π contributions to bonding in the AlO_6 coordination sphere.[195]

25.1.5.4 Oxyanion Ligands

The coordination chemistry of aluminum with oxyanion ligands is chiefly that of the more highly oxidized ions. Nitrites and phosphites are unknown. Arsenites, sulfites, selenites and tellurites are mentioned in the early literature but have received little recent attention.

25.1.5.4.1 Borate, carbonate and silicate

Aluminium borates $(Al_2O_3)_n\cdot B_2O_3$ are solids of some technological interest. Gaseous species AlBO and $AlBO_2$ are generated, along with B_2O_3, when aluminum borates are heated above 1000 °C.[196] $Al_2(CO_3)_3$ is unknown. Basic carbonates of Al are ill-defined materials precipitated from aqueous aluminate solutions in contact with CO_2 gas.[1] Aluminum silicate chemistry, which rests on the variety of structures possible for complex silicate anions, with the added possibility of replacing Si by Al in the lattice, encompasses many naturally occurring and synthetic materials, the structures of which are built up from SiO_4 tetrahedra and AlO_4 and/or AlO_6 units. This vast subject is beyond the scope of the present account.

25.1.5.4.2 Nitrate, phosphate and arsenate

From aqueous nitric acid solutions, aluminum gives hydrates $Al(NO_3)_3\cdot nH_2O$ (n = 6, 8 or 9) which contain the $[Al(OH_2)_6]^{3+}$ complex (Section 25.1.5.1). A basic nitrate produced when $Al(NO_3)_3\cdot 9H_2O$ is heated to 136 °C has the composition $Al(OH)_2NO_3\cdot 1.5H_2O$.[197] Further heating produces hydrated alumina. IR spectroscopy indicates that NO_3 groups are not coordinated to Al^{3+}, although there is Raman[198] and UV[199] evidence of interaction in aqueous solution. Anhydrous $Al(NO_3)_3$ is prepared with some difficulty from $AlBr_3 + ClNO_3$. $AlCl_3$ with liquid N_2O_4 yields the 'oxide nitrate' $Al_2O(NO_3)_4$ or adducts $Al_2O(NO_3)_4{\cdot}xN_2O_4$ which on heating afford $AlONO_3$.[200] Reactions involving N_2O_5 give $Al(NO_3)_3{\cdot}N_2O_5$, m.p. 85–98 °C (dec), distillable in vacuum, and believed to crystallize as $NO_2^+[Al(NO_3)_4]^-$.[201] $Et_4N[Al(NO_3)_4]$ is prepared by solvolysis of $AlCl_3 + Et_4NCl$ using excess N_2O_4. The aluminum complex migrates to the anode during electrolysis in nitromethane. Complex salts of the larger group I cations, $MAl(NO_3)_4$ and $M_2Al(NO_3)_5$ (M = Rb or Cs), have been isolated as hygroscopic crystals, stable at room temperature. The crystal structure of $Cs_2Al(NO_3)_5$ shows that the Al atom has distorted octahedral coordination, with the apices occupied by one bidentate and four unidentate nitrate groups.[202] Parameters of the discrete $[Al(NO_3)_5]^{2-}$ anion are as follows: $r(Al–O)_{bident}$ = 198 pm; $r(Al–O)_{unident}$ = 193 pm (mutually *trans*) and 189 pm (mutually *cis,* but *trans* to bidentate NO_3 linkages). For the unidentate ligands, the Al—O—N angles are close to 125 °. Nitrate groups are coordinated to Al in the trimethylaluminum complex ion $(Me_3AlNO_3)^-$ in which NO_3 is unidentate with Al—O distance of 193 pm and angle Al—O—N = 123.3° and $[(Me_3Al)_2NO_3]^-$ in which the nitrato group is bridging (Al–O distances of 195 and 201 pm).[203]

The chemistry of aluminum phosphates,[204,205] which are commercially important as inorganic binding agents, is centred on compounds with Al_2O_3:P_2O_5 molar ratios of 1:1, *i.e.* $AlPO_4$ and hydrates $AlPO_4{\cdot}xH_2O$, and 1:3, *i.e.* $Al(H_2PO_4)_3$. Various aluminum poly- and meta-phosphates have been characterized. Hydroxyphosphates of Al can be prepared by heating $Al(OH)_3$ with phosphoric acid. Crystalline $AlPO_4$ resembles SiO_2, with tetrahedral coordination around Al, and the structure can exist in quartz-like and other silica isomorphs. NMR (^{27}Al under magic angle spinning conditions) shows a clear distinction between the AlO_4 groups of $AlPO_4$, and the octahedral AlO_6 groups present in various crystalline phosphate complexes.[206] The crystal structure of aluminum trimetaphosphate, $Al(PO_3)_3$, shows it to be built up of infinite chains of PO_4 units, interconnected by AlO_6 octahedra in which the Al—O distance is 188.4 pm (average).[207] In $AlPO_4{\cdot}2H_2O$ the PO_4 tetrahedra share vertices with four $AlO_4(OH_2)_2$ octahedra.[208] The two water molecules are *cis* to one another, with Al—O distances of 189.2 and 195.3 pm. The other four Al—O bonds range from 185.9 to 188.8 pm (average 187.6 pm). Edge-sharing AlO_6 octahedra linked by PO_4 groups are found in the structure of $NaAl_3(PO_4)_2(OH)_2$.[209] The structure of $Al(H_2PO_4)_3$, which can be crystallized from solutions of alumina in phosphoric acid, contains AlO_6 octahedra which share corners with $PO_2(OH)_2$ groups. Several other acid phosphates have been identified in the Al_2O_3—P_2O_5—H_2O system.[210] The compounds $M(PO_2F_2)_3$, where M = Al or Ga, can be prepared from the trihalides and difluorophosphoric acid.[211] Phosphoric acid derivatives of the organometal cations R_2M^+ (M = Al, Ga, In or Tl) have cyclic structures in which the bifunctional M centres are linked through O—P—O groups.[212]

The Al^{3+} ion is strongly complexed by phosphoric acid in aqueous solutions. The complex $[Al(HPO_4)_3]^{3-}$ appears to be the major species present, but the high viscosity of concentrated solutions indicates that polymeric species are also formed. The presence of aquated $[Al(H_3PO_4)]^{3+}$, $[Al(H_2PO_4)]^{2+}$ and complex cations with several acid phosphate ligands has been inferred from ^{27}Al and ^{31}P NMR studies.[213] On addition of fluoride to the solutions, F enters the coordination sphere of Al forming a series of fluorophosphatoaluminum complexes. The solids $AlF(HPO_4){\cdot}nH_2O$ (n = 2 or 3) have subsequently been characterized.[214] Hydrates of the complex salt $(Me_4N)_3[Al(PO_4)_2]$ have been isolated from aqueous solutions believed to contain the $[Al(PO_4)_2]^{3-}$ anion.[1] Some ATP complexes of aluminum have been studied by NMR.[215] Much remains to be learned about the exact constitution of these solutions.

25.1.5.4.3 Sulfate, selenate and tellurate; tungstate and molybdate

Evaporation of solutions of alumina in sulfuric acid yields crystals of $Al_2(SO_4)_3{\cdot}16H_2O$. If a group I sulfate is also present, the products are alums $MAl(SO_4)_2{\cdot}12H_2O$, in which aluminum exists as $[Al(OH_2)_6]^{3+}$. Analogous selenates but not tellurates have been prepared. Basic salts, *e.g.* $Al_2(OH)_2(SO_4)_2{\cdot}10H_2O$, and solids obtained from aluminum sulfate solutions above 100 °C also contain complex ions in which the coordination sphere of Al^{3+} is occupied by OH groups and/or water molecules rather than sulfate ions. Thermal decomposition of the hydrate $Al_2(SO_4)_3{\cdot}16H_2O$ yields practically anhydrous $Al_2(SO_4)_3$ at 450 °C; further decomposition to

Al_2O_3 and SO_3 occurs above 600 °C.[216] Hydrogen sulfates $M(HSO_4)_3$ (M = Al or Ga) have been prepared (equation 8).[217] Excess water allows the hexahydrates to be formed.

$$M(SO_3Cl)_3 + 3H_2O \longrightarrow M(HSO_4)_3 + 3HCl \quad (8)$$

Anhydrous double sulfates ('alum anhydrides') $M^IM^{III}(SO_4)_2$ are formed by Al and Ga with NH_4 or the alkali metals (M^I).[218] Crystal structure determination establishes that Al^{3+} is coordinated by SO_4 groups, with r(Al—O) = 191 pm.[219] Complexes $M[AlCl_2(SO_4)]$ (M = Li, Na or K) have been prepared.[220] Chlorosulfates of aluminum with the same alkali metals, $M[Al(SO_3Cl)_4]$, are obtained by the solvolysis of $MAlCl_4$ in SO_2 at −15 °C.[221] All are very hygroscopic. The Li compound thermally decomposes to $Al_2(SO_4)_3$, whereas Na or K compounds give $MAl(SO_4)_2$. The sulfate ligand is found in bridging role in $(Me_2Al)_2SO_4$ (from $2Me_2AlCl + Na_2SO_4$), which is monomeric, D_{2d}, in solution but polymeric in the solid state. In aqueous solutions of $Al_2(SO_4)_3$ there is evidence from UV[222] and NMR[223] spectra of coordination of sulfate to the aluminum ion, as $[Al(SO_4)(OH_2)_5]^+$, at concentrations above 0.1 mol^{-1}.

Aluminum forms tungstates $Al_2(WO_4)_3$ and $MAl(WO_4)_2$.[224] Crystals of $Al(WO_4)_3$ are orthorhombic; as in the sulfate, Al^{3+} is octahedrally coordinated, with Al—O distances between 186 and 191 pm.[225] Aluminum can be introduced into the structures of polytungstates and molybdates to prepare $(AlW_{12}O_{40})^{5-}$ and $(AlMo_6O_{21})^{3-}$ for example.[226]

25.1.5.4.4 Chlorate, perchlorate, bromate, iodate and periodate

The salts of these anions are highly hydrated and all are likely to contain the $[Al(OH_2)_6]^{3+}$ complex, although the periodate $Al(IO_4)_3 \cdot 12H_2O$ is reported to decompose above 130 °C to yield the anhydrous iodate $Al(IO_3)_3$.[1] Hydrates $Al(ClO_4)_3 \cdot nH_2O$ (n = 3, 6, 9 or 15) lose water on heating and at 178 °C decompose to $Al(OH)(ClO_4)_2$. Anhydrous $Al(ClO_4)_3$ is obtained from $AlCl_3 + 3AgClO_4$ in benzene. Sparingly soluble perchlorato complexes $M_nAl(ClO_4)_{3+n}$ (M = group I or NH_4 cation; n = 1, 2 or 3) have recently been prepared.[227,228] IR spectra indicate that $[Al(ClO_4)_6]^{3-}$ possesses only unidentate ligands, whereas $[Al(ClO_4)_4]^-$ and $[Al(ClO_4)_5]^{2-}$ contain both uni- and bi-dentate ClO_4. By the same criteria, the compound $Al(ClO_4)_3$ contains bidentate groups, whilst $Al(ClO_4)_3 \cdot 3L$ (L = H_2O, THF or $MeNO_2$) has only unidentate ClO_4 ligands. $Al(ClO_4)_3$ dissolves in MeOH, HOAc or $(AcO)_2O$ with complete replacement of ClO_4 groups by solvent molecules, whereas the perchlorate ligands remain coordinated in MeCN, THF or $MeNO_2$. Mixed ligand complexes $[AlCl_n(ClO_4)_{4-n}]^-$ (n = 1, 2 or 3) have been detected by ^{27}Al NMR in solutions of $AlCl_3 + ClO_4^-$ ions in $CHCl_3$ or CH_2Cl_2, and there is evidence of perchlorate complexing of Al^{3+} in EtOH solutions.[229] Preparations of a salt $Me_4N[AlCl_3(ClO_4)]$,[230] and a molecular adduct $AlCl_2(ClO_4) \cdot 2DME$[231] of the ligand dimethoxyethane, have been reported; both appear to contain perchlorate coordinated to aluminum.

25.1.5.5 Carboxylate Ligands

Aluminum forms carboxylates $Al(O_2CR)_3$, and a range of basic carboxylates, including the mononuclear species $Al(OH)(O_2CR)_2$ and $Al(OH_2)(O_2CR)$, and polynuclear complexes such as $Al_3O(OH)(O_2CR)_6$ obtained from the above by elimination of water.[232] There are many uses in textile, paper and pharmaceutical industries which take advantage of the ability of Al centres to bind O ligands strongly.

Crystallization from concentrated solutions prepared by dissolving freshly precipitated $Al(OH)_3$ in formic acid yields $Al(OH)(O_2CH)_2 \cdot H_2O$. The hydrated triformate $Al(O_2CH)_3 \cdot 3H_2O$ can be obtained from solutions containing an excess of HCO_2H. Basic acetates $Al(OH)_n(O_2CMe)_{3-n}$ (n = 1 or 2) are obtained as white powders from aqueous acetate solutions on partial neutralization. The triacetate $Al(O_2CMe)_3$ is prepared by heating the metal or $AlCl_3$ with acetic acid and acetic anhydride. When heated, $Al(O_2CMe)_3$ loses acetic anhydride forming products with Al—O—Al linkages. The higher carboxylic acids form similar compounds. The trifluoroacetate $Al(O_2CCF_3)_3$ is obtained from Al/Hg and trifluoroacetic acid. A basic salt and the compounds $AlCl_n(O_2CCF_3)_{3-n}$ (n = 1 or 2) have been described.[233] Besides preparations from Al or the trihalides AlX_3, a number of routes to carboxylates employing alkoxides $Al(OR)_3$ or alkyls R_3Al are available.[232] By using the appropriate stoichiometry mixed alkoxycarboxylates and the compounds $R_2Al(O_2CR')$ and $RAl(O_2CR')_2$ can be obtained.[6]

25.1.5.5.1 Complex ions

Complex salts with a group I metal $MAl(O_2CR)_4$ can be prepared from carboxylate solutions or from $MAlR_4$ (R = various alkyl groups) by reaction with CO_2 under pressure at 200 °C. Acetate complexes of trimethylaluminum have also been prepared, *i.e.* $Me_4N(Me_3AlO_2CMe)$ and $Me_4N[(Me_3Al)_2O_2CMe]$. XRD study of the former shows it to be a four-coordinate aluminum complex, with r(Al—O) = 183 pm and the angle C—O—Al = 137 °.[234] The structures of several Al complexes with bridging acetate ligands have been determined.[235]

In aqueous solution Al^{3+} gives complex ions $[Al(O_2CR)]^{2+}$ and $[Al(O_2CR)_2]^+$ with formic, acetic, propionic and butyric acids, for which the formation constants have been measured.[236,237] Owing to the influence of chelation, complexes of higher stability are formed by dicarboxylic, amino and hydroxy acids, and both cationic and anionic complexes may be formed.[238,239] Data are available for oxalic, malonic, succinic and lactic, diglycolic, gallic, citric, salicylic, glutamic and aspartic acids.[240–247]

NMR spectra give evidence of polymerization in aqueous solutions of hydroxycarboxylic acid complexes. In lactate (or citrate) solutions the ^{27}Al NMR spectra show distinct peaks for AlL_2^+, AlL^{2+} and AlL_3 complexes.[248] Increased pH leads to substitution of H_2O in the coordination sphere of $Al(OH_2)_6^{3+}$ by lactate and eventual displacement of lactate by OH^-. The citrate ion can function as a uni-, bi- or tri-dentate ligand towards Al^{3+}.[244] A tris(oxalato) complex can be isolated as $K_3Al(O_4C_2)_3\cdot 3H_2O$.[249] In contrast to oxalato complexes of transition metals, $[Al(O_4C_2)_3]^{3-}$ is highly labile and has defeated attempts at the resolution of optical isomers. Studies of the interaction of Al^{3+} with the carboxylate groups of model membrane systems are leading to a better understanding of physiological processes, especially cumulative toxicity.[250]

25.1.5.6 Ethers and Other Oxygen Donor Ligands

25.1.5.6.1 Molecular complexes

Organic ligands which form O donor complexes with the aluminum trihalides, except AlF_3, include ethers, aldehydes, ketones, esters, nitro compounds, alkyl phosphates, sulfoxides, ureas, acid chlorides and amides.[4] 1:1 is the prevailing stoichiometry, although there are 1:2 AlX_3·ether adducts, some nitro complexes of this form (*e.g.* $AlCl_3\cdot 2PhNO_2$, $AlBr_3\cdot 2MeNO_2$), and a few instances of 2:1 type ($Al_2X_6\cdot L$). The ether complexes tend to decompose on heating, giving mixtures including AlX_2OR, and as a class the AlX_3 adducts are highly reactive and very susceptible to hydrolysis and nucleophilic attack. X-Ray studies have shown that the crystalline acetyl chloride complex of $AlCl_3$ is ionic, $MeCO^+AlCl_4^-$ (although it has a donor–acceptor structure in dichloromethane solution[251]), whereas the adducts with EtCOCl and PhCOCl are O-linked coordination compounds.[252] Bond lengths are Al—O = 182–185 pm and Al—Cl = 208–209 pm. A decrease in the ν(C=O) stretching frequency shows the carbonyl oxygen to be the donor site in these compounds, and also in the case of esters, amides and anhydrides. Vibrational spectra show that the complexes $AlX_3\cdot MeNO_2$ (X = Cl or Br) are simple molecular adducts.[253]

Several of the ether adducts have been investigated in detail. $AlCl_3\cdot 2THF$ has a molecular structure which is close to trigonal bipyramidal; the tetrahydrofuran ligands are *trans* diaxial with Al—O = 199.0 pm and Al—Cl = 215.3–216.4 pm.[254] There is spectroscopic proof that the ionic complex $AlCl_2(THF)_4^+AlCl_4^-$ can be obtained in the solid state, although it readily isomerizes to the molecular form. A molecular structure $AlX_3\cdot THF$ (X = Cl or Br) is found for the solid 1:1 adduct. These species can be detected in solution where the various forms are in equilibrium, and there is evidence of interconversion of *cis* and *trans* forms of $AlCl_3\cdot 2THF$.[255] The structure of $AlCl_3\cdot 2diox$ is also based on trigonal bipyramidal coordination.[256] Here, planar $AlCl_3$ units are bridged by 1,4-dioxane to form $(AlCl_3\cdot diox)_x$ chains; the second dioxane molecule is merely a solvate. The bond distances [Al—O = 201.7 pm and Al—Cl = 213.9 pm (average)] are close to the values in $AlCl_3\cdot 2THF$. Diethyl ether gives $AlX_3\cdot Et_2O$ (X = Cl, Br or I)[257] whereas Me_2O forms 1:1 and 1:2 types. The complexes $AlHCl_2\cdot Et_2O$ and $AlHBr_2\cdot Et_2O$

have been examined spectroscopically; ν(Al—O) was close to 330 cm^{-1} and shifted very little on changing Cl for Br.[258]

Coordination takes place through oxygen in the 1:1 and 2:1 sulfone complexes $AlBr_3{\cdot}R_2SO_2$ and $2AlBr_3{\cdot}R_2SO_2$ which have been studied by means of bromine NQR spectra.[259] A complex of aluminum trichloride with benzo-15-crown-5 is notable as another instance of a 2:1 type, effectively $Al_2Cl_6{\cdot}$ether.[260] It is obtained from benzene–diethyl ether solvent; on recrystallization from methanol the more stable 1:1 ($AlCl_3{\cdot}$ether) adduct is produced. Crystal structures of some related complexes with crown ethers, $(Me_3Al)_n{\cdot}$ether ($n = 2$ or 4) and $(AlCl_2{\cdot}\text{ether})^+$, have been reported.[261] These are useful agents for solubilizing ionic species M^+A^- in aromatic solvents, ultimately affording $M^+AlMe_3A^-$ and $M^+Al_2Me_6A^-$. A general review of etherates of organoaluminum compounds has been given.[262]

Measurements on systems $R_3Al{\cdot}L$, where L is one of the ethers mono-, di- or tri-glyme, yield low values (*ca.* 25 kJ mol^{-1}) for the activation energy of ligand exchange, suggesting that a bimolecular displacement rather than a dissociative mechanism is operative.[263] NMR measurements in solution confirm the existence of both $AlCl_3{\cdot}Me_2O$ and $AlCl_3{\cdot}2Me_2O$ and enable ligand exchange rates to be determined.[264] Competition experiments have shown the following order of ligand strength towards $AlCl_3$: $COCl_2 < MeNO_2 < PhNO_2 < PhCOCl < Me_2O < Ph_2CO < Et_2O < HCO_2Me < MeCN < Cl^-$.[265] Other aspects of the energetics of AlX_3–oxygen donor complexes are compared with those of amine adducts and of the AlX_4^- ions in appropriate sections. Ether adducts of other Al^{III} species, $AlH_3{\cdot}Et_2O$ and $Al(NCS)_3{\cdot}Et_2O$, are mentioned elsewhere in this chapter.

Of several inorganic ligands which form O-coordinated complexes with $AlCl_3$ or $AlBr_3$, phosphorus oxychloride has been the most fully investigated. The adduct $AlCl_3{\cdot}POCl_3$, m.p. 186 °C, is well known and a second adduct has been shown to be $2AlCl_3{\cdot}3POCl_3$ (m.p. 166 °C)[266] (not $AlCl_3{\cdot}2POCl_3$ as in early literature). Unstable crystals $AlCl_3{\cdot}6POCl_3$ decomposing at 42 °C have also been obtained. With other ligands there are reports of $AlCl_3{\cdot}COCl_2$, $2AlCl_3{\cdot}NO$, $AlCl_3{\cdot}NOCl$, $AlCl_3{\cdot}2NOCl$, $2AlCl_3{\cdot}SO_2$, $AlCl_3{\cdot}SO_2$, $2AlCl_3{\cdot}SOCl_2$, and a few analogous complexes of $AlBr_3$ and AlI_3.[1,4,264] All are not necessarily O-bound, for example the 1:1 complex of $AlCl_3$ with NOCl is better formulated as $NO^+AlCl_4^-$ (see Section 25.1.7.2). The aluminum trihalides do not react with carbon monoxide as such, but there are a number of complexes in which the AlX_3 Lewis acid group is attached to a carbonyl group of a transition metal complex,[267] and kinetic studies have shown that coordination to AlX_3 is an effective means of promoting carbonyl insertion reactions.[268] The structure of $[W(CO)_3Cp]_2Al{\cdot}3THF$ is a remarkable one; the Al atom is six-coordinate, being attached to O atoms of carbonyls and three THF ligands.[269]

25.1.5.6.2 Cationic complexes

There have been a number of investigations of the species formed by Al^{3+} in polar oxygen-containing solvents. Frequently these are complex cations AlL_6^{3+}; examples which have been characterized by NMR (^1H and ^{27}Al) and other measurements are $Al(DMSO)_6^{3+}$, $Al(DMF)_6^{3+}$, $Al(TMPA)_6^{3+}$, $Al(urea)_6^{3+}$ and $Al(EtOH)_6^{3+}$.[270] Kinetic and thermodynamic parameters have been determined for several of these systems. They show, for example, that $Al(DMF)_6^{3+}$ is a more labile complex than $Al(DMSO)_6^{3+}$, but that all seven species $Al(DMSO)_n(DMF)_{6-n}^{3+}$ ($n = 0$–6) are present at various DMSO:DMF ratios when both ligands are added to a solution of $Al(ClO_4)_3$ in nitromethane. The urea complex has been isolated as a crystalline salt, $Al(urea)_6(ClO_4)_3$, which has been studied crystallographically.[271]

Dimethyl sulfoxide systems have attracted most attention. Ligand exchange of DMSO with the complexes $M(DMSO)_6^{3+}$ (M = Al, Ga or In) has been shown to involve a dissociative mechanism, and rate and activation parameters have been determined for ligand displacement reactions of $Al(DMSO)_6^{3+}$ with H_2O, and with the stronger bases pyridine, bipyridyl and terpyridyl.[272] The halides $AlCl_3$ and $AlBr_3$ form the DMSO complexes $AlX_3{\cdot}1.5DMSO$ and $AlX_3{\cdot}6DMSO$. According to vibrational spectra, both these solids contain the DMSO-coordinated cation and should be formulated as $Al(DMSO)_6^{3+}3AlX_4^-$ and $Al(DMSO)_6^{3+}3X^-$, respectively.[273] Vibrational assignments are also available for the tetramethylene sulfoxide complex $Al(TMSO)_6^{3+}$.[274]

^1H NMR spectra of $AlCl_3$, in either MeOH or EtOH solutions, indicate a solvent coordination number of four which may imply that the species $AlCl_2(ROH)_4^+AlCl_4^-$ predominate.[275] In the cases of $Al(NO_3)_3$ and $Al(ClO_4)_3$ the solvation numbers are five and six, respectively. For $Al(ClO_4)_3$ in non-aqueous solvents, the addition of trimethyl phosphate produces $Al(TMP)_6^{3+}$, whereas hexamethylphosphoramide gives a tetrahedrally coordinated complex. The kinetics of ligand exchange show that an associative mechanism is followed by the four-coordinate species, which contrasts with the dissociative path to which the six-coordinate species are restricted. If water is also present, the octahedral species $AlL_n(OH_2)_{6-n}^{3+}$ (L = TMP or HMPA) are detected instead.[276]

25.1.6 SULFUR AND HEAVIER GROUP VI LIGANDS

25.1.6.1 Sulfides, Selenides and Tellurides

The compounds Al_2S_3, Al_2Se_3 and Al_2Te_3 are produced by heating the elements in 2:3 proportions.[277] All crystallize in the wurtzite hexagonal lattice of anions with Al^{3+} in two-thirds of the cation sites, although other forms exist, including some high-pressure phases based on a cubic lattice.[278] There are many ternary compounds M^IAlY_2 and $M^{II}Al_2Y_4$ (Y = S, Se or Te),[1,279,280] some of which are less susceptible to hydrolysis and other attack than the corresponding Al_2Y_3. Ternary group III compounds are also known, *e.g.* $InAlS_3$ in which Al is tetrahedrally coordinated by S atoms.[281]

Thiohalides AlSX (X = Cl, Br or I) can be prepared as crystalline compounds by heating Al_2S_3 with AlX_3.[1] They decompose on strong heating and are readily converted to the oxyhalides by moisture and to thioamides $AlSNH_2$ by ammonia. Selenium and tellurium form the analogous AlSeX and AlTeX which react similarly. The structure of AlSX (and AlSeX) deduced from IR/Raman spectra is basically a dimer with a planar Al_2S_2 ring and terminal Al—X.[282] The units are interlinked through halogen bridge bonding. Sulfide ions and $AlCl_4^-$ react in basic CsCl/$AlCl_3$ melts to produce chain-like complex anions $[Al_nS_{n-1}Cl_{2n+2}]^{n-1}$($n$ = 2–3 in dilute solution; higher when more concentrated). The polymeric compound $CsAlSCl_2$ can be isolated.[283] Raman spectra of the melts show features attributable to both bridging S and Cl ligands. In acid melts, dissociation to form solid AlSCl occurs.

25.1.6.2 Other Sulfur Donor Ligands

Dialkyl sulfide ligands interact with the halides AlX_3 to form the 1:1 complexes.[257] For $AlCl_3{\cdot}SMe_2$, rapid exchange of ligand with excess Me_2S is shown by NMR investigation.[264] $AlI_3{\cdot}SEt_2$, obtained by direct reaction of the compounds, has also been characterized by NMR. The dithiocarbamate $Al(S_2CNMe_2)_3$ can be prepared from $Al(NMe_2)_2$ by CS_2 insertion.[284] The structure is monomeric with six-fold coordination of Al by S (Al—S distances = 238.6–240.5 pm). In the case of $ClAl(NMe_3)_2$ the product is the dimer $[ClAl(S_2CNMe_2)_2]_2$. There are two bridging Cl ligands which, with the four S atoms, enable each of the Al atoms in the molecule to be six-coördinate. Various lithium tetrakis(thiocarbamato)aluminates $LiAl(YY'CNHR)_4$ (Y = S, Y′ = O or S) have been prepared and characterized by IR spectra.[285] Insertion of CS_2 into the Al—N bond in Me_2AlNPh_2 to yield $Me_2AlS_2CNPh_2$ occurs when the reactants are refluxed together in hexane.[286] The structure of the adduct $Me_3Al{\cdot}SMe_2$ has been determined by vapour phase ED which yields the Al—S bond length of 255 pm.[287] Reactions of Me_3Al with toluene-3,4-dithiol give a variety of cyclic and polymeric products with Al—S-bonded structures.[288] $(Me_2Al)_2S$, the sulfur analogue of tetramethylaluminoxane, has been reported; it is produced from trimethylaluminum by reaction with lead sulfide or liquid H_2S.[289]

25.1.7 HALOGEN LIGANDS

25.1.7.1 Fluoride Ligand

The AlF_3 molecule, trigonal planar, with Al—F distance of 163 pm, is produced in the gas phase when aluminum trifluoride sublimes (1272 °C at 1 atm).[1] Vibrational frequencies of the monomer, and the Al_2F_6 dimer, are available from matrix isolation studies,[290,291] as is also true for the OAlF molecule.[292]

Crystalline AlF_3 can be obtained from Al_2O_3 and HF at 700 °C. The lattice contains six-coordinate Al with Al—F = 179 pm. The hydrates $AlF_3{\cdot}nH_2O$ ($n = 3$ or 9) are dehydrated with difficulty *via* non-stoichiometric forms. Unlike the other halides, anhydrous AlF_3 is insoluble in inert solvents and only slightly soluble in water. It is more soluble in the presence of HF, forming the complexes AlF_6^{3-}, $AlF_n(OH_2)_{6-n}^{3-n}$ and $AlF_3(OH_2)_3$ detectable by NMR and other means.[293–296] Concentrated solutions deposit β-$AlF_3{\cdot}3H_2O$ on standing.

There are many complex fluoroaluminates, both natural and synthetic. Octahedral AlF_6 units are present, either as AlF_6^{3-} or within chain or layer structures where adjacent units share apices. Edge-sharing, which would produce short Al···Al distances, is avoided and Al—F—Al links are close to linear. The solids M_3AlF_6 assume cubic lattices at elevated temperature (the Li salt at 597 °C, Na at 560 °C, K at 300 °C).[1] Structures at ambient temperature are less regular. Recent structural studies include those of alkali metal hexafluoroaluminates, $Ba_3(AlF_6)_2$, $BaZnAlF_9$ (in which Al/ZnF_6 octahedra are linked into tetrameric anions)[297] and $Hg_2AlF_5{\cdot}2H_2O$ (in which AlF_6 octahedra are linked through *trans* fluorines into infinite chains).[298] In $KAlF_4$, layers of AlF_6 groups are connected by four corners. Al—F distances, measured by neutron diffraction, are 181.7 and 175.2 pm.[299]

Na_3AlF_6 'cryolite', which can be synthesized from $Al_2O_3 + HF + NaOH$, forms, with alumina and AlF_3, the basis of the electrolyte in the manufacture of aluminum. There have been critical studies of cryolite,[300] of the liquid phase dissociation $AlF_6^{3-} \rightarrow AlF_4^- + 2F^-$,[301,302] and of the vapour above the melt.[303] In the gas phase, the predominant species $NaAlF_4$ has the structure $Na(\mu\text{-}F)_2AlF_2$, and the $(NaAlF_4)_2$ dimer is also present. MO calculations have been used to compare various possible gas phase structures of $NaAlF_4$, $LiAlF_4$ and $MgAlF_5$.[304]

X-Ray studies[305] established the presence of a linear, symmetrical Al—F—Al bridge with Al—F 178.2 pm in the anion $Me_3Al(\mu\text{-}F)AlMe_3^-$ which is isoelectronic with disiloxane $(Me_3Si)_2O$. Me_2AlF is a cyclic trimer containing the Al_3F_3 ring.[306]

25.1.7.2 Chloride, Bromide, Iodide and Mixed Halide Ligands

Aluminum monohalides AlX are recognized in the high temperature gas phase from which they can be trapped by matrix isolation.[307] AlCl is significant in the chemical transport of aluminum through the equilibrium $Al_{(1)} + AlCl_{3(g)} \rightleftharpoons 3AlCl_{(g)}$. For the AlX_3 monomers, planar trigonal molecules, there are reports of bond distances, and of IR and Raman spectra in the gas phase, or as matrix-isolated species.[308,309] *Ab initio* MO calculations of the geometry of the $AlCl_3$ and Al_2Cl_6 molecules have been performed, and photoelectron spectroscopy has been used to investigate the equilibrium between MX_3 and the M_2X_6 dimers for M = Al or Ga; X = Cl, Br or I.[310] Structural data for the Al_2X_6 molecules are tabulated elsewhere,[1,2] as are the physical properties of the crystalline aluminum trihalides, so that only values which have been subsequently revised are cited here. For example, vibrational spectra of aluminum halide vapours have provided a basis for fresh NCA calculations on the AlX_3 and Al_2X_6 molecules,[308,311] and the Raman spectrum of Al_2Cl_6 in the melt has been reexamined.[312]

Aluminum trichloride is synthesized by heating the metal in chlorine or dry HCl. Various methods of purification have been employed.[1] Crystalline $AlCl_3$, a lattice containing six-coordinate Al, melts at 192 °C under 2.5 atm to give a liquid consisting of Al_2Cl_6 molecules. At 1 atm it sublimes at 150–180 °C, mainly as Al_2Cl_6, although the vapour then contains 1% of the trimer Al_3Cl_9. The dimer dissociates readily (ΔH for $Al_2Cl_6 \rightarrow 2AlCl_3$ is 170 kJ mol^{-1}),[313] and the monomer is the predominant form at 750 °C. Aluminum tribromide (m.p. 98 °C) and aluminum triiodide (m.p. 189 °C) are synthesized from Al and halogen. The standard enthalpies of formation have been redetermined, giving −495 and −280 kJ mol^{-1} for $AlBr_3$ and AlI_3, respectively.[314] The bromide consists of Al_2Br_6 molecules in solid and liquid states. In the vapour, MS detects clusters containing up to 30 atoms, $(AlBr_3)_n$, with strong peaks for $n = 2$, 4 and 6, besides the monomer formed by dissociation.[315] Crystalline AlI_3 has an infinite chain structure with the repeating unit —$AlI_2(\mu\text{-}I)$—.[316] The Al—I distances differ markedly, averaging 272.0 pm (bridging) and 245.3 pm (non-bridging). The Al—I—Al bridge angle is 119°. The liquid is molecular, Al_2I_6, as shown by the IR and Raman spectra.[1] As powerful Lewis acceptors, the trihalides form (with donor molecules) coordination compounds of the type D—AlX_3, and sometimes D_2AlX_3 or DAl_2X_6, discussed at various points in this chapter. $S_2N_2(AlCl_3)_2$ is an example of a compound in which a dibasic species, the S_2N_2 ring, coordinates to two separate $AlCl_3$ units.[317]

The importance of aluminum trichloride (and $AlBr_3$ which is more soluble in hydrocarbons) as a catalyst, particularly for Friedel–Crafts alkylation and acylation of aromatic compounds, has led to many investigations.[4,318–321] The prime function of $AlCl_3$ is to interact with the RCl or RCOCl molecule to generate cationic alkyl or acyl species with simultaneous production of $AlCl_4^-$ and/or $Al_2Cl_7^-$. With aromatic hydrocarbons (Ar) in the presence of HX, the halides can form $ArH^+Al_2X_7^-$ which is an effective catalyst. Anhydrous aluminum trihalides also interact with aromatic systems in the absence of HX, but the process is imperfectly understood. The alleged formation of a complex of Al_2Cl_6 with benzene has been refuted.[322] Mixtures of the bromide with benzene yield an inclusion compound Al_2Br_6·benzene, the NQR signals of which are similar to those of pure Al_2Br_6.[323] With toluene, Raman spectroscopy gives evidence of a complex throught to have the singly bridged structure $Br_3Al(\mu\text{-}Br)AlBr_2$·toluene,[324] and with mesitylene there is a π complex, $AlBr_3$·mesitylene.[1] Lamellar compounds including $C_9·AlBr_3·Br_2$ and $C_{24}·AlBr_3Br_{0.3}$ are formed when graphite is exposed to $AlBr_3$–Br_2 mixtures.[325]

Chemical transport of the chlorides of many other metals takes place in aluminum trichloride vapour, and this aspect of the Lewis acid behaviour of $AlCl_3$ has received considerable attention, focussing especially on the formulae of the gaseous complexes, their thermodynamic properties and possible applications.[313] The species concerned are molecular complexes which typically have $Al(\mu\text{-}X)_2M$ bridges, such that both Al (as AlX_4 or Al_2X_7) and the other metal, *e.g.* Be, Mg, Fe or Pd, attain a coordination number of four. In $NaAlCl_{4(g)}$ the Na atom is probably above one edge of the $AlCl_4$ tetrahedron, as in $NaAlF_4$.[326] With $CrCl_2$ or $NiCl_2$ the gas phase complex $MCl_2·2AlCl_3$ has the structure $ClAl(\mu\text{-}Cl)_3M(\mu\text{-}Cl)_3AlCl$ in which Al is four-coordinate and the transition metal is six-coordinate. The structures and properties of organoaluminum halides further demonstrate the coordination behaviour of halogen ligands towards Al. Compounds such as R_2AlCl and $RAlCl_2$ (R = Me or Et) are dimeric in the liquid state and tend to dissociate in the vapour.[327,328] The Al–Cl bridge bonds in the $Me_2Al(\mu\text{-}Cl)_2AlMe_2$ molecule (230 pm) are significantly longer than those in Al_2Cl_6 (225 pm).

25.1.7.2.1 Complex halide ions

Aluminum forms the four-coordinate anions AlX_4^- and the bridged species $Al_2X_7^-$, *i.e.* $X_3Al(\mu\text{-}X)AlX_3^-$, both of which are produced when aluminum halide is fused with a metal halide MX.[329] There are recent XRD and spectroscopic studies of the $AlCl_4^-$ ion as crystalline Li, Na, K, Tl and NH_4 salts at ambient and elevated temperatures.[329,330] The Al—Cl distances are 212–214 pm, and various distortions of $AlCl_4^-$ from tetrahedral symmetry are revealed. The corresponding bromides $MAlBr_4$ have also been characterized.[1,331] Near regular $AlCl_4^-$ ions with Al–Cl bond lengths of 212.9 pm occur in $Li(SO_2)_3AlCl_4$[332] and in $NOAlCl_4$[333] (from $AlCl_3 + NOCl$ in liquid SO_2). The $AlCl_4^-$ tetrahedra, which interact only weakly with the alkali metal cations, are linked through Cl bridges in the case of some transition metal chloroaluminates, *e.g.* $Cu(AlCl_4)_2$, $Ti(AlCl_4)_2$, $(\eta\text{-}C_6H_6)Ti(AlCl_4)_2$ and $(\eta\text{-}C_6H_6)V(AlCl_4)_3$.[334–336] NQR spectra are able to distinguish between compounds containing ionic AlX_4^- and those in which the groups interact with other metal centres, or form $Al_2X_7^-$.[337–339] The ready formation and low coordinating ability of the AlX_4^- ions has been instrumental in the use of aluminum halides to isolate compounds containing low-valent cluster cations or other unusual species, *e.g.* $Hg_3(AlCl_4)_2$, $Bi_8(AlCl_4)_2$,[340] $C_3X_3(AlX_4)$ (X = Cl or Br),[341] $SCl_3(AlCl_4)$,[342] $PI_4(AlI_4)$ and $P_2I_5(AlI_4)$.[343] In these last two compounds, AlI_4^- has near regular T_d symmetry with Al—I = 251.8 or 252.9 pm, respectively. A notable attempt to utilize $AlCl_4^-$ in isolating a silicon-based cation resulted instead in a compound with $AlCl_3$ coordinated to one of the N atoms of $(Me_2N)_3SiCl$.[344]

Much recent work on NaCl/$AlCl_3$ and related haloaluminate systems has concerned their behaviour as molten salts and non-aqueous electrolytes, and their use as aprotic reaction media.[345–352] $AlCl_3$ with *n*-butylpyridinium chloride, and allied low-melting combinations provide useful model systems for fused salt behaviour and have attracted considerable attention on this account.[353–356] There is evidence, particularly from ^{27}Al NMR spectra, that $Al_2Cl_7^-$ and also $Al_3Cl_{10}^-$ are present in $AlCl_3$-rich systems.[354,355]

The tetrahaloaluminates are hydrolyzed by water; however, the anion $AlCl_4^-$ has been characterized in solution as the complex acid $(Et_2O)_nH^+AlCl_4^-$.[357] This ion and solvated $AlCl^{2+}$ and $AlCl_2^+$ are formed when $AlCl_3$ dissociates in various solvents, such as ethers, $MeNO_2$ and MeCN.[358,359] NMR, IR and Raman spectra have identified the species and have enabled exchange processes between them to be monitored. The ions $[Me_nAlCl_{4-n}]^-$ (n = 1–3) have been investigated in benzene solution.[360] Vibrational analyses are available for $MeAlCl_3^-$ and

Me_3AlCl^-,[361] and the crystal structures of $K(MeAlCl_3)$, $Cs(Me_2AlI_2)$ and $Me_4N(Me_3AlI)$ have been determined.[362]

The heptachlorodialuminate ion has been obtained from the reaction of acetyl chloride and Al_2Cl_6 which yields $MeCO^+Al_2Cl_7^-$ in liquid HCl,[363] although the direct reaction yields $MeCO^+AlCl_4^-$.[364] (EtCOCl and $AlCl_3$ form a molecular adduct in which $AlCl_3$ is attached by the CO group of propionyl chloride.) Crystallographic studies of solids containing $Al_2Cl_7^-$, which has been observed in both eclipsed and staggered bent configurations, give the average Al—Cl distances 227.0 pm (bridging) and 208.4 pm (terminal) and a bridge bond angle Al—Cl—Al of 114 °.[365] The $Al_2Br_7^-$ anion is similar, having Al—Br = 243 pm (bridging) and 227 pm (terminal) with angle Al—Br—Al = 108 ° in the NH_4^+ salt.[366] $Et_4N(AlBr_4)$ and $Et_4N(Al_2Br_7)$ are obtained by reactions of $AlBr_3$ with Et_4NBr in acetonitrile.[367] Vibrational spectra of $Al_2X_7^-$ (X = Cl, Br or I in Me_4N^+, K^+ or Cs^+ salts, respectively) indicate higher symmetry with a linear Al—X—Al bridge in the liquid state.[368] In view of the experimental support for both linear and bent structures of $Al_2Cl_7^-$, it is interesting that *ab initio* calculations[369] favour a bent Al—Cl—Al bridge in the species $H_3Al(\mu\text{-Cl})AlH_3^-$ (in contrast, $Al_2H_7^-$ is predicted to be a linear).

25.1.7.2.2 Mixed halide species

The anions $[AlI_nX_{4-n}]^-$ (X = Cl or Br), readily identifiable by ^{27}Al NMR spectroscopy, are generated from AlI_3 and Pr^n_4NI in CH_2Cl_2 or CH_2Br_2, and the $[AlCl_nI_{4-n}]^-$ series has also been observed in molten salt media.[370] The series $[AlCl_nBr_{4-n}]^-$ and the ions AlX_2YZ^-, containing three different halogens, have been characterized similarly,[371] and equilibrium relationships between these various complexes have been estimated. The chemical shift range which separates $AlCl_4^-$ from AlI_4^- is much smaller than that which separates $GaCl_4^-$ from GaI_4^-, in line with differences in nuclear shielding of the Al and Ga mixed halide species. Raman and NMR spectra of mixtures containing $AlCl_nBr_{4-n}^-$ (n = 0–4) have been used to identify the individual ions and MO calculations have been applied to this system.[372] Mixed halide forms of the aluminum trihalide molecules are little known although the reaction of HCl with $AlBr_3$ is said to give mainly $AlCl_2Br$, and both $AlCl_2Br$ and $AlClBr_2$ have been identified by matrix isolation methods.[373]

25.1.8 HYDROGEN AND COMPLEX HYDRIDE LIGANDS

25.1.8.1 Hydride Ligand

Aluminum hydride as polymeric $(AlH_3)_x$, and the trimethylamine complex of alane, $AlH_3{\cdot}NMe_3$, were reported about 1940.[26] Synthesis of aluminum hydride, as a colourless solid, by the reaction of $LiAlH_4$ with $AlCl_3$, appeared in 1947.[374] A route which avoids residual chloride ligands uses the action of sulfuric acid on $LiAlH_4$ in THF.[375] A tested preparation of the complex $AlH_3(OEt_2)_{0.3}$ is available,[376] and some spectroscopic and X-ray data have been reported for this.[377] There are ν(Al—H) bands at 1620–1880 cm^{-1} and ν(Al–O) at 725 cm^{-1}; the crystal is cubic. AlH_3, free of ether, is obtained by running a fresh ethereal solution into pentane. At least six crystalline phases of AlH_3 have been identified.[378] In one modification, isostructural with AlF_3, the Al atoms are each surrounded by an octahedral array of hydrogens. The Al—H distance in this structure is 172 pm.[379] The monomeric AlH_3 and dimeric Al_2H_6 molecules can be generated as vapour phase species above 1000 °C. The Al—H bond energy in $AlH_{3(g)}$ is 340 kJ mol^{-1}, compared to the energy of the electron deficient bridge bond in Al_2H_6, estimated to be 116 kJ mol^{-1}.[380] Whereas AlH_3 is planar, the alane radical anion $^{\bullet}AlH_3^-$ has a pyramidal structure, demonstrated by a combination of matrix isolation and ESR techniques. The molecule $HAlCl_2$ has been generated from AlCl and HCl in an argon matrix under UV irradiation. The vibrational spectrum with ν(Al—H) at 1968 cm^{-1} and ν(Al—Cl) near 475 cm^{-1} favours a planar monomeric structure.[381]

Using *ab initio* procedure, the structure of the $(Me_2AlH)_2$ dimer has been calculated.[382] Similar calculations for $(Me_3Al)_2H^-$, and for the hypothetical species $Al_2H_7^-$, show that the lowest energy corresponds to a linear Al—H—Al bond.[383] The crystal structure of $Na(Me_3Al—H—AlMe_3)$ shows the Al—H distance to be 165 pm, compared to 168 pm in $(Me_2AlH)_2$, and that the Al—H—Al angle is accurately 180 °.[384] X-Ray analysis of $K(Me_3AlH)$ reveals a length of 173 pm for the non-bridging Al—H bond in this anion.[385] Secondary bridging through H and Cl occurs in the complexes formed by $AlH_3{\cdot}OEt_2$, or $AlH_3{\cdot}NMe_3$, with the yttrium chloride dimer $(Cp_2YCl)_2$.[386] This demonstrates both the tendency of the Al—H bond to form bridges, and also the residual unsaturation of the Al centre in $AlH_3{\cdot}L$ systems.

Important aluminum hydride derivatives are the adducts of AlH_3 with Lewis bases; the complex aluminohydrides, notably $LiAlH_4$; and a range of compounds AlH_nL_{3-n}. In this last category are halogenoalanes, AlH_nX_{3-n} (X = Cl, Br or I; $n = 1$ or 2); dialkylaminoalanes, $AlH_n(NR_2)_{3-n}$; iminoalanes, especially the polymeric substances $(—AlH(NR)—)_x$; phosphinoalanes, alkoxyalanes and mercaptoalanes. Aspects of the chemistry of these compounds and adducts of the type $AlH_3{\cdot}NR_3$ and $AlH_3{\cdot}OR_2$ are treated elsewhere in this chapter. There is a good review of work prior to 1970 which charts the early progress of aluminum hydride chemistry.[26]

25.1.8.1.1 Complex aluminohydrides

The principal complex hydrides are alkali metal salts, $MAlH_4$ and M_3AlH_6, especially lithium aluminum hydride. Early accounts described the initial development of these reagents.[387–389] The laboratory preparation and purification of $LiAlH_4$, its manufacture by the Schlesinger process, properties, uses and safe handling were covered. Characteristic reactions of aluminohydrides will not be described now, except for recent work, or where there are aspects of special relevance to aluminum coordination chemistry. Hydridoaluminates can react with metal–halogen, metal–alkoxy and metal–alkyl compounds to form the corresponding metal hydride. Al—H bonds are highly labile and species with such bonds react with active-hydrogen-containing compounds to release hydrogen gas.

Lithium aluminum hydride can be prepared from $AlCl_3$ plus 4LiH in diethyl ether. An alternative route, using $AlH_3{\cdot}NMe_3$ and LiH, illustrates the displacement of coordinated trimethylamine by the more powerful base, H^-. $NaAlH_4$ can be synthesized directly from the elements using THF solvent at 150 °C and 135 atm. The product is obtained by precipitation with toluene, and can be converted to $LiAlH_4$ by reaction with LiCl in Et_2O.

$LiAlH_4$ is stable in dry air below 120 °C, but highly reactive towards moisture and protic solvents. When heated to 100 °C, Li_3AlH_6 is produced by a reaction in which exactly 50% of the total hydrogen is liberated (equation 9).

$$3LiAlH_4 \longrightarrow Li_3AlH_6 + 2Al + 3H_2 \quad (9)$$

Preparative routes to Li_3AlH_6 use $LiAlH_4$ + butyllithium,[390] or $LiAlH_4$ + 2LiH in the presence of Et_3Al as catalyst.[391] $LiAl_2H_7$ was also reported but not substantiated by further work.[392] $NaAlH_4$ is more stable than the lithium compound, decomposing to Na_3AlH_6 at 220 °C under nitrogen. Above 245 °C further decomposition gives 3NaH + Al + hydrogen.[393] The crystal of $LiAlH_4$ contains tetrahedral AlH_4^- ions with an average Al—H distance of 155 pm. In $NaAlH_4$, the anion has a compressed tetrahedral geometry with the Al—H distance of 153.2 pm. The bond angles are 107.3° and 113.9°.[394] The structures and stabilities of $LiAlH_4$ and the series of complex hydrides $M^IM^{III}H_4$ (M^I = alkali metal, M^{III} = Al, Ga, In or Tl) have attracted recent theoretical as well as experimental attention.[395,396]

The complex hydrides $MAlH_4$ form solvates, *e.g.* $LiH_4{\cdot}nL$, where L = OEt_2 or THF and $n = 1$ or 2. $LiAlH_4{\cdot}2THF$ is soluble in benzene and is volatile with some loss of THF. Amine adducts of $LiAlH_4$ are also formed, and are more stable to heat than the parent compound. A hydridoaluminate $NaAlH(OMe)_2OEt$ is manufactured as the benzene-soluble reducing agent 'Red-Al'. Tested procedure is available to prepare a dihydroaluminate which is non-pyrophoric and soluble in aromatic hydrocarbons by the reaction shown in equation (10), using 2-methoxyethanol.[397] This compound probably owes its stability to six-coordinate Al with two H ligands and two chelating groups forming Al—O bonds. Related compounds $Na[AlH_n(OR)_{4-n}]$, where OR = $OC_2H_4NMe_2$ and $n = 1$ or 2 but not 3, are produced by the reaction of 2-dimethylaminoethanol with $NaAlH_4$. Mixed alkoxyhydrides $AlH_n(OPr^i)_{3-n}$ are formed by redistribution reactions of $AlH_3{\cdot}THF$ with $Al(OPr^i)_3$.[398]

$$Na + Al + 2MeOC_2H_4OH \xrightarrow[150\,°C]{H_2,\ 70\text{–}250\ atm} Na[AlH_2(OC_2H_4OMe)_2] \quad (10)$$

The structures of hydridoaluminates have been investigated in solution. Conductance measurements in ether solutions over a wide concentration range suggest that the ions of $LiAlH_4$ (and $NaAlH_4$) form solvent-separated ion pairs, whereas the ions of Bu_4NAlH_4 are in intimate contact.[399] A slight concentration dependence of the ^{27}Al NMR chemical shift, and the appearance of Al—H coupling below 0.15 mol l^{-1} is consistent with ion-cluster formation at higher concentations, with ion pairs at lower concentrations.[400] Rapid, intermolecular hydride exchange occurs at 25 °C in mono- or di-glyme solutions. No solvent coordination to AlH_4^- is detectable in IR/Raman spectra.[401] NCA of the vibrational data produces an order of M—H stretching force constants in which the aluminum value is less than that for gallium: $AlH_4^- < GaH_4^- < BH_4^-$.

Reaction of hydridoaluminates with $AlCl_3$ forms $AlHCl_2$ and AlH_2Cl as the etherates. The complexes $Li(AlHCl_3)$ and $Li(AlH_2Cl_2)$ also participate.[402] Excess $LiAlH_4$ with $AlCl_3$ in ether at −90 °C can furnish a hydridoaluminate of aluminum itself, $Al(AlH_4)_3{\cdot}OEt_2$, which decomposes to Al + Et_2O + hydrogen between 100 and 200 °C.[403]

25.1.8.2 Borohydride Ligand

Aluminum tris(tetrahydridoborate), $Al(BH_4)_3$, a colourless liquid, f.p. −64.5 °C, b.p. 44.5 °C (extrap.), is the most volatile known compound of aluminum.[1] It is formed by reactions of Me_3Al, AlH_3 or $LiAlH_4$ with diborane. An alternative synthesis is from $AlCl_3 + 3NaBH_4$. Thermal decomposition of $Al(BH_4)_3$ occurs above 70 °C with loss of diborane to form $HAl(BH_4)_2$. Although stable below 25 °C under vacuum, $Al(BH_4)_3$ is a hazardous substance, being oxidized explosively by air and reacting violently with water or any protic agent.

The vibrational spectra of $Al(BH_4)_3$ and $Al(BD_4)_3$ conform to a prismatic structure of D_{3h} symmetry which is maintained in gas, liquid and solid states.[404] NMR spectra show that rapid exchange of ligands is taking place such that bridging and outer protons are equivalent on the NMR timescale at ambient temperature. The dimensions of $Al(BH_4)_3$ have been determined in the vapour by ED.[405] Each BH_4 unit is linked to Al by two three-centre B—H—Al bonds. Similar bridging occurs in the trimethylamine adduct $Al(BH_4)_3{\cdot}NMe_3$, although the Al—H bonds are lengthened considerably from 180 pm in $Al(BH_4)_3$ to 197 pm in the adduct.[406] The Al—N coordinate bond length is 200 pm, a normal distance. A related structure, that of $Al(BH_4)_3{\cdot}NH_3$, has been investigated by single crystal XRD.[407]

Hydridoaluminum tetrahydridoborates $HAl(BH_4)_2$ and H_2AlBH_4 (as ether or amine adducts) are formed by exchange reactions of $Al(BH_4)_3$ with Et_2AlH,[408] or $AlH_3{\cdot}THF$ with B_2H_6,[409] in suitable media. Use of $LiAlH_4$ leads to mixtures of anionic species $[H_{4-n}Al(BH_4)_n]^-$ ($n = 1$–3), some of which appear to be solvated, raising the coordination number of Al above four. Various aluminoboranes of considerable thermal stability are produced by reactions of $Al(BH_4)_3$ with other boron hydrides, *e.g.* B_4H_{10} gives $Al(BH_4)(B_3H_8)_2$.[410] The skeletal structure of $Me_2AlB_3H_8$, determined by vapour phase ED,[411] is analogous to that of B_4H_{10}, the Al atom being linked to each of two boron atoms by a hydrogen bridge, Al—H—B. The Al—H distance is 190.6 pm. In the similar compound of gallium, Ga—H is 198.9 pm. Methylaluminum tetrahydridoborates $MeAl(BH_4)_2$ and $Me_2Al(BH_4)$ can be prepared from $Me_3Al + Al(BH_4)_3$, or from $Me_nAlCl_{3-n} + (3-n)LiBH_4$, where $n = 1$ or 2.[412] Like $Al(BH_4)_3$, they are volatile, monomeric compounds containing bidentate BH_4 ligands, with Al—H distances of 177 pm in four-coordinate $Me_2Al(BH_4)$ and 182 pm in five-coordinate $MeAl(BH_4)_2$.[413] A series of chloroaluminum tetrahydridoborates have been reported, molecular adducts $Cl_{3-n}Al(BH_4)_n{\cdot}OEt_2$ ($n = 1$ or 2) and complex anions, *e.g.* $ClAl(BH_4)_3^-$ as the K^+ salt, having been obtained.[414]

25.1.9 MIXED DONOR ATOM LIGANDS

25.1.9.1 Schiff Base Ligands

A series of pyridoxal amino acid Schiff base complexes have been prepared in which Al is trigonally coordinated by N and two O atoms.[415] These provide a model for the intermediate in the pyridoxal-catalyzed reactions of amino acids. Some Schiff base complexes produced in reactions of the bases with $Al(OPr^i)_3$ have been assigned a structure with five-coordinate aluminum.[416,417]

25.1.9.2 Ligands with N, O and Other Functions

A variety of chelate complexes of Al^{3+} with N and/or O donor atoms are known through stability constant data.[7,8] Their formation illustrates several aspects of Al coordination chemistry. Chelates and other multidentate ligand systems provide a means of regulating the reactivity of aluminum compounds. For example, Al alkoxides can be converted to amino alcohol derivatives to confer water solubility and a degree of hydrolytic stability on otherwise water-sensitive materials.

The edta complex is a particularly stable one: NMR spectroscopy shows that the formation of this complex in solution displaces aqua or hydroxo ligands from Al^{3+}. In other situations, H_2O molecules may be retained in the coordination sphere of the Al atom. A recent example is the six-coordinate complex $AlL(OH_2)_4$ given in solution by the tribasic acid 8-hydroxy-7-[(6′-sulfo-2′-naphthyl)azo]quinoline-5-sulfonic acid (H_3L).[418] The formation of this particular species demonstrates selective coordination to Al by the heterocyclic N and phenolic O atoms, in preference to the azophenol moiety. In another study of multidentate ligand interactions, IR spectra show that complex anions of Al, Ga, In or Tl with oxamic acid $M(NHCOCO_2)_3^{3-}$ are coordinated *via* the carboxyl oxygen and amidic nitrogen, after ionization of one of the hydrogens of the NH_2 group.[419] Similarly, the IR characterization of tris(diethylthiophosphato)-aluminum(III) indicates coordination of Al by both O and S atoms of the ligand.[420]

Molecules in which aluminum forms part of a heterocyclic ring system offer a synthetic challenge likely to lead to developments including Al—N analogues of the boron–nitrogen compounds. Several diazaphosphonia-aluminatacyclobutanes have been characterized in which the unit Al—N—P—N constitutes a four-membered ring.[421] A series of aluminaminophosphines, $R_2AlN(R')PR''_2$, which have been synthesized to act as amphoteric ligands, have been shown to facilitate CO transfer reactions in transition metal complexes. An intermediate has been isolated in which the ligand is part of a ring comprising the CPNAlO sequence.[422]

25.1.10 MACROCYCLIC LIGANDS

There have been several studies of aluminum porphyrins. Octaethylporphyrinatoaluminum hydroxide is readily acetylated and silylated at the OH group.[423] The complex has limited stability in acidic media. Tetraphenylporphinatoaluminum methoxide (TTPAlOMe) promotes the reaction of CO_2 with propylene oxide to produce propylene carbonate under very mild conditions,[424] while TTPAlEt itself reacts with CO_2 under visible light, but not in the dark, and holds interest for investigations of photosynthesis.[425] Phthalocyanine complexes of Al, Ga and In have attracted attention as photocatalysts.[426,427] The conductivity of polymeric materials $(P_cAlF)_n$ increases by factors as high as 10^9 when the structure is doped with iodine to compositions in the range $(P_cAlFI_x)_n$, where $x = 0.01$ to 3.4.

X-Ray studies of crown ether complexes of Al reveal some variation in Al—O coordination with ring size. In particular, 15-crown-5 forms $(Me_3Al)_4$·crown with Al—O = 200.5 pm, whilst dibenzo-18-crown-6 gives $(Me_3Al)_2$·crown with Al—O = 196.7 pm.[428] Other facets of crown ether coordination of aluminum are considered in Section 25.1.5.6.

25.1.11 GALLIUM: INTRODUCTION

Accounts of gallium which include coordination chemistry are those of Greenwood,[429] Sheka *et al.*,[430] and Wade and Banister.[1] Early work on the adducts of gallium trihalides with Lewis bases was documented in 1963,[4] and this area was thoroughly reviewed by Pidcock in 1972.[431] Other aspects of gallium coordination chemistry were examined by Tuck.[431] Stability constants have been tabulated for a number of Ga^{3+} complexes.[7,8]

Three oxidation states of gallium are treated in the following account. Besides the gallium(III) state, it considers the Ga^I state, in which Ga^+ retains the $4s^2$ pair of electrons, and the Ga^{II} state which is associated with the presence of a Ga—Ga bond, giving rise to the formal entity Ga_2^{4+} as the nucleus of Ga^{II} coordination compounds. The significance of metal–metal bonding in group III chemistry was explored in 1975.[432]

25.1.12 GALLIUM(I)

25.1.12.1 Oxide, Sulfide and Selenide Ligands

Gallium(I) oxide, Ga_2O, sublimes as a dark brown powder on heating Ga_2O_3 and Ga to 500 °C. It is stable in dry air, but reduces water to hydrogen and H_2SO_4 to H_2S. Molecular species Ga_2O and Ga_4O_2 are trapped when the vapour above heated gallium oxide is condensed in a low temperature matrix. In_2O, GaInO and $Ga_xIn_{4-x}O_2$ molecules were detected similarly.[433] ED studies of $Ga_2O_{(g)}$ give r(Ga—O) = 184 pm and Ga—O—Ga = 140 ± 10°.[1] Gallium or indium superoxide species MO_2 ($\equiv M^+O_2^-$) and M_2O_4 have been observed in frozen matrices.[434]

An apparent Ga^I coordination compound, $Ga(diox)_2Cl$, containing the 1,4-dioxane ligand was studied crystallographically some years ago.[1] This system is due for reinvestigation now that the related dioxane complex of Ga_2Cl_4 has been shown not to contain $Ga(diox)_2^+$, as previously supposed, but to be a molecular complex of Ga^{II} (Section 25.1.13.1).

Examination of the gallium–sulfur and gallium–selenium phase systems reveals Ga_2S and Ga_2Se, respectively.[1] Other compounds identified in these systems, *i.e.* GaS, Ga_4S_5 and GaSe, are materials with Ga—Ga bonds (see Section 25.1.13.2). Gallium(I) sulfide, Ga_2S, is obtained in amorphous form by reducing GaS with hydrogen at 925 °C.

25.1.12.2 Halide Ligands

All four halides GaX (X = F, Cl, Br or I) are known as vapour-phase species, but tend to disproportionate when condensed, producing elementary gallium. The solid once thought to be GaI has been shown to be Ga_2I_3 (below).

Gallium forms a number of mixed valence halides. In these compounds the Ga^+ cation is stabilized by the presence of a non-coordinating anion formed when the strong Lewis acid GaX_3 combines with halide ligands. Examples are the 'dihalides' $Ga^+GaX_4^-$,[435,436] and compounds Ga_3X_7, *i.e.* $Ga^+Ga_2X_7^-$ (X = Cl, Br or I).[437,438] Compounds of the type $Ga^+AlX_4^-$ are analogous to Ga_2X_4.[439] In the subhalides, Ga_2X_3, the stabilizing anion is $Ga_2X_6^{2-}$, the constitution being $(Ga^+)_2Ga_2X_6^{2-}$ (X = Cl,[437] Br[440] or I[441]).

The crystal structure of Ga_2Cl_4 shows Ga^+ surrounded by eight chlorines of the $GaCl_4^-$ counterions. The gallium(I)–chlorine distance is more than 300 pm, to be compared with the Ga–Cl bond length of 219 pm in $GaCl_4^-$, and implies that the Ga^+ cation is not coordinated. When Ga_2Cl_4 is vaporized, the vapour consists solely of GaCl and $GaCl_3$ molecules.[442] Recent structural analyses of Ga_2I_4 and Ga_2I_3 have directed attention to the influence on these structures of the $4s^2$ electron pair of the Ga^+ ion and confirm that iodide is coordinated only to the higher oxidation state centre.[441]

The halides Ga_2X_4 combine with various donor ligands to give products since shown to be molecular adducts of the metal–metal-bonded Ga^{II} state. There are reports of reactions of the halides Ga_2X_4 with alcohols,[443] water,[444] aqueous NaOH, HCl or HF, and H_2S gas.[445,446] Oxidative addition usually occurs, although there is evidence of the initial precipitation of Ga^I-containing solids in the reactions of Ga_2Cl_4 in benzene solution with H_2O or H_2S.

The crystalline solid Ga_2Cl_4 is soluble in benzene. Under regulated conditions, this solution yields crystals which prove to contain the complex $(C_6H_6)_2Ga^+GaCl_4^-$, and X-ray studies show the cation to be the η^6-bis(benzene) complex of Ga^I.[447] The $GaCl_4^-$ ions interact weakly with Ga^+ to provide bridging chlorine atoms between the Ga^I centres. With hexamethylbenzene the related compounds $Ga(C_6Me_6)^+GaX_4^-$ (X = Cl, Br or I) have been prepared.[448]

25.1.13 GALLIUM(II)

25.1.13.1 Nitrogen, Oxygen and Other Donor Ligands

Complexes derived from the 'gallium dihalides' $Ga^+GaX_4^-$ were prepared in 1959 by adding various N, P, As, O or S donor ligands to a solution of Ga_2X_4 in benzene. Originally these were thought to be Ga^I compounds of the form $GaL_4^+GaX_4^-$, where L_4 represents four monodentate, two bidentate or a single tetradentate ligand coordinated to Ga^+. The reaction with these donors has been reinvestigated and products shown to be molecular coordination complexes of the form $Ga_2X_4L_2$, with a covalent Ga—Ga bond.[449] In $Ga_2X_4(diox)_2$ (X = Cl or

Br),[450] 1,4-dioxane ligands function as monodentate ligands so that each Ga atom is four-coordinate, as in the case of the $Ga_2X_6^{2-}$ ions (Section 25.1.13.3). The molecules adopt a partially eclipsed conformation in the crystal structure. Attempts to form $Ga_2Br_4py_2$ by direct reaction of Ga_2Br_4 with pyridine fail owing to disproportionation to Ga metal and the Ga^{III} state, caused by the basic ligand. $Ga_2Br_4py_2$ can, however, be prepared successfully from the dioxane complex by ligand displacement.[451] The molecular structure resembles that of the dioxane complex, although in this case the *trans* conformation is observed.

Intense Raman bands, which are useful for recognizing these M—M-bonded species, are seen in the $Ga_2X_4(diox)_2$ series, at 238 cm^{-1} for X = Cl, 169 cm^{-1} for X = Br and 143 cm^{-1} for X = I.[449] Unstable species which are present in ether extracts from freshly prepared solutions of gallium in hydrobromic acid have been observed in this way. The species are now thought to be the complex acids $(H^+)_2Ga_2Br_6^{2-}$ and $H^+Ga_2Br_5OEt_2^-$, and the ether adduct $Ga_2Br_4(OEt_2)_2$.[432] As anticipated, treatment of the mixture of species with HBr displaces the ether ligands forming $(H^+)_2Ga_2Br_6^{2-}$.

In an interesting further development, the mixed metal compounds $InGaX_4L_2$ (X = various ligands, including py, THF and diox) have been obtained by carefully controlled reactions of $In^+GaX_4^-$ with the various ligands.[452] The salts $(Bu^n_4N)_2InGaX_6$ (X = Cl or Br) could also be prepared. Raman spectra indicate that these compounds contain In—Ga bonds, whilst bearing coordinated halogen and other ligands.

25.1.13.2 Sulfide, Selenide and Telluride Ligands

Gallium forms the golden yellow sulfide GaS, and analogous compounds GaSe and GaTe, which are of interest as semiconductor materials. The structures contain Ga atoms linked in pairs forming the Ga_2^{4+} entity which is bonded by S (Se or Te) ligands to form a layer lattice. The Ga atoms are tetrahedrally coordinated by three group VI atoms and the partnering Ga. There are several modifications of GaS which differ from one another in the stacking of layers and the interlayer spacing,[453–455] but all have essentially the same coordination within the Ga_2S_6 unit. The nature of metal–ligand bonding in GaS, GaSe and GaTe has been investigated by vibrational,[456,457] photoelectron[458] and NQR[459] spectroscopy. The selenogallate $Na_2Ga_2Se_3$, which can be synthesized by heating a 2:2:3 mixture of the elements, is structurally related to GaSe.[455] It contains linked Ga_2Se_6 units with Ga—Ga bonds of 240 pm, a distance slightly less than that in GaS (244.7 pm), GaSe (245.7 pm), GaTe (243.1 pm)[453] or $GaS_{0.6}Se_{0.4}$ (244.7 pm).[454]

25.1.13.3 Halide Ligands

Freshly prepared solutions of gallium metal in aqueous halogen acids act as powerful reducing agents, and this property proves to be due to Ga^{II} halide complexes. The species concerned are the anions $(X_3Ga—GaX_3)^{2-}$ of D_{3d} symmetry which have been isolated as crystalline salts with tetraalkylammonium cations,[460] Ph_4P^+ and Ph_3PH^+.[461] The gallium–halogen distances are similar to those of the GaX_4^- ions, while the Ga—Ga bond length (239 to 242 pm),[441,462] which is almost independent of change of halide, corresponds closely to twice the covalent radius of gallium as observed in the Ga^{III} state. The vibrational specra of all three $Ga_2X_6^{2-}$ ions have been assigned[460] and NQR spectra of $Ga_2Cl_6^{2-}$ and $Ga_2Br_6^{2-}$ have been studied.[463] The Ga NQR frequencies were found to be considerably lower than those of Ga^{III} compounds and the data were used to calculate electron distribution on Ga and halogen atoms as a means of comparing the bonding with that of Ga^{III} complexes. The formation of $Ga_2Cl_6^{2-}$ in gallium-rich chloride melts has been demonstrated by Raman spectroscopy,[437] and anions of this type have also been shown to occur in the low-valent compounds Ga_2X_3 formed in bromide and iodide systems.[437,440,441,464]

Some reactions of the hexahalogenodigallate(II) ions have been investigated.[432] Treatment of a solution of a $Ga_2X_6^{2-}$ salt in nitromethane with a different halide, as solid KY, results in replacement to form mixed halides of gallium(II), $Ga_2(X,Y)_6^{2-}$ which are characterized by strong bands in the 120–250 cm^{-1} region of Raman spectra. The analogous reaction of $Ga_2X_6^{2-}$ with potassium thiocyanate results in only partial replacement of halide by the pseudohalide. IR spectra show the NCS ligand to be coordinated to gallium through the nitrogen atom, as in the case of gallium(III) complexes. The Ga—Ga bond of $Ga_2X_6^{2-}$ ions is stable towards disproportionation, but readily undergoes scission in reactions with oxidizing agents, for example halogens, Ag^+ or Hg^{2+} ions, to yield gallium(III) complexes. The $Ga_2X_6^{2-}$ ions are slowly attacked by water. Hydrogen is evolved and hydroxo complexes of Ga^{III} are produced.

The formation of complex halides $(X_3GaInX_3)^{2-}$, containing the dinuclear $GaIn^{4+}$ entity,[452] has been mentioned in Section 25.1.13.1, and some dinuclear In^{II} compounds have also been synthesized. Thus, although the molecular form of M_2X_4 is not known in group III (except for boron), stable complexes of these halides can be prepared, either as anions $M_2X_6^{2-}$ or as molecular adducts $M_2X_4L_2$.

25.1.14 GALLIUM(III)

25.1.14.1 Group IV Ligands

25.1.14.1.1 Cyanide ligands

Neither gallium cyanide, $Ga(CN)_3$, nor binary complexes, $Ga(CN)_x^{3-x}$, appear to have been reported. The ion $Me_2Ga(CN)_2^-$ has been prepared as the potassium salt by the reaction of $K_2Hg(CN)_4$ with Me_3Ga.[465] The related complexes $(R_3Ga)_2CN^-$ (R = Me or Et) contain a bridging cyanide group, like the corresponding aluminum species.

25.1.14.1.2 Silicon and heavier group IV ligands

There has been little work on compounds in which gallium is bonded to silicon or a heavier group IV atom. Tris(trimethylsilyl)gallium is produced by treating $GaCl_3 + 3Me_3SiCl$ with lithium in THF, the initial adduct $Ga(SiMe_3)_3{\cdot}THF$ being converted to $Ga(SiMe_3)_3$ by vacuum sublimation.[466] $Ga(SiMe_3)_3$ is monomeric, with a planar $GaSi_3$ framework, characterized by vibrational bands $\nu(Ga—Si)_{sym} = 312\ cm^{-1}$ (Raman) and $\nu(Ga—Si)_{deg} = 349\ cm^{-1}$ (IR, Raman). Reaction of Me_2GaCl with $KGeH_3$ leads to $K(H_3GeGaMe_2Cl)$ and, by elimination of KCl, to Me_2GaGeH_3.[467] The compound $Li(Me_3GaSnMe_3)$ readily decomposes to $LiGaMe_4$ (*cf.* In).

25.1.14.2 Nitrogen Ligands

25.1.14.2.1 Ammonia and amine ligands

The gallane complex $GaH_3{\cdot}NH_3$ is unknown. Metal fluoride adducts $MF_3{\cdot}3NH_3$ (M = Ga or In) are presumably six-coordinate; above 100 °C step-wise loss of ammonia occurs to give MF_3. For chloride and bromide systems, adducts $GaX_3{\cdot}nNH_3$ (X = Cl, n = 1, 3, 5, 6, 7 and 14; X = Br, n = 1, 5, 7, 9 and 14) were reported in 1932.[4] In recent work, exposure of gallium trichloride or tribromide to excess ammonia yielded white powdery substances which proved difficult to characterize, but by restricting the amount of ammonia to one molar equivalent, glassy, low-melting solids, $GaX_3{\cdot}NH_3$, were obtained.[468] The vibrational spectra were fully assigned on the basis of C_{3v} symmetry and showed $\nu(Ga—NH_3)$ stretching bands at 498 cm^{-1} (X = Cl) or 480 cm^{-1} (Br). NCA yields Ga—N stretching force constants k(Ga—N) of 2.00 and 1.85 N cm^{-1} for $H_3N—GaCl_3$ and $H_3N—GaBr_3$, respectively. The $H_3N{\cdot}GaCl_3$ molecule can be observed in the gas phase where ED results have yielded the bond lengths Ga—N = 205.7 pm and Ga—Cl = 214.2 pm. The $GaCl_3$ unit has a rather flat pyramidal strucure suggesting a weaker donor–acceptor interaction than in the aluminum analogue. The Ga—N bond lengthens slightly to 208.1 pm in $H_3N{\cdot}GaBr_3$.[469] The enthalpies of dissociation of $GaX_3{\cdot}NH_3$ [134 kJ mol^{-1} (X = Cl) and 137 kJ mol^{-1} (Br)] are much smaller than those of other nitrogen bases.[470] The trimethylgallium adduct $Me_3Ga{\cdot}NH_3$ has been thoroughly characterized.[471] A cationic ammonia complex in which gallium appears to be four-coordinate occurs in $[Me_2Ga(NH_3)_2]Cl$.[472]

Amine adducts $GaH_3{\cdot}NR_3$ are produced by reaction of $LiGaH_4$ with the base hydrochloride, or by displacing a ligand, such as ether, from a gallane adduct $GaH_3{\cdot}L$. Trimethylamine forms $GaH_3{\cdot}NMe_3$, m.p. 70 °C, volatile without decomposition, and $GaH_3{\cdot}2NMe_3$, unstable above −60 °C. The second adduct is much weaker (ΔH for dissociation to $GaH_3{\cdot}NMe_3$ and Me_3N is 43 kJ mol^{-1}).[1] Structure and bonding in the $Me_3N—GaH_3$ molecules have been studied by

NMR,[473] XRD (Ga—N = 197 ± 9 pm),[1] and more recently by ED, microwave and vibrational spectroscopy.[474] These newer data furnish a length of 212 ± 1 pm and a force constant of 2.44 N cm^{-1} for the Ga—N bond. $GaH_3 \cdot NMe_3$ undergoes some ligand displacements. *e.g.* with dimethylamine forming $GaH_3 \cdot NHMe_2$ leading to $(H_2GaNMe_2)_2$ by loss of H_2. The reaction of $GaH_3 \cdot NMe_3$ with BF_3 apparently involves the initial formation of a five-coordinate complex, $Me_3N \cdot GaH_3 \cdot BF_3$, which decomposes to give $Me_3N \cdot BF_3 + Ga + H_2$.[475] The stabilities of gallane adducts decrease in the following order of bases:[1] $Me_2NH > Me_3N >$ pyridine $> Et_3N > Me_2PhN > Ph_3N$.

Gallium trihalides in non-polar solvents, or as ether adducts, react with amines usually in 1:1 stoichiometry. Recrystallization of GaX_3 from trimethylamine gives $GaX_3 \cdot NMe_3$ when X = Br or I.[476] The chloride initially forms $GaCl_3 \cdot 2NMe_3$ which easily releases one mole of NMe_3. Vibrational spectra have been assigned, and NCA for $Me_3N \cdot GaCl_3$ computes a substantial Ga—N force constant of 2.50 N cm^{-1}.[477] For $Me_3Ga \cdot NMe_3$, ED results yield r(Ga—N) = 220 pm and Raman spectral data provide k(Ga—N) = 1.6 N cm^{-1}.[478] In reaction with methylamines, the Lewis acid affinity of a series of gallium compounds including $GaCl_3$ decreases in the order $GaCl_3 > Bu^nGaCl_2 > Bu^n_2GaCl > Bu^n_3Ga$.[479] Calorimetric studies of the interaction of various amine bases with $GaCl_3$ have yielded bond dissociation energies $D(GaCl_3—L)$ for the coordinate link of about 290 kJ mol^{-1}. The value for oxygen donors is smaller: *ca.* 235 kJ mol^{-1}. $D(GaCl_3—Cl^-)$ is greater: *ca.* 365 kJ mol^{-1}. Detailed comparisons are available.[435]

In a recent study, $GaCl_3$ has been shown to form a 1:1 complex with hexamethylborazine, in which the conjugated system of the BN heterocycle is disrupted.[480] The ring is no longer planar and contains a tetracoordinate nitrogen linked to $GaCl_3$, with bond distances Ga—N = 206.4 pm and Ga—Cl = 215.3 pm (average). In solution at low temperature, the NMR spectrum indicates a different structure for the adduct, with a bidentate borazine ligand attached to five-coordinate gallium. Leaving aside the rather uncertain crystallographic measurement for $GaH_3 \cdot NMe_3$, coordinate Ga—N bond distances for amine donors are in the range 205–220 pm. The smallest values are for strong bases coordinated to $GaCl_3$.

25.1.14.2.2 N-Heterocyclic ligands

There are early reports from phase diagram studies of solids $GaX_3(py)_n$ (X = Cl, n = 1 or 2; X = Br, n = 1 or 3) and of some piperidine adducts.[1,3] Recent X-ray study shows that $GaCl_3 \cdot 2py$[481] has the ionic structure $GaCl_2(py)_4^+GaCl_4^-$. The six-coordinate cation adopts a *trans* octahedral structure with r(Ga—N) = 210.8 pm (average). The Ga—Cl distance in the cation is 231.2 pm, compared with 215.7 pm in the four-coordinate anion. Vibrational studies, making use of pyridine-d^5, locate two ν(Ga—py) stretching modes at 248 and 263 cm^{-1}. There is NMR evidence that the strongly basic heterocyclic ligands afford complex cations in MeCN solution, *e.g.* $GaX_2(py)_2^+$ and $GaX_2(phen)_2^+$ (X = Cl, Br or I).[482] Formation of the six-coordinate anions $GaX_3Y(py)_2^-$ (X = Cl or Br, Y = I′; X = I, Y = Cl or Br) has been claimed.[483] This is remarkable since gallium is usually restricted to a maximum coordination number of four in anionic halide complexes other than fluorides.

With bidentate 1,10-phenanthroline and 2,2′-bipyridyl ligands, gallium forms $Ga(ClO_4)_3 \cdot 3bipy$, $GaCl_3 \cdot phen$, $GaCl_3 \cdot 2phen$, $GaCl_3 \cdot bipy$ and $GaCl_3 \cdot 2bipy$.[484,485] A common feature is the presence of a complex cation, *e.g.* $Ga(bipy)_3^{3+}$ or $GaCl_2(bipy)_2^+$. The crystal structure of $GaCl_3 \cdot bipy$ shows it to be *cis*-$[GaCl_2(bipy)_2]^+GaCl_4^-$.[486] The Ga—N distances of 209.5 and 211.1 pm are very similar to those of the pyridine complex above. In the 1:2 (bidentate) adducts, Cl^- replaces $GaCl_4^-$ as the counteranion. The dimethylgallium derivative $Me_2GaCl(phen)$ is an example of a five-coordinate gallium compound.[487] Ionization to $Me_2Ga(phen)^+Cl^-$ occurs in aqueous solution. The terpyridyl adduct $GaCl_3 \cdot terpy$ is notable as a molecular complex; in it r(Ga—N) = 211.2 pm (mutually *trans*) and 203.4 pm (*trans* to Cl).[488] The compound $(D_2Ga \cdot pyrazol\text{-}1\text{-}yl)_2$ is also a molecular complex.[489] It contains the unit $D_2Ga(\mu\text{-}NN)_2GaD_2$ which assumes the boat conformation.

25.1.14.2.3 Amido and imido ligands

Potassium amidogallate, $K[Ga(NH_2)_4]$, is prepared by dissolving gallium in a solution of KOH in liquid ammonia. Heating this compound to 300 °C under vacuum removes two moles of ammonia, forming $KGa(NH)_2$.[1] Partial neutralization of the ammonia solution of $K[Ga(NH_2)_4]$ with NH_4Cl yields $Ga(NH_2)_3$. A reported amidogallane, $(H_2GaNH_2)_n$ appears to be polymeric.[1] The anion $[Ga(NH_2)_4]^-$ has tetrahedral symmetry, according to the single crystal Raman spectrum in which ν(Ga—N) appears near 550 cm^{-1}.[490] A more highly coordinated complex, $Na_2[Ga(NH_2)_5]$, has also been prepared. From evidence of IR spectra, this contains six-coordinate gallium, with a structure based on linked $Ga(NH_2)_6$ octahedra with both bridging and non-bridging NH_2 groups.[491]

Dimethylamido compounds $GaCl(NMe_2)_2$ and $Ga(NMe_2)_3$ are obtained from the reaction of $GaCl_3$ with $LiNMe_2$; an excess yields $Li[Ga(NMe_2)_4]$.[492] $Ga(NMe_2)_3$ dimerizes through bridging dimethylamido ligands, but this does not occur in the case of $Ga[N(SiMe_3)_2]_3$, presumably for steric reasons.[493] CS_2 inserts into all of the Ga—N bonds of $Ga(NMe_2)_3$, forming $Ga(S_2CNMe_2)_3$. Volatile compounds $(H_2GaNR_2)_n$ and $[H_2GaN(CH_2)_xCH_2]_n$ ($n = 2$ or 3) result from the reaction of amines or cyclic imines $CH_2(CH_2)_xNH$ with $GaH_3{\cdot}NMe_3$ at room temperature.[494] In $(H_2GaNMe_2)_2$ the Ga—N bond length is 202.7 pm and the central Ga_2N_2 ring is almost square,[495] as it is in related molecules where B, Al or In atoms are bridged by Me_2N ligands. An X-ray study of $[H_2GaN(CH_2)_2]_3$ reveals a Ga_3N_3 ring in the chair conformation, with Ga—N = 197 ± 9 pm.[496] Compounds $RGaCl_2$ (R = Me or Bu) react with two equivalents of $(Me_3Si)_2NH$ to afford $[R(Cl)GaNHSiMe_3]_2$ in which substituents on the Ga_2N_2 ring adopt the *trans-trans* conformation, according to IR and XRD evidence.[497] Reactions of Me_3Ga—base adducts eliminate methane to form $(Me_2GaNMe_2)_2$ from $Me_3Ga{\cdot}NHMe_2$, or $(Me_2GaNH_2)_2$ and $(MeGaNH)_x$ from $Me_3Ga{\cdot}NH_3$. Analogous compounds with P or As ligands in place of N bases are reviewed elsewhere.[498] The structure of an imido complex containing the Ga_8N_8 framework is shown in Table 1 (Section 25.1.3.3).

25.1.15.2.4 Azide, cyanate, thiocyanate and selenocyanate ligands

Azide complexes are formed by Ga^{3+} in aqueous solution.[499] The formation of $Ga(N_3)^{2+}$ is decidedly more exothermic than that of $In(N_3)^{2+}$. The dimethylgallium cation Me_2Ga^+ forms $[Me_2Ga(N_3)]_2$ and $Me_2Ga(N_3)_2^-$, in each of which the gallium atom is four-coordinate.[500] Similarly, Me_3Ga forms $Me_3Ga(N_3)^-$ and $(Me_3Ga)_2(N_3)^-$ (the latter with a bridging azido ligand), as well as the corresponding complexes with NCO, NCS, NCSe or CN ligands.

Little is known of gallium cyanates. There are reports of complex isocyanates $Ga(NCO)_3{\cdot}$bipy and $Ga(NCO)_3{\cdot}$phen and of an unstable compound $(Me_4N)_3Ga(NCO)_6$, thought from spectroscopic evidence to contain cyanate groups bonded through the O atom.[501] Thiocyanato complexes are formed in aqueous solution, *e.g.* $Ga(NCS)^{2+}$ for which the dissociation constant at 25 °C is 4.8×10^{-3}.[502] When excess NCS^- is present, anionic complexes are produced. Gallium can be solvent extracted from these solutions by ethers or by trilaurylamine as $(HB)Ga(NCS)_4$, where HB represents the protonated ether or amine, or by tributyl phosphate or a phosphonate in the form $Ga(NCS)_3L_3$.[503] The hydrate $Ga(NCS)_3{\cdot}3H_2O$ is soluble in ether and in polar organic solvents.[504] The coordinated water molecules can be replaced by diamines to yield $Ga(NCS)_3(H_2O){\cdot}$diam, $Ga(NCS)_3(\text{diam})_n$ ($n = 1.5$ or 2), and $(\text{Hdiam})Ga(NCS)_4{\cdot}$diam, where diam = bipy or phen.[505] Maintaining the first of these complexes under vacuum at 120 °C removes the molecule of water. IR spectra confirm Ga—NCS bonds and suggest that the bidentate amine is also coordinated to the metal. An ethylenediamine adduct of gallium thiocyanate has been formulated with a complex cation, $[Ga(en)_3](NCS)_3$, whereas the acetonitrile complex from $GaBr_3 + 4KNCS$ in MeCN is $K[Ga(NCS)_4{\cdot}MeCN]$[506] and may contain five-coordinate gallium.

Crystalline salts containing tetra-, penta- and hexa-isothiocyanate complex anions can be isolated from aqueous or methanolic solutions.[507] Frequencies ν(Ga—NCS) occur in IR spectra at 350 cm^{-1} in four-coordinate or 250 cm^{-1} in six-coordinate species. Although powder patterns have been reported, single-crystal X-ray data are lacking for gallium thiocyanates. Some gallium selenocyanide complexes $Ga(NCSe)_3L_3$ and $Ga(NCSe)_4^-$ have been prepared.[508] Attempts to obtain $Ga(NCS)_3$ or $Ga(NCSe)_3$ appear to have failed.

25.1.14.2.5 Nitrile ligands

As with aluminum, most recent work on nitrile complexes has been directed to understanding the constitution of solutions of GaX_3 in MeCN. ^1H NMR measurements have been interpreted in terms of the four-coordinate cations $[GaX_2(MeCN)_2]^+$ and GaX_4^- (X = Cl, Br or I).[509] Later ^{71}Ga NMR studies confirm the presence of $GaCl_4^-$ but favour $[Ga(MeCN)_6]^{3+}$ as the cationic species.[510] The equilibrium

$$4MCl_3(MeCN)_3 \rightleftharpoons M(MeCN)_6^{3+} + 3MCl_4^- + 6MeCN$$

lies further to the right for Ga than Al complexes and this can be attributed to the greater stability of the MCl_4^- ion formed by gallium. For various molecular complexes $GaCl_3{\cdot}RCN$, analysis of NQR data shows that approximately half of the charge transferred from the donor molecule is localized on the Ga atom.[511]

25.1.14.3 Phosphorus and Heavier Group V Ligands

Molecular adducts $GaH_3{\cdot}L$ have been prepared with Me_3P, Me_2PH, Ph_3P and Ph_2PH, but not with PH_3, by the reaction shown in equation (11). Gas phase displacement reactions indicate that Me_3P has about the same donor strength as Me_3N towards gallane.[512] Analysis of vibrational spectra of these adducts, however, suggests rather different force constants: $k(Ga—P) = 2.0\ N\ cm^{-1}$ in $GaH_3{\cdot}PMe_3$,[513] compared with $k(Ga—N) = 2.44\ N\ cm^{-1}$ in $GaH_3{\cdot}NMe_3$.[474]

$$LiGaH_4 + R_3P + HCl \longrightarrow GaH_3{\cdot}PR_3 + LiCl + H_2 \qquad (11)$$

Phosphine reacts with gallium trihalides to form the adducts $GaX_3{\cdot}PH_3$ (X = Cl or Br) as glassy solids. In contrast to the behaviour with ammonia, there is no tendency to combine with more than one mole of ligand. Gallium triiodide does not take up phosphine. Alkyl- and aryl-phosphines give 1:1 adducts with $GaCl_3$ and $GaBr_3$, and some GaI_3 adducts have also been prepared.[514] The C_{3v} molecular structure of the phosphine complexes $H_3P—GaX_3$, in solid or liquid states, is substantiated by Raman and IR spectra.[468] NCA applied to the $GaCl_3{\cdot}PH_3$ data yields $k(Ga—P) = 1.0\ N\ cm^{-1}$. As might be expected, the bond stretching force constant obtained for $GaCl_3{\cdot}PMe_3$ is larger, $k(Ga—P) = 1.75\ N\ cm^{-1}$.[515] NQR data also show Me_3P to be a very good donor towards $GaCl_3$, and in the same study a crystal structure determination has found the $Me_3P—GaCl_3$ molecule to have the unusual staggered conformation: Ga—P = 235.3 pm and Ga—Cl = 217.4 pm.[516]

Vibrational spectra of phenylphosphine adducts $GaCl_3{\cdot}PPh_3$ and $GaCl_3{\cdot}PPh_nH_{3-n}$ ($n = 1$ or 2) are fully in accord with the presence of a four-coordinate $PGaCl_3$ skeleton in solid and liquid states, and in non-polar solvents.[517–519] However, in acetone or nitromethane, conducting solutions are formed by $GaX_3{\cdot}PPh_3$ (X = Cl or Br) which have IR spectra consistent with GaX_4^- ions.[517] A plausible cation is $GaX_2(PPh_3)_2^+$ but confirmation is lacking. There have been a few studies of gallium halide adducts with bidentate P or As ligands. The vibrational spectra of solids $Ga_2X_6{\cdot}$depe (X = Cl or Br) are consistent with a structure in which each phosphorus atom carries a GaX_3 group.[518] The iodide has a markedly different spectrum and may be ionic $GaI_2(depe)^+GaI_4^-$. Nyholm had earlier reported $Ga_2Cl_6{\cdot}$diars and $GaX_3{\cdot}$diars (X = Br or I) and, on the basis of conductivity measurements, had formulated the latter as $GaX_2(diars)_2^+GaX_4^-$.[520] The existence of an apparently six-coordinate complex, with atoms as large as two Br or I and four As ligands, requires checking. Tris(bidentate) complexes $Ga(diars)_3^{3+}$ could not be obtained.

Some P and As donor complexes of organogallium compounds are known.[498] These include $Me_3Ga{\cdot}PH_3$, $Me_3Ga{\cdot}PMe_3$,[521] $Ph_3Ga{\cdot}PPh_3$ and $Ph_2ClGa{\cdot}PPh_3$[522] investigated by IR/Raman spectra. Pyrolysis of a mixture of Et_3Ga and $HPEt_2$ yields $(Et_2GaPEt_2)_3$, the structure of which is based on a six-membered Ga_3P_3 ring.[523] Me_2GaCl and $HPEt_2$ yield $[Me(Cl)GaPEt_2]_n$, where n indicates a mixture of oligomers, 2 and 3. Fragmentation of these compounds enables GaP to be deposited as a surface coating, significant as a type III–V semiconducting layer. The recently reported compounds $[R_2GaN(PPh_2)_2]_2$ (R = Me or Et)[524] appear to exist in solution as two isomers; one has a six-membered ring GaNPGaNP, the other is eight-membered GaPNPGaPNP. Further developments towards compounds with Ga—N, Ga—P and Ga—As networks can be expected.

25.1.14.4 Oxygen Ligands

25.1.14.4.1 Oxide, hydroxide and H_2O as ligands

Gallium hydroxide $Ga(OH)_3$ is precipitated as a gel from aqueous Ga^{3+} solutions by adding ammonia. It is also produced by the action of acids on solutions of gallates. On standing the hydroxide forms GaO(OH), which is obtained as crystalline material from $Ga(OH)_3$ at 110 °C.[1] The crystal structure is of the six-coordinate diaspore type. In another form, Ga is six-coordinate with the oxygens arranged in the shape of a distorted trigonal prism.[525] Heated above 300 °C, moist gallium hydroxide gives β-Ga_2O_3, the most stable of five oxide modifications.[1] Whereas α-Ga_2O_3 has the corundum structure with a Ga—O distance of 183.4 pm and octahedral coordination, β-Ga_2O_3 consists of vertex-sharing GaO_4 tetrahedra and edge-sharing octahedra, with Ga—O distances of 183 and 200 pm, respectively.[526]

On heating with other metal oxides, gallium oxide forms gallates M^IGaO_2, $M^{II}Ga_2O_4$ (spinels) and $M^{III}GaO_3$. There are oxides of Ga with aluminum or indium which form solid solutions, $M^{III}_{2-x}Ga_xO_3$. Mixed oxide stoichiometries other than 1:1 (O^{2-}:Ga_2O_3) are quite common. With sodium, there are complex gallates $NaGaO_2$, Na_3GaO_3, $Na_8Ga_2O_7$ and Na_5GaO_4.[527] The structures of these last two are built up from effectively isolated $Ga_2O_7^{8-}$ and GaO_4^{5-} units, the bonding within which has been explored by MO calculations.[528] Some hydrated forms have also been studied, such as $LiGaO_2{\cdot}6H_2O$, effectively $[Li(OH_2)_4]^+[GaO_2(OH_2)_2]^-$, and $LiGaO_2{\cdot}8H_2O$. Here the four-coordinate Ga—O distances are 185–189 pm.[529] $NaGa_{11}O_{16}(OH)_2$ has a complicated structure in which both GaO_4 tetrahedra and GaO_6 octahedra are present, with mean values *r*(Ga—O) of 185 and 200 pm, respectively.[530] Structures based on GaO_4 tetrahedral prevail among gallates of the divalent metals, *e.g.* $SrGa_2O_4$, $CaGa_2O_4$, $Ca_3Ga_4O_9$ and $Pb_9Ga_8O_{21}$.[531] In $CuGaInO_4$, the Ga atom has trigonal bipyramidal coordination, while that of In is distorted octahedral.[532] Gallium forms oxyhalides GaOX.[1] GaOF is inert to water; the solubility of the others, with extensive hydrolysis, increases in the order GaOCl < GaOBr < GaOI.

Gallium hydroxide is amphoteric, and is a much stronger acid than aluminum hydroxide. For $Ga(OH)_3$ the first acid dissociation constant is 1.4×10^{-7} [for $Al(OH)_3$ the value is 2×10^{-11}].[1] Polymerization occurs in aqueous Ga^{3+} solutions to which OH^- is added,[533] but this tendency is less than in the case of aluminum solutions (Section 25.1.5.1). The formation constants of mononuclear hydroxo complexes of Ga, including $Ga(OH)_4^-$, and the hydrolysis constants of gallium ions have been measured by a competing ligand technique.[534]

In acidic solutions, gallium exists as the hexaaqua cation $Ga(OH_2)_6^{3+}$, which occurs in hydrated crystals of oxyanion salts (Section 25.1.14.4.4). NMR studies employing ^{17}O give $Ga(OH_2)_6^{3+}$ as the hydrated complex.[535] Water exchange at 25 °C is characterized by the rate constant $k_1 = 1.8 \times 10^3\ s^{-1}$. This is much faster than the exchange rate of $Al(OH_2)_6^{3+}$ for which $k_1 = 0.17\ s^{-1}$, reflecting a lesser affinity of Ga^{3+} for oxygen as a ligand. A mechanism of S_N2 type involving $Ga(OH_2)_7^{3+}$ has been suggested.[536] The aquated gallium cation $Ga(OH_2)_6^{3+}$, with K_a of 2.5×10^{-3}, is a stronger acid than $Al(OH_2)_6^{3+}$ ($K_a = 1.1 \times 10^{-5}$).

The $Ga(OH_2)_6^{3+}$ complex is the usual reference state in ^{71}Ga NMR spectroscopy. Compared with this species, the Ga nucleus in $Ga(OH_4)^-$ exhibits a chemical shift of 192 p.p.m. and is more highly shielded.[537] Physical methods, particularly NMR, show that Ga^{3+}, in the form of the aqueous nitrate or perchlorate, continues to be hydrated by six H_2O molecules,[536] even in the presence of a large excess of acetone. Entry of a ligand species into the primary solvent shell can be induced by suitable conditions of temperature and solvent composition. For example, with solutions of gallium chloride in aqueous acetone at −60 °C, an increase in the amount of acetone (not itself coordinated) brings about a progressive fall in the hydration number from 6 to *ca.* 2. The resulting species may be the mixed ligand complex $GaCl_2(OH_2)_2^+$, Dimethylgallium hydroxide is a tetramer $(Me_2GaOH)_4$, and the presence of bridging OH ligands has also been established by X-ray analysis of a dimeric gallium chelate complex.[538]

25.1.14.4.2 Alkoxide ligands

Gallium alkoxides can be prepared by the methods employed for aluminum.[539] $Ga(OMe)_3$, a polymeric material, decomposes without melting but can be sublimed under vacuum. $Ga(OEt)_3$, m.p. 144 °C, $Ga(OPr^n)_3$ and $Ga(OBu^n)_3$ are more volatile than $Ga(OMe)_3$, and these are tetrameric in solution. $Ga(OPr^i)_3$, a liquid, and $Ga(OBu^t)_3$ have dimeric structures bridged by alkoxide ligands, according to NMR studies.[540] The bulky trimethylsilyl ligand results in $Ga(OSiMe_3)_3$ being monomeric, but does not prevent the formation of a complex anion $Ga(OSiMe_3)_4^-$. Aryloxides, *e.g.* $Ga(OPh)_3$, m.p. 192 °C, are obtained by prolonged heating of Ga metal with the phenol at 180–220 °C.[541] Like group III alkoxides generally, these compounds are Lewis acids and form 1:1 adducts with ammonia, pyridine and other bases. $Ga(OPh)_3$ reacts with boiling MeOH or EtOH to form the corresponding alkoxide.

Reaction of the halide GaX_3 (X = Cl or Br) with less than three moles of alcohol may initially give adducts $GaX_3 \cdot ROH$ but these are readily converted into products $GaX_n(OR)_{3-n}$ (n = 1 or 2). These compounds may be di-, tri- or tetra-meric: in all cases alkoxide ligands are bridging. If the halides Ga_2X_4 are used, the reactions involve oxidation of Ga^+ by alcohol (see equations 12 and 13).[443]

$$Ga_2X_4 + 3EtOH \xrightarrow{0\,^\circ C} GaX(OEt)_2 + GaX_3 \cdot EtOH + H_2 \quad (12)$$

$$Ga_2X_4 + 2EtOH \xrightarrow{25\,^\circ C} 2GaX_2(OEt) + H_2 \quad (13)$$

Like the Al compounds, gallium alkoxides are susceptible to hydrolysis giving ill-defined products, including $Ga_2O(OR)_4$ and GaO(OR). Methanolysis of Ga^{3+} ions produces $Ga(OMe)^{2+}$ and $Ga(OMe)_2^+$ in solvated form, as well as polymeric complexes.[542] In reactions with another metal alkoxide, four-coordinate complex anions of the type $Ga(OR)_4^-$ can be formed. Other routes, such as $MGaH_4 + 4ROH$, or $MGaR_4 + 2O_2$, are available. Structural data for gallium alkoxides are in short supply. There are Raman and IR spectra of MeGaCl(OMe) and R_2MOMe (R = Me, or Et; M = Ga or In). The Me_2Ga derivatives are trimeric with a puckered Ga_3O_3 ring.[543]

25.1.14.4.3 *β-Ketoenolate and related ligands*

Tris(acetylacetonato)gallium(III), m.p. 195 °C, exists in two forms differing slightly in density. Crystal structure determination shows the central GaO_6 octahedron of $Ga(acac)_3$ to be almost regular, with an average Ga—O distance of 195.2 pm.[544] $Ga(acac)_3$ is formed in aqueous solutions of Ga^{3+} and acetylacetonate ions, and is conveniently prepared by adding aqueous ammonia to a solution of $Ga_2(SO_4)_3$ + Hacac. However, at ligand to Ga ratios less than three, hydrolysis leading to the formation of hydroxo complexes must also be considered.[545]

Rates of ligand exchange have been measured for some β-diketonato complexes. Exchange between $Ga(acac)_3$ and ^{14}C-labelled Hacac in THF solution is much faster than in the case of $Al(acac)_3$.[546] A dissociative mechanism proceeds *via* a five-coordinate transition state in which one ligand is unidentate. Coalescence temperatures and activation energies for *cis–trans* rearrangements are conveniently measured by ^{19}F NMR spectroscopy, applied to tris(β-diketonates) with fluoro substituents.[188] The findings place gallium between Al and In in the usual order of a decrease in bond strength of Al—O > Ga—O > In—O, which is also consistent with Raman studies of $M(acac)_3$.[1] When $Ga(acac)_3$ is added to a solution of $Ga(ClO_4)_3$ in dimethylformamide the NMR spectrum shows new signals attributable to $[Ga(acac)(DMF)_4]^{2+}$ and $[Ga(acac)_2(DMF)_2]^+$.[547] The Al^{III} system behaves similarly and the two systems have been compared.

There is a dimethylgallium acetylacetonate $Me_2Ga(acac)$, m.p. 22 °C, and a dimethylgallium tropolonate, the structure of which proves to be dimeric with Ga—O bonds of 197.4 and 202.5 pm to the chelation ligand and distances of 246.3 pm within the central Ga_2O_2 ring of the dimer.[548]

25.1.14.4.4 *Oxyanion ligands*

(*i*) *Borate, carbonate, silicate and germanate*

Gallium borate $GaBO_3$ is unknown. Li_2GaBO_4, m.p. 896 °C (dec), is formed in the $LiBO_2$–$LiGaO_2$ system.[549] There is no carbonate $Ga_2(CO_3)_3$, but a basic carbonato complex $NH_4[Ga(OH)_2CO_3]$ has been characterized in which the CO_3 group is coordinated to Ga in bidentate fashion.[550] Ga_2O_3 and SiO_2 do not react when fused together, but may form ternary compounds in the presence of a third oxide. The products, gallosilicates, are similar to aluminosilicates, for example, the solids $M^{II}Ga_2Si_2O_8$ (M^{II} = Sr or Ba) are synthetic feldspars.[551] However, unlike the corresponding aluminum compounds, $NaGaSiO_4$ and $Na_4Ga_2Si_2O_9$ are soluble in water.[552] Soluble lithium gallosilicates are also known.[1,553] Gallium oxide combines with germanium oxide GeO_2 to give $Ga_6Ge_2O_{13}$, and various gallogermanates have been synthesized.[1,551]

(ii) Nitrate, phosphite, phosphate and arsenate

The hydrate $Ga(NO_3)_3{\cdot}9H_2O$, which presumably contains $Ga(OH_2)_6^{3+}$ ions, and a basic nitrate, $Ga_2(OH)_5(NO_3){\cdot}2H_2O$, can be prepared from aqueous solutions. The anhydrous nitrate $Ga(NO_3)_3$ and the complexes $Ga_2O(NO_3)_4$ and $MGa(NO_3)_4$ (M = Na, K or Cs) can be obtained from $GaBr_3/N_2O_5$ or $MBr/GaBr_3/N_2O_5$ mixtures.[554] Vibrational spectra indicate that the nitrate ligands are unidentate in $NO_2^+Ga(NO_3)_4^-$.[555] There are reports of gallium phosphites in early literature.[1] The phosphate $GaPO_4{\cdot}2H_2O$, a highly insoluble compound, is precipitated from aqueous Ga^{3+} solutions by NaH_2PO_4. Above pH 7, basic phosphates are produced. The crystal structure of $GaPO_4{\cdot}2H_2O$ shows Ga to have a coordination number of six: four PO_4 oxygens at a distance of 194 pm and two water molecules at 210 pm.[1] Above 200 °C this solid dehydrates to give $GaPO_4$, in which Ga is four-coordinate, and the Ga—O distance is smaller, *i.e.* 178 pm.

In acidic solutions of gallium phosphate there is evidence of $Ga(HPO_4)^+$, $Ga(H_2PO_4)^{2+}$, $Ga(H_3PO_4)^{3+}$ and possibly other complexes.[556,557] Gallium also forms a pyrophosphate, $Ga_4(P_2O_7)_3$, and various complex phosphates and pyrophosphates.[1,214] Gallium forms $GaAsO_4$ and $GaAsO_4{\cdot}2H_2O$.[1] The latter is precipitated when an acidified Ga^{3+} solution is neutralized in the presence of arsenate ions.

(iii) Sulfate, selenate and tellurate; tungstate and molybdate

Crystals of $Ga_2(SO_4)_3{\cdot}18H_2O$ lose water in stages, yielding $Ga_2(SO_4)_3$ above 165 °C. Various basic sulfates have been reported.[1] The structure of $H_3O[Ga_3(OH)_6(SO_4)_2]$ contains GaO_6 octahedra with SO_4 groups coordinated to gallium.[558] $Ga(SO_3F)_3$, from $GaCl_3 + S_2O_6F_2$, contains six-coordinate gallium with bridging fluorosulfato groups, according to IR spectra.[559] A chlorosulfate $Ga(SO_3Cl)_3$ and complexes $MGa(SO_3Cl)_4$ (M = Li or Na) have been prepared by the action of SO_3 on $MGaCl_4$.[560] Gallium selenate and tellurate resemble the sulfate.

$Ga_2(WO_4)_3$ is identified in the Ga_2O_3–WO_3 phase diagram. Several gallium tungstates and some polytungstate and molybdate complexes have recently been characterized.[561,562]

(iv) Chlorate, perchlorate, bromate and iodate

Gallium chlorate and perchlorate crystals obtained from aqueous solutions are hydrated, *e.g.* $Ga(ClO_4)_3{\cdot}6H_2O$. The aqueous solution shows evidence of polymerized cationic species: these are probably hydroxo complexes (Section 25.1.14.4.1) and are unlikely to contain ClO_4 ligands. The same is true of bromate and iodate solutions. $Ga(IO_3)_3{\cdot}2H_2O$ is sparingly soluble in water but readily dissolves in dilute acids. Anhydrous gallium perchlorate is obtained as very deliquescent crystals by the sequences in equation (14).[563] Vibrational spectra show that all three ClO_4 ligands are bidentate, forming a distorted octahedral GaO_6 unit.

$$GaCl_3 + 4Cl_2O_6 \longrightarrow ClO_2^+Ga(ClO_4)_4^- \xrightarrow[\text{vac}]{90\,^\circ\text{C}} Ga(ClO_4)_3 \qquad (14)$$

25.1.14.4.5 Carboxylate ligands

Gallium formate can be prepared from $Ga_2O_3 + HCO_2H$.[564] The acetate $Ga(O_2CMe)_3$ is obtained, *via* gallium chloroacetates, from $GaCl_3 + MeCO_2H$. Another preparative route is the slow reaction of the metal with anhydrous acetic acid. The benzoate and phenylacetate can be obtained similarly. Basic carboxylates result if aqueous acids are employed. The reaction of R_3Ga with an acid $R'CO_2H$ can also be applied to carboxylate synthesis. Structural data are lacking for simple formates or acetates of gallium. $MeGa(O_2CMe)_2$ contains both four- and five-coordinate Ga atoms, with Ga—O distances of 187.3 to 215.3 pm.[565] Gallium tris(trifluoroacetate), $Ga(O_2CCF_3)_3$, has been prepared from $GaCl_3 + CF_3CO_2H$ and shown to form 1:1 adducts with Lewis bases.[566] The complex $Cs[Ga(O_2CCF_3)_4]$ was also reported.

There have been a number of studies of gallium oxalate systems. $Ga(C_2O_4)_3 \cdot 4H_2O$ crystallizes from hot aqueous $Ga(NO_3)_3$ solution containing oxalic acid. On heating above 140 °C, a dihydrate and then anhydrous $Ga_2(C_2O_4)_3$ are produced.[1] The latter decomposes at 200 °C, forming Ga_2O_3. Complex salts containing $Ga(C_2O_4)_2^-$ and $Ga(C_2O_4)_3^{3-}$ anions have been prepared and the formation of these and the cation $Ga(C_2O_4)^+$, presumably hydrated, has been studied to obtain stability constants.[567] Basic complexes may appear in these aqueous solutions. Raman spectra of gallium oxalate solutions confirm the presence of $Ga(C_2O_4)_3^{3-}$. The mode $\nu(M—O)_{sym}$ appears at 573 cm^{-1} [585 cm^{-1} in $Al(C_2O_4)_3^{3-}$].[1] Gallium adopts the coordination number of four in bis(dimethylgallium)oxalate.[568]

There are long-standing reports of Ga complexes with hydroxy acids and related organic chelating agents.[1] The lactate, tartrate and citrate are being used in physiological studies (where Ga^{3+} has affinitities with Fe^{3+}) which involve gallium radioisotopes.[569] According to NMR studies (^1H, ^{13}C and ^{71}Ga) 1:1 Ga:citrate complexes predominate at low pH. Polymerized species exist in the range pH 2–6, and citrate ligands are displaced, forming $Ga(OH)_4^-$, under alkaline conditions.[570] Among other systems of Ga^{3+} and chelating organic acids for which solution equilibria have been studied are those of caprate,[571] various salicylates,[572] some phosphonates and a variety of multidentate ligands.[570,573] Analytical reagents which form complexes with Ga^{3+} include Tiron,[574] pyrocatechol violet,[575] *p*-nitrophenylfluorone,[576] thenoyltrifluoroacetone[577] and flavone kaempferol.[578] A point to emerge from these investigations is that basic complexes are often involved, and the hydrolyzed cation $Ga(OH)^{2+}$ shows enhanced reactivity compared with $Ga^{3+}_{(aq)}$.

The crystal structure of a dimethylgallium salicylaldehyde complex shows it to be dimeric. There is a central Ga_2O_2 ring; the Ga atoms are five-coordinate with distorted trigonal bipyramidal geometry.[579]

25.1.14.4.6 Ethers and other oxygen donor ligands

(*i*) *Molecular complexes*

Gallium compounds, especially the trihalides, form coordination complexes by reacting directly with a variety of oxygen donor ligands. Early examples to be characterized were the series $GaCl_3 \cdot L$ ($L = Me_2O$, Et_2O, Me_2CO, $POCl_3$ or $PhNO_2$) and $GaBr_3 \cdot L$ ($L = Me_2O$, Et_2O, $POCl_3$ or $POBr_3$).[4,431] The diethyl etherates are low-melting solids: $GaCl_3 \cdot Et_2O$, m.p. 16 °C; $GaBr_3 \cdot Et_2O$, m.p. 21 °C; $GaI_3 \cdot Et_2O$, m.p. 27 °C. Calorimetric measurements show that $GaCl_3$ is a slightly weaker Lewis acid than $AlCl_3$, although the former shows a preference for softer bases, notably S and P donor ligands.[435] Trends in donor bond strength within the $GaCl_3 \cdot L$ series can also be monitored by observing the NQR frequencies of Cl or Ga nuclei.[580] There is a linear correlation: the stronger the donor the lower the NQR frequency.

With the water molecule as ligand, both $GaCl_3$ and $GaBr_3$ form a monohydrate, and several other hydrates have been identified.[581] The highest of these, $GaBr_3 \cdot 15H_2O$, is probably a salt of the $Ga(OH_2)_6^{3+}$ cation with additional water of crystallization. The vibrational spectra of the $GaX_3 \cdot H_2O$ adducts are best assigned in terms of the local symmetry of the units X_3GaO and H_2O rather than the overall C_s symmetry of the molecules.[582] ν(Ga–O) bands are observed in the region 410–440 cm^{-1} and force constants for the coordinate Ga—O bond are computed to be 1.43 and 1.48 N cm^{-1} in chloride and bromide systems, respectively. The Ga–X stretching modes of these H_2O adducts are similar in frequency to those of complexes $GaX_3 \cdot L$, where $L = Et_2O$, Me_2S, Et_2S or an N donor ligand, whose constitution has been established earlier by vibrational and NMR spectroscopy.[1] Other 1:1 adducts to have been studied spectroscopically include $GaX_3 \cdot Ph_3PO$ (X = Cl, Br or I),[583] $Me_3Ga \cdot Me_2O$[584] and $GaX_3 \cdot pyO$ (X = Cl or Br).[585] The last of these studies encompassed several substituted pyridine *N*-oxides which exhibit decreases in ν(N—O) on coordination, showing that the ligand is bound to gallium *via* the oxygen atom. Debate about the structure of acyl chloride adducts $GaCl_3 \cdot RCOCl$ now favours the ionic form $RCO^+GaCl_4^-$ (Section 25.1.14.6.2).

ESR spectra have been used to study interaction of the free radical base 2,2,6,6-tetramethylpiperidine nitroxide (tempo), a weak Lewis base, with $GaCl_3$.[586] This system reveals the expected 1:1 complex $GaCl_3 \cdot$tempo. Only one bridging Cl bond in the Ga_2Cl_6 molecule has been broken, leaving a vacant site for the oxygen lone pair to coordinate weakly to gallium. $GaCl_3$·1,4-dioxane has recently been shown by XRD to contain five-coordinate Ga.[587] The structure consists of infinite chains in which planar $GaCl_3$ units are linked by dioxane molecules in chair conformation, with Ga—O bonds of length 221.0 pm. The involvement of six-coordinate species $Ga(NCS)_3L_3$, where L represents the O-bonded ligand, has been proposed to account for the solvent extraction of gallium from thiocyanate solutions by phosphonate or trialkylphosphate solvents.[503]

(ii) Cationic complexes

When gallium salts or the halides GaX_3 dissolve in polar solvents there is a pronounced tendency to ionization, and under these circumstances gallium generally exists as an octahedrally coordinated species. This occurs in aqueous solutions giving $Ga(OH_2)_6^{3+}$ (Section 25.1.14.4.1), in alcohols giving $Ga(ROH)_4^{3+}$ or $Ga(ROH)_6^{3+}$,[588] in dimethylformamide giving $Ga(DMF)_6^{3+}$ and in dimethyl sulfoxide giving $Ga(DMSO)_6^{3+}$.[537,589] In the DMSO complex a characteristic decrease in ν(S—O) confirms coordination through the oxygen. The symmetrical field about the metal nucleus in these species is favourable for Ga NMR spectroscopy, making this an important technique for studying the solution chemistry of these complexes.[537]

25.1.14.5 Sulfur and Heavier Group VI Ligands

25.1.14.5.1 Sulfide, selenide and telluride ligands

Gallium(III) sulfide Ga_2S_3 is prepared by passing H_2S over gallium at 950 °C and subliming the product under vacuum to obtain white crystals. When annealed at 1000 °C, the material is α-Ga_2S_3 which has the wurtzite lattice of close-packed sulfide ions, containing an ordered arrangement of Ga cations. Single crystal IR and Raman spectra of this phase have been assigned to vibrational modes of the GaS_4 groups.[590] Other forms, β- and γ-Ga_2S_3, have a random distribution of the Ga atoms in the wurtzite and zinc blende sulfide lattices, respectively. Ga_2S_3 decomposes slowly in moist air, evolving H_2S. It is soluble in acids, and in strong alkalis, to give gallate and thiogallate species. The compound $Na_2Ga(OH)_3S$ is reported to be formed when aqueous sodium gallate is treated with Na_2S or H_2S.[1]

There is a selenide Ga_2Se_3 and a telluride Ga_2Te_3, and materials can be synthesized which contain two of the group VI elements, *e.g.* Ga_2S_2Te. This compound contains GaS_3Te units linked to give parallel chains: Ga—S and Ga—Te distances are 230.7–235.3 and 255.6 pm, respectively.[591]

Gallium forms a variety of ternary sulfides and selenides. Common types are M^IGaS_2, $M^{II}Ga_2S_4$ and $M^{III}GaS_3$, and are akin to the mixed oxides. Owing to the unusual electrical properties of these and the corresponding indium compounds, there is an extensive literature concerned with their preparation, physical properties, growth as single crystals and X-ray studies.[592] Examples are $CsGaS_2$, $AgCaS_2$, $TlGaSe_2$, $CdGa_2S_4$ and $LaGaS_3$.[593,594] Compounds of other stoichiometries occur in mixed sulfide systems involving gallium, *e.g.* $Pb_2Ga_2S_4$ (or $2PbS·Ga_2S_3$).[595] Extended lattices composed of linked GaS_4 tetrahedra are a feature of these structures. In recent X-ray studies the Ga—S bond distances are within the range 218–236 pm. In $La_{3.33}Ga_6O_2S_{12}$ there are sheets of both GaS_4 and $GaOS_3$ tetrahedra.[596] An adamantane-like cluster anion $Ga_4S_{10}^{8-}$ is found in $K_8Ga_4S_{10}·14H_2O$, prepared by the action of aqueous potassium sulfide on Ga_2S_3.[597] The vibrational spectra have been assigned and X-ray structural analysis confirms near-ideal T_d symmetry. Ga—S distances are 228.9 (bridging) and 225.2 pm (terminal). The analogous seleno compound has been synthesized, as have the indium species. The structures have high thermal and solvolytic stability. Layers of Ga_4Se_{10} groups, linked through corners, comprise the structure of the ternary selenide $TlGaSe_2$, already mentioned.[593] Discrete $Ga_6Se_{14}^{10-}$ anions occur in $Cs_5Ga_3Se_7$: in this case six $GaSe_4$ tetrahedra are joined at their edges.[598] The Ga—Se distances range from 235.7 to 249.7 pm. $Cs_6Ga_2Se_6$ obtained as moisture-sensitive crystals, m.p. 685 °C, by reaction of Cs_2Se with Ga_2Se_3, contains the $Ga_2Se_6^{6-}$ anion which consists of just two tetrahedra sharing an edge. Here the Ga—Se distances are 247.0 and 250.5 pm (bridging), 238.5 and 240.7 pm (terminal).[599]

Gallium forms a series of compounds GaYX, where X = Cl, Br or I and Y = S, Se or Te, all of which are highly water sensitive. Crystallographic study of GaTeCl reveals a layer structure in which the Ga—Te framework resembles that in black phosphorus.[600] Recent investigations have shown that sulfur reacts with $GaCl_3$ in the medium of a CsCl–$GaCl_3$ melt to produce polymeric species $(GaSCl_2^-)_n$.[601] From Raman evidence these appear to have Ga—S—Ga linkages, like the corresponding aluminum complexes.

Alkylthiogallanes $Ga(SR)X_2$ and $Ga(SR)_4^-$ (R = Me or Et) are obtained by the reaction of GaX_3 with Me_3SiSR in benzene.[602] A dimeric structure with bridging thioalkyl ligands has been found. By varying the stoichiometry, $Ga(SR)_2Cl$, $Ga_2(SR)_3Br_3$ and $Ga(SR)_3$ can be prepared. A remarkable structure results from the reaction of Me_2S_2 with Ga_2I_4. The product, $Ga_4(SMe)_4S_2I_4$, is a cluster compound in which four tetrahedrally disposed GaI units are linked along the edges of the tetrahedron by sulfur ligands (four SMe groups and two S atoms). The result is a cage structure similar to that of $Ge_4S_6Br_4$, or to $Ga_4S_{10}^{8-}$ (above). Average bond distances are Ga—I = 251.6, Ga—S (Me) = 233.6 and Ga—S = 220.4 pm.[603]

25.1.14.5.2 Other sulfur-donor ligands

Gallium has a stronger tendency to react with sulfur ligands than has aluminum. This is evident in the chemistry of gallium sulfides and related species, and in the fact that Ga forms compounds with S ligands of which there are no counterparts in Al chemistry. There are adducts GaX_3SR_2 (X = Cl or Br; R = Me or Et), and a dithiolato complex $Ga(mnt)_3^{3-}$ (mnt = *cis*-1,2-$S_2C_2(CN)_2^{2-}$) which resemble those of In^{III} and Tl^{III}.[604] A series of tris(dithiocarbamates) $Ga(S_2CNR_2)_3$ and tris(xanthates) $Ga(S_2COR)_3$ have been prepared from aqueous solutions.[605] Thermal decomposition at 100 °C of two of these compounds is reported to proceed as follows, although the products are not well characterized (equations 15 and 16).

$$Ga(S_2CNH_2)_3 \longrightarrow Ga(NCS)_3 + 3H_2S \quad (15)$$

$$Ga(S_2CNHMe)_3 \longrightarrow Ga(SH)_3 + 3MeNCS \quad (16)$$

The crystal structure of $Ga(S_2CNEt_2)_3$ reveals a monomer, the GaS_6 core of which is closer to prismatic than regular octahedral symmetry. The Ga—S distances are 240.8 to 246.6 pm (average 243.6 pm).[606] The analogous indium compound is known. $(Ph_2GaSEt)_2$ has a structure in which gallium is four-coordinate with bridging EtS ligands.[607] The Ga—S distances are 237.3 and 238.4 pm. A monothio-β-diketonate of gallium has been obtained.[608] Crystal structure analysis to determine the geometry of the central GaO_3S_3 unit would be of considerable interest.

Gallium forms a thiophosphate $GaPS_4$. XRD measurements show that Ga atoms are four-coordinate.[609] The GaS_4 units are distorted tetrahedra, and the Ga—S distances, in the range 226.6–229.7 pm, are appreciably less than those of the GaS_6 group noted above.

25.1.14.6 Halogen Ligands

25.1.14.6.1 Fluoride ligand

The GaF_3 molecule is planar with a Ga—F distance of 188 pm from ED studies.[610] This species and the dimer $(GaF_3)_2$ have been examined in the vapour phase by mass spectrometry.[611] The OGaF molecule has been detected by matrix isolation.[612]

The compound gallium trifluoride GaF_3, colourless needles which are sparingly soluble in water, is produced by thermal decomposition of $(NH_4)_3GaF_6$ or $GaF_3{\cdot}3NH_3$,[613] excluding moisture which brings about the formation of $(NH_4)_2GaF_3(OH)_2$. Solutions of gallium in hydrofluoric acid yield $GaF_3{\cdot}3H_2O$ when evaporated. Stronger heating gives $GaF_2OH{\cdot}xH_2O$, and exposure to NH_3 forms $GaF_3{\cdot}3NH_3$. Aqueous gallium fluoride solutions contain the aquated complexes GaF^{2+}, GaF_2^+, GaF_3 and GaF_4^-;[614] probably all are six-coordinate, *e.g.* $GaF_4(OH_2)_2^-$. The octahedral anion GaF_6^{3-} exists in some fluorogallates for which vibrational analyses are available.[615,616] As with aluminum, there are group I fluorogallates, $MGaF_4$, M_2GaF_5, M_3GaF_6 and MGa_2F_7.[617] Other stoichiometries, *e.g.* $Na_5Ga_3F_{14}$, $BaGaF_5$, $Ba_3Ga_2F_{12}$ and $Ba_{11}Ga_4F_{34}$, are attained by the linking of GaF_6 groups into chains or sheets through bridging fluoride ligands.[618–620] NH_4GaF_4 forms a continuous series of solid solutions with NH_4AlF_4, as does γ-GaF_3 with γ-AlF_3.[621]

Some organogallium fluorides are known. These have cyclic structures $(R_2GaF)_n$, where $n = 3$ or 4,[622] and so differ from the chlorides and bromides $(R_2GaX)_2$ which are halogen-bridged dimers.

25.1.14.6.2 Chloride, bromide, iodide and mixed halide ligands

The trihalides $GaCl_3$ and $GaBr_3$ are usually prepared by burning the metal in a stream of Cl_2 (or Br_2) carried by nitrogen, in apparatus which includes several purification stages and uses all-glass or greaseless connections.[468] GaI_3 can be obtained by dissolving Ga in a refluxing solution of I_2 in CS_2. Gallium chloride and bromide have some advantages over the aluminum trihalides as Friedel–Crafts catalysts, particularly owing to their high solubility in organic solvents. All three halides are colourless, deliquescent, low-melting solids, whose physical properties have been tabulated.[1] In solid and liquid states, and in non-coordinating solvents, the gallium trihalide molecules are dimeric with bridging ligands spanning four-coordinate Ga atoms. In the vapour there is extensive dissociation of Ga_2X_6 to GaX_3. Molecular dimensions are known for monomers $GaBr_3$[623] and GaI_3[1] and dimers Ga_2Cl_6[1] and Ga_2I_6,[624] but precise values are lacking for $GaCl_3$ and Ga_2Br_6. Vibrational assignments are available for the GaX_3 monomers[623,625,626] (from gas phase and matrix isolation studies) and the Ga_2X_6 dimers.[627,628] NQR specra of Ga_2Cl_6 are in accordance with the symmetrical dimer of D_{2h} symmetry.[629]

Halogen-bridging, which enables Ga to be four-coordinate, is also a feature of the dimeric structures of organogallium halides, *e.g.* $(RGaX_2)_2$ where X = Cl, Br or I and R = various alkyl groups; or (Ph_nGaX_{3-n}) ($n = 1$ or 2; X = Cl, Br or I).[630] $(Me_3SiCH_2)_2GaCl$ appears to differ from the usual pattern and to be polymeric in the solid state.[631] In $(HGaCl_2)_2$, IR and Raman spectra establish a chlorine-bridged structure with the terminal ligands in the *cis* arrangement.[632] Structural and spectroscopic data for methylgallium halide anions $[Me_nGaX_{4-n}]^-$ ($n = 0$–4) have been analyzed.[633] A single halogen ligand links the R_3Ga groups in the anions $(R_3Ga)_2X^-$ (X = Cl or Br; R = Me or Et).[634]

$GaCl_3$ forms $GaCl_3{\cdot}BCl_3$,[635] and adducts with some chlorides of the later group elements, *e.g.* $GaCl_3{\cdot}TeCl_4$ and $GaCl_3{\cdot}PCl_5$, which have ionic structures including the $GaCl_4^-$ anion. In $GaCl_3{\cdot}SbCl_3$, XRD reveals a chain structure with Ga—Cl—Ga bridges.[636] The $GaCl_3$–$SnCl_2$ system forms $Sn(GaCl_4)_2$, whereas the $GaCl_3$–$SnCl_4$ system is eutectic and shows no intermediate compound.[637] Germanium is chemically transported in the vapour as $GeGaCl_5$, *i.e.* $Cl_2Ge(\mu\text{-}Cl)_2GaCl$.[638] Kindred vapour-phase species are present in gallium chloride–transition metal chloride systems.[639,640]

Being powerful acceptors, the gallium trihalides react with donor molecules L to form the adducts $GaX_3{\cdot}L$, and in some instances $GaX_3{\cdot}2L$, with a coordinate bond from the group V or VI ligand, which are treated in earlier sections.

(i) *Complex halide ions*

With chloride, bromide or iodide as ligand, gallium readily gives the tetrahedral anions GaX_4^-. These exist in aqueous solutions containing excess halide, from which they can be isolated as crystalline salts with a large cation, *e.g.* Bu_4NGaX_4, or by solvent extraction as complex acids $H(solv)^+GaX_4^-$.[357] Crystallographic data for solids containing the GaX_4^- ions are available.[1,486,641,642] Characteristic absorption spectra,[643] vibrational frequencies,[644] NMR chemical shifts[537] and NQR parameters[645] are well established. These properties can be used to study effects such as ligand exchange processes of $GaCl_4^-$ with Cl^-,[646] ion pairing in non-aqueous solutions of alkylammonium chlorogallates,[647] polymorphism of $KGaCl_4$[648] and solid state interactions in the crystalline solids $NO^+GaCl_4^-$[649] and $Ga^+GaCl_4^-$.[437] Bond dissociation energies $D(GaX_3\text{—}X^-)$ are 364 and 314 kJ mol^{-1} for X = Cl or Br, respectively.[650]

In dilute aqueous solutions the four-coordinate GaX_4^- ions dissociate, forming $[GaX_n]^{3-n+}$ (aq) ($n = 1, 2$ or 3) and finally $[Ga(OH_2)_6]^{3+}$. Stability constants are available for these equilibria.[7,8] When the solutions are frozen to the glassy state, Raman spectra show that the equilibria are displaced in favour of the aqua cation.[651] The complexes GaX_4^- are formed, together with $Ga_2X_7^-$ (see below), in the systems MX–GaX_3, where MX is a group I chloride or bromide, which have been studied in the solid state or as molten salts. The TlBr–$GaBr_3$ system reveals $TlGa_2Br_7$ and $TlGaBr_4$, whereas the NaBr–$GaBr_3$ system forms $Na_2Ga_2Br_7$, $NaGaBr_4$,

Na_2GaBr_5 and Na_3GaBr_6.[652] These last two probably contain $GaBr_4^-$ and Br^- ions, rather than complexes more highly coordinated than $GaBr_4^-$.

The halide dimers Ga_2X_6 react with MX in 1:2 stoichiometry to form $MGaX_4$, and in 1:1 stoichiometry yielding MGa_2X_7. The crystal structure of KGa_2Cl_7 reveals a bent conformation, the $Cl_3Ga(\mu\text{-}Cl)GaCl_3^-$ anion having C_{2v} symmetry. Vibrational spectra of the solid and molten salt phases have been assigned on this basis,[653] and the bridging and terminal chlorines give distinct NQR signals.[645] Raman studies of molten $CsCl–GaCl_3$ mixtures show the presence of Ga_2Cl_6, $GaCl_4^-$, $Ga_2Cl_7^-$ and possibly higher complexes.[654] The anions $Ga_2X_7^-$ (X = Cl or Br) are present in the mixed valence compounds Ga_3X_7.[437,655] In $MI–GaI_3$ systems, both $MGaI_4$ and MGa_2I_7 exist when M = Rb or Cs, but only $MGaI_4$ occurs when M = K. Interestingly, the $Ga_2I_7^-$ ion has a linear Ga—I—Ga bridge, vibrational spectra of melts indicating D_{3d} symmetry.[656] For the corresponding bromide systems,[657] there is evidence of a further gallium-rich compound MGa_3Br_{10} for which the Raman spectrum suggests the structure $Br_3Ga(\mu\text{-}Br)GaBr_2(\mu\text{-}Br)GaBr_3^-$. Nyholm and Ulm had prepared $[Ga(diars)_2Cl_2](Ga_3Cl_{10})$ in 1965,[658] although their proposed structure of $Ga_3Cl_{10}^-$, featuring a central six-coordinate Ga atom, can now be discounted.

(ii) Mixed halide species

The molecules $GaCl_2Br$ and $GaClBr_2$ have been observed in the gas phase by Raman spectroscopy.[442] However, attempts to study halogen bridging in mixed halide dimers, *e.g.* $Ga_2Cl_4Br_2$, or in complex anions, such as $Ga_2Cl_6Br^-$, have been defeated by the complexity of the spectra. Mixed halide species RGa_2X_4I have been obtained by the oxidative addition of an alkyl iodide to the halides Ga_2Cl_4 or Ga_2Br_4.[659]

Four-coordinate mixed halide complexes $[GaX_nY_{4-n}]^-$, where X and Y are two of Cl, Br or I, are readily generated in gallium halide systems in solution and solid states. Preparations of $Bu_4P(GaCl_3Br)$, $Ph_4As(GaClBr_3)$, $[(Et_2O)_2H](GaX_3Y)$ and $(py_2H)[GaX_3Y(py)_2]$ (X or Y = Cl, Br or I) have been described.[660,661] Vibrational assignments have been made for most of the ν(Ga—X) and ν(Ga—Y) stretching modes of the GaX_3Y^- anions.[660,662] ^{35}Cl NQR spectra suggest that the solids $Bu^n_4N(GaCl_nBr_{4-n})$ (n = 0–4) are individual compounds,[663] although these might have been expected to be disordered phases. The production of mixed halide anions in solution starting from GaX_4^- and GaY_4^- can be followed by Ga NMR spectroscopy and the chemical shifts of the binary mixed halide ions have been measured, as have those of $GaCl_2BrI^-$, $GaClBr_2I^-$ and $GaClBrI_2^-$.[664,665] Halide exchange is rapid in aqueous solution, but slow in CH_2Cl_2, and produces GaX_3Y^- and $GaXY_3^-$ ahead of $GaX_2Y_2^-$. The mechanism of exchange is thought to involve initial dissociation of GaX_4^- and the intermediacy of bridged species such as $X_3Ga—Y—GaX_3^-$.

25.1.14.7 Hydrogen and Complex Hydride Ligands

25.1.14.7.1 Hydride ligand

Gallium forms hydridogallates, $MGaH_4$ and M_3GaH_6, and gallane adducts, *e.g.* $GaH_3{\cdot}NMe_3$. These compounds have some advantages as reducing agents over the corresponding aluminum compounds; however, the chemistry of the gallium hydrides is much less well developed.

Lithium gallium hydride, $LiGaH_4$, a solid which slowly decomposes at 25 °C and a milder reducing agent than $LiAlH_4$, can be prepared in ether solution (equation 17). Tested methods are also available to synthesize Na, K and other salts, and to convert $LiGaH_4$ to $GaH_3{\cdot}NR_3$ by reaction with Me_3N (or Et_3N).[666,667] $LiGaH_4$ forms a dietherate, which is converted to the monoetherate $LiGaH_4{\cdot}OEt_2$ under vacuum. The Raman and IR spectra of alkali metal salts $MGaH_4$ confirm the tetrahedral symmetry of the anion and enable GaH_4^- to be compared with other hydrides.[668]

$$GaCl_3{\cdot}Et_2O + 4LiH \xrightarrow{-78\,^{\circ}C} LiGaH_4 + 3LiCl \qquad (17)$$

$GaH_3{\cdot}NMe_3$, a solid, is less sensitive to hydrolysis than $AlH_3{\cdot}NMe_3$. Recent measurements of the dimensions of the $Me_3N–GaH_3$ molecule in the vapour by ED include r(Ga—H) = 149.7 pm.[669] Similarly short terminal Ga—H bonds of 148.7 pm are found in the structure of $(Me_2NGaH_2)_2$.[495] $GaH_3{\cdot}NMe_3$ can undergo hydride displacement without rupture of the Ga—N bond (equation 18).

$$Me_3N{\cdot}GaH_3 + Me_3NHCl \longrightarrow Me_3N{\cdot}GaH_2Cl + Me_3N + H_2 \quad (18)$$

Following early claims of gallane synthesis, reinvestigation of the reaction of $GaH_3{\cdot}NMe_3$ with BF_3 provides no evidence for uncoordinated GaH_3.[475] MS experiments in which the metal is heated in a hydrogen atmosphere have detected small amounts of gallium hydride as GaH_3^+ (or InH_3^+ when indium is used),[670] but the amounts are much less than in the case of aluminum where both AlH_3^+ and $Al_2H_6^+$ are found. The gallane radical anion $^{\bullet}GaH_3^-$ has been generated in a low temperature matrix and has been shown by ESR study to be pyramidal, like $^{\bullet}AlH_3^-$ and $^{\bullet}SiH_3$.[671]

There are reports of unstable halogenogallanes, the compounds $GaHCl_2$ and $GaHBr_2$ having been prepared from $GaX_3 + Me_3SiH$. $GaHCl_2$ melts at 29 °C (dec), and decomposition to $Ga_2Cl_4 + H_2$ is complete at 150 °C. Evidence from IR/Raman spectra indicates that $GaHCl_2$ is a dimer, probably *cis*-$ClHGa(\mu\text{-}Cl)_2GaHCl$.[632]

25.1.14.7.2 Borohydride ligand

Reaction of $LiBH_4$ with $GaCl_3$ yields hydridogalliumbis(tetrahydroborate) $HGa[(\mu\text{-}H)_2BH_2]_2$. In the gas phase this molecule exhibits five-fold coordination by H atoms which form a distorted rectangularly based pyramid around the Ga atom.[672] As determined by ED, the length of bridge bonds averages 177.4 pm; the terminal Ga—H bond is much shorter, *i.e.* 156.5 pm. $Me_2Ga(BH_4)$, m.p. 5 °C, decomposing to $Ga + H_2 +$ methylated borons, can be obtained from $Me_2GaCl + LiBH_4$ (or from $Me_3Ga + B_2H_6$). Its structure shows four-coordinate gallium linked to bidentate BH_4 as $Me_2Ga(\mu\text{-}H)_2BH_2$ with a bridging Ga—H bond length of 179 pm.[673]

25.1.14.8 Mixed Donor Atom Ligands

25.1.14.8.1 Schiff base ligands

A few studies of Schiff base complexes of gallium have been made.[415,674] Frequencies ν(Ga—O) have been located in the range 600–685 cm^{-1}. Modes ν(C=N) are almost unaffected by complexation.

25.1.14.8.2 Amino acids, peptides and proteins

Complexing of Ga^{3+} by simple amino acids has attracted attention as the basis for investigations of more complicated systems.[675,676] IR evidence indicates that both N and O sites are coordinated to the metal. NMR studies of interactions of Ga^{3+} with the bleomycin antibiotics (glycopeptides used in tumour scanning as carriers for the ^{67}Ga radionuclide) have shown that the metal binds by displacing a proton from the α-amino group of the diaminopropionamide fragment of the drug.[677] The conformation of the gallium complex of parabactin has been determined by 300 MHz ^1H NMR spectroscopy, and the sequestering powers of several catechoylamide-type ligands for Ga^{3+} (and In^{3+}) have been investigated as a function of pH to assess the feasibility of the removal of gallium from association with the protein transferin.[678]

25.1.14.8.3 Ligands with N, O and other functions

In gallium complexes with multidentate ligands there is a marked tendency for the coordination number to increase from four to five or six. The reaction of Me_3Ga with ethanolamine yields Me_2GaL_2, where L_2 represents the chelating ethanolamine ligand. Bond distances within the five-membered ring are Ga—O = 191.7 pm, Ga—N = 206 pm.[679] If the ligand is changed to *N,N*-dimethylethanolamine, the resulting chelate dimerizes to form

$[Me_2Ga(OC_2H_4NMe_2)]_2$ with a central Ga_2O_2 ring such that the Ga atoms become five-coordinate. In *N,N'*-ethylenebis(salicylideneiminato)chlorogallium(III) the central Ga atom is again five-coordinate.[680] The geometry approximates to square pyramidal with Cl at the apex. The Ga atom is also five-coordinate in chlorobis(β-hydroxo-2-methylquinolinato)gallium(III) and in some 1:1 complexes of Ga^{3+} or $MeGa^{2+}$ with various O,N,O or O,N,S tridentate ligands.[681,682] Salicylaldimine complexes of Ga are lipophilic species of radiopharmaceutical interest. X-Ray study of one of these reveals the ligand occupying all six coordination sites around Ga, by bonding through three imino N atoms and the deprotonated phenolic O atoms.[683] Ga—N and Ga—O distances are slightly longer than those in four- or five-coordinate Schiff base complexes of gallium, but lie within the established range for octahedral Ga^{III} compounds. The crystal structure of the ethylenediaminetetraacetic acid complex $Ga(Hedta)\cdot H_2O$ shows the ligand bonding in a pentadentate fashion with the water molecule occupying a sixth position.[684] Average bond distances are Ga—N = 214 and Ga—O = 195 pm. The structure of a six-coordinate complex which bears Cl, a tridentate N donor and bridging OH ligands has also been determined.[538]

Stability constants have been evaluated for the complexes formed in aqueous solution between Ga^{3+} and a number of multidentate ligands.[685,686] Some other Ga chelates, including those of hydroxy acids, are treated in Section 25.1.14.4. Investigations by Storr, Trotter and co-workers into the structural chemistry of a series of transition metal complexes, which carry bi- or tri-dentate ligands based on MeGa or Me_2Ga groups (and so can be viewed alternatively as chelate complexes of gallium), are reviewed elsewhere.[687]

25.1.14.9 Macrocyclic Ligands

Gallium porphyrins resemble those of Al, In and transition metal M^{III} atoms. Examples in which the central atom of the GaN_4 kernel also bears a chlorine ligand[688] can be converted into Ga porphyrins with a Ga—C bond.[689] Certain polymeric fluorogallium and aluminum phthalocyanines are of importance on account of their high electrical conductivity (see Section 25.1.9). Crystallographic study of PcGaF[690] has revealed the orientation of the phthalocyanine rings and has established the following bond distances: Ga—F = 193 and Ga—N = 197 pm.

25.1.15 REFERENCES

1. K. Wade and A. J. Banister, in 'Comprehensive Inorganic Chemistry', ed. J. C. Bailar, Jr., H. J. Emeléus, R. Nyholm and A. F. Trotman-Dickenson, Pergamon, Oxford, 1972, vol. 1, p. 993.
2. H. J. Emeléus (ed.), 'MTP International Review of Science, Inorganic Chemistry Series Two', Butterworth, London, 1975, vol. 1, pp. 169, 219, 257.
3. R. Warncke (ed.), 'Gmelins Handbuch der Anorganischen Chemie', Formula Index, Springer Verlag, Berlin, 1975, vol. 1.
4. N. N. Greenwood and K. Wade, in 'Friedel–Crafts and Related Reactions', ed. G. A. Olah, Interscience, New York, 1963, vol. 1, p. 569.
5. T. Mole, *Organomet. React.*, 1970, **1,** 1.
6. J. J. Eisch, in 'Comprehensive Organometallic Chemistry', ed. G. Wilkinson, F. G. A. Stone and E. W. Abel, Pergamon, Oxford, 1982, vol. 1, p. 555.
7. L. G. Sillén and A. E. Martell, 'Stability Constants of Metal Ion Complexes', Special Publication, No. 17, 1964; No. 25, 1971, The Chemical Society, London.
8. R. M. Smith and A. E. Martell, 'Critical Stability Constants', Plenum, New York, vol. 1–4, 1974–77; vol. 5, 1982.
9. D. P. N. Satchell and R. S. Satchell, *Chem. Rev.*, 1969, **69,** 251.
10. G. K. Barker, M. F. Lappert, J. B. Pedley, G. J. Sharp and N. P. C. Westwood, *J. Chem. Soc., Dalton Trans.*, 1975, 1765.
11. C. J. Nicholls and D. S. Urch, *J. Chem. Soc., Dalton Trans.*, 1974, 901.
12. G. Wittig and H. Bille, *Z. Naturforsch., Teil B*, 1951, **6,** 226.
13. F. Weller and K. Dehnicke, *J. Organomet. Chem.*, 1972, **36,** 23.
14. K. Wallenfels and G. Bechtler, *Angew. Chem., Int. Ed. Engl.*, 1963, **2,** 515.
15. L. Rösch and G. Altnau, *Z. Naturforsch., Teil B*, 1980, **35,** 195.
16. L. Rösch and G. Altnau, *J. Organomet. Chem.*, 1980, **195,** 47.
17. L. Rösch, G. Altnau, C. Krüger and Y.-H. Tsay, *Z. Naturforsch., Teil B.*, 1983, **38,** 34.
18. L. Rösch, G. Altnau and W. H. Otto, *Angew. Chem., Int. Ed. Engl.*, 1981, **20,** 581, 583.
19. L. Rösch and W. Erb, *Angew. Chem., Int. Ed. Engl.*, 1978, **17,** 604.
20. W. C. Laughlin and N. W. Gregory, *Inorg. Chem.*, 1975, **14,** 1263.
21. M. Hargittai, J. James, M. Bihari and I. Hargittai, *Acta Chim. Acad. Sci. Hung.*, 1979, **99,** 127.
22. P. L. Baxter, A. J. Downs and D. W. H. Rankin, *J. Chem. Soc., Dalton Trans.*, 1984, 1755.

23. K. N. Semenenko, E. B. Lobkovskii, V. B. Polyakova, I. I. Korobov and O. V. Kravchenko, *Koord. Khim.*, 1978, **4,** 1649.
24. T. Østvold, E. Rytter and G. N. Papathedoru, *Spectrochim. Acta, Part A,* 1986, **41,** 1277.
25. K. N. Semenenko, I. I. Korobov, O. V. Kravchenko and S. P. Shilkin, *Russ. J. Inorg. Chem.* (*Engl. Transl.*), 1975, **20,** 1568.
26. S. Cucinella, A. Mazzei and W. Marconi, *Inorg. Chim. Acta Rev.*, 1970, **4,** 51.
27. W. Clegg, U. Klingebiel, J. Neeman and G. M. Sheldrick, *J. Organomet. Chem.*, 1983, **249,** 47.
28. A. M. Shirk and J. S. Shirk, *Inorg. Chem.*, 1983, **22,** 72.
29. J. W. Akitt and R. H. Duncan, *J. Chem. Soc., Dalton Trans.*, 1976, 1119.
30. E. C. Ashby, *J. Am. Chem. Soc.*, 1964, **86,** 1882.
31. V. S. Mastryukov, A. V. Golubinskii and L. V. Vilkov, *Zh. Strukt. Khim.*, 1979, **20,** 921.
32. K. R. Skillern and H. C. Kelly, *Inorg. Chem.*, 1974, **13,** 2802.
33. M. Boleslawski, J. Serwatowski and S. Pasynkiewicz, *J. Organomet. Chem.*, 1978, **161,** 279.
34. A. Sadownik, S. Pasynkiewicz, M. Boleslawski and H. Szachnowska, *J. Organomet. Chem.*, 1978, **152,** C49.
35. J. W. Wilson and I. J. Worrall, *J. Inorg. Nucl. Chem.*, 1968, **30,** 715; 1969, **31,** 1205, 1357.
36. J. W. Wilson and I. J. Worrall, *J. Chem. Soc.* (*A*), 1967, 392; 1968, 316, 2389.
37. P. Pullman, K. Hensen and J. W. Bats, *Z. Naturforsch., Teil B.*, 1982, **37,** 1312.
38. D. H. Brown, D. T. Stewart and D. E. H. Jones, *Spectrochim. Acta, Part A,* 1973, **29,** 213.
39. M. Dalibart, J. Derouault and M. T. Forel, *J. Mol. Struct.*, 1981, **70,** 199.
40. F. J. Arnáiz Garciá, *Inorg. Synth.*, 1980, **20,** 83.
41. J. G. Oliver and I. J. Worrall, *J. Inorg. Nucl. Chem.*, 1971, **33,** 1281.
42. G. J. Sutton, *Aust. J. Chem.*, 1963, **16,** 278, 1134.
43. K. H. Thiele and W. Bruesar, *Z. Anorg. Allg. Chem.*, 1966, **348,** 179.
44. J. Y. Corey and R. Lamberg, *Inorg. Nucl. Chem. Lett.*, 1972, **8,** 275.
45. S. Herzog, K. Geisler and H. Prackel, *Angew. Chem.*, 1963, **73,** 94.
46. W. Kaim, *J. Oganomet. Chem.*, 1981, **215,** 325, 337.
47. W. Kaim, *J. Am. Chem. Soc.*, 1984, **106,** 1712.
48. W. E. Dorogy and E. P. Schram, *Inorg. Chim. Acta,* 1983, **73,** 31.
49. G. Beran, K. Dymock, H. A. Patel, A. J. Carty and P. M. Boorman, *Inorg. Chem.*, 1972, **11,** 896.
50. G. Müller and C. Krüger, *Acta Crystallogr., Sect. C,* 1984, **40,** 628.
51. R. Caminiti, A. Lai, G. Saba and G. Crispani, *Chem. Phys. Lett.*, 1983, **97,** 180.
52. A. E. Findholt, C. Helling, V. Imhof, L. Nielson and E. Jacobson, *Inorg. Chem.*, 1963, **2,** 504.
53. P. Molinie, R. Brec, J. Rousel and P. Herpin, *Acta Crystallogr., Sect. B,* 1973, **29,** 925.
54. D. C. Bradley, *Adv. Inorg. Chem. Radiochem.*, 1972, **15,** 259.
55. R. G. Beach and E. C. Ashby, *Inorg. Chem.*, 1971, **10,** 1888.
56. G. M. Sheldrick and W. S. Sheldrick, *J. Chem. Soc.* (*A*), 1969, 2279.
57. M. J. Zaworotko and J. L. Atwood, *Inorg. Chem.*, 1980, **19,** 268.
58. H. M. M. Shearer, R. Snaith, J. D. Sowerby and K. Wade, *Chem. Commun.*, 1971, 1275.
59. K. Gosling and R. E. Bowen, *J. Chem. Soc., Dalton Trans.*, 1974, 1961.
60. E. C. Ashby and R. Kovar, *J. Organomet. Chem.*, 1970, **22,** C34.
61. U. Wannagat, T. Blumenthal, D. J. Brauer, and H. Bürger, *J. Organomet. Chem.*, 1983, **249,** 33.
62. K. Ouzounis, H. Riffel, H. Hess, U. Kohler and J. Weidlein, *Z. Anorg. Allg. Chem.*, 1983, **504,** 67.
63. M. Pelissier, J.-F. Labarre, L. V. Vilkov, A. V. Golubinsky and V. S. Mastryukov, *J. Chim. Phys.*, 1974, **71,** 702.
64. A. Ahmed, W. Schwarz and H. Hess, *Z. Naturforsch., Teil B,* 1978, **33,** 43.
65. G. M. McLaughlin, G. A. Sim and J. D. Smith, *J. Chem. Soc., Dalton Trans.*, 1972, 2197.
66. S. Amirkhalili, P. B. Hitchcock, A. D. Jenkins, J. Z. Nyathi and J. D. Smith, *J. Chem. Soc.* (*A*), 1971, 377.
67. O. T. Beachley and C. Tessier-Youngs, *Inorg. Chem.*, 1979, **18,** 3188.
68. K. Wakatsuki and T. Tanaka, *Bull. Chem. Soc. Jpn.*, 1975, **48,** 1475.
69. R. E. Bowen and K. Gosling, *J. Chem. Soc., Dalton Trans.*, 1974, 964.
70. S. Cucinella, G. Dozzi, A. Mazzei and T. Salvatori, *J. Organomet. Chem.*, 1975, **90,** 257.
71. H. Nöth and P. Wolfgardt, *Z. Naturforsch., Teil B,* 1976, **31,** 697.
72. H. Nöth and P. Konrad, *Chem. Ber.*, 1968, **101,** 3423; U. Thewalt and I. Kawada, *Chem. Ber.*, 1970, **103,** 2754.
73. G. Perego, G. Del Piero, M. Cesari, A. Zazzetta and G. Dozzi, *J. Organomet. Chem.*, 1975, **87,** 53.
74. S. Cucinella, T. Salvatori, C. Busetto, G. Perego and A. Mazzei, *J. Organomet. Chem.*, 1974, **78,** 185.
75. G. Perego, M. Cesari, G. Del Piero, A. Balducci and E. Cernia, *J. Organomet. Chem.*, 1975, **87,** 33.
76. G. Del Piero, S. Cucinella and M. Cesari, *J. Organomet. Chem.*, 1979, **173,** 263.
77. G. Perego and G. Dozzi, *J. Organomet. Chem.*, 1981, **205,** 21.
78. P. B. Hitchcock, J. D. Smith and K. M. Thomas, *J. Chem. Soc., Dalton Trans.*, 1976, 1433.
79. S. Amirkhalili, P. B. Hitchcock and J. D. Smith, *J. Chem. Soc., Dalton Trans.*, 1979, 1206.
80. S. Cucinella, T. Salvatori, C. Busetto and A. Mazzei, *J. Organomet. Chem.*, 1976, **108,** 13.
81. T. Salvatori, G. Dozzi and S. Cucinella, *Inorg. Chim. Acta,* 1980, **38,** 263.
82. I. Bousquet, J. J. Choury and P. Claudy, *Bull. Soc. Chim. Fr.*, 1967, 3852.
83. M. F. Lappert, *Adv. Inorg. Chem. Radiochem.*, 1966, **9,** 133.
84. E. Wiberg and H. Michaud, *Z. Naturforsch., Teil B,* 1954, **9,** 495, 504.
85. K. Dehnicke and N. Krüger, *Z. Anorg. Allg. Chem.*, 1978, **444,** 71.
86. J. Muller, *Z. Naturforsch., Teil B,* 1979, **34,** 531.
87. K. Dehnicke and N. Röder, *J. Organomet. Chem.*, 1975, **86,** 335.
88. J. L. Atwood and W. R. Newberry, *J. Organomet. Chem.*, 1975, **87,** 1.
89. F. Klanberg, *Z. Anorg. Allg. Chem.*, 1962, **316,** 197.
90. O. Schmitz-Dumont and B. Ross, *Angew. Chem., Int. Ed. Engl.*, 1964, **3,** 586.
91. S. K. Seale and J. L. Atwood, *J. Organomet. Chem.*, 1974, **64,** 57.
92. R. Shakir, M. J. Zaworotko and J. L. Atwood, *J. Organomet. Chem.*, 1979, **171,** 9.

93. R. Shakir, M. J. Zaworotko and J. L. Atwood, *J. Cryst. Mol. Struct.*, 1979, **9,** 135.
94. I. R. Beattie, P. J. Jones, J. A. K. Howard, L. E. Smart, C. J. Gilmore and J. W. Akitt, *J. Chem. Soc., Dalton Trans.*, 1979, 528.
95. J. L. Atwood, S. K. Seale and D. H. Roberts, *J. Organomet. Chem.*, 1973, **51,** 105.
96. K. Kawai and I. Kanesaka, *Spectrochim. Acta, Part A*, 1969, **25,** 263.
97. I. Y. Ahmed and C. D. Schmulbach, *Inorg. Chem.*, 1972, **11,** 228.
98. B. Dubois, L. Mocek, J. Pattyne and B. Vandorpe, *Bull. Chem. Soc. Fr, I,* 1984, 145.
99. J. W. Akitt, R. H. Duncan, I. R. Beattie and P. J. Jones, *J. Magn. Reson.*, 1979, **34,** 435.
100. E. W. Wehrli and R. Hoerdt, *J. Magn. Reson.*, 1981, **42,** 334; **44,** 197.
101. M. Dalibart, J. Derouault and P. Granger, *Inorg. Chem.*, 1981, **20,** 3975; 1982, **21,** 1040, 2241.
102. R. Höltje, *Z. Anorg. Allg. Chem.*, 1932, **209,** 244.
103. W. Levason and C. A. McAuliffe, *Coord. Chem. Rev.*, 1976, **19,** 173.
104. I. R. Beattie, G. A. Ozin and H. E. Blayden, *J. Chem. Soc. (A)*, 1969, 2535.
105. P. H. Bird and M. G. H. Wallbridge, *J. Chem. Soc.*, 1965, 3923; 1967, 664.
106. A. Almenningen, L. Fernholt, A. Haaland and J. Weidlein, *J. Organomet. Chem.*, 1978, **145,** 109.
107. C. C. Wang, M. Zaheeruddin and L. H. Spinar, *J. Inorg. Nucl. Chem.*, 1963, **25,** 326.
108. N. N. Greenwood, E. J. F. Ross and A. Storr, *J. Chem. Soc.*, 1965, 1400.
109. E. R. Alton, R. G. Montemayor and R. W. Parry, *Inorg. Chem.*, 1974, **13,** 2267.
110. L. Kolditz and W. Schmidt, *Z. Anorg. Allg. Chem.*, 1958, **296,** 188.
111. G. Fritz and R. Emül, *Z. Anorg. Allg. Chem.*, 1975, **416,** 19.
112. H. Schmidbaur, S. Lauteschläger and B. Milewski-Mahrla, *Chem. Ber.*, 1983, **116,** 1403.
113. C. Tessier-Youngs, C. Bueno, O. T. Beachley, Jr. and M. R. Churchill, *Inorg. Chem.*, 1983, **22,** 1054.
114. A. D. Norman, D. C. Wingleth and C. A. Heil, *Inorg. Synth.*, 1974, **15,** 178.
115. G. Fritz and H. Schäfer, *Z. Anorg. Allg. Chem.*, 1974, **406,** 167.
116. J. W. Anderson and J. E. Drake, *J. Chem. Soc. (A)*, 1971, 2246.
117. J. Crea and S. F. Lincoln, *Inorg. Chem.*, 1972, **11,** 1131.
118. T. Sato, *Z. Anorg. Allg. Chem.*, 1972, **391,** 69, 167.
119. M.-C. Stegmann, D. Vivien and C. Mazières, *J. Chim. Phys.*, 1974, **71,** 761.
120. N. Ishizawa, T. Miyata, I. Minato, F. Marumo and S. Iwai, *Acta Crystallogr., Sect. B,* 1980, **36,** 228.
121. H. L. Saalfeld and M. Wedde, *Z. Kristallogr.* 1974, **139,** 129.
122. D. E. Haycock, C. J. Nicholls, D. S. Urch, M. J. Webber and G. Wiech, *J. Chem. Soc., Dalton Trans.*, 1978, 1785.
123. L. M. Fraas, J. E. Moore and J. B. Salzberg, *J. Chem. Phys.*, 1973, **58,** 3585.
124. M. C. Saine, E. Husson and H. Brusset, *Spectrochim. Acta, Part A,* 1982, **38,** 25.
125. L. B. Alemany and G. W. Kirker, *J. Am. Chem. Soc.*, 1986, **108,** 6158.
126. M. G. Barker, P. G. Gadd and M. J. Begley, *J. Chem. Soc., Dalton Trans.*, 1984, 1139.
127. D. Müller, D. Hoebbel and W. Gessner, *Chem. Phys. Lett.*, 1981, **84,** 25.
128. B. N. Ivanov-Emin, G. Z. Kazier, V. I. Ivlieva and T. B. Aksenova, *Russ. J. Inorg. Chem. (Engl. Transl.),* 1981, **26,** 295.
129. L. S. Dent Glasser and R. Giovanoli, *Acta Crystallogr., Sect. B,* 1972, **28,** 519, 760.
130. C. Sabelli and R. T. Ferroni, *Acta Crystallogr., Sect. B,* 1978, **34,** 2407.
131. M. G. Vincent and J. W. Jeffrey, *Acta Crystallogr., Sect. B,* 1978, **34,** 1422.
132. A. P. M. Kentgens, K. F. Scholle and W. S. Veeman, *J. Phys. Chem.*, 1983, **87,** 4357.
133. D. Ginderow and F. Cesbron, *Acta Crystallogr., Sect. B,* 1979, **35,** 2499.
134. K. Hermansson, *Acta Crystallogr., Sect. C,* 1983, **39,** 925.
135. S. P. Best, R. S. Armstrong and J. K. Beattie, *J. Chem. Soc., Dalton Trans.*, 1982, 1655.
136. D. M. Adams and D. J. Hills, *J. Chem. Soc., Dalton Trans.*, 1978, 782.
137. A. V. Pankratov, A. N. Skachkov, O. N. Shalaeva and G. M. Kurbatov, *Russ. J. Inorg. Chem. (Engl. Transl.),* 1972, **17,** 47.
138. N. V. Kriutsov, G. N. Shirokova and V. Ya. Rosolovskii, *Russ. J. Inorg. Chem. (Engl. Transl.),* 1973, **18,** 503.
139. J. W. Akitt, *J. Chem. Soc., Dalton Trans.*, 1973, 42, 1177.
139a. R. C. Turner, *Can. J. Chem.*, 1976, **54,** 1528, 1910.
140. L.-O. Öhman and W. Forsling, *Acta Chem. Scand. Ser. A,* 1981, **35,** 795.
140a. N. I. Eremin, Yu. A. Volkhov and V. E. Mironov, *Russ. Chem. Rev. (Engl. Transl.),* 1974, **43,** 92.
141. J. W. Akitt and A. Farthing, *J. Chem. Soc., Dalton Trans.*, 1981, 1233, 1606, 1609, 1617, 1624.
142. J. W. Akitt and N. B. Milić, *J. Chem. Soc., Dalton Trans.*, 1984, 981.
143. D. L. Teagarden, N. L. White and S. L. Hem, *J. Pharm. Sci.*, 1981, **70,** 762.
144. D. N. Waters and M. S. Henty, *J. Chem. Soc., Dalton Trans.*, 1977, 243.
145. S. Schönter, H. Görz, R. Bertram, D. Müller and W. Gessner, *Z. Anorg. Allg. Chem.*, 1981, **476,** 188, 195; 1981, **483,** 153; 1983, **502,** 113, 132.
146. R. E. Mesmer and C. F. Baes, Jr., *Inorg. Chem.*, 1971, **10,** 2290.
147. R. J. Moolenaar, J. C. Evans and L. D. McKeever, *J. Phys. Chem.*, 1970, **74,** 3629.
148. S. G. Szabó, J. Wajand, I. Ruff and K. Burger, *Z. Anorg. Allg. Chem.*, 1978, **441,** 245.
149. J. W. Akitt and W. Gessner, *J. Chem. Soc., Dalton Trans.*, 1984, 147.
150. J. F. McIntyre, R. T. Foley and B. F. Brown, *Inorg. Chem.*, 1982, **21,** 1167.
151. H. Veillard, *J. Am. Chem. Soc.*, 1977, **99,** 7194.
152. R. J. Hill, G. V. Gibbs and R. C. Peterson, *Aust. J. Chem.*, 1979, **32,** 231.
153. O. Gropen, R. Johansen, A. Haaland and O. Stokkeland, *J. Organomet. Chem.*, 1975, **92,** 147.
154. R. H. Hange, J. W. Kaufman and J. L. Margrave, *J. Am. Chem. Soc.*, 1980, **102,** 6005.
155. D. C. Bradley, *Adv. Inorg. Chem. Radiochem.*, 1972, **15,** 259.
156. D. N. Bradley, J. W. Lorimer and C. Prevedorou-Demas, *Can. J. Chem.*, 1971, **49,** 2310.
157. J. G. Oliver and I. J. Worrall, *J. Chem. Soc. (A)*, 1970, 1389.
158. P. Maleki and M.-J. Schwing-Weill, *J. Inorg. Nucl. Chem.*, 1975, **37,** 435.

159. J.-P. Laussac and J.-P. Laurent, *J. Inorg. Nucl. Chem.*, 1976, **38,** 597.
160. J.-P. Laussac, R. Enjalbert, J. Galy and J.-P. Laurent, *J. Coord. Chem.*, 1983, **12,** 133.
161. M. F. Garbauskas, J. H. Wengrovius, R. C. Going and J. S. Kaspar, *Acta Crystallogr., Sect. C.*, 1984, **40,** 1536.
162. J. W. Akitt and R. H. Duncan, *J. Magn. Reson.*, 1974, **15,** 162.
163. N. Ya. Turova, V. A. Kozunov, A. I. Yanovskii, N. G. Bokii, Yu. T. Struchkov and B. L. Tarnopol'skii, *J. Inorg. Nucl. Chem.*, 1979, **41,** 5.
164. R. H. T. Bleyeveld, W. Fieggen and H. Gerding, *Recl. Trav. Chim. Pays-Bas,* 1972, **91,** 477.
165. K. S. Mazdiyasni, B. J. Schapter and L. M. Brown, *Inorg. Chem.*, 1971, **10,** 889.
166. A. Mehrotra and R. C. Mehrotra, *Inorg. Chem.*, 1972, **11,** 2170.
167. R. C. Mehrotra, S. Goel, A. B. Goel, R. B. King and R. C. Nainan, *Inorg. Chim. Acta,* 1978, **29,** 131.
168. E. Stumpp and U. Hillebrand, *Z. Naturforsch., Teil B,* 1979, **34,** 262.
169. R. C. Mehrotra, M. M. Agarwal and A. Mehrotra, *Synth. React. Inorg. Metal-Org. Chem.*, 1973, **3,** 181.
170. D. A. Horne, *J. Am. Chem. Soc.*, 1980, **102,** 6011.
171. N. Ya. Turova, N. I. Kozlova and M. I. Yanovkanya, *Koord. Khim.*, 1982, **8,** 148.
172. A. B. Goel, E. C. Ashby and R. C. Mehrotra, *Inorg. Chim. Acta,* 1982, **62,** 161.
173. S. Cucinella, G. Dozzi and G. Del Piero, *J. Organomet. Chem.*, 1982, **224,** 1.
174. M. Boleslawski and J. Serwatowski, *J. Organomet. Chem.*, 1983, **255,** 269.
175. J. L. Atwood and M. J. Zaworotko, *J. Chem. Soc., Chem. Commun.*, 1983, 302.
176. U. Thewalt and F. Stollmaier, *Angew. Chem., Int. Ed. Engl.*, 1982, **21,** 133.
177. D. A. Drew, A. Haaland and J. Weidlein, *Z. Anorg. Allg. Chem.*, 1973, **398,** 241.
178. M. Bonamico and G. Dessy, *J. Chem. Soc. (A)*, 1967, 1786.
179. R. D. Rogers and J. L. Atwood, *Organometallics,* 1984, **3,** 271.
180. J. R. Blackborow, C. R. Eady, E. A. Körner von Gustorf, A. Scrivanti and O. Wolfbeis, *J. Organomet. Chem.*, 1976, **108,** C32.
181. B. W. McLelland, *Acta Crystallogr., Sect. B,* 1975, **31,** 2496.
182. B. Nordén and I. Jonáš, *Inorg. Nucl. Chem. Lett.*, 1976, **12,** 33.
183. Y. Okamoto, E. Yashima and K. Hatada, *J. Chem. Soc., Chem. Commun.*, 1984, 1051.
184. M. L. Morris and R. L. Hildebrandt, *J. Mol. Struct.*, 1979, **53,** 69.
185. R. E. Hester and R. A. Plane, *Inorg. Chem.*, 1964, **3,** 513.
186. B. A. Kolesov and I. K. Igumenov, *Spectrochim. Acta, Part A,* 1984, **40,** 233.
187. A. J. Carty, R. H. Cragg, J. D. Smith and G. E. Toogood, *Annu. Rep. Prog. Chem., Sect. A, Phys. Inorg. Chem.*, 1976, **73,** 138; R. Jain, A. K. Rai and R. C. Mehrotra, *Inorg. Chim. Acta,* 1987, **126,** 99.
188. M. Das, D. T. Haworth and J. W. Beery, *Inorg. Chim. Acta,* 1981, **49,** 17.
189. M. Hirayama and K. Kitami, *J. Chem. Soc., Chem. Commun.*, 1980, 1030.
190. A. Pietrzykowski, S. Pasynkiewicz and A. Wolińska, *J. Organomet. Chem.*, 1980, **201,** 89.
191. C. M. Lukehart, *Acc. Chem. Res.*, 1981, **14,** 109.
192. M. M. Aly, *J. Inorg. Nucl. Chem.*, 1973, **35,** 537.
193. E. L. Muetterties and L. J. Guggenberger, *J. Am. Chem. Soc.*, 1972, **94,** 8046.
194. R. J. Irving and M. A. V. Ribiero da Silva, *J. Chem. Soc., Dalton Trans.*, 1975, 1257.
195. M. A. Al-Kadler and D. S. Urch, *J. Chem. Soc., Dalton Trans.*, 1984, 263.
196. P. E. Blackburn, A. Buchler and J. L. Stauffer, *J. Phys. Chem.*, 1966, **70,** 2469; N. I. Yanakiev and D. Trenafelov, *Dokl. Bolg. Akad. Nauk,* 1972, **25,** 1641.
197. D. J. Gardiner, R. E. Hester and E. Mayer, *J. Mol. Struct.*, 1974, **22,** 327.
198. D. W. James and R. L. Frost, *Aust. J. Chem.*, 1982, **35,** 1793.
199. J. F. McIntyre, R. T. Foley and B. F. Brown, *Inorg. Chem.*, 1982, **21,** 1167.
200. C. C. Addison and D. Sutton, *Prog. Inorg. Chem.*, 1967, **8,** 195.
201. G. N. Shirokova and V. Ya. Rosolovskii, *Russ. J. Inorg. Chem. (Engl. Transl.)*, 1971, **16,** 1699.
202. O. Ya. Dyachenko, L. O. Atovmyan, G. N. Shirokova and V. Ya. Rosolovskii, *J. Chem. Soc., Chem. Commun.*, 1973, 595.
203. J. L. Atwood, K. D. Crissinger and R. D. Rogers, *J. Organomet. Chem.*, 1978, **155,** 1.
204. J. H. Morris, P. G. Perkins, A. E. A. Rose and W. E. Smith, *Chem. Soc. Rev.*, 1977, **6,** 173.
205. C. Y. Shen and C. F. Callis, *Prep. Inorg. React.*, 1965, **2,** 137.
206. D. Müller, I. Grunze, E. Hallas and G. Ladwig, *Z. Anorg. Allg. Chem.*, 1983, **500,** 80.
207. H. van der Meer, *Acta Crystallogr., Sect. B,* 1976, **32,** 2423.
208. R. Kniep and D. Mootz, *Acta Crystallogr., Sect. B.* 1973, **29,** 2292.
209. B. M. Gatehouse and B. K. Miskin, *Acta Crystallogr., Sect. B,* 1974, **30,** 1311.
210. D. Brodalla, R. Kniep and D. Mootz, *Z. Naturforsch., Teil B,* 1981, **36,** 907.
211. A.-F. Shihada, B. K. Hassan and A. T. Mohammed, *Z. Anorg. Allg. Chem.*, 1980, **466,** 139.
212. B. Schaible, W. Haubold and J. Weidlein, *Z. Anorg. Allg. Chem.*, 1974, **403,** 289.
213. J. W. Akitt, N. N. Greenwood and G. D. Lester, *J. Chem. Soc. (A)*, 1971, 2450.
214. V. V. Pechkovskii, N. I. Gavrilyuk and R. Ya. Mil'nikova, *Russ. J. Inorg. Chem. (Engl. Transl.)*, 1980, **25,** 997; J. B. Parise, *Inorg. Chem.*, 1985, **24,** 4312.
215. S. J. Karlik, G. A. Elgavish, R. Pillai and G. L. Eichorn, *J. Magn. Reson.*, 1982, **49,** 164.
216. G. F. Knutsen and A. W. Searcy, *J. Electrochem. Soc.*, 1978, **125,** 327.
217. B. Vandorpe and M. Drache, *C. R. Hebd. Seances Acad. Sci., Ser. C,* 1973, **277,** 1211.
218. P. Rems, B. Orel and J. Šiftar, *Inorg. Chim. Acta,* 1971, **5,** 33; W. Franke and G. Henning, *Acta Crystallogr.*, 1965, **19,** 870; M. Cola, *Gazz. Chim. Ital.*, 1960, **90,** 220.
219. J. M. Manoli, P. Herpin and G. Pannetier, *Bull. Soc. Chim. Fr.*, 1970, 98.
220. B. Vandorpe and P. Barbier, *C. R. Hebd. Seances Acad. Sci., Ser. C,* 1968, **266,** 379.
221. M. Drache and J. Heubel, *Bull. Soc. Chim. Fr.*, 1980, 105.
222. H. Olapinski and J. Weidlein, *Z. Naturforsch., Teil B,* 1974, **29,** 114.
223. J. W. Akitt, N. N. Greenwood and B. L. Khandelwal, *J. Chem. Soc., Dalton Trans.*, 1972, 1226.
224. P. V. Klevtsov and V. A. Sinaiko, *Russ. J. Inorg. Chem. (Engl. Transl.)*, 1975, **20,** 1170.

225. J. J. De Boer, *Acta Crystallogr., Sect. B,* 1974, **30,** 1878.
226. J. W. Akitt and A. Farthing, *J. Chem. Soc., Dalton Trans.*, 1981, 1615.
227. Z. K. Nikitina and V. Ya. Rosolovskii, *Russ. J. Inorg. Chem.* (*Engl. Transl.*), 1977, **22,** 1458; 1978, **23,** 1293; 1979, **24,** 516, 779; 1980, **25,** 71.
228. T. Chausse, A. Potier and J. Potier, *J. Chem. Res.* (*S*), 1980, 316.
229. J. S. Martin and G. W. Stockton, *J. Magn. Reson.*, 1973, **12,** 218.
230. Yu. N. Matyuschin, T. S. Kon'kova, K. V. Titova, V. Ya. Rosolovskii and Yu. B. Lebedev, *Russ. J. Inorg. Chem.* (*Engl. Transl.*), 1981, **26,** 183.
231. M. Tropel, J.-C. Folest, C. Chevrot and J. Périchon, *Bull. Soc. Chim. Fr., Part 1,* 1978, 45.
232. R. C. Mehrotra and R. Bohra, 'Metal Carboxylates', Academic, London, 1983.
233. C. D. Garner and B. Hughes, *Adv. Inorg. Chem. Radiochem.*, 1975, **17,** 1.
234. J. L. Atwood, W. E. Hunter and K. D. Crissinger, *J. Organomet. Chem.*, 1977, **127,** 403.
235. J. Lamotte, O. Dideberg, L. Dupont and P. Durbut, *Cryst. Struct. Commun.*, 1981, **10,** 59.
236. P. H. Tedesco, V. B. de Rumi and J. A. Gonzalez-Quintana, *J. Inorg. Nucl. Chem.*, 1973, **35,** 285.
237. A. S. Kereichuk and L. M. Il'icheva, *Russ. J. Inorg. Chem.* (*Engl. Transl.*), 1975, **20,** 1291.
238. R. J. Motekaitis and A. E. Marvell, *Inorg. Chem.*, 1984, **23,** 18.
239. R. A. Hancock and S. T. Orszulik, *Polyhedron,* 1982, **1,** 313.
240. P. H. Tedesco and V. B. de Rumi, *J. Inorg. Nucl. Chem.*, 1975, **37,** 1833.
241. A. Napoli, *J. Inorg. Nucl. Chem.*, 1972, **34,** 1225.
242. L.-O. Öhman and S. Sjöberg, *Acta Chem. Scand., Ser. A,* 1981, **35,** 201; 1982, **36,** 47.
243. L.-O. Öhman and S. Sjöberg, *J. Chem. Soc., Dalton Trans.*, 1983, 2513; 1985, 2665.
244. M. A. Lopez-Quintella, W. Knoche and J. Veith, *J. Chem. Soc., Faraday Trans. 1,* 1984, **80,** 2313.
245. B. Perlmutter-Hayman and E. Tapuhi, *Inorg. Chem.*, 1979, **18,** 875.
246. L.-O. Öhman and S. Sjöberg, *Acta Chem. Scand., Ser. A,* 1983, **37,** 875.
247. M. K. Singh and M. N. Srivastava, *J. Inorg. Nucl. Chem.*, 1972, **34,** 567.
248. S. J. Karlik, E. Tarien, G. A. Elgavish and G. L. Eichhorn, *Inorg. Chem.*, 1983, **22,** 525.
249. D. Taylor, *Aust. J. Chem.*, 1978, **31,** 1455.
250. A. S. Tracey and T. L. Boivin, *J. Am. Chem. Soc.*, 1983, **105,** 4901.
251. J. Wilinski and R. J. Kurland, *J. Am. Chem. Soc.*, 1978, **100,** 2233.
252. T. Deeg and A. Weiss, *Ber. Bunsenges. Phys. Chem.*, 1976, **80,** 2.
253. M. Dalibart, J. Derouault and P. Granger, *Inorg. Chem.*, 1982, **21,** 2241.
254. A. H. Cowley, M. C. Cushner, R. E. Davis and P. E. Riley, *Inorg. Chem.*, 1981, **20,** 1179.
255. J. Derouault and M. T. Forel, *Inorg. Chem.*, 1977, **16,** 3207; 3214.
256. A. Boardman, R. W. H. Small and I. J. Worrall, *Acta Crystallogr., Sect. C,* 1983, **39,** 433.
257. P. J. Ogren, J. P. Cannon and C. F. Smith, Jr., *J. Phys. Chem.*, 1971, **75,** 282.
258. K. N. Semenenko, Yu. Ya. Kharitonov, V. B. Polyakova and V. N. Fokin, *Russ. J. Inorg. Chem.* (*Engl. Transl.*), 1974, **19,** 1437.
259. E. N. Gur'yanova, V. V. Puchkova and A. F. Volkov, *J. Struct. Chem.*, 1974, **15,** 374.
260. F. Wada and T. Matsuda, *Bull. Chem. Soc. Jpn.*, 1980, **53,** 421.
261. J. L. Atwood, D. C. Hrncir, R. Shakir, M. S. Dalton, R. D. Priester and R. D. Rogers, *Organometallics,* 1982, **1,** 1021; S. G. Bott, H. Elgamal and J. L. Atwood, *J. Am. Chem. Soc.*, 1985, **107,** 1796.
262. V. P. Mardykin, P. N. Gaponik and A. F. Popov, *Russ. Chem. Rev.* (*Engl. Transl.*), 1979, **48,** 487.
263. R. L. Shuler, R. A. de Marco and A. D. Berry, *Inorg. Chim. Acta,* 1984, **85,** 185.
264. B. Glavincevski and S. K. Brownstein, *Can. J. Chem.*, 1981, **59,** 3012.
265. D. E. H. Jones and J. L. Wood, *J. Chem. Soc.* (*A*), 1971, 3132, 3135.
266. P. H. Collons, *J. Appl. Crystallogr.*, 1975, **8,** 337.
267. M. R. Churchill, H. J. Wassermann, S. J. Holmes and R. R. Schrock, *Organometallics,* 1982, **1,** 766.
268. T. G. Richmond, F. Basolo and D. F. Shriver, *Inorg. Chem.*, 1982, **21,** 1272.
269. R. B. Peterson, J. J. Stezowski, C. Wan, J. M. Burlitch and R. E. Hughes, *J. Am. Chem. Soc.*, 1971, **93,** 3532.
270. J. J. Dechter, *Prog. Inorg. Chem.*, 1982, **29,** 285.
271. Yu. A. Buslaev and S. P. Petrosyants, *Polyhedron,* 1984, **3,** 265.
272. A. E. Merbach, P. Moore, O. W. Howarth and C. M. McAteer, *Inorg. Chim. Acta,* 1980, **39,** 129.
273. J. Meunier and M. T. Forel, *Can. J. Chem.*, 1972, **50,** 1157.
274. C. V. Berney and J. H. Weber, *Inorg. Chim. Acta,* 1971, **5,** 375.
275. D. Richardson and T. D. Alger, *J. Phys. Chem.*, 1975, **79,** 1733.
276. D. Canet, J.-J. Delpeuch, M. R. Khaddar and P. Rubini, *J. Magn. Reson.*, 1974, **15,** 325.
277. E. E. Hellstrom and R. A. Huggins, *Mater. Res. Bull.*, 1979, **14,** 127.
278. H. Haeuseler, A. Cansiz and H. D. Lutz, *Z. Naturforsch., Teil B,* 1981, **36,** 532.
279. H. J. Berthold and K. Köhler, *Z. Anorg. Allg. Chem.*, 1981, **475,** 45.
280. W. Klee and H. Schäfer, *Z. Anorg. Allg. Chem.*, 1981, **479,** 125.
281. M. Schulte-Kellinghaus and V. Krämer, *Acta Crystallogr., Sect. B,* 1979, **35,** 3016.
282. L. Zhengyan, H. Janssen, R. Mattes, H. Schnöckel and B. Krebs, *Z. Anorg. Allg. Chem.*, 1984, **513,** 67.
283. R. W. Berg, S. von Winbush and N. H. Bjerrum, *Inorg. Chem.*, 1980, **19,** 2688.
284. H. Nöth and P. Konrad, *Chem. Ber.*, 1983, **116,** 3552.
285. M. Mortag and K. Möckel, *Z. Chem.*, 1980, **20,** 183.
286. K. Wakatsuki, Y. Takeda and T. Tanaka, *Inorg. Nucl. Chem. Lett.*, 1974, **10,** 383.
287. L. Fernholt, A. Haaland, M. Hargittai, R. Seip and J. Weidlein, *Acta Chem. Scand., Ser. A,* 1981, **35,** 529.
288. A. A. Carey and E. P. Schram, *Inorg. Chim. Acta,* 1982, **59,** 75, 79, 83.
289. M. Boleslawski, S. Pasynkiewicz and A. Kunicki, *J. Organomet. Chem.*, 1974, **73,** 193.
290. J. W. Hastie, R. H. Hauge and J. L. Margrave, *J. Fluorine Chem.*, 1973, **3,** 285.
291. Y. S. Yang and J. S. Shirk, *J. Mol. Spectrosc.*, 1975, **24,** 39.
292. H. Schnöckel, *J. Mol. Struct.*, 1978, **50,** 267.
293. N. A. Matwiyoff and W. E. Wageman, *Inorg. Chem.*, 1970, **9,** 1031.

294. R. P. Agarwal and E. C. Moreno, *Talanta,* 1971, **18,** 873.
295. V. P. Vasil'ev and E. V. Kozlovskii, *Russ. J. Inorg. Chem. (Engl. Transl.),* 1975, **20,** 672.
296. Yu. A. Buslaev and S. P. Petrosyants, *Koord. Chim.,* 1981, **7,** 516.
297. T. Fleischer and R. Hoppe, *Z. Anorg. Allg. Chem.,* 1982, **492,** 83.
298. J. L. Fourquet, F. Plet and R. de Pape, *Acta Crystallogr., Sect. B,* 1981, **37,** 2136.
299. J. Nouet, J. Pannetier and J. L. Fourquet, *Acta Crystallogr., Sect. B,* 1981, **37,** 32.
300. J. Bondam, *Acta Chem. Scand.,* 1971, **25,** 3271.
301. B. Gilbert, G. Mamantov and G. M. Begun, *Inorg. Nucl. Chem. Lett.,* 1976, **12,** 415.
302. J.-J. Videau, J. Portier and B. Piriou, *Rev. Chim. Minér.,* 1979, **16,** 393.
303. M. Couzi, P. L. Loyzance, R. Mokhlisse, A. Bulou and J. L. Fourquet, *Ber. Bunsenges. Phys. Chem.,* 1983, **87,** 232.
304. L. A. Curtiss, *Inorg. Chem.,* 1982, **21,** 4100.
305. J. L. Atwood and W. R. Newberry, *J. Organomet. Chem.,* 1974, **66,** 15.
306. H. Schmidbaur, J. Weidlein, H.-F. Klein and K. Eiglmeier, *Chem. Ber.,* 1968, **101,** 2268.
307. H. Schnöckel, *Z. Naturforsch., Teil B,* 1976, **31,** 1291.
308. C. E. Sjøgren, P. Klaboe and E. Ritter, *Spectrochim. Acta, Part A,* 1984, **40,** 457.
309. M. Tranquille and M. Fouassier, *J. Chem. Soc., Faraday Trans., 2,* 1980, **76,** 26.
310. M. F. Lappert, J. B. Pedley, G. J. Sharp and M. F. Guest, *J. Chem. Soc., Faraday Trans. 2,* 1976, **72,** 539.
311. T. Tomita, C. E. Sjøren, P. Klaboe, G. N. Papatheodorou and E. Rytter, *J. Raman Spectrosc.,* 1983, **14,** 415.
312. K. Tanemoto, G. Mamantov and G. M. Begun, *Inorg. Chim. Acta,* 1983, **76,** L79.
313. H. Schäfer, *Adv. Inorg. Chem. Radiochem.,* 1983, **26,** 201.
314. M. E. Anthoney, A. Finch and P. J. Gardner, *J. Chem. Soc., Dalton Trans.,* 1973, 659.
315. T. P. Martin and J. Diefenbach, *J. Am. Chem. Soc.,* 1984, **106,** 623.
316. R. Kneip, P. Blees and W. Poll, *Angew. Chem., Int. Ed. Engl.,* 1982, **21,** 386.
317. U. Thewalt and M. Burger, *Angew. Chem., Int. Ed. Engl.,* 1982, **21,** 634.
318. B. Chevrier and R. Weiss, *Angew. Chem., Int. Ed. Engl.,* 1974, **13,** 1.
319. J. Wilinski and R. J. Kurland, *J. Am. Chem. Soc.,* 1978, **100,** 2233.
320. G. M. Kramer, *J. Org. Chem.,* 1979, **44,** 2616, 2619.
321. E. Mayer, *Z. Naturforsch., Teil B,* 1979, **34,** 548, 891.
322. J. Robinson and R. A. Osteryoung, *J. Am. Chem. Soc.,* 1979, **101,** 321.
323. T. Okuda, Y. Furukawa and H. Negita, *Bull. Chem. Soc. Jpn.,* 1972, **45,** 2245.
324. R. Schürmann and H. H. Perkampus, *Spectrochim. Acta, Part A,* 1979, **35,** 45.
325. T. Sasa, Y. Takahashi and T. Mukaibo, *Bull. Chem. Soc. Jpn.,* 1972, **45,** 2250, 2657.
326. U. Gesenhues and H. Wendt, *Z. Phys. Chem.,* 1982, **132,** 121.
327. K. Brendhaugen, A. Haaland and D. P. Novak, *Acta Chem. Scand., Ser. A,* 1974, **28,** 45.
328. J. E. House, Jr., *J. Organomet. Chem.,* 1984, **263,** 267.
329. M. Dalibart and J. Derouault, *Coord. Chem. Rev.,* 1986, **74,** 1.
330. E. Perenthaler, H. Schulz and A. Rabenau, *Z. Anorg. Allg. Chem.,* 1982, **491,** 259.
331. B. Vandorpe and B. Dubois, *C. R. Hebd. Seances Acad. Sci., Ser. C,* 1972, **275,** 487.
332. A. Simon, K. Peters, E.-M. Peters, H. Künnl and B. Koslowski, *Z. Anorg. Allg. Chem.,* 1980, **469,** 94.
333. S. Brownstein, A. Morrison and L. K. Tan, *Can. J. Chem.,* 1986, **64,** 265.
334. N. Kitajima, H. Shimanouchi, Y. Ono and Y. Sasada, *Bull. Chem. Soc. Jpn.,* 1982, **55,** 2064.
335. A. Justnes, E. Rytter and A. F. Andresen, *Polyhedron,* 1982, **1,** 393.
336. U. Thewalt and F. Stollmaier, *J. Organomet. Chem.,* 1982, **228,** 149.
337. D. J. Merryman, P. A. Edwards, J. D. Corbett and R. E. McCarley, *Inorg. Chem.,* 1974, **13,** 1471.
338. T. Okuda, K. Yamoda, H. Ishihara and H. Negita, *Chem. Lett.,* 1975, 785.
339. A. Fischner and A. Weiss, *Z. Naturforsch., Teil B,* 1980, **35,** 170.
340. B. Krebs, M. Hucke and C. J. Brendel, *Angew. Chem., Int. Ed. Engl.,* 1982, **21,** 445.
341. S. W. Tobey and R. West, *J. Am. Chem. Soc.,* 1966, **88,** 2481, 2489.
342. H. Grunze, *Z. Chem.,* 1983, **23,** 136.
343. S. Pohl, *Z. Anorg. Allg. Chem.,* 1983, **498,** 15, 20.
344. A. H. Cowley, M. C. Cushner and P. E. Riley, *J. Am. Chem. Soc.,* 1980, **102,** 624.
345. H. Ikeuchi and C. Krohn, *Acta Chem. Scand., Ser. A,* 1974, **28,** 48.
346. W. C. Laughlin and N. W. Gregory, *Inorg. Chem.,* 1975, **14,** 1263.
347. U. Anders and J. A. Plambeck, *J. Inorg. Nucl. Chem.,* 1978, **40,** 387.
348. A. I. Morozov, O. A. Solovkina and V. I. Evdokimov, *Russ. J. Inorg. Chem. (Engl. Trans.),* 1982, **27,** 1172.
349. J. P. Collman, J. I. Brauman, G. Tustin and G. S. Wann, III, *J. Am. Chem. Soc.,* 1983, **105,** 3913.
350. J. Hvistendahl, P. Klaeboe, E. Rytter and H. A. Øye, *Inorg. Chem.,* 1984, **23,** 706.
351. R. W. Berg, H. A. Hjuler and N. J. Bjerrum, *Inorg. Chem.,* 1984, **23,** 557.
352. M. Elam, I. Bahatt, E. Peled and E. Gileadi, *J. Phys. Chem.,* 1984, **88,** 1609.
353. S. Sahami and R. A. Osteryoung, *Inorg. Chem.,* 1984, **23,** 2511.
354. S. Takahashi, N. Koura, M. Murase and H. Ohno, *J. Chem. Soc., Faraday Trans. 2,* 1986, **82,** 49.
355. F. Taulelle and A. I. Popov, *J. Solution Chem.,* 1986, **15,** 463.
356. A. A. Fannin, Jr., L. A. King, J. A. Levisky and J. S. Wilkes, *J. Phys. Chem.,* 1984, **88,** 2609, 2614.
357. R. J. H. Clark, B. Crociani and A. Wasserman, *J. Chem. Soc. (A),* 1970, 2458.
358. H. Nöth, R. Rurländer and P. Wolfgardt, *Z. Naturforsch., Teil B,* 1982, **37,** 29.
359. M. Dalibart, J. Derouault and P. Granger, *Inorg. Chem.,* 1982, **21,** 1040, 2241.
360. S. Spange and G. Heublein, *Z. Chem.,* 1983, **23,** 337.
361. K. Kawai, I. Kanesaka and F. Ichimura, *Spectrochim. Acta, Part A,* 1970, **26,** 593, 2345.
362. R. Rogers and J. L. Atwood, *J. Cryst. Mol. Struct.,* 1980, **9,** 45.
363. M. E. Peach, V. L. Tracey and T. C. Waddington, *J. Chem. Soc. (A),* 1969, 336.
364. J.-M. Le Carpentier and R. Weiss, *Acta Crystallogr., Sect. B,* 1972, **28,** 1421; 1437.
365. F. Stollmaier and U. Thewalt, *J. Organomet. Chem.,* 1981, **208,** 327.

366. E. Rytter, B. E. D. Rytter, H. A. Øye and J. Krogh-Moe, *Acta Crystallogr., Sect. B,* 1975, **31,** 2177.
367. B. Dubois, P. Decock and B. Vandorpe, *C. R. Hebd. Seances Acad. Sci., Ser. C,* 1981, **292,** 517.
368. A. Manteghetti and A. Potier, *Spectrochim. Acta, Part A,* 1982, **38,** 141.
369. J. M. Howell, A. M. Sapse, E. Singman and G. Synder, *J. Am. Chem. Soc.*, 1982, **104,** 4758.
370. J. L. Gray and G. E. Maciel, *J. Am. Chem. Soc.*, 1981, **103,** 7147.
371. R. W. Berg, E. Kemnitz, H. A. Hjuler, R. Fehrmann and N. J. Bjerrum, *Polyhedron,* 1985, **4,** 457.
372. R. M. Canadine and D. E. H. Jones, *J. Chem. Soc., Faraday Trans. 2,* 1972, **68,** 1665.
373. R. G. S. Pong, A. E. Shirk and J. S. Shirk, *J. Chem. Phys.*, 1979, **71,** 525.
374. A. E. Finholt, A. C. Bond, Jr. and H. I. Schlesinger, *J. Am. Chem. Soc.*, 1947, **69,** 1199.
375. P. R. Oddis and M. G. H. Wallbridge, *J. Chem. Soc., Dalton Trans.*, 1978, 572.
376. D. L. Schmidt, C. B. Roberts and P. F. Reigler, *Inorg. Synth.*, 1973, **14,** 47.
377. K. N. Semenenko, Kh. A. Taisumov, A. P. Savachenkova and V. N. Surov, *Russ. J. Inorg. Chem.* (*Engl. Transl.*), 1971, **16,** 1104.
378. F. M. Brower, N. E. Matzek, P. F. Reigler, H. W. Rinn, C. B. Roberts, D. L. Schmidt, J. A. Snover and K. Terada, *J. Am. Chem. Soc.*, 1976, **98,** 2450.
379. J. W. Turley and H. W. Rinn, *Inorg. Chem.*, 1969, **8,** 18.
380. A. Okninski and S. Pasynkiewicz, *Inorg. Chim. Acta,* 1978, **28,** L125.
381. H. Schnöckel, *J. Mol. Struct.*, 1978, **50,** 275.
382. M. Pelissier, J. P. Malrieu, A. Serafini and J. F. Labarre, *Theor. Chim. Acta,* 1980, **56,** 175.
383. J. M. Howell, A. M. Sapse, E. Singman and G. Snyder, *J. Am. Chem. Soc.*, 1982, **104,** 4758.
384. J. L. Atwood, D. C. Hrncir, R. D. Rogers and J. A. K. Howard, *J. Am. Chem. Soc.*, 1981, **103,** 6787.
385. G. Hencken and E. Weiss, *J. Organomet. Chem.*, 1974, **73,** 35.
386. E. B. Lobkovsky, G. I. Soloveychik, B. M. Bulychev, A. B. Erofeev, A. I. Gusev and N. I. Kirillova, *J. Organomet. Chem.*, 1983, **254,** 167.
387. E. C. Ashby, *Adv. Inorg. Chem. Radiochem.*, 1966, **8,** 283.
388. J. S. Pizey, 'Lithium Aluminum Hydride', Horwood-Wiley, New York, 1977.
389. N. S. Gaylord, 'Reduction with Complex Metal Hydrides', Interscience, New York, 1956.
390. R. G. Beach and E. C. Ashby, *Inorg. Chem.*, 1971, **10,** 1888.
391. J. Mayet, J. Tranchant and S. Kovacevic, *Bull. Soc. Chim. Fr.*, 1973, 503, 506.
392. E. C. Ashby, J. J. Watkins and H. S. Prasad, *Inorg. Chem.*, 1975, **14,** 583.
393. V. A. Kuznetzov, N. D. Golubeva and K. N. Semenenko, *Russ. J. Inorg. Chem.* (*Engl. Transl.*), 1974, **19,** 669.
394. J. W. Lauher, D. Dougherty and P. J. Herley, *Acta Crystallogr., Sect. B,* 1979, **35,** 1454.
395. S. Wold and O. Strouf, *Acta Chem. Scand., Ser. A,* 1979, **33,** 463, 521.
396. A. I. Boldyrev and O. P. Charkin, *Russ. J. Inorg. Chem.* (*Engl. Transl*), 1980, **25,** 58.
397. B. Čásenský and J. Macháček, *Inorg. Synth.*, 1978, **18,** 149.
398. A. B. Goel, E. C. Ashby and R. C. Mehrotra, *Inorg. Chim. Acta,* 1982, **62,** 161.
399. E. C. Ashby, F. R. Dobbs and H. P. Hopkins, Jr., *J. Am. Chem. Soc.*, 1973, **95,** 2823.
400. H. Nöth, R. Rurländer and P. Wolfgardt, *Z. Naturforsch., Teil B,* 1981, **36,** 31.
401. A. E. Shirk and D. F. Shriver, *J. Am. Chem. Soc.*, 1973, **95,** 5904.
402. K. N. Semenenko, V. N. Fokin, A. P. Savchenkova and E. B. Lobkovskii, *Russ. J. Inorg. Chem.* (*Engl. Transl.*), 1973, **18,** 926.
403. P. Claudy, J. Étienne and R. Bonnetot, *Rev. Chim. Miner.*, 1972, **9,** 511.
404. D. A. Coe and J. W. Nibler, *Spectrochim. Acta, Part A,* 1973, **29,** 1789.
405. A. Almenningen, G. Gundersen and A. Haaland, *Acta Chem. Scand.*, 1968, **22,** 328.
406. N. A. Bailey, R. H. Bird and M. G. H. Wallbridge, *Inorg. Chem.*, 1968, **7,** 1575.
407. E. B. Lobkovskii, V. B. Polyakova, S. P. Shilkin and K. N. Semenenko, *J. Struct. Chem.* (*Engl. Transl.*), 1975, **16,** 66.
408. P. R. Oddy and M. G. H. Wallbridge, *J. Chem. Soc., Dalton Trans.*, 1978, 572.
409. H. Nöth and R. Rurländer, *Inorg. Chem.*, 1981, **20,** 1062.
410. F. L. Himpsl and A. C. Bond, *J. Am. Chem. Soc.*, 1981, **103,** 1098.
411. C. J. Dain, A. J. Downs and D. W. H. Rankin, *J. Chem. Soc., Dalton Trans.*, 1981, 2465.
412. P. R. Oddy and M. G. H. Wallbridge, *J. Chem. Soc., Dalton Trans.*, 1976, 869; 2076.
413. M. T. Barlow, C. J. Dain, A. J. Downs, P. D. P. Thomas and D. W. H. Rankin, *J. Chem. Soc., Dalton Trans.*, 1980, 1374.
414. K. N. Semenenko, Yu. Ya. Kharitonov, V. B. Polyakova and V. N. Fokin, *Russ. J. Inorg. Chem.* (*Engl. Transl.*), 1975, **20,** 19.
415. B. Szpoganicz and A. E. Martell, *Inorg. Chem.*, 1985, **24,** 2414.
416. A. Fratiello, D. D. Davis, S. Peak and R. E. Schuster, *Inorg. Chem.*, 1971, **10,** 1627.
417. M. Agrawal, J. P. Tandon and R. C. Mehrotra, *J. Inorg. Nucl. Chem.*, 1979, **41,** 1405.
418. K. Hayashi, K.-I. Okamoto, J. Hidaka and H. Einaga, *J. Chem. Soc., Dalton Trans.*, 1982, 1377.
419. J. K. Kouinis, J. M. Tsangaris and A. G. Galinos, *Z. Naturforsch., Teil B,* 1978, **33,** 987.
420. C. M. Mikulski, S. Chauhan, R. Rabin and N. M. Karayannis, *Inorg. Nucl. Chem. Lett.*, 1981, **17,** 195.
421. E. Niecke and R. Kröher, *Z. Naturforsch., Teil B,* 1979, **34,** 837.
422. J. A. Labinger and J. S. Miller, *J. Am. Chem. Soc.*, 1982, **104,** 6856, 6858.
423. J. W. Buchler, L. Puppe and H. H. Schneehage, *Liebigs Ann. Chem.*, 1971, **749,** 143.
424. N. Takeda and S. Inoue, *Bull. Chem. Soc. Jpn.*, 1978, **51,** 3564.
425. S. Inoue and N. Takeda, *Bull. Chem. Soc. Jpn.*, 1977, **50,** 984.
426. R. S. Nohr, P. M. Kuznesof, K. J. Wynne, M. E. Kenney and P. G. Sieberman, *J. Am. Chem. Soc.*, 1981, **103,** 4371.
427. A. B. P. Lever and P. C. Minor, *Inorg. Chem.*, 1981, **20,** 4015.
428. J. L. Atwood, D. C. Hrncir, R. Shakir, M. S. Dalton, R. D. Priester and R. D. Rogers, *Organometallics,* 1982, **1,** 1021.
429. N. N. Greenwood, *Adv. Inorg. Chem. Radiochem.*, 1963, **5,** 91.

430. I. A. Sheka, I. S. Chaus and T. T. Mityureva, 'The Chemistry of Gallium', Elsevier, New York, 1966.
431. H. J. Emeléus (ed.), 'MTP International Review of Science, Inorganic Chemistry, Series Two', Butterworths, London, 1975, vol. 1, p. 281, 331.
432. M. J. Taylor, 'Metal-to-Metal Bonded States of the Main Group Elements', Academic, London, 1975, chap. 3.
433. A. J. Hinchcliffe and J. S. Ogden, *J. Phys. Chem.*, 1973, **77**, 2537.
434. M. J. Zehe, D. A. Lynch, B. J. Kelsall and K. D. Carlson, *J. Phys. Chem.*, 1979, **83**, 656.
435. H. Ishihara and H. Negita, *Bull. Chem. Soc. Jpn.*, 1985, **58**, 2731.
436. J. C. Beamish, M. Wilkinson and I. J. Worrall, *Inorg. Chem.*, 1978, **17**, 2026.
437. M. J. Taylor, *J. Chem. Soc. (A)*, 1970, 2812.
438. E. Chemouni and A. Potier, *J. Inorg. Nucl. Chem.*, 1971, **33**, 2343.
439. P. I. Fedorov, N. S. Malova and I. Yu. Rodimtseva, *Russ. J. Inorg. Chem. (Engl. Transl.)*, 1978, **23**, 1376.
440. W. Hönle, G. Gerlach, W. Weppner and A. Simon, *J. Solid State Chem.*, 1986, **61**, 171.
441. G. Gerlach, W. Hönle and A. Simon, *Z. Anorg. Allg. Chem.*, 1982, **486**, 7.
442. I. R. Beattie and J. R. Horder, *J. Chem. Soc. (A)*, 1970, 2433.
443. J. G. Oliver and I. J. Worrall, *J. Chem. Soc. (A)*, 1971, 2315.
444. J. D. Corbett, *Inorg. Chem.*, 1963, **2**, 634.
445. F. M. Brewer, J. R. Chadwick, G. Garton and D. M. L. Goodgame, *J. Inorg. Nucl. Chem.*, 1963, **25**, 322; 324.
446. F. M. Brewer, G. S. Reddy and P. L. Goggin, *J. Indian. Chem. Soc.*, 1967, **44**, 179.
447. M. Uson-Finkenzeller, W. Bublak, B. Huber, G. Müller and H. Schmidbaur, *Z. Naturforsch., Teil B*, 1986, **41**, 346.
448. H. Schmidbaur, W. Bublak, B. Huber and G. Müller, *Organometallics*, 1986, **5**, 1647.
449. J. C. Beamish, A. Boardman, R. W. H. Small and I. J. Worrall, *Polyhedron*, 1985, **4**, 983.
450. R. W. H. Small and I. J. Worrall, *Acta Crystallogr., Sect. B*, 1982, **38**, 250.
451. R. W. H. Small and I. J. Worrall, *Acta Crystallogr., Sect. B*, 1982, **38**, 86; L. A. Woodward and M. J. Taylor, *J. Inorg. Nucl. Chem.*, 1965, **27**, 737.
452. I. Sinclair and I. J. Worrall, *Inorg. Nucl. Chem. Lett.*, 1981, **17**, 279.
453. S. Paashaus and R. Kniep, *Angew. Chem., Int. Ed. Engl.*, 1986, **25**, 752.
454. A. Kuhn, R. Chevalier, C. Desnoyers and J. C. J. M. Terhell, *Acta Crystallogr., Sect. B*, 1976, **32**, 1910.
455. J. Weiss and H. Schäfer, *Z. Naturforsch., Teil B*, 1976, **31**, 1341.
456. M. J. Taylor, *J. Raman Spectrosc.*, 1973, **1**, 355.
457. G. Lucazeau, *Solid State Commun.*, 1976, **18**, 917.
458. J. M. Thomas, I. Adams, R. H. Williams and M. Barber, *J. Chem. Soc., Faraday Trans. 2*, 1972, **68**, 755.
459. T. J. Bastow and H. J. Whitfield, *J. Magn. Reson.*, 1975, **20**, 1.
460. C. A. Evans, K.-H. Tan, S. P. Tapper and M. J. Taylor, *J. Chem. Soc., Dalton Trans.*, 1973, 988.
461. M. J. Taylor and D. G. Tuck, *Inorg. Synth.*, 1983, **22**, 135.
462. H. J. Cumming, D. Hall and C. E. Wright, *Cryst. Struct. Commun.*, 1974, **3**, 107.
463. T. Okuda, N. Yoshida, M. Hiura, H. Ishihara, K. Yamada and H. Negita, *J. Mol. Struct.*, 1982, **96**, 169.
464. L. G. Waterworth and I. J. Worrall, *J. Inorg. Nucl. Chem.*, 1973, **35**, 1535.
465. T. Eheman and K. Dehnicke, *J. Organomet. Chem.*, 1974, **64**, C33.
466. L. Rösch and H. Neumann, *Angew. Chem., Int. Ed. Engl.*, 1980, **19**, 55.
467. C. H. van Dyke, *Prep. Inorg. Reactions*, 1971, **6**, 157.
468. M. J. Taylor and S. Reithmiller, *J. Raman Spectrosc.*, 1984, **15**, 370.
469. M. Hargittai, I. Hargittai, V. P. Spiridonov and A. A. Ivanov, *J. Mol. Struct.*, 1977, **39**, 225.
470. V. I. Trusov, A. V. Suvorov and R. N. Abdukumova, *Russ. J. Inorg. Chem. (Engl. Transl.)*, 1975, **20**, 278.
471. J. R. Durig, C. G. Bradley and J. D. Odom, *Inorg. Chem.*, 1982, **21**, 1466.
472. D. F. Shriver and R. W. Parry, *Inorg. Chem.*, 1962, **1**, 835.
473. M. B. Dunn and C. A. McDowell, *Mol. Phys.*, 1972, **24**, 969.
474. J. R. Durig, K. K. Chatterjee, Y. S. Li, M. Jalilian, A. J. Zozulin and J. D. Odom, *J. Chem. Phys.*, 1980, **73**, 21.
475. A. E. Shirk and J. S. Shirk, *Inorg. Chem.*, 1983, **22**, 72.
476. I. R. Beattie, T. Gilson and G. A. Ozin, *J. Chem. Soc. (A)*, 1968, 1092.
477. J. R. Durig and K. K. Chatterjee, *J. Mol. Struct.*, 1982, **95**, 102.
478. J. R. Durig, C. B. Bradley, Y. S. Li and J. D. Odom, *J. Mol. Struct.*, 1981, **74**, 205.
479. R. A. Kovar, J. A. Johnson and R. L. Cook, *Inorg. Chem.*, 1980, **19**, 3264.
480. K. Anton and H. Nöth, *Chem. Ber.*, 1982, **115**, 2668.
481. I. Sinclair, R. W. H. Small and I. J. Worrall, *Acta Crystallogr., Sect. B*, 1981, **37**, 1290.
482. C. D. Schmulbach and I. Y. Ahmed, *Inorg. Chem.*, 1971, **10**, 1902.
483. D. Raptis, J. K. Kouinis, D. Kaminaris and A. G. Galinos, *Monatsh. Chem.*, 1981, **112**, 713.
484. A. J. Carty, *Can. J. Chem.*, 1970, **48**, 3524.
485. F. Ya. Kul'ba, V. L. Stolyarov and A. P. Zharkov, *Russ. J. Inorg. Chem. (Engl. Transl.)*, 1972, **17**, 57, 197.
486. R. Restivo and G. J. Palenik, *J. Chem. Soc., Dalton Trans.*, 1972, 341.
487. A. T. McPhail, R. W. Miller, C. G. Pitt, G. Gupta and S. G. Srivastava, *J. Chem. Soc., Dalton Trans.*, 1976, 1657.
488. G. Beran, A. J. Carty, H. A. Patel and G. J. Palenik, *Inorg. Chem.*, 1972, **11**, 896.
489. D. F. Rendle, A. Storr and J. Trotter, *J. Chem. Soc., Dalton Trans.*, 1973, 2252.
490. G. Lucazeau, A. Novak, P. Molinié and J. Rouxel, *J. Mol. Struct.*, 1974, **20**, 303.
491. P. Molinié, R. Brec and J. Rouxel, *C. R. Hebd. Seances Acad. Sci, Ser. C*, 1972, **275**, 487.
492. H. Nöth and P. Konrad, *Z. Naturforsch., Teil B*, 1975, **30**, 681.
493. H. Bürger, J. Cichon, U. Goetze, U. Wannagat and H. J. Wismar, *J. Organomet. Chem.*, 1971, **33**, 1.
494. A. Storr, B. S. Thomas and A. D. Penland, *J. Chem. Soc., Dalton Trans.*, 1972, 326.
495. P. L. Baxter, A. J. Downs, D. W. H. Rankin and H. Robertson, *J. Chem. Soc., Dalton Trans.*, 1985, 807.
496. W. Harrison, A. Storr and J. Trotter, *J. Chem. Soc., Dalton Trans.*, 1972, 1554.
497. W. R. Nutt, J. Anderson, J. D. Odom, M. W. Williamson and B. H. Rubin, *Inorg. Chem.*, 1985, **24**, 159.
498. D. G. Tuck, in 'Comprehensive Organometallic Chemistry', ed. G. Wilkinson, F. G. A. Stone and E. W. Abel, Pergamon, Oxford, 1982, vol. 1, p. 683.

499. E. Avsar, *Acta Chem. Scand., Ser. A,* 1982, **36,** 627.
500. K. Dehnicke and N. Röder, *J. Organomet. Chem.,* 1975, **86,** 335.
501. A. Yu. Tsivadze, G. V. Tsintsadze and Ts. L. Maknatadze, *J. Gen. Chem. USSR (Engl. Transl.),* 1974, **44,** 157.
502. Ya. I. Tur'yan, L. M. Makarova and V. N. Sirko, *Russ. J. Inorg. Chem. (Engl. Transl.),* 1974, **19,** 969.
503. J. P. Brunette, M. J. F. Leroy, B. Ceccaroli and J. Alstad, *Acta Chem. Scand., Ser. A,* 1978, **32,** 415.
504. S. J. Patel and D. G. Tuck, *Can. J. Chem.,* 1969, **47,** 229.
505. L. M. Mikeeva, A. I. Grigor'ev, A. I. Tarasova and N. V. Mikeev, *Russ. J. Inorg. Chem. (Engl. Transl.),* 1974, **19,** 1131, 1277.
506. B. Hajek, F. Petru and E. Holeckova, *Collect. Czech. Chem. Commun.,* 1967, **32,** 1988.
507. L. M. Mikeeva and A. I. Tarasova, *Russ. J. Inorg. Chem. (Engl. Transl.),* 1975, **20,** 202, 301.
508. V. V. Skopenko, G. V. Tsintsadze and E. I. Ivanova, *Russ. Chem. Rev. (Engl. Transl.),* 1982, **51,** 21.
509. I. Y. Ahmed and C. D. Schmulbach, *Inorg. Chem.,* 1972, **11,** 228.
510. S. F. Lincoln, *Aust. J. Chem.,* 1972, **25,** 2705.
511. L. A. Popkova, E. N. Guryanova and A. F. Volkov, *J. Mol. Struct.,* 1982, **83,** 341.
512. N. N. Greenwood, E. J. F. Ross and A. Storr, *J. Chem. Soc. (A),* 1965, 1400; 1966, 706.
513. J. D. Odom, K. K. Chatterjee and J. R. Durig, *J. Phys. Chem.,* 1980, **84,** 1843.
514. W. Levason and C. A. McAuliffe, *Coord. Chem. Rev.,* 1976, **19,** 173.
515. J. R. Durig and K. K. Chatterjee, *J. Mol. Struct.,* 1982, **81,** 167.
516. J. C. Carter, G. Jugie, R. Enjalbert and J. Galy, *Inorg. Chem.,* 1978, **17,** 1248.
517. A. Balls, N. N. Greenwood and B. P. Straughan, *J. Chem. Soc. (A),* 1968, 753.
518. A. J. Carty, *Can. J. Chem.,* 1967, **45,** 345; 3187.
519. M. J. Taylor, S. Riethmiller and D. S. Bohle, *J. Raman Spectrosc.,* 1984, **15,** 393.
520. R. S. Nyholm and K. Ulm, *J. Chem. Soc.,* 1965, 4199.
521. J. D. Odom, K. K. Chatterjee and J. R. Durig, *J. Mol. Struct.,* 1981, **74,** 193.
522. B. F. Helliwell and M. J. Taylor, *Aust. J. Chem.,* 1983, **36,** 385.
523. F. Maury and G. Constant, *Polyhedron,* 1984, **3,** 581.
524. H. Schmidbaur, S. Lauteschläger and B. Milewski-Mahrla, *Chem. Ber.,* 1983, **116,** 1403.
525. P. Vitse, J. Galy and A. Potier, *C. R. Hebd. Seances Acad. Sci., Ser. C,* 1973, **277,** 159.
526. D. Dohy and G. Lucazeau, *J. Molc. Struct.,* 1982, **79,** 419.
527. R. Hoppe and F. Griesfeller, *Z. Anorg. Allg. Chem.,* 1978, **440,** 74.
528. B. V. Shchegdev and M. F. Dyatkina, *J. Struct. Chem. (Engl. Transl.),* 1974, **15,** 302.
529. C. Caranoni, L. Capella, R. Haser and G. Pèpe, *Acta Crystallogr., Sect. B,* 1981, **37,** 15.
530. A. N. Christensen, *Acta Chem. Scand., Ser. A,* 1974, **28,** 145.
531. K.-B. Plötz and H. Müller-Buschbaum, *Z. Anorg. Allg. Chem.,* 1982, **484,** 153.
532. A. Roesler and D. Reinen, *Z. Anorg. Allg. Chem.,* 1981, **479,** 119.
533. J. D. Glickson, T. P. Pitner, J. Webb and R. A. Gams, *J. Am. Chem. Soc.,* 1975, **97,** 1679.
534. E. A. Biryuk and V. A. Nazarenko, *Russ. J. Inorg. Chem. (Engl. Transl.),* 1973, **18,** 1576.
535. J. P. Hunt and H. L. Friedman, *Prog. Inorg. Chem.,* 1983, **30,** 359.
536. A. Fratiello, *Prog. Inorg. Chem.,* 1972, **17,** 57.
537. J. P. Dechter, *Prog. Inorg. Chem.,* 1982, **29,** 285.
538. K. Dymock, G. J. Palenik and A. J. Carty, *J. Chem. Soc., Chem. Commun.,* 1972, 1218.
539. S. R. Bindal, V. K. Mathur and R. C. Mehrotra, *J. Chem. Soc. (A),* 1969, 863.
540. J. G. Oliver and I. J. Worrall, *J. Chem. Soc. (A),* 1970, 845.
541. H. Funk and A. Paul, *Z. Anorg. Allg. Chem.,* 1965, **337,** 142, 145; J. G. Oliver and I. J. Worrall, *J. Chem. Soc. (A),* 1971, 2315.
542. L. Asso, J. Haladjian, P. Bianco and R. Pilard, *J. Less-Common Met.,* 1974, **35,** 107.
543. G. Mann, H. Olapinski, R. Ott and J. Weidlein, *Z. Anorg. Allg. Chem.,* 1974, **410,** 195.
544. K. Dymock and G. J. Palenik, *Acta Crystallogr., Sect. B,* 1974, **30,** 1364.
545. J. Haladjian, P. Bianco and R. Pilard, *J. Chim. Phys.,* 1974, **71,** 1251.
546. C. Chatterjee, K. Matsuzawa, H. Kido and K. Saito, *Bull. Chem. Soc. Jpn.,* 1974, **47,** 2809.
547. W. G. Movius and N. A. Matwiyoff, *Inorg. Chem.,* 1969, **8,** 925.
548. I. Waller, T. Halder, W. Schwarz and J. Weidlein, *J. Organomet. Chem.,* 1982, **232,** 99.
549. G. K. Abdullaev, P. F. Rza-Zade and Kh. S. Mamedov, *Russ. J. Inorg. Chem. (Engl. Transl.),* 1981, **26,** 766.
550. T. D. Sidorenko, S. N. Grishaeva and Yu. G. Dorokhov, *Russ. J. Inorg. Chem. (Engl. Transl.),* 1979, **24,** 661.
551. H. Kroll, M. W. Phillips and H. Pentinghaus, *Acta Crystallogr., Sect. B,* 1978, **34,** 359.
552. T. M. Ponomareva, N. P. Tamilov and A. S. Berger, *Russ. J. Inorg. Chem. (Engl. Transl.),* 1979, **24,** 294.
553. N. P. Tamilov, V. E. Morozovka and A. S. Berger, *Russ. J. Inorg. Chem. (Engl. Transl.),* 1979, **24,** 1398.
554. B. V. Ivanov-Emin, Z. K. Odinets, V. A. Belosonov and B. E. Zaitsev, *Russ. J. Inorg. Chem. (Engl. Transl.),* 1973, **18,** 623; 1975, **20,** 843.
555. D. Bowler and N. Logan, *Chem. Commun.,* 1971, 582.
556. Yu. I. Mikhailyuk and V. I. Gordienko, *Russ. J. Inorg. Chem. (Engl. Transl.),* 1975, **20,** 1617.
557. R. E. Lenkinski, C. H. F. Chang and J. D. Glickson, *J. Am. Chem. Soc.,* 1978, **100,** 5383.
558. D. W. Kydon, M. Pintar and H. E. Petch, *J. Chem. Phys.,* 1968, **48,** 5348.
559. A. Storr, P. A. Yeats and F. Aubke, *Can. J. Chem.,* 1972, **50,** 452.
560. M. Drache, B. Vandorpe and J. Heubel, *Rev. Chim. Miner.,* 1973, **10,** 505.
561. O. W. Rollins and C. R. Skolds, *J. Inorg. Nucl. Chem.,* 1980, **42,** 371; 1360.
562. L. G. Maksimova, L. V. Tumurova, M. V. Mokosoev and S. R. Sambueva, *Russ. J. Inorg. Chem. (Engl. Transl.),* 1981, **25,** 1336.
563. M. Chaabouni, J.-L. Pascal, A. C. Pavia, J. Potier and A. Potier, *J. Chem. Res. (M),* 1977, 80.
564. J. D. Donaldson, J. F. Knighton and S. D. Ross, *Spectrochim. Acta,* 1964, **20,** 847.
565. H.-D. Hausen, K. Sille, J. Weidlein and W. Schwartz, *J. Organomet. Chem.,* 1978, **160,** 411.
566. P. Sartori, J. Fazekas and J. Schnackers, *J. Fluorine Chem.,* 1971, **1,** 463.
567. F. Ya. Kul'ba, N. A. Babkina and A. Zharkov, *Russ. J. Inorg. Chem. (Engl. Transl.),* 1974, **19,** 365.

568. H.-D. Hausen, K. Mertz and J. Weidlein, *J. Organomet. Chem.*, 1974, **67**, 7.
569. R. H. Cragg, J. D. Smith and G. E. Toogood, *Annu. Rep. Prog. Chem., Sect. A, Phys. Inorg. Chem.*, 1977, **74**, 139; G. A. Banta, S. J. Rettig, A. Storr and J. Trotter, *Can. J. Chem.*, 1985, **63**, 2545.
570. C. H. F. Changz, T. P. Pitner, R. E. Lenkinski and J. D. Glickson, *Bioinorg. Chem.*, 1978, **8**, 11.
571. H. Yamada and M. Tanaka, *J. Inorg. Nucl. Chem.*, 1973, **35**, 3307.
572. R. Corrigli, F. Secco and M. Venturini, *Inorg. Chem.*, 1982, **21**, 2992.
573. R. J. Motekaitis and A. E. Martell, *Inorg. Chem.*, 1980, **19**, 1646.
574. R. P. Guseva and V. N. Kumok, *Russ. J. Inorg. Chem.*, 1972, **17**, 1680.
575. R. Corrigli, F. Secco and M. Venturini, *Inorg. Chem.*, 1979, **18**, 3184.
576. L. S. Serdyuk and A. V. Fedin, *Russ. J. Inorg. Chem. (Engl. Transl.)*, 1971, **16**, 2159.
577. T. Selkine, Y. Komatsu and J.-I. Yumikura, *J. Inorg. Nucl. Chem.*, 1973, **35**, 3891.
578. B. S. Garg and R. P. Singh, *Talanta*, 1971, **18**, 761.
579. S. J. Rettig, A. Storr and J. Trotter, *Can. J. Chem.*, 1976, **54**, 1278.
580. D. A. Tong, *Chem. Commun.*, 1969, 790.
581. J. Burgess and J. Kijowski, *J. Inorg. Nucl. Chem.*, 1981, **43**, 2649.
582. J. Roziere, M.-T. Roziere-Bories, A. Manteghetti and A. Potier, *Can. J. Chem.*, 1974, **52**, 3274.
583. C. Perchard, J. Angenault and J.-C. Couturier, *Spectrochim. Acta, Part A*, 1977, **33**, 793.
584. J. D. Odom, F. M. Wasacz, J. F. Sullivan and J. R. Durig, *J. Raman Spectrosc.*, 1981, **11**, 469.
585. A. Sandez, J. S. Casas, J. Sordo and J. R. Masaguer, *J. Inorg. Nucl. Chem.*, 1978, **40**, 355, 357.
586. C. Hambly and J. B. Raynor, *J. Chem. Soc., Dalton Trans.*, 1974, 604.
587. A. Boardman, S. E. Jeffs, R. W. H. Small and I. J. Worrall, *Inorg. Chim. Acta*, 1984, **87**, L27.
588. Yu. A. Buslaev and S. P. Petrosyants, *Polyhedron*, 1984, **3**, 265.
589. A. J. Carty, H. A. Patel and P. M. Boorman, *Can. J. Chem.*, 1970, **48**, 492.
590. G. Lucazeau and J. Leroy, *Spectrochim. Acta, Part A*, 1978, **34**, 29.
591. A. Mazurier, S. Maneglier-Lacordaire, G. Ghemard and S. Jaulmes, *Acta Crystallogr., Sect. B*, 1979, **35**, 1046.
592. B. Krebs, *Angew. Chem., Int. Ed. Engl.*, 1983, **22**, 113.
593. D. Müller and H. Hahn, *Z. Anorg. Allg. Chem.*, 1978, **438**, 258.
594. M. Julien-Pouzol, S. Jaulmes and C. Dagron, *Acta Crystallogr., Sect. B*, 1982, **38**, 1566.
595. A. Mazurier, S. Jaulmes and M. Guittard, *Acta Crystallogr., Sect. B*, 1980, **36**, 1990.
596. A. Mazurier, M. Guittard and S. Jaulmes, *Acta Crystallogr., Sect. B*, 1982, **38**, 379.
597. B. Krebs, D. Voelker and K.-O. Stiller, *Inorg. Chim. Acta*, 1982, **65**, L101.
598. H.-J. Deiseroth and H. Fu-Son, *Angew. Chem., Int. Ed. Engl.*, 1981, **20**, 962.
599. H.-J. Deiseroth and H. Fu-Son, *Z. Naturforsch., Teil B*, 1983, **38**, 181.
600. R. Kniep and W. Welzel, *Z. Naturforsch., Teil B*, 1985, **40**, 26.
601. R. W. Berg and N. J. Bjerrum, *Polyhedron*, 1983, **2**, 179.
602. G. G. Hoffmann, *Chem. Ber.*, 1985, **118**, 3320; L. E. Maelia and S. A. Koch, *Inorg. Chem.*, 1986, **25**, 1896.
603. A. Boardman, S. E. Jeffs, R. W. H. Small and I. J. Worrall, *Inorg. Chim. Acta*, 1984, **83**, L39.
604. R. O. Fields, J. H. Watts and T. J. Bergendahl, *Inorg. Chem.*, 1971, **10**, 2808.
605. M. Delepine, *Ann. Chim. (Paris)*, 1951, **6**, 633, 645.
606. K. Dymock, G. J. Palenik, J. Slezak, C. L. Raston and A. H. White, *J. Chem. Soc., Dalton Trans.*, 1976, 28.
607. G. G. Hoffmann and C. Burschka, *J. Organomet. Chem.*, 1984, **267**, 229.
608. E. Uhleman and P. Thomas, *Z. Anorg. Allg. Chem.*, 1967, **356**, 71.
609. P. Buck and C. D. Carpentier, *Acta Crystallogr., Sect. B*, 1973, **29**, 1864.
610. W. J. Pietro, B. A. Levi, W. J. Hehre and R. F. Stewart, *Inorg. Chem.*, 1980, **19**, 2225.
611. D. H. Feather, A. Büchler and A. W. Searcy, *High Temp. Sci.*, 1972, **4**, 290.
612. H. Schnöckel and A. J. Göcke, *J. Mol. Struct.*, 1978, **50**, 281.
613. S. A. Polyschchuk, S. P. Kozerenko and Y. U. Gargarinsky, *J. Less-Common Met.*, 1972, **27**, 45.
614. Yu. A. Buslaev, S. P. Petrosyants and V. P. Tarasov, *J. Struct. Chem. (Engl. Transl)*, 1974, **15**, 187.
615. B. B. Srivastava, A. K. Dublish and A. N. Pandhey, *J. Mol. Struct.*, 1973, **15**, 421.
616. N. K. Sanyal and L. Dixit, *Spectrosc. Lett.*, 1973, **6**, 49.
617. J. Chassaign, *Rev. Chim. Miner.*, 1968, **5**, 1115.
618. R. Domesle and R. Hoppe, *Rev. Chim. Miner.*, 1978, **15**, 439.
619. A. de Kozat and M. Samouël, *Rev. Chim. Miner.*, 1981, **18**, 255.
620. R. Awadallah, J.-P. Laval and B. Frit, *C. R. Hebd. Seances Acad. Sci.*, 1982, **295**, 725.
621. W. Stoeger, A. Rabenau and H. M. Haendler, *Z. Anorg. Allg. Chem.*, 1974, **408**, 92.
622. H. Schmidbauer, J. Weidlein, H.-F. Klein and K. Eiglmeir, *Chem. Ber.*, 1968, **101**, 2268, 2278.
623. E. Rytter, M. A. Einarsrud and C. E. Sjøgren, *Spectrochim. Acta, Part A*, 1986, **42**, 1317.
624. R. Kniep, P. Blees and W. Poll, *Angew. Chem., Int. Ed. Engl.*, 1982, **21**, 386.
625. I. R. Beattie, H. E. Blayden, S. M. Hall, S. N. Jenny and J. S. Ogden, *J. Chem. Soc., Dalton Trans.*, 1976, 4171.
626. R. G. S. Pong, R. A. Stachnik, A. E. Shirk and J. S. Shirk, *J. Chem. Phys.*, 1975, **63**, 1525.
627. D. M. Adams and R. G. Churchill, *J. Chem. Soc. (A)*, 1970, 697.
628. E. Chemouni, *J. Inorg. Nucl. Chem.*, 1971, **33**, 2333.
629. H.-J. Deiseroth and H. Fu-Son, *Angew. Chem., Int. Ed. Engl.*, 1981, **20**, 962.
630. S. B. Miller and T. B. Brill, *J. Organomet. Chem.*, 1979, **166**, 293.
631. O. T. Beachley, Jr. and R. G. Simmons, *Inorg. Chem.*, 1980, **19**, 1021.
632. O. T. Beachley, Jr. and R. G. Simmons, *Inorg. Chem.*, 1980, **19**, 783.
633. A. Haaland and J. Weidlein, *Acta Chem. Scand., Ser. A*, 1982, **36**, 805.
634. I. L. Wilson and K. Dehnicke, *J. Organomet. Chem.*, 1974, **67**, 229.
635. R. R. Holmes, *J. Inorg. Nucl. Chem.*, 1960, **14**, 179.
636. G. Peylhard, P. Feulon and A. Potier, *Z. Anorg. Allg. Chem.*, 1981, **483**, 236.
637. P. I. Fedorov, *Russ. J. Inorg. Chem. (Engl. Transl.)*, 1979, **24**, 434.
638. H. Schäfer, *Z. Anorg. Allg. Chem.*, 1980, **461**, 29.
639. A. Anundskås, A. E. Mahgoub and H. A. Øye, *Acta Chem. Scand. Ser. A*, 1976, **30**, 193.

640. C. W. Schläpfer and C. Rohrbasser, *Inorg. Chem.*, 1978, **17,** 1623.
641. H.-D. Hausen, H. Binder and W. Schwarz, *Z. Naturforsch., Teil B,* 1978, **33,** 567.
642. I. Sinclair, R. W. H. Small and I. J. Worrall, *Acta Crystallogr., Sect. B,* 1981, **37,** 1290.
643. A. Bartecki and M. Sowińska, *Spectrochim. Acta, Part A,* 1979, **35,** 739.
644. S. P. Andrews, P. E. R. Badger, P. L. Goggin, N. W. Hurst and A. J. M. Rattray, *J. Chem. Res. (S),* 1978, 94.
645. T. Degg and A. Weiss, *Ber. Bunseges. Phys. Chem.*, 1975, **79,** 497.
646. S. F. Lincoln, A. C. Sandercock and D. R. Stranks, *J. Chem. Soc., Dalton Trans.*, 1975, 669.
647. R. A. Work and M. L. Good, *Spectrochim. Acta, Part A,* 1972, **28,** 1537.
648. P. Barbier, G. Mairesse, F. Wallart and J.-P. Wignacourt, *C. R. Hebd. Seances Acad. Sci.*, 1973, **C277,** 841.
649. E. Chemouni, *J. Inorg. Nucl. Chem.*, 1971, **33,** 2325.
650. R. C. Gearhart, Jr., J. D. Beck and R. H. Wood, *Inorg. Chem.*, 1975, **14,** 2413.
651. H. Kanno and J. Hiraishi, *Chem. Phys. Lett.*, 1979, **68,** 46.
652. A. K. Molodkin, M. Rabbani, A. G. Dudareva and A. I. Ezhov, *Russ. J. Inorg. Chem. (Engl. Transl.)*, 1980, **25,** 1244.
653. A. Manteghetti, D. Mascherpa-Corral and A. Potier, *Spectrochim. Acta, Part A,* 1981, **37,** 211.
654. J. H. von Barner, *Inorg. Chem.*, 1985, **24,** 1686.
655. P. Klinzing, W. Willing, U. Müller and K. Dehnieke, *Z. Anorg. Allg. Chem.*, 1985, **529,** 35.
656. D. Mascherpa-Corral and A. Potier, *J. Inorg. Nucl. Chem.*, 1976, **38,** 211.
657. D. Mascherpa-Corral and A. Potier, *J. Inorg. Nucl. Chem.*, 1977, **39,** 1519.
658. R. S. Nyholm and K. Ulm, *J. Chem. Soc.*, 1965, 4199.
659. W. Lind and I. J. Worrall, *J. Organomet. Chem.*, 1972, **36,** 35.
660. R. Rafaeloff and A. Silberstein-Hirsch, *Spectrochim. Acta, Part A,* 1975, **31,**. 183.
661. D. Raptis, J. K. Kouinis, D. Kaminaris and A. G. Galinos, *Monatsh. Chem.*, 1981, **112,** 713.
662. A. N. Grigor'ev, A. I. Grigor'ev, F. M. Spiridonov, L. M. Mikheeva and L. N. Komissarova, *Russ. J. Inorg. Chem. (Engl. Transl.),* 1977, **22,** 508.
663. I. M. Alymov, T. L. Khotsyanova, E. V. Bryukhova, A. N. Grigor'ev, L. M. Mikheeva and L. N. Komissarova, *Russ. J. Inorg. Chem. (Engl. Transl.)*, 1975, **20,** 1262.
664. R. Colton, D. Dakternieks and J. Hauenstein, *Aust. J. Chem.*, 1981, **34,** 949.
665. B. R. McGarvey, M. J. Taylor and D. G. Tuck, *Inorg. Chem.*, 1981, **20,** 2010.
666. A. E. Shirk and D. F. Shriver, *Inorg. Synth.*, 1977, **17,** 42; 45.
667. J. A. Dilts and W. R. Nutt, *Inorg. Synth.*, 1977, **17,** 48.
668. A. P. Kurbakova, L. A. Leites, V. V. Gavrilenko, Yu. N. Karaksin and L. I. Zakharkin, *Spectrochim. Acta, Part A,* 1975, **31,** 281.
669. P. L. Baxter, A. J. Downs and D. W. H. Rankin, *J. Chem. Soc., Dalton Trans.*, 1984, 1755.
670. P. Breisacher and B. Siegel, *J. Am. Chem. Soc.*, 1965, **87,** 4255.
671. J. C. Brand and B. P. Roberts, *J. Chem. Soc., Chem. Commun.*, 1984, 109.
672. M. T. Barlow, C. J. Dain, A. J. Downs, G. S. Laurenson and D. W. H. Rankin, *J. Chem. Soc., Dalton Trans.*, 1982, 597.
673. M. T. Barlow, A. J. Downs, P. D. P. Thomas and D. W. H. Rankin, *J. Chem. Soc., Dalton Trans.*, 1979, 1793.
674. M. Agrawal, J. P. Tandon and R. C. Mehrotra, *J. Inorg. Nucl. Chem.*, 1981, **43,** 1070.
675. I. A. Sheka and K. I. Arsenin, *J. Gen. Chem. USSR (Engl. Transl.),* 1975, **44,** 2286.
676. P. Bianco, J. Haladjian and R. Pilard, *J. Less-Common Met.*, 1975, **42,** 127.
677. R. E. Lenkinski, B. E. Peerce, J. L. Dallas and J. D. Glickson, *J. Am. Chem. Soc.*, 1980, **102,** 131.
678. R. J. Bergeron and S. J. Kline, *J. Am. Chem. Soc.*, 1984, **106,** 3089.
679. K. S. Chong, S. J. Rettig, A. Storr and J. Trotter, *Can. J. Chem.*, 1979, **57,** 586
680. K. S. Chong, S. J. Rettig, A. Storr and J. Trotter, *Can. J. Chem.*, 1981, **59,** 518.
681. A. Meller, W. Maringgele and R. Oesterle, *Monatsh. Chem.*, 1980, **111,** 1087.
682. L. Pellerito, R. Cefalu, G. Ruisi and M. T. Lo Guidice, *Z. Anorg. Allg. Chem.*, 1981, **481,** 218.
683. M. A. Green, M. J. Welch and J. C. Huffman, *J. Am. Chem. Soc.*, 1984, **106,** 3689.
684. C. H. L. Kennard, *Inorg. Chim. Acta,* 1967, **1,** 347.
685. C. H. Taliaferro and A. E. Martell, *Inorg. Chem.*, 1985, **24,** 2408.
686. M. J. Kappel, V. L. Pecoraro and K. N. Raymond, *Inorg. Chem.*, 1985, **24,** 2447.
687. A. J. Welch, *Annu. Rep. Prog. Chem., Sect. A, Phys. Inorg. Chem.*, 1981, **78,** 62.
688. S. S. Eaton, D. M. Fishwild and G. R. Eaton, *Inorg. Chem.*, 1978, **17,** 1524.
689. K. M. Kadish, B. Boisselier-Coclios, A. Coutsolelos, P. Mitaine and R. Guilard, *Inorg. Chem.*, 1985, **24,** 4521.
690. K. J. Wynne, *Inorg. Chem.*, 1985, **24,** 1339.

25.2

Indium and Thallium

DENNIS G. TUCK
University of Windsor, Ontario, Canada

25.2.1 INDIUM: INTRODUCTION

A review of the coordination of indium necessarily begins with reference to the general inorganic chemistry of the element,[1] and to other earlier general works.[2–4] More recent authors have reviewed the adducts of the trihalides and pseudohalides of gallium, indium and thallium,[5] other aspects of the inorganic chemistry of these elements,[6] the coordination chemistry of indium,[7] and the organometallic chemistry of indium.[8] A critical evaluation of the stability constants of indium complexes in solution has also appeared.[9] Each of these articles contains background material and references to earlier work.

25.2.2 INDIUM(I)

The coordination chemistry of indium(I) received little or no attention for many years, largely no doubt because of the lack of suitable starting materials. The indium(I) halides can be easily prepared,[1] but are intractable materials which readily undergo hydrolysis and disproportionation (but see Section 25.2.2.1). The only known organoindium(I) compounds are the cyclopentadienyl and methycyclopentadienyl derivatives,[8] and the former has proven a useful starting point for some synthetic work. Electrochemical methods have been used either to reduce higher states, or to oxidize the metal in the presence of suitable ligands, in order to produce indium(I) species.

25.2.2.1 Nitrogen Ligands

The neutral adducts of InBr and InI with ammonia, $InX{\cdot}nNH_3$ ($n = 1, 2$), are reported to be black insoluble substances which disproportionate in air or in the presence of moisture, and on heating, so that $InI{\cdot}2NH_3$ for example gives $InI_3{\cdot}5NH_3$, indium metal and ammonia.[10] An interesting new development which should open up indium(I) chemistry is that the monohalides InX (X = Cl, Br, I) dissolve in mixtures of aromatic solvents and organic bases.[11] These solutions are relatively stable at low temperature (−20 °C and below), and the dependence of solubility on TMEDA concentration suggests that in the case of InBr or InI, the solute species is InX·3TMEDA in dilute solution, although addition of petroleum ether to the cold solution precipitates InX·0.5TMEDA. The synthetic usefulness of such solutions is discussed below. Indium(I) halides do not otherwise dissolve in basic solvents and in any case disproportionation is always a major problem, but prolonged refluxing in either aniline or morpholine is reported to give products analyzing as $[In(aniline)_4]X$ and $[In(morpholine)_2]X$ (X = Cl, Br, I), whose formulation depends on the behaviour of the products as 1:1 electrolytes in nitrobenzene.[12] Some structural information on these unusual compounds would be most helpful.

25.2.2.2 Cyanate and Thiocyanate Ligands

Salts of $InCl_2^-$ and InI_3^{2-} (see Section 25.2.2.4) serve as the starting point for the synthesis of salts of $[In(NCS)_2]^-$, $[In(NCO)_2]^-$ and $[In(NCS)_3]^{2-}$ by metathesis in ethanol;[13] the IR spectrum shows that both ligands are N-bonded in structures which are apparently bridged homopolymers in the solid state. No other pseudohalide compounds of indium(I) have been reported.

25.2.2.3 β-Ketoenolates and Related Ligands

The unavailability of suitable starting materials has hindered work in this area, but the reaction of cyclopentadienylindium(I) in Et_2O/benzene mixtures with 4,4,4-trifluoro-1-(thien-2′-yl)butane-1,3-dionate (ttaH) gave the extremely hygroscopic In(tta), and similar derivatives of other bidentate ligands were obtained by this same process.[14] The reactions of the analogous In(oxine) (oxine = 8-hydroxyquinoline anion) show that such InL species are easily oxidized to indium(III) complexes (see Section 25.2.4.8). The insolubility of the product in the reaction mixture is presumably an important factor in this method.

25.2.2.4 Halogen Ligands

The solubility of the indium(I) halides in toluene/TMEDA was noted previously. Electron spectroscopy[15] shows that InCl forms covalent dimers in the gas phase. The *d*-shell electrons can be regarded as being part of the core and play no part in bonding.

Indium monohalides (InX; X = Cl, Br, I) react with the corresponding salt of *N,N′*-dimethyl-4,4′-bipyridylium to form $[Me_2bipy]^{2+}[InX_3]^{2-}$.[16,17] The anions have been formulated as mononuclear C_{3v} species on the basis of Raman spectra, supported by force constant calculations.[18] Electrochemical reduction of In^{III} in concentrated hydrochloric acid did not yield stable In^I species. The reaction of CpIn with HX in the presence of Et_4NX in non-aqueous media[13,19] gave the salts $Et_4N[InX_2]$ whose vibrational spectra indicate the presence of mononuclear C_{2v} anions. With the cation Me_2dppe^{2+} (= 1,2-bis(methyldiphenylphosphonio)-ethane), the salt $Me_2dppe[InI_3]$ was obtained. Metathesis yielded pseudohalide derivatives (see Section 25.2.2.2). The electrochemical oxidation of an indium anode in a non-aqueous solution of Et_4NI and I_2 provides an unusual route to $Et_4N[InI_2]$.[19,20] The oxidation of InX_2^- to $InX_2Y_2^-$ is discussed later.

Aqueous solutions of In^I in low concentration can be prepared by the anodic oxidation of the metal,[21] but the strong reducing properties of $In^+(aq)$ bar this as a route to complexes. The rate of disappearance of In^I in various media has been used to derive stability constants for In^I/X^- species.[22,23]

25.2.2.5 Macrocyclic Ligands

The indium(II) halides, whose solid state structures are $In^+[InX_4]^-$ for X = Br and I (see Section 25.2.2.5), have served as starting points for the preparation of $[In(crown)][InX_4]$(X = Cl, Br, I), $[In(crown)][AlCl_4]$ and $[In(cyclam)][InX_4]$(X = Br, I; crown = dibenzo-18-crown-6; cyclam = 1,4,8,11-tetraazacyclotetradecane).[24] The characteristic vibrational spectra of the InX_4^- anions were important in establishing the solid state structure of these salts of indium(I) cationic complexes. These macrocyclic ligands do not react with suspensions of InX (*cf.* thallium below).

25.2.3 INDIUM(II)

The structure of the few indium(II) compounds identified by early workers is one of the classical problems in group III chemistry, (*cf.* gallium, Section 25.3.3.3). It is now clear that both M—M-bonded compounds and ionic mixed valence species exist, but the factors which govern the relative stability of these species still remain to be elucidated.

25.2.3.1 Nitrogen Ligands

Neutral adducts of the indium dihalides are known with both mono- and bi-dentate nitrogen ligands. Condensation of ligand on to InX_2 (X = Cl, Br, I) cooled in liquid nitrogen gave InX_2L (L = piperidine, piperazine and morpholine). Raman emissions at *ca.* 173 (Cl), 143 (Br) or 105 (I) cm^{-1} were assigned as ν(In—In), leading to the formulation of these compounds as $X_2(L_2)InIn(L_2)X_2$, with the bases acting as monodentate ligands at five-coordinate indium(II)[25] (*cf.* gallium(II) compounds). The reaction between InX_2 (X = Br, I) and various ligands in benzene[26] yielded adducts only with TMEDA and Et_3P (see below); other ligands, and all reactions involving $InCl_2$, lead to disproportionation, either to $In^0 + InX_3$, or to $InX + InX_3$. A later X-ray crystallographic investigation[27] confirmed both the In—In-bonded structure of $In_2Br_3I{\cdot}2TMEDA$, for which r(In—In) = 2.775(2) Å, and five-coordination at indium. Possible reasons for the different behaviour of InX_2 in solution[26] and in the solid state[25] have been discussed[28] and are summarized below (Section 25.2.3.5).

Condensation of pyridine on to InX_2 gives the In—In-bonded compounds $X_2In_2X_4py_4$ (X = Br, I; X = Cl not obtained by this or other methods).[25] Adducts of InI_2 with $EtNH_2$ ($In_2I_4L_5$), and $EtNH_2$ and bipy ($In_2I_4{\cdot}2bipy{\cdot}4EtNH_2$, $In_2I_4{\cdot}Etbipy{\cdot}bipy{\cdot}8EtNH_2$, $In_2I_4{\cdot}Et_4bipy{\cdot}bipy{\cdot}8NH_3$) have also been reported,[29] but unfortunately no structural information is available on these unusual compounds.

25.2.3.2 Phosphorus Ligands

Bis adducts of Et_3P with In_2Br_4 and In_2I_4 have been prepared; vibrational spectra showed these to be $P(X_2)InIn(X_2)P$ species, with a staggered conformation in the solid state.[26]

25.2.3.3 Oxygen Ligands

Adducts of In_2Br_4 and In_2I_4 with THF, tetrahydropyran and DMSO have been prepared by condensing the ligand on to the dihalide.[25] The evidence from vibrational spectroscopy, and the obvious analogies with In_2X_4—nitrogen compounds and with Ga_2X_4 adducts, readily identify these as In—In species.

Two other types of unusual indium(II)–oxygen ligand complexes are known. The polarographic reduction of $In(acac)_3$, $In(tropolonate)_3$ or $In(pyronate)_3$ (pyronate = 3-hydroxy-2-methyl-4-pyronate anion) in non-aqueous solution shows a one-electron reduction in each case,[30] corresponding to the formation of a (formally) In^{II} complex. Similar behaviour was observed with $In(oxine)_3$, and in some cases a second one-electron reduction was also observed, implying the formation of In^I species, although in no case was the site of the reduction identified, so that the product may be either $[In^{II}(acac)_3]^-$ or $[In^{III}(acac)_2(acac^-)]$[6] (*cf.* sulfur ligands, Section 25.2.3.4).

Electrolysis in the cell $Pt/MeCO_2H:H_2O(trace)/In$ caused dissolution of the indium anode and formation of fine feathery crystals of the metal on the cathode. This material slowly reacted with the electrolyte to give crystals of $In(OAc)_2$ which could be converted to $In(OAc)_3$ on boiling with glacial acetic acid. It seems likely that the crystal lattice consists of $In^I + In^{III} + OAc^-$, but no detailed structure could be obtained.[31] This system warrants further investigation in view of work on the analogous thallium system (see below).

25.2.3.4 Sulfur and Heavier Group VI Ligands

The bis-tetrahydrothiophene adducts of In_2X_4 (X = Cl, Br, I) and the corresponding dimethyl sulfide complexes of In_2Br_4 and In_2I_4 have been prepared.[25]

The polarographic reduction[32] of a number of InL_2^{2-} complexes (L = toluene-3,4-dithiolate) $[InL_2D]^-$ (D = phen, bipy, *etc.*), InL_2^{2-} (L = 1,2-dicyanoethylene-1,2-dithiolate), *etc.*, in non-aqueous solution demonstrated a series of reversible one-electron changes corresponding formally to $In^{III} \rightarrow In^{II} \rightarrow In^{I} \rightarrow In^0$, in contrast to the $In^{III} \rightarrow In^0$ reduction observed in aqueous solution. As noted above (Section 25.2.3.3), the formal oxidation states are assigned to the metal on the assumption that the ligand itself is not reduced, and this remains to be confirmed experimentally.

Although not true coordination compounds of indium, the compounds In_6S_7 and In_6Se_7 should be noted, since they involve In_2^{4+} in the solid state [r(In—In) = 2.741(2), 2.760(5) Å respectively].[33,34] Similarly, In_4Se_3 and In_4Te_3 contain the catenated In_3^{5+} cation,[35] and might therefore prove useful starting materials for the synthesis of thio complexes of In^{II}. Another intriguing compound in this context is In_5Se_4,[36] in which indium is tetrahedrally coordinated in either $In(In')_4$ or $In(In'S_3)_4$.

25.2.3.5 Halogen Ligands

The solid state structure of In_2I_4 has been shown to be $In^+[InI_4]^-$ by X-ray crystallography,[37,38] isostructural with $GaCl_2$ (see Section 25.2.3.3). Structural work on $InBr_2$ indicates a similar structure,[39] and both this and InI_2 confirm previous assignments based on vibrational spectroscopy.[17] The structure of $InCl_2$ presents a much more complex problem. Vibrational spectra have demonstrated the presence of the $InCl_4^-$ anion in molten indium dichloride,[41,42] but the analogous studies of the room temperature phase are variously interpreted as confirming[43] or denying[41,42] the presence of the tetrahedral anion in the lattice. X-Ray powder studies of the solid are also claimed to support the presence of $InCl_4^-$ in both of two different phases of indium dichloride.[44,45] One explanation of these contradictions may be found in the work of Meyer[46,47] who has shown that the only mixed valence species in the $InCl/InCl_3$ system are In_3Cl_4, In_2Cl_3 and In_5Cl_9; the latter is formulated as $In_3^I[In_2Cl_9]$, with a confacial bioctahedral anion. This work removes much of the uncertainty from the problem, and the conclusion is in keeping with the known tendency of anionic chloro complexes of indium(III) to be six-coordinate, unlike the bromo and iodo compounds (see Section 25.2.4.6.2).

It is clear, then, that none of these molecules contains an In—In bond, and the same is apparently true of the only known indium(II) pseudohalide, $In(CN)_2$, formulated as $In[In(CN)_4]$ from vibrational spectroscopy.[48] The species formed by the coordination of electron donors to In_2X_4 on the other hand are clearly In—In molecules. The structures of the neutral adducts were noted earlier (Section 25.3.2.1). The anionic species $In_2X_6^{2-}$ (X = Cl, Br, I), prepared *via* InX_2,[49] are isostructural with the corresponding gallium(II) anions. In non-aqueous solution, ^{115}In NMR studies[50] show that these molecules disproportionate to $InX_2^- + InX_4^-$. Recent studies of the InX/InY_3 reaction[23] cast some light on these matters, and it appears that kinetic factors, and specifically intramolecular halide transfer, are at least as important as thermodynamic effects in the stability of both $In_2X_4L_2$ and $In_2X_6^{2-}$ species. The importance of phase is shown by the fact that the gas phase above molten InX_2 contains InX, InX_3 and In_2X_6 molecules.[51] Much clearly remains to be done in this area.

25.2.4 INDIUM(III)

As noted in the introduction to this chapter, the coordination chemistry of indium(III) is now firmly based and a number of reviews already exist. The following sections are therefore comprehensive, rather than totally inclusive, in scope.

25.2.4.1 Group IV Ligands

25.2.4.1.1 Cyanide ligands

Indium(III) cyanide is best prepared from InOI and cyanogen, although other methods have been described.[52] An ammonia adduct of $In(CN)_3$ may be formed in the reaction between $Hg(CN)_2$ and indium metal in liquid ammonia, but no other derivatives have been reported.

The efficient extraction of indium from aqueous hydrocyanic acid by long-chain amines implies the formation of anionic In^{III}/CN^- complexes,[53] but no stability constant results are available.[9] Organoindium cyanides have been prepared by reacting Me_3M (M = Ga, In) with Me_3GeCN, eliminating Me_4Ge and forming $[Me_2MCH]_4$[54] to give a structure believed to involve In—C≡N—In bridges.

25.2.4.1.2 Silicon, germanium and tin ligands

The reaction of $InCl_3$ with Me_3SiCl and lithium metal in THF at −30 °C yielded $(Me_3Si)_3In$; NMR and vibrational spectra, and vibrational force constants, have been reported.[55] Indium metal reacts with $(Et_3Ge)_2Hg$ over several days at 50–60 °C to form $(Et_3Ge)_3In$ in 50% yield,[56] and this molecule presumably resembles $(Me_3Si)_3In$ in having a trigonal planar InM'_3 kernel (M′ = Si, Ge). Analogous $In(SnR_3)_3$ compounds have not been prepared to date, but the reaction of $LiSnMe_3$ with Me_3In at low temperature has been shown by NMR methods to produce $Li[Me_3SnInMe_3]$ (*cf.* Al, Ga, Tl), which could not be isolated because of decomposition at room temperature (75% after 2 d).[57] Solutions of InX in toluene/TMEDA[58] do not react with Sn_2Ph_6 (see Section 25.2.4.5.2), but with Ph_3SnX the products are Ph_3SnInX_2·TMEDA (X = Cl, Br, I), in which indium is apparently five-coordinate with an $InSnX_3N_2$ kernel;[59] the anionic complex $Ph_3SnInX_3^-$ has also been prepared by treating such a molecule with R_4NX. The compound $L_2InSnPh_3$ (L = 2-[(dimethylamino)methyl]phenyl ligand) has been prepared[60] from L_2InCl and $NaSnPh_3$, so that a number of In—Sn bonded species are now known.

25.2.4.2 Nitrogen Ligands

25.2.4.2.1 Ammonia and amine ligands

The $[In(NH_3)_6]^{3+}$ cation has been identified in liquid ammonia solution,[61] but attempts to prepare the perchlorate salt were unsuccessful.[62] Mixed complexes such as $[In(NH_3)_5Br]^{2+}$ have been postulated in liquid ammonia,[63] but this has not been confirmed preparatively. Ethylenediamine yields the salt $[Inen_3](ClO_4)_3$ on reaction with $In(ClO_4)_{3(aq)}$ in ethanol,[62] and the analogous nitrate[64] and thiocyanate[65] salts have also been prepared.

Adducts of indium(III) halides or pseudohalides with ammonia have not been reported,[5] which is surprising given that the corresponding Me_3In compounds are known.[66] A number of amine adducts of InX_3 have been reported,[5,7] with stoichiometries InX_3L, InX_3L_2 or InX_3L_3, but structural information is scant; vibrational spectroscopy suggests that InX_3L species probably have D_{3h} symmetry. Adducts of the related bidentate nitrogen donors such as en or TMEDA have not been widely studied, which is unfortunate since here as with bidentate heterocyclic ligands, the question of the balance between neutral and ionic structures is interesting. For example, $InCl_3$/en gives rise[67] to $[Inen_3]Cl_3$ which on heating goes to $[Inen_2Cl_2]Cl$, with no neutral adducts known. With pyrazine, the adduct $InCl_3·L_{1.5}$ is polymeric, with bridging pyrazine and six-coordination at indium.[68] Similar problems have been identified with the corresponding adducts of indium thiocyanate (see Section 25.2.4.2.2). Earlier publications[5,7,8] have listed the adducts formed by inorganic and organic indium(III) species; recent information includes the structures of the compounds $EtInX_2$·TMEDA (X = Br, I) which are five-coordinate neutral mononuclear molecules in both solid and solution.[69]

Little work has been published on hydrazine adducts; the reported compounds $InCl_3·3N_2H_4$ and $InI_3·3N_2H_4$ presumably involve monodentate coordination of the ligand.[70]

25.2.4.2.2 N-heterocyclic ligands

Cationic indium(III) complexes could not be prepared with pyridine, but species such as $[In(bipy)_3]^{3+}$ and $[In(phen)_3]^{3+}$ are known as the perchlorate[62] or nitrate[64] salts, and $[In(terpy)_2](ClO_4)_3$ has also been reported.[67] Stability constants have been measured in the In^{3+}/bipy and In^{3+}/phen systems[71,72] and for some substituted pyridines.[9] The adducts of indium(III) halides and pseudohalides with pyridine and substituted pyridines have been reviewed previously.[5,7] The stoichiometries of the primary products are either InX_3L_3 or InX_3L_2, but thermal decomposition[73] can lead to loss of ligand yielding InX_3L_2 or InX_3L, depending on X and L (X = Cl, Br, I).

The interaction of ligands such as bipy or phen with InX_3 (X = Cl, Br, I, NCS, NCO) produces either the salts $[InL_3]X_3$[74] or compounds of stoichiometry $InX_3L_{1.5}$.[65,75,76] The structures of these latter have been the subject of several investigations, which have been reviewed previously.[7] In essence, the preparation[77] of salts such as PhAs[$InCl_4$(bipy)] and [$InCl_2(bipy)_2$]$PF_6 \cdot H_2O$, and the spectroscopic proof that such cationic and anionic indium species exist in solid $InCl_3 \cdot bipy_{1.5}$, have led to the formulation of the latter as the ionic dimer *cis*-$[InCl_2(bipy)]^+[InCl_4(bipy)]^-$ and it seems reasonable to generalize this conclusion to cover other $InX_3 \cdot L_{1.5}$ complexes. The reasons for the existence of these or similar structures, which are to be found in the chemistry of other M^{III} states (*e.g.* Ga, Tl, Ti), is clearly a subject for speculation, but in the absence of crystallographic and thermochemical data no useful arguments can be advanced at present. In addition to the neutral inorganic adducts discussed above, a number of compounds are known involving bipy and/or phen and organometallic compounds,[8] carboxylates and thiolene complexes (see Sections 25.2.4.4.5 and 25.2.4.5.2).

The compounds InX_3(terpy) (X = Cl, Br, I, NCS) are non-ionic,[78] and X-ray[79] studies have confirmed the $InCl_3N_3$ core for X = Cl.

25.2.4.2.3 Amido and imido ligands

The direct reaction of $InCl_3$ with $LiN(SiMe_3)_2$ gives $In[N(SiMe_3)_2]_3$, which is monomeric and has a planar InN_3 kernel;[80] an adduct with Me_3PO has been reported.[81] The established route to amido–indium species is *via* the thermal decomposition of $R'_3In \cdot NR_3$ adducts,[66,82] and in the case of $(Me_2InNMe_2)_2$, a crystal structure determination[83] has confirmed the presence of the In_2N_2 ring, which is also found in $[Me_2InN(Me)Ph]_2$. The isomerization of this compound has been studied in solution.[84] The product obtained from the reaction of $Me_2C{=}NCl$ and Me_3In, namely $[Me_2C{=}NInMe_2]_2$, has a similar In_2N_2 core,[85] being dimeric in both solution (benzene) and solid states. Given the detailed studies of such derivatives of the lighter group III elements, the information here is surprisingly sparse.

25.2.4.2.4 Azide, cyanate, thiocyanate and selenocyanate ligands

Neither neutral nor anionic inorganic azido complexes have been reported, but the organometallics R_2InN_3 are known. Reaction of Et_3In with ClN_3 gives $(Et_2InN_3)_2$, for which a structure based on the In_2N_2 core is proposed,[86] while with Me_3SiN_3 the product is $Me_3SiN_3InMe_3$ which on heating[87] goes to $(Me_2InN_2)_3$. The In_2N_2 bridging is obviously similar to that just discussed (Section 25.2.4.2.3), and the chemistry of such rings should prove interesting.

The parent compound $In(NCO)_3$ has not been isolated, since the reaction between $In_2(SO_4)_3$ and $Ba(NCO)_2$ gave only hydrolysis products, while reactions in alcoholic media between InI_3 and AgCNO produced AgI and an insoluble product which decomposed on heating to give an alkyl urethane derived from the solvent.[76] When adducts of InI_3 were used, the products were the corresponding $In(NCO)_3L_3$ (L = py, DMSO, Ph_3PO), $In(NCO)_3L_2$ (L = Ph_3PO) or $In(NCO)_3L_{1.5}$ (L = bipy, phen). Compounds with different stoichiometries (*e.g.* $In(NCO)_3py_2$, $In(NCO)_3$phen, $In(NCO)_3phen_3$, $In(NCO)_3bipy_3$) have also been reported.[88] The only anionic complex is $[In(NCO)_4]^-$, as the Me_4N^+ salt, and here both terminal and bridging ligands have been postulated on the basis of the vibrational spectrum.[88]

Indium(III) thiocyanate can be prepared by treating $In(CN)_3$ with sulfur,[52] from indium metal and $Hg(NCS)_2$,[52] from $InCl_3$ + NaNCS in ethanol,[52] or from $In_2(SO_4)_3 + Ba(NCS)_2$ in water.[65] The evidence from IR[65] and X-ray[52] results is that the solid state involves extensive In—NCS—In bridging, with six-coordinate indium. Adducts[65] with both monodentate (py, picoline, urea, DMSO, DMA, Ph_3PO, thiourea) and bidentate (bipy, phen, en) ligands are easily prepared. The monodentate ligand adducts are all of the type $In(NCS)_3L_3$, with the N bonding of thiocyanate shown by IR spectroscopy, while with bipy and phen, $In(NCS)_3L_{1.5}$ compounds are obtained; following the discussion above (Section 25.2.4.2.2), these are formulated as $[InL_2(NCS)_2][InL(NCS)_4]$. With en, the product is $[Inen_3](NCS)_3$, believed to involve the six-coordinate cationic complex described previously (see Section 25.2.4.2.1).

Organoindium thiocyanate derivatives do not appear to have been reported. The results of stability constant measurements for In^{3+}/NCS^- have been reviewed; the recommended values are similar to those for In^{3+}/Cl^- systems.[9]

The anionic thiocyanato complexes prepared[83,84] from $In(NCS)_3$ and different organic cations may be either $[In(NCS)_5]^{2-}$ or $[In(NCS)_6]^{3-}$, apparently depending on the properties of the cation. A subsequent X-ray structural investigation of $(Ph_4As)_3[In(NCS)_6]$ confirmed the N-bonding mode of the ligand, and the presence of an octahedral InN_6 kernel.[91]

A number of selenocyanate compounds formulated as the adducts $In(NCSe)_3L_n$ have been reported[5,92] with L = py ($n = 4$), DMF ($n = 3.5$), Pr_3PO ($n = 3$) and diantipyrylmethane ($n = 2$), but no structural details are available. The compounds $In(NCSe)_3L_3$ (L = bipy, phen) may well be salts of InL_3^{3+} and $NCSe^-$, while the anionic complexes $[In(NCSe)_6]^{3-}$ are reported to involve an N-bonded ligand and six-coordinate indium.[93,94]

25.2.4.2.5 Nitrile ligands

Surprisingly little work has been reported on nitrile derivatives of indium(III), and the neutral adduct $InCl_3 \cdot 2MeCN$ appears to be the only compound of its type subjected to spectroscopic investigation;[95] it seems reasonable to suppose that other such adducts of inorganic and organometallic indium(III) species could be obtained. Neutral adducts of InX_3 have been prepared by the direct electrochemical oxidation of indium metal, giving InX_3L_n (X = Cl, Br, $n = 3$; X = I, $n = 2$),[96] and with mixtures of MeCN and HBF_4/Et_2O the same method gave $[In(NCMe)_6](BF_4)_3$, the salt of a six-coordinate cationic complex.[97]

25.2.4.3 Phosphorus and Arsenic Ligands

The four-coordinate cationic complexes $[InL_4]^{3+}$ (L = Ph_3P, Ph_3As) and $[In(diphos)_2]^{3+}$ were prepared as the perchlorate salts,[62] but could not be obtained from nitrate media.[64] The change in coordination number between nitrogen ligands and the heavier analogues is striking.

Adducts of trialkyl- or triaryl-phosphines, triphenylarsine and diphosphines and diarsines[5,7] with indium trihalides are well known. The structures of $InCl_3 \cdot 2R_3P$ compounds have been generally shown to be trigonal bipyramidal with axial phosphines, and this has been confirmed by X-ray methods in the case of $InCl_3(Ph_3P)_2$.[98] With $InBr_3$ and InI_3, both InX_3L_2 and InX_3L compounds are known as monodentate donors, and interconversion by loss of ligands is often facile.[79] These halides also form 1:1 adducts with diphos, unlike indium trichloride,[5] although the latter does interact with diars to give a compound formulated as $[InCl_2(diars)_2][InCl_4]$.[99] Further crystal structure studies of these compounds would be helpful. Adducts of these group V donors with indium(III) pseudohalides are not known. An interesting reaction of coordinated Ph_3P is the *in situ* oxidation to Ph_3PO to give the Ph_3PO adduct of InX_3,[75] which occurs even in the absence of air; it appears that oxygen is abstracted from the solvent, so that the choice of solvent for the preparation of adducts may be critical.

Adducts of phosphorus and arsenic ligands with organoindium compounds have been reviewed elsewhere.[8] As with the amino derivatives (Section 25.2.4.2.3), one of the important reactions of these compounds involves alkane elimination. In the case of $Me_3In \cdot PH_3$, the product of thermal decomposition is the polymeric $(MeInPH)_n$,[100] while with Me_2PH, Et_2PH or Me_2AsH adducts of Me_3In, the decomposition products are polymeric glasses,[101] which can be sublimed, unlike the compounds derived from $PhPH_2$, $MeAsH_2$ or $PhAsH_2$. With Me_3In and $SbEt_3$ the 1:1 adduct is thermally unstable both to dissociation and decomposition,[102] finally yielding InSb. This route to InM (M = P, As, Sb) compounds is of considerable importance to the electronics industry.

25.2.4.4 Oxygen Ligands

25.2.4.4.1 Oxide, hydroxide and water as ligands

The earlier discussion of complexes of indium(III) with nitrogen ligands emphasized the importance of six-coordination, and in keeping with this, ^1H and ^{115}In NMR studies have identified $[In(H_2O)_6]^{3+}$ as the cation in aqueous solutions of perchloric acid.[103,104] Confirmatory evidence has been obtained from dilatometric[105] and X-ray diffraction[106] methods. Proton

NMR has also been used to demonstrate rapid exchange between coordinated water molecules and the bulk phase,[107] and to show the existence of appropriate mixtures of $[In(H_2O)_6]^{3+}$, $[InL_6]^{3+}$ and $[In(H_2O)_nL_{6-n}]^{3+}$ for L = DMF,[108] acetone,[109] trimethyl phosphate,[110] and also in trimethyl phosphate/acetone/water.[110]

Hydrolysis of the $In^{3+}(aq)$ ion to hydroxy species occurs readily[111] when the pH of the aqueous solution drops below ~3.0, and since the redissolution of any precipitated material may be slow[112] it is important to consider briefly the quantitative aspects of this problem. The first step in the hydrolysis of $In^{3+}(aq)$ is best represented as by equation (1) for which $-\log {}^*K_{1,1} = 4.42 \pm 0.05$ (3 M $NaClO_4$, $\log K_w = -14.22$). The values for ${}^*K_{1,2}$ *etc.* are still uncertain. Enthalpy and entropy changes for equation (1) have also been reported. Polymeric core-linked species generalized as $[In_{n+1}(OH)]^{(n+1)+}$ are believed to form as hydrolysis proceeds, with $In(OH)_3$ as the final product.[9,113,114] The ReO_3 solid state structure of $In(OH)_3$ is distorted by hydrogen bonding between OH ligands of the $M(OH)_6$ octahedra.[115,117] The solubility of the latter in water leads to $K_{sp} = -36.92 \pm 0.01$, although aging of the precipitate is a significant factor.[118] The substance is reported to be soluble in conc. aq. caustic soda, but the nature of the solute species is uncertain. Both $H_2InO_3^-$ and $In(OH)_4^-$ have been proposed[119,120] for the range up to ~1 M NaOH, but at higher OH^- concentrations the solubility of $In(OH)_3$ goes through a maximum (at 11.3 M) above which the solid phase[121] is hydrated $M_3[In(OH)_6]$ (M = Na, Rb). Further study of this topic would seem appropriate. When $In(OH)_3$ is heated with water at high pressure,[122] the product is In(O)OH, which has a deformed rutile structure.[123] The analogous InOX solids (X = F, Cl, Br, I)[1] have been prepared and the lattice parameters tabulated.[124] The possible existence of the related organoindium compounds $(Me_2In)_2O$, MeInO and $(Ph_2In)_2O$ has been discussed elsewhere.[8]

$$In^{3+}(aq) + H_2O \rightarrow In(OH)^{2+} + H^+ \tag{1}$$

Basic species involving both OH^- and halide ligand have been identified in the solid state and in solution. The best defined solution species is $[In(OH)Cl]^+$ for which stability constants have been reported,[9] while the solids[125] include $In(OH)_{1.5}Cl_{1.5}$ and $In(OH)_{1.75}Cl_{1.25}$. No structural work has been reported for these, but the analogous $In(OH)F_2$ has a distorted ReO_3 structure,[126] with octahedral In coordinated by four F and two OH, and this appears to provide a reasonable model for the chlorohydroxo species.

25.2.4.4.2 Alkoxide ligands

Indium(III) alkoxides are readily prepared by the treatment of trihalide with NaOR.[127] The chemistry of these compounds has been little studied, no doubt because they are insoluble materials in which the indium atom optimizes its coordination number by cross-linking *via* oxygen. The isopropoxides of aluminum and gallium react[128] with $In(OPr^i)_3$ to give $In[M(OPr^i)_4]_3$ (M = Al, Ga) which are volatile and monomeric in solution. The structure involves six-coordination at the smaller metal centre.

A number of organometallic alkoxides have been reported. The reaction of In with MeOH gives Me_2InOMe as a glassy polymeric solid, while Bu^tOH gives the dimeric $(Me_2InOBu^t)_2$;[66] with $(CF_3)_2CO$ the product[129] is the analogous $Me_2InOC(Me)(CF_3)_2$. The related compounds $[Me_2InOMMe_3]$ (M = C, Si) are obtained from $Me_3In \cdot OEt_2$ and Me_3MOH; the dimeric structure is believed to be based on a four-membered In_2O_2 ring,[130] similar to the In_2N_2 rings discussed earlier (Section 25.2.4.2.3). No structural details are available for the compound $K[Me_2In(OSiMe_3)_2]$, but analogous Al and Ga compounds are known.[131]

25.2.4.4.3 β-Ketoenolates and related ligands

Indium tris(pentane-1,2-dionate), $In(acac)_3$, was first prepared[132] in 1921 by the reaction of Hacac with freshly precipitated $In(OH)_3$, and this remains the essence of later preparations of related compounds, which include derivatives of dibenzoylmethane,[133] fluoro-4-6-octanedione,[134] and various $RCOCH_2COCF_3$ compounds (R = Me, Ph, naphthyl, thienyl, *etc.*).[135] Extraction into an organic solvent such as chloroform is a useful synthetic step in isolating the reaction product from aqueous solution. A direct route involving the electrochemical oxidation of indium metal in a solution of Hacac in methanol offers the advantage of a one-step synthesis.[136] Related compounds with InO_6 kernels include the tropolonate[137] and 3-oxo-2-methyl-3-pyronate derivatives[30] and it is also worth noting the anionic complex

$[In(tropolonate)_4]^-$, prepared as the Na^+ salt,[138] and apparently representing a rare example of eight-coordinate indium(III).

The IR[139,140] and Raman[141] spectra of $In(acac)_3$ and $In(tta)_3$ have been reported, and the ^1H NMR spectrum of $In(acac)_3$ has been discussed by a number of authors.[142–145] The diamagnetism allows a clear interpretation in terms of delocalized electron density in the OCCCO system, but it is not clear whether this bonding extends to the metal itself. The planarity and the InO_6 kernel implicit in these results have been confirmed by X-ray crystallography.[146,14] The preparation of $X_2In(acac)L$ (X = Cl, Br, I; L = 2py, bipy, phen) was achieved by the unexpected oxidation of InX in refluxing Hacac.[148] The crystal structure of $Cl_2In(acac)bipy$ shows that the molecule involves six-coordinate indium, with *cis* chlorides and an essentially planar acac ligand.[149] Dimethylindium acetylacetonate[66] and the corresponding $Me_2In[OC(R)CHC(CF_3)O]$ compounds[150] are easily prepared by the reaction of Me_3In with the parent diketone. These latter molecules are good acceptors, forming monomeric six-coordinate adducts with bidentate nitrogen and phosphorus donors.

The inter- and intra-molecular kinetic behaviour of $In(acac)_3$ and its analogues have been the subject of considerable study. Isomerization is too rapid to allow identification of the individual species in $In(tfa)_3$ ($tfa = CF_3COCHCOMe^-$) and it has generally been accepted that the mechanism involves a five-coordinate intermediate with one monodentate ligand.[151–155] A similar scheme has been used to explain the rapid ligand exchange between InL_3 and HL ($L = CF_3COCHCOR$),[156] and in exchange between Me_2InL and HL.[150] The kinetic orders and activation energies agree with easy change in the coordination number of indium ($6 \leftrightarrow 5$ and $4 \leftrightarrow 5$), which is a common and important feature of much of the coordination chemistry of this element.

The diketonates of indium(III) are thermodynamically stable complexes in non-aqueous solution,[9] and are sufficiently robust to allow studies of polarographic reduction, which has demonstrated one-electron reduction to the formally indium(II) and indium(I) compounds (see Section 25.2.3.3).[30]

25.2.4.4.4 Oxyanion ligands

Stability constant measurements have demonstrated the presence of indium(III) nitrite complexes in methanol, and the crystalline compound $In(NO_2)_3 \cdot MeOH$ has been isolated.[157]

Indium nitrate can be prepared from $InCl_3$ and N_2O_5;[158,159] the dissolution of indium metal in conc. nitric acid yields $In(NO_3) \cdot xH_2O$ ($x = 3, 4, 5$),[7] which has been used as the source of cationic and anionic complexes.[64] Salts of $[In(NO_3)_4]^-$ have been reported,[64,160] and Raman studies[161,162] have demonstrated the presence of In/NO_3^- complexes in solution; the stability constant results have been reviewed.[9]

Indium orthophosphate is essentially insoluble in water, but salts of $[In(PO_4)_2]^{3-}$ and $[In_2(PO_4)_4]^{6-}$ have been prepared,[163] and a solution phase complex with the $H_2PO_4^-$ anion has been identified.[164] The coordination chemistry of In^{3+} with the higher phosphates is at present ill defined. Stability constant measurements,[9] and preparative and spectroscopic studies[7] suggest that InO_6 is the central core of the compounds in the solid state. The same appears to be true for the complexes reported for L = fluorophosphate,[165] methylphosphonate,[166] dichlorophosphate,[167] dimethylthiophosphate,[168] diethylthiophosphinate[169,170] and diisopropylmethylphosphonate.[171]

Indium sulfates have been known for many years, and the literature includes a number of references to double salts which may be reformulated as anionic sulfato complexes.[1,2,7] The existence of indium sulfate complexes in solution has been demonstrated.[9] The solid state structures reveal that bridging by sulfate and/or H_2O is important. By analogy with $RbTl(SO_4)_2$, the salt $NH_4In(SO_4)_2$ was reported to involve infinite chains of $[In(SO_4)_2]^-$, with a distorted trigonal prismatic InO_6 kernel,[172] but in $(NH_4)_3In(SO_4)_3$ the infinite chains consist of InO_6 units in which each oxygen is supplied by a different SO_4^{2-} ligand.[173] Three hydrated salts have been studied by X-ray crystallography,[174,175] namely $NH_4In(SO_4)_2 \cdot 4H_2O$, $HIn(SO_4)_2 \cdot 4H_2O$ and $(H_3O)[In(H_2O)_2(SO_4)_2] \cdot 2H_2O$; all are based on *trans*-$In(OH_2)_2O_4$ units which are linked by sulfate groups. The compound $In(OH)SO_4 \cdot 2H_2O$, which crystallizes from aqueous ethanol solution, has bridged octahedral InO_6 kernels.[177] Here, as elsewhere, the importance of six-coordination of indium(III) by oxygen is a predominant feature of the coordination chemistry.

Indium(III) fluorosulfate, which can be prepared from $InCl_3$ and HSO_3F, is a Lewis acid, and forms both 1:1 and 1:2 adducts.[178] Indium(III) selenate octahydrate, acid salts and double salts are known, as is the case for tellurate and tellurite, but the only available structural information refers to $NH_4In(SeO_4)_2 \cdot 4H_2O$ which resembles the sulfate in having *trans*-$In(OH_2)_2O_4$ coordination at the metal.[179,180]

The weak complexing of indium(III) with chlorate, bromate and iodate has been studied, and stability constants reported, but no preparative work has been done. Perchlorate ion does not complex with indium(III).[9]

25.2.4.4.5 Carboxylate ligands

Beyond the preparation of many $In(O_2CR)_3$ compounds, the carboxylates have been little investigated. Preparation may be *via* the reaction of oxide or hydroxide with the appropriate acid,[1,31,181] or of $InMe_3$ with excess acid;[182] the latter route also yields $Me_2In(O_2CR)$ when smaller quantities of acid are used (see below). The mode of bonding of the carboxylate (mono, bidentate, bridging) in such compounds has not been settled. Indium(III) acetate, which apparently forms a polymeric lattice, acts as a weak acceptor with pyridine to give a 1:1 complex; with ethanediamine the product is the salt $[Inen_3][OAc]_3$, while R_4N^+ salts of OAc^- and ClO_4^- give the unusual $(Me_4N)_3In(OAc)_6$ and $(Et_4N)_3In(OAc)_3(ClO_4)_3$, whose structures have yet to be elucidated. Indium(III) formate forms a weak 1:1 adduct with pyridine.[31] Anhydrous $In(O_2CCF_3)_3$ is also a Lewis acid, forming 1:3 adducts with py, DMSO and DMF.[183] The bipy and phen adducts of $In(OAc)_3$ show an eight-coordinate InO_6N_2 kernel, with bidentate acetate.[184]

Other than the conclusions drawn from vibrational spectroscopy, there is little definitive structural information on these compounds. The X-ray structure of $Cl_2In(O_2CPh) \cdot 2py$, obtained as one of a number of products from the reaction of InCl with dibenzoyl peroxide, reveals the presence of an $InO_2N_2Cl_2$ kernel, with *cis* chloride and *trans* pyridine ligands;[185] the benzoate is formally bidentate. The structures of Me_2InOAc[186] and Et_2InOAc[187] again show that indium is six-coordinate, with the molecules stacked to produce $InC_2O_2(O')_2$ coordination. These compounds are prepared by the reaction of HOAc and R_3In,[31,188] but the $RIn(OAc)_2$ compounds have not been isolated, although Ph_3In in similar conditions yields Ph_2InOR and $PhIn(OR)_2$ depending on the conditions (R = OCMe, OCEt).[189] Another route is *via* the reaction of R_3M (R = Me, Et) with CO_2, when the product is the dimeric $[R_2M(O_2CR)]_2$, in which the carboxylate ligands bridge two metal atoms to give an $In_2O_2C_2$ ring,[190] so that a variety of modes of bonding of carboxylate to indium(III) must be considered.

The polymeric lattice of Me_2InOAc has important consequences for the coordination chemistry of this substance, which acts as an acceptor giving adducts with stoichiometries 1:1 (DMSO, py; en, diphos) or 2:1 (bipy, phen). The en complex is fluxional in solution. The 2:1 adducts are believed to involve the bidentate ligand bridging the metal atoms of a dimeric Me_2InOAc unit.[191] Similarly the reaction of Me_2InOAc with toluene-3,4-dithiol (H_2TDT) eliminates CH_4 in two separate solvent-dependent processes, and the first product is $[MeIn(OAc)HTDT]_2$, again derived from dimeric Me_2InOAc.[192]

The complex $[In(ox)_2(OH_2)_2]^-$ (ox^{2-} = oxalate anion) was postulated[193] as being the anion present in $NH_4[In(ox)_2(OH_2)_2]$, but recent X-ray studies have shown that the lattice has unusual eight-coordinate indium, with four bridging oxalate ligands giving an Archimedean antiprism.[194] Other solid species include $(NH_4)_3[In(ox)_3]$ and $MIn(ox)_2 \cdot xH_2O$ ($M = NH_4$, K, Cs; $x = 3$ or 6); the IR spectra suggest that the oxalate is not bidentate, but do not eliminate the possibility that bridging is again important.[195–197] The structure of $[In_2(ox)_3(H_2O)_4] \cdot 2H_2O$ is based on pentagonal bipyramidal seven-coordinate indium, oxalate-bridged into infinite chains.[198]

25.2.4.4.6 Ether and other oxygen donor ligands

A number of cationic complexes of the type InL_6^{3+} (L = monodentate oxygen donor) have been prepared,[7] and others are identified in solution by NMR studies (see Section 25.2.4.4.1). These are all readily explained as InO_6 species, with O_h symmetry in the core, again demonstrating the importance of this stereochemistry with hard donor ligands. With Ph_3PO, the product is $[InL_4]^{3+}$ and presumably steric interaction between ligands prevents further coordination.[62] The normal preparative route is by substitution of H_2O in a hydrated salt by excess ligand, but a direct electrochemical route to $[In(DMSO)_6](BF_4)_3$ has been reported.[96]

Adducts of the neutral indium trihalides with oxygen donors have been reviewed previously,[5,7] as have the corresponding organoindium adducts.[8] Stoichiometries of InX_3L_3, InX_3L_2 and InX_3L are known, but any detailed discussion is hindered by the lack of structural and thermodynamic information, although application of the cone angle concept of Tolman[199] has been successful in rationalizing some results.[7] A further complication is that some of the supposedly neutral adducts are in fact ionic dimers, so that $InI_3 \cdot 2DMSO$ is actually *cis*-$[InI_2(DMSO)_4][InI_4]$,[200] and here again there is no explanation of the factors which govern the choice of available structures. Mixed halo-aqua cations and anions are discussed below (Section 25.2.4.6).

25.2.4.5 Sulfur and Heavier Group VI Ligands

25.2.4.5.1 Sulfur and selenide ligands

Thioethers form adducts with indium trihalides,[5,7] but not apparently with organoindium compounds;[8] these compounds show the same range of stoichiometry as do the oxygen analogues. Thiourea complexes have also been studied, and cationic, neutral and anionic derivatives are known.[5,7] Indium tris-phenylthiolate $In(SPh)_3$ has recently been prepared from InX_3 and NaSPh, and adducts with neutral donors, and the anionic complexes $[In(SPh)_4]^-$ and $[XIn(SPh)_3]^-$ (X = Cl, Br, I) can be derived from $In(SPh)_3$.[201] There appears to have been no work on selenium analogues of these compounds. Indium(III) thiolate compounds can also be prepared by the oxidative insertion of InX (X = Cl, Br, I) into R_2S_2 (R = Me, Ph), which gives $(R_2S)_2InX$ and subsequently $(R_2S)_2InX \cdot py_2$ and $[(R_2S)_2InX_2]^-$. Selenide derivatives $(PhSe)_2InX \cdot py_2$ can also be synthesized, but Ph_2Te_2 does not react with InX under the conditions used.[58]

Interesting solid state structures are shown by some stoichiometrically simple In–sulfur compounds. Tetrahedral InS_4 is found in $InPS_4$,[202] but in $KInS_8$ both four- and six-coordinate indium are identified.[203] The basic unit in $Rb_6In_2S_6$ and $Rb_4In_2S_5$ is $In_2S_6^{2-}$, isoelectronic with In_4Cl_6, with both bridging and terminal sulfurs and further bridging gives $In_4S_{10}^{4-}$; indium is four-coordinate in each case.[204] The compounds $In_{2.01}X_3$ (X = S, Se, Te) have been formulated as $In^I[In^{III}oct]_2[In^{III}tet]_{73}X_{115}$, with InX_6 sharing edges with distorted InX_4 tetrahedra.[205] Some of these structures might prove useful in synthetic work (see also In_5S_4, Section 25.2.3.4).

25.2.4.5.2 Bidentate sulfur donor ligands

As part of the development of the coordination chemistry of bidentate sulfur ligands, studies of the appropriate indium(III) complexes have been reported. The In—S_6 analogue of $In(acac)_3$ has not been prepared[7], but a number of related neutral InL_3 chelates are known for L = dithiobenzoate,[206] dithiophenylacetate,[206] diphenyldithioarsenate,[207] *O,O*-diethyldithiophosphate,[208] pentamethylenedithiocarbamate,[209,210] trithiocarbonate and alkyl xanthates.[211,212] The InS_6 kernel has been confirmed by X-ray crystallography; both distorted octahedral[208,212] and trigonal prismatic[210] stereochemistries have been identified, but no doubt the bite of the ligand is a dominant factor. In $In(dithizonate)_3$, indium is five-coordinate (trigonal bipyramidal), with two bidentate and one monodentate ligand.[213] The only analogous In—Se_6 species is the tris-*O,O*-diethyldiselenophosphate complex.[214]

Complexes with anionic dithiolene ligands are prepared by the reaction of $InCl_3$ with the sodium salt of the appropriate ligand, such as MNT^{2-} (= maleonitriledithiolate), TDT^{2-} (= toluene-3,4-dithiolate) or *i*-MNT^{2-} (= 1,1-dicyanoethylene-2,2-dithiolate). The product anions show a range of In—S coordination numbers, from four ($[In(MNT)_2]^-$, $[In(TDT)_2]^-$ to six ($[In(MNT)_3]^{3-}$, $[In(i\text{-}MNT)_3^{3-}]$); the four-coordinate species readily give rise to six-coordinate adducts such as $[In(TDT)_2bipy]^-$. With *i*-MNT^{2-}, no four-coordinate anions were obtained, but $[In(i\text{-}MNT)_2X]^{2-}$ anions (X = Cl, Br, I) were prepared.[32,215] The reduction of these anions was discussed earlier (Section 25.2.3.4). The crystal structure of $(Et_4N)_3[In(MNT)_3]$ established the planarity of the InS_2C_2 ring system, and the distorted octahedral symmetry of the InS_6 kernel.[215,216] 1,2-Ethanediol (H_2EDT) gives the analogous $Et_4N[In(EDT)_2]$, but no six-coordinate species,[217] and the five-coordinate anion $[In(MNT)_2Cl]^-$ can be obtained from aqueous solution;[218] the structure of this anion is intermediate between trigonal bipyramidal and square-based pyramidal.[219]

The reaction of InX (X = Cl, Br, I) with 1,2-bis(trifluoromethyl)dithiete in non-aqueous solution[220] proceeds by insertion of indium into the S—S bond of the dithiete to give polymeric products containing an InS_2C_2 ring. Monomeric derivatives can be obtained by treatment with neutral donors to give adducts formulated as $[InL(DMSO)_4]X$ or $[InL(bipy)_2][InL_2bipy]$ ($L = (CF_3)_2C_2S_2^{2-}$). Similar reactions occur when the dithiete reacts with cyclopentadienylindium,[221] although in this case the neutral adducts are of the type CpInL·bipy. Another route to In—S coordination compounds involves alkane elimination,[222] so that Me_3In reacts with H_2TDT to give polymeric $[MeIn(TDT)]_2$ from which adducts can be obtained in the usual way.

25.2.4.6 Halogen Ligands

25.2.4.6.1 Fluoride ligands

Anhydrous indium trifluoride[223] is an insoluble substance whose lattice involves six-coordinate indium(III).[224] The hydrate $InF_3{\cdot}3H_2O$ is soluble in water,[225] but insoluble in common organic solvents. The structure has been reported to be based on InF_3O_3 octahedra,[226] but a later study showed sharing of apical H_2O between InF_4O_2 units, with the third H_2O in the lattice.[227] The hydrate is a rare example of an InF_3 adduct.[5] A series of ammonia adducts is known, with 1:1 and 1:3 stoichiometries;[228–230] the 1:1 adduct decomposes on heating to InN.[229] Mixed hydrazine/water adducts have also been prepared,[231] but have been formulated as salts (see below).

Earlier information on the preparation of anionic fluoro complexes from aqueous solutions or in high temperature phases has been reviewed elsewhere.[6,7] Stoichiometries corresponding to $MInF_4$, M_2InF_5 and M_3InF_6 have all been reported, as have the more unusual KIn_2F_7, $K_5In_5F_{14}$[232] and $Rb_2In_3F_{11}$.[233] Early X-ray studies[7] which identified octahedrally coordinated indium in K_3InF_6 and $(NH_4)_2InF_6$ have been confirmed by single crystal work[234] on $Cs_2Na[InF_6]$. Vibrational spectra also confirm this stereochemistry in other InF_6^{3-} salts, and it has been suggested that $MInF_4$ and M_2InF_5 consist of metal ions and InF_6 octahedra sharing edges.[235] In $Rb_2In_3F_{11}$, the $In_3F_{11}^{2-}$ anion has a distorted octahedral InF_6 bridging two pentagonal bipyramidal InF_5 units.[233]

Given the structure of $InF_3{\cdot}3H_2O$ (see above) and the tendency of indium(III) to be six-coordinate with hard ligands, it is reasonable that the mixed fluoro/aqua anionic complexes should also be $[InF_4(H_2O)_2]^-$ and $[InF_5(H_2O)]^{2-}$. This is compatible with the evidence from aqueous solution chemistry, where stability constants and thermochemical results indicate a series of $[In(H_2O)_{6-n}F_n]^{(3-n)+}$ species.[9] These complexes still await detailed systematic structural investigation.

25.2.4.6.2 Chloride, bromide and iodide ligands

Crystalline indium trihalides are readily prepared from the reaction of the metal and halogen[9] (sometimes in the presence of solvent), from metal + HX, or by the electrochemical oxidation of the metal in a non-aqueous solution of X_2.[20] Indium trichloride and tribromide have indium octahedrally coordinated by halide in the solid state, whereas the triiodide exists as the dimer $I_2InI_2InI_2$ in the solid or in solution; all three halides are dimeric in the vapor phase.[9] The many adducts of these compounds with neutral donors have been extensively reviewed previously.[5,7]

The dominant ion in the mass spectrum of the trihalides[236] is InX_2^+, isoelectronic with CdX_2, and the corresponding organoindium cation $InMe_2^+$ has been identified in solution[237] and in the solid state,[238] and there is also evidence for Ph_2In^+.[239] The symmetry of these R_2M^+ ions and their tendency to go to higher coordinate species have been discussed elsewhere.[8] The existence of cationic complexes such as $[InI_2(DMSO)_4]^+$ and $[InX_2bipy_2]^+$ has been noted earlier (Sections 25.2.4.2.2 and 25.2.4.4.5).

Anionic complexes have been known, if not recognized, for many years as mixed salts of both inorganic and organic cations.[7] In particular, Ekeley and Potratz[240] identified the stoichiometries $ClnX_4$, C_2InX_5, C_3InX_6 and C_4InX_7. The role of cation size, solution conditions and ligand properties in determining the coordination number of indium in such anions has been discussed earlier.[7,241,242] It is now clear that with X = I, the only anionic complex is the tetrahedral InI_4^-,[34,38,199] and at the opposite end of the scale that the crystalline C_4InX_7 compounds involve lattices of $4C^+ + InX_5^{3-} + X^-$ in which InX_6^{3-} (X = Cl, Br) is the octahedral complex,[243,244] also found in $(Me_3NH)_3InCl_6$[245] and in $Cs_2In^I[InCl_6]$.[246] The remaining InX_4^- anions are also tetrahedral in both solid state[247,248] and in non-aqueous solution.[249–251] The detailed symmetry of InX_5^{2-} has been a matter of discussion, but the description as a square-based pyramid will serve for most purposes.[252,253] Compounds of stoichiometry C_2InBr_5 appear to be based on a lattice of $2C^+ + InBr_4^- + Br^-$.[17,242] A meaningful treatment of the interrelationships between these anions is lacking.

The aqueous solution chemistry of indium(III)–halide species has been discussed on the basis of a scheme[9] involving the successive substitution of H_2O by X^- in $[In(H_2O)_6]^{3+}$, in keeping with the evidence on aqua/halogeno species (see below). Raman spectroscopy has supported this model for X = Cl.[254] A second type of coordination equilibria involves the formation of InX_4^- species in aqueous solution from $[InX_4(H_2O)_2]^-$, and appears to be favoured in the order I > Br > Cl[255] (see ref. 9 for a full discussion of this). Changes in the coordination number of indium(III) species in solution have been noted at several points elsewhere in this review, and clearly also play an important part in the chemistry of these halide complexes. The $InCl_5^{2-}/InCl_4^-$ equilibrium has been studied in non-aqueous solution,[256] and other examples have been reported.[241] The aqueous phase substitution scheme is supported by the existence of the anions *cis*-$[InX_4(H_2O)_2]^-$ (X = Cl, Br) and $[InCl_5(H_2O)]^-$, which are known from X-ray studies to be six-coordinate;[257–259] other similar mixed complexes known include $[InCl_4L_2]^-$ (L = urea, thiourea)[260] and $[InCl_4bipy]^-$ (see Section 25.2.4.2.2). Spectroscopic studies of the aqua/halogeno anions have been reported.[261,262]

25.2.4.6.3 Mixed halide ligands

The existence of the mixed halide complexes $InX_{4-n}Y_n^-$ in non-aqueous solution was demonstrated by ^{115}In NMR work for the whole range of $X \neq Y$ = Cl, Br, I, and the InX_2YZ^- anions were also identified.[50] The $InX_2Y_2^-$ anions can be prepared by the oxidation of InX_2^- by halogen Y_2, and InX_3Y^- from $In_3X_6^{2-} + Y_2$; detailed vibrational spectral analysis[263] and X-ray analysis of $InCl_3Br^-$ and $InBr_3Cl^-$ confirm the C_{3v} stereochemistry of these anions.[248] Organohalide anions have been reviewed elsewhere.[8]

25.2.4.7 Hydrogen and Hydride Ligands

25.2.4.7.1 Hydride ligands

There is no evidence to support the existence of neutral monomeric InH_3, although an unstable $(InH_3)_n$ is reported to be formed by the decomposition of $LiInH_4$, itself formed by treating $InBr_3$ with LiH ether.[1] No structural information is available. Trimethylindium reacts with MH (M = Li, Na, K) to give the unstable MMe_3InH, and MR_3InH (R = Me, Et) and $NaEt_2InH_2$ have also been described, but again decomposition has barred further work.[8]

25.2.4.7.2 Borohydride and related ligands

Trimethylindium and B_2H_6 yield a solution of $In(BH_4)_3$, which can be obtained as the THF adduct, which decomposes *in vacuo*. The aluminohydride analogue $In(AlH_4)_3$ is even more unstable,[1] and there seems to be little chance of extending the chemistry of indium(III)–hydrido systems.

25.2.4.8 Mixed Donor Atom Ligands

The largest group of mixed donor atom complexes is the tris chelates, of which In(oxine)$_3$ (oxine = 8-hydroxyquinolate) is typical. A number of spectroscopic studies have shown that such compounds contain InO_3N_3 kernels. The analogous InS_3N_3 compounds, such as the 8-mercaptoquinolate, have been used in analytical work and derivatives of monothiodibenzoylmethane are known.[8] Organoindium compounds with such ligands, and others, have been discussed elsewhere.[8] The structure of $(Ph_4As)_4[In_2(dto)_5]$ (dto = 1,2-dithiooxalate) has one dto ligand bonded to both indium atoms, while others act as bidentate (S_2) ligands to give InS_5O coordination.[264] Another important group of complexes, involving ethylenediaminetetraacetic acid and its analogues, has been extensively studied in solution, and thermodynamic results are available,[9] but the general coordination chemistry of these compounds adds little to what has been discussed previously.

25.2.4.9 Macrocyclic Ligands

The reaction of InX_3 (X = Cl, Br, I) with 18-crown-6 (= L) (*cf.* Section 25.2.2.5) yields In_2X_6L compounds, whose structure is written as $[InX_2L][InX_4]$ on the basis of vibrational spectroscopy.[22] The cation is believed to be involved in *trans*-X_2InL coordination. The corresponding cyclam compounds were also prepared, but the structure of the cations in such molecules is in doubt, since reaction may have occurred at the ring nitrogen atoms.

A series of earlier papers on complexes of indium(III) with porphyrins,[265] and phthalocyanines[266–268] has been summarized earlier.[7] The structures InLX (L = substituted porphyrin; X = Cl, OAc, OPh) imply a square pyramidal stereochemistry with the indium atom situated above the ligand N_4 system, and this has been confirmed by X-ray crystallography for X = Cl[269] and X = Me,[270] but for X = $MeOSO_2$, the indium atom is in the plane of the four nitrogen atoms.[271] A number of Schiff base derivatives have recently been described.[272]

25.2.5 THALLIUM: INTRODUCTION

As with indium, the literature contains a number of surveys of the general inorganic[1] and organometallic chemistry of thallium,[273] and the reviews of this element already cited are also concerned with the adducts of the halides and pseudohalides,[5] and with other aspects of the coordination chemistry.[6] A monograph by Lee provides a most useful treatment of the literature up to 1970.[274] The application of thallium compounds in organic synthesis has been discussed.[275,276]

25.2.6 THALLIUM(I)

One of the classical properties of the main group elements is that the stability of the lower oxidation states increases with atomic number, and the chemistry of thallium is a good example of this effect. In aqueous solution, the Tl^+ ion is stable with respect to oxidation by the solvent and there is accordingly an extensive chemistry of this oxidation state. The similarities between Tl^+ and the corresponding alkali metal cations have resulted in much interest in the use of this ion as a probe in biochemical systems, and the ease with which ^{205}Tl NMR spectra can be recorded has also had an impact on such studies.[277,278]

The similarities between Tl^+ and other M^+ ions have led to the synthesis of many salts and complexes in which Tl^+ serves as a balancing cation, but such compounds will not be included in the present review unless there is some feature of special interest. We may note immediately that the similarity of Tl^+ to, say, K^+ includes poor complexing (*i.e.* weak Lewis acid) behavior in both cases.

25.2.6.1 Group IV Ligands

25.2.6.1.1 Cyanide ligand

The salt TlCN can be prepared by the reaction of thallium(I) acetate and KCN in aqueous solution,[279] but there is no evidence for the formation of cyano complexes in aqueous solution.[280]

25.2.6.1.2 Other Group IV ligands

No compounds containing Tl—MR bonds are known for M = Si, Ge, Sn or Pb, which presumably reflects the dominant ionic character of Tl^+. By way of illustration of this, attempts to displace mercury from $Hg[Ge(C_6F_5)_2]_2$ by reaction with Tl^0 yielded the compound $Tl[Hg\{Ge(C_6H_5)_3\}_3]$ containing the Tl^+ cation.[281]

It is appropriate to note here some remarkable compounds formed by tin and thallium, although these are not formally thallium(I) compounds. Treatment of the alloy $NaSnTl_{1.5}$ with ethylenediamine gives $Na_5en_2(TlSn_8)$, containing the $TlSn_8^{5-}$ anion for which a *nido* structure is proposed.[282] Similarly, the reaction of KTlSn with en and [2.2.2]cryptate yields ([2.2.2]crypt-$K^+)_3(TlSn_8^{3-}$—$TlSn_8^{3-})0.5$en, in which the *closo* cluster anions have bicapped square antiprismatic and tricapped trigonal prismatic structures respectively.[283] The existence of such molecules suggests that other Tl—Sn-bonded species might exist.

25.2.6.2 Nitrogen Ligands

Given that the low polarizing power of the Tl^+ ion is similar to that of the heavier alkali metals, it is not surprising that no complexes have been reported with monodentate nitrogen donors such as ammonia. Thallium(I) hydroxide reacts with imidazole (RH) to give TlR, which is apparently a homopolymer linked by covalent Tl—N bonds, rather than a salt,[284] but in the benzotriazole derivative,[285] Tl^+ ions are well separated by the anions, with a distorted trigonal prismatic arrangement of six nitrogen atoms at distances ranging from 2.725 to 3.721 Å. Other than this, there appear to be no monodentate heterocyclic ligand derivatives. The bidentate heterocyclic donors phen and bipy form TlL_2^+ cations, identified as either nitrate or perchlorate salts, and the spectroscopic properties suggest a pseudotetrahedral TlN_4 coordination.[286] The extra stabilization inherent in chelate formation is clearly important here, and in the case of macrocyclic ligands (see Section 25.2.6.7).

Thallium(I) nitride is a highly explosive black solid,[1] but the yellow azide TlN_3 is more stable; the physical properties suggest that there is some covalent Tl—N bonding in the solid state. The mechanism of the decomposition has been reviewed.[287]

The pseudohalides and related compounds have been investigated by a number of workers. Thallium(I) cyanate and thiocyanate are both known, and the latter has been shown to be ionic in the solid state;[288] differing values have been reported for the stability constants in the Tl/NCS^- system in aqueous or mixed water/non-aqueous systems, but the overall evidence is that the complexes are extremely weak.[289] The ionic selenocyanate has been reported, but the chemistry has not been investigated.[1]

25.2.6.3 Oxygen Ligands

25.2.6.3.1 Oxide and hydroxide ligands

Thallium(I) oxide, Tl_2O, is a well-characterized compound and a useful source of Tl^+ salts.[1] Matrix isolation studies of the solid gave structural data in agreement with spectroscopic results.[290,291] The so-called peroxide TlO_2 is obtained electrochemically from aqueous Tl_2SO_4, but the properties include insolubility in water, alkali and some dilute acids, and treatment with dilute HCl yields oxygen. The stoichiometry suggests that the compound is in fact a superoxide, and this has been confirmed by spectroscopic studies on frozen matrices.[292,293]

The reaction of $Ba(OH)_2$ and Tl_2SO_4 in oxygen-free aqueous solution yields aqueous thallium(I) hydroxide, but the solid is best obtained from the metal and aqueous ethanol in the presence of oxygen. The yellow crystalline solid yields an alkaline solution. The mixed oxides MTlO (M = K, Rb) have been prepared by heating M_2O and Tl_2O, but these are apparently not thallate(I) species.[294]

25.2.6.3.2 Alkoxide ligands

Thallium(I) ethoxide has been known for many years, and a series of related alkoxides can be prepared by methods which include Tl + ROH, TlOH + ROH and Tl_2O + ROH.[295] The ethoxide,[296] and a number of other TlOR compounds,[297] are tetrameric in solution, and an X-ray crystallographic study of $(TlOMe)_4$ shows that the structure is based on that of cubane,[298] with no Tl—Tl bonding interactions. The bonding in such molecules has been discussed.[299] Phenoxide, and substituted phenoxides, can be prepared from PhOH + TlOH, while reaction between Me_3Tl and ROH gives Me_2TlOR (R = Et, Ph).[300–302] The Tl—O bond undergoes insertion by CS_2 and SO_2; the products are TlS_2COPh and TlO_2SOPh from TlOPh, while Me_2TlOR gives RSO_3TlMe_2.[301]

25.2.6.3.3 *β-Diketonate and related ligands*

Thallium(I) β-diketonates have been known for many years,[303] and are readily obtained by treating the appropriate acid with TlOH or Tl_2CO_3.[1,300,302] The dipole moments of Tl(acac), Tl(hfacac) and Tl(dibenzoylmethane) have been measured, and the two latter compounds shown to be monomeric in benzene.[304] The structure of Tl(hfacac) consists of chelated thallium(I) atoms (r = 2.62, 2.72 Å) which are 0.42 Å below the ligand O—C—C—C—O plane; chains of dimers then give pentacoordinate Tl^I, in which the presence of the inert pair is deduced.[305] The oxidative reactions of these compounds might prove interesting. The related Tl(oxinate) has been prepared,[302] but the thioxine analogue could not be obtained by standard methods.[306] Some unusual chelate derivatives of thallium(I) and $Me_3CNSi(Me_2)OCMe_3$ have been reported.[307,308]

25.2.6.3.4 *Oxyanion ligands*

As noted in the introduction, many thallium(I) salts of oxyacids are known. The preparation and properties of these have been reviewed previously.[1] Some recent structural information from X-ray crystallography includes the identification of two different Tl^I sites in Tl_2SO_3, prepared from $Tl_2CO_3 + SO_2$,[309] the demonstration that Tl^I is seven-coordinate in TlH_2PO,[310] but three-coordinate in Tl_6TlO_6,[311] and the proof of the importance of the inert pair effect in Tl_2HPO_4.[312] In $Tl_4P_4O_{12}$, Tl^+ ions are separated by $P_4O_{12}^{4-}$ rings.[313] The vapor phase of Tl_2SO_4 contains molecules of D_2 symmetry.[314] In a different approach, the shielding anisotropy of ^{205}Tl has been used to investigate solid state effects in $TlNO_3$.[315]

25.2.6.3.5 *Carboxylate ligands*

Many thallium(I) derivatives of carboxylic acids are known, and can be prepared by standard methods.[1] The electronic spectra[300] have been interpreted as indicating a significant covalent metal–ligand interaction, whereas the IR spectra have been taken to show that the compounds are ionic;[316,317] X-ray diffraction studies should help in setting this point. In Tl(ascorbate), the Tl atoms are far apart (4.05 Å) and the homopolymeric lattice involves oxygen bridging between ligands.[318] Thallium(I) carboxylates can act as Lewis acids, as in the formation of Tl(salicylate)-phen, which consists of dimeric units with pentagonal pyramidal TlN_2O_3 kernels, and again well-separated thallium(I) atoms; the inert pair effect is adduced in explaining this structure.[319] The structure of Tl formate in concentrated aqueous solution is found to be a cubane-like tetramer, with Tl—O—Tl and Tl—OCO—Tl bridging.[320]

Oxidation of TlOAc with bromide in acetic acid gives $Tl(OAc)Br_2$,[321] but as with the alkoxides such reactions have apparently not been investigated in detail. Thallium(I) carboxylates have found considerable application in organic synthesis.[274,275,322]

25.2.6.4 Sulfur and Heavier Group VI Ligands

Thallium(I) sulfide, selenide and telluride are all well-established compounds, and in the case of all three elements, binary compounds other than M^I and M^{III} species and double sulfides are also known, as are Tl salts of trithiocarbonate, azidothiocarbonate, *etc.*[1]

A series of dithiocarbamates can be prepared from a mixture of $R_2NH + CS_2 + Tl^+(aq)$ in alkaline solution. The structure of Pr_2NCS_2Tl consists of dimeric units bonded through a TlS_4Tl octahedron, with these dimers interacting to give chains.[323] In Bu_2NCS_2Tl, dimerization and chains are again found, and in this compound there is some evidence of a stereochemically significant inert pair.[324] Other known related compounds include monothiocarbamates, monothioxanthates,[325] xanthates[326] and *O,O*-substituted dithiophosphates. Derivatives of *i*-MNT^{2-} (see Section 25.2.8.5.2) such as Na_4Tl_2(*i*-MNT)$_3$, $NaTl_3$(*i*-MNT)$_2$ and Tl_2(*i*-MNT) are apparently ionic.[327] (See also Section 25.2.8.4).

The analogous selenium and tellurium chemistry has not been explored. The $[Tl_2Te_2]^{2-}$ anion has been obtained by reacting KTlTe with [2.2.2]cryptate in ethylenediamine to give $[K([2.2.2]crypt)]_2^+[Tl_2Te_2]^{2-}$ en, in which the anion is butterfly-shaped with a 50° fold.[328] The assignment of oxidation numbers is somewhat arbitrary in such systems (*cf.* Section 25.2.6.1.2), but it seems clear that thallium is in a low oxidation state.

Thiourea (tu) adducts of thallium(I) salts have been reported with stoichiometries such as TlX·4tu (X = Cl, Br, I, ClO_4, NO_3, NO_2, ClO_3, NCS), TlF·4tu·$2H_2O$, Tl_2CO_3·8tu and Tl_2SO_4·8tu.[329] The crystal structures of the halides show that these compounds are isostructural with the corresponding alkali halide adducts, while in TlX·4tu (X = NO_3, ClO_4) the coordination of sulfur to thallium involves distances [3.43(1) Å] which are greater than the sum of the ionic radii, so that Tl—S bonding is not an important factor.[330] A similar conclusion emerged from crystal structure studies of TlH_2PO_4·4tu and $Tl(C_6H_5CO_2)$·4tu.[331] A series of analogous selenourea (L) compounds TlX·4L (X = Cl, ClO_4, OAc, Br), TlX·3L (X = NCS, NO_3), and Tl_2SO_4·6L have been prepared and their mode of thermal decomposition studied.[332]

25.2.6.5 Halogen Ligands

25.2.6.5.1 Fluoride ligand

The physical properties of the thallium halides have been tabulated.[1] Crystalline TlF has a distorted rock-salt lattice, with two independent thallium sites.[333–335] The thermodynamic properties of the solid have been reassessed.[336] At high temperature, the vapor phase contains TlF and Tl_2F_2 molecules,[337] and matrix isolation studies have demonstrated the presence of both TlF and Tl_2F_2 in the solid; the dimers have linear symmetry.[338]

Stability constant studies show that in aqueous solution the anions TlF_2^-, TlF_3^{2-} and TlF_4^{3-} are formed;[289] some of these may be aquated. Many double salts have been reported,[1] but it seems that thallium is present in these as the Tl^+ cation rather than as $TlF_x^{(x+1)-}$ anionic complexes.

25.2.6.5.2 Chloride, bromide and iodide ligands

The crystal structures of the three heavier thallium(I) halides have been established[333,334] and lattice energies calculated;[339] the inert pair of Tl^+ is apparently insignificant in terms of the stability of the lattice. The enthalpy of hydration of the Tl^+ ion was also derived. Gas phase (TlBr, TlI) and matrix isolation studies (TlCl, TlBr, TlI) have shown that TlX and Tl_2X_2 species are important.[338,340] Complex formation[289] in aqueous solution decreases in the order Cl > Br > I, and as with the fluorides, the double salts show no evidence of anionic complex formation by Tl^I. Finally, it is important to note that TlI_3 is formulated as $Tl^+(I_3^-)$ from X-ray studies.[341,342] Thermal decomposition yields Tl_3I_4, whose structure has not been reported.

25.2.6.6 Hydrogen and Hydride Ligand

Thallium(I) hydride has been identified spectroscopically in the products of an electric discharge in hydrogen at a thallium cathode, or from thallium and atomic hydrogen.[343] An unstable polymeric solid has also been reported.[344] The borohydride $TlBH_4$ can be obtained by reacting Tl^I compounds with $LiBH_4$, and $TlAlH_4$ is formed similarly;[344] the borohydride, which has an ionic lattice, can also be prepared from KBH_4 and aqueous $TlNO_3$.[345]

25.2.6.7 Macrocyclic Ligands

As noted earlier, the similarities between Tl^+ and alkali metal cations have led to the use of the former as a probe in biological studies, including studies with various macrocyclic ligands, especially those with oxygen donor atoms. The thallium(I) cryptates behave kinetically like the potassium compounds,[346] and the binding constants to 18-crown-6 have been measured by ^{205}Tl NMR methods.[347] Several Tl^I compounds with crown ethers (L) have been prepared; in addition to TlXL compounds, $(TlL_2)X$ species were obtained, and these are formulated as sandwich structures. There is little evidence of an inert pair effect.[348] Strong interactions are detected between the metal and aromatic rings in the dibenzo-18-crown-6 derivative.[349]

25.2.7 THALLIUM(II)

The problems associated with the M^{II} state of the elements of group III have been noted previously (*cf.* Sections 25.1.13 and 25.2.3). The Tl^{2+} ion has been identified in pulse radiolysis decomposition of aqueous Tl_2SO_4,[350,351] and in γ-irradiated frozen aqueous solutions.[352] Thallium(II) has also been invoked as an intermediate in the photochemical reduction of aqueous Tl^{III} solutions,[353,354] but there is dispute as to its involvement in redox reactions.[355,356] It is safe to conclude that there is no equilibrium aqueous phase chemistry of Tl^{2+}.

The compound Tl_4O_3 is a mixed oxide lattice $(3Tl_2O + Tl_2O_3)$,[357] but the sulfide TlS and selenide TlSe can be formulated as mixed Tl^I/Tl^{III} species, with tetrahedral coordination at Tl^{III} in a $(TlS_2)_n$ chain; the Tl^I ions are then eight-coordinate.[358,359] The sulfide Tl_4S_3 consists of tetrahedral TlS_4^{5-} and Tl^+ ions,[360] while the structure of Tl_2S_5 is made up of Tl^+ and S_5^{2-}. The Tl^+ is coordinated by sulfur atoms from the anion to give pyramidal TlS_3 units, with the remaining site at Tl identified as being occupied by the inert pair.[361] There are both Tl—Te and Tl—Tl interactions in Tl_2Te_3 and Tl_5Te_2;[362,363] many of these compounds might be useful starting materials for synthetic work.

A range of subhalides has been reported in phase studies of TlF/TlF_3 mixtures,[364,365] but structural information is lacking. In the chloride system, the aqueous phase reaction of TlCl and $TlCl_3$ gives $Tl[TlCl_4]$, but the product of the analogous reaction in nitric acid is Tl_2Cl_3, whose structure has been identified as $Tl_3[TlCl_6]$, with $TlCl_6^{3-}$ octahedra [r(Tl—Cl) = 2.50–2.65 Å]; the Tl^+ ions are seven-, eight- or nine-coordinate with long (3.06–3.83 Å) Tl—Cl distances.[366] In the $Tl^I/Tl^{III}/Br^-$ system, $Tl[TlBr_4]$ and $Tl_3[TlBr_6]$ have been identified, and the structure of the former is confirmed by X-ray crystallography.[367] A dioxane (diox) adduct has been shown to be $[Tl(diox)][TlBr_4]$, in which the chains of Tl^+ are formed by dioxane bridging to give TlO_8 dodecahedra; the anion is tetrahedral[368] (*cf.* Section 25.2.8.5.2). The compound $Tl(CN)_2$, which has the structure $Tl[Tl(CN)_4]$, provides a further example of mixed oxidation states giving rise to TlX_2 stoichiometry.[1]

In addition to the halide systems there are a number of thallium(I) salts of thallium(III) complex anions in which the overall stoichiometry implies, incorrectly, the presence of thallium(II) species. These include the sulfate, selenate,[369] acetate[370] and oxalate[371] derivatives. An X-ray determination has confirmed that $Tl(OAc)_2$ is indeed $Tl[Tl(OAc)_4]$, in which the anions are linked through seven-coordinate Tl^+, with some evidence for a stereochemically active inert pair.[370] In $Tl[Tl(OH)(SO_4)_2]$, the structure consists of sheets of linked anions with Tl^+ ions.[372]

25.2.8 THALLIUM(III)

The oxidizing properties of thallium(III) in both aqueous and non-aqueous media obviously preclude the existence of a number of compounds analogous to those found for aluminum, gallium and indium in this oxidation state. Thus, while there is much richer chemistry for thallium in the +I state, there is a concomitant decrease in the detailed information on thallium(III) coordination chemistry.

25.2.8.1 Group IV Ligands

There is some evidence for the formation of Tl^{3+}/CN^- complexes in aqueous solution,[373] but no synthetic study of thallium(III) with this or other group IV ligands has apparently been reported, although the $[Tl(CN)_4]^-$ anion is known in $Tl(CN)_2$ (*cf.* Section 25.2.7). Organothallium(III) compounds are of course well known.[272]

25.2.8.2 Nitrogen and Other Group V Ligands

25.2.8.2.1 Ammonia and amine ligands

Ammoniates of thallium(III) are not stable species, although stability constant measurements suggest that $[Tl(OH)]_2^+$ forms strong complexes with ammonia in conc. ammonium nitrate solution.[374] A number of complexes of TlX_3 (X = Cl, Br, I) with ethylenediamine, 1,3-diaminopropane and similar ligands are known,[5,6] and in the case of X = Cl, Br, NO_3 or ClO_4, for example, the products are of the type $TlX_3{\cdot}3en$, suggesting that these are the ionic species $[Tlen_3]X_3$ with a six-coordinate Tl—N_6 complex cation. Spectroscopic results are available in some cases and have been reviewed elsewhere.[5,6]

25.2.8.2.2 N-heterocyclic ligands

As with amino ligands, adducts of TlX_3 (X = Cl, Br, I) with py, phen, bipy and terpy have been prepared.[5,6] The vibrational spectra of these and related adducts have been extensively reviewed; both neutral (*e.g.* $TlX_3{\cdot}py_2$) and ionic formulations (*e.g.* $[Tl_2X_2bipy_2][TlX_4]$) have been proposed.[375] The bipy and phen derivatives are very similar.[375,376] The dimeric structure of $TlCl_3{\cdot}phen$ has been shown to involve a TlN_2Cl_4 kernel, in which the phenanthroline nitrogen atoms, the thallium, and two chlorines are in a plane; dimerization involves donation from a neighbouring apical chlorine.[377] The related product $TlCl_3{\cdot}phen{\cdot}DMF$ apparently has the DMF ligand occupying this sixth coordination site.[376]

The compound $TlCl_3{\cdot}terpy$ is isomorphous with the gallium analogue (see Section 25.2.4.2.2),[375] although conductivity and spectroscopic results have been interpreted[378] in terms of the structure $[TlCl_2terpy][TlCl_4]$ (but see ref. 379). The problems implicit in using solution phase data to deduce solid state structures have been emphasized earlier, and the conflicts about this compound further illustrate this point. Anionic thallium(III) halogen species, such as $[TlCl_4(quin)_2]^-$ (quin = quinoline) and $[TlCl_3BrL_2]^-$ (L = pyridine, 3-picoline, quinoline) have been prepared,[380] as have the cationic complexes $[Tl(phen)_2X_2]^+$ (X = NO_3, Cl, Br, I).[381] In general, these compounds all point to the predominance of six-coordination in this aspect of the coordination chemistry of thallium(III).

25.2.8.2.3 Other nitrogen ligands

There is little to report here, since the oxidizing properties of thallium(III) prevent the formation of the neutral pseudohalides and similar compounds.[1] On the other hand, adducts of phen and bipy and $Tl(NCS)_3$ have been prepared (*e.g.* $Tl(NCS)_3bipy$) from the trichloride[382] suggesting that the presence of a neutral ligand may stabilize the +III state.

25.2.8.2.4 Other group V ligands

Oxidation of the ligand is again a severe problem in the synthesis of thallium(III) compounds with the heavier group V donors. The reaction of $TlCl_3{\cdot}py_2$ with Et_2PhP(= L) yields $TlCl_3{\cdot}L_{2.5}$, but attempts to extend this reaction to other phosphines gave thallium(I) species as the main product,[383] and it appears that the combination of oxidizing (Tl^{III}) and reducing (PR_3, AsR_3) reagents is such as to preclude preparative work in this area.

25.2.8.3 Oxygen Ligands

25.2.8.3.1 Oxide, hydroxide and H_2O as ligands

Thallium(III) oxide, Tl_2O_3, has the C—M_2O_3 rare earth sesquioxide structure, in which the metal is six-coordinate.[1] The corresponding hydroxide is apparently unknown. X-Ray studies have shown that $[Tl(OH_2)_6]^{3+}$ exists in solution (Tl—O = 2.235(5) Å),[384] and that the addition of Br^- brings about substitution to *trans*-$[TlBr_2(OH_2)_4]^+$ (Tl—Br = 2.481(2) Å), and $[TlBr_3(OH)_2]^{2+}$ which has a $TlBr_3O_2$ tbp kernel (Tl—O = 2.512(2) Å) (see also below). The solid $Tl(ClO_4)_3{\cdot}6H_2O$ also contains the exactly octahedral $[Tl(OH_2)_6]^{3+}$ cation (Tl—O =

2.23(2) Å),[385] while the tetrahydrates of $TlCl_3$ and $TlBr_3$ are based on distorted trigonal bipyramidal TlX_3O_2 units hydrogen bonded together through the remaining two water molecules.[386] The Tl—OH_2 bond lengths range from 2.428(15) to 2.602(16) Å.

Three oxide halides TlOX (X = F, Cl, Br) are known.[1] Preparations must be carried out at low temperatures[1,387] since the heavier oxyhalides in particular readily decompose. There is no structural information on either TlOCl or TlOBr, but X-ray crystallography shows that TlOF has a calcium fluoride structure, with Tl^{3+} ions in a cube of F^- or O^{2-} ions.[388]

25.2.8.3.2 Alkoxide, β-ketoenolates and related ligands

There appear to be no thallium(III) compounds with such ligands.

25.2.8.3.3 Oxyanion ligands

Thallium(III) nitrate, crystallized from Tl_2O_3 and conc. nitric acid as the hydrate, readily decomposes to $Tl_2O_3{\cdot}2H_2O$, and is a strong oxidizing agent,[389] having been used as such in organic synthesis. The crystal structure of $Tl(NO_3)_3{\cdot}3H_2O$ shows that the nitrate groups are bidentate and asymmetric (Tl—O = 2.299, 2.637 Å), but the water molecules are also bonded to the metal (Tl—O = 2.293 Å), giving a tricapped trigonal prismatic stereochemistry.[390] The anionic complex $NO[Tl(NO_3)_4]$ is formed in the reaction of N_2O_5 and $TlNO_3$.[391]

Thallium(III) orthophosphate, the only known phosphate of this oxidation state, is isostructural with $InPO_4$[392] with the metal atom coordinated by six oxygen atoms from six different PO_4^{3-} groups. The hydrate $TlPO_4{\cdot}2H_2O$ also involves six-coordination, with a $TlO_4(OH_2)_2$ kernel; again each oxygen is provided by a different phosphate.[393] The dichlorophosphate, $Tl(O_2PCl_2)_3$, prepared *via* $TlCl_3 + Cl_2O$ in $POCl_3$, is also believed to involve six-coordinate thallium with bidentate (O-bonded) chelating ligands.[394]

Solutions of Tl_2O_3 in dilute sulfuric acid eventually yield the sulfate $Tl_2(SO_4)_3{\cdot}7H_2O$, which at high pH goes to the basic $Tl_2(OH)_2(SO_4)_2{\cdot}4H_2O$ whose structure is based on chains of TlO_6 octahedra.[395] Crystallization from conc. sulfuric acid gives $Tl_2(SO_4)_2{\cdot}4H_2O$, which on heating reverts to $Tl_2(SO_4)_3$. A number of double sulfates have been reported, and many of these no doubt involve Tl^{III} sulfate complexes (*cf.* Section 25.2.4.4.4) but structural data are lacking at present. Thallium(III) selenate is similar to the sulfate. Thallium(III) tellurate, Tl_6TeO_{12}, consists of TeO_6 octahedra and Tl^{III} coordinated to seven oxygen atoms in a distorted monocapped trigonal prismatic stereochemistry.[311]

Thallium(III) perchlorate is obtained from $TlCl_3 + AgClO_4$, or by the anodic oxidation of $TlClO_4$. Complexing by perchlorate appears very unlikely and the same applies to chlorate, bromate or iodate.[1–3]

25.2.8.3.4 Carboxylate ligands

Thallium(III) carboxylates are readily formed by the reaction of $Tl(OAc)_3$ and the appropriate acid,[1] while $Tl(OAc)_3$ itself is prepared from Tl_2O_3 and glacial acetic acid.[396] The structure of $Tl(OAc)_3$ shows that the thallium atom is coordinated by three chelate acetate groups (Tl—O = 2.30 Å, av.), but two longer Tl—O bonds (2.57 Å) link adjacent molecules into chains of TlO_8 units.[397] The monohydrate is related to the anhydrate, but the eight-coordination is achieved by both acetate and water coordination.[398] The compound $PdTl(OAc)_5$ has palladium bonded directly to thallium but the metals are also bridged by four bidentate carboxylates.[399] There is an extensive literature on the use of thallium(III) carboxylates in organic synthesis.[274,275]

The hydrated double oxalate of rubidium and thallium has been formulated as $Rb[Tlox_2(H_2O)_2]{\cdot}2H_2O$. Water is lost on heating, and Tl^{III} undergoes reduction to Tl^I, so that final products are Tl^I oxide and Rb_2CO_3.[400]

25.2.8.3.5 Ethers and other oxygen donor ligands

Relatively few compounds of thallium(III) with simple organic oxygen donors are known, and in particular the adducts of TlX_3 do not appear to have been systematically investigated.[5]

Two recent crystal structures suggest that this situation should be rectified. The adduct $TlCl_3 \cdot L_2$ (L = 4-pyridinecarbonitrile *N*-oxide) has a trigonal bipyramidal structure, with axial oxygen donors,[401] while in $TlBr_3 \cdot$diox the same stereochemistry arises from planar $TlBr_3$ units bridged by 1,4-dioxane molecules in the chair conformation.[402]

Cationic complexes of Tl^{3+} with oxygen donors are known, and can be regarded as substitution products of $[Tl(OH_2)_6]^{3+}$ (*cf.* Section 25.2.8.3.1). The DMSO product has been characterized with various anions,[403] and there seems no reason to suppose that similar compounds cannot be prepared with donor ligands which are resistant to oxidation. The cation $[Tl(pyNO)_8]^{3+}$ (prepared as the perchlorate) may be an indication that eight-coordinate compounds are also accessible in thallium(III) chemistry.[404]

25.2.8.4 Sulfur and other Heavier Group VI Ligands

The oxidizing properties of Tl^{III} are held to be partly responsible for the absence of a binary sulfide Tl_2S_3, although Tl_2Se_3 and Tl_2Te_3 can be prepared by heating the elements,[1] which suggests that oxidation is not the only factor involved. The existence of TlS_4^{5-} tetrahedra in Tl_4S_3 was noted earlier, and the information on Tl_2Te_3 suggests that it is not a simple binary compound (see Section 25.2.7).

Compounds of thallium(III) with monodentate sulfur donors have not been reported, again no doubt because of the redox chemistry involved, but a number of compounds with bidentate sulfur ligands are known. With maleonitriledithiolate (= MNT^{2-}, *cf.* Section 25.2.4.5.2), the anions $[Tl(MNT)_2]^-$, $[Tl(MNT)_2Br]^{2-}$ and $[Tl(MNT)_3]^{3-}$ are all known,[218] implying four-, five- and six-coordination at the metal, although the Tl—S_4 species do not add neutral bidentate nitrogen donors to give Tl—S_4N_2 kernels.[405] Surprisingly, the reaction of TDT^{2-} (the dianion of toluenedithiol) with thallium(I) yields the thallium(III) anion $[Tl(TDT)_2]^-$, suggesting complex redox chemistry in such systems. Stepwise polarographic reduction was not observed for Tl—S_4 or Tl—S_6 complexes.[405] Other anionic complexes reported include $(Ph_4P)_3[Tl(S_2C{=}N{-}C{\equiv}N)_3]$[406] and $R_4N[TlL_2]$ (L = *cis*-ethanedithiolate or 4,5-dithiolate-2,3-xylene).[217] It seems that there are a number of interesting problems awaiting resolution in this area.

25.2.8.5 Halogen Ligands

25.2.8.5.1 Fluoride ligand

Crystalline TlF_3, obtained from the reaction of Tl_2O_3 with F_2, BrF_3 or SF_4,[1] forms orthorhombic crystals in which the Tl^{3+} ions have distorted trigonal prismatic coordination.[407] Stability constant measurements show that Tl^{3+} does not complex with fluoride ion in aqueous solution,[289] while in liquid HF, the equilibrium identified is $TlF_3(s) + F^-(soln) \rightleftarrows TlF_4^-$.[408] In keeping with these solution phase results, the solid phases in the system TlF/TlF_3 do not contain any Tl^{III}/F anionic complexes,[354] and the compounds $MTlF_4$ (M = Li, Na) are true double salts.[409] On the other hand, the compounds M_3TlF_6 (M = Na, K) contain MF_6^{3-} anions;[410,411] a number of other species were identified in these MF/TlF_3 phase studies, including $Na_3Tl_2F_3$, $Na_5Tl_9F_{32}$, $K_5Tl_3F_{14}$ and $K_2Tl_7F_{23}$.[411]

25.2.8.5.2 Chloride, bromide, iodide and mixed halide ligands

Thallium(III) chloride can be prepared by bubbling chlorine through a suspension of TlCl in acetonitrile.[412] The corresponding bromide is unstable in the solid state, and the iodide does not exist[1] (see Section 25.2.6.5.2), and here again the oxidizing properties of the +III state are critical. Hydrates of $TlCl_3$ and $TlBr_3$ can be prepared by oxidation of TlX with X_2; the stable room temperature species are $TlCl_3 \cdot 4H_2O$ and $TlBr_3 \cdot 4H_2O$. Non-aqueous solutions of TlX_3 (X = Cl, Br) can be obtained by similar methods. The structure of the hydrates was discussed earlier (Section 25.2.8.3.1).

Various adducts of TlX_3 have been noted earlier in this chapter, and a number of anionic complexes with coordination numbers of four, five or six have also been prepared. Salts of $TlCl_4^-$ and $TlCl_3X^-$ (X = Br, I) are readily formed from TlX_3 and (say) R_4NCl in non-aqueous media[412–415] and X-ray studies of a number of such salts[415–417] have confirmed the tetrahedral, or slightly distorted tetrahedral, structure predicted from vibrational[418] and electronic[414] spectroscopy. The five-coordinate anion $TlCl_5^{2-}$ has been identified in aqueous[419] and non-aqueous solution[256] and in the solid state, and is presumably isostructural with $InCl_5^{2-}$ (Section 25.2.4.6.2). The octahedral $TlCl_6^{3-}$ has been identified in a number of salts, although the stoichiometry M_3TlCl_6 is not a sure guide, and X-ray structural studies have been reported.[417,420] The Tl—Cl bond is significantly longer in $TlCl_6^{3-}$ (2.59 Å) than in $TlCl_4^-$ (2.43 Å). The structure of the salt $(pyH)_3Tl_2Cl_9$ is reported to involve $TlCl_4^-$ and $TlCl_5^{2-}$ anions on the basis of Raman and ^{205}Tl NMR spectroscopy,[421] although a previous spectroscopic and normal coordinate analysis of $Cs_3Tl_2Cl_9$ confirmed the $[Cl_3TlCl_3TlCl_3]^{3-}$ structure in this salt.[422] The use of ^{205}Tl NMR to identify $TlCl_5^{2-}$ and $TlCl_6^{3-}$ in both solid and solution should also be noted as an important use of this method.[421,423,424]

Salts of $TlBr_4^-$ can be prepared by the methods outlined above for $TlCl_4^-$. The vibrational spectra in the solid state,[424] and in aqueous[419] and non-aqueous solution,[425,426] support the expected tetrahedral structure, and this has been confirmed by later X-ray studies.[427,428] An interesting result from these experiments is that the salt $RbTlBr_4{\cdot}H_2O$ involves $TlBr_4^-$ rather than a five-coordinate hydrated anion. The $TlBr_5^{2-}$ anion has been identified, and the $TlBr_6^{3-}$ anion is assigned O_h symmetry from vibrational spectra.[424,429] The only reported anionic iodo complex is TlI_4^-, for which the tetrahedral geometry has been confirmed by vibrational spectroscopy,[424,425] and by X-ray studies of solid and solution.[430] The preparation of the mixed halogeno complexes $TlCl_3X^-$ was noted earlier and the electronic spectra of these and other $TlX_{4-n}Y_n^-$ have been discussed.[404]

25.2.8.6 Hydrogen and Hydride Ligands

It is likely that the previously reported hydrides TlH_3 and $(TlH_3)_n$ do not exist,[431,432] but the reaction of $TlCl_3$ and LiH in ether at −15 °C yields $LiTlH_4$ which decomposes above ~30 °C,[435] and is the least stable of the group III $LiMH_4$ compounds. The reaction between $TlCl_3$ and $LiMH_4$ (M = Al, Ga) at very low temperatures gives $Tl(MH_4)_3$ which decomposes above −115 °C (Al) or −95 °C (Ga). When $TlCl_3$ is heated with AlH_3 or $LiBH_4$ at −115 °C, the product is $TlCl(MH_4)_2$ (M = B, Al) which decomposes at −95 °C. In these circumstances, it is not surprising that there is no structural information on these very unstable substances.

25.2.8.7 Mixed Donor Atom Ligands

There is nothing significant to report under this heading.

25.2.8.8 Macrocyclic Ligands

Inorganic derivatives of thallium(I) and thallium(III) are not known, but the cationic complex $[TlMe_2(crown)]^+$ (crown = dibenzo-18-crown-6) has been prepared as the picrate salt.[434] X-Ray crystallography has demonstrated the existence of two isomers of the analogous complex with dicyclohexano-18-crown-6; the Me—Tl—Me group is orthogonal to the average ring plane.[435,436]

The porphyrin derivative of thallium(III) has been found[437] to have the metal atom above the N_4 plane by 0.9 Å. Electrochemical two-electron reduction results in reduction to Tl^IL^- (L = porphyrin), and subsequent protonation gives H_2L and Tl^+ in aqueous solution.[438]

25.2.9 REFERENCES

1. K. Wade and A. J. Banister, 'Comprehensive Inorganic Chemistry', ed. J. C. Bailer, Jr., H. J. Emeléus, R. Nyholm and A. F. Trotman-Dickenson, Pergamon, Oxford, 1973, vol. 1, p. 993.
2. J. W. Mellor, 'Comprehensive Treatise on Inorganic and Theoretical Chemistry', Longmans, London, 1930, vol. 5, p. 387.
3. Gmelins Handbuch der Anorganischen Chemie, System No. 37, Springer Verlag, Berlin, 1936.
4. P. Pascal, 'Nouveau Traité de Chimie Minérale', Masson, Paris, 1961, vol. 6.

5. A. Pidcock, in 'MTP International Review of Science, Inorganic Chemistry, Series 2', ed. M. F. Lappert, 1975, vol. 1, p. 281.
6. D. G. Tuck, in ref. 5, vol 1, p. 311.
7. A. J. Carty and D. G. Tuck, *Prog. Inorg. Chem.*, 1975, **19,** 245.
8. D. G. Tuck, in 'Comprehensive Organometallic Chemistry', ed. G. Wilkinson, F. G. A. Stone and E. W. Abel, Pergamon, Oxford, 1982, vol. 1, p. 683.
9. D. G. Tuck, *Pure Appl. Chem.*, 1983, **55,** 1477.
10. A. P. Kochetkova, V. G. Tronev and V. N. Gilyarov, *Dokl. Akad. Nauk SSSR,* 1962, **147,** 1086.
11. C. Peppe, D. G. Tuck and L. Victoriano, *J. Chem. Soc., Dalton Trans.*, 1982, 2165.
12. P. L. Goggin and I. J. McColm, *J. Inorg. Nucl. Chem.*, 1966, **28,** 2501.
13. J. J. Habeeb and D. G. Tuck, *J. Chem. Soc., Dalton Trans.*, 1976, 866.
14. J. J. Habeeb and D. G. Tuck, *J. Chem. Soc., Dalton Trans.*, 1975, 1815.
15. R. G. Edgell and A. F. Orchard, *J. Chem. Soc., Faraday Trans. 2,* 1978, 1179.
16. J. G. Contreras and D. G. Tuck, *Chem. Commun.*, 1971, 1552.
17. J. G. Contreras, J. S. Poland and D. G. Tuck, *J. Chem. Soc., Dalton Trans.*, 1973, 922.
18. J. G. Contreras and D. G. Tuck, *Inorg. Chem.*, 1972, **11,** 2967.
19. J. J. Habeeb and D. G. Tuck, *J. Chem. Soc., Chem. Commun.*, 1975, 601.
20. J. J. Habeeb and D. G. Tuck, *J. Chem. Soc., Chem. Commun.*, 1975, 808.
21. (a) R. S. Taylor and A. G. Sykes, *J. Chem. Soc. (A)*, 1969, 2419; (b) *J. Chem. Soc. (A)*, 1971, 1628.
22. A. N. Red'Kin, L. G. Dubovitskaya and V. A. Smirnov, *Zh. Neorg. Khim.*, 1982, **27,** 627.
23. V. A. Smirnov, *Zh. Neorg. Khim.*, 1983, **28,** 772.
24. M. J. Taylor, D. G. Tuck and L. Victoriano, *J. Chem. Soc., Dalton Trans.*, 1981, 928.
25. I. Sinclair and I. J. Worrall, *Can. J. Chem.*, 1982, **60,** 695.
26. M. J. Taylor, D. G. Tuck and L. Victoriano, *Can. J. Chem.*, 1982, **60,** 691.
27. M. A. Khan, C. Peppe and D. G. Tuck, *Can. J. Chem.*, 1984, **62,** 701.
28. C. Peppe and D. G. Tuck, *Can. J. Chem.*, 1984, **62,** 2793.
29. A. P. Kochetkova and O. N. Gilyarov, *Zh. Neorg. Khim.*, 1966, **11,** 1239.
30. D. G. Tuck and M. K. Yang, *J. Chem. Soc. (A)*, 1971, 3100.
31. J. J. Habeeb and D. G. Tuck, *J. Chem. Soc., Dalton Trans.*, 1973, 243.
32. D. G. Tuck and M. K. Yang, *J. Chem. Soc. (A)*, 1971, 214.
33. J. H. C. Hogg and W. J. Duffin, *Acta Crystallogr.*, 1967, **23,** 111.
34. J. H. C. Hogg, *Acta Crystallogr., Sect. B,* 1971, **27,** 1630.
35. J. H. C. Hogg, H. H. Sutherland and D. J. Williams, *Chem. Commun.*, 1971, 1568.
36. T. Wadsten, L. Arnberg and J. E. Berg, *Acta Crystallogr., Sect. B,* 1980, **36,** 2220.
37. H. P. Beck, *Z. Naturforsch., Teil B,* 1984, **39,** 310.
38. M. A. Khan and D. G. Tuck, *Inorg. Chim. Acta,* 1985, **97,** 73.
39. H. P. Beck, unpublished results.
40. J. J. Kenney and F. X. Powell, *J. Phys. Chem.*, 1968, **72,** 3094.
41. F. J. Brinkman and H. Gerding, *Recl. Trav. Chim. Pays-Bas,* 1969, **88,** 275.
42. J. H. R. Clarke and R. E. Hester, *Inorg. Chem.*, 1969, **8,** 1113.
43. A. W. Atkinson, J. R. Chadwick and E. Kinsella, *J. Inorg. Nucl. Chem.*, 1968, **30,** 401.
44. J. R. Chadwick, A. W. Atkinson and B. G. Huckstepp, *J. Inorg. Nucl. Chem.*, 1966, **28,** 1021.
45. A. W. Atkinson and B. O. Field, *J. Inorg. Nucl. Chem.*, 1968, **30,** 3177.
46. G. Meyer, *Z. Anorg. Allg. Chem.*, 1978, **445,** 140.
47. G. Meyer and R. Blachnik, *Z. Anorg. Allg. Chem.*, 1983, **503,** 126.
48. P. L. Goggin and I. J. McColm, *J. Less-Common Met.*, 1966, **11,** 292.
49. B. H. Freeland, J. L. Hencher, D. G. Tuck and J. G. Contreras, *Inorg. Chem.*, 1976, **15,** 2144.
50. B. R. McGarvey, C. O. Trudell, D. G. Tuck and L. Victoriano, *Inorg. Chem.*, 1980, **19,** 3432.
51. P. L. Radloff and G. N. Papatheodorou, *J. Chem. Phys.*, 1980, **72,** 992.
52. P. L. Goggin, I. J. McColm and R. Shore, *J. Chem. Soc. (A)*, 1966, 1314.
53. D. G. Tuck and E. J. Woodhouse, *J. Chem. Soc.*, 1964, 6017.
54. J. Muller, F. Schmock, A. Klopsch and K. Dehnicke, *Chem. Ber.*, 1975, **108,** 664.
55. H. Burger and U. Goetz, *Angew. Chem., Int. Ed. Engl.*, 1969, **8,** 140.
56. N. S. Vyazankin, G. A. Razuvaev and O. A. Kruglaya, *Organomet. Chem. Rev., Sect. A,* 1968, **3,** 323, 354.
57. A. T. Weibel and J. P. Oliver, *J. Am. Chem. Soc.*, 1972, **94,** 8590.
58. C. Peppe and D. G. Tuck, *Can. J. Chem.*, 1984, **62,** 2798.
59. T. Annan and D. G. Tuck, *J. Organomet. Chem.*, in press.
60. R. S. Steevensz, D. G. Tuck, H. A. Meinema and J. G. Noltes, *Can. J. Chem.*, 1985, **63,** 755.
61. P. Gans and J. B. Gill, *J. Chem. Soc., Dalton Trans.*, 1976, 779.
62. A. J. Carty and D. G. Tuck, *J. Chem. Soc.*, 1964, 6012.
63. R. Ochoa and H. M. Haendler, *Inorg. Chim. Acta,* 1969, **3,** 441.
64. D. G. Tuck, E. J. Woodhouse and P. Carty, *J. Chem. Soc. (A)*, 1966, 1077.
65. S. J. Patel, D. B. Sowerby and D. G. Tuck, *J. Chem. Soc. (A)*, 1967, 1187.
66. G. E. Coates and R. A. Whitcombe, *J. Chem. Soc.*, 1956, 3351.
67. (a) G. J. Sutton, *Aust. J. Chem.*, 1961, **14,** 37; (b) *Aust. Chem. Inst. J. Proc.*, 1948, **15,** 356.
68. R. A. Walton, *J. Chem. Soc. (A)*, 1967, 1485.
69. M. A. Khan, C. Peppe and D. G. Tuck, *J. Organomet. Chem.*, 1985, **280,** 17.
70. R. A. Walton, *Inorg. Chem.*, 1968, **7,** 640.
71. F. Ya. Kul'ba, Yu. A. Makashev and N. I. Fedyaev, *Zh. Neorg. Khim.*, 1971, **16,** 2352.
72. F. Ya. Kul'ba, V. L. Stolyarov and A. P. Zharkov, *Zh. Neorg. Khim.*, 1972, **17,** 378.
73. D. H. Brown and D. T. Stewart, *J. Inorg. Nucl. Chem.*, 1970, **32,** 3751.
74. (a) G. J. Sutton, *Aust. Chem. Inst. J. Proc.*, 1949, **16,** 115; (b) *Aust. J. Chem.*, 1961, **14,** 37; (c) *Aust. J. Chem.*, 1967, **20,** 1859.

75. A. J. Carty and D. G. Tuck, *J. Chem. Soc. (A)*, 1966, 1081.
76. S. J. Patel and D. G. Tuck, *J. Chem. Soc. (A)*, 1968, 1870.
77. R. A. Walton, *J. Chem. Soc. (A)*, 1969, 61.
78. G. Beran, K. Dymock, H. A. Patel, A. J. Carty and P. M. Boorman, *Inorg. Chem.*, 1970, **11**, 896.
79. K. Dymock and G. J. Palenik, unpublished results.
80. H. Burger, J. Cichon, U. Goetze, U. Wannagat and H. J. Wismar, *J. Organomet. Chem.*, 1971, **33**, 1.
81. D. C. Bradley and Y. C. Gao, *Polyhedron*, 1982, **1**, 307.
82. V. I. Scherbakov, S. F. Zhiltsov and O. N. Druzhkov, *Zh. Obshch. Khim.*, 1970, **40**, 1542.
83. K. Mertz, W. Schwarz, B. Eberwein, J. Weidlein, H. Hess and H. D. Hausen, *Z. Anorg. Allg. Chem.*, 1977, **429**, 99.
84. O. T. Beachley, C. Bueno, M. R. Churchill, R. B. Hallock and R. G. Simmons, *Inorg. Chem.*, 1981, **20**, 2423.
85. F. Weller and U. Mueller, *Chem. Ber.*, 1979, **112**, 2039.
86. J. Muller and K. Dehnicke, *J. Organomet. Chem.*, 1968, **12**, 37.
87. N. Roder and K. Dehnicke, *Chimia*, 1974, **28**, 349.
88. A. M. Golub, G. V. Tsintsadze and T. L. Makhatadze, *Soobshch. Akad. Nauk Gruz. SSR.*, 1961, **61**, 57 (*Chem. Abstr.*, 1971, **74**, 134 282).
89. J. J. Habeeb and D. G. Tuck, *J. Chem. Soc., Dalton Trans.*, 1973, 96.
90. H. H. Schmidtke and D. Garthoff, *Z. Naturforsch., Teil A*, 1969, **24**, 126.
91. F. W. B. Einstein, M. M. Gilbert, D. G. Tuck and P. L. Vogel, *Acta Crystallogr., Sect. B*, 1976, **32**, 2234.
92. V. F. Mikitchenko, V. V. Skopeno and G. V. Tsintsadze, *Ukr. Khim. Zh. (Russ. Ed.)*, 1971, **37**, 878 (*Chem. Abstr.*, 1972, **76**, 67 562).
93. V. V. Skopenko, V. F. Mikitchenko and G. V. Tsintsadze, *Zh. Neorg. Khim.*, 1969, **14**, 1790.
94. G. V. Tsintsadze, V. V. Skopenko, V. F. Mikitchenko and Z. Zhumabaev, *Dopo. Akad. Nauk. Ukr. RSR, Ser. B*, 1969, **31**, 130.
95. J. Reedjik and W. L. Groenveld, *Recl. Trav. Chim. Pays-Bas*, 1969, **87**, 552.
96. J. J. Habeeb, F. F. Said and D. G. Tuck, *J. Chem. Soc., Dalton Trans.*, 1980, 1161.
97. J. J. Habeeb, F. F. Said and D. G. Tuck, *J. Chem. Soc., Dalton Trans.*, 1981, 118.
98. M. V. Veidis and G. J. Palenik, *Chem. Commun.*, 1970, 1182.
99. R. S. Nyholm and K. Ulm, *J. Chem. Soc.*, 1965, 4199.
100. R. Didchenko, J. E. Alix and R. H. Toeniskoetter, *J. Inorg. Nucl. Chem.*, 1960, **14**, 35.
101. O. T. Beachley and G. E. Coates, *J. Chem. Soc.*, 1965, 3241.
102. L. M. Nemirovskii, B. I. Kozyrkin, A. F. Lanstov, B. G. Gribov, I. M. Skovortsov and I. A. Sredinakaya, *Dokl. Akad. Nauk SSSR*, 1974, **214**, 590.
103. T. H. Cannon and R. E. Richards, *Trans. Faraday Soc.*, 1966, **62**, 1378.
104. A. Fratiello, R. E. Lee, V. M. Nishida and R. E. Schuster, *J. Chem. Phys.*, 1968, **48**, 3705.
105. J. Celeda and D. G. Tuck, *J. Inorg. Nucl. Chem.*, 1974, **36**, 373.
106. M. Maeda and H. Ohtaki, *Bull. Chem. Soc. Jpn.*, 1977, **50**, 1893.
107. G. E. Glass, W. B. Shwabacher and R. S. Tobias, *Inorg. Chem.*, 1968, **7**, 2471.
108. A. M. Golub and V. M. Samoilenko, *Zh. Neorg. Khim.*, 1965, **10**, 328.
109. A. Fratiello, D. D. Davis, S. Peak and R. R. Schuster, *Inorg. Chem.*, 1971, **10**, 1627.
110. J. Crea and S. F. Lincoln, *Inorg. Chem.*, 1972, **11**, 1131.
111. T. Moeller, *J. Am. Chem. Soc.*, 1941, **63**, 2625.
112. N. Paskalev, *C. R. Acad. Bulg. Sci.*, 1967, **20**, 549 (*Chem. Abstr.*, 1967, **67**, 85 549s).
113. P. L. Brown, J. Ellis and R. N. Sylva, *J. Chem. Soc., Dalton Trans.*, 1982, 1911.
114. G. Biedermann and D. Ferri, *Acta Chem. Scand., Ser. A*, 1982, **36**, 611.
115. K. Schubert and A. Seitz, *Z. Anorg. Allg. Chem.*, 1948, **256**, 226.
116. A. N. Christensen, N. C. Broch, O. von Heidenstam and A. Nilsson, *Acta Chem. Scand.*, 1967, **21**, 1046.
117. D. F. Mullica, G. W. Beall, W. O. Milligan, J. D. Korp and I. Bernal, *J. Inorg. Nucl. Chem.*, 1979, **41**, 277.
118. N. V. Aksel'rud and V. B. Spirakovskii, *Zh. Neorg. Khim.*, 1959, **4**, 989.
119. L. C. A. Thompson and R. Pacer, *J. Inorg. Nucl. Chem.*, 1963, **25**, 1041.
120. N. V. Aksel'rud, *Dokl. Akad. Nauk SSSR*, 1960, **132**, 1067.
121. B. N. Ivanov-Emin, L. A. Nisel'son and Yu. Greksa, *Zh. Neorg. Khim.*, 1960, **5**, 1966.
122. R. Roy and M. W. Shafer, *J. Phys. Chem.*, 1954, **58**, 372.
123. A. N. Christensen, R. Gronback and S. E. Rasmussen, *Acta Chem. Scand.*, 1964, **18**, 1261.
124. B. Siegel, *Inorg. Chim. Acta*, 1968, **2**, 137.
125. N. V. Aksel'rud and V. B. Spivakovskii, *Zh. Neorg. Khim.*, 1959, **4**, 989.
126. H. E. Forsberg, *Acta Chem. Scand.*, 1957, **11**, 676.
127. S. Chatterjee, S. R. Bindal and R. C. Mehrotra, *J. Indian Chem. Soc.*, 1976, **53**, 867.
128. A. Mehrotra and R. C. Mehrotra, *Inorg. Chem.*, 1972, **11**, 2170.
129. H. C. Clark and A. L. Pickard, *J. Organomet. Chem.*, 1968, **13**, 61.
130. H. Schmidbaur and F. Schindler, *Chem. Ber.*, 1966, **99**, 2178.
131. H. Schmidbaur, B. Armer and M. Bergfeld, *Z. Chem.*, 1968, **8**, 254.
132. G. T. Morgan and H. D. K. Drew, *J. Chem. Soc.*, 1921, **119**, 1058.
133. A. Nikolovski, V. Guceva, I. Petrov and B. Soptrajanov, *Bull. Sci. Cons. Acad. Sci. Arts RSF Yugoslav., Sect. A*, 1971, **16**, 337.
134. R. E. Sievers and J. W. Connolly, *Inorg. Synth.*, 1970, **12**, 72.
135. G. T. Tanner, E. J. Wells and D. G. Tuck, *Can. J. Chem.*, 1972, **50**, 3950.
136. J. J. Habeeb and D. G. Tuck, *Inorg. Synth.*, 1979, **19**, 257.
137. B. E. Bryant, W. C. Fernelius and B. E. Douglas, *J. Am. Chem. Soc.*, 1953, **75**, 3784.
138. E. L. Muetterties and C. M. Wright, *J. Am. Chem. Soc.*, 1954, **86**, 5132.
139. L. A. Gribov, Yu. A. Zolotov and M. P. Noskova, *Zh. Strukt. Khim.*, 1968, **9**, 448.
140. M. P. Noskova, L. A. Gribov and Yu. A. Zolotov, *Zh. Strukt. Khim.*, 1969, **10**, 474.
141. R. E. Hester and R. A. Plane, *Inorg. Chem.*, 1964, **3**, 63.

142. R. H. Holm and F. A. Cotton, *J. Am. Chem. Soc.*, 1958, **80,** 5658.
143. J. A. S. Smith and J. D. Thwaites, *Discuss. Faraday Soc.*, 1962, **34,** 143.
144. A. J. Carty, D. G. Tuck and E. Bullock, *Can. J. Chem.*, 1965, **43,** 2559.
145. J. A. S. Smith and E. J. Wilkins, *J. Chem. Soc. (A)*, 1966, 1749.
146. J. G. Rodriguez, F. H. Cano and S. Garcia–Blanco, *Cryst. Struct. Commun.*, 1979, **8,** 53.
147. G. J. Palenik and K. R. Dymock, *Acta Crystallogr., Sect. B*, 1980, **36,** 2059.
148. J. G. Contreras and D. G. Tuck, *J. Chem. Soc., Dalton Trans.*, 1974, 1249.
149. J. G. Contreras, F. W. B. Einstein and D. G. Tuck, *Can. J. Chem.*, 1974, **52,** 3793.
150. H. L. Chung and D. G. Tuck, *Can. J. Chem.*, 1974, **52,** 3944.
151. R. C. Fay and T. S. Piper, *J. Am. Chem. Soc.*, 1963, **85,** 500.
152. R. C. Fay and T. S. Piper, *Inorg. Chem.*, 1964, **3,** 348.
153. M. Yamazaki and T. Takeuchi, *Nippon Kagaku Kaishi*, 1970, **91,** 965, 973.
154. M. Yamazaki and T. Takeuchi, *Kogyo Kagaku Zasshi*, 1970, **73,** 2634.
155. J. G. Gordon, M. J. O'Connor and R. H. Holm, *Inorg. Chim. Acta*, 1971, **5,** 381.
156. G. T. Tanner, D. G. Tuck and E. J. Wells, *Can. J. Chem.*, 1972, **50,** 3950.
157. A. M. Golub, R. Akmyradov and S. L. Uskova, *Izv. Vyssh. Uchebn. Zaved., Khim. Khim. Tekhnol.*, 1974, **17,** 1287 (*Chem. Abstr.*, 1975, **82,** 77 693d).
158. B. O. Field and C. J. Hardy, *J. Chem. Soc.*, 1964, 4428.
159. M. Schmeisser and G. Kohler, *Angew. Chem.*, 1965, **77,** 456.
160. B. N. Ivanov–Emin, Z. K. Odinets, V. A. Belonosov and B. E. Zaitsev, *Zh. Neorg. Khim.*, 1973, **18,** 1181.
161. R. E. Hester, R. A. Plane and G. E. Walrafen, *J. Chem. Phys.*, 1963, **38,** 249.
162. R. E. Hester and R. A. Plane, *Inorg. Chem.*, 1964, **3,** 769.
163. E. N. Deichman, G. V. Rodicheva and P. A. Chel'tsov, *Zh. Neorg. Khim.*, 1965, **10,** 89.
164. L. N. Filatova and T. N. Kurdyumova, *Zh. Neorg. Khim.*, 1974, **19,** 3190.
165. E. N. Deichmann and G. V. Rodicheva, *Zh. Neorg. Khim.*, 1970, **15,** 2081.
166. I. Yu. Manevich, Yu. A. Buslaev, K. A. Andrianov and G. I. Matrosov, *Izv. Akad. Nauk SSSR, Neorg. Mater.*, 1965, **1,** 1241.
167. H. Mueller and K. Dehnicke, *Z. Anorg. Allg. Chem.*, 1967, **350,** 231.
168. E. M. Wolterman and J. R. Wasson, *Inorg. Nucl. Chem. Lett.*, 1970, **6,** 475.
169. W. Kuchen and H. Hertel, *Angew. Chem., Int. Ed. Engl.*, 1967, **6,** 175.
170. W. Kuchen and H. Hertel, *Chem. Ber.*, 1968, **101,** 1991.
171. C. M. Mikulski, N. M. Karayannis, J. V. Minkiewicz, L. L. Pytlewski and M. M. Labes, *Inorg. Chim. Acta*, 1969, **3,** 523.
172. G. Pannetier, J. M. Manoli and P. Herpin, *Bull. Soc. Chim. Fr.*, 1972, 485.
173. B. Jolibois, G. Laplace, F. Abraham and G. Nowogrocki, *Acta Crystallogr., Sect. B*, 1980, **36,** 2517.
174. N. N. Mukhtarova, R. K. Rastsvetaeva and V. V. Ilyukhim, *Dokl. Akad. Nauk SSSR*, 1978, **239,** 322.
175. J. Tudo, B. Jolibois, G. Laplace, G. Nowogrocki and F. Abraham, *Acta Crystallogr., Sect. B*, 1979, **35,** 1580.
176. R. Caminiti, G. Marongiu and G. Paschina, *Cryst. Struct. Commun.*, 1982, **11,** 955.
177. G. Johansson, *Acta Chem. Scand.*, 1961, **15,** 1437.
178. R. C. Paul, R. D. Sharma, S. Singh and R. D. Verma, *J. Inorg. Nucl. Chem.*, 1981, **43,** 1919.
179. E. N. Soldatov, E. A. Kuzinin and N. V. Kadoshnikova, *Dokl. Akad. Nauk SSSR*, 1978, **240,** 362.
180. E. A. Soldatov, E. A. Kuzmin and V. V. Ilyukhin, *Zh. Strukt. Khim.*, 1979, **20,** 1081.
181. J. D. Donaldson, J. F. Knifton and S. D. Ross, *Spectrochim. Acta*, 1964, **20,** 847.
182. W. Lindel and F. Huber, *Z. Anorg. Allg. Chem.*, 1974, **408,** 167.
183. P. Sartori, J. Fazekas and J. Schnackers, *J. Fluorine Chem.*, 1972, **1,** 463.
184. H. Preut and F. Huber, *Z. Anorg. Allg. Chem.*, 1979, **450,** 120.
185. M. A. Khan, C. Peppe and D. G. Tuck, *Acta Crystallogr., Sect. C*, 1983, **39,** 1339.
186. F. W. B. Einstein, M. M. Gilbert and D. G. Tuck, *J. Chem. Soc., Dalton Trans.*, 1973, 248.
187. H. D. Hausen, *J. Organomet. Chem.*, 1972, **39,** C37.
188. Yu. A. Aleksandrov, N. V. Chikinova, G. I. Makin, N. V. Kornilova and V. I. Bregadze, *Zh. Obshch. Khim.*, 1978, **48,** 467.
189. I. M. Viktorova, N. I. Sheverdina, A. N. Rodionov and F. A. Kocheshkov, *Dokl. Akad. Nauk SSSR*, 1969, 315.
190. J. Weidlein, *Z. Anorg. Allg. Chem.*, 1970, **378,** 245.
191. J. J. Habeeb and D. G. Tuck, *Can. J. Chem.*, 1974, **52,** 3950.
192. J. J. Habeeb and D. G. Tuck, *J. Organomet. Chem.*, 1975, **101,** 1.
193. T. Moeller, *J. Am. Chem. Soc.*, 1940, **62,** 2444.
194. N. Bulc, L. Golic and J. Siftar, *Acta Crystallogr., Sect. C*, 1983, **39,** 176.
195. E. N. Deichmann, *Zh. Neorg. Khim.*, 1959, **4,** 2617.
196. E. N. Deichmann and G. V. Rodicheva, *Zh. Neorg. Khim.*, 1964, **9,** 807.
197. Ya. Ya. Kharitonov and E. N. Deichmann, *Zh. Neorg. Khim.*, 1965, **10,** 853.
198. N. Bulc and L. Golic, *Acta Crystallogr., Sect. C*, 1983, **39,** 174.
199. C. A. Tolman, *J. Am. Chem. Soc.*, 1970, **92,** 2956.
200. F. W. B. Einstein and D. G. Tuck, *Chem. Commun.*, 1970, 1182.
201. H. Mabrouk and D. G. Tuck, unpublished results.
202. R. Diehl and C.-D. Carpentier, *Acta Crystallogr., Sect. B*, 1978, **34,** 1097.
203. N. V. Porotnikov, O. I. Kondratov and K. I. Petrov, *Zh. Neorg. Khim.*, 1983, **28,** 105.
204. H. J. Dieseroth, *Z. Naturforsch., Teil B*, 1980, **35,** 953.
205. C. Svenson and J. Albertsson, *J. Solid State Chem.*, 1983, **46,** 46.
206. C. Furlani and M. L. Luciani, *Inorg. Chem.*, 1968, **7,** 1586.
207. A. Mueller and P. Werele, *Chem. Ber.*, 1971, **104,** 3782.
208. P. Coggon, J. D. Lebedda, A. T. McPhail and R. A. Palmer, *Chem. Commun.*, 1970, 78.
209. P. J. Hauser, A. F. Schreiner, J. D. Gunter, W. J. Mitchell and M. K. DeArmond, *Theor. Chim. Acta*, 1972, **24,** 78.

210. P. J. Hauser, J. Bordner and A. F. Schreiner, *Inorg. Chem.*, 1973, **12,** 1347.
211. G. J. Sutton, *Aust. Chem. Inst. J. Proc.*, 1950, **17,** 249.
212. B. F. Hoskins, E. R. T. Tiekink, R. Vecchiet and G. Winter, *Inorg. Chim. Acta,* 1984, **90,** 197.
213. J. MacB. Harrowfield, C. Pakowatchai and A. H. White, *J. Chem. Soc., Dalton Trans.*, 1983, 1109.
214. V. Krishnan and R. A. Zingaro, *Inorg. Chem.*, 1968, **8,** 2337.
215. F. W. B. Einstein, G. Hunter, D. G. Tuck and M. K. Yang, *Chem. Commun.*, 1968, 423.
216. F. W. B. Einstein and R. D. G. Jones, *J. Chem. Soc. (A).*, 1971, 2762.
217. E. Hoyer, W. Dietzsch, H. Mueller, A. Zschunke and W. Schroth, *Inorg. Nucl. Chem. Lett.*, 1968, **4,** 201.
218. R. O. Fields, J. H. Waters and T. J. Bergendahl, *Inorg. Chem.*, 1971, **10,** 2808.
219. R. O. Day and R. R. Holmes, *Inorg. Chem.*, 1982, **21,** 2379.
220. A. F. Berniaz, G. Hunter and D. G. Tuck, *J. Chem. Soc. (A)*, 1971, 3254.
221. A. F. Berniaz and D. G. Tuck, *J. Organomet. Chem.*, 1973, **51,** 113.
222. A. F. Berniaz and D. G. Tuck, *J. Organomet. Chem.*, 1972, **46,** 243.
223. O. Hannebohn and W. Klemm, *Z. Anorg. Allg. Chem.*, 1936, **229,** 337.
224. C. Hebecker and R. Hoppe, *Naturwissenschaften,* 1966, **53,** 104.
225. F. Ensslin and H. Dreyer, *Z. Anorg. Allg. Chem.*, 1942, **229,** 377.
226. S. P. Gabuda, S. A. Polischuk, L. M. Avkhutskii and Yu. V. Gagarinskii, *Zh. Strukt. Khim.*, 1969, **10,** 240.
227. P. Bukovec and V. Kaucic, *Inorg. Nucl. Chem. Lett.*, 1978, **14,** 79.
228. W. Klemm and H. Kilian, *Z. Anorg. Allg. Chem.*, 1939, **141,** 93.
229. R. J. Clark, E. Griswold and J. Kleinberg, *J. Am. Chem. Soc.*, 1958, **80,** 4764.
230. C. Mermant, C. Belinski and F. Lalan–Keraly, *C. R. Hebd. Seances Acad. Sci., Ser. C,* 1967, **265,** 1447.
231. P. Bokovec and J. Siftar, *Monatsh. Chem.*, 1970, **101,** 1184; 1971, **102,** 885.
232. J. Grannec, J. C. Champarnaud and J. Portier, *Bull. Soc. Chim. Fr.*, 1970, 3862.
233. J. C. Champarnaud–Mesjard and B. Frit, *Acta Crystallogr., Sect. B,* 1978, **34,** 736.
234. S. Schneider and R. Hoppe, *Z. Anorg. Allg. Chem.*, 1970, **376,** 277.
235. V. I. Sokol, E. N. Deichman and R. D. Yartseva, *Zh. Neorg. Khim.*, 1971, **16,** 951.
236. P. Dobud and D. G. Tuck, unpublished results.
237. C. W. Hobbs and R. S. Tobias, *Inorg. Chem.*, 1970, **9,** 1998.
238. J. S. Poland and D. G. Tuck, *J. Organomet. Chem.*, 1972, **42,** 315.
239. S. B. Miller, B. L. Jelus and T. B. Brill, *J. Organomet. Chem.*, 1975, **96,** 1.
240. J. B. Ekeley and H. A. Potratz, *J. Am. Chem. Soc.*, 1936, **58,** 907.
241. D. G. Tuck, *Proc. CNRS Conf.*, 1970, **191,** 159.
242. D. J. Gislason, M. H. Lloyd and D. G. Tuck, *Inorg. Chem.*, 1971, **10,** 1907.
243. H. V. Schlimper and M. L. Ziegler, *Z. Naturforsch., Teil B,* 1972, **27,** 377.
244. M. A. Khan and D. G. Tuck, *Acta Crystallogr., Sect. B,* 1981, **37,** 683.
245. J. G. Contreras, F. W. B. Einstein, M. M. Gilbert and D. G. Tuck, *Acta Crystallogr., Sect. B,* 1977, **33,** 1648.
246. G. Meyer, *Naturwissenschaften,* 1980, **67,** 143.
247. J. Trotter, F. W. B. Einstein and D. G. Tuck, *Acta Crystallogr., Sect. B,* 1969, **25,** 603.
248. M. A. Khan and D. G. Tuck, *Acta Crystallogr., Sect. B,* 1982, **38,** 803.
249. L. A. Woodward and P. T. Bill, *J. Chem. Soc.*, 1955, 1699.
250. L. A. Woodward and G. H. Singer, *J. Chem. Soc.*, 1958, 716.
251. L. A. Woodward and M. J. Taylor, *J. Chem. Soc.*, 1958, 4473.
252. D. S. Brown, F. W. B. Einstein and D. G. Tuck, *Inorg. Chem.*, 1969, **8,** 14.
253. G. Joy, A. P. Gaughan, I. Wharf, D. F. Shriver and T. A. Dougherty, *Inorg. Chem.*, 1975, **14,** 1795.
254. T. Jaarv, J. T. Bulmer and D. E. Irish, *J. Phys. Chem.*, 1977, **81,** 649.
255. M. P. Hanson and R. A. Plane, *Inorg. Chem.*, 1969, **8,** 746.
256. D. F. Shriver and I. Wharf, *Inorg. Chem.*, 1969, **8,** 2167.
257. M. L. Ziegler, H. V. Schlimper, B. Nuber, J. Weiss and G. Ertel, *Z. Anorg. Allg. Chem.*, 1975, **415,** 193.
258. J. P. Wignacourt, G. Mairesse and P. Barbier, *Acta Crystallogr. Sect. B,* 1980, **36,** 669.
259. J. P. Wignacourt, G. Mairesse and P. Barbier, *Cryst. Struct. Commun.*, 1976, **5,** 293.
260. D. G. Tuck and E. J. Woodhouse, *Chem. Ind. (London)*, 1964, 1363.
261. J. P. Wignacourt, A. Lorriaux–Rubbens, P. Barbier, G. Mairesse and F. Wallart, *Spectrochim. Acta, Part A,* 1980, **36,** 403.
262. J. P. Wignacourt, G. Mairesse, P. Barbier, A. Lorriaux–Rubbens and F. Wallart, *Can. J. Chem.*, 1982, **60,** 1747.
263. J. E. Drake, J. L. Hencher, L. N. Khasrou, D. G. Tuck and L. Victoriano, *Inorg. Chem.*, 1980, **19,** 34.
264. L. Golic, N. Bulc and W. Dietzsch, *Inorg. Chem.*, 1982, **21,** 3560.
265. C. Colaitis, *Bull. Soc. Chim. Fr.*, 1962, 32.
266. P. Muehl, *Krist. Tech.*, 1967, **2,** 431.
267. J. W. Buehler, G. Eikelmann, L. Puppe, K. Rohbock, H. H. Schneehage and D. Week, *Liebigs Ann. Chem.*, 1971, **745,** 135.
268. M. Bhatti, W. Bhatti and G. Mast, *Inorg. Nucl. Chem. Lett.*, 1972, **8,** 133.
269. R. G. Ball, K. M. Lee, A. G. Marshall and J. Trotter, *Inorg. Chem.*, 1980, **19,** 1463.
270. C. Lecomte, J. Protas, P. Cocolios and R. Guilard, *Acta Crystallogr., Sect. B,* 1980, **36,** 2769.
271. P. Cocolios, P. Fournari, R. Guilard, C. Lecomte, J. Protas and J. C. Boubel, *J. Chem. Soc., Dalton Trans.*, 1980, 2081.
272. S. Baskaran and R. N. Kapoor, *Synth. React. Inorg. Metal.-Org. Chem.*, 1984, **14,** 401.
273. H. Kurosawa, in 'Comprehensive Organometallic Chemistry', ed. G. Wilkinson, F. G. A. Stone and E. W. Abel, Pergamon, Oxford, 1982, vol. 1, p. 725.
274. A. G. Lee, 'The Chemistry of Thallium,' Elsevier, Amsterdam, 1971.
275. A. McKillop and E. C. Taylor, in 'Comprehensive Organometallic Chemistry', ed. G. Wilkinson, F. G. A. Stone and E. W. Abel, Pergamon, Oxford, 1982, vol. 7, p. 465.
276. S. Uemura, in 'Synthetic Reagents', ed. J. S. Pizey, Ellis Horwood, Chichester, 1983, p. 165.
277. J. P. Manners, K. G. Morallee and R. J. P. Williams, *Chem. Commun.*, 1970, 965.

278. J. J. Dechter and J. I. Zink, *J. Chem. Soc., Chem. Commun.*, 1974, 96.
279. R. A. Penneman and E. Staritzky, *J. Inorg. Nucl. Chem.*, 1958, **6**, 112.
280. R. O. Nilsson, *Årk. Kemi*, 1957, **10**, 363.
281. M. N. Bochkarev, N. I. Gur'ev, L. V. Pankratov and G. A. Razuvaev, *Inorg. Chim. Acta*, 1980, **44**, L59.
282. R. W. Rudolph, W. L. Wilson and R. C. Taylor, *J. Am. Chem. Soc.*, 1981, **103**, 2480.
283. R. C. Burns and J. D. Corbett, *J. Am. Chem. Soc.*, 1982, **104**, 2804.
284. A. G. Lee, *J. Chem. Soc. (A)*, 1971, 880.
285. J. Reedijk, G. Roelofsen, A. R. Siedle and A. L. Spek, *Inorg. Chem.*, 1979, **18**, 1947.
286. J. R. Hudman, M. Patel and W. R. McWhinnie, *Inorg. Chim. Acta*, 1970, **4**, 161.
287. P. Gray, *Q. Rev. Chem. Soc.*, 1963, **17**, 441.
288. S. J. Patel, *J. Inorg. Nucl. Chem.*, 1970, **32**, 3708.
289. 'Stability Constants of Metal-Ion Complexes', Special Publications Nos. 17 and 25, Chemical Society, London, 1964 and 1971.
290. A. J. Hinchcliffe and J. S. Ogden, *Chem. Commun.*, 1969, 1053.
291. V. F. Shevel'kov, N. A. Klyuev and A. A. Mal'tsev, *Vestn. Mosk. Univ., Khim.*, 1969, **24**, 32.
292. M. Tzentnershver and T. Trebaczkiewicz, *Z. Phys. Chem., Teil A*, 1933, **165**, 367.
293. B. J. Kelsall and K. D. Carlson, *J. Phys. Chem.*, 1980, **84**, 951.
294. H. Sabrowsky, *Z. Anorg. Allg. Chem.*, 1969, **365**, 146.
295. D. C. Bradley, R. C. Mehrotra and D. P. Gaur, 'Metal Alkoxides', Academic, London, 1978, pp. 4, 12, 24.
296. N. V. Sidgwick and L. E. Sutton, *J. Chem. Soc.*, 1930, 1461.
297. P. J. Burke, R. W. Matthews and D. G. Gillies, *J. Chem. Soc., Dalton Trans.*, 1980, 1439.
298. L. F. Dahl, G. L. Davis, D. L. Wampler and R. West, *J. Inorg. Nucl. Chem.*, 1962, **24**, 357.
299. S. F. A. Kettle, *Theor. Chim. Acta*, 1966, **4**, 150.
300. A. G. Lee, *J. Chem. Soc. (A)*, 1971, 2007.
301. A. G. Lee, *J. Chem. Soc. (A)*, 1970, 467.
302. N. S. Poonia and M. R. Truter, *J. Chem. Soc., Dalton Trans.*, 1972, 1791.
303. C. M. Fear and R. C. Menzies, *J. Chem. Soc.*, 1926, **129**, 937.
304. C. Z. Moore and W. H. Nelson, *Inorg. Chem.*, 1969, **8**, 143.
305. S. Tachiyashiki, H. Nakayama, R. Kuroda, S. Sato and Y. Saito, *Acta Crystallogr., Sect. B*, 1975, **31**, 1483.
306. V. A. Zakharov, O. A. Songina, I. M. Bessarabova and L. N. Lebedeva, *Zh. Anal. Khim.*, 1970, **25**, 879.
307. M. Veith and R. Rösler, *Angew. Chem., Int. Ed. Engl.*, 1982, **21**, 858.
308. M. Veith and R. Rösler, *J. Organomet. Chem.*, 1982, **229**, 131.
309. Y. Oddon, G. Pepe, J.-R. Vignalou and A. Tranquard, *J. Chem. Res. (S)*, 1978, 250.
310. Y. Oddon, A. Tranquard and G. Pepe, *Acta Crystallogr., Sect. B*, 1979, **35**, 542.
311. B. Frit, G. Roult and J. Galy, *J. Solid State Chem.*, 1983, **48**, 246.
312. Y. Oddon, J.-R. Vignalou, A. Tranquard and G. Pepe, *Acta Crystallogr., Sect. B*, 1979, **35**, 2525.
313. J. K. Fawcett, V. Kocman, S. C. Nyburg and R. J. O'Brien, *Chem. Commun.*, 1970, 1213.
314. K. P. Petrov, V. V. Ugarov and N. G. Rambidi, *Zh. Strukt. Khim.*, 1980, **21**, 159.
315. K. R. Metz and J. F. Hinton, *J. Am. Chem. Soc.*, 1982, **104**, 6206.
316. J. D. Donaldson, J. F. Knifton and S. D. Ross, *Spectrochim. Acta*, 1964, **20**, 847; 1965, **21**, 275.
317. T. V. Lysyak, S. L. Rusakov, I. S. Kolomnikov and Yu. Ya. Kharitonov, *Zh. Neorg. Khim.*, 1983, **28**, 756.
318. D. L. Hughes, *J. Chem. Soc., Dalton Trans.*, 1973, 2209.
319. D. L. Hughes and M. R. Truter, *J. Chem. Soc., Dalton Trans.*, 1972, 2214.
320. H. Ohtaki, K. Ozutsumi and A. Kusumegi, *23rd ICCC Abstracts*, 1984, p. 73.
321. B. Cocton and A. C. de Paulet, *Bull. Soc. Chim. Fr.*, 1966, 2947.
322. E. C. Taylor and A. McKillop, *Acc. Chem. Res.*, 1970, **3**, 338.
323. L. Nilson and R. Hesse, *Acta Chem. Scand.*, 1969, **23**, 1951.
324. H. Pritzkow and P. Jennische, *Acta Chem. Scand., Ser. A*, 1975, **29**, 60.
325. R. J. Magee and M. J. O'Connor, *Inorg. Chim. Acta*, 1971, **5**, 554.
326. J. W. Spanyer and J. P. Phillips, *Anal. Chem.*, 1956, **28**, 253.
327. M. Dräger and G. Gattow, *Z. Anorg. Allg. Chem.*, 1972, **387**, 281, 300.
328. R. C. Burns and J. D. Corbett, *J. Am. Chem. Soc.*, 1981, **103**, 2627.
329. G. Pfrepper, *Z. Anorg. Allg. Chem.*, 1966, **347**, 160.
330. J. C. A. Boeyens and F. H. Herbstein, *Inorg. Chem.*, 1967, **6**, 1408.
331. J. C. A. Boeyens, *Acta Crystallogr., Sect. B*, 1970, **26**, 1251.
332. V. M. Shul'man, L. N. Makeeva, E. V. Khylstunova, V. L. Varand and V. E. Fedorov, *Izv. Akad. Nauk SSSR, Ser. Khim.*, 1971, 1552.
333. N. W. Alcock, *Acta Crystallogr., Sect. A*, 1969, **25**, S101.
334. M. Barlow and C. C. Meredith, *Z. Kristallogr.*, 1969, **130**, 304.
335. N. W. Alcock and H. D. B. Jenkins, *J. Chem. Soc., Dalton Trans.*, 1974, 1907.
336. W. G. Lyon, E. F. Westrum and M. Chavret, *J. Chem. Thermodyn.*, 1971, **3**, 571.
337. F. J. Keneshea and D. Cubicciotti, *J. Phys. Chem.*, 1967, **71**, 1958.
338. J. M. Brom, Jr. and H. F. Franzen, *J. Chem. Phys.*, 1971, **54**, 2874.
339. M. F. C. Ladd and W. H. Lee, *Trans. Faraday Soc.*, 1970, **66**, 2767.
340. O. G. Polyachenok and L. D. Polyachenok, *Zh. Fiz. Khim.*, 1971, **45**, 1793.
341. A. G. Sharpe, *J. Chem. Soc.*, 1952, 2165.
342. A. C. Hazell, *Acta Crystallogr.*, 1963, **16**, 71.
343. B. Grundström and P. Valberg, *Z. Physik.*, 1938, **108**, 326.
344. E. Wiberg and E. Amberger, 'Hydrides of the Elements of the Main Groups I–IV', Elsevier, Amsterdam, 1971.
345. T. C. Waddington, *J. Chem. Soc.*, 1958, 4783.
346. R. Gresser, D. W. Boyd, A. W. Albrecht-Gary and J. P. Schwing, *J. Am. Chem. Soc.*, 1980, **102**, 651.
347. J. J. Dechter and J. I. Zink, *J. Am. Chem. Soc.*, 1976, **98**, 845.
348. M. E. Farago, *Inorg. Chim. Acta*, 1977, **23**, 211.

349. T. Platzner, J. K. Thomas and M. Grätzel, *Z. Naturforsch., Teil B,* 1978, **33,** 614.
350. M. Anbar, *Q. Rev., Chem. Soc.,* 1968, **22,** 578.
351. B. Cercek, M. Ebert and A. J. Swallow, *J. Chem. Soc. (A),* 1966, 612.
352. M. C. R. Symons and J. K. Yandell, *J. Chem. Soc. (A),* 1971, 760.
353. C. E. Burchill and G. G. Hickling, *Can. J. Chem.,* 1970, **48,** 2466.
354. C. E. Burchill and W. H. Wolodarsky, *Can. J. Chem.,* 1970, **48,** 2955.
355. D. R. Stranks and J. K. Yandell, *J. Phys. Chem.,* 1969, **73,** 840.
356. B. Falcinella, P. D. Felgate and G. S. Laurence, *J. Chem. Soc., Dalton Trans.,* 1974, 1367.
357. M. Tournoux, R. Marchand and M. Bouchama, *C. R. Hebd. Seances Acad. Sci., Ser. C,* 1969, **269,** 1201; 1970, **270,** 1007.
358. J. A. A. Ketelaar, W. H. t'Hart, M. Moerel and D. Polder, *Z. Kristallogr.,* 1939, **101,** 396.
359. F. I. Aliev and L. I. Tatarinova, *Kristallografiya,* 1966, **11,** 389.
360. B. Leclerc and M. Bailly, *Acta Crystallogr., Sect. B,* 1973, **29,** 2334.
361. B. Leclerc and T. S. Kabre, *Acta Crystallogr., Sect. B,* 1975, **31,** 1675.
362. S. Bhan and K. J. Schubert, *J. Less-Common Met.,* 1970, **20,** 229.
363. L. I. Man, R. M. Imamov and Z. G. Pinsker, *Kristallografiya,* 1971, **16,** 122.
364. J. Grannec and J. Portier, *C. R. Hebd. Seances Acad. Sci., Ser. C,* 1971, **272,** 942.
365. J. Grannec, L. Lozano, J. Portier and P. Hagenmuller, *Z. Anorg. Allg. Chem.,* 1971, **385,** 26.
366. R. Bohme, J. Rath, B. Grunwald and G. Thiele, *Z. Naturforsch., Teil B,* 1980, **35,** 1366.
367. A. C. Hazell, *J. Chem. Soc.,* 1963, 3459.
368. S. E. Jeffs, R. W. H. Small and I. J. Worrall, *Acta Crystallogr., Sect. C,* 1983, **39,** 1205.
369. J. Tudo and B. Jolibois, *J. Less-Common Met.,* 1980, **70,** 25.
370. I. D. Brown and R. Faggiani, *Acta Crystallogr., Sect. B,* 1980, **36,** 1802.
371. S. R. Sagi, M. S. P. Rao and K. V. Ramana, *J. Therm. Anal.,* 1981, **20,** 93.
372. F. Abraham, G. Nowogrocki, B. Jolibois and G. Laplace, *J. Solid State Chem.,* 1983, **47,** 1.
373. R. A. Horne, *J. Inorg. Nucl. Chem.,* 1958, **6,** 338.
374. B. I. Lobov, F. Ya. Kul'ba and V. E. Mironov, *Zh. Neorg. Khim.,* 1967, **12,** 334.
375. R. A. Walton, *J. Inorg. Nucl. Chem.,* 1970, **32,** 2875.
376. J. R. Hudman, M. Patcl and W. R. McWhinnie, *Inorg. Chim. Acta,* 1970, **4,** 161.
377. W. J. Baxter and G. Gafner, *Inorg. Chem.,* 1972, **11,** 176.
378. K. C. Malhotra and I. Balkrishnan, *J. Inorg. Nucl. Chem.,* 1977, **39,** 387; **39,** 1523.
379. R. A. Walton, *J. Inorg. Nucl. Chem.,* 1977, **39,** 549.
380. D. N. Sotiropoulos, J. K. Kouinis and A. G. Galinos, *Inorg. Nucl. Chem. Lett.,* 1981, **17,** 117.
381. F. Ya. Kul'ba, N. G. Yaroslavskii, L. V. Konovalov, A. V. Barsukov and V. E. Mironov, *Zh. Neorg. Khim.,* 1968, **13,** 153.
382. G. J. Sutton, *Aust. J. Sci. Res., Ser. A,* 1951, **4,** 654.
383. V. Ozimec, T. J. Smith and R. A. Walton, *J. Inorg. Nucl. Chem.,* 1977, **39,** 362.
384. J. Glaser and G. Johansson, *Acta Chem. Scand., Ser. A,* 1982, **36,** 125.
385. J. Glaser and G. Johansson, *Acta Chem. Scand., Ser. A,* 1981, **35,** 639.
386. J. Glaser, *Acta Chem. Scand., Ser. A,* 1979, **33,** 789.
387. J. R. Günter, *Z. Anorg. Allg. Chem.,* 1978, **438,** 203.
388. M. Vlasse, J. Grannec and J. Portier, *Acta Crystallogr., Sect. B,* 1972, **28,** 3426.
389. A. A. Noyes and C. S. Garner, *J. Am. Chem. Soc.,* 1936, **58,** 1268.
390. R. Faggiani and I. D. Brown, *Acta Crystallogr., Sect. B,* 1978, **34,** 1675.
391. D. W. Amos, *Chem. Commun.,* 1970, 19.
392. R. C. L. Mooney, *Acta Crystallogr.,* 1956, **9,** 113.
393. R. C. L. Mooney-Slater, *Acta Crystallogr.,* 1961, **14,** 1140.
394. S. Kauffmann and K. Dehnicke, *Z. Anorg. Allg. Chem.,* 1966, **347,** 318.
395. G. Johansson, *Acta Chem. Scand.,* 1959, **13,** 924.
396. J. K. Kochi and T. W. Bethea, *J. Org. Chem.,* 1968, **33,** 75.
397. R. Faggiani and I. D. Brown, *Acta Crystallogr., Sect. B,* 1978, **34,** 2845.
398. R. Faggiani and I. D. Brown, *Acta Crystallogr., Sect. B,* 1982, **38,** 2473.
399. A. F. M. J. van der Ploeg, G. van Koten and K. Vrieze, *Inorg. Chim. Acta,* 1980, **38,** 253.
400. S. R. Sagi, K. V. Ramana and M. S. P. Rao, *J. Therm. Anal.,* 1980, **18,** 291.
401. E. Gutierrez-Puebla, A. Vegas and S. Garcia-Blanco, *Acta Crystallogr., Sect. B,* 1980, **36,** 145.
402. S. E. Jeffs, R. W. H. Small and I. J. Worrall, *Acta Crystallogr., Sect. C,* 1983, **39,** 1628.
403. T. N. Srivastava and G. Mohan, *Indian J. Chem.,* 1970, **8,** 742.
404. J. Reedijk, *Recl. Trav. Chim. Pays-Bas,* 1969, **88,** 499.
405. G. Hunter and B. C. Williams, *J. Chem. Soc. (A),* 1971, 2554.
406. F. A. Cotton and J. A. McCleverty, *Inorg. Chem.,* 1967, **6,** 229.
407. C. Hebecker, *Z. Anorg. Allg. Chem.,* 1972, **393,** 223.
408. A. F. Clifford and E. Zamora, *Trans. Faraday Soc.,* 1961, **57,** 1963.
409. R. Hoppe and C. Hebecker, *Z. Anorg. Allg. Chem.,* 1965, **335,** 85; **341,** 336.
410. H. Bode and E. Voss, *Z. Anorg. Allg. Chem.,* 1957, **290,** 1.
411. J. Grannec, J. Portier and P. Hagenmuller, *J. Solid State Chem.,* 1971, **3,** 227.
412. F. A. Cotton, B. F. G. Johnson and R. M. Wing, *Inorg. Chem.,* 1965, **4,** 502.
413. R. A. Walton, *Inorg. Chem.,* 1968, **7,** 640; **7,** 1927.
414. R. W. Matthews and R. A. Walton, *J. Chem. Soc. (A),* 1968, 1639.
415. G. Thiele, B. Grunwald, W. Rink and D. Breitinger, *Z. Naturforsch., Teil B,* 1979, **34,** 1512.
416. J. Glaser, *Acta Chem. Scand., Ser. A,* 1980, **34,** 75.
417. J. Glaser, *Acta Chem. Scand., Ser. A,* 1982, **36,** 451.
418. D. M. Adams and D. M. Morris, *J. Chem. Soc. (A),* 1968, 694.
419. K. Schmidt, *J. Inorg. Nucl. Chem.,* 1970, **32,** 3549.

420. J. Glaser, *Acta Chem. Scand., Ser. A,* 1980, **34,** 141.
421. T. J. Bastow, B. D. James and M. B. Millikan, *J. Solid State Chem.*, 1983, **49,** 388.
422. I. R. Beattie, T. R. Gilson and G. A. Ozin, *J. Chem. Soc. (A),* 1968, 2765.
423. J. Glaser and U. Henricksson, *J. Am. Chem. Soc.*, 1981, **103,** 6642.
424. T. G. Spiro, *Inorg. Chem.*, 1965, **4,** 1290; 1967, **6,** 569.
425. M. J. Taylor, *J. Chem. Soc. (A),* 1968, 1780.
426. J. E. D. Davies and D. A. Long, *J. Chem. Soc. (A),* 1968, 2050.
427. J. Glaser, *Acta Chem. Scand., Ser. A,* 1980, **34,** 157.
428. H. W. Rotter and G. Thiele, *Z. Naturforsch., Teil B,* 1982, **37,** 995.
429. T. Barrowcliffe, I. R. Beattie, P. Day and K. Livingston, *J. Chem. Soc. (A),* 1967, 1810.
430. J. Glaser, P. L. Goggin, M. Sandström and V. Lutsko, *Acta Chem. Scand., Ser. A,* 1982, **36,** 55.
431. P. Breisacher and B. Siegel, *J. Am. Chem. Soc.*, 1965, **87,** 4255.
432. K. M. McKay, 'Hydrogen Compounds of the Metallic Elements', E. F. Spon, London, 1966.
433. E. Wiberg, O. Dittmann and M. Schmidt, *Z. Naturforsch., Teil B,* 1957, **12,** 60.
434. K. Henrick, R. W. Matthews, B. L. Podejma and P. A. Tasker, *J. Chem. Soc., Chem. Commun.*, 1982, 118.
435. D. L. Hughes and M. R. Truter, *J. Chem. Soc., Chem. Commun.*, 1982, 727.
436. D. L. Hughes and M. R. Truter, *Acta Crystallogr., Sect. C,* 1983, **39,** 329.
437. F. Brady, K. Henrick and R. W. Matthews, *J. Organomet. Chem.*, 1981, **210,** 281.
438. A. Giraudeau, A. Louati, J. H. Callot and M. Gross, *Inorg. Chem.*, 1981, **20,** 769.

26

Silicon, Germanium, Tin and Lead

PHILIP G. HARRISON
University of Nottingham, UK

26.1 INTRODUCTION

The metalloidal and metallic congeners of carbon in Group IVB exhibit what is probably the most varied coordination chemistry of any similar group of elements in the Periodic Table. Simple reasons for this observation are easy to find including (a) the change in electronegativity on proceeding down the group resulting in a transformation from the classical non-metallic element carbon to the respectably metallic elements, tin and lead, (b) the large increase in atomic radii leading to, *inter alia,* high coordination numbers for the heavier elements, and (c) the emergence of an 'inert pair' effect, leading to the possibility of lone pair stereochemical activity. However, although the principles of VSEPR can in general be applied successfully to predict the geometry of tetravalent and simple bivalent compounds, the factors underlying the stereochemistry of tin(II) and lead(II) compounds in the solid state can often be more subtle, and hence not so readily understood. The extremely wide scope of the structural chemistry in this area therefore only allows a panoramic overview to be described in this chapter, rather than a totally comprehensive treatise.

26.2 GENERAL TRENDS

26.2.1 Electronegativity

Estimated values of the electronegativities of the Group IVB elements are listed in Table 1. The more recently devised scales of Sanderson and Hargattai–Bliefelt (based on analysis of IR band intensities for $MHCl_3$ and MX_4 molecules and bond distance data in bivalent and tetravalent derivatives) concur with the alternating scale of Allred–Rochow, rather than the original Pauling scale.[1,2] It must be borne in mind, however, that the absolute electronegativity varies significantly upon the particular chemical environment in which the element atom finds itself. The concept of electronegativity is therefore only of utility as a general guide to relative behaviour.

26.2.2 Elemental Structure

Both silicon and germanium crystallize with the diamond structure, although with a substantially lower band gap [*e.g.* (kJ mol^{-1}): 106.8 (Si), 64.2 (Ge); *cf.* ~580 (C)], accounting

Table 1 Element Electronegativity Scales[a]

Element	*Pauling*	*Allred–Rochow*	*Sanderson*	*Hargittai–Bliefelt*
C	2.5	2.5	2.5	2.6
Si	1.8	1.7	1.7	1.9
Ge	1.8	2.0	2.3	2.5
Sn	1.8	1.7	2.0	2.3
Pb	1.9			

[a] Data taken from 'Inorganic Chemistry of the Main Group Elements', The Chemical Society, 1977, vol. 4, p. 145; I. Hargittai and C. Bliefelt, *Z. Naturforsch., Teil B.*, 1983, **38**, 1304; and N. N. Greenwood and A. Earnshaw, 'Chemistry of the Elements', Pergamon, Oxford, 1984, p. 431.

for their semiconducting properties. Tin forms two modifications; white or β-tin, which adopts a tetragonal lattice and is the stable form above the transition temperature of 13.2 °C, and grey or α-tin, which also has the diamond structure and is stable below this temperature.[3] Tetragonal modifications of silicon and germanium have been observed on the application of very high pressures.[4] Lead has a typically metallic face-centred cubic lattice.[3]

26.2.3 Elemental Radii

Estimates of the radii of the Group IV elements in various chemical situations are collected in Table 2. Values increase with increasing atomic number, so that the covalent radii of tin and lead are almost twice that of carbon. 'Ionic' radii are substantially smaller than the corresponding covalent radii, with the M^{II} 'ionic' radii being substantially larger than their M^{IV} counterparts.

Table 2 Elemental Radii (pm)

Element	*Non-bonded radius*[a]	M^{IV} *covalent radius*[b]	M^{IV} *'ionic' radius*[b]	M^{II} *'ionic' radius*[b]
C	125	77.2		
Si	155	117.6	40	
Ge	158	122.3	53	73
Sn	182	140.5	69	118
Pb		146	78	119

[a] Data taken from C. Glidewell, *Inorg. Chim. Acta*, 1979, **36**, 135.
[b] Data taken from N. N. Greenwood and A. Earnshaw, 'Chemistry of the Elements', Pergamon, Oxford, 1984, p. 431.

26.2.4 Bond Energy Data

Data for the single bond dissociation energy, D(M—X), are listed in Table 3. Two sets of values are given, one a general set and the other (in parentheses in Table 3) for trimethylmetal molecules, Me_3MX. Significant variation between the two sets of values occurs (by as much as 100 kJ mol^{-1}), and so only general trends should be noted. In addition, bond dissociation energies vary somewhat between individual molecules, as has been demonstrated for silicon compounds.[5] A steady decrease in bond energy usually accompanies an increase in size of both M and X. However, values for bonds between silicon and oxygen, nitrogen and the halogens are anomalously high, a phenomenon usually ascribed to a manifestation of $(p \rightarrow d)\pi$ reinforcement of the M—X σ bond. Various estimates of the π bond energy of the Si═C and Si═Si double bonds in silaalkenes and disilenes have appeared.[5,6] The most reliable estimate of the Si═C π bond energy in 1,1-dimethylsilaethylene is around 163 ± 21 kJ mol^{-1}.[5] The triple bond energy in the hypothetical species has been calculated to be 324 kJ mol^{-1}.[7] Experimental data afford a lower bound of 69 ± 11 kJ mol^{-1} for the Si═Si π bond energy for disilene, $H_2Si{=}SiH_2$, whereas *ab initio* calculations give a value of 94 ± 8 kJ mol^{-1}. Estimates for substituted disilenes are somewhat greater at around 108 ± 20 kJ mol^{-1}.[8] The Si═Cr bond dissociation energy in the silylene complex $[(OC)_5Cr{=}Si(OH)H]$ is calculated to be even higher at 128 kJ mol^{-1}.[9]

Table 3 Approximate Bond Energies, D(M—X) (kJ mol^{-1})[a]

M	*X* = *C*	*H*	*F*	*Cl*	*Br*	*I*	*O*	*S*	*N*	*M*
C	368	435	452	351	293	216	359		305	368
	(376)	(439)		(351)	(297)		(343)[b]	(297)[c]	(334)[e]	
Si	360	393	565	381	310	234	464		334	340
	(376)	(377)		(464)	(393)	(322)	(464)[b]	(414)[c]	(418)[e]	(339)
Ge	255	289	465	342	276	213	359		255	188
	(318)	(343)		(485)	(435)		(447)[b]			(276, 305)
Sn	226	251		320	272	187				151
	(272)	(309)		(422)	(555)		(351)[b]	(217)[d]	(172)[f]	(234)
Pb	130	205		244						
	(205)	(259)								(230)

[a] Data taken from N. N. Greenwood and A. Earnshaw, 'Chemistry of the Elements', Pergamon, Oxford, 1984, p. 436. Values in parentheses are for Me_3MX molecules taken from R. A. Jackson, *J. Organomet. Chem.*, 1979, **166**, 17; A. M. Doncaster and R. Walsh, *J. Phys. Chem.*, 1979, **83**, 578; R. Walsh, *Acc. Chem. Res.*, 1981, **14**, 246; and M. J. Almend, A. M. Doncaster, P. N. Noble and R. Walsh, *J. Am. Chem. Soc.*, 1982, **104**, 4717; J. C. Baldwin, M. F. Lappert, J. B. Pedley and J. S. Poland, *J. Chem. Soc., Dalton Trans.*, 1972, 1943.
[b] For Me_3MOEt derivatives.
[c] For Me_3MSBu^t derivatives.
[d] For Me_3SnSBu^n.
[e] For Me_3MNHMe derivatives.
[f] For Me_3SnNMe_2.

26.2.5 Coordination Geometries

The variety of coordination geometries exhibited by the Group IV elements is illustrated in Table 4, which contains examples of the principal types encountered. The list is extensive, with coordination numbers ranging from one to twelve, but by no means comprehensive. In particular, tin and lead exhibit many irregular coordination geometries of high coordination number. Indeed, except for some esoteric carbides and silicides, compounds with coordination number greater than six are exclusively confined to these metals, consistent with their large covalent radii. Stable, small multiply bonded molecules are formed only by carbon, although a few rather sophisticated molecules containing Si=C, Si=Si, Ge=Ge and Sn=Sn double bonds have been synthesized. Many other silicon and germanium species containing multiple bonds to oxygen, carbon, nitrogen and phosphorus have been postulated as transient intermediates. The only examples of multiple bonding between tin and lead and an electronegative element are the matrix-isolated species, MO and MO_2 (M = Sn, Pb).

Conversely, the stability of compounds of oxidation state +II increases dramatically as the atomic number of the element increases. Carbenes, CX_2, and silylenes, SiX_2, are well established as transient reaction intermediates, and structural data have been obtained in several cases either at high temperatures or by their generation in low-temperature matrices. However, only for germanium, tin and lead are compounds in this oxidation state stable under ordinary conditions. Compounds with the Group IV element in oxidation state +III, the formal oxidation state of the radical species, $R_3M^{\bullet}$, are also usually considered as unstable transients. However, when R is very bulky, these metal-centred radicals, such as for example $Sn[CH(SiMe_3)_2]_3$, have extremely long, perhaps indefinite, lifetimes in solution.

Silicon and germanium are generally tetrahedrally coordinated in their tetravalent derivatives. Although examples of five- and six-coordination are well documented, none are known which contain two or more silicon–carbon bonds. Tetrahedral four-coordination is only observed for tin and lead in R_4M (R = organic group) compounds and R_nMX_{4-n} derivatives which are either sterically crowded or in which the ligand X is weakly electronegative and a poor donor, and higher coordination numbers are achieved wherever possible. For tin and lead compounds with three metal–carbon bonds, coordination saturation invariably occurs at five, although one authenticated example of a six-coordinated complex, $Me_3Sn[(N_2C_3H_3\text{-}1)_3BH]$, has been reported (*vide infra*). For compounds with two or less metal–carbon bonds, coordination saturation is usually achieved at six in an octahedral or distorted octahedral geometry, but a few examples with coordination numbers of seven or eight have been

Table 4 Typical Coordination Geometries Exhibited by the Group IV Elements

Coordination number	*Geometry*	*Examples* *Carbon*	*Silicon*	*Germanium*	*Tin*	*Lead*
1	Diatomic	CO	SiO[a]	GeO[a]	SnO[a]	PbO[a]
2	Linear	CO_2, CS_2, alkynes, RN:C:NR, NCO^-	SiO_2,[a] SiOS[a]	GeO_2[a]	SnO_2[a]	PbO_2[a]
2	Bent	Carbenes, CX_2,[b] $Ph_3P{:}C{:}PPh_3$	Silylenes, SiX_2[b,c]	Germanium(II) halides,[a,c] $Ge[NC_5H_6Me_4]_2$, $Ge[OC_6H_2MeBu^t_2]_2$	Tin(II) halides,[a,c] $Sn[CH(SiMe_3)_2]_2$,[c] $Sn[OC_6H_2MeBu^t_2]_2$	Lead(II) halides,[a,c] $Pb[OC_6H_2MeBu^t_2]_2$, $Pb[N(SiMe_3)_2]_2$
3	Trigonal planar	Aromatic compounds, graphite, alkenes, OCX_2 (X = R, F, Cl, Br), CO_3^{2-}	Silaalkenes, $R_2Si{=}CH_2$,[b] silabenzene,[b] $SiOF_2$[a]	Germabenzene,[b] $GeOF_2$,[a] $(OC)_5Cr{:}Ge(SC_6H_2Me_3)_2$	$(OC)_5Cr{:}\leftarrow Sn[CH(SiMe_3)_2]_2$	
3	Pyramidal	Carbenes, CPh_3^-,[b] free radicals, $R_3C\cdot$[b]	SiH_3^-, silyl radicals $R_3Si\cdot$[b]	$GeCl_3^-$, GeI(acac), $GeCl(O_2PH_2)_2$, $GeCl_2\cdot$1,3-benzthiazole, germyl radicals, $R_3Ge\cdot$[b]	SnX_3^-, $SnX_2\cdot L$, stannyl radicals, $R_3Sn\cdot$,[b] SnS, $SnSO_4$	PbX_3^-, $PbX_2\cdot L$, $Pb(SPh_3)^-$, plumbyl radicals, $R_3Pb\cdot$[b]
4	Tetrahedral	Diamond, SiC, CX_4 and aliphatic sp^3 hybridized organic	$R_{4-n}SiX_n$ (n = 0–4), SiO_2 (quartz), silicates	$R_{4-n}GeX_n$ (n = 0–4), hexagonal GeO_2, germanates	$Me_{4-n}SnX_n$ (n = 0–4),[c,d] $Ph_{4-n}SnCl_n$ (n = 0–2), $Sn(NMe_2)_4$, Ph_5Sn_2	R_4Pb, Ph_3SAr, R_6Pb_2
4	Square pyramidal			Phthalocyaninato-germanium(II)	SnO, phthalo cyaninatotin(II)	PbO, phthalo cyaninatolead(II)
4	Pseudotrigonal bipyramidal[e]			GeF_2	$Sn[OCMeCHCPhO]_2$, $Sn(O_2CH)_2$, SnF_2, $Sn(S_2COMe)_2$, $Sn_6O_4(OMe)_4$, $SnCl_4^{2-}$	$[Pb_4(OSiPh_3)_6]O$
5	Trigonal bipyramidal	Me_6Al_2[f]	SiF_5^-, $Ph_2SiF_3^-$ 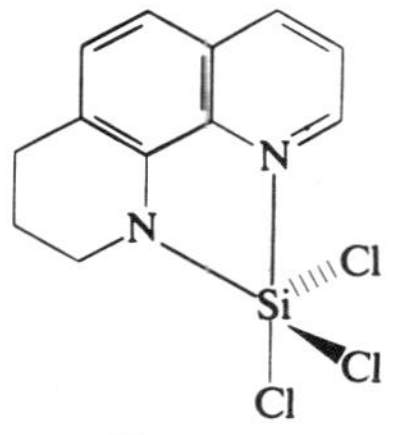 Silatranes	$FeF_4\cdot L$, $R_2GeX_3^-$ 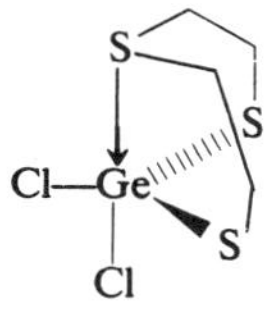 Germatranes	Me_3SnF, $Me_3SnCl\cdot py$, Me_3SnO_2CMe, $R_3SnX_2^-$, $R_2SnX_3^-$, $[Bu_3Sn(OH_2)_2]^+$	Me_3PbO_2CMe, Me_3PbN_3, Ph_3PbOH, Ph_3PbCl, R_3PbX_2

5	Square pyramidal	*closo*-1,6-$C_2B_4H_6$	$[(C_6H_4O_2)SiF]^-$, $[PhSi(O_2C_{10}H_6)]^-$	$[(C_6H_4O_2)_2GeF]^-$	$[(MeC_6H_3S_2)_2SnCl]^-$	
6	Octahedral	Metal carbides, MC (M = Ti, Zr, Hf, V, Nb, Ta, Mo, W, *etc.*)	SiF_6^{2-}, $Si(acac)_3^{2-}$, SiF_4·bipy, $Si(DMF)_6^{4+}$, SiO_2 (stishovite)	GeF_6^{2-}, $Ge(oxalate)_3^{2-}$, $Ge(acac)_3^{2-}$	SnX_6^{2-}, Me_2SnF_2, $Me_2Sn(acac)_2$, R_2SnX_2·2L, SnTe, $Me_2Sn(NCS)_2$·terpy	Ph_2PbCl_2, $Me_2Pb(OMe)_2$, $Ph_2PbX_4^{2-}$, Ph_2PbX_2·2L, $[Pb(antipyrine)_6]^{2+}$, PbS, PbSe, $[Ph_2Pb(O_2CMe)_2]_2$·H_2O
7	Pentagonal bipyramidal					$[Ph_2Pb(O_2CMe)_3]^-$,
8	Hexagonal bipyramidal					$Pb(SCN)_2(C_{12}H_{26}N_2O_4)$
8	Dodecahedral				$Sn(NO_3)_4$, $Sn(O_2CMe)_4$	$Pb(O_2CMe)_4$
8	Square antiprismatic	Be_2C	Mg_2Si		bis(phthalo cyaninato)tin(IV)	$Pb(O_2CC_6F_5)_2$·2MeOH, $\frac{3}{4}Pb(O_2CH)_2$·$4SC(NH_2)_2$
9	Trifacially capped trigonal prismatic				$SnCl_2$, $SnBr_2$, $Sn(SbF_6)_2(AsF_3)_2$	$PbCl_2$, $PbBr_2$
10	Various		$TiSi_2$, $CrSi_2$, $MoSi_2$			$Pb(SCN)_2(C_{18}H_{36}N_2O_6)$, $Pb_3(PO_4)_2$
12	Various					$Pb(NO_3)_2$, $PbSO_4$

[a] Matrix-isolated species.
[b] Unstable intermediates.
[c] In the gas phase.
[d] In solution.
[e] *i.e.* one equatorial coordination site occupied by the metal(II) lone pair.
[f] Precise geometry at aluminum not known.

characterized. Of the two five-coordinated geometries, square pyramidal and trigonal bipyramidal, the latter is strongly favoured.[10]

Two factors combine to lend a greater diversity in the stereochemistries exhibited by bivalent germanium, tin and lead compounds, the increased radius of M^{II} compared with that of M^{IV} and the presence of a non-bonding pair of electrons. When the non-bonding pair of electrons occupies the isotropic valence level s orbital, as in, for example, the complex cations $Pb[SC(NH_2)_2]_6^{2+}$ and $Pb[antipyrine]_6^{2+}$, or when they are donated to conductance band levels, as in the binary tin and lead selenides or tellurides or the perovskite ternary phases $CsMX_3$ ($M = Sn, Pb; X = Cl, Br, I$), then the metal coordination is regular. However, in the majority of compounds an apparent vacancy in the coordination sphere of the metal is observed, which is usually ascribed to the presence of the non-bonding pair of electrons in a hybrid orbital and cited as evidence for a 'stereochemically active lone pair'.

Examples of compounds exhibiting the two-coordinated 'bent' geometry are rather sparse and only when the two ligands are very bulky do they occur under ordinary conditions. The preferred coordination geometries for bivalent germanium and tin are the three-coordinate trigonal pyramid and the four-coordinate distorted pseudotrigonal bipyramid, where the stereochemically active lone pair occupies an equatorial coordination site. The latter stereochemistry appears to be favoured when the ligands are very electronegative. Similar three- and four-coordination is also found for bivalent lead, but much higher coordination numbers in a variety of geometries predominate. Although square antiprismatic eight-coordination is quite common, several other types, many of which are irregular, of coordination numbers from seven to twelve are also observed.

26.3 STRUCTURAL SURVEY[11]

26.3.1 Multiple Bonding Involving Silicon, Germanium and Tin

The reluctance of the heavier elements of Group IV to form stable compounds possessing multiple bonds is in sharp contrast to the behaviour of carbon, where $p\pi-p\pi$ multiple bond function is a major feature of its chemistry. The quest for multiply bonded compounds involving silicon is by no means new and dates back to the attempts of Frederick Stanley Kipping over 80 years ago to characterize compounds of the type $R_2Si{=}O$, which he termed silicones by analogy with organic ketones.[12] More recently, several multiply bonded silicon and germanium species have been generated as transient intermediates. The list of such intermediates is quite extensive and includes $Si{=}C$ (silaethylenes; also referred to as silaethenes, silaalkenes and silaolefins), $Si{=}O$ (silanones), $Si{=}S$ (silathiones), $Si{=}N$ (silaimines), $Si{=}Si$ (disilenes), $Ge{=}C$ (germaethylenes), $Ge{=}O$ (germanones), $Ge{=}S$ (germathiones), $Ge{=}N$ (germaimines) and $Ge{=}P$ (germaphosphimines). The evidence for many of these intermediates is indirect and lies only in the strategy of their generation and the nature of the reaction products. However, several have been characterized by physicochemical methods either in low-temperature matrices or at high temperatures. The formation and properties of these types of unstable multiply bonded intermediates were reviewed in 1979.[13] Later data may be found in the main group chemistry reviews.[14]

Various routes to silaethylenes have been devised:

Carbene rearrangement[15,16]

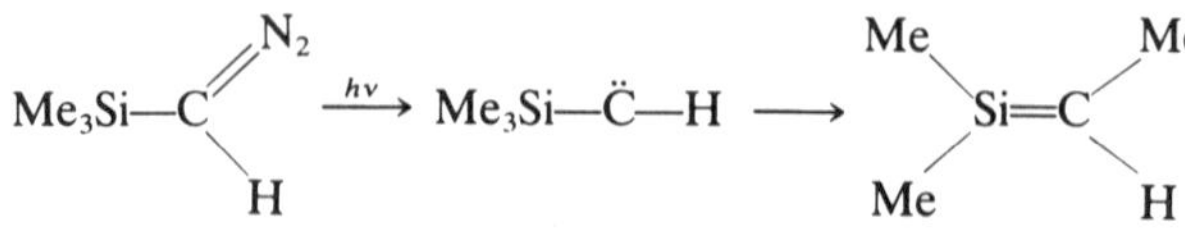

$$\tag{1}$$

(1)

Silacyclobutane pyrolysis or photolysis[17,18]

$$Me_2Si{=}CH_2 + C_2H_4 \tag{2}$$

(2)

Bridgehead extrusion from bicyclooctadienes[19]

$$\longrightarrow H_2Si{=}CH_2 + \quad (3)$$

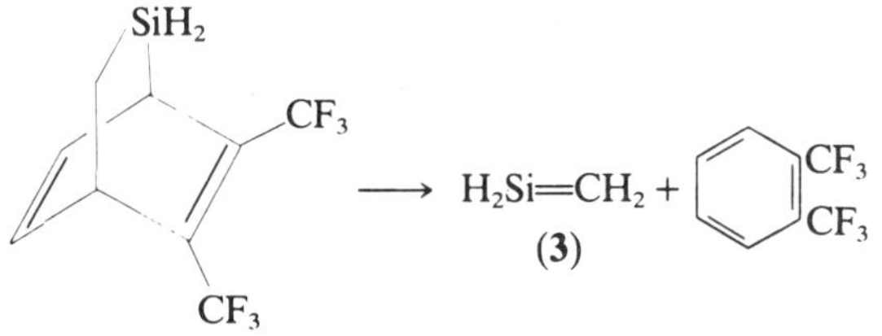

(3)

Sigmatropic rearrangement of acylpolysilanes[20]

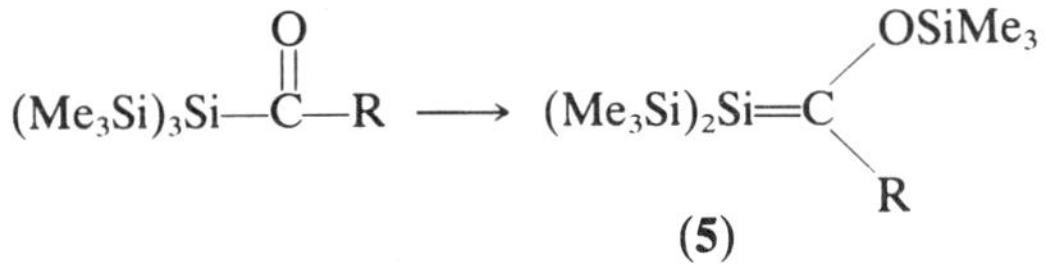

$$(Me_3Si)_3Si{-}\overset{\overset{\displaystyle O}{\|}}{C}{-}R \longrightarrow (Me_3Si)_2Si{=}C\overset{\displaystyle OSiMe_3}{\underset{\displaystyle R}{\big\langle}} \qquad (4)$$

(5)

R = Bu^t, adamantyl

Elimination of LiF[21]

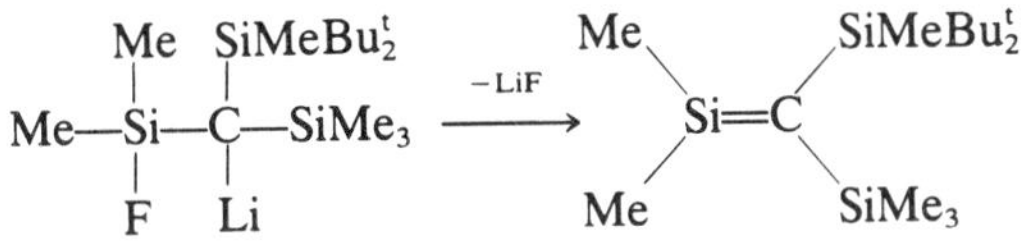

$$(5)$$

(6)

Small unsubstituted silaethylenes, such as **(1)**, **(2)** and **(3)** are stable only in low temperature matrices. On warming, 'head-to-tail' dimerization usually occurs. Steric crowding, however, has a dramatic stabilizing effect, and the sterically hindered silaethylenes, **(5)** and **(6)**, may be isolated as stable crystalline solids. Nevertheless, the Si$=$C double bond even in these compounds is very reactive, undergoing aerobic oxidation and many other reactions with both singly and multiply bonded reagents (Scheme 1).[22–26]

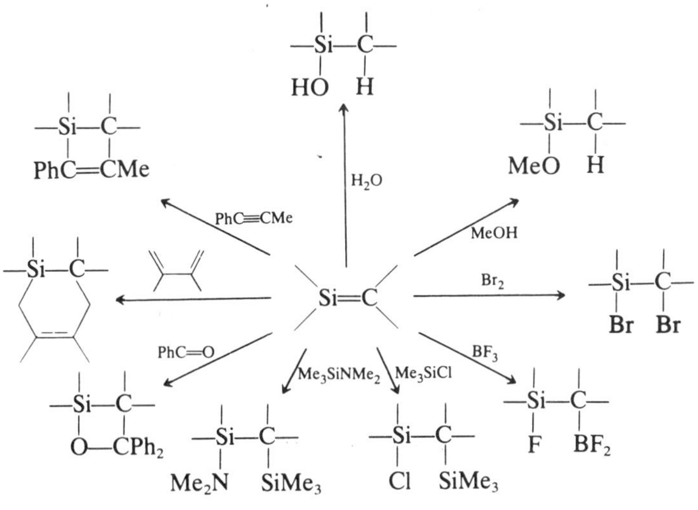

Scheme 1 Typical reactions of the Si$=$C double bond

As expected, experimental determinations of the Si$=$C double bond distance are lower than the corresponding single bond distance. In the electron diffraction study of $Me_2Si{=}CH_2$,[18] the Si$=$C double bond is 183(4) pm compared with a value for the Si$-$C single bond distance of 191(2) pm in the same molecule, and the more precise Si$-$C single bond distance of 186.7 pm derived for methylsilane from microwave data.[27] This value, however, is significantly higher than calculated values for silaethylene, $H_2Si{=}CH_2$, which fall in the range 163–175 pm.[28,29] That the discrepancy arises because of methyl-for-hydrogen substitution at silicon, proposed by Dewar[32] on the basis of MINDO/3 calculations, has been discounted since the more sophisticated double ζ SCF method predicts exactly the same Si$=$C distance (169.2 pm) for both $H_2Si{=}CH_2$ and $Me_2Si{=}CH_2$.[28] The double bond distances in the two sterically crowded crystalline silaethylenes, $(Me_3Si)_2Si{=}C(OSiMe_3)(C_{10}H_{15})$[20] and $Me_2Si{=}C(SiMe_3)(SiMeBu^t_2)\cdot$THF,[34] are 176.4 pm and 174.7(5) pm respectively, much closer to theoretical estimates. However, it should be noted that in both the silaethylene skeleton is not planar as would be expected for sp^2 hybridization at both silicon and carbon. In the former adamantyl-substituted silaethylene, the molecule is twisted 16° about the Si=C double bond most probably due to the bulky substituents. Such a motion would decrease the efficiency of the $p\pi{-}p\pi$ overlap and lead to a lengthening of the Si=C bond.[35] In crystals of the latter THF solvate, the THF molecule is coordinated to the silicon atom, which adopts a distorted tetrahedral geometry, although the planar geometry persists at carbon. Again the efficiency of $p\pi{-}p\pi$ bonding overlap will be reduced and the adduct is probably best regarded as a zwitterion in which the bonding is described in terms of the resonance formulation as shown in equation (6).

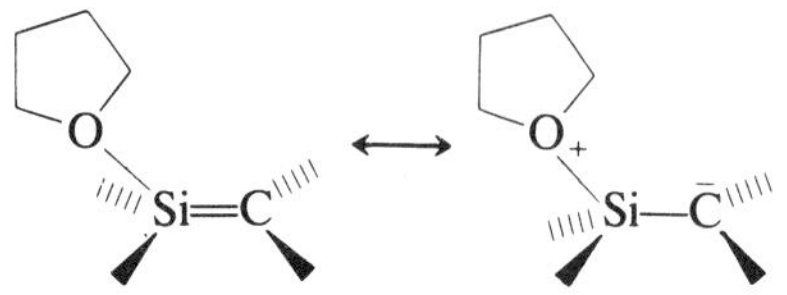

(6)

Silaacetylene, HSi≡CH, is predicted to have a *trans* bent equilibrium geometry, with a Si≡C bond distance of 163.5 pm, *i.e.* about 8 pm shorter than the best estimate for the Si=C distance. The bond order has been suggested to be intermediate between two and three.[36]

In the IR, the Si=C stretching vibration has been assigned to a band at 1002 cm^{-1} in the spectrum of $Me_2Si{=}CH_2$[37,38] and to one at 988 cm^{-1} for $MeHSi{=}CH_2$.[39] A later study of the isotopomeric molecules, $Me_2Si{=}CD_2$ and $(CD_3)_2Si{=}CD_2$, however, indicates that two bands, one around 880 cm^{-1} and the other around 1115 cm^{-1}, contribute to this vibration.[40] Significantly, Brook[22] considers the band at 1135 cm^{-1} in the spectrum of the adamantyl-substituted silaethylene (**5**; $R = C_{10}H_{15}$) to be characteristic of silaethylenes. ^{29}Si NMR chemical shifts for the sp^2-hybridized silicon atoms in silaethylenes lie in the rather small range 41–54 p.p.m. As expected, coupling constants involving sp^2-hybridized silicon are greater than those involving sp^3-hybridized silicon.[41]

The stabilization of a silicon-containing arene has not as yet been achieved. Nevertheless, silabenzene (**7**), 1-silatoluene (**8**) and hexamethyl-1,4-disilabenzene (**9**) have all been generated as transients.[42–47] Both silabenzene and 1-silatoluene are stable when condensed in argon matrices. The UV spectrum exhibits three bands at 212, 272 and 320 nm expected for a π-perturbed benzene. In its photoelectron spectrum, the lowest ionization energies are at 8.0, 9.3 and 11.3 eV (*cf.* calculated values derived by SCF methods of 8.2, 9.2 and 11.5 eV).[48]

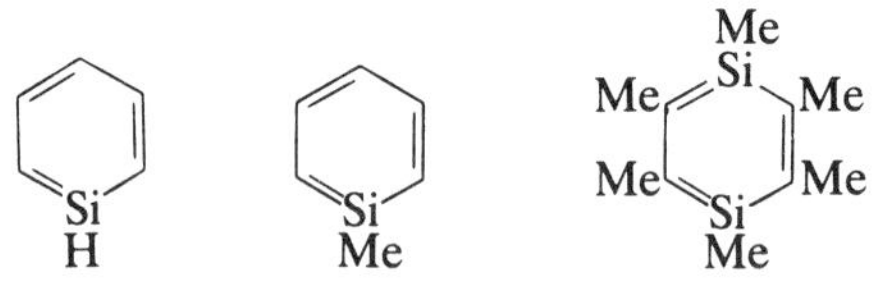

(**7**) (**8**) (**9**)

Disilenes, $R_2Si{=}SiR_2$, vary in stability from short-lived transients to crystalline solids and are indefinitely stable under inert atmospheres depending upon the steric crowding at the Si=Si double bond. Several methods have been employed in the generation of disilenes, including photolysis of linear and cyclic trisilanes,[49–51] photolysis of strained bridgehead molecules such as 7,7,8,8-tetra-*tert*-butyl-7,8-disilabicyclo[2.2.2]octa-2,5-diene[52] and the reduction of 1,2-dihalogenodisilanes using alkali metal naphthalides[53] or lithium wire with simultaneous ultrasonic irradiation.[54] These routes have been employed for the generation of tetraisopropyldisilene, tetra-*tert*-butyldisilene, tetrakis(1-ethylpropyl)disilene and tetraneopentyldisilenes, all of low stability and which may be trapped by the usual types of trapping reagents, and also tetramesityldisilene (**10**), *trans*-1,2-di-*tert*-butyl-1,2-dimesityldisilene (**11**) and tetrakis(2,6-diethylphenyl)disilene (**12**), respectively, which are all stable yellow crystalline solids. The physical and chemical properties of these derivatives indicate that the Si=Si double bond bears a close resemblance to, although distinctly weaker than, the C=C double bond. The most studied compound in this respect is tetramesityldisilene and some typical reactions are shown in Scheme 2. Corroboration of the double bond character comes from UV[49] and ^{29}Si MAS[55] spectra, and from structural data. The ^{29}Si MAS spectrum exhibits an anisotropy similar to that observed in the ^{13}C spectrum of ethylene (*cf.* the anisotropy in the spectrum of tetramesityldisilane which is much smaller as in the ^{13}C spectra of alkanes). As expected, the Si═Si bond distances in all three stable disilenes are distinctly shorter [216.0(1) pm (**10**), 214.3(1) pm (**11**) and 214.0(2) pm (**12**)] than typical single bond distances (*e.g.* H_3SiSiH_3, 245.3 pm). The two silicon and four carbon atoms in (**11**) are coplanar as in (**13**), but both (**10**) and (**12**) show some distortion. (**12**) exhibits a 10° twist about the Si═Si double bond, while retaining planarity about each silicon atom. The distortion in (**10**) involves a moderate antipyramidalization at the silicon atoms in addition to a slight twisting about the multiple bond as in (**14**).[56,57] In spite of these distortions, the observed Si═Si bond distances in these isolable disilene derivatives fall in the range predicted for disilene, $H_2Si{=}SiH_2$ (212.7–217 pm).[58–60] A further theoretical study has shown that the conformation of disilenes depends strongly upon the central π bond and also on the nature of the substituents.[61]

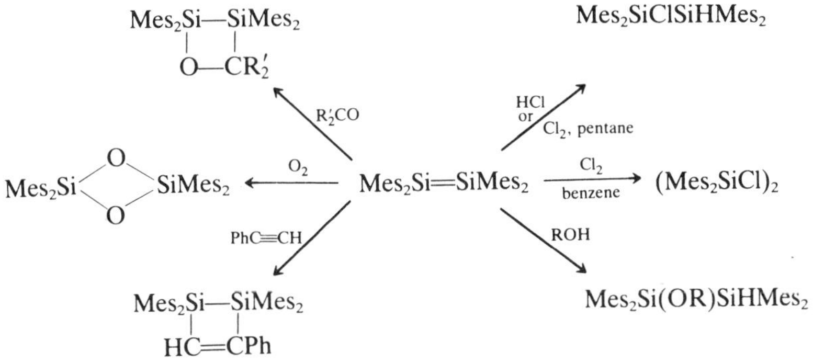

Scheme 2 Typical reactions of tetramesityldisilene (Mes = 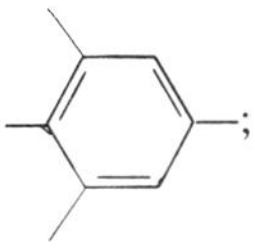 ; R = H, Et; R′ = Me, Ph)

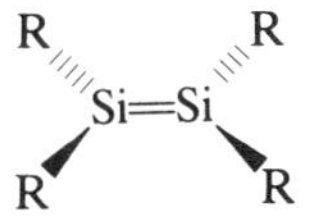

(**13**)

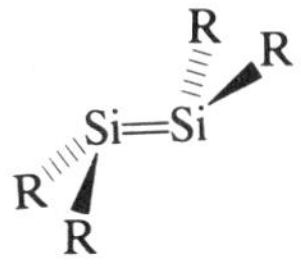

(**14**)

Silanones [$R_2Si{=}O$],[62–64] silaimines [$R_2Si{=}NR$],[65–67] siladiimides [$RN{=}Si{=}NR$],[68] silathiones [$R_2Si{=}S$],[69] germaalkenes [$R_2Ge{=}CR'R''$],[70,71] germaarenes [RC_6H_4GeMe-1,4],[72] germanones [$R_2Ge{=}O$],[73–77] germathiones [$R_2Ge{=}S$],[75–78] germaimines [$R_2Ge{=}NR$][74] and germaphosphimines [$R_2Ge{=}PR$][74–79] are known only as transients for which no structural data is available. However, double ζ SCF calculations predict planar structures for $H_2Si{=}O$, $H_2Ge{=}O$ and $H_2Ge{=}S$ with double bond distances of 148.5, 163 and 202 pm, respectively.[80,81]

Silicon monoxide is known to be stable at either very high or very low temperatures. The structure of the solid material is very complex. A regular cubic lattice may be formed above 1300 °C, but below this temperature the structure is irregular and comprises a mixture of different atomic arrangements.[82] Molecular SiO as well as several other analogues of multiply bonded carbon oxide molecules have been stabilized in low-temperature matrices. Structural and force constant data are collected in Table 5, from which it would appear that the M═O (M = Si, Ge, Sn) bonds have a generally similar electronic structure to the C═O double bond with bond orders of approximately two. Significantly, the force constants for the MO and MO_2 species for each element are similar, indicating that no further increase in bond order occurs in the two atom molecule (*cf.* CO and CO_2). All the species show a considerable shortening (and strengthening) of the M═O bond compared with typical values for M—O single bonds (*vide infra*). Likewise, the SiS force constant in matrix-isolated SiOS is consistent with a double bond, as in gaseous SiS ($f_{SiS} = 494\ N\ m^{-1}$).[91]

Table 5 Structural, Force Constant and Bond Order Data for Molecular MO, MO_2 and X_2MO Species

Species	Ground state geometry	Bond distance (pm)	MO force constant ($\times 10^2$ N m^{-1})	Bond order[a]
SiO	Diatomic	151.0[b]	9.26[b]	
		149.4[c]	11.16[c]	
SiO$_2$	Linear	148.8[c]	11.14[c]	
			9.2[d]	2.5[d]
SiOS[e]	Linear		SiO: 9.2	2.01
			SiS: 4.86	1.9
F$_2$SiO[f]	Planar		9.4	2.1
H$_2$SiO[g]	Planar	148.5	10.57	
GeO[h]	Diatomic	163	7.34	
		165[b]		
GeO$_2$[i]	Linear	163	7.32	2.0[d]
F$_2$GeO[j]	Planar		7.43	1.8
H$_2$GeO[k]	Planar	163.4	8.36	
SnO	Diatomic	183[l]	5.62[b]	
SnO$_2$[m]	Linear	181	5.36	1.9[d]

[a] Estimated using Siebert's rule. H. Siebert, 'Anwendungen der Schwingungsspektroskopie in der anorganischen Chemie', Springer-Verlag, Berlin, 1966.
[b] G. Herzberg, 'Molecular Spectra and Molecular Structure, Vol. I Spectra of Diatomic Molecules', Van Nostrand, New York, 1950.
[c] Calculated values. J. Pacansky and K. Hermann, *J. Chem. Phys.*, 1978, **69**, 963.
[d] H. Schnöckel, *Angew. Chem., Int. Ed. Engl.*, 1978, **17**, 616.
[e] H. Schnöckel, *Angew. Chem., Int. Ed. Engl.*, 1980, **19**, 323.
[f] H. Schnöckel, *J. Mol. Struct.*, 1980, **65**, 115.
[g] Calculated values. R. Jaquet, W. Kutzelnigg and V. Staemmler, *Theoret. Chim. Acta*, 1980, **54**, 205.
[h] J. S. Ogden and M. J. Ricks, *J. Chem. Phys.*, 1970, **52**, 352.
[i] A. Bos, J. S. Ogden and L. Ogree, *J. Phys. Chem.*, 1974, **78**, 1763.
[j] H. Schnöckel, *J. Mol. Struct.*, 1981, **70**, 183.
[k] Calculated values. G. Trinquier, M. Pelissier, B. Saint-Roch and H. Lavayssiere, *J. Organomet. Chem.*, 1981, **214**, 169.
[l] F. T. Loval and D. R. Lide, *Adv. High Temp. Chem.*, 1971, **3**, 177.
[m] A. Bos and J. S. Ogden, *J. Phys. Chem.*, 1973, **77**, 1513.

26.3.2 Silylenes, Germylenes and Stannylenes[92]

There has been considerable interest over the past decade in carbene analogues of silicon, germanium and tin. As with the multiply bonded derivatives of these elements, silylenes, germylenes and stannylenes are unisolable transient intermediates, although several have been studied either at high or low temperatures. Electronically, carbenes behave quite differently to their heavier analogues, and invariably have a triplet ($^3B_1$) ground state with the singlet ($^1A_1$) ground state to higher energy.[94,100–106] However, with dihalogenocarbenes,[107,108] silylenes,[109] germylenes[110,111] and stannylenes,[111] the opposite order of energy states is observed and the latter types of intermediate react as such.[112,113] Structurally, all are angular molecules, although the sterically crowded carbene, dimesitylcarbene, has a nearly linear skeleton (*cf.* other diarylcarbenes where the bond angle is *ca.* 150°).[114] Structural data, both calculated and experimental, are collected in Table 6. Consideration of the data confirm the singlet ground state in each case. Particularly diagnostic is the low value of the XMX bond angle, which is calculated to be about 15–25° higher in the triplet first excited state. In addition, the M—X bond distance is reduced in the triplet excited state, which would appear to indicate a greater p character for the bonds in the ground state. Vibrational spectra for mixed-isolated SiH$_2$,[115] SiMe$_2$[116] and halogenated silylenes,[117–120] germylenes and stannylenes[121] confirm the bent C_{2v} geometry for these molecules in this condensed phase. Irradiation of SiMe$_2$ with 450 nm light under these conditions promotes its photoconversion to 1-methylsilene (equation 7).[116]

$$\text{Me}_2\text{Si:} \xrightarrow[\text{argon matrix}]{450\,\text{nm}} \overset{\text{H}}{\underset{\text{Me}}{\diagdown}}\text{Si}{=}\text{CH}_2 \qquad (7)$$

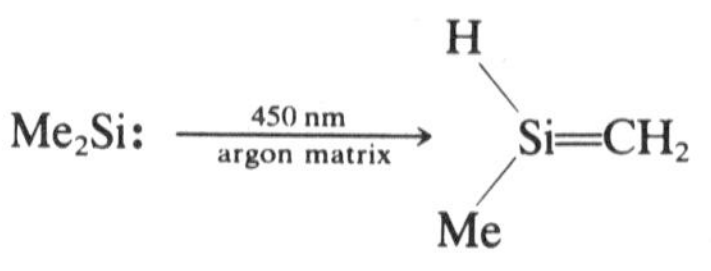

Table 6 Structural Data for Silylene, Germylene and Stannylene Molecules[a]

Molecule	*M—X* (pm)	⟨*XMX* (°)	*Molecule*	*M—X* (pm)	⟨*XMX* (°)	*Molecule*	*M—X* (pm)	⟨*XMX* (°)
SiH_2[b]	$^1A_1$: (151)	(93.5)	GeH_2[c]	$^1A_1$: (160)	(93)	SnH_2[d]	$^1A_1$: (175.6)	(92.7)
	$^3B_1$: (147)	(117.6)		$^3B_1$: (154.5)	(118.4)		$^3B_1$: (170.7)	(118.2)
SiHMe[e]	(151)[f]	(95.9)	$GeMe_2$[c]	$^1A_1$: (202)	(98)			
	(190)[g]			$^3B_1$: (200)	(117.5)			
SiF_2	159.01(1)[h]	100.76(2)[h]	GeF_2	173.21(1)[j]	97.16(1)[j]	SnF_2[k]		94(5)
	(159)[i]	(102.7)[i]		$^1A_1$: (176)[c]	(97)[c]			
				$^3B_1$: (175)[c]	(112)[c]			
$SiCl_2$[l]	208.3(4)	102.8(6)	$GeCl_2$[m]	218.3(4)	100.3(4)	$SnCl_2$[n]	234.7(7)	99(1)
$SiBr_2$[l]	224.3(5)	102.7(3)	$GeBr_2$[o]	233.7(13)	101.2(9)	$SnBr_2$[p]	255(2)	95[q]

[a] Values in parentheses are taken from *ab initio* calculations.
[b] J. H. Meadows and H. F. Schaeffer, *J. Am. Chem. Soc.*, 1976, **98,** 4383.
[c] J. C. Barthelat, B. Saint-Roch, G. Trinquier and J. Satgé, *J. Am. Chem. Soc.*, 1980, **102,** 4080.
[d] G. Olbrich, *Chem. Phys. Lett.*, 1980, **73,** 110.
[e] G. Trinquier and J. P. Malrieu, *J. Am. Chem. Soc.*, 1981, **103,** 6313.
[f] Si—H distance.
[g] Si—C distance.
[h] H. Shoji, T. Tanaka and E. Hirota, *J. Mol. Spectrosc.*, 1973, **47,** 268; V. M. Rao, R. F. Curl, P. L. Timms and J. L. Margrave, *J. Chem. Phys.*, 1965, **43,** 2557.
[i] B. Wirsam, *Chem. Phys. Lett.*, 1973, **22,** 360; C. Thomson, *Theor. Chim. Acta*, 1973, **32,** 93.
[j] H. Takeo, R. F. Curl and P. W. Wilson, *J. Mol. Spectrosc.*, 1971, **38,** 464; H. Takeo and R. F. Curl, *J. Mol. Spectrosc.*, 1972, **43,** 21.
[k] R. F. Hauge, J. W. Hastie and J. L. Margrave, *J. Mol. Spectrosc.*, 1973, **45,** 420.
[l] I. Hargittai, G. Schiltz, J. Tremmel, N. D. Kagramanov, A. K. Maltsev and O. M. Nefedov, *J. Am. Chem. Soc.*, 1983, **105,** 2895.
[m] G. Schultz, J. Tremmel, I. Hargittai, I. Berecz, S. Bohatka, N. D. Kagramanov, A. K. Maltsev and O. M. Nefedov, *J. Mol. Struct.*, 1979, **55,** 207.
[n] A. A. Ishchenko, L. S. Ivashkevich, E. Z. Zasorin, V. P. Spiridonov and A. A. Ivanov, Sixth Austin Symposium on Gas Phase Molecular Structure, Austin, Texas, 1976.
[o] G. Schultz, J. Tremmel, I. Hargittai, N. D. Kagramanov, A. K. Maltsev and O. M. Nefedov, *J. Mol. Spectrosc.*, 1982, **82,** 107.
[p] M. Lister and L. E. Sutton, *Trans. Faraday Soc.*, 1941, **37,** 393, 406.
[q] Assumed value.

26.3.3 Bivalent Germanium, Tin and Lead Halide Structures

The C_{2v} geometry characteristic of isolated molecules of bivalent germanium, tin and lead[122] halides in the vapour phase or low-temperature rare gas matrices (*vide supra*) is not preserved in their solid lattices. Rather, variants of close-packed lattices are formed with distortions arising because of the metal non-bonding pair of electrons. The basic structural unit is usually recognizable as a trigonal pyramid (**15**), where the metal forms three quite short bonds with angles of *ca.* 90°, but additional bonds or contacts are often present leading to distorted pseudotrigonal bipyramidal (**16**), octahedral (**17**) or facially capped trigonal prismatic (**18**) coordination geometries.

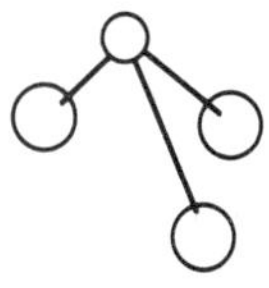
(**15**)

(**16**)

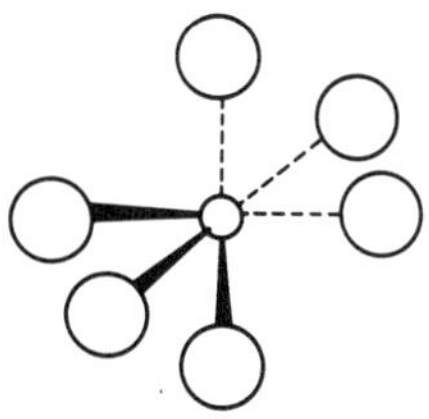
(**17**)

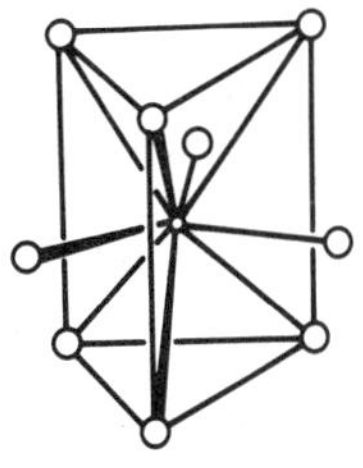
(**18**)

The structures of the binary metal(II) halides show a distinct trend to a more ionic type of lattice as the atomic numbers of the metal and halogen increase. Both germanium(II) and tin(II) fluorides have a strongly fluorine-bridged structure in which the structural unit is an $[MF_3]$ trigonal pyramid.[123–125] In GeF_2, these units are fluorine-bridged giving infinite spiral chains in which the germanium atoms have pseudotrigonal bipyramidal geometry (Figure 1a). Crystals of monoclinic SnF_2, however, comprise puckered eight-membered $[Sn_4F_4]$ rings, and each tin atom is coordinated by a highly distorted octahedron of fluorine atoms (Figure 1b). At normal temperatures, PbF_2 crystallizes with the cubic fluorite lattice,[126] but also forms an orthorhombic modification with the prototype $PbCl_2$-type lattice (**18**) exhibited by anhydrous $SnCl_2$, $SnBr_2$ and $PbBr_2$.[127–129] In this layer-type lattice, the metal atom is surrounded by nine halogen atoms from three different layers, six at the apices of a trigonal prism and three in facially capping positions. The coordination is, however, far from regular, and the metal forms three short bonds (the typical $[MX_3]$ structural unit) together with four additional close contacts. A further two halogen atoms are at even longer distances. Germanium(II) and lead(II) iodides crystallize with the hexagonal anti-CdI_2 lattice, in which the metal atom is surrounded octahedrally by six iodine neighbours, although both compounds exhibit a large number of modifications which differ in the layer sequence.[130,131] Crystals of tin(II) iodide possess a unique layer lattice with the metal atoms occupying two different sites. Two-thirds occupy sites based on the $PbCl_2$-type lattice, each metal atom being coordinated by six iodine atoms at the apices of the trigonal prism, and also by a seventh from a symmetry-related prism in a facial position (Figure 2). The remaining tin atoms are in $PdCl_2$-type chains, which interlock with the $PbCl_2$-type part of the structure to give almost perfect octahedral coordination.[132]

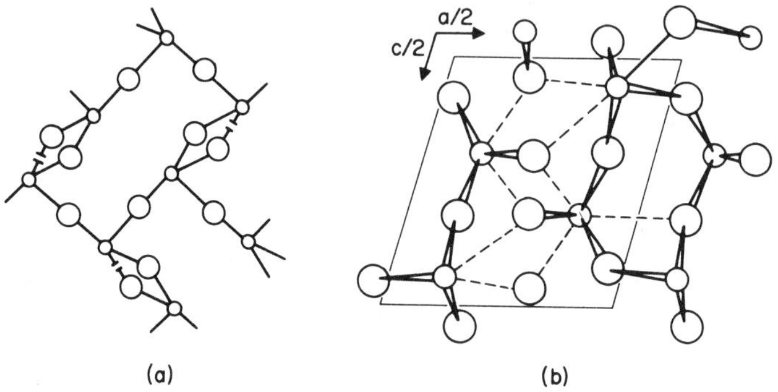

Figure 1 Crystal structures of (a) GeF_2 and (b) the monoclinic modification of SnF_2

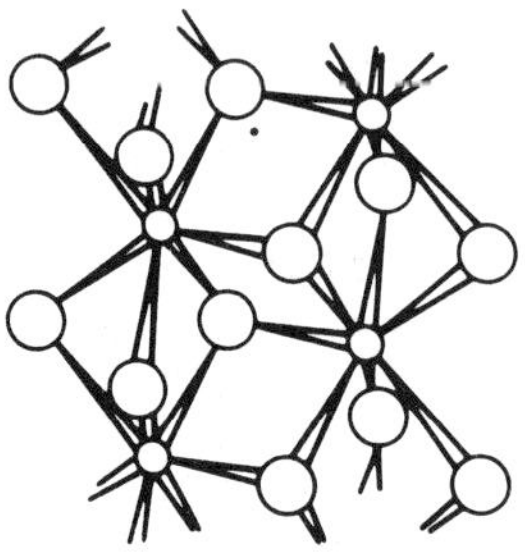

Figure 2 View of the trigonal prismatic sites in SnI_2

A similar range of coordination geometries is also observed for halide complexes and mixed halides. Although many complexes of stoichiometries $MX_2{\cdot}L$ and $MX_2{\cdot}2L$ have been synthesized, the structures of only a few are known. Crystals of the *N*-methylbenzimidazole complex of germanium(II) chloride comprise isolated pyramidal adduct molecules.[133] The basic unit of the thiourea complex of tin(II) chloride is again the pyramidal $[SnCl_2{\cdot}S{=}C(NH_2)_2]$ adduct, but in this case these units are tightly bound into chains by both chlorine and sulfur bridging.[134] In contrast, the structure of the 1,4-dioxane complex consists of $SnCl_2$ units linked by dioxane ligands to form polymeric chains, with pseudotrigonal bipyramidal geometry for tin (Figure 3).[135]

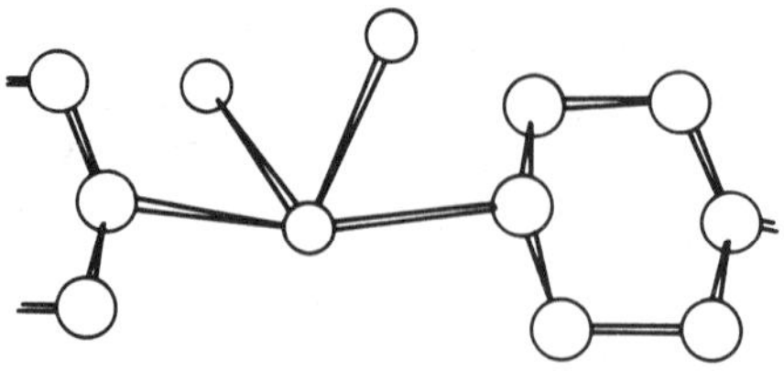

Figure 3 The stereochemistry at tin in *catena*-$SnCl_2{\cdot}$dioxane

The structures of the tin(II) and lead(II) bromide hydrates, $2SnBr_2{\cdot}H_2O$, $3SnBr_2{\cdot}H_2O$, $6SnBr_2{\cdot}5H_2O$ and $3PbBr_2{\cdot}2H_2O$, are based upon the trigonal prismatic geometry (**18**) of the anhydrous halides. The metal atoms are surrounded by a trigonal prism of bromine atoms, with bromine atoms, water molecules or vacancies in the facially capping sites. Different coordination types may occur within the same hydrate. For example, in $2SnBr_2{\cdot}H_2O$ the tin atoms may experience coordination by either eight bromines, seven bromines and a water molecule, or seven bromines.[136,137] Tin(II) chloride dihydrate has a markedly different structure. One of the water molecules is strongly coordinated to the tin atom forming $[SnCl_2{\cdot}H_2O]$ units with typical pyramidal geometry, which form double layers alternating with layers of water of crystallization. Layers are held together by a network of hydrogen bonds.[138,139] The structure of basic tin(II) chloride is very complex indeed. The correct composition is $Sn_{21}Cl_{16}(OH)_4O_6$, and not Sn(OH)Cl as first thought. The structure consists of infinite anionic and cationic nets of stoichiometry $[Sn_5(OH)_6Cl_6]_\infty^{2\infty-}$ and $[Sn_8O_3(OH)_4Cl_5]_\infty^{3\infty+}$, with $[Sn(OH)_3]$, $[Sn(OH)_2Cl]$, $[Sn(OH)_2Cl_2]$ and cubane $[Sn_4O_3(OH)]$ units and intercalated chloride ions.[140] The lead atom in basic lead chloride, Pb(OH)Cl, is surrounded by five chlorine atoms and three hydroxyl groups in a polyhedron best described as a strongly distorted square antiprism.[141]

Several examples of pyramidal $[MX_3]^-$ anions have been characterized, although most involve additional longer contacts. Discrete isolated $[GeCl_3]^-$ and $[SnCl_3]^-$ anions occur in $[C_{11}H_{17}N_2O_2][GeCl_3]H_2O$,[142] and red and green isomers of $[Co(Ph_2PCH_2CH_2PPh_2)Cl]$-$[SnCl_3]$[143] and $[Co(en)_3][SnCl_3]Cl_2$.[144] No interaction occurs between the two anions in the latter compound, whereas the closely related compound, $[Co(NH_3)_6][SnCl_4]Cl$, contains discrete pseudotrigonal bipryamidal $[SnCl_4]^{2-}$ anions.[144] An intermediate situation is present in the structure of $[NH_4]_2[SnCl_3]Cl{\cdot}H_2O$, where significant interaction between the chloride and trichlorostannate(II) ions occurs. However, neighbouring $[SnCl_3{\cdots}Cl]$ units are also connected by two longer $Sn{\cdots}Cl$ bridges forming a chain structure and completing a severely distorted octahedral environment about the tin.[145] A similar distorted octahedral coordination with three short and three much longer metal–halogen distances is also found in $[NH_4][SnF_3]$,[146,147] $RbGeCl_3$[148] and the low temperature forms of $CsGeCl_3$[149] and $CsSnCl_3$.[150] These two latter materials, along with other cesium trihalogenometallates(II), undergo phase transitions to modifications with the cubic perovskite lattice, in which the Group IV metal has ideal octahedral coordination.[151–155] The methylammonium salts, $[MeNH_3][SnBr_xI_{3-x}]$ ($x = 0$–3), $[MeNH_3][PbX_3]$ and $[MeNH_3][Pb_nSn_{1-x}X_3]$ (X = Cl, Br, I) also have the cubic perovskite lattice,[156] and like the alkali metal salts are intensely coloured and exhibit electrical conductivity due to population of lattice conduction bands by the non-bonding electrons of the Group IV metal. Regular octahedral coordination is very common with the larger halogens, presumably largely due to steric reasons, there being little room for metal(II) lone pair stereoactivity, and is found in crystals of orthorhombic $RbPbI_3$,[157] $CsSnI_3$[158] and the hydrates $KPbI_3{\cdot}2H_2O$, $NH_4PbI_3{\cdot}2H_2O$ and $RbPbI_3{\cdot}2H_2O$[159] which all comprise double chains of edge-sharing $[MI_6]$ octahedra held together by the cations and, in the case of the hydrates, water molecules.

Whereas crystals of $[NH_4][SnF_3]$ contain essentially isolated pyramidal $[SnF_3]$ anions,[146,147] the potassium salt $KSnF_3{\cdot}\frac{1}{2}H_2O$ consists of infinite $[SnF_3]_\infty^{\infty-}$ chains, in which the tin atoms have a distorted square pyramidal geometry (Figure 4a). Within the chains, each tin atom forms longer bonds to the bridging fluorines, whilst the lone pair is directed away from the basal plane.[160] $NH_4SnBr_3{\cdot}H_2O$ has an analogous structure, with water molecules and ammonium ions forming rows between the chains.[161] Intermediate between these two extremes are the $[Sn_2F_5]^-$ and $[Sn_3F_{10}]^{4-}$ anions. The former comprises two $[SnF_3]$ pyramids sharing a common fluorine atom (Figure 4(b)), although a fourth fluorine is located at a distance of only 253 pm from the tin and hence the structure may also be regarded as containing infinite $[Sn_2F_5]_\infty^{\infty-}$ chains.[162] The $[Sn_3F_{10}]^{4-}$ anion consists of three distorted square pyramidal $[SnF_4]$ units similar to those found in $KSnF_3{\cdot}\frac{1}{2}H_2O$ (Figure 4(c)).[163]

A number of materials have been characterized which contain cationic fluorotin(II) networks, including Sn_2F_3Cl,[164] Sn_3F_5Br,[165] Sn_2F_3I[166] and $[Sn_5F_9][BF_4]$.[167] In these, the tin atoms enjoy typical pyramidal $[SnF_3]$ or distorted square pyramidal $[SnF_4]$ coordination, with the halide or $[BF_4]^-$ anions occupying large holes within the networks, as illustrated in Figure 5 for Sn_3F_5Br. Closely related to these structures are those of the mixed-valence fluorides, Ge_5F_{12}[168] and Sn_3F_8,[169] where the cationic fluorometal(II) networks are cross-linked by *trans*-bridging octahedral $[M^{IV}F_6]$ units (Figure 6). In contrast, SnFCl[170] and SnF(NCS)[171] have fluorine-bridged ribbon-like structures with pendant chlorine atoms or isothiocyanate groups

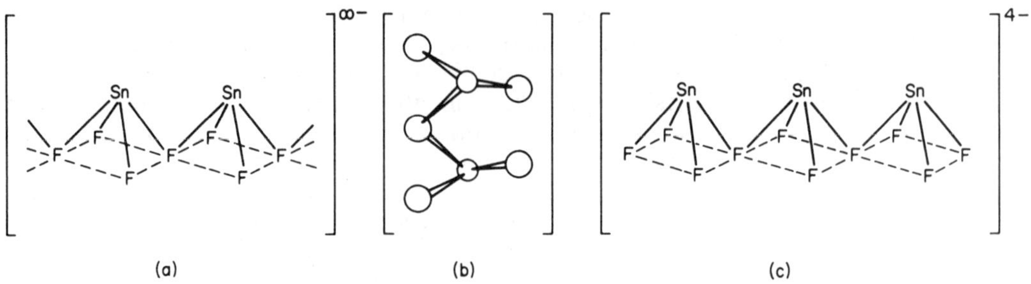

Figure 4 Representations of the structures of (a) the $[SnF_3]_\infty^{\infty-}$ anion, (b) the $[Sn_2F_5]^-$ anion, and (c) the $[Sn_3F_{10}]^{4-}$ anion

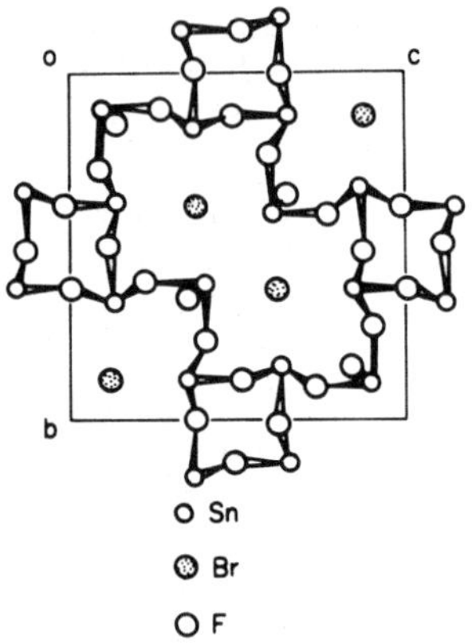

Figure 5 Projection of the structure of Sn_3F_5Br (reproduced by permission from *Acta Crystallogr., Sect. B,* 1978, **34,** 35)

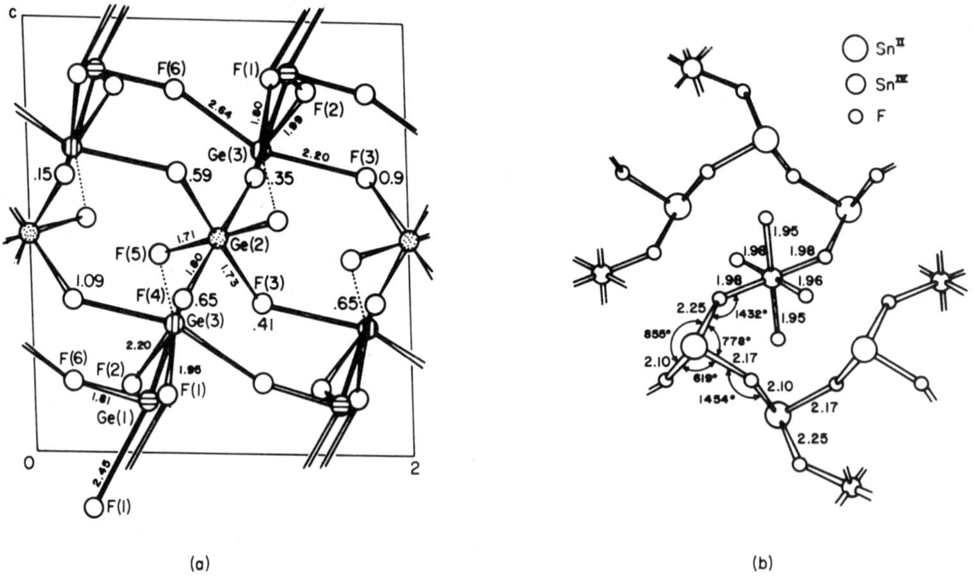

Figure 6 The structures of the mixed-valence fluorides, (a) Ge_5F_{12} and (b) Sn_3F_8 (reproduced by permission from *J. Am. Chem. Soc.,* 1973, **95,** 1834 and *J. Chem. Soc., Chem. Commun.,* 1973, 944)

from each four-coordinated tin. That of SnFCl is shown in Figure 7. The tin(II) and lead(II) chloride iodides are different still and have a distorted trigonal bipyramidal geometry.[172]

Although many complexes are known in which the trichlorostannate(II) anion functions as a donor to transition metals, very few to main group Lewis acids have been described and no strucural data are available. However, the complexes $B \rightarrow SnX_2 \rightarrow BF_3$ (X = Cl, Br, I; B = NMe_3, bipy, TMEDA and DMSO), in which the tin(II) halide functions simultaneously as both a Lewis acid and a Lewis base, have been reported.[173]

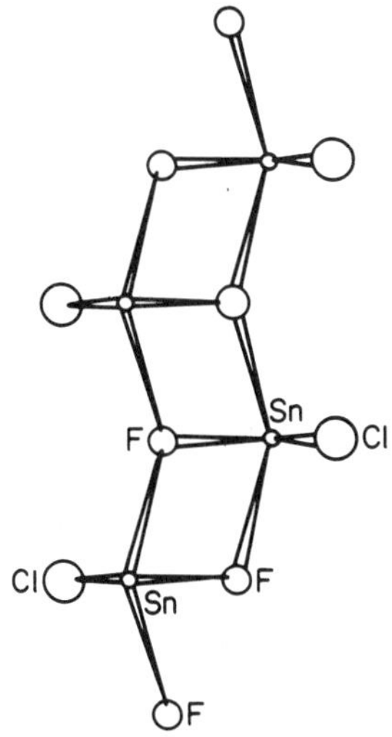

Figure 7 The ribbon-like structure of SnFCl (reproduced by permission from *Acta Crystallogr., Sect. B*, 1976, **32,** 3199)

26.3.4 Tetravalent Halides and Halide Complexes

Tetrahedral geometry dominates the structural chemistry of the halides of silicon and germanium, and all the tetrahalides MX_4, hydride halides MH_nX_{4-n} ($n = 1$–3) and organometallic halides, MR_nX_{4-n} ($n = 1$–3; R = alkyl, alkenyl, aryl), exhibit this stereochemistry in all three phases.[174] In contrast, although the analogous tin derivatives are also tetrahedral in the vapour and liquid phases,[196] the solid tin and lead halides show a preponderance for the formation of halogen-bridged lattices. The tetrahedral compounds warrant little comment, although it is notable that the metal–halogen distance increases whilst the metal–carbon distance decreases in the series Me_nMX_{4-n} as the halogen content of the molecule increases, which is consistent with a redistribution of *p* character in the bonds to the more electronegative ligands.

In contrast to the other tin(IV) halides which are tetrahedral molecules, tin(IV) fluoride forms a very strongly fluorine-bridged, two-dimensional sheet structure with octahedrally coordinated tin (**19**).[197] Dimethyltin fluoride[198] (and almost certainly methyltin trifluoride[199]) has a very similar sheet structure in which the non-bridging fluorine atoms in (**19**) are replaced by methyl groups. Fluorine bridging in trimethyltin fluoride results in a one-dimensional chain structure with a planar $[Me_3Sn]$ moiety and trigonal bipyramidal coordination at tin (**20**).[200] These polymeric materials are infusible and insoluble. Other organotin fluorides are similar and only when the organic groups are very large (*e.g.* in $(PhMe_2CCH_2)_3SnF$) does steric hindrance perhaps preclude association. Bridging by the other halogens is much weaker. A loosely associated one-dimensional chain structure similar to that of Me_3SnF has been observed for trimethyltin chloride at 135 K,[201] whereas crystals of Ph_3SnCl,[202] Ph_3SnBr,[203] $[(Me_3Si)_2CH_2]_3SnCl$[204] and $(Me_3SiCH_2)_3SnI$[205] contain discrete isolated molecules. The distortion away from tetrahedral geometry in tricyclohexyltin chloride suggests some intermediate state.[206] Association into one-dimensional chains occurs for dialkyltin dihalides when the alkyl group is very small. Two types of structure have been characterized. In crystals of dimethyltin dichloride[207] and diethyltin dichloride and dibromide,[208] the chains are formed by the two covalently bonded halogen atoms on each tin bridged to two different tin atoms in the chain as in (**21**), whereas in diethyltin diiodide[208] and bis(chloromethyl)tin dichloride[209] the two halogen atoms essentially chelate the next tin atom in the chain as in (**22**). In both, the geometry at tin is highly distorted, and is best regarded as intermediate between tetrahedral and octahedral. No intermolecular interaction appears to be present in crystals of diphenyltin dichloride[209,210] and bis(biphenyl-2)tin dichloride,[211] although both molecules are distorted from tetrahedral geometry. Bridging by chlorine and bromine is much more effective in the organolead halides. Triphenyllead chloride and bromide have an analogous structure to that of trimethyltin fluoride, although as with trimethyltin chloride, the two lead–halogen distances differ substantially.[212] Symmetrical chlorine bridging occurs in diphenyllead dichloride giving rise to polymeric chains with octahedrally coordinated lead (**23**).[213] Vibrational spectra indicate that the corresponding methyllead halides possess the same structures in the solid, but tetrahedral monomers occur in solution.[214,215]

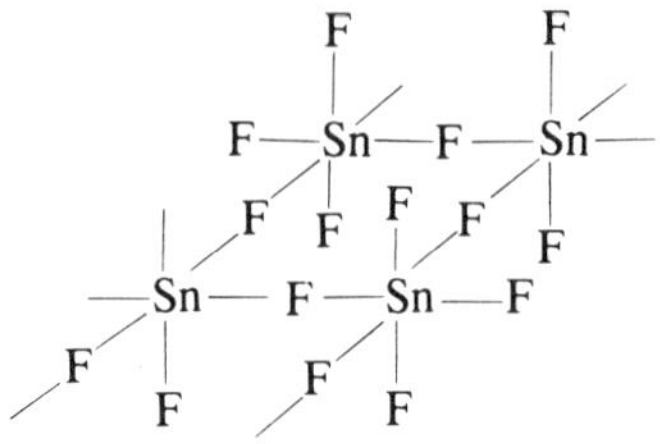

(**19**)

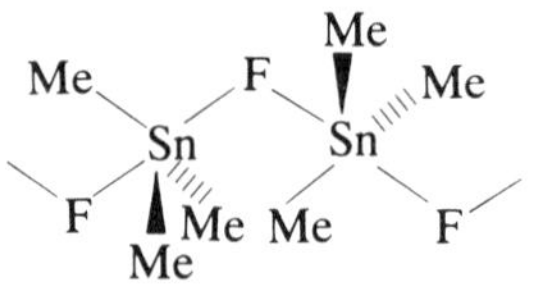

(**20**)

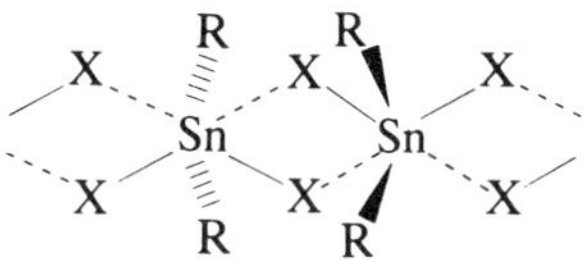

(**21**)

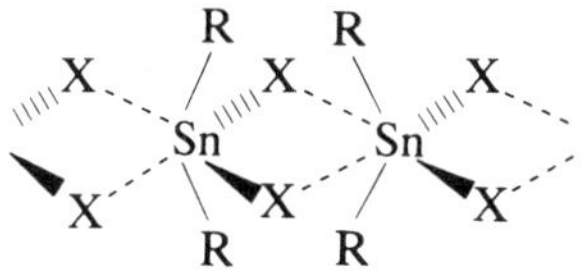

(**22**)

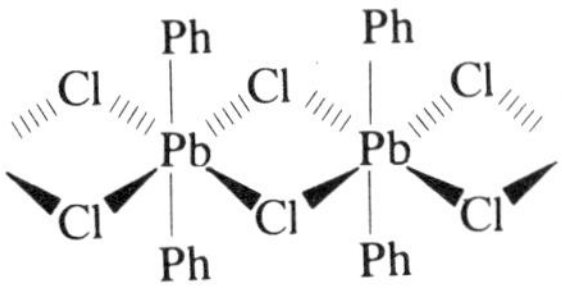

(**23**)

Higher coordination numbers in organotin chlorides can also arise by intramolecular coordination of a donor atom remote in the organic ligand such as in (**24**)–(**28**).[216–222] When two nitrogen atoms are present on the organic group, ionization of halide from tin can occur, as in (**29**).[223] The methylphenyltin homologues, (**27**; R = Me, R′ = Ph) and (**28**), contain chiral tin atoms and exhibit high optical stability since the intramolecular coordination blocks the pathways for stereoisomerization. However, dynamic processes do occur which have been examined by variable temperature NMR and involve Sn—N bond dissociation and inversion at the then uncoordinated nitrogen atom.[222,224] Analogous processes have been observed in the cationic species (**29**),[223] and also in the five-coordinated fluorosilanes (**30**).[225] Intramolecular coordination in the 2-benzoyloxyethylsilanes, $PhCO_2CH_2CH_2SiX_3$ (X = Cl, F), is much weaker than in the analogous tin derivatives such as (**25**), and these compounds can exist in solution as either an acyclic form with four-coordinated silicon or as an intramolecularly five-coordinated heterocycle depending on the concentration and the nature of the solvent.[226] Many other five-coordinated silicon halides have been characterized; (**31**),[227] (**32**)[228] and (**33**)[229] are recent examples illustrating the trigonal bipyramidal geometry typically adopted by such compounds.

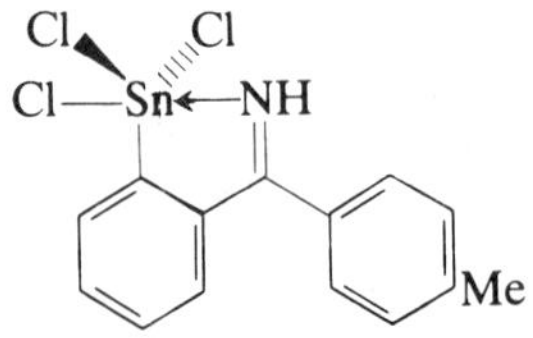

(**24**)

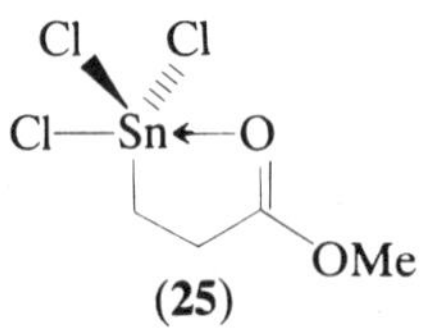

(**25**)

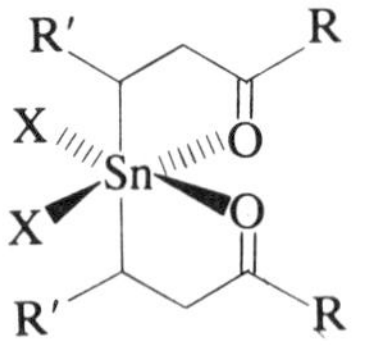

(**26**) X = Cl; R = OMe, NH_2; R′ = H
X = Br; R = OEt; R′ = CO_2Et

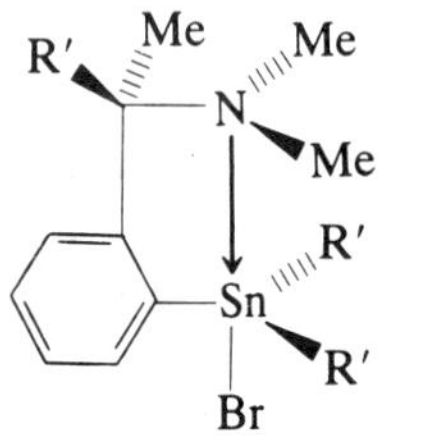

(**27**) R, R′ = Me, Ph; R″ = H, Ph

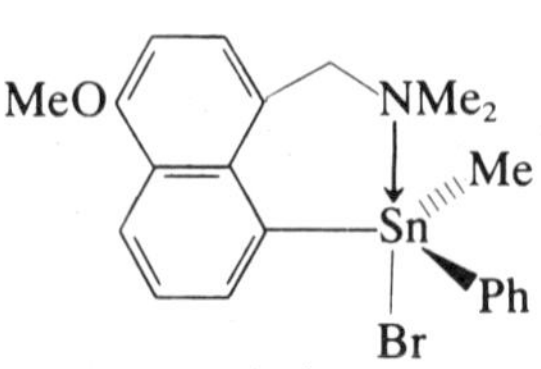

(**28**)

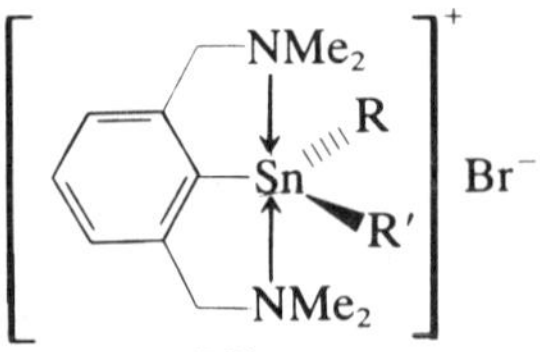

(**29**)

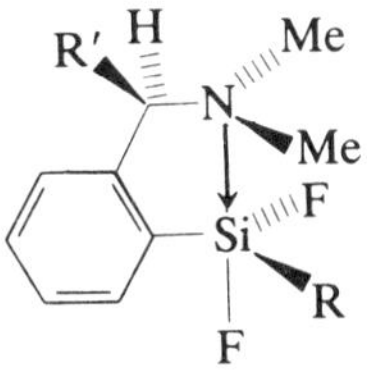

(30) R = F, Me

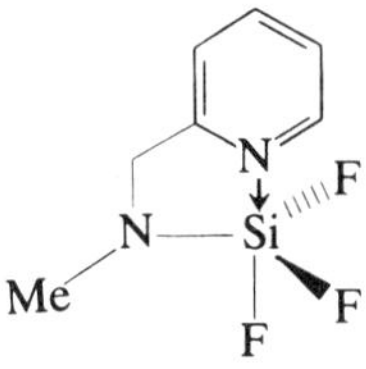

(31)

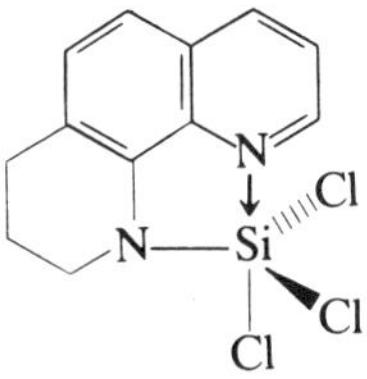

(32)

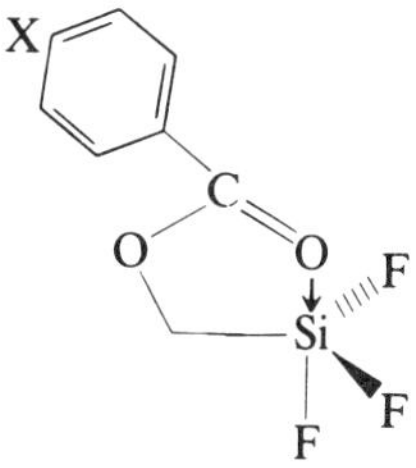

(33) X = F, Cl, Br

The silicon(IV), germanium(IV) and tin(IV) halides MX_4 and organometallic halides $R_{4-n}SnX_n$ are good Lewis acids and acceptors of halide and neutral donor molecules. Lewis acidity in these halides increases in the order Si < Ge < Sn < Pb, I < Br < Cl < F and $n = 1 < 2 < 3 < 4$, and is reflected in the stoichiometry and stability of the resulting complexes. Thus, whereas trimethylchlorosilane shows no tendency for complex formation with pyridine,[230] the corresponding tin adduct, $Me_3SnCl{\cdot}C_5H_5N$, was the first authenticated five-coordinated tin complex.[231] Coordination saturation appears to be reached in the trigonal bipyramidal arrangement (**34**) with a planar [Me_3M] moiety for the monohalides, but the dihalides, trihalides and tetrahalides can form 1:2 adducts with an octahedral geometry as well as 1:1 adducts with the trigonal bipyramidal geometry. In the latter, the organic groups always occupy equatorial sites.

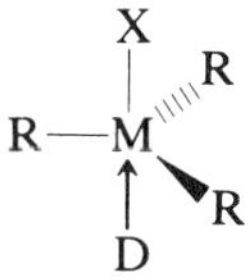

(34)

Typical examples of five- and six-coordinated complexes and complex anions are listed in Table 7. The 1:1 adduct of dimethyltin dichloride and diphenylcyclopropenone is unusual, and comprises a chlorine-bridged dimer with distorted octahedral geometry for tin (**35**),[232] rather than the usual trigonal bipyramidal geometry found in most other 1:1 complexes such as Ph_2SnCl_2·benzothiazole.[238] Two types of behaviour are observed with terdentate ligands. In some isolated cases, seven-coordinated adducts with a distorted pentagonal bipyramidal geometry such as (**36**) are formed.[234,235] but 2,2′,2″-terpyridine promotes displacement of halogen from tin and the formation of ionic complexes such as $[Me_2Sn(terpy)Cl]^+[Me_2SnCl_3]^-$ containing six-coordinate cations and five-coordinate anions.[236] Ionization of halide takes place more readily with bromine and iodine which are more weakly bonded to the Group IV element. Thus, whereas hexamethylphosphoric triamide forms a neutral, trigonal bipyramidal adduct with trimethyltin chloride, the ionic complex $[Me_3Sn(hmpt)_2]^+[Me_3SnBr_2]^-$ is formed with trimethyltin bromide.[237] Similarly, stable ionic 1:1 adducts of the type $[Me_3Si{\cdot}D]^+X^-$ (X = Cl, Br; D = pyridine, hmpt) have been isolated from the interaction of trimethylbromo- and trimethyliodo-silanes with the donor,[238] whilst all four Si—I bonds are cleaved on dissolution of silicon(IV) iodide in the same solvent, from which the complex $[Sn(hmpt)_6]^{4+}4I^-$ containing octahedrally coordinated silicon was isolated.[239]

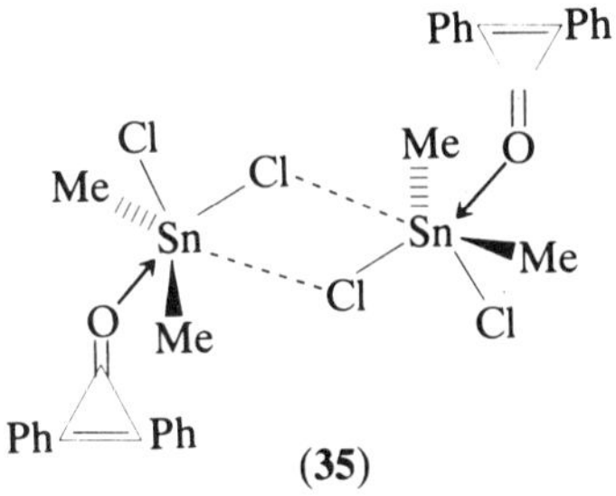

(35)

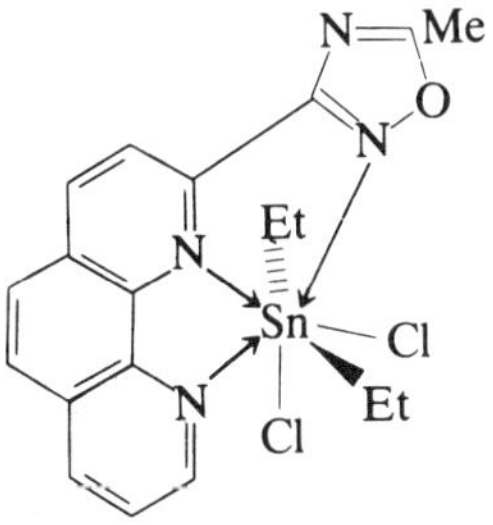

(36)

Various stereochemical permutations are observed for octahedral tin complexes. Adducts of tin(IV) chloride and bromide generally exhibit the *cis* geometry (**37**),[240–247] although some examples of the *trans* geometry are known.[248–252] Reasons for the preference for the *cis*

Table 7 Examples of Five- and Six-Coordinate Complexes and Complex Anions

Complex	*Ref.*	*Complex*	*Ref.*
$[SiF_5]^-$	a	$SnCl_4 \cdot 2MeCN$	u
$[SiF_6]^{2-}$	b, c	$[MeSnCl_4]^-$	v
$[PhSiF_5]^{2-}$	b	$[Me_2SnCl_3]^-$	w
$[Ph_2SiF_3]^-$	a, d, e	$[Me_2SnCl_4]^{2-}$	v
$[Ph_2SiF_4]^{2-}$	d, e	$[Me_3SnCl_2]^-$	w
$SiF_4 \cdot bipy$	f	$Me_3SnCl \cdot Ph_3PCHCOMe$	x
$[GeF_5]^-$	g	$Me_2SnCl_2 \cdot 2DMSO$	y, z
$[GeF_6]^{2-}$	h, i, j	$Ph_2SnCl_2 \cdot bipy$	aa
$[(CF_3)_2GeF_4]^{2-}$	k	$MeSnCl_3 \cdot 2py$	bb
$[SnCl_5]^-$	l, m	$[PbF_6]^{2-}$	cc
$[SnF_6]^{2-}$	n, o	$[PhPbCl_4]^-$	dd
$[SnCl_6]^{2-}$	p, q, r	$[PhPbCl_5]^{2-}$	dd
$[SnBr_6]^{2-}$	s	$Ph_3PbCl \cdot hmpt$	ee
$[SnCl_5(OPCl_3)]^-$	t	$Ph_2PbCl_2 \cdot 2hmpt$	ee
$SnF_4 \cdot bipy$	f	$Ph_2PbCl_2 \cdot hmpt$	ee

[a] D. Schumburg and R. Krebs, *Inorg. Chem.*, 1984, **23**, 1378.
[b] L. Tansjö, *Acta Chem. Scand.*, 1964, **18**, 465.
[c] R. A. J. Driessen, F. B. Hulsbergen, W. J. Vermin and J. Reedijk, *Inorg. Chem.*, 1982, **21**, 3594.
[d] K. Kuroda and N. Ishikawa, *J. Chem. Soc. Jpn.*, 1969, **90**, 322.
[e] R. Müller and C. Dathe, *Z. Anorg. Allg. Chem.*, 1966, **343**, 152.
[f] A. D. Adley, P. H. Bird, A. R. Fraser and M. Onyszchuk, *Inorg. Chem.*, 1972, **11**, 1402.
[g] T. E. Mallouk, B. Desbat and N. Bartlett, *Inorg. Chem.*, 1984, **23**, 3160.
[h] T. E. Mallouk, G. L. Rosenthal, G. Müller, R. Brusasco and N. Bartlett, *Inorg. Chem.*, 1984, **23**, 3167.
[i] B. Frlec, D. Gantar, L. Golic and I. Leban, *Acta Crystallogr., Sect. B*, 1981, **37**, 666.
[j] A. J. Edwards and K. O. Christie, *J. Chem. Soc., Dalton Trans.*, 1976, 175.
[k] D. J. Brauer, H. Bürger and R. Eujen, *Angew. Chem., Int. Ed. Engl.*, 1980, **19**, 836.
[l] R. F. Bryan, *J. Am. Chem. Soc.*, 1964, **86**, 744.
[m] A. G. Ginsburg, N. G. Bokii, A. I. Yanovskii, Yu. T. Struchkov, V. N. Setkina and D. N. Kursanov, *J. Organomet. Chem.*, 1977, **136**, 45.
[n] I. A. Baidina, V. V. Bakakin, S. V. Borisov and N. V. Podberezskaya, *J. Struct. Chem. (Engl. Transl.)*, 1976, **17**, 434.
[o] K. O. Christie, C. J. Schack and R. D. Wilson, *Inorg. Chem.*, 1977, **16**, 849.
[p] J. A. Lerbscher and J. Trotter, *Acta Crystallogr., Sect. B*, 1976, 2671.
[q] K. Nielsen and R. W. Berg, *Acta Chem. Scand., Sect. A*, 1980, **34**, 153.
[r] M. H. Ben Ghozlen, A. Daoud and J. W. Bats, *Acta Crystallogr., Sect. B*, 1981, **37**, 1415.
[s] K. B. Dillon, J. Halfpenny and A. Marshall, *J. Chem. Soc., Dalton Trans.*, 1983, 1091.
[t] A. J. Bannister, J. A. Durrant, I. Raymond and H. M. M. Shearer, *J. Chem. Soc., Dalton Trans.*, 1976, 928.
[u] M. Webster and H. E. Blayden, *J. Chem. Soc. (A)*, 1969, 2443.
[v] L. E. Smart and M. Webster, *J. Chem. Soc., Dalton Trans.*, 1976, 1924.
[w] A. J. Buttenshaw, M. Duchene and M. Webster, *J. Chem. Soc., Dalton Trans.*, 1975, 2231.
[x] J. Buckle, P. G. Harrison, T. J. King and J. A. Richards, *J. Chem. Soc., Dalton Trans.*, 1975, 1552.
[y] N. W. Isaacs and C. H. L. Kennard, *J. Chem. Soc. (A)*, 1970, 1257.
[z] L. A. Aslanov, V. M. Ionov, W. M. Attiya, A. B. Permin and V. S. Petrosyan, *J. Organomet. Chem.*, 1978, **144**, 39.
[aa] P. G. Harrison, T. J. King and J. A. Richards, *J. Chem. Soc., Dalton Trans.*, 1974, 1723.
[bb] L. A. Aslanov, V. M. Ionov, W. M. Attiya and A. B. Permin, *J. Struct. Chem. (Engl. Transl.)*, 1978, **19**, 166.
[cc] R. L. Davidovitch, T. F. Levchishina and T. A. Kaidalova, *Russ. J. Inorg. Chem. (Engl. Transl.)*, 1973, **18**, 325.
[dd] H. Lindemann and F. Huber, *Z. Anorg. Allg. Chem.*, 1972, **394**, 101.
[ee] I. Wharf, M. Onyszchuk, J. M. Miller and T. R. B. Jones, *J. Organomet. Chem.*, 1980, **190**, 417.

geometry are not obvious, but the complexes with the *trans* structure are those with rather bulky donor molecules. Similarly confusing are the structures adopted by the complexes of organotin trihalides and diorganotin dihalides. The two donors are *trans* in the 2:1 pyridine and hmpt complexes of methyltin trichloride as in (**39**), but the analogous DMF complex adopts the *cis* structure (**40**).[253,254] The two organic groups in $R_2SnX_2 \cdot 2L$ complexes are always (to date, anyway!) invariably mutually *trans*, but again the two halide and the two donor molecules can adopt either configuration, (**41**) or (**42**), depending upon the nature of the substituents.[253–262]

IR and NMR spectroscopy are useful techniques for the characterization of complexes which cannot be isolated and for studying solution equilibria. The matrix isolation IR technique has been employed to demonstrate the formation of anions such as SiF_5^-, $MeSiF_4^-$, $Me_2SiF_3^-$ and GeF_5^-, as well as neutral 1:1 adducts such as $SiF_4 \cdot NH_3$.[263–267] ^{19}F NMR shows that the three

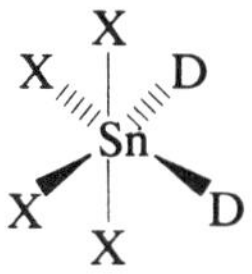

(37)

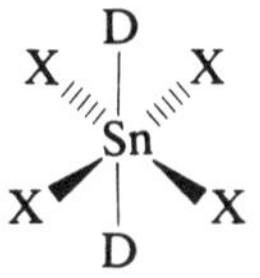

(38)

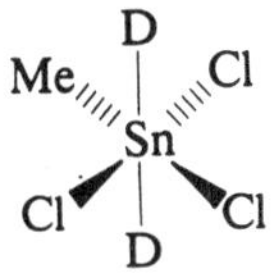

(39)

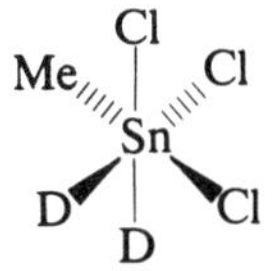

(40)

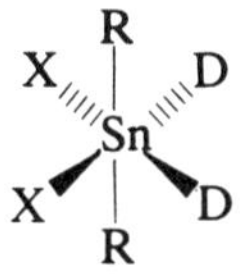

(41)

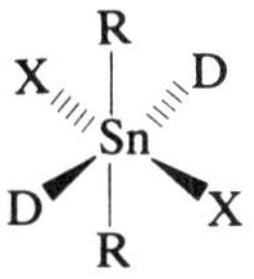

(42)

species, $[CF_3GeF_5]^{2-}$, *trans*-$[(CF_3)_2GeF_4]^{2-}$ and $[(CF_3)_3GeF_2]^-$, are present in solution when fluoride ion is added to the appropriate trifluoromethylgermanium fluoride $(CF_3)_nGeF_{4-n}$ ($n = 1$–3). Unusually, although the *trans* isomer of the $[(CCF_3)_2GeF_4]^{2-}$ anion exists in solution almost exclusively, crystallization affords the corresponding *cis* isomer (as the potassium salt).[268] ^{119}Sn NMR has been used to demonstrate the facile scrambling of halogen between different tin sites. Thus, all possible isomers occur in solution in mixtures of the tin(IV) halides, the anions SnX_5^- and SnX_6^{2-}, and neutral complexes such as $SnX_4{\cdot}2PBu_3$.[269,270] Formation constants may also be derived by an analysis of NMR chemical shift or coupling constant data.[271–273]

26.3.5 Binary and Ternary Chalcogenide Derivatives

Silica, SiO_2, exists in many different crystalline modifications, the principal ones being the α- and β-forms of quartz, tridymite and cristobalite. Other forms include the rather rare minerals, coesite and stishovite, synthetic forms such as keatite and W-silica, and vitreous silica as found in obsidian. All except for stishovite, which has the rutile structure with six-coordinated silicon, contain the $[SiO_4]$ tetrahedron as the basic structural unit. In α-quartz, the thermodynamically most stable form of silica at room temperature, and the other more common modifications, these tetrahedra are linked together by corner sharing forming infinite arrays. The structure of α-cristobolite is illustrated in Figure 8. Crystals of the two modifications of quartz comprise interlinked helical chains which may be either right- or left-handed so giving rise to the chirality of individual crystals. In the synthetic modification, W-SiO_2, adjacent $[SiO_4]$ tetrahedra are linked not by the corner sharing of single oxygen atoms but by the edge sharing of two oxygen atoms forming infinite parallel chains similar to those found in SiS_2 and $SiSe_2$ (Figure 9). Consistent with this structure, crystals of all three materials[274–277] are fibrous.

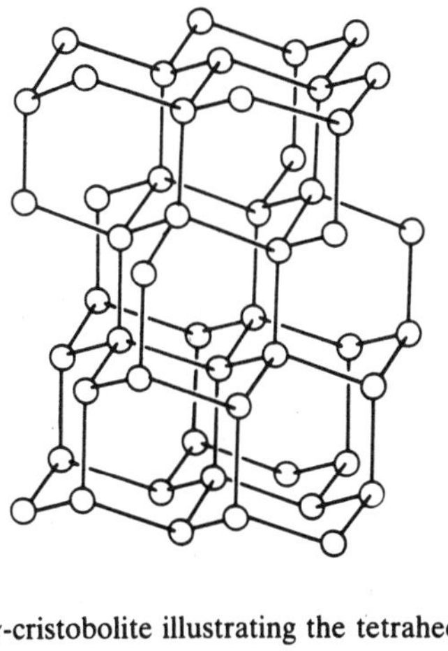

Figure 8 The structure of α-cristobolite illustrating the tetrahedral coordination for silicon.

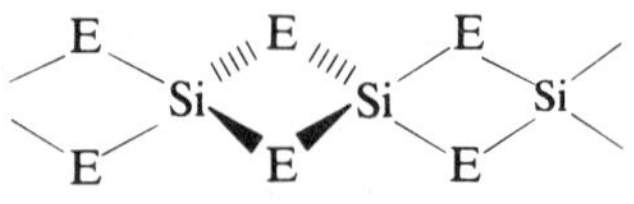

Figure 9 Schematic representation of the chain structures of W-SiO_2, SiS_2 and $SiSe_2$

Germanium(IV) oxide, GeO_2, forms two crystalline modifications. Hexagonal GeO_2 exhibits the β-quartz structure with four-coordinated germanium, whilst the tetragonal modification has a rutile-like structure with six-coordinated germanium and resembles stishovite.[278–280] Both tin(IV) and lead(IV) oxides possess the rutile structure (Figure 10), although another orthorhombic modification of PbO_2 is known with distorted octahedral coordination for lead.[279–283] Limited information is available for the mixed valence tin oxide Sn_3O_4,[284] but the lead analogue, Pb_3O_4, is well characterized and consists of chains of edge-sharing [$Pb^{IV}O_6$] octahedra linked by pyramidal [$Pb^{II}O_3$] units.[285–287]

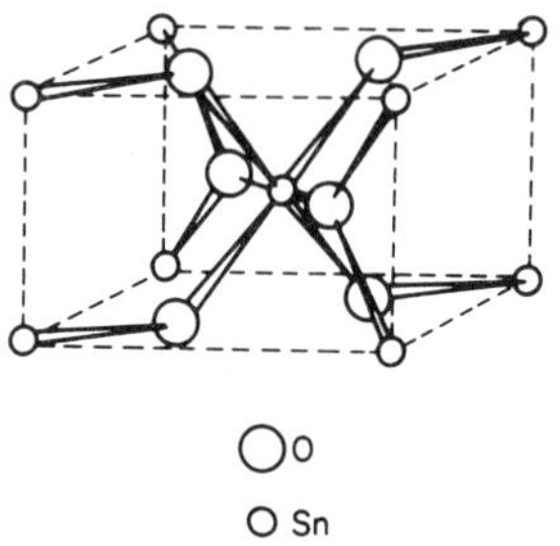

Figure 10 The unit cell of the rutile structure

Germanium(IV) sulfide exhibits two modifications, one with the chain structure illustrated in Figure 9, and the other, like the corresponding selenide, with a distorted CdI_2 lattice.[288] Tin(IV) sulfide and selenide both crystallize with the CdI_2 lattice.[274] The mixed valence tin sulfide, Sn_2S_3, consists of infinite double rutile strands of corner and edge sharing [SnS_6] octahedra connected by tin(II) atoms in typical pyramidal coordination.[289,290]

Tetragonal tin(II) and lead(II) oxides form layer-lattice structures with square pyramidal metal coordination and equal metal–oxygen distances (Figure 11).[291] The metastable orthorhombic modification of PbO has a similar, but more loosely bound layer structure with two short and two longer lead–oxygen distances.[292] Hydrolysis of tin(II) and lead(II) salts yields oxo-metal ions of many different types depending upon the conditions, particularly the pH, some of which have been isolated and characterized crystallographically. The basic tin(II) sulfate, [$Sn_3O(OH)_2$]SO_4, contains discrete [$Sn_3O(OH)_2$]$^{2+}$ cations in which two of the three tin(II) atoms have distorted square pyramidal four-coordination and the third pyramidal three-coordination.[293,294] The [$Pb_4(OH_4)$]$^{4+}$ cation, in which the lead atoms occupy the corners of a slightly distorted tetrahedron with the hydroxyl groups capping each face, is present in crystals of [$Pb_4(OH)_4$]$_3$(CO_3)(ClO_4)$_{10}$·$6H_2O$ and [$Pb_4(OH)_4$](ClO_4)$_4$.[295] The homologous [$Pb_6O(OH)_6$]$^{4+}$ cation is present in the α- and β-modifications of [$Pb_6O(OH)_6$](ClO_4)$_4$·H_2O. This cation is made up of three [Pb_4] tetrahedra sharing common faces, with the oxygen atom located at the centre of the central tetrahedron rather like in the structure of the neutral species $Pb_4O(OSiPh_3)_6$.[296] The six hydroxyl groups are located above each of the six unshared faces of the two terminal tetrahedra.[297,298] From neutral solutions, the hydrate oxides, $3MO{\cdot}H_2O$ (M = Sn, Pb), may be isolated which contain adamantane-like [M_6O_8] clusters,[299] very similar to those in [$Sn_6O_4(OMe)_4$].[300]

The influence of the non-bonding pair of electrons on the coordination geometry adopted by the binary metal chalcogenide decreases markedly with increase in size of other Group VI elements. Germanium(II) sulfide and selenide have very distorted NaCl-type lattices with three short and three long contacts.[301–303] Tin(II) sulfide has a structure with parallel zigzag —Sn—S—Sn—S—Sn—S chains, which are connected by short interchain Sn—S contacts giving a basic pyramidal [SnS_3] structural unit. Tin(II) selenide exhibits one phase which is isomorphous with the sulfide, and a second cubic phase with the NaCl lattice, which is also adopted by tin(II) telluride as well as lead(II) sulfide, selenide and telluride.[304–308]

Silicates, together with the closely related aluminosilicates, form what is certainly the largest

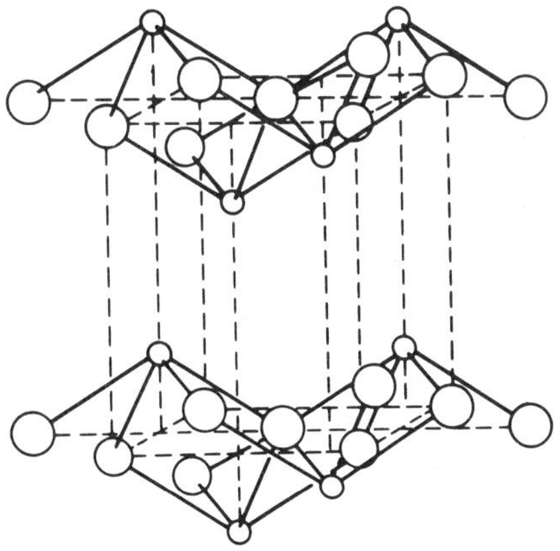

Figure 11 The layer lattice of tetragonal SnO and PbO

family of compounds known in inorganic chemistry. Many other excellent texts and treatises are available,[274,275,309–311] and hence no more than a cursory account is presented here. Despite the enormous variety of stoichiometries and structural types observed, all follow but a few structural principles: (i) silicon is almost without exception tetrahedrally coordinated by four oxygen atoms, (ii) the overall silicate lattice may be regarded as being made up from the close packing (or very nearly so) of oxygen atoms, in which silicon, aluminum and other metal atoms occupy interstices, (iii) small atoms (Li^{I}, Be^{II}, Al^{III}, Si^{IV}) can occupy tetrahedral interstices, whilst larger metal atoms (Na^{I}, K^{I}, Mg^{II}, Ca^{II}, Fe^{II}, Al^{III}, Ti^{IV}) occupy octahedral holes, (iv) charge balance is accomplished by hydroxyl-for-oxygen or by modification of the metal content of the lattice. Alternatively, silicate structures may be regarded as containing $[SiO_4]$ tetrahedra, which can be 'isolated' as in orthosilicates or *neso*-silicates such as in the olivine minerals or synthetic silicates such as Na_4SiO_4,[312] or form disilicates, chain and cyclic silicates, sheet silicates or infinite three-dimensional networks, by sharing one, two, three or all four oxygen atoms respectively, with other $[SiO_4]$ tetrahedra. Some typical examples are shown in Table 8.

Table 8 Principal Types of Silicate Structures

Nomenclature	*Description*	*Examples*
Orthosilicates or *neso*-silicates	Isolated $[SiO_4]$ tetrahedra	Olivine, $(Mg, Fe, Mn)(SiO_4)$ Zircon, $Zr(SiO_4)$
Disilicates or *soro*-silicates	Discrete $[Si_2O_7]$ units formed by two $[SiO_4]$ tetrahedra sharing a common oxygen	Thortveitite, $Sc_2(Si_2O_7)$ Hemimorphite, $Zn_4(OH)_2(Si_2O_7)H_2O$
Cyclic metasilicates or *cyclo*-silicates	Closed rings of $[SiO_4]$ tetrahedra sharing two oxygen atoms	Benitoite, $BaTi(Si_3O_9)$ Beryl, $Be_3Al_2(Si_6O_{18})$ Murite, $[Ba_{10}(Ca, Mn, Tl)_4(Si_8O_{24})$-$(Cl, OH, O)_{12}]\cdot 4H_2O$
Chain metasilicates or *ino*-silicates	Chains or ribbons of $[SiO_4]$ tetrahedra sharing two oxygen atoms	Jadeite, $NaAl(Si_2O_6)$ Wollastonite, $Ca_3(Si_3O_9)$ Tremolite, $Ca_2Mg_5(Si_4O_{11})_2(OH)_2$
Layer silicates or *phyllo*-silicates	Two-dimensional sheets of $[SiO_4]$ tetrahedra sharing three oxygen atoms	Kaolinite, $Al_4(OH)_8(Si_4O_{10})$ Talc, $Mg_3(OH)_2(Si_4O_{10})$ Muscovite, $KAl_2(OH)_2(AlSi_3O_{10})$
Three-dimensional or *tecto*-silicates	Three-dimensional frameworks of $[SiO_4]$ tetrahedra sharing all four oxygen atoms	Feldspars, zeolites, ultramarines

Despite being much less numerous, the structures of germanates bear a striking resemblance to their silicate analogues, and generally comprise lattices formed from basic $[GeO_4]$ tetrahedral units. Thus, the orthogermanates, $BeGeO_4$ and Zn_2GeO_4, contain isolated $[GeO_4]$ tetrahedra, the digermanate, $Sc_2Ge_2O_7$, the $[Ge_2O_7]$ unit, whilst the metagermanates, $BaTiGe_3O_9$ and $CaMgGe_2O_6$, possess cyclic $[Ge_3O_9]$ units and infinite $[GeO_3]_\infty$ chains, respectively. However, several germanates also exhibit octahedral $[GeO_6]$ structural units. For example, the structure of lead α-tetragermanate, α-$PbGe_4O_9$, is based on a mixed Ge—O lattice with the lead atoms situated in lattice vacancies formed by $[GeO_4]$ tetrahedra and $[GeO_6]$ octahedra,[313] whilst that of $K_2Ba[Ge_4O_9]_2$ consists of rings built up from $[GeO_4]$ tetrahedra, which are linked by $[GeO_6]$ octahedra to form a three-dimensional network.[314] Some germanate species are very unusual and do not have their counterparts in silicate chemistry. The $[Ge_5O_{14}]$ clusters (**43**) found in $Tl_8Ge_5O_{14}$ are formed from five $[GeO_4]$ tetrahedra, three of which share two oxygen atoms whilst the other two each share three.[315]

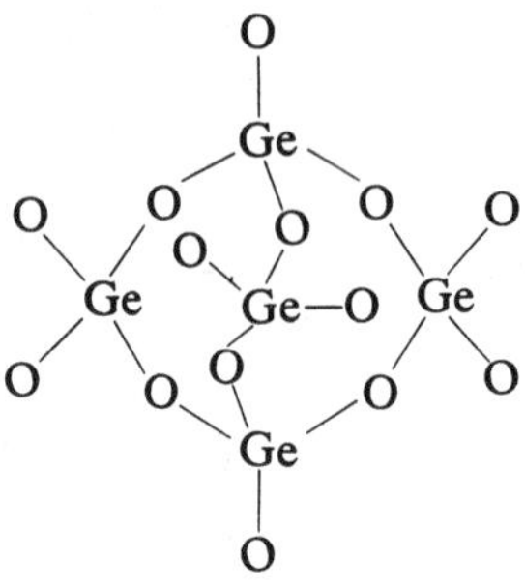

(**43**)

The structures adopted by the stannates are profoundly different to those of silicates and germanates, in spite of the similarity in stoichiometries. Tin generally is six-coordinated in a regular or slightly distorted octahedral fashion by oxygen atoms, although four- and five-coordination do occur. In addition, a number of hydroxystannate(II) species have been characterized. The sodium salts, $Na_2[Sn_2O(OH)_4]$ and $Na_4[Sn_4O(OH)_{10}]$, contain the $[Sn(OH)_3]^-$ and oxo-bridged $[(HO)_2SnOSn(OH)_2]^{2-}$ anions respectively,[316] whilst the barium salt, $Ba[SnO(OH)]_2$, contains the one-dimensional polyanion, $[SnO(OH)]_\infty^{\infty-}$.[317] All three exhibit typical pyramidal coordination for bivalent tin. Octahedral oxygen coordination about tin is present in all the hydroxystannates(IV), $M[Sn(OH)_6]$ (M = 2Na, 2K, Mg, Ca, Mn, Fe, Co, Ni, Cu, Zn, Cd).[318–326] Potassium orthostannate, K_4SnO_4, is isotypic with K_4GeO_4 and contains discrete, isolated $[SnO_4]$ tetrahedra,[327–329] but other materials of the stoichiometry M_2SnO_4 (M = Mg, Ca, Sr, Ba, Mn, Co, Cd, Zn) are not true orthostannates and contain $[SnO_6]$ octahedra which share common edges to give chain or layer structures.[330–335] Octahedrally coordinated tin is also present in the 'distannates', $M_2Sn_2O_7$ (M = lanthanide elements, Bi),[336–338] and most 'metastannates', $MSnO_3$ (M = Li, Tl, Mg, Ca, Sr, Ba, Mn, Cd, Pb).[339–345] The potassium and rubidium metastannates, however, have unusual square pyramidal five-coordination for tin, and the solid contains infinite $[SnO_3]_\infty^{\infty-}$ polyanions as shown in Figure 12.[346,347]

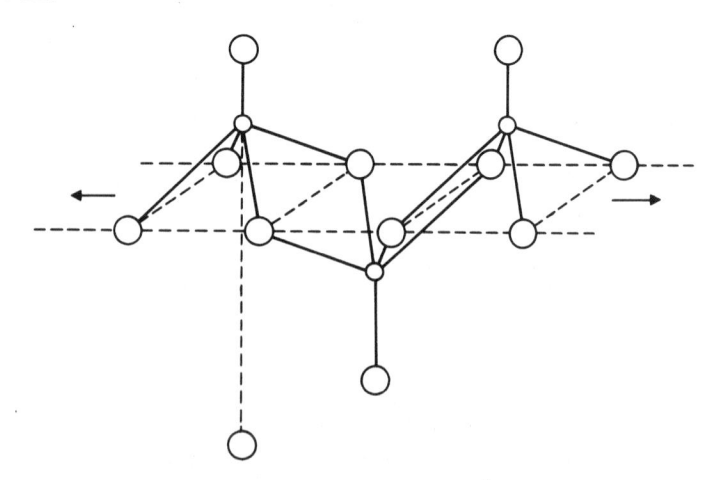

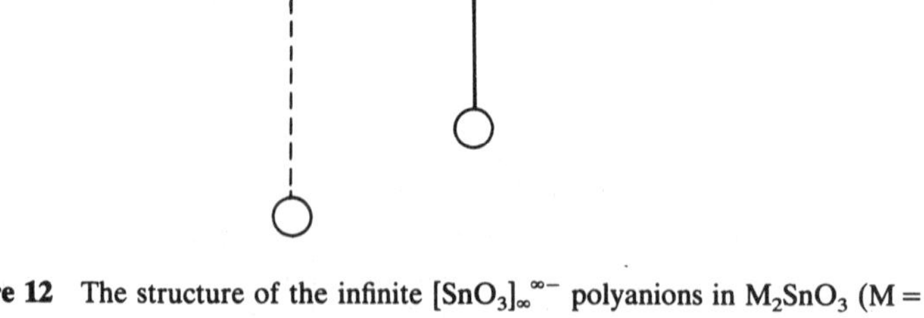

Figure 12 The structure of the infinite $[SnO_3]_\infty^{\infty-}$ polyanions in M_2SnO_3 (M = K, Rb)

Potassium, rubidium and cesium orthoplumbates M_4PbO_4 (M = K, Rb, Cs) all contain isolated $[PbO_4]$ tetrahedra, as does Rb_3NaPbO_4.[348–350] The structure of the 'metaplumbate', Li_2PbO_3, like Li_2SnO_3, is a variant of the NaCl lattice, with octahedrally coordinated lead,[351] whilst the diplumbate, $K_2Li_2(Pb_2O_8)$,[352] and the triplumbates, $M_2Li_{14}(Pb_3O_{14})$ (M = K, Rb, Cs),[353] contain the discrete $[Pb_2O_8]$ (**44a**) and $[Pb_3O_{14}]$ (**44b**) units respectively. Sodium plumbate(II), Na_2PbO_2, has a layer lattice,[354] although the analogous silver salt contains infinite $[PbO_2]_{\infty}^{2\infty-}$ chains.[355]

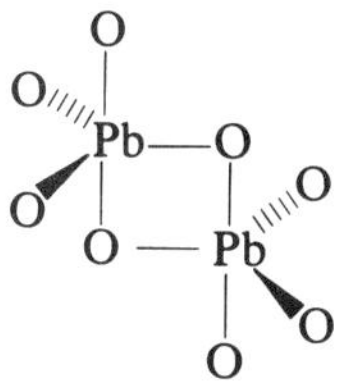

(**44a**)

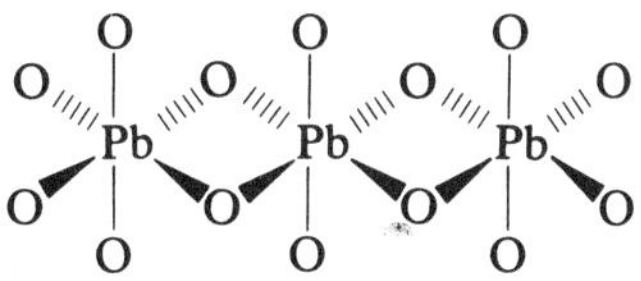

(**44b**)

The structures of heavier chalcogenide analogues of the silicates, germanates and stannates are largely based upon the tetrahedral $[MX_4]$ unit. Isolated orthothiosilicate anion, $[SiS_4]^{4-}$, and all six orthogermanate and orthostannate anions, $[MX_4]^{4-}$ (M = Ge, Sn; X = S, Se, Te), have been characterized.[356–363] Three types of binuclear anion are known. The $[M_2S_7]^{6-}$ (M = Si, Ge, Sn)[357,364,365] and $[M_2X_6]^{4-}$ (M = Ge, X = S, Se; M = Sn, X = S)[366,367] are formed by the vertex sharing (**45**) and edge sharing (**46**), respectively, of two $[MX_4]$ tetrahedra, but the tellurium derivatives, $[M_2Te_6]^{6-}$ (M = Si, Ge, Sn), contain the metal–metal-bonded arrangement (**47**) with a staggered conformation.[368–370] The metathiometallates, Na_2GeS_3[371] and $K_2SnS_3{\cdot}2H_2O$,[372] form infinite $[MS_3]^{2-}$ chains, whilst $Eu_3Ge_3S_9$[373] contains the cyclic $[Ge_3S_9]^{6-}$ anion. A two-dimensional sheet structure is present in La_2GeS_5,[374] formed as with the metathiometallates by the vertex sharing of tetrahedra. The tetranuclear anions, $[Ge_4S_{10}]^{4-}$, $[Ge_4Se_{10}]^{4-}$ and $[Si_4Te_{10}]^{4-}$, adopt a central adamantane-like framework, but may be more consistently described as being composed of four $[MX_4]$ tetrahedra each sharing two vertices with adjacent tetrahedra.[367,375–378] The $[Ge_4Te_{10}]^{8-}$ anion, however, has the cyclic structure $[Te(GeTe_2)_4Te]$.[379] Not unexpectedly, higher coordination numbers are observed with tin. In $Tl_2Sn_2S_5$, the tin has a five-coordinated distorted trigonal bipyramidal geometry,[380] and a six-coordinated distorted octahedral geometry is observed in the lanthanide thiostannates, La_2SnS_5,[381;382] Fe_2SnS_4[383] and Mn_2SnS_4.[384]

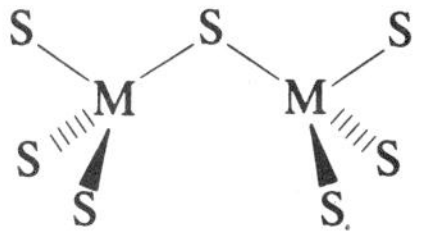

(**45**)

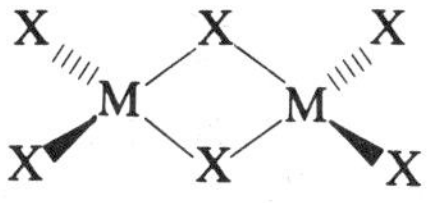

(**46**)

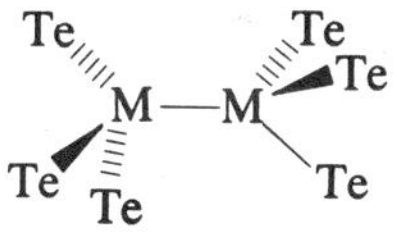

(**47**)

26.3.6 Tetravalent Compounds of Groups IV, V and VI

Compounds containing bonds to elements of Groups IV, V and VI illustrate the rather sharp distinction in the structural chemistries of silicon and germanium on the one hand and tin and lead on the other. The coordination chemistries of silicon and germanium are very similar indeed, and tetrahedral four-coordination is overwhelmingly the rule. Tin and lead compounds in contrast endeavour, wherever possible, to adopt structures containing coordination numbers higher than four, the most common being five and six, although examples of seven and eight are known as well.

Compounds containing four bonds to carbon are almost without exception four-coordinate and tetrahedral, although bond distances vary somewhat. Bonds to electronegative organic ligands are generally longer and weaker, and hence more susceptible to cleavage. For example, the Pb–C (cyclopentadienyl) distance in $Ph_3PbC_5H_5$ is 10 pm longer than the bonds to the phenyl groups, and the bond is easily cleaved by protic reagents;[385] whilst the Sn—C bond distances in $Sn[CF_3]_4$ are lengthened by 6 pm compared with those in $SnMe_4$.[386] When electron withdrawal from the metal is accompanied by the presence of a donor atom remote in the ligand, as in the ethyl ester of trimethylplumbylacetic acid, $Me_3PbC(N_2)CO_2Et$, weak intermolecular association with five-coordination for the metal can occur (**48**) in spite of the presence of four lead–carbon bonds.[387]

Similarly, compounds containing metal–metal bonds are four-coordinated, except in the case of the 1,2-bis(dicarboxylato)ditin compounds, $(RCO_2)R'_2SnSnR'_2(O_2CR)$, which have the structure (**49**) where the carboxylato groups span the tin–tin bond.[388,389]

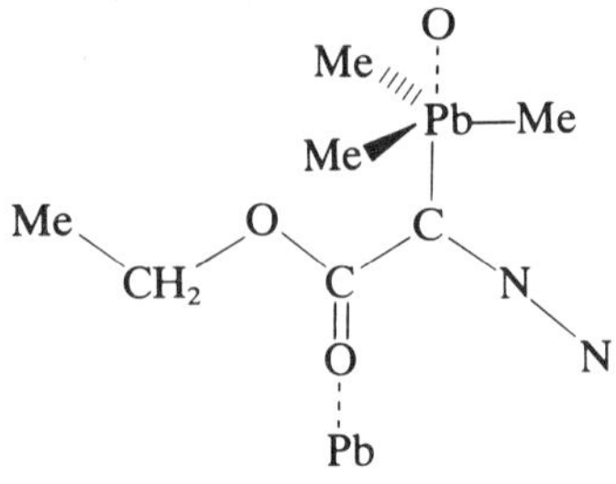

(**48**)

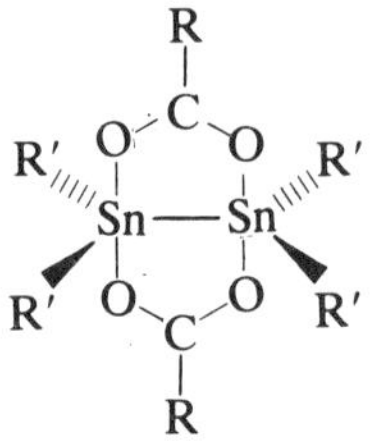

(**49**)

In contrast to other disiloxane derivatives, the SiOSi framework in $Ph_3SiOSiPh_3$ has unexpectedly been found to be linear[390] (*cf.* other disiloxanes and related compounds, *e.g.* $MeH_2SiOSiH_2Me$, $Me_2HSiOSiHMe_2$ and $Me_2SiOSiMe_3$,[391] $Ph_3GeOGePh_3$,[392] $Ph_3SnOSnPh_3$[393] and $Ph_3SiOPbPh_3$[394] which are bent at oxygen). The MOM frameworks in hexabenzyldigermoxane and hexabenzyldistannoxane are also linear, and the phenomenon in all three compounds was attributed to the low electronegativity of the metal coupled with the electron-donating character of the organic group.[395]

Both di-*tert*-butylsilanediol[396] and the analogous di-*tert*-germanediol[397] form hydrogen-bonded one-dimensional chain structures. Diorganodichlorosilanes and -germanes undergo hydrolysis to form cyclic compounds of composition $(R_2MO)_n$ where $n = 3, 4, 5, 6$, as well as polymeric material. All contain approximately tetrahedrally coordinated metalloid, although the exact conformation adopted by the heterocyclic rings depends upon the organic substituent and the ring size. The four-membered $[Si_2O_2]$ ring compound, tetramesityldisiloxane, obtained by the aerobic oxidation of tetramesityldisilene is rather intriguing. The most striking feature of the molecule is the transannular Si—Si distance (only 231 pm), which is actually shorter than a typical Si—Si bond distance (234 pm), and suggests the retention of some Si—Si bonding. Further, the two Si—O bond distances are longer than usual and unequal, and the SiOSi angle is highly constrained (86°). Further, treatment with lithium naphthalide followed by quenching with water yields the disilane-1,2-diol, $Mes_2Si(OH)$—$Si(OH)Mes_2$. Hence, the representation (**50**) is probably a truer description of the bonding than a conventional four-membered disiloxane ring (**51**).[398] No analogous transannular Si—Si interactions occur in other four-membered ring systems such as $Ph_2SiNPhSiPh_2NPh$,[399] $(Bu^tO)_2SiSSi(Bu^tO)S$,[400] $Me_2SiSSiMe_2S$,[401] $Cl_2SiCH_2SiCl_2CH_2$[402] and Ph_8Si_4.[403]

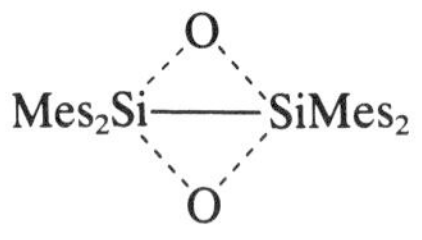

(**50**)

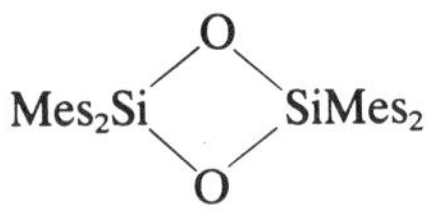

(**51**)

The sesquisiloxanes, $(RSiO_{1.5})_n$ ($n = 4, 6, 8, 10, 12$), adopt a variety of molecular cage structures (Figure 13).[404–408] The *tert*-butylgermanium sesquioxide, $(Bu^tGe)_6O_9$, has the same structure as its silicon analogue.[409]

The sharp antithesis between the molecular structures involving four-coordinated silicon and germanium and the diorgano-tin and -lead oxides, $(R_2MO)_n$ (M = Sn, Pb), and the stannonic and plumbonic acids, $[RM(OH)O]_n$, is astonishing for these latter materials are insoluble and infusible amorphous solids. No X-ray data are available, but the principal structural element in all of them is most probably the four-membered $[M_2O_2]$ ring. The tin compounds are much more studied than their lead counterparts. Indeed, of the plumbonic acids, only PhPb(OH)O has been characterized satisfactorily.[410] Spectroscopic data indicate a cross-linked, three-dimensional network such as (**52**) for the diorganotin oxides, where the stereochemistry of the tin atoms is close to trigonal bipyramidal with equatorial organic groups. The other materials most probably possess similar gross structures exhibiting the same major features. Cross-linking is, however, apparently prevented by bulky ligands, and di-*tert*-butyl- and bis[2,6-diethylphenyl]-tin oxides crystallize as trimeric molecules (**53**) with a planar $[Sn_3O_3]$ ring.[411,412]

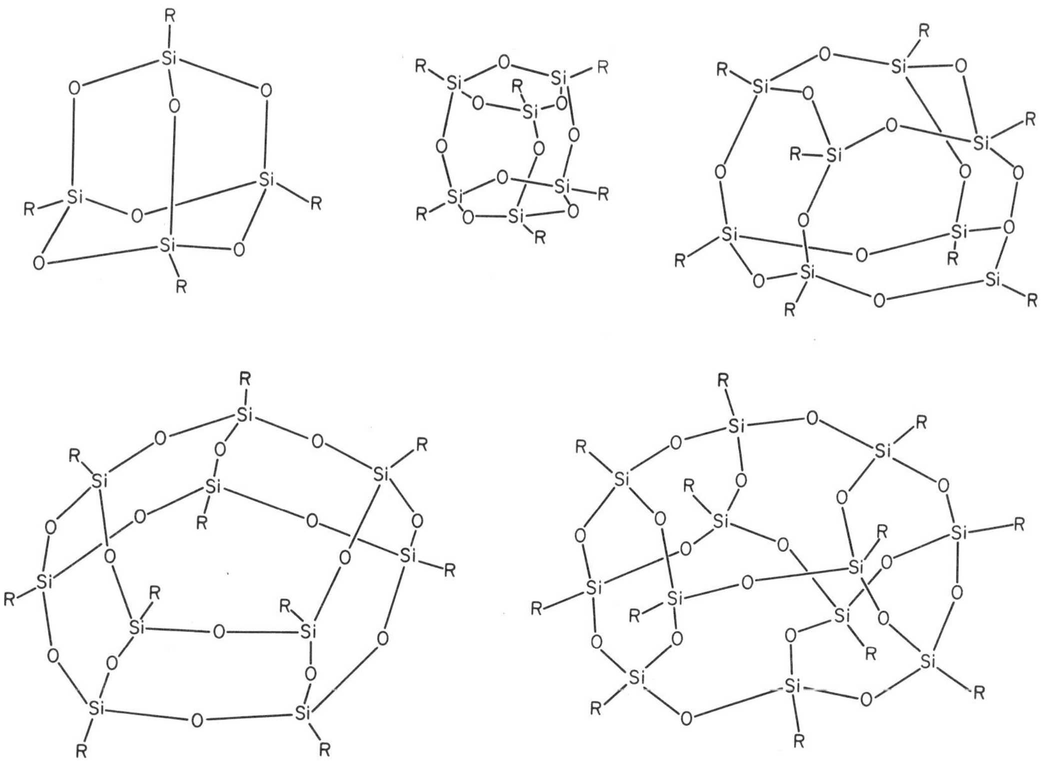

Figure 13 Cage structures formed by organosesquisiloxanes, $(RSiO_{1.5})_n$, where $n = 4$, 6, 8, 10 and 12

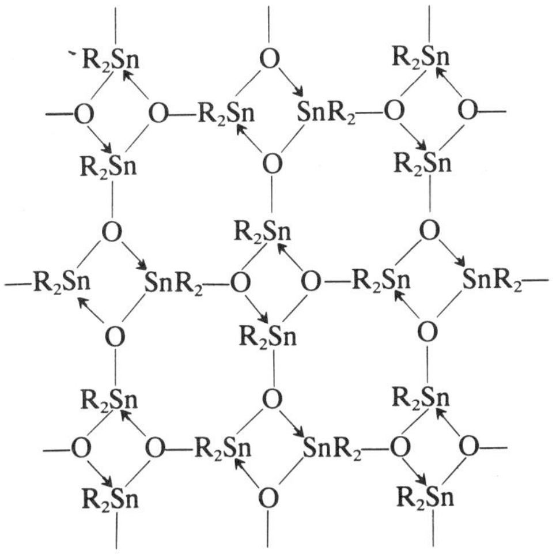

(52)

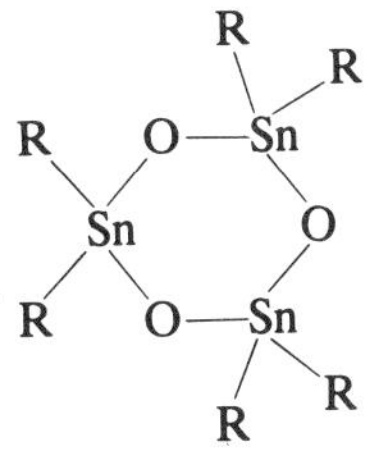

(53)

Much better characterized are the intermediates obtained from the hydrolysis of di- and mono-organotin halides or carboxylates. Hydrolysis of di-*tert*-butyltin dichloride gives initially di-*tert*-butyltin chloride hydroxide, $Bu^t_2Sn(OH)Cl$, which has the hydroxyl-bridged chain structure (**54**). Further hydrolysis yields a poorly defined dihydroxide, which readily dehydrates to the oxide. With less bulky organic groups, two intermediates can be isolated from the hydrolysis reaction, the 1,3-dichlorodistannoxanes, $[ClR_2SnOSnR_2Cl]_2$, and the 1-hydroxy-3-chlorodistannoxanes, $[(HO)R_2SnOSnR_2Cl]_2$. Both types are dimeric, with structures (**55**) and (**56**) respectively, and contain the central four-membered $[Sn_2O_2]$ ring and five-coordinated tin.[413,414] In the corresponding dimeric 1,3-dicarboxylatodistannoxanes, $[(RCO_2)R'_2SnOSn'_2(O_2CR)]_2$, one carboxylate group is unidentate, whilst the other functions as a bridging ligand connecting both types of tin atom (**57**).[415–417] Similar 'ladder'-type structures have also been characterized in $[ClPh_2SnOSnPh_2Cl]_2$,[418] $[ClPh_2SnOSnPh_2(OH)]_2$[418] and

$\{[(NCS)Me_2Sn]_2O\}_2$[419] and the tristannoxane, $\{(N_3)Bu_2SnOSnBu_2OSnBu_3(N_3)\}_2$.[420] Triorganotin and triorganolead hydroxides,[421,422] alkoxides[424–425] and oximes[426] show a marked tendency to associate into one-dimensional chain polymers in which planar $[R_3Sn]$ units are bridged by oxgyen atoms as in (**58**). Steric crowding at either the metal or oxgyen reduces this tendency towards association. Thus, whereas trimethyltin methoxide has the associated structure (**58**), tri-*n*-butyltin methoxide is an oil where association is negligible. Mössbauer spectroscopy may also be used to detect structural changes along homologous series. For example, in the series of cyclohexanone oxime derivatives, $R_3SnON{=}C_6H_{10}$, both the quadrupole splitting and the recoil-free fraction temperature coefficient decrease in the order $Me > Et > Pr^n > Ph$ as the group R increases in size, consistent with a change in structure from the linear polymer (**58**), characterized by crystallography for the trimethyltin derivative, through the more loosely associated structure (**59**) with non-planar $[R_3Sn]$ moieties for the ethyl and *n*-propyl homologues, to ultimately a lattice comprised isolated molecules (**60**) for the phenyl compound.[427,428] Similarly with lead, the triorganolead methoxides, R_3PbOMe, are postulated to have the associated structure (**58**),[424,425] but triphenyl(triphenylsiloxy)lead, $Ph_3PbOSiPh_3$, is monomeric.[429]

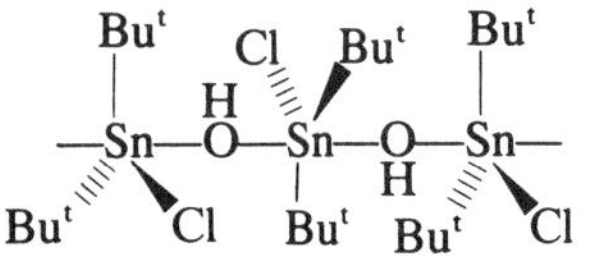

(**54**)

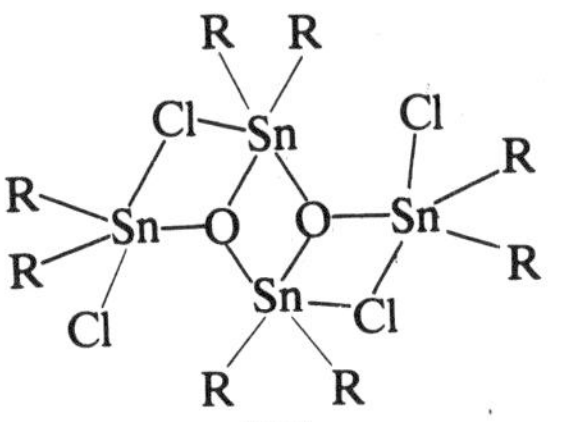

(**55**)

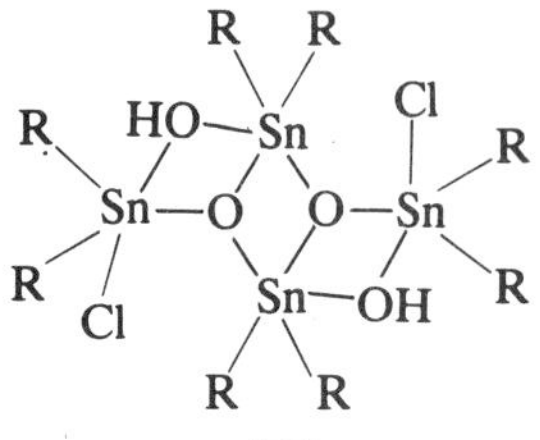

(**56**)

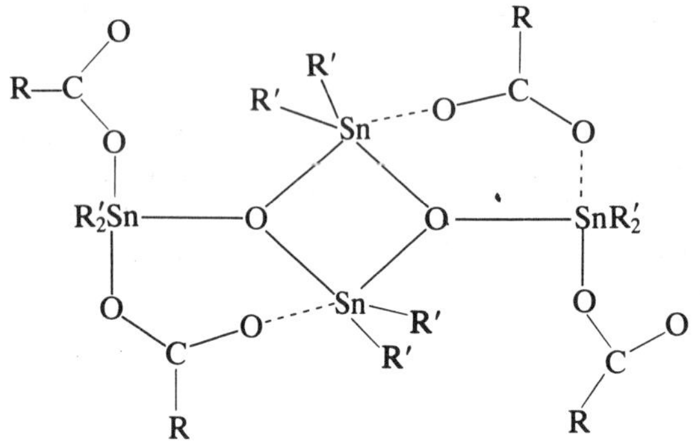

(**57**)

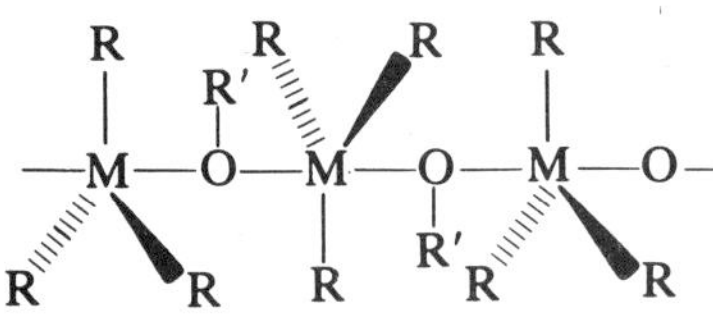

(**58**)

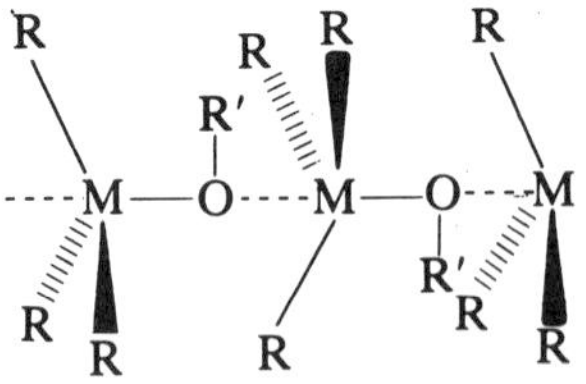

(**59**)

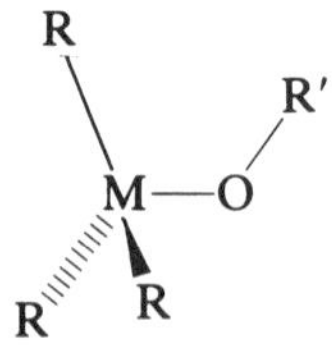

(**60**)

Similar polymeric chain structures are adopted by several other types of triorganotin and triorganolead derivatives including carboxylates, oxyacid derivatives and pseudohalides, and examples are listed in Table 9. Some of the chains are linear, others helical, but the most unusual, and at present unique, solid state structure of this type is the cyclic hexamer formed by triphenyltin diphenylphosphate, $Ph_3SnO_2P(OPh)_2$ (Figure 14).[430] Observed structures of triorganotin carboxylates fall into several categories. The majority, including trimethyl-, trivinyl-, triphenyl- and tribenzyl-tin carboxylates (and also trimethyl- and triphenyl-lead acetates), possess the rather tightly bound one-dimensional polymeric structure (**61**), with the carboxylato groups strongly bridging in the usual *syn,anti* fashion planar triorganotin residues. Tricyclohexyltin carboxylates, although having been previously misrepresented as examples of

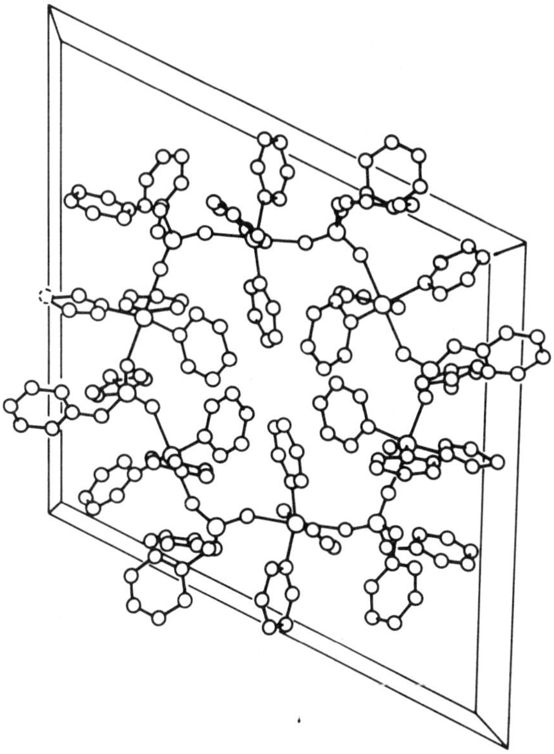

Figure 14 The hexameric structure of $Ph_3SnO_2P(OEt)_2$ (reproduced by permission from *Inorg. Chem.*, 1982, **21,** 960)

monomeric triorganotin carboxylates with unidentate carboxylato ligands and four-coordinated tin (as in **62**), have a similar one-dimensional structure, although the *syn,anti* bridging is longer and weaker. The only characterized example of a monomeric triorganotin carboxylate is the very sterically crowded, triphenyltin *o*-(2-hydroxy-5-methylphenylazo)benzoate, which adopts not a tetrahedral geometry (**63**) but the distorted trigonal bipyramidal structure (**64**) where the carboxylate group chelates *via* axial and equatorial sites.[431,432]

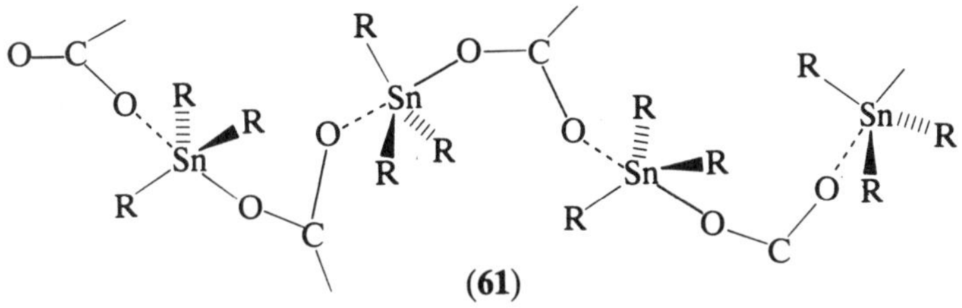

(61)

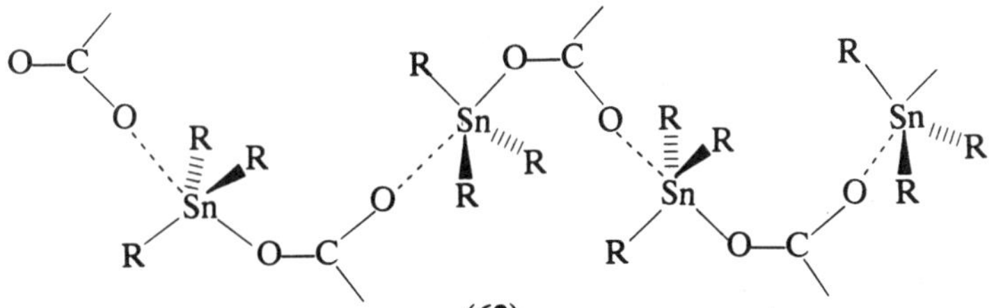

(62)

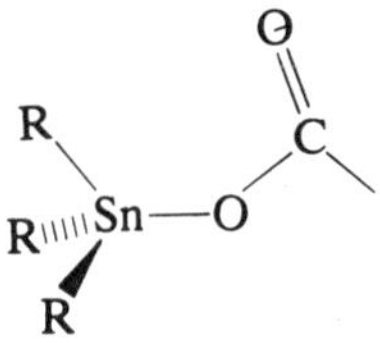

(63)

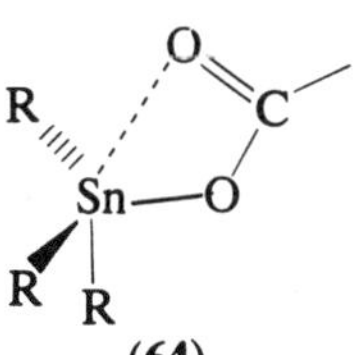

(64)

Table 9 Examples of Triorganotin and Triorganolead Compounds Containing Metal–Oxygen, –Sulfur or –Nitrogen bonds Exhibiting One-dimensional Associated Structures

Compound	*Ref.*	*Compound*	*Ref.*
Me_3SnOH	a	Me_3SnNO_3	m
Ph_3SnOH	b	Me_3SnO_2SMe	n, o
Ph_3PbOH	b	Me_3SnO_2SeMe	p
Me_3SnOMe	c	$Me_3SnO_2P(OH)Ph$	q
$Me_3SnON{=}C_6H_{10}$ Me_3SnO_2CMe $Me_3SnO_2CCF_3$	d e e	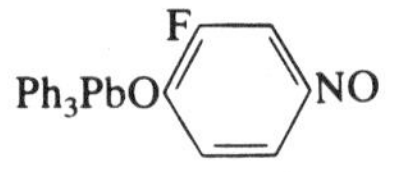	r
$(Me_3SnO_2C)CH_2$ $Me_3SnO_2CH_2NH_2$ Ph_3SnO_2CMe	f g h	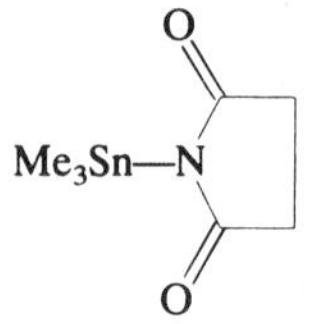	s
$(CH_2{=}CH)_3SnO_2CCl_3$ $(PhCH_2)_3SnO_2CMe$ Me_3PbO_2CMe Ph_3PbO_2CMe $Me_3SnN(CN)_2$	i j k l w	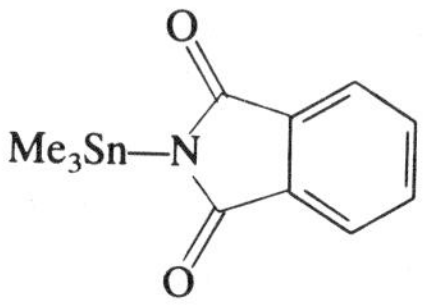	t
$Me_3SnN{:}C{:}NSnMe_3$	x	$Me_3SnNCO{\cdot}Me_3SnOH$	u
Me_3SnN_3	y	$Me_3SnN(Me)NO_2$	v
Me_3PbN_3	z	Ph_3SnNCO	cc
Me_3SnNCS	aa	Ph_3PbNCO	dd
Ph_3SnNCS	bb	$Me_3Sn{-}NSN{=}S{=}NSO_2$	ee
		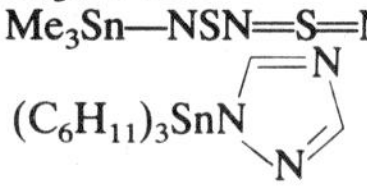	ff

[a] N. Kasai, K. Yasuda and R. Okawara, *J. Organomet. Chem.*, 1965, **3**, 172.
[b] C. Glidewell and D. C. Liles, *Acta Crystallogr., Sect. B*, 1978, **34**, 129.
[c] A. M. Domingos and G. M. Sheldrick, *Acta Crystallogr., Sect. B*, 1974, **30**, 519.
[d] P. F. R. Ewings, P. G. Harrison, T. J. King, R. C. Phillips and J. A. Richards, *J. Chem. Soc., Dalton Trans.*, 1975, 1950.
[e] H. Chih and B. R. Penfold, *J. Cryst. Mol. Struct.*, 1973, **3**, 285.
[f] U. Schubert, *J. Organomet. Chem.*, 1978, **155**, 285.
[g] B. Y. K. Ho, K. C. Molloy, J. J. Zuckerman, F. Reidinger and J. A. Zubieta, *J. Organomet. Chem.*, 1980, **187**, 213.
[h] K. C. Molloy, T. G. Purcell, K. Quill and I. W. Nowell, *J. Organomet. Chem.*, 1984, **267**, 237.
[i] S. Calogero, D. A. Clemente, V. Peruzzo and G. Tagliavini, *J. Chem. Soc., Dalton Trans.*, 1979, 1172.
[j] N. W. Alcock and R. E. Timms, *J. Chem. Soc. (A)*, 1968, 1873.
[k] G. M. Sheldrick and R. Taylor, *Acta Crystallogr., Sect. B*, 1975, **31**, 2740.
[l] C. Gaffney, Ph.D. Thesis, University of Nottingham, 1982.
[m] H. C. Clark, private communication quoted in D. Potts, H. D. Sharma, A. J. Carty and A. Walker, *Inorg. Chem.*, 1974, **13**, 1205.
[n] R. Hengel, U. Kunzo and J. Strähle, *Z. Anorg. Allg. Chem.*, 1976, **423**, 35.
[o] G. M. Sheldrick and R. Taylor, *Acta Crystallogr., Sect. B*, **33**, 135.
[p] U. Ansorge, E. Lindner and J. Strähle, *Chem. Ber.*, 1978, **111**, 3048.
[q] K. C. Molloy, M. B. Hossain, D. van der Helm, D. Cunningham and J. J. Zuckerman, *Inorg. Chem.*, 1981, **20**, 2402.
[r] N. G. Bokii, A. I. Udel'nov, Yu. T. Struchkov, D. N. Kravtsov and V. M. Pacherskaya, *Zh. Strukt. Khim.*, 1977, **18**, 1025.
[s] F. E. Hahn, J. S. Dory, C. L. Barnes, M. B. Hossain, D. van der Helm and J. J. Zuckerman, *Organometallics*, 1983, **2**, 969.
[t] E. V. Chuprunov, T. N. Tarkhova, Yu. T. Korallova and N. V. Belov, *Dokl. Akad. Nauk SSSR*, 1978, **242**, 606.
[u] J. B. Hall and D. Britton, *Acta Crystallogr., Sect. B*, 1972, **28**, 2133.
[v] A. M. Domingos and G. M. Sheldrick, *J. Organomet. Chem.*, 1974, **69**, 207.
[w] Y. M. Chow, *Inorg. Chem.*, 1971, **10**, 1938.
[x] R. A. Forder and G. M. Sheldrick, *J. Chem. Soc. (A)*, 1971, 1107.
[y] R. Allmann, R. Hohlfeld, A. Waskawska and J. Lorberth, *J. Organomet. Chem.*, 1980, **192**, 353.
[z] R. Allmann, A. Waskawska, R. Hohlfeld and J. Lorberth, *J. Organomet. Chem.*, 1981, **216**, 245.
[aa] R. A. Forder and G. M. Sheldrick, *J. Organomet. Chem.*, 1970, **21**, 115.
[bb] A. M. Domingos and G. M. Sheldrick, *J. Organomet. Chem.*, 1974, **67**, 257.
[cc] T. N. Tarkhova, E. V. Chuprunov, M. A. Simonov and N. V. Belov, *Kristallografiya*, 1977, **22**, 1004.
[dd] T. N. Tarkhova, E. V. Chuprunov, L. E. Nikola'eva, M. A. Simonov and N. V. Belov, *Kristallografiya*, 1978, **23**, 506.
[ee] H. W. Roesky, M. Witt, M. Diehl, J. W. Bats and H. Fuess, *Chem. Ber.*, 1979, **112**, 1372.
[ff] I. Hammann, K. H. Büchel, K. Bungartz and L. Born, *Pflanzenschutz-Nachr. Bayer (Engl. Ed.)*, 1978, **31**, 61.

Triorganotin acetylacetonates[433] and *N*-acylhydroxylaminates[434,435] exhibit contrasting behaviour, and have the *cis* trigonal bipyramidal structures, (**65**) and (**66**).

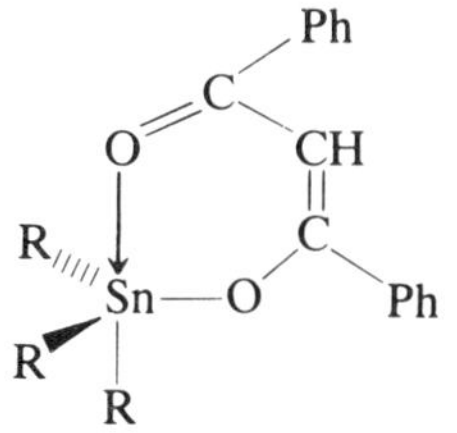

(**65**)

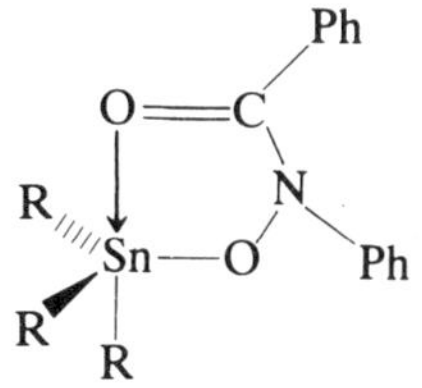

(**66**)

[Tris(pyrazolyl)borate](trimethyl)tin, $[HB(pz)_3]SnMe_3$, (**67**) is unique, being the only authenticated example amongst triorganotin compounds of six-coordination.[436]

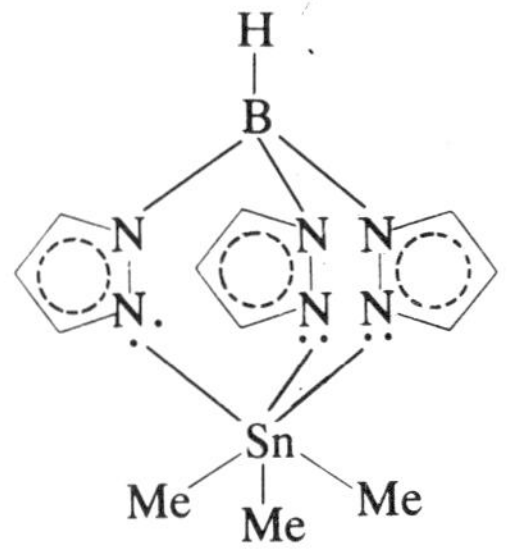

(**67**)

Few structural data are available for the corresponding diorgano-tin and -lead derivatives of these ligands, but their behaviour appears to be broadly similar. Carboxylato and oxyanion ligands form bridged structures, although the nitrate group functions as a chelating ligand, as do the acetylacetonato and *N*-acylhydroxylaminato groups. Thus, the anhydrous dicarboxylates are believed to possess the bridged polymeric structure (**68**) in the solid, which breaks down in solution to the monomeric species (**69**). Both have a *trans* $[SnC_2O_4]$ geometry. The structure of one diorganolead diacetate, $[Ph_2PbCO_2CMe)_2]_2{\cdot}H_2O{\cdot}C_6H_6$, has been determined, although this is probably unrepresentative of the general class of compound, but nevertheless notable. Crystals of this compound comprise binuclear units (Figure 15), in which both the lead atoms have a pentagonal bipyramidal geometry with the phenyl groups occupying axial sites. Each lead is chelated unsymmetrically by two acetate groups, one of which bridges both metals. The water molecule is coordinated to one lead and forms a hydrogen bond with an acetate group chelating the other.[437] Diphenyllead diacetate readily accepts further acetate ion to form the $[Ph_2Pb(O_2CMe)_3]^-$ anion, which has the hexagonal bipyramidal structure (**70**).[438] Dimethyltin bis(fluorosulfonate), $Me_2Sn(OSO_2F)_2$, has a polymeric sheet structure[439] and the dimethyltin phosphate, $(Me_2Sn)_3(PO_4)_2{\cdot}8H_2O$, an infinite 'ribbon' structure,[440] both with *trans* $[SnC_2O_4]$ geometry at tin.

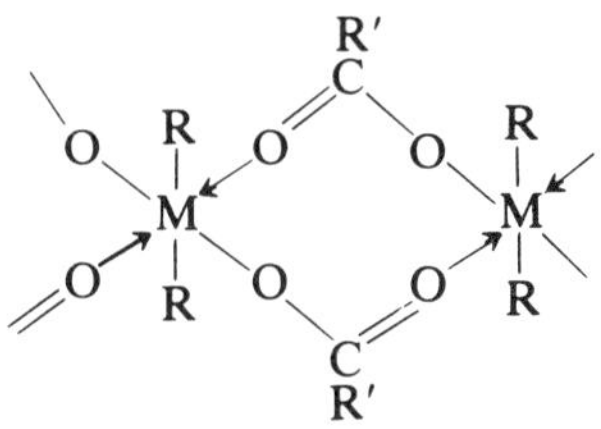

(**68**)

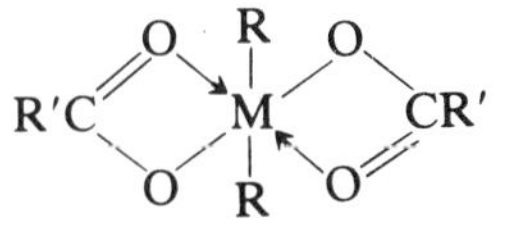

(**69**)

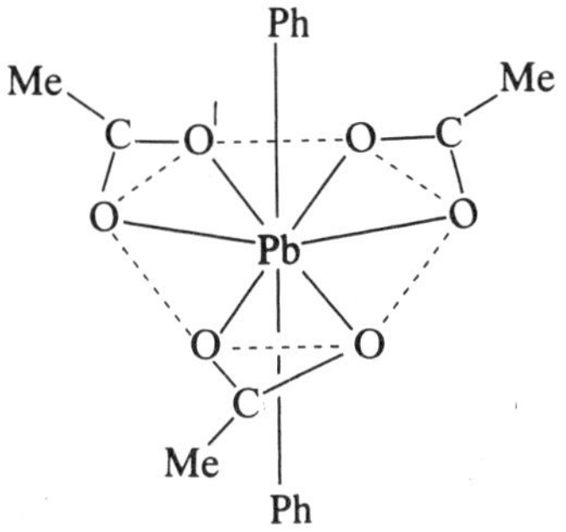

(**70**)

The stereochemistry of molecular six-coordinated dimethyltin derivatives can vary substantially from the *cis* $[SnC_2O_4]$ geometry (**71**), adopted by dimethyltin bis(oxinate)[441] and

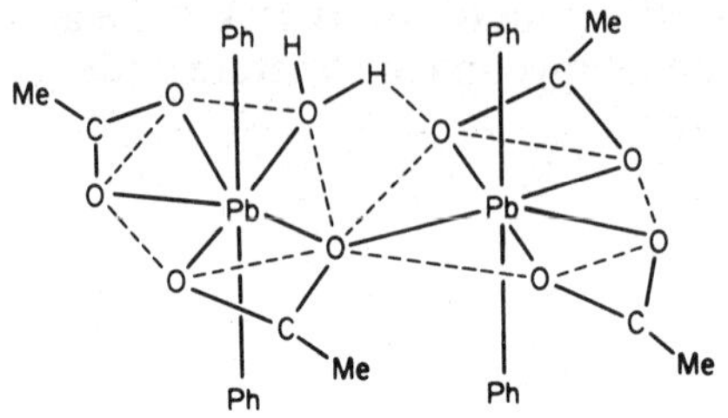

Figure 15 Schematic representation of the structure of $[Ph_2Pb(O_2CMe)_2]_2{\cdot}H_2O$

$Me_2Sn(ONHCOMe)_2$,[442] to the *trans* $[SnC_2O_4]$ geometry (**73**) of bis(acetylacetonato)dimethyltin,[443] in which the [CSnC] group is linear and the acetylacetonato group chelate the tin atom symmetrically. A number of compounds, however, exhibit a geometry (**72**) intermediate between these two extremes, including $Me_2Sn(NO_3)_2$,[444] bis{(salicylaldehyde)ethylenediiminato}dimethyltin,[445] $Me_2Sn(ONMeCOMe)_2$[446] and $Me_2Sn(ONHCOMe)_2{\cdot}H_2O$.[442] The difference in stereochemistry between $Me_2Sn(ONHCOMe)_2$ and its monohydrate is quite surprising, and illustrates the rather large effect that hydrogen bonding networks and crystal packing demands can have upon the stereochemistry adopted about the tin atom.

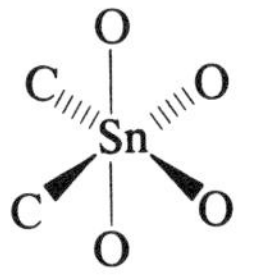

(**71**)

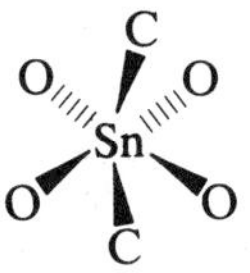

(**72**)

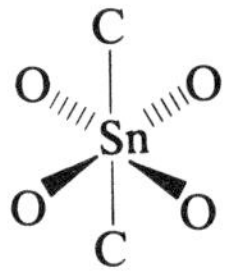

(**73**)

Little is known concerning the structural chemsitry of monoorganotin and monoorganolead derivatives. IR and colligative data suggest the seven-coordinate structure (**74**) for monoorganotintris(carboxylates),[447] similar to that determined for methyltin trinitrate (**75**).[448] Tin(IV) nitrate,[449] and tin(IV) and lead(IV) acetates[450,451] have an eight-coordinated dodecahedral structure (**76**).

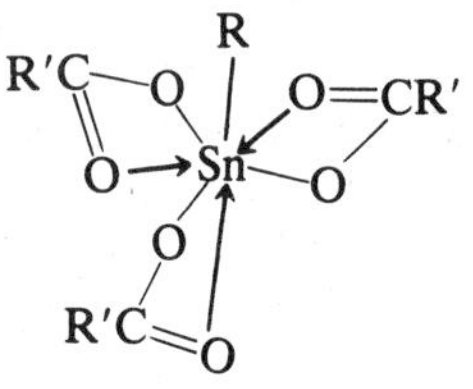

(**74**)

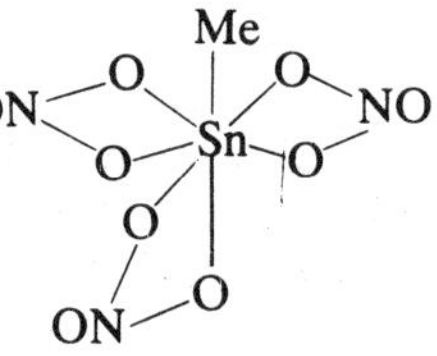

(**75**)

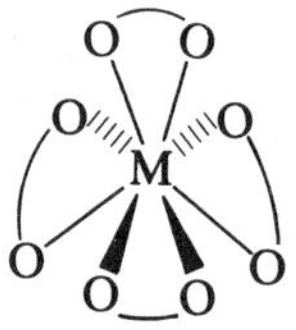

(**76**)

Silicon and germanium compounds with coordination numbers greater than four are much less abundant than for their heavier congeners. Silicon(IV) acetate, $Si(O_2CMe)_4$, is tetrahedral with unidentate acetate groups,[452] but the metalloids are six-coordinated in the anions, $[M(O_2CMe)_6]^{2-}$ (M = Si, Ge).[453] Unusually short intermolecular Si—O, Si---O═C and Si---S═C interactions have been detected in crystalline samples of silyl acetate, H_3SiO_2CMe, and silyl thioacetate, $H_3SiO(S)CMe$, at low temperatures, but under normal conditions silyl carboxylates, like the alkoxides, are monomeric.[454,455] Other examples of five- and six-coordinated compounds, which generally need to contain fewer than two bonds to carbon, include octahedral species such as the $[Si(OH)_2(bipy)_2]^{2+}$ cation,[456] and tris(oxalato)- and tris(acetylacetonato)-metal (2 −) anions,[457–463] and the neutral compounds (**77**)[464] and (**78**),[465] as well as mainly five-coordinated derivatives of chelating ligands. The most widely investigated class of five-coordinated compound are the silatranes and germatranes, which have a trigonal bipyramidal geometry as in (**79**) with short M←-N transannular interactions.[466–473] Electron diffraction studies on methylsilatrane, however, indicate that this type of bonding interaction is absent at higher temperatures.[469] Several other related compounds which exhibit transannular M---N interactions are known including [iminobis(ethyleneoxy)]diphenylsilane (**80**),[474] silatranones and germatranones (**81**)[473,475] and 2-carbagermatranes (**82**).[473] Stannatranes are similar, but are associated into the trimeric units (**83**) with octahedrally coordinated tin in both solution in apolar solvents and the solid phases.[476–480] In donor solvents, this association is broken down, and the mononuclear species (**84**) exist in water and DMSO. In contrast, 1-alkylthiostannatranes are monomeric in all solvents.[478]

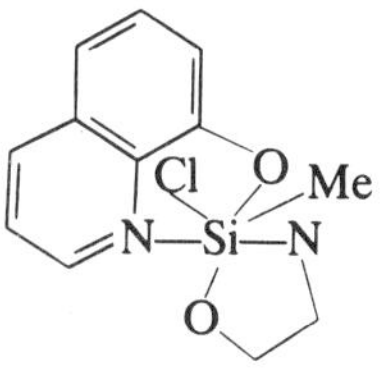

(**77**)

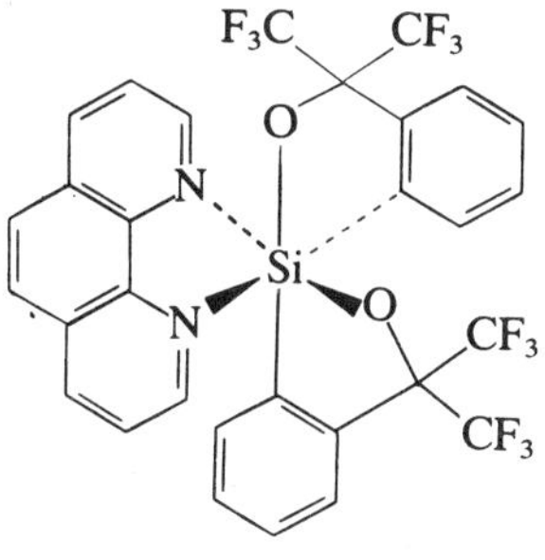

(**78**)

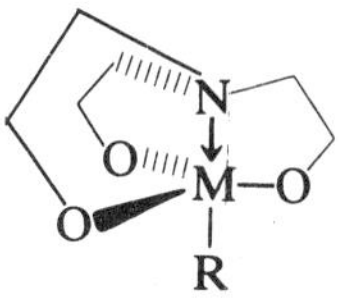

(**79**)

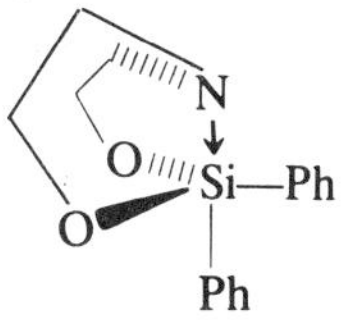

(**80**)

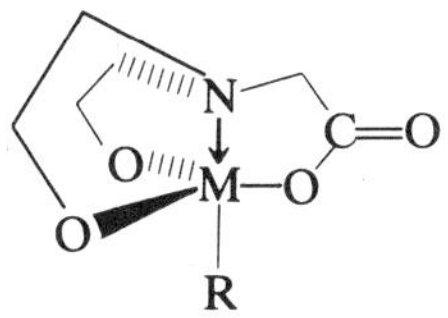

(**81**)

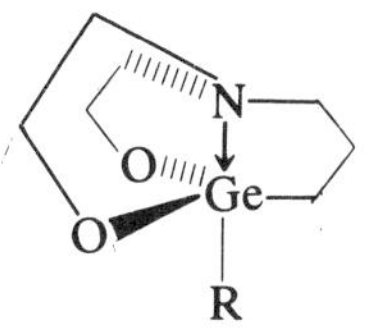

(**82**)

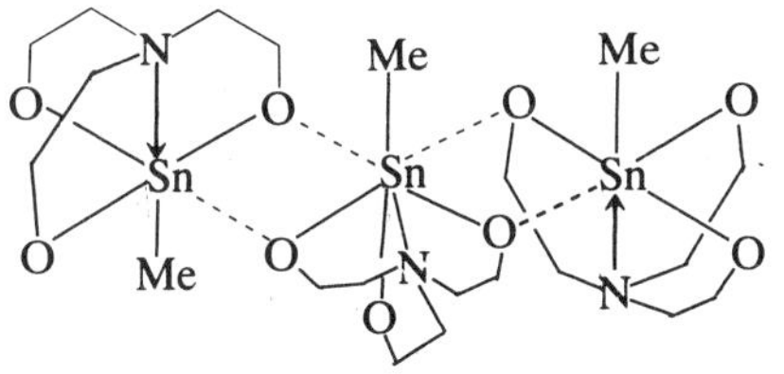

(**83**)

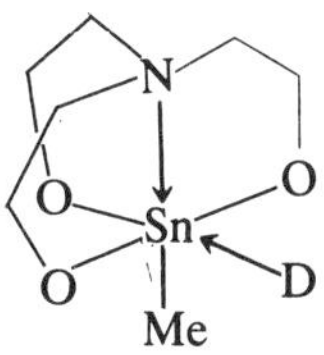

(**84**) D = H_2O, DMSO

The geometry of all the five-coordinated compounds described thus far is, or is very close to, a trigonal bipyramid. Examples of the alternative five-coordinated geometry, the square-based pyramid, are much rarer in Group IV chemistry, although several are known with geometries intermediate between the two extremes along the Berry pseudorotation coordinate.[481–484] Idealized square-based pyramidal geometry is favoured over the trigonal bipyramid by the presence of two large unsaturated chelating ligands with like atoms bonded to the central Group IV element. Thus, the bis(naphthalene-2,3-dioxy)phenylsilane(1 −) has an almost ideal square-based pyramidal structure, being 97.6% displaced along the Berry coordinate. Displacements along the Berry coordinate for other examples are shown in Table 10.

Associated structures are far less prevalent in compounds containing bonds to the heavier chalcogenides. Triorganotin and triorganolead thiolates, R_3MSR' (M = Sn, Pb), are monomeric and tetrahedral like their silicon and germanium counterparts.[485–490] Weak intermolecular Sn---N and Pb---Br interactions do, however, occur in $Ph_3SnSC_5H_4N$-4[491] and $Ph_3PbSC_6H_4Br$-2[489] respectively, causing severe distortion of the geometry at the metal towards a trigonal bipyramid. Likewise, bis(triphenylgermanium)sulfide,[492] and bis(triphenyltin)-sulfide[493] and -selenide[494] are monomeric molecules, which are bent at sulfur or selenium and contain tetrahedrally coordinated metal atoms. The corresponding diorganometal chalcogenides form ring structures (*cf.* the associated polymeric structures formed by diorganotin oxides), most commonly a cyclic trimer (**85**) with a twist-boat conformation.[495] However, with bulky organic substituents, a four-membered ring may be stabilized. Thus, di-*tert*-butyltin sulfide, selenide and telluride, $[Bu^t_2SnX]_2$ (X = S, Se, Te), exist as the dimers (**86**) with a planar central $[Sn_2X_2]$ ring.[496] Some diorganotin sulfides may also exist in a polymeric modification of lower solubility and higher melting than the trimeric form, and probably have the same linear polymeric structure (**87**) as that characterized for diisopropyltin sulfide.[497] The monoorganometal sesquisulfides generally crystallize with the adamantane skeleton (**88**).[498–500]

Table 10 Selected Examples of Five-coordinated Species Showing Significant Distortions towards the Square-based Pyramidal Geometry

Species	*Displacement*[a]	*Ref.*
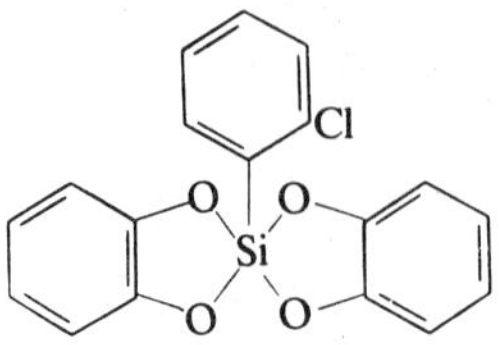	33.2	b
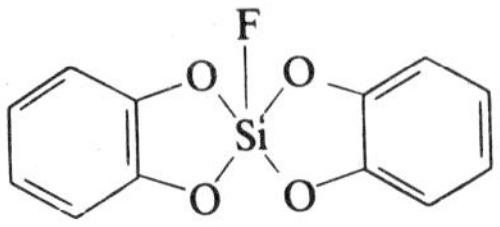	55 70	c
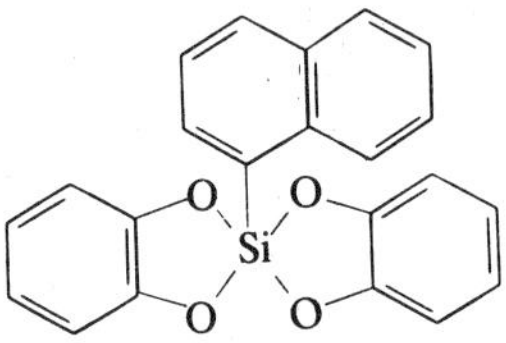	58.7	b
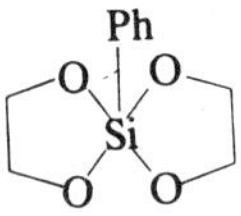	72	d
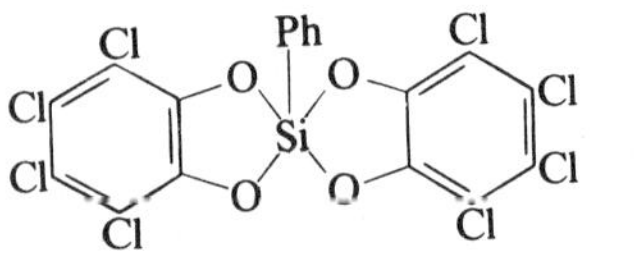	90	d
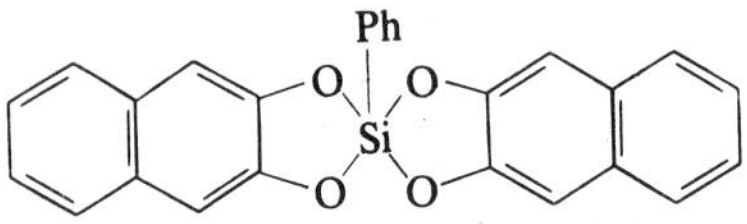	97.6	b
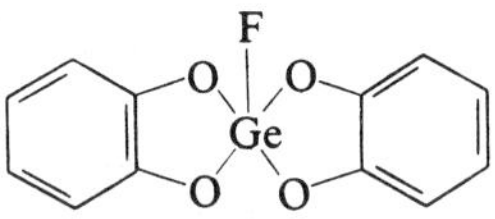	81	e
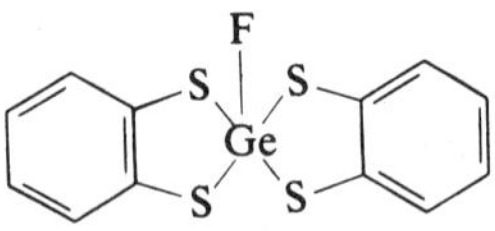	40	e
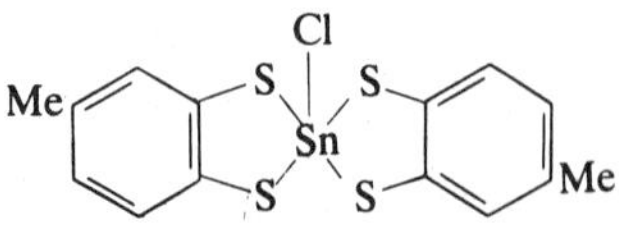	76.9	f

[a] Percentage displacement along the Berry pseudorotation coordinate.
[b] R. R. Holmes, R. O. Day, J. J. Harland and J. M. Holmes, *Organometallics*, 1984, **3**, 347.
[c] Two crystallographically independent species, J. J. Harland, R. O. Day, J. F. Vollano, A. C. Sau and R. R. Holmes, *J. Am. Chem. Soc.*, 1981, **103**, 5269.
[d] R. R. Holmes, R. O. Day, J. J. Harland, A. C. Sau and J. M. Holmes, *Organometallics*, 1984, **3**, 341.
[e] R. O. Day, J. M. Holmes, A. C. Sau and R. R. Holmes, *Inorg. Chem.*, 1982, **21**, 281.
[f] A. C. Sau, R. O. Day and R. R. Holmes, *Inorg. Chem.*, 1981, **20**, 3076.

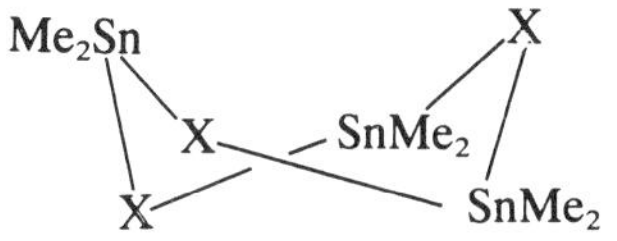

(85)

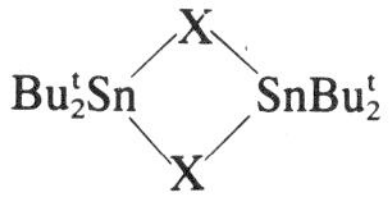

(86)

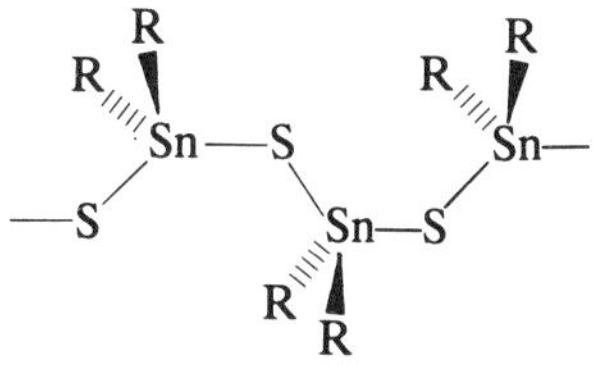

(87)

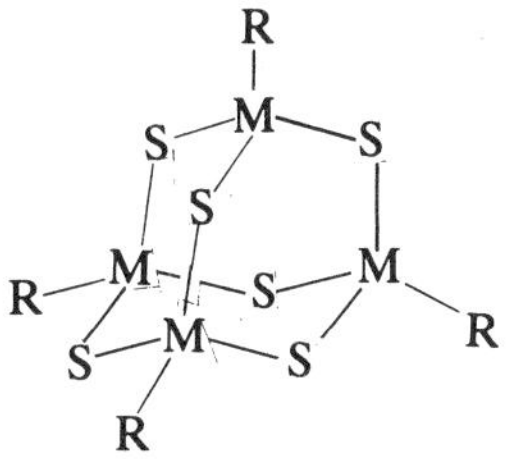

(88)

In contrast to a carboxylato and other oxyanion ligands, dithio ligands such as dithiocarbamate and dithiophosphate function as unidentate ligands in their triorganotin and triorganolead derivatives.[490,501–503] When two or less metal–carbon bonds are present, however, chelation by these types of ligands occurs. In the dimethyltin bis(dithiocarbamates), $Me_2Sn(S_2CNR_2)_2$ (R = Me, Et),[504,505] and the diphenyltin and diphenyllead dithiophosphoridates, $Ph_2Sn[S_2P(OEt)_2]_2$,[506] and $Ph_2Pb[S_2P(OPr^i)_2]_2$,[490] the ligands chelate in an anisobidentate fashion allowing the CMC bond angle to close significantly, but in $Ph_2Sn[S_2P(OPr^i)_2]_2$, the [CSnC] linkage is linear and the dithiophosphate groups chelate symmetrically.[507] Methyltin tri(diethyldithiocarbamate) has a distorted pentagonal bipyramidal geometry, in which two dithiocarbamato groups chelate *via* equatorial sites whilst the third spans the remaining equatorial and an axial position.[508]

The most significant feature in the structures of compounds containing bonds to nitrogen is the planar skeleton about nitrogen exhibited by some including, *inter alia,* trisilylamine, $(H_3Si)_3N$,[509] $(MeH_2Si)_3N$,[510] $(F_2P)(H_3Si)_2N$,[511] $(F_2P)_2(H_3Si)N$,[511] $(F_2P)_2GeNH_2$[511] and $(Me_3Sn)_3N$.[512] In these compounds, the metal–nitrogen bond is generally somewhat longer than in others where the geometry at nitrogen is pyramidal, *e.g.* $(Ph_2MeSi)_3N$.[513] In silylamines, $Me_2HSiNMe_2$, MeH_2SiNMe_2 and H_3SiNMe_2, the nitrogen atom has a shallow pyramidal configuration.[514] The similar silylphosphines, $H_3SiPH_nMe_{2-n}$ (n = 0–2), are pyramidal at phosphorus.[515–517] Planarity at nitrogen has been attributed to $(p \rightarrow d)\pi$ interactions between the filled p_z orbital on nitrogen and vacant metal d orbitals. Theoretical calculations have shown that such interactions are most important for silicon, but even in this case planarity at nitrogen was ascribed principally to electrostatic interactions.[518,519] Unlike other trimethylstannylamines, which according to Mössbauer spectroscopy are monomeric,[520] trimethyltin aziridine is associated in the solid (**89**) but is dimeric in solution (**90**).[521]

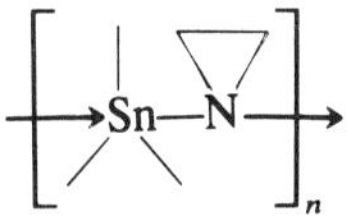

(89)

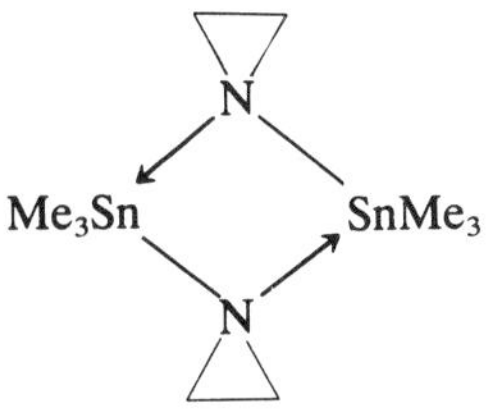

(90)

With phosphorus, a remarkably diverse range of ring and cage frameworks are formed. Three-membered $[MP_2]$ (M = Si, Ge, Sn) rings (**91**) are obtained by steric crowding at phosphorus and the Group IV element.[522–524] That in $Ph_2Si(PBu^t)_2$ is an almost perfect equilateral triangle. Both diastereoisomers of the tetraphospha-3-silaspiro[2.2]pentane (**92a**) and (**92b**) have been characterized.[525] Many different four-, five- and six-membered ring systems have been synthesized, although few structures have been determined. Examples of these, together with some examples of cage frameworks, are illustrated in Figure 16.[526–542] The dimethylsilylarsine derivative, $As_4(SiMe)_6$, has the adamantane structure (**93**).[543]

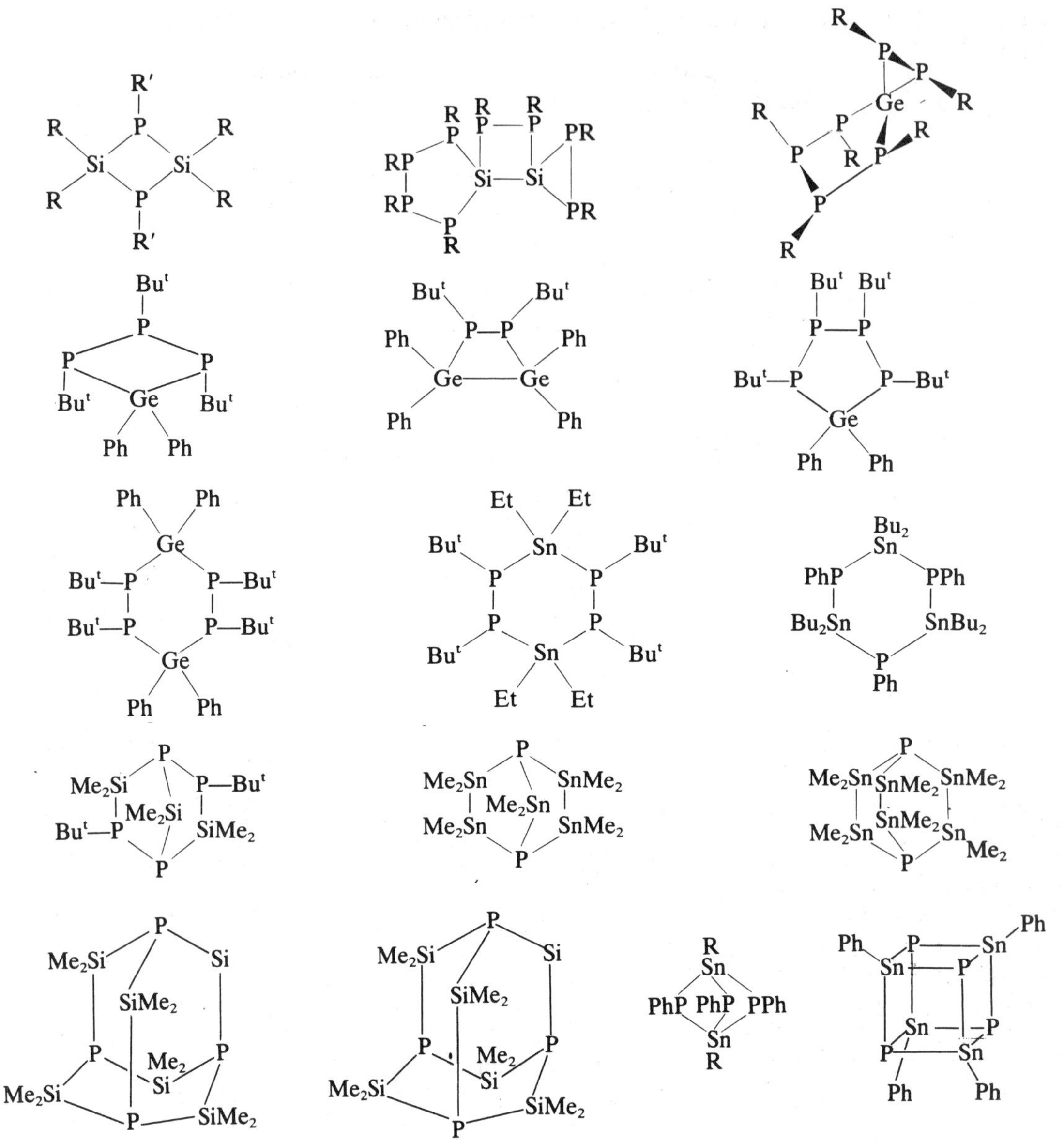

Figure 16 Examples of ring and cage structures formed by organo-silicon, -germanium and -tin phosphines

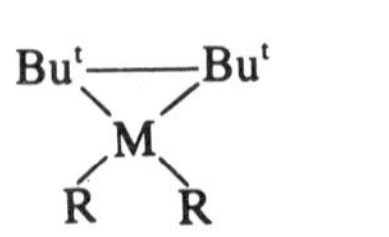

(91) M = Si, R = Ph
M = Ge, R = Et
M = Sn, R = Bu^t

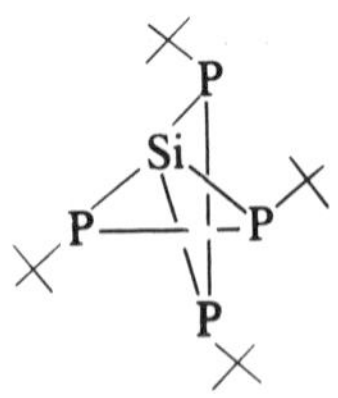

(92a)

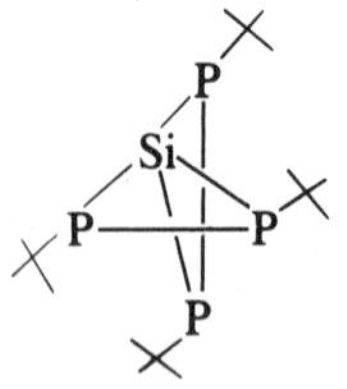

(92b)

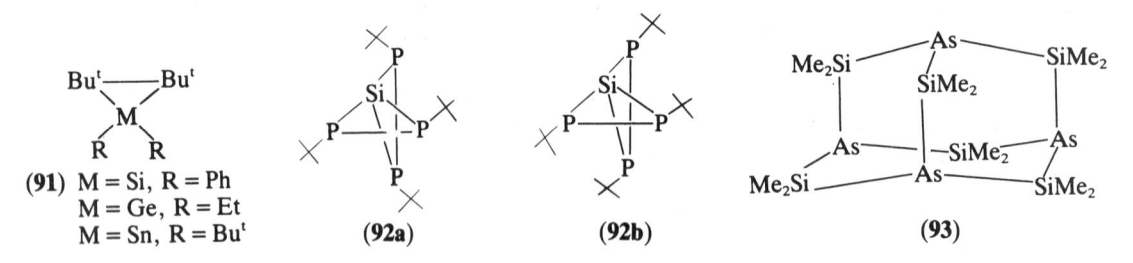

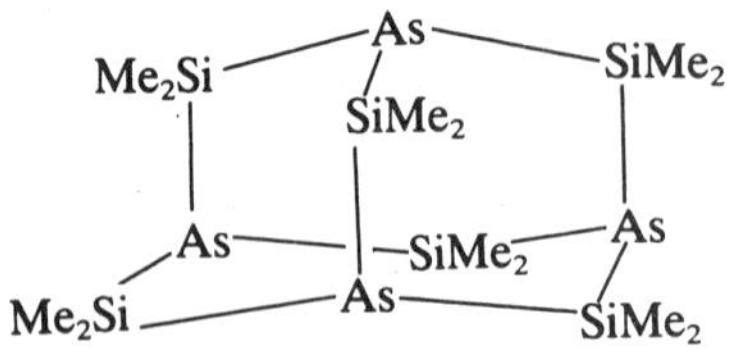

(93)

26.3.7 Bivalent Compounds Containing Bonds to Elements of Groups IV, V and VI

Simple σ-bonded bivalent dialkyl or diaryl compounds of germanium, tin and lead of the type R_2M (M = Ge, Sn, Pb) are not stable. Rather, attempts to generate such compounds always result in the formation of isomeric, metal–metal-bonded oligomers (**94**), which contain tetravalent metal atoms.[546–540] Predictably, this facile polymerization process may be precluded by employing a sterically crowded organic group, and bis[bis(trimethylsilyl)methyl-germanium(II) and -tin(II) are monomeric, angular compounds (**95**) in the vapour and in solution.[551–554] In the solid phase, however, dimerization to the *trans* bent structure (**96**) occurs.[551,552,555]

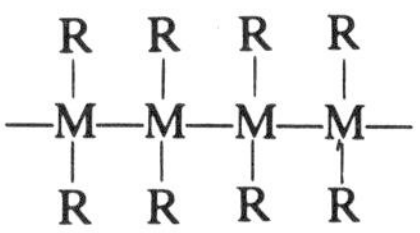

(94)

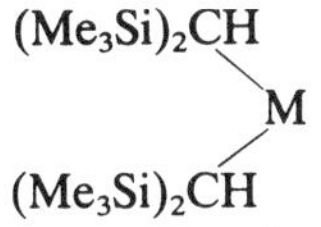

(95)

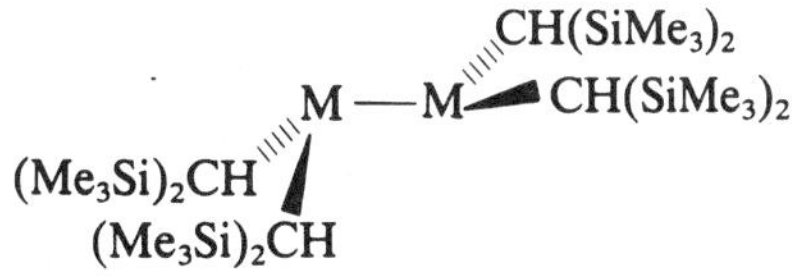

(96)

The nature of the metal–metal bond in these compounds is somewhat controversial. Initial interpretation of the bonding in terms of a double 'banana' bond are probably unrealistic,[556] albeit supported by theoretical calculations on the prototype dimetallenes, $H_2M{:}MH_2$ (M = Ge, Sn).[553,557] A more recent theoretical description of distannene, however, favours the singly bonded biradical structure (**97**), in which the unpaired electrons can interact conjugatively across space giving rise to a π or π-type bond, or σ-conjugatively *via* the Sn—Sn σ bond.[558] The corresponding lead compound, $Pb[CH(SiMe_3)_2]_2$, has also been synthesized, but its structure is as yet not known.[552]

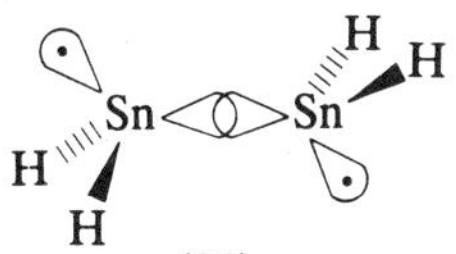

(97)

Cyclopentadienyl derivatives form the largest class of bivalent organometallic compounds and many are known (Table 11). All the bis(cyclopentadienyl)metal* compounds have the angular sandwich structure (**98**),[559–566] except for bis(pentaphenylcyclopentadienyl)tin[567] in which the two cyclopentadienyl rings are parallel and staggered (Figure 17). Stannocene itself has been the subject of extended Hückel,[559] SCF X_α scattered wave[568] and MNDO calculations,[558] which give generally good agreement between experimentally observed and calculated structural parameters for the angular sandwich geometry and place the non-bonding electron pair in a highly directional orbital *ca.* 2 eV more stable than the HOMO. Calculated metal–carbon distances for the (unstable) isomer with enforced D_{5d} symmetry are, however, substantially shorter (247 pm)[558] than those observed in decaphenylstannocene (268.6–220.5 pm)[567] indicating steric repulsion as the cause of the parallel ring structure in this case. Cyclopentadienylgermanium chloride, C_5H_5GeCl,[569] and pentamethylcyclopentadienylgermanium chloride, C_5Me_5GeCl,[566] also have angular structures in the gas phase, but additional Sn---Cl intermolecular contacts are present for cyclopentadienyltin chloride in the solid.[570]

Cleavage of the rings is very facile.[571–575] Treatment of pentamethylstannocene with tetrafluoroboric acid affords the pentamethylstannocenium salt, $C_5Me_5Sn^+BF_4^-$ (**99**),[559,576] analogous to stannocenium tetrachloroaluminate, $C_5H_5Sn^+AlCl_4^-$, obtained by the reaction of C_5H_5SnCl with aluminum(III) chloride.[577] The germanocenium[569] and stannocenium cations,[558,559] $C_5H_5M^+$ (M = Ge, Sn), too, have been the objects of theoretical study with reasonable agreement with observed data for $C_5Me_5Sn^+$. The two studies of $C_5H_5Sn^+$ do, however, disagree over the ordering of the molecular orbitals.

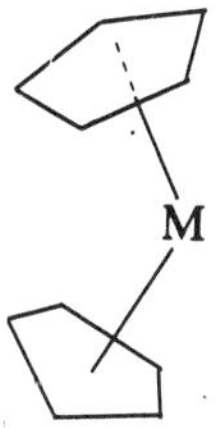

(98)

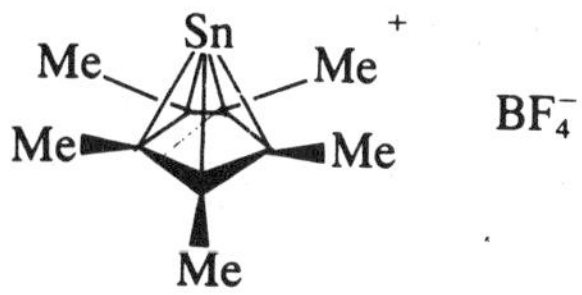

(99)

* The trivial nomenclature germanocene, stannocene and plumbocene appears to have been universally adopted for these compounds.

Table 11 Examples of Cyclopentadienyl Derivatives of Bivalent Germanium, Tin and Lead

Compound	*Ref.*	*Compound*	*Ref.*
$(C_5H_5)_2Ge$	a	$(C_5Me_5)_2Sn$	h
$(C_5H_4Me)_2Ge$	b	$(C_5Ph_5)_2Sn$	i
$(C_5Me_5)_2Ge$	c	C_5H_5SnCl	j
C_5Me_5GeCl	c	$[C_5Me_5Sn]^+BF_4^-$	h
$(C_5H_5)_2Sn$	d	$[(C_5Me_5)Sn(bipy)]^+O_3SCF_3$	k
$(C_5H_4Li)_2Sn$	e	$(C_5H_5)_2Pb$	l, m
$\{[(Pr^i_2N)_2P]C_5H_4\}_2Sn$	e	$(C_5Me_5)_2Pb$	d
$(C_5H_5)(C_5H_4COR)Sn$ (R = Me, OMe, OEt)	f	$[(Me_3Si)_nC_5H_{5-n}]_2Pb$ (n = 1–3)	n
$[1,2,4-(Me_3Si)_3C_5H_2]_2Sn$	g	C_5H_5PbCl	o

[a] M. Grenz, E. Hahn, W. W. DuMont and J. Pickardt, *Angew. Chem., Int. Ed. Engl.*, 1984, **23**, 61.
[b] J. Almlöf, L. Fernholt, K. Faegin, A. Haaland, B. E. R. Schilling, R. Seip and K. Taugbol, *Acta Chem. Scand., Ser. A*, 1983, **37**, 131.
[c] L. Fernholt, A. Haaland, P. Jutzi, F. X. Kohl and R. Seip, *Acta Chem. Scand., Ser. A*, 1984, **38**, 211.
[d] J. L. Atwood, W. E. Hunter, A. H. Cowley, R. A. Jones and C. A. Stewart, *J. Chem. Soc., Chem. Commun.*, 1981, 925.
[e] A. H. Cowley, J. G. Lasch, N. C. Norman, C. A. Stewart and T. C. Wright, *Organometallics*, 1983, **2**, 1691.
[f] T. S. Dory, J. J. Zuckerman and M. D. Rausch, *J. Organomet. Chem.*, 1985, **281**, C8.
[g] A. H. Cowley, P. Jutzi, F. X. Kohl, J. G. Lasch, N. C. Norman and E. Schlüter, *Angew. Chem., Int. Ed. Engl.*, 1984, **23**, 616.
[h] P. Jutzi, F. Kohl, P. Hofmann, C. Krüger and Y. H. Tsay, *Chem. Ber.*, 1980, **113**, 757.
[i] M. J. Heeg, C. Janiak and J. J. Zuckerman, *J. Am. Chem. Soc.*, 1984, **106**, 4259.
[j] K. D. Bos, E. J. Bulten and J. G. Noltes, *J. Organomet. Chem.*, 1975, **99**, 71.
[k] F. X. Kohl, E. Schlüter, P. Jutzi, C. Krüger, G. Wolmershauser, P. Hofmann and P. Stauffert, *Chem. Ber.*, 1984, **117**, 1178.
[l] A. Almenningen, A. Haaland and T. Motzfeld, *J. Organomet. Chem.*, 1967, **7**, 97.
[m] C. Panattoni, G. Bombieri and U. Groatto, *Acta Crystallogr.*, 1966, **21**, 823.
[n] P. Jutzi and E. Schlüter, *J. Organomet. Chem.*, 1983, **253**, 313.
[o] A. K. Holliday, P. H. Makin and R. J. Puddephatt, *J. Chem. Soc., Dalton Trans.*, 1976, 435.

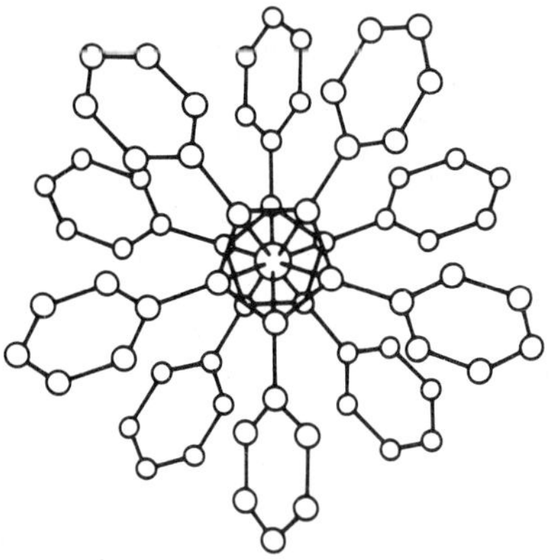

Figure 17 The structure of decaphenylstannocene, $(C_5Ph_5)_2Sn$, viewed along the five-fold axis (reproduced by permission from *J. Am. Chem. Soc.*, 1984, **106**, 4259)

Pentamethylgermanocenium and pentamethylstannocenium salts, $C_5Me_5M^+X^-$ (M = Ge, Sn; $X^- = CF_3SO_3^-$ or BF_4^-), react with nitrogen donors such as pyridine, pyrazine and 2,2′-bipyridine to form adducts. In both of the cations, $[C_5Me_5Sn{\cdot}L]^+$ (L = C_5H_5N, $C_{10}H_8N_2$), the bonding of the cyclopentadienyl ring to the tin atom relaxes from η^5 towards η^2/η^3. The gross structure of the pyridine complex, $[(C_5H_5N)Sn(C_5Me_5)]^+[O_3SCF_3]^-$, is a chain structure in which the anions link adjacent $[(C_5H_5N)Sn(C_5Me_5)]^+$ cations. In contrast, crystals of the corresponding bipyridine salt comprise isolated $[(C_{10}H_8N_2)Sn(C_5Me_5)]^+$ and $[CF_3SO_3]^-$ ions.[578,579]

Functionally substituted stannocene and plumbocene derivatives may be obtained either by ring substitution (equations 8 and 9),[580–582] or by metathesis (equations 10 and 11).[583,584]

Both the bis-alkyl derivatives, $[(Me_3Si)_2CH]_2M$ (M = Ge, Sn, Pb) and the metallocenes, $(C_5H_5)_2M$ (M = Sn, Pb), function as two-electron donors towards both transition metal and main group metal Lewis acid centres forming complexes of the types $[(Me_3Si)_2CH]_2M{:}\rightarrow$

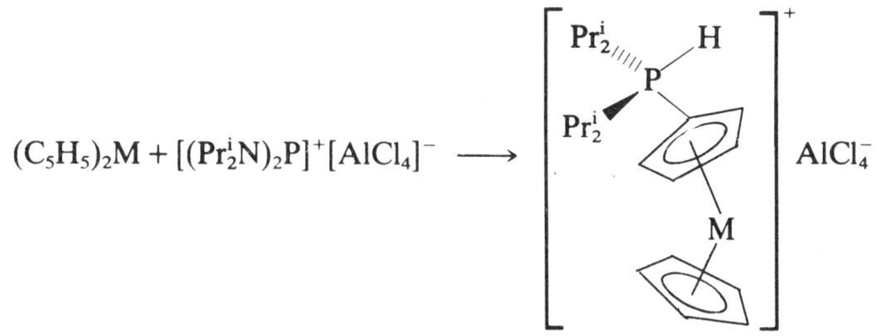

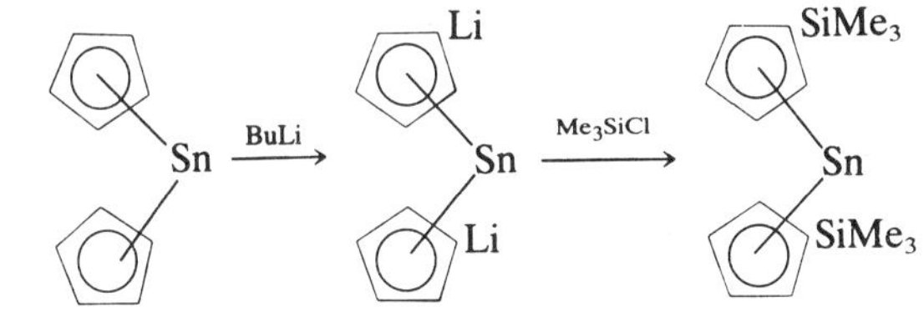

(9)

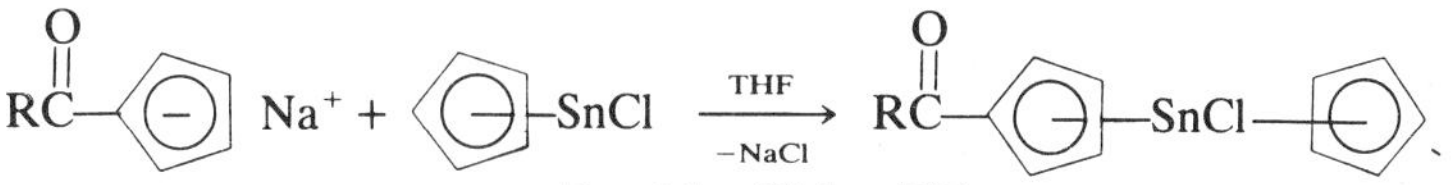

(10)

R = Me, OMe, OEt

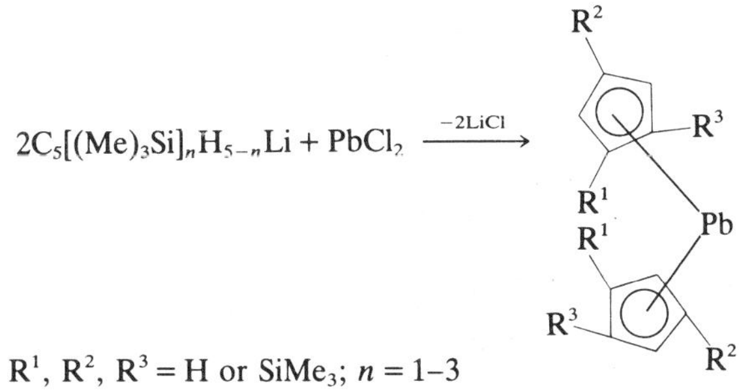

(11)

R^1, R^2, R^3 = H or $SiMe_3$; n = 1–3

$M^{VI}(CO)_5$ (M = Ge, Sn, Pb; M^{VI} = Cr, Mo, W),[551,552,555,585–588] $(C_5H_5)_2Sn{:}\rightarrow M^{VI}(CO)_5$,[589] $(C_5H_5)_2Sn{:}Mn(CO)_2(C_5H_4Me)$,[590] $[(C_5H_5)_2SnFe(CO)_4]_2$,[591] $(C_5H_4R)_2Sn{:}\rightarrow AlX_3$ (R = H, Me; X = Cl, Br)[592] and $(C_5H_5)_2Pb{:}\rightarrow BF_3$.[593] The constitution of the latter complex, however, must be placed in doubt since the analogous tin complex[594] has been shown by X-ray crystallography to have a very complex polymeric structure, perhaps best formulated as $\{[BF_4]^-(\mu\text{-}\eta^5\text{-}C_5H_5)_2Sn[\mu\text{-}\eta^5\text{-}C_5H_5Sn]^+THF\}_n$.[595]

Closely related to the cyclopentadienyls are carbaborane derivatives, where the bivalent Group IV metal occupies a vertex of the carbaborane framework. The *closo*-dodecacarboranes **(100)** (M = Ge, Sn, Pb) are very stable indeed, and may be sublimed unchanged *in vacuo* at high temperatures.[596] The smaller *C*-trimethylsilyl-substituted stannacarboranes **(101)** (R = H, Me, $SiMe_3$) are also stable enough to be sublimed,[597,598] but the similar plumbacarboranes **(102)** (R = H, Me) decompose at room temperature *in vacuo* although they can be stored indefinitely at 0 °C.[599] The stannacarboranes do not react with either $BH_3{\cdot}THF$ or BF_3 to form donor–acceptor complexes, indicating that the lone pair occupies an orbital of non-directional character. 1-Stanna-2,3-dimethyl-2,3-dicarba-*closo*-dodecacarborane, $Sn(Me_2)C_2B_9H_9$, can, however, function as a Lewis acid, forming a 2,2′-bipyridyl complex (Figure 18). As with the metallocene adducts, complexation is accompanied by a 'slippage' from η^5 bonding, and this complex may be regarded as an η^3-borallyl complex.

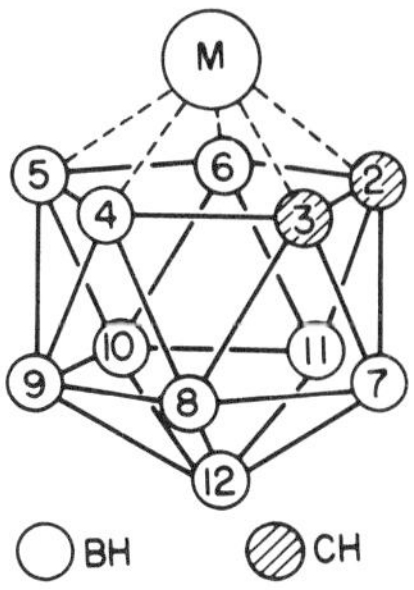

(100)

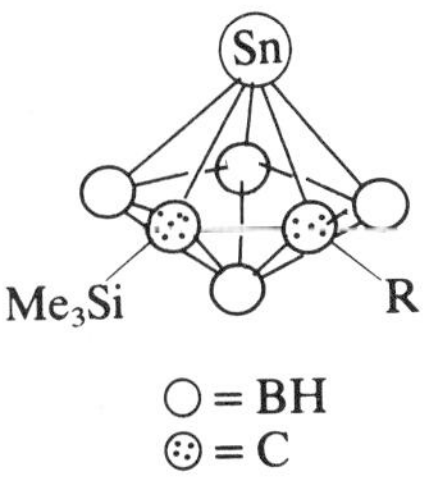

(101)

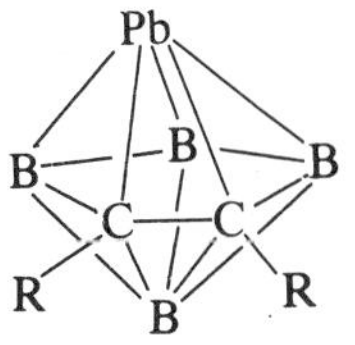

(102)

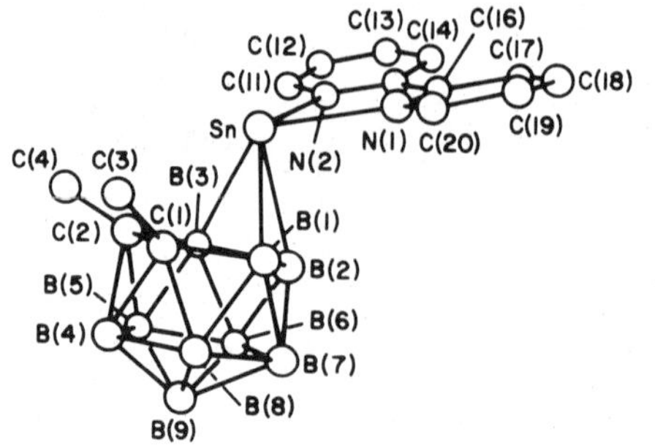

Figure 18 The structure of $Sn(Me)_2C_2B_9H_9$·bipy (reproduced by permission from *J. Chem. Soc., Chem. Commun.*, 1984, 1564)

Both tin and lead form η^6-arene complexes, although the arene is much more weakly bound to the metal than in similar complexes of Group VI transition metals. The tin and lead complexes, $[(C_6H_6)M{\cdot}(AlCl_4)_2{\cdot}C_6H_6]$ (M = Sn, Pb), are isomorphous, and comprise one-dimensional chain structures (Figure 19) with bridging tetrachloraluminate groups.[600,601] The structure of the related complexes, $[ArSnCl(AlCl_4)]_2$ (Ar = C_6H_6, $C_6H_4Me_2$-1,4), is similar, but contains a central four-membered $[Sn_2Cl_2]$ ring.[602,603]

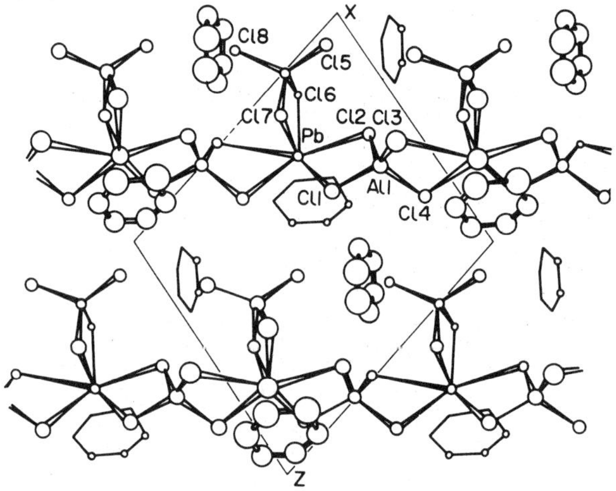

Figure 19 The structure of $[(C_6H_6)Pb(AlCl_4)_2]{\cdot}C_6H_6$ (reproduced by permission from *Inorg. Chem.*, 1974, **13**, 2429)

Monomeric amide, alkoxide and thiolate derivatives of germanium(II), tin(II) and lead(II) are only obtained with very sterically crowded ligands. Thus, bis(2,2,6,6-tetramethylpiperidyl)germanium,[604] the bis(trimethylsilyl)amides, $M[N(SiMe_3)_2]_2$ (M = Ge, Sn, Pb),[605] bis(aryloxides), $M(OC_6H_2Me\text{-}4\text{-}Bu^t_2\text{-}2,6)$ (M = Ge, Sn, Pb)[606] and the bis(arenethiolato)tin derivative $(SnSC_6H_2Bu^t_3\text{-}2,4,6)_2$[607] are all two-coordinated with an angular, 'V'-shaped geometry. Otherwise, for smaller ligands, association takes place *via* nitrogen, oxygen or sulfur bridging to give dimers, trimers or polymers depending upon the steric demands of the ligand. Thus, tin(II) bis(dimethylamido)tin(II) exists as the dimer (**103**),[608] whereas a polymeric structure such as (**104**) has been proposed for bis(aziridinyl)tin(II).[609] Crystals of the cyclic tin(II) amide, $[SnN(Bu^t)SiMe_2N(Bu^t)]_n$, comprise both monomeric and dimeric units.[610] Similarly, bis(*tert*-butoxy)-germanium(II)[561] and -tin(II),[611] and bis(triphenylsiloxy)-tin(II)[611] and -lead(II)[612] are dimeric, whilst the tin and lead bis(2,6-didisopropylbenzenethiolates) are trimeric, and small alkoxides, $M(OR)_2$ (M = Ge, Sn, Pb; R = Me, Et) are polymeric.[613,614] Bis(triphenylsiloxy)lead(II) is unstable, and readily eliminates hexaphenyldisiloxane affording the cage compound (**105**).[615] Other similar cage compounds such as $Sn_6O_4(OMe)_4$ (**106**) formed by hydrolysis of polymeric tin(II) methoxide,[300] and (**107**)[616] have also been characterized.

A relatively large number of tin–nitrogen cage structures have been characterized. The monomeric diazasilastannetidine, $\overline{SnN(Bu^t)Si(Me_2)N}Bu^t$, is cleaved quantitatively by *tert*-butylamine to give either the 'closed' $[Sn_2SiN_3]$ cage (**108**) or the 'open' cage (**109**). Thermolysis of the latter compound, however, results in conversion to the $[Sn_4N_4]$ cubane

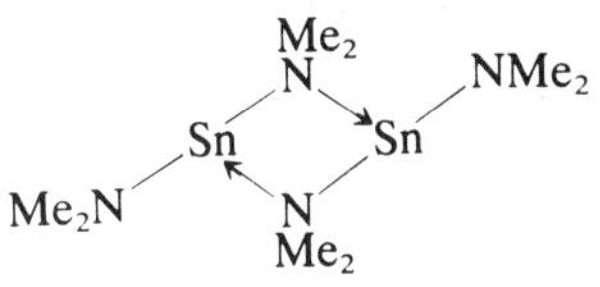

(103)

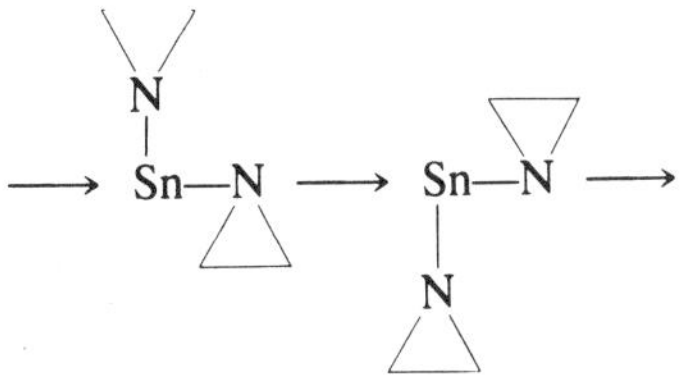

(104)

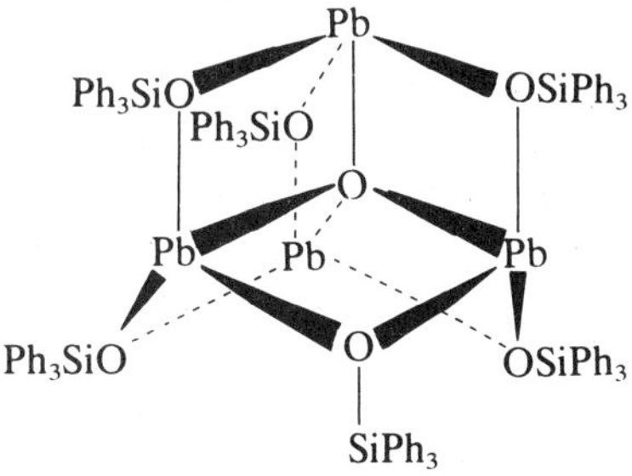

(105)

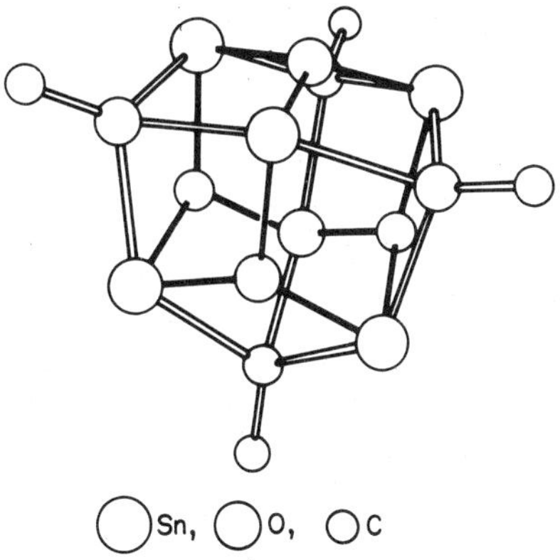

(106)

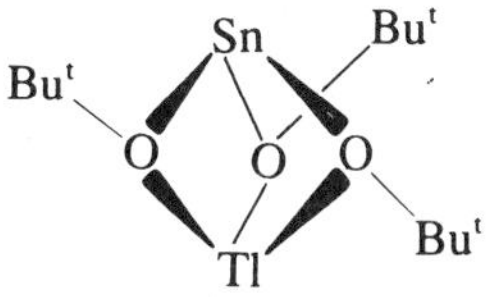

(107)

compound (**110**). The analogous $[Sn_4N_3O]$ cubane is obtained by hydrolysis of (**109**) (but not of **110**), and forms the adduct (**111**) with trimethylaluminum. The $[Sn_4N_4]$ core is also present in $(Me_2Si)(NMe)_5Sn_4$ (**112**), where one edge is enlarged by an [Me_2Si—NMe] bridge. The $[Sn_3N_2O_2]$ cage compound (**113**) has a crystallographic plane of symmetry, and may be regarded as a complex between the transient *tert*-butyliminostannanediyl, $[Sn(NBu^t)_2Sn]$, and tin(II) bis(*tert*-butoxide).[617–621]

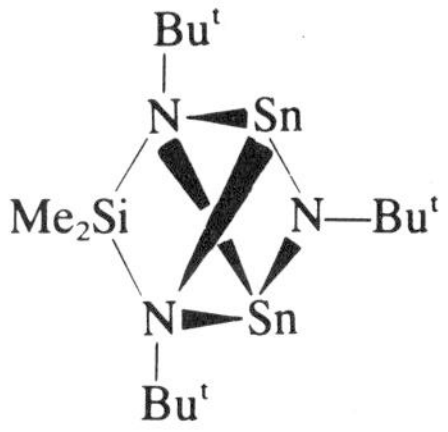

(108)

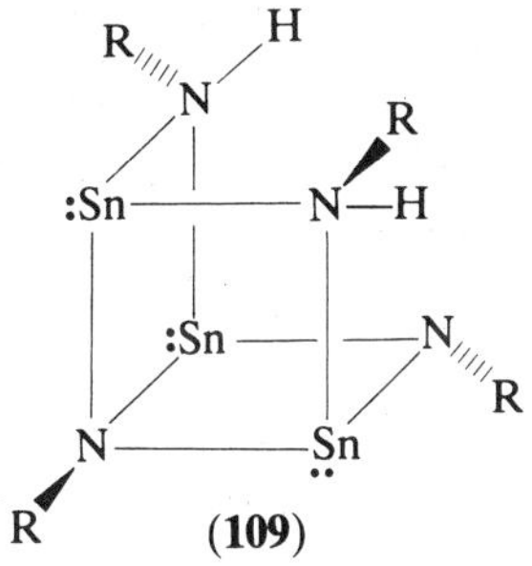

(109)

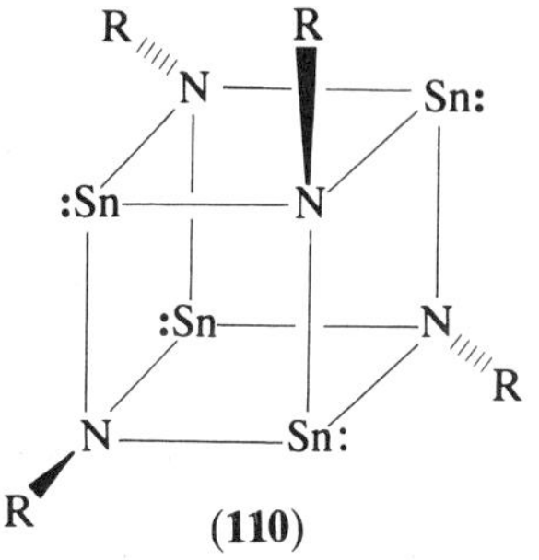

(110)

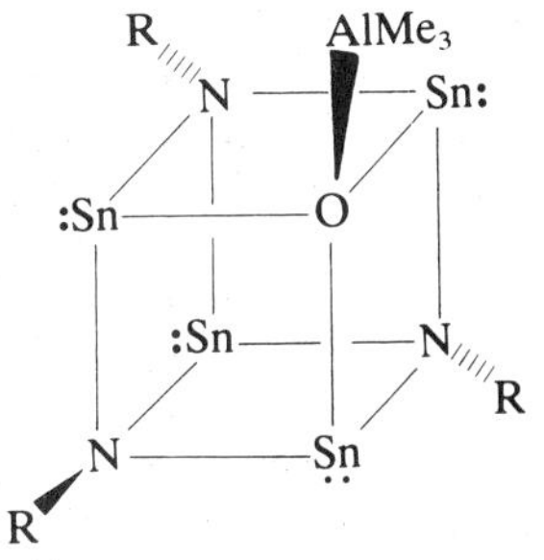

(111)

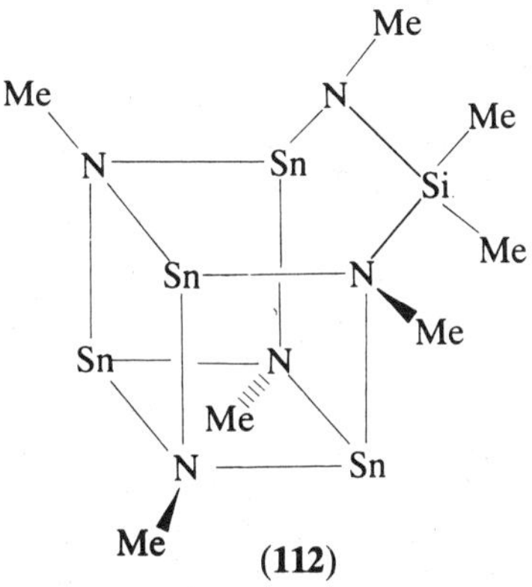

(112)

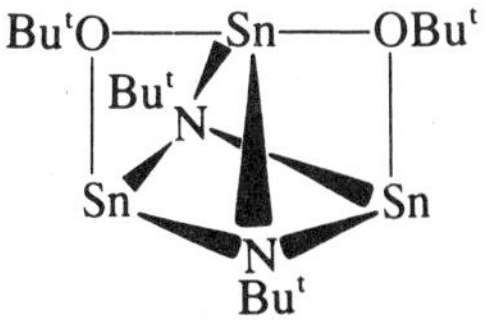

(113)

The proclivity of these types of compound to form associated structures derives from the propensity of bivalent germanium, tin and lead to achieve coordination numbers greater than two. When additional donor sites are available within the ligand itself, coordination saturation at the metal can be achieved either intramolecularly or intermolecularly resulting in the formation, respectively, of discrete monomeric species or polymers. Thus, GeI(acac),[622] $Ge(OCH_2CH_2)_2NMe$,[613] $Sn(SCH_2CH_2)_2E$ (E = Bu^tN, MeN, O, S),[623,624] $Sn(S_2COMe)_2$,[625] *N,N′*-ethylenebis(acetylideneimino)tin(II)[626] and tin(II) derivatives of 1,3-diones and related compounds[627–629] are all monomeric. Tin(II) and lead(II) carboxylates,[630–636] sulfates[637–639] and phosphates,[640–645] on the other hand, are polymeric in nature. Nevertheless, the coordination polyhedra about the two metals in these types of compound does vary substantially. With bivalent tin, only the usual three-coordinate pyramidal and four-coordinate pseudotrigonal bipyramidal geometries are found, and the anionic groups function solely as bridging ligands. With lead, however, simultaneous chelation and bridging occurs resulting in much higher coordination numbers. Values of six, seven, eight or nine are not atypical, whilst the particular stereochemistry adopted depends upon the individual compound. Thus, for example, whereas tin(II) formate has a two-dimensional sheet structure in which each formate group bridges two adjacent tin atoms (Figure 20),[630] lead(II) formate has a very complex three-dimensional polymeric structure with eight-coordinated lead, in which each formate group now links four different lead atoms (Figure 21).[636] Intermolecular association is much weaker in complexes involving dithio ligands. Thus, crystals of tin(II) and lead(II) dithiocarbonates (xanthates), $M[S_2COR]_2$ (M = Sn, Pb), and dithiocarbamates, $M[S_2CNR_2]_2$, generally comprise discrete molecules with chelating ligands.[646–651] Additional intermolecular interactions are weak or negligible. Lead(II) bis(diethyldithiophosphate), $Pb[S_2P(OEt)_2]_2$, is similar,[652] but the corresponding diisopropyldithiophosphate has a more strongly bonded chain structure.[653] Tin(II) and lead(II) bis(diphenyldithiophosphates) are quite unusual, and possess dimeric structures, where the two metal atoms are linked not only by bridging dithiophosphate groups, but also by an η^6 bonding interaction between the metal and one of the phenyl groups (Figure 22).[654,655]

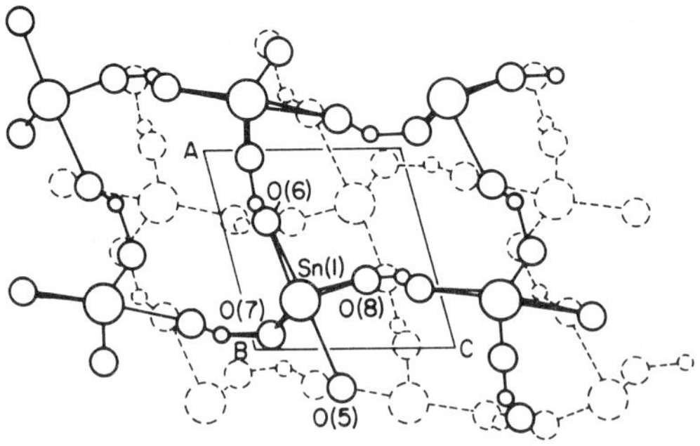

Figure 20 The layer structure of tin(II) formate (reproduced by permission from *J. Chem. Soc., Dalton Trans.*, 1978, 1274)

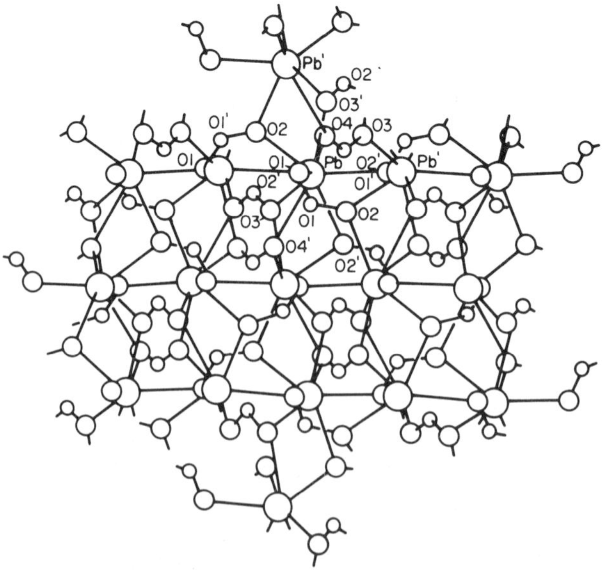

Figure 21 The three-dimensional lattice of lead(II) formate (reproduced by permission from *J. Organomet. Chem.*, 1982, **239**, 105)

The two mixed-valence tin carboxylates $[Sn(O_2CC_6H_4NO_2\text{-}o)_4O\cdot THF]_2$[656] and $[Sn_2(O_2CCF_3)_4O]_2\cdot C_6H_6$,[657] are quite unusual, and exhibit pentagonal pyramidal and square-based pyramidal coordination polyhedra, respectively, for bivalent tin. The former represents

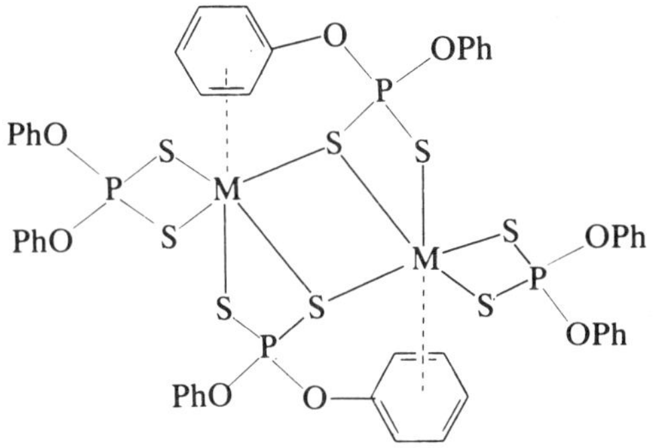

Figure 22 Schematic representation of the dimeric structure of tin(II) and lead(II) bis(O,O'-diphenyldithiophosphate)

the only example of this stereochemistry in Group IV. The tetravalent tin atoms enjoy typical, almost perfect, octahedral coordination.

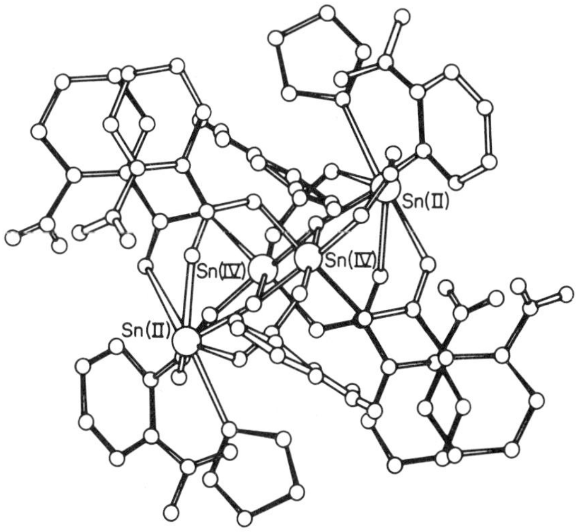

Figure 23 The structure of the mixed-valence tin carboxylate $[Sn_2(O_2CC_6H_4NO_2\text{-}o)_4O]_2{\cdot}THF$ (reproduced by permission from *J. Chem. Soc., Dalton Trans.*, 1976, 1602)

Bis(phosphino) and bis(arsino) derivatives of germanium, tin or lead are unknown. However, monophosphino- and monoarsino-germanium and tin halides, Bu^t_2EMX (M = Ge, Sn; E = P, As, X = Cl, Br), have been characterized as thermally stable, though easily oxidized solids, which are thought to be highly associated.[658–663]

26.4 REFERENCES

1. J. G. Chambers, T. E. Thomas, A. J. Barnes and W. J. Orville-Thomas, *Chem. Phys.*, 1975, **9,** 467.
2. I. Hargittai and C. Bliefelt, *Z. Naturforsch., Teil B,* 1983, **38,** 1304.
3. See A. F. Wells, 'Structural Inorganic Chemistry', 4th edn., Oxford University Press, Oxford, 1975.
4. J. C. Jamieson, *Science,* 1963, **139,** 762.
5. R. Walsh, *Acc. Chem. Res.*, 1981, **14,** 246.
6. L. E. Gusel'nikov and N. S. Nametkin, *Chem. Rev.*, 1979, **79,** 529.
7. P. H. Blustin, *J. Organomet. Chem.*, 1976, **105,** 161.
8. G. Olbrich, P. Potzinger, B. Riemann and R. Walsh, *Organometallics,* 1984, **3,** 1267.
9. H. Nakatsji, J. Ushio and T. Yonezawa, *J. Organomet. Chem.*, 1983, **258,** C1.
10. For an excellent discussion of five-coordination see R. R. Holmes, *Prog. Inorg. Chem.*, 1984, **32,** 119.
11. Useful review articles have appeared, germanium: K. C. Molloy and J. J. Zuckerman, *Adv. Inorg. Chem. Radiochem.*, 1983, **27,** 113; tin: J. Zubieta and J. J. Zuckerman, *Prog. Inorg. Chem.*, 1978, **24,** 251; M. Veith and O. Recktenwald, *Top. Curr. Chem.*, 1982, **104,** 1; P. A. Cusack, P. J. Smith, J. D. Donaldson and S. M. Grimes, International Tin Research Institute Publication No. 588; lead: P. G. Harrison, in 'Comprehensive Organometallic Chemistry', Pergamon, Oxford, 1982, vol. 2, p. 629; bivalent germanium, tin and lead: P. G. Harrison, *Coord. Chem. Rev.*, 1976, **20,** 1.
12. F. S. Kipping and L. L. Lloyd, *J. Chem. Soc., Trans.*, 1901, **79,** 449.

13. L. E. Gusel'nikov and N. S. Nametkin, *Chem. Rev.*, 1979, **79**, 529.
14. P. G. Harrison, *Coord. Chem. Rev.*, 1981, **34**, 154; 1982, **40**, 179; 1983, **49**, 193; 1984, **56**, 187; 1985, **66**, 190.
15. O. L. Chapman, C. C. Chang, J. Kole, M. E. Jung, J. A. Lowe, T. J. Barton and M. L. Turney, *J. Am. Chem. Soc.*, 1976, **98**, 7844.
16. M. R. Chedekel, M. Skoglund, R. L. Kreeger and H. Shechter, *J. Am. Chem. Soc.*, 1976, **98**, 7846.
17. S. Tokach, P. Boudjouk and R. D. Koob, *J. Phys. Chem.*, 1978, **82**, 1203.
18. P. G. Mahaffy, R. Gutowsky and L. K. Montgomery, *J. Am. Chem. Soc.*, 1980, **102**, 2854.
19. G. Maier, G. Mihm and H. P. Reisenauer, *Angew. Chem., Int. Ed. Engl.*, 1981, **20**, 597.
20. A. G. Brook, F. Abdesaken, B. Gutekunst, G. Gutekunst and R. K. Kallury, *J. Chem. Soc., Chem. Commun.*, 1981, 191; A. G. Brook, S. C. Nyburg, F. Abdesaken, B. Gutekunst, R. Krishna, M. R. Kalluny, Y. C. Poon, Y. M. Chang and W. Wong-Ng, *J. Am. Chem. Soc.*, 1982, **104**, 5667.
21. N. Wiberg and G. Wagner, *Angew. Chem., Int. Ed. Engl.*, 1983, **22**, 1005.
22. A. G. Brook, J. W. Harris, J. Lennon and M. El Sheikh, *J. Am. Chem. Soc.*, 1979, **101**, 83.
23. A. G. Brook, S. C. Nyburg, W. F. Reynolds, Y. C. Poon, Y. M. Chang, J. S. Lee and J. P. Picard, *J. Am. Chem. Soc.*, 1979, **101**, 6750.
24. N. Wiberg, G. Preiner and O. Schieda, *Chem. Ber.*, 1981, **114**, 2087, 3518.
25. N. Wiberg, G. Preiner, O. Schieda and G. Fischer, *Chem. Ber.*, 1981, **114**, 3505.
26. N. Wiberg and G. Wagner, *Angew. Chem., Int. Ed. Engl.*, 1983, **22**, 1005.
27. R. W. Kilb and L. Pierce, *J. Chem. Phys.*, 1957, **27**, 108.
28. H. F. Schaeffer, *Acc. Chem. Res.*, 1982, **15**, 283.
29. As Schaeffer points out,[28] the double-ζ SCF calculations of Hood and Schaeffer[30] and Alrichs and Heinzmann,[31] which predict Si=C bond distances of 171.5 pm and 169 pm, are probably the most reliable theoretical estimates. Note that these calculations were reported *before* the electron diffraction study of $Me_2Si{=}CH_2$ appeared.
30. D. M. Hood and H. F. Schaeffer, *J. Chem. Phys.*, 1978, **68**, 2985.
31. R. Alrichs and R. Heinzmann, *J. Am. Chem. Soc.*, 1977, **99**, 7452.
32. M. J. S. Dewar, D. H. Low and C. A. Ramsden, *J. Am. Chem. Soc.*, 1975, **97**, 1311.
33. Schaeffer[28] actually casts doubt on the interpretation of the electron diffraction data of Mahaffy, Gutowsky and Montgomery.[18]
34. N. Wiberg, G. Wagner, G. Mueller and J. Riede, *J. Organomet. Chem.*, 1984, **271**, 381.
35. This effect would reach a maximum at a twist angle of 90° when no π-bonding would occur. The predicted Si—C distance in the resulting diradical is 188 pm[30].
36. M. R. Hoffmann, Y. Yoshioka and H. F. Schaeffer, *J. Am. Chem. Soc.*, 1983, **105**, 1084.
37. L. E. Gusel'nikov, V. V. Volkova, V. G. Avakyan and N. S. Nametkin, *J. Organomet. Chem.*, 1980, **201**, 137.
38. O. M. Nefedov, A. K. Maltsev, V. N. Khabashesku and V. A. Korolev, *J. Organomet. Chem.*, 1980, **201**, 123.
39. C. A. Arrington, K. A. Klingensmith, R. West and J. Michl, *J. Am. Chem. Soc.*, 1984, **106**, 525.
40. A. K. Maltsev, V. N. Khabasheshku and O. M. Nefedov, *J. Organomet. Chem.*, 1984, **271**, 55.
41. A. G. Brook, F. Abdesaken, G. Gutenkunst and N. Plavac, *Organometallics*, 1982, **1**, 994.
42. T. J. Barton and G. J. Burns, *J. Am. Chem. Soc.*, 1978, **100**, 5246.
43. B. Solouki, P. Rosmus, H. Bock and G. Maier, *Angew. Chem., Int. Ed. Engl.*, 1980, **19**, 51.
44. G. Maier, G. Mihm and H. P. Reisenauer, *Angew. Chem., Int. Ed. Engl.*, 1980, **19**, 52.
45. C. L. Kreil, O. L. Chapman, G. T. Burns and T. J. Barton, *J. Am. Chem. Soc.*, 1980, **102**, 841.
46. J. D. Rich and R. West, *J. Am. Chem. Soc.*, 1982, **104**, 6884.
47. H. Bock, P. Rosmus, B. Solouki and G. Maier, *J. Organomet. Chem.*, 1984, **271**, 145.
48. H. Sakurai, Y. Nakadaira and T. Kobayashi, *J. Am. Chem. Soc.*, 1979, **101**, 487.
49. R. West and M. J. Fink, *Science*, 1981, **214**, 1343.
50. H. Watanabe, T. Okawa, M. Kato and Y. Nagai, *J. Chem. Soc., Chem. Commun.*, 1983, 781.
51. S. Masamune, H. Tobita and S. Murakami, *J. Am. Chem. Soc.*, 1983, **105**, 6524.
52. S. Masamune, S. Murakami and H. Tobita, *Organometallics*, 1983, **2**, 1464.
53. M. Wiedenbruch, A. Schäfer and K. L. Than, *Z. Naturforsch., Teil B*, 1983, **38**, 1695.
54. P. Boudjouk, B. H. Han and K. R. Anderson, *J. Am. Chem. Soc.*, 1982, **104**, 4992.
55. K. W. Zilm, D. M. Grant, J. Michl, M. J. Fink and R. West, *Organometallics*, 1983, **2**, 193.
56. S. Masamune, S. Murakami, J. T. Snow, H. Tobita and D. J. Williams, *Organometallics*, 1984, **3**, 333.
57. M. J. Fink, M. J. Michalczyk, K. J. Haller, R. West and J. Michl, *Organometallics*, 1984, **3**, 793.
58. K. J. Krogh-Jespersen, *J. Phys. Chem.*, 1982, **86**, 1492.
59. H. Lischka and H. J. Köhler, *Chem. Phys. Lett.*, 1982, **85**, 467; H. Lischka and H. J. Köhler, *J. Am. Chem. Soc.*, 1982, **104**, 5884.
60. R. A. Poiver and J. D. Goddard, *Chem. Phys. Lett.*, 1981, **80**, 37.
61. W. W. Schoeller and V. Staemmler, *Inorg. Chem.*, 1984, **23**, 3369.
62. H. Okinoshima and W. P. Weber, *J. Organomet. Chem.*, 1978, **149**, 279; **155**, 165.
63. D. Seyferth, T. F. O. Lim and D. P. Duncan, *J. Am. Chem. Soc.*, 1978, **100**, 1626.
64. T. J. Barton and W. D. Wulff, *J. Am. Chem. Soc.*, 1979, **101**, 2735.
65. N. Wiberg and G. Preiner, *Angew. Chem., Int. Ed. Engl.*, 1978, **17**, 362.
66. D. R. Parker and L. H. Sommer, *J. Am. Chem. Soc.*, 1976, **98**, 618.
67. S. A. Kazoura and W. P. Weber, *J. Organomet. Chem.*, 1984, **271**, 47.
68. W. Ande, H. Tsumaki and M. Ikenu, *J. Chem. Soc., Chem. Commun.*, 1981, 597.
69. C. W. Carlson and R. West, *Organometallics*, 1983, **2**, 1793.
70. T. J. Barton and S. K. Hoekman, *J. Am. Chem. Soc.*, 1980, **102**, 1584.
71. P. Riviere, A. Castel and J. Satgé, *J. Am. Chem. Soc.*, 1980, **102**, 5413.
72. G. Märkl, D. Rudnick, R. Schulz and A. Scheeif, *Angew. Chem., Int. Ed. Engl.*, 1982, **21**, 221.
73. H. Lavayssière, J. Barrau, G. Dousse, J. Satgé and M. Bouchaut, *J. Organomet. Chem.*, 1978, **154**, C9.
74. C. Couret, J. Satgé, J. D. Andriamizaka and J. Esudié, *J. Organomet. Chem.*, 1978, **157**, C35.
75. J. Barrau, M. Bouchaut, H. Lavayssière, G. Dousse and J. Satgé, *J. Organomet. Chem.*, 1983, **243**, 281.
76. J. Barrau, G. Rima, H. Lavayssière, G. Dousse and J. Satgé, *J. Organomet. Chem.*, 1983, **246**, 227.

77. G. Trinquier, M. Pelissier, G. Saint-Roch and H. Lavayssière, *J. Organomet. Chem.*, 1981, **214,** 169.
78. H. Lavayssière, G. Dousse, J. Barrau, J. Satgé and M. Bouchaut, *J. Organomet. Chem.*, 1978, **161,** C59.
79. J. Escudié, C. Couret, J. Satgé and J. D. Andriamizaka, *Organometallics,* 1982, **1,** 1262.
80. R. Jaquet, W. Kutzelnigg and V. Staemmler, *Theor. Chim. Acta,* 1980, **54,** 205.
81. G. Trinquier, M. Pelissier, B. Saint-Roch and H. Lavayssière, *J. Organomet. Chem.*, 1981, **214,** 169.
82. E. Hengge, *Top. Curr. Chem.*, 1974, **51,** 1.
83. J. W. Hastie, R. H. Hauge and J. L. Margrave, *Inorg. Chim. Acta,* 1969, **3,** 601.
84. H. Schnöckel, *Angew. Chem., Int. Ed. Engl.*, 1978, **17,** 616; 1980, **19,** 323.
85. H. Schnöckel, *J. Mol. Struct.*, 1980, **65,** 115; 1981, **70,** 183.
86. J. S. Ogden and M. J. Ricks, *J. Chem. Phys.*, 1970, **52,** 352.
87. A. Bos, J. S. Ogden and L. Ogree, *J. Phys. Chem.*, 1974, **78,** 1763.
88. J. S. Anderson, A. Bos and J. S. Ogden, *Chem. Commun.*, 1971, 1381.
89. A. Bos, A. T. Howe, B. W. Dale and L. W. Becker, *J. Chem. Soc., Chem. Commun.*, 1972, 730.
90. A. Bos and J. S. Ogden, *J. Phys. Chem.*, 1973, **77,** 1513.
91. G. Herzberg, 'Molecular Spectra and Molecular Structure', Van Nostrand, New York, 1950, vol. I, 'Spectra of Diatomic Molecules'.
92. The chemistry of the carbenes[93,94] and their analogues, also referred to as silanediyls,[95–97] germanediyls[97,98] and stannanediyls[97,99] (IUPAC nomenclature) have been reviewed extensively.
93. W. Kirmse, 'Carbene Chemistry', Academic, New York, 1971.
94. M. Jones and R. A. Moss, 'Carbenes', Wiley, New York, 1973/75, vols. I and II.
95. E. A. Chernyshev, N. G. Komalenkova and S. A. Bashkirova, *Usp. Khim.*, 1976, **45,** 1782.
96. E. A. Chernyshev, N. G. Komalenkova and S. A. Bashkirova, *J. Organomet. Chem.*, 1984, **271,** 129.
97. O. M. Nefedov, S. P. Kolesnikov and A. I. Ioffe, *J. Organomet. Chem.*, 1977, **5,** 181.
98. J. Satgé, M. Massol and P. Rivierè, *J. Organomet. Chem.*, 1973, **56,** 1.
99. W. P. Neumann, 'The Organometallic and Coordination Chemistry of Germanium, Tin and Lead', ed. M. Gielen and P. G. Harrison, Freund, Tel Aviv, 1978, 51.
100. B. O. Roos and P. M. Siegbahn, *J. Am. Chem. Soc.*, 1977, **99,** 7716.
101. C. W. Bauschlicher and I. Shavitt, *J. Am. Chem. Soc.*, 1978, **100,** 739.
102. S. Shih, S. D. Peyerimhoff, R. J. Buenker and M. Peric, *Chem. Phys. Lett.*, 1978, **55,** 206.
103. L. B. Harding and W. A. Goddard, *Chem. Phys. Lett.*, 1978, **55,** 217.
104. C. W. Bauschlicher, H. F. Schaefer and P. S. Bagus, *J. Am. Chem. Soc.*, 1977, **99,** 7106.
105. N. C. Baird and K. F. Taylor, *J. Am. Chem. Soc.*, 1978, **100,** 6930.
106. K. B. Eisenthal, N. J. Turro, M. Aikawa, J. A. Butcher, C. DuPuy, G. Hefferon, W. Hetherington, G. M. Korenowskii and M. J. McAuliffe, *J. Am. Chem. Soc.*, 1980, **102,** 6563.
107. S. P. So, *J. Chem. Soc., Faraday Trans. 2,* 1979, **75,** 820.
108. C. W. Bauschlicher, *J. Am. Chem. Soc.*, 1980, **102,** 5492.
109. J. H. Meadows and H. F. Schaefer, *J. Am. Chem. Soc.*, 1976, **98,** 4383.
110. J. C. Barthelat, B. Saint-Roch, G. Trinquier and J. Satgé, *J. Am. Chem. Soc.*, 1980, **102,** 4080.
111. G. Olbrich, *Chem. Phys. Lett.*, 1980, **73,** 110.
112. M. Schriewer and W. P. Neumann, *Angew. Chem., Int. Ed. Engl.*, 1981, **20,** 1019.
113. K. H. Scherping and W. P. Neumann, *Organometallics,* 1982, **1,** 1017.
114. A. S. Nazran, E. J. Gabe, Y. Le Page, D. J. Northcott, J. M. Park and D. Griller, *J. Am. Chem. Soc.*, 1983, **105,** 2912.
115. D. E. Milligan and M. E. Jacox, *J. Chem. Phys.*, 1970, **52,** 2594.
116. C. A. Arrington, K. A. Klingensmith, R. West and J. Michl, *J. Am. Chem. Soc.*, 1984, **106,** 525.
117. J. W. Hastie, R. H. Hauge and J. L. Margrave, *J. Am. Chem. Soc.*, 1969, **91,** 2536; G. Maase, R. H. Hauge and J. L. Margrave, *Z. Anorg. Allg. Chem.*, 1972, **392,** 295.
118. D. L. Perry, P. F. Meier, R. H. Hauge and J. L. Margrave, *Inorg. Chem.*, 1978, **17,** 1364.
119. Z. K. Ismail, L. Fredin, R. H. Hauge and J. L. Margrave, *J. Chem. Phys.*, 1982, **77,** 1626.
120. Z. K. Ismail, R. H. Hauge, L. Fredin, J. W. Kauffman and J. L. Margrave, *J. Chem. Phys.*, 1982, **77,** 1617.
121. G. A. Ozin and A. Vander Voet, *J. Chem. Phys.*, 1972, **56,** 4768.
122. I. Hargittai, J. Tremmel, E. Vadja, A. A. Ishchenko, A. A. Ivanov, L. S. Ivashkevich and V. P. Spiridonov, *J. Mol. Struct.*, 1977, **42,** 147.
123. J. Trotter, M. Akhtar and N. Bartlett, *J. Chem. Soc. (A),* 1966, 30.
124. J. D. Donaldson and R. Oteng, *Inorg. Nucl. Chem. Lett.*, 1967, **3,** 163.
125. R. C. McDonald, H. Ho-Kuen Hau and K. Eriks, *Inorg. Chem.*, 1976, **15,** 762.
126. A. Byström, *Ark. Kemi, Mineral. Geol.*, 1947, **24A,** No. 33.
127. I. Naray-Szabo, 'Inorganic Crystal Chemistry', Akademiai Kaido, Budapest, 1969, p. 215.
128. J. M. van den Berg, *Acta Crystallogr.*, 1961, **14,** 1002.
129. J. Anderson, *Acta Chem. Scand., Ser. A,* 1975, **29,** 956.
130. H. M. Powell and F. M. Brauer, *J. Chem. Soc.*, 1938, **21,** 197.
131. M. Chand and G. C. Trigunayat, *Acta Crystallogr., Sect. B,* 1975, **31,** 1222.
132. R. A. Howie, W. Moser and I. C. Moser, *Acta Crystallogr., Sect. B,* 1972, **28,** 2965.
133. P. Jutzi, H. J. Hoffman, D. J. Brauer and C. Kruger, *Angew. Chem., Int. Ed. Engl.*, 1973, **12,** 1002.
134. P. G. Harrison, B. J. Haylett and T. J. King, *Inorg. Chim. Acta,* 1983, **75,** 265.
135. E. Hough and D. G. Nicholson, *J. Chem. Soc., Dalton Trans.*, 1976, 1782.
136. J. Anderson, *Acta Chem. Scand.*, 1972, **26,** 1730, 2543, 3813.
137. J. Anderson and G. Lundgren, *Acta Chem. Scand.*, 1970, **24,** 2670.
138. N. Kamenar and D. Grdenic, *J. Chem. Soc.*, 1961, 3954.
139. H. Kiriyama, K. Kitahama, O. Nakamura and R. Kiriyama, *Bull. Chem. Soc. Jpn.*, 1973, **46,** 1389.
140. H. G. von Schnering, R. Nesper and H. Pelshenko, *Z. Naturforsch., Teil B,* 1981, **36,** 1551.
141. C. C. Venetopoulos and P. J. Rentzeperis, *Z. Kristallogr.*, 1975, **141,** 246.
142. S. Fregerslev and S. E. Rasmussen, *Acta Chem. Scand.*, 1968, **22,** 2541.

143. J. K. Stalik, P. W. R. Corfield and D. W. Meek, *Inorg. Chem.*, 1973, **12**, 1668.
144. H. J. Haupt, F. Huber and H. Preut, *Z. Anorg. Allg. Chem.*, 1976, **422**, 97, 255.
145. P. G. Harrison, B. J. Haylett and T. J. King, *Inorg. Chim. Acta*, 1983, **75**, 265.
146. G. Bergerhoff and L. Goost, *Acta Crystallogr., Sect. B*, 1973, **29**, 632.
147. G. Bergerhoff and H. Namgung, *Acta Crystallogr., Sect. B*, 1978, **34**, 699.
148. D. Messer, *Z. Naturforsch., Teil B*, 1978, **33**, 366.
149. A. N. Christensen and S. E. Rasmussen, *Acta Chem. Scand.*, 1965, **19**, 421.
150. F. R. Poulsen and S. E. Rasmussen, *Acta Chem. Scand.*, 1970, **24**, 150.
151. J. Barrett, S. R. A. Bird, J. D. Donaldson and J. Silver, *J. Chem. Soc. (A)*, 1971, 3105.
152. S. R. A. Bird, J. D. Donaldson and J. Silver, *J. Chem. Soc., Dalton Trans.*, 1972, 1950.
153. J. D. Donaldson and J. Silver, *J. Chem. Soc., Dalton Trans.*, 1973, 666.
154. J. D. Donaldson, D. Laughlin, S. D. Ross and J. Silver, *J. Chem. Soc., Dalton Trans.*, 1973, 1985.
155. J. D. Donaldson, J. Silver, S. Hadjiminolis and S. D. Ross, *J. Chem. Soc., Dalton Trans.*, 1975, 1500.
156. D. Weber, *Z. Naturforsch., Teil B*, 1978, **33**, 862, 1443; 1979, **34**, 939.
157. H. J. Haupt, F. Huber and H. Preut, *Z. Anorg. Allg. Chem.*, 1974, **408**, 209.
158. P. Mauersberger and F. Huber, *Acta Crystallogr., Sect. B*, 1980, **36**, 683.
159. D. Bedlivy and K. Mereiter, *Acta Crystallogr., Sect. B*, 1980, **36**, 782.
160. G. Bergerhoff, L. Goost and E. Schultze-Rhonhof, *Acta Crystallogr., Sect. B*, 1968, **24**, 803.
161. J. Anderson, *Acta Chem. Scand., Ser. A*, 1976, **30**, 229.
162. R. R. McDonald, A. C. Jarson and D. T. Cramer, *Acta Crystallogr.*, 1964, **17**, 1104.
163. G. Bergerhoff and L. Goost, *Acta Crystallogr., Sect. B*, 1970, **26**, 19.
164. G. Bergerhoff and L. Goost, *Acta Crystallogr., Sect. B*, 1974, **30**, 1362.
165. J. D. Donaldson and D. C. Puxley, *J. Chem. Soc., Chem. Commun.*, 1972, 289; S. Vilminot, W. Gramer and L. Cot, *Acta Crystallogr., Sect. B*, 1978, **34**, 35.
166. S. Vilminot, W. Gramer, Z. Al Oraibi and L. Cot, *Acta Crystallogr., Sect. B*, 1978, **34**, 3308.
167. J. Börisch and G. Bergerhoff, *Acta Crystallogr., Sect. C*, 1984, **40**, 2005.
168. J. C. Taylor and P. W. Wilson, *J. Am. Chem. Soc.*, 1973, **95**, 1834.
169. M. F. A. Dove, R. King and T. J. King, *J. Chem. Soc., Chem. Commun.*, 1973, 944.
170. C. Geneys, S. Vilminot, and L. Cot, *Acta Crystallogr., Sect. B*, 1976, **32**, 3199.
171. S. Vilminot, W. Gramer, Z. Al Oraibi and L. Cot, *Acta Crystallogr., Sect. B*, 1978, **34**, 3306.
172. S. Vilminot, W. Gramer, Z. Al Oraibi and L. Cot, *Acta Crystallogr., Sect. B*, 1980, **36**, 1537.
173. C. C. Hsu and R. A. Geanangel, *Inorg. Chem.*, 1980, **19**, 110.
174. More recently determined data include those for SiF_4,[175] $MeSiF_3$,[176] $CH_2{:}CH{\cdot}CH_2SiF_3$,[177] $PhSiF_3$,[178] $ClCH_2SiCl_3$,[179] $F_2H_3C_3SiCl_3$,[180] $C_5H_5SiCl_3$,[181] $CX_2(SiCl_3)_2$ (X = H, Cl),[182] Me_2SiCl_2,[183] $(CH_2{:}CH)Me_2SiCl$,[184] Ph_3SiCl,[185] $MeGeF_3$,[186] Me_2GeF_2,[186] $MeGeH_2F$,[187] Cl_3CGeCl_3,[188] $MeGeCl_3$,[189] Me_3GeCl,[190] $GeBr_4$,[191] $MeGeBr_3$,[186] Me_2GeBr_2,[186] Ph_3GeBr,[192] GeH_3Br,[193] GeH_2X_2 (X = Cl, Br),[194] GeD_3X and $GeHD_2X$ (X = F, Cl, Br, I).[195]
175. B. Beagley, D. P. Brown and J. M. Freeman, *J. Mol. Struct.*, 1973, **18**, 337.
176. J. R. Durig, Y. S. Li and C. C. Tong, *J. Mol. Struct.*, 1972, **14**, 255.
177. T. M. Kuznetsova, N. V. Alekseev, and N. N. Veniaminov, *J. Struct. Chem. (Engl. Transl.)*, 1979, **20**, 281.
178. T. M. Il'enko, N. N. Veniaminov and N. V. Alekseev, *J. Struct. Chem. (Engl. Trans.)*, 1975, **16**, 270.
179. E. Vadja, T. Szekely and M. Hargittai, *J. Mol. Struct.*, 1981, **73**, 243.
180. T. M. Kuznetsova, N. V. Alekseev and N. V. Veniaminov, *J. Struct. Chem. (Engl. Transl.)*, 1979, **20**, 823.
181. A. H. Cowley, E. A. V. Ebsworth, S. K. Mehrotra, D. W. H. Rankin and M. D. Walkinshaw, *J. Chem. Soc., Chem. Commun.*, 1982, 1099.
182. E. Vadja, M. Kolonits, B. Rozsondai, G. Fritz and E. Matern, *J. Mol. Struct.*, 1982, **95**, 197.
183. V. S. Mastryukov, A. V. Golubinskii and L. V. Vilkov, *J. Struct. Chem. (Engl. Transl.)*, 1980, **21**, 37.
184. Q. Shan, *J. Mol. Struct.*, 1982, **95**, 215.
185. E. B. Lobkovskii, V. N. Fokii and K. N. Semenenko, *J. Struct. Chem. (Engl. Transl.)*, 1982, **22**, 603.
186. J. E. Drake, R. T. Hemmings, J. L. Hencher, F. M. Mustoe and Q. Shen, *J. Chem. Soc., Dalton Trans.*, 1976, 394, 811.
187. R. F. Roberts, R. Varma and J. F. Nelson, *J. Chem. Phys.*, 1976, **64**, 5053.
188. E. Vadja, I. Hargittai, A. K. Maltsev and O. M. Nefedov, *J. Mol. Struct.*, 1974, **23**, 417.
189. J. R. Durig, P. J. Cooper and Y. S. Li, *J. Mol. Spectrosc.*, 1975, **57**, 169.
190. J. R. Durig and K. L. Hellams, *J. Mol. Spectrosc.*, 1975, **57**, 349.
191. G. B. G. Souza and J. D. Wieser, *J. Mol. Struct.*, 1975, **25**, 442.
192. H. Preut and F. Huber, *Acta Crystallogr., Sect. B*, 1979, **35**, 83.
193. L. C. Krisher and S. N. Wolf, *J. Chem. Phys.*, 1973, **58**, 396.
194. B. Beagley, D. P. Brown and J. M. Freeman, *J. Mol. Struct.*, 1973, **18**, 335.
195. S. Cradock, D. C. McKean and M. W. Mackenzie, *J. Mol. Struct.*, 1981, **74**, 265.
196. J. Zubieta and J. J. Zuckerman, *Prog. Inorg. Chem.*, 1978, **24**, 268–270.
197. R. Hoppe and W. Dähne, *Naturwissenschaften*, 1962, **49**, 254.
198. E. O. Schlemper and W. C. Hamilton, *Inorg. Chem.*, 1966, **5**, 995.
199. L. E. Levchuk, J. R. Sams and F. Aubke, *Inorg. Chem.*, 1972, **11**, 43.
200. H. C. Clark, R. J. O'Brien and J. Trotter, *J. Chem. Soc.*, 1964, 2332.
201. M. B. Hossain, J. L. Lefferts, K. C. Molloy, D. van der Helm and J. J. Zuckerman, *Inorg. Chim. Acta*, 1979, **36**, L409.
202. N. G. Bokii, G. N. Zakharova and Yu. T. Struchkov, *J. Struct. Chem.*, 1970, **11**, 828.
203. H. Preut and F. Huber, *Acta Crystallogr., Sect. B*, 1979, **35**, 744.
204. M. J. S. Gyrane, M. F. Lappert, S. J. Miles, A. J. Carty and N. J. Taylor, *J. Chem. Soc., Dalton Trans.*, 1977, 2009.
205. L. N. Zakharov, B. I. Petrov, V. A. Lebedev, E. A. Kuz'min and N. V. Belov, *Kristallografiya*, 1978, **23**, 1049.
206. S. Calogero, P. Ganis, V. Peruzzo and G. Tagliavini, *J. Organomet. Chem.*, 1979, **179**, 145.

207. A. G. Davies, H. J. Milledge, D. C. Puxley and P. J. Smith, *J. Chem. Soc. (A)*, 1970, 2862.
208. N. W. Alcock and J. F. Sawyer, *J. Chem. Soc., Dalton Trans.*, 1977, 1090.
209. N. G. Bokii, Yu. T. Struchkov and A. K. Prokof'ev, *J. Struct. Chem. (Engl. Transl.)*, 1972, **13,** 619.
210. P. T. Greene and R. F. Bryan, *J. Chem. Soc. (A)*, 1971, 2549.
211. J. L. Baxter, E. M. Holt and J. J. Zuckerman, *Organometallics,* 1985, **4,** 255.
212. H. Preut and F. Huber, *Z. Anorg. Allg. Chem.*, 1977, **435,** 234.
213. M. Marumi, V. Busetti and A. Del Pra, *Inorg. Chim. Acta,* 1967, **1,** 419.
214. R. J. H. Clark, A. G. Davies and R. J. Puddephatt, *Inorg. Chem.*, 1969, **8,** 457.
215. R. J. H. Clark, A. G. Davies and R. J. Puddephatt, *J. Am. Chem. Soc.*, 1968, **90,** 6923.
216. P. G. Harrison, T. J. King and M. A. Healy, *J. Organomet. Chem.*, 1979, **182,** 17.
217. T. Kimura, T. Ueki, N. Yasuoka, N. Kasai and M. Kakudo, *Bull. Chem. Soc. Jpn.*, 1969, **42,** 2479.
218. M. Yoshida, T. Ueki, N. Yasuoka, N. Kasai, M. Kakudo, I. Omae, S. Kikkawa and S. Matsuda, *Bull. Chem. Soc. Jpn.*, 1968, **41,** 1113.
219. B. W. Fitzsimmons, D. G. Othen, H. M. M. Shearer, K. Wade and G. Whitehead, *J. Chem. Soc., Chem. Commun.*, 1977, 215.
220. G. van Koten, J. G. Noltes and A. L. Spek, *J. Organomet. Chem.*, 1976, **118,** 183.
221. G. van Koten, J. T. B. H. Jastrzebskii, J. G. Noltes, W. M. G. F. Pontenagel, J. Kroon and A. L. Spek, *J. Am. Chem. Soc.*, 1978, **100,** 5021.
222. G. van Koten, J. T. B. H. Jastrzebskii, J. G. Noltes, G. J. Verhoeck, A. L. Spek and J. Kroon, *J. Chem. Soc., Dalton Trans.*, 1980, 1352.
223. G. van Koten, J. T. B. H. Jastrzebskii, J. G. Noltes, A. L. Spek and J. C. Schoone, *J. Organomet. Chem.*, 1978, **148,** 233.
224. G. van Koten and J. G. Noltes, *J. Am. Chem. Soc.*, 1976, **98,** 5393.
225. R. J. P. Corriu, A. Kpoton, M. Poirier, G. Royo and J. Y. Corey, *J. Organomet. Chem.*, 1984, **277,** C25.
226. M. G. Voronkov, L. I. Gubanova, Y. L. Frolov, N. K. Chernov, G. A. Gavrilova and N. N. Chipanina, *J. Organomet. Chem.*, 1984, **271,** 164.
227. G. Klebe, M. Nix and K. Hensen, *Chem. Ber.*, 1984, **117,** 797.
228. G. Klebe, J. W. Bats and H. Fuess, *J. Am. Chem. Soc.*, 1984, **106,** 5202.
229. E. A. Zel'bst, V. E. Shklorov, Yu. T. Struchkov, Yu. L. Frolov, A. A. Kashaev, L. I. Gubanova, V. M. D'yakov and M. G. Voronkov, *J. Struct. Chem. (Engl. Transl.)*, 1981, **22,** 377.
230. K. Hensen and R. Busch, *Z. Naturforsch., Teil B,* 1982, **37,** 1174.
231. R. Hulme, *J. Chem. Soc.*, 1963, 1524.
232. S. W. Ng, C. L. Barnes, M. B. Hossain, D. van der Helm, J. J. Zuckerman and V. G. Kumar Das, *J. Am. Chem. Soc.*, 1982, **104,** 5359.
233. P. G. Harrison and K. C. Molloy, *J. Organomet. Chem.*, 1978, **152,** 63.
234. C. Pelizzi, G. Pelizzi and P. Tarasconi, *Polyhedron,* 1983, **2,** 145.
235. R. Graziani, U. Casllato, R. Eltone and G. Plazzogna, *J. Chem. Soc., Dalton Trans.*, 1982, 805.
236. F. W. B. Einstein and B. R. Penfold, *J. Chem. Soc. (A)*, 1968, 3019.
237. L. A. Aslanov, W. M. Attiya, V. M. Ionov, A. B. Permin and V. S. Petrosyan, *J. Struct. Chem.*, 1977, **18,** 884.
238. J. Chojnowski, M. Cypryk and J. Michalski, *J. Organomet. Chem.*, 1978, **161,** C31; K. Hensen, T. Zengerly, P. Pickel and G. Klebe, *Angew. Chem., Int. Ed. Engl.*, 1983, **22,** 725.
239. B. Deppisch, B. Gladrow and D. Kummer, *Z. Anorg. Allg. Chem.*, 1984, **519,** 42.
240. A. D. Adley, P. H. Bird, A. R. Fraser and M. Onyszchuk, *Inorg. Chem.*, 1972, **11,** 1402.
241. C. I. Brandon, *Acta Chem. Scand.*, 1963, **17,** 759.
242. Y. Hermodsson, *Acta Crystallogr.*, 1960, **13,** 656; Y. Hermodsson, *Ark. Kemi,* 1970, **31,** 73.
243. P. Domiano, A. Musatti, M. Nardelli, C. Pelizzi and G. Predieri, *Inorg. Chim. Acta,* 1980, **38,** 9.
244. M. Webster and H. E. Blayden, *J. Chem. Soc. (A)*, 1969, 2443.
245. D. M. Barnhart, C. N. Caughlan and M. Ul-Haque, *Inorg. Chem.*, 1968, **7,** 1135.
246. J. C. Barnes and T. J. R. Weakley, *J. Chem. Soc., Dalton Trans.*, 1976, 1786.
247. H. W. Roesky, M. Kuhn and J. W. Bats, *Chem. Ber.*, 1982, **115,** 3025.
248. M. M. Olmstead, K. A. Williams and W. K. Musker, *J. Am. Chem. Soc.*, 1982, **104,** 5567.
249. L. A. Aslanov, V. M. Ionov, W. M. Attia, A. B. Permin and V. S. Petrosyan, *J. Organomet. Chem.*, 1978, **144,** 39.
250. I. R. Beattie, M. Milne, M. Webster, H. E. Blayden, P. J. Jones, R. C. G. Killean and J. L. Lawrence, *J. Chem. Soc. (A)*, 1969, 482.
251. G. G. Mather, G. M. McLaughlin and A. Pidcock, *J. Chem. Soc., Dalton Trans.*, 1973, 1823.
252. I. R. Beattie, R. Hulme and L. Rule, *J. Chem. Soc.*, 1965, 1581.
253. L. A. Aslanov, V. M. Ionov, W. M. Attiya and A. B. Permin, *J. Struct. Chem. (Engl. Transl.)*, 1978, **19,** 166.
254. L. A. Aslanov, V. M. Ionov, M. W. Attiya, A. B. Permin and V. S. Petrosyan, *J. Organomet. Chem.*, 1978, **144,** 39.
255. E. A. Blom, B. R. Penfold and W. T. Robinson, *J. Chem. Soc. (A)*, 1969, 913.
256. L. Randaccio, *J. Organomet. Chem.*, 1973, **55,** C58.
257. N. W. Isaacs and C. H. L. Kennard, *J. Chem. Soc. (A)*, 1970, 1257.
258. S. L. Chadha, P. G. Harrison and K. C. Molloy, *J. Organomet. Chem.*, 1980, **202,** 247.
259. P. G. Harrison, T. J. King and J. A. Richards, *J. Chem. Soc., Dalton Trans.*, 1974, 1723.
260. L. Coghi, C. Pelizzi and G. Pelizzi, *Gazz. Chim. Ital.*, 1974, **104,** 873.
261. P. Ganis, V. Peruzzo and G. Valle, *J. Organomet. Chem.*, 1983, **256,** 245.
262. L. Prasad, Y. Le Page and F. E. Smith, *Inorg. Chim. Acta,* 1983, **68,** 45.
263. B. S. Ault, *Inorg. Chem.*, 1979, **18,** 3339; 1981, **20,** 2817.
264. B. S. Ault and U. Tandoc, *Inorg. Chem.*, 1981, **20,** 1937.
265. T. J. Lorenz and B. S. Ault, *Inorg. Chem.*, 1982, **21,** 1758.
266. A. M. McNair and B. S. Ault, *Inorg. Chem.*, 1982, **21,** 1762, 2603.
267. A. M. Walther and B. S. Ault, *Inorg. Chem.*, 1984, **23,** 3892.

268. D. J. Brauer, H. Bürger and R. Eujen, *Angew. Chem., Int. Ed. Engl.*, 1980, **19,** 836.
269. R. Cotton, D. Dakternicks and C. A. Harvey, *Inorg. Chim. Acta*, 1982, **61,** 1.
270. R. Cotton and D. Dakternicks, *Inorg. Chim. Acta*, 1983, **71,** 101.
271. H. C. Clark and B. I. Swanson, *J. Am. Chem. Soc.*, 1979, 1604.
272. M. Zeldin, P. Mehta and W. D. Vernon, *Inorg. Chem.*, 1979, **18,** 463.
273. V. Lucchini and P. R. Wells, *J. Organomet. Chem.*, 1980, **199,** 217.
274. A. F. Wells, 'Structural Inorganic Chemistry', 4th edn, Oxford University Press, Oxford, 1975.
275. N. N. Greenwood and A. Earnshaw, 'Chemistry of the Elements', Pergamon, Oxford, 1984, pp. 393–418.
276. J. Peters and B. Krebs, *Acta Crystallogr., Sect. B*, 1982, **38,** 1270.
277. A. J. Leadbetter, T. W. Smith and A. F. Wright, *Nature (London), Phys. Sci.*, 1973, **244,** 125.
278. A. W. Laubengayer and D. S. Morton, *J. Am. Chem. Soc.*, 1932, **54,** 2303.
279. W. H. Baur and A. A. Khan, *Acta Crystallogr., Sect. B*, 1971, **27,** 2133.
280. K. J. Seifert, H. Nowotny and E. Hauser, *Monatsch. Chem.*, 1971, **102,** 1006.
281. W. H. Baur, *Acta Crystallogr.*, 1956, **9,** 575.
282. A. Byström, *Ark. Kemi, Mineral. Geol.*, 1945, **20A,** No. 11.
283. A. I. Zaslavsky, Yu. D. Kandrashev and S. Stolkachev, *Dokl. Akad. Nauk SSSR*, 1950, **75,** 559.
284. F. Lawson, *Nature (London)*, 1967, **215,** 955.
285. S. J. Gross, *J. Am. Chem. Soc.*, 1941, **65,** 1107.
286. M. Straumanis, *Z. Phys. Chem.*, 1942, **B52,** 127.
287. A. Byström and A. Westgren, *Ark. Kemi, Mineral. Geol.*, 1943, **16B,** No. 14; 1947, **25A,** No. 13.
288. G. Dittmar and H. Schafer, *Acta Crystallogr., Sect. B*, 1975, **31,** 2060; 1976, **32,** 1188, 2226.
289. D. Mootz and H. Puhl, *Acta Crystallogr.*, 1967, **23,** 471.
290. R. Kniep, D. Mootz, U. Severin and H. Wunderlich, *Acta Crystallogr., Sect. B*, 1982, **38,** 2022.
291. W. J. Moore and L. Pauling, *J. Am. Chem. Soc.*, 1941, **63,** 1392.
292. A. Byström, *Ark. Kemi*, 1943, **17B,** No. 8.
293. S. Grimvall, *Acta Chem. Scand.*, 1973, **27,** 1447.
294. C. G. Davies, J. D. Donaldson, D. R. Laughlin, R. A. Howie and R. Beddoes, *J. Chem. Soc., Dalton Trans.*, 1975, 2241.
295. S. H. Hong and A. Olin, *Acta Chem. Scand.*, 1973, **27,** 2304; 1974, **28,** 233.
296. C. Gaffney, P. G. Harrison, and T. J. King, *J. Chem. Soc., Chem. Commun.*, 1980, 1251.
297. T. G. Spiro, D. H. Templeton and A. Zalkin, *Inorg. Chem.*, 1969, **8,** 856.
298. A. Olin and R. Sönderquist, *Acta Chem. Scand.*, 1972, **26,** 3505.
299. R. A. Howie and W. Moser, *Nature (London)*, 1968, **219,** 372.
300. P. G. Harrison, B. J. Haylett and T. J. King, *J. Chem. Soc., Chem. Commun.*, 1978, 112.
301. W. H. Zachariasen, *Phys. Rev.*, 1932, **40,** 917.
302. A. Okazaki, *J. Phys. Soc. Jpn.*, 1958, **13,** 1151.
303. C. R. Kannewurf, A. Kelly and R. J. Cashman, *Acta Crystallogr.*, 1960, **13,** 449.
304. W. Hoffman, *Z. Kristallogr.*, 1935, **92,** 161.
305. R. E. Rundle and D. H. Olsen, *Inorg. Chem.*, 1964, **3,** 596.
306. A. Okazaki and I. Ueda, *J. Phys. Soc. Jpn.*, 1956, **11,** 470.
307. E. Zeipel, *Ark. Mat., Astron. Fys.*, 1935, **25A,** 1.
308. J. W. Earley, *Am. Mineral.*, 1950, **35,** 338.
309. W. A. Deer, R. A. Howie and J. Zussman, 'An Introduction to Rock-forming Minerals', Longmans, London, 1966.
310. F. Liebau, 'Silicon, element 14', in 'Handbook of Geochemistry', ed. K. H. Wedepohl, vol. II-2.
311. B. Mason and L. G. Berry, 'Elements of Mineralogy', ed. W. H. Freeman, San Francisco, 1968.
312. M. G. Barker and P. G. Good, *J. Chem. Res. (S)*, 1981, 274.
313. A. Yu. Shashov, V. A. Efremov, I. Matsichek, N. V. Rannev, Yu. N. Venevtsev and V. K. Trunov, *Russ. J. Inorg. Chem. (Engl. Trans.)*, 1981, **26,** 316.
314. O. Baumgartner and H. Völlenkle, *Monatsch. Chem.*, 1978, **109,** 1145.
315. M. Toubovl and Y. Feutelais, *Acta Crystallogr., Sect. B*, 1979, **35,** 810.
316. H. G. von Schnering, R. Nesper and H. Pelshenke, *Z. Anorg. Allg. Chem.*, 1983, **499,** 117.
317. R. Nesper and H. G. von Schnering, *Z. Anorg. Allg. Chem.*, 1983, **499,** 109.
318. C. O. Björling, *Ark. Kemi, Mineral Geol.*, 1941, **15B,** 1.
319. H. Strune and B. Contag, *Acta Crystallogr.*, 1960, **13,** 601.
320. I. Morgenstern-Badarau, P. Poix and A. Michel, *C. R. Hebd. Seances Acad. Sci.*, 1964, **258C,** 3036.
321. I. Morgenstern-Badarau, Y. Billiet, P. Poix and A. Michel, *C. R. Hebd. Seances Acad. Sci.*, 1965, **260,** 3668.
322. C. Cohen-Addad, *Bull. Soc. Fr. Mineral. Cristallogr.*, 1967, **90,** 32.
323. A. N. Christensen and R. G. Hazell, *Acta Chem. Scand.*, 1969, **23,** 1219.
324. I. Morgenstern-Badarau, *J. Solid State Chem.*, 1976, **17,** 399.
325. I. Morgenstern-Badarau, C. Levy-Clement and A. Michel, *C. R. Hebd. Seances Acad. Sci.*, 1969, **268,** 696.
326. E. Dubler, R. Hess and H. R. Oswald, *Z. Anorg. Allg. Chem.*, 1976, **421,** 61.
327. R. Marchand, Y. Piffard and M. Tournoux, *Acta Crystallogr., Sect. B*, 1975, **31,** 511.
328. R. Olacuaga, J. M. Reau, M. Devalette, G. Le Flem and P. Hagenmuller, *J. Solid State Chem.*, 1975, **13,** 275.
329. B. Nowitzki and R. Hoppe, *Z. Anorg. Allg. Chem.*, 1983, **505,** 105.
330. P. Poix, *Ann. Chim.*, 1965, **10,** 49.
331. M. Trömel, *Z. Anorg. Allg. Chem.*, 1969, **371,** 237.
332. R. Weiss and R. Faivre, *C. R. Hebd. Seances Acad. Sci.*, 1959, **248,** 106.
333. G. Wagner and H. Binder, *Z. Anorg. Allg. Chem.*, 1959, **298,** 12.
334. M. Nogues and P. Poix, *Ann. Chim.*, 1968, **3,** 335.
335. L. Siegel, *J. Appl. Crystallogr.*, 1978, **11,** 284.
336. F. Brisse and O. Knop, *Can. J. Chem.*, 1968, **46,** 859.
337. C. G. Whinfrey and A. Tauber, *J. Am. Chem. Soc.*, 1961, **83,** 755.

338. G. Vetter, F. Queyroux and J. C. Gilles, *Mater. Res. Bull.*, 1978, **13,** 211.
339. G. Kreuzburg, F. Stewner and R. Hoppe, *Z. Anorg. Allg. Chem.*, 1970, **379,** 242.
340. A. Verbaere, M. Dion and M. Tournoux, *J. Solid State Chem.*, 1974, **11,** 184.
341. C. Levy-Clement, I. Morgenstern-Badarau, Y. Billier and A. Michel, *C. R. Hebd. Seances Acad. Sci., Ser. C,* 1967, **265,** 585.
342. B. Durand and H. Loiseleur, *J. Appl. Crystallogr.*, 1978, **11,** 289.
343. R. S. Roth, *J. Res. Nat. Bur. Stand.*, 1957, **58,** 75.
344. G. Wagner and H. Binder, *Z. Anorg. Allg. Chem.*, 1959, **298,** 12.
345. R. D. Shannon, J. L. Gillson and R. J. Bouchard, *J. Phys. Chem. Solids,* 1977, **38,** 877.
346. B. M. Gatehouse and D. J. Lloyd, *J. Solid State Chem.*, 1970, **2,** 410.
347. R. Hoppe and K. Seeger, *Z. Anorg. Allg. Chem.*, 1970, **375,** 264.
348. B. Brazel and R. Hoppe, *Z. Anorg. Allg. Chem.*, 1983, **499,** 161; **505,** 99.
349. B. Nowitzki and R. Hoppe, *Z. Anorg. Allg. Chem.*, 1983, **505,** 111.
350. K. P. Martens and R. Hoppe, *Z. Anorg. Allg. Chem.*, 1980, **471,** 64.
351. B. Brazel and R. Hoppe, *Z. Naturforsch., Teil B,* 1983, **38,** 661.
352. B. Brazel and R. Hoppe, *Z. Anorg. Allg. Chem.*, 1983, **497,** 176.
353. B. Brazel and R. Hoppe, *Z. Anorg. Allg. Chem.*, 1982, **493,** 93; 1983, **498,** 167.
354. P. Panek and R. Hoppe, *Z. Anorg. Allg. Chem.*, 1973, **400,** 219.
355. A. Byström and L. Evers, *Acta Chem. Scand.*, 1950, **4,** 613.
356. H. Vincent, E. F. Bertaut, W. H. Baur and R. D. Shannon, *Acta Crystallogr., Sect. B,* 1976, **32,** 1749.
357. J. Mandt and B. Krebs, *Z. Anorg. Allg. Chem.*, 1976, **420,** 31.
358. B. Krebs and H. J. Jacobson, *Z. Anorg. Allg. Chem.*, 1976, **421,** 97.
359. N. Ryanck, P. Larvelle and A. Katty, *Acta Crystallogr., Sect. B,* 1976, **32,** 692.
360. W. Schiwy, S. Pohl and B. Krebs, *Z. Anorg. Allg. Chem.*, 1973, **402,** 77.
361. R. A. Beskrovnaya, L. D. Dyatlova and V. V. Serebrennikov, *Tr. Tomsk. Univ.*, 1971, 403 (*Chem. Abstr.*, 1973, **78,** 131 485).
362. B. Krebs and H. U. Hürter, *Z. Anorg. Allg. Chem.*, 1980, **462,** 143.
363. B. Eisenmann, H. Schäfer and H. Schrod, *Z. Naturforsch., Teil B,* 1983, **38,** 921.
364. B. Krebs and W. Schiwy, *Z. Anorg. Allg. Chem.*, 1973, **398,** 63.
365. J. C. Jumas, J. Olivier-Fourcade, F. Vermont-Gaud-Daniel, M. Ribes, E. Philippot and M. Maurin, *Rev. Chim. Miner.*, 1974, **11,** 13.
366. B. Krebs, S. Pohl and W. Schiwy, *Z. Anorg. Allg. Chem.*, 1972, **393,** 241.
367. B. Krebs and H. Müller, *Z. Anorg. Allg. Chem.*, 1983, **496,** 47.
368. G. Dittmar, *Acta Crystallogr., Sect. B,* 1978, **34,** 2390.
369. G. Dittmar, *Z. Anorg. Allg. Chem.*, 1979, **453,** 68.
370. B. Eisenmann, H. Schwerer and H. Schäfer, *Z. Naturforsch., Teil B,* 1981, **36,** 2538.
371. J. Oliver-Fourcade, E. Philippot, M. Ribes and M. Maurin, *C. R. Hebd. Seances Acad. Sci.*, 1972, **274C,** 1185.
372. W. Schiwy, C. Blutau, D. Gathje and B. Krebs, *Z. Anorg. Allg. Chem.*, 1975, **412,** 1.
373. G. Bugli, D. Carre and S. Barnier, *Acta Crystallogr., Sect. B,* 1978, **34,** 3186.
374. A. Mazurier and J. Etienne, *Acta Crystallogr., Sect. B,* 1973, **29,** 817.
375. B. Krebs and S. Pohl, *Z. Naturforsch., Teil B,* 1971, **26,** 853.
376. S. Pohl and B. Krebs, *Z. Anorg. Allg. Chem.*, 1976, **424,** 265.
377. B. Eisenmann and H. Schäfer, *Z. Anorg. Allg. Chem.*, 1982, **491,** 67.
378. G. Eulenberger, *Z. Naturforsch., Teil B,* 1981, **36,** 521.
379. B. Eisenmann, H. Schäfer and H. Schwerer, *Z. Naturforsch., Teil B,* 1983, **38,** 924.
380. G. Eulenberger, *Z. Naturforsch., Teil B,* 1981, **36,** 687.
381. S. Jaulmes, *Acta Crystallogr., Sect. B,* 1974, **30,** 2283.
382. M. Julien-Pouzol and S. Jaulmes, *Acta Crystallogr., Sect. B,* 1979, **35,** 2672.
383. J. C. Jumas, E. Philippot and M. Maurin, *Acta Crystallogr., Sect. B,* 1977, **33,** 3850.
384. J. C. Jumas, M. Ribes, M. Maurin and E. Philippot, *Ann. Chim. (Paris),* 1978, **3,** 125.
385. C. Gaffney and P. G. Harrison, *J. Chem. Soc., Dalton Trans.*, 1982, 1057.
386. R. Eujen, H. Bürger and H. Oberhammer, *J. Mol. Struct.*, 1981, **71,** 109.
387. M. Birkhahn, E. Glozbach, W. Massa and J. Lorberth, *J. Organomet. Chem.*, 1980, **192,** 171.
388. G. Bandoli, D. A. Clemente and C. Pannatoni, *J. Chem. Soc., Chem. Commun.*, 1971, 311.
389. R. Faggiani, J. P. Johnson, I. D. Brown and T. Birchall, *Acta Crystallogr., Sect. B,* 1978, 3742.
390. C. Glidewell and D. C. Liles, *Acta Crystallogr., Sect. B,* 1978, **34,** 125.
391. D. W. H. Rankin and H. E. Robertson, *J. Chem. Soc., Dalton Trans.*, 1983, 265; B. Csakvan, Z. Wagner, P. Gömöny, F. C. Mijlhoff, B. Rozsondai and I. Hargittai, *J. Organomet. Chem.*, 1976, **107,** 287.
392. L. G. Kuz'mina, *Zh. Strukt. Khim.*, 1972, **13,** 946.
393. C. Glidewell and D. C. Liles, *Acta Crystallogr., Sect. B,* 1978, **34,** 1693.
394. P. G. Harrison, T. J. King, J. A. Richards and R. C. Phillips, *J. Organomet. Chem.*, 1976, **116,** 307.
395. C. Glidewell and D. C. Liles, *J. Chem. Soc., Chem. Commun.*, 1979, 93; *J. Organomet. Chem.*, 1979, **174,** 275; *Acta Crystallogr., Sect. B,* 1979, **35,** 1689.
396. O. Graalmann, U. Klingebiel, W. Clegg, M. Haase and G. M. Sheldrick, *Chem. Ber.*, 1984, **117,** 2988.
397. H. Puff, S. Franken, W. Schuh and W. Schwab, *J. Organomet. Chem.*, 1983, **244,** C41; 1983, **254,** 33.
398. M. J. Fink, K. J. Haller, R. West and J. Michl, *J. Am. Chem. Soc.*, 1984, **106,** 822.
399. L. Parkanyi, M. Dunaj-Jurco, L. Bihatsi and P. Henesei, *Cryst. Struct. Commun.*, 1980, **9,** 1049.
400. W. Wojnowski, K. Peters, D. Weber and H. G. von Schnering, *Z. Anorg. Allg. Chem.*, 1984, **519,** 134.
401. M. Yokoi, T. Nomura and K. Yamasaki, *J. Am. Chem. Soc.*, 1955, **77,** 4484.
402. L. V. Vilkov, M. M. Kusakov, N. S. Nametkin and V. D. Oppengeim, *Akad. Nauk SSSR Proc.*, 1968, **181,** 1038.
403. L. Parkanyi, K. Sasvan and I. Barta, *Acta Crystallogr., Sect. B,* 1978, **34,** 883.
404. M. A. Hossain, M. B. Hursthouse and K. M. A. Malik, *Acta Crystallogr., Sect. B,* 1979, **35,** 2258.

405. V. E. Shkolev, Yu. T. Struchkov, N. N. Makarova and K. A. Andrianov, *J. Struct. Chem.* (*Engl. Transl.*), 1978, **19,** 944.
406. W. Clegg, G. M. Sheldrick and N. Vater, *Acta Crystallogr., Sect. B,* 1980, **36,** 3162.
407. I. A. Baidina, N. V. Podberezskaya, V. I. Alekseev, T. N. Marktynova, S. V. Borisov and A. N. Kanov, *J. Struct. Chem.* (*Engl. Transl.*), 1979, **20,** 550.
408. I. A. Baidina, N. V. Padberezskaya, S. V. Borisov, V. I. Alekseev, T. N. Martynova and A. N. Kanov, *J. Struct. Chem.* (*Engl. Trans.*), 1980, **21,** 352.
409. H. Puff, S. Franken and W. Schuk, *J. Organomet. Chem.*, 1983, **256,** 23.
410. H. Lindemann and F. Huber, *Z. Anorg. Allg. Chem.*, 1972, **394,** 101.
411. H. Puff, W. Schuk, R. Sievers and R. Zimmer, *Angw. Chem., Int. Ed. Engl.*, 1981, **20,** 591.
412. S. Masamune, L. R. Sita and D. J. Williams, *J. Am. Chem. Soc.*, 1983, **105,** 630.
413. P. G. Harrison, M. J. Begley and K. C. Molloy, *J. Organomet. Chem.*, 1980, **186,** 213.
414. H. Puff, I. Bung, E. Friedrichs, and A. Jansen, *J. Organomet. Chem.*, 1983, **254,** 23.
415. C. D. Garner, B. Hughes and T. J. King, *Inorg. Nucl. Chem. Lett.*, 1976, **12,** 859.
416. R. Graziani, G. Bambieri, E. Forsellini, P. Furlan, V. Peruzzo and G. Tagliavini, *J. Organomet. Chem.*, 1977, **125,** 43.
417. R. Faggiani, J. P. Johnson, I. D. Brown and T. Birchall, *Acta Crystallogr., Sect. B,* 1978, **34,** 3743.
418. J. F. Vollano, R. O. Day and R. R. Holmes, *Organometallics,* 1984, **3,** 745.
419. Y. M. Chow, *Inorg. Chem.*, 1971, **10,** 673.
420. H. Matsada, A. Kashina, S. Matsuda, N. Kasai and K. Jitsumori, *J. Organomet. Chem.*, 1972, **34,** 341.
421. N. Kasai, K. Yasuda and R. Okawara, *J. Organomet. Chem.*, 1965, **3,** 172.
422. C. Glidewell and D. C. Liles, *Acta Crystallogr., Sect. B,* 1978, **34,** 129.
423. A. M. Domingos and G. M. Sheldrick, *Acta Crystallogr., Sect. B,* 1974, **30,** 579.
424. A. G. Davies and R. J. Puddephatt, *J. Chem. Soc.* (*C*), 1967, 2663.
425. E. Amberger and R. Hönigschmidt-Grossich, *Chem. Ber.*, 1965, **98,** 3795; 1969, **102,** 3589.
426. P. F. R. Ewings, P. G. Harrison, T. J. King, R. C. Phillips and J. A. Richards, *J. Chem. Soc., Dalton Trans.*, 1975, 1950.
427. P. G. Harrison, *Inorg. Chem.*, 1970, **9,** 175.
428. P. G. Harrison, R. C. Phillips and E. W. Thornton, *J. Chem. Soc., Chem. Commun.*, 1977, 603.
429. P. G. Harrison, T. J. King, J. A. Richards and R. C. Phillips, *J. Organomet. Chem.*, 1976, **116,** 307.
430. K. C. Molloy, F. A. K. Nasser, C. L. Barnes, D. van der Helm and J. J. Zuckerman, *Inorg. Chem.*, 1982, **21,** 960.
431. P. G. Harrison, K. Lambert, T. J. King and B. Majee, *J. Chem. Soc., Dalton Trans.*, 1983, 363; *et loc. cit.*
432. K. C. Molloy, T. G. Purcell, K. Quill and I. W. Nowell, *J. Organomet. Chem.*, 1984, **267,** 237.
433. G. M. Bancroft, B. W. Davies, N. C. Payne and T. K. Sham, *J. Chem. Soc., Dalton Trans.*, 1975, 973.
434. T. J. King and P. G. Harrison, *J. Chem. Soc., Dalton Trans.*, 1974, 2298.
435. P. G. Harrison, T. J. King and K. C. Molloy, *J. Organomet. Chem.*, 1980, **185,** 199.
436. B. K. Nicholson, *J. Organomet. Chem.*, 1984, **265,** 153.
437. C. Gaffney and P. G. Harrison, *J. Chem. Soc., Dalton Trans.*, 1982, 1061.
438. N. W. Alcock, *J. Chem. Soc., Dalton Trans.*, 1972, 1189.
439. F. H. Allen, J. A. Lerbscher and J. Trotter, *J. Chem. Soc.* (*A*), 1971, 2507.
440. J. P. Ashmore, T. Chivers, K. A. Kerr and J. H. G. van Roode, *Inorg. Chem.*, 1977, **16,** 191.
441. E. O. Schlemper, *Inorg. Chem.*, 1967, **6,** 2012.
442. P. G. Harrison, T. J. King and R. C. Phillips, *J. Chem. Soc., Dalton Trans.*, 1976, 2317.
443. G. A. Miller and E. O. Schlemper, *Inorg. Chem.*, 1973, **12,** 677.
444. J. J. Hilton, E. K. Nunn and S. C. Wallwork, *J. Chem. Soc., Dalton Trans.*, 1973, 173.
445. M. Calligaris, G. Nardin and L. Randaccio, *J. Chem. Soc., Dalton Trans.*, 1972, 2003.
446. P. G. Harrison, T. J. King and J. A. Richards, *J. Chem. Soc., Dalton Trans.*, 1975, 826.
447. H. H. Anderson, *Inorg. Chem.*, 1964, **3,** 912.
448. G. S. Brownlee, A. Walker, S. C. Nyburg and J. J. Szymanski, *J. Chem. Soc., Chem. Commun.*, 1971, 1073.
449. C. D. Garner, P. Sutton and S. C. Wallwork, *J. Chem. Soc.* (*A*), 1967, 1949.
450. N. W. Alcock and V. L. Tracy, *Acta Crystallogr., Sect. B,* 1979, 80.
451. B. Kamenar, *Acta Crystallogr., Sect. A,* 1963, **16,** 34.
452. B. Kamenar and M. Bruro, *Z. Kristallogr.*, 1975, **141,** 97.
453. N. W. Alcock, V. M. Tracy and T. C. Waddington, *J. Chem. Soc., Dalton Trans.*, 1976, 2243.
454. M. J. Barrow, S. Cradock, E. A. V. Ebsworth and D. W. H. Rankin, *J. Chem. Soc., Dalton Trans.*, 1981, 1988.
455. M. J. Barrow, E. A. V. Ebsworth, C. M. Huntley and D. W. H. Rankin, *J. Chem. Soc., Dalton Trans.*, 1982, 1131.
456. G. Sawitzki, H. G. von Schering, D. Kummer and T. Seshadri, *Chem. Ber.*, 1978, **111,** 3705.
457. G. Schott and D. Lange, *Z. Anorg. Allg. Chem.*, 1972, **391,** 27.
458. A. Nagasawa and K. Saito, *Bull. Chem. Soc. Jpn.*, 1974, **47,** 131.
459. T. Inone and K. Saito, *Bull. Chem. Soc. Jpn.*, 1973, **46,** 2417.
460. T. B. Shkodina, E. I. Krylov and V. A. Sharov, *Russ. J. Inorg. Chem.* (*Engl. Transl.*), 1974, **19,** 1595.
461. A. Nagasawa and K. Saito, *Bull. Chem. Soc. Jpn.*, 1978, **51,** 2015.
462. N. Jorgensen and T. J. R. Weakley, *J. Chem. Soc., Dalton Trans.*, 1980, 2051.
463. T. Ho, K. Toriumi, F. B. Ueno and K. Saito, *Acta Crystallogr., Sect. B,* 1980, **36,** 2998.
464. W. Farnham and J. F. Whitney, *J. Am. Chem. Soc.*, 1984, **106,** 3992.
465. G. Klebe and D. T. Qui, *Acta Crystallogr., Sect. B,* 1984, **40,** 476.
466. L. Parkanyi, K. Simon and J. Nagy, *Acta Crystallogr., Sect. B,* 1974, **30,** 2328.
467. L. Parkanyi, J. Nagy and K. Simon, *J. Organomet. Chem.*, 1975, **101,** 11.
468. A. A. Kemme, Y. Y. Bleidelis, V. M. D'yakov and M. G. Voronkov, *J. Struct. Chem.* (*Engl. Transl.*), 1975, **16,** 847.
469. O. Shen and R. L. Hilderbrandt, *J. Mol. Struct.*, 1980, **64,** 257.

470. S. N. Gurkova, A. I. Gusev, I. R. Segel'man, N. V. Alekseev, T. K. Gar and N. Yu. Khromova, *J. Struct. Chem. (Engl. Transl.)*, 1981, **22,** 461.
471. S. N. Gurkova, A. I. Gusev, N. V. Alekseev, I. R. Segel'man, T. K. Gar and N. Yu Khromova, *J. Struct. Chem. (Engl. Transl.)*, 1981, **22,** 984.
472. A. Tzschach, K. Jurkschat and C. Mügge, *Z. Anorg. Allg. Chem.*, 1982, **492,** 135.
473. S. N. Gurkova, A. I. Gusev, V. A. Sharapov, N. V. Alekseev, T. K. Gar and N. Yu Khromova, *J. Organomet. Chem.*, 1984, **268,** 119.
474. J. J. Daly, *J. Chem. Soc., Dalton Trans.*, 1974, 2051.
475. L. Parkanyi, P. Hencsei and E. Popowski, *J. Organomet. Chem.*, 1980, **197,** 275.
476. M. Zeldin and J. Ochs, *J. Organomet. Chem.*, 1975, **86,** 369.
477. A. Tzschach and K. Pörricke, *Z. Anorg. Allg. Chem.*, 1975, **413,** 136.
478. K. Jurkschat, C. Mügge, A. Tszchach, A. Zschunke and G. W. Fischer, *Z. Anorg. Allg. Chem.*, 1980, **463,** 123.
479. A. Tzschach, K. Jurkschat and C. Mügge, *Z. Anorg. Allg. Chem.*, 1982, **492,** 135.
480. R. G. Swisher, R. O. Day and R. R. Holmes, *Inorg. Chem.*, 1983, **22,** 3692.
481. R. R. Holmes, R. O. Day, J. J. Harland, A. C. Sau and J. M. Holmes, *Organometallics,* 1984, **3,** 341.
482. R. R. Holmes, R. O. Day, J. J. Harland and J. M. Holmes, *Organometallics,* 1984, **3,** 347.
483. R. O. Day, J. M. Holmes, A. C. Sau and R. R. Holmes, *Inorg. Chem.*, 1982, **21,** 281.
484. A. C. Sau, R. O. Day and R. R. Holmes, *Inorg. Chem.*, 1981, **20,** 3076.
485. G. D. Andreeth, G. Bocelli, G. Calestani and P. Sgarabotto, *J. Organomet. Chem.*, 1984, **273,** 31.
486. N. G. Bokii, Yu. T. Struchkov, D. N. Kravtsov and E. M. Rokhlina, *J. Struct. Chem. (Engl. Transl.)*, 1973, **14,** 258.
487. M. E. Cradwick, R. D. Taylor and J. L. Wardell, *J. Organomet. Chem.*, 1974, **66,** 43.
488. P. L. Clarke, M. E. Cradwick and J. L. Wardell, *J. Organomet. Chem.*, 1973, **63,** 279.
489. N. G. Furmanova, A. S. Batsanov, Yu. T. Struchkov, D. N. Kravtsov and E. M. Rokhlina, *J. Struct. Chem. (Engl. Transl.)*, 1979, **20,** 245.
490. M. J. Begley, C. Gaffney, P. G. Harrison and A. T. Steele, *J. Organomet. Chem.*, 1985, **289,** 281.
491. N. G. Bokii, Yu. T. Struchkov, D. N. Kravtsov and E. M. Rokhlina, *J. Struct. Chem. (Engl. Transl.)*, 1973, **14,** 458.
492. B. Krebs and H. J. Jacobsen, *J. Organomet. Chem.*, 1979, **179,** 13.
493. O. A. D'yachenko, A. B. Zolotoi, L. O. Atovmyan, R. G. Mirskov and M. G. Voronkov, *Dokl. Phys. Chem. (Engl. Transl.)*, 1977, **237,** 1142.
494. B. Krebs and H. J. Jacobsen, *J. Organomet. Chem.*, 1979, **178,** 301.
495. B. Menzbach and P. Blackmann, *J. Organomet. Chem.*, 1975, **91,** 291.
496. H. J. Jacobsen and B. Krebs, *J. Organomet. Chem.*, 1977, **136,** 333.
497. M. Dräger, A. Blecher, H. J. Jacobsen and B. Krebs, *J. Organomet. Chem.*, 1978, **161,** 319.
498. H. Puff, R. Gattermeyer, R. Hundt and R. Zimmer, *Angew. Chem., Int. Ed. Engl.*, 1977, **16,** 547.
499. H. Puff, A. Bongartz, R. Sievers and R. Zimmer, *Angew. Chem., Int. Ed. Engl.*, 1978, **17,** 939.
500. R. H. Benno and C. J. Fritchie, *J. Chem. Soc., Dalton Trans.*, 1973, 543.
501. S. Pohl, *Angew. Chem., Int. Ed. Engl.*, 1976, **15,** 162.
502. D. Kobett, E. F. Paulus and H. Scherer, *Acta Crystallogr., Sect. B.*, 1972, **28,** 2323.
503. G. M. Sheldrick and W. S. Sheldrick, *J. Chem. Soc. (A)*, 1970, 490.
504. G. M. Sheldrick, W. S. Sheldrick, R. F. Dalton and K. Jones, *J. Chem. Soc. (A)*, 1970, 493.
505. K. C. Molloy, M. B. Hossain, D. van der Helm, J. J. Zuckerman and I. Haiduc, *Inorg. Chem.*, 1979, **18,** 3507.
506. T. Kimura, N. Yasuoka, N. Kasai and M. Kakudo, *Bull. Chem. Soc. Jpn.*, 1972, **45,** 1649.
507. J. S. Morris and E. O. Schlemper, *J. Cryst. Mol. Struct.*, 1980, **9,** 13.
508. B. W. Liebich and M. Tomassini, *Acta Crystallogr., Sect. B,* 1978, **34,** 944.
509. K. C. Molloy, M. B. Hossain, D. van der Helm, J. J. Zuckerman and I. Haiduc, *Inorg. Chem.*, 1980, **19,** 2041.
510. J. S. Morris and E. O. Schlemper, *J. Cryst. Mol. Struct.*, 1978, **8,** 295.
511. M. J. Barrow and E. A. V. Ebsworth, *J. Chem. Soc., Dalton Trans.*, 1964, 563.
512. E. A. V. Ebsworth, E. K. Murray, D. W. H. Rankin and H. E. Robertson, *J. Chem. Soc., Dalton Trans.*, 1981, 1501.
513. G. S. Laurenson and D. W. H. Rankin, *J. Chem. Soc., Dalton Trans.*, 1981, **425,** 1047.
514. L. S. Khaikin, A. V. Belyakov, G. S. Koptev, A. V. Golubinskii, L. V. Vilkov, N. V. Girbasova, E. T. Bogoradovskii and V. S. Zargorodnii, *J. Mol. Struct.*, 1980, **66,** 191.
515. J. J. Daly and F. Sanz, *Acta Crystallogr., Sect. B,* 1974, **30,** 2766.
516. G. Gundersen, R. A. Mayo and D. W. H. Rankin, *Acta Chem. Scand., Ser. A,* 1984, **38,** 579.
517. C. Glidewell, P. M. Pinder, A. G. Robiette and G. M. Sheldrick, *J. Chem. Soc., Dalton Trans.*, 1972, 1402.
518. R. Varma, K. R. Ramaprasad and J. F. Nelson, *J. Chem. Phys.*, 1975, **63,** 915.
519. J. R. Durig, M. Jalilian, Y. S. Li and R. O. Cater, *J. Mol. Struct.*, 1979, **55,** 177.
520. L. Noodleman and N. L. Paddock, *Inorg. Chem.*, 1979, **18,** 354.
521. P. Livant, M. L. McKee and S. D. Worley, *Inorg. Chem.*, 1983, **22,** 895.
522. P. G. Harrison and J. J. Zuckerman, *J. Organomet. Chem.*, 1973, **55,** 261.
523. M. E. Bishop and J. J. Zuckerman, *Inorg. Chim. Acta,* 1976, **19,** L1.
524. K. F. Tebbe, *Z. Anorg. Allg. Chem.*, 1980, **468,** 202.
525. K. F. Tebbe and B. Freckmann, *Acta Crystallogr., Sect. C,* 1984, **40,** 254.
526. M. Baudler and H. Suchomel, *Z. Anorg. Allg. Chem.*, 1983, **505,** 39.
527. K. F. Tebbe and T. Heinlein, *Z. Anorg. Allg. Chem.*, 1984, **575,** 7.
528. W. Clegg, M. Haase, U. Klingebeil and G. M. Sheldrick, *Chem. Ber.*, 1983, **116,** 146.
529. U. Klingebeil, N. Vater, W. Clegg, M. Haase and G. M. Sheldrick, *Z. Naturforsch., Teil B,* 1983, **38,** 1557.
530. M. Baudler, T. Pontzen, U. Schings, K. F. Tebbe and M. Feher, *Angew. Chem., Int. Ed. Engl.*, 1983, **22,** 775.
531. M. Baudler and H. Suchomel, *Z. Anorg. Allg. Chem.*, 1983, **503,** 7.
532. M. Baudler and H. Suchomel, *Z. Anorg. Allg. Chem.*, 1983, **506,** 22.
533. K. F. Tebbe and R. Fröhlich, *Z. Anorg. Allg. Chem.*, 1983, **506,** 27.

534. M. Baudler, L. de Riese-Meyr and U. Schings, *Z. Anorg. Allg. Chem.*, 1984, **519**, 24.
535. M. Baudler and H. Suchomel, *Z. Anorg. Allg. Chem.*, 1983, **505**, 39.
536. H. Schumann and H. Benda, *Angew. Chem., Int. Ed. Engl.*, 1968, **7**, 812.
537. W. Hönle and H. G. von Schnering, *Z. Anorg. Allg. Chem.*, 1978, **442**, 107.
538. W. Hönle and H. G. von Schnering, *Z. Anorg. Allg. Chem.*, 1978, **442**, 91.
539. A. R. Dahl, A. D. Norman, H. Shenav and R. Schaeffer, *J. Am. Chem. Soc.*, 1975, **97**, 6364.
540. B. Mathiasch and M. Dräger, *Angew. Chem., Int. Ed. Engl.*, 1978, **17**, 767.
541. B. Mathiasch, *J. Organomet. Chem.*, 1979, **165**, 295.
542. M. Dräger and B. Mathiasch, *Angew. Chem., Int. Ed. Engl.*, 1981, **20**, 1029.
543. H. Schumann and H. Banda, *Angew. Chem., Int. Ed. Engl.*, 1969, **8**, 989.
544. H. Schumann and H. Banda, *Angew. Chem., Int. Ed. Engl.*, 1968, **7**, 813,.
545. G. Becker, G. Gutenkunst and H. J. Wessely, *Z. Anorg. Allg. Chem.*, 1980, **462**, 113.
546. L. Ross and M. Dräger, *Z. Naturforsch., Teil B*, 1983, **38**, 665.
547. M. Dräger, B. Mathiasch, L. Ross and M. Ross, *Z. Anorg. Allg. Chem.*, 1983, **506**, 99.
548. D. H. Olsen and R. E. Rundle, *Inorg. Chem.*, 1963, **2**, 1310.
549. W. P. Neumann and J. Fu, *J. Organomet. Chem.*, 1984, **273**, 295.
550. L. C. Willemsens and G. J. M. van der Kerk, 'Investigations in the Field of Organolead Chemistry', International Lead Zinc Research Organization Inc., 1965.
551. D. E. Goldberg, D. H. Harris, M. F. Lappert and K. M. Thomas, *J. Chem. Soc., Chem. Commun.*, 1976, 261.
552. P. J. Davidson, D. H. Harris and M. F. Lappert, *J. Chem. Soc., Dalton Trans.*, 1976, 2268.
553. T. Fjeldberg, A. Haaland, B. E. R. Schilling, H. V. Volden, M. F. Lappert and A. J. Thorne, *J. Organomet. Chem.*, 1984, **276**, C1.
554. T. Fjeldberg, A. Haaland, M. F. Lappert, B. E. R. Schilling, R. Seip and A. J. Thorne, *J. Chem. Soc., Chem. Commun.*, 1982, 1407.
555. P. B. Hitchcock, M. F. Lappert, S. J. Miles and A. J. Thorne, *J. Chem. Soc., Chem. Commun.*, 1984, 480.
556. L. Pauling, *Proc. Natl. Acad. Sci. USA*, 1983, **80**, 3871.
557. G. Trinquier, J. P. Malneu and P. Rivière, *J. Am. Chem. Soc.*, 1982, **104**, 4529.
558. M. J. S. Dewar, G. L. Grady, D. R. Kutin and K. M. Merz, *J. Am. Chem. Soc.*, 1984, **106**, 6773.
559. P. Jutzi, F. Kohl, P. Hofmann, C. Kruger and Y. H. Tsay, *Chem. Ber.*, 1980, **113**, 757.
560. J. L. Atwood, W. E. Hunter, A. H. Cowley, R. A. Jones and C. A. Stewart, *J. Chem. Soc., Chem. Commun.*, 1981, 925.
561. M. Grenz, E. Hahn, W. W. du Mont and W. W. Pickardt, *Angew. Chem., Int. Ed. Engl.*, 1984, **23**, 61.
562. C. Panattoni, G. Bambieri and U. Croatto, *Acta Crystallogr.*, 1966, **21**, 823.
563. A. Almennigen, A. Haaland and T. Motzfeldt, *J. Organomet. Chem.*, 1967, **7**, 97.
564. J. Almlöf, L. Fernholt, K. Faergn, A. Haaland, B. R. Schilling, R. Seip and K. Taugbol, *Acta Chem. Scand., Ser. A*, 1983, **37**, 131.
565. A. H. Cowley, J. G. Lasch, N. C. Norman, C. A. Stewart and T. C. Wright, *Organometallics*, 1983, **2**, 1691.
566. L. Fernholt, A. Haaland, P. Jutzi, F. X. Kohl and R. Seip, *Acta Chem. Scand., Ser. A*, 1984, **38**, 211.
567. M. J. Heeg, C. Janiak and J. J. Zuckerman, *J. Am. Chem. Soc.*, 1984, **106**, 4259.
568. S. G. Baxter, A. H. Cowley, J. G. Lasch, M. Lattman, W. P. Sharum and C. A. Stewart, *J. Am. Chem. Soc.*, 1982, **104**, 4064.
569. A. Haaland and B. E. R. Schilling, *Acta Chem. Scand., Ser. A*, 1984, **38**, 217.
570. K. D. Bos, E. J. Bultan and J. G. Noltes, *J. Organomet. Chem.*, 1975, **99**, 71.
571. P. G. Harrison and S. R. Stobart, *J. Chem. Soc., Dalton Trans.*, 1973, 940.
572. P. G. Harrison and S. R. Stobart, *Inorg. Chim. Acta*, 1973, **7**, 306.
573. P. F. R. Ewings, P. G. Harrison and D. E. Fenton, *J. Chem. Soc., Dalton Trans.*, 1975, 821.
574. P. F. R. Ewings and P. G. Harrison, *J. Chem. Soc., Dalton Trans.*, 1975, 1717.
575. A. B. Cornwell and P. G. Harrison, *J. Chem. Soc., Dalton Trans.*, 1975, 1722.
576. P. Jutzi and F. Kohl, *J. Organomet. Chem.*, 1979, **164**, 141.
577. P. G. Harrison and J. A. Richards, *J. Organomet. Chem.*, 1976, **108**, 35.
578. P. Jutzi, F. Kohl, C. Krüger, G. Wolmershäuser, P. Hofmann and P. Stauffert, *Angew. Chem., Int. Ed. Engl.*, 1982, **21**, 70.
579. F. X. Kohl, E. Schlüter, P. Jutzi, C. Krüger, G. Wolmershäuser, P. Hofmann and P. Stauffert, *Chem. Ber.*, 1984, **117**, 1178.
580. A. H. Cowley, R. A. Kemp and C. A. Stewart, *J. Am. Chem. Soc.*, 1982, **104**, 5239.
581. A. H. Cowley, J. G. Lasch, N. C. Norman, C. A. Stewart, and T. C. Wright, *Organometallics*, 1983, **2**, 1691.
582. A. H. Cowley, P. Jutzi, F. X. Kohl, J. G. Lasch, N. C. Norman and E. Schlüter, *Angew. Chem., Int. Ed. Engl.*, 1984, **23**, 616.
583. P. Jutzi and E. Schlüter, *J. Organomet. Chem.*, 1983, **253**, 313.
584. T. S. Dory, J. J. Zuckerman and M. D. Rausch, *J. Organomet. Chem.*, 1985, **281**, C8.
585. J. D. Cotton, P. J. Davidson, M. F. Lappert, J. D. Donaldson and J. Silver, *J. Chem. Soc., Dalton Trans.*, 1976, 2286.
586. J. D. Cotton, P. J. Davidson, D. E. Goldberg, M. F. Lappert and K. M. Thomas, *J. Chem. Soc., Dalton Trans.*, 1974, 893.
587. J. D. Cotton, P. J. Davidson and M. F. Lappert, *J. Chem. Soc., Dalton Trans.*, 1976, 2275.
588. M. F. Lappert, S. J. Miles, P. P. Power, A. J. Carty and N. J. Taylor, *J. Chem. Soc., Chem. Commun.*, 1977, 458.
589. A. B. Cornwell, P. G. Harrison and J. A. Richards, *J. Organomet. Chem.*, 1976, **108**, 47.
590. A. B. Cornwell and P. G. Harrison, *J. Chem. Soc., Dalton Trans.*, 1976, 1054.
591. P. G. Harrison, T. J. King and J. A. Richards, *J. Chem. Soc., Dalton Trans.*, 1975, 2097.
592. P. G. Harrison and J. A. Richards, *J. Organomet. Chem.*, 1976, **108**, 35.
593. A. K. Holliday, P. H. Makin and R. J. Puddephatt, *J. Chem. Soc., Dalton Trans.*, 1976, **228**, 435.
594. P. G. Harrison and J. J. Zuckerman, *J. Am. Chem. Soc.*, 1970, **92**, 2577.

595. T. S. Dory, J. J. Zuckerman and C. L. Barnes, *J. Organomet. Chem.*, 1985, **281,** C1.
596. R. W. Rudolph, R. L. Voorhees and R. E. Cochoy, *J. Am. Chem. Soc.*, 1970, **92,** 3351.
597. N. S. Hosmane, N. N. Sirmokadam and R. H. Herber, *Organometallics,* 1984, **3,** 1665.
598. A. H. Cowley, P. Galow, N. S. Hosmane, P. Jutzi and N. C. Norman, *J. Chem. Soc., Chem. Commun.*, 1984, 1564.
599. K. S. Wong and R. N. Grimes, *Inorg. Chem.*, 1977, **16,** 2053.
600. H. Lüth and E. L. Amma, *J. Am. Chem. Soc.*, 1969, **91,** 7515.
601. A. G. Gash, P. F. Rodesiler and E. L. Amma, *Inorg. Chem.*, 1974, **13,** 2429.
602. M. S. Weininger, P. F. Rodesiler, A. G. Gash and E. L. Amma, *J. Am. Chem. Soc.*, 1972, **94,** 2135.
603. M. S. Weininger, P. F. Rodesiler and E. L. Amma, *Inorg. Chem.*, 1979, **18,** 751.
604. M. F. Lappert, M. J. Slade, J. L. Atwood and M. J. Zaworatko, *J. Chem. Soc., Chem. Commun.*, 1980, 621.
605. T. Feldberg, H. Hope, M. F. Lappert, P. P. Power and A. J. Thorne, *J. Chem. Soc., Chem. Commun.*, 1983, 639.
606. B. Cetinkay, I. Gümrükgu, M. F. Lappert, J. L. Atwood, R. D. Rogers and M. J. Zawototko, *J. Am. Chem. Soc.*, 1980, **102,** 2088.
607. P. B. Hitchcock, M. F. Lappert, B. J. Samways and E. L. Weinberg, *J. Chem. Soc., Chem. Commun.*, 1983, 1492.
608. M. M. Olmstead and P. P. Power, *Inorg. Chem.*, 1984, **23,** 413.
609. K. C. Molloy, M. P. Bigwood, R. H. Herber and J. J. Zuckerman, *Inorg. Chem.*, 1982, **21,** 3709.
610. M. Veith, *Z. Naturforsch., Teil B,* 1978, **33,** 7.
611. W. W. du Mont and M. Grenz, *Z. Naturforsch., Teil B,* 1983, **38,** 113.
612. W. W. du Mont and M. Grenz, *Chem. Ber.*, 1981, **114,** 1180.
613. L. D. Silverman and M. Zeldin, *Inorg. Chem.*, 1980, **19,** 272.
614. B. J. Haylett, Ph.D. Thesis, University of Nottingham, 1981.
615. C. Gaffney, P. G. Harrison and T. J. King, *J. Chem. Soc., Chem. Commun.*, 1980, 1251.
616. M. Veith and R. Rosler, *Angew. Chem., Int. Ed. Engl.*, 1982, **21,** 858.
617. M. Veith, M. L. Sommer and D. Jäger, *Chem. Ber.*, 1979, **112,** 2581.
618. M. Veith and H. Lange, *Angew. Chem., Int. Ed. Engl.*, 1980, **19,** 401.
619. M. Veith, M. Grossev and O. Recktenwald, *J. Organomet. Chem.*, 1981, **216,** 27.
620. M. Veith and O. Recktenwald, *Z. Naturforsch., Teil B,* 1981, **36,** 144.
621. M. Veith and W. Frank, *Angew. Chem., Int. Ed. Engl.*, 1984, **23,** 158.
622. S. R. Stobart, M. R. Churchill, F. J. Hollander and W. J. Youngs, *J. Chem. Soc., Chem. Commun.*, 1979, 911.
623. A. Tzschach, M. Scheer, K. Jurkschat, A. Zschunke and C. Mugge, *Z. Anorg. Allg. Chem.*, 1983, **502,** 158.
624. A. Tzschach, M. Scheer and K. Jurkschat, *Z. Anorg. Allg. Chem.*, 1984, **515,** 147.
625. P. F. R. Ewings, P. G. Harrison and T. J. King, *J. Chem. Soc., Dalton Trans.*, 1976, 1399.
626. P. F. R. Ewings, P. G. Harrison and A. Mangia, *J. Organomet. Chem.*, 1976, **114,** 35.
627. P. F. R. Ewings, P. G. Harrison and D. E. Fenton, *J. Chem. Soc., Dalton Trans.*, 1975, 821.
628. P. F. R. Ewings, P. G. Harrison and T. J. King, *J. Chem. Soc., Dalton Trans.*, 1975, 1455.
629. A. B. Cornwell and P. G. Harrison, *J. Chem. Soc., Dalton Trans.*, 1975, 1722.
630. P. G. Harrison and E. W. Thornton, *J. Chem. Soc., Dalton Trans.*, 1978, 1274.
631. A. D. Christie, R. A. Howie and W. Moser, *Inorg. Chim. Acta,* 1979, **36,** L447.
632. N. Kh. Dzhafarov, I. R. Amiraslanov, G. N. Nadzhafov, E. M. Movsumov, F. R. Kerimova and K. S. Mamedov, *J. Struct. Chem. (Engl. Transl.)*, 1981; **22,** 245.
633. I. R. Amiraslanov, N. Kh. Dzhafarov, G. N. Nadzhafov, Kh. S. Mamedov, E. M. Movsumov and B. T. Usubaliev, *J. Struct. Chem. (Engl. Transl.)*, 1980, **21,** 109.
634. N. Kh. Dzafarov, I. R. Amiraslanov, G. N. Nadzhafov, E. M. Movsumov and Kh. S. Mamedov, *J. Struct. Chem. (Engl. Transl.)*, 1981, **22,** 242.
635. I. R. Amiraslanov, N. Kh. Dzhafarov, G. N. Nazhafov, Kh. S. Mamedov, E. M. Movsumov and B. J. Usubakiev, *J. Struct. Chem. (Engl. Transl.)*, 1980, **22,** 104.
636. P. G. Harrison and A. T. Steel, *J. Organomet. Chem.*, 1982, **239,** 105.
637. G. Lundgren, C. Wernfors and T. Yamaguchi, *Acta Crystallogr., Sect. B,* 1982, **38,** 2357.
638. J. D. Donaldson and S. M. Grimes, *J. Chem. Soc., Dalton Trans.*, 1984, 1301.
639. J. D. Donaldson, S. M. Grimes, A. Nicolaides and P. J. Smith, *Polyhedron,* 1985, **4,** 391.
640. R. Herak, B. Prelesnik, M. Curic and P. Vasic, *J. Chem. Soc., Dalton Trans.*, 1978, 566.
641. T. R. J. Weakly and W. W. L. Watt, *Acta Crystallogr., Sect. B,* 1979, **35,** 3023.
642. R. C. McDonald and K. Eriks, *Inorg. Chem.*, 1980, **19,** 1237.
643. T. H. Jordan, B. Dickens, L. W. Schroeder and W. E. Brown, *Inorg. Chem.*, 1980, **19,** 2551.
644. M. Hala, F. Maruno, S. I. Iawi and H. Aoki, *Acta Crystallogr., Sect. B,* 1980, **36,** 2128.
645. P. Vasic, B. Prelesnik, R. Herak and M. Giric, *Acta Crystallogr., Sect. B,* 1981, **37,** 660.
646. H. Hagihara and S. Yamashita, *Acta Crystallogr., Sect. B,* 1966, **21,** 350.
647. H. Hagihara, Y. Watanabe and S. Yamashita, *Acta Crystallogr., Sect. B,* 1968, **24,** 960.
648. J. Potenza and D. Mastropaolo, *Acta Crystallogr., Sect. B,* 1973, **29,** 1830.
649. H. Iwasaki and H. Hagihara, *Acta Crystallogr., Sect. B,* 1972, **28,** 507.
650. H. Iwasaki, *Acta Crystallogr., Sect. B,* 1980, **36,** 2138.
651. M. Ho and H. Iwasaki, *Acta Crystallogr., Sect. B,* 1980, **36,** 443.
652. T. Ho and H. Iwasaki, *Acta Crystallogr., Sect. B,* 1972, **28,** 1034.
653. S. L. Lawton and G. Kokotailo, *Inorg. Chem.*, 1972, **11,** 363.
654. J. L. Lefferts, K. C. Molloy, M. B. Hossain, D. van der Helm and J. J. Zuckerman, *Inorg. Chem.*, 1982, **21,** 1410.
655. A. T. Steel, Ph.D. Thesis, University of Nottingham, 1983.
656. P. F. R. Ewings, P. G. Harrison, A. Morris and T. J. King, *J. Chem. Soc., Dalton Trans.*, 1976, 1602.
657. T. Birchall and J. J. Johnson, *J. Chem. Soc., Dalton Trans.*, 1981, 69.

658. W. W. DuMont and H. Schumann, *J. Organomet. Chem.*, 1975, **85,** C45.
659. W. W. DuMont and H. Schumann, *Angew. Chem., Int. Ed. Engl.*, 1975, **14,** 368.
660. W. W. DuMont and B. Neudert, *Z. Anorg. Allg. Chem.*, 1978, **441,** 86.
661. W. W. DuMont, *Inorg. Chim. Acta,* 1978, **29,** L195.
662. W. W. DuMont and G. Rudolph, *Inorg. Chim. Acta,* 1979, **35,** L341.
663. W. W. DuMont, *Z. Anorg. Allg. Chem.*, 1979, **450,** 57.

27

Phosphorus

REINHARD SCHMUTZLER
Technische Universität Carolo-Wilhelmina, Braunschweig, Federal Republic of Germany

Because of ill health and other factors beyond the editors' control, the manuscript for this chapter was not available in time for publication. However, it is anticipated that the material will appear in the journal *Polyhedron* in due course as a Polyhedron Report.

28

Arsenic, Antimony and Bismuth

CHARLES A. McAULIFFE
University of Manchester Institute of Science and Technology, UK

28.1 INTRODUCTION

This chapter will be divided into three main sections, dealing with the chemistry of arsenic, antimony and bismuth individually, since their interest is intrinsic rather than comparative. However, the Periodic Table trends are important and their main features will be outlined.

The triad As, Sb, Bi has an important place in the history of chemistry, the isolation of the elements being one of the achievements of alchemists in the pre-Scientific Age. An interesting history of their discovery can be found in ref. 1.

28.2 ARSENIC

The thirty-third member of the Periodic Table has a sinister reputation. Since its isolation in the Middle Ages the main use of arsenic compounds has been the termination of lifeforms. Its use for homicide was widened to include *inter alia* herbicide, fungicide and bactericide (see Section 28.7). Various arsenic compounds have been used from ancient times as pigments, *e.g.* lemon-yellow orpiment, As_2S_3, and red realgar, As_4S_4.

The common oxidation states are −3, +3 and +5, but the simple ions As^{3-}, As^{3+} and As^{5+} are not known. The krypton electronic structure could be attained by gain of three electrons, but there is a high energy requirement. Equally, the loss of five valence electrons to form As^{5+} is unrealizable because of the high ionization enthalpy.

The most common structures of arsenic compounds are tetrahedral and pyramidal, which are similar when the sterically active lone pair is counted. Tetrahedral symmetry holds the potential for chirality and indeed many chiral organoarsenic compounds have been prepared. Arsenic may also use *d* orbitals for $(d\text{–}d)\pi$ bonding and for hybridization with s^2 and p^3 orbitals, resulting in trigonal bipyramidal or octahedral structures. In the former the more electronegative substituents occupy the apical position.

28.3 ARSENIC–GROUP 4 BONDS

28.3.1 Carbon Ligands

28.3.1.1 Organoarsenic compounds

There is a comprehensive review of this area[2] so only a few recent developments will be mentioned here. Organoarsenic intramolecular coordination compounds, *e.g.* (**1**), have also recently been reviewed,[3] and organoarsenic chemistry is reviewed annually.[4] There is a review containing 102 references on arsonium ylides.[5] The canonical structures of the arsenic ylides are

$$Ph_3As{=}CRR' \longleftrightarrow Ph_3\overset{+}{As}{-}\overset{-}{C}RR'$$

The overlap of carbon *p* orbitals with arsenic *d* orbitals is less effective than with the *d* orbitals of phosphorus, and so the covalent canonical structure is expected to make less of a contribution to the hybrid structure. This has been confirmed in an X-ray study of 2-acetyl-3,4,5-triphenylcyclopentadienetriphenylarsorane.[6] Yamamoto and Schmidbaur[7] found (^{13}C NMR) that the bonding in arsenic ylides was probably sp^3 (*cf.* phosphorus, which changes from $sp^3 \rightarrow sp^2$), resulting in arsenic pseudotetrahedral geometry (*cf.* phosphorus ylides, which are planar).

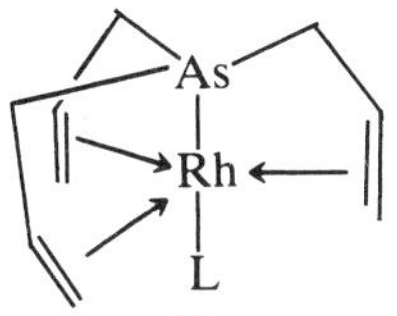

(**1**)

28.3.2 Silicon Ligands

The parent compound for this series is H_3SiAsH_2, which was first produced in 1962 in low yield in the gas phase reaction between silane, SiH_4, and arsine, AsH_3, promoted by silent electric discharge.[8] This volatile pyrophoric compound is better prepared from lithium tetraarsinoaluminate and bromosilane.[9] It is also possible to prepare $H_3SiSiH_2AsH_2$ (59%), $MeSiH_2AsH_2$ (57%) and $(Me_3Si)_3As$ (77%) by this route. Trisilylarsine has been shown to have a pyramidal structure by electron diffraction;[10] it is prepared from bromosilane and potassium dihydrogenarsenide at low temperature (−122 to −38 °C).[11]

Higher homologues are coveniently prepared from 'sodium potassium arsenide' and chlorosilanes. The arsenide is a complex mixture formed by mixing a sodium–potassium alloy with powdered arsenic suspended in 1,2-dimethoxyethane (equation 1).[36]

$$ClSiMe_3 + Na_3As/K_3As \longrightarrow As(SiMe_3)_3 \quad (80\text{–}90\%) \qquad (1)$$

Tris(trimethylsilyl)arsine is a convenient starting material for the production of $LiAs(SiMe_3)_2 \cdot 2L$ (L = coordinating solvent); see, for example, equation (2) and structure (**2**).

$$As(SiMe_3)_3 + LiMe \xrightarrow{THF} LiAs(SiMe_3)_2 \cdot 2THF + SiMe_4 \qquad (2)$$

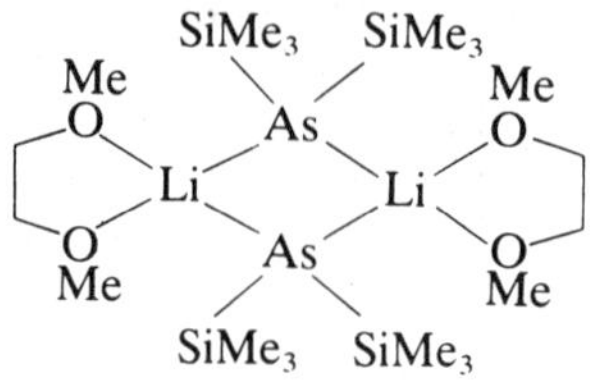

(**2**)

28.3.3 Other Group 4 Ligands

While many compounds of the type $X_nAs(ER_3)_{3-n}$ ($n = 0$–2; X = any other group, *e.g.* H, Hal, R *etc.*; R = H, alkyl, aryl; E = Ge,[263] Sn,[264] Pb[265]) have been produced, they are mainly of interest to theoreticians and synthesists. They have the expected pyramidal geometry of AsX_3 compounds, and their interest to coordination chemists lies in their behaviour as ligands, which is dealt with elsewhere (Chapter 12.2); there is some special interest in cluster coordination compounds involving As—Ge bonding.

28.4 ARSENIC–GROUP 5 BONDS

Arsenic forms a large number of compounds with other Group 5 members. These are dealt with element by element down the group. However, in the case of arsenic only its catenated and π-bonded compounds are dealt with.

28.4.1 Nitrogen Ligands

The arsenic–nitrogen bond is not as strong as the As—C, As—O, or As—S bonds and use is made of this in the synthetic chemistry of arsenic. Amongst the wide range of As—N compounds are open-chain aminoarsines, cyclic and macrocyclic compounds of various types, fused ring compounds and three-dimensional fused ring systems of the adamantine type. The calculated As—N bond length from Pauling's covalent single bond radii, and applying the Schomaker–Stevenson correction, is 1.87 Å. This compares with values given for As—N sp^3 bonds from X-ray data.

28.4.1.1 Arsenic(III)

(i) Alkylaminoarsines

These are readily prepared by adding amines to trihaloarsines in cooled ether solutions. The reaction proceeds stepwise, halogens being successively replaced by alkylamino groups (equations 3–5). Some examples of dialkylaminoarsines are given in Table 1.

$$AsCl_3 + 2Me_2NH \xrightarrow[0\,^\circ C]{Et_2O} Me_2NAsCl_2 + Me_2NH_2^+Cl^- \qquad (3)$$

$$Me_2NAsCl_2 + 2Me_2NH \longrightarrow (Me_2N)_2AsCl + Me_2NH_2^+Cl^- \qquad (4)$$

$$(Me_2N)_2AsCl + 2MeNH \longrightarrow (Me_2N)_3As + Me_2NH_2^+Cl^- \qquad (5)$$

The As—N bond is labile and this has been widely exploited in synthetic arsenic chemistry. Some idea of the versatility[16a] can be seen from Schemes 1 and 2. Refluxing secondary amines with tris(dimethylamino)arsine effects transamination (equation 6). These tris(dialkylamino)arsines undergo the general reactions in Scheme 1, enabling ready access to a wide variety of compounds, many of them finding use as ligands in transition metal complexes (see Chapter 14 of this work).

$$As(NMe_2)_3 + 3HNR_2 \xrightarrow[reflux]{C_6H_6} As(NR_2)_3 + 3HNMe_2 \qquad (6)$$

Table 1 Dialkylaminoarsines

Compound	*B.p.* (°C/Torr)	d^{20}	n_D^{20}	*Ref.*
Me_2NAsCl_2	73/25	1.6560	1.5532	12
$(Me_2N)_2AsCl$	69–70/10		1.5427	13
		1.3460		12
$(Me_2N)_3As$	42–43/3.5	1.1248	1.4848	12
	60/13			14
Et_2NAsCl_2	83–84/13	1.4727	1.5335	12
$(Et_2N)_2AsCl$	118–120/17	1.2225	1.5098	12
$(Et_2N)_3As$	91–93/2	1.2418	1.4825	12
Me_2NAsBr_2	70/3		1.6428	13
$(Me_2N)_2AsBr$	72/3		1.5660	13
Me_2NAsF_2	30/15			15
Et_2NAsF_2	35/15			15
$(Bu^n_2N)_3As$	159–161/0.01		1.4789	173
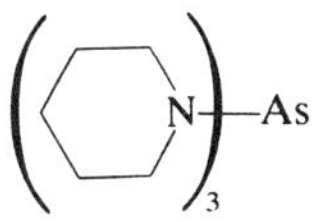	139–142/0.01 (m.p. 49–51°C)		1.4839	173
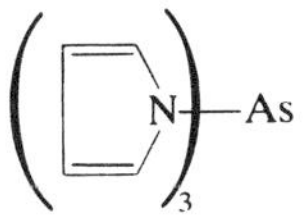	125–132/1			17
$(Me_3Si)_2NAsCl_2$	96–98/4			18
$[(Me_3Si)_2N]_2AsCl$	140–142/3			18
$Me_2NAs(NHBu^t)_2$	67/0.01			173

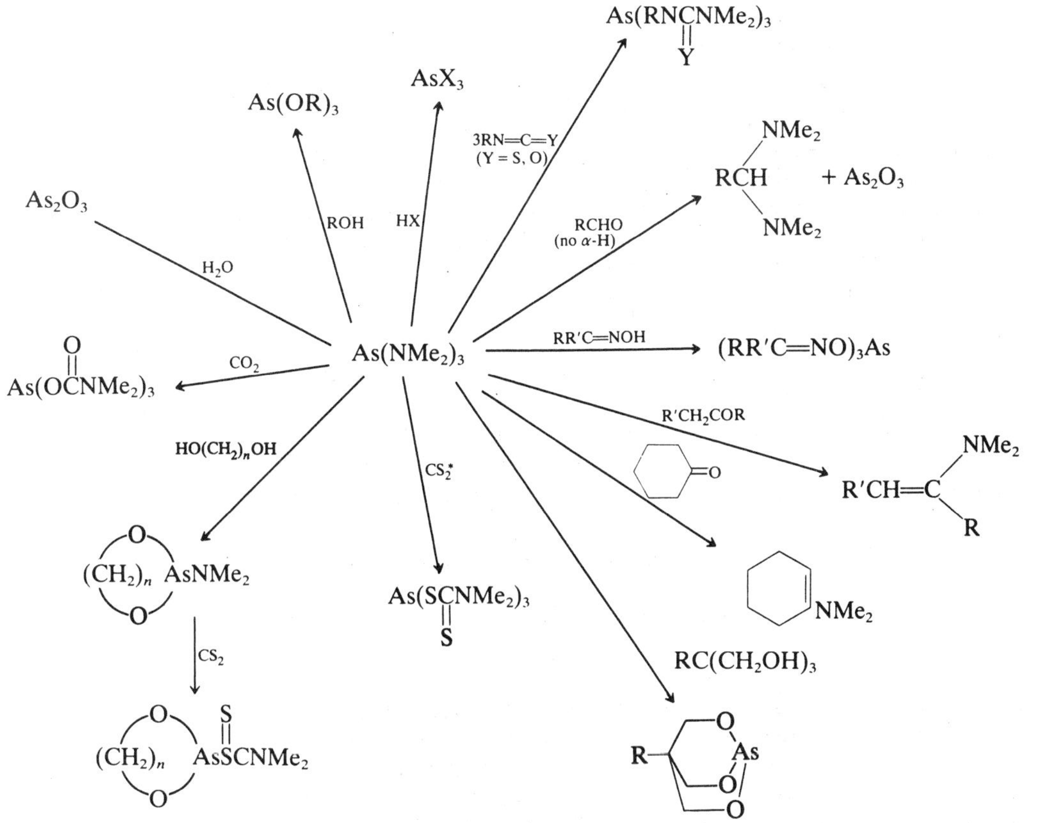

* Reaction with CS_2 is general for tris(dialkylamino)arsines and the tris(dialkyldithiocarbamidato)arsanes may be used for characterization.

Scheme 1

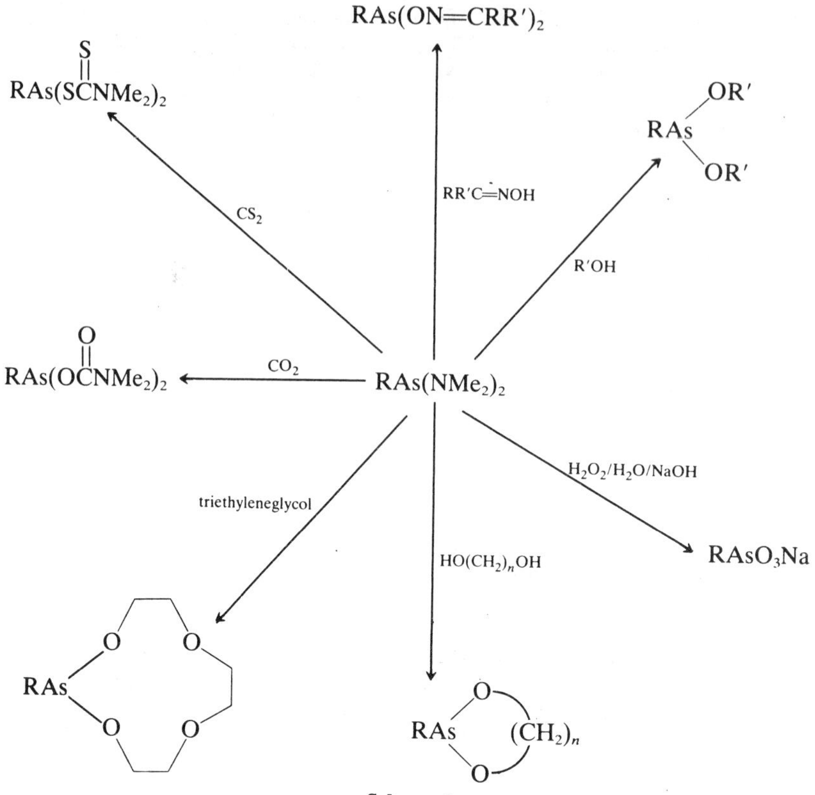

Scheme 2

(*ii*) *Secondary amines and* AsF_3

Dialkylaminodifluoroarsines are formed from the reaction of R_2NH with AsF_3. Interestingly, the fluorination of dialkylaminodichloroarsines is unsuccessful using SbF_3, although AsF_3 works.[15]

28.4.1.2 Arsenic(V)

Roesky and co-workers[19,20] have produced an interesting series of compounds which illustrate some general points of arsenic coordination chemistry. When $[(CF_3)_2As\{N(SiMe_3)_2\}]$ is treated in a strict 1:1 ratio with chlorine, $[(CF_3)_2AsCl_2\{N(SiMe_3)_2\}]$ is obtained. This is a trigonal bipyramidal molecule (**3**) with chlorine atoms in axial positions. When $[(CF_3)_2As\{N(SiMe_3)_2\}]$ is treated with an excess of chlorine, the dimer $[(CF_3)_2As(Cl)\{N(SiMe_3)\}]_2$ (**4**) is formed, the first reported example of five-coordinated arsenic

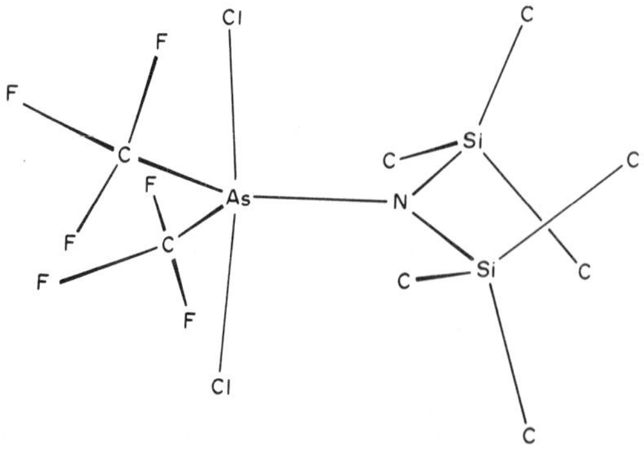

(**3**)

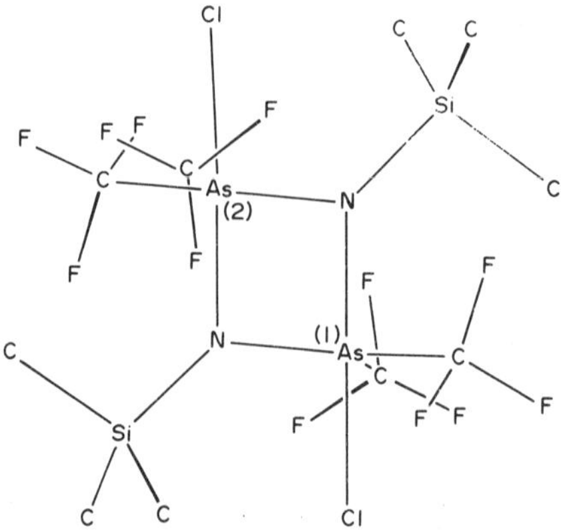

(**4**)

in a four-membered ring. The ligands are disposed about arsenic in a distorted trigonal bipyramidal geometry, the longer As—N bonds being axial and the short bonds being equatorial. Once more the chlorine occupies the other axial position instead of one of the CF_3 groups.

When $[(CF_3)_2As(Cl)\{N(SiMe_3)\}]_2$ is refluxed in an inert solvent or pyrolyzed *in vacuo*, trimethylsilyl chloride is derived and expanded heterocycles are formed. The CN of arsenic drops from five to four and the geometry from distorted trigonal bipyramidal to distorted tetrahedral (equation 7). In (**5**) the As and N lie alternately at ±38 and ±63 pm, respectively, from the mean plane through the ring atoms. There are two types of As—N bond, *viz.* at 1.716 Å for bonds on the same side of the plane and the longer 1.732 Å bonds across the plane. The As—N—As angles average 123.5° and the N—As—N angles average 125.0°. The difference in bond lengths, Δ = 0.016 Å, is much less than that found by Glick,[21] Δ = 0.12 Å, for alternating N—As (1.67 Å and 1.79 Å) bonds in the eight-membered ring $[Ph_2AsN]_4$.

$$[(CF_3)_2As(Cl)\{N(SiMe_3)\}]_2 \longrightarrow \underset{\text{six-membered, four-coordinate}}{[(CF_3)_2AsN]_3} + \underset{(\mathbf{5})}{[(CF_3)_2AsN]_4} \quad (7)$$

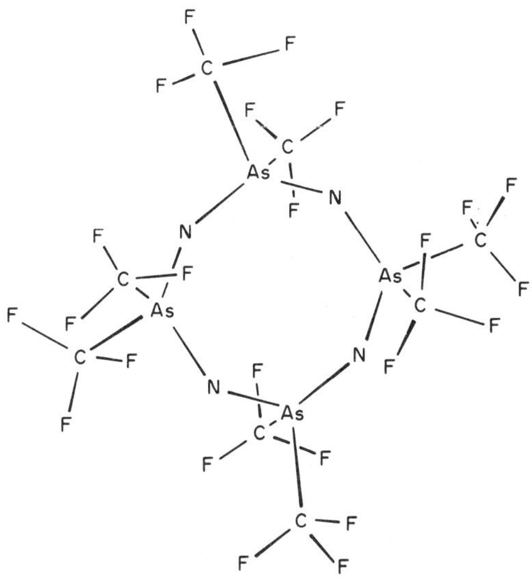

(**5**)

28.4.2 Phosphorus Ligands

Phosphorus as a ligand to arsenic is discussed in several sections of this work. There is no merit in repeating here under a separate heading that which is more naturally dealt with elsewhere.*

Sufficient to say that, of all the members of Group 5, the analogies between phosphorus and arsenic are closest. When used as coreactant with arsenic compounds, phosphorus compounds may react interchangeably. The recent preparation of P=As compounds by Escudie *et al.*[22] relied on this scrambling of arsenic and phosphorus reactants (equation 8).

$$(Me_3Si)_3CPCl_2 + (Me_3Si)_3CAsCl_2 \xrightarrow{\text{BuLi}} \underset{(35)\%}{(Me_3Si)_3CP{=}PC(SiMe_3)_3} + \underset{(30\%)}{(Me_3Si)_3CP{=}AsC(SiMe_3)_3} \quad (8)$$

* Antimony and bismuth bonds to arsenic are dealt with in Sections 28.10.2 and 28.16.2 under antimony and bismuth, respectively.

The interchange of As and P is not necessarily symmetrical. In a study[23] of Group 5 adducts of the type $R_nMX_{3-n}\cdot yM'R'_3$ (X = halogen; R = alkyl or aryl; M, M′ = P, As, Sb; y = 1, 2) it was found that complexes were formed when M′ = P and M = As but not when M′ = As and M = P. A few examples of the complexes are given in Table 2. The adducts are probably quaternary salts $[R_2AsPR_2]X$, similar to those of Wolfsberger, $[Me_3P{=}NPMe_2AsMe_2]Cl$.[24]

Table 2 Arsenic–Phosphorus Complexes

Adduct	*Yield* (%)	*M.p.*[a] (°C)	*Conductivity*[b] (10^{-3} S cm^{-1})
$MeAsCl_2\cdot PMe_3$	83	d	37.8
$AsI_3\cdot 2PMe_3$	—	d	117
$MeAsI_2\cdot 2PMe_3$	87	d	124
$EtAsCl_2\cdot PMe_3$	85	d	—
$PhAsCl_2\cdot PMe_3$	90	115 (d)	38.3
$PhAsI_2\cdot PMe_3$	59	120–8	57.2
$Me_2AsCl\cdot PMe_3$	82	—	—

[a] d = decomposed. [b] In $MeNO_2$.

28.4.3 Arsenic Ligands

28.4.3.1 Open-chain arsenic compounds

The first established As—As bond in cacodyl, $Me_2AsAsMe_2$, was also among the earliest of arsenic compounds[25] to be isolated at the beginning of the modern chemical era.[26] However, the building of arsenic chains has not progressed. Arsenic does not appear to form small molecular chain compounds because the thermodynamically stable configurations for three to six arsenic atoms are rings. However, it may be unwise to speculate that such chain compounds will never be produced, since until recently it was thought that As═As systems would never be prepared. The only known exception presently is a red-black polymeric substance of empirical formula $(MeAs)_n$ characterized by Rheingold *et al.*[27] and Daly *et al.*[28]

28.4.3.2 Coordinated open-chain arsenic compounds

While unpolymerized open-chain arsenic compounds are unknown to date, a number of such systems are known when coordinated to transition metals.[29–31] They are prepared by the reaction between transition metal carbonyls and *cyclo*-$(RAs)_n$ (n = 5, 6). The reactions go by a variety of mechanisms leading to either the coordination of the unchanged ring,[32] an expanded ring,[31] fragmentamtion of the ring,[33] ring opening and coordination of the open chain without loss of arsenic,[30] or a ring opening and lengthening of the coordinated chain.[31]

When $Fe(CO)_5$ reacts with *cyclo*-$(AsMe)_5$, the ring is opened and the open chain of *four* arsenic atoms is stabilized by ligation (**6**):[29] As(1)—As(2) and As(3)—As(4) bond lengths are 2.453 Å, and the mid-chain As(2)—As(3) bond length is 2.391 Å. The non-bonding arsenics, As(1)· · ·As(4), are 2.888 Å apart. A similarly stabilized open chain of eight arsenic atoms occurs in $[Mo_2(CO)_6(AsPr^n)_8]$ (**7**), formed from $Mo(CO)_6$ and *cyclo*-$(AsPr^n)_5$, *i.e.* chain lengthening (5→8) has occurred. The two terminal As atoms bridge the Mo atoms (Mo—As = 2.554 Å) and two, As(4) and As(5), are only coordinated to one of each of the Mo atoms (Mo—As = 2.623 Å), leaving four uncoordinated As chain atoms, As(2)—As(3) = 2.434 Å. The Mo atoms are seven-coordinate, each lying at the centre of a distorted edge-shared octahedron.

In the case of $\{[CpMo(CO_2)]_2$ [μ-*catena*-$(MeAs)_5$]} simple chain opening and subsequent coordination occurred in the reaction between $[Mo(Cp)(CO)_3]_2$ and *cyclo*-$(AsMe)_5$. The mid-chain As—As distances are 2.4335 Å (average) and the terminal As—As distances are 2.449 Å (average) with non-bonding separation between the two terminal arsenics of As(1)· · ·As(5) of 2.835 Å; Mo—As is 2.6285 Å (average).

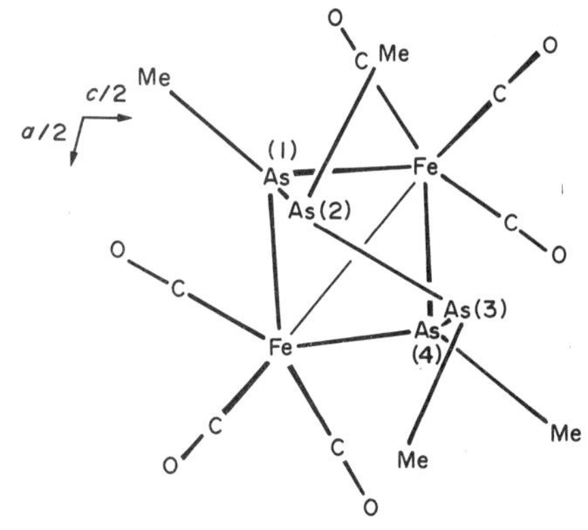

(**6**) [Fe(CO)$_3$]$_2$(AsMe)$_4$ (viewed down the *b* axis)

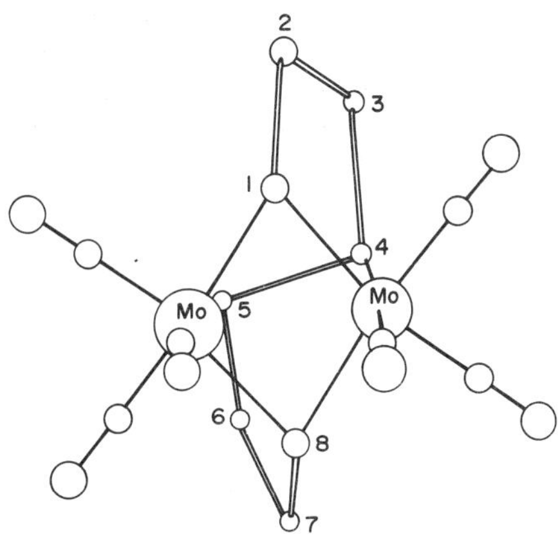

(**7**) [Mo$_2$(CO)$_6$(AsPr^n)$_8$] (*n*-propyl groups are omitted)

28.5 ARSENIC–GROUP 6 LIGANDS

28.5.1 Oxygen Ligands

The chemistries of oxygen, sulfur, nitrogen and halogen ligands of arsenic are quite deeply enmeshed with each other. In this presentation the divisions chosen are not to be thought of as 'the best' but as the most convenient. The oxyhalides are discussed under halogen ligands (Section 28.6). The mixed As/O/S compounds, including oxathia acids, will be included in the sulfur ligand section (28.5.2).

28.5.1.1 Arsenic acid esters and other oxaorganic compounds

Simple esters of arsenic acid, OAs(OR)$_3$, have tetragonal structures; some examples are listed in Table 3. Structurally more interesting are the spiroarsanes (**8**). Goldwhite and Teller[34] initiated the study of the variation in geometry of these complexes from trigonal bipyramidal to rectangular pyramidal. Their findings have been confirmed and the series of structures between the two extremes has been bridged by Holmes and co-workers.[35] The Berry exchange mechanism for a five-coordinate pyramidal system proceeds *via* a rectangular pyramidal intermediate.

Table 3 Esters of Arsenic Acid, H_3AsO_4[202]

Compound	*B.p.* (°C/Torr)	d^{20}	n_D^{20}
$(MeO)_3AsO$	97–9/13	1.5572	1.4367
$(EtO)_3AsO$	118–120/15.5 235–238/760	1.3023	1.4343
$(PrO)_3AsO$	144–146/14	1.1945	1.4381
$(BuO)_3AsO$	169–171/12	1.1257	1.4421
$(C_6H_{13}O)_3AsO$	189–191/2	1.0496	1.4488

Spiroarsoranes corresponding to many degrees of the transition states are known (**9–11**). Holmes and co-workers studied the variations in substituents and their effect on the percentage displacement from TBP (Table 4).

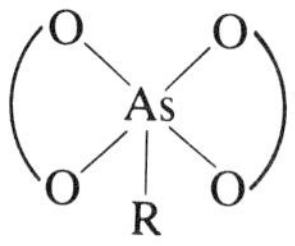

(**8**) R = alkyl, aryl, OH

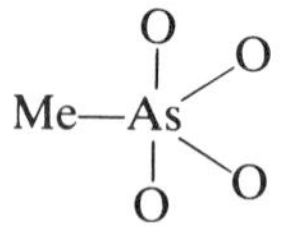

(**9**) Distorted trigonal bipyramid (TBP)

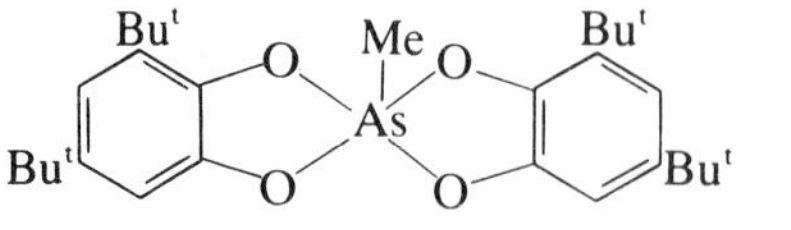

(**10**) Almost 45% displaced from TBP to RP

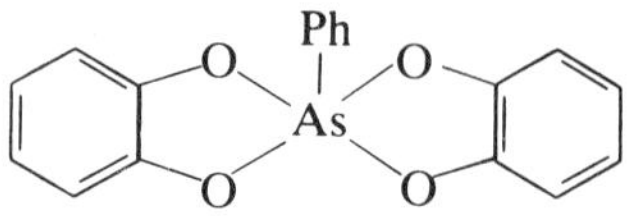

(**11**) Distorted regular pyramid (RP)

Table 4 Spirocyclic Arsoranes: As—O Bond Lengths and TBP→RP Distortions

Compound	*Bond lengths* (Å)	*TBP distortion towards RP* (%)	*Ref.*
$PhAs(O_2C_2Me_4)_2$	*d*(As—O) 1.779 (av) *d*(As—C) 1.908	16	34
$HOAs(O_2C_2H_4)_2$	*d*(As—O) 1.779 (av) *d*(As—OH) 1.704	60	34
$MeAs(O_2C_2Me_4)_2$	*d*(As—O) 1.789 (av) *d*(As—C) 1.910 *d*(As—O) 1.1819 (av)[a] *d*(As—O) 1.759 (av)[b]	11	35
$MeAs(O_2C_6H_2Bu^t_2)_2$	*d*(As—O) 1.835[a] *d*(As—O) 1.785[b] *d*(As—C) 1.93	20	35
$HOAs(O_2C_6H_2Bu^t_2)_2$	*d*(As—O) 1.788 (av) *d*(As—O) 1.808 (av)[a] *d*(As—O) 1.769 (av)[b] *d*(As—OH) 1.704	45	35

[a] Axial. [b] Equatorial.

28.5.1.2 Arsenic sulfates

Arsenic does not form true metallic-like salts with sulfuric acid of the type $As_2(SO_4)_3$. However, As_2O_3 dissolves in oleum and crystalline compounds are formed, the compositions of which depend on the concentration of SO_3, *e.g.* $(As_2O_2)(SO_4)_2$, $(As_2O)(SO_4)_2$ and $As(SO_4)_3$.[37] X-Ray studies are listed in Tables 5 and 6. The crystalline compounds consist of AsO_3 pyramids and SO_4 tetrahedra linked by one, two or three oxygen bridges (**12–14**).

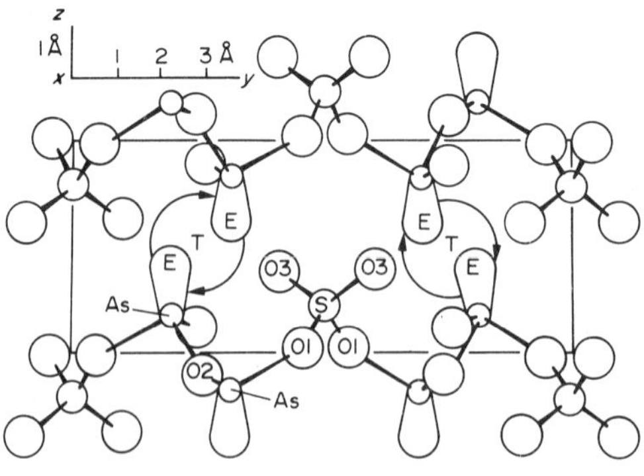

(**12**) $(As_2O_2)(SO_4)$

The lone pair on arsenic is sterically active and the arsenic moiety is considered to be AsO_3E (E = lone pair) rather than merely AsO_3. Analysis of the X-ray data[37] enables the definition of

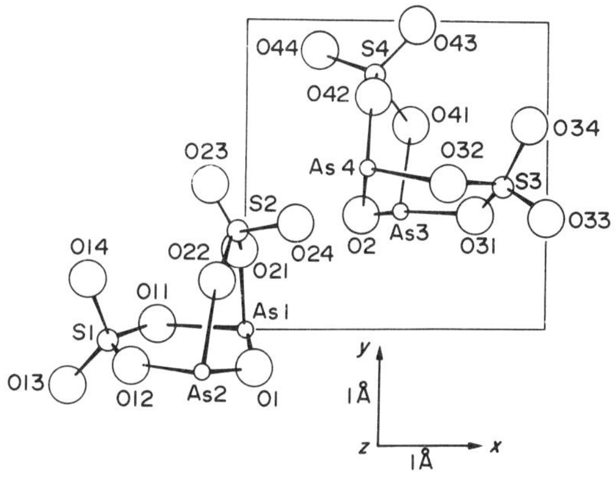

(13) $(As_2O)(SO_4)_2$

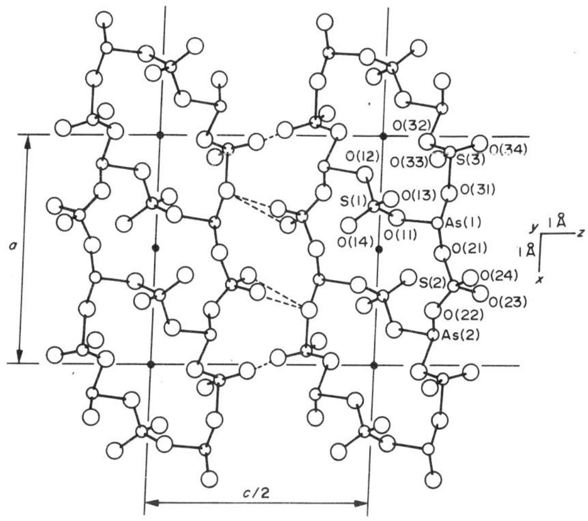

(14) $As_2(SO_4)_3$

Table 5 Arsenic(III) Sulfates

Compound				
Sulfate formula	*Mixed oxide formula*	*Crystal form*	*D*	*No of SO_4 units in contact with AsO_3*
$(As_2O_2)(SO_4)$	$As_2O_3{\cdot}SO_3$	Orthorhombic	—	1
$(As_2O)(SO_4)_2$	$As_2O_3{\cdot}2SO_3$	Monoclinic	3.23	2
$As_2(SO_4)_3$	$As_2O_3{\cdot}3SO_3$	Monoclinic	3.05	3

Table 6 Arsenic(III) Sulfates, Bond Lengths (Å) and Angles (°)

Compound	*d[As—O(As)]*	*d[(As—O(S)]*	*d(S—O$_{br}$)*[a]	*d(S—O$_t$)*[b]	*d(As—E)*[c]	*O—As—O*
$(As_2O_2)(SO_4)$	1.78, 1.75	1.89	1.51	1.44	1.15	—
$(As_2O)(SO_4)_2$	1.75 (av)	1.86 (av)	1.53	1.42 (av)	1.25	95 (av)
$As_2(SO_4)_3$	—	1.83 (av)	1.54	1.42	1.23	90

[a] br = bridging. [b] t = terminal. [c] E = lone pair.

the lone pairs in the structures, including estimates of d(As—E) (Table 6). This was considered especially important in the case of $(As_2O_2)(SO_4)$, where the chain structure of the As_2O_2 enables the lone pairs to form tunnels running along the chains and between the sheets. The possibility of utilizing these tunnels for intercalation complexes for electrical applications is being currently investigated.

28.5.2 Sulfur Ligands

28.5.2.1 Thioarsenious acid derivatives

The free acid, $As(SH)_3$, is unknown, but the series of salts M_3AsS_3 is well known; many are minerals. A few examples are listed in Table 7.

Table 7 Some Thioarsenate Compounds

Compound	*d(As—O)* (Å)	*d(As—S)* (Å)	*Ref.*
K_3AsS_4	—	2.165	247
$Na_3AsS_4\cdot 8D_2O$	—	2.145–2.173	248
$Na_3AsO_2S_2\cdot 11H_2O$	1.69	2.14	249
$Na_3AsO_3S\cdot 7H_2O$	1.67	2.14	250
$(NH_4)_3AsS_4$	—	2.20–2.26	251
Cu_3AsS_4	—	2.18	252

The thioarsenite ion is pyramidal[38] in Na_3AsS_3: d(As—S) 2.25 Å, S—As—S 109.9°. The AsS_3^- are stacked to give six sulfur neighbours for Na ions of variable bond distances.

The metathioarsenite ion in $NaAsS_2$ has been shown[39] to be a linear corner-sharing pyramidal polymeric ion (**15**). Similar polymeric structures have been shown for $PbAs_2S_4$ and $TlAsS_2$.

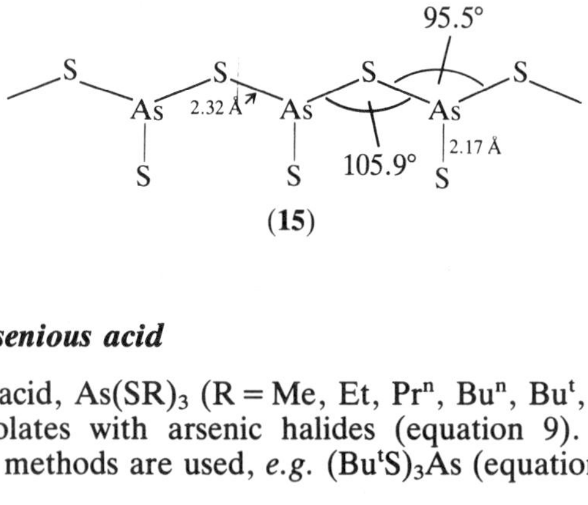

(**15**)

28.5.2.2 Esters of thioarsenious acid

Esters of thioarsenious acid, $As(SR)_3$ (R = Me, Et, Pr^n, Bu^n, Bu^t, Ph), may be obtained by reacting alkali metal thiolates with arsenic halides (equation 9). In difficult cases (steric hindrance) more vigorous methods are used, *e.g.* $(Bu^tS)_3As$ (equation 10).

$$3RSNa + AsX_3 \longrightarrow (RS)_3As + 3NaX \qquad (9)$$

$$AsF_3 + 3Me_3SiSBu^t \longrightarrow (Bu^tS)_3As + 3Me_3SiF \qquad (10)$$

Structurally, the esters are pyramidal and an interesting X-ray study of $(PhS)_3As$ has revealed[40] the structure shown in (**16**). The sulfur atoms are in a plane above the arsenic, which

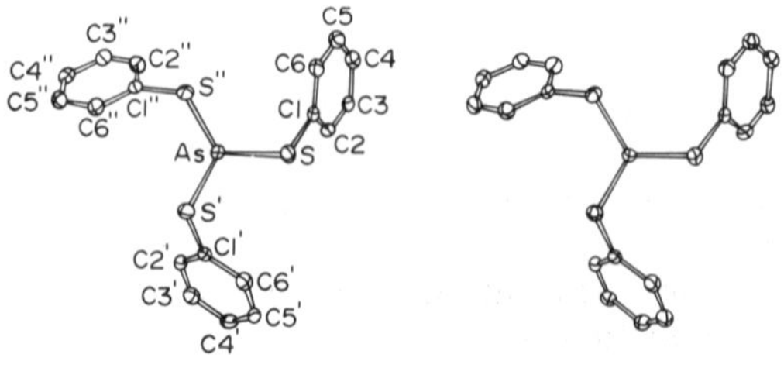

(**16**) Stereoscopic *ORTEP* drawing of $(PhS)_3As$

together with the phenyl rings shield the arsenic atoms (**17**). In solution or in liquid phase the arsenic would be more exposed and available for ligand bond formation with metals, because of thermal inversions at both S and As.

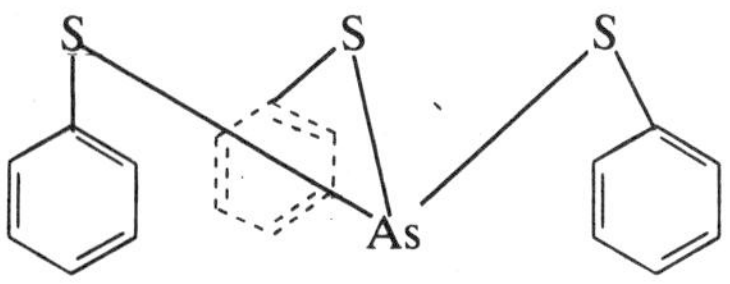

(**17**) View at right angles to sulfur plane

The structure of 2-chloro-1,3,6,2-trithioarsocane shows two interesting features.[41] There is a transannular S–As interaction resulting in the stabilizing of one of the ring conformations. This also results in a pseudotrigonal bipyramidal structure (**18**). Axial bond lengths are As—Cl 2.36 Å and As···S 2.72 Å; equatorial bond lengths are As—S 2.25 and 2.26 Å. Electron donation from the S lone pairs interact with the As *d* orbitals, resulting in sp^3d hybridization and TBP geometry. Raston and White[42] have shown that $As(S_2CNEt_2)_3$ is a six-coordinate arsenic(III) compound (**19**). The crystals are monoclinic and the geometry is grossly distorted towards C_3 symmetry. The mean As—S bond lengths are 2.84 and 2.34 Å. In free Et_2NCS_2 the C—S bonds are of equal length, but when coordinated to As the C—S bonds are 1.68 and 1.76 Å, respectively.

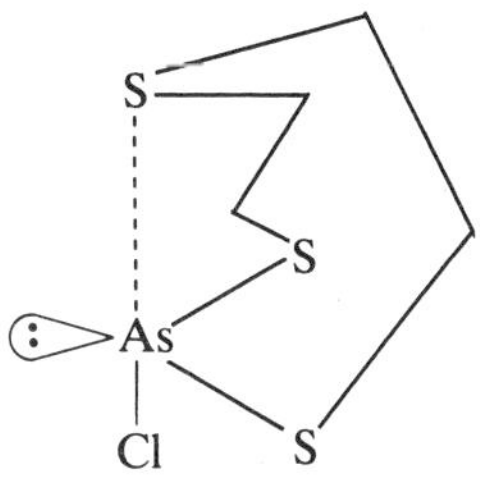

(**18**)

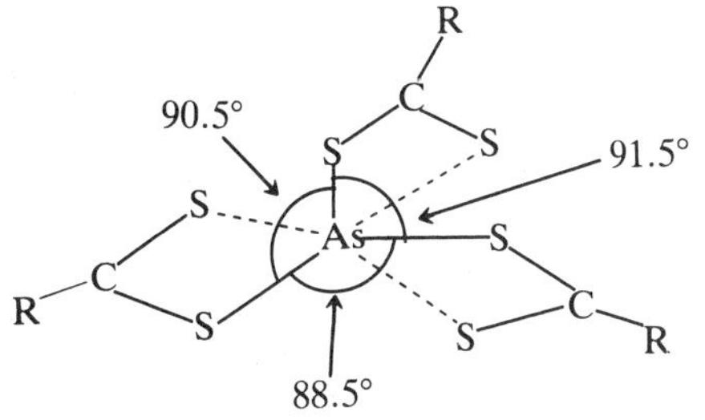

(**19**)

28.5.2.3 Trithioarsenate salts (As^V)

The parent acid H_3AsS_4 has not been isolated, but the partly oxidized acids $H_3AsO_2S_2$ and H_3AsO_3S have been isolated[43] at low temperatures. The thioarsenate ion has the expected four-coordinate tetrahedral geometry. The $(AsO_{4-n}S_n)^{3-}$ anions are well known.

28.5.3 Selenium Ligands

28.5.3.1 Selenoarsenous acid and its salts

Selenoarsenous acid, $As(SeH)_3$, exists only as its salts. Matsumoto and co-workers[44] solved the X-ray crystal structure of silver orthoselenoarsenite: the As atom is pyramidally linked to three Se in $AsSe_3$ units. The silver atoms are bonded to Se making —Se—Ag—Se—Ag— infinite spiral chains; for the $AsSe_3$ pyramids *d*(As—Se)(av) is 2.411 Å, Se—As—Se are 9.81° (av); regarding silver *d*(Se—Ag) (av) are 2.548 Å, 2.527 Å, As—Se—Ag is 107.5°, 97.4°, Se—Ag—Se is 158.8°, Ag—Se—Ag is 79.9°.

28.5.3.2 Selenoarsenous acid esters

The esters of $As(SeH)_3$ are more difficult to prepare than the oxygen or sulfur analogues. They are all either solids or high boiling oils, and their structures are almost certainly pyramidal. Darmadi *et al.*[45] prepared the trifluoromethyl ester using $Hg(SeCF_3)_2$, which is a yellow oil (m.p. −31 °C) (equation 11).

$$3Hg(SeCF_3)_2 + 2AsBr_3 \longrightarrow 2As(SeCF_3)_3 + 3HgBr_2 \qquad (11)$$

Anderson *et al.* used Me₃SiSeMe as the selenating reagent to give high yields (94–98%) of the arsenic esters (equation 12).

$$R_nAsCl_{3-n} + (3-n)Me_3SiSeMe \longrightarrow R_nAs(SeMe)_{3-n} + (3-n)Me_3SiCl \qquad (12)$$

28.5.3.3 Six-coordinate arsenic–selenium compounds

Manoussakis *et al.*[46] have synthesized a series of selenocarbamate six-coordinate complexes of arsenic (**20** and Table 8; *cf.* As—S analogues). By reacting the piperidyl derivative with halogens dissolved in CCl₄ a series of haloselenato carbamate complexes are produced which are probably pentacoordinate (equation 13).

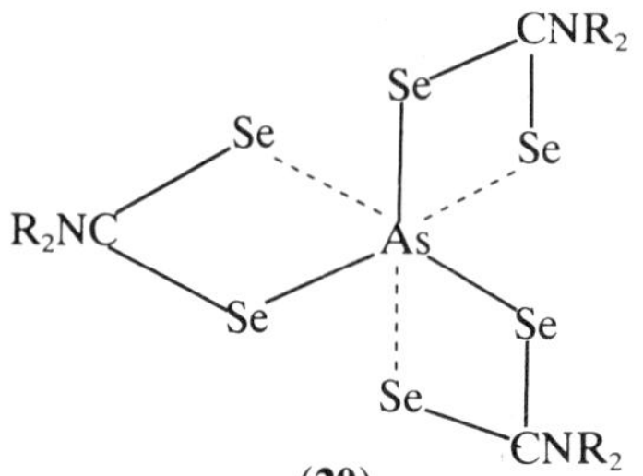

(20)

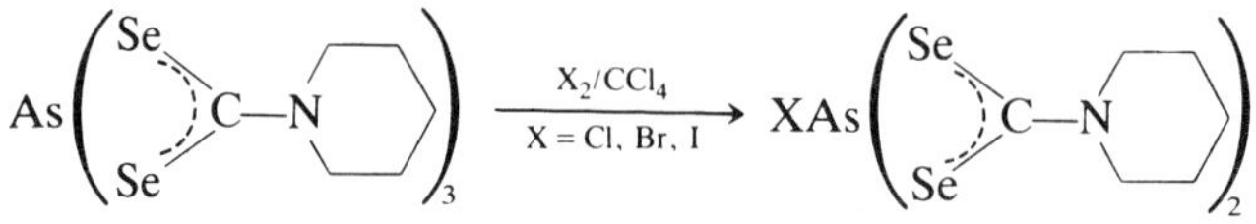

$$As\left(\underset{Se}{\overset{Se}{\diagup}}\hspace{-0.3em}C{-}N\right)_3 \xrightarrow[X = Cl, Br, I]{X_2/CCl_4} XAs\left(\underset{Se}{\overset{Se}{\diagup}}\hspace{-0.3em}C{-}N\right)_2 \qquad (13)$$

Table 8   Diselenocarbamate Complexes of Arsenic(III)

Compound	Structure	Colour	M.p.[a] (°C)
Tris(diethyldiselenocarbamato)arsine	$As\left(\overset{Se}{\underset{Se}{\diagup}}CNEt_2\right)_3$	Yellow	205 (d)
Tris(diisobutyldiselenocarbamato)arsine	$As\left(\overset{Se}{\underset{Se}{\diagup}}CNBu^i_2\right)_3$	Orange	157–9(d)
Tris(dibenzyldiselenocarbamato)arsine	$As\left(\overset{Se}{\underset{Se}{\diagup}}CN(CH_2Ph)_2\right)_3$	Yellow	192–4 (d)
Tris(piperidylselenocarbamato)arsine	$As\left(\overset{Se}{\underset{Se}{\diagup}}CN\bigcirc\right)_3$	Pale yellow	234 (d)

[a] d = decomposed.

28.6 ARSENIC–HALOGEN COMPOUNDS

28.6.1 Three-coordinate Arsenic(III) Halides

Various transient species containing arsenic in low CN with halogens have been recorded spectroscopically, *e.g.* AsI₂ from the photodissociation of AsI₃.[47] The only stable species is the red crystalline iodide, As₂I₄ (m.p. 260 °C),[48] which is decomposed by air or moisture. The

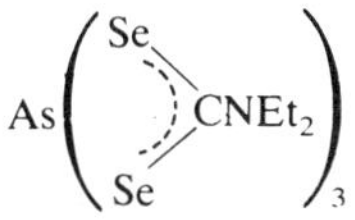

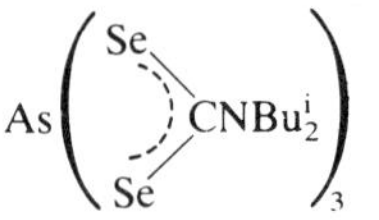

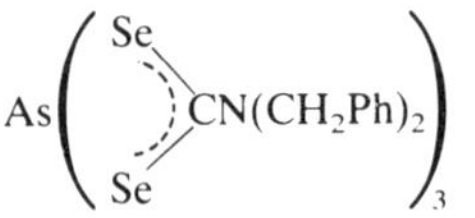

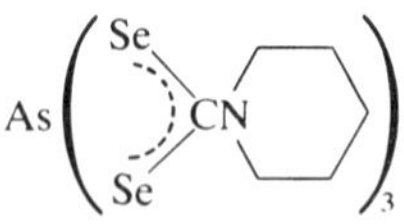

sub-iodide disproportionates when heated above 200 °C (equation 14).[49]

$$3As_2I_4 \xrightleftharpoons{300-350\,^{\circ}C} 4AsI_3 + 2As \quad (14)$$

The arsenic trihalides have been shown by a variety of techniques (Table 9) to be pyramidal molecules in the gas phase. The X—As—X angles increase from X = F 96° to X = I 100°. The *d*(As—X) values increase with increase of temperature,[53] and the X—As—X angle changes in proportion.

Table 9 Arsenic–Halogen Bond Lengths and Angles

Compound	*d(As—X)* (Å)	*d(As—E)* (Å)	*X—As—X* (°)	*Ref.*
AsF_3(g)	1.706		96.2	49
$MeAsF_2$(g)	1.74		96 ± 4	50
AsF_5(g)	1.711 (ax) 1.656 (eq)			49
$AsCl_3$(g)	2.161		98.4	51
$AsBr_3$(g)	2.324		99.8	52
$AsBr_3$(g), 373 K	2.324		99.64	53
$AsBr_3$(g), 466 K	2.328		99.59	
$AsBr_3$(cryst)	2.36		97.7	54
AsI_3(g), 503 K	2.577		100.2	55
AsI_3(cryst)	2.591 (intramol) 3.467 (intermol)		99.67	56
$KAsF_6$	1.719 (av)		90.0 (av)	57
$[S_8]^{2+}[AsF_6]_2^-$	1.67 (av), range 1.59–1.72		171.4–178.8 86.1–92.0	58
$AsF_5{\cdot}N(Me)SOF_2$	1.695 (av)	(As–N) 1.985	174.3 91.5 (av) S—N—As, 124	59
$M_2(As_2F_8O_2)$ (M = K, Rb, Cs)	1.739 (av)	(As—O) 1.80	O—As—O, 84.9 (av)	60a
$Me_3N{\cdot}AsCl_3$	2.385 (ax) 2.187 (eq)	(As—N) 2.286	96.1 (av) N—As—Cl (ax), 178.7 N—As—Cl (eq), 88.4	60b
$[PPh_4][As_2SCl_5]$	2.71 (av) (Cl_{br}) 2.32 (av) (ax) (Cl_t) 2.20 (av) (eq) (Cl_t)	(As—S) 2.23 (av)	170 (av) 93.9 (av) Cl—As—S, 85–101 As—S—As, 104.0	61
$[PPh_4][As_2Cl_6]{\cdot}$ $C_2H_4Cl_2$	2.81 (av) (br) 2.31 (av) (Cl_t)	(As—S) 2.26 (av)	173 (av) 90 (av) Cl—As—S, 84–95 As—S—As, 93.5	61
$M_2(As_2F_{10}O)$ (M = K, Rb, Cs)	1.72 (av)	(As—O) 1.72, 1.77	As—O—As, 136.5 (av)	62
$[AsCl_4][AsF_6]$	(As—Cl) 2.00 (As—F) 1.72		Cl—As—Cl, 109.6 F—As—F, 90	63
$[Me_3AsBr]Br$	2.15	(As—C) 1.99		64
Me_3AsCl_2	2.45	(As—C) 1.93		64
Ph_3AsF_2	1.834	(As—C) 1.93	F—As—F, 177.9 C—As—C, 120.2 (av)	65
$[AsCl_4]^+[SbCl_6]^-{\cdot}$ $AsCl_3$	(As^V) 2.05 (av) (As^{III}) 2.075 (av)	(Sb—Cl) 2.36	Cl—As^V—Cl, 109.5 (av) Cl—As^{III}—Cl, 97, 106	66

Crystalline arsenic triiodide consists of discrete AsI_3 molecules, which are stacked such that the iodine atoms are hexagonally close packed. The arsenic atoms are octahedrally surrounded by iodine, but with the arsenic off-centre resulting in two sets of *d*(As—I) (Table 9); three bonds of 2.59 Å and three of 3.47 Å (*cf.* AsI_3 vapour, *d*(As—I) 2.56 Å). Arsenic(III) triastatine has not yet (mid-1986) been reported, although it would be surprising if the synthesis has not been attempted.

28.6.2 Reactions of Arsenic(III) Halides

Arsenic(III) halides are not salt-like compounds but typically non-metallic halides comparable to, for example, $GeCl_4$ and SCl_2 besides their predecessors in Group 5, PX_3. However $AsCl_3$ is not as reactive to nucleophilic attack as is PCl_3.

Arsenic trichloride is a Lewis acid, shown by its formation of adducts with bases, *e.g.* $Me_3N \cdot AsCl_3$ (**21**), a trigonal bipyramid with a stereoactive lone pair on arsenic.[60b] The *d*(As—N) is exceptionally long and may be considered to have some ionic character.

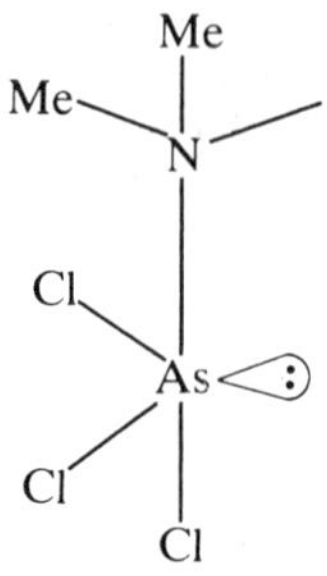

(**21**)

28.6.3 Arsenic(III) Pseudohalides

The three known examples of arsenic(III) pseudohalides are $As(CN)_3$,[67] $As(NCO)_3$[68] and $As(NCS)_3$;[69] the thiocyanate is unstable at room temperature. The structure of $As(CN)_3$ (**22**) shows the CN group to be carbon bonded to form a pyramidal $As(CN)_3$ unit. These pyramids are linked together by long As· · ·N 'bonds' to form infinite chains.

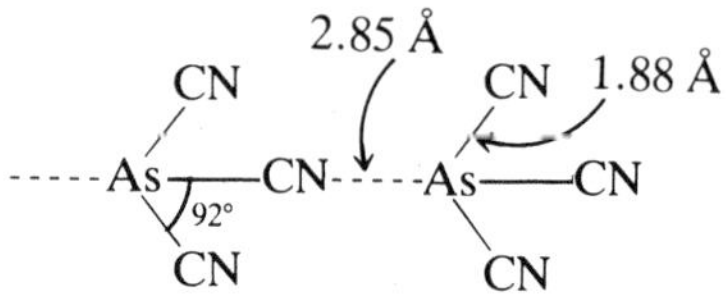

(**22**)

28.6.4 Arsenic(V) Halides

28.6.4.1 Arsenic pentafluoride

Arsenic pentafluoride (b.p. −52.8 °C) is the only stable well-characterized arsenic(V) halide. Electron diffraction[49] shows it to be a trigonal bipyramidal molecule, with equatorial *d*(As—F) 1.656 Å and axial *d*(As—F) 1.711 Å; ^{19}F NMR shows that the five fluorines are equivalent. The apparent inconsistency displayed by the two techniques is due to the differing time scales of the measured effects. On the slower NMR time scale the axial–equatorial bonds 'rotate' *via* a square pyramidal intermediate resulting in the observed equivalence (pseudorotation).

Arsenic pentafluoride is a Lewis acid capable of forming adducts with basic molecules, *e.g.* (**23**), the adduct with *N*-methyl-*S*,*S*-difluorosulfoximine.[59] Neglecting the methyl group there is a mirror plane through F(ax),As,N,S,O. The methyl C atom lies *ca.* 0.05 Å out of this plane;

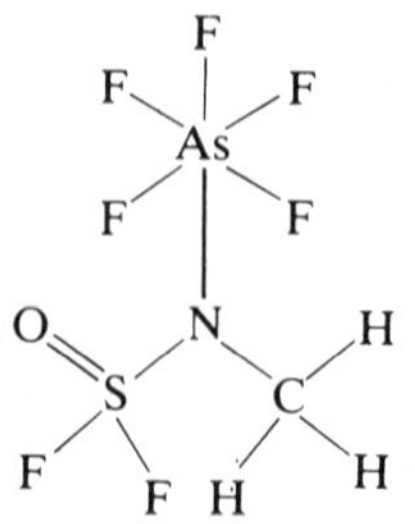

(**23**)

the d(As—N) of 1.985 Å is longer than the normal d(As—N) (see Section 28.4.1), but the bond is shorter than that in the $AsCl_3{\cdot}NMe_3$ adduct, d(As—N) 2.286 Å.

Arsenic pentafluoride, although outclassed as a 'superacid' component by SbF_5, has still been widely used.[70] Arsenic pentafluoride is a powerful halide acceptor and this property has been utilized in stabilizing various cationic species, *e.g.* $[ClO_2]^+[AsF_6]^-$,[71] $[S_8]^{2+}[AsF_6]_2^-$[58] and $[NS]^+[AsF_6]^-$.[72] The arsenic in AsF_6^- is octahedral, but this is often distorted by the cation field in the crystal unit cell. In the case of $[S_8][AsF_6]_2$ the average d(As—F) is 1.67 Å, *cf.* $KAsF_6$ value of 1.72 Å,[57] but the range of d(As—F) is 1.59–1.72 Å in the four AsF_6^- ions in the crystallographic unit cell. The preparation of several $[AsF_6]^-$ compounds, including the oxidizing agent nitrosyl hexafluoroarsenate $[NO]^+[AsF_6]^-$, is described in detail in 'Inorganic Synthesis', volume 24 (1986).

28.6.4.2 Arsenic pentachloride

While both phosphorus and antimony pentachlorides are stable, $AsCl_5$ is unstable. This anomalous behaviour of arsenic is attributed to d-block contraction; the highest valencies of the p-block elements of the post 3d transition elements are, in general, unstable with respect to other members of their group. In the case of arsenic trichloride the lowering of the energy of the 4s orbital makes the promotion of the $4s^2$ electrons, for further As—Cl bonds to form $AsCl_5$, more difficult. The As—Cl bond strength *per se* in $AsCl_5$ is not thought to be the cause of the molecule's instability.[73] Because of this inherent instability $AsCl_5$ was thought to be one of the so-called non-existent compounds, until the successful proof in 1976 by Seppelt.[74] $AsCl_3$ in liquid chlorine was irradiated with UV light at −150 °C to give $AsCl_5$, a pale yellow solid (m.p. *ca.* −50 °C with decomposition) which was shown by Raman spectroscopy to have a trigonal bipyramidal structure (equation 15).

$$AsCl_3 + Cl_2 \underset{-50\,^{\circ}C}{\overset{-150\,^{\circ}C,\ h\nu}{\rightleftharpoons}} AsCl_5 \qquad (15)$$

28.6.4.3 Arsenic(V) mixed halides

In the solid state[63] $AsCl_2F_3$ is, in fact, $[AsCl_4]^+[AsF_6]^-$; the $[AsCl_4]^+$ tetrahedron has d(As—Cl) of 2.00 Å and Cl—As—Cl of 109.3° and 109.8°, very near the ideal tetrahedral angles. The $[AsF_6]^-$ bond lengths vary, *viz.* 1.71, 1.68, 1.78 Å, with F—As—F of 90.5° and 89.5°. The arsenic centres in the polyhedra are between 4.84 and 6.03 Å apart.

When $[AsCl_4][AsF_6]$ is pyrolyzed, the mixed halide $AsClF_5$ is formed in preparative quantities.[75] IR studies indicate the chlorine to be in an equatorial position, as expected from its lower electronegativity.

The chlorination of $SbCl_5$ in $AsCl_3$ solution at 80 °C results in the crystallization of a mixed As^V–Sb^V chloride (equation 16). This has been shown by X-ray analysis[66] to be $[AsCl_4]^+[SbCl_6]^-{\cdot}AsCl_3$. The $AsCl_3$ 'solvate' is slightly distorted, Cl—As—Cl of 97° and 106° (*cf.* 98.5° in free $AsCl_3$). The arsenic tetrahedron and the antimony octahedron are only slightly distorted from the ideal angles (Table 9).

$$SbCl_5 + AsCl_3 + Cl_2 \xrightarrow{AsCl_3} AsCl_5{\cdot}SbCl_5{\cdot}AsCl_3 \qquad (16)$$

28.6.4.4 Complex arsenic(V) chloride anions

Although arsenic(V) chloride is unstable, the derivative anions are well known, though not as widely studied or used as the corresponding fluorides. In $Cs_3As_2Cl_9$ the highly distorted $AsCl_6$ octahedron has two types of As—Cl bonds, d(As—Cl) at 2.25 and 2.75 Å, the latter suggesting an almost ionic bond.[217]

In the thioarsenic halide anions $[As_2SCl_5]^-$ and $[As_2SCl_6]^{2-}$ the As atoms are bridged by S and Cl.[61] In the former the chlorine bridge forms an axial bond with each arsenic atom, which has trigonal pyramidal geometry **(24)**.

In $[As_2SCl_6]^{2-}$ the arsenic atoms are bridged by two chlorines and the sulfur ligand; the arsenic atoms are five-coordinate and have distorted octahedral geometry (**25**). The stereoview shows the unit cell packing of $[Ph_4P][As_2SCl_5]$ and the trigonal bipyramidal disposition of the arsenic ligands is well displayed (**26**).

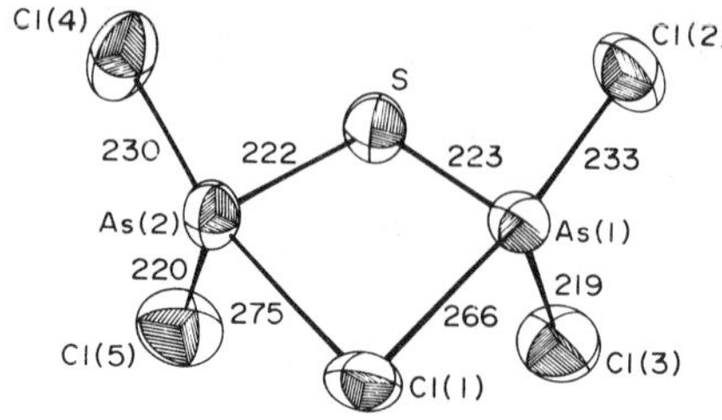

(**24**)

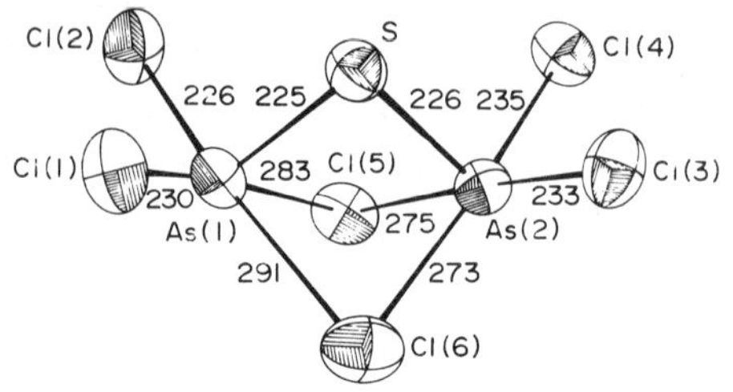

(**25**)

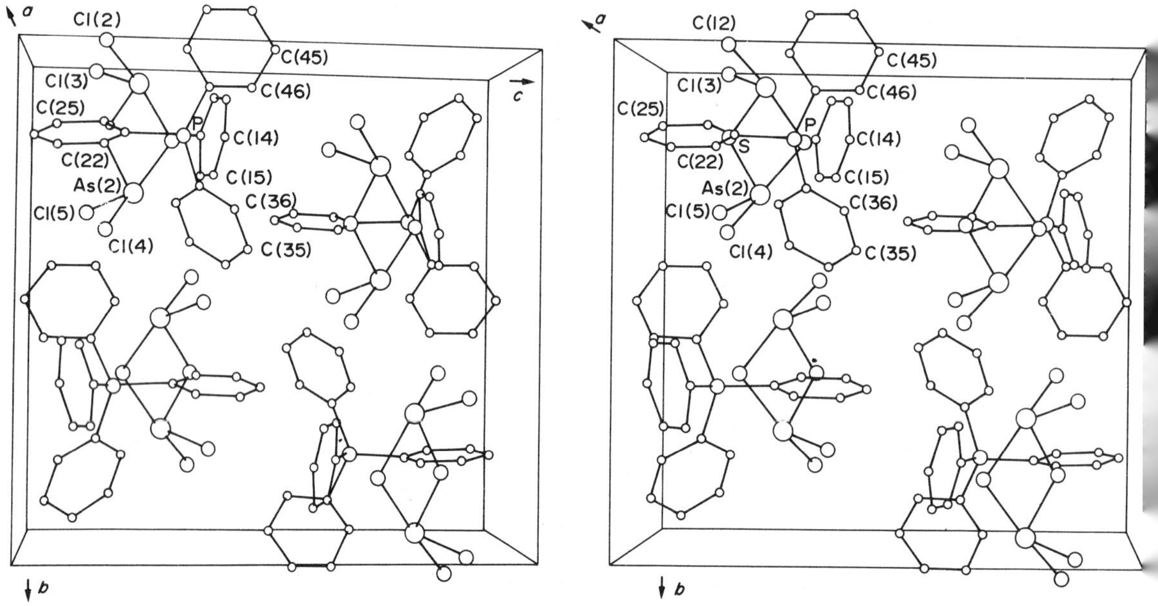

(**26**)

28.6.4.5 *Arsenic(V) oxyfluorides*

A number of oxygen-bridged arsenic(V) fluoride anions have been studied. In $[As_2F_{10}O]^{2-}$ two AsF_5 groups are bridged[62] by oxygen with an As—O—As angle of 138° and d(As—O) 1.75 Å (av) (**27**). There are small changes in bond lengths and angles when the cation is changed (Table 10).

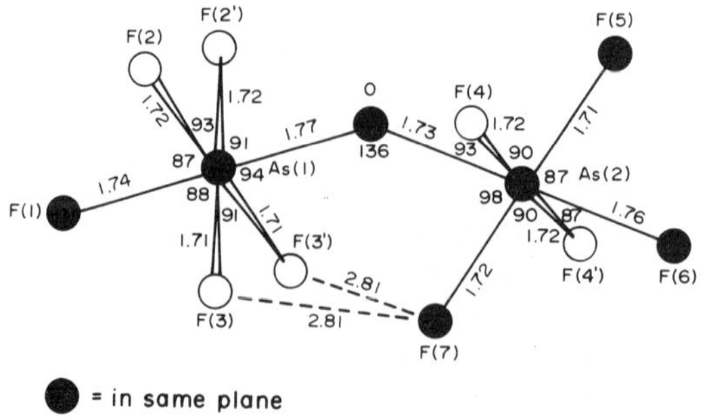

(**27**) $(As_2F_{10}O)^{2-}$anion in $Rb_2(As_{10}F_{10}O)$

In the $[As_2F_8O_2]^{2-}$ anion[60] there are two bridging oxygens forming a four-membered $(As—O)_2$ ring (**28**) with As—O—As of 96° and O—As—O of 84.9° (av), d(As—O) 1.82 Å (av) (Table 10).

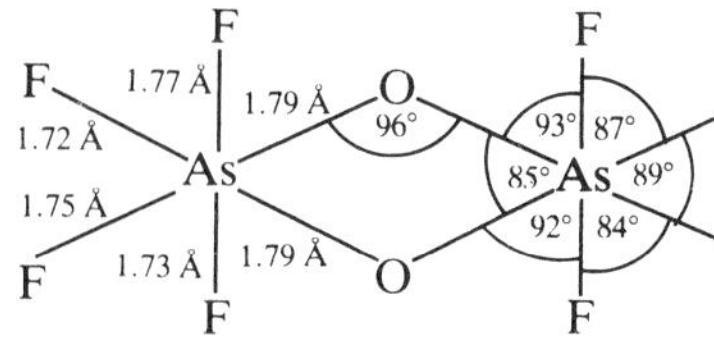

(**28**)

Table 10 Crystallographic Data for Some $[As_2F_{10}O]^{2-}$ and $[As_2F_8O_2]^{2-}$ Derivatives

Compound	*d(As—F)* (Å)	*d[As(1)—O]* (Å)	*d[As(2)—O]* (Å)	*As—O—As* (°)
$Cs_2[As_2F_{10}O]$	1.74	1.77	1.68	138.8
$Rb_2[As_2F_{10}O]\cdot H_2O$	1.72	1.77	1.73	136.1
$K_2[As_2F_{10}O]\cdot H_2O$	1.70	1.76	1.71	136.9
Average	1.72	1.77	1.71	137.3
				O—As—O (Å)
$Cs_2[As_2F_8O_2]$	1.74	1.79	—	83.7
$Rb_2[As_2F_8O_2]$	1.73	1.81	—	84.9
$K_2[As_2F_8O_2]$	1.74	1.81	—	86.0
Average	1.739	1.80	—	84.9

28.7 BIOLOGICAL ASPECTS OF ARSENIC

28.7.1 Toxicity of Arsenic

Trivalent arsenic compounds are generally more toxic than the pentavalent compounds; the former compounds inactivate enzymes containing —SH groups. However, arsenate, AsO_4^{3-}, is an analogue of phosphate, PO_4^{3-}, and if present in the body acts as an uncoupler in the oxidation of glyceraldehyde 3-phosphate by forming the highly labile acyl arsenate, 1-arseno-3-phosphoglycerate.[76] Because of its high toxicity the maximum acceptable concentration for an 8 h period has been set at 0.05 p.p.m. for arsine, AsH_3.[77] The lethal human dose is less than 0.25 g As.

In a notorious case in Manchester in 1900–1901 6000 people were affected by drinking beer containing 15 p.p.m. arsenic; 70 people died as a result. The arsenic originated from sulfuric acid containing 1.4% arsenious acid, produced from As-containing iron pyrites. This sulfuric acid was used in a process to manufacture glucose for the brewers, and the glucose contained several hundred parts per million arsenic.

28.7.2 Arsenic in the Environment

Arsenic compounds occur naturally and ubiquitously in nature. The primary sources result from volcanic action and stream waters from arsenic rock strata.

Severe local arsenic pollution can occur adjacent to industrial enterprises producing the various arsenicals used in timber preservation and as agricultural pesticides. Concern expressed by environmental groups has led to a reduction in the use of arsenicals in agriculture over the last decade, and this trend is likely to continue. Waste disposal is the main problem in industrial arsenic production and this problem has been responsible for closing several plants in Western Europe.

28.7.3 Arsenic in Peace and War

Earlier in this century the toxicity of arsenic was utilized in medicine and, later, in warfare. In 1909 Ehrlich developed salvarsan, the first effective antisyphilis agent and one of the first systematically screened chemotherapeutic agents. While salvarsan was replaced in clinical use by the antibiotics in the 1940s, some specialist arsenical drugs, *e.g.* glycobiarsol (**29**), are still the treatment of choice for amoebic infections.

Amongst the toxic gases, liquids and powders used in World War I there were a few arsenicals. Diphenylchloroarsine, Ph_2AsCl, and diphenylcyanoarsine, Ph_2AsCN, were used as sternutators (irritants to nose and throat, inducing feeling of choking); 10-chloro-5,10-dihydrophenarsazine (**30**), an 'improvement' came 'too late' for use in World War I. Also 'too late' was the vesicant Lewisite, *trans*-ClCH=CHAsCl$_2$.

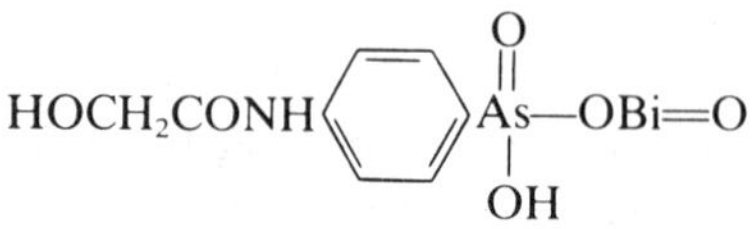

(**29**)

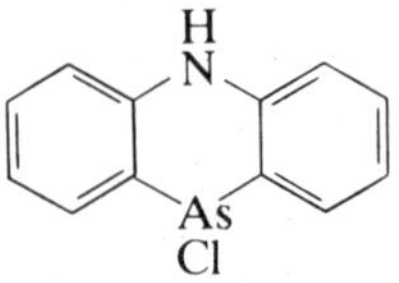

(**30**)

28.7.4 Arsenic in Life

Until recently there was only one word to describe arsenic in relation to life and that was 'lethal'. However, recent work has revealed a new class of bioelements. Beside the bulk elements C, H, O, Ca, P, *etc.* and the trace elements such as Fe, Zn, Br, *etc.*, a new group, the ultratrace elements, has been defined.[78] Among this new class of essential elements is arsenic. Arsenic deficiency signs in experimental animals (fed on rigorously arsenic free diets) include impairment of growth, reproductive and heart function.[79] The detailed role of ultramicrotrace elements is one of the challenges now being addressed by biochemists. It will not be without significance that certain lower marine life forms exhibit a high tolerance for arsenic: up to 175 p.p.m. for prawns, and trimethylarsine has been detected as an off-flavour component of prawns.[80] Arsenic is accumulated in crustacea mainly as the non-toxic arsenobetaine,[81] but this is not the source of the trimethylarsine.

28.8 ANTIMONY

Antimony, the fourth member of the pnictide family, is the only one to exhibit substantial natural isotope variability. The electronic structure of antimony is $1s^22s^22p^63s^23p^63d^{10}4s^24p^64d^{10}5s^25p^3$; the Group 5 outer electronic structure of $5s^25p^3$ presages its main valencies of three and five. The CN in its compounds is variable, with six becoming much more common. Simple theory predicts $5p^3$ bonding for Sb^{III} compounds with pyramidal bonds at 90°. In fact three-coordinate antimony(III) compounds have variable angles, usually >90°, indicating a stereoactive lone pair and involving some *s* character in the *p* bonding. The trigonal pyramidal shape for SbX_3 compounds has been shown frequently by X-ray crystallographic studies. Antimony(V) compounds usually involve sp^3d-hybridized orbitals, resulting in trigonal bipyramidal geometry in its pentacoordinate complexes. In six-coordinate compounds sp^3d^2 hybridization is generated, *e.g.* in SbF_6^-. In 'square pyramidal' coordination a stereochemically active lone pair occupies an octrahedral axial position.

28.9 ANTIMONY–GROUP 4 BONDS

The chemistry of antimony is generally less well developed than that of arsenic and particularly in respect of Group 4 ligands. For the purposes of this discussion organoantimony compounds will be dealt with first and those of silicon, germanium, tin and lead will be taken together.

28.9.1 Carbon Ligands

The synthetic routes developed in arsenic chemistry usually also apply to antimony, but they have not been as fully investigated. A good outline of organoantimony chemistry has been given by Wardell[82] and an excellent compilation of data and synthetic methods may be found in Gmelin (now in English).[83] For German readers Houben–Weyl will prove useful.[84]

28.9.1.1 Antimony(III) compounds

Stibines, SbR_3, are trigonal pyramidal molecules with C—Sb—C ≈ 90°, but this angle is sensitive to electronic and steric factors, *e.g.* in $Sb(CF_3)_3$, *d*(Sb—C) is 2.202 Å and C—Sb—C is 100°.[85]

Trivinylstibine has been shown to be trigonal pyramidal by IR and Raman spectroscopy,[86] and the bis(stibine) $(Me_2Sb)_2CH_2$ has been shown[87] by NMR to be a mixture of structures with C_s and C_{2v} symmetries (**31**). The halostibine Ph_2SbF[88] is of interest because it exemplifies the tendency of antimony to increase its CN, in this case by bridging Sb···F bonds (**32**). Antimony adopts a trigonal bipyramidal geometry involving a stereoactive lone pair.

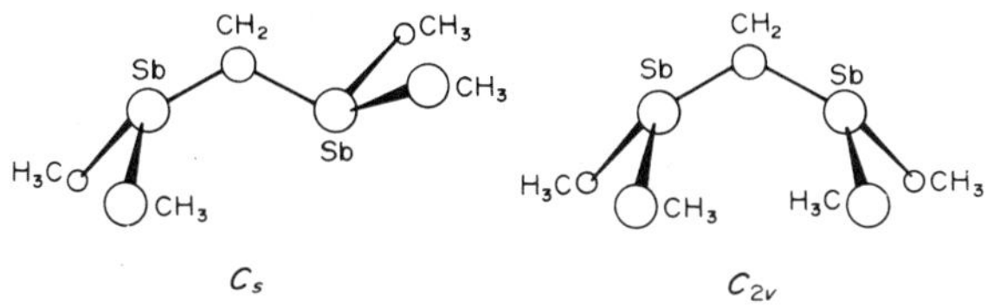

(31)

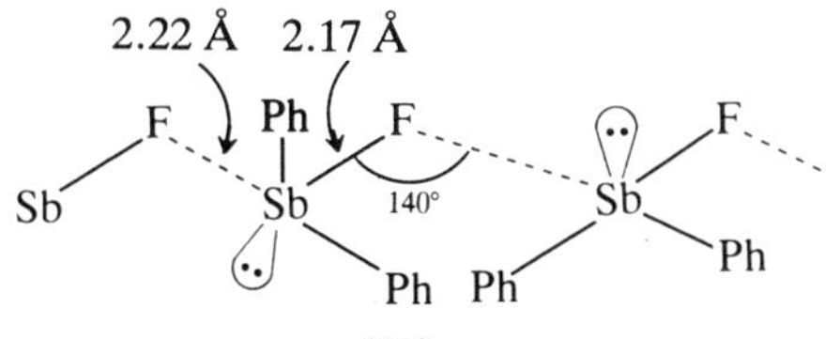

(**32**)

28.9.1.2 Antimony(V) organohalogen compounds

Stibines may be halogenated by the use of X_2 or by a halogenating agent, *e.g.* trivinylstibine may be chlorinated using sulfuryl chloride (equation 17),[86] and this may be fluorinated using potassium fluoride (equation 18). Both of these compounds are trigonal bipyramidal with apical halogens. Similarly $(Me_2Sb)_2CH_2$, when chlorinated using SO_2Cl_2, afforded the tetrachloro derivative, m.p. 175 °C.[87]

$$(CH_2{=}CH)_3Sb \xrightarrow{SO_2Cl_2} (CH_2{=}CH)_3SbCl_2 \tag{17}$$

$$(CH_2{=}CH)_3SbCl_2 \xrightarrow{KF} (CH_2{=}CH)_3SbF_2 \tag{18}$$

The trigonal bipyramidal geometry around antimony (**33**) has been shown by X-ray analysis. Some bond lengths are given in Table 11; Cl—Sb—Cl 176.75° (av), Sb—C—Sb 118.2° (rather large for a td angle about carbon).

A halogen atom may be abstracted from haloorganoantimony(V) compounds to form a cation (equations 19 and 20),[86,92] but note the reaction shown in equation (21).[92] The compound $Me_3SbCl_2{\cdot}SbCl_3$ is non-ionic, and an X-ray study has shown the $SbCl_3$ moiety to be chlorine-bridged to three Me_3SbCl_2 moieties, resulting in CNs of six and five, respectively.[92]

$$(CH_2{=}CH)_3SbCl_2 + SbCl_5 \longrightarrow [(CH_2{=}CH)_2SbCl]^+[SbCl_6]^- \tag{19}$$

$$Me_3SbCl_2 + SbCl_5 \longrightarrow [Me_3SbCl]^+[SbCl_6]^- \tag{20}$$

$$Me_3SbCl_2 + SbCl_3 \longrightarrow Me_3SbCl_2{\cdot}SbCl_3 \tag{21}$$

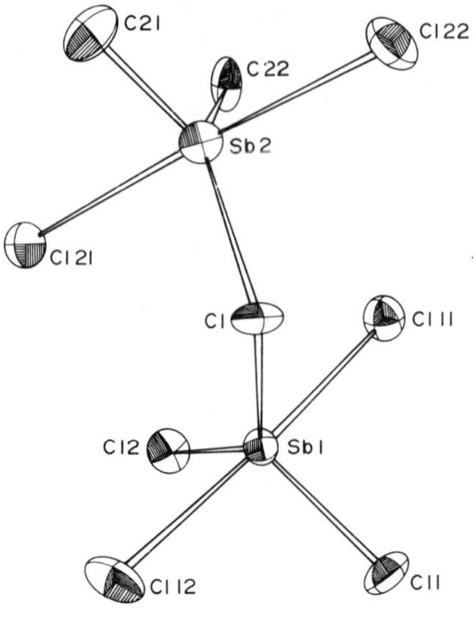

(33) $(Me_2Cl_2Sb)_2CH_2$

Table 11 Bond Lengths and Angles for Some Organoantimony Compounds

Compound	*d(Sb—C)* (Å)	*d(Sb—E)* (Å)	*C—Sb—C* (°)	*Other* (°)	*Ref.*
$Sb(CF_3)_3$	2.202	—	100	—	85
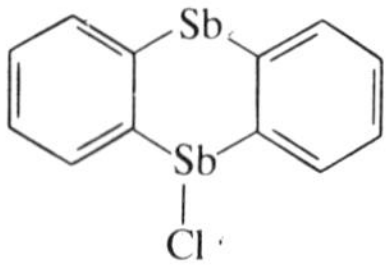	2.141	2.390 (Cl)	100.2	—	89
$CH_2(SbCl_2Me_2)_2$	2.114 (CH_2) 2.109 (Me)	2.479 (Cl)	90.1 (av)	—	87
$Sb(SiH_3)_3$[a]	—	2.557 (Si)	—	88.6[b]	95

[a] *d*(Si—H) 1.394 Å. [b] Si—Sb—Si

The tendency for antimony to achieve CN six is shown by the behaviour of organohaloantimony(V) compounds to substitution by oxo ligands. Jha *et al.*[93] have employed the bidentate ligands acetylacetone, 8-hydroxyquinoline, salicylaldehyde, *o*-hydroxyacetophenone and 2-hydroxynaphthaldehyde HL (equation 22).

$$R_3SbBr_2 + NaL \xrightarrow[R=Ph,\ Me]{C_6H_6,\ MeOH} R_3Sb(OMe)L + 2NaBr \qquad (22)$$

The complexes produced have been assigned *fac* structures, *e.g.* (**34**). The increasing stability of higher coordination and valency of antimony is shown in the organotrihaloantimony(V) compounds R_2SbCl_3 formed by the oxidative chlorination of monochlorostibines (equation 23). These compounds can be used to prepare pure dihalostibines by thermolysis (equation 24). In the case of Me_2SbCl_3, in solution the methyl groups are equatorial (C_{2v} symmetry), but they become *trans* on reaction with NMe_4Cl.[94]

$$R_2SbCl_2 + Cl_2 \longrightarrow R_2SbCl_2 \qquad (23)$$

$$R_2SbCl_3 \longrightarrow RSbCl_2 + RCl \qquad (24)$$

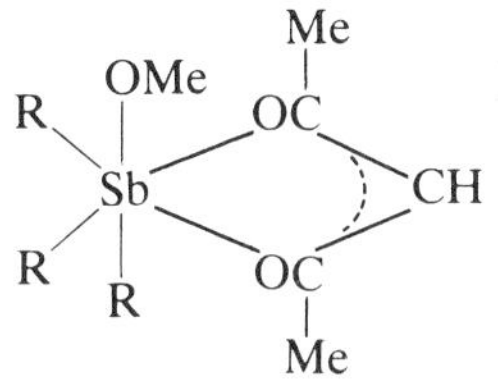

(**34**)

28.9.1.3 Other Group 4 ligands (Si, Ge, Sn and Pb)

Silicon forms a series of compounds with antimony analogous to silane. Trisilylstibine, $Sb(SiH_3)_3$, has been shown to be pyramidal by electron diffraction.[95] There was no evidence of bond angle widening, nor bond shortening, thus suggesting the absence of $(p \rightarrow d)\pi$ bonding. The silylstibines are prepared by conventional methods (equation 25).

$$Li_3Sb + 3Me_3SiCl \longrightarrow (Me_3Si)_3Sb \quad (88\%) \qquad (25)$$

The silylstibines may be used to prepare germylstibines (equation 26).

$$3Et_3GeH + (Et_3Si)_3Sb \longrightarrow 3Et_3SiH + (Et_3Ge)_3Sb \qquad (26)$$

Stannylstibines are prepared[96] by conventional routes (equation 27).

$$3Ph_3SnLi + SbCl_3 \longrightarrow (Ph_3Sn)_3Sb + 3LiCl_3 \qquad (27)$$

Tris(organoplumbyl)stibines have been prepared by condensing organolead chlorides with stibine (equation 28).[97]

$$3Me_3PbCl + 3Et_3N + SbH_3 \longrightarrow (Me_3Pb)_3Sb + 3Et_3NHCl \qquad (28)$$

The general coordination chemistry of stibines as donors has been reviewed.[91,98,99]

28.10 ANTIMONY–GROUP 5 BONDS

28.10.1 Nitrogen Ligands

The chemistry of antimony and nitrogen is not as rich as that of arsenic and nitrogen. The N—Sb bond mainly exists in compounds where N is but one of a number of donor atoms. These compounds, where significant, will be dealt with under various other headings (mainly oxo and halogen ligands). Some Sb—N bond data are assembled in Table 12.

Antimony trichloride does not react with dimethylamine to give dimethylaminostibines (compare arsenic trichloride). The more vigorous reagent Me_2NLi, however, gives a good yield[100] of tris(dimethylamino)stibine, a colourless, air- and moisture-sensitive liquid (Table 13; equations 29 and 30[103]).

$$SbCl_3 + 3LiNMe_2 \longrightarrow Sb(NMe_2)_3 + 3LiCl_3 \quad (>60\%) \qquad (29)$$

$$Me_{3-n}SbCl_n + nMe_3SiNMeLi \longrightarrow \left(\begin{matrix} Me_3Si \\ Me \end{matrix} \!\! \right\rangle \! N \Bigg)_n SbMe_{3-n} \quad (50\text{–}70\%) \qquad (30)$$

The cyclic aminochlorostibine was synthesized by Wannagat and Schlingmann[104] using milder conditions and triethylamine as HCl acceptor (equation 31).

Scherer *et al.*[105] synthesized the novel bicyclic Sb—N—P ring system (**35**; equation 32). The

Table 12 Structural Data for Antimony–Nitrogen Compounds

Compound	*CN*	*d(Sb—N)* (Å)	*d(Sb—E)* (Å)	*N—Sb—N* (°)	*N—Sb—E* (°)	*Ref.*
$Ph_3Sb(NCO)_2$	5	—	—	—	—	253
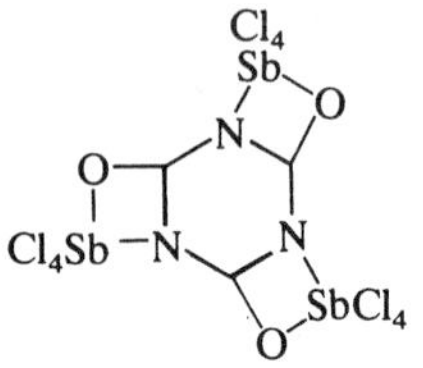	6	2.23	2.17 (O) 2.299 (Cl)	—	—	163
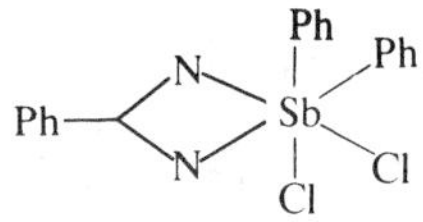	6	—	—	—	—	255
$(N_3SbPh_3)_2O$	5	2.236 (ax)	1.985 (O)	139.8[a]	178.3 (O)	256
$PhNH_2{\cdot}SbCl_3$		2.53 (ax)	2.52 (Cl, ax)	—	166.3 (Cl)	257
$C_{16}H_{36}ClN_4P_2Sb$		2.413 (ax) 2.100 (eq) 2.089 (eq)	2.492 (Cl, ax)	—	147.2 (Cl)	258

[a] Sb—O—Sb

Table 13 Aminostibines

Compound	*B.p.* (°C/Torr)	*M.p.* (°C)	*Ref.*
$Sb(NMe_2)_3$	55–61/5 32–34/0.45 mm Hg	—	100
$Sb(NEt_2)_3$	59–62/0.02	—	101
$Sb(NPr_2)_3$	107–111/0	—	101
$Me_2SbN(Me)SiMe_3$	44–46/1.0	—	102
$MeSb[N(Me)SiMe_3]_2$	59–61/0.1	—	102
$Sb[N(Me)SiMe_3]_3$	78–79/0.1	9.11	102
$Bu^t_2SbNMe_2$	25/0.3	−78	103
$Bu^t_2SbN(SiMe_3)_2$	73–75/0.5	—	103
$Me_2SiN(Et)SbN(Et)SiMe_2$	86/0.2	$d_4^{20}/1.2944$	104

X-ray analysis showed the structure to be distorted trigonal bipyramidal about antimony with axial Cl—Sb—N 147.2°.

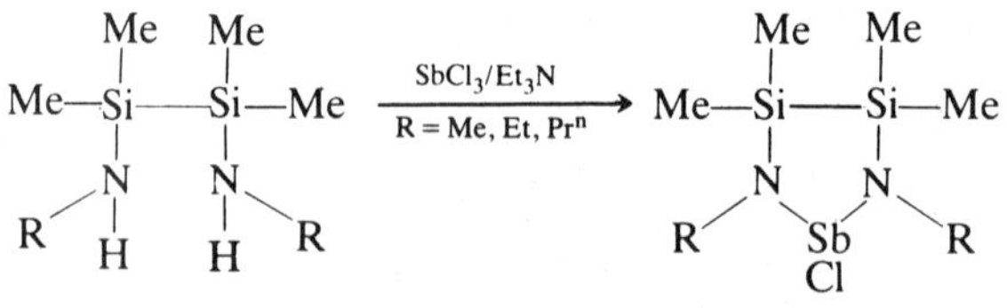

(31)

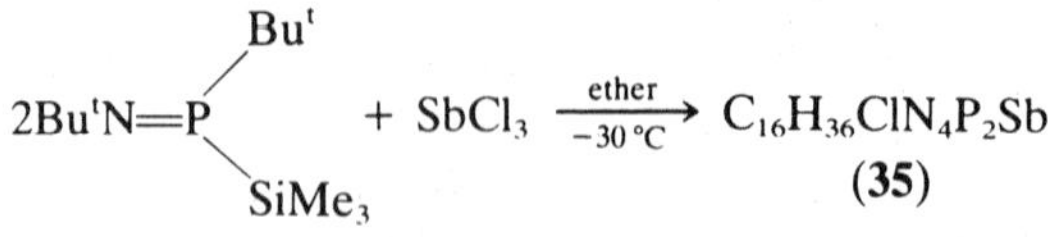

(32)

Tris(dialkyl)stibines react with sulfur dioxide (equation 33), affording tetraalkylsulfurous diamides and an unidentified antimony/nitrogen compound.[101]

$$Sb(NR_2)_3 + SO_2 \longrightarrow (R_2N)_2SO + 1/n(R_2NSbO)_n \qquad (33)$$

There are in the literature some interesting features of antimony chemistry, but which need some crystallographic confirmation, for example the reaction (34). This implies a two-

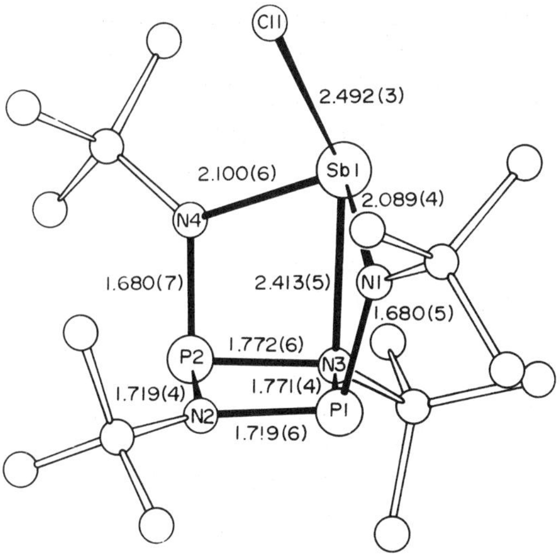

(35)

coordinate antimony.[106] Another unusual reaction is shown in equation (35).[107]

$$R_3Sb + R'SO_2N_3 \longrightarrow RSb{=}NSO_2R' \qquad (34)$$

$$SbI_3 + RNH_2 \xrightarrow[\text{reflux}]{\text{MeCN}} RNSbI \qquad (35)$$

$$R = Pr, Bu, PhCH_2, Ph$$

28.10.2 Other Group 5 Ligands (P, Sb and Bi)

Simple antimony–antimony compounds are known, but are less common than those with higher members of Group 5; the Sb—Sb bond energy is 142 kJ mol^{-1}, which is even weaker than the As—As value of 180 kJ mol^{-1}. Several compounds R_2SbSbR_2 (R = Me, CF_3, Et, Bu^t, Ph) are characterized, and all are air sensitive, although the recently prepared $(C_6H_8Sb)_2$ (Table 14) is air stable. This compound[108] is an intensely coloured irridescent purple-blue by reflected light and red by transmitted light. X-Ray analysis revealed the molecular structure to be as expected, *d*(Sb—Sb) 2.835 Å, but with a short intermolecular Sb···Sb distance, 3.625 Å, which is much less than the calculated van der Waals' distance of 4.40 Å. The intermolecular Sb···Sb distance in Ph_2Sb—$SbPh_2$ is 4.88 Å.[90] The intermolecular chains and weak Sb···Sb bond are thought to be responsible for the unusual colouring.

Breunig[109] observed the unusual coupling reaction of tris(trimethylstannyl)stibine when oxidized by air (equation 36).

$$(Me_3Sn)_3Sb \xrightarrow[\text{oxidation}]{\text{air}} (Me_3Sn)_2Sb{-}Sb(SnMe_3)_2 + (Me_3Sn)_2O \qquad (36)$$

Antimony–element double bonds are still uncommon. Cowley *et al.*[110] found the stability of P=E (E = P, As, Sb, Bi) falls down the group, *i.e.* P > As > Sb > Bi (equation 37; the Bi analogue could not be prepared). The dark orange solution of the P=Sb compound slowly decomposes to give RP—PR (R = mesityl), making isolation impossible. The arsenic analogue has been characterized by X-ray analysis.[111] The ^{31}P NMR spectrum of the P=Sb compound exhibits an extremely deshielded signal (δ = 620.0 p.p.m.), among the lowest ever recorded.

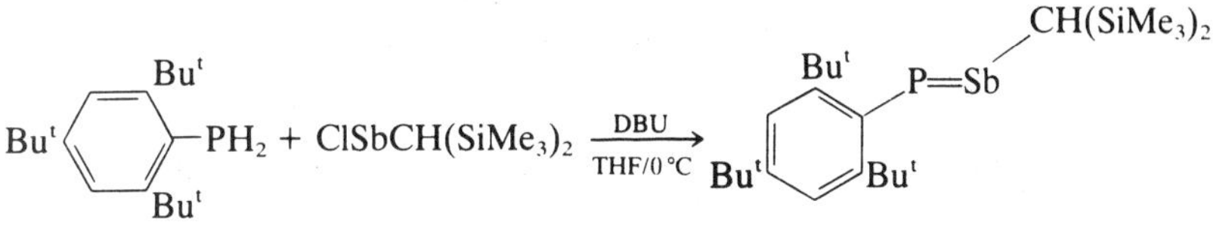

(37)

28.10.3 Antimony Catenation

Compared to other members of Group 5 antimony displays a marked preference for polymeric structures over cyclic structures (equations 38 and 39[112]). Alkyl primary stibines break down at −78 °C to afford brown/black solids (equation 40).

$$PhEH_2 + PhEI \longrightarrow (PhE)_6 + 6HI \quad (38)$$

$$E = P, As$$

$$PhSbH_2 + PhSbI_2 \longrightarrow (PhSb)_x + xHI \quad (39)$$

$$RSbH_2 \longrightarrow (RSb)_x + H_2 \quad (40)$$

This clear preference for polymeric primary organostibine structures indicates this to be the thermodynamically stable state. Rheingold *et al.*[113] have produced a polymeric material $(MeSb)_x$ by reacting $MeSbH_2$ with Me_2SiCl_2 at −78 °C in toluene. It is thought that the weak Sb—Sb bond and lower steric restrictions because of the size of the Sb atom, together with a lack of transannular stabilizing interactions, significantly favour chain formation over ring formation when alternative reaction pathways are available.

28.10.4 Cyclic Antimony(III) Compounds

Although antimony–antimony compounds tend to form as polymers rather than as cyclic structures, several *cyclo*-antimony(III) compounds have recently been characterized. Ellermann and Veit[114] have synthesized the Sb_3R_4 analogues of the well known three-membered arsenic ring compounds (equation 41).

$$MeC(CH_2SbCl_2)_3 \xrightarrow{Na/THF}$$ 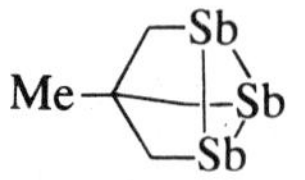 (41)

The four-membered Sb ring with bulky Bu^t stabilizing groups, $(SbBu^t)_4$, has been shown to have the all-*trans* structure (**36**).[115]

Issleib and Balszuweit[116] synthesized a yellow-orange compound thought to be $(PhSb)_6{\cdot}C_6H_6$, and Breunig *et al.*[117] have succeeded in isolating $(PhSb)_6{\cdot}C_4H_8O_2$ (**37**; equation 42; Table 14).

$$6PhSb(SiMe_3)_2 + 3O_2 + C_4H_8O_2 \longrightarrow (\mathbf{37}) + 6(Me_3Si)_2O \quad (42)$$

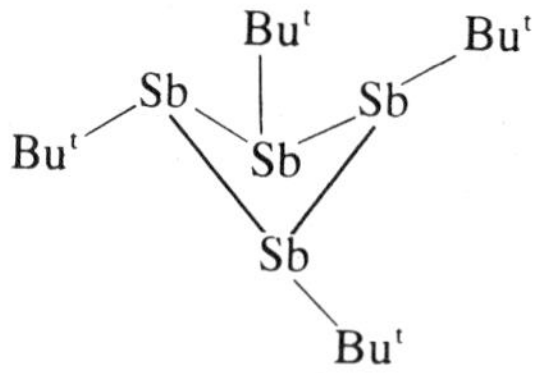

(**36**)

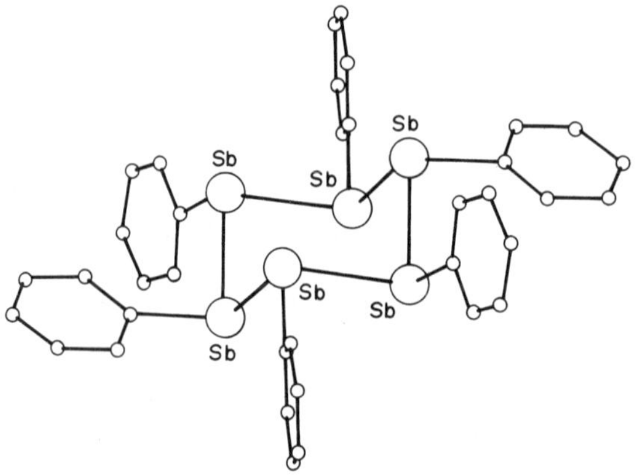

(**37**)

Table 14 Sb—Sb Bond Lengths and Angles

Compound	*d(Sb—Sb)* (Å)	*d(Sb—C)* (Å)	*Sb—Sb—Sb* (°)	*Sb—Sb—C* (°)	*Ref.*
$Ph_2SbSbPh_2$	2.837[a]	—	—	—	90
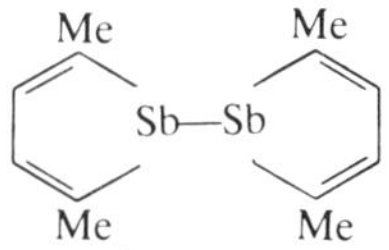	2.835 3.625[a]	2.138 (av)	—	—	108
$(SbBu^t)_4$	2.82	2.21	85 (av)	99 (av)	115
$(SbPh)_6$	2.836–2.839	2.16–2.17	86.8–93.6	92.8–99.7	117

[a] Intermolecular Sb · · · Sb distance.

28.11 ANTIMONY-GROUP 6 LIGANDS

28.11.1 Antimony(III) Compounds with Oxoacids

Antimony(III) forms a variety of compounds with oxoacids, some with empirical formulae that may be written in the form of a metal salt, *e.g.* $Sb_2(HPO_3)_3$[118] and $Sb_2(SO_4)_3$,[119] and others as 'basic' salts or mixed oxides, *e.g.* $SbO(H_2PO_4){\cdot}H_2O$[120] and $Sb_2O(SO_4)_2$.[121] When the antimony lone pair is involved, the difference between pyramidal and trigonal bipyramidal geometry about Sb is a small change in position (Figure 1 and Table 15). Table 15 was compiled by Bovin[122] and lists typical trigonal bipyramids at the top and typical pyramidal structures (tetrahedral with lone pair) at the bottom. The mean equatorial distance in the first seven compounds listed in Table 15 (Sb-1 and Sb-2) is 2.01 Å and axial (Sb-3 and Sb-4) is 2.22 Å, with mean angles of 92.0° and 148.0°.

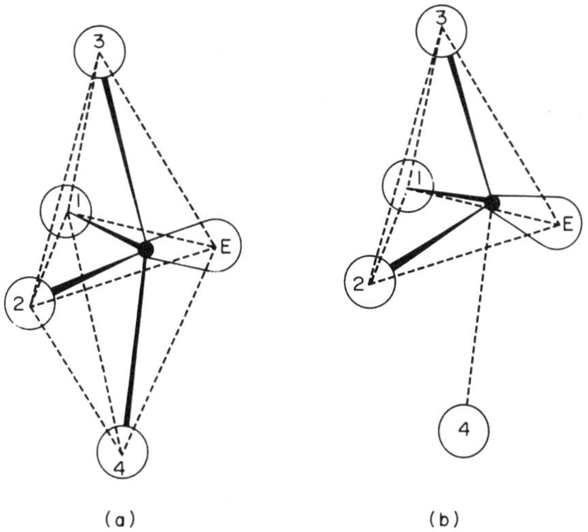

Figure 1 The coordination polyhedra of antimony(III). (a) and (b) show idealized fourfold and threefold coordination, respectively.

The structures of these oxoacid compounds can be complex, *e.g.* $Sb_6O_7(SO_4)_2$,[122] the structure of which consists of distorted tetrahedra (lone pair at the one corner) and *d*(Sb—O) of 1.994–2.207 Å. In these units three different SbO_3 polyhedra share corners and edges to give cylindrical units with SO_4^{2-} tetrahedra between the cylinders. In contrast, antimony(III) phosphite, $Sb_2(HPO_3)_3$, is a fairly simple structure with Sb^{III} coordinated to four isolated HPO_3^{2-} tetrahedra (**38**). The geometry about antimony is trigonal bipyramidal with O(ax)—Sb—O(ax) of 159.4°.[118]

Table 15 Distances and Angles in Some Antimony(III)–Oxygen (Fluorine) Complexes

Compound[a]	*Sb—O(F) bond distances* (Å) *Sb—1*	*Sb—2*	*Sb—3*	*Sb—4*	*Angles* (°) *1—Sb—2*	*3—Sb—4*	*Remarks*
L-SbOF	1.956 (18)	1.993 (17)	2.162 (5)	2.162 (5)	96.1 (7)	142.9 (8)	1 = F^-
$SbPO_4$	1.983 (10)	2.035 (12)	2.181 (10)	2.181 (10)	87.9 (5)	164.8 (5)	
β-Sb_2O_4	2.032 (9)	2.032 (9)	2.218 (9)	2.218 (9)	87.9 (4)	148.1 (4)	
Sb(pyrogallate)·H_2O	2.021 (8)	2.021 (8)	2.244 (4)	2.244 (4)	91.9 (5)	149.1 (1)	
α-Sb_2O_4	2.011 (33)	2.043 (38)	2.209 (42)	2.259 (33)	88 (1)	148 (1)	
$Sb_4O_4(OH)_2(NO_3)_2$	2.019 (6)	2.020 (6)	2.236 (6)	2.265 (6)	98.3 (2)	141.6 (2)	
$SbO(H_2PO_4)\cdot H_2O$	1.970 (8)	1.977 (8)	2.145 (9)	2.291 (10)	94.4 (3)	156.3 (3)	
M-SbOF	2.006 (13)	2.012 (13)	2.038 (13)	2.304 (15)	98.4 (5)	146.2 (4)	3 = F^-
$Sb_6O_7(SO_4)_2$	1.994 (17)	2.066 (11)	2.207 (18)	2.327 (14)	85.2 (5)	147.9 (5)	
$SbNbO_4$	2.008 (9)	2.035 (9)	2.134 (9)	2.331 (9)	92.1 (4)	150.7 (4)	
$Sb_6O_7(SO_4)_2$	2.025 (10)	2.128 (12)	2.167 (15)	2.331 (13)	87.1 (7)	141.1 (6)	
$Sb_4O_5(OH)ClO_4\cdot\frac{1}{2}H_2O$	1.952 (9)	2.030 (9)	2.145 (9)	2.350 (9)	93.7 (4)	141.3 (3)	
M-SbOF	1.952 (12)	1.987 (15)	2.062 (12)	2.364 (13)	89.9 (6)	144.0 (6)	1 = F^-
$Sb_4O_5(OH)ClO_4\cdot\frac{1}{2}H_2O$	1.952 (9)	2.009 (9)	2.085 (9)	2.378 (8)	91.2 (4)	144.5 (3)	
$Sb_6O_7(SO_4)_2$	2.011 (16)	2.014 (17)	2.154 (10)	2.382 (19)	94.0 (2)	146.8 (5)	
$Sb_4O_5Cl_2$	1.902 (1)	2.061 (5)	2.144 (6)	2.469 (6)	91.8 (2)	139.0 (2)	
$Sb_2O_3\cdot 2SO_3$	1.894 (5)	2.122 (10)	2.161 (10)	2.490 (11)	91.3 (4)	156.0 (3)	
$Sb_4O_5OH)ClO_4\cdot\frac{1}{2}H_2O$	1.975 (9)	1.997 (8)	2.027 (8)	2.572 (10)	94.8 (4)	142.8 (3)	
Sb_2O_3 (orthorhombic)	1.977 (7)	2.023 (4)	2.019 (6)	2.619 (6)	98.1 (2)	139.3 (2)	
Sb_5O_7I	1.935 (19)	1.978 (18)	2.091 (15)	2.782 (19)	94.6 (8)	148.0 (6)	
$Sb_4O_4(OH)_2(NO_3)_2$	1.942 (7)	2.067 (6)	2.052 (6)	2.745 (9)	82.9 (3)	152.7 (3)	
Sb_5O_7I	1.955 (17)	1.965 (18)	2.063 (14)	2.794 (17)	91.3 (7)	150.2 (6)	
$Sb_4O_5Cl_2$	1.931 (6)	2.015 (5)	2.080 (6)	2.905 (3)	94.8 (3)	153.9 (2)	4 = Cl^-
$Fe_4Ti_3SbO_{13}(OH)$	2.006 (3)	2.009 (4)	2.006 (3)	2.918 (3)	92.5 (1)	155.4 (1)	
Sb_2O_3(cubic)	1.977 (1)	1.977 (1)	1.977 (1)	2.918 (2)	96.0 (3)	160.9 (9)	
Sb_5O_7I	1.962 (19)	1.988 (16)	2.001 (18)	2.988 (16)	94.1 (7)	134.3 (6)	
Sb_5O_7I	1.959 (16)	1.961 (17)	2.024 (16)	3.042 (18)	89.4 (7)	149.2 (6)	
Sb_5O_7I	1.949 (17)	2.000 (19)	2.016 (17)	3.086 (17)	89.5 (7)	146.8 (6)	

[a] See ref. 122 for literature references.

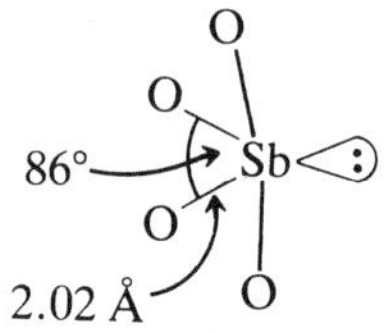

(38)

In the series $Me_3Sb(X)OSb(X)Me_3$ (X = N_3, Cl, ClO_4) trigonal bipyramidal structures were found with X and O in the axial positions. The variation of *d*(Sb—O) with X is given in Table 16.

Table 16 Axial Bond Lengths in $(Me_3SbX)_2O$

Compound	*d(Sb—O)* (Å)	*Axial group*	*d(Sb—X)* (Å)	*Electronegativity of X (Pauling)*
$(Me_3SbCl)_2O$	1.907	Cl	2.628	3.0
$(Me_3SbN_3)_2O$	1.872	N (N_3)	2.344	3.0
$[Me_3Sb(OClO_3)]$	1.862	O (ClO_4)	2.601	3.5

28.11.2 Antimony Oxyhalides

Antimony forms polymeric oxyhalides, and not metallic as in BiOCl. The fluoride, SbOF, has been prepared in two forms: '*L*', with a ladder structure, and '*M*', which has a layered structure. Both forms have a trigonal bipyramidal structure about antimony with three oxygens, one fluorine and one lone pair. Structural parameters are given in Table 15, from which it can be seen that *L*-SbOF heads the table as the nearest to an ideal fit for trigonal bipyramidal geometry.

The oxychlorides SbOCl and $Sb_4O_5Cl_2$ have sheets of Sb—O layers, with chlorines between the layers. The SbOCl layer has one third of the Sb in SbO_4 and two thirds in SbO_2Cl units. In

$Sb_4O_5Cl_2$ one half of the Sb is in SbO_3 coordination and half in SbO_4.[123] Antimony oxoiodides, Sb_3O_4I, $Sb_8O_{11}I_2$ and Sb_5O_7I, also have layered structures.

28.11.3 Antimony(V) Oxide and Derivatives

28.11.3.1 Antimony oxoacids and their salts

Antimonic acid is poorly characterized and reported examples are hydrated Sb_2O_5 of unknown structure. It is known in its salts, however, and these are of the two main types, based on either $[Sb(OH)_6]^-$ or $[SbO_6]^{n-}$ octahedra. The oxyions, *e.g.* $[SbO_4]^{3-}$, do not exist *per se*.

28.11.3.2 Antimonates containing [Sb(OH)₆]⁻ ions

The X-ray crystal structures of sodium pyroantiomonate revealed that the salt should be written $Na[Sb(OH)_6]$, and the silver salt is $Ag[Sb(OH)_6]$. This has led to the reformulation of the older notation (Table 17).

Table 17 Revision of Antimonate Formulae

Old formula	New formula
$Na_2H_2Sb_2O_7 \cdot 5H_2O$	$Na[Sb(OH)_6]$
$Mg(SbO_3)_2 \cdot 12H_2O$	$[Mg(H_2O)_6][Sb(OH)_6]_2$
$Cu(NH_3)_3(SbO_3)_2 \cdot 9H_2O$	$[Cu(NH_3)_3(H_2O)_3][Sb(OH)_6]_2$
$LiSbO_3 \cdot 3H_2O$	$Li[Sb(OH)_6]$
$HaF \cdot SbOF_3 \cdot H_2O$	$Na[SbF_4(OH)_2]$

28.11.4 Sulfur Ligands

28.11.4.1 Thioantimonates

There are a large number of minerals and synthetic materials based on the antimony–sulfur bond. The structures are diverse and are based on trigonal pyramidal, tetragonal, trigonal bipyramidal (usually distorted) or octagonal units. Frequently several of these types of structural unit are used in one structure, *e.g.* Livingstonite, $HgSb_4S_8$, although cases of simple trigonal pyramidal structures are known, *e.g.* Samsonite, $MnAg_4Sb_2S_6$.

28.11.4.2 Organosulfur ligands

These molecules are usually formed by reacting antimony halides with an organosulfur compound (equations 43,[124] 44[126] and 45[125]).

$$2SbCl_3 + 3Pb(SC_6Cl_5)_2 \xrightarrow[\text{reflux}]{C_6H_6} Sb(SC_6Cl_5)_3 + 3PbCl_2 \qquad (43)$$
$$(73\%)$$

$$SbF_3 + 3Me_3SiBu^t \longrightarrow Sb(SBu^t)_3 + 3Me_3SiF \qquad (44)$$
$$(100\%)$$

$$3 \underset{Me}{\diamond\!\!-\!\!NCS_2Na} + SbCl_3 \longrightarrow \left(\underset{Me}{\diamond\!\!-\!\!NCS_2} \right)_3 Sb + 3NaCl \qquad (45)$$

In the case of ligands with larger bite, *i.e.* chains between the sulfur atoms, cyclic disulfides may result, for example 5-chloro-1-oxa-4,6-dithia-5-stibocene,[127] the structure of which consists of antimony in a distorted trigonal bipyramid with Cl and O in axial positions (**39**). A range of crystallographic data is contained in Table 18.

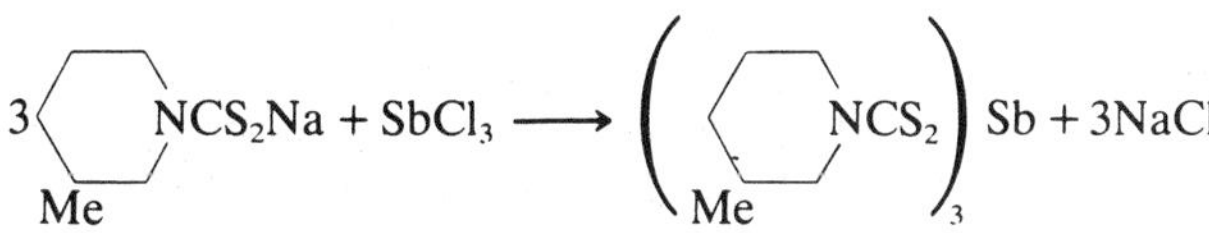

Table 18 Sulfur–Antimony Bond Lengths and Angles

Compound	*(Sb—S)* (Å)	*(Sb—E)* (Å)	*Sb—S—E* (°)	*Sb—S—E* (°)	*S—E—S* (°)	*Ref.*
$(Et_2NCS_2)_3Sb$	3 × (2.487–2.631) 3 × (2.886–2.965)				118	128
$(Pr^n_2NCS_2)_3Sb$	3 × 2.630 (av) 3 × 2.815 (av)				117.7	129
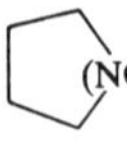 $(NCS_2)_3Sb$	3 × 2.607 (av) 3 × 2.858 (av)				120	130
$(Ph_2PS_2)_3Sb$	3 × 2.55 (av) 3 × (2.923–3.187)		S_{apex}—Sb—S_{base} 92.5 (av)			131
$[(MeO)_2PS_2]_3Sb$	3 × 2.529 (av) 3 × 3.005 (av)				112.3	132
$Sb(exa)(oxin)_2$	3.155 (av)	(Sb—N) 2.371 (av) (Sb—O) 2.105 (basal) (Sb—O) 2.018 (ax)	O_{ax}—Sb—E 85.9 × 4 (av) O_{ax}—Sb—$N_{same\ ligand}$ 74.2		124.4	133
$[(Pr^iO)_2PS_2]_3Sb$	3 × 2.523 (av) 3 × 3.015 (av)				111.2	132
$(EtOCS_2)_3Sb$ distorted octahedron	3 × 2.511 (av) 3 × 3.002 (av)				123.8	133
Compound	*(Sb—S)* (Å)	*(Sb—E)* (Å)	*E—Sb—E* (°)	*Other* (°)		*Ref.*
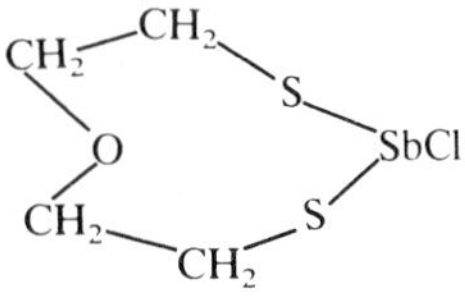 distorted trig bipyr	(eq) 2.435 (av)	(Sb—Cl) (ax) 2.448 (Sb · · · O) (ax) 2.529	S—Sb—S 107.8	O—Sb—Cl 152		127
$Ph_2SbS(2,6\text{-}Me_2C_6H_3)$	2.439	(Sb—C) 2.145 (av)	S—Sb—C 97.3 93.4	C—Sb—C 94.1		134
Ph_3SbS	2.244	(Sb—C) 2.098 (av)	S—Sb—C 112.4 (av)	C—Sb—C 106.4 (av)		266

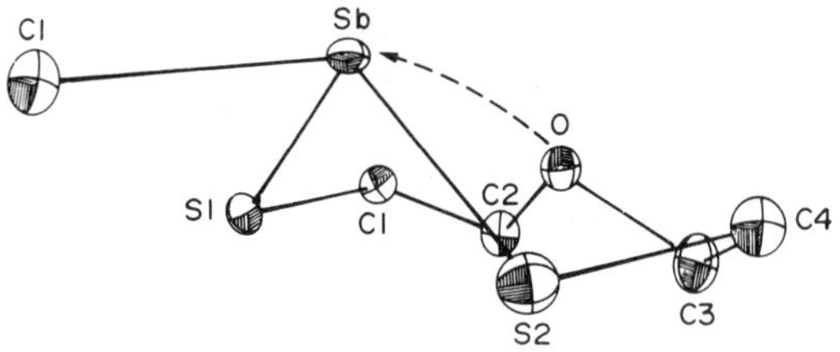

(**39**)

Antimony(III) complexes with bidentate sulfur ligands form an interesting series of structural types. In tris(diethyldithiocarbamate)antimony(III) the sulfur atoms fall into two groups. Firstly, there are three *fac* sulfurs with short *d*(Sb—S) 2.487–2.631 Å bonds, one from each ligand, and then there are three longer bonds *d*(Sb—S) 2.886–2.965 Å (**40**). There is a large gap in the coordination sphere and this has been ascribed[128] to a sterically active lone pair. There is a weak Sb· · ·S 3.389 Å interaction between molecules within van der Waals' radii. Similar results have been obtained for $Sb(S_2CNPr^n_2)$ and tris(1-pyrrolidinecarbodithioato)antimony(III).[129,130] Hoskins *et al.*[133] used the bidentate $EtOCS_2^-$ to prepare analogous SbL_3 species, and found once again that there are three *fac* short Sb—S 2.511 Å bonds and three longer 3.002 Å bonds. The triangles formed by each of these groups of three sulfur atoms (projected into a plane) are rotated at an angle of 41.6° relative to one another (**41**), (*cf.* ideal octahedral geometry of 60°). The thrust of the lone pair on Sb is considered to be in the direction through the centre of the triangle formed by the longer-bonded sulfurs, accounting for the 'splaying out' of that triangle when compared to the short- or primary-bonded sulfur atoms. The S—C—S angle is 123° (*cf.* $(Et_2NCS_2)_3Sb$ S—C—S 118°) and this subtle change in bite angle of the ligand may be a factor in the resulting distorted octahedral geometry rather than the pyramidal geometry of $(Et_2NCS_2)_3Sb$. Hoskins *et al.*[133] also adventitiously isolated the mixed ligand complex (*O*-ethylxanthato)bis(quinolin-8-olato)antimony(III), which is pentagonal bipyramidal about antimony (**42**). The oxygen atom of one of the hydroxyquinoline ligands is the apical donor with a short *d*(Sb—O) 2.018 Å. The pyramidal structure is indicated by the angles the other ligand atoms make *via* antimony with this axial oxygen; average angle is 86° (ideal structure, 90°).

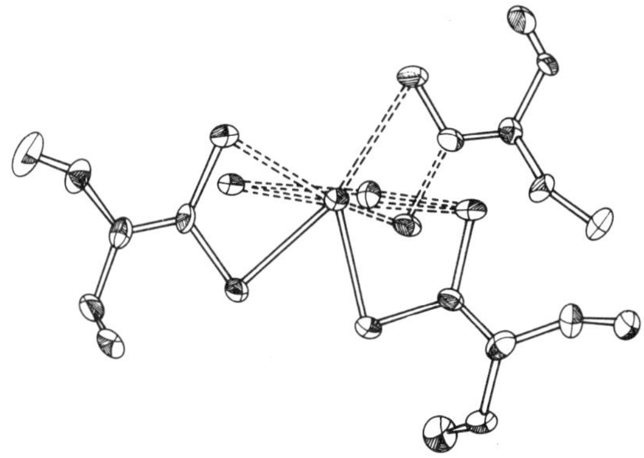

(40)

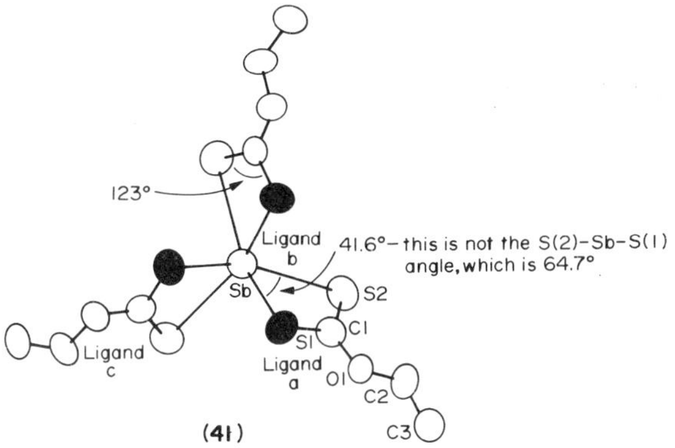

(41)

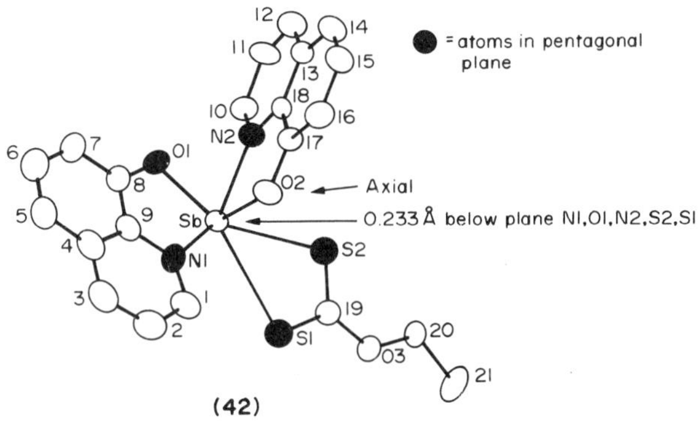

(42)

In $Sb(PhCOS)_3$ the shorter bonds are with the sulfur, *d*(Sb—S) 2.493 Å, and the weaker with oxygen, *d*(Sb—O) 2.802 Å. Interestingly, the three oxygen atoms and the antimony are nearly coplanar (**43**). This gives a grossly distorted octahedral geometry which may be thought of as an extreme extension of the type of distortion in $(EtOCS_2)_3Sb$ (**41**) at the threshold of pentagonal bipyramidal geometry.

The geometry about antimony in tris(diphenyldithiophosphinato)antimony(III) is pentagonal pyramidal (**44**) with the lone pair thought to be occupying one of the apical positions in a pentagonal bipyramid. This is similar to the structure revealed by Russel *et al.*[135] for the $M_3[Sb(C_2O_4)_3]$ complex ($M = NH_4$, K; **45**). Sowerby *et al.*[132] have produced $Sb\{S_2P(OR)_2\}_3$ (equation 46).

$$SbCl_3 + 3(NH_4)S_2P(OR)_2 \xrightarrow{EtOH} Sb\{S_2P(OR)_2\}_3 + 3NH_4Cl \tag{46}$$

$$R = Me, Et, Pr^i, Bu^n, Bu^i, Bu^s$$

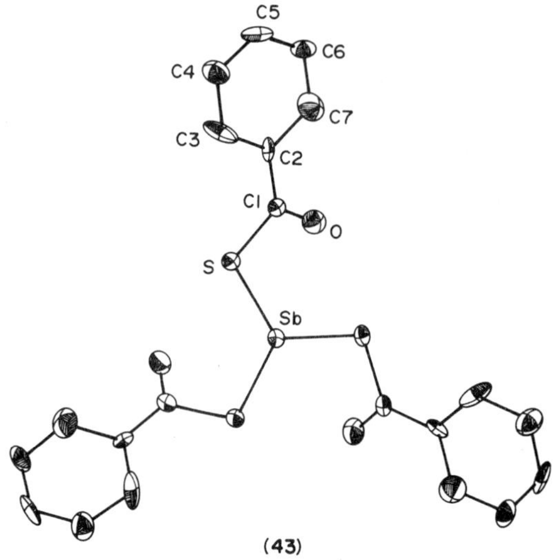

(43)

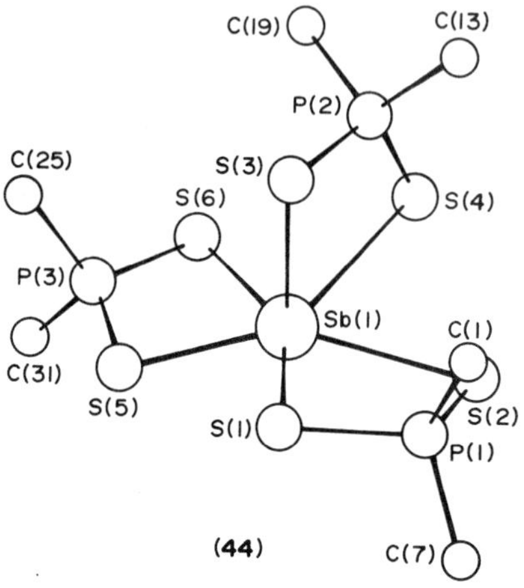

(44)

A perspective view of the $[Sb(C_2O_4)_3]^{3-}$ion

A projection of the $[Sb(C_2O_4)_3]^{3-}$ ion approximately perpendicular to the equatorial plane

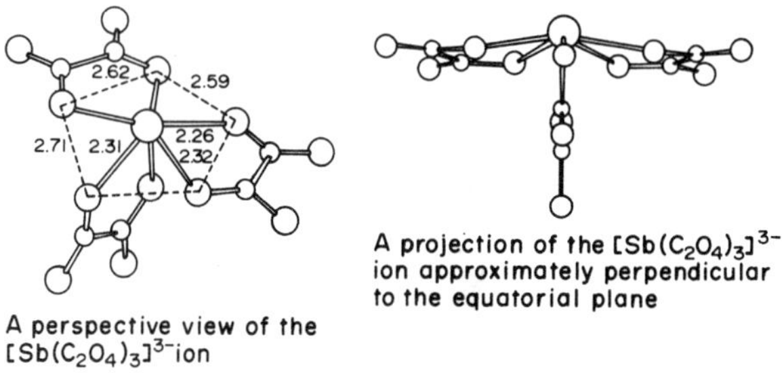

(45)

X-Ray structures were produced for the Me (**46**) and Pr^i (**47**) derivatives, both being described in terms of distorted octahedral geometry; the twist angles between the two planes of sulfur atoms are *ca.* 45° in both cases (ideal octahedral twist angle is 60°).

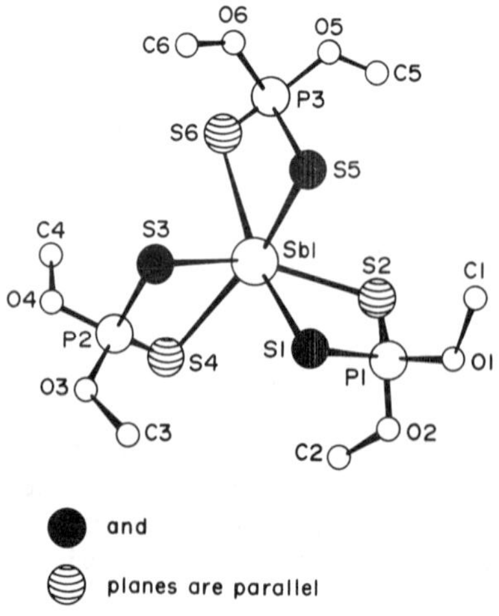

(46)

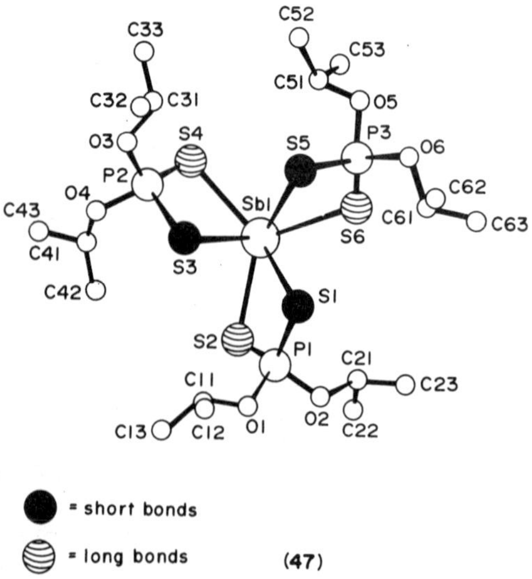

(47)

In a formal sense in these compounds with sulfur the central antimony atom has a total of seven pairs of electrons. The lone pair then has to occupy a capping position over a triangular face of the octahedron (**48**). The longer Sb—S bonds are then rationalized by the interaction of the lone pair with a group of three *fac* sulfur atoms, say *x, y, z*. However, it could equally well be argued that other geometric imperatives imposed, for example, by small bite ligands or inductive factors by adjacent atoms or groups, or the requirement to minimize intermolecular interaction in the crystal, are responsible. The balance of interactions and their relative importance has not yet been fully worked out.

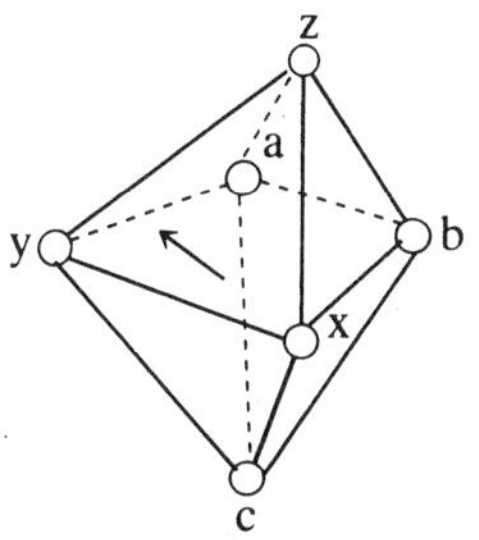

(**48**)

Some interest is being shown in some Sb—S compounds for their pharmacological effect. For example Alonzo *et al.*[136] have prepared $SbL_3 \cdot H_2O$ (LH = L-cysteine), but the structure is not yet known).

28.11.4.3 *Organoantimony–sulfur compounds*

The known types are well reviewed in Gmelin up to *ca.* 1980, but some representatives will be mentioned here.

Thioesters are prepared from stibonous esters (equation 47).[137] Thiostibinous esters may be made from the oxo compounds (equation 48).[138]

$$MeSb(OEt)_2 + 2EtSH \xrightarrow[\text{fractionate}]{\text{EtOH, reflux}} MeSb(SEt)_2 \quad (47)$$

b.p. 100 °C/13 mmHg

$$(R_2Sb)_2O + 2R'SH \xrightarrow[\text{reflux}]{C_6H_6} R_2SbSR' \quad (48)$$

These are compounds of pharmaceutical interest possessing bactericidal and fungicidal properties and finding use in, for example, paint formulations.

28.11.5 Some Pentacoordinate Molecules

Spiroarsoranes[139] (see p. 245) are distorted towards square pyramidal geometry, so it is surprising that 2-phenyl-4,4,4′4′,5,5,5′5′-octamethyl-2,2′-spirobi(1,3,2)^5-dioxastibolan is found to have a distorted trigonal bipyramidal geometry (**49**).[140]

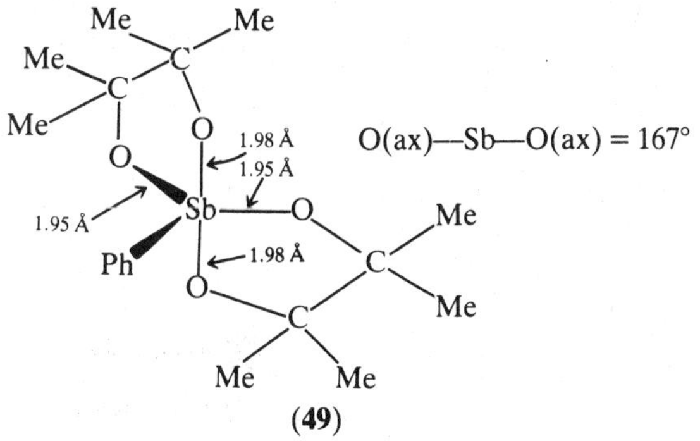

(**49**)

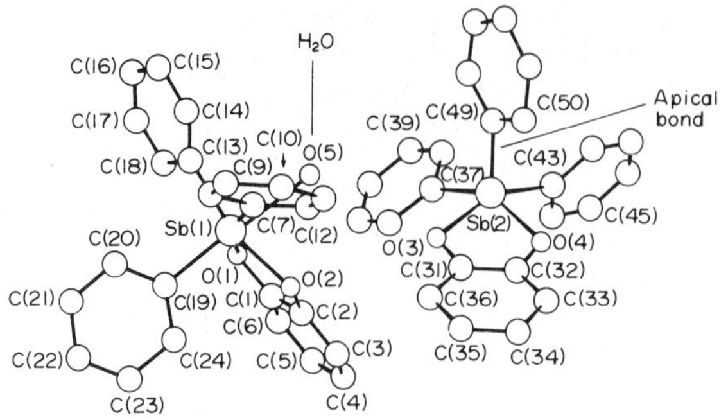

(50) $(Ph_3SbO_2C_6H_4)_2 \cdot H_2O$

Perhaps the most remarkable five-coordinate compound is bis(triphenylantimony-catecholate)hydrate,[141] which displays six-coordinate octagonal geometry and square pyramidal geometry (Table 19). As seen in (**50**) one molecule of Ph_3Sb(catecholate) is bonded to a water molecule; this bonding is very weak with *d*(Sb—O) 2.512 Å, corresponding to a bond order of 0.17. The oxygens are *fac*. The other molecule of the pair has a base of two carbons, *d*(Sb—C) 2.143, 2.125 Å, two oxygens and an apical *d*(Sb—C) 2.099 Å.

Table 19 Bond Lengths in Some Five-Coordinated Sb^V Compounds

Compound	*Geometry*[a]		
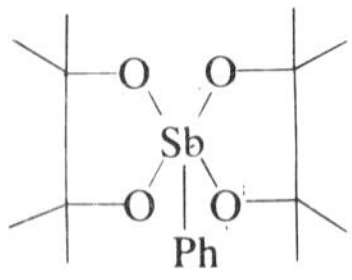	Trigonal pyramidal	Sb—O(ax) 1.98 Sb—O(eq) 1.95 Sb—C(eq) 2.11	O(ax)—Sb—O(ax) 167°
$(Ph_3SbO_2C_6H_4)_2 \cdot H_2O$	(i) Octahedral	Sb—C 2.140 (av) Sb—O 2.049 (chelate) Sb—O 2.512 (H_2O)	
	(ii) Square pyramidal	Sb—C 2.099 (apical) Sb—C 2.143, 2.125 (base) Sb—O 2.060, 2.013 (base)	

[a] Bond lengths in Å.

28.12 ANTIMONY–HALOGEN COMPOUNDS

28.12.1 Antimony(III) Halides

The structure of SbX_3 molecules is trigonal bipyramidal in the vapour phase (Table 20) and this symmetry is retained in the crystal state with some variations in bond lengths. In the case of the iodide the *d*(Sb—I) is 2.72 Å (I—Sb—I 99°); the crystal consists of hexagonal close packed assemblies of iodine atoms with antimony in the octahedral interstices, but off-centre, giving two different *d*(Sb—I) distances of 2.87 and 3.32 Å.

Antimony(III) halides are chemically reactive, but less so than their phosphorus or arsenic analogues. Antimony(III) chloride forms a clear solution with water, and there is no evidence for Sb^{3+} ions; dilution results in precipitation of insoluble oxychlorides of various compositions, *e.g.* SbOCl, $Sb_4O_5Cl_2$, $Sb_8O_{11}Cl_2$. Some reactions of $SbCl_3$ are shown in Scheme 3. Antimony(III) fluoride is an important fluorinating agent.

Antimony(III) halides are Lewis acids and will form adducts with Lewis bases. An example is provided by Lipka and Mootz,[149] who prepared an adduct of diphenyl with $SbCl_3$, $2SbCl_3 \cdot PhPh$ (**51**). A molecule of $SbCl_3$ is bonded to each of the biphenyl rings, but there are two differently bonded antimony atoms. One which forms the chain is six-coordinate: there are three chlorine atoms at a normal distance, 2.36 Å (av), and two at the longer distances of 3.443

Table 20 Antimony–Halogen Bond Lengths

Compound	*d(Sb—X)* (Å)	*d(Sb—E)* (Å)	*E—Sb—X* (°)	*E—X—Sb* (°)	*Ref.*
SbF_3	1.92		87.3		142
$SbCl_3$(vap)	2.33		97.2		143
$SbCl_3$(crystal)	1 × 2.340 2 × 2.368		1 × 91 2 × 95.7		144
$SbBr_3$(vap)	2.49		98		145
$SbBr_2$(crystal)	2.50		95		146
SbI_3(vap)	2.719		99.1		147
SbI_3(crystal)	3 × 2.686, 3 × 3.316		95.8		148
$SbCl_5$(vap)	2.43 (ax) 2.31 (eq)				267
$SbCl_5$(crystal)	2.34 (ax) 2.29 (eq)				268
[$2SbCl_3$·Ph·Ph]	2 × 2.36 (av) 2 × 3.45 (av)	Sb · · · Ph 3.26 Sb · · · Ph 3.08	91–94		149
SbF_5 (tetramer)	2.025 (br) 1.81 (t)		2 × 141 2 × 170		150
[SbF_5·SO_2]	1.85				151
[SbF_5·$POCl_3$]	2.33				152
[$SbCl_5$·OC(H)NMe_2]	2.34				153
$SbCl_3$·$PhNH_2$	2.516 (ax) 2.329 (av) (eq)	2.525 (ax)	1.663		154
$(SbCl_3F_2)_4$	2.07 (br) (Sb—F) 1.97 (t) (Sb—F) 2.25 (av) (Sb—Cl)				155
$(SbCl_4OMe)_2$	2.328 (av)	2.05, 2.15		110.3	156
$[MeCO]^+[SbF_6]^-$	1.851 (av)		89.9 (av) 178 (ax)		157
$(Me_4N)_3(Sb_2Br_9{\cdot}Br_2)$	3.043 (br) 2.625 (t)		174.0 (5)		158
$(NH_4)_4(Sb^{III}Br_6)(Sb^{V}Br_6)$	2.795 (Sb^{III}) 2.564 (Sb^{V})				159
$[XeF_3]^+[Sb_2F_{11}]^-$	2.01 (br) 1.85 (av of rest)				160
Fe(TPP)($FSbF_5$)·PhF	1.905 (br) 1.844 (av of rest)	2.105 (Fe—F)			161
$(SbCl_4N_3)_2$	2.317 (av)	2.186 (av)	69.1	110.8	162
$(SbCl_4NCO)_3$	2.299 (av)	2.17 (O) 2.23 (N)			163
$(Sb_2F_4)^{2+}$	2.07 (av) (br)		148		164
$NH_4(SbF)PO_4{\cdot}H_2O$	1.934				165
$Na(SbF)PO_4{\cdot}nH_2O$ (n = 2–4)	1.923				165

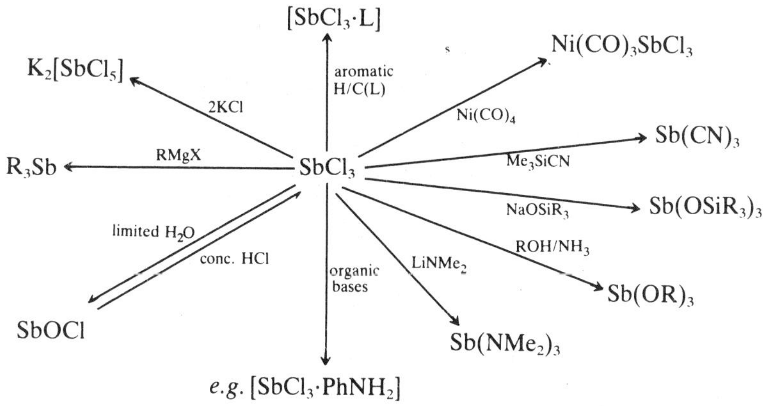

Scheme 3 Representative reactions of antimony(III) chloride

and 3.456 Å; the phenyl ring is the sixth ligand. The other type of Sb only binds to one extra chlorine (3.442 Å) and the other phenyl (3.08 Å), resulting in pentacoordination.

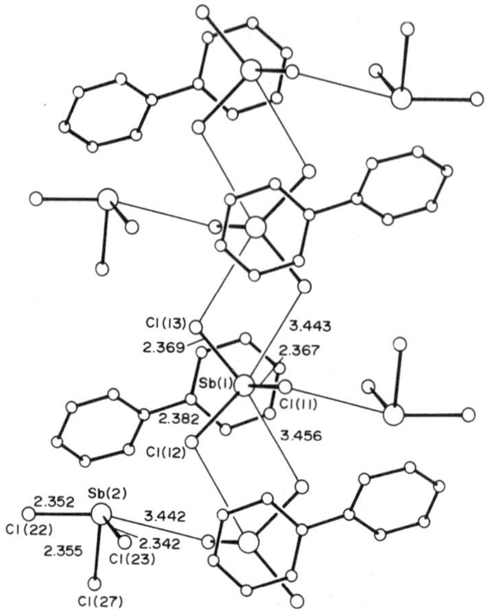

(51)

Placing an NH group between the phenyl rings, *i.e.* diphenylamine, resulted in a similar structure, but because of the greater flexibility of the ligand the Sb···Ph distances for the two antimony atoms are approximately the same, 3.08 and 3.09 Å (**52**). However, when the $[PhNH_2]^+Cl^-$ salt is used as the ligand, a completely different structure resulted. Both antimony atoms are in the same environment of a distorted octahedron of six chlorine atoms, and the amine ion is a bridging position between chlorine atoms of the antimony–chlorine-bridged tubular chain (**53**).

These adducts may be usefully compared to that formed by Scruton and Hulme[154] in $SbCl_3{\cdot}PhNH_2$. In this case the bonding is with nitrogen and the structure is trigonal bipyramidal with N and Cl apical and two chlorines and the lone pair equatorial. The balance between the factors of the Lewis basicity of the amine *versus* that of the aromatic system together with crystal packing energies must be finely drawn in these systems.

Antimony(III) halides are powerful halide acceptors and form complexes with, for example, monovalent halides to give a variety of interesting structures, *e.g.* (**54**)–(**56**). The tetramethylammonium nonabromodiantimonate(III) dibromine is of interest since the crystal structure contains bromine molecules which bridge the $[Sb_2Br_9]^{3-}$ anions (**56**).[158]

Gillespie and coworkers[164] isolated a 1:1 adduct of SbF_3 and SbF_5 from a reaction mixture of SbF_5 and S_4N_4 in liquid SO_2, the crystal structure of which shows it to be $[Sb_2F_4]^{2+}[SbF_6]_2^-$ (**57**). Similar $[SbF_2]^+$ cations have been described in $M(SbF_2)SO_4$ (M = Rb, Cs)[166] and $K(SbF_2)HPO_4$.[167]

A crystallographically characterized species, $[SbF]^{2+}$, has been described by Mattes and Holz.[165] On mixing 0.1–0.3 M solutions of SbF_3 and MH_2PO_4 the compounds shown in Scheme 4 are isolated. Compound (I) is the same as that produced in $K(SbF_2)HPO_4$ (above), but complexes (II) and (III) have the interesting structure (**58**).

28.12.2 Antimony(V) Halides

Two simple antimony(V) halides are known: SbF_5 (m.p. 8.3 °C, b.p. 141 °C, *d* 3.11 g cm^{-3} at 20 °C) and $SbCl_5$ (m.p. 40 °C, b.p. 140 °C, *d* 2.35 g cm^{-3} at 21 °C). The former is a syrupy viscous liquid (*ca.* 850 centipoise, 20 °C, which is a similar value to glycerol), due probably to the polymeric chains of *cis*-bridged SbF_6 octahedra.[168] The viscosity is greatly reduced by the addition of small amounts of $SbCl_5$, which breaks the Sb—F—Sb bridges. This is of industrial

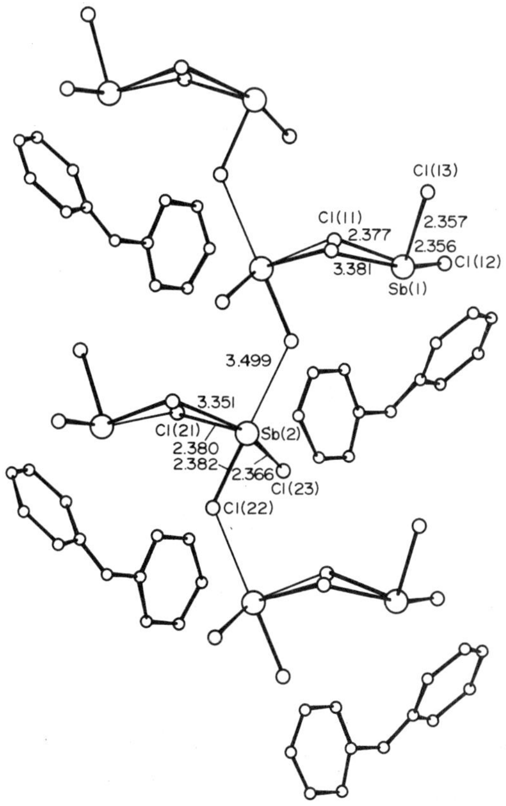
Cl(13)
2.357
Cl(11)
2.377
2.356
Cl(12)
3.381
Sb(1)
3.499
3.351
Cl(21)
Sb(2)
2.380
2.382
2.366
Cl(23)
Cl(22)

(52)

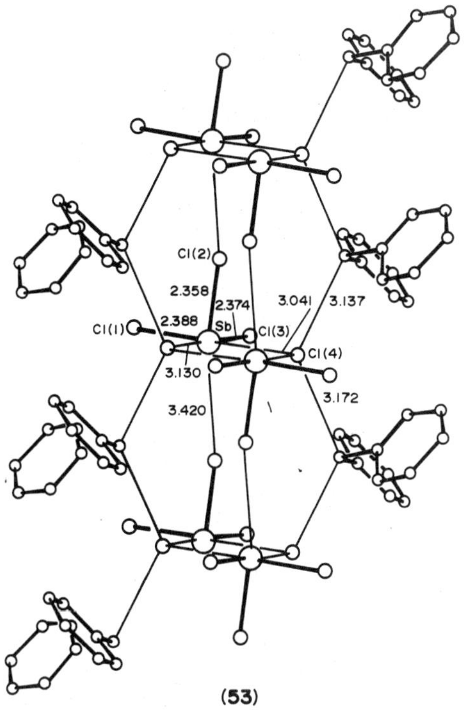
Cl(2)
2.358
2.374
3.041
3.137
2.388
Sb
Cl(1)
Cl(3)
Cl(4)
3.130
3.420
3.172

(53)

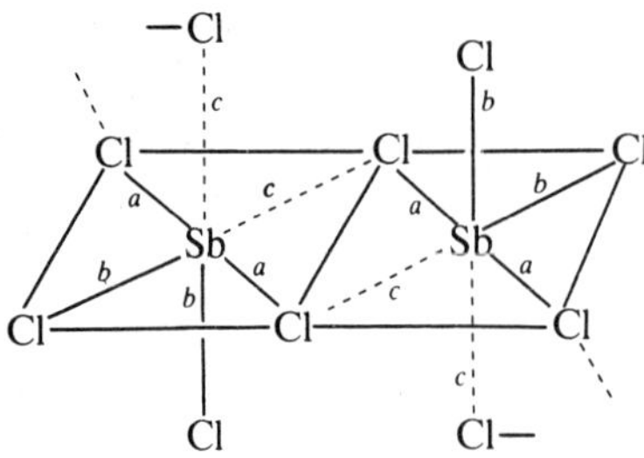

(54) $a = 2.64$ Å; $b = 2.38$ Å; $c = 3.12$ Å; all angles ≈ 90°

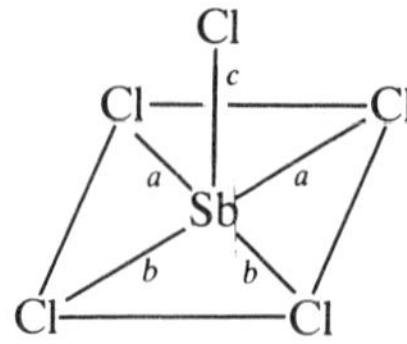

(55) $a = 2.58$ Å; $b = 2.69$ Å; $c = 2.36$ Å

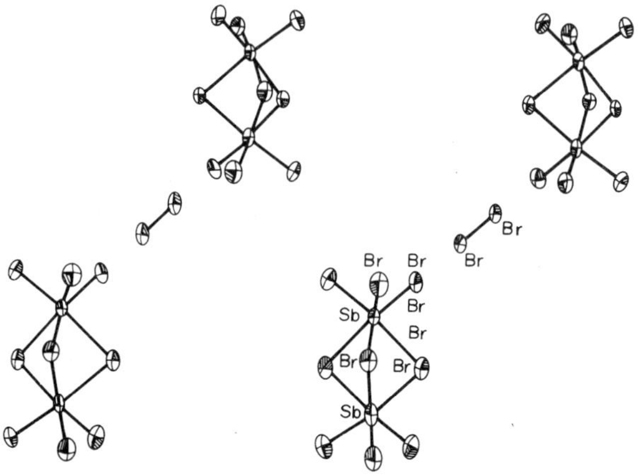

(56)

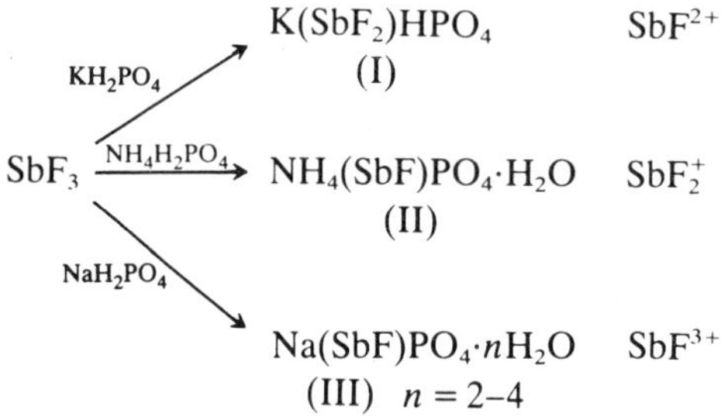

Scheme 4

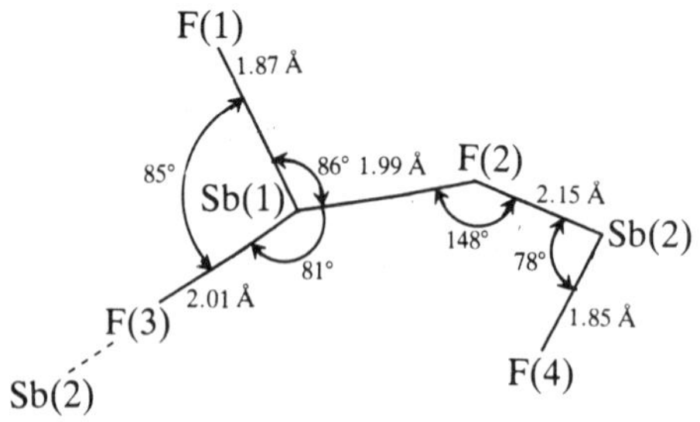

(57)

importance in processes such as the Swarts reactions. In the crystalline state SbF_5 is a tetramer, (**59**).[150]

A variety of mixed antimony(V) chloride/fluorides have been described, in the main arising from the fluorination of $SbCl_5$; the product of fluorination is strongly dependent on the fluorinating agent. Bridging fluorides exist in the cyclic tetramer $SbCl_4F$, similar to the $(SbF_5)_4$ (**59**).[168] One of the products from the HF reaction, $SbCl_3F_2$, is also a tetramer with bridging fluorine, *d*(Sb—F) 2.07 Å.[155] The other HF fluorination product, $[SbCl_4][Sb_2Cl_2F_9]$, contains the unit $[ClF_4Sb—F—SbF_4Cl]$, with a distorted octahedral geometry about antimony, Sb—F—Sb 163°.[169]

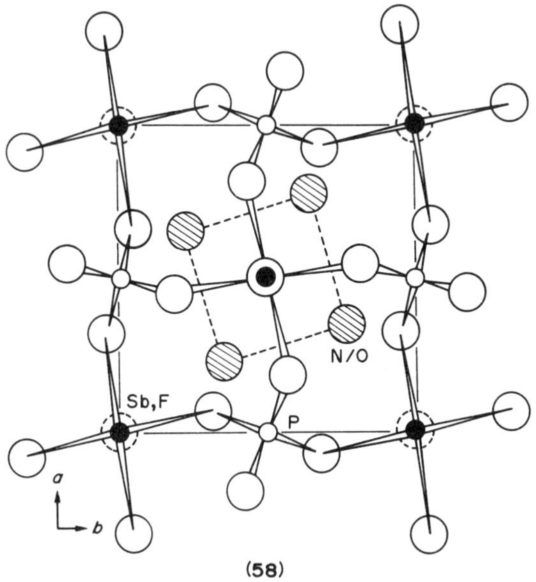

(58)

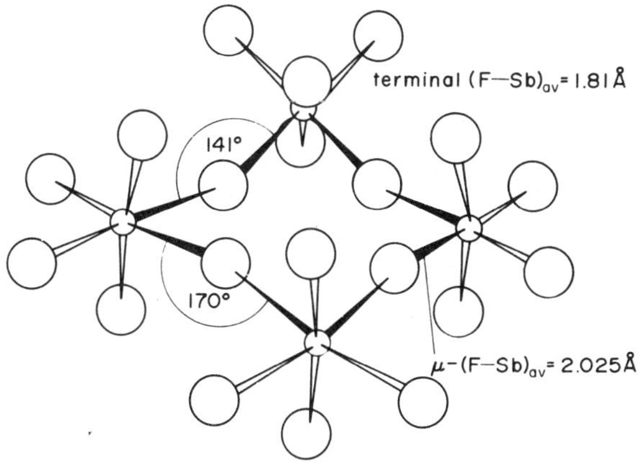

(59)

Birchall and co-workers[155] found that the fluorination of $SbCl_4F$ gave a product best described as a tetramer of $SbCl_3F_2$ with partial replacement of the fluorine by chlorine and corresponding to the general formula $Sb_4Cl_{12+x}F_{8-x}$ ($x = 0$–4). In the case where $x = 1$ (equation 49) the substitution results in a compound with a disordered crystal structure.

$$SbCl_4F \xrightarrow[SO_2(l)]{SbF_5} Sb_4Cl_{13}F_7 \qquad \text{m.p. 45–54 °C} \tag{49}$$

One of the more notable features of antimony(V) halide chemistry is the tendency to achieve a CN of six, thus resulting in the facile formation of complex anions, particularly with halide donors (Table 21); the d(Sb—F) depends on the nature of the cation. Their structures are related to the F—Sb···F interactions between crystal structure units, which is dependent upon the potential field of the cation.

This powerful halide-abstracting ability results in SbF_5 being the strongest Lewis acid with which HSO_3F forms the strongest 'superacid', the so-called 'magic acid'.[70] The Sb^V halides are excellent Friedel–Crafts catalysts.[269]

Although XeF_4 is a poor F-donor, such is the strength of SbF_5 as a Lewis acid that reaction between them produces $[XeF_3]^+[Sb_2F_{11}]^-$ (**60**), but it should bc noted that the nearest approach of an F atom of the anion to the cation is 2.50 Å.[160]

Care must be exercised in interpreting initial data when using SbF_5 to generate $[SbF_6]^-$ anions. An example from the recent literature is that of Fe(TPP)($FSbF_5$)·PhF.[161] It had been supposed, not unreasonably, that the SbF_6 moiety was an anion, but careful X-ray analysis shows that this moiety is, in fact, a ligand covalently bound to iron (**61**).

Table 21 X-Ray Crystal Data for $M[SbF_6]$

Compound	*Crystal type*	*d(Sb—F)* (Å)	*Ref.*
$Li[SbF_6]$	Rhombohedral	1.877	259
$Na[SbF_6]$	Cubic (NaCl type)	1.78	260
$K[SbF_6]$	Cubic (CsCl type)	1.77	261
$pyH[SbF_6]$	Rhombohedral	1.87	262

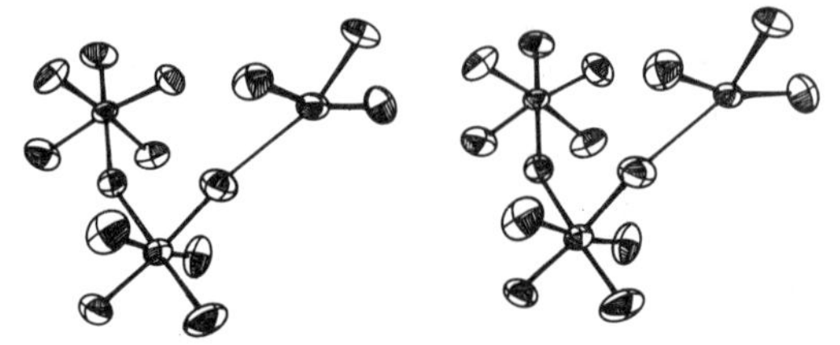

(60) Stereogram of the $[XeF_3^+][Sb_2F_{11}^-]$ structural unit

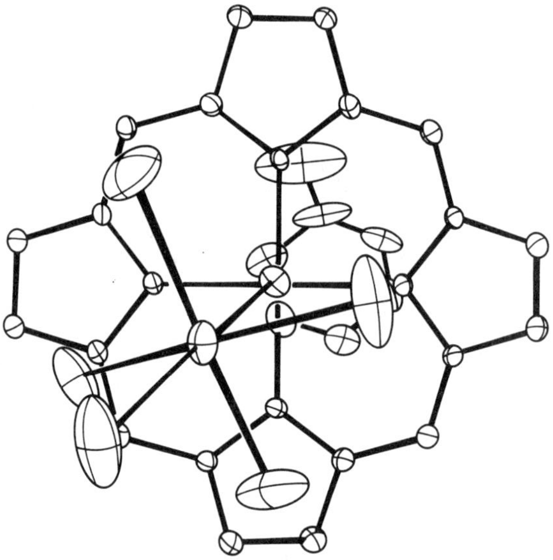

(61) View of $Fe(TPP)(FSbF_5)\cdot PhF$ showing the orientation of the hexafluoroantimonate ligand and the disposition of the fluorobenzene solvate. The porphinato core is in the plane of the paper.

Antimony(V) chloride is also a reactive compound, readily chlorinating and oxidizing a wide variety of compounds. It has a trigonal bipyramidal structure in both the vapour and crystalline forms.[148] Substitution of one of the chlorines by a methoxy group results in an oxygen-bridged dimer (**62**).[156] Two interesting examples of $SbCl_5$ adduct with Lewis bases can be cited. Weber[170] first prepared $SbCl_5\cdot SeOCl_2$ in 1865, but its structure was only solved 100 years later (**63**).[171] Cyclooctasulfur monoxide is an unstable material, but in the orange $SbCl_5$ adduct (**64**), it is stabilized.[172]

28.12.3 Antimony Pseudohalides

The pseudohalides are discussed in the halogen ligand section rather than in the nitrogen ligand section because in the majority of the examples studied the pseudohalide is the minor ligand, *i.e.* one among many—most studied are antimony halide pseudohalides.

Antimony pentachloride reacts with alkali pseudohalides (equation 50). Addition of further KX results in dimer formation, $[SbCl_4X]_2$. The crystal structure of $[SbCl_4N_3]_2$ shows a distorted

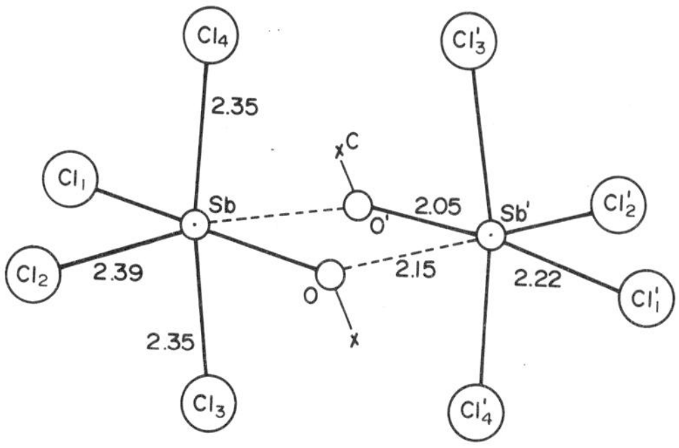

(62)

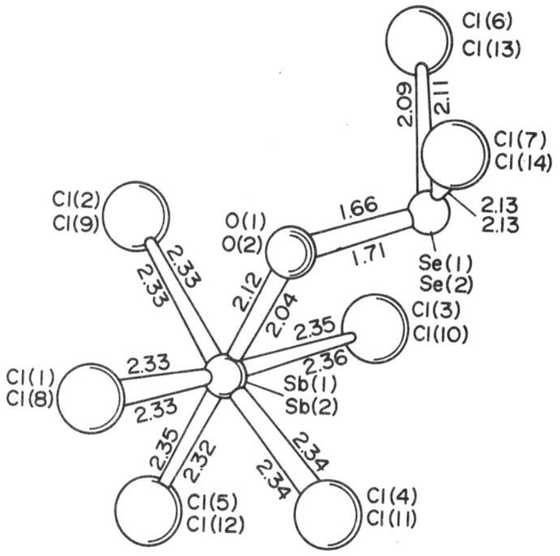

(63)

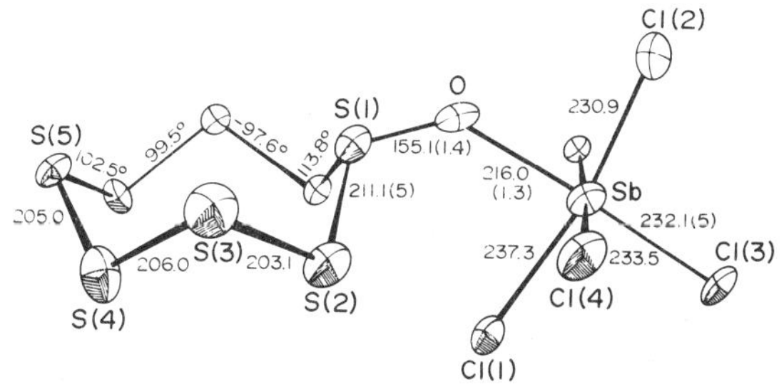

(64) Molecular structure, bond lengths (pm) and torsional angles of $S_8O \cdot SbCl_5$

octahedral environment about antimony (**65**).[162]

$$SbCl_5 + KX \longrightarrow K[SbCl_5X] \quad (50)$$
$$X = CN, N_3, NCO, NCS$$

In the case of $[SbCl_4(NCO)]_2$, however, the material decomposes at 20 °C to a trimer, $[SbCl_4(NCO)]_3$, which is a cyanuric acid derivative (**66**).[163]

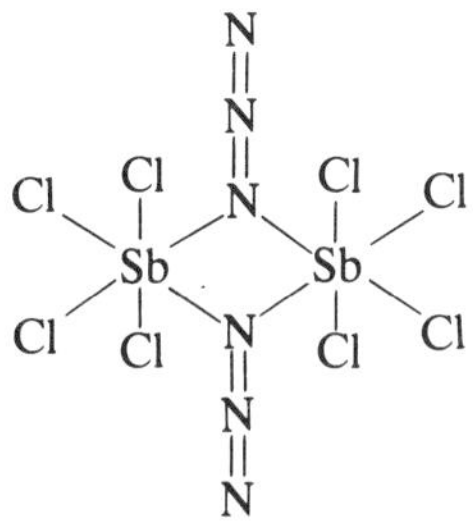

(65)

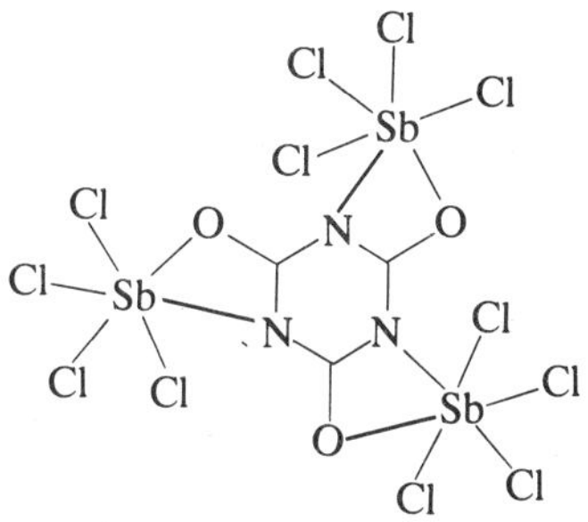

(66)

28.13 BIOLOGICAL ASPECTS OF ANTIMONY

28.13.1 Toxicology

Prudence requires that all antimony compounds should be treated with caution and, until proved otherwise, be considered toxic. While it is true that some studies have shown that, in the short term, antimony oxides may be considered non-hazardous, in the longer term they have been shown to be carcinogenic in rats. Dusts from the 'natural' antimony oxides are more dangerous than chemically pure Sb_2O_3 because the minerals are often associated with arsenic and lead.

'Antimony spots' are temporary pustular skin eruptions that afflict workers exposed to antimony compounds. Prolonged or acute exposure results in the build up of antimony in the tissues, especially in liver, kidney, adrenals and thyroid. Antimony(III) is considered to be more toxic than antimony(V) because it is relatively slowly excreted. Long term exposures (up to 28 years) have resulted in pneumoconiosis and emphysema, but even after a few months exposure (5–10 mg m^{-3}) to fumes from antimony smelting pathological symptoms were observed, *e.g.* rhinitis, pharyngitis and tracheitis.[174]

Antimony pentachloride, in a case of acute poisoning, caused severe pulmonary oedema in the three victims, two of whom died.[175] Antimony compounds form complexes with thiol groups (see Section 28.11.4), and in enzymes with thiol groups then these processes are blocked.

Evidence is accumulating that antimony is teratogenic and carcinogenic; 30 μl dm^{-3} cause mutations in *Bacillus subtilis*.[176] Antimony has also caused mutations in plants and animals.[177]

Antimony does not appear to be an essential ultratrace element and does not appear to undergo metabolic reactions analogous to those of arsenic; for example, no biomethylation of antimony has been detected.[178] Antimony trioxide is now suspected to be carcinogenic to humans.

28.13.2 Pharmacology

The hostility of antimony to life has been harnessed by chemical technology in a range of bactericides and fungicides, some of which were mentioned in Section 28.11.4. Since the advent of a rational chemotherapy a rich pharmacology based on antimony has developed. Because of the essential toxicity of antimony, its medical use is of necessity a trade off between the resistance of the patient and the pathogen to antimony poisoning. These drugs are not 'magic bullets'. Nevertheless, the antimony drugs are still the most effective treatment for a number of diseases. The reduction in mortality from 95% to less than 5% in the case of Kala-Azar is due to antimony therapy. Some of the drugs are listed in Table 22.

Table 22 Some Antimony Compounds Used in Medicine

Compound	*Pathogen*	*Human toxicity*	LD_{50} (mg kg^{-1})	*Disease*
Antimony potassium tartrate	*Schistosoma japonicum*	Toxic	48.8 (mouse interperitoneal)	Schistosomiasis (bilharziasis)
Antimony(V) sodium tartrate	*Schistosoma japonicum*	Less toxic than K salt	257 (mouse intravenous)	Schistosomiasis
Sodium antimony(III)bispyro-chatecol-3,5-disulfonate)	*Schistosoma haematobium*		80 (rabbit intravenous)	Schistosomiasis
(Stibophen)	*Schistosoma mansoni*		320 (mouse subcutaneous)	Leishmaniasis
Antimony(III) sodium gluconate		Less toxic than tartrates	172 (mouse intravenous)	Schistosomiasis
Sodium antimony(III) dimercaptosuccinate (Stibocaptate)				Schistosomiasis
'Ethylstibamine'[a]				Kala Azar Espudia
Antimony(V) sodium gluconate				Cutaneous leishmania

[a] A mixture of *p*-aminophenylstibonic acid, *p*-acetylaminophenylstibonic acid, antimonic(V) acid and diethylamine.

The structure of crystalline tartar emetic, dipotassium di-μ-(+)-tartrato(4)-bisantimony(III) trihydrate, was first elucidated by X-ray crystallography in China as the (±) salt by H.-C. Mu[179] and further refined by Jacobson and Gress[180] as the (+) isomer by three dimensional X-ray and white radiation neutron diffraction. The structure of $[Sb_2\{(+)\text{-}C_4H_2O_6\}_2]^{2-}$ is shown in (**67**); the coordination about antimony is that of a highly distorted trigonal bipyramid. The axial oxygen atoms are of the carboxylate group and the equatorial is the hydroxyl oxygen (**68**). The bond lengthening between Sb—O(ax) and Sb—O(eq) is on average 0.17 Å. Between the dimers there is a weak Sb—O interaction involving the uncoordinated carboxylate oxygen atom. The dimers are bound loosely to clusters of three water molecules, forming sheets. Potassium ions lie between the sheets holding them in place in the crystal.

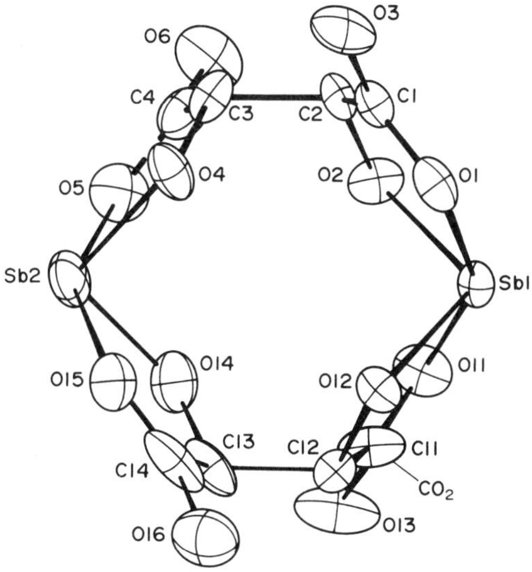

(67)

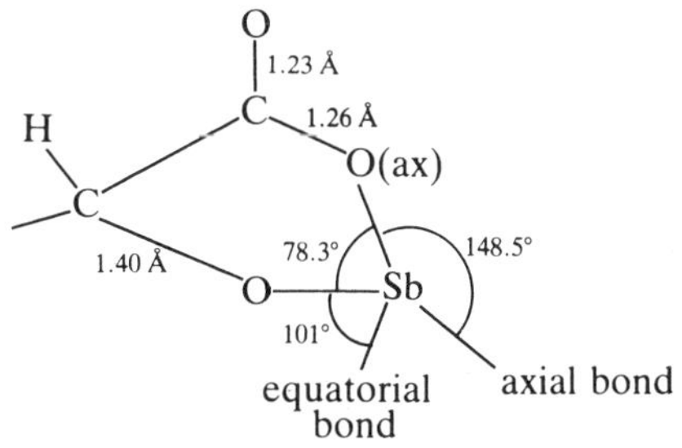

(**68**)

Finally, an antimony analogue of choline, $HOCH_2CH_2NMe_3^+X^-$, has been synthesized (Scheme 5). The compound acted as a substrate for choline kinase.[181]

$$Me_3Sb + ICH_2CH_2OH \xrightarrow{\text{diglyme}} Me_3SbCH_2CH_2OH^+I^-$$

$$\downarrow \text{ion exchange}$$

$$Me_3SbCH_2CH_2OOMe^+Cl^- \xleftarrow{Ac_2O} Me_3SbCH_2CH_2OH^+Cl^-$$

Scheme 5

28.14 BISMUTH

Bismuth was established as an element in 1739 by a Mr Potts. It had been known since the Middle Ages, but had been thought to be a type of lead. It is the heaviest member of Group 5 and has the electronic configuration $[Xe]4f^{14}5d^{10}6s^26p^3$.

28.15 BISMUTH–GROUP 4 BONDS

28.15.1 Carbon Ligands

Tetraphenylbismuthonium diphenylbis(trifluoroacetato)bismuthate (equation 51)[182] contains a distorted tetrahedral cation and an anion (**69**), the structure of which has been interpreted in terms of a stereochemically active lone pair. This appears to be the only four-coordinate trigonal bipyramidal bismuth structure established to date; in antimony chemistry these types form polymeric polyanionic structures.

$$Ph_5Bi + PhBi(CF_3CO_2)_2 \xrightarrow{Et_2O} [Ph_4Bi][Ph_2Bi(CF_3CO_2)_2] \quad (51)$$
$$86\%$$

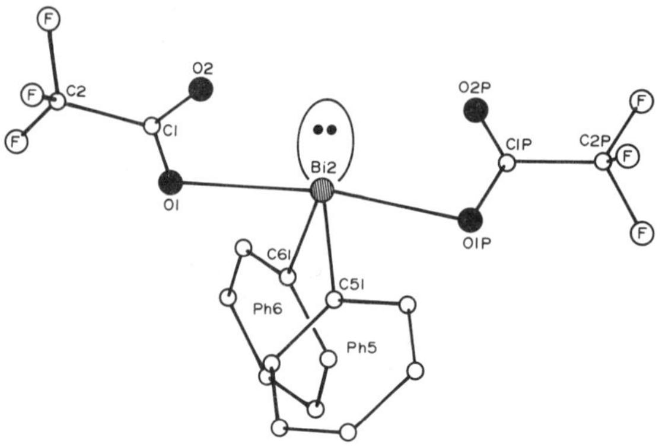

(69)

X-Ray analysis has shown Ph_3BiCl_2 to have a trigonal bipyramidal structure with axial chlorines.[183] 8-Hydroxyquinoline can displace one chlorine to give $Ph_3Bi(8\text{-}HQ)Cl$, which also has a trigonal bipyramidal structure (**70**). The nucleophilic N interaction with bismuth, *d*(Bi—N) 2.807 Å, causes the twisting of the two nearest phenyl rings.[182]

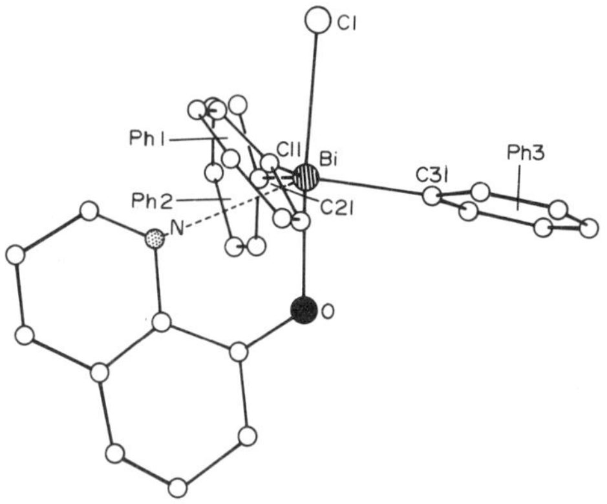

(70)

28.15.2 Silicon, Germanium and Tin Ligands

28.15.2.1 Silicon ligands

The silicon–bismuth bond has been demonstrated in a number of compounds such as $Bi(SiF_3)_3$, a volatile crystalline compound[184] which decomposes on heating; it is air and moisture sensitive. Germanium–bismuth compounds can be prepared by the use of $Bi(SiEt_3)_3$ (equations 52 and 53).[185]

$$3Et_3SiH + Et_3BI \xrightarrow[15\ h]{165\ °C} (Et_3Si)_3Bi + 3S_2H_6 \quad (52)$$
$$33\%$$

$$(Et_3Si)_3Bi + Et_3GeH \xrightarrow[16\ h]{180\ °C} (Et_3Ge)_3Bi + Et_3SiH \quad (53)$$
$$73\%$$

The physical properties of some bismuth complexes containing Si, Ge and Sn ligands are listed in Table 23.

Table 23 Some Physical Properties of Si-, Ge- and Sn-containing Bismuthines

Compound	*B.p.* (°C/Torr)	*M.p.*[a] (°C)	*Density* (g cm^{-3})	*Ref.*
$(Et_3Si)_3Bi$	145–6/1		1.273	185
$(Ph_3Si)_3Bi$		216–218 (d)		186
$(Me_3Ge)_3Bi$	114–6/0.1			187
$(Et_3Ge)_3Bi$	167–8/2.5		1.586	188
$(Et_3Ge)_2BiEt$	132–6/2			188
$(Ph_3Ge)_3Bi$		225–7		186
$[(C_6F_5)_3Ge]_2BiEt$		148–150		189
$(Et_3Sn)_3Bi$		160–170 (d)	1.743	184
$(Ph_3Sn)_3Bi$		138–142 (d)		190

[a] d = decomposed.

28.15.2.2 Germanium ligands

Germanium–bismuth compounds have been prepared by transmetallation reactions,[185,191] as shown in the previous section (28.15.2.1).

Triethylbismuthine reacts with $(C_6F_5)_3GeH$ to produce[189] mono- and di-substituted germanium compounds, but not the trisubstituted species. If disubstituted germanes are used, an interesting Bi_2Ge_3 ring system is produced (Scheme 6).

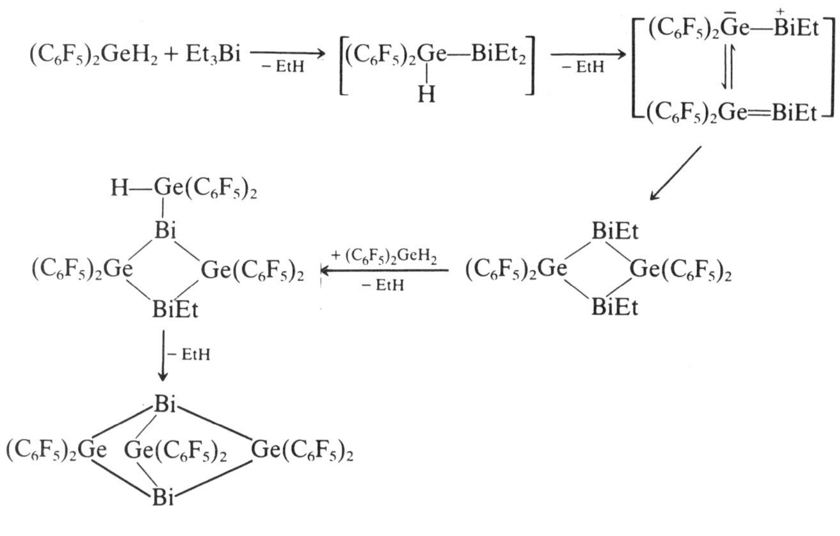

Scheme 6

X-Ray analysis has shown the structure of $[(C_6F_5)_2Ge]_3Bi_2$ to be a trigonal bipyramid (**71**) in which Bi has trigonal pyramidal geometry.[192] The average *d*(Bi–Ge) is 2.739 Å and average angles are Ge—Bi—Ge 72.1° and Bi—Ge—Bi 93.9°.

28.15.2.3 Tin ligands

Tin bismuthines have been prepared in a similar manner to the germanium bismuthines (equation 54).[185,193] The tin bismuthines decompose on heating and are air and moisture sensitive. They have, as yet, found little use.

$$Et_3Bi + Me_3SnH \xrightarrow[5\,h]{100\,°C} (Me_3Sn)_3Bi + C_2H_6 \qquad (54)$$

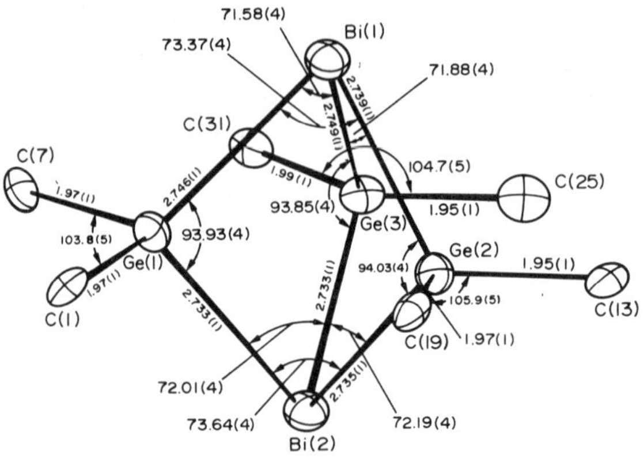

(71)

28.16 BISMUTH–GROUP 5 BONDS

28.16.1 Nitrogen Ligands

There are a large number of compounds which contain bismuth–nitrogen bonds, but very few in which such bonding represents the main framework of the bonding. There does exist $BiCl_3$ coordination compounds such as the stable colourless $[BiCl_3(NH_3)_3]$ or the unstable red $[(BiCl_3)_2(NH_3)]$. These compounds are described in more appropriate sections, except where a point about the Bi—N bond is being made.

28.16.1.1 Bismuth(III) compounds

A number of aminobismuthines have been described (Table 24) and are prepared by similar methods to the aminoarsines (equations 55 and 56).[102,194]

$$(Me_3Si)(Me)NLi + Me_{3-n}BiBr_n \longrightarrow [(Me_3Si)(Me)N]_nBiMe_{3-n} \quad (55)$$

$$BiCl_3 + 3LiNR_2 \xrightarrow{THF} 3LiCl + Bi(NR_2)_3 \quad (56)$$

Table 24 Some Physical Properties of Alkylaminobismuthines

Compound	*B.p.* (°C/Torr)	*M.p.* (°C)	n_{20}^{D}	*Ref.*
$(Me_2N)_3Bi$	46–49/0.16	40–40.5		194
$(Et_2N)_3Bi$	80–82/0.09		1.5322	194
$(Pr^n_2N)_3Bi$	110–112/0.02		1.4086	194
$[(Me_3Si)(Me)N]_3Bi$	90–92/0.1	27–29		102
$[(Me_3Si)(Me)N]_2BiMe$	70–71/0.1			102
$(Me_3Si)(Me)N—BiMe_2$	31–32/0.1			102

The bismuth–nitrogen bonds in these compounds are labile and decomposition slowly occurs at room temperature; they may be stored at −78 °C. Some reactions are shown in Scheme 7.

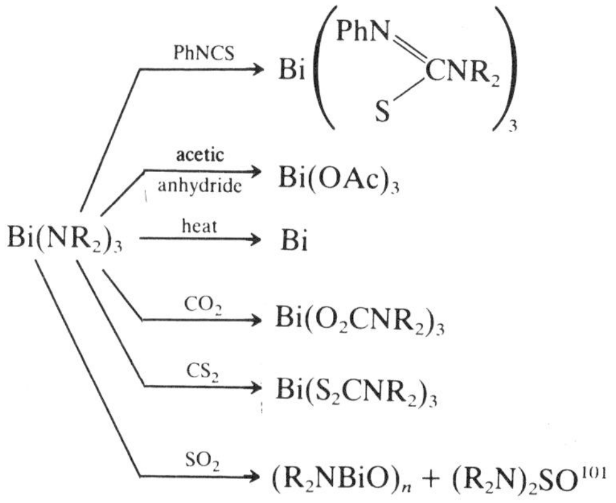

Scheme 7 Some reactions of alkylaminobismuthines

The bismuth–azide bond has received some attention. The synthesis is often surprisingly facile (equation 57).[195] The organobismuthine azides are more unstable, but the azide group displaces an alkyl group from trialkylbismuthine (equation 58).[196]

$$Bi(NO_3)_3(aq) + NaN_3(aq) \longrightarrow BiON_3 \quad (57)$$

$$Me_3Bi + HN_3 \xrightarrow{Et_2O} Me_2BiN_3 + CH_4 \quad (58)$$
$$\text{(m.p. 150 °C dec.)}$$

A series of mixed donor ligand complexes involving Bi—N bonds has been developed by Alonzo (equation 59).[197]

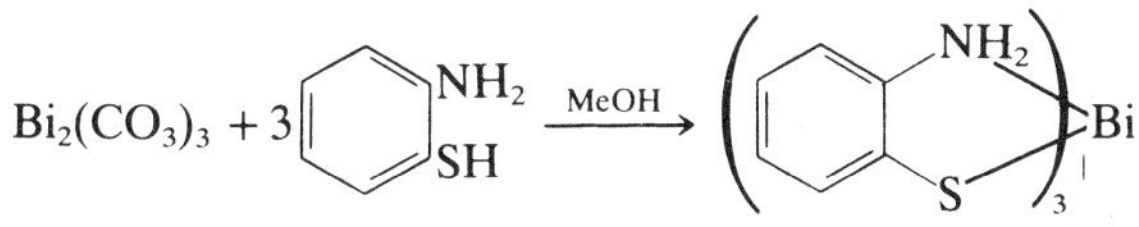

(59)

28.16.1.2 Bismuth(V) compounds

The nitrogen–bismuth bond in bismuth(V) compounds is usually found in organobismuth compounds, *e.g.* in the formation of $Ph_3Bi(N_3)_2$, remarkably from aqueous solution, as white crytals, m.p. 95 °C (dec.)[198] (equation 60). The material is metastable at room temperature, slowly decomposing. On heating the diazide decomposes into phenyl azide and the bismuth(III) azide (equation 61). Ph_2BiN_3 is itself unstable with respect to Ph_3Bi and continual heating results in formation of Ph_3Bi.

$$Ph_3BiCl_2 + NaN(aq) \longrightarrow Ph_3Bi(N_3)_2 + 2NaCl \quad (60)$$

$$Ph_3Bi(N_3)_2 \longrightarrow Ph_2BiN_3 + PhN_3 \quad (61)$$

28.16.2 Other Group 5 Bonds

Interpnictide bonding is not a well-developed field, with the exception of N—P compounds, although in recent years there have been notable achievements, such as the synthesis of RE═ER compounds. Developments in bismuth–pnictide bonding have not been as successful. However, it may confidently be predicted that analogous compounds will be developed in the near future.

The remarkable five-membered ring (MoP_2Bi_2) was prepared by Stelzer *et al.* (equation 62).[199]

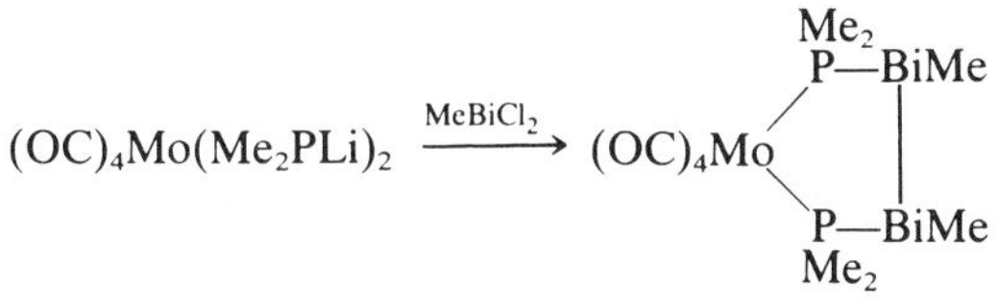

(62)

28.16.2.1 Bismuth–bismuth bonding

There are no well-characterized polymeric bismuth catenates, although a black ether-insoluble substance, described as $(PhBi)_x$, results from the reduction of phenylbismuth halides with $LiAlH_4$.[200] A polymeric bismuthine is thought to result from the decomposition of 1,1′-bibismolane.[201] To date, bismuth catenation is limited to Bi_2 in the case of free molecules and to some complex anionic and cationic clusters.

(i) Dibismuthines

Tetrakis(trimethylsilyl)dibismuthine has a *trans*-staggered conformation, shown to good effect in the stereoview **(72)**.[203] The intramolecular d(Bi—Bi) is 3.035 Å and the intramolecular d(Bi···Bi) is 3.804 Å. The Bi—Bi distance averages 2.68 Å. The molecules are linked into chains in the crystal, allowing electronic excitations along the chain, accounting for the green metallic lustrous appearance of these crystals. The compound is red in the liquid state.

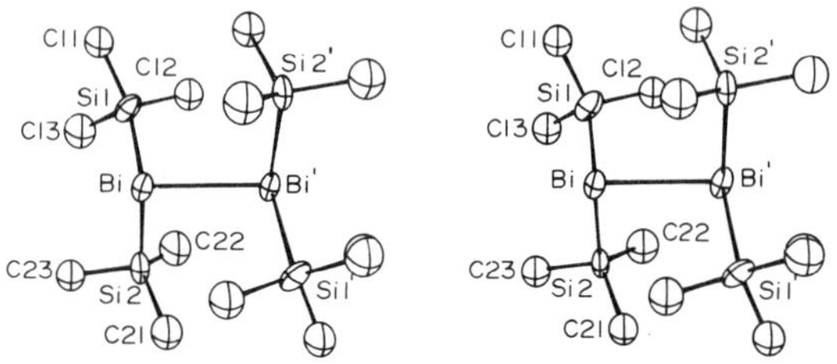

(72)

28.17 BISMUTH–GROUP 6 LIGANDS

28.17.1 Oxygen Ligands

Bismuth trioxide is a basic oxide and dissolves in mineral acids to give salts such as $Bi_2(SO_4)_3$, $Bi(NO_3)_2{\cdot}5H_2O$ and $BiPO_4$. The hydroxide, $Bi(OH)_3$, may be precipitated by the addition of a base; its composition is variable and compounds such as BiO(OH) have been described, although no structures have been confirmed. Before precipitation takes place complex cations are formed, and at one time it was thought that these were $[Bi_6(OH)_{12}]^{6+}$, but it has subsequently been shown[204] that the most common cation present is $[Bi_6O_4(OH)_4]^{6+}$. An X-ray crystal structure of $Bi_6O_4(OH)_4(NO_3)_6{\cdot}H_2O$ has shown the six Bi atoms to be at the apices of an octahedron at non-bonding separations; the octahedron is face capped by O and OH, forming Bi—O—Bi bridges. A similar structure has been demonstrated for $Bi_6O_4(OH)_4(ClO_4)_6{\cdot}7H_2O$. The lone pairs on the bismuth atoms are directed away from the cage. Each bismuth is bonded to two equatorial oxygen atoms and two axial hydroxide oxygen atoms, resulting in a distorted trigonal bipyramidal structure about bismuth; the bismuth atoms are also weakly bonded to anions in the crystal lattice.

28.17.2 Sulfur Ligands

A truly remarkable bismuth–sulfur compound was recently reported by Muller *et al.* (equation 63).[205] The dark brown compound contains an anion which is essentially two S_7 rings joined by a spiro bismuth atom, and two of these bismuth groups are joined by an S_6 chain **(73)**. This bismuth coordination polyhedron is a highly distorted square pyramid, the apex of which is the sulfur of the S_6 chain, $d(Bi{-}S_{apical})$ 2.683 Å (av) and $d(Bi{-}S_{eq})$ 2.817 Å.

$$Ph_4AsCl + S_8 + BiCl_3 + H_2S \xrightarrow[NH_3]{MeCN} \underset{60\%}{[AsPh_4]_4[Bi_2S_{34}]} \quad (63)$$

An eight-membered ring containing bismuth is formed by the condensation of diethoxybismuthines with dithiols (Table 25).[206]

Table 25 Bismuth–Sulfur Bond Lengths and Angles

Compound	*d(Bi—S)* (Å)	*d(Bi—E)* (Å)	*Bond angles* (°)		*Ref.*
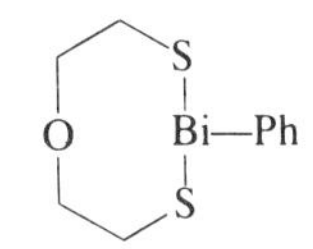	2.581 (av)	2.25 (C) 2.97 (O)	S—Bi—S C—Bi—S	104.2 88.75 (av)	206
$Rb[Bi(NCS)_4]$	2.785 (av)	2.74 (· · · N)	S(ax)—Bi—S(ax)	158.2	207
Isopropylxanthogenato-diphenylbismuthine	2.66 (ax) 3.23 (ax)	2.25 [C(eq)] (av)	S(ax)—Bi—S(ax)	161.9	208
$MeBi(S_2CNEt_2)_2$	2 × (2.688) (av) 2 × (2.955) (av)	2.24 [C(apex)] (av)	S—Bi—C	84.5, 95.1	209

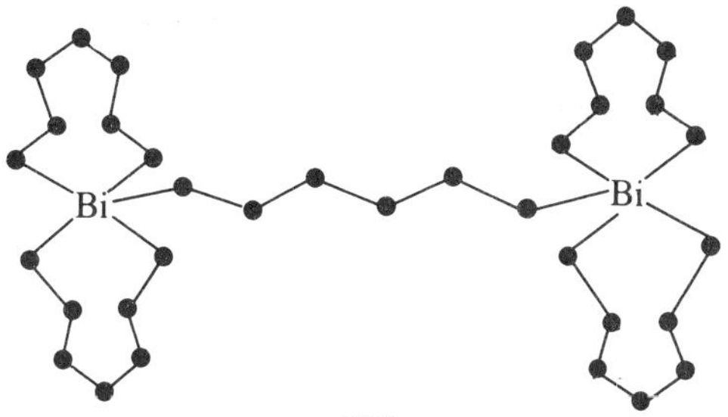

(73)

An X-ray study of 5-phenyl-1-oxa-4,6-dithia-5-bismocane revealed structure (**74**). The Bi—O interaction ensures the conformation shown, in the crystal at any rate. The bismuth coordination sphere is completed by intermolecular sulfur ligands (**75**) and the intermolecular stacking is shown in (**76**).

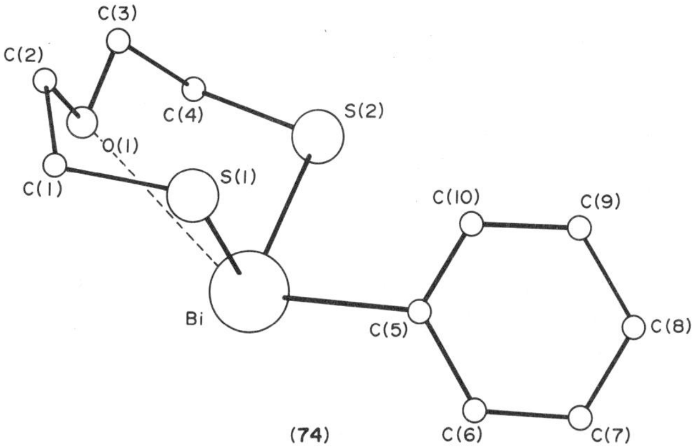

(74)

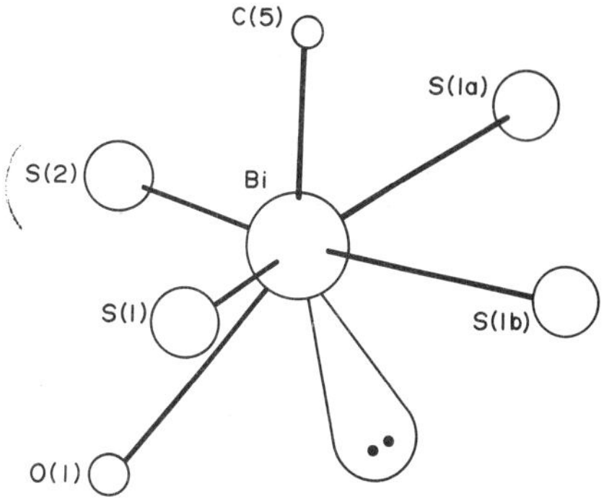

(75)

In the case of ambidentate ligands such as thiocyanate, bismuth has a preference for the 'softer' sulfur ligand. This is demonstrated in the X-ray study of $Rb[Bi(NCS)_4]$.[207] Four SCN^- groups coordinate *via* sulfur, *d*(Bi—S) 2.785 Å (av), in a trigonal bipyramidal arrangement, the lone pair occupying an equatorial site. The distortion is shown in the S(ax)—Bu—S(ax) angle of 158.2° (**77**). A weak interaction is suggested by the *d*(Bi· · ·N) of 2.74 Å.

28.17.2.1 Organosulfur ligands

The preference of bismuth for sulfur donor atoms is demonstrated by their method of synthesis. A frequent method used is to displace alkoxide groups by thio-alkyls or -aryls (equation 64). However, one of the problems in syntheses of thiobismuthines is a tendency in some cases to disproportionate, for example the reaction (65) does not occur; instead, the dithiobismuthine is obtained (equation 66).[210] However, in the case of R groups with

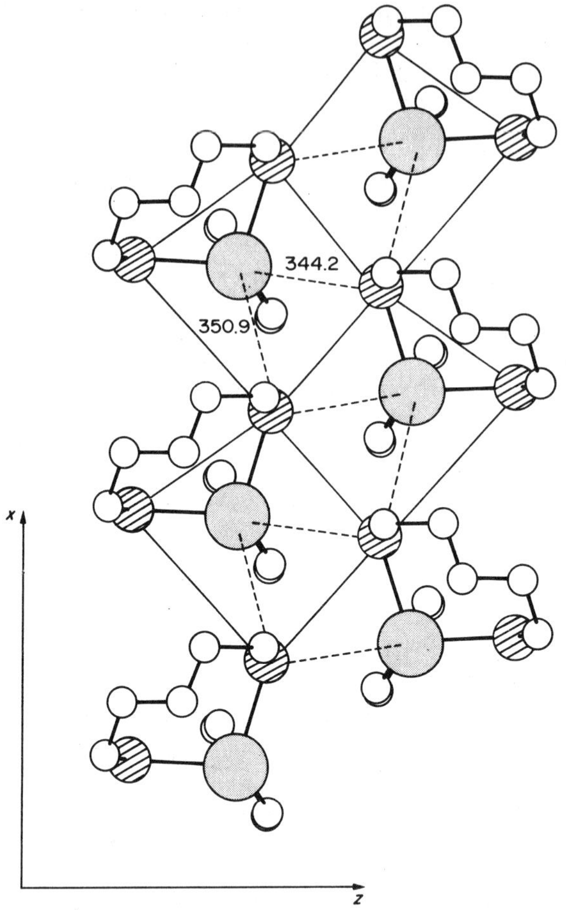

(76)

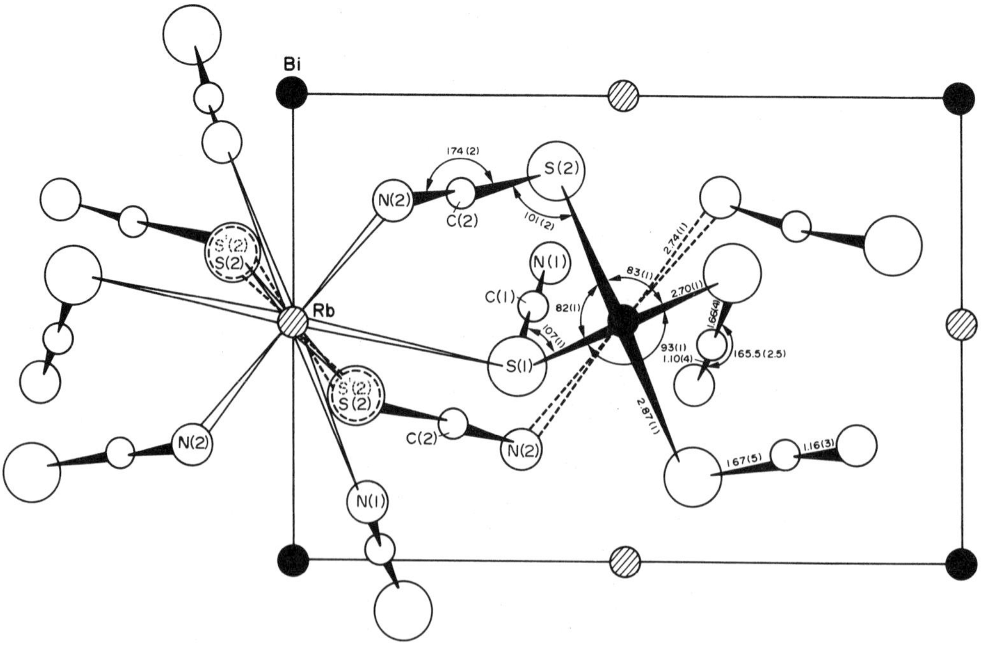

(77) The structure of $Rb[Bi(SCN)_4]_2$ projection along the (001) axis

a 1,3-diaza-2-yl heterocyclic structure, *e.g.* pyrimidin-2-yl, the monosulfide is obtained (equation 67).

$$MeBi(OEt)_2 + 2HSR \longrightarrow MeBi(SR)_2 + 2HOEt \qquad (64)$$

$$Me_2BiBr + NaSR \not\longrightarrow Me_2BiSR + NaBr \qquad (65)$$

$$Me_2BiBr + NaSR \longrightarrow Me_3Bi + MeBi(SR)_2 \qquad (66)$$

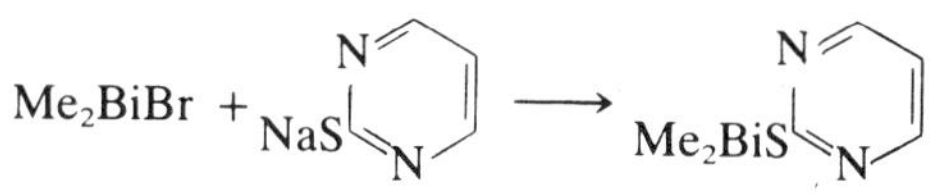

(67)

Examples of $(RS)_3Bi$, $(RS)_2BiR'$ and $RSBiR'_2$ are given in Tables 26 and 27 together with some physical data.

Table 26 Bis(organothio)bismuthines, $RBi(SR')_2$

Compound	*Colour*	*M.p.* (°C)	*Ref.*
$PhBi(SPh)_2$	Yellow plates	170	211, 212
p-$MeC_6H_4Bi(SPh)_2$	—	155	211
p-$ClC_6H_4Bi(SPh)_2$	Yellow	170	211
α-NaphthylBi$(SPh)_2$		240	211
$MeBi(SPh)_2$	Yellow crystals	131	213
$MeBi(SEt)_2$	Yellow crystals	91	213
$MeBi(SCH_2Ph)_2$	Yellow crystals	90–92	213
$MeBi(SCH_2CH_2OH)_2$	Pale yellow crystals	86	213
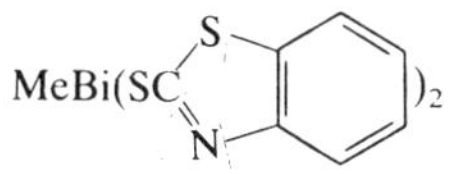	Yellow	156	213
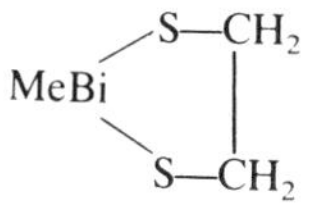	Yellow crystals	100	213
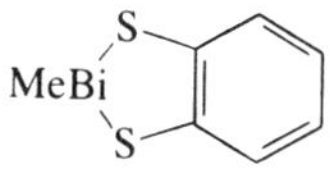	Orange-red crystals	185	213
$PhBi(SEt)_2$	Yellow crystals	74	212
$PhBi(SCH_2Ph)_2$	Large yellow crystals	90	212
$PhBi(SCH_2CH_2OH)_2$	Pale-yellow crystals	57	212
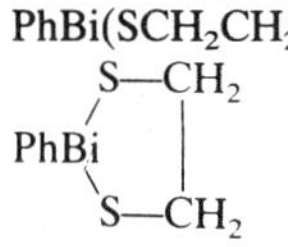	Pale-yellow crystals	137	212

Table 27 Some Tris(organothio)bismuthines, $(RS)_3Bi$

Compound	*Colour*	*M.p.*[a] (°C)	*Ref.*
$(PhS)_3Bi$	Orange-red needles	90–91 98	211 212
$(Bu^tS)_3Bi$		150 (d)	214
$(C_6Cl_5S)_3Bi$			215
$(EtS)_3Bi$	Deep-yellow crystals	80–82	212
$(PhCH_2S)_3Bi$	Deep-yellow crystals	88–90	212
$(HOCH_2CH_2S)_3Bi$	Pale-yellow crystals	77 (d)	212

[a] d = decomposed.

Bidentate organosulfur ligands have been synthesized in a similar manner to the simple organosulfur compounds, *e.g.* equation (68).

$$\mathrm{Ph_2BiBr + NaS\underset{\underset{S}{\|}}{C}OR \longrightarrow Ph_2BiS\underset{\underset{S}{\|}}{C}OR} \qquad (68)$$

Some examples of alkylxanthogenato and organodithiocarbamatobismuthines are given in Tables 28 and 29. The crystal structure of isopropylxanthogenatodiphenylbismuth has been solved (**78**).[208] The intermolecular sulfur ligand makes the bismuth coordination polyhedron a trigonal bipyramid (Table 25). In the case of one organo group on bismuth and two bidentate sulfur ligands, *e.g.* bis(diethylthiocarbamato)methylbismuthine,[209] a pentagonal pyramidal structure is adopted in the crystalline state. The compound is monomeric in benzene solution, but crystallizes as a dimer (**79**). The four basal sulfur ligands and the bismuth atom are approximately coplanar, with an apical methyl group (Table 25). It is assumed that the bismuth lone pair is *trans* to the methyl group.

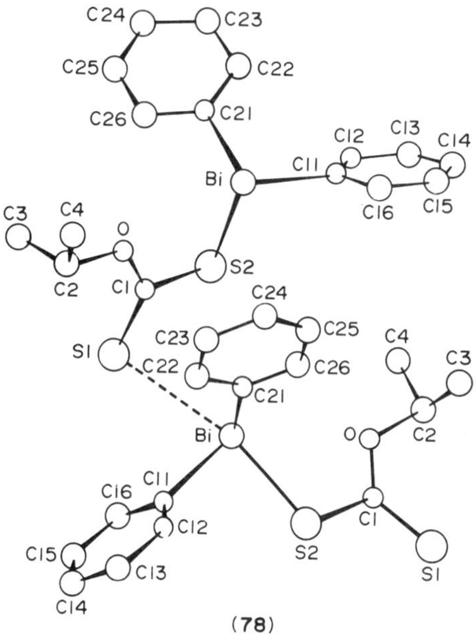

(78)

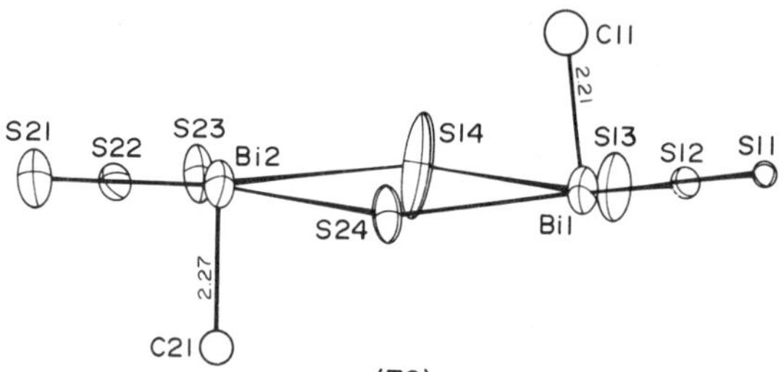

(79)

In the case of three bidentate sulfur ligands, *e.g.* $[Bi(pedt)_3]_2 \cdot EtOH$ (pedt = 1-pyrrolidinecarbondithioate), more complex structures are formed. The dimer contains two differently coordinated bismuth atoms: one is trigonal prismatic with an extra position on a rectangular face and the other bismuth atom is pentagonal pyramidal. The overall molecular structure is (**80**).[270]

Table 28 Some Bis(diorganodithiocarbamato)organobismuthines, $RBi(S_2CNR'_2)_2$[216]

R	*R'*	*Colour*	*M.p.* (°C)
Me	Me	Pale yellow	185
Me	Et	Bright yellow	141
Me	Piperidyl[a]	Pale yellow	192
Ph	Me	Yellow	219
Ph	Et	Yellow	137
Ph	Piperidyl[a]	Yellow	228

[a] Piperidyl = R'_2.

Table 29 Some Alkylxanthogenatodimethyl (or Phenyl) Bismuthines, $(R_2BiS)C(=S)(OR')$

R	*R'*	*Colour*	*M.p.* (°C)	*Ref.*
Me	Me	Pale yellow	56	271
Me	Et	Pale yellow	99	271
Me	Pr^n	Pale yellow	84	271
Me	Pr^i	Pale yellow	107	271
Me	Bu^n	Colourless	68	271
Me	Bu^i	Colourless	78	271
Ph	Me	Pale yellow	96	208
Ph	Et	Pale yellow	77	208
Ph	Pr^n	Pale yellow	86	208
Ph	Pr^i	Pale yellow	135	208
Ph	Bu^n	Pale yellow	72	208
Ph	Bu^i	Pale yellow	90	208

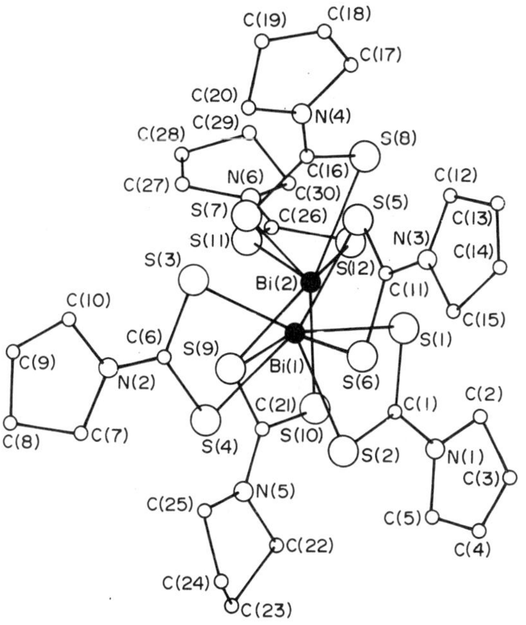

(80)

28.17.3 Selenium and Tellurium Ligands

28.17.3.1 Organobismuth–selenium and –tellurium bonds

Selenium–bismuth bonds are formed in the mutual cleavages of dibismuthines and diselenides (equation 69).[218] This particular compound is stable at −30 °C, unlike the tellurium analogue. At room temperature the compound disproportionates to give the diseleno derivative (equation 70). Similar behaviour was noted in the analogous sulfur compounds (Section 28.17.2.1).

$$Me_2BiBiMe_2 + Ph_2Se_2 \longrightarrow Me_2BiSePh \qquad (69)$$

$$Me_2BiSePh \xrightarrow{\text{r.t.}} Me_3Bi + MeBi(SePh)_2 \qquad (70)$$

28.18 BISMUTH–HALOGEN COMPOUNDS

28.18.1 Bismuth(III) Halides

28.18.1.1 Bismuth trifluoride

Reference has been made to the similarity of Bi^{3+} to Y^{3+} and rare earth M^{3+} in properties and structures of their compounds. In this respect BiF_3 is a good example, since its structure is similar to that of YF_3, being a nine-coordinate distorted tricapped trigonal prism with essentially ionic bonding.[219] But, the eight near neighbours are at 2.217–2.502 Å, and the ninth fluorine 'ligand' is at 3.100 Å, as determined by neutron diffraction;[219] BiF_3 is thus best described as eight-coordinate. There are indications, *e.g.* an anomalously large unit cell, that the lone pair is structurally active in BiF_3.

28.18.1.2 Bismuth trichloride

In the gas phase $BiCl_3$ is a pyramidal molecule with *d*(Bi—Cl) 2.48 Å and Cl—Bi—Cl 100°.[220] In the crystalline state there are five additional near neighbours giving a bicapped trigonal prism for the bismuth atom (Table 30).

Bismuth trichloride readily forms addition compounds, *e.g.* $BiCl_3(NH_3)_3$, $BiCl_3(NO)$, $BiCl_3(NO_2)_2$, $BiCl_2(NOCl)$. These complexes are stable in dry air, but the ligands are displaced by water. It dissolves in organic solvents such as ether, probably forming etherates, $BiCl_3(Et_2O)_n$. Complex halides such as $Cs_2Na(BiCl_6)$ and $(Et_2NH_2)(BiCl_4)$ are known (Table 30).

The ligand 3,4,5,6-tetrahydropyrimidine-2(1*H*)-thione (THPT) forms an S-bonded complex, $BiCl_3(THPT)_3$.[225] The crystal structure (**81**) shows a *fac* octahedron. Interestingly, with bismuth perchlorate $Bi(ClO_4)_3(THPT)_5$ is formed.

As is well known, simple stoichiometry of a compound is not a good guide to the coordination number of the central atom. This is well shown in $(BiCl_3)_3{\cdot}7tu$ (tu = thiourea), where bismuth is octahedrally coordinated to thiourea *via* the sulfur atom.[240] The structure was found to consist of bismuth cations and anions, $[Bi_2Cl_4tu_6]^{2+}[BiCl_5tu]^{2-}$ (**82**).

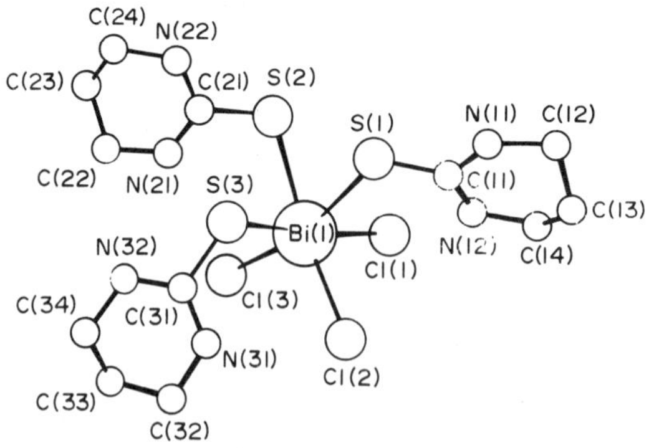

(81)

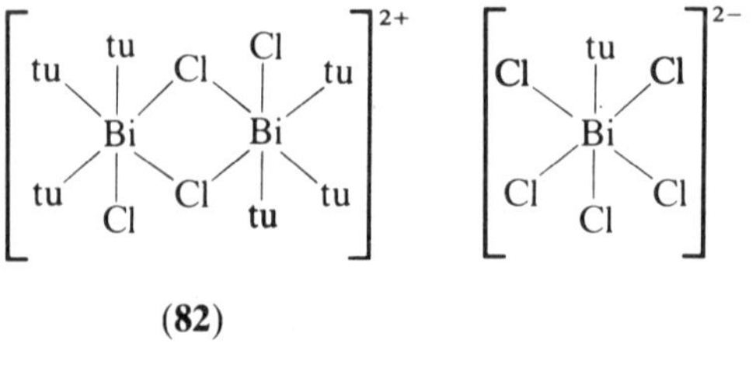

(82)

Table 30 Some Crystal Structure Data for the Bismuth Halides

Compound	*d(Bi—X)* (Å)	*d(Bi—E)* (Å)	*Bond angles* (°)		*Ref.*
$Bi_{12}Cl_{14}$ $[BiCl_5]^{2-}$	2.617–3.100				221
$[Bi_2Cl_8]^{2-}$	2.605–3.017				
$BiCl_3$(vap)	2.48		Cl—Bi—Cl	100.0	220
$BiCl_3$(cryst)	2.50		Cl—Bi—Cl	84, 94	222
$Cs_2Na[BiCl_6]$ (reg octahed)	6 × 2.66				223
$[Et_2NH_2][BiCl_4]$ ($BiCl_6$octahedra)	2 × 2.526 (t) (av) 2 × 2.718 (μ) (av) 2 × 2.961 (μ) (av)				224
$BiCl_3$·3THPT	2 × 2.737 (av)	3 × 2.799 (av) (S)	Cl—Bi—Cl S—Bi—S	90.6–99.5 74.3–84.0	225
BiF_3	8 × 2.217–2.502				219
BiF_5 (BiF_6 octahedra)	4 × 1.90 2 × 2.11				226
$[NH_4][BiF_4]$ (BiF_9 polyhedron)	9 × 2.19–2.86				227
$Bi_{12}Br_{14}$ $[BiBr_5]^{2-}$	2.798–2.148				228
$[Bi_2Br_8]^{2-}$	2.742–3.066				
α-$BiBr_3$	3 × 2.66 3 × 3.32				229
β-$BiBr_3$	6 × 2.82				229
$BiBr_3$(vap)	2.63		Br—Bi—Br	100 ± 4	220
$K_4[Bi_2Br_{10}]·4H_2O$	2.993 (μ) (av) 2.805 (t) (av)		Bi—Br—Bi	98.3–99.5	230
[2-picH]$[BiBr_4]$ ($BiBr_6$ octahedra)	2.64(t) 2.97–3.27(μ)				231
$[Ph_4P]_3[Bi_2Br_9]$ ($μ_3$-$BiBr_6$)	2.74 (t) (av) 3.046 (μ) (av)		Bi—Br—Bi	83.2	232
$[Et_2NH_2]_3[BiBr_6]$ (dist octah)	3 × 2.749 (Br′) 3 × 3.006 (Br″)		Br′—Bi—Br′ Br″—Bi—Br″	92.8 89.6	233
$Cs_3[Bi_2Br_9]$	2.713 (t) 2.979 (μ)				234
$Rb_3[BiBr_6]$	6 × 2.849 (av)				235
$[Hg_7(HgBr)_2As_4][Bi_2Br_{10}]$ (see text)	5 × 2.855 ($BiBr_6$) 4 × 2.777 ($BiBr_5$) 2 × 2.970 (μ-Br)	Bi · · · Br 3.258 Br—Hg 2.502			236
BiI_3 (BiI_6 octahedra)	6 × 3.07		I—Bi—I	90.0	237
$Cs_3[Bi_2I_9]$ (face-sharing octahed)	2.923 (t) 3.249 (μ)				238
$Rb_5I(I_3)(BiI_6)(H_2O)_2$	3.073 (av)		I—Bi—I	89.0–91.0	239

28.18.1.3 Bismuth tribromide

Bismuth tribromide is a pryramidal molecule with d(Bi—Br) 2.66 Å. The golden-yellow crystalline $BiBr_3$ occurs in two structural modifications (equation 71). Both modifications are monoclinic. The α-$BiBr_3$ is trigonal pyramidal with three near neighbours, 3 × 2.66 Å, and three more distant bromine ligands, 3 × 3.32 Å. Above 158 °C the structure is an $AlCl_3$ type, *i.e.* polymeric with six bromine near neighbours, 6 × 2.82 Å. Thus, $BiBr_3$ is the only Group 5 trihalide which can assume both a molecular (α-$BiBr_3$) and a polymeric (β-$BiBr_3$) structure.

$$\alpha\text{-BiBr}_3 \xrightleftharpoons{158\,^\circ\text{C}} \beta\text{-BiBr}_3 \qquad (71)$$

The tribromide is prepared in a similar fashion to the chloride (Section 28.18.1.2), and it readily forms complexes. These may be simple octahedra $BiBr_6$, *e.g.* in the case of [2-picH]$[BiBr_6]$, doubly bridged dimers as in the case of $K_4[Bi_2Br_{10}]·nH_2O$ (**83**) and triply bridged dimers as in the case of $[Ph_4P][Bi_2Br_9]$ (**84**).

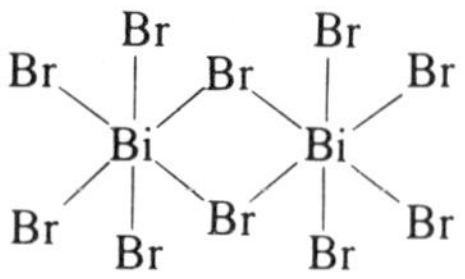

(83)

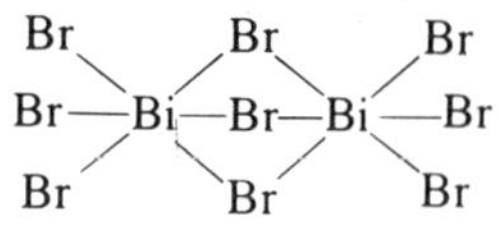

(84)

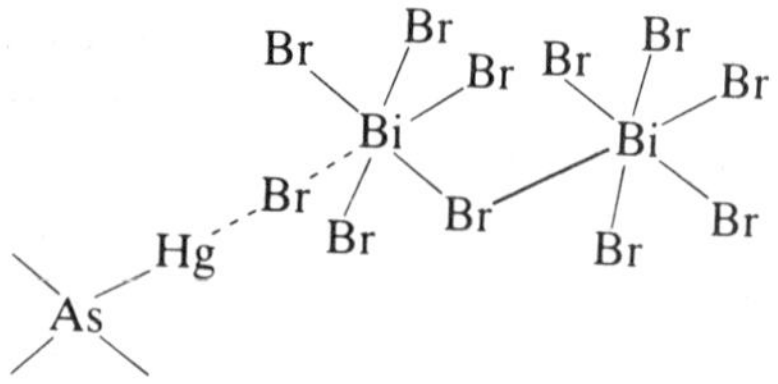

(85)

In the case of the singly bridged dimer anion $[Bi_2Br_{11}]^{5-}$, a bromine ligand is 'lost' to the polymeric cation, forming $[Hg_7(HgBr)_2As_4]^{4+}[Bi_2Br_{10}]^{4-}$ (**85**).[236] The anion thus contains bismuth in octahedral and tetragonal pyramidal coordination, and is quite distinct from the $[Bi_2Br_{10}]^{4-}$ anion in $K_4[Bi_2Br_{10}]\cdot 4H_2O$.

28.18.1.4 Bismuth triiodide

The greenish-black crystals of BiI_3 are composed of a hexagonal close packed array of iodine atoms with bismuth atoms in the interstitial spaces, with six nearest neighbours at 3.07 Å. It is prepared by heating the elements in stoichiometric quantities, and purified by subliming under reduced pressure.

Bismuth triiodide forms similar complexes to its lighter congeners, *e.g.* $KBiI_4$ and K_3BiI_6. The interesting compound $Rb_5I(I_3)(BiI_6)(H_2O)_2$ was formed adventitiously when preparing Rb_3BiI_6 (Table 30).[239]

28.18.2 Bismuth Pentafluoride

The only stable bismuth(V) halide is BiF_5, a white crystalline material, the structure[241] of which consists of BiF_6 octahedra in infinite *trans* chains in the manner of α-UF_5. It reacts explosively with water to form OF_2, O_3 and a brown solid (Bi^V oxyfluoride?) and it is a powerful fluorinating agent.

Bismuth pentafluoride is a considerably weaker Lewis acid than is SbF_5, as is seen in the structure of the adduct formed with XeF_4 (**86**).[242] The antimony analogue is described as ionic, $[XeF_3]^+[SbF_6]^-$, on the basis of the 'bridging' fluorine ligand at d(Bi—F) 2.49 Å.[243] In $[XeF_3][BiF_6]$ the μ-F distance, d(Bi—F) 2.25 Å, is shorter and clearly has much greater covalent character than that of the antimony analogue.

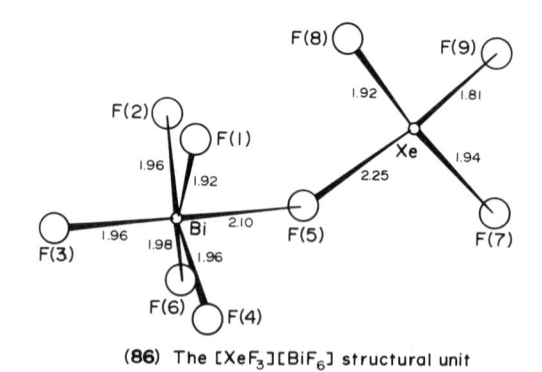

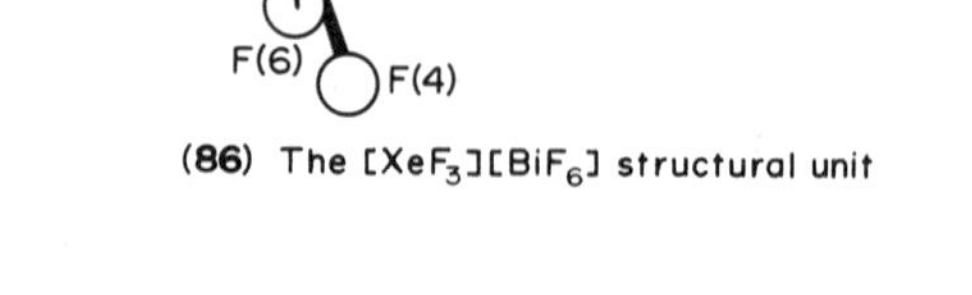

(**86**) The $[XeF_3][BiF_6]$ structural unit

28.19 BIOLOGICAL ASPECTS OF BISMUTH

28.19.1 Pharmacology

Bismuth and its compounds have traditionally been widely used in medicine and veterinary practice. Finely divided bismuth was injected intramuscularly as an aqueous suspension in the treatment of syphilis and yaws. Its advantage was the slow absorption, possibly as a protein bound complex, giving continuous therapeutic action. The 1982 edition of 'Martindale' lists 18 bismuth compounds used in medical practice,[244] although most of these are not 'drugs of choice', and are mainly used out of traditional sentiment. This tradition has been summarized by Jaenicke.[245]

Trichlorotris(3-sulfanilamido-6-methoxypyridazine)bismuth(III) is a pharmaceutical which contains three nitrogen–bismuth bonds and three weak oxygen–bismuth links (**87**): d(Bi—Cl), 2.529 Å, is similar to that in other coordination compounds; d(Bi—N), 2.90 Å, is longer than the sum of the covalent or ionic radii; d(Bi—O) is long at 3.09 Å. The polyhedron has been described as a distorted octahedron (three chlorine, three nitrogen donors) with three extra positions occupied by the oxygen atoms. The oxygen and nitrogen atoms form a hexagon in a chair configuration (**88**).[272]

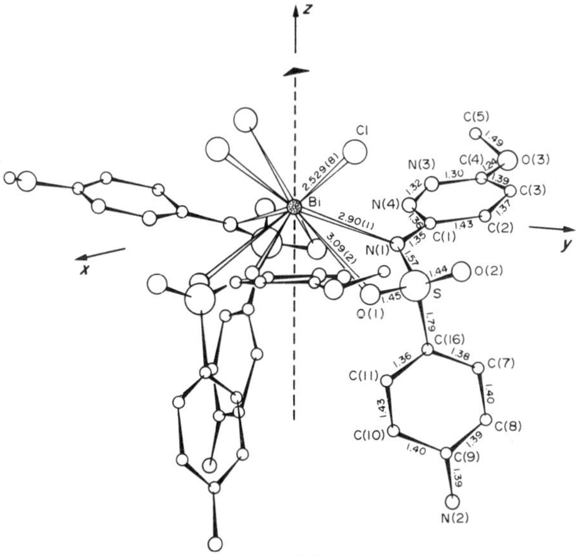

(**87**)

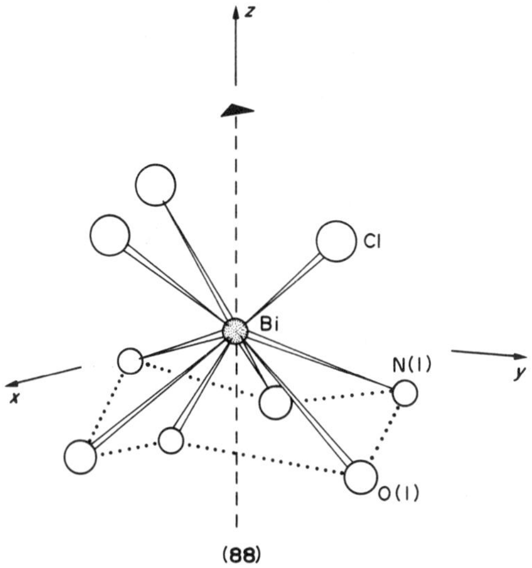

(**88**)

28.19.2 Toxicology

Bismuth is the least toxic of the As, Sb, Bi triad, which is unusual, since toxicity normally increases down a group, as shown by elements to the left of bismuth, *i.e.* Pb, Tl, Hg. However, it must be emphasized that this is not to say that bismuth is not toxic and experimentalists should avoid contact and ingestion of its compounds.

Table 31 Toxicity of Bismuth Compounds

Compound	*Mammal*	LD_{50}(mg kg^{-1})	LDL_0(mg kg^{-1})	*Method of administration*
Bi metal	Man		221	Not stated
BiOCl	Rat	22		Not stated
$Bi(NO_3)_3$	Mouse		21	Intravenous
$Bi(NO_3)_3$	Mouse		2500	Intraperitoneal
Bi trisodiumthioglycollate	Rabbit	21		Intravenous
Bi potassium sodium tartrate	Rabbit	55		Intravenous
Bi tin oxide	Mouse	178		Intravenous
$BiMe_3$	Dog		233	Oral
	Dog	12		Intravenous
$BiPh_3$	Dog	180		Intravenous

The effects of acute bismuth intoxication include gastrointestinal disturbance, anorexia, headache, malaise, skin reactions, discolouration of the mucous membranes and mild jaundice.[244]

There have been no fatalities in industry attributed to bismuth and it is regarded as relatively non-toxic for a heavy metal.[246] The toxic problems which have been recorded have in the main been iatrogenic illnesses. A characteristic blue-black line on the gums, the 'bismuth line', which may persist for years, is a feature of bismuth overdosage. Soluble salts are excreted *via* urine and may cause mild kidney damage. Less soluble salts may be excreted in the faeces, which may be black in colour due to the presence of bismuth sulfide. Table 31 contains some toxicity data.

28.20 REFERENCES

1. M. E. Weeks, H. M. Leicester and F. B. Dains, 'Discovery of the Elements', Journal of Chemical Education, Washington, DC, 1968.
2. J. L. Wardell, in 'Comprehensive Organometallic Chemistry', ed. G. Wilkinson, F. G. A. Stone and E. W. Abel, Pergamon, Oxford, 1982, vol. 2, chap. 13, p. 681.
3. I. Omae, 'Organometallic Intramolecular Coordination Compounds', Elsevier, Amsterdam, 1986.
4. Annual reviews appear in The Royal Society of Chemistry Specialist Periodical Reports on Organometallic Chemistry and in Coordination Chemistry Reviews.
5. H. Yaozeng and S. Yanchang, *Adv. Organomet. Chem.*, 1982, **20,** 157.
6. G. Ferguson and D. F. Rendle, *Chem. Commun.*, 1971, 1647.
7. Y. Yamamoto and H. Schmidbaur, *J. Chem. Soc., Chem. Commun.*, 1975, 668.
8. J. E. Drake and W. L. Jolly, *Chem. Ind. (London)*, 1962, 1470.
9. J. W. Anderson and J. E. Drake, *J. Chem. Soc. (A)*, 1970, 3131.
10. B. Beagley, A. G. Robiette and G. M. Sheldrick, *J. Chem. Soc. (A)*, 1968, 3006.
11. E. Amberger and H. Boeters, *Angew. Chem., Int. Ed. Engl.*, 1962, **1,** 268.
12. G. Kamai and Z. L. Khisamova, *J. Gen. Chem. USSR (Engl. Transl.)*, 1956, **26,** 125.
13. H. J. Vetter and H. Noth, *Z. Anorg. Allg. Chem.*, 1964, **330,** 233.
14. K. Modritzer, *Chem. Ber.*, 1959, **92,** 2637.
15. F. Kober, *J. Fluorine Chem.*, 1972, **2,** 247.
16. (a) F. Kober, *Synthesis,* 1982, 173; (b) H. J. Vetter and H. Noth, *Angew. Chem., Int. Ed. Engl.*, 1962, **1,** 663.
17. J. L. Atwood, A. H. Cowley, W. E. Hunter and S. K. Mehrotra, *Inorg. Chem.*, 1982, **21,** 1354.
18. M. J. S. Gynane, A. Hudson, M. F. Lappert, P. P. Power and H. Goldwhite, *J. Chem. Soc., Dalton Trans.*, 1980, 2428.
19. H. W. Roesky, *J. Organomet. Chem.*, 1985, **281,** 69.
20. H. W. Roesky, R. Bohra, J. Lucas, M. Noltemeyer and G. M. Sheldrick, *J. Chem. Soc., Dalton Trans.*, 1983, 1011.
21. M. D. Glick, *Diss. Abstr.*, 1965, **25,** 5546.
22. J. Escudie, C. Couret, J. Ranaivonjatova and J.-G. Wolf, *Tetrahedron Lett.*, 1983, 3625.
23. J. C. Summers and H. H. Sisler, *Inorg. Chem.*, 1970, **9,** 862.
24. W. Wolfsberger, *Z. Naturforsch., Teil B,* 1974, **29,** 35.
25. (a) L. Cadet de Gassicourt, 'Memoires de Mathématique et Physique des Savants Etrangers', 1760, **3,** 623; (b) R. Bunsen, *Ann. Chem. Pharm.*, 1841, **37,** 1.
26. A. V. Baeyer, *Ann. Chem. Pharm.*, 1858, **107,** 257.
27. A. L. Rheingold, J. E. Lewis and J. M. Bellama, *Inorg. Chem.*, 1973, **12,** 2845.
28. J. J. Daly and F. Sanz, *Helv. Chim. Acta,* 1970, **53,** 1879.
29. B. M. Gatehouse, *Chem. Commun.*, 1969, 948.
30. A. L. Rheingold and M. R. Churchill, *J. Organomet. Chem.*, 1983, **243,** 165.
31. P. S. Elms, B. M. Gatehouse, D. J. Lloyd and B. O. West, *J. Chem. Soc., Chem. Commum.*, 1974, 953.
32. P. S. Elms and B. O. West, *Aust. J. Chem.*, 1970, **23,** 2247.

33. P. S. Elms, P. Leverett and B. O. West, *Chem. Commun.*, 1971, 747.
34. H. Goldwhite and R. G. Teller, *J. Am. Chem. Soc.*, 1978, **100**, 5357.
35. C. A. Poutasse, R. O. Day, J. H. Holmes and R. R. Holmes, *Organometallics*, 1985, **4**, 708.
36. G. Becker, G. Gutekunst and H. J. Wessely, *Z. Anorg. Allg. Chem.*, 1980, **462**, 113.
37. R. Mercier and J. Douglade, *Acta Crystallogr., Sect. B*, 1982, **38**, 720, 896, 1731.
38. H. Sommer and R. Hoppe, *Z. Anorg. Allg. Chem.*, 1977, **430**, 199.
39. M. Pallazzi and S. Jaulmes, *Acta Crystallogr., Sect. B*, 1977, **33**, 908; G. D. McLeod, *US Pat.* 2 768 192 (*Chem. Abstr.*, 1957, **51**, 5816); G. Kamai, *Zh. Obshch. Khim.*, 1966, **36**, 704 (*Chem. Abstr.*, 1966, **65**, 10 618).
40. G. C. Pappalardo, *Acta Crystallogr., Sect. C*, 1983, **39**, 1618.
41. M. Drager, *Chem. Ber.*, 1974, **107**, 2601.
42. C. L. Raston and A. H. White, *J. Chem. Soc., Dalton Trans.*, 1975, 2425.
43. M. Schmidt and M. Wieber, *Z. Anorg. Allg. Chem.*, 1963, **326**, 182; W. Lauer, M. Becke-Goehring and K. Sommer, *Z. Anorg. Allg. Chem.*, 1969, **371**, 193.
44. K. Sakai, T. Koide and T. Matsumoto, *Acta Crystallogr., Sect. B*, 1978, **34**, 3326.
45. A. Darmadi, H. Haas and M. Kaschani-Motlagh, *Z. Anorg. Allg. Chem.*, 1979, **448**, 35; *Inorg. Chem.*, 1973, **12**, 3015.
46. G. E. Manoussakis, C. A. Tsipsis and A. G. Christophides, *Z. Anorg. Allg. Chem.*, 1975, **417**, 235.
47. M. Kawasaki and R. Bersohn, *J. Chem. Phys.*, 1978, **68**, 2105.
48. M. Baudler and H. J. Stassen, *Z. Anorg. Allg. Chem.*, 1966, **343**, 244.
49. R. Hillel, J. Bouix and R. Avre, *Bull. Soc. Chim. Fr.*, 1975, 2458.
50. L. J. Nugent and C. D. Cornwell, *J. Chem. Phys.*, 1962, **37**, 523.
51. P. Kisliuk and C. H. Townes, *J. Chem. Phys.*, 1950, **18**, 1109.
52. A. G. Robiette, *J. Mol. Struct.*, 1976, **35**, 81.
53. S. Samdal, D. M. Barnhart and K. Hedberg, *J. Mol. Struct.*, 1976, **35**, 67.
54. J. Trotter, *Z. Kristallogr., Kristallgeom., Kristallphys., Kristallchem.*, 1965, **122**, 230.
55. Y. Morino, T. Ukaji and T. Ito, *Bull. Soc. Chem. Jpn.*, 1966, **39**, 71.
56. R. Enjalbert and J. Galy, *Acta Crystallogr., Sect. B*, 1980, **36**, 914.
57. G. Gafner and G. J. Kruger, *Acta Crystallogr., Sect. B*, 1974, **30**, 250.
58. R. J. Gillespie, C. G. Davies, J. J. Park and J. Passmore, *Inorg. Chem.*, 1971, **10**, 2781.
59. S. Bellard, A. V. Rivera and G. M. Sheldrick, *Acta Crystallogr., Sect. B*, 1978, **34**, 1034.
60. (a) W. Haase, *Chem. Ber.*, 1974, **107**, 1009; (b) N. Webster and S. Keats, *J. Chem. Soc.* (*A*), 1971, 836.
61. A. T. Mohammed and U. Muller, *Z. Anorg. Allg. Chem.*, 1985, **523**, 45.
62. W. Haase, *Acta Crystallogr., Sect. B*, 1974, **30**, 1722.
63. H. Preiss, *Z. Anorg. Allg. Chem.*, 1971, **380**, 45.
64. M. B. Hursthouse and I. A. Steer, *J. Organomet. Chem.*, 1971, **27**, C11.
65. A. Augstine, G. Ferguson and F. C. March, *Can. J. Chem.*, 1975, **53**, 1647.
66. H. Preiss, *Z. Anorg. Allg. Chem.*, 1971, **380**, 71.
67. K. Emerson and D. Britton, *Acta Crystallogr.*, 1963, **16**, 113.
68. H. C. Fielding and I. M. Pollock, *Br. Pat.* 931 240 (1963). (*Chem. Abstr.*, 1963, **59**, 8384a).
69. D. B. Sowerby, *J. Inorg. Nucl. Chem.*, 1961, **22**, 205.
70. G. A. Olah, G. K. S. Prakash and J. Sommer, 'Superacids', Wiley, New York, 1985.
71. K. O. Christie and W. Sawodny, *Inorg. Chem.*, 1969, **8**, 212, 2489.
72. R. Mews, *Angew. Chem., Int. Ed. Engl.*, 1976, **15**, 691.
73. K. Seppelt, *Z. Anorg. Allg. Chem.*, 1977, **434**, 5.
74. K. Seppelt, *Angew. Chem., Int. Ed. Engl.*, 1976, **15**, 377.
75. F. Claus, M. Glaser, V. Wolfel and R. Minkwitz, *Z. Anorg. Allg. Chem.*, 1984, **517**, 207.
76. L. Stryer, 'Biochemistry', 2nd edn., W. H. Freeman, San Francisco, 1981, p. 275.
77. 'Arsenic and Its Inorganic Compounds', US National Safety Council Data Sheet 499, Washington, DC, 1961.
78. E. Frieden (ed.), 'Biochemistry of the Essential Ultratrace Elements', Plenum, New York, 1984.
79. E. Frieden, *J. Chem. Educ.*, 1985, **62**, 917.
80. F. B. Whitfield, D. J. Freeman and K. J. Shaw, *Chem. Ind.* (*London*), 1983, 786.
81. J. S. Edmonds and K. A. Francesconi, *Nature* (*London*), 1977, **265**, 436.
82. J. L. Wardell, in 'Comprehensive Organometallic Chemistry', ed. G. Wilkinson, F. G. A. Stone and E. W. Abel, Pergamon, Oxford, 1982, vol. 2, chap. 13, p. 681.
83. 'Gmelin Handbook of Inorganic Chemistry'; Organoantimony(III) Compounds, Parts 1 and 2 (1981); Organoantimony(V) Compounds, Part 3 (1982), Springer-Verlag, Berlin, 1981, 1982.
84. S. Samaan, 'Methoden zur Herstellung und Unwandlung von Organo-arsen, -antimon, und -wismut Verbindung', in 'Methoden der Organischen Chemie (Houben-Weyl)', vol. XIII/8, ed. E. Muller, Verlag, Stuttgart, 1978.
85. H. J. M. Bowen, *Trans. Faraday Soc.*, 1954, **50**, 463.
86. K. Sille, J. Weidlein and A. Haaland, *Spectrochim. Acta, Part A*, 1982, **38**, 475.
87. W. Kolondra, W. Schwarz and J. Weidlein, *Z. Anorg. Allg. Chem.*, 1983, **501**, 137.
88. S. P. Bone and D. B. Sowerby, *J. Chem. Soc., Dalton Trans.*, 1979, 1430.
89. A. W. Cordes, W. T. Pennington, J. C. Graham and Y. W. Jung, *Acta Crystallogr., Sect. C*, 1983, **39**, 709.
90. K. von Deuten and D. Rehder, *Cryst. Struct. Commun.*, 1980, **9**, 167.
91. C. A. McAuliffe (ed.), 'Transition Metal Complexes of Phosphorus, Arsenic and Antimony', Macmillan, London, 1973.
92. J. Werner, W. Schwarz and A. Schmidt, *Z. Naturforsch., Teil B*, 1981, **36**, 556.
93. N. K. Jha and D. M. Joshi, *Polyhedron*, 1985, **4**, 2083.
94. N. Bertazzi, G. Alonzo and F. DiBianca, *Spectrochim. Acta, Part A*, 1982, **38**, 527.
95. B. Beagley, D. W. H. Rankin, A. G. Robiette, G. M. Sheldrick and T. G. Hewitt, *J. Inorg. Nucl. Chem.*, 1969, **31**, 2351.
96. H. Schumann and M. Schmidt, *Angew. Chem., Int. Ed. Engl.*, 1964, **3**, 316.

97. H. Schumann, A. Roth, O. Stelzer and M. Schmidt, *Inorg. Nucl. Chem. Lett.*, 1966, **2,** 311.
98. C. A. McAuliffe and W. Levason, 'Phosphine, Arsine and Stibine Complexes of the Transition Elements', Elsevier, Amsterdam, 1979.
99. W. Levason and C. A. McAuliffe, *Acc. Chem. Res.*, 1978, **11,** 363.
100. K. Moedritzer, *Angew. Chem., Int. Ed. Engl.*, 1964, **3,** 609.
101. F. Ando, J. Koketsu and Y. Ishi, *Bull. Chem. Soc. Jpn.*, 1978, **51,** 1481.
102. O. J. Scherer, P. Hornig and M. Schmidt, *J. Organomet. Chem.*, 1966, **6,** 259.
103. O. J. Scherer and W. Janssen, *Chem. Ber.*, 1970, **103,** 2784.
104. U. Wannagat and M. Schlingmann, *Z. Anorg. Allg. Chem.*, 1976, **424,** 87.
105. O. J. Scherer, G. Wolmershauser and J. Conrad, *Angew. Chem., Int. Ed. Engl.*, 1983, **22,** 404.
106. Z. I. Kuplennik, Zh. N. Belaya and A. M. Pincgiuk, *Zh. Obshch. Khim.*, 1981, **51,** 2711.
107. D. Hass, *Z. Chem.*, 1964, **4,** 31.
108. A. N. Ashe, W. Butler and T. R. Diephouse, *J. Am. Chem. Soc.*, 1981, **103,** 207.
109. H. J. Breunig, *Z. Naturforsch., Teil B*, 1984, **39,** 111.
110. A. H. Cowley, J. G. Lasch, N. C. Norman, M. Pakulski and B. R. Whittlesey, *J. Chem. Soc., Chem. Commun.*, 1983, 881.
111. A. H. Cowley, J. E. Kilduff, J. G. Lasch, S. K. Mehrotra, N. C. Norman, M. Pakulski, B. R. Whittlesey and J. L. Atwood, *Inorg. Chem.*, 1984, **23,** 2582.
112. E. Wiberg and K. Moedritzer, *Z. Naturforsch., Teil B*, 1957, **12,** 131.
113. See refs in A. L. Rheingold, 'Homoatomic Rings, Chains and Macromolecules of Main Group Elements', Elsevier, New York, 1977.
114. J. Ellermann and A. Veit, *Angew. Chem., Int. Ed. Engl.*, 1982, **21,** 375.
115. O. Mundt, G. Becker, J. H. Wessley, H. J. Breuning and H. Kischkel, *Z. Anorg. Allg. Chem.*, 1982, **486,** 70.
116. K. Issleib and A. Balszuweit, *Z. Anorg. Allg. Chem.*, 1976, **419,** 87.
117. H. J. Breunig, K. Haberle, M. Drager and T. Severengiz, *Angew. Chem., Int. Ed. Engl.*, 1985, **24,** 72.
118. J. Loub and H. Paulus, *Acta Crystallogr., Sect. B*, 1981, **37,** 1106.
119. R. Mercier, J. Douglade and J. Bernard, *Acta Crystallogr., Sect. B*, 1976, **32,** 2787.
120. C. Sarnstrand, *Acta Chem. Scand., Ser. A*, 1974, **28,** 275.
121. R. Mercier, J. Douglade and F. Theobald, *Acta Crystallogr., Sect. B*, 1975, **31,** 2081.
122. J. O. Bovin, *Acta Crystallogr., Sect. B*, 1976, **32,** 1771.
123. C. Sarnstrand, *Acta Crystallogr., Sect. B*, 1978, **34,** 2402.
124. C. R. Lucas and M. E. Peach, *Can. J. Chem.*, 1970, **48,** 1869.
125. P. Zannini, A. C. Fabretti, A. Giusti, C. Preti and G. Tosi, *Polyhedron*, 1986, **5,** 871.
126. A. F. Janzen, O. C. Vaidya and C. J. Wallis, *J. Inorg. Nucl. Chem.*, 1981, **43,** 1469.
127. M. Drager, *Z. Anorg. Allg. Chem.*, 1976, **424,** 183.
128. A. H. White and C. L. Raston, *J. Chem. Soc., Dalton Trans.*, 1976, 791.
129. C. A. Kavounis, S. C. Kokkou, P. J. Rentzeperis and P. Karagiannidis, *Acta Crystallogr., Sect. B*, 1982, **38,** 2686.
130. C. A. Kavounis, S. C. Kokkou, P. J. Rentzeperis and P. Karagiannidis, *Acta Crystallogr., Sect. B*, 1980, **36,** 2954.
131. D. B. Sowerby and M. J. Begley, *J. Chem. Soc., Chem. Commun.*, 1980, 64.
132. D. B. Sowerby, I. Haiduc, A. Barbul-Rusu and M. Salajan, *Inorg. Chim. Acta*, 1983, **68,** 87.
133. B. F. Hoskins, E. R. T. Tiekink and G. Winter, *Inorg. Chim. Acta*, 1985, **97,** 217.
134. L. G. Kuzmina, N. G. Bokii, T. V. Timofeeva, Yu. T. Struchkov, D. N. Kravtson and S. I. Pombruk, *J. Struct. Chem. (Engl. Transl.)*, 1978, **19,** 279.
135. D. R. Russel and M. C. Poore, *Chem. Commun.*, 1971, 18.
136. G. Alonzo, N. Bertazzi and M. Consiglio, *Inorg. Chim. Acta*, 1984, **85,** L35.
137. N. Baumann and M. Wieber, *Z. Anorg. Allg. Chem.*, 1974, **408,** 261.
138. *US Pat.* 3 530 158 (1971) (*Chem. Abstr.*, 1971, **75,** 6106).
139. R. O. Day, J. M. Holmes, A. C. San, J. R. Devillers, R. R. Holmes and J. A. Deiters, *J. Am. Chem. Soc.*, 1982, **104,** 2127.
140. M. Wieber, N. Baumann, H. Wunderlich and H. Rippstein, *J. Organomet. Chem.*, 1977, **133,** 183.
141. M. Hall and D. Sowerby, *J. Am. Chem. Soc.*, 1980, **102,** 628.
142. A. J. Edwards, *J. Chem. Soc. (A)*, 1970, 2751.
143. S. Konaka and M. Kimura, *Bull. Chem. Soc. Jpn.*, 1973, **46,** 404.
144. A. Lipka, *Acta Crystallogr., Sect. B*, 1979, **35,** 3020.
145. S. Konaka and M. Kimura, *Bull. Chem. Soc. Jpn.*, 1973, **46,** 413.
146. D. W. Cushen and R. Hulme, *J. Chem. Soc.*, 1964, 4162.
147. A. Almenningen and T. Bjorvatten, *Acta Chem. Scand.*, 1963, **17,** 2573.
148. J. Trotter and T. Zobel, *Z. Kristallogr.*, 1966, **123,** 67.
149. A. Lipka and D. Mootz, *Z. Anorg. Allg. Chem.*, 1978, **440,** 217, 224, 231.
150. A. J. Edwards and P. Taylor, *Chem. Commun.*, 1971, 1377.
151. J. W. Moore, H. W. Baird and H. B. Miller, *J. Am. Chem. Soc.*, 1968, **90,** 1358.
152. C. I. Branden and I. Lindqvist, *Acta Chem. Scand.*, 1963, **17,** 353.
153. L. Brun and C. I. Branden, *Acta Crystallogr.*, 1966, **20,** 749.
154. J. C. Scruton and R. Hulme, *J. Chem. Soc. (A)*, 1968, 2448.
155. J. G. Ballard, T. Birchall and D. R. Slim, *J. Chem. Soc., Dalton Trans.*, 1977, 1469; 1979, 62.
156. H. Preiss, *Z. Anorg. Allg. Chem.*, 1971, **380,** 65.
157. F. P. Boer, *J. Am. Chem. Soc.*, 1968, **90,** 6706.
158. C. R. Hubbard and R. A. Jacobson, *Inorg. Chem.*, 1972, **11,** 2247.
159. S. L. Lawton and R. A. Jacobson, *J. Am. Chem. Soc.*, 1966, **88,** 616.
160. D. E. McKee, A. Zalhin and N. Bartlett, *Inorg. Chem.*, 1973, **12,** 1713.
161. K. Shelly, T. Bartczak, W. R. Scheidt and C. A. Reed, *Inorg. Chem.*, 1985, **24,** 4325.

162. U. Muller, *Z. Anorg. Allg. Chem.*, 1972, **388,** 207.
163. U. Muller, *Z. Anorg. Allg. Chem.*, 1976, **422,** 134.
164. R. J. Gillespie, D. R. Slim and J. E. Vekris, *J. Chem. Soc., Dalton Trans.*, 1977, 971.
165. R. Mattes and K. Holz, *Angew. Chem., Int. Ed. Engl.*, 1983, **22,** 11.
166. R. Fourcade, M. Bourgault, B. Bonnet and B. Ducourant, *J. Solid State Chem.*, 1982, **43,** 81.
167. S. Hurter, R. Mattes and D. Ruhl, *J. Solid State Chem.*, 1983, **46,** 204.
168. T. K. Davies and K. C. Moss, *J. Chem. Soc. (A)*, 1970, 1054.
169. H. Preiss, *Z. Anorg. Allg. Chem.*, 1972, **389,** 254.
170. R. Weber, *Pogg. Ann. Phys.*, 1865, **125,** 325.
171. Y. Hermudsson, *Acta Chem. Scand.*, 1967, **21,** 1313.
172. R. Steudel, T. Sandow and J. Steidel, *J. Chem. Soc., Chem. Commun.*, 1980, 180.
173. H. J. Vetter, H. Strametz and H. Nöth, *Angew. Chem., Int. Ed. Engl.*, 1963, **2,** 218.
174. H. U. Wolf, in 'Ullmann's Encyclopedia of Industrial Chemistry', ed. W. Gerhartz, VCH, Weinheim, 1985, chap. 13.
175. E. M. Cordasco and F. D. Stone, *Chest,* 1973, **64,** 182.
176. N. Kanematsu, M. Hara and T. Kada, *Mutat. Res.*, 1980, **77,** 109.
177. C. P. Flessel, *Adv. Exp. Med. Biol.*, 1977, **91,** 117.
178. J. S. Thanger and F. E. Brinckmann, *Adv. Organomet. Chem.*, 1982, **20,** 313.
179. H.-C. Mu, *K'o Hseuh T'ung Pao*, 1966, **17,** 502 (*Chem. Abstr.*, 1967, **66,** 69 731).
180. M. E. Gress and R. A. Jacobson, *Inorg. Chim. Acta,* 1974, **8,** 209.
181. E. M. Meyer, D. L. White, S. B. Andrews, R. J. Barrnett and J. R. Cooper, *Biochem. Pharmacol.*, 1981, **30,** 3003.
182. D. H. R. Basrton, B. Charpiot, E. T. H. Dau, W. B. Motherwell, C. Pascard and C. Pichon, *Helv. Chim. Acta,* 1984, **67,** 586.
183. G. Ferguson and D. M. Hanley, *J. Chem. Soc. (A)*, 1968, 2539.
184. S. David, B. Estramareix, J. C. Fischer and M. Therisod, *J. Am. Chem. Soc.*, 1981, **103,** 7341.
185. N. S. Vyazankin, G. A. Razuvaev, O. A. Kruglaya and G. S. Semchikova, *J. Organomet. Chem.*, 1966, **6,** 474.
186. N. S. Vyazankin, G. S. Kalinina, O. A. Kruglaya and G. A. Ruzuvaev, *J. Gen. Chem. USSR (Engl. Transl.)*, 1969, **39,** 1964.
187. I. Schumann-Ruidisch and H. Blass, *Z. Naturforsch., Teil B,* 1967, **22,** 1081.
188. O. A. Kruglaya, N. S. Vyazankin and G. A. Razuvaev, *J. Gen. Chem. USSR (Engl. Transl.)*, 1965, **35,** 392.
189. M. N. Bochkarev, N. I. Gurev and G. A. Razuvaev, *J. Organomet. Chem.*, 1978, **162,** 289.
190. H. Schumann and M. Schmidt, *Angew. Chem., Int. Ed. Engl.*, 1964, **3,** 316.
191. H. Schumann, *Angew. Chem., Int. Ed. Engl.*, 1969, **8,** 937.
192. M. N. Bochkarev, G. A. Razuvaev, L. N. Azkharov and Yu. T. Struchkov, *J. Organomet. Chem.*, 1980, **199,** 205.
193. H. Schumann and H. J. Breunig, *J. Organomet. Chem.*, 1975, **87,** 83.
194. F. Ando, T. Hayashi, K. Ohashi and J. Koketsu, *J. Inorg. Nucl. Chem.*, 1975, **37,** 2011.
195. H. Dehnicke, *Z. Anorg. Allg. Chem.*, 1974, **409,** 311.
196. J. Muller, *Z. Anorg. Allg. Chem.*, 1971, **381,** 103.
197. G. Alonzo, *Inorg. Chim. Acta,* 1983, **73,** 141.
198. R. G. Goel and H. S. Prasad, *J. Organomet. Chem.*, 1973, **50,** 129.
199. O. Stelzer, E. Unger and V. Wray, *Chem. Ber.*, 1977, **110,** 3430.
200. E. Wiberg and K. Moedritzer, *Z. Naturforsch., Teil B,* 1957, **12,** 132.
201. A. J. Ashe, E. G. Ludwig and J. Oleksyszyn, *Organometallics,* 1983, **2,** 1859.
202. K. I. Kuz'min and G. Kamai, *Sb. Statei Obshch. Khim.* 1953, **1,** 223 (*Chem. Abstr.*, 1955, **49,** 841).
203. G. Becker, M. Rossler, O. Mundt and E. Witthauer, *Z. Anorg. Allg. Chem.*, 1983, **506,** 42.
204. B. Sundrall, *Acta Chem. Scand., Ser. A,* 1979, **33,** 219.
205. A. Muller, M. Zimmermann and H. Bogge, *Angew. Chem., Int. Ed. Engl.*, 1986, **25,** 273.
206. M. Drager and B. M. Schmidt, *J. Organomet. Chem.*, 1985, **290,** 133.
207. Z. Galdecki, M. L. Glonka and B. Golinski, *Acta Crystallogr., Sect. B,* 1976, **32,** 2319.
208. M. Wieber, H. G. Rudling and C. Burschka, *Z. Anorg. Allg. Chem.*, 1980, **470,** 171.
209. C. Burschka and M. Wieber, *Z. Naturforsch., Teil B,* 1979, **34,** 1037.
210. H. G. Rudling and M. Wieber, *Z. Anorg. Allg. Chem.*, 1983, **505,** 147.
211. H. Gilman and H. L. Yale, *J. Am. Chem. Soc.*, 1951, **73,** 2880.
212. M. Wieber and U. Baudis, *Z. Anorg. Allg. Chem.*, 1976, **423,** 47.
213. M. Wieber and U. Baudis, *Z. Anorg. Allg. Chem.*, 1976, **423,** 40.
214. A. G. Jansen, O. C. Vaidya and C. J. Willis, *J. Inorg. Nucl. Chem.*, 1981, **43,** 1469.
215. C. R. Lucas and M. E. Peach, *Can. J. Chem.*, 1970, **48,** 1869.
216. M. Wieber and A. Basel, *Z. Anorg. Allg. Chem.*, 1979, **448,** 89.
217. M. Zalewics and B. Ptaszynski, *Pol. J. Chem.*, 1979, **53,** 1437 (in English) (*Chem. Abstr.*, 1979, **91,** 199 552).
218. M. Wieber and I. Sauer, *Z. Naturforsch., Teil B,* 1984, **39,** 1668.
219. A. K. Cheetham and N. Norma, *Acta Chem. Scand., Ser. A,* 1974, **28,** 55.
220. H. A. Skinner and L. E. Sutton, *Trans. Faraday Soc,* 1940, **36,** 681.
221. A. Hershaft and J. D. Corbett, *Inorg. Chem.*, 1963, **2,** 979.
222. S. G. Nyburg, G. A. Ozin and J. T. Szymanski, *Acta Crystallogr., Sect. B,* 1971, **27,** 2298; 1972, **28,** 2855.
223. L. R. Mors and W. R. Robinson, *Acta Crystallogr., Sect. C,* 1972, **28,** 653.
224. B. Blazic and F. Lazarini, *Acta Crystallogr., Sect. C,* 1985, **41,** 1619.
225. U. Praeckel, F. Huber and H. Preut, *Z. Anorg. Allg. Chem.*, 1982, **494,** 67.
226. C. Hebecker, *Z. Anorg. Allg. Chem.*, 1971, **384,** 111.
227. B. Aurivillius and C.-I. Lindlum, *Acta Chem. Scand.*, 1964, **18,** 1554.
228. H. von Benda, A. Simon and W. Bauhoffer, *Z. Anorg. Allg. Chem.*, 1978, **438,** 53.
229. H. von Benda, *Z. Kristallogr.*, 1980, **151,** 271.

230. F. Lazarini, *Acta Crystallogr., Sect. B*, 1977, **33,** 1954.
231. B. K. Robertson, W. G. McPherson and E. A. Mayers, *J. Phys. Chem.*, 1967, **71,** 3531.
232. F. Lazarini, *Acta Crystallogr., Sect. B*, 1977, **33,** 2686.
233. F. Lazarini, *Acta Crystallogr., Sect. C*, 1985, **41,** 1617.
234. F. Lazarini, *Acta Crystallogr., Sect. B*, 1977, **33,** 2961.
235. F. Lazarini, *Acta Crystallogr., Sect. B*, 1978, **34,** 2288.
236. H. Puff, M. Gronke, B. Kilger and P. Moltgen, *Z. Anorg. Allg. Chem.*, 1984, **518,** 120.
237. J. Trotter and T. Zobel, *Z. Kristallogr., Kristallgeochem., Kristallphys., Kristallchem.*, 1966, **123,** 67.
238. O. Lindqvist, *Acta Chem. Scand.*, 1968, **22,** 2943.
239. F. Lazarini, *Acta Crystallogr., Sect. B.*, 1977, **33,** 1957.
240. L. P. Battaglia, A. B. Corradi, G. Pellizzi and M. E. V. Tani, *Cryst. Struct. Commun.*, 1975, **4,** 399.
241. C. Hebecker, *Z. Anorg. Allg. Chem.*, 1971, **384,** 111.
242. R. J. Gillespie, D. Martin, G. J. Schrobilgen and D. R. Slim, *J. Chem. Soc., Dalton Trans.*, 1977, 2234.
243. P. Boldrini, R. J. Gillespie, P. R. Ireland and G. J. Schrobilgen, *Inorg. Chem.*, 1974, **13,** 1690.
244. J. E. F. Reynolds (ed.) 'Martindale, The Extra Pharmacopoeia', 20th edn., The Pharmaceutical Press, London, 1982.
245. L. Jaenicke, 'Bismuth, A Survey of Pharmacological and Medical Uses', Mining and Chemical Products, London, 1950.
246. N. I. Sax, 'Dangerous Properties of Industrial Materials', Van Nostrand and Rheinhold, New York, 1984.
247. M. Palazzi, S. Jaulmes and P. Laruelle, *Acta Crystallogr., Sect. B*, 1974, **30,** 2378.
248. K. Mereiter, A. Preisinger, O. Baumgartner, G. Heger, W. Mikenda and H. Steidl, *Acta Crystallogr., Sect. B*, 1982, **38,** 401.
249. S. Jaulmes and M. Palazzi, *Acta Crystallogr., Sect. B*, 1976, **32,** 2119.
250. M. Palazzi, *Acta Crystallogr., Sect. B*, 1976, **32,** 516.
251. H. Schafter, G. Schafer and A. Weiss, *Z. Naturforsch., Teil B*, 1963, **18,** 665.
252. G. Adiwidjaja and J. Lohn, *Acta Crystallogr., Sect. B*, 1970, **26,** 1878.
253. G. Ferguson, R. G. Goel and D. R. Ridley, *J. Chem. Soc., Dalton Trans.*, 1975, 1288.
254. U. Muller, *Z. Anorg. Allg. Chem.*, 1976, **422,** 134, 141.
255. J. Pebler, K. Hartke, K. Dehnicke and H. M. Wolff, *Z. Anorg. Allg. Chem.*, 1982, **486,** 61.
256. G. Ferguson, R. G. Goel, F. C. Marsch, D. R. Ridley and H. S. Prasat, *Chem. Commun.*, 1971, 1547.
257. R. Hulme and J. C. Scruton, *J. Chem. Soc. (A)*, 1968, 2448.
258. O. J. Scherer, G. Wolmershäuser and H. Conrad, *Angew. Chem., Int. Ed. Engl.*, 1983, **22,** 404.
259. J. H. Burns, *Acta Crystallogr.*, 1962, **15,** 1098.
260. G. Tuefer, *Acta Crystallogr.*, 1956, **9,** 539.
261. H. Bode and E. Voss, *Z. Anorg. Allg. Chem.*, 1951, **264,** 144.
262. R. F. Copeland, S. H. Conner and E. A. Meyers, *J. Phys. Chem.*, 1966, **70,** 1288.
263. J. W. Anderson and J. E. Drake, *J. Chem. Soc. (A)*, 1970, 3131.
264. J. W. Anderson and J. E. Drake, *J. Inorg. Nucl. Chem.*, 1973, **35,** 1032.
265. H. Schumann, *Angew. Chem., Int. Ed. Engl.*, 1969, **8,** 937.
266. J. Pebler, F. Weller and K. Dehnicke, *Z. Anorg. Allg. Chem.*, 1982, **492,** 139.
267. M. Rouault, *Ann. Phys. (Leipzig)*, 1940, **14,** 78.
268. S. M. Oldberg, *J. Am. Chem. Soc.*, 1959, **81,** 811.
269. G. Yakobson and G. G. Furin, *Synthesis*, 1980, 345.
270. L. P. Battaglia and A. B. Corradi, *J. Chem. Soc., Dalton Trans.*, 1986, 1513.
271. M. Wieber, D. Wirth and K. Hess, *Z. Anorg. Allg. Chem.*, 1983, **505,** 150.
272. M. B. Ferrari, L. C. Capacchi and G. F. Gasparri, *Acta Crystallogr., Sect. B*, 1972, **28,** 1169.

29

Sulfur, Selenium, Tellurium and Polonium

FRANK J. BERRY
University of Birmingham, UK

29.1 INTRODUCTION

Sulfur, selenium, tellurium and polonium constitute the heavier elements of group VIB of the periodic table and are sometimes referred to as the chalcogens, chalcogenins, chalcogenides or chalconides. Developments in the understanding and interest in the chemistry of these elements have been reviewed at appropriate intervals during the past 20 years.[1–8]

29.1.1 Bonding, Valence and Geometry

The elements have six electrons in their valence shells and show essentially non-metallic covalent character except for polonium and, to a lesser extent, tellurium. Although they may complete the noble gas configuration by forming the chalcogenide ions S^{2-}, Se^{2-} and Te^{2-} it is pertinent to note that these species only exist in the salts of the most electropositive elements. Indeed this divalent state involving two covalent bonds and two lone pairs of electrons is unstable for selenium and tellurium, especially the dihalides, and these species tend to disproportionate to quadrivalent compounds and the elemental state. The elements may also achieve the noble gas configuration by forming two electron pair bonds, *e.g.* hydrogen sulfide, sulfur dichloride and dimethyl sulfide, with the hydrides H_2X decreasing in stability as the group VI element series is descended. The noble gas configuration may also be attained by the formation of ionic species with one bond and one negative charge, *e.g.* RS^-, or three bonds and one positive charge, *e.g.* R_3S^+.

In addition to the divalent species, the elements also form compounds in the formal oxidation state of IV involving two, three, four, five or six bonds. Although the quadrivalent polonium halides are thermally unstable and give the divalent compounds at relatively low temperatures, the quadrivalent state is generally stable for the heavier elements of group VIB particularly in the oxides, oxo acids and halides where, for example, the formation of the hexacoordinated anionic complexes of the type $XHal_6^{2-}$ is increasingly favoured as group VIB is descended. Indeed, six is the characteristic coordination number for tellurium.

The formation of compounds in the formal oxidation state of VI is well established for all four elements, for example, the sexivalent fluorides and the TeF_8^{2-} ion. However, oxidation to this highest valency state becomes progressively more difficult as group VIB is descended since the inert pair effect causes the elements to behave as though two of their valence electrons are absent. Some examples of the compounds formed by the group VIB elements and their stereochemistries are described in Table 1.

Table 1 Compounds of Group VI Elements and their Stereochemistries

Valence	*Number of bonds*	*Geometry*	*Examples*
II	2	Angular	Me_2S, H_2Te, S_n
	3	Pyramidal	Me_3S^+
	4	Square	$Te[SC(NH_2)_2]_2Cl_2$
IV	2	Angular	SO_2
	3	Pyramidal	SF_3^+, OSF_2, SO_3^{2-}, Me_3TeBPh_4
		Trigonal planar	$(SeO_2)_n$
	4	ψ-Trigonal bipyramidal	SF_4, RSF_3, Me_2TeCl_2
		Tetrahedral	Me_3SO^+
	5	ψ-Octahedral (square pyramidal)	$SeOCl_2py_2$, SF_5^-, TeF_5^-
	6	Octahedral	$SeBr_6^{2-}$, PoI_6^{2-}, $TeBr_6^{2-}$
	7	Distorted pentagonal bipyramidal	$TePh(S_2CNEt_2)_3$
	8	Distorted octahedral	$Te(S_2CNEt_2)_4$
VI	3	Trigonal planar	$SO_3(g)$, S(NCR)
	4	Tetrahedral	SeO_4^{2-}, SO_3 (s), SeO_2Cl_2
	5	Trigonal bipyramidal	SOF_4
	6	Octahedral	RSF_5, SeF_6, $Te(OH)_6$
	8 (?)	?	TeF_8^{2-} (?)

Reproduced by permission from F. A. Cotton and G. Wilkinson, 'Advanced Inorganic Chemistry', 4th edn., Wiley-Interscience, New York, 1980.

The accessibility of the elements from sulfur to polonium to higher oxidation states and coordination numbers, as compared with the four coordination and divalency of oxygen, is a reflection of the availability of *d* orbitals for the formation of σ bonds. It is also notable that sulfur and selenium appear to use $d\pi$ orbitals in the formation of multiple bonds. The strong tendency of sulfur to catenation, which gives rise to compounds with no oxygen, selenium, tellurium or polonium analogues must also be noted, although selenium rings and unbranched elemental chains of selenium and tellurium are an important feature of the chemistry of these elements, as is their tendency to form secondary bonds in crystalline solid compounds.

29.1.2 Group Trends

The greater differences between the chemistry of oxygen and sulfur, as compared with differences between the other group VIB elements, may be associated with a number of features including the lower electronegativities of the elements sulfur to polonium. This tends to lessen the ionic character of compounds which are analogous to those of oxygen, as well as alter the relative stabilities of some bonds and reduce the importance of hydrogen bonding.

Although there are similarities between the chemistry of the chalcogenide elements, the properties of selenium and tellurium clearly lie between those of non-metallic sulfur and metallic polonium. The enhancement in metallic character as the group is descended is illustrated in the emergence of cationic properties by polonium, and marginally by tellurium, which are reflected in the ionic lattices of polonium(IV) oxide and tellurium(IV) oxide and the formation of salts with strong acids.

Finally it is worth noting that, unlike sulfur, selenium and tellurium, neither stable nor long-lived isotopes of polonium are known in nature. The only readily accessible isotope of polonium is polonium-210 which is the penultimate member of the radium decay series and which decays by α-emission with a half-life of 138.4 days.

29.2 COORDINATION COMPOUNDS OF THE GROUP VIB ELEMENTS

A review of the complexes of sulfur and tellurium[9] in 1964 was followed[1] by an extensive survey of the heavy chalcogen complexes in a comprehensive text on the chemistry of selenium, tellurium and polonium in 1966. The subject of coordination in organoselenium and organotellurium chemistry,[10] together with the structural chemistry of complexes of selenium and tellurium with sulfur- and selenium-containing ligands,[11] has received recent attention in conference proceedings. The group VIB complexes may be conveniently considered within five categories:[4] the anionic halo complexes, the anionic complexes formed by oxo acids, and finally the oxygen, nitrogen and sulfur donor complexes.

29.2.1 Halo Complexes

The complexes of sulfur tetrafluoride with several inorganic fluorides were initially considered[12] in terms of SF_4 acting as a Lewis base giving addition compounds formulated as donor–acceptor complexes of the type $F_4S{\cdot}BF_3$. The stability of these types of compounds was found to decrease in the sequence $SbF_5 > AsF_5 > IrF_5 > BF_3 > PF_5 > AsF_3$. Later investigations of these materials by IR spectroscopy[13] resulted in them being formulated as $SF_3^+BF_4^-$, suggesting that SF_4 acts as a fluoride ion lone-pair donor as opposed to a sulfur lone-pair donor.

The electron acceptor properties of sulfur tetrafluoride are well illustrated by the formation of the stable 1:1 complex with pyridine, $C_5H_5NSF_4$, which is thought to adopt pseudooctahedral or square pyramidal geometry, and the reaction of SF_4 with CsF at 125 °C and Me_4NF at −20 °C to give[14] $CsSF_5$ and $[NMe_4]^+[SF_5]^-$. The square pyramidal structure of the SF_5^- ion in $CsSF_5$ (Figure 1) was deduced from vibrational spectra.[15]

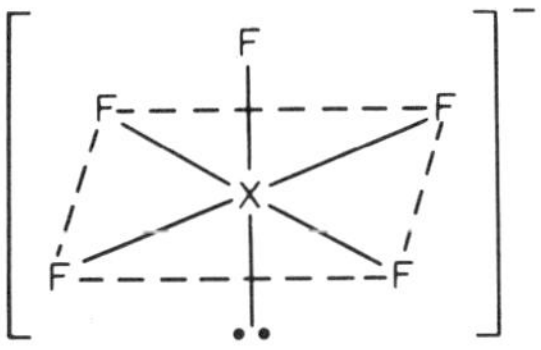

Figure 1 The structure of the SF_5^- ion in $CsSF_5$ (reproduced by permission from *Inorg. Chem.*, 1972, **11**, 1679)

The fluorides of the heavier group VIB elements also form a significant grouping within the halo complexes category, for example the quadripositive elements form complexes of the type MXF_5. For selenium the alkali metal and thallous pentafluoroselenates(IV) are formed as white solids from the dissolution of the metal fluorides in selenium tetrafluoride, but dissociate quite readily at room temperature to the component fluorides.[16] The corresponding pentafluorotellurates(IV) were initially prepared[17] by the addition of cesium fluoride to a solution of tellurium tetrafluoride in hydrogen fluoride and were isolated as colourless crystals which were formulated as $CsTeF_5$. Other pentafluorotellurates(IV), described as $NH_4TeF_5{\cdot}H_2O$, $KTeF_5$ and $Ba(TeF_5)_2{\cdot}2H_2O$, have been isolated from reactions of alkali metal fluoride and tellurium dioxide in hydrofluoric acid[18] or selenium tetrafluoride.[19] The square pyramidal geometry of the pentafluorotellurate(IV) anion has been identified from IR and Raman spectroscopy,[18] and by an X-ray structural determination of the potassium salt.[20]. It is pertinent to note that although the solubility of polonium(IV) hydroxide increases with the concentration of solvent hydrofluoric acid, as might be expected if complex formation were occurring, no complexes have yet been isolated.[21]

The fluoro complex of tellurium with pyridine, $[C_5H_5NH]TeF_5$, has been obtained[22] as a pale green solid by treatment of the pyridine complex $TeF_4{\cdot}py$ with aqueous hydrofluoric acid, whilst a white modification has been reported[18] to be isolated from a mixture of tellurium dioxide and pyridine in hydrofluoric acid.

The nitronium salts of pentafluoroselenate(IV) and pentafluorotellurate(IV) of composition NO_2XF_5 have been prepared[23] from the reaction of selenium or tellurium dioxide with nitryl fluoride.

Other pentahalo anions of group VIB elements have also been identified. For example, studies of solutions of selenium(IV) oxide and tellurium(IV) oxide in hydrobromic acid by Raman spectroscopy[24] have indicated the likely presence of the pentabromo anions, and in non-aqueous solutions there is some evidence that $TeCl_5^-$ and $TeBr_5^-$ might be formed.[25] Although a chloride of this type has been reported[26] a more complex chloro compound $(C_2H_5)_4NTeCl_{13}$, which was obtained from non-aqueous solution and examined by IR and Raman spectroscopy, has been described as $[(C_2H_5)_4N^+(TeCl_3^+)_3]^{4+}Cl_4$. A series of halogeno complexes with urea of the type $[CON_2H_5]TeX_5$, where X = Cl, Br, I, have been prepared[27] and the stability of the $[TeX_5]^-$ anion reported to increase from the chloro to the iodo complex. Indeed, the chloro complexes were readily converted into the bromo and iodo complexes when treated with the appropriate acid. It is also pertinent to note that the vibrational spectra of the adduct $TeCl_4{\cdot}PCl_5$ are consistent[28] with the formulation $PCl_4^+TeCl_5^-$, whereas those of $TeCl_4{\cdot}AlCl_3$ are indicative of the formulation $TeCl_3^+AlCl_4^-$.

It is interesting to note the compound $HTeI_5{\cdot}2oxin{\cdot}10H_2O$, which has been reported[29] to be isolated from a solution of oxine in hydroiodic acid containing tellurium. The instability constants for some tellurium halo complexes[30,31] and the polonium iodo complexes,[32] including the POI_5^- species, have been reported.

It is also interesting to note the reported[33,34] formation of hydrogen halogeno acids of composition $TeCl_4{\cdot}HCl{\cdot}5H_2O$, $TeBr_4{\cdot}HBr{\cdot}5H_2O$, and $TeI_4{\cdot}HI{\cdot}8H_2O$. The hygroscopic needle-shaped prismatic yellow chloride, red bromide and black iodide were found to lose the hydrogen halide when heated. It is also pertinent to note the identification of the oxide fluorides $TeF_4{\cdot}3TeO_2{\cdot}6H_2O$, $TeF_4{\cdot}TeO_2{\cdot}2H_2O$ and $TeF_4{\cdot}TeO_2{\cdot}H_2O$ and the proposed[33,35] existence of the complex anion $[TeO_2F_2]^{2-}$. In these respects it is also relevant to record that selenium tetrachloride and tellurium tetrachloride both form addition compounds with ammonia and several metal and non metal chlorides,[1] whilst polonium tetrachloride is known[1] to form complexes with nitrosyl chloride, tributyl phosphate and edta.

Selenium oxychloride, $SeOCl_2$, disssolves many metal chlorides, and a number of solvates, for example $FeCl_3{\cdot}2SeOCl_2$, have been reported.[1] An addition compound of $SeOCl_2$ with pyridine, $SeOCl_2{\cdot}2py$, is also known[36] in which the coordination polyhedron around the central selenium atom is a tetragonal pyramid (Figure 2). In this structure two chlorine and two nitrogen atoms (from the pyridine) are bonded in the basal plane with atoms of the same kind in *trans* positions and oxygen atoms at the apex. A tendency to dimerization involving chlorine atoms to give a distorted octahedron around the selenium atoms is observed. It is also relevant to note that selenium dioxide forms addition compounds with hydrogen chloride and hydrogen bromide.[37]

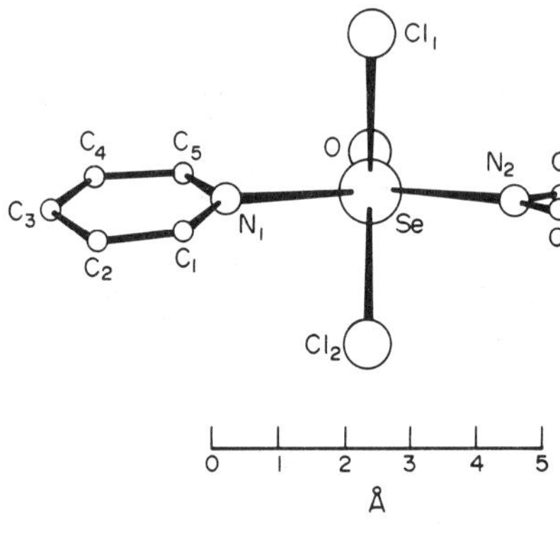

Figure 2 The structure of the addition compound $SeOCl_2{\cdot}2C_5H_5N$ (reproduced by permission from *Acta Crystallogr.*, 1959, **12,** 638)

Tellurium tetrabromide is amphoteric in fused arsenic tribromide, from which complex bromides such as $[(Et)_4N]_2TeBr_6$ and ammine adducts such as $TeBr_4{\cdot}2PhNMe_2$ have been obtained. Addition compounds with pyridine, tetramethylthiourea and piperazine have also been reported.[38,39] Alkali metal salts such as $TeBr_4{\cdot}2CsCl$ and the full brominated complexes of the type M_2TeBr_6 and M_2PoBr_6 are also known.

The tellurium dichlorobromide, $TeCl_2Br_2$, formed by reaction of bromine with tellurium dichloride gives a cream coloured addition compound, $TeCl_2Br_2{\cdot}2py$, with pyridine in ether.[1] The complex dissolves in hot concentrated hydrochloric acid, from which pale orange needles of $(pyH)_2TeCl_4Br_2$ separate on cooling. This compound, as well as the original pyridine complex, dissolves in hot concentrated hydrobromic acid and red needles of the hexabromotellurate(IV), $(pyH)_2TeBr_6$, crystallize on cooling. This is an example of the stability of the hexahalogen complexes (*vide infra*).

A number of hexahalogenotellurates(IV) have been investigated. A few rather unstable hexafluoro complexes are known, for example the nitrosonium compounds of composition $(NO)_2XF_6$ have been obtained[40] from the reaction of nitrosulfuryl fluoride, $NOSO_2F$, with selenium or tellurium tetrachloride. The selenium compound loses nitrosyl fluoride at room temperature to leave the pentafluoroselenate(IV), whereas the tellurium analogue remains stable to 100 °C. Although the pyridinium hexafluoro complex has been reported[22] to result from the treatment of tellurium dioxide with pyridine in concentrated hydrofluoric acid it must be recalled that (*vide supra*) the pentafluoro complex has been made[18] from the same reaction mixture and that the existence of the hexafluoro complex is therefore a matter of some uncertainty. It is pertinent to note that the fluoro complexes are the least stable of the tellurium(IV) halide complexes.[22]

Tellurium hexafluoride itself has been found to form complexes with potassium, rubidium and cesium fluorides. The cesium compound has a stochiometry approaching Cs_2TeF_8 and, in these types of complexes, the stability of the lattice appears to be inversely related to the polarizing power of the cation and the compounds are only stable in the solid state.[41] Several heptafluorotellurates(VI) have been prepared from the reaction of tellurium with chlorine trifluoride in anhydrous hydrofluoric acid followed by addition of the metal fluoride.[42]

Several hexachloro, hexabromo and hexaiodo complexes of the group VIB elements have been reported.[1] For example, solutions of selenium tetrachloride in concentrated hydrochloric acid contain the hexachloro anion $SeCl_6^{2-}$ from which the ammonium and alkai metal salts can be readily obtained with the cesium salts being the least soluble. The hexahalogenotellurates(IV), such as the chlorides, bromides and iodides of potassium, rubidium and cesium and the chlorides and bromides of ammonium and thallium, have received significant attention.[1] These complexes are generally prepared by precipitation of the complex salt from a solution of the chalcogen dioxide in the concentrated halogen acid. This reaction has also been found[43] to be very effective for the preparation of hexaiodoselenates(IV). Complexes of this type are generally of the form A_2XY_6 in which the anions are octahedrally coordinated. The IR spectra of the selenium and tellurium hexahalogeno anions have been reported.[43] The tellurium-125 Mössbauer chemical isomer shifts, recorded[44] from the hexahalogenotellurates(IV), have been found to be characteristic of the anion and independent of the cation. The data have been associated with the stereochemically inactive lone pair of non-bonding electrons which may be presumed to have considerable *s* electron character and to be highly contracted towards the tellurium nucleus.

The stability of compounds containing $[TeX_6]^{2-}$ ions apparently depends to a considerable extent on the degree of polarization caused by the cation. Hence strongly polarizing hydrogen, lithium and sodium ions may be envisaged as making the $[TeX_6]^{2-}$ anion unstable and thereby mitigate against the isolation of either the free hexahalogeno acids or the lithium and sodium derivatives.

It is pertinent to record that the differences in the properties of compounds of selenium and tellurium in hydrochloric acid solution and the formation of several types of complex ion by these elements are the basis of extraction and ion-exchange methods of separating selenium and tellurium.[1]

29.2.2 Pseudohalide Complexes

The group VIB cyanides, thiocyanates and selenocyanates and their complexes with species such as thiourea have been described.[1,45] For example, the tellurium dithiocyanate complex has been prepared[45] by treatment of tellurium dichloride or tellurium dibromide with ammonium thiocyanate. It seems that little information exists on the preparation of tellurocyanates and there is a sparsity of data on polonium derivatives. Indeed, the only known cyanide of polonium is probably a salt of the quadrivalent element.[1]

29.2.3 Oxo Acid Complexes

The oxo acid complexes of the group VIB elements are largely restricted to those of tellurium and polonium and the information on these types of materials has not changed significantly in recent years.[4]

Tellurium(IV) sulfato complexes of composition $2(2TeO_2{\cdot}SO_3){\cdot}MHSO_4{\cdot}2H_2O$ have been reported,[46] from which the anhydrous compounds were obtained by calcination. Carboxylic acids have also been found to form anionic complexes with tellurium(IV) and polonium(IV). For example, the silver salts of the citrato- and tartrato-tellurates(IV) have been described[47] as insoluble in water but soluble in nitric acid.

Studies of the solubility of polonium(IV) in formic, acetic, oxalic and tartaric acids have provided evidence of complex formation,[48] with the acetato complex emerging as more stable than the hexachloro anion. Other studies of the solubility of polonium(IV) hydroxide in carbonate[49] and nitrate[50] solution, together with investigations[51] of the ion exchange behaviour of polonium(IV) at high nitrate ion concentration, have been discussed in terms of the formation of anionic complex species.

29.2.4 Oxygen Donor Complexes

There is a sparsity of information on oxygen donor complexes of the group VIB elements. Although sulfur tetrafluoride reacts with oxygen atoms in several species to give materials containing oxygen–sulfur bonds,[8] *e.g.* $SF_4(OSF_5)_2$, the characterization of many of the species as genuine coordination compounds is difficult to assess. Selenium trioxide has been reported[52] to form 1:1 adducts with ether and dioxane. The 1:1 complex of tellurium tetrafluoride with dioxane has been examined[53] by IR spectroscopy and described as $[(dioxane)_2TeF_3]^+[TeF_5]^-$. The 1:1 complex of tellurium tetrachloride with pyridine *N*-oxide (PNO) has similarly been formulated[54] as $[(PNO)_2TeCl_3]^+[TeCl_5]^-$.

Although tellurium tetrachloride has been reported[55] to form a 1:2 complex with acetamide, there appears to be little other information available on the complex. Similarly, the reported tributyl phosphate complex of polonium tetrachloride[56] has received little attention. The formation of a polonium(IV) perchlorate complex with tributyl phosphate has been suggested[57] in the solvent extraction of polonium from perchloric acid.

29.2.5 Nitrogen Donor Complexes

The study of sulfur–nitrogen compounds is currently a very active area of inorganic chemistry research with many new cyclic and acyclic compounds being synthesized which have unusual structures and bonding properties.[8] Indeed, in many cases their description as coordination complexes is difficult to justify without further evidence on their bonding characteristics. The donor properties of nitrogen to sulfur have been referred to previously (Section 29.2.1), *e.g.* during the discussion of the 1:1 stable adduct between pyridine and sulfur tetrafluoride and the formation of $[NMe_4]^+[SF_5]^-$ from the reaction between Me_4NF and SF_4. It is also relevant to note that pyridine forms an adduct with sulfur trioxide.

The selenium tetrachloride complex with pyridine, $SeCl_4{\cdot}2py$, has been discussed[58] in terms of octahedral $[SeCl_3{\cdot}2py]^+$ ions in which the sixth position around selenium is occupied by a lone pair of electrons. It is therefore relevant to recall that the complex $SeOCl_2{\cdot}2py$[36] (Section 29.2.1, Figure 2) may also be described as octahedral with the chlorine atoms and pyridine rings occupying *trans* positions and in which the oxygen atom is *trans* to the lone pair of electrons. It is also pertinent to note that $SeCl_4{\cdot}2py$ is isomorphous with $SnCl_4{\cdot}2py$, which also has a *trans* octahedral configuration, and that conductivity measurements[59] suggest that the complex reacts with the solvent or with impurities. Selenium monobromide dissolved in carbon disulfide has been reported[60,61] to form simple nitrogen donor complexes of composition $Se_2Br_2{\cdot}2L$ with a variety of amines and heterocyclic bases. Although several of the nitrogen donor complexes of the chalcogen tetrahalides have compositions of the type $MX_4{\cdot}2L$, where L is a monodentate ligand, adducts of selenium or tellurium tetrachloride with four molecules of ammonia or two molecules of ethylenediamine which may involve six-coordination, *e.g.* $[(SeCl_2en_2)Cl_2]$, are also well established.[62]

The pentahalo anions which are found in complexes with nitrogen (Section 29.2.1) deserve a little more comment. For example, both mono- and bi-dentate nitrogen donor complexes[53] of composition $TeF_4{\cdot}L$ and $2TeF_4{\cdot}L$ have been described as $[L_2TeF_3]^+[TeF_5]^-$ or $[L'TeF_3]^+[TeF_5]^-$. The conductivity data recorded[54] from the analogous selenium and tellurium tetrachloride complexes suggest that they are 1:1 electrolytes in non-aqueous solvents and similar evidence recorded[54] from complexes of composition $MY_4{\cdot}2L$, in which the anion may be a simple halide ion, suggests that these are likely to be of a similar nature. Tellurium hexafluoride has been found[63] to form amine complexes of the type $TeF_6{\cdot}2R_3N$ which have been described as eight-coordinate species.

29.2.6 Sulfur and Selenium Donor Complexes

Complexes of group VIB elements with sulfur or selenium donor atoms are mainly concerned with complexes of selenium(II), tellurium(II) and tellurium(IV). The structural chemistry of complexes of selenium and tellurium with sulfur- and selenium-containing ligands has been well reviewed in the recent past.[11]

During work on nucleophilic substitution at divalent sulfur,[64,65] a series of tellurium derivatives with sulfur-containing ligands were prepared in a straightforward manner by addition of a warm aqueous solution of, for example, thiourea to tellurium dioxide dissolved in warm hydrochloric or hydrobromic acid, and which involved the reduction of tellurium(IV) to tellurium(II).

Two types of three-coordinate selenium(II) complexes, isolated as the triselenocyanate and the triselenourea salts, and which have essentially similar structures, have been identified.[66] Several three-coordinate tellurium(II) complexes have also been obtained[67] when one of the ligands has a very strong *trans* effect, for example a phenyl group, as in phenylbis(thiourea)tellurium(II) chloride.[68]

Several types of four-coordinate selenium(II) complexes with monodentate ligands have been identified. The isomorphous *cis*-dichloro- and *cis*-dibromo-bis(thiourea) selenium(II) complexes have recently been reported,[69] whilst the dimeric selenium triselenocyanate and trithiocyanate ions have been known for some time.[70–72] The distorted square planar coordination of the $[SeBr_2(tmtu)]$ complex[73] and the selenium dithiocyanate and diselenocyanate complexes[74] is characteristic of the four-coordinate selenium(II) complexes with monodentate ligands. Numerous four-coordinate tellurium(II) complexes with monodentate ligands have been reported,[64] including a large number of square planar complexes with TeS_4, TeS_2S_2 and TeS_4 coordination spheres, some of which may be dimeric. The tellurium tris(ethylenethiourea) cation, $[Te_2(etu)_6]^{4+}$, in the perchlorate salt (Figure 3) is illustrative[75] of the distorted square planar tellurium coordination in these types of complexes. A few square planar complexes with TeS_2X_2 and $TeSX_3$ coordination spheres are also known.[11] The IR and Raman spectra of the tellurium dichloride and dibromide complexes with thiourea have been reported and it appears that the Te–S bond in these compounds is relatively weak.[76]

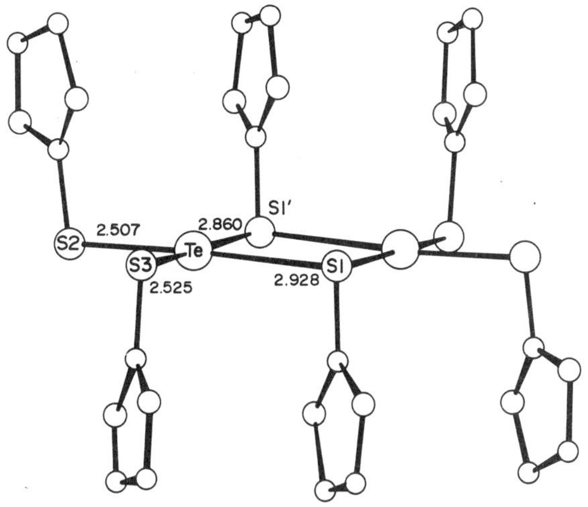

Figure 3 The structure of hexakis(ethylenethiourea)ditellurium(II) perchlorate (reproduced by permission from *Inorg. Chem.*, 1980, **19**, 1050)

Several four-coordinate selenium(II) complexes with bidentate ligands, all of which involve highly distorted square planar SeS_4 or $SeSe_4$ coordination spheres, and some tellurium analogues have been described.[11]

It is pertinent to note the preparation and structure[77] of the tris(ethylxanthato)tellurium(II) anion in its tetraethylammonium salt, $Et_4N^+[Te(EtOCS_2)_3]^-$ (Figure 4), in which the five-coordinate tellurium(II) is coordinated to three ethyl xanthate ligands, one of which is monodentate whilst the other two are anisobidentate. The resulting TeS_5 coordination sphere is pentagonal planar.

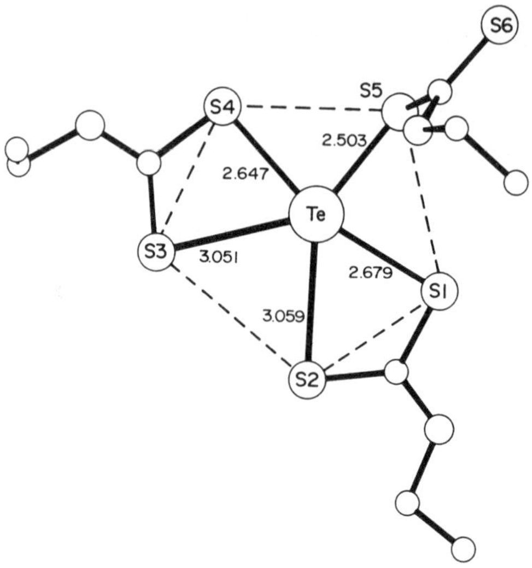

Figure 4 The structure of the tris(ethylxanthato)tellurium(II) anion in its tetraethylammonium salt (reproduced by permission from *Aust. J. Chem.*, 1976, **29**, 2337)

Complexes of both selenium(IV) and tellurium(IV) with sulfur-containing ligands are known although the structural properties appear to have been determined only for the tellurium(IV) species. However, due to the smaller size of selenium(IV) and its poorer acceptor properties, as compared with tellurium(IV), its maximum coordination number appears to be six whilst it has been found to be eight for tellurium.[78] Furthermore, the greater tendency of tellurium to form secondary bonds may cause the structures of similar selenium(IV) and tellurium(IV) compounds to be different.[79] It is also pertinent to note that the reduction of tellurium(IV) to

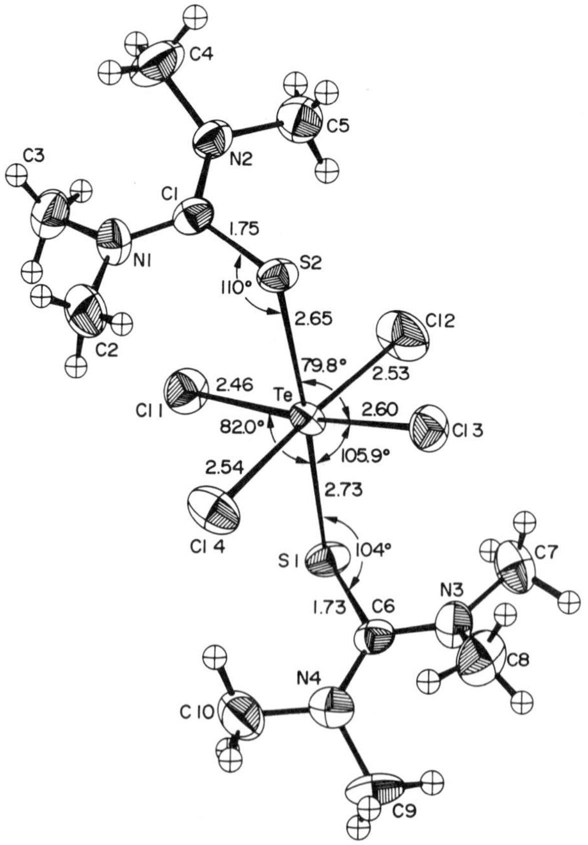

Figure 5 The structure of the monoclinic form of *trans*-tetrachlorobis(tetramethylthiourea)tellurium(IV) (reproduced by permission from *Acta Chem. Scand., Ser. A*, 1975, **29**, 141)

tellurium(II) by sulfur- or selenium-containing ligands does not always occur when strong donor ligands are used.[80,81]

A few four- and five-coordinate tellurium(IV) complexes with sulfur- or selenium-containing ligands are known, but despite the data on the hexahalogenotellurates(IV) (*vide supra*), six-coordinate tellurium(IV) complexes with other ligands are relatively rare. The six-coordinate complexes generally have a TeX_4S_2 coordination sphere, with both *cis* and *trans* isomers, and are considerably distorted from perfect octahedral geometry. The structure[82] of the monoclinic form of *trans*-tetrachlorobis(tetramethylthiourea)tellurium(IV), $[TeCl_4(tmtu)_2]$ (Figure 5) is illustrative of this common distortion from octahedral coordination which may be related to the stereochemical activity of the lone pair of electrons.

Some seven- and eight-coordinate tellurium(IV) complexes with bidentate dialkyldithiocarbamate ligands are known. The eight-coordinate complexes all have dodecahedral D_{2d} structures in which each dodecahedron may be seen as two planar trapezoidal $Te(R_2NCS_2)_2$ groups fused at right angles at the central tellurium ligand. The structure[83] of tetrakis(4-morpholinecarbodithioato)tellurium(IV) (Figure 6) shows the central tellurium atoms to be bonded to all eight sulfur atoms in a slightly distorted dodecahedral configuration.

Complexes of polonium with sulfur donor atoms appear to be less common, although there is some evidence that polonium is complexed by thiourea[84] and of the formation[85] of a volatile polonium diethyldithiocarbamate.

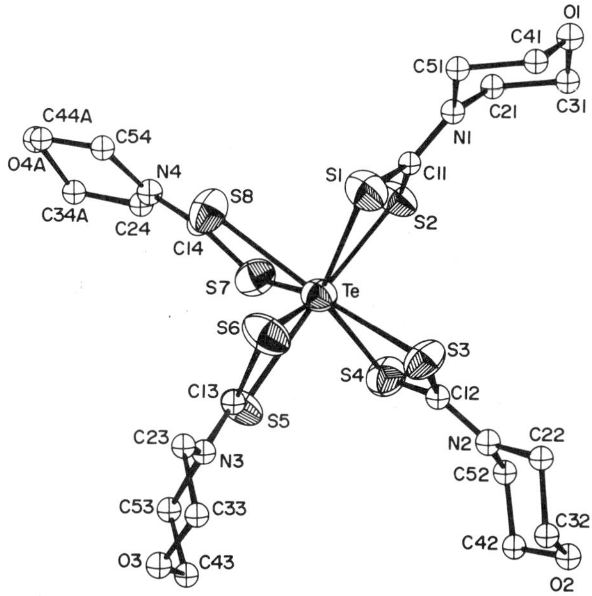

Figure 6 Structure of tetrakis(4-morpholinecarbodithioato)tellurium(IV) (reproduced by permission from *Acta Chem. Scand., Ser. A*, 1975, **29,** 185)

29.3 BIOLOGICAL PROPERTIES

The health hazards and toxicity of selenium and tellurium compounds have been known for some time.[1,2,5] Most selenium compounds are regarded as being toxic and great care has been recommended[1] for the handling of all selenides, organic selenium compounds and soluble selenites and selenates. Ingested selenides are reported[1] to be only slowly released from the body and cause the subject to suffer offensive-smelling breath and perspiration for a long time. Tellurium compounds are also regarded as harmful, although they appear to be rapidly reduced in the body to elemental tellurium which is ultimately excreted in the form of unpleasant smelling organotellurium compounds. Polonium is far more dangerous than selenium or tellurium because of its radioactivity. It appears that all three elements are absorbed in the kidneys, spleen and liver which, in the case of polonium, become the subject of irreversible radiation damage.

22.4 REFERENCES

1. K. W. Bagnall, 'The Chemistry of Selenium, Tellurium and Polonium', Elsevier, Amsterdam, 1966.
2. W. C. Cooper (ed.), 'Tellurium', Van Nostrand Reinhold, New York, 1971.
3. M. Schmidt and W. Siebert, in 'Comprehensive Inorganic Chemistry', ed. J. C. Bailar, H. J. Emeléus, R. Nyholm and A. F. Trotman-Dickenson, Pergamon, Oxford, 1973, p. 795.
4. K. W. Bagnall, in 'Comprehensive Inorganic Chemistry', ed. J. C. Bailar, H. J. Emeléus, R. Nyholm and A. F. Trotman-Dickenson, Pergamon, Oxford, 1973, p. 935.
5. R. A. Zingaro and W. C. Cooper (eds.), 'Selenium', Van Nostrand Reinhold, New York, 1974.
6. F. A. Cotton and G. Wilkinson, 'Advanced Inorganic Chemistry', 4th edn., Wiley-Interscience, New York, 1980, p. 502.
7. A. F. Wells, 'Structural Inorganic Chemistry', Clarendon, Oxford, 1984, p. 699.
8. N. N. Greenwood and A. Earnshaw, 'Chemistry of the Elements', Pergamon, Oxford, 1984, p. 757.
9. D. I. Ryabchikov and I. I. Nazarenko, *Russ. Chem. Rev.* (*Engl. Transl.*), 1964, **33,** 55.
10. W. R. McWhinnie, in 'Proceedings of the Fourth International Conference on the Organic Chemistry of Selenium and Tellurium', ed. F. J. Berry and W. R. McWhinnie, University of Aston, Birmingham, 1983, p. 3.
11. S. Husebye, in 'Proceedings of the Fourth International Conference on the Organic Chemistry of Selenium and Tellurium', ed. F. J. Berry and W. R. McWhinnie, University of Aston, Birmingham, 1983, p. 298.
12. N. Bartlett and P. L. Robinson, *Chem. Ind.* (*London*), 1956, 1351.
13. F. Seel and O. Detner, *Z. Anorg. Allg. Chem.*, 1959, **301,** 113.
14. L. Muettertis, *J. Am. Chem. Soc.*, 1960, **82,** 1082.
15. K. O. Christie, E. C. Curtis, C. J. Chack and D. Philipovich, *Inorg. Chem.*, 1972, **11,** 1679.
16. E. E. Aynsley, R. D. Peacock and P. L. Robinson, *J. Chem. Soc.*, 1952, 1231.
17. H. Wells and J. Willis, *Am. J. Sci.*, 1901, **12,** 190.
18. N. N. Greenwood, A. C. Sarma and B. P. Straughan, *J. Chem. Soc.*, 1966, 1446.
19. A. J. Edwards, M. A. Mouty, R. D. Peacock and A. J. Suddens, *J. Chem. Soc.*, 1964, 4087.
20. A. J. Edwards and M. A. Mouty, *J. Chem. Soc.* (*A*), 1969, 703.
21. K. W. Bagnall, J. H. Freeman, D. S. Robertson, P. S. Robinson and M. A. A. Stewart, UKAEA Unclassified Report CIR 2566, 1958.
22. E. E. Aynsley and G. Hetherington, *J. Chem. Soc.*, 1953, 2802.
23. E. E. Aynsley, G. Hetherington and P. L. Robinson, *J. Chem. Soc.*, 1954, 1119.
24. P. J. Hendra and Z. Jovic, *J. Chem. Soc.* (*A*), 1968, 600.
25. J. Dobrowolski, *Zesz. Nauk. Politech. Gdansk., Chem.*, 1966, **11,** 3 (*Chem. Abstr.*, 1968, **68,** 45 817s).
26. J. A. Creighton and J. H. S. Green, *J. Chem. Soc.* (*A*), 1968, 808.
27. E. E. Aynsley and W. Campbell, *J. Chem. Soc.*, 1957, 832.
28. I. R. Beattie and H. Chudzynska, *J. Chem. Soc.* (*A*), 1967, 984.
29. K. Geiersberger, *Z. Anal. Chem.*, 1952, **11,** 135.
30. R. Korewa and Z. Szponar, *Rocz. Chem.*, 1965, **39,** 349.
31. V. I. Marashova, *J. Anal. Chem. USSR* (*Engl. Transl.*), 1966, **21,** 303.
32. K. W. Bagnall, R. W. M. D'Eye and J. H. Freeman, *J. Chem. Soc.*, 1956, 3385.
33. R. Metzer, *C. R. Hebd. Seances Acad. Sci.*, 1879, **125,** 23.
34. 'Gmelins Handbuch der Anorganischen Chemie', 8 Auflage, Syst. -Nr. 11, Tellurium, Verlag-Chemie, Berlin, 1940, pp. 334, 343, 349.
35. E. Prideaux and O. Millot, *J. Chem. Soc.*, 1929, 2703.
36. I. Lindqvist and G. Nahringbauer, *Acta Crystallogr.*, 1959, **12,** 638.
37. S. Witekowa, *Z. Chem.*, 1962, **2,** 315.
38. E. Montignie, *Z. Anorg. Allg. Chem.*, 1960, **306,** 234.
39. E. Montignie, *Z. Anorg. Allg. Chem.*, 1961, **307,** 109.
40. F. Seel and H. Massett, *Z. Anorg. Allg. Chem.*, 1955, **280,** 186.
41. E. L. Muetterties, *J. Am. Chem. Soc.*, 1957, **79,** 1004.
42. A. F. Clifford and A. G. Morris, *J. Inorg. Nucl. Chem.*, 1957, **5,** 71.
43. N. N. Greenwood and B. P. Straughan, *J. Chem. Soc.* (*A*), 1966, 962.
44. T. L. Gibb, R. Greatrex, N. N. Greenwood and A. C. Sarma, *J. Chem. Soc.* (*A*), 1970, 212.
45. V. Gutmann, *Monatsh. Chem.*, 1951, **82,** 280.
46. R. Metzner, *Ann. Chim. Phys.*, 1898, **15,** 203.
47. B. Brauner, *J. Chem. Soc.*, 1889, **55,** 382.
48. K. W. Bagnall and J. H. Freeman, *J. Chem. Soc.*, 1957, 2161.
49. H. V. Moyer (ed.), USAEC Unclassified Document TID-5221, 1956.
50. K. W. Bagnall, D. S. Roberston and M. A. A. Stewart, *J. Chem. Soc.*, 1958, 3633.
51. E. Orban, USAEC Unclassified Report MLM-973, 1954.
52. M. Schmidt and I. Wilhelm, *Chem. Ber.*, 1964, **97,** 872.
53. N. N. Greenwood, A. C. Sarma and B. P. Straughan, *J. Chem. Soc.* (*A*), 1968, 1561.
54. D. A. Couch, P. S. Elmes, J. E. Fergusson, M. L. Greenfield and C. J. Wilkins, *J. Chem. Soc.* (*A*), 1967, 1813.
55. R. C. Paul and R. Dev, *Indian J. Chem.*, 1965, **3,** 315.
56. K. W. Bagnall and D. S. Robertson, *J. Chem. Soc.*, 1957, 509.
57. N. I. Ampelogova, *Radiokhimiya*, 1963, **5,** 626.
58. A. W. Cordes and T. V. Hughes, *Inorg. Chem.*, 1964, **3,** 1640.
59. I. R. Beattie, M. Milne, M. Webster, H. E. Blaydon, P. J. Jones, R. C. G. Killean and J. L. Lawrence, *J. Chem. Soc.* (*A*), 1969, 482.
60. S. Prased and B. L. Khandelwal, *J. Indian Chem. Soc.*, 1960, **37,** 645.
61. S. Prased and B. L. Khandelwal, *J. Indian Chem. Soc.*, 1961, **38,** 107.
62. V. G. Tronev and A. N. Grigorovich, *Zh. Neorg. Khim.*, 1957, **2,** 2400.
63. E. L. Muetterties and W. D. Phillips, *J. Am. Chem. Soc.*, 1957, **79,** 2957.

64. O. Foss, *Pure Appl. Chem.*, 1970, **24,** 31.
65. O. Foss and S. Hauge, *Acta Chem. Scand.*, 1959, **13,** 1252.
66. S. Hauge, Ph.D. Dissertation, University of Bergen, 1977.
67. O. Foss, S. Husebye and K. Maroy, *Acta Chem. Scand.*, 1963, **17,** 1806.
68. S. Hauge, O. Johannessen and O. Vikane, *Acta Chem. Scand., Ser. A,* 1978, **32,** 901.
69. S. V. Bjornevag and S. Hauge, *Acta Chem. Scand., Ser. A,* 1983, **37,** 235.
70. S. Hauge, *Acta Chem. Scand., Ser. A,* 1975, **29,** 771.
71. S. Hauge, *Acta Chem. Scand., Ser. A,* 1975, **29,** 843.
72. S. Hauge and P. A. Henriksen, *Acta Chem. Scand., Ser. A,* 1975, **29,** 778.
73. K. J. Wynne, P. S. Pearson, M. G. Newton and J. Golen, *Inorg. Chem.* 1972, **11,** 1192.
74. S. Hauge, *Acta Chem. Scand., Ser. A,* 1979, **33,** 313.
75. A. Foust, *Inorg. Chem.*, 1980, **19,** 1050.
76. P. J. Hendra and Z. Jovic, *J. Chem. Soc. (A)*, 1967, 735.
77. B. F. Hoskins and C. D. Pannan, *Aust. J. Chem.*, 1976, **29,** 2337.
78. S. Esperas, S. Husebye and S. E. Svaeren, *Acta Chem. Scand.*, 1971, **25,** 3539.
79. N. W. Alcock, *Adv. Inorg. Chem. Radiochem.*, 1972, **15,** 1.
80. O. Foss, *Acta Chem. Scand.*, 1953, **7,** 226.
81. G. S. Nikolov, N. Jordanov and I. Havezov, *J. Inorg. Nucl. Chem.*, 1970, **33,** 1055.
82. S. Esperas, J. W. George, S. Husebye and O. Mikalsen, *Acta Chem. Scand., Ser. A,* 1975, **29,** 141.
83. S. Esperas and S. Husebye, *Acta Chem. Scand., Ser. A,* 1975, **29,** 185.
84. H. Mabuchi, *Bull. Chem. Soc. Jpn,* 1958, 245.
85. K. Kimura and T. Ishimori, in 'Proceedings of the 2nd International Conference on Peaceful Uses of Atomic Energy', United Nations, New York, 1958, p. 151.

30

Halogenium Species and Noble Gases

ANTHONY J. EDWARDS
University of Birmingham, UK

30.1 INTRODUCTION

This chapter deals with coordination complexes formed by the elements of both the halogen and noble gas groups, groups VII and 0, of the periodic table. It is convenient and useful to consider these two sets of elements together, since they have many similarities arising from their high electronegativities as the central atoms in complexes, and hence with the rather restricted range of ligands which form such complexes, particularly in the case of the noble gases. Recently[1] the statement has been made, 'It is now widely accepted that noble gas chemistry, which in most cases means nothing but xenon chemistry, must be viewed in relation to the high valency state chemistry of the neighbour elements in the periodic system, especially iodine'.[1]

The general reviews available of the chemistry of the halogens[2b,3–5] and of the noble gases[2c,3–5] are usually concerned with the complete chemistry of the elements and deal with coordination complexes only as part of this chemistry. This is largely due to the difficulty of defining such complexes for these elements. Thus, on the most general definition, almost all compounds containing the central atom in a positive oxidation state could be included. This review will be selective, concentrating on the relationships between the elements of the two groups.

The halogens will be restricted to chlorine, bromine and iodine since fluorine, as the most electronegative element, does not function as the central atom in a complex and astatine has only short-lived, radioactive isotopes, so that very little of its coordination chemistry has been investigated.[2b]

As stated above, noble gas chemistry is almost restricted to that of xenon, with a few krypton compounds of lower stability and with the chemistry of radon largely unexplored, due to the short half-lives of its isotopes. Two reviews[1,6] give good coverage of more recent noble gas chemistry.

30.2 OXIDATION STATES AND COORDINATION GEOMETRY

For both groups of elements the normal oxidation states always differ by two. Thus the halogens form compounds with the odd positive oxidation states I, III, V and VII, while noble gases correspondingly exhibit the even oxidation states II, IV, VI and VIII. These oxidation states vary in their accessibility with different ligands and, as expected, the higher the oxidation state, the more electronegative the ligands needed for stabilization. Thus the only examples of nitrogen ligands bonded to a noble gas atom are found for xenon(II), and these only for a ligand with electron-withdrawing groups attached to the nitrogen atom.

For the halogens the ease of attainment of the higher oxidation states increases as the group is descended from chlorine to iodine, and conversely the lower oxidation states become less accessible down the group.

The coordination geometry of complexes of these elements is most usefully correlated by reference to the valence shell electron pair repulsion (VSEPR) theory.[7] This bonding model, although based on a theoretically doubtful premise, is simple to apply and predicts correctly structures for most of the complexes considered. Only for coordination numbers greater than six does the model run into difficulties, as will be seen later for XeF_6 and $[XeF_8]^{2-}$. Since the structures predicted depend on the number of electron pairs, isostructural arrangements are found for corresponding, isoelectronic, neutral noble gas and anionic halogen species, and for cationic noble gas and neutral halogen species. Some examples of such relationships are shown in Table 1, where the various coordination geometries are listed. The diatomic interhalogens and isoelectronic noble gas cations have not been included.

Table 1 Coordination Geometries Based on VSEPR Theory

Valence shell electrons	*Theoretical arrangement*	*Number of bonds*	*Shape*	*Examples*
8	Tetrahedron	2	Angular	$[BrF_2]^+$
10	Trigonal bipyramid	2	Linear	XeF_2, $[ICl_2]^-$
		3	T-shaped	BrF_3, $[XeF_3]^+$
		4	Distorted square pyramid	$[IF_4]^+$
12	Octahedron	4	Square planar	XeF_4, $[ICl_4]^-$
		5	Square pyramid	IF_5, $[XeF_5]^+$
		6	Octahedral	$[IF_6]^+$
14	Pentagonal bipyramid?	6	Distorted octahedron	XeF_6, $[BrF_6]^-$
		7	Pentagonal bipyramid (fluxional)	IF_7
16	Square antiprism or dodecahedron	7	—	$[XeF_7]^-$
		8	—	$[IF_8]^-$
18	Tricapped trigonal prism?	8	Square antiprism	$[XeF_8]^{2-}$

30.3 COORDINATION COMPOUNDS

No strict definition of coordination compound will be followed for the compounds of these two groups, but coverage will be concentrated on those compounds where inter-group relationships are most obvious. In particular, the halogen oxides and derived oxo ions will not be considered.

The order in which the ligands are discussed will deviate from that generally followed in this series. Complexes with halogen ligands cover such an extensive range of oxidation states and provide so many structural examples for other ligands that they will be considered first. They also provide the largest number of examples of coordination compounds for chlorine and bromine, and the only examples for krypton. Xenon forms some compounds with other ligands, almost all linked to oxygen, and halogen complexes of other ligands are mainly restricted to those of iodine, often with halogen ligands as part of the coordination sphere to enhance stability.

30.3.1 Halogen Ligands

Coordination compounds containing halogen ligands (of which the great majority are fluorides) form the largest group of compounds for these elements. They are best considered as based on the neutral species, with cations and anions, which can be theoretically derived (and usually practically prepared) from the formal self-ionization of the neutral compound.

Considering bromine trifluoride as an example, the self-ionization is shown in equation (1). Dissolution of fluoride acceptors, such as antimony pentafluoride, gives compounds formally containing the difluorobromine(III) cation, and correspondingly, dissolution of fluoride donors, such as potassium fluoride, gives compounds containing the tetrafluorobromate(III) anion.

$$2BrF_3 \rightleftharpoons [BrF_2]^+ + [BrF_4]^- \quad (1)$$

Although the formulation as ionic species appears to correlate well with the known crystal structures for the anionic complexes, the situation is much less clear for the cations, and secondary bonding[8] between the central atom of the cation and atoms of the anion is important. Thus in the structure of difluorobromine(III) hexafluoroantimonate(V) the bromine atom has two fluorine atom neighbours at 1.69 Å and two more, from $[SbF_6]^-$ units, at 2.29 Å, completing a distorted, square planar arrangement of fluorine atoms around the bromine atom.[9]

Similar fluorine bridging in the structure[10] of trifluoroxenon(IV) hexafluorobismuthate(V) leads again to a distorted, square planar arrangement, with three fluorine atoms at 1.81, 1.92 and 1.94 Å from the xenon atom, and the fourth, from the $[BiF_6]^-$ unit, at 2.25 Å.

The interhalogens, noble gas fluorides and derived ions will be treated systematically. The diatomic interhalogens will not be discussed, but the isoelectronic $[XeF]^+$ and $[KrF]^+$ cations will be included due to their important secondary bonding to fluoride to form an asymmetric, two-coordinate arrangement.[6]

30.3.1.1 Halogen(I) and noble gas(II) compounds

These lowest oxidation states can be associated with a coordination number of two and a linear arrangement of the ligands about the central atom. On VSEPR theory these are 10-electron species, with a trigonal bipyramidal arrangement of three lone pairs of electrons in equatorial positions and two halogen ligands in axial positions.

The halogen compounds are the anions derived from the diatomic interhalogens, usually involving the same halogen, symmetrically bonded, as ligands, but, for example with $[BrICl]^-$, the different I—Br (2.51 Å) and I—Cl (2.91 Å) distances giving an asymmetric, linear ion.[11] Generally these anions are prepared by the interaction of the appropriate interhalogen compound with the appropriate halide, with reaction conditions tailored to the properties of reactants and products.

The noble gas compounds include the two difluorides of xenon and krypton and the derived cations $[XeF]^+$ and $[KrF]^+$, if for these latter species the interaction with fluoride in the anion is included. The simplest example of this arrangement occurs in the ions $[Xe_2F_3]^+$ and $[Kr_2F_3]^+$, prepared by interaction of excess difluoride with an acceptor pentafluoride, which can best be described as assemblies of two $[NgF]^+$ ions (Ng = Xe or Kr) with an F^- anion between them, symmetrically placed[12] for Xe, but possibly asymmetrically[13] for Kr. This gives a linear arrangement for Ng and an angular coordination for the central F^-. This asymmetric coordination by two fluorine ligands is also found in the structures of salts of the $[XeF]^+$ cation, investigated by X-ray crystallography,[14–16] and is also indicated for corresponding $[KrF]^+$ salts from spectroscopic measurements.[6] Some representative examples of this coordination arrangement are given in Table 2.

30.3.1.2 Halogen(III) and noble gas(IV) compounds

The species considered for these oxidation states are associated with coordination numbers two, three and four. The corresponding geometrical arrangements are angular for two-coordination in the cationic halogen species, T-shaped for the three-coordinate interhalogens and cationic noble gas species, and square-planar for the anionic halogen complexes and the neutral noble gas fluorides. The situation is complicated by secondary bonding. Thus the

Table 2 Linear Two-coordinate Complexes

	A—X (Å)	*X—A—Y* (°)	*Ref.*
$K^+[ICl_2]^-$	2.55 × 2	180	17
$Cs^+[IBr_2]^-$	2.62, 2.78	178	18
$Cs^+[Br_3]^-$	2.440, 2.698	177.5	19
$[NH_4]^+[BrICl]^-$	2.51(Br), 2.91	180	11
XeF_2	1.983 × 2	180	35
$[XeF]^+[AsF_6]^-$	1.873, 2.212	179	14
$[XeF]^+[RuF_6]^-$	1.87, 2.18	177	15
$[XeF]^+[Sb_2F_{11}]^-$	1.84, 2.35	~180	16
$[Xe_2F_3]^+[AsF_6]^-$	1.90, 2.14	178	12

angular cations for halogens and the T-shaped cations for noble gases can approach the square planar arrangement, as described earlier in this section.

On VSEPR theory the angular cations are eight-electron species with a tetrahedral arrangement of two lone pairs of electrons and two halogen ligands. The T-shaped complexes are 10-electron species, with two lone pairs of electrons in equatorial positions and the halogen ligands in the remaining equatorial and the axial positions. The square planar arrangement is associated with 12-electron species, with two lone pairs *trans* to each other in an octahedral geometry.

The triatomic cations are usually prepared from the corresponding interhalogen, by reaction with a halogen acceptor molecule. This is impossible for some of the more exotic species which have no interhalogen precursor. For example, the $[Br_3]^+$ cation has been characterized[20] in $[Br_3]^+[AsF_6]^-$, formed from the interaction of arsenic pentafluoride and a bromine/bromine pentafluoride mixture and the $[Cl_2F]^+$ cation results from the interaction of antimony pentafluoride with chlorine monofluoride.[21]

Secondary bonding interactions have been shown to give a distorted square planar arrangement for these ions and the variation of this interaction can be correlated with both the acceptor strength of the anion involved and the acceptor strength of the cationic species itself. Thus with the $[SbF_6]^-$ anion the interactions[9] of $[BrF_2]^+$ are stronger than those[22] of $[ClF_2]^+$, and $[ICl_2]^+$ shows[23] one of the shortest iodine to fluorine secondary bonds as yet reported (Table 3).

Table 3 Angular Two-coordinate Halogen Complexes

	Average A—X (Å)	*Average A · · · X* (Å)	*Ref.*
$[ClF_2]^+[SbF_6]^-$	1.57	2.38	22
$[BrF_2]^+[SbF_6]^-$	1.69	2.29	9
$[ICl_2]^+[SbF_6]^-$	2.268	2.650	23
$[IBr_2]^+[Sb_2F_{11}]^-$	2.423	2.81	24
$[I_3]^+[AsF_6]^-$	2.665	3.04	25
$[ICl_2]^+[SbCl_6]^-$	2.31	2.92	26

The T-shaped geometry is established for the trifluorides of chlorine and bromine in the vapour phase and for[27] solid ClF_3. In the solid state for[28] BrF_3 there is a Br · · · F contact at 2.49 Å, compared with the Br—F distances of 1.72, 1.84 and 1.85 Å, which, although coplanar with the three bonds, is offset from the symmetrical square position. It can be considered as symmetrically placed above the centre of a triangular face, including the two lone pairs of electrons, of the trigonal bipyramidal arrangement. This interaction can be considered too weak to force a rearrangement to the pseudooctahedral form. In contrast, iodine trichloride[29] achieves a close approach to the square planar arrangement by dimerization, and only the longer bridge distances lead to distortion (Table 4).

Although the T-shaped $[XeF_3]^+$ cation forms[10] a coplanar, symmetric secondary bond with the $[BiF_6]^-$ ion, through a very strong contact (Bi · · · F = 2.25 Å), which results in a distorted square planar geometry, the $[SbF_6]^-$ [30] and $[Sb_2F_{11}]^-$ [31] anions interact more weakly to give the asymmetric arrangement described above for solid BrF_3.

The square planar arrangement found in the anionic halogen complexes and in xenon tetrafluoride is usually regular, although small distortions have been attributed to the influence of the countercations in the halogen compounds. For example, in the compound $[SCl_3]^+[ICl_4]^-$ lengthening of I—Cl bonds[32] can be correlated with the number of secondary bonds to sulfur formed by the particular chlorine atom (Table 4).

Table 4 Planar Four-coordinate Complexes

	A—X (Å)	*X—A—X* (°)	*Ref.*
$K^+[BrF_4]^-$	1.89 × 4	90.9 × 2, 89.1 × 2	33
$K^+[ICl_4]^-H_2O$	2.42, 2.47, 2.53, 2.60	89.2, 89.5, 90.6, 90.7	34
$[SCl_3]^+[ICl_4]^-$	2.439, 2.469, 2.543, 2.603	88.0, 89.5, 91.1, 91.4	32
XeF_4	1.952 × 4	90.0 × 4	35
I_2Cl_6	2.38 × 2, 2.70 × 2	94, 84, 91 × 2	29

30.3.1.3 Halogen(V) and noble gas(VI) compounds

Complexes for these oxidation states are restricted to fluorine ligands, and to xenon for the noble gases, with associated coordination numbers ranging from four to eight. The geometrical arrangements are distorted square pyramidal for four-coordination in the cationic halogen species, square pyramidal for the five-coordinate interhalogens and the cationic xenon species, and distorted octahedral for six-coordinate xenon hexafluoride. For the hexafluoroanions of the halogens, and the heptafluoroxenate(VI) ion, no crystallographic results are available, but the octafluoroxenate(VI) ion has been shown to have a square antiprismatic arrangement.[36]

VSEPR theory correlates the four-coordinate species with a 10-electron, trigonal bipyramidal arrangement of one lone pair of electrons in an equatorial position and four fluorine ligands. The square pyramidal complexes are based on a 12-electron arrangement with one lone pair of electrons. For higher numbers of valence-shell electrons, the VSEPR treatment is less satisfactory. The effect of the lone pair of electrons is certainly seen in the structure of monomeric XeF_6, but this is a fluxional molecule, and the geometry is therefore hard to define. The situation with $[XeF_8]^{2-}$ is difficult to rationalize in terms of VSEPR theory, since the square antiprismatic coordination is only slightly distorted, showing no clear position for the lone pair of electrons.

The tetrafluorohalogen(V) cations are prepared by interaction of the parent fluoride with a good fluoride acceptor molecule, typically SbF_5. The geometry in the solid state has been established for $[BrF_4]^+$ in the $[Sb_2F_{11}]^-$ salt,[37] although the accuracy of the results is low, for $[IF_4]^+$ in the $[SbF_6]^-$ salt,[38] which has a disordered arrangement, and for $[IF_4]^+$ in the $[Sb_2F_{11}]^-$ salt.[39] In the latter, the F—I—F angle between equatorial fluorine ligands is 92.4° and that between axial ligands is 160.3°, as expected from VESPR theory, but there are two pairs of secondary bonds averaging 2.56 and 2.87 Å, grouped around the lone pair position.

The halogen pentafluorides and the $[XeF_5]^+$ cation have a square pyramidal configuration and any weak secondary bonds are found below the base of the pyramid and situated to avoid the axial, lone pair position. These contacts are much more significant for the xenon compounds than for the interhalogens, where they are so weak as to be virtually indistinguishable from normal intermolecular contacts, as seen in the structure of IF_5.[40]

Salts of the $[XeF_5]^+$ cation are synthesized by interaction of XeF_6 with fluoride acceptor molecules in the correct proportions. Excess XeF_6 will produce compounds containing the $[Xe_2F_{11}]^+$ cation. For $[XeF_5]^+$ in combination with $[MF_6]^-$ or $[MF_6]^{2-}$ anions there are either three[41] or four [15] secondary bonds, symmetrically arranged around the lone pair position, with $Xe\cdots F$ distances in the range 2.5–3.0 Å. In solid XeF_6, which exists as both tetramers, best represented as $[(XeF_5)^+F^-]_4$, and the corresponding hexamers,[42] the distances are 2.56 Å in the hexamer and 2.23 and 2.60 Å in the tetramer. The $[Xe_2F_{11}]^+$ cation can be represented as $[XeF_5^+\cdots F^-\cdots XeF_5^+]$ and in the $[AuF_6]^-$ salt[43] has an $Xe\cdots F^-$ distance of 2.23 Å and a secondary $Xe\cdots F$ bond distance to $[AuF_6]^-$ of 2.64 Å In comparison the overall average Xe—F bond distance in $[XeF_5]^+$ compounds is 1.85 Å.

Compounds with six fluorine ligands have provided structural difficulties. Xenon hexafluoride exists as the monomer in the vapour state and has been shown largely by electron diffraction experiments to be non-octahedral, and also fluxional.[44] Although VSEPR theory might have predicted a pentagonal bipyramidal arrangement, including the lone pair of electrons, the lone pair is thought to take up a position capping a face of the octahedron of fluorine ligands, and interchanging positions, by moving through the midpoint of an edge. The ions BrF_6^- and IF_6^- result from the interaction of the pentafluorides with alkai metal fluorides as fluoride donors,[45,46] but no definitive crystal structure has been determined. Spectroscopic results indicate a non-octahedral arrangement for the ions.

Interaction of XeF_6 with fluoride donors produces compounds formally containing $[XeF_7]^-$ and $[XeF_8]^{2-}$. The latter anion has been characterized[36] in the compound $[NO]_2^+[XeF_8]^{2-}$, by X-ray crystallography. The Xe—F distances are in the range 1.95–2.10 Å and the structural arrangement is a slightly distorted, square antiprism with no definite position for the lone pair of electrons.

30.3.1.4 Halogen(VII) and noble gas(VIII) compounds

Although in theory the Xe^{VIII} oxidation state might be attainable, no such compounds containing only fluorine ligands have yet been reported. The halogens are represented by iodine heptafluoride and its related $[IF_6]^+$ cation and $[IF_8]^-$ anion. $[ClF_6]^+$ and $[BrF_6]^+$ cations have also been prepared, although lacking their neutral precursors, using the powerful oxidizing properties of platinum hexafluoride or the fluorokrypton(II) cation. Thus interaction of ClF_5 and PtF_6 in a sapphire reactor produces $[ClF_6]^+[PtF_6]^-$ [47a] and warming a solution of $[KrF]^+[AsF_6]^-$ or $[Kr_2F_3]^+[AsF_6]^-$ in liquid BrF_5 or ClF_5 produces $[BrF_6]^+$ [48] $[AsF_6]^-$ or the $[ClF_6]^+$ salt.[47b]

The three hexafluorohalogen(VII) cations have been shown to have octahedral symmetry by spectroscopic methods. This undistorted octahedral geometry is predicted by VSEPR theory, since no lone pairs of electrons are involved.

The heptafluoride has been reported not to have a pentagonal bipyramidal arrangement in the solid state,[49] although the considerable difficulties involved in the crystal structure determination lead to less accurate results than could be desired.[50] In the vapour state electron diffraction results[51] show that the pentagonal bipyramidal arrangement is present but the molecule is non-rigid, with deformation from D_{5h} symmetry, and this may account for the solid state structure.

The octafluoroiodate(VII) ion has been reported[52] to result from the interaction of the heptafluoride with the fluoride donors CsF or NOF, but its structure has not been determined.

30.3.2 Carbon Ligands

This section is concerned almost exclusively with halogen compounds, since although there is an extensive chemistry of halogens in higher oxidation states bonded to carbon in ligands, there is no well-characterized noble gas compound of this type. Mass spectrometer experiments involving *in situ* decay of organic iodides containing ^{131}I have given evidence for the existence of the C—$^{131}Xe^+$ bond,[53] but only one compound has been described on a macroscopic scale. The reaction of XeF_2 with CF_3 free radicals, generated in a plasma, produced a waxy, colourless solid,[54] which was more volatile than XeF_2, and decomposed at room temperature with a half-life of 30 minutes. Handling difficulties and lack of a suitable solvent precluded conclusive identification of the compound.

An extensive review[55] of the chemistry of halogen complexes containing carbon ligands, the 'hypervalent halogen compounds', has been published recently. This is written from an organic chemical viewpoint and includes discussion of the organic chemistry involved in reactions of these compounds. Only aspects directly related to coordination chemistry will be mentioned here.

The review also discusses in detail the bonding in these compounds, in terms of hypercovalent bonds. The gross stereochemistry of these compounds will be related here to the previously described VSEPR model, sufficient for the present discussion.

Although there are a few examples of iodine bonded only to carbon ligands, most complexes contain other, more electronegative ligands in the coordination sphere. There are few examples of chlorine and bromine complexes, and, as yet, no examples even with iodine of oxidation state VII. Most of these complexes, and especially the more stable examples, have aromatic rather than aliphatic carbon ligands. Classification of these compounds has been in terms of the oxidation state of the halogen and the number of halogen–carbon bonds.

30.3.2.1 Halogen(I) compounds

This class of compound is represented by the dicyanoiodate(I) anion, involving a pseudohalide rather than an alkyl or aryl ligand. The linear ion, with I—C distance of 2.302 Å, has been characterized in the compound $K[I(CN)_2]C_6H_{12}N_2O_2$, resulting from reaction of an aqueous solution of KCN with an ethanolic solution of ICN. The diimino oxalic acid diethyl ester molecule apparently stabilizes the structure.[56]

30.3.2.2 Halogen(III) compounds

This is the largest group of compounds and, although many examples have been prepared since the original report of phenyliodine(III) dichloride a century ago,[57] few crystal structures have been determined. The compounds have been classified as shown in Table 5.

Table 5 Classification of Organohalogen(III) Compounds

Number of ligands	*Type of ligands*	*Example*
2	Two carbon	PhBr(cyim)[a]
2	One carbon, one heteroatom	PhIO
3	Three carbon	Ph_3I
3	Two carbon, one heteroatom	Ph_2ICl
3	One carbon, two heteroatom	$PhICl_2$

[a] cyim = 3,4-dicyanoimidazolylide.

The compounds with two carbon ligands are halonium ylides, with a formal positive charge on the halogen atom. On VESPR theory they are eight-electron species, with two lone pairs of electrons and an overall tetrahedral arrangement. The C—I—C angles determined are 97.8° in 2-(*o*-chlorophenyliodonium) 4-nitro-6-chlorophenoxide[58] and 98.7° in phenyliodonium 2,5-dicarboethoxy-3,4-dicyanocyclopentadienylide.[59] The C—Br—C angle in the structure of the only bromine compound structurally characterized, phenylbromonium 3,4-dicyanoimidazolylide,[60] is 99.3°. When one carbon atom is replaced by a heteroatom (in practice a multiply bonded oxygen atom) a neutral species of the same stereochemical type results, since the oxygen ligand accounts for the extra pair of electrons in its multiple bond. These compounds are unstable with respect to disproportionation and have not been definitively characterized structurally.

The compounds with three ligands should have the T-shaped stereochemistry characteristic of 10-electron species on VSEPR theory. For three carbon ligands, triphenyliodine, prepared from diphenyliodine(III) iodide and phenyllithium at −80 °C in ether,[61] decomposes at −10 °C, and even more stable compounds, with cyclic structures involving the iodine atom,[62] still decompose over relatively short periods. No crystal structures have been determined, although NMR studies support the structural assignment.[62]

The diphenyliodine(III) halides have dimeric structures in the solid state,[63] directly comparable with that of ICl_3.[29] Thus, these compounds have a planar coordination of the iodine atom by two carbon and two bridging halogen atoms. This corresponds to the 12-electron, octahedral arrangement. The I—Cl bonds in the chloride, of 3.085 Å, are longer than those of 2.70 Å in the trichloride and presumably reflect the significance[63] of the ionic form $\{[Ph_2I]^+[Cl]^-\}_2$.

The characteristic T-shaped coordination is found for compounds with one carbon and two heteroatom ligands. The simplest structure is that of phenyliodine(III) dichloride[64] with I—Cl bond distances of 2.54 Å and Cl—I—C angles of 86°. In general, in the other examples for which structures have been determined,[55] the corresponding angles are less than 90°. Secondary bonding is important in the overall packing of these compounds and has been fully analyzed for two examples.[65]

Recent synthetic work has achieved the preparation of two important bromine(III) compounds. The interaction of BrF_3 and $C_6F_5SiF_3$ produces[66] $C_6F_5BrF_2$, a thermally stable, but reactive solid. Interaction of this compound with trifluoroacetic anhydride gives the stable carboxylate, $C_6F_5Br(O_2CCF_3)_2$. The much more stable compound (**1**) results from treatment[67] of the diol precursor with BrF_3. This latter compound, which melts at 153–154 °C, is unreactive towards water, dilute hydrochloric acid and dilute sodium hydroxide.

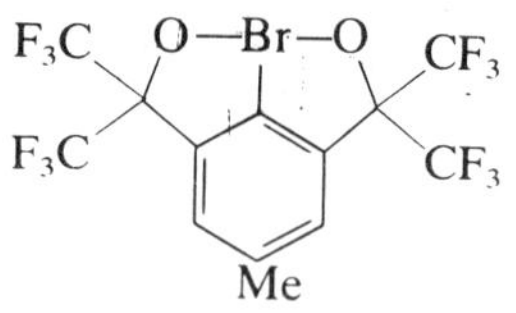

(**1**)

30.3.2.3 Halogen(V) compounds

These compounds are much less common than the halogen(III) compounds and only examples with iodine have been isolated. They have been classified as shown in Table 6. There is a paucity of detailed structural information, with three significant crystal structures and NMR data on some fluoro derivatives.

Table 6 Classification of Organohalogen(V) Compounds

Number of ligands	*Type of ligands*	*Example*
3	One carbon, two heteroatom	$PhIO_2$
4	One carbon, three heteroatom	$PhIOF_2$
4	Two carbon, two heteroatom	Ph_2IOF
5	Two carbon, three heteroatom	Ph_2IF_3
5	One carbon, four heteroatom	CF_3IF_4

The three-ligand class is represented structurally by *p*-chlorophenyliodine(V) dioxide (*p*-chloroiodylbenzene) and the unsubstituted phenyl compound. VSEPR theory would predict pyramidal geometry for these compounds, with multiple bonds to oxygen, and this is found approximately[68] for the *p*-chloro compound, although the accuracy of the determination is low. The angles around iodine are 94, 95 and 103° for the two C—I—O and the O—I—O angles respectively. For the phenyl compound[69] the corresponding angles are 91.1, 76.3 and 147.8° and the distortion in geometry is linked to the effect of the three strong secondary bonds to the iodine atom, which are grouped around the lone-pair position.

Although there is little structural information on four-ligand compounds with two carbon ligands, the structure of '*o*-iodoxybenzoic acid', $C_6H_4CO_2I(O)OH$, provides data for the class with a single carbon ligand.[70] The carboxylate oxygen forms a strong intermolecular bond to iodine of 2.324 Å, to give an arrangement based on a trigonal bipyramid, with the bonds to carbon and the multiply bonded oxygen (I—O = 1.784 Å) in equatorial positions, with the lone pair of electrons, and the intermolecular bond and the I—OH bond of 1.895 Å in axial positions.

For the five-ligand compounds there are no crystallographic data. These compounds are expected to have square pyramidal coordination, based on a 12-electron arrangement. ^{19}F NMR studies have provided some evidence in support of the square pyramidal arrangement. Thus diphenyliodine(V) trifluoride [diphenyl (difluoroiodosyl) fluoride] in nitromethane solution gives a spectrum consisting of two singlets, assigned to the fluorine atom opposite a phenyl ligand (more weakly bonded) and the two equivalent fluorine ligands opposite one another.[71] The lone pair is assumed to occupy the sixth coordination position opposite a phenyl group. Similarly[72] phenyliodine(V)tetrafluoride(tetrafluoroiodobenzene) shows just one singlet in its spectrum, corresponding to the four equivalent fluorine atoms. The corresponding trifluoromethyliodine(V) tetrafluoride[73] [(tetrafluoroiodo) trifluoromethane] has a spectrum typical of an A_3X_4 spin system and shows F—C—I—F coupling, which is consistent with the square pyramidal arrangement.

30.3.3 Nitrogen Ligands

The most significant aspect of chemistry in this section recently has been the characterization of the xenon–nitrogen bond. The carefully controlled reaction between XeF_2 and $HN(SO_2F)_2$ was reported to yield[74,75] the compounds $FXeN(SO_2F)_2$ for 1:1 proportions, and $Xe[N(SO_2F)_2]_2$ for 1:2 proportions, with elimination of HF. This has been confirmed by the subsequent X-ray crystallographic structure determination[76] of $FXeN(SO_2F)_2$ which shows a linear F—Xe—N arrangement with Xe—F and Xe—N distances of 1.967 and 2.200 Å respectively. Reaction of this compound with arsenic pentafluoride forms a bright yellow solid,[77,75] unstable at room temperature, which appears to be $[(FO_2S)_2NXe]^+[AsF_6]^-$ and which decomposes under vacuum to form the pale yellow $[\{(FO_2S)_2NXe\}_2F]^+[AsF_6]^-$, with the cation analogous to $[Xe_2F_3]^+$.

Halogen–nitrogen bonds are rather more common, although research in this area has not been extensive. Most compounds reported are halogen(I) cations, with a linear coordination geometry. Thus, dissolution of I_2, IBr or ICl in pyridine leads[78] to formation of $[py_2I]^+$ cations, and the crystal structure determination[79] of the solid, empirically $py(I_2)_2$, formed from iodine in ethanol solution, has shown the presence of planar $[py_2I]^+$ cations with I_3^- anions and I_2 molecules. The linear $[N—I—N]^+$ unit has an I—N distance of 2.16 Å, the linear arrangement corresponding on VSEPR theory to a 10-electron species.

This linear coordination, but with unequal I—N bonds, has been found in the compound $(bipy)(IN_3)_2$[80] where the I—N(bipy) distance averages 2.44 Å and I—N(azide) averages 2.17 Å, reflecting the difference in the bonding to the two ligands. The two bromine(I) examples studied both show asymmetry of the N—Br—N linear system, although the two ligands are equivalent. Thus in bis(quinuclidine)bromine(I) tetrafluoroborate[81] the Br—N distances are 2.120 and 2.156 Å and in bis(quinoline)bromine(I) perchlorate[82] they are 2.100 and 2.165 Å. The reason for the asymmetry is not obvious.

Although the interaction of pyridine and $ClNO_3$ has been reported[83] to yield the $[py_2Cl]^+$ cation, there is as yet no crystallographic determination on a chlorine(I) compound.

30.3.4 Oxygen Ligands

Apart from the very few examples mentioned previously, oxygen is the only element other than fluorine to form bonds to xenon. For the halogens in positive oxidation states the oxides and oxoanions provide some of the most familiar compounds of these elements, with a very extensive chemistry, adequately reviewed elsewhere.[2] Similarly the oxides and oxoanions of xenon include the simplest examples of compounds of xenon(VIII), with the tetraoxide and the xenate(VIII) (perxenate) ion having the expected, undistorted tetrahedral and octahedral coordination arrangements respectively. In this chapter attention will be concentrated on ligands containing oxygen as the donor atom, rather than the multiply bonded oxide ligand itself. These are mainly oxoacid salts, with earlier work concentrating on such ligands as nitrates and perchlorates and more recent work on the more electronegative species such as the fluorosulfates and particularly the pentafluoroselenates(VI) and pentafluorotellurates(VI). It has been suggested that these latter ligands may be more electronegative than fluorine[84] and the pentafluorotellurate(VI) gives the most thermally stable noble gas compounds of this type.

In general, apart from those involving the more electronegative ligands, the compounds containing Xe—O bonds tend to have limited thermal stability and high chemical reactivity and often to involve explosions. Thus, relatively little crystallographic structure determination has been achieved and spectroscopic techniques have been used to show structural analogies to corresponding interhalogen compounds.

30.3.4.1 Halogen(I) and noble gas(II) compounds

By analogy with the halogen ligand compounds discussed previously, these oxidation states are associated with coordination number two and a linear arrangement of ligands about the central atom. The anionic halogen compounds have been little studied. An IR spectroscopic study[85] of the dinitrato-bromate(I) and -iodate(I) anions was concerned primarily with the denticity of the nitrate ligands, which was not unambiguously determined. Although

$Ag[I(OClO_3)_2]$ is among the products of the low temperature reaction[86] of I_2 with $AgClO_4$ it has not been obtained pure, and no spectroscopic results have been reported.

More interest has been shown in the noble gas compounds, since the majority of these compounds have been prepared for this oxidation state. Compounds are derived from XeF_2 by substitution of one or both fluorine atoms. The order of stability based on the reported decomposition of the mono- and di-substituted species[1] is $OTeF_5 > OSeF_5 > OSO_2F > OPO_2F > OClO_3 > OSO_2CF_3$, OSO_2Me, $OCOCF_3$, OCOMe. The last four ligands lead to explosive products.

The characteristic linear coordination of xenon has been verified crystallographically for two examples. The structure of $FXeOSO_2F$ shows[87] a linear arrangement with Xe—F and Xe—O distances of 1.94 Å and 2.16 Å respectively. Although the structure determination[88] of $Xe(OSeF_5)_2$ shows a disordered arrangement, the O—Xe—O group is linear with an Xe—O distance of 2.12 Å.

The novel cation $[(FXeO)_2SOF]^+$ has been characterized crystallographically as the $[AsF_6]^-$ salt.[89] The bridging fluorosulfate group is analogous to the bridging fluoride in the $[Xe_2F_3]^+$ cation. The Xe—F and Xe—O distances are 1.86 and 2.21 Å respectively.

30.3.4.2 Halogen(III) and noble gas(IV) compounds

The halogen compounds are known for neutral, cationic and anionic species. These should be three, two and four coordinate, with T-shaped, angular and square planar coordination arrangements respectively. No crystallographic work has been reported on the simple species and the spectroscopic results are not always conclusive.

Thus iodine(III) perchlorate[90] and bromine(III) and iodine(III) fluorosulfates[91] are apparently polymeric in the solid state, although the degree of polymerization has not been established. The $[Br(OSO_2F)_2]^+$ and $[I(OSO_2F)_2]^+$ cations have been reported in combination with[92] $[Sn(OSO_2F)_6]^{2-}$ and[93] $[Au(OSO_2F)_4]^-$ anions, and IR and Raman spectra support the formulation as ions, but do not give structural details. The spectra of the anionic species in[91] $K[Br(OSO_2F)_4]$, $K[I(OSO_2F)_4]$, $Cs[I(OClO_3)_4]$[90] and $Rb[Br(OSeF_5)_4]$[94] show features characteristic of the $[BrO_4]$ and $[IO_4]$ square planar units.

A recent structure determination[95] has shown that the compound $I(OSO_2F)_2I$ has the characteristic T-shaped coordination with the I^{III}—I distance of 2.678 Å in the equatorial position and two I—O axial distances of 2.09 and 2.26 Å with O—I—I angles of 91.0° and 85.3°. A secondary I · · · O bond of 3.05 Å completes a coordination similar to that in the $[XeF_3]^+$ ion described earlier.[30] The only noble gas compound of this class is $Xe(OTeF_5)_4$, prepared from the interaction of XeF_4 and $B(OTeF_5)_3$, for which the Raman spectrum supports the square planar arrangement.[96]

30.3.4.3 Halogen(V) and noble gas(VI) compounds

Although there are some examples of halogen(V) compounds containing oxygen ligands, together with fluoride or multiply bonded oxide, examples with five oxygen ligands are rare and often ill characterized. For example the interaction[97] of IF_5 with $Me_3Si(OMe)$ or $Me_2Si(OMe)_2$ produces the compounds $IF_{(5-n)}(OMe)_n$ for $n = 1$–4, with only tentative identification of $I(OMe)_5$. However, $I(OTeF_5)_5$ has been characterized[84] as well as the substituted species $F_nI(OTeF_5)_{5-n}$ for $n = 1$–4 and $OI(OTeF_5)_3$. The pentakis(pentafluorotellurate) could not be prepared directly from IF_5 but resulted from the oxidation of $I(OTeF_5)_3$ with $Xe(OTeF_5)_2$. The mixed fluoro species were characterized by NMR spectroscopy, all having fluorine in the axial position of the square pyramidal arrangement.

Pentafluorotellurate compounds of xenon(VI) similar to the iodine(V) compounds above have also been prepared.[96] Thus $OXe(OTeF_5)_4$ results from the interaction of $OXeF_4$ with $B(OTeF_5)_3$ and there is NMR evidence from a solution of the pentafluorotellurate in $OXeF_4$ for the species $OXeF_n(OTeF_5)_{4-n}$ for $n = 1$–3. The structures of these compounds are square pyramidal like the iodine compounds above, with the multiply bonded oxide in the apical position. $Xe(OTeF_5)_6$ has been prepared from XeF_6 using the versatile $B(OTeF_5)_3$ reagent at low temperatures. Although it decomposes above −10 °C, powder X-ray diffraction has shown it to have a similar, but significantly different, unit cell to that of $Te(OTeF_5)_6$.[98] From this difference, a distorted octahedral configuration for xenon has been inferred, by comparison with the undistorted [TeO_6] central unit in the tellurium compound,[96] and the lone pair appears to have some stereochemical effect.

30.3.5 Sulfur and Selenium Ligands

There are no examples of xenon bonded to sulfur or selenium and few for the halogens. The reported compounds involve the thiocyanate[99,100] and selenocyanate[100] ligands acting as pseudohalides, in the anions $I(SCN)_2^-$ and $I(SeCN)_2^-$, for which IR and Raman spectra show the thiocyanate to have a linear S—I—S arrangement but give no definitive results for the selenocyanate, and two examples with more complex thio ligands for which crystal structures have been determined.

For the bis(thiourea) iodine(I) cation,[101] in the iodide salt, there is a linear S—I—S arrangement with the I—S distance of 2.629 Å. The complex cation[102] $[\{(en)_2Co(SCH_2CH_2NH_2)\}_2I]^{5+}$, with the two octahedrally coordinated cobalt complexes linked through an S—I—S bridge, has an S—I—S angle of 173° and an I—S distance of 2.619 Å.

30.4 REFERENCES

1. K. Seppelt and D. Lentz, *Prog. Inorg. Chem.*, 1982, **29,** 167.
2. (a) J. C. Bailar, H. J. Emeleus, R. Nyholm and A. F. Trotman-Dickenson (eds.), 'Comprehensive Inorganic Chemistry', Pergamon, Oxford, 1973; (b) A. J. Downs and C. J. Adams, 'Chlorine, Bromine, Iodine and Astatine', in ref. 2(a), vol. 2, p. 1107; (c) N. Bartlett and F. O. Sladky, 'The Chemistry of Krypton, Xenon and Radon', in ref. 2(a), vol. 1, p. 213.
3. 'Gmelins Handbuch der Anorganischen Chemie', 8th edn., and Supplement vols., Springer-Verlag, Berlin, System-Nos. 6, 'Chlorine', 7, 'Bromine', 8 'Iodine' and 1, 'Noble Gases'.
4. J. W. Mellor, 'A Comprehensive Treatise on Inorganic and Theoretical Chemistry', Supplement II, Part I, F, Cl, Br, I, Longmans, London, 1956.
5. (a) F. A. Cotton and G. Wilkinson, 'Advanced Inorganic Chemistry', 4th edn., Wiley-Interscience, New York, 1980; (b) K. F. Purcell and J. C. Kotz, 'Inorganic Chemistry', Holt–Saunders, Philadelphia, 1977; (c) N. N. Greenwood and A. Earnshaw, 'Chemistry of the Elements', Pergamon, Oxford, 1984.
6. H. Selig and J. H. Holloway, *Top. Curr. Chem.*, 1984, **124,** 33.
7. R. J. Gillespie, 'Molecular Geometry', van-Nostrand Reinhold, London, 1972.
8. N. W. Alcock, *Adv. Inorg. Chem. Radiochem.*, 1972, **15,** 1.
9. A. J. Edwards and G. R. Jones, *J. Chem. Soc. (A)*, 1969, 1467.
10. R. J. Gillespie, D. Martin, G. J. Schrobilgen and D. R. Slim, *J. Chem. Soc., Dalton Trans.*, 1977, 2234.
11. T. Migchelsen and A. Vos, *Acta Crystallogr.*, 1967, **22,** 812.
12. N. Bartlett, B. G. DeBoer, F. J. Hollander, F. O. Sladky, D. H. Templeton and A. Zalkin, *Inorg. Chem.*, 1974, **13,** 780.
13. R. J. Gillespie and G. J. Schrobilgen, *Inorg. Chem.*, 1976, **15,** 22.
14. A. Zalkin, D. L. Ward, R. N. Bagioni, D. H. Templeton and N. Bartlett, *Inorg. Chem.*, 1978, **17,** 1318.
15. N. Bartlett, M. Gennis, D. D. Gibler, B. K. Morrell and A. Zalkin, *Inorg. Chem.*, 1973, **12,** 1717.
16. V. M. McRae, R. D. Peacock and D. R. Russell, *Chem. Commun.*, 1969, 62.
17. S. Soled and G. B. Carpenter, *Acta Crystallogr., Sect. B*, 1973, **29,** 2104; S. Gran and C. Rømming, *Acta Chem. Scand.*, 1968, **22,** 1686.
18. S. Soled and G. B. Carpenter, *Acta Crystallogr., Sect. B*, 1973, **29,** 2556; J. E. Davies and E. K. Nunn, *Chem. Commun.*, 1969, 1374.
19. G. L. Breneman and R. D. Willett, *Acta Crystallogr., Sect. B*, 1969, **25,** 1073.
20. O. Glemser and A. Smalc, *Angew. Chem., Int. Ed. Engl.*, 1969, **8,** 517.
21. R. J. Gillespie and M. J. Morton, *Inorg. Chem.*, 1970, **9,** 811.
22. A. J. Edwards and R. J. C. Sills, *J. Chem. Soc. (A)*, 1970, 2697.
23. T. Birchall and R. D. Myers, *Inorg. Chem.*, 1981, **20,** 2207.
24. T. Birchall and R. D. Myers, *Inorg. Chem.*, 1983, **22,** 1751.
25. J. Passmore, G. Sutherland and P. S. White, *Inorg. Chem.*, 1981, **20,** 2169.
26. C. G. Vonk and E. H. Wiebenga, *Acta Crystallogr.*, 1959, **12,** 859.
27. R. D. Burbank and F. N. Bensey, *J. Chem. Phys.*, 1953, **21,** 602.
28. R. D. Burbank and F. N. Bensey, *J. Chem. Phys.*, 1957, **27,** 982.
29. K. H. Boswijk and E. H. Wiebenga, *Acta Crystallogr.*, 1954, **7,** 417.
30. P. Boldrini, R. J. Gillespie, P. R. Ireland and G. J. Schrobilgen, *Inorg. Chem.*, 1974, **13,** 1690.
31. D. E. McKee, A. Zalkin and N. Bartlett, *Inorg. Chem.*, 1973, **12,** 1713.
32. A. J. Edwards, *J. Chem. Soc., Dalton Trans.*, 1978, 1723.
33. A. J. Edwards and G. R. Jones, *J. Chem. Soc. (A)*, 1969, 1936.
34. R. J. Elema, J. L. deBoer and A. Vos, *Acta Crystallogr.*, 1963, **16,** 243.
35. J. H. Burns, P. A. Agron and H. A. Levy, *Science*, 1963, **139,** 1208; H. A. Levy and P. A. Agron, *J. Am. Chem. Soc.*, 1963, **85,** 241.
36. S. W. Patterson, J. H. Holloway, B. A. Coyle and J. M. Williams, *Science*, 1971, **173,** 1238.
37. M. D. Lind and K. O. Christe, *Inorg. Chem.*, 1972, **11,** 608.

38. H. W. Baird and H. F. Giles, *Acta Crystallogr., Sect. A,* 1969, **25,** S115.
39. A. J. Edwards and P. Taylor, *J. Chem. Soc., Dalton Trans.,* 1975, 2174.
40. R. D. Burbank and G. R. Jones, *Inorg. Chem.,* 1974, **13,** 1071.
41. K. Leary, D. H. Templeton, A. Zalkin and N. Bartlett, *Inorg. Chem.,* 1973, **12,** 1726.
42. R. D. Burbank and G. R. Jones, *J. Am. Chem. Soc.,* 1974, **96,** 43.
43. K. Leary, A. Zalkin and N. Bartlett, *Inorg. Chem.,* 1974, **13,** 775.
44. R. M. Gavin, Jr. and L. S. Bartell, *J. Chem. Phys.,* 1968, **48,** 2460; L. S. Bartell and R. M. Gavin, Jr., *J. Chem. Phys.,* 1968, **48,** 2466; K. S. Pitzer and L. S. Bernstein, *J. Chem. Phys.,* 1975, **63,** 3849.
45. R. Bougon, P. Charpin and J. Soriano, *C. R. Hebd. Seances Acad. Sci., Ser. C,* 1971, **272,** 565.
46. K. O. Christie, *Inorg. Chem.,* 1972, **11,** 1215.
47. (a) K. O. Christie, *Inorg. Chem.,* 1973, **12,** 1580; (b) K. O. Christe, W. W. Wilson and E. C. Curtis, *Inorg. Chem.,* 1983, **22,** 3056.
48. R. J. Gillespie and G. J. Schrobilgen, *Inorg. Chem.,* 1974, **13,** 1230.
49. R. D. Burbank and F. N. Bensey, *J. Chem. Phys.,* 1957, **27,** 981; R. D. Burbank, *Acta Crystallogr.,* 1962, **15,** 1207.
50. J. Donohue, *Acta Crystallogr.,* 1965, **18,** 1018.
51. W. J. Adams, H. B. Thompson and L. S. Bartell, *J. Chem. Phys.,* 1970, **53,** 4040.
52. C. J. Adams, *Inorg. Nucl. Chem. Lett.,* 1974, **10,** 831.
53. T. A. Carlson and R. M. White, *J. Chem. Phys.,* 1962, **36,** 2883.
54. L. J. Turbini, R. E. Aikman and R. J. Lagow, *J. Am. Chem. Soc.,* 1979, **101,** 5833.
55. G. F. Koser, in 'The Chemistry of Functional Groups', Supplement D: The Chemistry of Halides, Pseudohalides and Azides, ed. S. Patai and Z. Rappoport, Wiley, 1983, p. 721.
56. K.-F. Tebbe and R. Fröhlich, *Z. Anorg. Allg. Chem.,* 1983, **505,** 7.
57. C. Willgerodt, *J. Prakt. Chem.,* 1886, **33,** 154.
58. S. W. Page, E. P. Mazzola, A. D. Mighell, V. L. Himes and C. R. Hubbard, *J. Am. Chem. Soc.,* 1979, **101,** 5858.
59. U. Drück and W. Littke, *Acta Crystallogr., Sect. B,* 1978, **34,** 3092.
60. J. L. Attwood and W. A. Sheppard, *Acta Crystallogr., Sect. B,* 1975, **31,** 2638.
61. G. Wittig and M. Rieber, *Liebigs Ann. Chem.,* 1949, **562,** 187; G. Wittig and K. Clauss, *Liebigs Ann. Chem.,* 1952, **578,** 136.
62. H. J. Reich and C. S. Cooperman, *J. Am. Chem. Soc.,* 1973, **95,** 5077.
63. N. W. Alcock and R. M. Countryman, *J. Chem. Soc., Dalton Trans.,* 1977, 217.
64. E. M. Archer and T. G. van Schalkwyk, *Acta Crystallogr.,* 1953, **6,** 88.
65. N. W. Alcock, R. M. Countryman, S. Esperas and J. F. Sawyer, *J. Chem. Soc., Dalton Trans.,* 1979, 854.
66. H. J. Frohn and M. Giesen, *J. Fluorine Chem.,* 1984, **24,** 9.
67. T. T. Nguyen and J. C. Martin, *J. Am. Chem. Soc.,* 1980, **102,** 7382.
68. E. M. Archer, *Acta Crystallogr.,* 1948, **1,** 64.
69. N. W. Alcock and J. F. Sawyer, *J. Chem. Soc., Dalton Trans.,* 1980, 115.
70. J. Z. Gougoutas, *Cryst. Struct. Commun.,* 1981, **10,** 489.
71. V. V. Lyalin, V. V. Orda, L. A. Alekseeva and L. M. Yagupol'skii, *J. Org. Chem. USSR (Engl. Transl.),* 1972, **8,** 215.
72. L. M. Yagupol'skii, V. V. Lyalin, V. V. Orda and L. A. Alekseeva, *J. Gen. Chem. USSR (Engl. Transl.),* 1968, **38,** 2714.
73. G. Oates and J. M. Winfield, *J. Chem. Soc., Dalton Trans.,* 1974, 119.
74. R. D. LeBlond and D. D. DesMarteau, *J. Chem. Soc., Chem. Commun.,* 1974, 555.
75. D. D. DesMarteau, R. D. LeBlond, S. F. Hossain and D. Nothe, *J. Am. Chem. Soc.,* 1981, **103,** 7734.
76. J. F. Sawyer, G. J. Schrobilgen and S. J. Sutherland, *J. Chem. Soc., Chem. Commun.,* 1982, 210.
77. D. D. DesMarteau, *J. Am. Chem. Soc.,* 1978, **100,** 6270.
78. S. G. W. Ginn and J. L. Wood, *Trans. Faraday Soc.,* 1966, **62,** 777.
79. O. Hassell and H. Hope, *Acta Chem. Scand.,* 1961, **15,** 407.
80. H. Dörner, K. Dehnicke, W. Massa and R. Schmidt, *Z. Naturforsch., Teil B,* 1983, **38,** 437.
81. L. K. Blair, K. D. Parris, P. S. Hii and C. P. Brock, *J. Am. Chem. Soc.,* 1983, **105,** 3649.
82. N. W. Alcock and G. B. Robertson, *J. Chem. Soc., Dalton Trans.,* 1975, 2483.
83. M. Schmeisser and K. Brändle, *Angew. Chem.,* 1961, **73,** 388.
84. D. Lentz and K. Seppelt, *Angew. Chem., Int. Ed. Engl.,* 1978, **17,** 355; *Z. Anorg. Allg. Chem.,* 1980, **460,** 5.
85. M. Lustig and J. K. Ruff, *Inorg. Chem.,* 1966, **5,** 2124.
86. N. W. Alcock and T. C. Waddington, *J. Chem. Soc.,* 1962, 2510.
87. N. Bartlett, M. Wechsberg, G. R. Jones and R. D. Burbank, *Inorg. Chem.,* 1972, **11,** 1124.
88. L. K. Templeton, D. H. Templeton, K. Seppelt and N. Bartlett, *Inorg. Chem.,* 1976, **15,** 2718.
89. R. J. Gillespie, G. J. Schrobilgen and D. R. Slim, *J. Chem. Soc., Dalton Trans.,* 1977, 1003.
90. K. O. Christe and C. J. Schack, *Inorg. Chem.,* 1972, **11,** 1682.
91. H. A. Carter, S. P. L. Jones and F. Aubke, *Inorg. Chem.,* 1970, **9,** 2485.
92. P. A. Yeats, B. Landa, J. R. Sams and F. Aubke, *Inorg. Chem.,* 1976, **15,** 1452.
93. K. C. Lee and F. Aubke, *Inorg. Chem.,* 1980, **19,** 119.
94. K. Seppelt, *Chem. Ber.,* 1973, **106,** 1910.
95. M. J. Collins, G. Dénès and R. J. Gillespie, *J. Chem. Soc., Chem. Commun.,* 1984, 1296.
96. E. Jacob, D. Lentz, K. Seppelt and A. Simon, *Z. Anorg. Allg. Chem.,* 1981, **472,** 7.
97. G. Oates, J. M. Winfield and O. R. Chambers, *J. Chem. Soc., Dalton Trans.,* 1974, 1380.
98. D. Lentz, H. Pritzkow and K. Seppelt, *Angew. Chem., Int. Ed. Engl.,* 1977, **16,** 729; *Inorg. Chem.,* 1978, **17,** 1926.
99. C. Long and D. A. Skoog, *Inorg. Chem.,* 1966, **5,** 206.
100. G. A. Bowmaker and D. A. Rogers, *J. Chem. Soc., Dalton Trans.,* 1981, 1146.
101. H. Hope and G. H.-Y. Lin, *Chem. Commun.,* 1970, 169.
102. D. L. Nosco, M. J. Heeg, M. D. Glick, R. C. Elder and E. Deutsch, *J. Am. Chem. Soc.,* 1980, **102,** 7784.

31
Titanium

CHARLES A. McAULIFFE and DAVID S. BARRATT
University of Manchester Institute of Science and Technology, UK

31.1 INTRODUCTION

Titanium is a reasonably common element which has been known for over 170 years. The average titanium content of the earth's crust is 0.63% by weight, which makes it the ninth most abundant element in the earth's crust, 20 times more abundant than carbon, and only outranked by oxygen, silicon, aluminum, iron, magnesium, calcium, sodium and potassium. It is only really in this century that elemental titanium has developed any industrial potential, partly because of difficulties associated with its refinement.

The element was first obtained pure in 1910 by Hunter[1] *via* reduction of the tetrachloride with sodium. In 1925, van Arkel and de Boer[2] obtained a very pure form of the metal by dissociation of the tetraiodide. Nonetheless, the titanium metal industry really dates from the publication of the Kroll process[3] in 1940, which involves reduction of the tetrachloride with magnesium. The principal properties of titanium which recommend it to metallurgists and engineers are its low density, strength and resistance to corrosion.

The most important industrial use of titanium is in the form of the dioxide (rutile or anatase) as a white pigment in the paint industry. Its outstanding property is its capacity or covering power; this, with its relative chemical inertness, has resulted in its almost complete replacement of all other white pigments. Related uses arise in the paper, plastics, rubber, textile and vitreous enamel industries.

31.1.1 Discovery

Titanium was discovered in 1791 by William Gregor, an English clergyman and amateur chemist. He identified it in a black sand (now known to be ilmenite) sent to him for analysis from the Menacchan Valley in Cornwall. Four years later, the famous German chemist Klaproth rediscovered the element in the ore rutile, one form of titanium dioxide. He gave it the name titanium after the Titans who in Greek mythology were the sons of Earth.

Titanium is widely distributed on the earth's surface, and also occurs in the sun, in stars and in meteorites. In the earth's crust the principal ores are ilmenite ($FeTiO_3$) and rutile (TiO_2). An important deposit of ilmenite occurs in the Ilmen mountains in Russia, from which the ore gets its name. Some other minerals are titanomagnetite [$Fe_3O_4(Ti)$], ramsayite (lorenzenite) ($Na_2(TiO)_2Si_2O_7$), titanite ($CaTiSiO_5$), lamprophyllite [$(Ba, Sr, K)Na(Ti, Fe)TiSi_2(O, OH, F)_9$], benitoite [$BaTi(Si_3O_9)$], warwickite [$(Mg, Fe)_3TiB_2O_8$], osbornite (TiN) and perovskite ($CaTiO_3$). Crystallographic details for all these have been tabulated.[4]

31.1.2 Preparation of the Metal

It is not possible to obtain titanium metal by the usual method of reduction of the dioxide with carbon, because a very stable carbide is formed. Moreover the metal is reactive towards nitrogen and oxygen at elevated temperatures, and hydrogen at 900 °C only reduces TiO_2 as far as Ti_3O_5. Reduction of the dioxide with most reducing metals, *e.g.* Na, Al, Ca or Mg, seldom seems to yield a pure product; the most common contaminants, and in some cases the principal products, are lower oxides of titanium. Reduction of the tetrachloride is, therefore, the basis of the preferred methods.

A successful laboratory method is reduction of the dioxide with an excess of calcium hydride in a molybdenum boat. The reaction is carried out at 900 °C in a vacuum or in an atmosphere of hydrogen[5] (equation 1). Most of the hydrogen present can be removed by heating the metal in a vacuum at 1100 °C.

$$TiO_2 + 2CaH_2 \longrightarrow Ti + 2CaO + 2H_2 \qquad (1)$$

The first general method involved reduction of the tetrachloride with sodium (equation 2).[1]

The reaction, which is very exothermic, is carried out at 700–800 °C in a steel bomb under an inert atmosphere. The method, which requires very careful control of the temperature, is capable of yielding titanium with a purity of 99.9%. It is also possible to prepare the metal by hydrogen reduction[6] of the tetrachloride at approximately 2000 °C (equation 3). The difficulties of removing hydrogen from the metal, and the high temperatures required, make the method impracticable industrially.

$$TiCl_4 + 4Na \longrightarrow Ti + 4NaCl \tag{2}$$

$$TiCl_4 + 2H_2 \longrightarrow Ti + 4HCl \tag{3}$$

Another important way of producing the metal is the Kroll process, in which magnesium is the reducing agent (equation 4). The magnesium is contained in a molybdenum-coated iron crucible in an atmosphere of argon at 850–950 °C; the $TiCl_4$ is added drop by drop. Towards the end of the addition the temperature is slowly raised to a maximum of 1180 °C, *i.e.* to slightly above the boiling point of magnesium. After cooling the crucible in argon, the magnesium and magnesium chloride are extracted with water and dilute HCl; alternatively, these materials may be removed by volatilization above 1000 °C. The process is clearly expensive, because it involves the prior preparation of the tetrachloride from ilmenite or rutile, followed by fractionation of the $TiCl_4$ to remove any $FeCl_3$.

$$TiCl_4 + 2Mg \longrightarrow Ti + 2MgCl_2 \tag{4}$$

Extremely pure titanium can be made on the laboratory scale by the van Arkel–de Boer method[2] in which TiI_4[7] that has been carefully purified is vaporized and decomposed on a hot wire in a vacuum.[8]

31.1.3 Properties of the Metal

Titanium is dimorphic with a transformation temperature of 882 ± 5 °C. Below this temperature, the metal is hexagonal (D_{6h}—$6/mmc$, $Z = 2$)[4] with lattice parameters at 25 °C of $a = 2.9504$ Å and $c = 4.6833$ Å (α form).[4,9] The second (β) form is cubic (O_h—$Im3m$, $Z = 2$),[9] with the cell parameter[10] of 3.3065 Å at 900 °C. The two forms are isomorphous with the corresponding forms of zirconium.

Titanium is a silvery, ductile metal with important industrial uses because it is less dense than iron, much stronger than aluminum and almost as corrosion resistant as platinum. Although it is unlikely ever to be as inexpensive as steel, its rare combination of properties make it ideal for a variety of uses, particularly in engines, aircraft frames, some marine equipment, in industrial plants and in laboratory equipment. Certain properties may be improved by alloying it with aluminum.

The metal has an enormous affinity for dioxygen, dinitrogen and dihydrogen at elevated temperatures, but only in the case of dihydrogen is the absorption reversible. Moreover, titanium dust is explosive in air, and even the fresh surface of the metal will ignite spontaneously in dioxygen at 25 atm. Traces of oxygen or nitrogen render the metal very brittle and hence of little utility. The kinetics of oxidation of the metal in the range 850–1050 °C have been studied.[11] It is not attacked by mineral acids in the cold or by hot aqueous alkali, but dissolves in HF, hot HCl, hot HNO_3 and hot concentrated H_3PO_4. It reacts with many metals and non-metals forming intermetallic or interstitial compounds, *e.g.* with Al, Sb, Be, Cr, Fe and B.

31.1.3.1 Isotopes of titanium

Titanium has five naturally occurring isotopes with mass numbers ranging from 46 to 50; their percentage isotopic abundances are given in Table 1, with the atomic masses of the individual nucleides.

The atomic weight of titanium is 47.88 (IUPAC; 1971; based on $^{12}C = 12.000$) and was originally determined[12] by the ratio $TiCl_4/4Ag$ and $TiBr_4/4Ag$. Slight separation of the titanium isotopes is possible in the extraction of thiocyanic acid complexes of titanium by ether from aqueous sulfate solutions. The separation factors[13] are $^{50}Ti:^{48}Ti = 1.006$, $^{48}Ti:^{46}Ti = 1.008$ and $^{50}Ti:^{46}Ti = 1.014$.

Table 1 Nuclides of Titanium

Nuclide	*%*	*Isotopic mass*	*Nuclear spin*	*Half-life*
^{43}Ti	—	42.96850	—	0.6 s
^{44}Ti	—	43.95957	—	48.2 y[a]
^{45}Ti	—	44.95813	7/2	3.08 h
^{46}Ti	7.93	45.952633	—	—
^{47}Ti	7.28	46.95176	5/2	—
^{48}Ti	73.94	47.947948	—	—
^{49}Ti	5.51	48.947867	7/2	—
^{50}Ti	5.34	49.944789	—	—
^{51}Ti	—	50.94662	—	5.9 min
^{52}Ti	—	—	—	12 min

[a] P. E. Moreland and D. Heymann, *J. Inorg. Nucl. Chem.*, 1965, **27**, 493.

31.1.4 General Properties of the Group IV Elements

Titanium is the first member of the 3*d* transition series and has four valence electrons, $3d^24s^2$. The most stable and most common oxidation state, +4, involves the loss of all these electrons. However, the element may also exist in a range of lower oxidation states, most importantly as Ti(III), (II), (0) and −(I). Zirconium shows a similar range of oxidation states, but the tervalent state is much less stable relative to the quadrivalent state than is the case with titanium. The chemistry of hafnium closely resembles that of zirconium; in fact, the two elements are amongst the most difficult to separate in the periodic table.

The ionization energies of the group IV elements are given in Table 2, together with the sum of the first four ionization energies in each case. There is a clear alternation in the ionization energies on passing down the group, this being associated with the alternation in the screening constants and hence the effective nuclear charges (Z^*) of the different atoms. In particular, the third member of a group (Ge in this case) first suffers the transition metal contraction associated with the poor screening of the underlying d^{10} shell, and the fifth member (Pb) first suffers the double contraction associated with the even poorer screening of the underlying f^{14} shell. Consequently, these elements have higher ionization energies than might otherwise have been expected. The elements in group IVA (Ti, Zr, Hf) have uniformly lower ionization energies than those of group IVB, and as a consequence it is believed that the former achieve high coordination numbers with greater facility than the latter. The radii of the elements show these alternations in inverse, as the ionization energy is inversely related to the radius.

Table 2 Ionization Energies (eV) of Group IV Elements[a]

	I_1	I_2	I_3	I_4	$\Sigma_1^4 I_n$
C	11.256	24.376	47.871	64.476	147.98
Si	8.149	16.34	33.46	45.13	103.8
Ge	7.88	15.94	34.21	45.7	103.72
Sn	7.342	14.628	30.49	40.72	93.18
Pb	7.415	15.028	31.93	42.31	96.68
Ti	6.82	13.57	27.47	43.24	91.10
Zr	6.84	13.13	22.98	34.33	77.28
Hf	7.00	14.90	23.20[b]	33.30[b]	78.40

[a] C. E. Moore, Atomic Energy Levels, Circular No. 467, National Bureau of Standards, Washington, 1949, 1952, 1958 and appendices.
[b] P. F. A. Klinkenberg, T. A. M. van Kleef and P. E. Norman, *Physica*, 1961, **27**, 151.

The energy required for the removal of the four valence electrons of titanium is so high that the Ti^{IV} ion does not exist as such; nor do the Zr^{IV} and Hf^{IV} ions. Titanium(IV) compounds are, in general, covalent and bear some resemblance to the corresponding compounds of Si, Ge, Sn and Pb, especially Sn (partly because of the similarities between the ionic and covalent radii of Ti and Sn). For instance TiO_2 (rutile) is isomorphous with SnO_2 (cassiterite), and both become yellow on heating; both $TiCl_4$ and $SnCl_4$ are colourless, distillable liquids which are rapidly hydrolyzed; both behave as Lewis acids, forming a wide variety of addition compounds. Both $SiCl_4$ and $GeCl_4$ are less powerful acceptor molecules than $TiCl_4$ and $SnCl_4$.

In the lower oxidation states the chemistry of titanium has little or no counterpart in the chemistries of the group IVB elements. The only lower oxidation state of these elements is two, for which the stability order is Ge < Sn < Pb. However, both zirconium(III) and hafnium(III) are similar to if less stable (towards oxidation) than titanium(III) and have comparable although less extensively investigated chemistries.

The most common coordination number of titanium is six (recognized for all oxidation states of the metal), although compounds are known in which the coordination number is four, five, seven or eight. The common oxidation states of titanium with the associated coordination numbers and stereochemistries are summarized in Table 3. The properties of these molecules will be discussed in the appropriate sections. In brief, however, titanium compounds in the +III or lower oxidation states are readily oxidized to the +IV state. Furthermore, titanium compounds can usually be hydrolyzed to compounds containing Ti—O linkages.

Table 3 Oxidation States and Stereochemistry of Titanium

Oxidation state	*Coordination number*	*Geometry*	*Examples*
Ti^{-I}	6	Octahedral	$[Ti(bipy)_3]^-$
Ti^0	6	Octahedral	$[Ti(bipy)_3]$
Ti^{II}, d^2	4	Dist. tetrahedral	$(\eta^5\text{-}C_5H_5)_2Ti(CO)_2$
	6	Octahedral	$TiCl_2$
Ti^{III}, d^1	3	Planar	$Ti\{N(SiMe_3)_2\}_3$
	5	Tbp	$TiBr_3(NMe_3)_2$
	6[a]	Octahedral	TiF_6^{3-}, $TiCl_3(THF)_3$
Ti^{IV}, d^0	4[a]	Tetrahedral	$TiCl_4$
		Dist. tetrahedral	$(\eta^5\text{-}C_5H_5)TiCl_2$
	5	Dist. tbp	$K_2Ti_2O_5$
		Square planar	TiO(porphyrin)
	6[a]	Octahedral	TiF_6^{2-}, $Ti(acac)_2Cl_2$
	7	ZrF_7^{3-} type	$[Ti(O_2)F_5]^{3-}$
		Pentagonal bipyramidal	$Ti_2(ox)_3\cdot 10H_2O$
	8	Dist. dodecahedral	$TiCl_4(diars)_2$
			$Ti(S_2CNEt_2)_4$

[a] Most common state.

31.1.5 Sources of Information

The main source of information on titanium and its chemistry remains the important review by Clark.[14] References to phosphine and analogous complexes are contained in volumes by McAuliffe and Levason,[15–17] and the update reviews which appear from time to time in *Coordinated Chemistry Reviews.*[18–21] In order to minimize the number of references quoted, much of the early chemistry is referred to in these review sources.

31.2 NIROGEN LIGANDS

31.2.1 Low Valent Titanium Complexes

Although the coordination chemistry of titanium(IV), and to a much lesser extent titanium(III), has been investigated relatively well, few complexes are known in the lower oxidation states −II, −I, O, +I, +II, and all are almost invariably unstable to oxidation.[14,22] Reduction of $TiCl_4$ with excess of dilithium 2,2′-bipyridyl in THF with excess bipyridyl gives $Li[Ti(bipy)_3]\cdot 3.7THF$, as shown in equations (5) and (6). Further reduction to $Li_2[Ti(bipy)_3]\cdot 5.7$ is possible.[14]

$$TiCl_4 + 2Li_2bipy + bipy \xrightarrow{THF} [Ti(bipy)_3] + 4LiCl \quad (5)$$

$$TiCl_4 + 2.5Li_2bipy + 0.5bipy \xrightarrow{THF} Li[Ti(bipy)_3]\cdot 3.7THF + 4LiCl \quad (6)$$

The titanium(II) complexes are somewhat better understood. The deeply coloured $TiX_2(MeCN)_2$ (X = Cl, Br) complexes react with other ligands to yield $TiCl_2L_2$ (L = C_4H_8O, $C_5H_{10}O$, py, $\frac{1}{2}$bipy, $\frac{1}{2}$phen) and $TiBr_2$(bipy), the new species reacting with water to liberate dihydrogen.[23] Reaction of Na_2TiCl_4 with NMe_3 results in oxidation of the halide and the formation of $TiCl_3(NMe_3)_2$.[24] Dialkylamides of titanium(II) have been prepared[25] by the volatility-controlled disproportionation, as shown in equation (7).

$$2Ti^{III} \rightleftharpoons Ti^{II} + Ti^{IV} \qquad (7)$$

31.2.2 Titanium(III) Complexes

31.2.2.1 Monodentate ligands

Compounds of many stoichiometries are known, *e.g.* $[TiL_6]^{3+}$, $[ML_4X_3]$ and $[ML_2X_4]^-$. A survey of the range of these compounds and their electronic spectral and magnetic properties is given in Clark's review.[14] Fowles *et al.* have isolated a range of titanium(III) iodide species: TiI_3L_2 (L = NMe_3, α-picoline), TiI_3L_3 (L = py, α-picoline) and $TiI_3(MeCN)_4$, the last being assigned the formula $[TiI_2(MeCN)_4]I$. The magnetic properties of $TiBr_3(NMe_3)_2$[7] and its single crystal X-ray structure have been studied,[28] the latter showing it to be monomeric and basically trigonal bipyramidal. It is possible to use electronic spectroscopy to distinguish the pentacoordinate MX_3L_2 (L = NMe_3, α-picoline) species from the hexacoordinate MX_3L_3 species, since the former have characteristic absorption bands in the near-IR,[29] whereas ESR spectroscopy has been used to show that the $TiCl_3(py)_2$ complex is dimeric and hexacoordinate *via* chloro bridges in pyridine at 77 K.[30]

The reactive Ti^{III}—N complexes of the dialkylamides are useful synthetic reagents, readily reacting with protic reagents such as alcohols (to give alkoxides), unsaturated compounds such as isocyanates to yield ureido derivatives, or metal hydrides to give Ti—M-bonded compounds.[25] Rare three-coordination occurs in $Ti\{N(SiMe_3)_2\}_3$.[31] The He PE spectra of a number of three-coordinate metal amides $M\{N(SiMe_3)_2\}_3$ (M = Sc, Ti, Cr, Fe, Ga, In) show that, whereas the d^0 scandium compound exhibits two bands at IE < 9 eV (assigned to lone-pair orbitals), the spectra of the open-shell compounds are very similar, and it was concluded that bands corresponding to ionization from the metal 3*d* orbitals lie at IE > 9 eV, or are masked by bands due to ionization from nitrogen lone-pair orbitals (>8.1 eV). This was attributed to the strong −I effect of the ligand which, in contrast to compounds of the type $V(NMe_2)_4$, is not counterbalanced by an $\overset{\curvearrowright}{\ddot{N}\text{—}M}$ π interaction because of strong competing $\overset{\curvearrowright}{\ddot{N}\text{—}Si}$ $(p \rightarrow d)$ π bonding.

31.2.2.2 Bidentate and polydentate ligands

The blue en and pn complexes have the formula $[Ti(ligand)_3]^{3+}$, whereas for bipy and phen $[TiCl_2(ligand)_2][TiCl_4(ligand)]$ salts are obtained;[14] the salts $[TiI_2(phen)_2]I$ and $[TiI_2(bipy)_2][TiI_4(bipy)]$ are also known.[26] Whereas most transition metal complexes of tris(2-dimethylaminoethyl)amine (Me_6tren) normally involve all four nitrogens coordinated, in $TiCl_3(Me_6tren)$ only three amines are bound. With terpy all three nitrogens are bound in the octahedral monomeric $TiCl_3$(terpy).[22]

31.2.3 Magnetic Properties of Titanium(III) Complexes

Nearly all titanium(III) complexes exhibit room temperature magnetic moments close to 1.73 BM, and at lower temperatures the fall in value is related not only to the temperature but also to the asymmetry of the ligand field.[14,22,27] Most complexes contain a ground term $^2T_{2g}$ split by ~500 cm^{-1}; the orbital angular momentum is reduced by a factor *k*, which is usually 0.65–0.95. The ESR signals of titanium(III) complexes are almost invariably near the spin-only value of 2.0023.[14,30]

31.2.4 Titanium(IV) Complexes

31.2.4.1 Monodentate ligands

There are a large number of complexes TiX_4 with monodentate nitrogen ligands. For TiF_4 both 1:2 and 1:1 complexes are known, the former being pseudooctahedral and the latter, because of the high Ti—F bond strength, are fluoro-bridge polymers.[14] With $TiCl_4$ and $TiBr_4$ six-coordinate monomers are the commonest adducts, *viz.* TiX_4L_2. The area has been well reviewed.[14] However, reactions between $TiCl_4$ and amines can, unlike reaction with oxygen donor ligands, give rise to $TiCl_3$ derivatives.[14]

More recent work will be described here. Adducts of $TiOCl_2$ with MeCN and NMe_3, for example, form $TiOCl_2L_2$, but with py the adduct $TiOCl_2py_{2.5}$ forms. Only in the case of $TiOCl_2(\alpha\text{-pic})_2$ and $TiOCl_2(NMe_3)_2$ do IR spectra indicate the presence of a terminal Ti═O moiety (978 cm^{-1}), the other complexes presumably having a Ti—O—Ti bridge (*ca.* 890 cm^{-1}), the other complexes presumably having a Ti—O—Ti bridge (*ca.* 890 cm^{-1}). The potentially bidentate *o*-allylaniline complexes only *via* the nitrogen in TiX_4L (X = Cl, Br) to form pentacoordinate monomers; the hexacoordinate $TiBr_4L_2$ is also known.[37]

Titanium(IV) chloride and bromide form $TiX_4(Me_2C{=}N{-}N{=}CMe_2)_2$ in the presence of excess of acetone azine, but when the halides are in excess the products formed are $(TiCl_4)_4(\text{ligand})_3$ and $TiBr_4(\text{ligand})$.[38] Stable azides $[Ti(OPr^i)_2(N_3)_2]$ and $[TiCl(OPr^i)_2(N_3)]$ complexes are known; the azido group can be displaced by reaction with protic reagents and HN_3 is eliminated. The complexes also react with phosphines to form phosphiniminato complexes with N_2 elimination.[39] Mono- and bis-azido complexes have been prepared as shown in equations (8) and (9).[39]

$$[TiCl(OCHMe_2)_3] + Me_3Si(N_3) \longrightarrow [TiCl(OCH_2Me_2)_2(N_3)] + Me_3SiOCHMe_2 \quad (8)$$

$$[TiCl(OCHMe_2)_3] + 2Me_3Si(N_3) \longrightarrow [Ti(OCH_2Me_2)_2(N_3)_2] + Me_3SiOCHMe_2 + Me_3SiCl \quad (9)$$

An X-ray study of $[TiCl_4(N_3)_2]^{2-}$ reveals that the azide groups occupy *trans* positions.[40]

Electrolytic reduction of $TiCl_4$ in the presence of N_2 in DMSO produces a 1:1$TiCl_3$:N_2 complex.[41] Two previously reported products from $[ReCl(N_2)(PPhMe_2)_4]$ and $TiCl_3$ have been shown to be derived from $TiCl_4$. $[\{ReCl(N_2)(PPhMe_2)_4\}_2TiCl_4]$ and $[\{ReCl(N_2)(PPhMe_2)_4\}\text{-}TiCl_4(THF)]$ can be obtained from $TiCl_4$, the former being the first fully characterized example of a trinuclear complex containing two bridging dinitrogen ligands. Titanium tetrachloride in excess reacts with $[ReCl(N_2)(PPhMe_2)_4]$ in Et_2O to yield $[\{ReCl(N_2)PPhMe_2)_4\}(Ti_2Cl_6O)(Et_2O)]$, which exhibits a very low $\nu(N_2)$ at 1622 cm^{-1}.[42]

The X-ray crystal structure of the product of the reaction of $TiCl_4$ with $N(SiMe_3)_3$ shows it to contain planar four-membered $(Ti{-}N)_2$ rings with planar geometry at N apparently with considerable π bonding in the rings. These are linked by chlorine bridges to give five-coordinate Ti atoms in approximately trigonal bipyramidal geometry.[43] The reaction of $TiCl_4$ with an excess of $Li[N(SiMe_3)_2]$ yields $[TiCl\{N(SiMe_3)_2\}_3]$. The resistance to complete chlorine substitution suggests considerable steric crowding due to the bulky $[N(SiMe_3)_2]^-$ ligands. Both X-ray crystallography and ^1H NMR spectroscopy confirm this premise. In the solid state $[TiCl\{N(SiMe_3)_2\}_3]$ has crystallographically imposed C_3 symmetry and a distorted tetrahedral geometry. The ^1H NMR spectrum, which exhibits two equally intense resonances below the coalescence temperature of 34 °C, has been interpreted in terms of restricted rotation about the Ti—N bonds. At low temperatures the three silylamide ligands remain equivalent, but the $SiMe_3$ groups proximal and distal to chlorine are non-equivalent.[44]

The TiF_4 and Pr^i_2NH, Et_2NH, py or NMe_3 systems have been shown to be quite complex by Drago and coworkers.[45] For example, in TiF_4-rich solutions in acetone, high conductance has been explained by equation (10).

$$3TiF_4 + R_2NH \longrightarrow TiF_3(R_2NH)^+ + Ti_2F_9^- \quad (10)$$

31.2.4.2 Bidentate and polydentate ligands

A number of dinitrile ligands (succinonitrile, glutaronitrile, adiponitrile) form $TiCl_4(\text{ligand})$ complexes in which the nitriles bridge two titanium atoms.[46] Reactions of $CH_2(NMe_2)_2$ or $SiMe_2(NMe_2)_2$ with $TiCl_4$ give both chelated $TiCl_4(\text{ligand})$ complexes and $TiCl_3(NMe_2)$ species.[47]

The five-coordinate *meso*-tetraphenylporphyrinato complex [TiF(TPP)] has been synthesized by reduction of $[TiF_2(TPP)]$ with zinc amalgam. Electrochemical reduction of $[TiF_2(TPP)]^-$, or chemical reduction with sodium anthracenide, gives $[TiF_2(TPP)]^-$. Consistent with its coordinative unsaturation, [TiF(TPP)] adds ligands L (L = THF, PBu_3, pyridine, *N*-methylimidazole, or *N*-methylpyrrolidone) yielding [TiF(L)(TPP)], and adds F^- affording $[TiF_2(TPP)]^-$. All these Ti^{III} complexes have been characterized in solution by EPR spectroscopy; the unpaired electron is believed to reside in the $3d_{xy}$ orbital. Oxidation of [TiF(TPP)] yields a mixture of $[TiF_2(TPP)]$ and the peroxo complex $[Ti(O_2)(TPP)]$. EPR evidence suggests that this reaction proceeds *via* the dioxygen adduct $[TiF(O_2)(TPP)]$, as shown in equations (11) and (12), a species which could be formulated as a superoxotitanium(IV) complex but which appears to have a significant amount of peroxo porphyrin cation radical titanium(IV) character.[48]

$$[TiF(TPP)] + O_2 \rightleftharpoons [TiF(O_2)(TPP)] \quad (11)$$

$$[TiF(TPP)] + [TiF(O_2)(TPP)] \longrightarrow [TiF_2(TPP)] + [Ti(O_2)TPP)] \quad (12)$$

31.3 OXYGEN LIGANDS

31.3.1 Titanium(III) Complexes

31.3.1.1 Electronic spectra

As pointed out by Clark, the $Ti(H_2O)_6^{3+}$ ion has an important historical connection with ligand field theory. In a perfect octahedral field of O_h symmetry, the ^2D ground term of titanium(III) is split into a lower $^2T_{2g}$ term and an upper ^2E_g term. The broad, quite asymmetric band at *ca.* 20 100 cm^{-1} in the visible spectrum of $Ti(H_2O)_6^{3+}$ arises from the $^2E_g \leftarrow {}^2T_{2g}$ transition. The energy of this transition for a complex is a direct measure of the ligand field strength. The splitting of the ^2E_g term is a consequence of the Jahn–Teller effect.

Table 4 is a compilation of some spectral parameters for six-coordinate titanium(III) complexes. There are only a few TiL_6^{3+} complexes of neutral oxygen donor ligands. Addition of chloride to the purple $Ti(H_2O)_6^{3+}$ solution in water causes a continuous shift of the $^2E_g \leftarrow {}^2T_{2g}$ band to lower energy, from 20 100 to almost 14 500 cm^{-1} and probably reflects the formation of such species as $[TiCl(H_2O)_5]^{2+}$, $[TiCl_2(H_2O)_4]^+$, $[TiCl_3(H_2O)_3]$, *etc.* Moreover, the electronic spectra can give a hint to quite subtle structural information. This is illustrated for the compound $NH_4Ti(C_2O_4)_2{\cdot}2H_2O$ which exhibits the $^2E \leftarrow {}^2T_g$ band at 12 800 cm^{-1}, pointing to a weak ligand field generated by bridging oxalate groups, (**1**).[49]

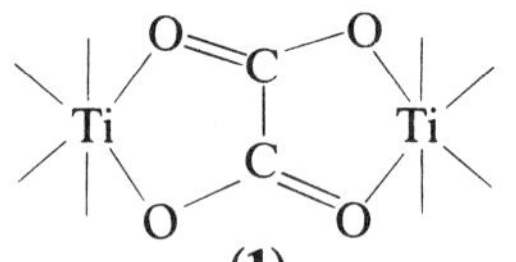

(**1**)

Early workers assumed that alcohol ligands formed $[TiL_6]Cl_3$ complexes, analogous to the hexaaqua species. However, electronic spectra have pointed to these complexes as being of the type $[TiCl_2L_4]Cl$ (Table 4). For THF, dioxane, DMF, MeCN and some other nitrogen ligands the $[TiL_3Cl_3]$ complexes are known, but all of these are very unstable to oxidation.

31.3.1.2 Monodentate and bidentate ligands

Apart from the early interest in monodentate oxygen donors most subsequent work has been with bidentate systems, particularly the oxalates. However, one TiO_6^{3+} system which has been studied is $[Ti(urea)_6]I_3$, the structure having rigorous D_3 point symmetry and essentially linear arrays of iodide ions parallel to the crystallographic threefold axis. The Ti—O distance is 2.014 Å, and there are significant distortions of the coordination polyhedron from the idealized O_h symmetry and from the D_{3d} model often assumed for trigonally distorted octahedral complexes. The three O—Ti—O angles are 90.5, 86.4 and 92.8·. The stability of this compound to aerial oxidation apparently arises from intramolecular hydrogen bonding involving the urea molecules and also hydrogen bonds involving the iodide ions.[50]

Fowles and co-workers have investigated the complexes $KTi(ox)_2{\cdot}2H_2O$, $NH_4Ti(ox)_2{\cdot}2H_2O$

Table 4 Electronic Spectra of Six-coordinate Titanium(III) Complexes

Complex	$^2E_g \longleftarrow$	^2T_g (cm^{-1})	Splitting (cm^{-1})	Method[a]	Ref.
$K_3Ti(CN)_6 \cdot 2KCN$	22 300 —	18 900 —	3400	A	1
$[Ti(H_2O_6]Cl_3$	20 100 (5.7)	17 000 (3.9)	3100	A	2
$CsTi(SO_4)_2 \cdot 12H_2O$	19 900 (4.2)	18 000 (2.2)	1900	C	3, 4
$Ti^{III}Al_2O_3$[77 K]	20 300	18 450	1850	C	5
$[Ti(OCN_2H_4)_6]I_3$	17 550 (10)	16 000 (10)	1550	A, B, C	6
$[Ti(OC_3H_6N_2)]I_3$	17 050 —	15 400 —	1650	B	7
$(NH_4)_3[TiF_6]$	19 000 —	15 100 —	3900	D	8
$Na_2K[TiF_6]$	18 900 —	16 000 —	2900	D	8
$NaK_2[TiF_6]$	18 900 —	16 000 —	2800	D	8
$Bu^n_3[Ti(NCS)_6]$	18 400 (51)	asym —	—	A, B	9
α-$TiCl_3$	13 800 —	12 000 (sh)	1800	B	10
$(PhN)_3[TiCl_6]$	12 750 —	10 800 (sh)	1750	B	7
$(Li,K)_3[TiCl_6]$	13 000 (3.3)	10 000 (2.4)	3000	E	11
$(PhN)_3[TiBr_6]$	11 400 —	9650 —	1750	B	7
$[Ti(MeOH)_4Cl_2]Cl$	16 800 (4.3)	14 700 (3.8)	2100	A	2
$[Ti(EtOH)_4Cl_2]Cl$	16 800 (4.0)	14 700 (3.5)	2100	A	2
$[TiPr^iOH)_4Cl_2]Cl$	16 700 (9.1)	asym —	—	A, B	12
$[Ti(Bu^sOH)_4Cl_2]Cl$	16 100 (9.1)	asym —	—	A, B	12
$[Ti(C_6H_{11}OH)_4Cl_2]Cl$	16 100 (7.8)	asym —	—	A, B	12
$[TiCl_3 \cdot 3MeCN]$	17 100 (22)	14 700 (13)	2400	A, B	13
$[TiCl_3 \cdot 3EtCN)$	17 200 —	14 700 —	2500	B	14
$[TiCl_3 \cdot 3n$-$PrCN]$	17 200 —	14 700 —	2500	B	14
$[TiCl_3 \cdot 3C_5H_5N]$	16 600 —	asym —	—	A, B	15
$[TiCl_3 \cdot 3\gamma$-pic]	16 750	asym —	—	A, B	15
$[TiCl_3 \cdot 3C_4H_9ON]$	15 400	13 300 (sh)	2100	B	16
$[TiCl_3 \cdot 3C_4H_8O]$	14 700	13 500 —	1200	B	13
$[TiCl_3 \cdot 3MeCOMe]$	15 400 (37)	13 300 (28)	2100	A	13
$[TiCl_3 \cdot 3C_4H_8O_2]$	15 150 —	13 400 (sh)	1750	B	16
$[TiBr_3 \cdot 3MeCN]$	16 300 (50)	asym —	—	B	14, 17
$[TiBr_3 \cdot 3C_5H_5N]$	16 000 —	14 500 —	1500	B	18
$[TiBr_3 \cdot 3\gamma$-pic]	16 000 (br)	— —	—	B	8, 18, 17
$[TiBr_3 \cdot 3C_4H_8O_2]$	14 100 —	12 500 (sh)	1600	B	17
$[TiBr_3 \cdot 3C_5H_{11}NO]$	14 250 —	— —	—	B	17
$Et_4N[TiCl_4(MeCN)_2]$	15 050 —	14 300 (sh)	750	B	10
$Et_4N[TiBr_4(MeCN)_2]$	14 800 (sh)	12 750 —	2050	B	10
$(C_5H_6N)_3[TiCl_4Br_2]$	12 350 —	10 250 (sh)	2100	B	10
$[Ti_2Cl_6(NMe_3)_3]$	17 700 —	10 300 —	—	D	
$[KTi(Ox)_2 \cdot 2H_2O]$	12 800	— —			19

[a] A = solution absorption spectrum; B = diffuse reflectance spectrum; C = crystal transmission spectrum; D = transmission spectrum as KCl pellet; E = absorption spectrum of melt at 400 °C maximum; extinction coefficient in parentheses following the band maxima.

1. H. L. Schläfer and R. Gotz, *Z. Anorg. Allg. Chem.*, 1961, **309**, 104.
2. H. Hartmann, H. L. Schläfer and K. H. Hansen, *Z. Anorg. Allg. Chem.*, 1956, **284**, 153.
3. H. L. Schläfer, in 'Proceedings of the Conference on the Theory and Structure of Complex Compounds, Wroclaw', Pergamon, Oxford, 1962, p. 181.
4. H. Hartmann and H. L. Schläfer, *Z. Naturforsch., Teil A*, 1951, **6**, 754.
5. D. S. McClure, *J. Chem. Phys.*, 1962, **36**, 2757.
6. H. Hartmann, H. L. Schläfer and K. H. Hansen, *Z. Anorg. Allg. Chem.*, 1957, **289**, 40.
7. M. Blažeková and H. L. Schläfer, *Z. Anorg. Allg. Chem.*, 1968, **360**, 169.
8. H. D. Bedon, S. M. Horner and S. Y. Tyree, *Inorg. Chem.*, 1964, **3**, 647.
9. R. J. H. Clark, *J. Chem. Soc.*, 1964, 417.
10. G. W. A. Fowles and B. J. Russ, *J. Chem. Soc. (A)*, 1967, 517.
11. D. M. Gruen and R. L. McBeth, in 'Plenary Lectures, 7th International Conference on Coordination Chemistry', Butterworths, London, 1963, p. 23.
12. H. L. Schläfer and R. Gotz, *Z. Anorg. Allg. Chem.*, 1964, **328**, 1.
13. R. J. H. Clark, J. Lewis, D. J. Machin and R. S. Nyholm, *J. Chem. Soc.*, 1963, 379.
14. M. W. Duckworth, G. W. A. Fowles and R. A. Hoodless, *J. Chem. Soc.*, 1963, 5665.
15. G. W. A. Fowles and R. A. Hoodless, *J. Chem. Soc.*, 1963, 33.
16. G. W. A. Fowles, R. A. Hoodless and R. A. Walton, *J. Chem. Soc.*, 1963, 5873.
17. G. W. A. Fowles and R. A. Walton, *J. Chem. Soc.*, 1964, 4953.
18. G. W. A. Fowles and R. A. Walton, *J. Less-Common Met.*, 1965, **9**, 457.
19. D. J. Eve and G. W. A. Fowles, *J. Chem. Soc. (A)*, 1966, 1183.

and $Ti_2(ox)_3 \cdot 10H_2O$ (ox = oxalate).[51] The crystal structure of the decahydrate, (**2**), contains a centrosymmetric dimer $Ti_2(ox)_3(H_2O)_6$ and also four non-bonded water molecules. Each metal atom is a seven-coordinate pentagonal bipyramid with two molecules in axial positions and one in an equatorial plane, together with one bridging and one non-bridging oxalate ligand. Further work on the anionic oxalato complexes of titanium(III) has led to the isolation of $Cs[Ti(ox)_2(H_2O)_3] \cdot 2H_2O$, which contains a discrete anion involving again a pentagonal bipyramidal seven-coordinate titanium atom with two non-bridging oxalate chelating ligands in the equatorial plane and the three water molecules occupying the remaining equatorial site and the axial site.[52] Although good crystallographic data are not available for the $K[Ti(ox)_2] \cdot 2H_2O$ complex, an eight-coordination sphere has been proposed, (**3**). Magnetic measurements indicate, not surprisingly, that the bridging oxalate provides an effective path for magnetic exchange between the two titanium(III) atoms; for example in $Ti_2(ox)_3 \cdot 6H_2O$ the intramolecular exchange parameter is rather large, $J = -60\ cm^{-1}$.[54]

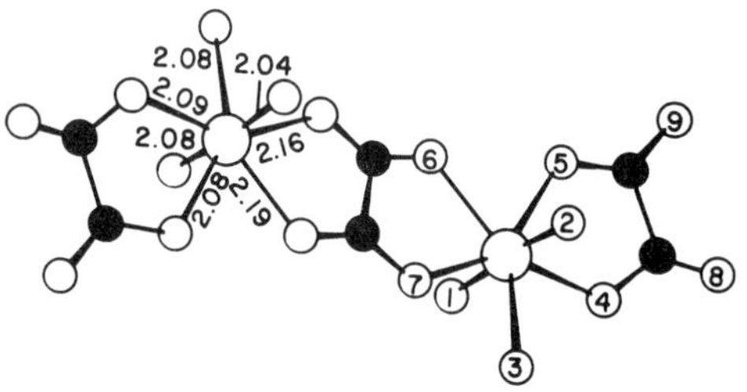

(**2**)

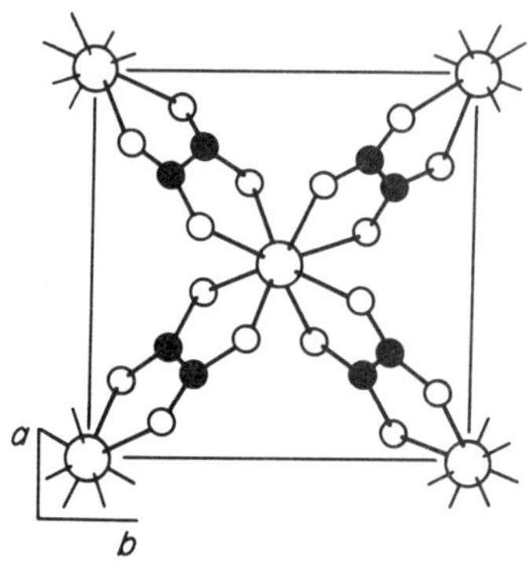

(**3**) Proposed structure for $[Ti(ox)_2]_n^{n-}$; ○ Ti; ○ Oxygen; ● Carbon.

Reaction of $TiCl_3(THF)_3$ with β-diketones gives the deeply coloured $Ti(diketonate)Cl_2(THF)_2$, the ESR spectra of which suggests structure (**4**).[55,56] These complexes react with $LiC_6H_4CH_2NMe_2$, $LiCH_2C_6H_4NMe_2$ and $LiN(SiMe_3)_2$ to give novel air-sensitive organometallic compounds, *e.g.* (**5**).[56] The coordinated THF may readily be displaced by neutral ligands such as PMe_3 or bipy. Magnetic susceptibility measurements on $TiCl_3(THF)_3$ and its thermal decomposition products $TiCl_3(THF)_2$ and $TiCl_3(THF)$ are consistent with monomeric structures for the first two compounds and a polymeric species for $TiCl_3(THF)$.[57]

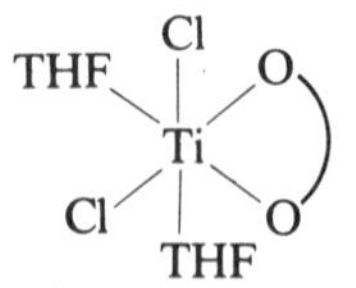

(**4**)

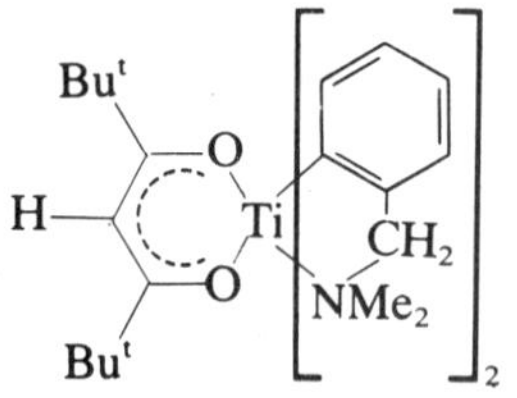

(**5**)

Polymeric $Ti(ligand)_3$ complexes of diisopropyl phosphate (dipp) and di-*n*-propyl phosphate (dnpp) have been compared to assess steric and inductive effects of alkyl substituents on the strength of the M—O bond in metal complexes with organophosphate ligands. In comparison with the dnpp complex, $Ti(dipp)_3$ exhibits a lower ν(Ti—O), lower ligand field parameters and a higher magnetic moment. These data point to a weaker Ti—O bond in the dipp complex, which is attributable to the steric effect of the $CHMe_2$ substituents, since inductive effects of $CHMe_2$ and Pr groups are about the same. These complexes may have a linear, chain-like, triply bridged $\text{+}M(OP(R)_2O_3\text{+}_n$ structure.[58]

The tridentate ether ligand diglyme forms $[TiCl_3\{O(CH_2CH_2OMe)_2\}]$ and crystals contain two molecules in the asymmetric unit. In both, the titanium(III) atoms are six-coordinate with a *fac* arrangement of ligands. The metals are bonded to three chlorine atoms (2.31–2.35 Å) and to the three oxygen atoms of the trioxanonane ligand (2.115–2.182 Å) and the ligand conformation differs in the two molecules, being $\delta\delta$ in one and $\delta\lambda$ in the other, (**6**).[59]

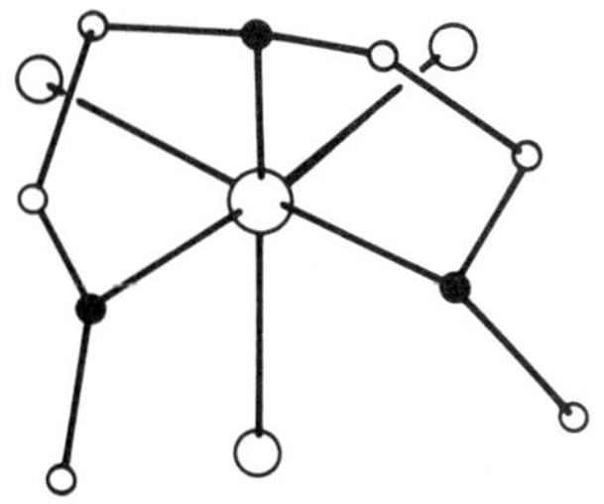

(**6**)

31.3.2 Titanium(IV) Complexes

31.3.2.1 Monodentate ligands

(i) Alkoxides

Perhaps the best studied group of titanium(IV) complexes is the alkoxides. The metal alkoxides generally have received a great deal of attention because of their ease of hydrolysis and reactivity with hydroxylic molecules, and their tendency to increase the coordination number of the metal which is opposed by the steric effect of the alkyl group. These properties result in materials, the characteristics of which range from polymeric solids to volatile liquids. The definitive review of this area is that by Bradley.[60]

The original method of preparation involved reaction of the sodium alkoxide and titanium tetrachloride in the appropriate alcohol, as shown by equation (13). However, this is slow and only works for those alcohols which readily form sodium alkoxides. An improved method is shown by equation (14). If a proton-accepting reagent, such as ammonia, is not employed, however, the reaction proceeds only as far as the $TiCl_2(OR)_2$ derivative.

$$TiCl_4 + 4NaOR \longrightarrow Ti(OR)_4 + 4NaCl \tag{13}$$

$$TiCl_4 + 4ROH + 4NH_3 \longrightarrow Ti(OR)_4 + 4NH_4Cl \tag{14}$$

Compounds such as $Ti(OMe)_4$ are usually insoluble polymers, although a soluble tetramer has been isolated, $[Ti(OMe)_4]_4$ (**7**). This complex has the same M_4O_{16} framework as $Ti_4(OEt)_{16}$ and $Ti_4(OMe)_4(OEt)_{12}$. All of these frameworks are centrosymmetric and contain six-coordinate titanium in a group of four edge-sharing octahedra. Each tetrameric unit contains two triply bridging (Ti_3OR), four doubly bridging (Ti_2OR) and ten terminal alkoxy groups. There are two types of Ti atoms: two are coordinated by three bridging and three terminal oxygens and two are coordinated by two bridging and two terminal oxygens. Bradley[60] had discussed the contributions from σ and π bonding in this complex. There is another form of $Ti(OMe)_4$, which is more insoluble and has a different X-ray powder diffraction pattern from the tetramer.[61] Moreover, there are quite large differences in the IR spectra of both species. The insoluble form may well contain no triply bridged OMe groups and thus, of necessity, be more polymeric in nature.

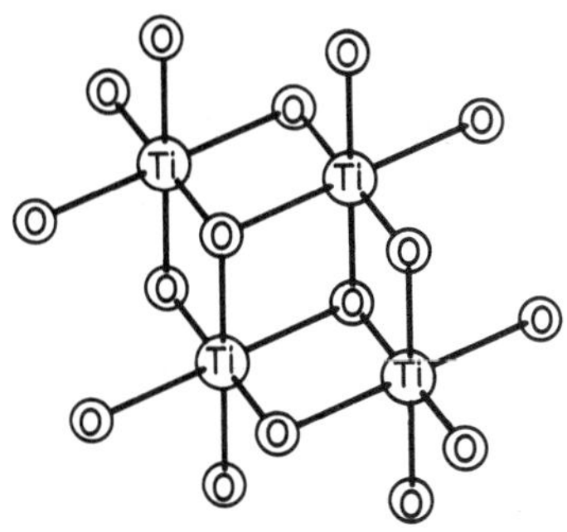

(**7**)

For $Ti(OEt)_4$ the same tetrameric structure obtains in the solid state. The Ti—O bond lengths reflect the nature of the coordinated oxygen atom, *viz.* Ti_3OEt (2.23 Å), Ti_2OEt (2.03 Å), TiOEt (1.77 Å). This compound is very soluble in organic solvents and may be initially obtained as a supercooled liquid by vacuum distillation. However, over some months of standing at room temperature it solidifies to the tetrameric form. In solution it is probably predominantly trimeric. From ebullioscopic molecular weight measurements a whole series $[Ti_{3(x+1)}O_{4x}(OEt)_{4(x+3)}]$ ($x = 1, 2, \ldots, \infty$) has been postulated for the hydrolysis products of $Ti_3(OEt)_{12}$. From solution, crystalline $[Ti_6O_4(OEt)_{16}]$ (*i.e.* $x = 1$) has been obtained.

As mentioned above, the ability of $TiCl_{4-n}(OR)_n$ to function as a Lewis acid decreases as n increases. Few adduct complexes of $Ti(OR)_4$ are known, although $Ti_2(OEt)_8$·en, $Ti_2(OPr^i)_8$·en and $Ti(OPr^i)_4$·en have been reported,[14] but these complexes almost completely dissociate in solution. However, $Ti(OPh)_4$ readily forms 1:1 adducts with PhOH, NH_3, NH_2Me, $NHMe_2$, NMe_3, py, dioxane, $PhNH_2$, Me_2CO. These complexes have been the subject of a review.[62] The difference in the behaviour of the $Ti(OPh)_4$ to the $Ti(OR)_4$ (R = Me, Et) towards Lewis bases is undoubtedly related to the monomeric, and thus the coordinatively unsaturated, nature of the phenoxides in solution.

A range of $TiX_{4-n}(OR)_n$ (R = alkyl, cycloalkyl, alkenyl, aryl; X = halogen; $n = 1, 2, 3$) are known and are best prepared by the direct reaction of the $Ti(OR)_4$ complex with the correct molar ratio of the TiX_4 salt in an inert solvent such as benzene. The nature of the complexes depends upon the degree of replacement of the alkoxy groups by the halide, *e.g.* $[TiCl_3(OBu)_3]_3$, $[TiCl_2(OBu)_2]_2$ and $[TiCl_3(OBu)]$. Both $TiCl_2(OPh)_2$ and $TiCl_2(OEt)_2$ have been shown by X-ray crystallography to be trigonal bipyramidal *via* alkoxy bridges in the solid state, the former having Ti—O_t (1.744 Å), Ti—O_{br} (1.910, 2.122 Å), Ti—Cl (2.209, 2.219 Å)[63] and the latter Ti—O_t (1.77 Å), Ti—O_{br} (1.96 A) and Ti—Cl (2.19, 2.20 Å).[64] Variable temperature ^{13}C NMR studies of $Ti(OR)_4$ (R = Et, Pr^n, Bu^n, Pr^i, Bu^i, Bu^t) suggest the coexistence of trimers and trimer aggregates for the straight chain alkoxides, and monomer–dimer–trimer equilibria for the branched chain derivatives.[65] Recently, $[Ti(Cp)(OPh)_3]$ and $[Ti(Cp)(OPh)_2Cl]$ have been reported.[66] The X-ray crystal structure of $Ti(OPh)_4$·HOPh reveals it to be a dimer of octahedrally coordinated alkoxytitanium molecules (**8**).[67]

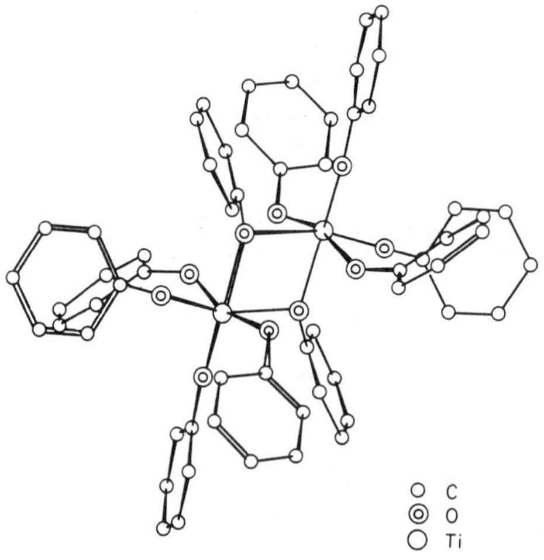

(**8**) A view of the dimer projected along the *a* axis

There have been very few reports of alkoxy complexes derived from unsaturated alcohols. A vinoxytitanium derivative can be obtained from the reaction of tetraisopropoxytitanium and acetaldehyde (equation 15).[68] More recently, the dimer $[Ti(OCH_2CHCH_2)_4]_2$ has been isolated from reaction (16).[69]

$$Ti(OCHMe_2)_4 + 2MeCHO \longrightarrow Ti(OCHCH_2)_2(OCHMe_2)_2 + 2Me_2CHOH \quad (15)$$

$$Ti(OCH_2Me)_4 + 4CH_2CHCH_2OH \longrightarrow Ti(OCH_2CHCH_2)_4 + 4Me_2CHOH \quad (16)$$

(*ii*) *Other monodentate ligands*

Titanium tetrachloride reacts with trimethylsilanol in the presence of ammonia (equation 17). This complex is a colourless liquid, and the method can be used to produce complexes of the general type $Ti(OSiR_3)_4$ and $Ti(OSiR_nR'_{3-n})_4$. These complexes have received particular attention because the products of their hydrolysis have potential as polymers.[70]

$$TiCl_4 + 4Me_3SiOH + 4NH_3 \longrightarrow Ti(OSiMe_3)_4 + 4NH_4Cl \qquad (17)$$

The fact that trifluoromethanesulfonic acid is known to liberate HCl from sodium chloride suggested the reaction of CF_3SO_3H with $TiCl_4$. At 25 °C this results in the formation of a hygroscopic lemon yellow solid $TiCl_3(SO_3CF_3)$. On heating *in vacuo* above 60 °C this complex decomposes, as shown by equation (18), giving another lemon yellow solid which may also be prepared by the reaction of $TiCl_4$ with excess CF_3SO_3H at 25 °C.[71] No higher trifluoromethanesulfonate was isolated. Complexes containing the 2,2,2-trifluoroethoxy, $OCH_2CF_3(OR_f)$ are known, *viz.* $TiCl_2(OR_f)_2$, $TiCl(OR_f)_3$, $Ti(OPr_f)(OR_f)_2$ and $Ti(OPr^i)_3(OR_f)$.[72]

$$2TiCl_3(SO_3CF_3) \longrightarrow TiCl_4 + TiCl_2(SO_3CF_3)_2 \qquad (18)$$

Reaction of a purple suspension of $TiCl_3$ in THF with two mol equivalent of HMPA (HMPA = hexamethylphosphoramide) under argon resulted in the formation of a turquoise suspension attributed to $TiCl_3(HMPA)_2$. This suspension was stable to argon, but exposure to air or dioxygen caused the formation of a crystalline yellow solid, (**9**). This contains Ti^{IV}, and a Raman band at 970 cm^{-1} is assigned to the O—O stretch.[73]

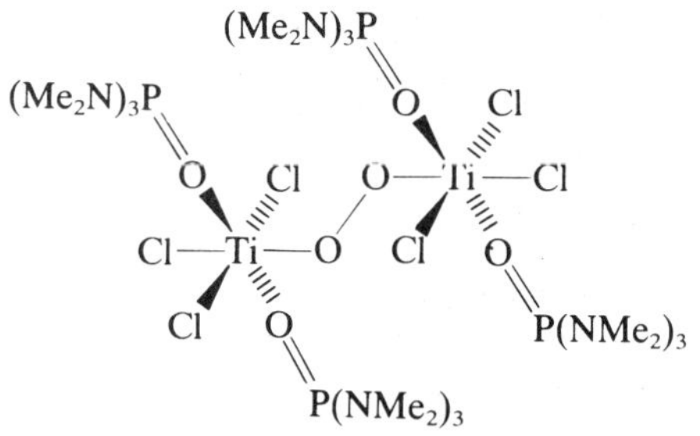

(**9**)

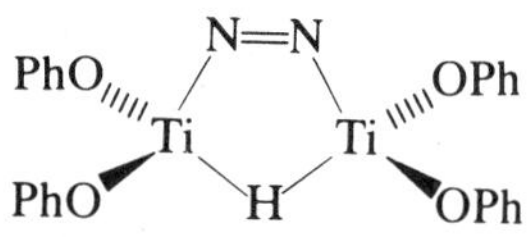

(**10**)

Before leaving this section some special mention should be made of the interesting chemistry developed by Cole-Hamilton and Wilkinson on dichlorodiphenoxotitanium(IV) (Scheme 1),[74] and interesting compounds such as (**10**) were isolated.

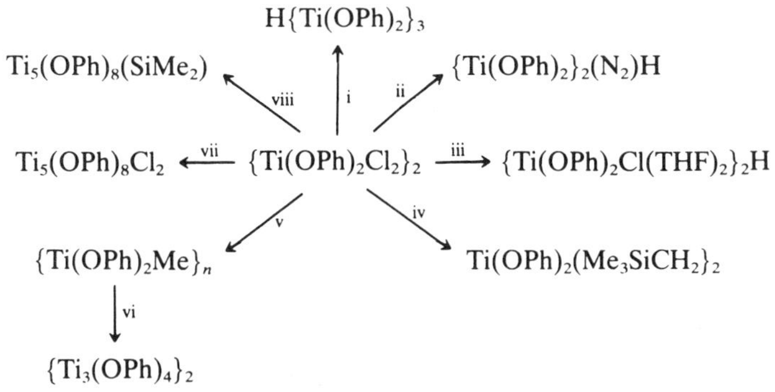

Reactions of $(Ti(OPh)_3Cl_2)_2$ i, 2 g-atom of K under argon, toluene, 25 °C; ii, excess of Na/Hg under N_2, THF 25 °C; iii, excess of $NaBH_4$, THF, 25 °C. iv, $(Me_3SiCH_2)_2Mg$, diethyl ether, −10 °C. v, Me_3Mg, diethyl ether, −30 °C. vi, Warm to 25 °C. vii, excess of Zn, toluene, 25 °C. viii, 2 mole equiv. $LiN(SiMe_3)_2$, petroleum, 25 °C.

Scheme 1

31.3.2.2 Bidentate ligands

By far the greatest number of reports have been on the β-diketonate (LH = β-diketone) complexes, and these provide a large series of types TiX_3L, $TiX_4(LH)$, TiX_2L_2, $[TiL_3]^+$, TiOXL, $[TiOL]_2$ (see Table 5 and structures **11–15**).

Table 5 Some β-Diketonate Complexes of Titanium(IV)

Complex	*x*	*Technique/comments*	*Ref.*
$TiCl_3(acac)$		X-Ray, dimeric (**11**)	1
$TiX_4(C_7H_{12}O_2)$	X = Cl, Br, I	^1H NMR	2
$TiX_3(acac)$	X = F, Cl, Br	IR, X-ray, X = Cl, dimeric (**12**)	3
$TiX_3(bzac)$	X = F, Cl, Br	IR	3
$[Ti(acac)_3][FeCl_4]$		Mössbauer	4
$[\{Ti(acac)_2\}_2O]\cdot CHCl_3$		X-Ray, dimeric (**13**)	5
$TiX_2(bzac)$	X = F, Cl, Br	X-Ray powder	6
$TiX_2(bzbz)$	X = F, Cl, Br	^{19}F NMR	6
cis-$TiX_2(acac)$	X = OMe, OEt, OCH_2CF_3, OPr^i, OBu^n, OBu^t, Cl	^1H NMR	7
cis-$TiX_2(C_6H_9O_2)_2$	X = F, Cl, Br	Monomeric	8
$TiX_3(C_6H_9O_2)$	X = Cl, Br		8
$[Ti(C_6H_9O_2)_3]Y$	Y = $FeCl_4$, $SbCl_6$		8
$TiOCl(acac)$		Dimer in $CHCl_3$	9
$[\{TiO(acac)\}_2]$		X-Ray, dimeric (**14**)	10
$[\{TiO(acac)\}_2]\cdot 2C_4H_8O_2$		X-Ray, dimeric	10
$TiCl(OR)(acac)$		^1H NMR	11
$TiI_2(acac)_2$		50% *cis*, 40% *trans*, 10% Ti^{III}	12
cis-$TiX_2(acac)_2$	X = F, Cl, Br	Bonding discussion	13
$Ti(C_{12}H_{17}O)_2(acac)_2$		X-Ray, (**15**)	14
$TiX_2(acac)_2$	X = F, OEt	Ligand-exchange equilibrium	15
$TiX_2(acac)_2$	X = NCO, NCS	^1H NMR	16

1. N. Serpone, P. H. Bird, D. G. Brickley and D. W. Thompson, *J. Chem. Soc., Chem. Commun.*, 1972, 217.
2. A. L. Allred and D. W. Thompson, *Inorg. Chem.*, 1968, **7**, 1196.
3. N. Serpone, P. H. Bird, A. Somogyvari and D. G. Bickley, *Inorg. Chem.*, 1977, **16**, 2381.
4. R. J. Woodruff, J. L. Marini and J. P. Fackler, Jr., *Inorg. Chem.*, 1964, **3**, 687.
5. K. Watenpaugh and C. N. Caughlan, *Inorg. Chem.*, 1967, **6**, 963.
6. N. Serpone and R. C. Fay, *Inorg. Chem.*, 1967, **6**, 1835.
7. D. C. Bradley and C. E. Holloway, *J. Chem. Soc. (A)*, 1969, 282.
8. D. W. Thompson, W. A. Somers and M. O. Workman, *Inorg. Chem.*, 1970, **9**, 1252.
9. R. E. Collis, *J. Chem. Soc. (A)*, 1969, 1895.
10. C. D. Smith, C. N. Caughlan and J. A. Campbell, *Inorg. Chem.*, 1972, **11**, 2989.
11. D. W. Thompson, W. R. C. Munsey and T. V. Morris, *Inorg. Chem.*, 1973, **12**, 2190.
12. R. C. Fay and R. N. Lowry, *Inorg. Chem.*, 1970, **9**, 2048.
13. H.-H. Schmidtke and U. Voets, *Inorg. Chem.*, 1981, **20**, 2766.
14. P. H. Bird, A. R. Fraser and C. F. Lau, *Inorg. Chem.*, 1973, **12**, 1322.
15. R. C. Fay and R. N. Lowry, *Inorg. Chem.*, 1974, **13**, 1309.
16. A. F. Lindmark and R. C. Fay, *Inorg. Chem.*, 1975, **14**, 282.

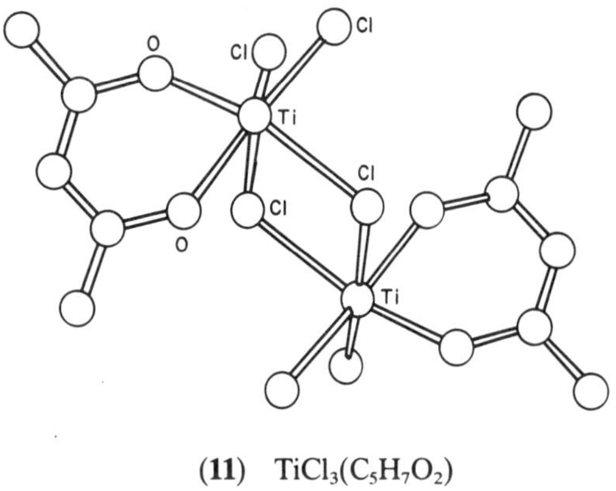

(**11**) $TiCl_3(C_5H_7O_2)$

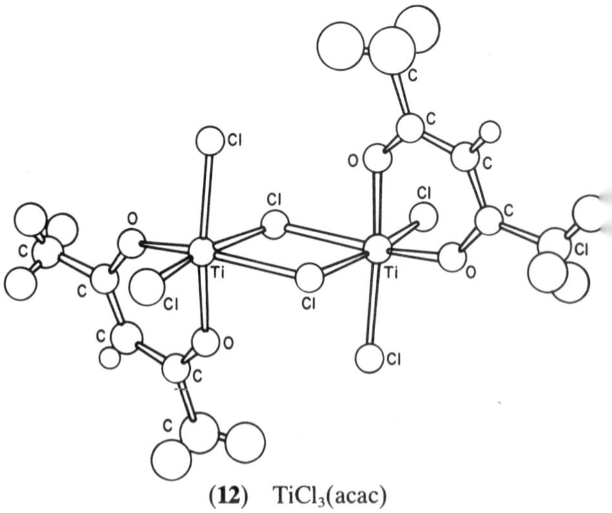

(**12**) $TiCl_3(acac)$

Both bis-oxalato, $[TiO(ox)_2]$,[75] and tris-oxalato, $[Ti(ox)_3]^{2-}$,[76] complexes are known. It is also observed that 1,2-dimethoxyethane (DME) forms chelates $TiX_4(DME)$ (X = F, Cl).[77] The nitrate group is also bidentate in anhydrous $Ti(NO_3)_4$. This complex crystallizes with four $Ti(NO_3)_4$ in each monoclinic unit cell of dimensions $a = 7.80$, $b = 13.57$, $c = 10.34$ Å, $\beta = 125.0°$, and space group $P2_1/c$. The structure, (**16**), consists of individual molecules in each of which four symmetrically bidentate nitrato groups are arranged according to D_{2d} symmetry

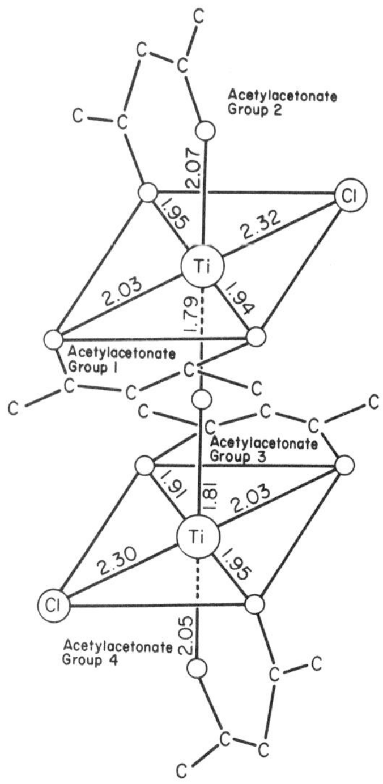

(**13**) $\{TiCl(acac)_2\}_2O$

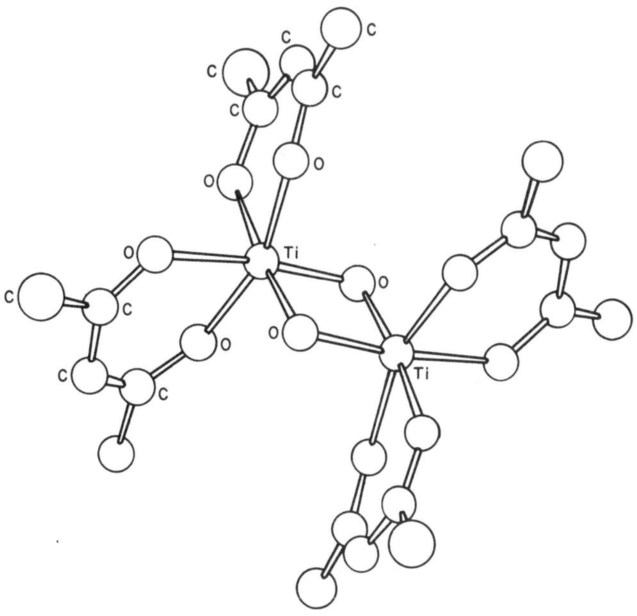

(**14**)

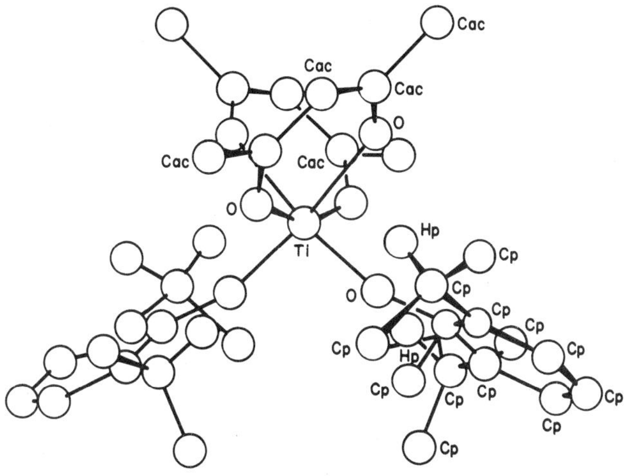

(**15**)

in a flattened tetrahedral manner round the central titanium atom.[78] The dimensions of the nitrato groups differ considerably from those of a typical nitrate ion, having average N—O distances of 1.292 Å for bonds adjacent to the titanium atom and 1.185 Å for the outer bonds. These indicate a high contribution of the bond structure (**17**) to the resonance form of the molecule, consistent with the unusually high IR stretching frequency, 1635 cm^{-1}.

The reaction of titanium(IV) alkoxides with 2-methyl-2,4-pentanedione, pyrocatechol, salicyclic acid or *P,P'*-diphenylmethylenediphosphinic acid (LH_2) yields chelates of type $Ti(OR)_2L$ (R = Et, Pr^i), but instead $Ti(OC_3H_6O)_2$ and $Ti(OR)_4$ are recovered. Replacement of the alkoxy groups in the pyrocatecholate derivatives gives the monomeric $Ti(o\text{-}C_6H_4O_2)(C_5H_7O_2)_2$.[79] The potentially bidentate but-3-en-1-ol forms $TiCl_3(OCH_2CH_2CHCH_2)$, which probably does not contain a Ti–alkene bond. With $TiCl_4$ and 4-methyoxybut-1-ene, a chloride-bridged dimer containing six-coordinate titanium, $[TiCl_4(MeOCH_2CH_2CHCH_2)]_2$, forms.

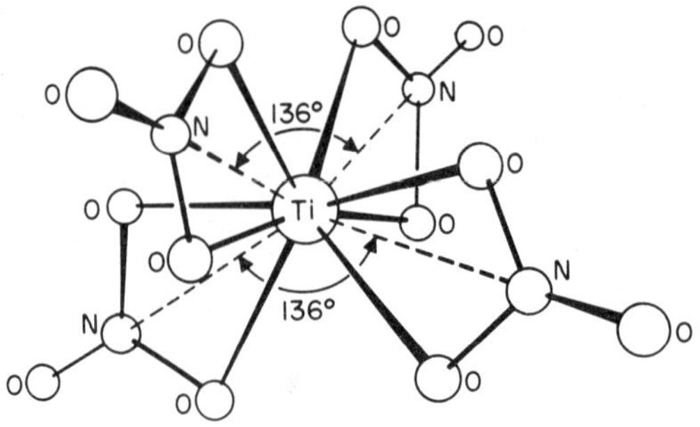

(16)

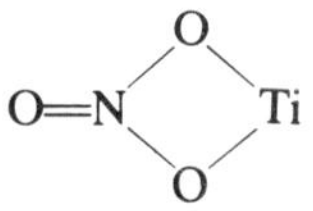

(17)

31.4 SULFUR LIGANDS

31.4.1 Neutral Donors

Sulfur complexes of titanium have been only poorly investigated but, nonetheless, there has been some interest in this area. Titanium(IV) must definitely be predicted to be a 'hard' acid, whereas titanium(III) might be expected to show intermediate 'hard/soft' behaviour. However, contrary to expectation, titanium(IV) has been shown to prefer sulfur to oxygen as a donor atom as evidenced by the existence of the Ti—S bonds in the 1,4-thiozen complexes $TiX_4(C_4H_8OS)_2$.[81] Indeed, titanium(IV) halides form stable adducts with a variety of thioethers. Even more puzzling is the fact that this ligand is both sulfur- and oxygen-bound in the titanium(III) derivative $TiCl_3(C_4H_8OS)_x$.[82] Titanium(III) halides also form 1:2 adducts with dimethyl sulfide and tetrahydrothiophene, TiX_3L_2. The chloro complexes are probably dimeric with a direct Ti—Ti bond, whereas the bromo and iodo derivatives are likely to be six-coordinate halogen-bridged dimers.[81] The bidentate 1,2-dimethylthioethane, 1,2-diethylthioethane, 1,2-diphenylthioethane and *cis*-dimethylthiomaleonitrile yield 1:1 adducts TiX_4L (X = Cl, Br).[83]

31.4.2 Anionic Ligands

Attempts to prepare mercaptide complexes of $TiCl_4$ by the use of either thiol and ammonia or sodium mercaptide have not proved successful (equation 19). However, by employing titanium dialkylamides as precursors substitution did occur (equation 20). The insoluble, non-volatile tetramercaptides usually had weakly bound thiol or amine associated with them, *viz.* $Ti(SR')_4{\cdot}(R'SH)_x(R_2NH)_y$. The tetramercaptides are thought to be six-coordinate polymers *via* mercaptide bridges. More controlled addition of ethanethiol or 1-methylethanethiol to $Ti(NR)_4$ gave soluble compounds, *e.g.* $Ti(NMe_2)_3(SEt)$, $Ti(NMe_2)_2(SEt)_2$, $Ti(NMe_2)_3(SPr^i)$ and $Ti(NMe_2)_2(SPr^i)_2$.[84] Titanium tetrachloride reacts with difluorodithiophosphinic acid to give $TiCl_3(S_2PF_2) + HCl$, but no $TiCl_2(S_2PF_2)_2$ could be isolated.[85]

$$TiCl_4 + 4RSH + 4NH_3 \longrightarrow Ti(SR)_4 + 4NH_4Cl \quad (19)$$

$$Ti(NR_2)_4 + 4RSH \longrightarrow Ti(SR')_4 + 4R_2NH \ (R = Me, Et, Pr^i, R' = Me, Et) \quad (20)$$

There has been tremendous interest in metal complexes of 1,2-dithiolenes[86] and 1,1-dithiolato[87] complexes, arising from both their industrial applications and the bonding in the complexes. The interest in dialkyldithiocarbamate complexes of titanium(IV) was initially because X-ray analysis of $Ti(S_2CNEt_2)_4$ showed it to contain discrete eight-coordinate (dodecahedral) molecules.[88] However, even seven-coordination has been shown[89] to exist in $TiCl(S_2CNMe_2)_3$. Prior to this discovery the existence of seven-coordinate titanium(IV) was, apart from the oxo-bridged dinuclear peroxodipicolinato complex $[TiO_{0.5}(O_2)(C_7H_3O_4N)(H_2O)]_2^-$, not certain. However, $Ti(tropolonate)_3Cl$, $Ti(oxine)_3Cl$ and $TiCl_4(triars)$ are thought to be seven-coordinate (see refs. quoted in ref. 89). Thus, Fay and co-workers[89] prepared complexes of the type $Ti(S_2CNR_2)_nCl_{4-n}$ ($n = 2, 3, 4$; R = Me, Pr^i, Bu^i; or, when $n = 3$, R = Et) by the reaction of $TiCl_4$ with anhydrous sodium, *N,N*-dialkyldithiocarbamates in refluxing CH_2Cl_2 or benzene. A number of physical techniques indicate that those complexes are monomeric non-electrolytes in which all of the dithiocarbamate ligands are bidentate. Thus, titanium exhibits coordination numbers six, seven and eight when $n = 2$, 3 and 4, respectively. Coordination number seven has been confirmed for $TiCl(S_2CNMe_2)_3$ (**18**) by X-ray diffraction; the molecule has a pentagonal bipyramidal structure with the chlorine atom in an axial position. Fay and co-workers[90] have examined the low-temperature ^1H NMR of the eight-coordinate monothiocarbamate complexes $Ti(R_2mtc)_4$ (R = Et, Pr^i), which exhibit four methyl resonances of approximately equal intensity, consistent with the dodecahedral *mmmm*-C_{2v} stereoisomer found[91] for $Ti(Et_2mtc)_4$ (**19**) by X-ray diffraction. This isomer has an all-*cis* structure with four sulfur atoms on one side of the coordination group and the four oxygen atoms on the other side. The dipole moment in benzene (4.49 D) is in accord with the highly polar *mmmm*-C_{2v} stereoisomer being the principal species in solution. Variable temperature NMR spectra afford evidence for two distinct kinetic processes: (1) a low-temperature process that involved metal-centred rearrangement, and (2) a high-temperature process that involves hindered rotation about the C—N bond in the ligand. Thus, this complex is sterically rigid on the NMR timescale. Both ν(Ti—O), 558 cm^{-1}, and ν(Ti—S), 313 cm^{-1}, bands have been identified.

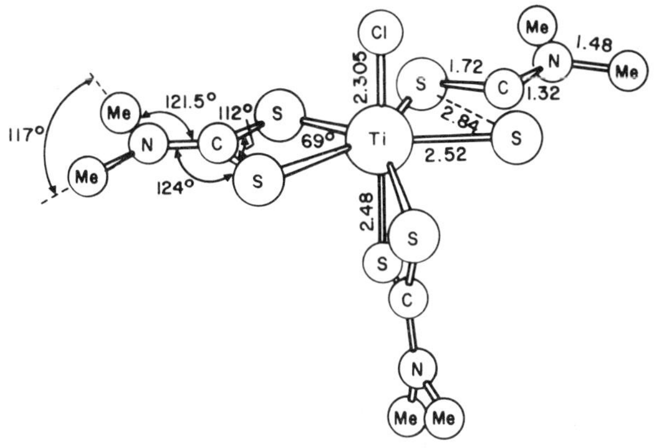

(**18**)

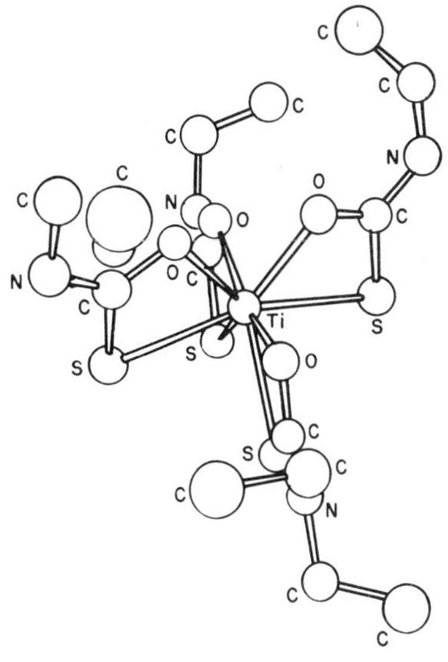

(**19**)

During the period 1965–1975 the chemistry of the 1,2-dithiolene complexes of the transition metals was the subject of considerable study.[86,87,91–98] However, during this period of great activity few complexes of the early transition metals were reported aside from those of vanadium. The problem had much to do with synthetic procedures, since reaction of, say, the anhydrous metal chlorides with the dithiolene or its sodium salt did not prove successful. However, the use of metal dialkylamides[99] did result in clean reactions (*e.g.* equation 21).

$$M(NR_2)_x + (6-x)NaHS_2C_6H_3Me + (x-3)(HS)_2C_6H_3Me \longrightarrow Na_{6-x}M(S_2C_6H_3Me)_3$$
$$M = Ti, Zr, Hf; x = 4 \qquad M = Nb, Ta; x = 5 \tag{21}$$

Much of the interest in the bidentate sulfur donor ligands was the discovery that trigonal prismatic as well as octahedral coordination could occur. Thus, X-ray studies of $Re(S_2C_2Ph_2)_3$ and $V(S_2C_2Ph_2)_3$ showed trigonal prismatic geometry. However, Gray and co-workers[100] have concluded from electronic spectral measurements that $Ti(mnt)^{2-}$ (mnt = maleonitriledithiolate) has a severely distorted octahedral structure.

In order to gain some insight into the possible role of π bonding in eight-coordinate complexes involving four symmetrical bidentate ligands, the complexes of dithiobenzoic acid (dtbH) with titanium(IV) and vanadium(V) were examined. A range of complexes $VO(dtb)_3$ and $Ti(dtb)_nCl_{4-n}$ (n = 2, 3, 4) were isolated and coordination numbers of six, seven and eight were assigned for the species containing two, three and four dtb ligands. By comparing the electronic spectra of these complexes with the known dithiocarboxylato and dithiocarbamato eight-coordinate species with d^0, d^1, d^2 configurations it was concluded that π bonding interactions of dtb with d^0 systems are much less effective than with d^1 and d^2 systems.

31.5 MIXED DONOR LIGANDS

31.5.1 Schiff Base Ligands

These ligands have been a constant source of interest for inorganic chemists. Both the bidentate type (**20**) (for which, for example, control of tetrahedral–square planar equilibria by the steric effect of the *N*-alkyl group has been studied in bis(chelate)nickel(II) complexes, and for which oxygen bridging has been observed in some *N*-methyl derivatives) and the tetradentate type (**21**) (since the backbone of the ligand, —Y—, can be varied giving either flexibility or rigidity) have been extensively studied.

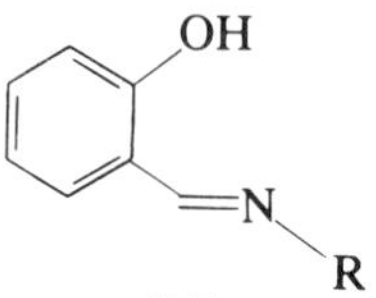

(**20**)

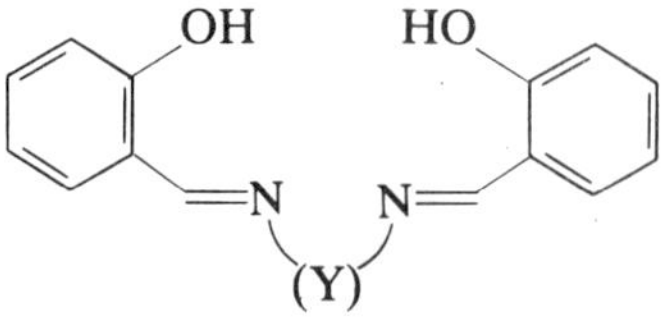

(**21**)

However, despite the potential of these ligands there has been very little work done on titanium systems. Bradley and co-workers[107] were interested to see if four bidentate ligands could bind to titanium(IV) to give an eight-coordination species. They prepared their tetrakis derivatives using Bradley's dialkylamide precursors, as shown in equation (22). The ^1H NMR spectrum (R = Et) in $CDCl_3$ solution exhibited broadened methylene, methyl and methine proton resonances at room temperature, which at low temperature split into double peaks suggesting an exchange between two types of ligand in equal proportions. This could mean an eight-coordinate tetrachelated species with a configuration containing the ligands in two distinguishable environments, or a six-coordinate complex containing two bidentate and two monodentate ligands. In the solid state, X-ray crystallography shows that the latter alternative (**22**) obtains. The fact that six-coordination occurs is not surprising, but that two of the Schiff base ligands coordinate as monodentate is wholly surprising. The adoption of the *cis* unidentate configuration recalls the similar behaviour in $Ti(acac)_2(OR)_2$ complexes.[102]

$$Ti(NMe)_4 + 4HOC_6H_4CHNR \longrightarrow Ti(OC_6H_4CHNR)_4 + 4HNMe_2$$
$$R = Me, Et, Bu^t \qquad (22)$$

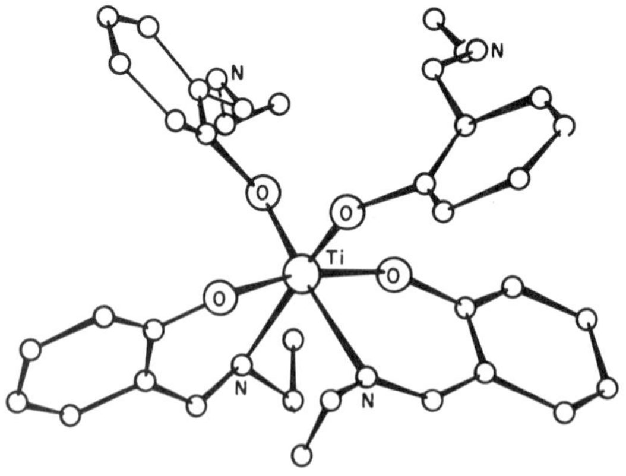

(**22**)

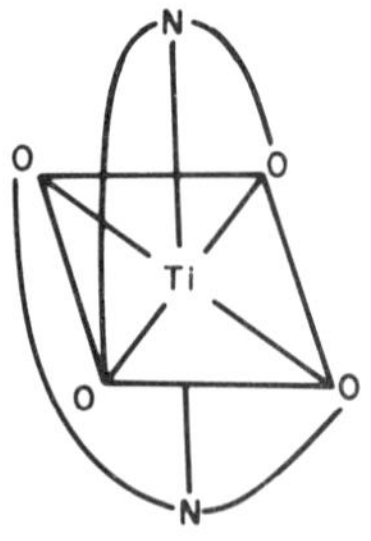

(**23**)

Some complexes with tridentate O_2N donor Schiff base ligands form from the reaction of $OTi(ClO_4)_2$ with H_2L in aqueous methanol. A structure containing four oxygens in the plane of the octahedron and *trans* axial azomethine groups has been assigned (**23**). There is no evidence for the Ti═O group (which is reported to occur in the IR spectrum at *ca.* 1280 cm^{-1}).[103]

Thus far we have seen that Schiff base ligands tend to promote six-coordination, but an unusual reaction has been found to occur between $TiCl_2(salen)$ and borohydride. No reduction at titanium is seen, but rather, addition of BH_4^- to the imino function of the ligand occurs to give a dimeric seven-coordinate titanium(IV) compound, (**24**), containing amine–boranes as donor groups. There is no IR band at *ca.* 1610–1630 cm^{-1}, confirming the reduction of the imino groups.[104] Thus, there is here observed the sexidentate ligand $(salen)(BH_4)_2$ having an $H_2N_2O_2$ set of donor atoms.

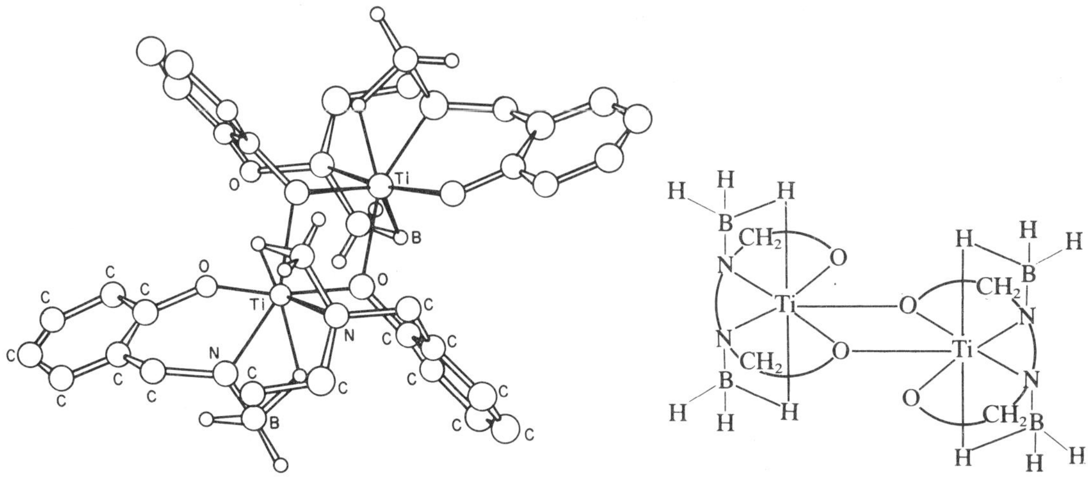

(24)

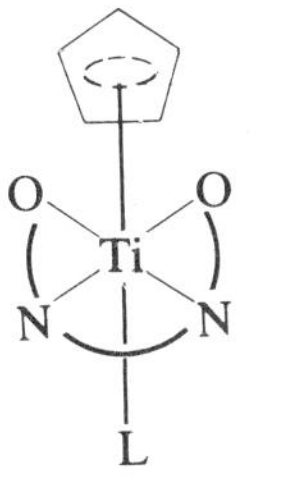

(25)

Employing silyated quadridentate Schiff bases, *e.g.* *O,O'*-bis(trimethylsilyl)-*N,N'*-ethylenebis(salicylideneimine),[105] a range of titanium(IV) complexes $Ti(\eta^5\text{-}C_5H_5)(salen)L$ (**25**; L = Cl, OMe, SMe, NMe_2, $SnPh_3$, acac) have been prepared.

Bowden and Ferguson[106] have obtained a range of interesting complexes starting with $TiCl_3(THF)_3$ and potentially quadridentate Schiff bases. Species (**26**), a titanium(III) complex, initially green, slowly turns red (**27**) on stirring in THF. Complex (**29**) was obtained from the filtrate resulting from the isolation of (**28**). Partial hydrogenation of the Schiff base ligand has occurred. In this complex there occurs another example of a reduced imine function binding to titanium; however, no external reductant was used in producing this species.

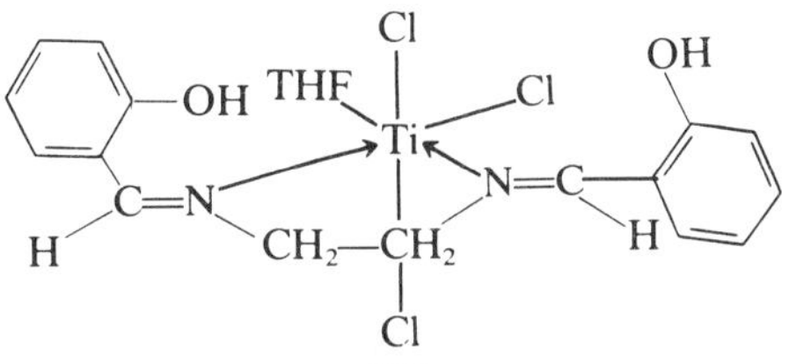

(26)

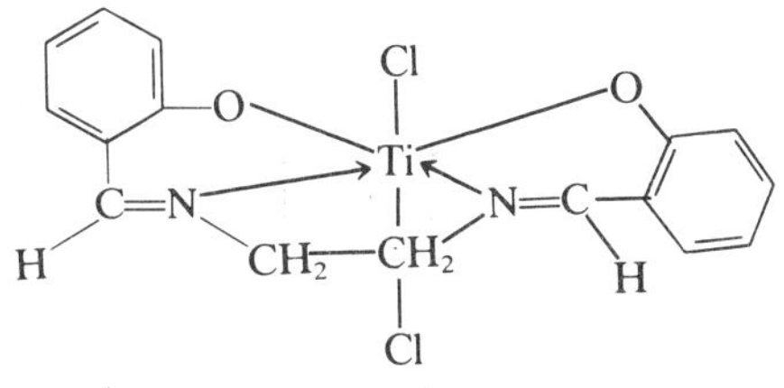

(27)

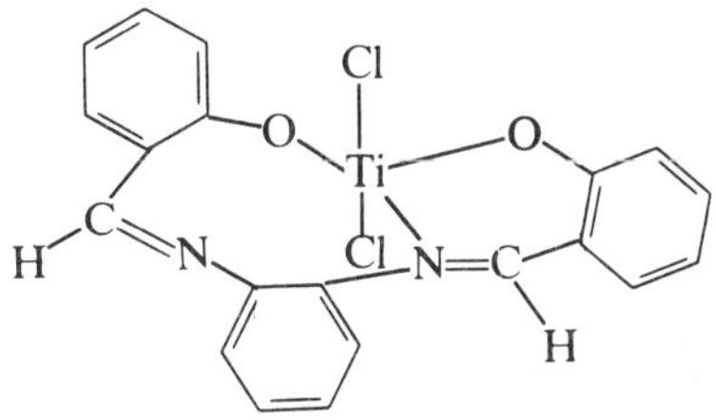

(28)

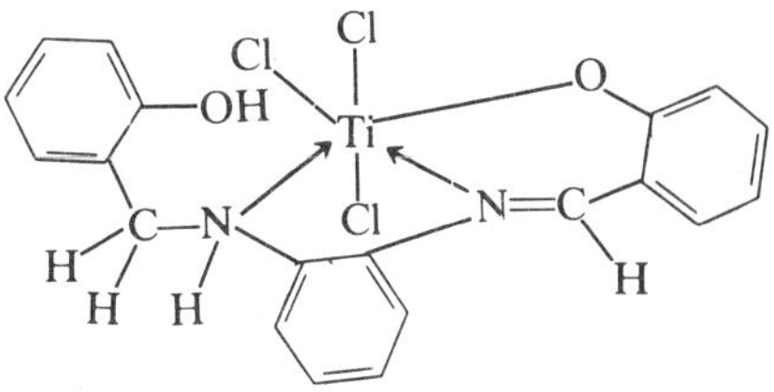

(29)

Reduction of the titanium(III) species [$TiCl_2$(salen)] with zinc dust in THF gives a deep blue solution from which dimeric [{Ti(salen)$(THF)_2$}][$ZnCl_4$] can be isolated. The reduced paramagnetism of this complex (μ_{eff} = 0.96 BM, 293 K) suggests a strong titanium–titanium interaction (**30a** or **30b**). However, treating this blue solution with pyridine leads to the isolation of the very air-sensitive [TiCl(salen)(py)]·THF, whose magnetic moment (μ_{eff} = 1.79 BM, 293 K) corresponds to that expected for a d^1 system. X-Ray analysis of this complex (**31**) confirms the pseudooctahedral coordination around the metal, with salen in the equatorial plane and Cl and py mutually *trans*.[107]

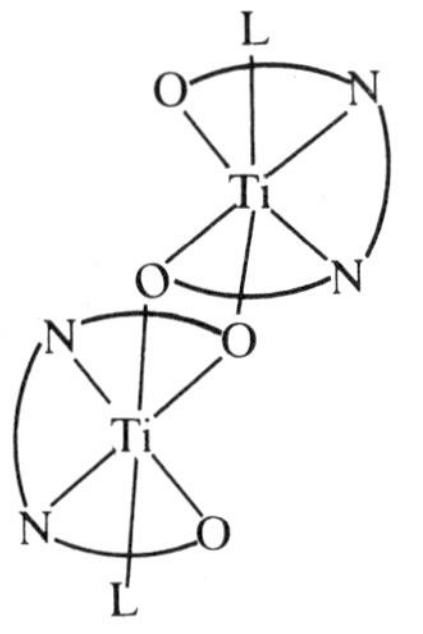

(30a)

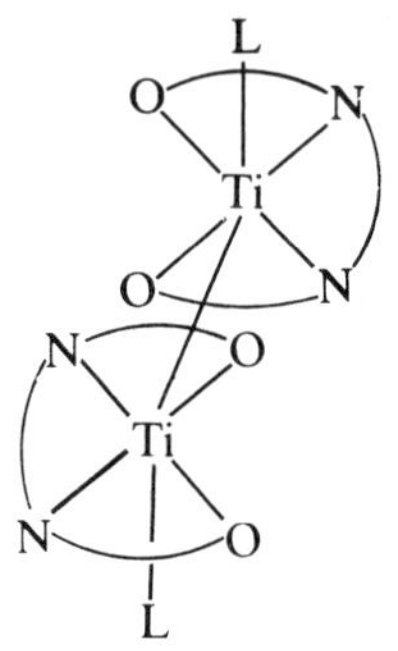

(30b)

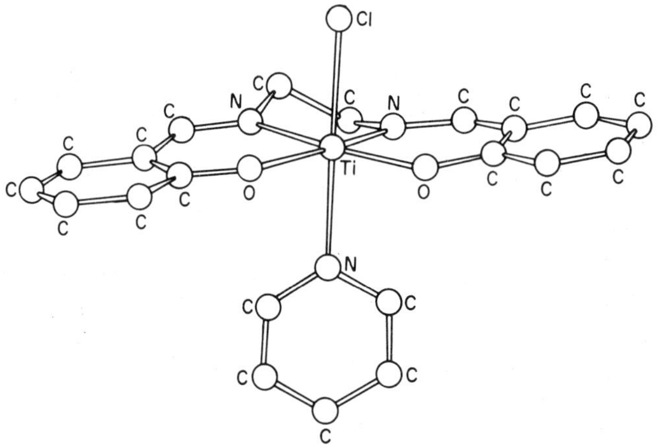

(31)

31.5.2 Aminoacidate Ligands

Recent interest in the peroxo complexes of the early transition metal has resulted in the report of the formation of orange 1:1 peroxo–titanium(IV) species from acidic perchlorate solutions of titanium(IV) and H_2O_2,[108] and Schwarzenbach's group have characterized[109] a number of monomeric and dimeric peroxo–titanium(IV) complexes containing chelating ligands such as pyridine-2,6-dicarboxylate (dipic) and nitrilotriacetate (NTA). Wieghardt and coworkers[110] have isolated [TiO(dipic)(OH_2)]·$3H_2O$, [TiO(IDA)(OH_2)]·$2H_2O$ (IDA = iminodiacetate) and Cs_4[TiO(NTA)]$_4$·$6H_2O$, which are thought to be precursors to Schwarzenbach's species. The last complex is a tetramer, (**32**), containing four asymmetric μ-oxo bridges, and the titanium atoms are in a distorted octahedral environment. All the above complexes react with H_2O_2 in acid perchloric media to form 1:1 peroxo–titanium(IV) species. Schwarzenbach has isolated the potassium salt of the dipicolinic acid derivative of titanium(IV) in the presence of H_2O_2, *viz.* $K_2Ti_2O_5(dipic)_2$·$5H_2O$. This dinuclear complex has C_2 symmetry, and the titanium atoms are coordinated approximately pentagonal bipyramidally sharing an apical μ-oxygen, (**33**). In each bipyramid the peroxy group and the dipic ligand occupy the equatorial sites and the remaining apex is occupied by water. The Ti—O—Ti angle (178.1°) is not significantly different from 180°. The peroxy group, O(5)–O(6), is attached symmetrically to each titanium. The heptacoordination about titanium is similar to that found for other transition metal peroxide complexes.[112] Three of the corners of the pentagon are occupied by the dipic ligand with angles N—Ti—O(3) and N—Ti—O(4) of 72.1 ± 0.3°. The pentagon is distorted mainly because of the short O(5)—O(6) distance. The non-chelated binuclear peroxytitanium complex thus presumably has the composition $[\{Ti(O_2)(H_2O)_4\}_2O]^{2+}$. The Ti—O(7) distance is very short (1.825 Å) but agrees well with that found in the binuclear acac complex of titanium(IV).[113] The O(5)—O(6) distance of 1.453 Å in the peroxy group is equal to the distance found by Busing and Levy in H_2O_2.[114]

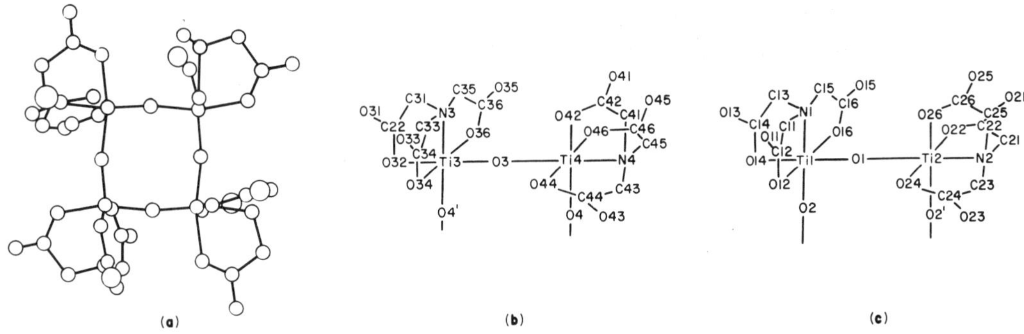

(32) (a) Perspective view of the complex $([TiO(NTA)]_4)^{4-}$ anion; (b) and (c) atomic numbering schemes for the two crystallographically independent halves of the tetrameric anions. The open circles represent centres of symmetry.

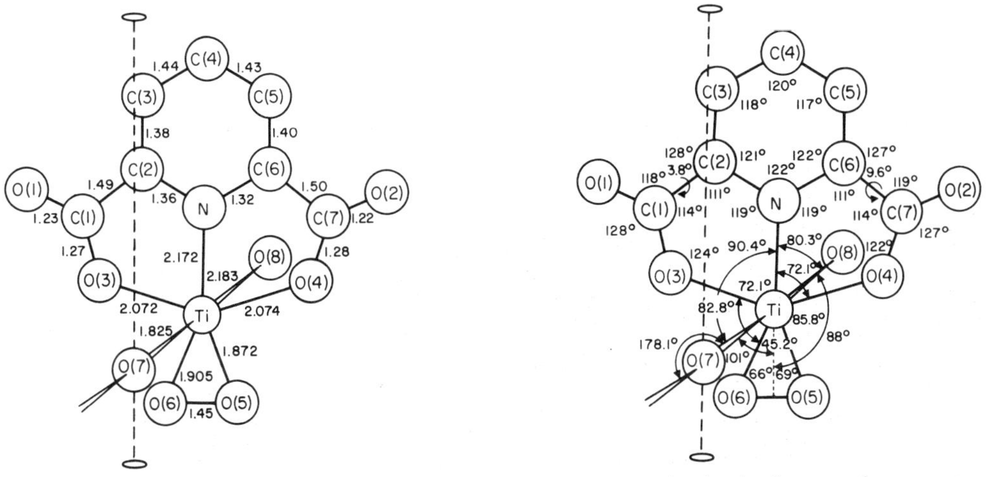

Distances in the complex

Angles in the complex

(33)

The peroxy complex of titanium(IV), $Ti(O_2)(edta)^{2-}$, has been used as part of an analytical method for either titanium(IV) or H_2O_2. More recently the structural nature of $TiO(edta)^{2-}$ in solution and the reaction of $TiO(edta)^{2-}$ with H_2O_2 to form $Ti(O_2)(edta)^{2-}$ have been investigated over the pH range 2.0–5.2. Two distinctly different primary coordination spheres in terms of spectral properties and reactivities towards H_2O_2 are observed for $TiO(edta)^{2-}$ solutions in this pH range. As shown in Scheme 2, these are the aquated forms (A), $TiO(edtaH_n)(H_2O)^{n-2}$, having two coordinated carboxylates and two amines and one coordinated water, and the fully chelated forms (B), $TiO(edtaH_n)^{n-2}$, with three carboxylates and two amines coordinated to titanium(IV).[115]

31.5.3 Mixed Ligands Containing Sulfur Donors

Clark and McAlees[116] have prepared a large number of pseudooctahedral complexes of methyltitanium trichloride with symmetrical and unsymmetrical bidentate ligands, (**34**). It is known that oxidation of $MeTiCl_3$ itself in a hydrocarbon solvent gives a product formulated as $TiCl_3OMe{\cdot}TiCl_2(OMe)_2$, and the oxidation of (**34**) gave the corresponding complexes of methoxytitanium trichloride $TiCl_3OMe(XCH_2CH_2Y)$ (X, Y = OMe, NMe_2, SMe).

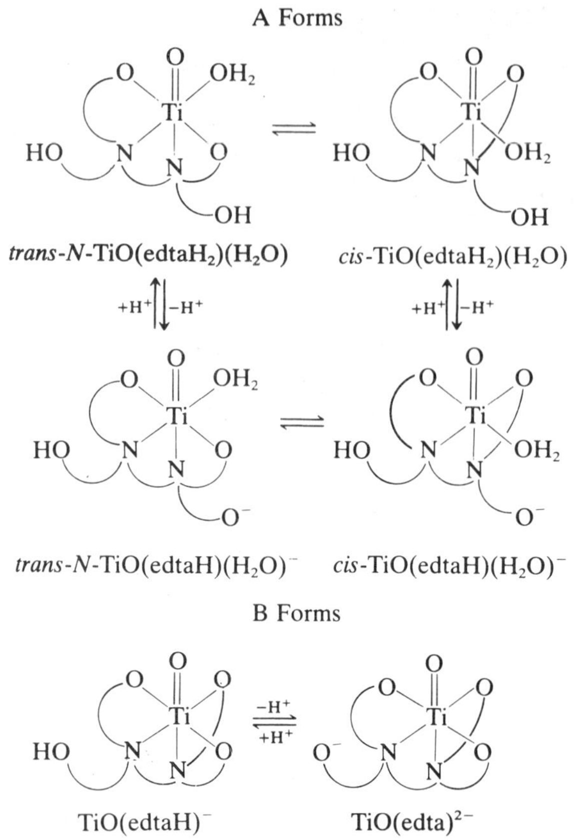

Scheme 2

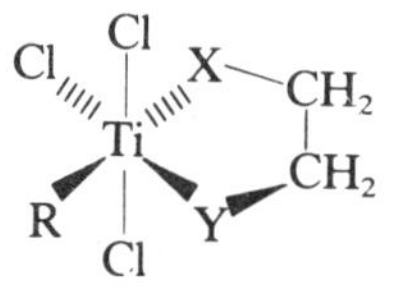

(34a)

X = Y = OMe
X = OMe, Y = NMe_2

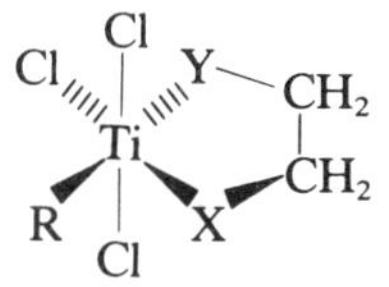

(34b)

X = Y = NMe_2
X = OMe, Y = SMe

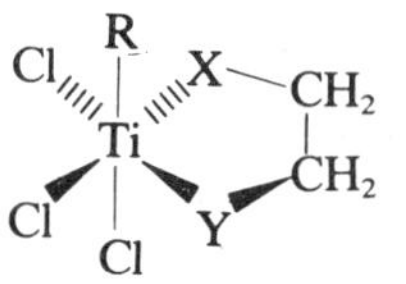

(34c)

X = Y = SMe
X = NMe_2, Y = SMe

McAlees[117] has also examined some potentially terdentate tripod ligands MeC(X)(Y)(Z) (X = Y = Z = CH_2OMe; X = Y = CH_2OMe, Z = CH_2SMe and CH_2NMe_2; X = CH_2OMe, Y = CH_2SMe, Z = CH_2NMe_2). The precipitates obtained on mixing hexane solutions of equimolar quantities of titanium tetrachloride with these ligands consist of mixtures of products which may include the corresponding 1:1 complex, products in which oxygen–methyl cleavage has occurred to give a titanium alkoxide, and a 3:2 complex. However, the reaction in chloroform medium does give 1:1 products, as it appears that reactions leading to the other types of products take place much more slowly. Variable-temperature NMR shows that all four ligands act only as bidentates towards titanium and that exchange can take place between free and coordinated ligand atoms. The binding to titanium is stronger for nitrogen than for oxygen or sulfur, and oxygen coordination is preferred over sulfur when nitrogen is already bound, whereas sulfur binding is preferred to oxygen coordination when an oxygen is already bound. These observations led McAlees to reexamine the published data[118] on the analogous arsine complex $TiCl_4 \cdot MeC(CH_2AsMe_2)_2$. It had previously been concluded that all three arsenic atoms are coordinated since all the arsenic methyl groups resonate downfield from their resonance position in the free ligand. However, the reinvestigation suggests that one arsenic is uncoordinated.[117]

31.5.4 Other Oxygen–Nitrogen Systems

A number of mixed monodentate ligands have been coordinated to titanium(III), *viz.* $TiCl_3(L)(L')$, $TiCl_3(L)_2(L')$ (L = THF, L′ = MeCN),[119] and some titanium(IV) alkoxy derivatives of 2-pyrrolidinone have been isolated: $Ti(OR)_3L$, $TiCl(OR)_2L$, $TiCl_2(OR)_2{\cdot}LH$, $TiCl_2(OR)_2{\cdot}2LH$, $TiCl_4{\cdot}LH$ and $TiCl_4{\cdot}2LH$.[120] The starting point for the preparation of the 2-pyrrolidinone complexes is $Ti(OR)_4$ (R = Et, Pr^i), and thus these complexes represent the unusual example of a Ti—O bond being replaced by a Ti—N bond. Moreover, it was found possible to substitute NR_2 by pyrrolidinyl for, on reaction with $Ti(OR)_3(NR)_2$, HNR_2 was evolved. The generalized reaction is shown by equation (23).

$$XTi(OR)_3 + \overline{(CH_2)_3C(O)NH} \longrightarrow \overline{(CH_2)_3C(O)N}Ti(OR)_3 + XH \qquad (23)$$

$$(= LH)$$

$$X = OR,\ NR_2,\ Me,\ \text{but not Cl}$$

Analogous to the work on dipicolinic acid reported in Section 31.5.2, a range of titanium(IV) peroxo complexes (**35**) have been prepared which encompass an oxygen–nitrogen anionic chelating ligand and hexamethylphosphortriamide.[121] The crystal structure of peroxobis(picolinato)(hexamethylphosphortriamide)titanium(IV) (**36**) reveals that it is another example of a deformed pentagonal pyramid similar to that found for (**32**). The equatorial positions are occupied by the nitrogen atoms of the pyridine-2-carboxylate groups, the oxygen of HMPT and the oxygen atoms of the peroxo group. The axial positions are occupied by one oxygen atom of each pyridine-2-carboxylate group. The peroxo group is, as usual, π bonded to the titanium and the Ti—O(8) and Ti—O(7) bonds are equivalent, and are in the range found by Schwarzenbach[111] and by Guilard in a peroxotitanium(IV) porphyrin adduct.[122] However, the O—O bond is significantly shorter, 1.419 Å, and thus the O—Ti—O angle is smaller, 45.2°. The nitrogen *trans* to the HMPT group (Ti—N, 2.207 Å) is shorter than the Ti—N *trans* to the peroxo group (2.340 Å). Of the complexes prepared, (**35**), all proved to be unusually stable and were reluctant to react with simple alkenes, allylic alcohols and cyclic ketones. Complex (**35a**) did react with tetracyanoethylene (TCNE) to produce '$Ti(O_2)(TCNE)(C_6H_4NO_2)_2(HMPT)CH_2Cl_2$', but as the Ti—$O_2$ IR vibrations disappear the reaction is best viewed according to equation (24).

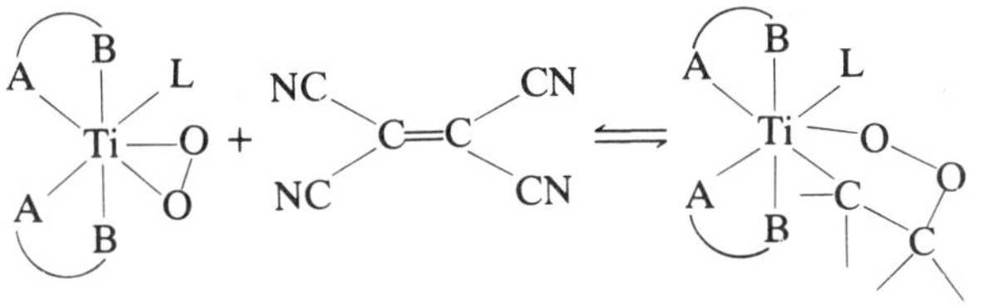

(24)

Titanium tetrachloride reacts with 8-quinolinol (oxineH) in 1:2 mol ratio to form the adduct $TiCl_4$(oxineH), from which two moles of HCl can be eliminated to produce $TiCl_2(\text{oxine})_2$. This compound is a member of a large group of compounds of general formula $MX_2(L)_2$ (M = Ti, Ge, Sn; X = F, Cl, Br, I; L = oxine, acac, salen). The $TiCl_2(\text{oxine})_2$ is only one member of this group, others being $TiCl_4$(oxineH), $TiCl_4(\text{oxineH})_2$, $TiCl(\text{oxine})_3$ and $TiCl_3$(oxine).[123] The crystal structure of $TiCl_2(\text{oxine})_2$ shows it to be six-coordinate, (**37**), with *cis* chlorines (Ti—Cl; 2.283 Å), *cis* nitrogens (Ti—N = 2.200 Å) and *trans* oxygens (Ti—O = 1.888 Å). From the reaction of $TiCl_2(\text{oxine})_2$ with oxineH in the presence of NEt_3, $Ti(\text{oxine})_4$ can be isolated. $Ti(\text{sal})_4$ (sal = salicyldehydato) and $Ti(\text{oxine})_2(\text{sal})_2$ can be similarly prepared. The $Ti(\text{oxine})_4$ is probably eight-coordinate.[125,126]

Dimethylglyoxime, $DMGH_2$, forms $TiCl_4(DMGH_2)_n$ (n = 1, 2) in light petroleum. Warming to 40 °C results in the loss of HCl to give $TiCl_3$(DMGH). Benzil monoxime adducts, $TiCl_4(BMOH)_n$ (n = 1, 2) can also be obtained. The dimethylglyoxime complexes decompose violently on heating, but the benzil monoxime species undergo a Beckmann fragmentation at 120 °C to give phenyl cyanide.[127]

31.6 PHOSPHORUS, ARSENIC AND ANTIMONY COMPLEXES

There are two major reference works to metal complexes containing these donor atoms.[16,17] The chemistry of these donor ligands with titanium is sparse, and clearly there is much scope for exploitation in this area.

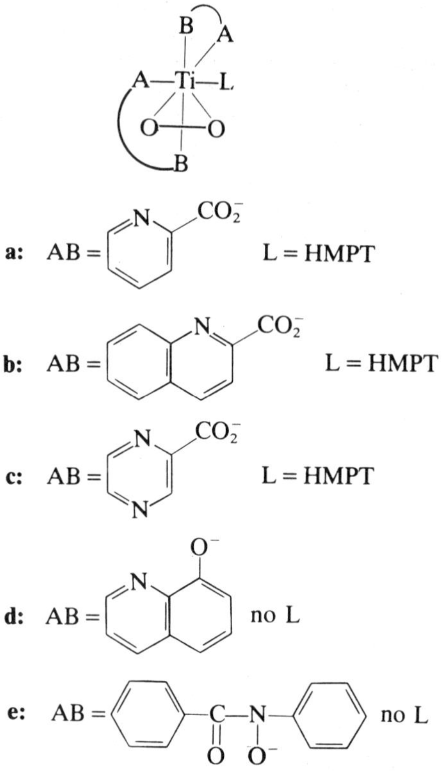

(35)

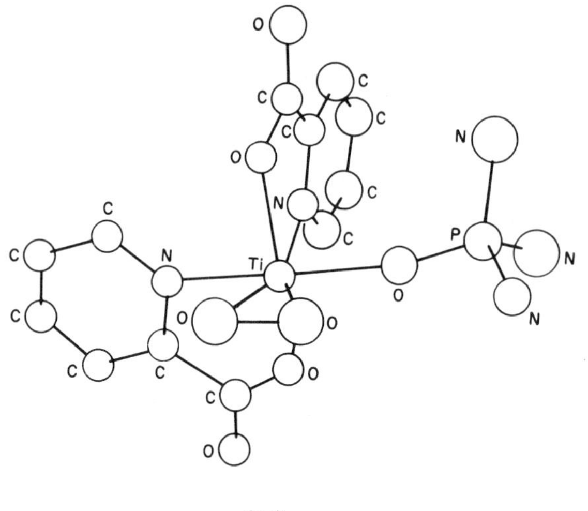

(36)

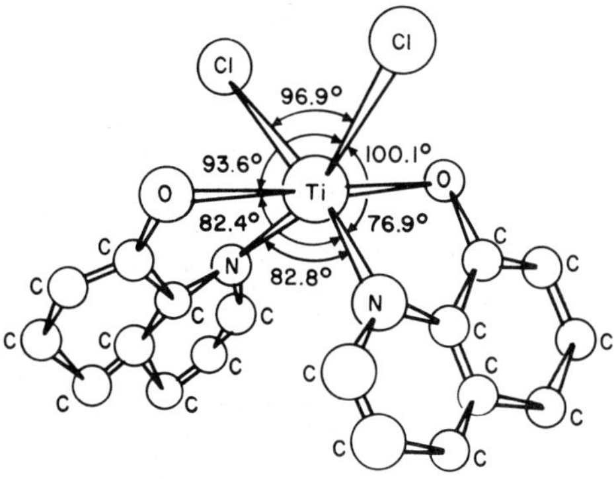

(37)

31.6.1 Titanium(II) Complexes

Issleib and Weuschuh prepared $Ti(PCy)_2$ by reaction of $TiCl_3 \cdot THF$ or $TiCl_4 \cdot 2THF$ with $LiPCy_2$. The dark brown product ($\mu_{eff} = 0.60$ BM) was pyrophoric, was decomposed by water and reacted with iodine in benzene to give $TiI_2(PCy_2)_2$ and $TiI_3(PCy_2)_2$.[128] Although somewhat outside the scope of this treatise, an important organometallic complex has recently been prepared $[(\eta^5\text{-}C_5H_5)_2Ti(PMe_3)_2]$.[129] Monomeric titanocene, Cp_2Ti, has always eluded isolation as a discrete compound, since *in statu nascendi* it immediately reacts to form bi- or multi-nuclear species. However, $Cp_2Ti(PMe_3)_2$, synthesized as shown by equation (25), makes accessible the Cp_2Ti unit, since the two phosphine ligands may be replaced by a very large number of substrates under very mild conditions (CO, SPh, C_4H_4, C_4H_8, dmpe, MeCN). The crystal structure of this complex, (**38**), indicates that two crystallographically independent molecules are present, but both molecules possess the same dimensions. The Ti—P distances (2.524, 2.527 Å) are in excellent agreement with the recently reported $(MeC_5H_4)_2Ti(dmpe)$.[130]

$$Cp_2TiCl_2 + Mg + 2PMe_3 \xrightarrow{THF} Cp_2Ti(PMe_3)_2 + MgCl_2 \quad (25)$$

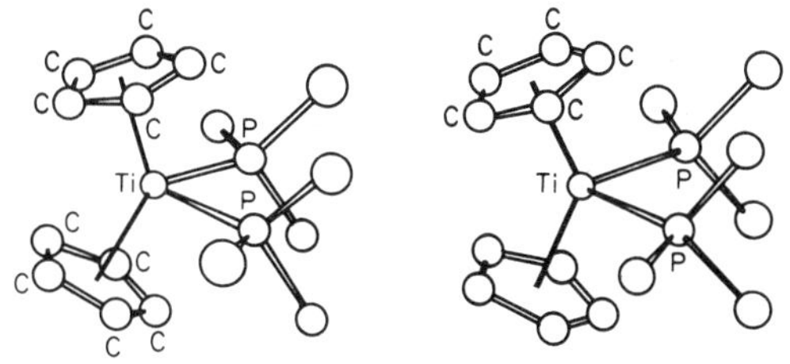

(38)

31.6.2 Titanium(III) Complexes

When titanium and phosphorus are heated together in the presence of iodine TiI_4 is formed, but rapidly decomposes. The elemental Ti thus created reacts rapidly with phosphorus to give either TiP or TiP_2 (depending on stoichiometry). Iodine is a catalyst in this sequence.[131]

The first titanium(III) complex, $TiCl_3(PEt_3)\cdot C_2H_4$, was prepared by Chatt and Hayter,[132] but more recently monomeric $TiCl_3(PR_3)_2$ ($R_3 = MeH_2$, Me_2H, Me_3, Et_3) have been isolated from direct combination of the components in toluene at 85 °C. At −20 °C there is ESR evidence for $TiCl_3(PR_3)_3$ species for solutions of $TiCl_3$ and $3PR_3$ in toluene.[133] The grey-green $TiCl_3(PEt_3)_2$ is monomeric in solution, and by analogy, the red-brown complexes with the other phosphines are also assumed to be pentacoordinate. The weaker donors, PPh_3 and PCl_3, do not complex with titanium(III).

31.6.3 Titanium(IV) Complexes

31.6.3.1 Monodentate ligands

The isolation of $TiCl_4(PH_3)$ was first reported in 1832;[134] $TiCl_4(PH_3)_2$ has also been prepared.[135] A large number of $TiCl_4(PR_3)_2$ complexes are now known,[16,17] but only quite recently have $TiBr_4(PPh_3)$ and $TiBr_4(PPh_3)_2$ been prepared, the former having C_{3v} symmetry and the latter a *cis* octahedral structure.[136] There is some evidence for the probable existence of binuclear species of the type $Ti_2Cl_8(PR_3)$.[16,17] Titanium tetrachloride reacts with $Pt(PPh_3)_n$ ($n = 3, 4$) to yield emerald green $(TiCl_4)_2Pt(TiCl_4\cdot PPh_3)_3$, which, on heating *in vacuo* or treatment with more PPh_3, produces $Pt(TiCl_4\cdot PPh_3)_3$. It is thought[137] that the former species is pentacoordinate with two axial $TiCl_4$ and three equatorial $TiCl_4\cdot PPh_3$ moieties bonded to platinum *via* the titanium atoms.

Although $TiCl_4$ and $AsPh_3$ were known to complex as early as 1924, the adduct $TiCl_4(AsPh_3)$ has only been investigated relatively recently.[16,17] The reports of a bis adduct $TiCl_4(AsPh_3)_2$ must be viewed with some suspicion.[138]

The reaction between $SbPh_3$ and $TiCl_4$ in dry benzene initially gives a deep violet solution and ultimately a gummy black solid. It is likely that reduction of $TiCl_4$ has occurred.[139] Although there is a paucity of well-characterized arsine and stibine complexes, a surprisingly large number of patents have dealt with catalyst systems employing these compounds (Table 6) which often contain Lewis acids in addition to Lewis bases. The function of the Lewis acid is often not clearly delineated, but its presence has a pronounced effect on the catalytic activity as does the Lewis base.

31.6.3.2 Bidentate ligands

With 1,2-bis(diphenylarsino)ethane (dpae), $TiCl_4$ forms only a 2:3 complex $[(TiCl_4)_2(dpae)_3]$ although diphos forms 1:1 and 3:2 adducts.[139] Tetramethyldiphosphine and $TiCl_4$ react to form a deep red paramagnetic product of unknown composition, the paramagnetism suggesting that partial reduction to titanium(III) has occurred. With $TiCl_3(THF)_3$ an insoluble, moisture- and air-sensitive complex $(TiCl_3)_2(Me_2PPMe_2)_3$ is produced, for which structure (**39**) has been proposed. Tetraethyldiphosphine does not yield any isolable complex with $TiCl_3(THF)_3$.[17]

Investigations of the diars complexes of the titanium(IV) halides resulted in the first

Table 6 Catalyst Systems Containing Titanium Salts[16]

Titanium Salt	*Components Group VB*	*Other*	*Use*[a]
$TiCl_4$	R_3As	Al	P
$TiCl_3$	R_3As, R_3Sb	$RAlX_2$	P
$TiCl_3$	Ph_3As, Ph_3Sb	Et_3Al	P
$TiCl_3$	Ph_3As	CaH_2	P
$TiCl_3$	$(CH_2{=}CH)_3Sb$	Et_3Al	P
$TiCl_3$	$Et_3Sb\cdot 2ZnCl_2$	Et_3Al	P
$TiCl_3$	$Et_3Sb\cdot ZnCl_2$	Et_3Al	P
[b]	R_3As, R_3Sb	None	X
$TiCl_3$	Et_3Sb, Et_3Bi	None	M
$TiCl_3$	$(C_6H_{13})_3As$	$NaAlEt_4$	P
π-Allylic compound	R_3As,	Lewis acid	P, O
$3TiCl_3{-}AlCl_3$	R_3As, R_3Sb	$RAlX_2$	P
$TiCl_4$	Et_3Sb	$LiAlH_4$	A
$Ti(OBu^n)_4$	Et_3As, Et_3Sb	$EtAlCl_2$	A
Ti halide	R_3As, R_3Sb	Na, Li, K, Mg, Zn	A
[c]	R_3As, R_3Sb	Na, Li, K, Al, Mg, Zn, Cu	P
$Ti(BH_4)_3$	Bu_3Sb	None	P
$TiCl_4$	Ph_2BiH, $(BuO)_2SbH$	None	P
$TiCl_3$, $(BuO)_2TiCl_2$	R_3As, R_3Sb	$RAlCl_2$	P

[a] P = monoalkene polymerization; O = monoalkene oligomerization; X = oxo process; D = alkadiene polymerization; M = methyl methacrylate polymerization; A = 1-alkene polymerization.
[b] Complex unspecified.
[c] Halide, alkyl halide or acetylacetonate of Ti or Zr.

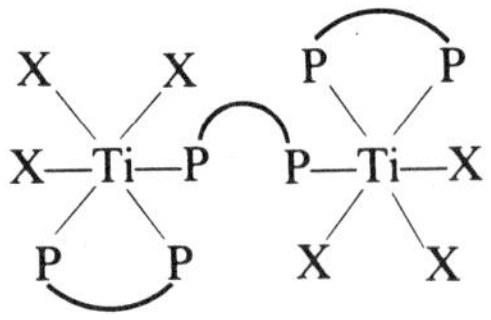

(39)

examples of eight-coordination in the first transition series.[16,17] In $[TiCl_4(diars)_2]$ the dodecahedron contains two diars forming an elongated tetrahedron and four chlorine atoms in a flattened tetrahedron with respect to the fourfold inversion axis, (**40**).[140] Although numerous attempts have been made to prepare similar complexes with a range of bidentate ligands, only those with the *o*-phenylene or similar backbone have proved successful. Thus 1,2-bis(dimethylarsino)-3-fluoro- or 4-fluoro-benzene form only the hexacoordinate $TiCl_4$(ligand) complexes; presumably the fluorine-substituted ligands are poorer σ donors. Eight-coordination is also not achieved with 1,2-bis(dimethylarsino)-3,3,4,4-tetrafluorocyclobutene ($f_4$fars) or 1,2-bis(dimethylarsino)-3,3,4,4,5,5-hexafluorocyclopentene ($f_6$fars), probably due to

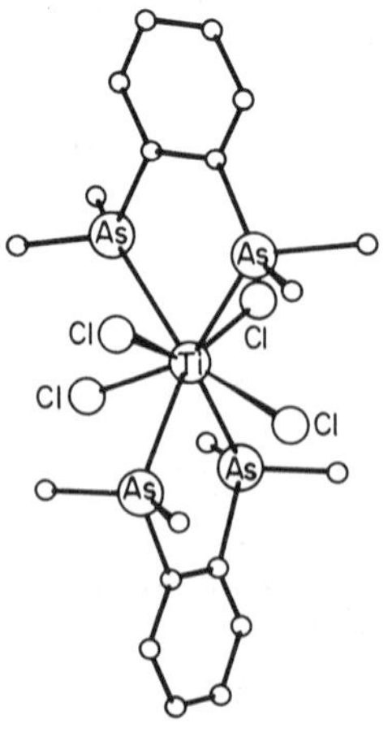

(40)

the larger As—As 'bite' of the ligands, nor with *rac-o*-phenylenebis(phenylmethylarsine), due to steric hindrance. However, 1,8-bis(dimethylarsino)naphthalene, which has a short As—As separation, readily forms brown $TiCl_4(ligand)_2$. These results reinforce earlier conclusions about the importance of steric factors in promoting eight-coordination, but also suggest that the electronic properties of the ligands are not negligible.[17]

A number of $[TiX_4(diars)]$ (X = Cl, Br), $[(TiF_4)_2(diars)]$ and $[TiI_4(diars)_2]$ are known.[16,17] No authentic complexes of titanium(III) are known, save for $[TiX_3(diars)(H_2O)]$ (X = Cl, Br).

31.6.3.3 Multidentate ligands

The triarsines $(o\text{-}Me_2AsC_6H_4)_2AsMe$ and $MeC(CH_2AsMe_3)_3$ are reported to form seven-coordinate adducts of type $TiCl_4L$, on the basis that they are monomeric and non-electrolytes.[142] However, the complex of the latter ligand has been shown, by NMR studies of McAlees and coworkers,[117] to contain one unbound and two coordinated arsine groups. It is probable that both complexes are hexa-coordinate. Another triarsine, $MeAs\{(CH_2)_3AsMe_2\}_2$, is reported to form a series of complexes $[TiF_4]_3 \cdot L$, $[TiCl_4]_2 \cdot L$, $[TiBr_4] \cdot L$ and $[TiI_3 \cdot L]I$.[143] All need further characterization.

Cyclophosphazenes $(RNPX)_n$, especially the diazadiphosphetidines ($n = 2$), have been shown to be capable of acting both as bridging and chelating bidentates through the two phosphorus atoms. It is interesting to see that, once again, the supposed 'hard' acceptor $TiCl_4$ chooses to bind this multidentate ligand *via* the two phosphorus atoms in the *cis* configuration, (**41**).[144] A ν(Ti—P) band is assigned at 435 cm^{-1}, very close to those found for $[TiCl_4(diphos)]$ (453, 414 cm^{-1}).

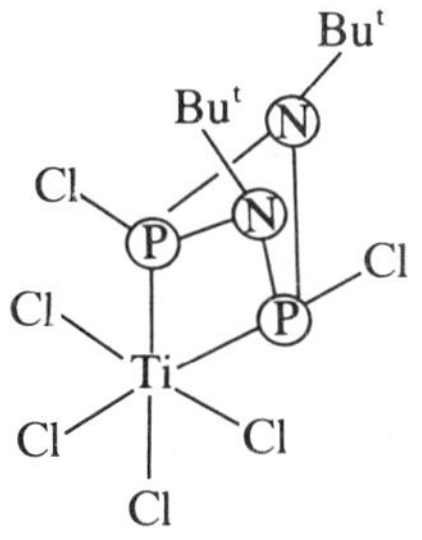

(**41**)

31.7 MACROCYCLIC LIGANDS

31.7.1 Porphyrin Complexes

Although it is just 50 years since Treibs obtained vanadyl deoxophylloerythroetioporphyrin, the major metalloporphyrin in petroleum and shales,[145] it is only relatively recently that Tsutsui obtained the first titanium porphyrin, (*meso*-DME)Ti=O.[146] A review of this general area is available;[147] the nomenclature is given in Table 7.

31.7.1.1 Titanium(III) complexes

Reaction of $(por)TiX_2$ with zinc amalgam gives the reduced (por)TiX (por = TPP; X = F, Cl, I, SMe) and some hexa-coordinate $[(por)TiF_2]^-$ and [(por)TiF(1-MeIm)] have also been synthesized.[148] For (TPP)Ti(OMe) (**42**), the Ti—N distances (2.12 Å) are very close to those found in analogous titanium(IV) derivatives; in each case the Ti atom lies 0.62 Å out of the plane of the ring.[149] When (TPP)TiF is exposed to dioxygen the reaction shown in equation (26) occurs, and a superoxide intermediate, $(TPP)Ti(O_2)F$, is thought to exist, from ESR measurements.[148]

$$2(TPP)TiF + O_2 \longrightarrow (TPP)TiF_2 + (TPP)Ti(O_2) \quad (26)$$

Table 7 Porphyrin Nomenclature

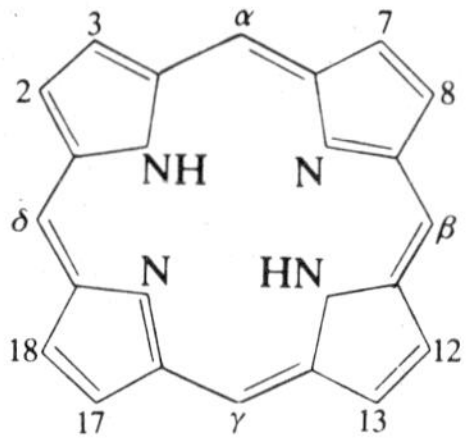

Abbreviation	*Name*	*Substituents* 2	3	7	8	12	13	17	18	α, β, γ, δ
P	Porphine	H	H	H	H	H	H	H	H	H
OMP	Octamethylporphyrin	Me	Me	Me	Me	Me	Me	Me	Me	H
OEP	Octaethylporphyrin	Et	Et	Et	Et	Et	Et	Et	Et	H
TPP	Tetraphenylporphyrin	H	H	H	H	H	H	H	H	Ph
TmTP	Tetra-*m*-tolylporphyrin	H	H	H	H	H	H	H	H	*m*-Tol
TpTP	Tetra-*p*-tolylporphyrin	H	H	H	H	H	H	H	H	*p*-Tol
Etio-I	Etioporphyrin-I	Me	H	Me	H	Me	P^H	P^H	Me	H
meso-DME	*meso*-Porphyrin dimethyl ester	Me	Et	Me	Et	Me	P^{Me}	P^{Me}	Me	H

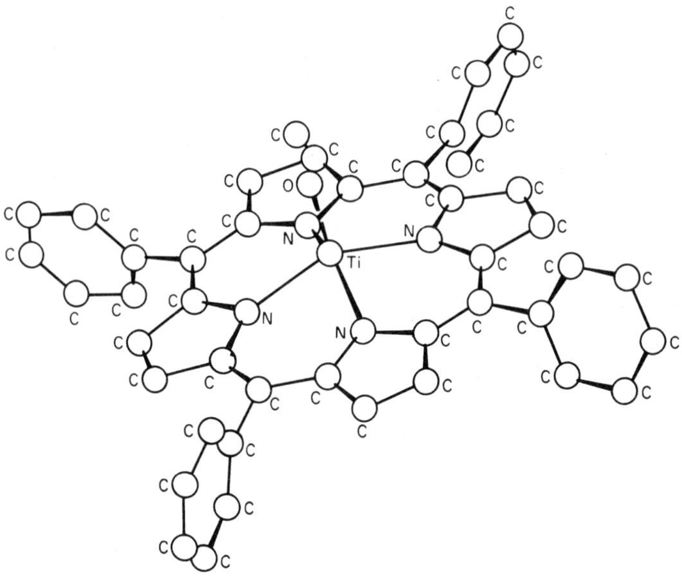

(42)

31.7.1.2 Titanyl derivatives

The (por)Ti=O species can be prepared either by the reaction of $(acac)_2TiO$ in phenol at 180–240 °C, or $TiCl_4$ in a variety of solvents with the porphyrin. The characteristic ν(Ti=O) occurs in the range 1050–950 cm^{-1}.

A number of crystal structures are known, *viz.* (OEP)TiO, (TPP)TiO, $(OEPMe_2)TiO$, (**43**).[147] For all these complexes titanium has four nitrogen donors and a double-bonded oxygen, resulting in a classical square pyramid. The information obtained from X-ray diffraction and EXAFS is in excellent agreement.

31.7.1.3 Peroxotitanium(IV) derivatives

Earlier it has been noted that there is a good deal of interest in peroxo derivatives of titanium(IV). It has been shown that (por)TiO can be converted to $(por)Ti(O_2)$ by reaction with either H_2O_2 or benzoyl peroxide.[150] The crystal structure of $(OEP)Ti(O_2)$ (**44**) reveals O—O = 1.445 Å, characteristic of a peroxo moiety, and the Ti—O_2 moiety is generally the same as found in other peroxo complexes (Table 8).

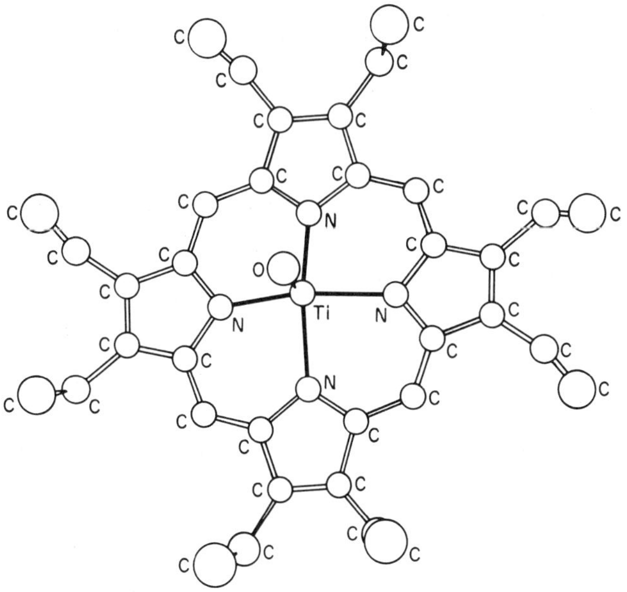

(43)

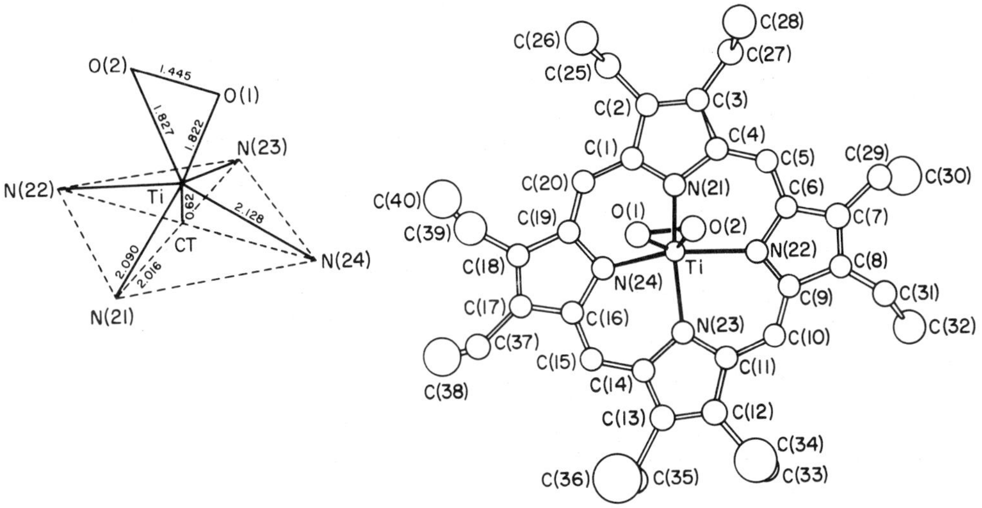

(44)

Table 8 Crystallographic Data for some Titanium(IV) Peroxo Complexes

Complex	*Ti—0(1)*	*Ti—0(2)*	*0(1)—0(2)*	*Ti—0(1)—0(2)*	*0(1)—(02)—Ti*	*0(1)—Ti—0(2)*
$[H_2O(dipic)Ti(O_2)]_2O$	1.872(7)	1.905(7)	1.45(1)			45.2(4)
$(H_2O)_2(dipic)Ti(O_2)$	1.834(2)	1.858(2)	1.464(2)	67.50(6)	67.78(6)	46.72(6)
triclinic	1.842(1)	1.862(1)	1.469(1)			
$(H_2O)_2(dipic)Ti(O_2)$	1.833(1)		1.458(2)	66.56(2)		46.88(6)
orthorhombic	1.844		1.477			
$[F_2(dipic)Ti(O_2)]^{2-}$	1.846(4)	1.861(4)	1.463(5)	67.3(2)	66.2(2)	46.5(2)
$[\{(NTA)Ti(O_2)\}_2O]^{4-}$	1.892(2)	1.889(2)	1.469(2)	67.02(6)	67.26(6)	45.72(6)
	1.897	1.898	1.481			
$(OEP)Ti(O_2)$	1.827(4)	1.822(4)	1.445(5)	66.5(2)	66.9(2)	46.7(2)

Fluxional behaviour for (OEP)Ti(O_2) and (TPP)Ti(O_2) has been shown from ^1H and ^{13}C NMR studies. At temperatures down to −50 °C the peroxo ligand undergoes fast exchange between two equivalent sites; the barrier to rotation of the peroxo ligand at the coalescence temperature is 43.9 ± 2.1 kJ mol^{-1} for (TPP)Ti(O_2).

31.7.1.4 Titanium(IV) dihalide derivatives

Oxotitanium(IV) porphyrin complexes react with HX (X = F, Cl, Br) to yield (por)TiX_2. The crystal structure of (TPP)TiBr_2 (**45**) has been determined.[151]

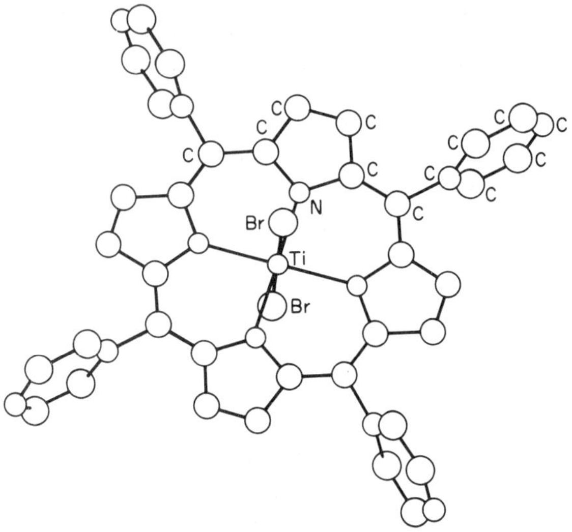

(45)

31.7.2 Phthalocyanine Derivatives

Very few complexes have been examined. It is known that (Pc)TiCl can be converted to (Pc)TiO.[152] Dichloro(phthalocyaninato)titanium(IV) can be synthesized *via* equation (27). This reaction is a convenient one, inasmuch as the titanium(III) furnishes the electrons necessary to convert phthalonitrile to the phthalocyanine anion.[153]

$$2TiCl_3 + 4C_6H_4(CN)_2 \longrightarrow (Pc)TiCl_2 + TiCl_4 \quad (27)$$

31.8 GROUP IV DERIVATIVES

Heintz[154] has reported a dark blue octahedral complex cyanide of titanium(III), $K_3Ti(CN)_6$, as well as a compound formulated as $K_3Ti(CN)_6KCN$. These complexes were isolated from aqueous solutions of $TiCl_3$ and KCN under conditions that might be expected to produce only titanium(III) hydroxide. Nicholls *et al.*[155] have repeated these preparations and find the only product of these reactions to contain the dark-blue hydroxide, $[Ti(OH)(H_2O)_5]^{2+}$. However, these workers have isolated the grey-green $K_5Ti(CN)_8$ from the reaction of $TiBr_3$ and KCN in liquid ammonia; this species is probably a seven-coordinate $K_4Ti(CN)_7 \cdot KCN$ moiety.

Nicholls and Ryan have produced other complex cyanides of titanium: $Rb_5Ti(CN)_8$, $Cs_4Ti(CN)_7$ and $K_3Ti(CN)_6$. The $K_5Ti(CN)_8$, $Rb_5Ti(CN)_8$ and $Cs_4Ti(CN)_7$ are believed to contain seven-coordinate titanium, while $K_3Ti(CN)_6$ is probably six-coordinate. A titanium(II) cyanide, $K_2Ti(CN)_4$, was obtained by reduction of $K_3Ti(CN)_6$ with one equivalent of potassium in liquid ammonia.[156] The extremely reactive titanium(0) cyanide complex, $K_4Ti(CN)_4$, is prepared by potassium reduction of a mixture of $TiBr_3$ and KCN in liquid ammonia. The $Ti(CN)_4^{4-}$ is a powerful reducing agent, converting bipy into bipy$^-$, leaving a brown residue of $KTi(NH)_2$. Nicholls and co-workers have also produced evidence for the formation of $Ti(CN)_3$.[157]

31.9 TITANIUM ALLOY HYDRIDES

Titanium absorbs dihydrogen reversibly above 300 °C to give, eventually, TiH_2, which is a grey, air-stable powder. Complete desorption occurs at *ca.* 1000 °C. A useful sourcebook is available.[158]

There are usually quite large differences in stability among the protides, deuterides and tritides of a given metal or alloy. Recently, Tanska *et al.*[159] decided to study a series of alloys formed between a single hydride-forming metal and a number of other metals which did not themselves form stable hydrides. Titanium was chosen as the constant constituent since it forms intermetallic compounds or solid solutions with almost every transition element, and many of these alloys are known to form hydrides. The particular interest was the periodic trend of the hydrogen isotope effect as a function of the transition metal alloyed with titanium. The titanium distribution coefficient was defined as in equation (28), and the results obtained are given in Table 9. To place into comparison alloys with stoichiometries other than 1:1, such as $TiMo_2$ and TiCrMn, the electron/atom (e/a) ratio may be plotted *vs.* α, (Figure 1). All the alloys which show significant inverse isotope ratios fall in the range e/a 4.5–5.5. This is not too surprising, in that it is known that the heat of solution of dihydrogen is dependent on the electronic structure of the metal.

$$\alpha = (T/H)\text{solid}/(T/H)\text{ gas} \tag{28}$$

Table 9 Tritium Isotope Effects

Material equilibrated	*Max H*	*Temperature (°C)*	*Activity balance (%)*	α *(cpm of solid/ cpm of gas)*	*Structure of alloy*
TiH_2	H_2	350	105.4	0.67	
$TiVH_{4.15}$	$H_{4.0}$	40	93	1.18	
$TiCrH_{2.35}$	$H_{2.92}$	40	88	1.54	X-Ray pattern similar to that of TiMo
$TiCr_2H_{1.68}$	$H_{2.66}$	−20	90	2.01	fcc $MgCu_2$ $a = 6.943$ Å at high temp., hexagonal $MgZn_2$ $a = 4.932$, $c = 7.961$ Å
$TiCr_2H_{1.48}$	$H_{2.66}$	0	97	2.03	
$TiCr_3H_{1.76}$	$H_{2.91}$	−20	98	1.68	$TiCr_2$ + Cr
$TiCr_3H_{1.36}$	$H_{2.91}$	0	100	1.59	
$TiCr_3H_{0.875}$	$H_{2.91}$	+20	95	1.69	
$TiCrMnH_{2.19}$	$H_{2.91}$	−20	99	2.05	$TiCr_2$ pattern
$TiCrMnH_{1.28}$	$H_{2.91}$	0	105	1.80	
$Ti_2MoH_{4.77}$	$H_{\sim 4.94}$	40	100.8	1.61	Solid solution bcc
$TiMoH_{2.99}$	$H_{2.99}$	40	100	1.61	Solid solution
$TiMo_2H_{1.10}$	$H_{1.31}$	−20	96.8	1.87	Solid solution bcc
$TiMnH_{1.99}$	$H_{2.21}$	40	93	1.37	
$FeTiH_{1.88}$	$H_{1.97}$	0	95	0.92	CsCl, $a = 2.976$ Å, Elliott 438
$FeTiH_{1.69}$	$H_{1.97}$	0	97	0.95	
$FeTiH_{1.21}$	$H_{1.97}$	40	105	1.0	
$Fe_{0.6}TiMn_{0.2}H_{1.67}$	$H_{1.75}$	40	101	1.0	
$TiCoH_{1.44}$	$H_{1.46}$	40	98	0.852	CsCl, $a = 2.987$ to 2.944 Å
$TiNiH_{1.44}$	$H_{1.50}$	40	99	0.74	CsCl, $a = 3.013$ to 3.015 Å

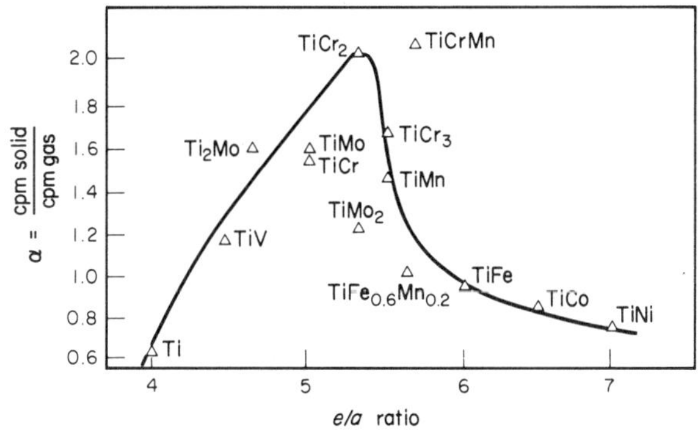

Figure 1 Tritium–protium separation factors for alloys of varying electron-to-atom ratio

31.10 COMPLEX HALIDES

For general properties of the halides and oxyhalides of titanium, reference to the work by Clark is recommended.[14] The chemistry of the complex halides will be treated here.

31.10.1 Complex Fluorides

The most common form of complex fluoride is that of the TiF_5^- moiety. One method of preparation is to dissolve TiF_4, and to add $[NPr^n_2H_2]F$ and a stoichiometric amount of the donor; a second method involves adding a stoichiometric amount of $[NPr^n{}_2H_2]_2[TiF_6]$ to a solution of TiF_4L_2.[160] In the ^{19}F NMR spectra of the solutions made by the first method which did not form $[TiF_5L]^-$, a triplet appears at *ca.* −20 ppm with a coupling constant of between 43 and 45 Hz, assigned to $[Ti_2F_{11}]^{3-}$ (**46**). The complexes $[TiF_5L]^-$ and $[Ti_2F_{11}]^{3-}$ are listed in Table 10.

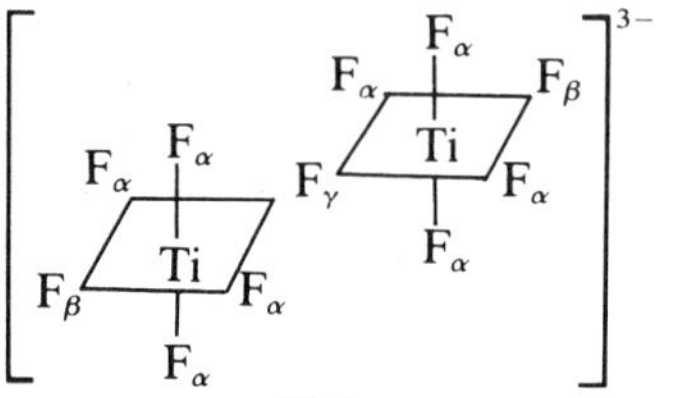

(**46**)

Table 10 Formation of $[TiF_5]^-\cdot D$ and $[Ti_2F_{11}]^{3-}$ Related to some of the Physical Properties of the Donor Molecules

Donor	*Solvent*	$[TiF_5]^-\cdot D$ *formed*	$[Ti_2F_{11}]^{3-}$ *formed*	*Dipole moment of donor*[a] (D)	*Dielectric constant of donor*[b]	$D_{II.I}$ (cm^{-1})
H_2O	EtOH	Yes	No	1.85	78.5	5490
MeOH	MeOH	Yes	No	1.70	32.6	4420
py	MeCN	Yes	No	2.19	12.3	4390
EtOH	EtOH	Yes	No	1.69	24.3	4260
SMe_2O	MeCN	Yes	No	3.96	47.6	4180
	$ClCH_2CN$					
$Me_2NC(O)H$	$ClCH_2CN$	Yes	No	3.82	37.6	3940
$(NH_2)_2CO$	$ClCH_2CN$	Yes	Yes			
4-NCpy	MeCN	Yes	Yes			
3-NCpy	MeCN	Yes	Yes			
2-MeOpy	MeCN	No	Yes			
2-Mepy	MeCN	No	Yes			
$MeO(CH_2)_2OMe$	$MeO(CH_2)_2OMe$	No	No			
Me_2CO	Me_2CO	No		2.88	20.7	
$(MeCO)_2O$	$ClCH_2CN$	No	Yes		20.7(20)	
$(Me_2N)_2CS$	$ClCH_2CN$	No	Yes			
MeCN	MeCN	No	Yes	3.92	37.5(20)	2600
$ClCH_2CN$	$ClCH_2CN$	No	Yes			
$MeNO_2$	$MeNO_2$	No	No	3.46	35.9(20)	2260
$CHCl_3$	$CHCl_3$	No	No	1.01	4.8(20)	1940

[a] In the gas phase; ID ≈ 3.33×10^{-30} C m (from 'Handbook of Chemistry and Physics', ed. R. C. Weast, Chemical Rubber Co., Cleveland, 1973).
[b] In the liquid phase at 25 °C, except where indicated in parentheses (from 'Handbook of Analytical Chemistry', ed. L. Meites, McGraw-Hill, New York, 1963).

A procedure for the fluorination of 1 mm thick wafers of rutile has been developed by Wold and co-workers.[161] The products are $TiO_{2-x}F_x$, and the value of x is an essentially linear function of the fluorination temperature over the range 575–700 °C. For example, $x = 0.002$ for the preparation at 700 °C. Samples with the smallest value of x exhibit the largest photocurrent on irradiation despite having the highest resistance. Under irradiation from a xenon arc source, rutile wafers fluorinated at 575 °C produced almost twice the photocurrent of rutile reduced at the optimum temperature (600 °C).

31.10.2 Complex Chlorides

The $TiCl_6^{3-}$ anion is believed to form when titanium(III) is fused in chloride melts, but it can be isolated from the direct reaction of $TiCl_3$ with an excess of anhydrous pyHCl or by the reaction of $TiCl_3(MeCN)_3$ with pyHCl in chloroform/acetonitrile. However, when Et_4NCl and Et_4NBr are used in the second method the expected hexahalogeno salts are not formed, but $[TiCl_4(MeCN)_2]$ and $(Et_4N)[TiCl_3Br(MeCN)_2]$ were produced. The reaction of $TiBr_3py_3$ with liquid anhydrous hydrogen chloride gives the mixed $(pyH)_3[TiCl_4Br_2]$.[162] Thermal decomposition of $[TiCl_4(MeCN)_2]^-$ gives $[TiCl_4]^-$ species which are six-coordinate polymers and not tetrahedral.[163] Vibrational spectra suggest that $(Et_4N)TiCl_5$ and $(Et_4N)Ti_2Cl_9$ (**47**) have trigonal bipyramidal and chlorine-bridged structures, respectively; ν(Ti—Cl) occurs at 324 cm^{-1} for $TiCl_6^{2-}$.[164] Kyber and Schramm[165] have very usefully collated titanium–chlorine stretching frequency data (Table 11). The $[Ti_2Cl_{10}]^{2-}$ (**48**) species are also known.[166] Adducts of the type $(PCl_5)_n \cdot mTiCl_4$ (n, m = integer) have been known since 1867. Treatment of a PCl_3–$TiCl_4$ solution with chlorine gas led to the complex $PCl_5 \cdot TiCl_4$. A more recent investigation has shown that the reaction of PCl_5 with $TiCl_4$ in $POCl_3$ gives $[PCl_4]_2[Ti_2Cl_{10}]$ (**48**) and in $SOCl_2$ gives $[PCl_4][Ti_2Cl_9]$ (**47**). These anions are unusual examples of face-shared bioctahedra containing transition metal ions in the formal oxidation state +4.[167] The titanium(III) analogue of (**47**) $Cs_3Ti_2Cl_9$ is also known.[168] Optical spectra show evidence for both titanium(III) and titanium(II) in $AlCl_3$–KCl melts ($1.00 \geqslant X_{AlCl_3} \geqslant 0.49$) at temperatures 471–894 K and in the LiCl–KCl eutectic melt at 658–1185 K. Titanium(III) is found to be octahedrally coordinated in pure $AlCl_3$(1) and in $AlCl_3$–KCl melts with $X_{AlCl_3} \geqslant 0.67$, whilst octahedral-tetrahedral equilibria are established at lower $AlCl_3$ contents. Titanium(II) has octahedral coordination at compositions $1.00 \geqslant X_{AlCl_3} \geqslant 0.51$, but a disproportionation equilibrium is observed for $X_{AlCl_3} \leqslant 0.60$, culminating at $X_{AlCl_3} = 0.49$, where titanium(II) is unstable and the spectrum of titanium(III) in an octahedral–tetrahedral coordinated equilibrium is found.[169]

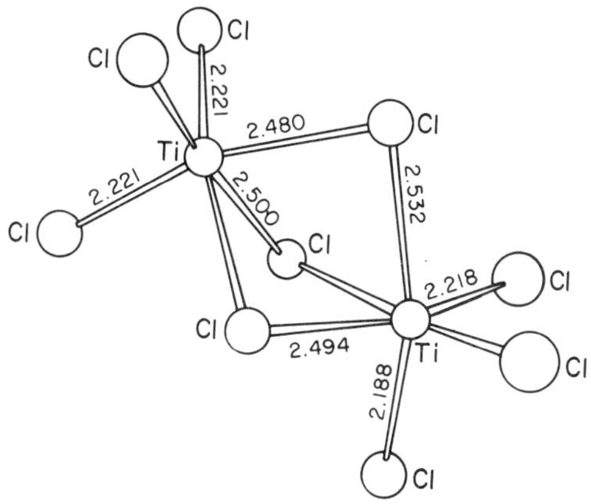

(**47**)

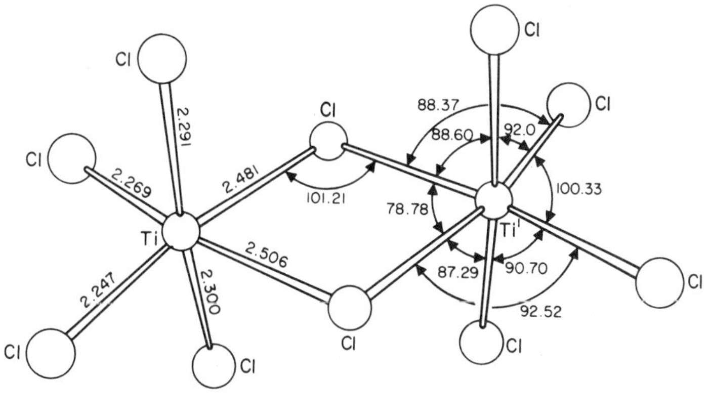

(**48**)

Table 11 Titanium–Chlorine Vibrations for some Selected Compounds—a criterion for Coordination Number

Compound	Phase	CN	r(MCl) (Å)	Ref.	ν(MCl) (cm^{-1})	Symmetry	Ref.
$TiCl_4$	Gas	4	2.18	a	506vs (t_2)	T_d	b
$TiCl_4 \cdot NMe_3$	Solid	5	—	c	456s (*e*), 376s (*e*), 344m (a_1)	C_3	c
$TiCl_3 \cdot 2NMe_3$	Solid	5	—	d	387s, b, 330m, 254m	D_{3h}?	e
$TiBr_3 \cdot 2NMe_3$	Solid	5	—	d	323s (a_2''), 238ms (*e*)	D_{3h}	e
$TiCl_3 \cdot (2SMe_2)$	Solid	5	—	d, f	390s, b, 232m, 280b	C_{4v}	d
$TiCl_3 \cdot 2SC_4H_4$	Solid	5	—	d, f	390s, 324m, 290m	C_{4v}	d
$[Et_2NH_2]TiCl_5$	Solid	5	—	g	385vs, 346vs, 212m, 170m (bending)	D_{3h}?	g
$[Et_2NH_2]_2TiCl_6$	Solid	6	*ca.* 2.34	i	325vs, b (t_{1u})	O_h	h
$(NH_4)_2TiCl_6$	Aq. soln.	6	*ca.* 2.34	i	330 (ν_3), 193 (ν_4)	O_h	j
$[Me_4N]_2TiCl_6$	Solid	6	*ca.* 2.34	i	324vs (ν_3), 183m (ν_4)	O_h	g
$[Et_4N]_2TiCl_6$	Solid	6	*ca.* 2.34	i	321vs (ν_3), 182m (ν_4)	O_h	g
$TiCl_4 \cdot 2THF^m$	Solid	6	*ca.* 2.34	k	371s, 298m	C_{2v}	k
$TiCl_4 \cdot 1,4\text{-}D^m$	Solid	6	*ca.* 2.34	k	389s, 340w	C_{2v}	k
$TiCl_4 \cdot 2A^m$	Solid	6	*ca.* 2.34	k	387s, b, 318w	C_{2v}	k
$TiCl_4 \cdot 2py^m$	Solid	6	*ca.* 2.34	k	368, 280	C_{2v}	k
$TiCl_4 \cdot bipy^m$	Solid	6	—	—	384s, 366sh, 307w	C_{2v}	k
$TiCl_4phen^m$	Solid	6	—	—	379s, 356sh, 303w	C_{2v}	k
$TiCl_3 \cdot 3THF$	Solid	6	—	—	360s, 331s, 300m	—	k
$TiCl_3 \cdot 3A$	Solid	6	—	—	365s, *ca.*334m, b	—	k
$\alpha\text{-}TiCl_3$	Solid	6	*ca.* 3.6	l	289 (bridge)	—	k
$TiCl_4 \cdot 2B^m$	Solid	8	2.44	k	317 (b_2), 325sh (*e*)	D_{2d}	k

[a] By electron diffraction: M. Lister and L. E. Sutton, *Trans. Faraday Soc.*, 1941, **37**, 393.
[b] M. F. A. Dove, J. A. Creighton and L. A. Woodward, *Spectrochim. Acta*, 1962, **18**, 267.
[c] The IR spectra of the solid is identical with that in C_6H_6 solution: I. R. Beattie and T. Gilson, *J. Chem. Soc.* 1965, 6596. Complex is monomeric in C_6H_6: G. W. A. Fowles and R. A. Hoodless, *J. Chem. Soc.*, 1963, 33.
[d] By X-ray diffraction, *i.e.* $TiCl_3 \cdot 2NMe_3$ is isomorphous (powder pattern) to $TiBr_3 \cdot 2NMe_3$ for which a *trans* trigonal bipyramidal structure was found *via* a single crystal X-ray study: B. J. Russ and J. S. Wood, *Chem. Commun.*, 1966, 745.
[e] G. W. A. Fowles, T. E. Lester and R. A. Walton, *J. Chem. Soc. (A)*, 1968, 198.
[f] Both complexes were considered to be five-coordinate because ν_{as}(MCl) shifts to lower energy when the complexes dissolve in excess base; the first charge-transfer band in the UV spectrum of both complexes shifts to lower energy upon dissolving in excess base. These shifts are consistent with the presence of five-coordinate monomers in the solid state which dissolve in excess base as six-coordinate complexes.
[g] J. A. Creighton and J. H. S. Green, *J. Chem. Soc. (A)*, 1968, 808.
[h] R. A. Walton and B. J. Brisdon, *Spectrochim. Acta, Part A.* 1967, **23**, 2222.
[i] By X-ray diffraction, *e.g.*, for Cs_2TiCl_6 (2.35 Å) and Rb_2TiCl_6 (2.33 Å): G. Engel, *Z. Kristallogr.*, 1935, **90**, 341.
[j] D. M. Adams and D. C. Newton, *J. Chem. Soc. (A)*, 1968, 2262.
[k] R. J. H. Clark, *Spectrochim. Acta*, 1965, **21**, 955.
[l] X-Ray diffraction: G. Natta, P. Corradini and G. Allegra, *J. Polym. Sci.*, 1961, **51**, 399.
[m] Abbreviations: THF, tetrahydrofuran; D, dioxane; A, acetonitrile; py, pyridine; bipy, 2,2′-bipyridyl; phen, *o*-phenanthroline; B, *o*-phenylenebis(dimethylarsine). For *cis* (C_{2v}) complexes four ν(MCl) are allowed in the IR spectra ($2A_1 + B_1 + B_2$) whereas for *trans* (D_{4h}) complexes only one ν(MCl) mode is allowed (E_u); these complexes are considered to have the *cis* (C_{2v}) structure.

Schram and co-workers have produced a range of unusual titanium chloride species. Treatment of $B_2[NMe_2]_4$ with a six-fold molar excess of $TiCl_4$ afforded $\{[Me_2N]_2BCl\}_2\{TiCl_4\}_3$, small amounts of Me_2NTiCl_3 and $Me_2NBCl \cdot Ti_2Cl_6$.[170] The last complex contains coordinated dimethylamino moieties and the pentacoordinate structure (**49**) is proposed.[165] Treatment of this compound with HCl produced $\{[Me_2NH]_2BCl_2\}^+\{Ti_2Cl_7\}^-$, and pyrolysis of this complex affords Me_2NBCl_2 and $\{Me_2NH_2\}\{Ti_2Cl_7\}$.[171] As well as IR criteria for coordination numbers (Table 11), Kyker and Schram have also proposed that charge-transfer spectra can also be used as an indication of coordination number (Table 12).

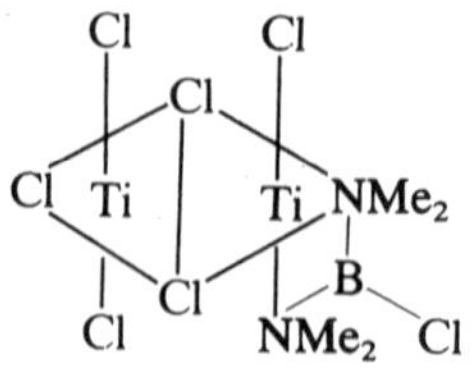

(**49**)

The reaction of $TiCl_4$ with $Pt(PPh_3)_n$ ($n = 3, 4$) was mentioned in Section 31.6.3.1, the product being $(TiCl_4)_2Pt(TiCl_4 \cdot PPh_3)_3$, from which $Pt(TiCl_4 \cdot PPh_3)_3$ can be obtained by heating *in vacuo* or addition of further PPh_3.[137] Schram has also isolated the $Ti_3Cl_{11}^-$ ion in the

Table 12 Charge-transfer Spectral Data for some Selected Titanium(III) and Titanium(IV) Complexes — a Possible Criterion for Coordination Numbers

Complex	*Phase*	*CN*	*Ref.*	*Frequency (kK)*[a]	*Ref.*
$TiCl_4$	Gas	4	1, 2	35.65, 43.15	1
$TiCl_4{\cdot}NMe_3$	Solid	5	3	30.22, 46.05	4
$TiCl_3{\cdot}2NMe_3$	Solid	5	5	30.0, 37.0, 44.9	6
K_3TiCl_6	Solid	6	7	23.4, 30.0, 34.5, 37.8sh, *ca.* 43sh, 48.1	7
α-$TiCl_3$	Solid	6	8	18.9, 27.5	9
β-$TiCl_3$	Solid	6	8	24.12, 29.20, 38.20	10
γ-$TiCl_3$	Solid	6	8	19.00, 27.50	11
$TiCl_4{\cdot}2B^b$	Solid	8	12	19.6	13

[a] $1\ kK = 1000\ cm^{-1}$.
[b] B = *o*-phenylenebis(dimethylarsine).

1. D. S. Alderice, *J. Mol. Spectrosc.*, 1965, **15**, 509.
2. By electron diffraction: M. Lister and L. E. Sutton, *Trans. Faraday Soc.*, 1965, 393.
3. Vibrational spectroscopy: I. R. Beattie and T. Gilson, *J. Chem. Soc.*, 1965, 6559.
4. Data obtained in this work.
5. By X-ray diffraction; isomorphous to $TiBr_3{\cdot}2N(CN_3)_3$ which has been shown to have a *trans* trigonal bipyramidal structure *via* a single-crystal X-ray study: B. J. Russ and J. S. Wood, *Chem. Commun.*, 1966, 745.
6. M. W. Duckworth, G. W. A. Fowles and P. T. Green, *J. Chem. Soc. (A)*, 1967, 1592.
7. *Via* IR spectroscopy; B. J. Brisdon, *Spectrochim. Acta. Part A*, 1967, **23**, 1969.
8. *Via* X-ray diffraction; G. Natta, P. Corradini and G. Allegra, *J. Polym. Sci.*, 1961, **51**, 399.
9. C. Dijkgraaf and J. P. G. Rousseau, *Spectrochim. Acta, Part A*, 1967, **23**, 1267.
10. R. J. H. Clark, *J. Chem. Soc.*, 1964, 417.
11. C. Dijkgraaf, *Nature (London)*, 1964, **201**, 1121.
12. R. J. H. Clark, *Spectrochim. Acta*, 1964, **21**, 955.
13. R. J. H. Clark, J. Lewis and R. S. Nyholm, *J. Chem. Soc.*, 1962, 2460.

$[(Ph_3P)_3PtCl]^+[Ti_5Cl_{11}]^-$ salt. Little is known about the structure of this mixed oxidation state cluster anion; addition of $TiCl_4$ affords $[(Ph_3P)_3PtCl]^+[Ti_5Cl_{19}]^-$.[172]

31.10.3 Complex Bromides

Concomitant with his work on the chloride system[165,170,171] Schram has produced $\{[Me_2N]_2BBr\}_2\{TiBr_4\}_3$, proposing the following type of coordination:[173]

$$Br_4Ti \longleftarrow N(Me)_2B(Br)(Me)_2N \longrightarrow TiBr_4 \longleftarrow N(Me)_2B(Br)(Me)_2N \longrightarrow TiBr_4$$

Treatment of $TiBr_4$ with $B_2[NMe_2]_4$ produces $B_2Br_2[NMe_2]_2{\cdot}TiBr_3$, for which the dinuclear structure (**50**) is proposed.[174] Complexes $M_2[TiBr_5(H_2O)]$ (M = Cs, Rb, pyH) are known.[175]

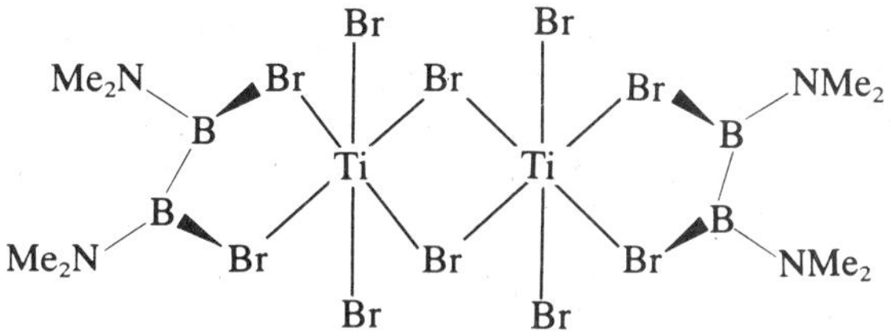

(**50**)

31.10.4 Complex Iodides

It would be predicted that TiI_6^{2-}, analogous to TiX_6^{2-} (X = Cl, Br), would not be a very stable complex, and because of the electronic spectra of the latter species, it would also be predicted that TiI_6^{2-} would be black. When $[Et_4N]_2TiCl_6$ is treated with liquid HI a very black material, resembling carbon, is produced. This material is very unstable in air and changes in colour to the dark red-brown of I_3^- in the time required to pour it quickly from one container to another. The complex is unstable in almost all solvents; it was only possible to obtain the reflectance spectrum of $[Et_4N]_2TiI_6$ (Table 13).

Table 13 Electron-transfer Spectra of TiX_6^{2-} [a]

$TiCl_6^{2-}$	(25.0) [1300], (29.6) [~17 000], 31.8 [22 000], 43.8 [30 000]
$TiBr_6^{2-}$	(17.7) [~700], (21.7) [~8 000], 24.9 [16 000], (27) [~12 000], 36.1 [20 000]
$[Et_4N]_2TiI_6$	(9.5) [0.3], 12.1 [0.65], 14.3 [0.65], (18.7) [0.8], 23.2 [1.0]

[a] Wavenumbers in kilokaisers with shoulders in parentheses. For $TiCl_6^{2-}$ and $TiBr_6^{2-}$ values in brackets are molar extinction coefficients and for $[Et_4N]_2TiI_6$ values are intensities relative to the strongest listed transition taken equal to 1.0.

31.11 ZIEGLER–NATTA CATALYSIS

Ziegler and Natta were jointly awarded the Nobel Prize for Chemistry in 1963. It was Ziegler[177] who first showed that ethylene can be induced to polymerize under ambient conditions in the presence of a mixture of $TiCl_4$ and $AlEt_3$, and Natta[178] demonstrated that by suitably modifying these catalysts stereoregular polymers of alkenes can be produced. Prior to these discoveries, the polymerization of ethylene involved high temperatures and pressures. The most likely mechanism is shown in Scheme 3,[179] involving (i) replacement of a coordinated chloride by an ethyl group derived from $AlEt_3$; (ii) a vacant coordination site on the surface and an ethylene molecule becoming bound; (iii) '*cis* insertion' as the ethylene migrates to the ethyl group; and (iv) another ethylene becoming bound at the vacant coordination site.

Processes (ii) and (iii) are repeated.

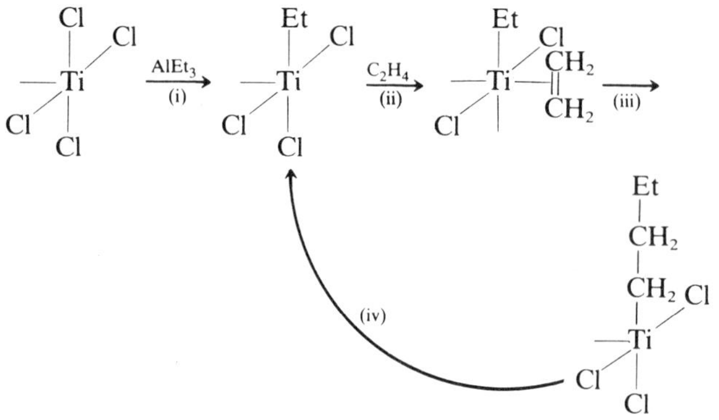

Scheme 3

Quite apart from the economic advantage of being able to polymerize alkenes under ambient conditions, another advantage is that with propylene, for example, the polymerization is regular and not atactic.

31.12 MISCELLANEOUS

This chapter has been devoted to the coordination chemistry of titanium and has made no attempt to describe the more basic chemistry of this element. References to alloys, to the simple halides and oxyhalides, the oxides, sulfides, selenides, tellurides, nitrides, azides, phosphides, arsenides and antimonides are well reviewed by Clark,[14] and the recent text by Greenwood and Earnshaw[180] contains a good section on titanium.

31.13 REFERENCES

1. M. A. Hunter, *J. Am. Chem. Soc.*, 1910, **32**, 330.
2. A. E. Van Arkel and J. H. de Boer, *Z. Anorg. Allg. Chem.*, 1925, **148**, 345.
3. W. Kroll, *Trans. Electrochem. Soc.*, 1940, **78**, 35; *J. Less-Common Met.*, 1965, **8**, 361, and refs. therein.
4. P. Pascal, 'Nouveau Traité de Chimie Minerale', Tome IX, Masson, Paris, 1963, p. 7.
5. W. Freundlich and M. Bichara, *C. R. Hebd. Seances Acad. Sci.*, 1954, **238**, 1324.
6. A. Münster and W. Ruppert, *Z. Elektrochem.*, 1953, **57**, 558.
7. E. G. M. Tornqvist and W. F. Libby, *Inorg. Chem.*, 1979, **18**, 1792.

8. I. E. Campbell, R. I. Jaffee, J. M. Bolcher, J. Gurlard and B. W. Gonser, *Trans. Electrochem. Soc.*, 1948, **93,** 271.
9. H. T. Clark, *J. Met.*, 1949, **1,** 588.
10. D. S. Eppelsheimer and R. R. Penman, *Nature (London)*, 1950, **166,** 960.
11. J. Stringer, *J. Less-Common Met.*, 1964, **6,** 207.
12. G. P. Baxter and A. Q. Butler, *J. Am. Chem. Soc.*, 1926, **48,** 3117.
13. E. M. Kuznetsova, N. V. Zakurim and O. T. Nikiti, *Russ. J. Inorg. Chem. (Engl. Transl.)*, 1962, **7,** 343.
14. R. J. H. Clark, in 'Comprehensive Inorganic Chemistry', ed. J. C. Bailar, Jr., H. J. Emeleus, R. S. Nyholm and A. F. Trotman-Dickenson, Pergamon, Oxford, 1973, p. 355.
15. W. Levason and C. A. McAuliffe, *Adv. Inorg. Chem. Radiochem.*, 1972, **14,** 173.
16. C. A. McAuliffe (ed.), 'Transition Metal Complexes of Phosphorus, Arsenic and Antimony Ligands', Macmillan, London, 1973.
17. C. A. McAuliffe and W. Levason, 'Coordination Complexes Containing Phosphorus, Arsenic and Antimony Ligands', Elsevier, Amsterdam, 1979.
18. R. C. Fay, *Coord. Chem. Rev.*, 1981, **37,** 9.
19. R. C. Fay, *Coord. Chem. Rev.*, 1982, **45,** 9.
20. J. R. Ufford and N. Serpone, *Coord. Chem. Rev.*, 1984, **57,** 301.
21. N. Serpone, M. A. Jamieson, F. iSalvio, P. A. Takats, L. Yeretsian and J. R. Ufford, *Coord. Chem. Rev.*, 1984, **58,** 87.
22. R. J. H. Clark, 'The Chemistry of Titanium and Vanadium', Elsevier, Amsterdam, 1968.
23. G. W. A. Fowles and T. E. Lester, *Chem. Commun.*, 1967, 47.
24. G. W. A. Fowles, T. E. Lester and R. A. Walton, *J. Chem. Soc. (A)*, 1968, 198.
25. M. F. Lappert and A. R. Sanger, *J. Chem. Soc. (A)*, 1971, 874.
26. G. W. A. Fowles, T. E. Lester and B. J. Russ, *J. Chem. Soc. (A)*, 1968, 805.
27. D. J. Machin, K. S. Murray and R. A. Walton, *J. Chem. Soc. (A)*, 1968, 195.
28. P. T. Greene, B. J. Russ and J. S. Wood, *J. Chem. Soc. (A)*, 1971, 3636.
29. P. C. Crouch, G. W. A. Fowles and R. A. Walton, *J. Chem. Soc. (A)*, 1968, 2172.
30. S. G. Carr and T. D. Smith, *J. Chem. Soc., Dalton Trans.*, 1972, 1887.
31. D. C. Bradley and R. G. Copperthwaite, *Chem. Commun.*, 1971, 764.
32. M. F. Lappert, J. B. Pedley, G. J. Sharp and D. C. Bradley, *J. Chem. Soc., Dalton Trans.*, 1976, 1737.
33. J. Hughes and G. R. Willey, *Inorg. Chim. Acta*, 1975, **13,** L1.
34. G. D. McDonald, M. Thompson and E. M. Larsen, *Inorg. Chem.*, 1968, **7,** 648.
35. W. Giggenbach and C. H. Brubaker, Jr., *Inorg. Chem*, 1968, **7,** 129.
36. G. W. A. Fowles, D. F. Lewis and R. A. Walton, *J. Chem. Soc. (A)*,. 1968, 1468.
37. D. A. Baldwin and R. J. H. Clark, *J. Chem. Soc. (A)*, 1971, 1725.
38. F. King and D. Nicholls, *Inorg. Chim. Acta*, 1978, **28,** 55.
39. R. Choukroun and D. Gervais, *J. Chem. Soc., Dalton Trans.*, 1980, 1800.
40. W.-M. Dyck, K. Dehnicke, F. Weller and U. Muller, *Z. Anorg. Allg. Chem.*, 1980, **470,** 89.
41. T. C. Franklin and R. C. Byrd, *Inorg. Chem.*, 1970, **9,** 986.
42. R. Robson, *Inorg. Chem.*, 1974, **13,** 475.
43. N. W. Alcock, M. Pierce-Butler and G. R. Willey, *J. Chem. Soc., Dalton Trans.*, 1976, 707.
44. C. Airoldi, D. C. Bradley, H. Chudzynska, M. B. Hursthouse, K. M. Abdul Malik and P. R. Raithby, *J. Chem. Soc., Dalton Trans.*, 1980, 2010.
45. J. A. Chandler and R. S. Drago, *Inorg. Chem.*, 1962, **1,** 356; J. A. Chandler, J. E. Wuller and R. S. Drago, *Inorg. Chem.*, 1962, **1,** 65.
46. M. Kubota and S. R. Schulze, *Inorg. Chem.*, 1964, **3,** 853.
47. S. R. Wade and G. R. Willey, *J. Chem. Soc., Dalton Trans.*, 1981, 1264.
48. J.-M. Latour, J.-C. Marchon and M. Nakajima, *J. Am. Chem. Soc.*, 1979, **101,** 3974.
49. D. J. Eve and G. W. A. Fowles, *J. Chem. Soc. (A)*, 1966, 1183.
50. P. H. Davis and J. S. Wood, *Inorg. Chem.*, 1970, **9,** 1111.
51. M. G. B. Drew, G. W. A. Fowles and D. F. Lewis, *Chem. Commun.*, 1969, 876.
52. M. G. B. Drew and D. J. Eve, *Acta Crystallogr., Sect. B*, 1977, **33,** 2919.
53. M. G. B. Drew and D. J. Eve, *Inorg. Chim. Acta*, 1977, **25,** L111.
54. J. T. Wrobleski and D. B. Brown, *Inorg. Chim. Acta*, 1980, **38,** 227.
55. F. L. Bowden and D. Ferguson, *Inorg. Chim. Acta*, 1978, **26,** 251.
56. L. E. Manzer, *Inorg. Chem.*, 1978, **17,** 1552.
57. M. Zikmund, M. Kohutova, M. Handlovic and D. Miklos, *Chem. Zvesti*, 1979, **33,** 180 (*Chem. Abstr.*, 1979, **91,** 150 242).
58. C. M. Mikulski, T. Tran, L. L. Pytlewski and N. N. Karayannis, *J. Inorg. Nucl. Chem.*, 1979, **41,** 1671.
59. M. G. B. Drew and J. A. Hutton, *J. Chem. Soc., Dalton Trans.*, 1978, 1176.
60. D. C. Bradley, *Adv. Inorg. Chem. Radiochem.*, 1972, **15,** 259.
61. G. Winter and G. A. Kakos, *Aust. J. Chem.*, 1968, **21,** 793.
62. R. Masthoff, H. Köhler, H. Böhland and F. Schmeil, *Z. Chem.*, 1965, **5,** 122.
63. K. Watenpaugh and C. N. Caughlan, *Inorg. Chem.*, 1966, **5,** 1782.
64. W. Haase and H. Hoppe, *Acta Crystallogr., Sect. B*, 1968, **24,** 281.
65. C. E. Holloway, *J. Chem. Soc., Dalton Trans.*, 1976, 1050.
66. S. R. Wade, M. G. H. Wallbridge and G. R. Willey, *J. Chem. Soc., Dalton Trans.*, 1982, 271.
67. G. W. Svetich and A. A. Voge, *Chem. Commun.*, 1971, 676.
68. J. H. Haslam, *US Pat.* 2 708 205 (1955) (*Chem. Abstr.*, 1956, **54,** 4211g).
69. P. N. Kapoor, S. K. Mehrotra, R. C. Mehrotra, R. B. King and K. C. Nainan, *Inorg. Chim. Acta*, 1975, **12,** 273.
70. D. C. Bradley, *Coord. Chem. Rev.*, 1967, **2,** 299.
71. R. E. Noftle and G. H. Cady, *Inorg. Chem.*, 1966, **5,** 2182.
72. M. Basso-Bert and D. Gervais, *Inorg. Chim. Acta.*, 1979, **34,** 191.

73. D. P. Bauer and R. S. Macomber, *Inorg. Chem.*, 1976, **15**, 1985.
74. A. Flamini, D. J. Cole-Hamilton and G. Wilkinson, *J. Chem. Soc., Dalton Trans.*, 1978, 454.
75. F. Brisse and M. Haddad, *Inorg. Chim. Acta*, 1977, **24**, 173.
76. P. A. W. Dean, D. F. Evans and R. F. Phillips, *J. Chem. Soc. (A)*, 1969, 363.
77. R. S. Borden, P. A. Loeffler and D. S. Dyer, *Inorg. Chem.*, 1972, **11**, 2481.
78. C. D. Garner and S. C. Wallwork, *J. Chem. Soc. (A)*, 1966, 1496.
79. G. H. Dahl and B. P. Block, *Inorg. Chem.*, 1966, **5**, 1394.
80. R. J. H. Clark and M. A. Coles, *J. Chem. Soc., Dalton Trans.*, 1974, 1462.
81. G. W. A. Fowles, T. E. Lester and R. A. Walton, *J. Chem. Soc. (A)*, 1968, 198.
82. R. A. Walton, *Inorg. Chem.*, 1966, **5**, 643.
83. R. J. H. Clark and W. Errington, *Inorg. Chem.*, 1966, **5**, 650.
84. D. C. Bradley and P. A. Hammersley, *J. Chem. Soc. (A)*, 1967, 1894.
85. R. G. Cavell and A. R. Sanger, *Inorg. Chem.*, 1972, **11**, 2016.
86. R. P. Burns and C. A. McAuliffe, *Adv. Inorg. Chem. Radiochem.*, 1979, **22**, 303.
87. R. P. Burns, F. P. McCullough and C. A. McAuliffe, *Adv. Inorg. Chem. Radiochem.*, 1980, **23**, 211.
88. D. C. Bradley, M. Colapietro, M. B. Hursthouse, I. F. Rendall and A. Vaciago, *Chem. Commun.*, 1970, 743.
89. A. N. Bhat, R. C. Fay, D. F. Lewis, A. F. Lindmark and S. H. Strauss, *Inorg. Chem.*, 1974, **13**, 886.
90. S. L. Hawthorne, A. H. Bruder and R. C. Fay, *Inorg. Chem.*, 1983, **22**, 3368.
91. W. L. Steffen and R. C. Fay, *Inorg. Chem.*, 1978, **17**, 2120.
92. H. B. Gray, *Transition Met. Chem*, 1965, **1**, 239.
93. J. A. McCleverty, *Prog. Inorg. Chem.*, 1968, **10**, 49.
94. G. N. Schrauzer, *Transition Met. Chem.*, 1968, **4**, 299.
95. G. N. Schrauzer, *Acc. Chem. Res.*, 1969, **2**, 72.
96. R. Eisenberg, *Prog. Inorg. Chem.*, 1970, **12**, 295.
97. E. Hoyer, W. Dietzsch and W. Schroth, *Z. Chem.*, 1971, **11**, 41.
98. G. N. Schrauzer, *Adv. Chem. Ser.*, 1972, **110**, 73.
99. J. L. Martin and J. Takats, *Inorg. Chem.*, 1975, **14**, 73.
100. E. I. Stiefel, L. E. Bennett, Z. Dori, T. H. Crawford, C. Simo and H. B. Gray, *Inorg. Chem.*, 1970, **9**, 281.
101. D. C. Bradley, M. B. Hursthouse and I. F. Rendall, *Chem. Commun.*, 1969, 672.
102. D. C. Bradley and C. E. Holloway, *J. Chem. Soc. (A)*, 1969, 282.
103. N. S. Biradar, V. B. Mahale and V. H. Kulkarni, *Inorg. Chim. Acta*, 1973, **7**, 267.
104. G. Fachinetti, C. Floriani, M. Mellini and S. Merlino, *J. Chem. Soc., Chem. Commun.*, 1976, 300; G. Dell'Amico, F. Marchetti and C. Floriani, *J. Chem. Soc., Dalton Trans.*, 1982, 2197.
105. K. Dey, D. Koner, A. K. Biswas and S. Ray, *J. Chem. Soc., Dalton Trans.*, 1982, 911.
106. F. L. Bowden and D. Ferguson, *J. Chem. Soc., Dalton Trans.*, 1974, 460.
107. M. Pasquali, F. Marchetti, A. Landi and C. Floriani, *J. Chem. Soc., Dalton Trans.*, 1978, 545.
108. M. Orhanovic and R. G. Wilkins, *J. Am. Chem. Soc.*, 1967, **89**, 278.
109. D. Schwarzenbach and K. Girgis, *Helv. Chim. Acta*, 1975, **58**, 2391.
110. K. Wieghardt, U. Quilitzsch, J. Weiss and B. Nuber, *Inorg. Chem.*, 1980, **19**, 2514.
111. D. Schwarzenbach, *Inorg. Chem.*, 1970, **9**, 2391.
112. J. A. Connor and E. A. V. Ebsworth, *Adv. Inorg. Chem. Radiochem.*, 1964, **6**, 279.
113. K. Watenpaugh and C. N. Caughlan, *Inorg. Chem.*, 1967, **6**, 963.
114. W. R. Busing and H. A. Levy, *J. Chem. Phys.*, 1965, **42**, 3054.
115. F. J. Kristine, R. E. Shepherd and S. Siddiqui, *Inorg. Chem.*, 1981, **20**, 2571.
116. R. J. H. Clark and A. J. McAlees, *J. Chem. Soc. (A)*, 1972, 640; 1970, 2026.
117. A. J. McAlees, R. McCrindle and A. R. Woon-Fat, *Inorg. Chem.*, 1976, **15**, 1065.
118. R. J. H. Clark, M. L. Greenfield and R. S. Nyholm, *J. Chem. Soc. (A)*, 1966, 1254.
119. G. R. Hoff and C. H. Brubaker, Jr., *Inorg. Chem.*, 1971, **10**, 2063.
120. R. Choukroun and D. Gervais, *Inorg. Chim. Acta*, 1978, **26**, 17.
121. H. Mimoun, M. Postel, F. Casabianca, J. Fischer and A. Mitschler, *Inorg. Chem.*, 1982, **21**, 1303.
122. R. Guilard, M. Fontesse, P. Fournari, C. Lecomte and J. Protas, *J. Chem. Soc., Chem. Commun.*, 1976, 161.
123. M. J. Frazer and Z. Goffer, *J. Chem. Soc. (A)*, 1966, 544.
124. B. F. Studd and A. G. Swallow, *J. Chem. Soc. (A)*, 1968, 1961; A. G. Swallow and B. F. Studd, *Chem. Commun.*, 1967, 1197.
125. D. Cunningham, I. Douek, M. J. Frazer, W. E. Newton and B. Rimmer, *J. Chem. Soc. (A)*, 1969, 2133.
126. M. J. Frazer and B. Rimmer, *J. Chem. Soc. (A)*, 1968, 69.
127. J. Charalambous and M. J. Frazer, *J. Chem. Soc. (A)*, 1968, 2361.
128. K. Issleib and E. Wenschuh, *Chem. Ber.*, 1964, **97**, 715.
129. L. B. Kool, M. D. Rausch, H. G. Alt, M. Herberhold, U. Thewalt and B. Wolf, *Angew. Chem., Int. Ed. Engl.*, 1985, **24**, 394.
130. G. S. Girolami, G. Wilkinson, M. Thornton-Pett and M. B. Hursthouse, *J. Chem. Soc., Dalton Trans.*, 1984, 2347.
131. S. V. Muchnik, K. A. Lynchak, V. B. Chernogorenko and A. N. Rafal, *Inorg. Mater. (Engl. Transl.)*, 1979, **15**, 472.
132. J. Chatt and R. G. Hayter, *J. Chem. Soc.*, 1963, 1343.
133. C. D. Schmulbach, C. H. Kolich and C. C. Hinckley, *Inorg. Chem.*, 1972, **11**, 2841.
134. H. Rose, *Poggendorfs Ann. Phys.*, 1832, **24**, 141, 259.
135. R. Höltje, *Z. Anorg. Allg. Chem.*, 1930, **190**, 241.
136. C. M. F. Rae and K. R. Seddon, *Inorg. Chim. Acta.*, 1976, **17**, L35.
137. J. F. Plummer and E. P. Schram, *Inorg. Chem.*, 1975, **14**, 1505.
138. A. K. Anagnostopolous, *Chem. Chron.*, 1966, **31**, 141 (*Chem. Abstr.*, 1967, **66**, 82 009).
139. A. D. Westland and L. Westland, *Can. J. Chem.*, 1965, **43**, 426.
140. R. J. H. Clark, J. Lewis, R. S. Nyholm, P. Pauling and G. B. Robertson, *Nature (London)*, 1961, **192**, 222.

141. G. J. Sutton, *Aust. J. Chem.*, 1959, **12,** 122.
142. R. J. H. Clark, M. L. Greenfield and R. S. Nyholm, *J. Chem. Soc. (A)*, 1966, 1254.
143. G. A. Barclay, I. K. Gregor and S. B. Wild, *Chem. Ind.*, 1964, 1710.
144. L. S. Jenkins and G. R. Willey, *J. Chem. Soc., Dalton Trans.*, 1979, 777.
145. A. Treibs, *Angew. Chem.*, 1936, **49,** 682.
146. M. Tsutsui, R. A. Velapolidi, K. Suzuki and T. Koyano, *Angew. Chem.*, 1968, **80,** 914.
147. R. Guilard and C. Lecomte, *Coord. Chem. Rev.*, 1985, **65,** 87.
148. J. M. Latour, J. C. Marchon and M. Nakajima, *J. Am. Chem. Soc.*, 1979, **101,** 3974.
149. C. J. Boreham, G. Buisson, E. Duee, J. Jordanov, J. M. Latour and J. C. Marchon, *Inorg. Chim. Acta*, 1983, **70,** 77.
150. R. Guilard, M. Fontesse, P. Fournari, C. Lecomte and J. Protas, *J. Chem. Soc., Chem. Commun.*, 1976, 161.
151. C. Lecomte, J. Protas, J. C. Marchon and M. Nakajima, *Acta Crystallogr., Sect. B*, 1978, **34,** 2858.
152. R. Taube, *Z. Chem.*, 1963, **3,** 194.
153. B. P. Block and E. G. Meloni, *Inorg. Chem.*, 1965, **4,** 111.
154. E. A. Heintz, *Nature (London)*, 1963, **197,** 690.
155. D. Nicholls, T. A. Ryan and K. R. Seddon, *J. Chem. Soc., Chem. Commun.*, 1974, 635.
156. D. Nicholls and T. A. Ryan, *Inorg. Chim. Acta*, 1980, **41,** 233.
157. E. S. Dodsworth, J. P. Eaton, D. Nicholls and T. A. Ryan, *Inorg. Chim. Acta*, 1979, **35,** L355.
158. V. A. Livanov, A. A. Bukhanova and B. A. Kolachev, 'Hydrogen in Titanium', Israel Programme for Scientific Translations, Jerusalem, 1965.
159. J. Tanaka, R. H. Wiswall and J. J. Reilly, *Inorg. Chem.*, 1978, **17,** 498.
160. H. G. Lee, D. S. Dyer and R. O. Ragsdale, *J. Chem. Soc., Dalton Trans.*, 1976, 1325.
161. S. N. Subbarao, Y. H. Yun, R. Kershaw, K. Dwight and A. Wold, *Inorg. Chem.*, 1979, **18,** 488.
162. B. J. Russ and G. W. A. Fowles, *Chem. Commun.*, 1966, 19.
163. G. W. A. Fowles and B. J. Russ, *J. Chem. Soc. (A)*, 1967, 517.
164. J. A. Creighton and J. H. S. Green, *J. Chem. Soc. (A)*, 1968, 808.
165. G. S. Kyker and E. P. Schram, *Inorg. Chem.*, 1969, **8,** 2313.
166. C. S. Creaser and J. A. Creighton, *J. Chem. Soc., Dalton Trans.*, 1975, 1402.
167. T. J. Kistenmacher and G. D. Stucky, *Inorg. Chem.*, 1971, **10,** 122.
168. B. Briat, O. Kahn, I. Morgenstern-Badaran and J. C. Rivoal, *Inorg. Chem.*, 1981, **20,** 4193.
169. M. Sorlie and H. A. Oye, *Inorg. Chem.*, 1981, **20,** 1384.
170. G. S. Kyker and E. P. Schram, *Inorg. Chem.*, 1969, **8,** 2306.
171. G. S. Kyker and E. P. Schram, *Inorg. Chem.*, 1969, **8,** 2319.
172. S. Wongnawa and E. P. Schram, *Inorg. Chim. Acta*, 1979, **36,** 45.
173. M. R. Suliman and E. P. Schram, *Inorg. Chem.*, 1973, **12,** 920.
174. M. R. Suliman and E. P. Schram, *Inorg. Chem.*, 1973, **12,** 923.
175. S. E. Adnitt, D. W. Barr, D. Nicholls and K. R. Seddon, *J. Chem. Soc., Dalton Trans.*, 1974, 644.
176. J. L. Ryan, *Inorg. Chem.*, 1969, **8,** 2058.
177. K. Ziegler, E. Holzkamp, H. Breiland and H. Martin, *Angew. Chem.*, 1955, 67, 541.
178. G. Natta, *J. Polym. Sci.*, 1955, **16,** 143.
179. R. Feld and P. L. Cowe, 'The Organic Chemistry of Titanium', Butterworths, London, 1965, pp. 149–167.
180. N. N. Greenwood and A. Earnshaw, 'Chemistry of the Elements', Pergamon, Oxford, 1984.

32

Zirconium and Hafnium

ROBERT C. FAY
Cornell University, Ithaca, NY, USA

32.1 INTRODUCTION

The coordination chemistry of zirconium and hafnium is dominated by oxidation state IV and by higher coordination numbers, especially coordination number eight. Oxidation states and coordination geometries are summarized in Table 1.

Only a small number of zirconium(III) and hafnium(III) complexes are known. Nearly all of these are metal trihalide adducts with simple Lewis bases, and few are well characterized. Just one zirconium(III) complex has been characterized structurally by X-ray diffraction, the chlorine-bridged dimer $[\{ZrCl_3(PBu_3)_2\}_2]$. Although a number of reduced halides and organometallic compounds are known in which zirconium or hafnium exhibits an oxidation state less than III, coordination compounds of these metals in the II, I or 0 oxidation states are unknown, except for a few rather poorly characterized Zr^0 and Hf^0 compounds, *viz.* $[M(bipy)_3]$, $[M(phen)_3]$ and $M'Zr(CN)_5$ (M = Zr or Hf; M′ = K or Rb).

The preference of zirconium(IV) and hafnium(IV) for higher coordination numbers follows from the relatively large size and high charge of the +4 ions. Because these ions have electronic configuration d^0, there are no stereochemical preferences arising from a partly filled *d* shell. Consequently, Zr^{IV} and Hf^{IV} exhibit a great variety of coordination geometries. The three idealized coordination polyhedra for the most common coordination number, coordination number eight, are shown in Figure 1; polyhedral edges and vertices are labeled according to the nomenclature of Hoard and Silverton[1] and Porai-Koshits and Aslanov.[2] The eight vertices of the D_{4d} square antiprism are equivalent, while the D_{2d} dodecahedron has two sets of inequivalent vertices (A and B) and the C_{2v} bicapped trigonal prism has three sets of vertices (M, N and the capping sites, C). In Table 1, and throughout this chapter, the stereochemistry of tetrakis chelate complexes is indicated by specifying the polyhedral edges that are spanned by the bidentate ligands. The most common stereoisomer for tetrakis chelates is the dodecahedral *mmmm* isomer. The square antiprismatic *ssss* isomer is less common, and the bicapped trigonal prismatic $t_1t_1p_2p_2$ stereoisomer and dodecahedral D_2-*gggg*, S_4-*gggg* and *abmg* stereoisomers are rare. Of the three eight-coordination polyhedra, the dodecahedron is preferred.

A survey of the ligands which coordinate to zirconium and hafnium indicates that complexes with metal–oxygen bonds are most abundant. Complexes with metal–halogen and metal–nitrogen bonds follow in decreasing order of abundance. The chemistries of zirconium and hafnium are closely similar, more similar than for any other pair of elements. Because of the lanthanide contraction, zirconium–ligand and hafnium–ligand bond lengths are nearly identical. Based on a limited amount of thermochemical data, hafnium–ligand bonds appear to be slightly stronger than zirconium–ligand bonds.

Zirconium and hafnium complexes are highly fluxional. NMR studies indicate stereochemical nonrigidity, facile exchange of terminal and bridging ligands, and rapid intermolecular ligand exchange.

This chapter reviews the literature up to approximately January 1984. Previous comprehensive reviews of the chemistry of zirconium and hafnium are available,[3–6] and annual surveys of the inorganic and coordination chemistry of these elements have been published since 1981.[7–9] The present treatment emphasizes the chemistry of discrete, isolable complexes, although some attention has been given to the extremely complex aqueous solution chemistry of Zr^{IV} and Hf^{IV}. In general, zirconium forms slightly more stable complexes in solution than does hafnium; stability constants may be found in standard compilations.[10–13] In cases where the structural chemistry of discrete complexes is closely related to that of chain, layer or extended three-dimensional structures, the discussion of the discrete complexes has been set in the broader context; the section on fluorometallate complexes is a case in point. Organometallic compounds of zirconium and hafnium[14] have been excluded almost entirely.

32.2 ZIRCONIUM(0) AND HAFNIUM(0)

32.2.1 Group IV Ligands

Air- and moisture-sensitive complex cyanides of composition $M'_5Zr(CN)_5$ (M′ = K or Rb) have been prepared by reaction at −50 °C in anhydrous ammonia of zirconium(III) bromide with an excess of either KCN or RbCN. In the case of the reaction involving KCN, the 1:3 distribution of zirconium between the gray, ammonia-insoluble $K_5Zr(CN)_5$ and the soluble zirconium(IV) products is in accord with the disproportionation shown in equation (1). The

Table 1 Oxidation States and Stereochemistry of Zirconium and Hafnium

Oxidation state	*Coordination number*	*Coordination polyhedron*	*Ligand wrapping pattern*	*Examples (M = Zr or Hf)*
Zr^{o}, Hf^{0}	6	Octahedron (?)		$[[M(bipy)_3]$, $[M(phen)_3]$ (?)
Zr^{III}, Hf^{III}	5	?		$[ZrCl_3\{(CF_2)_2C_2(AsMe_2)_2\}]$
	6	Octahedron		$[\{ZrCl_3(PBu_3)_2\}_2]$
Zr^{IV}, Hf^{IV}	4	Tetrahedron		$MCl_4(g)$, $[Zr(NEt_2)_4]$, $[MCl\{N(SiMe_3)_2\}_3]$, $[HfCl\{2,6\text{-}OC_6H_3(CMe_3)_2\}_3]$, $[\{Zr(NCMe_3)(NMe_2)_2\}_2]$
	5	Trigonal bipyramid		$[ZrCl_3\{OC(CMe_3)_3\}\cdot Li(OEt_2)_2]$, $[ZrCl_5]^-$ (?)
		?		$[ZrCl_4(POCl_3)]$, $[Zr(acac)(OCMe_3)_3]$
	6	Octahedron		$Li_2[ZrF_6]$, *cis*-$[MCl_4(MeCN)_2]$, $[MCl_6]^{2-}$ *cis*-$[M(acac)_2Cl_2]$, $[Zr(\eta^1\text{-}NC_4H_4)_6]^{2-}$, $[M(MoO_4)_6]^{8-}$, $[\{ZrCl_4(SOCl_2)\}_2]$, $[\{ZrCl_4(N_3)\}_2]^{2-}$
	6	Octahedron/trigonal prism		$[Zr(S_2C_6H_4)_3]^{2-}$
	7	Pentagonal bipyramid		$[Zr(acac)_3Cl]$, $[NH_4]_3[ZrF_7]$. $K_2[Cu(H_2O)_6][Zr_2F_{12}]$, $[NMe_4]_2[Zr_2F_{10}(H_2O)_2]$
		Monocapped trigonal prism		$[H_3NC_2H_4NH_3]_3[ZrF_7]_2\cdot 2H_2O$, $Ba_2[Zr_2F_{12}]$, $Na_5[Zr_2F_{13}]$
	8	Dodecahedron		$Li_6[BeF_4][ZrF_8]$, $[Hf(SO_4)_4(H_2O)_2]^{4-}$, $[Zr(edta)(H_2O)_2]$, $[Zr_2F_8(H_2O)_6]$, $[Zr_2(OH)_2(CO_3)_6]^{6-}$, $[Zr_4(OH)_8(H_2O)_{16}]^{8+}$, K_2ZrF_6
			mmmm	$[M(C_2O_4)_4]^{4-}$, $[Zr\{ON(Ph)NO\}_4]$, $[Zr(SOCNEt_2)_4]$, $[Zr(acac)_2(NO_3)_2]$
			D_2-*gggg*	$[Zr(8\text{-}O\text{-}quin)_4]$
			S_4-*gggg*	$[Hf\{ON(Ph)COPh\}_4]$, $[Zr(2\text{-}OC_6H_4CH{=}NEt)_4]$
			abmg	$[Zr(acac)_3(NO_3)]$
		Square antiprism		$[Cu(H_2O)_6]_2[ZrF_8]$, $[Cu(H_2O)_6][Cu_2(H_2O)_{10}][Zr_2F_{14}]$, $[Zr(NCS)_4(bipy)_2]$, $[M(OEP)(O_2CMe_2)_2]$
			ssss	$[Zr(acac)_4]$, $[Zr(Sacac)_4]$
		Bicapped trigonal prism		$TlZrF_5$, $(N_2H_6)ZrF_6$
			$t_1t_1p_2p_2$	$[Hf\{MeC_6H_4C(S)N(O)Me\}_4]$

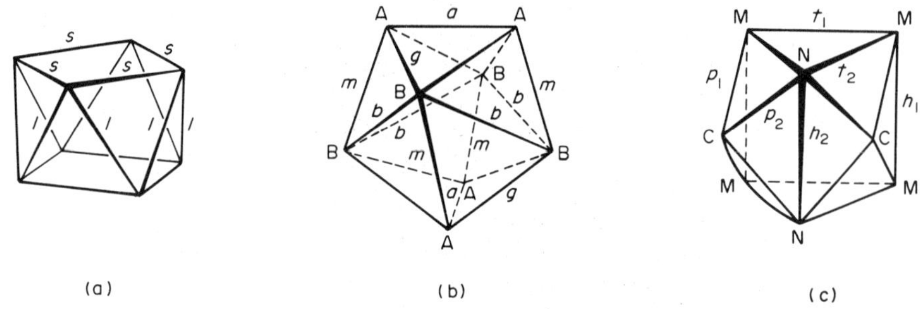

Figure 1 Perspective views of (a) the D_{4d} square antiprism, (b) the D_{2d} dodecahedron and (c) the C_{2v} bicapped trigonal prism

magnetic moment of $K_5Zr(CN)_5$ is 2.69 BM at room temperature and its ESR spectrum shows a signal with $g = 1.986$. The absence of free KCN in the product has been confirmed by X-ray powder diffraction. The IR spectrum of $K_5Zr(CN)_5$ exhibits at least five bands in the C≡N stretching region at 2195, 2173, 2112, 2097 and 2073 cm^{-1}, suggesting a rather low symmetry structure.[15]

$$4ZrBr_3 + 5KCN + 6NH_3 \longrightarrow K_5Zr(CN)_5(s) + 3ZrBr_3(NH_2) + 3NH_4Br \quad (1)$$

32.2.2 Nitrogen Ligands

The tris(2,2′-bipyridyl) and tris(1,10-phenanthroline) complexes, $[M(bipy)_3]$ and $[M(phen)_3]$ (M = Zr or Hf), can be synthesized in ~80% yield by sodium amalgam reduction of $ZrCl_4$ or $HfCl_4$ in THF solutions that contain a stoichiometric amount of the ligand. These compounds are dark colored, extremely air-sensitive, monomeric and diamagnetic in solution.[16] $[Zr(bipy)_3]$ has also been prepared by using $Li_2(bipy)$ as the reducing agent. This procedure afforded copper-colored crystals that have an effective magnetic moment of 0.31 BM.[17] Further reduction of $[Zr(bipy)_3]$ in THF gives $Li[Zr(bipy)_3]\cdot 4THF$ (dark violet crystals, $\mu_{eff} = 1.70$ BM) and $Li_2[Zr(bipy)_3]\cdot 8THF$ (dark colored crystals, $\mu_{eff} = 1.10$ BM).[18] The formal oxidation state of zirconium in the $[Zr(bipy)_3]^{n-}$ ($n = 0$, 1 or 2) could be regarded as 0, −I or −II, respectively; however, because of the likelihood of electron delocalization on to the aromatic ligands, there are the usual problems here in assigning formal oxidation states.

32.3 ZIRCONIUM(III) AND HAFNIUM(III)

The coordination chemistry of zirconium(III) is an ill-defined and relatively unexplored area.[19] Only one magnetically dilute zirconium(III) complex has been isolated, $[ZrCl_3(tfars)]\cdot MeCN$ [tfars = 1, 2-bis(dimethylarsino)-3,3,4,4-tetrafluorocyclobutene $(CF_2)_2C_2(AsMe_2)_2$][20] and the first structurally characterized zirconium(III) complex $[\{ZrCl_3(PBu_3)_2\}_2]$ has just recently been reported.[21] The known zirconium(III) complexes and their physical properties are listed in Table 2. Even less is known about the coordination chemistry of hafnium(III). Just one complex, $HfCl_3(py)_4$, has been reported.[22]

32.3.1 Nitrogen Ligands

The ammonia adduct $ZrBr_3(NH_3)_6$ has been obtained as a dark brown crystalline solid upon treatment of $ZrBr_3$ with liquid ammonia.[15]

Fowles *et al.*[23,24] have prepared a number of adducts of ZrX_3 (X = Cl, Br or I) with pyridine, 2,2′-bipyridyl, 1,10-phenanthroline and acetonitrile (Table 2). These highly colored, air-sensitive compounds have unusual stoichiometries and low magnetic moments. The metal-to-ligand ratio is 1:2 for $ZrX_3(py)_2$ (X = Cl, Br or I) and $ZrBr_3(bipy)_2$, 1:1.5 for $Zr_2X_6(bipy)_3$ (X = Cl or I) and $Zr_2Cl_6(phen)_3$, and 1:2.5 for $Zr_2X_6(MeCN)_5$ (X = Cl, Br or I). The magnetic properties of these compounds suggest dinuclear or oligonuclear structures with some electronic interaction between neighboring Zr atoms. For example, halogen-bridged dimeric

Table 2 Zirconium(III) and Hafnium(III) Complexes

Complex	*Color*	μ_{eff}[a] (BM)	Λ_M[b] (Ω^{-1} cm^2 mol^{-1})	*Spectral data*[c]	*Ref.*
$ZrBr_3(NH_3)_6$	Dark brown				1
$ZrCl_3(py)_2$	Chocolate	1.29	42	IR: 418s, 293br, 270sh UV-vis (solid): 27.6br, 33.9m, 37.6m, 44.7s; (MeCN): 13.25sh (12), 29.2sh (1800), 31.3 (1900), 39.4br (50 000)	2, 3
$ZrCl_3(py)_3$		0.39			4
$ZrCl_3(3,5\text{-lutidine})_2$	Maroon			IR: 340s Vis (solid): 21.1s, 11.1m	5
$ZrCl_3(2,4\text{-lutidine})_{1.2-1.3}$	Orange			IR: 321s Vis (solid): 20.4s, 11.1m	5
$ZrBr_3(py)_2$	Red-brown	1.24	24	IR: 417br, 382w, 220br UV-vis (solid): 25.6br, 30.0sh, 37.6m, 44.8s; (MeCN): 13.0sh (6), 28.25sh (1500), 40.0br (50 000)	2, 3
$ZrI_3(py)_2$	Yellow	1.16	145	IR: 472br, 417s, 320s UV–vis (solid): 13.4sh, w, 24.2sh, 27.5m, 36.8sh, 45.0s; (MeCN): 28.6 (1400), 34.3 (1500), 39.4br (50 000)	2, 3
$Zr_2Cl_6(bipy)_3$	Chocolate	1.35	28	IR: 434sh, 416s, 403sh, 347sh, 315sh, 292s, 278sh UV-vis (solid): 14.5sh, w, 21.4sh, w, 31.5br, 37.9sh, 45.5s; (MeCN): 14.3 (7), 20.0sh (200), 25.0 (800), 31.7sh, 35.5 (1100), 42.2 (1800)	2, 3
$ZrBr_3(bipy)_2$	Dark brown	1.20	65	UV-vis (solid): 14.0sh, w, 25.7sh, w, ~32.0br, 37.1sh, 45.3s; (MeCN): 31.9sh, 35.5 (1700), 40.9 (1800), 42.2 (2200)	2, 3
$Zr_2I_6(bipy)_2$	Yellow	0.96	199	UV-vis (solid): 12.8sh, w, 26.0br, 37.0sh, 45.1s; (MeCN): 13.2sh, w, 14.0sh, w, 31.6 (1510), 40.5 (1900)	2, 3
$Zr_2Cl_6(phen)_3$	Purple-brown	1.27	122	IR: 410s, 360w, 302s, 265m, 252sh UV-vis (solid): 21.2sh, 24.4m, 29.4sh, 32.4sh, ~36.3br, 45.0s; (MeCN): 25.4sh, 26.2 (500), 37.9 (1300), 43.4 (1800)	2, 3
$Zr_2Cl_6(MeCN)_5$	Red-brown	0.46	41	IR: 440br, 294s UV-vis (solid): 15.5w, ~22.0sh, 29.5br, 37.4sh, 46.3s; (MeCN): 15.4 (13), 21.9 (220), 31.1 (1300), ~36.8br, sh (1800), 46.6 (3000)	2, 3

Table 2 *(continued)*

Complex	*Color*	μ_{eff} [a] (BM)	Λ_M [b] (Ω^{-1} cm^2 mol^{-1})	*Spectral data* [c]	*Ref.*
$Zr_2Br_6(MeCN)_5$	Brown	0.29	57	IR: 461br, 395s, 296w, 223br UV-vis (solid): 12.4sh, ~27.6br, ~37.2br, sh, 44.8s; (MeCN): 38.8sh (2000), 40.9 (2200)	2, 3
$Zr_2I_6(MeCN)_5$	Yellow-brown	1.10	143	IR: 459br, 395m UV-vis (solid): 12.9sh, w, 29.7m, 35.5sh, 45.1s; (MeCN): 14.3 (2), 29.2sh (800), 34.0 (1060), 40.5 (2100)	2, 3
$[\{ZrCl_3(PMe_3)_2\}_2]$	Green			^1H NMR (CH_2Cl_2, −20 °C): δ 1.61 (*d*, $^2J(HP)$ 6 Hz)	6
$[\{ZrCl_3(PEt_3)_2\}_2]$	Green			$^{31}P\{^1H\}$ NMR (C_6D_6): δ −3.82br, s	7
$[\{ZrCl_3(PPr_3)_2\}_2]$					7
$[\{ZrCl_3(PBu_3)_2\}_2]$	Green			$^{31}P\{^1H\}$ NMR (toluene-d_8, −30 °C): δ −8.73s	7
$[ZrCl_3(tfars)]\cdot MeCN$	Brown	1.73	30	UV-vis (solid): 24.0, bands at higher energy	8
K_2ZrF_5	White				9
$HfCl_3(py)_4$		diamag.			4

[a] At room temperature. [b] 10^{-3} M in MeCN. [c] IR frequencies in cm^{-1}. UV-vis frequencies in 10^3 cm^{-1}; (solid), reflectance spectra; (MeCN), solution spectra with ε in parentheses.

1. D. Nicholls and T. A. Ryan, *Inorg. Chim. Acta*, 1977, **21,** L17.
2. G. W. A. Fowles, B. J. Russ and G. R. Willey, *Chem. Commun.*, 1967, 646.
3. G. W. A. Fowles and G. R. Willey, *J. Chem. Soc. (A)*, 1968, 1435.
4. D. A. Wasmund, Ph.D. Thesis, Southern Illinois University, 1969, University Microfilms, Inc., Ann Arbor, Michigan; *Diss. Abstr. B*, 1970, **30,** 4549.
5. E. M. Larsen and T. E. Henzler, *Inorg. Chem.*, 1974, **13,** 581.
6. J. H. Wengrovius and R. R. Schrock, *J. Organomet. Chem.*, 1981, **205,** 319.
7. J. H. Wengrovius, R. R. Schrock and C. S. Day, *Inorg. Chem.*, 1981, **20,** 1844.
8. D. L. Kepert and K. R. Trigwell, *Aust. J. Chem.*, 1975, **28,** 1359.
9. P. Ehrlich, *Angew. Chem.*, 1952, **64,** 617.

structures, $[(py)_2X_2ZrX_2ZrX_2(py)_2]$ and $[(MeCN)_3X_2ZrX_2ZrX_2(NCMe)_2]$, have been proposed for the pyridine and acetonitrile adducts. However, because the complexes are too insoluble for molecular weight measurements, structural proposals are necessarily speculative. Conductance measurements (Table 2) indicate that the iodide complexes and $Zr_2Cl_6(phen)_3$ behave as 1:1 electrolytes in acetonitrile; species such as $[ZrI_2(py)_2(MeCN)_2]^+I^-$, $[ZrI_2(bipy)_2]^+[ZrI_4(bipy)]^-$ and $[ZrCl_2(phen)_2]^+[ZrCl_4(phen)]^-$ may be formed.

In a thesis, Wasmund[22] has reported the reduction of $ZrCl_4$ and $HfCl_4$ with alkali metals in pyridine to give MCl_3–pyridine adducts having stoichiometries $ZrCl_3(py)_3$ and $HfCl_3(py)_4$, respectively. The zirconium complex is weakly paramagnetic ($\mu_{eff} = 0.39$ BM) and the hafnium compound is diamagnetic, but ESR spectra of both complexes in pyridine are said to indicate the existence of 'a paramagnetic species of unknown character.'

Larsen and Henzler[25] have described the preparation of $ZrCl_3(3,5\text{-lutidine})_2$ and $ZrCl_3(2,4\text{-lutidine})_{1.2-1.3}$ by slow reaction of crystalline $ZrCl_3$ with benzene solutions of the appropriate lutidine at room temperature over periods of up to 30 days. These investigators view the lutidine adducts as intercalation compounds, in which the lutidine is inserted into the solid between the chains of $ZrCl_6$ octahedra present in crystalline $ZrCl_3$. The frequencies of the ν(Zr—Cl) IR band and visible absorption bands of the lutidine complexes (Table 2) are appreciably shifted from the values of 333 cm^{-1} and 17 400 cm^{-1} found for $ZrCl_3$. The magnetic susceptibilities of the complexes are nearly independent of temperature, but they are very dependent on field strength. It is interesting that the more sterically hindered 2,6-lutidine did not react with $ZrCl_3$.

Olver and co-workers[26,27] have studied the polarographic reduction of $ZrCl_4$ and $HfCl_4$ in acetonitrile with 0.1 M $[NEt_4][ClO_4]$ as the supporting electrolyte. Three plateaus were interpreted in terms of reduction to the III, II and metallic states. However, controlled potential electrolysis of $ZrCl_4$ led to soluble $[ZrCl_6]^{2-}$ ions and poorly characterized precipitates, and no evidence for stable concentrations of Zr^{III} complexes could be found. Evidently, the Zr^{III} is rapidly oxidized by perchlorate, solvent or residual water.

32.3.2 Phosphorus and Arsenic Ligands

Dinuclear zirconium(III) complexes of the type $[\{ZrCl_3(PR_3)_2\}_2]$ (R = Et, Pr or Bu) have been prepared in high yield by reduction of $[ZrCl_4(PR_3)_2]$ with one equivalent of sodium amalgam.[21] The less soluble methyl derivative $[\{ZrCl_3(PMe_3)_2\}_2]$ was obtained by photolysis of $[Zr(CH_2CMe_3)_2Cl_2(PMe_3)_2]$.[28] Molecular weight measurements suggest that the $[\{ZrCl_3(PR_3)_2\}_2]$ complexes undergo reversible phosphine dissociation according to equation (2), and ^1H and ^{31}P NMR spectra indicate that coordinated and free phosphine exchange at a rate on the order of the NMR time scale. NMR data are included in Table 2. A weak ESR signal for toluene solutions of $[\{ZrCl_3(PR_3)_2\}_2]$ and free phosphine, which becomes more intense when excess phosphine is added, has been interpreted in terms of formation of the mononuclear complex $[ZrCl_3(PR_3)_3]$ (equation 3). $[\{ZrCl_3(PBu_3)_2\}_2]$ has a centrosymmetric chlorine-bridged structure (**1**), with distortions from regular octahedral geometry that include bending back of the four terminal Zr—Cl bonds away from the central $Zr(\mu\text{-}Cl)_2Zr$ unit; (terminal Cl)—Zr—(terminal Cl) = 165.1°. The Zr—Zr bond length is 3.182(1) Å, close to that in zirconium metal (3.1789 Å) and somewhat longer than that in β-$ZrCl_3$ (3.07 Å). Thus the Zr—Zr bond is long and presumably weak. Reactions of the $[\{ZrCl_3(PR_3)_2\}_2]$ complexes with ethylene, propylene and butadiene have been studied.[21]

$$[Zr_2Cl_6(PR_3)_4] \rightleftharpoons [Zr_2Cl_6(PR_3)_3] + PR_3 \qquad (2)$$

$$0.5[\{ZrCl_3(PR_3)_2\}_2] \rightleftharpoons [ZrCl_3(PR_3)_2] \underset{-PR_3}{\overset{PR_3}{\rightleftharpoons}} [ZrCl_3(PR_3)_3] \qquad (3)$$

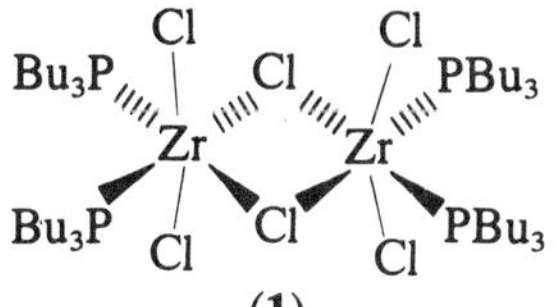

(**1**)

Prolonged reaction of $ZrCl_4$ with 1,2-bis(dimethylarsino)3,3,4,4-tetrafluorocyclobutene,

$(CF_2)_2C_2(AsMe_2)_2$ (tfars), in acetonitrile (sealed tube, eight days at 100 °C, six months at room temperature) yields $[ZrCl_3(tfars)]\cdot MeCN$ as a brown solid. A weak IR band at 2255 cm^{-1} indicates that the acetonitrile is not coordinated to the metal. This compound is a nonelectrolyte in acetonitrile and exhibits a magnetic moment at room temperature of 1.73 BM. It appears to be the first magnetically dilute zirconium(III) complex to have been reported.[20]

32.3.3 Oxygen Ligands

No zirconium(III) complexes with oxygen donor ligands have been isolated. However, the electronic absorption spectra of aqueous solutions of ZrI_3 have been interpreted in terms of the formation of aqua complexes (equation 4).[29] The spectrum of a freshly prepared solution of ZrI_3 exhibits a band at 24 400 cm^{-1}, which decays over a period of 40 minutes, and a shoulder at 22 000 cm^{-1}, which decays more rapidly. The 24 400 cm^{-1} band has been assigned to $[Zr(H_2O)_6]^{3+}$, and the 22 000 cm^{-1} shoulder has been attributed to an unstable intermediate iodo–aqua complex. If it is assumed that the absorption band of $[Zr(H_2O)_6]^{3+}$ is due to the $^2T_{2g} \rightarrow {}^2E_g$ ligand-field transition, the value of Δ is 24 400 cm^{-1}. This corresponds to a Δ value of 20 300 cm^{-1} for $[Ti(H_2O)_6]^{3+}$ [30] and 17 400 cm^{-1} for the octahedral $ZrCl_6$ chromophore in zirconium(III) chloride.[25]

$$ZrI_3 \xrightarrow{\text{fast}} [ZrI_n(H_2O)_{6-n}]^{3-n} \xrightarrow{\text{slow}} [Zr(H_2O)_6]^{3+} \xrightarrow[\text{oxidation}]{\text{slow}} Zr^{IV} \qquad (4)$$

32.3.4 Halogen Ligands

K_2ZrF_5 has been mentioned by Ehrlich.[31] It is described as a white solid which can be sublimed and which exhibits slight temperature-independent paramagnetism.

Reduction of Zr^{IV} in ZrX_4–Al_2X_6 (X = Cl, Br or I) melts, with either zirconium or aluminum metal, yields soluble blue species. These species were not isolated, but they have been formulated as $Zr(X_2AlX_2)_3$ complexes on the basis of their electronic spectra. Absorption bands at 17 400 (X = Cl), 16 000 (X = Br) and 14 000 cm^{-1} (X = I) may be assigned to the $^2T_{2g} \rightarrow {}^2E_g$ transition of Zr^{III} in an octahedral environment of halide ions. The solid zirconium(III) halides exhibit ligand-field bands at closely similar frequencies.[32]

32.4 ZIRCONIUM(IV) AND HAFNIUM(IV)

32.4.1 Group IV Ligands

32.4.1.1 Cyanides

Very little is known about cyanide complexes of zirconium(IV) or hafnium(IV). Just one compound has been reported. $[Hf(CN)\{N(SiMe_3)_2\}_3]$ (m.p. 170–172 °C dec.; ν(CN) 2160 cm^{-1}) has been prepared by reaction of $[HfCl\{N(SiMe_3)_2\}_3]$ with Me_3SiCN in toluene at reflux.[33]

32.4.1.2 Tin-containing ligands

$Zr(SnPh_3)_4$ has been obtained as a yellow solid (m.p. 70–73 °C; sintering at 65 °C) by reaction of $Zr(NEt_2)_4$ with four equivalents of Ph_3SnH. The related compound containing nine metal atoms, $(Ph_3Sn)_3ZrSnPh_2Zr(SnPh_3)_3$, was prepared *via* the sequence of reactions in equations (5)–(7). The transamination with *N*-phenylformamide (equation 6) is necessary in order to prevent diethylamine-catalyzed decomposition of Ph_2SnH_2 in the next reaction step. After purification by chromatography, the product was obtained in 22% yield as a yellow solid which melted at 100 °C (sintering at 85 °C).[34]

$$Zr(NEt_2)_4 + 3Ph_3SnH \longrightarrow (Ph_3Sn)_3Zr(NEt_2) + 3Et_2NH \qquad (5)$$

$$(Ph_3Sn)_3Zr(NEt_2) + HNPhC(O)H \longrightarrow (Ph_3Sn)_3ZrNPhC(O)H + Et_2NH \qquad (6)$$

$$2(Ph_3Sn)_3ZrNPhC(O)H + Ph_2SnH_2 \longrightarrow (Ph_3Sn)_3ZrSnPh_2Zr(SnPh_3)_3 + 2HNPhC(O)H \qquad (7)$$

32.4.2 Nitrogen Ligands

32.4.2.1 Ammonia and amines

The reactions of ammonia and aliphatic amines with zirconium(IV) chloride and hexachlorozirconate(IV) salts have been reviewed by Fowles.[35] Early work on ammonia adducts of zirconium tetrahalides is summarized in the monograph by Blumenthal.[3] Compounds of the type $ZrCl_4 \cdot nNH_3$ ($n = 8$, 4 or 2)[36] and $ZrBr_4 \cdot 4NH_3$[37] were prepared more than 80 years ago by reaction of the tetrahalides with gaseous ammonia. Thermal decomposition of $ZrCl_4 \cdot 8NH_3$ in argon gives stepwise formation of $ZrCl_4 \cdot 6NH_3$, $ZrCl_4 \cdot 4NH_3$ and finally $ZrCl_4$.[38] Tensimetric studies indicate that the $ZrCl_4 \cdot 8NH_3$ obtained upon reaction of $ZrCl_4$ with liquid ammonia is a mixture of $ZrCl_3(NH_2) \cdot 6NH_3$ and NH_4Cl.[39] The ammonolysis product can be freed of NH_4Cl by washing with liquid ammonia, and the resulting $ZrCl_3(NH_2) \cdot nNH_3$ is converted to $ZrCl_3(NH_2) \cdot NH_3$ by heating *in vacuo* at 100 °C.[40] The same ammonolysis products are obtained upon reaction of liquid ammonia with M'_2ZrCl_6 ($M' = NH_4$, Rb or Cs).[41] The reaction of gaseous $ZrCl_4$ with ammonia at 300 °C yields $ZrCl_4 \cdot 2NH_3$, which has a cubic unit cell with $a = 10.13$ Å.[42]

Zirconium(IV) and hafnium(IV) halides form adducts with a variety of monodentate and polydentate amines. Typical complexes are listed in Table 3. In general, these compounds are air- and moisture-sensitive white solids (yellow when the halide is iodide), thermally stable and insoluble in most organic solvents.

Table 3 Complexes of Zirconium(IV) and Hafnium(IV) Halides with Amines

Complex[a]	*Physical method*[b]	*Ref.*
cis-$[ZrCl_4(NMe_3)_2]$	IR, UV, NMR, χ, M	1
cis-$[ZrBr_4(NMe_3)_2]$	IR, UV, NMR, χ	1
trans-$[ZrI_4(NMe_3)_2]$	IR, UV, NMR, χ	1
cis-$[HfCl_4(NMe_3)_2]$	IR, NMR	2
trans-$[HfBr_4(NMe_3)_2]$	IR, NMR	3
trans-$[HfI_4(NMe_3)_2]$	IR, NMR	3
$ZrCl_4(NEt_3)$		4
$ZrCl_4\{(Me_2NCH_2C_5H_4)Fe(C_5H_5)\}$	IR	5
$[ZrCl_4(Et_2NCH_2CN)]$	IR	6
$ZrCl_4(NHR_2)_2$ (R = Me, Et, Pr)		4
$MrBr_4(en)$		7
$[MCl_4(TMEDA)]$	IR	8
$[ZrBr_4(TMEDA)]$	IR	8
$[ZrCl_4\{(NMe_2)_2SiMe_2\}]$	IR, NMR	9
$[HfCl_4\{(NMe_2)_2SiMe_2\}]$	IR	9
$[MCl_4\{Ti(NMe_2)_4\}]$	IR	10
$MCl_4 \cdot Me_6tren$	IR, NMR	11

[a] M = Zr or Hf. [b] IR = infrared spectrum; UV = electronic spectrum; NMR = ^{1}H NMR spectrum; χ = magnetic susceptibility; M = molecular weight.

1. J. Hughes and G. R. Willey, *Inorg. Chim. Acta*, 1976, **20**, 137.
2. G. R. Willey, *Inorg. Chim. Acta*, 1977, **21**, L12.
3. S. R. Wade and G. R. Willey, *J. Less-Common Met.*, 1979, **68**, 105.
4. J. E. Drake and G. W. A. Fowles, *J. Chem. Soc.*, 1960, 1498.
5. S. C. Jain and R. Rivest, *J. Inorg. Nucl. Chem.*, 1970, **32**, 1579.
6. S. C. Jain and R. Rivest, *Inorg. Chem.*, 1967, **6**, 467.
7. J. J. Habeeb, F. F. Said and D. G. Tuck, *Can. J. Chem.*, 1977, **55**, 3882.
8. T. C. Ray and A. D. Westland, *Inorg. Chem.*, 1965, **4**, 1501.
9. S. R. Wade and G. R. Willey, *J. Chem. Soc., Dalton Trans.*, 1981, 1264.
10. S. R. Wade, M. G. H. Wallbridge and G. R. Willey, *J. Chem. Soc., Dalton Trans.*, 1982, 271.
11. S. R. Wade and G. R. Willey, *Inorg. Nucl. Chem. Lett.*, 1978, **14**, 363.

Tertiary amines yield MX_4L_2 or MX_4L complexes, depending on the steric bulk of the substituents on the nitrogen atom. Trimethylamine gives well-characterized monomeric $[MX_4(NMe_3)_2]$ (M = Zr or Hf; X = Cl, Br or I) complexes that have been assigned an octahedral *cis* or *trans* structure on the basis of the number of ν(M—X) bands in their IR spectra. On the other hand, 1:1 adducts were obtained upon reaction of $ZrCl_4$ with triethylamine, *N,N*-dimethylaminomethylferrocene or diethylaminoacetonitrile. The IR spectrum of $[ZrCl_4(Et_2NCH_2CN)]$ and a molecular weight measurement on the analogous titanium complex suggest that Et_2NCH_2CN behaves as a bidentate ligand, coordinating to the metal through the amine nitrogen atom and the triple bond of the nitrile group.

While zirconium(IV) chloride forms simple adducts with tertiary and secondary amines, aminolysis takes place with primary amines yielding amonobasic chlorides such as $ZrCl_2(NHMe)_2 \cdot NH_2Me$ and $ZrCl_3(NHR) \cdot NH_2R$ (R = Et, Pr or Bu).[43] The reactions of amines (A = Me_3N, Et_3N, Me_2NH, Et_2NH or $EtNH_2$) with the corresponding alkylammonium hexachlorozirconates $[AH]_2[ZrCl_6]$ have also been studied.[44] No reaction was observed with the tertiary amines; very slow aminolysis occurred with the secondary amines; and relatively rapid aminolysis of one Zr—Cl bond took place with ethylamine (equation 8).

$$[EtNH_3]_2[ZrCl_6] + 2EtNH_2 \longrightarrow [EtNH_3]_2[ZrCl_5(NHEt)] + [EtNH_3]Cl \quad (8)$$

Diamines such as ethylenediamine, tetramethylethylene diamine and $SiMe_2(NMe_2)_2$ give 1:1 adducts with zirconium(IV) and hafnium(IV) halides. The ν(M—X) region of IR spectra of $[MX_4(TMEDA)]$ (X = Cl or Br) and $[MCl_4\{(NMe_2)_2SiMe_2\}]$ complexes is consistent with an octahedral *cis* (C_{2v}) structure. The presence of two *N*-methyl resonances in ^1H NMR spectra of $[ZrCl_4\{(NMe_2)_2SiMe_2\}]$ suggests the presence of a non-planar ZrN_2Si chelate ring; the two resonances coalesce to a broad singlet at 100 °C owing to a rapid ring-inversion process. Pale yellow solid adducts, $[MCl_4\{Ti(NMe_2)_4\}]$ (M = Zr or Hf), have been prepared by displacement of trimethylamine from *n*-pentane suspensions of $[MCl_4(NMe_3)_2]$. IR spectra of the $[MCl_4\{Ti(NMe_2)_4\}]$ complexes suggest an octahedral *cis* structure about zirconium or hafnium in which $Ti(NMe_2)_4$ behaves as a bidentate chelating ligand (**2**).

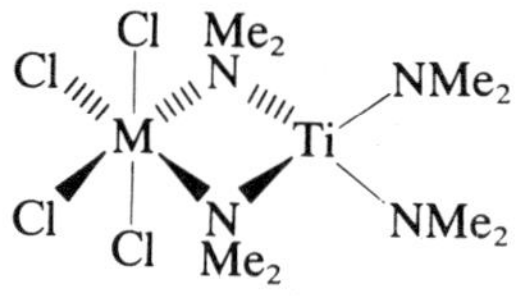

(2)

The tripod ligand tris(2-dimethylaminoethyl)amine (Me_6tren) also displaces trimethylamine from $[MCl_4(NMe_3)_2]$ yielding 1:1 adducts $MCl_4 \cdot Me_6tren$ (M = Zr or Hf). On the basis of IR and ^1H NMR data these compounds have been formulated as $[MCl_2\{Me_6tren\}]Cl_2$ salts in which Me_6tren behaves as a tetradentate ligand. Unfortunately, conductance studies proved unreliable.

32.4.2.2 N-heterocyclic ligands

Monodentate N-heterocyclic amines such as pyridine, substituted pyridines, benzoquinolines, imidazoles and indazoles, and bidentate amines such as 2,2′-bipyridyl and 1,10-phenanthroline react with zirconium(IV) and hafnium(IV) halides to give moisture-sensitive, diamagnetic adducts that are generally insoluble in most organic solvents. Typical complexes are listed in Table 4. They may be prepared by prolonged stirring of a solution of the ligand and a suspension of the metal halide in an inert solvent such as benzene or dichloromethane. Alternatively, the adducts may be precipitated by mixing solutions of the ligand and metal halide in a donor solvent such as acetonitrile, ether or ethyl acetate.

Octahedral 1:2 adducts $[MX_4L_2]$ are the usual products of the reaction of zirconium(IV) or hafnium(IV) halides with monodentate N-heterocyclic amines. However, there are exceptions. Both 1:2 and 1:3 adducts have been reported for X = Cl and L = pyridine; the $MCl_4(py)_3$ (M = Zr or Hf) complexes are converted to $[MCl_4(py)_2]$ upon heating *in vacuo* for seven hours at 65–75 °C.[45] While 3,5-lutidine and 2,4-lutidine react with ZrX_4 (X = Cl, Br or I) to give the expected 1:2 adducts, the more sterically hindered 2,6-lutidine yields 1:1 adducts. IR and ^1H NMR spectra of $ZrCl_4$(2,6-lutidine) suggest that this complex may have a symmetrical, chlorine-bridged dimeric structure.[25] 5,6-Benzoquinoline also gives a 1:1 adduct with zirconium(IV) chloride, whereas 3,4-benzoquinoline yields the expected 1:2 complex.

X-Ray structures of these complexes are not yet available. However, the $[MX_4L_2]$ adducts have been assigned octahedral *cis* or *trans* structures on the basis of the number of metal–halogen stretching bands in IR spectra. *cis*-$[MCl_4L_2]$ complexes exhibit two or more ν(M—Cl) bands in the region 275—360 cm^{-1},[45] while *trans*-$[MCl_4L_2]$ compounds show a single ν(M—Cl) band at about 335–350 cm^{-1}.[25] Both *cis* and *trans* isomers are observed (Table 4), and the preferred configuration is not easily rationalized in terms of the structure of the ligand.

Table 4 Complexes of Zirconium(IV) and Hafnium(IV) Halides with N-heterocyclic Ligands

Ligand	*Complex*[a]	*Physical method*[b]	*Structure (ligand donor atom)*	*Ref.*
Pyridine	$[MCl_4(py)_2]$	P, IR, ΔH	*cis*	1–7
	$MCl_4(py)_3$	IR		4
	$[ZrBr_4(py)_2]$	P, IR		1, 2, 4
Picolines	$[ZrF_4(4\text{-}MeC_5H_4N)_2]$			8
	$[ZrCl_4(n\text{-}MeC_5H_4N)_2]$ ($n = 2, 3, 4$)	ΔH		7
Lutidines	$[ZrX_4(3,5\text{-}Me_2C_5H_3N)_2]$ (X = Cl, Br, I)	IR	*trans* (X = Cl)	9
	$[ZrX_4(2,4\text{-}Me_2C_5H_3N)_2]$ (X = Cl, Br, I)	IR	*trans* (X = Cl)	9
	$ZrX_4(2,6\text{-}Me_2C_5H_3N)$ (X = Cl, Br, I)	IR, NMR		9
	$[ZrCl_4(2,6\text{-}Me_2C_5H_3N)_2]$	ΔH		7
3-Cyanopyridine	$[MCl_4(3\text{-}NCC_5H_4N)_2]$	IR	*cis* (pyridine N)	10
2-Acetylpyridine	$[MCl_4(2\text{-}MeCOC_5H_4N)_2]$	IR	*cis* (pyridine N)	11
Nicotinamide	$[ZrCl_4(3\text{-}NH_2COC_5H_4N)_2]$	IR	*cis* (pyridine N)	12, 13
3,4-Benzoquinoline	$[ZrCl_4(C_{13}H_9N)_2]$	IR	*trans*	14
5,6-Benzoquinoline	$ZrCl_4(C_{13}H_9N)$	IR		15
Imidazoles	$[MCl_4L_2]$	IR	(unsaturated N)	16, 17
Indazole	$[ZrCl_4(C_7H_6N_2)_2]$	IR	*trans* (pyrrole N)	18
5-Aminoindazole	$[ZrCl_4(C_7H_7N_3)_2]$	IR	(pyrrole N)	19
Pyrazine	$ZrX_4(C_4H_4N_2)_2$ (X = Cl, Br)	IR, Λ		20
2,2′-Bipyridyl	$[ZrF_4(bipy)_2]$	IR, Λ	D_{2d}	21
	$[MCl_4(bipy)]$	IR, Λ, χ	C_{2v}	3, 4, 22
	$MCl_4(bipy)_{1.5}$	IR, ΔH, χ	$[MCl_2(bipy)_3][MCl_6]$(?)	7, 21
	$[ZrBr_4(bipy)]$	IR, Λ, χ	C_{2v}	3, 4, 22
1,10-Phenanthroline	$[MCl_4(phen)]$	IR, Λ, χ		4, 22
	$[ZrBr_4(phen)]$	IR, χ		4, 22

[a] M = Zr or Hf. [b] P = dissociation pressure measurements; IR = infrared spectrum; ΔH = thermochemical data; NMR = ^{1}H NMR spectrum; Λ = electrical conductance; χ = magnetic susceptibility.

1. H. J. Emeleus and G. S. Rao, *J. Chem. Soc.*, 1958, 4245.
2. G. S. Rao, *Z. Anorg. Allg. Chem.*, 1960, **304**, 176.
3. I. R. Beattie and M. Webster, *J. Chem. Soc.*, 1964, 3507.
4. T. C. Ray and A. D. Westland, *Inorg. Chem.*, 1965, **4**, 1501.
5. W. M. Graven and R. V. Peterson, *J. Inorg. Nucl. Chem.*, 1969, **31**, 1743.
6. A. D. Westland and V. Uzelac, *Can. J. Chem.*, 1970, **48**, 2871.
7. Ts. B. Konunova and M. F. Frunze, *Russ. J. Inorg. Chem. (Engl. Transl.)*, 1973, **18**, 951.
8. E. L. Muetterties, *J. Am. Chem. Soc.*, 1960, **82**, 1082.
9. E. M. Larsen and T. E. Henzler, *Inorg. Chem.*, 1974, **13**, 581.
10. S. C. Jain, *J. Inorg. Nucl. Chem.*, 1973, **35**, 505.
11. S. C. Jain, *Labdev, Part A*, 1970, **8**, 169.
12. R. C. Paul, H. Arora and S. L. Chadha, *Inorg. Nucl. Chem. Lett.*, 1970, **6**, 469.
13. R. C. Paul, H. Arora and S. L. Chadha, *Indian J. Chem.*, 1971, **9**, 698.
14. S. A. A. Zaidi, T. A. Khan and N. S. Neelam, *Indian J. Chem., Sect A*, 1979, **18**, 461.
15. S. A. A. Zaidi, T. A. Khan and N. S. Neelam, *Indian J. Chem., Sect A*, 1980, **19**, 169.
16. V. T. Panyushkin, A. D. Garnovskii, O. A. Osipov, A. L. Sinyavin and A. F. Pozharskii, *J. Gen. Chem. USSR (Engl. Transl.)*, 1967, **37**, 292.
17. A. D. Garnovskii, V. T. Panyushkin, L. I. Kuznetsova, O. A. Osipov, V. I. Minkin and V. I. Martynov, *J. Gen. Chem. USSR (Engl. Transl.)*, 1968, **38**, 1806.
18. K. S. Siddiqi, N. S. Neelam, T. A. Khan and S. A. A. Zaidi, *Bull. Soc. Chim. Fr.*, 1980, I-360.
19. K. S. Siddiqi, N. S. Neelam, F. R. Zaidi and S. A. A. Zaidi, *Croat. Chem. Acta*, 1981, **54**, 421.
20. G. W. A. Fowles and R. A. Walton, *J. Chem. Soc.*, 1964, 4330.
21. R. J. H. Clark and W. Errington, *J. Chem. Soc. (A)*, 1967, 258.
22. G. W. A. Fowles and R. A. Walton, *J. Less-Common Met.*, 1963, **5**, 510.

IR spectra are also useful in identifying the ligand donor atom in the case of complexes that contain ambidentate or potentially bidentate ligands. Thus, 3-cyanopyridine, 2-acetylpyridine and nicotinamide behave as monodentate ligands in $[MCl_4L_2]$ complexes, coordinating to the metal only through the pyridine nitrogen atom. Imidazoles bind to metal tetrachlorides at the unsaturated nitrogen atoms, while indazoles are attached at the pyrrole nitrogen atom. The absence of a band at 960 cm^{-1} in the IR spectra of $ZrX_4(pyrazine)_2$ (X = Cl or Br) complexes indicates that, at least in the solid state, both pyrazine nitrogen atoms are coordinated. Evidently pyrazine bridges between two metal atoms. The molar conductance of a $\sim 10^{-3}$ M acetonitrile solution of $ZrCl_4(pyrazine)_2$ is close to that expected for a nonelectrolyte; the small conductance observed may be due to slight dissociation (equation 9). Unfortunately, the low solubility of the $ZrX_4(pyrazine)_2$ complexes precluded molecular weight measurements.

$$ZrCl_4(pyrazine)_2 + MeCN \rightleftharpoons [ZrCl_3(MeCN)(pyrazine)_2]^+ + Cl^- \quad (9)$$

Zirconium(IV) fluoride reacts with 2,2′-bipyridyl to give $[ZrF_4(bipy)_2]$, whereas zirconium(IV) and hafnium(IV) chlorides yield $[MCl_4(bipy)]$ or $MCl_4(bipy)_{1.5}$, depending on the stoichiometry of the reaction mixture. Only the 1:1 adduct $[ZrBr_4(bipy)]$ is known in the case of the bromide. $[ZrF_4(bipy)_2]$ has been assigned an eight-coordinate structure on the basis of IR spectra and conductance measurements. This compound exhibits $\nu(Zr—F)$ bands at 483 and 504 cm^{-1}, consistent with the two IR-active Zr—F stretching modes ($b_2 + e$) expected for a D_{2d} dodecahedral structure.[46] An ionic structure, $[MCl_2(bipy)_3][MCl_6]$, has been suggested for the isomorphous $MCl_4(bipy)_{1.5}$ (M = Zr or Hf) complexes, but attempts to isolate the eight-coordinate cation by preparation of the perchlorate or tetrafluoroborate salts were unsuccessful.[46] The $\nu(M—X)$ region of IR spectra of the $[MX_4(bipy)]$ (M = Zr or Hf, X = Cl; M = Zr, X = Br) complexes is consistent with an octahedral *cis* (C_{2v}) structure. The structure of the $[MX_4(phen)]$ analogues may be more complex since IR spectra of the $[MCl_4(phen)]$ complexes exhibit five $\nu(M—Cl)$ bands.[45] Molar conductances of the $[ZrX_4L]$ (L = bipy or phen) complexes in acetonitrile are much less than expected for a 1:1 electrolyte, but slight dissociation in this solvent (equation 10) is indicated.[47]

$$[ZrX_4L] + MeCN \rightleftharpoons [ZrX_3L(MeCN)]^+ + X^- \quad (10)$$

Oxozirconium(IV) complexes of the type $ZrOX_2L_n$ (L = pyridine, α-picoline, 2-aminopyridine, 2,4-lutidine, 2,6-lutidine or quinoline; $n = 4$ when X = Cl, Br, I or ClO_4; $n = 2$ when X = NO_3 or NCS) have been reported.[48–50] The corresponding 2,2′-bipyridyl and 1,10-phenanthroline complexes have stoichiometries $ZrO(ClO_4)_2L_3$, $ZrOX_2L_2$ (X = Cl, Br or I) and $ZrOX_2L$ (X = NO_3 or NCS). All of these compounds are monomeric nonelectrolytes in nitrobenzene except for the perchlorate derivatives, which behave as 1:2 electrolytes. The structures of these complexes are unknown. However, several conclusions have been drawn on the basis of IR studies: (i) the nitrate ligands are bidentate; (ii) the NCS ligands bond to zirconium through the nitrogen atom; and (iii) the perchlorate compounds contain uncoordinated ClO_4^- ions. A weak IR band in the 900–980 cm^{-1} region has been assigned to $\nu(Zr{=}O)$,[50] but the presence of the zirconyl group needs to be confirmed by X-ray diffraction since this species rarely occurs[51–54] (see Section 32.4.4.1).

Zirconium and hafnium cyanate, thiocyanate, and selenocyanate complexes that contain N-heterocyclic ligands are discussed in Sections 32.4.2.5.iii, 32.4.2.5.iv and 32.4.2.5.v.

32.4.2.3 Hydrazine and nitro ligands

The reaction of zirconium(IV) fluoride with excess anhydrous hydrazine yields the 1:2 adduct $ZrF_4(N_2H_4)_2$, which loses one molecule of hydrazine at 230–245 °C. $ZrF_4(N_2H_4)_2$ and $ZrF_4(N_2H_4)$ have been characterized by their IR spectra.[55]

Hydrazine and substituted-hydrazine complexes of the type $ZrOX_2(NH_2NHR)_4$ (R = H, Me or Ph; X = Cl, Br, I, NCS or NO_3) have been synthesized by addition of six equivalents of the hydrazine to an ethanol solution of the appropriate zirconyl salt. An IR band at ~970 cm^{-1} assigned to $\nu(N—N)$ indicates the presence of bidentate chelating or bridging hydrazine ligands; these complexes may be polymeric since they are insoluble in common organic solvents.[56]

Substituted phenylhydrazido(1−) complexes $[XC_6H_4NHNH_3]_2[ZrCl_{6-n}(NHNHC_6H_4X)_n]$ (X = Cl, Br, I or NO_2; $n = 1, 2, 3, 4$ or 6) are formed upon reaction in alcohol or dioxane of $ZrCl_4$ with $XC_6H_4NHNH_2$ or the corresponding hydrazine hydrochloride. The thermal decomposition of these compounds has been studied.[57–59]

Nitro complexes $[M(NO_2)_4(DMF)_2]$ (M = Zr or Hf) have been prepared by reaction of $ZrCl_4$ with four equiv. of $AgNO_2$ in dimethylformamide–acetonitrile. Addition of 2,2′-bipyridyl or 1,10-phenanthroline to solutions of $[M(NO_2)_4(DMF)_2]$ yields the corresponding $[M(NO_2)_4(bipy)]$ or $[M(NO_2)_4(phen)]$ complexes. The reaction of $[Zr(NO_2)_4(DMF)_2]$ with $(NBu_4)Br$ gives $(NBu_4)_2[Zr(NO_2)_6]$. IR spectra of these complexes indicate that the nitro group is coordinated through the nitrogen atom.[60]

32.4.2.4 *Dialkylamido-, bis(trimethylsilylamido)- and bis(amido)-silane ligands*

(i) Dialkylamido ligands

Metal dialkylamides were reviewed by Bradley in 1972.[61] A comprehensive list of all zirconium and hafnium amides reported prior to 1979 may be found in the book by Lappert *et al.*[62]

Dialkylamido complexes of the type $M(NR_2)_4$ (M = Zr or Hf; R = Me, Et, Pr, Bu, $CHMe_2$ or CH_2CHMe_2; $R_2 = NC_5H_{10}$, NC_5H_9Me or $NC_5H_8Me_2$) may be prepared by reaction of the metal tetrachloride with the appropriate lithium dialkylamide (equation 11) or by transamination between $M(NR_2)_4$ (R = Me or Et) and a higher boiling secondary amine (equation 12).[63–66] The transamination approach may also be used to prepare mixed ligand complexes such as $[Zr(NMe_2)_2\{N(CHMe_2)_2\}_2]$ or $[Zr(NEt_2)_3\{N(CHMe_2)_2\}]$.[63] The $M(NR_2)_4$ complexes are moisture-sensitive, volatile liquids or low-melting solids that can be distilled or sublimed at ~0.1 mmHg. Originally, some of these compounds were said to be green or blue,[63] but it now appears that the pure compounds are generally colorless or yellow.[6,64–66] Molecular weight measurements in boiling benzene indicate that $Zr(NMe_2)_4$ is partly polymerized; the degree of association is 1.22 in benzene, and ^1H NMR studies in dichloromethane show that the degree of association increases with decreasing temperature.[63,65] $[Zr(NEt_2)_4]$ is monomeric in boiling benzene, and complexes that contain higher amides are presumably monomeric as well.

$$4MCl_4 + 4LiNR_2 \longrightarrow M(NR_2)_4 + 4LiCl \qquad (11)$$

$$M(NR_2)_4 + 4R'_2NH \longrightarrow M(NR'_2)_4 + 4R_2NH \qquad (12)$$

He^I photoelectron spectra of $[M(NR_2)_4]$ (M = Zr or Hf; R = Me or Et) have been recorded, and the lowest energy band has been assigned to ionization from molecular orbitals having nitrogen 'lone pair' character. Splitting of this band indicates that free rotation about the M—N bond is not occurring in these compounds.[67] Calorimetric measurements on alcoholysis reactions of $[Zr(NR_2)_4]$ (R = Me or Et) and $[Hf(NEt_2)_4]$ have afforded the following values for the mean thermochemical bond energies (kJ mol^{-1}): Zr—N, 377; Hf—N, 397.[68] IR spectra of the $[M(NR_2)_4]$ complexes exhibit a single M—N stretching frequency (at 530–680 cm^{-1}), consistent with the local T_d symmetry of the MN_4 core.[65]

Zirconium and hafnium dialkylamides are highly reactive compounds. They undergo (i) protolytic substitution reactions with reagents such as alcohols, cyclopentadiene and bis(trimethylsilyl)amine;[63,64] (ii) insertion reactions with CO_2, CS_2, COS, nitriles, phenyl isocyanate, methyl isothiocyanate, carbodiimides and dimethyl acetylenedicarboxylate;[69–72] and (iii) addition reactions with metal carbonyls.[73] These reactions are summarized with reference to $Zr(NMe_2)_4$ in Scheme 1.

Transamination between $[Zr(NEt_2)_4]$ and primary amines does not give stable $Zr(NHR)_4$ complexes; the reaction of $[Zr(NEt_2)_4]$ with RNH_2 (R = Et, Pr, Bu or Ph) in hydrocarbon solvents at 0–5 °C yields precipitates of composition $Zr(NR)(NHR)_2$.[74] The dinuclear *t*-butylimido derivatives $[\{M(NCMe_3)(NMe_2)_2\}_2]$ (M = Zr or Hf) **(3)** have been prepared by heating a hexane solution of $M(NMe_2)_4$ under reflux with one equivalent of Me_3CNH_2, followed by removal of the solvent and sublimation of the residue at 150 °C *in vacuo*. The dimeric structure of **(3)** has been confirmed by an X-ray study of the zirconium complex. The geometry about the Zr atom is roughly tetrahedral, with the Zr—N bonds to the nearly planar dimethylamide ligands (average 2.060 Å) and the symmetrically bridging *t*-butylimide ligands (average 2.066 Å) being nearly identical. The molecule is centrosymmetric, and the Zr_2N_2 ring is planar.[75]

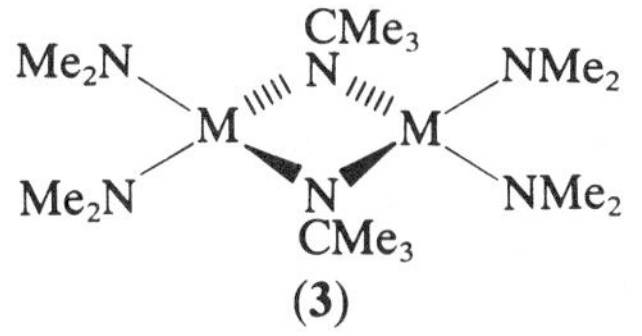

(3)

Acylamido complexes of the type $Zr(NHCOR)_n(OCHMe_2)_{4-n}$ (n = 1, 2, 3 or 4; R = Me, Ph or C_5H_4N) have been prepared in quantitative yield by reaction of $Zr(OCHMe_2)_4{\cdot}HOCHMe_2$ with stoichiometric amounts of acetamide, benzamide or nictotinamide. These compounds are insoluble in common organic solvents, nonvolatile and probably polymeric. IR spectra show

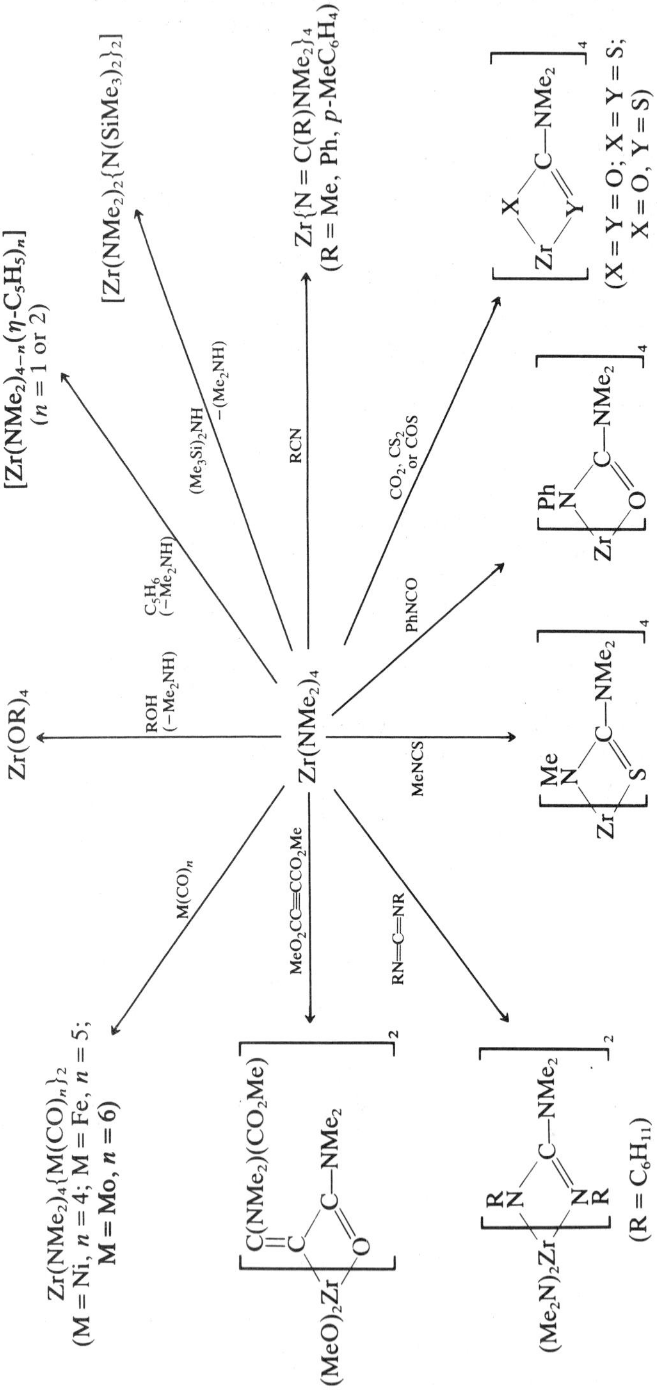
Zr(NMe2)4
Zr(OR)4
ROH (−Me2NH)
[Zr(NMe2)4−n(η-C5H5)n] (n = 1 or 2)
C5H6 (−Me2NH)
[Zr(NMe2)2{N(SiMe3)2}2]
(Me3Si)2NH −(Me2NH)
Zr{N=C(R)NMe2}4 (R = Me, Ph, p-MeC6H4)
RCN
CO2, CS2 or COS
(X = Y = O; X = Y = S; X = O, Y = S)
PhNCO
MeNCS
RN=C=NR
(R = C6H11)
MeO2CC≡CCO2Me
M(CO)n
Zr(NMe2)4{M(CO)n}2 (M = Ni, n = 4; M = Fe, n = 5; M = Mo, n = 6)

Scheme 1

unperturbed ν(C═O) vibrations, indicating that the NHCOR ligands coordinate through the nitrogen atom.[76]

Reaction of an excess of imidazole with tetrabenzylzirconium in THF at 52 °C yields $Zr(C_3N_2H_3)_4$ as a yellow crystalline solid.[77] The pyrrolyl complex $[Na(THF)_6]_2[Zr(\eta^1\text{-}NC_4H_4)_6]$ has been prepared by reaction of $[ZrCl_2(\eta\text{-}C_5H_5)_2]$ with (pyrrolyl)sodium in THF at reflux temperature. The $[Zr(\eta^1\text{-}NC_4H_4)_6]^{2-}$ ion is octahedral (crystallographic point group symmetry $\bar{3}$) with N—Zr—N bond angles near 90°. Short Zr—N bond lengths (2.198 Å) and Zr—N—(centroid η^1-NC_4H_4) angles of 179° are indicative of considerable d_π–p_π character in the Zr—N bonds.[78]

(*ii*) *Bis(trimethylsilylamido) ligands*

Chlorobis(trimethylsilyl)amido complexes $[MCl\{N(SiMe_3)_2\}_3]$ and $[MCl_2\{N(SiMe_3)_2\}_2]$ (M = Zr or Hf) have been prepared by reaction of the metal tetrachloride with a stoichiometric amount of lithium or sodium bis(trimethylsilyl)amide.[79–81] The $[MCl\{N(SiMe_3)_2\}_3]$ complexes are surprisingly inert to substitution reactions; they do not react with dioxygen, water, dilute mineral acids, aqueous $AgNO_3$, $LiCH_2SiMe_3$, $LiBH_4$ or excess $Na[N(SiMe_3)_2]$. However, they do react with methyllithium, yielding $[M\{N(SiMe_3)_2\}_3(Me)]$,[79] and $[HfCl\{N(SiMe_3\}_3]$ is conveniently converted to $[HfX\{N(SiMe_3)_2\}_3]$ (X = Br, I, N_3 or CN) by reaction for 12 hours in refluxing toluene with the appropriate trimethylsilyl halide or pseudohalide (equation 13).[33]

$$[HfCl\{N(SiMe_3)_2\}_3] + Me_3SiX \longrightarrow [HfX\{N(SiMe_3)_2\}_3] + Me_3SiCl \qquad (13)$$

The corresponding $[MCl_2\{N(SiMe_3)_2\}_2]$ (M = Zr or Hf) complexes are considerably more reactive than the $[MCl\{N(SiMe_3)_2\}_3]$ analogues. The dichlorobis(silylamides) are air- and moisture-sensitive. They can be converted to $[MCl(BH_4)\{N(SiMe_3)_2\}_2]$ (M = Zr or Hf) and $[HfCl(OSiMe_3)\{N(SiMe_3)_2\}_2]$ derivatives, and they serve as starting materials for the preparation of a variety of alkyl complexes of the type $[M\{N(SiMe_3)_2\}_2R_2]$, $[Hf\{N(SiMe_3)_2\}_2RR']$ and $[HfCl\{N(SiMe_3)_2\}_2R$.[81,82]

Physical and spectroscopic properties of the inorganic bis(trimethylsilylamido) complexes are listed in Table 5. These compounds are relatively low-melting, colorless or white crystalline solids; they have been characterized by IR, ^{1}H NMR and $^{13}C\{^1H\}$ NMR spectroscopy.

Table 5 Bis(trimethylsilylamido) Complexes of Zirconium(IV) and Hafnium(IV)

Compound	*M.p. (°C)*	*ν(M—N)*[a]	*ν(M—X)*[a]	*^{1}H NMR*[b]	*$^{13}C\{^1H\}$ NMR*[b]	*Ref.*
$[ZrCl\{N(SiMe_3)_2\}_3]$	182–183	408, 400	348	0.67	6.15	1
	183–188	408, 400	349	0.38[c]		2
$[ZrCl_2\{N(SiMe_3)_2\}_2]$	53–55	420	388	0.57	4.46	3
$[ZrCl(BH_4)\{N(SiMe_3)_2\}_2]$	42–44	415	355	0.50		3
$[HfCl\{N(SiMe_3)_2\}_3]$	180–181	404, 388	338	0.62	6.38	1
	186–189	397, 395	330	0.39[c]		2
$[HfBr\{N(SiMe_3)_2\}_3]$	65–67	415, 372	230	~0.5		4
$[HfI\{N(SiMe_3)_2\}_3]$	92–96	415, 370		~0.5		4
$[Hf(N_3)\{N(SiMe_3)_2\}_3]$	150 (dec.)	400, 385	355	~0.5		4
$[Hf(CN)\{N(SiMe_3)_2\}_3]$	170–172 (dec.)	395	295	~0.5		4
$[HfCl_2\{N(SiMe_3)_2\}_2]$	44–45	418	405	0.62	4.79	3
$[HfCl(BH_4)\{N(SiMe_3)_2\}_2]$	55–57			0.54		3
$[HfCl(OSiMe_3)\{N(SiMe_3)_2\}_2]$	82–86	415, 383	350	0.50	5.10	5

[a] IR frequencies (cm^{-1}); X = halide, azide or cyanide. [b] δ values relative to $SiMe_4$ for the $N(SiMe_3)_2$ group. In benzene unless indicated otherwise. [c] In toluene-d_8.

1. R. A. Andersen, *Inorg. Chem.*, 1979, **18**, 1724.
2. C. Airoldi, D. C. Bradley, H. Chudzynska, M. B. Hursthouse, K. M. Abdul Malik and P. R. Raithby, *J. Chem. Soc., Dalton Trans.*, 1980, 2010.
3. R. A. Andersen, *Inorg. Chem.*, 1979, **18**, 2928.
4. R. A. Andersen, *Inorg. Nucl. Chem. Lett.*, 1980, **16**, 31.
5. R. A. Andersen, *J. Organomet. Chem.*, 1980, **192**, 189.

X-Ray and NMR studies[80] of the $[MCl\{N(SiMe_3)_2\}_3]$ (M = Zr or Hf) complexes indicate the presence of considerable steric crowding due to the bulky $N(SiMe_3)_2$ ligands. In the solid state, these compounds have a distorted tetrahedral structure (**4**) of crystallographically imposed C_3 symmetry [Zr—Cl = 2.394(2), Zr—N = 2.070(3), Hf—Cl = 2.436(5), Hf—N = 2.040(10) Å].

Crowding is evidenced by the Cl—M—N angles being 10–13° less than the N—M—N angles, and by the M—N—Si angles proximal to Cl being 8–10° less than the M—N—Si angles distal to Cl. ^{1}H NMR spectra, which exhibit two equally intense resonances below the coalescence temperature of 4–5 °C, are consistent with structure (**4**), and the observed line broadening has been interpreted in terms of restricted rotation about the M—N bonds. The unusual stability of the [MCl{N(SiMe$_3$)$_2$}$_3$] complexes is no doubt due to the steric bulk of the bis(trimethylsilyl)amide ligands, which protect the metal atom from attack by potential reactants.

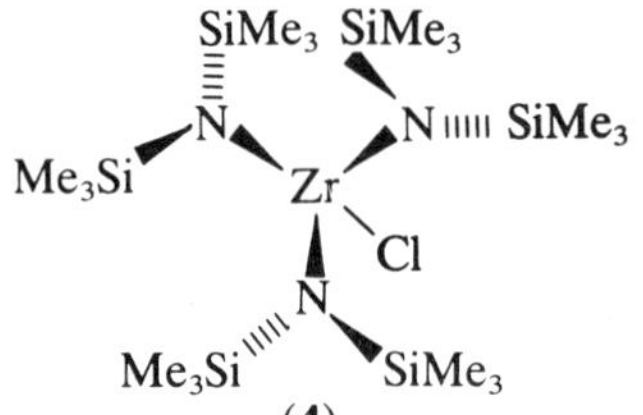

(**4**)

(iii) Bis(amido)silane ligands

Reaction of zirconium or hafnium tetrachlorides with the lithium-substituted bis(amido)silanes Me$_2$Si{N(Li)R}$_2$, {SiMe$_2$N(Li)Me}$_2$ or Y{SiMe$_2$N(Li)Me}$_2$ yields the spirobicyclic amides (**5**)–(**11**).[83–87] These compounds are colorless or light yellow, low-melting, volatile solids. They have been characterized by mass spectra, IR and Raman spectra, and ^{1}H and ^{29}Si NMR spectra. An X-ray study of compound (**5**) has established a spirobicyclic structure of approximate D_{2d} symmetry with planar {ZrN$_2$Si} rings (Zr—N = 2.053 Å, N—Zr—N bite angle = 77.9°).[83] Compounds (**9**)–(**11**) are monomeric in freezing benzene.[87]

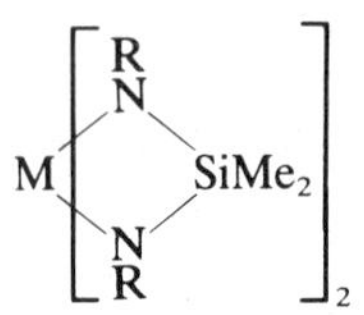

(**5**) M = Zr, R = CMe$_3$
(**6**) M = Zr, R = SiMe$_3$
(**7**) M = Hf, R = SiMe$_3$

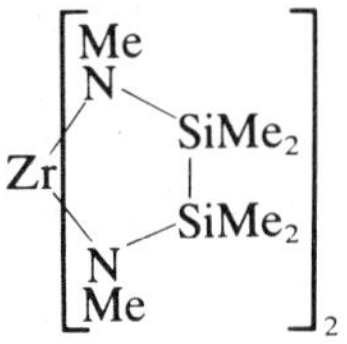

(**8**)

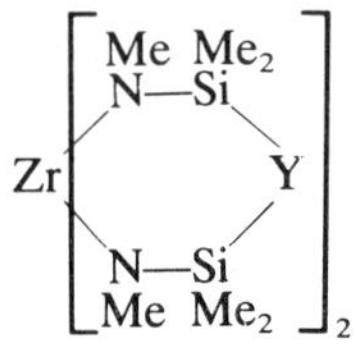

(**9**) Y = NMe
(**10**) Y = O
(**11**) Y = CH$_2$

The spirobicyclic compounds are cleaved by ZrCl$_4$ or HfCl$_4$, yielding the corresponding dichloro-monocyclic compounds (equations 14 and 15).[84,87] The dichloro compounds are oligomeric or polymeric; IR spectra suggest chlorine-bridged structures, and cryoscopic molecular weight measurements on ZrCl$_2${(NMeSiMe$_2$)$_2$O} in benzene indicate a degree of association of ~4.

$$[M\{(NSiMe_3)_2SiMe_2\}_2] + MCl_4 \longrightarrow 2MCl_2\{(NSiMe_3)_2SiMe_2\} \quad (14)$$
(**6**) or (**7**) (M=Zr or Hf)

$$[Zr\{(NMeSiMe_2)_2Y\}_2] + ZrCl_4 \longrightarrow 2ZrCl_2\{(NMeSiMe_2)_2Y\} \quad (15)$$
(**9**), (**10**) or (**11**) (Y=NMe, O or CH$_2$)

32.4.2.5 Triazenes, azides, cyanates, thiocyanates, selenocyanates and related ligands

(i) Triazenes

The dark red tetrakis(1,3-diphenyltriazenido)zirconium(IV), [Zr(dpt)$_4$] (m.p. 230 °C), has been prepared by reaction of ZrCl$_4$ and Ag(dpt) in diethyl ether.[88] The analogous, orange 1,3-dimethyltriazenido derivative, [Zr(dmt)$_4$] (m.p. 120 °C), was synthesized from ZrCl$_4$ and Mg(dmt)I.[89] Both compounds are monomeric in benzene and are presumably eight-coordinate (**12**).

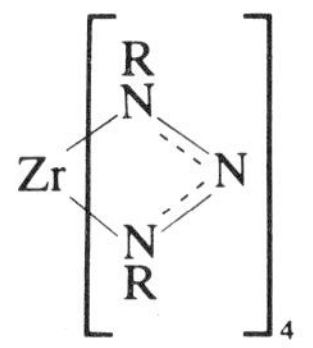

(**12**) R = Me or Ph

(*ii*) *Azides*

Highly explosive $ZrCl_3(N_3)$ is obtained from the reaction of $ZrCl_4$ with iodine azide in dichloromethane (equation 16). IR spectra indicate that this compound is polymeric, having both azide and chlorine bridges. The reaction of $ZrCl_4$ with one and two equivalents of $[PPh_4][N_3]$ yields the thermally and mechanically stable complexes $[PPh_4]_2[\{ZrCl_4(N_3)\}_2]$ and $[PPh_4]_2[ZrCl_4(N_3)_2]$, respectively. $[PPh_4]_2[\{ZrCl_4(N_3)\}_2]$ contains centrosymmetric dimeric anions (**13**) in which the zirconium atoms are linked by the α-nitrogen atoms of the nearly linear azide groups; average bond lengths and angles are: Zr—Cl = 2.407 Å, Zr—N = 2.203 Å, N—Zr—N = 66.7°, Zr—N—Zr = 113.3°. IR spectra suggest that $[ZrCl_4(N_3)_2]^{2-}$ has an octahedral *trans* structure.[90]

$$ZrCl_4 + IN_3 \longrightarrow ZrCl_3(N_3) + ICl \quad (16)$$

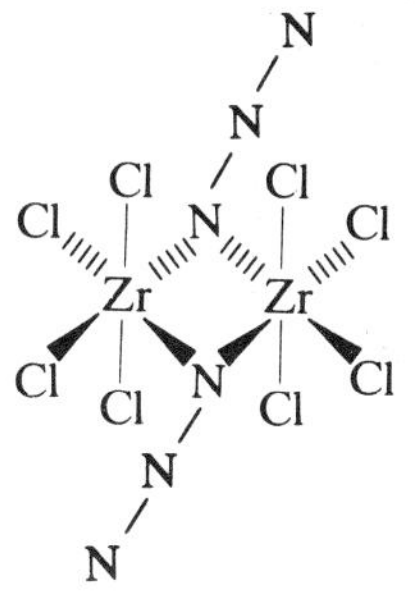

(**13**)

(*iii*) *Cyanates*

Three kinds of cyanate complexes have been reported: (i) $ZrCl_3(NCO)$;[91] (ii) $[NEt_4]_2[M(NCO)_6]$ (M = Zr or Hf);[92] and (iii) $M(OCN)_4L_2$ (M = Zr or Hf; L = DMSO, py, 0.5bipy or 0.5phen).[93] These compounds have been prepared from the metal tetrahalides by the reactions outlined in equations (17)–(21).

$$ZrCl_4 + IOCN \xrightarrow[-30\,^\circ C]{CH_2Cl_2} ZrCl_3(NCO) + ICl \quad (17)$$

$$MCl_4 + 2[NEt_4][NCO] \xrightarrow{MeCN} [NEt_2]_2[MCl_4(NCO)_2] \quad (18)$$

$$[NEt_4]_2[MCl_4(NCO)_2] + 4AgNCO \xrightarrow{MeCN} [NEt_4]_2[M(NCO)_6] + 4AgCl \quad (19)$$

$$MCl_4 + 2Pb(NCO)_2 \xrightarrow{THF} M(OCN)_4 + 2PbCl_2 \quad (20)$$

$$M(NCO)_4 + 2L \xrightarrow{THF} M(OCN)_4L_2 \quad (21)$$

An N-bonded attachment of the cyanate ligands in $ZrCl_3(NCO)$ and $[M(NCO)_6]^{2-}$ has been established by vibrational spectra, which exhibit a shift of the ν(C—O) mode to higher frequency upon complex formation.[91,92] IR spectra of the $M(OCN)_4L_2$ complexes suggest that the cyanate ligands in these compounds may be coordinated through the oxygen atom;[93] however, this needs to be confirmed by structural studies. Conductance measurements show that the $M(OCN)_4L_2$ complexes dissociate according to equation (22) upon dissolution in DMF.[93]

$$M(OCN)_4L_2 + DMF \longrightarrow [M(OCN)_3L_2(DMF)]^+ + [NCO]^- \quad (22)$$

(*iv*) *Thiocyanates*

Two classes of salts containing complex thiocyanate anions have been isolated from aqueous solutions: (i) $A_2[M(NCS)_6]\cdot nH_2O$ (M = Zr or Hf; A = tetraalkylammonium or pyridinium cation)[94–98] and (ii) $A[MO(NCS)_3(H_2O)]\cdot H_2O$ (A = NH_4, C_5H_5NH, K, Rb or Cs when M = Zr; A = Cs when M = Hf).[97–100] The related oxohafnium complexes, $[C_5H_5NH][HfO(NCS)_3(H_2O)]$ and $(C_5H_5NH]_3[(HfO)_2(NCS)_7]\cdot H_2O$, have also been reported.[98,100] All of these compounds exhibit a ν(C—S) IR band at higher frequency (787–879 cm^{-1}) than in KSCN (749 cm^{-1}), indicative of an N-bonded attachment of the thiocyanate ligands. The oxometal complexes display a characteristic band at 913–927 cm^{-1} (M = Zr) or 934–940 cm^{-1} (M = Hf), which is not present in spectra of the $[C_5H_5NH]_2[M(NCS)_6]$ salts. This band may be due to the M=O stretching mode[99,100] (see Section 32.4.4.1).

Oxozirconium complexes of the type $ZrO(NCS)_2L_2$, where L is an N or O donor ligand, are discussed in Sections 32.4.2.2, 32.4.2.3 and 32.4.4.1.

The complexes $[M(NCS)_4(DMF)_4]$ and $K_2[M(NCS)_5]\cdot n$DMF ($n = 6$ when M = Zr; $n = 4$ when M = Hf) have been prepared by mixing saturated DMF solutions of KSCN and the metal tetrachloride. IR spectra indicate that the thiocyanate ligands are N-bonded and DMF is coordinated through the carbonyl oxygen atom in the eight-coordinate $[M(NCS)_4(DMF)_4]$ complexes.[101] Related $[M(NCS)_4(bipy)_2]$ and $M(NCS)_4(phen)_3$ complexes have been reported,[102] and the X-ray structure of $[Zr(NCS)_4(bipy)_2]$ has been determined.[103] The structure (**14**) features an N-bonded attachment of the linear, monodentate thiocyanate ligands (Zr—NCS = 2.182 Å) and a square antiprismatic coordination polyhedron of crystallographically imposed D_2 symmetry. The 2,2′-bipyridyl ligands (Zr—N = 2.412 Å) span an opposite pair of lateral edges. The coordination polyhedron is slightly distorted in the direction of a D_{2d} dodecahedron; the corresponding dodecahedral stereoisomer would have the bipy ligands on the *a* edges and the NCS groups in the B sites.

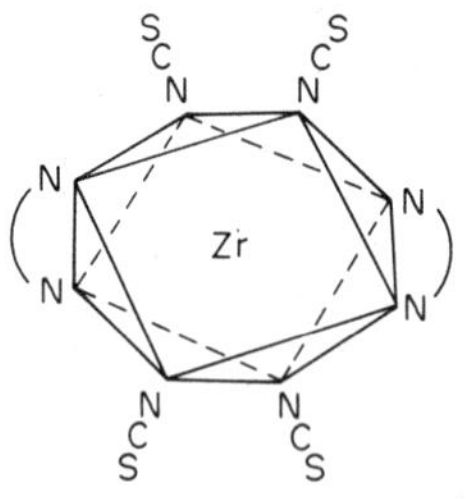

(**14**)

In aqueous solution, zirconium(IV) and hafnium(IV) form complexes $M(NCS)_n^{4-n}$, where $n = 1$–8.[104] Selective extraction of hafnium thiocyanate complexes from acidic aqueous solution by methyl isobutyl ketone is a widely used industrial method for the separation of zirconium and hafnium. Separation methods have been reviewed by Vinarov.[105]

(*v*) *Selenocyanates*

Tan colored $K_2[M(NCSe)_6]$ (M = Zr or Hf) complexes have been prepared by extracting the metal tetrachlorides into an acetonitrile solution of KSeCN. IR spectra indicate an N-bonded attachment of the NCSe ligands.[106]

Treatment of a THF solution of $K_2[M(NCSe)_6]$ with 2,2′-bipyridyl or 1,10-phenanthroline yields the complexes $M(NCSe)_4(bipy)_2$ and $M(NCSe)_4(phen)_3(THF)$ (M = Zr or Hf).[107] Radial distribution curves for the hafnium compounds based on X-ray powder data show maxima assigned to Hf···Se at 4.8–4.85 Å. These data are consistent with coordination of the NCSe ligand through the nitrogen atom.[102]

The reaction of $ZrCl_4$ with freshly prepared $(SeCN)_2$ in CS_2 at −60 °C yields the substitution product $ZrCl_3(NCSe)$ (equation 23). IR spectra suggest that the SeCN group acts as a bridging ligand, coordinating through the nitrogen and selenium atoms.[108]

$$ZrCl_4 + (SeCN)_2 \longrightarrow ZrCl_3(NCSe) + ClSeCN \qquad (23)$$

(*vi*) *Halogenoids, interhalogen–halogenoids and interhalogenoids*

Zirconium(IV) chloride reacts with freshly prepared oxycyanogen $(OCN)_2$ in dichloromethane at −70 °C to give the 1:1 adduct $ZrCl_4(NCO)_2$. IR spectra suggest that the NCO—OCN ligand is attached to the metal through the nitrogen atoms. The nitrogen atoms appear to occupy *cis* positions in the $ZrCl_4N_2$ coordination group,but it is not known whether this compound is a monomer or a polymer.[109]

Cyanogen iodide forms a 1:1 adduct with $ZrCl_4$ in which ICN is N-bonded; a dimeric C_{2h} structure (**15**) has been proposed on the basis of IR and Raman spectra.[110] X-ray powder patterns show that the 1:2 adducts $[MCl_4(NCCl)_2]$ (M = Zr or Hf) are isostructural with $[ZrCl_4(NCMe)_2]$, and the ν(C≡N) and ν(M—Cl) regions of IR spectra confirm an octahedral *cis* geometry. The $[MCl_4(NCCl)_2]$ complexes lose one molecule of ClCN at 95–125 °C to give $MCl_4(NCCl)$ complexes, which are stable up to 190 °C.[111]

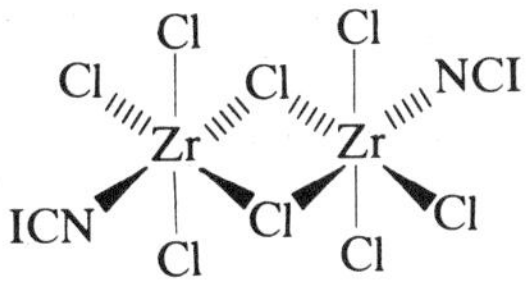

(**15**)

Vibrational spectra indicate that $ZrCl_4$ adducts with $S(CN)_2$ and $Se(CN)_2$ adopt rather different structures. The 1:2 adduct $[ZrCl_4\{S(CN)_2\}_2]$ has been assigned an octahedral *cis* structure in which the $S(CN)_2$ ligands are S-bonded. $[ZrCl_4\{Se(CN)_2\}]$, a 1:1 adduct, appears to contain an N-bonded $Se(CN)_2$ ligand, and a chelate structure (**16**) has been suggested.[112] However, oligomeric structures in which $Se(CN)_2$ acts as an N-bonded bridging ligand may be more likely in view of the geometry of $Se(CN)_2$.[113]

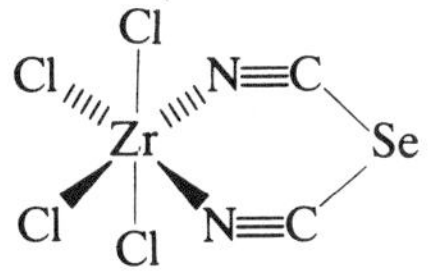

(**16**)

32.4.2.6 *Poly(pyrazolyl)borates*

Poly(pyrazolyl)borate complexes of the type $[ZrCl_3(RBPz_3)]$ (R = H, $CHMe_2$, Bu or Pz) and $[ZrCl_3\{HB(3,5\text{-}Me_2Pz)_3\}]$ have been prepared by reaction of sodium or potassium salts of the ligands (**17**) and (**18**) with two equivalents of $ZrCl_4$.[114,115] Treatment of these complexes with one equivalent of $KOCMe_3$ in toluene yields the mono-*t*-butoxide derivatives $[ZrCl_2(OCMe_3)(RBPz_3)]$ and $[ZrCl_2(OCMe_3)\{HB(3,5\text{-}Me_2Pz)_3\}]$. The $[ZrCl_2(OCMe_3)(RBPz_3)]$ complexes are fluxional on the NMR time scale ($\Delta G^‡ = 56 \pm 2$ kJ mol^{-1} for R = Bu), while the sterically more crowded $[ZrCl_2(OCMe_3)\{HB(3,5\text{-}Me_2Pz)_3\}]$ is stereochemically rigid at temperatures up to 140 °C. These rearrangements appear to take place by a trigonal twist mechanism.[115] $[ZrCl_2(OCMe_3)\{HB(3,5\text{-}Me_2Pz)_3\}]$ serves as the starting material for preparation of stable alkyl derivatives $[Zr(OCMe_3)R_2\{HB(3,5\text{-}Me_2Pz)_3\}]$ (R = Me, CH_2Ph or C≡CMe).[116] Di- and tri-*t*-butoxide derivatives $[ZrCl_{3-n}(OCMe_3)_n\{HB(3,5\text{-}Me_2Pz)_3\}]$ (n = 2 or 3) have also been reported.[115,116]

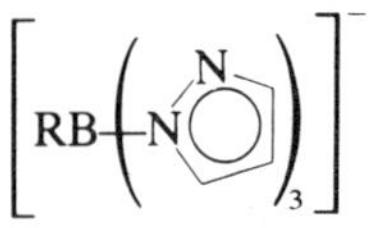

(**17**) $[RBPz_3]^-$

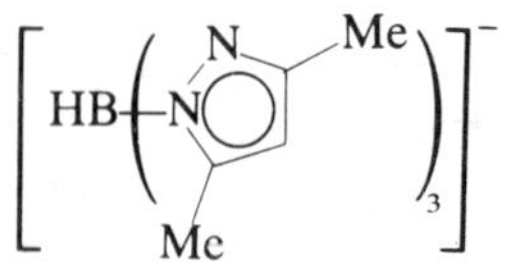

(**18**) $[HB(3,5\text{-}Me_2Pz)_3]^-$

Asslani *et al.*[117] have described the synthesis of the tetrakis(pyrazolyl)borate complexes $[ZrCl_2(BPz_4)_2]$, $[ZrBr_2(BPz_4)_2]$ and $[Zr(BPz_4)_4]$. Analytical and ^1H NMR data were reported for these compounds, but Reger and Tarquini[115] have been unable to reproduce the NMR data.

32.4.2.7 Nitriles

Zirconium(IV) and hafnium(IV) chlorides and bromides form 1:2 adducts of the type $[ZrX_4(RCN)_2]$ (R = Me, Et, Pr or Ph; X = Cl or Br) and $[HfX_4(MeCN)_2]$ (X = Cl or Br).[118–124] These complexes may be prepared by (i) direct reaction of the metal tetrahalide with an excess of the nitrile[120,123] or (ii) electrochemical oxidation of zirconium or hafnium metal in the presence of a solution of chlorine or bromine in acetonitrile.[118] The adducts are moisture-sensitive, white solids, insoluble in nonpolar solvents, but soluble in acetonitrile. In the later solvent, $[ZrBr_4(MeCN)_2$ behaves as a nonelectrolyte.[122]

IR (4000–180 cm^{-1}) and Raman (4000–10 cm^{-1}) spectra of $[MCl_4(MeCN)_2]$ and $(MCl_4(CD_3CN)_2]$ (M = Zr or Hf) have been reported, and normal coordinate calculations have been performed using a modified Urey–Bradley force field. Four ν(M—Cl) frequencies were observed in the region 375–310 cm^{-1}, consistent with an octahedral *cis* structure.[124] A shift of the $\nu(C{\equiv}N)$ band to higher frequencies upon complexation seems to be characteristic of coordination through the lone pair of electrons on the nitrile nitrogen atom.[120,121] UV spectra of the $[ZrX_4(RCN)_2]$ (R = Me or Et; X = Cl or Br) complexes exhibit intense bands that have been attributed to halogen→metal and nitrile→metal charge-transfer transitions.[122]

IR spectra suggest different structures for the $ZrCl_4$ complexes with diethylcyanamide and diethylaminoacetonitrile. $[ZrCl_4(Et_2NCN)_2]$ is a typical octahedral *cis* complex in which Et_2NCN coordinates through the nitrile nitrogen atom.[125] However, in $[ZrCl_4(Et_2NCH_2CN)]$, a 1:1 adduct that exhibits an 80 cm^{-1} shift of $\nu(C{\equiv}N)$ to lower frequency upon complexation, Et_2NCH_2CN appears to behave as a bidentate ligand, coordinating to zirconium through the amine nitrogen atom and the triple bond of the nitrile group.[126] IR spectra of $ZrCl_4\{(NCCH_2C_5H_4)Fe(C_5H_5)\}$ indicate that cyanomethylferrocene coordinates to zirconium through the nitrile nitrogen atom.[127]

The dinitrile complexes $[ZrCl_4(NCCH_2CN)_2]$ and $ZrCl_4\{NC(CH_2)_nCN\}$ ($n = 4$ or 8) were obtained as white solids from the reaction of excess nitrile with a suspension of $ZrCl_4$ in dichloromethane. These complexes exhibit high-frequency shifts of $\nu(C{\equiv}N)$ upon complexation, characteristic of coordination through the nitrile nitrogen atoms. The $ZrCl_4\{NC(CH_2)_nCN\}$ adducts were formulated as chelate compounds or coordination polymers, while $[ZrCl_4(NCCH_2CN)_2]$ is an ordinary 1:2 adduct in which chelation is precluded by the short length of the chain between the two nitrile groups.[128] The IR spectrum of the pale yellow phthalonitrile adduct $[ZrCl_4\{C_6H_4(CN)_2\}_2]$ suggests that phthalonitrile behaves as a monodentate ligand.[129]

32.4.2.8 Ureas

A few zirconium tetrahalide adducts with urea and its derivatives have been reported. Known compounds include $[ZrF_4\{(Me_2N)_2CO\}]$,[130] $[ZrCl_4\{(H_2N)_2CO\}_2]$[131] and $[ZrCl_4\{RC(O)NHC(O)NH_2\}]$ (R = Me or Ph).[132] IR studies indicate that urea coordinates to zirconium through the carbonyl oxygen atom.[131] The acylureas coordinate through both oxygen atoms, giving a chelate structure.[132]

32.4.3 Phosphorus and Arsenic Ligands

32.4.3.1 Ligands containing only P or As donor atoms

Zirconium and hafnium tetrahalides form six-, seven- and eight-coordinate adducts with polydentate phosphine and arsine ligands. The 1:1 adducts [MX_4(dppe)] and [MX_4(dpae)] (M = Zr, X = Cl or Br; M = Hf, X = Cl; dppe = $Ph_2PCH_2CH_2PPh_2$, dpae = $Ph_2AsCH_2CH_2AsPh_2$) have been prepared in benzene media. IR spectra of the [MCl_4(dppe)] complexes exhibit four ν(M—Cl) modes in the 353–278 cm^{-1} region, consistent with a six-coordinate octahedral structure.[45]

Eight-coordinate 1:2 adducts [$MX_4(diars)_2$] (M = Zr or Hf; X = Cl or Br) are obtained from the reaction of the metal tetrahalides with *o*-phenylenebis(dimethylarsine).[133,134] These compounds are isomorphous with the corresponding titanium complex [$TiCl_4(diars)_2$], which has a D_{2d} dodecahedral structure with the chlorine atoms in the B sites and the bidentate diarsine ligands spanning the polyhedral *a* edges.[135] The analogous [$ZrCl_4(dmpe)_2$] (dmpe = $Me_2PCH_2CH_2PMe_2$) complex has been reported recently.[136]

Reaction of the tritertiary arsine ligand methylbis(3-dimethylarsinopropyl)arsine, $MeAs\{(CH_2)_3AsMe_2\}_2$ (tas), with $ZrBr_4$ in acetonitrile yields the 1:1 complex [$ZrBr_4$(tas)], which is believed to be seven-coordinate on the basis of IR and conductance evidence. The methyl rocking modes at 890 and 852 cm^{-1} in the complex are at higher frequencies than in the free ligand, consistent with a tridentate attachment of the arsine ligand, and the molar conductance of a 10^{-3} M solution of [$ZrBr_4$(tas)] in nitrobenzene indicates little dissociation of bromide ions. However, conductance studies in nitrobenzene, nitromethane, acetone and acetonitrile indicate increasing dissociation upon dilution (equation 24).[137]

$$[ZrBr_4(tas)] + solvent \rightleftharpoons [ZrBr_3(tas)(solvent)]^+ + Br^- \qquad (24)$$

Zirconium(IV) chloride reacts with $P(NCO)_3$ in benzene and with $P(CN)_3$ in benzene–diethyl ether to give the substitution products $ZrCl_3\{P(NCO)_2\}$ and $ZrCl_3\{P(CN)_2\}$, respectively (equation 25). These compounds have been obtained as somewhat impure, very hygroscopic solids. On the basis of IR evidence, the $P(NCO)_2$ and $P(CN)_2$ ligands are believed to be coordinated through the phosphorus atom.[138]

$$ZrCl_4 + PX_3 \longrightarrow ZrCl_3(PX_2) + ClX \qquad (25)$$
$$(X = CN \text{ or } NCO)$$

32.4.3.2 Ligands containing mixed donor atoms

The reaction of $LiN(SiMe_2CH_2PR_2)_2$ with either $ZrCl_4$ or $HfCl_4$ yields [$MCl_2\{N(SiMe_2CH_2PR_2)_2\}_2$] (M = Zr or Hf; R = Me or Ph) as colorless, air- and moisture-sensitive crystalline solids. ^1H and $^{31}P\{^1H\}$ NMR spectra of these complexes are consistent with a chiral, octahedral structure (**19**) in which the bulky $N(SiMe_2CH_2PR_2)_2$ ligands are bidentate, with the two uncoordinated $SiMe_2CH_2PR_2$ groups being disposed to opposite sides of the equatorial plane. Structure (**19**), of approximate C_2 symmetry, has been established for [$ZrCl_2\{N(SiMe_2CH_2PMe_2)_2\}_2$] by a single-crystal X-ray study; averaged bond distances are Zr—P = 2.798, Zr—N = 2.096 and Zr—Cl = 2.460 Å.[139]

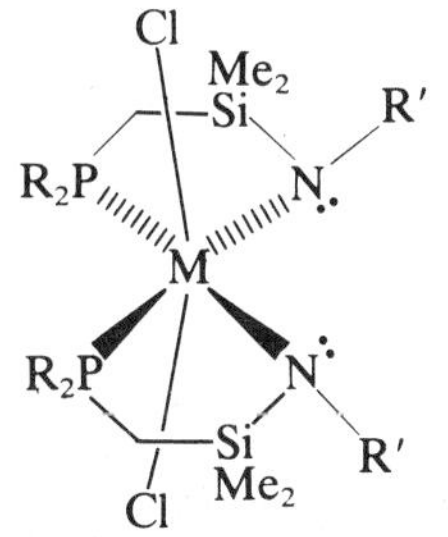

(**19**) R′ = $SiMe_2CH_2PR_2$

The [$MCl_2\{N(SiMe_2CH_2PR_2)_2\}_2$] complexes are unreactive towards nucleophiles. However,

$[HfCl_2\{N(SiMe_2CH_2PMe_2)_2\}_2]$ disproportionates in the presence of excess $HfCl_4$, yielding the 'mono' amide complex $[HfCl_3\{N(SiMe_2CH_2PMe_2\}_2\}\cdot HfCl_4]_x$, which can be converted to the monomeric complex $[Hf(BH_4)_3\{N(SiMe_2CH_2PMe_2)_2\}]$ and $[Hf(BH_4)_4]$ upon reaction with excess $LiBH_4$. Treatment of $[Hf(BH_4)_3\{N(SiMe_2CH_2PMe_2)_2\}]$ with Lewis bases (NEt_3 or PMe_3) generates the dinuclear trihydride $[\{Hf[N(SiMe_2CH_2PMe_2)_2]\}_2(H)_3(BH_4)_3]$.[140]

32.4.4 Oxygen Ligands

32.4.4.1 Water, hydroxide and oxide

The extremely complicated aqueous solution chemistry of Zr^{IV} and Hf^{IV} has been reviewed by Larsen,[5] Solovkin and Tsvetkova,[51] and Clearfield.[52] Zirconium(IV) and hafnium(IV) ions undergo extensive hydrolysis, and the predominant solution species are polynuclear, even in dilute ($>10^{-4}$–10^{-3} M) solutions of high acidity (1–2 M). Spectrophotometric,[141] ultracentrifugation,[142–144] and light scattering[145] studies point to trinuclear and tetranuclear hydrolysis products; complexes such as $[M_3(OH)_4]^{8+}$ and $[M_4(OH)_8]^{8+}$ have been suggested. The mononuclear ions M^{4+} are predominant solution species only at trace metal ion concentrations ($<\sim 10^{-4}$ M). A recent potentiometric determination of hydrolysis constants, $K_h = [M(OH)^{3+}][H^+]/[M^{4+}]$, for Zr^{4+} and Hf^{4+} in perchlorate media ($[H^+] = 0.5$–4 M) confirms that Zr^{4+} ($K_h = 0.28 \pm 0.05$) is more strongly hydrolyzed than Hf^{4+} ($K_h = 0.08 \pm 0.03$).[146]

Direct evidence for the existence of tetranuclear hydroxo complexes comes from X-ray diffraction studies of the zirconyl halides, $ZrOX_2\cdot 8H_2O$ (X = Cl or Br).[147,148] These isomorphous compounds contain the $[Zr_4(OH)_8(H_2O)_{16}]^{8+}$ cation (Figure 2) in which four Zr atoms at the corners of a slightly puckered square are connected by double hydroxo bridges. Each zirconium atom is attached, in addition, to four water molecules, giving an eight coordination polyhedron best described as a distorted dodecahedron. Averaged Zr—O bond distances are Zr—OH_2 = 2.272 and Zr—OH = 2.142 Å.[148] Subsequent studies[149,150] of X-ray scattering from concentrated aqueous solutions of zirconium(IV) and hafnium(IV) halides and zirconium(IV) perchlorate are consistent with a predominant tetranuclear solution species, $[M_4(OH)_8(H_2O)_{16}]^{8+}$ (M = Zr or Hf), with a structure similar to that of the cation in the crystalline zirconyl halides.

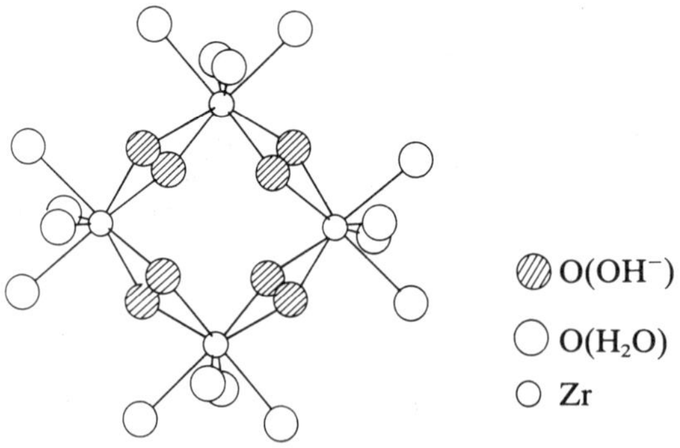

Figure 2 Structure of the $[Zr_4(OH)_8(H_2O)_{16}]^{8+}$ cation

^1H NMR spectra of zirconyl perchlorate in water–acetone mixtures at −70 °C indicate that Zr^{IV} has an average hydration number of ~four.[151] This is in accord with $[Zr_4(OH)_8(H_2O)_{16}]^{8+}$ if it is assumed that only the bound water molecules are observed; the lack of an NMR signal for the hydroxy protons could be due to rapid proton exchange.

Raman spectra of acidic aqueous solutions of $MOCl_2\cdot 8H_2O$ (M = Zr or Hf) have been examined in an effort to detect the zirconyl and hafnyl ions, $M{=}O^{2+}$. No evidence for a $\nu(M{=}O)$ Raman band in the region 800–1000 cm^{-1} was found.[152–154] A broad IR band centered at ~1000 cm^{-1} is observed in the multiple attenuated internal reflectance spectrum of a concentrated $ZrOCl_2$ solution.[154] However, this band shifts to lower frequency upon deuteriation, indicating that the solution species contains hydroxo bridges, and not zirconyl groups. Bands due to δ(ZrOH) in the 910–1030 cm^{-1} region of the IR spectrum of crystalline

$ZrOCl_2{\cdot}8H_2O$ exhibit similar isotopic shifts in the deuteriated analogue.[154,155] Raman spectra of 2.5 M $ZrOCl_2$ in 5 M HCl show a polarized band at 580 cm^{-1}, a depolarized band at 450 cm^{-1} and a shoulder at ~420 cm^{-1}. The former two bands are assigned to totally symmetric and non-totally symmetric ν(Zr—OH) vibrations of the $[Zr_4(OH)_8(H_2O)_{16}]^{8+}$ ion, and the 420 cm^{-1} band is tentatively assigned to a $\nu(Zr—OH_2)$ vibration. Upon addition of NaOH, the 420 cm^{-1} band decreases in intensity and a new band appears at ~530 cm^{-1}, probably due to ν(Zr—OH) vibrations involving terminal hydroxyl groups of a more highly hydrolyzed species such as $[Zr_4(OH)_8(OH)_4(H_2O)_{12}]^{4+}$.[154]

When perchloric acid solutions of Zr^{IV} are diluted from $[Zr^{IV}] \simeq 10^{-2}$ M to $[Zr^{IV}] \leq 10^{-4}$ M and the $HClO_4$ concentration is increased from 0.5 M to 2–4 M, $[Zr_4(OH)_8(H_2O)_{16}]^{8+}$ is converted to the mononuclear ion, presumably $[Zr(H_2O)_8]^{4+}$. The kinetics of this process, abbreviated $Zr_4 \rightarrow Zr$, follow the simple rate law $k_H[H^+][Zr_4]$; at ionic strength 2.0 M ($NaClO_4$): $k_H(25\,°C) = 0.95 \times 10^{-3}\ M^{-1}\,s^{-1}$, $\Delta H^{\ddagger} = 55.2\ kJ\,mol^{-1}$, and $\Delta S^{\ddagger} = -117 \pm 8\ J\,K^{-1}\,mol^{-1}$. Acids HX, such as H_3PO_4, $H_2C_2O_4$ and HF, also induce $Zr_4 \rightarrow Zr$ conversion, giving the rate law $k_H[H^+][Zr_4] + k_{HX}[HX][Zr_4]$.[156]

X-Ray diffraction studies have established the presence of coordinated water molecules and bridging hydroxo ligands in normal and basic Zr (Hf) salts and in certain neutral and anionic complexes. Water- and hydroxo-containing compounds that consist of discrete complexes are listed in Table 6. Averaged metal–oxygen bond lengths to coordinated water molecules and bridging hydroxo ligands are in the ranges 2.20–2.29 and 2.10–2.14 Å, respectively. Structures of these complexes, mostly dodecahedral, are discussed in later sections dealing with the other ligands.

Table 6 Structural Data for Aqua and Hydroxo Complexes of Zirconium(IV) and Hafnium(IV)

Structural formula	*Coordination polyhedron*[a]	*Averaged bond lengths* (Å) *Zr—OH₂*	*Zr—OH*	*Ref.*
$[Zr_4(OH)_8(H_2O)_{16}]Cl_8{\cdot}12H_2O$	DD	2.272	2.142	1
$Na_4[Hf(SO_4)_4(H_2O)_2]$	DD	2.26		2
$[NH_4]_4[Hf(SO_4)_4(H_2O)_2]{\cdot}2H_2O$	DD	2.196		3
$Na_6[Hf(SO_4)_5(H_2O)]$		2.28		4
α-$[(H_2O)_4(SO_4)Zr(SO_4)_2Zr(SO_4)(H_2O)_4]{\cdot}2H_2O$	DD	2.21		5
β-$[(H_2O)_4(SO_4)Zr(SO_4)_2Zr(SO_4)(H_2O)_4]{\cdot}2H_2O$	DD	2.24		6
$[(H_2O)_4(SO_4)Zr(SO_4)_2Zr(SO_4)(H_2O)_4]{\cdot}6H_2O$	DD	2.23		7
$K_4[(H_2O)_2(SO_4)_2Zr(SO_4)_2Zr(SO_4)_2(H_2O)_2]$	DD	2.29		8
	BTP	2.22		9
$K_6[(CO_3)_3Zr(OH)_2Zr(CO_3)_3]{\cdot}6H_2O$	DD		2.10	10
$[NH_4]_6[(CO_3)_3Zr(OH)_2Zr(CO_3)_3]{\cdot}4H_2O$	DD		2.11	11
$[H_2O)_3F_3ZrF_2ZrF_3(H_2O)_3]$	DD	2.283		12
$[NMe_4]_2[(H_2O)F_4ZrF_2ZrF_4(H_2O)]$	PBP	2.271		13
$[Zr(edta)(H_2O)_2]{\cdot}2H_2O$	DD	2.27		14

[a] DD = dodecahedron; BTP = bicapped trigonal prism; PBP = pentagonal bipyramid.

1. T. C. W. Mak, *Can. J. Chem.*, 1968, **46**, 3491.
2. D. L. Rogachev, L. M. Dikareva, V. Ya. Kuznetsov and G. G. Sadikov, *J. Struct. Chem.* (*Engl. Transl.*), 1981, **22**, 470.
3. D. L. Rogachev, L. M. Dikareva, V. Ya. Kuznetsov, V. V. Fomenko and M. A. Porai-Koshits, *J. Struct. Chem.* (*Engl. Transl.*), 1982, **23**, 765.
4. D. L. Rogachev, V. Ya. Kuznetsov, L. M. Dikareva, G. G. Sadikov and M. A. Porai-Koshits, *J. Struct. Chem.* (*Engl. Transl.*), 1980, **21**, 118.
5. I. J. Bear and W. G. Mumme, *Acta Crystallogr.*, *Sect. B*, 1969, **25**, 1566.
6. I. J. Bear and W. G. Mumme, *Acta Crystallogr.*, *Sect. B*, 1969, **25**, 1572.
7. I. J. Bear and W. G. Mumme, *Acta Crystallogr.*, *Sect. B*, 1969, **25**, 1558.
8. W. G. Mumme, *Acta Crystallogr.*, *Sect. B*, 1971, **27**, 1373.
9. M. A. Porai-Koshits, V. I. Sokol and V. N. Vorotnikova, *J. Struct. Chem.* (*Engl. Transl.*), 1972, **13**, 815.
10. Yu. E. Gorbunova, V. G. Kuznetsov and E. S. Kovaleva, *J. Struct. Chem.* (*Engl. Transl.*), 1968, **9**, 815.
11. A. Clearfield, *Inorg. Chim. Acta*, 1970, **4**, 166.
12. F. Gabela, B. Kojic-Prodic, M. Sljukic and Z. Ruzic-Toros, *Acta Crystallogr.*, *Sect. B*, 1977, **33**, 3733.
13. R. L. Davidovich, M. A. Medkov, V. B. Timchenko and B. V. Bukvetskii, *Bull. Acad. Sci. USSR, Div. Chem. Sci.*, 1983, 2181.
14. A. I. Pozhidaev, M. A. Porai-Koshits and T. N. Polynova, *J. Struct. Chem.* (*Engl. Transl.*), 1974, **15**, 548.

Very little X-ray data is available for zirconium or hafnium complexes that contain oxo ligands. The only X-ray evidence for the existence of the zirconyl or hafnyl group, $M{=}O^{2+}$, comes from a structural study of K_2ZrO_3.[157] This compound contains chains of ZrO_5 square pyramids that share basal edges. The Zr—O distance to the apical oxygen atom (1.92 Å) is appreciably shorter than the sum of ionic radii (2.20 Å), the sum of the covalent radii (2.22 Å) and the Zr—O distances to the four basal, bridging oxygen atoms (2.13 Å). In his 1973 review of the structural chemistry of zirconium compounds, MacDermott[54] identifies the apical Zr—O bond in K_2ZrO_3 as the only probable example of a zirconium–oxygen double bond yet reported. He adds, 'There is no doubt that the 'zirconyl' group as a persistent species in solution is mythical.'

Planalp and Anderson[158] have recently reported the structure of $[\{ZrME[N(SiMe_3)_2]_2\}_2O]$, a centrosymmetric oxo-bridged dinuclear complex having a linear Zr—O—Zr bridge and a short Zr—O bond distance (1.950(1) Å). μ-Oxo compounds of the type $[\{MX(\eta\text{-}C_5H_5)_2\}_2O]$ (M = Zr or Hf; X = Cl, Me or SPh) are common in organometallic chemistry.[14] These compounds exhibit a characteristic IR band associated with the M—O—M unit at ~750–800 cm^{-1} and M—O bond distances in the neighborhood of 1.95 Å. The Zr—O bond lengths (average 1.959(3) Å) in the cyclic trinuclear complex $[\{ZrO(\eta\text{-}C_5H_5)_2\}_3]$ are similar to those in the dinuclear complexes, indicative of an appreciable amount of zirconium–oxygen $d\pi$–$p\pi$ bonding.[159]

Numerous attempts have been made to identify the zirconyl or hafnyl group in solid compounds on the basis of vibrational spectroscopy. Barraclough, Lewis and Nyholm[160] found that compounds known to have an M═O bond exhibit a strong, sharp IR band in the region 900–1100 cm^{-1} while polymers with an —M—O—M—O— chain structure give a broad band at somewhat lower frequency. Kharitonov and Zaitsev[161] suggested the range ~800–1000 cm^{-1} for ν(Zr═O) and ν(Hf═O), and noted that δ(MOH) deformation modes can give rise to strong bands in the same region. Thus, study of deuteriated compounds is necessary in order to avoid incorrect assignments. A case in point is $ZrOCl_2{\cdot}4H_2O$. This compound exhibits a sharp IR band at 918 cm^{-1}, assigned to ν(Zr═O) by Spasibenko and Goroshchenko.[162] However, Powers and Gray[155] showed that the 918 cm^{-1} band is displaced to 697 cm^{-1} in the spectrum of $ZrOCl_2{\cdot}4D_2O$, and therefore it cannot be assigned to the zirconyl group.

The IR spectrum of anhydrous $ZrOCl_2$, prepared by reaction of $ZrCl_4$ with Cl_2O, exhibits an intense band at 877 cm^{-1} and a broad band at 538 cm^{-1}. The 877 cm^{-1} band was assigned to the ZrO^{2+} cation, and the 538 cm^{-1} band to the —Zr—O—Zr—O— chain of a polymeric $\{ZrOCl_4^{2-}\}_n$ anion.[163] The same bands have been observed in the IR spectrum of an amorphous form of anhydrous $ZrOCl_2$ obtained from the reaction of $ZrCl_4$ with Sb_2O_3.[164] On the other hand, anhydrous $ZrOCl_2$ prepared by (i) thermal decomposition of $Zr_2OCl_6{\cdot}4MeCN$[165] or (ii) partial dehydration of $ZrOCl_2{\cdot}8H_2O$ with thionyl chloride followed by heating *in vacuo* at 200 °C[166] showed no bands that could be assigned to the zirconyl group.

Several oxochloro complexes of zirconium have been reported. These compounds are listed in Table 7 along with frequencies of IR bands that have been assigned to zirconium–oxygen stretching modes. The acetonitrile and dioxane adducts of Zr_2OCl_6, $[Zr_2OCl_8]^{2-}$ and $[Zr_2OCl_{10}]^{2-}$ are believed to contain Zr—O—Zr bridging units. The strong, sharp band at 1104 cm^{-1} in the spectrum of $Zr_2OCl_6{\cdot}4POCl_3$ has been cited as evidence for a Zr═O group, and both Zr═O groups and —Zr—O—Zr—O— chains have been suggested for $(C_5H_5NH)_2ZrOCl_4$.

Table 7 Zirconium—Oxygen Stretching Frequencies for Oxo–Chloro Complexes of Zirconium(IV)

Compound	ν(*ZrO*) (cm^{-1})	*Ref.*
$Zr_2OCl_6{\cdot}4MeCN$	800	1
$Zr_2OCl_6{\cdot}5C_4H_8O_2$	740	1
$Zr_2OCl_6{\cdot}4POCl_3$	1104	1
$[NO]_2[Zr_2OCl_8]$	605, 584	2
$[NEt_4]_4[Zr_2OCl_{10}]$	750	3
$(C_5H_5NH)_2ZrOCl_4$[a]	1000, 770	4

[a] Isolated from methanol. Another form of $(C_5H_5NH)_2ZrOCl_4$, isolated from carbon tetrachloride, shows a ν(ZrO) band at ~920 cm^{-1} (ref. 5).

1. A. Feltz, *Z. Anorg. Allg. Chem.*, 1965, **335**, 304.
2. H. Prinz, K. Dehnicke and U. Müller, *Z. Anorg. Allg. Chem.*, 1982, **448**, 49.
3. A. Feltz, *Z. Anorg. Allg. Chem.*, 1968, **358**, 21.
4. G. M. Toptygina, I. B. Barskaya and I. Z. Babievskaya, *Russ. J. Inorg. Chem.* (*Engl. Transl.*), 1972, **17**, 1104.
5. G. M. Toptygina, I. B. Barskaya and I. Z. Babievskaya, *Russ. J. Inorg. Chem.* (*Engl. Transl.*), 1971, **16**, 686.

Kharitonov and coworkers have presented IR evidence for the existence of the M═O group in two classes of zirconyl (hafnyl) compounds: (i) oxyfluorides such as $ZrOF_2$, $KZrOF_3 \cdot 2H_2O$, $Zr_4F_{10}O_3$, $HfOF_2$, $Hf_4F_{14}O$ and $Hf_4F_{12}O_2$;[167] (ii) isothiocyanates such as $A[ZrO(NCS)_3(H_2O)] \cdot H_2O$ ($A = NH_4$, C_5H_5NH, K, Rb or Cs), $A[HfO(NCS)_3(H_2O)] \cdot H_2O$ ($A = C_5H_5NH$ or Cs) and $[C_5H_5NH]_3[(HfO)_2(NCS)_7] \cdot H_2O$.[99,100] However, Selbin[53] has been unable to duplicate the oxyfluoride work and Clearfield[52] has suggested that the intense, but rather broad, IR band reported for the oxyfluorides in the region 833–896 cm^{-1} could be assigned to an M—O—M vibration instead of an M═O mode. The isothiocyanate complexes display a sharp, rather intense band at 913–927 cm^{-1} (M = Zr) or 934–940 cm^{-1} (M = Hf),[99,100] and this has been confirmed by Selbin[53] for $A[ZrO(NCS)_3(H_2O)] \cdot H_2O$ (A = Cs or NMe_4). However, the corresponding deuteriated compounds have not been studied, and therefore it is possible that the 913–927 and 934–940 cm^{-1} band arises from a δ(MOH) vibration rather than a ν(M═O) mode.[161] Lipatova and Semenova[168] have reported that this band disappears when $Cs[MO(NCS)_3(H_2O)] \cdot H_2O$ (M = Zr or Hf) is dissolved in acetone, and reappears when the solvent is evaporated. This observation was interpreted in terms of a reversible hydrolysis of the M═O group of the solid compound, giving an $M(OH)_2$ moiety in solution. The mode of attachment of the oxygen atom in the $A[MO(NCS)_3(H_2O)] \cdot H_2O$ complexes remains uncertain, and X-ray studies are needed to establish the structure of these compounds.

The zirconyl group has also been said to be present in oxo zirconium(IV) complexes of the type $[ZrOX_2L_2]$ (X = Cl, Br, NO_3 or NCS), $[ZrOL_4]I_2$, and $[ZrOL_6][ClO_4]_2$, where L is an oxygen donor ligand such as diphenyl sulfoxide,[169,170] tetramethylene sulfoxide,[171] pyridine *N*-oxide,[172] 2-methylpyridine *N*-oxide,[173] 2,6-lutidine *N*-oxide,[172] triphenylphosphine oxide[174] or hexamethylphosphoramide[175] and in the analogous 2,2′-bipyridine 1,1′-dioxide complexes $[ZrOX_2(bipyO_2)] \cdot 2H_2O$ (X = Cl, Br, NO_3 or NCS), $[ZrO(bipyO_2)_2]I_2 \cdot 2H_2O$ and $[ZrO(bipyO_2)_3][ClO_4]_2 \cdot 2H_2O$.[176,177] In general, the chloro, bromo, nitrato and isothiocyanato complexes are monomeric nonelectrolytes in ionizing organic solvents, while the iodide and perchlorate compounds behave as 1:2 electrolytes. IR spectra of these compounds show that (i) the nitrate ligands are bidentate, (ii) the NCS ligands are attached to zirconium through the nitrogen atom, and (iii) the perchlorate compounds contain uncoordinated ClO_4^- ions. A weak IR band in the 900–980 cm^{-1} region has been assigned to the ν(Zr═O) mode, but the low intensity of this band argues against such an assignment. On the other hand, the molecular weight and conductance data are consistent with mononuclear structures that contain a Zr═O group. Unfortunately, no X-ray structural data are available. Other compounds of unknown structure that may possibly contain a Zr═O group include oxozirconium(IV) complexes with N-heterocyclic ligands (Section 32.4.2.2), tridentate Schiff bases (Section 32.4.9.1.i) and 1,10-phenanthroline mono-*N*-oxide (Section 32.4.9.3).

Previous reviewers [51–54] have maintained a healthy skepticism concerning the existence of the $Zr{=}O^{2+}$ and $Hf{=}O^{2+}$ ions. There is no evidence for the existence of these species in aqueous solution and IR evidence for the presence of Zr═O and Hf═O groups in solid compounds is equivocal. These structural units appear to be rare, and only X-ray studies can provide really convincing evidence for their existence.

Relatively insoluble, high-melting oxometal complexes are likely to have polymeric structures based on —M—O—M—O chains. IR bands assignable to ν(Zr—O—Zr) vibrations suggest structures of this type for $ZrO(O_2CR)_2 \cdot nRCO_2H$ (R = Me, Et, Pr, CH_2Cl or Ph),[178] $ZrO(O_2SeC_6H_4X)_2$ (X = 4-Cl, 3-Cl or 3-Br), and $ZrO(O_2SeC_6H_4\text{-}3\text{-}NO_2)_2 \cdot 2H_2O$.[179] The carboxylato complexes exhibit a broad band of medium intensity in the 800–865 cm^{-1} region, and the benzeneseleninato complexes show a medium or strong band at 739–750 cm^{-1}.

32.4.4.2 Peroxide and hydroperoxide

A variety of peroxo and hydroperoxo complexes of zirconium(IV) and hafnium(IV) have been isolated from aqueous or aqueous methanolic hydrogen peroxide solutions that contain additional ligands such as sulfate, oxalate or fluoride. Examples of recently reported complexes are listed in Table 8 along with characteristic vibrational frequencies and the pH employed in the aqueous preparations. Earlier work on peroxo compounds has been reviewed by Connor and Ebsworth[180] and by Larsen.[5]

The peroxo complexes are white solids, soluble in mineral acids, but insoluble in common organic solvents. They are relatively stable at room temperature; however, they lose active oxygen at elevated temperatures. Several of the compounds in Table 8 were isolated with

Table 8 IR and Raman Data and Synthesis Conditions (pH) for Peroxo and Hydroperoxo Complexes of Zirconium(IV) and Hafnium(IV)

Compound		$\nu(O—O)$ (cm^{-1})	$\nu_{as}(MO_2)$ (cm^{-1})	$\nu_s(MO_2)$ (cm^{-1})	*pH*	*Ref.*
$[Zr_2(O_2)_3(SO_4)(H_2O)_4]\cdot(4–6)H_2O$	IR	865m, 840vw			0.2–0.3	1
	R	872s, 840s				
$[Zr(O_2)(C_2O_4)(H_2O)_2]\cdot(3–4)H_2O$	IR	862m	600s	500w	0.2–0.3	2, 3
	R	867s		495s		
$[NH_4]_3[Zr(O_2)F_5]$	IR	839w	480s, bd	555w	7	4
$K_2Zr_2(O_2)_2F_6\cdot(4–5)H_2O$	IR	840				5, 6
$(NH_4)_3Zr_2(O_2)_2F_7\cdot(0.6–2)H_2O$	IR	850m	600s			7
$K_3Zr_2(O_2)_2F_7\cdot(0.6–2)H_2O$	IR	840m	590s			5, 7
	R	850s		536m		
$Rb_3Zr_2(O_2)_2F_7\cdot(0.6–2)H_2O$	IR	850m	590s			7
	R	850s		536m		
$Cs_3Zr_2(O_2)_2F_7\cdot(0.6–2)H_2O$	IR	850m	600s			7
	R	850s		538m		
$Zr(O_2)F_2\cdot 2H_2O$	IR	870vw			2–3	8
	R	867vs, 846s		500m		
$(NH_4)_3Zr_2(O_2)_2F_7$	IR	862w			6–7	8
	R			565s		
$K_3Zr_2(O_2)_2F_7$	IR	840vw, 845vw			5–6	8
	R	861m, 840vs		528m		
$[NH_4]_3[ZrF_6(O_2H)]$	IR	844m			6–7	8
	R	837vs				
$[NH_4]_3[HfF_6(O_2H)]$					7–8	8
$K_2ZrF_5(O_2H)$	IR	835m			5–6	8
	R	838vs				
$K_2HfF_5(O_2H)$	IR	840m			7	8

1. G. V. Jere and G. D. Gupta, *J. Inorg. Nucl. Chem.*, 1970, **32**, 537.
2. G. D. Gupta and G. V. Jere, *Indian J. Chem.*, 1968, **6**, 54.
3. G. D. Gupta and G. V. Jere, *Indian J. Chem.*, 1972, **10**, 102.
4. W. P. Griffith and T. D. Wickins, *J. Chem. Soc. (A)*, 1968, 397.
5. M. T. Santhamma and G. V. Jere, *Synth. React. Inorg. Metal-Org. Chem.*, 1977, **7**, 413.
6. G. V. Jere, G. D. Gupta, V. Raman and M. T. Santhamma, *Indian J. Chem., Sect. A*, 1978, **16**, 435.
7. G. V. Jere and M. T. Santhamma, *Inorg. Chim. Acta*, 1977, **24**, 57.
8. S. O. Gerasimova, Yu. Ya. Kharitonov, S. A. Polishchuk and L. M. Avkhutskii, *Sov. J. Coord. Chem. (Engl. Transl.)*, 1981, **7**, 267.

variable amounts of lattice-held water of crystallization. The distinction between lattice-held and coordinated water molecules was made on the basis of thermogravimetric analysis. These complexes exhibit characteristic IR and Raman bands in the regions 840–880 and 500–600 cm^{-1} (Table 8). The band at 840–880 cm^{-1}, observed in both IR and Raman spectra, has been assigned to the O—O stretching mode of the bidentate peroxo ligand in the triangularly linked MO_2 group (**20**).[181–184] Two bands in the 500–600 cm^{-1} region have been attributed to the symmetric and asymmetric metal–oxygen stretching vibrations, $\nu_s(MO_2)$ and $\nu_{as}(MO_2)$, although this assignment is somewhat uncertain for the peroxofluoro complexes owing to the presence of ν(M—F) modes in the same frequency region.[181,185]

(20)

Jere and Gupta[182] have reported two ν(O—O) bands for $[Zr_2(O_2)_3(SO_4)(H_2O)_4]\cdot(4–6)H_2O$, a band at ~870 cm^{-1}, fairly strong in both IR and Raman spectra, which has been assigned to a triangularly linked bidentate peroxo ligand, and a band at 840 cm^{-1}, strong in the Raman spectrum but very weak in the IR, which was attributed to a bridging peroxo group. In principle, one should be able to distinguish these two modes of attachment since ν(O—O) of a triangular MO_2 group is both IR and Raman active, while ν(O—O) of an M—O—O—M group is active in the Raman spectrum but inactive (C_i symmetry) or only weakly allowed (C_2 symmetry) in the IR spectrum. However, Gerasimova *et al.*[185] have noted that ν(O—O) of a triangular MO_2 group may also be very weak in the IR, and thus distinguishing the two modes of peroxide attachment may be problematic.

Structures have been suggested for some of the peroxo complexes on the basis of the vibrational spectra.[182–184] However, no X-ray structures are yet available.

The hydroperoxo complexes $K_2MF_5(O_2H)$ (M = Zr or Hf) may be distinguished from the peroxo complexes on the basis of a unique IR band of moderate intensity at ~1450 cm^{-1}. This band has been assigned to $\delta(O_2H)$.[185] X-Ray powder patterns indicate that $[NH_4]_3[MF_6(O_2H)]$ and $[NH_4]_3[MF_7]$ (M = Zr or Hf) are isostructural. Since $[MF_7]^{3-}$ has a pentagonal bipyramidal structure, the $[MF_6(O_2H)]$ ions may be assigned an analogous seven-coordinate structure, with the O_2H^- group behaving as a monodentate ligand.[185]

32.4.4.3 Alkoxides, trialkysilyloxides and aryloxides

Metal alkoxides have been extensively reviewed by Bradley[61,186–188] and by Mehrotra.[188–191] Zirconium and hafnium tetraalkoxides may be prepared by (i) reaction of alcohols with the anhydrous metal chlorides or pyridinium hexachlorometallates in the presence of ammonia (equations 26 and 27), (ii) alcoholysis or alcohol interchange (equation 28), (iii) transesterifications (equation 29) or (iv) reaction of alcohols with metal dialkylamides or metal acetylacetonates (equations 30 and 31). A preferred method for preparation of many $M(OR)_4$ (M = Zr or Hf) compounds in a high state of purity involves alcoholysis of $M(OCHMe_2)_4 \cdot HOCHMe_2$ in benzene solution; the isopropanol solvate can be purified by recrystallization from isopropanol, and the isopropanol produced in the alcoholysis is conveniently removed by fractional distillation of the benzene–isopropanol azeotrope. This approach, however, is inefficient for synthesis of the *t*-butoxides because of the small (~0.2 °C) difference in boiling points for isopropanol and *t*-butanol. The *t*-butoxides are better prepared from the tetrakis(diethylamide) (equation 30) or by transesterification between $M(OCHMe_2)_4 \cdot HOCHMe_2$ and *t*-butyl acetate.

$$MCl_4 + 4ROH + 4NH_3 \longrightarrow M(OR)_4 + 4NH_4Cl \quad (26)$$

$$[C_5H_5NH]_2[MCl_6] + 4ROH + 6NH_3 \longrightarrow M(OR)_4 + 6NH_4Cl + 2C_5H_5N \quad (27)$$

$$M(OR)_4 + 4R'OH \longrightarrow M(OR')_4 + 4ROH \quad (28)$$

$$M(OR)_4 + 4MeCOOR' \longrightarrow M(OR')_4 + 4MeCOOR \quad (29)$$

$$Zr(NEt_2)_4 + 4Me_3COH \longrightarrow Zr(OCMe_3)_4 + 4Et_2NH \quad (30)$$

$$Zr(acac)_4 + 4ROH \longrightarrow Zr(OR)_4 + 4Hacac \quad (31)$$

The reaction of alcohols with $ZrCl_4$ in the absence of ammonia yields chloride alkoxides $ZrCl_3(OR) \cdot nROH$ and $ZrCl_2(OR)_2 \cdot nROH$. Mixed alkoxides such as $Zr(OR)(OCMe_3)_3$ and $Zr(OR)_2(OCMe_3)_2$ (R = Me or Et) have been obtained from alcoholysis reactions between (i) $Zr(OR)_4$ and *t*-butanol or (ii) $Zr(OCMe_3)_4$ and ROH. The monomethoxy complex is dimeric, $[\{Zr(OMe)(OCMe_3)_3\}_2]$, and probably has a methoxy-bridged structure.

Boiling points and molecular association data for the tetraalkoxides are listed in Table 9. The methoxides are insoluble, polymeric solids of low volatility; $Zr(OMe)_4$ sublimes in a molecular still at 280 °C. $Ti(OMe)_4$ is a tetramer, but the structures of $Zr(OMe)_4$ and $Hf(OMe)_4$ remain unknown. Most of the other tetraalkoxides are volatile liquids, which are soluble in organic solvents. Two forms of zirconium isopropoxide (liquid b.p. 160 °C/0.1 mmHg and crystalline m.p. 135 °C) have been described.[192] For tetraalkoxides that contain the same number of carbon atoms, the boiling point and the degree of association (n) in boiling benzene decrease with increasing branching of the alkyl chain. Thus, in the case of the eight isomeric zirconium amyloxides, the primary amyloxides have values of n in the range 3.7–2.4, while the secondary amyloxides are dimers and the tertiary amyloxide $Zr(OCMe_2Et)_4$ is a monomer. The tertiary alkoxides are monomeric because steric hindrance prevents alkoxide bridging, which causes oligomerization of the primary and secondary alkoxides. Careful vapor pressure measurements on $M(OCMe_3)_4$ and $M(OCMe_2Et)_4$ have shown that volatility increases as M varies in the order Ti < Zr < Hf; the increase in volatility with increasing molecular weight has been attributed to an entropy effect.[193,194] $Ti(OCMe_2Et)_4$ and $Hf(OCMe_2Et)_4$ have been separated by gas chromatography, but attempts to separate $Zr(OCMe_2Et)_4$ and $Hf(OCMe_2Et)_4$ were unsuccessful.[195]

^1H NMR spectra of the trimeric isopropoxides $M(OCHMe_2)_4$ (M = Zr or Hf) indicate rapid exchange of terminal and bridging isopropoxide groups; ^1H NMR spectra of mixtures of

Table 9 Boiling Points and Degree of Association for Zirconium(IV) and Hafnium(IV) Alkoxides and Trialkylsilyloxides

Compound	*B.p.* (°C/mmHg)	n^a	*Ref.*	*Compound*	*B.p.* (°C/mmHg)	*M.p.*	n^a	*Ref.*
$Zr(OMe)_4$	280[b]		1	$Hf(OMe)_4$				2
$Zr(OEt)_4$	180–200/0.1	3.6	2, 3	$Hf(OEt)_4$	180–200/0.1			2
$Zr(OPr)_4$	208/0.1		3					
$Zr(OCHMe_2)_4$	160/0.1	3.0, 3.6[c]	4, 5	$Hf(OCHMe_2)_4$	170/0.35		3.3[c]	2, 5
$Zr(OBu)_4$	243/0.1	3.4	3					
$Zr(OCHMeEt)_4$	164/0.1	2.5	4					
$Zr(OCMe_3)_4$	89.1/5.0	1.0	6	$Hf(OCMe_3)_4$	90/6.5			2
$Zr(OCH_2CH_2CH_2CH_2Me)_4$	276/0.1	3.2	3	$Zr(OSiMe_3)_4$	135/0.1[b]	152	2.0	12
$Zr(OCH_2CH_2CHMe_2)_4$	247/0.1	3.3	7					
$Zr(OCH_2CHMeEt)_4$	238/0.1	3.7	7					
$Zr(OCH_2CMe_3)_4$	188/0.2	2.4	7					
$Zr(OCHEt_2)_4$	181/0.1	2.0	4					
$Zr(OCHMePr)_4$	178/0.1	2.0	4					
$Zr(OCHMeCHMe_2)_4$	176/0.1	2.0	4					
$Zr(OCMe_2Et)_4$	95/0.1	1.0	7	$Hf(OCMe_2Et)_4$	92/0.1			2
$Zr(OCHMeBu)_4$	190/0.1		4	$Zr(OSiMe_2Et)_4$	105/0.1	105	1.2	12
$Zr(OCHMeCMe_3)_4$	128/0.1	1.0	4					
$Zr(OCMeEt_2)_4$	171.4/5.0	1.0	6	$Zr(OSiMeEt_2)_4$	120/0.1	30	1.0	12
$Zr(OCMe_2Pr)_4$	161.6/5.0		6	$Zr(OSiMe_2Pr)_4$	103/0.05	60	1.1	12
$Zr(OCMe_2CHMe_2)_4$	133/0.1		6	$Zr(OSiMe_2CHMe_2)_4$	110/0.1		1.0	12
$Zr(OCHPr_2)_4$	163/0.1	1.0	4					
$Zr\{OCH(CHMe_2)_2\}_4$	158/0.1	1.0	4					
$Zr(OCEt_3)_4$	180/0.4	1.0	6	$Zr(OSiEt_3)_4$	147/0.1		1.0	12
$Zr(OCMe_2CMe_3)_4$	168/0.5		6					
$Zr(OCMeEtCHMe_2)_4$	194/0.7		6					
$Zr(OCH_2CF_3)_4$	146/0.8	3.0	8					
$Zr\{OCH(CF_3)_2\}_4$	110/0.8	1.8	9, 10	$Hf\{OCH(CF_3)_2\}_4$				10
$Zr(OCMe_2CCl_3)_4$		1.0	11					

[a] Degree of association, in boiling benzene unless indicated otherwise. [b] Sublimes. [c] In freezing benzene.

1. D. C. Bradley and M. M. Factor, *Nature (London)*, 1959, **184,** 55.
2. D. C. Bradley, R. C. Mehrotra and W. Wardlaw, *J. Chem. Soc.*, 1953, 1634.
3. D. C. Bradley, R. C. Mehrotra, J. D. Swanwick and W. Wardlaw, *J. Chem. Soc.*, 1953, 2025.
4. D. C. Bradley, R. C. Mehrotra and W. Wardlaw, *J. Chem. Soc.*, 1952, 5020.
5. D. C. Bradley and C. E. Holloway, *J. Chem. Soc. (A)*, 1968, 1316.
6. D. C. Bradley, R. C. Mehrotra and W. Wardlaw, *J. Chem. Soc.*, 1952, 4204.
7. D. C. Bradley, R. C. Mehrotra and W. Wardlaw, *J. Chem. Soc.*, 1952, 2027.
8. P. N. Kapoor and R. C. Mehrotra, *Chem. Ind. (London)*, 1966, 1034.
9. P. N. Kapoor, R. N. Kapoor and R. C. Mehrotra, *Chem. Ind. (London)*, 1968, 1314.
10. K. S. Mazdiyasni, B. J. Schaper and L. M. Brown, *Inorg. Chem.*, 1971, **10,** 889.
11. D. C. Bradley, R. N. P. Sinha and W. Wardlaw, *J. Chem. Soc.*, 1958, 4651.
12. D. C. Bradley and I. M. Thomas, *J. Chem. Soc.*, 1959, 3404.

$M(OCMe_3)_4$ and *t*-butanol evidence rapid intermolecular exchange of *t*-butoxide ligands.[196] IR spectra of zirconium and hafnium tetraalkoxides exhibit a strong ν(C—O) band near 1000 cm^{-1} and one or more ν(M—O) bands in the region 520–590 cm^{-1}.[197,198] Calorimetric measurements have afforded the following values for mean thermochemical bond energies (kJ mol^{-1}) in $Zr(OCHMe_2)_4$ and $Hf(OCHMe_2)_4$: Zr—O, 552; Hf—O, 573.[68] No X-ray structural data are available for the tetraalkoxides except for a preliminary report of a study of the isopropanol solvates $M(OCHMe_2)_4{\cdot}HOCHMe_2$ (M = Zr or Hf). These complexes have a dimeric structure in which two MO_6 octahedra share an edge.[199]

Zirconium and hafnium tetraalkoxides are highly reactive compounds. They react with water, alcohols, silanols, hydrogen halides, acetyl halides, certain Lewis bases, aryl isocyanates and other metal alkoxides. With chelating hydroxylic compounds HL, such as β-diketones, carboxylic acids and Schiff bases, they give complexes of the type $ML_n(OR)_{4-n}$; these reactions are discussed in the sections dealing with the chelating ligand.

Partial hydrolysis of zirconium alkoxides yields polymeric oxozirconium alkoxides such as $Zr_3O_4(OR)_4$ (R = Et or CMe_3), $Zr_2O_3(OCHMe_2)_2$ and $Zr_2O_3(OCMe_3)_2{\cdot}4HOCMe_3$;[200,201] complete hydrolysis gives hydrated zirconium dioxide. An X-ray study[202] of an oxo zirconium methoxide hydrolysis product has shown that the crystals contain molecules of $[Zr_{13}O_8(OMe)_{36}]$. This complex has a $\bar{3}m(D_{3d})$ structure (Figure 3) consisting of a cubic close packed arrangement of 13 Zr atoms, three above and three below a plane of seven. The central

Zr atom is surrounded by a slightly distorted cube of eight oxygen atoms (at 2.19 and 2.23 Å), each of which is attached to three of the other Zr atoms. The 12 peripheral Zr atoms are seven-coordinate, being bonded to two bridging oxygen atoms (at 2.15–2.19 Å), four bridging methoxide ligands (at 2.12–2.21 Å) and one terminal methoxide ligand (at 2.00 and 2.03 Å). An alternate route to oxozirconium alkoxides is the reaction of alcohols with $ZrOCl_2{\cdot}2MeCO_2H$ in the presence of piperidine. This approach has been used to prepare compounds of the type $ZrO(OR)_2{\cdot}nROH$ and $ZrOCl(OR){\cdot}nROH$ (R = Me, Et, $CHMe_2$ or CMe_3; $n = 0.5–2.0$.[203]

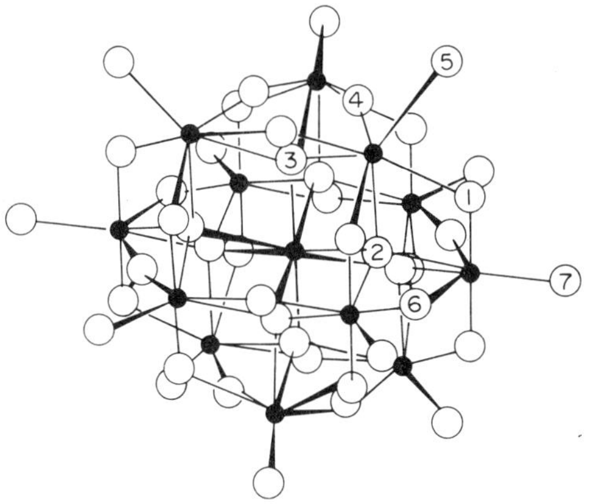

Figure 3 The arrangement of zirconium atoms (solid circles) and oxygen atoms (open circles) in molecules of $[Zr_{13}O_8(OMe)_{36}]$; O(2) and O(3) are bridging oxygen atoms while the other oxygen atoms are part of methoxide groups (reproduced by permission from ref. 202)

The synthetic utility of alcoholysis reactions of $Zr(OCHMe_2)_4{\cdot}HOCHMe_2$ has been mentioned already. Additional examples are the preparation of glycoxides of the type $Zr\{O(CR_2)_xO\}$ $(OCHMe_2)_2$, $Zr\{O(CR_2)_xO\}_2$ and $Zr\{O(CR_2)_xO\}\{O(CR_2)_xOH\}_2$ (R = H or Me)[204] and *N*-methylaminoalkoxides of the type $Zr(OR)_n(OCHMe_2)_{4-n}$ (R = CH_2CH_2NHMe, $CH_2CH_2NMe_2$ or $CHMeCH_2NMe_2$; n = 1, 2, 3 or 4)[205] from stoichiometric amounts of $ZrO(OCHMe_2)_4{\cdot}HOCHMe_2$ and the appropriate glycol or aminoalcohol. The *N*-methylaminoalkoxides are monomeric in boiling benzene except for $Zr(OCH_2CH_2NHMe)_n(OCHMe_2)_{4-n}$ (n = 1 or 2), which are dimeric.

Silanolysis of $Zr(OR)_4$ (R = $CHMe_2$ or CMe_3) (equation 32) gives the corresponding trialkylsilyloxides $Zr(OSiR'_3)_4$ (Table 9). These compounds are colorless liquids or white solids that can be distilled or sublimed under reduced pressure. They are less susceptible to hydrolysis and possess greater thermal stability than the corresponding *t*-alkoxides. It is interesting to note that the trialkylsilyloxides are slightly more associated than the *t*-alkoxides, presumably because steric effects of the R_3SiO group are smaller that those of the R_3CO group; $Zr(OSiMe_3)_4$ is a dimer in boiling benzene.

$$Zr(OR)_4 + 4R'_3SiOH \longrightarrow Zr(OSiR'_3)_4 + 4ROH \qquad (32)$$

Halide alkoxides $ZrX_n(OR)_{4-n}$ (X = Cl or Br) are formed in reactions of zirconium alkoxides with hydrogen halides or acetyl halides. Thus $Zr(OCHMe_2)_4{\cdot}HOCHMe_2$ reacts with acetyl chloride in 1:1, 1:2 and 1:3 mole ratios yielding isopropyl acetate and $ZrCl(OCHMe_2)_3{\cdot}HOCHMe_2$, $ZrCl_2(OCHMe_2)_2{\cdot}HOCHMe_2$ and $ZrCl_3(OCHMe_2)$, respectively. When an excess of acetyl chloride is used, the product is $ZrCl_4{\cdot}2MeCO_2CHMe_2$.[206]

Relatively few coordination compounds are known in which zirconium or hafnium alkoxides bond to Lewis bases. Zirconium isopropoxide forms 1:1 adducts with isopropanol, pyridine, hydrazine and ethylenediamine, but it does not bond to donors such as diethyl ether, dimethylamine, triethylamine, 2,2′-bipyridyl or thiourea.[207,208] Halogen substitution in the alkoxide ligand appears to increase the Lewis acid character; thus $Zr(OCH_2CCl_3)_4$ forms a 1:2 adduct with acetone. Chloride alkoxides $ZrCl_n(OR)_{4-n}$ (n = 1, 2 or 3) and ZrOCl(OR) are better Lewis acids than the corresponding $Zr(OR)_4$ and $ZrO(OR)_2$ compounds. $ZrOCl(OCHMe_2)_2{\cdot}2HOCHMe_2$ reacts in dichloromethane with a variety of Lewis bases yielding adducts of the type ZrOCl $(OCHMe_2){\cdot}L$ (L = py, bipy, quinoline, DMF, dimethylacetamide or phthalimide) and $ZrOCl(OCHMe_2){\cdot}3DMSO$, whereas the corresponding dialkoxides $ZrO(OR)_2{\cdot}ROH$ (R = Me, Et or $CHMe_2$) failed to react.[209]

Zirconium tetraalkoxides $Zr(OR)_4$ (R = Pr, $CHMe_2$ or CMe_3) react with stoichiometric amounts of phenyl or naphthyl isocyanate to give the insertion products $Zr\{NR'C(O)OR\}_n(OR)_{4-n}$ ($R' = Ph$ or $C_{10}H_7$; $n = 1, 2, 3$ or 4).[210]

Reaction of zirconium and hafnium alkoxides with alkali metal, alkaline earth metal, aluminum and gallium alkoxides yields bimetallic alkoxides of the type: $M'Zr(OCMe_3)_5$, $M'M_2(OR)_9$ and $M'_2M_3(OR)_{14}$ ($M' =$ alkali metal; $M = Zr$ or Hf; $R =$ alkyl); $M''\{M_2(OCHMe_2)_9\}_2$ and $M''M_3(OCHMe_2)_{14}$ ($M'' = Mg$, Ca, Sr or Ba); $M'''M(OCHMe_2)_7$ and $M'''_2M(OCHMe_2)_{10}$ ($M''' = Al$ or Ga).[188,190] The bimetallic alkoxides are volatile, covalent compounds that can be sublimed or distilled under reduced pressure. Most of these compounds are monomeric in boiling benzene, though some tend to dimerize. Structures have been suggested on the basis of molecular weight data, ^1H NMR spectra, and the products of alcoholysis reactions, but no X-ray structural data are yet available. A recent study of solubility in the $Al(OCHMe_2)_3$–$Zr(OCHMe_2)_4$–L (L = THF or $HOCHMe_2$) system has confirmed the existence of $AlZr(OCHMe_2)_7$, but failed to provide evidence for $Al_2Zr(OCHMe_2)_{10}$.[211] The existence of bimetallic alkoxides $M'''Zr_3(OCHMe_2)_{15}\cdot 3HOCHMe_2$ has been established in the $M'''(OCHMe_2)_3$–$Zr(OCHMe_2)_4$–$HOCHMe_2$ ($M''' = Y$ or La) systems.[212,213]

Compounds of composition $[C_5H_5NH]_2[MCl_4(OMe)_2]$ ($M = Zr$ or Hf), $[C_5H_5NH]_2$-$[ZrCl_5(OEt)]$ and $[C_5H_5NH]_2[ZrCl_2(OEt)_4]$ have been prepared by reaction of the metal tetrachlorides with methanol or ethanol in the presence of pyridine.[214]

The reaction of $ZrCl_4$ with Li(tritox) in diethyl ether yields $[ZrCl_3(tritox)_2\cdot Li(OEt_2)_2]$, where tritox is the sterically bulky tri-*t*-butylmethoxide ligand, $(Me_3C)_3CO$. An X-ray study of this compound revealed trigonal bipyramidal coordination about the zirconium atom, with the two tritox ligands in equatorial sites and three chlorine atoms in the remaining sites. The equatorial chlorine atom and one of the axial chlorine atoms bridge to the lithium atom, which is also attached to two diethyl ether molecules. The Zr—O bond lengths (average 1.895 Å) are exceptionally short and the Zr—O—C angles (average 169°) are nearly linear, indicative of strong $p\pi$–$d\pi$ bonding. It has been suggested that the neutral $(Me_3C)_3CO\cdot$ ligand may function as a five-electron donor, like a cyclopentadienyl group. Extraction of $[ZrCl_3(tritox)_2\cdot Li(OEt_2)_2]$ with hexane yields $ZrCl_2(tritox)_2$, which can be converted to $ZrMe_2(tritox)_2$ by reaction with MeLi.[215]

Conductometric titration of $ZrCl_4$ with KOPh in nitrobenzene indicates the formation of $Zr(OPh)_4$, $ZrCl_3(OPh)$ and $ZrCl_2(OPh)_2$, which have been isolated and characterized by chemical analysis. The Lewis acidity of $Zr(OPh)_4$ and $ZrCl_2(OPh)_2$ has been demonstrated by isolation of adducts with py, bipy, phen and their *N*-oxides.[216]

$ZrCl_4$ and $HfCl_4$ react with LiOAr (OAr = 2,6-di-*t*-butylphenoxide) in benzene or diethyl ether to give the aryloxide complexes $[MCl(OAr)_3]$ ($M = Zr$ or Hf). Attempts to prepare the corresponding $MCl_2(OAr)_2$ compounds were unsuccessful. $[HfCl(OAr)_3]$ has a sterically congested, tetrahedral structure with bond distances Hf—Cl = 2.365(1) Å, Hf—O = 1.938(3), 1.925(2) and 1.917(3) Å, and Hf—O—C bond angles in the range 152–159°. Low-temperature ^1H NMR spectra exhibit inequivalent *t*-butyl groups proximal and distal to the chlorine atom owing to a slowing at ~ -60 °C of rotation about the Hf—O—Ar bonds. Because the sterically bulky *t*-butyl groups tend to protect the metal atom from nucleophilic attack, the $[MCl(OAr)_3]$ complexes are not easily alkylated though $[HfCl(OAr)_3]$ does react with MeLi to give $[HfMe(OAr)_3]$.[217]

32.4.4.4 β-Ketoenolates, tropolonates, catecholates and quinones; ethers, aldehydes, ketones and esters

(*i*) *β-Ketoenolates*

Tetrakis(acetylacetonato) complexes of zirconium and hafnium were reported in 1904[218] and 1926,[219] respectively, and a large number of β-diketonate derivatives have been described subsequently (Table 10). These compounds are of the type $[M(dik)_4]$, $[M(dik)_3X]$, $[M(dik)_2X_2]$, $M(dik)X_3$ and $[M(dik)_3]Y$ (dik = β-diketonate anion; X = Cl, Br, I, NO_3, or alkoxide; $Y = [FeCl_4]$, $[AuCl_4]$, $\frac{1}{2}[PtCl_6]$ or $\frac{1}{2}[Zr(SO_4)_3]$). Additional β-diketonate compounds include the anionic complex $[NEt_4][Zr(bzbz)F_4]$ (bzbz = dibenzoylmethanate)[220] and the 1:1 $ZrCl_4$–diketone adducts $[ZrCl_4(MeCOCR_2COMe)]$ (R = H or Me).[221,222] Early work on zirconium and hafnium β-diketonates has been reviewed by Larsen,[5] Bradley and Thornton,[6] Fackler[223] and Mehrotra *et al.*[224]

The following methods have been employed for preparation of zirconium and hafnium

Table 10 β-Diketonate Complexes of Zirconium(IV) and Hafnium(IV), $[M(RCOCR'COR'')_nX_{4-n}]$

M	*R*	*R'*	*R''*	*X*	*M.p. (°C)*	*Method of preparation*[a]	*Physical method*[b]	*Ref.*
M(dik)$_4$								
Zr	Me	H	Me		194.5–195 (dec.)	A, E	XD, IR, R, UV, H-NMR, C-NMR, MS, PES, E, M	1–13
	Me	H	CF_3		130–131; subl. 115°/0.05 mmHg	A, C	R, UV, H-NMR, F-NMR, MS, GC	5–7, 14–17
	Me	H	CMe_3		93–94 (dec.)	C	H-NMR, M	15
	Me	H	2-Furyl		198–201	A	UV	6
	Me	H	2-Thienyl		244–245	A	UV	6
	Me	H	Ph		200–202	D, E	IR, M	2, 18
	Me	H	4-Biphenyl		275–277	D		19
	CHF_2	H	2-Thienyl		187 (dec.)	A	C-NMR, F-NMR	20
	CHF_2	H	Ph		182 (dec.)	A	C-NMR, F-NMR	20
	CF_3	H	CF_3		39–42[22] 48.5–49[23] 152–154[24]	C, E	IR, H-NMR, F-NMR, MS, GC, TA, M	21–24
	CF_3	H	CMe_3		188	A	H-NMR, F-NMR, M	15
	CF_3	H	2-Pyrrolyl		184–185	B	UV	6
	CF_3	H	2-Furyl		199–201	B	UV	6
	CF_3	H	2-Thienyl		225–226	A, B, E	XD, IR, UV, C-NMR, F-NMR, D, TA, M	6, 20, 21, 25
	CF_3	H	Ph		168–169	A, C, E	C-NMR, F-NMR, MS, D, M	16, 20, 21
	CF_3	H	4-C_6H_4F		169–170	A	C-NMR, F-NMR	20
	CF_3	H	4-C_6H_4Me		184–185	A	C-NMR, F-NMR	20
	CF_3	H	2-Naphthyl		150–152	A	C-NMR, F-NMR	20
	C_2F_5	H	2-Thienyl		201–203	A	C-NMR, F-NMR, D, M	20
	C_2F_5	H	Ph		157–158	A	C-NMR, F-NMR, D, M	20
	C_3F_7	H	CMe_3		170.5–171.5	C	MS	16, 26
	C_3F_7	H	2-Thienyl		135–136	A	C-NMR, F-NMR, D, M	20
	C_3F_7	H	Ph		129–131	A	C-NMR, F-NMR, D, M	20
	$CHMe_2$	H	$CHMe_2$		>250	C	H-NMR	15
	CMe_3	H	CMe_3		346–350	C, D	H-NMR, MS, TA	15, 23, 27–29
	Ph	H	Ph		245–246	C, D, E	XD, IR	2, 19, 30, 31
	Ph	H	4-Biphenyl		254–255	E		19
	1-Naphthyl	H	1-Naphthyl		276–277	A, E		19
	2-Naphthyl	H	2-Naphthyl		278–279	E		19
	4-Biphenyl	H	4-Biphenyl		302–303	C, E		19
	Me	Br	Me		169–170		IR, UV	32
	Me	NO_2	Me		205–206 (dec.)		IR, UV	33
	Me	CN	Me		273–275	C	IR, H-NMR	34
	Me	Bu	Me		131.5–133 (dec.)	D		31
	Me	Mesityl	Me			A	IR, H-NMR, MS	35

Table 10 *(continued)*

M	*R*	*R'*	*R''*	*X*	*M.p.* (°C)	*Method of preparation*[a]	*Physical method*[b]	*Ref.*
	Ph	Ph	Ph		210–220	C		19
Hf	Me	H	Me		193–195	A	XD, IR, R, UV, H-NMR, C-NMR, TDPAC	4, 6–8, 36, 37
	Me	H	CF_3		128–129; subl. 115 °C/0.05 mmHg	A, C	UV, H-NMR, F-NMR, GC	6–8, 14, 17
	Me	H	2-Furyl		200–202	A	UV	6
	Me	H	2-Thienyl		239–242	A	UV	6
	Me	H	Ph		205.5–207.5	D		38
	CF_3	H	CF_3		47.5–49.5	C	IR, H-NMR, F-NMR, MS, GC, TA	22
	CF_3	H	2-Pyrrolyl		185–186	B	UV	6
	CF_3	H	2-Furyl		195–197	B	UV	6
	CF_3	H	2-Thienyl		220–223	A, B	XD, IR, UV, TA	6, 25
	C_3F_7	H	CMe_3		170.5–171.5	C	MS	26
	CMe_3	H	CMe_3			C		39
	Ph	H	Ph		238–239	D		19
M(dik)$_3$*X*								
Zr	Me	H	Me	Cl	159.5–161[40] 156–158[31] 134–136[44] 101–102[45]	C, F, G	XD, IR, R, H-NMR, D, M, Λ	4, 18, 31, 40–45
				Br	162.5–164	C	XD, IR, R, H-NMR, D, M, Λ	4, 40–42
				I	178–183 (dec.)	C, F	XD, IR, H-NMR, D, M, Λ	4, 40–42
				NO_3	150.5–153	A, F	XD, IR, H-NMR, M, Λ	15, 46, 47
				$OCHMe_2$	B.p. 124–130/0.6–2 mmHg	E	IR, M	2, 48
				$OCMe_3$	B.p. 125–128/0.6–1.0 mmHg		M	2
	Me	H	Ph	Cl	146–147.5[18] 126–127[45]	C, D, F, G	M, Λ	18, 44, 45
				NO_3	152–153.5	F		46
				$OCHMe_2$		E	IR, M	2, 48
	Me	H	4-Biphenyl	Cl	255 (dec.)	C, E		19
	CF_3	H	CF_3	$OCHMe_2$		E	IR, M	21
	CF_3	H	2-Thienyl	$OCHMe_2$		E	IR	21
	CF_3	H	Ph	$OCHMe_2$		E	IR, M	21
	Ph	H	Ph	Cl	262	A, C, E	M, Λ	19, 44, 45
				Br	279.5 (dec.)	C		19, 49
				I	233	A		19
				NO_3	237	F		19
				$OCHMe_2$		E	IR, M	2
				$OCMe_3$			M	2
				$[FeCl_4]$	170		M, Λ	44, 45

				[$AuCl_4$]	155–156 (dec.)		M, Λ	44, 45
				0.5[$PtCl_6$]				45
				0.5[$Zr(SO_4)_3$]	>350			19
	Me	CN	Me	Cl	238–240 (dec.)	C	IR, H-NMR	34
Hf	Me	H	Me	Cl	161.5–162.5	C, F, G	XD, IR, R, H-NMR, D, M, Λ	4, 38, 40, 42, 44
				Br	161.5–163.5	C	XD, IR, R, H-NMR, D, M, Λ	4, 40, 42
				NO_3	157–157.5	F		38
	Me	H	Ph	Cl	126–127	C, F, G	M, Λ	18, 44
	Ph	H	Ph	Cl	261–262	A, C, D	M, Λ	19, 38, 44
				[$FeCl_4$]			M, Λ	44
M(dik)$_2$*X*$_2$								
Zr	Me	H	Me	Cl	180.5–182[40]	C, F	XD, IR, R, UV, H-NMR, D, M, Λ	4, 40–42, 44, 50, 51
					189–193[50]			
					236[44]			
				Br	185–186.5	C	XD, IR, R, H-NMR, D, M, Λ	4, 40, 42
				NO_3	149–151	A	XD, IR, H-NMR, M, Λ	15, 46, 52
				$OCHMe_2$		E	IR, M	2, 48
				OSi[$OCMe_3$]$_3$	69–70 (dec.)		IR, H-NMR	53
	Me	H	Ph	Cl	232–234	C		54
				NO_3	185–186	A, D		46
				$OCHMe_2$		E	M	2
	CF_3	H	CF_3	$OCHMe_2$		E	M	21
	CF_3	H	2-Thienyl	$OCHMe_2$		E		21
	CF_3	H	Ph	$OCHMe_2$		E	M	21
	Ph	H	Ph	Cl	231–234 (dec.)[51]	F		31, 51
					271–273 (dec.)[31]			
				Br	289.5–290			49
				NO_3	254–256	A		19
				$OCHMe_2$		E	IR, M	2
				$OCMe_3$			M	2
	Me	CN	Me	Cl	236–238 (dec.)	C	IR, H-NMR	34
	Me	Bu	Me	Cl	133.5–135	F		31
	Me	CN	Ph	Cl	192–194	C	IR	34
Hf	Me	H	Me	Cl	186–188	C	XD, IR, R, H-NMR, D, M, Λ	4, 40, 42
				Br	190–192	C	XD, IR, R, H-NMR, D, M, Λ	4, 40, 42
				NO_3	148–148.5	A		38
	Me	H	Ph	NO_3	189.5–190	A		38
	Ph	H	Ph	Br	294.5–295.5			49
				NO_3	237–238	A, D		38
M(dik)X$_3$								
Zr	Me	H	Me	Cl[c]	152–154	F		31
				OPr				55
				$OCHMe_2$	B.p. 149 °C/0.4 mmHg	E	M	2
				$OCMe_3$			M	2
	Me	H	Ph	$OCHMe_2$		E	M	2

Table 10 *(continued)*

M	*R*	*R'*	*R"*	*X*	*M.p.* (°C)	*Method of preparation*[a]	*Physical method*[b]	*Ref.*
	CF_3	H	CF_3	$OCHMe_2$		E	M	21
	CF_3	H	2-Thienyl	$OCHMe_2$		E		21
	CF_3	H	Ph	$OCHMe_2$		E	M	21
	Ph	H	Ph	$OCHMe_2$		E	M	2
	Me	Bu	Me	Cl^c	113–114 (dec.)	F		31
Hf	Me	H	Me	Cl^c	152–156	F		38

[a] See text. [b] XD = X-ray diffraction; IR = infrared spectrum; R = Raman spectrum; UV = ultraviolet spectrum; H-NMR = ^{1}H NMR spectrum; C-NMR = ^{13}C NMR spectrum; F-NMR = ^{19}F NMR spectrum; MS = mass spectrum; PES = photoelectron spectrum; E = electric polarization and dielectric loss measurements; D = dipole moment measurements; TDPAC = time differential perturbed angular correlation measurements; GC = gas chromatography; TA = thermal analysis; M = molecular weight; Λ = electrical conductance. [c] Isolated as the THF adduct $M(dik)Cl_3 \cdot C_4H_8O$.

1. R. C. Young and A. Arch, *Inorg. Synth.*, 1946, **2,** 121.
2. U. B. Saxena, A. K. Rai, V. K. Mathur, R. C. Mehrotra and D. Radford, *J. Chem. Soc. (A)*, 1970, 904.
3. J. V. Silverton and J. L. Hoard, *Inorg. Chem.*, 1963, **2,** 243.
4. R. C. Fay and T. J. Pinnavaia, *Inorg. Chem.*, 1968, **7,** 508.
5. C. J. Wiedenheft, *Inorg. Nucl. Chem. Lett.*, 1971, **7,** 439.
6. E. M. Larsen, G. Terry and J. Leddy, *J. Am. Chem. Soc.*, 1953, **75,** 5107.
7. T. J. Pinnavaia and R. C. Fay, *Inorg. Chem.*, 1966, **5,** 233.
8. C. A. Wilkie and D. T. Haworth, *J. Inorg. Nucl. Chem.*, 1978, **40,** 195.
9. C. G. Macdonald and J. S. Shannon, *Aust. J. Chem.*, 1966, **19,** 1545.
10. I. Fragala, G. Condorelli, A. Tondello and A. Cassol, *Inorg. Chem.*, 1978, **17,** 3175.
11. M. Casarin, E. Ciliberto, I. Fragala and G. Granozzi, *Inorg. Chim. Acta,* 1982, **64,** L247.
12. A. E. Finn, G. C. Hampson and L. E. Sutton, *J. Chem. Soc.*, 1938, 1254.
13. E. N. DiCarlo, E. Watson, Jr., C. E. Varga and W. J. Chamberlain, *J. Phys. Chem.*, 1973, **77,** 1073.
14. M. L. Morris, R. W. Moshier and R. E. Sievers, *Inorg. Synth.*, 1967, **9,** 50.
15. R. C. Fay and J. K. Howie, *J. Am. Chem. Soc.*, 1979, **101,** 1115.
16. S. Tsuge, J. J. Leary and T. L. Isenhour, *Anal. Chem.*, 1973, **45,** 198.
17. R. E. Sievers, B. W. Ponder, M. L. Morris and R. W. Moshier, *Inorg. Chem.*, 1963, **2,** 693.
18. R. Kh. Friedlina, E. M. Brainina and A. N. Nesmeyanov, *Bull. Acad. Sci. USSR, Div. Chem. Sci.*, 1957, 45.
19. L. Wolf and C. Troltzsch, *J. Prakt. Chem.*, 1962, **17,** 78.
20. M. Das, J. W. Beery and D. T. Haworth, *Synth. React. Inorg. Metal-Org. Chem.*, 1982, **12,** 671.
21. P. C. Bharara, V. D. Gupta and R. C. Mehrotra, *Indian J. Chem.*, 1975, **13,** 725.
22. S. C. Chattoraj, C. T. Lynch and K. S. Mazdiyasni, *Inorg. Chem.*, 1968, **7,** 2501.
23. T. J. Pinnavaia, M. T. Mocella, B. A. Averill and J. T. Woodward, *Inorg. Chem.*, 1973, **12,** 763.
24. M. L. Morris, R. W. Moshier and R. E. Sievers, *Inorg. Chem.*, 1963, **2,** 411.
25. Y. Baskin and N. S. Krishna Prasad, *J. Inorg. Nucl. Chem.*, 1963, **25,** 1011.
26. J. J. Leary, S. Tsuge and T. L. Isenhour, *Anal. Chem.*, 1973, **45,** 1269.
27. M. Balog, M. Scheiber, M. Michman and S. Patai, *Thin Solid Films,* 1977, **47,** 109.
28. J. D. McDonald and J. L. Margrave, *J. Less-Common Met.*, 1968, **14,** 236.
29. K. J. Eisentraut and R. E. Sievers, *J. Inorg. Nucl. Chem.*, 1967, **29,** 1931.
30. H. K. Chun, W. L. Steffen and R. C. Fay, *Inorg. Chem.*, 1979, **18,** 2458.
31. E. M. Brainina and R. Kh. Friedlina, *Bull. Acad. Sci. USSR, Div. Chem. Sci.*, 1964, 1331.
32. E. M. Brainina and L. I. Strunkina, *Bull. Acad. Sci. USSR, Div. Chem. Sci.*, 1977, **26,** 1925.
33. P. R. Singh and R. Sahai, *Aust. J. Chem.*, 1967, **20,** 649.
34. G. A. Lock and D. W. Thompson, *J. Chem. Soc., Dalton Trans.*, 1980, 1265.
35. G. Brill and H. Musso, *Justus Liebigs Ann. Chem.*, 1979, 803.
36. G. von Hevesy and M. Logstrup, *Chem. Ber.*, 1926, **59,** 1890.
37. L. M. Lowe, W. V. Prestwich and H. Zmora, *Can. J. Phys.*, 1975, **53,** 1327.
38. E. M. Brainina and E. I. Mortikova, *Bull. Acad. Sci. USSR, Div. Chem. Sci.*, 1967, 2418.
39. M. Balog, M. Schieber, M. Michman and S. Patai, *Thin Solid Films,* 1977, **41,** 247.
40. T. J. Pinnavaia and R. C. Fay, *Inorg. Chem.*, 1968, **7,** 502.
41. T. J. Pinnavaia and R. C. Fay, *Inorg. Synth.*, 1970, **12,** 88.
42. N. Serpone and R. C. Fay, *Inorg. Chem.*, 1969, **8,** 2379.
43. R. B. VonDreele, J. J. Stezowski and R. C. Fay, *J. Am. Chem. Soc.*, 1971, **93,** 2887.
44. M. Cox, J. Lewis and R. S. Nyholm, *J. Chem. Soc.*, 1964, 6113.
45. G. T. Morgan and A. R. Bowen, *J. Chem. Soc.*, 1924, **125,** 1252.
46. E. M. Brainina, R. Kh. Freidlina and A. N. Nesmeyanov, *Bull. Acad. Sci. USSR, Div. Chem. Sci.*, 1958, 910.
47. E. G. Muller, V. W. Day and R. C. Fay, *J. Am. Chem. Soc.*, 1976, **98,** 2165.
48. D. M. Puri, *J. Indian Chem. Soc.*, 1970, **47,** 535.
49. E. M. Brainina and M. Kh. Minacheva, *Bull. Acad. Sci. USSR, Div. Chem. Sci.*, 1975, **24,** 1482.
50. H.-H. Schmidtke and U. Voets, *Inorg. Chem.*, 1981, **20,** 2766.
51. R. Kh. Freidlina, E. M. Brainina and A. N. Nesmeyanov, *Dokl. Akad. Nauk. SSSR, Chem. Sect. (Engl. Transl.)*, 1961, **138,** 628.
52. V. W. Day and R. C. Fay, *J. Am. Chem. Soc.*, 1975, **97,** 5136.
53. Y. Abe, K. Hayama and I. Kijima, *Bull. Chem. Soc. Jpn.*, 1972, **45,** 1258.
54. W. Dilthey, *J. Prakt. Chem.*, 1925, **111,** 147.
55. E. M. Brainina and R. Kh. Friedlina, *Bull. Acad. Sci. USSR, Div. Chem. Sci.*, 1961, 1489.

β-diketonates, and the application of these methods to specific compounds is documented in Table 10: (A) reaction of a hydrated metal salt and the β-diketone in aqueous or nonaqueous media, often in the presence of a base such as carbonate or triethylamine; (B) extraction of a hydrochloric acid solution of the metal ion with a benzene solution of the diketone; (C) reaction of a metal tetrahalide with the diketone (or diketonate anion) in an organic solvent at reflux (equation 33); (D) exchange of diketonate ligands between a metal diketonate and a less volatile β-diketone (*e.g.* equation 34); (E) reaction of a metal alkoxide with a stoichiometric amount of the diketone in benzene at reflux (*e.g.* equation 35); (F) ligand exchange between metal complexes, *e.g.* $M(dik)_4$ and MX_4 (equation 36) or $M(dik)_4$ and $M(dik)_2X_2$ (equation 37); (G) conversion of $[M(dik)_4]$ to $[M(dik)_3Cl]$ by reaction with acetyl chloride. The physical methods that have been used to characterize zirconium and hafnium diketonates are also listed in Table 10.

$$MX_4 + n\text{Hdik} \longrightarrow [M(\text{dik})_nX_{4-n}] + n\text{HX} \quad (n = 2, 3 \text{ or } 4) \tag{33}$$

$$[Zr(acac)_4] + 4\text{Hbzbz} \xrightarrow[25\ \text{mm Hg}]{100\ ^\circ\text{C}} [Zr(bzbz)_4] + 4\text{H(acac)} \tag{34}$$

$$Zr(OCHMe_2)_4{\cdot}HOCHMe_2 + n\text{Hdik} \xrightarrow[\text{reflux}]{C_6H_6} Zr(\text{dik})_n(OCHMe_2)_{4-n} + (n+1)HOCHMe_2 \quad (n = 1, 2, 3 \text{ or } 4) \tag{35}$$

$$3[Zr(acac)_4] + ZrI_4 \xrightarrow{\text{THF}} 4[Zr(acac)_3I] \tag{36}$$

$$[Zr(acac)_4] + [Zr(acac)_2(NO_3)_2] \xrightarrow[\text{reflux}]{C_6H_6} 2[Zr(acac)_3(NO_3)] \tag{37}$$

(*a*) *Tetrakis(diketonates)*, $[M(dik)_4]$. X-Ray crystal structures have been reported for $[Zr(acac)_4]$[225] and $[Zr(bzbz)_4]$.[226] Both complexes have an eight-coordinate square antiprismatic structure (Figure 4) in which the bidentate β-diketonate ligands span *s* edges of the antiprism to give the D_2-*ssss* stereoisomer. The antiprism is slightly distorted, in the direction of a C_{2v} bicapped trigonal prism in the case of $[Zr(bzbz)_4]$, and in the direction of a D_{2d} dodecahedron in the case of $[Zr(acac)_4]$. These distortions have been analyzed in some detail by Chun *et al.*[226] The Zr—O bonds in $[Zr(bzbz)_4]$ fall into two symmetry-inequivalent sets in accord with the idealized D_2 symmetry of the coordination group. The Zr—O bonds to the 'outer' sites (1, 3, 6 and 8 in Figure 4) (average 2.153 Å) are significantly shorter than the bonds to the 'inner' sites (average 2.192 Å). The averaged length of all eight Zr—O bonds in $[Zr(bzbz)_4]$ (2.172 Å) is 0.026 Å less than the averaged Zr—O bond length in $[Zr(acac)_4]$ (2.198 Å). X-ray powder patterns have shown that $[Hf(acac)_4]$ is isostructural with $[Zr(acac)_4]$, and time differential perturbed angular correlation measurements on $[Hf(acac)_4]$ are consistent with a D_2-*ssss* square antiprismatic structure.[227]

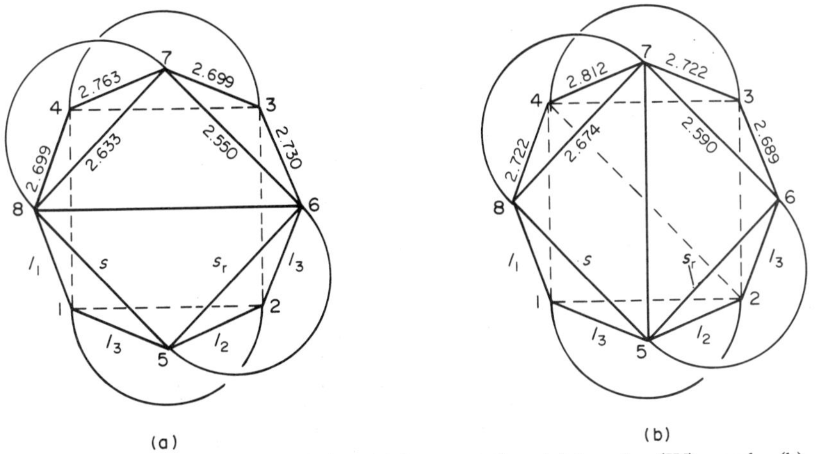

Figure 4 Structures of (a) tetrakis(1,3-diphenyl-1,3-propanedionato)zirconium(IV) and (b) tetrakis(2,4-pentanedionato)zirconium(IV). Dimensions of the coordination polyhedra are averaged in accord with approximate D_2-222 symmetry. Distortions of the square antiprisms involve slight folding of the quadrilateral faces about the indicated face diagonals (reproduced by permission from reference 226)

^1H and ^{19}F NMR spectra of Zr^{IV} and Hf^{IV} tetrakis(β-diketonates) have been investigated in $CHClF_2$ solution at temperatures down to −170 °C, and several complexes have been identified which become stereochemically rigid on the NMR time scale at temperatures in the range −115 to −170 °C.[228] Spectra of $[Zr(acac)_4]$ and $[Hf(acac)_4]$ in the slow-exchange limit exhibit two methyl proton resonances of equal intensity, consistent with the D_2-*ssss* stereoisomer found in the solid state. ^{19}F spectra of $[Zr(tfacac)_4]$ (tfacac = $MeCOCHCOCF_3$) show that more than one geometric isomer is present in solution. The kinetics of intramolecular rearrangement of several of these complexes have been studied by NMR line-shape analysis; kinetic data are summarized in Table 11. Tetrakis(diketonates) are highly fluxional; for $[Zr(acac)_4]$, the extrapolated rate constant at 25 °C for exchange of methyl groups between the two inequivalent sites of the D_2-*ssss* stereoisomer is $4.7 \times 10^5\ s^{-1}$ and the rearrangement barrier $\Delta H^\ddagger$ is only 17 kJ mol^{-1}.

Table 11 Kinetic Data for Methyl or *t*-Butyl Group Exchange in Zr^{IV} β-Diketonate Complexes[a]

	$[Zr(acac)_4]$[b]	$[Zr(dmh)_4]$[c]	$[Zr(acac)_2(NO_3)_2]$[b]
T_c (°C)	−145	−116	−144
$k_{25\,°C}$ (s^{-1})	4.7×10^5	1.7×10^5	1.4×10^6
$k_{-125\,°C}$ (s^{-1})	2.0×10^2	6.8	2.9×10^2
$\Delta G^\ddagger$ (−125 °C) (kJ mol^{-1})	28.9 ± 0.3	33.1 ± 0.1	28.5 ± 0.2
$\Delta H^\ddagger$ (kJ mol^{-1})	17.2 ± 1.3	23.0 ± 2.1	18.8 ± 1.3
$\Delta S^\ddagger$ (J mol^{-1} K^{-1})	−78.2 ± 10.5	−67.8 ± 12.1	−63.6 ± 10.5
E_a (kJ mol^{-1})	18.4 ± 1.3	24.3 ± 2.1	20.1 ± 1.3
log *A*	8.8 ± 0.6	9.4 ± 0.6	9.5 ± 0.5

[a] In $CHClF_2$ solution. [b] Methyl group exchange. [c] *t*-Butyl group exchange; dmh = $MeCOCHCOCMe_3$.

Intermolecular exchange of β-diketonate ligands is also a rapid process. Ligand exchange between $[M(acac)_4]$ and $[M(tfacac)_4]$ (M = Zr or Hf) is fast on the ^1H NMR time scale at temperatures above ~70 °C.[229] ^1H and ^{19}F NMR spectra of $[Zr(acac)_4]$–$[Zr(tfacac)_4]$ mixtures indicate the presence in solution of all five $[Zr(acac)_n(tfacac)_{4-n}]$ (n = 0, 1, 2, 3 or 4) complexes, with the mixed-ligand complexes being favored at the expense of the parent complexes.[229,230] It is interesting to note that equilibrium constants K_n for formation of the mixed-ligand complexes (equation 38) depend strongly on the difference in the degree of fluorination of the two diketonate ligands dik and dik′. Thus, observed values of K_n for the $[Zr(acac)_4]$–$[Zr(bzbz)_4]$ system are slightly less than the statistical values for random scrambling of ligands ($K_1 = K_3 = 4$; $K_2 = 6$), while values of K_n for the $[Zr(acac)_4]$–$[Zr(hfacac)_4]$ and $[Zr(dpm)_4]$–$[Zr(hfacac)_4]$ (dpm = $Me_3CCOCHCOCMe_3$) systems are ~10^3–10^5 times the statistical values.[231] Values of K_n for the $[Zr(acac)_4]$–$[Zr(tfacac))_4]$ system are intermediate, ~4–7 times the statistical values. In the $[Zr(acac)_4]$–$[Zr(tfacac)_4]$ system, deviations from random scrambling of ligands are due to entropy effects,[230] while in the $[Zr(acac)_4]$–$[Zr(hfacac)_4]$ system, enthalpy effects are the principal causes of the nonstatistical behavior.[231] Minacheva and Brainina[232] have reported the isolation of a number of mixed-ligand complexes, including $[M(acac)_2(bzbz)_2]$ (M = Zr or Hf).

$$\frac{n}{4}[M(dik)_4] + \frac{4-n}{4}[M(dik')_4] \overset{K_n}{\rightleftharpoons} [M(dik)_n(dik')_{4-n}] \qquad (38)$$

(n = 1, 2 or 3)

The kinetics of ligand exchange between $[M(dik)_4]$ (M = Zr or Hf; dik = acac or tfacac) and the corresponding free β-diketones have been studied by Adams and Larsen.[233] $[Zr(acac)_4]$ exchanges ligands with a polystyrene resin functionalized with acetylacetonate groups to give a polymer-bound zirconium complex.[234]

$[M(acac)_4]$ (M = Zr or Hf) complexes react with *N*-bromosuccinimide yielding the 3-bromo-2,4-pentanedionate derivatives $[M(acac)(3\text{-}Bracac)_3]$ and $[M(3\text{-}Bracac)_4]$; the products depend on the reaction conditions and often are isolated as succinimide adducts.[235] $[Zr(bzac)_4]$ (bzac = MeCOCHCOPh) and $[Zr(bzbz)_4]$ are not brominated under the same reaction conditions. The reaction of bromine with $[M(acac)_4]$ results in both electrophilic substitution on the chelate ring and replacement of two acetylacetonate ligands; the products $M(3\text{-}Bracac)_2Br_2 \cdot nH_2O$ (n = 1 or 2) were isolated as hydrates due to the presence of traces of moisture.[236] Only ligand replacement occurs in reactions of bromine with $[M(bzbz)_4]$; products are [$M(bzbz)_2Br_2$] (M = Zr or Hf) and [$Zr(bzbz)_3Br$].[237] With iodine, [$Zr(acac)_4$] forms a 1:1 charge-transfer complex, which has been studied in solution by spectrophotometric methods.[238] IR studies indicate extensive hydrogen bonding between [$Zr(acac)_4$] and $CDCl_3$ in CCl_4 solution, presumably involving the acetylacetonate oxygen atoms.[239]

Many [$M(dik)_4$] complexes are volatile, especially those that contain fluorinated diketonate ligands. Mass spectra and gas chromatographic behavior of several of these complexes have been studied (see Table 10). Isenhour and coworkers[240,241] have employed fluorinated diketonates in mass spectrometric procedures for determination of Zr and Zr/Hf ratios in geological samples. The most intense peak in mass spectra of [$M(dik)_4$] complexes is [$M(dik)_3$]$^+$. Sievers *et al.*[242] have used gas chromatography of metal trifluoroacetylacetonates to separate Zr from Al, Cr and Rh. However, attempts to separate [$Zr(tfacac)_4$] and [$Hf(tfacac)_4$] by gas chromatography were unsuccessful. Zirconium and hafnium can be separated by solvent extraction procedures that employ fluorinated diketones.[105] [$M(dik)_4$] (M = Zr or Hf; dik = acac, dpm, tfacac or hfacac) have been used as volatile source materials for chemical vapor deposition of thin films of the metal oxides.[243,244]

(*b*) *Tris(diketonates)*, [*M*(*dik*)$_3$*X*]. Known compounds of this type have X = halide, nitrate or alkoxide. The methods used to prepare these complexes are indicated in Table 10.

IR and Raman spectra, dipole moment data, and molecular weight and conductance measurements point to seven-coordinate structures for the halo complexes [$M(acac)_3X$] (M = Zr or Hf, X = Cl or Br; M = Zr, X = I). The vibrational spectra exhibit metal–halogen stretching bands (Table 12) and indicate that all of the carbonyl groups are coordinated.[245] Dipole moments of ~5–6 D in benzene solution are in good agreement with the values expected for a seven-coordinate structure.[246] The [$M(acac)_3X$] (X = Cl or Br) complexes are monomeric nonelectrolytes in ionizing solvents,[247,248] but [$Zr(acac)_3I$] is ~10% dissociated in 1,2-dichloroethane and ~26% dissociated in nitrobenzene.[247] The dissociated species of [$Zr(acac)_3I$] in 1,2-dichloroethane are markedly stabilized by tetrahydrofuran, probably due to coordination by the THF (equation 39). Crystallization of [$Zr(acac)_3I$] from THF yields a $Zr(acac)_3I\cdot THF$ adduct, which loses THF upon heating for 5 h at 80 °C *in vacuo*.

$$[Zr(acac)_3I] + THF \rightleftharpoons [Zr(acac)_3(THF)]^+ + I^- \qquad (39)$$

Table 12 Selected IR and Raman Vibrational Frequencies (cm^{-1}) for $M(acac)_4$, $M(acac)_3X$ and $M(acac)_2X_2$ (X = Cl, Br or I)

Compound		$\nu_s(C{\cdots}O)$	$\nu_{as}(C{\cdots}O)$	$\nu_s(M—O)$	$\nu_{as}(M—O)$	$\nu(M—X)$
$Zr(acac)_4$	IR	1591	1397		421	
	R	a		441p[b]	416	
$Hf(acac)_4$	IR	1592	1397		422	
	R	a		446p	417dp[b]	
$Zr(acac)_3Cl$	IR	1581, 1568	1381	449	432	314
	R	a		448p	431	
$Zr(acac)_3Br$	IR	1580, 1566	1381	449	434	164
	R	a		449p	~430	
$Zr(acac)_3I$	IR	1562	1380	452	438	93
$Hf(acac)_3Cl$	IR	1579, 1570	1387	454	432	293
	R	a		452p	433	~296
$Hf(acac)_3Br$	IR	1574, 1565	1386	454	433	166
	R	a		452p	~430	
$Zr(acac)_2Cl_2$	IR	1575, 1555	1338	460	450	335, 325
	R	a		458p	~450	338p
$Zr(acac)_2Br_2$	IR	1570, 1551	1333	460	450	213
	R	a		459p	~450	212p
$Hf(acac)_2Cl_2$	IR	1579, 1556	1341	463	446	322, 313
	R	a		462p	~444	323p
$Hf(acac)_2Br_2$	IR	1576, 1553	1337	463	447	207
	R	a		460p	~445	208p

[a] This region obscured by solvent absorption. [b] p = polarized; dp = depolarized.

A seven-coordinate structure for [Zr(acac)$_3$Cl] has been confirmed by X-ray diffraction.[249] This complex has distorted pentagonal bipyramidal geometry with the chlorine atom in an axial position (Zr—Cl = 2.472(6) Å). The axial Zr—O bond is ~0.06 Å shorter than the averaged length (2.14 Å) for the five, quite uniform equatorial Zr—O bonds. Because of close O···O contacts in the equatorial plane, the pentagonal girdle of oxygen atoms is appreciably puckered. In solution, [Zr(acac)$_3$Cl] is stereochemically nonrigid on the NMR time scale at temperatures down to −130 °C.[247]

It is interesting to note that [Zr(acac)$_3$Cl] can be recrystallized unchanged from hot acetylacetone.[250] An NMR study[247] has shown that substitution of the chlorine atom is very slow under anhydrous conditions; heating [Zr(acac)$_3$Cl] with neat acetylacetone for 24 h at 80 °C yields a product which is only ~20% [Zr(acac)$_4$]. In the presence of water, [Zr(acac)$_3$Cl] is readily converted to [Zr(acac)$_4$],[250,251] evidently due to preliminary hydrolysis of the Zr—Cl bond. In anhydrous media, the ease of complete substitution is dependent on the nature of the diketone; certain aryl-[252] and trifluoromethyl-substituted[253] diketones readily react with $ZrCl_4$ to give the tetrakis(diketonates).

The dibenzoylmethanate complex [Zr(bzbz)$_3$Cl] reacts with iron(III) chloride, gold(III) chloride or chloroplatinic acid in chloroform or glacial acetic acid yielding [Zr(bzbz)$_3$]Y (Y = [FeCl$_4$], [AuCl$_4$] or 0.5[PtCl$_6$]).[248,250] [Hf(bzbz)$_3$][FeCl$_4$] is also known.[248] The tetrachloroferrate and tetrachloroaurate salts behave as 1:1 electrolytes in nitromethane or nitrobenzene.

IR spectra of the 3-cyano-2,4-pentanedionate complex (Zr(3-CNacac)$_3$Cl suggest that one of the three cyano groups is coordinated to a second Zr atom, thus increasing the coordination number of zirconium to eight. Coordination of one of the cyano groups also occurs in the dichloro complex Zr(3-CNacac)$_2$Cl$_2$.[254]

Several [M(dik)$_3$(NO$_3$)] complexes are known (Table 10). [Zr(acac)$_3$(NO$_3$)] has an eight-coordinate structure in which the bidentate nitrate ligand spans an *a* edge and the three acetylacetonate ligands span *b*, *m* and *g* edges, respectively, of a distorted D_{2d} dodecahedron. This unusual *abmg* ligand wrapping pattern (Figure 5) is preferred over the commonly observed *mmmm* wrapping pattern because the bite of the acetylacetonate ligand is too large to permit two acac ligands to be located on the same trapezoid of a ZrO_8 dodecahedron. The acac ligand bite shows significant variation with the type of polyhedral edge it spans: m(2.618(2) Å) < g(2.689(2) Å) < b(2.786(2) Å). Zr—O bonds to the acac ligands (average 2.141(1) Å) are systematically shorter than Zr—O bonds to the nitrate ligands (average 2.366(2) Å).[255] In solution, [Zr(acac)$_3$(NO$_3$)] is a monomeric nonelectrolyte which is stereochemically nonrigid on the NMR time scale at temperatures down to at least −144 °C.[228,255] The monoalkoxy complexes [Zr(dik)$_3$(OR)] (Table 10) are monomeric in solution and presumably are seven-coordinate.

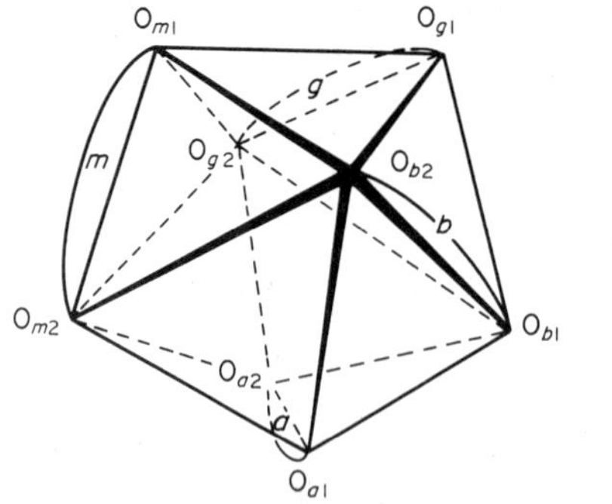

Figure 5 Ligand wrapping pattern in [Zr(acac)$_3$(NO$_3$)]. The ligands are labeled *a, b, m* or *g* according to the dodecahedral edge which the ligand spans (reproduced by permission from ref. 255)

(*c*) *Bis*(*diketonates*), [*M*(*dik*)$_2$*X*$_2$]. The best characterized compounds in this class (Table 10) are [M(acac)$_2$X$_2$] (M = Zr or Hf; X = Cl or Br) and [Zr(acac)$_2$(NO$_3$)$_2$]. The [M(acac)$_2$X$_2$] complexes are monomeric nonelectrolytes in nitrobenzene.[247] In benzene solution, they possess dipole moments of ~8 D.[246] Vibrational spectra of the [M(acac)$_2$X$_2$] complexes exhibit coincident metal–ligand stretching bands in IR and Raman spectra (Table 12), and the [M(acac)$_2$Cl$_2$] complexes display two ν(M—Cl) IR bands.[245] These data point to an octahedral *cis* structure. In vibrational spectra of the series of compounds [M(acac)$_2$X$_2$], [M(acac)$_3$X] and [M(acac)$_4$] (Table 12), the ν(M—O) and ν(M—X) bands exhibit systematic shifts to lower frequency, and the ν(C⋯O) bands show corresponding shifts to higher frequency, as the coordination number of the metal increases from six to seven to eight. The $[M(acac)_2X_2]$ complexes are highly fluxional in solution; a single time-averaged methyl proton resonance was observed for $[Zr(acac)_2Cl_2]$ at temperatures down to −130 °C.[247] X-Ray powder patterns indicate that the $[Zr(acac)_2X_2]$ and $[Hf(acac)_2X_2]$ complexes which contain the same halogen are isomorphous.

$[Zr(acac)_2Cl_2]$ reacts with two equivalents of PhMgBr, two equivalents of PhLi or one equivalent of Ph_2M (M = Mg, Cd or Hg) in THF at low temperatures yielding $Zr(acac)_2ClPh$. This compound was formulated as a chlorine-bridged dimer, but no molecular weight data were reported.[256] Treatment of $[Zr(acac)_2Cl_2]$ with four equivalents of PhLi results in phenyl substitution of the chelate ring (equation 40).

$$[Zr(acac)_2Cl_2] + 3PhLi \longrightarrow Zr(3\text{-}Phacac)_2ClPh + LiCl + 2LiH \qquad (40)$$

$[Zr(acac)_2(NO_3)_2]$ has an eight-coordinate structure in which bidentate acetylacetonate and bidentate nitrate ligands span the *m* edges of distorted D_{2d} dodecahedron.[257] The observed stereoisomer is the C_2-*mmmm* isomer having one acetylacetonate ligand and one nitrate ligand on each BAAB trapezoid. Averaged Zr—O bond distances are Zr—O(acac) = 2.096 Å and Zr—O(nitrate) = 2.295 Å. Within a particular chelate ring, Zr—O bond lengths involving dodecahedral A sites exceed those involving B sites by 0.015–0.051 Å (5σ–17σ) and these differences appear to be propagated in the C—O and N—O bond lengths in the ligands. $[Zr(acac)_2(NO_3)_2]$ is a monomeric nonelectrolyte in solution, and retention of coordination number eight is suggested by the similarity of solid-state and solution-state IR spectra. ^1H NMR spectra in $CHClF_2$ solution[228] indicate that $[Zr(acac)_2(NO_3)_2]$ becomes stereochemically rigid on the NMR time scale at temperatures below −144 °C, and the spectra in the slow-exchange limit are consistent with the C_2-*mmmm* structure found in the solid state. Kinetic data for exchange of acetylacetonate methyl groups between the two inequivalent sites of the C_2-*mmmm* stereoisomer are included in Table 11.

(*d*) *Mono(diketonates), M(dik)X*$_3$. A few $M(dik)Cl_3$ complexes (Table 10) have been isolated as 1:1 adducts with THF following ligand exchange reactions between $[M(dik)_4]$ and MCl_4 in THF.[258,259] Several $Zr(dik)(OCHMe_2)_3$ compounds have been prepared by reaction of equimolar amounts of $Zr(OCHMe_2)_4{\cdot}HOCHMe_2$ and the β-diketone. The complexes that contain fluorinated diketonate ligands (Table 10) appear to be monomeric[260] and presumably five-coordinate, while those that contain acac, bzac or bzbz ligands have a tendency to dimerize.[261] A structure having six-coordinate zirconium and two isopropoxy bridges has been suggested for the dimer. Reaction of $Zr(acac)(OCHMe_2)_3$ with *t*-butanol affords $Zr(acac)(OCMe_3)_3$, which has only a small tendency to dimerize.

(*e*) *Miscellaneous β-diketonates and related compounds*. Heterocyclic tetrakis(β-diketonates) of the type $Zr(dik)_n(OCHMe_2)_{4-n}$, where dik is the anion of a 4-acyl-3-methyl-1-phenyl-2-pyrazol-5-one (**21**) and n = 1, 2, 3 or 4, have been prepared by reaction of stoichiometric amounts of $Zr(OCHMe_2)_4{\cdot}HOCHMe_2$ and the pyrazolone in benzene at reflux.[262] These compounds are monomeric in boiling benzene, except for the $Zr(dik)(OCHMe_2)_3$ complexes, which tend to dimerize. The $Zr(dik)_n(OCHMe_2)_{4-n}$ complexes exchange alkoxide ligands with *t*-butanol yielding the corresponding $Zr(dik)_n(OCMe_3)_{4-n}$ complexes, and they undergo insertion reactions with phenyl isocyanate giving $Zr(dik)_n\{NPhC(O)OCHMe_2\}_{4-n}$ (n = 1, 2 or 3). Additional synthetic routes have been reported for preparation of $[Zr(dik)_4]$, where dik is the anion of (**21**; R = Ph or CF_3).[263,264]

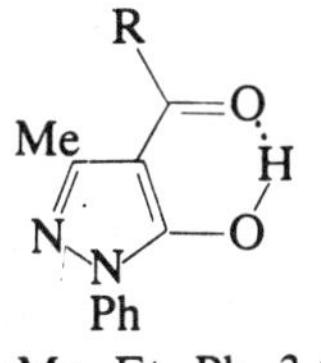

(**21**) R = Me, Et, Ph, 3-ClC_6H_4

Tetrakis(salicylaldehydato)zirconium(IV), tetrakis(3-methoxysalicylaldehydato)zirconium(IV) and tetrakis(ethyl acetoacetato)zirconium(IV) have been synthesized by reaction of $[Zr(acac)_4]$ with salicylaldehyde, 3-methoxysalicylaldehyde or ethyl acetoacetate, respectively.[265,266] The reaction of $ZrCl_4$ with salicylaldehyde in diethyl ether gives the addition compound $ZrCl_4 \cdot 2\{2\text{-}OC_6H_4CHO\}$ at −15 °C, and the substitution product $[Zr\{2\text{-}OC_6H_4CHO\}_2Cl_2]$ at reflux temperature.[221] Salicylaldehyde reacts with $Zr(OCHMe_2)_4 \cdot HOCHMe_2$ in benzene at ambient temperature yielding $Zr\{2\text{-}OC_6H_4CHO\}_n(OCHMe_2)_{4-n}$ ($n = 1$ or 2). When these reactions are carried out in refluxing benzene, the products undergo Meerwein–Ponndorf reduction yielding compounds such as $Zr_2\{2\text{-}OC_6H_4CH_2O\}\{2\text{-}OC_6H_4CHO\}(OCHMe_2)_5$ and $Zr_2\{2\text{-}OC_6H_4CH_2O\}_3(2\text{-}OC_6H_4\text{-}CHO\}(OCHMe_2)$.[267] Several β-keto ester complexes have been synthesized from $Zr(OCHMe_2)_4 \cdot HOCHMe_2$; these include $Zr(PhCOCHCO_2Et)_n(OCHMe_2)_{4-n}$ ($n = 1$, 2, 3 or 4)[261] and $[Zr(MeCOCHCO_2Me)_3(OCHMe_2)]$.[268] The reaction of $Zr(OCHMe_2)_4 \cdot HOCHMe_2$ with bis(acetoacetates), $MeCOCH_2CO_2(CH_2)_nO_2CCH_2COMe$ ($n = 2$, 4 or 5) yields oligonuclear products.[269]

Neutral bis-chelate complexes of the type $[\{fac\text{-}(OC)_3Re(MeCO)_3\}_2Zr]$ and $[\{fac\text{-}(OC)_3Re(MeCO)_2(RCO)\}_2Hf]$ (R = Me, $CHMe_2$ or $PhCH_2$) have been prepared by reaction of $[RC(O)Re(CO)_5]$ with MeLi in THF, followed by addition of a THF solution of $ZrCl_4$ or $HfCl_4$. The $[fac\text{-}(OC)_3Re(MeCO)_3]^{2-}$ dianion is the formal, metalla analogue of the triacetylmethanate anion. The Zr and Hf complexes were obtained in rather poor yields as pale yellow air-sensitive solids. IR and ^1H NMR spectra are consistent with tridentate coordination (**22**) of the triacetylmetallate ligands.[270]

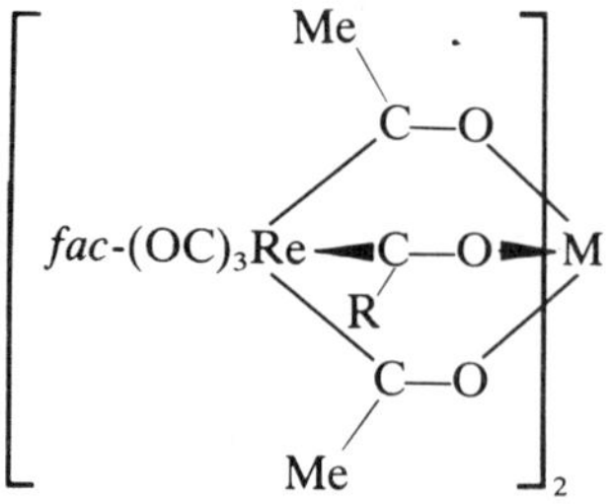

(**22**)

(ii) Tropolonates

Tetrakis(tropolonato) complexes $[MT_4]$ (M = Zr or Hf; T = $C_7H_5O_2$) have been prepared by reaction of tropolone with (i) $ZrCl_4$ or $HfCl_4$ in chloroform,[271] or (ii) a Zr^{IV} or Hf^{IV} salt in 50% water–methanol.[272] The $[MT_4]$ complexes tend to form solvates when crystallized from polar organic solvents. X-Ray diffraction studies of $[ZrT_4] \cdot 2.25CHCl_3$[273] and $[HfT_4] \cdot DMF$[274] have established that both crystals contain eight-coordinate $[MT_4]$ complexes. The bidentate tropolonate ligands span the *m* edges of a D_{2d} dodecahedron. As is often observed in dodecahedral complexes, averaged bond lengths to the dodecahedral A sites are slightly longer than averaged bond lengths to the B sites ($Zr—O_A = 2.194$, $Zr—O_B = 2.174$, $Hf—O_A = 2.188$, $Hf—O_B = 2.176$ Å). A very strong band at 288.0 cm^{-1} in the IR spectrum of $[^{90}ZrT_4]$, which shifts to 283.9 cm^{-1} in $[^{94}ZrT_4]$, has been assigned to Zr—O stretching; the corresponding ν(Hf—O) mode in $[HfT_4]$ is at 245 cm^{-1}.[272] Quadrupole coupling parameters for $[HfT_4]$, obtained from time differential perturbed angular correlation measurements, have been interpreted in terms of a square antiprismatic or dodecahedral structure of D_2 symmetry.[275,276] However, more recent measurements on $[HfT_4]$, $[HfT_4] \cdot CHCl_3$ and $[HfT_4] \cdot DMF$[277] have shown that the quadrupole coupling parameters are strongly influenced by intermolecular interactions between $[HfT_4]$ and dipolar solvent molecules.

^1H NMR studies of the corresponding α-, β- and γ-isopropyl-tropolonato complexes, $[M(C_{10}H_{11}O_2)_4]$ (M = Zr or Hf)[278] indicate that these complexes are stereochemically nonrigid on the NMR time scale at −85 to −100 °C. The $[M(C_{10}H_{11}O_2)_4]$ complexes and the free ligand undergo intermolecular ligand exchange at elevated temperatures; the approximate rates for $[M(\gamma\text{-}C_{10}H_{11}O_2)_4]$ (M = Zr or Hf) are 50 s^{-1} at 140 °C in tetrachloroethane. These ligand-exchange rates are slower than those for analogous β-diketonate complexes.[233]

(iii) Catecholates and quinones

Colorless crystals of $Na_4[Hf(O_2C_6H_4)_4]\cdot 21H_2O$ have been isolated following addition of an O_2-free aqueous solution of hafnium tetrachloride to an O_2-free alkaline solution of catechol. The crystals contain discrete eight-coordinate $[Hf(O_2C_6H_4)_4]^{4-}$ anions having $\bar{4}$ site symmetry. The coordination polyhedron is a D_{2d} dodecahedron, and the bidentate catecholate ligands span the *m* polyhedral edges; Hf—O_A = 2.220(3) and Hf—O_B = 2.194(3) Å.[279]

The $[Hf(O_2C_6H_4)_3]^{2-}$ anion has been prepared in THF by reaction of $[Hf(NEt_2)_4]$ with one equivalent of catechol and two equivalents of the sodium salt $NaOC_6H_4OH$. The complex was isolated from acetonitrile–ether as $[NEt_4]_2[Hf(O_2C_6H_4)_3]$, which behaves as a 2:1 electrolyte in acetonitrile and exhibits a ligand-to-metal charge-transfer band at 33 900 cm^{-1}.[280] Aqueous solutions that contain zirconyl chloride, catechol and excess ammonia yield precipitates of composition $(NH_4)Zr(OH)(O_2C_6H_4)_2$, $[NH_4]_2[Zr(O_2C_6H_4)_3]$ or $[NH_4]_2[Zr(O_2C_6H_4)_3]\cdot C_6H_4(OH)_2\cdot 1.5H_2O$, depending on the catechol/Zr^{IV} ratio in the initial solution.[281] Anhydrous $ZrCl_4$ reacts with molten catechol yielding a colorless crystalline powder of composition $H_2Zr(O_2C_6H_4)_3$. Related compounds of the type $ZrL_2(amine)_2$ ($L = C_{12}H_8O_2$, $C_{20}H_{12}O_2$ or $C_{10}H_6O_2$; amine = Et_3N, Et_2NH or C_5H_5N) have been prepared by reaction in methanol of $ZrCl_4$ with 2,2′-dihydroxybiphenyl, 2,2′-dihydroxybinaphthyl, or 1,8-dihydroxynaphthalene followed by addition of the appropriate amine. The amine-free compound $Zr(C_{12}H_8O_2)_2$ is obtained by reaction of $ZrCl_4$ with molten 2,2′-dihydroxybiphenyl.[282]

Formation of zirconium and hafnium complexes with polyhydroxyaromatic compounds in aqueous solution has been studied by potentiometric, spectrophotometric and ion exchange methods. Much of this work has been reviewed by Larsen.[5] A potentiometric and light-scattering study of the zirconium(IV)–Tiron (Tiron = disodium 1,2-dihydroxybenzene-3,5-disulfonate) system provides evidence for mixed hydroxo–Tiron chelates at Tiron/Zr^{IV} ratios less than 2.5, but at higher Tiron/Zr^{IV} ratios the unhydrolyzed complexes $[Zr_2L_5]^{12-}$ and $[ZrL_4]^{12-}$ (L = tetranegative anion of 1,2-dihydroxybenzene-3,5-disulfonic acid) are present.[283,284]

Quinones such as alizarin red S (1,2-dihydroxyanthraquinone-3-sulfonate) and chloroanilic acid have long been used as analytical reagents for spectrophotometric determination of zirconium.[285,286] These reagents form 1:1 complexes with Zr^{IV} in strongly acidic solutions.[287–289] The binding site in alizarin red S has been a matter of some controversy,[288–290] but it now appears, on the basis of voltametric and spectrophotometric evidence, that Zr^{IV} coordinates to alizarin red S through the two adjacent phenolic oxygen atoms.[288,289]

(iv) Ethers, aldehydes, ketones and esters

Zirconium and hafnium tetrahalides react with ethers, aldehydes, ketones and esters in solvents such as carbon tetrachloride, dichloromethane, benzene, acetyl chloride or liquid sulfur dioxide to give moisture-sensitive solid adducts of the type MX_4L and $[MX_4L_2]$ (Table 13). The more common 1:2 adducts are obtained when two equivalents or an excess of the Lewis base is used, while synthesis of the 1:1 adducts generally requires strict adherence to 1:1 stoichiometry. Procedures for preparation of $[MCl_4(THF)_2]$ (M = Zr or Hf) have been published in *Inorganic Syntheses*.[291] The adducts are thermally stable at room temperature but decompose at elevated temperatures. They are insoluble in most organic solvents, though they show some solubility in benzene and in nitrobenzene in which they behave as nonelectrolytes. Complex formation is accompanied by characteristic changes in the ν(C—O—C) frequency of the ethers and the ν(C=O) frequency of the aldehydes, ketones and esters; the IR bands due to these modes are shifted to lower frequency in the adducts.

Many of the 1:2 adducts have been assigned an octahedral *cis* structure on the basis of dipole moment data[292–297] (Table 13), and this has been confirmed for $[ZrCl_4(MeCO_2Et)_2]$ by IR spectroscopy.[298] On the other hand, $[HfCl_4(THF)_2]$[299] and $[ZrCl_4L_2]$ [L = naphthaldehyde[300] or coumarin[301] (**23**)] have been assigned an octahedral *trans* structure on the basis of the number of ν(M—O) and/or ν(M—Cl) bands in IR spectra.

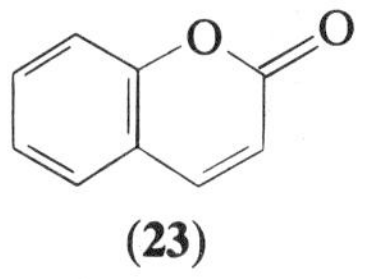

(23)

Table 13 Complexes of Zirconium(IV) and Hafnium(IV) Halides with Ethers, Aldehydes, Ketones and Esters

Complex	*M.p.* (°C)	μ^a (D)	$-\Delta H^b$ (kJ mol^{-1})	*Physical method*[c]	*Ref.*
Ether complexes					
$ZrCl_4(Me_2O)$				IR	1
$[ZrCl_4(Me_2O)_2]$	148–153			IR	1, 2
$[HfCl_4(Me_2O)_2]$	160–165 (dec.)			IR	2
cis-$[ZrCl_4(Et_2O)_2]$	87–93, 95	4.87		IR	2, 3
$[HfCl_4(Et_2O)_2]$	105–108			IR	2
$[ZrCl_4(THF)_2]$	170–171 (dec.)		141	IR, NMR	4–6
$[ZrBr_4(THF)_2]$			151	NMR	5
trans-$[HfCl_4(THF)_2]$	187		148	IR, NMR	4, 5, 7
$ZrCl_4$(1,4-dioxane)	220 (dec.)			IR	2, 8
cis-$[ZrCl_4$(1,4-dioxane)$_2]$		5.94			9
$HfCl_4$(1,4-dioxane)	260 (dec.)			IR	2
Aldehyde complexes					
$[ZrCl_4(PhCHO)_2]$			79		10
trans-$[ZrCl_4$(naphthaldehyde)$_2]$	250			IR, M, Λ	11
$[ZrCl_4$(furanaldehyde)$_2]$				IR, Λ	11
cis-$[ZrCl_4\{1,2\text{-}C_6H_4(CHO)_2\}]$	250			IR, Λ	11
Ketone complexes					
$ZrCl_4(Me_2CO)$				IR	12
cis-$[ZrCl_4(Me_2CO)_2]$	113–115[2], 125[3]	6.63[9], 7.67[3]		IR, Λ	2, 3, 9, 13
$[HfCl_4(Me_2CO)_2]$	131–134			IR	2
$[ZrCl_4(MeCOPh)_2]$	250			IR, Λ	13
$[ZrCl_4(PhCOPh)_2]$				IR, Λ	13
$[ZrCl_4(MeCOCN)_3]$	200			IR	14
$[HfCl_4(MeCOCN)_3]$	200			IR	14
$[ZrCl_4$(benzanthrone)$_2]$	>300			IR	15
$[ZrBr_4$(benzanthrone)$_2]$	>325			IR	15
$[ZrCl_4$(anthrone)$_2]$	>325			IR	15
$[ZrCl_4$(10-nitroanthrone)$_2]$	251			IR	15
$[ZrCl_4$(10-benzalanthrone)$_2]$	266–267			IR	15
cis-$[ZrCl_4(MeCOCMe_2COMe)]$	163–166			IR, NMR	16
Ester complexes					
$[ZrCl_4(HCO_2Me)_2]$	75–77			IR	2
$[HfCl_4(HCO_2Me)_2]$	82–84			IR	2
$ZrCl_4(HCO_2Et)$	94	4.15	86		17, 18
$ZrBr_4(HCO_2Et)$	108	4.50	82	M	18, 19
cis-$[ZrCl_4(HCO_2Et)_2]$		7.63	100	M	18, 20, 21
cis-$[ZrBr_4(HCO_2Et)_2]$		7.85	105	M	18, 19
cis-$[ZrI_4(HCO_2Et)_2]$		4.24			22
$ZrCl_4(HCO_2CHMe_2)$		4.18			17, 23
$HfCl_4(HCO_2CHMe_2)$					23
cis-$[ZrCl_4(HCO_2CHMe_2)_2]$		7.76			20
$ZrCl_4(MeCO_2Et)$	170	3.89	86	IR	17, 18
$ZrBr_4(MeCO_2Et)$	164	3.15	80	M	18, 19
$ZrI_4(MeCO_2Et)$		3.17			22
cis-$[ZrCl_4(MeCO_2Et)_2]$		6.75	100	IR, M	18, 20, 21, 24
cis-$[ZrBr_4(MeCO_2Et)_2]$		3.40	98	M	18, 19
cis-$[ZrI_4(MeCO_2Et)_2]$		3.49		M	22
$ZrCl_4(MeCO_2CHMe_2)$	129	3.85			17
cis-$[ZrCl_4(MeCO_2CHMe_2)_2]$		6.73			20
cis-$[ZrBr_4(MeCO_2CHMe_2)_2]$		3.97			19
cis-$[ZrI_4(MeCO_2CHMe_2)_2]$		3.93			22
cis-$[ZrCl_4(MeCO_2CH_2CHMe_2)_2]$		6.74			20
cis-$[ZrBr_4(MeCO_2CH_2CHMe_2)_2]$		4.21			19
cis-$[ZrI_4(MeCO_2CH_2CHMe_2)_2]$		4.23			22
cis-$[ZrCl_4(MeCO_2CH_2Ph)_2]$		6.54			20
$ZrCl_4(PrCO_2Et)$	163	3.53	86		17, 18
$ZrBr_4(PrCO_2Et)$	157	3.57	84		18, 19
cis-$[ZrCl_4(PrCO_2Et)_2]$		5.23	99	M	18, 20, 21
cis-$[ZrBr_4(PrCO_2Et)_2]$		3.80	97		18, 19
cis $[ZrI_4(PrCO_2Et)_2]$		3.07			22
$[ZrCl_4(PhCO_2Me)_2]$			115	M	25, 26
$[HfCl_4(PhCO_2Me)_2]$			101	M	25, 26
$[ZrCl_4(PhCO_2Et)_2]$			118	M	25, 26

Table 13 *(continued)*

Complex	M.p. (°C)	μ^a (D)	$-\Delta H^b$ (kJ mol^{-1})	Physical method[c]	Ref.
$[HfCl_4(PhCO_2Et)_2]$			108	M	25, 26
$[ZrCl_4(PhCO_2Ph)_2]$			38	M	25
$ZrCl_4(NCCH_2CO_2Et)$	180 (dec.)			IR	27
trans-$[ZrCl_4(coumarin)_2]$	137–138			IR, Λ	28
$ZrCl_4\{(CO_2Et)_2\}$		7.52		IR, M, Λ	29, 30
$ZrBr_4\{CO_2Et)_2\}$				IR, M, Λ	29
$ZrCl_4\{CH_2(CO_2Et)_2\}$		9.21		IR, M, Λ	29, 30
$ZrBr_4\{CH_2(CO_2Et)_2\}$				IR, M, Λ	29
$ZrX_4\{[(CH_2)_nCO_2Et]_2\}$ (X = Cl, Br; n = 1, 2)				IR, M, Λ	29
$\{ZrX_4\}_2\{[(CH_2)_nCO_2Et]_2\}$ (X = Cl, Br; n = 1, 2)				IR, M, Λ	29
$[ZrCl_4\{1,2\text{-}C_6H_4(CO_2Et)_2\}]$			76	M	31
$[HfCl_4\{1,2\text{-}C_6H_4(CO_2Et)_2\}]$			74	M	31

[a] Dipole moment in benzene solution at 20 °C. [b] ΔH is the enthalpy of the reaction in equation (42). [c] IR = infrared spectrum; NMR = ^{1}H NMR spectrum; M = molecular weight; Λ = electrical conductance.

1. G. Rossmy and H. Stamm, *Liebigs Ann. Chem.*, 1958, **618,** 59.
2. W. M. Graven and R. V. Peterson, *J. Inorg. Nucl. Chem.*, 1969, **31,** 1743.
3. O. A. Osipov and Yu. B. Kletenik, *J. Gen. Chem. USSR (Engl. Transl.)*, 1961, **31,** 2285.
4. L. E. Manzer, *Inorg. Synth.*, 1982, **21,** 135.
5. F. M. Chung and A. D. Westland, *Can. J. Chem.*, 1969, **47,** 195.
6. A. D. Westland and V. Uzelac, *Can. J. Chem.*, 1970, **48,** 2871.
7. S. R. Wade and G. R. Willey, *J. Less-Common Met.*, 1979, **68,** 105.
8. E. C. Alyea and E. G. Torrible, *Can. J. Chem.*, 1965, **43,** 3468.
9. O. A. Osipov, V. I. Minkin and V. A. Kogan, *Russ. J. Phys. Chem. (Engl. Transl.)*, 1962, **36,** 468.
10. M. H. Dilke and E. D. Eley, *J. Chem. Soc.*, 1949, 2601.
11. R. C. Paul, H. R. Singal and S. L. Chadha, *Indian J. Chem.*, 1971, **9,** 995.
12. P. T. Joseph and W. B. Blumenthal, *J. Org. Chem.*, 1959, **24,** 1371.
13. R. C. Paul and S. L. Chadha, *J. Inorg. Nucl. Chem.*, 1969, **31,** 1679.
14. S. C. Jain and B. P. Hajela, *Indian J. Chem.*, 1974, **12,** 843.
15. R. C. Paul, R. Parkash and S. S. Sandhu, *J. Inorg. Nucl. Chem.*, 1966, **28,** 1907.
16. A. L. Allred and D. W. Thompson, *Inorg. Chem.*, 1968, **7,** 1196.
17. Yu. B. Kletenik, O. A. Osipov and E. E. Kravtsov, *Russ. J. Inorg. Chem. (Engl. Transl.)*, 1959, **29,** 13.
18. O. A. Osipov and Yu. B. Kletenik, *Russ. J. Inorg. Chem. (Engl. Transl.)*, 1960, **5,** 1076.
19. Yu. B. Kletenik and O. A. Osipov, *J. Gen. Chem. USSR (Engl. Transl.)*, 1959, **29,** 1398.
20. O. A. Osipov and Yu. B. Kletenik, *J. Gen. Chem. USSR (Engl. Transl.)*, 1957, **27,** 2953.
21. O. A. Osipov and Yu. B. Kletenik, *J. Gen. Chem. USSR (Engl. Transl.)*, 1959, **29,** 1351.
22. O. A. Osipov and Yu. B. Kletenik, *J. Gen. Chem. USSR (Engl. Transl.)*, 1959, **29,** 2085.
23. Yu. A. Lysenko and L. I. Khokhlova, *Russ. J. Inorg. Chem. (Engl. Transl.)*, 1974, **19,** 690.
24. M. F. Lappert, *J. Chem. Soc.*, 1962, 542.
25. W. S. Hummers, S. Y. Tyree, Jr. and S. Yolles, *J. Am. Chem. Soc.*, 1952, **74,** 139.
26. F. W. Chapman, W. S. Hummers, S. Y. Tyree and S. Yolles, *J. Am. Chem. Soc.*, 1952, **74,** 5277.
27. S. C. Jain and R. Rivest, *Can. J. Chem.*, 1964, **42,** 1079.
28. R. C. Paul and S. L. Chadha, *Indian J. Chem.*, 1970, **8,** 739.
29. K. C. Malhotra and S. M. Sehgal, *Indian J. Chem.*, 1975, **13,** 599.
30. O. A. Osipov, V. M. Artemova, V. A. Kogan and Yu. A. Lysenko, *J. Gen. Chem. USSR (Engl. Transl.)*, 1962, **32,** 1354.
31. R. V. Moore and S. Y. Tyree, *J. Am. Chem. Soc.*, 1954, **76,** 5253.

IR spectra also provide information about the coordination site in these adducts. In $[ZrCl_4L_2]$ (L = furanaldehyde[300] or coumarin[301]), only the carbonyl oxygen atom is attached to the Zr atom. All three carbonyl groups, and none of the nitrile groups, appear to be coordinated in the pyruvonitrile complexes $[MCl_4(MeCOCN)_3]$ (M = Zr or Hf), suggesting that these compounds are seven-coordinate.[302] Among the compounds in Table 13, the $[MCl_4(MeCOCN)_3]$ complexes are the only 1:3 adducts to have been reported. IR spectra indicate that both the carbonyl oxygen atom and the nitrile nitrogen atom are coordinated in the 1:1 ethyl cyanoacetate adduct, $ZrCl_4(NCCH_2CO_2Et)$. A structure having a six-membered chelate ring has been suggested,[303] but because of the geometry of the ligand, a polymeric structure with ethylcyanoacetate bridges seems more likely.

Not all of the compounds listed in Table 13 have been isolated in the solid state. The existence of some of the ester adducts has been inferred from electric polarization and cryoscopic molecular weight measurements on 1:1 and 1:2 stoichiometric mixtures of the metal tetrahalide and the ester in benzene solution. The concentration dependence of the polarizations and molecular weights[294–297,304] indicates that the 1:1 complexes tend to dimerize in the more concentrated solutions and the 1:2 complexes tend to dissociate (equation 41) in the more dilute solutions. Dissociation of the $[ZrX_4(RCO_2R')_2]$ complexes increases as X varies in the order Cl < Br < I. Thus, estimated equilibrium constants for dissociation of the $[ZrX_4(MeCO_2Et)_2]$ complexes according to equation (41) are $\sim 5\times10^{-4}$ (X = Cl), 2×10^{-2} (X = Br) and 3.5×10^{-1} (X = I). Dimerization of the $[ZrX_4(RCO_2R']$ complexes increases as X varies in the reverse order (I < Br < Cl). The ease of dissociation of the $[ZrX_4(RCO_2R')_2]$ complexes also depends on the size of the R group. Thus, the ethyl formates $[ZrX_4(HCO_2Et)_2]$ (X = Cl or Br; $\mu = 7.63$ and 7.85 D, respectively) are hardly dissociated, while the ethyl butyrates $[ZrX_4(PrCO_2Et)_2]$ (X = Br or I; apparent $\mu = 3.80$ and 3.07 D, respectively) are essentially completely dissociated to $[ZrX_4(PrCO_2Et)]$ and $PrCO_2Et$. These dissociation reactions complicate interpretation of the dipole moment data in Table 13. The polarization and molecular weight measurements provide no evidence for dissociation of the 1:1 complexes, and they indicate that the 1:2 complexes do not bind additional ester molecules in the presence of an excess of the ester.

$$[ZrX_4(RCO_2R')_2] \rightleftharpoons [ZrX_4(RCO_2R')] + RCO_2R' \quad (41)$$

Table 13 includes heats of formation for some of the adducts; values of ΔH, determined calorimetrically, refer to equation (42). In formation of the $ZrX_4(RCO_2Et)_n$ (R = H, Me or Pr, $n = 1$ or 2) complexes, ΔH for addition of the first ester molecule is much larger than ΔH for addition of the second ester molecule,[305] consistent with the ease of dissociation of the 1:2 complexes.

$$MX_4(s) + nL(l) \longrightarrow MX_4L_n(\text{s or l}) \quad (42)$$
$$(n = 1 \text{ or } 2)$$

Also included in Table 13 are zirconium and hafnium tetrahalide adducts with diethers, dialdehydes, diketones and diesters. 1,4-Dioxane gives both 1:1 and 1:2 adducts; the 1:1 complexes may be polymeric.[120,306] Phthalaldehyde, 3,3-dimethylacetylacetone, and diethyl phthalate yield 1:1 adducts with MCl_4 (M = Zr or Hf); IR and molecular weight data suggest that these complexes have an octahedral *cis* structure in which the Lewis base behaves as a chelating ligand. ZrX_4 (X = Cl or Br) form 1:1 adducts with diethyl oxalate and diethyl malonate and both 1:1 and 2:1 (*i.e.* $\{ZrX_4\}_2\{\text{diester}\}$) adducts with diethyl succinate and diethyl adipate.[307] Apparently conflicting molecular weight data have been reported for the 1:1 complexes[307,308] and the structure of these compounds remains uncertain.

32.4.4.5 Oxoanions as ligands

(*i*) *Nitrate*

Anhydrous zirconium(IV) nitrate has been prepared by reaction of $ZrCl_4$ with liquid dinitrogen pentoxide at 30 °C (equation 43). This compound is a covalent, hygroscopic, colorless crystalline solid which sublimes at 100 °C *in vacuo* and is extremely soluble in polar organic solvents.[309] IR and Raman spectra of $[Zr(NO_3)_4]$ have been interpreted[310] in terms of the eight-coordinate, D_{2d}-dodecahedral *mmmm* structure that has been established by X-ray diffraction for $[Ti(NO_3)_4]$[311] and $[Sn(NO_3)_4]$.[312] Nitrogen–oxygen stretching bands in the 1576–1650, 1223–1280 and 986–1028 cm^{-1} regions have been assigned to $\nu(N{=}O)$, $\nu_{as}(NO_2)$ and $\nu_s(NO_2)$, respectively. The $\nu(N{=}O)$ and $\nu_{as}(NO_2)$ modes arise from splitting of the ν_3 (E') mode of the D_{3h} $[NO_3]^-$ ion, and the large difference between the frequencies of these two modes is consistent with bidentate coordination of the nitrate ligands.[313] Hafnium(IV) nitrate has been prepared by reaction of hydrated hafnium nitrate with liquid N_2O_5. Sublimation of the reaction product at 100 °C *in vacuo* afforded the addition complex $Hf(NO_3)_4{\cdot}N_2O_5$.[314]

$$ZrCl_4 + 4N_2O_5 \longrightarrow [Zr(NO_3)_4] + 4NO_2Cl \quad (43)$$

The reaction of $[NMe_4]_2[MCl_6]$ (M = Zr or Hf) with N_2O_4 in acetonitrile at room temperature yields white, moisture-sensitive hexanitrato complexes $[NMe_4]_2[M(NO_3)_6]$.[315] A similar reaction between $Cs_2[ZrCl_6]$ and liquid N_2O_5 gives a mixture of $CsNO_3$ and $[Zr(NO_3)_4]$. The $[NMe_4]_2[M(NO_3)_6]$ complexes are thermally unstable, slowly evolving nitrogen oxides at 130 °C (M = Zr) or 230 °C (M = Hf). They exhibit $\nu(N{-}O)$ IR bands in the regions 1517–1600,

1269–1294 and 981–1021 cm^{-1}. The splitting between the bands arising from the ν_3 (E') mode of $[NO_3]^-$ is substantial, though not as large as for $[Zr(NO_3)_4]$. The $[NMe_4]_2[M(NO_3)_6]$ complexes probably contain monodentate nitrate ligands, but their structures have not yet been determined.

Quite large splittings of the ν_3 (E′) mode have been observed for the hydrated tetranitrates $M(NO_3)_4(H_2O)_4$ (M = Zr or Hf)[166,316] and the basic nitrates $MO(NO_3)_2(H_2O)_2$, $MO(NO_3)_2(H_2O)_6$, $M_2O_3(NO_3)_2(H_2O)_5$ (M = Zr or Hf)[317] and $Zr(OH)(NO_3)_3(H_2O)_3$,[166] indicative of strong nitrate coordination; however, the mode of attachment of the nitrate ligands is unknown. Bidentate attachment has been proposed for the $[ZrO(NO_3)_2L_2]$ (L = N-heterocyclic or O-donor ligand) complexes mentioned in Sections 32.4.2.2 and 32.4.4.1, and this has been established by X-ray diffraction for the β-diketonate complexes $[Zr(acac)_3(NO_3)]$[255] and $[Zr(acac)_2(NO_3)_2]$[257] discussed in Section 32.4.4.4.i.

(ii) *Dichlorophosphate*

Dichlorophosphates of the type $ZrCl_3(O_2PCl_2)\cdot POCl_3$ and $MOCl(O_2PCl_2)\cdot POCl_3$ (M = Zr or Hf) have been synthesized by reaction of the metal tetrachlorides with dichlorine monoxide in $POCl_3$ (equations 44 and 45). IR and Raman spectral studies suggest that $ZrCl_3(O_2PCl_2)\cdot POCl_3$ has a centrosymmetric dimeric structure (**24**), while the very insoluble $MOCl(O_2PCl_2)\cdot POCl_3$ compounds are assumed to have an oxo-bridged polymeric structure.[318]

$$ZrCl_4 + 2POCl_3 + Cl_2O \longrightarrow ZrCl_3(O_2PCl_2)\cdot POCl_3 + 2Cl_2 \quad (44)$$

$$ZrCl_4 + 2POCl_3 + 2Cl_2O \longrightarrow MOCl(O_2PCl_2)\cdot POCl_3 + 4Cl_2 \quad (45)$$

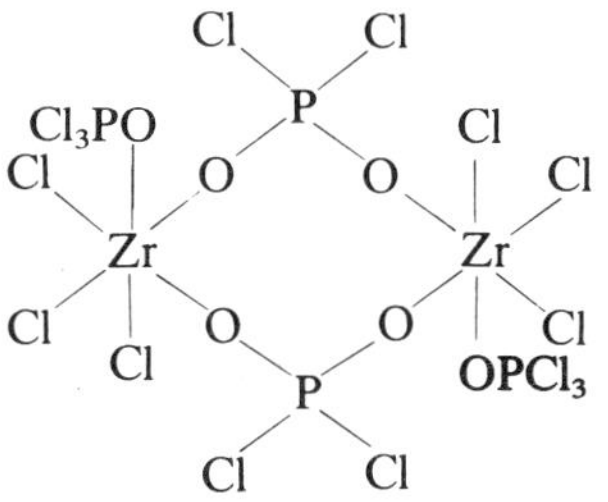

(**24**)

(iii) *Sulfate, fluorosulfate and selenate*

Zirconium(IV) and hafnium(IV) form a plethora of crystalline sulfates. The normal sulfates include five anhydrous phases, four monohydrates, a tetrahydrate, three pentahydrates and a heptahydrate. The structural relationships and transformations among these compounds have been reviewed by Bear and Mumme.[319,320] Of the normal zirconium sulfates whose structures have been determined thus far, only α-$Zr(SO_4)_2(H_2O)_5$,[321] β-$Zr(SO_4)_2(H_2O)_5$[322] and $Zr(SO_4)_2(H_2O)_7$[323] contain discrete complexes. All three of these compounds contain the centrosymmetric dinuclear $[Zr_2(SO_4)_4(H_2O)_8]$ molecule in which each Zr atom is dodecahedrally coordinated by one bidentate chelating sulfate ligand, two bridging sulfate ligands and four water molecules; Zr—O bond lengths are in the ranges 2.16–2.28 Å (chelating sulfate), 2.06–2.18 Å (bridging sulfate), and 2.16–2.27 Å (water). $Zr(SO_4)_4(H_2O)_4$,[324] α-$Zr(SO_4)_2(H_2O)$[325] and γ-$Zr(SO_4)_2(H_2O)$[326] have layer structures, while anhydrous α-$Zr(SO_4)_2$[327,328] has an extended three-dimensional structure. The Zr atom exhibits square antiprismatic eight-coordination in the tetrahydrate and tetragonal base–trigonal base seven-coordination in the monohydrates and the anhydrous sulfate. The sulfate ligands in these compounds are variously bidentate chelating, bidentate bridging, tridentate bridging or tetradentate bridging. X-Ray data indicate that the corresponding hafnium sulfates[329] are isostructural with the zirconium compounds.

Basic zirconium sulfates[5,6] are unlikely to contain discrete complexes. $Zr(OH)_2SO_4$[330] and $Zr_2(OH)_2(SO_4)_3(H_2O)_4$[331] have three-dimensional framework structures that feature hydroxo and sulfato bridges.

Numerous complex sulfates of the type $M'_2M(SO_4)_3(H_2O)_x$, $M'_4M(SO_4)_4(H_2O)_x$ and

$M'_6M(SO_4)_5(H_2O)_x$ (M′ = Na, K, Rb, Cs, or NH_4; M = Zr or Hf) have been isolated from aqueous sulfuric acid solutions,[332–340] and related oxalato-, carbonato- and fluoro-containing mixed-ligand complexes such as $(NH_4)_2Zr(C_2O_4)(SO_4)_2(H_2O)_5$,[341] $(NH_4)_6Zr_2(CO_3)_6(SO_4)(H_2O)_6$,[341] $M'_2ZrF_2(SO_4)_2(H_2O)_x$ and $M'_2ZrF_4(SO_4)(H_2O)_x$ (M′ = K, Rb, Cs or NH_4)[342] have also been reported. Recent X-ray studies have shown that the complex sulfates can have mononuclear, dinuclear or polynuclear structures. The tetrasulfato complexes $Na_4[Hf(SO_4)_4(H_2O)_2]$[343] and $[NH_4]_4[Hf(SO_4)_4(H_2O)_2]\cdot 2H_2O$[344] contain discrete mononuclear anions in which the Hf atom is dodecahedrally coordinated by eight oxygen atoms from two bidentate sulfate ligands, two monodentate sulfate ligands and two water molecules. In the pentasulfato complex $Na_6[Hf(SO_4)_5(H_2O)]$,[345] the Hf atom is attached to two bidentate sulfate groups, three monodentate sulfate groups and one water molecule; the geometry of the HfO_8 coordination group in this complex does not closely approximate that of any of the common eight-coordination polyhedra. Hf—O bond distances in these complexes are in the ranges 2.15–2.33 Å (bidentate sulfate), 2.07–2.17 Å (monodentate sulfate) and 2.18–2.33 Å (water). The trisulfato complex $K_2Zr(SO_4)_3(H_2O)_2$ contains centrosymmetric $[Zr_2(SO_4)_6(H_2O)_4]^{4-}$ anions in which the eight-coordinate Zr atoms are held together by two bidentate bridging sulfate ligands; the ZrO_8 coordination polyhedron has been variously described as dodecahedral[346] or bicapped trigonal prismatic.[347] Still another type of structure has been found for the isostructural complexes $Na_2M(SO_4)_3(H_2O)_3$ (M = Zr[348] or Hf[349]). In these compounds the metal atom is attached to two water molecules, two bidentate chelating sulfate ligands, and two bidentate bridging sulfate ligands, giving infinite spiral chains in which the repeating unit is $[M(SO_4)_3(H_2O)_2]^{2-}$. The chains are held together by Na^+ cations and water molecules. $Cs_2Hf(SO_4)_3(H_2O)_2$[350] has a similar chain structure.

Cation exchange, solvent extraction and Raman spectroscopic studies of highly acidic zirconium sulfate solutions suggest that these solutions contain a mixture of solution species having from one to four sulfate ligands per Zr atom.[153,351] The sulfato–zirconium complexes appear to be somewhat more stable than the analogous hafnium complexes.[351]

Zirconium(IV) fluorosulfates $Zr(SO_3F)_4$, $ZrO(SO_3F)_2$, $Zr(O_2CMe)_2(SO_3F)_2$ and $Zr(O_2CMe)_3(SO_3F)$ have been prepared by reaction of fluorosulfuric acid with $Zr(O_2CCF_3)_4$, anhydrous $ZrOCl_2$ or $Zr(O_2CMe)_4$. IR spectra suggest that the fluorosulfate ligands are bidentate in the first three compounds and tridentate (C_{3v} symmetry) in $Zr(O_2CMe)_3(SO_3F)$. All four compounds are good Lewis acids and form complexes with pyridine, 2,2′-bipyridyl and triphenylphosphine oxide.[352]

Mixed-ligand selenato complexes of the type $K_2MF_2(SeO_4)_2(H_2O)_3$, $K_2MF_2(SO_4)(SeO_4)(H_2O)_3$ and $M'_2M_2F_8(SeO_4)(H_2O)_3$ (M = Zr or Hf; M′ = K or Rb) have been isolated from aqueous selenic acid solutions. Possible polymeric and dimeric structures have been proposed for these compounds on the basis of IR spectra.[353,354]

(*iv*) *Molybdate*

Alkali metal zirconium and hafnium molybdates have been isolated from high-temperature melts, and X-ray structures of representative compounds have been determined. The isostructural $K_8[M(MoO_4)_6]$ (M = Zr or Hf) complexes[355] contain discrete $[M(MoO_4)_6]^{8-}$ anions that are linked together by K^+ cations. The Zr (Hf) atoms are octahedrally coordinated by six oxygen atoms (mean Zr—O = 2.08, mean Hf—O = 2.05 Å) which are corner-shared with six MoO_4 tetrahedra. $Cs_8[Zr(MoO_4)_6]$[356] crystallizes in a different space group but contains the same complex anions (mean Zr—O = 2.11 Å). In the structure of $Cs_2Hf(MoO_4)_3$,[357] each Hf atom is again surrounded octahedrally by six MoO_4 tetrahedra. However, the molybdate groups bridge between adjacent Hf atoms giving infinite columns of HfO_6 octahedra and MoO_4 tetrahedra with repeating unit $[Hf(MoO_4)_3]^{2-}$.

(*v*) *Perchlorate*

Anhydrous zirconium(IV) and hafnium(IV) perchlorates have been prepared by reaction of $ZrCl_4$ or $HfCl_4$ with a solution of Cl_2O_7 in anhydrous perchloric acid at −10 °C.[358,359] $[Zr(ClO_4)_4]$ (m.p. 95.5–96.0 °C) and $[Hf(ClO_4)_4]$ (m.p. 104.5–105.0 °C) are volatile, hygroscopic, colorless crystalline solids. The vapor pressure of $[Zr(ClO_4)_4]$ varies from 0.8 mmHg at 25 °C to 210.8 mmHg at 95.7 °C,[360] while that of the even more volatile $[Hf(ClO_4)_4]$ ranges from 2.2 mmHg at 30.5 °C to 37.9 mmHg at 60.9 °C.[361] Above 100 °C, the $[M(ClO_4)_4]$ complexes decompose to $MO(ClO_4)_2$ (M = Zr or Hf). IR and Raman spectra of $[Zr(ClO_4)_4]$[358] and $[Hf(ClO_4)_4]$[359] have been interpreted in terms of an eight-coordinate structure that contains bidentate chelating perchlorate ligands. A large frequency difference between the terminal and bridging ν(Cl—O) bands indicates a considerable amount of covalent character in the metal–perchlorate bonds. IR spectra of gaseous $[M(ClO_4)_4]$ (M = Zr[360] or Hf[361]) are not significantly different from spectra of crystalline $[M(ClO_4)_4]$, indicating no significant change in molecular structure upon vaporization.

$[Zr(ClO_4)_4]$ and $[Hf(ClO_4)_4]$ react with cesium or rubidium perchlorates in $HClO_4$–Cl_2O_7 yielding (perchlorato)metallates of the type $Cs_n[Zr(ClO_4)_{4+n}]$ (n = 1, 2, 3 or 4),[362] $Cs_n[Hf(ClO_4)_{4+n}]$ (n = 1, 2 or 3)[363] and $Rb_n[Hf(ClO_4)_{4+n}]$ (n = 1 or 2).[363] IR spectra suggest that, as n increases, the number of bidentate perchlorate ligands decreases and the number of monodentate perchlorate groups increases, thus maintaining a coordination number of eight for the metal atom.

32.4.4.6 Carboxylates, carbamates, carbonates and oxalates

(i) Carboxylates

Zirconium carboxylates have long been known,[3] but these compounds are still not very well characterized and no X-ray structural data are available. Known compounds include normal carboxylates $M(O_2CR)_4$, mixed-ligand compounds $Zr(O_2CR)_nX_{4-n}$ (X = Cl or $OCHMe_2$) and a variety of basic carboxylates such as $MO(O_2CR)_2(H_2O)_x$, $Zr(OH)_3(O_2CR)$, $Zr_2O(O_2CR)_6$ and $M_2O(OH)(O_2CR)_5(H_2O)_x$ (M = Zr or Hf).

The following normal carboxylates have been prepared by reaction of $ZrCl_4$ or $HfCl_4$ with an excess of hot, anhydrous carboxylic acid: $Zr(O_2CR)_4$ (R = H, Me, Et, Pr, $CHMe_2$, $C_{15}H_{31}$, $C_{17}H_{35}$, CH_2Ph, Ph, CH_2Cl, $CHCl_2$ or CCl_3)[221,364–371] and $M(O_2CCF_3)_4$ (M = Zr or Hf).[372] Temperature control appears to be an important factor in these syntheses since intermediate products are obtained at low temperatures and thermal decomposition occurs at high temperatures. Thus Ludwig and Schwartz[366] reported that the reaction of $ZrCl_4$ with RCO_2H at −15 °C in carbon tetrachloride yields the addition compounds $ZrCl_4(RCO_2H)_2$ (R = Et, Pr or $CHMe_2$). In xylene solution, the reaction products are the mixed-ligand complexes $Zr(O_2CR)_2Cl_2(RCO_2H)_2$ at 60–65 °C, and the tetrakis(carboxylates) $Zr(O_2CR)_4$ (R = Me, Et, Pr or $CHMe_2$) at 125 °C. Paul *et al.*[365] have synthesized $Zr(O_2CR)_4$ (R = Me, Et, CH_2Cl, $CHCl_2$ or CCl_3) by heating at reflux a mixture of $ZrCl_4$ and an excess of the pure carboxylic acid under reduced pressure at temperatures below 80 °C. If the reactants are allowed to reflux at 1 atm pressure, the thermal decomposition products $ZrO(O_2CR)_2$ are obtained.

Brainina *et al.*[373] have prepared $Zr\{O_2C(CH_2)_nX\}_4$ (n = 4, 5, 6, 11, 15 or 17 when X = H; n = 10 when X = Cl) from $Zr(acac)_4$ and the fatty acid (equation 46), the acetylacetone being removed by heating the reaction mixture in a stream of dry dinitrogen at reduced pressure. A similar approach has been employed for synthesis of $M(O_2CR)_4$ (M = Zr or Hf; R = Pr, Ph, 4-$C_6H_4NO_2$, or 4-$C_6H_4NH_2$).[374–377]

$$Zr(acac)_4 + 4RCO_2H \longrightarrow Zr(O_2CR)_4 + 4Hacac \qquad (46)$$

The tetrakis(carboxylates) rapidly hydrolyze in moist air and thermally decompose to oxocarboxylato compounds at elevated temperatures. They are nonelectrolytes in nitrobenzene and acetonitrile. Because of their insolubility in most organic solvents, they have been assumed to be polymeric,[365] and this is supported by molecular weight data for $M(O_2CPr)_4$ (M = Zr or Hf).[374] On the other hand, more recent molecular weight measurements on $Zr(O_2CMe)_4$ in antimony(III) chloride,[367] $Zr(O_2CEt)_4$ in propionic acid, and $Zr(O_2CPr)_4$ and $Zr(O_2CCHMe_2)_4$ in the corresponding butyric acids and in carbon tetrachloride[366] indicate that all four compounds are monomeric. Moreover, IR spectra of $Zr(O_2CMe)_4$, which exhibit $\nu_{as}(CO_2)$ and $\nu_s(CO_2)$ bands at 1540 and 1455 cm^{-1}, respectively, suggest that the acetate ligands are chelating rather than bridging.[378] Therefore, the tetrakis(carboxylates) may be discrete eight-coordinate chelate complexes; this needs to be confirmed by X-ray diffraction. $Zr(O_2CCCl_3)_4$ behaves as a Lewis acid, forming $Zr(O_2CCCl_3)_4L_2$ adducts with pyridine, pyridine *N*-oxide and triphenylphosphine oxide.[371]

Mixed-ligand compounds of the type $Zr(O_2CCH_2Ph)_nCl_{4-n}$ (n = 2 or 3)[369] and

$Zr(O_2CR)_n(OCHMe_2)_{4-n}$ ($R = C_{15}H_{31}$ or $C_{17}H_{35}$; $n = 1$ or 2)[368] have been prepared by reaction of stoichiometric amounts of the carboxylic acid with $ZrCl_4$ or $Zr(OCHMe_2)_4 \cdot HOCHMe_2$, respectively. Similar reactions of $Zr(OCHMe_2)_4 \cdot HOCHMe_2$ with furan-2-carboxylic acid, indole-3-acetic acid or indole-3-butyric acid yield the heterocyclic carboxylato complexes $Zr(O_2CR)_n(OCHMe_2)_{4-n}$ ($R = C_4H_3O$, C_9H_8N or $C_{11}H_{12}N$; $n = 1$, 2 or 3); IR spectra of these complexes exhibit an 80–120 cm^{-1} separation between the $\nu_{as}(CO_2)$ and $\nu_s(CO_2)$ bands, suggesting a symmetrical bidentate attachment of the carboxylate ligands.[379] The reaction of $Zr(OCHMe_2)_4 \cdot HOCHMe_2$ with fatty acids present in excess in refluxing benzene yields the basic carboxylates $Zr_2O(O_2CR)_6$ ($R = C_9H_{19}$, $C_{11}H_{23}$, $C_{15}H_{31}$ or $C_{17}H_{35}$).[368]

Basic carboxylates are generally obtained from reactions of $MOCl_2 \cdot 8H_2O$ (M = Zr or Hf) with carboxylic acids or acyl chlorides.[178,380–383] Compounds of the type $MO(O_2CEt)_2(H_2O)$ and $M_2O(OH)(O_2CR)_5(H_2O)_n$ (R = H or Me; $n = 1$ or 2) have been studied extensively by Russian workers,[381–386] and some, as yet unconfirmed, tetranuclear and octanuclear structures have been proposed on the basis of IR and molecular weight data. Dicarboxylic acids precipitate Zr^{IV} from aqueous zirconyl chloride solutions at pH 2–3 yielding polymeric complexes of composition $Zr(OH)_3(O_2CCH{=}CHCO_2H)$, $Zr(OH)_3(O_2CC_6H_4CO_2H)$ and $Zr(OH)_3\{O_2C(CH_2)_nCO_2H\}$ ($n = 1$, 2 or 4).[387,388]

Amino acid complexes of the type $[M_4(OH)_8(O_2CCHRNH_3)_8]X_8 \cdot 12H_2O$ (M = Zr or Hf; R = H or Me; X = Cl, Br or NO_3) have been prepared in aqueous solution from glycine or α-alanine and the appropriate zirconyl (hafnyl) salt. IR and X-ray photoelectron spectra indicate that (i) the amino acid ligands are in the zwitterionic form, $^-O_2CHRNH_3^+$, and are coordinated through the carboxylate oxygen atoms, and (ii) the X^- ions are not coordinated. The tetrameric structure of $[Zr_4(OH)_8(H_2O)_{16}]Cl_8 \cdot 12H_2O$ is assumed to be preserved in the amino acid complexes.[389–392]

(*ii*) *Carbamates*

Tetrakis(*N,N*-dimethylcarbamato)zirconium(IV), an air- and moisture-sensitive white crystalline solid, has been prepared by reaction of CO_2 with $Zr(NMe_2)_4$ in toluene.[70] This compound is monomeric in benzene, and IR studies indicate that the carbamato ligands are bidentate. An eight-coordinate dodecahedral structure like that found for the analogous monothiocarbamato complex $[Zr(SOCNEt_2)_4]$ (Section 32.4.5.2) is likely. ^1H NMR spectra of $[Zr(O_2CNMe_2)_4]$ at −120 °C in 1:1 dichloromethane–Freon 11 exhibit only a single methyl resonance. In the absence of accidental degeneracy, this observation would indicate that $[Zr(O_2CNMe_2)_4]$ is stereochemically nonrigid at −120 °C.

(*iii*) *Carbonates*

Carbonato complexes of the type $K_4[M(CO_3)_4] \cdot nH_2O$, $[C(NH_2)_3]_4[M(CO_3)_4] \cdot H_2O$ and $[Co(NH_3)_6]_4[Zr(CO_3)_4]_3 \cdot 11H_2O$ (M = Zr or Hf; $n = 2$ or 6) have been isolated from aqueous solutions that contain $MOCl_2 \cdot 8H_2O$ and appropriate carbonate salts.[393,394] The potassium and guanidinium salts exhibit molar conductances in aqueous solution that are close to the value expected for a 4:1 electrolyte.[393] IR spectra show strong bands in the regions 1580–1620, 1280–1360 and 1012–1060 cm^{-1} due to stretching vibrations of bidentate carbonato ligands.[395] The $[M(CO_3)_4]^{4-}$ ions are assumed[393–395] to have an eight-coordinate dodecahedral structure like that of $[Zr(C_2O_4)_4]^{4-}$ (Section 32.4.4.6.iv).

IR and conductance data also suggest the likelihood of mononuclear eight-coordinate anions in the mixed-ligand complexes $[C(NH_2)_3]_6[Zr(CO_3)_4(C_2O_4)] \cdot 1.5H_2O$, $[C(NH_2)_3]_6[Zr(CO_3)_3(C_2O_4)_2] \cdot 7H_2O$, $K_6[Zr(CO_3)(C_2O_4)_4] \cdot 7.5H_2O$,[396] $[NH_4]_4[Zr(CO_3)_2F_4]$[341] and $K_4[M(CO_3)(C_2O_4)_2F_2] \cdot 5H_2O$ (M = Zr or Hf).[354] In the $[Zr(CO_3)_n(C_2O_4)_{5-n}]^{6-}$ ($n = 1$, 3 or 4) complexes, some of the ligands are bidentate and some are monodentate; IR spectra indicate that the carbonato–metal chelate rings are more stable than the oxalato–metal chelate rings.

Pospelova and Zaitsev[393] have reported a series of basic carbonato complexes of zirconium in which the metal-to-carbonate ratio varies from 1:1 to 1:3. Typical compositions are: $M'_nZr_2O(OH)_{6-n}(CO_3)_n(H_2O)_x$ ($n = 2$, 3, 4 or 6) and $M'_4Zr_2O(CO_3)_5(H_2O)_x$ (M′ = Na, K, NH_4, $C(NH_2)_3$, *etc.*). Analogous carbonato-oxalato complexes such as $M'_6Zr_2O(CO_3)_n(C_2O_4)_{6-n}(H_2O)_x$ ($n = 2$ or 4) are also known.[396] IR studies have shown that the basic carbonato complexes contain bidentate carbonate ligands,[395] but only the 1:3 complexes of composition $K_6Zr_2O(CO_3)_6(H_2O)_7$ and $(NH_4)_6Zr_2O(CO_3)_6(H_2O)_5$ have been characterized structurally by X-ray diffraction.[397,398] These complexes, properly formulated as $K_6[Zr_2(OH)_2(CO_3)_6]\cdot 6H_2O$ and $[NH_4]_6[Zr_2(OH)_2(CO_3)_6]\cdot 4H_2O$, both contain a centrosymmetric dimeric anion (Figure 6) in which each Zr atom is dodecahedrally coordinated by two bridging hydroxo groups (Zr—O = 2.07–2.13 Å) and three bidentate carbonate ligands (Zr—O = 2.16–2.29 Å). The two dodecahedra share an *m* edge, and the carbonate ligands span the remaining six *m* edges. Analogous hafnium complexes, $K_6[Hf_2(OH)_2(CO_3)_6]\cdot 6H_2O$ and $[C(NH_2)_3]_6[Hf_2(OH)_2(CO_3)_6]\cdot 2H_2O$, have been reported by Dervin *et al.*[394]

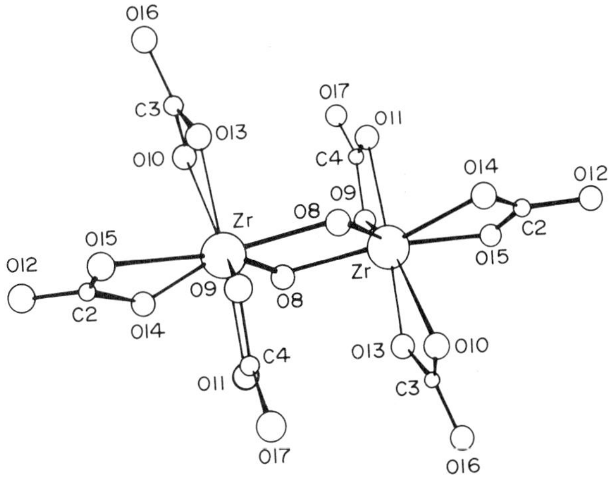

Figure 6 Structure of the centrosymmetric anion in $[NH_4]_6[Zr_2(OH)_2(CO_3)_6]\cdot 4H_2O$ (reproduced by permission from ref. 398)

(*iv*) *Oxalates*

Oxalato complexes of zirconium have been known for many years.[3,399] The best characterized is the $[Zr(C_2O_4)_4]^{4-}$ ion, which has been isolated in alkali,[399] alkaline earth,[400,401] cadmium and lead[401] salts. High-yield procedures for preparation of $K_4[M(C_2O_4)_4]\cdot 5H_2O$ (M = Zr or Hf) have appeared in *Inorganic Syntheses*.[402] The preparations involve addition of an aqueous solution of $MOCl_2\cdot 8H_2O$ to a solution containing potassium oxalate and oxalic acid, followed by addition of ethanol to precipitate the product. X-Ray studies of the isostructural $Na_4[M(C_2O_4)_4]\cdot 3H_2O$ (M = Zr or Hf) salts[403] and the isostructural $K_4[M(C_2O_4)_4]\cdot 5H_2O$ (M = Zr or Hf) salts[404,405] have shown that the $[M(C_2O_4)_4]^{4-}$ ion has a slightly distorted D_{2d} dodecahedral structure in which the virtually flat, bidentate oxalate ligands span the *m* polyhedral edges. Averaged metal–oxygen bond lengths (Table 14) to the dodecahedral A sites are 1.6–2.9% longer than those to the B sites. Perturbed angular correlation measurements of the electric field gradient at the metal site in $M'_4[Hf(C_2O_4)_4]\cdot nH_2O$ (M′ = Na, K, Rb, Cs, or NH_4) complexes indicate significant ligand→metal π bonding involving the oxygen atoms in the dodecahedral B sites and the metal $d_{x^2-y^2}$ orbital.[406] Metal–oxygen stretching bands in IR spectra of $K_4[Zr(C_2O_4)_4]\cdot 5H_2O$ ($321.5\ cm^{-1}$) and $K_4[Hf(C_2O_4)_4]\cdot 5H_2O$ ($280\ cm^{-1}$) have been assigned on the basis of metal isotopic substitution studies [$\nu(^{90}Zr—O) = 324.0\ cm^{-1}$; $\nu(^{94}Zr—O) = 316.6\ cm^{-1}$].[407] The lability of $[M(C_2O_4)_4]^{4-}$ (M = Zr or Hf) complexes with respect to ligand exchange has been demonstrated by studies of aqueous solutions of $K_4[M(C_2O_4)_4]\cdot 5H_2O$ and ^{14}C-labeled oxalic acid; complete exchange of oxalate ligands occurs within the few minutes required to complete the experiment.[408]

Tris(oxalato) and pentakis(oxalato) complexes of the type $M'_2Zr(C_2O_4)_3\cdot nH_2O$ (M′ = NH_4 or K), $(NH_4)_3H_3Zr(C_2O_4)_5\cdot nH_2O$ and $(NH_4)_6Zr(C_2O_4)_5\cdot nH_2O$ have also been reported.[399] $(NH_4)_6Zr(C_2O_4)_5\cdot nH_2O$ was obtained by passing ammonia over the acid salt $(NH_4)_3H_3Zr(C_2O_4)_5\cdot nH_2O$. The other compounds were isolated from aqueous and/or alcoholic solutions of zirconyl chloride that contain various amounts of $H_2C_2O_4$ or $[HC_2O_4]^-$. The $M'_2Zr(C_2O_4)_3\cdot nH_2O$ salts are 2:1 electrolytes,[399] and IR spectra indicate that all of the oxalate ligands are bidentate.[409] It has been suggested, based on the structure of $[Zr(C_2O_4)_4]^{4-}$, that the tris(oxalato) complexes contain an eight-coordinate $[Zr(C_2O_4)_3(H_2O)_2]^{2-}$ anion.[409] In addition to the IR bands characteristic of bidentate oxalate ligands, $(NH_4)_6Zr(C_2O_4)_5\cdot nH_2O$ shows a band at $\sim 1600\ cm^{-1}$,

Table 14 Averaged Metal–Oxygen Bond Lengths (Å) in Tetrakis(oxalato) Complexes of Zirconium(IV) and Hafnium(IV)

Compound	*M—O*	*M—O*$_A$	*M—O*$_B$	*M—O*$_A$/*M—O*$_B$	*Ref.*
$Na_4[Zr(C_2O_4)_4]\cdot 3H_2O$	2.199	2.230	2.168	1.029	1
$K_4[Zr(C_2O_4)_4]\cdot 5H_2O$	2.193	2.210	2.176	1.016	2
$K_4[Hf(C_2O_4)_4]\cdot 5H_2O$	2.190	2.209	2.171	1.018	3

1. G. L. Glen, J. V. Silverton and J. L. Hoard, *Inorg. Chem.*, 1963, **2**, 250.
2. B. Kojic-Prodic, Z. Ruzic-Toros and M. Sljukic, *Acta Crystallogr., Sect. B*, 1978, **34**, 2001.
3. D. Tranqui, P. Boyer, J. Laugier and P. Vulliet, *Acta Crystallogr., Sect. B*, 1977, **33**, 3126.

which has been assigned to the $\nu_{as}(CO_2)$ mode of 'uncombined CO_2' groups. A structure for $[Zr(C_2O_4)_5]^{6-}$ having three bidentate and two monodentate oxalate ligands is favored.[409]

Mixed-ligand sulfato-oxalato and carbonato-oxalato complexes were mentioned in Sections 32.4.4.5.iii and 32.4.4.6.iii. A variety of fluoro-oxalato complexes have also been reported. Typical compositions are: $M'_4M(C_2O_4)_3F_2(H_2O)_n$, $M'_3M(C_2O_4)_2F_3(H_2O)_3$, $M'_2Zr(C_2O_4)F_4(H_2O)_n$, $M'_2Zr_2(C_2O_4)F_8(H_2O)_4$ and $K_{14}M_4(C_2O_4)_5F_{20}(H_2O)_2$ (M′ = alkali metal; M = Zr or Hf).[341,410,411]

Zaitsev and coworkers[412–414] have prepared basic oxalates such as $M'_2ZrO(C_2O_4)_2\cdot nH_2O$ (M′ = K, NH_4, or 0.5 Ba) and $M'Zr(OH)_3(C_2O_4)\cdot nH_2O$ (M′ = Na, K, NH_4, or 0.5 Ba) and have suggested that these complexes may have tetranuclear structures based on the $\{Zr_4(OH)_8^{8+}\}$ unit found in $[Zr_4(OH)_8(H_2O)_{16}]Cl_8\cdot 12H_2O$. In recent years, divalent metal zirconyl oxalates have been of interest as precursors for the preparation of ceramic materials; $M''ZrO(C_2O_4)_2\cdot nH_2O$ complexes decompose at elevated temperatures yielding metal zirconates $M''ZrO_3$ (M″ = alkaline earth, Zn, Cd or Pb).[415,416]

32.4.4.7 *Aliphatic and aromatic hydroxy acids*

α-Hydroxycarboxylic acids RR′C(OH)CO_2H such as glycolic (R = R′ = H), lactic (R = H; R′ = Me), malic (R = H; R′ = CH_2CO_2H), tartaric (R = H; R′ = CH(OH)CO_2H), trihydroxyglutaric (R = H, R′ = CH(OH)CH(OH)CO_2H), and citric (R = R′ = CH_2CO_2H) acids form stable complexes with Zr^{IV} and Hf^{IV}, even in strongly acidic solutions.[417–419] The greater complexing ability of the hydroxy acids relative to that of the corresponding carboxylic acids is attributed to the formation of five-membered chelate rings (**25**). A recent study by Hala and Londyn[419] indicates that Hf^{IV} in 1.0–3.0 M $HClO_4$ media forms 1:1 and 1:2 complexes with glycolic, tartaric and citric acids and 1:1, 1:2, and 1:3 complexes with lactic and malic acids. The stability of the 1:1 complex increases in the order tartrate < malate < glycolate < citrate < lactate. Earlier, Ryabchikov *et al.*[417,418] found that the Hf^{IV} complexes are somewhat less stable than the Zr^{IV} analogues.

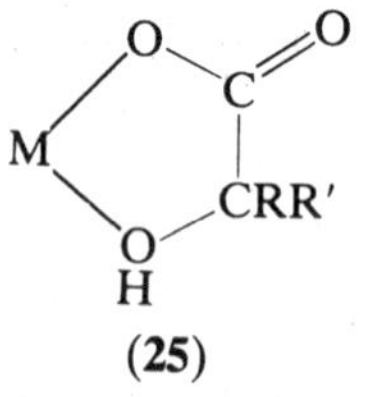

(**25**)

Hydroxo(glycolato) complexes of composition $M_2O(OH)_2(CH_2OHCO_2)_4\cdot nH_2O$ and $M(OH)(CH_2OHCO_2)_3\cdot nH_2O$ (M = Zr or Hf; n = 3–7) have been isolated by mixing 1 M solutions of $MOCl_2\cdot 8H_2O$ and glycolic acid in 1:1, 1:2, 1:3 and 1:4 molar ratios.[420] IR spectra indicate that the carboxylate and hydroxyl oxygen atoms are attached to the metal atom, but the detailed structures of these compounds are unknown. The lactato complexes $Zr(MeCHOHCO_2)_4$ and $Zr(OH)(MeCHOHCO_2)_3$ have been obtained from 3–5 M HCl solutions; $Zr(MeCHOHCO_2)_4$ is very easily hydrolyzed.[421]

Ermakov *et al.*[422] have isolated compounds of composition $Zr(C_4H_2O_6)\cdot 2H_2O$ and $Zr(C_5H_4O_7)\cdot 3H_2O$ from 0.5 M $ZrOCl_2$ solutions that were treated with one equivalent of tartaric acid or trihydroxyglutaric acid, respectively. A similar experiment with citric acid gave a mixture of $Zr(C_6H_4O_7)\cdot 4H_2O$ and $Zr_2(OH)_4(C_6H_4O_7)\cdot 2H_2O$. IR studies of the tartrate and citrate complexes suggest that complex formation involves the carboxylate and the hydroxyl groups and is accompanied by loss of the hydroxy protons.

Mandelic acid and its derivatives have long been used as analytical reagents for determination of zirconium and hafnium.[285,286] The tetrakis(mandelates) $M\{(\pm)\text{-}PhCHOHCO_2\}_4$ are precipitated when a solution of (±)-mandelic acid is added to a hot (80–90 °C) 5 M hydrochloric acid solution of the metal oxychloride. Larsen and Homeier[421] have proposed that the $M\{(\pm)\text{-}PhCHOHCO_2\}_4$ complexes have a polymeric structure in which eight-coordinate metal atoms are linked by (±) pairs of bridging mandelate ligands (**26**). The stability of this structure was attributed to short, double-minimum hydrogen bonds involving the hydroxyl protons and the carboxylate oxygen atoms, and it was suggested that these hydrogen bonds can only form when a (±) pair of ligands is present in the bridge. Evidence for this polymeric structure includes: (i) the very low solubility and lack of volatility of the $M\{(\pm)\text{-}PhCHOHCOO)\}_4$ complexes, (ii) the presence of broad ν(O—H···O) IR bands centered at 2600, 2300 and 1850 cm^{-1}, and (iii) the fact that the tetrakis(mandelato)-complexes could only be prepared when the ligand was (±)-mandelate; attempts to prepare $Zr\{(+)\text{-}PhCHOHCHO_2\}_4$ and $Zr\{(-)\text{-}PhCHOHCO_2\}_4$ yielded basic mandelates $Zr(OH)_x(PhCHOHCO_2)_{4-x}$. The presence of broad, low-frequency ν(O—H···O) bands in IR spectra of $M\{(\pm)\text{-}PhCHOHCO_2\}_4$ (M = Zr or Hf) has been confirmed by Karlysheva *et al.*,[423] although these workers have suggested a monomeric tetrakis(chelate) structure.

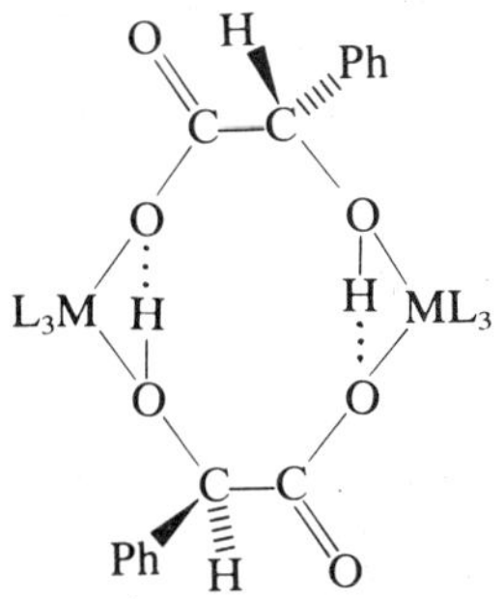

(**26**) $L = PhCHOHCO_2^-$

$M\{(\pm)\text{-}PhCHOHCO_2\}_4$ precipitates dissolve in solutions of weak bases such as ammonia, organic amines and sodium bicarbonate. The dissolution reaction is consistent with enhanced acidity of the OH groups in a chelate structure and has been attributed by Hahn and coworkers[424,425] to salt formation, *e.g.* equation (47). Larsen and Homier,[421] however, have questioned the identity of Hahn's reaction products and have suggested that the dissolution reaction involves attack by bases on the hydrogen bonds of a polymeric structure.

$$M\{\pm\text{-}PhCHOHCO_2\}_4 + 4NH_3 \longrightarrow [NH_4]_4[M\{\pm\text{-}PhCH(O)CO_2\}_4] \quad (47)$$

The reaction of mandelic acid with $ZrCl_4$ in acetonitrile yields a mixture of $[Zr\{(\pm)\text{-}PhCHOHCO_2\}_3Cl]$ and $[Zr\{(\pm)\text{-}PhCHOHCO_2\}_4]$.[421] The latter soluble product is believed to be a monomeric, seven-coordinate isomer of the polymeric tetrakis(mandelate) obtained in aqueous media; it exhibits an IR band at 1700 cm^{-1}, which is attributed to ν(C=O) of a monodentate mandelate ligand.

Mandelic acid reacts with $Zr(OCHMe_2)_4\cdot HOCHMe_2$ in refluxing benzene to give $Zr\{PhCH(O)CO_2\}(OCHMe_2)_2$, $Zr(PhCHOHCO_2)\{PhCH(O)CO_2\}(OCHMe_2)$ or $Zr(PhCHOHCO_2)_4$, depending on the stoichiometry of the reaction mixture. $Zr(PhCHOHCO_2)\{PhCH(O)CO_2\}(OCHMe_2)$ loses isopropanol when heated under reduced pressure, yielding $Zr\{PhCH(O)CO_2\}_2$, and is converted to $Zr(PhCHOHCO_2)_2(OCHMe_2)_2$ when treated with isopropanol in benzene. Reaction of $Zr(PhCHOHCO_2)\{PhCH(O)CO_2\}(OCHMe_2)$ with 1-butanol in refluxing benzene gives the butoxide derivative $Zr(PhCHOHCO_2)\{PhCH(O)CO_2\}(OBu)$.[426]

Reactions between salicylic acid and $Zr(OCHMe_2)_4\cdot HOCHMe_2$ in benzene yield similar products; $Zr(2\text{-}OC_6H_4CO_2)(OCHMe_2)_2$, $Zr(2\text{-}HOC_6H_4CO_2)(2\text{-}OC_6H_4CO_2)(OCHMe_2)$, $Zr(2\text{-}OC_6H_4CO_2)_2$ and $Zr(2\text{-}HOC_6H_4CO_2)_2(2\text{-}OC_6H_4CO_2)$ have been isolated, depending on the reaction temperature and the stoichiometry of the reaction mixture. The reaction of $Zr(OCHMe_2)_4 \cdot HOCHMe_2$ with excess methyl salicylate gives the tetrakis complex $Zr(2\text{-}OC_6H_4CO_2Me)_4$.[427] In aqueous media, salicylic acid and its derivatives precipitate Zr^{IV} in the form of basic salicylates.[286,428,429]

The reactions of $Zr(OCHMe_2)_4 \cdot HOCHMe_2$ with the β-hydroxy esters $PhCR(OH)CH_2CO_2Et$ (R = H or Me) and $C_6H_{10}(OH)CH_2CO_2Et$ in 1:1, 1:2, 1:3 and 1:4 molar ratios in benzene at reflux afford monomeric, viscous liquid complexes of the type (**27**) and (**28**). IR spectra and molecular weight measurements indicate that the Zr atom has a coordination number of five, six, seven or eight when *n* is 1, 2, 3 or 4, respectively. The complexes that contain $OCHMe_2$ groups can be converted to the corresponding *t*-butoxide derivatives by reaction of (**27**) and (**28**) ($n = 1$, 2 or 3) with Me_3COH in benzene at reflux.[430] Chloro(hydroxyester) complexes of the type $Zr\{MeCH(O)CO_2Et\}_2Cl_2$, $Zr\{PhCH(O)CO_2Me\}_2Cl_2$ and $Zr(2\text{-}OC_6H_4CO_2Me)_nCl_{4-n}$ ($n = 2$ or 3) have been known for many years.[221]

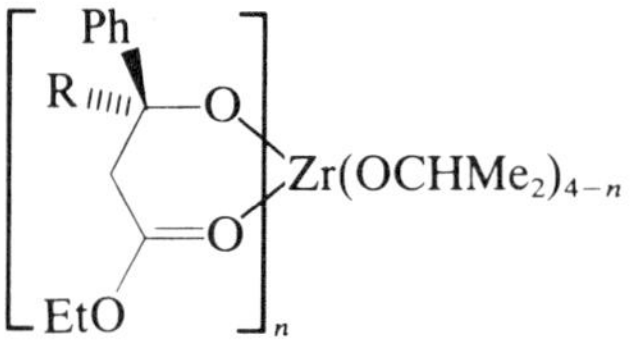

(**27**) $n = 1$, 2, 3 or 4

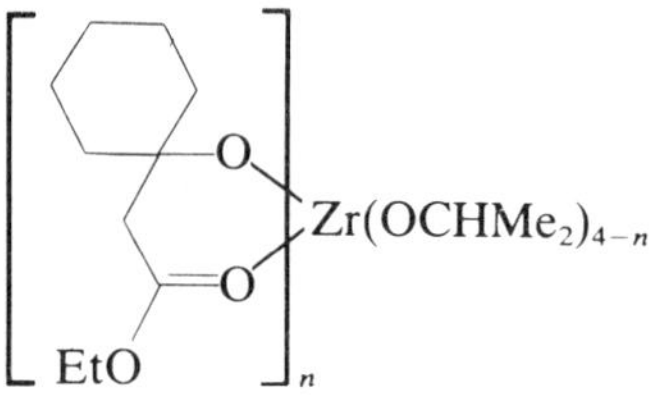

(**28**) $n = 1$, 2, 3 or 4

32.4.4.8 Sulfoxides, selenoxides, amides, N-oxides, P-oxides and related ligands

Zirconium(IV) and hafnium(IV) halides react with sulfoxides, selenoxides, amides, *N*-oxides, *P*-oxides and related ligands to give moisture-sensitive solid adducts that are generally insoluble in common organic solvents. Typical examples are listed in Table 15. Use of two equivalents or an excess of Lewis base ordinarily yields octahedral 1:2 adducts $[MX_4L_2]$, some of which have been assigned *cis* or *trans* configurations on the basis of the number of metal–halogen stretching bands in vibrational spectra. For example, IR and Raman spectra and normal coordinate calculations[431] have established a *cis* structure for $[MCl_4(POCl_3)_2]$ (M = Zr or Hf) complexes, which exhibit four ν(M—Cl) modes in the region 370–305 cm^{-1}.

A few 1:1 adducts have been reported. For example, thermal decomposition of $[MCl_4L_2]$ (M = Zr or Hf; L = $POCl_3$ or $POFCl_2$) under reduced pressure yields solid MCl_4L complexes[432] that have been shown to be monomeric in nitrobenzene when L = $POCl_3$.[433] Vapor pressure measurements in the $ZrCl_4$–$POCl_3$ system have shown that $[ZrCl_4(POCl_3)]$ exists in the gas phase as well.[434] On the other hand, the thionyl chloride adduct $[\{ZrCl_4(SOCl_2)\}_2]$ is a centrosymmetric dimer (**29**) in which each Zr atom is octahedrally coordinated by two bridging chlorine atoms (at 2.58 ± 0.01 Å), three terminal chlorine atoms (at 2.34 ± 0.02 Å) and the thionyl chloride oxygen atom (at 2.273(13) Å).[435] Phthalimide forms a 1:1 adduct with $ZrCl_4$, which is a monomeric nonelectrolyte in nitrobenzene but which may be polymeric in the solid state since IR spectra suggest that both carbonyl groups are coordinated to the metal.[436]

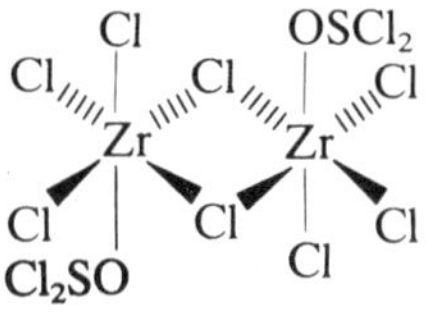

(**29**)

The reactions of zirconium and hafnium tetrahalides with an excess of dimethyl sulfoxide or a primary amide yield MX_4L_n compounds having values of *n* as large as 10 (see Table 15). IR studies of $MCl_4(DMSO)_8$[437] and $ZrCl_4(HCONH_2)_4$[438] indicate the presence of both coordinated and noncoordinated Lewis base ligands; the formamide complex has been formulated as $[ZrCl_4(HCONH_2)_2] \cdot 2HCONH_2$. On the other hand, IR spectra of the 2,6-lutidine *N*-oxide complex $[ZrCl_4(2,6\text{-}Me_2C_5H_3NO)_3]$ suggest that all three *N*-oxide ligands are attached to a

Table 15 Complexes of Zirconium(IV) and Hafnium(IV) Halides with Sulfoxides, Selenoxides, Amides, *N*-Oxides, *P*-Oxides and Related Ligands

Complex[a]	*Physical method*[b]	*Ref.*
Sulfoxide and selenoxide complexes		
cis-$[ZrF_4(DMSO)_2]$	IR	1, 2
$ZrCl_4(DMSO)_n$ ($n = 2, 3, 8, 9$)	IR, ΔH, Λ	3–5
$HfCl_4(DMSO)_n$ ($n = 2, 8, 9$)	IR, ΔH	3–5
$MBr_4(DMSO)_n$ ($n = 2, 10$)	IR, ΔH	3, 4
$[ZrCl_4(Ph_2SeO)_2]$	IR	6
$[\{ZrCl_4(SOCl_2)\}_2]$	XD	7
$[ZrCl_4(SeOCl_2)_2]$		8
Amide and related complexes		
$ZrCl_4(HCONH_2)_n$ ($n = 4, 8, 10$)	IR	9
$ZrCl_4(MeCONH_2)_4$		9
$ZrF_4(DMF)$		1
$[MCl_4(DMF)_2]$	IR	9, 10
$[ZrBr_4(DMF)_2]$	NMR	11
$[ZrCl_4(MeCONMe_2)_2]$		9
$[ZrCl_4(NCCH_2CONH_2)_2]$	IR, M, Λ	12
$[ZrCl_4$(1-methyl-2-pyrrolidinone$)_2]$	IR	13
trans-$[ZrCl_4$(antipyrine$)_2]$	IR	14
trans-$[ZrCl_4$(4-aminoantipyrine$)_2]$	IR	14
$ZrCl_4$(phthalimide)	IR, M, Λ	15
N-*Oxide and related complexes*		
$[ZrCl_4(Me_3NO)_2]$	IR	16
$[ZrF_4(4\text{-}XC_5H_4NO)_2]$ ($X = NO_2$, Cl, H, Me, OMe)	IR	17
$[ZrCl_4(2,6\text{-}Me_2C_5H_3NO)_3]$	IR	18
cis-$[HfCl_4(2,6\text{-}Me_2C_5H_3NO)_2]$	IR	18
cis-$[ZrCl_4$(benzofuroxane$)_2]$	IR, R	19
P-*Oxide and related complexes*		
$[ZrCl_4(R_3PO)_2]$ (R = Me, Bu, CH_2Ph, Ph)	IR	16, 20, 21
$[ZrCl_4\{R_2P(O)(OR')\}_2]$ (R, R′ = alkyl or aryl)	IR, Λ	21
$[ZrCl_4\{RP(O)(OR')_2\}_2]$ (R, R′ = alkyl or aryl)	IR, Λ	21
$[ZrCl_4\{(RO)_3PO\}_2]$ (R = Me, Et, Bu)	IR, Λ	21
$[ZrCl_4\{R_2P(O)H\}_2]$ (R = alkyl or aryl)	IR	22
$ZrCl_4(\{Ph_2P(O)CH_2\}_2P(O)R)$ (R = Ph, OEt, OH)	IR, Λ	23
$[MCl_4(POCl_3)]$	XD, M, P	24–27
$MCl_4(POF_nCl_{3-n})$ ($n = 1, 2, 3$)		24
cis-$[MCl_4(POCl_3)_2]$	XD, IR, R, M, P	24–29
$[MCl_4(POFCl_2)_2]$		24
$MCl_4\{PO(NCS)_3\}_2$	XD, IR	30
trans-$[ZrCl_4\{PO(NMe_2)_3\}_2]$	IR, Λ	31
$[ZrBr_4\{PO(NMe_2)_3\}_2]$	IR, Λ	31
$ZrI_4\{PO(NMe_2)_3\}$	IR	31

[a] M = Zr or Hf. [b] IR = infrared spectrum; R = Raman spectrum; NMR = ^{1}H NMR spectrum; ΔH = thermochemical data; M = molecular weight; Λ = electrical conductance; XD = X-ray diffraction; P = vapor pressure measurements.

1. E. L. Muetterties, *J. Am. Chem. Soc.*, 1960, **82,** 1082.
2. R. J. H. Clark and W. Errington, *J. Chem. Soc. (A)*, 1967, 258.
3. R. Makhija and A. D. Westland, *Inorg. Chim. Acta*, 1978, **29,** L269.
4. H. L. Schlafer and H.-W. Willie, *Z. Anorg. Allg. Chem.*, 1965, **340,** 40.
5. A. M. Golub, T. P. Lishko and P. F. Lozovskaya, *Russ. J. Inorg. Chem. (Engl. Transl.)*, 1974, **19,** 13.
6. R. Paetzold and P. Vordank, *Z. Anorg. Allg. Chem.*, 1966, **347,** 294.
7. R. K. Collins and M. G. B. Drew, *J. Chem. Soc. (A)*, 1971, 3610.
8. V. Gutmann and R. Himml, *Z. Anorg. Allg. Chem.*, 1956, **287,** 199.
9. A. Clearfield and E. J. Malkiewich, *J. Inorg. Nucl. Chem.*, 1963, **25,** 237.
10. W. M. Graven and R. V. Peterson, *J. Inorg. Nucl. Chem.*, 1969, **31,** 1743.
11. S. J. Kuhn and J. S. McIntyre, *Can. J. Chem.*, 1965, **43,** 375.
12. R. C. Paul and S. L. Chadha, *Aust. J. Chem.*, 1969, **22,** 1381.
13. S. C. Jain and G. S. Rao, *J. Indian Chem. Soc.*, 1976, **53,** 25.
14. S. A. A. Zaidi, T. A. Khan and N. S. Neelam, *Synth. React. Inorg. Metal-Org. Chem.*, 1979, **9,** 243.
15. R. C. Paul and S. L. Chadha, *J. Inorg. Nucl. Chem.*, 1969, **31,** 2753.
16. S. H. Hunter, V. M. Langford, G. A. Rodley and C. J. Wilkins, *J. Chem. Soc. (A)*, 1968, 305.
17. F. E. Dickson, E. W. Baker and F. F. Bentley, *J. Inorg. Nucl. Chem.*, 1969, **31,** 559.
18. N. M. Karayannis, A. N. Speca, L. L. Pytlewski and M. M. Labes, *J. Less-Common Met.*, 1970, **22,** 117.
19. V. Fernandez and C. Muro, *Z. Anorg. Allg. Chem.*, 1980, **466,** 209.
20. A. K. Majumdar and R. G. Bhattacharyya, *J. Inorg. Nucl. Chem.*, 1967, **29,** 2359.
21. B. Kautzner and P. C. Wailes, *Aust. J. Chem.*, 1969, **22,** 2295.
22. A. A. Muratova, E. G. Yarkova, V. P. Plekhov, R. G. Zagetova and A. N. Pudovik, *J. Gen. Chem. USSR (Engl. Transl.)*, 1972, **42,** 966.
23. P. Bronzan and H. Meider, *J. Less-Common Met.*, 1978, **59,** 101.

24. E. M. Larsen, J. Howatson, A. M. Gammill and L. Wittenberg, *J. Am. Chem. Soc.*, 1952, **74**, 3489.
25. E. M. Larsen and L. J. Wittenberg, *J. Am. Chem. Soc.*, 1955, **77**, 5850.
26. B. A. Voitovich, A. S. Barabanova, V. P. Klochkov and E. V. Sharkina, *Sov. Prog. Chem. (Engl. Transl.)*, 1966, **32**, 130.
27. A. V. Suvorov and E. K. Krzhizhanovskaya, *Russ. J. Inorg. Chem. (Engl. Transl.)*, 1969, **14**, 434.
28. E. K. Krzhizhanovskaya and A. V. Suvorov, *Russ. J. Inorg. Chem. (Engl. Transl.)*, 1971, **16**, 1355.
29. Y. Hase and O. L. Alves, *Spectrochim. Acta, Part A*, 1981, **37**, 957.
30. V. V. Skopenko, A. I. Brusilovets and A. V. Sinkevich, *Dopo. Akad. Nauk Ukr. RSR, Ser. B*, 1982, 44 (*Chem. Abstr.*, 1982, **97**, 65 411).
31. R. J. Masaguer, Midel Carmen Touza, J. Sordo, J. S. Casas and M. R. Bermejo, *Acta Cient. Compostelana*, 1977, **14**, 313 (*Chem. Abstr.*, 1977, **87**, 210 395).

seven-coordinate Zr atom.[439] It is interesting to note that the reaction of excess 2,6-lutidine *N*-oxide with $ZrCl_4$ yields a 1:3 complex, while the analogous reaction with $HfCl_4$ gives the 1:2 adduct *cis*-$[HfCl_4(2,6\text{-}Me_2C_5H_3NO)_2]$.

Formation of the adducts in Table 15 is accompanied by characteristic low-frequency shifts in the IR bands of ligand vibrations that involve stretching of bonds to the oxygen atom; typical low-frequency shifts upon coordination to MCl_4 (M = Zr or Hf) are $\sim 100\ cm^{-1}$ for $\nu(SO)$ in DMSO,[437] $\sim 60\ cm^{-1}$ for $\nu(SeO)$ in Ph_2SeO,[440] $\sim 15\ cm^{-1}$ for $\nu(C{=}O)$ in DMF[120] and $115\text{–}155\ cm^{-1}$ for $\nu(P{=}O)$ in phosphine oxides, phosphinates, phosphonates and phosphates.[441] These data indicate that the Lewis bases are coordinated through the oxygen atom. Coordination through the carbonyl oxygen atom in $[ZrBr_4(DMF)_2]$ is supported by the presence of two methyl resonances in the ^1H NMR spectrum.[442] The C≡N group is not coordinated in the cyanoacetamide complex $[ZrCl_4(NCCH_2CONH_2)_2]$; this compound is a monomeric nonelectrolyte in nitrobenzene, and IR spectra are consistent with six coordination, the $NCCH_2CONH_2$ being attached to zirconium through the carbonyl oxygen atom.[443]

Several lines of evidence suggest that the Lewis acid strength of zirconium and hafnium tetrahalides toward oxygen donor ligands increases in the order $ZrX_4 < HfX_4$: (i) metal–oxygen stretching force constants in *cis*-$[MCl_4(POCl_3)_2]$ are 0.344 mdyn $Å^{-1}$ (M = Zr) and 0.434 mdyn $Å^{-1}$ (M = Hf);[432] (ii) $\Delta H°_{298}$ values ($kJ\ mol^{-1}$) for conversion of $MCl_4(POCl_3)_2(s)$ to $MCl_4(POCl_3)(s)$ and $POCl_3(g)$ are 67 ± 8 for M = Zr and 88 ± 8 for M = Hf;[444] and (iii) ΔH°_{298} values ($kJ\ mol^{-1}$) for formation of $MX_4(DMSO)_2(s)$ from $MX_4(g)$ and 2DMSO(g) are: −379 ($MX_4 = ZrCl_4$), −392 ($HfCl_4$), −367 ($ZrBr_4$), −399 ($HfBr_4$).[445]

Oxozirconium(IV) complexes of the type $[ZrOX_2L_2]$ (X = Cl, Br, NO_3 or NCS), $[ZrOL_4]I_2$ and $[ZrOL_6][ClO_4]_2$, where L is sulfoxide, *N*-oxide or *P*-oxide, have been discussed in Section 32.4.4.1. In all of these compounds, the Lewis base is coordinated to the metal through the oxygen atom. Additional complexes of this type include $[ZrO(DMSO)_6][ClO_4]_2 \cdot n$DMSO ($n = 0$ or 2)[446,447] and $[ZrOL_6][ClO_4]_2$ (L = pyridine *N*-oxide or quinoline *N*-oxide).[448,449]

Paul *et al.*[450] have prepared amide complexes of the type $ZrOCl_2(HCONH_2)_4$ and $ZrOCl_2L_2$ (L = DMF, $MeCONH_2$, MeCONHMe, MeCONHPh, $MeCONMe_2$, or $PhCONH_2$) by addition of the amide to an acetone solution of anhydrous $ZrOCl_2$. Kharitonov *et al.*[451] have isolated DMF complexes of composition $MOCl_2(DMF)_2(H_2O)_2$, $MO(OH)(NO_3)(DMF)$, $M(SO_4)_2(DMF)_4$ and $MO(SO_4)(DMF)(H_2O)$ (M = Zr or Hf) from DMF solutions of hydrated zirconium and hafnium salts; they have suggested mononuclear, tetranuclear and polynuclear structures for some of these compounds on the basis of IR spectra. Savant *et al.*[452] have reported $ZrO(antipyrine)_3(ClO_4)_2$ and $Zr(antipyrine)_6(ClO_4)_4$; these complexes are electrolytes in $PhNO_2$, $MeNO_2$ and DMSO, with the extent of dissociation depending on the solvent. For all of these compounds, IR spectra indicate that the amide or antipyrine ligand is coordinated through the carbonyl oxygen atom.

$MOCl_2 \cdot 8H_2O$ (M = Zr or Hf) and $ZrO(ClO_4)_2 \cdot 8H_2O$ react with the potentially tridentate phosphine oxide, phosphinate and phosphinic acid ligands L = $\{Ph_2P(O)CH_2\}_2P(O)R$ (R = Ph, OEt or OH) yielding complexes of composition $M(OH)_2LCl_2(H_2O)_8$ and $Zr(OH)_2L(ClO_4)_2(H_2O)_8$, respectively. The perchlorate compounds are electrolytes in $PhNO_2$, while the chloro complexes are nonelectrolytes. Reaction of $ZrCl_4$ with L gives the $ZrCl_4(L)$ complexes (Table 15), which are also nonelectrolytes in $PhNO_2$. Low-frequency shifts of $60\text{–}90\ cm^{-1}$ in $\nu(P{=}O)$ upon complex formation indicate that the phosphorus ligands are coordinated through the phosphoryl oxygen atoms.[453] Phosphine oxides and organic phosphates, particularly tri-*n*-octylphosphine oxide and tri-*n*-butyl phosphate, are important complexing agents in solvent extraction procedures for separation of zirconium and hafnium from other metals.[286]

32.4.4.9 Hydroxamates, cupferron and related oxygen ligands

N-Benzoyl-*N*-phenylhydroxylamine and cupferron (the ammonium salt of *N*-nitroso-*N*-phenylhydroxylamine) have long been used as analytical reagents for gravimetric determination of zirconium.[286] Addition of these reagents to moderately acidic aqueous solutions of Zr^{IV} or Hf^{IV} precipitates the tetrakis chelates (**30**) and (**31**). Bhatt and Agrawal[454] have prepared $[Zr\{ON(Ph)COPh\}_4]$ at pH 5.5, but Karlysheva and Sheka[455] report that basic cupferrates of variable composition are obtained if the acid concentration is less than 1 N. $[M\{ON(Ph)COPh\}_4]$ (M = Zr or Hf) complexes have also been isolated from benzene solution following extraction of Zr^{IV} or Hf^{IV} from 1–3 M $HClO_4$ or 1–5 M HCl; extraction from 6–9 M HCl affords the chloro complexes $[M\{ON(Ph)COPh\}_3Cl]$.[456]

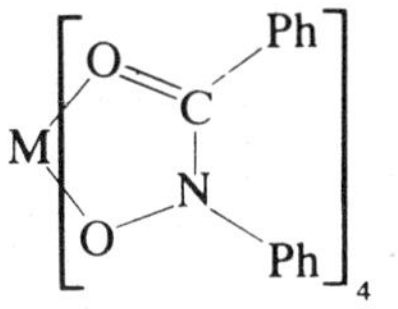

(**30**) M = Zr or Hf

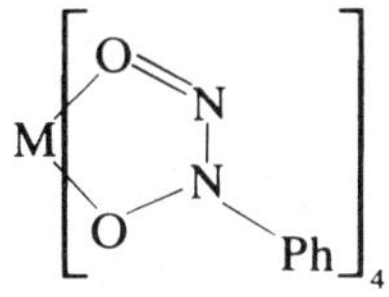

(**31**) M = Zr or Hf

An X-ray structural determination[457] has shown that $[Hf\{ON(Ph)COPh\}_4]$ (**30**; M = Hf) exists in the solid state as the unusual S_4-*gggg* dodecahedral stereoisomer. Because the four *g* edges spanned by the ligands (average 2.527 Å) are appreciably shorter than the other four *g* edges (2.869 Å), the BAAB trapezoids of the dodecahedron are somewhat twisted; the ligating oxygen atoms are displaced from the mean trapezoidal planes by 0.08–0.20 Å. The PhCO oxygen atoms occupy the dodecahedral A sites and the PhNO oxygen atoms take the B sites, in accord with Orgel's rule for d^0 complexes.[458,459] As expected, the averaged Hf—O_A bond distance (2.258 Å) is longer than the Hf—O_B distance (2.116 Å), but the ratio Hf—O_A/Hf—O_B (1.067) is unusually large.

Zirconium(IV) cupferrate (**31**; M = Zr) exists in the solid state as the more usual D_{2d}-*mmmm* dodecahedral stereoisomer.[460] In this structure, the PhNO oxygen atoms are located in the A sites and the nitroso oxygen atoms are found in the B sites; averaged bond lengths are Zr—O_A = 2.168 Å, Zr—O_B = 2.212 Å, Zr—O_A/Zr—O_B = 0.980. If the PhNO oxygen atom carries more of the ligand's negative charge than the nitroso oxygen atom, this structure would appear to be a violation of Orgel's rule. However, the two N—O bond lengths within the chelate ring are equal (1.30 Å), suggesting that the charge on the two oxygen atoms may not be significantly different.

The difference in ligand wrapping pattern for $[Hf\{ON(Ph)COPh\}_4]$ and $[Zr\{ON(Ph)NO\}_4]$ is interesting. Tranqui *et al.*[457] have noted the close interpenetration of adjacent molecules in the crystal structure of $[Hf\{ON(Ph)COPh\}_4]$ and have suggested that the molecular structure may be influenced by crystal packing effects. Another factor that may be of importance is a difference in the normalized bites of the hydroxamate and cupferrate ligands, 1.155 in (**30**; M = Hf) and 1.108 in (**31**; M = Zr); the D_{2d}—*mmmm* wrapping pattern becomes less stable as the normalized bite of the ligand increases.[461]

Time differential perturbed angular correlation (TDPAC) measurements on the solid compounds (**30**; M = Hf) and (**31**; M = Hf) have been interpreted[275] in terms of two different hafnium environments, a conclusion which is not in accord with the X-ray crystal structures.[457,460] A more recent TDPAC study[462] provides information about π bonding involving the hafnium $d_{x^2-y^2}$ orbital and the oxygen atoms in the dodecahedral B sites. The charge transferred into $d_{x^2-y^2}$ has been estimated to be ~20% of the overall charge donated by the ligands in (**30**; M = Hf) and ~16% in (**31**; M = Hf). The overall charge donated by the ligands into the metal valence orbitals is in the range −0.7 to −1.2 e; thus the Hf—O bonds are still quite highly ionic.

Mixed-ligand hydroxamato complexes of the type $Zr(ONHCOR)_n(OCHMe_2)_{4-n}$ (R = Ph or CH_2Ph; *n* = 1, 2, 3 or 4) have been prepared by reaction of the appropriate hydroxamic acid with $Zr(OCHMe_2)_4 \cdot HOCHMe_2$ in benzene at reflux; the value of *n* depends on the stoichiometry of the reaction mixture.[463] Zirconium(IV) alkyls ZrR_2Cl_2 (R = CH_2SiMe_3) or ZrR_4 (R = CH_2SiMe_3 or CH_2CMe_3) react with nitric oxide to give the six- or eight-coordinate *N*-alkyl-*N*-nitrosohydroxylaminato complexes $[Zr(ONRNO)_2Cl_2]$ or $[Zr(ONRNO)_4]$, respectively.[464]

Zirconium(IV) forms a tetrakis chelate with the hydroxamate-related 1-oxy-2-pyridonate ligand (**32**); the chloroform solvate $Zr(C_5H_4NO_2)_4 \cdot CHCl_3$ was isolated from chloroform–ethanol.[465]

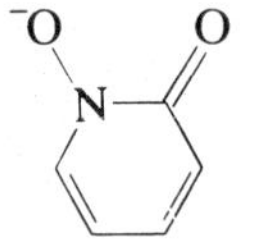

(**32**)

32.4.4.10 Miscellaneous oxygen donor ligands

The diethylammonium salt of the tris(tetraphenyldisiloxanediolato)zirconate(IV) anion (**33**) has been obtained in very low yield from the reaction of $[Zr(NEt_2)_4]$ with diphenylsilanediol in THF. The ligand results, under the basic reaction conditions, from condensation of two $Ph_2Si(OH)_2$ molecules. Anion (**33**) has an octahedral tris-chelate structure with a twist angle (46.9°) that indicates some distortion towards trigonal prismatic geometry.[466]

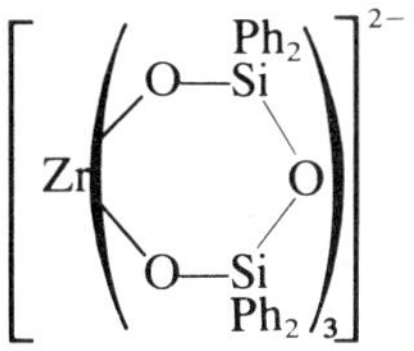

(**33**)

Tetrakis(perfluoroalkanesulfonates) $Zr(R_FSO_3)_4$ ($R_F = CF_3$ or C_2F_5) have been prepared by reaction of $ZrCl_4$ with an excess of the perfluoroalkanesulfonic acid.[467] The mode of attachment of the perfluoroalkanesulfonate ligands is unknown.

Bis(benzeneseleninato)oxozirconium(IV) complexes, $ZrO(O_2SeC_6H_4X)_2$ (X = 4-Cl, 3-Cl or 3-Br) and $ZrO(O_2SeC_6H_4\text{-}3\text{-}NO_2)_2 \cdot 2H_2O$ have been synthesized in aqueous solution by reaction of a 1:4 molar ratio of $Zr(NO_3)_4$ and the sodium salt of the ligand. IR spectra indicate an *O,O′*-bidentate attachment of the $O_2SeC_6H_4X$ ligands and an oxo-bridged polymeric structure. The presence of just one ν(Zr—O—Zr) IR band (739–750 cm^{-1}) and two ν(Zr—O) bands for the bonds to the $O_2SeC_6H_4X$ ligands (405–504 cm^{-1}) suggests an octahedral trans environment for the Zr atom.[179]

$Zr(acac)_4$ reacts with four equivalents of diphenylphosphinic acid at 300 °C yielding $Zr\{OP(O)Ph_2\}_4$, which is dimeric at 100 °C in 1,2-dichlorobenzene. IR spectra of this compound exhibit ν(P=O) bands due to coordinated and uncoordinated P=O groups, suggesting the presence of bridging and monodentate diphenylphosphinate ligands. $Zr\{OP(O)Ph_2\}_4$ has remarkable thermal stability, not melting up to 490 °C. A polymeric, diphenylphosphinate-bridged compound $ZrCl_2\{OP(O)Ph_2\}_2$ has been obtained by reaction of $ZrCl_4$ with two equivalents of diphenylphosphinic acid in *n*-butanol.[468] Di-*n*-butylphosphates of the type $Zr(NO_3)_2\{OP(O)(OBu)_2\}\{HOPO(OBu)_2\}_n$ ($n = 0$, 2 or 4) are also known.[469]

32.4.5 Sulfur Ligands

32.4.5.1 Thiolates and thioethers

Only one zirconium thiolate coordination compound is known; the blue $Zr(SPh)_4$ has been prepared by reaction of $ZrCl_4$ with $Al(SPh)_3(Et_2O)$.[470] The dithiolates $Zr(SCH_2CH_2S)(OCHMe_2)_2$ and $Zr(SCH_2CH_2S)_2$ have been obtained as white solids from the reaction of stoichiometric amounts of $Zr(OCHMe_2)_4 \cdot HOCHMe_2$ and ethanedithiol in benzene at reflux.[471]

The following metal halide–thioether adducts have been reported: $[ZrCl_4(Me_2S)_2]$,[472] $[MCl_4(THT)_2]$ and $[ZrBr_4(THT)_2]$ (M = Zr or Hf; THT = tetrahydrothiophene),[473] $[ZrCl_4(MeSCH_2CH_2SMe)_2]$[474] and $ZrCl_4$(1,4-oxathiane).[306] These relatively insoluble compounds are generally prepared by prolonged stirring of a mixture of the metal halide and the thioether in an inert solvent such as benzene or cyclohexane.

Thermochemical measurements[473] indicate that $[ZrBr_4(THT)_2]$ is less stable than $[ZrCl_4(THT)_2]$, presumably because of steric hindrance. Although $[ZrCl_4(THF)_2]$ and

$[HfCl_4(THF)_2]$ exhibit closely similar heats of adduct formation, $[HfCl_4(THT)_2]$ is more stable than $[ZrCl_4(THT)_2]$, indicating that Hf^{IV} has more class b character than Zr^{IV}.

IR spectra show that the conformation of 1,2-dimethylthioethane changes from predominantly *trans* to *gauche* on formation of $[ZrCl_4(MeSCH_2CH_2SMe)_2]$.[474] This is consistent with an eight-coordinate structure in which $MeSCH_2CH_2SMe$ coordinates through both sulfur atoms. A high coordination number for zirconium is also indicated by the low frequencies of the ν(Zr—Cl) and ν(Zr—S) IR bands, 305 and 275 cm^{-1}, respectively. In the six-coordinate, probably polymeric compound $ZrCl_4$(1,4-oxathiane),[306] ν(Zr—Cl) is at 345 cm^{-1}. The IR spectra indicate that 1,4-oxathiane is in the chair form and both the sulfur and the oxygen atoms are coordinated to zirconium atoms.

32.4.5.2 Dithiocarbamates and monothiocarbamates

Eight-coordinate *N,N*-dialkyldithiocarbamato complexes $[Zr(S_2CNR_2)_4]$ (R = Me, Et, Pr) have been prepared by reaction of $[Zr(NR_2)_4]$ with CS_2 in cyclohexane.[69] The isopropyl derivative $[Zr\{S_2CN(CHMe_2)_2\}_4]$ was obtained from the reaction of $ZrCl_4$ with $Na\{S_2CN(CHMe_2)_2\}$ in the presence of triethyl phosphite.[475] $[Zr(S_2CNEt_2)_4]$ is isomorphous with $[Ti(S_2CNEt_2)_4]$, which has been shown by X-ray diffraction to have a D_{2d}-*mmmm* dodecahedral structure.[476] The mass spectrum of $[Zr(S_2CNEt_2)_4]$ exhibits a peak for the parent ion; the base peak is $[Zr(S_2CNEt_2)_3]^+$.[477] Low-temperature ^{1}H, ^{13}C and ^{19}F NMR spectra of $[Zr(S_2CNR_2)_4]$ (R = Et or Pr), $[Zr\{S_2CN(Me)(Pr)\}_4]$ and $[Zr\{S_2CN(Me)(C_6F_5)\}_4]$ indicate that these complexes undergo very rapid intramolecular rearrangement. No evidence of stereoisomers or inequivalence of A-site and B-site *N*-alkyl groups was observed at temperatures down to −140 °C.[478,479] Intermolecular ligand exchange between $[Zr(S_2CNPr_2)_4]$ and $[Ti(S_2CNPr_2)_4]$ is slow on the NMR time scale at 80 °C.[478]

Variable temperature ^{1}H NMR spectra of $[Zr\{S_2CN(CHMe_2)_2\}_4]$ (**34**) show that the isopropyl groups are inequivalent owing to hindered rotation about the C—N single bonds. Rate constants for this process range from 20 s^{-1} at −25.0 °C to 670 s^{-1} at 21.0 °C; $\Delta H^\ddagger = 45 \pm 1$ kJ mol^{-1} and $\Delta S^\ddagger = -39 \pm 5$ J mol^{-1} deg^{-1}.[475]

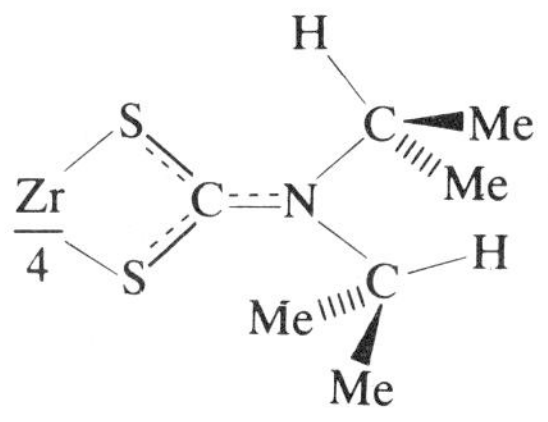

(**34**)

Five-, six-, seven- and eight-coordinate complexes of the type $[M(S_2CNRR')_nCl_{4-n}]$ (n = 1, 2, 3 or 4; M = Zr or Hf; R = Et, R′ = 3-C_6H_4Me; R = H, R′ = cyclopentyl or cycloheptyl) have been synthesized by reaction of stoichiometric amounts of MCl_4 and anhydrous $Na(S_2CNRR')$ in dichloromethane at reflux. These complexes are monomeric nonelectrolytes in solution, and IR spectra indicate a bidentate attachment of the dithiocarbamate ligands.[480,481] The same ligands react with aqueous solutions of $MOCl_2$ (M = Zr or Hf) to give compounds of composition $MO(S_2CNRR')_2 \cdot 2H_2O$; these complexes contain bidentate dithiocarbamate ligands, but their structures are unknown.[482] Six-coordinate complexes, $[Zr(S_2CNR_2)_2(OPh)_2]$ and $[Zr(S_2CNRR')_2(OPh)_2]$ (R = Me, Et or $CHMe_2$; R′ = cyclohexyl), have been prepared from $Zr(OPh)_2Cl_2$ and $Na(S_2CNR_2)$ or $Na(S_2CNRR')$ in THF at reflux.[483]

ZrX_4 (X = F or Cl) reacts with the sodium salt of piperazine-1,4-dicarbodithioate to give complexes of composition $ZrF_2\{S_2CN(CH_2CH_2)_2NC(S)SH\}_2$ and $Zr_2Cl_6\{S_2CN(CH_2CH_2)_2NCS_2\}$ respectively.[484,485] On the basis of IR evidence, the fluoride complex has been assigned an octahedral *trans* structure in which one CS_2^- group is bidentate while the other CS_2^- group is attached to a proton.[484] IR spectra indicate that the piperazine-1,4-dicarbodithiolate behaves as a tetradentate ligand in the chloride complex.[485]

Two recent papers on $[(\eta\text{-}C_5H_5)M(S_2CNMe_2)_3]$ (M = Zr or Hf)[486] and $[(\eta\text{-}C_5H_5)_2ZrCl(S_2CNR_2)]$ (R = Me or Et)[487] are mentioned in passing to provide an entry into the current literature on organo-zirconium and -hafnium dithiocarbamato complexes.

The moisture-sensitive monothiocarbamato complexes $[Zr(SOCNR_2)_4]$ (R = Me, Et or

$CHMe_2$) may be prepared by reaction of $ZrCl_4$ with $Na(SOCNR_2)$ in dry acetonitrile.[488] $[Zr(SOCNMe_2)_4]$ has also been synthesized by insertion of carbonyl sulfide into the Zr–N bonds of $[Zr(NMe_2)_4]$.[70] An X-ray study[489] has established that $[Zr(SOCNEt_2)_4]$ exists in the solid state as the dodecahedral stereoisomer that has an *mmmm* ligand wrapping pattern and a C_{2v} arrangement of the sulfur and oxygen atoms. This isomer is one of six possible stereoisomers having the *mmmm* ligand wrapping pattern (Figure 7). In the C_{2v} isomer, two of the sulfur atoms occupy A sites and two occupy B sites so as to give a highly polar all-*cis* arrangement with the four sulfur atoms clustered on one side of the coordination polyhedron; $Zr—S_A = 2.689(1)$, $Zr—S_B = 2.669(1)$, $Zr—O_A = 2.200(2)$, $Zr—O_B = 2.180(2)$ Å. The dipole moment of $[Zr(SOCNEt_2)_4]$ (3.61 ± 0.16 D) is relatively large, indicating that the C_{2v} isomer is the predominant solution species as well.[488] This isomer may be stabilized by a *trans* effect;[489] one of the two D_{2d} stereoisomers (Figure 7) would have been expected on the basis of Orgel's rule.[458,459]

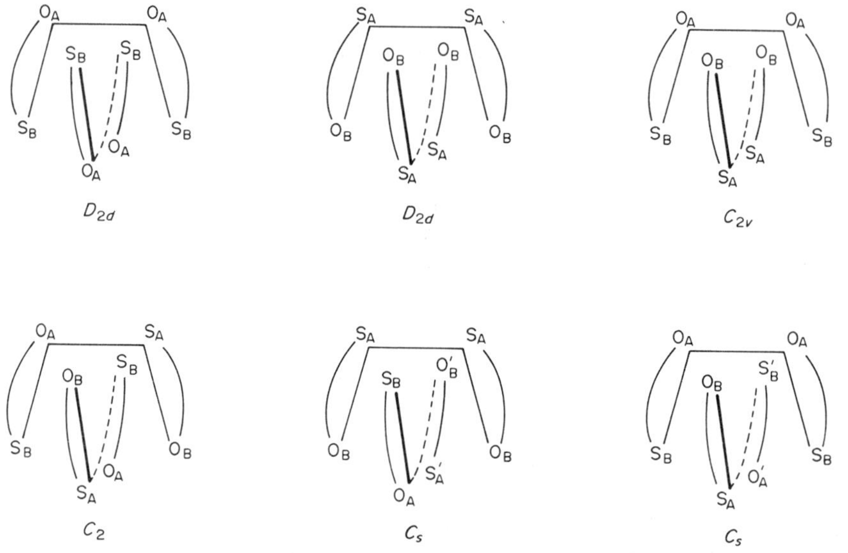

Figure 7 Schematic representation of the six possible dodecahedral stereoisomers of a $[Zr(SOCNR_2)_4]$ complex. The two mutually perpendicular trapezoids of the D_{2d} dodecahedron are outlined, and the A- and B-site occupancy of the sulfur and oxygen atoms is shown (reproduced by permission from ref. 488)

The low-temperature ^1H NMR spectrum of $[Zr(SOCNMe_2)_4]$ can be interpreted in terms of four overlapping methyl resonances, a spectrum consistent with that expected for the C_{2v} stereoisomer. Variable-temperature ^1H NMR spectra[490] exhibit evidence of two distinct kinetic processes: (i) a low-temperature process (LTP) that involves metal-centered rearrangement and (ii) a high-temperature process (HTP) that involves rotation about the C⋯N partial double bond in the monothiocarbamate ligand. Coalescence temperatures for the LTP and HTP are −53.5 and +78.2 °C, respectively. Activation parameters (kJ mol^{-1} or J mol^{-1} deg^{-1}) are: $\Delta G^{\ddagger}(-53.5\,°C) = 47.1 \pm 0.2$, $\Delta H^{\ddagger} = 40 \pm 1$, and $\Delta S^{\ddagger} = -34 \pm 5$ for the LTP; $\Delta G^{\ddagger}(78.2\,°C) = 82.7 \pm 0.5$, $\Delta H^{\ddagger} = 99 \pm 5$, and $\Delta S^{\ddagger} = 47 \pm 15$ for the HTP. The greater stereochemical rigidity of $[Zr(SOCNMe_2)_4]$ in comparison with $[Ti(SOCNMe_2)_4]$ and analogous $[M(S_2CNR_2)_4]$ (M = Ti or Zr) complexes is evidence for a polytopal rearrangement mechanism for the metal-centered process.

32.4.5.3 Dithiolenes and related ligands

The tris(dithiolene) complexes $[M(S_2C_6H_3R)_3]^{2-}$ (M = Zr, R = H; M = Zr or Hf, R = Me) have been synthesized in THF solution by reaction of the metal diethylamide with a mixture of benzene-1,2-dithiol (or toluene-3,4-dithiol) and its sodium salt (equation 48). The complexes were isolated as moisture-sensitive, red or orange $[NEt_4]_2[M(S_2C_6H_3R)_3]$ salts following addition of $(NEt_4)Br$ to the THF solution of the sodium salt. The $[NEt_4]_2[M(S_2C_6H_3R)_3]$ complexes exhibit a strong $\nu(M—S)$ IR band at 311–321 cm^{-1}, and they show an intense ligand $(\pi_v) \rightarrow$ metal $(d_{x^2-y^2}, d_{xy})$ charge-transfer transition in electronic spectra at 20 000–27 800 cm^{-1}.

^1H and ^{13}C NMR spectra of $[NEt_4]_2[Zr(S_2C_6H_3Me)_3]$ gave only a single, sharp (time-averaged?) methyl resonance. Polarographic reduction of $[Zr(S_2C_6H_3Me)_3]^{2-}$ in MeCN was highly irreversible.[491]

$$[M(NEt_2)_4] + 2NaHS_2C_6H_3R + (HS)_2C_6H_3R \longrightarrow Na_2[M(S_2C_6H_3R)_3] + 4HNEt_2 \quad (48)$$

The $[Zr(S_2C_6H_4)_3]^{2-}$ anion in $[NMe_4]_2[Zr(S_2C_6H_4)_3]$ has a structure (Figure 8) that is intermediate between octahedral and trigonal prismatic; the trigonal twist angle is 37°. Averaged distances are Zr—S = 2.543, S···S (intraligand) = 3.265, S···S (interligand) = 3.584 Å. The averaged S—C bond length (1.765 Å) is close to that expected for an S—C single bond, indicating that $[Zr(S_2C_6H_4)_3]^{2-}$ is best formulated as a dithiolatozirconium(IV) complex. It is interesting to note that the isoelectronic complexes $[Nb(S_2C_6H_4)_3]^-$ and $[Mo(S_2C_6H_4)_3]$ are trigonal prismatic. The change in geometry as the metal *d*-orbital energies decrease in the series $Zr^{IV} > Nb^{V} > Mo^{VI}$ has been rationalized in terms of increasing ligand $(\pi_v) \rightarrow$ metal $(d_{x^2-y^2}, d_{xy})\pi$ bonding.[492]

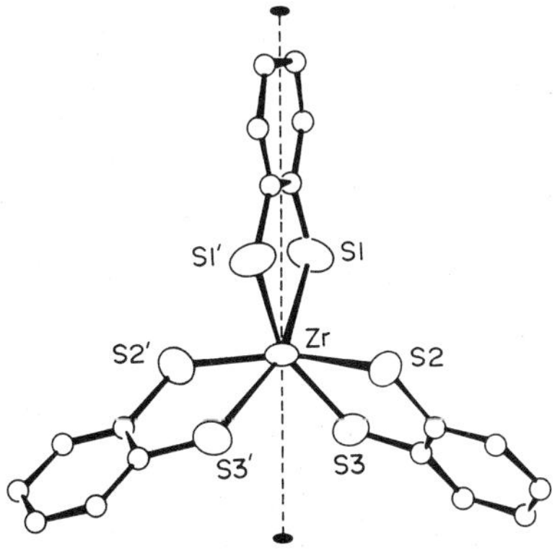

Figure 8 A perspective view of the $[Zr(S_2C_6H_4)_3]^{2-}$ anion. The complex is located on a crystallographic twofold axis

Tris(*o*-mercaptophenolato) complexes, $[NEt_4]_2[M(SOC_6H_4)_3]$ (M = Zr or Hf),[280] have been prepared by a method similar to that used for the analogous dithiolates. A compound of composition $Zr(SOC_6H_4)(HSOC_6H_4)_2$ has been obtained from the reaction of $ZrCl_4$ with three equivalents of *o*-mercaptophenol.[493]

32.4.5.4 Other sulfur donor ligands

Reaction of $Na\{S_2P(OEt)_2\}$ with $ZrCl_4$ in toluene affords the tetrakis(diethyldithiophosphate) $[Zr\{S_2P(OEt)_2\}_4]$.[494] An X-ray study[495] has established that the corresponding diisopropyl derivative $[Zr\{S_2P(OCHMe_2)_2\}_4]$ has an eight-coordinate D_{2d}-*mmmm* dodecahedral structure, with Zr—S_A = 2.746(3) and Zr—S_B = 2.653(3) Å. This structure is somewhat surprising[461] in view of the rather large S···S bite (3.178(5) Å) of the dithiophosphate ligand; the normalized bite is 1.177 Å.

Zirconium and hafnium tetrachlorides form 1:1 and 1:2 adducts with thiourea. IR spectra indicate that the 1:1 adducts contain terminal and bridging chlorine atoms and S-bonded thiourea ligands. A chlorine-bridged dimeric structure $[\{H_2N)_2CS\}Cl_3MCl_2MCl_3\{SC(NH_2)_2\}]$ has been proposed.[496] The IR spectra and calorimetric measurements suggest that only one thiourea ligand is coordinated to the metal in the 1:2 adducts; the second ligand appears to be attached to the complex *via* hydrogen bonds.[496,497]

The reaction of $ZrCl_4$ with triphenylphosphine sulfide in benzene at reflux affords a 1:1 adduct, which has been assigned a μ_2-chlorine-bridged structure on the basis of molecular weight, conductance and IR data.[498] An insoluble red-brown thiophene adduct of composition $ZrCl_4(C_4H_4S)_2$ has been obtained by prolonged shaking of a 4:1 mixture of thiophene and $ZrCl_4$ in benzene.[292]

32.4.6 Selenium Ligands

The reaction of $ZrCl_4$ with $Al(SePh)_3$ in ether–benzene yields the phenylselenolate $Zr(SePh)_4$ as a turquoise, microcrystalline product.[499]

32.4.7 Halogens as Ligands

32.4.7.1 Tetrahalides

Zirconium(IV) and hafnium(IV) halides are high-melting crystalline solids, which do not contain discrete complexes. α-ZrF_4[500] and β-ZrF_4[501] have extended three-dimensional structures in whch ZrF_8 dodecahedra (α form) or square antiprisms (β form) are linked together by sharing fluorine atoms. Crystalline $ZrCl_4$ consists of infinite zigzag chains of $ZrCl_6$ octahedra which share non-opposite edges; $ZrBr_4$, $HfCl_4$ and $HfBr_4$ are isostructural with $ZrCl_4$.[502] In zirconium(IV) iodide, ZrI_6 octahedra share non-opposite edges so as to give a helical chain structure with an identity period of six octahedra.[503] Hafnium(IV) iodide has yet another type of chain structure in which sharing of non-opposite edges of HfI_6 octahedra results in folded chains with an identity period of four octahedra.[504] Much useful information on the preparation and properties of zirconium(IV) and hafnium(IV) halides may be found in previous reviews[5,6] and in the monograph by Canterford and Colton.[505]

In the gas phase, the tetrahalides exist as monomeric regular tetrahedral molecules. This has been established by electron diffraction and vibrational spectroscopy. Bond lengths and vibrational frequencies are listed in Table 16. Force constants for M—X bond stretching are slightly larger for the hafnium tetrahalides than for the zirconium analogues; the force constants decrease appreciably as X varies in the order Cl > Br > I.[506]

Table 16 Bond Lengths (Å) and Vibrational Frequencies[a] (cm^{-1}) in Gaseous Zirconium(IV) and Hafnium(IV) Halides

Compound	*Bond length*	$\nu_1\ (a_1)$	$\nu_2\ (e)$	$\nu_3\ (t_2)$	$\nu_4\ (t_2)$	*Ref.*
ZrF_4	1.902(4)	633 (ED)[b]	185 (ED)	668 (IR)	187 (ED)	1–3
HfF_4	1.909(5)	659 (ED)	187 (ED)	645 (IR)	179 (ED)	1–3
$ZrCl_4$	2.32(1)	377	98	418	113	3–8
	2.32(2)	380.5		423 (IR)		
				421 (IR)		
$HfCl_4$	2.316(5)	382	101.2	390	112	3, 5, 6, 9
	2.33(2)			393 (IR)		
$ZrBr_4$	2.465(4)	225.5	60	315	72	6, 10
$HfBr_4$	2.450(4)	235.5	63	273	71	6, 10
ZrI_4	2.660(10)	158	43	254	55	6, 10
HfI_4	2.662(8)	158	55	224	63	6, 10

[a] Raman frequencies, unless indicated otherwise. [b] Calculated from electron diffraction data and the $\nu_3\ (t_2)$ IR frequency.

1. V. M. Petrov, G. V. Girichev, N. I. Giricheva, O. K. Shaposhnikova and E. Z. Zasorin, *J. Struct. Chem. (Engl. Transl.)*, 1979, **20**, 110.
2. G. V. Girichev, N. I. Giricheva and T. N. Malysheva, *Russ. J. Phys. Chem. (Engl. Transl.)*, 1982, **56**, 1120.
3. A. Büchler, J. B. Berkowitz-Mattuck and D. H. Dugre, *J. Chem. Phys.*, 1961, **34**, 2202.
4. M. Kimura, K. Kimura, M. Aoki and S. Shibata, *Bull. Chem. Soc. Jpn.*, 1956, **29**, 95.
5. V. P. Spiridonov, P. A. Akishin and V. I. Tsirel'nikov, *J. Struct. Chem. (Engl. Transl.)*, 1962, **3**, 311.
6. R. J. H. Clark, B. K. Hunter and D. M. Rippon, *Inorg. Chem.*, 1972, **11**, 56.
7. W. Brockner and A. F. Demiray, *J. Raman Spectrosc.*, 1978, **7**, 330.
8. J. K. Wilmshurst, *J. Mol. Spectrosc.*, 1960, **5**, 343.
9. G. V. Girichev, V. M. Petrov, N. I. Giricheva, A. N. Utkin and V. N. Petrova, *J. Struct. Chem. (Engl. Transl.)*, 1981, **22**, 694.
10. G. V. Girichev, E. Z. Zasorin, N. I. Giricheva, K. S. Krasnov and V. P. Spiridonov, *J. Struct. Chem. (Engl. Transl.)*, 1977, **18**, 34.

Metal tetrahalide adducts with Lewis bases (*cf.* Tables 3, 4, 13, 15 and 23) are discussed in other sections. The tetrafluorides form stable hydrates $MF_4(H_2O)_n$ ($n = 1$ or 3), in contrast to the other tetrahalides which react with water to give oxyhalides. It is interesting to note that $ZrF_4(H_2O)_3$ and $HfF_4(H_2O)_3$ have rather different structures. The zirconium compound[507,508] consists of discrete, centrosymmetric $[Zr_2F_8(H_2O)_6]$ complexes in which two dodecahedral $ZrF_5(H_2O)_3$ units are joined by sharing an F···F *m* edge. The three water molecules (Zr—O = 2.263–2.323 Å) and a bridging fluorine atom (Zr–F = 2.214 Å) occupy the dodecahedral A sites, while the other bridging fluorine atom (Zr—F = 2.118 Å) and the three terminal fluorines (Zr—F = 1.996–2.032 Å) take the B sites. Occupation of the B sites by the better π-donor ligand is in accord with Orgel's rule.[458,459] In the hafnium compound,[509] $HfF_6(H_2O)_2$ units share two F···F polyhedral edges giving a chain structure with repeating unit $HfF_4(H_2O)_2$; the uncoordinated water molecule is held between the chains by hydrogen bonds. The monohydrate $ZrF_4(H_2O)$ has a three-dimensional network structure[510] in which $ZrF_6(H_2O)_2$ dodecahedra share six corners (four fluorine atoms and two water molecules) with six adjacent dodecahedra.

32.4.7.2 Fluorometallates

Zirconium and hafnium form a vast array of complex fluorides, $M'_xM_yF_z$, where *M* is Zr or Hf, and M′ is an alkali, alkaline earth, divalent transition metal, or lanthanide cation, or the conjugate acid of a nitrogen base. The number of fluorine atoms per Zr (Hf) atom is commonly eight, seven, six or five, but non-integral ratios are also found, as in $Na_3Zr_2F_{11}$, $Na_5Zr_2F_{13}$, $Li_3Zr_4F_{19}$, $Rb_5Zr_4F_{21}$ and $Na_7Zr_6F_{31}$. The variety of alkali fluorometallates is illustrated in Table 17.

Table 17 Alkali Fluorometallates

Compound	*Alkali cation (M′)*
M'_4ZrF_8	Li[1]
M'_3ZrF_7	Li,[2] Na,[3–5] K,[5,6] NH_4,[7–10] Rb,[5,11] Cs[12,13]
M'_2ZrF_6	Li,[1,2] Na,[3–5] K,[5,6,11] NH_4,[7–9] Rb,[5] Cs,[5,11–13] Tl[14]
$M'ZrF_5$	Na,[3] K,[6] NH_4,[7–9,15,16] Cs,[12,13] Tl[17]
$M'ZrF_5\cdot H_2O$	Na,[5] K,[5] NH_4,[15,16] Rb,[5] Cs[5]
$M'_3Zr_2F_{11}$	Na,[3] K[6]
$M'_5Zr_2F_{13}$	Na,[3–5] K[6]
$M'_3Zr_4F_{19}$	Li,[2] Na[3]
$M'_5Zr_4F_{21}$	Rb[18]
$M'_7Zr_6F_{31}$	Na,[3,4] K[6]
M'_3HfF_7	Na,[4,19] K,[19,20] NH_4,[21] Rb,[20] Cs[13]
M'_2HfF_6	Na,[4,19] K,[19,20,22] NH_4,[21,23] Rb,[20] Cs,[13,20,24] Tl[14]
$M'HfF_5$	Na,[4,19] K,[19,22] NH_4,[21,23] Cs,[13,24] Tl[17]
$M'HfF_5\cdot H_2O$	K,[20] Rb,[20] Cs[20]
$M'_5Hf_2F_{13}$	Na[1]
$M'_7Hf_6F_{31}$	Na[4]

1. Yu. M. Korenev and A. V. Novoselova, *Inorg. Mater.* (*Engl. Transl.*), 1965, **1,** 546.
2. R. E. Thoma, H. Insley, H. A. Friedman and G. M. Hebert, *J. Nucl. Mater.*, 1968, **27,** 166.
3. C. J. Barton, W. R. Grimes, H. Insley, R. E. Moore and R. E. Thoma, *J. Phys. Chem.*, 1958, **62,** 665.
4. I. D. Ratnikova, A. A. Kosorukov, Yu. M. Korenev and A. V. Novoselova, *Russ. J. Inorg. Chem.* (*Engl. Transl.*), 1975, **20,** 782.
5. I. V. Tananaev and L. S. Guzeeva, *Russ. J. Inorg. Chem.* (*Engl. Transl.*), 1966, **11,** 590.
6. A. V. Novoselova, Yu. M. Korenev and Yu. P. Simanov, *Dokl. Chem.* (*Engl. Transl.*), 1961, **139,** 771.
7. H. M. Haendler, C. M. Wheeler, Jr. and D. W. Robinson, *J. Am. Chem. Soc.*, 1952, **74,** 2352.
8. B. Gaudreau, *Rev. Chim. Min.*, 1965, **2,** 1.
9. H. Hull and A. G. Turnbull, *J. Inorg. Nucl. Chem.*, 1967, **29,** 951.
10. H. M. Haendler, F. A. Johnson and D. S. Crocket, *J. Am. Chem. Soc.*, 1958, **80,** 2662.
11. D. S. Crocket and H. M. Haendler, *J. Am. Chem. Soc.*, 1960, **32,** 4158.
12. G. D. Robbins, R. E. Thoma and H. Insley, *J. Inorg. Nucl. Chem.*, 1965, **27,** 559.
13. A. A. Kosorukov, V. Ya. Frenkel', Yu. M. Korenev and A. V. Novoselova, *Russ. J. Inorg. Chem.* (*Engl. Transl.*), 1973, **18,** 1025.
14. H. Bode and G. Teufer, *Z. Anorg. Allg. Chem.*, 1956, **283,** 18.
15. H. M. Haendler and D. W. Robinson, *J. Am. Chem. Soc.*, 1953, **75,** 3846.
16. H. Hull and A. G. Turnbull, *J. Inorg. Nucl. Chem.*, 1967, **29,** 2903.
17. D. Avignant, I. Mansouri, R. Chevalier and J. C. Cousseins, *J. Solid State Chem.*, 1981, **38,** 121.
18. G. Brunton, *Acta Crystallogr., Sect. B*, 1971, **27,** 1944.
19. I. N. Sheiko, V. T. Mal'tsev and G. A. Bukhalova, *Sov. Progr. Chem.*, 1966, **32,** 978.
20. I. V. Tananaev and L. S. Guzeeva, *Russ. J. Inorg. Chem.* (*Engl. Transl.*), 1966, **11,** 587.
21. B. Gaudreau, *C.R. Hebd. Seances Acad. Sci., Ser. C*, 1966, **263,** 67.
22. A. A. Opalovskii, T. F. Gudimovich, M. Kh. Akhmadeev and L. D. Ishkova, *Russ. J. Inorg. Chem.*, 1982, **27,** 664.
23. N. P. Sazhin, B. V. Shchepochkin and G. A. Yagodin, *Bull. Acad. Sci. USSR, Div. Chem. Sci.*, 1965, 1097.
24. A. A. Opalovskii, T. F. Gudimovich, N. Kh. Akhmadeev and L. D. Ishkova, *Russ. J. Inorg. Chem.* (*Engl. Transl.*), 1983, **28,** 1065.

Table 18 summarizes structural data for fluorozirconates whose structures have been determined by diffraction methods. In some cases, the analogous fluorohafnates are known to be isomorphous with the fluorozirconates. The compounds are grouped according to the coordination number of the Zr atom (six, seven or eight); individual structures are discussed in the following sections. In general, there is no simple relation between structure and stoichiometry. Fluorine bridging between Zr atoms is common, resulting in oligonuclear, chain, layer and extended three-dimensional structures in which the coordination number of the metal exceeds that expected on the basis of the simplest formula. A variety of coordination polyhedra have been found in fluorozirconate structures; the more common ones are the octahedron for coordination number six, the pentagonal bipyramid and C_{2v} monocapped trigonal prism for coordination number seven, and the D_{2d} dodecahedron, D_{4d} square antiprism, and C_{2v} bicapped trigonal prism for coordination number eight. Averaged Zr—F bond lengths (Table 18) are in the ranges 1.98–2.04, 2.04–2.08 and 2.10–2.13 Å for coordination numbers six, seven and eight, respectively. In compounds that contain Zr—F—Zr bridges, the averaged Zr—F (bridging) bond length is as much as 0.1 Å larger than these values, and the averaged Zr—F (terminal) bond length is correspondingly smaller. The following discrete anions have been characterized structurally: $[ZrF_6]^{2-}$, $[ZrF_7]^{3-}$, $[ZrF_8]^{4-}$, $[Zr_2F_{12}]^{4-}$, $[Zr_2F_{10}(H_2O)_2]^{2-}$, $[Zr_2F_{13}]^{5-}$, $[Zr_2F_{14}]^{6-}$ and $[Zr_4F_{24}]^{8-}$.

Table 18 Structural Data for Complex Fluorozirconates

Compound	*Structural description*[a]	*Average Zr—F bond length* (Å)	*Ref.*
Coordination number 6			
$Li_2[ZrF_6]$	Isolated $[ZrF_6]^{2-}$ octahedra	2.016	1
$Rb_2[ZrF_6]$	Isolated $[ZrF_6]^{2-}$ octahedra	2.04	2
$Cs_2[ZrF_6]$	Isolated $[ZrF_6]^{2-}$ octahedra	2.04	2
$[Cu(H_2O)_4][ZrF_6]$	$[ZrF_6]^{2-}$ octahedra	1.996	3
$FeZrF_6$	Three-dimensional network; ZrF_6 and FeF_6 octahedra share corners	1.991 (cubic) 1.995 (trigonal)	4
$CuZrF_6$	Three-dimensional network; ZrF_6 and CuF_6 octahedra share corners	1.995 (cubic)	5
$SmZrF_7$	Three-dimensional network; ZrF_6 octahedra and SmF_8 polyhedra share corners	2.006	6
Coordination number 7			
$[NH_4]_3[ZrF_7]$	Isolated $[ZrF_7]^{3-}$ PBP; dynamically disordered	2.02	7, 8
$[H_3NC_2H_4NH_3]_3[ZrF_7]_2{\cdot}2H_2O$	Isolated $[ZrF_7]^{3-}$ CTP	2.059	9
$[NMe_4]_2[Zr_2F_{10}(H_2O)_2]$	Centrosymmetric dimer; two $ZrF_6(H_2O)$ PBPs share an equatorial F···F edge	2.045	10
$[H_3NC_2H_4NH_3]_2[Zr_2F_{12}]$	Centrosymmetric dimer; two ZrF_7 PBPs share an equatorial edge	2.062	11
$K_2[Cu(H_2O)_6][Zr_2F_{12}]$	Centrosymmetric dimer; two ZrF_7 PBPs share an equatorial edge	2.063	12
α-$BaZrF_6$	Centrosymmetric $[Zr_2F_{12}]^{4-}$; two ZrF_7 CTPs share an edge	2.082	13
$Na_5[Zr_2F_{13}]$	C_{2h} symmetry; two ZrF_7 CTPs share the capping site	2.06	14
γ-Na_2ZrF_6	ZrF_7 irregular polyhedra share two F atoms	2.065	15
Coordination number 8			
$Li_6[BeF_4][ZrF_8]$	Isolated $[ZrF_8]^{4-}$ DD	2.10	16
$[Cu(H_2O)_6]_2[ZrF_8]$	Isolated $[ZrF_8]^{4-}$ SAP	2.103	17
$[Cu(H_2O)_6][Cu_2(H_2O)_{10}][Zr_2F_{14}]$	Centrosymmetric dimer; two ZrF_8 SAPs share an *s* edge	2.109 2.105	18 19
K_2ZrF_6	Chain structure; ZrF_8 DDs share *m* edges	2.114	20
$MnZrF_6{\cdot}5H_2O$	Chain structure; ZrF_8 DDs share *a* edges	2.12	21, 22
β-$BaZrF_6$	Chain structure; ZrF_8 DDs share *a* edges	2.126	23
$PbZrF_6$	Isostructural with β-$BaZrF_6$	2.10	24
$(N_2H_6)ZrF_6$	Zigzag chain structure; ZrF_8 BTPs share t_1 edges	2.12	25
$TlZrF_5$	Layers of $(ZrF_5^-)_n$; ZrF_8 BTPs share edges and corners	2.112	26
$Mn_2ZrF_8{\cdot}6H_2O$	Layer structure; ZrF_8 SAPs and $MnF_4(H_2O)_3$ PBPs share F···F edges	2.11	27
$Cd_2ZrF_8{\cdot}6H_2O$	Isostructural with $Mn_2ZrF_8{\cdot}6H_2O$	2.114	28
$Na_7Zr_6F_{31}$	Three-dimensional network; ZrF_8 SAPs share corners and edges	2.113	29
Coordination numbers 6, 7 and 8			
$Rb_5Zr_4F_{21}$	Cross-linked structure containing:		30
	ZrF_6 octahedra	1.98	
	ZrF_7 polyhedra (CO/PBP)	2.06	
	ZrF_7 polyhedra (CO/PBP)	2.08	
	ZrF_8 irregular antiprisms	2.13	

[a] PBP = pentagonal bipyramid; CTP = C_{2v} monocapped trigonal prism; CO = monocapped octahedron; DD = dodecahedron; SAP = square antiprism; BTP = bicapped trigonal prism.

1. G. Brunton, *Acta Crystallogr., Sect. B*, 1973, **29**, 2294.
2. H. Bode and G. Teufer, *Z. Anorg. Allg. Chem.*, 1956, **283**, 18.
3. J. Fischer and R. Weiss, *Acta Crystallogr., Sect. B*, 1973, **29**, 1955.
4. P. Kohl, D. Reinen, G. Decher and B. Wanklyn, *Z. Kristallogr., Kristallgeom., Kristallphys., Kristallchem.*, 1980, **153**, 211.
5. V. Propach and F. Steffens, *Z. Naturforsch, Teil B*, 1978, **33**, 268.

Table 18 *(footnotes continued)*

6. M. Poulain, M. Poulain and J. Lucas, *J. Solid State Chem.*, 1973, **8,** 132.
7. H. J. Hurst and J. C. Taylor, *Acta Crystallogr., Sect. B,* 1970, **26,** 417.
8. H. J. Hurst and J. C. Taylor, *Acta Crystallogr., Sect. B,* 1970, **26,** 2136.
9. I. P. Kondratyuk, B. V. Bukvetskii, R. L. Davidovich and M. A. Medkov, *Sov. J. Coord. Chem. (Engl. Transl.),* 1982, **8,** 113.
10. R. L. Davidovich, M. A. Medkov, V. B. Timchenko and B. V. Bukvetskii, *Bull. Acad. Sci. USSR, Div. Chem. Sci.*, 1983, **32,** 2181.
11. I. P. Kondratyuk, M. F. Eiberman, R. L. Davidovich, M. A. Medkov and B. V. Bukvetskii, *Sov. J. Coord. Chem. (Engl. Transl.),* 1981, **7,** 549.
12. J. Fischer and R. Weiss, *Acta Crystallogr., Sect. B,* 1973, **29,** 1958.
13. J.-P. Laval, R. Papiernik and B. Frit, *Acta Crystallogr., Sect. B,* 1978, **34,** 1070.
14. R. M. Herak, S. S. Malcic and L. M. Manojlovic, *Acta Crystallogr.*, 1965, **18,** 520.
15. G. Brunton, *Acta Crystallogr., Sect. B,* 1969, **25,** 2164.
16. D. R. Sears and J. H. Burns, *J. Chem. Phys.*, 1964, **41,** 3478.
17. J. Fischer, R. Elchinger and R. Weiss, *Acta Crystallogr., Sect. B,* 1973, **29,** 1967.
18. J. Fischer and R. Weiss, *Acta Crystallogr., Sect. B,* 1973, **29,** 1963.
19. T. S. Chernaya, L. E. Fykin, V. A. Sarin, N. I. Bydanov, N. M. Laptash, B. V. Bukvetskii and V. I. Simonov, *Sov. Phys. Crystallogr. (Engl. Transl.),* 1983, **28,** 393.
20. R. Hoppe and B. Mehlhorn, *Z. Anorg. Allg. Chem.*, 1976, **425,** 200.
21. L. P. Otroshchenko, R. L. Davidovich and V. I. Simonov, *Sov. J. Coord. Chem. (Engl. Transl.),* 1978, **4,** 1079.
22. L. P. Otroschenko, V. I. Simonov, R. L. Davidovich, L. E. Fykin, V. Ya. Duderov and S. P. Solov'ev, *Sov. Phys.-Crystallogr. (Engl. Transl.),* 1980, **25,** 416.
23. B. Melhorn and R. Hoppe, *Z. Anorg. Allg. Chem.*, 1976, **425,** 180.
24. J.-P. Laval, D. Mercurio-Lavaud and B. Gaudreau, *Rev. Chim. Min.*, 1974, **11,** 742.
25. B. Kojic-Prodic, S. Scavnicar and B. Matkovic, *Acta Crystallogr., Sect. B,* 1971, **27,** 638.
26. D. Avignant, I. Mansouri, R. Chevalier and J. C. Cousseins, *J. Solid State Chem.*, 1981, **38,** 121.
27. L. P. Otroshchenko, R. L. Davidovich, V. I. Simonov and N. V. Belov, *Sov. Phys. Crystallogr. (Engl. Transl.),* 1981, **26,** 675.
28. M. F. Eiberman, T. A. Kaidalova, R. L. Davidovich, T. F. Levchishina and B. V. Bukvetskii, *Koord. Khim.*, 1980, **6,** 1885 (*Chem. Abstr.*, 1981, **94,** 112794).
29. J. H. Burns, R. D. Ellison and H. A. Levy, *Acta Crystallogr., Sect. B,* 1968, **24,** 230.
30. G. Brunton, *Acta Crystallogr., Sect. B,* 1971, **27,** 1944.

(*i*) *Alkali metal salts*

Alkali fluorometallates may be obtained by crystallization of MF_4—M′F melts (M = Zr or Hf; M′ = alkali metal), or by reaction of a suitable Zr^{IV} or Hf^{IV} compound with an alkali fluoride in hydrofluoric acid, water, acetic acid or methanol. References to the preparation of specific compounds are given in Table 17.

Phase equilibria in MF_4–M′F systems are rather complex, and polymorphism is common. For example, the following compounds have been identified in the ZrF_4–NaF system: $NaZrF_5$, $Na_3Zr_2F_{11}$, $Na_3Zr_4F_{19}$, $Na_7Zr_6F_{31}$, two polymorphic forms of Na_3ZrF_7 and $Na_5Zr_2F_{13}$, and at least three polymorphs of Na_2ZrF_6.[511,512]

From aqueous MF_4–M′F systems, the usual products are M'_3MF_7, M'_2MF_6 or $M'MF_5{\cdot}H_2O$ depending on the stoichiometry of the reaction mixture. The hexa- and penta-fluorometallates can be recrystallized from water, but crystallization of the heptafluorometallates requires the presence of excess alkali fluoride. For the aqueous MF_4–CsF system, the highest fluorometallate obtained, even in the presence of excess CsF, is Cs_2MF_6.[513,514] Synthesis of the cesium heptafluorometallates Cs_3MF_7 is accomplished by crystallization of a 1:3 MF_4—CsF melt. The pentafluorozirconate that crystallizes from aqueous ZrF_4—NH_4F—HF solutions depends on temperature and the concentration of HF; at 20 °C and HF concentrations less than 5 wt%, $(NH_4)ZrF_5{\cdot}H_2O$ is obtained, while at higher HF concentrations or at elevated temperatures, anhydrous $(NH_4)ZrF_5$ is formed.[515] $(NH_4)ZrF_5$ may also be prepared by thermolysis of $(NH_4)_3ZrF_7$[516,517] and by reaction of $(NH_4)_2ZrF_6$ with XeF_2.[518] Other routes to anhydrous fluorometallates include (i) reaction of HfF_4 with M′F in anhydrous acetic acid,[519,520] (ii) reaction of $ZrBr_4$ with M′F in methanol[521,522] and (iii) reaction of ZrO_2 or HfO_2 with NH_4F at 130 °C.[517,523]

The thermodynamic properties of ammonium fluorozirconates have been determined[515,516] from heat of solution measurements; values of ΔH_f° (kJ mol^{-1}) at 298 K are -3386 ± 3 for $(NH_4)_3ZrF_7$, -2918 ± 3 for α-$(NH_4)_2ZrF_6$, -2217 ± 3 for $(NH_4)ZrF_5$ and -2728 ± 3 for $(NH_4)ZrF_5{\cdot}H_2O$. The heat of dehydration for the pentafluorozirconate (equation 49) is 54 ± 2 kJ mol^{-1}, and the equilibrium water vapor pressure associated with this reaction is 16.1 mmHg at 20 °C. These values indicate very weak bonding of the water. Thermal decomposition of ammonium fluorozirconates[524] occurs in a stepwise manner, with the decomposition temperatures depending on pressure; the steps in equation (50) have been identified at 1 atm pressure.

$$(NH_4)ZrF_5{\cdot}H_2O(s) \longrightarrow (NH_4)ZrF_5(s) + H_2O(g) \qquad (49)$$

$$(NH_4)_3ZrF_7 \xrightarrow{297\,^\circ C} (NH_4)_2ZrF_6 \xrightarrow{357\,^\circ C} (NH_4)ZrF_5 \xrightarrow{410\,^\circ C} ZrF_4 \qquad (50)$$

Raman spectra of aqueous solutions of $(NH_4)_3HfF_7$ and $(NH_4)_2HfF_6$[525] are identical (ν_1 589, ν_5 230 cm^{-1}), and are similar to spectra of crystalline Cs_2MF_6 (M = Zr or Hf) salts, which are known to contain $[MF_6]^{2-}$ anions. Therefore, $[HfF_7]^{3-}$ dissociates in solution into $[HfF_6]^{2-}$ and F^-. Paper chromatographic studies[526] have shown that $[ZrF_7]^{3-}$ is similarly dissociated. Raman spectra of ZrO_2 and HfO_2 in 5 M HF indicate that the $[MF_6]^{2-}$ anions are the predominant species in solution.[152]

^{19}F NMR spectra of 1 M aqueous solutions of $(NH_4)_2ZrF_6$ and $(NH_4)_2HfF_6$[525] exhibit a single, rather broad resonance (line width ~8 Hz for $[ZrF_6]^{2-}$ and ~40 Hz for $[HfF_6]^{2-}$), which broadens and shifts upfield upon addition of excess F^-. These observations have been interpreted in terms of rapid fluoride exchange, probably proceeding *via* an $[MF_7]^{3-}$ intermediate. For a solution 0.25 M in $[ZrF_6]^{2-}$ and 1.5 M in F^-, which gave a ^{19}F linewidth of 1550 Hz, the average lifetime of a fluorine nucleus in the $[ZrF_6]^{2-}$ and F^- sites was estimated to be ~1.4×10^{-5} s. The ^{91}Zr NMR spectrum of $(NH_4)_2ZrF_6$ in D_2O[527] also consists of a single, broad line (minimum linewidth 50 Hz); no Zr—F coupling is observed, in accord with rapid fluoride exchange.

Because K_2HfF_6 is ~1.7 times more soluble in water than K_2ZrF_6,[528] zirconium and hafnium can be separated by fractional crystallization of the hexafluorometallates. This approach is used on an industrial scale in the USSR.[286] Conductance measurements on aqueous solutions of M'_2MF_6 (M′ = K, Rb, or Cs) indicate very little hydrolysis of the $[MF_6]^{2-}$ ions.[529] Alkaline hydrolysis of potassium and ammonium fluorozirconates yields crystalline $M'ZrF_3(OH)_2{\cdot}H_2O$ complexes, which are easily dehydrated to $M'ZrF_3(OH)_2$.[530]

The coordination number of the Zr (Hf) atom in the crystalline hexafluorometallates varies from six to eight depending on the alkali cation. Li_2ZrF_6[531] and the isomorphous rubidium and cesium salts M'_2MF_6[532] contain the octahedral $[MF_6]^{2-}$ anion. The Zr atom in γ-Na_2ZrF_6 is surrounded by an irregular polyhedron of seven fluorine atoms, two of which are bridging.[533] The isomorphous K_2MF_6 salts[532] contain chain anions which are built up from MF_8 dodecahedra that share opposite *m* edges, i.e. all four bridging fluorine atoms are located on one trapezoid of the dodecahedron (Zr—F terminal = 2.038, 2.077; Zr—F bridging = 2.111, 2.231 Å).[534] Still another structure (space group *Pmma*) is adopted by the isomorphous $(NH_4)_2MF_6$ and Tl_2MF_6 salts. In these compounds, the Zr (Hf) atom may have a pentagonal bipyramidal environment; however, the locations of the fluorine atoms are not well defined by the X-ray data.[532]

The isomorphous, cubic heptafluorometallates $(NH_4)_3MF_7$ and K_3ZrF_7 contain dynamically disordered $[MF_7]^{3-}$ ions. Both capped octahedral[535] and pentagonal bipyramidal[536] models have been suggested, and the latter is supported by single-crystal X-ray and neutron diffraction studies of $(NH_4)_3ZrF_7$.[537,538] The averaged Zr—F bond length in $(NH_4)_3ZrF_7$ (2.02 Å) appears to be a bit too short in comparison with Zr—F bond lengths in other seven-coordinate structures (Table 18). The sodium salts Na_3MF_7 crystallize in the tetragonal system (space group *I4/mmm*) and are also dynamically disordered; the Zr (Hf) atom appears to be surrounded by a cube of F-atom sites, one of which is vacant.[539] The dynamics of F-atom motion in alkali heptafluorometallates have been studied by ^{19}F NMR spectroscopy[540,541] and by perturbed angular correlation measurements.[542,543]

The coordination environment of the Zr atom in Li_4ZrF_8 is unknown, but the $[ZrF_8]^{4-}$ anion has been found in Li_6BeZrF_{12}. This compound, which is obtained from a LiF—BeF_2—ZrF_4 melt of stoichiometric composition, contains tetrahedral $[BeF_4]^{2-}$ and dodecahedral $[ZrF_8]^{4-}$ anions (Zr—F_A = 2.16, Zr—F_B = 2.05 Å).[544] Eight-coordinate Zr atoms are also found in the pentafluorozirconate $TlZrF_5$. In this compound, ZrF_8 bicapped trigonal prisms share one edge and four corners so as to give $(ZrF_5^-)_n$ layers that are held together by Tl^+ cations.[545]

The sodium salt $Na_5Zr_2F_{13}$ contains $[Zr_2F_{13}]^{5-}$ anions of C_{2h} symmetry in which two ZrF_7 monocapped trigonal prisms share the F atom that occupies the capping site (Zr—F terminal = 2.00–2.10, Zr—F bridging = 2.10 Å).[546]

$Rb_5Zr_4F_{21}$ has a complex cross-linked chain structure that contains four crystallographically independent Zr atoms having three different coordination numbers. The structure is built up from edge- and/or corner-shared ZrF_6 octahedra, ZrF_8 irregular antiprisms and two independent ZrF_7 polyhedra.[547] Application of the dihedral angle criteria of Muetterties and Guggenberger[548] indicates that both ZrF_7 polyhedra are midway between a capped octahedron and a pentagonal bipyramid.

$Na_7Zr_6F_{31}$ contains octahedral arrays of six ZrF_8 square antiprisms which share corners in such a way as to give Zr_6F_{36} units. These units in turn share external edges with six neighboring Zr_6F_{36} units, giving an extended three-dimensional anion with formula $(Zr_6F_{30}^{6-})_n$. The 'extra'

F^- ion is located within the cuboctahedral cavity surrounded by the six ZrF_8 square antiprisms.[549]

Mixed-ligand fluorometallates that contain peroxide, sulfate, selenate, carbonate or oxalate ligands are discussed in Sections 32.4.4.2, 32.4.4.5.iii, 32.4.4.6.iii and 32.4.4.6.iv.

(*ii*) *Hydrazinium, alkylammonium and guanidinium salts*

Hydrazinium(2+) salts, $(N_2H_6)MF_6$ and $(N_2H_6)_3M_2F_{14}$ (M = Zr or Hf), have been isolated from hydrofluoric acid solutions of the metal oxides and hydrazinium(2+) fluoride.[550,551] $(N_2H_6)ZrF_6$ contains infinite zigzag chains of ZrF_8 bicapped trigonal prisms which share the two t_1 edges on the quadrilateral face.[552] The reaction of $(N_2H_6)MF_6$ with XeF_6 at room temperature yields thermally unstable compounds $nXeF_6 \cdot MF_4$ ($n \leqslant 6$) which decompose *in vacuo* at room temperature to white, moisture-sensitive solids of composition $XeF_6 \cdot MF_4$. Vibrational spectra suggest that these compounds contain $[XeF_5]^+$ cations and polymeric $(MF_5^-)_n$ anions.[553] A hydrazinium(1+) salt of composition $(N_2H_5)_3Zr_4F_{19} \cdot 3H_2O$ has been prepared by reaction of ZrF_4 and hydrazinium(1+) fluoride in aqueous solution.[55]

Large, colorless, prismatic crystals of $[NMe_4]_2[Zr_2F_{10}(H_2O)_2]$ have been isolated from dilute HF solutions of $ZrF_4(H_2O)_3$ and tetramethylammonium fluoride. The $[Zr_2F_{10}(H_2O)_2]^{2-}$ anion is a centrosymmetric dimer in which two $ZrF_6(H_2O)$ pentagonal bipyramids share an equatorial F···F edge (Zr—F terminal = 1.972–1.997; Zr—F bridging = 2.143, 2.184 Å); the water molecule occupies an equatorial site (Zr—O = 2.271 Å). The following metal–ligand stretching bands are observed in the IR spectrum: ν(Zr—F terminal) 520, 480, 420; ν(Zr—F bridging) 325; ν(Zr—O) 385 cm^{-1}.[554] The hexafluorometallates $[NEt_4]_2[MF_6] \cdot 2H_2O$ have been prepared by direct electrochemical oxidation of zirconium or hafnium anodes in PhCN solutions that contain $[NEt_4]F \cdot 3HF$.[555]

The ethylenediammonium fluorozirconates $(H_3NC_2H_4NH_3)(ZrF_5)_2 \cdot H_2O$, $(H_3NC_2H_4NH_3)ZrF_6$ and $(H_3NC_2H_4NH_3)_3(ZrF_7)_2 \cdot 2H_2O$ have been prepared in aqueous solution by reaction of various molar ratios of $(H_3NC_2H_4NH_3)F_2 \cdot HF$ and $H_2ZrF_6 \cdot 2H_2O$. The presence of ν(Zr—F terminal) and ν(Zr—F bridging) in the 600–380 cm^{-1} region of IR spectra of the pentafluorozirconate indicates that this compound probably contains polymeric $(ZrF_5^-)_n$ anions.[556] Crystals of $(H_3NC_2H_4NH_3)ZrF_6$ contain centrosymmetric $[Zr_2F_{12}]^{4-}$ ions in which two pentagonal bipyramidal ZrF_7 groups share a common equatorial edge (Zr—F terminal = 1.960–2.073; Zr—F bridging = 2.146, 2.154 Å),[557] while $(H_3NC_2H_4NH_3)_3(ZrF_7)_2 \cdot 2H_2O$ contains two crystallographically independent monocapped trigonal prismatic $[ZrF_7]^{3-}$ ions.[558]

The following guanidinium and aminoguanidinium fluorozirconates have been isolated from aqueous media and have been characterized by IR spectroscopy and X-ray diffraction: $(CN_3H_6)_2ZrF_6$, $(CN_3H_6)_3ZrF_7$, $(CN_4H_7)_2ZrF_6$, $(CN_4H_8)ZrF_6 \cdot 0.5H_2O$ and $(CN_4H_8)ZrF_6 \cdot H_2O$. The IR and X-ray studies indicate the presence of $[ZrF_6]^{2-}$ in $(CN_4H_7)_2ZrF_6$, $[Zr_2F_{12}]^{4-}$ in $(CN_3H_6)_2ZrF_6$ and $[Zr_4F_{24}]^{8-}$ ions in $(CN_4H_8)ZrF_6 \cdot 0.5H_2O$. Comparison of IR spectra of $(CN_3H_6)_3ZrF_7$ and $(CN_4H_8)ZrF_6 \cdot H_2O$ with spectra of compounds of known structure suggests that the former compound contains $[ZrF_7]^{3-}$ ions while the latter probably contains $(ZrF_6^{2-})_n$ chain anions like those in K_2ZrF_6. Thermal decomposition of the $(CN_4H_8)ZrF_6 \cdot nH_2O$ compounds involves loss of water at 100–120 °C and elimination of HF at 130–190 °C; the latter process yields the pentafluorozirconate $(CN_4H_7)ZrF_5$.[559]

(*iii*) *Divalent metal fluorometallates*

Several types of divalent transition metal fluorometallate hydrates have been obtained by slow, room-temperature evaporation of aqueous HF solutions of MO_2 or MF_4 (M = Zr or Hf) and the divalent metal fluoride $M''F_2$. A 1:1 molar ratio of $M''F_2$ and MO_2 gives well-formed crystals of $M''MF_6 \cdot 6H_2O$ (M″ = Fe, Co, Ni or Zn), $MnMF_6 \cdot 5H_2O$ or $CuMF_6 \cdot 4H_2O$, while a 2:1 ratio of $M''F_2$ and MF_4 in water, slightly acidified with HF, yields crystals of $M''_2MF_8 \cdot 12H_2O$ (M″ = Co, Ni, Cu or Zn). From HF solutions of CdF_2 and MO_2, crystals of $Cd_2MF_8 \cdot 6H_2O$ are obtained, regardless of the molar ratio of CdF_2 and MO_2.[560,561]

The $M''MF_6 \cdot 6H_2O$ compounds are isostructural with $FeSiF_6 \cdot 6H_2O$ (space group $R\bar{3}m$), which is known to contain octahedral $[Fe(H_2O)_6]^{2+}$ and $[SiF_6]^{2-}$ ions in a slightly distorted CsCl structure.[560,561] The structure of $CuZrF_6 \cdot 4H_2O$ consists of square planar $[Cu(H_2O)_4]^{2+}$ and octahedral $[ZrF_6]^{2-}$ ions (Zr—F = 1.982–2.007 Å), which are arranged in infinite

$\cdots[ZrF_6]^{2-}\cdots[Cu(H_2O)_4]^{2+}\cdots[ZrF_6]^{2-}\cdots$ chains. Thus the coordination polyhedron about copper is a tetragonally distorted octahedron, with two rather long (2.246 Å) *trans* Cu···F interactions.[562] The structure of $MnZrF_6{\cdot}5H_2O$ contains infinite chains of ZrF_8 dodecahedra, which are threaded on the twofold (*c*) axis of the monoclinic crystal (Figure 9). Each ZrF_8 dodecahedron shares the two *a* edges with neighboring ZrF_8 dodecahedra and shares two B vertices with two $MnF_2(H_2O)_4$ octahedra. An additional uncoordinated water molecule is present in the lattice. Averaged bond lengths are Zr—F_A = 2.21 and Zr—F_B = 2.03 Å.[563,564]

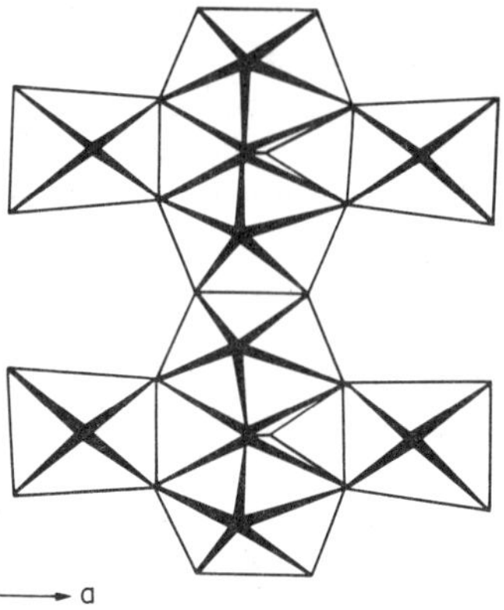

Figure 9 Projection of part of the structure of $MnZrF_6{\cdot}5H_2O$ on the *ac* plane showing ZrF_8 dodecahedra and $MnF_2(H_2O)_4$ octahedra

The isostructural octafluorozirconates $M''_2ZrF_8{\cdot}6H_2O$ (M″ = Mn or Cd) have a layer structure in which ZrF_8 square antiprisms and $M''F_4(H_2O)_3$ pentagonal bipyramids are linked together by sharing F···F polyhedral edges.[565,566] On the other hand, $Cu_2ZrF_8{\cdot}12H_2O$ contains isolated square antiprismatic $[ZrF_8]^{4-}$ anions and $[Cu(H_2O)_6]^{2+}$ cations.[567]

The heptafluorozirconate $Cu_3(ZrF_7)_2{\cdot}16H_2O$, isolated from 40% hydrofluoric acid, contains $[Cu(H_2O)_6]^{2+}$, $[Cu_2(H_2O)_{10}]^{4+}$, and $[Zr_2F_{14}]^{6-}$ ions, which are held together by a three-dimensional network of hydrogen bonds. The $[Zr_2F_{14}]^{6-}$ ion is a centrosymmetric dimer in which two ZrF_8 square antiprisms are joined together by sharing an *s* polyhedral edge (Zr—F terminal = 2.048–2.107; Zr—F bridging = 2.171, 2.185 Å).[568,569] The structure of $K_2Cu(ZrF_6)_2{\cdot}6H_2O$ consists of K^+, $[Cu(H_2O)_6]^{2+}$, and centrosymmetric $[Zr_2F_{12}]^{4-}$ ions. In $[Zr_2F_{12}]^{4-}$, two ZrF_7 pentagonal bipyramids share an equatorial edge (Zr—F terminal = 1.967–2.064; Zr—F bridging = 2.156, 2.161 Å).[570]

Anhydrous alkaline earth and transition metal hexafluorometallates $M''MF_6$ are generally prepared from $M''F_2$ and MF_4 by crystallization of a melt or by prolonged annealing of the metal fluorides. Most of these compounds exist in two polymorphic forms, depending on the temperature. The low temperature (α) form of $BaZrF_6$ contains centrosymmetric $[Zr_2F_{12}]^{4-}$ anions in which two ZrF_7 monocapped trigonal prisms share the F···F edge that joins the capping site to an adjacent vertex (Zr—F terminal = 1.999–2.079; Zr—F bridging = 2.155, 2.254 Å).[571] The structure of the high-temperature (β) form of $BaZrF_6$ contains infinite chains of *a*-edge-shared ZrF_8 dodecahedra, similar to the chains in $MnZrF_6{\cdot}5H_2O$ (Figure 9); bond lengths in β-$BaZrF_6$ are Zr—F_A = 2.242 and Zr—F_B = 2.010 Å.[572] α-$SrZrF_6$, α-$EuZrF_6$, and $PbZrF_6$ are isostructural with β-$BaZrF_6$; a single-crystal study of the lead compound gave Zr—F_A = 2.19 and Zr—F_B = 2.00 Å.[573] In studies of phase equilibria for the SrF_2–MF_4 (M = Zr or Hf) systems, the compounds α- and β-Sr_2MF_8 and Sr_3MF_{10} have been identified in addition to α- and β-$SrMF_6$.[574]

$M''MF_6$ (M″ = Mn, Fe, Co, Ni or Cu) exist in several crystalline forms, the most common being a high-temperature, cubic ordered ReO_3 structure and a low-temperature, trigonal $LiSbF_6$ structure. Both modifications consist of octahedra which share corners in three dimensions and which are centered alternately by M″ or M.[575–577] The isostructural lanthanide fluorometallates $LnMF_7$[578] also have an extended three-dimensional structure; an X-ray study of $SmZrF_7$[579] has shown that the structure is built on corner sharing of ZrF_6 octahedra and irregular SmF_8 polyhedra.

(iv) Vibrational spectra of fluorometallates

The vibrational spectra of alkali fluorometallates have been examined in considerable detail. In the hexafluorometallates Li_2ZrF_6 and M'_2MF_6 (M′ = Rb or Cs; M = Zr or Hf) the octahedral $[MF_6]^{2-}$ anion occupies a site of D_{3d} symmetry,[531,532] and consequently the triply degenerate normal modes [$\nu_3\,(t_{1u})$, $\nu_4\,(t_{1u})$, $\nu_5\,(t_{2g})$ and $\nu_6\,(t_{2u})$ in O_h] are split into *a* and *e* components. Vibrational frequencies (Table 19) have been assigned on the basis of polarized Raman[580,581] and polarized IR reflectance[581] spectra of single crystals, and assignments have been confirmed by normal coordinate analyses.

Table 19 Vibrational Frequencies for Solid Alkali Hexafluorometallates

Compound	ν_1 a_{1g}	ν_2 e_g	ν_3 e_u	ν_3 a_{2u}	ν_4 e_u	ν_4 a_{2u}	ν_5 a_{1g}	ν_5 e_g	ν_6 e_u	*Ref.*
Li_2ZrF_6	585	580 (?)	560	498	246	187	303	251		1
Rb_2ZrF_6	589		537	522	240	192	258	244		2
Cs_2ZrF_6	579	585 (?)	535	502	229	190	245	233	142 (?)	1
Cs_2HfF_6	572		498	490	217	187	257	247		2

1. L. M. Toth and J. B. Bates, *Spectrochim. Acta, Part A*, 1974, **30**, 1095.
2. I. W. Forrest and A. P. Lane, *Inorg. Chem.*, 1976, **15**, 265.

Far IR spectra of solid K_2MF_6, $(NH_4)_3ZrF_7$, $(NH_4)_2ZrF_6$, $(NH_4)ZrF_5$ and $(NH_4)ZrF_5{\cdot}H_2O$ exhibit broad band systems in the region 550–150 cm^{-1}. These spectra are more complex and less well defined than spectra of $[MF_6]^{2-}$. The broad bands have been attributed to metal–fluorine vibrations of complex anions of low site symmetry.[582,583] The close similarity between the spectra of $(NH_4)ZrF_5$ and $(NH_4)ZrF_5{\cdot}H_2O$ in the Zr—F stretching region strongly suggests that the water molecule is not attached to the Zr atom.[583]

Attempts have been made to correlate the frequency of the Raman-active symmetric stretching mode of crystalline fluorozirconates of known structure with the coordination number of the Zr atom.[584,585] The data in Table 20 show that, for the alkali metal salts, ν_s(Zr—F) decreases with increasing coordination number, although the interpretation of this trend is complicated by fluorine-bridging in the seven- and eight-coordinate compounds. No correlation exists when all of the data are considered; values of ν_s(Zr—F) for the isostructural eight-coordinate β-$BaZrF_6$, $PbZrF_6$ and α-$SrZrF_6$ vary considerably among themselves and are more similar to ν_s(Zr—F) frequencies for the six- and seven coordinate fluorozirconates of alkali metals. Raman spectra of LiF—NaF—ZrF_4 melts have been examined, and composition-dependent systematic variations in Raman frequencies have been interpreted in terms of an equilibrium distribution of $[ZrF_8]^{4-}$, $[ZrF_7]^{3-}$, $[ZrF_6]^{2-}$ and possibly species in which the Zr atom has a coordination number less than six.[584]

Table 20 Coordination Numbers of the Zirconium Atom and ν_s (Zr—F) Raman Frequencies for Crystalline Fluorozirconates

Coordination number	*Compound*	ν_s *(Zr—F)* (cm^{-1})	*Ref.*
6	Li_2ZrF_6	585	1
	Rb_2ZrF_6	589	2
	Cs_2ZrF_6	579	2
7	Na_3ZrF_7	556	1
	α-$BaZrF_6$	560	3
8	K_2ZrF_6	525	1
	$Na_7Zr_6F_{31}$	548	1
	β-$BaZrF_6$	562	3
	$PbZrF_6$	567	3
	α-$SrZrF_6$	578	3

1. L. M. Toth, A. S. Quist and G. E. Boyd, *J. Phys. Chem.*, 1973, **77**, 1384.
2. I. W. Forrest and A. P. Lane, *Inorg. Chem.*, 1976, **15**, 265.
3. Y. Kawamoto and F. Sakaguchi, *Bull. Chem. Soc. Jpn.*, 1983, **56**, 2138.

An IR spectral criterion for identification of dinuclear anions having pentagonal bipyramidal coordination about zirconium has been proposed by Davidovich *et al.*[554] These workers have noted that the $[Zr_2F_{12}]^{4-}$ anion in $[H_3NC_2H_4NH_3]_2[Zr_2F_{12}]$ and $[CN_3H_6]_4[Zr_2F_{12}]$ and the $[Zr_2F_{10}(H_2O)_2]^{2-}$ anion in $[NMe_4]_2[Zr_2F_{10}(H_2O)_2]$ exhibit an intense ν(Zr—F bridging) band in the 300–400 cm^{-1} region and a characteristic triplet of ν(Zr—F terminal) bands in the 400–600 cm^{-1} region; the central band of this triplet, located at 470–490 cm^{-1}, is more intense than the other two.

32.4.7.3 Chlorometallates, bromometallates and iodometallates

In contrast to the great variety of fluorometallates, nearly all of the known chloro-, bromo- and iodo-metallates are salts that contain the simple, octahedral $[MX_6]^{2-}$ anion. The best characterized alkali metal salts are listed in Table 21. The hexachloro complexes M'_2MCl_6 may be prepared (i) by fusing a stoichiometric mixture of M′Cl and MCl_4 at elevated temperatures,[586,587] (ii) by heating M′Cl in the presence of MCl_4 vapor at ~1 atm pressure[588,589] or (iii) by reaction of stoichiometric amounts of M′Cl and MCl_4 or $MOCl_2 \cdot 8H_2O$ in concentrated hydrochloric acid saturated with hydrogen chloride gas.[41,590,591] The recently reported M'_2MX_6 (X = Br or I) complexes were obtained by fusing or sintering a stoichiometric mixture of M′X and MX_4. Analogous ammonium salts $(NH_4)_2MCl_6$ have been isolated from hydrochloric acid solutions of NH_4Cl and $MOCl_2 \cdot 8H_2O$.[590,591]

Table 21 Physical Properties of Alkali Hexachloro, Hexabromo and Hexaiodo Complexes of Zirconium(IV) and Hafnium(IV)

Compound	*M.p.* (°C)	ΔH_F[a] (kJ mol^{-1})	T_{dec}[b] (°C)	ΔH_c[c] (kJ mol^{-1})	ΔH_f^0 (298 K) (kJ mol^{-1})	a[d] (Å)
Li_2ZrCl_6	535[1]	39 ± 7[1]	501[2]			
Li_2HfCl_6	557[1]	37 ± 5[1]	513[2]			
Na_2ZrCl_6	646[3]	17 ± 5[1]	634[2]	−32 ± 2[4]	−1835 ± 4[4]	
Na_2HfCl_6	660[5]			−46 ± 3[4]	−1857 ± 12[4]	
K_2ZrCl_6	799[3]	23 ± 8[1]	831[2]	−79 ± 2[4]	−1932 ± 4[4]	10.082(3)[3]
K_2HfCl_6	808[6]			−96 ± 3[4]	−1957 ± 12[4]	10.036(3)[7]
Rb_2ZrCl_6	768[8]		904[2]			10.178(4)[9]
Rb_2HfCl_6	795[10]					
Cs_2ZrCl_6	809[2]	73 ± 2[11]	1040[2]	−145 ± 2[4]	−1992 ± 4[4]	10.407(5)[9]
Cs_2HfCl_6	826[2]		953[2]			10.395(4)[7]
K_2ZrBr_6				−59 ± 5[12]	−1603 ± 6[12]	10.62(1)[12]
K_2HfBr_6						10.58(1)[12]
Cs_2ZrBr_6				−116 ± 2[12]	−1665 ± 3[12]	10.91(1)[12]
Cs_2HfBr_6				−167 ± 4[12]	−1793[12]	10.91(1)[12]
K_2ZrI_6	674[13]					
Rb_2ZrI_6	737[14]					
Cs_2ZrI_6	788[14]					11.613(3)[15]
Cs_2HfI_6						11.609(3)[15]

[a] Enthalpy of fusion. [b] Decomposition temperature at which the equilibrium pressure of MCl_4(g) reaches 1 atm. [c] Enthalpy of the complexing reaction: $2M'X(s) + MX_4(s) \rightarrow M'_2MX_6(s)$. [d] Cubic lattice parameter.

1. J. E. Dutrizac and S. N. Flengas, *Can. J. Chem.*, 1967, **45,** 2313.
2. D. A. Asvestas, P. Pint and S. N. Flengas, *Can. J. Chem.*, 1977, **55,** 1154.
3. R. L. Lister and S. N. Flengas, *Can. J. Chem.*, 1964, **42,** 1102.
4. P. Gelbman and A. D. Westland, *J. Chem. Soc., Dalton Trans.*, 1975, 1598.
5. I. S. Morozov and S. In'-Chzhu, *Russ. J. Inorg. Chem.* (*Engl. Transl.*), 1959, **4,** 307.
6. G. J. Kipouros and S. N. Flengas, *Can. J. Chem.*, 1978, **56,** 1549.
7. G. J. Kipouros and S. N. Flengas, *Can. J. Chem.*, 1983, **61,** 2183.
8. G. M. Toptygina and I. B. Barskaya, *Russ. J. Inorg. Chem.* (*Engl. Transl.*), 1965, **10,** 1226.
9. G. Engel, *Z. Kristallogr., Kristallgeom., Kristallphys., Kristallchem.*, 1935, **90,** 341.
10. I. B. Barskaya and G. M. Toptygina, *Russ. J. Inorg. Chem.* (*Engl. Transl.*), 1967, **12,** 14.
11. A. S. Kucharski and S. N. Flengas, *Can. J. Chem.*, 1974, **52,** 946.
12. R. Makhija and A. D. Westland, *J. Chem. Soc., Dalton Trans.*, 1977, 1707.
13. V. V. Chibrikin, A. V. Nevzorov, Z. B. Mukhmetshina, V. P. Seleznev and G. A. Yagodin, *Russ. J. Inorg. Chem.* (*Engl. Transl.*), 1981, **26,** 1216.
14. V. V. Chibrikin, Yu. V. Shabaev, Z. B. Mukhametshina, V. P. Seleznev and G. A. Yagodin, *Russ. J. Inorg. Chem.* (*Engl. Transl.*), 1981, **26,** 1376.
15. D. Sinram, C. Brendel and B. Krebs, *Inorg. Chim. Acta*, 1982, **64,** L131.

The M'_2MX_6 compounds are moisture-sensitive, high-melting, white (X = Cl or Br) or yellow (X = I) crystalline solids that decompose at elevated temperatures to solid M'X and gaseous MX_4. The thermal stability of the M'_2MCl_6 compounds, as measured by the decomposition temperature T_{dec} (Table 21), increases with increasing size of the alkali cation. Values of ΔH_c, the enthalpy of the complexing reaction in equation (51), indicate that $[HfCl_6]^{2-}$ is slightly more stable than $[ZrCl_6]^{2-}$, but $[HfBr_6]^{2-}$ is appreciably more stable than $[ZrBr_6]^{2-}$, thus illustrating the enhanced class b character of Hf^{IV}. The thermodynamic data in Table 21 and data for related Nb^{IV} and Sn^{IV} complexes suggest that halogen $p\pi \rightarrow$ metal $d\pi$ bonding is important in $[ZrX_6]^{2-}$ and $[HfX_6]^{2-}$ complexes.[587,592]

$$2M'X(s) + MX_4(s) \longrightarrow M'_2MX_6(s) \qquad (51)$$

X-Ray powder patterns indicate that M'_2MCl_6 (M' = K, Rb, Cs, or NH_4),[591] M'_2MBr_6 (M' = K or Cs),[592] and Cs_2MI_6[593] have the cubic K_2PtCl_6 structure (space group *Fm*3*m*), which contains regular octahedral anions. The powder data afford Zr—Cl bond lengths of 2.44(10) Å for Rb_2ZrCl_6 and 2.45(10) Å for Cs_2ZrCl_6.[594] An averaged Hf—Cl bond length of 2.446 Å in the octahedral $[HfCl_6]^{2-}$ ion has been obtained from a single-crystal study of $Bi^+[Bi_9^{5+}][HfCl_6^{2-}]_3$.[595] The Hf—I distance in Cs_2HfI_6 is 2.829(2) Å.[593] Interestingly, the Na_2MCl_6 complexes are not cubic; X-ray powder data for Na_2HfCl_6 have been indexed assuming a tetragonal cell with $a = 15.99$ and $c = 13.21$ Å.[596]

Quite a number of hexachlorozirconates and some hexachlorohafnates, hexabromozirconates and hexabromohafnates of nitrogen bases have been reported. These compounds are of the type A_2MX_6 (A = R_4N, R_3NH, R_2NH_2, RNH_3 or C_5H_5NH; R = alkyl or aryl). They may be prepared by (i) reaction of MX_4 with a substituted ammonium halide in thionyl chloride,[597–599] acetyl chloride,[600] benzoyl chloride[601,602] or acetonitrile;[603] (ii) reaction of $[MX_4(EtCN)_2]$ with the ammonium halide in chloroform;[604] or (iii) reaction of MX_4 with an amine in ethanol saturated with hydrogen chloride gas.[44,46,605] $[NEt_4]_2[MBr_6]$ complexes have been obtained by electrochemical oxidation of zirconium or hafnium metal in the presence of bromine and $[NEt_4]Br$ in benzene; however, analogous oxidation in the presence of chlorine and $[NPr_4]Cl$ in thionyl chloride gave the pentachlorometallates $[NPr_4][MCl_5]$.[118] $[C_5H_5NH]_2[ZrCl_6]$ reacts with NaOEt in ethanol at reflux yielding the substitution product $[C_5H_5NH]_2[ZrCl_5(OEt)]$.[605] Aminolysis reactions of alkylammonium hexachlorozirconates were mentioned in Section 32.4.2.1.

Vibrational spectra of $Cs_2(MCl_6)$,[599] $[NEt_4]_2[MX_6]$,[599,606] and $[Et_2NH_2]_2[MX_6]$[603] (X = Cl or Br) have been reported. Raman (ν_1, ν_2 and ν_5) and IR (ν_3 and ν_4) bands have been assigned and force constants determined. Frequencies for the NEt_4^+ salts, taken from the most complete study to date,[599] are listed in Table 22. Far IR spectra of M'_2ZrI_6 (M' = Li, Na, K, Rb or Cs) display a strong band in the region 195–175 cm^{-1}; the analogous M'_2HfI_6 complexes exhibit a similar band at 165–145 cm^{-1}.[607] Electronic spectra of $[Et_2NH_2]_2[ZrX_6]$ (X = Cl or Br) have been recorded, and UV bands have been assigned to halogen $p\pi \rightarrow$ metal *d* charge-transfer transitions.[604]

Table 22 Vibrational Spectral Data for Solid $[NEt_4]_2[MX_6]$ (X = Cl or Br) Complexes[1]

Compound	ν_1 (a_{1g})	ν_2 (e_g)	ν_3 (t_{1g})	ν_4 (t_{1u})	ν_5 (t_{1g})
$[NEt_4]_2[ZrCl_6]$	321s	250w, sh	293s	152s	151s
$[NEt_4]_2[HfCl_6]$	326s	257w, sh	275s	145s	156s
$[NEt_4]_2[ZrBr_6]$	194s	144w, sh	223s	106m	99s
$[NEt_4]_2[HfBr_6]$	197s	142w, sh	189s	102m	101s

1. W. van Bronswyk, R. J. H. Clark and L. Maresca, *Inorg. Chem.*, 1969, **8**, 1395.

Other hexachlorometallates that have been characterized by vibrational spectroscopy include $[SbPh_4]_2[ZrCl_6]$,[608] $[NO]_2[MCl_6]$ (M = Zr or Hf),[609] and $[SCl_3]_2[ZrCl_6]$.[610] The $[NO]_2[MCl_6]$ complexes were prepared by treating a concentrated solution of the metal tetrachloride in thionyl chloride at 0 °C with NOCl gas; their IR and Raman spectra exhibit the bands expected for $[MCl_6]^{2-}$ and show a strong $\nu(NO^+)$ band at 2191–2199 cm^{-1}. $[SCl_3]_2[ZrCl_6]$ was obtained by reaction of sulfur, chlorine and $ZrCl_4$ in a sealed tube; Raman frequencies of $[ZrCl_6]^{2-}$ in this compound were identified at 324 (ν_1), 258 (ν_2) and 154 cm^{-1} (ν_5).

Raman spectra of $ZrCl_4$–PCl_5 mixtures reveal the presence of at least one chlorozirconate species in addition to $[ZrCl_6]^{2-}$. The new Raman frequencies have been attributed to trigonal bipyramidal $[ZrCl_5]^-$, which is formed as a result of the equilibrium in equation (52).[611] IR evidence for the presence of solvated $[ZrCl_5]^-$ in acetonitrile solutions of $ZrCl_4$ and $[NEt_4]Cl$ has been presented by Olver and Bessette,[27] and $[NEt_4][ZrCl_5(MeCN)]$ has been isolated from acetonitrile by Feltz.[612] This compound loses acetonitrile *in vacuo* at 90 °C yielding $[NEt_4][ZrCl_5]$.

$$[ZrCl_6]^{2-} + [PCl_4]^+ \rightleftharpoons [ZrCl_5]^- + PCl_5 \quad (52)$$

The reaction of $ZrCl_4$ with *t*-butyl chloride and phosphorus(III) chloride in carbon disulfide gives $[Me_3CPCl_3][Zr_2Cl_9]$ as a moisture-sensitive, insoluble solid. IR and Raman spectra of the anion are consistent with a D_{3h} structure in which two $ZrCl_6$ octahedra share a common face; ν(Zr—Cl terminal) 350–387, ν(Zr—Cl bridging) 240–315 cm^{-1}.[613]

32.4.8 Hydroborates and Hydrides as Ligands

Zirconium(IV) and hafnium(IV) tetrakis(tetrahydroborates) $M(BH_4)_4$ are of interest as extremely volatile, covalent complexes that contain tridentate BH_4^- ligands and exhibit rapid intramolecular exchange of bridging and terminal hydrogen atoms.[614,615] These compounds were prepared initially from $NaMF_5$ (M = Zr or Hf) and excess $Al(BH_4)_3$ (equation 53),[616] but they are obtained more conveniently from the reaction of the anhydrous metal tetrachloride with excess lithium tetrahydroborate (equation 54), either in the solid state[617,618] or in the presence of a small amount of diethyl ether.[619]

$$NaMF_5 + 2Al(BH_4)_3 \longrightarrow M(BH_4)_4 + 2AlF_2(BH_4) + NaF \quad (53)$$

$$MCl_4 + 4LiBH_4 \longrightarrow M(BH_4)_4 + 4LiCl \quad (54)$$

$Zr(BH_4)_4$ and $Hf(BH_4)_4$ have very similar properties. Both are low-melting (m.p. 29 °C), colorless crystalline solids that exhibit a vapor pressure of 15 mmHg at 25 °C. They are highly reactive, spontaneously inflaming in dry air, detonating on contact with water, and slowly decomposing with liberation of hydrogen gas on standing at 25 °C. The solid complexes are very soluble in nonpolar solvents.

The chemical reactions of $M(BH_4)_4$ complexes have received very little attention. Both compounds behave as Lewis acids in reactions with tetraalkylammonium and lithium tetrahydroborates, yielding $[(C_8H_{17})_3NPr][M(BH_4)_5]$, $[NBu_4][M(BH_4)_5]$ and $LiM(BH_4)_5$.[620] $Zr(BH_4)_4$ reacts with $LiAlH_4$ in ether to give $Zr(AlH_4)_4$, an unstable white solid, which decomposes within several hours at room temperature to a pyrophoric black solid.[617]

Tridentate coordination of the BH_4^- ligand in $M(BH_4)_4$ complexes was first established by a low-temperature (−160 °C) single-crystal X-ray study of $Zr(BH_4)_4$.[621] A very recent single-crystal neutron diffraction study (at −163 °C) of the isostructural $Hf(BH_4)_4$ analogue[622] has afforded accurate positions for the bridging hydrogen atoms. The regular tetrahedral $Hf(BH_4)_4$ molecule (Figure 10) occupies a site of crystallographic $\bar{4}3m(T_d)$ symmetry, and the BH_4^- ligands adopt the conformation in which the Hf—H bonds to one BH_4^- ligand are staggered with respect to the Hf—B bonds to the other three BH_4^- ligands. Bond lengths are Hf—B = 2.281(8), Hf—H_b = 2.130(9), B—H_b = 1.235(10) and B—H_t = 1.150(19) Å, where H_b and H_t denote the bridging and terminal hydrogen atoms, respectively. An electron diffraction study[623] has afforded the following bond lengths for the gaseous $Zr(BH_4)_4$ molecule: Zr—B = 2.308 ± 0.010, Zr—H_b = 2.21 ± 0.04, B—H_b = 1.27 ± 0.05 and B—H_t = 1.18 ± 0.12 Å. The electron diffraction data suggest that the gaseous molecule may have only T symmetry, with the tridentate BH_4^- ligands being rotated out of the exactly staggered conformation (Figure 10) by an averaged torsion angle of 22°.

IR and Raman spectra of $Zr(BH_4)_4$ and $Hf(BH_4)_4$ have been assigned in terms of both T_d and T models,[624–627] and assignments have been supported by deuterium isotopic substitution studies[625–627] and normal coordinate analyses.[626–629] The vibrational spectra have not provided unequivocal evidence for a choice between these two closely related point groups. However, the spectra strongly support a tridentate attachment of the BH_4^- ligands.[615,630] Characteristic IR and Raman bands for $Zr(BH_4)_4$ and $Hf(BH_4)_4$ occur in the following frequency regions:

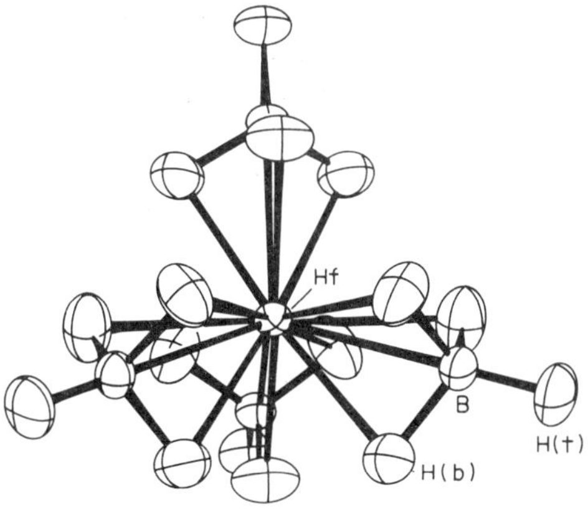

Figure 10 The structure of $Hf(BH_4)_4$ (reproduced by permission from ref. 622)

$\nu(B—H_t)$ 2560–2580, $\nu(B—H_b)$ 2100–2200, bridge deformation 1200–1300 and $\nu(M—BH_4)$ 480–560 cm^{-1}. The Raman studies[626] and normal coordinate analyses[626,627] suggest that direct metal–boron bonding interactions may be important.

He(I) and He(II) photoelectron spectra of $M(BH_4)_4$ (M = Zr or Hf) have been reported. Assignments were made on the basis of qualitative molecular-orbital models[631,632] and an LCAO–HFS(Xα) calculation on $Zr(BH_4)_4$.[632]

^1H and ^{11}B NMR spectra of $M(BH_4)_4$ (M = Zr or Hf)[618,633–636] and ^{91}Zr NMR spectra of $Zr(BH_4)_4$[636] at 25 °C indicate the equivalence of all of the hydrogen atoms owing to rapid intramolecular exchange between the bridging and terminal positions. ^1H and ^{11}B spectra of the zirconium and hafnium compounds are closely similar. The ^1H spectra consist of a quartet of equally intense lines due to spin–spin splitting by boron-11 ($I = \frac{3}{2}$; $J(^{11}B—H) = 90$ Hz), with some additional weaker lines due to splitting by boron-10 ($I = 3$; $J(^{10}B—H) = 30$ Hz). The ^{11}B spectra exhibit a 1:4:6:4:1 quintet ($J(^{11}B—H) = 90$ Hz). The inner nine lines of the anticipated 17-line spectrum have been observed in the ^{11}B-decoupled ^{91}Zr spectrum of $Zr(BH_4)_4$; the averaged coupling constant $J(Zr—H)$ is 28 Hz.

On cooling solutions of $M(BH_4)_4$ to −80 °C, the ^1H resonance lines are markedly broadened. This effect was originally attributed to a slowing of the rate process that exchanges hydrogen atoms between the bridging and terminal sites.[634] However, it was later shown that the observed line-shape changes are due to quadrupolar effects and that the hydrogen-atom exchange rate is still rapid (probably $> 10^4\ s^{-1}$) at −80 °C.[635]

Variable temperature, solid-state ^1H NMR studies of $M(BH_4)_4$[637] have provided evidence for two intramolecular motional processes. The higher temperature process has been assigned to exchange of bridging and terminal hydrogen atoms; for M = Zr, $E_a = 21.8 \pm 1.3$ and $\Delta G^{\ddagger}$ (−73 °C) = 30.5 kJ mol^{-1}; for M = Hf, $E_a = 35.1 \pm 1.3$ and $\Delta G^{\ddagger}$ (−59 °C) = 33.9 kJ mol^{-1}. The lower temperature process has been interpreted in terms of rotation of the BH_4^- ligands about the threefold $(M—B—H_t)$ axis; for M = Zr, $E_a = 22.6 \pm 0.8$ and $\Delta G^{\ddagger}$ (−149 °C) = 18.0 kJ mol^{-1}; for M = Hf, $E_a = 19.2 \pm 0.8$ and $\Delta G^{\ddagger}$ (−140 °C) = 19.2 kJ mol^{-1}.

The methyltrihydroborato complex $Zr(BH_3Me)_4$ has been prepared by reaction of $ZrCl_4$ and $LiBH_3Me$ in chlorobenzene. It has a tetrahedral structure, like $Zr(BH_4)_4$, with tridentate attachment of the BH_3Me^- ligands (Zr—B = 2.335(3), average $Zr—H_b = 2.06$, average $B—H_b = 1.14$ Å).[638]

The mixed-ligand complexes $[MCl(BH_4)\{N(SiMe_3)_2\}_2]$ (M = Zr or Hf) and $[Hf(BH_4)_3\{N(SiMe_2CH_2PMe_2)_2\}]$ were mentioned in Sections 32.4.2.4.ii and 32.4.3.2 respectively. IR and ^1H NMR studies of the $[MCl(BH_4)\{N(SiMe_3)_2\}_2]$ complexes indicate that the BH_4^- ligand is tridentate and exchange of bridging and terminal hydrogen atoms is fast on the NMR time scale at −80 °C.[81] On the other hand, $[Hf(BH_4)_3\{N(SiMe_2CH_2PMe_2)_2\}]$ contains bidentate BH_4^- ligands as evidenced by two $\nu(B—H_t)$ (2470 and 2420 cm^{-1}) and two $\nu(B—H_b)$ (2110 and 1950 cm^{-1}) IR bands. ^1H and ^{11}B NMR spectra of this complex indicate rapid exchange of inequivalent BH_4^- ligands as well as rapid exchange of bridging and terminal hydrogen atoms.[140]

Treatment of $[Hf(BH_4)_3\{N(SiMe_2CH_2PMe_2)_2\}]$ with Lewis bases (NEt_3 or PMe_3) yields the dinuclear trihydride $[\{Hf[N(SiMe_2CH_2PMe_2)_2]\}_2(H)_3(BH_4)_3]$. This complex is fluxional; ^1H and selectively decoupled ^{31}P NMR spectra show the presence of three equivalent hydrides coupled to four equivalent phosphorus nuclei.[140]

32.4.9 Mixed Donor Atom Ligands

32.4.9.1 *Open polydentate ligands*

Zirconium(IV) forms a rather large number of complexes with ONO- and ONS-tridentate Schiff base ligands, ONNO- and SNNS-tetradentate Schiff base ligands, and related species such as diacyl hydrazines. Relatively few hafnium analogues have been reported thus far.

(*i*) *ONO- and ONS-Tridentate Schiff base ligands*

$Zr(OCHMe_2)_4{\cdot}HOCHMe_2$ reacts in benzene at reflux with a variety of dibasic tridentate Schiff bases H_2L in 1:1 and 1:2 mole ratios to give complexes of the type $Zr(OCHMe_2)_2(L)$ and $Zr(L)_2$, respectively. The Schiff bases that undergo these reactions include aldimines and ketamines derived from carbonyl compounds and hydroxyalkyl amines (**35**)[639–641] or mercaptoalkyl amines (**36**),[642] azines (**37**),[643] semicarbazones (**38**),[644] thiosemicarbazones (**39**),[645] and Schiff bases derived from *S*-alkyldithiocarbazates (**40**);[646] one example of each type of ligand is shown in (**35**)–(**40**). The carbonyl compounds from which these Schiff base ligands are derived are commonly salicylaldehyde, 2-hydroxy-1-naphthaldehyde, acetylacetone or 2-hydroxyacetophenone. With but a few exceptions, the $Zr(L)_2$ complexes are monomeric in solution, while the $Zr(OCHMe_2)_2(L)$ analogues are dimeric. IR spectra indicate that $(L)^{2-}$ behaves as an ONO- or ONS-tridentate ligand, and six-coordinate structures have been suggested, for example (**41**) and (**42**). The $Zr(OCHMe_2)_2(L)$ complexes undergo alkoxide exchange reactions with alcohols such as *t*-butanol, 2-methylpentane-2,4-diol ($C_6H_{14}O_2$), or benzene-1,2-diol ($C_6H_6O_2$) yielding the $[Zr(OCMe_3)_2(L)]$, $[Zr(C_6H_{12}O_2)(L)]$ and $[Zr(C_6H_4O_2)(L)]$ derivatives, respectively; these complexes are monomeric in solution and presumably five-coordinate.

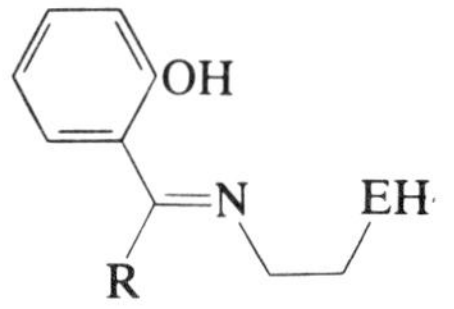

(**35**) H_2L; R = H or Me, E = O
(**36**) H_2L; R = H or Me, E = S

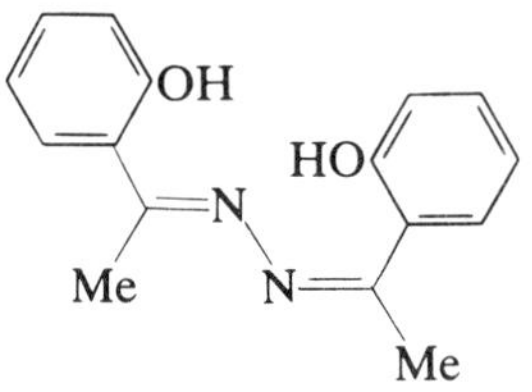

(**37**) H_2L

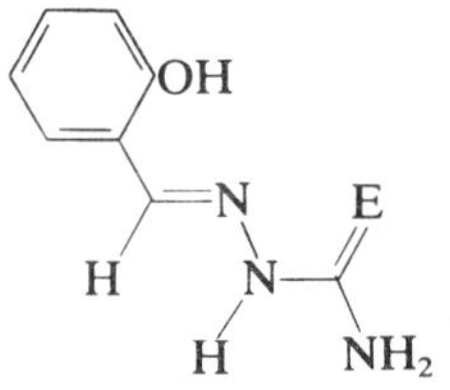

(**38**) H_2L; E = O
(**39**) H_2L; E = S

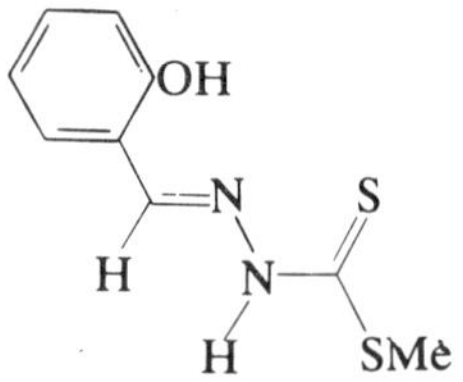

(**40**) H_2L

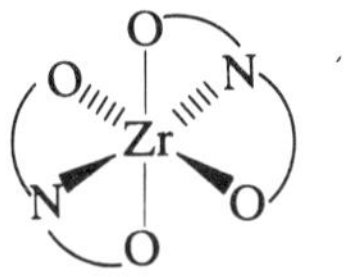

(**41**)

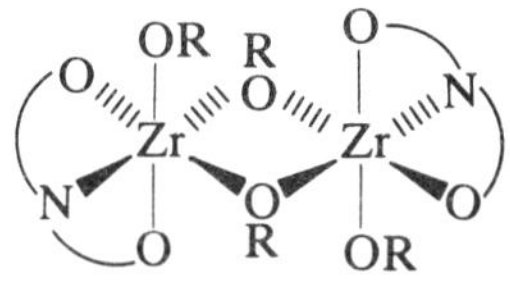

(**42**) R = $CHMe_2$

Mixed-ligand complexes [Zr(L)(L′)] that contain dinegative anions of two different ONO-tridentate Schiff base ligands have been prepared in benzene at reflux by reaction of a 1:1:1 mole ratio of $Zr(OCHMe_2)_4{\cdot}HOCHMe_2$, H_2L and H_2L'.[647] Several $[Zr(L)_2]$, $[\{Zr(OCHMe_2)_2(L)\}_2]$, $[Zr(C_6H_{12}O_2)(L)]$ and $[Zr(C_6H_4O_2)(L)]$ complexes that contain the dinegative anions of ONS-tridentate Schiff bases have been synthesized from benzothiazolines.[648]

The reaction of $Hf(OCHMe_2)_4{\cdot}HOCHMe_2$ with stoichiometric amounts of the *S*-methyldithiocarbazate (**40**) or the dibasic benzoyl hydrazones (**43**) yields analogous hafnium complexes $Hf(L)_2$ and $Hf(OCHMe_2)_2(L)$.[649,650] Only the disubstitution products $HfCl_2(L)$ could be obtained from the reaction of $HfCl_4$ with (**40**) or (**43**). The $HfCl_2(L)$ complexes are moisture-sensitive, insoluble, yellow solids and are probably polymeric.[651,652] On the other hand, a monomeric trigonal bipyramidal structure with *trans* chlorine atoms has been suggested for thiosemicarbazone complexes of the type $ZrCl_2(L)$; these compounds exhibit a single ν(Zr—Cl) IR band at ~290 cm^{-1}.[653] Unfortunately, both the $HfCl_2(L)$ and $ZrCl_2(L)$ complexes are too insoluble for molecular weight measurements.

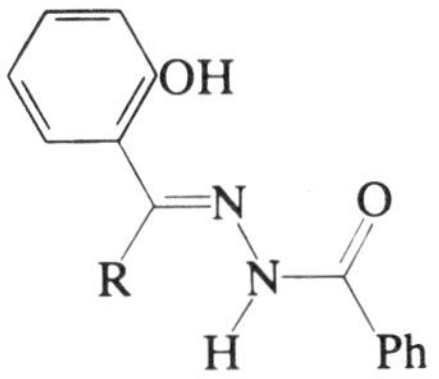

(**43**) H_2L; R = H or Me

Zirconyl chloride (or acetate) reacts in methanol at reflux with tridentate Schiff bases prepared by condensation of substituted salicylaldehydes with ethanolamine,[654] *o*-hydroxybenzylamine,[655] *o*-aminobenzyl alcohol[656] or salicylhydrazide[657] to give 1:2 metal:ligand complexes of the type $[ZrO(HL)_2]$. IR spectra indicate that the Schiff bases behave as monobasic ONO- tridentate ligands; for example, (**35**) coordinates to zirconium through the phenolate oxygen, azomethine nitrogen and hydroxylic oxygen atoms. The complexes are monomeric nonelectrolytes in solution, and they exhibit a medium intensity IR band in the region 880–945 cm^{-1}, which has been attributed to ν(Zr=O). If these compounds do contain a Zr=O group, the Zr atom would be seven coordinate.

Dibasic tridentate Schiff bases derived from salicylaldehydes and 2-aminobenzoic acid[658] or 1-amino-2-mercaptobenzene[659] react with aqueous zirconium nitrate to give monomeric complexes of the type $[Zr(OH)_2(L)(H_2O)]$. IR spectra of these compounds support an ONO- or ONS-tridentate attachment of the $(L)^{2-}$ ligands.

(*ii*) *ONNO- and SNNS-Tetradentate Schiff base ligands*

The eight-coordinate complex $[Zr(dsp)_2]$, where $(dsp)^{2-}$ is the dianion of the tetradentate Schiff base (**44**), has been prepared by condensation of tetrakis(salicylaldehydato)zirconium(IV) $[Zr(sal)_4]$ with *o*-phenylenediamine, and its structure has been determined by X-ray diffraction.[266] This complex has a dodecahedral coordination polyhedron with nitrogen atoms in the A sites and oxygen atoms in the B sites. The tetradentate ligands span the *mam* polyhedral edges, and the trapezoidal planes are very nearly perpendicular (dihedral angle = 89.2°). The Zr—N bonds (mean 2.43 Å) are appreciably longer than the Zr—O bonds (mean 2.10 Å). The corresponding $[Zr(dspeb)_2]$ complex, where $(dspeb)^{2-}$ is the dianion of (**45**), is of interest as a possible starting material for synthesis of coordination polymers.[660] A glossy red polymer of approximate molecular weight 20 000–40 000 has been prepared by condensation of $[Zr(sal)_4]$ and 1,2,4,5-tetraaminobenzene.[661]

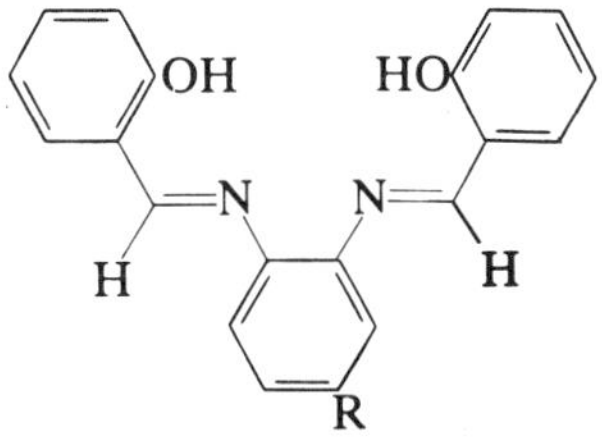

(**44**) H_2dsp; R = H
(**45**) H_2dspeb; R = CO_2Et

The reaction of $Zr(OCHMe_2)_4{\cdot}HOCHMe_2$ with dibasic ONNO-tetradentate Schiff base ligands derived from diaminoalkanes and salicylaldehyde, 2-hydroxy-1-naphthaldehyde, acetylacetone or 2-hydroxyacetophenone yields dimeric $[\{Zr(OCHMe_2)_2(L)\}_2]$ and monomeric $[Zr(L)_2]$ complexes,[662] analogous to the ONO-tridentate Schiff base complexes discussed in the previous section. Monomeric $[Zr(OCHMe_2)_2(L)]$ and $[Zr(L)_2]$ complexes, where $(L)^{2-}$ is the dianion of the SNNS-tetradentate Schiff bases (**46**) have been prepared from $Zr(OCHMe_2)_4{\cdot}HOCHMe_2$ and the appropriate benzothiazoline.[648] A few compounds of composition $ZrCl_2(L)$[663,664] and $ZrO(L)(H_2O)_3$[665] have been synthesized in methanol by reaction of ONNO-tetradentate Schiff bases and zirconyl chloride.

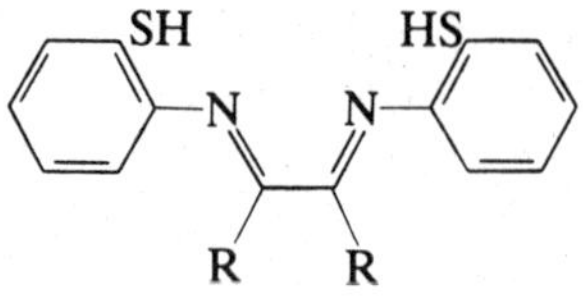

(**46**) H_2L; R = H or Me

(iii) Diacyl hydrazines

With diacyl hydrazines, zirconium tetrafluoride and tetrachloride form 2:1, 1:1 and 1:2 adducts of the type $(ZrF_4)_2$(RCONHNHCOR') (R = 4-C_5H_4N, R' = Me or Ph),[666] ZrF_4(RCONHNHCOR') (R = R' = H or Me; R = Me, R' = Ph),[667] and $ZrCl_4$-$(MeCONHNHCOPh)_2$.[668] In addition, zirconium tetrachloride undergoes substitution reactions yielding $ZrCl_2$(RCONNCOR') (R = R' = H, Me or Ph; R = Me, R' = Ph; R = 4-C_5H_4N, R' = Me or Ph).[666,668] All of these compounds may be polymeric since they have high decomposition temperatures and are insoluble in common organic solvents. IR spectra suggest that the diacyl hydrazines function as ONNO-tetradentate ligands.

32.4.9.2 *Aminopolycarboxylates*

Zirconium(IV) and hafnium(IV) form numerous stable chelate complexes with the anions of aminopolycarboxylic acids. Among the compounds that have been isolated in the solid state are $K_2[Zr(nta)_2]$,[669] $[M(edta)(H_2O)_2]\cdot 2H_2O$[670] and $H[Zr(dtpa)]\cdot 3H_2O$[671] (M = Zr or Hf; nta = nitrilotriacetate; edta = ethylenediaminetetraacetate; dtpa = diethylenetriaminepentaacetate). These compounds are prepared by concentration of hot, acidic aqueous solutions that contain stoichiometric amounts of an M^{IV} salt and the chelating ligand. Solid complexes of the type $M'Hf(OH)(edta)\cdot 2H_2O$ and $M'Hf(OH)(cdta)\cdot 2H_2O$ (M' = Na or NH_4; cdta = 1,2-cyclohexanediaminetetraacetate) are also known.[672]

Eight-coordinate dodecahedral structures have been established for $[Zr(nta)_2]^{2-}$[673] and $[Zr(edta)(H_2O)_2]$[674] by single-crystal X-ray studies. $[Zr(nta)_2]^{2-}$ exhibits a ligand wrapping pattern (Figure 11) in which the two nitrogen atoms occupy dodecahedral A sites, in accord with Orgel's rule; the three glycinate groups of each tetradentate nta ligand span *a, g* and *m* polyhedral edges. In comparison with cubic, square antiprismatic, and other dodecahedral stereoisomers, the observed stereoisomer is ideally suited for minimizing nonbonded repulsions and chelate-ring strain. Zr—O bonds to the dodecahedral A sites (2.251(7) Å) are significantly longer than Zr—O bonds to the B sites (2.124(9), 2.136(8) Å), and the Zr—N bonds are extraordinarily long (2.439(9) Å). In $[Zr(edta)(H_2O)_2]$ (Figure 11), the two nitrogen atoms and the water molecules occupy the dodecahedral A sites, while the four oxygen atoms of the hexadentate edta ligands take the B sites, again in accord with Orgel's rule (Zr—N = 2.43(2); Zr—O = 2.12(2), 2.14(2); Zr—OH_2 = 2.27(2) Å).

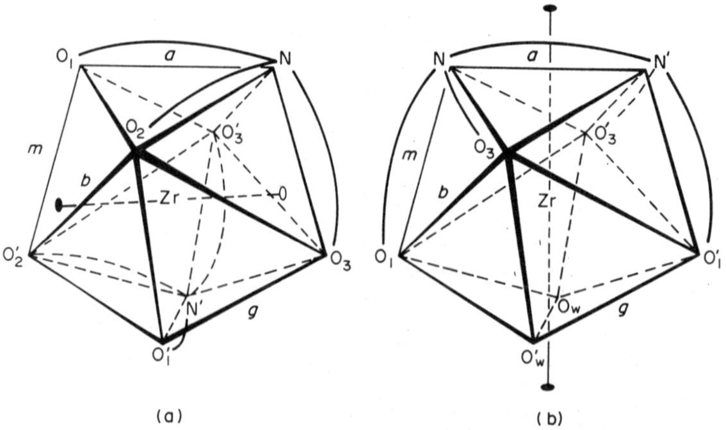

Figure 11 Ligand wrapping pattern in (a) $[Zr(nta)_2]^{2-}$ and (b) $[Zr(edta)(H_2O)_2]$. Primed and unprimed symbols are related by the indicated crystallographic twofold axes

The existence of a wide variety of zirconium(IV)-aminopolycarboxylate complexes in solution was established by the potentiometric titration studies of Intorre and Martell.[675–677] This work and related solution studies have been reviewed by Larsen.[5] Zirconium(IV) forms 1:1 and 1:2 complexes with tetradentate aminopolycarboxylates, 1:1 complexes with hexadentate and octadentate aminopolycarboxylates, and mixed ligand complexes with aminopolycarboxylates and various bidentate ligands, for example 1:1:1 Zr–edta–Tiron and 1:1:2 Zr—nta–Tiron complexes (Tiron = 1,2-dihydroxybenzene-3,5-disulfonate). The stoichiometries of these complexes are consistent with the tendency of zirconium to achieve a coordination number of eight.

^1H NMR spectra of the 1:1 Zr–edta and Hf–edta complexes in the pH range 1–3.5 exhibit an AB pattern for the inequivalent glycinate protons and a single resonance for the ethylene protons.[678,679] The spectra are not as complex as might have been expected on the basis of the low (C_2) symmetry of the dodecahedral $[Zr(edta)(H_2O)_2]$ molecule; evidently the inequivalent glycinate groups undergo rapid exchange, but the rate of Zr—N bond rupture is slow on the NMR time scale. Changes in the NMR spectra above pH 3.5 have been interpreted in terms of hydrolysis and polymerization.

Thermodynamic parameters for formation of M(edta) and $M(nta)^+$ (M = Zr or Hf) from M^{4+} and $(edta)^{4-}$ or $(nta)^{3-}$ have been determined at 15, 25 and 35 °C from calorimetric measurements.[680–683] Results for Zr(edta) at 25 °C are $\Delta G° = -187.2 \pm 0.3$ kJ mol^{-1}, $\Delta H° = -2.6 \pm 0.6$ kJ mol^{-1} and $\Delta S° = 618.8 \pm 2.5$ J mol^{-1} deg^{-1}. The large positive entropy change is associated with the effects of dehydration of the highly charged reacting ions.

32.4.9.3 NO-, NS- and OS-bidentate ligands

Zirconium(IV) and hafnium(IV) halides from a variety of 1:1 and 1:2 adducts with NO- and NS-bidentate ligands (Table 23). These compounds are generally prepared by mixing solutions or suspensions of the components in a polar organic solvent at room temperature. They are moisture-sensitive, white or yellow solids, insoluble in most organic solvents. Structures of the adducts have not been established; however the mode of coordination of the ligands is suggested by frequency shifts in characteristic IR bands upon complexation. Most of the ligands in Table 23 behave as NO- or NS-bidentate ligands, but 8-quinolinol and the *N*-arylsalicylaldimines appear to coordinate only through the nitrogen atom. IR spectra suggest that only one of the two benzaldehyde thiosemicarbazone ligands is attached to the metal in $MCl_4(PhCH{=}NNHCSNH_2)_2$ (M = Zr or Hf). Coordination of the hydroxylic oxygen atom of 3- and 4-aminophenol in the 1:4 adducts $MCl_4(H_2NC_6H_4OH)_4$ is uncertain.

Anhydrous zirconyl chloride forms insoluble, yellow 1:2 adducts, $ZrOCl_2(2\text{-}HOC_6H_4CH{=}NC_6H_4R)_2$ (R = H, 4-Me, 4-Cl or 4-OMe), with *N*-arylsalicylaldimines. Five-coordinate structures in which the salicylaldimines behave as N-bonded monodentate ligands have been suggested on the basis of IR spectra.[684]

The reaction of $ZrCl_4$ with two equivalents of ethanolamine or diethanolamine in ethyl acetate yields the substitution products $ZrCl_2(OC_2H_4NH_2)_2$ or $ZrCl_2(OC_2H_4NHC_2H_4OH)_2$, respectively, while the analogous reaction with triethanolamine gives the 1:4 adduct $ZrCl_4\{N(C_2H_4OH)_3\}_4$. IR spectra indicate that $(OC_2H_4NH_2)^-$ behaves as an NO-bidentate ligand.[685]

The following moisture-sensitive substitution products have been prepared by prolonged heating (in refluxing chloroform, THF or dichloromethane) of stoichiometric amounts of $ZrCl_4$ or $HfCl_4$ and benzoyl hydrazine, 4-pyridinecarboxylic acid hydrazide, benzoylhydrazones or Schiff bases derived from *S*-methyldithiocarbazate: $ZrCl_2(PhCONNH_2)_2$,[668] $ZrCl_3(4\text{-}NC_5H_4CONNH_2)$,[666] $HfCl_3(L)$ and $HfCl_2(L)_2$ (L = RR'C=NNC(O)Ph or RR'C=NNC(S)-SMe).[651,652] IR spectra indicate that these ligands act as monobasic NO- or NS-bidentate ligands, forming five-membered chelate rings upon complexation. The complexes are insoluble in most organic solvents, and they may be polymeric. The *N*-phenylsalicylaldiminato derivative $ZrCl_2(2\text{-}OC_6H_4CH{=}NPh)_2$ has been synthesized by condensation of aniline and dichlorobis(salicylaldehydato)zirconium(IV).[686]

$ZrCl_4$ and $HfCl_4$ react with 8-quinolinol (eight equivalents) in THF at reflux or in the absence of solvent at 140 °C to give the hydrolytically and thermally stable, isomorphous tetrakis chelates [M(8-O-quin)$_4$] (M = Zr or Hf). These compounds can be sublimed at ~350 °C/0.05 mmHg.[687] Precipitation of Zr^{IV} from aqueous solutions with 8-quinolinol often gives products of variable composition, but pure [Zr(8-O-quin)$_4$] can be obtained from hot

Table 23 Complexes of Zirconium(IV) and Hafnium(IV) Halides with NO- and NS-Bidentate Ligands

Ligand	*Complex*[a]	*Ligand donor atom(s)*	*Ref.*
Semicarbazide	$[MCl_4(H_2NNHCONH_2)]$	O and hydrazine pri-amino N	1
Thiosemicarbazide	$[MCl_4(H_2NNHCSNH_2)]$	S and hydrazine pri-amino N	2
Acetone thiosemicarbazone	$[MCl_4(Me_2C{=}NNHCSNH_2)]$	S and azomethine N	3
Benzaldehyde thiosemicarbazone	$[MCl_4(PhCH{=}NNHCSNH_2)]$	S and azomethine N	4
	$MCl_4(PhCH{=}NNHCSNH_2)_2$	S and azomethine N	4
Benzoyl hydrazine	$ZrF_4(PhCONHNH_2)_2$	O and pri-amino N	5
	$ZrCl_4(PhCONHNH_2)_2$	O and pri-amino N	6
4-Pyridinecarboxylic acid hydrazide	$ZrF_4(4\text{-}NC_5H_4CONHNH_2)$	O and pri-amino N	7
Aminopyrine	$[ZrCl_4(C_{13}H_{17}N_3O)]$	O and dimethylamino N	8
2-Pyrazine carboxamide	$[ZrCl_4(C_5H_7N_3O)]$	O and pyrazine N	9
8-Quinolinol	ZrF_4(8-HO-quin)	N	10
	MCl_4(8-HO-quin)$_2$	N	10
N-Arylsalicylaldimines	$MCl_4(2\text{-}HOC_6H_4CH{=}NC_6H_4R)_2$ (R = H, 4-Me, 4-NO_2)	Azomethine N	11
Aminophenols	$MCl_4(2\text{-}H_2NC_6H_4OH)_4$	O and N	12, 13
	$MCl_4(3\text{-}H_2NC_6H_4OH)_4$	N, O (?)	12, 13
	$MCl_4(4\text{-}H_2NC_6H_4OH)_4$	N, O (?)	12, 13

[a] M = Zr or Hf.

1. Ts. B. Konunova, A. Yu. Tsivadze, A. N. Smirnov and S. A. Kudritskaya, *Russ. J. Inorg. Chem. (Engl. Transl.)*, 1982, **27**, 807.
2. Ts. B. Konunova, A. V. Ablov, S. A. Kudritskaya and V. D. Brega, *Sov. J. Coord. Chem. (Engl. Transl.)*, 1976, **2**, 589.
3. Ts. B. Konunova, A. Yu. Tsivadze, A. N. Smirnov and S. A. Kudritskaya, *Russ. J. Inorg. Chem. (Engl. Transl.)*, 1983, **28**, 1282.
4. Ts. B. Konunova, A. V. Ablov, S. A. Kudritskaya and V. D. Brega, *Sov. J. Coord. Chem. (Engl. Transl.)*, 1979, **5**, 658.
5. R. C. Aggarwal, B. N. Yadav and T. Prasad, *Indian J. Chem.*, 1972, **10**, 671.
6. R. C. Aggarwal, B. N. Yadav and T. Prasad, *J. Inorg. Nucl. Chem.*, 1973, **35**, 653.
7. R. C. Aggarwal, T. Prasad and B. N. Yadav, *J. Inorg. Nucl. Chem.*, 1975, **37**, 899.
8. B. P. Hajela and S. C. Jain, *Indian J. Chem., Sect. A*, 1982, **21**, 530.
9. S. C. Jain, M. S. Gill and G. S. Rao, *J. Indian Chem. Soc.*, 1976, **53**, 537.
10. M. J. Frazer and B. Rimmer, *J. Chem. Soc. (A)*, 1968, 2273.
11. V. A. Kogan, V. P. Sokolov and O. A. Osipov, *Russ. J. Inorg. Chem. (Engl. Transl.)*, 1968, **13**, 1195.
12. Ts. B. Konunova, T. N. Popova, Z. P. Burnasheva and V. D. Brega, *Sov. J. Coord. Chem. (Engl. Transl.)*, 1978, **4**, 341.
13. Ts. B. Konunova, T. N. Popova, Z. P. Burnasheva and V. D. Brega, *Sov. J. Coord. Chem. (Engl. Transl.)*, 1977, **3**, 278.

solutions that contain an excess of oxalate.[688] The structure of [Zr(8-O-quin)$_4$] is dodecahedral with the nitrogen atoms in the A sites and oxygen atoms in the B sites, in accord with Orgel's rule (Zr—N = 2.405(8), Zr—O = 2.106(6) Å). The molecule adopts the rather uncommon D_2-*gggg* ligand wrapping pattern, presumably to avoid steric crowding of *o*-hydrogen atoms along the *a* edges of the more common D_{2d}-*mmmm* stereoisomer.[689]

Tetrakis(*N*-ethylsalicylaldiminato)zirconium(IV) $[Zr(2\text{-}OC_6H_4CH{=}NEt)_4]$ has a closely related dodecahedral structure with nitrogen atoms in the A sites and oxygen atoms in the B sites, but in this case the bidentate ligands span the *g* edges so as to give the very rare S_4-*gggg* stereoisomer. The Zr—N bond length (2.539(9) Å) is extraordinarily long, and the Zr—O bond length (2.055(7) Å) is unusually short. $[Zr(2\text{-}OC_6H_4CH{=}NEt)_4]$ and the *N*-isopropyl and *N*-*t*-butyl analogues were prepared by reaction of the appropriate salicylaldimine with $Zr(NMe_2)_4$.[690]

The *N*,*N*-diethylhydroxylamido(1−) complex $[Zr(ONEt_2)_4]$ has been prepared from $Zr(OCHMe_2)_4{\cdot}HOCHMe_2$ and excess Et_2NOH. An X-ray study of the analogous titanium(IV) compound shows that the complex exists as the expected dodecahedral D_{2d}-*mmmm* stereoisomer with nitrogen atoms in the A sites and oxygen atoms in the B sites. Coalescence of the NMR resonances of the diastereotopic methylene protons of $[Zr(ONEt_2)_4]$ at 42 °C indicates rapid cleavage of the Zr—N bonds ($\Delta G^{\ddagger} = 69 \pm 6$ kJ mol^{-1}).[691]

Numerous NO-bidentate Schiff base complexes and a few NS-bidentate Schiff base complexes of the type $M(OCHMe_2)_{4-n}(L)_n$ (M = Zr or Hf; n = 1, 2, 3 or 4) have been synthesized by reaction of stoichiometric amounts of $M(OCHMe_2)_4{\cdot}HOCHMe_2$ and the Schiff base (HL) in refluxing benzene. The known zirconium complexes contain ligands such as *N*-arylsalicylaldimines, *N*-arylnaphthylaldimines[692,693] and *N*-(2-pyridyl)salicylaldimine.[694,695] In some cases, only two or three of the isopropoxy groups could be replaced by the Schiff base ligand. The $Zr(OCHMe_2)(L)_3$ and $Zr(L)_4$ complexes are monomeric in solution, but association, presumably involving isopropoxy bridges, is sometimes observed for the $Zr(OCHMe_2)_3(L)$ and $Zr(OCHMe_2)_2(L)_2$ complexes. Benzoyl hydrazone and *S*-methyldithiocarbazate Schiff base complexes of hafnium, $Hf(OCHMe_2)_{4-n}\{RCH{=}$

NNC(O)Ph$\}_n$ and Hf$(OCHMe_2)_{4-n}$\{RCH=NNC(S)SMe$\}_n$ (R = Ph or C_4H_3O), have been reported by Verma *et al.*[649,650]

Oxozirconium(IV) *N*-arylsalicylaldiminato complexes of composition ZrOCl(L) and $ZrO(L)_2$ (L = 2-OC_6H_4CH=NAr) have been prepared by reaction of 1:1 and 1:2 mole ratios of $ZrOCl_2{\cdot}8H_2O$ and the Schiff base in MeOH–Et_2O. The ZrOCl(L) complexes can also be obtained from a 1:1:1 molar mixture of $ZrOCl_2{\cdot}8H_2O$, salicylaldehyde and aromatic amine.[696]

Treatment of a hot methanol solution of appropriate zirconyl salts with 1,10-phenanthroline mono-*N*-oxide (phenNO) yields the oxozirconium(IV) complexes [ZrO(phenNO)X_2] (X = Cl, Br, NO_3 or NCS), [ZrO$(phenNO)_2$]I_2 or [ZrO$(phenNO)_3$][ClO_4]$_2$.[697] These compounds are analogues of the 2,2′-bipyridine-1,1′-dioxide complexes discussed in Section 32.4.4.1.

Insertion of phenyl isocyanate or methyl isothiocyanate into the metal-nitrogen bonds of $M(NMe_2)_4$ gives the tetrakis (NO- or NS-bidentate) chelates [M\{NPhC(O)$NMe_2\}_4$] (M = Zr or Hf) or [Zr\{NMeC(S)$NMe_2\}_4$].[72] Analogous insertion into the Zr—C bonds of $Zr(CH_2Ph)_4$ yields [Zr\{NPhC(O)$CH_2Ph\}_4$] or [Zr\{NMeC(S)$CH_2Ph\}_4$],[698] while insertion of phenyl or naphthyl isocyanate into the Zr—O bonds of zirconium alkoxides gives Zr\{NR′C(O)OR$\}_n(OR)_{4-n}$ (R′ = Ph or $C_{10}H_7$; R = Pr, $CHMe_2$ or CMe_3; n = 1, 2, 3 or 4).[210]

A few complexes have been reported in which zirconium(IV) or hafnium(IV) is attached to an OS-donor bidentate ligand. Alkyl thioglycolates, $Zr(OCHMe_2)_2(L)$ and $Zr(L)_2$ (L = [SCH=C(OR)O]$^{2-}$; R = Me or Et), have been synthesized from stoichiometric amounts of $Zr(OCHMe_2)_4$ and the alkyl thioglycolate, H_2L. The isopropoxy groups of $Zr(OCHMe_2)_2(L)$ are replaced by *t*-butoxy groups when the isopropoxy complex is heated under reflux with an excess of Me_3COH in benzene. Molecular weight measurementts and IR spectra indicate that the $Zr(OR)_2(L)$ complexes have alkoxy-bridged dimeric structures.[699] Related $Zr(OCHMe_2)_2(L)$ and $Zr(L)_2$ (L = $[SCH_2CH_2O]^{2-}$, $[SCH_2CH_2CO_2]^{2-}$ or $[2\text{-}SC_6H_4CO_2]^{2-}$) complexes have been reported.[471,700] Reactions of stoichiometric amounts of zirconium isopropoxide and 2-mercaptopropionic acid yield $Zr(OCHMe_2)(L)(HL)$ and $Zr(L)_2$.[471]

Tetrakis(thioacetylacetonato)zirconium(IV) [Zr$(Sacac)_4$] has been prepared by reaction of stoichiometric amounts of $ZrCl_4$ and Na(Sacac) in dichloromethane. [Zr$(Sacac)_4$] has a square antiprismatic structure (**47**) in which the ligands span the *s* polyhedral edges in such a way as to cluster the sulfur atoms in all-*cis* positions. The observed C_2-*ssss* stereoisomer is distorted in the direction of the dodecahedral C_1-*mmgg* and bicapped trigonal prismatic C_1-$t_1t_1p_2p_2$ stereoisomers. Consistent with the former distortion, the averaged Zr—O and Zr—S bond lengths fall into two classes (Zr—O_A = 2.185, Zr—O_B = 2.132, Zr—S_A = 2.724, Zr—S_B = 2.665 Å).[701]

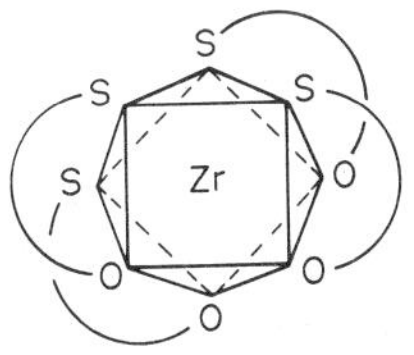

(**47**)

The sulfur atoms are also clustered in all-*cis* positions in the $t_1t_1p_2p_2$ bicapped trigonal prismatic structure of tetrakis(*N*-methyl-*p*-thiotolylhydroxamato)hafnium(IV) [Hf\{MeC_6H_4-C(S)N(O)Me$\}_4$]. Averaged metal–ligand bond distances are Hf—O = 2.150 and Hf—S = 2.678 Å.[702]

Monothiocarbamato complexes are discussed in Section 32.4.5.2.

32.4.10 Multidentate Macrocyclic Ligands

32.4.10.1 Porphyrins

Octaethylporphinato complexes of the type [M(OEP)$(O_2CMe)_2$] (M = Zr or Hf) have been prepared by reaction of H_2OEP and [M$(acac)_4$] in molten phenol at 210–240 °C followed by crystallization from pyridine–acetic acid–water mixtures. The complexes have been characterized by IR and mass spectra[703,704] and by electronic absorption and emission spectra.[705] They have an eight-coordinate square antiprismatic structure (**48**) in which the four porphinato nitrogen atoms occupy the coordination sites on one square face and the two bidentate acetate ligands span *s* edges on the opposite (rectangular) face. Bond lengths in the hafnium complex of C_{2v} symmetry are Hf—O = 2.278(3), Hf—N_1 = 2.266(5) and Hf—N_2 = 2.248(3) Å; the zirconium and hafnium complexes are nearly isodimensional.[706]

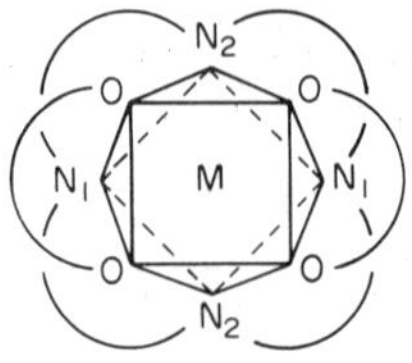

(48)

The red-violet mixed-ligand complex [Zr(OEP)(acac)(OPh)] has been isolated from the reaction of H_2OEP and [Zr(acac)$_4$] in molten phenol, and has been converted to bis(β-diketonato) derivatives [Zr(OEP)(dik)$_2$] (dik = acac or bzbz) by reaction with an excess of the β-diketone in boiling pyridine. The hafnium analogues [Hf(OEP)(dik)$_2$] (dik = acac or bzbz) were prepared from the β-diketone and [Hf(OEP)($O_2CMe)_2$]. ^{1}H NMR spectra of the [M(OEP)(bzbz)$_2$] complexes at 0 °C (M = Zr) or −40 °C (M = Hf) exhibit two triplets and two quartets for the porphinato ethyl groups, consistent with a C_{2v} structure like (**48**).[707]

The *meso*-tetraphenylporphyrin complex [Zr(TPP)Cl$_2$] has been prepared by reaction of H_2TPP with $ZrCl_4$ in boiling benzonitrile. The electronic spectrum of [Zr(TPP)Cl$_2$] and the facile kinetics of TPP dissociation in acidic media suggest that the zirconium atom lies considerably out of the plane of the TPP ligand with both chlorine atoms being located on the same side of the TPP plane.[708]

32.4.10.2 Phthalocyanines

The bluish chlorophthalocyanine complexes [M(PcCl)Cl$_2$] (M = Zr or Hf; PcCl = [$C_{32}H_{15}ClN_8]^{2-}$) have been synthesized by reaction of the metal tetrachlorides with *o*-phthalonitrile at 280 °C (equation 55).[709] Treatment of the [M(PcCl)Cl$_2$] complexes with boiling acetic acid gives the dichroic red-violet/dark blue acetato analogues [M(PcCl)($O_2CMe)_2$],[709] while treatment of the [M(PcCl)$_2$] complexes with ethanol and then warm water yields the blue hydroxo derivatives M(PcCl)(OH)$_2\cdot$2H$_2$O.[710] These complexes exhibit enormous thermal stability. [Zr(PcCl)($O_2CMe)_2$] can be sublimed in high vacuum at 500 °C.[709] [Hf(Pc)$_2$] (Pc = [$C_{32}H_{16}N_8]^{2-}$) sublimes at 540–560 °C and does not decompose until ~686 °C.[711]

$$ZrCl_4 + 4C_8H_4N_2 \longrightarrow [Zr(PcCl)Cl_2] + HCl \tag{55}$$

32.4.10.3 Other polyaza macrocycles

$ZrOCl_2\cdot 8H_2O$ reacts with 2,6-dipicolinoyl dihydrazine (H_2dpdh) yielding a 1:1 complex that has been formulated as ZrO(dpdh)($H_2O)_2$ or Zr(OH)$_2$(dpdh)(H_2O) (**49**). The free NH_2 groups of compound (**49**) condense with the carbonyl groups of β-diketones affording compounds (**50**) that are believed to contain 12-membered macrocyclic rings.[712]

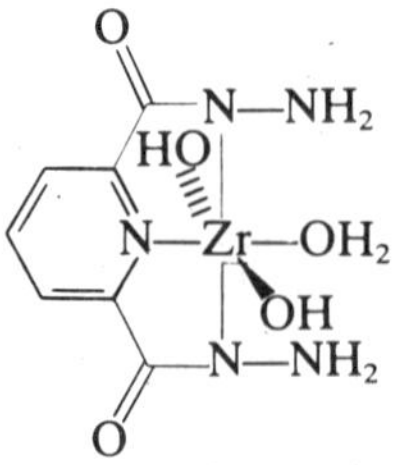

(49)

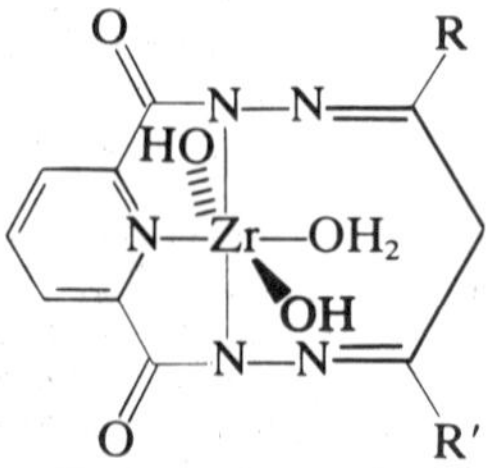

(50)

32.5 REFERENCES

1. J. L. Hoard and J. V. Silverton, *Inorg. Chem.*, 1963, **2,** 235.
2. M. A. Porai-Koshits and L. A. Aslanov, *J. Struct. Chem.* (*Engl. Transl.*), 1972, **13,** 244.
3. W. B. Blumenthal, 'The Chemical Behavior of Zirconium', Van Nostrand, Princeton, 1958.
4. P. Pascal, 'Nouveau Traité de Chimie Minérale', tome IX, Masson, Paris, 1963.
5. E. M. Larsen, *Adv. Inorg. Chem. Radiochem.*, 1970, **13,** 1.
6. D. C. Bradley and P. Thornton, in 'Comprehensive Inorganic Chemistry', ed. J. C. Bailar, Jr., H. J. Eméleus, R. S. Nyholm and A. F. Trotman-Dickenson, Pergamon, Oxford, 1973, vol. 3, pp. 419–490.
7. R. C. Fay, *Coord. Chem. Rev.*, 1981, **37,** 41.
8. R. C. Fay, *Coord. Chem. Rev.*, 1982, **45,** 41.
9. R. C. Fay, *Coord. Chem. Rev.*, 1983, **52,** 285.
10. L. G. Sillén and A. E. Martell, 'Stability Constants of Metal–Ion Complexes', Special Publication No. 17, The Chemical Society, London, 1964.
11. L. G. Sillén and A. E. Martell, 'Stability Constants of Metal–Ion Complexes', Supplement No. 1, Special Publication No. 25, The Chemical Society, London, 1971.
12. E. Hogfeldt, 'Stability Constants of Metal-Ion Complexes. Part A: Inorganic Ligands', IUPAC Chemical Data Series, No. 21, Pergamon, Oxford, 1982.
13. D. D. Perrin, 'Stability Constants of Metal–Ion Complexes. Part B: Organic Ligands', IUPAC Chemical Data Series, No. 22, Pergamon, Oxford, 1979.
14. D. J. Cardin, M. F. Lappert, C. L. Raston and P. I. Riley, in 'Comprehensive Organometallic Chemistry', ed. G. Wilkinson, F. G. A. Stone and E. W. Abel, Pergamon, Oxford, 1982, vol. 3, pp. 549–646.
15. D. Nicholls and T. A. Ryan, *Inorg. Chim. Acta,* 1977, **21,** L17.
16. J. Quirk and G. Wilkinson, *Polyhedron,* 1982, **1,** 209.
17. S. Herzog and H. Zühlke, *Z. Naturforsch., Teil B,* 1960, **15,** 466.
18. S. Herzog and H. Zühlke, *Z. Chem.*, 1966, **6,** 382.
19. D. A. Miller and R. D. Bereman, *Coord. Chem. Rev.*, 1972, **9,** 107.
20. D. L. Kepert and K. R. Trigwell, *Aust. J. Chem.*, 1975, **28,** 1359.
21. J. H. Wengrovius, R. R. Schrock and C. S. Day, *Inorg. Chem.*, 1981, **20,** 1844.
22. D. A. Wasmund, Ph.D. Thesis, Southern Illinois University, 1969, University Microfilms, Inc., Ann Arbor, Michigan; *Diss. Abstr. B,* 1970, **30,** 4549.
23. G. W. A. Fowles, B. J. Russ and G. R. Willey, *Chem. Commun.*, 1967, 646.
24. G. W. A. Fowles and G. R. Willey, *J. Chem. Soc.* (*A*), 1968, 1435.
25. E. M. Larsen and T. E. Henzler, *Inorg. Chem.*, 1974, **13,** 581.
26. J. W. Olver and J. W. Ross, Jr., *J. Inorg. Nucl. Chem.*, 1963, **25,** 1515.
27. J. W. Olver and R. R. Bessette, *J. Inorg. Nucl. Chem.*, 1968, **30,** 1791.
28. J. H. Wengrovius and R. R. Schrock, *J. Organomet. Chem.*, 1981, **205,** 319.
29. W. A. Baker, Jr. and A. R. Janus, *J. Inorg. Nucl. Chem.*, 1964, **26,** 2087.
30. C. K. Jorgansen, 'Absorption Spectra and Chemical Bonding in Complexes', Pergamon, Oxford, 1962, p. 284.
31. P. Ehrlich, *Angew. Chem.*, 1952, **64,** 617.
32. E. M. Larsen, J. W. Moyer, F. Gil-Arnao and M. J. Camp, *Inorg. Chem.*, 1974, **13,** 574.
33. R. A. Andersen, *Inorg. Nucl. Chem. Lett.*, 1980, **16,** 31.
34. H. M. J. C. Creemers, F. Verbeek and J. G. Noltes, *J. Organomet. Chem.*, 1968, **15,** 125.
35. G. W. A. Fowles, *Prog. Inorg. Chem.*, 1964, **6,** 1.
36. J. M. Matthews, *J. Am. Chem. Soc.*, 1898, **20,** 815.
37. J. M. Matthews, *J. Am. Chem. Soc.*, 1898, **20,** 839.
38. V. P. Orlovskii, N. V. Rudenko and B. N. Ivanov-Emin, *Russ. J. Inorg. Chem.* (*Engl. Transl.*), 1967, **12,** 1217.
39. G. W. A. Fowles and F. H. Pollard, *J. Chem. Soc.*, 1953, 4128.
40. J. E. Drake and G. W. A. Fowles, *J. Less-Common Met.*, 1960, **2,** 401.
41. J. E. Drake and G. W. A. Fowles, *J. Less-Common Met.*, 1961, **3,** 149.
42. A. Yajima, Y. Segawa, R. Matsuzaki and Y. Saeki, *Bull. Chem. Soc. Jpn.*, 1983, **56,** 2638.
43. J. E. Drake and G. W. A. Fowles, *J. Chem. Soc.*, 1960, 1498.
44. J. E. Drake and G. W. A. Fowles, *J. Inorg. Nucl. Chem.*, 1961, **18,** 136.
45. T. C. Ray and A. D. Westland, *Inorg. Chem.*, 1965, **4,** 1501.
46. R. J. H. Clark and W. Errington, *J. Chem. Soc.* (*A*), 1967, 258.
47. G. W. A. Fowles and R. A. Walton, *J. Less-Common Met.*, 1963, **5,** 510.
48. R. K. Agarwal, B. S. Tyagi, M. Srivastava and A. K. Srivastava, *Thermochim. Acta,* 1983, **61,** 241.
49. T. N. Srivastava, S. K. Tandon and N. Bhakru, *J. Inorg. Nucl. Chem.*, 1978, **40,** 1180.
50. A. K. Srivastava, M. Srivastava and R. K. Agarwal, *Croat. Chem. Acta,* 1982, **55,** 429.
51. A. S. Solovkin and S. V. Tsvetkova, *Russ. Chem. Rev.* (*Engl. Transl.*), 1962, **31,** 655.
52. A. Clearfield, *Rev. Pure Appl. Chem.*, 1964, **14,** 91.
53. J. Selbin, *Angew. Chem., Int. Ed. Engl.*, 1966, **5,** 712.
54. T. E. MacDermott, *Coord. Chem. Rev.*, 1973, **11,** 1.
55. P. Galvic, A. Bole and J. Slivnik, *J. Inorg. Nucl. Chem.*, 1975, **37,** 1316.
56. A. K. Srivastava, R. K. Agarwal, M. Srivastava, V. Kapur, S. Sharma and P. C. Jain, *Transition Met. Chem.*, 1982, **7,** 41.
57. M. S. Novakovskii and V. R. Timofeeva, *J. Gen. Chem. USSR* (*Engl. Transl.*), 1973, **43,** 1407.
58. V. R. Timofeeva and M. S. Novakovskii, *J. Gen. Chem. USSR* (*Engl. Transl.*), 1979, **49,** 1046.
59. M. S. Novakovskii and V. R. Timofeeva, *J. Gen. Chem. USSR* (*Engl. Transl.*), 1982, **52,** 2136.
60. A. M. Golub and T. P. Lishko, *Russ. J. Inorg. Chem.* (*Engl. Transl.*), 1974, **19,** 1292.
61. D. C. Bradley, *Adv. Inorg. Chem. Radiochem.*, 1972, **15,** 259.
62. M. F. Lappert, A. R. Sanger, R. C. Srivastava and P. P. Power, 'Metal and Metalloid Amides,' Ellis Horwood, Chichester, 1980, pp. 513–516.

63. D. C. Bradley and I. M. Thomas, *J. Chem. Soc.*, 1960, 3857.
64. G. Chandra and M. F. Lappert, *J. Chem. Soc. (A)*, 1968, 1940.
65. D. C. Bradley and M. H. Gitlitz, *J. Chem. Soc. (A)*, 1969, 980.
66. C. Airoldi and D. C. Bradley, *Inorg. Nucl. Chem. Lett.*, 1975, **11**, 155.
67. S. G. Gibbins, M. F. Lappert, J. B. Pedley and G. J. Sharp, *J. Chem. Soc., Dalton Trans.*, 1975, 72.
68. M. F. Lappert, D. S. Patil and J. B. Pedley, *J. Chem. Soc., Chem. Commun.*, 1975, 830.
69. D. C. Bradley and M. H. Gitlitz, *J. Chem. Soc. (A)*, 1969, 1152.
70. M. H. Chisholm and M. W. Extine, *J. Am. Chem. Soc.* 1977, **99**, 782.
71. D. C. Bradley and M. C. Ganorkar, *Chem. Ind. (London)*, 1968, 1521.
72. G. Chandra, A. D. Jenkins, M. F. Lappert and R. C. Srivastava, *J. Chem. Soc. (A)*, 1970, 2550.
73. D. C. Bradley, J. Charalambous and S. Jain, *Chem. Ind. (London)*, 1965, 1730.
74. R. K. Bartlett, *J. Inorg. Nucl. Chem.*, 1966, **28**, 2448.
75. W. A. Nugent and R. L. Harlow, *Inorg. Chem.*, 1979, **18**, 2030.
76. K. R. Nahar, A. K. Solanki and A. M. Bhandari, *Indian J. Chem., Sect. A*, 1980, **19**, 69.
77. K. Andrä and S. Büttner, *Z. Chem.*, 1981, **21**, 33.
78. R. V. Bynum, W. E. Hunter, R. D. Rogers and J. L. Atwood, *Inorg. Chem.*, 1980, **19**, 2368.
79. R. A. Andersen, *Inorg. Chem.*, 1979, **18**, 1724.
80. C. Airoldi, D. C. Bradley, H. Chudzynska, M. B. Hursthouse, K. M. Abdul Malik and P. R. Raithby, *J. Chem. Soc., Dalton Trans.*, 1980, 2010.
81. R. A. Andersen, *Inorg. Chem.*, 1979, **18**, 2928.
82. R. A. Andersen, *J. Organomet. Chem.*, 1980, **192**, 189.
83. D. J. Brauer, H. Bürger, E. Essig and W. Geschwandtner, *J. Organomet. Chem.*, 1980, **190**, 343.
84. H. Bürger, W. Geschwandtner and G. R. Liewald, *J. Organomet. Chem.*, 1983, **259**, 145.
85. M. Schlingmann and U. Wannagat, *Z. Anorg. Allg. Chem.*, 1976, **419**, 115.
86. H. Bürger, M. Schlingmann and G. Pawelke, *Z. Anorg. Allg. Chem.*, 1976, **419**, 121.
87. K. Wiegel and H. Bürger, *Z. Anorg. Allg. Chem.*, 1976, **426**, 301.
88. F. E. Brinckman and H. S. Haiss, *Chem. Ind. (London)*, 1963, 1124.
89. F. E. Brinckman, H. S. Haiss and R. A. Robb, *Inorg. Chem.*, 1965, **4**, 936.
90. W. M. Dyck, K. Dehnicke, G. Beyendorff-Gulba and J. Strähle, *Z. Anorg. Allg. Chem.*, 1981, **482**, 113.
91. M. S. Delgado and V. Fernandez, *Z. Anorg. Allg. Chem.*, 1979, **455**, 119.
92. E. J. Peterson, A. Galliart and T. M. Brown, *Inorg. Nucl. Chem. Lett.*, 1973, **9**, 241.
93. A. M. Golub, T. P. Lishko and V. P. Chistoplyasova, *Russ. J. Inorg. Chem. (Engl. Transl.)*, 1974, **19**, 823.
94. R. A. Bailey, T. W. Michelsen and A. A. Nobile, *J. Inorg. Nucl. Chem.*, 1970, **32**, 2427.
95. M. A. Shashkin and V. S. Kravtsov, *Russ. J. Inorg. Chem. (Engl. Transl.)*, 1977, **22**, 140.
96. M. A. Shashkin and V. S. Kravtsov, *Sov. J. Coord. Chem. (Engl. Transl.)*, 1977, **3**, 21.
97. I. V. Tananaev and I. A. Rozanov, *Russ. J. Inorg. Chem. (Engl. Transl.)*, 1962, **7**, 957.
98. I. V. Tananaev, I. A. Rozanov and A. G. Kolgushkina, *Russ. J. Inorg. Chem. (Engl. Transl.)*, 1963, **8**, 522.
99. Yu. Ya. Kharitonov and I. A. Rozanov, *Bull. Acad. Sci. USSR, Div. Chem. Sci.*, 1962, **3**, 373.
100. Yu. Ya. Kharitonov, I. A. Rozanov and I. V. Tananaev, *Bull. Acad. Sci. USSR, Div. Chem. Sci.*, 1963, **4**, 538.
101. A. M. Golub and T. P. Lishko, *Russ. J. Inorg. Chem. (Engl. Transl.)*, 1970, **15**, 783.
102. A. M. Golub, T. P. Lishko, G. V. Tsintsadze and V. V. Skopenko, *Russ. J. Inorg. Chem. (Engl. Transl.)*, 1972, **17**, 1100.
103. E. J. Peterson, R. B. Von Dreele and T. M. Brown, *Inorg. Chem.*, 1976, **15**, 309.
104. A. M. Golub and V. N. Sergun'kin, *Russ. J. Inorg. Chem. (Engl. Transl.)*, 1966, **11**, 419.
105. I. V. Vinarov, *Russ. Chem. Rev. (Engl. Transl.)*, 1967, **36**, 522.
106. A. Galliart and T. M. Brown, *J. Inorg. Nucl. Chem.*, 1972, **34**, 3568.
107. A. M. Golub, T. P. Lishko and V. V. Skopenko, *Sov. Prog. Chem. (Engl. Transl.)*, 1971, **37**, 83.
108. M. S. Delgado and V. Fernandez, *Z. Anorg. Allg. Chem.*, 1981, **474**, 226.
109. M. S. Delgado and V. Fernandez, *Z. Anorg. Allg. Chem.*, 1981, **476**, 149.
110. M. S. Delgado and V. Fernandez, *Inorg. Nucl. Chem. Lett.*, 1978, **14**, 505.
111. J. MacCordick and A. Westland, *Bull. Soc. Chim. Fr.*, 1975, 1117.
112. M. S. Delgado and V. Fernandez, *Z. Anorg. Allg. Chem.*, 1979, **455**, 112.
113. K.-H. Linke and F. Lemmer, *Z. Anorg. Allg. Chem.*, 1966, **345**, 211.
114. D. L. Reger and M. E. Tarquini, *Inorg. Chem.*, 1982, **21**, 840.
115. D. L. Reger and M. E. Tarquini, *Inorg. Chem.*, 1983, **22**, 1064.
116. D. L. Reger, M. E. Tarquini and L. Lebioda, *Organometallics*, 1983, **2**, 1763.
117. S. Asslani, R. Rahbarnoohi and B. L. Wilson, *Inorg. Nucl. Chem. Lett.*, 1979, **15**, 59.
118. J. J. Habeeb, F. F. Said and D. G. Tuck, *Can. J. Chem.*, 1977, **55**, 3882.
119. H. J. Emeleus and G. S. Rao, *J. Chem. Soc.*, 1958, 4245.
120. W. M. Graven and R. V. Peterson, *J. Inorg. Nucl. Chem.*, 1969, **31**, 1743.
121. G. S. Rao, *Z. Anorg. Allg. Chem.*, 1960, **304**, 351.
122. G. W. A. Fowles and R. A. Walton, *J. Chem. Soc.*, 1964, 2840.
123. D. Brown, G. W. A. Fowles and R. A. Walton, *Inorg. Synth.*, 1970, **12**, 225.
124. Y. Hase and O. L. Alves, *Spectrochim. Acta, Part A*, 1981, **37**, 711.
125. S. C. Jain and R. Rivest, *J. Inorg. Nucl. Chem.*, 1970, **32**, 1117.
126. S. C. Jain and R. Rivest, *Inorg. Chem.*, 1967, **6**, 467.
127. S. C. Jain and R. Rivest, *J. Inorg. Nucl. Chem.*, 1970, **32**, 1579.
128. S. C. Jain and R. Rivest, *Can. J. Chem.*, 1963, **41**, 2130.
129. G. W. A. Fowles and R. A. Walton, *J. Chem. Soc.*, 1964, 4330.
130. E. L. Muetterties, *J. Am. Chem. Soc.*, 1960, **82**, 1082.
131. R. C. Paul and S. L. Chadha, *Spectrochim. Acta, Part A*, 1967, **23**, 1243.
132. R. C. Paul, S. Sood and S. L. Chadha, *J. Inorg. Nucl. Chem.*, 1971, **33**, 2703.
133. R. J. H. Clark, J. Lewis and R. S. Nyholm, *J. Chem. Soc. (A)*, 1962, 2460.

134. R. J. H. Clark, W. Errington, J. Lewis and R. S. Nyholm, *J. Chem. Soc. (A)*, 1966, 989.
135. R. J. H. Clark, J. Lewis, R. S. Nyholm, P. Pauling and G. B. Robertson, *Nature (London)*, 1961, **192,** 222.
136. S. Datta, S. S. Wreford, R. P. Beatty and T. J. McNeese, *J. Am. Chem. Soc.*, 1979, **101,** 1053.
137. G. A. Barclay, I. K. Gregor, M. J. Lambert and S. B. Wild, *Aust. J. Chem.*, 1967, **20,** 1571.
138. M. S. Delgado and V. Fernandez, *Z. Anorg. Allg. Chem.*, 1980, **467,** 225.
139. M. D. Fryzuk, H. D. Williams and S. J. Rettig, *Inorg. Chem.*, 1983, **22,** 863.
140. M. D. Fryzuk and H. D. Williams, *Organometallics*, 1983, **2,** 162.
141. A. J. Zielen and R. E. Connick, *J. Am. Chem. Soc.*, 1956, **78,** 5785.
142. K. A. Kraus and J. S. Johnson, *J. Am. Chem. Soc.*, 1953, **75,** 5769.
143. J. S. Johnson, K. A. Kraus and R. W. Holmberg, *J. Am. Chem. Soc.*, 1956, **78,** 26.
144. J. S. Johnson and K. A. Kraus, *J. Am. Chem. Soc.*, 1956, **78,** 3937.
145. R. L. Angstadt and S. Y. Tyree, *J. Inorg. Nucl. Chem.*, 1962, **24,** 913.
146. B. Noren, *Acta Chem. Scand.*, 1973, **27,** 1369.
147. A. Clearfield and P. A. Vaughan, *Acta Crystallogr.*, 1956, **9,** 555.
148. T. C. W. Mak, *Can. J. Chem.*, 1968, **46,** 3491.
149. G. M. Muha and P. A. Vaughan, *J. Chem. Phys.*, 1960, **33,** 194.
150. M. Aberg, *Acta Chem. Scand., Ser. B*, 1977, **31,** 171.
151. A. Fratiello, G. A. Vidulich and F. Mako, *Inorg. Chem.*, 1973, **12,** 470.
152. W. P. Griffith and T. D. Wickins, *J. Chem. Soc. (A)*, 1967, 675.
153. F. G. Baglin and D. Breger, *Inorg. Nucl. Chem. Lett.*, 1976, **12,** 173.
154. K. A. Burkov, G. V. Kozhevnikova, L. S. Lilich and L. A. Myund, *Russ. J. Inorg. Chem. (Engl. Transl.)*, 1982, **27,** 804.
155. D. A. Powers and H. B. Gray, *Inorg. Chem.*, 1973, **12,** 2721.
156. D. H. Devia and A. G. Sykes, *Inorg. Chem.*, 1981, **20,** 910.
157. B. M. Gatehouse and D. J. Lloyd, *Chem. Commun.*, 1969, 727.
158. R. P. Planalp and R. A. Andersen, *J. Am. Chem. Soc.*, 1983, **105,** 7774.
159. G. Fachinetti, C. Floriani, A. Chiesi-Villa and C. Guastini, *J. Am. Chem. Soc.*, 1979, **101,** 1767.
160. C. G. Barraclough, J. Lewis and R. S. Nyholm, *J. Chem. Soc.*, 1959, 3552.
161. Yu. Ya. Kharitonov and L. M. Zaitsev, *Russ. J. Inorg. Chem. (Engl. Transl.)*, 1968, **13,** 476.
162. T. P. Spasibenko and Ya. G. Goroshchenko, *Russ. J. Inorg. Chem. (Engl. Transl.)*, 1969, **14,** 757.
163. K. Dehnicke and K.-U. Meyer, *Z. Anorg. Allg. Chem.*, 1964, **331,** 121.
164. A. I. Morozov and E. V. Karlova, *Russ. J. Inorg. Chem. (Engl. Transl.)*, 1971, **16,** 12.
165. A. Feltz, *Z. Anorg. Allg. Chem.*, 1965, **335,** 304.
166. C. J. Hardy, B. O. Field and D. Scargill, *J. Inorg. Nucl. Chem.*, 1966, **28,** 2408.
167. Yu. Ya. Kharitonov and Yu. A. Buslaev, *Bull. Acad. Sci. USSR, Div. Chem. Sci.*, 1962, 365.
168. I. P. Lipatova and E. I. Semenova, *Bull. Acad. Sci. USSR, Div. Chem. Sci.*, 1967, 3.
169. R. K. Agarwal and S. C. Rastogi, *Thermochim. Acta*, 1983, **63,** 113.
170. V. V. Savant and C. C. Patel, *J. Inorg. Nucl. Chem.*, 1969, **31,** 2319.
171. R. K. Agarwal and A. K. Srivastava, *Inorg. Nucl. Chem. Lett.*, 1980, **16,** 311.
172. A. K. Srivastava, V. Kapur, R. K. Agarwal and T. N. Srivastava, *J. Indian Chem. Soc.*, 1981, **58,** 279.
173. R. K. Agarwal, P. C. Jain, M. Srivastava, A. K. Srivastava and T. N. Srivastava, *J. Indian Chem. Soc.*, 1980, **57,** 374.
174. R. K. Agarwal, A. K. Srivastava and T. N. Srivastava, *Indian J. Chem., Sect. A*, 1979, **18,** 459.
175. A. K. Srivastava, R. K. Agarwal and T. N. Srivastava, *Croat. Chem. Acta*, 1981, **54,** 329.
176. R. K. Agarwal, V. Kapoor, A. K. Srivastava and T. N. Srivastava, *Natl. Acad. Sci. Lett. (India)*, 1979, **2,** 447.
177. S. K. Madan and A. M. Donohue, *J. Inorg. Nucl. Chem.*, 1966, **28,** 1303.
178. R. C. Paul, S. K. Gupta, R. D. Sharma and S. K. Vasisht, *Z. Naturforsch., Teil B*, 1977, **32,** 543.
179. C. Preti, G. Tosi and P. Zannini, *Transition Met. Chem.*, 1980, **5,** 200.
180. J. A. Connor and E. A. V. Ebsworth, *Adv. Inorg. Chem. Radiochem.*, 1964, **6,** 279.
181. W. P. Griffith and T. D. Wickins, *J. Chem. Soc. (A)*, 1968, 397.
182. G. V. Jere and G. D. Gupta, *J. Inorg. Nucl. Chem.*, 1970, **32,** 537.
183. G. D. Gupta and G. V. Jere, *Indian J. Chem.*, 1972, **10,** 102.
184. G. V. Jere and M. T. Santhamma, *Inorg. Chim. Acta*, 1977, **24,** 57.
185. S. O. Gerasimova, Yu. Ya. Kharitonov, S. A. Polishchuk and L. M. Avkuhutskii, *Sov. J. Coord. Chem. (Engl. Transl.)*, 1981, **7,** 267.
186. D. C. Bradley, *Prog. Inorg. Chem.*, 1960, **2,** 303.
187. D. C. Bradley, *Prep. Inorg. React.*, 1965, **2,** 169.
188. D. C. Bradley, R. C. Mehrotra and D. P. Gaur, 'Metal Alkoxides,' Academic, New York, 1978.
189. R. C. Mehrotra, *Inorg. Chim. Acta Rev.*, 1967, **1,** 99.
190. R. C. Mehrotra and A. Mehrotra, *Inorg. Chim. Acta Rev.*, 1971, **5,** 127.
191. R. C. Mehrotra, *Adv. Inorg. Chem. Radiochem.*, 1983, **26,** 269.
192. N. I. Kozlova and N. Ya. Turova, *Russ. J. Inorg. Chem. (Engl. Transl.)*, 1980, **25,** 1188.
193. D. C. Bradley and J. D. Swanwick, *J. Chem. Soc.*, 1959, 748.
194. D. C. Bradley and J. D. Swanwick, *J. Chem. Soc.*, 1959, 3773.
195. L. M. Brown and K. S. Mazdiyasni, *Anal. Chem.*, 1969, **41,** 1243.
196. D. C. Bradley and C. E. Holloway, *J. Chem. Soc. (A)*, 1968, 1316.
197. C. G. Barraclough, D. C. Bradley, J. Lewis and I. M. Thomas, *J. Chem. Soc.*, 1961, 2601.
198. C. T. Lynch, K. S. Mazdiyasni, J. S. Smith and W. J. Crawford, *Anal. Chem.*, 1964, **36,** 2332.
199. S. A. Imam, *Nature (London)*, 1965, **206,** 1146.
200. D. C. Bradley and D. G. Carter, *Can. J. Chem.*, 1961, **39,** 1434.
201. D. C. Bradley and D. G. Carter, *Can. J. Chem.*, 1962, **40,** 15.
202. B. Morosin, *Acta Crystallogr., Sect. B*, 1977, **33,** 303.
203. R. C. Paul, S. K. Gupta, V. Goyal and S. K. Vasisht, *Monatsh. Chem.*, 1977, **108,** 1019.

204. U. B. Saxena, A. K. Rai and R. C. Mehrotra, *Inorg. Chim. Acta,* 1973, **7,** 681.
205. P. C. Bharara, V. D. Gupta and R. C. Mehrotra, *Synth. React. Inorg. Metal-Org. Chem.*, 1977, **7,** 537.
206. D. C. Bradley, F. M. Abd-el Halim, R. C. Mehrotra and W. Wardlaw, *J. Chem. Soc.*, 1952, 4609.
207. M. S. Bains and D. C. Bradley, *Can. J. Chem.*, 1962, **40,** 1350.
208. M. S. Bains and D. C. Bradley, *Can. J. Chem.*, 1962, **40,** 2218.
209. S. K. Vasisht and S. K. Gupta, *Monatsh. Chem.*, 1979, **110,** 1197.
210. R. C. Mehrotra, V. D. Gupta and P. C. Bharara, *Indian J. Chem.*, 1975, **13,** 156.
211. N. Ya. Turova, S. I. Kucheiko and N. I. Kozlova, *Sov. J. Coord. Chem. (Engl. Transl.)*, 1983, **9,** 513.
212. N. Ya. Turova, N. I. Kozlova and A. V. Novoselova, *Russ. J. Inorg. Chem. (Engl. Transl.)*, 1980, **25,** 1788.
213. N. Ya. Turova and N. I. Kozlova, *Sov. J. Coord. Chem. (Engl. Transl.)*, 1982, **8,** 346.
214. G. M. Toptygina, I. B. Barskaya and I. Z. Babievskaya, *Russ. J. Inorg. Chem. (Engl. Transl.)*, 1972, **17,** 1102.
215. T. V. Lubben, P. T. Wolczanski and G. D. Van Duyne, *Organometallics,* 1984, **3,** 977.
216. K. C. Malhotra, G. Mehrotra and S. C. Chaudry, *Natl. Acad. Sci. Lett. (India)*, 1980, **3,** 21.
217. L. Chamberlain, J. C. Huffman, J. Keddington and I. P. Rothwell, *J. Chem. Soc., Chem. Commun.*, 1982, 805.
218. W. Biltz and J. A. Clinch, *Z. Anorg. Allg. Chem.*, 1904, **40,** 218.
219. G. von Hevesy and M. Logstrup, *Chem. Ber.*, 1926, **59,** 1890.
220. E. M. Brainina, V. A. Shreider and I. N. Rozhkov, *Izv. Akad. Nauk SSSR, Ser. Khim.*, 1980, 1920 (*Chem. Abstr.*, 1980, **93,** 196 861).
221. G. Jantsch, *J. Prakt. Chem.*, 1927, **115,** 7.
222. A. L. Allred and D. W. Thompson, *Inorg. Chem.*, 1968, **7,** 1196.
223. J. P. Fackler, Jr., *Prog. Inorg. Chem.*, 1966, **7,** 361.
224. R. C. Mehrotra, B. Bohra and D. P. Gaur, 'Metal β-Diketonates and Allied Derivatives', Academic, New York, 1978.
225. J. V. Silverton and J. L. Hoard, *Inorg. Chem.*, 1963, **2,** 243.
226. H. K. Chun, W. L. Steffen and R. C. Fay, *Inorg. Chem.*, 1979, **18,** 2458.
227. L. M. Lowe, W. V. Prestwich and H. Zmora, *Can. J. Phys.*, 1975, **53,** 1327.
228. R. C. Fay and J. K. Howie, *J. Am. Chem. Soc.*, 1979, **101,** 1115.
229. A. C. Adams and E. M. Larsen, *Inorg. Chem.*, 1966, **5,** 228.
230. T. J. Pinnavaia and R. C. Fay, *Inorg. Chem.*, 1966, **5,** 233.
231. T. J. Pinnavaia, M. T. Mocella, B. A. Averill and J. T. Woodard, *Inorg. Chem.*, 1973, **12,** 763.
232. M. Kh. Minacheva and E. M. Brainina, *Bull. Acad. Sci. USSR, Div. Chem. Sci.*, 1973, **22,** 672.
233. A. C. Adams and E. M. Larsen, *Inorg. Chem.*, 1966, **5,** 814.
234. S. Bhaduri, H. Khwaja and V. Khanwalkar, *J. Chem. Soc., Dalton Trans.*, 1982, 445.
235. E. M. Brainina and L. I. Strunkina, *Bull. Acad. Sci. USSR, Div. Chem. Sci.*, 1977, **26,** 1925.
236. E. M. Brainina, M. Kh. Minacheva, Z. S. Klimenkova and B. V. Lokshin, *Bull. Acad. Sci. USSR, Div. Chem. Sci.*, 1976, **25,** 380.
237. E. M. Brainina and M. Kh. Minacheva, *Bull. Acad. Sci. USSR, Div. Chem. Sci.*, 1975, 1482.
238. P. R. Singh and R. Sahai, *Aust. J. Chem.*, 1970, **23,** 269.
239. T. S. Davis and J. P. Fackler, Jr., *Inorg. Chem.*, 1966, **5,** 242.
240. J. J. Leary, S. Tsuge and T. L. Isenhour, *Anal. Chem.*, 1973, **45,** 1269.
241. S. Tsuge, J. J. Leary and T. L. Isenhour, *Anal. Chem.*, 1974, **46,** 106.
242. R. E. Sievers, B. W. Ponder, M. L. Morris and R. W. Moshier, *Inorg. Chem.*, 1963, **2,** 693.
243. M. Balog, M. Schieber, M. Michman and S. Patai, *Thin Solid Films,* 1977, **41,** 247.
244. M. Balog, M. Schieber, M. Michman and S. Patai, *Thin Solid Films,* 1977, **47,** 109.
245. R. C. Fay and T. J. Pinnavaia, *Inorg. Chem.*, 1968, **7,** 508.
246. N. Serpone and R. C. Fay, *Inorg. Chem.*, 1969, **8,** 2379.
247. T. J. Pinnavaia and R. C. Fay, *Inorg. Chem.*, 1968, **7,** 502.
248. M. Cox, J. Lewis and R. S. Nyholm, *J. Chem. Soc.*, 1964, 6113.
249. R. B. VonDreele, J. J. Stezowski and R. C. Fay, *J. Am. Chem. Soc.*, 1971, **93,** 2887.
250. G. T. Morgan and A. R. Bowen, *J. Chem. Soc.*, 1924, **125,** 1252.
251. E. M. Brainina and R. Kh. Freidlina, *Bull. Acad. Sci. USSR, Div. Chem. Sci.*, 1961, 1489.
252. L. Wolf and C. Tröltzsch, *Z. Prakt. Chem.*, 1962, **17,** 78.
253. M. L. Morris, R. W. Moshier and R. E. Sievers, *Inorg. Synth.*, 1967, **9,** 50.
254. G. A. Lock and D. W. Thompson, *J. Chem. Soc., Dalton Trans.*, 1980, 1265.
255. E. G. Muller, V. W. Day and R. C. Fay, *J. Am. Chem. Soc.*, 1976, **98,** 2165.
256. G. A. Razuvaev, L. I. Vyshinskaya and A. M. Rabinovich, *J. Gen. Chem. USSR (Engl. Transl.)*, 1980, **50,** 1349.
257. V. W. Day and R. C. Fay, *J. Am. Chem. Soc.*, 1975, **97,** 5136.
258. E. M. Brainina and R. Kh. Freidlina, *Bull. Acad. Sci. USSR, Div. Chem. Sci.*, 1964, 1331.
259. E. M. Brainina and E. I. Mortikova, *Bull. Acad. Sci. USSR, Div. Chem. Sci.*, 1967, 2418.
260. P. C. Bharara, V. D. Gupta and R. C. Mehrotra, *Indian J. Chem.*, 1975, **13,** 725.
261. U. B. Saxena, A. K. Rai, V. K. Mathur, R. C. Mehrotra and D. Radford, *J. Chem. Soc. (A)*, 1970, 904.
262. S. Saxena, Y. P. Singh and A. K. Rai, *Synth. React. Inorg. Metal-Org. Chem.*, 1983, **13,** 855.
263. M. Y. Mirza and F. Ik. Nwabue, *J. Inorg. Nucl. Chem.*, 1981, **43,** 1817.
264. E. C. Okafor, *Z. Naturforsch., Teil B,* 1981, **36,** 213.
265. R. Kh. Freidlina, E. M. Brainina and A. N. Nesmeyanov, *Bull. Acad. Sci. USSR, Div. Chem. Sci.*, 1957, 45.
266. R. D. Archer, R. O. Day and M. L. Illingsworth, *Inorg. Chem.*, 1979, **18,** 2908.
267. R. C. Mehrotra, V. D. Gupta and P. C. Bharara, *Indian J. Chem.*, 1973, **11,** 814.
268. D. M. Puri, *J. Indian Chem. Soc.*, 1970, **47,** 535.
269. U. B. Saxena, A. K. Rai and R. C. Mehrotra, *Indian J. Chem.*, 1971, **9,** 709.
270. D. T. Hobbs and C. M. Lukehart, *Inorg. Chem.*, 1979, **18,** 1297.
271. E. L. Muetterties and C. M. Wright, *J. Am. Chem. Soc.*, 1965, **87,** 4706.
272. B. Hutchinson, D. Eversdyk and S. Olbricht, *Spectrochim. Acta, Part A,* 1974, **30,** 1605.
273. A. R. Davis and F. W. B. Einstein, *Acta Crystallogr., Sect. B,* 1978, **34,** 2110.

274. D. Tranqui, A. Tissier, J. Laugier and P. Boyer, *Acta Crystallogr., Sect. B*, 1977, **33**, 392.
275. P. Boyer, A. Tissier, J. I. Vargas and P. Vulliet, *Chem. Phys. Lett.*, 1972, **14**, 601.
276. P. Boyer, A. Tissier and J. I. Vargas, *Inorg. Nucl. Chem. Lett.*, 1972, **8**, 813.
277. A. Baudry, P. Boyer and A. Tissier, *Mol. Phys.*, 1978, **36**, 1037.
278. E. L. Muetterties and C. W. Alegranti, *J. Am. Chem. Soc.*, 1969, **91**, 4420.
279. S. R. Sofen, S. R. Cooper and K. N. Raymond, *Inorg. Chem.*, 1979, **18**, 1611.
280. J. L. Martin and J. Takats, *Can. J. Chem.*, 1975, **53**, 572.
281. R. N. Kapoor and R. C. Mehrotra, *Z. Anorg. Allg. Chem.*, 1957, **293**, 100.
282. K. Andra, *Z. Anorg. Allg. Chem.*, 1968, **361**, 254.
283. F. G. R. Gimblett and B. H. Massey, *Inorg. Chem.*, 1970, **9**, 2038.
284. B. H. Massey and F. G. R. Gimblett, *Inorg. Chem.*, 1970, **9**, 2043.
285. S. V. Elinson and K. I. Petrov, 'Analytical Chemistry of Zirconium and Hafnium', ed. D. Slutzkin, Academy of Science of the USSR, (transl. N. Kaner, Israel Program for Scientific Translations, Jerusalem, 1965).
286. A. K. Mukherji, 'Analytical Chemistry of Zirconium and Hafnium', Pergamon, Oxford, 1970.
287. B. J. Thamer and A. F. Voigt, *J. Am. Chem. Soc.*, 1951, **73**, 3197.
288. H. E. Zittel and T. M. Florence, *Anal. Chem.*, 1967, **39**, 320.
289. T. M. Florence, Y. J. Farrar and H. E. Zittel, *Aust. J. Chem.*, 1969, **22**, 2321.
290. E. M. Larsen and S. T. Hirozawa, *J. Inorg. Nucl. Chem.*, 1956, **3**, 198.
291. L. E. Manzer, *Inorg. Synth.*, 1982, **21**, 135.
292. O. A. Osipov and Yu. B. Kletenik, *J. Gen. Chem. USSR (Engl. Transl.)*, 1961, **31**, 2285.
293. O. A. Osipov, V. I. Minkin and V. A. Kogan, *Russ. J. Phys. Chem. (Engl. Transl.)*, 1962, **36**, 468.
294. O. A. Osipov and Yu. B. Kletenik, *J. Gen. Chem. USSR (Engl. Transl.)*, 1957, **27**, 2953.
295. O. A. Osipov and Yu. B. Kletenik, *J. Gen. Chem. USSR (Engl. Transl.)*, 1959, **29**, 1351.
296. Yu. B. Kletenik and O. A. Osipov, *J. Gen. Chem. USSR (Engl. Transl.)*, 1959, **29**, 1398.
297. O. A. Osipov and Yu. B. Kletenik, *J. Gen. Chem. USSR (Engl. Transl.)*, 1959, **29**, 2085.
298. M. F. Lappert, *J. Chem. Soc.*, 1962, 542.
299. S. R. Wade and G. R. Willey, *J. Less-Common Met.*, 1979, **68**, 105.
300. R. C. Paul, H. R. Singal and S. L. Chadha, *Indian J. Chem.*, 1971, **9**, 995.
301. R. C. Paul and S. L. Chadha, *Indian J. Chem.*, 1970, **8**, 739.
302. S. C. Jain and B. P. Hajela, *Indian J. Chem.*, 1974, **12**, 843.
303. S. C. Jain and R. Rivest, *Can. J. Chem.*, 1964, **42**, 1079.
304. Yu. B. Kletenik, O. A. Osipov and E. E. Kravtsov, *J. Gen. Chem. USSR (Engl. Transl.)*, 1959, **29**, 13.
305. O. A. Osipov and Yu. B. Kletenik, *Russ. J. Inorg. Chem. (Engl. Transl.)*, 1960, **5**, 1076.
306. E. C. Alyea and E. G. Torrible, *Can. J. Chem.*, 1965, **43**, 3468.
307. K. C. Malhotra and S. M. Sehgal, *Indian J. Chem.*, 1975, **13**, 599.
308. O. A. Osipov, V. M. Artemova, V. A. Kogan and Yu. A. Lysenko, *J. Gen. Chem. USSR (Engl. Transl.)*, 1962, **32**, 1354.
309. B. O. Field and C. J. Hardy, *Proc. Chem. Soc.*, 1962, 76.
310. J. Weidlein, U. Müller and K. Dehnicke, *Spectrochim. Acta, Part A*, 1968, **24**, 253.
311. C. D. Garner and S. C. Wallwork, *J. Chem. Soc. (A)*, 1966, 1496.
312. C. D. Garner, D. Sutton and S. C. Wallwork, *J. Chem. Soc. (A)*, 1967, 1949.
313. C. C. Addison, N. Logan, S. C. Wallwork and C. D. Garner, *Chem. Soc. Rev.*, 1971, **25**, 289.
314. B. O. Field and C. J. Hardy, *J. Chem. Soc.*, 1964, 4428.
315. K. W. Bagnall, D. Brown and J. G. H. du Preez, *J. Chem. Soc., Suppl. 1*, 1964, 5523.
316. J. R. Ferraro, *J. Mol. Spectrosc.*, 1960, **4**, 99.
317. Yu. Ya. Kharitonov, L. I. Yuranova, V. E. Plyushchev and V. G. Pervykh, *Russ. J. Inorg. Chem. (Engl. Transl.)*, 1965, **10**, 399.
318. G. Münninghoff, E. Hellner, M. El Essawi and K. Dehnicke, *Z. Kristallogr., Kristallgeom., Kristallphys., Kristallchem.*, 1978, **147**, 231.
319. I. J. Bear and W. G. Mumme, *J. Solid State Chem.*, 1970, **1**, 497.
320. I. J. Bear and W. G. Mumme, *Rev. Pure Appl. Chem.*, 1971, **21**, 189.
321. I. J. Bear and W. G. Mumme, *Acta Crystallogr., Sect. B*, 1969, **25**, 1566.
322. I. J. Bear and W. G. Mumme, *Acta Crystallogr., Sect. B*, 1969, **25**, 1572.
323. I. J. Bear and W. G. Mumme, *Acta Crystallogr., Sect. B*, 1969, **25**, 1558.
324. J. Singer and D. T. Cromer, *Acta Crystallogr.*, 1959, **12**, 719.
325. I. J. Bear and W. G. Mumme, *Acta Crystallogr., Sect. B*, 1970, **26**, 1131.
326. I. J. Bear and W. G. Mumme, *Acta Crystallogr., Sect. B*, 1970, **26**, 1125.
327. I. J. Bear and W. G. Mumme, *Acta Crystallogr., Sect. B*, 1970, **26**, 1140.
328. D. L. Rogachev, A. S. Antsyshkina and M. A. Porai-Koshits, *J. Struct. Chem. (Engl. Transl.)*, 1969, **10**, 259.
329. I. J. Bear and W. G. Mumme, *J. Inorg. Nucl. Chem.*, 1970, **32**, 1159.
330. D. B. McWhan and G. Lundgren, *Acta Crystallogr., Sect. A*, 1963, **16**, 36.
331. D. B. McWhan and G. Lundgren, *Inorg. Chem.*, 1966, **5**, 284.
332. I. G. Atanov and L. M. Zaitsev, *Russ. J. Inorg. Chem. (Engl. Transl.)*, 1967, **12**, 188.
333. D. L. Motov and T. P. Spasibenko, *Russ. J. Inorg. Chem. (Engl. Transl.)*, 1969, **14**, 1321.
334. D. L. Motov and M. P. Ritter, *Russ. J. Inorg. Chem. (Engl. Transl.)*, 1969, **14**, 1325.
335. L. I. Fedoryako and I. A. Sheka, *Sov. Prog. Chem. (Engl. Transl.)*, 1971, **37**, 6.
336. Yu. P. Sozinova, D. L. Rogachev, D. L. Motov and S. A. Kobycheva, *Russ. J. Inorg. Chem. (Engl. Transl.)*, 1976, **21**, 375.
337. D. L. Motov and Yu. P. Sozinova, *Russ. J. Inorg. Chem. (Engl. Transl.)*, 1976, **21**, 1054.
338. Yu. P. Sozinova and D. L. Motov, *Russ. J. Inorg. Chem. (Engl. Transl.)*, 1978, **23**, 609.
339. Yu. P. Sozinova and D. L. Motov, *Russ. J. Inorg. Chem. (Engl. Transl.)*, 1979, **24**, 756.
340. D. L. Motov and Yu. P. Sozinova, *Russ. J. Inorg. Chem. (Engl. Transl.)*, 1982, **27**, 1804.
341. L. M. Zaitsev, L. A. Pospelova, I. G. Atanov and V. N. Kokunova, *Russ. J. Inorg. Chem.*, 1968, **13**, 689.

342. R. L. Davidovich, M. A. Medkov and S. B. Ivanov, *Bull. Acad. Sci. USSR, Div. Chem. Sci.*, 1974, **23**, 1152.
343. D. L. Rogachev, L. M. Dikareva, V. Ya. Kuznetsov and G. G. Sadikov, *J. Struct. Chem. (Engl. Transl.)*, 1981, **22**, 470.
344. D. L. Rogachev, L. M. Dikareva, V. Ya. Kuznetsov, V. V. Fomenko and M. A. Porai-Koshits, *J. Struct. Chem. (Engl. Transl.)*, 1982, **23**, 765.
345. D. L. Rogachev, V. Ya. Kuznetsov, L. M. Dikareva, G. G. Sadikov and M. A. Porai-Koshits, *J. Struct. Chem. (Engl. Transl.)*, 1980, **21**, 118.
346. W. G. Mumme, *Acta Crystallogr., Sect. B*, 1971, **27**, 1373.
347. M. A. Porai-Koshits, V. I. Sokol and V. N. Vorotnikova, *J. Struct. Chem. (Engl. Transl.)*, 1972, **13**, 815.
348. I. J. Bear and W. G. Mumme, *Acta Crystallogr., Sect. B*, 1971, **27**, 494.
349. D. L. Rogachev, L. M. Dikareva, V. P. Nikolaev and V. Ya. Kuznetsov, *J. Struct. Chem. (Engl. Transl.)*, 1981, **22**, 473.
350. V. P. Chalyi, I. A. Sheka, L. I. Fedoryako and V. V. Fomenko, *Ukr. Khim. Zh. (Russ. Ed.)*, 1978, **44**, 532 (*Chem. Abstr.*, 1978, **89**, 68 936).
351. B. Noren, *Acta Chem. Scand.*, 1969, **23**, 379.
352. S. Singh, M. Bedi and R. D. Verma, *J. Fluorine Chem.*, 1982, **20**, 107.
353. R. L. Davidovich and M. A. Medkov, *Sov. J. Coord. Chem. (Engl. Transl.)*, 1975, **1**, 1355.
354. M. A. Medkov and R. L. Davidovich, *Bull. Acad. Sci. USSR, Div. Chem. Sci.*, 1978, **27**, 231.
355. R. F. Klevtsova, L. A. Glinskaya and N. P. Pasechnyuk, *Sov. Phys.—Crystallogr. (Engl. Transl.)*, 1977, **22**, 677.
356. R. F. Klevtsova, E. S. Zolotova, L. A. Glinskaya and P. V. Klevtsov, *Sov. Phys.—Crystallogr. (Engl. Transl.)*, 1980, **25**, 558.
357. R. F. Klevtsova, A. A. Antonova and L. A. Glinskaya, *Sov. Phys.—Crystallogr. (Engl. Transl.)*, 1980, **25**, 92.
358. V. P. Babaeva and V. Ya. Rosolovskii, *Bull. Acad. Sci. USSR, Div. Chem. Sci.*, 1977, **26**, 445.
359. V. P. Babaeva and V. Ya. Rosolovskii, *Russ. J. Inorg. Chem. (Engl. Transl.)*, 1979, **24**, 206.
360. A. V. Dudin, V. P. Babaeva and V. Ya. Rosolovskii, *Russ. J. Inorg. Chem. (Engl. Transl.)*, 1983, **28**, 950.
361. A. V. Dudin, V. P. Babaeva and V. Ya. Rosolovskii, *Russ. J. Inorg. Chem. (Engl. Transl.)*, 1981, **26**, 1568.
362. V. P. Babaeva and V. Ya. Rosolovskii, *Russ. J. Inorg. Chem. (Engl. Transl.)*, 1978, **23**, 527.
363. V. P. Babaeva, L. S. Skogareva and V. Ya. Rosolovskii, *Russ. J. Inorg. Chem. (Engl. Transl.)*, 1982, **27**, 51.
364. R. C. Paul, O. B. Baidya and R. Kapoor, *Z. Naturforsch., Teil B*, 1976, **31**, 300.
365. R. C. Paul, O. B. Baidya, R. C. Kumar and R. Kapoor, *Aust. J. Chem.*, 1976, **29**, 1605.
366. J. Ludwig and D. Schwartz, *Inorg. Chem.*, 1970, **9**, 607.
367. V. L. Pavlov, Yu. A. Lysenko and A. A. Kalinichenko, *Russ. J. Inorg. Chem. (Engl. Transl.)*, 1972, **17**, 1768.
368. R. N. Kapoor and R. C. Mehrotra, *J. Chem. Soc.*, 1959, 422.
369. K. L. Jaura, H. S. Banga and R. L. Kaushik, *J. Indian Chem. Soc.*, 1962, **39**, 531.
370. K. C. Malhotra and R. G. Sud, *J. Inorg. Nucl. Chem.*, 1976, **38**, 393.
371. K. C. Malhotra, A. Kumar and S. C. Chaudhry, *Indian J. Chem., Sect. A*, 1979, **18**, 423.
372. P. Sartori and M. Weidenbruch, *Chem. Ber.*, 1967, **100**, 2049.
373. E. M. Brainina, R. Kh. Freidlina and A. N. Nesmeyanov, *Bull. Acad. Sci. USSR, Div. Chem. Sci.*, 1961, 560.
374. Z. N. Prozorovskaya, L. N. Komissarova and V. I. Spitsyn, *Russ. J. Inorg. Chem. (Engl. Transl.)*, 1968, **13**, 369.
375. Z. N. Prozorovskaya, L. N. Komissarova and V. A. Smirnov, *J. Gen. Chem. USSR (Engl. Transl.)*, 1969, **39**, 1284.
376. V. A. Smirnov, Z. N. Prozorovskaya and L. N. Komissarova, *Russ. J. Inorg. Chem. (Engl. Transl.)*, 1970, **15**, 178.
377. V. A. Smirnov, Z. N. Prozorovskaya and L. N. Komissarova, *Russ. J. Inorg. Chem. (Engl. Transl.)*, 1970, **15**, 649.
378. N. W. Alcock, V. M. Tracy and T. C. Waddington, *J. Chem. Soc., Dalton Trans.*, 1976, 2243.
379. K. R. Nahar, A. K. Solanki and A. M. Bhandari, *Synth. React. Inorg. Metal–Org. Chem.*, 1982, **12**, 805.
380. S. Cartlidge and H. Sutcliffe, *J. Fluorine Chem.*, 1981, **17**, 549.
381. L. N. Komissarova, S. V. Krivenko, Z. N. Prozorovskaya and V. E. Plyushchev, *Russ. J. Inorg. Chem. (Engl. Transl.)*, 1966, **11**, 146.
382. L. N. Komissarova, Z. N. Prozorovskaya and V. I. Spitsyn, *Russ. J. Inorg. Chem. (Engl. Transl.)*, 1966, **11**, 1089.
383. Z. N. Prozorovskaya, L. N. Komissarova and V. I. Spitsyn, *Russ. J. Inorg. Chem. (Engl. Transl.)*, 1967, **12**, 1348.
384. V. I. Spitsyn, L. N. Komissarova, Z. N. Prozorovskaya and V. F. Chuvaev, *Russ. J. Inorg. Chem. (Engl. Transl.)*, 1967, **12**, 785.
385. L. N. Komissarova, M. V. Savel'eva and V. E. Plyushchev, *Russ. J. Inorg. Chem. (Engl. Transl.)*, 1963, **8**, 27.
386. Z. N. Prozorovskaya, K. I. Petrov and L. N. Komissarova, *Russ. J. Inorg. Chem. (Engl. Transl.)*, 1968, **13**, 505.
387. H. Bilinski, N. Brnicevic and Z. Konrad, *Inorg. Chem.*, 1981, **20**, 1882.
388. H. Bilinski and N. Brnicevic, *Croat. Chem. Acta*, 1983, **56**, 53.
389. L. N. Pankratova and G. S. Kharitonova, *Russ. J. Inorg. Chem. (Engl. Transl.)*, 1972, **17**, 1389.
390. G. S. Kharitonova, V. I. Nefedov and L. N. Pankratova, *Russ. J. Inorg. Chem. (Engl. Transl.)*, 1972, **17**, 1649.
391. G. S. Kharitonova, V. I. Nefedov, L. N. Pankratova and V. L. Pershin, *Russ. J. Inorg. Chem. (Engl. Transl.)*, 1974, **19**, 469.
392. G. S. Kharitonova and L. N. Pankratova, *Russ. J. Inorg. Chem. (Engl. Transl.)*, 1977, **22**, 379.
393. L. A. Pospelova and L. M. Zaitsev, *Russ. J. Inorg. Chem. (Engl. Transl.)*, 1966, **11**, 995.
394. J. Dervin, J. Faucherre and H. Pruszek, *Rev. Chim. Min.*, 1974, **11**, 372.
395. Yu. Ya. Kharitonov, L. A. Pospelova and L. M. Zaitsev, *Russ. J. Inorg. Chem. (Engl. Transl.)*, 1967, **12**, 1390.
396. L. A. Pospelova and L. M. Zaitsev, *Russ. J. Inorg. Chem. (Engl. Transl.)*, 1968, **13**, 1535.
397. Yu. E. Gorbunova, V. G. Kuznetsov and E. S. Kovaleva, *J. Struct. Chem. (Engl. Transl.)*, 1968, **9**, 815.
398. A. Clearfield, *Inorg. Chim. Acta*, 1970, **4**, 166.
399. G. S. Bochkarev, L. M. Zaitsev and V. N. Kozhenkova, *Russ. J. Inorg. Chem. (Engl. Transl.)*, 1963, **8**, 1177.
400. V. P. Shvedov and N. A. Susorova, *Russ. J. Inorg. Chem. (Engl. Transl.)*, 1963, **8**, 458.

401. A. A. Grinberg and V. I. Astapovich, *Russ. J. Inorg. Chem. (Engl. Transl.)*, 1961, **6,** 164.
402. F. A. Johnson and E. M. Larsen, *Inorg. Synth.*, 1966, **8,** 40.
403. G. L. Glen, J. V. Silverton and J. L. Hoard, *Inorg. Chem.*, 1963, **2,** 250.
404. B. Kojic-Prodic, Z. Rusic-Toros and M. Sljukic, *Acta Crystallogr., Sect. B*, 1978, **34,** 2001.
405. D. Tranqui, P. Boyer, J. Laugier and P. Vulliet, *Acta Crystallogr., Sect. B*, 1977, **33,** 3126.
406. P. Vulliet, A. Baudry, P. Boyer and A. Tissier, *Chem. Phys. Lett.*, 1978, **55,** 297.
407. B. Hutchinson, A. Sunderland, M. Neal and S. Olbricht, *Spectrochim. Acta, Part A*, 1973, **29,** 2001.
408. F. A. Johnson and E. M. Larsen, *Inorg. Chem.*, 1962, **1,** 159.
409. Yu. Ya. Kharitonov, G. S. Bochkarev and L. M. Zaitsev, *Russ. J. Inorg. Chem. (Engl. Transl.)*, 1964, **9,** 745.
410. R. L. Davidovich and M. A. Medkov, *Bull. Acad. Sci. USSR, Div. Chem. Sci.*, 1978, **27,** 227.
411. R. L. Davidovich, I. G. Dergacheva and S. B. Ivanov, *Bull. Acad. Sci. USSR, Div. Chem. Sci.*, 1975, **24,** 2539.
412. L. M. Zaitsev and G. S. Bochkarev, *Russ. J. Inorg. Chem. (Engl. Transl.)*, 1962, **7,** 802.
413. Yu. Ya. Kharitonov, L. M. Zaitsev, G. S. Bochkarev and O. N. Evstaf'eva, *Russ. J. Inorg. Chem. (Engl. Transl.)*, 1964, **9,** 876.
414. L. M. Zaitsev, *Russ. J. Inorg. Chem. (Engl. Transl.)*, 1964, **9,** 1279.
415. A. K. Sharma, S. Kumar and N. K. Kaushik, *Thermochim. Acta*, 1981, **47,** 149.
416. A. K. Sharma and N. K. Kaushik, *Thermochim. Acta*, 1982, **56,** 221.
417. D. I. Ryabchikov, A. N. Ermakov, V. K. Belyaeva and I. N. Marov, *Russ. J. Inorg. Chem. (Engl. Transl.)*, 1960, **5,** 505.
418. D. I. Ryabchikov, I. N. Marov, A. N. Ermakov and V. K. Belyaeva, *J. Inorg. Nucl. Chem.*, 1964, **26,** 965.
419. J. Hala and P. Londyn, *Collect. Czech. Chem. Commun.*, 1981, **46,** 1764.
420. K. I. Petrov and L. N. Komissarova, *Russ. J. Inorg. Chem. (Engl. Transl.)*, 1972, **17,** 1545.
421. E. M. Larsen and E. H. Homeier, *Inorg. Chem.*, 1972, **11,** 2687.
422. A. N. Ermakov, I. N. Marov and L. P. Kazanskii, *Russ. J. Inorg. Chem. (Engl. Transl.)*, 1967, **12,** 1437.
423. K. F. Karlysheva, A. V. Koshel', I. A. Sheka and G. S. Semenova, *Russ. J. Inorg. Chem. (Engl. Transl.)*, 1975, **20,** 521.
424. R. B. Hahn and L. Weber, *J. Am. Chem. Soc.*, 1955, **77,** 4777.
425. R. B. Hahn and P. T. Joseph, *J. Am. Chem. Soc.*, 1957, **79,** 1298.
426. R. N. Kapoor and R. C. Mehrotra, *J. Am. Chem. Soc.*, 1958, **80,** 3569.
427. R. N. Kapoor and R. C. Mehrotra, *J. Am. Chem. Soc.*, 1960, **82,** 3495.
428. K. F. Karlysheva, L. A. Malinko and I. A. Sheka, *Russ. J. Inorg. Chem. (Engl. Transl.)*, 1967, **12,** 1076.
429. C. S. Pande and G. N. Misra, *Indian J. Chem.*, 1973, **11,** 292.
430. M. Pal and R. N. Kapoor, *Inorg. Chim. Acta*, 1980, **40,** 99.
431. Y. Hase and O. L. Alves, *Spectrochim. Acta, Part A*, 1981, **37,** 957.
432. E. M. Larsen, J. Howatson, A. M. Gammill and L. Wittenberg, *J. Am. Chem. Soc.*, 1952, **74,** 3489.
433. E. M. Larsen and L. J. Wittenberg, *J. Am. Chem. Soc.*, 1955, **77,** 5850.
434. A. V. Suvorov and E. K. Krzhizhanovskaya, *Russ. J. Inorg. Chem. (Engl. Transl.)*, 1969, **14,** 434.
435. R. K. Collins and M. G. B. Drew, *J. Chem. Soc. (A)*, 1971, 3610.
436. R. C. Paul and S. L. Chadha, *J. Inorg. Nucl. Chem.*, 1969, **31,** 2753.
437. A. M. Golub, T. P. Lishko and P. F. Lozovskaya, *Russ. J. Inorg. Chem. (Engl. Transl.)*, 1974, **19,** 13.
438. A. Clearfield and E. J. Malkiewich, *J. Inorg. Nucl. Chem.*, 1963, **25,** 237.
439. N. M. Karayannis, A. N. Speca, L. L. Pytlewski and M. M. Labes, *J. Less-Common Met.*, 1970, **22,** 117.
440. R. Paetzold and P. Vordank, *Z. Anorg. Allg. Chem.*, 1966, **347,** 294.
441. B. Kautzner and P. C. Wailes, *Aust. J. Chem.*, 1969, **22,** 2295.
442. S. J. Kuhn and J. S. McIntyre, *Can. J. Chem.*, 1965, **43,** 375.
443. R. C. Paul and S. L. Chadha, *Aust. J. Chem.*, 1969, **22,** 1381.
444. E. K. Krzhizhanovskaya and A. V. Suvorov, *Russ. J. Inorg. Chem. (Engl. Transl.)*, 1971, **16,** 1355.
445. R. Makhija and A. D. Westland, *Inorg. Chim. Acta*, 1978, **29,** L269.
446. V. Krishnan and C. C. Patel, *J. Inorg. Nucl. Chem.*, 1964, **26,** 2201.
447. V. Krishnan and C. C. Patel, *J. Inorg. Nucl. Chem.*, 1965, **27,** 244.
448. P. R. Ramamurthy and C. C. Patel, *Can. J. Chem.*, 1964, **42,** 856.
449. V. Krishnan and C. C. Patel, *Can. J. Chem.*, 1966, **44,** 972.
450. R. C. Paul, S. L. Chadha and S. K. Vasisht, *J. Less-Common Met.*, 1968, **16,** 288.
451. Yu. Ya. Kharitonov, M. M. Bekhit and S. F. Belevskii, *Sov. J. Coord. Chem. (Engl. Transl.)*, 1980, **6,** 763.
452. V. V. Savant, P. Ramamurthy and C. C. Patel, *J. Less-Common Met.*, 1970, **22,** 479.
453. P. Bronzan and H. Meider, *J. Less-Common Met.*, 1978, **59,** 101.
454. K. Bhatt and Y. K. Agrawal, *Synth. React. Inorg. Met.-Org. Chem.*, 1972, **2,** 175.
455. K. F. Karlysheva and I. A. Sheka, *Russ. J. Inorg. Chem. (Engl. Transl.)*, 1962, **7,** 666.
456. K. F. Fouché, H. J. le Roux and F. Phillips, *J. Inorg. Nucl. Chem.*, 1970, **32,** 1949.
457. D. Tranqui, J. Laugier, P. Boyer and P. Vulliet, *Acta Crystallogr., Sect. B*, 1978, **34,** 767.
458. L. E. Orgel, *J. Inorg. Nucl. Chem.*, 1960, **14,** 136.
459. W. D. Bonds, Jr., R. D. Archer and W. C. Hamilton, *Inorg. Chem.*, 1971, **10,** 1764.
460. W. Mark, *Acta Chem. Scand.*, 1970, **24,** 1398.
461. D. G. Blight and D. L. Kepert, *Inorg. Chem.*, 1972, **11,** 1556.
462. A. Baudry, P. Boyer, A. Tissier and P. Vulliet, *Inorg. Chem.*, 1979, **18,** 3427.
463. M. Pal and R. N. Kapoor, *Indian J. Chem., Sect. A*, 1980, **19,** 912.
464. A. R. Middleton and G. Wilkinson, *J. Chem. Soc., Dalton Trans.*, 1980, 1888.
465. P. E. Riley, K. Abu-Dari and K. N. Raymond, *Inorg. Chem.*, 1983, **22,** 3940.
466. A. Hossain and M. B. Hursthouse, *Inorg. Chim. Acta*, 1980, **44,** L259.
467. M. Schmeisser, P. Sartori and B. Lippsmeier, *Chem. Ber.*, 1970, **103,** 868.
468. J. J. Pitts, M. A. Robinson and S. I. Trotz, *J. Inorg. Nucl. Chem.*, 1969, **31,** 3685.
469. E. G. Teterin, N. N. Shesterikov, P. G. Krutikov and A. S. Solovkin, *Russ. J. Inorg. Chem. (Engl. Transl.)*, 1971, **16,** 77.

470. H. Funk and M. Hesselbarth, *Z. Chem.*, 1966, **6**, 227.
471. S. Chatterjee, D. Sukhani, V. D. Gupta and R. C. Mehrotra, *Indian J. Chem.*, 1970, **8**, 362.
472. I. R. Beattie and M. Webster, *J. Chem. Soc.*, 1964, 3507.
473. F. M. Chung and A. D. Westland, *Can. J. Chem.*, 1969, **47**, 195.
474. J. B. Hamilton and R. E. McCarley, *Inorg. Chem.*, 1970, **9**, 1339.
475. A. F. Lindmark and R. C. Fay, *Inorg. Chem.*, 1983, **22**, 2000.
476. M. Colapietro, A. Vaciago, D. C. Bradley, M. B. Hursthouse and I. F. Rendall, *Chem. Commun.*, 1970, 743.
477. D. C. Bradley, I. F. Rendall and K. D. Sales, *J. Chem. Soc., Dalton Trans.*, 1973, 2228.
478. E. L. Muetterties, *Inorg. Chem.*, 1973, **12**, 1963.
479. E. L. Muetterties, *Inorg. Chem.*, 1974, **13**, 1011.
480. S. Kumar and N. K. Kaushik, *Synth. React. Inorg. Met.-Org. Chem.*, 1982, **12**, 159.
481. S. Kumar and N. K. Kaushik, *Indian J. Chem., Sect. A*, 1982, **21**, 182.
482. S. Kumar and N. K. Kaushik, *Inorg. Nucl. Chem. Lett.*, 1980, **16**, 389.
483. H. S. Sangari, G. S. Sodhi, N. K. Kaushik and R. P. Singh, *Synth. React. Inorg. Met.-Org. Chem.*, 1981, **11**, 373.
484. S. A. A. Zaidi, T. A. Khan and B. S. Neelam, *J. Inorg. Nucl. Chem.*, 1980, **42**, 1525.
485. S. A. A. Zaidi, T. A. Khan and N. S. Neelam, *Indian J. Chem., Sect. A*, 1979, **17**, 419.
486. R. C. Fay, J. R. Weir and A. H. Bruder, *Inorg. Chem.*, 1984, **23**, 1079.
487. M. E. Silver, O. Eisenstein and R. C. Fay, *Inorg. Chem.*, 1983, **22**, 759.
488. S. L. Hawthorne, A. H. Bruder and R. C. Fay, *Inorg. Chem.*, 1978, **17**, 2114.
489. W. L. Steffen and R. C. Fay, *Inorg. Chem.*, 1978, **17**, 2120.
490. S. L. Hawthorne, A. H. Bruder and R. C. Fay, *Inorg. Chem.*, 1983, **22**, 3368.
491. J. L. Martin and J. Takats, *Inorg. Chem.*, 1975, **14**, 73.
492. M. Cowie and M. J. Bennett, *Inorg. Chem.*, 1976, **15**, 1595.
493. K. Andra and W. Rollka, *Z. Anorg. Allg. Chem.*, 1972, **391**, 19.
494. R. N. McGinnis and J. B. Hamilton, *Inorg. Nucl. Chem. Lett.*, 1972, **8**, 245.
495. V. V. Tkachev, S. A. Shchepinov and L. O. Atovmyan, *J. Struct. Chem. (Engl. Transl.)*, 1977, **18**, 823.
496. Ts. B. Konunova, A. Yu. Tsivadze, A. N. Smirnov and S. A. Kudritskaya, *Russ. J. Inorg. Chem. (Engl. Transl.)*, 1983, **28**, 515.
497. Ts. B. Konunova and S. A. Kudritskaya, *Russ. J. Inorg. Chem. (Engl. Transl.)*, 1983, **28**, 447.
498. K. C. Malhotra, G. Mehrotra and S. C. Chaudhry, *Indian J. Chem., Sect. A*, 1979, **17**, 421.
499. K. Andrä, *Z. Anorg. Allg. Chem.*, 1970, **373**, 209.
500. R. Papiernik, D. Mercurio and B. Frit, *Acta Crystallogr., Sect. B*, 1982, **38**, 2347.
501. R. D. Burbank and F. N. Bensey, US Atomic Energy Commission, rep. K-1280, 1956.
502. B. Krebs, *Z. Anorg. Allg. Chem.*, 1970, **378**, 263.
503. B. Krebs, G. Henkel and M. Dartmann, *Acta Crystallogr., Sect. B*, 1979, **35**, 274.
504. B. Krebs and D. Sinram, *J. Less-Common Met.*, 1980, **76**, 7.
505. J. H. Canterford and R. Colton, 'Halides of the Second and Third Row Transition Metals', Wiley-Interscience, London, 1968.
506. R. J. H. Clark, B. K. Hunter and D. M. Rippon, *Inorg. Chem.*, 1972, **11**, 56.
507. T. N. Waters, *Chem. Ind. (London)*, 1964, 713.
508. F. Gabela, B. Kojic-Prodic, M. Sljukic and Z. Ruzic-Toros, *Acta Crystallogr., Sect. B*, 1977, **33**, 3733.
509. D. Hall, C. E. F. Rickard and T. N. Waters, *J. Inorg. Nucl. Chem.*, 1971, **33**, 2395.
510. B. Kojic-Prodic, F. Gabela, Z. Ruzic-Toros and M. Sljukic, *Acta Crystallogr., Sect. B*, 1981, **37**, 1963.
511. C. J. Barton, W. R. Grimes, H. Insley, R. E. Moore and R. E. Thoma, *J. Phys. Chem.*, 1958, **62**, 665.
512. I. D. Ratnikova, A. A. Kosorukov, Yu. M. Korenev and A. V. Novoselova, *Russ. J. Inorg. Chem. (Engl. Transl.)*, 1975, **20**, 782.
513. I. V. Tananaev and L. S. Guzeeva, *Russ. J. Inorg. Chem. (Engl. Transl.)*, 1966, **11**, 590.
514. I. V. Tananaev and L. S. Guzeeva, *Russ. J. Inorg. Chem. (Engl. Transl.)*, 1966, **11**, 587.
515. H. Hull and A. G. Turnbull, *J. Inorg. Nucl. Chem.*, 1967, **29**, 2903.
516. H. Hull and A. G. Turnbull, *J. Inorg. Nucl. Chem.*, 1967, **29**, 951.
517. B. Gaudreau, *Rev. Chim. Min.*, 1965, **2**, 1.
518. J. Slivnik, B. Druzina and B. Zemva, *Z. Naturforsch, Teil B*, 1981, **36**, 1457.
519. A. A. Opalovskii, T. F. Gudimovich, M. Kh. Akhmadeev and L. D. Ishkova, *Russ. J. Inorg. Chem. (Engl. Transl.)*, 1982, **27**, 664.
520. A. A. Opalovskii, T. F. Gudimovich, N. Kh. Akhmadeev and L. D. Ishkova, *Russ. J. Inorg. Chem. (Engl. Transl.)*, 1983, **28**, 1065.
521. H. M. Haendler, F. A. Johnson and D. S. Crocket, *J. Am. Chem. Soc.*, 1958, **80**, 2662.
522. D. S. Crocket and H. M. Haendler, *J. Am. Chem. Soc.*, 1960, **82**, 4158.
523. B. Gaudreau, *C. R. Hebd. Seances Acad. Sci., Ser. C*, 1966, **263**, 67.
524. H. M. Haendler, C. M. Wheeler, Jr. and D. W. Robinson, *J. Am. Chem. Soc.*, 1952, **74**, 2352.
525. P. A. W. Dean and D. F. Evans, *J. Chem. Soc. (A)*, 1967, 698.
526. L. Kolditz and A. Feltz, *Z. Anorg. Allg. Chem.*, 1961, **310**, 195.
527. B. G. Sayer, N. Hao, G. Dénes, D. G. Bickley and M. J. McGlinchey, *Inorg. Chim. Acta*, 1981, **48**, 53.
528. O. I. Egerev and A. D. Pogorelyi, *J. Appl. Chem. USSR (Engl. Transl.)*, 1966, **39**, 866.
529. R. H. Schmitt, E. L. Grove and R. D. Brown, *J. Am. Chem. Soc.*, 1960, **82**, 5292.
530. L. Kolditz and A. Feltz, *Z. Anorg. Allg. Chem.*, 1961, **310**, 204.
531. G. Brunton, *Acta Crystallogr., Sect. B*, 1973, **29**, 2294.
532. H. Bode and G. Teufer, *Z. Anorg. Allg. Chem.*, 1956, **283**, 18.
533. G. Brunton, *Acta Crystallogr., Sect. B*, 1969, **25**, 2164.
534. R. Hoppe and B. Mehlhorn, *Z. Anorg. Allg. Chem.*, 1976, **425**, 200.
535. G. C. Hampson and L. Pauling, *J. Am. Chem. Soc.*, 1938, **60**, 2702.
536. W. H. Zachariasen, *Acta Crystallogr.*, 1954, **7**, 792.
537. H. J. Hurst and J. C. Taylor, *Acta Crystallogr., Sect. B*, 1970, **26**, 417.

538. H. J. Hurst and J. C. Taylor, *Acta Crystallogr., Sect. B,* 1970, **26,** 2136.
539. L. A. Harris, *Acta Crystallogr.,* 1959, **12,** 172.
540. V. P. Tarasov and Yu. A. Buslaev, *Inorg. Mater.* (*Engl. Transl.*), 1969, **5,** 978.
541. V. P. Tarasov and Yu. A. Buslaev, *J. Struct. Chem.* (*Engl. Transl.*), 1969, **10,** 816.
542. P. da R. Andrade, A. Vasquez, J. D. Rogers and E. R. Fraga, *Phys. Rev. B,* 1970, **1,** 2912.
543. P. Boyer, O. de O. Damasceno, J. D. Fabris, J. R. F. Ferreira, A. L. de Oliveira and J. de Oliveira, *J. Phys. Chem. Solids,* 1976, **37,** 1019.
544. D. R. Sears and J. H. Burns, *J. Chem. Phys.,* 1964, **41,** 3478.
545. D. Avignant, I. Mansouri, R. Chevalier and J. C. Cousseins, *J. Solid State Chem.,* 1981, **38,** 121.
546. R. M. Herak, S. S. Malcic and L. M. Manojlovic, *Acta Crystallogr.,* 1965, **18,** 520.
547. G. Brunton, *Acta Crystallogr., Sect. B,* 1971, **27,** 1944.
548. E. L. Muetterties and L. J. Guggenberger, *J. Am. Chem. Soc.,* 1974, **96,** 1748.
549. J. H. Burns, R. D. Ellison and H. A. Levy, *Acta Crystallogr., Sect. B,* 1968, **24,** 230.
550. J. Slivnik, A. Smalc, B. Sedej and M. Vilhar, *Vestn. Slov. Kem. Drus.,* 1964, **11,** 53.
551. J. Slivnik, B. Jerkovic and B. Sedej, *Monatsh. Chem.,* 1966, **97,** 820.
552. B. Kojic-Prodic, S. Scavnicar and B. Matkovic, *Acta Crystallogr., Sect. B,* 1971, **27,** 638.
553. B. Zemva, S. Milicev and J. Slivnik, *J. Fluorine Chem.,* 1978, **11,** 545.
554. R. L. Davidovich, M. A. Medkov, V. B. Timchenko and B. V. Bukvetskii, *Bull. Acad. Sci. USSR, Div. Chem. Sci.,* 1983, **32,** 2181.
555. V. A. Shreider and I. N. Rozhkov, *Bull. Acad. Sci. USSR, Div. Chem. Sci.,* 1979, **28,** 236.
556. M. A. Medkov, M. D. Davidovich, M. D. Rizaeva, I. P. Kondratyuk and B. V. Bukvetskii, *Bull. Acad. Sci. USSR, Div. Chem. Sci,* 1980, **29,** 1185.
557. I. P. Kondratyuk, M. F. Eiberman, R. L. Davidovich, M. A. Medkov and B. V. Bukvetskii, *Sov. J. Coord. Chem.* (*Engl. Transl.*), 1981, **7,** 549.
558. I. P. Kondratyuk, B. V. Bukvetskii, R. L. Davidovich and M. A. Medkov, *Sov. J. Coord. Chem.* (*Engl. Transl.*), 1982, **8,** 113.
559. R. L. Davidovich, M. A. Medkov, M. D. Rizaeva and B. V. Bukvetskii, *Bull. Acad. Sci. USSR, Div. Chem. Sci.,* 1982, **31,** 1291.
560. R. L. Davidovich, T. F. Levchishina, T. A. Kaidalova and Yu. A. Buslaev, *Inorg. Mater.* (*Engl. Transl.*), 1970, **6,** 433.
561. R. L. Davidovich, T. F. Levchischina and T. A. Kaidalova, *Inorg. Mater.* (*Engl. Transl.*), 1971, **7,** 1773.
562. J. Fischer and R. Weiss, *Acta Crystallogr., Sect. B,* 1973, **29,** 1955.
563. L. P. Otroshchenko, R. L. Davidovich and V. I. Simonov, *Sov. J. Coord. Chem.* (*Engl. Transl.*), 1978, **4,** 1079.
564. L. P. Otroshchenko, V. I. Simonov, R. L. Davidovich, L. E. Fykin, V. Ya. Duderov and S. P. Solov'ev, *Sov. Phys. Crystallogr.* (*Engl. Transl.*), 1980, **25,** 416.
565. L. P. Otroshchenko, R. L. Davidovich, V. I. Simonov and N. V. Belov, *Sov. Phys. Crystallogr.* (*Engl. Transl.*), 1981, **26,** 675.
566. M. F. Eiberman, T. A. Kaidalova, R. L. Davidovich, T. F. Levchishina and B. V. Bukvetskii, *Koord. Khim.,* 1980, **6,** 1885 (*Chem. Abstr.,* 1981, **94,** 112 794).
567. J. Fischer, R. Elchinger and R. Weiss, *Acta. Crystallogr., Sect. B,* 1973, **29,** 1967.
568. J. Fischer and R. Weiss, *Acta Crystallogr., Sect. B,* 1973, **29,** 1963.
569. T. S. Chernaya, L. E. Fykin, V. A. Sarin, N. I. Bydanov, N. M. Laptash, B. V. Bukvetskii and V. I. Simonov, *Sov. Phys. Crystallogr.* (*Engl. Transl.*), 1983, **28,** 393.
570. J. Fischer and R. Weiss, *Acta Crystallogr., Sect. B,* 1973, **29,** 1958.
571. J.-P. Laval, R. Papiernik and B. Frit, *Acta Crystallogr., Sect. B,* 1978, **34,** 1070.
572. B. Mehlhorn and R. Hoppe, *Z. Anorg. Allg. Chem.,* 1976, **425,** 180.
573. J.-P. Laval, D. Mercurio-Lavaud and B. Gaudreau, *Rev. Chim. Min.,* 1974, **11,** 742.
574. I. D. Ratnikova, Yu. M. Korenev and A. V. Novoselova, *Russ. J. Inorg. Chem.* (*Engl. Transl.*), 1980, **25,** 452.
575. D. Reinen and F. Steffens, *Z. Anorg. Allg. Chem.,* 1978, **441,** 63.
576. P. Köhl, D. Reinen, G. Decher and B. Wanklyn, *Z. Kristallogr., Kristallgeom., Kristallphys., Kristallchem.,* 1980, **153,** 211.
577. V. Propach and F. Steffens, *Z. Naturforsch., Teil B,* 1978, **33,** 268.
578. Yu. M. Korenev, P. I. Antipov and A. V. Novoselova, *Russ. J. Inorg. Chem.* (*Engl. Transl.*), 1980, **25,** 698
579. M. Poulain, M. Poulain and J. Lucas, *J. Solid State Chem.,* 1973, **8,** 132.
580. L. M. Toth and J. B. Bates, *Spectrochim. Acta, Part A,* 1974, **30,** 1095.
581. I. W. Forrest and A. P. Lane, *Inorg. Chem.,* 1976, **15,** 265.
582. A. P. Lane and D. W. A. Sharp, *J. Chem. Soc.* (*A*), 1969, 2942.
583. P. W. Smith, R. Stoessiger and A. G. Turnbull, *J. Chem. Soc.* (*A*), 1968, 3013.
584. L. M. Toth, A. S. Quist and G. E. Boyd, *J. Phys. Chem.,* 1973, **77,** 1384.
585. Y. Kawamoto and F. Sakaguchi, *Bull. Chem. Soc. Jpn.,* 1983, **56,** 2138.
586. I. S. Morozov and S. In'-Chzhu, *Russ. J. Inorg. Chem.* (*Engl. Transl.*), 1959, **4,** 307.
587. P. Gelbman and A. D. Westland, *J. Chem. Soc., Dalton Trans.,* 1975, 1598.
588. R. L. Lister and S. N. Flengas, *Can. J. Chem.,* 1964, **42,** 1102.
589. D. A. Asvestas, P. Pint and S. N. Flengas, *Can. J. Chem.,* 1977, **55,** 1154.
590. G. M. Toptygina and I. B. Barskaya, *Russ. J. Inorg. Chem.* (*Engl. Transl.*), 1965, **10,** 1226.
591. I. B. Barskaya and G. M. Toptygina, *Russ. J. Inorg. Chem.* (*Engl. Transl.*), 1967, **12,** 14.
592. R. Makhija and A. D. Westland, *J. Chem. Soc., Dalton Trans.,* 1977, 1707.
593. D. Sinram, C. Brendel and B. Krebs, *Inorg. Chim. Acta,* 1982, **64,** L131.
594. G. Engel, *Z. Kristallogr., Kristallgeom., Kristallphys., Kristallchem.,* 1935, **90,** 341.
595. R. M. Friedman and J. D. Corbett, *Inorg. Chem.,* 1973, **12,** 1134.
596. G. J. Kipouros and S. N. Flengas, *Can. J. Chem.,* 1983, **61,** 2183.
597. S. S. Sandhu, B. S. Chakkal and G. S. Sandhu, *J. Indian Chem. Soc.,* 1960, **37,** 329.
598. D. M. Adams, J. Chatt, J. M. Davidson and J. Gerratt, *J. Chem. Soc.,* 1963, 2189.

599. W. van Bronswyk, R. J. H. Clark and L. Maresca, *Inorg. Chem.*, 1969, **8,** 1395.
600. K. Goyal, R. C. Paul and S. S. Sandhu, *J. Chem. Soc.*, 1959, 322.
601. V. Gutmann and H. Tannenberger, *Monatsh. Chem.*, 1957, **88,** 292.
602. R. C. Paul, K. Chander and G. Singh, *J. Indian Chem. Soc.*, 1958, **35,** 869.
603. B. J. Brisdon, G. A. Ozin and R. A. Walton, *J. Chem. Soc. (A)*, 1969, 342.
604. B. J. Brisdon, T. E. Lester and R. A. Walton, *Spectrochim. Acta, Part A*, 1967, **23,** 1969.
605. D. C. Bradley, F. M. Abd-el Halim, E. A. Sadek and W. Wardlaw, *J. Chem. Soc.*, 1952, 2032.
606. D. M. Adams and D. C. Newton, *J. Chem. Soc. (A)*, 1968, 2262.
607. S. B. Muchamecina and H. Hobart, *Z. Chem.*, 1982, **22,** 273.
608. H. K. Sharma, S. N. Dubey and D. M. Puri, *Proc. Indian Acad. Sci. (Chem. Sci.)*, 1981, **90,** 555.
609. J. MacCordick, C. Devin, R. Perrot and R. Rohmer, *C. R. Hebd. Seances Acad. Sci.*, 1972, **274,** 278.
610. F. W. Poulsen, *Inorg. Nucl. Chem. Lett.*, 1980, **16,** 355.
611. A. F. Demiray and W. Brockner, *Monatsh. Chem.*, 1980, **111,** 21.
612. A. Feltz, *Z. Anorg. Allg. Chem.*, 1968, **358,** 21.
613. J. I. Bullock, F. W. Parrett and N. J. Taylor, *Can. J. Chem.*, 1974, **52,** 2880.
614. B. D. James and M. G. H. Wallbridge, *Prog. Inorg. Chem.*, 1970, **11,** 99.
615. T. J. Marks and J. R. Kolb, *Chem. Rev.*, 1977, **77,** 263.
616. H. R. Hoekstra and J. J. Katz, *J. Am. Chem. Soc.*, 1949, **71,** 2488.
617. W. E. Reid, Jr., J. M. Bish and A. Brenner, *J. Electrochem. Soc.*, 1957, **104,** 21.
618. B. D. James, R. K. Nanda and M. G. H. Wallbridge, *J. Chem. Soc. (A)*, 1966, 182.
619. B. D. James and B. E. Smith, *Synth. React. Inorg. Metal-Org. Chem.*, 1974, **4,** 461.
620. M. Ehemann and H. Nöth, *Z. Anorg. Allg. Chem.*, 1971, **386,** 87.
621. P. H. Bird and M. R. Churchill, *Chem. Commun.*, 1967, 403.
622. R. W. Broach, I.-S. Chuang, T. J. Marks and J. M. Williams, *Inorg. Chem.*, 1983, **22,** 1081.
623. V. Plato and K. Hedberg, *Inorg. Chem.*, 1971, **10,** 590.
624. B. D. James, B. E. Smith and H. F. Shurvell, *J. Mol. Struct.*, 1976, **33,** 91.
625. N. Davies, M. G. H. Wallbridge, B. E. Smith and B. D. James, *J. Chem. Soc., Dalton Trans.*, 1973, 162.
626. T. A. Keiderling, W. T. Wozniak, R. S. Gay, D. Jurkowitz, E. R. Bernstein, S. J. Lippard and T. G. Spiro, *Inorg. Chem.*, 1975, **14,** 576.
627. B. E. Smith, H. F. Shurvell and B. D. James, *J. Chem. Soc., Dalton Trans.*, 1978, 710.
628. J. C. Whitmer and S. J. Cyvin, *J. Mol. Struct.*, 1978, **50,** 21.
629. B. N. Cyvin, S. J. Cyvin and J. C. Whitmer, *J. Mol. Struct.*, 1979, **53,** 281.
630. T. J. Marks, W. J. Kennelly, J. R. Kolb and L. A. Shimp, *Inorg. Chem.*, 1972, **11,** 2540.
631. A. J. Downs, R. G. Edgell, A. F. Orchard and P. D. P. Thomas, *J. Chem. Soc., Dalton Trans.*, 1978, 1755.
632. A. P. Hitchcock, N. Hao, N. H. Werstiuk, M. J. McGlinchey and T. Ziegler, *Inorg. Chem.*, 1982, **21,** 793.
633. V. V. Volkov, E. V. Sobolev, E. E. Zaev and O. I. Shapiro, *J. Struct. Chem. (Engl. Transl.)*, 1967, **8,** 407.
634. N. A. Bailey, P. H. Bird, N. Davies and M. G. H. Wallbridge, *J. Inorg. Nucl. Chem.*, 1970, **32,** 3116.
635. T. J. Marks and L. A. Shimp, *J. Am. Chem. Soc.*, 1972, **94,** 1542.
636. B. G. Sayer, J. I. A. Thompson, N. Hao, T. Birchall, D. R. Eaton and M. J. McGlinchey, *Inorg. Chem.*, 1981, **20,** 3748.
637. I.-S. Chuang, T. J. Marks, W. J. Kennelly and J. R. Kolb, *J. Am. Chem. Soc.*, 1977, **99,** 7539.
638. R. Shinomoto, E. Gamp, N. M. Edelstein, D. H. Templeton and A. Zalkin, *Inorg. Chem.*, 1983, **22,** 2351.
639. P. Prashar and J. P. Tandon, *J. Less-Common Met.*, 1968, **15,** 219.
640. P. Prashar and J. P. Tandon, *Z. Anorg. Allg. Chem.*, 1971, **383,** 81.
641. R. K. Sharma, R. V. Singh and J. P. Tandon, *Proc. Indian Acad. Sci. (Chem. Sci.)*, 1981, **90,** 461.
642. R. K. Sharma, R. V. Singh and J. P. Tandon, *Curr. Sci.*, 1980, **49,** 89.
643. R. K. Sharma and J. P. Tandon, *J. Prakt. Chem.*, 1980, **322,** 161.
644. R. K. Sharma, R. V. Singh and J. P. Tandon, *J. Prakt. Chem.*, 1980, **322,** 508.
645. R. K. Sharma, R. V. Singh and J. P. Tandon, *J. Inorg. Nucl. Chem.*, 1980, **42,** 463.
646. R. V. Singh, R. K. Sharma and J. P. Tandon, *Synth. React. Inorg. Met.-Org. Chem.*, 1981, **11,** 139.
647. R. K. Sharma, R. V. Singh and J. P. Tandon, *Indian J. Chem., Sect. A*, 1979, **18,** 360.
648. R. K. Sharma, R. V. Singh and J. P. Tandon, *J. Inorg. Nucl. Chem.*, 1980, **42,** 1267.
649. V. Verma, S. Kher and R. N. Kapoor, *Synth. React. Inorg. Metal-Org. Chem.*, 1983, **13,** 943.
650. V. Verma, S. Kher and R. N. Kapoor, *Transition Met. Chem.*, 1983, **8,** 34.
651. V. Verma, S. Kher and R. N. Kapoor, *Acta Chim. Acad. Sci. Hung.*, 1982, **111,** 371.
652. V. Verma, S. Kher and R. N. Kapoor, *Croat. Chem. Acta*, 1983, **56,** 97.
653. N. S. Biradar and A. L. Locker, *J. Inorg. Nucl. Chem.*, 1974, **36,** 1915.
654. A. Syamal and D. Kumar, *J. Less Common Met.*, 1980, **71,** 113.
655. A. Syamal and D. Kumar, *Synth. React. Inorg. Met.-Org. Chem.*, 1980, **10,** 63.
656. A. Syamal and D. Kumar, *Indian J. Chem., Sect. A*, 1980, **19,** 1018.
657. A. Syamal and D. Kumar, *Pol. J. Chem.*, 1984, **55,** 1747.
658. A. K. Agarwal, R. Kumar, N. Singh and B. D. Kansal, *J. Chin. Chem. Soc. (Taipei)*, 1980, **27,** 31.
659. A. K. Agarwal, S. Agrawal, N. Agrawal and B. D. Kansal, *J. Chin. Chem. Soc. (Taipei)*, 1982, **29,** 55.
660. M. L. Illingsworth and R. D. Archer, *Polyhedron*, 1982, **1,** 487.
661. R. D. Archer, W. H. Batschelet and M. L. Illingsworth, *J. Macromol. Sci., Chem.*, 1981, **A16,** 261.
662. R. K. Sharma, R. V. Singh and J. P. Tandon, *Curr. Sci.*, 1979, **48,** 886.
663. C. G. Macarovici, E. Perte and E. Motiu, *Rev. Roum. Chim.*, 1966, **11,** 59.
664. N. S. Birdadar and A. L. Locker, *J. Inorg. Nucl. Chem.*, 1975, **37,** 1308.
665. A. N. Sundar Ram and C. P. Prabhakaran, *Indian J. Chem., Sect. A*, 1977, **15,** 980.
666. R. C. Aggarwal, T. Prasad and B. N. Yadav, *J. Inorg. Nucl. Chem.*, 1975, **37,** 899.
667. R. C. Aggarwal, B. N. Yadav and T. Prasad, *Indian J. Chem.*, 1972, **10,** 671.
668. R. C. Aggarwal, B. N. Yadav and T. Prasad, *J. Inorg. Nucl. Chem.*, 1973, **35,** 653.
669. E. M. Larsen and A. C. Adams, *Inorg. Synth.*, 1967, **10,** 7.

670. H. G. Langer, *J. Inorg. Nucl. Chem.*, 1964, **26,** 59.
671. R. E. Sievers and J. C. Bailar, Jr., *Inorg. Chem.*, 1962, **1,** 174.
672. A. Marques-Netto and J. C. Abbe, *J. Inorg. Nucl. Chem.*, 1975, **37,** 2235.
673. J. L. Hoard, E. W. Silverton and J. V. Silverton, *J. Am. Chem. Soc.*, 1968, **90,** 2300.
674. A. I. Pozhidaev, M. A. Porai-Koshits and T. N. Polynova, *J. Struct. Chem.* (*Engl. Transl.*), 1974, **15,** 548.
675. B. J. Intorre and A. E. Martell, *J. Am. Chem. Soc.*, 1960, **82,** 358.
676. B. J. Intorre and A. E. Martell, *J. Am. Chem. Soc.*, 1961, **83,** 3618.
677. B. J. Intorre and A. E. Martell, *Inorg. Chem.*, 1964, **3,** 81.
678. Y. O. Aochi and D. T. Sawyer, *Inorg. Chem.*, 1966, **5,** 2085.
679. N. A. Kostromina, V. P. Shelest, T. V. Ternovaya and Ts. B. Konunova, *Russ. J. Inorg. Chem.* (*Engl. Transl.*), 1977, **22,** 1659.
680. V. P. Vasil'ev, V. P. Lymar and A. I. Lytkin, *Russ. J. Inorg. Chem.* (*Engl. Transl.*), 1978, **23,** 29.
681. V. P. Vasil'ev, V. P. Lymar and A. I. Lytkin, *Russ. J. Inorg. Chem.* (*Engl. Transl.*), 1978, **23,** 525.
682. V. P. Vasil'ev, V. P. Lymar and A. I. Lytkin, *Russ. J. Inorg. Chem.* (*Engl. Transl.*), 1978, **23,** 683.
683. V. P. Vasil'ev, V. A. Borodin, V. P. Lymar and A. I. Lytkin, *Russ. J. Inorg. Chem.* (*Engl. Transl.*), 1981, **26,** 1306.
684. N. S. Biradar, A. L. Locker and V. H. Kulkarni, *Rev. Roum. Chim.*, 1974, **19,** 45.
685. Ts. B. Konunova, E. I. Toma and V. D. Brega, *Sov. J. Coord. Chem.* (*Engl. Transl.*), 1976, **2,** 800.
686. V. A. Kogan, V. P. Sokolov and O. A. Osipov, *J. Gen. Chem. USSR* (*Engl. Transl.*), 1970, **40,** 292.
687. M. J. Frazer and B. Rimmer, *J. Chem. Soc.* (*A*), 1968, 2273.
688. A. V. Vincogradov and V. S. Shpinel', *Russ. J. Inorg. Chem.* (*Engl. Transl.*), 1961, **6,** 687.
689. D. F. Lewis and R. C. Fay, *J. Chem. Soc., Chem. Commun.*, 1974, 1046.
690. D. C. Bradley, M. B. Hursthouse and I. F. Rendall, *J. Chem. Soc., Chem. Commun.*, 1970, 368.
691. K. Wieghardt, I. Tolksdorf, J. Weiss and W. Swiridoff, *Z. Anorg. Allg. Chem.*, 1982, **490,** 182.
692. S. R. Gupta and J. P. Tandon, *Bull. Acad. Pol. Sci., Ser. Sci. Chim.*, 1973, **21,** 911.
693. J. Uttamchandani and R. N. Kapoor, *Monatsh. Chem.*, 1979, **110,** 841.
694. M. Singh, B. L. Mathur, K. S. Gharia and R. Mehta, *Inorg. Nucl. Chem. Lett.*, 1981, **17,** 291.
695. M. Singh, *Pol. J. Chem.*, 1982, **56,** 393.
696. A. Kriza and M. Nasta, *Rev. Roum. Chim.*, 1981, **26,** 693.
697. A. K. Srivastava, R. K. Agarwal, M. Srivastava and T. N. Srivastava, *J. Inorg. Nucl. Chem.*, 1981, **43,** 2144.
698. J. F. Clarke, G. W. A. Fowles and D. A. Rice, *J. Organomet. Chem.*, 1974, **74,** 417.
699. D. Bhatia, A. M. Bhandari, R. N. Kapoor and S. K. Sahni, *Synth. React. Inorg. Met.-Org. Chem.*, 1979, **9,** 95.
700. M. Hasan, B. S. Sankhla and R. N. Kapoor, *Aust. J. Chem.*, 1968, **21,** 1651.
701. M. E. Silver, H. K. Chun and R. C. Fay, *Inorg. Chem.*, 1982, **21,** 3765.
702. K. Abu-Dari and K. N. Raymond, *Inorg. Chem.*, 1982, **21,** 1676.
703. J. W. Buchler, G. Eikelmann, L. Puppe, K. Rohbock, H. H. Schneehage and D. Weck, *Liebigs Ann. Chem.*, 1971, **745,** 135.
704. J. W. Buchler and K. Rohbock, *Inorg. Nucl. Chem. Lett.*, 1972, **8,** 1073.
705. M. Gouterman, L. K. Hanson, G.-E. Khalil, J. W. Buchler, K. Rohbock and D. Dolphin, *J. Am. Chem. Soc.*, 1975, **97,** 3142.
706. N. Kim, J. L. Hoard, J. W. Buchler and R. Rohbock, unpublished results cited by J. L. Hoard in 'Porphyrins and Metalloporphyrins', ed. K. M. Smith, Elsevier, Amsterdam, 1975, pp. 347–348.
707. J. W. Buchler, M. Folz, H. Habets, J. van Kaam and K. Rohbock, *Chem. Ber.*, 1976, **109,** 1477.
708. B. D. Berezin and T. N. Lomova, *Russ. J. Inorg. Chem.* (*Engl. Transl.*), 1981, **26,** 203.
709. P. Mühl, *Z. Chem.*, 1967, **7,** 352.
710. V. E. Plyushchev, L. P. Shklover and I. A. Rozdin, *Russ. J. Inorg. Chem.* (*Engl. Transl.*), 1964, **9,** 68.
711. P. N. Moskalev, V. Ya. Mishin, E. M. Rubtsov and I. S. Kirin, *Russ. J. Inorg. Chem.* (*Engl. Transl.*), 1976, **21,** 1243.
712. S. Kher, S. K. Sahni, V. Kumari and R. N. Kapoor, *Synth. React. Inorg. Metal-Org. Chem.*, 1980, **10,** 431.

33
Vanadium

LUIS VILAS BOAS and JOAO COSTA PESSOA
Instituto Superior Tecnico, Lisbon, Portugal

33.1 GENERAL INTRODUCTION

The element was recognized in 1831, when N. G. Sefström was able to isolate and characterize the oxide; the name vanadium was derived from Vanadis, a goddess in Scandinavian mythology. The beautiful colours of vanadium compounds had been observed as early as 1801, by A. M. del Rio in his experiments with a new element he called erythronium because of the red colour after treatment with acid.[1]

Some tribulations in the experimental work and the variety of colours observed in early vanadium chemistry can now be explained as due to the formation of various coordination compounds, early symptoms of a rich and challenging chemistry.

There are vanadium compounds with formal oxidation states from −3 up to +5 with the exception of −2. Under ordinary conditions, the most stable states are +4 and +5. Suitable sources of information for the early chemistry of vanadium are Mellor's[2] and Sidgwick's[3] books. More recent work is covered by several reviews.[4–9]

The coordination chemistry of vanadium is strongly influenced by the oxidizing/reducing properties of the metallic centre, and the chemistry of vanadium ions in aqueous solution is limited to oxidation states +2, +3, +4 and +5, although V^{2+} can reduce water. Redox potentials are given in Table 1 and an *E vs.* pH diagram is shown in Figure 1.

The oxidation state −3 has been observed only in organometallic compounds, and until now there have been no reports of compounds with oxidation state −2, although there is no reason

Table 1 Standard Potentials of Vanadium Couples in Aqueous Solutions at 25 °C

Couple	*Standard potentials* (V) *in* Strongly acid solutions	Weakly acid solutions	Neutral and basic solutions
V^{V}–V^{IV}	1.000	0.723	0.991
V^{IV}–V^{III}	0.337	0.481	0.542
V^{III}–V^{II}	−0.255	−0.082	−0.486
V^{II}–V^{0}	−1.13	−1.13	−0.820

Data from Y. Israel and L. Meites, in 'Standard Potentials in Aqueous Solution', ed. A. J. Bard, R. Parsons and J. Jordan, Dekker, New York, 1985.

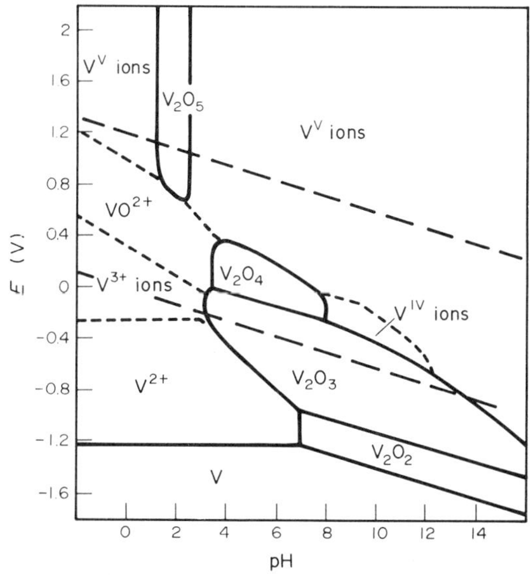

Figure 1 Potential *versus* pH diagram for the vanadium–water system at 25 °C. The dashed lines indicate the domains of relative predominance of the dissolved forms of the metal, but the various dissolved forms for each oxidation state are not explicit. The solid lines correspond to saturated solutions with a total vanadium concentration of 0.51 g dm^{-3}. The long dashed lines correspond to oxidation and reduction of water (for $E°$ values of 1.23 and 0.00 V respectively) (adapted from E. Deltombe, N. Zoubov and M. Pourbaix, in 'Atlas d'Equilibres Electrochimiques', ed. M. Pourbaix, Gauthier-Villars, Paris, 1963)

to believe that such compounds cannot exist. Coordination compounds with formal oxidation states from −1 up to +2 are usually octahedral; for some it is not possible to ascertain a formal oxidation state as the bonding scheme may alternatively be explained by coordination of the reduced ligand. The ability of V^{2+} to reduce water and the instability of most of the complexes in these low oxidation states explains the small number of compounds. On the other hand, the reducing properties of vanadium(II) ions and their complexes have been used in many preparative methods, and among such reactions, the dinitrogen reduction deserves special attention.

Although many vanadium(III) complexes are unstable towards air, there are quite a few compounds in this oxidation state, most with octahedral geometry. However, remarkable seven-coordinate complexes were also characterized.

High concentration of vanadium in the blood of some tunicates has been a long-standing problem of biochemistry. That vanadium(III) ions are part of the respiratory pigment has been ruled out recently. The efficient mechanism used to concentrate vanadium from sea water is now understood but the utility of vanadium for these living organisms is still an intriguing question.

The +4 oxidation state is the most stable under ordinary conditions and the majority of vanadium(IV) compounds contain the VO^{2+} unit (vanadyl ion), which can persist through a variety of reactions. Its complexes typically have square pyramidal or bipyramidal geometry with the vanadyl oxygen apical and the V atom lying above the plane defined by the equatorial ligands. Trigonal bipyramidal complexes are also known.

The VO^{2+} entity bonds most effectively to the more electronegative atoms, *e.g.* F, Cl, O, N (and also S and P).

Many vanadyl complexes are not air stable and generally they can be easily hydrolyzed and/or oxidized to the +5 oxidation state.

Table 2 gives examples of the coordination geometry adopted by each oxidation state.

Table 2 Oxidation States and Stereochemistries of Vanadium Complexes

Oxidation state	*Coordination number*	*Stereochemistry*	*Examples*
−3	6		$(Et_4N)_2[(Ph_3Sn)V(CO)_5]$
	7	Monocapped octahedral	$(Et_4N)[(Ph_3Sn)_2V(CO)_5]$
−1	6	Octahedral	$[V(bipy)_3]^-$
0	6	Octahedral	$[V(bipy)_3]$
1	6	Octahedral	$[V(bipy)_3]^+$
Low but not defined	6	Trigonal prismatic	$[V(S_2C_2Ph_2)_3]$
2	6	Octahedral	$[V(CN)_6]^{4-}$, $[V(H_2O)_6]^{2+}$
3	3	Planar	$[V\{N(SiMe_3)_2\}_3]$
	4	Tetrahedral	$[VCl_4]^-$
	5	Trigonal bipyramidal	*trans*-$[VCl_3(NMe_3)_2]$
	6	Octahedral	$[V(H_2O)_6]^{3+}$, $[V(C_2O_4)_3]^{3-}$
	7	Pentagonal bipyramidal	$K_4[V(CN)_7]\cdot 2H_2O$
4	4	Tetrahedral	$[VCl_4]$
	5	Trigonal bipyramidal	$[VOCl_2(NMe_3)_2]$
		Square pyramidal	$[VO(acac)_2]$, $[VS(acac)_2]$
	6	Octahedral	$[VO(acac)_2(dioxane)]$, $[V(catecholate)_3]^{2-}$, $[V(acac)_2Cl_2]$
	8	Dodecahedral	$[V(MeCS_2)_4]$, $[VCl_4(diars)_2]$
5	4	Tetrahedral	$[VOCl_3]$, $[V(Me_3SiO)_3(adamantylimido)]$
	5	Trigonal bipyramidal	VF_5 (g), VCl_5 (g)
		Square pyramidal	$[VOF_4]^-$
	6	Octahedral	$[VO(OMe)_3]$, $[VO_2(ox)_2]^{3-}$, $[VOCl_3(benzonitrile)_2]$
	7	Pentagonal bipyramidal	$[VO(NO_3)_3(MeCN)]$, $[V(dipic)(H_2NO)(NO)(H_2O)]^-$
	8	Dodecahedral	$[V(O_2)_4]^{3-}$

Many vanadium(V) compounds are oxo complexes containing the VO^{3+} or the VO^{2+} entity, and the *cis* geometry in dioxo complexes has been confirmed by structure determinations. A great number of oxo complexes containing halides, alkoxides, peroxide, hydroxamates and aminocarboxylate have been characterized. The oxidation of ligands by vanadium(V) prevents the isolation of a larger number of complexes. On the other hand, the oxidizing properties of vanadium(V) compounds are useful for many preparative reactions, namely for the catalysis of oxidations. Important examples are catalysts used for the oxidation of SO_2 to SO_3 in the industrial production of sulfuric acid, and there is great concern over the presence of small concentrations of vanadium in fuels (coal and petroleum) because of this ability of vanadium to catalyze the oxidation of sulfur dioxide.

Some similarities between vanadate and phosphate explain the biological activity of the metal in the oxidation state +5. The most remarkable example is the strong inhibition of the Na^+ pump (Na^+, K^+-ATPase). This and other problems in the bioinorganic chemistry of vanadium are likely to continue as areas of active research.

The great recent development in electrochemical techniques will certainly be helpful for the study of redox processes of a metal which can occur in so many oxidation states. Multinuclear NMR spectrometers will allow increased use of ^{51}V resonance as a routine method for the characterization of complexes in solution. Other recent developments are the study of polynuclear complexes, metal clusters (homo and hetero-nuclear) and mixed valence complexes, and it can be anticipated that these topics will soon become important areas of vanadium coordination chemistry, although the isolation of compounds with such complex structures is again symptomatic of the difficulties that may be found in the chemistry of this metal.

33.2 LOW OXIDATION STATES (BELOW +2)

33.2.1 Introduction

The low oxidation states (from −3 to +1) are less usual; ligands in these complexes are those capable of stabilizing low oxidation states of other metals.

The derivatives of the pentacarbonylvanadate(-III) ion have the lowest formal oxidation state known for vanadium. Solutions of $[Na(diglyme)_2][V(CO)_6]$ are reduced by sodium to $[V(CO)_5]^{3-}$. Various salts were isolated by using the alkali iodide or onium halide;[10] Rb^+ and Cs^+ salts are rather thermally stable. $[V(CO)_5]^{3-}$ with R_3EX (R = alkyl or aryl; E = Sn, Pb; X = halide) leads to $[(R_3E)V(CO)_5]^{2-}$ and $[(R_3Sn)_2V(CO)_5]^-$. The IR spectra of crystalline $(Et_4N)[(Ph_3Sn)_2V(CO)_5]$ and its solutions are inconsistent with a pentagonal bipyramidal geometry and X-ray methods showed[11] that the anion structure is approximately a monocapped octahedron: the Sn—V lengths are 2.757(3) and 2.785(3) Å with ∠SnVSn 137.9(1)°. The corresponding compounds with oxidation number −2 have not been isolated.

33.2.2 Cyanide Complexes

VBr_3, potassium and potassium cyanide react in liquid ammonia.[12] The brown product, $K_2V(CN)_2 \cdot 0.5NH_3$, was paramagnetic (1.86 BM) and extremely pyrophoric. ν(CN) at 1905 cm^{-1} is consistent with a low oxidation state; the UV reflectance spectrum was recorded: 45.0, 36.0 sh, 28.0 sh, 20.0 br ($cm^{-1} \times 10^3$).

33.2.3 Nitrogen Ligands

33.2.3.1 Bipyridyl, o-phenanthroline and terpyridyl

Complexes with formal oxidation states from +2 to −3 were reported by Herzog and co-workers in classic experiments; those for the bipyridyl complexes are in Scheme 1. Solutions of $[V(bipy)_3]I_2$ in 50% methanol on reduction by magnesium or zinc yield $[V(bipy)_3]$.[13] This had μ_{eff} = 1.9 BM and a two-dimensional X-ray study[14] indicated that the nitrogen atoms were at the corners of a distorted octahedron with V—N 2.10 ± 0.03 Å and ∠NVN 73.6 ± 1.5°. ESR spectra[15] also suggested trigonal symmetry and an electronic structure analogous to that for $[VS_6C_6Ph_6]$ (Section 33.2.5).

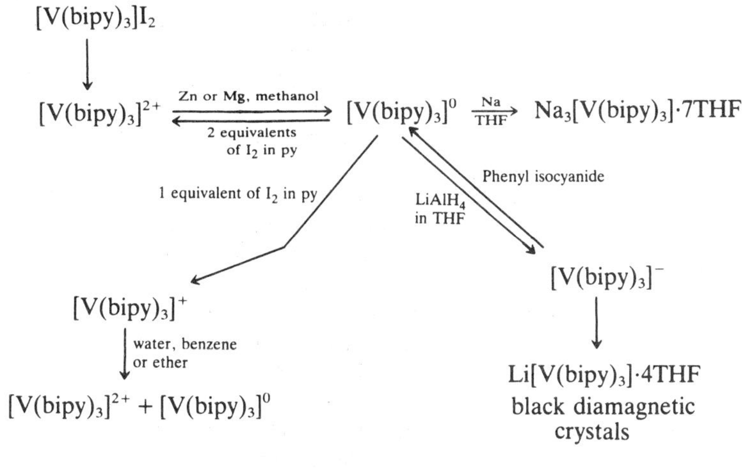

Scheme 1 Preparation and interconversions of bipy complexes in low oxidation states

Solutions of $[V(bipy)_3]^+$ can be obtained by partial oxidation of $[V(bipy)_3]$ with iodine but the solid obtained with empirical formula $[V(bipy)_3]I\cdot\frac{1}{2}py$ was suggested[16] to be a mixed crystal of the similar compounds of vanadium(0) and vanadium(II). $Na_3[V(bipy)_3]\cdot 7THF$ was obtained[17] on reducing the tris(bipy) complex with sodium. This air-sensitive compound has $\mu_{eff} = 2.76$ BM. The corresponding tris(phenanthroline) complexes were obtained[18] on reducing a $[V(phen)_3]I_2$ solution in THF with Li_2benzophenone to $[V(phen)_3]$ which could be reoxidized with two equivalents of I_2 back to $[V(phen)_3]^{2+}$. Further reduction of $[V(phen)_3]$ with Li_2benzophenone led to $[V(phen)_3]^-$ and $Li[V(phen)_3]\cdot 3.5THF$ was isolated. Similarly, black $[V(terpy)_2]$ was isolated when $[V(terpy)_2]I_2$ was reduced with magnesium or $LiAlH_4$.[19]

The reduction of VCl_3 by sodium amalgam in the presence of a suitable ligand has been proposed[20] as an improved method for the preparation of several zerovalent complexes VL_3 (L = 2,2′-dipyridyl disulfide, 2,2′-dipyridylamine, di-*t*-butyldiimine). The magnetic moments were 1.82, 1.71 and 1.79 BM. Such similarity of magnetic properties is in favour of similar bonding. An alternative description considers that the electrons of the reduced species are located in π^* orbitals of the ligand. Reduced forms of bipy ($bipy^-$ and $bipy^{2-}$) are known.

The UV-vis spectra of $[V(bipy)_3]$ were interpreted[21] as indicating the existence of partially reduced ligand ($bipy^-$ or $bipy^{2-}$) at least in the excited states, but were also interpreted[22] as indicating V^0 bonded to neutral bipy. By comparing IR spectra of $[V(bipy)_3]^z$ ($z = 2+, 0$) and Li(bipy), $[V(bipy)_3]$ should be considered[23] as a complex of the reduced ligand $bipy^-$ or a complex of collectively reduced ligands.[24]

The ESR spectrum of $[VL_3]$ [L = benzilbis(*N*-phenylimine)] shows[25] hyperfine splitting due to ^{51}V, corresponding to a d^5 low-spin configuration with 80–90% of the charge of the unpaired electron at the metal.

Polarography and cyclic voltammetry[26] of $[V(bipy)_3]^z$ ($z = 2+, 1+, 0, 1-, 2-, 3-$) gave $E_{1/2}$ values of −0.8, −1.01, −1.11, −1.55 and −2.23 V for these five steps, and smaller polarographic waves at −2.10 and −2.6 V were attributed to the reduction of free bipy. From comparison of the $E_{1/2}$ values for the bipy complexes with those for 4,4′-dimethyl-2,2′-bipy and 5,5′-dimethyl-2,2′-bipy, the electron added (or removed) in the reactions of $[V(L)_3]^z$ ($z = 2+, 1+, 0$) was at t_{2g} orbitals.[26] On the other hand, for $V(bipy)_3/V(bipy)_3^-$ and $V(bipy)_3^-/V(bipy)_3^{2-}$ the added electron would occupy orbitals with a predominant π^* character. However, $[V(bipy)_3]$ is diamagnetic, and an intermediate character has been proposed for it.

33.2.3.2 Nitrosyl

Using $[Co(X)(NO)_2]_2$ as a nitrosylating agent, Rehder and co-workers synthesized various nitrosyl and dinitrosyl complexes as depicted in Scheme 2.[27,28] An extensive study of ^{51}V NMR of solutions of nitrosylvanadium species has been published.[27] THF molecules in *cis*-$[V(NO)_2(THF)_4]$ are easily replaced by other L leading to *cis*-$[V(NO)_2(THF)_{4-n}L_n]X$ (L = several ligands such as py, phen, MeCN, $OPEt_3$, thiophene, acetone; $n = 2$ to 4; X = Cl, Br, I). For L = CNR (R = cyclohexyl, Pr^i, Bu^t), $[VX(NO)_2L_3]$ are also formed and the presence of coordinated X was established[27] by the normal halogen dependence (Cl < Br < I) of ^{51}V shielding in ^{51}V NMR.

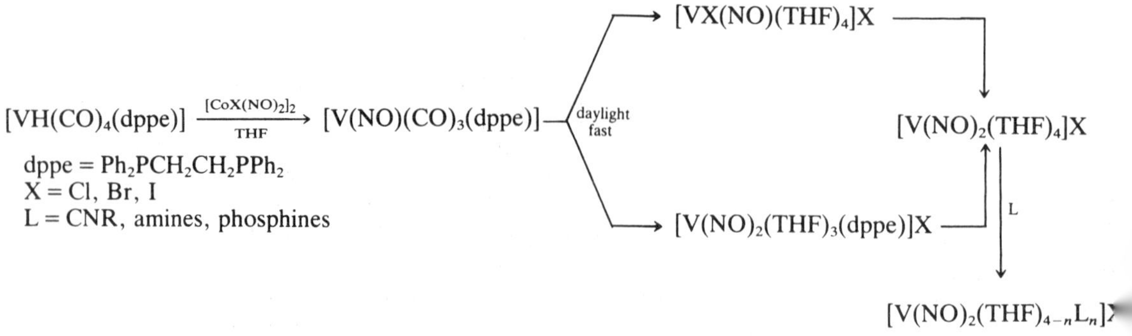

Scheme 2 Preparation of dinitrosyl complexes

A cyanonitrosyl complex, $K_3[V(NO)(CN)_5]\cdot 2H_2O$, was obtained[29] from VO_4^{3-} and CN^- in alkaline medium using hydroxylamine as the source of NO. Its structure[30] is tetragonally distorted octahedral: the mean V—C distance is 2.17 Å, there is a short V—N bond (1.66 Å) and the VCN and VNO linkages are almost linear. By conducting nitrosylation in H_2S,[31] $K_4[V(NO)(CN)_6]\cdot H_2O$ was isolated and $K_4[V(NO)(CN)_6]\cdot \frac{1}{2}KOH\cdot \frac{1}{2}H_2O$ studied by diffraction.[32] $[V(CN)_6(NO)]^{4-}$ has approximately pentagonal bipyramidal geometry with the nitrosyl group axial. The mean V—C bond length is 2.15 Å and the V—N bond length is 1.68 Å. These complexes have been used to prepare $K_2[V(NO)(CN)_4L]$ from $K_4[V(NO)(CN)_6]$ with L = bipy or phen[33] and $[V(NO)_2(L)_2]^+$ derivatives from $K_3[V(NO)(CN)_5]$.[34]

Using an excess of hydroxylamine in the reductive nitrosylation of VO_4^{3-} forms the dinitrosyl anion $[V(NO)_2(CN)_4]^{2-}$, and on treatment with bipy (or phen), complexes of formula $[V(NO)_2(CN)_2L]\cdot H_2O$ were isolated.[35]

Magnetic moments (μ_{eff} = 1.5 and 1.6 BM for bipy and phen derivatives) confirmed the existence of one unpaired electron and ESR (no $^{51}V-^{15}N$ or $^{51}V-^{13}C$ superhyperfine splitting) indicated that it is located in a metal orbital. Two strong ν(N—O) bands at 1650 and 1520 cm^{-1} confirm the *cis* NO groups.

A brown polynuclear dichlorotrinitrosylvanadium(−I) was isolated[36] from a solution of $[VCl_4]$ in carbon tetrachloride by reaction with dry NO, and Beck and co-workers reported[37] that the same product was obtained from $[V(CO)_6]$ with NOCl. A trigonal bipyramidal arrangement with terminal and bridging chloro ligands was proposed for this compound $[V(NO)_3Cl_2]_n$, which reacts with solvents losing NO and forming mononitrosyls $[V(NO)L_4Cl_2]$ (L = MeCN, Ph_3PO, $\frac{1}{2}$phen). Herberhold and Trampisch[38] used sodium amalgam or zinc powder to prepare $[V(NO)_2L_2Cl]_x$ (L = acetonitrile, methyl isonicotinate) and $[V(NO)_2(Bu^tNC)_4]PF_6$. IR and ^{51}V NMR indicate that a single *cis* dinitrosyl species is the intermediate. The extensive work with ^{51}V NMR is summarized in Figure 2.

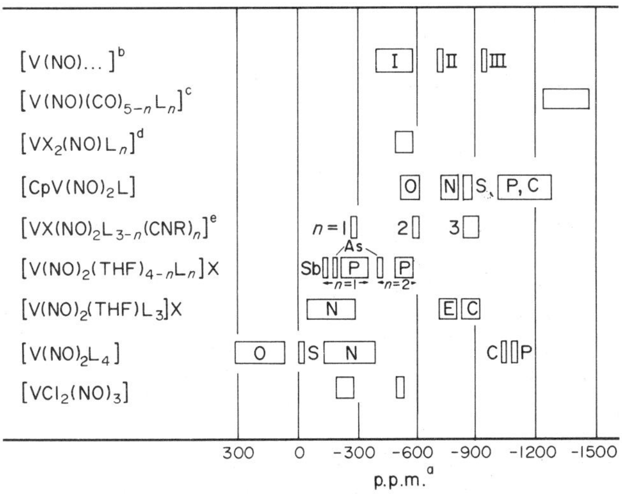

Figure 2 ^{51}V NMR shift ranges of nitrosyl vanadium complexes. Ligand abbreviations: N = NR_3, NCR; O = OCR_2, OR_2, OPR_3; P = PR_3, As = AsR_3, Sb = SbR_3, E = P, As, Sb; S = SR_2; X = Cl, Br, I, PF_6. (a) ^{51}V shielding increases from left to right; shift values are measured taking $VOCl_3$ as standard. (b) I = $[V(nitrilotripropanolato)(NO)L]^-$ (L = O ligand); II = $[V(dipicolinato)(NO)(ONH_2)(H_2O)]^-$; III = $[V(CN)_4(NO)(ONH_2)]^{3-}$. (c) L = P, As or CO. (d) L = O ligand or acetonitrile. (e) L = THF or acetonitrile. (f) I and II are respectively the di- and mono-nitrosyl derivatives formed in acetonitrile solution; III, resonance measured in CD_3NO_2 solution (data taken from refs. 27 and 38)

33.2.3.3 Dinitrogen complexes

Complexes of vanadium(0) with dinitrogen were obtained by co-condensations. $[V(N_2)_6]$ has been prepared[39] in frozen matrices; this formulation was based on the similarity of IR and UV-vis spectra with the corresponding $[V(CO)_6]$. By increasing the vanadium concentration, changes in the IR were explained as due to a dimer, $[V_2(N_2)_{12}]$.[39] Dinitrogen is acting as a weak ligand compared with CO, *i.e.* a poorer σ donor and a poorer π acceptor.[40]

33.2.4 Phosphorus Ligands

The dropwise addition of sodium naphthalenide to THF containing vanadium trichloride and 1,2-bis(dimethylphosphino)ethane (dmpe) causes changes of colours suggesting a stepwise reduction from +2 to 0. Brown $[V(dmpe)_3]$ was isolated[41] ($\mu_{eff} = 2.10$ BM) and IR data suggest octahedral coordination. The same complex was synthesized by a metal vapour technique.[42] The ESR was that expected and the unit cell is cubic with $a = 11.041(3)$ Å.

$[Et_4N][V(PF_3)_6]$ was prepared[43,44] by UV irradiation of $[Na(diglyme)_2][V(CO)_6]$ in diglyme saturated with PF_3 and studied[45] by multinuclear (^{19}F, ^{31}P, ^{51}V) NMR. The well-resolved spectra confirmed a (non-ideal) cubic geometry for the complex and V—PF_3 is weaker than V—CO.

Other V—P bonds may be found in phosphinehydrido complexes. The heptacoordinate $[V(H)(PF_3)_6]$ was prepared[43] from $[V(PF_3)_6]$ on heating with phosphoric acid. Reaction of a THF slurry of $[V_2(\mu\text{-}Cl)_3(THF)_6]_2Zn_2Cl_6$ with $PMePh_2$ followed by addition of $LiBH_4$ yielded a dark red air-sensitive material, a Zn_2V_2 aggregate with a V=V double bond and hydrogen bridging V to Zn atoms (Figure 3).[46]

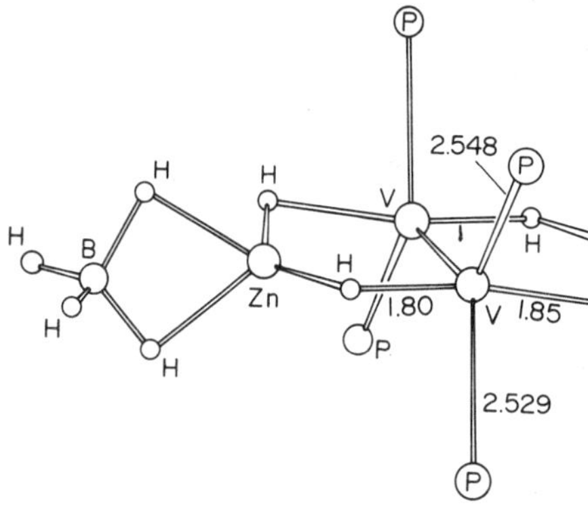

Figure 3 Structure of the phosphinohydrido complex $V_2Zn_2H_4(BH_4)_2(PMePh_2)_4$. The V—V distance of 2.400(2) Å was interpreted as corresponding to a double bond[46]

33.2.5 Tris(dithiolene) and Tris(diselenolene) Complexes

'Dithiolene complexes' was the name suggested by McCleverty[47] for complexes of unsaturated 1,2-dithiols without implying any particular ligand structure or valence formalism. The dithiolene complexes described differ from the vanadium(III) dithiolates (Section 33.4.6) reported by three independent groups.[48]

The charge at the vanadium in the usual dithiolene complexes may be estimated[49] as less than 2, and these complexes are therefore included in the section on low valence states. Various tris(dithiolene) complexes have been isolated and one-electron oxireductions studied by polarographic and voltammetric techniques. Table 3 summarizes methods of preparation and electrochemical behaviour. Schrauzer and co-workers[50] correlated electrochemical data with Taft's constants: the observed linear correlation reflected the ligand π orbital origin of the orbitals involved in the redox process.

$[V(S_2C_2Ph_2)_3]$ is trigonal prismatic, slightly distorted by a trigonal twist of 8.5°.[51] $[V(S_2C_2Ph_2)_3]^-$ was also trigonal prismatic as its electronic spectrum is similar to the isoelectronic $[Cr(S_2C_2Ph_2)_3]$,[51] which was isomorphous with $[V(S_2C_2Ph_2)_3]$. $(Me_4N)_2[V(S_2C_2(CN)_2)_3]$ was studied by diffraction[52] and a structure between octahedral and trigonal prismatic was proposed for the anion $[V(S_2C_2(CN)_2)_3]^{2-}$.

The C=C stretching frequencies of the $S_2C_2Ph_2$ and $S_2C_2H_2$ complexes increase with the negative charge of the complexes, *i.e.* the sulfur ligands become more 'dithiolate' in character.[53] The dithiolene complexes have several intense transitions in the visible, whose assignment is not yet complete. Using previous work as well as the resonance-Raman spectra for $[V(S_2C_2Ph_2)_3]^{0,1-}$ and $[V(S_2C_2(CN)_2)_3]^{2-}$, a set of assignments was produced (Figure 4).[54]

The ESR of $[V\{S_2C_2(CN)_2\}_3]^{2-}$ doped in single crystals of isomorphous

Table 3 Preparation and Electrochemical Studies of Tris(dithiolene) and Tris(diselenolene) Complexes

Method of preparation of isolated solid complex	*Formula*	*Z*	*R*	$E_{1/2}$ *values* (V)	*Ref.*
	$[V(S_2C_2R_2)_3]^Z$	−2	H		
				−0.722[a]	1
$Na_2S_2C_2H_2$		−1			2
				+0.25[a]	1
$Na_2S_2C_2H_2$ + VO(acac)$_2$		0			3
		−3	CN		
				−0.49[b]	4
$Na_2S_2C_2(CN)_2 + VCl_3$		−2			4
				+0.66[b]	4
(z = −2 complex) + $[Mo\{S_2C_2(CF_3)_2\}_3]$		−1			5
		−3	CF_3		
				−1.10[d]	
Na(diglyme) + $[V(CO)_6] + S_2C_2(CF_3)_2$		−2			5
				+0.47,[b] +0.08[d]	5
$[AsPh_4][V(CO)_6] + S_2C_2(CF_3)_2$		−1			5
		0			
Isolation as $(NEt_4)_2[V(S_2C_2Ph_2)_3]$		−2	Ph		3
				−0.71[b]	5
(Benzoin + P_4S_{10}) + (VCl_3 + Et_4NBr)		−1			5
				+0.30[b]	5
(z = −1 complex) + iodine		0			5
		−2	p-MeC_6H_4		
				−0.745[a]	1
Acyloin + P_4S_{10} method		−1			3
				+0.323[a]	1
		0			
		−2	p-$MeOC_6H_4$		
				−0.783[a]	1
Acyloin + P_4S_{10} method		−1			3
				+0.260[a]	1
		0			
	$[V(S_2C_6H_3R)_3]^Z$	−3			
				−0.12[b]	5
$Na_2(S_2C_6H_4) + VCl_3$		−2			5
		−3	Me		
				−0.18[b]	5
$Na_2S_2C_6H_3Me + VCl_3$		−2			5
	$[V(Se_2C_2(CF_3)_2)_3)]^Z$	−3			6
				−1.05[d]	6
(z = −1 complex) + $[AsPh_4]Cl$		−2			6
				+0.07[d]	6
$Se_2C_2(CF_3)_2 + [AsPh_4][V(CO)_6]$		−1			6

[a] Potentials measured in DMF solutions *vs.* Ag/AgCl electrode in an aqueous 0.10 M LiCl solution.[c]
[b] Potentials measured in acetonitrile solutions at room temperature *vs.* an aqueous calomel electrode saturated with NaCl.[c]
[c] The difference in potentials measured with techniques (a) and (b) was estimated as ~0.03 V, *i.e.* potentials in DMF may be estimated by adding 0.03 V to the corresponding values measured in acetonitrile.
[d] Potentials measured in CH_2Cl_2 *vs.* saturated calomel electrode.

1. D. C. Olson, V. P. Mayweg and G. N. Schrauzer, *J. Am. Chem. Soc.*, 1966, **88**, 4876.
2. G. N. Schrauzer, V. P. Mayweg and W. Heinrich, *Chem. Ind. (London)*, 1965, 1464.
3. G. N. Schrauzer and V. P. Mayweg, *J. Am. Chem. Soc.*, 1966, **88**, 3235.
4. A. Davison, N. Edelstein, R. H. Holm and A. H. Maki, *J. Am. Chem. Soc.*, 1964, **86**, 2799.
5. A. Davison, N. Edelstein, R. H. Holm and A. H. Maki, *Inorg. Chem.*, 1965, **4**, 55.
6. A. Davison and E. T. Shawl, *Inorg. Chem.*, 1970, **9**, 1820.

$[AsPh_4]_2[Mo\{S_2C_2(CN)_2\}_3]$ was fitted by an axially symmetric spin Hamiltonian with $g_{\parallel}$ = 2.000, $g_{\perp}$ = 1.974, $A_{\parallel}$ = 10 G and $A_{\perp}$ = −100 G, interpreted in terms of a $^2A_1$ ground state in D_3 symmetry.[49] The unpaired electron has predominantly metal d_{z^2} character. A molecular orbital scheme (Figure 4) was proposed for the dithiolene complexes.[55] This scheme has been used in the discussion of ESR results,[49] and electronic and resonance-Raman spectra.[54]

Tris(diselenolene) complexes derived from $Se_2C_2(CF_3)_2$ have been prepared by similar methods.[56]

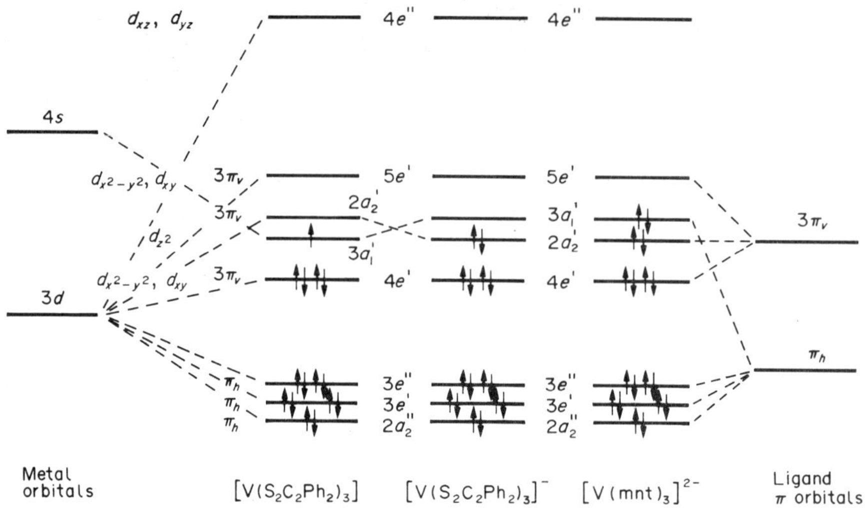

Figure 4 Electronic configurations of vanadium tris(1,2-dithiolene) complexes (from ref. 54, based on the scheme in ref. 55)

33.2.6 Hydrido Complexes

Hydridophosphino complexes have been described previously (Section 33.2.4). The bond dissociation energy of 209 kJ mol^{-1} has been measured[57] by ion beam mass spectrometry using the reaction in equation (1). The bond energies between V^+ and other atoms were compared with the corresponding carbon bonds: the energy of the bond to vanadium is about one half that of the corresponding carbon bond.

$$V^+ + H_2 \longrightarrow VH^+ + H \qquad (1)$$

33.3 VANADIUM(II)

33.3.1 Introduction

The small number of vanadium(II) compounds may be explained by (i) difficulties in keeping the compounds (they are very often easily oxidizable) and (ii) difficulties in characterization (because of contamination by higher oxidation state(s)). Although the easy oxidation may be inconvenient in the isolation and handling of compounds of vanadium(II), there are many examples of applications of vanadium(II) as a reducing agent for a large variety of both organic and inorganic species (see Section 33.3.8), usually behaving as a one-electron reducing agent.

Vanadium(II) compounds are usually prepared by reduction of acidic solutions of higher oxidation states. The usual geometry is octahedral and the electronic spectra often show three *d–d* spin-allowed bands. These spectra, as well as those of tetrahedral and linear complexes, have been reviewed.[58] For this d^3 configuration, a spin-only magnetic moment of 3.87 BM is expected. Slow substitution can be predicted for vanadium(II) complexes but such reactions are not as slow as those for the isoelectronic Cr^{III} ions.

33.3.2 Cyanides and Isocyanides

A solution of VCl_3 reduced with zinc and added to saturated KCN gave yellow $K_4[V(CN)_6]$. The structure[59] is almost a regular octahedron with V—C having a mean value of 2.161(4) Å and C—N a mean value of 1.153(7) Å, comparable to other cyano complexes of vanadium. The value 3.78 BM[60] for the room temperature magnetic moment is consistent with vanadium(II) and the spectrum showed[60] two *d–d* bands at 22.3×10^3 and 27.7×10^3 cm^{-1}. A purple powder[14] $K_2V(CN)_4$ was obtained by reaction of $K_2V(CN)_2{\cdot}0.5NH_3$ with NH_4CN in liquid ammonia. Its low magnetic moment (3.2 BM) and high ν(CN) were due to a polymeric cyanide-bridged structure.

$[V(CNBu^t)_6]^{2+}$ has been prepared by reduction of $V(CNBu^t)_3Cl_3$ or oxidation of $[V(CO)_6]^-$ in the presence of an excess of *t*-butyl isocyanide. In red $[V(CNBu^t)_6][V(CO)_6]_2$,[61] the octahedral cation has V—C 2.10(1) Å comparable to 2.161(4) Å for the hexacyanide.

33.3.3 Nitrogen Ligands

33.3.3.1 *Ammonia and amines*

Vanadium(II) chloride does not react with liquid ammonia at −33 °C, but at room temperature, it reacts slowly giving purplish brown $V(NH_3)_{5.4}Cl_{1.9}$ with a magnetic moment of 3.70 BM.[62] $[V(NH_3)_6]Br_2$ was isolated[63] from solutions of $[V(H_2O)_6]Br_2$ in ethanol by bubbling dry NH_3, or by condensing ammonia on to the hexahydrate. The magnetic moment is ~3.9 BM and the powder photograph has been indexed in terms of a face-centered cubic unit cell of side 10.5 Å, isomorphous with $[Ni(NH_3)_6]Br_2$. The three bands in the diffuse reflectance spectrum (32 000, 21 200 and 14 800 cm^{-1}) have been interpreted in terms of octahedral symmetry. Methylamine and VCl_2 in a Carius tube at 160 °C formed purple-brown $V(MeNH_2)_6Cl_2$.[64]

Complexes $[VL_3]X_2$ (L = en, 1,2-diaminopropane, 1,3-diaminopropane; X = Cl, Br, I) and $[V(dien)_2]X_2$ were prepared by mixing ethanolic solutions of amine and of vanadium(II) halide. The magnetic moments (3.66–3.91 BM) and electronic spectra were typical of octahedral vanadium(II) and there was no halogen coordination.[65]

33.3.3.2 *N-heterocyclic ligands*

Vanadium(II) alcoholates in ethanol[66] or hydrated vanadium(II) halides in *n*-butanol[67] were used to prepare complexes $[VL_6]X_2$ and $[VL_4X_2]$ (L = pyrazole, imidazole, dimethylimidazole, 1-methylimidazole, 2-methylimidazole, benzimidazole and isoquinoline; X = Cl, Br, I). $[V(benzimidazole)_2Cl_2]\cdot0.5$butanol has been obtained[67] and the reflectance spectrum was consistent with an essentially octahedral environment. The magnetic moment (2.41 BM at 298 K decreasing with temperature) also suggests a polymeric structure. On heating the green $V(1\text{-methylimidazole})_6Cl_2$ under vacuum, violet compounds formulated as V(1-methylimidazole)$_n$Cl$_2$ (n = 4, 5) were obtained.

(*i*) *Polypyrazolyl ligands*

Mani and co-workers[68] synthesized complexes of polypyrazolylborates $H_nB(pz)^-_{4-n}$ (n = 0, 1, 2) with vanadium(II). For n = 0 or 1, $[VL_2]$ were obtained with electronic spectra (and magnetic moments) consistent with an octahedral geometry and $HB(pz)_3^-$ and $B(pz)_4^-$ as tridentate ligands. For n = 2, $K[V(H_2B(pz)_2)_3]\cdot EtOH$ was isolated: the anion has a slightly distorted octahedral geometry with a mean V—N of 2.169(19) Å.[68] The isoelectronic poly(1-pyrazolyl)methane ligands have been used to synthesize vanadium(II) complexes.[69] The potentially bidentate ligands $H_2C(pz)_2$ and $H_2C(Me_2pz)_2$ react with vanadium(II) salts giving different stoichiometries, depending on the coordinating power of the anions: VL_2X_2 (L = $H_2C(pz)_2$, X = Cl, Br, I, NSC; L = $H_2C(Me_2pz)_2$, X = Br, NCS) and $[VL_2Cl]Y$ (L = $H_2C(pz)_2$, Y = BPh_4; L = $H_2C(Me_2pz)_2$, Y = PF_6, BPh_4).

They were octahedral with the exception of $[V(H_2C(Me_2pz)_2)_2Cl]Y$ (Y = PF_6 or BPh_4): these two have moments of 3.82–3.83 BM, virtually independent of temperature, and are monomeric and five-coordinate. The potentially tridentate $HC(pz)_3$ formed $[VL_2]Y_2$ (Y = Br, PF_6) and $[VL_2]_2[V(NCS)_6]$, with octahedral coordination.

Complexes of tris(1-pyrazolylethyl)amines (shown in Figure 5 and abbreviated as tpea) have been characterized. Complexes $[V(NCS)_2L]$ (L = tpea, R = H,[70] Me[71]) possess a slightly distorted octahedral geometry, the complex $[VClL]BPh_4$ (L = tpea, R = H) (μ_{eff} = 3.54 BM at 295 K and 2.93 at 90 K) has a polymeric structure with bridging chlorine atoms.

(*ii*) *Pyridine and related ligands*

$[V(py)_6][V(CO)_6]_2$ precipitated by adding solid $[V(CO)_6]$ to a toluene solution containing an excess of pyridine. By dissolving it in CH_2Cl_2, an isocarbonyl-bridged complex

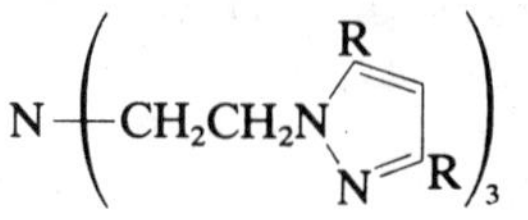

Figure 5 General formula of the tris(1-pyrazolylethyl)amine ligands

$[V(py)_4][V(CO)_6]_2$ is formed.[72] $[VL_4][V(CO)_6]_2$ (L = β- or γ-picoline) were obtained by reaction of $V(CO)_6$ and the ligand in benzene.[73]

$V(py)_4Cl_2$ has a *trans* structure with a mean V—N of 2.189 Å.[74] On heating it at 170 °C under vacuum, red-brown $V(py)_2Cl_2$ was detected and finally $V(py)Cl_2$ is formed. The formulae of complexes with pyridine, β-picoline and γ-picoline are listed in Table 4 with magnetic moments and thermal decomposition data.

Table 4 Complexes of Vanadium(II) with Pyridine and Picolines

Compound	μ_{eff} (BM) (300/90 *K*)	*Product*	*Thermal decomposition Temperature range* (K)	ΔH (kJ mol^{-1})
$V(py)_4Cl_2$	3.85/3.73	$V(py)_2Cl_2$	435–470	97 ± 2
$V(py)_2Cl_2$		$V(py)Cl_2$	475–575	59 ± 1
$V(py)_4Br_2$	3.92/3.75	$V(py)_2Br_2$	430–475	104 ± 2
$V(py)_2Br_2$		$V(py)Br_2$	505–550	71 ± 4
$V(py)_4I_2$		$V(py)_2I_2$	340–425	120 ± 2
$V(\beta\text{-pic})_4Cl_2$	3.91/3.86	$V(\beta\text{-pic})_2Cl_2$	410–470	133 ± 2
$V(\beta\text{-pic})_2Cl_2$		VCl_2	495–555	70 ± 3
$V(\beta\text{-pic})_4Br_2$	3.84/3.78	$V(\beta\text{-pic})_2Br_2$	420–500	166 ± 4
$V(\beta\text{-pic})_2Br_2$	3.41/2.28	$V(\beta\text{-pic})Br_2$	505–550	71 ± 4
$V(\gamma\text{-pic})_4Br_2$		$V(\gamma\text{-pic})Br_2$	450–560	200 ± 10
$V(\gamma\text{-pic})_4I_2$	3.94/3.85	VI_2	470–567	194 ± 2

Data taken from M. M. Khamar, L. F. Larkworthy, K. C. Patel, D. J. Phillips and G. Beech, *Aust. J. Chem.*, 1974, **27**, 41.

Complexes like $[V(L)_3]I_2$ (L = bipy, phen) and $[V(terpy)_2]I_2$ (already mentioned as starting materials for low-valent compounds) were isolated from solutions of VSO_4 by adding the corresponding ligand and iodide. The spectrum of $V(bipy)_3(ClO_4)_2$ shows[21] three bands due to *d–d* transitions with much metal-to-ligand charge-transfer character. The fine structure was due to vibrational interactions, evidence for electron delocalization within the excited states. $[V(aminoquinoline)_3]X_2{\cdot}2H_2O$ (X = Cl, Br, I) were synthesized[65] and showed spectra resembling those of the bipy complex.

Pyridine (and derivative) complexes $[VL_4(NCS)_2]$ have been isolated from aqueous ethanol; IR spectra indicated *trans* structures. On the other hand, $[VL_2(NCS)_2]$ with L = phen or bipy exhibit split CN absorptions suggestive of *cis* structures.[75]

33.3.3.3 Isothiocyanates

$K_4[V(NCS)_6]{\cdot}EtOH$ was prepared from VSO_4 solution in the presence of an excess of potassium thiocyanate,[76] and other counterions (NMe_4^+, NEt_4^+, Hpy^+) have been reported.[77] Thiocyanato groups are N-bonded and the moments are near the spin-only value. The electronic spectra have been interpreted on the basis of an octahedral d^3 ion although the splitting observed in the ν_2 band makes the assignment uncertain. On the other hand, X-ray powder photographs showed[77] that $[NEt_4]_4[V(NCS)_6]$ is isomorphous with the nickel complex, suggesting an octahedral geometry. In aqueous solutions, $[V(H_2O)_6]^{2+}$ reacts with the NCS^- ion;[78] the results from the thermodynamic and kinetic studies of this equilibrium have been summarized in Table 5.

33.3.3.4 Acetonitrile

The $[V(NCMe)_6]^{2+}$ ion was first obtained[73] in a solid $[V(NCMe)_6][V(CO)_6]_2$ as a product of the disproportionation of $[V(CO)_6]$. More recently, the structure of the

Table 5 Data on the Complexation of Vanadium(II) with Thiocyanate in Water at 25 °C

			Ref.
Composition of the complex[a]			
$V(H_2O)^{2+} + NCS^- \rightleftarrows V(H_2O)_5NCS^+$			1
Thermodynamic data			
$K = 21 \pm 3$,[b] 27 ± 8[c]			2[d]
$\Delta H = 22 \pm 3$ kJ mol^{-1}, $\Delta S = 46 \pm 13$ J K^{-1} mol^{-1}			1[e]
Kinetic data			
	Forward reaction (M^{-1} s^{-1})	Reverse reaction (s^{-1})	
k	22.3 ± 1.3	1.06 ± 0.09	2[f]
	15 ± 2	1.4 ± 0.2	3[g]
$\Delta H^{\neq}$ (kJ mol^{-1})	66.6 ± 1.6	62.7 ± 2.6	2
$\Delta S^{\neq}$ (J K^{-1} mol^{-1})	4.9 ± 5.4	-35.0 ± 9.4	2
$\Delta V^{\neq}$ (cm^3 mol^{-1})	-2.1 ± 0.8	-11.5 ± 0.9	2

Mechanism of reaction and estimated volume changes (cm^3 mol^{-1})

$$[V(H_2O)_6]^{2+} + NCS^- \underset{}{\overset{+3.2}{\rightleftarrows}} [V(H_2O)_6,NCS]^+ \xrightarrow{-5.3} \left\{ \begin{matrix} H_2O \\ SCN \end{matrix} V(H_2O)_5 \right\}^+ \xrightarrow{+11.5} [V(H_2O)_5NCS]^+ + H_2O \qquad 2$$

[a] Job's method of continuous variations. [b] From kinetic data. [c] From spectrophotometric data. [d] Ionic strength 0.5 M ($LiClO_4$). [e] Ionic strength 0.84 M ($LiClO_4$). [f] Stopped flow method. [g] Temperature jump method, 24 °C, ionic strength 1 M ClO_4^-/Cl^- medium.

1. J. M. Malin and J. H. Swinehart, *Inorg. Chem.*, 1968, **7**, 250.
2. P. J. Nichols, Y. Ducommun and A. E. Merbach, *Inorg. Chem.*, 1983, **22**, 3993.
3. W. Kruse and D. Thusius, *Inorg. Chem.*, 1968, **7**, 464.

hexakis(acetonitrile)vanadium(II) ion has been determined[79] in $[V(NCMe)_6][ZnCl_4]$ obtained from the reduction of VCl_3 with $ZnEt_2$ in acetonitrile. The vanadium is surrounded by the six acetonitrile nitrogen atoms in near octahedral geometry (with mean V—N of 2.11 Å) and the VNC bond angle (173.4° av.) shows a slight deviation from linearity.

33.3.4 Phosphines

Bis(triethylphosphine)vanadium(II) chloride may be prepared by the zinc reduction of $VCl_3(PEt_3)_2$ or from '$VCl_2(THF)_2$' upon addition of triethylphosphine. The solid has a low value of 1.98 BM consistent with a polymeric structure and μ_{eff} remained unchanged in benzene; cryoscopy in benzene showed a dimer.[80] The spectra of such solutions contain three *d–d* transition bands at 10 040 ($\varepsilon = 21$), 15 120 ($\varepsilon = 19$) and 22 500 cm^{-1} ($\varepsilon = 30$).

$[V_2(\mu\text{-}Cl)_3(THF)_6]_2[Zn_2Cl_6]$ was used to synthesize other dinuclear compounds with bridging chlorine atoms, bridging diphosphine ligands and bidentate tetrahydroborates: $[V(\mu\text{-}Cl)(\mu\text{-}diphos)BH_4]_2$ (diphos = bis(diphenylphosphino)methane[81] or bis(dimethylphosphinomethane[82]). The structures of these two were very similar; the bisdimethylphosphine complex is represented in Figure 6. The V—V distances (respectively 3.112(3) Å and 3.124(2) Å) are too long to suggest any significant metal–metal bonding.

Using 1,2-bis(dimethylphosphino)ethane (dmpe) mononuclear *trans*-$VX_2(dmpe)_2$ (X = Cl or Me) were obtained[83] and the *trans* structure of the dichloro complex was confirmed: the mean V—Cl and V—P were 2.44 and 2.50 Å respectively.

33.3.5 Oxygen Ligands

33.3.5.1 Water

A hydrolysis constant of 10^{-7} has been assumed[84] from the colour changes in VCl_2 titrated with base, and a pK value of 6.49 was determined from pH measurements in solutions of

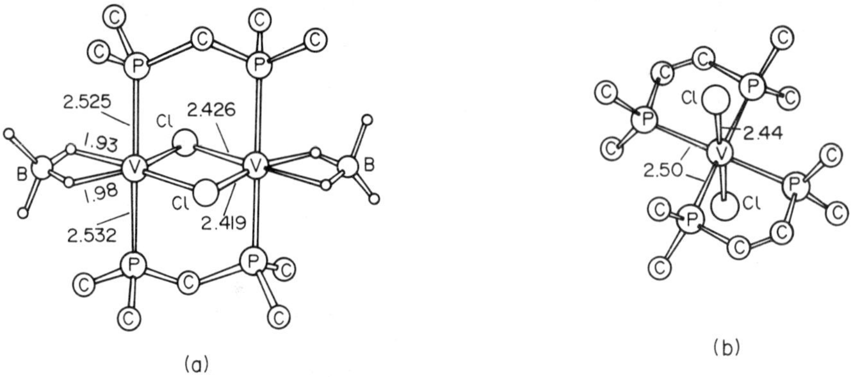

Figure 6 Structure of the phosphine complexes. (a) The complex with bis(dimethylphosphino)methane. A similar structure is found in the corresponding complex with bis(diphenylphosphino)methane.[82] (b) The complex with bis(diphenylphosphino)ethane[83]

VSO_4.[85] Polarography suggested dimeric species.[86] Baes and Mesmer[87] criticized the value for the hydrolysis constant, since it is difficult to avoid oxidation of the metal by water. Contamination with higher oxidation states would lead to high values of the constant. Molecular orbital studies[88] of hexahydrated divalent metals show that p$K > 10$ comparable to manganese(II) ions should be expected. Exchange in $[V(H_2O)_6]^{2+}$ has been studied by ^{17}O NMR:[89] $k_m^{298} = 87 \pm 4\ s^{-1}$, $\Delta H^* = 61.8 \pm 0.7$ kJ mol^{-1}, $\Delta S^* = -0.4 \pm 1.9$ J mol^{-1} K^{-1} and $\Delta V^* = -4.1 \pm 0.1$ cm^3 mol.$^{-1}$ This negative volume of activation was evidence of an associative interchange mechanism.

On reducing V_2O_5 in sulfuric acid by electrolysis, sodium amalgam or zinc, $[V(H_2O)_6]SO_4$ may be isolated, isomorphous with zinc sulfate. Tutton's salts are double sulfates $M_2[V(H_2O)_6](SO_4)_2$ ($M = NH_4^+$, K^+, Rb^+, Cs^+); in $(NH_4)_2[V(H_2O)_6](SO_4)_2$, the $[V(H_2O)_6]^{2+}$ have mean V—O of 2.15 Å.[90]

Several hydrated halides have been prepared[91] by dissolving V_2O_5 in acid (hydrochloric, hydrobromic or hydroiodic) and reducing by electrolysis: $[V(H_2O)_6]X_2$ (X = Br or I), $[V(H_2O)_4X_2]$ (X = Cl, Br or I) and $[V(H_2O)_2X_2]$ (X = Cl or Br). The hexa- and tetra-hydrates are magnetically dilute, with moments near the spin-only value and invariant with temperature. On the other hand, the dihydrates have moments which indicate a polymeric structure.

33.3.5.2 Alcohols and ethers

Green $[V(MeOH)_6][V(CO)_6]_2$ was prepared by reacting methanol with $[V(CO)_6]$.[92] Vanadium(II) chloride methanolates were prepared by electrolytic reduction of VCl_3 in methanol.[93] The hexasolvated vanadium(II) complex was converted to tetrakis and bis methanolates by careful heating. $[VCl_2(MeOH)_2]$ has μ_{eff} lower than the spin-only value and a polymeric structure (with Cl bridges) may be assumed.[94] In $[V(MeOH)_6]Cl_2$, the VO_6 unit has almost ideal octahedral geometry with average V—O 2.132(3) Å.[95] The electronic spectrum confirms a slightly weaker field for methanol compared to water. An alternative route[66] for vanadium(II) alcoholate synthesis involved treatment of ethanolic solutions of hydrated vanadium(II) salts with triethyl orthoformate. Several solids were isolated ($VBr_2{\cdot}6EtOH$, $VCl_2{\cdot}4EtOH$, $VCl_2{\cdot}2EtOH$).

The V^{II} cation in $[V(THF)_4][V(CO)_6]_2$ was shown to have octahedral coordination:[96] four THF molecules are in a planar array with V—O 2.170 Å. The axial positions are occupied by oxygen atoms of the two hexacarbonylvanadate(−I) anions with V—O(CO) 2.079 Å.

33.3.5.3 β-Diketones

A basic solution with a large excess of acetylacetone is strong blue. Complexes have been studied ($\log \beta_1 = 5.383$, $\log \beta_2 = 10.189$, $\log \beta_3 = 14.704$). A potential of −1.0 V *vs.* SHE was measured for the reduction of the V^{III} tris complex.[84] Mixed complexes $V(L)_2(py)_2$ (L = acac, trifluoroacetylacetonato or dibenzoylmethanato) isolated from solutions containing VSO_4 and the appropriate ligand[97] show intense absorption in the region 700–300 nm.

33.3.5.4 *Oxalate and carbonate*

$V(C_2O_4)\cdot 2H_2O$ may be obtained[98] from VCl_2 solution and sodium oxalate or oxalic acid; its thermal decomposition under a CO_2 atmosphere (~3000 bar) was shown[99] to occur at 262 ± 2 °C and by 280 °C the conversion into V_2O_3 is complete. The vanadium(II) carbonate was prepared from a VSO_4 aqueous solution and Na_2CO_3.[99] Amorphous $VCO_3\cdot 2H_2O$ was isolated. When this was heated in CO_2 (~3500 bar), the water content was lowered and crystalline products were obtained.

33.3.6 Halides

A compound obtained from reduction of VCl_3 by zinc metal in THF was shown to be $[V_2(\mu\text{-Cl})_3(THF)_6]_2[Zn_2(\mu\text{-Cl})_2Cl_4]$. Its structure has been determined at −105 °C[100,101] and −160 °C[102] avoiding the extensive and detrimental disorder in the THF ligands. Figure 7 shows the $[V_2(\mu\text{-Cl})_3(THF)_6]^+$ ion. At −105 °C, the mean V—Cl and V—O distances are 2.478(3) and 2.147(7) Å. The V—V (2.99 Å) corresponds to some interaction, and antiferromagnetic behaviour was observed with an exchange integral $-J = 75\ cm^{-1}$. The visible spectra in dichloromethane with a few drops of THF show a strong sharp doublet at *ca.* 25 000 cm^{-1},[101] assigned as a double spin-flip transition of the coupled pair of V^{2+} ions.

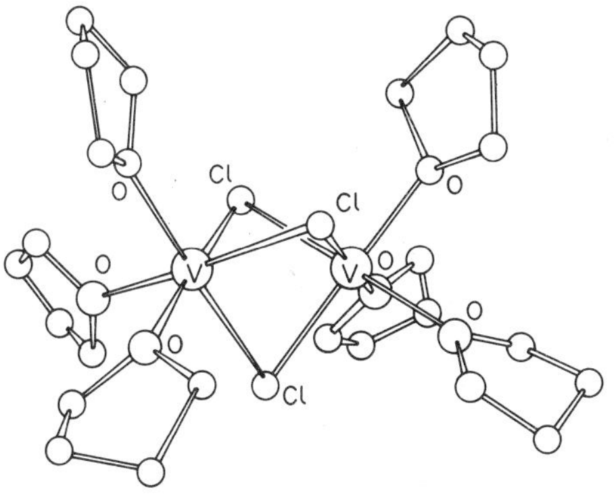

Figure 7 Structure of the $[V_2(\mu\text{-Cl})_3(THF)_6]^+$ cation[101–102]

Vanadium(II) dihalides and complex halides are prepared by high temperature dry routes. The preparation, structure and properties were reviewed.[103] Blue crystals of VF_2 have a rutile-type structure: the V—F distances (2.073 and 2.092 Å) are near the sum of the ionic radii.[104,105] Complex fluorides MVF_3 (M = Na, K or Rb) have been obtained from VF_2 and the appropriate fluoride by heating at 1200 °C and separating the crystals by means of a microscope[106] or heating at 800 °C and treating the crystals with dilute acid solution.[107] $NaVF_3$ has a distorted perovskite-type structure[106,107] and KVF_3 and $RbVF_3$ have a cubic structure of the ideal perovskite type.[107] By replacing some V^{2+} ions by V^{3+} in KVF_3, new compounds having the general formula $K_xV^{II}_xV^{III}_{1-x}F_3$ or K_xVF_3 may be isolated,[108,109] with structures similar to the potassium tungsten bronzes.

VCl_2 can be prepared by such methods as reduction of VCl_3 with hydrogen at 675 °C,[110] or the mixing of stoichiometric quantities of VCl_3 and V powder in a quartz tube under vacuum and slowly heating to 800 °C.[111] Crystals have a CdI_2-type structure.[112]

Using dry high temperature procedures, several complexes have been isolated: $KVCl_3$,[98,113] $CsVCl_3$,[98,113] NH_4VCl_3[114] and $TlVCl_3$.[114] $RbVCl_3$ has been obtained by careful dehydration of $RbVCl_3\cdot 6H_2O$.[115] All these have $CsNiCl_3$-type structures[98,113–115] with a slightly distorted octahedron of chloride ions.

VBr_2 can be prepared by reduction of VBr_3 with hydrogen[116] and VI_2 by reaction of stoichiometric amounts of the elements at 160 °C.[116] The diiodide may form black crystals by using an excess of iodine but red crystals are obtained by sublimation at 750 °C with vanadium in excess.[117] The dibromide[118] and the red form of vanadium diiodide[117] have structures of the CdI_2 type while the black form of VI_2 exhibits a $CdBr_2$-type structure.[117] The complexes $CsVBr_3$[113] and $CsVI_3$,[119] prepared by heating stoichiometric amounts of dihalide and the Cs halide at 700 °C in sealed quartz tubes, have structures similar to the chloride complexes (*i.e.*

$CsNiCl_3$-type). Table 6 summarizes results on electronic spectra. The three spin-allowed transitions are usually observed and extensive vibrational contribution has been observed in single crystal spectra.[120,121] Intense spin-forbidden transitions may also be observed[122] due to exchange interactions in the solid phase. The diiodide[123,124] and the complex halides[121] show trigonal distortion.

Table 6 The Absorption Spectra of Vanadium Dihalides and Complex Halides

Compound	ν_1 (cm^{-1})	ν_2 (cm^{-1})	ν_3 (cm^{-1})	*Dq* (cm^{-1})	*B* (cm^{-1})	*Ref.*
VF_2	10 300	16 450		1030	687	1
VCl_2	9200	14 270	22 220	920	615	2
	8980	13 920	21 840	800	755	3
$KVCl_3$				1010	640	4
$RbVCl_3$	7300 to 9700	13 200 to 14 900	21 900 to 25 600	1029		5
$CsVCl_3$				950	640	6
VBr_2	8600	13 330	20 330	860	530	2
$CsVBr_3$				870	625	6
VI_2	7870	12 270	18 920	790	510	2
	7870	12 280	20 330	787	540	7
$CsVI_3$	7700	12 000	18 700	770	515	8
				790	590	6

1. M. W. Shafer, *Mater. Res. Bull.*, 1969, **4**, 905; *B* value calculated in C. Cros, *Rev. Inorg. Chem.*, 1979, **1**, 163.
2. W. E. Smith, *J. Chem. Soc., Dalton Trans.*, 1972, 1634.
3. S. S. Kim, S. A. Reed and J. W. Stout, *Inorg. Chem.*, 1970, **9**, 1584.
4. W. E. Smith, *J. Chem. Soc. (A)*, 1969, 2677.
5. C. Cros, *Rev. Inorg. Chem.*, 1979, **1**, 163.
6. A. Hauser and H. U. Güdel, *Chem. Phys. Lett.*, 1981, **82**, 72.
7. W. van Erk and C. Haas, *Phys. Stat. Solidi B*, 1975, **71**, 537.
8. G. L. McPherson, L. J. Sindel, H. F. Quarls, C. B. Frederik and C. Doumit, *Inorg. Chem.*, 1975, **14**, 1831.

The magnetic properties have been studied by means of magnetic susceptibility and specific heat measurements; some results are in Table 7. In VF_2 there are strong exchange interactions (antiferromagnetic) between the vanadium atoms.[104,125] In the lamellar dihalides (VCl_2, VBr_2 and VI_2), the predominant interactions take place between atoms in the same plane. There are relatively strong antiferromagnetic interactions at low temperature, which decrease from the chloride to the iodide.[118] In the complex fluorides there are strong antiferromagnetic interactions due to superexchange couplings between the vanadium ions. The vanadium(II) complex halides of the $CsNiCl_3$ type (chloride, bromide and iodide) do not obey the Curie–Weiss law. The molar susceptibilities are weak and increase significantly from chloride to iodide. The strongly negative antiferromagnetic interactions in these halides result from either direct t_{2g}–t_{2g} or superexchange t_{2g}–t_{2g}–t_{2g} and e_g–s–e_g coupling.[113,126]

Table 7 Magnetic Moments and Curie–Weiss Law Parameters of Vanadium(II) Dihalides and Complex Halides

Halide	μ_{exp} (BM)	C_{exp} ($cm^3\ mol^{-1}$ K)	θ (K)	*Ref.*
VF_2	3.79	1.79	−90	1
$NaVF_3$	3.70	1.72	−52.5	2
	3.82	1.82	−59	3
KVF_3	3.94	1.94	−220	2
$RbVF_3$	3.93	1.93	−190	2
VCl_2	3.96	1.96	−437	4
VBr_2	4.07	2.07	−335	4
VI_2	4.07	2.07	−143	4

1. C. Cros, R. Feurer and M. Pouchard, *J. Fluorine Chem.*, 1975, **5**, 457.
2. C. Cros, R. Feurer and M. Pouchard, *Mater. Res. Bull.*, 1976, **11**, 117.
3. R. W. Williamson and W. O. J. Boo, *Inorg. Chem.*, 1977, **16**, 646.
4. M. Niel, C. Cros, G. Le Flem, M. Pouchard and P. Hagenmuller, *Nouv. J. Chim.*, 1977, **1**, 127; *Physica B + C (Amsterdam)*, 1977, **86–88B**, 702.

33.3.7 Macrocyclic Ligands

By reacting the appropriate halide with the ligand in anhydrous DMF, VLX_2 (L = Me_2[14]aneN_4, X = Cl; L = Me_6[14]aneN_4, X = Cl, Br, I) were obtained.[127] All have moments close to spin-only, the reflectance spectra can be interpreted assuming a tetragonally distorted octahedral structure, and the IR spectra confirm that the ligands adopt a planar arrangement of the four nitrogen atoms, the halides occupying the *trans* axial positions.

The chemistry of vanadium porphyrins has been reviewed.[128] Reduction of dihalogenovanadium(IV) porphyrins produced the corresponding vanadium(II) complexes *trans*-[V(por)L_2] (por = OEP, TPP, TpTP, TmTP; L = THF, $PPhMe_2$). The structures of [V(OEP)$(THF)_2$] and [V(OEP)$(PPhMe_2)_2$] have been determined.[129] In both, the vanadium is in the plane defined by the four porphyrin nitrogens and V—N are 2.051(4) Å in [V(OEP)$(PPhMe_2)_2$] and 2.046(4) Å in [V(OEP)$(THF)_2$]. V—O in the THF complex is 2.174(4) Å and the V—P bond in the phosphine complex is 2.523(1) Å, comparable with lengths of V—O in other THF complexes (Sections 33.3.5.2 and 33.3.6) and V—P in complexes with diphosphine ligands (Section 33.3.4).

33.3.8 Vanadium(II) as Reducing Agent

Vanadium(II) has been the reducing agent for a large variety of organic and inorganic substrates and both aqueous and nonaqueous solutions have been used. The solutions of vanadium(II) can be conveniently obtained by electrolysis (Section 33.3.5), from reduction of V_2O_5 with zinc metal[130] and by reduction of VCl_3 with lithium tetrahydroaluminate in anhydrous THF.[131]

The reduction of organic compounds by low valent metals has been reviewed[132] and Table 8 shows examples for vanadium(II). Vanadium acts as a one-electron reducing agent and usually the reactions involve the formation of organic radicals. For instance,[133] [$VCl_2(py)_4$] was rapidly oxidized by several aryl and alkyl halides RX to vanadium(III) and an almost quantitative yield of R—R obtained. Initial attack on the V^{II} complex by the aralkyl halide *via* an outer-sphere mechanism probably forms a radical which rapidly reacts with a further V^{II} yielding an aralkyl vanadium(III) intermediate. The coupled products are formed by oxidation of this by another molecule of RX. The reduction of organic radicals has also been studied: the rapid reduction of methyl radical by vanadium(II) was observed[134] in the reaction of V^{2+} with hydrogen peroxide and Me_2SO: methane (~25% based on the amount of H_2O_2) was formed with traces of ethane (<2%). The formation of methane was assumed to occur according to Scheme 3.

The oxidation of hexaaquavanadium(II) by aliphatic radicals has been studied by kinetic competition based on the homolytic decomposition of organopentaaquachromium(III) cations.[135] $[V(H_2O)_6]^{2+}$ is oxidized to V^{3+} by organic radicals such as $\cdot CMe_2OH$ or $\cdot$CHMeOEt with formation of Me_2CHOH and Et_2O respectively. The rate constants measured for the reactions were 2.1×10^5 (3.6×10^4 in 92% D_2O) and $5.9 \times 10^4\ M^{-1}\ s^{-1}$. Two mechanisms have been suggested:[134,135] either the formation of a seven-coordinate organovanadium intermediate or the direct abstraction of a hydrogen atom from a water molecule coordinated to a vanadium(II) ion.

33.3.8.1 The reduction of N_2 and related compounds

Shilov and collaborators discovered[136] considerable yields of hydrazine and ammonia when dinitrogen reacts with a suspension of freshly prepared V^{II} and Mg^{II} hydroxides. The magnesium is essential and yields of hydrazine and ammonia depend on temperature, pH, dinitrogen pressure and solvent (Scheme 4). The system has been used to reduce acetylene to ethylene and ethylene to ethane.[137] Shilov and collaborators assumed a multinuclear structure for the centre where N_2 reduction occurs and suggested the mechanism shown in Scheme 5.

An alternative explanation by Schrauzer and collaborators[138,139] assumes mononuclear vanadium(II) as a two-electron reductant: dinitrogen would be reduced in a first step to N_2^{2-}. N_2 and hydrazine would be obtained by disproportionation of N_2H_2 (diazene), inside the $Mg(OH)_2$ lattice, which would protect the diazene from rapid decomposition to the elements. On reducing acetylene in deuterated media, the resulting ethylene was exclusively *cis*.[139]

Solutions of vanadium(II) complexes of catechol (1,2-dihydroxybenzene) can reduce

Table 8 Examples of Reductions of Organic Compounds by Vanadium(II) Ions

Vanadium reagent	*Organic reagent*	*Product*	*Ref.*
$V(OH)_2$ or $V(catecholate)_3^-$	Acetylene or ethylene	Ethylene and ethane	1
$V(ClO_4)_2$	Aliphatic radicals	Aliphatic compounds	2
VCl_2	Aryl azides	Amines	3
VCl_2	Oximes	Ketones	4
VSO_4	Nitro aromatics	Arylamines	5
VCl_2	Alcohols	Dimers	6
	Aldehydes and aromatic aldehydes	Hydrobenzoins	7
	Quinones	Hydroquinones	8
	Benzils	Benzoins	8
	Maleic or fumaric acid	Succinic acid	9
$V(ClO_4)_2$	α-Keto acids	Tartaric acid derivatives	10
VCl_2	Sulfoxides	Sulfides	11
$V(py)_4Cl_2$	Aralkyl halides	Biaralkyls	12
VCl_2	α-Bromo or chloro ketones	Ketones	13
$V(ClO_4)_2$	Mercaptoacetic acid	Succinic acid	14
$V(acac)_2$	Epoxides	Alkenes	15
VCl_2	Tropyl cation	Dimer	16

1. L. A. Nikonova, S. A. Isaeva, N. I. Pershikova and A. E. Shilov, *J. Mol. Catal.*, 1975, **1**, 367 and references therein; G. N. Schrauzer and M. R. Palmer, *J. Am. Chem. Soc.*, 1981, **103**, 2659 and references therein.
2. J.-T. Chen and J. H. Espenson, *Inorg. Chem.*, 1983, **22**, 1651.
3. T.-L. Ho, M. Henninger and G. A. Olah, *Synthesis*, 1976, 815.
4. G. A. Olah, M. Arvanaghi and G. K. Surya Prakash, *Synthesis*, 1980, 220.
5. P. C. Banerjee, *J. Indian Chem. Soc.*, 1942, **19**, 30.
6. J. B. Conant and A. W. Sloan, *J. Am. Chem. Soc.*, 1925, **47**, 572; J. B. Conant, L. F. Small and B. S. Taylor, *J. Am. Chem. Soc.*, 1925, **47**, 1959; L. H. Slaugh and J. H. Raley, *Tetrahedron*, 1964, **20**, 1005.
7. J. B. Conant and H. B. Cutter, *J. Am. Chem. Soc.*, 1926, **48**, 1016.
8. T.-L. Ho and G. A. Olah, *Synthesis*, 1976, 815.
9. E. Vrachnou-Astra, P. Sakellaridis and D. Katakis, *J. Am. Chem. Soc.*, 1970, **92**, 811.
10. N. Katsaros, E. Vrachnou-Astra, J. Konstantatos and C. I. Stassinopoulou, *Tetrahedron Lett.*, 1979, **44**, 4319; J. Konstantatos, E. Vrachnou-Astra, N. Katsaros and D. Katakis, *J. Am. Chem. Soc.*, 1980, **102**, 3035.
11. G. A. Olah, G. K. Surya Prakash and T.-L. Ho, *Synthesis*, 1976, 810.
12. T. A. Cooper, *J. Am. Chem. Soc.*, 1973, **95**, 4158.
13. T.-L. Ho and G. A. Olah, *Synthesis*, 1976, 807.
14. J. Konstantatos, G. Kalatzis, E. Vrachnou-Astra and D. Katakis, *Inorg. Chem.*, 1983, **22**, 1924.
15. Y. Hayasi and J. Schwartz, *Inorg. Chem.*, 1981, **20**, 3473.
16. G. A. Olah and T.-L. Ho, *Synthesis*, 1976, 798; M. Pomerantz, G. L. Combs, Jr. and R. Fink, *J. Org. Chem.*, 1980, **45**, 143.

$$V^{2+} + H_2O_2 \longrightarrow V^{3+} + OH\cdot + OH^-$$
$$OH\cdot + (CH_3)_2SO \longrightarrow CH_3^{\cdot} + CH_3SO_2H$$
$$V^{2+} + CH_3^{\cdot} + H^+ \longrightarrow V^{3+} + CH_4$$

Scheme 3 Mechanism of formation of methane in the reaction of V^{2+} with hydrogen peroxide and DMSO

$$4V(OH)_2 + N_2 + 4H_2O \longrightarrow 4V(OH)_3 + N_2H_4$$
$$2V(OH)_2 + N_2H_4 + 2H_2O \longrightarrow 2V(OH)_3 + 2NH_3$$
$$6V(OH)_2 + N_2 + 6H_2O \longrightarrow 6V(OH)_3 + 2NH_3$$

Scheme 4 Dinitrogen reduction by vanadium(II) hydroxide

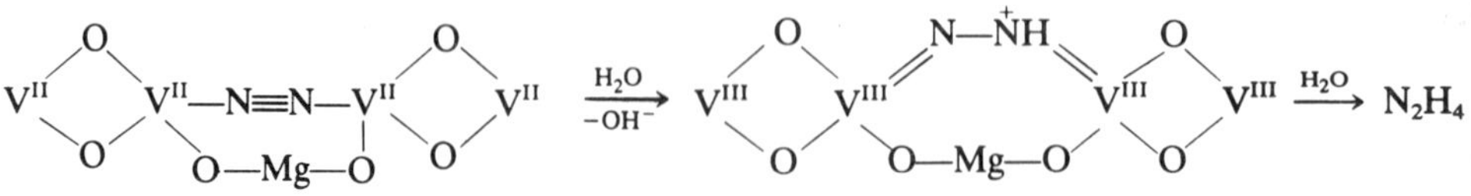

Scheme 5 Mechanism suggested for the reduction of dinitrogen by the suspension of freshly prepared V^{II} and Mg^{II} hydroxide

dinitrogen to ammonia in the pH range 8.5–13.5.[140] There is parallel reduction of solvent with dihydrogen evolution. The yield of NH_3 increases with the dinitrogen pressure up to 75% of the reducing agent: the other 25% are used in the simultaneous reduction of the solvent. In the limit, the stoichiometry of the reaction is as shown in equation (2), which is remarkable as the same stoichiometry is found in the biological fixation of dinitrogen (catalyzed by nitrogenase). This vanadium system has other similarities with the biological systems: (i) carbon monoxide is also reduced[141] and inhibits the formation of ammonia but not of dihydrogen;[142] (ii) acetylene is reduced to ethylene, and *cis*-deuteroethylene is obtained from C_2D_2;[143] (iii) if the reaction is stopped, there is an intermediate from which hydrazine can be obtained by acid.

$$8V^{2+} + N_2 + 8H_2O \longrightarrow 8V^{3+} + 2NH_3 + H_2 + 8OH^- \tag{2}$$

$$\frac{d[V^{(II)}]}{dt} = k_1[V^{(II)}]^2[N_2] + k_2[V^{(II)}]^{1/2} \tag{3}$$

A spectrophotometric method was used to study[144] the rate equation (3). The first term corresponds to the reduction of dinitrogen to ammonia and the second to the reduction of the protons of solvent to H_2.

The stoichiometry (equation 2) requires eight electrons transferred from vanadium(II) ions to dinitrogen atoms (six electrons) and hydrogen atoms (two electrons) and this kinetic equation (3) was interpreted as evidence for a polynuclear structure of the transition state during the reduction of dinitrogen. An alternative mechanism was suggested.[145]

For several α,ω-dicarboxylates and polymeric carboxylates the highest yield of hydrazine was found for $—O_2C(CH_2)_nCO_2—$ with $n = 5$ or 6.[146]

The most important contributions to studies of these reactions came from the groups of Schrauzer[147] and Shilov;[148] there has been a long controversy about the mechanism.

Aprotic media containing vanadium as the reducing species have also been proposed for dinitrogen reduction. Combinations of $[VO(acac)_2]$ with organometallic compounds ($EtMgBr$, Bu^nLi, Bu^i_3Al) react with dinitrogen.[149] On reacting VCl_3 in THF with Mg under N_2, nitrogen-containing complexes were found and it was suggested that the reduction occured *via* Scheme 6.[150] The reducing species is divalent. However, for $[VCl_3]$ or $[VCl_4]$ in THF with an excess of $Li(naphthalenide)$[151,152] and $Li_2(naphthalenide)$,[153] ESR experiments suggested that V^I or V^0 may be responsible. $[VCl_4]$ with Mg under dinitrogen gave a reaction similar to that with $[VCl_3]$.[150] After the dinitrogen fixation, the atmosphere was replaced by carbon dioxide and one molecule was fixed per vanadium: $[(THF)_3Cl_2Mg_2OV(NCO)]$ precipitated.[154] The reaction of trimethylchlorosilane with lithium in THF in the presence of VCl_3 under an N_2 atmosphere leads to tris(triethylsilyl)amine.[155]

$$\{V\}_2 + N_2 \rightleftharpoons \{VN_2V\}$$

$$\{VN_2V\} \xrightarrow{Mg} \{VN_2V\} \xrightarrow{3Mg} \begin{cases} V_2N_2Mg_4Cl_4(THF)_2 \\ \text{or} \\ VNMgCl_2(THF) \end{cases}$$
$$\qquad\qquad\quad \downarrow \atop Mg$$

Scheme 6 Mechanism suggested for the reduction of dinitrogen by VCl_3 and Mg in THF

33.3.8.2 *The reduction of H₂O, H₂O₂ and O₂*

$V(OH)_2$ is a strong reducing agent and freshly prepared $V(OH)_2$ reacts with water with dihydrogen evolution. In acidic solution, reduction of water may be induced by UV radiation.[138] From solutions containing complexes with catechol there is evolution of dihydrogen with simultaneous oxidation of the metal to vanadium(III). The reaction is first order in vanadium(II) and autocatalytic (Scheme 7).[145]

$$V^{II}(pyrocat) + H^+ \longrightarrow \left[\begin{array}{c} | \diagup \\ —V—H \\ \diagup | \end{array} \right] \xrightarrow{H^+} \left[\begin{array}{c} | \diagup \\ —V—H\text{---}H^+ \\ \diagup | \end{array} \right] \longrightarrow V^{IV}(pyrocat) + H_2$$

$$V^{IV}(pyrocat) + V^{II}(pyrocat)_n \longrightarrow 2V^{III}(pyrocat)$$

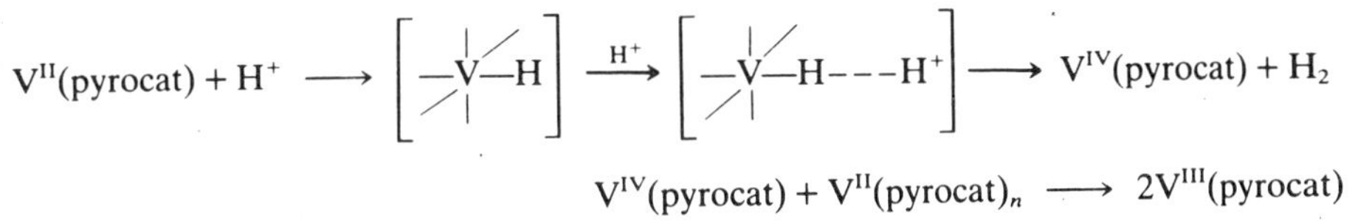

Scheme 7 Mechanism suggested for the reduction of water by $V^{(II)}$ pyrocathecolate complexes

Dihydrogen evolves from vanadium(II)-cysteine at pH 6.0–9.5. This reduction is first order in V^{II} and independent of pH in the range 7.5–8.5. If cysteamine or cysteine methyl ester is used, dihydrogen is still evolved. The reaction with serine is 1000 times slower than with cysteine even though the half wave potentials are comparable.[156] This reaction may be explained by a hydride pathway similar to that proposed for catechol complexes or alternatively Scheme 8.

$$H^+ + H^-{-}V^{IV}(cys)_n \xrightarrow{\text{fast}} H_2 + V^{IV}(cys)_n$$

$$V^{II}(cys)_m + V^{IV}(cys)_n \longrightarrow 2V^{III}(cys)_n$$

Scheme 8 Alternative mechanism suggested for the reduction of water by V^{II} cysteinato complexes

The reduction of H_2O_2 was studied[157] by stopped flow confirming evidence for a one-electron path.[158] A Fenton-type mechanism is likely.[134]

The reaction with oxygen was studied[157] by stopped flow; the results and mechanism are in Scheme 9.

$$V^{2+} + O_2 \xrightarrow{k_3=2\times10^3\ M^{-1}\ s^{-1}} (V{\cdot}O_2)^{2+} \underset{-V^{2+},\ k_{-5}\approx 20\ s^{-1}}{\overset{V^{2+},\ k_5\approx 3.7\times10^3\ M^{-1}\ s^{-1}}{\rightleftharpoons}} (V{\cdot}O_2{\cdot}V)^{4+}$$

$$(V{\cdot}O_2)^{2+} \underset{H_2O}{\xrightarrow{k_4\approx 100\ s^{-1}}} VO^{2+} + H_2O_2 \qquad (V{\cdot}O_2{\cdot}V)^{4+} \xrightarrow{k_6\approx 35\ s^{-1}} 2VO^{2+}$$

Scheme 9 Kinetics and mechanism of oxidation of vanadium(II) by molecular oxygen

A polymeric Schiff base complex gave reversible dioxygen addition:[159] the oxygen could be liberated in vacuum and above 200 °C.

33.3.8.3 The reduction of metal complexes

There are examples of outer and inner sphere mechanisms.[160] Many cobalt(III) complexes have been substrates. For halogenopentaamminecobalt(III), the rate constants ($k_F = 3.95$, $k_{Cl} = 10.2$, $k_{Br} = 30.0$, $k_I = 127\ M^{-1}\ s^{-1}$) show a systematic trend.[161] Comparisons with the outer-sphere reductant hexaammineruthenium(II) (k_{Ru})[162] or with $Cr(bipy)_3^{2+}$ (k_{Cr})[161] show linear correlation (equations 4 and 5).

$$\log k_V = 0.89 \log k_{Ru} - 0.64 \tag{4}$$

$$\log k_V = 1.18 \log k_{Cr} - 6.1 \tag{5}$$

The reduction of carboxylatopentamminecobalt(III) complexes by vanadium(II) is predominantly inner-sphere.[163] From $[Co(en)(C_2O_4)_2]^-$ and $[Co(C_2O_4)_3]^{3-}$, $V(C_2O_4)^-$ and V^{3+} ions are primary products.[164] For N-bonded glycinato- and β-alaninato-pentaamminecobalt(III), there is an intermediate with V^{II} attached to the unprotonated carboxylate group.[165]

The reduction of diamminebis(dimethylglyoximato)cobalt(III) is a substitution-controlled inner-sphere process while the monohalogeno complexes (amminebromo and amminechloro) react by an outer-sphere mechanism.[166] The activation entropies are -46.0 ± 1.7, -129.7 ± 1.7 and $-137.7 \pm 1.7\ J\ K^{-1}\ mol^{-1}$. In the cationic cobaloximes $[Co(HDMG)_2L_2]^+$ ($L = EtNH_2$, $PhNH_2$, py or 3,5-Me_2py), a plot of $\Delta G^{\neq}$ *vs.* $E_{1/2}$ illustrates progression from inner- to outer-sphere mechanism (Figure 8). $[Co(HDMG)_2(NH_3)X]$ ($X = N_3^-$, NCS) have $\Delta S^{\neq}$ values of -73.0 ± 3.0 and $48.6 \pm 4.0\ J\ K^{-1}\ mol^{-1}$.[167]

Several azidoammines react *via* inner-sphere pathways, except *trans*-$Co(NH_3)_4(H_2O)(N_3^-)^{2+}$ ($k \approx 100\ M^{-1}\ s^{-1}$).[168]

Some reactions with cobalt(III) complexes show autocatalysis by the unbound ligand released during the primary reaction,[169] *via* reduction to a radical, Lig·, which reduces another cobalt(III) and regenerates the ligand (Scheme 10).

Tris(picolinato)vanadate(II) reduces $[Co(edta)]^-$, $[Co(NH_3)_6]^{3+}$, $[Co(en)_3]^{3+}$ and $[Co(sep)]^{3+}$ (sep = sepulchrate) with second-order rate constants of 7.7×10^4, 8.0×10^3, 8.2×10^2 and 8.4×10^4.[170] $k_{11} = 3.1 \times 10^6\ M^{-1}\ s^{-1}$ for self exchange of $V(pic)_3^{0/-}$,[170] eight orders

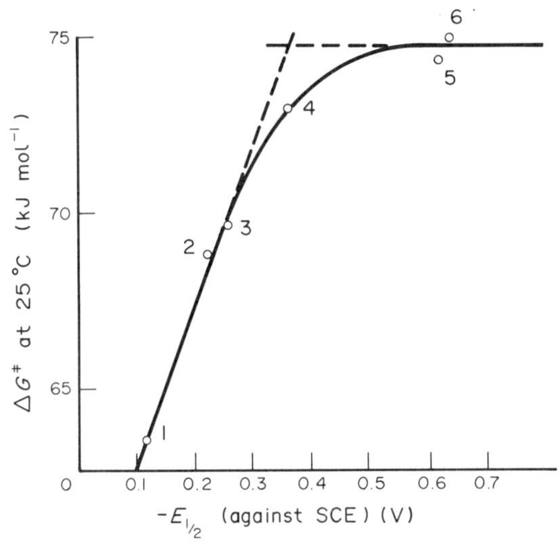

Figure 8 Free energy for the V^{2+}(aq) reduction of cobaloximes $[Co(HDMG)_2LL']$ compared with their polarographic reduction

$$\text{Lig} + V^{2+} \underset{k_{-1}}{\overset{k_1}{\rightleftharpoons}} \text{Lig}\cdot + V^{3+}$$

$$\text{Lig}\cdot + \text{Co}^{\text{III}}\text{ complex} \xrightarrow{k_2} \text{Lig} + \text{Co}^{\text{II}}\text{ complex}$$

Scheme 10 Mechanism of the autocatalysis observed in the reduction of cobalt(III) complexes by vanadium(II)

of magnitude larger than for $[V(H_2O)_6]^{3+/2+}$ but comparable to $[V(bipy)_3]^{3+/2+}$; a more difficult rearrangement is associated with electron transfer for $[V(H_2O)_6]^{3+/2+}$.

The reduction of platinum(IV) complexes involves two one-equivalent changes $Pt^{IV} \rightarrow Pt^{III} \rightarrow Pt^{II}$; the first step is rate-determining. Second-order rate constants with vanadium(II) (k_V) and with hexaammineruthenium(II) (k_{Ru}) show linear correlation (equation 6).[171]

$$\log k_V = 0.89 \log k_{Ru} - 1.68 \quad (6)$$

A wide variety of microorganisms synthesize powerful iron(III) chelators (siderophores). Their reduction (ferroxamine B, ferrichrome and ferrichrome A) has been studied.[172]

33.3.9 Biological Dinitrogen Fixation

The isolation of a vanadium nitrogenase suggests that vanadium may have a more important role in the nitrogen cycle than was recognized.[173] This vanadium enzyme may have similarities with the chemical systems described in Section 33.3.8.1.

33.4 VANADIUM(III)

33.4.1 Introduction

Vanadium(III) compounds are usually prepared from V^V or V^{IV} by electrolytic reduction; many complexes were prepared using VCl_3 as starting material. Usually, vanadium(III) complexes are unstable towards air or moisture and have to be prepared under an inert atmosphere. Many have octahedral coordination and their electronic spectra are typical of d^2 ions with three spin-allowed d–d transitions:

$$\nu_1 = {}^3T_{2g}\,(F) \longleftarrow {}^3T_{1g}\,(F)$$
$$\nu_2 = {}^3T_{1g}\,(P) \longleftarrow {}^3T_{1g}\,(F)$$
$$\nu_3 = {}^3A_{2g}\,(F) \longleftarrow {}^3T_{1g}\,(F).$$

Usually, the third band is not observed as it is weak and frequently masked by charge-transfer transitions.

33.4.2 Cyanide and Isocyanide

Several procedures for the preparation of cyanide complexes have been reported. The heptacyanovanadate(III) ion has been shown to be a pentagonal bipyramid from a crystallographic study of $K_4V(CN)_7{\cdot}2H_2O$.[174,175] The averaged V—C distance is 2.15(1) Å, and there is no significant difference between the equatorial and the axial V—C distances. The coordination number seven is unusual in complexes of monodentate ligands. The crystal field stabilization energy cannot explain the pentagonal bipyramidal geometry as a cubic configuration has a nearly equivalent stabilization energy.[176]

The observed structure has been rationalized in terms of the 'nine-orbital' or the '18-electron' rule.[175] The two electrons of the paramagnetic $[V(CN)_7]^{4-}$ ion ($\mu_{eff} = 2.8$ BM) will occupy two degenerate orbitals of the metal ($3d_{xz}$ and $3d_{yz}$) and the seven orbitals of the ligands containing the donated electrons achieve the 'nine orbitals' of the rare gas krypton. A molecular orbital treatment of this species also suggested large interactions between orbitals on adjacent equatorial ligands not encountered in octahedral symmetry.[177]

A monohydrate[60] of $K_4[V(CN)_7]$ and the preparation of the anhydrous material have also been reported.[178] The geometry of the anions in all cases has been suggested to be the same.[179] The IR spectrum of the dihydrate shows two bands (2100 and 2070 cm^{-1}) similar to those in solution (2104 and 2027 cm^{-1}). Therefore the anion in solution must be $[V(CN)_7]^{4-}$ with a pentagonal bipyramidal geometry. Attempted studies[179,180] of the ESR spectrum have been unsuccessful due to the disproportionation products: $[VO(CN)_5]^{3-}$ and $[V(CN)_6]^{4-}$.

The energy ordering of the 3*d* metal orbitals is $d_{xz}, d_{yz} < d_{x^2-y^2}, d_{xy} < d_{z^2}$ predicted in a crystal field approach[181] or by a molecular orbital treatment.[176] $K_4[V(CN)_7]{\cdot}2H_2O$ in a Nujol mull at liquid nitrogen temperature shows[185] three absorption bands at 20 600, 22 800 and 28 600 cm^{-1},[181] assigned as $^3A_2' \rightarrow {}^3E_2''$, $^3A_2' \rightarrow {}^3E_1[(e_1'')(e_2')]$ and $^3A_2' \rightarrow {}^3A_1''[(e_1'')(a_1')]$.

VCl_3 treated with Bu^tNC produces *mer*-$[VCl_3(CNBu^t)_3]$. The structure was determined by X-ray diffraction.[182] The ^1H NMR spectra in $CDCl_3$ showed that the free ligand exchanges more rapidly with the isocyanide *trans* to chloride than with the *trans* isocyanide groups. Therefore, the kinetic *trans* effect of chloride ion is larger than that of isocyanide.

33.4.3 Nitrogen Ligands

33.4.3.1 Ammonia, amines and amides

On reaction of ammonia with vanadium(III) halides there is ammonolysis of one V—X bond. The amidohalides $VX_2(NH_2){\cdot}nNH_3$ (X = Cl[62,183] or Br,[183,184] $n = 4$–5) have an ammonia content that depends on the temperature. A similar compound $V(NCS)_2(NH_2){\cdot}4NH_3$ has been prepared from $V(NCS)_3$ in THF with NH_3.[185] These amidohalides are polymeric.[6,185]

VCl_3 with primary amines gave similar derivatives containing one coordinated amide and one or more molecules of amine,[64] but ethylenediamine[186,187] and propylenediamine[186] give $[V(L)_3]Cl_3$ (L = en, pn). These are isomorphous with the chromium(III) complexes; the chlorine atoms are ionic, and the cations are tris-bidentate.

Tetrahedral blue-black compounds $[VL_4](SCN)_3$ (L = en or Me_4en) were prepared[188] from $K_3[V(NCS)_6]$ and their electronic spectra are comparable to other tetrahedral vanadium(III) complexes with substituted pyridines (see Section 33.4.3.2).

$VCl_3(NMe_3)_2$ reacts in benzene with lithium bis(trimethylsilyl)amide to give $[V\{N(SiMe_3)_2\}_3]$.[189] Similarity of spectroscopic properties led to the suggestion of a trigonal structure similar to that of the iron(III) complex.[190] The complex is paramagnetic ($\mu_{eff} =$ 2.38 BM) but shows no electron paramagnetic resonance. Bands at 19 200, 15 900 and 12 000 cm^{-1} have been assigned to *d–d* transitions.[191]

On reacting vanadium(IV) chloride or vanadium(III) bromide with trimethylamine $VX_3(NMe_3)_2$ (X = Cl or Br) were obtained.[192] The driving force for the reduction of $[VCl_4]$ is the formation of $VCl_3(NMe_3)$.[193] This chocolate brown compound has been isolated from the solution remaining after extraction of the pink-mauve $[VCl_3(NMe_3)_2]$ whose structure has been determined and as shown in Figure 9 is a trigonal bipyramid.[194]

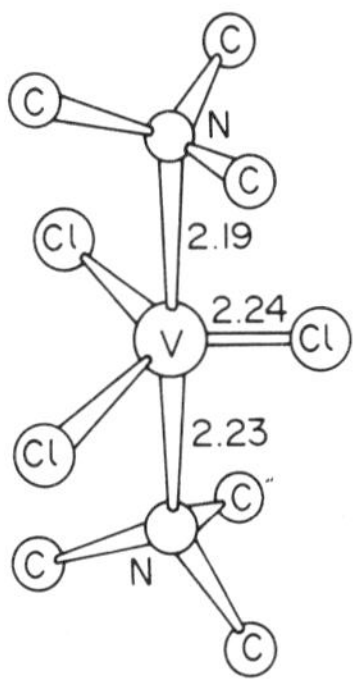

Figure 9 Structure of $[VCl_3(NMe_3)_2]$[194]

The electronic spectra of these bipyramidal species are more complex than the related octahedral chromophores and two characteristic spin-allowed bands are predicted in the near IR region: ${}^3A_2' \rightarrow {}^3A_1'', {}^3A_2''$ and ${}^3A_2' \rightarrow {}^3E''$ respectively at 4900 and 7100 cm^{-1} in $[VCl_3(NMe_3)_2]$.[195] In benzene, it decomposes irreversibly to $V_2Cl_6(NMe)_3$ with a second-order rate constant of 0.0166 $l\,mol^{-1}\,s^{-1}$.[196] The IR spectrum shows V—Cl terminal and bridging bands and the magnetic moment (2.55 BM) is significantly smaller than for a magnetically dilute vanadium(III) complex.

33.4.3.2 N-heterocyclic ligands

Complexes $V(L)Cl_2 \cdot 2EtOH$ (LH = guanine)[197] and $V(LH)L_2Cl \cdot EtOH$ (LH = purine or adenine)[198] have been isolated after refluxing VCl_3 with the ligand in a mixture of ethanol and triethyl orthoformate.

$VCl_3(py)_3$ has been prepared[187,199] and binding energies measured by ESCA:[200] 514.1 (for V $2p^{3/2}$ level), 196.7 (for Cl 2*s*) and 398.0 eV (for N 1*s*). On reacting $K_3[V(NCS)_6]$ in Pr^iOH with pyridine or alkylpyridines, tetrahedral blue-black compounds $[VL_4](SCN)_3$ have been isolated although they were unstable,[188] forming the corresponding octahedral complexes by liberating amine on rearrangement. Their spectra have been studied.[201] For example, $[V(py)_4](NCS)_3$ has three bands at 9500, 16 200 and 25 000 cm^{-1} assigned as ${}^3A_2 \rightarrow {}^3T_2$, ${}^3A_2 \rightarrow {}^3T_1\,(F)$ and ${}^3A_2 \rightarrow {}^3T_1\,(P)$.

A similar preparative procedure was used to isolate tetrahedral complexes of bidentate donors (L = bipy or phen) $[VL_2](SCN)_3$. However, complexes of vanadium(III) chloride or bromide with these same bidentate donors have stoichiometries depending on conditions.[199] With excess of bipy or phen, complexes $[VX_2L_2]X$ were obtained. In EtOH or in MeCN two series of complexes were isolated:[202] VCl_3LL' (L = bipy or phen; L′ = EtOH or MeCN) and $[VCl_2L_2]^+ + [VCl_4L]^-$. In the reaction of vanadium(III) with phenanthroline[202,203] or bipyridine[204] there is evidence for extensive hydrolysis. Spectrophotometry was used to estimate the equilibrium constant ($K_{eq} = 0.43$) for the reaction in equation (7).[203]

$$V(OH)^{2+} + 2phenH^+ \rightleftharpoons [V(OH)(phen)_2]^{2+} + 2H^+ \tag{7}$$

For $[V] > 2 \times 10^{-3}$ M, an increase in absorptivity indicates dimerization. The nature of the dimer has been discussed.[204]

33.4.3.3 Thiocyanate, selenocyanate and azide

From spectrophotometry, the formation constant of $V(NCS)^{2+}$ is $\sim 10^2$.[205] Complexes containing $[V(NCS)_6]^{3-}$ have been isolated and the structure of $K_3[V(NCS)_6] \cdot 12H_2O$ determined by X-ray diffraction,[206] confirming that thiocyanate coordinates *via* the N atom (the mean V—N distance in this complex is 2.044(2) Å).

When VCl_3 in acetonitrile reacts with ammonium thiocyanate, a progressive exchange occurs.[207] Complexes with various stoichiometries were isolated: $VCl_2(NCS) \cdot 3MeCN$, $VCl(NCS)_2 \cdot 3MeCN$, $V(NCS)_3 \cdot 5MeCN$ and $Ct_3[V(NCS)_6] \cdot 2MeCN$ ($Ct = K^+$, NH_4^+, pyH^+, piperidinium, Ph_4P^+, $MePh_3As^+$ or Me_3S^+). Similar complexes containing bromide and thiocyanate were also isolated as well as a mixed complex of NCS^- and $NCSe^-$ formulated as $M_3[V(NCS)_3(NCSe)_3]\cdot 2MeCN$ ($M = K^+$ or Na^+). Complexes $Ct_3[V(NCSe)_6]$ ($Ct = Ph_4As^+$, Me_4N^+, Et_4N^+ or K^+) have also been prepared.[208,209]

Several azido species have been detected in acetonitrile: $V(N_3)^{2+}$, $V(N_3)_3$ and $V(N_3)_6^{3-}$.[210] $(Bu_4N)_3[V(N_3)_6]$ has been isolated.[208] As expected for octahedral V^{III}, the magnetic moments are slightly lower than spin-only values and there is a marked temperature dependence at low temperatures.[211]

33.4.3.4 Nitriles

Anhydrous vanadium trichloride or tribromide reacts with acetonitrile to yield green or red-brown solutions. Complexes $[VX_3(MeCN)_3]$ (X = Cl or Br) crystallize with varying amounts of acetonitrile in the lattice,[212] where the IR spectra show the free (2253.5 cm^{-1}) and coordinated (2292 cm^{-1}) acetonitrile.

In acetonitrile, vanadium trichloride reacts with large organic chlorides to yield cream-yellow complexes of the type $Ct[VCl_4(MeCN)_2]$ ($Ct = Et_4N^+$, $MePh_3As^+$ or Ph_4As^+). Bromide-containing complexes have also been isolated: $(Et_4N)[VBr_4(MeCN)_2]$ and $(Et_4N)[VCl_3Br(MeCN)_2]$. The acetonitrile molecules of $(Ph_4As)[VCl_4(MeCN)_2]$ may be removed at 100 °C under vacuum.[213] Acrylonitrile and propionitrile complexes have been described.[214]

33.4.3.5 Silazanes

Hexamethylcyclotrisilazane, $(Me_2SiNH)_3$, and octamethylcyclotetrasilazane, $(Me_2SiNH)_4$ (shown in Figure 10 and abbreviated as HMT and OMT respectively), form octahedral complexes. VCl_3 reacted with HMT in THF to form $VCl_3(HMT)$,[215] which dissolves in dichloromethane giving a nonconducting solution. The ligand is terdentate and two V—Cl bands (at 358 and 326 cm^{-1}) confirm the facial geometry for the chlorine atoms. On reacting $[VCl_3(THF)_3]$ with OMT in benzene $[(VCl_3)_2(OMT)(THF)_2]$ was obtained,[216] with both coordinated and uncoordinated NH groups and with bridging chlorine atoms.

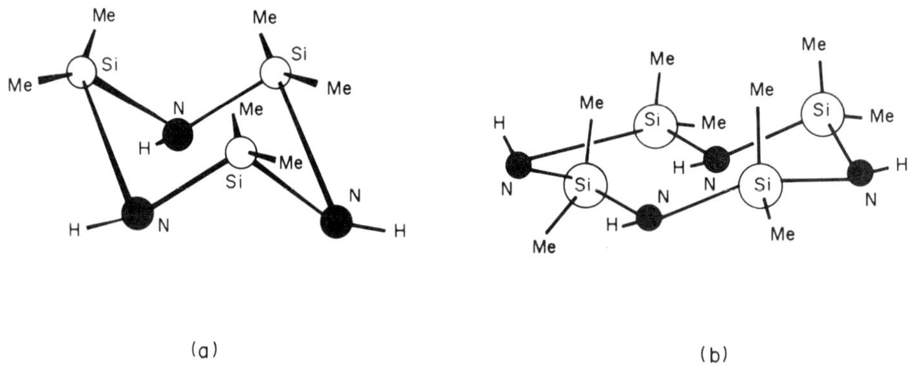

Figure 10 Formulae of silazanes: (a) hexamethylcyclotrisilazane (HMT), (b) octamethylcyclotetrasilazane (OMT)

33.4.4 Phosphines and Arsines

VCl_3 reacted with Et_3P or Pr_3P in THF to form trigonal bipyramidal $[VCl_3L_2]$.[217] $[VCl_3(Me_2PCH_2CH_2PMe_2)(THF)]$ was prepared by reaction of $[VCl_3(THF)_3]$ with ligand.[218] Five-coordinate $[VCl_3(PMePh_2)_2]$ has been synthesized from $[VCl_3(THF)_3]$ and phosphine upon reflux in benzene.[219] X-Ray diffraction has shown that the crystalline solid contains two independent but nearly identical molecules, each of them trigonal bipyramidal (Figure 11).[219]

On refluxing in THF, VCl_3 reacted with triarsines to yield red-brown complexes $[VCl_3(TAS)]$ (TAS = bis(dimethylarsinophenyl)methylarsine or 1,1,1-tris(dimethylarsinomethyl)ethane).[220] These are non-electrolytes, and the ligand is terdentate. Both have a facial structure.

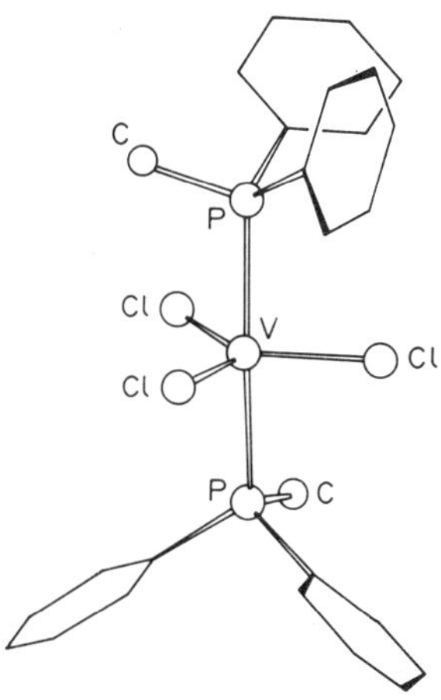

Figure 11 Structure of $[VCl_3(PMePh_2)_2]$. The V—Cl distance to the underlined atom (2.17 Å) is shorter than the average of the other V—Cl bonds (2.25 Å). Another distortion of the trigonal bipyramid is the observed P—V—P bending (166.6°). The small difference in V—P bond lengths (2.547 and 2.525 Å) has no chemical significance[219]

33.4.5 Oxygen Ligands

33.4.5.1 *Aqua species,* OH^-

The characterization of $[V(H_2O)_6]^{3+}$ has been done on the cesium alum $CsV(SO_4)_2{\cdot}12H_2O$[221] and on $[V(H_2O)_6](H_5O_2)(CF_3SO_3)_4$;[222] the VO_6 unit has an octahedral geometry. When hydrogen atoms of the coordinated water molecules are considered, the symmetry is lowered from O_h to D_{3d}. Two remarkable features are underlined in Figure 12: (i) the water molecules are nearly flat, *i.e.* for each coordinated water the vanadium centre, the oxygen and the two hydrogen atoms are lying in the same plane; (ii) the six hydrogen atoms of three mutually *cis* water molecules are essentially coplanar (the deviations from the mean plane are less than 0.1 Å with an average absolute deviation of 0.056 Å).

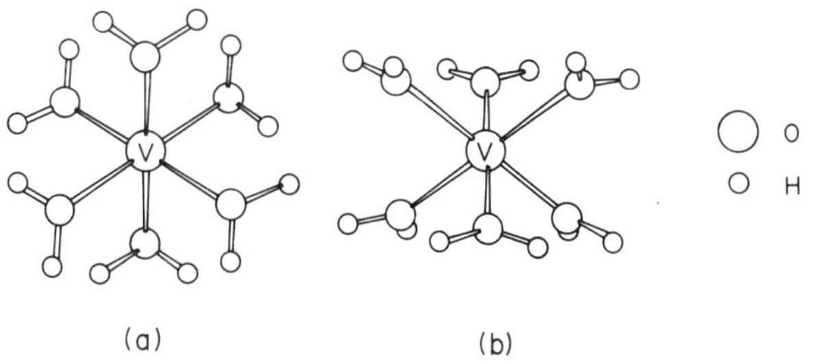

Figure 12 Structure of the $[V(H_2O)_6]^{3+}$ ion. The oxygen atoms are octahedrally coordinated with an average V—O distance of 1.995(4) Å. (a) The ion viewed down the C_3 axis; (b) a view perpendicular to the C_3 axis[222]

In the IR spectrum of the cesium alum, two bands were assigned to antisymmetric metal–ligand stretch ν_3 (532 cm^{-1}) and antisymmetric metal–ligand bend ν_4 (607 cm^{-1}),[223] and in the oriented single-crystal Raman spectrum, the totally symmetric stretching mode of the cation occurs at 525 cm^{-1}.[224]

The electronic spectra of the cesium alum and of vanadium(III) in perchloric acid solutions are very similar ($\nu_1 = 17\,800$ and $\nu_2 = 25\,700\ cm^{-1}$) and it was concluded that vanadium(III) existed as the hexaaqua ion in such solutions.[225] The coordination number six was confirmed by NMR.[226] Water exchange has been studied by ^{17}O NMR spectroscopy as a function of temperature (255–413 K) and pressure (up to 250 MPa):[227] $k_{ex} = (5.0 \pm 0.3) \times 10^2\ s^{-1}$ (at 298 K), $\Delta H^{\neq} = (49.4 \pm 0.8)\ kJ\ mol^{-1}$, $\Delta S^{\neq} = -(28 \pm 2)\ J\ K^{-1}\ mol^{-1}$, $\Delta V^{\neq} = -(8.9 \pm 4)\ cm^3\ mol^{-1}$ and the compressibility coefficient of activation $\Delta\beta^{\neq} = -(1.1 \pm 0.3) \times 10^{-2}\ cm^3\ mol^{-1}\ MPa^{-1}$.

Table 9 Data on Hydrolysis and Oxoreduction of Vanadium(III)

Reaction		*Ref.*
$V(H_2O)_6^{3+} \rightarrow V(H_2O)_5(OH)^{2+} + H^+$	$\log \beta = -3.07 \pm 0.03$	1
$2V(H_2O)_6^{3+} \rightarrow V_2(H_2O)_{10}(OH)_2 + 2H^+$	$\log \beta = -3.94 \pm 0.02$	1
$2V(H_2O)_6^{3+} \rightarrow V_2(H_2O)_9(OH)_3 + 3H^+$	$\log \beta = -7.87 \pm 0.09$	1
$VO^{2+} + \frac{1}{2}H_2 + H^+ \rightarrow$ 'V^{3+}'	$\log \beta = 6.192 \pm 0.004$	1
	$E^\circ_{4,3} = 366.3 \pm 0.3$ mV	
$V(H_2O)_6^{3+} + e^- \rightarrow V(H_2O)_6^{2+}$	$E_{1/2} = -475$ mV (*vs.* SCE)	2
	$\Delta S^\circ = 155$ J K^{-1} mol^{-1}	

1. F. Brito and J. M. Gonçalves, *An. Quím.*, 1982, **78,** 104; solutions 3 M in KCl at 25 °C with total vanadium concentration in the range 0.006–0.160 M and $0 < \bar{n} < 0.85$.
2. E. L. Yee, R. J. Cave, K. L. Guyer, P. D. Tyma and M. J. Weaver, *J. Am. Chem. Soc.*, 1979, **101,** 1131; solution 0.2 M $LiClO_4$ (pH 1–1.8) at 25 °C.

The hydrolysis has been studied by potentiometric[228] and spectrophotometric[229] techniques and below pH 3.2 may be described by the reactions and equilibrium constants of Table 9.

The magnetic susceptibilities of solutions of hydrolysis products were consistent with the binuclear complex in such solutions having a double hydroxide bridge $[V(OH)_2V]^{4+}$ rather than an oxo bridge $[V—O—V]^{4+}$.[230]

The solids from systems $VX_3 + MX + HX + H_2O$ (X = Cl, Br; M = K, NH_4, Rb, Cs) depend on the concentration of the halide and on the temperature during precipitation.[231]

There was evidence for the existence of green $[V(H_2O)_4X_2]^+$ in $[V(H_2O)_4X_2]X \cdot 2H_2O$ (X = Cl, Br), $Rb[VCl_2(H_2O)_4]Cl_2 \cdot 2H_2O$ and $Cs_2[V(H_2O)_4X_2]X_3 \cdot nH_2O$ (X = Br, $n = 1$; X = Cl, $n = 0$). The red $[VCl_5(H_2O)]^{2-}$ may be found in $M_2[VCl_5(H_2O)]$ (M = Cs, Rb, K, NH_4) prepared by dehydration of the corresponding double salts containing the $[VCl_2(H_2O)_4]^+$ ion. The structure of *trans*-$[V(H_2O)_4X_2]^+$ has been confirmed by X-ray diffraction in $[V(H_2O)_4X_2]X \cdot 2H_2O$ (X = Br or Cl),[232] $Cs_2[V(H_2O)_4X_2]X_3$ (X = Br or Cl)[233] and $Cs_3[VCl_2(H_2O)_4]Cl_4$.[234] The low temperature crystal electronic and IR spectra of $Cs_3[VCl_2(H_2O)_4] \cdot Cl_4$ have been reported.[234]

33.4.5.2 Dioxygen

Dioxygen oxidizes VCl_3 or VCl(salen) to vanadium(IV) with the intermediate formation of dioxygen adducts,[235] with equilibrium constants of the order of 10^2; they are involved in the catalysis of oxygenation of 3,5-di-*t*-butylpyrocatechol.[236]

33.4.5.3 Alcohols and ethers

VCl_3 reacted with several alcohols to form $[VCl_2(ROH)_4]Cl$ or $[VCl_3(ROH)_3]$ depending on the preparation and on the alcohol.[237] The complexes dissolved in parent alcohol behave as 1:1 electrolytes; electronic spectra are consistent with the $[VCl_2(ROH)_4]^+$ ion. The IR spectra show two V—Cl frequencies and a *cis* geometry has been suggested.[237] The fluorescence and photochemistry of the charge-transfer band in alcoholic solutions have been reported.[238]

THF and 1,2-dimethoxyethane react with VX_3 (X = Cl or Br) to form $[VX_3(THF)_3]$ and $VX_3 \cdot 1.5C_4H_{10}O_2$.[239]

A *trans* structure proposed for $[VCl_3(THF)_3]$ from the IR spectrum was confirmed by X-ray diffraction.[100] The dimethoxyethane complex loses dimethoxyethane on heating to give $VX_3 \cdot C_4H_{10}O_2$. Both complexes of dimethoxyethane were polymeric.[239]

33.4.5.4 β-Ketoenolates and catecholate

For vanadium(III) β-diketonates the recommended preparation is reduction with the dithionite of compounds such as oxovanadium(IV) sulfate, vanadium(V) oxide or ammonium vanadate followed by the addition of the β-diketone in alcohol.[240]

$[V(acac)_3]$ crystallizes in two forms.[241] The vanadium atom is surrounded by an octahedron of oxygen atoms with V—O lengths of 1.979 and 1.982 Å for the α and β forms. There is a slight trigonal distortion of these octahedra: the intraring OVO angles are 88.0 and 87.3° respectively.

The *d–d* transitions in the electronic spectrum (~18 200 and 21 700 cm^{-1} in solution in ethanol) are difficult to locate as they occur as shoulders on a strong charge-transfer band.[242] Molecular orbital calculations were used in the interpretation of the spectra of $[V(acac)_3]$;[243] charge-transfer and Rydberg transitions may contribute at energies higher than 40 000–50 000 cm^{-1}. The He-I photoelectron spectra of $[V\{(CF_3CO)_2CH\}_3]$ have been studied to elucidate the energies of the metal *d* and ligand orbitals.[244] Based on core binding energies,[245] the metal *d* orbitals are not involved in bonding.

Raman spectra of $[V(acac)_3]$ suggest that the contributions of dative π bonding should be small.[246] In IR spectra, the shifts of ν(M—O) bands were related to the electronic effects of the substituents.[247]

NMR techniques are suitable for the study of solutions containing vanadium(III) complexes. Ligand exchange is catalyzed by acids and inhibited by bases.[248] The facial and meridional isomers of complexes with unsymmetrically substituted ligands (*e.g.* 1,1,1-trifluoroacetylacetone) could be detected.[249] The same technique was used in the study of the stereochemistry and rearrangement of tris(β-diketonato) complexes[250] as well as those of the related ligands dehydroacetic acid, methyl acetoacetate and ethyl acetopyruvate.[251] In many systems, measured *fac/mer* ratios were smaller than the statistical 1:3. The rearrangement (isomerization) in a number of diketonate complexes is intramolecular, but the NMR data are consistent with twisting or bond rupture.[248] However, a bond breaking mechanism should be catalyzed by acid and such catalysis was not observed.

The β-diketonates are usually thermally stable and may be fused or volatilized with little or no decomposition. These complexes were tested by gas–liquid chromatography without evidence for on-column degradation.[252] However, some retention of fluorodiketonato complexes was due to reaction in the stationary phase or with active sites.[253] The sublimation enthalpy of tris(trifluoroacetylacetonate)vanadium(III) is 118 ± 2 kJ mol^{-1} and the evaporation enthalpy is 77.8 ± 0.8 kJ mol^{-1}.[254] At 300–470 °C there is decomposition of $[V(acac)_3]$; acetone, CO and CO_2 are the products.[255]

$K_3[V(catecholate)_3]\cdot 1.5H_2O$ has been synthesized from VCl_3 and catechol in ammonia solution, on adding a solution of KOH. The vanadium is surrounded by six oxygen atoms of catecholate ions: the average V—O bond length is 2.013(9) Å.[256] A tris(*o*-semiquinone)vanadium(III) complex[257] from $V(CO)_6$ and quinone was better formulated as bis(catecholate)(*o*-semiquinone)vanadium(V).[256]

33.4.5.5 Oxoanions

Several complexes $[V(OR_2P{=}O)_3]_x$, $[V\{O(RO)R'P{=}O\}_3]_x$ and $[V\{O(RO)_2P{=}O\}_3]_x$ (R = alkyl or aryl) were prepared from VCl_3 and phosphinates ($R_2(RO)P{=}O$),[258] phosphonates ($R(RO)_2P{=}O$)[259] and phosphate alkyl esters ($(RO)_3P{=}O$).[260]

Tris(organoseleninate) complexes have been synthesized and, in DMF or nitromethane, are non-conducting.[261] The ligands act as bidentate *O,O'*-seleninate and the complexes have an octahedral configuration with D_3 symmetry.

The electronic spectrum of vanadium(III) ions in concentrated sulfuric acid with λ_{max} at 13 600 ($\varepsilon = 7.8$), 21 800 ($\varepsilon = 11$) and 29 000sh cm^{-1} has been interpreted as due to octahedral coordination.[262] The solid methyl sulfate $V(MeSO_4)_3$ has been isolated ($\mu_{eff} = 2.68$ BM) as well as complexes of methyl sulfate containing three molecules of pyridine and one or two molecules of bipy.[263]

Sodium sulfinates $NaRSO_2$ (R = Me, Ph or MeC_6H_4) reacted in ethanol with VCl_3 and tris(*O,O'*-sulfinato) complexes were obtained.[264] The IR spectra are consistent with the sulfinato ions as bidentate. The electronic spectra show two *d–d* transitions: ν_1 in the range 14 700–14 930 cm^{-1} and $\nu_2 = 22\,990$ cm^{-1}.

33.4.5.6 Carboxylic acids

There have been several reports on monocarboxylate complexes although different formulae and various structures have been produced. After boiling VCl_3 with anhydrous formic acid, a solid $V(O_2CH)_3 \cdot HCO_2H$ was isolated and at 180 °C green $[V(O_2CH)_3]$ was obtained.[265] VCl_3 and acetic acid under reflux gave a solid $V_3(OH)(O_2CMe)_8$ or $H[V_3O(O_2CMe)_8]$.[266] Later, a different formulation was proposed, $[V_3O(O_2CMe)_7(HOCMe)]_n$, *i.e.* a polymeric structure with a bridging acetate joining the $[V_3O(O_2Me)_6]$ clusters.

$[V_3O(O_2CCH_2Cl)_6(H_2O)_3]ClO_4 \cdot 3H_2O$ contains the planar V_3O group and there is no metal–metal bonding in this oxo-centered cluster.[267]

There was an earlier claim for the synthesis of $[V_2(O_2CR)_6]$ (R = Me or Ph) using vanadium diboride as starting material (equation 8).[268]

$$2VB_2 + 6RCO_2H \longrightarrow V_2(O_2CR)_6 + 3H_2 + 4B \quad (8)$$

$[VCl_3(THF)_3]$ and sodium benzoate in dichloromethane produced green $[V_3O_3(THF)(O_2CPh)_6]$, characterized by X-ray diffraction.[269]

It is likely that the oxidation to vanadium(IV) occurred during the crystallization, with intermediate trimeric vanadium(III) benzoate complexes.

Another example of a complicated polynuclear complex is an octanuclear basic benzoate containing four vanadium(III) and four zinc(II) atoms: $[VZnO(O_2CPh)_3(THF)]_4 \cdot 2THF$.[270] The crystal structure shows that the distances between these four vanadium atoms are just below 3 Å and therefore some V—V bonding may occur.

Values for the formation constants of $[V(C_2O_4)]^+$ and $[V(C_2O_4)_2]^-$ were respectively 1.6×10^7 and 3.5×10^{12}.[271] By irradiation of these solutions at 254 nm, there is a photodecomposition of the oxalate sensitized by the oxalato vanadium(III) complex; the resulting products were equimolar amounts of CO and CO_2. $K_3[V(C_2O_4)_3] \cdot 3H_2O$ can be easily crystallized from aqueous solutions containing ammonium vanadium alum and potassium oxalate. The measured magnetic moment is 2.78 BM and this compound obeys the Curie law quite closely.[272] The thermal decomposition has been studied.[273]

By adding the resolving ions $(+)$-$[Co(en)_3]^{3+}$ and $(-)$-$[Rh(en)_3]^{3+}$ to solutions of the labile $[V(C_2O_4)_3]^{3-}$ ion, the diastereoisomeric salts $[M(en)_3][V(C_2O_4)_3]$ (M = Co or Rh) have been isolated and the CD spectra were measured as dispersions in CsCl discs.[274] The ORD and CD spectra of solutions containing $[V(C_2O_4)_3]^{3-}$ could be observed by adding $(+)$-cinchonine hydrochloride.[275] Although $[V(C_2O_4)_3]^{3-}$ is labile, partial resolution occurs. The structure of the *trans*-diaquabis(oxalato)vanadate(III) ion has been determined in $Cs[V(C_2O_4)_2(H_2O)_2] \cdot 4H_2O$ and $(MeNH_3)[V(C_2O_4)_2(H_2O)_2] \cdot 4.5H_2O$.[276] In both, the oxalate groups are coplanar, with V—O bond lengths of 2.005(2) and 2.017(4) Å respectively.

$K_3[V(malonate)_3]$ may be prepared by a procedure similar to that of the trisoxalato complex.[272] The kinetics of ligand exchange have been studied by NMR line broadening:[277] $k_{ex} = 50\ s^{-1}$ at 70 °C.

33.4.5.7 Hydroxy acids

Vanadium(III) and tartaric acid (H_4t) in aqueous solution form 1:1 and 1:2 complexes with various degrees of protonation:[278] $V(H_2t)^+$, $V(H_2t)_2^-$, $V(Ht)^0$, $V(Ht)_2^{3-}$ and $V(Ht)_3^{6-}$. The formation constants of complexes 1:1, 1:2 and 1:3 were estimated by NMR relaxation in solutions containing vanadium(III) and lactic ($\beta_1 = 1.1 \times 10^2$, $\beta_2 = 2.0 \times 10^4$ and $\beta_3 = 1.8 \times 10^6$) or mandelic acids ($\beta_1 = 1.2 \times 10^2$, $\beta_2 = 2.2 \times 10^4$ and $\beta_3 = 1.5 \times 10^6$).[279] The reactions between vanadium(III) and salicylic, *p*-aminosalicylic and 5-nitrosalicylic have been studied.[280]

33.4.5.8 Dimethyl sulfoxide and amides

Dimethyl sulfoxide saturated with SO_2 reacts with vanadium[281] or V_2O_5[282] to form $[V(DMSO)_6](S_2O_7)_{3/2}$ which loses DMSO up to 170 °C giving $V(S_2O_7)_{3/2}$. DMSO coordinates through oxygen while the disulfate is present as an anion.[283] Earlier work reported the $[V(DMSO)_6]^{3+}$ perchlorate.[284]

Vanadium was allowed to react with bromine in DMF; $V(DMF)_6(Br_3)_3 \cdot 0.3(DMF)$ was isolated,[285] for which $10Dq = 17\,300\ cm^{-1}$.

$[V(urea)_6]I_3$ crystallizes from a solution of the sulfate on adding potassium iodide. The oxygen atoms of the urea molecules form an octahedron twisted 7° about the three-fold axis with V—O lengths of 1.98 Å.[286] The magnetic susceptibility of the iodide and perchlorate salts of $[V(urea)_6]^{3+}$ ion and the magnetic anisotropy of the iodide have been measured. The electronic spectrum of the perchlorate has been reinterpreted.[287]

33.4.6 Sulfur Ligands

33.4.6.1 Thioethers

Vanadium trichloride or tribromide reacts with thioethers giving $[VX_3L_2]$ (X = Cl or Br; L = SMe_2, tetrahydrothiophene or SEt_2).[288,289] The complex with di-*n*-propyl sulfide could not be isolated.[288] These compounds are oxidized very easily. Solubility, molecular weights and conductance show that they are monomeric and non-ionic. Dipole moments, IR and electronic spectra are consistent with *trans* trigonal bipyramidal structures.

The similarity of properties for $[VX_3{\cdot}(SEt_2)_2]$ in solution or solid indicates that the solid is also five-coordinate bipyramidal.[289]

33.4.6.2 Dithiocarbamates

$[V(S_2CNR_2)_3]$ (R = Me, Et, Pr^n or Bu^i) were obtained by oxidation of $VBr_2{\cdot}6H_2O$ with diethylammonium dialkyldithiocarbamates in anaerobic ethanol/triethyl orthoformate.[290] Complexes $[VL_3]$ and $[VClL_2]$ of cyclic substituted dithiocarbamates L (L = piperidino, morpholino, thiomorpholino and *N*-methylpiperazino) were reported;[291] their properties are consistent with bidentate dithiocarbamate.

The diethyldithiocarbamato complex is isomorphous with the iron(III) complex and must have an approximate octahedral symmetry. As expected for octahedral vanadium(III), two bands (ν_1 and ν_2) appeared in the diffuse reflectance spectra. $[V(S_2CNRR')_3]$ (RR' = MePh or BzBz) were prepared from VCl_3 and a sodium or lithium salt of the dithiocarbamate; IR spectra in CS_2 confirm the tris(bidentate chelate) coordination. By NMR,[292] two processes can be observed for $[V(S_2CNMePh)_3]$ in CD_2Cl_2: *cis–trans* isomerization causing line broadening between −89 and −61 °C and metal-centred optical inversion without *cis–trans* isomerization causing broadening below −97 °C.

33.4.6.3 Dithiolate and dithiophosphinates

Although dithiolene complexes have been known for some time (see Section 33.2.5), the first 'pure' dithiolate complexes of vanadium have only recently been characterized.[48] Structures of $(Et_4N)_2[V_2(SCH_2CH_2S)_4]$, $(Ph_4P)_2[V_2(SCH_2CH_2S)_4]$ and $(Ph_4P)_2[V_2(SCH_2CH_2S)_4]{\cdot}4MeOH$ were published almost simultaneously. The three independent reports agree on the structure of the $[V_2(SCH_2CH_2S)_4]^{2-}$ ion (Figure 13). The vanadium atoms are bridged by the four S atoms of two dithiolate ions and the V—V distance (2.60 ± 0.02 Å) suggests metal–metal bonding, confirmed by the observed diamagnetic behaviour.

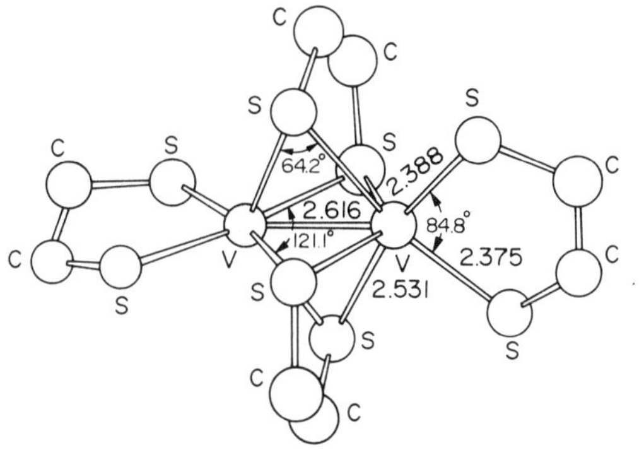

Figure 13 Structure of $[V_2(edt)_4]^{2-}$ with selected bond distances and angles

Dithiophosphinate complexes $[V(S_2PX_2)_3]$ (X = F, CF_3, Me, Ph, OEt) were reported[293] and the structure where X = OEt has been determined.[294] The six sulfur atoms form a trigonally distorted octahedron around vanadium with the average V—S distance 2.45 ± 0.02 Å. Visible spectra (in CH_2Cl_2 or an oriented crystal) show four bands: a charge-transfer band is observed at 22 700 and 23 300 cm^{-1} in the solution and crystal. The others at 12 820, 18 080 and 26 400 cm^{-1} (13 200, 18 400 and 26 400 in the crystal) are the expected transitions for octahedral V^{III}.

33.4.7 Halides

Several solvated vanadium(III) halides have been described in the previous sections and those with water, alcohol, acetonitrile or THF have been useful starting materials for the synthesis of vanadium(III) complexes.

A great number of anionic complexes containing four, five or six halide ions have been characterized. However, in aqueous solutions containing V^{III} ions and F^-, the potentiometric technique using a fluoride selective electrode has shown no indication of any complex other than VF^{2+} (stability constant: $\log \beta = 5.00 \pm 0.03$ at 25 °C in 1.0 M $NaClO_4$ solution).[295]

Salts of hexafluorovanadate can be obtained by high temperature techniques or from solutions containing hydrofluoric acid. For instance, X-ray patterns and DTA were used for the characterization of 17 double fluorides obtained by solid state reactions in the systems $VF_3 + MF$ (M = Li, Na, K, Rb, Cs, Tl) and seven double fluorides in the systems $VF_3 + MF_2$ (M = Ca, Sr, Ba, Pb),[296] and the lattice constants[297,298] and magnetic properties[298] of $A_2B[VF_6]$ (A, B = Cs, Rb, Tl, K, Na, Li) were also reported. The structure of the high temperature β phase of Li_3VF_6 has been determined and compared with the cryolite-type stable α form. The vanadium atoms have an octahedral coordination.[299]

K_3VF_6 can be obtained by reacting V_2O_5 and KI with hydrofluoric acid on a steam bath[300a] or by heating at 800 °C powdered vanadium and potassium hydrogen fluoride;[300b] the magnetic moment at 20 °C was 2.79 BM.[300a] X-Ray powder data have been given for its three solid forms.[301] Predicted transition energies[302] were compared with those observed at 10 200, 14 800 and 23 250 cm^{-1}.[303]

The standard enthalpy of formation of K_3VF_6(c) is −3026 kJ mol^{-1}.[304] The calculated ligand field stabilization has been compared with those of the elements from scandium to gallium. The ligand field effect has also been discussed for the IR spectrum of $[Cr(NH_3)_6][VF_6]$ and isomorphous $[Cr(NH_3)_6][MF_6]$ (M = Sc, Ti, Cr, Mn or Fe).[305]

Hydrazinium fluorovanadates $(N_2H_5)_3VF_6$[306] and $(N_2H_6)VF_5$[307] and their thermal decomposition were reported.[308]

$(NH_4)_3VF_6$ can be crystallized from $VF_3{\cdot}3H_2O$ and ammonium fluoride in water.[309] It has a moment of 2.79 BM[300a] and on reaction with XeF_2 forms a V^V complex with coordination of more than six fluoride ions.[309] On thermal decomposition the final product is VF_3 at ~450 °C. There was no intermediate formation of NH_4VF_4 but a method for the preparation of this has been reported.[310]

Several studies[311,312] appeared on the phase transitions at 135, 413 and 482 K in the perovskite-type $RbVF_4$, in which vanadium is octahedrally coordinated. In $Rb[VF_4(H_2O)_2]$ the water molecules are *trans* and hydrogen bonds form chains of VF_4O_2 octahedra.[313] Complexes $M_2VF_5{\cdot}H_2O$ (M = K, Rb, Cs) were precipitated from solutions containing $VF_3{\cdot}3H_2O$, HF and KF, and on heating in vacuum, M_2VF_5 (M = K, Rb, Cs) formed.[314]

The chloride complexes have also been prepared with various stoichiometries. Green $MVCl_4{\cdot}6H_2O$ (M = K, Rb, Cs) precipitated from cold solutions (−30 to −10 °C) of VCl_3 and MCl saturated with HCl.[315] From such solutions at high temperature (100–120 °C) red $M_2VCl_5{\cdot}H_2O$ (M = K, Rb, Cs) are obtained.[316]

In both solid compounds (containing large cations each as Ph_4P^+, Ph_4As^+) and in solutions $[VCl_4]^-$ has distorted tetrahedral symmetry as suggested by spectra in benzene:[317] 8300 and 9250 cm^{-1} assigned as ${}^3A_2 \rightarrow {}^3E$, 10 000 cm^{-1} as ${}^3A_2 \rightarrow {}^3A$ and 15 030 cm^{-1} ($\varepsilon = 260$) as ${}^3A_2 \rightarrow {}^3T_1(P)$. X-Ray powder patterns show the tetrachlorovanadates to be isomorphous with other known tetrahedral MCl_4 species and IR data suggest a distorted tetrahedral symmetry.[318] Magnetic data show antiferromagnetic interaction of the vanadium atoms; these magnetically dilute species are completely different from $(NEt_4)VCl_4$ which is similar to $(NEt_4)_3V_2Cl_9$. Therefore a polymeric structure was suggested for $(NEt_4)VCl_4$ with octahedral vanadium. Its electronic spectrum is very different from the tetrahedral complexes,[319] consistent with a distorted octahedral geometry.

$Cs_3V_2Cl_9$ was prepared by heating a stoichiometric mixture of VCl_3 and CsCl.[320] The similarity of the powder pattern to that of the corresponding Cr complex suggests similar structures: $V_2Cl_9^{3-}$ consists of octahedra sharing a trigonal face. The moment at 300 K (2.74 BM) is consistent with little metal–metal interaction.

Green crystals of $Ct_2VBr_5 \cdot 5H_2O$ (Ct = Rb, Cs, Hpy) were prepared in hydrobromic acid[321] and, on thermal decomposition, successively dehydrated *via* $Ct_2VBr_5 \cdot 3H_2O$ and $Ct_2VBr_5 \cdot H_2O$ to the corresponding anhydrous Ct_2VBr_5 (Ct = Rb, Cs, Hpy).

33.4.8 Mixed Donor Atom Ligands

33.4.8.1 Schiff bases

By allowing salen to react with anhydrous VCl_3 in pyridine under high purity nitrogen, a brown solid formulated as [VCl(salen)(py)] precipitated.[235] The geometry is probably octahedral with Cl^- and py *trans*.[322] Conductance showed it to be a 1:1 electrolyte. Oxidation with O_2 resulted in the corresponding VO^{2+} compound, [VO(salen)(py)], *via* a dioxygen adduct (Section 33.4.5.2). Complexes with ligands (**1**) and (**2**) and with the structurally related (**3**) have been prepared by adding the THF adduct of VCl_3 to solutions of the corresponding ligands, with subsequent addition of potassium *t*-butoxide or in the presence of triethylamine.[323] They are red-brown monomers [VCl(SB)] (SB = Schiff bases (**1–3**). IR indicates coordination by phenolic oxygen and azomethine nitrogen and the visible spectra and magnetic moments (2.5–2.7 BM) are consistent with pseudooctahedral V^{III}.[323] Oxidation of [VCl(SB)] (SB = **1**) by O_2 produces the VO^{2+} compound.[323] If $TlBF_4$ is added to [VCl(SB)] (SB = **3**) in acetone, TlCl precipitates and an orange solid is isolated, formulated as $[V(SB)(acetone)]BF_4$ from displacement of chloride by O-coordinated acetone.[323] This needs confirmation.

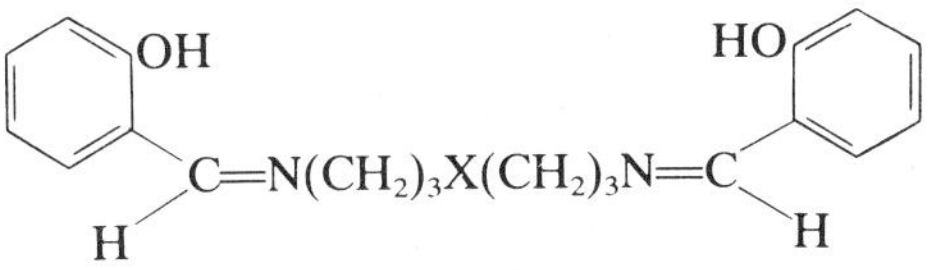

(**1**) X = N
(**2**) X = P

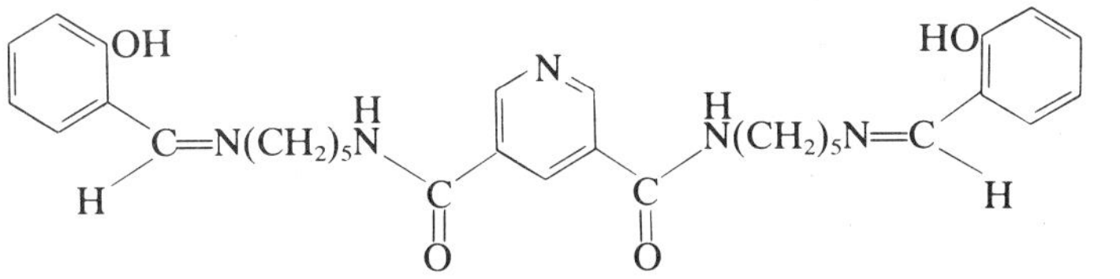

(**3**)

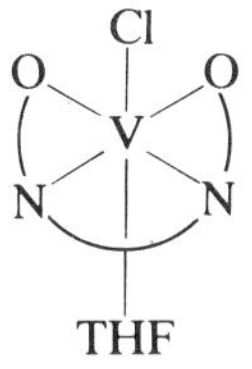

(**4**)

In one synthesis of [V(acen)Cl(THF)] (**4**), a THF suspension of Na_2acen is allowed to react with $[VCl_3(THF)_3]$.[324] A second method involves reductive deoxygenation of [VO(acen)] (equation 9); once [VO(acen)] has been synthesized, it is a straightforward procedure.

$$[VO(acen)] + TiCl_3(THF)_3 \xrightarrow[-TiOCl_2]{\text{reflux in THF}} (\mathbf{4}) \tag{9}$$

The chloride in (**4**) can be easily replaced and several derivatives [V(acen)(X)(L)] (X = OPh, SPh, NCS^-; L = solvent molecule, *e.g.* THF, py) have been prepared.[325] Complex (**4**) has been used for synthesis of organometallic derivatives.[325] The structures of several ($X = Cl^-$ (**4**), OPh^- (**5**) and NCS^- (**6**)) have been determined,[324] and consist of discrete molecules containing the same [V(acen)(THF)] unit. The coordination geometry can be described as a square bipyramid whose equatorial plane is formed by the N_2O_2 atoms of the tetradentate ligand, with oxygen from THF at one apex, the other apex being occupied by Cl^-, O from OPh^- or N from NCS^- (Scheme 11). The N_2O_2 core is not exactly planar, and the V atom is displaced out of the N_2O_2 'plane' towards X (0.143, 0.151 and 0.082 Å for **4**, **5** and **6**).

Complex (**7**) may be either monomeric or dimeric (as observed for formally five-coordinate V^{III} in $[V(acen)(CH_2Ph)]_2$).[325]

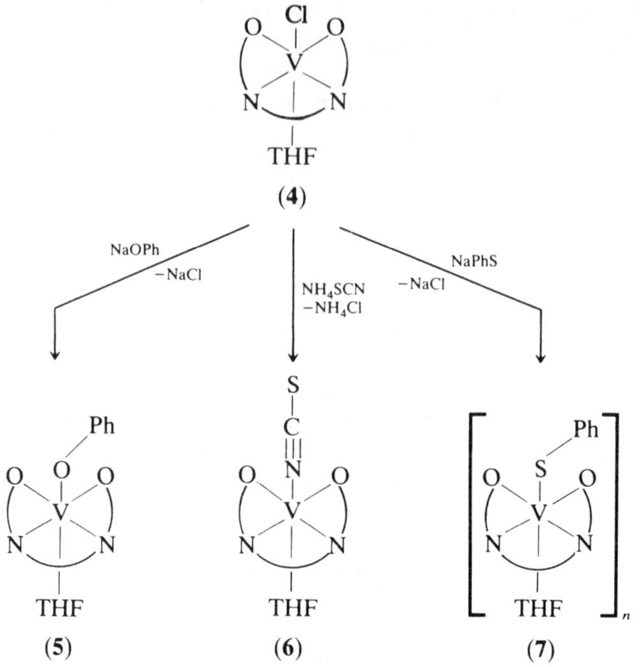

Scheme 11

When reaction (**9**) is started at room temperature for a few minutes, then cooled to −30 °C and later to −70 °C, a different complex (**8**) may be isolated, whose structure consists of binuclear [V^{IV}(Cl)(acen)O{$TiCl_3(THF)_2$}] in which an O atom bridges the [VCl(acen)] and $TiCl_3(THF)_2$ units.[324] The V—Cl distance (2.352 Å) is similar to that of (**4**). For the bridging O atom, d(V—O)(1.937 Å) is longer than in other μ-oxo vanadium complexes.[324]

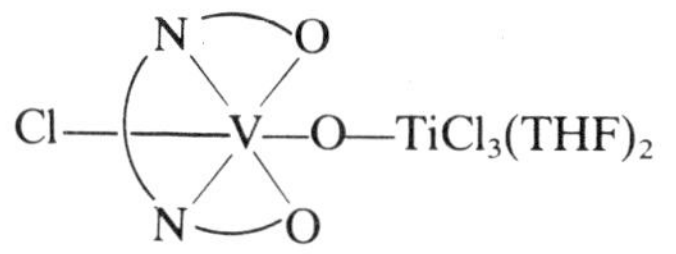

(**8**)

33.4.8.2 Amino acids

Simple and mixed complexes of V^{III} with aspartic and glutamic acids and with methionine have been reported[326] although their elemental analysis cannot be considered entirely satisfactory. These green, water soluble compounds were prepared from VCl_3 solutions (normally obtained *in situ* by reduction of V_2O_5 by Zn), addition of the amino acid(s) and precipitation with acetone. Complexes with L-Ser, L-Thr and L-Leu have been isolated and studied by IR, UV-vis, CD and thermogravimetric measurements.[327] The location of ν(V—Cl) bands in the region 350–395 cm^{-1}, the magnetic moments and other considerations led to formulations $VCl(Ser)_2 \cdot H_2O$ (μ_{eff} = 2.79 BM), $VCl_2(Thr)(ThrOH) \cdot 2H_2O$ and $VCl_2(Leu)(LeuOH) \cdot 3H_2O$ (μ_{eff} = 2.31 and 2.39 BM, respectively), with polymeric structures (note that low μ_{eff} values can result from VO^{2+} impurities).[327] [$VCl_3(Gly)_3$][328] and [$V(H\text{-}Had)_3$] (H-Had = *N*-formyl-*N*-hydroxylglycine[329]) have also been prepared. Further structural studies are needed.

Preparation of complexes with amino acids in the pH range 4–7 is difficult as V^{III} is partially hydrolyzed and can be easily oxidized. This also makes solution studies difficult. At low pH (1.5–3) complexes $[VL(H_2O)_5]^{2+}$ are believed to form,[327–331] with, at higher pH, $[V(Gly)_2]$ and $[V(Glu)_2]^-$ (pH *ca.* 3.5 and 4–4.5, respectively).[332] From potentiometry, log β_1 of 8.8, 8.4 and 9.0 and log β_2 of 15.4, 14.8 and 16.3 were determined for Ser, Thr and Leu. With cysteine, complexation of V^{III} is faster than that of V^{II} by three orders of magnitude (same pH).[156]

Enthalpies of activation are comparable but entropies differ. The maximum number of ligands is two, and the formation constants determined (equation 10) were $\beta_1 = 10^{13}$ and $\beta_2 = 10^{20}$.[156]

$$V^{(III)}(aq) + n\overset{+}{N}H_3CH(CO_2^-)CH_2SH \overset{\beta_n}{\rightleftharpoons} [V(NH_2CH(CO_2^-)CH_2S)_n] + 2nH^+ \quad (10)$$

33.4.8.3 Complexones

Complexes of NTA formulated as $[V(NTA)(H_2O)_2]\cdot 5H_2O$ and $Cs_3[V(NTA)_2]$ have been prepared.[333] The pK_a of the coordinated water molecule in the former is 6.2. Although $Cs_3[V(NTA)_2]$ exhibits no tendency to form hydroxo complexes in aqueous solution, the second NTA^{3-} is markedly more weakly bound than the first (log $\beta_1 = 13.4$ and log $\beta_2 = 22.1$ at 20 °C and $I = 0.1$ M).

Schwarzenbach *et al.*[334] prepared $Na[V(edta)]\cdot 4H_2O$ by reduction of the VO^{2+} complex with sodium amalgam and determined the formation constant of $[V(edta)]^-$ as $10^{25.9}$. The formation constant for reaction **(11)** in Scheme 12 ($L = edta^{4-}$ and $n = 2$) was $10^{4.4}$. In the V^{III} dimer $(H_2en)[V(Hnhet)]_2\cdot 2H_2O$, the anion **(9)** displays several uncommon features.[335] Each V^{III} was seven-coordinate, having a distorted pentagonal bipyramidal configuration. The bridging atoms are the alkoxy oxygens formed by deprotonation of the *N*-hydroxyethyl functionality of the parent $nhet^{3-}$ ligand ($nhet^{3-}$ = *N*-(hydroxyethyl)ethylenediaminetriacetate). The V—V distance is 3.296 Å and the angles of the $\{V_2O_2\}$ core (VOV = 108°; OVO = 72°) suggest little strain is present.

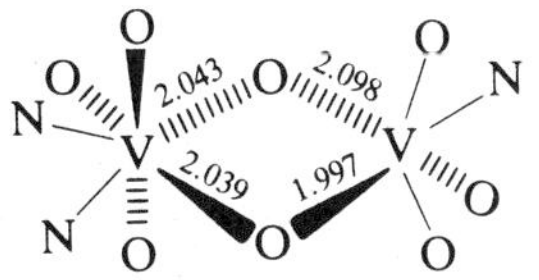

(9)

The binuclear complex **(9)** may be produced from monomeric [V(nhet)] by two paths: (i) by combination of two molecules of $[V(nhet)(OH)]^-$ or $[V(nhet)(OH)]^-$ and $[V(nhet)(H_2O)]$,[336] or (ii) by a cross-redox reaction between $[V^{IV}O(nhet)]^-$ and $[V^{II}(nhet)]^-$ (see also Section 33.5.9.3).[337,338] The unusually intense spectral features of $[V^{II}V^{IV}O(nhet)_2]^{2-}$ originate in oxo bridging between V^{II} and V^{IV} in the cross reaction.[336–338] This intermediate has a short lifetime (~25 ms), but it is unusually long by comparison with other inner-sphere systems.

The cross-redox reaction discussed above and formation of the V^{III} dimers has also been reported in the case of the edta complexes. Scheme 12 summarizes part of the equilibria proposed to explain potentiometric and spectroscopic data at $t = 25$ °C and $I = 0.20$ M in $NaClO_4$.[336–338] The polarographic behaviour of V^{III} complexes of edta and 'analogous' ligands has also been studied.[339]

$$[VL(H_2O)]^{1-n} + H_2O \overset{K_a}{\rightleftharpoons} [VL(OH)]^{n-} + H_3O^+ \quad (11)$$

$$2[VL(H_2O)]^{1-n} + 2H_2O \overset{K_D}{\rightleftharpoons} [V^{III}\,dimer] + 2H_3O^+ \quad (12)$$

$$2[VL(OH)]^{n-} \overset{K_d}{\rightleftharpoons} [V^{III}\,dimer] + H_2O \quad (13)$$

If $L = edta^{4-}$ then $n = 2$, p$K_a = 10.16$, p$K_D = 16.71$ and p$K_d = -3.62$; if $L = Hedta^{3-}$ then $n = 1$, p$K_a = 6.4$, p$K_D = 9.05$ and p$K_d = -3.7$[335]

Scheme 12

Reaction **(14)** has been studied.[340,341] The equilibrium constant K_R is of the order 10^{-10} so that comparable concentrations of the reactant and product coexist at pH ~5. The reaction goes to completion at higher pH values (5.4–7.1) and is quantitative in the reverse direction at pH 2.5–4.0, where mixtures of the Fe^{III} and V^{III} complexes are stable in the absence of O_2.

$$[Fe^{III}(edta)]^- + [V^{III}(edta)]^- \overset{K_R}{\rightleftharpoons} [Fe^{II}(edta)]^{2-} + [V^{IV}O(edta)]^{2-} + 2H^+ \quad (14)$$

In the redox reaction between [Ti(nhet)(H_2O)] and [VO(nhet)]$^-$, a binuclear species containing Ti^{IV} and V^{III} appears by a substitution after the electron transfer step.[342] The orange [$Ti^{IV}V^{III}$(nhet)$_2$] forms in the pH range 2.5–7.5, and can also be generated by combination of [V(nhet)(H_2O)] and [TiO(nhet)]$^-$.

33.4.9 Vanadium in Tunicates

The sea squirts or tunicates are fascinating marine creatures, their name being derived from the tunic made of cellulosic material that surrounds the body of the animal. In 1911, Henze discovered vanadium in the blood of *Phallusia mammillata* C.[343] He later found the same with other ascidians (a class of tunicates). In vanadium-accumulating species, most vanadium is located in the vacuoles—vanadophores—of certain types of blood cells—the vanadocytes. The concentration in the vanadophore can be as high as ~1 M and this value must be compared with concentrations of the order of 2×10^{-8} M for vanadium in sea water.[344] Kustin *et al.* have reviewed the work done to understand the efficient accumulation and the possible biological roles of the metal.[345]

Although local concentrations of the metal may be very high, isolation is difficult due to the sizes of the vanadophores (2 μm) and the vanadocytes (8 μm). It was thought that vanadium was bound to a protein named hemovanadin[346] but later it was found that this was an artefact during hemolysis and isolation. Proton NMR spectra of living blood cells show that the vanadium(III) concentration range of the vanadophores is sharply limited and the spectra were consistent with a labile vanadium(III) aqua complex.[347] X-Ray absorption spectroscopy of the vanadocytes has shown that only a small amount (<10%) of the vanadium is present as VO^{2+} in the vanadocyte and that the vanadium(III) ions are surrounded by a single shell of six oxygen atoms at an average distance of 1.99 Å, ruling out N ligands such as porphyrin or imidazole.[348] The combined NMR and EXAFS were consistent with vanadophores containing a simple vanadium(III) aqua complex probably with another small oxygen ligand like sulfate that exists in high concentration in the vanadocyte. ESR confirmed the existence of concentrations of vanadium(IV) of the order of 10^{-2} M,[349] compared with the estimated total of vanadium in the vanadophore (1.4 M). Again, the data were consistent with most vanadium existing as $V(SO_4)(H_2O)_{4-5}{}^+$.

For the vanadophore region, the sulfate concentration has been estimated to be of the order of 1.3 M and pH ~2.[349] With no evidence for a stable complex within the vanadocyte, Figure 14 depicts an interesting mechanism proposed for vanadium and sulfate accumulation.[350] Anionic vanadium(V) (as HVO_4^{2-}) and sulfate ions enter the cell. Provided that within the vanadophore there is a strong reducing agent, vanadium can be reduced to vanadium(IV) and vanadium(III), cationic at the low pH in the vanadophore. If the vanadophore membrane is permeable to anions but not cations, the reduced vanadium remains trapped.

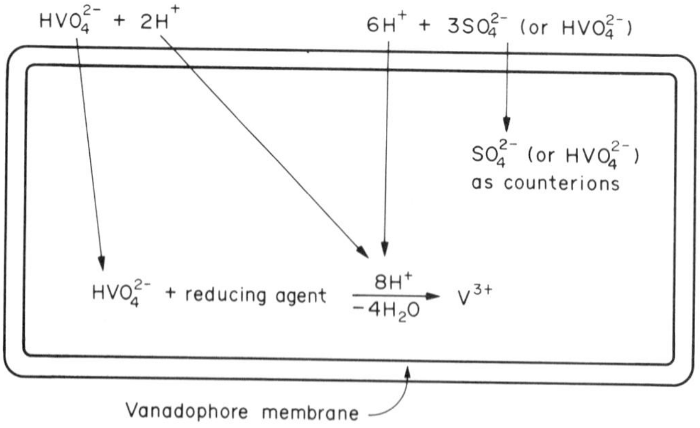

Figure 14 Mechanism of vanadium accumulation in the vanadophore assuming a membrane permeable only to anions

Recently, a reducing blood pigment named tunichrome B-1 (**10**) was isolated from the tunicate *Ascidia nigra* L. and characterized.[351] It has a structure derived from three (3,4,5-trihydroxy)phenylalanine units and the trihydroxyphenyl group is suitable for reducing the metal.

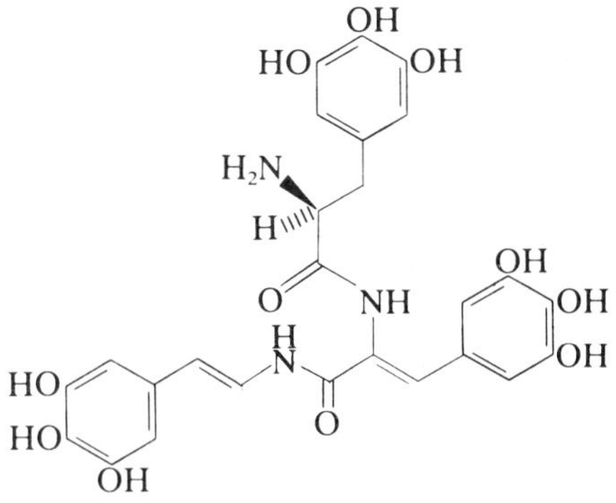

(**10**)

The uses of the metal for the living creature have been the subject of much speculation but the possibility that vanadium(III) is involved in oxygen transport has been ruled out experimentally.[352] The role of the high concentrations of vanadium in tunicates remains tantalizing.

33.5 VANADIUM(IV)

33.5.1 Introduction

This oxidation state is the most stable under ordinary conditions.[353] Important compounds contain the VO^{2+} unit (vanadyl(IV) ion or oxovanadium(IV) ion), which can persist through a variety of reactions and is considered the most stable oxycation of the first row transition ions.[354] It forms stable anionic, cationic and neutral complexes with all types of ligands, and has one coordination position occupied by the vanadyl oxygen (Figure 15).

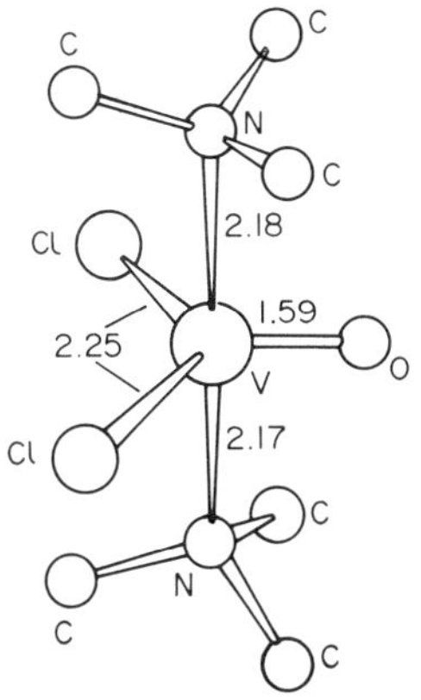

Figure 15 Square bipyramidal geometry of vanadyl(IV) complexes; more common ranges of bond lengths (Å) are indicated. The V atom lies above the plane of the four L ligands

The complexes typically have square pyramidal (**11**) or bipyramidal structure (**12**) with vanadyl oxygen apical, the vanadium atom lying 0.035–0.055 Å above the plane defined by the equatorial ligands.[354] There have been reports of five-coordinate 'trigonal bipyramidal' complexes (**13**) with structures determined by X-ray diffraction techniques.

(**11**)

(**12**)

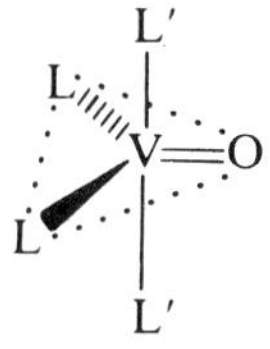

(**13**)

Only three general reviews on vanadium(IV) complexes have been given;[355–357] others deal with subnormal magnetic complexes,[358] ESR spin probes,[354] vanadium(IV) halides,[6] optical spectra and phosphorus donor ligands,[359] vanadium porphyrins[128] and annual reviews for 1979, 1980 and 1981.

Many publications discuss ordering of molecular orbitals based on ESR or optical spectra. Although assignment of the bands has been a matter of controversy,[355,360–375] most interpret them in terms of the energy level scheme proposed by Balhausen and Gray (Figure 16).[364] Although developed for $VO(H_2O)_5^{2+}$ (assuming C_{4v} symmetry), it has been applied to other complexes. Its universality has been questioned because the order of orbitals will probably vary from complex to complex and will depend on symmetry. In this model, the unpaired electron resides in a largely nonbonding d_{xy} orbital (the z axis is taken as the vanadyl bond) with lobes pointing between the equatorial ligands; ESR measurements are generally consistent with this.[354] The interaction of d_{xz},d_{yz} with the ligand may be π acceptor in character and the relative order of the B_2 and E orbitals could be inverted.[374] Usually three optical bands are observed in the visible,[356] assigned (C_{4v} symmetry) to (I)$^2B_2\,(d_{xy})\rightarrow{}^2E\,(d_{xz},d_{yz})$; (II)$^2B_2\,(d_{xy})\rightarrow{}^2B_1\,(d_{x^2-y^2})$; and (III)$^2B_2\,(d_{xy})\rightarrow{}^2A_1\,(d_{z^2})$ (or a charge-transfer transition). For some complexes four bands are observed, possibly arising from splitting of the d_{xz},d_{yz} levels in low symmetry (Figure 16c) and band III is often obscured by intense absorption tailing in from the UV. Transition III($d_{xy}\rightarrow d_{z^2}$) may be at 125–200 nm.[362,370] However, it will most likely be at 330–400 nm if not obscured by charge transfer. This is discussed by Lever[368] and further evidence is given by weak bands at ~360 nm in the CD spectra of complexes with (R)-lactate[372] and (S)-peida[373] (peida = N-[1-(2-pyridyl)ethyl] iminodiacetate).

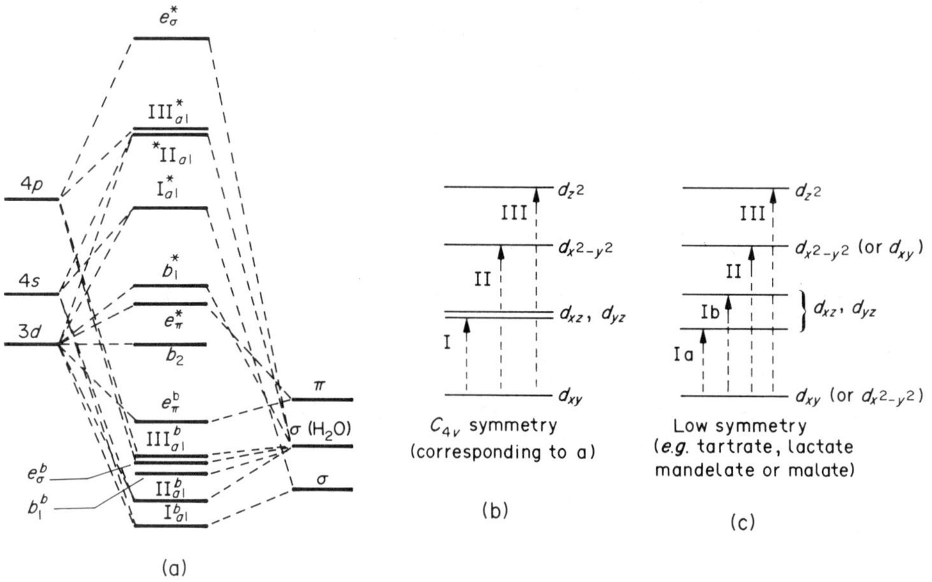

Figure 16 (a) Molecular orbital scheme for [$VO(H_2O)_5^{2+}$] (C_{4v} symmetry) according to Balhausen and Gray;[364] (b), (c) splitting of the vanadium d levels (b_2, e and b_1 considered mainly as V orbitals)

Mabbs and co-workers studied the single crystal polarized spectra of pentacoordinate VO^{2+} compounds at 298 and 5 K and of poly(dimethylsiloxane) mulls of six VO^{2+} compounds (five- and six-coordinate complexes).[369] Ligand-field strengths ranged from that of cyanide to that of bromide. Transitions I and II were assigned to $3d_{xy}\rightarrow 3d_{x^2-y^2}$ and to $3d_{xy}\rightarrow 3d_{xz},3d_{yz}$. Spectra of six-, five- and four-coordinate compounds of vanadium(IV), *e.g.* $[VCl_6]^{2-}$, $[VCl_4]^+\cdot[VCl_5]^-$ and VCl_4, are in Lever's textbook.[368]

According to the Ballhausen and Gray model,[364] the vanadium–oxygen bond is triple and it has been explicitly considered as such by several authors.[376–378] In most textbooks and publications, including this one, the structures of oxovanadium(IV) compounds are reported with a double bond between vanadium and oxygen.

ESR spectroscopy is a particularly useful technique as it provides information on the stereochemistry, ligand type and degree of covalency of V^{IV} complexes. The fundamental ESR properties of VO^{2+} complexes have been reviewed.[354] Petrakis and co-workers have shown that the ESR spectra of various vanadyl complexes yield A_0 (isotropic hyperfine splitting constant) and g_0 (isotropic g) values in reasonably well-defined ranges characteristic of the types of metal–ligand bonds, *i.e.* g_0 increases and A_0 decreases in the series VOO_4, VON_2O_2, VOS_2O_2, VON_4, VOS_4.[379] ENDOR studies of complexes randomly oriented in a matrix can provide valuable data.[380–383] The hyperfine and quadrupole tensor components of ligand nuclei identify ligands and give more insight than that provided by ESR studies alone.

The V=O stretching frequency is an important characteristic of oxovanadium(IV) complexes, generally observed at $985 \pm 50\ cm^{-1}$.[353,355] It is quite sensitive to the nature of the ligands. Donors that increase the electron density of the metal reduce its acceptor properties towards O, the V—O multiple bond character and the stretching frequency. For complexes $VOL_5^{n\pm}$, $\nu(V{=}O)$ falls in the order $L = H_2O > NCS^- > CN^- > DMSO \sim F^-$; this series bears no obvious relation to the spectrochemical and nephelauxetic series.[353]

Although most oxovanadium(IV) complexes are blue, some Schiff base complexes may vary from yellow to maroon. Earlier suggestions that such colours, together with reductions in the V—O frequencies from 950–1000 to 800–850 cm^{-1}, indicate polymerization or VO $\cdots$ VO $\cdots$ VO interaction are erroneous. Even for coordination environments which are very similar, the V=O lengths may be very different, the vanadyl oxygens being exposed to the influence of the neighbouring molecules in the crystal.[384]

VO^{2+} bonds most effectively to electronegative atoms, *e.g.* F, Cl, O and N, and bonds to S and P are also known.[359] Fluoro and oxygen complexes are specially stable. Complexes with oxygen donors follow the expanded Irving–Williams series:[354] $VO^{2+} > Cu^{2+} > Ni^{2+} > Co^{2+} > Fe^{2+} > Mn^{2+}$. The presence of one or two nitrogen donors places VO^{2+} lower in the series.

Vanadyl compounds hydrolyze much more easily than any other metal chelates of the first transition series, and in some conditions the vanadyl 1:1 chelates prefer combination with hydroxyl to a second ligand molecule.[385] Unless otherwise stated formation constants β_{xyz} presented in this review correspond to the reaction shown in equation (15).

$$x\,VO^{2+} + y\,L + z\,H^+ \rightleftharpoons (VO)_x(L)_y(H)_z \qquad \beta_{xyz} = \frac{[(VO)_x(L)_y(H)_z]}{[VO^{2+}]^x[L]^y[H^+]^z} \tag{15}$$

33.5.2 Group IV Ligands

Most complexes of vanadium(IV) with group IV ligands are normally organometallic compounds, which have been reviewed.[9]

In cyanide complexes, several bonding modes are available but CN^- will show more basic character when coordinating through the carbon lone pair.[386] No stable cyanide complexes exist in acid solution, but complexes $M_3[VO(CN)_5]$ ($M = Et_4N^+$, Me_4N^+ or Cs^+) have been isolated; apparently the large cation provides the necessary crystal stability. They are prepared from aqueous $VOSO_4$ and NaCN (6 M) on addition of salts of the corresponding counterion.[387] $\nu(V{=}O)$ was in the range 952–956 cm^{-1} (with a shoulder at 965–968 cm^{-1}).[388]

An old compound, $K_2V(CN)_6$,[389] was reformulated as $K_3[VO(CN)_5]$,[390] based on a band at 935 cm^{-1} assigned to $\nu(V{=}O)$. Mabbs and co-workers assigned $\nu(V{=}O)$ at 930 cm^{-1}, which probably reflects high electron density in the vanadium from the CN^-;[369] ESR studies of $[VO(CN)_5^{3-}]$ anion indicate significant covalency.[356] UV-visible spectra of solutions containing VO^{2+} and cyanide do not support the existence of $V(CN)_6^{2-}$.[391]

33.5.3 Nitrogen Ligands

33.5.3.1 Ammonia and aliphatic amines

Complexes $[VOCl_2L_n]$ (L = monodentate or bidentate neutral ligand; $n = 1$, 2 or 3) are made by methods including controlled hydrolysis of vanadium(IV) chloride adducts,[392,393] controlled oxidation of vanadium(III) chloride adducts,[394,395] reaction of ligand with aqueous $VOCl_2$,[396,397] with $VOCl_3$,[398,399] $VOBr_3$[400,401] or with $VOCl_2(MeOH)_3$ in methanol.[402] Methods for the related $Ct_2[VOCl_4]$ (Ct = monopositive cation) include reaction of $[VOCl_2(diox)_2]$ (diox = 1,4-dioxane) with CtCl in liquid SO_2,[403,404] thermal decomposition of hydrates or ethanenitrile adducts,[403–407] reaction of $VOCl_2(MeCN)_2(diox)_{0.5}$ with CtCl in ethanenitrile[403,404] and direct reaction between $VOCl_2$ and CtCl in $CHCl_3$.[408] A general synthetic procedure for

$[VOCl_2L_n]$ and $Ct_2[VOCl_4]$ is more reproducible and gives fewer problems from solvolysis and thermolysis.[409] The method consists of ligand displacements in solutions of $[VOCl_2(MeCN)_2]$ (very simple to prepare on a large scale) in THF or ethanenitrile according to equations (16) and (17). $[VOCl_2(Me_3N)_2]$ was prepared by this method and bands at 404 and 383 cm^{-1} were assigned to ν(V—Cl).[410] Two strong IR bands in $[VOCl_2L_2]$ indicate geometry of type (**13**);[409] when there is only one ν(V—Cl) band, the geometry is of type (**11**). For $VOBr_2L_2$,[401] one class gives two strong ν(V—Br) stretches (the higher in the region 360–340 cm^{-1}); the other shows only one ν(V—Br), 340–320 cm^{-1}.

$$VOCl_2(MeCN)_2 \xrightarrow[\text{THF or MeCN}]{L} VOCl_2L_n \quad (16)$$

$$VOCl_2(MeCN)_2 \xrightarrow[\text{MeCN}]{\text{CtCl}} Ct_2[VOCl_4] \quad (17)$$

Wood and co-workers prepared $[VOCl_2(Me_3N)_2]$ by reduction of $VOCl_3$ with Me_3N under strictly anhydrous conditions.[398] After redissolution in trimethylamine, blue crystals suitable for X-ray work formed. The geometry around the vanadium atom is trigonal bipyramidal (**13**) (see Figure 17), probably due to steric effects of the bulky Me_3N. Several similar compounds have been prepared by reduction of $VOCl_3$ with the appropriate amine (see Table 10). With methylamine, the reduction is slower than with the others; the silvery grey product is probably polymeric. A compound formulated as $VOCl_2 \cdot 5NH_3$ had a room temperature moment of 1.62 BM and obeyed the Curie–Weiss law with $\theta = 60$ K.[411]

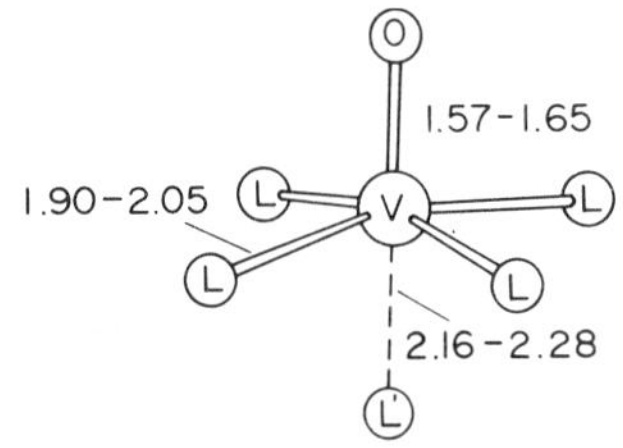

Figure 17 A perspective view of $[VOCl_2(Me_3N)_2]$. The bond lengths are in Å. The angles in the equatorial plane are 120°[398]

Table 10 $VOCl_2 \cdot n$(amine) Complexes

Amine	ν(V=O) (cm^{-1})	$\mu^{r.t.}$ (BM)	ν(V—Cl) (cm^{-1})	*Ref.*
NMe_3 ($n = 2$)	990	1.73	404, 383	1, 2
$NHMe_2$ ($n = 2$)	1000	1.68		1
NH_2Me ($n = 4$)	972	1.72		1

1. K. L. Baker, D. A. Edwards, G. W. A. Fowles and R. G. Williams, *J. Inorg. Nucl. Chem.*, 1967, **29**, 1881.
2. J. Cave, P. R. Dixon and K. R. Seddon, *Inorg. Chim. Acta*, 1978, **30**, 349.

The tetrahalides $[VX_4]$ are Lewis acids and produce adducts with a variety of donors, namely aliphatic amines or amides. The common coordination number is six, although this and the precise geometry have not often been determined. VCl_4 reacts with amines and, in rigorously anhydrous conditions, gives satisfactory products like VCl_4NHMe_2, $VCl_2(NHMe)_2 \cdot 4NH_2Me$, VCl_4NMe_3 and $VCl_3 \cdot 2NMe_3$.[412] A compound formulated $VCl_4 \cdot 2EtNH_2$ has been obtained in a vapor-phase reaction;[413] $VCl_4 \cdot 2L$ (L = MeCN, $PhCH_2CN$, EtCN) have been prepared,[6] and VCl_4(dibenzylamine)$_2$.[414] An adduct formulated $VCl_4(MeCOCN)_2$ was prepared by reaction of VCl_4 with an excess of ligand in petrol.[415] With VCl_4 in excess, $(VCl_4)_2(MeCOCN)_3$ was obtained. With benzoyl cyanide, only $(VCl_4)_2(PhCOCN)_3$ was obtained pure. IR spectra suggest nitrogen coordination. VCl_4 normally forms 1:2 adducts with unidentate ligands. Octahedral as well as tetragonal structures have been proposed. Solutions of such adducts are not common, but $[VCl_4(RCN)_2]$ give brown solutions in benzene and molecular-weight measurements indicate extensive dissociation.[6] $[VF_4NH_3]$ (probably polymeric, with $\mu_{eff}^{r.t.} =$ 1.83 BM) readily gives a brown aqueous solution.[416]

The chloride of $[VCl_4]$ may be substituted by amide. With primary and secondary amines at room temperature, aminolysis stops at the dichlorodiamides. The ammonia-insoluble, presumably polymeric $[V(NH_2)_2Cl_2]$ dissolves in NH_4Cl in liquid NH_3; complex anions are formed.[6] The green $VCl_2(RHN)_2{\cdot}4RH_2N$ (R = Me, Et, Pr^n, Bu^n), insoluble in benzene and the parent amines, are probably polymeric.[412,417] They release two molecules of amine at 60 °C, leaving purple $[VCl_2(RHN)_2{\cdot}2RH_2N]$. Products from secondary amines $[VCl_2(R_2N)_2]$ (R = Me, Et, Pr^n) are more soluble and molecular weights in benzene indicated a mixture of monomer and dimer.[417]

Complete replacement of halogen can be achieved using lithium dialkylamides. $[VCl(Et_2N)_3]$[418] and the dark green liquid $[V(Et_2N)_4]$ have been obtained as well as $[V(Me_2N)_4]$.[418,419] $[V(Et_2N)_4]$ obeys the Curie–Weiss law with small θ; $\mu_{eff} = 1.71$ BM. ESR and absorption spectra are consistent with monomeric $[V(Et_2N)_4]$.[419] The metal–nitrogen bond in these complexes is sensitive to protic reagents HL (where L = OH, OR, $OSiR_3$, halide, HNR, NR_2, SR, C_5H_5, C≡CR, *etc.*) and this has been put to synthetic use, originally by Bradley and Thomas,[420–422] who prepared vanadium(IV) alkoxides and trialkylsilyloxides (equation 18).

$$V(R_2N)_4 + yHL \longrightarrow VL_y(R_2N)_{4-y} + yNHR_2 \qquad (18)$$

$$\underset{\text{(Slight excess)}}{VCl_4 + \text{amine}} \xrightarrow[\text{ii, vacuum dry, } CaCl_2]{\text{i, } CHCl_3\text{, } \sim 0\,°C} VCl_4{\cdot}(\text{amine})_n \qquad (19)$$

The complex of tetrakis(aminomethyl)methane was studied by ESR at 77 K[423] and the triplet state spectrum at ~1500 G, due to $\Delta M_I = 2$, interpreted assuming dimeric oxovanadium(IV) with V—V of 5.7 Å.

From the reaction (equation 19) of anilines and naphthylamines, $[VCl_4L_2]$ were obtained,[414,424] and with diamines (**14**)–(**16**), compounds VCl_4L' result. They are hygroscopic, insoluble in $CHCl_3$, CCl_4, benzene and nitrobenzene and slightly soluble in acetone and dioxane.

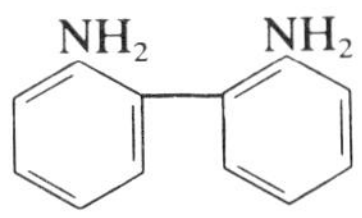

(**14**)

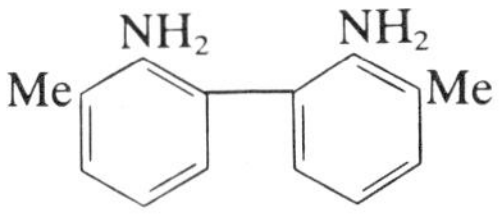

(**15**)

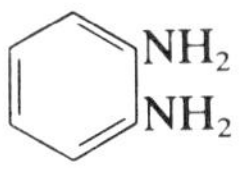

(**16**)

Ethylenediamine oxovanadium(IV) complexes in water have been reported[355] and dark brown '$[V(en)_3]F_4$' has been prepared by adding VF_5 to the diamine at low temperature.[425] Further studies are required to sort out whether the fluorides really are only counterions.

$VOSO_4{\cdot}5H_2O$ with 1,4,7-triazacyclononane (9-aneN_3) (1:1) in H_2O at 50 °C yields blue acidic solutions (equation 20), from which a blue perchlorate was precipitated, by $NaClO_4$. With NaI, a blue iodide was obtained for which ν(V=O) was at 950 cm^{-1};[426] absorptions at 1210, 1140 and 1030 cm^{-1} were assigned to μ-sulfato.

Attempts to recrystallize (**17**) from aqueous solutions resulted in the release of sulfate and formation of species (**18**) (equation 21). Upon addition of $NaClO_4$, NaBr or NaI, grey salts of (**18**) were obtained. From the temperature dependence of χ_M, $J = -177$ cm^{-1}, indicating strong antiferromagnetic coupling of the two V^{IV}; ν(V=O) = 950 cm^{-1}, and a new band at 540 cm^{-1} was assigned to ν(V—O) of the μ-hydroxo bridges. The crystal and molecular structure of the bromide of (**18**) consists of discrete binuclear cations with the V^{IV} centres in a distorted octahedral environment. The $V(\mu\text{-}OH)_2V$ four-membered ring is planar and V···V is 3.033 Å. The oxo groups are mutually *trans*; VO is 1.603 Å, much shorter than the V—OH bond lengths of ~1.96 Å. The vanadyl O atoms influence the *trans* V—N bond (~0.15 Å longer than the equatorial V—N distances).

For the dimer (**18**), at 200 K, the ESR powder spectrum shows an intense central signal and a strong half-field absorption.[427] Below 200 K, well-resolved hyperfine structure appears. The hyperfine constant for the half-field transition, 84 G, is approximately half of the A_z value for the monomers. From single crystal ESR,[427] the exchange integral $J_{xy,x^2-y^2} \approx -410$ cm^{-1}, compared to the value from magnetic susceptibility: $J_{xy,x^2-y^2} \approx 354$ cm^{-1}.[426]

$2VO(H_2O)_5^{2+} + SO_4^{2-} + (9\text{-aneN}_3) \longrightarrow$ 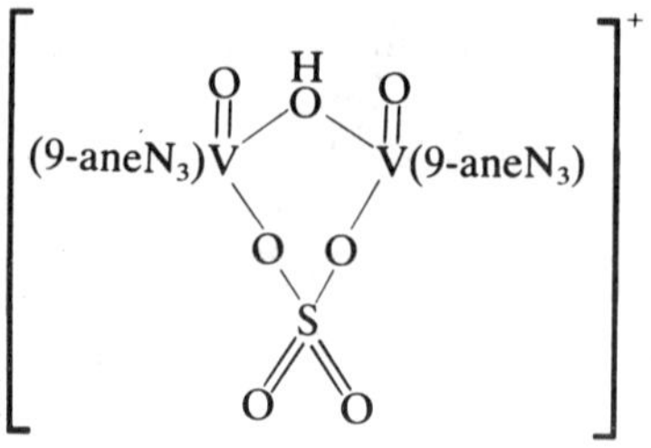 (20)

(17)

$(\mathbf{17}) + H_2O \longrightarrow$ 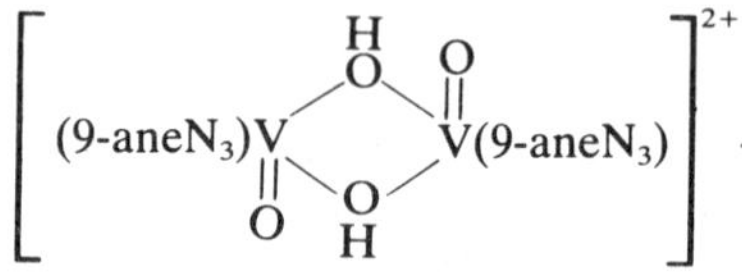 $+ H^+ + SO_4^{2-}$ (21)

(18)

33.5.3.2 N-heterocyclic ligands

Adducts of $VO(acac)_2$ with pyridines have been extensively studied (see Section 33.5.5.4.ii). Adducts $VOF_4{\cdot}py$[416] and $VOCl_4{\cdot}L_n$ with L = py, PhN=NPh, pyrrole, pyrazine, 2,6-dimethylpyrazine, bipy and phen have been discussed.[6] A $[VCl_4L_2]$ complex (L = carbazole) has been reported.[414] Again, the rule is the formation of 1:2 adducts with monodentate nitrogen donors and 1:1 adducts with bidentate donors. IR and solubility data have been used to decide whether bidentate ligands are chelated or bridging.

The stoichiometry of products from $VOCl_2{\cdot}3MeOH$ with aliphatic and heterocyclic nitrogen donors[402] depends on pK_a, bulk of the ligand and the reaction conditions (see also Section 33.5.5.3). The ligands included N-heterocyclic and other nitrogen bases (*e.g.* Et_2NH, Et_3N and *N,N,N′,N′*-tetramethyl-1,2-diaminoethane). The complexes isolated have the stoichiometries $VOCl_2{\cdot}xL$ ($x = 2$ and 3), $VOCl_2{\cdot}xL{\cdot}yMeOH$ ($x = 1, 2$; $y = 1, 2$), $VOCl_2{\cdot}B$, $2VOCl_2{\cdot}3B'$, $(LH)[VOCl_2(OMe)]$, $(BH_2)[VOCl_2(OMe)]_2$, $(B'H)[VOCl_2(OMe)]$, $VO(OMe)_2{\cdot}2LHCl$, $VO(OMe)_2{\dots}{\cdot}BH_2Cl_2$, $VOCl(OMe){\cdot}xL$ ($x = 1$ and 2) and $VOCl(OMe){\cdot}2L{\cdot}L'{\cdot}HCl$ where L and L′ are monodentate (ligands), and B and B′ are bidentate chelating and bridging ligands respectively. $VOCl_2{\cdot}xL$, $VOCl_2{\cdot}B$ and $(Ct)_2[VOCl_4]$ (Ct = Hpy^+, L = py ($x = 2$, 3), bipy and phen) have been prepared by the general procedure in equations (16) and (17).[409] When $(LH)[VOCl_2(OMe)]$ (L = 2,6-lutidine or *N,N*-dimethylaniline, bulky ligands with low pK_a values) react with py, a simple adduct $VOCl_2{\cdot}2py$ forms. In contrast, reactions of complexes where the base has a high pK_a (*e.g.* Et_3N) give $VOCl(OMe){\cdot}2py{\cdot}LHCl$.[402]

VOF_2 combines readily with water or fluoride. It has less tendency to combine with nitrogen or sulfur, but many complexes of these donors and oxygen donors were easily prepared by reaction (**22**).[428] The products containing N-heterocyclic donors are listed in Table 11, with some properties.

$$VOF_2{\cdot}H_2O + nL \xrightarrow{\text{EtOH}} VOF_2(L)_n(H_2O)_m \qquad (22)$$

$$n = 1 \text{ or } 2,\ m = 0 \text{ or } 1$$

Table 11 Some Vanadium(IV) Complexes Containing N-heterocyclic Ligands

Complex	*ν(V=O) or ν(V—Cl)*[a] (cm^{-1})	μ_{eff} (BM)	*Ref.*
$[VOF_2(phen)]$	968 (964, 928)	1.71 (306 K)	1–3
$[VOCl_2(phen)]$[b]	366[a]		4
$[VOCl(phen)_2]Cl$	982 (960, 908)		1, 2
$[VOBr(phen)_2]Br{\cdot}H_2O$	981, 984 (908)		1, 2
$[VOSO_4(phen)]$	978		1
$[VOSO_4(phen)_2]$	960 (923, 911, 900)		1, 2
$[VO(phen)_2](ClO_4)_2$	987		1, 2
$2 \times [VO(NCS)_2(phen)_{1.5}]$[c]	973 (863)		1, 2
$[VO(NCS)_2(BuOH)(phen)]$			5

Table 11 (*continued*)

Complex	*ν(V=O) or ν(V—Cl)*[a] (cm^{-1})	μ_{eff} (BM)	*Ref.*
[VO(oxalate)(phen)]	989 (959)		1, 2
[VO(malonate)(phen)]·H_2O	981 (972)	1.63	6
[VO(malonate)(phen)]·$3H_2O$	980 (973)	1.81	6
[VO(lut)(phen)]		1.86	7
[VO(coll)(phen)]		1.82	7
[VO(imda)(phen)]			7
[VO(chel)(phen)]		1.84	7
[VCl_2(phen)$_2$]			8
[VBr_2(phen)$_2$]			8
[VCl_4phen]			4
[VCl_4·2phen]			9
[VOF_2(bipy)]	968 (957, 949)	1.71 (306 K)	1–3
[$VOCl_2$(bipy)][b]	371[a]		4
[VOCl(bipy)$_2$]Cl			1
[VOBr(bipy)$_2$]Br·H_2O			1
[VO(SO_4)(bipy)]	979		2
[VO(bipy)$_2$](ClO_4)$_2$	979 (923)		1, 2
2 × [VO(NCS)$_2$(bipy)$_{1.5}$][c]	977 (967)		1, 2
[VO(oxalate)(bipy)]	979		1, 2
[VO(malonate)(bipy)]·$2H_2O$	981 (973)	1.69	6
[VO(maleate)(bipy)]·H_2O	980 (975)	1.85	6
[VO(bipy)$_2$](PF_6)$_2$	980		10
[VO(bipy)(lut)]		1.83	7
[VO(bipy)(coll)]		1.84	7
[VOF_2(py)$_2$]		1.70 (306 K)	3
[VCl_4(bipy)]			4
[$VOCl_2$·2py]	373[a]		11, 12
[$VOCl_2$·3py]			13
[$VOCl_2$·5py] (?)			12
(Hpy)$_2$[$VOCl_4$] (blue)	399,[a] 356[a]		11
(Hpy)$_2$[$VOCl_4$] (green)	345[a]		11
[VO(NCS)$_2$(py)$_2$]			14
[$VOBr_2$(py)$_2$]	940, 334[a]	1.61	15
[$VOBr_2$(py)$_3$]	965, 959, 307,[a] 257[a]	1.69	15
[$VOBr_2$(bipy)]	876, 355,[a] 345 sh[a]	1.68	15
[$VOBr_2$(quin)$_2$]	1012, 993 (?), 342,[a] 307[a]	1.63	15
[VOF_2(4-pic)$_2$]		1.71 (306 K)	3
[VOF_2(2,6-lut)$_2$]		1.74 (306 K)	3
[VOF_2(4,6-lut)$_2$]		1.73 (306 K)	3
[VO(benz)$_4H_2O$]SO_4		1.63 (308 K)	16
[VO(benz)$_2H_2O$]			17
[VO(Hgubenz)H_2O]SO_4		1.50 (303 K)	16
[VO(gubenz)$_2$]		1.59 (308 K)	16
[VO(benzc)$_2$]·H_2O		1.60 (306 K)	16
[VO(benzt)$_2H_2O$]			17

H_2lut = pyridine-2,6-dicarboxylic acid; H_3coll = pyridine-2,4,6-tricarboxylic acid; imda = iminodiacetic acid; H_2chel = 4-hydroxypyridine-2,6-dicarboxylic acid; quin = quinoline; benz = benzimidazole; Hgubenz = 2-guanidinobenzimidazole; Hbenzc = benzimidazole-2-carboxylic acid; Hbenzt = 1*H*-benzotriazole

[a] Numbers with a superscript 'a' correspond to ν(V—Cl) stretching frequencies; all the others were assigned as vanadyl IR bands.

[b] Complexes with a *cis* structure.

[c] From conductance measurements in nitrobenzene and IR spectroscopy, Selbin and Holmes suggested the formulation [VO(NCS)(AA)$_2$][VO(NCS)$_3$(AA)] (AA = phen or bipy).[1]

1. J. Selbin and L. H. Holmes, Jr., *J. Inorg. Nucl. Chem.*, 1962, **24**, 1111.
2. J. Selbin, L. H. Holmes, Jr. and S. P. McGlynn, *J. Inorg. Nucl. Chem.*, 1963, **25**, 1359.
3. M. C. Chakravorti and A. R. Sarkar, *J. Inorg. Nucl. Chem.*, 1978, **40**, 139.
4. R. J. H. Clark, *J. Chem. Soc.*, 1963, 1377.
5. V. P. R. Rao, D. Satyanarayana and Y. Anjaneyulu, *Proc. Chem. Symp, 2nd*, 1970, **1**, 151 (*Chem. Abstr.*, 1972, **76**, 158 938*r*).
6. J. Sala-Pala and J. E. Guerchais, *Bull. Soc. Chim. Fr.*, 1971, 2444.
7. R. L. Dutta and S. Ghosh, *J. Inorg. Nucl. Chem.*, 1966, **28**, 247.
8. A. Jezierski and J. B. Raynor, *J. Chem. Soc., Dalton Trans.*, 1981, 1.
9. W. W. Brandt, F. P. Dwyer and E. C. Cyarfas, *Chem. Rev.*, 1954, **54**, 959.
10. J. Selbin, H. R. Manning and G. Cessac, *J. Inorg. Nucl. Chem.*, 1963, **25**, 1253.
11. J. Cave, P. R. Dixon and K. R. Seddon, *Inorg. Chim. Acta*, 1978, **30**, 349.
12. J. Meyer and W. Taube, *Z. Anorg. Allg. Chem.*, 1935, **222**, 167.
13. H. Funk, W. Weiss and M. Zeising, *Z. Anorg. Allg. Chem.*, 1958, **296**, 36.
14. G. H. Ayres and L. E. Scroggie, *Anal. Chim. Acta*, 1962, **26**, 470.
15. P. Nicholls and K. R. Seddon, *J. Chem. Soc., Dalton Trans.*, 1973, 2751.
16. R. L. Dutta and S. Lahiry, *J. Indian Chem. Soc.*, 1964, **41**, 62.
17. R. L. Dutta and S. Lahiry, *Sci. Cult. (Calcutta)*, 1960, **26**, 139.

At room temperature, the reaction (equation 16) between Hpy·Cl and $[VOCl_2(MeCN)_2]$ in MeCN gives the expected green $(Hpy)_2[VOCl_4]$ with spectra similar to other square pyramidal $Ct_2[VOCl_4]$ compounds.[409] However, at −30 °C, a blue compound of identical composition forms, with different diffuse reflectance typical of a trigonal bipyramidal structure. The compound undergoes temperature-dependent ($C_{2v} \leftrightarrow C_{4v}$) reversible transformation like that of $(Et_4N)_2[VOBr_4]$) (Section 33.5.8.1, equations 52).[400] The C_{4v} green form is the stable isomer at room temperature and the blue form is stable at room temperature for only a few days.

Polycrystalline $[VOBr_2(py)_2]$ and $[VOCl_2(MeCN)_2]$ give single-line ESR spectra as does $(Hpy)_2[VOCl_4]$, while the spectrum of $(Et_4N)_2[VOCl_4]$ shows fine structure.[409] In nonaqueous solvents at room temperature, ESR spectra of $[VOX_2L_n]$ show the occurrence of ligand dissociation and displacement; for example, the following reaction shown in equation (23) occurs.[409] Even in non-coordinating solvents, then, the electronic spectra of $[VOX_2L_n]$ may be open to doubt in the absence of a careful study of the species present in solution. $[VOBr_2L_n]$ have been prepared by reduction of $VOBr_3$ (see Table 11).[401] The reactions liberate Br_2 when 1:2 ratios of $VOBr_3$ to ligand are used, so cyclohexene was used as solvent.

$$VOBr_2(py)_3 \underset{}{\overset{\text{toluene}}{\rightleftharpoons}} VOBr_2(py)_2 + py \qquad (23)$$

Solutions from the reactions of $[VO(L)SO_4]$, $[VO(L)_2SO_4]$ (L = bipy or phen) and $[VO(bzim)_2]^{2+}$ (bzim = benzimidazol-2-olate) with $SOCl_2$ or $SOBr_2$ gave similar ESR spectra after several hours at reflux. The deoxygenated vanadium(IV) compounds formed (see Section 33.5.5.7) are almost certainly *cis*-dihalogeno dichelates because of steric hindrance due to the α protons of the chelates.[429]

Imidazole (im) (**19**) complexing oxovanadium(IV) has been studied by ESR and proton and nitrogen ENDOR spectra.[383] Changes in orientation of the im plane, imposed by geometric constraints of multidentate ligands, such as in histidine (**20**), are reflected in the proton as well as the nitrogen ENDOR spectra. Four im ligands are bound to VO^{2+} and are oriented in a similar fashion, all making the same angle with the equatorial plane.

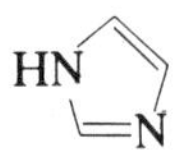

(**19**)

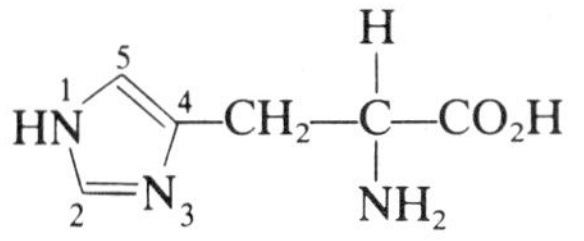

(**20**)

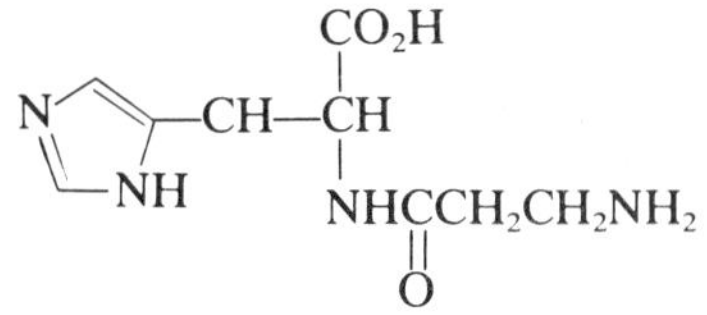

(**21**)

33.5.3.3 Hydrazine and related ligands

$[VOX_2L_4]$ (X = Cl, Br, I; L = N_2H_4, $PhNHNH_2$, Me_2NNH_2) exhibit subnormal moments (1.26–1.36 BM) and may be polymeric but further structural studies are needed.[431] VO^{2+} and 2,6-dipicolinoyl dihydrazine (DPH) (**23**) give $[VO(DPH)_2]$ and $[VO(DPH)X_2]$ (X = Cl^- or $\frac{1}{2}SO_4^{2-}$) (*cf.* Section 33.5.9.4.i).[432] They react with β-diketones according to Scheme 13. The β-diketones were acac, Prac (R = Et, R′ = Me) and Bzac (R = Ph, R′ = Me). (**25**) is obtained when $VOCl_2$ is added to the condensation product of DPH and acac (Scheme 14). VO^{2+} induces the cyclization; this template effect is 'kinetic'.[433] Complexes of (**26**) and (**27**) have been prepared,[434] for use in treatment of arteriosclerosis and to lower blood pressure.

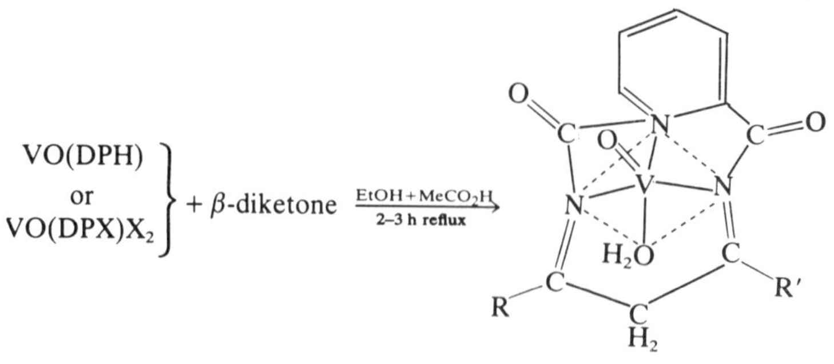

(**22**)

Scheme 13

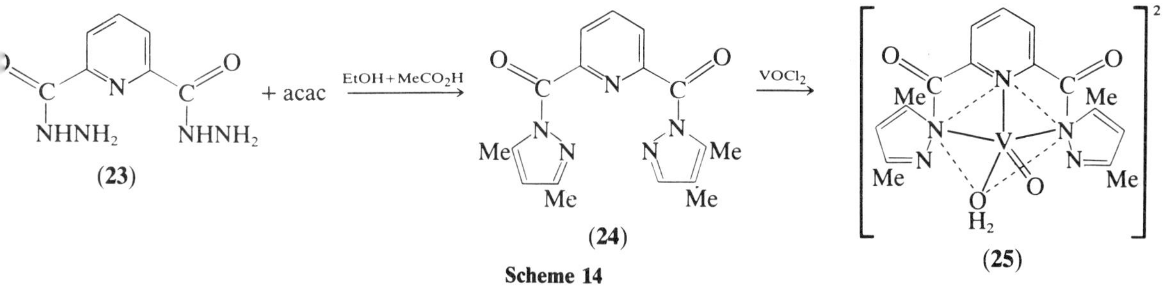

Scheme 14

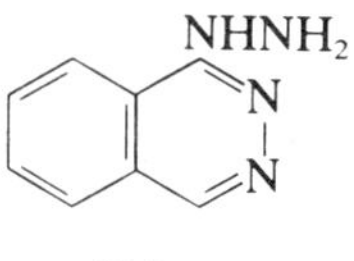

(26)

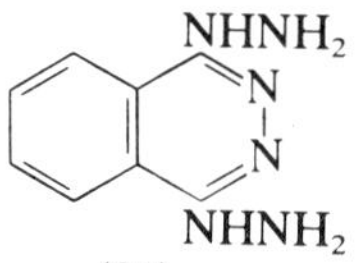

(27)

33.5.3.4 Thiocyanate

Thiocyanato oxovanadium(IV) complexes have been extensively studied. Table 12 summarizes the isolated complexes containing NCS^-. In the thiocyanato thiourea complexes, IR shows that all ligands are coordinated through nitrogen.[435] The thiocyanato complexes $(Ct)_2[VO(NCS)_4(H_2O)]\cdot4H_2O$ ($Ct = K^+$, NH_4^+) have been studied by X-ray techniques.[436] Prepared by the addition of the appropriate thiocyanate to aqueous $VOSO_4$, the complex is extracted with ethyl acetate. The geometry around the vanadium is shown in Figure 18. The NCS^- coordinates as isothiocyanate and the structure may be regarded as chains of anions linked by the two ammonium ions *via* the vanadyl oxygen and one water. The bond distance to the axial H_2O in $[VO(NCS)_4(H_2O)]^{2-}$(2.22 Å) is shorter than in $[VO(H_2O)_5]^{2+}$(2.4 Å).[356] Probably the stronger ligand NCS^- weakens V=O sufficiently to allow a stronger *trans* attachment by H_2O. ν(V=O) values support this idea, with values of 982 and 963 cm^{-1} for the thiocyanato complex and 1000 and 975 cm^{-1} for the aqua ion.[388]

Table 12 Some Vanadium(IV) Complexes Containing NCS^- as a Ligand

Complex	*ν(V=O)* (cm^{-1})	μ_{eff} (BM)	*Ref.*
$(Et_4N)_3[VO(NCS)_5]$	997, 979		1
$(Me_4N)_3[VO(NCS)_5]$	987, 979		1
$K_3[VO(NCS)_5]$	982, 969		2
$(Me_4N)[VO(NCS)_4(H_2O)]$			3
$(NH_4)_2[VO(NCS)_4(H_2O)]\cdot4H_2O$	1003, 975	1.72	4
$K_2[VO(NCS)_4(H_2O)]\cdot4H_2O$			4
$KVO(NCS)_3(1.5EtOH)$			5
$(Ph_4As)[VO(NCS)_4]$			3
$VO(NCS)_2$(2-dioxane)			5
$[VO(NCS)_2(py)_2]\cdot py$			5
$VO(NCS)_2\{(Me_2N)_3PO\}$			6
$VO(NCS)_2(L)_2$			7
(L = tu, *p*-Clptu, *p*-Brptu, dptu, dcytu)[a]		1.70–1.76	
$(VO)(NCS)_2(phen)_{1.5}$	973, 963		1
$(VO)(NCS)_2(bipy)_{1.5}$	977, 967		1
[VO(Apacac)NCS]NCS[a]			8
$[VO(NCS)_2(9\text{-ane}N_3)]$	950	1.63	9

[a] tu = thiourea; *p*-Clptu = *p*-chlorophenylthiourea; *p*-Brptu = *p*-bromophenylthiourea; dbtu = 1,3-dibenzylthiourea; dcytu = 1,3-dicyclohexylthiourea; Apacac = 2-pyN=C(Me)CHC(Me)=Npy-2; 9-aneN_3 = 1,4,7-triazacyclononane.

1. J. Selbin and L. H. Holmes, Jr., *J. Inorg. Nucl. Chem.*, 1962, **24**, 1111.
2. J. Selbin and L. H. Holmes, Jr., *J. Inorg. Nucl. Chem.*, 1963, **25**, 1359.
3. D. Collison, B. Gahan, C. D. Garner and F. E. Mabbs, *J. Chem. Soc., Dalton Trans.*, 1980, 667.
4. A. C. Hazell, *J. Chem. Soc.*, 1963, 5745.
5. J. Selbin, *Chem. Rev.*, 1965, **65**, 153.
6. M. Ziegler, H. Winkler and L. Ziegeler, *Naturwissenschaften*, 1965, **52**, 302.
7. A. H. Jamkhandi and A. S. R. Murty, *Curr. Sci.*, 1980, **49**, 615.
8. V. R. Rana, D. P. Singh and M. P. Teotia, *Transition Met. Chem. (Weinheim, Ger.)*, 1981, **6**, 189.
9. K. Wieghardt, U. Bossek, K. Volckmar, W. Swiridoff and J. Weiss, *Inorg. Chem.*, 1984, **23**, 1387.

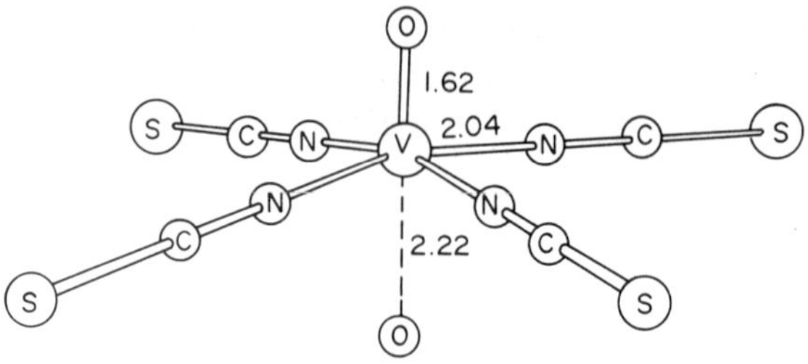

Figure 18 Geometry and internuclear distances (Å) in the complex $(NH_4)[VO(NCS)_4(H_2O)]\cdot 4H_2O$[436]

Smith and Martell tabulate stability constants ($\log \beta_{110} = 3.32$ and $\log \beta_{120} = 3.68$, at 25 °C)[437] and Che and Kustin determined the rate constant for NCS^- complexation ($k_1 \approx 1 \times 10^3\ M^{-1}\ s^{-1}$, at 25 °C and $I = 0.05$ M KNO_3).[438] Thiocyanate complexes have high formation constants, and find use for photometric determination of vanadium.[435,439–441]

33.5.3.5 Nitriles

The exchange of ligands having different π donor/acceptor properties with $[VOCl_2(MeCN)_2]$ and $[VO(DMSO)_4](ClO_4)_2$ was studied by ESR in MeCN;[442] coordinated MeCN could be substituted readily but large excesses of fairly strong donors (*e.g.* DMSO) were required to substitute the chlorine atoms. $VOBr_3$ and acetonitrile in cyclohexene give $[VOBr_2(MeCN)_3]$; with pyridinium bromide in acetonitrile $(Hpy)[VOBr_3(MeCN)_2]$ is obtained.[400] $[VOF_2(MeCN)(H_2O)]$ was prepared by the procedure described in Section 33.5.3.2 for $[VOF_2L_n]$.[429]

For malonodinitrile (maldn) and succinodinitrile (sucdn), $VO(maldn)SO_4\cdot 2H_2O$ and $VO(sucdn)SO_4\cdot 2H_2O$ were prepared according to reaction (24).[443]

$$VOSO_4 + \begin{matrix} \text{maldn} \\ \text{or} \\ \text{sucdn} \end{matrix} \xrightarrow[\text{5 h reflux}]{\text{DMF}} \text{complex} \qquad (24)$$

33.5.3.6 Biguanide and related ligands

Complexes of biguanide and substituted derivatives (big = $R^1R^2C({=}NH)NHC({=}NH)NH_2$) formulated as $[VO(big)_2(H_2O)]$ have been prepared from $VOSO_4$ and an excess of biguanide in alkaline solution.[444] The bluish green to light green crystalline products (see Table 13) are hydrolyzed by water. Room temperature moments are near the spin-only value. Syamal studied the ESR, IR and electronic spectra of polycrystalline complexes of biguanide, several dibiguanide derivatives and of *o*-methyl-1-amidinourea (Table 13).[445] The pure samples exhibited a single line at $g \approx 1.97$ but magnetic dilution in the corresponding diamagnetic Ni^{II} complexes resolved the hyperfine splitting, an axial spectrum (eight parallel and eight perpendicular lines) being obtained. The low values of $\nu(V{=}O)$ (~940 cm^{-1}) are consistent with the presence of strong out-of-plane $d_\pi - p_\pi$ bonding. The electronic spectra were recorded in mulls and bands at ~590, 430 and 400 nm were assigned to bands I, II and III (see Figure 16b).[74] Complexes of guanylalkylureas appear as silky bluish green to rose-violet precipitates (Table 13). They were formulated as $[VO(gua)_2]$ complexes involving nitrogen coordination.[446]

33.5.4 Phosphorus and Arsenic Ligands

Only a few non-organometallic complexes with V—P bonds have been studied. The first oxovanadium(IV) phosphine complexes obtained pure were the green compounds $VOX_2\cdot L\cdot H_2O$ (X = Cl, Br; L = $Ph_2PCH_2CH_2PPh_2$) and $VOBr_2\cdot L'\cdot H_2O$ (L′ = $Ph_2PCH_2PPh_2$),[447] prepared from $VOCl_2$ or $VOBr_2$ in carefully degassed solvents. In spite of precautions, some phosphine oxide compounds were formed. Table 14 summarizes some properties. $[VOBr_2(Ph_3P)_2]$ was prepared by substitution in THF from $[VOBr_2(MeCN)_3]$.[401]

ESR spectra of powdered $[VOCl_2(Ph_3P)_2]$ show some fine structure.[409] Spectra for $VOCl_2$

Table 13 Some Oxovanadium(IV) Biguanide, Guanylurea and Dibiguanide Complexes

Ligand (H_nL)	*Formulation*	$\mu_{eff}^{r.t.}$	g_{av}, A_{av} [a]	*Ref.*
$n = 1$				
Biguanide	$VO(L)_2(H_2O)$	1.75	1.97, 87	1, 2
N-Methylbiguanide	$VO(L)_2$	1.62		1
N-Ethylbiguanide	$VO(L)_2$	1.45		1
N,N-Dimethylbiguanide	$VO(L)_2$	1.60		1
N-(*n*-Hexyl)biguanide	$VO(L)_2(H_2O)$	1.81		1
N,N-Ethylenedibiguanide (R^1 = H and R^2 = $(CH_2)_2$biguanide)	$VO(L)(H_2O)\cdot 0.5H_2O$	1.94		1
N,N-Hexamethylenedibiguanide (R^1 = H and R^2 = $(CH_2)_6$biguanide)	$VO(L)(H_2O)\cdot 1.5H_2O$	1.50		1
Guanylmethylurea	$VO(L)_2$	1.67		3
Guanylethylurea	$VO(L)_2$	1.68		3
Guanylbutylurea	$VO(L)_2$	1.67		3
Guanylethoxyethylurea	$VO(L)_2$	—		3
Guanylbutoxyethylurea	$VO(L)_2$	1.69		3
$n = 2$				
o-Methyl-1-amidinourea	$VO(omau)_2(H_2O)$	—	1.97, 87	2
Ethylenedibiguanide	$VO(endibig)(H_2O)$	—	1.97, 88	2
Trimethylenedibiguanide	$VO(tdibig)(H_2O)$	—	1.97, 88	2
Tetramethylenedibiguanide	$VO(ttdibig)(H_2O)$	—	1.97, 87	2
Hexamethylenedibiguanide	$VO(hxdibig)(H_2O)$	—	1.97, 87	2
Piperazinedibiguanide	$VO(pipdibig)(H,_2O)$	—	1.97, 87	2

[a] A_{av} in 10^{-4} cm^{-1} units.

1. P. Ray, *Chem. Rev.*, 1961, **61,** 313.
2. R. L. Dutta and S. Lahiry, *J. Indian Chem. Soc.*, 1963, **40,** 863.
3. A. Syamal, *Indian J. Pure Appl. Phys.*, 1983, **21,** 130.

Table 14 Phosphine Complexes of Oxovanadium(IV)

L in $[VOX_2L_n]$	$\nu(V{=}O)$ (cm^{-1})	$\nu(V{-}X)$ (cm^{-1})	*Ref.*
$n = 1$, $X = Cl^-$			
$Ph_2PCH_2CH_2PPh_2$	1003		1
$Ph_2PCH_2PPh_2$	1000	394	1, 2
$n = 2$, $X = Cl^-$			
Ph_3P		412, 356	2
Ph_2MeP		409, 360	2
$n = 1$, $X = Br^-$			
$Ph_2PCH_2CH_2PPh_2$	992		1, 2
$n = 2$, $X = Br^-$			
Ph_3P	988	347 s, 313 m	3

1. J. Selbin and G. Vigee, *J. Inorg. Nucl. Chem.*, 1968, **30,** 1644.
2. J. Cave, P. R. Dixon and K. R. Seddon, *Inorg. Chim. Acta,* 1978, **30,** L349.
3. D. Nicholls and K. R. Seddon, *J. Chem. Soc.* (*A*), 1973, 2751.

solubilized in toluene by an excess of phosphine R_3P show that each of the eight lines of the $^{51}V^{IV}$ nucleus is split into a 1:2:1 triplet.[448] This structure can be attributed to two equivalent phosphorus nuclei and was thus ascribed to $[VOCl_2(R_3P)_2]$. Varying the bulk of R, the spectra show increasing line width with increasing molecular volume and it was suggested that the trigonal bipyramid is the more probable configuration for $[VOCl_2(R_3P)_2]$.[448] This agrees with a suggestion that $[VOCl_2L_2]$ with trigonal bipyramidal structure should give two strong ν(V—Cl) IR-active bands;[409] in fact these were observed for $[VOCl_2L_2]$ (L = Ph_3P and Ph_2MeP; Table 14). As in $[VOX_2L_n]$ where L is a nitrogen ligand, ESR spectra with phosphines in non-aqueous solvents provide evidence for the occurrence of ligand dissociation and displacement.[409] For example, reaction (25) proceeds even in non-coordinating solvents and this again emphasizes the need for examination of species in solution when interpreting the electronic spectra (and other properties) of $[VOX_2L_2]$ complexes.

$$VOBr_2(Ph_3P)_2 \underset{Ph_3P}{\overset{THF}{\rightleftharpoons}} VOBr_2(THF)_2 + 2Ph_3P \quad (25)$$

A dicyclohexylphosphide (Cy_2P) complex, $[Li(DME)][V(Cy_2P)_4]$ (DME = 1,2-dimethoxyethane) was isolated by the addition of a THF solution of $LiCy_2P$ to a suspension of a vanadium halide THF adduct at −80 °C, addition of DME and cooling to −30 °C.[449] The detailed structure was not determined; oxidation with $[Cp_2Fe](BF_4)$ gives the thermally unstable $[V(Cy_2P)_4]$.

The first compound with arsenic donors was orange $[VCl_4(diars)_2]$.[450] With the similar ligand detars (**28**), brown-black $[VCl_4(detars)]$ was obtained;[451] with tritertiary arsines (**29**) and (**30**) $[VCl_4L]$ were obtained.[452] The complexes were prepared by adding the arsines to VCl_4 in CCl_4; extreme care must be taken to avoid formation of the vanadyl species. Hydrolysis of (**28**) forms no complexes with definite composition, unlike the case of bipy and phen which form $[VOCl_2L]$.[451] However, with (**29**) and (**30**), similar VO^{2+} complexes were obtained by hydrolysis,[452] but were not well characterized. $[VCl_4(detars)]$ has $\mu_{eff} \sim 1.76$ BM (~293 K) and IR and solubility suggest that detars acts as a bidentate chelate (not a bridge).[6] A dodecahedral structure with bidentate arsines was suggested[6] for $[VCl_4(diars)_2]$ as it is isostructural with the corresponding titanium complex. When diars reacts with VCl_4 the only product is the eight-coordinate $[VCl_4(diars)_2]$, while with detars Clark[451] reported that $[VCl_4L]$ is obtained and in no circumstances was another molecule added to the six-coordinate complex. The tritertiary arsines (**29**) and (**30**) in $[VCl_4L]$ were originally considered terdentate;[452] thus V^{IV} would be seven-coordinate. Further structural studies are needed.

(**28**)

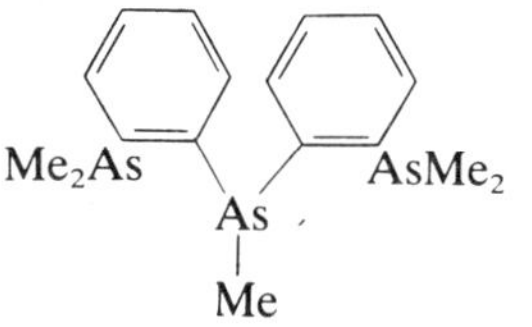

(**29**)

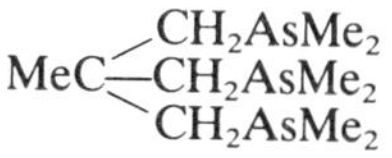

(**30**)

Attempts to prepare $[VOCl_2L_n]$ (L = Ph_3As and Ph_3Sb) were unsuccessful and only the As-containing compound $(Ph_4As)_2[VOCl_4]$ was obtained.[409] The ESR and the single ν(V—Cl) band (346 cm^{-1}) suggest a square pyramidal structure.[409]

33.5.5 Oxygen Ligands

33.5.5.1 Oxovanadium(IV) in aqueous solutions

(*i*) *The vanadyl ion. Rate of exchange*

In moderately acidic solutions, oxovanadium(IV) exists as VO^{2+} (or $VO(H_2O)_5^{2+}$) commonly called 'vanadyl ion' (Figure 19). The four water molecules of the tetragonal bipyramid have long residence time (see Table 15), and the axial water exchanges rapidly with the solvent. The 'labilization' of the *trans* water has been explained as a result of π bonding.[454] Water exchange is slow compared to other doubly charged ions.[455] The high charge of vanadium is important not only in the formation of the vanadyl group but also in the coordination of the water molecules.

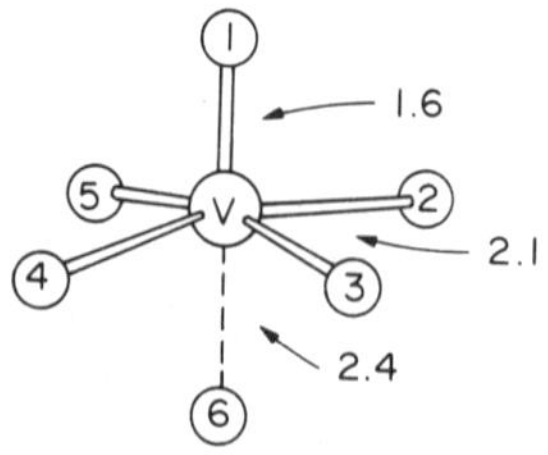

Figure 19 Schematic representation of $[VO(H_2O)_5^{2+}]$ and average internuclear distances (Å) from electron spin echo modulation (ESEM), electron nuclear double resonance (ENDOR) and X-ray diffraction studies[453]

Proton and deuteron relaxation show that relatively slow hydrogen exchange occurs in neutral solutions. The exchange is, however, acid catalyzed. Rivkind interpreted increased spin–spin relaxation rate with increasing acidity as due to protonated vanadyl and estimated

Table 15 Residence Times and Rate of Exchange of Water Molecules and Oxygen Coordinated to VO^{2+} at 25 °C

'Type' of water molecule (*Figure* 15)	Residence time (s)	Rate of exchange (s^{-1})	Method used	Ref.
2, 3, 4, 5 (equatorial)	1.35×10^{-3}	5×10^2	^{17}O NMR	1, 2
6 (axial)	$\sim 10^{-11}$	$>5 \times 10^8$	^{17}O NMR	1, 2
1		<20		2
1 (0 °C)	$t_{1/2} \sim 400$ min	2.9×10^{-5}	F^- precipitation and H_2O sampling	3

1. J. Reuben and D. Fiat, *Inorg. Chem.*, 1967, **6**, 579.
2. K. Wüthrich and R. E. Connick, *Inorg. Chem.*, 1967, **6**, 583.
3. R. K. Murmann, *Inorg. Chim. Acta*, 1977, **25**, L43.

the constant $\beta_p = [VOH^{3+}]/([VO^{2+}][H^+])$.[456] However, the similarity of the activation parameters for the proton-catalyzed proton exchange of VO^{2+} and that of Ni^{2+} and Cr^{3+} is a strong argument against Rivkind's interpretation.[457] The hydrogen lifetime in the first coordination shell of VO^{2+} ions is determined by chemical exchange of the equatorial water.[458,459] Nágypál *et al.* studied equilibrium (**26**) by spectrophotometric methods at constant $[ClO_4^-]$ (6 M) and interpreted NMR relaxation at 2.5 and 100 MHz taking into account both chemical environments and reaction pathways.[460] The UV spectrum of solutions cannot be explained by protonation of the vanadyl oxygen: the protonation site is a coordinated water, contrary to Rivkind's interpretation. The protonation constant for reaction (**26**) is 0.23 ± 0.05. Variations in absorption spectra have also been explained through anation by the acid (including ClO_4^- [461]). For high acid concentrations, V^{IV} may be oxidized. Ballhausen and Gray explained the resistance of VO^{2+} to protonation in terms of the M.O. bonding scheme.[364]

$$VO^{2+} + H^+ \rightleftharpoons VOH^{3+} \quad (26)$$

Considering values in Table 15 and other kinetic data, Nishizawa and Saito classified the rates of substitutions into five categories (Figure 20):[462] (1) very rapid apical substitution, (2) fairly rapid intramolecular rearrangement from apical to basal, (3) rather slow direct basal substitution, (4) slow chelate formation of a bidentate ligand coordinated as a unidentate and (5) very slow exchange of the oxo ligand. According to this view, regioselectivity between *trans* (apical) and *cis* (basal) sites on substitution reaches 10^9.[462]

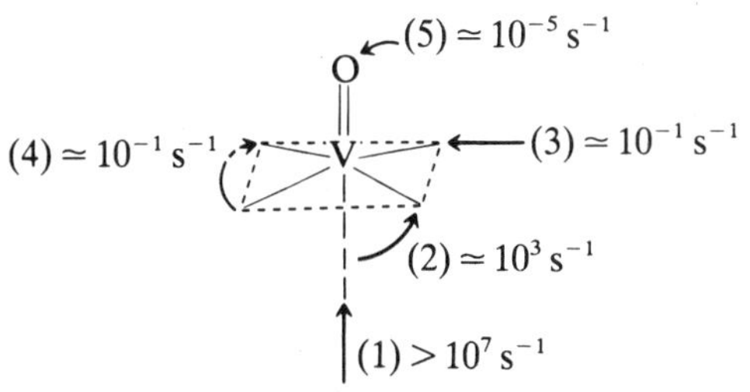

Figure 20 Different substitution rates at various coordination sites of oxovanadium(IV) ion (see text), with approximate first-order rate constants at room temperature

(*ii*) *Hydrolysis in solution*

From potentiometry, calorimetry (Table 16) and ESR spectroscopy,[463–465] there is agreement about the species in moderately acidic solutions. On the other hand, the nature of species present in neutral or basic solutions is still controversial.

In moderately acid solutions, oxovanadium(IV) exists mainly as $[VO(H_2O)_5]^{2+}$. As pH

Table 16 Hydrolysis of Oxovanadium(IV) at 25 °C

Medium	*log* β_{11}	*log* β_{22}	*log* β_{12}	*log* β_{21}	*Ref.*
$NaClO_4$ (3 M)	−6.0	−6.88	—	—	1
$NaClO_4$ (3 M)	−5.77	−6.90	—	—	2
$NaClO_4$ (3 M)	−6.1	−6.91	−10.05	—	3
$NaClO_4$ (3 M)	−5.67	−6.67	—	—	4
$NaClO_4$ (1 M)	−6.07	−6.59	—	—	5
KNO_3 (0.1 M)	−5.44	−7.18	—	—	6
$LiClO_4$ (0.1 M)	−5.04	−6.72	—	—	7
KCl (3 M)	−6.4	−7.45	−10.00	—	8
$MgSO_4$ (1 M)[a]	—	−7.5	—	3.75	9

[a] Temperature is probably 25 °C.
1. F. J. C. Rossotti and H. S. Rossotti, *Acta Chem. Scand.*, 1955, **9**, 1177.
2. F. Brito, *An. Quim.*, 1966, **62B**, 123.
3. F. Brito, *An. Quim.*, 1966, **62B**, 193.
4. R. P. Henry, P. C. H. Mitchell and J. E. Prue, *J. Chem. Soc., Dalton Trans.*, 1973, 1156.
5. I. Fábyán and I. Nagypál, *Inorg. Chim. Acta*, 1982, **62**, 193.
6. M. M. Taqui Khan and A. E. Martell, *J. Am. Chem. Soc.*, 1968, **90**, 6011.
7. A. Komura, M. Hayashi and H. Imanaga, *Bull. Chem. Soc. Jpn.*, 1977, **50**, 2927.
8. S. Mateo and F. Brito, *An. Quim.*, 1971, **68**, 37.
9. A. Diaz, Dr. Ph. Thesis, Facultad de Ciencias, UCU, Caracas, 1966.

increases (but pH $\lesssim$ 4), evidence for the formation of $[VO(OH)]^+$ and $[(VO)_2(OH)_2]^{2+}$ is strong. Table 16 summarizes the relevant constants (β_{np} is defined according to equation 27).

$$nVO^{2+} + pH_2O \rightleftarrows (VO)_n(OH)_p^{2n-p} + pH^+ \qquad \beta_{np} = \frac{[(VO)_n(OH)_p^{2n-p}]}{[VO^{2+}]^n[H^+]^p} \tag{27}$$

Oxovanadium(IV) is amphoteric but only a few quantitative studies exist on vanadium(IV) in neutral or basic solutions. Ducret explained his observations by the existence at pH $\geq$ 8–9 of $[(VO)_2(OH)_5]^-$.[466] Britton and Welford[468] also suggested $[(VO)_4(OH)_{10}]^{2-}$ to explain titration in basic solutions and Lingane and Meites[469] interpreted polarographic data in terms of $[(VO)_2(OH)_6]^{2-}$ and $[(VO)_4(OH)_{10}]^{2-}$. Ostrowetsky also assumed the formation of $[(VO)_4(OH)_{10}]^{2-}$ in neutral and basic solutions and obtained brown reddish crystals, $(NH_4)_2V_4O_9$.[470] Chasteen remarked that species in the pH range 7–11 are ESR silent[471] and suggested that polymeric $[VO(OH)_3]_n^-$ exist at pH ~ 7.[354]

Virtually all amphoteric oxides are converted to monomeric anions in sufficiently strong basic media. Rieger and co-workers studied strongly basic oxovanadium(IV) solutions by ESR, optical and Raman spectroscopy.[472] Raman absorption at 987 cm^{-1} confirms the presence of the vanadyl group in solution, and from ESR spectra and titration, the authors concluded that vanadium(IV) exists as a monomer $[VO(OH)_3]^-$ for pH $\gtrsim$ 12.

In Figures 21(a) and 21(b), speciations of solutions with two different total oxovanadium(IV) concentrations are presented assuming that the predominant soluble species in the pH range *ca.* 6–11 is $[(VO)_2(OH)_5]^-$. Further research is required to sort out the species and equilibria existing in neutral and basic solutions.

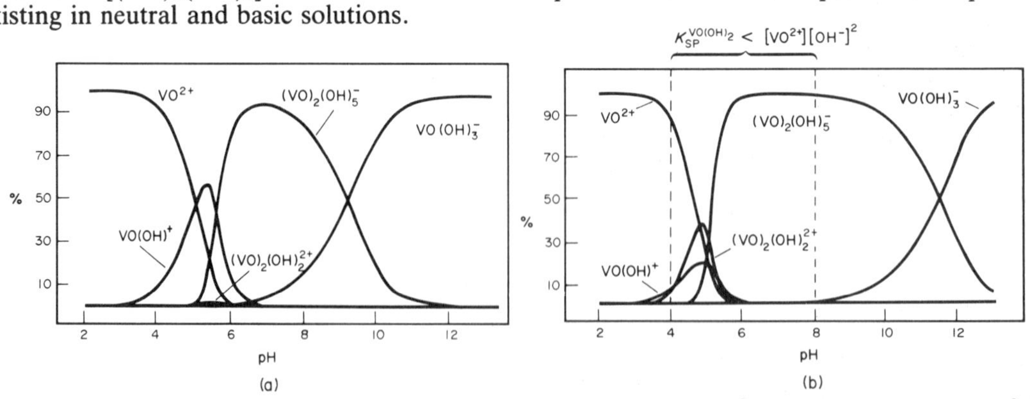

Figure 21 Speciation in oxovanadium(IV) aqueous solutions: (a) $C_{VO} = 10^{-5}$ M and (b) $C_{VO} = 2 \times 10^{-3}$ M. Formation constants β_{11}, β_{22} and β_{25} are those reported by Komura *et al.* for 0.1 M $LiClO_4$ and 25 °C.[467] For pH $\gtrsim$ 5 these speciations should only be used as a guide in order to give an idea of what may probably be the relative importance of the total concentrations of species $[(VO)_2(OH)_5^-]_n$ and $[(VO)_2(OH)_6^{2-}]_m$. For pH $\gtrsim$ 12 apparently only $[VO(OH)_3^-]$ predominates. Note that the ratio $C_{monomeric} : C_{polymeric}$ increases as the C_{VO} decreases and that in (b) $K_{sp} \lesssim [VO^{2+}][OH^-]^2$ in the pH region *ca.* 4–8. In the calculations β_{13} was assumed to be 1×10^{-18}

(iii) Redox behaviour of oxovanadium(IV) solutions

Because of the multiple and slow equilibria, the elucidation of all the species present and the hydrolysis and redox equilibria is a difficult task. Redox behaviour of oxovanadium(IV) will not be discussed here: detailed information is given by Pourbaix,[474] Hepler and co-workers,[475] 'Gmelins Handbuch'[357] or Baes and Mesmer.[87]

The most important half-cell reactions are shown in equations (28) and (29).[476] Oxovanadium(IV) can be oxidized by O_2, more easily as pH increases; in acidic solutions, the blue VO^{2+} solutions are stable to air. Air oxidation in alkaline solutions (0.006–3.8 M NaOH) is rapid at 15 °C.[477] The initial rate is proportional to $[OH^-]$. Fe^{III} catalyzes the reaction while Cr^{III} inhibits it. With $[V] > 0.002$ M, the rate is controlled by diffusion of oxygen. When VO^{2+} is chelated, oxidation may be considerably retarded and—as a rule—the more stable the complexes, the more pronounced this 'retardation effect'. However, in 1:1 solutions of oxovanadium(IV) and tartrate at pH ~ 8, the oxidation is rapid.[478] The dissolved oxygen was depleted within a few seconds.

$$VO_2^+(aq) + 2H^+ + e^- \rightleftharpoons VO^{2+}(aq) + H_2O \qquad E^0 = 1.00\ V \tag{28}$$

$$VO^{2+}(aq) + 2H^+ + e^- \rightleftharpoons V^{3+}(aq) + H_2O \qquad E^0 = 0.359\ V \tag{29}$$

A study of several VO^{2+}–complexone–aqua complexes with a few outer-sphere oxidants proved the importance of an equatorial H_2O or OH^- ligand.[479] The products certainly have a hexacoordinate *cis*-dioxo structure. Presumably the aqua ligand can change into a new oxo ligand without changing its site, and the other donors are retained without rearrangement or partial dissociation. Further, significant regioselectivity exists:[479] *cis*-oxo-hydroxo complexes are much more rapidly converted into *cis*-dioxo complexes (V^V compounds) than *cis*-oxo-aqua complexes or those without hydroxo or aqua *cis* to the vanadyl oxygen.

(iv) Mixed-valence species

Ostrowetsky studied solutions containing V^V and V^{IV} and concluded:[481] (i) at pH ≳ 10, no mixed-valence complex forms; (ii) at 8 ≲ pH ≲ 10, one forms with a V^{IV}/V^V ratio of 2, and at least six vanadium atoms; a solid, $Na_3[V_2^VV_4^{IV}O_{15}H]$, was isolated; (iii) at 4 ≲ pH ≲ 6.5, six different mixed-valence complexes form but only those with V^{IV}/V^V ratios of 3/7 and 7/3 are stable; (iv) a compound containing eight V^V and two V^{IV} transforms into the more stable 7/3 compound in a few minutes; (v) all these compounds contain a V_{10} group in the molecule.

Deep violet crystalline $[Et_4N]_2[V_5O_{14}H_2]$ was prepared from CH_2Cl_2 (or similar) solutions containing $[VO(acac)_2]$, Et_4NCl and $[Cu(acac)_2]$.[359,482] Why it forms is not understood. It was formulated as $[Et_4N]_4[V_8^VV_2^{IV}O_{28}H_4]$. Two types of mixed-valence vanadium compounds are known: the very inert bronzes and the aqueous decavanadates. This compound appears intermediary, somehow related to both types.[482] Its colour, hardness and metallic lustre are similar to the bronzes. Yet it is soluble in water, producing a green solution with a spectrum similar to one of the 7/3 decavanadates described by Ostrowetsky.[481] Robin and Day also discuss the formation of the stable 7/3 species proposed by Ostrowetsky and assume that all these anions almost certainly retain the $V_{10}O_{28}$ skeleton.[483]

33.5.5.2 Peroxide ligand

Orange-red oxoperoxovanadates(IV) $(Ct)[VO(O_2)F(H_2O)_2]$ ($Ct = NH_4^+$, K^+, Rb^+, Cs^+) were synthesized from $(Ct)[VOF_4]$ with H_2O_2 in the molar ratio 1:12 (pH ~ 4).[484] Insoluble in common organic solvents, they decompose in water. Redox titrations indicate the presence of O_2^{2-} coordinated to the V^{IV} of VO^{2+}. The $\mu_{eff}^{r.t.}$ are 1.70–1.75 BM and the optical spectra showed three absorptions at *ca.* 476 (or lower), 565 and 855 nm, the first obscured by strong charge transfer.[484] ν(V═O) of compounds $(Ct)[VO(O_2)F(H_2O)_2]$ was in the range 950–970 cm^{-1}.

33.5.5.3 Alcohols

Alcohols and phenols cleave two bonds of VCl_4 forming dichloroalkoxides. Those formed by aliphatic alcohols are dark green and dimeric in boiling benzene.[6] At 150 °C (0.1 mmHg) they yield $[V_2OCl_3(OR)_3]$. The *t*-butyl and *t*-amyl alkoxide chlorides are made by alcohol exchange on the isopropoxides.[6] $[VCl_2(OR)_2{\cdot}ROH]$, where R = *n*- or *t*-butyl, give well-resolved ESR spectra with $g_{av} = 1.95$.[485] While alkali metal alkoxides do not usually effect complete cleavage of V—Cl bonds, pale blue $[V(OSiPh_3)_4]$ may be obtained in the reaction of VCl_4 with sodium triphenylsilanolate.[486] The liquid analogue $[V(OSiEt_3)_4]$ forms from $[V(NEt_2)_4]$ and triethylsilanol in benzene.[420]

Complete replacement of the halogen atoms in VCl_4 can be achieved using lithium dialkylamides. As described in Section 33.4.3.1.i, VCl_4 with $LiNEt_2$ gives the dark green liquid $[V(NEt_2)_4]$. This and $[V(NMe_2)_4]$ react vigorously with alcohols to give moisture-sensitive tetraalkoxides $[V(OR)_4]$.[420] Molecular weight measurements in benzene show that the tertiary alkoxide complexes are monomeric liquids, the secondary are predominantly monomeric and the primary are associated.[487] An eight-line partially resolved ESR spectrum was obtained with pure $V(OBu^t)_4$ at 303 K.[487,419] The ESR parameters from room temperature[487] and frozen solutions[420,488] indicate distorted D_{2d} symmetry with a $d_{x^2-y^2}$ ground state; the compound is magnetically dilute with $\mu_{eff} = 1.69$ BM independent of temperature. Solid vanadium tetramethoxide gave a very broad ESR spectrum. The results are consistent with a trimeric species $[V_3(OMe)_{12}]$. It shows significant temperature variation of μ_{eff}.[419] Dimeric $[V_2(OEt)_8]$ also gave a broad ESR spectrum consistent with an orbital singlet ground state with magnetic dipole interaction. NMe_4^+ and NH_4^+ salts of $[VCl_5(OR)]^{2-}$ (R = Me, Et, Pr^n, Bu^n) gave μ_{eff} values near the spin-only value for a d^1 system with negligible temperature variation.[489]

$VOCl_2{\cdot}3MeOH$ may be prepared when VCl_4 reacts with MeOH[490] (*cf.* Section 33.5.3.2). Several other complexes $[VO(L)_n(X)_m]$ where L is a relatively strongly bound ligand (*e.g.* Cl^-) and X is alkoxide or alcohol have been included in early reviews,[6,355,356] *e.g.* $[NMe_4][VOCl_4(EtOH)]$. Some are discussed in other sections according to the nature of L. Glycerol inhibits V^{IV} oxidation in alkaline solutions and prevents precipitation in the whole pH range. Absorption spectra indicate coordination of glycerol.[491]

The stability constants with ascorbic (**31**) and 5,6-*O,O'*-isopropylideneascorbic acids (**32**) have been estimated and are almost independent of the nature of the ligands,[492] which are unidentate, the hydroxyl groups in the lactone ring being involved in the coordination. A vanadyl ascorbate has been synthesized, formulated $[VO(C_6H_7O_6OH)]{\cdot}2H_2O$.[493]

33.5.5.4 Catecholate, polyketonate, hydroxyaldehydate and hydroxyketonate ligands

(i) Catecholates

Aromatic polyalcohols act as strong coordinating agents and Table 17 summarizes reported formation constants. The complexes are quite stable; this behaviour has been used for the qualitative and quantitative determination of vanadium (*e.g.* refs. 494 and 495). At pH 3–4, an initial vanadyl catechol complex slowly converts to a tris complex.[496] In fact complexes with 1:3 metal–ligand stoichiometry have been isolated (see below), but since in the equilibrium (30) no protons are consumed or liberated, $[VO(cat)_2]^{2-}$ and $[V(cat)_3]^{2-}$ are not distinguishable by potentiometric studies.

$$VO(cat)_2^{2-} + H_2cat \rightleftharpoons V(cat)_3^{2-} + H_2O \quad (30)$$

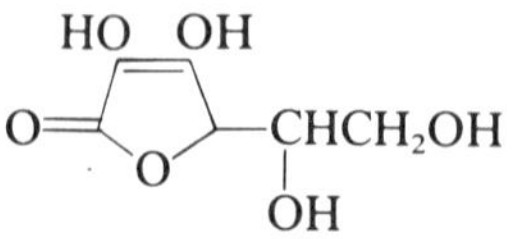

(**31**)

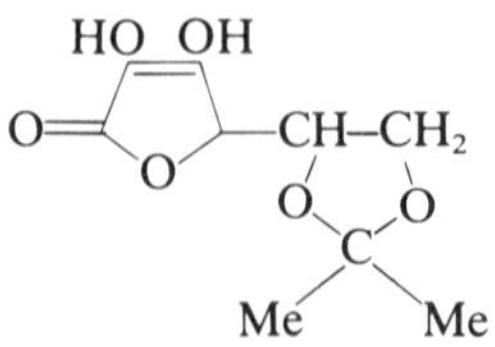

(**32**)

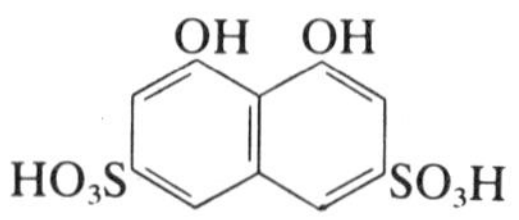

(**33**)

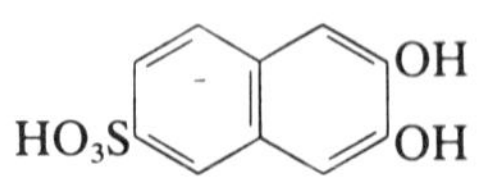

(**34**)

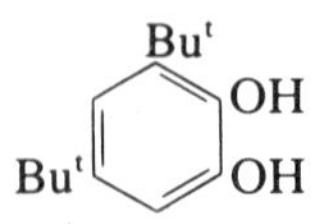

(**35**)

Table 17 Formation Constants β_{xyz} of Oxovanadium(IV) Complexes with Aromatic Polyalcohols

Ligand	*Log* β_{110}	*Log* β_{120}	*Log* β_{11-1}	*Log* β_{22-2}	*Medium*	*Ref.*
Catechol	15.76	29.42	—	—	25 °C, 0.06 M $NaClO_4$	1
(1,2-dihydroxybenzene)	16.85	31.48	—	—	35 °C, 0.2 M $NaClO_4$	2
	16.69[a]	30.78[a]	10.83[a]	—	25 °C, $I = 0$	3
	17.7	33.5	—	—	20 °C, 0.1 M KNO_3	4, 5
1,2-Dihydroxybenzene-4-sulfonic acid	16.7	31.2	—	—	20 °C, 0.1 M KNO_3	4, 5
1,2-Dihydroxybenzene-	16.8	31.2	—	—	20 °C, 0.1 M KNO_3	4, 5
3,5-Disulfonic acid (Tiron)	16.74	32.79	10.44	25.18	25 °C, 0.1 M KNO_3	5, 6
Pyrogallol	15.0	28.7	—	—	20 °C, 0.1 M KNO_3	4
	15.15	28.57	—	—	35 °C, 0.2 M $NaClO_4$	2
(**33**)	17.6	—	6.0[b]	—		5, 7
(**34**)	17.31	32.66	—	—	25 °C, $I = 0$	8
2,5-Dihydroxy-1,4-benzoquinone (= H_2D)	—	21.27[c]	—	—	25 °C, and $I = 1.0$ M	9
3,6-Dibromo-H_2D	—	18.98[c]	—	—	(acetate)	9
3,6-Dichloro-H_2D	—	18.82[d]	—	—		9

[a] The constants were reported in terms of H_2L. They were converted to β_{xyz} using the pK_a values for $I = 0.1$ M KNO_3.
[b] The value corresponds to $[VO(L)]/([VO(OH)L][H^+])$.
[c] The values correspond to β_{122} studied in the pH range 5–7.
[d] The value corresponds to β_{122} studied in the pH range 6.5–8.5.
1. R. Trujillo, F. Brito and J. Cabrera, *An. Fis. Quim.*, 1956, **52B**, 589.
2. G. M. Husain and P. K. Bhattacharya, *J. Indian Chem. Soc.*, 1969, **46**, 875.
3. R. P. Henry, P. C. H. Mitchell and J. E. Prue, *J. Chem. Soc., Dalton Trans.*, 1973, 1156.
4. J. Zelinka and M. Bartušek, *Collect. Czech. Chem. Commun.*, 1971, **36**, 2615.
5. R. M. Smith and A. E. Martell, 'Critical Stability Constants', Plenum, New York, 1976.
6. G. E. Mont and A. E. Martell, *J. Am. Chem. Soc.*, 1966, **88**, 1387.
7. S. Chaberek, Jr., R. L. Gustafson, R. C. Courtney and A. E. Martell, *J. Am. Chem. Soc.*, 1959, **81**, 515.
8. H. J. L. Lajunen and S. Parhi, *Finn. Chem. Lett.*, 1979, 140 (*Chem. Abstr.*, 1979, **91**, 217 742g).
9. J. F. Verchere and J. M. Poirier, *J. Inorg. Nucl. Chem.*, 1980, **42**, 1514.

Although 1:2 complexes may be isolated, formation of 1:3 complexes and displacement of the vanadyl oxygen seem to be favoured. This effect is probably due to the exceptional chelating ability of 1,2-dihydroxybenzenes which appear particularly good σ and π donors, an effect that stabilizes metal and ligand without a change in formal change. An interesting comparison can be made for $K_3[V^{III}(cat)_3]$,[497] $(Et_3NH)_2[V^{IV}(cat)_3]$[256] and $Na[V^V(DTBcat)_3]$[256] (cat = $C_6H_4O_2^{2-}$, $H_2DTBcat$ = **35**); $d(V—O)_{av}$ are 2.013, 1.942 and 1.91 Å, and the ionic radii are 0.78, 0.72 and 0.68 Å, respectively.

From equation (30), whether a tris chelate is formed depends on the donor ability of the ligand. Several vanadium(IV) complexes with 1,2-dihydroxybenzenes formed from solutions of the ligand, $VOSO_4$ and thallium(I) acetate.[498] The compounds are of two types: (I) $Tl_2[V(RC_6H_3O_2)_3]$ (R = H, 3-Me, 4-Me, 3-MeO) and (II) $Tl_2[VO(RC_6H_3O_2)_2]$ (R = CHO).

The crystal and molecular structures of $(Et_3NH)_2[V(cat)_3]\cdot MeCN$ (**37**) and $K_2[VO(cat)_2]\cdot EtOH\cdot H_2O$ (**36**) were determined. Complex (**36**) has a square pyramidal geometry; $d(V{=}O) = 1.616$ Å (longer than usual in complexes with geometry of type (**41**)) and $d(V—O(cat)) = 1.956$ Å. Complex (**37**) has an approximately octahedral coordination geometry. $(Et_3NH)_2[VO(L)_2]\cdot 2MeOH$ (L = **35** and **37**) were prepared by displacement using $[VO(acac)_2]$ and adding the appropriate catechol.[256] A similar method was used for $Na_2[VO(DTBcat)_2]$ (**38**).[500] Its absorption spectrum in CH_2Cl_2 is dominated by intense change transfer which results in a bronze-coloured solution. When one equivalent of $SOCl_2$ is added to one of (**38**) in CH_2Cl_2, the solution immediately changes to blue. The resulting blue $Na_2[VCl_2(DTBcat)_2]$ (**39**) can also be formed by the same procedure as for the synthesis of (**38**), but with $VCl_2(acac)_2$ in place of $[VO(acac)_2]$.

In contrast with catechol[498,501,502] and norepinephrine[501] complexes where V^{IV} seems to be the stable oxidation state, V^{III} and V^{IV} DTBcat complexes are oxidized to V^V by protic solvents.[500] However, observation of V^{IV} ESR does not necessarily imply complete reduction of V^V by catechol;[256] either partial reduction or even adventitious V^{IV} could yield a similar result.

Redox chemistry of vanadium–catechol systems is complicated; References 256, 497 and 499–508 discuss this subject in detail. In complexes, the metal centre may be in the +5, +4, +3 (and +2) formal oxidation state and quinones complex in three localized electronic forms:

catecholate, semiquinolate and quinolate. A complex containing mixed charged ligands, $[V^VO(DTBSQ)(DTBcat)]_2$ (DTBSQ = semiquinonate of **35**), has been characterized structurally and the formation of $[V(Cl_4SQ)(Cl_4Q)_2]$[497] (Cl_4Q = tetrachloroquinonato and Cl_4SQ = tetrachlorosemiquinonato) has also been suggested.

(*ii*) *β-Diketonato ligands*

In addition to the very stable, commercially available, crystalline blue-green $[VO(acac)_2]$, several other $[VO(\beta\text{-diketonato})_2]$ complexes have been prepared;[509,510] they vary from yellow to green and brown. The $\mu_{eff}^{r.t.}$ values are in the range 1.71–1.76 BM and normally decrease slightly in the range 300–380 K.[509]

$[VO(acac)_2]$ may be prepared by several methods (*e.g.* by one-electron reduction of V^V by ethanol or mixing the diketone with $VOSO_4$) and its crystal and molecular structure has been determined.[511] It consists of discrete molecules of $[VO(acac)_2]$ in which vanadium presents a nearly square pyramidal geometry (Figure 22). Some internuclear distances are in Table 18. The crystal structure of $[VO(bzac)_2]$ (bzac = 1-phenyl-1,3-butanedionato) has also been determined.[512a] The ligands are arranged in a square pyramidal geometry with a *cis* configuration and each phenyl ring is nearly coplanar with its chelate ring. Some distances are in Table 18. IR and Raman spectra of $[VO(acac)_2]$ were studied.[512b]

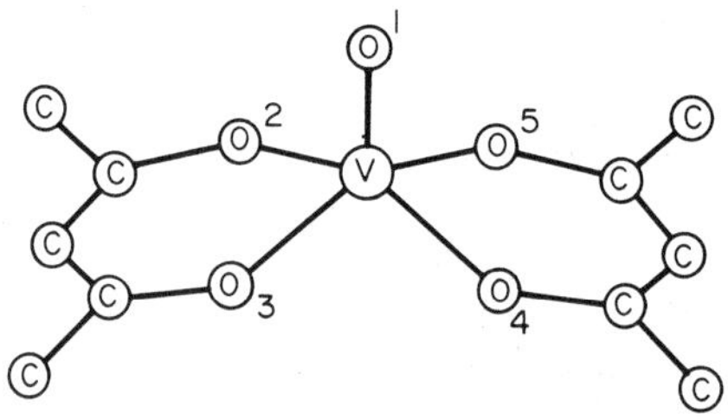

Figure 22 Perspective view of $VO(acac)_2$ molecule.[511] Internuclear distances are given in Table 18

Table 18 Internuclear Distances for Several Oxovanadium(IV) β-Diketonato Complexes

Compound	V—O(1)	V—O(2)	V—O(3)	V—O(4)	V—O(5)	V—X	Ref.
			Internuclear distances (Å)				
$[VO(acac)_2]$	1.56	1.97	1.96	1.96	1.98	—	1
	1.571	1.974	1.955	1.962	1.983	—	2
$[VO(bzac)_2]$	1.612	1.952	1.986	1.982	1.946	—	3
$[VO(acac)_2(4\text{-Phpy})]$	1.58	a	a	a	a	a	4
$[VO(acac)_2(dioxane)]$	1.62	~1.99	~1.99	~1.99	~1.99	a	5

[a] Not reported.
1. R. P. Dodge, D. H. Templeton and A. Zalkin, *J. Chem. Phys.*, 1961, **35**, 55.
2. P. K. Hon, R. L. Belford and C. E. Pfluger, *J. Chem. Phys.*, 1965, **43**, 3111.
3. P. K. Hon, R. L. Belford and C. E. Pfluger, *J. Chem. Phys.*, 1965, **43**, 1323.
4. M. R. Caira, J. M. Haigh and L. R. Nassimbeni, *Inorg. Nucl. Chem. Lett.*, 1972, **8**, 109.
5. K. Dichmann, G. Hamer, S. C. Nyburg and W. F. Reynolds, *Chem. Commun.*, 1970, 1295.

In many solvents, particularly water and alcohols, $[VO(\beta\text{-diketonato})_2]$ are slowly oxidized by atmospheric O_2,[513] but very rapidly with H_2O_2.[355]

The constant for reaction (**31**) for β-diketones in tautomeric keto (HK) and enol (HE) forms, can be partitioned between K_K and K_E (reactions **32** and **33**). At equilibrium, equations (34) and (35) hold, and values obtained are given in Table 19.[514] Kinetics showed that the reaction occurs through the enol form by parallel acid-independent and inverse-acid paths (Scheme 15).[514]

$$VO^{2+} + HL \overset{K_1}{\rightleftharpoons} VO(L)^+ + H^+ \quad (31)$$

$$VO^{2+} + HK \overset{K_K}{\rightleftharpoons} VO(E)^+ + H^+ \quad (32)$$

$$VO^{2+} + HE \overset{K_E}{\rightleftharpoons} VO(E)^+ + H^+ \quad (33)$$

$$\frac{1}{K_1} = \frac{1}{K_E} + \frac{1}{K_K} \tag{34}$$

$$\frac{[HE]}{[HK]} = \frac{K_K}{K_E} \tag{35}$$

Table 19 Equilibrium Constants for Formation of Mono-β-diketone Complexes of VO^{2+} in Aqueous Solution at 25 °C

	$K_1{}^a$	$K_K{}^a$	$K_E{}^a$	$log\,\beta_{110}$	$log\,\beta_{120}$	*Ref.*
acac[b]	0.55	0.67	3.14	8.7	15.8	1, 2
tfacac[b]	0.32	0.32	73.05	—	—	1
tftacac[b]	0.11	0.11	10.39	—	—	1
bzac[c]	—	—	—	10.5	20.5	3

tfacac = 1,1,1-trifluoropentane-2,4-dione; tftacac = 4,4,4-trifluoro-1-(2-thienyl)-butane-1,3-dione.

[a] Constants K_1, K_K and K_E are defined by equations (31)–(33).
[b] Except for $log\,\beta_{110}$, β_{120}, $I = 1.0$ M $NaClO_4$.
[c] $I = 0.1$ M.

1. M. J. Hynes and B. D. O'Regan, *J. Chem. Soc., Dalton Trans.*, 1980, 7.
2. R. Trujillo and F. Brito, *Anal. Fis. Quim.*, 1956, **52B**, 407.
3. N. V. Melchakova, N. A. Krasnyanskaya and V. M. Peshkova, *J. Anal. Chem. USSR (Engl. Transl.)*, 1970, **25**, 1756.

$$VO^{2+} + HE \underset{k_{-a}}{\overset{k_a}{\rightleftarrows}} VO^{2+} + E^- + H^+$$

$$k_{-ai}\Big\uparrow\Big\downarrow k_{ai} \qquad\qquad k_{-ad}\Big\uparrow\Big\downarrow k_{ad}$$

$$VO(HE)^{2+} \overset{K}{\rightleftarrows} VO(E)^+ + H^+$$

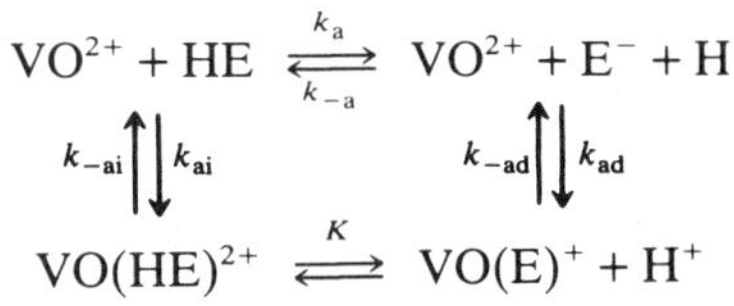

Scheme 15

In dichloroethane the ligand exchange of $[VO(acac)_2]$ with Hacac $[^{14}C]$ occurs without side reactions and the rate may be expressed by $R = k_2[VO(acac)_2][Hacac]$, where [Hacac] is the concentration of the free acac in enol form.[515]

Electrochemistry of $[VO(acac)_2]$ in DMSO has been studied by cyclic voltammetry and controlled-potential coulometry at a platinum electrode.[516] $[VO(acac)_2]$ is irreversibly reduced by one electron at -1.9 V *vs.* SCE (saturated calomel electrode) to a stable V^{III} product. In the presence of an excess of ligand, $[VO(acac)_2]$ is reduced by two electrons to $[V(acac)_3]^-$ with the V^{III} species mentioned above and $[V(acac)_3]$ as intermediates. The one-electron oxidation of $[VO(acac)_2]$ at $+0.81$ V *vs.* SCE gives a V^V product.

The room temperature ESR spectrum of $[VO(acac)_2]$ in toluene consists of the eight lines expected for an atom of spin $\frac{7}{2}$, with $g = 1.971$ and an average coupling constant of 120 G.[517,518] In frozen (77 K) toluene solution, molecular tumbling ceases, different molecular axes of the axially symmetric $[VO(acac)_2]$ are oriented along the applied field and the spectrum splits into two overlapping eight-line spectra. The single crystal spectra of *cis*-$[VO(bzac)_2]$ show anomalous fine structure.[519]

Solvent effects and adduct formation of $[VO(acac)_2]$ and other $[VO(\beta\text{-diketonato})_2]$ complexes have been studied by several methods[361,366,520–532] and in coordinating solvents $[VO(acac)_2]$ is known to add a sixth ligand according to equation (36). Older reports include a spectrophotometric and calorimetric study of $[VO(acac)_2]$ and $[VO(tfacac)_2]$ adducts.[521] With $[VO(acac)_2]$ in nitrobenzene, the enthalpy change for reaction (36) ranges from 44.3 kJ mol^{-1} for *n*-decylamine to 24.3 kJ mol^{-1} for methanol. Equilibrium constants K were between 1000 and ~0.6. $[VO(acac)_2]$ is not a sensitive indicator of relative base strength.[521] For $[VO(acac)_2]$, A_0 decreases by ~3 G (and g_0 increases by ~0.0004) as amine adducts are formed.[522]

$$VO(acac)_2 + ligand \overset{K}{\rightleftarrows} VO(acac)_2(ligand) \tag{36}$$

Equilibrium constants for reaction (36) have been estimated from spectrophotometry. In some studies, visible spectra of solutions containing $[VO(acac)_2]$ at constant concentration and base in variable concentration show an isosbestic point, which suggests but does not prove that

Table 20 Adduct Formation Constants of $[VO(acac)_2]$ (Equation 36) with Amides, Amines, Sulfoxides and Pyridine Derivatives

Amide ligand[a,1]	*K*	*Amine ligand*[a,2]	*K*	*Sulfoxide*[b,3]	*K*	*Substituted pyridine*[c,4]	*K*
N,N-Dimethylacetamide	35.4	Isopropylamine	9600	Dimethyl sulfoxide	4.3	Pyridine	67.9
N,N-Diethylacetamide	38.7	*n*-Butylamine	8960	Diethyl sulfoxide	4.8	2-Methylpyridine	10.6
N,N-Di(*n*-propyl)acetamide	38.7	*t*-Butylamine	8515	Dipropyl sulfoxide	4.8	3-Methylpyridine	79.5
N,N-Di(isopropyl)acetamide	41.6	Cyclohexylamine	9995	Dibutyl sulfoxide	5.0	4-Methylpyridine	130.0
N,N-Di(*n*-butyl)acetamide	27.4	Diethylamine	162	Tetramethylene sulfoxide	6.4	2,4-Dimethylpyridine	19.1
N,N-Di(isobutyl)acetamide	24.4	Diisopropylamine	184	Phenyl methyl sulfoxide	2.4	2,6-Dimethylpyridine	1.3
N-Methyl-*N*-phenylacetamide	27.0	Diisobutylamine	169	*p*-Tolyl methyl sulfoxide	2.5	3,4-Dimethylpyridine	140.7
N-Ethyl-*N*-phenylacetamide	26.6	Dicyclohexylamine	107			3,5-Dimethylpyridine	97.3
N-Methyl-*N*-benzylacetamide	26.4	Morpholine	4360				
N,N-Dicyclohexylacetamide	125.2	Piperidine	12970				
N-Acetylpiperidine	40.5	Quinoline	15.3				
N-Acetylmorpholine	26.4	Isoquinoline	828				
N,N,N',N'-Tetramethylurea	41.8	Pyridine	867				
N,N-Dimethylbenzamide	9.4	2-Picoline	10.5				
N,N-Dimethylchloroacetamide	24.2	3-Picoline	1550				
		4-Picoline	1910				
		2,4-Lutidine	41.7				
		3,5-Lutidine	1810				
		2-Hydroxypyridine	646				
		2-Cyanopyridine	16.6				
		3-Cyanopyridine	170				
		4-Cyanopyridine	159				
		4-Ethylpyridine	1830				
		2-Aminopyridine	78.1				
		3-Aminopyridine	1623				
		4-Aminopyridine	4540				
		3-Methyl-2-aminopyridine	551				
		4-Methyl-2-aminopyridine	571				
		5-Methyl-2-aminopyridine	552				
		6-Methyl-2-aminopyridine	19.3				

[a] In CH_2Cl_2 at 20 °C.
[b] In CH_2Cl_2 at 25 °C.
[c] In 1,2-dichloroethane (DCE) at 25 °C.

1. E. Kwiatkowski and J. Trojanowski, *J. Inorg. Nucl. Chem.*, 1975, **37,** 979.
2. E. Kwiatkowski and J. Trojanowski, *J. Inorg. Nucl. Chem.*, 1976, **38,** 131.
3. E. K. Jaffee and A. P. Zipp, *J. Inorg. Nucl. Chem.*, 1978, **40,** 839.
4. M. J. Macazaga, J. A. Garcia-Vazquez, R. M. Medina and J. R. Masaguer, *Transition. Met. Chem.* (*Weinheim, Ger.*), 1983, **8,** 215.

there are only two absorbing species. However, in all cases the results were interpreted according to equation (36).

Table 20 summarizes results from several spectroscopic equilibrium studies of adduct formation $[VO(acac)_2]$.[525,532–534]

In early work, solvent effects were explained by coordination at the vacant axial position. This was questioned by Guzy *et al.* who measured spectra of $[VO(acac)_2]$ in numerous solvents and pointed out that, especially for hydrogen-bonding solvents, there might be interaction with vanadyl oxygen.[361] Selbin concluded that for alcohols (and $CHCl_3$ and CH_2Cl_2), the H-bonding effect is greater with the vanadyl oxygen than with the acac oxygens.[359] The interaction of $[VO(acac)_2]$ with solvents has been measured by the ^{51}V hyperfine coupling in mixtures of ethanol and CCl_4 as a function of solvent composition.[529]

IR studies with pyridine derivatives[535,536] showed the existence of adducts with the sixth ligand bound *cis* rather than *trans* to the oxygen; an X-ray study of the 4-phenylpyridine (Phpy) adduct confirmed this (Figure 23a).[537] Interaction of ligands (solvent or a dissolved molecule) with $[VO(acac)_2]$ may involve (i) coordination in a sixth position *trans* to the vanadyl oxygen, (ii) coordination in a sixth position *cis* to the vanadyl oxygen, and (iii) hydrogen bonding to the oxygen (or atoms of the ligand).

While type (ii) was observed with the 4-phenylpyridine adduct, interaction of type (i) exists between $[VO(acac)_2]$ and 1,4-dioxane (diox) or 4-picoline. With diox, a substantial change in the $\nu(V{=}O)$ stretch is observed when it is added to $[VO(acac)_2]$ in $CHCl_3$[538] but only small effects for the visible spectra in nitrobenzene.[521] In the complex $[VO(acac)_2(diox)]$, from X-ray analysis, the oxygen atoms of diox are coordinated *trans* to the vanadyl oxygen (Figure 23b).[539] The crystal structure of $[VO(acac)_2(4\text{-picoline})]$ consists of discrete molecules with 4-picoline in the *trans* position and a very short V=O of 1.557(17) Å.[540]

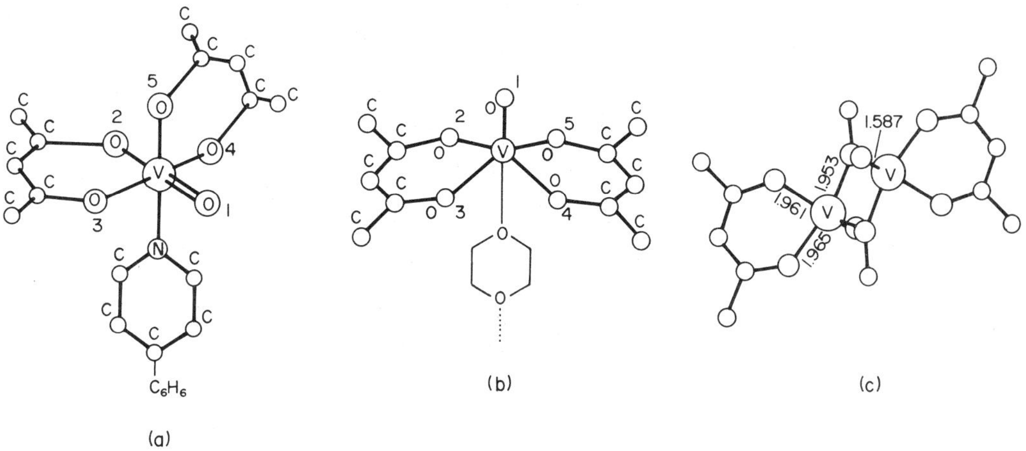

Figure 23 Schematic representation of the molecular structure of (a) $[VO(acac)_2(4\text{-phenylpyridine})]$,[537] (b) $[VO(acac)_2(1,4\text{-dioxane})]$[539] (a dioxane molecule bridges two $[VO(acac)_2]$ groups) and (c) $[VO(OMe)(acac)]_2$.[553] Some internuclear distances are included in Table 18

Besides the structure of $[VO(acac)_2(Phpy)]$[537] and IR evidence,[535,536] ENDOR studies (see Section 33.5.1) suggest formation of *cis* adducts and/or the existence of *cis* and *trans* isomers.[527] For $[VO(acac)_2]$ with methanol and substituted pyridines in frozen $CHCl_3$/toluene solutions, there must be a well-defined binding site for $CHCl_3$, most likely *via* a hydrogen bond to the vanadyl oxygen, and methanol is at the vacant axial position. For $[VO(acac)_2]$ and substituted pyridines in $CHCl_3$, complexation induced constant shifts of $31 \pm 3\ cm^{-1}$ and/or $51 \pm 3\ cm^{-1}$ (Table 21)[535] attributed to *cis* ($31\ cm^{-1}$ shift) and/or *trans* ($51\ cm^{-1}$ shift) adducts. However, the IR of solid $[VO(acac)_2(\text{substituted py})]$ (Table 22) were interpreted on the reverse assumption that *cis* coordination produces the larger shift of $\nu(V{=}O)$.[536] Kirsk and van Willigen consider that the group of pyridines that coordinate in the *cis* position induces a larger shift in $\nu(V{=}O)$;[527] accepting this, they find a good correlation between the solution IR data[535] and their own ENDOR. In a number of instances, the isomer that is present in the solid state[536] may not predominate in liquid[535] and frozen solution. Other investigations emphasize subtle factors controlling equilibrium between different conformation.[526,531,541] For example, Mannix and Zipp[526] studied the visible spectra of $[VO(acacX)_2]$ (X = Cl or Br) in several solvents with a wide range of donor properties and obtained a linear relationship between the energy differences of the first two visible bands, $D_{1,2}$, and the solvent donor number.[542] However, the $[VO(acacX)_2]$ compounds were found to be poorer acceptors than $[VO(acac)_2]$; this is surprising.

Table 21 $\nu(V{=}O)$ (cm^{-1}) in Solutions Containing $[VO(acac)_2]$ and Substituted Pyridines (1:5) in $CHCl_3$ at Room Temperature

Substituent	*ν(V=O) bands and relative intensity*	
No adduct		1003
3-CN		975
4-CN	951	<975
3-Cl	955	~972
3-Br	955	~974
3-CHO	953	>972
4-Br	953	>969
4-Cl	953	>969
4-CHO	953	≫971
3-OH	950	
H	955	
3-Me	953	
4-Me	954	≫971
$3-NH_2$	952	
$4-NH_2$	949	
2-Me	954	~975

J. J. R. Fraústo da Silva and R. Wootton, *Chem. Commun.*, 1969, 421.

Table 22 $\nu(V{=}O)$ (cm^{-1}) of $[VO(acac)_2L]$ (L = Substituted Pyridines)

Substituent	*ν(V=O)*
No adduct	999
$4-CO_2Et$	955
4-COPh	957
4-Ph	955
$3,5-Me_2$	959
4-Et	959
$3-NH_2$	959
$4-Bu^t$	962
$4-NMe_2$	953
$4-NH_2$	958
4-CN	974
H	973
$4-Pr^n$	971
4-Me	971
$3,4-Me_2$	975

M. R. Caira, J. M. Haigh and L. R. Nassimbeni, *J. Inorg. Nucl. Chem.*, 1972, **34**, 3171.

A rather different study of the kinetics of decomposition of solid complexes of $[VO(dbm)_2(L)]$ (dbm = dibenzoylmethanato, L = py and several methyl-, dimethyl- and amino-pyridines) used differential scanning calorimetry (DSC).[531] Using the temperature that corresponds to the loss of the molecule L in equation (37), a linear relationship was found between it and the basicity of L, except for 4-amino- and 4-methyl-pyridine.

$$VO(dbm)_2L \longrightarrow VO(dbm)_2 + L \qquad (37)$$

In the coordination of 4-picoline *N*-oxide (MPNO) to a series of $[VO(\beta\text{-diketonato})_2]$

complexes, IR spectra and equilibrium constants for adduct formation followed the order: tftacac > tfacac > dbm > bzac > acac > dpm (dipivaloylmethanato). A similar order was found with several other neutral bases.[524] For [VO(acac)$_2$], the MPNO coordinates *trans* but for [VO(bzac)$_2$] and others the coordination is *cis*. A similar trend was found in compounds [VO(β-diketonato)$_2$(L)] with acac or tfacac and L = py, phen, α- and β-naphthoquinoline (α- and β-nq).[543] [VO(acac)$_2$(phen)]·2MeOH was prepared from an excess of phen with [VO(acac)$_2$] in methanol. The same reaction in CH$_2$Cl$_2$ results in V^{IV} oxidation and [VO$_2$(acac)(phen)] is obtained,[544] while [VO(tfacac)$_2$(phen)] is produced in both solvents if the starting material is [VO(tfacac)$_2$]. In the two [VO(β-diketonato)$_2$(phen)] complexes, phen was said to be unidentate and a *cis* octahedral structure was proposed for [VO(tfacac)$_2$(L)] complexes (L = py, phen, α-nq, β-nq).[543]

Early publications on [VO(β-diketonato)$_2$] have been reviewed.[355] More recently, complexes with benzoyl *m*-nitroacetanilide, benzoyl acetanilide[545] and 1,1'-(1,3-phenylene)-bis(butane-1,3-dione)[546] have been synthesized. Other [VO(β-dik)$_2$] adducts have been isolated, for example [VO(acac)$_2$] adducts with a series of pyridine *N*-oxides[547] and several pyridine carboxamides,[548] and [VO(bzac)$_2$] adducts with pyridine, methylamine, isoquinoline and 4-picoline.[549] Equilibrium constants of 1:1 and 2:1 adducts of pyrazine with [VO(tfacac)$_2$] have been determined (equation 38).[550] In the 2:1 complex, the pyrazine bridge between two equatorial sites of adjacent vanadium atoms promotes a weak exchange interaction. The nitroxide radical 2,2,6,6-tetramethylpiperidinyl *N*-oxide also forms an adduct with [VO(hfacac)$_2$] in which there is a strong interaction between the electrons on the metal and nitroxide.[551]

$$\text{VO(tfacac)}_2 + \text{pz} \underset{}{\overset{K=1.5\times10^6}{\rightleftharpoons}} \text{VO(tfacac)}_2\text{pz} \underset{}{\overset{K\approx2\times10^3}{\rightleftharpoons}} [\{\text{VO(tfacac)}_2\}_2\text{pz}] \qquad (38)$$

Oxidation of alkenes, sulfides, sulfoxides and amines by alkyl hydroperoxides (ROOH) is catalyzed by [VO(acac)$_2$] (equations 39–42),[552] and mechanisms involving association of ROOH with [VO(acac)$_2$] forming V^V compounds have been suggested.[552] The reactions of phenoxyl, iminoxyl, nitroxyl, peroxyl and alkoxyl radicals with [VO(acac)$_2$] in solution were studied by kinetic ESR spectroscopy[552] and the net reaction was found to be catalytic reduction of the radical, probably also involving initial formation of a V^V compound.

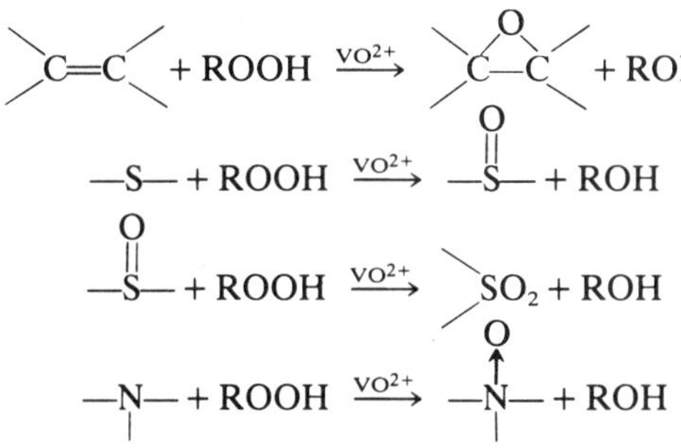

$$\text{C=C} + \text{ROOH} \xrightarrow{\text{VO}^{2+}} \text{C-C (epoxide)} + \text{ROH} \qquad (39)$$

$$-\text{S-} + \text{ROOH} \xrightarrow{\text{VO}^{2+}} -\overset{\text{O}}{\underset{}{\text{S}}}- + \text{ROH} \qquad (40)$$

$$-\overset{\text{O}}{\underset{}{\text{S}}}- + \text{ROOH} \xrightarrow{\text{VO}^{2+}} \text{SO}_2 + \text{ROH} \qquad (41)$$

$$-\text{N-} + \text{ROOH} \xrightarrow{\text{VO}^{2+}} -\overset{\text{O}}{\underset{}{\text{N}}}- + \text{ROH} \qquad (42)$$

The reaction in alcoholic medium of a β-diketone, a base and VOCl$_2$ in 1:2:1 molar ratios leads to compounds [VO(alkoxy)(β-diketonato)]$_2$.[553] This is an interesting result since [VO(β-diketonato)$_2$] form if the reaction with the same species is carried out in molar ratios of 2:2:1 or 1:1:1. Therefore, the stoichiometry in which the reagents are mixed will drive the reaction towards one product or the other. This behaviour was rationalized assuming Scheme 16 where an intermediate [VO(OH)L] (not isolated) forms and is irreversibly transformed to the corresponding [VO(OR)L] compound.

$$\text{VOCl}_2 + \text{HCl} + 2\text{OH}^- \underset{}{\overset{-2\text{Cl}^-, -\text{H}_2\text{O}}{\rightleftharpoons}} \text{VO(OH)L} \overset{-\text{Cl}^-}{\rightleftharpoons} \tfrac{1}{2}\text{VOCl}_2 + \tfrac{1}{2}\text{VOL}_2 + \text{OH}^-$$

$$\downarrow \substack{\text{ROH} \\ -\text{H}_2\text{O}}$$

$$\text{VO(OR)L} \longrightarrow [\text{VO(OR)L}]_2$$

Scheme 16

Several complexes [VO(OR)$_2$(β-dik)]$_2$ were prepared including [VO(OMe)(acac)]$_2$; for its crystal structure,[580] see Figure 23(c).[553] It consists of dimeric units with a chair-like arrangement and the geometry around the vanadium atom was best described as tetragonal bipyramidal. The vanadium lies 0.60 Å above the plane of two β-diketone oxygens and two bridging alkoxy oxygen atoms. The $\mu_{eff}^{r.t.}$ is 1.80 BM, *i.e.* magnetic exchange is much greater in the Cu^{II} analogues. V—V is 3.102 Å. Although the compositions, homogeneity and molecular weights of the polymers obtained using [VO(OR)(β-dik)]$_2$ complexes as Ziegler–Natta catalysts are almost the same as those obtained using the commercial catalysts [V(acac)$_3$] and [$VOCl_3$], the latter show higher activities.

A few [VO(OMe)(β-diketonato)]$_2$ and [VO(L)]$_2$ complexes (L = polydentate O-coordinating ligands) also have a high catalytic activity for the oxygenation of catechols (namely 3,5-di-*t*-butyl-1,2-dihydroxybenzene)[554] and this reaction was considered similar to the one catalyzed by pyrocatechase.

A different group includes vanadium^{IV} hexacoordinate complexes with no vanadyl bond. Complexes [VO(L)$_n$] (L = tetradentate or two bidentate ligands) react under mild conditions with $SOCl_2$ or $SOBr_2$ producing [V(L)$_n$Cl$_2$] or [V(L)$_m$Br$_2$] hexacoordinate V^{IV} complexes.[429] The same deoxygenation can be accomplished in dioxane by PCl_5, which is even more reactive than $SOCl_2$.[555] Complexes [V(β-diketonato)X$_2$] (X = Cl, Br) have been studied with acac, bzac, dbm, tfacac: *cis* and *trans* isomers may form, in relative abundances which are solvent dependent. In non-polar solvent, the abundance of the *trans*-dichloro isomer decreases in the order tfacac > acac > bzac > dbm.[429] All these dark compounds are rapidly hydrolyzed in air to vanadyl complexes.

(*iii*) *Hydroxyaldehydato and hydroxyketonato ligands*

Several VO^{2+} salicylaldehydato (and derivative) complexes [VO(L)$_2$] have been prepared although practically no work is recent. The early work has been reviewed[355,357b] and includes salicylaldehyde (Hsal) and the derivatives 5-Cl-, 5-Br-, 3,5-Br$_2$-, 3,5-Cl$_2$-, 4-phenyl- and 4-amino-salicylaldehyde. The oxovanadium(IV) complex of β-hydroxynaphthaldehyde was also reported.[556,559] More recent studies include the synthesis of compounds formulated as [VO(L)$_2$] with L = (**46**; R = 3-MeO, 5-Cl),[557,558] (**47**), 3,7-dimethyl-7-hydroxyoctan-1-al[560] and 4,6-dihydroxycoumaran-3-one (**40**).[561] IR spectra of the last suggests that coordination involves the 3- and 4-oxygen atoms. No magnetic data are given and ν(V=O) = 965 cm^{-1}.

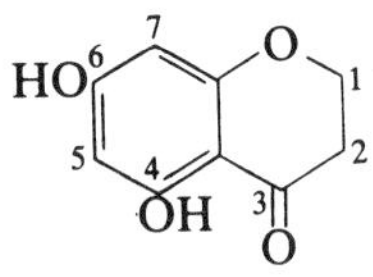

(**40**)

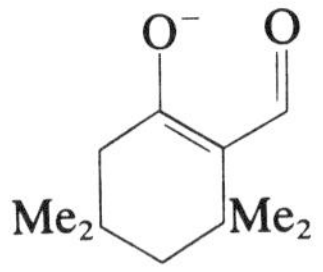

(**41**)

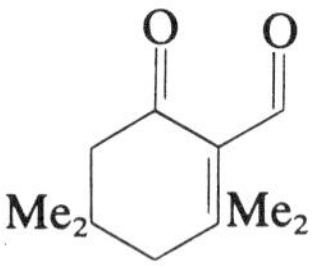

(**42**)

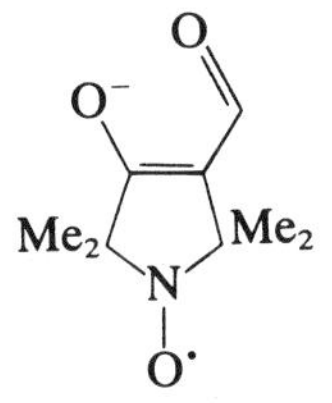

(**43**)

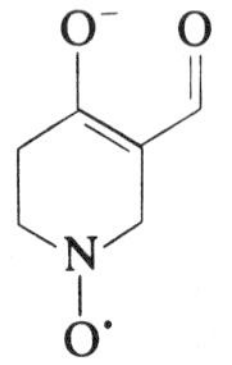

(**44**)

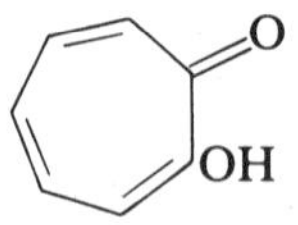

(**45**)

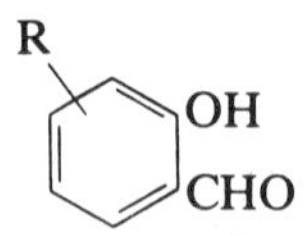

(**46**)

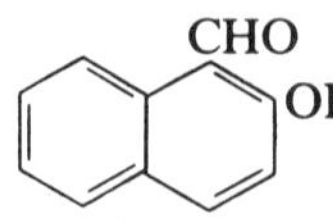

(**47**)

After mixing for a few hours a solution of (**42**) in CH_2Cl_2 with a solution of $VOSO_4$ (pH 3) in H_2O under N_2, a chelate with the β-ketoaldehyde (**41**), formulated as [VOL$_2$], was obtained from the organic phase.[562] ESR showed A_0 = 105 G. Similarly formulated complexes with the nitroxides (**43**) and (**44**) were observed in solution. [VO(tropolonato)$_2$] (tropolone = **45**) and the deoxygenated compounds obtained by reaction with $SOCl_2$ and $SOBr_2$ (Section 33.5.5.4.ii), [VX$_2$(tropolone)$_2$] (X = Cl, Br), have been studied.[429] A trigonal prismatic structure is proposed with the chelate ligands occupying two vertical edges.

33.5.5.5 Oxoanions as ligands

(i) Sulfate

Hydrated sulfates $VOSO_4{\cdot}xH_2O$ are one of the most common starting materials for the preparation of oxovanadium(IV) compounds. Several forms have been characterized; some are in Table 23, which shows that V=O is 1.58–1.59 Å and is longer in α- and β-$VOSO_4$ where the VO^{2+} group is involved in $\cdots$V=O$\cdots$V=O$\cdots$ chains; the H_2O molecules are either free or coordinated to the vanadium atom and the oxygens of VO_6 octahedra belong to water molecules, to SO_4 groups or to other VO_6 octahedra.

Table 23 Anhydrous and Hydrated Forms of Vanadyl Sulfates $VOSO_4{\cdot}xH_2O$: X-Ray Diffraction Data

	V—O internuclear distances (Å)				
x	*Vanadyl oxygen*	*Equatorial oxygens*	*Axial oxygen*	*Type of vanadium environment*	*Ref.*
0 (β form) (unstable)	1.594	2.056, 2.005 2.05, 2.016	2.284	Three-dimensional network with chains of VO_6 distorted octahedra connected pairwise by an SO_4 group *via* corner sharing	1
0 (α form)	1.63	2.04[a] (× 4)	2.47	A VO_6 octahedron has four SO_4 tetrahedra in the equatorial plane	2
				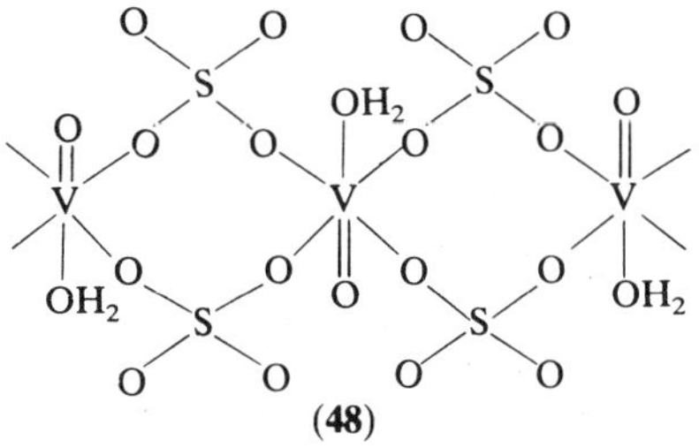 **(48)**	3
3	1.576	2.033, 2.048 2.003[a], 2.001[a]	2.280	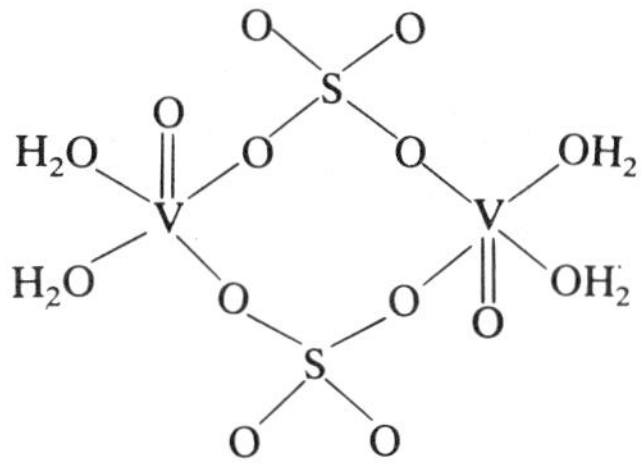 **(49)**	4, 5
5	1.591	2.048, 2.037 2.035, 1.983[a]	2.223	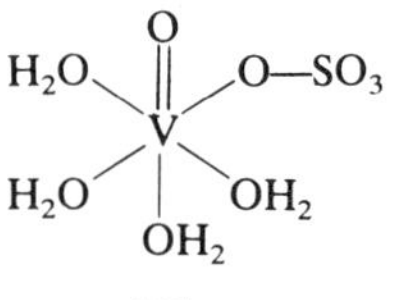 **(50)**	6
5	1.584	2.008, 2.019 2.006, 1.994[a]	2.181		7
5 (β form) unstable)	1.591	2.031, 2.031 2.018, 2.018	2.218	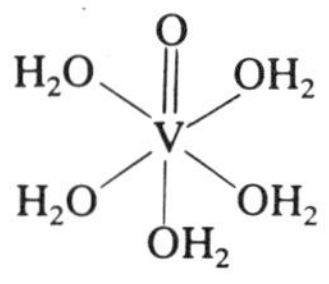 **(51)**	8
6	1.586	2.023, 2.021 2.160, 2.004	2.029		9

Oxygen atoms of SO_4 groups.

. P. Kierkegaard and J. M. Longo, *Acta Chem. Scand.*, 1965, **19**, 1906.
. J. M. Longo and R. J. Arnott, *J. Solid State Chem.*, 1970, **1**, 394.
. G. Ladwig, *Z. Anorg. Allg. Chem.*, 1969, **364**, 225.
. F. Théobald and J. Galy, *Acta Crystallogr., Sect. B*, 1973, **29**, 2732.
. M. Tachez and F. Théobald, *Acta Crystallogr., Sect. B*, 1980, **36**, 2873.
. C. J. Ballhausen, B. F. Djurinskij and K. J. Watson, *J. Am. Chem. Soc.*, 1968, **90**, 3305.
. M. Tachez, F. Théobald, K. J. Watson and R. Mercier, *Acta Crystallogr., Sect. B*, 1979, **35**, 1545.
M. Tachez and F. Théobald, *Acta Crystallogr., Sect. B*, 1980, **36**, 1757.
. M. Tachez and F. Théobald, *Acta Crystallogr., Sect. B*, 1980, **36**, 249.

Neutron powder data on $VOSO_4{\cdot}3D_2O$ improved the location of hydrogen in the structure.[563] In $2VOSO_4{\cdot}H_2SO_4$, X-ray techniques show VO_6 octahedra linked by monodentate SO_4 tetrahedra into polymeric $[VOSO_4]_\infty$ layers joined together by sulfuric acid molecules.[564] Several $(VO)_2H_2(SO_4)_3$ hydrates have been reported.[357a]

Oxovanadium(IV) sulfate solutions were studied by Ducret who concluded that $[VO(SO_4)]$ and $[VO(SO_4)_2]^{2-}$ form and reported $[VOSO_4]/([VO^{2+}][SO_4^{2-}]) = 63$.[466] Others obtained 2.4×10^2.[565] Pressure jump relaxation techniques on $VOSO_4$ solutions were explained according to equation (43): $K = (3.0 \pm 0.5) \times 10^2$.[566] Grigor'eva studied vanadium(IV) solutions over a wide range of H_2SO_4 and SO_4^{2-} concentrations.[567]

$$VO^{2+} + SO_4^{2-} \underset{\text{very fast}}{\rightleftharpoons} \underset{\text{(outer sphere complex)}}{VO(H_2O)SO_4} \overset{k_d \approx 10^3}{\rightleftharpoons} \underset{\text{(inner sphere complex)}}{VOSO_4} \qquad K = \frac{[VO(H_2O)SO_4] + [VOSO_4]}{[VO^{2+}][SO_4^{2-}]} \tag{43}$$

(*ii*) *Phosphoric acids*

There are many old reports of solid vanadyl phosphates.[355] Complex formation in aqueous acidic ($pH \leq 3$) vanadyl phosphate solutions was explained by formation of $[VOHPO_4]$, $[HVO(HPO_4)_2]^-$ and $[VO(HPO_4)_2]^{2-}$.[568] Orthophosphate coordinates as bidentate HPO_4^{2-}. The most important is $[VOHPO_4]$. The formation of 1:3 complexes was considered unlikely. These[568] and spectrophotometric results[569] include a total of five monomeric species and were later discussed on the basis of Scheme 17.[570] $[VO(H_2PO_4)_2]$ is dominant in 1 M $H_2PO_4^-$ solution at pH 2,[570] and the rate of loss of $H_2PO_4^-$ was determined from ^{31}P NMR. Aquation and anation rate constants at 25 °C were estimated to be $2.5 \times 10^4\ s^{-1}$ and $2 \times 10^6\ mol^{-1}\ s^{-1}$, respectively. The anation rate constant indicates that coordinated water in $[VO(H_2PO_4)]^+$ is more labile than the equatorial H_2O molecules in the complex $VO(H_2O)_5^{2+}$.

$$VO^{2+} + H_2PO_4^- \rightleftharpoons VO(H_2PO_4)^+$$
$$VO(H_2PO_4)^+ + H_2PO_4^- \rightleftharpoons VO(H_2PO_4)_2$$
$$VO(H_2PO_4)^+ \rightleftharpoons VO(HPO_4) + H^+$$
$$VO(H_2PO_4)_2 \rightleftharpoons VO(H_2PO_4)(HPO_4)^- + H^+$$
$$VO(H_2PO_4)(HPO_4)^- \rightleftharpoons VO(HPO_4)_2^{2-} + H^+$$

Scheme 17

The 22-line ESR spectra of oxovanadium(IV) pyrophosphate ($5 \leq pH \leq 6$) suggested a pyrophosphate trimer.[572] $Na_6(VO)_3(P_2O_7)_3{\cdot}18H_2O$ had $\mu = 3.6$ BM and its solution ESR also exhibits a 22-line signal. Osmometry supported the trimeric nature.[573] Addition of base to its solution causes a decrease in the intensity of the ESR and at $pH \geq 11$ no signal was detected. The explanation assumed (i) at pH 7–8, disproportionation of trimer to monomeric $[VO(P_2O_7)_2]^{6-}$ and an anionic species $[(VO)_4(OH)_{10}]^{2-}$ and (ii) at pH 9.5–10.5, base hydrolysis of the monomeric complex to $[(VO)_4(OH)_{10}]^{2-}$.[572]

In vanadium phosphorus oxides, the oxidation state and the phosphorus:vanadium ratio can be varied. They act as heterogeneous catalysts in the oxidation of *n*-butane and *n*-butene to maleic anhydride.[574–576] The best catalysts have vanadium in the +4 state and a P:V ratio of ~1. Understanding mechanisms requires crystal structures. That of $[(VO)_2P_2O_7]$ is built of double chains of VO_6 octahedra that share edges across the chain.[577] Pyrophosphates link the double chains into a three-dimensional network by sharing oxygen corners with vanadium. This compound forms at 400 °C from $[(VO)_2H_4P_2O_9]$, was studied by several techniques[575,576] and characterized as $VO(HPO_4){\cdot}0.5H_2O$.[578] Vanadyl hydrogen phosphate layers stacked along the *c* axis are held together by H bonding. The layers contain face-sharing VO_6 octahedra; one face-shared O atom is from an H_2O molecule that bridges the two vanadyl groups. The four remaining O atoms of each octahedron are shared with phosphate tetrahedra. Each tetrahedron has three oxygens shared with vanadium and one OH group. $(VO)_2P_2O_7$ contains double chains of $V^{(IV)}$ cations, and $VO(HPO_4)\cdot 0.5H_2O$ contains isolated $V^{(IV)}$—$V^{(IV)}$ dimers.[575]

(iii) *Perchlorate–oxovanadium(IV) solutions and salts*

Solutions of the perchlorate have been prepared by several methods (*e.g.* ref. 357a, p. 250): (i) precipitate VO_2 in the absence of O_2 and dissolve the solid with perchloric acid;[579] (ii) start with $VOSO_4$ and precipitate with $Ba(ClO_4)_2$;[580] (iii) reduce $V^{(V)}$ compounds, *e.g.* reduce NH_4VO_3 with HCl and KI and precipitate halides with $AgClO_4$;[460] (iv) reduce a slurry of V_2O_5 in perchloric acid at a platinum cathode.[581] SO_2 has also been used.[582] Oxovanadium(IV) in stock solutions has been determined mainly by oxidation by $KMnO_4$ but several have used spectrophotometry.[460] Frausto da Silva and co-workers used titrations of 1:1 VO^{2+} and edta,[377,378] Tapscott and Belford titrated 1:2 vanadyl perchlorate–sodium oxalate[583] and others[584,585] used Gran's method.[586]

The blue, very hygroscopic $[VO(H_2O)_5](ClO_4)_2$ has been shown to involve no ClO_4^- coordination in solution or solid[587] and has $\mu = 1.70$ BM[587] (see also ref. 357a, p. 250). It is easier to get solids with ligands other than water, for example $[VO(DMSO)_5]\cdot(ClO_4)_2$.[355]

Benali-Baitich studied spectra of VO^{2+}–perchloric acid ($0 \leqslant [HClO_4] \leqslant 13.3$ M).[461] For high acid concentrations, V^{IV} is oxidized; others observed rapid oxidation in solutions containing ~2.2 M $NaClO_4$ for pH $\geqslant 2.5$.[588] From the visible spectra, ClO_4^- coordinates.[461] However, such changes may arise from VOH^+ (see Section 33.5.5.1.i), not ClO_4^- coordination.[460]

(iv) *Nitrate–oxovanadium(IV) solutions*

Although dilute solutions of nitrate may be prepared, the instability of concentrated solutions and the reaction of HNO_3 with VO^{2+} have been reported.[582,588] Oxidation of $VOSO_4$ in acid at 80 °C by nitrate is rapid, after a long induction which resembles an autocatalytic reaction.[582]

(v) *Miscellaneous anions*

A few sulfite, arsenate, selenite and selenate compounds were reported[355,357] but should be reinvestigated. VO^{2+} forms a deep purple complex with phosphotungstic acid,[589] in contrast with the yellow complex with V^V. Spectrophotometrically, the formation constant is 1.3×10^5. The kinetics with 12-tungstovanadophosphates were analogous to those with 12-molybdotungstophosphates.[590]

33.5.5.6 Carboxylates

(i) *Monocarboxylates*

Table 24 summarizes most known $VO(RCO_2)_2$ complexes. The magnetic properties of formate hydrates differ from those of higher carboxylates. $\mu_{eff}^{r.t.}$ values are normally 1.45–1.80 BM and these complexes obey the Curie–Weiss law.[358,595] A magnetically normal hydrated formate, $VO(HCO_2)_2\cdot H_2O$, is known[596] and the compound $K_2[VO(HCO_2)_4]$ consists of infinite zigzag chains in which the V atoms are linked in an axial–equatorial manner by formate bridges (Figure 24).[597] The geometry is pseudo-octahedral, the V atom lying 0.27 Å out of the equatorial plane. $\mu_{eff}^{r.t.} = 1.79$ BM, indicating little or no magnetic interaction between the V atoms. Formation constant studies have been performed.[591–594]

Subnormal magnetic oxovanadium(IV) complexes that include this type of ligand have been reviewed by Syamal (Table 24).[358] $\nu(V{=}O)$ values appear ~100 cm^{-1} lower than commonly found. Vanadyl acetate and halogenoacetates $VO(CH_{3-n}X_nCO_2)_2$, ($n = 1, 2, 3$; X = Cl or Br) are characterized by one $\nu(V{=}O)$ at 895 ± 5 cm^{-1}, by very low $\mu_{eff}^{r.t.}$ (Table 24) and by antiferromagnetic spin–spin interactions. These results have been interpreted on the basis of a polymeric structure (**52**).[358,598] The same characteristics and interactions exist in many complexes of other aliphatic and aromatic carboxylic acids.[358] [VO(salnpn)],[599] polymeric with

Table 24 Vanadyl Monocarboxylate Complexes $VO(RCO_2)_2$

Ligand (*R in* RCO_2^-)	$\mu_{eff}^{r.t.}$ (BM)	$\nu(V{=}O)$ (cm^{-1}) 300 *K*	100 *K*	*Ref.*	*Ligand* (*R in* RCO_2^-)	$\mu_{eff}^{r.t.}$ (BM)	*Ref.*
H (sesquihydrate)	1.70			1	CH_2Cl	1.16	1
H (monohydrate)	1.65			1		0.93	4
H (hemihydrate)	1.45			1	$CHCl_2$	1.16	1
Me	1.24	902	891	2		1.20	4
	1.19			1	CCl_3	1.20	1
Et	1.24	893	885	3		1.24	4
	1.20			1	CH_2Br	1.27	4
Pr^n	1.25	891	883	3	$CHBr_2$	1.25	4
	1.20			1	CBr_3	1.02	4
Pr^i	1.33	891	883	3	$Me(CH_2)_{16}$	—	5
	1.23			1	$Me(CH_2)_7CH{=}CH(CH_2)_7$	—	5
Bu^n	1.21			1	$PhCH{=}CH$	1.30	1
Bu^s	1.23	893	884	3	Ph_2CH	1.18	1
	1.26			1	*m*-ClC_6H_4	1.33	3
Bu^t	1.22			1	*o*-IC_6H_4	1.25	3
$Me(CH_2)_4$	1.25			1			
$Me(CH_2)_5$	1.23			1			
$Me(CH_2)_6$	1.25			1			
$Me(CH_2)_7$	1.25			1			
Ph	1.38			1			
	1.32	893	886	3			
$PhCH_2$	1.25			1			
	1.19	893	886	3			

1. V. T. Kalinnikov, V. V. Zelentsov, O. N. Kuz'micheva and T. G. Aminov, *Russ. J. Inorg. Chem.* (*Engl. Transl.*), 1970, **15**, 341.
2. A. T. Casey and J. R. Thackeray, *Aust. J. Chem.*, 1969, **22**, 2549.
3. A. T. Casey, B. S. Morris, E. Sinn and J. R. Thackeray, *Aust. J. Chem.*, 1972, **25**, 1195.
4. J. P. Walter, D. Dartiguenave and Y. Dartiguenave, *J. Inorg. Nucl. Chem.*, 1973, **35**, 3207.
5. S. Prasad and K. N. Upadhyaya, *J. Indian Chem. Soc.*, 1961, **38**, 163.

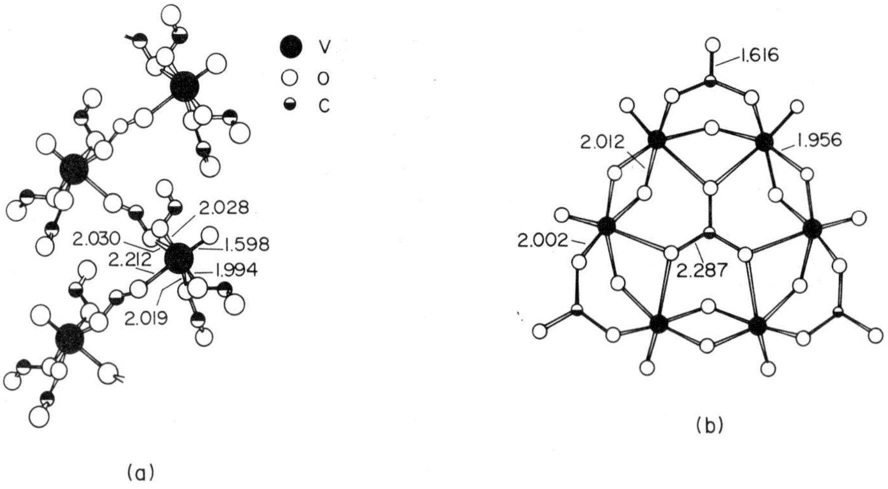

Figure 24 Schematic representation of the molecular structure of (a) the anion of $K_2[VO(HCO_2)_4]$ showing the polymeric structure[597] and (b) the $[(VO)_6(CO_3)_4(OH)_9]^{5-}$ anion.[604] Some internuclear distances are indicated

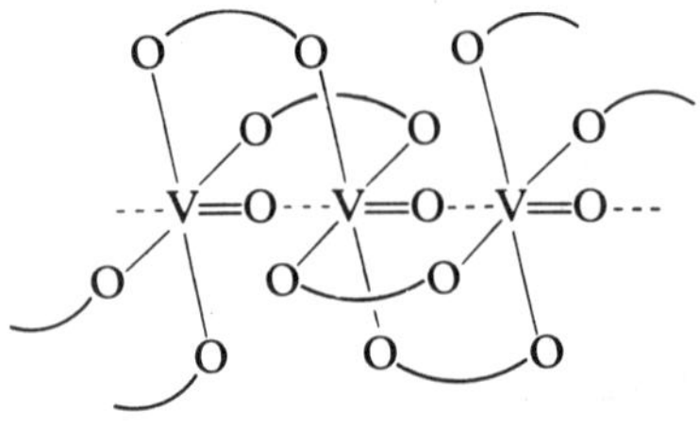

(52)

V=O···V=O··· interactions (see Figure 30), has small ferromagnetic interaction in contrast to that in polymeric $[VO(MeCO_2)_2]$.

ESR studies on $[VO(MeCO_2)_2]$ and its adducts with ethanol, acetic acid, phen, bipy and 2-picoline have been reported:[600,601] Different monocarboxylate compounds have been prepared (Table 25).

Table 25 Monocarboxylate Compounds of Vanadium(IV) with Molecular Structures Determined by X-Ray Diffraction Studies

	Internuclear distances (Å)			
Basic units	*Vanadyl oxygen*	*Oxygen (trans)*	*Equatorial monocarboxylate oxygens*	*Ref.*
$K_2[VO(HCO_2)_4]$ (polymeric)	1.598	2.212	2.018 (av)	1
$[V_3O_3(PhCO_2)_6(THF)]$	1.626	2.186 (THF)	1.978 (av)	2
(oxo-centred)	1.582	2.344	2.002 (av)	
	1.568	2.452	1.980 (av)	
$Na_4[(VO)_2(CF_3CO_2)_8(THF)_6(H_2O)_2]$	1.951	2.290 (THF)	2.004 (av)	3
(cyclic centrosymmetric units)	(ν(V=O) = 981 cm^{-1})			

1. T. R. Gilson, I. M. Thom-Postlethwaite and M. Webster, *J. Chem. Soc., Dalton Trans.*, 1986, 895.
2. F. A. Cotton, G. E. Lewis and G. N. Mott, *Inorg. Chem.*, 1982, **21**, 3127.
3. F. A. Cotton, G. E. Lewis and G. N. Mott, *Inorg. Chem.*, 1983, **22**, 1825.

Revision of early work[603] on carbonato complexes formulated the violet NH_4^+ salt obtained at pH ~3.5 as $V_{12}O_{28}H_{14}(CO_3)_8(NH_4)_{10}\cdot 23H_2O$ and the purple K^+ salt obtained at pH ~8.5 as $V_{12}O_{28}H_{14}(CO_3)_8K_{10}\cdot 16H_2O$.[602] ν(V=O) occurs at 950 cm^{-1} in both and ν(V—O—V) at 525 and 515 cm^{-1} for the NH_4^+ and K^+ salts.[602] Deep violet crystals of the NH_4^+ salt made by reacting $VOCl_2$ with NH_4HCO_3 under CO_2 had the formula $(NH_4)_5[(VO)_6(CO_3)_4(OH)_9]\cdot 10H_2O$.[604] The anion (Figure 24b) consists of crown-shaped hexanuclear aggregates consolidated by bridging hydroxo and carbonato groups, the latter functioning in a μ_6 mode as well as in the more common μ_2 mode. The six V atoms, each octahedral, are within 0.07 Å of their mean plane in a pseudo-hexagonal arrangement.

(ii) *Polycarboxylates*

Oxovanadium(IV) forms stable complexes with polycarboxylic acids and many formation constants have been determined, including mixed complexes (Section 33.5.10). Potentiometry and NMR led to analysis of the paramagnetic contribution to relaxation as a function of $[VO^{2+}]$, [VO(oxal)] and $[VO(oxal)_2]^{2-}$ for the oxalate system and $[VO^{2+}]$, $[VO(Hmalon)]^+$, [VO(malon)] and $[VO(malon)_2]^{2-}$ for the malonate system.[584] Table 26 summarizes their rate constants. Proton exchange rate constants are higher than those for the free ion ($k = 1.7\times10^5$ s^{-1},[584] 6.2×10^{-4} s^{-1},[459] 8×10^4 s^{-1} [605] and 1.8×10^5 s^{-1} [606]). Similar faster exchange was found in other systems and may be explained assuming that the coordination of the ligand weakens the VO—OH_2 bond, increasing the exchange rate of the water molecule, reflected also in the proton exchange. The surprising proton exchange rate for $[VOL_2]^{2-}$ was explained by an intramolecular rearrangement (**53**↔**54**).[584]

Table 26 First-order Rate Constants of the Proton Exchange Between the Bulk Water and the Different Paramagnetic Species (I = 1 M $NaClO_4$ and T = 25 °C)

Species	$k\times10^{-5}$ (s^{-1})
[VO(oxalate)]	1.85
$[VO(oxalate)_2]^{2-}$	2.30
[VO(malonate)]	1.85
$[VO(malonate)_2]^{2-}$	1.55
$[VO(Hmalonate)]^+$	1.43

I. Nagypál and I. Fábyán, *Inorg. Chim. Acta*, 1982, **61**, 109.

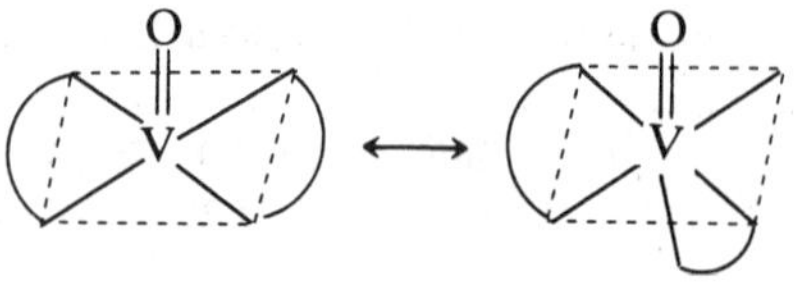

(53) **(54)**

Che and Kustin studied complexation;[438] results for oxalic and malonic acids are in Table 27. From previous relaxation data[607] and their own results, they concluded that the rate constants are more consistent with a 'normal' dissociative pathway if VOL formation from $[VO(OH)]^+$ is assumed.

Table 27 Forward Rate Constants for Vanadyl Complexation ($I = 0.5$ M KNO_3 and $T = 25$ °C)

Ligand	$k_1 \times 10^{-3}$ ($mol^{-1}\,s^{-1}$)	*Ref.*
Malonate	~20	1
	2.18[a]	2
Oxalate	~40	1
	~1[a]	2

[a] Including $[VO(OH)^+]$ in the overall reaction mechanism.
1. H. Hoffmann and W. Ulbricht, *Ber. Bunsenges. Phys. Chem.*, 1972, **76,** 1052.
2. T. H. Che and K. Kustin, *Inorg. Chem.*, 1980, **19,** 2275.

Many formation constants involve polycarboxylates; Table 28 summarizes the data. Nágypál and Fábián's report on the oxalic and malonic systems seems the most complete as hydrolysis of both metal ion and complexes has been included.[584] A concentration distribution of the complexes in the malonic system is shown in Figure 25. The order of basicities is succinic > citraconic > itaconic > maleic > malonic acid and $\log \beta_{110}$ should follow the same order. However, from Table 28, the order of stabilities is citraconic > malonic > maleic > itaconic > succinic acid.[608]

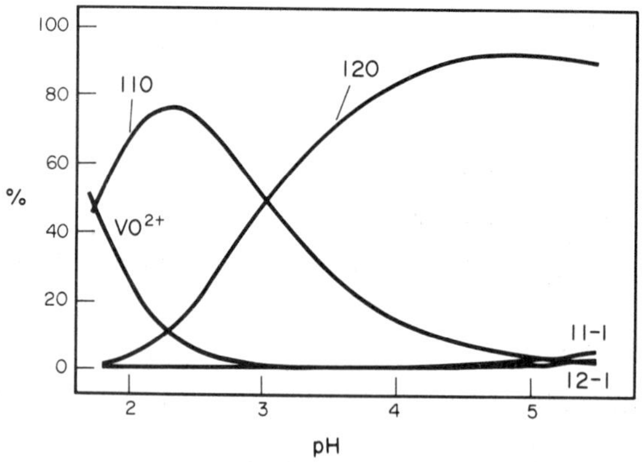

Figure 25 Speciation in oxovanadium(IV) + malonic acid solutions calculated assuming Nágypál *et al.*[584] formation constants ($I = 1.0$ M $NaClO_4$ and 25 °C) for $C_{VO} = 0.02$ M and $C_{mal}/C_{VO} = 2.0$

Many solids have been isolated and Table 29 summarizes most. Solid oxalate complexes have been prepared from vanadium(IV) salts and from V_2O_5. They form readily and are mainly of two types: $(M)_2[VO(oxal)_2]\cdot xH_2O$ and $(M)_2[(VO)_2(oxal)_3]\cdot xH_2O$. $VO(malon)\cdot 4H_2O$ and $VO(maleate)\cdot 2H_2O$ were prepared from vanadyl hydroxide,[609] and bis malonate and maleate complexes have been synthesized from vanadium(IV) and vanadium(V) salts. Several mixed-ligand complexes containing polycarboxylates as one ligand have also been prepared (see Section 33.5.10).

Complexes with succinate, glutarate and adipate have subnormal magnetic properties that indicate V···V interactions.[358] Oxalato, malonato and maleato 1:1 complexes appear monomeric as indicated by $\nu(V{=}O)$ and $\mu_{eff}^{r.t.}$, and ease of formation, stability constants and decrease in $\nu(V{=}O)$ follow the order: oxalate > malonate > maleate. For the 1:2 complexes

Table 28 Formation Constants β_{xyz} of Oxovanadium(IV) Polycarboxylate Complexes

Ligand (H_nL)	*Log* β_{110}	*Log* β_{120}	*Log* β_{111}	*Log* $\beta_{11\text{-}1}$	*Log* $\beta_{12\text{-}1}$	*Ref.*
Malonic acid	5.59	9.48	6.20	0.52	2.56	1[a]
	5.23	8.85	—	—	—	2[b]
	6.10	10.60	—	—	—	3[c]
	3.80	—	—	—	—	4[d]
Oxalic acid	6.45	11.78	—	—	—	2[b]
	4.65	9.35	—	1.07	—	5[e]
	f	12.0	—	1.75	—	6
	—	9.76	—	—	—	7[g]
	—	9.76	—	—	—	8[h]
Maleic acid	5.19	—	—	—	—	3[c]
	4.41	—	—	−0.33	—	5[e]
Diglycollic acid	4.05	6.68	—	—	—	9[i]
	3.94	6.43	—	—	—	9[j]
	3.84	6.26	—	—	—	9[k]
	4.06	6.70	—	—	—	9[l]
	5.01	—	—	—	—	10[m]
Dithiodiacetic acid	3.96	6.78	—	—	—	9[i]
	3.82	6.48	—	—	—	9[j]
	3.66	6.28	—	—	—	9[k]
	3.98	6.81	—	—	—	9[l]
Glutaric acid	3.18	—	—	—	—	11[m]
Succinic acid	3.65	—	—	—	—	3[c]
	3.66	—	—	—	—	5[e]
Adipic acid	3.48	—	—	—	—	5[e]
Itaconic acid	3.91	—	—	—	—	3[c]
Citraconic acid	6.33	—	—	—	—	3[c]
Thiodiacetic acid	3.14	—	—	—	—	10[m]
Ethylenedithioacetic acid[n]	2.68	—	—	—	—	11[m]
Phthalic acid	—					12[o]

[a] I = 1 M $NaClO_4$ and T = 25 °C. [b] I = 1 M $NaClO_4$ and T = 20 °C. [c] I = 0.1 M $NaClO_4$ and T = 30 °C. [d] T = 20 °C. [e] I = 0.1 M KNO_3 and T = 30 °C. [f] Evidence for formation. [g] I = 0.5 M $NaClO_4$ and T = 18 °C. [h] I = 0.05 M $NaClO_4$ and T = 18 °C. [i] I = 0.1 M $NaClO_4$ and T = 25 °C. [j] I = 0.2 M $NaClO_4$ and T = 25 °C. [k] I = 0.3 M and T = 25 °C. [l] I = 0.1 M $NaClO_4$ and T = 35 °C. [m] I = 0.5 M and T = 25 °C. [n] Comparing the stabilities of the glutaric, thiodiacetic and ethylenedithiodiacetic acids, Napoli (ref. 11) referred to the basicity of the ligand computing $\log \beta_{110}/\log \beta_{012}$ and concluded that sulfur is probably also involved in the coordination of the ethylenedithiodiacetate ion. [o] Room temperature ESR spectra of aqueous VO^{2+}–phthalic acid (H_2Pta) solutions at pH < 5 were interpreted by McBride (ref. 12) in terms of VO^{2+} and [VO(Pta)] only.

1. I. Nagypál and I. Fábián, *Inorg. Chim. Acta*, 1982, **61**, 109.
2. A. A. Ivakin, E. M. Voronova and L. V. Chaschina, *Russ. J. Inorg. Chem. (Engl. Transl.)*, 1970, **15**, 1839.
3. K. J. Narasimhulu and V. V. Seshaiah, *Indian J. Chem., Sect. A*, 1980, **19**, 1027.
4. P. K. Bhattacharya and S. N. Banerji, *Curr. Sci.*, 1960, **29**, 147.
5. S. P. Singh and J. P. Tandon, *Acta Chim. Acad. Sci. Hung.*, 1974, **80**, 425.
6. L. P. Ducret, *Ann. Chim. (Paris)*, 1951, S12, **6**, 705.
7. R. Trujillo and F. Torres, *An. Fis. Quim.*, 1956, **52B**, 157.
8. R. Trujillo, F. Torres and J. Ascanio, *An. Fis. Quim.*, 1956, **52B**, 669.
9. R. K. Baweja, S. N. Dubey and D. M. Puri, *J. Indian Chem. Soc.*, 1980, **57**, 244.
10. A. Napoli, *J. Inorg. Nucl. Chem.*, 1973, **35**, 3360.
11. A. Napoli, *Monatsh. Chem.*, 1981, **112**, 1347.
12. M. B. McBride, *Soil Sci. Soc. Am. J.*, 1980, **44**, 495.

the stability and bond strength (as indicated by ν(V=O)) follow the same order.[610] The ease of dissociation follows the inverse order.

$(NH_4)[VO(oxal)_2(H_2O)]\cdot H_2O$ has been studied by X-ray diffraction and involves the equatorial coordination of one water molecule (see Figure 26a).[611] (Hpy)$[VO(oxal)(F)(H_2O)_2]$ is monomeric with a pseudo-octahedral geometry (see Figure 26b).[612] In $(H_2Tmen)[VO(malon)_2(H_2O)]\cdot 2H_2O$, the vanadium is octahedrally coordinated to four malonato oxygens, to vanadyl oxygens and to a water molecule (*trans* to V=O).[613] Comparison with $[VOCl_2(tmn)_2]$ (tmn = *N,N,N',N'*-tetramethylurea) suggests that a weak extended exchange interaction of magnitude comparable to the nuclear hyperfine splitting is responsible for the ESR.[614]

33.5.5.7 Hydroxycarboxylates

(i) Tartrate

Salts of oxovanadium(IV) tartrate anions have been prepared, *e.g.* by reduction of V_2O_5 with an excess of either the optically active or racemic acid followed by addition of MOH.[583]

Table 29 Oxovanadium(IV) Polycarboxylic Acid Complexes

Complex	$\mu_{eff}^{r.t.\ b}$ (BM)	$\nu(V{=}O)^b$ (cm^{-1})	*Ref.*
$VO(oxal)\cdot nH_2O$ (n = 2, 3 or 4)	1.71–1.73	986–988	1, 2
$[VO(oxal)(DMSO)_2]$	1.74	986	3
$[VO(oxal)(antipyrine)_2]$	1.71	980	3
$VO(oxal)\cdot H_2oxal\cdot 3H_2O$	—	973, 988	4
$4\times VO(oxal)\cdot H_2oxal\cdot H_2O$	—	—	5
$(Hpy)[VO(oxal)(F)(H_2O)_2]$	—	—	6
$(Hmorph)[VO(oxal)(H_2O)]$	1.83		7
$(NH_4)[VO(oxal)(F)(H_2O)_2]$	1.82		7
$K_2[VO(oxal)(F)_2(H_2O)]\cdot H_2O$	1.82		7
$(NH_4)_3[VO(oxal)(F)_3]\cdot H_2O$	1.82		7
$(Ct)_2[VO(oxal)_2]\cdot xH_2O$ (Ct = Li, Rb^+, Cs^+, Hpy, $\frac{1}{2}H_2en$, Hmorph)	1.91–1.99	[940–980]	8
$(Me_4N)_2[(VO)_2(oxal)_3]\cdot 3.5H_2O$	1.23	[940–980]	8
$(2\text{-Hpicoline})[(VO)_2(oxal)_2]\cdot 6H_2O$	1.14	[940–980]	8
$[Co(NH_3)_6]_2[(VO)_2(oxal)_5]\cdot 5H_2O$	1.25	[940–980]	8
$(NH_4)_2[VO(oxal)_2(H_2O)]\cdot H_2O$	—	—	9
$(Ct)[(VO)_2(oxal)_3]\cdot nH_2O$ (n = 0, 4, 6 or 8.5 and Ct = NH_4^+, Na^+ or K^+)	1.65–1.70	990–1000	3, 10–13
$(NH_4)_2[(VO)_2(oxal)_2]\cdot 2H_2O$	1.72	988, 976	10, 12
$VO(malon)\cdot 4H_2O$	1.78	990	14
$(Ct)_2[VO(malon)_2]\cdot nH_2O$ (n = 0–5; Ct = H^+, NH_4^+, Na^+, K^+, Rb^+, Cs^+, Tl^+; $\frac{1}{2}$Ct = Ca^{2+}, Sr^{2+}, Ba^{2+}, K^+H^+)	1.72–1.87	967–988	12, 15–17
$(H_2tmen)[VO(malon)_2(H_2O)]\cdot H_2O$			18
$VO(maleate)\cdot 2H_2O$	1.72	1000	14
$(NH_4)_2[VO(maleate)_2]\cdot 3H_2O$	0.72	1003	16
[VO(succinate)]	1.22	—	15
[VO(glutarate)]	1.29	—	15
[VO(adipate)]	1.23	—	15

[a] H_2tmen = *N,N,N',N'*-tetramethylethylenediamidium; Hmorph = morpholinium.
[b] When the line of the table corresponds to more than one complex, the values presented for $\mu_{eff}^{r.t.}$ and $\nu(V{=}O)$ are the lowest and highest reported in the original publications. For many compounds no data are available.

1. D. N. Sathyanarayana and C. C. Patel, *J. Inorg. Nucl. Chem.*, 1965, **27**, 297.
2. K. C. Satapathy, R. Parmar and B. Sahoo, *Indian J. Chem.*, 1963, **1**, 402.
3. D. N. Sathyanarayana and C. C. Patel, *J. Inorg. Nucl. Chem.*, 1966, **28**, 2277.
4. J. Selbin and L. H. Holmes, Jr., *J. Inorg. Nucl. Chem.*, 1962, **24**, 1111.
5. Gmelins Handbuch der Anorganischen Chemie, 8. Auflage, Vanadium, Teil B, Lieferung 1, 1967, p. 349.
6. A. J. Edwards, D. R. Slim, J. Sala-Pala and J. E. Guerchais, *Bull. Soc. Chim. Fr.*, 1975, 2015.
7. A. K. Sengupta, B. B. Bhaumik and R. K. Chattopadhyay, *Indian J. Chem., Sect. A*, 1980, **19**, 914.
8. B. B. Bhaumik and R. K. Chattopadhyay, *Indian J. Chem., Sect. A*, 1981, **20**, 417.
9. R. E. Oughtred, E. S. Raper and H. M. M. Shearer, *Acta Crystallogr., Sect. B*, 1976, **32**, 82.
10. D. N. Sathyanarayana and C. C. Patel, *J. Inorg. Nucl. Chem.*, 1965, **27**, 2549.
11. J. Selbin, L. H. Holmes, Jr. and S. P. McGlynn, *J. Inorg. Nucl. Chem.*, 1963, **25**, 1359.
12. C. G. Barraclough, J. Lewis and R. S. Nyholm, *J. Chem. Soc.*, 1959, 3552.
13. Gmelins Handbuch der Anorganischen Chemie, 8. Auflage, Vanadium, Teil B, Lieferung 2, 1967, pp. 408 and 441.
14. D. N. Sathyanarayana and C. C. Patel, *Bull. Chem. Soc. Jpn.*, 1967, **40**, 794.
15. V. T. Kalinnikov, V. V. Zelentsov, O. N. Kuz'micheva and T. G. Aminov, *Russ. J. Inorg. Chem. (Engl. Transl.)*, 1970, **15**, 341.
16. D. N. Sathyanarayana and C. C. Patel, *Indian J. Chem.*, 1968, **6**, 523.
17. Ref. 13, pp. 408, 441, 475, 491, 499, 517, 523, 530, 551.
18. D. Collison, B. Gahan and F. E. Mabbs, *J. Chem. Soc., Dalton Trans.*, 1983, 1705.

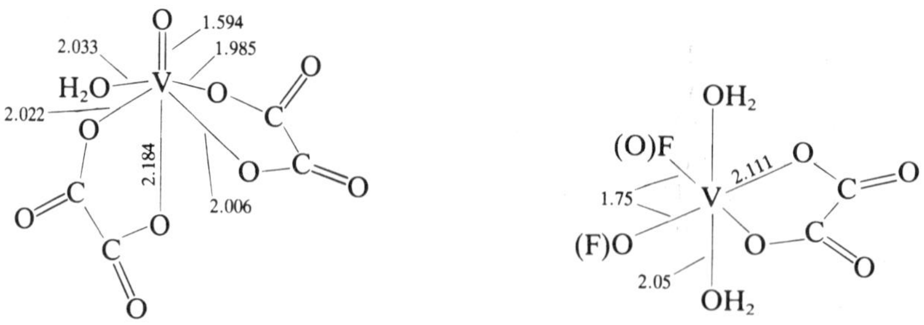

Figure 26 Perspective diagrams of the complex anions (a) $[VO(oxal)_2(H_2O)]^{2-}$ [611] and (b) $[VO(F)(oxal)(H_2O)_2]^-$;[612] interatomic distances are in Å. In (b) the H_2O and F^- are distributed in a 'disordered fashion' and the 1.75 Å distance may be considered an average of the distances expected for a V—F bond (1.9 Å) and a V=O bond (1.6 Å)

According to M (Na^+, K^+, NH_4^+, NMe_4^+, Rb^+, Ba^{2+}, NEt_4^+), $M_n[(VO)_2\{(+)\text{-tart}\}\{(-)\text{-tart}\}]\cdot xH_2O$ or $M_n[(VO)_2\{(+)\text{- or }(-)\text{-tart}\}_2]\cdot xH_2O$ have been obtained.[583,615] Several methyl tartrate derivatives have been used as ligands and compounds isolated. Table 30 summarizes products. The (+, +) [or (−, −)] isomer requires a *trans* configuration about vanadyl and the ± a *cis* configuration. Compounds include $(NH_4)_4[(VO)_2\{(+)\text{-tart}\}_2]\cdot H_2O$,[615] $Na_4[(VO)_2\{(+)\text{-tart}\}\{(-)\text{-tart}\}]\cdot 12H_2O$,[616] $(NEt_4)[(VO)_2\{(+)\text{-tart}\}\{(-)\text{-tart}\}]\cdot 8H_2O$ (V—V distance = 3.895 Å), $Na_4[(VO)_2\{(+)\text{-dmt}\}\{(-)\text{-dmt}\}]\cdot xH_2O$ (x = 6,[617] 12[618]) and $Na_4[(VO)_2\{(+)\text{-}threo\text{-mmt}\}\{(-)\text{-}threo\text{-mmt}\}]\cdot 14H_2O$.[619] The geometries of the dimeric anions are those of Figure 27 and only slight changes occur with the counterion.[620]

Table 30 Some Oxovanadium(IV) Tartrate Compounds

Compound[a]	$\nu(V{=}O)$ *or* $\mu_{eff}^{r.t.}$ (300 K)	*Band maxima* (μm)[b] *I*	*II*	*III*
$Na_4[(VO)_2\{(+)\text{-tart}\}\{(-)\text{-tart}\}]\cdot 12H_2O$	1.72 BM	1.33	1.88	2.34
$Na_4[(VO)_2\{(+)\text{-tart}\}_2]\cdot 6H_2O$	925 cm^{-1}, 1.75 BM	1.10	1.70, 1.95	2.51
$(NH_4)_2[(VO)_2\{(+)\text{-tart}\}_2]\cdot 2H_2O$		1.20	1.70, 1.96	2.55
$Na_4[(VO)_2\{(+)\text{-mmt}\}\{(-)\text{-mmt}\}]\cdot 14H_2O$	924 cm^{-1}	1.39	1.90	2.36
$Na_4[VO)_2\{(+)\text{-mmt}\}\{(-)\text{-mmt}\}]\cdot 10H_2O$		1.40	1.90	2.38
$Na_4[(VO)_2\{(+)\text{-mmt}\}_2]\cdot 8H_2O$		1.10	1.65, 1.87	2.43
$Na_4[(VO)_2\{(+)\text{-dmt}\}\{(-)\text{-dmt}\}]\cdot 12H_2O$	939 cm^{-1}	1.20	1.75	2.74
$Na_4[(VO)_2\{(+)\text{-dmt}\}\{(-)\text{-dmt}\}]\cdot 6H_2O$	937 cm^{-1}	1.26	1.88	2.70
$Na_4[(VO)_2\{(\pm)\text{-dmt}\}_2]\cdot 14H_2O$		1.06	1.72, 1.89	2.52
$Na_4[(VO)_2\{(+)\text{-dmt}\}_2]\cdot 12H_2O$		1.10	1.72, 1.90	2.45
$Ba_2[(VO)_2\{(\pm)\text{-dmt}\}_2]\cdot 12H_2O$	924 cm^{-1}	1.13	1.71, 1.91	2.58

[a] tart = tartrate^{4-} anion, mmt = monomethyltartrate^{4-} anion (*threo* isomer) and dmt = dimethyltartrate^{4-} anion.
[b] Positions of band maxima in visible spectra of solid complex salts as Nujol mulls.
1. S. K. Hahs, R. B. Ortega, R. E. Tapscott, C. F. Campana and B. Morosin, *Inorg. Chem.*, 1982, **21**, 664.

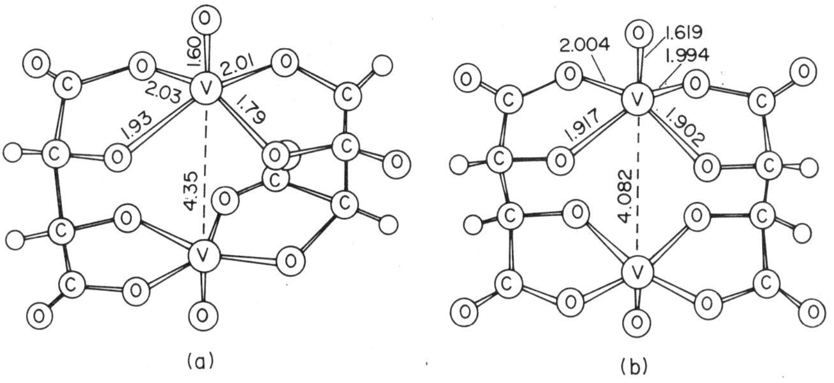

Figure 27 Schematic representation of $[(VO)_2(tart)_2]^{4-}$ units in tartrate complexes. Internuclear distances are given for (a) $(NH_4)_4[(VO)_2\{(+)\text{-tart}\}_2]\cdot H_2O$[615] and (b) $Na_4[(VO)_2\{(+)\text{-tart}\}\{(-)\text{-tart}\}]\cdot 12H_2O$[616]

ESR suggested that the methyl-substituted dimers exhibit a shortened V—V distance.[621] In $Na_4[(VO)_2\{(+)\text{-dmt}\}\{(-)\text{-dmt}\}]\cdot 12H_2O$ the V—V distance is 3.429 Å and the vanadium distance to the basal plane is 0.394 Å, 0.60 and 0.13 Å shorter than for (±)-tartrate^{4-}.[619]

Tartaric acid is dibasic (H_2tart), carboxyl protons being ionized by pH 5. The hydroxyl groups are not ionized ($pK_{a3} \approx 14$ and $pK_{a4} \approx 15.6$).[622] The formation constants of (+)-, (−)-, (±)- and *meso*-tartaric acids with VO^{2+} confirm binuclear complexes of optically active and, to a lesser extent, *meso* acids with VO^{2+}.[623]

Electronic and CD spectra[624] of the solid and solutions of oxovanadium(IV) (+)-tartrate are identical, as are their IR spectra,[615] good evidence that the dimeric species in solution is the same as that found in the crystals. An unusually well-defined triplet state ESR has been recorded for liquid and frozen solutions.[625,626]

Racemic and optically active solutions behave differently, with large stereoselective effects.[627] The reaction corresponding to equation (44) has $K = 16^{623}$ and a reaction enthalpy of $(6.9 \pm 0.4) \times 10^3$ J mol^{-1}.[628,629] The kinetics were also studied.[629]

$$\tfrac{1}{2}[(VO)_2\{(+)\text{-tart}\}_2]^{4-} + \tfrac{1}{2}[(VO)_2\{(-)\text{-tart}\}_2]^{4-} \rightleftharpoons [(VO)_2\{(+)\text{-tart}\}\{(-)\text{-tart}\}]^{4-} \quad (44)$$

(ii) Other hydroxycarboxylate ligands

Hydroxycarboxylate systems have been extensively studied. Solution spectra of most VO^{2+} complexes contain, at most, three bands in the visible region that may be assigned to *d–d* transitions; however, alkaline solutions of most α-hydroxycarboxylates exhibit four bands[365] ascribed to a reduced symmetry of the ligand field by *trans* coordination of two strongly bound ionized hydroxyl oxygens[478] (perhaps accompanied by distortion towards a trigonal bipyramidal geometry).[630] Table 31 summarizes part of the reported spectra: in several cases, the four bands suggest that α-hydroxycarboxylato complexes have *trans* coordination. In $Na(Et_4N)[VO(benzilato)_2]\cdot 2PrOH$, vanadium is five-coordinate and distorted square pyramidal, and the bidentate ligands are mutually *trans*.[630] The solution, single crystal and polycrystalline optical spectra exhibit four bands similar to those of the *trans*-vanadyl–(+)-tartrate system (Table 31).

Table 31 Spectral Bands Observed in Optical Spectra of Oxovanadium(IV) Hydroxycarboxylates in Aqueous Solution

Ligand	pH	λ/nm (ε/1 mol^{-1} cm^{-1}) I	IIA	IIB	III	Ref.
(±)-Tartrate	~8	737 (25.4)	536 (25.4)		421 (52.9)	1
(+)-Tartrate	~8	902 (22.6)	590 (22.4)	533 (27.7)	399 (46.4)	1
Benzilate	8.5	847 (31)	599 (26)	541 (27)	417 (36)	1
Mandelate	8.5	826	599	535	413	1
Lactate	>7	813	599	529	407	1
Malate	>7	815	608	530	410	1
Citrate	>9	775	671	555		2
	8	1052	649	571		2

1. Collected in R. M. Holland and R. E. Tapscott, *J. Coord. Chem.*, 1981, **11**, 17.
2. B. M. Nikolova and G. S. Nikolov, *J. Inorg. Nucl. Chem.*, 1967, **29**, 1013.

Complexes with (*R*)-lactate have been studied by ESR[631] and CD.[580,632] Lactate is near water in the spectrochemical series and its low pK_a helps reduce error due to hydrolysis and oxidation. Table 32 summarizes the data. The pK_a of the R—OH is quite high (especially for the aliphatic acids). Many have considered coordination in VO^{2+} hydroxycarboxylate to involve the CO_2^- and OH groups and assumed the dissociation of the OH. As structures (**55**) and (**56**) or (**57**) and (**58**) are not pH-metrically distinguishable, if the pK_a^{OH} is not known, other techniques must establish the species present.

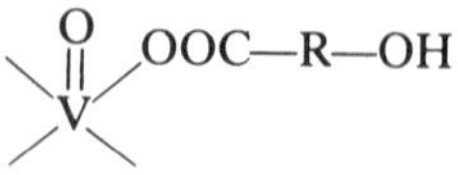

(**55**)

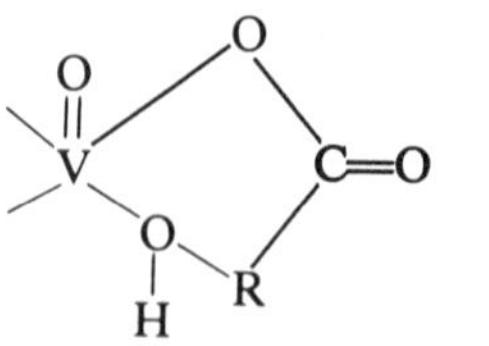

(**56**)

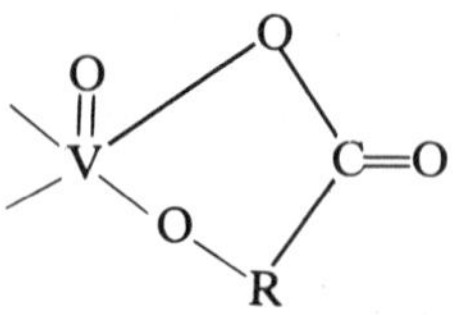

(**57**)

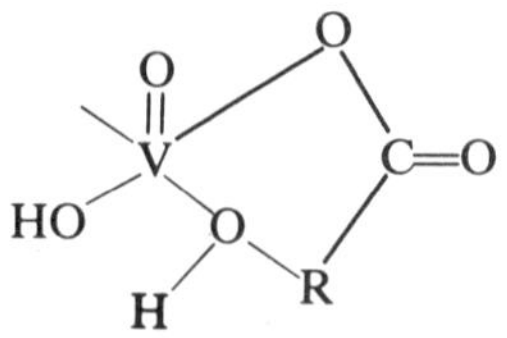

(**58**)

From ESR, for (*R*)-lactic acid (H_2lact), five species were required: $[VO(Hlact)]^+$, $[VO(Hlact)_2]_2$, [VO(lact)], $[VO(lact)(Hlact)]^-$ and $[VO(lact)_2]^{2-}$; for glycolic acid (H_2glyc): [VO(glyc)] and $[VO(glyc)_2]^{2-}$.[631] Another ESR study assumed only $[VO(glyc)_2]^{2-}$ formation.[626]

With salicylic acid (H_2SA) and 5-sulfosalicylic acid (H_2SSA), titrations have been interpreted in terms of coordination with CO_2H and OH ionized, although the pK_a values of the OH groups are high ($pK_a^{OH} \approx 11.4$ for $HSSA^-$ and ≳13 for HSA).[633] ESR spectra of solutions containing salicylic acid (H_2SA) at pH ≲ 5 were interpreted in terms of VO^{2+}, [VO(SA)] and $[VO(SA)_2]^{2-}$ species.[634] A different model involves VO^{2+}, $[VO(HSA)^+]$ and

Table 32 Formation Constants of Hydroxycarboxylate Oxovanadium(IV) Complexes

Ligand	pK_a^{OH}	$Log\,\beta_{110}$	$Log\,\beta_{120}$	$Log\,\beta_{111}$	$Log\,\beta_{210}$	$Log\,\beta_{11\text{-}1}$	$Log\,\beta_{12\text{-}1}$	$Log\,\beta_{12\text{-}2}$	$Log\,\beta_{22\text{-}2}$	*Medium conditions*	*Ref.*
(*R*)-Lactate[a]	—	2.68	4.83	—	—	—	—	—	—	20 °C, 1 M $NaClO_4$	1
	—	~2.48	~4.08	—	—	~−1.77	~0.58	~−4.42	—	25 °C, 0.47 M $NaClO_4$	2
Citrate[a]	—	8.8	—	—	11.5	—	—	—	—	20 °C, 1 M NaCl	3
Mandelate[a]	—	—	—	—	—	—	—	−2.21[c]	—	25 °C, 0.5 M KNO_3	4
Vanillomandelate[a]	—	—	—	—	—	—	—	−2.36[c]	—	25 °C, 0.5 M KNO_3	4
Glycolate	—	—	—	—	−1.40[b]	—	—	−1.52[d]	—	25 °C, 0.47 M $NaClO_4$	2
	—	2.56[a]	4.22[a]	—	—	—	—	—	—	25 °C, 1 M $NaClO_4$	5
Salicylate[b]	13.90	13.38	—	—	—	—	—	—	—	25 °C, 0.1 M KNO_3	6
(H_2SA)	13.1	12.7	22.4	—	—	—	—	—	—	20 °C, 0.1 M KNO_3	7
	13.6	13.18	—	—	—	8.55	—	—	—	30 °C, 0.1 M KNO_3	8
	13.0	12.68	—	14.68	—	6.54	—	—	—	25 °C, 0.10 M $NaClO_4$	9[e]
	13.0	12.52	—	14.88	—	6.07	—	—	—	25 °C, 0.40 M $NaClO_4$	9[e]
	13.0	12.56	—	15.25	—	6.62	—	—	—	25 °C, 0.70 M $NaClO_4$	9[e]
Sulfosalicylate[b]	12.0	12.0	20.6	—	—	—	—	—	—	20 °C, 0.1 M KNO_3	7
(H_2SSA)	11.67	11.71	—	—	—	4.49	—	—	14.3	25 °C, 0.1 M KNO_3	6
	11.42	11.37	—	—	—	7.03	—	—	—	30 °C, 0.1 M KNO_3	8
Salicyl-2-phosphate[b]	6.61	5.81	—	—	—	0.11	—	—	2.5	25 °C, 0.1 M KNO_3	6
Gallate[b]	—	11.37	20.95	—	—	—	—	—	—	35 °C, 0.2 M $NaClO_4$	10
Protocatechuate	—	9.00	7.47	—	—	—	—	—	—	35 °C, 0.2 M $NaClO_4$	10
3,4-Dihydroxybenzoate[b]	8.69, 12.5	16.63	—	—	20.83	—	—	—	—	25 °C, 0.10 M $NaClO_4$	9[e]
	8.59, 12.5	16.34	—	—	20.65	—	—	—	—	25 °C, 0.40 M $NaClO_4$	9[e]
	8.62, 12.5	16.24	—	—	20.76	—	—	—	—	25 °C, 0.70 M $NaClO_4$	9[e]

[a] Constants β_{mnp} refer to the reaction $m\mathrm{VO}^{2+} + n\mathrm{HOXCO_2^-} + p\mathrm{H}^+ = [(\mathrm{VO})_m(\mathrm{HOXCOO}^-)_n(\mathrm{H}^+)_p]$.
[b] Constants β_{mnp} refer to reaction $x\mathrm{VO}^{2+} + y\mathrm{OXCO_2^{2-}} + z\mathrm{H} = [(\mathrm{VO})_x(\mathrm{OXCO_2^{2-}})_y(\mathrm{H}^+)_z]$.
[c] Complexes correspond to $[\mathrm{VO(L^{2-})_2}]^{2-}$. As K_a^{OH} is not known, $[\mathrm{VO(L^{2-})_2}]^{2-}$ and $[\mathrm{VO(HL^-)(OH^-)_2}]$ are not pH-metrically distinguishable; the constant corresponds to $\beta_{12\text{-}2}$ in terms of $\mathrm{HL^-}$.
[d] A value defined as $K2 * K_{a2}$ is reported; it is not equal to the defined $\beta_{12\text{-}2}$.
[e] Other species may be present in minor amounts in the pH range studied (2.5–4).

1. K. M. Jones and E. Larsen, *Acta Chem. Scand.*, 1965, **19**, 1205.
2. R. R. Reeder and P. H. Rieger, *Inorg. Chem.*, 1971, **10**, 1258.
3. B. M. Nikolova and G. S. Nikolov, *J. Inorg. Nucl. Chem.*, 1967, **29**, 1013.
4. T. M. Che and K. Kustin, *Inorg. Chem.*, 1980, **19**, 2275.
5. A. Lorenzotti, D. Leonesi and A. Cingolani, *Inorg. Chim. Acta*, 1981, **52**, 149.
6. G. E. Mont and A. E. Martell, *J. Am. Chem. Soc.*, 1966, **88**, 1387.
7. J. Zelinka and M. Bartušek, *Collect. Czech. Chem. Commun.*, 1971, **36**, 2615.
8. S. P. Singh and J. P. Tandon, *Monatsh. Chem.*, 1975, **106**, 271.
9. M. L. Gonçalves and A. M. Mota, *Talanta*, in press.
10. R. M. Smith and A. E. Martell, 'Critical Stability Constants', Plenum, New York, 1976.

[VO(SA)] (see Table 32).[635] Early reports include $(Ct)_2[VO(SA)_2]\cdot 3H_2O$ ($Ct = K^+$, NH_4^+ and $\frac{1}{2}Ca^{2+}$) and $Tl_2[VO(SA)_2]$.[3,357b]

DMF solutions of chelates of 1-hydroxycyclohexanecarboxylate and mandelate exhibit the half field ESR at ~1500 G due to $\Delta M = 2$ ($g \simeq 4$),[358] as does aqueous VO^{2+} citrate at pH 6–9, indicating V···V interactions.[636] $[(VO)_3(citrato)_2]\cdot 4H_2O$ has $\mu_{eff}^{r.t.} = 1.51$ BM, like VO(trihydroxyglutarato)$\cdot 0.5H_2O$ ($\mu_{eff}^{r.t.} = 1.57$ BM).[637] Dimeric 2:2 complexes with citrate and malate explain proton relaxation in the pH range 3–10;[638] comparative dialysis and spectrophotometry at pH 2–7 suggested the existence of 1:3, 1:2, 1:1 and polymeric 1:1 species.[639] Early reports include $(Ct)_4[(VO)_2(citrato)_2]\cdot nH_2O$ ($Ct = Na^+$, K^+, NH_4^+ with $n = 12$, 6 and 0 respectively).[357b]

$(NH_4)_2[VO(benzilato)_2]$, $(NH_4)_2[VO(fsa)(OH)_2H_2O]$ and $(NH_4)_2[VO(fsa)_2]\cdot 2H_2O$ (fsa = 3-formylsalicylato) have $\mu_{eff}^{r.t.}$ 1.67–1.73 BM,[640] as well as $[VO(3,5\text{-dinitro})SA)_2]\cdot 2H_2O$.[641] Some of these have high antimicrobial activity.[640]

The effect of exchange of lactic, mandelic and sulfosalicylic acids on the relaxation of solvent protons gave rate constants (k) of exchange from 1.73 to 0.70 l mol^{-1} s^{-1}.[642] Kinetics of complex formation with mandelic (HMDA) and vanillomandelic acids (HVMDA) gave rate constants (1.09×10^3 and 1.13×10^3 mol^{-1} s^{-1} for MDA^- and $VMDA^-$) consistent with a dissociative (Eigen) mechanism.[438] As in the case of oxalic and malonic acids (Section 33.5.5.5.ii; Table 27), species with coordinated hydroxyl are labilized.

33.5.5.8 *Other oxygen ligands*

(*i*) *Sulfoxides*

Complexes with DMSO as one of several ligands have been reported frequently and early ones include $[VO(X)_2(DMSO)_3]$ where $X = \frac{1}{2}SO_4^{2-}$, Cl^-, Br^- [387,643] and $[VO(L)_2(DMSO)]$ where L is a bidentate ligand (*e.g.* acac).[355] $[VOBr_2(DPSO)_3]$ (DPSO = diphenyl sulfoxide) and $[VO(DMSO)_5]Br_2$ were prepared by the reaction of $VOBr_3$ with the sulfoxides in cyclohexene.[401] $[VOCl_2(DMSO)_3]$ and $[VOCl_2(DPSO)_3]$ were prepared by the method in equation (16);[409] two V—Cl IR-active bands were observed for both compounds (353, 321 cm^{-1} and 312, 282 cm^{-1}, respectively), suggesting a trigonal bipyramidal (**13**) structure. For adducts of VOX_2 ($X = Cl$, $\frac{1}{2}SO_4^{2-}$) with aromatic and aliphatic sulfoxides, the sulfoxide groups are coordinated through the oxygen.[644] In $[VO(DMSO)_5](ClO_4)_2$, d(V═O) = 1.591, d(V—O(equatorial)) = 2.034 (av) and d(V—O(axial)) = 2.18 Å.[387,645] 'Adducts' of $VOBr_2$ and VOI_2 with DMSO and DMF have been isolated; in $[VO(DMSO)_5]I_2$ the distances are slightly shorter (*ca.* 0.01 Å) than in $[VO(DMSO)_5](ClO_4)_2$.[646]

(*ii*) N-*Oxides and related ligands*

$[VO(C_5H_5NO)_5](ClO_4)_2$, $[VOCl_2(Ph_3PO)_2]$, $[VO(Ph_3PO)_4]$, $[VO(Ph_3AsO)_4](ClO_4)_2$ and $[VOCl_2(DMSO)_2]$ have been prepared by the reaction of $VCl_3\cdot 6H_2O$ with the ligand in ethanol.[643] Perchlorates are obtained if $LiClO_4$ is added. In $[VOCl_2(Ph_3PO)_2]$, vanadium is pentacoordinate with the Ph_3PO ligands *trans*; their O atoms and two Cl atoms comprise the base of a square pyramid completed by an apical multiply bonded O atom.[647] Internuclear distances are as follows: d(V═O) = 1.584, d(V—O) = 1.986 and 2.002, and d(V—Cl) = 2.312 and 2.304 Å. This and $[VOCl_2(hmpa)_2]$ (hmpa = hexamethylphosphoramide) have also been prepared by equation (16) and have only one IR-active V—Cl band each at 384 and 385 cm^{-1}. Seddon's suggestion would imply square pyramidal structures.[409] However, the structure determined by X-ray is based on a distorted trigonal bipyramid.[648] In $[VOCl_2(dimp)_2]$ (dimp = Pr^i_2MePO), ν(V—Cl) is at 431 cm^{-1}.[649]

From the single crystal ESR of $[VOCl_2(Ph_3PO)_2]$, no magnetic exchange interactions exist; the spectrum consists of eight lines from interaction of the electron with the nuclear spin of ^{51}V ($I = \frac{7}{2}$).[650] In poly(dimethylsiloxane) mulls of $[VOCl_2(Ph_3PO)_2]$ and $[VOCl_2(hmpa)_2]$, optical bands were assigned according to a crystal field orbital splitting different from that of Ballhausen and Gray (Section 33.5.1, Figure 16).[364]

(iii) Miscellaneous oxygen ligands

Complexes with *N,N,N',N'*-tetramethylurea (tmu) have been extensively studied. In $[VOCl_2(tmu)_2]$, the vanadium is pentacoordinate and square pyramidal (see Figure 28).[651] The single crystal ESR of $[VOX_2(tmu)_2]$ (X = Br, Cl) gave unusual fine structure at 170–120 K, which was interpreted in terms of an electronic exchange interaction, within a two-dimensional layer.[614] ESR for $[VO(pftc)_2]$ (pftc = perfluoropinacolate) and the deoxygenated $[VCl_2(pftc)_2]$ obtained by its reaction with $SOCl_2$[429] suggests a trigonal prismatic structure for $[VCl_2(pftc)_2]$. The reaction between $VCl_3 \cdot 3THF$ and three equivalents of 2-hydroxy-6-methylpyridine (Hmhp; **59**) in CH_2Cl_2 in air produces bright blue $[V_2O_2Cl_4(\mu\text{-Hmhp})_3]$ (**60**) characterized by crystallography, ESR ($g = 1.976$) and IR,[652] in the region $1000 \pm 50\ cm^{-1}$. The structure consists of discrete dinuclear units.

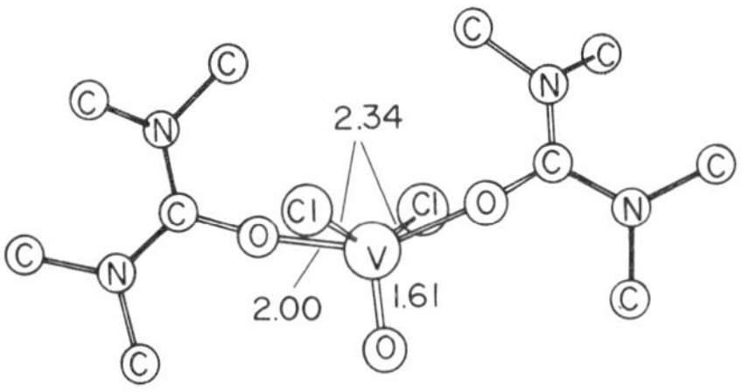

Figure 28 Molecular structure of $[VOCl_2(tetramethylurea)_2]$. Some internuclear distances (in Å) are indicated[651]

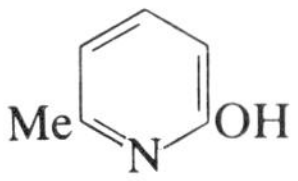

(59)

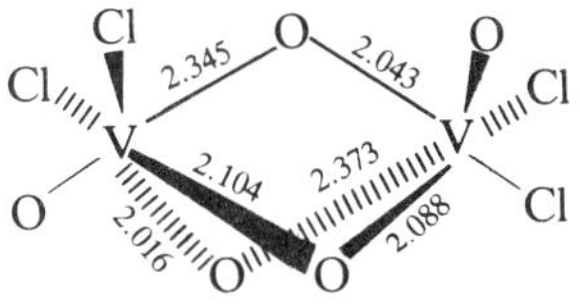

(60)

Several complexes which contain halogen ligands (with and without the vanadyl bond) with oxygen donors such as diethyl ether, $MeOCH_2CH_2OMe$, benzaldehyde, benzophenone, THF, dioxane and pyridine *N*-oxide derivatives are included in 'Gmelins Handbuch'.[357b] $[VOF_2L_n]$ complexes[428] (equation 22) include $[VOF_2L_2]$ with L = urea and formamide (fmd) and $[VOF_2(H_2O)L]$ with L = DMF and dioxane.

33.5.6 Sulfur Ligands

33.5.6.1 Thiovanadyl complexes

The first complexes of the thiovanadyl ion, VS^{2+}, were made by the reaction of [VO(salen)] or [VO(acen)] (H_2salen = *N,N'*-ethylenebis(salicylideneamine), H_2acen = *N,N'*-ethylenebis(acetylacetonylideneamine)) with the powerful oxophile B_2S_3 in CH_2Cl_2 with rigorous exclusion of oxygen and moisture (equations 45 and 46).[653,654] The magenta [VS(salen)] is relatively stable to oxidation or hydrolysis and shows no colour change in air for months, although it then smells of H_2S and IR shows the presence of V=O. In sharp contrast, on swirling in air a solution in CH_2Cl_2, it quickly changed to green, forming [VO(salen)] quantitatively with the liberation of H_2S.[654] The magenta [VS(acen)] is also relatively stable but — in contrast to [VS(salen)] — solutions in organic solvents are decomposed to [VO(acen)] only after a few hours of exposure to the atmosphere. IR of (**61**) and (**62**) showed no V=O stretch, and new medium-to-strong bands at 543 (**61**) and 556 cm^{-1} (**62**), were assigned to ν(V=S). (**61**) and (**62**) exhibit eight-line isotropic solution ESR; g and A decrease in magnitude on substitution of oxygen by sulfur, consistent with greater covalency in the VS bond.[653] [VS(acen)] has square pyramidal geometry with the sulfur atom at the apex and the coordinated atoms of acen comprising the basal plane.[655] V=S was 2.061 Å, shorter than V—S single bonds (Section 33.5.6.2 and 33.5.6.3). The vanadyl analogue (Figure 32, p. 536) also has square pyramidal geometry around the V atom; the difference between V=S and V=O is 0.47 Å.

$$VO(salen) + B_2S_3 \xrightarrow[48\ h\ reflux]{CH_2Cl_2} \xrightarrow{hexane} VS(salen) \quad (45)$$

(61)

$$VO(acen) + B_2S_3 \xrightarrow[24\ h\ at\ 25\ °C]{CH_2Cl_2} \xrightarrow{hexane} VS(acen) \quad (46)$$

(62)

By allowing a slurry of $(PPh_4)Na[VO(edt)_2]\cdot 2EtOH$ (edt^{2-} = ethane-1,2-dithiolate) in MeCN to react with $(Me_3Si)_2S$, orange-brown crystals were obtained.[656] They contain $[VS(edt)_2]^{2-}$, whose structure is essentially identical with $[VO(edt)_2]^{2-}$ (see Section 33.5.6.3). The difference between V=O and V=S is 0.462 Å.

In $(NMe_4)Na[VO(edt)_2]\cdot 2EtOH$, ν(V=O) was at 930 cm^{-1} and the V=O distance (1.625 Å) is rather long but is similar to that in $[VO(pdt)_2]^{2-}$ (pdt = propane-1,3-dithiolate) (1.628Å).[656] On O/S substitution, the band at 930 cm^{-1} disappears and a strong, sharp band at 502 cm^{-1} appears, ν(V=S). This change (428 cm^{-1}) is similar to that between VE(acen) and VE(salen) (E = O, S), paralleling changes in V=E bond length.

Electrochemical oxidations of [VS(acen)] and [VO(acen)] are reversible at 0.45 and 0.66 V *vs.* SCE, with one electron involved.[657] In contrast to $[VO(acen)]^+$, which is quite stable, the $[VS(acen)]^+$ product readily reacts with residual water and is converted to $[VO(acen)]^+$ and H_2S. Vanadyl complexes with O/N-based ligands have oxidation potentials in the range 0.6–1.1 V.[516,657] The potential for $[VO(edt)_2]^{2-}$ is more negative by ~1 V.[656] On O/S substitution, the potential becomes even more negative by 0.27 V, behaviour also present in VE(acen) (E = O, S) (~0.26 V). Again the basal ligands govern the absolute magnitude of a property whereas changes in O/S substitution are essentially constant.[656]

By treatment with $(Me_3Si)_2S$ the conversion of $[VO(edt)_2]^{2-}$ to $[VS(edt)_2]^{2-}$ gave an intermediate $[V(OSiMe_3)(edt)_2]^-$ as the PPh_4^+ salt.[658] It contains a discrete, square pyramidal V^{IV} with d(V—O) = 1.761 Å, mean d(V—S) = 2.322 Å and a V—O—Si angle of 140.8°.

Changes in ESR and electronic absorption from $[VO(edt)_2]^{2-}$ to $[VS(edt)_2]^{2-}$ are similar to those in the series [VE(acen)] and [VE(salen)] (E = O, S).[654] The spectroscopic results for the $[VE(edt)_2]^{2-}$ complexes have been interpreted assuming a $d_{x^2-y^2}$ ground state.[658]

33.5.6.2 CS_2-based and related ligands

(i) Dithiocarboxylates

The interesting reaction of oxovanadium(IV) with dithiocarboxylates RCS_2^- cleaves the vanadyl bond, in water or ethanol, with formation of stable $[V(RCSS)_4]$. When pure, they are not air-sensitive.[661] They are insoluble in polar solvents and generally soluble in nonpolar ones; however, the solutions decompose rapidly. Various stereochemistries are possible but molecular weights, IR, electronic and ESR spectra are consistent with eight-coordinate geometry,[661–664] with $d_{x^2-y^2}$ lowest-lying (D_{2d} symmetry). Several [V(dithiocarboxylato)$_4$] complexes have vanadium coordinated to eight sulfur atoms with a dodecahedral geometry.[663,665,666] The V—S distances are of two kinds and for $[V(MeCS_2)_4]$, nonequivalent atoms have (i) a regular or (ii) a distorted dodecahedral configuration. Some data are in Table 33.

(ii) Dithiocarbamates

Dithiocarbamato (dtc) complexes are prepared by mixing $VOSO_4$ and the ligand in water–ethanol solutions. Some of the isolated complexes are in Table 34; they are susceptible to air oxidation, especially in solution, and are generally monomeric.

$[VO(Et_2NCS_2)_2]$ has discrete molecules with five-coordinate square pyramids, with the oxygen at the apex with d(V=O) = 1.59 Å.[667] The vanadium atom lies approximately 0.75 Å above the 'plane' of the four sulfur atoms and the average V—S distance is 2.402 Å, shorter than for most vanadium dithiolate complexes;[667] IR, electronic and ESR spectra are consistent

Table 33 Some Properties and X-Ray Data of [V(dithiocarboxylato)$_4$] Complexes

	R in *[V(RCSS)$_4$]*	*Colour*	$\mu_{eff}^{r.t.}$ (BM)[a]	*V—S average internuclear distances* (Å) *V—S$_A$*	*V—S$_B$*	*ν(V—S)* (cm^{-1})	*Ref.*
1	Me	Red-brown	1.74[b]	2.50	2.46	—	1, 2
2	Ph	Red	1.70–1.79	2.53	2.47	—	3, 4
3	$PhCH_2$	Red	1.71	2.56	2.45	—	3
4	*p*-MeC_6H_4	Red	1.69–1.75	—	—	—	1
5	C_5H_8N[c]	Red-brown	—	—	—	460	5

[a] The magnetic behaviour of compounds 1–4 in the temperature range 77–300 K was studied (ref. 2), but the authors considered that ferromagnetic impurities, undetectable by chemical analysis, are the origin of part of the anomalous behaviour of some of the complexes.
[b] Determined by the Faraday method.
[c] Hacda, 2-amino-1-cyclopentenedithiocarboxylate.

1. O. Piovesana and C. Furlani, *Chem. Commun.*, 1971, 256.
2. O. Piovesana and G. Cappuccilli, *Inorg. Chem.*, 1972, **11**, 1543.
3. N. Bonamico, G. Dessy, V. Fares and L. Scaramuzza, *J. Chem. Soc., Dalton Trans.*, 1974, 1258.
4. M. Bonamico, G. Dessy, V. Fares, P. Porta and L. Scaramuzza, *Chem. Commun.*, 1971, 365.
5. R. D. Bereman and J. R. Dorfman, *Polyhedron*, 1983, **2**, 1013.

Table 34 Oxovanadium(IV) Dithiocarbamate Complexes [VO(dtc)$_2$]

	R in RCSS$^-$, or R′, R″ in R′R″NCSS	*ν(V=O)* (cm^{-1})	$\mu_{eff}^{r.t.}$ (BM)	*Ref.*
1	Me	982	1.69	1, 2
2	Et	984	1.73	1–4
3	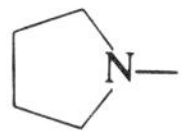	992	1.74	1, 2, 5
4	Pr[i]	995	1.77	2, 5
5	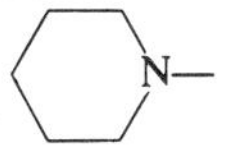	985	[1.56]	3, 6, 7
6	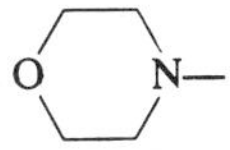	985	—	7
7	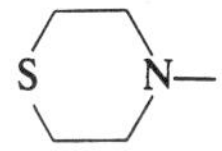	945	[1.75]	7
8		965	—	7
9	Et, Ph	970	—	3, 6
10	Me, Cy	—	—	3, 6
11	Cy, Cy	968	—	3, 6
12	H, cyclopentyl	940	1.70–1.72	8
13	H, cycloheptyl	942	1.70–1.72	8
14	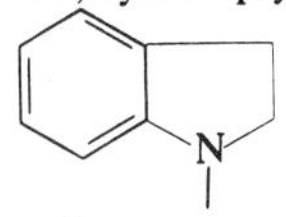	~985	—	8
15	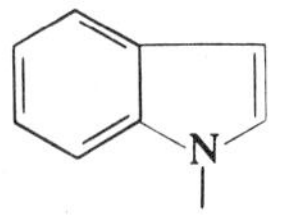	~985	—	5
16	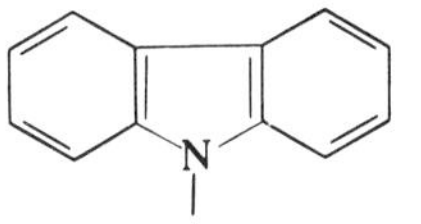	~985	—	5

1. B. J. McCormick, *Inorg. Nucl. Chem. Lett.*, 1967, **3**, 293.
2. B. J. McCormick, *Inorg. Chem.*, 1968, **7**, 1965.
3. G. Vigee and J. Selbin, *J. Inorg. Nucl. Chem.*, 1969, **31**, 3187.
4. D. C. Bradley, I. F. Rendall and K. D. Sales, *J. Chem. Soc., Dalton Trans.*, 1973, 2228.
5. R. D. Bereman and D. Nalewajek, *J. Inorg. Nucl. Chem.*, 1978, **40**, 1313.
6. J. Selbin, *Coord. Chem.*, Proc. John C. Bailar, Jr., Symp., 1969, 248.
7. F. Forghieri, G. Graziosi, C. Petri and G. Tosi, *Transition Met. Chem.* (*Weinheim, Ger.*), 1983, **8**, 372.

with a square pyramidal structure for other $[VO(dtc)_2]$ complexes. The structures are undoubtedly similar to $[VO(acac)_2]$ (Figure 22) and so it is not surprising that adducts form in solution[668,669] and in the solid state.[670] The tendency of $[VO(dtc)_2]$ to bind a sixth ligand is lower than for $[VO(acac)_2]$ and no correlation between $\nu(V{=}O)$ and the ligands seems to exist.[669] The studies are more difficult than with $[VO(acac)_2]$ owing to the sensitivity to oxygen.

Although some $[V(dtc)_4]$ complexes have been isolated,[671] oxovanadium(IV) compounds are usually obtained by the direct reaction of oxovanadium(IV) salts with dtc bases. Several $[V(dtc)_4]$ complexes have also been prepared by insertion of CS_2 into V—N bonds of vanadium tetraamides (equation 47).[419,672]

$$V(NR_2)_4 + 4CS_2 \xrightarrow[\text{(oxygen-free conditions)}]{\text{cyclohexane}} V(R_2NCS_2)_4 \tag{47}$$

Table 35 shows some of the $[V(dtc)_4]$ complexes prepared. Compounds 2–7 in Table 35 were prepared from $VOSO_4$ and in some cases in the presence of water. This therefore is one more example where the vanadyl bond is broken; others include $SOCl_2$ (and $SOBr_2$), catechols and dithiocarboxylates (these forming eight-coordinate complexes). However, $[V(dtc)_4]$ complexes are not stable towards conversion back to vanadyl. Thus, these ligands seem to be intermediate in behaviour between dithiocarboxylates and other dithiocarbamates; this probably reflects π-bonding capabilities.[671] ESR spectra of compounds 4–7 (Table 35) show well-resolved hyperfine structure and the rather low $A_{\parallel}$ and $A_{\perp}$ values ($\sim$117 and $\sim 37 \times 10^{-4}\ cm^{-1}$) indicate both a significantly covalent system and an eight-coordinate geometry.[671] The ESR spectra of $[VO(dtc)_2]$ complexes with the same ligands show much higher $A_{\parallel}$ and $A_{\perp}$ values ($\sim$151 and $\sim 49 \times 10^{-4}\ cm^{-1}$).[673] In addition to the reaction of $[V(Et_2NCS_2)_4]$ with oxygen producing vanadium(V) (equation 48), two other reactions may occur upon heating (equations 49 and 50).[672]

$$2V(Et_2NCS_2)_4 + O_2 \longrightarrow 2VO(Et_2NCS_2)_3 + (EtNCS_2)_2 \tag{48}$$

$$V(Et_2NCS_2)_4 \xrightarrow{\Delta} V(Et_2NCS_2)_3 + \tfrac{1}{2}(Et_2NCS_2)_2 \tag{49}$$

$$VO(Et_2NCS_2)_3 \xrightarrow{\Delta} VO(Et_2NCS_2)_2 + \tfrac{1}{2}(Et_2NCS_2)_2 \tag{50}$$

Table 35 Some [V(dithiocarbamato)$_4$] Complexes

	R in RCSS	*g isotropic*	*Colour*	$\mu_{eff}^{r.t.}$ (BM)	*Solvent*	*Ref.*
1	Me_2N	—	Brown	1.72	c	1
2	Et_2N	1.975[a]			C_6H_6	2, 3
3	$(PhCH_2)_2N$	—	Red-brown	1.68	$CHCl_3/C_6H_6/H_2O$	4
4	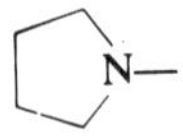	1.973,[a] 1.976[b]	Red-brown	—	$MeCN/H_2O$	5
5	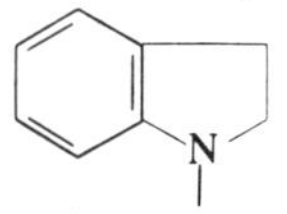	1.979[a], 1.976[b]	Red-brown	—	$MeCN/H_2O$	5
6	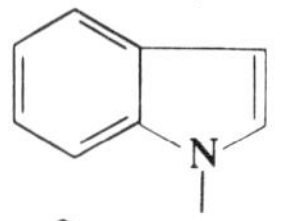	1.976,[a], 1.979[b]	Red-brown	—	$MeCN/H_2O$	5
7	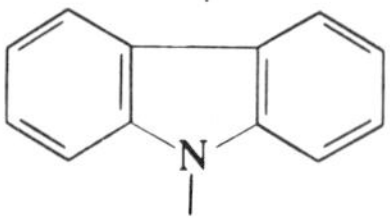	1.975,[a] 1.980[b]	Red-brown	—	$MeCN/H_2O$	5

[a] In toluene solution. [b] In CH_2Cl_2 solution. [c] CS_2 insertion reaction into $[V(NMe)_4]$.

1. E. C. Alyea and D. C. Bradley, *J. Chem. Soc. (A)*, 1969, 2330.
2. D. C. Bradley, I. F. Rendell and K. D. Sales, *J. Chem. Soc., Dalton Trans.*, 1973, 2228.
3. M. B. Dehkordy, B. Crociani, M. Nicolini and R. L. Richards, *J. Organomet. Chem.*, 1979, **181**, 69.
4. G. Soundararajan and M. Subbaiyan, *Indian J. Chem., Sect. A*, 1983, **22**, 454.
5. R. D. Bereman and D. Nalewajek, *J. Inorg. Nucl. Chem.*, 1978, **40**, 1309.

Treatment of tris(tricaprylylmethylammonium)tetrathiovanadate(V) solution with solid tetraisobutylthiuram causes an immediate change from violet to brown. Yellow-brown crystals were obtained and characterization showed that a dimer (**63**) formed *via* induced internal redox (equation 51).[674]

$$2VS_4^{3-} + 2(Bu^i_2NCS_2)_2 \longrightarrow \underset{(\mathbf{63})}{V_2(\mu\text{-}S_2)_2(Bu^i_2NCS_2)_4} + S_x^{2-} \quad (51)$$

(**63**) consists of discrete dimers where the V atoms are bridged by two symmetry-related μ-η^2-S_2 ligands, forming an $M_2(\mu\text{-}\eta^2\text{-}S_2)_2$ core, the octahedral coordination of vanadium being completed by two bidentate dithiocarbamates.[673] d(V—S) = 2.405 (av) and d(V—V) = 2.851 Å;[674] the V—V distance and the observed diamagnetism are consistent with a V—V bond.

(*iii*) *Dithiophosphate and related ligands*

ESR spectra of complexes of dithiophosphate (dtpo, $P(OR)_2SS^-$), dithiophosphinate (dtpi, PR_2SS^-), $PR'(OR)SS^-$ and dithioarsinate are of particular interest since they exhibit appreciable phosphorus-31 superhyperfine splitting which arises from V—P interactions over distances of about 3 Å or more. For example, in solutions containing (**64**), phosphorus causes the vanadium hyperfine lines to be further split into three components, intensitities 1:2:1. A total of 24 lines are predicted; however, owing to overlap, only 17 were observed.[675,676] With the cyclohexanol (R′O) ethyl (R) derivative, 23 of the expected lines are resolved whereas only 17 are for the corresponding cyclohexanol phenyl complex (**65**).[677] The superhyperfine splittings have been attributed to direct transannular V–P interactions involving mainly the $3d_{x^2-y^2}$ vanadium orbital and 3s (and 3p) phosphorus orbitals.[676,677] Wasson and co-workers[677] studied $[VO\{S_2PR(OR')\}_2]$ and discussed $[VO(dtpo)_2]$ and $[VO(dtpi)_2]$.[678,679] The general trend of the phosphorus superhyperfine splitting constants is $S_2P(OR)_2 > (S_2PR(OR') > S_2PR_2$. The ESR of $[VO(S_2PCy_2)_2]$,[676] $[VO(dtc)_2]$, $[VO(dtpi)_2]$ and $[VO(dtc)(dtpo)]$ have been reported,[680,681] as has coordination of py, DMF and hexamethylphosphoramide to five-coordinate $[VO(S_2PMe_2)]$, $[VO(S_2PPh_2)]$ and $[VO\{S_2P(OEt)_2\}]$,[682] using hyperfine interactions between ^{51}V and ^{31}P.

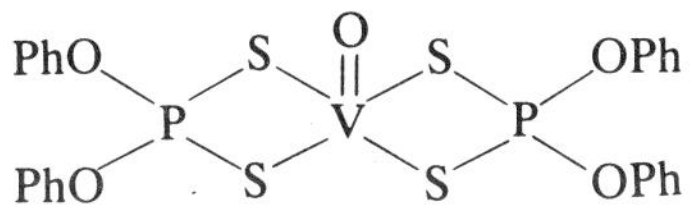

(**64**)

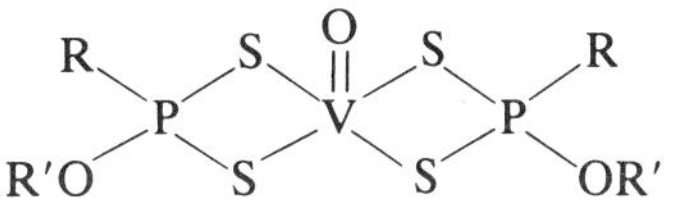

(**65**)

Most solid compounds (see Table 36) were prepared with $VOSO_4$ as starting material ($VOCl_3$ for the fluoro complex) and can be kept indefinitely under nitrogen or vacuum. The ethoxy, fluoro and trifluoromethyl complexes are rapidly decomposed by air. $[VO(S_2PMe_2)_2]$ consists of monomeric square pyramidal chelate complexes.[679] However, the clearly different properties of the F- and Me-substituted complexes, with anomalously high $\mu_{eff}^{r.t.}$ (~2.0 BM) and ν(V═O) at 860–870 cm^{-1}, suggest V···V interactions. In solution, the 860–870 cm^{-1} bands disappear and $\mu_{eff}^{r.t.}$ are close to 1.75 BM, indicating absence of magnetic exchange. The inverse dependence of susceptibility on field strength (H), and the slight increase of μ_{eff} with decrease in temperature suggest ferromagnetic interaction.[358] However, molecular weights suggest a monomeric nature and no half-field ESR spectrum was observed.

The compound $[VO(dtas)_2]$,[683] prepared by the reaction of $VOCl_2$ with dimethyldithioarsinate (dtas), has ν(V═O) = 986 cm^{-1} and the electronic spectrum is similar to those for $[VO(dtc)_2]$ and $[VO(dtpi)_2]$ complexes. Its isotropic ESR spectrum shows at least 27 discernible lines, from interaction with at least one ^{75}As atom.

$[VO(dpd)_2]$ (Hdpd = $(EtO)_2P(O)NHCSSH$) has been reported.[684]

(*iv*) *Xanthates*

Xanthato compounds have been prepared by the addition of aqueous $VOSO_4$ to a solution of the ligand at 0–5 °C;[685] green products precipitate and were recrystallized from CH_2Cl_2, $CHCl_3$ or n-hexane (or mixtures). Some properties are in Table 37. They were studied by electronic

Table 36 Some [VO(dithiophosphinate)$_4$] and [VO(dithiophosphate)$_4$] Complexes

R in [$VO(S_2PR_2)_4$]	$\mu_{eff}^{r.t.}$ (BM)[a]	θ(K) *from Curie–Weiss law*[a]	$\nu(V{=}O)$[b] (cm^{-1})	*Ref.*
EtO	1.66	0.7	960 s	1
Me	1.72	−1.5	991 s	1
Ph	1.76	−5.1	998 s	1
F	2.25	17	1025 w, br, 860 s	1
CF_3	2.13	10	1020 w, br, 870 ms	1
Cy	—	—	ν(V—S) = 325 (?)	2

[a] With no TIP corrections. [b] ν(V=O) is indicated unless otherwise stated.
1. R. G. Cavell, E. D. Day, W. Byers and P. M. Watkins, *Inorg. Chem.*, 1972, **11**, 1591.
2. H. J. Stoklosa, G. L. Seebach and J. R. Wasson, *J. Phys. Chem.*, 1974, **78**, 962.

spectroscopy, thermogravimetry and differential thermogravimetry.[685] [VO(Pr^{i}OCSS)$_2$] with pyridine and piperidine produces the last compounds in Table 37, which are diamagnetic and have no visible band. These vanadium(V) compounds result from oxidation during the long reflux times used to prepare them. In fact, in most reactions with long reflux times, oxidation can be avoided only by taking extreme care to avoid O_2 participation, or in the presence of reducing agents.[686,687]

Table 37 Some [VO(xanthate)$_2$] Compounds

Complex	$\nu(V{=}O)$ (cm^{-1})	$\nu(V{-}S)$ (cm^{-1})	$\mu_{eff}^{r.t.}$ (BM)
[VO(EtOCSS)$_2$]	1020, 990 sh	370	1.62
[VO(Pr^{i}OCSS)$_2$]	1010, 980 sh	385	1.68
[VO(Bu^{i}OCSS)$_2$]	970, 930 sh	370	1.66
[VO(i-C_5H_{11}OCSS)$_2$]	1030, 970 sh	390	1.67
[VO(C_7H_7OCSS)$_2$]	980	340	1.63
[VO(Pr^{i}OCSS)(py)$_2$]·5H_2O	950	380	Diamagnetic
[VO(Pr^{i}OCSS)(pip)$_2$]·5H_2O	1000	380	Diamagnetic

[a] pip = piperidine.
A. D. Villarejo, A. Doadrio, R. Lozano and V. Ragel, *An. Quim.*, 1981, **77**, 50.

33.5.6.3 Dithiolates

By mixing stoichiometric amounts of $VOSO_4 \cdot 5H_2O$ and $NaS_2CCN \cdot 3DMF$ at 20–50 °C for ~1.5 h an olive green maleonitrile dithiolate (mnt) $(Ph_3PMe)_2$[VO(mnt)$_2$] was obtained.[688] If heated for 3–5 h at about 80 °C, deoxygenation occurs and $[V(mnt)_3]^{2-}$ forms. The vanadyl complex is monomeric and $\nu(V{=}O) = 963\ cm^{-1}$. In M(1,2-dithiolate)$_3$, the more oxidized species usually have a trigonal prismatic coordination, whereas highly reduced species possess near octahedral structures.[49] A detailed ESR study of magnetically dilute oriented crystals (with diamagnetic hosts, $(Ph_4P)_2$[Mo(mnt)$_3$] for $[V(mnt)_3]^{2-}$ and $(MePPh_3)_2[TiO(C_2O_4)_2]$ for $[VO(mnt)_2]^{2-}$) led for the vanadyl complex to a scheme $d_{x^2-y^2} < d_{xz},\ d_{yz} < d_{xy} < d_{z^2}$ ($d^1_{x^2-y^2}$ ground state).[689] Kwik and Stiefel[49] also studied the ESR spectrum of $[V(mnt)_3]^{2-}$ doped in crystals of $(Ph_4As)_2$[Mo(mnt)$_3$]. While the experimental results agree, the interpretations differ. A third assignment differs again.[690]

(Ct)$_2$[V(mnt)$_3$] (Ct = Me_4N^+, Et_4N^+, Bu_4N^+, Ph_4N^+, Ph_3PMe^+, Ph_4As^+) have been prepared from $VOCl_3$,[52,688,691] and in $(Me_4N)_2$[V(mnt)$_3$] the six sulfur atoms are around the vanadium at an average distance of 2.36 ± 0.01 Å (Figure 29a), in distorted D_{3h} symmetry.[52]

Reaction of [VO(acac)$_2$] ethanolic solutions containing an excess of sodium ethane-1,2-dithiolate (Na_2edt) under inert atmosphere yields deep green solutions. NMe_4Cl slowly deposits green prisms of (NMe_4)Na[VO(edt)$_2$]·2EtOH (Figure 29b);[656,692] the vanadium is square pyramidal with $d(V{=}O) = 1.625$ Å and $d(V{-}S) = 2.378$ Å (av). The V atom lies 0.668 Å above the sulfurs. (NMe_4)Na[VO(edt)$_2$] in DMSO exhibits rather similar $g_\parallel$ and $g_\perp$ (1.981 and 1.984) and relatively small values of $A_\parallel$ and $A_\perp$ (133 and $42 \times 10^{-4}\ cm^{-1}$) when compared with [VO(acac)$_2$] for example. This suggests significant π interactions between the sulfur lone pairs and the metal d orbitals.[692] $[VO(pdt)_2]^{2-}$ (pdt = propane-1,3-dithiolate) is similar to $[VO(edt)_2]^{2-}$.[656]

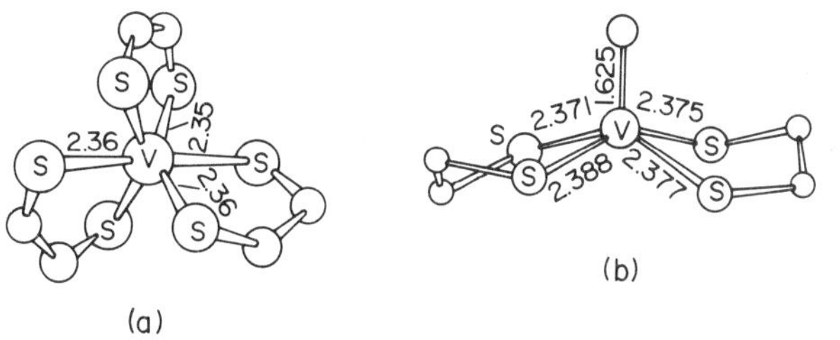

Figure 29 Molecular structure of (a) the $[V(mnt)]_3^{2-}$ anion[52] (the cyano groups are not included) and (b) the $[VO(SCH_2CH_2S)_2^{2-}]$ anion[692]

33.5.6.4 Other S-containing ligands

Early publications include the vapor-phase reaction of VCl_4 with thiophene to yield $VCl_4{\cdot}C_4H_4S$.[693] Reduction of $VOCl_3$ by sulfides has long been known and fairly pure $[VOCl_2L_2]$ (L = Me_2S and Et_2S) complexes have been obtained.[399]

$[VOCl_2(tmtu)_2]$ (tmtu = tetramethylthiourea) had $\nu(V{=}O) = 990\ cm^{-1}$.[369] In $[VOF_2(tu)_2]$ (tu = thiourea), prepared by the method in equation (22), the IR indicates a V—S bond[428] in contrast with $[VO(NCS)_2L_2]$ (L = tu or *N*-substituted thioureas), with nitrogen bonding (see also Table 12).

33.5.7 Selenium Ligands

Early reports include 1:1 $[VCl_4]$ adducts with $MeSe(CH_2)_3SeMe$ and Pr^iSeMe, purple-grey decomposing at 114 and 130 °C.[694]

From iterative extended Hückel molecular orbital (EHMO) calculations, $[VX]^{n+}$ (X = S, Se; $n = 0, 1, 2$) ions and their tetrachloro complexes $[VXCl_4]^{m-}$ ($m = 4, 3, 2$) decrease in stability O > S > Se and with increasing charge of the anion.[695] The $[VXCl_4]^{2-}$ and $[VXCl_4]^{4-}$ complexes are expected to be $S = \frac{1}{2}$ systems while $[VXCl_4]^{3-}$ species are expected to be diamagnetic.[695]

33.5.8 Halogen Ligands

Only complexes $[VOX_n]^{2-n}$ and $[VX_m]^{4-m}$ will be included. Gmelins Handbuch discusses work in this area up to 1967.[357a]

33.5.8.1 Oxovanadium(IV) Complexes

Stability of the VO^{2+} complexes will decrease $F^- \gg Cl^- > Br^- > I^-$, with complexes formed by the three heavier halides being weak. Fluoride and chloride have been studied in aqueous solutions.[696–698] For fluoride, the following stability constants have been determined (at 20 °C and $I = 1$ M $NaClO_4$): $\beta_1 = 2.0 \times 10^3$, $\beta_2 = 3.7 \times 10^5$, $\beta_3 = 1.3 \times 10^7$ and $\beta_4 = 6 \times 10^7$,[698] much stronger than chloride ($\beta_1 \approx 1\ M^{-1}$). The fluoride complexes are more stable than those of Cu^{2+}, Zn^{2+} and Ni^{2+} but are weaker than the UO_2^{2+} complexes.[697] VOF_2 is known as well as some hydrated forms $VOF_2{\cdot}xH_2O$.[355,428] Fluoro complexes have been mainly studied with alkali and ammonium ions[356] and stoichiometries $[VOF_3]^-$, $[VOF_4]^{2-}$, $[VOF_5]^{3-}$, $[V_2O_2F_7]^{3-}$ and $[V_2O_2F_6(H_2O)_2]$ have been reported.[699,700] The basic structural units are $[VOF_5]^{3-}$ octahedra, found isolated in $(Ct)_3[VOF_5]$[701,702] and $[M(NH_3)_6][VOF_5]$,[699] connected in infinite chains linked by *cis*-bridging F atoms in $K_2[VOF_4]$[703] and $(NH_4)_2[VOF_4]$[704] or to complex chains in $\{Cs[VOF_3]\}_2{\cdot}H_2O$,[705] where two VOF_5 octahedra are linked *via* an F—F edge; the V—F distances span the range 1.881–2.205 Å and V=O (1.583 and 1.595 Å) are shorter than in $[VOF_4(H_2O)]^{2-}$ ($d(V{=}O) = 1.602$ and $d(V{-}O(\textit{trans})) = 2.268$ Å).[706] In $Cs_3[V_2O_2F_7]$, the anion consists of two octahedra sharing a face.[707] In $(NMe_4)_2[V_2O_2F_6(H_2O)_2]$, dimeric units of $[V_2O_2F_6(H_2O)_2]^{2-}$ are linked in chains by short hydrogen bonds, V—F distances ranging from 1.894 to 2.173 Å.[708] $d(V{-}V) = 3.292$ and $d(V{-}OH_2) = 2.074$ Å. Blue crystalline

$(N_2H_5)[VOF_3]$ has been synthesized from V_2O_5 and an excess of hydrazine hydrate in aqueous HF.[700] The hydrazine hydrate not only acts as a reducing agent but also provides the $N_2H_5^+$ cation. The Ct(Na^+, K^+ and NH_4^+) salts were prepared by metathesis between $(N_2H_5)[VOF_3]$ and CtF. Their $\mu_{eff}^{r.t.}$ are in the range of 1.51–1.53 BM.[700] Spin-Hamiltonian parameters are in Table 38.

Table 38 Spin-Hamiltonian Parameters for Vanadyl Complexes

Complex	$g_{\parallel}$	$g_{\perp}$	$A_{\parallel}$ ($\times 10^{-4}$ cm^{-1})	$A_{\perp}$	*Ref.*
$[VO(H_2O)_5]^{2+}$	1.993	1.981	182.8	72.0	1
$[VOF_3]^{-}$ a	1.937	1.978	177.2	68.5	2
$[VOF_4]^{2-}$ b	1.932	1.973	182.0	66.7	3
$[VOF_5]^{3-}$ c	1.937	1.977	178.5	64.0	4
$[VOCl_4]^{2-}$ d	1.948	1.979	168.8	62.8	5
$[VOCl_5]^{3-}$ e	1.945	1.985	173.0	63.8	6
$[VOCl_4]^{2-}$ f	1.959	1.984	172.0	70.7	7

a Aqueous solutions of $(N_2H_5)[VOF_3]$ at 100 K.
b $[VOF_4]^{2-}$ in single crystals of $(NH_4)_2SbF_5$.
c $[VOF_5]^{3-}$ in single crystals of $(NH_4)_3AlF_6$.
d $[VOCl_4]^{2-}$ in single crystals of $(NH_4)_2SbCl_5$.
e $[VOCl_5]^{3-}$ (or $[VO(Cl_4)(H_2O)]^{2-}$?) in single crystals of $(NH_4)_2[In(Cl_5)(H_2O)]$.
f Ethanenitrile solution of $(Ph_4As)_2[VOCl_4]$ at 138 K.

1. C. J. Ballhausen and H. B. Gray, *Inorg. Chem.*, 1962, **1,** 111.
2. M. K. Chandhuri, S. K. Ghosh and J. Subramanian, *Inorg. Chem.*, 1984, **23,** 4439.
3. K. K. Sunil and M. T. Rogers, *Inorg. Chem.*, 1981, **20,** 3283.
4. P. T. Manoharan and M. T. Rogers, *J. Chem. Phys.*, 1968, **49,** 3912.
5. J. M. Flowers, J. C. Hempel, W. E. Hatfield and D. H. Dearman, *J. Chem. Phys.*, 1973, **58,** 1479.
6. K. DeArmond, B. B. Garrett and H. S. Gutowsky, *J. Chem. Phys.*, 1965, **42,** 1019.
7. K. R. Seddon, Report 1980, EOARD-TR-80-13 (from *Gov. Rep. Announce. Index (U.S.)*, 1980, **80** (26), 5628; *Chem. Abstr.*, 1981, **94,** 112 126t).

Low ν(V=O) values are expected for fluoride complexes. In fact, in $(NMe_4)_2[V_2O_2F_6(H_2O)_2]$ the ν(V—O) was at 968 cm^{-1} (980 cm^{-1} Raman); in $(NH_4)_3[VOF_5]$ it was at 937 and 947 cm^{-1},[388] in $Ni[VOF_4]\cdot 7H_2O$ at 950 cm^{-1},[709] in (Ct)$[VOF_3]$ (Ct = $N_2H_5^+$, NH_4^+, Na^+, K^+) at 970–980 cm^{-1},[700] while in $(M(NH_3)_6)[VOF_5]$ (M = Cr, Co or Rh) ν(V=O) was at 907–913 cm^{-1}.[699] ν(V=O) in hexamminemetal(III) pentafluorooxovanadates(IV) reflected multiple hydrogen bonds.[699]

Simple $(Ct)_2[VOCl_4]$ are green (hydrates may have other colours) and are extremely air and moisture sensitive, due to the ease with which V—Cl bonds are hydrolyzed. In many $(Ct)_2[VOCl_4]$, $[VOCl_4]^{2-}$ probably has a square pyramidal structure as they present only one strong IR-active band (ν(V—Cl) = 340–350 cm^{-1})[409] (Sections 33.5.3.1 and 33.5.3.2) and the ESR spectra of $[VOCl_4]^{2-}$ and $[VOCl_5]^{3-}$ have been interpreted according to this (Table 38). (Ct)$[VOCl_3]$ have also been reported.[712]

Oxobromovanadates(IV) were first characterized in 1973, of the types (Ct)$[VOBr_3]$, $(Ct)_2[VOBr_4]$ and $(Ct)_3[VOBr_5]$ as well as several $[VOBr_2]$ adducts.[400,401] Most $[VOBr_2]$ adducts were included in a special report.[711] Normally, the oxobromovanadates(IV) are prepared by reduction of $VOBr_3$ by bromides in acetonitrile and/or nitromethane. When V^V is reduced by Br^-, free bromine is liberated and unless conditions are carefully regulated, perbromides form, giving preparative difficulties.[400] Generally, ν(V—F) have been assigned in the region 460–520 cm^{-1},[699,700,708] ν(V—Cl) bands in the region 340–400 cm^{-1} [409] and ν(V—Br) in the region 320–360 cm^{-1}.[400,711] The adducts reveal the trend for ν(V=O): anionic > five-coordinate > six-coordinate > bidentate.[711] The far IR of $(NEt_4)_2[VOBr_4]$ shows a complete and reversible transformation within the temperature range −30 to 35 °C, due to a reversible equilibrium (equation 52) between two forms of $[VOBr_4]^{2-}$, a C_{4v} form being stable at room temperature with ν(V—Br) at 296 cm^{-1}, a C_{2v} form being stable at lower temperatures with two ν(V—Br) bands at 274 and 325 cm^{-1}.[400] A similar transformation has been suggested for $[VOF_4]^{-}$ [713] as well as for $[VOCl_4]^{2-}$ (Section 33.5.3.2).

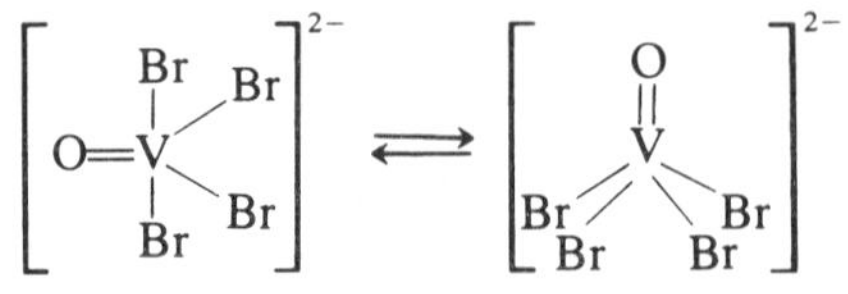

(52)

C_{2v} (−30 °C) $\quad$ C_{4v} (35 °C)

Anhydrous VOI_2 has not yet been prepared and most reports of iodides of V^{IV} are not firmly based. I^- can reduce vanadium(IV).

33.5.8.2 Vanadium(IV) complexes

Potassium hexafluorovanadate(IV) can be prepared from $[VF_4]$ and KF in selenium tetrafluoride(III)[416] or by fluorination of K_2VF_5 for example. K^+, Rb^+ and Cs^+ salts obey the Curie–Weiss law with high values of θ. $[VF_4]$ reacts with py and NH_3 forming $[VF_4L]$, probably fluorine-bridged polymers.[416]

Chloro complexes of V^{IV} have been reported ($[VCl_5]^-$, $[VCl_6]^{2-}$, $[VCl_7]^{3-}$ and $[VCl_8]^{4-}$),[714] but the last two at least need to be confirmed. In the reaction of PCl_5 with $VOCl_3$ in hot $POCl_3$ or CH_2Cl_2 and of $[VOCl_2]$ with PCl_5 in $AsCl_3$ or $POCl_3$, $(PCl_4^+)[VCl_5]^-$ may be precipitated.[714] The anion is trigonal bipyramidal and exhibits weak, presumably *d–d* transitions at 6200, 8100 and 16 000 cm^{-1}.[715] By the reaction of $[VCl_4]$ or its ethyl cyanide adduct with alkylammonium chloride, $(Ct)_2[VCl_6]$ ($Ct^+ = Et_3NH^+$ and $Et_2NH_2^+$) may be isolated.[716] Other salts were prepared by the reaction of $SOCl_2$ with $(Ct)_2[VOCl_4]$.[717] $[VCl_4]$ may form adducts with oxygen, sulfur, selenium, nitrogen, phosphorus and arsenic donors.[6,718,720] With unidentate ligands, 1:2 complexes are normal (*e.g.* with $S(CN)_2$ or $Se(CN)_2$).[719] Thermal decomposition often results in V^{III} compounds.[6] With bidentate ligands, 1:1 complexes normally form.[6] Besides additions, another method has been used. [VO(chelate)$_2$] complexes react with $SOCl_2$, PCl_5 and $SOBr_2$ producing [VX_2(chelate)$_2$] (X = Cl or Br) hexacoordinate V^{IV} complexes[428,555,721] where the two halide ligands may add either *cis* or *trans* (*cf.* Section 33.5.5.4.ii, where the chelate is acac, or to other sections according to the chelate ligand involved, *e.g.* salen, oxine, *etc.*).

33.5.9 Mixed Donor Atom Ligands

33.5.9.1 Schiff base ligands

No general review has been published since 1967.[357b] Reviews by Syamal[358] and others[722] cover only part of the published work on oxovanadium(IV) Schiff base complexes.

(i) Vanadium(IV) complexes with Schiff bases derived from the reaction of diamines and salicylaldehyde and derivatives

The complexes with Schiff bases (SBs) derived from diamines and salicylaldehydes (**66**) or β-diketones (**67**) normally have a 1:1 stoichiometry.

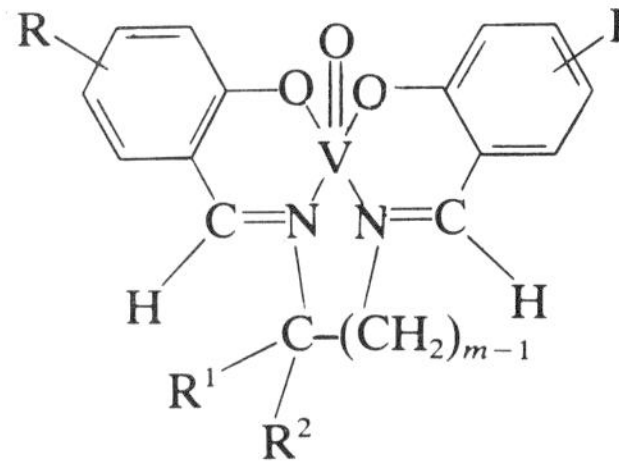

(**66**)

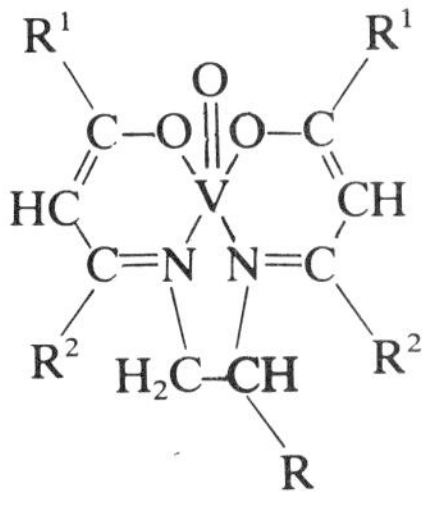

(**67**)

[VO(salnpn)] (**66**; $R^1 = R^2 = R = H$; $n = 3$), prepared simply by mixing vanadyl salts and the SB, even in the presence of DMSO or py, as sparingly soluble yellow-orange needles, consists of molecules packed so that the vanadyl oxygen of one occupies the sixth position of the V in a neighbour.[723] The result is an infinite chain of molecules about a two-fold screw axis (Figure 30). d(V=O) (1.633 Å) is one of the longest reported to date, consistent with ν(V=O) = 854 cm^{-1}, and the bridging character of the vanadyl oxygen.

The magnetic moment was 1.78 BM at 295 K and it obeys the Curie–Weiss law over the range 95–295 K with $\theta = -7$ K.[599b] Other authors conclude that not only do magnetic spin–spin

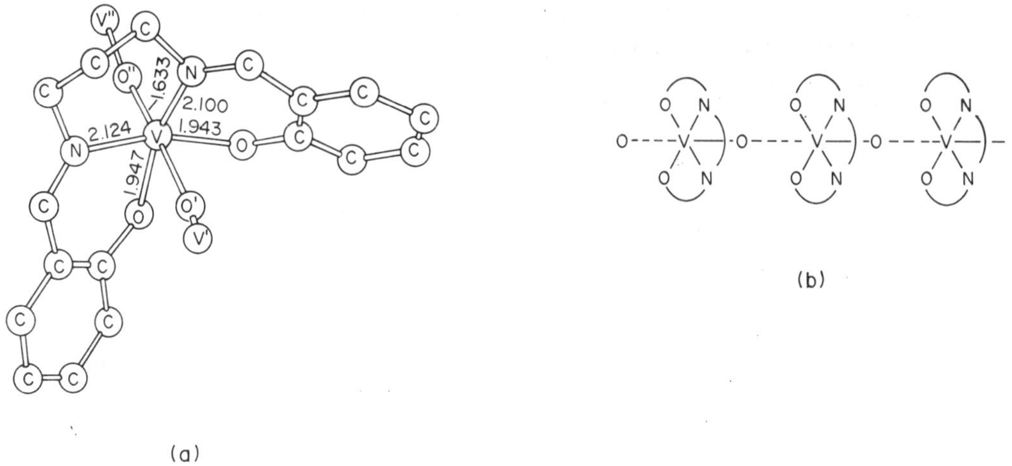

Figure 30 (a) A view of [VO(salnpn)] molecule with some internuclear distances (in Å);[723] (b) schematic representation of the chain structure

interactions exist in this compound but also that a small ferromagnetic interaction is indeed found.

Because the vanadyl ion is lopsided, its chelates have the possibility of forming a greater number of isomers than monatomic ions. For example, for an N,O-coordinating chelate (*e.g.* glycine), two enantiomers, (**68a**) and (**69a**), may form (S = solvent).

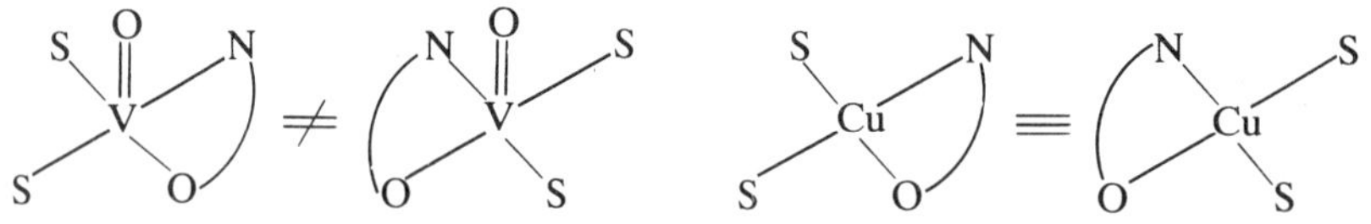

(**68a**) (**69a**) (**68b**) (**69b**)

In the case of dissymmetric SB (**66**), VO^{2+} has the possibility of forming twice the number of isomers for each *R* or *S* configuration of the diamine chain (Scheme 18). For example, in the case of [VO(sal(*S*)pn)], where sal(*S*)pn is the SB derived from Hsal and (*S*)-(+)-propane-1,2-diamine, the preferred conformation does not always depend on the absolute configuration of the diamine; structure (**70**) represents the favoured form since it involves the least steric interaction between the vanadyl oxygen and the (*S*)-pn methyl group.[724,725] The actual situation existing in solution is probably an equilibrium of the isomers with a predominance of the λ conformation since the barrier for ring interconversion from λ to δ is small.[724] In some complexes, as in [VO(sal(*SS*)chxn)] ((*SS*)chxn = (S,S)-(+)-cyclohexane-1,2-diamine), the ligand is stereospecifically coordinated so that the chelate ring is locked in a δ conformation.

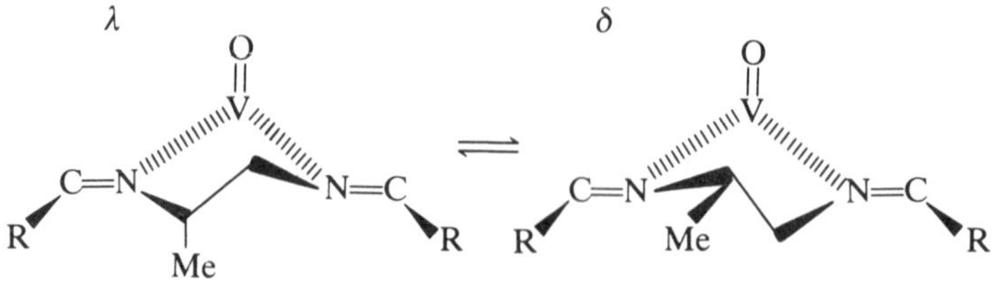

(**70**)

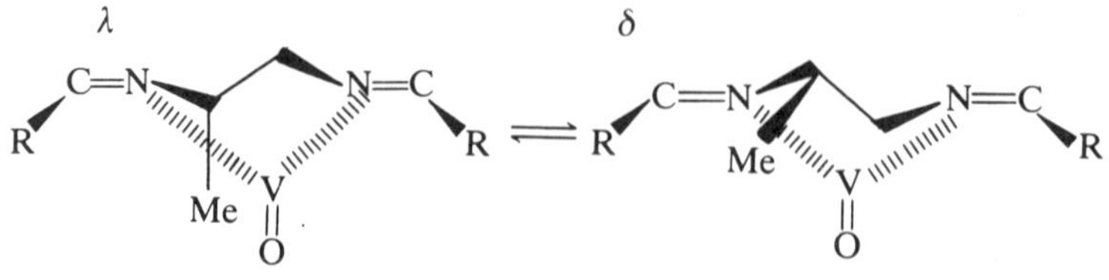

Scheme 18

Dissymmetry may occur if the ligand adopts a preferred flattened tetrahedral geometry about the metal ion. CD spectra of [VO(sal(*S*)pn)] and [VO(sal(*SS*)chxn)] exhibit evidence of Δ (**71**) and Λ (**72**) configurations.[724] From CD spectra, for complexes with tetradentate SBs,[726] the geometry around VO^{2+} has a slight pseudotetrahedral distortion.

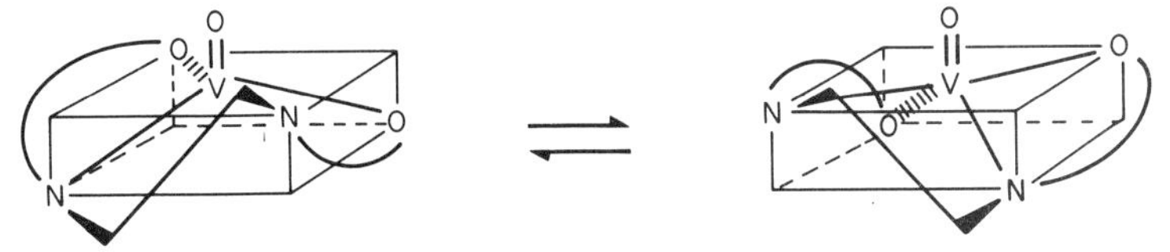

λ, Δ   δ, Λ

(71)   **(72)**

Many investigations have involved vanadyl complexes of type (**66**),[599a,723–726] particularly with 1,2-diamines. In the solid state $\mu_{eff}^{r.t.}$ values are normally near the spin-only value.[725,727]

Complexes of SB (**73**) are five-coordinate in solution and normally show no tendency to coordinate a basic ligand to form six-coordinate compounds.[725] The formation of adducts depends on the chelate ring, as in (**70**). With electronegative groups in the aromatic ring, *e.g.* 5-NO_2salen (**73**; $R^1 = R^2 = H$, $R = 5\text{-}NO_2$), electronic and CD spectra show hexacoordinate adducts, and solid adducts were obtained with py and DMSO.[725]

Expansion of the 'central chelate ring' in tetradentate (**66**) from $n = 2$ to $n = 3$ favours six-coordinate species, as [VO(salnpn)] (Figure 30).[726]

Complexes (**66**; R = H and $n = 2$) were five-coordinate square pyramidal monomers; in py, six-coordinate species form.[728] Compound (**66**; $R^1 = R^2 = R = H$), polymeric in the solid state (Figure 30), was considered to break down to five-coordinate monomers in $CHCl_3$ and six-coordinate in py. Changes in the spectrum of (**66**; $R^1 = R^2 = R = H$, $n = 9$) in $CHCl_3$ on dilution indicate that the complexes with $n \geq 4$ are not monomers. Compounds (**66**; R = 3-MeO) were five-coordinate monomers except when $R^1 = R^2$ and $n = 3$.[557] Compounds (**66**; R = 5-Cl) tend to be polymeric solids. Complexes with (**74**) appear monomeric in the solid state except when $n = 3$.[559]

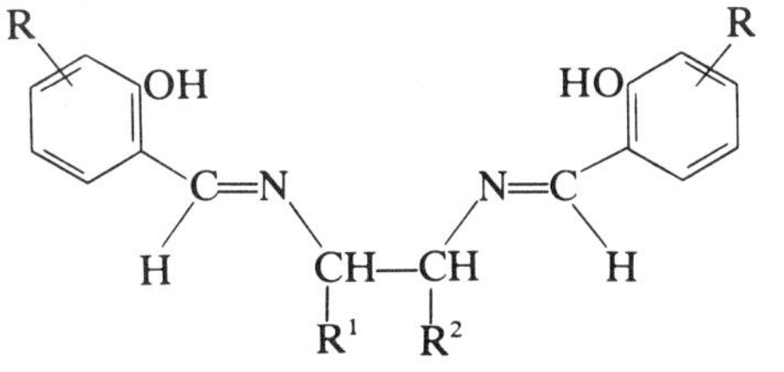

(73)

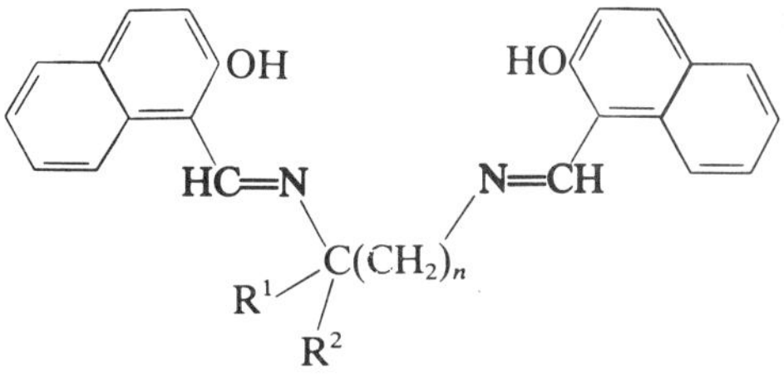

(74)

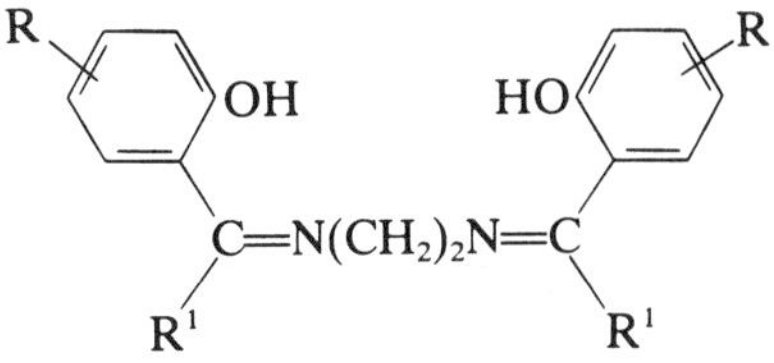

(75)

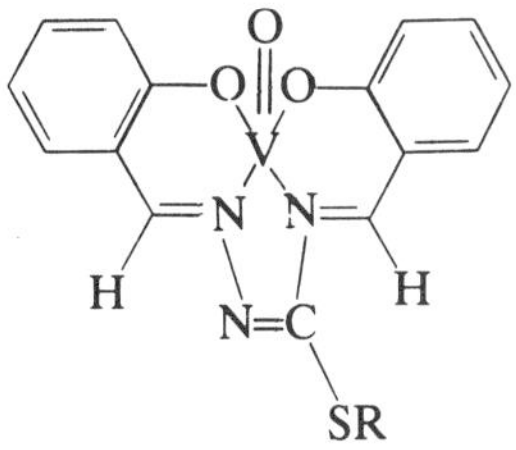

(76)

For (**66**; $R^1 = R^2 = H$, R = 3-MeO), the changes in the ordering of *d* orbitals as *n* varies are schematically represented in Figure 31. Complexes with (**75**; R′ = Me or Ph) have been reported.[729,730]

Alkylation of the thiocarbonyl sulfur of thiosemicarbazone derivatives induces not only complexation through the terminal amino group but also enough acidic character for it to

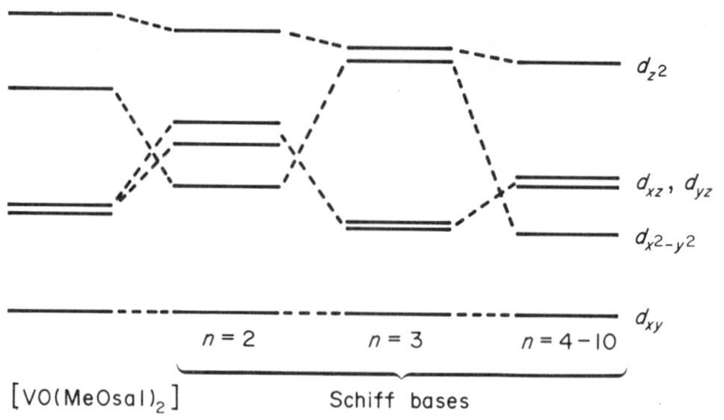

Figure 31 Changes in the ordering of the oxovanadium(IV) *d* orbitals proposed for complexes with Schiff bases (**66**) with $R^1 = R^2 = H$ and R = 5-MeO[557] (not to scale)

function as a monoacidic ligand.[731] In the presence of VO^{2+} (and Ni^{2+}, Cu^{2+}) salts, these ligands are capable of condensing at the terminal amino nitrogen atom through another aldehyde or ketone producing quadridentate ligands and forming (**76**).[732] In (**76**) and similar Ni^{2+} and Cu^{2+} complexes (R = H, NH_4, Na, K), the set of coordinating atoms and the geometry appear similar to those with (**73**).[733–735] The VO^{2+} compounds react with alkyl iodides producing complexes with R = Me, Et, Pr. Their electronic spectra in pyridine differ from the spectra in $CHCl_3$;[736] a decrease in intensity with parallel decrease of the ESR shows that VO^{2+} is oxidized to vanadium(V),[736] surprising compared with examples of V^V complexes being reduced to V^{IV}[385,686,687] and even to V^{III} compounds.[720] Other oxidations in the presence of py are in Table 37.

When [VO(salen)], only slightly soluble in THF, acetone and acetonitrile, is added to $NaBPh_4$ in these solvents, a green solution is produced from which the sodium cation complex (**77**) crystallizes as a light green solid, stable in air (equation 53).[737] [VO(saloph)] (saloph = *N,N'*-(*o*-phenylene)bis(salicylidenimine)) gives $([(VO(saloph)]_2Na)BPh_4$.[737] Compound (**77**) contains Na^+ trapped in a coordination cage provided by six oxygen atoms. In $([VO(salen)]_2Na)BPh_4$, all the oxygens of [VO(salen)] are involved in coordination to the Na^+, giving a cationic polymeric structure. The geometry around Na^+ is pseudo-octahedral; the six oxygens are provided by four different [VO(salen)] units: two vanadyl O at an average of 2.368 Å and two pairs of O from salen units at an average of 2.405 Å. Vanadium is five-coordinate square pyramidal and is 0.606 Å above the plane of the O_2N_2 atoms of the salen unit. The mean V=O is 1.584 Å.

$$Na^+BPh_4^- + 2VO(salen) \longrightarrow [\{VO(salen)\}_2Na]BPh_4 \quad (53)$$
(**77**)

Treatment of a basic aqueous solution of EHPG (EHPG = ethylenebis(*o*-hydroxyphenyl glycine), protonated and deprotonated forms will not be specified) with $VOSO_4$ afforded light blue $(NH_4)[VO(EHPG)]\cdot H_2O\cdot EtOH$;[738] the VO^{2+} is bonded to five atoms of the $EHPG^{3-}$ ligand with one amine nitrogen, one phenolate oxygen and two carboxylate oxygens coordinated equatorial and an amine nitrogen coordinated *trans* to the vanadyl oxygen. In contrast, $VOSO_4$ and EHPG in aqueous ethanol yield a neutral dark blue product characterized as the V^V complex $[V^VO(EHPG)]$,[738–741] stable as a solid. On allowing warm solutions to stand, green [VO(salen)] forms.[338,339] This corresponds to an oxidative decarboxylation in which two molecules of CO_2 are lost. The reaction, rapid in DMF, proceeds slowly in alcohols with $[V^VO(EHGS)]$ (EHGS = *N,N'*-ethylene((*o*-hydroxyphenyl)glycine)) as monodecarboxylated intermediate, isolated as the sodium salt $Na[VO(EHGS)]\cdot 1.5H_2O\cdot MeOH$, which may be prepared by exhaustive electrolysis of the V^V analogue. It is six-coordinate with the two nitrogens and phenolic oxygen binding equatorial and an oxygen of carboxylate *trans* to the vanadyl oxygen.[738,739] The final product of decarboxylation is formed by discrete molecules of [VO(salen)];[738,739] all aspects of the structure agree with those for $([VO(salen)]_2Na)BPh_4$ (**77**).[737] The electrochemistry of $[V^{IV}O(EHPG)]^-$ and *meso*-$[V^VO(EHPG)]$ has been studied by cyclic voltammetry.[741]

In $[VO(EHPG)]^-$, $[VO(EHGS)]^-$ and [VO(salen)], *v*(V=O) correlates well with axial (V=O) and equatorial bond lengths.[738]

Complexes with SBs of general formula (**78**) form by adding a $VOCl_2$ solution to a solution containing the appropriate derivative and substituted ethylenediamine, followed by the addition of $NaMeCO_2$ and heating under reflux for 10–60 min.[742] According to the nature of R and R′, two types may be obtained: (i) complexes [VO(XsalenN(R)R′)$_2$] in which the vanadium is linked to two SBs, and (ii) complexes [VO(Xsal)(XsalenN(R)R′)] in which the vanadium is linked to only one SB and to a salicylaldehyde molecule. When the R and R′ groups have an electron-releasing effect, the β nitrogen becomes a strong donor, the SB acts as a tridentate ligand, and six-coordinate mixed-ligand complexes of type (ii) form. However, while with Ni^{2+}, the bidentate ligand is an SB, with VO^{2+} it is a sal$^-$ ion; this occurs even if the reaction is carried out using stoichiometric amounts of preformed SB (**78**) and the vanadyl salt. Type (i) complexes are green, and monomeric, with structure (**79**); the pyridine readily adds to vanadium in solution.[742] In type (ii) complexes, the vanadium is coordinatively saturated and the electronic spectra in inert solvents and in py are very similar; this and other observations led to a proposed structure (**80**).[742]

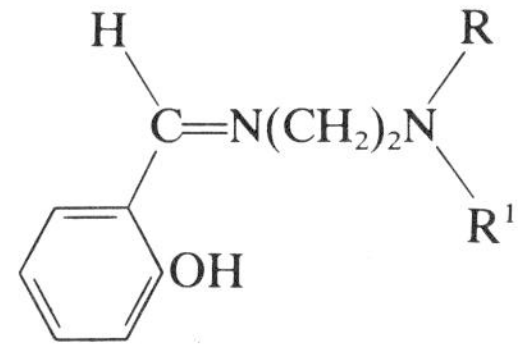

(**78**)

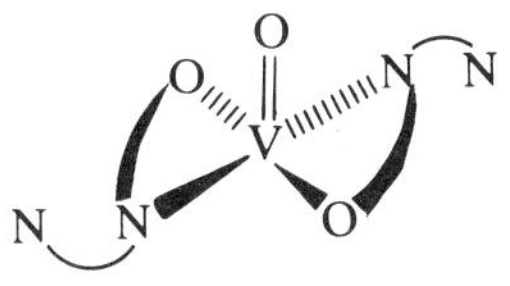

(**79**)

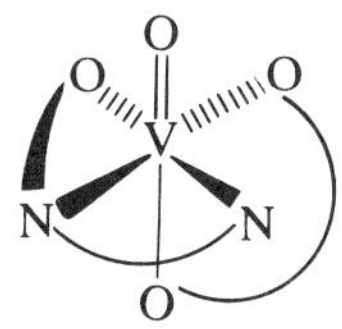

(**80**)

(ii) Vanadium(IV) complexes with Schiff bases derived from the reaction of diamines and β-diketones

Several complexes with Schiff bases derived from the reaction of β-diketones and diamines (**67**) have been prepared and many publications on them have appeared.[523,725,743–748] Almost all are monomeric (no V···V interactions) and five-coordinate square pyramidal, as in (**67**). This structure has been confirmed in [VO(acen)] (acen = **67**; $R^1 = R^2 = Me$, R = X = H).

Complexes (**67**) are often prepared by chelate exchange (equation 54), using [VO(acac)$_2$] as the starting product. In the case of H_2tfen (tfen = **67**; $R^1 = CF_3$, $R^2 = R = X = H$) this yields [VO(acen)], and [VO(tfen)] was obtained starting with [VO(tfacac)$_2$] instead of [VO(acac)$_2$].[745] The complexes may be recrystallized from benzene, $CHCl_3$, *etc.*, but, as in the case of (**66**), solvates are often obtained.

$$VO(acac)_2 + \beta\text{-diketone·diamine} \xrightarrow{\Delta} VO(\beta\text{-diketone·diamine}) + 2acac \quad (54)$$

The separation of the four possible isomers of [VO(acpn)] (**67**; $R^1 = R^2 = R = Me$, X = H) $S\lambda$, $S\delta$, $R\lambda$, $R\delta$ (compare with Scheme 18) was claimed.[743,744] The O_2N_2 chromophore is not planar but has a slight tetrahedral distortion and the CD spectra suggest that the corresponding chirality has Δ absolute configuration for diamines of *S* or *S,S* absolute configuration.[725]

In [VO(acen)], the geometry around the vanadium is square pyramidal (see Figure 32),[746,747] but β-ketimine VO^{2+} complexes do not readily add a ligand to form six-coordinate complexes, except if electron-withdrawing groups are attached to the acac portion of the SB. For example, Cotton effects appear in the *d–d* transition of [VO(tfen)] in the presence of (–)-1-phenyl-1-aminoethane in acetonitrile,[725] and, also, though a solid [VO(tfen)]·DMSO adduct forms, the sulfoxide is lost on standing overnight in the air.

In (**67**; $R^1 = R^2 = Me$, R = H) with X = Br green or yellow complexes were obtained, according to the solvent used, with ν(V=O) at *ca.* 980 and 900 cm^{-1}, respectively.[748] When X = Cl, only the green-type complex may be obtained. The authors assume five-coordinate square pyramidal geometry for the green complexes and a chain structure ··V=O···V=O···, similar to [VO(salnpn)], for the yellow complex. Solid adducts were isolated with 1-methylimidazole.

Allowing [VO(acen)] to react with $SOCl_2$, [VCl_2(acen)] was isolated,[749] which is soluble in acetone, DMSO and CH_2Cl_2. This is less stable as a solid than [VCl_2(salen)] but its decomposition product from reaction with air is not [VO(acen)].

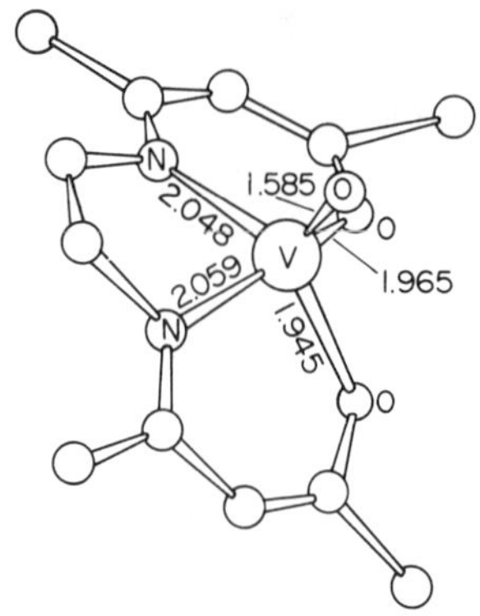

Figure 32 The molecular structure of [VO(acen)].[747] Some internuclear distances are indicated. The 'en' portion of the ligand assumes a configuration approximately halfway between *gauche* and eclipsed. The V atom is 0.58 Å above the plane of the SB coordinating atoms

(iii) Bidentate Schiff bases

Most of the compounds with SBs expected to have structure of type (**81**) or (**82**) are in Table 39. Compounds 1–6 of this table were included in Gmelins Handbuch.[357] Preparation consists simply in adding an aqueous-alcoholic (EtOH or MeOH) solution of the corresponding amine and $NaMeCO_2$ to an alcoholic solution of $VOCl_2$ (or $VOSO_4$) and the salicylaldehyde derivative. The complexes are often purified by recrystallization from ethanol. Compounds 7–13 are brownish grey-green and complexes 14–21 are green;[750] they are soluble in $CHCl_3$ and CH_2Cl_2 but less soluble in methanol, ethanol and benzene. These complexes show electronic spectra in pyridine which are different from the spectra of the same complexes in the solid state and in $CHCl_3$; this indicates, in pyridine, six-coordinate solvates.[750]

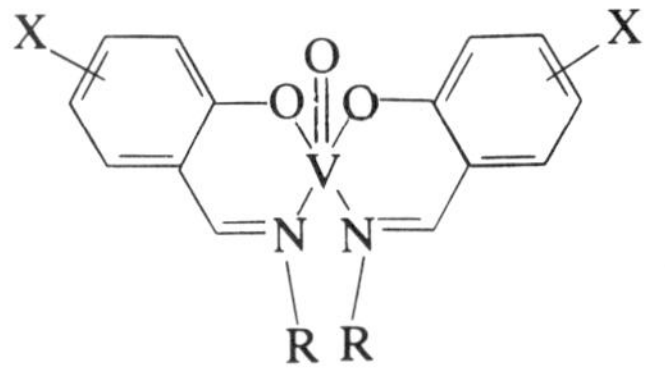

(**81**)

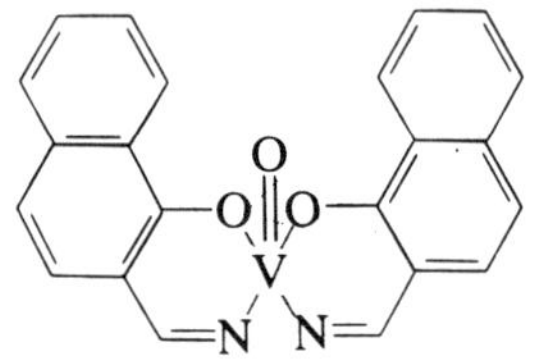

(**82**)

For several SB complexes ranging from yellow to green and maroon (22–30 in Table 39), poor solubility with failure to obtain six-coordinate species initially suggested a polymer.[721] However, they are soluble in hydrocarbons, and cryoscopy in benzene suggests a monomeric nature. For yellow bis(*N*-(4-chlorophenyl)salicylideneiminato)oxovanadium(IV) (22 in Table 39, $\nu(V{=}O) = 885\ cm^{-1}$), the unit cell comprises monomers of C_2 symmetry and the coordination was distorted trigonal bipyramidal (**13**). The three oxygens and vanadium are in the equatorial plane and the two nitrogen donors axial. V=O (1.615 Å) is longer than normal.

Syamal and Kale[751–753] reinvestigated previous reports[727,754,755] where a dimeric structure was proposed for compounds 33–36 in Table 39 and concluded that all exhibit moments in the range 1.73–1.75 BM at 295 K and obey the Curie–Weiss law with $\theta = 0$–4 K, indicating the absence of magnetic ordering in the temperature range 83–296 K. Osmometry indicates a monomeric nature. ESR spectra in $CHCl_3$ give the 'normal' eight-line spectra with $g = 1.98$ and no triplet state spectra ($\Delta M_s = \pm 2$ transition) were detected around 1600 G. On the basis of this evidence, complexes with SBs (**81**) and (**82**) with R = Ph are monomers and have square pyramidal structures. Although the compounds are not strictly the same, this conclusion disagrees in part with the work of Pasquali *et al.*[721] where a distorted trigonal bipyramidal structure was established for compound 22 in Table 39.

Oxovanadium(IV) complexes with SBs derived from Hsal or substituted Hsal and 2-aminoethanethiol (**83**) and 3-aminothiophenol (**84**) were reported by Syamal[756–758] and some properties are summarized in Table 40. The corresponding ligands with oxygen atoms instead of sulfur form tridentate complexes with antiferromagnetic properties (see Section 33.5.9.1.iv)

Table 39 Oxovanadium(IV) Complexes $VO(SB)_2$ with Schiff Bases of Type (**81**)

	X	*R*	*Colour*	$\mu_{eff}^{r.t.}$ (BM)	$\nu(V{=}O)$ (cm^{-1})	*Ref.*
1	5-Cl	H				1
2	3,5-Cl_2	H				1
3	5-Br	H		1.78		2
4	H	Me	Brown			3
5	H	Ph	Greenish yellow	1.75	988	3–5
6	5-Br	Ph		1.74	986	2–5
7	H	Et	Grey brown			6
8	H	Pr	Green			6
9	H	Pr^i	Green			6
10	H	Bu	Green	1.77	975, 895[a]	6, 7
11	H	Bu^s	Green			6
12	H	Bu^i	Green			6
13	H	Cy	Green			6
14	3-MeO	Me	Green			6
15	3-MeO	Et	Green			6
16	3-MeO	Pr	Green			6
17	3-MeO	Pr^i	Green			6
18	3-MeO	Bu	Green			6
19	3-MeO	Bu^s	Green			6
20	3-MeO	Bu^i	Green			6
21	3-MeO	Cy	Green			6
22	H	C_6H_4Cl-4	Golden yellow	1.81	885	7
23	H	C_6H_4Cl-3	Yellow green	—	940	7
24	H	$C_6H_4NO_2$-4	Light green	1.76	875	7
25	H	$C_6H_4NO_2$-3	Light green	1.66	980	7
26	H	C_6H_4OMe-4	Yellow green		970	7
27	H	$PhCH_2$	Green	1.76	992	7
28	H	C_6H_4OMe-2	Yellow green		970	7
29	H	$C_{12}H_{25}$	Beige maroon		970, 985[a]	7
30	H	$C_{16}H_{33}$	Beige maroon		975, 985[a]	7
31	H	C_6H_4Cl-4	Greenish yellow	1.67	1028 (?)	8
32	H	C_6H_4OMe-4	Greenish yellow			8
33	H	$C_6H_4SO_2NH_2$-4	Greenish yellow	1.54		8
				1.70		9
34	5-Cl	Ph	Greenish yellow	1.75	984	1, 4, 5, 8
35	5,6-Benzo	Ph	Greenish yellow	1.54	978	8, 10
				1.69		9
36	5,6-Benzo	C_6H_4Cl-4	Greenish yellow	1.54		8
37	5,6-Benzo	C_6H_4OMe-4	Greenish yellow	1.72		8
38	4-OH	C_6H_4Cl-4	Greenish yellow	1.68	990	8
39	4-OH	C_6H_4OMe-4		1.70	994	8
40	5,6-Benzo	C_6H_4Cl-4		1.71		9
41	H	C_6H_4Me-2	Cinnamon			11
42	H	C_6H_4Me-3	Brown			11
43	H	C_6H_4Me-4	Cinnamon			11
44	H	$C_6H_3Me_2$-2,3	Cinnamon			11
45	H	$C_6H_3Me_2$-2,4	Cinnamon			11
46	H	$C_6H_2Me_3$-2,4,6	Greenish yellow			11
47	H	C_6H_4OMe-3	Green			11
48	H	C_6H_4OH-4	Green			11
49	H	C_6H_4Br-4	Yellowish green			11

[a] In CS_2.

1. A. S. Pesis and E. V. Sokolova, *Zh. Neorg. Khim.*, 1963, **8,** 2518.
2. V. V. Zelentsov, I. A. Savich and V. I. Spitsyn, *Proc. Acad. Sci. USSR, Chem. Sect.*, 1958, **118–123,** 649.
3. L. Sacconi and U. Campigli, *Inorg. Chem.*, 1966, **5,** 606.
4. A. Syamal and K. S. Kale, *Curr. Sci.*, 1977, **46,** 258.
5. A. Syamal and K. S. Kale, *J. Mol. Struct.*, 1977, **38,** 195.
6. S. Yamada and Y. Kuge, *Bull. Chem. Soc. Jpn.*, 1969, **42,** 152.
7. M. Pasquali, F. Marchetti, C. Floriani and S. Merlino, *J. Chem. Soc., Dalton Trans.*, 1977, 139.
8. R. L. Dutta and G. P. Sengupta, *J. Indian Chem. Soc.*, 1971, **48,** 34.
9. A. Syamal, *Bull. Chem. Soc. Jpn.*, 1977, **50,** 557.
10. S. I. Gusev, V. I. Kumov and E. V. Sokolova, *J. Anal. Chem. USSR* (*Engl. Transl.*), 1960, **15,** 205.
11. V. A. Kogan, L. E. Lempert, O. A. Osipov, G. V. Nemirov and A. N. Nyrkova, *Zh. Neorg. Khim.*, 1967, **12,** 1400 (*Chem. Abstr.*, 1967, **67,** 60 501g).

However, (**83**) and (**84**) behave as bidentate monobasic ligands and form complexes $[VO(SB)_2]$; a square pyramidal structure was suggested with coordination through the oxygen atom and nitrogen of the azomethine group. The $[VO(SB)_2]$ complex with (**83**; R = H) was also studied by IR and cyclic voltammetry.[759] Although the presence of the protonated SH group could not be confirmed, the coordination probably involves N and O donors. Schiff bases (**85**) may behave as tridentate ONS ligands forming $VOL{\cdot}nH_2O$ complexes (Section 33.5.9.1.iv); however, if heated, these SBs are capable of cyclization and may then form $[VO(SB)_2]$ with (**86**) bidentate.[760]

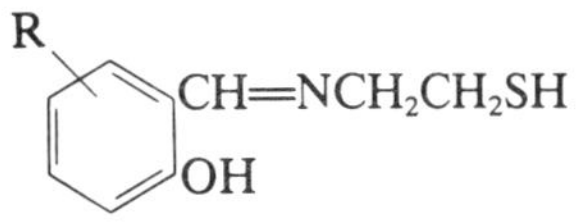

(**83**)

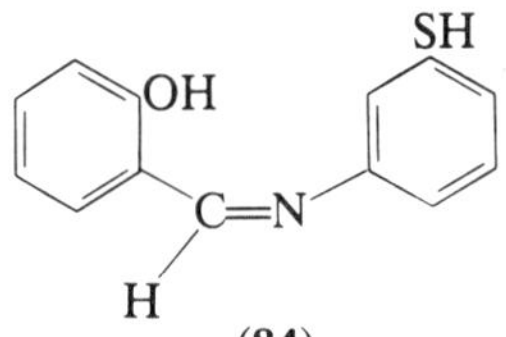

(**84**)

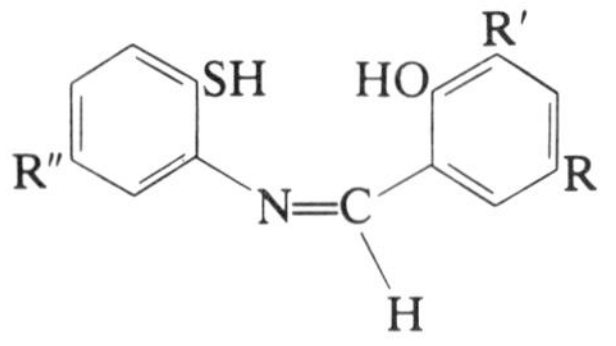

(**85**)

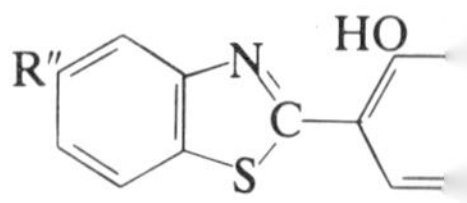

(**86**)

Table 40 Some Oxovanadium(IV) Complexes Derived from Salicylaldehyde or Substituted Salicylaldehyde with 2-Aminoethanethiol (**83**), 3-Aminothiophenol (**84**) or 2-Aminothiophenol (**86**) (on reflux)

	Schiff base (**83**)[1]		*Schiff base* (**84**)[2]			*Schiff base* (**86**)[3]	
R	$\mu_{eff}^{r.t.}$ (BM)	$\nu(V{=}O)$ (cm^{-1})	$\mu_{eff}^{r.t.}$ (BM)	$\nu(V{=}O)$ (cm^{-1})		$\mu_{eff}^{r.t.}$ (BM)	$\nu(V{=}O)$ (cm^{-1})
H	1.71	980	1.73	970	R, R′, R″ = H	1.86	914
5-Cl	1.71	980	1.72	980	R = Cl; R′, R″ = H	1.86	900
5-Br	1.72	980	1.71	980	R = NO_2; R′, R″ = H	1.76	910 (900)
5-MeO	1.75	975			R′ = NO_2; R, R″ = H	1.69	920
4-MeO	1.72	970	1.74	980	R″ = Cl; R, R′ = H	1.78	895
3-MeO	1.75	970					
5-NO_2	1.72	950	1.71	940			
2-OH	1.72	985	1.73	980			

1. A. Syamal, *Transition Met. Chem.* (*Weinheim, Ger.*), 1978, **3**, 259.
2. A. Syamal, *Transition Met. Chem.* (*Weinheim, Ger.*), 1978, **3**, 297.
3. C. C. Lee, A. Syamal and L. J. Theriot, *Inorg. Chem.*, 1971, **10**, 1669.

(*iv*) *Tridentate Schiff bases*

With bidentate SBs, monomeric $[VO(SB)_2]$ complexes are normally formed. On the contrary, with tridentate dibasic SBs, dimeric species often form. It appears that the dibasic character of the ligands forces the VO^{2+} ion to dimerize leading to anomalous magnetic properties. Proposed structures have not been verified owing to their non-crystalline nature. Insolubility precluded molecular weight determinations and ESR and electronic spectral studies have been carried out in the solid state only.

Solutions containing VO^{2+} and dinegative, tridentate ligands of the type $^-O{-}N{-}O^-$ are easily oxidized to the corresponding V^V complexes and possibly V^{IV} complexes are stable only if insoluble. Evidence for dimeric structures based on single-temperature subnormal magnetic moments should be treated with caution, as small amounts of a V^{IV} contaminant in a V^V complex may be the cause of the observed magnetic susceptibility.

Complexes with the tridentate dibasic SBs (**87**)–(**90**) have subnormal magnetic properties. Part of the published work on these type of complexes was reviewed by Syamal[358] and Casellato *et al.*[722] and many of the reported compounds are included in Table 41. They are normally obtained by mixing alcoholic (MeOH or EtOH) solutions of the appropriate SB, often prepared *in situ*, and an alcoholic solution of a vanadyl salt. Zelentsov first characterized such complexes,[762] prepared from SB (**87**) derived from Hsal or substituted Hsal and *o*-aminophenol. Ginsberg *et al.* prepared these as well as some others and studied the magnetic properties in detail from 1.4 to 300 K.[763] The moments of the complexes (with the exception of R = NO_2, R′ = NO_2) decrease considerably as the temperature is lowered and the temperature dependence of the susceptibility is characteristic of intramolecular antiferromagnetic exchange. The authors suggested a dimeric structure (**91**) and calculated the exchange integral, *J*, according to the Bleaney–Bowers equation,[764] concluding that substituents at R and R′

positions influence J in a different manner in oxovanadium(IV) and copper(II) complexes of the same ligands.[764] This was attributed to a difference in the exchange mechanism operating in VO^{2+} complexes resulting from the orbitals containing the unpaired electron not having the same symmetry. A tetrametallic structure (**92**) with V═O···V interaction was proposed for complex 5 in Table 41 to explain its higher J value and IR spectra.

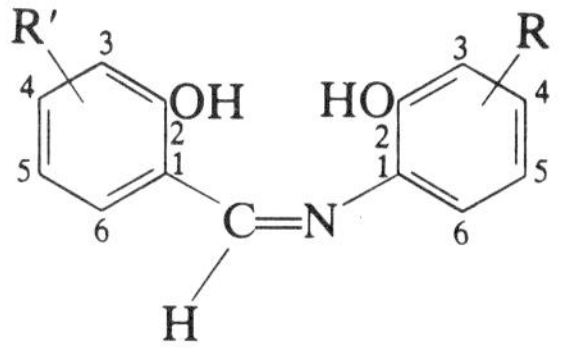

(**87**)

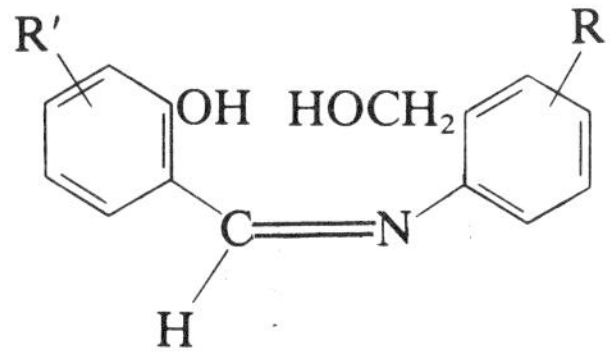

(**88**)

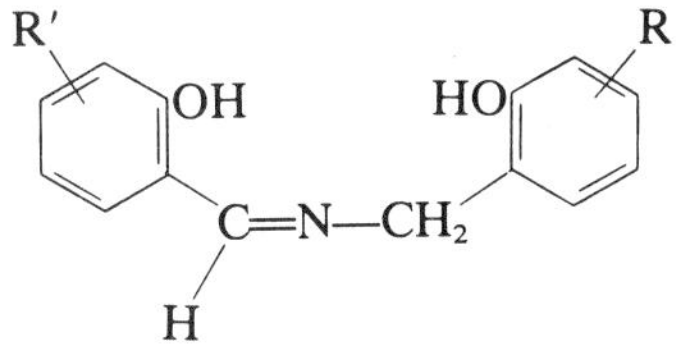

(**89**)

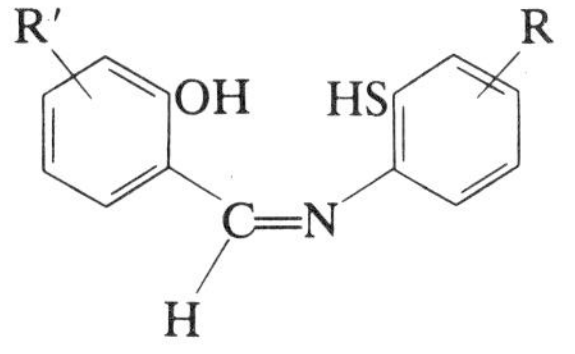

(**90**)

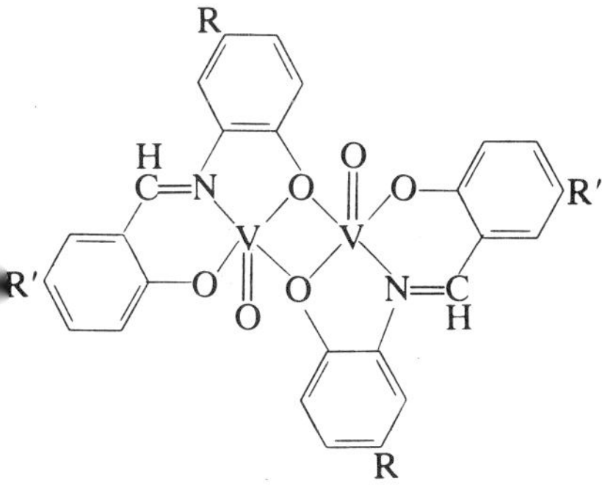

(**91**)

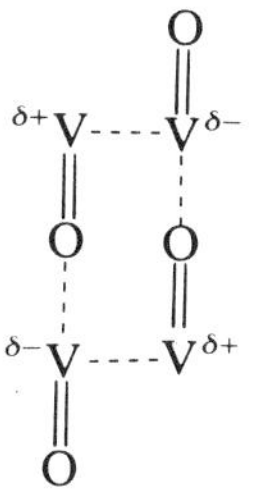

(**92**)

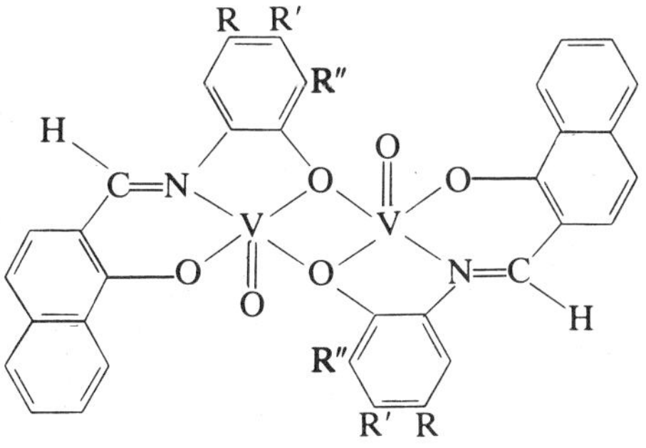

(**93**)

The complex with $R = NO_2$, $R' = H$ (7 in Table 41) has only a small TIP (temperature independent paramagnetism) term and does not follow the single–triplet susceptibility curve; the spins are completely coupled and only the diamagnetic single state ($S = 0$) is populated. The IR spectrum provides no evidence for bridging through the vanadyl oxygen and the complex probably has a different structure.[763] The complex with $R = R' = 5\text{-}NO_2$ (9) obeys the Curie–Weiss law with $\theta = 1.4$ K and $\mu = 1.62$ BM and the authors suggest a five-coordinate monomer in which a molecule of solvent occupies a coordination position.[763]

On treatment of (**91**; $R = R' = H$) with a strong chelating agent, such as phen, the dimer is broken, with the formation of a mononuclear mixed-ligand complex [VO(SB)(phen)].[765] The dimeric structure is also broken on treatment of complex (**91**; $R = H$, $R' = 5\text{-Cl}$) with pyridine; a monopyridine adduct forms which obeys the Curie–Weiss law with $\theta = 2$ K and $\mu_{eff}^{r.t.} = 1.75$ BM.[763]

Syamal reported the ESR spectrum of complex (**91**; $R = R' = H$) in polycrystalline solids; it exhibits a single-line spectrum with $g_{av} = 1.99$ at room temperature and 77 K, with no half-field band or hyperfine splitting.[358]

Subnormal VO^{2+} complexes (**93**) of SBs derived from 2-hydroxynaphthaldehyde and *o*-aminophenol and some substituted derivatives have also been obtained.[765] The ESR spectra of the complexes (**93**; $R = R' = R'' = H$; $R = Cl$, $R' = R'' = H$) in polycrystalline solids at room temperature exhibit two parallel lines, two perpendicular lines and a broad line around 1600 G due to the $\Delta M_s = 2$ transition. At 77 K, both high- and low-field spectra exhibit hyperfine splittings. Several complexes of type (**93**) were also studied by Carlisle *et al.* (Table 41, 27–34).[766,767] ν(V═O) values of complexes (**93**) lie in the range 955–1000 cm^{-1} and the compounds exhibit only one *d–d* band in the region 620–700 nm. As in the case of (**91**; $R = R' = H$), on treatment of compound (**93**; $R = R' = R'' = H$) with phen the dimer is broken and a mixed-ligand complex is obtained.[765]

Table 41 Oxovanadium(IV) Complexes with Tridentate Schiff Bases (**87**)–(**90**)

	R	*R'*	*R''*	$\mu_{eff}^{r.t.}$ (BM)	$\lvert J\rvert$ (cm^{-1})	$\nu(V{=}O)$ (cm^{-1})	*Ref.*
Schiff bases (**87**)							
1	H	H		1.48	125	990	1, 2
2	H	5-Me		1.46	118	993	2
3	H	5-Cl		1.54	90	992	1, 2
4	H	5-Br		1.50	115	991	1, 2
5	H	5-NO_2		1.30	218	900 (1010)	2
6	5-Cl	H		1.47	120	1000, 989	2
7	5-NO_2	H		—			2
8	5-Cl	5-Cl		1.52	132	983, 993	2
9	5-NO_2[a]	5-NO_2[a]				999	2
10	H	5-Clmonopyridinate		1.75		978	2
11	5-NO_2	5-Br					1
Schiff bases (**88**)							
12	H	H		1.52	126	910–985	3, 4
13	H	5-Cl		1.18	245		3, 4
14	H	5-Br		1.43	198		3, 4
15	H	3-MeO		1.57	121		3, 4
16	H	3-EtO		1.42	201		3, 4
Schiff bases (**89**)							
17	H	H		1.11	298	910	5, 6
18	H	5-Cl		1.19	266	900	5, 6
19	H	5-Br		1.18	264	910	5, 6
20	H	4-MeO		1.23	241	910	5, 6
21	H	3, 5-Cl_2		1.11	307	900	5, 6
22	H	hydrox[e]		1.10	274	910	5, 6
Schiff bases (**90**)							
23	H	5-Cl[b]		1.29		1005, 995	7
24	H	5-Br		1.34		900	7
25	H	5-NO_2		1.27		914, 904	7
26	H	hydrox[e]		1.52		960	7
V^{IV} *Schiff base complexes* (**93**)							
	R	*R'*	*R''*				
27	H	H	H	1.54[c]	151	1000	8
				1.84, 1.78			9, 10
28	Cl	H	H	1.44, 1.51	182	990	8, 10
29	NO_2	H	H	1.34, 1.60[c]	201	955	8, 10
30	H	H	NO_2	1.71		989	9
31	H	H	Cl	1.48		990	9
32	NO_2	H	NO_2	1.70,[d] 1.65[d]		992	9, 10
33	H	NO_2	H	1.49, 1.44			9, 10
34	Ph	H	H	1.46, 1.47			9, 10

Unless otherwise indicated, complexes are of anhydrous $[VOL]_2$ type.
[a] Formulated as $[VOL]\cdot 0.5H_2O\cdot EtOH$. [b] Formulated as $[VO(L)(H_2O)]$. [c] Formulated as $[VOL]\cdot EtOH$. [d] Formulated as $[VOL]\cdot H_2O$. [e] hydrox = 2-hydroxy-1-naphthaldehyde.

1. V. V. Zelentsov, *Dokl. Akad. Nauk SSSR*, 1961, **139**, 1110.
2. A. P. Ginsberg, E. Koubek and H. J. Williams, *Inorg. Chem.*, 1966, **5**, 1656.
3. A. Syamal and K. S. Kale, *Indian J. Chem. Sect. A*, 1980, **19**, 225.
4. A. Syamal, K. S. Kale and S. Banerjeee, *J. Indian Chem. Soc.*, 1979, **56**, 320.
5. A. Syamal and K. S. Kale, *J. Indian Chem. Soc.*, 1978, **55**, 606.
6. A. Syamal and K. S. Kale, *Inorg. Chem.*, 1979, **18**, 992.
7. C. C. Lee, A. Syamal and L. J. Theriot, *Inorg. Chem.*, 1971, **10**, 1669.
8. A. Syamal and L. J. Theriot, *J. Coord. Chem.*, 1973, **2**, 193.
9. G. O. Carlisle, D. A. Crutchfield and M. D. McKnight, *J. Chem. Soc., Dalton Trans.*, 1973, 1703.
10. G. O. Carlisle and D. A. Crutchfield, *Inorg. Nucl. Chem. Lett.*, 1972, **8**, 443.

With (**88**) and (**89**), several VO^{2+} complexes were prepared (Table 41).[768–771] High melting or decomposition temperatures (>250 °C) and insolubility in common noncoordinating solvents suggest a dimeric or polymeric nature. $\nu(V{=}O)$ occurs at 910–985 cm^{-1} for complexes of (**88**) and at 900–910 cm^{-1} for those of (**89**); this argues against the presence of a $\cdots V{=}O\cdots V{=}O\cdots$ chain structure. IR suggests a phenolic oxygen bridge in (**89**) complexes[769] and enolic oxygen bridges in (**88**) complexes.[771] $\mu_{eff}^{r.t.}$ values of the complexes are remarkably less than the spin-only value (Table 41) and the magnetic moments decrease significantly as the temperature is lowered, suggesting antiferromagnetic exchange interaction of neighbouring VO^{2+} ions. The ground state is $S = 0$ (singlet state) and the complexes exhibit no dependence on the magnetic field strength.

The J values of the VO^{2+} complexes with (**89**) are higher than those with (**88**); this suggests that in complexes with (**88**) and (**89**), bridging does not involve the phenolic O atoms of Hsal. If this was the case one should expect similar J values in the two series of complexes[769,771] since the electronic environment in the chelate ring would be the same.

Complexes with (**90**) involving ONS coordination apparently similar to the ONO coordination in (**87**) were reported (Table 41).[760] ν(V=O) lies in the range 900–1005 cm^{-1} and some exhibit splitting of the V=O band. $\mu_{eff}^{r.t.}$ values are in the range 1.27–1.52 BM and the temperature dependence indicates exchange-coupled antiferromagnetism. A dimeric oxygen-bridged structure, similar to (**91**), which provides an appropriate symmetry for the $3d_{xy}$ orbitals to overlap and form a strong σ V—V bond, was suggested.[760]

The synthesis and characterization of VO^{2+} complexes of hydrazones (**94**) and (**95**) derived from benzoylhydrazine and *o*-hydroxy aromatic aldehydes and ketones were also reported[772] and molecular weight determinations, IR, electronic spectra and magnetic properties were explained assuming a dimeric structure (**96**) with each unit having a five-coordinate square pyramidal geometry.

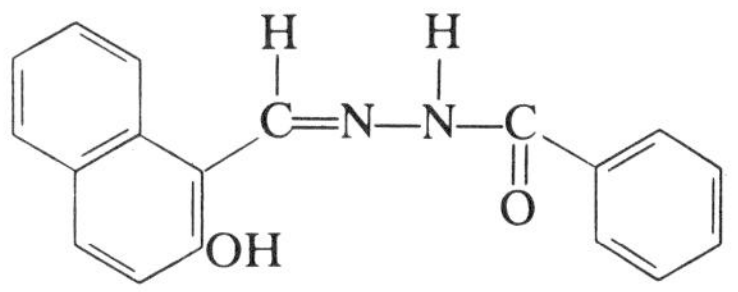

(**94**)

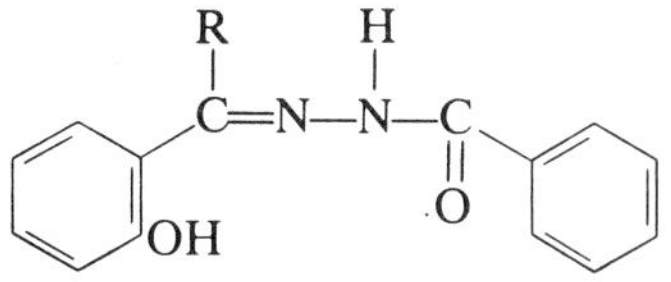

(**95**)

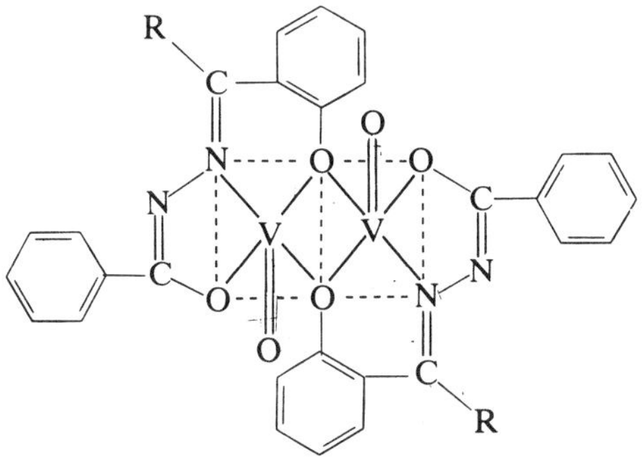

(**96**)

Oxovanadium(IV) complexes with (**97**) and (**98**) derived from alkyl aminoalcohols and Hsal or *o*-hydroxyacetophenone (and substituted derivatives), respectively, have been prepared as well as VO^{2+} complexes with (**99**) derived from 2-hydrazinoethanol and Hsal and substituted Hsal. Most of these and some properties are summarized in Table 42. Generally the complexes were prepared by adding an alcoholic solution of the SB, either previously obtained as a solid or prepared *in situ*, to alcoholic vanadyl acetate. The mixture is normally refluxed for several hours and the precipitates are collected by filtration; as they are not soluble they are not recrystallized.

Compounds obtained with SBs (**97**)–(**99**) have been formulated as $[VOL]_2$ compounds with structure (**100**). This is based on magnetic properties and IR spectra.[773–776] They have subnormal $\mu_{eff}^{r.t.}$ and the dependence of χ_M^{corr} on temperature is characteristic of antiferromagnetic exchange. For at least some (15–23 in Table 42), it was reported[773,774] that the complexes exhibit no dependence of χ_M^{corr} on magnetic field strength, indicating absence of ferromagnetic interaction. ESR studies on polycrystalline samples at room temperature normally give g_{av} ~1.99 and the half-field spectrum corresponding to the $\Delta M_s = 2$ is often observed. At 77 K, the high-field spectra often exhibit hyperfine splittings similar to those observed in VO^{2+} tartrates (Section 33.5.5.7.i).

The spectra of complexes with the tridentate SBs (**97**)–(**99**) generally exhibit three ligand field bands: (i) *ca.* 715–770, (ii) *ca.* 625 and (iii) *ca.* 500–550 nm, assigned to *d–d* transitions according to the Vanquickenborne and McGlynn MO scheme,[360] which corresponds to an ordering of levels as in the Ballhausen and Gray model (Figure 16a): $d_{xy} < d_{xz}, d_{yz} < d_{x^2-y^2} <$

Table 42 Oxovanadium(IV) Complexes with Schiff Bases (**97**)–(**99**)[a]

	R	*R'*	$\mu_{eff}^{r.t.}$ (BM)	$\|J\|$ (cm^{-1})	$\nu(V{=}O)$ (cm^{-1})	*Colour*	*Ref.*
1	$(CH_2)_2$	H	1.42			Brown	1
			1.06				2
			1.41	215	971		3
2	$(CH_2)_3$	H	1.36			Green	2
			1.53	141	970		3
3	$(CH_2)_3$	5-MeO	1.17	398	998	Yellow green	3
4	$(CH_2)_2$	5-Cl	1.26	330	995	Brown	3
5	$(CH_2)_2$	5-Br	1.27	321	987	Brown	3
6	$(CH_2)_2$	5-NO_2	1.19	339	916	Reddish brown	3
7	$(CH_2)_2$	3-NO_2	0.94	486	995	Green	3
8	$(CH_2)_2$	3-MeO	0.89	519	976, 995	Grey	3
9	$(CH_2)_3$	5-Cl	1.52	153	980	Green	3
10	$(CH_2)_3$	5-Br	1.48	166	968	Green	3
11	$(CH_2)_3$	5-NO_2	1.34	294	880	Yellow	3
12	$(CH_2)_3$	3-NO_2	1.54	147	963	Green	3
13	$(CH_2)_3$	5-MeO	1.58	117	963	Green	3
14	$(CH_2)_3$	3-MeO	1.54	137	974, 986	Green	3
			0.87				2
15	$CH_2C(Me)H$	H	1.01	369			4
16	$CH_2C(Me)H$	5-Cl	0.88	470			4
17	$CH_2C(Me)H$	3,5-Cl_2	0.81	450	965–995		4
18	$CH_2C(Me)H$	5-Br	0.98	350			4
19	$CH_2C(Me)H$	3-MeO	0.83	385			4
20	$CH_2C(Me)H$	hydrox[b]	1.27	—			4
21*	$(CH_2)_2$	H	1.40	380			5
22*	$(CH_2)_3$	H	1.50	350	965–985		5
23*	$CH_2C(Me)H$	H	1.40	380			5
24†	$(CH_2)_2$	H	1.32		970		6
25†	$(CH_2)_2$	5-Cl	1.28		980		6
26†	$(CH_2)_2$	5-Br	1.27	245–336	970		6
27†	$(CH_2)_2$	5-NO_2	1.35		970		6
28†	$(CH_2)_2$	3-EtO	1.30		960		6
29†	$(CH_2)_2$	hydrox[b]	1.39		970		6

[a] Complexes with Schiff bases of type (**98**) are indicated with an asterisk (*), those with Schiff bases of type (**99**) by a dagger (†).
[b] hydrox = 2-hydroxynaphthaldehyde.

1. S. N. Poddar, K. Dey, J. Haldar and S. C. Nathsarkar, *J. Indian Chem. Soc.*, 1970, **47**, 743.
2. Y. Kuge and S. Yamada, *Bull. Chem. Soc. Jpn.*, 1970, **43**, 3972.
3. A. Syamal, E. F. Carey and L. J. Theriot, *Inorg. Chem.*, 1973, **12**, 245.
4. A. Syamal and K. S. Kale, *Indian J. Chem., Sect. A*, 1977, **15**, 431.
5. A. Syamal and K. S. Kale, *Indian J. Chem., Sect. A*, 1980, **19**, 486.
6. A. Syamal, S. Ahmed and O. P. Singhal, *Transition Met. Chem. (Weinheim, Ger.)*, 1983, **8**, 156.

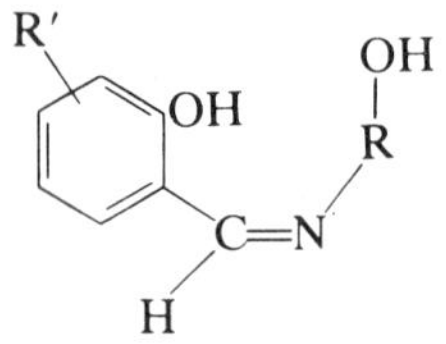

(**97**)

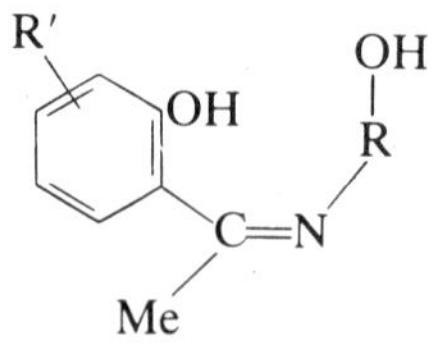

(**98**)

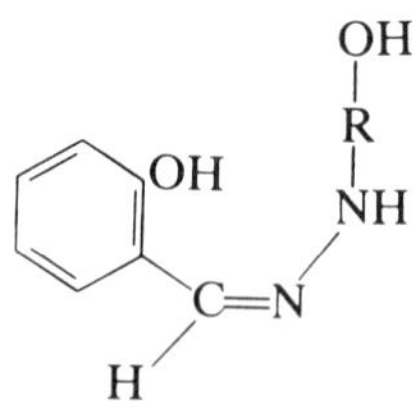

(**99**)

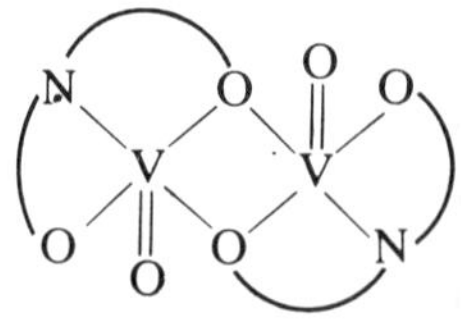

(**100**)

d_{z^2}.[364] Some of these bands are not well developed and this gave rise to ambiguity in their assignment. No band characteristic of VO^{2+}–VO^{2+} interaction has been observed in complexes with these SBs.

In Table 42, in complexes with SBs (**97**)–(**99**), the $|J|$ values of compounds with $n = 2$ (n in $R = (CH_2)_n$) are larger than with $n = 3$. This suggests that the antiferromagnetic exchange interaction is stronger with $n = 2$, and this has been attributed to the difference in the chelate ring (five- or six-membered).[775] For $n = 2$, the J values also depend appreciably on the nature and position of the Hsal substituent; for $n = 3$, they are relatively insensitive to ring substituents.[358]

In an early publication, Zelentsov reported the preparation of a complex with subnormal magnetic properties ($\mu_{eff}^{r.t.} = 0.87$ BM) with the SB (**101**).[777] The compound was prepared from dibenzoylmethane and *o*-aminophenol. The VO^{2+} complex with (**102**) has also been shown to be involved in antiferromagnetic exchange with $J = -139\ cm^{-1}$. Revenko and Gerbeleu prepared several VO^{2+} complexes of (**103**) derived from Hsal and thiosemicarbazide which were reported to involve ONS coordination.[778] The measured $\mu_{eff}^{r.t.}$ were 1.27, 1.21 and 1.33 BM for R′ = H, Cl and Br, respectively. The complexes are not anhydrous and a dimeric structure (**104**) was proposed.

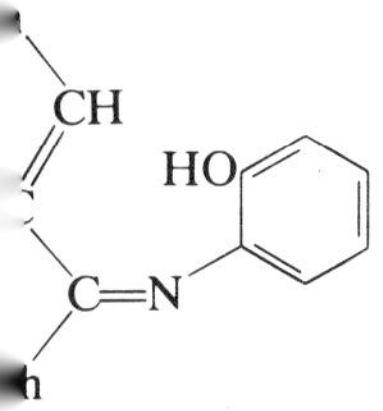
(**101**)

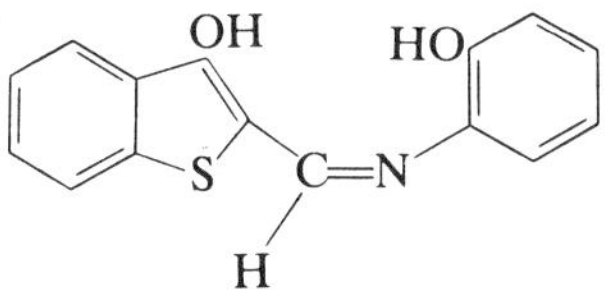
(**102**)

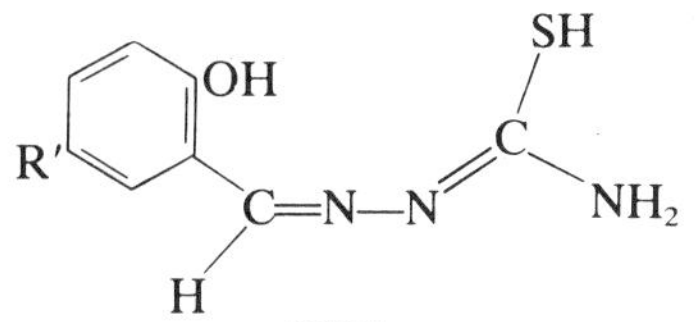
(**103**)

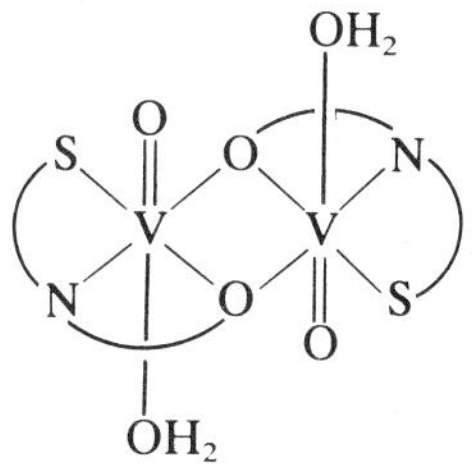
(**104**)

Several complexes were prepared with (**105**; R′ = H or Me). All decompose above 300 °C and are insoluble in common nonpolar solvents. $\mu_{eff}^{r.t.}$ values, in the range 1.32–1.58 BM, suggest exchange-coupled antiferromagnetism. The authors formulated the complexes as $[VO(SB)(Cl)]_2$ and proposed a dimeric structure (**106**) but report no χ_M^{corr} temperature dependence results. In the IR, ν(V=O) and ν(V—Cl) were in the range 975–985 and 360–385 cm^{-1}.[779] Electronic spectra in Nujol mulls had three bands: (i) ~395–412, (ii) 640–670 and (iii) 745–800 nm. Structure (**106**) is probably not correct.

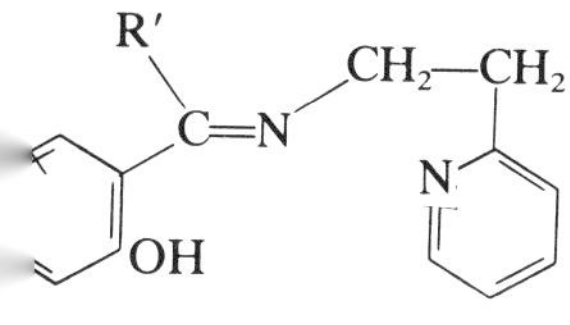
(**105**)

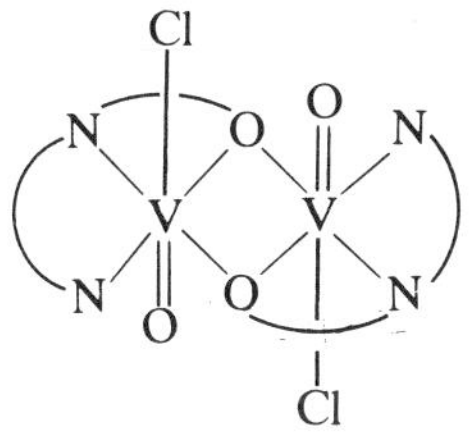
(**106**)

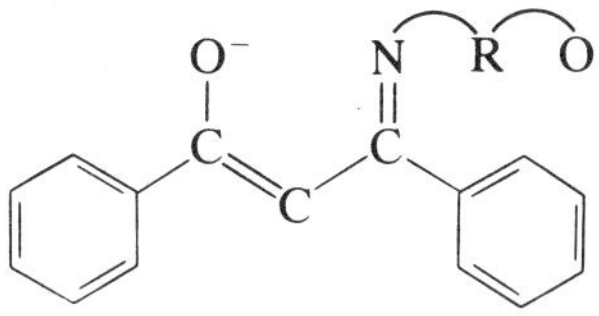
(**107**)

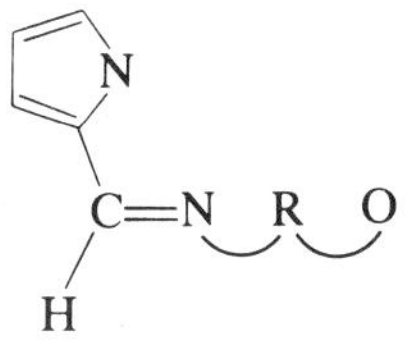
(**108**)

Several VO^{2+} complexes with (**107**) and (**108**) derived from dibenzoylmethane or pyrrole-2-carboxyaldehyde and several amines (taurine, anthranilic acids, β-alanine) were formulated as $[VO(SB)(H_2O)_2]$;[780–783] it was proposed that these SBs behave as tridentate ligands and assume a monomeric structure on the basis of $\mu_{eff}^{r.t.}$ and ebulliometric measurements in dioxane. A pyridine adduct $[VO(PE)(py)_2]$ was also obtained where PE is the SB (**108**; R—O = $CH_2CH_2SO_3^-$).[782]

Benzoylhydrazine and $[VO(acac)_2]$ react in methanol under dry N_2 to produce violet $[V(abh)_2]$ where abh is the dianion of acetylacetone benzoylhydrazone (**109**; R = H).[784] In air, a smaller yield of $[V(abh)_2]$ is obtained, with V^V complexes containing abh. The violet complex has $\mu_{eff}^{r.t.} = 1.87$ BM and the ESR shows the 'normal' eight peaks with $g = 1.972$; the structure has trigonal prismatic coordination in which the tridentate abh ligand forms both five- and six-membered chelate rings (**110**). Depending on the relative amounts of the reactants, chelate exchange between $[VO(acac)_2]$ and Schiff bases (**109**) provides (i) complexes with subnormal $\mu_{eff}^{r.t.}$, probably dimeric, if a 1:1 stoichiometry is used, and (ii) vanadium bis complexes, in which abstraction of vanadyl oxygen occurs, if 1:2 $[VO(acac)_2]$:SB stoichiometry is used.[785]

Complexes with SBs from Hsal and α-amino acids (glycine, L- and DL-alanine, L-methionine, L-valine, L-leucine, L- and DL-phenylalanine) were prepared and characterized.[786] The complexes are bluish-grey and there is no appreciable interaction between V atoms. The compounds were $[VO(SB)(H_2O)]$ (**111**), confirmed in an X-ray study of (**111**; R = Me, L-Ala derivative).[787] ν(V=O) was at ~1000 cm^{-1}; the compounds are not soluble in noncoordinating solvents, but when dissolved in py ν(V=O) shifts to ~970 cm^{-1}; several orange py adducts (**112**) were isolated.[686] Complexes (**111**) have electronic spectra that resemble those of

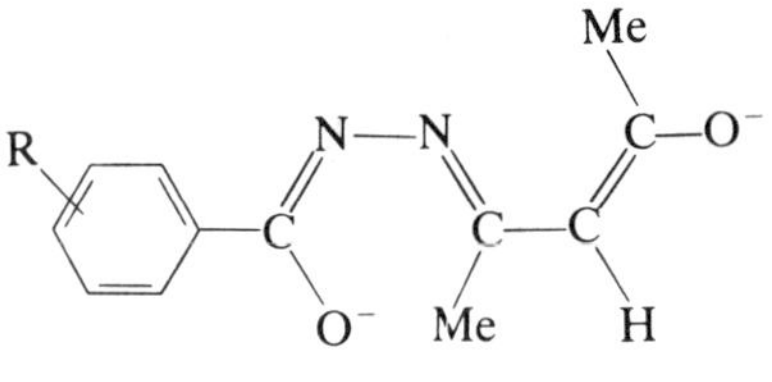

(109)

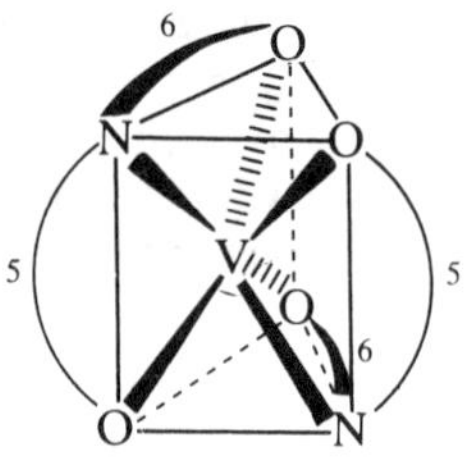

(110)

[VO(acac)$_2$] and were considered as having a square pyramidal structure, although a trigonal bipyramidal structure was not precluded.[786] Detailed IR studies of [VO(sal-Gly)(H_2O)] were reported.[788,789]

In complexes (**111**), one water molecule occupies a coordination position and 'prevents' the dimerization of the compounds. Dimers were obtained on dissolving the monomer (**111**) in absolute methanol, when equilibrium (ii) in Scheme 19 takes place. The dimeric nature of the isolated complexes[790] is supported by the marked differences from the mononuclear complexes, by the low $\mu_{eff}^{r.t.}$ obtained and by molecular weights; ν(V═O) was around 980 cm^{-1}. From dry methanol the orange products are the vanadium(V) complexes (**114**) (probably dimeric); the oxidizing agent seems to be dissolved O_2.[686] In wet methanol, complexes (**115**) were obtained as red needles. Further evidence for V^V arises from the fact that on treating (**115**) or (**114**) with a reducing agent (*e.g.* NaSH·$x$$H_2O$), bluish-grey complexes (**111**) are regenerated. Reaction of (**115**) with py gives (**112**), corroborating other reports of the reducing action of pyridine on vanadium(V).[395]

Dimeric oxovanadium(IV) complexes (NBu_4)[(VO)(SB)]$_2$·$n$$H_2O$ have also been prepared with the SB from salicylaldehyde and L, D,L-serine and *L*-cysteine.[791] Magnetic moments are 1.1–1.4 BM, higher than for the dimeric SB derived from the L-amino acids: Ala, Val, Leu, Met or Phe (0.5–0.8 BM) which are probably best described as V^V complexes (**114**).[790]

Green complexes [VO(SB)(H_2O)], where H_2SB = 2-hydroxynaphthylidene amino acids, react with py or phen to form [VO(SB)(py)$_2$] or [VO(SB)(phen)], which exhibit $\mu_{eff}^{r.t.}$ near the spin-only value.[765] In methanol, these gave brown solutions from which the monomeric brown V^V complexes [VO(OMe)(SB)(MeOH)] were isolated. The treatment of these brown complexes with CH_2Cl_2 followed by n-hexane gave green precipitates of [VO(OH)(SB)]$_2$;[765] in CH_2Cl_2, the complexes are dimers.

(*v*) *Other polydentate Schiff bases*

Several polydentate SBs may act as binucleating ligands; see Section 33.5.12.1. Only a few miscellaneous polydentate SBs will be included here. A complex with the SB obtained in the reaction of two moles of benzoyl hydrazide with acetylacetone had ν(V═O) = 995 cm^{-1}, $\mu_{eff}^{r.t.}$ = 1.7 BM and was monomeric square pyramidal (**117**).[792] A few complexes with the ethylenediamine derivatives of 2,2′-dihydroxychalcones (**118**) were prepared in ethanol.[793] The stoichiometry is 1:1 and the atoms involved in the coordination were two deprotonated oxygens and two nitrogen atoms. ν(V═O) (970–985 cm^{-1}) and $\mu_{eff}^{r.t.}$ (1.73–1.83 BM) suggest a monomer; however, molecular weights in nitrobenzene suggest a dimer.

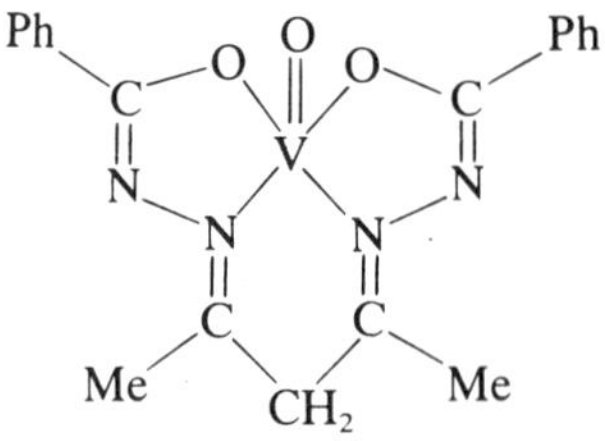

(117)

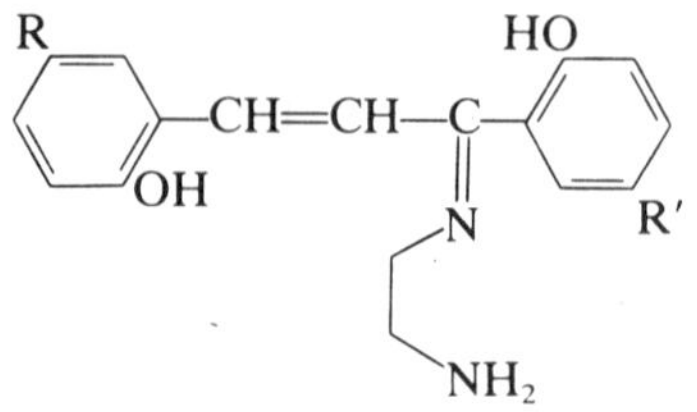

(118)

33.5.9.2 Amino acids

The equilibria in solutions containing amino acids have been studied by different methods.[383,581,794–808] Relatively high pH is necessary to achieve an appropriate free ligand

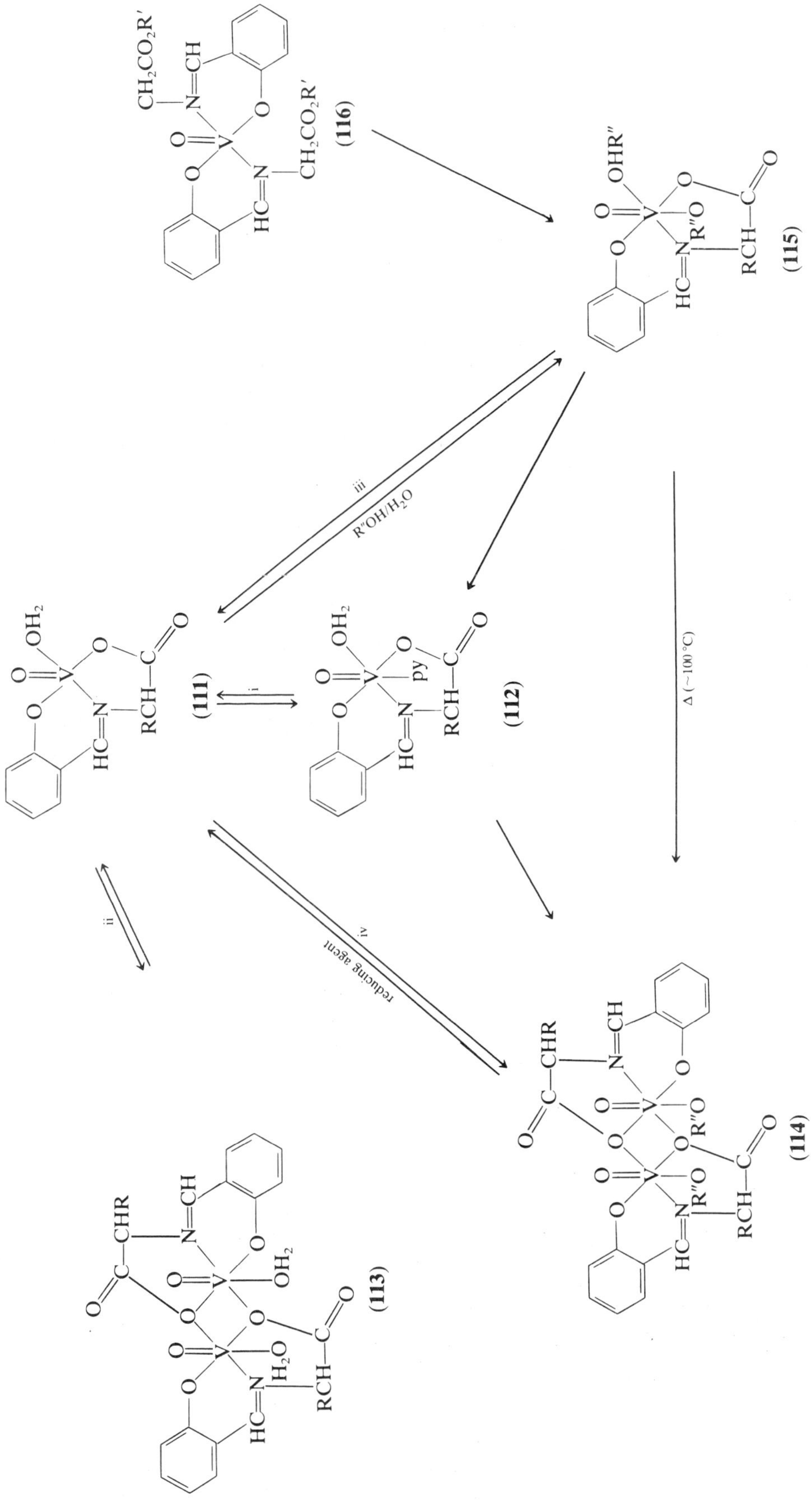
(111)
(112)
(113)
(114)
(115)
(116)
i
ii
iii
R″OH/H2O
iv
reducing agent
Δ (~100 °C)

Scheme 19

concentration for complex formation. At pH ≥ 4, however, hydrolysis of VO^{2+} takes place and oxidation to V^V may also be significant. Hydrolysis of VO^{2+} in the pH range 5–12 is not well understood (see Section 33.5.5.1.ii).

For the glycine system, models I and II in Table 43 have been suggested.[581] But (*cf.* Section 33.5.5.1.ii), for pH ≥ 5 other VO^{2+} hydrolysis products should be included.[585,588] For the L-alanine system (certainly similar to glycine), results were explained according to model III included in Table 43. $[VO(HAlaO)^{2+}]$ (**119**) must be included to explain the CD spectra. The inclusion of (**124**) gives a better fit of the data.

Table 43 Species Included in Equilibrium Models proposed for the Oxovanadium(IV)–Glycine and –L-Alanine Systems at 25 °C with High Ligand/Metal Ratios

Glycine (1 *M NaClO$_4$ solutions*)		L-*Alanine* (2.2 *M NaNO$_3$ solutions*)[3]
Model I	*Model II*	*Model III*
Tomiyasu and Gordon[1]	*Fábyán and Nagypál*[2]	*Gillard, Pessoa and Vilas-Boas*[3,4]
H_2GlyO^+, HGlyO, $GlyO^-$	H_2GlyO^+, HGlyO, $GlyO^-$	H_2AlaO^+, HAlaO, $AlaO^-$
$[VO^{2+}]$, $[VO(OH)]^+$, $[(VO)_2(OH)_2]^{2+}$	$[VO^{2+}]$, $[VO(OH)]^+$, $[(VO)_2(OH)_2]^{2+}$	$[VO^{2+}]$, $[VO(OH)]^+$, $[(VO)_2(OH)_2]^{2+}$, $[(VO)_2(OH)_5^-]_n$[a] (and $[VO(OH)_3^-]$)
$[VO(HGlyO)]^{2+}$ (**119**)	$[VO(HGlyO)]^{2+}$	$[VO(HAlaO)]^{2+}$
$[VO(GlyO)]^+$ (**120**)	$[VO(GlyO)]^+$	$[VO(HAlaO)_2]^{2+}$ (**124**)
$[VOGlyO]^+$ (**121**)	$[VO(GlyO)(HGlyO)]^+$ (**123**)	$[VO(AlaO)^+]$
$[VO(GlyO)_2]$ (**122**)	$[VO(GlyO)_2]$	$[VO(AlaO)(HAlaO)]^+$
	$[VO(GlyO)(OH)]$	$[VO(AlaO)_2]$
	$[(VO)_2(GlyO)_2(OH)_2]$	$[VO(AlaO)_2(OH)]^-$
	$[(VO)(GlyO)_2(OH)]^-$	$[(VO)_2(AlaO)_2(OH)_2]$
	$[VO(GlyO)(OH)_2]^-$	$[(VO)(AlaO)(OH)]$ (?)
	$[(VO)_2(GlyO)_2(OH)_3]^-$ [b]	$[VO(AlaO)(OH)_2]^-$ [c]
	$[(VO)_2(GlyO)_3(OH)_2]^-$ [b]	$[(VO)_2(AlaO)_2(OH)_3]^-$
	$[(VO)_2(GlyO)(OH)]_2^-$ [b]	$[(VO)_2(AlaO)_3(OH)]_2^-$ [c]

[a] A value of *n* of 1 was 'arbitrarily' assumed. [b] The presence of these species was assumed in order to get an acceptable fit of the experimental data. [c] The existence of these species cannot be definitely precluded but their inclusion is not necessary to explain the experimental results for pH < 8.0.

1. H. Tomiyasu and G. Gordon, *J. Coord. Chem.*, 1973, **3,** 47.
2. I. Fábyán and I. Nagypál, *Inorg. Chim. Acta*, 1982, **62,** 193.
3. R. D. Gillard, J. C. Pessoa and L. Vilas-Boas, Comunicação C31.9, 5° Encontro Nacional de Quimica, Porto 1982.
4. R. D. Gillard, J. C. Pessoa and L. Vilas-Boas, Comunicação PA54, 6° Encontro Nacional de Quimica, Aveiro 1983.

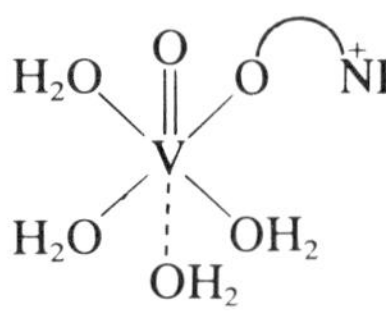

(**119**)

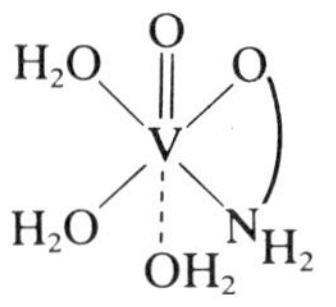

(**120**)

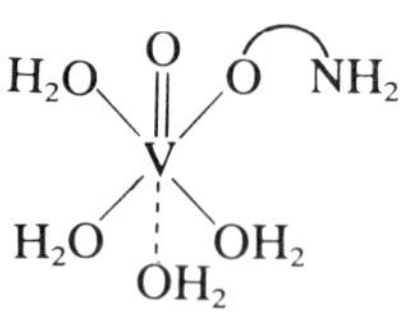

(**121**)

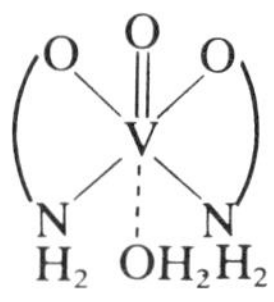

(**122**)

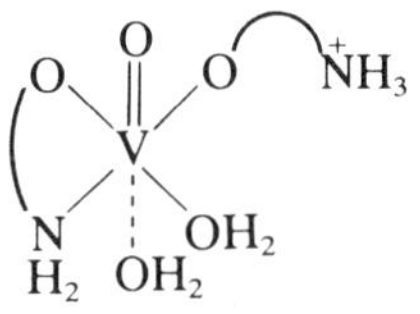

(**123**)

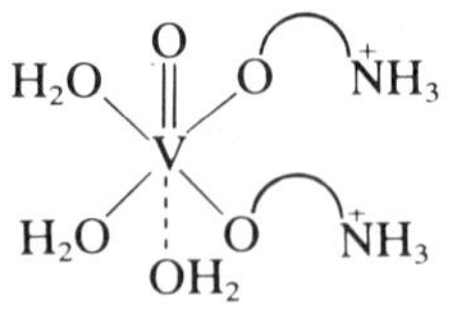

(**124**)

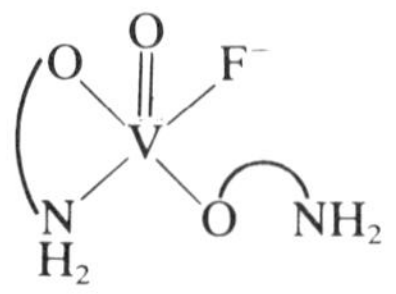

(**125**)

The kinetics of processes involving glycinato–VO^{2+} are summarized in Table 44. Only one species is formed from the addition of F^- ion to a $[VO(GlyO)_2]$ solution (equation 55; calculated $K = 17\ mol^{-1}$), with structure (**125**).[800] From ^{19}F NMR, relaxation is controlled by the exchange reaction (56), which is first order in F^-.

Cysteine reduces vanadium(V);[803] the kinetics have been studied.[804] VO^{2+}–cysteine compounds form in the reduction of vanadium(V) with an excess of cysteine[805] and a Chinese publication reports $K[VO(HCysO)(H_2O)]SO_4$ and *cis-* and *trans*-$K_2[VO(CysO)_2]$ (H_2CysO = cysteine), and ESR as a function of pH.[806] A purple complex from a 5:1 mixture of cysteine

Table 44 Some Kinetic Data for the Oxovanadium(IV)–System at 25 °C

Rates of reaction for VO^{2+} *substitution reactions*[1] *Reaction*	*Method used*	*k at 25 °C* (s^{-1})
$[VO(HGlyO)]^{2+}$ formation	Stopped flow	1.3×10^3 [a]
$[VO(HGlyO)]^{2+}$ dissociation	Calculated	4.6×10^2
HGlyO exchange on $[VO(HGlyO)]^{2+}$	NMR[b]	$<2.5 \times 10^3$
HGlyO exchange on $[VO(GlyO)_2]$	NMR[b]	3.6×10^2
Ring closure of $[VO(HGlyO)]^{2+}$	Stopped flow, temperature jump[2]	35

First-order rate constants of the proton exchange between bulk water and the different paramagnetic species assumed in Model II[3] (*Table* 43)

$[VO(GlyO)]^+$	$(1.0 \pm 0.5) \times 10^5\ s^{-1}$
$[VO(GlyO)_2]$	$(1.2 \pm 0.1) \times 10^5\ s^{-1}$
$[VO(HGlyO)]^{2+}$	$(3.5 \pm 0.3) \times 10^5\ s^{-1}$
$[VO(GlyO)(GlyOH)]^+$	$(4.1 \pm 0.3) \times 10^5\ s^{-1}$
$[VO(GlyO)_2(OH)]^-$	$(1.6 \pm 0.03) \times 10^6\ s^{-1}$

[a] Second-order rate constant ($mol^{-1}\ s^{-1}$). [b] Process considered to be zero order in glycine.
1. H. Tomiyasu, K. Dreyer and G. Gordon, *Inorg. Chem.*, 1972, **11**, 2409.
2. H. Tomiyasu and G. Gordon, *Inorg. Chem.*, 1976, **15**, 870.
3. I. Fábyán and I. Nagypál, *Inorg. Chim. Acta*, 1982, **62**, 193.

$$VO(GlyO)_2 + F^- \overset{K}{\rightleftharpoons} VOF(GlyO)GlyO^- \quad (55)$$

$$VOF(GlyO)GlyO^- + {}^*F \rightleftharpoons VO{}^*F(GlyO)GlyO^- + F^- \quad (56)$$

methyl ester and $VOSO_4$ in borate buffer stirred in air for 5–6 hours showed a typical isotropic eight-line ESR spectrum and a second component in the ESR, IR and Raman.[807] The two species were considered to be the *trans* and *cis* isomers (**126**) and (**127**) but further work is needed. The reduction of A_{iso} for histidine and sulfhydryl side chain residues is large enough to be of diagnostic utility for VO^{2+} protein labelling studies.[798]

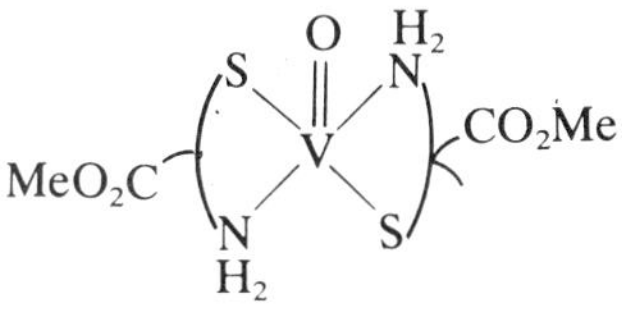

(**126**)

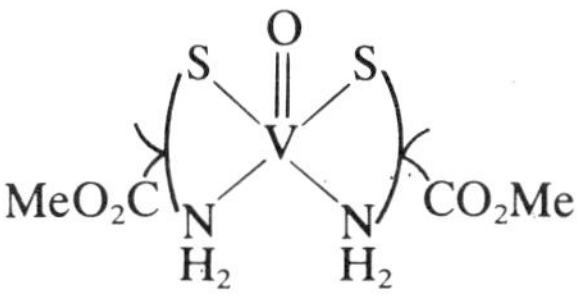

(**127**)

Van Willigen and co-workers reported an ENDOR study of frozen solutions containing VO^{2+} and histidine (**20**), carnosine (**21**) and imidazole (**19**) (*cf.* Section 33.5.3.2).[383] The complex formed between VO^{2+} and an excess of histidine in aqueous solutions of pH ~7 showed no axial H_2O lines. Coordination involved the N-3 nitrogen atom of the im ring of histidine, giving a structure in which the cation is shielded from water. According to the ENDOR spectra,[383] the geometry of the VO^{2+}–im moiety in the $[VO(im)_4]^{2+}$ and VO^{2+} carnosine complexes is similar but a distinct structure appears to exist for the VO^{2+} histidine complex.

33.5.9.3 Complexones

pH titrations show that in solutions containing VO^{2+} and ttha (triethylenetetraminehexaacetic acid) (2:1), only one major complex is formed in the pH range *ca.* 4–10.[809,810] The ESR spectrum was attributed to a binuclear chelate.[809] $Na_2[(VO)_2ttha]\cdot 10H_2O$ is composed of discrete $[(VO)_2ttha]^{2-}$ units and the V atom is octahedrally coordinated to the oxo ligand and three O and two N atoms of the ttha ligand (Figure 33a). The V atom is displaced 0.39 Å towards the oxo ligand. The molecular structure of the solid gives no explanation for the ESR spectrum observed. Reduction of O_2 with dinuclear $[V_2^{III}O(ttha)]^{2-}$ may produce $[V^{III},V^{IV}]$ and/or $[V^{IV},V^{IV}]$ complexes.[942]

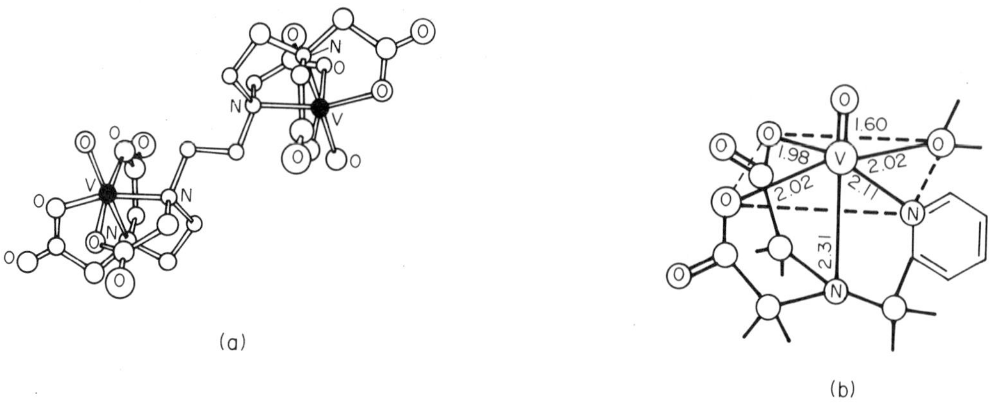

Figure 33 Schematic representation of the molecular structure of (a) $Na_2[(VO)_2ttha]\cdot 10H_2O$[810] and (b) $[VO(pmida)(H_2O)]\cdot 2H_2O$[811]

In $[VO(pmida)(H_2O)]\cdot 2H_2O$ (pmida = *N*-(2-pyridylmethyl)iminodiacetic acid), vanadium deviates by 0.39 Å from the equatorial plane of two *cis* carboxylate oxygens, a water molecule and the pyridine nitrogen (see Figure 33b).[811]

The structures of $[VO(S\text{-peida})(H_2O)]\cdot 2H_2O$ (**128**), $Na[V_2O_3(S\text{-peida})_2]\cdot NaClO_4\cdot H_2O$ (**129**) and $Li[VO_2(S\text{-peida})]\cdot 2MeOH$ (**130**) (*S*-peida = (*S*)-((1-(2-pyridyl)ethyl)imino)diacetic acid) were also determined (Figure 34).[812] In the mixed-valence binuclear ion (**129**), two octahedra share the equatorial oxo ligand. In the $O{=}V_A{-}O_b{-}V_B{=}O$ segment, the $V{-}O_b{-}V$ angle is ~180° and the $V_A{-}O_b$ bond (1.875 Å) differs from the $V_B{-}O_b$ bond (1.763 Å). Furthermore, the volumes of the octahedra around V_A and V_B are close to those of the V^{IV} and V^V complexes, (**128**) and (**129**) respectively. This and other features indicate that V_A and V_B are quadri- and quinque-valent.

Other compounds containing the $V_2O_3^{3+}$ core, $H[V_2O_3(pmida)_2]\cdot 4H_2O$ and $K_7[V_9O_{16}(bdta)_4]\cdot 27H_2O$ (btda = butanediaminetetraacetate), were prepared and their molecular structures were determined.[813] Both also contain $V_2O_3^{3+}$ with both terminal O atoms *cis* to the bridge but *trans* to each other. Contrary to (**129**), no systematic variation in bond lengths around the two V atoms was detected. The $[V_9O_{16}(btda)_4]^{7-}$ anion exhibits a quite unusual structure. It is organized around a central tetrahedral V^VO_4 group from which four mixed-valence binuclear groups spread, each containing the $V_2O_3^+$ core. The mixed-valence groups are linked together by the four bdta ligands.[813] Contrary to $[V_2O_3(pmida)_2]^-$, the two V atoms are in different environments. As in (**129**), V^{IV} is more displaced than V^V with respect to the equatorial plane.

In the spectra of $[V_2O_3(pmida)_2]^-$ and $[V_2O_3(nta)_2]^{3-}$, a new band around 1000 nm was assigned to the intervalence transition.[814] The spectrum of $[V_9O_{16}(btda)_4]^{7-}$ is the same in solution and in the solid.[813] The band at 900 and a shoulder at 1020 nm corresponded to two components of the intervalence band.

Mixed-valence species containing V^{II} and V^{IV} bridged by an oxygen atom were intermediate in electron transfer between $[V(Hedta)]^-$ and $[VO(Hedta)]^-$, and $[V(edta)]^{2-}$ and $[VO(edta)]^{2-}$.[815] Similarly bridged species were assumed in the reaction of $[V(H_2O)_6]^{2+}$ with $[VO(H_2O)_5]^{2+}$. Vanadium(III) dimeric complexes are the reaction products (see Section 33.4.9.3).

Compounds $[VOL_2]$ (L = ida and idpa; see Table 45) involved addition of $VOSO_4$ to the barium salt of the complexones. IR spectra suggest both coordinated and noncoordinated carboxyl groups and structure (**131**; R = H) was proposed.[816]

Complexes of polyaminocarboxylic acids (**132**)–(**134**) were investigated by Felcman and Fraústo da Silva.[377] Their results and other data are in Table 45. Log K_{VOL} values are almost equal to the values for Ni^{II} and a linear correlation (log K_{VOL} *vs.* log K_{NiL}) exists.

Fraústo da Silva and co-workers proposed structure (**131**; R = OH) for the species predominant in the pH range 4–7, which corresponds to the structure normally assumed for 'amavadin' which occurs in the toadstool *Amanita muscaria*.[377,378]

Edta, egta and dtpa (diethylenetriaminepentaacetic acid) all form 1:1 chelates and give rise to a broadened ESR signal.[809]

When equimolar solutions of $[V^{IV}O(nta)(H_2O)]^-$ and $[V^VO_2(nta)]^{2-}$ are mixed, a deep blue appears and the product was obtained crystalline as the Na^+ and NH_4^+ salts.

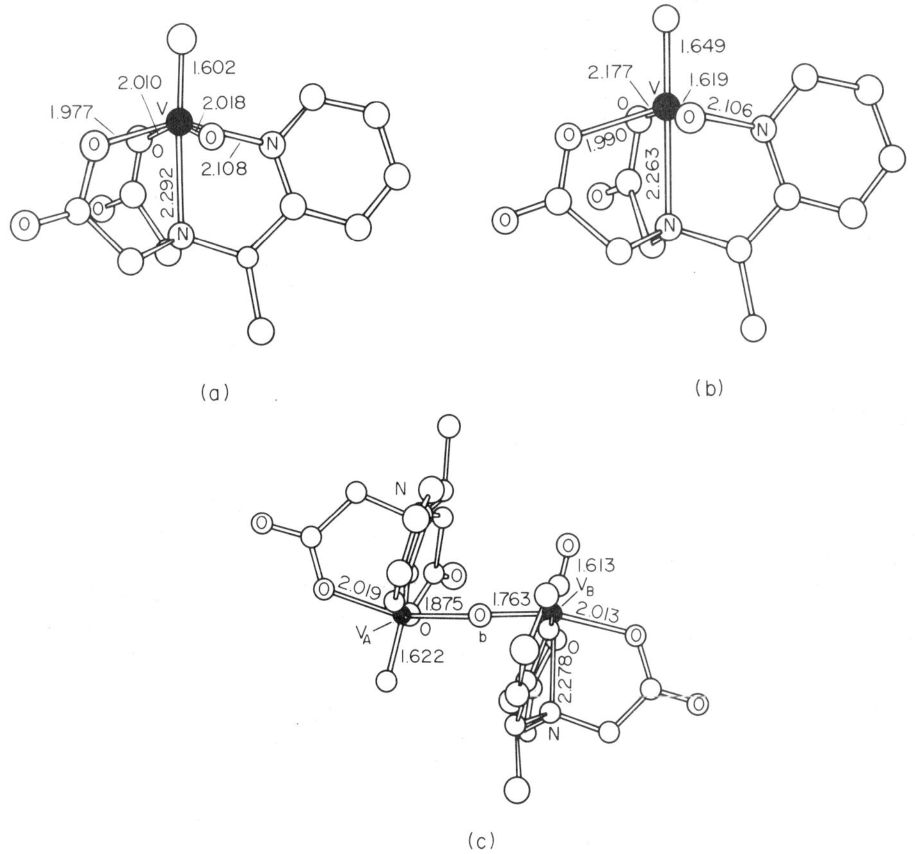

Figure 34 The molecular structure of (a) [VO(*S*-peida)(H_2O)] (**128**), (b) [VO_2(*S*-peida)]$^-$ (**129**) and (c) [V_2O_3(*S*-peida)$_2$]$^-$ (**130**). Internuclear distances are in Å[812]

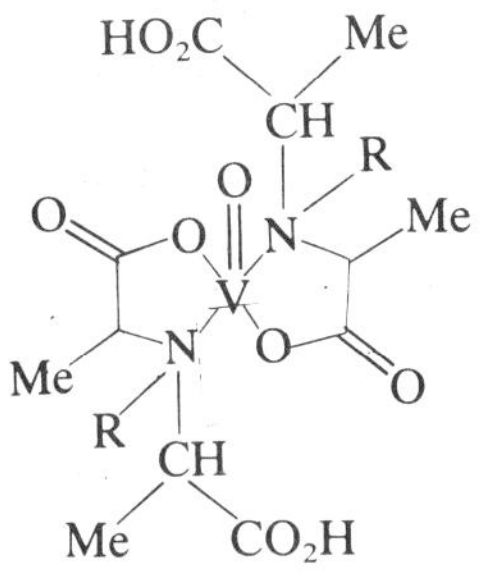

(**131**)

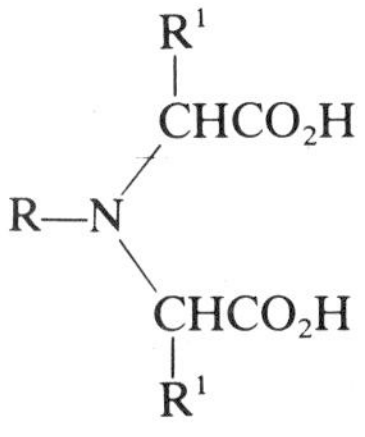

(**132**)

(**133**)

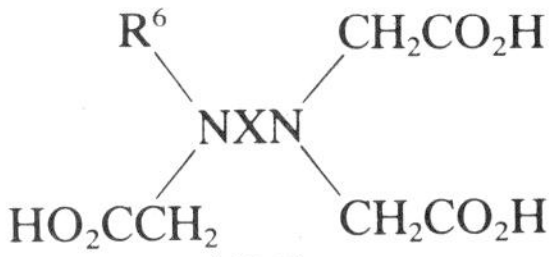

(**134**)

$(NH_4)_3[V_2O_3(nta)_2]\cdot 3H_2O$ is a binuclear species containing the $V_2O_3^{3+}$ core.[817,818] Near IR and electronic spectra of the solution and of the crystals indicated the establishment of an equilibrium (equation 57) with $K = 20\,\text{mol}^{-1}$. The binuclear complex gives very large extinctions in the visible and near IR region, and ESR spectroscopy also indicates an almost equal distribution of unpaired electrons around the two vanadium atoms. The two vanadium ions are equivalent, both in the crystalline state and in solution.

$$V^{IV}O(nta)(H_2O)^- + V^{V}O_2(nta)^{2-} \rightleftharpoons (nta)\overset{\overset{O}{\|}}{V}{-}O{-}\overset{\overset{O}{\|}}{V}(nta) + H_2O \qquad (57)$$

Saito and coworkers prepared the *cis*-aquaoxovanadium(IV) complex

Table 45 Stability Constants for Oxovanadium(IV) Complexes of Polyaminocarboxylic Acids (**132**)–(**134**) at 25 °C[a]

Ligand	*Ligand abbreviation*	*Log* β_{110}	$-Log\ K^{H}_{VOL}$ [c]	*Log* $\beta_{11\text{-}1}$ [c]	*Log* $\beta_{22\text{-}2}$	*Log* β_{111}	*Log* β_{120}
R = H	ida	8.98[1] (9.30[3])	5.8[1]	3.2	8.9[2] (8.21[3])	11.25[3]	—
R = Me	mida	9.56[1] (9.44[5])	5.9[1] (5.76[5])	3.7	10.1[2] (9.83[5])	—	—
R = NH_2	hda	7.61[1]	5.3[1]	2.3	7.2[2]	—	—
R = CH_2CH_2OH	eida	9.60[1] (9.26[5])	5.2[1] (5.11[5])	4.4	12.2	—	—
R = CH_2CO_2H	nta	11.47[1] (12.3[5])	7.06[1] (7.15[5])	4.4	—	—	—
R = $CH(Me)CO_2H$	mnta	11.77[1]	7.30[1]	4.5	—	—	—
R = $CH_2C_5H_5N$	pmida	11.30[1]	6.61[1] (6.45[6])	3.1	—	—	—
R = $C_6H_4(CO_2H)$	cfida	9.49[1]	6.97[1]	2.9	—	—	—
R = $C_3H_2N_2O_3$	uda	>14[1]	7.18[1]	6.8	—	—	—
R = OH	hida	7.16[9] (9.26[5])	5.0[9] (5.11[5])	2.2[9]	7.9[9]	—	13.26 (21.9[10])
R^1 = Me	idpa	9.54[9]	6.1[9]	3.44[9]	9.88[9]	—	—
R = OH, R^1 = Me	hidpa	7.34[9]	5.0[9]	2.3[9]	8.1[9]	—	12.85 (23.4[10])
R^2, R^3, R^4, R^5 = H	edda	13.4[1]	7.29[2]	6.1[2]	—	—	
R^2, R^4 = Me	eddpa	13.34[1]	7.50[2]	5.8[2]	—	—	
R^2, R^3, R^4, R^5 = Me	eddiba	12.23[1]	7.88[2]	4.4[2]	—	—	
$R^6 = CH_2CO_2H$, $X = CH_2CH_2$	edta	18.77[7]	—	—	—	3.0[7]	
$R^6 = CH_2CO_2H$, $X = (CH_2)_2O(CH_2)_2O(CH_2)_2$	egta	10.48[1]	5.58[6]	—	—	—	
$R^6 = (CH_2)_2CH_2CO_2H$, $X = C_6H_{10}$	dcta	20.1[6]	—	—	—	—	
$R^6 = CH_2CH_2OH$, $X = CH_2CH_2$	hedta	17.12[1] (20.1[6])	—	—	—	—	
N,N-Bis(hydroxyethyl)glycine		6.35 and 11.55[8 d]					
Triethylenetetraminehexaacetic acid	ttha					8.95, 4.36, 2.32[4,e]	

[a] Unless otherwise stated β_{xyz} are according to equation (15).
[b] Unless otherwise stated the R groups are H atoms.
[c] K^{H}_{VOL} corresponds to the reaction $VOL(H_2O) \rightleftarrows VOL(OH) + H^+$ and $\beta_{11\text{-}1} = \beta_{110} \times K^{H}_{VOL}$.
[d] The values correspond to log β_{110} and log β_{120} respectively at 20 °C and $I = 0.1$ M $NaClO_4$.
[e] The values correspond to the protonation constants K^{H}_{VOHL}, $K^{H}_{VOH_2L}$ and $K^{H}_{VOH_3L}$, respectively, at 25 °C and $I = 0.5$ M $NaClO_4$.

1. J. Felcman and J. J. R. Fraústo da Silva, *Talanta*, 1983, **30**, 565 ($I = 0.1$ M KNO_3).
2. J. Felcman, Dissertation, Pontifícia Universidade Católica de Rio de Janeiro, 1983 ($I = 0.1$ M KNO_3).
3. A. Napoli and L. Pontelli, *Gazz. Chim. Ital.*, 1973, **103**, 1219 ($I = 0.5$ M $NaClO_4$).
4. A. Napoli, *Gazz. Chim. Ital.*, 1976, **106**, 597 ($I = 0.5$ M $NaClO_4$).
5. A. Napoli, *J. Inorg. Nucl. Chem.*, 1977, **39**, 463 ($I = 0.5$ M $NaClO_4$).
6. (a) L. G. Sillen and A. E. Martell, 'Stability Constants of Metal Ion Complexes', The Chemical Society, London, 1964; (Special Publ. no. 17) and 1971 (Special Publ. no. 25); (b) R. M. Smith and A. E. Martell, 'Critical Stability Constants', Plenum, New York, 1976 and 1982 (1st suppl.).
7. G. Schwarzenbach, R. Gut and G. Anderegg, *Helv. Chim. Acta*, 1954, **37**, 936 ($I = 0.1$ M KNO_3 and $t = 20$ °C).
8. R. S. Saxena and A. K. Gupta, *J. Electrochem. Soc. India*, 1980, **29**, 275 ($I = 0.1$ M $NaClO_4$).
9. J. Felcman, M. C. T. A. Vaz and J. J. R. Fraústo da Silva, *Inorg. Chim. Acta*, 1984, **93**, 101 ($I = 0.10$ M KNO_3).
10. G. Anderegg, E. Koch and E. Bayer, *Inorg. Chim. Acta*, 1987, **127**, 183.

$Na_2[VO(pida)(H_2O)]\cdot 3.5H_2O$ ($pida^{4-}$ = ((phosphonomethyl)imino)diacetate), which undergoes protonation at the phosphonate oxygen with $pK_{a_1} = 3.8$ and deprotonation from the coordinated water with $pK_{a_2} = 8.4$.[819] All kinetic parameters were considered consistent with a hydrogen-bonded aqua ligand in $[VO(pida)(H_2O)]^{2-}$ having an intermediate character between aqua and hydroxo ligands.

33.5.9.4 *Other polydentate ligands*

(*i*) *Polydentate ligands involving nitrogen and oxygen coordination*

Several complexes with pyridinecarboxylic acid derivatives were isolated. Two forms of VO^{2+} lutidinato tetrahydrate were obtained. One is green and orthorhombic (see Figure 35), the other is blue and triclinic. Both have a square pyramid of O donors around V^{IV} and nitrogen coordination in the sixth position.[820]

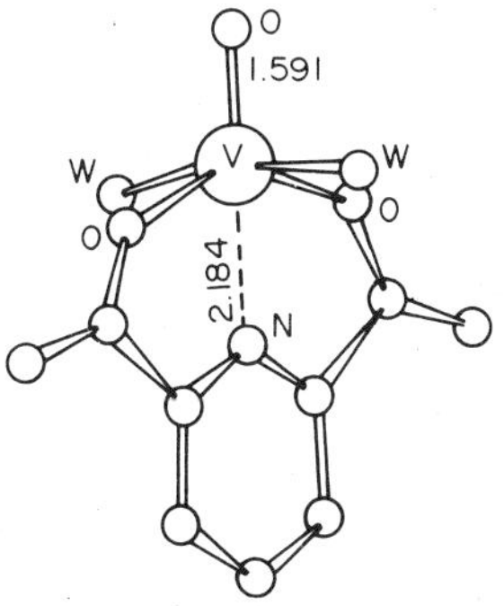

Figure 35 Perspective view of the green (orthorhombic) form of $[VO(lut)(H_2O)]\cdot 2H_2O$ (H_2lut = lutidinic acid). The V atom is six-coordinate in a distorted octahedral site which closely approximates C_{2v} symmetry; the remaining two water molecules (w) link the chelate molecules by hydrogen bonding. Some other internuclear distances are V—O(carboxylate) = 2.017 and V—O(water) = 2.027 Å

With the hydrazides (**135**)–(**137**) from lutidinic acid, monomeric compounds were obtained[432] by reaction (58). For all complexes, the IR spectra suggest that the ligands (**135**)–(**137**) are pentadentate.

$$\begin{array}{c} VOSO_4 \\ \text{or} \\ VOCl_2 \end{array} + \text{hydrazides (L)} \xrightarrow[\text{water–acetone}]{\text{water–ethanol or}} \begin{array}{c} VOLX_2 \\ \text{or} \\ VOL \end{array} \qquad (58)$$

(**135**)–(**137**)

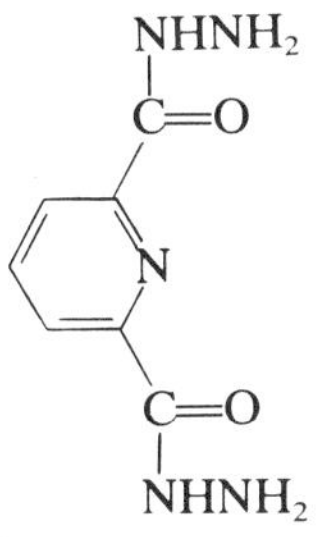

(**135**)

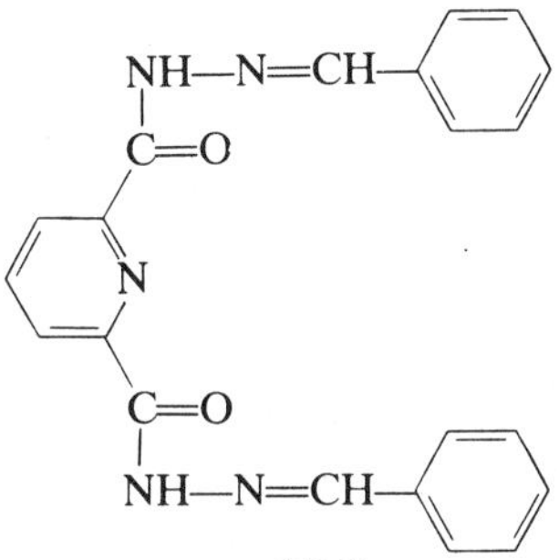

(**136**)

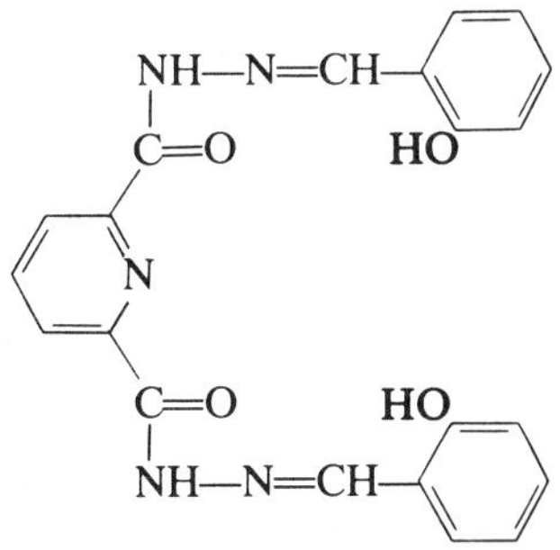

(**137**)

The stability constants[821] for the 8-hydroxyquinoline (oxine) complexes have been questioned owing to the ease with which the V^{IV} complexes undergo oxidation, even by atmospheric O_2.[355,687] However, adducts $[VO(oxine)_2L]$ (L = py or other nitrogen bases) were more stable towards oxidation. The V^{IV} oxine complex gradually turns magenta black when extracted into a nonpolar solvent, indicating formation of the V^V oxine complex in air, but in the presence of a nitrogen base (*e.g.* py), the extract remains yellow. The adducts can be schematically represented as in equation (59).[687]

$$\mathrm{VO(oxine)_2 + base \rightleftharpoons VO(oxine)_2(base)} \qquad (59)$$

$$\left.\begin{array}{c}\mathrm{VO(OMe)(oxine)_2}\\ \text{or}\\ \mathrm{VO(OH)(oxine)_2}\end{array}\right. + \mathrm{VO(oxine)_2 + OH^-} \longrightarrow \mathrm{[(V^VO)(oxine)_2O(V^{IV}O)(oxine)_2]^-} + \begin{array}{c}\mathrm{MeOH}\\ \text{or}\\ \mathrm{H_2O}\end{array} \qquad (60)$$

An equimolar mixture of [VO(OMe)(oxine)$_2$], [VO(oxine)$_2$] and base in methanol results in the binuclear complex (equation 60). Likewise, an equimolar mixture of [VO(OH)(oxine)$_2$], [VO(oxine)$_2$] and base in methanol yields the same product.[822] This V^{IV}–V^V dimer was isolated as the tetra-*n*-butylammonium salt.

Early solid complexes include [VO(oxine)$_2$], [VO(oxine)$_2$py] and [VO(5,7-X$_2$oxine)$_2$] (X = Br, I).[357] [VOX(oxine)] (X = Cl, Br, OH) were also prepared with magnetic moments dependent on temperature.[823] On the contrary, for brown [VO(oxine)$_2$], prepared from $VOSO_4$ in aqueous solution,[387] $\mu_{\text{eff}}^{\text{r.t.}}$ is near the spin-only value and $\nu(\mathrm{V{=}O}) = 979\ \mathrm{cm^{-1}}$.[595] These values indicate a monomeric structure and this was confirmed for its 2-methyl derivative by X-ray diffraction (Figure 36).[824]

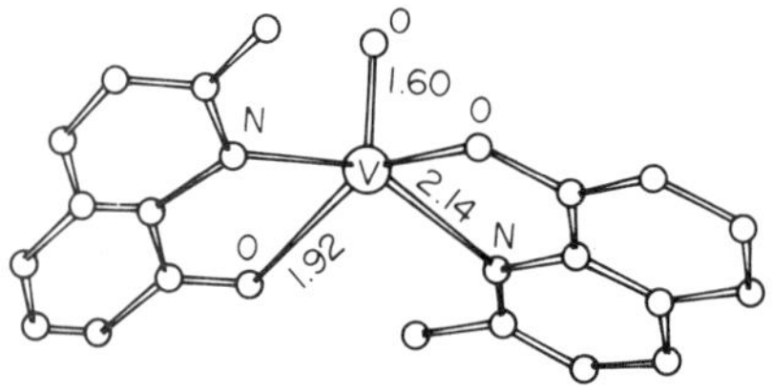

Figure 36 A perspective view of the molecular structure of [VO(2-methyl-8-quinolinolato)$_2$] with some internuclear distances (in Å)[824]

Attempts to prepare adducts directly from [VO(oxine)$_2$] (**138**) were unsuccessful and a trigonal bipyramidal structure was suggested for this compound as for [VO(2-Me-oxine)$_2$] (**139**).[825] The [VO(oxine)$_2$X] compounds were prepared by addition of $VOSO_4$ to an aqueous solution containing 8-hydroxyquinoline and the base X. For the adducts [VO(oxine)$_2$X], a correlation was found between $\nu(\mathrm{V{=}O})$ and the pK_a of the X ligand, except with bases with substituents where steric hindrance may be operating, and no evidence was obtained for the existence of both *cis* and *trans* isomers, unlike the case of [VO(acac)$_2$] adducts (Section 33.5.5.4.ii). For the tris complexes [VO(oxine)$_3$]$^-$, an octahedral structure in which one oxine ligand is unidentate and bonded to V was proposed.[825]

[VO(oxine)$_2$] (**138**) and [VO(oxine)$_2$py]·C_6H_6 (**140**), prepared by a ligand displacement reaction (equation 61), had $\nu(\mathrm{V{=}O}) = 963$ and $945\ \mathrm{cm^{-1}}$ and $\mu_{\text{eff}} = 1.75$ and 1.64 BM, respectively. When 8-quinolinol is replaced by 2-methyl-8-quinolinol in reaction (61), [VO(2-Me-oxine)$_2$] is obtained, with $\nu(\mathrm{V{=}O}) = 970\ \mathrm{cm^{-1}}$. Complexes (**138**) and (**140**) react readily with O_2 and a diamagnetic dinuclear μ-oxo V^V species [(oxine)$_2$—OV—O—VO(oxine)$_2$] is produced.[826] Nitrobenzene and 2-nitrosobiphenyl also react in a similar manner in THF giving the same complex. With *p*-benzoquinone a dinuclear diamagnetic V^V complex [(oxine)$_2$OVOC$_6$H$_4$OVO(oxine)$_2$] is obtained.

$$\mathrm{VO(acac)_2 + 2Hoxine \underset{C_6H_6}{\rightleftharpoons} VO(oxine)_2 + 2Hacac} \qquad (61)$$

The preparation of [VCl$_2$(oxine)$_2$] by the reaction of [VO(oxine)$_2$] with $SOCl_2$ in benzene or in 1,4-dioxane with PCl_5 (Section 33.5.5.4.ii) was found to be much easier than from [VCl$_4$][555] as it proceeds at room temperature in a few minutes.

Early publications with N,O donor ligands include oxovanadium(IV) and (V) complexes with a large number of substituted hydroxytriazenes.[828,829] The vanadyl complexes are of the [VOL$_2$] type with $\mu_{\text{eff}}^{\text{r.t.}} = 1.6$–2.1 BM and cryoscopic measurements in freezing benzene suggest a monomeric nature. In basic medium containing pyridine, 2-picoline or aniline, the complexes are oxidized and produce diamagnetic V^V complexes formulated as [VO$_2$LX] compounds.

Several complexes with hydroxyoximes (especially aromatic) have been prepared. Typically, an ethanolic solution of ligand is added to a solution of a vanadyl salt, the pH and temperature are adjusted and the solid precipitates. Most complexes are of $[VOL_2]$ type and include ligands L with HL = 2-hydroxybenzaldoxime[830,831] and derivatives,[830] several 2-hydroxyalkylphenone oximes and derivatives (**141**),[832–838] salicylaldoxime (**142**),[831,839] 2-hydroxynaphthaldoxime[830,840] and *o*-vanillin oxime (**143**).[839]

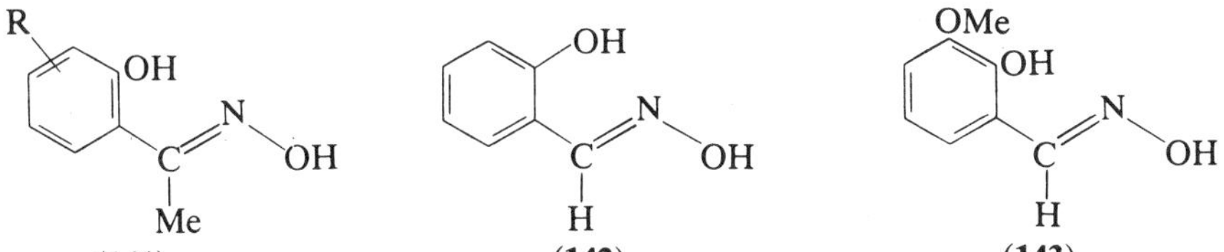

(**141**) (**142**) (**143**)

Other complexes with N,O donor ligands include the dimeric compound (**144**) and monomeric (**145**).[841] The ESR spectra of the vanadyl dimer (**144**) in frozen DMSO and DMF exhibited only zero-field splitting and the V—V distance estimated from this was 8.66 Å.

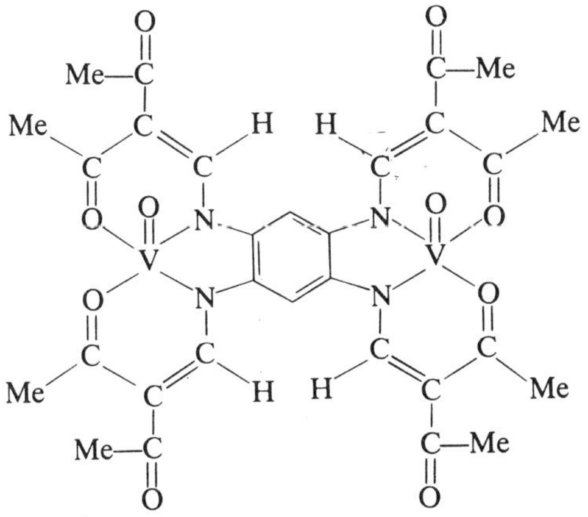

(**144**)

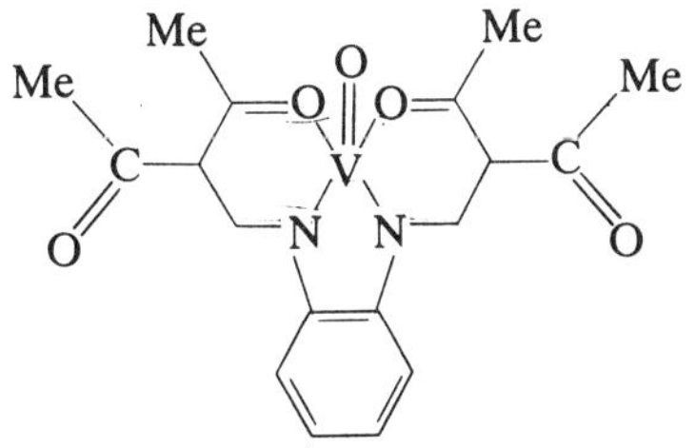

(**145**)

$[VOL_2]$ (HL = salicylaldehyde hydrazone (salhy) and salicylaldehyde phenylhydrazone (salpheny)) are monomeric ($\mu_{eff}^{r.t.}$ = 1.76 and 1.74 BM).[842,843] Reaction of $SOCl_2$ in DMF gave the corresponding deoxygenated $[VCl_2L_2]$.[843] $\mu_{eff}^{r.t.}$ values are 1.78 and 1.79 BM for L = salhy and salpheny, respectively, suggesting a d^1 configuration but $[VCl_2(salhy)_2]$ in DMF at 77 K gives no ESR signal.[843] As in the case of the porphyrinato vanadium(IV) dihalide, the absence of an ESR spectrum was attributed to a degenerate ground state.[843] The reaction of $[VOCl_2(salhy)_2]$ with several nitrogen bases such as aniline, 1,4-diaminobenzene, aminopyridine and adenine proceeds according to equation (62).[843]

$$VCl_2L_2 + 2Hamine \longrightarrow VL_2(amine)_2 + 2HCl \quad (62)$$

(L = salicylaldehyde hydrazone and salicylaldehyde phenyl hydrazone)

Complexes with thiocarbohydrazones (**146**; R = H, Me, OMe, Cl) have been prepared in an ethanolic medium.[844] They have 1:1 stoichiometry, and $\mu_{eff}^{r.t.}$ are 1.73–1.94 BM; ν(V=O) was at 980 cm^{-1}.

Oxovanadium(IV) complexes have also been prepared with the carbohydrazones and thiocarbohydrazones formed in the condensation of carbohydrazide or thiocarbohydrazide with benzaldehyde, *o*-nitrobenzaldehyde, anisaldehyde, cinnamaldehyde, acetophenone and 2-acetylpyridine.[845] $\mu_{eff}^{r.t.}$ values are 1.68–1.72 BM and ν(V=O) is at 960–980 cm^{-1}. The compounds, $[VO(L)_2(H_2O)]$ (**147**), were suggested to be of structure (**12**) (p. 487).[845]

(*ii*) *Polydentate ligands involving oxygen, sulfur and/or nitrogen coordination*

A complex with 8-mercaptoquinoline $[VO(thiooxine)_2]$ is magnetically dilute.[823,358] Complexes with Cl and Br derivatives of 8-mercaptoquinoline have also been reported.[357] In

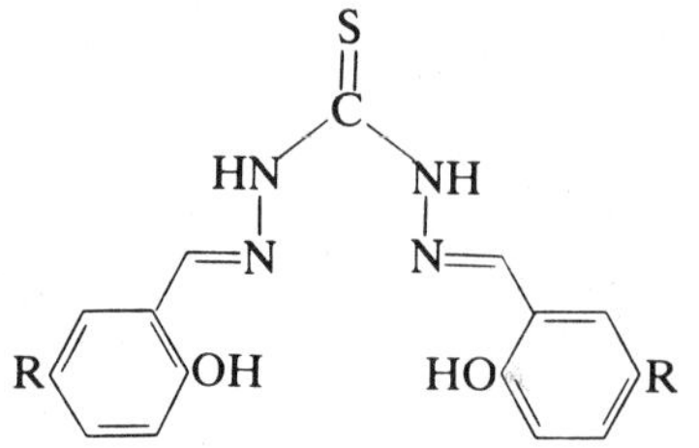

(146)

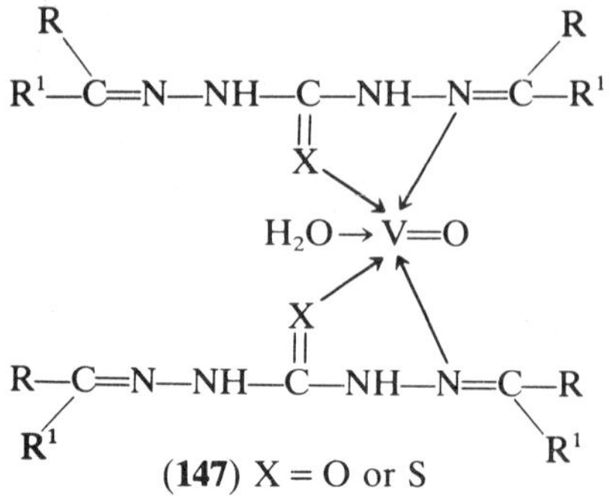

(147) X = O or S

$[VO(QT)_2]$ (HQT = **148**), with $\mu_{eff}^{r.t.} = 1.85$ BM, coordination involves sulfur (as S^{-2}) and nitrogen.[846]

Hodge *et al.* synthesized complexes containing the N,S and O,S bidentate ligands **(149)**–**(152)**.[847] These were of $[VOL_2]$ type with a square pyramidal geometry (except $[VO(pto)_2]$) (pto = **152**); all attempts to prepare *N,N*-dialkyl-2-aminoethanethiol VO^{2+} complexes failed. $[VO(pto)_2py]$ appears more stable than $[VO(abt)_2py]$, and may exist as a mixture of *cis* and *trans* isomers (in relation to the oxo group). The reaction of $[VO(abt)_2py]$ with a variety of 4-substituted pyridine *N*-oxides, where the steric factor is expected to be approximately constant, resulted in a systematic change in the visible spectra, and the changes correlated with the Hammett σ values of the bases.[847]

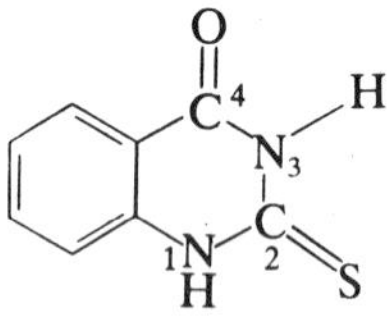

(148)

$NH_2CH_2CH_2SH$
(149) Haet

$C_{10}N_{21}NHCH_2CH_2SH$
(150) Hdaet

(151) Habt

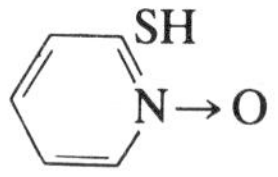

(152) Hpto

With 3-hydroxyquinazoline-4(3*H*)-thiones (HQ) (Scheme 20; Z = S, Y = NOH), the ESR spectra were explained assuming equilibria (63).[848] Compounds isolated, $[VOQ_2]$, involved sulfur and oxygen atoms in a square pyramidal geometry.

$$VO^{2+} \overset{HQ}{\rightleftharpoons} VOQ^+ \overset{HQ}{\rightleftharpoons} VOQ_2 + 2H^+ \qquad (63)$$

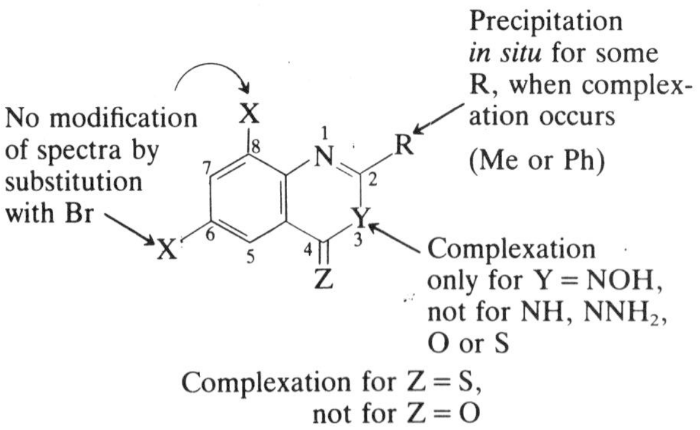

Scheme 20

Bozis and McCormick prepared several green complexes with monothio-β-diketones **(153)**–**(155)**, formulated as monomeric square pyramidal $[VOL_2]$.[849] With **(156)** and **(157)**, the complexes have a golden yellow colour, diffuse reflectance spectra differ from those with **(153)**–**(155)** and ν(V=O) was at 868 and 876 cm^{-1} for $[VO(ttmbs)_2]$ and $[VO(nebas)_2]$, respectively.[850] The solid complexes probably have a structure that involves —V—O—V— bridging, as found for [VO(salnpn)] (Figure 30).

Early reports on tetrasulfur tetranitride (S_4N_4 = L) suggest the formation of $[VCl_4L]$

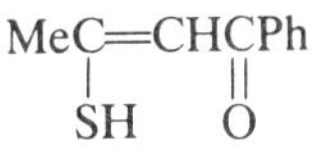

$$MeC{=}CHCPh$$
$$\underset{\text{SH}}{|} \quad \underset{\text{O}}{||}$$
(153) Hbas

$$PhC{=}CHCPh$$
$$\underset{\text{SH}}{|} \quad \underset{\text{O}}{||}$$
(154) Hdbms

$$PhC{=}CHCOEt$$
$$\underset{\text{SH}}{|} \quad \underset{\text{O}}{||}$$
(155) Hebas

$$p\text{-}O_2NC_6H_4C{=}CHCOEt$$
$$\underset{\text{SH}}{|} \quad \underset{\text{O}}{||}$$
(156) Hnebas

$$C_4H_3SC{=}CHCCF_3$$
$$\underset{\text{SH}}{|} \quad \underset{\text{O}}{||}$$
(157) Httmbs

adducts.[851] More recently, it was found that S_4N_4 reacts with excess of VCl_4 in CH_2Cl_2 forming a mixture of $VCl_2(S_2N_3)$ and $S_2N_2{\cdot}VCl_4$. Sublimation at 110 °C gives black needles $[VCl_2(S_2N_3)]$, chlorine-bridged dimers linked by additional weak V—N interactions to form polymeric chains (Figure 37).[852] Thus, only nitrogen coordination is involved. The six-membered VNSNSN ring is planar.

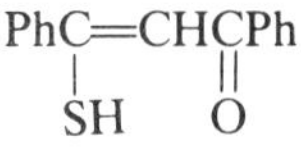

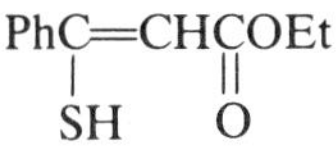

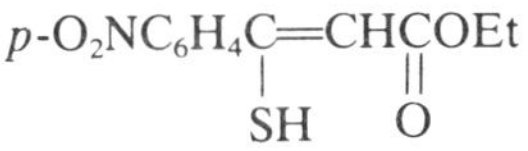

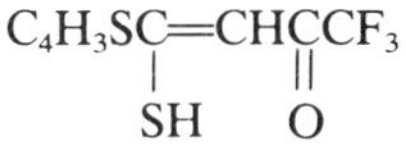

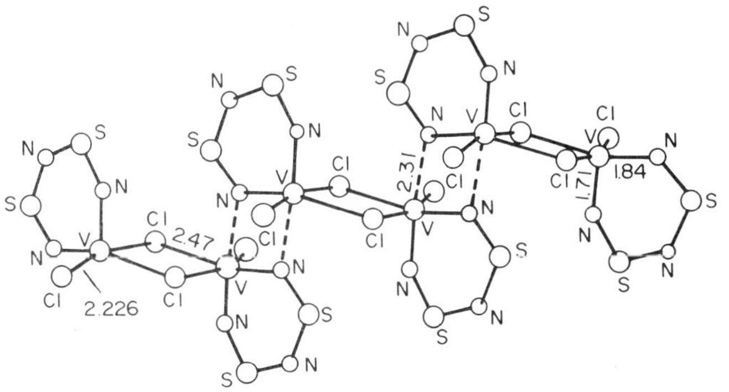

Figure 37 Three-dimer section of the polymeric chain of the compound $[VCl_2(S_2N_3)]$.[852] The longer V—N 'interactions' which link the dimers are indicated

33.5.10 Mixed-ligand Complexes

Strictly speaking, all oxovanadium(IV) complexes should be considered of mixed-ligand type. Of course, such criteria have not been used and this section concerns complexes with at least two different chelating ligands.

The overall reaction for the formation of a 1:1:1 mixed-ligand chelate in aqueous solution may be written as in equation (64) where A^{n-} is considered the 'primary ligand' and L the 'secondary ligand'. Ternary complexes will form only if complex $[VOAL]^{(2-m-n)}$ is more stable than the binary complexes of VO^{2+} with A^{n-} or L^{m-}. Table 46 summarizes most of the data available. For a given primary ligand, the relative stabilities of [VOAL] follow the order of the relative basicities of the secondary ligand, although steric factors might also be operating in some cases; steric interference increases with the increasing size of the ligand and invariably decreases the stability of the chelate.[633,676,853–855]

$$VO^{2+} + A^{n-} + L^{m-} \rightleftharpoons VOAL^{(2-m-n)} \tag{64}$$

$$\beta_{VOAL} = \frac{[VOAL]}{[VO^{2+}][A^{n-}][L^{m-}]} \tag{65}$$

Early mixed-ligand complexes $[VO(A)(L)]$[856–858] with A potentially tridentate (or bidentate) (lutidinic, chelidamic, collidinic, iminodiacetic and oxalic acids) and L a bidentate ligand (phen, bipy, acac, glycine, oxine, salicylaldoxime and others) appear to be monomeric and were prepared from the parents $[VO(A)(X)(H_2O)]$, where X is a monodentate ligand (H_2O, py, 2-picoline, aniline, *etc.*), by displacement. One such, $[VO(ida)(bipy)]{\cdot}2H_2O$, consists of monomeric units with two loosely bound water molecules of crystallization.[859] The geometry around V is octahedral ($d(V{=}O) = 1.596$ Å). Several $[VO(A)(L)]$, A being the anions of oxalic, malonic and maleic acids and L = phen, were prepared from $[VO(A)(H_2O)_2]$.[860] The ease of preparation follows the order: oxalate > malonate > maleate.

Table 46 Stability Constants β_{VOAL} of Some Mixed-Ligand Chelates of Oxovanadium(IV)[a]

H_nA	H_mL	Log β_{VOAL}	Ref.	H_nA	HmL	Log β_{VOAL}	Ref.
Oxalic acid	Phthalic	9.70	1	Picolinic acid	Oxalic	10.40	2
	Maleic	10.56	1		Phthalic	9.48	2
	SA	17.87	1		Maleic	10.53	2
	SSA	15.83	1		SA	18.63	3
	HQSA	14.45	1		SSA	16.44	3
Anthranilic acid	Oxalic	9.29	4	Hippuric acid	SA	14.78	5
	Malonic	10.62	4		SSA	13.54	5
	Phthalic	9.40	4		HQSA	12.55	5
	Maleic	10.24	4				
	SA	10.62	4				
	SSA	9.83	4				
HQSA	Phthalic	13.88	6	Catechol[d]	Succinic	8.68	8
	Maleic	15.96	6		Adipic	8.24	8
	SA	23.21	6		Itaconic	9.15	8
	SSA	21.22	6		Phthalic	8.79	8
	Tiron	26.04	7		3,5-DinitrosoSA	12.61	8
	Chromotropic	26.36	7		SSA	23.16	8
Pyridine-2-aldoxime	Oxalic	12.83	9	bipy[c]	Oxalic	8.88	11
	Phthalic	12.63	9		Malonic	11.34	11
	Maleic	13.65	9		Phthalic	10.97	11
	SA	21.63	10		Maleic	11.44	11
	SSA	19.52	10		Glycolic	23.26	11
	HQSA	18.73	10		SA	27.78	11
					SSA	22.93	11
phen	SA	19.92	12	bipy	SA	18.51	12
	SSA	18.12	12		SSA	16.87	12
	HQSA	16.75	12				
phen[d]	*o*CA	19.22	13	bipy[d]	*o*CA	18.53	13
	*m*CA	18.90	13		*m*CA	18.17	13
	*p*CA	18.18	13		*p*CA	18.36	13
Succinic acid[d]	Malonic	8.70	14	Iminodiacetic acid[e]	Oxalic	7.66	15
	Phthalic	8.34	14		Malonic	11.85	15
	SSA	11.29	14		SA	18.35	15
	3,5-DinitrosoSA	9.40	14		SSA	18.35	15
					Mandelic	17.52	15
Adipic acid[d]	Malonic	9.68	14		Glycolic	17.96	15
	SSA	11.67	14				
	3,5-DinitrosoSA	9.36	14				
Itaconic acid[d]	SSA	11.39	14				
	3,5-DinitroSA	9.68	14				
SSA[d]	3,5-DinitroSA	11.86	14				
	4-HydroxySA	30.05	14				
3,5-DinitrosoSA[d]	4-HydroxySA	14.30	14				

[a] Unless otherwise stated, the β_{VOAL} values were determined in 0.1 M KNO_3 solutions at $T = 30$ °C by pH-potentiometric titrations.
[b] SA, SSA and HQSA = salicylic, 5-sulfosalicylic and 8-hydroxy-quinoline-5-sulfonic acids, respectively; *o*CA, *m*CA and *p*CA = 2-hydroxy-*n*-methylbenzoic acid ($n = 3$, 4 and 5, respectively).
[c] Values determined in 0.1 M KNO_3 solutions at $T = 35$ °C.
[d] Values determined in 0.1 M $NaClO_4$ solutions at $T = 30$ °C.
[e] Values determined in 0.1 M KNO_3 solutions at $T = 25$ °C.

1. S. P. Singh and J. P. Tandon, *Bull. Acad. Pol. Sci., Ser. Sci. Chim.*, 1974, **22,** 725.
2. S. P. Singh and J. P. Tandon, *Acta Chim. Acad. Sci. Hung.*, 1975, **84,** 419.
3. H. S. Verma, S. P. Singh and J. P. Tandon, *Talanta*, 1979, **26,** 982.
4. A. K. Jain and G. K. Chaturvedi, *J. Indian Chem. Soc.*, 1979, **56,** 850.
5. H. S. Verma, B. P. Singh and S. P. Singh, *J. Indian Chem. Soc.*, 1981, **58,** 315.
6. S. P. Singh and J. P. Tandon, *Monatsh. Chem.*, 1975, **106,** 271.
7. S. P. Singh and J. P. Tandon, *Acta Chim. Acad. Sci. Hung.*, 1976, **88,** 335.
8. R. G. Deshpande, D. N. Shelke and D. V. Jahagirdar, *Indian J. Chem., Sect. A*, 1977, **15,** 320.
9. S. P. Singh, H. S. Verma and J. P. Tandon, *J. Indian Chem. Soc.*, 1978, **55,** 539.
10. S. P. Singh, H. S. Verma and J. P. Tandon, *J. Inorg. Nucl. Chem.*, 1979, **41,** 587.
11. A. K. Jain, V. Kumari and G. K. Chaturvedi, *J. Indian Chem. Soc.*, 1978, **55,** 1251.
12. M. M. Islam, R. S. Singh and B. G. Bhat, *Indian J. Chem., Sect. A*, 1982, **21,** 751.
13. M. M. Islam, R. S. Singh and B. G. Bhat, *Trans. SAEST*, 1983, **18,** 32.
14. D. N. Shelke and D. V. Jahagirdar, *J. Inorg. Nucl. Chem.*, 1977, **39,** 2223.
15. A. K. Jain, R. C. Sharma and G. K. Chaturvedi, *Pol. J. Chem.*, 1978, **52,** 2259 (*Chem. Abstr.*, 1978, **88,** 198 690h).

Compounds [VO(acac)(2-aminobenzoato)]·nH_2O ($n = 0$ or 1) were prepared with rather low $\mu_{eff}^{r.t.}$ (1.49 and 1.34 BM for $n = 1$ and 0).[861] In $CHCl_3$, the ESR spectrum shows splitting of $A_\perp$ into A_x and A_y components ascribed to asymmetry of the ligands.[861] In [VO(oxine)(2-aminobenzoato)]·nH_2O ($n = 0$ or 1), a $\mu_{eff}^{r.t.}$ value of 1.33 BM was reported for $n = 1$.[862]

33.5.11 Multidentate Macrocyclic Nitrogen Ligands

33.5.11.1 Porphyrins

Much interest in environmental vanadium stems from its presence in fossil fuels, largely as porphyrins and related compounds which pose problems in refining and burning of high-vanadium-content fuels. The metal porphyrins in crude oils and oil shales have been extensively studied since their discovery by Treibs, who postulated that the principal compound in geological samples is vanadyl deoxophylloerythroetioporphyrin (DPEP) and developed a detailed scheme to show that this compound was formed from chlorophyll.[863,864] Based on Treibs' observations, metal porphyrins in crude oils show their biological origin. The literature is extensive and will be very briefly covered. Although the geoporphyrin system is more complex than Treibs' original scheme,[865] vanadyl porphyrins found in crude oils and bitumens are believed to be the end product of the diagenesis of chlorophyll. The vanadyl analogue of chlorophyll, [VO(DPEP)] (**158**), provides some insight into some of chlorophyll's properties, including its resistance to oxidation.[866] XANES/EXAFS experiments on pulverized coal gave arguments discounting the presence of vanadyl porphyrins.[867,868] More recently, the identification, including X-ray structure, of a vanadyl porphyrin isolated from a geological sample unlikely to have been chemically altered by the extraction procedure was reported.[869] Further, using asphaltenic fractions, from EXAFS, XANS, ESR and UV-vis spectroscopy vanadium is predominantly porphyrinic.[870] Several chemical and structural studies of vanadium porphyrins emphasize low valent metalloporphyrins and related systems with sulfur and selenium.[128] Two general methods can be used to prepare vanadyl porphyrins: (1) using $VO(acac)_2$ in phenol at reflux temperatures, or (2) using VCl_4 or VCl_3 in various solvents. In boiling quinoline, 100% incorporation is obtained in 2 h for *meso*-tetraphenylporphyrin (TPP) and octaethylporphyrin (OEP).[871,872]

ESR has been used to characterize vanadium compounds in crude oil.[354,873] In ESR spectra of [VO(TTP)], extrahyperfine structure from interactions with N atoms was detected in CS_2 and $CHCl_3$ glasses and the splitting constants $A_\parallel^N$, $A_\perp^N$ were calculated (2.9 ± 0.05 and 2.8 ± 0.05 G).[517]

In [VO(DPEP)] (**158**), the four nitrogens are coplanar and the vanadium lies 0.48 Å out of plane.[874] In [VO(OEP)] (**159**) (OEP = 2,3,7,8,12,13,17,18-octaethylporphyrinato), the four V—N distances average 2.10 Å, the same as for [VO(DPEP)] (excepting V—N(c)).[875] The vanadium is 0.543 Å from the mean plane of the four N atoms, in the range 0.48–0.72 Å predicted by EHMO calculations.[876] For [VO(TTP)] (**160**), [VO(ETP)] (**161**) and $[VO(ETP)_2](H_2Q)$ (**162**) (ETP = etioporphyrin and H_2Q = 1,4-dihydroxybenzene) (Figure 38), the vanadium is five-coordinate.[877] One interesting feature in (**162**) is the hydrogen bond between the quinol (H_2Q) and the VO group. The O···O distance is 2.82 Å and the V═O···O and C—O···O angles are 159.3 and 118.9°.

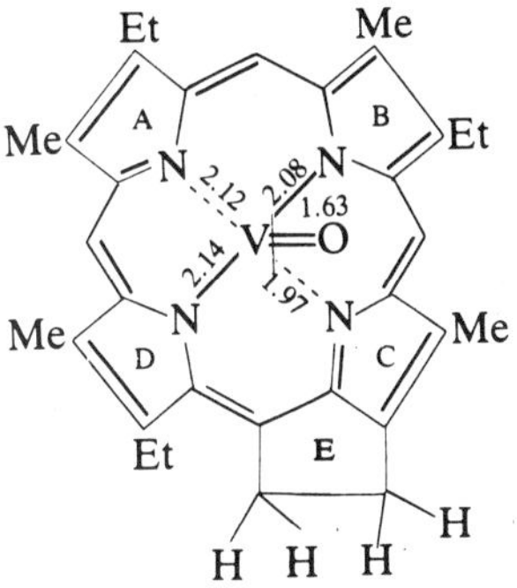

(**158**)

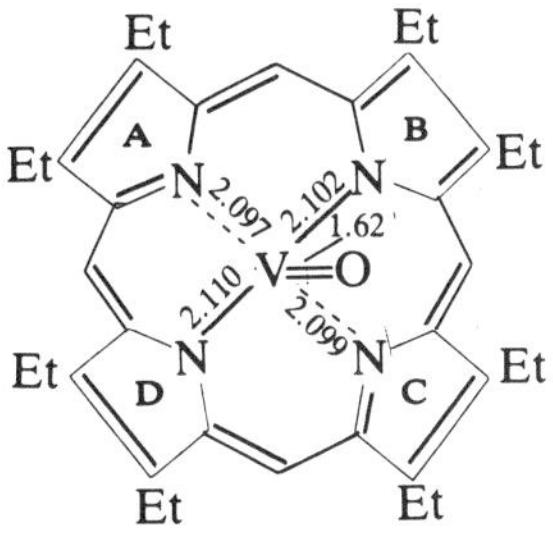

(**159**)

EXAFS with [VO(OEP)] and [VO(DPEP)] gave V═O and V—N bond lengths almost equal

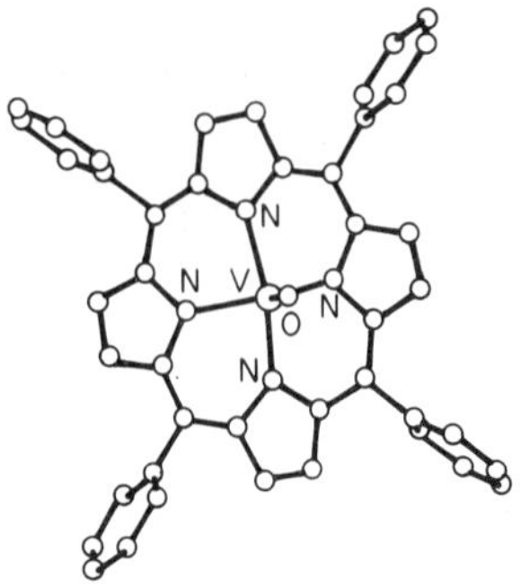

(160)

(a)

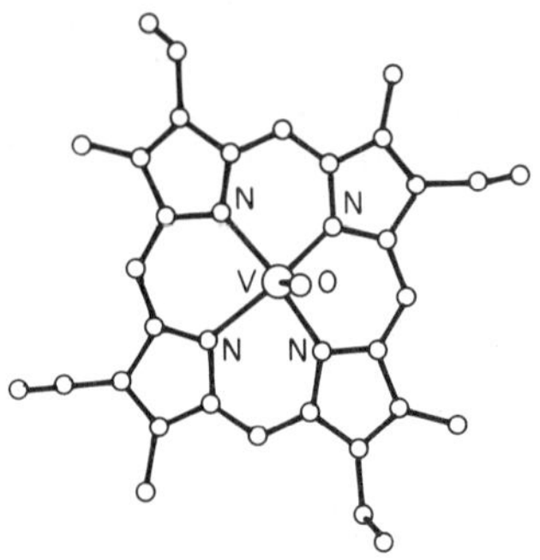

(161)

(b)

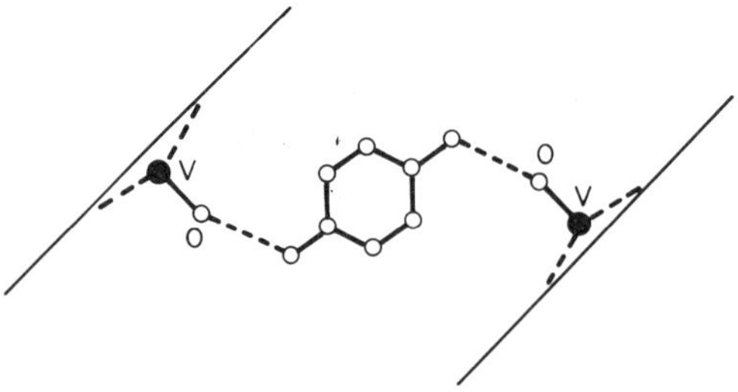

(162)

(c)

Figure 38 Schematic representation of the molecular structures of (a) [VO(TPP)], (b) [VO(ETP)] and (c) $[VO(ETP)]_2(H_2O)$ (side view).[877] The V=O bond distances are 1.625, 1.599 and 1.614 Å, respectively, and the vanadium 'N_4 plane' distances are 0.53, 0.49 and 0.51 Å, respectively. The V—N bond lengths are all in the range 2.04–2.09 Å

to the X-ray results,[878] showing that this is a useful tool for investigating the coordination sphere of metalloporphyrins; it is often difficult to obtain good quality crystals.

In solution, vanadyl porphyrins [VOP] can form adducts [VOPX] in the presence of a large excess of X (X = py or piperidine) but the equilibrium constants are rather small.[879–881]

For vanadyl uroporphyrin^{I}, spectra in alkaline medium are different from those in organic solvents or crystalline.[882] If a six-coordinate dihydroxide species $[V(OH)_2UroP]$ really forms, it would be the first example of a deoxygenated V^{IV} complex bound to two OH^- groups. However, the arguments presented leave the possibility of $[VO(OH)UroP]^-$ in alkaline solution.[882]

The action of SOX_2 or COX_2 (X = Cl or Br) on several VO^{2+} porphyrins gives the dihalogenovanadium(IV) porphyrins.[883] These are extremely reactive and act as precursors in the synthesis of low-valent porphyrins. $[VBr_2(OEP)]$ was studied by EXAFS and the two Br atoms were coordinated *trans*. In reactions of porphyrins with halides MX_n ($TiCl_4$, $SnCl_4$, $AlCl_3$, $AlBr_3$ and also $SbCl_5$ and $SOCl_2$), deoxygenation of the vanadyl occurs, octahedral $[VX_2P]$ being formed.[884] In the intermediate stage, the authors suggest the formation of a bimetallic complex of vanadium porphyrin with halide. In air, the products are dihalogen complexes of V^{IV} with mono- and di-cation porphyrin ligands. All are entirely converted to the original [VOP] when small amounts of water are added.

Reduction by zinc amalgam of a dry and oxygen-free THF solution containing $[VCl_2(TPTP)]$ (TPTP = *meso*-tetra-*p*-tolylporphyrinato) and addition of elemental sulfur yields [VS(TPTP)].[659] Similar thiovanadyl complexes were prepared with OEP, TPP and TMTP (TMTP = *meso*-tetra-*m*-tolylporphyrinato). Analytical results and mass spectra agree with the formulation [VS(por)], and ν(V=S) was at 550–565 cm^{-1} (see Section 33.5.6.1). FT-EXAFS of [VO(OEP)] and [VS(OEP)] were recorded and the difference FT spectrum yielded approximate values for the interatomic distances: $V{=}O \approx 1.6$ and $V{=}S \approx 2.06 \pm 0.02$ Å, which agrees with 2.061 Å for [VS(acen)].[659]

A convenient procedure for the isolation of selenovanadium(IV) porphyrins has been reported and involves the use of Cp_2TiSe_5 (instead of elemental selenium) (Scheme 21).[885,886] The IR spectra exhibit a medium-to-strong band at 434–447 cm^{-1} assigned to $\nu(V{=}Se)$. EXAFS with VSe(OEP) suggests that the V—Se bond is ~2.19 Å and that the V atom is ~0.47 Å above the plane of the four nitrogen ligands.[128]

$$VCl_2(TPTP) \xrightarrow[\text{THF, 24 h}]{\text{Zn amalgam}} V^{II}(TPTP)(THF)_2 \begin{array}{l} \xrightarrow{S_8} VS(TPTP) \\ \xrightarrow[\text{THF}]{Cp_2TiSe_5} VSe(TPTP) \end{array}$$

Scheme 21

33.5.11.2 Phthalocyanines

The VO^{2+} phthalocyanine [VO(pc)] has been the subject of patents as it is useful in photoelectrophoretic and xerographic imaging.[887] It can exist in at least three solid phases[888] and the structure of so-called phase II was determined.[889] It is composed of sheets of parallel and overlapping [VO(pc)] molecules and no discrete dimer pair exists. The vanadium is five-coordinate and square pyramidal, and is 0.575 Å above the 'plane' of the four nitrogens. The V—N distances do not differ significantly with a mean of 2.026 Å; the V=O distance is 1.580 Å.

ESR measurements of aqueous solutions of the VO^{2+} complex of 4,4'4,″,4‴-tetrasulfophthalocyanine (tspc) show that it is polymeric.[890] In dilute solutions (10^{-5} M), the dimeric form predominates and higher polymers increase with concentration.[891] The formation of adducts prevents polymerization and in solutions containing DMF most of the chelate is dimeric with a V—V distance of ~4.5 Å.

33.5.11.3 Tetraaza[14]annulenes and other macrocycles

Complexes with the tetradentate tetraaza[14]annulenes (163) and (164) ($= H_2L$), studied by spectroscopy,[892] were formulated as [VOL], probably similar to the complexes with porphyrins. $\nu(V{=}O)$ was at 967 and 940 cm^{-1} for (163) and (164). In the ESCA spectra, the nitrogen donors are equivalent.[892]

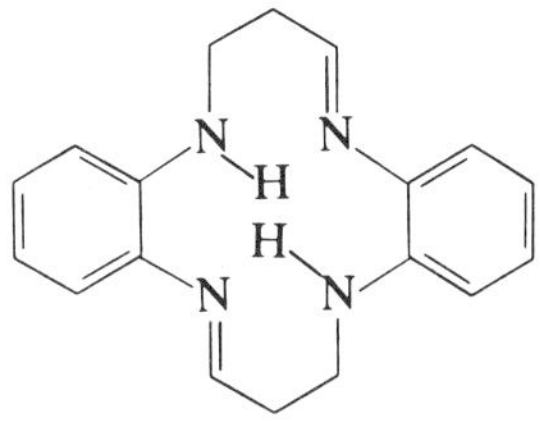

(163)   (164)

The tetramethyltetraaza[14]annulene (165) reacts with vanadyl acetate to yield (166) similar to complexes of (163) and (164).[660] It was the starting material for several reactions shown in Scheme 22, and exhibits a reversible one-electron oxidation wave at +0.245 V (*vs.* SCE) but reacts irreversibly with mild oxidants (O_2, Ag^+, Ce^{4+} or I_2) to yield (167). With Et_3N, (167) yielded a complex with properties similar to (166) but which mass spectral data indicate is a dimer. Structure (170) was proposed.

Reaction of (166) with anhydrous HCl deoxygenates the VO^{2+} compound, forming (168).[660] Its ESR parameters are those expected for the substitution by two chloride ligands. (168) with H_2S and base yielded the thiovanadyl analogue of (166) for which $\nu(V{=}S)$ was at 545 cm^{-1}. Hydrolysis of (169) regenerates (166).

On allowing $V(acac)_3$ to react with di(am)sar (172) for 3 days under N_2, a deep red-brown developed.[893] After separation (by cation exchange chromatography) of the red-brown product, its crystalline chloride was converted into the dithionate [V^{IV}(di(amH)sar-

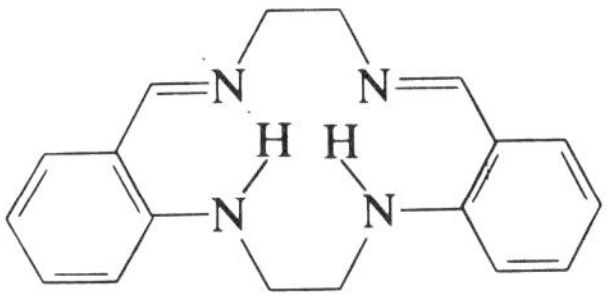

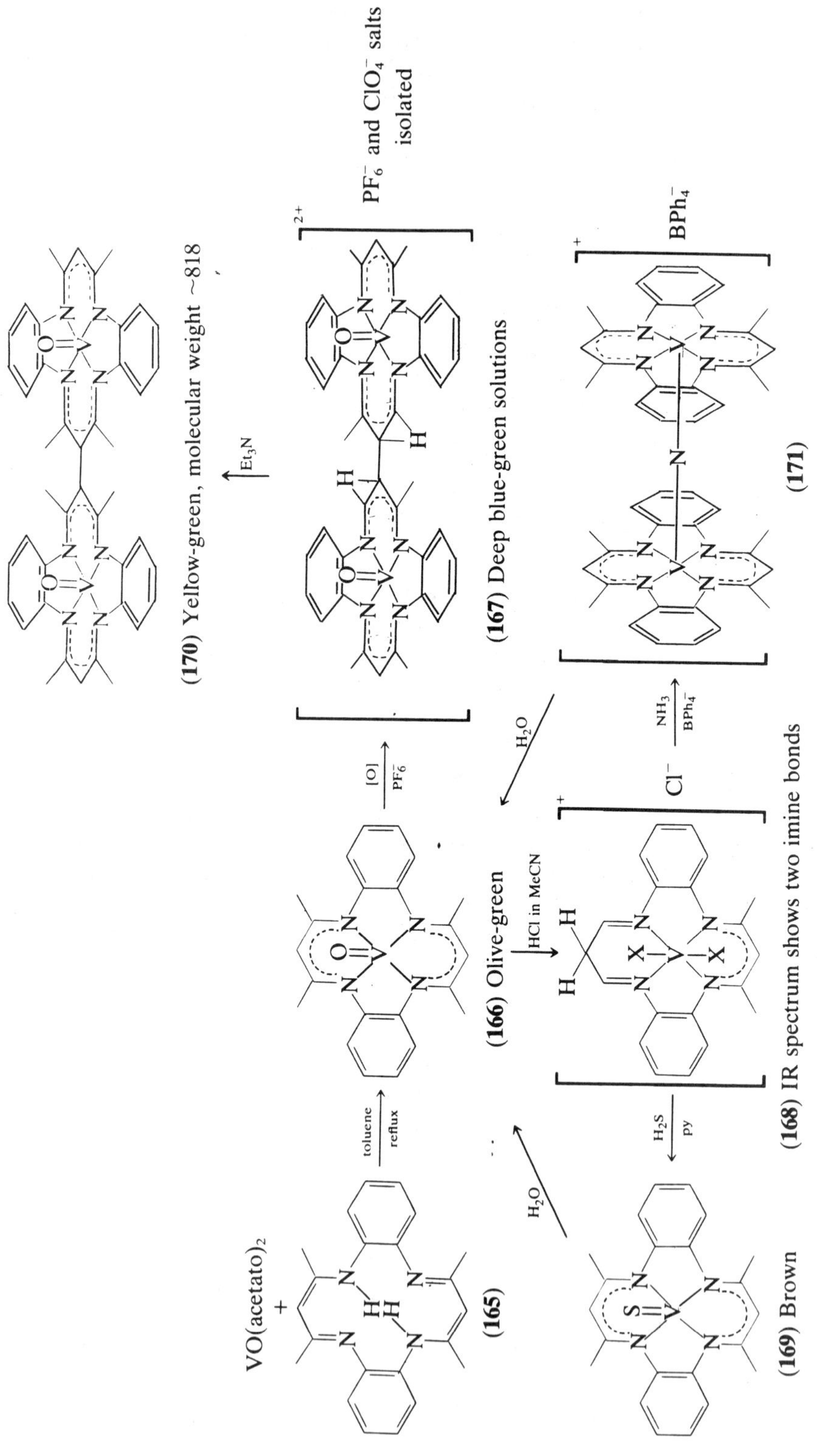
VO(acetato)$_2$
+
(165)
toluene
reflux
(166) Olive-green
[O]
PF_6^-
2+
PF_6^- and ClO_4^- salts isolated
(167) Deep blue-green solutions
Et_3N
(170) Yellow-green, molecular weight ~818
HCl in MeCN
Cl^-
(168) IR spectrum shows two imine bonds
H_2O
H_2S
py
(169) Brown
NH_3
BPh_4^-
(171)

Scheme 22

$2H)](S_2O_6)_2 \cdot 2H_2O$ whose geometry is close to trigonal prismatic with V—N in the range 2.06–2.10 Å. It has $\mu_{eff}^{r.t.} = 1.80$ BM and is stable in the pH range 1–10. Spectroscopic techniques have been used to study the V^{IV} di(am)sar complex[943] and other macrocyclic complexes.[944]

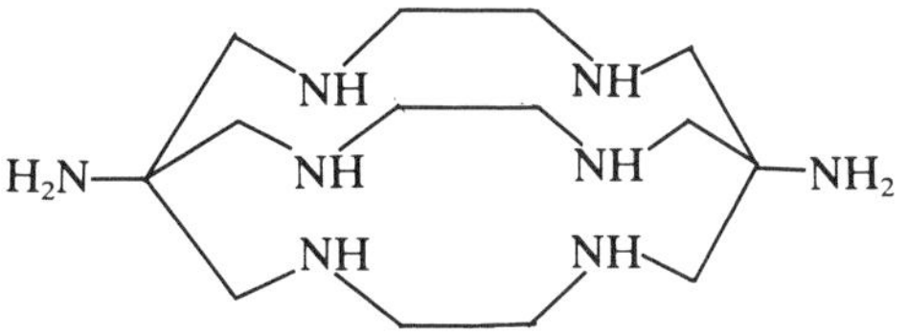

(**172**) di(am)sar

33.5.12 Polynuclear Complexes

33.5.12.1 Binucleating and compartmental ligands

Mononuclear VO^{2+} and heterodinuclear VO^{2+} complexes (with Cu^{2+}, Ni^{2+}, Mn^{2+}, Zn^{2+}, Pd^{2+} as the second metal ion) have been reviewed.[894] In this section, we shall mainly discuss more recent work. Many complexes were with Schiff bases from the reaction of β-triketones or β-ketophenols with ethylenediamine. One important aim has been to compare the magnetic, spectral and chemical properties for the metal ion M_a in $[M_a(diken)]$ and M_b in $[M_b(diketonato)_2]$, with the corresponding properties in the binuclear complexes formed (see equation 66).

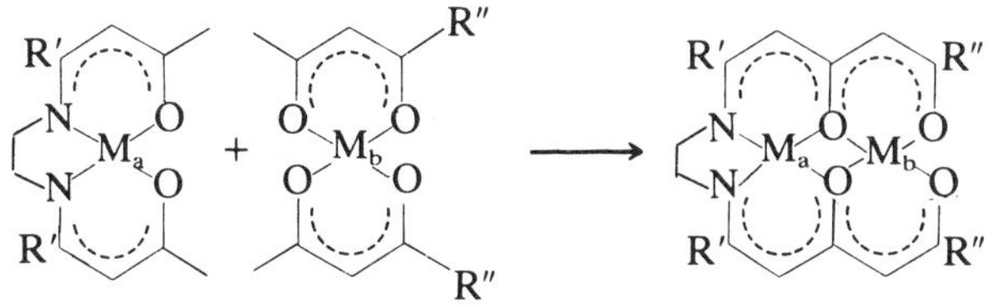

M_a(diken) M_b(diketonato)$_2$ (**173**) (66)

In mononuclear complexes with ligands of type (**173**), VO^{2+} has a strong preference for the O_2O_2 environment as clearly demonstrated by the single crystal molecular structure determinations with $[VO(H_2baaen)]$ (**174**; $R = (CH_2)_2$, $R' = Me$, $R'' = Ph$),[893,895,896] $[VO(H_2daaen)]$ (**174**; $R = (CH_2)_2$, $R' = R'' = Me$)[897] and $[VO(H_2daanpn)] \cdot CHCl_3$ (**174**; $R = (CH_2)_3$, $R' = R'' = Me$).[897] In the case of ligands (**178**; R = Me, Et), the outside O_2O_2 donor set shows a low tendency to coordinate and VO^{2+} coordinates in the N_2O_2 site as was found for the Ni^{2+} complex.[898]

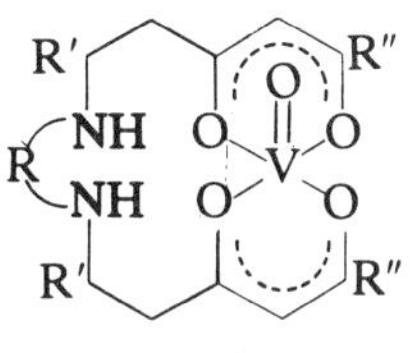

(**174**)

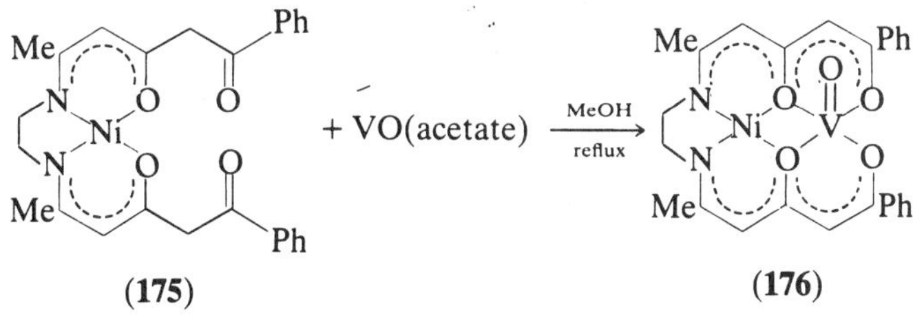

(**175**) (**176**) (67)

To prepare pure heterobinuclear complexes, it is important to minimize the possibility of producing homobinuclear chelates and to avoid the problem of the metal ions changing positions, *i.e.* positional isomers. $[Ni(H_2baaen)]$ (**175**) was the starting material for binuclear complexes including [NiVO(baeen)] (**176**).[896] As Ni^{2+} in (**175**) has a relatively strong selectivity to N_2O_2, it can be easily prepared as a pure mononuclear N_2O_2 positional isomer and used for the synthesis of binuclear complexes (equation 67). In (**176**), the Ni^{2+}, coordinated in a square planar geometry, is diamagnetic with the entire paramagnetism arising from VO^{2+};[899] μ_{eff} values at 297 and 77 K are 1.71 and 1.73 BM.[896] The vanadium is 0.616 Å above the O_2O_2 'plane' in a square pyramidal geometry; the Ni—V distance is relatively short (2.991 Å). [CuVO(daaen)]·H_2O (**179**) and [NiVO(daaen)]·H_2O came from the parent Cu and Ni complexes by a reaction similar to equation (67), and [$(VO)_2$(baaen)] from [VO(H_2baaen)]·Me_2CO.[900] The crystal structures of (**179**) and of the mononuclear VO^{2+} complexes [VO(H_2daaen)] and [VO(H_2daanpn)] were determined (Figure 39).[897] Cu—V distance in (**179**) is 2.985 Å and [VO(H_2daaen)] crystallizes in two forms (monoclinic and triclinic), but the corresponding molecular structures are indistinguishable. The molecular structure of [$(VO)_2$(daaen)], only a small proportion of the molecules in the structure of [VO(H_2daaen)], was proposed and is shown in Figure 39(d);[897] one important feature is that both vanadyl bonds point in the same direction and thus the vanadium atoms are probably in good positions for relatively strong V···V interaction.

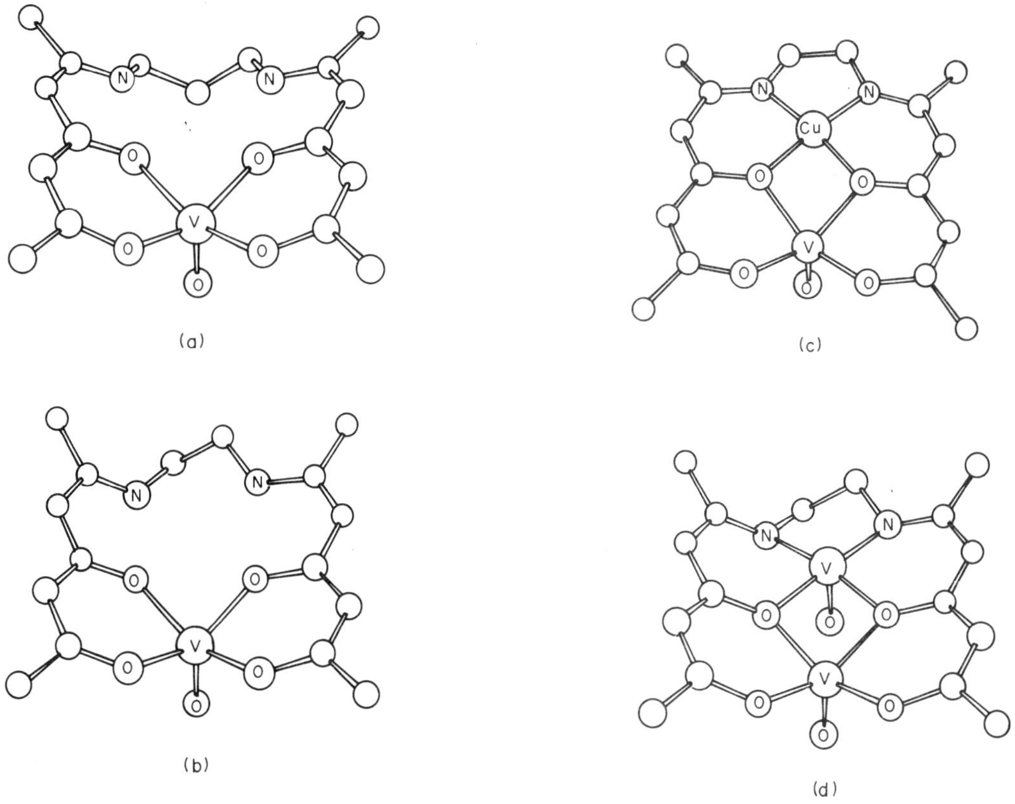

Figure 39 The molecular structure of (a) [VO(H_2daaen)], (b) [VO(H_2daanpn)], (c) [CuVO(daaen)]·H_2O (**179**) and (d) [$(VO)_2$(daaen)]. In all four molecules, the vanadium atom is coordinated in a square pyramidal geometry and is displaced above the O_2O_2 'plane' by *ca.* 0.56 Å (for (d), no data are given)[897]

As mentioned above, in the heterobinuclear preparations the presence of both mono- and homobi-nuclear VO^{2+} complexes was detected;[897] two alternative explanations were suggested: (i) any HCl present in the $CHCl_3$ used for recrystallizing could strip the metals from the ligand and allow subsequent incorporation of VO^{2+} into the outer compartment; (ii) during the synthesis, the mechanism depicted in Scheme 23 could take place. These observations suggest that in solutions of the heterobinuclear complexes, several species are probably present and the product isolated is a function of the insolubility of the individual species.

When $VOCl_2$ was reacted with the β,δ-triketones 1,5-bis(*p*-methoxyphenyl)1,3,5-pentanetrione (H_2dmba), 1,5-diphenyl-1,3,5-pentanetrione (H_2dba), 2,4,6-heptanetrione (H_2daa) or the β-diketophenols 2-acetoacetylphenol (H_2aap) and 2-benzoylacetylphenol (H_2bap) in the presence of 2 moles of LiOH, mononuclear complexes were normally obtained.[901] With cations other than VO^{2+}, *e.g.* Ni^{2+} and Co^{2+}, binuclear complexes were the usual result. With ketophenols, mononuclear chelates can be obtained but are only stable at

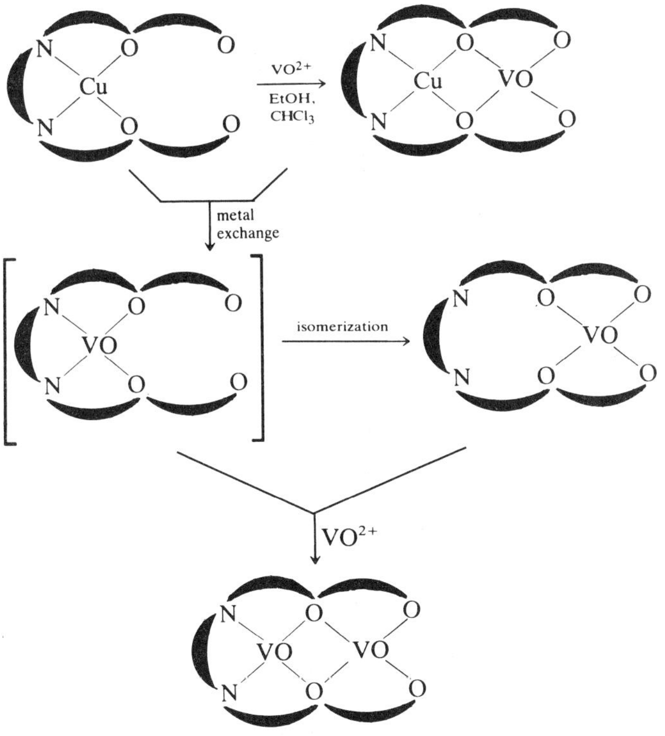

Scheme 23

low temperatures. With the exception of $[VO(Hbap)_2(OH)]$, which is yellow and diamagnetic (it is a V^V compound), $\mu_{eff}^{r.t.}$ values of the complexes obtained (in the range 1.65–1.73 BM) are those expected for magnetically dilute VO^{2+} complexes.

While attempts to prepare heterobinuclear complexes starting from the mononuclear chelates were unsuccessful, green or red oxovanadium(IV) homobinuclear complexes may be obtained (depending on the reaction conditions) with ligands dba and dmba, while only one type of complex is obtained with all other ligands. The main difference in the IR spectra is that the green compounds show only one ν(V═O) at 985–990 cm^{-1}, while the red ones have an additional strong band at 890–900 cm^{-1} suggesting V═O · · · V═O interactions.[901] The $\mu_{eff}^{r.t.}$ of the binuclear complexes are lower than the spin-only values except for $[(VO)_2(bap)_2OH]$; $\mu_{eff}^{r.t.} = 1.98$ BM, suggesting the presence of a VO^V and a VO^{IV} in the molecule.[901] The authors also report results of polymerization tests with these VO^{2+} complexes as Ziegler–Natta catalysts. The features of the copolymer obtained are similar to those from widely used industrial catalysts but while the mononuclear complexes showed catalytic activities comparable to those of $[V(acac)_3]$ and $[VOCl_3]$, dinuclear compounds show much lower activities.

The complex $[(VO)_2(dana)_2]$ (dana = 1,5-bis(*p*-methoxyphenyl)-1,3,5-pentanetrionato) was prepared and temperature dependent χ_M measurements show antiferromagnetic behaviour with $J = -80$ cm^{-1}.[902] The $|J|$ value is much lower than with $[M_2(dana)_2(py)_2]$ (M = Co, Cu) and this probably results from a different spatial orientation of the exchanging electrons: if the unpaired electron is considered to be initially in a d_{xy} orbital, a direct metal–metal interaction may be possible; as the V—V distance is large (*ca.* 3.0–3.2 Å; V atoms are probably 0.5–0.6 Å out of the plane and possibly one above and one below the ligand plane[902]), one would expect a weak exchange for the direct V—V interaction.

A VO^{2+} complex with the compartmental ligand (**177**) was obtained by the dropwise addition of a $VOSO_4$ solution to an aqueous acetone solution containing LiOH and the ligand.[903] The complex is monomeric and the coordination involves the four O atoms. With Ni^{2+} and Cu^{2+} both mononuclear positional isomers N_2O_2 and O_2O_2 may be prepared according to conditions, but with VO^{2+} only the O_2O_2 positional isomer is obtained.

From the reaction between isophthalaldehyde (Hfsal) and substituted derivatives and aminophenol in boiling methanol, the ligand $H_3fsal(ap)_2$ (**180**) may be obtained, which in its trianionic form may act as a binucleating agent. Several VO^{2+} complexes with SBs (**180**; R′ = Me and R variable) were studied by Okawa *et al.* and were formulated as

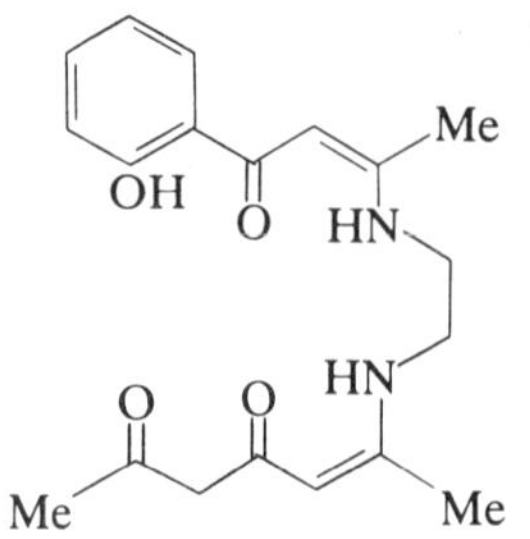

(177)

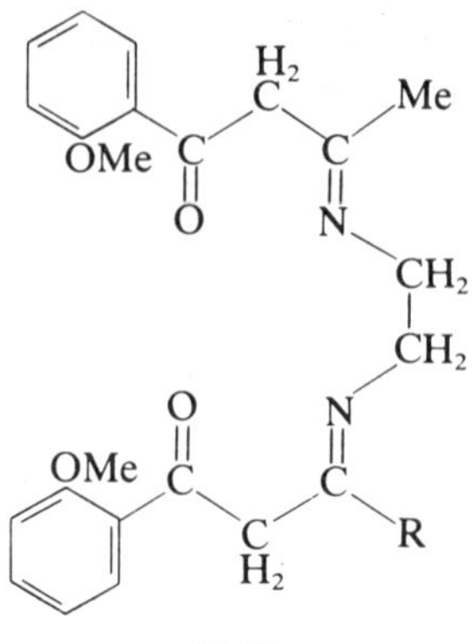

(178)

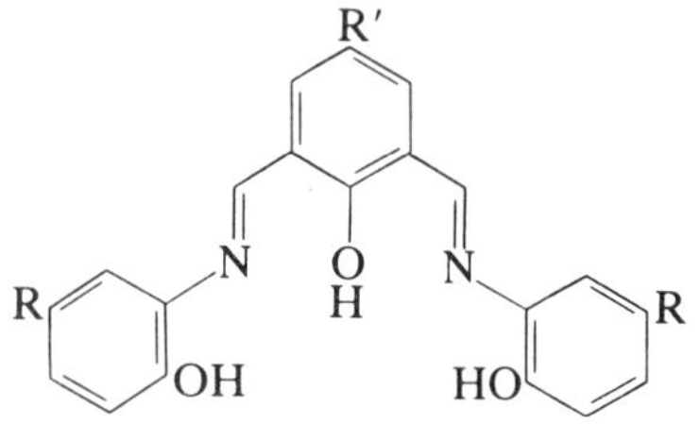

(180)

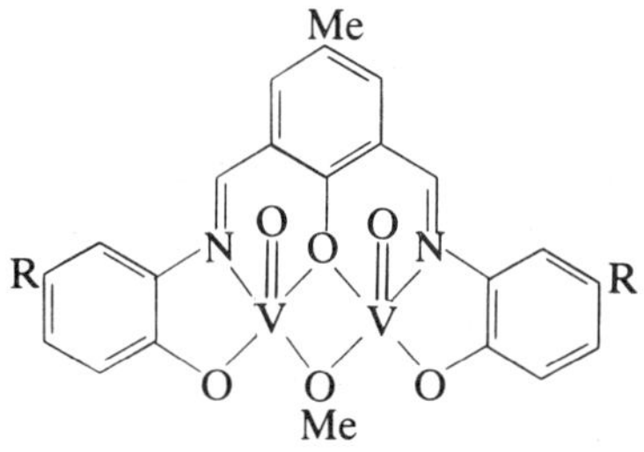

(181)

[$(VO)_2$\{fsal(ap)$_2$\}(OMe)] with a structure (**181**).[904] ν(V=O) of the complexes is at about 990 cm^{-1} except for (**181**; R = NO_2) for which ν(V=O) = 900 cm^{-1}. All the ESR spectra are comparable and show bands at about 3200 and 1600 G. From magnetic susceptibility data, the $|J|$ values were obtained and increase with the increasing order of electron-withdrawing abilities of the substituent R. Since direct coupling between d_{xy} orbitals seems to be the determining factor for the spin-exchange interaction in the binuclear VO^{2+} complexes, one may expect that the overlapping of these orbitals will increase in the order R′ = Me > H > Cl; in fact the $|J|$ values increase in the same order.[722]

Oxovanadium(IV) complexes with SBs (**182**; n = 1, 2, 4) were prepared and formulated as [$(VO)_2(SB)(H_2O)_2$] having coordinating sites at the azomethine nitrogen and enolic and phenolic O atoms.[905] $\mu_{eff}^{r.t.}$ values are in the range 1.62–1.70 BM, close to the spin-only value, but χ_M measurements as a function of temperature were not performed.

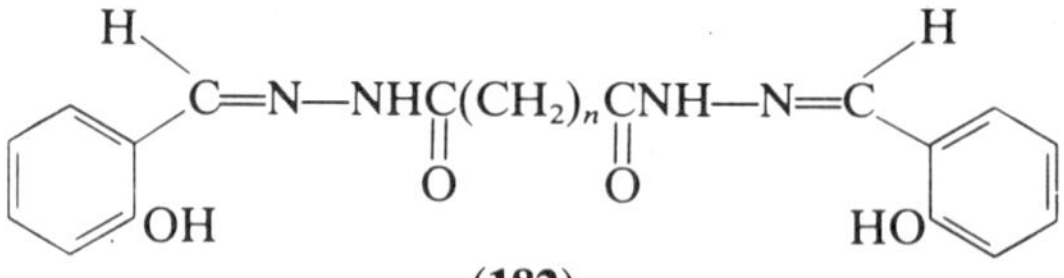

(182)

Selbin and Ganguly obtained complex (**183**) with $R^1 = R^2 = R^3$ = H by reacting [VO(salen)] and a Cu^{II} halide.[906] The $\mu_{eff}^{r.t.}$ of 1.69–1.90 BM per complex molecule was considered as indicating the presence of strong antiferromagnetic interaction between the Cu^{2+} and VO^{2+} ions. However, Okawa and Kida[907] argue that the large g value determined from ESR measurements is not compatible with the low $\mu_{eff}^{r.t.}$ and prepared Selbin's compounds and four other new complexes (R^1, R^2 or R^3 = Me, X = Br or Cl). They obtained much lower $\mu_{eff}^{r.t.}$ than the values reported by Selbin and from χ_M measurements at several temperatures[907] concluded that most of these compounds are diamagnetic, the paramagnetism being attributed to a paramagnetic impurity such as monomeric Cu^{2+} and VO^{2+} complexes. In order to explain the diamagnetism of the present complexes in terms of the antiferromagnetic spin-exchange interaction between VO^{2+} and Cu^{2+}, one would have to consider a strong spin pairing between the two metal ions and this seems unlikely since their ground state configurations, $(d_{xy})^1$ for VO^{2+} and $(d_{x^2-y^2})^1$ for Cu^{2+}, are not appropriate for strong interaction (see also Section 33.5.12.2). A different explanation, suggested by Okawa and Kida for the diamagnetism of complexes (**183**) is that they are composed of V^V and Cu^I, and thus the metal ions have no unpaired electrons at all.

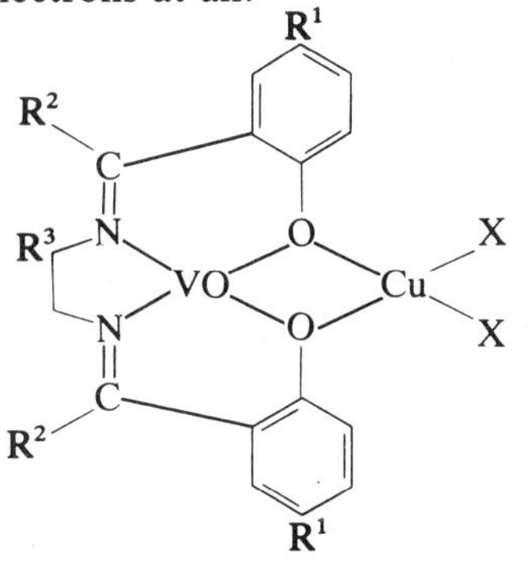

(**183**) X = Cl, Br

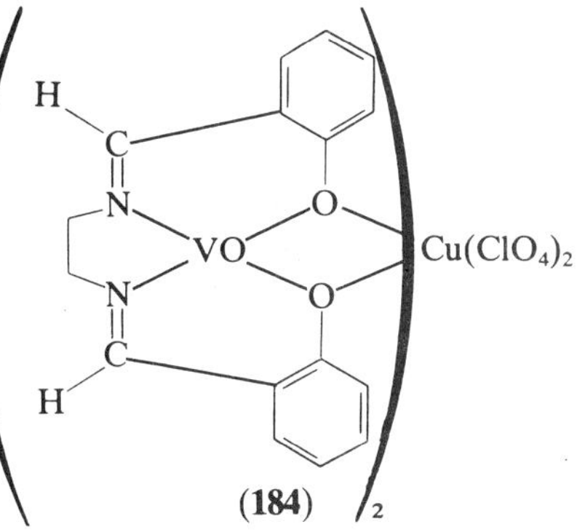

(**184**)

Selbin and Ganguly also reported the preparation of the heterotrinuclear complex (**184**) with a $\mu_{eff}^{r.t.}$ value of 3.93 BM,[906] and the authors infer the absence of magnetic exchange in the complex although one would expect a value of only 3.0 BM if this were the case. Orbital contribution, especially from the Cu atom, may be a partial explanation, but magnetic exchange is also probably involved.[358]

Polynuclear complexes with more than two metal ions have been prepared with polyketone ligands but no such complexes have been reported with VO^{2+}. With the tetraketone 1,1′-(2,6-pyridyl)-bis-1,3-butanedione (H_2L), several polynuclear complexes have been prepared including $[(VO)_2(L)_2]$ for which $\mu_{eff}^{r.t.} = 1.21$ BM was obtained.[908]

33.5.12.2 Heteropolynuclear complexes

Studies involving heteropolynuclear complexes containing V^{IV} have been discussed throughout this review. Magnetic interactions (VO^{2+}–VO^{2+} and VO^{2+}–Cu^{2+}) have been mentioned. A study comparing the interaction in Cu^{2+}–Cu^{2+}, VO^{2+}–VO^{2+} and Cu^{2+}–VO^{2+} pairs through the same oxalato bridging ligand was reported.[909] Two new bimetallic complexes were synthesized, namely $[(acac)VO(C_2O_4)VO(acac)]\cdot 4H_2O$ and $[(tmen)Cu(C_2O_4)VO(C_2O_4)]\cdot 3H_2O$ (tmen = *N,N,N′,N′*-tetramethylethylenediamine) and the crystal and molecular structure of $[(tmen)(H_2O)Cu(C_2O_4)Cu(H_2O)(tmen)](ClO_4)_2$ was determined (Figure 40). The energy gaps arising from the intramolecular interaction, determined from magnetic data, are 385.4, 5.75 and $<1\ cm^{-1}$ in the CuCu, VOVO and CuVO compounds, respectively. The authors rationalized the results according to an orbital model in which the 'magnetic orbitals' are defined as the singly occupied molecular orbitals for each monomeric fragment (Φ_{VO} and Φ_{Cu}), made up by the metal centre surrounded by the nearest neighbour ligands. The 'magnetic orbital' Φ_{VO} around VO^{2+} in the VOO_5 chromophore, schematized in Figure 40(c), transforms as a' (referring to C_s site symmetry) and is delocalized towards the oxygen atoms of the bridge with π metal–oxygen overlaps. Since the π overlaps are weaker than the σ ones, the delocalization towards the bridge is less pronounced in Φ_{VO} than in Φ_{Cu}. In the VOVO compound the interaction between the two 'magnetic orbitals' is weaker than in the CuCu compound since the delocalization of Φ_{VO} towards the nearest neighbour oxygen atom is less pronounced. In the CuVO compound the two magnetic orbitals are strictly orthogonal but as they do not give any region of strong overlap density this explains why the triplet and singlet states are found to be almost degenerate.

A rather different situation is found in the complex $[CuVO\{(fsa)_2en\}(MeOH)]$ (**185**) where $(fsa)_2en^{4-}$ denotes the binucleating ligand derived from the SB *N,N′*-(2-hydroxy-3-carboxybenzylidene)-1,2-diaminoethane. The crystal structure is made up of heterobinuclear units (Figure 41a) with the MeOH molecule weakly bound to the Cu atom.[910,911] The thermal variation of χ_M shows that upon cooling from 300 K to around 50 K, $\chi_M T$ increases and this clearly demonstrates that the triplet state ($S = 1$) is actually the lowest in energy. Thus, the intramolecular coupling is ferromagnetic and the separation between the triplet and singlet states was found to be $118\ cm^{-1}$. The nature of the intramolecular interaction was explained by the orthogonality of the 'magnetic orbitals' Φ_{Cu} and Φ_{VO}, centred on the Cu^{2+} and VO^{2+} ions (see Figure 41b). The overlap integral of Φ_{Cu} and Φ_{VO} is identically zero and this means that

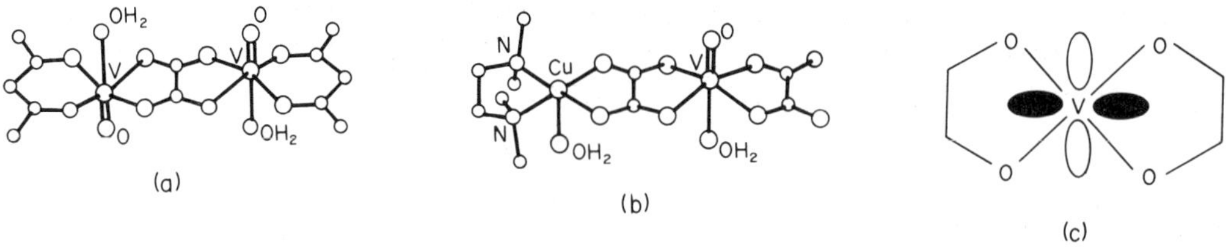

Figure 40 Schematic structures proposed for (a) $[(acac)VO(C_2O_4)VO(acac)]\cdot4H_2O$ and (b) $[(tmen)(H_2O)Cu(C_2O_4)VO(C_2O_4)]\cdot3H_2O$[909] (only the structure of $[(tmen)(H_2O)Cu(C_2O_4)Cu(tmen)(H_2O)]^{2+}$ was determined by X-ray diffraction); (c) schematic representation of the 'magnetic orbital Φ_{VO}' in the VOO_5 chromophore[909]

there is no interaction between the orbitals favouring the pairing of electrons on a molecular orbital of low energy. In order to explain why J is so large, the authors emphasize that the entire molecular symmetry is important in deciding the nature of the coupling and not only the local symmetry for each A—O—B linkage.[911] In the Cu^{2+}–VO^{2+} pair, the strict orthogonality which leads to ferromagnetic interaction occurs only if the phases at both bridging oxygen atoms are taken into account.

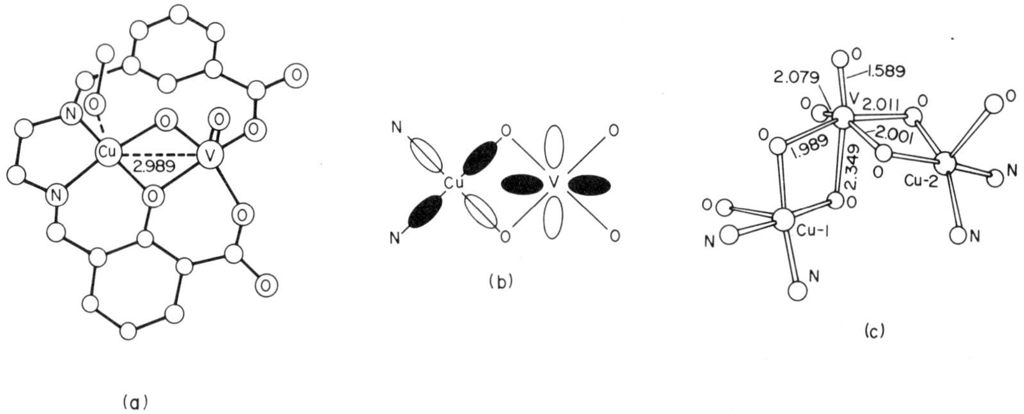

Figure 41 (a) schematic representation of the molecular structure of $[CuVO((fsa)_2en)(MeOH)]$ (**185**);[911] (b) relative symmetries of the 'magnetic orbitals';[911] (c) schematic representation of the geometry around VO^{2+} in the trinuclear cation $[VO(CuHAPen)_2]^{2+}$ (**186**)[914]

Copper complexes can act as bidentate ligands towards transition metal ions.[912,913] On mixing a solution of CuHAPen in $CHCl_3$ (CuHAPen = N,N'-ethylenebis(o-hydroxyacetophenone-iminato)copper(II)) with an ethanol solution of $VO(ClO_4)$, a precipitate formed. After recrystallization from nitroethane the compound $[VO(CuHAPen)_2](ClO_4)_2\cdot3H_2O\cdot\frac{1}{2}EtNO_2$ (**186**) was obtained and its crystal structure determined.[914] The two CuHAPen complex ligands bind to a central hexacoordinate VO^{2+} occupying nonequivalent positions in a distorted octahedron forming a dipositive trinuclear cation (see Figure 41c). The V atom lies 0.38 Å above the 'equatorial plane'. The Cu(1)—V, Cu(2)—V and Cu(1)—Cu(2) distances are 3.132, 3.015 and 4.558 Å, respectively. The temperature dependence of χ_M of (**186**) was studied in the range 4.2–300 K;[914] the ground state of the trinuclear cation is a spin doublet, with the quartet separated by $\sim3\,cm^{-1}$, and a second doublet at $\sim85\,cm^{-1}$ $(=J_{12})$. The large ferromagnetic constant J_{12} is associated with the coupling between Cu-2 and V, based (i) on comparisons with other similar copper complexes and with (**185**) and (ii) on the geometry of the Cu—V bridges and the symmetry of the 'magnetic orbitals' of Cu-1, Cu-2 and V.

Oxovanadium(IV) complexes were also prepared by the reaction of the phosphonate anions (**187**) and (**189**) with $[VO(acac)_2]$ (equations 68 and 69), and the products were characterized by elemental analysis and mass spectra.[915] In contrast to the reaction of (**187**) with $[VO(acac)_2]$ which gives (**188**) with $\mu_{eff}^{r.t.} = 2.03$ BM, in the reaction of (**189**) only one acac ligand is displaced by the organometallic chelate.

Allowing $TiCl_3(THF)_3$ and $[VO(acen)]$ to react in THF at room temperature, a reductive deoxygenation of the VO^{2+} occurs and a red solid precipitates, characterized as $[V^{III}$

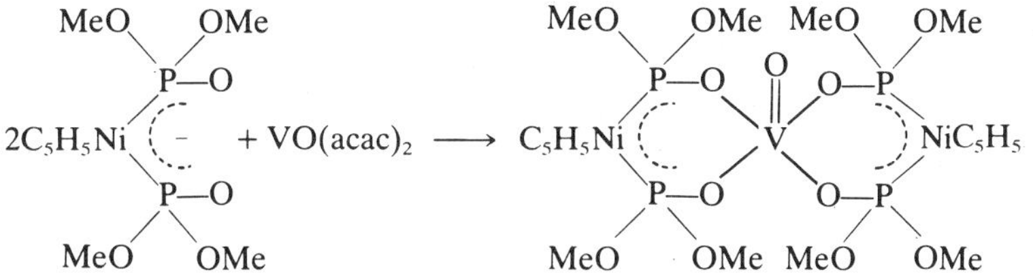

(187) (188)

(68)

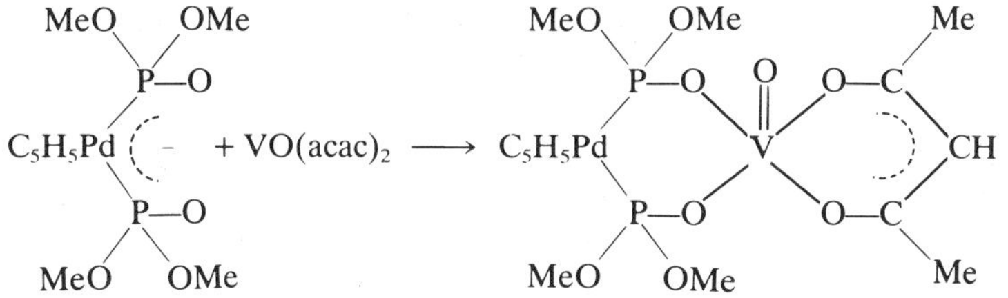

(189) (190)

(69)

(acen)(Cl)(THF)] (**4**) (Section 33.4.8.1).[916] However, if the reaction is carried out at low temperature (*ca.* −70 °C), a red crystalline solid precipitates which was shown to be the bimetallic vanadium(IV)–titanium(IV) μ-oxo complex $[(Cl)(acen)VOTi(Cl)_3(THF)_2]$. This is paramagnetic, with one unpaired electron and is probably an intermediate in the reductive deoxygenation of [VO(acen)] to yield (**4**). X-Ray data show that the μ-O atom bridges the [V(Cl)(acen)] and $[Ti(Cl)_3(THF)_2]$ units with a rather long V—O bridging distance: 1.937 Å (or 1.973 Å?); the Cl^- is coordinated *trans* to the μ-O and the V atom lies approximately in the plane of the N_2O_2 core of the acen ligand, its displacement from it being 0.042 Å towards chlorine.[916]

33.5.13 Naturally Occurring Ligands

This subject covers a wide range of compounds and many of them have been included in previous sections. Most investigations concerning complex formation between VO^{2+} and proteins have been carried out with VO^{2+} acting as a physicochemical marker of the binding sites. In recent years, there has been growing interest in employing VO^{2+} in protein studies because of its chemical and spectroscopic properties. Vanadyl forms hundreds of stable complexes with almost all kinds of ligands; thus, it may coordinate with the functional groups encountered in proteins. Since vanadium(IV) complexes are d^1 systems, the interpretation of optical spectra is relatively simple. Further, oxovanadium(IV) has one position occupied by the vanadyl oxygen and the IR stretching frequency of the VO bond can be used as an additional probe of metal-binding sites in crystalline proteins where the VO^{2+} concentration is sufficiently large.[917]

As vanadyl complexes consistently give sharp ESR spectra both in frozen and room temperature solutions and ^{51}V is nearly 100% abundant, ESR is particularly useful and there has been a growing interest in employing VO^{2+} as an ESR spin probe. ENDOR studies have also been performed and future work using this technique and electron spin echo (ESE) will possibly be of great value in giving an insight into VO^{2+} binding site structure. The use of VO^{2+} as an ESR spin probe was reviewed by Chasteen.[354] A few studies have appeared since, *e.g.* ESR studies of the VO^{2+} apoferritin complex[918] and the VO^{2+}–*S*-adenosylmethionine synthetase interaction,[919] a CD, ESR and acid–base study of VO^{2+} binding to bleomycin,[920] and redox kinetics and complexation of V^{IV} and V^V with serum proteins.[945] Some equilibria in solution, likely coordination geometries and ligand contributions to the observed hyperfine splitting A_0 were also discussed.[798]

Since the discovery by Cantley *et al.* that vanadate is a powerful and allosteric inhibitor of Na^+,K^+-ATPase,[921] vanadium began to interest biochemists; and Chasteen has reviewed work published up to 1983.[471] Particularly interesting new studies involve vanadium peroxidases[946] and nitrogenases.[947]

Waltermann *et al.* studied the nonenzymatic hydrolysis of ATP by VO^{2+} and VO^{2+} coupled with H_2O_2 and concluded that oxidation of V^{IV} to V^{V} is an important factor in enhancement of ATP hydrolysis.[922] The authors also observed that ATP protects the VO^{2+} from oxidation by H_2O_2 and from ESR spectra they concluded that binding occurs between VO^{2+} and ATP in aqueous solution. Sakurai *et al.* studied the formation of complexes between ATP and VO^{2+} by potentiometric titrations, optical and ESR spectra and by ^{31}P and ^{13}C NMR spectrometry.[923,924] They concluded that as pH increases three different complexes form and proposed detailed coordination geometries for each of them. At acidic and neutral pH, two different 1:1 species predominate and in alkaline solutions a green 2:1 complex forms. The authors argue that ^{31}P and ^{13}C NMR are very effective for determining coordination sites in compounds containing organic phosphate and that the detailed structures for the ATP–VO^{2+} complexes can thus be determined.[924]

Complexes formulated as $[VOCl_2(HL)]$ [HL = purine (**191**), adenine (**192**) and guanine (**193**)] were prepared by refluxing the ligand and $VOCl_2$ mixtures in ethanol–triethyl orthoformate.[925,926] Characterization was based on elemental analysis and spectral and magnetic studies. In the visible spectra (Nujol mulls) the *d–d* transitions were at ~695, ~500 and ~430 nm, and in the IR, ν(V=O) was assigned at ~1005 cm^{-1}; $\mu_{eff}^{r.t.} \approx 1.7$ BM. These $VOCl_2$ adducts are said to be polymeric with a linear, chain-like structure (**194**) involving single bridges of bidentate ligands between adjacent VO^{2+} ions, the bridging probably being through N-7 and N-9 of imidazole.[925,926] This suggestion is based on IR evidence and by comparison with the molecular structure of $[Cu(HL)(H_2O)_4](SO_4)\cdot 2H_2O$ (HL = **191**).[927]

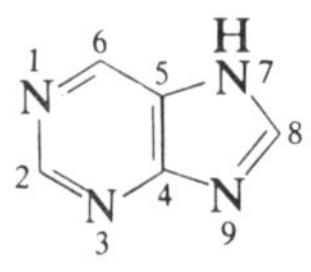
(**191**)

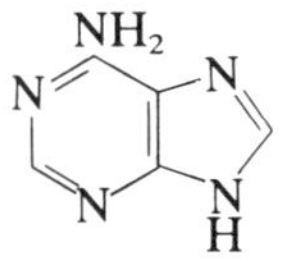
(**192**)

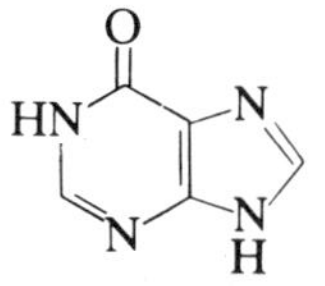
(**193**)

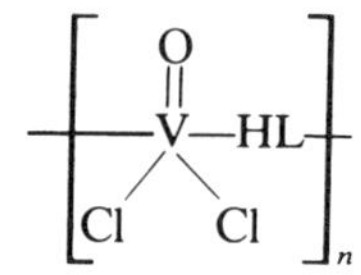
(**194**)

Several VO^{2+} complexes with nucleotides have been prepared by the addition of aqueous $VOSO_4$ to an aqueous solution of the nucleotide.[928] The compounds were formulated as $[VOL]\cdot nH_2O$ with a suggested polymeric structure. As in the case of the complexes with the purines (**191**)–(**193**), further work is needed to sort out the molecular structure of these compounds.

Amavadin, a natural compound which occurs in the toadstool *Amanita muscaria* has been said to be a 1:2 VO^{2+} complex of Hidpa (**131**; R = OH).[378,816,929–934,948,949] Fraústo da Silva and co-workers studied the formation of complexes of VO^{2+} with ligands of type (**132**) by pH potentiometric methods and concluded that with a ligand-to-metal ratio of 2, only 1:1 complexes form (see Table 45) except for the *N*-hydroxy derivatives (**132**; R = OH). With these ligands (*e.g.* Hidpa), the pK_a associated with the N group drops ~10 000 times and at pH ~7 the 2:1 complexes predominate. However, the formation constants calculated[378,932] seem low to explain the existence of amavadin in the toadstool and its 'intact' isolation after several extraction and chromatographic procedures, and further studies are needed to establish the detailed structure of amavadin and understand why VO^{2+} binds *N*-hydroxyiminodiacetate ligands so strongly. Different formation constants have recently been estimated.[948]

Several studies have concerned interaction of VO^{2+} with soil humic and fulvic acids. Strongly complexed VO^{2+} occurs naturally in the organic fractions isolated both from lignite and ball clay[935] and from soil.[936,937] Since the ESR spectra of VO^{2+} provide information on the environment of the metal, much can be learned about the interactions between vanadium and fulvic and humic acids using VO^{2+} as a paramagnetic probe, and several studies have appeared with VO^{2+} synthetically introduced to probe the metal-binding sites of these materials.[938–94] The ESR parameters indicate that in most cases the ligands are oxygen donors, probably from carboxylate and phenolate groups. Nitrogen coordination has also been suggested[938] but, as noted by Chasteen,[354] catechol coordination complexes have VO^{2+} ESR parameters which are comparable to those obtained with nitrogen chelates. McBride[941] studied the influence of pH and metal ion content on VO^{2+}-fulvic acid interactions and concluded that very low pH values prevent VO^{2+} from bonding with fulvic acid and that insignificant quantities of free VO^{2+} remain in solution at pH values of 4 or higher. Very high pH caused all the VO^{2+} associated with fulvic acid to be oxidized except for a very small quantity apparently strongly bound to four nitrogen ligands, possibly porphyrin.

VO^{2+} bound to wet humic acids exhibits a rigid limit ESR spectrum indicative of a highly immobilized probe ion.[939] The ESR parameters suggest that bonding is mainly ionic. Room temperature solution ESR spectra of VO^{2+} complexes with podzol fulvic acid exhibit reduced intensity due to VO^{2+}–VO^{2+} magnetic dipole interactions within the same fulvic acid molecule[940] and VO^{2+} binding facilitates aggregation of fulvic acid to form a soluble polymeric system.

33.6 REFERENCES

1. M. E. Weeks and H. M. Leicester, 'Discovery of the Elements', 7th edn., Chemical Education Publishing, Easton, PA, 1968, p. 351.
2. J. W. Mellor, 'Comprehensive Treatise on Inorganic and Theoretical Chemistry', vol. IX, Longmans, Green and Co., New York, 1929.
3. N. V. Sidgwick, 'The Chemical Elements and their Compounds', Oxford University Press, London, 1950.
4. 'Gmelins Handbuch der Anorganischen Chemie', vol. 48, Verlag Chemie, Weinheim, 1967/1968.
5. (a) R. J. H. Clark, 'The Chemistry of Titanium and Vanadium', Elsevier, London, 1968; (b) R. J. H. Clark and D. Brown, 'The Chemistry of Vanadium, Niobium and Tantalum', Pergamon, Oxford, 1973; (c) R. J. H. Clark, in 'Comprehensive Inorganic Chemistry', ed. J. C. Bailar, Jr., H. J. Emeléus, R. Nyholm and A. F. Trotman-Dickenson, Pergamon, Oxford, 1973, vol. 3, p. 491.
6. D. Nicholls, *Coord. Chem. Rev.*, 1966, **1,** 379.
7. D. L. Kepert, 'The Early Transition Metals', Academic, London, 1972, p. 142.
8. D. A. Rice, *Coord. Chem. Rev.*, 1981, **37,** 61; 1982, **45,** 67; E. M. Page, *Coord. Chem. Rev.*, 1984, **57,** 237.
9. N. G. Connelly, in 'Comprehensive Organometallic Chemistry', ed. G. Wilkinson, F. G. A. Stone and E. W. Abel, Pergamon, Oxford, 1982, vol. 3, p. 647.
10. J. E. Ellis and M. C. Palazzotto, *J. Am. Chem. Soc.*, 1976, **98,** 8264; J. E. Ellis, K. L. Fjare and T. G. Hayes, *J. Am. Chem. Soc.*, 1981, **103,** 6100.
11. J. E. Ellis, T. G. Hayes and R. E. Stevens, *J. Organomet. Chem.*, 1981, **220,** 45.
12. E. S. Dodsworth and D. Nicholls, *Inorg. Chim. Acta,* 1982, **61,** 9.
13. S. Herzog, *Z. Anorg. Allg. Chem.*, 1958, **294,** 155.
14. G. Albrecht, *Z. Chem.*, 1963, **3,** 182.
15. A. Davison, N. Edelstein, R. H. Holm and A. H. Maki, *Inorg. Chem.*, 1965, **4,** 55.
16. E. König, H. Fischer and S. Herzog, *Z. Naturforsch., Teil B,* 1963, **18,** 432.
17. S. Herzog and U. Grimm, *Z. Chem.*, 1967, **7,** 432.
18. S. Herzog and U. Grimm, *Z. Chem.*, 1964, **4,** 32; 1966, **6,** 380.
19. S. Herzog and H. Aul, *Z. Chem.*, 1966, **6,** 382.
20. J. Quirk and G. Wilkinson, *Polyhedron,* 1982, **1,** 209.
21. E. König and S. Herzog, *J. Inorg. Nucl. Chem.*, 1970, **32,** 601.
22. Y. Kaizu, T. Yazaki, Y. Torii and H. Kobayashi, *Bull. Chem. Soc. Jpn.*, 1970, **43,** 2068.
23. Y. Saito, J. Takemoto, B. Hutchinson and K. Nakamoto, *Inorg. Chem.*, 1972, **11,** 2003.
24. E. König and E. Lindner, *Spectrochim. Acta, Part A,* 1972, **28,** 1393.
25. R. Kirmse, J. Stach and D. Walther, *Z. Anorg. Allg. Chem.*, 1983, **496,** 147.
26. T. Saji and S. Aoyagui, *J. Electroanal. Chem. Interfacial Electrochem.*, 1975, **63,** 405.
27. F. Näumann, D. Rehder and V. Pank, *Inorg. Chim. Acta,* 1984, **84,** 117.
28. J. Schiemann, E. Weiss, F. Näumann and D. Rehder, *J. Organomet. Chem.*, 1982, **232,** 219.
29. W. P. Griffith, J. Lewis and G. Wilkinson, *J. Chem. Soc.*, 1959, 872.
30. S. Jagner and N.-G. Vannerberg, *Acta Chem. Scand.*, 1970, **24,** 1988.
31. A. Müller, P. Werle, E. Diemann and P. J. Aymonino, *Chem. Ber.*, 1972, **105,** 2419.
32. M. G. B. Drew and C. F. Pygall, *Acta Crystallogr., Sect. B,* 1977, **33,** 2838.
33. R. Srivastava and S. Sarkar, *Transition Met. Chem.* (*Weinheim, Ger.*), 1980, **5,** 122.
34. S. Sarkar, R. Maurya, S. C. Chaurasia and R. Srivastava, *Transition Met. Chem.* (*Weinheim, Ger.*), 1976, **1,** 49.
35. R. Bhattacharyya, P. S. Roy and A. K. Dasmahapatra, *J. Organomet. Chem.*, 1984, **267,** 293.
36. W. Beck, K. Lottes and K. H. Schmidtner, *Angew. Chem., Int. Ed. Engl.*, 1965, **4,** 151.
37. W. Beck, H. G. Fick, K. Lottes and K. H. Schmidtner, *Z. Anorg. Allg. Chem.*, 1975, **416,** 97.
38. M. Herberhold and H. Trampisch, *Inorg. Chim. Acta,* 1983, **70,** 143.
39. H. Huber, T. A. Ford, W. Klotzbücher and G. A. Ozin, *J. Am. Chem. Soc.*, 1976, **98,** 3176.
40. A. B. P. Lever and G. A. Ozin, *Inorg. Chem.*, 1977, **16,** 2012.
41. J. Chatt and H. R. Watson, *J. Chem. Soc.*, 1962, 2545.
42. F. G. N. Cloke, P. J. Fyne, M. L. H. Green, M. J. Ledoux, A. Gourdon and C. K. Prout, *J. Organomet. Chem.*, 1980, **198,** C69.
43. T. Kruck and H. U. Hempel, *Angew. Chem.*, 1974, **86,** 233.
44. H. Schmidt and D. Rehder, *Transition Met. Chem.* (*Weinheim, Ger.*), 1980, **5,** 214.
45. D. Rehder, H.-C. Bechthold and K. Paulsen, *J. Magn. Reson.*, 1980, **40,** 305.
46. R. L. Bansemer, J. C. Huffman and K. G. Caulton, *J. Am. Chem. Soc.*, 1983, **105,** 6163.
47. J. A. McCleverty, *Prog. Inorg. Chem.*, 1968, **10,** 49.
48. (a) J. R. Dorfman and R. H. Holm, *Inorg. Chem.*, 1983, **22,** 3179; (b) R. W. Wiggins, J. C. Huffman and G. Christou, *J. Chem. Soc., Chem. Commun.*, 1983, 1313; (c) D. Szeymies, B. Krebs and G. Henkel, *Angew. Chem., Int. Ed. Engl.*, 1983, **22,** 885.
49. W.-L. Kwik and E. I. Stiefel, *Inorg. Chem.*, 1973, **12,** 2337.
50. D. C. Olson, V. P. Mayweg and G. N. Schrauzer, *J. Am. Chem. Soc.*, 1966, **88,** 4876.
51. R. Eisenberg and H. B. Gray, *Inorg. Chem.*, 1967, **6,** 1844.
52. E. I. Stiefel, Z. Dori and H. B. Gray, *J. Am. Chem. Soc.*, 1967, **89,** 3353.
53. G. N. Schrauzer and V. P. Mayweg, *J. Am. Chem. Soc.*, 1966, **88,** 3235.

54. R. J. H. Clark and P. C. Turtle, *J. Chem. Soc., Dalton Trans.*, 1978, 1714.
55. E. I. Stiefel, R. Eisenberg, R. C. Rosenberg and H. B. Gray, *J. Am. Chem. Soc.*, 1966, **88**, 2956.
56. A. Davison and E. T. Shawl, *Inorg. Chem.*, 1970, **9**, 1820.
57. N. Aristov and P. B. Armentrout, *J. Am. Chem. Soc.*, 1984, **106**, 4065.
58. A. B. P. Lever, 'Inorganic Electronic Spectroscopy', 2nd edn., Elsevier, New York, 1984, p. 444.
59. S. Jagner, *Acta Chem. Scand., Ser. A*, 1975, **29**, 255.
60. B. G. Bennett and D. Nicholls, *J. Chem. Soc. (A)*, 1971, 1204.
61. L. D. Silverman, P. W. R. Corfield and S. J. Lippard, *Inorg. Chem.*, 1981, **20**, 3106.
62. G. W. A. Fowles, P. G. Lanigan and D. Nicholls, *Chem. Ind. (London)*, 1961, 1167.
63. L. F. Larkworthy and J. M. Tabatabai, *Inorg. Nucl. Chem. Lett.*, 1980, **16**, 509.
64. G. W. A. Fowles and P. G. Lanigan, *J. Less-Common Met.*, 1964, **6**, 396.
65. M. M. Khamar, L. F. Larkworthy, K. C. Patel, D. J. Phillips and G. Beech, *Aust. J. Chem.*, 1974, **27**, 41.
66. L. F. Larkworthy and M. W. O'Donoghue, *Inorg. Chim. Acta*, 1983, **71**, 81.
67. M. Ciampolini and F. Mani, *Inorg. Chim. Acta*, 1977, **24**, 91.
68. P. Dapporto, F. Mani and C. Mealli, *Inorg. Chem.*, 1978, **17**, 1323.
69. F. Mani, *Inorg. Chim. Acta*, 1980, **38**, 97.
70. F. Mani, *Inorg. Nucl. Chem. Lett.*, 1981, **17**, 45.
71. F. Mani, *Inorg. Chim. Acta*, 1982, **65**, L197.
72. T. G. Richmond, Q.-Z. Shi, W. C. Trogler and F. Basolo, *J. Am. Chem. Soc.*, 1984, **106**, 76.
73. W. Hieber, J. Peterhans and E. Winter, *Chem. Ber.*, 1961, **94**, 2572.
74. D. J. Brauer and C. Krueger, *Cryst. Struct. Commun.*, 1973, **2**, 421.
75. L. F. Larkworthy and B. J. Tucker, *Inorg. Chim. Acta*, 1978, **33**, 167.
76. G. Trageser and H. H. Eysel, *Inorg. Chem.*, 1977, **16**, 713.
77. L. F. Larkworthy and B. J. Tucker, *J. Chem. Soc., Dalton Trans.*, 1980, 2042.
78. P. J. Nichols, Y. Ducommun and A. E. Merbach, *Inorg. Chem.*, 1983, **22**, 3993.
79. P. Chandrasekhar and P. H. Bird, *Inorg. Chim. Acta*, 1985, **97**, L31.
80. V. M. Hall, C. D. Schmulbach and W. N. Soby, *J. Organomet. Chem.*, 1981, **209**, 69.
81. F. A. Cotton, S. A. Duraj and W. J. Roth, *Inorg. Chem.*, 1984, **23**, 4113.
82. F. A. Cotton, S. A. Duraj, L. R. Falvello and W. J. Roth, *Inorg. Chem.*, 1985, **24**, 4389.
83. G. S. Girolami, G. Wilkinson, A. M. R. Galas, M. Thornton-Pett and M. B. Hursthouse, *J. Chem. Soc., Dalton Trans.*, 1985, 1339.
84. W. P. Schaefer, *Inorg. Chem.*, 1965, **4**, 642.
85. J. Podlaha and J. Podlahova, *Collect. Czech. Chem. Commun.*, 1964, **29**, 3164.
86. W. J. Biermann and W.-K. Wong, *Can. J. Chem.*, 1963, **41**, 2510.
87. C. F. Baes and R. E. Mesmer, 'The Hydrolysis of Cations', Wiley-Interscience, New York, 1976.
88. N. J. Fitzpatrick and M. A. McGinn, *Inorg. Chim. Acta*, 1984, **90**, 137.
89. Y. Ducommun, D. Zbinden and A. E. Merbach, *Helv. Chim. Acta*, 1982, **65**, 1385.
90. H. Montgomery, R. V. Chastain, J. J. Natt, A. M. Witkowska and E. C. Lingafelter, *Acta Crystallogr.*, 1967, **22**, 775.
91. L. F. Larkworthy, K. C. Patel and D. J. Phillips, *J. Chem. Soc. (A)*, 1970, 1095.
92. W. Hieber, E. Winter and E. Schubert, *Chem. Ber.*, 1962, **95**, 3070.
93. H. J. Seifert and T. Auel, *Z. Anorg. Allg. Chem.*, 1968, **360**, 50.
94. H. J. Seifert and T. Auel, *J. Inorg. Nucl. Chem.*, 1968, **30**, 2081.
95. F. A. Cotton, S. A. Duraj, L. E. Manzer and W. J. Roth, *J. Am. Chem. Soc.*, 1985, **107**, 3850.
96. M. Schneider and E. Weiss, *J. Organomet. Chem.*, 1976, **121**, 365.
97. Y. Torii, H. Iwaki and Y. Inamura, *Bull. Chem. Soc. Jpn.*, 1967, **40**, 1550.
98. H. J. Seifert and B. Gerstenberg, *Z. Anorg. Allg. Chem.*, 1962, **315**, 56.
99. H. Ehrhardt, H. C. Schober and H. Seidel, *Z. Anorg. Allg. Chem.*, 1980, **465**, 83.
100. F. A. Cotton, S. A. Duraj, M. W. Extine, G. E. Lewis, W. J. Roth, C. D. Schmulbach and W. Schwotzer, *J. Chem. Soc., Chem. Commun.*, 1983, 1377.
101. F. A. Cotton, S. A. Duraj and W. J. Roth, *Inorg. Chem.*, 1985, **24**, 913.
102. R. J. Bouma, J. H. Teuben, W. R. Beukema, R. L. Bansemer, J. C. Huffman and K. G. Caulton, *Inorg. Chem.*, 1984, **23**, 2715.
103. C. Cros, *Rev. Inorg. Chem.*, 1979, **1**, 163.
104. J. W. Stout and W. O. J. Boo, *J. Appl. Phys.*, 1966, **37**, 966.
105. W. H. Baur, S. Guggenheim and J.-C. Lin, *Acta Crystallogr., Sect. B*, 1982, **38**, 351.
106. R. F. Williamson and W. O. J. Boo, *Inorg. Chem.*, 1977, **16**, 646.
107. C. Cros, R. Feurer and M. Pouchard, *J. Fluorine Chem.*, 1976, **7**, 605.
108. Y. S. Hong, R. F. Williamson and W. O. J. Boo, *J. Chem. Educ.*, 1980, **57**, 583.
109. C. Cros, R. Feurer, M. Pouchard and P. Hagenmuller, *Mater. Res. Bull.*, 1975, **10**, 383.
110. R. E. Young and M. E. Smith, *Inorg. Synth.*, 1953, **4**, 128.
111. A. Hauser, U. Falk, P. Fischer and H. U. Guedel, *J. Solid State Chem.*, 1985, **56**, 343.
112. P. Ehrlich and H. J. Seifert, *Z. Anorg. Allg. Chem.*, 1959, **301**, 282.
113. M. Niel, C. Cros, M. Pouchard and J. P. Chaminade, *J. Solid State Chem.*, 1977, **20**, 1.
114. M. Niel, C. Cros, G. Le Flem and M. Pouchard, *C. R. Hebd. Seances Acad. Sci.*, 1975, **280**, 1093.
115. I. E. Grey and P. W. Smith, *Chem. Commun.*, 1968, 1525.
116. G. Brauer, 'Handbook of Preparative Inorganic Chemistry', 2nd edn., Academic, New York, 1965.
117. V. D. Juza, D. Giegling and H. Schäfer, *Z. Anorg. Allg. Chem.*, 1969, **366**, 121.
118. W. Klemm and L. Grimm, *Z. Anorg. Allg. Chem.*, 1941, **249**, 198; M. Niel, C. Cros, G. Le Flem, M. Pouchard and P. Hagenmuller, *Nouv. J. Chim.*, 1977, **1**, 127.
119. Tuig-I Li, G. D. Stucky and G. L. McPherson, *Acta Crystallogr., Sect. B*, 1973, **29**, 1330.
120. S. S. Kim, S. A. Reed and J. W. Stout, *Inorg. Chem.*, 1970, **9**, 1584.
121. A. Hauser and H. U. Güdel, *Chem. Phys. Lett.*, 1981, **82**, 72.

122. G. L. McPherson, L. J. Sindel, H. F. Quarls, C. B. Frederik and C. J. Doumit, *Inorg. Chem.*, 1975, **14,** 1831.
123. W. E. Smith, *J. Chem. Soc., Dalton Trans.*, 1972, 1634.
124. W. van Erk and C. Haas, *Phys. Stat. Solidi B*, 1975, **71,** 537.
125. J. W. Stout and H. Y. Lau, *J. Appl. Phys.*, 1967, **38,** 1472.
126. M. Niel, C. Cros, G. Le Flem, M. Pouchard and P. Hagenmuller, *Physica B + C (Amsterdam)*, 1977, **86–88B,** 702.
127. A. Dei and F. Mani, *Inorg. Chim. Acta*, 1976, **19,** L39.
128. R. Guilard and C. Lecomte, *Coord. Chem. Rev.*, 1985, **65,** 87.
129. J. L. Poncet, J. M. Barbe, R. Guilard, H. Oumous, C. Lecomte and J. Protas, *J. Chem. Soc., Chem. Commun.*, 1982, 1421; *J. Chem. Soc., Dalton Trans.*, 1984, 2677.
130. M. Pomerantz, G. L. Combs, Jr. and N. L. Dassanayake, *Inorg. Synth.*, 1982, **21,** 185.
131. T.-L. Ho and G. A. Olah, *Synthesis*, 1977, 170.
132. T.-L. Ho, *Synthesis*, 1979, 1.
133. T. A. Cooper, *J. Am. Chem. Soc.*, 1973, **95,** 4158.
134. V. Gold and D. L. Wood, *J. Chem. Soc., Dalton Trans.*, 1981, 2462.
135. J.-T. Chen and J. H. Espenson, *Inorg. Chem.*, 1983, **22,** 1651.
136. N. T. Denisov, O. N. Efimov, N. I. Shuvalova, A. K. Shilova and A. E. Shilov, *Zh. Fiz. Khim.*, 1970, **44,** 2694.
137. A. G. Ovcharenko, A. E. Shilov and L. A. Nikonova, *Izv. Akad. Nauk SSSR, Ser. Khim.*, 1975, 534.
138. S. I. Zones, T. M. Vickrey, J. G. Palmer and G. N. Schrauzer, *J. Am. Chem. Soc.*, 1976, **98,** 7289.
139. S. I. Zones, M. R. Palmer, J. G. Palmer, J. M. Doemeny and G. N. Schrauzer, *J. Am. Chem. Soc.*, 1978, **100,** 2113.
140. L. A. Nikonova, A. G. Ovcharenko, O. N. Efimov, V. A. Avilov and A. E. Shilov, *Kinet. Katal.*, 1972, **13,** 1602.
141. S. A. Isaeva, L. A. Nikonova and A. E. Shilov, *Nouv. J. Chim.*, 1981, **5,** 21.
142. L. A. Nikonova, S. A. Isaeva, N. I. Pershikova and A. E. Shilov, *Kinet. Katal.*, 1977, **18,** 1606.
143. L. A. Nikonova, S. A. Isaeva, N. I. Pershikova and A. E. Shilov, *J. Mol. Catal.*, 1975, **1,** 367.
144. N. P. Luneva, L. A. Nikonova and A. E. Shilov, *React. Kinet. Catal. Lett.*, 1976, **5,** 149.
145. G. N. Schrauzer and M. R. Palmer, *J. Am. Chem. Soc.*, 1981, **103,** 2659.
146. B. Folkesson and R. Larsson, *Acta Chem. Scand., Ser. A*, 1979, **33,** 347.
147. G. N. Schrauzer, in 'New Trends in the Chemistry of Nitrogen Fixation', ed. J. Chatt, L. M. C. Pina and R. L. Richards, Academic, London, 1980, p. 103.
148. A. E. Shilov, in 'New Trends in the Chemistry of Nitrogen Fixation', ed. J. Chatt, L. M. C. Pina and R. L. Richards, Academic, London, 1980, p. 121.
149. M. E. Vol'pin, M. A. Ilatovskaya, E. I. Larikov, M. L. Khidekel, Y. A. Shvetsov and V. B. Shur, *Dokl. Akad. Nauk SSSR*, 1965, **164,** 331.
150. A. Yamamoto, S. Go, M. Ookawa, M. Takahashi, S. Ikeda and T. Keii, *Bull. Chem. Soc. Jpn.*, 1972, **45,** 3110.
151. G. Henrici-Olivé and S. Olivé, *Angew. Chem., Int. Ed. Engl.*, 1967, **6,** 873.
152. T. P. M. Beelen and W. Van Erk, *Recl. Trav. Chim. Pays-Bas*, 1971, **90,** 1197.
153. T. P. M. Beelen and W. Van Erk, *Recl. Trav. Chim. Pays-Bas.*, 1979, **98,** 8.
154. P. Sobota, B. Jezowska-Trzebiatowska and Z. Janas, *J. Organomet. Chem.*, 1976, **118,** 253.
155. K. Shiina, *J. Am. Chem. Soc.*, 1972, **94,** 9267.
156. G. Kalatzis, J. Konstantatos, E. Vrachnou-Astra and D. Katakis, *J. Am. Chem. Soc.*, 1983, **105,** 2897; *J. Chem. Soc., Dalton Trans.*, 1985, 2461.
157. J. D. Rush and B. H. J. Bielski, *Inorg. Chem.*, 1985, **24,** 4282.
158. J. H. Swinehart, *Inorg. Chem.*, 1965, **4,** 1069.
159. W. Sawodny, R. Grünes and H. Reitzle, *Angew. Chem., Int. Ed. Engl.*, 1982, **21,** 775.
160. R. G. Linck, in 'MTP International Review of Science, Series One, Inorganic Chemistry', ed. M. L. Tobe, Butterworths, London, 1972, vol. 9, p. 303; Series Two, 1974, p. 173.
161. M. R. Hyde, R. S. Taylor and A. G. Sykes, *J. Chem. Soc., Dalton Trans.*, 1973, 2730.
162. T. D. Hand, Ph.D. Thesis, University of Leeds, 1976, p. 29 (quoted in ref. 171).
163. F. F. Fan and E. S. Gould, *Inorg. Chem.*, 1974, **13,** 2647.
164. B. Grossman and A. Haim, *J. Am. Chem. Soc.*, 1971, **93,** 6490.
165. C. S. Glennon, J. D. Edwards and A. G. Sykes, *Inorg. Chem.*, 1978, **17,** 1654.
166. R. H. Prince and M. G. Segal, *J. Chem. Soc., Dalton Trans.*, 1975, 330; 1975, 1245.
167. P. N. Balasubramanian and V. R. Vrjayaraghavan, *Indian J. Chem., Sect. A*, 1981, **20,** 892.
168. K. W. Hicks, D. L. Toppen and R. G. Linck, *Inorg. Chem.*, 1972, **11,** 310.
169. Y.-T. Fanchiang, J. C. Thomas, V. D. Neff, J. C. K. Heh and E. S. Gould, *Inorg. Chem.*, 1977, **16,** 1942, and references therein.
170. A. M. Lannon, A. G. Lappin and M. G. Segal, *Inorg. Chem.*, 1984, **23,** 4167.
171. C. S. Glennon, T. D. Hand and A. G. Sykes, *J. Chem. Soc., Dalton Trans.*, 1980, 19.
172. S. A. Kazmi, A. L. Shorter and J. V. McArdle, *Inorg. Chem.*, 1984, **23,** 4332.
173. R. L. Robson, R. R. Eady, T. H. Richardson, R. W. Miller, M. Hawkins and J. R. Postgate, *Nature (London)*, 1986, **322,** 388.
174. R. A. Levenson and R. L. R. Towns, *J. Am. Chem. Soc.*, 1972, **94,** 4345.
175. R. L. R. Towns and R. A. Levenson, *Inorg. Chem.*, 1974, **13,** 105.
176. R. K. Boggess and W. D. Wiegele, *J. Chem. Educ.*, 1978, **55,** 156.
177. R. A. Levenson, *Chem. Phys. Lett.*, 1973, **22,** 293.
178. R. Nast and D. Rehder, *Chem. Ber.*, 1971, **104,** 1709.
179. R. A. Levenson, R. J. G. Dominguez, M. A. Willis and F. R. Young, III, *Inorg. Chem.*, 1974, **13,** 2762.
180. A.-M. Soares and W. P. Griffith, *J. Chem. Soc., Dalton Trans.*, 1981, 1886.
181. R. A. Levenson and R. J. G. Dominguez, *Inorg. Chem.*, 1973, **12,** 2342.
182. L. D. Silverman, J. C. Dewan, C. M. Giandomenico and S. J. Lippard, *Inorg. Chem.*, 1980, **19,** 3379.
183. H. Remy and I. May, *Naturwissenschaften*, 1961, **48,** 524.
184. D. Nicholls, *J. Inorg. Nucl. Chem.*, 1962, **24,** 1001.

185. H. Boehland, *Z. Chem.*, 1972, **12,** 148.
186. R. J. H. Clark and M. L. Greenfield, *J. Chem. Soc. (A)*, 1967, 409.
187. H. A. Rupp, *Z. Anorg. Allg. Chem.*, 1970, **377,** 105.
188. B. Hackel-Wenzel and G. Thomas, *J. Less-Common Met.*, 1971, **23,** 185.
189. D. C. Bradley and R. G. Copperthwaite, *Inorg. Synth.*, 1978, **18,** 117.
190. E. C. Alyea, D. C. Bradley and R. G. Copperthwaite, *J. Chem. Soc., Dalton Trans.*, 1972, 1580.
191. E. C. Alyea, D. C. Bradley, R. G. Copperthwaite and K. D. Sales, *J. Chem. Soc., Dalton Trans.*, 1973, 185.
192. M. W. Duckworth, G. W. A. Fowles and P. T. Greene, *J. Chem. Soc. (A)*, 1967, 1592.
193. L. S. Jenkins and G. R. Willey, *Inorg. Chim. Acta*, 1978, **29,** L201.
194. P. T. Greene and P. L. Orioli, *J. Chem. Soc. (A)*, 1969, 1621.
195. P. C. Crouch, G. W. A. Fowles and R. A. Walton, *J. Chem. Soc. (A)*, 1968, 2172.
196. J. Hughes and G. R. Willey, *Inorg. Chim. Acta*, 1977, **24,** L81.
197. C. M. Mikulski, L. Mattucci, L. Weiss and N. M. Karayannis, *Inorg. Chim. Acta*, 1984, **92,** L29.
198. C. M. Mikulski, S. Cocco, N. de Franco and N. M. Karayannis, *Inorg. Chim. Acta*, 1983, **80,** L61.
199. G. W. A. Fowles and P. T. Greene, *J. Chem. Soc. (A)*, 1967, 1869.
200. B. Horvath, J. Stutz, J. Geyer-Lippmann and E. G. Horvath, *Z. Anorg. Allg. Chem.*, 1981, **483,** 181.
201. E. König and G. Thomas, *J. Inorg. Nucl. Chem.*, 1972, **34,** 1173.
202. A. T. Casey and R. J. H. Clark, *Transition Met. Chem. (Weinheim, Ger.)*, 1977, **2,** 76.
203. N. L. Babenko and M. Sh. Blokh, *Russ. J. Inorg. Chem., (Engl. Transl.)*, 1980, **25,** 1481.
204. K. S. Murray and R. M. Sheahan, *J. Chem. Soc., Dalton Trans.*, 1973, 1182.
205. S. C. Furman and C. S. Garner, *J. Am. Chem. Soc.*, 1951, **73,** 4528.
206. C. Brattas, S. Jagner and E. Ljungström, *Acta Crystallogr., Sect. B*, 1978, **34,** 3073.
207. H. Böhland and P. Malitze, *Z. Anorg. Allg. Chem.*, 1967, **350,** 70.
208. H. H. Schmidtke and D. Garthoff, *Z. Naturforsch., Teil A*, 1969, **24,** 126.
209. (a) G. V. Tsintsadze, V. V. Skopenko and E. I. Ivanova, *Soobshch. Akad. Nauk Gruz. SSR*, 1968, **51,** 107 (*Chem. Abstr.*, 1969, **70,** 15 539b); (b) V. V. Skopenko and E. I. Ivanova, *Zh. Neorg. Khim.*, 1969, **14,** 742 (*Chem. Abstr.*, 1969, **70,** 102 599g).
210. V. Gutmann, O. Leitmann, A. Scherhaufer and H. Czuba, *Monatsh. Chem.*, 1967, **98,** 188.
211. J. J. Salzmann and H. H. Schmidke, *Inorg. Chim. Acta*, 1969, **3,** 207.
212. R. J. H. Clark, R. S. Nyholm and D. E. Scaife, *J. Chem. Soc. (A)*, 1966, 1296.
213. A. T. Casey, R. J. H. Clark, R. S. Nyholm and D. E. Scaife, *Inorg. Synth.*, 1971, **13,** 165.
214. R. J. Kern, *J. Inorg. Nucl. Chem.*, 1963, **25,** 5.
215. G. R. Willey, *J. Am. Chem. Soc.*, 1968, **90,** 3362.
216. J. Hughes and G. R. Willey, *J. Am. Chem. Soc.*, 1973, **95,** 8758.
217. K. Issleib and G. Bohn, *Z. Anorg. Allg. Chem.*, 1959, **301,** 188.
218. J. Nieman, J. H. Teuben, J. C. Huffman and K. G. Caulton, *J. Organomet. Chem.*, 1983, **255,** 193.
219. R. L. Bansemer, J. C. Huffman and K. G. Caulton, *Inorg. Chem.*, 1985, **24,** 3003.
220. R. J. H. Clark, M. L. Greenfield and R. S. Nyholm, *J. Chem. Soc. (A)*, 1966, 1254.
221. J. K. Beattie, S. P. Best, B. W. Skelton and A. H. White, *J. Chem. Soc., Dalton Trans.*, 1981, 2105.
222. F. A. Cotton, C. K. Fair, G. E. Lewis, G. N. Mott, F. K. Ross, A. J. Schultz and J. M. Williams, *J. Am. Chem. Soc.*, 1984, **106,** 5319.
223. S. P. Best, R. S. Armstrong and J. K. Beattie, *Inorg. Chem.*, 1980, **19,** 1958.
224. S. P. Best, J. K. Beattie and R. S. Armstrong, *J. Chem. Soc., Dalton Trans.*, 1984, 2611.
225. H. Hartmann and H. L. Schläfer, *Z. Naturforsch., Teil A*, 1951, **6,** 754.
226. A. M. Chmelnick and D. Fiat, *J. Magn. Reson.*, 1972, **8,** 325.
227. A. D. Hugi, L. Helm and A. E. Merbach, *Helv. Chim. Acta*, 1985, **68,** 508.
228. L. Pajdowski, *Rocz. Chem.*, 1963, **37,** 1363.
229. L. Pajdowski, *J. Inorg. Nucl. Chem.*, 1966, **28,** 433.
230. L. Pajdowski and B. Jezowka-Trzebiatowska, *J. Inorg. Nucl. Chem.*, 1966, **28,** 443.
231. L. P. Podmore and P. W. Smith, *Aust. J. Chem.*, 1972, **25,** 2521.
232. W. F. Donovan and P. W. Smith, *J. Chem. Soc., Dalton Trans.*, 1975, 894.
233. W. F. Donovan, L. P. Podmore and P. W. Smith, *J. Chem. Soc., Dalton Trans.*, 1976, 1741.
234. P. J. McCarthy, J. C. Laufenburger and M. M. Schreiner, *Inorg. Chem.*, 1981, **20,** 1571.
235. J. H. Swinehart, *Chem. Commun.*, 1971, 1443; D. J. Halko and J. H. Swinehart, *J. Inorg. Nucl. Chem.*, 1979, **41,** 1589.
236. Y. Tatsuno, M. Tatsuda and S. Otsuka, *J. Chem. Soc., Chem. Commun.*, 1982, 1100.
237. A. T. Casey and R. J. H. Clark, *Inorg. Chem.*, 1969, **8,** 1216.
238. Y. Doi and M. Tsutsui, *J. Am. Chem. Soc.*, 1978, **100,** 3243.
239. G. W. A. Fowles, P. T. Greene and T. E. Lester, *J. Inorg. Nucl. Chem.*, 1967, **29,** 2365.
240. S. Dilli and E. Patsalides, *Aust. J. Chem.*, 1976, **29,** 2389.
241. B. Morosin and H. Montgomery, *Acta Crystallogr., Sect. B*, 1969, **25,** 1354.
242. D. W. Barnum, *J. Inorg. Nucl. Chem.*, 1961, **21,** 221; 1961, **22,** 183.
243. L. S. Lussier, C. Sandorfy, A. Goursot, E. Penigault and J. Weber, *J. Phys. Chem.*, 1984, **88,** 5492.
244. S. Evans, A. Hamnet, A. F. Orchard and D. R. Loyd, *Faraday Discuss. Chem. Soc.*, 1972, **54,** 227; S. Evans, A. Hamnet and A. F. Orchard, *J. Coord. Chem.*, 1972, **2,** 57.
245. T. F. Schaaf, S. C. Avanzino, W. L. Jolly and R. E. Sievens, *J. Coord. Chem.*, 1976, **5,** 157.
246. B. A. Kolesov and I. K. Igumenov, *Koord. Khim.*, 1985, **11,** 485.
247. C. A. Fleming and D. A. Thornton, *J. Mol. Struct.*, 1973, **17,** 79.
248. A. J. C. Nixon and D. R. Eaton, *Can. J. Chem.*, 1978, **56,** 1005; 1978, **56,** 1012; 1978, **56,** 1928.
249. F. Röhrscheid, R. E. Ernst and R. H. Holm, *Inorg. Chem.*, 1967, **6,** 1315.
250. J. G. Gordon, II, M. J. O'Connor and R. H. Holm, *Inorg. Chim. Acta*, 1971, **5,** 381.
251. K. Hiraki, M. Onishi, Y. Nakashima and Y. Obayashi, *Bull. Chem. Soc. Jpn.*, 1979, **52,** 625.
252. S. Dilli and E. Patsalides, *Aust. J. Chem.*, 1976, **29,** 2369; 1976, **29,** 2381.

253. S. Dilli and E. Patsalides, *J. Chromatogr.*, 1979, **176**, 305.
254. N. Matsubara and T. Kuwamoto, *Inorg. Chem.*, 1985, **24**, 2697.
255. L. M. Dyagileva, E. I. Tsyganova and V. P. Mar'in, *Zh. Obshch. Khim.*, 1982, **52**, 2024 (*Chem. Abstr.*, 1982, **97**, 203 738x).
256. S. R. Cooper, Y. B. Koh and K. N. Raymond, *J. Am. Chem. Soc.*, 1982, **104**, 5092.
257. R. M. Buchanan, H. H. Downs, W. B. Shorthill, C. G. Pierpont, S. L. Kessel and D. N. Hendrickson, *J. Am. Chem. Soc.*, 1978, **100**, 4318.
258. C. M. Mikulski, J. Uuruh, R. Rabin, F. J. Iaconnianni, L. L. Pytlewski and N. M. Karayannis, *Transition Met. Chem. (Weinheim, Ger.)*, 1981, **6**, 79.
259. C. M. Mikulski, P. Sanford, N. Harris, R. Rabin and N. M. Karayannis, *J. Coord. Chem.*, 1983, **12**, 187, and references therein.
260. C. M. Mikulski, N. M. Karayannis, M. J. Strocko, L. L. Pytlewski and M. M. Labes, *Inorg. Chem.*, 1970, **9**, 2053.
261. I. P. Lorenz, *Inorg. Nucl. Chem. Lett.*, 1979, **15**, 127; G. Graziosi, C. Preti and G. Tosi, *Transition Met. Chem. (Weinheim, Ger.)*, 1982, **7**, 267.
262. J. A. Duffy and W. J. D. Macdonald, *J. Chem. Soc., Dalton Trans.*, 1970, 977.
263. S. K. Sharma, R. K. Mahajan, B. Kapila and V. P. Kapila, *Polyhedron*, 1983, **2**, 973.
264. E. Lindner, I. P. Lorenz and G. Vitzhum, *Chem. Ber.*, 1970, **103**, 3182; E. König, E. Lindner, I. P. Lorenz, G. Ritter and H. Gausmann, *J. Inorg. Nucl. Chem.*, 1971, **33**, 3305.
265. H. J. Seifert, *J. Inorg. Nucl. Chem.*, 1965, **27**, 1269.
266. H. Funk, C. Muller and A. Paul, *Z. Chem.*, 1966, 227; B. G. Bennett and D. Nicholls, *J. Inorg. Nucl. Chem.*, 1972, **34**, 673.
267. T. Glowiak, M. Kubiak and B. Jezowska-Trzebiatowska, *Bull. Acad. Pol. Sci., Ser. Sci. Chim.*, 1977, **25**, 359.
268. N. N. Greenwood, P. V. Parish and P. Thornton, *J. Chem. Soc. (A)*, 1966, 320.
269. F. A. Cotton, G. E. Lewis and G. N. Mott, *Inorg. Chem.*, 1982, **21**, 3127.
270. A. Matsumoto, H. Kumafuji and J. Shiokawa, *Inorg. Chim. Acta*, 1980, **42**, 149.
271. B. N. Figgis, J. Lewis and F. Mabbs, *J. Chem. Soc.*, 1960, 2480.
272. F. A. Cotton, S. A. Duraj and W. J. Roth, *Inorg. Chem.*, 1984, **23**, 4042.
273. K. Nagase, *Bull. Chem. Soc. Jpn.*, 1972, **45**, 2166; 1973, **46**, 144.
274. R. D. Gillard, D. J. Shepherd and D. A. Tarr, *J. Chem. Soc., Dalton Trans.*, 1976, 594.
275. N. Ahmad and S. Kirshner, *Inorg. Chim. Acta*, 1975, **14**, 215.
276. I. E. Grey, I. C. Madsen, K. Sirat and P. W. Smith, *Acta Crystallogr., Sect. C*, 1985, **41**, 681.
277. Y. Ikeda, S. Soya, H. Tomiyasu and H. Fukutomi, *Bull. Chem. Soc. Jpn.*, 1981, **54**, 3768.
278. A. N. Glebov, Yu. I. Sal'nikov and N. E. Zhuravleva, *Zh. Neorg. Khim.*, 1981, **26**, 2089.
279. A. A. Popel, A. N. Glebov, Yu. I. Sal'nikov and N. E. Zhuravleva, *Zh. Neorg. Khim.*, 1980, **25**, 2175.
280. B. Perlmutter-Hayman and E. Tapuhi, *J. Coord. Chem.*, 1979, **9**, 177; 1980, **10**, 219; *Inorg. Chem.*, 1979, **18**, 2872.
281. W. D. Harrison, J. B. Gill and D. C. Goodall, *J. Chem. Soc., Chem. Commun.*, 1976, 540; *J. Chem. Soc., Dalton Trans.*, 1979, 847.
282. B. Jeffreys, J. B. Gill and D. C. Goodall, *J. Chem. Soc., Dalton Trans.*, 1985, 99.
283. N. K. Graham, J. B. Gill and D. C. Goodall, *Aust. J. Chem.*, 1983, **36**, 1991.
284. C. H. Langford and F. M. Chung, *Can. J. Chem.*, 1970, **48**, 2969.
285. R. R. Windolph and A. J. Leffler, *Inorg. Chem.*, 1972, **11**, 594.
286. B. N. Figgis and L. G. B. Wadley, *J. Chem. Soc., Dalton Trans.*, 1972, 2182.
287. J. Baker and B. N. Figgis, *Aust. J. Chem.*, 1980, **33**, 2377.
288. M. W. Duckworth, G. W. A. Fowles and P. J. Greene, *J. Chem. Soc. (A)*, 1967, 1592.
289. R. J. H. Clark and G. Natile, *Inorg. Chim. Acta*, 1970, **4**, 533.
290. L. F. Larkworthy and M. W. O'Donoghue, *Inorg. Chim. Acta*, 1983, **74**, 155.
291. F. Forghieri, G. Graziosi, C. Preti and G. Tosi, *Transition Met. Chem. (Weinheim, Ger.)*, 1983, **8**, 372.
292. L. Que, Jr. and L. H. Pignolet, *Inorg. Chem.*, 1974, **13**, 351.
293. R. G. Cavell, E. D. Day, B. Byers and P. M. Watkins, *Inorg. Chem.*, 1971, **10**, 2716.
294. C. Furlani, A. A. G. Tomlinson, P. Porta and A. Sgamellotti, *J. Chem. Soc. (A)*, 1970, 2929.
295. R. D. Hancock, F. Marsicano and E. Rudolph, *J. Coord. Chem.*, 1980, **10**, 223.
296. J. C. Cretenet, *Rev. Chim. Miner.*, 1973, **10**, 399.
297. D. Babel, R. Haegele, G. Pausewang and F. Wall, *Mat. Res. Bull.*, 1973, **8**, 1371.
298. E. Alter and R. Hoppe, *Z. Anorg. Allg. Chem.*, 1975, **412**, 110.
299. W. Massa, *Z. Kristallogr.*, 1980, **153**, 201.
300. (a) R. S. Nyholm and A. G. Sharpe, *J. Chem. Soc.*, 1952, 3579; (b) B. M. Wanklyn, *J. Inorg. Nucl. Chem.*, 1965, **27**, 481.
301. K. Koyama, Y. Hashimoto, S. Omori and K. Terawaki, *Bull. Chem. Soc. Jpn.*, 1984, **57**, 2311.
302. D. R. Armstrong, R. Fortune and P. G. Perkins, *J. Chem. Soc., Dalton Trans.*, 1976, 753.
303. C. J. Ballhausen and F. Winther, *Acta Chem. Scand.*, 1959, **13**, 1729.
304. P. G. Nelson and R. V. Pearse, *J. Chem. Soc., Dalton Trans.*, 1983, 1977.
305. H. Siebert, *Z. Naturforsch., Teil B*, 1972, **27**, 1101.
306. J. Slivnik, J. Pezdic and B. Sedej, *Monatsh. Chem.*, 1967, **98**, 204.
307. J. Slivnik, *J. Inorg. Nucl. Chem.*, 1976, **38**, 2317.
308. J. Slivnik, J. Macek, A. Rahten and B. Sedej, *Thermochim. Acta*, 1980, **39**, 21.
309. J. Slivnik, B. Druzina and B. Zemva, *Z. Naturforsch., Teil B*, 1981, **36**, 1457.
310. P. Bukovec and J. Siftar, *Monatsh. Chem.*, 1974, **105**, 510.
311. M. Hidaka, K. Inoue, B. J. Garrard and B. M. Wanklyn, *Phys. Stat. Solidi A*, 1982, **72**, 809; M. Hidaka, H. Fujii, B. J. Garrard and B. M. Wanklyn, *Phys. Stat. Solidi A*, 1984, **86**, 75.
312. R. Deblieck, J. Van Landuyt and S. Amelinckx, *J. Solid State Chem.*, 1985, **59**, 379.
313. B. V. Bukvetskii, L. A. Muradyan, R. L. Davidovich and V. I. Simonov, *Koord. Khim.*, 1976, **2**, 1129.

314. W. Liebe, E. Weise and W. Klemm, *Z. Anorg. Allg. Chem.*, 1961, **311**, 281.
315. S. M. Horner and S. Y. Tyree, *Inorg. Chem.*, 1964, **3**, 1173.
316. H. J. Seifert and H. W. Loh, *Inorg. Chem.*, 1966, **5**, 1822.
317. E. F. King and M. L. Good, *Spectrochim. Acta, Part A*, 1973, **29**, 707.
318. D. E. Scaife, *Aust. J. Chem.*, 1970, **23**, 2205.
319. A. T. Casey and R. J. H. Clark, *Inorg. Chem.*, 1968, **7**, 1598.
320. R. Saillant and R. A. D. Wentworth, *Inorg. Chem.*, 1968, **7**, 1606.
321. D. Nicholls and D. N. Wilkinson, *J. Chem. Soc. (A)*, 1969, 1232.
322. R. H. Holm, G. W. Everett and A. Chakravorty, *Prog. Inorg. Chem.*, 1965, **7**, 83.
323. T. J. McNeese and T. E. Mueller, *Inorg. Chem.*, 1985, **24**, 2981.
324. M. Mazzanti, C. Floriani, A. Chiesi-Villa and C. Guastini, *Inorg. Chem.*, 1986, **25**, 4158.
325. S. Gambarotta, M. Mazzanti, C. Floriani, A. Chiesi-Villa and C. Guastini, *J. Chem. Soc., Chem. Commun.*, 1985, 829 and references therein.
326. I. Grecu, R. Sandulescu and M. Neamtu, *An. Quim.*, 1983, **79**, 18.
327. D. Kovala-Demertzi, M. Demertzis and J. M. Tsangaris, *Bull. Soc. Chim. Fr.*, 1986, 558.
328. M. Castillo and E. Ramirez, *Transition Met. Chem. (Weinheim, Ger.)*, 1984, **9**, 268.
329. H. P. Fritz and O. Stetten, *Z. Naturforsch., Teil B*, 1973, **28**, 772.
330. L. Pajdowski and Z. Karwecka, *Rocz. Chem.*, 1970, **44**, 1957; 1970, **14**, 2055.
331. Z. Karwecka and L. Pajdowski, *Adv. Mol. Relaxation Processes*, 1973, **5**, 45.
332. M. F. Grigoreva, N. I. Babenko and I. A. Tserkovnitskaya, *Instrum. Khim. Metody Anal.*, 1973, 92.
333. J. Podlaha and P. Petras, *J. Inorg. Nucl. Chem.*, 1970, **32**, 1963.
334. G. Schwarzenbach and J. Sandera, *Helv. Chim. Acta*, 1953, **36**, 1089.
335. R. E. Shepherd, W. E. Hatfield, D. Ghosh, C. D. Stout, F. J. Kristine and J. R. Ruble, *J. Am. Chem. Soc.*, 1981, **103**, 5511.
336. F. J. Kristine and R. E. Shepherd, *J. Am. Chem. Soc.*, 1977, **99**, 6562.
337. F. J. Kristine, D. R. Gard and R. E. Shepherd, *J. Chem. Soc., Chem. Commun.*, 1976, 994.
338. F. J. Kristine and R. E. Shepherd, *J. Am. Chem. Soc.*, 1978, **100**, 4398.
339. N. Kato, K. Nakano and K. Tanaka, *J. Electroanal. Chem. Interfacial Electrochem.*, 1983, **149**, 83.
340. J. A. M. Ahmad and W. C. E. Higginson, *J. Chem. Soc., Dalton Trans.*, 1983, 1449.
341. J. A. M. Ahmad and W. C. E. Higginson, *J. Chem. Soc., Dalton Trans.*, 1984, 451.
342. F. J. Kristine and R. E. Shepherd, *Inorg. Chem.*, 1981, **20**, 215.
343. M. Henze, *Hoppe-Zeyler's Z. Physiol. Chem.*, 1911, **72**, 494; 1932, **213**, 125.
344. A. W. Morris, *Deep-Sea Res.*, 1975, **22**, 49.
345. K. Kustin, G. C. McLeod, T. R. Gilbert and L. B. R. Briggs, *Struct. Bonding (Berlin)*, 1983, **53**, 139.
346. L. Califano and P. Caselli, *Pubbl. Stn. Zool. Napoli*, 1948, **21**, 261; L. Califano and E. Boeri, *J. Exp. Biol.*, 1950, **27**, 253.
347. R. M. K. Carlson, *Proc. Natl. Acad. Sci. USA*, 1975, **72**, 2217.
348. T. D. Tullius, W. O. Gillum, R. M. K. Carlson and K. O. Hodgson, *J. Am. Chem. Soc.*, 1980, **102**, 5670.
349. P. Frank, R. M. K. Carlson and K. O. Hodgson, *Inorg. Chem.*, 1986, **25**, 470.
350. I. G. Macara, G. C. McLeod and K. Kustin, *Biochem. J.*, 1979, **181**, 457.
351. R. C. Bruening, E. M. Oltz, J. Furukawa and K. Nakanishi, *J. Am. Chem. Soc.*, 1985, **107**, 5298.
352. I. G. Macara, G. C. McLeod and K. Kustin, *Comp. Biochem. Physiol.* **A**, 1979, **62**, 821.
353. F. A. Cotton and G. Wilkinson, 'Advanced Inorganic Chemistry', 3rd edn., Wiley-Interscience, New York, 1972.
354. N. D. Chasteen, 'Biological Magnetic Resonance', Plenum, New York, 1981, vol. 3, p. 54.
355. J. Selbin, *Chem. Rev.*, 1965, **65**, 153.
356. J. Selbin, *Coord. Chem. Rev.*, 1966, **1**, 293.
357. (a) Ref. 4, 8. Auflage, Teil B, Lieferung 1; (b) Ref. 4, 8. Auflage, Teil B, Lieferung 2.
358. A. Syamal, *Coord. Chem. Rev.*, 1975, **16**, 309.
359. J. Selbin, *Coord. Chem. Rev., Proc. John Bailar Jr. Symp.*, 1969, 248.
360. L. G. Vanquickenborne and S. P. McGlynn, *Theor. Chim. Acta*, 1968, **9**, 390.
361. C. M. Guzy, J. B. Raynor and M. C. R. Symons, *J. Chem. Soc. (A)*, 1969, 2791.
362. J. R. Wasson and H. J. Stoklosa, *J. Inorg. Nucl. Chem.*, 1974, **36**, 227.
363. H. J. Stoklosa, J. R. Wasson and B. J. McCormick, *Inorg. Chem.*, 1974, **13**, 592.
364. C. J. Ballhausen and H. B. Gray, *Inorg. Chem.*, 1962, **1**, 111.
365. J. Selbin and L. Morpurgo, *J. Inorg. Nucl. Chem.*, 1965, **27**, 673.
366. J. Selbin, G. Maus and D. L. Johnson, *J. Inorg. Nucl. Chem.*, 1967, **29**, 1735.
367. S. Radhakrishna, B. V. R. Chowdari and M. Salagram, *J. Chem. Phys.*, 1980, **72**, 1908.
368. A. B. P. Lever, 'Inorganic Electronic Spectroscopy' 2nd edn., Elsevier, Amsterdam, 1984, pp. 384–391 and references therein.
369. D. Collison, B. Gahan, C. D. Garner and F. E. Mabbs, *J. Chem. Soc., Dalton Trans.*, 1980, 667.
370. H. J. Stoklosa and J. R. Wasson, *J. Inorg. Nucl. Chem.*, 1974, **38**, 667.
371. R. J. H. Clark, *J. Chem. Soc.*, 1963,. 1377.
372. K. M. Jones and E. Larsen, *Acta Chem. Scand.*, 1965, **19**, 1210.
373. K. Okazaki and K. Saito, *Bull. Chem. Soc. Jpn.*, 1982, **55**, 785.
374. J. R. Wasson and H. J. Stoklosa, *J. Chem. Educ.*, 1973, **50**, 186.
375. H. Johansen and K. Tanaka, *Chem. Phys. Lett.*, 1985, **116**, 155.
376. Ref. 5c, page 518.
377. J. Felcman and J. J. R. Fraústo da Silva, *Talanta*, 1983, **30**, 565.
378. J. Felcman, M. C. T. A. Vaz and J. J. R. Fraústo da Silva, *Inorg. Chim. Acta*, 1984, **93**, 101.
379. F. E. Dickson, C. J. Kunesh, E. L. McGinis and L. Petrakis, *Anal. Chem.*, 1972, **44**, 978.
380. H. van Willigen, *Chem. Phys. Lett.*, 1979, **65**, 490.
381. H. van Willigen, *J. Phys. Chem.*, 1981, **85**, 1220.
382. C. F. Mulks and H. van Willigen, *J. Phys. Chem.*, 1981, **85**, 1220.

383. C. F. Mulks, B. Kirste and H. van Willigen, *J. Am. Chem. Soc.*, 1982, **104,** 5906.
384. M. Pasquali, F. Marchetti, C. Floriani and S. Merlino, *J. Chem. Soc., Dalton Trans.*, 1977, 139.
385. G. E. Mont and A. E. Martell, *J. Am. Chem. Soc.*, 1966, **88,** 1387.
386. MTP International Review of Science, Inorganic Chemistry, Series One, 'Transition Elements', Butterworths, London, 1972, vol. 5, parts 1 and 2.
387. J. Selbin and L. H. Holmes, Jr., *J. Inorg. Nucl. Chem.*, 1962, **24,** 1111.
388. J. Selbin, L. H. Holmes, Jr. and S. P. McGlynn, *J. Inorg. Nucl. Chem.*, 1963, **2,** 1315.
389. A. Yakimach, *C. R. Hebd. Seances Acad. Sci.*, 1930, **191,** 789 (Chem. Abstr., 1931, **25,** 2935).
390. V. V. Dovgei, L. I. Pavlenko and D. I. Zubritskaya, *Russ. J. Inorg. Chem.* (*Engl. Transl.*), 1980, **25,** 1349.
391. J. Selbin, T. R. Ortolano and F. J. Smith, *Inorg. Chem.*, 1963, **2,** 1315.
392. A. Feltz, *Z. Anorg. Allg. Chem.*, 1967, **354,** 225.
393. R. J. H. Clark, *J. Chem. Soc.*, 1963, 1377.
394. R. J. Kern, *J. Inorg. Nucl. Chem.*, 1962, **24,** 1105.
395. H. Funk, W. Weiss and M. Zeising, *Z. Anorg. Allg. Chem.*, 1958, **296,** 36.
396. Z. O. Dzhamalova, O. F. Khodzhaev and N. A. Parpiev, *Russ. J. Inorg. Chem.* (*Engl. Transl.*), 1974, **19,** 1658.
397. H. J. Seifert and W. Sanerteig, *Z. Anorg. Allg. Chem.*, 1970, **376,** 245.
398. J. E. Drake, J. Vekris and J. S. Wood, *J. Chem. Soc.* (*A*), 1968, 1000.
399. K. L. Baker, D. A. Edwards, G. W. A. Fowles and R. G. Williams, *J. Inorg. Nucl. Chem.*, 1967, **29,** 1881.
400. D. Nicholls and K. R. Seddon, *J. Chem. Soc., Dalton Trans.*, 1973, 2747.
401. D. Nicholls and K. R. Seddon, *J. Chem. Soc., Dalton Trans.*, 1973, 2751.
402. G. W. A. Fowles, D. A. Rice and J. D. Wilkins, *Inorg. Chim. Acta,* 1973, **7,** 642.
403. A. Feltz, *Z. Chem.*, 1967, **7,** 23.
404. A. Feltz, *Z. Anorg. Allg. Chem.*, 1967, **355,** 120.
405. I. S. Morosov and A. Aimorozov, *Izv. Akad. Nauk SSSR, Neorg. Mater.*, 1967, **3,** 925;
406. P. A. Kilty and D. Nicholls, *J. Chem. Soc.* (*A*), 1966, 1175.
407. D. J. Machin and K. S. Murray, *J. Chem. Soc.* (*A*), 1967, 1330.
408. J. E. Drake, J. Vekris and J. S. Wood, *J. Chem. Soc.* (*A*), 1969, 345.
409. J. Cave, P. R. Dixon and K. R. Seddon, *Inorg. Chim. Acta,* 1978, **30,** 1349.
410. G. W. A. Fowles, D. F. Lewis and R. A. Walton, *J. Chem. Soc.* (*A*), 1968, 1468.
411. L. V. Kobets, L. P. Dmitrieva, N. I. Vorob'ev and V. V. Pechkovskii, *Dokl. Akad. Nauk BSSR,* 1971, **15,** 713 (*Chem. Abstr.*, 1971, **75,** 135 052n).
412. G. W. Fowles and C. M. Pleass, *J. Chem. Soc.*, 1957, 1674.
413. D. Cozzi and S. Cecconi, Ric. Sci., 1953, **23,** 609.
414. S. Prasad and R. C. Srivastava, *Z. Anorg. Allg. Chem.*, 1965, **340,** 325.
415. D. Nicholls, A. C. Thomson and A. J. Varney, *Inorg. Chim. Acta,* 1986, **115,** L13.
416. R. G. Cavell and H. C. Clark, *J. Chem. Soc.*, 1962, 2692.
417. M. W. Duckworth and G. W. A. Fowles, *J. Less-Common Met.*, 1962, **4,** 338.
418. H. O. Frohlich and H. Kacholdt, *Z. Chem.*, 1975, **15,** 233.
419. E. C. Alyea and D. C. Bradley, *J. Chem. Soc.* (*A*), 1969, 2330.
420. I. M. Thomas, *Can. J. Chem.*, 1961, **39,** 1368.
421. D. C. Bradley and I. M. Thomas, *Proc. Chem. Soc.*, 1959, 225.
422. D. C. Bradley and I. M. Thomas, *J. Chem. Soc.*, 1960, 3857.
423. S. G. Carr, P. D. W. Boyd and T. D. Smith, *J. Chem. Soc., Dalton Trans.*, 1972, 1491.
424. S. Prasad and R. C. Srivastava, *Z. Anorg. Allg. Chem.*, 1965, **337,** 221.
425. R. G. Cavell and H. C. Clark, *J. Inorg. Nucl. Chem.*, 1961, **17,** 257.
426. K. Wieghart, U. Bossek, K. Vockmar, W. Swiridoff and J. Weiss, *Inorg. Chem.*, 1984, **23,** 1387.
427. A. Ozarowski and D. Reinen, *Inorg. Chem.*, 1986, **25,** 1704.
428. M. C. Chakravorti and A. R. Sarkar, *J. Inorg. Nucl. Chem.*, 1978, **40,** 139; A. H. Jamkhandi and A. S. R. Murty, *Curr. Sci.*, 1980, **49,** 615.
429. A. Jezierski and J. B. Raynor, *J. Chem. Soc., Dalton Trans.*, 1981, 1.
430. R. J. Sundberg and R. B. Martin, *Chem. Rev.*, 1974, **74,** 471.
431. A. K. Srivastava, R. K. Agarwal, M. Srivastava, V. Kapur, S. Sharma and P. C. Jain, *Transition Met. Chem.* (*Weinheim, Ger.*), 1982, **7,** 41.
432. V. B. Rana, S. K. Sahni and S. K. Sangal, *J. Inorg. Nucl. Chem.*, 1979, **41,** 1498.
433. S. Kher, S. K. Sahni, V. Kumari and R. N. Kapoor, *Inorg. Chim. Acta,* 1979, **37,** 121.
434. Ciba Ltd., *Belg. Pat.* 613 432 (1962) (*Chem. Abstr.*, 1963, **58,** 11 374e).
435. N. A. Verdizade and Z. B. Ragimova, *Azerb. Khim. Zh.*, 1979, 86 (*Chem. Abstr.*, 1980, **92,** 103 734w).
436. A. C. Hazell, *J. Chem. Soc.*, 1963, 5745.
437. (a) R. M. Smith and A. E. Martell, 'Critical Stability Constants', Plenum, New York, 1976, vols. 1–4; (b) A. E. Martell and R. M. Smith, 'Critical Stability Constants', Plenum, New York, 1982, vol. 5, 1st suppl.
438. T. M. Che and K. Kustin, *Inorg. Chem.*, 1980, **19,** 2275.
439. G. H. Ayres and L. E. Scroggie, *Anal. Chim. Acta,* 1962, **26,** 470.
440. V. Rao, R. Pandu, D. Satyanarayana and Y. Anjaneyulu, *Proc. Chem. Symp., 2nd.*, 1970, **1,** 151 (*Chem. Abstr.*, 1972, **76,** 158 938r).
441. T. Sato, H. Watanabe, S. Kotani and M. Yamamoto, *Anal. Chim. Acta,* 1976, **84,** 397.
442. A. A. Konovalova, G. M. Larin and V. V. Kolalev, *Koord. Khim.*, 1981, **7,** 548.
443. M. Y. Al-Janabi and N. J. Ali, *Z. Naturforsch., Teil A,* 1976, **31,** 1696.
444. P. Ray, *Chem. Rev.*, 1961, **61,** 313.
445. A. Syamal, *Indian J. Pure Appl. Phys.*, 1983, **21,** 130 (*Chem. Abstr.*, 1983, **99,** 61 075u).
446. R. L. Dutta and S. Lahiry, *J. Indian Chem. Soc.*, 1963, **40,** 863.
447. J. Selbin and G. Vigee, *J. Inorg. Nucl. Chem.*, 1968, **30,** 1644.
448. G. H. Olive and S. Olive, *J. Am. Chem. Soc.*, 1971, **93,** 4154.
449. R. T. Baker, P. J. Krusic, T. H. Tulip, J. C. Calabrese and S. S. Wreford, *J. Am. Chem. Soc.*, 1983, **105,** 6763.

450. R. J. H. Clark, J. Lewis and R. S. Nyholm, *J. Chem. Soc.*, 1962, 2460.
451. R. J. H. Clark, *J. Chem. Soc.*, 1965, 5699.
452. R. J. H. Clark, M. H. Greenfield and R. S. Nyholm, *J. Chem. Soc.*, 1966, 1254.
453. L. Kevan, *J. Phys. Chem.*, 1984, **88,** 327.
454. F. A. Cotton and G. Wilkinson, 'Advanced Inorganic Chemistry', 4th edn., Wiley, New York, 1980, p. 716.
455. K. Wuthrich and R. E. Connick, *Inorg. Chem.*, 1967, **6,** 583.
456. A. J. Rivkind, *Dokl. Akad. Nauk SSSR,* 1962, **142,** 137 (*Chem. Abstr.*, 1962, **57,** 126h).
457. T. J. Swift, T. A. Stephenson and G.-R. Stein, *J. Am. Chem. Soc.*, 1967, **89,** 1611.
458. J. Reuben and D. Fiat, *Inorg. Chem.*, 1969, **8,** 1821.
459. J. Reuben and D. Fiat, *Inorg. Chem.*, 1967, **6,** 579.
460. I. Nágypál, I. Fábián and R. E. Connick, *Acta Chim. Acad. Sci. Hung.*, 1982, **110,** 447.
461. O. Benalti-Baitich and E. Wendling, *Rev. Chim. Miner,* 1975, **12,** 223.
462. M. Nishizawa and K. Saito, *Inorg. Chem.*, 1978, **17,** 3676.
463. J. Francavilla and N. D. Chasteen, *Inorg. Chem.*, 1975, **14,** 2860.
464. M. M. Iannuzzi, C. P. Kubiak and P. H. Rieger, *J. Phys. Chem.*, 1976, **80,** 541.
465. N. F. Albanese and N. D. Chasteen, *J. Phys. Chem.*, 1978, **82,** 910.
466. L. P. Ducret, *Ann. Chim. (Paris),* 1951, **6,** S12, 705.
467. A. Komura, M. Hayashi and H. Imanaga, *Bull. Chem. Soc. Jpn.*, 1977, **50,** 2927.
468. H. T. S. Britton and G. Welford, *J. Chem. Soc.*, 1940, 748.
469. J. J. Lingane and L. Meites, *J. Am. Chem. Soc.*, 1967, **69,** 1882.
470. S. Ostrowetsky, *Bull. Soc. Chim. Fr.*, 1964, 1012.
471. N. D. Chasteen, *Struct. Bonding (Berlin),* 1983, **53,** 105.
472. M. M. Iannuzzi and P. H. Rieger, *Inorg. Chem.*, 1975, **14,** 895.
473. G. K. Johnson and E. O. Schlemper, *J. Am. Chem. Soc.*, 1978, **100,** 3645.
474. M. Pourbaix, 'Atlas d'Equilibres Electrochimiques', Gauthier-Villards, Paris, 1963, sect. 9.1.
475. J. O. Hill, I. G. Worsley and L. G. Hepler, *Chem. Rev.*, 1971, **71,** 127.
476. D. A. Skoog and D. M. West, 'Fundamentals of Analytical Chemistry', 4th edn., Holt-Saunders, Philadelphia, PA, 1982.
477. G. A. Dean and J. F. Herrinshaw, *Talanta,* 1963, **10,** 793.
478. R. M. Holland and R. E. Tapscott, *J. Coord. Chem.*, 1981, **11,** 17.
479. K. Saito, in 'Coordination Chemistry—20', ed. D. Banerjea, Pergamon, Oxford, 1980, p. 173.
480. M. Nishizawa, Y. Sasaki and K. Saito, *Inorg. Chem.*, 1985, **24,** 767.
481. S. Ostrowetsky, *Bull. Soc. Chim. Fr.*, 1964, 1018.
482. C. Heitner-Wirguin and J. Selbin, *J. Inorg. Nucl. Chem.*, 1968, **30,** 3181.
483. M. B. Robin and P. Day, *Adv. Inorg. Chem. Radiochem.*, 1967, **10,** 247.
484. M. K. Chaudhuri and S. K. Ghosh, *J. Chem. Soc., Dalton Trans.*, 1984, 507.
485. J. C. W. Chien and C. R. Boss, *J. Am. Chem. Soc.*, 1961, **83,** 3767.
486. M. M. Chamberlain, G. A. Jabs and B. B. Wayland, *J. Org. Chem.*, 1961, **311,** 281.
487. G. F. Kokoszka, H. C. Allen and G. Gordon, *Inorg. Chem.*, 1966, **5,** 91.
488. D. C. Bradley, R. H. Moss and K. D. Sales, *Chem. Commun.*, 1969, 1255.
489. R. D. Bereman and C. H. Brubaker, *Inorg. Chem.*, 1969, **8,** 2480.
490. H. Funk, G. Mohaupt and A. Paul, *Z. Anorg. Allg. Chem.*, 1959, **302,** 199.
491. S. A. Jimeno, M. L. A. Bartolome and J. A. G. Calzon, *Afinidad,* 1983, **40,** 52.
492. G. T. Kurbatova, E. E. Kriss, Y. E. Alekseev and T. P. Sudareva, *Russ. J. Inorg. Chem.*, (*Engl. Transl.*), 1980, **25,** 1503.
493. G. T. Kurbatova, E. E. Kriss, Y. E. Alekseev and V. A. Tyumenev, *Russ. J. Inorg. Chem.*, (*Engl. Transl.*), 1979, **24,** 1047.
494. A. M. Nardillo and J. A. Catoggio, *Anal. Chim. Acta,* 1975, **74,** 85.
495. Y. Shijo, T. Shimizu and K. Sakai, *Bull. Chem. Soc. Jpn.*, 1981, **54,** 700.
496. P. K. Battacharya and S. N. Banerji, *Z. Anorg. Allg. Chem.*, 1962, **315,** 118.
497. M. E. Cass, N. R. Gordon and C. G. Pierpont, *Inorg. Chem.*, 1986, **25,** 3962.
498. R. P. Henry, P. C. H. Mitchell and J. E. Prue, *J. Chem. Soc. (A),* 1971, 3392.
499. G. A. l'Heureux and A. E. Martell, *J. Inorg. Nucl. Chem.*, 1966, **28,** 481.
500. P. J. Bosserman and D. T. Sawyer, *Inorg. Chem.*, 1982, **21,** 1545.
501. S. Y. Shaiderman, A. N. Demidovskaya and V. G. Zaletov, *Russ. J. Inorg. Chem.* (*Engl. Transl.*), 1972, **17,** 348.
502. J. Zelinka, M. Bartusek and A. Okac, *Collect Czech. Chem. Commun.*, 1974, **39,** 83.
503. J. P. Wilshire and D. T. Sawyer, *J. Am. Chem. Soc.*, 1978, **100,** 3972.
504. M. E. Cass, D. L. Greene, R. M. Buchanan and C. G. Pierpont, *J. Am. Chem. Soc.*, 1986, **105,** 2680.
505. K. Kustin, S. T. Liu, C. Nicolini and D. L. Toppen, *J. Am. Chem. Soc.*, 1974, **96,** 7410.
506. K. Kustin, C. Nicolini and D. L. Toppen, *J. Am. Chem. Soc.*, 1974, **96,** 7416.
507. E. Pelizzetti, E. Mentasti, E. Pramauro and G. Saini, *J. Chem. Soc., Dalton Trans.*, 1974, 1940.
508. J. H. Ferguson and K. Kustin, *Inorg. Chem.*, 1979, **18,** 3349.
509. K. S. Patel, *J. Inorg. Nucl. Chem.*, 1981, **43,** 667.
510. D. A. Johnson and A. B. Waugh, *Polyhedron,* 1983, **2,** 1323.
511. R. P. Dodge, D. H. Templeton and A. Zalkin, *J. Chem. Phys.*, 1961, **35,** 55.
512. (a) P. K. Hon, R. L. Belford and C. E. Pluger, *J. Chem. Phys.*, 1965, **43,** 1323; (b) B. Vlčková, B. Strauch and M. Horák, *Collect. Czech. Chem. Commun.*, 1987, **52,** 686.
513. C. P. Stewart and A. L. Porte, *J. Chem. Soc., Dalton Trans.*, 1972, 1661.
514. M. J. Hynes and B. D. O'Regan, *J. Chem. Soc., Dalton Trans.*, 1980, 7.
515. M. Nishizawa and K. Saito, *Bull. Chem. Soc. Jpn.*, 1978, **51,** 483.
516. M. Nawi and T. L. Riechel, *Inorg. Chem.*, 1981, **20,** 1974.
517. D. Kivelson and S. K. Lee, *J. Chem. Phys.*, 1964, **41,** 1896.
518. A. Serianz, J. R. Shelton, F. L. Urbach, R. C. Dunbar and R. F. Kopczewski, *J. Chem. Educ.*, 1976, **53,** 394.

519. G. D. Simpson, R. L. Belford and R. Biagioni, *Inorg. Chem.*, 1978, **17,** 2424.
520. I. Bernal and P. H. Rieger, *Inorg. Chem.*, 1963, **2,** 256.
521. R. L. Carlin and F. A. Walker, *J. Am. Chem. Soc.*, 1965, **87,** 2128.
522. F. A. Walker, R. L. Carlin and P. H. Rieger, *J. Phys. Chem.*, 1966, **45,** 4181.
523. R. G. Garvey and R. O. Ragsdale, *Inorg. Chim. Acta,* 1968, **2,** 191.
524. N. Adachi, Y. Fukuda and K. Sone, *Bull. Chem. Soc. Jpn.*, 1977, **50,** 401.
525. E. K. Jaffé and A. P. Zipp, *J. Inorg. Nucl. Chem.*, 1978, **40,** 839.
526. C. E. Mannix and A. P. Zipp, *J. Inorg. Nucl. Chem.*, 1979, **41,** 59.
527. B. Kirste and H. van Willigen, *J. Phys. Chem.*, 1982, **86,** 2743.
528. B. M. Sawant, A. L. W. Shroyer, G. R. Eaton and S. S. Eaton, *Inorg. Chem.*, 1982, **21,** 1093.
529. N. M. Atherton, P. J. Gibbon and M. C. B. Shohoji, *J. Chem. Soc., Dalton Trans.*, 1982, 2289.
530. A. Urbanczyk and M. K. Kalinowski, *Montash. Chem.*, 1983, **114,** 1311.
531. R. Lozano, J. Martinez, A. Martinez and A. Doadrio-Lopez, *Polyhedron,* 1983, **2,** 977.
532. M. J. Macazaga, J. A. Garcia-Vasquez, R. M. Medina and J. R. Masaguer, *Transtiion Met. Chem.*, (*Weinheim, Ger.*), 1983, **8,** 215.
533. E. Kwiatkowski and J. Trojanowski, *J. Inorg. Nucl. Chem.*, 1975, **37,** 979.
534. E. Kwiatkowski and J. Trojanowski, *J. Inorg. Nucl. Chem.*, 1976, **38,** 131.
535. J. J. R. Fraústo da Silva and R. Wootton, *Chem. Commun.*, 1969, 421.
536. M. R. Caira, J. M. Haig and L. R. Nassimbeni, *J. Inorg. Nucl. Chem.*, 1972, **34,** 3171.
537. M. R. Caira, J. M. Haigh and L. R. Nassimbeni, *Inorg. Nucl. Chem. Lett.*, 1972, **8,** 109.
538. J. Selbin, H. R. Manning and G. Cessac, *J. Inorg. Nucl. Chem.*, 1963, **25,** 1253.
539. K. Dichmann, G. Hamer, S. C. Nyburg and W. F. Reynolds, *Chem. Commun.*, 1970, 1295.
540. M. Shao, L. Wang and Z. Zhang, *Huaxue Xuebao,* 1983, **41,** 985.
541. N. S. Al-Niaimi, A. R. Al-Karaghouli, S. M. Aliwi and M. G. Jalhoom, *J. Inorg. Nucl. Chem.*, 1974, **36,** 283.
542. V. Guttmann, 'Coordination Chemistry in Non-Aqueous Solution', Springer-Verlag, Vienna, 1963.
543. K. Isobe, Y. Nakamura and S. Kawaguchi, *J. Inorg. Nucl. Chem.*, 1978, **40,** 607.
544. K. Isobe, S. Ooi, Y. Nakamura, S. Kawaguchi and H. Kuroya, *Chem. Lett.*, 1975, 35.
545. A. Syamal and K. S. Kale, *Indian J. Chem., Sect. A,* 1979, **17,** 518.
546. D. E. Fenton, C. M. Regan, U. Casellato, P. A. Vigato and M. Vidali, *Inorg. Chim. Acta,* 1980, **44,** L105.
547. C. J. Popp, J. H. Nelson and R. O. Ragsdale, *J. Am. Chem. Soc.*, 1969, **91,** 610.
548. R. C. Aggawal and D. S. S. Narayana, *Indian J. Chem., Sect. A,* 1982, **21,** 318.
549. M. M. Jones, *J. Am. Chem. Soc.*, 1954, **76,** 5995.
550. M. S. Haddad, D. N. Hendrickson, J. P. Cannady, R. S. Drago and D. S. Bieksza, *J. Am. Chem. Soc.*, 1979, **101,** 898.
551. R. S. Drago, T. C. Kuechler and M. Kroeger, *Inorg. Chem.*, 1979, **18,** 2337.
552. J. A. Howard, J. C. Tait, T. Yamada and J. H. B. Chenier, *Can. J. Chem.*, 1981, **59,** 2184 and references therein.
553. M. M. Musiani, F. Milani, R. Graziani, M. Vidali, U. Casellato and P. A. Vigato, *Inorg. Chim. Acta,* 1982, **61,** 115.
554. U. Casellato, S. Tamburini, P. A. Vigato, M. Vidali and D. E. Fenton, *Inorg. Chim. Acta,* 1984, 101 and references therein.
555. M. Pasquali, F. Marchetti and C. Floriani, *Inorg. Chem.*, 1979, **18,** 2401.
556. A. S. Pesis and E. V. Sokolova, *Zh. Neorg. Khim.*, 1963, 2518 (*Chem. Abstr.*, 1964, **60,** 5067).
557. K. S. Patel and G. A. Kolawole, *J. Inorg. Nucl. Chem.*, 1981, **43,** 3107.
558. G. A. Kolawole and K. S. Patel, *J. Coord. Chem.*, 1982, **12,** 121.
559. K. S. Patel and G. A. Kolawole, *J. Coord. Chem.*, 1982, **12,** 231.
560. M. Nagar and R. K. Baslas, *J. Indian Chem. Soc.*, 1980, **57,** 848.
561. R. Raina and M. L. Dhar, *J. Indian Chem. Soc.*, 1980, **57,** 673.
562. J. C. Espi, R. Ramasseul, A. Rassat and P. Rey, *Bull. Soc. Chim. Fr., Part 2,* 1981, 33.
563. M. Tachez, F. Theobald and A. W. Hewat, *Acta Crystallogr., Sect. B,* 1982, **38,** 1807.
564. M. Tachez and F. Theobald, *Acta Crystallogr., Sect. B,* 1981, **37,** 1978.
565. B. Behr and H. Wendt, *Z. Electrochem.*, 1962, **66,** 223.
566. H. Strehlow and H. Wendt, *Inorg. Chem.*, 1963, **2,** 6.
567. M. F. Grigor'eva, N. I. Slesar and I. A. Tserkovnitskaya, *Vestn. Leningr. Univ. Fiz. Khim.*, 1976, 133 (*Chem. Abstr.*, 1976, **84,** 186 611).
568. J. E. Salmon and D. Whyman, *J. Chem. Soc.* (*A*), 1966, 980.
569. A. A. Ivakin, Z. M. Voronova and L. D. Koorbatova, *Zh. Neorg. Khim.*, 1975, **20,** 1246.
570. W. C. Copenhafer, M. W. Kendig, T. P. Russell and P. H. Rieger, *Inorg. Chim. Acta,* 1976, **17,** 167.
571. M. F. Grigor'eva, V. M. Zakharenko, I. A. Tserkovnitskaya, E. V. Kos'yanova and E. F. Strizhev, *Russ. J. Inorg. Chem.* (*Engl. Transl.*), 1982, **27,** 809.
572. C. C. Parker, R. R. Reeder, L. B. Richards and P. H. Rieger, *J. Am. Chem. Soc.*, 1970, **92,** 5230.
573. A. Hasegawa, S. Yamada and M. Miura, *Bull. Chem. Soc. Jpn.*, 1969, **42,** 846.
574. E. Bordes and P. Courtine, *J. Catal.*, 1979, **57,** 236.
575. J. W. Johnson, D. C. Johnston, A. J. Jacobson and J. F. Brody, *J. Am. Chem. Soc.*, 1984, **106,** 8123 and references therein.
576. D. C. Johnston and J. W. Johnson, *J. Chem. Soc., Chem. Commun.*, 1985, 1720 and references therein.
577. Y. E. Gorbunova and S. A. Linde, *Sov. Phys.-Dokl.* (*Engl. Transl.*), 1979, **24,** 138.
578. C. C. Torardi and J. C. Calabrese, *Inorg. Chem.*, 1984, **23,** 1310.
579. R. P. Henry, P. C. H. Michell and J. E. Prue, *J. Chem. Soc., Dalton Trans.*, 1973, 1156.
580. K. M. Jones and E. Larsen, *Acta Chem. Scand.*, 1965, **19,** 1205.
581. H. Tomiyasu and G. Gordon, *J. Coord. Chem.*, 1973, **3,** 47.
582. J. T. Kummer, *Inorg. Chim. Acta,* 1983, **76,** L291.
583. R. E. Tapscott and L. Belford, *Inorg. Chem.*, 1967, **6,** 735.

584. I. Nágypál and I. Fábián, *Inorg. Chim. Acta,* 1982, **61,** 109.
585. R. D. Gillard, J. Costa Pessoa and L. Vilas-Boas, Comunicação PA54, 6° Encontro Nacional de Química, Aveiro, 1983.
586. (a) G. Gran, *Analyst,* 1952, **77,** 661; (b) F. J. C. Rossotti and H. Rossotti, *J. Chem. Educ.*, 1965, **42,** 375.
587. D. N. Sathyanarayana and C. C. Patel, *Indian J. Chem.*, 1965, **3,** 486.
588. R. D. Gillard, J. Costa Pessoa and L. Vilas-Boas, Comunicação C31.9, 5° Encontro Nacional de Química, Porto, 1982.
589. R. Trujillo and F. Brito, *Anales Fis. Quím.*, 1956, **52B,** 151.
590. I. V. Kozhevnikov, O. A. Kholdeeva, M. A. Fedotov and K. I. Matveev, *Izv. Akad. Nauk SSSR, Ser. Khim.*, 1983, 717 (*Chem. Abstr.*, 1983, **99,** 11 599x).
591. A. Lorenzotti, D. Leonesi, A. Gingolani and A. Turolla, *Gazz. Chim. Ital.*, 1983, **113,** 229.
592. A. Lorenzotti, D. Leonesi and A. Cingolani, *Inorg. Chim. Acta,* 1981, **52,** 149.
593. A. Lorenzotti, D. Leonesi, A. Cingolani and P. Bernardo, *J. Inorg. Nucl. Chem.*, 1981, **43,** 737.
594. M. F. Grigor'eva, G. V. Chernyauskaya and I. A. Tserkavnitskaya, *Russ. J. Inorg. Chem.* (*Engl. Transl.*), 1976, **21,** 230.
595. V. T. Kalinnikov, V. V. Zelentsov, O. N. Kuz'micheva and T. G. Aminov, *Russ. J. Inorg. Chem.* (*Engl. Transl.*), 1970, **15,** 441.
596. H. J. Seifert, *Z. Anorg. Allg. Chem.*, 1962, **317,** 123.
597. T. R. Gilson, I. M. Thom-Postlethwaite and M. Webster, *J. Chem. Soc., Dalton Trans.*, 1986, 895.
598. A. Rachavan and M. Santappa, *J. Inorg. Nucl. Chem.*, 1973, **35,** 3363.
599. (a) R. F. Drake, V. H. Crawford, W. E. Hattfield, G. D. Simpson and G. O. Carlisle, *J. Inorg. Nucl. Chem.*, 1975, **37,** 291; (b) D. M. L. Goodgame and S. V. Waggett, *Inorg. Chim. Acta,* 1970, **5,** 155.
600. A. T. Casey and J. R. Thackeray, *Aust. J. Chem.*, 1969, **22,** 2549.
601. G. M. Larin, V. T. Kalinnikov, V. V. Zelentsov and M. E. Dyatkina, *Teor. Eksp. Khim.*, 1970, **6,** 213 (*Chem. Abstr.*, 1970, **73,** 93 451t).
602. D. Labonnette and B. Taravel, *J. Chem. Res.* (*S*), 1984, 34.
603. P. Koppel, R. Goldmann and A. Kaufmann, *Z. Anorg. Allg. Chem.*, 1905, **45,** 345.
604. T. C. W. Mak, P. Li, C. Zheng and K. Huang, *J. Chem. Soc., Chem. Commun.*, 1986, 1597.
605. T. J. Swift, T. A. Stephenson and G. R. Stein, *J. Am. Chem. Soc.*, 1967, **89,** 1611.
606. K. Wüthrich and R. E. Connick, *Inorg. Chem.*, 1968, **7,** 1377.
607. H. Hoffman and W. Ulbricht, *Ber. Bunsenges. Phys. Chem.*, 1972, **76,** 1052.
608. K. J. Narasimhulu and U. V. Seshaiah, *Indian J. Chem., Sect. A,* 1980, **19,** 1027.
609. D. N. Sathyanarayana and C. C. Patel, *Bull. Chem. Soc. Jpn.*, 1967, **40,** 794.
610. D. N. Sathyanarayana and C. C. Patel, *Indian J. Chem.*, 1968, **6,** 523.
611. R. E. Oughtred, E. S. Raper and H. M. M. Shearer, *Acta Crystallogr., Sect. B,* 1976, **32,** 82.
612. A. J. Edwards, D. R. Slim, J. Sala-Sala and J. E. Guerchais, *Bull. Soc. Chim. Fr.*, 1975, 2015.
613. D. Collison, B. Gahan and F. E. Mabbs, *J. Chem. Soc., Dalton Trans.*, 1983, 1705.
614. B. Gahan and F. E. Mabbs, *J. Chem. Soc., Dalton Trans.*, 1983, 1695.
615. J. G. Forrest and C. K. Prout, *J. Chem. Soc.* (*A*), 1967, 1312.
616. R. E. Tapscott, R. L. Belford and I. C. Paul, *Inorg. Chem.*, 1968, **7,** 356.
617. H. D. Beeson, R. E. Tapscott and E. N. Duesler, *Inorg. Chim. Acta,* 1985, **102,** 5.
618. S. K. Hahs, R. B. Ortega, R. E. Tapscott, C. F. Campana and B. Morosin, *Inorg. Chem.*, 1982, **21,** 664.
619. R. B. Ortega, R. E. Tapscott and C. F. Campana, *Inorg. Chem.*, 1982, **21,** 672.
620. R. B. Ortega, C. F. Campana and R. E. Tapscott, *Acta Crystallogr., Sect. B,* 1980, **36,** 1786.
621. M. E. McIlvain, R. E. Tapscott and W. F. Coleman, *J. Magn. Reson.*, 1977, **26,** 35.
622. M. T. Beck, B. Csaszar and P. Szarvas, *Nature* (*London*), 1960, **188,** 846.
623. L. D. Pettit and J. L. M. Swash, *J. Chem. Soc., Dalton Trans.*, 1978, 286.
624. R. D. Gillard and S. H. Laurie, *J. Chem. Soc.* (*A*), 1970, 59.
626. R. L. Belford, N. D. Chasteen, H. So and R. E. Tapscott, *J. Am. Chem. Soc.*, 1969, **91,** 4675.
627. J. H. Dunlop, D. F. Evans, R. D. Gillard and G. Wilkinson, *J. Chem. Soc.* (*A*), 1966, 1260.
628. R. E. Tapscott, L. D. Hansen and E. A. Lewis, *J. Inorg. Nucl. Chem.*, 1975, **37,** 2517.
629. T. M. Anaya and R. E. Tapscott, *Inorg. Chim. Acta,* 1981, **49,** 11.
630. N. D. Chasteen, R. L. Belford and I. C. Paul, *Inorg. Chem.*, 1969, **8,** 408.
631. P. Reeder and P. H. Rieger, *Inorg. Chem.*, 1971, **10,** 1258.
632. K. M. Jones and E. Larsen, *Acta Chem. Scand.*, 1965, **19,** 1210.
633. S. P. Singh and J. P. Tandon, *Monatsh. Chem.*, 1975, **106,** 271.
634. M. B. McBride, *J. Soil Sci. Soc. Am.*, 1980, **44,** 495.
635. M. L. Goncalves and A. M. Mota, *Talanta,* in press.
636. R. H. Dunhill and T. D. Smith, *J. Chem. Soc.* (*A*), 1968, 2189.
637. V. T. Kalinnikov, V. V. Zelentsov, O. D. Ubozhenko and T. G. Aminov, *Dokl. Akad. Nauk SSSR,* 1972, **206,** 627 (*Chem. Abstr.*, 1973, **78,** 21 723).
638. A. N. Glebov, Yu. I. Sal'nikov, N. I. Zhuravleva, V. V. Cheleva and P. A. Vasil'ev, *Russ. J. Inorg. Chem.* (*Engl. Transl.*), 1982, **27,** 2146.
639. Y. K. Tselinskii, L. V. Shevchenko and I. I. Kusel'man, *Ukr. Khim. Zh.*, 1980, **46,** 656 (*Chem. Abstr.*, 1980, **93,** 81 541x).
640. K. Dey, R. K. Maiti and K. S. Roy, *Indian J. Chem., Sect. A,* 1983, **22,** 580.
641. B. D. Heda, S. G. Kaskedikar and P. V. Khadikar, *Indian J. Hosp. Pharm.*, 1980, **17,** 39.
642. A. N. Glebov, Y. I. Sal'nikov, A. V. Zakharov, Z. A. Saprykova and E. L. Gogolashvili, *Zh. Neorg. Khim.*, 1983, **28,** 1443.
643. S. M. Horner, S. Y. Tyree and D. L. Venezky, *Inorg. Chem.*, 1962, **1,** 844.
644. C. A. L. Filgueiras and M. B. G. Lima, *Inorg. Nucl. Chem. Lett.*, 1981, **17,** 31.

645. T. S. Khodashova, M. A. Porai-Koshits, U. S. Sergienko, L. A. Butman, Y. A. Buslaev and A. A. Kuznetsova, *Koord. Khim.*, 1978, **4,** 1909.
646. H. J. Seifert and J. Vebach, *Z. Anorg. Allg. Chem.*, 1981, **479,** 32.
647. M. R. Caira and B. J. Gellatly, *Acta Crystallogr., Sect. B*, 1980, **36,** 1198.
648. M. Laing, C. Nicholson and T. Ashworth, *J. Cryst. Mol. Struct.*, 1976, **5,** 423 (*Chem. Abstr.*, 1977, **86,** 10 915s).
649. N. M. Karayannis, L. L. Pytlewski and C. Owens, *J. Inorg. Nucl. Chem.*, 1980, **42,** 675.
650. B. Gahan and F. E. Mabbs, *J. Chem. Soc., Dalton Trans.*, 1983, 1713.
651. J. Coetzer, *Acta Crystallogr., Sect. B*, 1970, **26,** 872.
652. F. A. Cotton, G. E. Lewis and G. N. Mott, *Inorg. Chem.*, 1983, **22,** 378.
653. K. P. Callahan, P. J. Durand and P. H. Rieger, *J. Chem. Soc., Chem. Commun.*, 1980, 75.
654. K. P. Callahan and P. J. Durand, *Inorg. Chem.*, 1980, **19,** 3211.
655. M. Sato, K. M. Miller, J. H. Enemark, C. E. Strouse and K. P. Callahan, *Inorg. Chem.*, 1981, **20,** 3571.
656. J. K. Money, J. C. Huffman and G. Christou, *Inorg. Chem.*, 1985, **24,** 3297.
657. R. Seangprasertkij and T. L. Riechel, *Inorg. Chem.*, 1984, **23,** 991.
658. J. K. Money, K. Folting, J. C. Huffman, D. Collison, J. Temperley, F. E. Mabbs and G. Christou, *Inorg. Chem.*, 1986, **25,** 4583.
659. J. L. Poncet, R. Guilard, P. Friant and J. Goulon, *Polyhedron,* 1983, **2,** 417.
660. V. I. Goedken and J. A. Ladd, *J. Chem. Soc., Chem. Commun.*, 1981, 910.
661. O. Piovesana and G. Cappuccilli, *Inorg. Chem.*, 1972, **11,** 1543.
662. O. Piovesana and C. Furlani, *Chem. Commun.*, 1971, 256.
663. M. Bonamico, G. Dessi, V. Fares and L. Scaramuzza, *J. Chem. Soc., Dalton Trans.*, 1974, 1258.
664. R. D. Bereman and J. R. Dorfman, *Polyhedron,* 1983, **2,** 1013.
665. M. Bonamico, G. Dessy, V. Fares, P. Porta and L. Scaramuzza, *Chem. Commun.*, 1971, 365.
666. L. Fanfani, A. Nunzi, P. F. Zanazzi and A. R. Zanzari, *Acta Crystallogr., Sect. B,* 1972, **28,** 1298.
667. K. Henrick, C. L. Raston and A. H. White, *J. Chem. Soc., Dalton Trans.*, 1976, 26.
668. B. J. McCormick and E. M. Bellott, Jr., *Inorg. Chem.*, 1970, **9,** 1779.
669. B. J. McCormick, *Inorg. Chem.*, 1968, **7,** 1965.
670. B. J. McCormick, *Can. J. Chem.*, 1969, **47,** 4283.
671. R. D. Bereman and D. Nalewajek, *J. Inorg. Nucl. Chem.*, 1978, **40,** 1309.
672. D. C. Bradley, I. F. Rendall and K. D. Sales, *J. Chem. Soc., Dalton Trans.*, 1973, 2228.
673. R. D. Bereman and D. Nalewajek, *J. Inorg. Nucl. Chem.*, 1978, **40,** 1311.
674. T. R. Halbert, L. L. Hutchings, R. Rhodes and E. I. Stiefel, *J. Am. Chem. Soc.*, 1986, **108,** 6437.
675. L. L. Borer, *J. Chem. Educ.*, 1982, **59,** 1065.
676. H. J. Stoklosa, G. L. Seebach and J. R. Wasson, *J. Phys. Chem.*, 1974, **78,** 962.
677. D. R. Lorenz, D. K. Johnson, H. J. Stoklosa and J. R. Wasson, *J. Inorg. Nucl. Chem.*, 1974, **36,** 1184.
678. J. R. Wasson, *Inorg. Chem.*, 1971, **10,** 1531.
679. R. G. Cavell, E. D. Day, W. Byers and P. M. Watkins, *Inorg. Chem.*, 1972, **11,** 1591.
680. B. S. Prabhananda, S. J. Shah and B. Venkataraman, *Proc. Chem. Symp., 2nd,* 1970, **1,** 195 (*Chem. Abstr.*, 1972, **76,** 133 861x).
681. B. S. Prabhananda, S. J. Shah and B. Venkataraman, *Indian J. Pure Appl. Phys.*, 1972, **10,** 365 (*Chem. Abstr.*, 1972, **77,** 145 938m).
682. G. A. Miller and R. E. D. McLung, *Inorg. Chem.*, 1973, **12,** 2552.
683. B. J. McCormick, J. L. Featherstone, H. J. Stoklosa and J. R. Wasson, *Inorg. Chem.*, 1973, **12,** 692.
684. G. S. Sodhi and N. K. Kaushik, *Proc. Indian Acad. Sci.*, 1983, **92,** 73.
685. A. D. Villarejo, A. Doadrio, R. Lozano and V. Ragel, *An. Quim.*, 1981, **77,** 50.
686. J. J. R. Fraústo da Silva, R. Wootton and R. D. Gillard, *J. Chem. Soc. (A)*, 1970, 3369.
687. B. S. Sundar, B. S. R. Sarma and V. P. R. Rao, *J. Inorg. Nucl. Chem.*, 1981, **43,** 404.
688. N. M. Atherton, J. Locke and J. A. McCleverty, *Chem. Ind. (London)*, 1965, 1300.
689. N. M. Atherton and C. J. Winscom, *Inorg. Chem.*, 1973, **12,** 383.
690. R. J. H. Clark and P. C. Turtle, *J. Chem. Soc., Dalton Trans.*, 1978, 1714.
691. A. Davison, N. Edelstein, R. H. Holm and A. H. Maki, *J. Am. Chem. Soc.*, 1964, **86,** 2799.
692. R. W. Wiggins, J. C. Huffman and G. Christou, *J. Chem. Soc., Chem. Commun.*, 1983, 1313.
693. D. Cozzi and S. Cecconi, *Ric. Sci.*, 1953, **23,** 609 (*Chem. Abstr.*, 1954, **48,** 3891b).
694. E. E. Aynsley, N. N. Greenwood and J. B. Leach, *Chem. Ind. (London)*, 1966, 379.
695. J. R. Wasson, J. W. Hall and W. E. Hatfield, *Transition Met. Chem., (Weinheim, Ger.)*, 1978, **3,** 195.
696. Ref. 107, p. 765.
697. S. Ahrland and B. Noren, *Acta Chem. Scand.*, 1958, **12,** 1595.
698. J. Rydberg, *Acta Chem. Scand.*, 1961, **15,** 1723.
699. A. Demšar and P. Bukovec, *Inorg. Chim. Acta*, 1977, **25,** L121.
700. M. K. Chaudhuri, S. K. Ghosh and J. Subramanian, *Inorg. Chem.*, 1984, **23,** 4439.
701. G. Pausewang, *Z. Anorg. Allg. Chem.*, 1971, **381,** 189.
702. G. Pausewang and W. Rudorff, *Z. Anorg. Allg. Chem.*, 1969, **364,** 69.
703. K. Waltersson and B. Karlsson, *Cryst. Struct. Commun.*, 1978, **7,** 459.
704. P. Bukovec and L. Golič, *Acta Crystallogr., Sect. B.*, 1980, **36,** 1925.
705. K. Waltersson, *J. Solid State Chem.*, 1979, **28,** 121.
706. K. Waltersson, *J. Solid State Chem.*, 1979, **29,** 195.
707. K. Waltersson, *Cryst. Struct. Commun.*, 1978, **7,** 507.
708. P. Bukovec, S. Milicev, A. Demšar and L. Golič, *J. Chem. Soc., Dalton Trans.*, 1981, 1802.
709. A. Klein, P. Carroll and G. Mitra, *Inorg. Synth.*, 1976, **16,** 87.
710. K. Wieghart and J. Weiss, *Acta Crystallogr., Sect. B,* 1972, **28,** 529.
711. K. R. Seddon, *Gov. Rep. Announce. Index (U.S.)*, 1980, **80** (26), 5628 (*Chem. Abstr.*, 1981, **94,** 112 126t) and references therein.

712. P. A. Kilty and D. Nicholls, *J. Chem. Soc.*, 1966, 1175.
713. J. A. S. Howell and K. C. Moss, *J. Chem. Soc. (A)*, 1971, 270.
714. I. M. Griffiths, D. Nicholls and K. R. Seddon, *J. Chem. Soc. (A)*, 1971, 2513, and references therein.
715. M. L. Ziegler, B. Nuber, K. Wiedenhammer and G. Hoch, *Z. Naturforsch., Teil B*, 1977, **32,** 18.
716. G. W. A. Fowles and R. A. Walton, *J. Inorg. Nucl. Chem.*, 1964, **27,** 735.
717. P. A. Kilty and D. Nicholls, *J. Chem. Soc.*, 1965, 4915.
718. D. L. Kepert, 'The Early Transition Elements', Academic, London, 1972.
719. M. S. Delgado, F. Alguacil and V. Fernandez, *Z. Anorg. Allg. Chem.*, 1982, **493,** 187.
720. B. E. Bridgland, G. W. A. Fowles and R. A. Walton, *J. Inorg. Nucl. Chem.*, 1965, **27,** 383.
721. M. Pasquali, A. Torres-Filho and C. Floriani, *J. Chem. Soc., Chem. Commun.*, 1975, 534.
722. U. Casellato, P. A. Vigato and M. Vidali, *Coord. Chem. Rev.*, 1977, **23,** 31.
723. M. Mathew, A. J. Carty and G. J. Palenik, *J. Am. Chem. Soc.*, 1970, **92,** 3197.
724. R. L. Farmer and F. L. Urbach, *Inorg. Chem.*, 1970, **9,** 2562.
725. A. Pasini and M. Gullotti, *J. Coord. Chem.*, 1974, **3,** 319.
726. R. L. Farmer and F. L. Urbach, *Inorg. Chem.*, 1974, **13,** 587.
727. R. L. Dutta and G. P. Sengupta, *J. Indian Chem. Soc.*, 1971, **48,** 34.
728. A. Kolawole and K. S. Patel, *J. Chem. Soc., Dalton Trans.*, 1981, 1241.
729. S. N. Poddar, K. Dey and S. C. N. Sarkar, *J. Indian Chem. Soc.*, 1963, **40,** 489.
730. M. M. Patel, M. R. Patel, M. N. Patel and R. P. Patel, *Indian J. Chem., Sect. A*, 1981, **20,** 623.
731. V. M. Leovac, N. V. Gerbeleu and V. D. Canic, *Russ. J. Inorg. Chem. (Engl. Transl.)*, 1982, **27,** 514.
732. N. V. Gerbeleu and F. K. Zhovmir, *Russ. J. Inorg. Chem. (Engl. Transl.)*, 1982, **27,** 309, and references therein.
733. N. V. Gerbeleu and M. D. Revenko, *Zh. Neorg. Khim.*, 1971, **16,** 1046.
734. N. V. Gerbeleu, M. D. Revenko and A. V. Ablov, *Zh. Neorg. Khim.*, 1972, **17,** 136.
735. N. V. Gerbeleu and M. D. Revenko, *Zh. Neorg. Khim.*, 1972, **17,** 2176.
736. N. V. Gerbeleu and M. D. Revenko, *Russ. J. Inorg. Chem. (Engl. Transl.)*, 1973, **18,** 1268.
737. M. Pasquali, F. Marchetti, C. Floriani and M. Cesari, *Inorg. Chem.*, 1980, **19,** 1198.
738. P. E. Riley, V. L. Pecoraro, C. J. Carrano and J. A. Bonadies, *Inorg. Chem.*, 1986, **25,** 154.
739. V. Pecoraro, J. A. Bonadies, C. A. Marrese and C. J. Carrano, *J. Am. Chem. Soc.*, 1984, **106,** 3360.
740. J. A. Bonadies and C. J. Carrano, *J. Am. Chem. Soc.*, 1986, **108,** 4088.
741. J. A. Bonadies and C. J. Carrano, *Inorg. Chem.*, 1986, **25,** 4358; J. A. Bonadies, W. M. Butler, V. L. Pecoraro and C. J. Carrano, *Inorg. Chem.*, 1987, **26,** 1218.
742. L. Sacconi and U. Campigli, *Inorg. Chem.*, 1966, **5,** 606.
743. K. Ramaiah and D. F. Martin, *J. Inorg. Nucl. Chem.*, 1965, **27,** 1663.
744. D. F. Martin and K. Ramaiah, *J. Inorg. Nucl. Chem.*, 1965, **27,** 2027.
745. L. J. Boucher and T. F. Yen, *Inorg. Chem.*, 1968, **7,** 2665.
746. L. J. Boucher, D. E. Bruins, T. F. Yen and D. L. Weaver, *Chem. Commun.*, 1969, 363.
747. D. Bruins and D. L. Weaver, *Inorg. Chem.*, 1970, **9,** 130.
748. K. Kasuga, T. Nagahara and Y. Yamamoto, *Bull. Chem. Soc. Jpn.*, 1982, **55,** 2665.
749. R. Seangprasertkij and T. L. Riechel, *Inorg. Chem.*, 1986, **25,** 4268.
750. S. Yamada and Y. Kuge, *Bull. Chem. Soc. Jpn.*, 1969, **42,** 152.
751. A. Syamal, *Bull. Chem. Soc. Jpn.*, 1977, **50,** 557.
752. A. Syamal and K. S. Kale, *Curr. Sci.*, 1977, **46,** 258.
753. A. Syamal and K. S. Kale, *J. Mol. Struct.*, 1977, **38,** 195.
754. R. L. Dutta and G. P. Sengupta, *J. Indian Chem. Soc.*, 1967, **44,** 738.
755. Y. Kuge and S. Yamada, *Bull. Chem. Soc. Jpn.*, 1972, **45,** 799.
756. A. Syamal, *Curr. Sci.*, 1978, **47,** 759.
757. A. Syamal, *Transition Met. Chem. (Weinheim, Ger.)*, 1978, **3,** 297.
758. A. Syamal, *Transition Met. Chem. (Weinheim, Ger.)*, 1978, **3,** 259.
759. R. Seangprasertkij and T. L. Riechel, *Inorg. Chem.*, 1985, **24,** 1115.
760. C. C. Lee, A. Syamal and L. J. Theriot, *Inorg. Chem.*, 1971, **10,** 1669.
761. A. A. Diamantis, J. M. Frederiksen, M. A. Salam, M. R. Snow and E. R. T. Tiekink, *Aust. J. Chem.*, 1986, **39,** 1081.
762. V. V. Zelentsov, *Dokl. Akad. Nauk SSSR*, 1961, **139,** 1110 (*Chem. Abstr.*, 1962, **56,** 1442c).
763. A. P. Ginsberg, E. Koubek and H. J. Williams, *Inorg. Chem.*, 1966, **5,** 1656.
764. B. Bleaney and K. D. Bowers, *Proc. R. Soc. London, Ser. A*, 1952, **214,** 451.
765. A. Syamal and L. J. Theriot, *J. Coord. Chem.*, 1973, **2,** 193.
766. G. O. Carlisle, D. A. Crutchfield and M. D. McKnight, *J. Chem. Soc., Dalton Trans.*, 1973, 1703.
767. G. O. Carlisle and D. A. Cruchfield, *Inorg. Nucl. Chem. Lett.*, 1972, **8,** 443.
768. A. Syamal and K. S. Kale, *J. Indian Chem. Soc.*, 1978, **55,** 606.
769. A. Syamal and K. S. Kale, *Inorg. Chem.*, 1979, **18,** 992.
770. A. Syamal, K. S. Kale and S. Banerjee, *J. Indian Chem. Soc.*, 1979, **56,** 320.
771. A. Syamal and K. S. Kale, *Indian J. Chem. Sect. A*, 1980, **19,** 225.
772. D. K. Rastogi, S. K. Sahni, V. B. Rana, K. Dua and S. K. Dua, *J. Inorg. Nucl. Chem.*, 1979, **41,** 21.
773. A. Syamal and K. S. Kale, *Indian J. Chem., Sect. A*, 1977, **15,** 431.
774. A. Syamal and K. S. Kale, *Indian J. Chem., Sect. A*, 1980, **19,** 486.
775. A. Syamal, E. F. Carey and L. J. Theriot, *Inorg. Chem.*, 1973, **12,** 245.
776. A. Syamal, S. Ahmed and O. P. Singhal, *Transition Met. Chem. (Weinheim, Ger.)*, 1983, **8,** 156.
777. V. V. Zelentsov, *Russ. J. Inorg. Chem., (Engl. Transl.)*, 1962, **7,** 670.
778. M. D. Revenko and N. V. Gerbeleu, *Russ. J. Inorg. Chem. (Engl. Transl.)*, 1972, **17,** 529.
779. D. K. Rastogi and P. C. Pachauri, *J. Inorg. Nucl. Chem.*, 1977, **39,** 151.
780. R. P. Matur, P. Matur and R. K. Mehta, *Curr. Sci.*, 1983, **52,** 481.
781. K. G. Sharma, R. P. Matur and R. K. Mehta, *Proc. Indian Natl. Sci. Acad.*, 1981, **47,** 331.
782. C. P. Gupta, N. K. Sankhla and R. K. Mehta, *J. Inorg. Nucl. Chem.*, 1979, **41,** 1392.

783. K. G. Sharma, C. P. Gupta and R. K. Mehta, *J. Indian Inst. Sci.*, 1981, **63,** 149.
784. A. Diamantis, M. R. Snow and J. A. Vanzo, *J. Chem. Soc., Chem. Commun.*, 1976, 264.
785. R. L. Dutta and M. M. Hossain, *Indian J. Chem., Sect. A*, 198b3, **22,** 204.
786. L. J. Theriot, G. O. Carlisle and H. J. Hu, *J. Inorg. Nucl. Chem.*, 1969, **31,** 2841.
787. R. Hamalainen and U. Turpeinen, *Acta Crystallogr., Sect. C*, 1985, **41,** 1726.
788. J. B. Hodgson and G. C. Percy, *Spectrosc. Lett.*, 1975, **8,** 975.
789. J. B. Hodgson and G. C. Percy, *Spectrochim. Acta, Part A*, 1976, **32,** 1291.
790. L. J. Theriot, G. O. Carlisle and H. J. Hu, *J. Inorg. Nucl. Chem.*, 1969, **31,** 3303.
791. P. O'Brien, J. A. L. Silva, A. L. Vieira and L. Vilas-Boas, Comunicação 5CP8, 9° Encontro Nacional de Química, Coimbra, 1986.
792. R. L. Dutta and M. M. Hossain, *Indian J. Chem., Sect. A*, 1982, **21,** 746.
793. S. N. Hiremath and V. H. Kulkarni, *Acta Chim. Acad. Sci. Hung.*, 1982, **111,** 225.
794. O. Farooq, A. U. Malik, N. Ahmad and S. M. F. Rahmao, *J. Electroanal. Chem., Interfacial Electrochem.*, 1970, **24,** 464.
795. M. V. Chidambaram and P. K. Bhattacharya, *Indian J. Chem., Sect. A*, 1970, **8,** 941.
796. R. C. Tewari and M. N. Srivastava, *Talanta*, 1973, **20,** 133.
797. H. Tomiyasu and G. Gordon, *Inorg. Chem.*, 1976, **15,** 870.
798. C. R. Johnson and R. E. Shepherd, *Bioinorg. Chem.*, 1978, **8,** 115.
799. R. S. Saxena and S. K. Dhawan, *Trans. SAEST*, 1981, **16,** 209.
800. O. Yokoyama, H. Tomiyasu and G. Gordon, *Inorg. Chem.*, 1982, **21,** 1136.
801. R. S. Saxena and S. K. Dhawan, *J. Electrochem. Soc. India*, 1982, **31,** 37.
802. I. Fábián and I. Nágypál, *Inorg. Chim. Acta*, 1982, **62,** 193.
803. F. Bergel, K. R. Harrap and A. M. Scott, *J. Chem. Soc.*, 1962, 1101.
804. J. P. Paumard and M. Cadiot, *Bull. Soc. Chim. Fr., Part 1*, 1981, 442.
805. H. Sakurai and S. Shimomura, *Inorg. Chim. Acta*, 1981, **55,** L67.
806. Y. Xu, X. Li, L. Yu, C. Liu, J. Lu and W. C. Liu, *Jiegon Huaxe*, 1982, **1,** 55 (*Chem. Abstr.*, 1983, **98,** 136 555r).
807. H. Sakurai, Y. Hamada, S. Shimomura and S. Yamashita, *Inorg. Chim. Acta*, 1980, **46,** L119.
808. R. D. Gillard, R. Marques, J. C. Pessoa and L. Vilas-Boas, 9° Encontro Nacional de Química, Comunicação 5CP12, Coimbra, 1986.
809. T. D. Smith, J. F. Boas and J. R. Pilbrow, *Aust. J. Chem.*, 1974, **27,** 2535.
810. G. D. Fallon and B. M. Gatehouse, *Acta Crystallogr., Sect. B*, 1976, **32,** 71.
811. S. Ooi, M. Nishizawa, K. Matsumoto, H. Kuroya and K. Saito, *Bull. Chem. Soc. Jpn.*, 1979, **52,** 452.
812. A. Kojima, K. Okazaki, S. Ooi and K. Saito, *Inorg. Chem.*, 1983, **22,** 1168.
813. J. P. Launay, Y. Jeannin and M. Daoudi, *Inorg. Chem.*, 1985, **24,** 1052.
814. F. Babonneau, C. Sanchez, J. Livage, J. P. Launay, M. Daoudi and Y. Jeannin, *Nouv. J. Chim.*, 1982, **6,** 353.
815. F. J. Kristine, D. R. Gard and R. E. Shepherd, *J. Chem. Soc., Chem. Commun.*, 1976, 994.
816. M. A. Nawi and T. L. Riechel, *Inorg. Chim. Acta*, 1984, **93,** 131.
817. M. Nishizawa and K. Saito, *Inorg. Chem.*, 1980, **19,** 2284.
818. M. Nishizawa, K. Hirotsu, S. Ooi and K. Saito, *J. Chem. Soc., Chem. Commun.*, 1979, 707.
819. B. Wang, Y. Sasaki, K. Okazaki, K. Kanesato and K. Saito, *Inorg. Chem.*, 1986, **25,** 3745.
820. B. H. Bersted, R. L. Belford and I. C. Paul, *Inorg. Chem.*, 1968, **7,** 1557.
821. R. Trujillo and F. Brito, *Ann. R. Soc. Esp. Fis. Quim., Ser. B*, 1957, **53,** 313.
822. T. L. Riechel and D. T. Sawyer, *Inorg. Chem.*, 1975, **14,** 1869.
823. G. M. Klesova, V. V. Zelentsov and V. I. Spitsyn, *Zh. Obshch. Khim.*, 1973, **43,** 454 (*Chem. Abstr.*, 1973, **79,** 12 978u).
824. M. Shiro and Q. Fernando, *Chem. Commun.*, 1971, 63.
825. R. P. Henry, P. C. Mitcell and J. E. Prue, *Inorg. Chim. Acta*, 1972, **7,** 125.
826. M. Pasquali, A. Landi and C. Floriani, *Inorg. Chem.*, 1979, **18,** 2397.
827. A. Doadrio and J. Martinez, *An. Quim.*, 1970, **66,** 325.
828. R. L. Dutta and S. Lahiry, *J. Indian Chem. Soc.*, 1963, **40,** 857.
829. R. L. Dutta and S. Lahiry, *J. Indian Chem. Soc.*, 1963, **40,** 67.
830. N. S. Biradar and S. D. Angadi, *Rev. Roum. Chim.*, 1975, **20,** 755.
831. H. J. Bielig and H. Mollinger, *Liebigs Ann. Chem.*, 1957, **605,** 117.
832. N. S. Biradar, S. D. Angadi and V. B. Mahale, *Curr. Sci.*, 1975, **44,** 539.
833. D. B. Ingle and D. D. Khanolkar, *J. Indian Chem. Soc.*, 1973, **50,** 25.
834. R. N. Saksena and K. K. Panday, *J. Indian Chem. Soc.*, 1973, **50,** 609.
835. J. Singh and S. P. Gupta, *Acta Cienc. Indica*, 1974, **1,** 107.
836. K. Lal and S. R. Malhotra, *An. Quim.*, 1983, **79B,** 56.
837. K. Lal, J. Singh and S. P. Gupta, *J. Inorg. Nucl. Chem.*, 1978, **40,** 359.
838. K. R. Reddy, P. N. Rao and K. A. Reddy, *J. Indian Chem. Soc.*, 1982, **59,** 422.
839. I. Rani, K. B. Pandeya and R. P. Singh, *Indian J. Chem., Sect. A*, 1982, **21,** 502.
840. S. I. Gusev, V. I. Kumovand and E. V. Sokolova, *J. Anal. Chem. USSR, (Engl. Transl.)*, 1960, **15,** 205.
841. E. F. Hasty, T. J. Colburn and D. N. Hendrickson, *Inorg. Chem.*, 1973, **12,** 2414.
842. R. C. Aggarwal, N. K. Singh and R. P. Singh, *Inorg. Chim. Acta*, 1979, **32,** L87.
843. M. Tirant and T. D. Smith, *Inorg. Chim. Acta*, 1984, **90,** 111.
844. S. A. Patil and V. H. Kulkarni, *Inorg. Chim. Acta*, 1984, **95,** 195.
845. O. P. Pandey, *Polyhedron*, 1986, **5,** 1587.
846. L. D. Dave, C. Mathew and V. Oommen, *Indian J. Chem., Sect. A*, 1982, **21,** 645.
847. A. Hodge, K. Nordquest and E. L. Blinn, *Inorg. Chim. Acta*, 1972, **6,** 491.
848. D. Chaigne, J. F. Hemidy, L. Legrand and D. Cornet, *J. Chem. Res. (S)*, 1977, 198.
849. R. A. Bozis and J. McCormick, *Inorg. Chem.*, 1970, **9,** 1541.
850. B. J. McCormick and R. A. Bozis, *Inorg. Chem.*, 1971, **10,** 2806.

851. Ref. 123 in ref. 7: D. Neubauer and J. Weiss, *Z. Anorg. Allg. Chem.*, 1960, **303**, 28; D. Neubauer, J. Weiss and M. Becke-Goehring, *Z. Naturforsch., Teil B*, 1959, **14**, 284.
852. H. W. Roesky, J. Anhaus, H. G. Schmidt, G. M. Sheldrick and M. Noltemeyer, *J. Chem. Soc., Dalton Trans.*, 1983, 1207.
853. A. K. Jain, U. Kumari and G. K. Chaturvedi, *J. Indian Chem. Soc.*, 1978, **55**, 1251.
854. A. K. Jain and G. K. Chaturvedi, *J. Indian Chem. Soc.*, 1979, **56**, 850.
855. S. P. Singh, H. S. Verma and J. P. Tandon, *J. Inorg. Nucl. Chem.*, 1979, **41**, 587.
856. R. L. Dutta and S. Ghosh, *J. Inorg. Nucl. Chem.*, 1966, **28**, 247.
857. R. L. Dutta and S. Ghosh, *J. Indian Chem. Soc.*, 1967, **44**, 296.
858. R. L. Dutta and S. Ghosh, *J. Indian Chem. Soc.*, 1967, **44**, 306.
859. M. Ghosh and S. Ray, *Acta Crystallogr., Sect. C*, 1983, **39**, 1367.
860. J. Sala-Pala and J. E. Guerchais, *Bull. Soc. Chim. Fr.*, 1971, 2444.
861. R. C. Aggarwal, R. Bala and R. L. Prasad, *Indian J. Chem., Sect. A*, 1983, **22**, 568.
862. R. G. Aggarwal, R. A. Rai and T. R. Rao, *Indian J. Chem., Sect. A*, 1983, **22**, 255.
863. A. Triebs, *Annalen*, 1934, **509**, 103; 1935, **517**, 172.
864. A. Triebs, *Angew. Chem.*, 1936, **49**, 682.
865. E. W. Baker and S. E. Palmer, in 'The Porphyrins', ed. D. Dolphin, Academic, New York, 1978, vol. 1, part A.
866. E. W. Baker, A. H. Corwin, E. Klesper and P. E. Wei, *J. Org. Chem.*, 1968, **33**, 3144.
867. J. Wong, F. W. Lytle, R. P. Messmer and D. H. Maylotte, 'EXAFS and Near Edge Structure', Springer-Verlag, Berlin, 1983, 130.
868. D. H. Maylotte, J. Wong, R. L. Stpeters, F. W. Lytle and R. B. Greegor, *Science*, 1981, **214**, 554.
869. A. Ekstrom, C. J. R. Fookes, T. Hambley, H. J. Loeh, S. A. Miller and J. C. Taylor, *Nature (London)*, 1983, **306** (5939), 173.
870. J. Goulon, A. Retournard, P. Priant, C. Goulon-Ginet, C. Berthe, J. F. Muller, J. L. Poncet, R. Guilard, J. C. Escalier and B. Neff, *J. Chem. Soc., Dalton Trans.*, 1984, 1095.
871. S. Bencosme, M. Labady and C. Romero, *Inorg. Chim. Acta*, 1986, **123**, 15.
872. J. W. Buchler, G. Eikelman, L. Puppe, K. Rohbock, H. Schneehage and D. Weck, *Liebigs Ann. Chem.*, 1971, **745**, 135.
873. B. B. Garrett, *J. Chem. Soc., Faraday Trans.*, 1983, **79**, 1733; J. Subramanian, In 'Porphyrins and Metalloporphyrins', ed. K. M. Smith, Elsevier Scientific Publ. Co., Amsterdam, 1975, p. 567; W. C. Lin, in 'The Porphyrins', ed. D. Dolphin; Academic Press, Vol. IV, New York, 1979, p. 355–377; T. G. Brown and B. M. Hoffman, *Mol. Phys.*, 1980, **39**, 1073; K. N. More, S. S. Eaton and G. R. Eaton, *J. Am. Chem. Soc.*, 1981, **103**, 1087.
874. R. C. Petersen and L. E. Alexander, *J. Am. Chem. Soc.*, 1968, **90**, 3873.
875. F. S. Molinaro and J. A. Ibers, *Inorg. Chem.*, 1976, **15**, 2278.
876. M. Zerner and M. Gouterman, *Inorg. Chem.*, 1966, **5**, 1699.
877. M. G. B. Drew, P. C. H. Mitchell and C. E. Scott, *Inorg. Chim. Acta*, 1984, **82**, 63.
878. J. Goulon, A. Retournard, P. Friand, C. Goulon-Ginet, C. Berthe, J. F. Muller, J. F. Poncet, R. Guilard, J. C. Escalier and B. Neff, *J. Chem. Soc., Dalton Trans.*, 1984, 1095.
879. F. A. Walker, E. Hui and J. M. Walker, *J. Am. Chem. Soc.*, 1975, **97**, 2390.
880. J. G. Erdman, V. G. Ramsey, N. W. Kalenda and W. E. Hanson, *J. Am. Chem. Soc.*, 1956, **78**, 5844.
881. C. S. Bencosme, C. Romero and S. Simoni, *Inorg. Chem.*, 1985, **24**, 1603.
882. J. A. Shelnutt and M. M. Dobry, *J. Phys. Chem.*, 1983, **87**, 3012.
883. P. Richard, J. L. Poncet, J. M. Barbe, R. Guilard, J. Goulon, D. Rinaldi, A. Cartier and P. Tola, *J. Chem. Soc., Dalton Trans.*, 1982, 1451.
884. G. E. Selyutin and A. A. Shklyaev, *Bull. Acad. Sci. USSR*, 1982, 18.
885. J. L. Poncet, R. Guilard, P. Friant and J. Goulon, *Polyhedron*, 1983, **5**, 417.
886. J. L. Poncet, R. Guilard, P. Friant, C. Goulon-Ginet and J. Goulon, *Nouv. J. Chim.*, 1984, **8**, 583.
887. (a) R. J. Gruber and B. Grushkin, *US Pat.* 3 825 422 (1974) (*Chem. Abstr.*, 1975, **82**, 37 302u); (b) J. Y. C. Chu, R. L. Schank and S. Tutsihashi, *US Pat.* 4 181 772 (1979) (*Chem. Abstr.*, 1980, **92**, 119 706z); (c) Fuji Xerox Co., *Jpn Pat.* 57 146 255 (1982) (*Chem. Abstr.*, 1984, **100**, 15 282a); (d) Fuji Xerox Co., *Jpn Pat.* 57 148 747 (1982) (*Chem. Abstr.*, 1984, **100**, 15 285d).
888. G. H. Griffiths, M. S. Walker and P. Goldstein, *Mol. Cryst. Liq. Cryst.*, 1976, **33**, 149.
889. R. F. Ziolo, C. H. Griffiths and J. M. Troup, *J. Chem. Soc., Dalton Trans.*, 1980, 2300.
890. P. D. W. Boyd and T. D. Smith, *J. Chem. Soc., Dalton Trans.*, 1972, 839.
891. D. M. Wagnerova, P. Stopka, H. Rein and J. Friedrich, *Collect. Czech. Chem. Commun.*, 1981, **46**, 457.
892. K. Sakata, M. Hashimoto, N. Tagami and Y. Murakami, *Bull. Chem. Soc. Jpn.*, 1980, **53**, 2262.
893. B. Tomlonovic, R. L. Hough, M. D. Glick and R. L. Lintvedt, *J. Am. Chem. Soc.*, 1975, **97**, 2925; P. Comba and A. M. Sargenson, *Aust. J. Chem.*, 1986, **39**, 1029.
894. U. Casellato, P. A. Vigato, D. E. Fenton and M. Vidali, *Chem. Soc. Rev.*, 1978, **7**, 199.
895. P. Comba, L. M. Engelhard, J. MacB. Harrowfield, G. A. Lawrance, L. L. Martin, A. M. Sargeson and A. H. White, *J. Chem. Soc., Chem. Commun.*, 1985, 174.
896. R. L. Lintvedt, M. D. Glick, B. K. Tomlonovic and D. P. Gavel, *Inorg. Chem.*, 1976, **15**, 1646.
897. H. Adams, N. A. Bailey, D. A. Fenton, M. S. L. Gonzalez and C. A. Phillips, *J. Chem. Soc., Dalton Trans.*, 1983, 371.
898. U. Casellato, P. A. Vigato, R. Graziani and M. Vidali, *Inorg. Chim. Acta*, 1981, **49**, 173.
899. M. D. Glick, R. L. Lintvedt, D. P. Gavel and B. Tomlonovic, *Inorg. Chem.*, 1976, **15**, 1654.
900. D. E. Fenton and S. E. Gayda, *J. Chem. Soc., Dalton Trans.*, 1977, 2109.
901. P. A. Vigato, U. Casellato, S. Tamburini, M. Vidali, F. Milani and M. M. Musiani, *Inorg. Chim. Acta*, 1982, **61**, 89.
902. M. J. Heeg, J. L. Mack, M. D. Glick and R. L. Lintvedt, *Inorg. Chem.*, 1981, **20**, 833.
903. J. P. Costes and D. E. Fenton, *J. Chem. Soc., Dalton Trans.*, 1983, 2235.
904. H. Okawa, I. Ando and S. Kida, *Bull. Chem. Soc. Jpn.*, 1974, **47**, 3041.

905. R. Chandra and R. N. Kapoor, *Ann. Chim.* (*Rome*), 1982, **72,** 309.
906. J. Selbin and L. Ganguly, *Inorg. Nucl. Chem. Lett.*, 1969, **5,** 815.
907. H. Okawa and S. Kida, *Inorg. Chim. Acta,* 1977, **23,** 253.
908. D. E. Fenton, J. R. Tate, U. Casellato, S. Tamburini, P. A. Vigato and M. Vidali, *Inorg. Chim. Acta,* 1984, **83,** 23.
909. M. Julve, M. Verdaguer, M. F. Charlot, O. Khan and R. Claude, *Inorg. Chim. Acta,* 1984, **82,** 5.
910. O. Khan, P. Tola, J. Galy and H. Condanne, *J. Am. Chem. Soc.*, 1978, **100,** 3931.
911. O. Khan, J. Galy, Y. Journaux, J. Jaud and Y. Morgenstern-Badarau, *J. Am. Chem. Soc.*, 1982, **104,** 2165.
912. E. Sinn and C. H. Harris, *Coord. Chem. Rev.*, 1969, **4,** 391.
913. K. A. Leslie, R. S. Drago, G. D. Stucky, D. J. Kitto and J. A. Broee, *Inorg. Chem.*, 1979, **18,** 1885.
914. A. Bencini, C. Benelli, A. Dei and D. Gatteschi, *Inorg. Chem.*, 1985, **24,** 695.
915. H. Werner and T. N. Khac, *Inorg. Chim. Acta,* 1978, **30,** L347,.
916. M. Mazzanti, C. Floriani, A. Chiesi-Villa and C. Guastini, *Inorg. Chem.*, 1986, **25,** 4158.
917. N. D. Chasteen, R. J. Dekoch, B. L. Rogers and M. W. Hanna, *J. Am. Chem. Soc.*, 1973, **95,** 1301.
918. N. D. Chasteen and E. C. Theil, *J. Biol. Chem.*, 1982, **257,** 7672.
919. G. D. Markham, *Biochemistry,* 1984, **23,** 470.
920. L. Banci, A. Dei and D. Gatteschi, *Inorg. Chim. Acta,* 1982, **67,** L53.
921. L. C. Cantley, L. Josephson, R. Warner, M. Yanagisawa, C. Lechene and G. Guidotti, *J. Biol. Chem.*, 1977, **252,** 7421.
922. G. M. Wolterman, R. A. Scott and G. P. Haight, Jr., *J. Am. Chem. Soc.*, 1974, **96,** 7569.
923. H. Sakurai, T. Goda, S. Shimomura and T. Yoshimura, *Biochem. Biophys. Res. Commun.*, 1982, **104,** 1421.
924. H. Sakurai, T. Goda and S. Shimomura, *Biochem. Biophys. Res. Commun.*, 1982, **108,** 474.
925. C. M. Mikulski, S. Cocco, N. Franco and N. M. Karayannis, *Inorg. Chim. Acta,* 1983, **78,** 125.
926. C. M. Mikulski, S. Mattucci, L. Weiss and N. M. Karayannis, *Inorg. Chim. Acta,* 1984, **92,** 275.
927. P. I. Vesnes and E. Sletten, *Inorg. Chim. Acta,* 1981, **52,** 269.
928. N. Katsaros, *Transition Met. Chem.* (*Weinheim, Ger.*), 1982, **7,** 72.
929. E. Bayer and H. Kneifel, *Z. Naturforsch., Teil B,* 1972, **27,** 207.
930. H. Kneifel and E. Bayer, *Angew. Chem., Int. Ed. Engl.*, 1973, **12,** 508.
931. R. D. Gillard and R. J. Lancashire, *Phytochemistry,* 1984, **23,** 179.
932. G. Bemski, J. Felcman, J. J. R. Fraústo da Silva, I. Moura, J. J. Moura, M. Candida Vaz and L. F. Vilas-Boas, *Rev. Port. Quím.*, 1985, **27,** 418.
933. H. Kneifel and E. Bayer, *J. Am. Chem. Soc.*, 1986, **108,** 3075.
934. E. Koch, H. Kneifel and E. Bayer, *Z. Naturforsch., Teil B,* 1986, **41,** 359.
935. P. L. Hall, B. R. Angel and J. Braven, *Chem. Geol.*, 1974, **13,** 97.
936. M. V. Cheshire, M. L. Berrow, B. A. Goodman and C. M. Mundie, *Geochim. Cosmochim. Acta,* 1977, **41,** 1131.
937. A. L. Abdul-Halim, J. C. Evans, C. C. Rowlands and J. H. Thomas, *Geochim. Cosmochim. Acta,* 1981, **45,** 481.
938. B. A. Goodman and M. V. Cheshire, *Geochim. Cosmochim. Acta.*, 1975, **39,** 1711.
939. M. B. McBride, *Soil Sci.*, 1978, **126,** 200.
940. G. D. Templeton and N. D. Chasteen, *Geochim. Cosmochim. Acta,* 1980, **44,** 741.
941. M. B. McBride, *Can. J. Soil Sci.*, 1980, **60,** 145.
942. T. K. Myser and R. E. Shepherd, *Inorg. Chem.*, 1987, **26,** 1544.
943. P. Comba and A. M. Sargenson, *Aust. J. Chem.*, 1986, **39,** 1029.
944. H. D. S. Yadava, S. K. Sengupta and S. C. Tripathi, *Inorg. Chim. Acta,* 1987, **128,** 1.
945. N. D. Chasteen, J. K. Grady and C. E. Holloway, *Inorg. Chem.*, 1986, **25,** 2754.
946. E. de Boer, Y. van Kooyk, M. G. M. Tromp, H. Plat and R. Wever, *Biochim. Biophys. Acta,* 1986, **869,** 48; E. de Boer, M. G. M. Tromp, H. Plat, B. E. Krenn and R. Wever, *Biochim. Biophys. Acta,* 1986, **872,** 104; H. Vilter and D. Rehder, *Inorg. Chim. Acta,* 1987, **136,** L7.
947. R. L. Robson, R. R. Eady, T. H. Richardson, R. W. Miller, M. Hawkins and J. R. Postgate, *Nature* (*London*), 1986, **322,** 388; J. M. Arber, B. R. Dobson, R. R. Eady, P. Stevens, S. S. Hasnain, C. D. Garner and B. E. Smith, *Nature* (*London*), 1987, **325,** 372.
948. G. Anderegg, E. Koch and E. Bayer, *Inorg. Chim. Acta,* 1987, **127,** 183.
949. M. A. Nawi and T. L. Rieckel, *Inorg. Chim. Acta,* 1987, **136,** 33.

34

Niobium and Tantalum

LILIANE G. HUBERT-PFALZGRAF, MICHÈLE POSTEL and JEAN G. RIESS
Université de Nice, France

34.1 SURVEY

Niobium (formerly called columbium) and tantalum, when compared with the neighboring early transition metals, are seen to share with titanium and zirconium an affinity for dioxygen and a reluctance to be reduced, but to differ from them in their much stronger tendency to form metal–metal bonds and in their rich cluster chemistry. With molybdenum and tungsten they share a propensity to yield oxo species and an ability to give multiple bonds with non-metals, as in carbenes, alkylidenes, nitrenes, *etc.*, and to form M_6-based clusters, although these are mainly of the $[M_6X_8]^{q+}$ type for molybdenum and of the $[M_6X_{12}]^{q+}$ type for niobium or tantalum. Carbonyls are generally much less stable for the group V than for the group VI metals. Roughly, it may be said that niobium and tantalum have chemistries more comparable to those of titanium and zirconium in their higher oxidation states, but closer to those of molybdenum and tungsten in their lower ones.

Compared to vanadium, the higher oxidation states are much more frequently and the lower ones much less frequently encountered in niobium and tantalum. Niobium and tantalum are also much more prone to extended metal–metal bonding in their lower oxidation states; V_6-based clusters, for example, remain undiscovered. Little resemblance is found to the group VB elements phosphorus and arsenic.

The chemistry of niobium and tantalum ranges from oxidation state +V to −III, but with no species of oxidation state −II presently known (Table 1). The largest number of molecular compounds, by far, is found for oxidation state V. The very reactive pentahalides provide the most convenient entry to the molecular chemistry of these metals.

Table 1 Oxidation States and Common Stereochemistries of Nb and Ta Compounds

Oxidation state	*Coordination number*	*Idealized geometry*	*Examples*
M^V	4	Tetrahedron	$[NbO_4]^{3-}$, $[NbO\{N(SiMe_3)_2\}_3]$
	5	Trigonal bipyramid	$[Ta(NEt_2)_5]$
		Square-based pyramid	$[Nb(NMe_2)_5]$, $[NbOCl_4]^-$, $[NbSCl_3(Ph_3PS)]$
	6	Octahedron	$[M_2X_{10}]$, $[MOX_3]$, $[NbAlCl_8]$, $[Nb(OMe)_5]_2$, $[MCl_3(OR)_2]_2$
		Trigonal prism	$[M(S_2C_6H_4)_3]^-$
	7	Pentagonal bipyramid	$[NbO(C_2O_4)_3]^{3-}$, $[NbO(R_2dtc)_3]$, $[MS(R_2dtc)_3]$, $[NbOF_6]^{3-}$, $[M(O_2)F_5]^{2-}$
		Capped trigonal prism	$[MF_7]^{2-}$
	8	Bicapped trigonal prism	$[TaF_8]^{3-}$, $[Ta(PS_4)S_2]$
		Dodecahedron	$[MCl_4(diars)_2]^+$, $[M(O_2)_4]^{3-}$, $[M(O_2)_3(phen)]^-$, $[M(O_2)_2(C_2O_4)_2]^{3-}$
	9	?	$[\{NbCl_2(Cp)_2\}_2O]$
	10	?	$[MCl_5(OxH)_5]$
M^{IV}	4	D_{2d} tetrahedron	$[Nb(NR_2)_4]$
	5	Trigonal bipyramid	$[MH_2(OSiBu^t_3)_2]_2$
	6	Octahedron	$[MbX_4L_2]$
	7	Pentagonal bipyramid	$[NbF_7]^{3-}$
	8	Dodecahedron	$[Nb\{S_2P(OR)_2\}_4I]$, $[NbCl_4(diars)_2]$
		Square antiprism	$[Nb(dpm)_4]$, $[TaCl_4(dmpe)_2]$, $[Nb_2Cl_4(PhPMe_2)_4(\mu\text{-}Cl)_4]$
	9	?	$[Nb(acac)_4(dioxane)]$
M^{III}	6	Octahedron	$[NbCl_3(py)_3]$, $[Nb_2X_9]^{3-}$, $[M_2Cl_4L_4(\mu\text{-}Cl)_2]$
	7	Pentagonal bipyramid	$[Ta(PMe_3)_3(\eta^2\text{-}CH_2PMe_2)(\eta^2\text{-}CHPMe_2)]$, $[TaH(PPh_2)_2(dmpe)_2]$
	8	Dodecahedron (solid)	$[Nb(CN)_8]^{5-}$
		Square antiprism (solution)	
M^{II}	6	Octahedron	$[MCl_2(PMe_3)_4]$
	8	Dodecahedron	$[TaH_2Cl_2(PMe_3)_4]$
		Square antiprism	$[TaH_2Cl_2(dmpe)_2]$
M^I	5	Pentagonal bipyramid	$[TaL_2R(C_2H_4)_2]$
	6	?	$[M(\eta^4\text{-}C_4H_6)_2(\eta^3\text{-}Meallyl)]$
	7	Capped octahedron	$[TaH(CO)_2(dmpe)_2]$, $[Ta(CO)_3(PMe_3)_4]^+$
		Capped trigonal prism	$[TaX(CO)_2(dmpe)_2]$, $[MCl(CO)_3(PMe_3)_3]$
		'Four-legged pianostool'	$[Nb(CO)_4(Cp)]$
M^0	6	Octahedron	$[M(dmpe)_3]$
		Sandwich	$[M(arene)_2]$
M^{-I}	6	Octahedron	$[M(CO)_6]^-$
M^{-III}	5	?	$[M(CO)_5]^{3-}$
	6	?	$[M(CO)_3(Cp)]^{2-}$

Until recently little attention was paid to oxidation states lower than III, with the exception of the octahedral clusters of type $[M_6X_{12}]^{q+}$, which are predominant in aqueous solutions. Preparation of molecular compounds in these oxidation states was for a long time more or less accidental unless cyclopentadienyl groups were present. This lack of development is mainly due to the paucity until recently of suitable precursors (halides or carbonyls), as well as to the high sensitivity to dioxygen and to moisture of products in such intermediate oxidation states as +III. Chemical reduction of pentavalent or tetravalent derivatives under strictly controlled conditions has since proven able to provide convenient starting materials for developing the chemistry of niobium and tantalum in their lower oxidation states. Powerful reductants such as sodium amalgam or naphthalenide are required for access to oxidation states lower than III. The least -documented chemistry of the formally positive oxidation states of these metals is that of oxidation state +II.

Among the most valuable starting materials now available for access to oxidation states IV to II are the tetrahalides and some of their more soluble molecular adducts such as $[NbCl_4(THF)_2]$, $[NbCl_4(MeCN)_2]\cdot MeCN$ and $[MCl_4(dmpe)_2]$, adducts of the trihalides, in particular $[M_2Cl_6(PR_3)_4]$, and their derived hydrides, $[M_2Cl_6L_3]$ ($L = SMe_2$, THT) and $[MCl_2(dmpe)_2]$. Pentavalent alkylidenes also offer attractive entries into the chemistry of oxidation states +III and +I.

Oxidation state −I is mainly represented by $[M(CO)_6]^-$ and $[M(PF_3)_6]^-$. The first compounds isolated with a metal in the −III oxidation state were $Cs_3[M(CO)_5]$ and $Cs_2[M(CO)_3(Cp)]$ with M = Nb or Ta.

Niobium and tantalum compounds form adducts with virtually all types of neutral and anionic donors. The coordination chemistry of the higher halides is widely developed, and their activity as Friedel–Crafts catalysts is another manifestation of their Lewis acidity. The strong acceptor capacity of the high valent metal compounds tends to favor the formation of dimers, and sometimes of higher condensation products, which competes with coordination with other donor molecules. Numerous simple anionic or heteropolyanionic species, but little cationic chemistry, and no simple metal salts, are known.

Specific to the lower oxidation states is the tendency to metal bonding found in various arrangements, including isolated dimers (with single or double metal–metal bonds), polymers with M—M··· M—M··· bond alternation, triangular M_3 units, and, the most abundant, octahedral M_6-derived clusters. These clusters, with non-integral oxidation states ranging from +2.67 for $[M_6X_{12}]^{q+}$ to +1.67 for $[Nb_6I_8]^{q+}$ exhibit complex redox chemistry that may be used in photocatalysis.

Being among the largest early transition metals, niobium and tantalum can easily accommodate high coordination numbers. Although coordination number six is the most frequent, stable compounds with coordination numbers of seven (especially in the lower oxidation states), eight, nine and even ten are found; in contrast, coordination numbers below five are extremely rare. The rate relatively large metal radii also account for the less tight coordination sphere, and hence for the stereodynamic character often encountered in the complexes.

Reactions other than Lewis acid–base associations/dissociations are frequently observed with donor molecules, leading notably to solvolysis, oxygen or sulfur abstraction, insertion reactions and carbon–carbon coupling reactions. The tendency to form metal–element multiple bonds is remarkable; in this respect the activation of dinitrogen by tantalum or niobium is unique. The formation and chemistry of constrained reactive metallacycles open another promising fast-developing area, on the frontier with organometallic chemistry.

Owing to lanthanide contraction, niobium and tantalum have virtually identical atomic radii (1.47 Å) and close ionization energies (Nb 6.67, Ta 7.3 eV), and usually display very similar chemical behavior. Some definite differences can however be noted; these can usually be traced to the lower sensitivity of tantalum to reduction and to its higher affinity for dioxygen. The tantalum–element multiple bonds are usually stabler, while M_6X_8 arrangements are so far known only for niobium.

The ^{93}Nb nucleus has 100% abundance, $I = 9/2$, and the third highest sensitivity (after ^{1}H and ^{19}F and before ^{27}Al) in NMR; the use of this technique is therefore likely to expand. ^{181}Ta (99.9877%), $I = 7/2$, is less promising in view of its much lower sensitivity and its tenfold larger quadrupolar coupling. The magnetic moment in the Nb^{IV} derivatives usually ranges between 1.6 and 1 BM (the spin only value expected for d^1 compounds being 1.73 BM). The magnetic moments of Ta^{IV} are usually lower than 1 BM as a result, at least in part, of a larger spin–orbit coupling ($\lambda = 750\ cm^{-1}$ for Nb^{4+} and $1400\ cm^{-1}$ for Ta^{4+}). The magnetic susceptibilities usually exhibit a Curie law dependence upon reciprocal temperature, which strongly supports an orbital singlet ground state well separated from the next excited state. Typical isotropic ESR spectra display 10 lines for niobium and eight lines for tantalum derivatives, the line width being larger for the latter, probably owing to non-resolved quadrupolar coupling. The g values of the niobium compounds are closer to 2 than are those of tantalum, which is no surprise as the latter has a higher spin–orbit coupling constant. Magnetic data on the other oxidation states are still scarce.

This review intends to cover essentially the molecular non-organometallic chemistry of niobium and tantalum. When organometallic compounds are mentioned, it is in principle only in so far as they are involved in some coordination process. The literature has been covered to the end of 1983; thanks to their authors, a few more recent preprints were able to be analyzed. Emphasis is given to the more recent material. Some of the presently most active areas such as those involving multiple bonding, dinitrogen activation and nitrenes are highlighted by subtitles. Earlier reviews on the subject are those of Fairbrother (1967),[1] Kepert (1972),[2] Walton (1972),[3] Gmelin (1973),[4] Brown (1973)[5] and Mehrotra (1976).[6] The first review of the chemistry of the d^1 metal compounds of early transition metals was published by Bereman in 1972.[7] For the organometallic chemistry of these elements see Labinger (1982).[8]

The following abbreviations and conventions will be used throughout this chapter: A: cation; Z: anion, M = Nb, Ta; M′ ≠ Nb, Ta; X: halogen; X and Y′: halogens when mixed; Y = S, Se, Te; E = N, P, As, Sb or Bi; LL = bidentate; ax: axial; eq: equatorial; b: bridging; t: terminal. Ligands: bzac: benzoylacetone; bipyO$_2$: bipyridyl *N,N′*-dioxide; cat: catechol; Cp′: η^5-C$_5$R$_5$; dbmH: dibenzoylmethane; DEF: diethylformamide; DMA: dimethylacetamide; dpac: 1,2-bis(diphenylarsino)ethane; dmpe: 1,2-bis(dimethylphosphino)ethane; dpmH: dipivaloylmethane; dth: 2,5-dithiahexane; edta: ethylenediaminetetraacetate; oxH: 8-quinolinol (8-hydroxyquinoline); pic: picoline; pyO: pyridine *N*-oxide; R$_2$dtc: dialkyldithiocarbamate; SB: Schiff's base; T: tropolone; THP: tetrahydropyran; THT: tetrahydrothiophene; TPO: triphenylphosphine oxide; triars: triarsine; TU: thiourea.

34.2 OXIDATION STATE +V

34.2.1 Pentahalides

The pentahalides of niobium and tantalum are predominantly covalent. Their volatility decreases from the fluorides to the iodides. Nb and Ta belong to the few elements which form stable pentaiodides. Selected physical and thermodynamic properties for the halides are listed in Table 2. All are sensitive to moist air, water or hydroxylic solvents.

Table 2 Selected Thermodynamic Properties for the Pentahalides[a,b]

	M.p. (°C)	*B.p.* (°C)	*Density*[c] (g cm^{-3})	*Viscosity*[c] (cP)	*Conductivity* (mho)	ΔH_{form}	ΔH_{vap}	ΔH_{sub}	ΔH_{fus}	ΔS_{form}	ΔS_{vap}	ΔS_{sub}
[NbF$_5$]	79	234	2.6995	91.41	1.63×10^{-5}	−1810	60	94.5	12.2	397.15	—	222
[TaF$_5$]	97	229	3.8800	70.31	1.56×10^{-5}	−1902	54	—	—	378	—	—
[NbCl$_5$]	203.4	247.4	2.75	0.921	2.2×10^{-7}	−796	63.5	89[e]	38	245	101.5	180[e]
[TaCl$_5$]	215.9	232.9	3.68	1.003	3.0×10^{-7}	−857	55	85.4[e]	41.5	248	108	172.6[e]
[NbBr$_5$]	254	365	—	—	—	−556	93.5	116	—	305	132	198
[TaBr$_5$]	256	344	—	—	—	−598[d]	61.5	106	45.5	305	99.5	184
[NbI$_5$]	—	—	—	—	—	−270	—	—	—	343	—	—
[TaI$_5$]	496	543	—	—	—	−293	75.5	82	6.7	—	—	—

[a] From D. Brown, 'Comprehensive Inorganic Chemistry', ed. J. C. Bailar, Jr., H. J. Emeléus, R. Nyholm, and A. F. Trotman-Dickenson, Pergamon, Oxford, 1973, p. 553, unless specified.
[b] ΔH values: kJ mol^{-1}; ΔS values: J mol^{-1} K^{-1}; all values measured at 298 K.
[c] At m.p.
[d] A. D. Westland, *Can. J. Chem.*, 1979, **57**, 2665.
[e] A. D. Westland and D. Lal, *Can. J. Chem.*, 1972, **50**, 1604.

34.2.1.1 Pentafluorides

The pentafluorides are best prepared by direct fluorination of the metals at 250 °C.[1,2] Alternative methods include reactions of the metals with HF, ClF$_3$ or BrF$_3$, or SnF$_2$. They have also been prepared from MCl_5 and HF or ZnF_2. The synthesis of high purity NbF_5 by competitive Lewis acid–base reactions between $[SbF_5]$ and $[MF_6]^-$ in HF has also been reported.[9]

The MF_5 molecules crystallize as tetramers in which the metal atoms are approximately octahedrally surrounded by fluorine atoms, and are linked together by linear bridging fluorine atoms (M—F_b: 2.06 Å; M—F_t: 1.77 Å).[10] It was suggested, on the basis of the high viscosity of the melts and the high Trouton constants, that this polymerization state is retained in the liquid.[2] Molecular weight data in the vapor, obtained through mass spectrometry and vapor density measurements, were interpreted to imply the simultaneous presence of monomeric, dimeric and trimeric species.[11–13] However, a different model, which assumes mixtures of monomers and tetramers, has since been reported to fit the vapor density data better.[14] Electron diffraction investigations also showed that the gaseous MF_5 molecules are associated at lower temperatures;[1] thus, at the nozzle temperature of about 45 °C, TaF_5 was found to be mostly trimeric (Ta—F_b: 2.062(2) to 2.076(3) Å; Ta—F_t: 1.81(6) to 1.894(6) Å).[15] The molten pentafluorides possess measurable electrical conductivities that were attributed to slight (<1%) self-ionization.[2]

34.2.1.2 Pentachlorides

The pentachlorides have been prepared by a variety of methods,[2,3] chief of which are the direct chlorination of the metal and the action of chlorinating agents on the pentoxides.

The crystal structure of $NbCl_5$ consists of Nb_2Cl_{10} dimers (Nb—Cl_b: 2.56 Å; Nb—Cl_t: 2.25 and 2.30 Å).[2] $TaCl_5$ was presumed to have the same structure; $TaCl_5$ and $NbCl_5$ form a complete series of mixed crystals. The dimeric structure of M_2Cl_{10} was confirmed by IR and Raman spectroscopy, and normal coordinate analyses were performed.[16] Raman spectroscopy shows a gradual dissociation of Ta_2Cl_{10} to monomeric $TaCl_5$ between 220 and 350 °C.[17] Halogen and metal NQR transitions of M_2X_{10} (M = ^{93}Nb, ^{181}Ta; X = ^{35}Cl, ^{81}Br, ^{127}I) indicate that coordination about M becomes closer to octahedral symmetry as the halogen is varied from Cl to Br and I. Terminal halogen atom resonances were reported to indicate M—X π bonding.[18] MX_5 appears to remain dimeric in non-complexing solvents,[1,3] while monomeric adducts are formed in donor solvents. In the vapor state MCl_5 is monomeric, with a trigonal bipyramidal structure.

Despite the drastic solvent limitations, polarographic data could be obtained for MCl_5 in non-aqueous solvents (Nb^V/Nb^{IV}: $E° = -0.52$ V; Nb^{IV}/Nb^{II}: $E° = -0.88$ V; Ta^V/Ta^{III}: $E° = -0.88$ V in MeCN). Data on molten salts are also available for both metals.[1] The reduction of Nb^V in aqueous concentrated HCl solutions proceeds through Nb^{IV}, which disproportionates to Nb^V and Nb^{III} (assumed to be $[NbCl_6]^{3-}$). The Nb^{IV} species may be stabilized by addition of ethylene glycol. Further reduction to Nb^{II} has been suggested.[4]

34.2.1.3 Pentabromides

The orange-yellow $NbBr_5$ and yellow $TaBr_5$ are prepared in the same way as the pentachlorides.[1,3,19] $NbBr_5$ has also been obtained by halogen exchange between $NbCl_5$ and BBr_3 at room temperature.

α-MBr_5 is isomorphous with $NbCl_5$ and consists of M_2Br_{10} dimers.[1] When $NbBr_5$ was mixed with sulfur in a closed tube, crystals of a new modification, β-$NbBr_5$, which show a one-dimensional disorder, were isolated.[20] β-$NbBr_5$ is isostructural with TaI_5 and consists of layers of Nb_2Br_{10} molecules (Nb—Br_b: 2.462(2) and 2.403(3) Å); the bromine atoms have a hexagonal-close-packed arrangement. Electron diffraction studies showed that the gaseous bromides consist of trigonal bipyramids.[3]

34.2.1.4 Pentaiodides

MI_5 species contrast in their preparation and properties with the other pentahalides, and differ from one another.[1,3] TaI_5 is stable to at least 500–600 °C, and can be sublimed under vacuum, while NbI_5 must be prepared and sublimed in the presence of an excess pressure of iodine vapor; when heated under vacuum, it loses iodine at temperatures as low as 200 °C, and is the most air sensitive of these halides.

TaI_5 consists of iodine atoms in a hexagonal-close-packed arrangement in which 2/5 of the octahedral holes are randomly occupied by tantalum atoms.[21] The Ta_2I_{10} molecules are arranged in layers (Ta—I_b: 2.932 Å; Ta—I_{ax}: 2.704 Å; Ta—I_{eq}: 2.619 Å). Triclinic NbI_5 is isostructural with β-UCl_5.[22]

34.2.1.5 Mixed pentahalides

MCl_4F was obtained by the action of AsF_3 on the $[MCl_5(PCl_5)]$ adducts in $AsCl_3$.[3,23] The compounds are tetrameric, with bridging fluorines (M—F: 2.10–2.11 Å; M—Cl: 2.26–2.31 Å). A series of mixed $[M_4Cl_xF_{20-x}]$ derivatives was isolated.[24,25] Reactions of mixtures of $AlBr_3$ and AlI_3 with Ta_2O_5 have been reported to yield $[TaBr_4I]$ and $[TaBrI_4]$.[3]

Crystalline $[NbAlCl_8]$ was prepared directly from Al_2Cl_6 and Nb_2Cl_{10}:[26] the crystal structure consists of an $AlCl_4$ tetrahedron sharing an edge with an $NbCl_6$ octahedron (Nb—Cl_{ax}: 2.288(1) Å; Nb—Cl_{eq}: 2.210(1) Å; Nb—Cl_b: 2.643(1) Å). Raman spectra of $SbCl_5$–$NbCl_5$ mixtures were interpreted in terms of formation of the $[NbSbCl_{10}]$ dimer.[27,28] The interaction of MCl_5 with chlorides of other elements has been described; $TaCl_5$ was reported to exhibit no chlorinating ability, and $NbCl_5$ to chlorinate only $SnCl_2$.[29]

34.2.1.6 Applications of the pentahalides

MF_5 and MCl_5 are strongly electrophilic, and catalyze Friedel–Crafts type reactions, as first shown by Fairbrother.[30] The alkylation of aromatic compounds[31] and of alkenes[32] was reported. Among the group of extremely strong acid catalysts, the HF/TaF_5 system has interesting characteristics. It is very stable to reduction, especially by hydrogen and organic ions, and this has been exploited in the selective acid-catalyzed isomerization and hydrogenolysis of cycloalkanes.[33] It is also a liquid phase hydrogenation catalyst for aromatics,[34,35] alkylaromatics and heteroaromatics.[36] Synthesis of high polymers with simple Nb^V and Ta^V halides as catalysts has been reported;[37–40] disubstituted alkynes which, owing to steric hindrance, cannot be polymerized by Ziegler–Natta type catalysts were polymerized, whereas monosubstituted alkynes selectively gave cyclotrimers.

34.2.2 Halo Adducts

34.2.2.1 Anionic halo complexes

(i) Anionic fluoro complexes

Several species, including $[MF_6]^-$, $[MF_7]^{2-}$, $[MF_8]^{3-}$, $[MF_9]^{4-}$ and $[M_2F_{11}]^-$, have been detected in HF solutions of the metals, using a selective fluoride ion electrode,[41–43] or by ^{19}F NMR studies.[42,44–46] These complexes can be extracted by organic solvents; IR showed that solutions of trioctylamine in To or CCl_4 extract M^V from 1–12 M HF solutions as $[MF_6]^-$.[47]

MF_5 forms double salts with most ionic fluorides (Table 3). These can be obtained by direct mixing of the constituents in the molten state, or more usually by addition of an ionic fluoride to a solution of M_2O_5 in aqueous HF. Oxofluoro species were also formed; as expected, this occurred more readily for Nb than for Ta. Thus the addition of the appropriate metal carbonate to M_2O_5 solutions in HF yielded $[NbOF_5]^{2-}$ and $[TaF_7]^{-2}$ salts.[48]

The structure of $K_2[NbF_7]$ was first determined by X-ray diffraction,[49] then confirmed by neutron diffraction.[50] The niobium atom is at the center of a slightly distorted capped trigonal prism of six fluorine atoms, the seventh capping one of the rectangular faces of the prism (Nb—F: 1.940–1.978 Å); the same structure is found for $K_2[TaF_7]$. The $[TaF_8]^{3-}$ ion consists of a square antiprism.[51] The standard heats of crystallization of $A_2[MF_7]$ were evaluated from the heats of dissolution of the complexes in 3 M H_2SO_4 (2970 kJ mol^{-1} for $(NH_4)_2[TaF_7]$ to -3215 kJ mol^{-1} for $K_2[TaF_7]$).[52] The fused salt baths that are used in electrolytic production of the metals were shown to consist mainly of $[MF_7]^{2-}$ and $[MF_8]^{3-}$.[52,54]

$(NH_4)(HF_2)$[55] reacts exothermically with powdered Nb and Ta with intermediate formation

Table 3 Fluorometallates

Anion	*Cation*	*Ref.*
$[MF_6]^-$	Li^+, Na^+, K^+	1
	Rb^+, Cs^+, Tl^+, $(NH_4)^+$	2
	Cu^{2+}	3
	$(N_2H_6)^{2+}$	4
	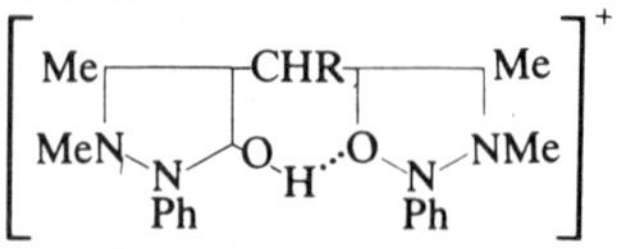	5, 6
$[MF_7]^{2-}$	Li^+, Na^+, K^+, $(NH_4)^+$	1, 7, 8
	Rb^+, Cs^+	9
	Mn^{2+}, Co^{2+}, Ni^{2+}, Zn^{2+}, Cd^{2+}	10
	$(N_2H_6)^{2+}$	4
$[MF_8]^{3-}$	Li^+, Na^+, K^+, $(NH_4)^+$	1, 4
	Ba^{2+}	8
	$(N_2H_6)^{2+}$	4

1. D. Bizot and M. Malek-Zadeh, *Rev. Chim. Miner.*, 1974, **11,** 710.
2. F. Fairbrother, 'Halides of Niobium and Tantalum', in 'Halogen Chemistry', ed. V. Gutmann, Academic, London, 1967, p. 123.
3. A. C. Baxter, J. H. Cameron, A. McAuley, F. M. McLaren and J. M. Winfield, *J. Fluorine Chem.*, 1977, **10,** 289.
4. B. Frlec and M. Vilhar, *J. Inorg. Nucl. Chem.*, 1971, **33,** 4069.
5. I. A. Arapova, N. Bausova and V. A. Zhilyaev, *Chem. Abstr.*, 1977, **87,** 15 314.
6. V. K. Akimov, A. I. Busev and U. A. Murachashvili, *Zh. Neorg. Khim.*, 1971, **16,** 680.
7. L. K. Marinina, E. G. Rakov, B. V. Gromov and O. V. Markina, *Zh. Fiz. Khim.*, 1971, **45,** 1592.
8. A. I. Nikolaev, V. Ya. Kuznetsov and A. G. Babkin, *Zh. Neorg. Khim.*, 1977, **22,** 2380.
9. D. Bizot, *J. Fluorine Chem.*, 1978, **11,** 497.
10. R. L. Davidovich, T. F. Levchishina, T. A. Kaidalova and V. I. Sergienko, *J. Less-Common Met.*, 1972, **27,** 35.

of metal hydrides to yield $(NH_4)_2[MF_7]$. Fluoride ion transfer reactions to the MF_5 Lewis acids were found to occur quite readily (Table 4).

The $[MF_5(SeF_4)]$ adducts are isostructural; the structure of the niobium compound, solved by X-ray diffraction, led to the ionic formulation $[SeF_3][NbF_6]$.[56] Strong bridging was found to occur between the ions, with four formulae units linked together in such a way that the Se and Nb atoms alternate at the corners of a distorted cube. The atomic arrangement found for $[(NbF_5)_2(SeF_4)]$[57] indicated the ionic structure $[SeF_3][Nb_2F_{11}]$.

The Lewis acidity of BF_3 and SbF_5 towards F^- was found to be higher than that of TaF_5 and NbF_5 (Table 4). $[NbF_5(SbF_5)]$ consists of an endless zigzag chain arrangement with alternating Sb and Nb atoms linked by *cis*-bridging fluorines.[58] Both metal atoms show a distorted octahedral coordination by F atoms consistent with a substantial contribution from the ionic $[NbF_4][SbF_6]$ form; four Nb—F distances were found at 1.81 Å and two at 2.17 Å.

While $K[MoF_6]$ was found to donate fluorine to MF_5, the ^{19}F NMR spectra measured on solutions of $[MF_6]^-$ and MoF_5 indicated the formation of $[M_2F_{11}]^-$ and $[MoMF_{11}]^-$.[59] The $[NbTaF_{11}]^-$ heteroanion was prepared from $[NbF_6]^-$ and TaF_5.[60] The $(O_2)[M_2F_{11}]$ salts[61] were obtained by heating the powdered metals with an F_2/O_2 mixture at 300–500 °C. When an excess of NOF was allowed to react with these salts, $(NO)[MF_6]$ and $(NO)_2[MF_7]$ were formed (equation 1).

$$\begin{aligned}(O_2)[M_2F_{11}] + 2NOF &\longrightarrow 2(NO)[MF_6] + O_2 + 0.5F_2\\ (NO)[NbF_6] + NOF &\longrightarrow (NO)_2[NbF_7]\end{aligned} \qquad (1)$$

MF_5 was reported to combine with brown solutions of iodine in IF_5 to give intense blue paramagnetic solutions, but only in the case of Ta was a stable solid of composition $I[TaF_6]$ isolated.[62] KrF_2, when allowed to react with NF_3 in the presence of NbF_5 in HF solution, gave $(NF_4)[NbF_6]$.[63]

Table 4 F Transfer Reactions Between Fluorides

Fluoride interacting with MF_5	*Compounds and comments*		*Ref.*
SeF_4	$[SeF_3][MF_6]$	X-Ray structure	1
	$[SeF_3][M_2F_{11}]$	X-Ray structure	2
$ClOF_3$	$(ClO_2F)[MF_6]$	X-Ray powder data	3
IO_2F_3	$[MF_4(IO_2F_4)]$	Oxygen-bridged polymers	4
XeF_2	$(XeF)[MF_6]$	Xe--F--M bridges (Raman)	5, 6
	$(XeF)[M_2F_{11}]$	^{93}Nb NQR	7
	$(Xe_2F_3)[MF_6]$	11% dissociation at m.p.	8
XeF_6	$(XeF_5)[MF_6]$	X-Ray powder data	9
	$(Xe_2F_{11})[MF_6]$		10
KrF_2	$(KrF)[MF_6]$, $(KrF)[M_2F_{11}]$, $(Kr_2F_3)[MF_6]$		11
BF_3	$[BF_3] + [MF_6]^- \leftrightarrows [BF_4]^- + [MF_5]$	^{19}F NMR	12, 13
	$[MF_5] + [MF_6]^- \leftrightarrows [M_2F_{11}]^-$	^{19}F NMR	
	$[BF_3] + [BF_4]^- \leftrightarrows [B_2F_7]^-$	^{19}F NMR	
SbF_5	$[NbF_4][SbF_6]$	X-Ray structure, ^{19}F NMR	14, 15
$[MoF_6]^-$	$[MoF_6]^- + [MF_5] \xrightarrow{CS_2} [MoF_5] + [MF_6]^-$		16
$[WF_6]^-$	$2[WF_6]^- + 2[TaF_5] \xrightarrow{CS_2} [WF_6] + [TaF_6]^- + [TaWF_{10}]^-$		
UF_4O	$[(MF_5)_3(UF_4O)]$	Small ionic contribution from $[UF_2O][MF_6][M_2F_{11}]$	17

1. A. J. Edwards and G. R. Jones, *J. Chem. Soc. (A)*, 1970, 1891.
2. A. J. Edwards and G. R. Jones, *J. Chem. Soc. (A)*, 1970, 1491.
3. R. Bougon, T. Bui Huy, A. Cadet, P. Charpin and R. Rousson, *Inorg. Chem.*, 1974, **13**, 690.
4. R. J. Gillespie and J. P. Krasznai, *Inorg. Chem.*, 1977, **16**, 1384.
5. B. Frlec and J. H. Holloway, *J. Chem. Soc., Dalton Trans.*, 1975, 535.
6. R. J. Gillespie and B. Landa, *Inorg. Chem.*, 1973, **12**, 1383.
7. J. C. Fuggle, D. A. Tong, D. W. A. Sharp, J. M. Winfield and J. H. Holloway, *J. Chem. Soc., Dalton Trans.*, 1974, 205.
8. J. Fawcett, B. Frlec and J. H. Holloway, *J. Fluorine Chem.*, 1976, **8**, 505.
9. B. Zemva and J. Slivnik, *J. Fluorine Chem.*, 1976, **8**, 369.
10. J. Aubert and G. H. Cady, *Inorg. Chem.*, 1970, **9**, 2600.
11. B. Frlec and J. H. Holloway, *Inorg. Chem.*, 1976, **15**, 1263.
12. M. E. Ignatov, E. G. Il'in and Yu. A. Buslaev, *Dokl. Akad. Nauk SSSR*, 1979, **245**, 604.
13. S. Brownstein, *J. Inorg. Nucl. Chem.*, 1973, **35**, 3567.
14. A. J. Edwards, *J. Chem. Soc., Dalton Trans.*, 1972, 2325.
15. P. A. W. Dean and R. J. Gillespie, *Can. J. Chem.*, 1971, **49**, 1736.
16. S. Brownstein, *Can. J. Chem.*, 1973, **51**, 2530.
17. J. H. Holloway, D. Laycolk and R. Bougon, *J. Chem. Soc., Dalton Trans.*, 1983, 2303.

(*ii*) *Anionic complexes of the higher halogens*

MCl_5 and MBr_5 show much less affinity for Cl^- or Br^- than does MF_5 for F^-. MCl_5 when allowed to react with ACl ($A = NH_4^+$, alkali metal) in closed systems at *ca.* 300 °C gave $A[MCl_6]$.[64–66] The Ta derivatives are thermally more stable than those of Nb.[79] $(NH_4)[TaCl_6]$ was also obtained through the reaction of $ClNH_2$ with MCl_5 in CCl_4.[65] The $[MX_5(RCN)]$ adducts (X = Cl, Br), when allowed to react with $(NEt_4)X$, afforded $(NEt_4)[MX_6]$.[67] The ^{93}Nb spectra measured on MeCN solutions of MX_5 (X = Cl, Br) indicate the presence of $[NbX_6]^-$. Passing HX′ through these solutions gave the $[NbX_6']^-$ anions (X′ = F or Cl).[68] $[MCl_6]^-$ can be extracted from 10 M HCl solutions in chloroform by amines.[69,70] Reactions of $K[HB(Me_2pz)_3]$ with MCl_5 yielded the interesting new compounds $[MCl_4][HB(Me_2pz)_3]$ and $[HB(Me_2pz)_3BH][NbCl_6]$;[71] the structure of the novel dibora cation was established.

Halogen transfer reactions were observed in the $Bu^tCl + RPCl_2 + MCl_5$ (R = Cl or Me) systems, yielding complexes of type $(Bu^tRPCl)[MCl_6]$.[72] The compounds $(EPh_4)[MX_6]$ (E = P, As) were obtained from the reaction of MX_5 (X = Cl, Br) with $(EPh_4)X$ in CH_2Cl_2 solution.[73,74] The $[MCl_5(PCl_5)]$ adducts are ionic and consist of discrete tetrahedral $(PCl_4)^+$ cations and octahedral $[MCl_6]^-$ anions.[75] A 1:1 mixture of $NbCl_5$ and $GaCl_3$ was reported to react in the melt to give molecular $[NbGaCl_8]$ built of an octahedron and tetrahedron sharing an edge.[76]

The X-ray powder patterns of $A[MX_6]$ (A = Na, K, Rb, Cs, Tl)[65,66,77] showed the low temperature modifications, with the exception of $K[NbCl_6]$, to be isotypic with $Cs[WCl_6]$, and the high temperature forms to crystallize in a cubic-face-centered arrangement. The Raman spectra were taken to indicate O_h symmetry of the $[MCl_6]^-$ ions (A = Li, Na, K, Rb, Cs, Tl),[64] while other authors interpreted the spectra measured on $K[NbCl_6]$ and $Na[TaCl_6]$ as supporting a formulation such as $K_2[Nb_2Cl_{10}]Cl_2$ with isolated edge-sharing bioctahedra.[65] The electron transfer spectra of $(Et_4N)[MX_6]$ (X = Cl, Br) were investigated.[3,78]

The structures of $A[NbCl_6]$ (A = $(PCl_4)^+$,[75] $(PPh_4)^+$ [73]), of $(PPh_4)[NbBr_6]$,[74] of $A[TaCl_6]$ (A = $(PCl_4)^+$,[75] $[HB(Me_2pz)_3BH]^+$ [71]) and of $[TaBr_4(diars)_2][TaBr_6]$[80] have been reported. The metal displays octahedral coordination, with M—Cl distances ranging from 2.29 to 2.37 Å and M—Br distances from 2.48 to 2.51 Å.

(*iii*) *Mixed halo complexes*

MeCN solutions containing MCl_5[81] or MBr_5[82] and HF in a ratio of *ca.* 1:3 were found by ^{19}F NMR to contain the $[MF_6]^-$, $[MF_5X]^-$, *cis*- and *trans*-$[MF_4X_2]^-$, *mer*-$[MF_3X_3]^-$, and *cis*- and *trans*-$[MF_2X_4]^-$ ions. The ^{93}Nb NMR spectra of mixtures of $NbCl_5$ and $NbBr_5$ in MeCN show the presence of all seven possible mixed halides $[NbCl_xBr_{6-x}]^-$ with close to random halogen distributions.[83]

Solutions of mixed anions containing three halides were prepared.[84] Of the 46 possible octahedral $[MF_xCl_yBr_z]^-$ complex anions, the formation in solution of 38 tantalum complexes, of which 22 were bromochlorofluoro species, was recognized. The redistribution of the halides did not occur randomly, and the differences in free energy changes, $\Delta G = \Delta G_{exp} - \Delta G_{random}$, showed that the stability of the fluorochloro complexes decreases in the series: *trans*-$[TaF_4Cl_2]^- >$ *trans*-$[TaF_2Cl_4]^- > [TaF_5Cl]^- >$ *fac*-$[TaF_3Cl_3]^- >$ *cis*-$[TaF_4Cl_2^- >$ *cis*-$[TaF_2Cl_4]^- >$ *mer*-$[TaF_3Cl_3]^-$.[85]

34.2.2.2 Neutral adducts of the pentahalides

Characteristic of the Lewis acid behavior of MX_5 are their reactions with donor molecules. Differences are found, however, between the reactions of MF_5 and those of the other pentahalides. For example, $NbCl_5$ reacts with L = DMSO or HMPA to give $[NbOCl_3L_2]$, and the reaction of $TaCl_5$ with py under appropriate conditions gives $[TaCl_4(py)_2]$, while MF_5 reacts with DMSO and py to give $[MF_5L_n]$ complexes ($n = 1, 2$). $NbCl_5$ behaves both as a softer and as a weaker Lewis acid than $TaCl_5$ with respect to ethers and sulfides, for example.[86] The acceptor power of $[NbCl_{5-x}Me_x]$ decreases with x.[87,88]

The pentahalides form stable 1:1 adducts with a variety of donors. These are octahedral monomeric non-electrolytes, except for the adduct of TaF_5 with PBu_3,[89] which was shown by NMR to be the ionic $[TaF_4(PBu_3)_2]F$. NMR has been thoroughly used to study their structures in solution, as well as their relative stabilities and dynamic behavior. Relative stability scales for adducts with nitriles,[90] dialkyl chalcogenides[86,91] and phosphine oxides[92,93] have been established. The stability of the nitrile and phosphine oxide adducts was found to be controlled by inductive factors, while it was concluded that steric factors were determinant in the ether and sulfide adducts. The ligand exchange (equation 2) was reported to take place through a dissociative mechanism for L = RCN and OPR_3,[92,95] while an associative mechanism was postulated for L = Me_2Y (Y = S, Se, Te)[96] in non-coordinating solvents. The exchange rates obey the sequences $MCl_5 \ll MBr_5$ for the dissociative mechanism, and $MF_5 > MCl_5 > MBr_5$ for the associative one.[97] NMR studies of $[TaCl_4(dppe)(Cp')]$ showed exchange between the free and monodentate dppe. The rates of intramolecular exchange between equatorial and axial fluorine atoms in $[MF_5L]$[95] have been evaluated by ^{19}NMR for L = OMe_2 and $OP(NMe_2)_{3-n}(OMe)_n$, where $n = 0$–3. NQR spectra have been measured for $[MCl_5L]$, where L = nitriles, ethers and phosphine oxides.[98] The electron transfer spectra of MX_5 adducts of several series of ligands (nitriles, dialkyl chalcogenides, phosphoryl and thiophosphoryl ligands, phosphines)[99] have been reported. Intermolecular charge transfer spectra indicate that MCl_5 interacts with aromatic hydrocarbons;[100] the equilibrium constants showed these interactions, however, to be very weak. The donor capacity of the solvent towards $NbCl_5$ was reported to decrease in the order: $Me_2NCOH > MeCN > RCOOR' > RCOR' > R_2O \approx C_4H_8O_2 > C_6H_6$.[101]

$$[MCl_5L] + L^* \rightleftharpoons [MCl_5L^*] + L \qquad L = RCN, OPR_3, Me_2Y \quad (Y = S, Se, Te) \tag{2}$$

$[MX_5L_2]$ has also been isolated for many O, S, N, P and As donors. The MF_5 derivatives consist of ionic $[MF_4L_4][MF_6]$ in solution, while molecular structures with seven-coordinate metals appear to be equally likely in the solid state.[102,103] The analogous chlorides and bromides seem to be essentially molecular.

(*i*) *Oxygen donors*

MF_5 reacts with aliphatic ethers to yield volatile adducts of the $[MF_5(OR_2)]$ type; DMSO forms the bis adducts $[MF_5(DMSO)_2]$. MCl_5 and MBr_5 form numerous $[MX_5(ether)]$ adducts. Abstraction of oxygen from the ethers occurs at around 100 °C to produce oxo trihalides and alkyl halides.[1] The ability of crown ethers to coordinate and reduce MCl_5 has been explored;[104] $[(MCl_5)_2L]$ and $[MCl_5L]$ were isolated with L = dibenzo-18-crown-16, 18-crown-6 and 15-crown-5. NMR data indicate that these compounds disproportionate.

MCl_5, MBr_5 and acetophenone or benzophenone yield 1:1 complexes.[105] $TaCl_5$, when allowed to react with ethyl methyl ketone or diethyl ketone in a 1:1 molar ratio, gave 1:1 adducts, whereas oxygen abstraction occurred in the case of $NbCl_5$, resulting in the formation of $[(NbCl_4)_2O(R_2CO)_2]$;[106] with excess ketone, $[MOCl_3(R_2CO)_2]$ was obtained in both cases. Ketone complexes $[NbCl_5L]$ were, however, isolated from benzene solutions.[107] The reactions of $[MCl_{5-x}Me_x]$ ($x = 1, 2$) with bulky ketones yielded $[MCl_{5-x}Me_x(RR'CO)]$ for R = R' = cyclohexyl; R = Me, R' = Bu^t; R = Me, R' = neopentyl, while addition of the methylmetal group to the carbonyl was found to occur with less bulky ketones and with benzaldehyde.[108] Dialkyl esters RCOOR' were allowed to react with $NbCl_5$[109] to yield 1:1 and 2:1 complexes; the former, partially ionic in solution, were formulated as $[NbCl_4(RCOOR')_2][NbCl_6]$.

In the $[MX_5L_n]$ adducts formed with amides (X = Cl, $n = 1, 3, 4$; X = Br, $n = 1, 3$; L = $RCONH_2$, R = H, Me or Ph), the amide is bonded to the metal through its oxygen only.[110] Urea complexes $[MX_5L]$[111] (X = Cl, Br) have been prepared from MX_5 and tetramethylurea in CH_2Cl_2; coordination was shown by IR to be through oxygen only. On the other hand, oxathiane and oxaselenane gave $[NbCl_5(C_4H_8OY)]$ (Y = S, Se) in which, on the basis of IR and NMR data, they were proposed to be S- or Se-bonded, and not O-bonded, to the metal.

(*ii*) *Adducts of phosphine or arsine oxides, sulfides and selenides*

MF_5 forms $[MF_5L]$ and $[M_2F_{10}L]$ complexes with phosphoryl ligands.[93,112,113] In the dimeric compounds, the ligand is *cis* to the bridging fluorine.[113] With PSX_3 and $PSeX_3$, molecular pseudooctahedral $[MF_5L]$ complexes were obtained in which the ligands are bound to the metal through S or Se.[114] Atmospheric oxygen reacts with $[MF_5(SPX_3)]$ or $[MF_5(SePX_3)]$ to give $[MF_5(OPX_3)]$ along with sulfur or selenium precipitates.[114] Oxygen coordination took place when MF_5 was allowed to react with $R_2P(O)SR$ to give $[MF_5L]$ and $[M_2F_{10}L]$.[113] A thione–thiol rearrangement was observed when thiophosphoric esters $(EtO)_2P(S)OMe$ were added to a solution of TaF_5,[115] resulting in the formation of $[TaF_5\{OP(OEt)_2SMe\}]$; in the presence of traces of oxygen $[TaF_5\{OP(OEt)_2OMe\}]$ was also detected. The same rearrangement occurred in a solution of TaF_5 and $Ph_2P(S)OEt$, but at slower rates, so that both $[TaF_5\{SP(OEt)Ph_2\}]$ and $[TaF_5\{OP(SEt)Ph_2\}]$ complexes could be identified along with (in the presence of air) $[TaF_5\{OP(OEt)Ph_2\}]$. Solutions of TaF_5 and PBu_3 yielded $[TaF_5(OPBu_3)]$ upon aging.[89] ^{19}F NMR showed that equimolar solutions of MF_5 and of R_3AsY (Y = O, S; R = Bu, Ph) contain molecular $[MF_5L]$ together with ionic $[MF_4L_2][MF_6]$ compounds.[116] The following donor capacities towards TaF_5 were derived from ^{19}F NMR chemical shifts: $R_3AsO > R_3PO > R_3AsS \approx R_3PS$.

Phosphine oxide complexes $[MCl_5(R_3PO)]$ (R = Ph, Me, NMe_2)[117–119] and $[MCl_5(R_2R'PO)]$ (R = Ph; R' = Bz) have been prepared. Nb^V forms an oxochloro adduct in the presence of excess $OPPh_3$.[1,120] When conducted in a non-basic solvent in the absence of MeCN, the action of HMPA on Nb_2Cl_{10} led to oxygen abstraction and formation of $[NbOCl_3(HMPA)_2]$,[117] while the 1:1 adduct $[NbCl_5(HMPA)]$ formed exclusively when a mixture of MeCN and CH_2Cl_2 was used as solvent, indicating the initial formation of a stable monomeric six-coordinate adduct $[NbCl_5(MeCN)]$. The O-coordinated $[MCl_5(OPCl_3)]$ adducts are monomeric in benzene but dissociate in nitromethane.[121] The existence of the niobium adduct in the gas phase has also been established.[122] The enthalpy of formation of the crystalline adduct from its components was evaluated.[123] $TaCl_5$ and $NbCl_5$ were found to react readily with triphenylphosphole oxide to yield 1:1 adducts.[124]

MX_5L (X = Cl, Br; L = $SPPh_3$ and $SePPh_3$) was obtained; the formation of 2:1 complexes was not observed with these ligands. The pentabromides, but not the pentachlorides, gave 1:1 complexes with $Ph_2P(S)CH_2P(S)Ph_2$.[125]

(*iii*) *Sulfur, selenium and tellurium donors*

$[MX_5L]$ complexes (X = F, L = SMe_2, SEt_2; X = Cl, L = SMe_2, SEt_2, SPr_2, SC_4H_8, SC_4H_{10}, SC_4H_8O, SeC_4H_8O; X = Br, L = SMe_2, SEt_2, SC_4H_8, SC_5H_{10}, SC_4H_8) were easily prepared; they could all be either distilled or sublimed unchanged, except for $[NbBr_5(SEt_2)]$, which decomposed on heating. Alternatively, the ether ligands of $[MCl_5(OEt_2)]$ and $[MCl_5(OPr_2)]$ were readily replaced by their sulfide analog, while $[MCl_5(OMe_2)]$ remained unaffected by SMe_2. IR, Raman and NMR studies of the complexes formed with ambidentate O,S ligands showed bonding to occur *via* sulfur exclusively and the complexes to be mononuclear. With SMe_2 and SC_4H_8 evidence was found for $[NbX_5L_2]$ adducts.[126]

In $[NbCl_5(dth)]$, the dth = $MeSCH_2CH_2SMe$ ligand was shown by IR to be in the *gauche* chelating form. Bidentate ligands having bulky end groups, *e.g.* $Bu^tSCH_2CH_2Bu^t$ and $PhSCH_2CH_2SPh$, adopt a *trans* conformation and bridge the two metal atoms in the $[(MCl_5)_2L]$ adducts isolated.[126] Reactions of MCl_5 with cyclic polysulfides yield dimeric complexes in which the two MCl_5 units are bridged by the ligand; such $[(MCl_5)_2L]$ complexes were isolated with L = 1,4-dithiane[127] and various S_4 and S_6 cyclic polysulfides.[128] A 1:1 complex $[TaCl_5L]$ was, however, obtained with 1,3,5-trithiane.[127] The structure of $[(NbCl_5)_2(C_{10}H_{20}S_4)]$ (Figure 1) shows that the ligand exists with the most unusual 'inside out' conformation.[129,130] MCl_5 yielded S-coordinated 1:1 complexes when allowed to react with tetramethylthiourea.[111] A sulfur abstraction reaction occurred however when the reaction was followed by exposure to an intense light source, leading to $[(Me_2N)_2CCl][NbSCl_4]$.

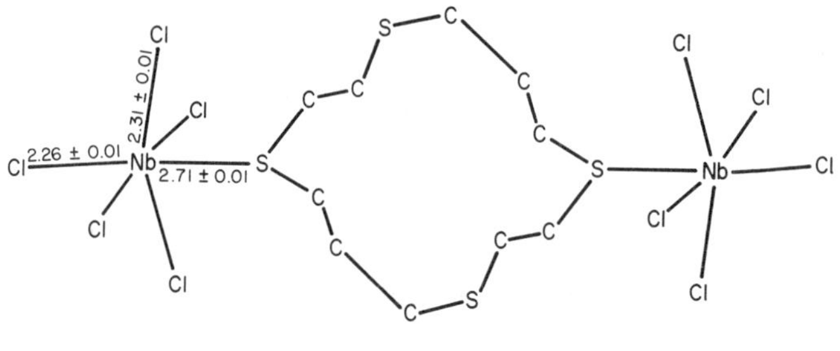

Figure 1 Molecular structure of $[(NbCl_5)_2(C_{10}H_{20}S_4)]$ (reproduced from ref. 128 with permission)

$NbCl_5$ and $TaCl_5$ were found to form 1:1 adducts with MeNCS and MeSCN;[131,132] the presence of both S- and N-bonded isomers was assumed on the basis of IR. Sulfur abstraction reactions occurred when $TaCl_5$ was allowed to react with PhNCS, to yield $[TaSCl_3(PhNCCl_2)]$. Reactions of $[MCl_{5-x}Me_x]$ (x = 1, 2) with MeSCN afforded $[MCl_4Me(MeSCN)]$ and $[MCl_3Me_2(MeSCN)]$,[133] which consist of mixtures of N- and S-bonded isomers.

(*iv*) *Nitrogen donors*

Niobium and tantalum halides form adducts with various nitrogen donor ligands including aliphatic and aromatic amines; nitriles, Schiff's bases and imidazoles (Table 5). The reactions of MX_5 with pyridine and related ligands such as bipy or phen depend critically on the reaction conditions. With py at low temperature MX_5 (X = Cl, Br) yielded 1:1 adducts that are rapidly reduced to $[MX_4(py)_2]$ on increasing the temperature, with formation of 1-(4-pyridyl)pyridinium halide. Similarly, bipy and phen reduced the metal in MeCN to oxidation state +IV and formed monoadducts of type $[MX_4(bipy)]$ at room temperature, while at 0 °C the same reactions yielded $[NbCl_5(bipy)(MeCN)]$ and $[TaX_5(bipy)(MeCN)]$ (X = Cl or Br). $NbBr_5$ and TaI_5 formed $[MX_5(bipy)_2]$, which were formulated as the eight-coordinate $[MX_4(bipy)_2]X$.[1] Reduction of the metal can however be prevented, even at room temperature,

Table 5 Complexes Formed by MX_5 with Nitrogen Donors

Ligand	*Compounds and comments*	*Ref.*
Amines		
NH_2R (R = H, Et)	$[NbF_5L_2]$	1, 2
$NHEt_2$	$[NbF_5L]$, $[MCl_5L]$; from C_6H_6	1, 2, 3
NR_3 (R = Me, Et)	$[NbF_5L]$	1, 2
	$[NbCl_5(NEt_3)(MeCN)]$; from MeCN	4
Aromatic amines		
py	$[NbX_5(py)]$; rapidly reduced to $[NbX_4(py)_2]$	
	$[TaX_5(py)]$, $[TaX_5(py)_2]$; X = Cl, Br	1
bipy	$[MX_5(bipy)]$ M = Nb; X = F, Cl, Br; M = Ta; X = Cl, Br	5
	$[MX_4(bipy)_2]X$; X = Br, I	1
	$[TaCl_2Me_3(bipy)]$; X-ray structure	6
terpy	$[MX_5(terpy)]$; X = Cl, M = Nb; X = Br; M = Nb, Ta	7
	$[Ta_2Cl_{10}(terpy)]$	
phen	$[TaCl_5(phen)]$; ionic	8
Nitriles		
RCN	$[MF_5(MeCN)]$, $[MF_4(MeCN)_4][MF_6]$	9, 10
	$[MCl_5(RCN)]$; R = Me, CD_3, Et	11, 12
	R = H; X-ray structure	13
	$[MBr_5(MeCN)]$	14
	$[MX_5(CH_2{=}CHCN)]$; X = Br, Cl	15
	$[MCl_5(PhCN)]$, $[MCl_4Me(PhCN)]$, $[TaCl_4Me(CH_2{=}CHCN)]$	16
	$[NbCl_3Me_2(CH_2ICN)]$, $[NbCl_3Me_2(CH_2ClCN)]$, $[NbCl_3Me_2(CH_2{=}CHCN)]$	
	$[MCl_5(RCN)]$; R = Me, Bu^t, CH_2F, CH_2Cl, CH_2I, Ph, C_6H_4Cl (relative stabilities; L exchange)	17
Dinitriles		
$CN(CH_2)_nCN$	$[MX_5L]$; n = 1–4, X = Cl, Br	18
	$[(MX_5)_2L]$	19
Schiff's bases		
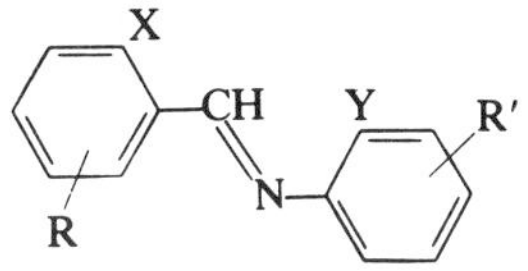	$[NbCl_5L]$, $[NbCl_5L_2]$, $[NbCl_5L_3]$; R = H, 4-NO_1, 3-NO_2; X = OH, H; R′ = H, 4′-Me, 4′-Br, 4′-NO_2, 3′-NO_2, 4′-I; Y = H, OH	20
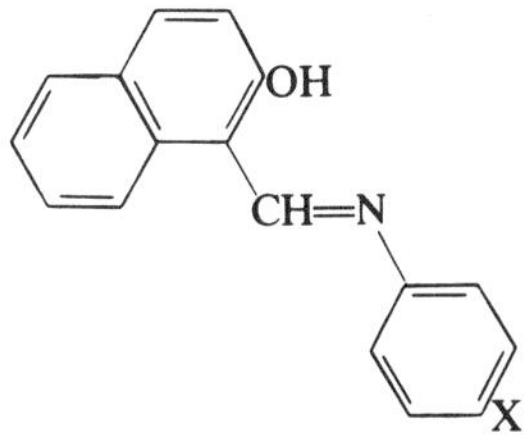	X = H, Me, Br, I, NO_2	21
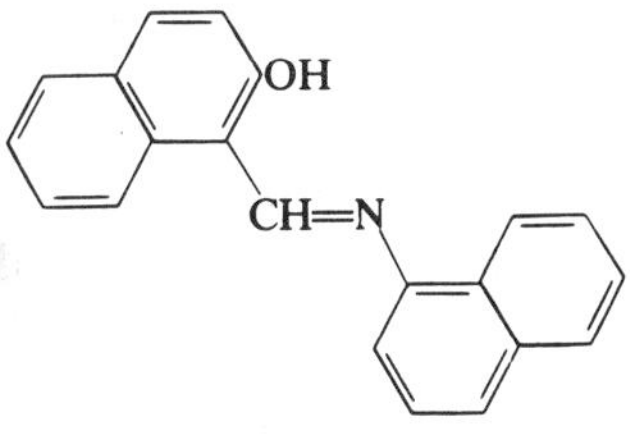		
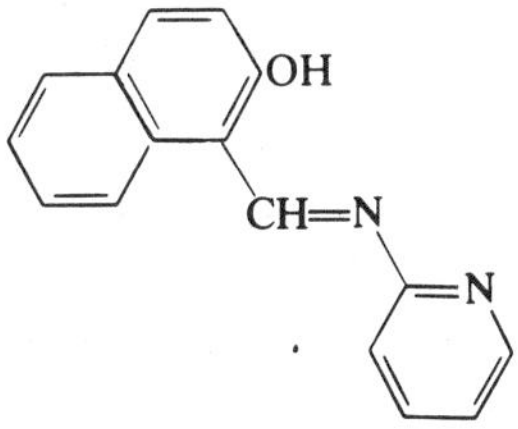		

Table 5 (*Continued*)

Ligand	*Compounds and comments*	*Ref.*
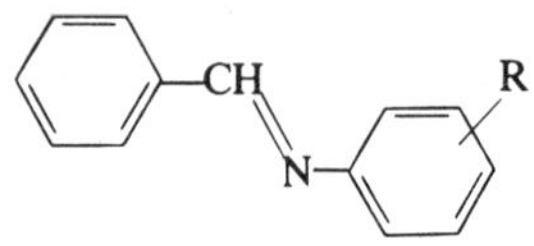	R = H, 4-Me, 4-OMe, 4-Cl, 4-Br, 4-NO_2, 3-Me, 3-NO_2, 3-OMe, 2-Cl, 2-OMe, 2,4,6-Br, 2,4,6-Me	22
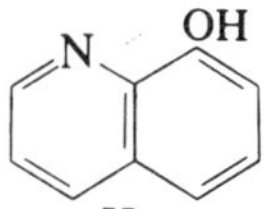 oxH	$[MX_5(oxH)_n]$; $n = 2, 4, 6, 8$ or 10; X = Cl, Br	23, 24
Imidazoles		
3,5-Dimethylpyrazole		
Benzimidazole	$[NbCl_5L_n]$ $n = 1, 2, 3$	25
Benzothiazole		
Benzazole		
R-benzimidazoles	$[NbCl_5L_4]$	
	R = 2-Me, 2-Et, 2-Ph, 2-(o-OHC_6H_4)	26
	R = 2-(o-aminophenyl), 2-(β-pyridyl)	27

1. D. L. Kepert, 'The Early Transition Metals', Academic, London, 1972.
2. E. G. Il'in, M. M. Ershova, M. A. Glushkova and Yu. A. Buslaev, *Dokl. Akad. Nauk SSSR*, 1978, **243**, 1459.
3. Yu. A. Buslaev, M. M. Ershova and M. A. Glushkova, *Koord. Khim.*, 1979, **5**, 532.
4. Yu. A. Buslaev, M. A. Glushkova, N. A. Chumaevslii, M. M. Ershova and L. V. Kumelevskaya, *Koord. Chim.*, 1982, **8**, 457.
5. C. Djordjevic and V. Katovic, *J. Chem. Soc. (A)*, 1970, 3382.
6. M. G. B. Drew and J. D. Wilkins, *J. Chem. Soc., Dalton Trans.*, 1973, 1830.
7. B. Begolli, V. Valjak, V. Allegretti and V. Katovic, *J. Inorg. Nucl. Chem.*, 1981, **43**, 2785.
8. L. A. Ugalava, N. I. Pirtskhalava, V. A. Kogan, A. S. Egorov and A. O. Osipov, *Chem. Abstr.*, 1975, **83**, 184 471.
9. K. C. Moss, *J. Chem. Soc. (A)*, 1970, 1224.
10. J. A. S. Howell and K. C. Moss, *J. Chem. Soc. (A)*, 1971, 2483.
11. D. L. Kepert and R. S. Nyholm, *J. Chem. Soc.*, 1965, 2871.
12. G. A. Ozin and R. Walton, *J. Chem. Soc. (A)*, 1970, 2236.
13. C. Chavant, G. Constant, Y. Jeannin and R. Morancho, *Acta Crystallogr., Sect. B*, 1975, **31**, 1823.
14. D. Brown, G. W. A. Fowles and R. A. Walton, *Inorg. Synth.*, 1970, **12**, 225.
15. G. W. A. Fowles and K. F. Gadd, *J. Chem. Soc. (A)*, 1970, 2232.
16. J. D. Wilkins, *J. Organomet. Chem.*, 1975, **92**, 27.
17. A. Merbach and J. C. Bünzli, *Helv. Chim. Acta*, 1971, **54**, 2543.
18. M. S. Gill, H. S. Ahuja and G. S. Rao, *Inorg. Chim. Acta*, 1973, **7**, 359.
19. M. S. Gill, H. S. Ahuja and G. S. Rao, *J. Inorg. Nucl. Chem.*, 1974, **36**, 3731.
20. L. V. Surpina, O. A. Osipov and V. A. Kogan, *Zh. Neorg. Khim.*, 1971, **16**, 685.
21. L. V. Surpina, G. F. Litovchencko, S. M. Artamonova, A. D. Garnovskii and O. A. Osipov, *Zh. Obshch. Khim.*, 1978, **48**, 1830.
22. L. A. Ugulava, N. I. Pirtskhalava, V. A. Kogan, A. S. Egorov and O. A. Osipov, *Zh. Obshch. Khim.*, 1975, **45**, 1575.
23. M. J. Frazer, B. G. Gillespie, M. Goldstein and L. I. B. Haines, *J. Chem. Soc. (A)*, 1970, 703.
24. A. V. Leshchenko and O. A. Osipov, *Zh. Obshch. Khim.*, 1977, **47**, 2581.
25. L. V. Surpina, A. D. Garnovskii, Yu. A. Kolodyazhnyi and O. A. Osipov, *Zh. Obshch. Khim.*, 1971, **41**, 2279.
26. N. S. Biradar, T. R. Goudar and V. H. Kulkarni, *J. Inorg. Nucl. Chem.*, 1974, **36**, 1181.
27. N. S. Biradar and T. R. Goudar, *J. Inorg. Nucl. Chem.*, 1977, **39**, 358.

when the reaction is conducted in dry benzene;[134] a series of isomorphous heptacoordinate compounds of type $[MX_5(bipy)]$ have thus been obtained.

The structure of $[TaCl_2Me_3(bipy)]$[135] shows tantalum to have distorted capped trigonal prismatic environment, with one chlorine atom in the capping position (Ta—Cl: 2.54(1) Å); the nitrogen atoms (Ta—N: 2.29(2) Å), one chlorine and one methyl group form the capped quadrilateral face, with two methyl groups (Ta—C: 2.24(4) and 2.16(6) Å) occupying the remaining edge.

Reactions of MX_5 (X = Cl, Br) with 2,2′,2″-terpyridyl[136] yielded the poorly soluble and moisture sensitive $[MBr_5(terpy)]$, $[NbCl_5(terpy)]$ and $[Ta_2Cl_{10}(terpy)]$. IR data suggest that all three pyridyl groups are coordinated.

Alkanenitriles readily gave 1:1 and 2:1 adducts with MF_5;[137,103] the 2:1 adducts are ionic and should be formulated as $[MF_4L_4][MF_6]$, while the 1:1 adducts are neutral $[MF_5L]$. The penta-chlorides, -bromides and -iodides afforded the monomolecular $[MX_5(RCN)]$ (Table 5).[138]

In $[NbCl_5(HCN)]$,[139] the niobium atom is octahedrally surrounded by five chlorines and by the nitrogen atom of HCN; the Nb—Cl bond *trans* to nitrogen is the shortest (2.243(1) as compared to an average 2.313(1) Å for the other Nb—Cl distances); the N≡C distance of 1.090(4) Å is shorter than in free HCN. The reactions of $[MCl_{5-x}Me_x]$ ($x = 1$–3) with nitriles yield simple 1:1 adducts (Table 5), except for the formation of a small quantity of insertion product in the reaction of $[TaCl_4Me]$ with CCl_3CN.[140]

Complexes of MX_5 with dinitriles $CN(CH_2)_nCN$ ($n = 1$–4) have been prepared.[141,142] Both $[MX_5L]$ and $[MX_5L_2]$ have been isolated in the case of malonitrile, succinonitrile and glutaronitrile, whereas only $[(MX_5)_2L]$ formed with adiponitrile. In the case of succinonitrile and glutaronitrile, *trans* conformers have been assigned to the 2:1 adducts, while a *gauche* conformation was assumed in 1:1 complexes; the more stable *gauche–trans–gauche* conformation for adiponitrile was proposed in its 1:2 adducts.

The reactions of 8-quinolinol (oxH) with NbX_5 (X = Cl, Br) produced $[MX_5(oxH)_n]$ ($n = 2$, 4, 5, 6 or 10).[143,144] The interaction of $NbCl_5$ with Schiff's bases led to isolation of crystalline, sparingly soluble compounds having 1:1, 2:1 and 3:1 stoichiometries (Table 5); the ligands act as monodentates through the azomethine nitrogen atom. $NbCl_5$ was reported to form 4:1 adducts with substituted imidazoles,[145–147] which are 1:1 electrolytes in DMF. IR data indicate that coordination takes place through the unsaturated nitrogen of the imidazole ring in an eight-coordinated $[NbCl_4L_4]^+$ cation.

$TaCl_5$ and S_4N_4 were reported to form $[TaCl_5(S_4N_4)]$,[148,149] the S_4N_4 group being bonded to the metal through one of its nitrogen atoms (Figure 2).

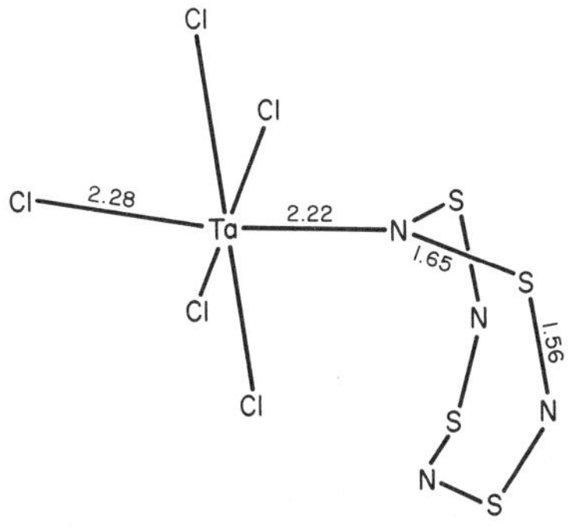

Figure 2 Molecular structure of $[TaCl_5(S_4N_4)]$ (reproduced from ref. 149 with permission)

Complexes $[NbCl_4L]Cl$, where L = RNHCSNHpy, were obtained;[150] their IR spectra indicate bidentate behavior of L *via* sulfur and the heterocyclic nitrogen; they exhibit semiconductor properties. Similar behavior and properties were reported for the adducts formed with *N,N′*-diaryl-substituted formamidino-*N″*-aryl-substituted carbamides and thiocarbamides.

v) Phosphorus and arsenic donors

The group Vb triphenyls form $[NbCl_5(EPh_3)]$ (E = N, P, As, Sb or Bi) in carbon tetrachloride, *n*-hexane or cyclohexane, while $[(NbCl_5)_2(EPh_3)]$ was obtained in benzene.[1] Table 6 illustrates the tendency toward increasing coordination numbers in the series $PPh_3 < PPh_2Me < P(OPh)_3 < PMe_2Ph \approx$ dppe < dmpe, which follows the decrease in the ligand's cone angle

Addition of diars to MCl_5 immediately precipitated the isomorphous red $[NbCl_5(diars)]$ and yellow $[TaCl_5(diars)]$;[151] being diamagnetic, they were reported to be seven-coordinate. X-Ray structure determinations,[152] however, indicated the presence of $[TaCl_4(diars)_2][TaCl_5(OEt)]$, $[NbCl_4(diars)_2][NbOCl_4]$ and $[NbCl_4(diars)_2][NbCl_3O_2]$, with an eight-coordinate dodecahedral cation. Similarly the compound first formulated as $[TaBr_5(diars)]$[151] was shown to be $[TaBr_4(diars)_2][TaBr_6]$;[80] the cation is a crystallographically imposed dodecahedron, (Ta—Br: .583(10); Ta—As: 2.765(1) Å). Fluorinated ditertiary arsines afforded $[MX_5(diarsine)]$[153] (X = Cl, Br), for which the exact formulation could not be established.

Table 6 Complexes with Phosphorus and Arsenic donors

Compounds	*Comments*	*Ref.*
With monodentates		
$[MCl_5(PPh_3)]$	Obtained from CCl_4, *n*-hexane, cyclohexane	1, 2
$[MCl_5(AsPh_3)]$		1, 2
$[(NbCl_5)_2(EPh_3)]$	E = P or As; obtained from benzene	1
$[TaCl_5(PMe_3)]$	Unstable at room temperature	3
$[TaCl_5(PPr^i_3)]$	Stable only at −60 °C in solution	3
$[MCl_5(PBu_3)]$	Obtained from CCl_4	2
$[MCl_5(PMe_2Ph)_2]$	$[MCl_5(PMe_2Ph)]$ also present in solution	4
$[NbCl_5\{P(OPh)_3\}_2]$		4
$[NbCl_4Me\{P(NMe_2)_3\}]$		5
$[NbCl_{5-x}Me_x(PPh_3)]$	$x = 1–3$	5, 6, 7
$[TaCl_4(PMe_3)(Cp)]$		8
$[TaCl_4\{P(OMe)_3\}(Cp)]$		8
With bidentates		
$[MCl_5(dppe)]$	Seven-coordinate compound	9
$[MCl_5(dmpe)]$	Seven-coordinate compound	4
$[MCl_5(dpae)]$	Seven-coordinate compound	4
$[MCl_{5-x}Me_x(dppm)]$	$x = 3$	7
$[MCl_{5-x}Me_x(dppe)]$	$x = 1–3$	5, 6, 7
$[NbBr_4Me(dppe)]$		5
$[TaCl_4(dppe)(Cp)]$	Monodentate phosphine	8
$[TaBr_4(diars)_2][TaBr_6]$	X-Ray structure	10
$[MCl_4(diars)_2]^+$	X-Ray structure	11
$[MX_5(diarsine)]$[a]	Either seven-coordinate, or mixed six–eight-coordinate ionic formulation	12

[a] Diarsine =

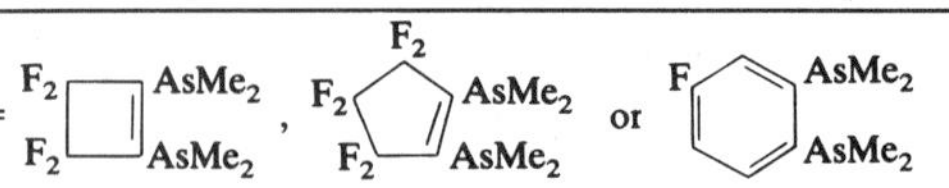

1. D. L. Kepert, 'The Early Transition Metals', Academic, London, 1972.
2. M. A. Glushkova, M. M. Ershova, N. A. Ovchinnikova and Yu. A. Buslaev, *Zh. Neorg. Khim.*, 1972, **17,** 147.
3. M. Valloton and A. E. Merbach, *Helv. Chim. Acta,* 1975, **58,** 2272.
4. G. Jamieson and W. E. Lindsell, *Inorg. Chim. Acta,* 1978, **28,** 113.
5. C. Santini-Scampucci and J. G. Riess, *J. Chem. Soc., Dalton Trans.*, 1973, 2436.
6. G. W. A. Fowles, D. A. Rice and J. D. Wilkins, *J. Chem. Soc., Dalton Trans.*, 1972, 2313.
7. G. W. A. Fowles, D. A. Rice and J. D. Wilkins, *J. Chem. Soc., Dalton Trans.*, 1974, 1080.
8. R. D. Sanner, S. T. Carter and W. J. Bruton, Jr., *J. Organomet. Chem.*, 1982, **240,** 157.
9. J. D. Wilkins, *J. Inorg. Nucl. Chem.*, 1975, **37,** 2095.
10. M. G. B. Drew, A. P. Wolters and J. D. Wilkins, *Acta Crystallogr., Sect. B,* 1975, **31,** 324.
11. J. C. Dewan, D. L. Kepert, C. L. Raston and A. H. White, *J. Chem. Soc., Dalton Trans.*, 1975, 2031.
12. D. L. Kepert and K. R. Trigwell, *Aust. J. Chem.*, 1976, **29,** 433.

34.2.3 Solvolysis Products of the Pentahalides

34.2.3.1 Alkoxides

(i) Synthesis

Equation (3) represents the reactions of MCl_5 with alcohols and phenols. They can be driven to completion by using ammonia as proton acceptor in the case of primary and secondary alcohols (Table 7).[154–156] $[Ta(OBu^t)_5]$ was obtained in the presence of pyridine, while under the same conditions $NbCl_5$ yielded $[NbO(OBu^t)_3]$. $[Nb(OBu^t)_5]$ could, however, be prepared from $[Nb(NR_2)_5]$.[157]

$$\begin{aligned} MCl_5 + 3ROH &\longrightarrow [MCl_2(OR)_3] + 3HCl \\ MCl_5 + 5PhOH &\longrightarrow [M(OPh)_5] + 5HCl \end{aligned} \quad (3)$$

Transesterification reactions have been extensively used for the preparation of further alkoxides.[158–165] Mixed alkoxides $[M(OR)_{5-x}(OR')_x]$ were obtained from the same alcoholysis reactions.[166,167] Exchange reactions between $[Nb(OEt)_5]$ or $[Nb(OPr^i)_5]$ and organic acetates have also been exploited for the preparation of higher alkoxides.

Trialkylsiloxo derivatives were prepared by the reaction of the pentaethoxides with

Table 7 Pentaalkoxides $M(OR)_5$

R	*Comments*	*Ref.*
Me	Dimeric both in the solid and in solution	1, 2
	^1H NMR: dynamic intramolecular exchange	3, 4, 5
	Solvation studies	6
	X-Ray structure	7
Et, Pr, Bu, *n*-pentyl	Dimeric in benzene; monomeric in alcohol	1, 2, 3
Bu^t, Pr^i		8, 9
CH_2CF_3	Dimeric in benzene	10
SiR_3		8, 11
$MeCH{=}CHCH_2$	Dimeric in benzene	12
$CH_2{=}CHC(Me)_2$—	Monomeric in benzene	13
$CH_2{=}CHCH_2$	Dimeric in benzene	14
$MeCH{=}CHCH(Me)$	Average association in benzene: 1.5	15
$CH_2{=}CHCH_2(Me)$	Monomeric in benzene	16
$MeCH{=}CHCH(Et)$	Monomeric in benzene	17
$(Ph)_2CH{=}CHCH_2$	Monomeric in benzene	18
Ph	Dimeric in benzene	19
Naphthyl	Dimeric in benzene	19

1. D. C. Bradley, B. N. Chakravarty and W. Wardlaw, *J. Chem. Soc.*, 1956, 2381.
2. D. C. Bradley, W. Wardlaw and A. Whitley, *J. Chem. Soc.*, 1955, 726.
3. D. C. Bradley and C. E. Holloway, *J. Chem. Soc. (A)*, 1968, 219.
4. J. G. Riess and L. G. Pfalzgraf, *Bull. Soc. Chim. Fr.*, 1968, 2401.
5. L. G. Pfalzgraf and J. G. Riess, *Bull. Soc. Chim. Fr.*, 1968, 4348.
6. J. G. Riess and L. G. Hubert-Pfalzgraf, *J. Chim. Phys. Phys. Chim. Biol.*, 1973, **70**, 646.
7. A. A. Pinkerton, D. Schwarzenbach, L. G. Hubert-Pfalzgraf and J. G. Riess, *Inorg. Chem.*, 1976, **15**, 1196.
8. I. M. Thomas, *Can. J. Chem.*, 1961, **39**, 1386.
9. R. C. Mehrotra and P. N. Kapoor, *J. Less-Common Met.*, 1964, **7**, 98.
10. P. N. Kapoor and R. C. Mehrotra, *Chem. Ind. (London)*, 1966, 1034.
11. D. C. Bradley and I. M. Thomas, *Chem. Ind. (London)*, 1958, 1231.
12. S. C. Goel and R. C. Mehrotra, *Z. Anorg. Allg. Chem.*, 1978, **440**, 281.
13. S. C. Goel, V. K. Singh and R. C. Mehrotra, *Z. Anorg. Allg. Chem.*, 1978, **447**, 253.
14. P. N. Kapoor, S. K. Mehrotra, R. C. Mehrotra, R. B. King and K. C. Nainan, *Inorg. Chim. Acta*, 1975, **12**, 273.
15. S. C. Goel, V. K. Singh and R. C. Mehrotra, *Synth. React. Inorg. Metal-Org. Chem.*, 1979, **9**, 459.
16. S. C. Goel and R. C. Mehrotra, *Synth. React. Inorg. Metal-Org. Chem.*, 1981, **11**, 35.
17. S. C. Goel, *Synth. React. Inorg. Metal-Org. Chem.*, 1983, **13**, 725.
18. S. C. Goel, S. K. Mehrotra and R. C. Mehrotra, *Synth. React. Inorg. Metal-Org. Chem.*, 1977, **7**, 519.
19. K. C. Malhotra and R. C. Martin, *J. Organomet. Chem.*, 1982, **239**, 159.

trialkylsilanols,[158] or trialkylsilyl acetates, or by that of dialkylamides with triethylsilanol.[157] Mixed alkoxo–siloxo derivatives were obtained when $M(OEt)_5$ was allowed to react with Me_3SiOAc in various molar ratios.[168]

Double alkoxides with alkali metals $A[M(OR)_6]$ (Table 8) were formed from the reaction of the pentaalkoxides with alkali metal alkoxides.[6] The double alkoxides of Mg, Ca, Sr and Ba with Nb and Ta have been synthesized in the presence of $MgCl_2$ as a catalyst.[169] Refluxing $M(OPr^i)_5$ and $A(OPr^i)_3$ (A = Al or Ga) in isopropyl alcohol afforded double isopropoxides of the type $[MA(OPr^i)_8]$ and $[MA_2(OPr^i)_{11}]$[170]. $[NbTa(OMe)_{10}]$ appears to be the first mixed transition metal alkoxide isolated.[171] NMR showed it to be in dynamic equilibrium with the symmetrical $M_2(OMe)_{10}$ dimers in solution, with close to random distribution of the three species.

(*ii*) *Structural aspects*

$Nb(OMe)_5$ appears to be the only one of these pentaalkoxides whose structure has been solved.[172] It revealed the presence of two crystallographically different dimers with distinct conformations in the unit cell. Both are centrosymmetric and consist of two approximately octahedral units with a shared edge, but with either a *cis* or a *trans* arrangement of the terminal methoxo groups on each metal atom with respect to the equatorial plane (Figure 3).

The volatility of the alkoxides increases, and their molecular complexity decreases, with increased ramification of the alkyl chains (Tables 7 and 8). Variations observed in the degree of association in different solvents were assigned to the donor capacity of the solvent rather than to differences in dielectric constants.[154,155] The mean dissociation energies of the M—OR bonds

Table 8 Double and Mixed Alkoxides

Compounds	*Comments*	*Ref.*
$[NbTa(OMe)_{10}]$	Dynamic equilibrium with $[Nb_2(OMe)_{10}]$ and $[Ta_2(OMe)_{10}]$ in solution	1
$A[M(OR)_6]$	A = Li, Na, K	2
$Mg[Nb_2(OPr^i)_{12}]$		3
$[MA(OPr^i)_8]$	A = Al or Ga	4
$[MA_2(OPr^i)_{11}]$	A = Al or Ga	
$[M(OEt)(OR)_4]$	Molecular complexity: R = Pr: 2.05 (Ta) R = Bu: 1.96 R = Pr^i: 1.93 R = pentyl: 1.15 (Ta); 1.13 (Nb)	5
$[M(OBu^t)(OR)_4]$	Molecular complexity: R = Me, 2.05 (Nb); 2.07 (Ta) R = Et: 2.02 (Nb); 2.03 (Ta) R = Pr^i: 1.15 (Nb); 1.21 (Ta)	5
$[M(OEt)_{5-x}(OSiMe_3)_x]$	$x = 1–4$	6

1. L. G. Hubert-Pfalzgraf and J. G. Riess, *Inorg. Chem.*, 1975, **14**, 2854.
2. R. C. Mehrotra, A. K. Rai, P. N. Kapoor and R. Bohra, *Inorg. Chim. Acta*, 1976, **16**, 237.
3. S. Govil, P. N. Kapoor and R. C. Mehrotra, *J. Inorg. Nucl. Chem.*, 1976, **38**, 172.
4. S. Govil, P. N. Kapoor and R. C. Mehrotra, *Inorg. Chim. Acta*, 1975, **15**, 43.
5. R. C. Mehrotra and P. N. Kapoor, *J. Less-Common Met.*, 1966, **10**, 354.
6. R. C. Mehrotra and P. N. Kapoor, *J. Indian Chem. Soc.*, 1967, **44**, 345.

were estimated to be 429 ± 10 and $440 \pm 12\,kJ\,mol^{-1}$ for the Nb and Ta alkoxides respectively.[173]

Proton NMR studies have shown that $M(OR)_5$ species (R = Me, Et, Bu, Pr)[6,174–178] have dimeric bioctahedral structures in solution, and that the bridging and the two distinct types of terminal groups rapidly exchange positions. When R = Me the terminal–terminal alkoxide exchange was found to occur at a significantly faster rate than the terminal–bridging alkoxide exchange; hence these proceed through distinct mechanisms. The activation energy (E_a) for the intramolecular scrambling of terminal and bridging alkoxo groups was found to be fairly constant (42 to 50 kJ mol^{-1}), while the activation free energy $\Delta G^\ddagger$ increased by small amounts as the degree of chain-branching in the alkoxo ligand increased.[177] The presence of a monomer–dimer equilibrium was proven in the case of the pentaisopropoxides and the enthalpy of dissociation of the dimer was evaluated.[178] The monomeric *t*-butoxides showed a rapid intramolecular exchange down to −90 °C. Alcohol interchange was slower with dimeric species than with monomeric ones.

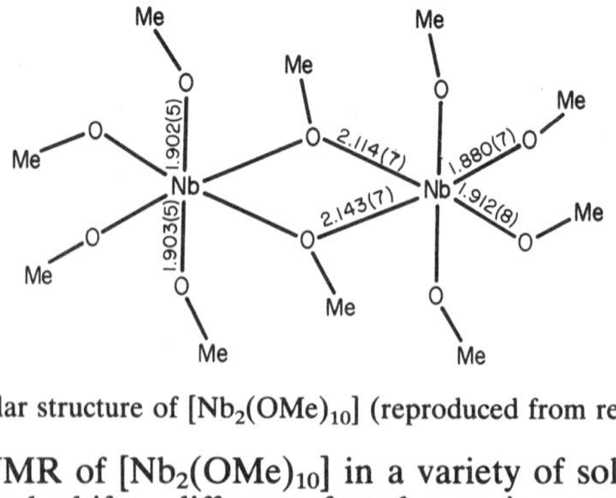

Figure 3 Molecular structure of $[Nb_2(OMe)_{10}]$ (reproduced from ref. 172 with permission)

Low temperature ^1H NMR of $[Nb_2(OMe)_{10}]$ in a variety of solvents has revealed significant variations of the chemical shifts, different for the various non-equivalent methoxo sites, particularly where bulky polar solvents are concerned; these were interpreted in terms of preferential solvation.[176]

(*iii*) *Chemical properties*

$M(OR)_5$ species do not readily form coordination complexes but prefer to autoassociate into dimers, trimers, *etc.* The coordination chemistry of $M(OMe)_5$ has been investigated by low temperature NMR, and rationalized in terms of frontier orbitals and charge distribution on both the ligand and alkoxide.[176,179–184]

The electro-attractive alkoxo groups render metal alkoxides very sensitive to nucleophilic attack.[6] They are extremely susceptible to hydrolysis, which under controlled conditions yields oxo alkoxides; they require strictly anhydrous conditions of handling.

$M(OR)_5$ reacts with β-diketones, thio-β-diketones, β-keto esters and β-hydroxy esters (Table 9). Only three of the five OR groups could be replaced by these bidentate monoanionic O͡OH ligands. Surprisingly, the reaction of tropolone (HT) with $Nb(OEt)_5$ was reported to yield $[Nb(OEt)T_4]$,[185] whose structure remains unknown.

A more versatile behavior was observed with dianionic HO͡OH ligands (Table 9): diols act as bidentates,[186,187] while α-hydroxycarboxylic acids behave as monodentate monoanionic ligands, *e.g.* in cinnamates,[188] or as bidentate dianionic ligands[189,190] in alkoxycarboxylates. In the case of catechol, ionic species were shown to exist in solution for both metals, but were isolated for tantalum only.[191] The bidentate monoanionic diphenylphosphinate ligand can replace from one to three alkoxo groups in $Ta(OMe)_5$ to yield low molecular weight polymers (Table 9).[192]

The alkoxo groups on Nb^V and Ta^V are not readily replaced by nitrogen ligands. Thus, with one equivalent of ethanolamine only the hydroxylic group was substituted.[193,194] Substitution by the amino group could however be forced by using a two- or three-fold excess of ethanolamine. Diethanolamine behaved as a tridentate trianionic ligand, while triethanolamine yielded metallatranes. The oximates and oxyaminates of Nb and Ta have been isolated.[185,195–198] The reactions of $M(OR)_5$ with both monofunctional bidentate and bifunctional tridentate Schiff's bases have been studied (Table 10).[199–203]

Insertion of unsaturated isocyanates into the M—O bond of $M(OR)_5$ has been reported[6,204,205] to give mono- to penta-substituted products (equation 4). No insertion into the Nb—O bond was observed with phenyl isothiocyanate.

$$[M(OR)_5] + x\,PhNCO \longrightarrow [M(OR)_{5-x}\{N(Ph)COOR\}_x] \quad (4)$$

$M(OMe)_5$ undergoes peroxidation[206] on treatment with H_2O_2. The use of labelled $H_2O_2^*$ indicated the fixation of the labeled oxygen between M and OMe.

$M(OEt)_5$, when allowed to react with MeCOX (X = Cl, Br)[207] in 1:1, 1:2 and 1:3 molar ratios, formed $[MX(OEt)_4]$, $[MX_2OEt)_3]$ and $[MX_3(OEt)_2(MeCOOEt)]$, respectively; molar ratios of 1:4 or higher yielded $[MX_4(OEt)(MeCOOEt)]$, while the reaction with acetyl bromide afforded $[MOBr_3(MeCOOEt)]$. $Nb(OPh)_5$, when refluxed with excess acetyl bromide, gave $[NbBr_5(MeCOOPh)]$.[183] Redistribution reactions were shown to occur between TaI_5 and $[Ta(OPh)_5]$ in MeCN, and $[TaI_4(OPh)(MeCN)]$ crystallized out. Attempts to synthesize mixed $[NbI_{5-x}(OR)_x]$ derivatives from similar reactions have so far failed.[208]

34.2.3.2 Halide alkoxides and related compounds

(i) Halide alkoxides

Direct reaction of MCl_5 with ROH yielded the mixed halide alkoxides $[MCl_x(OR)_{5-x}]$ (R = Me or Et, $x = 2$ or 3;[6] R = $MeOCH_2CH_2$, $x = 2$;[209] R = Ph, $x = 1$; R = β-naphthyl, $x = 2$; R = anthranyl, $x = 2$;[210] R = $CHF_2CF_2CH_2$, $x = 3$[6]). Stepwise alcoholysis of MX_5 in CCl_4 afforded $[MX_x(OR)_{5-x}]$ ($x = 0$–4 when X = Cl, R = Ph or naphthyl;[211] $x = 1$ or 2 for X = Br and R = Ph[212]). The reactions of $M(OR)_5$ with acetyl halides afforded easy access to $[MX_x(OEt)_{5-x}]$ (X = Cl, Br; $x = 1$ or 2[207]) and to $[TaI(OPh)_4]$.[208] Substitution of chlorine for a third or a fourth alkoxo group proceeded with coordination of the corresponding ester to the metal, while substitution by a fourth bromine resulted in the formation of oxo derivatives. Treatment of $TaCl_5$ with an excess of LiOAr (OAr = 2,6-di-*t*-butylphenoxide) in benzene led to $[TaCl_3(OAr)_2]$.[213] $TaCl_5$, when allowed to react with the chelating ether $MeOCH_2CH_2OMe$, gave $[TaCl_4(OCH_2CH_2OMe)]$, through cleavage of an alkyl–oxygen bond.[214] Neopentylidene complexes of the type $[TaX_3(CHCMe_3)(PR_3)_2]$ reacted with $[WO(OBu^t)_4]$ to yield $[TaX_3(OBu^t)_2]_2$.[215] Alkoxo fluorides were synthesized from the appropriate chloro methoxide with KF, or by the reaction of $M(OEt)_5$ with MF_5.[6]

Treatment of the halide alkoxides with $(NEt_4)Cl$ afforded crystalline $(NEt_4)[MCl_x(OR)_{6-x}]$.[216] With 1:2 molar ratios, the oxo derivative $(NEt_4)_2[NbOCl_5]$ was

Table 9 Reactions of the Alkoxides with O-containing Reagents

Reagent	*Compounds and comments*	*Ref.*
β-Diketones: *R′C(O)CH$_2$C(O)R″*		
	Monomeric species in boiling benzene	
acacH	$[M(OR)_{5-x}(acac)_x]$; $x = 1$–3; R = Et, Bu^t	1
bzacH	$[M(OR)_{5-x}(bzac)_x]$; $x = 1$–3; R = Et, Bu^t	2
R′ = Me, R″ = *p*-ClC$_6$H$_4$ ClbzacH	$[M(OR)_{5-x}(Clbzac)_x]$; $x = 1$–3; R = Et, Bu^t	3, 4
R′ = Me, R″ = *p*-FC$_6$H$_4$ FbzacH	$[M(OR)_{5-x}(Fbzac)_x]$; $x = 1$–3; R = Et, Bu^t	3, 4
R′ = CF$_3$, R″ = thenoyl	$M[(OR)_{5-x}L_x]$; $x = 1$–3; R = Et, Bu^t	3, 4
Thio-β-diketones: *R′C(S)CH$_2$C(O)R″(HL)*		
R′ = Me, Ph and R″ = Ph, Me	$M[(OR)_4L]$; R = Me, Et, Pr^i; R′ = R″ = Me, Ph;	5
β-Keto esters: *MeC(O)CH$_2$COOR′(HL)*		
R′ = Me, Et	$[Ta(OR)_{5-x}L_x]$, $x = 1$–3; R = Et, Bu^t;	6, 7
β-Hydroxy esters (HL)		
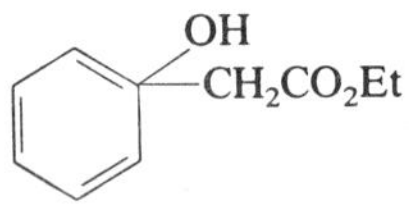 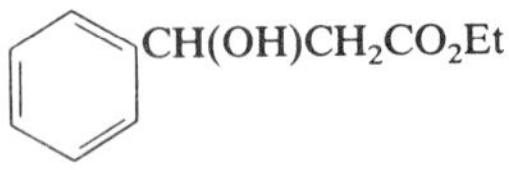 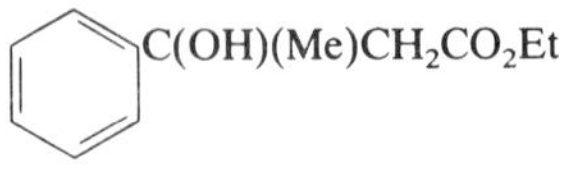	$M[(OR)_{5-x}L_x]$; $x = 1$–3; R = Et, Bu^t	8
Tropolone (HT)	$[Nb(OEt)T_4]$	9
	$[NbT_4]^+$; X-ray structure	10
Diols (H$_2$L)		
Propane-1,2-diol	$[M(OEt)_{5-2x}L_x]$; $x = 1, 2$; dimeric compounds	
Propane-1,3-diol	$[ML_2(HL)]$: monomeric and dimeric species in solution	11, 12
Butane-2,3-diol		
Hexane-1,6-diol	$[M_2L_5]$	
Pinacol		
Catechol (H$_2$L)		
	$[Ta(OR)_3L]$; $K[Ta(OR)_4L]$; $K[Ta(OR)_2L_2]$; $K_2[Ta(OR)L_3]$	2
α-Hydroxycarboxylic acids (H$_2$L)		
Lactic, mandelic or salicyclic acids	$[M(OR)_{5-2x}L_x]$; $x = 1$–2, monomeric; $[ML_2(HL)]$	13
Benzylic acid	$[M(OR)_{5-2x}L_x]$, $x = 1$–2, dimeric	14
trans-Cinnamic acid or dihydrocinnamic acid	$[M(OR)_{5-x}L_x]$; $x = 1$–3; monodentate cinnamate	15
$[OPPh_2O]^-$	$[Ta(OMe)_{5-x}(OPPh_2O)_x]$; $x = 1$–3	16

1. P. N. Kapoor and R. C. Mehrotra, *J. Less-Common Met.*, 1965, **8**, 339.
2. R. Gut, H. Buser and E. Schmid, *Helv. Chim. Acta*, 1965, **48**, 878.
3. A. K. Narula, R. R. Goyal and R. N. Kapoor, *Synth. React. Inorg. Metal-Org. Chem.*, 1983, **13**, 1.
4. A. K. Narula, B. Singh, P. N. Kapoor and R. N. Kapoor, *Synth. React. Inorg. Metal-Org. Chem.*, 1983, **13**, 887.
5. R. K. Kanjolia and V. D. Gupta, *Inorg. Nucl. Chem. Lett.*, 1980, **16**, 449.
6. R. C. Mehrotra and P. N. Kapoor, *J. Less-Common Met.*, 1964, **7**, 453.
7. R. C. Mehrotra and P. N. Kapoor, *J. Less-Common Met.*, 1964, **7**, 176.
8. A. K. Narula, B. Singh, P. N. Kapoor and R. N. Kapoor. *Transition Met. Chem.*, 1982, **7**, 325.
9. S. K. Mehrotra, R. N. Kapoor, J. Uttamchandani and A. M. Bhandri, *Inorg. Chim. Acta*, 1977, **22**, 15.
10. A. R. Davis and F. W. B. Einstein, *Inorg. Chem.*, 1975, **14**, 3030.
11. R. C. Mehrotra and P. N. Kapoor, *J. Less-Common Met.*, 1966, **10**, 237.
12. R. C. Mehrotra and P. N. Kapoor, *J. Less-Common Met.*, 1965, **8**, 419.
13. S. Prakash and R. N. Kapoor, *Inorg. Chim. Acta*, 1971, **5**, 372.
14. A. K. Narula, B. Singh, P. N. Kapoor and R. N. Kapoor, *J. Indian Chem. Soc.*, 1982, **59**, 195.
15. A. K. Narula, B. Singh, P. N. Kapoor and R. N. Kapoor, *Transition Met. Chem.*, 1983, **8**, 195.
16. H. D. Gillman, *J. Inorg. Nucl. Chem.*, 1975, **37**, 1909.

Table 10 Reactions of the Alkoxides with Schiff's Bases

Schiff's base	*Compounds and comments*	*Ref.*
H_2L 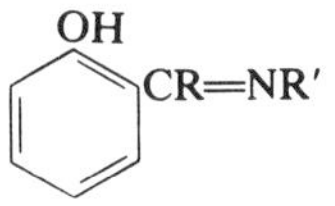 (R = H, Me; R' = CH_2CH_2OH, $CH_2CH(OH)Me$)	$[M(OPr^i)_{5-2x}L_x]$ $x = 1$: hexacoordinated monomer $x = 2$: solution equilibrium between hepta-coordinated monomer and octacoordinated dimer (OPr^ibridges)	1, 2, 3, 4
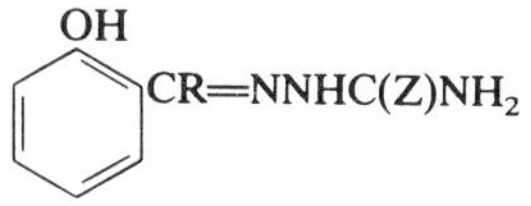 **(R = H, Me; Z = O, S)** 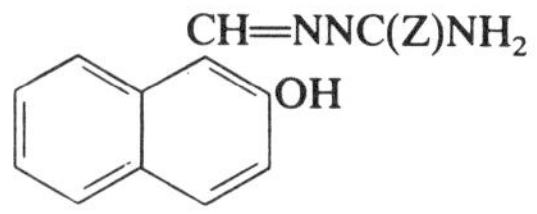 (Z = O, S)	$[ML_2(HL)]$ octacoordinate monomer	
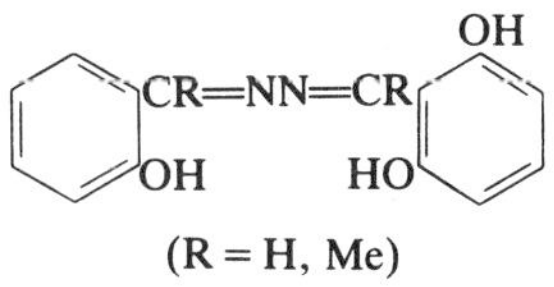 (R = H, Me) 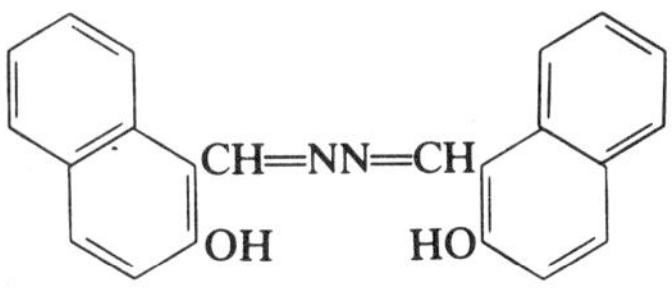	$[M(OR)_{5-2x}L_x]$, R = Pr^i, Bu^t, $x = 1, 2$ $[ML_2(HL)]$	5
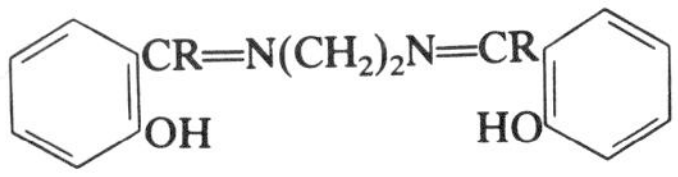 (R = Me, Ph)	$[M(OPr^i)_3L]$ heptacoordinated monomer	6
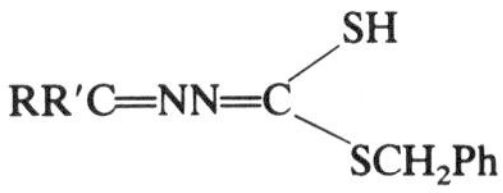 (R = PhOH, R' = H; R = $MeCOCH_2$, R' = Me)	$[Nb(OPr^i)L_2]$ solution equilibrium between monomer and dimer $[Nb_2L_5]$	7
HL 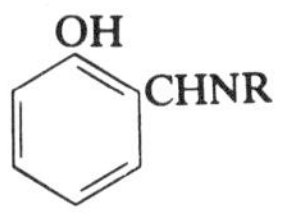 (*R* = *o*-toluidine, aniline or 2-pyridyl) 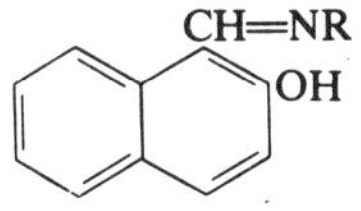 PhCN=NR (R=CH_2CH_2OH, $CH_2CH(Me)OH$, PhOH)	$[M(OPr^i)_{5x}L_x]$ $x = 1$–5 Monomeric hexacoordinated ($x = 1$) or heptacoordinated ($x = 2$) species in boiling benzene	8, 9

1. P. Prashar and J. P. Tandon, *Z. Naturforsch., Teil B*, 1971, **36,** 11.
2. P. Prashar and J. P. Tandon, *Bull. Chem. Soc. Jpn.*, 1970, **43,** 1244.
3. M. N. Mookerjee, R. V. Singh and J. P. Tandon, *Ann. Soc. Sci. Bruxelles, Ser. 1,* 1980, **94,** 207.
4. M. N. Mookerjee, R. V. Singh and J. P. Tandon, *Indian J. Chem.*, 1981, **20,** 246.
5. M. N. Mookerjee, R. V. Singh and J. P. Tandon, *Gazz. Chim. Iral.*, 1981, **111,** 109.
6. J. P. Tandon, S. R. Gupta and R. N. Prasad, *Acta Chim. Acad. Sci. Hung.*, 1975, **86,** 33 (*Chem. Abstr.*, 1975, **83,** 171 946).
7. M. N. Mookerjee, R. V. Singh and J. P. Tandon, *Synth. React. Inorg. Metal-Org. Chem.*, 1982, **12,** 621.
8. S. R. Gupta and J. P. Tandon, *Ann. Soc. Sci. Bruxelles, Ser. 1,* 1974, **88,** 251.
9. J. Uttamchandani, S. K. Mehrotra, A. M. Bhandari and R. N. Kapoor, *Transition Met. Chem.*, 1976, **1,** 249.

formed. The reaction between $(NEt_4)[MCl_6]$ and methanol yielded $(NEt_4)[MCl_4(OMe)_2]$, while $(NEt_4)[TaCl_x(OCH_2CH_2OMe)_{6-x}]$ ($x=4$, 5) was obtained with 2-methoxyethanol.[209] NaOEt and pyHCl, when allowed to react with $[NbCl_4(OEt)]$ and $[NbCl(OEt)_4]$, yielded $Na[NbCl_5(OEt)]$, $Na[NbCl_4(OEt)_2]$, $(pyH)[NbOCl_4]$ and $Na[Nb(OEt)_6]$.[6] The thermal decomposition of $[NbCl_2(OMe)_3]$[217] in an argon or oxygen atmosphere gave Nb_2O_5 and ethylene as the main products.

The ^{19}F and ^{93}Nb NMR spectra of $MF_5/M(OEt)_5$ or $M(OEt)_5/HF$ mixtures established the formation of neutral, anionic and cationic complexes.[218–221]

The structure of $[TaCl_3(OCH_2CH_2OMe)_2]$[222] showed the metal atom to have an octahedral environment, one ligand being bidentate and the other monodentate, with three different Ta—O bond lengths (1.80(3) Å for the monodentate; 1.91(3) and 2.23 Å for the chelating ligand).

(*ii*) *Adducts of the halide alkoxides and related compounds*

The halide alkoxides behave as Lewis acids; they react with O, S, N and P donor ligands to give the adducts listed in Table 11. Haloalkoxo adducts of the type $MX_3(OR)_2L$ were obtained in anhydrous alcohol.[223,224] Unexpectedly, $[NbCl_3(OR)_2(HMPA)]$ was also obtained from the reaction of $[Nb(OR)_3Cl_2]_2$ with HMPA in CH_2Cl_2 or Et_2O.[225] Mixing of $M(OR)_5$ and MX_5 in MeCN gave $[NbCl(OEt)_4(MeCN)]$,[6] $[NbBr_4(OR)(MeCN)]$,[212] where R = Me, Et, Pr, Bu, CH_2Me_3, CMe_3, $[NbI_{5-x}(OR)_x(MeCN)]$, where $x = 0$–4 and R = Me, Et, Pr, Ph, and $[TaI_4(OR)(MeCN)]$, where R = Et, Ph.[208] $[NbCl_3(OEt)_2(MeCOOEt)]$ was isolated by allowing MeCOCl to react with $[NbCl_2(OEt)_3]_2$ and further afforded $[NbCl_3(OEt)_2(py)]$ by addition of pyridine.[207] These adducts are moisture sensitive, monomeric and predominantly non-ionic. The crystal of $[NbCl_3(OPr^i)_2(HMPA)]$ is built up of the meridional isomer.[225] The same isomer was always predominant in fresh solutions, while oxo species were detected in aged solutions.

MCl_5 when allowed to react with *N*-arylsalicylaldimines [(salR)H] in methanol or ethanol[226] afforded $[MCl_2(OR)_3(salR)H]$, in which the neutral Schiff's base behaves as a unidentate ligand bound through the oxygen atom. When chlorophenoxo derivatives were allowed to react with (salR)H, $[MCl_{4-x}(OPh)_x(salR)]$ ($x = 1$–4) was obtained.[227] When MCl_5 was treated in CCl_4 with a slight excess of monoanionic HL or dianionic H_2L' Schiff's bases derived from salicylaldehyde, vanillin, *p*-phenylazoaniline or *p*-aminodiphenylamine, $[NbCl_3L_2]$, $[NbCl_3L']$ and $[NbClL'_2]$[228] were obtained, in which Nb is heptacoordinated. The mass spectra and fragmentation paths of $[MCl(OR)_2L]$ were examined.[229]

Halide–8-quinolinol adducts $[MX_5oxH)_n]$ lose HX upon heating:[143] $[MX_x(ox)_{5-x}]$ complexes ($x = 1$–4) were prepared for all combinations of M, X and x except for $[NbBr_2(ox)_3]$. $[MBr(ox)_4]$ complexes were found to be 1:1 electrolytes, while conductance measurements and IR indicated the others to be non-electrolytes, with metal coordination numbers of nine, eight, seven and six for $x = 1$ to 4. Complexation of $NbCl_5$ by 1,6-dihydroxyphenazine yielded a brown insoluble linear coordination polymer formulated as $[M_2Cl_8(C_{12}H_6N_2O_2)_n]$.[230] With bidentate hydroxy esters (HL), MCl_5 gave $[MCl_{5-x}L_x]$ ($x = 1$, 2).[231]

Complexes of the type $[MX_x(OR)_{4-x}(\beta\text{-diket})]$, where X = Cl or Br, R = Me or Et, β-diket = acac, bzac (Table 11), have been obtained from MX_5 and the β-diketones in the parent alcohol as solvent. Chlorophenoxo niobium and tantalum derivatives react with benzoin (HL) to yield $[MCl(OPh)_3L]$ and $[MCl_2(OPh)_2L]$ in which L acts as a chelating ligand,[232] while MCl_5 were found to form $[MCl_4L]$ and $[MCl_3L_2]$ with benzoin.[233] The preparation of relatively stable Nb^V complexes containing *N,N*-dialkyldithiocarbamate ligands (R_2dtc) has been reported; the structure of $[NbCl(OMe)_2(Et_2dtc)_2]$[234] shows the metal's environment to be a pentagonal bipyramid with the halide and four sulfur atoms forming a plane containing the metal; the two OMe groups occupy the apical sites.

(*iii*) *Halide carboxylates and related compounds*

Chlorocarboxylato complexes were obtained from the reaction of $TaCl_5$ with RCOOH;[235] dimeric $[MCl_4(RCOO)]_2$ compounds were isolated for R = H, Me, Et, Pr, Pr^i, Bu^t, CF_3, while $[TaCl_3(Bu^tCOO)_2]$ was the only compound obtained with two carboxylato groups. Both types of compound were proposed to have sevenfold coordination. The reaction of $NbCl_5$ with dicarboxylic acids[236] gave $[NbCl_3\{X(COO)_2\}]$, where X = $(CH_2)_n$ ($n = 0$–4), CH(OH)CH(OH)

Table 11 Adducts of the Halide Alkoxides

Compounds	*Comments*	*Ref.*
$[NbCl_3(OR)_2(phen)]$	R = Me, Et, Pr, Bu	1
$[MCl_3(OEt)_2(PPh_3)]$		2
$[MX_3(OR)_2L]$	X = Cl, Br; R = Me, Et; L = $OPPh_3$, $OAsPh_3$, HMPA, $OSPh_2$	3
$[NbCl_3(OPr^i)_2(HMPA)]$	X-Ray structure	4
$[NbCl_3(OEt)_2(MeCN)]$		5
$[NbCl_3(OEt)_2(py)]$		5
$[MCl_3(OPh)_2L]$	L = $OPCl_3$, $OPPh_3$, HMPA, pyO	6
$[MCl_3(OR)_2\{(salR')H\}]$	R = Me, Et; R′ = Ph, *p*-MeC_6H_4	7
$[MCl_4(OPh)L]$	L = py, α-pic	8
	L = $OPCl_3$, $OPPh_3$, HMPA, pyO	6
	L = α-, β- and γ-picO, HCONHMe, DMF, MeCONHMe, $MeCONMe_2$	9
	L = MePhCO, Ph_2CO	10
$[MCl(OMe)_3(acac)]$		11
$[NbX(OMe)_3(Et_2dtc)]$	X = Cl, Br; X-ray structure	12
$[MCl(OPh)_3L]$	HL = benzoin	17
$[MCl(OPh)_3(sal)]$	salH = salicylaldehyde	13
$[MX_2(OR)_2(acac)]$	X = Cl, Br; R = Me, Et	14
$[MX_2(OMe)_2(bzac)]$		14
$[MCl_2(OMe)_2L]$	HL = salicylaldehyde, acetoacetanilide, benzoylacetanilide, 2-hydroxyacetophenone, *O*-hydroxybenzophenone, benzoylacetophenone	15
$[MCl_2(OPh)_2L]$	HL = benzoin	17
$[MCl_2(OPh)_2(sal)]$		13
$[MCl_2(OR)_2L]$	R = Me, Et; HL = 2,4-pentaline, salicyaldehyde	16
$[MCl_3(OPh)(sal)]$		13
$[MCl_x(OPh)_{4-x}L]$	HL = benzoin	

1. K. M. Sharma and S. K. Anand, *J. Inst. Chem. (India)*, 1979, **51**, 208 (*Chem. Abstr.*, 1980, **92**, 156 979).
2. M. A. Glushkova, M. M. Ershova, N. A. Ovchinnikova and Yu. A. Buslaev, *Zh. Neorg. Khim.*, 1972, **17**, 147.
3. K. Behzadi and A. Thompson, *J. Less-Common Met.*, 1977, **56**, 9.
4. L. G. Hubert-Pfalzgraf, A. A. Pinkerton and J. G. Riess, *Inorg. Chem.*, 1978, **17**, 663.
5. R. C. Mehrotra and P. N. Kapoor, *J. Less-Common Met.*, 1966, **10**, 348.
6. K. C. Malhotra, U. K. Banerjee and S. C. Chaudhry, *Transition Met. Chem.*, 1982, **7**, 14.
7. K. Yamanouchi and S. Yamada, *Inorg. Chim. Acta*, 1976, **18**, 201.
8. K. C. Malhotra, U. K. Banerjee and S. C. Chaudhry, *J. Indian Chem. Soc.*, 1980, **57**, 868.
9. K. C. Malhotra, U. K. Banerjee and S. C. Chaudhry, *Natl. Acad. Sci. Lett. (India)*, 1979, **2**, 175 (*Chem. Abstr.*, 1980, **92**, 68 793).
10. K. C. Malhotra, U. K. Banerjee and S. C. Chaudhry, *Acta Cienc. Indica (Ser. Chem.)*, 1980, **6**, 236 (*Chem. Abstr.*, 1981, **95**, 53 985).
11. R. Gut, H. Buser and E. Schmid, *Helv. Chim. Acta*, 1965, **94**, 878.
12. J. W. Moncrief, D. C. Pantaleo and N. E. Smith, *Inorg. Nucl. Chem. Lett.*, 1971, **7**, 255.
13. K. C. Malhotra, U. K. Banerjee and S. C. Chaudhry, *J. Indian Chem. Soc.*, 1979, **56**, 754.
14. C. Djordjevic and V. Katovic, *J. Inorg. Nucl. Chem.*, 1963, **25**, 1099.
15. A. Syamal, D. H. Fricks, D. C. Pantaleo, P. G. King and R. C. Johnson, *J. Less-Common Met.*, 1969, **19**, 141.
16. D. Stefanovic, J. Stambolija and V. Katovic, *Org. Mass Spectrom.*, 1973, **7**, 1357.
17. K. C. Malhotra, U. K. Banerjee and S. C. Chaudhry, *Indian J. Chem., Sect. A*, 1978, **16**, 987.

and *o*-C_6H_4. ^{19}F NMR data on solutions of $[MF_5]$ and $(CF_3COO)^-$ indicated that $[TaF_5(CF_3COO)]^-$ and $[TaF_4(CF_3COO)_2]^-$ had formed[237] with monodentate trifluoroacetate; a *cis* structure was assigned to the dicarboxylato derivative.

Reactions between MCl_5 and hydrogen phosphonates (HL = $(RO)_2P(O)H$; R = Et, Pr^i, Bu, Ph) afforded $[MCl_{5-x}L_x]$ (x = 1–3), which were reported to be polymeric through —O—P—O bridges.[238] NbX_5 when allowed to react with $Na[S_2P(OR)_2]$ (R = Me, Et, cyclohexyl) gave $[Nb(OMe)_2Cl\{S_2P(OR)_2\}_2]$;[239] recrystallization from the appropropriate alcohol afforded complexes having other alkoxo groups. The reaction of dithiophosphates with TaX_5 did not afford the analogous tantalum derivatives.

34.2.3.3 *Thiolato and dithiocarbamato derivatives*

Dithioalkoxo- and dithiophenoxo-niobium(V) trichlorides $[MCl_3(SR)_2]$ were obtained from $NbCl_5$ and thiols and benzenethiol respectively.[240] The synthesis of tris(benzenedithiolato) complexes was accomplished by reaction of the appropriate metal amide with a mixture of the dithiol and its sodium salt in THF. The structures of $(AsPh_4)[M(S_2C_6H_4)_3]$[241–243] showed the metal to be surrounded by six sulfur atoms in a close to trigonal prismatic coordination, with the dithiolene ligands radiating from the metal in the usual 'paddle-wheel' arrangement (Figure 4). The coordination sphere of $[Nb(SCH_2CH_2S)_3]^-$ is midway between trigonal prismatic and octahedral.[243b]

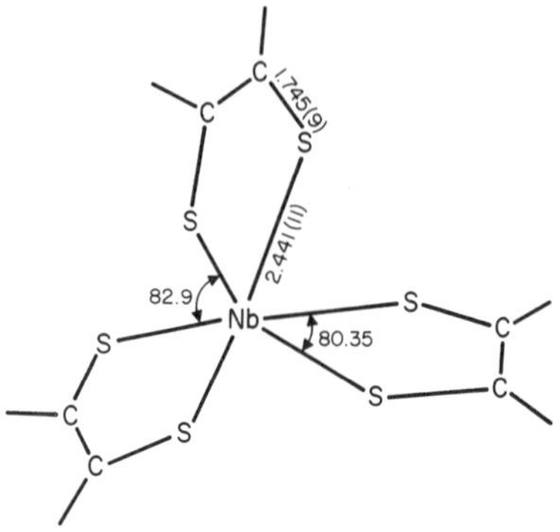

Figure 4 Structure of $[Nb(S_2C_6H_4)_3]^-$ (reproduced from ref. 243 with permission)

The trinuclear $[Ta(Cp)_2(\mu\text{-}SMe)_2Pt(\mu\text{-}SMe)_2Ta(Cp)_2]^{2-}$ anion was obtained by reaction of $[Ta(SMe)_2(Cp)_2]$ and $[PtCl_2(PhCN)_2]$.[244] The X-ray diffraction analysis[245] showed the arrangement around the Pt atom to be an almost regular tetrahedron, a situation hitherto unknown for Pt^{II}. The compound was therefore assumed to formally contain Ta^V and Pt^O, although the short Pt—Ta contacts of 2.788(7) and 2.809(8) Å are indicative of metal–metal bonds.

$[Ta(R_2dtc)_5]$, where $(R_2dtc) = (S_2CNR_2)^-$, were reported to result from insertion of CS_2 into $Ta(NR_2)_5$; in the case of niobium, the final product was $[Nb(R_2dtc)_4]$.[246] $[Ta(R_2dtc)_5]$[247] and $[MX_x(R_2dtc)_{5-x}]$ ($x = 1$–3) have been obtained in the reaction between MCl_5 and $Na(R_2dtc)$ in oxygen free solvents, while $[MX(OR)_2(R_2dtc)_2]$ was formed in alcohol.[239] $[Nb_2Br_3(R_2dtc)_5]$ was isolated from $NbBr_5$ in benzene.[250] Niobium, in acidic media in the presence of ClO_4^- ions, was found to form $[Nb(R_2dtc)_4](ClO_4)$ with pyrrolidine- and hexamethylene-dithiocarbamates.[251] $[Nb(R_2dtc)Me(Cp)_2](R_2dtc)$ was obtained when $[NbMe_2(Cp)_2]$ was allowed to react with $R_2NC(S)SC(S)NR_2$.[252]

In all these air sensitive complexes the dithiocarbamate ligand is bidentate through the sulfur atoms. The $[MX_x(R_2dtc)_{5-x}]$ adducts were found to be ionic, hence must be formulated as $[M(R_2dtc)_4]X$, $[\{Nb(R_2dtc)_3\}_2Cl_3]Cl$ and $[M(R_2dtc)_4][MX_6]$.[249] The structures of $[Ta(S_2CNMe_2)_4]^+$ in the Cl^- [253] and $[TaCl_6]^-$ [254] salts showed the eight-coordinate $[Ta(R_2dtc)_4]^+$ cation to be at the center of a D_{2d} dodecahedron, the bidentate ligand spanning the AB edges. Surprisingly, this cation is stereochemically rigid on the NMR time scale and appears to be the first eight-coordinate chelate for which this is observed.

34.2.3.4 Amido derivatives

Several reviews have dealt with the chemistry of transition metal–nitrogen derivatives,[255–257] including compounds with metal–nitrogen multiple bonds,[258,259] especially those encountered in high oxidation states.

Solvolysis of MCl_5 by ammonia was studied as early as 1924, but the products were not conclusively characterized.[260,261] Halo monoalkylamides[262] and dialkyl amides were prepared. The monoalkylamides undergo facile α-H abstraction reactions to nitrenes, especially when labile substitutents are present (Section 34.2.3.5); dialkylamides (Table 12) have been extensively studied by Bradley.[263–265]

(i) Synthesis and structure

Homoleptic Nb^V and Ta^V dialkylamides have been obtained (equation 5).[263,264] Extensive reduction, increasing with the length of the alkyl chain, was observed with niobium, the predominant products then being $[Nb(NR_2)_4]$ (R = Et, Pr^n, Bu^n; Section 34.3.5.4). Oxidation

Table 12 Niobium(V) and Tantalum(V) Dialkylamides

Compound	*Comments*	*Ref.*
$[MF_{5-x}(NEt_2)_x]$ ($x = 1, 2$)	Insoluble	1
$[MF_{5-x}(NEt_2)_xL]$	Isolated for L = py, γ-pic, unstable for L = MeCN, OEt_2	1
$[TaCl_3(NMe_2)_2]_2$	Impure, contaminated by $[TaCl_2(NMe_2)_3]_2$	4
$[MX_3(NR_2)_2(NHR_2)]$ (X = Cl, Br; R = Me, Et)	X-Ray structure for $[TaCl_3(NMe_2)_2(NHMe_2)]$	2, 3, 4
$[TaCl_2(NMe_2)_3]_2$	X-Ray structure, unstable towards redistribution	4
$[TaCl_2(NMe_2)_3L]$	Isolated for L = Me_2NH, observed in solution for L = py, γ-pic	4
$[M(NMe_2)_5]$	X-Ray structure, M = Nb tetragonal pyramid; PES M = Ta: isomorphous with the niobium analog	5, 6, 7
$[Ta(NR_2)_5]$ (R = Et, Pr^n)	R = Et, trigonal bipyramid	5, 8
$[Ta(NMe_2)_5(NHMe_2)]$	Observed in solution	9
$[M(NMeBu)_5]$		5
$[Nb(NC_5H_{10})_5]$	X-Ray structure, tetragonal pyramid	5, 6
$[Nb(NMe_2)_{5-x}(NEt_2)_x]$ ($x = 2, 3$)		5
$[Ta(NMe_2)_2Me_3(Cp')]$	Unstable, evolution to nitrene at room temperature	10
$[Ta(NMe_2)_2Me_3]$		11

1. J. C. Fuggle, D. W. A. Sharp and J. M. Winfield, *J. Chem. Soc., Dalton Trans.*, 1972, 1766.
2. P. J. H. Carnell and G. W. A. Fowles, *J. Less-Common Met.*, 1962, **4,** 40; G. W. A. Fowles and C. M. Pleass, *J. Chem. Soc.*, 1957, 2078.
3. P. J. H. Carnell and G. W. A. Fowles, *J. Chem. Soc. (A)*, 1959, 4113.
4. M. H. Chisholm, J. C. Huffman and L. S. Tan, *Inorg. Chem.*, 1981, **20,** 1859.
5. D. C. Bradley and I. M. Thomas, *Proc. Chem. Soc.*, 1959, 225; *Can. J. Chem.*, 1962, **40,** 449, 1355; D. C. Bradley and M. H. Gitlitz, *J. Chem. Soc. (A)*, 1969, 980.
6. C. E. Heath and M. B. Hursthouse, *J. Chem. Soc., Chem. Commun.*, 1971, 143.
7. M. H. Chisholm, A. H. Cowley and M. Lattman, *J. Am. Chem. Soc.*, 1980, **102,** 46.
8. R. J. Smallwood, Ph.D. Thesis, London, 1975.
9. M. H. Chisholm and M. W. Extine, *J. Am. Chem. Soc.*, 1977, **99,** 792.
10. J. M. Mayer, C. J. Curtis and J. E. Bercaw, *J. Am. Chem. Soc.*, 1983, **105,** 2651.
11. C. Santini-Scampucci and G. Wilkinson, *J. Chem. Soc., Dalton Trans.*, 1976, 807.

state +V was retained for tantalum, and either the amides $[Ta(NR_2)_5]$ or the mixed nitrene amides $[Ta(NR)(NR_2)_3]$ (R = Et, Pr^n; Section 34.2.3.5) were isolated.

$$MCl_5 + 5LiNR_2 \longrightarrow [M(NR_2)_5] + 5LiCl \quad R = Me, Et, Pr^n, Bu^n \tag{5}$$

The coordination polyhedron of the dialkylamides is largely determined by the steric requirements of the NR_2 groups and by their ability to form σ and π bonds.[256] Even the NMe_2 group is sterically demanding, and the homoleptic Nb and Ta amides are all monomers. Dimeric halogen-bridged structures were observed for mixed dialkylamido compounds, as, for example, in $[TaCl_2(NMe_2)_3]_2$. For pentacoordinated d^0 compounds, a trigonal bipyramidal geometry is expected to be the most stable.[266] $[Ta(NEt_2)_5]$ indeed exhibits this geometry,[256] but $[Nb(NMe_2)_5]$ (Figure 5) and $[Nb(NC_5H_{10})_5]$ approach a square pyramidal geometry.[267] Although all nitrogen atoms have approximately planar geometry, only the axial Nb—N bond appears to have some π character on the basis of its shorter length (single Nb—N bond lengths are expected to be in the 2.04 to 2.08 Å range); the planarity found for nitrogen has been interpreted to result from the rather tight packing of the ligands around the metal. $[M(NMe_2)_5]$ species were found by photoelectron spectrometry to be isomorphous in the vapor phase, but their exact geometry has not been established.[268]

Mixed halodialkylamido derivatives are accessible by several methods.[262,269] The chlorodimethylamidotantalum(V) derivatives have received special attention from Chisholm and coworkers.[270] Reaction of the pentachloride with an excess of dimethylamine provided $[TaCl_3(NHMe_2)]$ as the major product (85%), accompanied by $[TaCl_2(NMe_2)_3(NHMe_2)]$ (6%) and by a dinuclear oxo derivative $[\{TaCl_2(NMe_2)_2(NHMe_2)\}_2O]$ (2%), whose origin is undetermined. A common feature of these products is their fairly strong Ta—N bond. Metathesis reactions between $[Ta(NMe_2)_5]$ and Me_3SiCl led to close to random distributions of the halodialkylamido complexes, from which $[TaCl_2(NMe_2)_3]_2$ was easily isolated from pentane

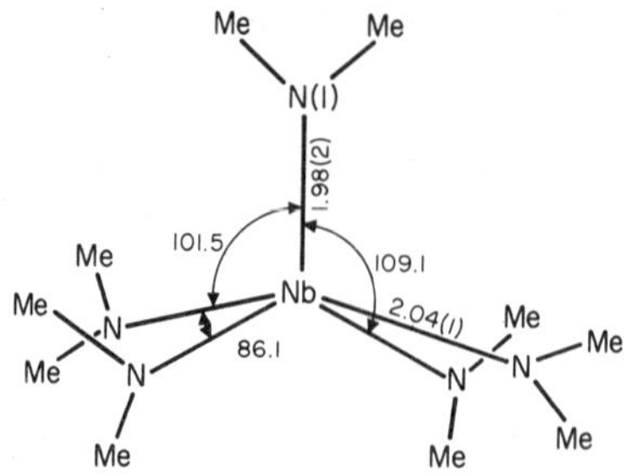

Figure 5 Coordination polyhedron of $[Nb(NMe_2)_5]$ (reproduced from ref. 267 with permission)

and a mixture of $[TaCl_2(NMe_2)_3]_2$ and $[TaCl_3(NMe_2)_2]_2$ from toluene solutions. The dimeric $[TaCl_2(NMe_2)_3]_2$ is readily cleaved by nitrogen donors to form $[TaCl_2(NMe_2)_3L]$ adducts ($L = Me_2NH$, γ-pic). Attempts to isolate $[TaCl(NMe_2)_4]$ were unsuccessful.

The structures of $[TaCl_3(NMe_2)_2(NHMe_2)]$ and $[TaCl_2(NMe_2)_3]_2$ have been solved.[270] Tantalum adopts a distorted octahedral geometry in all structures. The central $TaCl_3N_3$ core of the trichloro adduct has a meridional arrangement of the chlorine atoms with *cis* dimethylamido ligands. The two Ta—NMe_2 bonds are comparable in length (1.958 Å av.), and much shorter than the Ta—$NHMe_2$ distance (2.370(6) Å). The centrosymmetric dimeric structure of $[TaCl_2(NMe_2)_3]_2$ displays two *fac*-TaN_2Cl_3 arrangements with bridging chlorines (Ta—Cl_b: 2.586(1) and 2.635(2) Å). The amido $TaNC_2$ units are all planar. This, together with the short Ta—N distances and the close to 90° dihedral planes between these units, provides evidence for strong Ta—N d_π–p_π bonding. The Me_2N ligand in each of these molecules was considered as a four-electron donor ($\sigma^2 + \pi^2$), leading to 18- or 16-electron valence shells in $[TaCl_2(NMe_2)_3]_2$ and $[TaCl_3(NMe_2)_2(NHMe_2)]$, respectively. A *fac*-$TaN_3$ arrangement is a common feature of all these structures, as ligands having a strong *trans* influence tend to prefer a mutual *cis* configuration, irrespective of their π acceptor or π donor properties.[271] A *cis* arrangement of the dimethylamido groups was therefore assumed for $[TaCl_3(NMe_2)_2]_2$ and $[TaCl_2(NMe_2)_3(NHMe_2)]$.

Pentavalent silylamido derivatives are limited to Ta, as reduction occurs with Nb.[272,273] The bulkiness of the $[N(SiMe_3)_2]^-$ ligand precluded complete chlorine substitution as well as formation of dimers. $[TaCl_3\{N(SiMe_2)_2\}_2]$ has a trigonal bipyramidal structure with short equatorial Ta—N bonds (1.930 Å av.).

The first hydrazido(−1) compounds, $[M_2Cl_8(HNNRR')_2]$ ($R = Ph$, $R' = H$ or Me) and $[TaCl_3(HNNHPh)_2(H_2NNHPh)]$, were obtained from the reaction of the pentachlorides with the hydrazines.[274] Spectroscopic data on $[TaCl_3(HNNHPh)_2(H_2NNHPh)]$ are consistent with an equilibrium between neutral and ionic structures, as also observed for molybdenum hydrazido derivatives.

(*ii*) *Reactivity*

Niobium and tantalum dialkylamides can easily undergo substitution or insertion reactions, as well as C—H activation. Substitution reactions have been used to obtain dialkylamides[262] or dithiolenes.[275] Aminolysis reactions, for instance, were assumed to occur *via* a simple associative mechanism (equation 6).[276]

$$M{-}NR_2 + NHR'_2 \rightleftharpoons \left[\begin{array}{c} R'_2N{-}H \\ \vdots \quad \vdots \\ M{-}NR_2 \end{array}\right] \rightleftharpoons M{-}NR'_2 + HNR_2 \qquad (6)$$

The M—N bond is susceptible to insertion reactions, the Ta—N being even more reactive than the Ta—C bond; thus in $[Ta(NMe_2)_2Me_3]$, insertion of CS_2 was found to occur only in the former bond.[277] Insertion of CO_2, CS_2 and COS in $[M(NR_2)_5]$ has been observed.[276,278] In the case of niobium, the insertion of CS_2 was accompanied by reduction, giving $[NB(S_2CNR_2)_4]$ (Section 34.3.5.3.i).[279] The *N,N*-dialkylcarbamato ligand, O_2CNR_2, is much less prone to oxidation and the pentavalent state was retained for both niobium and tantalum. The synthesis of monothiocarbamates is limited to $[Ta(OSCNMe_2)_5]$.[278]

The mechanism of the CO_2 exchange and insertion reactions has been extensively studied on $[M(NMe_2)_5]$ using kinetic and labeling experiments. The insertion is catalyzed by amines (equations 7 and 8). It is rapid and quantitative, and proceeds through $[M(NMe_2)_{5-x}L_x]$ intermediates, the Ta derivatives being more stable than those of Nb.[272] None of the mixed species is favored. $[M(NMe_2)_2(O_2CNMe_2)_3]$ formed for both metals, but could be isolated only for tantalum. The formation of $[M(NMe_2)_3(O_2CNMe_2)_2]$ was not observed.

$$HNMe_2 + CO_2 \rightleftharpoons HO_2CNMe_2 \quad (7)$$

$$MNMe_2 + LH \rightleftharpoons ML + HNMe_2 \qquad L = O_2CNMe_2 \quad (8)$$

The coordination spheres of $[M(O_2CNMe_2)_5]$ and $[Ta(NMe_2)_2(O_2CNMe_2)_3]$ contain both mono- and bi-dentate dialkylcarbamato ligands. The magnitudes of the $\Delta\nu(^{12}C-^{3}C)$ and $\Delta\nu(^{16}O-^{18}O)$ ($\Delta^{18}O$) shifts in the IR were rationalized by assuming that $\nu(O_2CN)$ is mainly C═N in character for a bidentate ligand ($\Delta^{18}O < 5\ cm^{-1}$), while a significant contribution from the C═O moiety is observed for a monodentate ligand ($\Delta^{18}O$ *ca.* $20\ cm^{-1}$).

X-Ray diffraction data for $[Nb(O_2CNMe_2)_5]$[278] show that niobium is eight coordinated, with three bidentate and two monodentate *cis* $(Me_2NCO_2)^-$ ligands; The NbO_8 moiety does not correspond to any idealized MX_8 polyhedron. The Nb—O bond distances are much shorter for the monodentate (av. Nb—O = 1.91(2) Å) than for the bidentate ligands, the latter being asymmetrically bonded (av. Nb—O = 2.08(4) and 2.19(2) Å). In $[Ta(NMe_2)_2(O_2CNMe_2)_3]$[280] the metal's coordination polyhedron can be described as a distorted pentagonal bipyramid with two bidentate carbamato ligands in the equatorial plane and one monodentate ligand in an axial site. This geometry is particularly propitious to Ta—N π bonding (Ta—N: 1.970 Å av.), to attain a saturated valence shell configuration, which accounts for the kinetic and thermodynamic stability of the compound. It should be noted that the bite of the carbamato ligand (2.15(5) Å) compares with that of NO_3^- in $[Ti(NO_3)_4]$ and is considerably smaller than that of the dithiocarbamato ligand (2.84(3) Å), which may explain why oxidation state +V is maintained for niobium.

In solution these compounds are highly fluxional. Whereas the methyl groups are all stereochemically inequivalent in solid $[Nb(O_2CNMe_2)_5]$, they appear equivalent in solution, even at −80 °C, on the ^{13}C NMR time scale. The mono- and bi-dentate carbamato substituents can, however, be distinguished in solution by IR. A facile exchange between coordinated and uncoordinated oxygens of the monodentate carbamato ligand was also observed in $[Ta(NMe_2)_2(O_2CNMe_2)_3]$. Hindered rotation about the Ta—N linkage was detected at −95 °C by ^{13}C NMR.[280]

Although cyclometallation reactions have seldom been observed with d^0 metal compounds, intramolecular metallation involving early transition metal dialkylamides with formation of azametallacyclopropane, which occurs by β H abstraction, has been reported.[281]

On distillation, $[Ta(NEt_2)_5]$ undergoes substantial decomposition (equation 9) to give ethyliminoethyl(*C,N*)tris(diethylamido)tantalum (**1**) and ethylimidotris(dimethylamido)tantalum (**2**).[282,283] The IR and NMR data of the methine carbon of (**1**) [^{13}C: δ = 64 p.p.m., J(C–H) = 147 Hz (d)] indicate a notable C—N double bond character. Compound (**1**) reacts easily with methyl isocyanate, according to equation (10). The metallacycle (**1**) was shown to be an intermediate in the formation, above 100 °C, of the ethylimido complex (**2**) and ethylene. The reaction is first order with respect to the azametallacycle; the activation parameters were evaluated. The formation of ethylimino (*C,N*) groups (**3**) appears to be a general feature in the thermolysis of 'overcrowded' early transition metal diethylamido derivatives. It has also been observed for niobium, together with $[Nb(NEt_2)_4]$ as the major product. A similar three-membered metallacycle could not be isolated in the case of $[Ta(NMe_2)_5]$, but was presumed to form as an intermediate in the facile and reversible incorporation of deuterium from $^2HNMe_2$ into the methyl groups (equation 11). The intermediate azametallacyclopropane (**4**) can apparently be trapped by terminal alkenes, resulting in the catalytic aminomethylation of the alkenes (Scheme 1). $[Ta(NMe_2)_5]$ was recovered unchanged after the reaction, while the niobium analog was reduced to an as yet unidentified active species with evolution of methane.

Hydrogen abstraction reactions from the $(NR_2)^-$ ligand have been observed even at room temperature (equation 12).[284] With diisopropylamidolithium, the reaction is more complex; the only product isolated (≈10%) was the unusual methylene- and hydride-bridged compound (**7**; equation 13). Labeling experiments gave evidence that the hydride originates from the

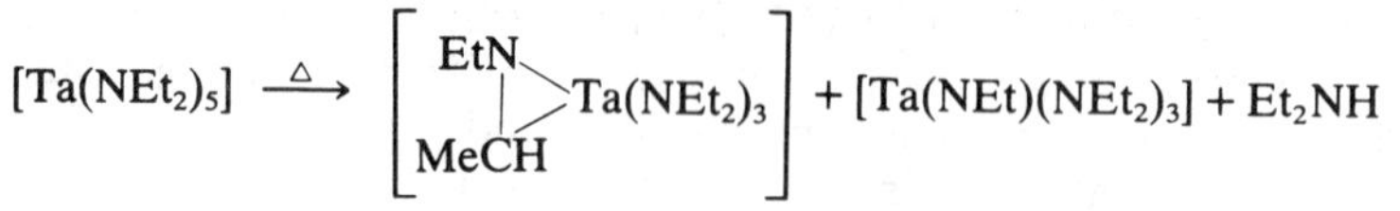

(9)

(1) 22% **(2)** 31%

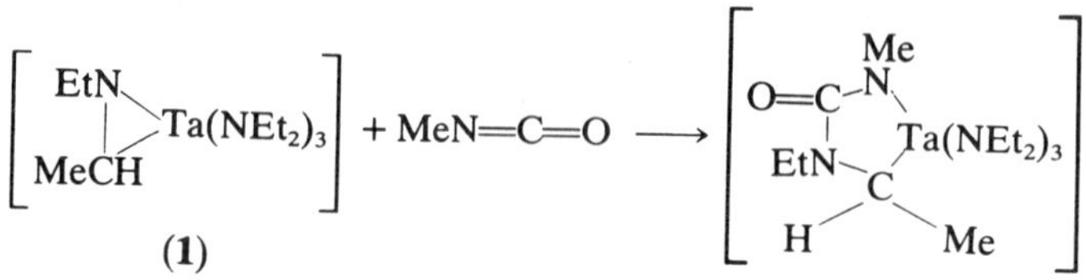

(10)

(1)

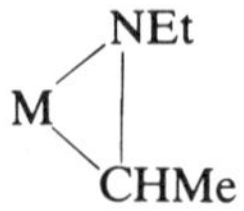

(3)

$$(CD_3)_2ND \xrightleftharpoons{[M(NMe_2)_5]\,140\text{–}180\,^{\circ}C} (Me)(CH_2D)NH \qquad (11)$$

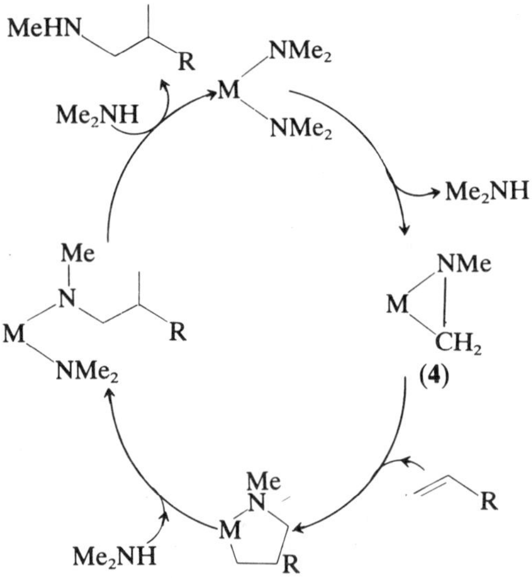

Scheme 1

diisopropylamide. It was suggested that the formation of a Ta—N bond is prevented by the steric bulk of the NPr^i group; hydride donation then becomes competitive. These spontaneous or thermally induced reactions may all be viewed as unimolecular β hydrogen abstraction or elimination reactions. With monoalkylamides, α hydrogen transfer processes appear to be more facile than β hydrogen migrations, and open a route to nitrenes (equation 14). The rate of the hydrogen transfer process depends on the nature of the group (OR, NR_2, CR_3) bonded to tantalum; caution should be used in extrapolating the observed rate order, since it is based on compounds with the $TaMe_3(Cp')$ moiety.

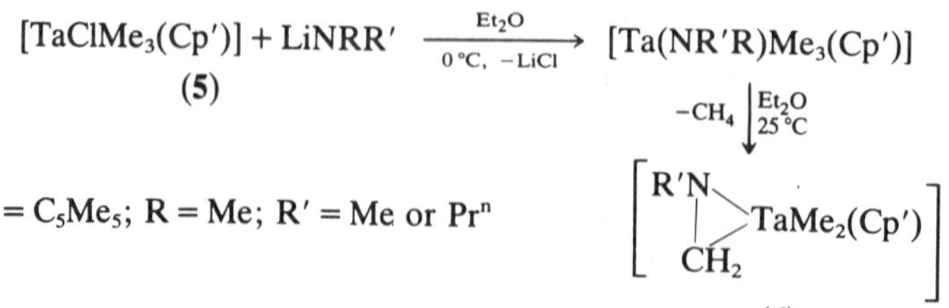

$$[TaClMe_3(Cp')] + LiNRR' \xrightarrow[0\,^{\circ}C,\ -LiCl]{Et_2O} [Ta(NR'R)Me_3(Cp')] \qquad (12)$$

(5)

$\downarrow$ $-CH_4$, Et_2O, 25 °C

η^5-Cp' = C_5Me_5; R = Me; R' = Me or Pr^n

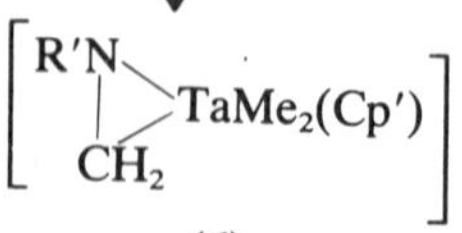

(6)

$$[TaClMe_3Cp')] \xrightarrow[To]{LiN(CHMe_2)_2} [(Cp')Me_2Ta(\mu\text{-}CH_2)(\mu\text{-}H)_2TaMe_2(Cp')] \quad (13)$$

(5) (7)

$$[TaClMe_3(Cp')] + LiNHR \xrightarrow{Et_2O} [Ta(NHR)Me_3(Cp')] \longrightarrow [Ta(NR)Me_2(Cp')] + CH_4 \quad (14)$$

(5)

$R = Me, Bu^t, CH_2CMe_3$

34.2.3.5 Nitrenes

Coordinated nitrenes RN^{2-} have been thoroughly investigated in recent years[285,286] owing to their formal similarity to carbenes and to their potential importance, *e.g.* as catalysts in the Haber ammonia process. Theoretical studies have suggested that imido ligands may also prove effective in promoting alkene metathesis.[287]

(i) Synthesis and structure

The first tantalum nitrene was obtained in 1959 by thermolysis of $[Ta(NEt_2)]_5$.[288] This class of compounds is presently accessible by several routes, including hydrogen abstraction from the mono- or di-alkylamides, reaction of metallacarbenes with organic imines, oxidation of low valent species by organic azides, or reductive coupling of nitriles (Table 13). The tantalum derivatives are usually stabler than those of niobium.

As stated above, α H abstraction is favored with respect to β H transfer, and the unstable monoalkylamides are converted to nitrenes. $[Ta(NBu^t)(NMe_2)_3]$ (**8**) has been prepared by two

Table 13 Niobium(V) and Tantalum(V) Nitrene Derivatives (without Cyclopentadienyl Ligands)

Compounds	*Comments*	*Ref.*
$[M(NR)X_3L_2]$ (R = Me, Bu^t, Ph; X = Cl, Br; L = THF R = Ph, $NSiMe_3$; L = THF, PR_3)	^{15}N NMR; X-Ray structure for *mer*-$[Ta(NPh)Cl_3(PEt_3)(THF)]$	1, 2, 3
$[Ta(NMe)Cl_3]$	Analytical data only	4
$[M(NPh)Cl_3(SMe_2)]_2$	X-Ray structure, M=M and N=N; metathesis	5
$[Ta(NPh)Np_3(THF)]$	Stable with respect to neopentylidene formation	1
$[Ta(NBu^t)Cl_2X(NH_2Bu^t)]_2$ (X = NHBu^t, OEt)	X-Ray structures; for related compounds see Scheme 2	6
$[Ta(NSiMe_3)Cl(CHCMe_3)(PMe_3)_2]$		1
$[M(NR)(NR'_2)_3]$ (M = Nb, Ta; R = Bu^t; R = Me; R = Bu; R'_2 = BuMe; M = Ta; R = R' = Et, Pr)	X-Ray structure for $[Ta(NBu^t)(NMe_2)_3]$, linear Ta=NC required by symmetry; alternate synthesis	7, 8, 9
$[Ta(NBu^t)(NMe_2)(OCPh_2NMe_2)_2]$	No further insertion of Ph_2CO	7
$[M(NBu^t)(O_2CNMe_2)_3]$		7
$[M(NBu^t)(R_2dtc)_3]$ M = Ta; R = Bu^t; R' = Me, Et	X-Ray structure for $[Nb(Np\text{-tolyl})(Et_2dtc)_3]$; M = Nb, various R and R', see equation (17)	10
$[Nb(NPh)_2Cl(dmpe)]$		5

1. S. M. Rocklage and R. R. Shrock, *J. Am. Chem. Soc.*, 1980, **102,** 7808; 1982, **104,** 3077.
2. M. R. Churchill and H. J. Wasserman, *Inorg. Chem.*, 1982, **21,** 223.
3. J. R. Dilworth, S. J. Harrison, R. A. Henderson and D. R. M. Walton, *J. Chem. Soc., Chem. Commun.*, 1984, 176.
4. P. J. H. Carnell and G. W. A. Fowles, *J. Less-Common Met.*, 1962, **4,** 40.
5. L. G. Hubert-Pfalzgraf and G. Aharonian, *Inorg. Chim. Acta*, 1985, **100,** L22; F. A. Cotton, S. A. Duraj and W. J. Roth, *J. Am. Chem. Soc.*, 1984, **106,** 4749.
6. T. C. Jones, A. J. Nielson and C. E. F. Rickard, *J. Chem. Soc., Chem. Commun.*, 1984, 205.
7. W. A. Nugent and R. L. Harlow, *J. Chem. Soc., Chem. Commun.*, 1978, 579; W. A. Nugent, *Inorg. Chem.*, 1983, **22,** 965.
8. D. C. Bradley and I. M. Thomas, *Proc. Chem. Soc.*, 1959, 225; *Can. J. Chem.*, 1962, **40,** 449, 1355.
9. D. C. Bradley and M. H. Gitlitz, *J. Chem. Soc. (A)*, 1969, 980.
10. L. S. Tan, G. V. Goeden and B. L. Haymore, *Inorg. Chem.*, 1983, **22,** 1744.

routes (equations 15 and 16).[289] Base-promoted deprotonation reactions of primary amido groups have also been used (Scheme 2).[290]

$$[Ta(NMe_2)_5] + Bu^tNH_2 \longrightarrow [Ta(NBu^t)(NMe_2)_3] + 2Me_2NH \tag{15}$$
$$(\mathbf{8})$$

$$TaCl_5 + LiNHBu^t + 4LiNMe_2 \longrightarrow [Ta(NBu^t)(NMe_2)_3] + Me_2NH + 5LiCl \tag{16}$$
$$(\mathbf{8})$$

$$M_2Cl_{10} \xrightarrow{C_6H_6} \begin{cases} \xrightarrow[3PMe_3]{Me_3SiNHBu^t} [M(NBu^t)Cl_3(PMe_3)_2] + PMe_3 \cdot HCl \\[2em] \xrightarrow[]{2Me_3SiNHBu^t} [M(NBu^t)Cl_3(NH_2Bu^t)]_n + 2Me_3SiCl \\[2em] \xrightarrow[]{8Bu^tNH_2} [M(NBu^t)Cl_2(NHBu^t)(NH_2Bu^t)]_n \end{cases}$$

$$(\mathbf{9}) \quad M = Ta$$

Scheme 2

The structure of $[TaCl(NBu^t)(NHBu^t)(NH_2Bu^t)(\mu - Cl)]_2$ (**9**) was shown to be dimeric with nearly symmetrical, long chloro bridges (Figure 6). Compound (**9**) is remarkable as imido, amido and amino ligands are present simultaneously on the same metal. Also of interest is the very short Ta—N(amido) bond and the large corresponding TaNC bond angle. The lengthening of the Ta—Cl bond *trans* to it has been considered as a confirmation of significant π contribution to the Ta—N(amido) bond. A close intermolecular approach to the metal (Ta $\cdots$ H—N: 2.0 Å), resulting in pseudo seven-coordination, was observed for amido protons; a two-electron three-centered bond has been suggested.

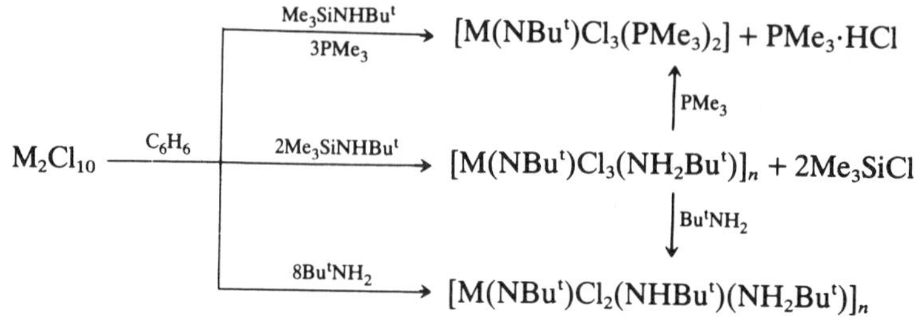

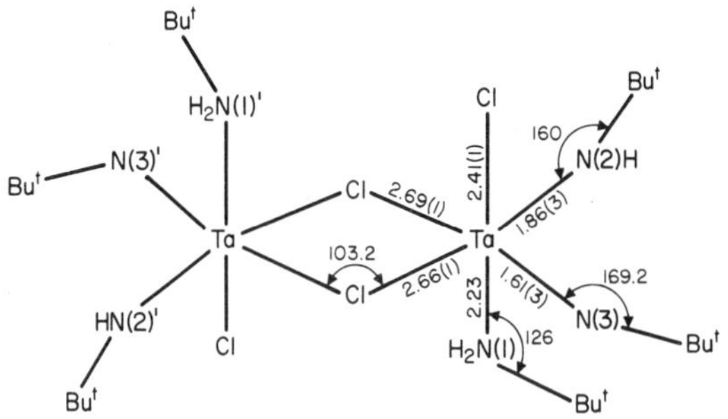

Figure 6 Molecular structure of $[TaCl(NBu^t)(NHBu^t)(NH_2Bu^t)]_2^-$ (**9**) (reproduced from ref. 290 with permission)

More complicated sequences involving successive substitutions of the chlorine on the metal have also been used successfully to obtain nitrenes (equation 17).[291] A comparable approach led to the first Nb hydrazido(−2) compounds: $[Nb(NNR_2)(R_2'dtc)_3]$ ($R_2 = Me_2$; PhMe; $R' = Me$ or Et) and $[Nb\{NN(CH_2)_5\}(Et_2dtc)_2]$.

$$NbCl_5 + 3RNH_2 + 3Me_3Si(R_2'dtc) \longrightarrow [Nb(NR)(R_2'dtc)_3] + 3Me_3SiCl + 2RNH_3Cl$$

$$R = Me, Pr^n, Bu^i, Bu^t, Ph, p\text{-tolyl}, p\text{-NC}_6H_4OMe, \quad R' = Et; \tag{17}$$

$$R = Bu^t, p\text{-tolyl}, p\text{-C}_6H_4OMe, \quad R' = Me$$

Metatheses between neopentylidene complexes and organic imines are an attractive simple general and selective entry to alkylimido compounds.[292] The experimental results can be rationalized by the formation of an intermediate with an MC_2N ring (equation 18). They also provide a straightforward route to μ-dinitrogen compounds, using $PhCH=NN=CHPh$ (Section 34.2.3.6). Alkenylimides were prepared by nucleophilic attack of nitriles on the α

carbon of alkylidenes (equation 19). A similar reaction with acetonitrile was used to trap the unstable [TaCl(CHCMe$_3$)(CH$_2$CMe$_3$)$_2$] at −78 °C.[293]

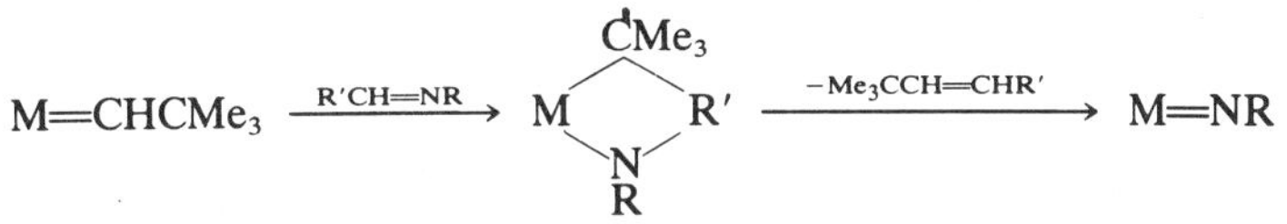

$$M{=}CHCMe_3 \xrightarrow{R'CH=NR} M \overset{\overset{\displaystyle \overset{\,}{C}Me_3}{\diagdown}}{\underset{\underset{\displaystyle R}{N}}{\diagup}} R' \xrightarrow{-Me_3CCH=CHR'} M{=}NR \qquad (18)$$

$$[M({=}CHCMe_3)(CH_2CMe_3)_3] \xrightarrow{RCN} \left[(Me_3CCH_2)_3M{=}N \underset{R}{\diagdown} C{=}CHCMe_3 \right] \qquad (19)$$
$$M = Ta;\ R = Me,\ Ph$$
$$M = Nb;\ R = Me$$

Oxidative addition of organic azides to low valent species, mainly in oxidation state +III, was found to be a general method for obtaining pentavalent mono- or di-nuclear nitrene derivatives.[294] Applied to Nb^I compounds, it led to the first dinitrene compound.

Niobium(III) and tantalum(III) adducts with sulfur or nitrile ligands undergo remarkable reactions involving reductive coupling of nitriles and oxidation of the metal to pentavalent nitrenes.[295,296,297] [Ta$_2$Cl$_6$(THT)$_3$] (10) (THT = C$_4$H$_8$S) can serve as a convenient source of Ta^{III}; its reaction (equation 20) with nitriles is rapid and quantitative. As the reaction was not observed with pivalonitrile or benzonitrile, it appears to be sensitive to steric hindrance. [Ta$_2$Cl$_6$(THF)$_2$(μ-C$_6$H$_4$N$_2$)] (11; Figure 7)[297] and the [Nb$_2$Cl$_8$(MeCN)$_2$(μ-C$_6$H$_4$N$_2$)]$^{2-}$ anions (12)[295] are centrosymmetric with octahedral arrangements around the metal atoms. Most remarkable is the new bridging 1,2-dimethyl-1,2-diimodoethylene ligand, which derives from two MeCN molecules by a process formally represented in equation (21). The Ta—N and Nb—N bond distances (1.75(1) Å for both 11 and 12) imply a close to triple bond character (a distance of 1.70 Å is expected for an M≡N bond). The MNC arrangements are essentially linear, the organic skeleton being planar with a *trans* configuration. The central C—C distances are slightly longer than the standard C=C distance of 1.335 Å. A notable lengthening, of *ca.* 0.20 Å, is seen in (11) for the Ta—O distance *trans* to the multiple Ta—N bond.

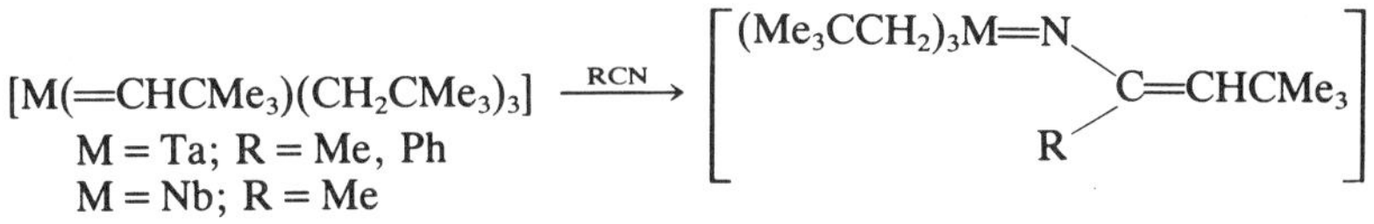

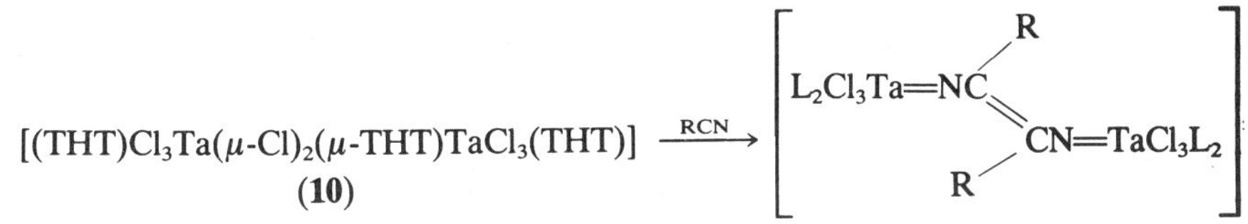

$$[(THT)Cl_3Ta(\mu\text{-}Cl)_2(\mu\text{-}THT)TaCl_3(THT)] \xrightarrow{RCN} \left[\begin{array}{c} L_2Cl_3Ta{=}NC\underset{R}{\overset{R}{\diagdown}}{=}\underset{}{\overset{}{C}}\diagup{CN{=}TaCl_3L_2} \end{array} \right] \qquad (20)$$
$$(10)$$

$$L = RCN;\ R = Me,\ Et,\ Pr^n,\ Bu^i$$

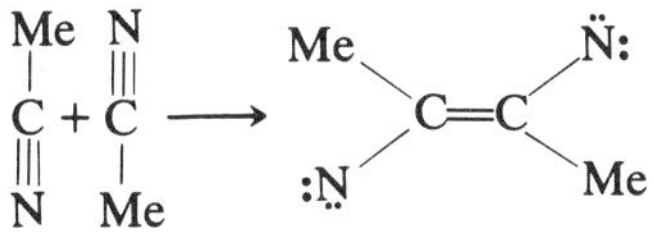

$$\underset{\underset{N}{\overset{\displaystyle |}{\underset{|||}{C}}}}{\overset{\overset{\displaystyle Me\ \ N}{\overset{|}{}}}{}} + \underset{\underset{Me}{\overset{\displaystyle |}{\underset{|}{C}}}}{\overset{\overset{\displaystyle N}{\overset{|||}{}}}{}} \longrightarrow \underset{:N}{\overset{Me}{\diagdown}}C{=}C\underset{Me}{\overset{\ddot{N}:}{\diagup}} \qquad (21)$$

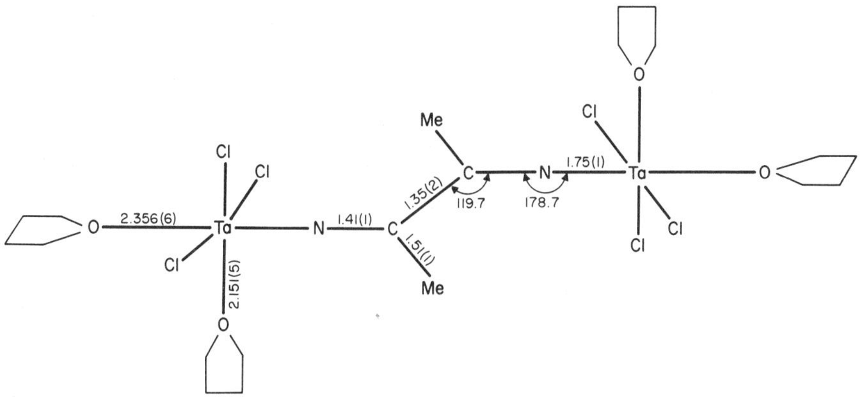

Figure 7 X-Ray structure of [Ta$_2$Cl$_6$(THF)$_2$(μ-C$_6$H$_4$N$_2$)] (11) (reproduced from ref. 297 with permission)

X-Ray data are also available for monomeric nitrene derivatives such as $[Ta(NBu^t)(NMe_2)_3]$ (**8**),[290] $[Ta(NPh)Cl_3(PEt_3)(THF)]$ (**13**)[298] and $[Nb(p\text{-}NC_6H_4Me)(Et_2dtc)_3]$ (**14**).[291] M—N bonds compare well with those reported above, and the MNC angles are close to 180°, in all but (**14**) in which the metal is heptacoordinated. The metal–nitrene bond was therefore described as an M≡NR triple bond, and the imido ligand as a four-electron donor. The oxo and organoimido ligands are isoelectronic. The Nb≡NR bond in (**14**) is indeed just 0.04 Å longer—the expected difference in radii between N and O—than the oxo bond (1.74(1) Å) in $[NbO(Et_2dtc)_3]$. These observations are consistent with those made for tantalum on $[TaO(NPr^i_2)_3]$[285] and $[Ta(NBu^t)(NMe_2)_3]$.[290]

All these compounds have high thermal stability, and are more soluble than their oxo analogs. They show a vibration around 1350 cm^{-1} in the IR, which shifts to lower frequency upon ^{15}N labeling ($\Delta\nu(^{14}N\text{–}^{15}N) = 22\ cm^{-1}$).[299,292] This fairly high energy absorption, whose frequency is only slightly affected by the ancillary ligands, has been interpreted as a combination of the M—N and N—C stretching modes.[291] ^{15}N NMR data have been reported for both Ta^V and Ta^{III} nitrenes, the latter appearing at higher fields.[292] In the ^1H NMR, the key feature is a downfield shift of *ca.* 1 p.p.m. of the hydrogen on the α carbon of the NR group.[285] The ^{13}C chemical shift difference, $\Delta = \delta C_\alpha - \delta C_\beta$, was used as a probe for evaluating the electron density on nitrogen.[300]

(*ii*) *Reactivity*

Information on the chemical reactivity of these nitrenes is limited; the NR groups are tightly bound to the pentavalent metal and not readily displaced.

The Ta^V nitrenes formed by reductive coupling of acetonitrile appear to be relatively resistant to protic attack.[296] $[M(NR)(R'_2dtc)_3]$ undergoes reversible protonation by strong acids, but the site of attack, as well as the reaction products, could not be determined.[291] It is of interest that hydrogenation of $[Ta(NR)Me_2(Cp')]$ afforded the imido hydrides $[TaH_2(NR)L(Cp')]$ (**15**) and not the amide complexes, even in the presence of an excess of phosphine (equation 22).[284] Compounds (**15**) appear to be, with the Re^V derivatives $[ReH(NR)Cl_2(PPh_3)_2]$ and $[ReH(NR)Cl(OR')(PPh_3)_2]$, the only mononuclear transition metal imide hydride complexes described so far. Their hydrogens are 'hydridic'; they reduce acetone to the isopropoxo group (equation 23). This reaction and the metathesis of imidochloride derivatives by alkaline reagents[292] constitute alternatives for producing imido alkoxides. The action of alcohols removes the imido ligand (equation 24). $Ta(OR)_5$ was obtained in high yields from $[Ta(NPr)(NPr_2)_3]$, although the NR group is more resistant to alcohols than the NR_2 group.[157]

$$[Ta(NR)Me_2(Cp')] + L \xrightarrow{H_2(\sim 4\ atm)} \underset{(\mathbf{15})}{[TaH_2(NR)L(Cp')]} + 2CH_4 \qquad (22)$$

$$R = Bu^t, CH_2CMe_3; \quad L = PMe_3, PPhMe_2$$

$$\underset{(\mathbf{15})}{[TaH_2(NR)L(Cp')]} + 2O{=}CMe_2 \longrightarrow [Ta(NR)(OCHCMe_2)(Cp')] + L \qquad (23)$$

$$R = Bu^t, CH_2CMe_3; \quad L = PMe_3, PPhMe_2$$

$$\underset{(\mathbf{15})\quad L = PMe_3}{[TaH_2(NR)L(Cp')] + MeOH} \longrightarrow [Ta(OMe)_4(Cp')] + RNH_2 + L \qquad (24)$$

The various Nb and Ta nitrenes were converted to oxo derivatives by an excess of carboxylic compounds.[291,296] This seems characteristic of the nucleophilic reactivity of nitrenes, but water is sometimes required (equation 25).

$$[M(NBu^t)(Me_2dtc)_3] + PhCHO \underset{dry\ (\times)}{\overset{H_2O}{\rightleftharpoons}} [M(O)(Me_2dtc)_3] + Bu^tN{=}CHPh \qquad (25)$$

Apart from the reactivity involving the NR group, the above compounds display the reactivity of their ancillary ligands. Compound (**8**) undergoes the typical reactions of the NR_2 ligands with electrophiles (ROH, R_2CO, CO_2, CS_2 or RI). The insertion observed with ketones (equation 26) is noteworthy, both for its reversibility and for its sensitivity to steric restrictions.[289] Alkoxo substituents allow the reactions of an alkylidene ligand to give metathesis products with alkenes.[301] An imidoalkylidene analog $[Ta(NSiMe_3)Cl(CHCMe_3)(PMe_3)_2]$ reacted with ethylene to give rearrangement but no metathesis products.[292]

$$\underset{(\mathbf{8})}{[Bu^tN{=}Ta(NMe_2)_3]} \xrightarrow{Ph_2C{=}O(\text{excess})} Bu^tN{=}Ta(OCPh_2NMe_2)_2(NMe_2) \xrightarrow{CO_2\ \text{excess}} [Bu^tN{=}Ta(O_2CNMe_2)_3] \qquad (26)$$

34.2.3.6 Dinitrogen activation

The reaction of niobium and tantalum derivatives in low oxidation states with N_2 led to $[\{NbBu(Cp)_2\}_2(N_2)(O_2)]$ (equation 27),[302] $[\{NbCl(dmpe)_2\}_2(\mu\text{-}N_2)]$[303] (equation 28) and $[Nb_2Cl_6(Bu^t_3P)_3(N_2)]$ (equation 29),[304] which were only poorly characterized owing to their low stability. IR investigations of the interaction between the terminal N_2 ligand of *trans*-$[ReCl(N_2)(PhPMe_2)_4]$ and a number of acceptor transition metal halides showed that $NbCl_5$ produces the greatest shift of the N_2 stretching frequency ($292\ cm^{-1}$), but the nature of the polynuclear species formed was not reported.[305] The development of this exciting area is mainly due to Schrock and coworkers, who were able to obtain stable tantalum μ-N_2 compounds in high yields, by reducing neopentylidene complexes in the presence of N_2 at normal pressure (equation 30).[292,306]

$$[NbCl_2(Cp)_2] + BuLi \xrightarrow[0.1\%\ N_2,\ 0.1\%\ O_2]{Ar} [\{NbBu(Cp)_2\}_2(N_2)(O_2)] \qquad (27)$$

$$[NbCl_4(dmpe)_2] + Mg\ (\text{excess}) \xrightarrow[N_2\ (1\ atm)]{CH_2Cl_2} [\{NbCl(dmpe)_2\}_2(\mu\text{-}N_2)] + 3/2MgCl_2 \qquad (28)$$

$$2NbCl_5 + 3PBu^t_3 + Mg\ (\text{excess}) \xrightarrow[N_2\ (1\ atm)]{CH_2Cl_2} [Nb_2Cl_6(PBu^t_3)_3(N_2)] + MgCl_2 \qquad (29)$$

$$[TaCl_3(CHCMe_3)(PMe_3)_2] \xrightarrow{2Na/Hg} \begin{cases} \xrightarrow{N_2} \underset{(\mathbf{16})}{[\{TaCl(CHCMe_3)(PMe_3)\}_2(\mu\text{-}N_2)]} \\ \xrightarrow{Ar} [TaCl(CHCMe_3)(PMe_3)_4] \end{cases} \qquad (30)$$

The structure of $[\{Ta(CHCMe_3)(CH_2CMe_3)(PMe_3)_2\}_2(\mu\text{-}N_2)]$ (**17**) revealed an unusually high activation of the N_2 ligand, with formation of a close to linear Ta(μ-N_2)Ta bridge, formulated as a Ta═N—N═Ta system.[307] This 'diimido' character is confirmed by the fact that similar compounds are accessible in high yields by reaction of tantalum or niobium alkylidenes with a diimine reagent, PhCH═NN═CHPh (equation 31).[292,308] The molecular structure of the centrosymmetric $[\{TaCl_3(PBz_3)(THF)\}_2(\mu\text{-}N_2)]$ (**19**) is depicted in Figure 8. The μ-N_2 ligand is the sole bridge between the two metals, which present distorted octahedral environments with a meridional arrangement of the chlorine atoms.[309] The Ta—N bonds (1.796(5) Å) are much

$$2\ \textit{cis,mer}\text{-}[MCl_3(CHCMe_3)(THF)_2] \xrightarrow[-2PhCH{=}CHCMe_3]{+PhCH{=}NN{=}CHPh} \underset{(\mathbf{18})}{[\{MCl_3(THF)_2\}_2(\mu\text{-}N_2)]} \xrightarrow[-2THF]{2PBz_3} [\{TaCl_3(PBz_3)(THF)\}_2(\mu\text{-}N_2)] \qquad (31)$$

shorter than the usual Ta^V—N single bond (*ca.* 1.97 Å), and close to the Ta═N distance of 1.765(5) Å found in [$Ta(NPh)Cl_3(PEt_3)(THF)$] (**13**).[298] It is also notable that in the alkylidene (**17**), the Ta—N bonds (av. 1.839(8) Å) are shorter than the Ta═C bonds (av. 1.935(9) Å). These data support the description of the Ta—N linkages in (**17**) and (**19**) as double bonds. The N—N bond length of 1.282(6) Å in (**19**) is significantly longer than in free dinitrogen (N≡N: 1.0976 Å), suggesting a bond order between one (N—N ≈ 1.45 Å) and two (N═N ≈ 1.24 Å). The production of such μ-dinitrogen compounds with M═N—N═M units could become a common feature in niobium and tantalum chemistry (Table 14).

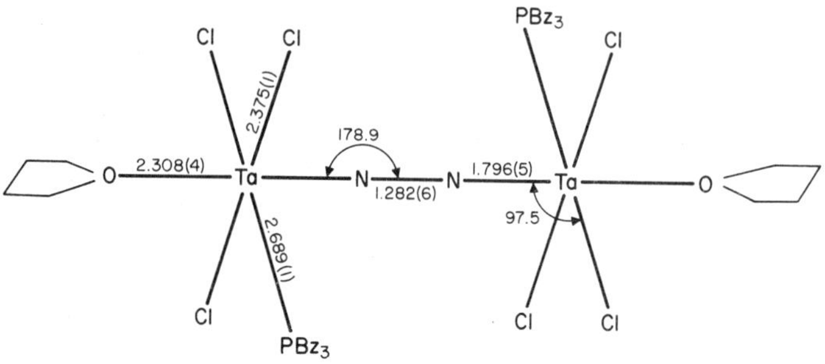

Figure 8 Molecular structure of [$\{TaCl_3(PBz_3)(THF)\}_2(\beta\text{-}N_2)$] (**19**) (reproduced from ref. 309 with permission)

These μ-N_2 complexes contain the longest most 'activated' N_2 groups reported to date for simple M_2N_2 systems (longer N—N bonds, *ca.* 1.35 Å, are known for complexes such as [$\{(PhLi)_3Ni\}_2(N_2)(Et_2O)_2]_2$ and related species, but in which N_2 interacts with more than two metallic centers [310]). They contrast sharply with the μ-N_2 complexes reported for group IVA metals, where these bridges are almost linear and relatively short (N—N bonds ranging from 1.15 to 1.18 Å for [$\{M'(N_2)(Cp')_2\}_2(\mu\text{-}N_2)$] (M′ = Ti[311a] or Zr[311b]), the corresponding metal–nitrogen bonds being relatively long (2.00 to 2.09 Å).

If one considers the metal to be Ta^V, the Ta═N bond then implies a combination of ligand to metal σ and π donation from a formal N_2^{4-} ligand, with unit N—N bond order. The experimental N—N bond lengths suggest bond orders between 1 and 2, consistent with donation of electrons from the π^* 2*p* orbitals of N_2^{4-} into the empty *d* orbitals of Ta^V.[320] This description also accounts for the eclipsed conformation of the equatorial ligands in (**19**), although the staggered one should be the more favored sterically.[309]

Compounds (**17**) and (**19**) are formally 14-electron species. By contrast, the μ-dinitrogen complexes [$\{M'(N_2)(Cp')_2\}_2(\eta\text{-}N_2)$] (M′ = Ti, Zr) are respectively 16-electron and 18-electron species, and the low oxidation state of the metal is preserved. The spectroscopic data also illustrate the differences between the tantalum and these Ti^{II} or Zr^{II} μ-dinitrogen compounds. The ^{15}N spectrum of (**16**) shows a singlet at 414 p.p.m.,[306] a chemical shift comparable to that

Table 14 Niobium and Tantalum μ-Dinitrogen Compounds

Compounds	*Comments*	*Ref.*
[$\{MCl_3(THF)_2\}_2(\mu\text{-}N_2)$]	Insoluble for M = Ta	1, 2, 3
[$\{TaCl_3L_2\}_2(\mu\text{-}N_2)$] (L = PEt_3, $PhPMe_2$, $PhPEt_2$, PPr^n_3)	$\nu(Ta\text{—}N_2) = 855\ cm^{-1}$; *trans–mer* isomer	1, 2
[$\{TaCl_3(PBz_3)(THF)\}_2(\mu\text{-}N_2)$]	X-Ray structure, linear Ta═NN═Ta linkage	2, 4
[$\{M(CH_2CMe_3)_3(THF)\}_2(\mu\text{-}N_2)$]		2
[$\{Ta(OBu^t)_3(THF)\}_2(\mu\text{-}N_2)$]		2
[$\{M(Et_2dtc)_3\}_2N_2$]	Fluxional	3
[$\{Ta(CHCMe_3)ClL_2\}_2(\mu\text{-}N_2)$] (L = PMe_3)	^{15}N NMR = 414 p.p.m. (NH_3)	1, 5
[$\{Ta(CH_2CMe_3)(CHCMe_3)L_2\}_2(\mu\text{-}N_2)$] (L = PMe_3)	X-Ray structure linear, Ta═NN═Ta linkage, trigonal bipyramidal coordination	1, 6

1. S. M. Rocklage and R. R. Schrock, *J. Am. Chem. Soc.*, 1980, **102**, 7809.
2. S. M. Rocklage and R. R. Schrock, *J. Am. Chem. Soc.*, 1982, **104**, 3077.
3. J. R. Dilworth, S. J. Harrison, R. A. Henderson and D. R. M. Walton, *Chem. Commun.*, 1984, p. 176.
4. M. R. Churchill and H. J. Wasserman, *Inorg. Chem.*, 1982, **21**, 218.
5. S. M. Rocklage, H. W. Turner, J. D. Fellmann and R. R. Schrock, *Oganometallics*, 1982, **1**, 703.

found in imido compounds such as $[Ta(NPh)Cl_3(THF)_2]$ ($\delta = 350$–370 p.p.m.), but different from the 679 p.p.m. observed for the bridging dinitrogen in $[\{Zr(N_2)(Cp')\}_2(\mu\text{-}N_2)]$.[311b] ^{15}N labeling experiments have established that the Ta═N—N═Ta group is characterized by an absorption around 845 cm^{-1} in the IR. This description of the Ta═N—N═Ta bridge is also supported by the reactivity of these μ-N_2 adducts, which behave as imido complexes. Thus the addition of an excess of acetone produced dimethylketazine, $Me_2C{=}N{-}N{=}CMe_2$. Several μ-N_2 complexes also reacted readily with excess HCl to give hydrazine (as $N_2H_4{\cdot}2HCl$) only; no ammonia was detected.[292,306,308]

34.2.3.7 *Azides* (Table 15)

Both neutral and ionic niobium and tantalum chloride azides are known. The neutral azides $[MN_3Cl_4]_2$ were obtained from MCl_5 and ClN_3.[312] They both explode spontaneously, although the tantalum derivative is somewhat stabler. The more stable ionic $[MN_3X_5]^-$ (X = Cl, Br) species were synthesized from the hexahalometallates(V) and iodoazide or from the pentahalides and phosphonium or arsonium azides.[73,258]

Table 15 Azido and Phosphineiminato compounds

Compounds	*Comments*	*Ref.*
(a) Azido compounds		
$[MCl_4N_3]_2$	$MCl_5 + ClN_3$ or NaN_3; explosive; X-ray structure for M = Ta	1
$(PPh_4)[MX_5N_3]$ (X = Cl, Br)	$MX_5 + N_3PPh_3$; non-explosive; X-ray structure for M = Nb and X = Cl (disorder of the azido group)	2, 3
$(AsPh_4)[TaCl_5N_3]$		2
$(Et_4N)[NbCl_5N_3]$	$NbCl_5 + Et_4NN_3$; non-explosive; analysis only	4
$[Ta\{N(SiR_3)_2\}_2(N_3)_3]$	$[TaCl_3\{N(SiMe_3)_2\}_2] + Me_3SiN_3$	5
(b) Phosphaneiminato compounds		
$[MCl_4(NPPh_3)]_2$	X-Ray structure for both M = Nb and Ta	6, 7
$(PPh_4)[MX_5(NPPh_3)]$ (X = Cl, Br)		3, 2
$(AsPh_4)[NbCl_5(NAsPh_3)]$	Photochemical synthesis	3
$(PPh_4)[TaCl_4(NPPh_3)_2]$	$TaCl_4N_3 + PPh_3$; X-ray structure, *trans* isomer	6
$[(NHPPh_3)TaCl_4(NPPh_3)]$	Proton abstraction from the solvent ($C_2H_4Cl_2$) by $[TaCl_4][TaCl_4(NPPh_3)_2]$	7

1. J. Strähle, *Z. Anorg. Allg. Chem.*, 1974, **405,** 139.
2. R. Dubgen, U. Muller, F. Weller and K. Dehnicke, *Z. Anorg. Allg. Chem.*, 1980, **471,** 89.
3. U. Muller, R. Dubgen and K. Dehnicke, *Z. Anorg. Allg. Chem.*, 1981, **473,** 115.
4. M. Kasper, R. D. Bereman, *Inorg. Nucl. Chem. Lett.*, 1974, **10,** 443.
5. R. A. Andersen, *Inorg. Nucl. Chem. Lett.*, 1980, **16,** 31.
6. H. Bezler and J. Strähle, *Z. Naturforsch., Teil B*, 1979, **34,** 1199.
7. H. Bezler and J. Strähle, *Z. Naturforsch., Teil B*, 1983, **38,** 317.

The structures of $[TaN_3Cl_4]_2$[312] and $(PPh_4)[NbN_3Cl_5]$[73] have been determined. Two independent dimeric molecules, of C_{2h} and D_{2h} symmetry, were found for the former with azido groups bridging the two metals in a slightly asymmetric manner, in a dimer of C_{2h} symmetry (Figure 9). The $(Ta{-}\mu\text{-}N)_2$ rings are planar, and the Ta—N linkages are all longer than expected for single bonds (2.08 Å). The azido compounds are characterized by stretching frequencies around 2100 $[\nu_{as}(N_3)]$, 1350 $[\nu_s(N_3)]$ and 420 $[\nu(M{-}N)]$ cm^{-1} in the IR.

Both the neutral and the ionic azides undergo reactions analogous to the Staudinger reaction, giving compounds containing the phosphineiminato M═N═P group (equation 32)[313,314] The reaction between $[TaN_3Cl_4]_2$ and PPh_3 provided $[TaCl_4(NPPh_3)]_2$ and $[TaCl_4(NPPh_3)(NHPPh_3)]$.[314] The latter was suggested to result from the reaction of the initially formed $[TaCl_4][TaCl_4(NPPh_3)_2]$ salt with the solvent $C_2H_4Cl_2$. As a result of the higher stability of the tantalum azides, the reactions with $[TaN_3X_5]^-$ must be activated photochemically. Attempts to extend the Staudinger reaction to $AsPh_3$ were successful only with $[NbN_3Cl_5]^-$.[74]

$$[NbN_3Cl_4]_2 + 2PPh_3 \longrightarrow [NbCl_4(NPPh_3)]_2 + 2N_2 \qquad (32)$$

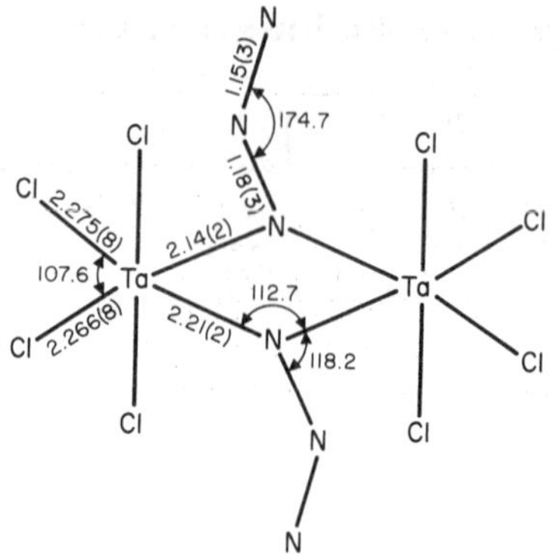

Figure 9 X-Ray structure of the $[TaCl_4N_3]_2$ dimer of C_{2h} symmetry (reproduced from ref. 312 with permission)

Centrosymmetric dimers with terminal phosphineiminato groups were found in the solid for $[MCl_4(NPPh_3)]_2$ (Figure 10).[313,314] The asymmetry of the bridge was interpreted as resulting from the strong *trans* effect (0.34 Å for M = Ta) of the phosphineiminato ligand, which also leads to a very long Ta—Cl bridge. The PNM sequence is linear, and the distances consistent with P═N and M═N double bonds. The Ta—N bond is longer (1.97 Å) in *trans*-$[TaCl_4(NPPh_3)_2]^-$, while the N—P distance is only 1.56 Å and the TaNP arrangement is slightly bent (162°). Thus the Ta—N linkage exhibits almost single bond character, probably as a result of the unfavorable *trans* position of the phosphineiminato groups.[316] Although no X-ray data are presently available for the arsineiminato $[NbCl_5(NAsPh_3)]^-$ compound, the IR data suggest a lower π character of the As═N bond with respect to the P═N bond.

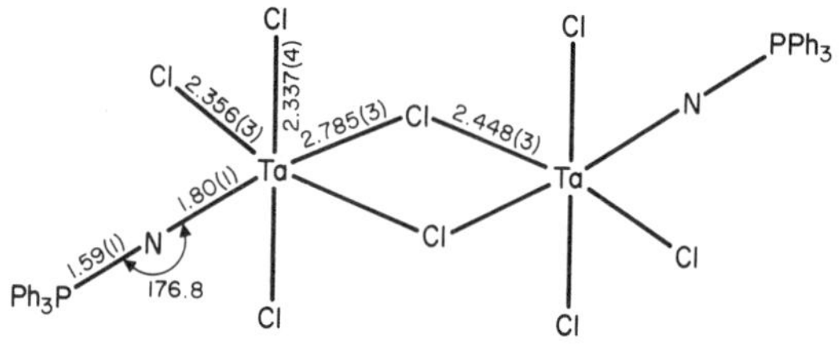

Figure 10 Structure of the phosphineiminato complex $[TaCl_4(NPPh_3)]_2$ (reproduced from ref. 313 with permission)

34.2.3.8 Nitrides (Table 16)

The obtention of nitrides by thermal decomposition of the azides was only successful for the synthesis of $[MNBr_4]^{2-}$ from $[MN_3Br_5]^-$. Thermolysis of MX_5 with ammonium salts afforded a more general access to nitride halides in oxidation state V, or IV if higher temperatures were used.[316,317] Oxo nitrides were obtained if moisture was admitted during the process.[318]

The only well-characterized nitrides are $[Ta_2NBr_{10}]^{3-}$ [316] and the oxo nitrides MON.[318] The structure of the former is similar to that of $[Ta_2OF_{10}]^{2-}$ (Section 34.2.6.1.i), with a symmetrical linear TaNTa arrangement and a Ta—N bond length (1.849(2) Å) consistent with double bond character. The ligands *cis* to the bridge are eclipsed; the lengthening of the axial Ta—Br bond (0.18 Å) reflects a strong *trans* effect. A polymeric ribbon type structure with nitrido and halo bridges has been assumed for all nitride halides $[MNX_2]_n$ (X = F, Cl, Br).[319,320] Their stability decreases from the fluoride to the bromide, as does the bond order of the M═N linkage, as reflected for instance by the diminution of the Nb═N frequency in the IR from 800 for X = F to 720 cm^{-1} for X = Br.

34.2.3.9 Miscellaneous

(*i*) *Porphyrin and related complexes*

Porphyrin derivatives of Nb^V and Ta^V were first reported in 1972 and have since aroused increasing interest. Fluoride complexes MF_3L were obtained from MCl_5 and the octaethylporphyrin in the presence of HF (L = OEP).[321,322] Niobium chloride and bromide

Table 16 Pentavalent Niobium and Tantalum Nitrides

Compounds	*Comments*	*Ref.*
$(NH_4)_3[M_2NBr_{10}]$	$NH_4Br + NH_4MBr_6$ at 400 °C; X-ray structure	1
$(PPh_4)_2[MNBr_4]$	Decomposition of $[MBr_5N_3]^-$, $\nu(M{\equiv}N) \approx 900\ cm^{-1}$	2
$[(TaNCl_2)_n]$	$TaCl_5 + NH_4Cl$ at 350 °C; polymeric (ribbons)	3
$[(NbNX_2)_n]$ (X = F, Cl, Br)	$NbX_5 + NH_4X$ at 210 °C (X = Cl, Br); $NbNCl_2 + F_2$ at room temperature	4
[MON]	$(NH_4)[MCl_6]$ heated in the presence of water or $MOCl_3 + NH_3$; X-ray structure, heptacoordination	5
$[Ta_3N_5]$	$TaCl_5 + NH_4Cl$ at 850 °C, X-ray structure	6

1. K. P. Frank, J. Strähle and J. Weidlein, *Z. Naturforsch., Teil B,* 1980, **35,** 300; M. Hörner, K. P. Frank and J. Strähle, *Z. Naturforsch., Teil B,* 1986, **41,** 423.
2. R. Dubgen, U. Muller, F. Weller and K. Dehnicke, *Z. Anorg. Allg. Chem.*, 1980, **471,** 89.
3. M. A. Glushkova, M. M. Ershova and Y. A. Buslaev, *Russ. J. Inorg. Chem. (Engl. Transl.)*, 1965, **10,** 1290.
4. E. M. Shustorovich, *J. Struct. Chem. (Engl. Transl.)*, 1962, **3,** 204; S. M. Sinitsyna, *Russ. J. Inorg. Chem. (Engl. Transl.)*, 1977, **22,** 402.
5. W. Weishaupt and J. Strähle, *Z. Anorg. Allg. Chem.*, 1977, **429,** 261.
6. J. Strähle, *Z. Anorg. Allg. Chem.*, 1973, **402,** 47.

complexes have been prepared by treatment of $[NB_2O_3L_2]$ with dry gaseous HX for L = TPP (*meso*-tetraphenylporphyrinato), TMTP (*meso*-tetra-*m*-tolylporphyrinato) and TPTP (*meso*-tetra-*p*-tolylporphyrinato).[323,324] The kinetics of the solvoprotolytic dissociation of $[NbCl_3(TPP)^-]$ in $MeCOOH/H_2SO_4$ mixtures have been studied.[325]

The reaction of porphyrins with $NbCl_5$ afforded after hydrolysis the $[Nb_2O_3L_2]$ dimers,[326] with L = OEP, TPP, TPTP and OMP (octamethylporphyrinato). An unusual geometry was found for these dimers, as illustrated in Figure 11 in the case of L = TPP.[327,330] The Nb atoms are heptacoordinated with the two Nb—TPP units linked together through three μ-oxo bridges; Nb atoms lie 1.00 Å from the 4N planes of the porphyrinato ligands. The geometry of this d^0–d^0 $[(NbL)_2O_3]$ compound was rationalized in terms of optimal Nb—O π interactions.[331]

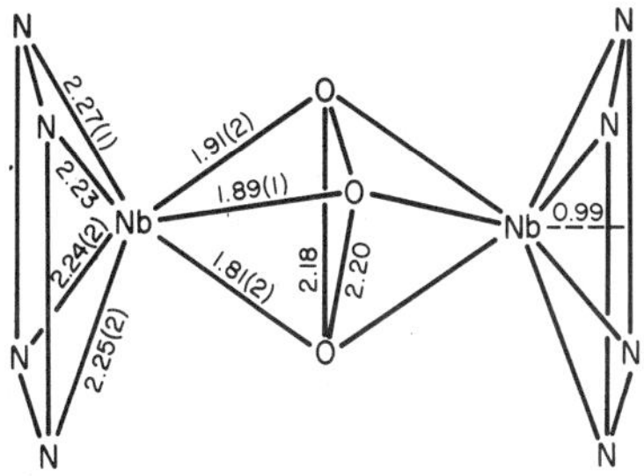

Figure 11 Coordination polyhedron of $[Nb_2(TPP)_2O_3]$ (reproduced from ref. 330 with permission)

Bubbling HF through solutions of the $[Nb_2O_3L_2]$ dimer in anhydrous benzene afforded [NbOFL] (L = OEP, TPP, OMP, TPTP).[321,332] The structure with L = OEP showed a *cis* geometry on the six-coordinate niobium. The metal–macrocycle distance of 0.905 Å is shorter than in seven-coordinated complexes, probably as a result of decreased steric hindrance (Nb═O: 1.7620(6) Å; Nb—F: 1.888 Å). Addition of HI to $[Nb_2O_3(OEP)_2]$ in CH_2Cl_2 solutions yielded $[NbOI_3(OEP)]$, which on heating under vacuum liberated iodine and [NbOI(OEP)].[322] [NbO(TPP)(acac)] was prepared by reaction of $[Nb_2O_3(TPP)_2]$ with excess acacH.[324]

Recrystallization of $[Nb_2O_3L_2]$ from acetic acid gave [NbOL(MeCOO)(MeCOOH)] (L = OEP, TPP, TPTP and OMP).[326] The structure of the L = TPP derivative[332,333] showed the niobium atom to be seven-coordinated by the four nitrogen atoms of the porphyrin and by the three oxygen atoms, which are *cis* to each other; the metal is displaced by 0.99(2) Å from the porphyrinato plane towards the oxygens. A high resolution NMR study of [NbOL(MeCOO)(MeCOOH)] and $[Nb_2O_3(OEP)_2]$ revealed the anisochrony of the methylenic protons of the porphyrin, induced by the out-of-plane position of the metal atom.

The redox chemistry of [NbO(TPP)(MeCOO)] was investigated by means of cyclic voltammetry and controlled potential electrolysis; the reduction of Nb^V to Nb^{III} was found to occur prior to the reduction of the ligand. The redox potentials were measured for the $[Nb^VO(TPP)(MeCOO)]/[Nb^{IV}O(TPP)]/[Nb^{III}(TPP)]^+/[Nb^{II}(TPP)]$ systems and found to be −0.94, 1.1 and −1.48 V respectively.[334]

. Irreversible oxygenation of the low valent [NbX$_2$L] porphyrins by O$_2$ yielded [Nb(O$_2$)X$_2$L] which, from ESR and IR data, was described as a superoxide.[323]

Phthalonitrile and phthalocyaninedilithium react with MX$_5$ to yield [MX$_3$(PC)].[335,336] The reaction between [TaCl$_2$Me$_3$] and Nasalen yielded [Ta(salen)Me$_3$].[277]

(ii) Pyrazolylborate derivatives

Tantalum pyrazolylborate derivatives [TaClMe$_3$\{HB(pz)$_3$\}] (**20**) and [TaClMe$_3$\{H$_2$B(3,5-Me$_2$pz)$_2$\}] (**21**) (pz = pyrazolyl) have been reported and characterized by X-ray and NMR studies.[337] Complex (**21**) adopts a capped octahedral structure in the solid with a close contact between tantalum and one hydrogen, which was described as a three-centered B—H—Ta bond, filling the seventh coordination site (Figure 12). The Ta—B distance is short (2.90(1) Å; usual M' – – – B distances are *ca.* 3.8 Å) probably as a consequence of extreme puckering of the six-membered ring. A methyl group is found to cap the face formed by C(5), C(6) and Cl(4); there are two geometrical isomers (one chiral) in the crystal. The existence of the Ta—H—B interaction was also shown by IR (v(B–H) = 2013, 2070 cm^{-1}) and by ^{11}B NMR in solution, where it is preserved up to at least 110 °C. Dynamic behavior, which equilibrates the three isomers present in solution, was observed.

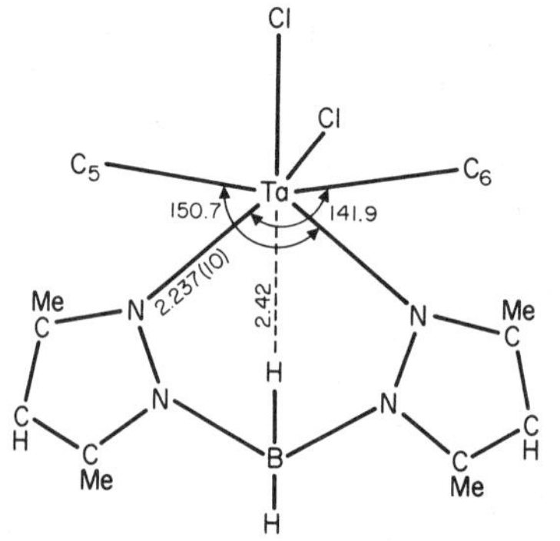

Figure 12 Molecular structure of one isomer of [TaClMe$_3$\{H$_2$B(3,5-Me$_2$pz)$_2$\}] (**21**) (reproduced from ref. 337 with permission)

Reactions of K[HB(Me$_2$pz)$_3$] with MCl$_5$ in equimolar proportions yielded [MCl$_4$\{HB(Me$_2$pz)$_3$\}] and [HB(Me$_2$pz)$_3$BH][MCl$_6$].[71] The presence of the novel dibora cation was confirmed by X-ray diffraction in the case ot tantalum. Acid-catalyzed cleavage of B—N(pyrazolyl) bonds also takes place and generates, for instance, [NbCl\{HB(Me$_2$pz)$_3$\}(Me$_2$pz)$_3$] with pyrazolate ligands. Reactions between NbCl$_5$ and K[HB(pz)$_3$] and K[H$_2$B(pz)$_2$] have led to ready reduction to Nb^{IV} products (Section 34.3.5.4.iii); the Nb^V salts K[NbCl$_5$\{HB(pz)$_3$\}] and K[NbCl$_5$\{H$_2$B(pz)$_2$\}] could however be isolated from low temperature reactions.[338]

(iii) Nitrogen-containing insertion products (Table 17)

Reactions of [MCl$_{5-x}$Me$_x$] (x = 1, 2 or 3) with heterocumulenes have led to insertions into the M—C bond.[8] *N,N'*-Dialkylacetamidinato derivatives[339] were obtained (equation 33). *N*-Acetamides or *N*-thioacetamides[340] were formed with isocyanates or isothiocyanates, respectively (equation 34). The rate of insertion varied in the order [MCl$_4$Me] > [MCl$_3$Me$_2$] > [MCl$_2$Me$_3$], following the established order of acceptor properties of the methylmetal chlorides, the isocyanates being more reactive than the isothiocyanates. Hexa- or hepta-coordinated species were obtained.

$$[MCl_2Me_3] + 2Pr^iN{=}C{=}NPr^i \longrightarrow Cl_2MeM\left(\begin{array}{c} Pr^i \\ N \\ \diagdown \\ CMe \\ \diagup \\ N \\ Pr^i \end{array}\right)_2 \qquad (33)$$

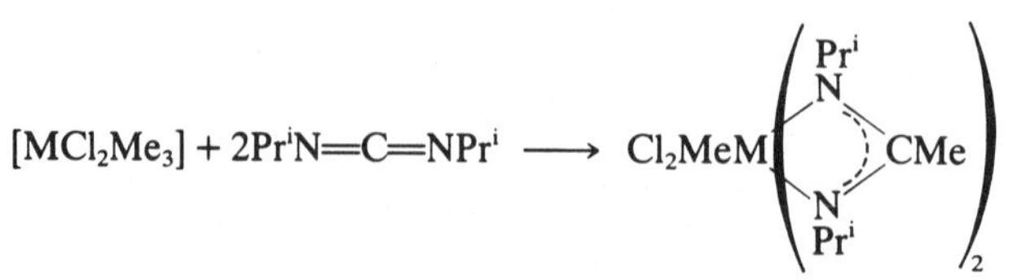

Table 17 Niobium and Tantalum Nitrogen-containing Insertion Products (Acetamidinato, Acetamido, Thioacetamido and Nitrosohydroxylaminato Derivativea)

Compounds	Comments	Ref.
$[MCl_3X'\{NRC(Me)NR\}]$ ($R = Pr^i$, Cy, *p*-tolyl; $X' = Cl$, Me)	Six coordinated	1
$[MCl_2Me_2\{NRC(Me)NR\}]$ ($R = Pr^i$, Cy; $M = Ta$; $R = p$-tolyl)	Six coordinated, Me probably *cis*	1
$[TaCl_2X'\{NRC(Me)NR\}_2]$ ($R = Pr^i$, Cy, $X' = Cl$, Me; $R = p$-tolyl, $X' = Me$)	X-Ray structures: pentagonal bipyramid for $X' = Cl$, $R = Pr^i$ (orthorhombic and triclinic)	1, 2, 3
$[MCl_2X'\{NRC(O)Me\}_2]$ ($R = Me$, Ph; $X' = Cl$, Me)	Bidentate acetamide, $X' = Me$ unstable for $M = Nb$	4
$[TaCl_2\{NMeC(O)Me\}_3]$	Probably seven coordinated	4
$[MCl_4\{NRC(S)Me\}]$	$M = Nb$, $R = Me$; $M = Ta$, $R = Ph$	5
$[TaCl_3Me\{NRC(S)Me\}]$ ($R = Me$, Ph)	Disproportionate rapidly for $M = Nb$	5
$[NbCl_3\{NMeC(S)Me\}_2]$	X-Ray structure: pentagonal bipyramid	5, 6
$[MCl_2X'\{ON(Me)NO\}_2]$ ($X' = Cl$, Me; $X' = OMe$, $M = Ta$)	$M = Ta$, $X' = Me$; X-ray structure: pentagonal bipyramid	7
$[NbMe_2\{ON(Me)NO\}_3]$	Eight coordinated	8
$[TaMe_3\{ON(Me)NO\}_2]_2$		8
$[Nb(O)\{ON(Me)NO\}(Cp)_2]_n$	Final product from $[NbMe_2(Cp)_2]$ + NO	8

1. J. D. Wilkins, *J. Organomet. Chem.*, 1975, **80**, 349.
2. M. G. B. Drew and J. D. Wilkins, *J. Chem. Soc., Dalton Trans.*, 1974, 1579, 1973; 1975, 2611.
3. M. G. B. Drew and J. D. Wilkins, *Acta Crystallogr., Sect. B*, 1975, **31**, 177, 2642.
4. J. D. Wilkins, *J. Organomet. Chem.*, 1974, **67**, 269.
5. J. D. Wilkins, *J. Organomet. Chem.*, 1974, **65**, 383.
6. M. G. B. Drew and J. D. Wilkins, *J. Chem. Soc., Dalton Trans.*, 1974, 198.
7. J. D. Wilkins and M. G. B. Drew, *J. Organomet. Chem.*, 1974, **69**, 111.
8. A. R. Middleton and G. Wilkinson, *J. Chem. Soc., Dalton Trans.*, 1980, 1888.

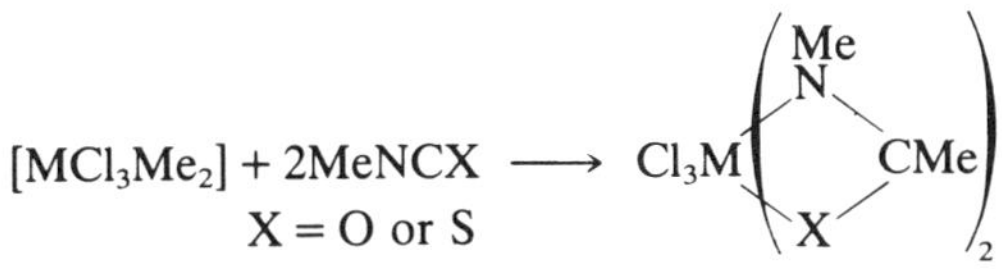

$$[MCl_3Me_2] + 2MeNCX \longrightarrow Cl_3M\left(\begin{array}{c} Me \\ N \\ \diagdown \quad CMe \\ X \end{array}\right)_2 \qquad X = O \text{ or } S \tag{34}$$

The coordination polyhedron of the metal in its N,N-dialkylacetamidinato derivatives is generally a distorted pentagonal bipyramid with two chlorines in the apical positions.$[TaCl_2Me\{CyNC(Me)NCy\}_2]$ adopts a different geometry, with one bidentate ligand in equatorial sites, while the other spans axial and equatiorial positions, thus maximizing the intramolecular distances between the cyclohexyl group.[341]

Comparable insertions occurred when methylmetal derivatives were exposed to NO, yielding N-alkyl-N-nitrosohydroxylaminato compounds (equation 35).[342] Nitric oxide may however also behave as an oxidant, and the reactions are sometimes more complicated (Section 34.2.6.3).[343]

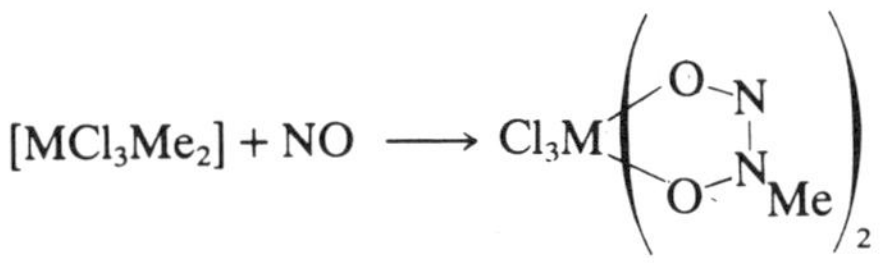

$$[MCl_3Me_2] + NO \longrightarrow Cl_3M\left(\begin{array}{c} O\diagdown N \\ \mid \\ O\diagdown N \\ Me \end{array}\right)_2 \tag{35}$$

34.2.4 Pseudohalo Derivatives

34.2.4.1 Cyanides and isocyanides

$NbCl_5$ and HCN yielded $[NbCl_4(CN)(Et_2O)]$ in Et_2O and $[NbCl_5(HCN)]$ in CCl_4. The addition of NEt_3 to the latter afforded $(HNEt_3)[NbCl_5(CN)]$; the corresponding bromo complex was also obtained.[344]

No isocyanide complex of Nb^V or Ta^V appears to have been reported. The insertion of isocyanides into M—C bonds of $[TaCl_2Me_3]$[340] and into the M—X bonds of MX_5[345–347] has however been described. $[MX_4\{CX(NR)\}(RNC)]$ was isolated ($R = Me$, Bu^t). In the case of $R = Me$,[345] further insertion occurred, yielding $[MCl_3\{CCl(NMe)\}]_2\cdot(MeNC)$.

$[TaBr_2\{Br(NMe)\}_3(MeNC)]$ could even be obtained by the reaction of Ta_2Br_{10} with excess MeNC in CH_2Cl_2; in Et_2O, the same reactants afforded an oxo species, $[TaOBr_2\{CBr(NMe)\}(MeNC)]$.[347]

The monoinserted compounds $[MX_4\{CX(NMe)\}(MeNC)]$ reacted with PPh_3 to form, through a second insertion, $[MX_3\{CX(NMe)\}_2(PPh_3)]$.[345] $[TaBr_3\{Cbr(NMe)\}(dppe)]Br$ was prepared directly from $[TaBr_5(dppe)]$ and MeNC. Attempts to induce further insertion under the action of phosphines when $R = Bu^t$ instead of Me resulted in the displacement of the terminal isocyanide ligand to yield $[MX_4\{CX(NBu^t)\}L]$, where $L = PPh_3$, PMe_2Ph, $PMePh_2$, dppe.[347]

34.2.4.2 Thio- and seleno-cyanates (Table 18)

The synthesis of $M(NCS)_5$ complexes was first effected by allowing KSCN to react with $TaCl_5$ in MeOH.[348] They were isolated from MeCN solutions as $A[M(NCS)_6(MeCN)]$. Complexes of the type $[MCl_x(NCS)_{5-x}L]$ ($x = 0$–4) were obtained on varying the KSCN/MCl_5 stiochiometry in MeCN or Et_2O. $[TaX_5L]$ (X = NCS or NCSe) formed with L = bipy or phen. $[Nb_2Cl_{10}\{\mu\text{-}C_2H_4(SCN)_2\}]$ was obtained from Nb_2Cl_{10} and the organic thiocyanate in CCl_4.[349] ^{93}Nb NMR data on MeCN solutions of $NbCl_5$ in the presence of KSCN[350] were taken to indicate the presence of $[Nb(NCS)_xX_{6-x}]^-$ (X = Cl or Br) and $[Nb(NCS)_7]^{2-}$.

Table 18 Isothio- and Isoseleno-cyanides

Compounds	*Comments*	*Ref.*
$A[M(NCS)_6]$	A = Li, Na, K, NH_4, $AsPh_4$	1–4
$A[M(NCSe)_6]$	A = K, $AsPh_4$	5
$[MCl_x(NCS)_{5-x}S]$	$x = 1$–4; S = MeCN, Et_2O	1, 6, 7
$[M(NCS)_5]_2$		5
$[Ta(NCS)_5L]$, $[Ta(NCSe)(bipy)]$	L = py, bipy	8
$[M(NCS)_2(OR)_3(bipy)]$	R = Me, Et; octacoordinated dimer	9
$[M(NCS)_2(OR)_2(dbm)]$	R = Me, Et, Pr, Pr^i, Bu, Bu^i	10
	X-Ray structure (R = Et): hexacoordination	11
$[M(NCS)(OR)_3(dbm)]$	R = Me, Et, Pr^i	10
	X-Ray structure (R = Pr^i): hexacoordination	12

1. H. Böhland and E. Tiede, *J. Less-Common Met.*, 1967, **13**, 224.
2. T. M. Brown and G. F. Knox, *J. Am. Chem. Soc.*, 1967, **89**, 5296.
3. G. F. Knox and T. M. Brown, *Inorg. Chem.*, 1969, **8**, 1401.
4. G. F. Knox and T. M. Brown, *Inorg. Synth.*, 1971, **13**, 226.
5. T. M. Brown and E. Zenker, *J. Less-Common Met.*, 1971, **25**, 397.
6. H. Böhland and E. Zenker, *J. Less-Common Met.*, 1968, **14**, 397.
7. H. Böhland, E. Tiede and E. Zenker, *J. Less-Common Met.*, 1968, **15**, 89.
8. J. N. Smith and T. M. Brown, *Inorg. Chem.*, 1972 **11**, 2697.
9. N. Vuletic and C. Djordjevic, *J. Chem. Soc., Dalton Trans.*, 1972, 2322.
10. R. Kergoat, M.-C. Senechal-Tocquer, J. E. Guerchais and F. Dahan, *Bull. Soc. Chim. Fr.*, 1976, 1203.
11. F. Dahan, R. Kergoat, M.-C. Tocquer and J. E. Guerchais, *Acta Crystallogr., Sect. B*, 1976, **32**, 1038.
12. F. Dahan, R. Kergoat, M.-C. Senechal-Tocquer and J. E. Guerchais, *J. Chem. Soc., Dalton Trans.*, 1976, 2202.

Alkoxoisothiocyanato complexes $[M(NCS)_2(OR)_2(dbm)]$ or $[M(NCS)(OR)_3(dbm)]$ have been synthesized from $NbCl_5$, KSCN and dibenzoylmethane (dbmH) in the appropriate alcohol.[351] The $[M(NCS)_2(OR)_3(bipy)]$ adducts were obtained from $[MCl_2(OR)_3]$.[352]

$[M_2(NCS)_{10}]$ reacts with ammonia and with primary and secondary aliphatic amines to yield $[M(NCS)(NH_2)_2(NH)]_2$, $[M(NCS)_3(NHBu)_2(H_2NBu)]$ or $[M(NCS)_3(NEt_2)_2(HNEt_2)]$.[353] Similar reactions with $[M(NCS)_5(MeCN)]$ afforded, through aminolytic cleavage of the M—NCS bonds, $[M(NCS)\{NC(NH_2)Me\}_4\{MeC(NH)NH_2\}]_n$ and $[M(NCS)_3\{NC(NHBu)Me\}\{MeC(NH)NHBu\}]$ complexes.[353]

The isothiocyanate groups are N bonded, in agreement with the hard and soft acid–base model. The structures of $[Nb(NCS)_2(OEt)_2(dbm)]$[355] and $[Nb(NCS)(OPr^i)_3(dbm)]$[356] showed the niobium atom to be located at the center of a distorted octahedron (Nb—N: 2.088(5), 2.101(5) and 2.176(7) Å respectively).

Bonding of the thiocyanate ligand through its sulfur atom has however been proposed[357] on the basis of ^{93}Nb NMR investigations of acetonitrile solutions containing $[Nb(NCS)_x(SCN)_yCl_{6-(x+y)}]^-$; $[Nb(SCN)_6]^-$, the first example of a homoleptic complex in which the thiocyanate ion is S bonded to a class a metal ion, was shown to be present in these solutions at concentrations comparable to those of $[Nb(NCS)_6]^-$.

34.2.5 Oxo, Thio and Seleno Halides

34.2.5.1 Oxo halides

The synthetic routes available for the preparation of the known oxyhalides, listed in Table 19, range from the halogenation of the oxides to the carefully controlled reaction of the halides with dioxygen.

Table 19 Anhydrous Oxo Halides of Niobium(V) and Tantalum(V)

Fluorides		*Chlorides*		*Bromides*		*Iodides*	
$[NbOF_3]$	$[TaOF_3]$	$[NbOCl_3]$	$[TaOCl_3]$	$[NbOBr_3]$	$[TaOBr_3]$	$[NbOI_3]$	—
—	—	—	$[Ta_2O_3Cl_4]$	—	—	—	—
$[NbO_2F]$	$[TaO_2F]$	$[NbO_2Cl]$	$[TaO_2Cl]$	—	$[TaO_2Br]$	$[NbO_2I]$	$[TaO_2I]$
$[Nb_3O_7F]$	$[Ta_3O_7F]$	$[Nb_3O_7Cl]$	—	—	—	—	—
$[Nb_5O_{12}F]$	—	—	—	—	—	—	—
$[Nb_{17}O_{42}F]$	—	—	—	—	—	—	—
$[Nb_{31}O_{77}F]$	—	—	—	—	—	—	—

(*i*) *Oxo fluorides*

MOF_3, in contrast to the other oxo trihalides, has not been fully characterized but merely described as an intermediate formed during the hydrolysis of MF_5, or the thermal decomposition at 600 °C or above of MO_2F.[358] The IR on the vapor above NbO_2F at 600–800 °C shows a strong band at 1030 cm^{-1}, which was assigned to the $\nu(NbO)$ stretching vibration of $NbOF_3$; on cooling this species was reported to decompose into NbO_2F and NbF_5.[359]

NbO_2F and TaO_2F were obtained by allowing the oxides to react with HF or F_2.[360,361] They show an ReO_3 type structure in which the fluorine and oxygen atoms are randomly distributed in octahedral positions around the metal atoms.[360] The thermal decomposition of TaO_2F was reported to result in the formation of gaseous TaF_5 and $TaOF_3$ and in the release of solid Ta_3O_7F and Ta_2O_5.[358]

Nb_3O_7F and $Nb_5O_{12}F$ were prepared by heating mixtures of Nb_2O_5 and NbO_2F in sealed platinum tubes at 800 °C. Nb_3O_7F crystallized in the orthorhombic system[359] with the U_3O_8 structure type.[362]

(*ii*) *Oxo chlorides, bromides and iodides*

Of the known oxo halides (Table 19), $NbOCl_3$ is the most extensively studied, partly because it forms during the preparation of $NbCl_5$ from which, owing to their similar volatilities, it is extremely difficult to separate. $NbOCl_3$ can however be conveniently prepared by the reaction of oxygen on $NbCl_5$ at 300–500 °C, or of anhydrous Nb_2O_5 with an excess of $NbCl_5$ vapor.[3] The activation energy of 93.5 kJ mol^{-1} found for the formation of $NbOCl_3$ from solid Nb_2O_5 and molten $NbCl_5$ is relatively low, and this redistribution has been shown to be mainly diffusion controlled.[363] Partial hydrolysis of $NbCl_5$ by $H_2{}^{18}O$ has been utilized to prepare $Nb^{18}OCl_3$.[364] $NbOCl_3$ was also formed when $NbCl_5$–R_2O adducts were heated, as well as from the reaction of $NbCl_5$ with oxygen-containing ligands such as sulfoxides, phosphine oxides or arsine oxides.[3] Similar reactions were observed with substituted halides.

The structure of $NbOCl_3$ consists of chains of dimeric Nb_2Cl_6 units linked together by bridging oxygen atoms, the coordination around each niobium being approximately octahedral.[3] A strong band at 770 cm^{-1} in the IR was attributed to the Nb—O—Nb stretching vibration. The Raman spectra of gaseous $NbOCl_3$ was shown to be consistent with a monomeric C_{4v} molecule, and the vibration found at 997 cm^{-1} is characteristic of an Nb=O double bond.

$TaOCl_3$ has been much less extensively studied. Unlike $NbOCl_3$ it can apparently not be prepared by partial hydrolysis of $TaCl_5$.[3]

Disporportionation of $MOCl_3$ at 300–400 °C yielded MO_2Cl and MCl_5.[3] Nb_3O_7Cl has been prepared from Nb_2O_5 and $NbOCl_3$ at 600 °C; its structure shows octahedrally coordinated metal atoms. A tantalum analog has been mentioned.

$NbOBr_3$ was obtained by methods analogous to those used for $NbOCl_3$.[365] A broad complex absorption around 740 cm^{-1} in the IR, characteristic of Nb—O—Nb chains, and in the NQR two groups of lines with markedly different frequencies, indicating the presence of both terminal and bridging bromine atoms, attest that the structure of $NbOBr_3$ is analogous to that of $NbOCl_3$. $TaOBr_3$ and TaO_2Br have also been obtained.[3]

$NbOI_3$ was conveniently prepared by the reaction of Nb_2O_5 with NbI_5, or from a mixture of niobium and iodine.[3] It decomposes to $NbOI_2$. MO_2I has been obtained as red needles by heating a 1:2:6 molar ratio of $M:M_2O_5:I_2$ at *ca.* 500 °C.[3] By contrast with the oxo triiodides, they are air stable.

34.2.5.2 Thio and seleno halides and their complexes

The thio and seleno derivatives reported so far for Nb^V and Ta^V appear to be limited to the $[M_6S_{17}]^{4-}$,[366] $[MY_4]^{3-}$ (Y = S or Se),[367] $[NbOS_3]^{3-}$ [368] and $[NbO_2S_2]^{3-}$ [369] anions (Section 34.2.8.2.i), to $[MYX_3]$ species and to some of their adducts and derivatives.

The thio and seleno halides $[MYX_3]$ were prepared by the reaction of Sb_2Y_3 with MCl_5 at room temperature in CS_2.[370] $MSCl_3$ was formed when B_2S_3 was allowed to react with MCl_5.[371] MYX_3, with Y = S, Se, and X = Cl, Br, were obtained through thermal decomposition of MY_2X_2.[372]

$[NbSX_3(THT)_2]$ was formed from $NbSX_3$ and THT in solution; a crystal structure (X = Br)[373] showed the Nb atom to be six-coordinated (Nb=S: 2.09 Å). The reaction between $NbSCl_3$ and Ph_3PS afforded a 1:1 adduct[374,375] whose unit cell contains two identical five-coordinate monomers and a centrosymmetric six-coordinate dimer with halogen bridges, the coexistence of which is rather unusual. The niobium atom is displaced from the plane formed by the three chlorine and the ligand's sulfur atoms toward the sulfido sulfur atom (0.533 Å monomer, 0.397 Å dimer); this displacement is larger than in $NbOCl_3$ adducts. $[TaSCl_3(PhNCCl_2)]$ was obtained, as the result of sulfur abstraction, when Ta_2Cl_{10} was allowed to react with PhNCS.[132] It was reported to react with alkali metal thiocyanates to give $[TaSCl_x(NCS)_{3-x}(PhNCCl_2)]$ (x = 1 or 2) and $[TaS(NCS)_4(PhNCCl_2)]^-$.[376] $[MS(Et_2dtc)_3]$[249] was isolated from the reaction of MCl_5 and NaS_2CNEt_2 in non-aqueous solvents. The Ta compound[377] is monomeric with seven sulfur atoms bound to Ta at the corners of a distorted pentagonal bipyramid (Ta=S: 2.181 Å).

34.2.6 Complexes of the Oxo Halides

34.2.6.1 Anionic oxohalo complexes

(*i*) *Oxofluoro anions*

Oxofluoro complexes are formed much more readily for niobium than for tantalum. This difference, indeed, is the basis of the classical method of separation of the two metals. When, for example, a 2–3% HF solution of the oxides is concentrated, K_2TaF_7 precipitates first and $K_2(NbOF_5)\cdot H_2O$ only later.[1]

(*a*) $[MOF_6]^{3-}$. Guanidinium oxohexafluoroniobate was synthesized by adding CN_3H_6F to a solution of Nb_2O_5 in HF.[378] The crystals consist of $[NbOF_6]^{3-}$ and guanidinium ions linked together into chains through NH···F hydrogen bonds. The structure determination for α-$Na_3[NbOF_6]$[379] shows niobium to be located at the center of a slightly distorted pentagonal bipyramid and displaced from the equatorial plane toward the oxygen atom (Nb—F_{eq}: 2.016–2.050(3) Å; Nb—F_{ax}: 2.084(3) Å; Nb=O: (1.738(3) Å).

(*b*) $[MOF_5]^{2-}$. Oxofluoroniobates $A[MOF_5]$ (A = Mn, Co, Ni, Cu, Zn, Cd) were obtained from Nb_2O_5, HF and the carbonates of the bivalent metals; in the same conditions Ta_2O_5 gave fluorotantalates.[48] The strength of the M=O bond depends on the nature of the counterion, as illustrated by the Nb=O bond strength constants shown in Table 20. In the hydrazinium salt the $[NbOF_5]^{2-}$ anion has a distorted octahedral configuration (Nb=O: 1.75 Å).[380]

(*c*) $[M_2OF_{10}]^{2-}$. The formation of $[Ta_2OF_{10}]^{2-}$ from $[TaF_6]^-$ and Et_4NOH was proposed to

Table 20 Calculated Nb═O bond strength in A[$NbOF_5$][1]

A	Mn	Co	Ni	Cu	Zn	Cd	Rb
$k \times 10^8$ (N Å^{-1})	7.24	7.31	6.87	6.78	7.36	7.17	6.50[2]

1. R. L. Davidovich, T. F. Levchishina, T. A. Kaidalova and V. I. Sergienko, *J. Less-Common Met.*, 1972, **27**, 35.
2. L. Surendra, D. N. Sathyanarayana and G. V. Jere, *J. Fluorine Chem.*, 1983, **23**, 115.

proceed through an intermediate hydroxo compound $[TaF_5OH]^-$ followed by a condensation reaction.[381] Its X-ray analysis[382] showed the two tantalum atoms to be bridged by an oxygen atom, the overall symmetry being approximately D_{4h}. The linearity of the M—O—M bridge (M—O = 1.875 Å) may be interpreted as resulting from π-bonding interactions between tantalum and oxygen. This is further supported by the high value of $\nu_{as} = 880\ cm^{-1}$ and low value of $\nu_s = 270\ cm^{-1}$ found for ν(Ta—O—Ta) in the IR and Raman spectra.[381]

(*d*) $[MOF_4]^-$. $[NbOF_4]^-$ was first obtained in 1866 by C. Marignac, by adding NH_4F to an HF solution of Nb_2O_5.[1] It contains Nb—O—Nb bridges.[383]

(*e*) *Miscellaneous.* $(NEt_4)_4[Ta_4O_6F_{12}]$ was obtained by hydrolysis of $(NEt_4)_2[(TaF_5)_2O]$.[384] The anion has an adamantane type Ta_4O_6 skeleton comparable to that of P_4O_6, each metal atom being further coordinated by three fluorine atoms in a *fac* octahedral arrangement (Figure 13). ^{19}F NMR data indicated the presence of the $[Nb_2O_2F_9]^{3-}$, $[NbOF_5]^{2-}$, $[NbOF_6]^{3-}$, $[TaOF_6]^{3-}$ and $[Ta_2O_3F_7]^{3-}$ anions in aqueous solutions.[385–388] The stepwise formation constants of the $[NbOF_x]^{(3-x)}$ complexes with $2 < x < 6$, which exist in solution for a range of concentrations in acid, metal and fluoride ion, were also determined.[41]

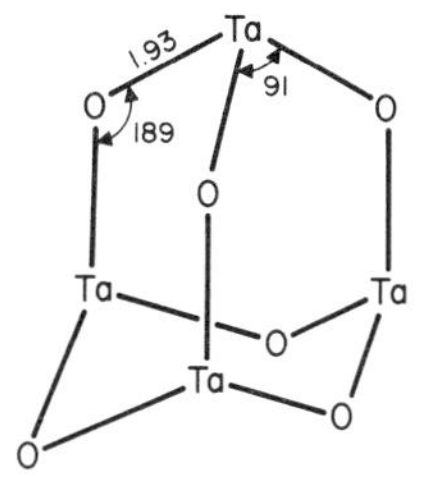

Figure 13 The Ta_4O_6 skeleton of $[Ta_4F_{12}O_6]$ (reproduced from ref. 384 with permission)

(ii) *Oxo anions of the higher halides*

(*a*) $[MOX_5]^{2-}$ ($X = Cl, Br$). $NbCl_5$, $NbOCl_3$ and Nb_2O_5 dissolve in concentrated HCl, from which $A_2[MOCl_5]$ salts ($A = NH_4^+$) were obtained by addition of an excess of ACl.[318,389–392] $A_2[NbOCl_5]$ salts also formed as intermediates during the oxygenation of $A[NbCl_6]$ (A = K, Rb, Cs).[393] $A_2[TaOCl_5]$ was prepared by heating the appropriate hexachlorometallate(V) with a stoichiometric amount of Sb_2O_3 in chlorine or under vacuum.[5] Analytical methods for extracting niobium(V) from HCl solutions involve the use of A = amines,[394] salicylhydroxamic acid[395] or benzohydroxamic acid;[396] the substances entering the organic phase were shown to be $(AH)_2[NbOCl_5]$. $(AsPh_4)_2[NbOCl_5]$ was formed in the reaction of $NbOCl_3$ with $(AsPh_4)Cl$.[397] The structure shows an octahedral arrangement for $[NbOCl_5]^{2-}$ with a very long Nb═O distance of 1.97 Å. Nb_2O_5 forms oxobromo complexes $A_2[NbOBr_5]$ (A = Rb, Cs) in concentrated HBr solutions of the alkali metal bromides.[5]

(*b*) $[Ta_2Cl_{10}O]^{2-}$ *and related species.* The compound $(PMe_3Ph)_2[Ta_2Cl_{10}O]$ was obtained 'serendipitously' by passing oxygen through a toluene suspension of $TaCl_5$ in the presence of PMe_2Ph and Na/Hg amalgam.[398] The structure revealed a linear Ta^V—O—Ta^V unit (Ta—O: 1.880(1) Å) comparable to that found in the $[F_5Ta—O—TaF_5]^{2-}$ ion (Section 34.2.6.1.i).[382] $(PPh_4)_2[Ta_2Cl_{10}O]$ was prepared from $NO[TaOCl_4]$ and PPh_4Cl.[399] $[\{TaCl_2(NMe_2)_2(HNMe_2)\}_2O]$, where three chlorine atoms on each tantalum of the parent $[Ta_2Cl_{10}O]^{2-}$ ion are replaced by nitrogen atoms, was obtained as a minor product (2%) from the reaction of $TaCl_5$ and $HNMe_2$.[270] The central Ta—O—Ta bonds are almost equal (Ta—O: 1.928(6) and 1.917(6) Å) and significantly longer than those found in $[Cl_5Ta—O—Cl_5]^{2-}$ [398] and

$[F_5Ta—O—TaF_5]^{2-}$ (1.880 and 1.875(1) Å respectively).[382] The Ta—Cl bonds *trans* to the NMe_2 groups are longer than in the decachloro anions, which can be accounted for by the stronger tendency of these groups, as compared to bridging oxygen, to be involved in π bonding to the metal.

(*c*) $[M_2Cl_9O]^-$. $[\{TaCl_2(Cp')\}_2(\mu\text{-}Cl)_3]_2[Ta_2Cl_9O)_2]$[400] was isolated as a by-product in the reaction of $[SnBu_3(C_5Me_5)]$ with $TaCl_5$. $(C_4ClPh_4)_2[Nb_2Cl_9O]_2$ [401] was obtained from $NbCl_5$ and PhC≡CPh in the presence of water. The M—O—M bridges are asymmetrical in both $[M_2Cl_9O]_2^{2-}$ anions, as are the two central chlorine bridges that link the two $[M_2Cl_9O]$ units together (Figure 14).

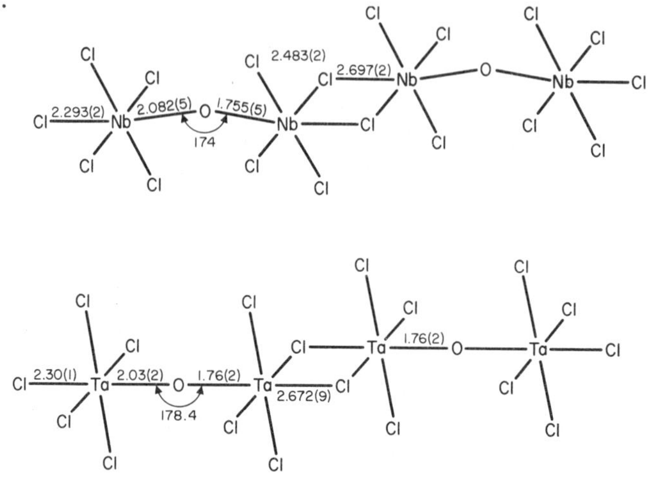

Figure 14 The $[M_2Cl_9O]_2^{2-}$ anions (reproduced from refs. 400 and 401 with permission)

(*d*) $[MOCl_4]^-$ *and* $[MO_2Cl_3]^{2-}$. $A[MOCl_4]$ (A = alkali metal) salts were obtained by heating the appropriate hexachlorometallate(V) with the stoichiometric amount of Sb_2O_3 in chlorine or under vacuum.[420] $(NO)[NbOCl_4]$ was prepared from NOCl and $NbOCl_3$. $[NbOCl_4](pyH)$[403,404] and $(R_3NH)[NbOCl_4]$[405] were precipitated from HCl solutions of $NbCl_5$ or $NbOCl_3$ and the appropriate amine; IR indicated the presence of —NbONbONb— chains.

The addition of diars to $NbCl_5$ in dry CCl_4 caused the precipitation of $NbCl_5$(diars); attempts to recrystallize it yielded $[NbCl_4(diars))_2][NbOCl_4]$ and $[NbCl_4(diars)_2]_2[NbO_2Cl_3]$. $[NbOCl_4]^-$ is basically a square pyramid with the oxygen at the apex (Nb═O: 1.70(2) Å); one of the Nb—Cl bonds is outstandingly long (2.438(4) Å) with the chlorine atom pointing to the sixth potential coordination site of the metal of a neighboring $[NbOCl_4]^-$ anion, with a very short contact of 3.011(6) Å.[152]

(*e*) $[MO_2Cl_2]^-$. Compounds of composition $A[TaO_2Cl_2(H_2O)]$, where A = K, Rb, Cs or NH_4, were isolated by saturating solutions of $TaCl_5$ and of the alkali or ammonium chloride with HCl.[406] The IR absorptions found in the 830–875 cm^{-1} region were assigened to tantanyl groups joined by asymmetrical Ta═O—Ta bridges. Under the same conditions, $NbCl_5$ led to $A_2[NbOCl_5]$ (see above).

34.2.6.2 Neutral adducts of the oxo halides and related compounds

Numerous 1:1 and 1:2 adducts are known for MOX_3, but few for $NbOF_3$, and none yet for $NbOI_3$. The adducts of the oxo trihalides of Nb^V and to a lesser extent of Ta^V were often obtained from the pentahalides through oxygen abstraction from the ligand or the solvent. Others were prepared directly from the oxo trihalides.

(*i*) *Oxygen, sulfur and selenium donors*

Oxygen was abstracted from phosphine oxides, phosphonates,[407] arsine oxides and sulfoxides by MX_5 as well as by $[MX_4Me]$ (X = Cl, Br),[408] and more easily by the Nb than by the Ta derivatives, to yield MOX_3 adducts (equation 36; Table 21). IR indicated the following sequence for the metal's acceptor character: $MCl_5 > MOCl_3 > MOCl_2Me$. NMR showed that in the presence of an excess of oxo ligand $[NbCl_4Me]_2$ is first and rapidly converted to the 1:1 adduct $[NbCl_4MeL]$; the oxygen *vs.* chlorine exchange then occurs at a slower rate, which depends, in the case of phosphoryl ligands, on the π character of the phosphoryl bond (Table 22).[408]

$$MX_5 + 3O{=}E \longrightarrow [MOX_3(O{=}E)_2] + ECl_2 \qquad E = PR_3, AsR_3, SR_2; \quad X = Cl, Br \tag{36}$$

Table 21 Typical Adducts of MOX_3 with Oxygen Donors

L	*Compounds*	*Comments*	*Ref.*
HMPA	$[MOCl_3L_2]$	X-Ray structure (M = Nb)	1–4
$[OP(NMe_2)_2]_2O$	$[MOCl_3LL]$	Monomeric	2, 3
$[OP(C_6H_4X)_3]$	$[MOCl_3L_2]$	X = H, *p*-NMe_2, *p*-OMe, *p*-Me, *p*-Cl, *p*-Br, *m*-Cl, *m*-NO_2	5–7
	$[NbOF_3L_2]$	$\Delta\nu(P{-}O)_X = 1168 + 7.6\sigma_X^+\ cm^{-1}$	
$(OAsPh_3)$	$[MOX_3L_2]$	X = F, M = Nb; X = Cl, Br, M = Nb, Ta	4, 6, 7
diars	$[MOX_3LL]$	$\nu(Nb{-}O)$: 967 cm^{-1} (X = Cl); 951 cm^{-1} (X = Br)	8
	$[(MCl_4LL)_2O]$	$\nu(Nb{-}O{-}Nb)$: 765 cm^{-1}	
OSR_2	$[NbOX_3L_2]$	X = F, Cl; R = Me, Ph	7, 9, 10
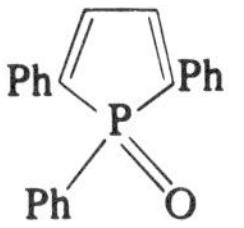	$[MOCl_3L]$		11

1. L. G. Hubert-Pfalzgraf and A. A. Pinkerton, *Inorg. Chem.*, 1977, **16**, 1895.
2. R. J. Dorschner, *J. Inorg. Nucl. Chem.*, 1972, **34**, 2665.
3. L. G. Hubert-Pfalzgraf, R. C. Muller, M. Postel and J. G. Riess, *Inorg. Chem.*, 1976, **15**, 40.
4. D. Brown, J. F. Easey and J. G. H. Du Preez, *J. Chem. Soc. (A)*, 1966, 258.
5. E. G. Amarskii, A. A. Shvets and O. A. Osipov, *Zh. Obshch. Khim.*, 1975, **45**, 898.
6. D. B. Copley, F. Fairbrother and A. Thompson, *J. Less-Common Met.*, 1965, **8**, 256.
7. J. Sala-Pala, J. Y. Calves and J. E. Guerchais, *J. Inorg. Nucl. Chem.*, 1975, **37**, 1294.
8. R. J. H. Clark, D. L. Kepert and R. S. Nyholm, *J. Chem. Soc.*, 1965, 2877.
9. D. B. Copley, F. Fairbrother, K. H. Grundy and A. Thompson, *J. Less-Common Met.*, 1964, **6**, 407.
10. J. G. Riess, R. C. Muller and M. Postel, *Inorg. Chem.*, 1974, **8**, 1802.
11. D. Budd, R. Churchman, D. G. Holah, A. N. Hughes and B. C. Hui, *Can. J. Chem.*, 1972, **50**, 1008.

Table 22 Dependence of the Oxygen *vs.* Chlorine Exchange Reaction Rate upon the π Character of the P=O Bond[a]

L	*P$\pi\pi'$ character*	*ν(P=O)* (cm^{-1})	*Reaction time in CH_2Cl_2* (min)
$OPMe_3$	0	1182	30
HMPA	0.2	1210	30
$O[OP(NMe_2)_2]_2$	—	1230	30
$OPPh_3$	0.370	1195	180
$OP(OMe)(NMe_2)_2$	0.390	1220	60
$OP(OMe)_2(NMe_2)$	0.77	1260	180
$OP(OMe)_3$	1.079	1278	b

[a] From C. Santini-Scampucci and J. G. Riess, *J. Chem. Soc., Dalton Trans.*, 1974, 1433.
[b] No apparent reaction after 6 months in CH_2Cl_2.

Although the P=O bond is weaker in the phosphole oxides (TPPO) than in Ph_3PO, only traces of $[NbOCl_3(TPPO)]$ were obtained from $NbCl_5$ and excess phosphole.[124] Furthermore, oxygen was picked up at phosphorus and not at the metal, to form $[NbX_5(TPPO)]$, when NbX_5 was allowed to react with TPPSe in a non-rigorously-oxygen-free atmosphere.

(*ii*) *Nitrogen donors* (Table 23)

Niobium and tantalum oxo halides react with amines in stoichiometric amounts to form polymeric 1/1 materials containing MOMO chains. Monomolecular $[MOX_3(NR_3)_2]$ derivatives were obtained in the presence of excess amine.[365,408–411]

Table 23 Nitrogen Donor Adducts of MOX_3

L	*Compounds*	*Comments*	*Ref.*
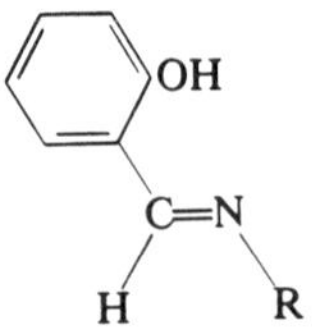	$[MOCl_3L_2]$	R = Me, Et	1
$RC_6H_4N{=}CHC_6H_4R'$	$[MOCl_3L]$	R = H; R′ = H, *p*-Br, *p*-NO_2, *m*-NO_2, *o*-OH	2
	$[MOCl_3L_2]$	R = *o*-OH; R′ = H, *p*-Br, *p*-NO_2, *p*-NMe_2	
RCN	$[MOCl_3L_2]$	X-Ray structure (M = Nb; R = Me)	3
bipy	$[NbOX_3L]$	X = Cl, Br	4–7
	$[Nb_2OCl_8L_2]$		4
	$[Ta_4OCl_{18}L_4]$		4
phen	$[NbOBr_3L]$		7
NR_3	$[NbOX_3L]_n$	X = Cl, Br; R = Me, Et, Bu, octyl	8–11
	$[NbOX_3L_2]$	R = Me, Et, Bu	10
$NPh(Et)_2$	$[NbOX_3L]_n$		12
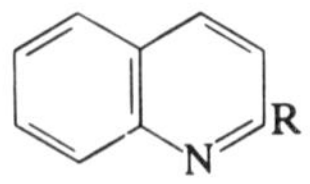	$[NbOX_3L]_n$	R = Me, Ph	12
$NHEt_2$; NH_2Bu	$[NbOCl_3L]_n$		11

1. K. Yamanouchi and S. Yamada, *Inorg. Chim. Acta*, 1976, **18**, 201.
2. A. V. Leshchenko, L. V. Orlova, A. D. Garnovskii and O. A. Osipov, *Zh. Obshch. Khim.*, 1969, **39**, 1843.
3. C. Chavant, J. C. Daran, Y. Jeannin, G. Constant and R. Morancho, *Acta Crystallogr., Sect. B*, 1975, **31**, 1828.
4. C. Djordjevic and V. Katovic, *Chem. Commun.*, 1966, 224.
5. D. Djordjevic and V. Katovic, *J. Chem. Soc. (A)*, 1970, 3382.
6. V. Katovic and C. Djordjevic, *Inorg. Chem.*, 1970, **9**, 1720.
7. N. Brnicevic and C. Djordjevic, *J. Less-Common Met.*, 1967, **13**, 470.
8. S. M. Sinitsyna, T. M. Gorlova, V. F. Chistyakov, *Zh. Neorg. Khim.*, 1973, **18**, 2114.
9. S. M. Sinitsyna, V. G. Khlebodarov and M. A. Bukhtereva, *Z. Neorg. Khim.*, 1974, **19**, 1532.
10. S. M. Sinitsyna, V. I. Sinyagin and Yu. A. Buslaev, *Izv. Akad. Nauk SSSR, Neorg. Mater.*, 1969, **5**, 605.
11. Yu. A. Buslaev, S. M. Sinitsyna, V. I. Sinyagin and M. A. Polikarpova, *Zh. Neorg. Khim.*, 1970, **15**, 2324.
12. A. V. Leshchenko, V. T. Panyushkin, A. D. Garnovskii and O. A. Osipov, *Zh. Neorg. Khim.*, 1966, **11**, 2156.

Reactions of bipy with MX_5 and $NbOCl_3$ [134,364,412–414] yielded $[MX_5(bipy)]$ and $[NbOCl_3(bipy)]$ in benzene, while mixtures formulated as $[NbOCl_3(bipy)][NbCl_5(bipy)]$ and $[TaOCl_3(bipy)][TaCl_5(bipy)]_3$ were isolated in ether. Solutions of $NbCl_5$ and bipy in alcohols containing a small, controlled amount of water yielded $[NbOCl_2(OR)(bipy)]$.

A series of monomeric, six-coordinate niobium(V) adducts with neutral Schiff's bases was synthesized from $NbOCl_3$ [410] or $NbOCl_3(OPPh_3)_2$.[207]

(*iii*) *Structures*

In spite of the large number of oxo trihalide adducts reported, there is still little structural information available on them. X-Ray diffraction studies are limited to those of $[NbOCl_3(MeCN)_2]$,[419] $[NbOCl_3(HMPA)_2]$[416] and $[NbOCl_3(OPCl_3)]_4$.[235] $[NbOCl_3(HMPA)_2]$ is monomeric in the solid (Figure 15) and in solution. In the solid the metal is octahedrally surrounded; the two HMPA ligands are *cis* to each other, one of them being *trans* to the oxo bond; the same isomer always predominates in solution. The Nb═O distance is comparable to that found in $[NbOCl_3(MeCN)_2]$ (1.68(2) Å); the lesser steric requirement of the MeCN ligands allows for a larger distortion of the coordination polyhedron and for shorter Nb—Cl bonds. A considerable *trans* effect was found for the metal–ligand bonds in both the HMPA and the MeCN adducts. In solution $[NbOCl_3(HMPA)_2]$ undergoes drastic changes and gives complex solvent dependent mixtures, as depicted in Scheme 3. All the species in solution are in dynamic equilibrium on the NMR time-scale. By evaporation the initial crystalline *trans* isomer of $[NbOCl_3(HMPA)_2]$ was integrally recovered. Similar solution behavior was observed for $[NbOCl_3(DMSO)_2]$, $[NbOCl_3(OMPA)]$, $[NbOCl_2(OEt)(HMPA)_2]$ and $[NbOCl_2Me(HMPA)_2]$.[408]

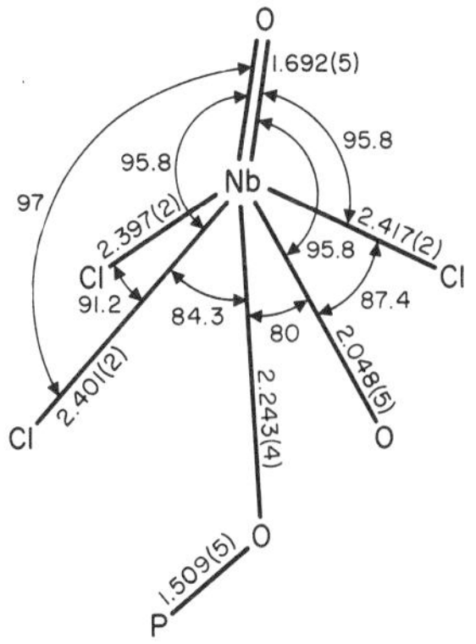

Figure 15 Coordination polyhedron of $[NbOCl_3(HMPA)_2]$ (reproduced from ref. 410 with permission)

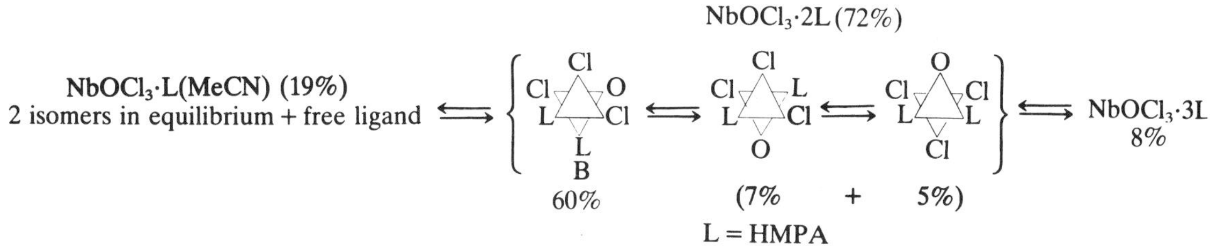

Scheme 3

The oxygen-bridged tetrameric Nb_4O_4 unit in $[NbOCl_3(OPCl_3)]_4$ [235] is almost planar with two different values for the Nb—O distances in agreement with a (—Nb═O—)$_4$ structure.

34.2.6.3 Miscellaneous oxygen abstraction reaction products

Oxygen abstraction reactions from alcohols, ketones, acyl halides and other oxygen-containing compounds by MCl_5 or related derivatives also afforded a route to relatively simple complexes of the oxo halides (Table 24). Thus Nb_2Cl_{10} was found to abstract oxygen from higher ketones to yield, after hydrolysis, alkanes, chloroalkanes or arylalkanes.[414]

Table 24 Typical Oxygen Abstraction Reactions

O species	*Reactions*	*Compounds*	*Ref.*
RCOX (X = Cl, Br)	$Nb(OR')_5$ + RCOX (excess)	$[NbOX_3(RCOOR')]$	1
Et_2O	$NbCl_5 + NbCl_5(Et_2O)$ + 2bipy	$[Nb_2OCl_8(bipy)_2]$	2
	$3TaCl_5 + TaCl_5(Et_2O)$ + 4bipy	$[Ta_4OCl_{18}(bipy)_4]$	2
Me_2CO	$NbCl_3Me_2 + 2Me_2CO$	$[NbOCl_2(OCMe_3)]$	3
RR'CO(R = R' = Et;	MCl_5 + RR'CO	$[\{MCl_4(RR'CO)\}_2O]$	4
R = Me, R' = Et)	MCl_5 + RR'CO (excess)	$[MOCl_3(RR'CO)_2]$	4

1. R. C. Mehrotra and P. N. Kapoor, *J. Less-Common Met.*, 1966, **10**, 348.
2. C. Djordjevic and V. Katovic, *J. Chem. Soc. (A)*, 1970, 3382.
3. J. D. Wilkins, *J. Organomet. Chem.*, 1974, **80**, 357.
4. M. S. Gill, H. S. Ahuja and G. S. Rao, *J. Indian Chem. Soc.*, 1978, **55**, 875.

Oxygen abstraction from $SOCl_2{\cdot}H_2O$ by $[NbCl_2(RCp)(R'Cp)]$ led to a series of new $[NbOCl(Cp)(Cp')]$ compounds.[417] This reaction, which allowed the synthesis of diastereoisomeric pairs of optically active species, could be reversed upon heating (Scheme 4).

The unexpected formation of $[NbOCl(Cp)_2]$ as a by-product in the synthesis of $[NbCl_3(Cp)_2]$[418] can probably be attributed to the presence of traces of oxygen or moisture. When $[NbCl_2(Cp)_2]$ was allowed to react with oxygen, $[NbOCl(Cp)_2]$[302,419,420] or $[\{NbCl_2(Cp)_2\}_2O]$[421] was obtained. The latter compound shows anticancer activity, lower than

$$[Nb^{IV}Cl_2(Cp)_2] \xrightarrow{SOCl_2 \cdot H_2O} [Nb^VOCl(Cp)_2]$$

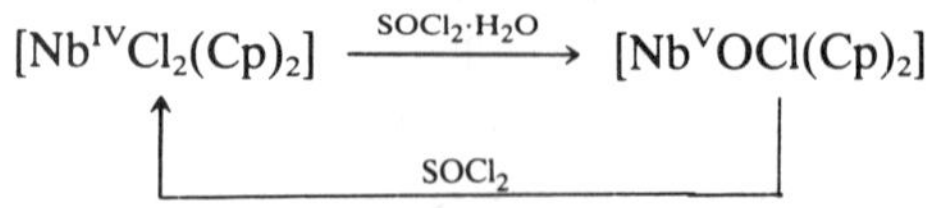

Scheme 4

that of $[NbCl_2(Cp)_2]$ but paralleled by a significant reduction of the toxic side-effects. The ionic $[\{NbCl(Cp)_2\}_2O][BF_4]^{422}$ derivative was obtained from $[NbCl_2(Cp)_2]$ by oxygen abstraction from water and addition of HBF_4; its Nb—O—Nb bridge is significantly non-linear (169°) but the Nb—O bonds (1.88(1) Å) are relatively short, indicating a certain degree of π bonding.[423] The reaction between $[NbMe_2(Cp)_2]$ and an excess of NO proceeded *via* a nitrosyl adduct, probably with a bent MNO arrangement ($v(NO)$: $1670\,cm^{-1}$) and gave a methyloxo derivative which further inserted NO (equation 37).[343]

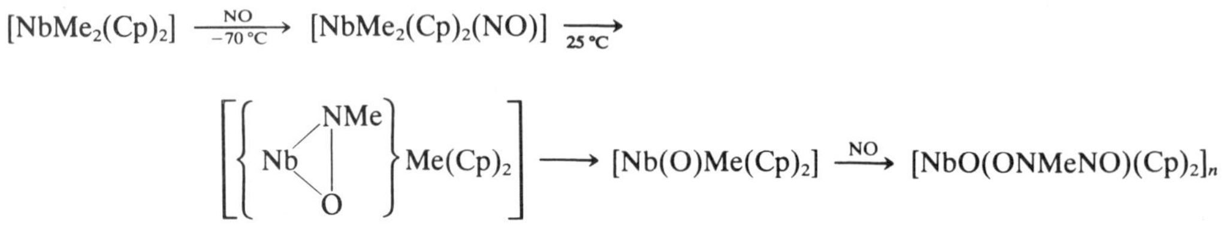

$$[NbMe_2(Cp)_2] \xrightarrow[-70\,°C]{NO} [NbMe_2(Cp)_2(NO)] \xrightarrow{25\,°C}$$

(37)

$$\left[\left\{Nb\overset{NMe}{\underset{O}{\big|}}\right\}Me(Cp)_2\right] \longrightarrow [Nb(O)Me(Cp)_2] \xrightarrow{NO} [NbO(ONMeNO)(Cp)_2]_n$$

Catalytic oxidation cycles involving the niobium oxide cations NbO^+ and NbO_2^+ in the gas phase have been reported.[424]

34.2.7 Solvolysis Products of the Oxo Trihalides

34.2.7.1 From Oxygen Compounds

(i) Oxoalkoxo compounds (Table 25)

Synthetic routes to well-characterized Nb^V and Ta^V oxoalkoxides remain surprisingly scarce. The direct alkoxylation of MOX_3 is far from providing a general route to these compounds. The alternative exchange of oxygen *vs.* chlorine between $[NbCl_2(OR)_3]_2$ and an oxo ligand often results in redistribution products.

The controlled hydrolysis of MCl_5 or $[Ta(OR)_5]_2$ (R = Et, $SiMe_3$) led to polymeric oxo alkoxides. On the other hand, the hydrolysis of $[NbCl_2(OR)_3]_2$ in the presence of bipyridyl led to $[NbOCl_2(OR)(bipy)]$ (R = Et, Pr^i) adducts.[134,364,412] An X-ray structure for R = Et showed a distorted coordination octahedron, with the two chlorine atoms *trans* to each other (Nb=O: 1.71(3) Å). The Nb—OR bond is only 0.16 Å longer than the Nb=O oxo bond, and was presumed to have considerable double bond character, which is also consistent with the very open NbOC bond angle of 149°.

MCl_5 when allowed to react with acidic aqueous solutions of tropolone (HT) gave the tetrakis(tropolonate) cations $[MT_4]^+$. Hydrolysis of the niobium derivative afforded $[NbOT_3]$;[425] an oxygen-labeling experiment established that the oxo atom is abstracted from the displaced tropolone molecule.

N-Benzoylphenylhydroxylamine (BPHA) reacts with Nb_2O_5 in bioling acidic aqueous solutions to yield a monomeric seven-coordinated niobium compound $[NbO(BPHA)_3]$.[426] The hydroxylamine ligand could be displaced by tropolone or 8-quinolinol, and the rate-determining step was shown to be associative (equation 38).[427] Oxoniobium and tantalum(V) 8-quinolinolates have also been reported. The optimal conditions for the formation and extraction of the $[NbO(C_6H_4O_2)_3]^{3-}$ ion formed in the niobium(V)–pyrocatechol system, have been investigated.[428]

$$[NbO(BPHA)_3] \underset{L=HT,\,oxH}{\overset{HL}{\rightleftarrows}} [NbO(BPHA)_3(HL)] \longrightarrow products$$

(38)

(ii) Oxo-β-diketonato derivatives

The action of acetylacetone on MOX_3 (X = Cl, Br) was reported to yield $[NbOX_2(acac)]$,[413,429] while the analogous Ta^V compounds were not formed. Ionic β-diketonates $[NbOCl_3(RCOCHCOR')]^-$ were obtained from oxotetrachloroaquaniobates.[430]

Table 25 Oxoalkoxo Derivatives

Compounds	*Comments*	*Ref.*
$[NbOCl_2(OR)bipy)]$	R = Et, Pr; monomeric	1, 4
$[NbOCl_2(OMe)(HMPA)_2]$	Solutions consist of complex mixtures of monomeric species in dynamic equilibrium	5
$[NbOCl_2(OMe)(MeOH)]_2$	Dimeric (Nb—O—Nb) compound	6
$[TaOCl_2(OMe)(DMSO)_2]$	Monomeric; NMR: various monomeric isomers in dynamic equilibrium	7
$[NbOCl(OCH_2CCl_3)_2(DMSO)_2]$	Monomeric; NMR: various monomeric isomers in dynamic equilibrium	7
$[Nb_4O_7Cl_4(OEt)_2(bipy)_2(H_2O)_2]$	M—O—M bridges	8
$[Ta_6O_{11}Cl_6(OEt)_2(bipy)_2(H_2O)_{10}]$	M—O—M bridges	8
$[NbO(OCH_2CF_3)_3(MeCN)]_2$	Dimeric; NMR: dynamic equilibrium between μ-alkoxo and μ-oxo isomers	6
$[NbO(OEt)_3]_2$	Dimeric, pentacoordinated Nb	6
$[NbO(OMe)(OC_6H_4CHO)_2]_2$	Nb—O—Nb	6
$[NbO(OMe)_2(OC_6H_4CHO)(MeCN)]$	NMR: various monomeric hexacoordinated species in dynamic equilibrium	6
$[NbO(OR)_3]$, $[Nb_2O(OR)_8]$	R = Bu^t	9
$[TaO_x(OSiMe_3)_{5-2x}]_n$	Polymeric compounds (Ta—O—Ta)	10
$[NbOT_3]$	HT = tropolone	11
$[NbO(BPHA)_3]$	BPHA = $[PhC(O)N(O)Ph]^-$; monomer	12
$[MO(ox)_3(oxH)]$	oxH = 8-hydroxyquinoline	13
$[NbO(cat)_3]^{3-}$, $H[NbO(cat)_2]$	cat = catechol	14

1. C. Djordjevic and V. Katovic, *Chem. Commun.*, 1966, 224.
2. C. Djordjevic and V. Katovic, *J. Chem. Soc. (A)*, 1970, 3382.
3. V. Katovic and C. Djordjevic, *Inorg. Chem.*, 1970, **9**, 1720.
4. B. Kamenar and C. K. Prout, *J. Chem. Soc. (A)*, 1970, 2379.
5. L. G. Hubert-Pfalzgraf, R. Muller, M. Postel and J. G. Riess, *Inorg. Chem.*, 1976, **15**, 40.
6. L. G. Hubert-Pfalzgraf and J. G. Riess, *Inorg. Chim. Acta*, 1980, **41**, 111.
7. L. G. Hubert-Pfalzgraf and J. G. Riess, *Inorg. Chim. Acta*, 1981, **47**, 7.
8. C. Djordjevic and V. Katovic, *J. Less-Common Met.*, 1970, **21**, 325.
9. D. C. Bradley, B. N. Charkravarty and W. Wardlaw, *J. Chem. Soc.*, 1956, 4439.
10. D. C. Bradley and C. Prevedoroudemas, *J. Chem. Soc. (A)*, 1966, 1139.
11. E. L. Muetterties and C. W. Alegranti, *J. Am. Chem. Soc.*, 1969, **91**, 4420.
12. F. I. Lobanov, V. M. Fes'kova and I. M. Gibalo, *Zh. Neorg. Khim.*, 1971, **16**, 776.
13. H. A. Szymanski and J. H. Archibald, *J. Am. Chem. Soc.*, 1958, **80**, 1811.
14. F. I. Lobanov, O. D. Savrova and I. M. Gibalo, *Zh. Neorg. Khim.*, 1973., **18**, 408.

Their solutions consisted of mixtures of the three possible monomeric octahedral isomers; these are rapidly interconverting on the NMR time-scale through an intramolecular mechanism.[431] The crystal structure of the $[NBOCl_3\{C_4H_3SC(O)CHC(O)CF_3\}]^-$ ion established that the isomer present in the solid state has the thenoyl groups *trans* to the Nb═O bond.[432] The niobium atom displays a distorted octahedral arrangement, and is displaced toward the oxo oxygen by 0.295 Å (Nb═O: 1.70(3) Å). Of the two chelate Nb—O distances, 2.044(34) and 2.285(3) Å, the longer is *trans* to the oxo group.

(ii) *Oxocarboxylato*

Oxocarboxylato derivatives were shown to form by treatment of NbX_5 or $Nb(OR)_5$ with an excess of carboxylic acids[235,433,434] or anhydrides[435] (equations 39a and 39b).

$$[MCl_5]_2 + 4RCOOH \longrightarrow 2[MOCl_2(RCOO)] + 4HCl + 2RCOCl \tag{39a}$$
$$M = Ta;\quad R = H, Me, Et, Pr^i, Bu^t$$
$$M = Nb;\quad R = Me, Et, Pr^i, Bu^t, CH_2Cl, CH_2F, Ph$$

$$[M(OEt)_5] + 4RCOOH \longrightarrow [MO(RCOO)_3] + RCOOEt + 4EtOH \tag{39b}$$
$$M = Nb, Ta;\quad R = Me, Et, Pr$$

The reaction of niobioporphyrin $[Nb_2O_3(TPP)_2]$ with acetic acid gave $[NbO(TPP)(MeCOO)]$ n which the metal is seven-coordinated and lies within a polyhedron of symmetry close to C_s, with a square based defined by the four nitrogen atoms of the pyrrole rings and a nearly parallel triangular base formed by the three oxygen atoms (Figure 16).[330]

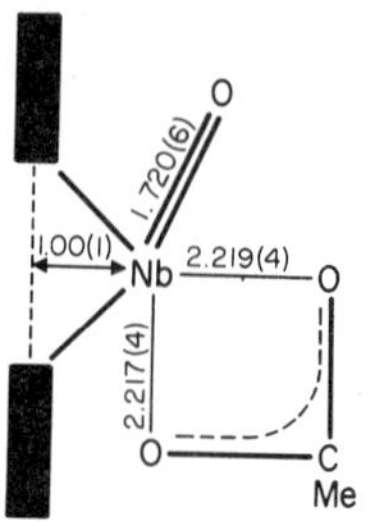

Figure 16 Structure of [NbO(TPP)(MeCOO)] (reproduced from ref. 339 with permission)

Ascorbic acid with niobium(V) in acid solution forms an intensely colored complex $[NbO(OH)_2(C_6H_7O_6)]$, which was used for the spectrophotometric determination of Nb.[436] Malato complexes of niobium(V) were obtained from aqueous racemic malic acid solutions of Nb_2O_5.[437] Niobic and tantalic acids dissolve in solutions of oxalic acid,[1] and mono-, bis- or tris-oxalatooxo niobium(V) compounds were reported to exist in these solutions, depending on pH and concentrations.[438–440] X-Ray diffraction studies on $[NbO(C_2O_4)_3]^{3-}$,[441,442] $[NbO(C_2O_4)_2(OH)(H_2O)]^{2-}$ [443] and $[NbO(C_2O_4)_2(H_2O)_2]^{-}$ [444] have shown niobium to be seven-coordinated. The oxo atom is at one of the apices of a pentagonal bipyramid and the metal is displaced toward the oxo atom by 0.18, 0.21 and 0.245 Å (Nb═O: 1.71, 1.66 and 1.69 Å respectively). The water molecules in $[NbO(C_2O_4)_2(H_2O)_2]^{-}$ exchange with oxygen donors; mixed ligand complexes with sulfoxides,[445] phosphine and arsine oxides[446] have been reported. Mixed complexes with the 4-(2-pyridylazo)resorcinol (PAR) N donor ligand are used in analytical chemistry.[447,448]

Oxohalooxalato niobium anions $[NbO(C_2O_4)_2F_2]^{3-}$ [449] and $[NbO(C_2O_4)_3Br]^{4-}$ [450] have been obtained from the parent tris(oxotrioxalato) anion. Complexation of the oxalato ligand as a bis(bidentate) bridge in $[Nb_2O_2X_6(C_2O_4)]^{2-}$ (X = F, Cl, Br) has been established on the basis of IR.[451]

The solubility of Ta_2O_5 in aqueous oxalic acid solutions is much lower than that of Nb_2O_5 and the resulting solutions are more complex owing to the presence of polymeric species.[452]

34.2.7.2 *From sulfur compounds*

$[NbO(R_2dtc)_3]$ has been prepared for R = ethyl[453,454], cyclopentyl and cycloheptyl.[455] All are monomers, with sulfur-bonded dithiocarbamates. The structure of the *N*-diethyl derivative[456] showed the niobium atom at the center of a pentagonal bipyramid with the oxygen at one of the apices. Two of the chelating ligands lie approximately in the equatorial plane, while the third spans an axial and an equatorial site. The axial M—S distance (2.753(4) Å) is longer than the equatorial ones (2.47–2.598 Å). The metal atom lies above the best plane passing through the equatorial sulfur atoms on the side of the oxygen (Nb═O: 1.74(1) Å).

34.2.7.3 *From nitrogen compounds: oxo amides and oxo isothiocyanates*

The aminolysis of $NbOCl_3$ [410] or $NbOBr_3$ [365] by an excess of primary amine was shown to affect only one of the halogen atoms, to give $[NbOX_2(NHR)]$. Excess of secondary amines with short alkyl chains displaces two bromine atoms. With increasing chain length, the tendency of a second halide atom to be substituted decreases and the second molecule of amine rather coordinates to the initial aminolysis product to yield $[NbOX_2(NR_2)(NHR_2)]$. Trialkylamines gave the addition products $[NbOX_3(NR_3)]$. Monomeric $[NbO\{N(SiMe_3)_2\}]$ was obtained from $NbCl_4(THF)_2$.[458]

$[NbO(NCS)]^{2-}$ salts were precipitated from acidic solutions of Nb^V with excess thiocyanate.[459] The structure[457] of its $(Ph_4As)^+$ salt showed the metal to be at the center of a distorted octahedron (Nb═O: 1.70 Å), and to be displaced by 0.21 Å out of the nitrogen's plane toward the oxygen. The Nb—N bond *trans* to the oxo ligand is longer (2.27 Å) than the others (2.03 to 2.16 Å).

Addition of heterocyclic nitrogen bases to acid solutions of Nb and KSCN gave the mixed monomeric complexes listed in Table 26. $[TaOCl_3(MeCN)_2]$ and $[TaSCl_3(PhNCCl_2)]$ react with alkali metal thiocyanates in MeCN to yield $[TaOCl_{3-x}(NCS)_x(MeCN)_2]$, $A[TaO(NCS)_4]$, $[TaSCl_{3-x}(NCS)_x(PhNCCl_2)]$ or $A[TaS(NCS)_4(PhNCCl_2)]$ ($x = 1$–3; $A = NH_4$, K, Na).[376] $[NbO(NCS)_3L]$, where L is $C_6H_{10}O$ or $C_9H_{14}O$, was isolated from $NbCl_5$ and KNCS solutions in acetone.[460]

Table 26 Mixed Oxo–Isothiocyanato complexes[1]

Compounds	*Medium*	$\nu(Nb{=}O)$ (cm^{-1})
$(HA)[NbO(NCS)_3(OH)H_2O)]$[a]	H_2SO_4	914
$(HA)[NbOCl_2(NCS)_2(H_2O)]$[a]	HCl	922
$(HDMP)[NbOCl_3(NCS)(H_2O)]$[b]	HCl	935
$(HColl)[NbO(NCS)_4(Coll)]$[c]	H_2SO_4	938

[a] A = phen.
[b] DMP = 2,3-dimethylphen.
[c] Coll = (*S*)-collidine.

1. From F. N. Lobanov, V. M. Zatonskaya and I. M. Gibalo, *Zh. Neorg. Khim.*, 1980, **25**, 3003.

34.2.8 Oxide- and Sulfide-derived Compounds

34.2.8.1 Oxides and peroxides

The ultimate products of the oxidation of the metals are the pentoxides. Reduction of M_2O_5 by H_2 yields MO_2 and MO. These are beyond the scope of this chapter. Numerous orthometallates $[MO_4]^{3-}$, metametallates $[MO_3]^-$ and polymetallates are known; their chemistry is reviewed in Chapter 38.

(*i*) *Peroxo acids and salts*

Orthoperoxometallates $A_3[M(O_2)_4]$ precepitate from alkaline solutions of niobates or tantalates on additon of H_2O_2.[461–467] The four O_2 groups are linked side-on to the metal in $[M(O_2)_4]^{3-}$, which releases oxygen progressively in mild conditions, and explosively at 80 °C.[468] The $[Nb(O_2)_4]^{3-}$ ion has nearly D_{2d} symmetry; the M—O distances range from 1.99 to 2.07 Å, and the O—O mean distance is 1.50 Å.[466]

On treatment with H_2SO_4, the orthoperoxo salts gave the corresponding acids HMO_4.[461,469–472] The niobium derivative was formulated as $[Nb_2O_2(O_2)_2(OH)_2]\cdot xH_2O$ on the basis of IR data.[461] Metaperoxo salts AMO_5 (A = Na, K, Rb, Cs) have also been obtained.[1,473] Oxidation of the coordinated O_2^{2-} moiety by Ce^{IV} or OH radicals yielded complexes of Nb^V of type $[Nb(O_2)(OH)_2]$ containing the O_2^- radical ion, as evidenced by EPR.[474,475]

(*ii*) *Peroxo complexes*

The di- and tri-peroxo compounds (Table 27), in which the metal is bound to the oxalato,[476,477] phen,[478] or edta ligands, were prepared from alkali metal niobates or tantalates(V) in hydrogen peroxide solutions. $[Nb(O_2)_3phen]^-$ is to date the only example of a triperoxo complex authenticated by X-ray crystallography (Figure 17). The metal, at the center of a distorted dodecahedron, is located 0.34 Å above the mean plane containing two peroxo groups and one nitrogen atom, on the same side as the third peroxo group. In $[Nb(O_2)_2(C_2O_4)_2]^{3-}$,[441] the two peroxo groups are *cis*, which is unusual (O—O mean distance 1.48 Å); the coordination polyhedron is a distorted dodecahedron.

Table 27 Di- and Tri-peroxo Complexes

Compounds	*Comments*	*Ref.*
$A_3[Nb(O_2)_3(C_2O_4)]$	A = NH_4, K, Rb, Cs	1
$A_3[Nb(O_2)_2(C_2O_4)_2]$	A = NH_4, K, Rb, Cs; X-ray structure (NH_4)	1, 2, 3
$A_3[M(O_2)_3(edta)]$	A = K, NH_4	4
$K_3[Nb(O_2)_3(phen)]$	X-Ray structure	5

1. R. N. Shchelokov, E. N. Traggeim, M. A. Michnik and K. I. Petrov, *Russ. J. Inorg. Chem. (Engl. Transl.)*, 1972, **17**, 1270.
2. J. E. Guerchais and B. Spinner, *Bull. Soc. Chim. Fr.*, 1965, 1122.
3. G. Mathern and R. Weiss, *Acta Crystallogr., Sect. B*, 1971, **27**, 1572.
4. N. Vuletic, E. Prcic and C. Djordjevic, *Z. Anorg. Allg. Chem.*, 1979, **450**, 67.
5. G. Mathern and R. Weiss, *Acta Crystallogr., Sect. B*, 1971, **27**, 1582.

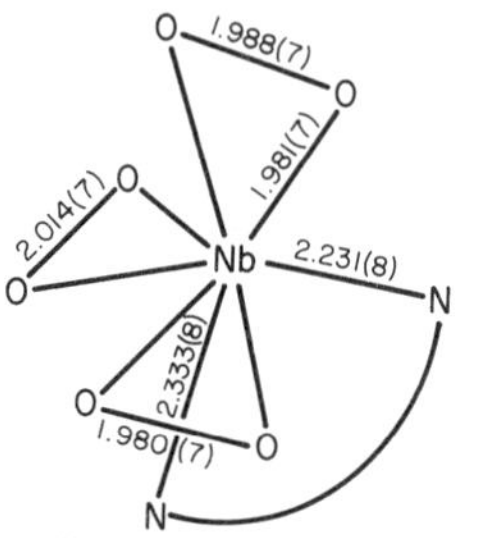

Figure 17 Structure of the $[Nb(O_2)_3(phen)]^-$ anion (reproduced from G. Mathern and R. Weiss, *Acta Crystallogr., Sect. B*, 1971, **27**, 1582 with permission)

(*iii*) *Peroxohalo derivatives* (Table 28)

Peroxofluoro complex anions $[Nb(O_2)F_6]^{3-}$, $[M(O_2)F_5]^{2-}$, $[M(O_2)F_4L]^-$, $[M(O_2)_2F_4]^{3-}$, $[M(O_2)_2F_3]^{2-}$ and $[Ta_3(O_2)_3OF_{13}(H_2O)]^{6-}$ were reported. Their structures often proved difficult to solve, due to disorder in the crystal and to facile decomposition of the peroxo moiety. The structure of the $[M(O_2)F_5]^{2-}$ anion has been repeatedly investigated; pentagonal bipyramidal arrangements have been found with the peroxo group in the equatorial plane (O—O: 1.45–1.47 Å for M = Nb; 1.389–1.443 Å for M = Ta at 170 K).

The $[Ta(O_2)F_4(C_6H_7NO)]^-$ anion also exhibits a seven-coordinate tantalum atom at the center of a pentagonal bipyramid, the two apical positions being occupied by fluorine (O—O: 1.43–1.67 Å). Salts of $[M(O_2)_2F_4]^{3-}$ were isolated and characterized by IR; a disordered structure has been reported for $[Ta(O_2)_2F_4]^{3-}$. $[Ta_3(O_2)_3OF_{13}]^{6-}$ consists of $[Ta(O_2)_2F_5]^{2-}$ and of μ-oxo-bridged $[Ta_2O(O_2)_2F_8]^{4-}$ anions in the same unit cell, with very different O—O distances in the various peroxo groups. The Ta—O—Ta bridge in the dimeric anion is asymmetrical (Figure 18).

A series of $[M(O_2)F_3(diket)]^-$ complexes has been prepared by the reaction of β-diketones with $[M(O_2)F_4(H_2O)]^-$. ^1H and ^{19}F NMR spectra indicate that the products exist as a mixture of geometrical isomers in MeCN. The reaction of oxalic acid with $[Nb(O_2)F_4(H_2O)]^-$ yielded $[Nb_2(O_2)_2F_6(C_2O_4)]^{2-}$. A bridging oxalato group and a coordination number of seven for the metal were proposed. $[M(O_2)F_3(phen)]$, $[M_2O(O_2)F_4(bipy)_2]$ and $[M(O_2)F_3(OAsPh_3)_2]$ were obtained when the neutral ligands were allowed to react with $[M(O_2)F_4(H_2O)]^-$.

A few peroxochloro derivatives of the $[M(O_2)Cl_5]^{2-}$ [391] and $[M(O_2)Cl_4(H_2O)]^-$ [431] anions are known. Addition of β-diketones to solutions of the latter afforded $[Nb(O_2)Cl_3(diket)]^-$; the stereorigidity of these heptacoordinated species in solution was assigned to the bidentate peroxo moiety.[431]

$[NbCl_2(RCp)_2]$ (R = H, Me) complexes react with H_2O_2, affording $[Nb(O_2)Cl(RCp)_2]$. In the complex with R = H the two oxygen atoms (O—O: 1.47(1) Å) and the chlorine atom are in the plane which bisects the $Nb(Cp)_2$ bent sandwich system; the coordination around niobium is thus pseudotetrahedral. In the presence of H_2O_2, $[Nb(O_2)Cl(Cp)_2]$ catalytically converts cyclohexene to its epoxide.

Table 28 Peroxo Halo Derivatives

Compounds	*Comments*	*Ref.*
$A_2[M(O_2)F_5]$	A = Na, K, NH_4, oxH_2	1
	X-Ray structure	2–6
$A_3[M(O_2)_2F_4]$	A = Na, K, NH_4	1
	X-Ray structure	7
$A[M(O_2)F_4L]$	L = H_2O, 2-MepyO; ^{19}F NMR	8–9
	X-Ray structure (A = NEt_4)	10
$K_6[Ta_3(O_2)_3OF_{13}(H_2O)]$	X-Ray structure: $[Ta(O_2)F_5]^-$ and $[\{Ta(O_2)F_4\}_2O]^{4-}$ in the same cell	12
$A_2[M(O_2)_2F_3]$		11
$(NEt_4)[M(O_2)F_3(RCOCHCOR')]$	R = R' = Me; R = R' = Ph; R = Ph, R' = CF_3 R = C_4H_3S, R' = CF_3; R = R' = CF_3	8–9
$(NEt_4)[\{Nb(O_2)F_3\}_2(C_2O_4)$	Heptacoordinated Nb; bridging oxalato	9
$A_2[Nb(O_2)Cl_5]$	A = K, Rb, Cs, NH_4	13, 14
$(NEt_4)[M(O_2)Cl_4(H_2O)]$	NMR: stereorigidity	15
$(NEt_4)[M(O_2)Cl_3(RCOCHCOR')]$	Heptacoordinated	15
$[M(O_2)F_3L_2]$	L = $OAsPh_3$; L_2 = phen	16
$[M_2O(O_2)_2F_4(bipy)_2]$	M—O—M bridges	16
$[Nb(O_2)Cl(Cp)_2]$	X-Ray structure	17

1. N. Vuletic and C. Djordjevic, *J. Less-Common Met.*, 1976, **45,** 85.
2. R. Stomberg, *Acta Chem. Scand., Ser. A*, 1981, **35,** 489.
3. R. Stomberg, *Acta Chem. Scand., Ser. A*, 1981, **35,** 389.
4. R. Stomberg, *Acta Chem. Scand., Ser. A*, 1982, **36,** 423.
5. R. Stomberg, *Acta Chem. Scand., Ser. A*, 1982, **36,** 101.
6. R. Stomberg, *Acta Chem. Scand., Ser. A*, 1983, **37,** 523.
7. R. Schmidt, G. Pausewang and W. Massa, *Z. Anorg. Allg. Chem.*, 1982, **488,** 108.
8. J. Y. Calves and J. E. Guerchais, *J. Fluorine Chem.*, 1974, **4,** 47.
9. J. Y. Calves and J. E. Guerchais, *J. Less-Common Met.*, 1973, **32,** 155.
10. J. C. Dewan, A. J. Edwards, J. Y. Calves and J. E. Guerchais, *J. Chem. Soc., Dalton Trans.*, 1977, 981.
11. Yu. A. Buslaev, E. G. Ilyin, V. D. Kopanev and V. P. Tarasov, *Zh. Strukt. Khim.*, 1972, **13,** 930.
12. W. Massa and G. Pausewang, *Z. Anorg. Allg. Chem.*, 1979, **456,** 169.
13. J. E. Guerchais, B. Spinner and R. Rohmer, *Bull. Soc. Chim. Fr.*, 1965, 55.
14. J. Dehand, J. E. Guerchais and R. Rohmer, *Bull. Soc. Chim. Fr.*, 1966, 346.
15. J. Y. Calves, J. E. Guerchais, R. Kergoat and N. Kheddar, *Inorg. Chim. Acta*, 1979, **33,** 95.
16. J. Y. Calves, J. Sala-Pala, J. E. Guerchais, A. J. Edwards and D. R. Slim, *Bull. Soc. Chim. Fr.*, 1975, 517.
17. J. Sala-Pala, J. Roue and J. E. Guerchais, *J. Mol. Catal.*, 1980, **7,** 141.
18. I. Bkouche-Waksman, C. Bois, J. Sala-Pala and J. E. Guerchais, *J. Organomet. Chem.*, 1980, **195,** 307.

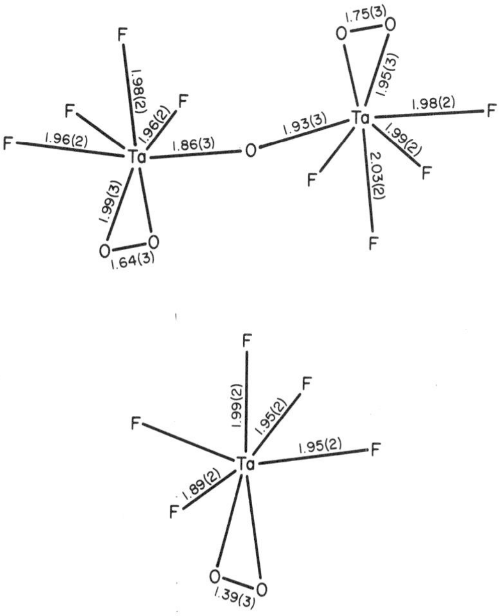

Figure 18 Structure $[Ta_3(O_2)_3OF_{13}(H_2O)]^{6-}$ showing the $[Ta(O_2)F_5]^{2-}$ and $[Ta_2O(O_2)F_8]^{4-}$ anions (reproduced from W. Massa and G. Pausewang, *Z. Anorg. Allg. Chem.*, 1979, **456,** 169 with permission)

34.2.8.2 Sulfides, selenides and related compounds

$[MY_4]^{3-}$ ($Y = S$ or Se)[367] and $[NbOS_3]^{3-}$ [368] are known only as insoluble salts synthesized from the elements at high temperatures; $[NbO_2S_2]^{3-}$ [369] was obtained from strongly basic solutions of Nb_2O_5 saturated with H_2S. The salts $(NEt_4)_4[M_6S_{17}]$,[366] sensitive to O_2 and H_2O, were prepared from $[M(OEt)_5]$, $(Me_3Si)_2S$ and Et_4NCl (1:6:3) in MeCN. Single crystal X-ray analysis (Figure 19) revealed 10 non-planar M_2S_2 rhombs in convex fusion forming a crown-shaped M_6S_{10} cage. Each metal atom is additionally bonded to a terminal S atom and to the bridging S atom inside the cage, thus achieving pentagonal seven-coordination for top M(1), and tetragonal six-coordination for the other metal atoms.

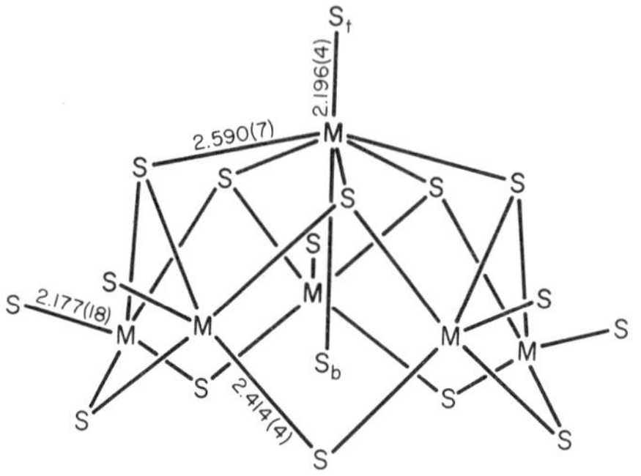

Figure 19 Structures of $[M_6S_{17}]^{4-}$ (reproduced from ref. 366 with permission)

Nb^V and Ta^V complexes with the S_2 and Se_2 ligands, formally similar to O_2, appear to be limited to bis(cyclopentadienyl) complexes. The reaction of $[NbCl_2(OH)(Cp)_2]$ with H_2S in methanol, in the presence of ionic halides or pseudohalides (KCl, KBr, NH_4I, KSCN), afforded the monomeric $[Nb(S_2)X(Cp)_2]$ (X = Cl, Br, I, SCN) complexes.[418] When X = Cl or Br, a second compound of the same formula was obtained and assumed to be polymeric with disulfide bridging groups. Surprisingly, when X = SCN, a cyano complex, $[Nb(S_2)CN(Cp)_2]$, was formed. $[NbMe_2(Cp)_2]$ was reported to be oxidized by S_8 to yield $[Nb(S_2)Me(Cp)_2]$.[479] Similar procedures with $[Nb(\eta^1\text{-}Cp)_2(Cp)_2]$, $[NbH(CH_2{=}CHEt)(Cp)_2]$ and $[TaMe_2(Cp)_2]$ gave the $[M(S_2)R(Cp)_2]$ complexes, where $R = \eta^1$-Cp, Bu and Me, respectively.[480] The reaction of $[NbMe_2(Cp)_2]$ with red selenium afforded $[Nb(Se_2)Me(Cp)_2]$.[480] $[Nb(S_2)\{SP(S)(OR)_2\}(Cp)_2]$ complexes (R = Me, Et, Pr^i, Bu^t) were obtained from $[NbCl_2(Cp)_2]$ and P_4S_{10}.[481] The R = Et and Pr^i derivatives were reported to form from S_8 and $[Nb\{SP(S)(OR)_2\}_2(Cp)_2]$.[252] The structures of $[Nb(S_2)X(Cp)_2]$ were solved for X = Cl,[482] CN[482] and Me[483] (S—S: 2.03(1) Å). $[Nb(S_2)Me(Cp)_2]$ was shown to transfer sulfur to a phosphine.[480]

34.2.8.3 Other oxygen- or sulfur-substituted metal compounds

(i) Sulfates

The interest in sulfates arises mainly from the development of processes using sulfuric acid in the treatment of mineral concentrates. MCl_5 reacts with SO_3 in SO_2Cl_2 to form the insoluble $[Nb_2O(SO_4)_4]$ and $[Ta_2(SO_4)_5]$, respectively. Whereas $[Nb_2O(SO_4)_4]$, $[Nb_2O_2(SO_4)_3]$, $[Nb_2O_3(SO_4)_2]$ and $[Nb_2O_4(SO_4)]$ could be isolated from the Nb_2O_5–SO_3–H_2O system, only $[Ta_2(SO_4)_5]$ and $[Ta_2O_4(SO_4)(H_2O)]$ were obtained in the case of Ta^V.[1] The hydrated oxo sulfates $[Ta_2O_3(SO_4)_2(H_2O)_n]$ ($n = 3$, 4 or 5) were also reported. In H_2SO_4, M_2O_5 and $MOCl_3$ form $H[MO(HSO_4)_4]$, while MX_5 gives $H[M(HSO_4)_6]$.[484] The more soluble NH_4,[1] Na,[485] Cs[486,487] and Rb[488] double sulfates have also been described.

(ii) Nitrates, phosphates, arsenates and related compounds

The reaction between MCl_5 and liquid N_2O_5 resulted in the moisture sensitive tris(nitrato)oxo compounds $[MO(NO_3)_3]$.[1] The $(NMe_4)[NbO(NO_3)_4]$ complexes were prepared by addition of liquid N_2O_5 to the appropriate hexachloro complexes with covalently bonded nitrato groups. Solvated polymeric nitratodioxides, $[MO_2NO_3]$, were obtained from MCl_5 and N_2O_4 in ionizing solvents.

Metaphosphates, $[MOPO_4]$ or $[M_2O_5(P_2O_5)]$, have been reported. Single crystals of $NbOPO_4$ were prepared by heating together H_3PO_4 and Nb_2O_5 in a gold capsule at 600 °C under 1800 bar.[489] A compound of composition $[Nb_9O_{25}P]$ was obtained by a similar procedure at 750 °C under 2000 bar.[490] $[TaOPO_4(H_2O)]$, $[TaOPO_4]$, $[TaOAsO_4]$ and $[Ta_9AsO_{25}]$ are also known.[491] The orthophosphates $[M_3(PO_4)_5]$ were prepared by reaction of excess H_3PO_4 with MX_5 in the molten state.[1]

$NbOPO_4$ has tetragonal symmetry, and its structure is made up of chains of corner-sharing NbO_6 octahedra with non-equivalent Nb—O bonds (1.78 and 2.32 Å), linked together by PO_4 tetrahedra. In the hydrated crystalline niobium phosphates, the individual NbO_6 octahedra are no longer connected to each other, and display a layer structure capable of hosting various molecules such as H_2O[492,493] or H_3PO_4.[494]

$NbCl_5$ and $TaCl_5$ abstract oxygen from diisopropylmethylphosphonate to yield polymeric complexes $[Mo(IMP)_3]_n$ (IMP = isopropoxymethylphosphonate),[407] which contain M=O groups and both chelating and bridging phosphonato groups.

The methylphosphinates $[M_2O(MePO_3)_4]$ precipitate when methylphosphinic acid and the pentachlorides are heated together.[1] $[Ta(OR)_{5-x}(Ph_2PO_2)_x]$ (x = 1–3) were prepared from $Ta(OR)_5$ and $Ph_2P(O)OH$ as low molecular weight polymers. The polymeric oxides $[TaO_2(Ph_2PO)]_n$, $[Ta_2O_3(Ph_2PO)_2]_n$ and $[TaO(Ph_2PO)_3]_n$ have been obtained by suspending the methoxide or ethoxide in boiling water.

The sodium salts of dialkyldithiophosphate (ddtp) react with NbX_5 (X = Cl or Br) in methanol to yield $[MX(OMe)_2(ddtp)_2]$; recrystallization from the appropriate alcohol afforded $[MX(OR)_2(ddtp)_2]$.[239] The P—S stretching vibrations in the IR around 600–750 cm^{-1} suggest that the ligands coordinate through both sulfur atoms. The surprising formation of the diamagnetic $[Nb(S_2)\{SP(S)(OR)_2\}(Cp)_2]$ from P_4S_{10} and $[NbCl_2(Cp)_2]$ in ROH illustrates the ease with which Nb^{IV} is oxidized in such complexes;[481] the dialkyldithiophosphate ligand appears to be monodentate; the origin of the S_2 group is as yet not understood.

The reactions of F_2PS_2H with MCl_5 gave the chloro(difluorophosphinato) complexes $[MCl_3(S_2PF_2)_2]$,[495] in which both $S_2PF_2^-$ ligands are bidentate.

34.2.9 Miscellaneous Metal–Element Derivatives

Zwitterionic four-membered PN_2Nb heterocycles were obtained in the reaction of η^5-phosphorus–nitrogen ylides with $NbCl_5$ according to equation (40).[496]

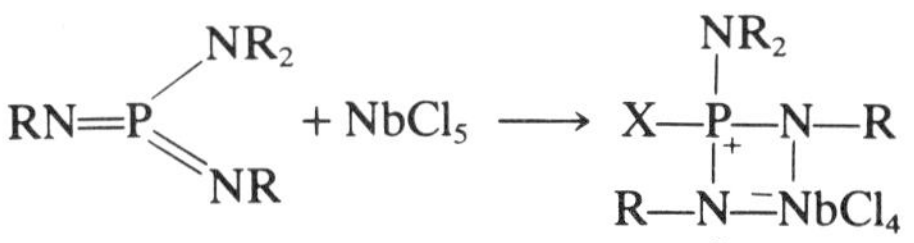

$$
\begin{array}{ccc}
RN{=}P\Big\langle{\!\!\begin{array}{l} NR_2 \\ NR \end{array}} + NbCl_5 & \longrightarrow & X{-}\underset{\underset{R{-}N{=}NbCl_4}{|^+}}{\overset{\overset{NR_2}{|}}{P}}{-}N{-}R
\end{array} \tag{40}
$$

$$R = SiMe_3$$

In the course of an investigation of the Ta/P/S system, formation of a compound of overall composition $TaPS_6$ was noted.[497] Its structure showed each tantalum atom to have eight nearest sulfur neighbors arranged in a bicapped triangular prism. Two such prisms, sharing a common face, form Ta_2S_{12} units; these are linked together by PS_4 tetrahedra into endless chains. The compound can be formally considered as containing two cations, Ta^{5+} and P^{5+}, and two anions, $(S_2)^{2-}$ and S^{2-}, and can be formulated as $[Ta(PS_4)S_2]$.

Compounds containing Nb—Zn or Ta—Zn bonds have been obtained by the reaction of $[MH_3(Cp)_2]$ with $[Zn(Cp)_2]$. The structures of $[NbH_2(Cp)_2Zn(Cp)]$ and $[TaH(Cp')_2\{Zn(Cp)\}_2]$ were determined.[499] The reaction of $[NbH_3(Cp)_2]$ with HgX_2 afforded the first complexes with Nb—Hg bonds;[500] $[Nb(Cp)_2(HgS_2CNEt_2)_3]$ has been characterized by X-ray diffraction analysis (Nb—Hg: mean 2.790 Å).

34.3 OXIDATION STATE +IV

34.3.1 Tetrahalides

The niobium and tantalum tetrahalides are all known, except TaF_4, whose non-existence is consistent with the fact that $[TaF_5]_4$ should be the least reducible halide of the group VA elements.[1,2,4,5] They are dark solids, obtainable by reduction of the pentahalides by H_2, Al, Nb or Ta at high temperatures. The most popular preparation involves the thermal gradient method, with a metal as reductant. NbI_4 is easily obtained on heating NbI_5 to 300 °C.[501] The reduction of TaI_5 is slower, and an easier way to prepare TaI_4 is to allow TaI_5 to react with an excess of pyridine, giving $[TaI_4(py)_2]$ which, upon heating, loses the ligand.[502] NbI_4 and TaI_4 are trimorphic and dimorphic, depending on the synthetic procedure. All tetrahalides except NbF_4 disproportionate thermally. These reactions may be suppressed by maintaining a sufficiently high partial pressure of the pentahalides during the high temperature synthesis.

NbF_4 is a paramagnetic solid in which the Nb atoms lie at the center of octahedra that share four corners in an infinite sheet structure.[2] The other tetrahalides are essentially diamagnetic, and often orthorhombic. $NbCl_4$ [503] and NbI_4 [504] form one-dimensional polymers consisting of *trans* edge-sharing polyoctahedra, with very weak interchain coupling. These $[Nb_2X_4(\mu\text{-}X)_4]_\infty$ polymers exhibit bond alternation (Figure 20), with metal–metal pairs. Powder data suggest that $NbBr_4$ and TaX_4 (X = Cl, Br, I) are isomorphous with $NbCl_4$ and α-NbI_4. $NbBr_4$ is extremely sensitive to moist air, while $NbCl_4$ is relatively air stable.

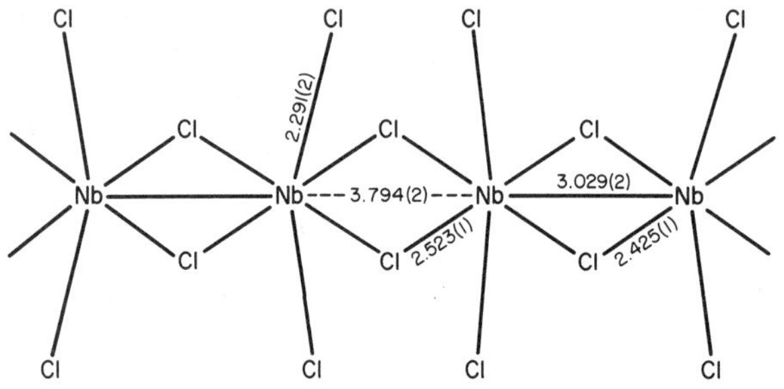

Figure 20 Section of the linear chain of $(Nb_2Cl_8)_n$ showing the bond alternation (reproduced from ref. 503 with permission)

Recent interest in low dimensional conducting materials has led to various experimental and theoretical studies on NbX_4.[505,506] Their structure, with two types of metal-bridging halide bonds on unequal lengths alternating along the chain, is responsible for their semiconducting properties. Their electrical conductivities increase sharply with temperature or pressure, and NbI_4 was shown to exhibit metallic behavior under high pressure.[507] The activation energy for conduction decreases from $NbCl_4$ to NbI_4, despite the increase in metal–metal distance. This trend has been accounted for by a diminution of the valence band and band gap energies (~0.1 and 1.0 eV in $NbCl_4$; ~0.3 and 0.8 eV in NbI_4).

34.3.2 Halo Complexes

34.3.2.1 Halometallates(IV)

The hexahalometallates obtained by electroreduction of NbX_6^- or from MX_4 have been isolated as their alkali or tetraalkylammonium salts $A_2[MX_6]$ (M = Nb, X = Cl, A = K, Rb, Cs, NH_4; X = Br, Et_4N, A = Cs and M = Ta, X = Cl).[7,508] Their thermal stability, higher for tantalum, is in marked contrast with that of the corresponding neutral tetrahalides (ΔH°_{298} in the −1614 to −1719 kJ mol^{-1} range, compared to −694 kJ mol^{-1} for $NbCl_4$).[508] $K_3[NbF_7]$ is isostructural with $K_3[NbOF_6]$ and $(NH_4)_3[ZrF_7]$; it approaches D_{5h} symmetry.[510]

34.3.2.2 Neutral adducts of tetrahalides

(i) Synthesis

Numerous adducts of MX_4 with nitrogen, phosphorus, arsenic, oxygen and sulfur donors have been obtained, mainly for niobium with X = Cl or Br[7] (Tables 29–33). The absence of MF_4 adducts, except for $[NbF_4(py)_2]$,[511] may be due less to their instability than to a lack of investigations. An ill-defined fluoro trichloride complex, $[NbCl_3F(MeCN)_{2.65}]$, has also been reported.[512]

These adducts were synthesized by extraction of solid $[M_2X_8]_n$ by the ligand, or by reduction of MX_5 (X = Cl, Br, I) with an excess of ligand. Such reductions occur most readily with monodentate, or, even better, with bidentate or heterocyclic nitrogen donors (equation 41),[502,515] although it was also observed with macrocyclic ethers, phosphines[513] and diarsines.[514] Longer reaction times or higher temperatures are required with the tantalum pentahalides. As a result, the adducts of Nb^{IV} and Ta^{IV} with nitrogen donors (nitriles and heterocycles) are predominant (Table 29). Partial solvolysis of NbX_4 was often observed with aliphatic amines (Et_2NH, en, *etc.*) and their adducts are confined to $[NbCl_4(NEt_3)]_n$. Adducts with nitriles are limited to $[NbCl_4(RCN)_2]$ (R = Me, Ph) and acrylonitrile $[MX_4(C_2H_3N)_2]$ (X = Cl, Br) adducts.[516] Adducts with various phosphines have been isolated (Table 30), while the adducts with arsenic donors are limited to those of the bidentate *o*-phenylenebisdimethylarsine (diars) and related ligands (Table 31). A few mixed ligand adducts have also been reported.

$$\begin{aligned} 2NbBr_5 + 7py &\longrightarrow [NbBr_4(py)_2] + pyH^+Br^- + (py)_2^{2+}Br \\ NbI_5 + 3py &\longrightarrow [NbI_4(py)_2] + pyI_2 \end{aligned} \qquad (41)$$

The complexes with group VIB donors are collected in Table 32. No TaX_4 adducts with monodentate oxygen donors have so far been isolated, probably as a consequence of the strong O abstraction properties of Ta^{IV}; the reaction between $TaCl_4$ and dioxane led, for instance, to $[TaOCl_3(diox)_2]$.[517] Such reactions are less favored for Nb^{IV} and require powerful oxo ligands such as DMSO;[512] they are not observed with dry DMF.[519] $[NbCl_4(THF)_2]$ is of interest because of the lability of the THF molecules, which makes it a useful starting material in Nb^{IV} chemistry.[519] So far, only crown ethers have been able to achieve O complexation of $TaCl_4$, providing $[(TaCl_4)_2(18\text{-}CRW\text{-}6)]$.[513] Sulfide adducts are listed in Table 33.

Table 29 Adducts of Tetrahalides with Amines

Compounds	*Comments*	*Ref.*
$[NbF_4(py)_2]$	*Trans* isomer (IR)	1
$[NbCl_4(Rpy)_2]$[a]	R = H isolated, powder X-ray data; R ≠ H not isolated, ESR evidence, *trans* isomer X = Cl, μ = 1.37 or 1.53 BM	2, 3, 4
$[NbX_4(py)_2]$ (X = Br, I)	Powder X-ray data; X = Br interconvertible green and red isomers; X = I, unstable in solution (I_3^- + Nb^{III}?)	2, 4
$[TaX_4(py)_2]$ (X = Cl, Br, I)	X-Ray powder data, X = Cl *trans* isomer; X = I loses py on heating	2, 5
$[MX_4(\gamma\text{-pic})_2]$ (X = Cl, Br)	X = Cl, M = Nb μ = 1.18 BM; M = Ta μ = 0.77 BM	6, 4
$[MX_4(LL)]$ (X = Cl, Br; LL = bipy, *o*-phen)	M = Nb, X = Cl, LL = bipy: ionic in MeCN; LL = *o*-phen, X = Cl, M = Nb μ = 1.38 BM; M = Ta μ = 0.67 BM	5, 6, 7
$[MCl_4(terpy)]$	Seven coordinate species	8
$[NbCl_4(terpy)(py)_3]$	Tridentate terpy, free py; reduction to Nb^{III} (E° = −0.51 V) by cyclic voltammetry	8
$[M_2Br_8(terpy)]$	Electrolyte in Me_2CO	8
$[NbCl_4(NEt_3)]_n$	Diagmagnetic	9
$[NbX_4(N_2H_{12}C_6)]$[b] (X = Cl, Br, I)	Partial reduction	9

[a] Rpy = 3-RC_5H_4N (R = Me, Et, NH_2, Br); 4-RC_5H_4N (R = Me, Et, Ph, NH_2).
[b] $C_6H_{12}N_2$ = N,N,N′,N′-tetramethylethylenediamine.

1. F. E. Dickson, R. A. Hayden and W. G. Fateley, *Spectrochim. Acta, Part A*, 1969, **25**, 1875.
2. R. E. McCarley, B. G. Hughes, J. C. Boatman and B. A. Torp, *Adv. Chem. Ser.*, 1963, **37**, 243; R. E. McCarley and B. A. Torp, *Inorg. Chem.*, 1963, **2**, 540.
3. D. P. Johnson, J. Wilkinski and R. D. Bereman, *J. Inorg. Nucl. Chem.*, 1973, **35**, 2365.
4. D. J. Machin and J. F. Sullivan, *J. Less-Common Met.*, 1969, **19**, 405.
5. R. E. McCarley and J. C. Boatman, *Inorg. Chem.*, 1963, **2**, 547.
6. M. Allbutt, K. Feenan and G. W. A. Fowles, *J. Less-Common Met.*, 1964, **6**, 299.
7. G. W. A. Fowles, D. J. Tidmarsh and R. A. Walton, *Inorg. Chem.*, 1969, **8**, 631.
8. B. Begolli, V. Valjak, V. Allegretti and V. Katovic, *J. Inorg. Nucl. Chem.*, 1981, **47**, 2785.
9. M. T. Brown and G. S. Newton, *Inorg. Chem.*, 1966, **5**, 1117.

Table 30 Adducts of Niobium and Tantalum Tetrahalides with Phosphorus Donors

Compounds	*Comments*	*Ref.*
$[MCl_4(PEt_3)_2]$	M = Ta, X-ray structure: *trans* isomer; M = Nb unstable at room temperature	1, 2, 3
$[TaX_4(PMe_2Ph)_2]$(X = Cl, Br)	X-Ray structure: *cis* isomer, X = Br, C—H activation of P—CH_3	4, 2, 13
$[NbCl_4(PEtPh_2)_2]$	X-Ray structure: *trans* isomer	1
$[NbCl_4(PR_3)_2]$ (R = Bu^n, Bu^i)	Not isolated, ESR data (*trans* isomer)	5
$[TaCl_4(PMe_3)_3]$	X-Ray structure, capped octahedron (C_{3v} symmetry)	2
$[Nb_2Cl_4(PR_3)_4(\mu\text{-}Cl)_4]$ (R_3 = Me_3, Me_2Ph)	X-Ray structures, monomer–dimer equilibrium in solution	1, 6
$[Nb_2Cl_8(PR_3)_x]$ (R = Cy, x = 4, 3; R = Bu^t, x = 3, 2)	Diamagnetic, soluble species, unstable with respect to Nb^{III}	7
$[Nb_2Cl_8(PR_3)_3]_n$ (R = Ph, Bz)	Diamagnetic, insoluble	8
$[NbCl_4(PPh_3)(MeCN)_2]$	Analysis only	9
$[NbCl_4(PBu^i_3)(THF)]$	Not isolated, ESR, data (*trans* geometry)	10
$[NbCl_4(dppe)]$	Analysis only	9
$[MCl_4(dmpe)_2]$	M = Ta; X-ray structure, approximate square antiprism	3
$[NbCl_4(depe)_2]$		11, 12

1. F. A. Cotton, S. A. Duraj and W. J. Roth, *Inorg. Chem.*, 1984, **23,** 3592.
2. F. A. Cotton, S. A. Duraj and W. J. Roth, *Inorg. Chem.*, 1984, **23,** 4046.
3. L. E. Manzer, *Inorg. Chem.*, 1977, **16,** 525.
4. L. G. Hubert-Pfalzgraf, M. Tsunoda and J. G. Riess, *Inorg. Chim. Acta,* 1981, **52,** 231.
5. G. Labauze, E. Samuel and J. Livage, *Inorg. Chem.*, 1980, **19,** 1384.
6. F. A. Cotton and W. J. Roth, *Inorg. Chem.*, 1984, **23,** 945.
7. J. L. Morançais and L. G. Hubert-Pfalzgraf, *Transition Met. Chem.*, 1984, **9,** 130.
8. D. J. Machin and J. F. Sullivan, *J. Less-Common Met.*, 1969, **19,** 405.
9. R. Gut and W. Perron, *J. Less-Common Met.*, 1972, **26,** 369.
10. E. Samuel, G. Labauze and J. Livage, *Nouv. J. Chim.*, 1977, **1,** 93.
11. S. Datta and S. S. Wreford, *Inorg. Chem.*, 1977, **16,** 1134.
12. F. A. Cotton, L. R. Falvello and R. C. Najjar, *Inorg. Chem.*, 1983, **22,** 770.
13. N. Hovnanian, Ph.D. Thesis, University of Nice, 1986.

Table 31 Tetrahalide Adducts with Bidentate Arsines

Compounds	*Comments*	*Ref.*
$[NbX_4(diars)_2]$ (X = Cl, Br, I)	X-Ray structure for X = Cl (dodecahedron), iodide not isomorphous (μ = 1.8 BM; X = Cl, I)	1, 2, 3
$[TaCl_4(diars)_2]$	Elusive, impure material, X-ray powder data	2
$[TaCl_4(diars)]_n$	Low magnetic moment (μ = 0.4 BM)	2
$[NbX_4(Etdiars)_2]$ (X = Cl, Br, I)	X = Br or I, partially dissociated in MeCN and completely in py	2
$[TaCl_4(Etdiars)]_n$	Diamagnetic	2
$[NbCl_4(Rdiars)_2]$	R = Me, F	2, 3
$[NbCl_4(naars)_2]$[a]	Dodecahedron (UV)	4

[a] naars = 1,8-bis(dimethylarsino)naphthalene

1. J. H. Clark, D. L. Kepert, J. Lewis and R. S. Nyholm, *J. Chem. Soc. (A)*, 1965, 2865.
2. R. L. Deutscher and D. L. Kepert, *Inorg. Chem.*, 1970, **9,** 2305.
3. D. L. Kepert, B. W. Skelton and A. H. White, *J. Chem. Soc., Dalton Trans.*, 1981, 652.
4. D. L. Kepert and K. R. Trignell, *Aust. J. Chem.*, 1975, **28,** 1245.

(*ii*) *Structure*

The solubility of the MX_4 adducts is generally poor, but increases from the chlorides to the iodides. In contrast to the tetrahalides themselves, most of their adducts are monomeric and paramagnetic. Characterization by ESR was sometimes achieved. The absence of notable paramagnetism was usually taken to support the existence of oligomeric or polymeric structures. X-Ray data remain scarce, other than for a number of phosphine adducts.

With monodentate ligands, paramagnetic pseudooctahedral diadducts $[MX_4L_2]$ of D_{4h} or C_{2v} symmetry are most frequently formed. X-Ray structure determinations are available on *trans*-$[MCl_4(PEt_3)_2]$, *cis*-$[TaCl_4(PMe_2Ph)_2]$,[520] *cis*-$[NbBr_4(MeCN)_2]$[521] and *cis*-$[NbCl_4(MeCN)_2]\cdot$MeCN.[522] The stereochemistry of most $[MX_4L_2]$ compounds has however been

Table 32 Tetrahalide Adducts with Oxygen Donors

Compounds	*Comments*	*Ref.*
$[NbX_4(O_2C_4H_8)_2]$ ($X = Cl, Br$)	Dioxane monodentate, *cis* stereochemistry in the solid (IR), *trans* in solution (ESR)	1, 2
$[NbX_4(THF)_2]$ ($X = Cl, Br$)	THF very labile, *cis* stereochemistry in the solid (IR), *trans* in solution (ESR)	3, 1
$[NbX_4(THP)_2]$ ($X = Cl, Br$)	*Cis* stereochemistry in the solid (IR), $X = Cl$; $\mu = 1.54$ BM	1
$[NbCl_4(DME)]$	Not isolated, ESR evidence	2
$[NbX_4(DMF)_2]$ ($X = Cl, Br$)	*Trans* isomer, $X = Br$ ionization to $[NbBr_{4-m}(DMF)_n]^{m+}$ in solution	4, 2
$[NbI_4(DMF)_8]$	Extensive ionization to $[NbI_{4-m}(DMF)_n]^{m+}$ even in the solid	4
$[NbCl_4(DEF)_2]$	Not isolated, ESR evidence (*trans*)	2
$[NbCl_4(DMA)_2]$	Not isolated, ESR evidence (*trans*)	2
$[NbCl_4(HMPA)]$	Not isolated, ESR evidence $\langle g \rangle = 1.8910$, $\langle A \rangle = 183$ G	2
$[NbCl_4(HMPA)_2]$	Not isolated, ESR evidence (*trans*) $\langle g \rangle = 1.8869$,[a] $\langle A \rangle = 199.9$ G[a]	2
$[(NbCl_4)_3(18\text{-CRW-}6)_2]$	Diamagnetic	5
$[(NbCl_4)_3(15\text{-CRW-}5)_2]$	Diamagnetic	5
$[(NbCl_4)_2(15\text{-CRW-}5)]$	Diamagnetic	5
$[(TaCl_4)_2(18\text{-CRW-}6)]$	Diamagnetic	5

[a] Estimated assuming $A_\perp$ and $g_\perp$ are the same for the mono and bis adducts.

1. G. W. A. Fowles, D. J. Tidmarsh and R. A. Walton, *Inorg. Chem.*, 1969, **8,** 631.
2. D. P. Johnson and R. D. Bereman, *J. Inorg. Nucl. Chem.*, 1972, **34,** 2957.
3. L. E. Manzer, *Inorg. Chem.*, 1977, **16,** 525.
4. K. Kirksey and J. B. Hamilton, *Inorg. Chem.*, 1972, **11,** 1945.
5. L. G. Hubert-Pfalzgraf and M. Tsunoda, *Inorg. Chim. Acta*, 1980, **38,** 43.

Table 33 Tetrahalide Adducts with Sulfur Donors

Compounds	*Comments*	*Ref.*
$[NbX_4(SR_2)]$ ($X = Cl, Br, I$; $R = Me$)	*Cis* isomer probable, unstable with respect to the monoadducts	1, 2
$[NbX_4(SR_2)]_2$ ($X = Cl, Br, I$; $R = Me, Et$)	Dimer–monomer equilibrium in C_6H_6	1, 2
$[TaX_4(SMe_2)_2]$ ($X = Cl, Br$)	*Cis* isomer (IR)	1
$[NbCl_4(THT)_2]$	*Cis* isomer (IR)	1, 2
$[NbBr_4(THT)_2]$	Two products (α and β) of different solubilities and structures	1
$[TaX_4(THT)_2]$ ($X = Cl, Br$)	*Cis* isomer (IR)	1
$[NbX_4(dth)_2]$ ($X = Cl, Br, I$)	ESR for $X = Cl, Br$ (dodecahedron)	3, 4
$[NbCl_4(dth)]$	$[NbCl_4] + [NbCl_4(dth)_2]$ (very low yield) six-coordinated metal	3
$[NbCl_4(TU)_2]$	S-bonded, soluble and extremely air sensitive	5
$[NbI_4(TU)_3]$	S-bonded, as $[Nb(TU)_3I_3]I$ in MeCN	5

1. G. W. A. Fowles, D. J. Tidmarsh and R. A. Walton, *J. Inorg. Nucl. Chem.*, 1969, **31,** 2373.
2. J. B. Hamilton and R. E. McCarley, *Inorg. Chem.*, 1970, **9,** 1333.
3. J. B. Hamilton and R. E. McCarley, *Inorg. Chem.*, 1970, **9,** 1339.
4. B. L. Wilson and J. B. Hamilton, *Inorg. Nucl. Chem. Lett.*, 1976, **12,** 59.
5. D. J. Machin and J. F. Sullivan, *J. Less-Common Met.*, 1969, **19,** 405.

deduced from IR or ESR data. Distorted d^1 octahedral MX_4L_2 complexes generally exhibit $g_\parallel > g_\perp$ except for $[NbCl_6]^{2-}$, where all ligands have similar electron-donating properties. The hyperfine coupling parameters decrease with increasing ligand basicity (Table 34).

A few monoadducts have been reported for Nb^{IV}. These are favored by bulky ligands; thus tetrahydrothiophene formed diadducts with $NbCl_4$, while the bulkier methyl and ethyl sulfides provided monoadducts and unstable diadducts.[523] Monoadducts were also obtained with HMPA, NEt_3 and PBu^t_3; they are only slightly paramagnetic, and halogen-bridged bioctahedral structures with M—M bonds have therefore been postulated. The monoadducts are stabler for the chlorides than for the iodides, illustrating the importance of M—M bonding.

Table 34 ESR Data for Nb^{IV} and Ta^{IV} Derivatives (Frozen Solutions)

Compounds	$\langle g \rangle$	$g_{\parallel}$	$g_{\perp}$	$\langle A \rangle_M$	$A_{\parallel}$	$A_{\perp}$	$a_{\parallel P}$[a]	*Comments*	*Ref.*
$[NbCl_6]^{2-}$	1.892	1.919	1.951	177	—	—	—	—	1,2
$[NbOCl_4(H_2O)]^{2-}$	1.894	1.916	1.883	179	277	131	—	Silent at room temperature	3
$[NbO(acac)_2]$	1.906	1.903	1.964	159	249	108	—	C_{4v} symmetry	4
$[NbCl_4(THF)_2]$	1.892	1.913	1.894	177	270	122	—	*Cis* isomer	3
$[NbCl_4(\gamma\text{-pic})_2]$	1.945	1.924	1.955	183	269	139	—	*Trans* isomer	5
$[NbCl_4(PEt_3)_2]$	1.927	1.959	1.912	141	218	103	23.6	*Trans* isomer	6
$[TaCl_4(PEt_3)_2]$	1.740	1.831	1.695	211	285	175	25	*Trans* isomer	6
$[NbCl_4(dth)_2]$	1.997	1.931	2.029	138	187	112	—	Dodecahedron (D_{2d})	7
$[NbBr_4(dth)_2]$	2.008	1.990	2.018	131	187	101	—	Dodecahedron (D_{2d})	7
$[Nb(CN)_8]^{4-}$	1.984	1.969	1.991	101	151	76	—	Solid: dodecahedron (D_{2d})	8
	1.984	2.002	1.976	99	62	117	—	Solution: square antiprism (D_{4d})	8
$[Nb(NEt_2)_4]$	1.952	1.923	1.966	107	201	60	—	D_{2d} distortion of the tetrahedron	9

[a] Hyperfine splittings in gauss.

1. S. Maniv, W. Low and A. Gabay, *Phys. Lett. A.*, 1969, **29**, 536.
2. M. Lardon and H. H. Gunthard, *J. Chem. Phys.*, 1966, **44**, 2010.
3. D. P. Johnson and R. D. Bereman, *J. Inorg. Nucl. Chem.*, 1972, **34**, 2957 and 679.
4. I. F. Gainullin, N. S. Garifyanov and B. M. Kozyrev, *Dokl. Akad. Nauk SSSR*, 1968, **180**, 858 (*Chem. Abstr.*, 1968, 72 714).
5. D. P. Johnson, J. Wilinski and R. D. Bereman, *J. Inorg. Nucl. Chem.*, 1973, **35**, 2365.
6. E. Samuel, G. Labauze and J. Livage, *Nouv. J. Chim.*, 1977, **1**, 93.
7. B. L. Wilson and J. B. Hamilton, *Inorg. Nucl. Chem. Lett.*, 1976, **12**, 59.
8. P. M. Kiernan and W. P. Griffith, *J. Chem. Soc., Dalton Trans.*, 1975, 2489.
9. D. C. Bradley and M. H. Chisholm, *J. Chem. Soc. (A)*, 1971, 1511.

The systematic study of the phosphine adducts of MCl_4 offered a series of $[MCl_4(PR_3)_n]$ compounds. Of particular interest are the dinuclear $[Nb_2Cl_8(PR_3)_4]$ ($R_3 = Me_3$, Me_2Ph), which display two square antiprismatic $[NbCl_4(PR_3)_2]$ units sharing a square Cl_4 face (Figure 21).[524] Strongly basic phosphines such as PBu^t_3 or PCy_3 also favor metal–metal interactions to give soluble dinuclear $[Nb_2Cl_8(PBu^t_3)_2]$ and $[Nb_2Cl_8(PR_3)_3]$ ($R = Bu^t$, Cy) complexes.[304] The latter stoichiometry is rather unusual, and triply-chloride-bridged structures, comparable to those found in $[Nb_2Cl_9]^{3-}$ [525] or $[Nb_2Cl_3(CO)_8]^-$,[526] have been assumed. Tantalum is apparently less prone to adopt dinuclear structures, and the seven-coordinate $[TaCl_4(PMe_3)_3]$ (a capped octahedron of C_{3v} symmetry) and *cis*-$[TaCl_4(PhPMe_2)_2]$ have been obtained.

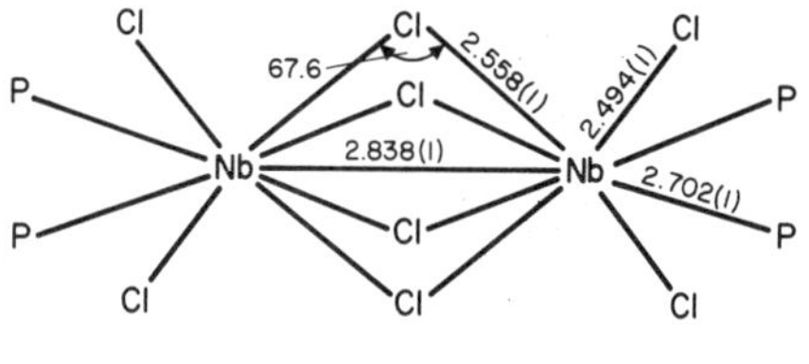

Figure 21 Drawing of the central core of the $[Nb_2Cl_8(PPhMe_2)_4]$ molecule (reproduced from ref. 524 with permission)

Multidentate ligands tend to stabilize higher coordination numbers, generally giving monomeric paramagnetic structures, although distorted octahedral complexes were obtained with some bidentate ligands such as bipy and phen. Heptacoordination is seldom encountered with bidentates, but occurs in $[MCl_4(terpy)]$. Octacoordination is more common, as illustrated by the formation of $[MCl_4(dmpe)_2]$,[519] $[NbX_4(dth)_2]$[527] and $[NbX_4(diars)_2]$ (X = Cl, Br, I),[528] in which the coordination polyhedra were assumed to be dodecahedral on the basis of UV or ESR data. It has been found that usually $g_{\parallel} > g_{\perp}$ for an archimedian antiprism, while $g_{\parallel} < g_{\perp}$ for a triangular dodecahedron (Table 34). The latter geometry was established by X-ray diffraction for $[NbCl_4(diars)_2]$,[530] while $[TaCl_4(dmpe)_2]$[529] was found to be a square antiprism. The bidentate *o*-phenylenebisdimethylarsine displays a different behavior, since only the monoadduct $[TaCl_4(diars)]$ was found to be stable, but its structure remains unknown. The $[MCl_4(dmpe)_2]$ (**22**) complexes have been widely used as precursors to lower oxidation states.

The tetrahalide adducts are stable to disproportionation, except for '$TaX_4(MeCN)_2$' (X = Cl, Br), which gives Ta^V and Ta^{III} species, the latter undergoing reductive coupling with MeCN. These observations corroborate the results obtained through polarography in MeCN; $TaCl_5$ is reduced directly to Ta^{III}.[531]

Spontaneous reduction to niobium(III) was observed for the crowded dinuclear PCy_3 and

PBu^t_3 adducts. Reduction to undefined products was also noted for $[NbI_4(py)_2]$[515] and $[NbX_4(LL)]$, where LL = *N,N,N',N'*-tetramethylethylenediamine.[532]

A poorly characterized borohydride derivative $[Nb(BH_4)_2Cl_2(MeCN)_2]$ has been reported.[533]

34.3.3 Pseudohalo Complexes (Table 35)

A number of stable niobium(IV) isothiocyanates and less stable isoselenocyanates are known. Rigorous exclusion of water is required to avoid the formation of selenium during the synthesis of the latter.[534]

Table 35 Niobium^{IV} Pseudohalide Derivatives

Compounds	*Comments*	*Ref.*
(a) Thiocyanates and selenocyanates		
$A_2[Nb(NCS)_6]$ (A = K, $AsPh_4$)	N-bonded in the solid (IR)	1
$K_2[Nb(NCSe)_6]$	N-bonded in the solid, unstable thermally, dissociation in dilute MeCN solution	2
$[Nb(NCS)_4(py)_2]$	*Trans* isomer (IR)	3
$[Nb(NCS)_4(LL)_2]$ (LL = bipy, dmbipy)^a	X-Ray structure for LL = bipy (square antiprism), air stable	3, 4
$[Nb(NCSe)_4(LL)_2]$ (LL = bipy, dmbipy)^a	N-bonded in the solid (IR), air stable	3
$[NbCl(NCS)_3]_n$	Not fully characterized, probably dinuclear *via* NCS bridges	5
$[Nb(SCN)_2(Cp)_2]$	Isolated, ESR	6
$[NbCl(SCN)(Cp)_2]$	Not isolated, ESR evidence	6
(b) Cyanides		
$A_4[Nb(CN)_8]\cdot xH_2O$ (A = Na, $x = 4$; A = K, Cs, $x = 2$; A = Ti, $PrNH_3$, $x = 0$)	X-Ray structure for A = K, (D_{2d} dodecahedron); D_{4d} in solution (ESR); stable in aqueous solution only in the dark	7, 8, 9
$[NbCl_3(CN)(MeCN)_2]_n$	Not fully characterized	5
$[Nb(CN)_2(Cp)_2]$	Isolated, ESR	6
$[NbCl(CN)(Cp)_2]$	Not isolated, ESR evidence	6
(c) Cyanates		
$[Nb(CNO)_3Cl]_n$	Not fully characterized, N-bonded (IR)	5
$[Nb(CNO)_2(Cp)_2]$	Isolated, ESR	6

^{a} dmbipy = 4,4'-dimethyl-2,2'-bipyridyl.

1. T. M. Brown and G. F. Knox, *J. Am. Chem. Soc.*, 1967, **89**, 5296; G. F. Knox and T. M. Brown, *Inorg. Chem.*, 1969, **8**, 1401.
2. T. M. Brown and B. L. Bush, *J. Less-Common Met.*, 1971, **25**, 397.
3. J. N. Smith and T. M. Brown, *Inorg. Chem.*, 1972, **11**, 2697.
4. E. J. Peterson, R. B. Von Dreele and T. M. Brown, *Inorg. Chem.*, 1976, **15**, 309.
5. D. J. Machin and J. F. Sullivan, *J. Less-Common Met.*, 1969, **19**, 413.
6. C. P. Stewart and A. L. Porte, *J. Chem., Dalton Trans.*, 1973, 722.
7. P. M. Kierman, J. F. Gibson and W. P. Griffith, *J. Chem. Soc., Chem. Commun.* 1973, 816.
8. P. M. Kierman and W. P. Griffith, *J. Chem. Soc., Dalton Trans.*, 1975, 2489.
9. M. Laing, G. Gafner, W. P. Griffith and P. M. Kierman, *Inorg. Chim. Acta*, 1979, **33**, L119.

Neutral six- or eight-coordinated adducts were obtained by allowing $[Nb(NCS)_6]^{2-}$ or $[Nb(NCSe)_6]^{2-}$ to react with pyridine or bipyridyl, respectively. An alternative route to the paramagnetic pyridine adducts is the reduction of $[Nb(NCS)_6]^-$ by the ligand. $[Nb(NCS)_4(bipy)_2]$ adopts a nearly perfect D_{4d} square antiprismatic structure, with the bidentate ligand bridging the square faces.[535] N coordination has generally been established for the ambidentate thiocyanate ligand in the solid.

Paramagnetic Nb^{IV} octacyanide anions have been isolated as alkali metal or tetraalkylammonium salts. $K_4[Nb(CN)_8]$ has been obtained in low yield by air- or hydrogen-peroxide-oxidation of $K_5[Nb(CN)_8]$ prepared by electrolytic reduction of $NbCl_5$ in an MeOH–KCN medium.[536] $K_4[Nb(CN)_8]\cdot 2H_2O$ disproportionated photolytically to restore the Nb^{III} derivative. The $[Nb(CN)_8]^{4-}$ anion is dodecahedral in the solid; all Nb—C distances are equal

(2.255 Å).[537] ESR, IR and Raman spectroscopies support a configurational change from D_{2d} symmetry in the solid to D_{4d} (antiprism) in solution.

The known cyanato compounds are summarized in Table 35.

34.3.4 Oxo, Thio and Seleno Halides

34.3.4.1 Oxo halides and oxo compounds

Apart from the fluorides, all six MOX_2 compounds have been reported.[1,2,5] Attempts to prepare $NbOF_2$ yielded a single phase between the stoichiometric limits NbO_2F and $[NbO_{1.25}F_{1.75}]$ (average metal oxidation state of 4.25), with an ReO_3 type structure, and it was concluded that $NbOF_2$ does not exist.

The oxo halides are generally obtained through high temperature reduction of M_2O_5 by the metal in the presence of the halogen. $NbOI_2$ may also be prepared by decomposition of $NbOI_3$. They are diamagnetic, stable in air as well as thermally, and the structure of $NbOCl_2$ is closely related to that of $NbCl_4$.

Some Nb^{IV} oxohalo anions have been identified (Table 36), but only the slightly paramagnetic dioxofluoroniobate salts $A[NbO_2F]$ (A = Li, Na, K) were isolated. $[NbOCl_4(L)]^{2-}$ (L = H_2O or EtOH) and $[NbOCl_5]^{3-}$ were prepared by reduction of MX_5 by Zn in aqueous HCl; they result from the hydrolysis of $[NbCl_6]^{2-}$, and have been characterized *in situ* by ESR. Extraction with acetylacetone yielded $[NbO(acac)_2]$, whose structure is related to that of $[VO(acac)_2]$.[538]

Table 36 Nb^{IV} Oxo Halide Derivatives

Compounds	*Comments*	*Ref.*
$A[NbO_2F]$ (A = Li, Na, K)	NbO_2 + excess AF (600–800 °C)	1
$[NbOF_4]^{2-}$	Not isolated, ESR data	2
$[NbOCl_4L]^{2-}$ (L = H_2O, EtOH)	Not isolated, ESR data	2, 3
$[NbOCl_5]^{3-}$	Not isolated, unstable $[Cl^-] < 12$ N, ESR	3
$[NbO(acac)_2]$	Not isolated, $NbCl_5$ + Zn + HCl, acacH, ESR	2
$[NbOCl_2(dpm)]_n$	Isolated, $Nb(dpm)_4 + Cl_2$ in oxygen free CCl_4, not fully characterized	4
$[Nb_2OCl_4(OEt)_2(THF)_2]$	X-Ray structure, metathesis, from Nb^{III}	5
$[NbOCl_2(MeCN)_2(NO)]$	Isolated, hydrolysis of $[NbCl_4(MeCN)_2(NO)]$, not fully characterized	6

1. W. Ruedorff and D. Krug, *Z. Anorg. Allg. Chem.*, 1964, **329,** 211.
2. J. F. Gainullin, N. S. Garifyanov and B. M. Kozyrev, *Dokl. Akad. Nauk. SSSR*, 1968, **180,** 858 (*Chem. Abstr.*, 1968, **69,** 72 714).
3. D. P. Johnson and R. D. Bereman, *J. Inorg. Nucl. Chem.*, 1972, **34,** 679 and 2957.
4. G. Podolsky, *Diss. Abs. B*, 1973, **33,** 5199 (*Chem. Abstr.*, 1973, **79,** 487 845).
5. F. A. Cotton, M. P. Diebold and W. J. Roth, *Inorg. Chem.*, 1985, **24,** 3509.
6. Yu. A. Buslaev, M. A. Glushkova and N. A. Ovchinnikova, *Koord. Khim.*, 1975, **1,** 319 (*Chem. Abstr.*, 1975, **83,** 52 565).

Neutral adducts are limited to the oxo porphyrin derivatives (Section 34.3.5.4.i) and to the uncharacterized $[NbOCl_2(NO)(MeCN)_2]$ obtained by hydrolysis of $[NbCl_4(NO)(MeCN)_2]$.[539] This high stability of the MOX_3 (X = Cl, Br) adducts toward reduction probably hinders easy access to such M^{IV} adducts. The reduction of $NbSX_3$ (X = Cl, Br), on the contrary, allowed the development of sub-thiohalide coordination chemistry.

34.3.4.2 Thio, seleno and telluro halides

The sulfido, seleno and telluro halides of the transition metals have been reviewed recently.[540–542] Numerous Nb^{IV} chalcogenide halides have been reported. For tantalum, only $[Ta(S_2)Cl_2]$ and $[Ta(S_2)_mCl_n]$ (m = 1–3, n = 1–4), of unknown structure, obtained from $TaCl_5$ and S_8, have so far been reported.[543]

$[Nb(Y)_2X_2]$ (Y = S or Se, X = Cl, Br, I) was prepared from the elements at 400–500 °C by chemical transport reactions. The obtention of pure $[Nb(Y)_2I_2]$ required an excess of iodine, and it must be stored under a pressure of NbI_4 even at room temperature. The chlorine derivatives are also available (equations 42–43). A similar reaction between tantalum and S_2Cl_2 led to impure TaS_3. $[Nb_2(Y)_2Br_6]$ (Y = Se, Te) and $[Nb_2(Te)_2I_6]$ were obtained by heating the elements to 780 °C.

$$3Nb + 10Y + 2NbCl_5 \xrightarrow{500\text{–}480\,°C} 5NbY_2Cl_2 \quad (42)$$

$$Y = S, Se$$

$$Nb + S_2Cl_2 \xrightarrow{500\text{–}480\,°C} NbS_2Cl_2 \quad (43)$$

The isomorphous $[Nb(Y_2)X_2]$ compounds are monoclinic[544,545] at room temperature and triclinic at low temperature. The structure of $[Nb(S_2)Cl_2]_2$ (Figure 22) exhibits the pairwise connection of two niobium atoms by metal–metal bonds specific of most of the Nb^{IV} halides. The molecules contain cage-shaped $Nb_2(Y_2)_2$ units of approximate D_{2h} symmetry consisting of a pair of Nb atoms lying perpendicular to the plane of the two Y_2 groups. The Y—Y distances in $[Nb(Y_2)Cl_2]$ (Y = S, Se) correspond to single bonds, and the structures may be described as $^2_\infty[Nb_2(Y_2)_2X_{8/2}]$. The occurrence of metal pairs explains their diamagnetic and semiconducting properties. The IR and Raman spectra of $[Nb(S_2)X_2]$ (X = Cl, Br, I) have been extensively studied.[546] The vibrations of the S_2 groups, at 588 cm^{-1} for $[Nb(S_2)Cl_2]$, decrease from the chloride to the iodide; they are higher than those of the thioniobyl group.

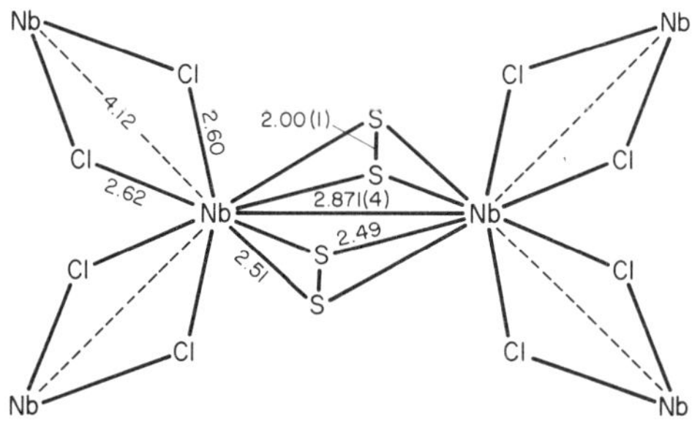

Figure 22 Drawing showing the $^2[Nb_2(S_2)Cl_{8/2}]$ structure of NbS_2Cl_2 (reproduced from ref. 545 with permission)

$[Nb_2(Y_2)X_6]$ is best described as chains of type $^1_\infty[Nb_2(Y_2)X_4X_{4/2}]$ with short Nb—Nb and Y—Y distances. The Nb_2 and Y_2 pairs (Nb—Nb: 2.832(4); Se—Se: 2.305 Å for $[Nb_2(Se_2)Br_6]$) interact to form a quasitetrahedral cluster. The structural and magnetic properties support the formal oxidation states Nb^{4+} and Y^{1-}.[547]

Isotypic compounds of stoichiometry $[Nb_3Se_5X_7]$ (X = Cl, Br) have been obtained from the corresponding $[Nb(Se_2)X_2]$ derivatives and $NbCl_4$ in the case of the chlorine compounds, or in the case of bromine by decomposition of $[Nb(Se_2)Br_2]$ in the presence of $[NbSeBr_3]$.[548] The structure of $[Nb_3Se_5Cl_7]$ (Figure 23) shows two types of niobium atoms; Nb(1) which form pairs, while Nb(2) is isolated. Similarly, the atoms Se(1) and Se(2) form an Se_2 group (2.30 Å) corresponding to a single bond, while Se(3) is coordinated to one metal atom only, at a remarkably short distance, which suggests the existence of a selenoniobyl group. X-Ray and ESCA data established that $[Nb_3Se_5Cl_7]$ is a mixed valence compound that should be formulated as $[Nb^{IV}_2Nb^{V}(Se_2)_2SeCl_7]$, Nb(2) being in oxidation state +V. The overall structure consists of chains of composition $[Nb^{IV}_2(Se_2)_2Cl_5]$ to which side chains $[Nb^{V}SeCl_2]$ are attached. Semiconducting behavior was found for this compound.

34.3.4.3 Adducts of the thio and seleno halides

A synthesis of $[NbSCl_2]$ from a solid state reaction between $NbCl_4$ and Sb_2S_3 at 200 °C has been claimed,[549] but attempts to reproduce it were unsuccessful.[522] By contrast, the reaction between NbX_4 (X = Cl, Br) and Sb_2S_3 in MeCN at 50 °C easily provided adducts of type $[NbSX_2(MeCN)_2]_2$ as highly air sensitive solids. The chlorine adducts were obtained as brown or green materials, depending on the number of MeCN groups. They are the only adducts of $NbSX_2$ that have been obtained directly from tetravalent metal derivatives. The dimers have D_{2h} symmetry; the niobium atoms are bridged by two sulfur atoms (Nb—S: 2.338(8)–

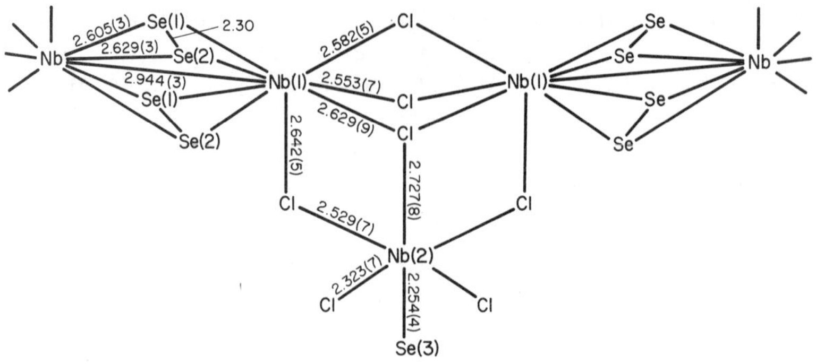

Figure 23 Section of the structure of the mixed valence $[Nb_2^{IV}Nb^V(Se)_2SeCl_7]$ (reproduced from ref. 548 with permission)

2.349(7) Å), and bear two chlorine atoms in apical and two acetonitrile ligands in equatorial positions. The octahedral environment of the metal is distorted as a result of niobium–niobium interactions (2.862(2)–2.872(3) Å).

$[Nb_2S_3X_4(THT)_4]$ (X = Cl, Br) results from a complex disproportionation in neat THT (equations 44 and 45), two S^{2-} ions being formally oxidized to $(S_2)^{2-}$. Sulfur abstraction reactions from $[NbSBr_3]$ have also been suggested.[373] Reactions between $[NbSCl_3]$ and THT, and those between $[NbSX_3]$ (X = Cl, Br) and Me_2S follow similar patterns. Analog studies with $[TaSX_3]$ and THT gave no evidence for Ta^{IV} compounds. Crystal structure determination led to the formulation $[Nb_2X_4(THT)_4(\mu\text{-}S)(\mu\text{-}S_2)]$ of approximate C_{2v} symmetry (Figure 24), the most salient feature being the $Nb(\mu\text{-}S)Nb(\mu\text{-}S_2)$ ring. The Nb—Nb distances suggest a single bond; the three sulfur atoms form isosceles triangles (3.78(1), 3.76(1) and 2.01(1) Å). The S—S distances are slightly shorter than in ionic $(S_2)^{2-}$. No significant *trans* effect was observed.

$$[NbSBr_3] + 2THT \longrightarrow [NbSBr_3(THT)_2] \quad (44)$$

$$[NbSBr_3(THT)_2] \longrightarrow [Nb_2S_3Br_4(THT)_4] + [NbBr_5(THT)_n] \quad (45)$$

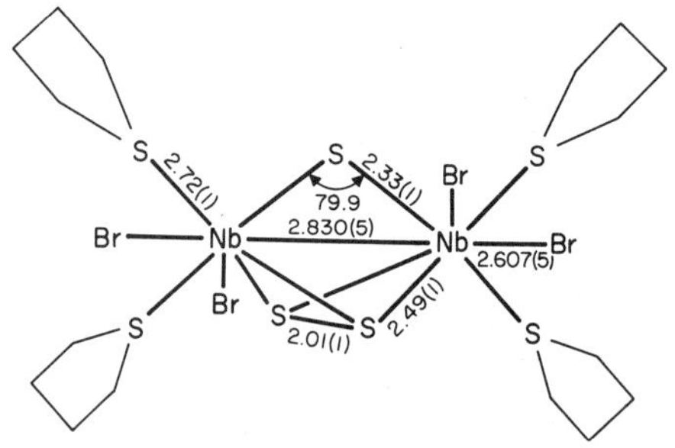

Figure 24 Molecular structure of $[(THT)_2Br_2Nb(\mu\text{-}S)(\mu\text{-}S_2)NbBr_2(THT)_2]$ (reproduced from ref. 373 with permission)

Attempts to synthesize selenium analogs of $[NbSX_2(MeCN)_2]_2$ from $[NbSeBr_2]$ led to reduction of the metal and formation of a tetranuclear cluster (equation 46). Formally two Se^{2-} ions form an $(Se_2)^{2-}$ ion with concomitant release of two electrons, which reduce two Nb^{IV} to Nb^{III} atoms, and formation of the mixed valence tetramer.[550] The latter (Figure 25) is best described as consisting of a central $(MeCN)_2Br_2Nb(\mu\text{-}Se_2)(\mu\text{-}Se)NbBr_2(MeCN)_2$ fragment linked by long bromide bridges to two *fac*-$NbBr_3$ units. Se(3) is bound to all three metal atoms. No metal–metal interaction is found between the two Nb^{III} centers, as also observed for $[NbCl_3(PhC{\equiv}CPh)]_4$ (Section 34.4.2.2.ii). The sample was ESR silent down to −196 °C, and diamagnetic, which was considered to be consistent with the two Nb^{III} atoms' being in an approximately C_{3v} environment and having a low-lying doubly occupied dz^2 orbital, while the two Nb^{IV} atoms exhibit a metal–metal interaction.

$$4[NbBr_4(MeCN)_2] + Sb_2Se_3 \xrightarrow[50\,°C,\,7\,d]{MeCN} [Nb_4Br_{10}(Se_2)(Se)(MeCN)_4] + 2SbBr_3 \quad (46)$$

(23)

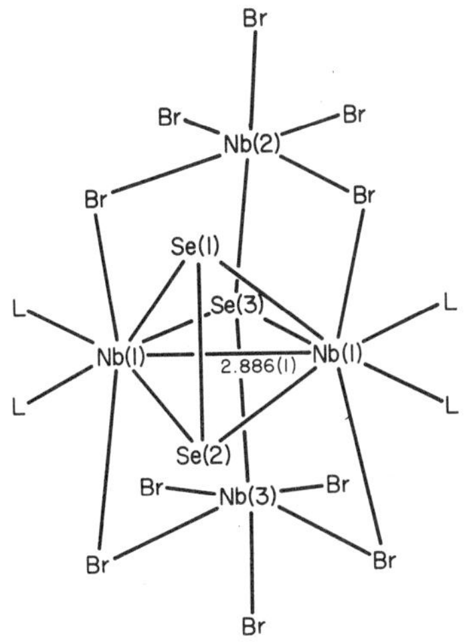

Figure 25 Molecular structure of $[Nb_4Br_{10}(Se_2)(Se)(MeCN)_4]$ (**23**) (reproduced from ref. 550 with permission)

34.3.5 Solvolysis of the Tetrahalides

34.3.5.1 Alkoxides (Table 37)

Several Nb^{IV} alkoxo compounds have been obtained, by miscellaneous methods, but were often only poorly characterized. A series of salts containing the $[NbCl_5(OR)]^{2-}$ anion (R = Me, Et, Pr^i) has been prepared by electrolytic reduction of $NbCl_5$ in saturated HCl–ROH solutions.[7] A different experimental treatment of the initial electrolytic solution led to the diamagnetic $[NbCl(OEt)_3py]_2$, which was further converted to the extremely moisture- and oxidation-sensitive diamagnetic $[Nb(OEt)_4]_n$ (Scheme 5).

Table 37 Niobium(IV) and Tantalum(IV) Alkoxides

Compounds[a]	*Comments*	*Ref.*
$A_2[NbCl_5(OR)]$ (R = Me, Et, Pr, A = quiH, pyH; R = Me, A = picH, Et_2NH; R = Me, Et, A = Me_4N)	Electrolytic reduction of ROH–HCl–$NbCl_5$, addition of ACl, A = Me_4N alternate synthesis: solvolysis of $NbCl_4(MeCN)_3$	1, 2
$K[Nb_2(OMe)_9]$	$[NbCl_4(MeCN)_2$, MeCN] + KOMe, analysis only	3
$[Nb(OEt)_4]_n$	Sublimable, diamagnetic	4
$[Nb(Rado)_4]$ (R = H, OMe)	Alcoholysis of $[Nb(NEt_2)_4]$, monomeric; R = OMe adduct $0{\cdot}5Et_2NH$, diamagnetic in the solid	5
$[M_2Cl_2(OR)_6(L)_2]$ (M = Nb, R = OEt, L = py, EtOH; M = Ta, R = OPr^i, L = γ-pic)	M = Nb electrolytic reduction, diamagnetic; X-ray structure for L = EtOH diamagnetic	4, 6
$Nb_2Cl_4(OMe)_4(MeOH)_2$	X-Ray structure, metathesis from Nb^{III}	7
$[Nb_2Cl_5(OEt)_3(bipy)]$	Electrolytic reduction	8
$[NbCl_3(OR)(bipy)]$ (R = Me, Et, Bu^t, Pr)	$[Nb_2Cl_4(OR)_6]$ + Libipy; unidentified Nb^{III} species if excess Libipy	9
$[NbCl_2(OMe)_2(MeOH)_2]$	$[NbCl_4(MeCN)_2$, MeCN] + MeOH, analysis only	3
$[Nb_2Cl_4(o\text{-cat})_2(THF)_2]$	Diamagnetic	6
$[Nb_2Cl_2(o\text{-cat})_3(MeCN)]_2$	Diamagnetic	6

[a] Qui = quinuclidine; ado = adamanto

1. R. A. D. Wentworth and C. H. Brubaker, *Inorg. Chem.*, 1963, **2,** 551.
2. R. A. D. Wentworth and C. H. Brubaker, *Inorg. Chem.*, 1962, **1,** 971.
3. R. Gut and W. Perron, *J. Less-Common Met.*, 1972, **26,** 369.
4. R. A. D. Wentworth and C. H. Brubaker, *Inorg. Chem.*, 1964, **3,** 47.
5. M. Bochmann, G. Wilkinson, G. B. Young, M. B. Hursthouse and K. M. A. Malik, *J. Chem. Soc., Dalton Trans.*, 1980, 901.
6. L. G. Hubert-Pfalzgraf and P. Laurent, to be published.
7. F. A. Cotton, M. P. Diebold and W. J. Roth, *Inorg. Chem.*, 1985, **24,** 3509.
8. C. Djordjevic and V. Katovic, *J. Chem. Soc. (A)*, 1970, 3382.
9. N. Vuletic and C. Djordjevic, *J. Chem. Soc., Dalton Trans.*, 1973, 550.

$$[NbCl(OEt)_3(py)_2] \xrightarrow[EtOH]{NaOEt} [Nb(OEt)_4]$$

$$\underset{EtOH}{py, HCl} \searrow \qquad \swarrow \underset{EtOH}{py, HCl}$$

$$(pyH)_2[NbCl_5(OEt)]$$

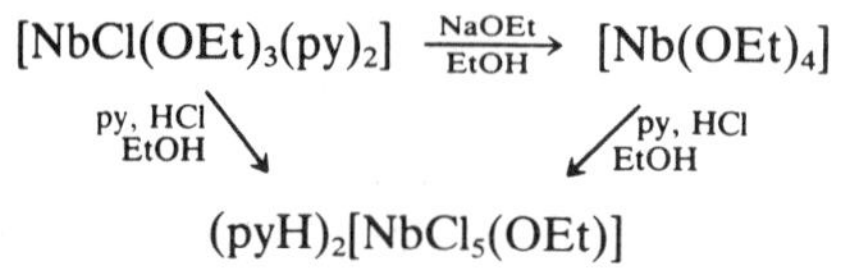

Scheme 5

The synthesis of $M(OEt)_4$ shows that, contrary to previous observations, tetravalent alkoxides are stable in the presence of alcohol, at least for a short time. The alcoholysis of $[Nb(NR_2)_4]$ usually led to $[Nb(OR')_5]$,[263] but oxidation state +IV was retained with bulky alcohols, *e.g.* for tetra(1-adamanto)niobium and tetra(1-adamantylmethoxo)niobium, illustrating the close dependence of the metal's oxidation state on steric requirements. $[Nb(1\text{-}ado)_4]$ was reported to be monomeric, but the absence of ESR signals down to $-150\,^{\circ}C$, as well as its reluctance to form complexes (Me_2NH, PMe_3, THT) is puzzling.[551] Dinuclear and tetranuclear diamagnetic niobium chlorocatecholates have also been obtained. The Ta^{IV} alkoxides are limited to some diamagnetic chloro alkoxide adducts.

34.3.5.2 β-Diketonates, carboxylates and related chelates

$NbCl_4$ reacts with a variety of β-ketoenols or oximes in the presence of a base to give moderately soluble eight-coordinate tetrakischelates (Table 38). These have one unpaired electron and are not isostructural with the related Zr and Hf compounds. Dodecahedral structures have generally been proposed on the basis of UV or ESR data.[552] However, the larger bite of the dipivaloylmethane ligand ($b = 1.28$ Å) allows the stabilization of a square antiprism. $[Nb(dpm)_4]$ corresponds to an idealized $D_4(1111)$ isomer, and was the first example of this geometry to be reported.[553] Evidence for nine-coordination was found in $[Nb(acac)_4(dioxane)]$.[557] $TaCl_4$ generally reacts by abstraction of oxygen, and tetrakischelated compounds are limited to $[Ta(dbm)_4]$. Some chloro-β-diketonato compounds, which result from reactions conducted in the absence of base, are also described. $[Ta_2Cl_4(dpm)_4]$ was derived from the tantalum(III) adduct $[Ta_2Cl_6(\gamma\text{-pic})_4]$.[544]

Table 38 Tetravalent β-Diketonates and Related Chelates

Compounds	Comments	Ref.
$[Nb(acac)_4]$	ESR (dodecahedron), no complexation with THF	1, 2
$[Nb(\beta\text{-diket})_4\text{dioxane}]$	β-Diket = acac, bzac, monodentate dioxane in the solid (nine coordination), for bzac free dioxane in solution	1
$[Nb(dpm)_4]$	X-Ray structure, square antiprism	3, 4
$[M(dbm)_4]$	M = Nb, ESR (dodecahedron); M = Ta, not isomorphous, low μ (0.53 BM)	1
$[NbCl_2(\beta\text{-diket})_2]$	β-Diket = acac, dbm; analysis only; dpm: X-ray	1, 5, 6
$[TaCl_2(dpm)_2]_2$	Diamagnetic, soluble, obtained from Ta^{III}	7
$[Nb(LL)_4]$	LL = benzoyltrifluoroacetone (bta), tetrathenoyltrifluoroacetone (tta)	1
$[NbT_4]_n$	Polymeric ($\mu = 0.74$ BM)	1
$[NbClT_3]_n$	Polymeric, ionic species in MeCN	1
$[Nb(8\text{-hdq})_4]$		1

1. D. L. Kepert and R. L. Deutscher, *Inorg. Chim. Acta*, 1970, **4**, 645.
2. R. L. Deutscher and D. L. Kepert, *Chem. Commun.*, 1969, 121.
3. T. J. Pinnavaia, G. Podolski and P. W. Godding, *J. Chem. Soc., Chem. Commun.*, 1973, 242.
4. T. J. Pinnavaia, B. L. Barnett, G. Podolski and A. Tulinsky, *J. Am. Chem. Soc.*, 1975, **97**, 2712.
5. R. Gut and W. Perron, *J. Less-Common Met.*, 1972, **26**, 369.
6. F. A. Cotton, M. P. Diebold and W. J. Roth, *Polyhedron*, 1985, **4**, 1485.
7. J. L. Morançais, L. G. Hubert-Pfalzgraf and P. Laurent, *Inorg. Chim. Acta*, 1983, **71**, 119.

The O dealkylation reactions often observed between carboxylic acids and group VA halides could be controlled by the use of bulky substituted acids. Stepwise substitution and oxidation of $[Nb(BH_4)(Cp)_2]$ *via* $[Nb(O_2CBu^t)(Cp)_2]$ and $[Nb(O_2CBu^t)_2(Cp)_2]$ thus provided $[Nb(O_2CBu^t)_4]$ in good yield.[555] The latter is paramagnetic, the metal being octacoordinated. It is a rare example of a monomeric Nb^{IV} compound soluble in non-polar hydrocarbons, and may therefore constitute an interesting starting material for Nb^{IV} chemistry, but its reactivity does

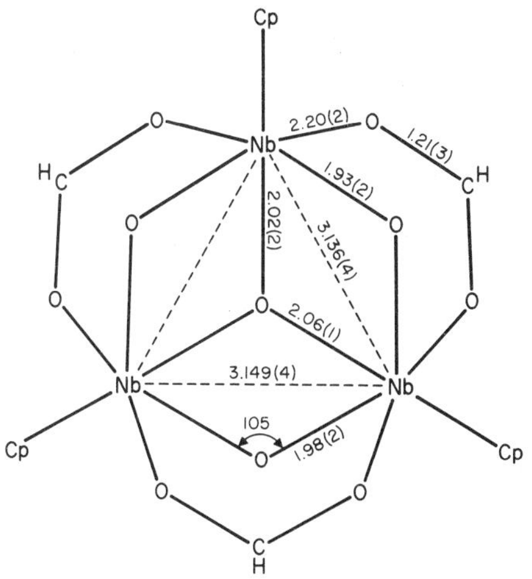

Figure 26 Molecular geometry of [{(Cp)Nb(μ-O_2CH)}$_3$(μ-OH)$_2$(μ_3-O)]. The nature of μ_2 groups is not known with certainty (reproduced from ref. 556 with permission)

not seem to have been investigated. [Nb(O_2CCR_3)$_2$(Cp)$_2$] (R = F, Ph) was formed in similar reactions. The only Ta^{IV} carboxylate known is the diamagnetic [{Ta_2Cl_4(O_2CBu^t)$_4$}2Bu^tCO_2H].

A trinuclear compound [{Nb(O_2CH)(Cp)}$_3$(OH)$_2$(O)$_2$] was isolated from the reaction between [Nb(CO)$_3PPh_3$(Cp)] and HCOOH, which involves decarboxylation of the formate group.[556] It has an almost regular triangular structure (Figure 26) with Nb - - - Nb distances close to the usual single Nb—Nb bond length, but the magnetic properties (one unpaired electron) and the large angles around the bridging oxygen argue against direct metal–metal bonding.

34.3.5.3 Thiolato and related derivatives

The dialkylthiocarbamates (Table 39) were initially made by insertion of CS_2 into the M—N bond of [M(NR_2)$_5$], which, in the case of niobium, proceeds with reduction to Nb^{IV} and

Table 39 Tetravalent Dialkyldithiocarbamates and Related Compounds

Compounds	*Comments*	*Ref.*
(*a*) *Dithiocarbamates*		
[Nb(S_2CNR_2)$_4$] (R = Me, Et)	Insertion of CS_2 into the Nb^V—NR_2 bond (R = Me, Et) or metathesis of $NbCl_4$ (R = Et); octacoordinated (IR)	1, 2, 3
[Nb{$S_2CN(CH_2)_4$}$_4$]	Metathesis of $NbCl_4$, octacoordinated (IR)	4
[Nb_2Br_3(S_2CNEt_2)$_5$]	Electrolyte [Nb_2Br_2(S_2CNEt_2)$_5$]Br in $MeNO_2$	3
[NbCl(S_2CNEt_2)$_2${C(Cl)NBu^t}$_2$]$_2$		5
[Nb(S_2CNEt_2)(Cp)$_2$]Z·H_2O (Z = PF_6, BPh_4)	Metathesis of [$NbCl_2$(Cp)$_2$], not fully characterized	6
(*b*) *Dithiocarboxylates and dithiolates*		
[Nb(S_2CR)$_4$] (R = Me, Ph)	Single crystal ESR studies, octacoordinated	7
[Nb($S_2C_2H_4$)$_3$]$^{2-}$	Not isolated, ESR (trigonal prism)	8
[M($S_2C_6H_4$)$_3$]$^{2-}$	Polarographic evidence (M = Nb, $E_{1/2}$ = −0.38 V; M = Ta, $E_{1/2}$ = −0.71 V)	9
[$NbCl_2$($S_2C_6H_4$)]$_2$	Isolated, diamagnetic	10

1. D. C. Bradley and M. H. Gitlitz, *J. Chem. Soc. (A)*, 1969, 1152.
2. J. N. Smith and T. M. Brown, *Inorg. Nucl. Chem. Lett.*, 1970, **6**, 44.
3. D. J. Machin and J. F. Sullivan, *J. Less-Common Met.*, 1969, **19**, 413.
4. T. M. Brown and J. N. Smith, *J. Chem. Soc., Dalton Trans.*, 1972, 1614.
5. M. Behnam-Dehkordy, B. Crociani, M. Nicolini and R. L. Richards, *J. Organomet. Chem.*, 1979, **181**, 69.
6. J. Amaudrut, J. E. Guerchais and J. Sala-Pala, *J. Organomet. Chem.*, 1978, **157**, C10.
7. D. Attanasio, C. Bellitto and A. Flamini, *Inorg. Chem.*, 1980, **19**, 3419.
8. J. Stach, R. Kirmse, W. Dietzsch, I. N. Markov and V. K. Belyaeva, *Z. Anorg. Allg. Chem.*, 1980, **466**, 36.
9. J. L. Martin and J. Takats, *Inorg. Chem.*, 1975, **14**, 73.
10. L. G. Hubert-Pfalzgraf, to be published.

oxidation of the organic moiety to tetraalkylthiuran disulfides.[279] In the case of tantalum, the pentavalent state is retained, and no Ta^{IV} dithiocarbamates are presently known. IR has been used as a criterion for establishing the coordination mode of the dithiocarbamato ligand, which appears to act as a bidentate in all these compounds. Nb^{IV} dithiocarboxylates have been prepared, and were studied by single crystal ESR. Their geometries are intermediate between a triangular dodecahedron and a square antiprism.[557]

Tris(maleonitriledithiolato)niobate $[Nb(mnt)_3]^{2-}$ (mnt = $S_2C_2H_4^{2-}$) has been obtained in solution by reduction of $NbCl_5$ by Zn in an HCl–MeOH solution followed by *in situ* complexation of the Nb^{4+} species.[558] The ESR data ($\langle g \rangle = 1.988$, $\langle A \rangle = 94.4$ G) are consistent with a nearly trigonal prismatic arrangement. They exclude the formation of an oxo species such as $[Nb(mnt)_2]^{2-}$ whose ESR parameters should compare with those of $[NbO(acac)_2]$ ($\langle g \rangle = 1902$, $\langle A \rangle = 142$ G). Polarographic results indicated that tris(dithiolene) Nb^V and Ta^V undergo one-electron transfers. $[NbCl_2(S_2C_6H_4)]_2$ appears to be the only dithiolene adduct isolated so far. Compounds with thiolato groups are unstable, especially in solution, but may be stabilized as $[M(SR)_2(Cp)_2]$ derivatives (M = Nb; R = Me, Pr, Ph;[8,559] M = Ta, R = Me). Selenium analogs are known for niobium, and have been used as ligands toward Fe(NO)(CO) and $Co(CO)_2$ moieties.[560]

Tetrakis *O,O'*-dialkylthiophosphates have been obtained for niobium starting from oxidation states +V or +IV (Table 40). These chelates display a distorted dodecahedral geometry in the solid (X-ray) as well as in solution (ESR).[561] An unusual cation, $[\{Nb(Cp)_2\}_2(PS_4)]^+$, in which a PS_4^{3-} moiety is doubly bridging the terminal $NbCp_2$ units, has been obtained from $[NbCl_2(Cp)_2]$ and P_4S_{10}. The metal has a distorted tetrahedral environment of approximate D_{2d} symmetry, while symmetry at phosphorus is reduced to C_{2v}.

Table 40 Niobium(IV) Thiophosphate Derivatives

Compounds	*Comments*	*Ref.*
$[Nb\{S_2P(OR)_2\}_4]$ (R = Et, Pr^i)	X-Ray structure, ESR (dodecahedron)	1, 2
$[Nb_2X_5\{S_2P(OEt)_2\}_3]$ (X = Cl, Br, I)	Analysis only; $[Nb_2X_4\{S_2P(OEt)_2\}_4]$ also compatible	1
$[NbCl_3(S_2PF_2)]$		3
$[Nb\{S_2P(OR)_2\}(Cp)_2](PF_6)$ (R = Me, Et, Pr^i)	ESR	4, 5
$[\{Nb(Cp)_2\}_2(\mu\text{-}PS_4)](PF_6)$	X-Ray structure	7
$[Nb\{SP(S)(OR)_2\}_2(Cp)_2]$ (R = Et, Pr^i)	Monodentate thiophosphate	6

1. R. N. McGinnis and J. B. Hamilton, *Inorg. Nucl. Chem. Lett.*, 1972, **8**, 245.
2. V. V. Tkachev, S. Shchepinov and L. O. Atovmyan, *J. Struct. Chem. (Engl. Transl.)*, 1977, **18**, 823.
3. R. G. Cavell and A. R. Sanger, *Inorg. Chem.*, 1972, **11**, 2016.
4. B. Viard, J. Sala-Pala, J. Amaudrut, J. E. Guerchais, C. Sanchez and J. Livage, *Inorg. Chim. Acta*, 1980, **39**, 99.
5. C. Sanchez, D. Vivien, J. Livage, J. Sala-Pala, B. Viard and J. E. Guerchais, *J. Chem. Soc., Dalton Trans.*, 1981, 64.
6. J. Sala-Pala, J. L. Migot, J. E. Guerchais, L. Le Gall and F. Grosjean, *J. Organomet. Chem.*, 1983, **248**, 299.
7. B. Viard, J. Sala-Pala, J. E. Guerchais, R. Mercier and B. Douglade, to be published.

34.3.5.4 Alkylamides and related compounds

(i) Dialkylamides and dialkylphosphides

A series of homoleptic Nb^{IV} dialkylamido compounds has been obtained as highly air sensitive liquids by spontaneous reduction of $[Nb(NR_2)_5]$ (R = Et, Pr^n, Bu^n; Table 41). Aminolysis of either Nb^V or Nb^{IV} dialkylamides provided mixed species.[256] The existence of quadrivalent tantalum dialkylamides is still questionable. $[Ta(NEt_2)_4]$ was reported to have been detected during the thermolysis of $[Ta(NEt_2)_5]$,[263] which led predominantly to the Ta^V nitrene (**2**; Section 34.2.3.4.ii), but other workers could not find any evidence for its formation in a similar experiment.[282] $[Ta(NMeBu)_4]$ has been reported to be formed in the decomposition of the corresponding pentavalent dialkylamide.

The monomeric, paramagnetic alkylamides constitute rare examples of magnetically dilute

Table 41 Niobium(IV) Alkylamides and Silylamides

Compounds	*Comments*	*Ref.*
$[Nb(NR_2)_4]$ (R = Et, Pr^n, Bu^n)	Monomeric, ESR ($g_{\parallel} < g_{\perp}$: D_{2d} symmetry)	1, 2
$[M(NMeBu^n)_4]$	From thermolysis of $[M(NMeBu^n)]_5$	
$[Nb(NC_5H_{10})_4]$	Analytical data only	1
$[Nb(NEt_2)_{4-x}(NC_5H_{10})_x]$ ($x = 1, 2$)	Analytical data only	1
$[Nb(NMe_2)_3(Cp)]_n$	Diamagnetic, structure unknown	3

1. D. C. Bradley and I. M. Thomas, *Can. J. Chem.*, 1962, **40**, 449.
2. D. C. Bradley and M. H. Chisholm, *J. Chem. Soc.* (*A*), 1971, 1511.
3. A. D. Jenkins, M. F. Lappert and R. C. Srivastava, *J. Organomet. Chem.*, 1970, **23**, 165.

tetrahedral Nb^{IV} compounds. The ESR data ($g_{\perp} > g_{\parallel}$; Table 34) are consistent with the unpaired electron residing in the $d_{x^2-y^2}$ orbital in a field of D_{2d} symmetry, with strong metal–nitrogen covalent character.[279]

The alkylamides should constitute valuable starting materials for the synthesis of other Nb^{IV} derivatives, but studies of their reactivity have so far been limited to some substitution (with ROH, R_2NH, RLi) and CS_2 insertion reactions. Compounds of type $[Nb(NR_2)_{4-x}R'_x]$ ($x = 1$, 2; R = Me, Et; R′ = Bu^t, Np, Bu, Me_3SiCH_2) have proven useful for removing small amounts of oxygen from gases and as components of catalysts for the polymerization of ethylene.[562] Although solvolysis of NbX_4 by primary and secondary aliphatic amines has been shown to occur, no reaction products were identified.[532] A product of puzzling formula, $[NbCl_4(salen)]$, was formed by the reaction of $NbCl_4$ with *N,N′*-bis(salicylidene)ethylenediammine.[563]

The unstable neutral $[Nb(PCy_2)_4]$ has recently been obtained through oxidation of an anionic Nb^{III} precursor $[Nb(PCy_2)_4][Li(DME)_2]$.[565] The spectral data were consistent with a distorted tetrahedral structure similar to that of $[Mo(PCy_2)_4]$. It is the first group Va member of a potentially large class of analogs of the dialkylphosphides. Mixed halodialkylphosphides were also obtained by oxidative addition of phosphines to M^{II} adducts.[564]

(*ii*) *Porphyrin and phthalocyanine adducts*

Zinc amalgam, or electrochemical reduction of porphyrinatoniobium(V) oxide acetates led to $[Nb^{IV}OL]$ (L = OEP, TPP, TPTP).[566] The $[NbOLF]^-$ anions have been obtained by controlled potential electrolysis of (i) the corresponding Nb^V oxo acetate in the presence of PF_6^-, and (ii) the appropriate Nb^V oxo fluoride. Reduction by zinc amalgam of $[NbLX_3]$ (X = Cl, Br) gave $[Nb^{IV}LX_2]$;[327] the ESR spectra of the six-coordinate complexes in frozen THF solution are consistent with *trans* coordination.

Reaction between $NbCl_5$ and phthalonitrile in quinoline at 100 °C led to $[NbCl_2(PC)]$ (PC = phthalocyanine); the metal is displaced by 0.98 Å from the central C_8N_8 core.[567]

(*iii*) *Miscellaneous derivatives with M—N bonds*

A number of dinuclear, diamagnetic, but rather unstable Nb^{IV} poly(1-pyrazolyl)borates have been obtained from either $NbCl_5$ or $NbCl_4$ with $[KH_xB(pz)_{4-x}]$ ($x = 1$, 2).[338] $[HB(pz)_3]$ generally acted as a bidentate ligand. Bridging through the chlorine or the poly(1-pyrazolyl)borate anion was postulated. Formation of dinuclear species even with bulky ligands such as $HB(pz)_3$, $N(SiMe_3)_2$ or Cp [568] illustrates the tendency towards metal–metal bonding.

The stronger reducing character of $PhMeNNH_2$ as compared to $PhNHNH_2$ gave, directly from Nb^V, diamagnetic $[Nb_2Cl_6(HNNMePh)_2(MeCN)_2]$.[274]

Nb^{IV} iminoethyl derivatives have been obtained by insertion of Bu^{t}NC into the Nb—Cl bond of $[NbCl_4(THF)_2]$ and substitution of the remaining chlorine. The products

$[NbCl_3\{C(Cl)NBu^t\}(CNBu^t)]_2$, $[NbCl_2\{C(Cl)NBu^t\}(ox)]_2$ and $[NbCl\{C(Cl)NBu^t\}(S_2CNEt_2)]_2$ are dinuclear, but only the last two are diamagnetic.[569]

Diamagnetic polymeric nitrides, $[TaNBr]_n$[324] and $[NbNI]_n$,[325] have been obtained by high temperature thermolysis of MX_5 with ammonium salts.

34.3.6 Chalcogenides

The chalcogenides MY_3 and MY_2 ($Y = S$, Se) can be prepared by various methods, including heating the elements together, thermal decomposition of the $[Nb(Y_2)X_2]$ halides, and action of CS_2 on the oxides at about 900–1300 °C.[1,542]

The structures of $[NbS_3(py)]$ and $NbSe_3$[570] show trigonal prismatic chain arrangements in which the chalcogenide atoms form approximate isosceles triangles. It was suggested that they should formally be considered as $[Nb^{IV}(Y^{2-})\{(Y_2)^{2-}\}]$ ($Y = S$ or Se), an idea which is supported by studies on the electronic structure of $NbSe_3$.[571] The presence of Nb—Nb pairs (3.04 Å) in NbS_3 is responsible for the observed diamagnetism and semiconducting properties; by contrast no Nb—Nb pairs were found in $NbSe_3$, which indeed exhibits metallic properties. The remarkable electrical properties of TaS_3 may also be explained by Ta—Ta pairing.[572]

A number of nearly two-dimensional intercalates of formula $[MY_2(L)_x]$ ($Y = S$, Se; L = N donors; $x = 0.5$ for L = py, Et_2NH, Et_3N, piperdine, *etc.*) with promising superconducting properties have been described.[573] Niobium disulfide is an intercalation electrode in lithium batteries, whose effectiveness is close to that of titanium disulfide.[574]

34.3.7 Activation of small molecules

The short-lived $[MH_2(Cp)_2]$ and $[TaH_4(dmpe)_2]$ have been obtained from the M^V hydrides using photogenerated *t*-butoxy radicals, and were characterized by low temperature ESR.[575] On the other hand, thermally stable, well-defined dinuclear or mononuclear M^{IV} hydrides have been prepared by oxidative addition of H_2 to dinuclear M^{III} or mononuclear M^{II} halide phosphine adducts, respectively. They constitute attractive entries to lower oxidation state compounds, and will be reviewed in Sections 34.4.3.1.i and 34.6.1.2.i.

$[TaHCl_2(Cp'')]_2$ (**24**) ($Cp'' = \eta^5$-C_5Me_4Et) reacts easily with CO (equation 47) to give $[\{Ta(Cl_2(Cp'')\}_2(\mu\text{-}H)(\mu\text{-}CHO)]$ (**25**).[576] Crossover experiments have established that the dimer (**24**) is not split by CO. X-Ray diffraction data show that the hydride and formyl ligands are bridging two skewed $TaCl_2(Cp'')$ fragments; the Ta—Ta distance was considered too long for a full single bond, but consistent with some interaction (Figure 27).[577] The C—O distance is unusually long; this activation is in agreement with its being cleft by PMe_3 (equation 48) to give (**26**) for which a Ta—Ta bonding has been proposed (2.992(1) Å).[578] The formyl ligand of (**25**) is apparently stabilized by side-on bonding between the two metals, and (**25**) does not react readily with CO or H_2. It is, however, susceptible to electrophilic or nucleophilic attack; methanol is produced when excess aqueous HCl is added to (**25**) in protic solvents.

$$[TaHCl_2(Cp'')]_2 + CO \xrightarrow{1\ atm,\ -78\,^\circ C} [\{TaHCl_2(Cp'')\}_2(CO)]$$
(**24**)

$$\xrightarrow{0\,^\circ C} [\{TaCl_2(Cp'')\}_2(\mu\text{-}H)(\mu\text{-}CHO)] \quad (47)$$
(**25**)

$$[\{TaCl_2(Cp'')\}_2(\mu\text{-}H)(\mu\text{-}CHO)] + PMe_3 \longrightarrow [\{TaCl_2(Cp'')\}_2(\mu\text{-}H)(\mu\text{-}O)(\mu\text{-}CHPMe_3)] \quad (48)$$
(**25**) (**26**)

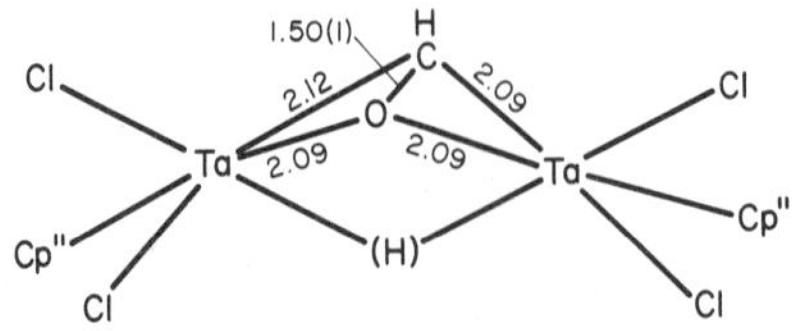

Figure 27 Drawing of the structure of $[\{TaCl_2(Cp'')\}_2(\mu\text{-}H)(\mu\text{-}CHO)]$ (**25**) (reproduced from ref. 576 with permission)

$[TaHCl_2(Cp'')]_2$ (**24**) reacts rapidly with one and only one equivalent of acetonitrile to give $[\{TaCl_2(Cp'')(\mu\text{-H})(\mu\text{-Cl})(\mu\text{-}\eta^1\text{-}N,\eta^2\text{-}C,N\text{-NCHMe})\}TaCl(Cp'')]$ (**27**),[579] whose molecular structure resembles those of (**25**) and (**26**). The $TaCl_2(Cp'')$ and $TaCl(Cp'')$ fragments are asymmetrically triply bridged by a chloride ligand, a hydride ligand and the NCHMe ligand derived from acetonitrile by hydrogenation. Multiple bond character has been suggested for Ta(1)—N (1.901(8) Å), but not for Ta(2)—N (2.059(8) Å).

The paramagnetic $[\{Nb(Bu^n)(Cp)_2\}_2(N_2)(O_2)]$ (equation 27) is the first example of simultaneous fixation of N_2 and O_2 by a niobium complex ($\nu(N_2) = 1740\ cm^{-1}$).

The reaction between NO and $[NbCl_4(MeCN)_2]\cdot MeCN$ led to $[NbCl_4(NO)(MeCN)_2]$.[539] Ligand exchange reactions provided $[NbCl_4(NO)(MeCN)(LL)]$ (LL = bipy or phen), $[NbCl_4(NO)(MeCN)(DMF)_3]$ and $[NbCl_4(NO)(MeCN)(Ph_2PO)]$, this last resulting from the oxidation of the PPh_3 ligand by NO. The diamagnetism of these products suggests polymeric structures in the solid. The phenanthroline and dimethylformamide adducts behave as electrolytes in solution. For the reaction between $[NbMe_2(Cp)_2]$ and excess NO, see Section 34.2.6.3.

34.4 OXIDATION STATE +III

34.4.1 Survey

The chemistry of niobium and tantalum in their lower oxidation states is expanding rapidly. The first structurally characterized molecular Nb^{III} derivative was reported in 1970,[525] while Nb^{III} and Ta^{III} halide adducts were described in 1973[580] and 1978, respectively.[581]

That this now fertile ground had long remained untilled is essentially attributable to the lack of convenient starting materials (in particular those without cyclopentadienyl ligands). Indeed, the stoichiometric trihalides MX_3 (X = Cl, Br) are only incidental compositions in homogeneous phases.[1,2,5] They can be obtained by reduction of the pentahalides by the metal, or, in the case of tantalum, by disproportionation of the tetrahalides. Reduction to the trichloride can be used as a means of separating tantalum from niobium. Stoichiometric NbI_3 appears as a residue in the thermal decomposition of the higher iodides; it is also the main product of the reaction of the metal with iodine at about 500 °C. NbI_3 is, however, unstable and disproportionates at 513 °C to α-NbI_4 and Nb_3I_8.[501]

The existence of trifluorides has been claimed, but the products isolated invariably contained oxygen, and the existence of pure MF_3 remains doubtful.[5,582]

The 'trihalides' are thermally unstable, dark-colored, polymeric, unreactive solids, and their chemical properties have received little attention.

34.4.2 Halo Adducts

34.4.2.1 Anionic halo complexes

The niobium(III) salts $Cs_3[Nb_2X_9]$ (X = Cl, Br, I) and $Rb_3[Nb_2Br_9]$ have been prepared by disproportionation reactions (equation 49). Other alkali metals gave rise to different reaction products.[525] There is as yet no evidence for the corresponding tantalum anions $[Ta_2X_9]^{3-}$. These salts were reported to be isotypic with $Cs_3[Cr_2X_9]$. The magnetic susceptibility of $Cs_3[Nb_2X_9]$ corresponds to two unpaired electrons ($\mu = 2.68$ BM, $\theta = -210$ °C, for X = Br), which was considered to be in agreement with a formal Nb═Nb double bond and a D_{3h} symmetry.

$$3Nb_3X_8 + 12AX \xrightarrow{600\text{–}690\,^\circ C} 4A_3[Nb_2X_9] + Nb \quad (49)$$
$$A = Cs, Rb \quad X = Cl, Br, I$$

The reaction between $[Nb_2Cl_6(THT)_3]$ and Et_4NCl in various stoichiometries provided the ionic $(Et_4N)_2[Nb_2Cl_8(THT)]$ and $(Et_4N)_3[Nb_2Cl_9]$.[580]

34.4.2.2 Neutral adducts of the trihalides

(i) Synthesis

Activity in this area has stemmed from the discovery that reduction of MX_5 or their complexes to yield molecular M^{III} trihalide adducts can be achieved rather easily by using reductants such as sodium amalgam, sodium–potassium alloy and magnesium in the presence of ligands.

The first report on this approach was on the synthesis of $[Nb_2X_6(THT)_3]$ (X = Cl, Br, I) by McCarley and Maas, using a one-electron procedure (equation 50).[525] The soluble Nb^{III} adducts formed were easily separated from the sodium derivative and isolated in high yields (60–70%). The tantalum analogues $[Ta_2X_6(THT)_3]$ (**10**; X = Cl, Br) were similarly prepared, but from TaX_5.[581]

$$2[NbX_4(THT)_2] + 2Na/Hg \xrightarrow{C_6H_6} [Nb_2X_6(THT)_3] + 2NaX + THT \quad (50)$$
$$X = Cl, Br, I$$

Similar reducing procedures—the commonest using Na/Hg in toluene, or Mg in methylene chloride—were applied to the pentahalides in the presence of a large variety of monodentate ligands (Table 42). Dinuclear diamagnetic compounds of different stoichiometries, such as $[M_2Cl_6(SMe_2)_3]$ (**28**),[583–585] $[M_2X_6(PhPMe_2)_4]$ (**29**; M = Nb, X = Cl; M = Ta, X = Cl),[586] $[M_2Cl_6(PMe_3)_4]$[587,588] and $[Ta_2Cl_6(\gamma\text{-pic})_4]$ (**30**)[554] were thus obtained in variable yields. $[NbCl_6(diox)_2]$ was obtained by reduction of $NbCl_5$ with Mg in the presence of dioxane.[584] Side-reactions were often observed when oxygen donors were used. Reduction of $NbCl_5$ by Mg also occurred in the presence of THF, but no pure Nb^{III} trihalide adduct could be isolated, as a result of THF degradation reactions.[589]

The nature of the adducts formed with phosphines depends on the experimental procedure. The diamagnetic $[Ta_2Cl_6(PMe_3)_4]$ (**31**) was obtained through reduction of $TaCl_5$ by sodium amalgam in toluene,[587] while the paramagnetic $[TaCl_3(PMe_3)_3]$ was isolated when the reduction was achieved in diethyl ether in the presence of an excess of ligand, or by an exchange reaction from $[Ta_2Cl_6(THT)_3]$ at 60 °C.[306] Alkylidene derivatives may also be converted to trivalent phosphine adducts (equations 51 and 52).[590] $[TaCl_3(PMe_3)_3]$ converts readily in solution to the dimer (**31**). Puzzling results were obtained when $TaBr_5$ was reduced by Mg in the presence of $PhPMe_2$.[586] A very clean one-step reduction of $NbCl_5$ was observed with diphenylacetylene, which acts both as the reductant and as the ligand (equation 53).[592]

$$2[NbCl_3(CHCMe_3)(PMe_3)_2] \xrightarrow[Et_2O]{C_2H_4\ (3\text{–}4\ atm)} [Nb_2Cl_6(PMe_3)_4] \quad (51)$$

$$[TaI_3(CHCMe_3)(PMe_3)_2] \xrightarrow[To]{C_2H_4\ (2\ atm)} [TaI_3(C_2H_4)(PMe_3)_2] \quad (52)$$
$$C_2H_4\ (2\ atm),\ To \uparrow\downarrow PMe_3,\ To$$
$$mer\text{-}[TaI_3(PMe_3)_3]$$

$$4NbCl_5 + 8PhC{\equiv}CPh \xrightarrow{CH_2Cl_2} [NbCl_3(PhC{\equiv}CPh)]_4 + 4t\text{-}PhClC{=}CClPh \quad (53)$$
$$75\%$$

Although some adducts with monodentate ligands, including $[NbCl_3L_3]$ (L = py, γ-pic, 3,5-Me_2py)[593] and $[TaCl_3(PMe_3)_3]$ have been obtained from $[M_2Cl_6L_3]$ (L = SMe_2 or THT) through ligand exchange reactions, this procedure was mainly employed to prepare adducts with bidentate donors, yielding $[M_2Cl_6(LL)_2]$ [LL = dppe,[594] bipy, dto (dto = dithiaoctane)], $[Nb_2Cl_6(LL)_2]$ [LL = diars, triars (triars = 1,1,1-tris(dimethylarsinomethyl)ethane)] and $[Ta_2Cl_6(dmpe)_2]$.[595] The latter compound could not be prepared directly from $TaCl_5$ with Mg, since the reduction stopped at $[TaCl_4(dmpe)_2]$,[557] while the use of Na/Hg offered $[TaCl_2(dmpe)_2]$,[596] without evidence for a Ta^{III} intermediate. Mixed ligand adducts such as $[Ta_2Cl_6(SMe_2)\{P(NMe_2)_3\}_2]$,[597] $[Nb_4Cl_{12}(SMe_2)_4(dppm)]$[598] and $[Nb_4Cl_{12}(PhPMe_2)_6\{MeN(PF_2)_2\}]$[599] were also obtained by the ligand exchange procedure.

Table 42 Neutral Adducts of the Niobium and Tantalum Trihalides with Monodentate Donors

Compounds	*Comments*	*Ref.*
$[M_2X_6(THT)_3]$ (M = Nb, X = Cl, Br, I; M = Ta, X = Cl, Br)	X-Ray structure for M = Nb, X = Br and M = Ta, X = Cl, Br; NQR	1, 2, 3
$[M_2Cl_6(SMe_2)_3]$	X-Ray structure for M = Ta (Ta—Ta = 2.691(1) Å)	4, 5, 6, 7
$[Ta_2Cl_6(SMe_2)\{P(NMe_2)_3\}_2]$	X-Ray structure (Ta—Ta = 2.704(1) Å)	8
$[NbCl_3L_3]$ (L = py, γ-pic; 3,5-Me_2py)	ESR at room temperature: $\langle g\rangle = 2.04$, $\langle A\rangle = 179.1$ G for L = γ-pic	9
$[Ta_2Cl_6L_4]$ (L = py, γ-pic)	Diamagnetic, dissociation in solution for L = γ-pic	9, 10
$[TaX_3(PMe_3)_3]$ (X = Cl, Br, I)	ESR active in frozen solution only: $\langle g\rangle = 1.92$, $\langle A\rangle = 240$ G; unstable towards dimerization, oxidative addition of H_2	11, 12
$[M_2Cl_6(PR_3)_4]$ (R_3 = Me_3, $PhMe_2$)	X-Ray structure for M = Ta; oxidative addition of H_2	13, 14, 15
$[NbBr_3(PhPMe_2)_3]$	X-Ray structure	16
$[M_2Cl_6(PR_3)_x]$ (M = Nb, R = Bu^t, $x = 2$; R = Cy, $x = 3$, 4; M = Ta, R = Cy, Bu^t, $x = 3$)	Activation of N_2(1 atm) during the reductive synthesis with Mg; $[M_2Cl_6(PCy_3)_4]$ in solution by addition of PCy_3	17
$[TaCl_3L_2(C_2H_4)]$ (L = PMe_3, THF)		11, 18
$[TaCl_3LL'(C_2H_4)]$	L = L' = $PhPMe_2$; L = PMe_3, L' = $PhPMe_2$ not isolated, identified by ^{31}P NMR	18

1. R. E. McCarley and E. T. Maas, *Inorg. Chem.*, 1973, **12,** 1096.
2. J. L. Templeton, W. C. Dorman, J. C. Clardy and R. E. McCarley, *Inorg. Chem.*, 1978, **17,** 1263.
3. J. L. Templeton and R. E. McCarley, *Inorg. Chem.*, 1978, **17,** 2293.
4. A. D. Allen and S. Naito, *Can. J. Chem.*, 1976, **54,** 2948.
5. L. G. Hubert-Pfalzgraf, M. Tsunoda and J. G. Riess, *Inorg. Chim. Acta*, 1980, **41,** 283.
6. M. Tsunoda and L. G. Hubert-Pfalzgraf, *Inorg. Synth.*, 1983, **21,** 16.
7. F. A. Cotton and R. C. Najjar, *Inorg. Chem.*, 1981, **20,** 2716.
8. F. A. Cotton, L. R. Falvello and R. C. Najjar, *Inorg. Chim. Acta*, 1982, **63,** 107.
9. M. E. Clay and M. T. Brown, *Inorg. Chim. Acta*, 1982, **58,** 1.
10. J. L. Morançais, L. G. Hubert-Pfalzgraf and P. Laurent, *Inorg. Chim. Acta*, 1983, **71,** 119.
11. S. M. Rocklage, H. W. Turner, J. D. Fellmann and R. R. Schrock, *Organometallics*, 1982, **1,** 703; H. W. Turner, R. R. Schrock, J. D. Fellmann and S. J. Holmes, *J. Am. Chem. Soc.*, 1983, **105,** 4942.
12. J. R. Wilson, A. P. Sattelberger and J. C. Huffman, *J. Am. Chem. Soc.*, 1982, **104,** 858.
13. A. P. Sattelberger, R. B. Wilson and J. C. Huffman, *Inorg. Chem.*, 1982, **21,** 2392, 4179.
14. L. G. Hubert-Pfalzgraf and J. G. Riess, *Inorg. Chim. Acta*, 1978, **29,** L251.
15. L. G. Hubert-Pfalzgraf, M. Tsunoda and J. G. Riess, *Inorg. Chim. Acta*, 1981, **52,** 231; N. Hovnanian and L. G. Hubert-Pfalzgraf, to be published.
16. F. A. Cotton, M. P. Diebold and W. J. Roth, *Inorg. Chim. Acta*, 1985, **105,** 41.
17. J. L. Morançais and L. G. Hubert-Pfalzgraf, *Transition Met. Chem.*, 1984, **9,** 130.
18. S. M. Rocklage, J. D. Fellmann, G. A. Rupprecht, L. W. Messerle and R. R. Schrock, *J. Am. Chem. Soc.*, 1981, **103,** 1440.

(*ii*) *Structure and characterization*

The isolated monomeric trivalent adducts are limited to $[NbCl_3L_3]$ (L = py, γ-pic), $[TaX_3(PMe_3)_3]$ (X = Cl, Br, I) and $[NbBr_3(PhPMe_2)_3]$. The existence of magnetically dilute d^2 compounds was confirmed by magnetic susceptibility ($\mu = 2.03$ BM for $[NbCl_3(\gamma\text{-pic})_3]$ at 20 °C), and ESR data.[593] The ESR signal of $[TaCl_3(PMe_3)_3]$ could be detected only in frozen solutions, and ^{1}H NMR indicates a meridional arrangement of the PMe_3 ligands.[306] $[NbBr_3(PhPMe_2)_3]$ is the only monomeric structurally characterized adduct.[591]

The formation of diamagnetic compounds involving metal–metal bonding is more frequent. Four different geometries—for dinuclear and tetranuclear compounds—have so far been found by X-ray crystal studies. The metal center is generally heptacoordinated. The adducts with monodentate ligands were characterized by ^{1}H or ^{31}P NMR techniques. Usually no ligand dissociation reactions were observed. The adducts formed with bidentate ligands are often of poor solubility.

^{1}H NMR data on the diamagnetic $[M_2X_6L_3]$ (L = SMe_2 or THT[583,580]) show one bridging neutral ligand for two terminal ones. Their crystal structures display confacial bioctahedra of C_{2v} symmetry with one bridging sulfur atom *trans* to both terminal ones (Figure 28). The short

M—M distances (Nb—Nb: 2.728(5); Ta—Ta: 2.710(2) Å) found for [{MBr_2(THT)}(μ-Br)$_2$(μ-THT){MBr_2(THT)}] were considered to reflect strong M—M bonding,[600] and to be in agreement with a formal double bond. The M—S_b bonds are markedly shorter than the M—S_t bonds; the S_t atoms appear to be hindered from closer approach to the metal by the repulsion exerted by the four *cis* halogens in the plane (short S_t—Br contact of 3.51 Å in [$Ta_2Br_6(THT)_3$]). An unusually close approach (3.30 Å) of the two bridging bromines has been observed, and was explained, on the basis of MO considerations, by the stereochemical activity of the metal–metal π-bonding electrons. The low C_{2v} symmetry of these [$M_2X_6L_3$] adducts was considered to be responsible for their diamagnetic behavior, compared to the paramagnetic behavior of $[Nb_2X_9]^{3-}$.

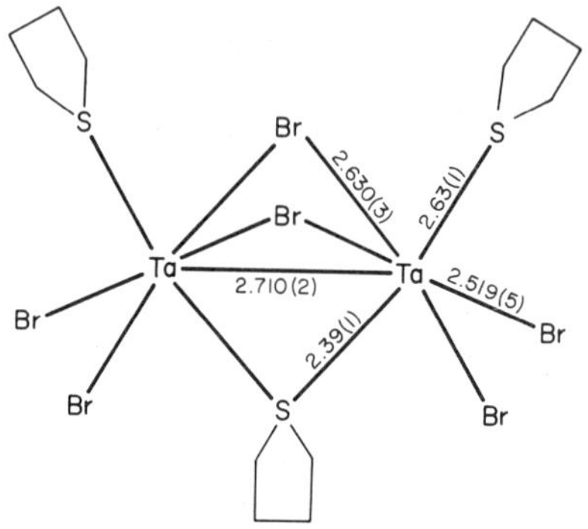

Figure 28 Molecular structure of [(THT)$TaBr_2$(μ-Br)$_2$(μ-THT)$TaBr_2$(THT)] (**10**) (reproduced from ref. 580 with permission)

A thorough NQR study (^{93}Nb, ^{35}Cl, ^{81}Br, ^{127}I) reflects the environmental difference between bridging and terminal halides. The latter resonate at lower frequencies, which was considered to mean that the relative amount of π bonding from halide to metal is much larger for X_t than for X_b.

A single crystal X-ray study of [Ta_2Cl_4{P(NMe_2)$_3$}$_2$(μ-Cl)$_2$(μ-SMe_2)], the only one available on a mixed ligand adduct in oxidation state +III, showed the structure to be related to that of [$Ta_2Cl_4(SMe_2)_2$(μ-Cl)$_2$(μ-SMe_2)].[597]

The isolation of [$M_2Cl_6(PR_3)_3$] (Table 42) shows that adducts of that stoichiometry can also be stabilized by crowded phosphines;[304] their structure is probably related to that of $[Nb_2Cl_9]^{3-}$ with triple chlorine bridges.

The [$M_2Cl_6L_4$] stoichiometry is the most commonly encountered with phosphines, although it has also been observed for M = Ta and L = py, γ-pic. X-Ray data are limited to (**31**), which consists of two somewhat distorted octahedra sharing a common edge (Figure 29). The short Ta—Ta distance and acute Ta—Cl_b—Ta angle indicate strong metal–metal bonding. Considerable distortions involving the axial substituents and non-bonded H · · · Cl interactions are observed. The structure with axial and equatorial phosphines was reported to be favored for electronic reasons.[587] According to the MO calculations of Hoffmann *et al.* for bridged M_2L_{10} complexes,[601] the ordering of the low-lying *d*-orbitals is $\sigma(a_g) < \pi(b_{1u}) < \sigma^*(a_u) < \delta(b_{3g}) < \pi^*(b_{2g}) < \sigma^*(b_{3u})$. This means a formal bond order of 2 for the d^2–d^2 system. The ^{31}P NMR of (**31**) offered a rare example of P–P coupling across a metal–metal bonded dimer ($J(P–P') = \pm 2.46$ Hz).

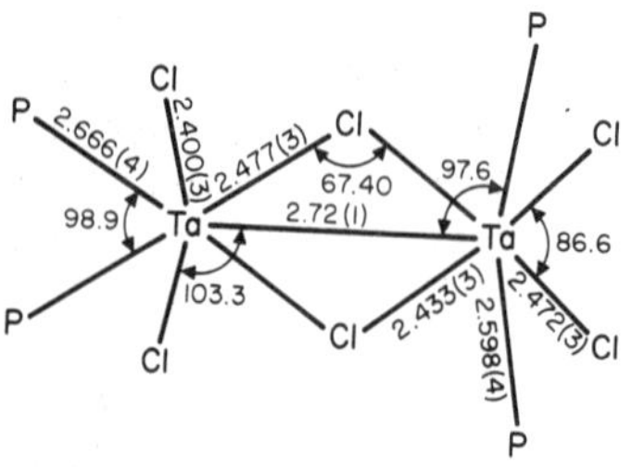

Figure 29 Coordination polyhedron of: [$Ta_2Cl_4(PMe_3)_4$(μ-Cl)$_2$] (**31**) (reproduced from ref. 587 with permission)

The adducts involving bidentate ligands are all of type $[M_2Cl_6(LL)_2]$. The structures of those with bidentate phosphines (M = Nb, LL = dppe;[602] M = Ta, LL = dmpe[595]) are related to that of (**31**), but with the chelate rings spanning two equatorial positions. The metal–metal distances are consistent with either a double or a single bond (Nb—Nb: 2.712(2) Å; Ta—Ta: 2.710(1) Å).

In contrast to the diamagnetism—with small temperature independent paramagnetic contributions—of the adducts formed with monodentate ligands, those with bidentate ligands exhibit temperature dependent paramagnetism and produce easily observable ESR spectra even at room temperature.[594] Although the magnetic moments (0.54 to 0.87 BM at room temperature for the niobium adducts) are lower than those generally observed for the tetravalent adducts, it has been suggested that the metals are only singly bonded, the magnetism being presumed to be lowered by spin–orbit coupling and/or antiferromagnetic interactions. A complex hyperfine structure was observed for the bipyridyl adducts. The IR data suggest that the bipy ligand is negatively charged with electron delocalization into the ligand ring system.

^{1}H NMR data on the niobium(III) adducts with triphos and triars established that these ligands behave as bidentates only.[583,603]

$[NbCl_3(PhC{\equiv}CPh)]_4$, the only structurally characterized tetranuclear compound (Figure 30),[592] can be regarded as a centrosymmetric distorted double hexahedron with two missing corners. The Nb atoms are heptacoordinated, linked by two triply bridging chlorines, but do not exhibit metal–metal bonding. The alkyne C—C distances and C—C—C angle are consistent with its formulation as a metallacyclopropene (Section 34.4.3.2.i).

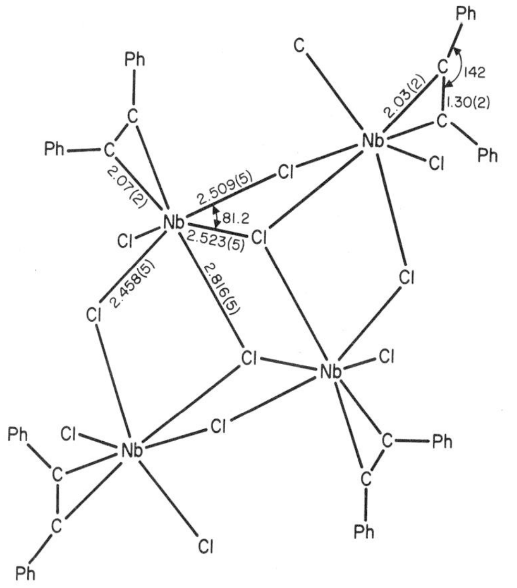

Figure 30 Molecular structure of $[NbCl_3(PhC{\equiv}CPh)]_4$ (reproduced from ref. 592 with permission)

The trivalent adducts display a wide range of colors, but electronic absorption spectra have been reported only for $[M_2X_6(THT)_3]$ (X = Cl, Br), $[Nb_2I_6(THT)_3]$, $(Et_4N)_2[Nb_2Cl_8(THT)]$ and $(Et_4N)_3[Nb_2Cl_9]$;[580,599] the magnitude of the extinction coefficients supports strong metal–metal interactions. The similarity between the spectra of $[Nb_2Cl_6(THT)_3]$ and $(Et_4N)_2[Nb_2Cl_8(THT)]$ incidates that the bridging THT ligand is retained.

(*iii*) *Reactivity*

Investigations on the reactivity of the M^{III} trihalide derivatives have so far mainly been limited to that of $[Ta_2Cl_6(PMe_3)_4]$, $[M_2Cl_6(PhPMe_2)_4]$, $[Ta_2Cl_6(\gamma\text{-pic})_4]$ and $[M_2Cl_6L_3]$ (L = SMe_2 or THT). Oxidative additions occur easily, and these compounds are therefore suitable precursors of M^V and dinuclear M^{IV} derivatives. Metathesis reactions may lead to other trivalent derivatives. The dinuclear unit is generally retained, although condensation to tetranuclear species is sometimes observed. The only monomeric substitution products reported so far, $[Nb(NCS)_3(py)_3]$ and $[Nb(Et_2dtc)_3]$, derive from $[NbCl_3(py)_3]$. Chemical reduction of the chloro adducts appears difficult, and seems favored for monomeric M^{III} adducts (Section 34.6). Electrochemical reduction offers possible alternatives.[604]

34.4.3 Activation Reactions Involving Niobium(III) and Tantalum(III) Adducts

34.4.3.1 Activation of small molecules

(i) Dihydrogen and related molecules

The reactivity of the red $[Ta_2Cl_4(PMe_3)_4(\mu\text{-}Cl)_2]$ (**31**) is of interest, because it leads to Ta^{IV} and Ta^{I} compounds (Scheme 6). Complex (**31**) reacts cleanly with H_2 under mild conditions to produce the air stable emerald green diamagnetic $[Ta_2Cl_4(PMe_3)_4(\mu\text{-}H)_2(\mu\text{-}Cl)_2]$ (**32**), one of the very few compounds with a metal–metal bond bridged by four ligands (Ta—Ta: 2.621(1) Å). It is also the first example of direct interaction of H_2 with a metal–metal multiple bond.[605–607]

The terminal and bridging groups are in a mutually staggered arrangement, and the coordination polyhedron about each tantalum is roughly a square antiprism. The equatorial phosphines, but not the axial ones, are chemically equivalent. The presence and location of the hydrides were established by spectroscopic techniques ($\delta_H = 8.5$(m) p.p.m.; ν(Ta—H—Ta) $\approx$ 1280 cm^{-1}). The Ta—Ta separation (2.621(1) Å) is shorter than the Ta=Ta bond in the original Ta^{III} adduct.

Compound (**32**) is the precursor of the air sensitive $[Ta_2Cl_4(PMe_3)_4(\mu\text{-}H)_2]$ (**33**), characterized by X-ray diffraction at -160 °C. The $Ta_2Cl_4(PMe_3)_4$ substructure resembles that of the quadruply bonded dimers $[M'_2Cl_4(PMe_3)_4]$ (M′ = Mo or W) with the ligands in eclipsed conformation and retention of the Ta=Ta bond (2.545(1) Å). Rapid rotation of the end groups and/or bridging hydrides is required to account for the apparent magnetic equivalences in ^1H and ^{31}P NMR. Complex (**33**) offers a convenient entry into dinuclear Ta^{IV} chemistry. Its ether solutions react readily with H_2, HCl or Cl_2 to give the quadruply bridged Ta^{IV} compounds (**34**), (**35**) or (**32**) (Scheme 6). Alternative syntheses of $[Ta_2Cl_4L_4(\mu\text{-}H)_4]$ (**34**) are the hydrogenation of the $[TaHCl_2(CHCMe_3)L_3]$ alkylidene or the thermolysis of $[TaH_2Cl_2(PMe_3)_4$ (Section 34.6.1.2.i, Scheme 9).

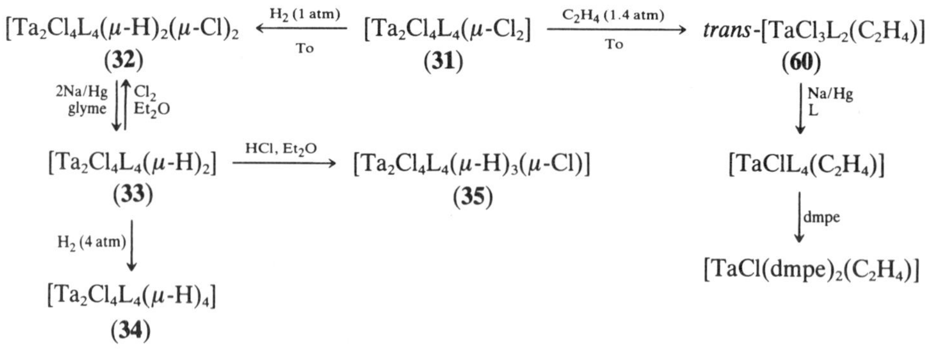

Scheme 6 (L = PMe_3)

Compound (**34**) has an imposed D_{2d} symmetry and a very short Ta—Ta length (2.511(2) Å)—the shortest reported to date; the bridging hydrogens were located in the solid (Ta—H: 1.81(2) Å). Unlike its tantalum(III) precursor (**33**), it is static on the NMR time-scale. According to the predictions of Hoffmann,[601] all the metal μ type orbitals are engaged in the Ta—H—Ta bonding, and a substantial rotational barrier arises from the overlap of orbitals of π symmetry.

The oxidative addition of H_2, HCl and Cl_2 to the multiply bonded metal–metal complex (**31**) occurs without gross structural rearrangements, and has no equivalent in binuclear chemistry. Oxidative addition of H_2 under mild conditions with formation of diamagnetic M^{IV} hydrides appears to be a general feature for the dinuclear M^{III} adducts with monodentate phosphines. Bridging or terminal hydrides are found.[557] $[Ta_2Cl_6(dmpe)_2]$ and $[Nb_2Cl_6(dppe)_2]$, however, are inert toward H_2 even under severe conditions.[595] Monomeric diamagnetic $[TaH_2Cl(PMe_3)_4]$

(**36**), was obtained by reduction of Ta^{IV} monomeric hydride, resulting from the oxidative addition of Ta^{II} derivatives (Section 34.6.1.2.i).

(ii) Dioxygen and dinitrogen

Admission of oxygen into a solution obtained by reduction of $TaCl_5$ by two equivalents of Na/Hg in the presence of PMe_2Ph led to the linear oxo derivative $(PMe_3Ph)_2[Cl_5TaOTaCl_5]$ (Section 34.2.6.1.ii).

Although $[\{TaCl_3(PMe_3)_2\}_2(\mu\text{-}N_2)]$ could be obtained *via* the 'azine route' (Section 34.2.3.6), this product did not form when $[TaCl_3(PMe_3)_3]$ was exposed to dinitrogen, even under 68 atm. The dimer (**31**) probably forms too rapidly from $[TaCl_3(PMe_3)_3]$, and does not react readily with dinitrogen.[306]

Activation of N_2 under mild conditions (25 °C, 1 atm) was observed during the reduction of the MCl_5 by magnesium (equation 29).[304] Unfortunately, the products were unstable and only a compound of empirical formula $[Nb_2Cl_6(PBu^t_3)_3N_2]$ could be isolated. IR data ($\nu(N_2) \approx 1720\ cm^{-1}$) suggest a mode of bonding different from those found in Schrock's compounds.

34.4.3.2 Reactions with unsaturated substrates

(i) Alkynes—three-membered metallacycles

The tetrahydrothiophene adducts $[M_2Cl_6(THT)_3]$ (**10**) react with a variety or alkynes $RC{\equiv}CR'$, but the result depends considerably on R and R′ (Scheme 7). With diphenylacetylene the initial product, presumed to be $[TaCl_3(THT)_2(PhC{\equiv}CPh)]$, converted to $(pyH)[TaCl_4(py)(PhC{\equiv}CPh)]$ by recrystallization in the presence of pyridine.[608] The most remarkable feature of this seven-coordinate anion is the very strong symmetrical binding of tolan to Ta^V. The Ta—C and C—C distances were interpreted to mean that the alkyne acts as a four-electron donor, or as a two-electron donor in a simple bent bond mode.[609] A comparable metallacyclopropane–Ta^V moiety was obtained when *t*-butylmethylacetylene was used, giving $[Ta_2Cl_4(THT)_2(Bu^tC{\equiv}CMe)_2(\mu\text{-}Cl_2)]$.[610] Evidence for strong interaction with the alkyne is given by the substantial asymmetry of the chlorine bridges, the Ta—Cl(1) bond *trans* to the alkyne being the longest. The Ta—Ta distance of 4.144(1) Å is too long to support any bonding interaction. Di-*t*-butylacetylene afforded $[Ta_2Cl_4(THT)_2(\mu\text{-}Cl)_2(\mu\text{-}Bu^tC{\equiv}CBu^t)]$,[611] whose structure derives from that of (**10**) with THF molecules in place of the terminal THT ligands and a bridging alkyne substituted for the THT bridge. The short metal–metal distance indicates the retention of the double bond.

Adducts (**10**) catalyze the polymerization of substituted alkynes: into cyclic trimers for the primary alkynes $RC{\equiv}CH$ (R = Ph, Et, Pr^i), into high polymers for the internal alkynes $RC{\equiv}CR'$ (R = Ph, R′ = Pr; R = Me, R′ = Bu^5).[612] Metallacyclopropene complexes might be the first step in cyclotrimerization. Quite different results were obtained by allowing the unusual alkyne $Ph_2PC{\equiv}CPPh_2$ to react with $[Ta_2Cl_6(SMe_2)_3]$.[613] Two products were obtained: $[Cl_4Ta{-}CH(PPh_2)C(PPh_2){=}C(PPh_2)CH(PPh_2){-}TaCl_4]$ (**37**) and $[(Me_2S)Cl_3Ta{=}C(PPh_2)C(PPh_2){=}C(PPh_2)Cl(PPh_2){=}TaCl_3(SMe_2)]$ (**38**; Scheme 8). In these two centrosymmetric compounds, the starting alkyne has dimerized to form a four-carbon chain, which acts as a bridging ligand (equation 54), while the tantalum atom's oxidation state has increased to +V. Each metal atom belongs to an unusual three-membered metallacycle, Ta—C—P for (**37**) and Ta=C—P for (**38**). In (**37**), one Cl^- ligand is replaced by Me_2S in a heptacoordinated environment. Complex (**37**) appears to contain the first example of an M—C—P ring reported for an early transition metal, while the ring with the Ta=C double bond of (**38**) is entirely unprecendented.

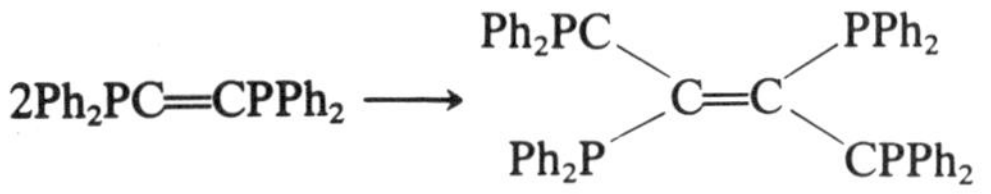

(54)

Comparable three-membered metallacycles were also obtained by oxidative addition of C—H bonds. Reduction of $TaCl_5$ by sodium sand in PMe_3 as a solvent afforded $[Ta(PMe_3)_3(\eta^2\text{-}CH_2PMe_2)(\eta^2\text{-}CHPMe_2)]$ (**39**; 7% yield), whose structure (Figure 31) shows that C(1) is bonded

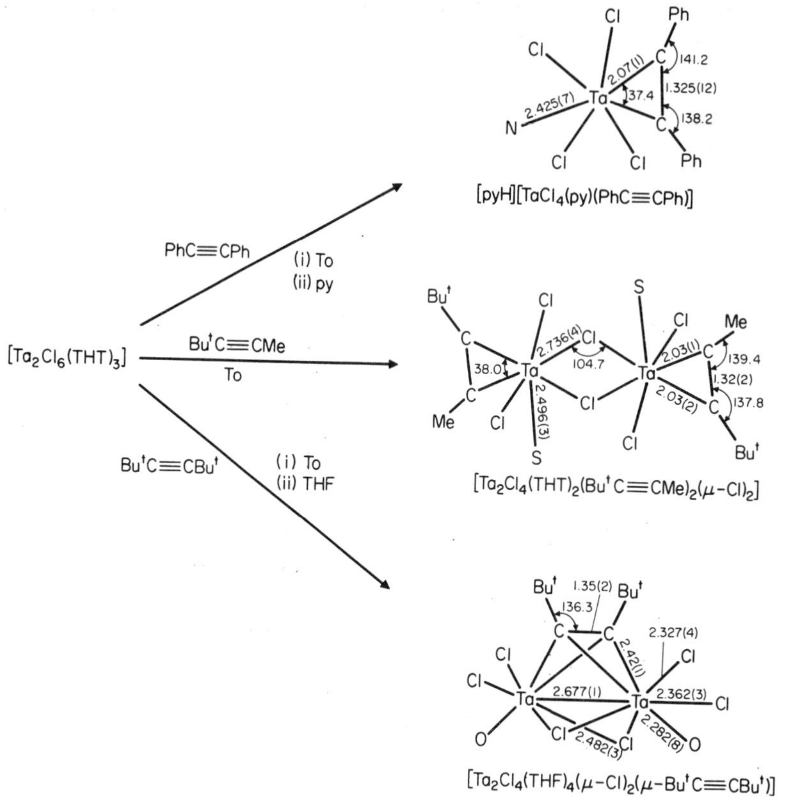

Scheme 7

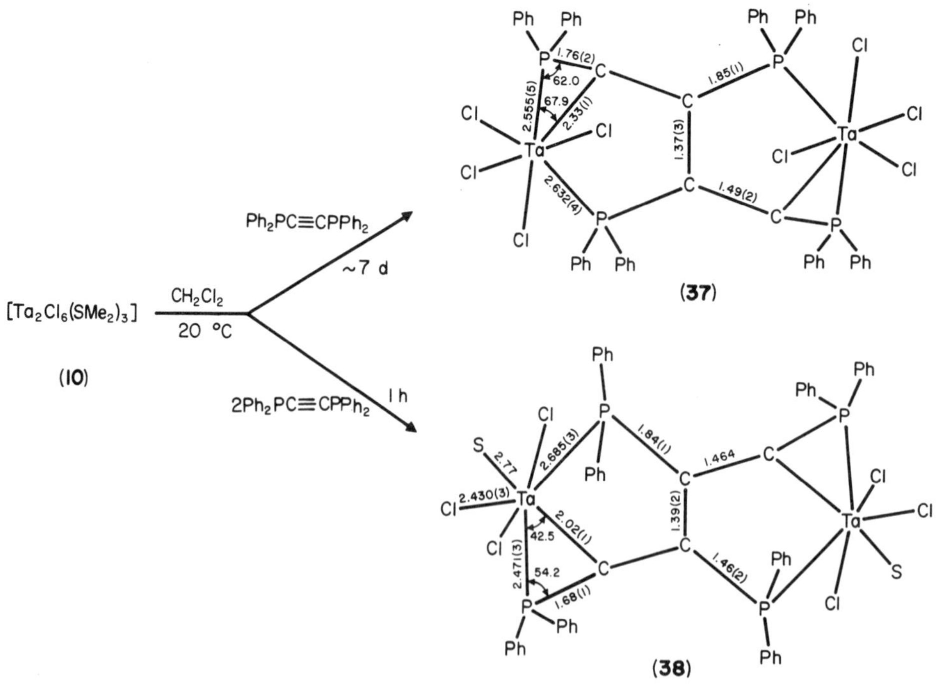

Scheme 8

to only one hydrogen atom, which is coplanar with the Ta—C—P triangle, the whole group affording the first example of an M(η-CHPMe$_2$) moiety. Complex (**39**) is an example of a seven-coordinate, pentagonal bipyramidal, 18-electron Ta^{III} compound.[614]

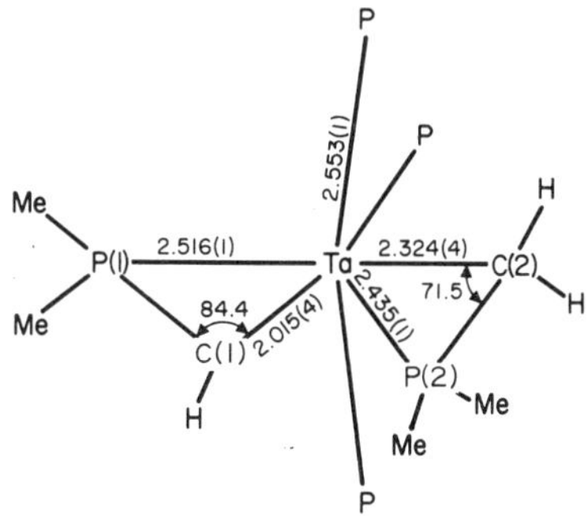

Figure 31 Crystal structure of [Ta(PMe$_3$)$_3$(η^2-CH$_2$PMe$_2$)(η^2-CHPMe$_2$)] (**39**) (reproduced from ref. 614 with permission)

(*ii*) *Reactions with other unsaturated substrates*

Reductive coupling reactions with C—C bond formation of unsaturated ligands such as nitriles (Section 34.2.3.5.i), alkynes and isocyanides were often observed for the [M$_2$Cl$_6$L$_3$] (L = THT, SMe$_2$) adducts. Thus the reaction between (**28**) and an excess of *t*-butyl isocyanide led to binuclear and trinuclear products (Section 34.5.2.2; equation 55).[615] In the dinuclear species (**40**), the bridging ligand arises from the *cis* coupling of two isocyanides (Figure 32). The average C—N distance (1.35(2) Å), the C—C distances and the planarity at the nitrogen atoms suggest a delocalized π system with C—C and C—N bond orders higher than one. The molecule is fluxional. Oxidation state +IV for each niobium, or +III for one and +V for the other, may account for the observed diamagnetism.

$$\underset{(\mathbf{28})}{[Nb_2Cl_6(SMe_2)_3]} \xrightarrow{Bu^tNC} \underset{(\mathbf{40})\ 60\%}{[Nb_2Cl_6(Bu^tNC)_4(\mu\text{-}Bu^tNCCNBu^t)]} + \underset{(\mathbf{41})\ 20\%}{[Nb_3Cl_8(Bu^tNC)_5]} \qquad (55)$$

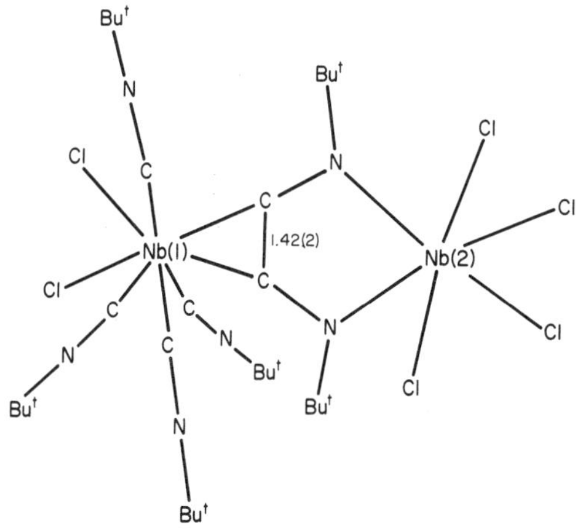

Figure 32 [Nb$_2$Cl$_6$(Bu^{t}NC)$_4$(μ-Bu^{t}NCCNBu^t)] (**40**) (reproduced from ref. 615 with permission)

34.4.4 Solvolysis Products of the Trihalides

Products that formally derive from the trihalides by solvolysis have been obtained (i) by metathesis of M^{III} trichloride adducts; (ii) from reactions of [TaCl$_2$(dmpe)$_2$] (**23**); and (iii) by reductive synthesis starting from M^{V} compounds. The second and third procedures usually give mononuclear species, while di- or poly-nuclear compounds are generally obtained by solvolysis of M^{III} adducts.

34.4.4.1 Alkoxides and related compounds

Diamagnetic $[Ta_2Cl_2(OR)_4]$, $[Ta_2Cl_2(OR)_4(\gamma\text{-pic})_2]$ (**42**; R = Pr^i, Bu^t) and $[Ta_2Cl_2(OBu^t)_4]_2$ have been obtained from $[Ta_2Cl_6(\gamma\text{-pic})_4]$ (**30**) and the lithium alkoxides.[554,616] Complex (**42**) may coordinate an additional molecule of γ-picoline to give the paramagnetic $[TaCl(OR)_2(\gamma\text{-pic})_3]$. Reaction of (**42**) with excess acetone generated various pinacolates by reductive coupling of acetone. Carbon–carbon coupling was also observed with $MeCO_2C{\equiv}CCO_2Me$. $[Nb_2Cl_5(OPr^i)(Pr^iOH)]$ has been characterized by X-ray diffraction.[616b] Monomeric chloro butoxides $[TaCl_2(OBu^t)(PMe_3)_2(C_2H_4)]$ and $[MCl(OBu^t)_2(PMe_3)_2(C_2H_4)]$ were also shown to form by reduction of alkylidenes in ethylene metathesis experiments (equation 56).[301]

$$[TaCl_2(OBu^t)(CHCMe_3)(PMe_3)_2] + C_2H_4 \longrightarrow [TaCl_2(OBu^t)(PMe_3)_2(C_2H_4)] + \underset{24\%}{Me_3CCH{=}CHMe} + \underset{65\%}{Me_3CCH_2CH{=}CH_2} \quad (56)$$

The reaction between (**28**) and $o\text{-}C_6H_4(OSiMe_3)_2$ offered tetranuclear chloro catecholates $[Nb_4Cl_8(cat)_2(SMe_2)_2]$ or $[Nb_4Cl_4(cat)_4(SMe_2)_2]$ depending on the atmosphere—Ar or NO—used. A diamagnetic complex $[Nb(SQ)_3]$ (SQ = semiquinone of 3,5-di-*t*-butyl-1,2-benzoquinone) of unknown structure has been synthesized from $NbCl_5$.[617] Displacement of the alcoholic ligands from impure '$NbCl_3(C_4H_8O)_3$' by salicylaldehyde provided $[Nb_2Cl_4(OC_6H_4CHO)_2(THF)_4]$.[584]

$[Ta_2Cl_4(dmpe)_2(\mu\text{-O})(\mu\text{-SMe}_2)]$, whose formula was established by X-ray diffraction, was isolated as a minor product in the preparation of $[Ta_2Cl_6(dmpe)_2]$. By contrast with $[Ta_2Cl_4(dmpe)_2(\mu\text{-Cl})_2]$, the chelating ligand spans axial and equatorial positions. The Ta—Ta distance (2.726(1) Å) has been taken to support the assignment of oxidation state +III to the metal.[618]

34.4.4.2 β-Diketonates and carboxylates

M^{III} trihalide adducts display different types of behavior towards β-diketones. With nobium, a careful control of the reaction conditions allowed the isolation of the diamagnetic, dinuclear, soluble Nb^{III} chlorodiketonato complexes $[Nb_2Cl_3(dpm)_3]$, $[Nb_2Cl_5(dpm)(PhPMe_2)_2]$ and $[Nb_2Cl_4(acac)_2(acacH)_2(PhPMe_2)_2]$.[619] No Ta^{III} chloro diketonates have so far been stabilized, even with the bulky dpmH ligand; oxidation to $[Ta_2Cl_4(dpm)_4]$ was observed with (**30**), while oxygen abstraction reactions, with formation of $[TaOCl_2(dpm)]_2$ are favored for (**29**). A paramagnetic ($\mu = 1.39$ BM) compound formulated as $[Ta(dbm)_3]$ has been obtained by addition of Et_3N and dbmH to the solution which results from refluxing $TaCl_4$ in MeCN.[620] Its formulation is however questionable, because in these conditions $[Ta_2Cl_6(MeCN)_2(\mu\text{-}C_4H_6N_2)]$ would be expected to form instead of $[TaCl_3(MeCN)_3]$.

Adducts $[Nb_3O_2(O_2CR)_6(THF)_3]^+$ (R = Ph, Bu^t) have been prepared by allowing (**28**) to react with sodium carboxylates (Section 34.5.2.3). $[Nb_2Cl_2(OAc)_5(THF)]$ was obtained with NMe_4OAc.[621] The reaction between (**30**) and excess pivalic acid afforded $[Ta_2Cl_4(Bu^tCO_2)_4]\cdot 2Bu^tCO_2H$ as a diamagnetic soluble species. The reaction between $Ta_2Cl_6(SMe_2)_3$ and lithium pivolate gave $Ta_2Cl_5(O_2CBu^t)(SMe_2)(THF)_2$.[621] A niobium trilactate has been isolated by adding lactic acid to the Nb^{III} species generated electrochemically from $NbCl_5$ in DMF–HCl.[622] $[Nb(O_2CBu^t)(Cp)_2]$ with an anomalously high $\nu(CO_2)$ coordination shift (347 cm^{-1}) has also been described, but chelating behavior was found by X-ray diffraction.[623]

34.4.4.3 Thiolates and dithiocarbamates

The arenethiolates $[M(SR)_3]$ (R = phenyl or naphthyl) were easily obtained as monomeric soluble products by reductive elimination of organic disulfides from the pentathiolates

$[M(SR)_5]$.[624] The selenophenolates were formed by a similar procedure. Monomeric $[Nb(Et_2dtc)_3]$ has been prepared (equation 57); dimerization was observed in non-polar solvents in the presence of dithiocarbamates.[593] Poorly characterized pyrrolidinedithiocarbamate, $[Nb(S_2NC_5H_8)_3]$, has also been reported.[625]

$$[NbCl_3(py)_3] + 3NaEt_2dtc \xrightarrow{MeCN} [Nb(Et_2dtc)_3] + 3NaCl + 3py \qquad (57)$$

34.4.4.4 *Dialkylphosphides*

Attempts to obtain M^{III} amido derivatives have apparently so far been restricted to the reactions between $[Nb_2Cl_6(SMe_2)_3]$ or $Nb_2Cl_6(PhPMe_2)_4$ and $LiN(SiMe_3)_2$, but these reactions are of poor selectivity. Related M^{III} dialkylphosphides are better known. A diamagnetic phosphido compound, $[TaH(PPh_2)_2(dmpe)_2]$, was formed from $[TaCl_2(dmpe)_2]$, apparently by hydrogen abstraction from the solvent (equation 58). It was the first structurally characterized example of a terminal phosphide, and consists of a pentagonal bipyramid, with apical PPh_2 moieties.[626] Salient features are the monodentate diphenylphosphido ligands and the tilting of two of their phenyl rings toward the hydride ion. The geometry about the diphenylphosphido centers is nearly trigonal planar, as often observed in the structurally related dialkylamides. The metal achieves a closed shell electronic configuration if the Ta—P(PPh_2) bond orders are assumed to be 1.5.

An ionic homoleptic Nb^{III} dialkylphosphide, $[Li(DME)_2][Nb(PCy_2)_4]$, has been obtained directly from $NbCl_5$.[565] Terminal phosphido groups, stabilized by interaction of two of the phosphorus atoms with lithium, have been postulated.

$$[TaCl_2(dmpe)_2] + 2KPPh_2(dioxane)_2 \xrightarrow[0\,°C]{THF} \underset{56\%}{[TaH(PPh_2)_2(dmpe)_2]} + 2KCl + 2\,dioxane \qquad (58)$$

34.4.4.5 *Nitrenes*

Ta^{III} nitrenes have been prepared by reduction of Ta^V nitrenes under argon (equation 59). The lability of the PMe_3 ligands allowed the production of $[Ta(NPh)ClL_3(alkene)]$ (alkene = ethylene or styrene).

$$[Ta(NR)Cl_3(THF)_2] \xrightarrow[4L,\ Ar,\ THF]{2Na/Hg} trans\text{-}[Ta(NR)ClL_4] \qquad (59)$$

$$R = Me, Bu^t, Ph \quad L = PMe_3, 0.5\ dmpe$$

A diimido Ta^{III} complex (**43**) analogous to the Ta^V derivatives (Section 34.2.3.6) was formed by reduction of $[TaCl_3(PMe_3)_2(C_2H_4)]$ (equation 60). Dissociation of the PMe_3 ligand *trans* to the μ-N_2 unit was observed by NMR.

trans-$[TaCl_3(PMe_3)_2(C_2H_4)]$ $\xrightarrow[Et_2O,\ THF;\ N_2\ (1\ atm)]{2Na/Hg}$ 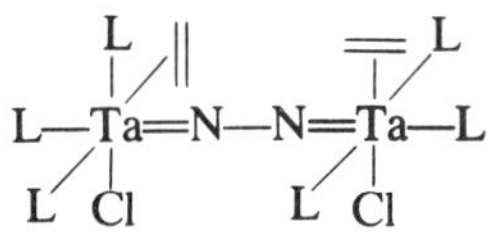 (60)

(**43**)

34.4.4.6 *Organometallics*

A few salient results that have appeared since the publication of 'Comprehensive Organometallic Chemistry' have been collected here.

Low-valent tantalum–bromide–$PhPMe_2$ adducts undergo facile P—Me bond cleavage reactions spontaneously or in the presence of strong bases, giving tantalum σ-methyl and tantalum methylphosphido derivatives.[627]

Smooth substitution of chlorine in $[M_2Cl_6(PhPMe_2)_4]$ using Me_3SnCp easily yielded $[Ta_2Cl_4(PhPMe_2)_2(Cp)_2]$, while the poorly soluble $[Nb_2Cl_5(PhPMe_2)_3(Cp)]$, is favored for niobium, although the disubstituted compound may be obtained after long reaction times. The reaction of $[TaCl_2(dmpe)_2]$ with $[NaCp(DME)]$ apparently proceeds through disproportionation, giving $[Ta(dmpe)(Cp)_2]Cl$ and $[TaCl(dmpe)_2(Cp)]Cl$.[528]

$[MX_2(CO)_3(Cp)]$ has been obtained by several routes.[8] Carbonylative reduction at atmospheric pressure of $[MCl_4(Cp)]$ was achieved (equation 61).[629] The tantalum derivative was obtained only as an impure material in poor yield (<10%). The niobium compound is a chlorine-bridged dimer (**44**).[564] $[NbI_2(CO)_3(Cp)]$ (**45**) was prepared by oxidative addition of iodine to $[Nb(CO)_4(Cp)]$.[630] Its carbonyl groups can be removed under vacuum, resulting in a carbonyl free material that reversibly absorbs CO to give a compound suggested to be a dimer comparable to (**44**). Complex (**45**) could be converted quantitatively to $[NbI_2(CO)(dmpe)(Cp)]$.

$$3[MCl_4(Cp)] + 9CO + 2Al \xrightarrow[THF]{HgCl_2} \underset{(\mathbf{44})}{3[MCl_2(CO)_3(Cp)]} + 2AlCl_3 \qquad (61)$$

A thermally robust dihapto CO_2 complex was easily obtained according to equation (62).[631] Complex (**46**) is the first authentic (X-ray diffraction) CO_2 complex of a hard oxophilic early transition element. The CO_2 (C—O bond lengths: 1.283(8) and 1.216(8) Å) lies close to the alkyl group, but its orientation is such that its insertion into the metal–alkyl bond would produce the unfavored metallacarboxylate ester instead of a carboxylato complex.

$$[NbCl(CH_2SiMe_3)(CpR)_2] \xrightarrow[CO_2\ (1\ atm)\ 30\ mm]{Na/Hg,\ THF} \underset{(\mathbf{46})\quad 34\%}{[Nb(CH_2SiMe_3)(\eta^2\text{-}CO_2)(CpR)_2]} +$$

$$[Nb(CH_2SiMe_3)(CO)(CpR)_2] + [Nb(CH_2SiMe_3)(CO_3)(CpR)_2] \qquad (62)$$

$$R = Me$$

$[NbX(Cp)_2L]$ (L = PMe_2Ph, $P(OMe)_3$, PhC≡CPh, PhC≡CH for X = Cl or Br; L = HC≡CH, CNCy, CNPh for X = Cl; L = CO for X = Br) has been obtained by reduction of $[NbX_2(Cp)_2]$ with Na/Hg in the presence of the ligands;[632,633] the remarkable stability of the alkyne adducts contrasts with that of the derivatives with phosphorus donors. Formation of strong metal–alkyne bonds seems to be a common feature of the mono- and dicyclopentadienyl derivatives. $[NbX(Cp)_2]$ were obtained for both X = Cl and X = Br,[8] but they display different behavior toward π-acid ligands. A dinuclear structure has been suggested for $[NbBr(Cp)_2]$ on the basis of its inertness toward CO or $PhPMe_2$.

34.4.5 Pseudohalo Derivatives

The monomeric $[NbCl_3(py)_3]$ has been used for the preparation of $[Nb(NCS)_3(py)_3]$.[593] $K_5[Nb(CN)_8]$, obtained by electrosynthesis,[536] is the only X-ray-characterized monomeric homoleptic Nb^{III} compound. A dodecahedral D_{2d} geometry with two sets of Nb—C distances was found in the solid.[635] IR data suggest a change to a D_{4d} geometry in solution. ^{13}C NMR data are compatible with either a D_{4d} or a fluxional D_{2d} polyhedron.

34.4.6 Miscellaneous

34.4.6.1 Chemistry in aqueous solution: sulfato, oxalato and related compounds

Reduction in aqueous solutions under argon of M^V sulfato complexes by Zn in the presence of sulfuric acid for Nb, by zinc amalgam or electrolytically in the presence of an oxalate for Ta, provided $K[Nb(SO_4)_2]\cdot 4H_2O$[636] or $K_5[Ta(C_2O_4)_4]$ and $[(PhCH_2)_3NH]_5[Ta(C_2O_4)_4]\cdot 4H_2O$[637] as diamagnetic, oxygen sensitive solids. A distorted dodecahedral geometry was assumed for the highly labile Ta^{III} oxalate.

The redox potential of the Nb^V–Nb^{III} couple has been determined in 3 N H_2SO_4 ($E'^\circ \approx -0.275$ V).[638] Its reducing power is comparable to that of Ti^{III}. Solutions of $K[Nb(SO_4)_2]\cdot 4H_2O$ are stable under CO_2, and were used for the titration of Cu^{II}, Fe^{III}, Tl^{III}, Mo^{VI}, V^V and U^{VI}.[639] Strongly reducing behavior of the Nb^{III} solutions towards NO, giving ammonia, was also observed. The reducing power of the Ta^V–Ta^{III} couple is close to that of the $Cr^{3+}(aq)/Cr^{2+}(aq)$ couple.

The potassium disulfatoniobate(III) tetrahydrate was used as starting material for the synthesis of $[Nb(oxin)_3]$ (oxin = oxinate) and of $[Nb_2O(PO_4)_2]\cdot 6H_2O$. The two products are weakly paramagnetic ($\mu = 0.76$ and 0.68 BM, respectively). $[Nb(oxin)_3]$ is quantitatively oxidized to $[NbO(oxin)_3]$ in air. Potentiometric titration of the phosphato derivative showed that it should be regarded as a niobyl phosphoniobate(III) of formula $[Nb^VONb^{III}(PO_4)_2]\cdot 6H_2O$.[640]

34.4.6.2 *Other products*

Chalcogeno halides NbYX (Y = S, Se; X = Br, I) were formed according to equation (63). Their structure contains $Nb_2(Y_2)_2$ fragments and is best described as $[Nb_2(Y_2)_2X_{8/2}]$.[641] Reduction of $NbCl_5$ with excess $LiAlH_4$ in Et_2O at room temperature led to a product presumed to be $[LiNb_2(AlH_4)_7]$.[642]

$$Nb_3X_8 + 5Nb + 8Y \xrightarrow{700\text{–}800\,^{\circ}C} 8NbYX \qquad (63)$$
$$X = Br, I; \quad Y = S, Se$$

34.5 NON-INTEGRAL OXIDATION STATES

34.5.1 Survey

The lower halides of niobium and tantalum consist of tightly bound clusters of metal atoms, with metal–metal distances close to those found in the metal. They contain ions with average oxidation numbers between +III and +I (Table 43). Their size depends on the valence electron concentrations (VEC) that are available on the metal atoms for M—M bonding, and on the halide-metal ratio.[644] Several reviews have been devoted to the clusters of early transition metals.[3,643]

Table 43 Various Niobium and Tantalum Clusters with Non-integral Oxidation States

Average oxidation number of the metal	*Close-packed structures*	*Cluster strucures*
3.00	$[Cs_3Nb_2X_9]$(X = Cl, Br, I)	—
2.67	$[Nb_3X_8]$(X = Cl, Br, I)	$(M_6X_{12})^{4+}$ } X = Cl, Br
2.50	$[CsNb_4X_{11}]$(X = Cl, Br)	$(M_6X_{12})^{3+}$ } X = Cl, Br
2.33	—	$(M_6X_{12})^{2+}$ } X = Cl, Br
1.83	—	$(Nb_6I_8)^{3+}$
1.67	—	$(Nb_6I_8)^{2+}$

34.5.2 Triangular Clusters

The triangular clusters, the simplest ones, have been obtained under a variety of conditions, including ligand exchange and metathesis reactions of molecular Nb^{III} derivatives. The three basic structural types[645]—without, with one, and with two μ_3-ligands—are known, and compounds with seven or eight valence electrons on the M_3 core have been described.

34.5.2.1 *Binary halides*

Trimeric clusters of formula M_3X_8 (X = Cl, Br, I) are known for Nb but not for Ta.[1,5,643] They were usually prepared by thermal decomposition of higher halides or by reduction of the pentahalides by niobium. The light green Nb_3Cl_8 and black Nb_3Br_8 are the Nb rich limits of the homogeneous phases that extend over the composition ranges $NbCl_{2.67}$ to $NbCl_{3.13}$ and $NbBr_{2.67}$ to $NbBr_{3.03}$.

Both the α and β forms contain Nb_3 triangles, with metal–metal distances ranging from 2.81 (for Nb_3Cl_8) to 3.00 Å (in β-Nb_3I_8). In β-Nb_3Br_8, for instance, the niobium atoms are bridged by three bromine atoms located below the plane of the triangle, while a fourth is located above the plane and bridges all three metal atoms.[5] The remaining halogen atoms are involved in intercluster bridging, and the structure is described as $^{2}_{\infty}[(Nb_3X^{i}_{4})\,|\,X^{o-o}_{6/2}X^{o-o}_{3/3}]$, where i and o stand for the halogen inside and outside the core, respectively.[646] The metallic centers are crystallographically and chemically equivalent in all the Nb_3X_8 compounds. The α and β forms differ mainly in the arrangement of the successive layers of the X—Nb—X atoms. Only the β form of Nb_3I_8, which is stable in air for several days, is stoichiometric.[501] Nb_3Cl_8, β-Nb_3Br_8 and β-Nb_3I_8 are paramagnetic. The low magnetic moment (1.95 BM at 573 K falling to 0.5 BM at 90 K for Nb_3Cl_8) indicates that there is some magnetic exchange between distinct Nb_3 clusters.

The Nb_3 binary halides with seven electrons available for metal–metal bonding within the triangle are 'electron poor' clusters. A simplified MO scheme suggests that the HOMO and LUMO possess metal *d* character.[645] One unpaired electron is expected to be present. The existence of a low-lying metal cluster orbital is also indicated by the electronic spectra.

34.5.2.2 *Adducts of the neutral halides*

The reactivity of the Nb_3X_8 clusters has received little attention as yet. The metal–metal bonds were however found to be susceptible to disruption, leading to dinuclear,[525] tetranuclear[647] and hexanuclear species.[648]

Several complexes of the M_3X_8 clusters have been obtained, generally during reduction procedures promoted by the ligands; thus, the reaction between $[Nb_2Cl_6(SMe_2)_3]$ and Bu^tNC (equation 55) yielded $[Nb_3Cl_8(Bu^tNC)_5]$ (**41**). The molecule has an approximate symmetry plane.[615] The Nb atoms define an isosceles triangle, with a bridging chlorine atom located below each edge and an unprecedented type of multiply bridging isocyanide ligands above the triangle (Figure 33). The short Nb(1)—Nb(2) distance suggests a two-electron bond, while the interpretation of the two larger Nb—Nb distances is more problematical. It is difficult to assign oxidation numbers unambiguously; ESR data at room temperature ($\langle g \rangle \approx 1.95$ and $\langle A \rangle =$ 130 G) are consistent with the unpaired electron localized on Nb(3). Two pairs of π electrons of the bridging Bu^tNC ligand are probably involved in bonding with Nb(1) and Nb(2), while C(1) forms a strong donor bond to Nb(3). A formally similar situation is found for one of the CO ligands in $[Nb_3(CO)_7(Cp)_3]$ (**47**).[649]

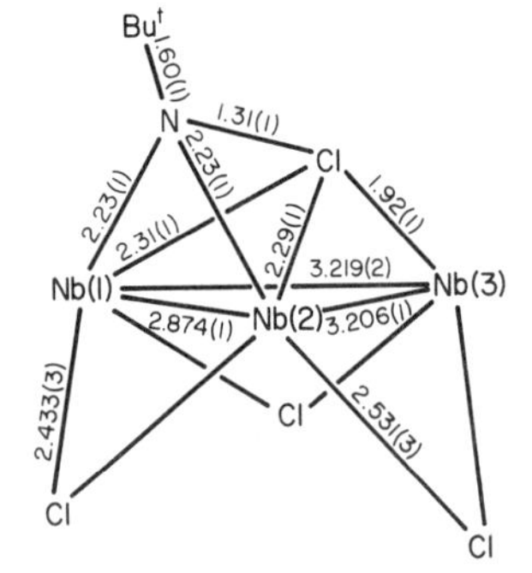

Figure 33 Central core of the $[Nb_3Cl_8(Bu^tNC)_5]$ (**41**) molecule. Each Nb is surrounded by two Cl and one Bu^tNC ligand (reproduced from ref. 615 with permission)

A compound of formula $[Nb_3Cl_8(SMe_2)_2(OEt_2)]$, but of unknown structure, has been observed to form during the synthesis of $[Nb_2Cl_6(SMe_2)_3]$.[584] Another, formulated as $[Ta_3Br_8(SMe_2)_2]$, resulted from a low yield reduction of $TaBr_4$.[650]

34.5.2.3 *Solvolysis products*

Compounds of formula $[M_3X_2B_6L_3]$, where B is a bridging anionic donor and L a neutral one, may be considered as formally derived from M_3X_8 adducts by solvolysis. The reduction of Nb^V in acid solutions has been used for its qualitative and quantitative determination, but the exact nature of the reduced species remains unknown. The red-brown salt obtained as early as 1912 by reduction of Nb^V in aqueous H_2SO_4 has been formulated as $[K_4(H_5O_2)][Nb_3O_2(SO_4)_6(H_2O)_3]\cdot 5H_2O$ (**48**) on the basis of a recent X-ray structure determination.[651] Complex (**48**) has nearly D_{3h} symmetry and is the first example of a cluster having bridging sulfate ligands (Figure 34). Its overall geometry, with two μ_3-ligands, is reminiscent of the structural and coordination type $[M_3'X_2(O_2CR)_6L_3]$, mainly found for Mo and W. The $[H_5O_2]^+$ ion exhibits an O···H···O distance of 2.42(1) Å, comparable to those found in other compounds containing the diaquahydrogen ion. The 5− charge of the cluster leads to a non-integral oxidation number of +3.66 for the metal atoms, and suggests that four electrons are involved in Nb—Nb bonding. According to MO calculations, the bond order should be 2/3, in agreement with the experimental M—M distance.

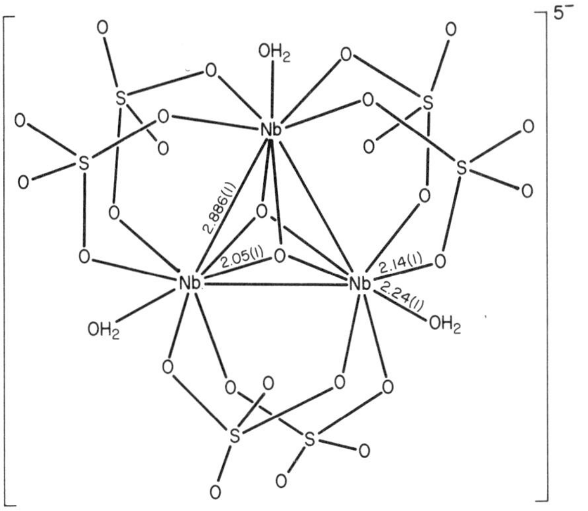

Figure 34 Structure of the $[Nb_3O_2(H_2O)_3(\mu\text{-}SO_4)_6]^{5-}$ (**48**) anion. The distances reported correspond to average values (reproduced from ref. 651 with permission)

The structurally related niobium complexes $[Nb_3O_2(O_2CR)_6(THF)_3]^+$ (Z = $[NbOCl_4(THF)]^-$ for R = Ph; Z = BPh_4^- for R = Bu^t) have been prepared from $[Nb_2Cl_6(SMe_2)_3]$ and sodium carboxylates.[621] The triangular array of Nb atoms (Nb—Nb: 2.824(3) Å for R = Bu^t) is capped by two oxygen atoms, bridged by six carboxylato ligands, and coordinated in the axial position by the THF molecules. Unlike (**48**), they are diamagnetic. $[Nb_3(\mu_3\text{-}S)(\mu_2\text{-}O)(NCS)_9]^{6-}$ further illustrates the tendency of Nb to form oxo trinuclear species, and has been considered to derive from the green $[Nb_3O_4(H_2O)_9]^{n+}$ cation.[651b]

34.5.2.4 Clusters with the hexamethylbenzene ligand

Trimetallic cations $[M_3(\mu\text{-}X)_6(C_6Me_6)_3]^+$ (M = Nb, X = Cl, Br; M = Ta, X = Cl) have been obtained from hexamethylbenzene, aluminum powder, $AlCl_3$ and MX_5 in the molten state.[8,652] The same approach, applied to $TaBr_5$, led to $[TaBr_2(C_6Me_6)]_n$, where the metal is in the formal oxidation state +II. The diamagnetic $[Nb_2(\mu\text{-}Cl)_6(C_6Me_6)]Cl$ has C_{3h} symmetry. Each metal is coordinated in an approximately square planar fashion and bears an η^6-C_6Me_6 ligand parallel to this plane. The Nb—Nb distances (3.334(6) Å) are consistent with the three metals' sharing two electrons.[653]

The $[M_3(\mu\text{-}X)_6(C_6Me_6)_3]^+$ derivatives show an unusual, reversible redox behavior.[654] Their oxidation by ammonium hexanitratocerate(IV) yields a species of stoichiometry $[Nb_3Cl_6(C_6Me_6)_3]^{2+}$, whose diamagnetic behavior implies dimerization to $[Nb_6Cl_{12}(C_6Me_6)_6]^{4+}$. This cluster is related to those of type $[(M_6X_{12})X_6]^{(q-6)}$ from which they derive by substitution of the six terminal halides by the C_6Me_6 ligand acting formally as a two-electron donor. The oxidation reaction (equation 64), analogous to that of $[(M_6Cl_{12})Cl_6]^{4-}$ to $[(M_6Cl_{12})Cl_6]^{2-}$ (Section 34.5.4.1.iii), has been accomplished by Ce^{IV}, NBS, iodine and even air in acidic solutions. The water soluble hexanitratocerate(IV) salt was converted to the insoluble $[Nb_6Cl_{12}(C_6Me_6)_6]Z_4$ (Z = PF_6^-, BPh_4^-, SCN^-). The related tantalum chloride and niobium bromide clusters display comparable behavior, but their derivatives are generally less stable.

$$2[Nb_3Cl_6(C_6Me_6)_3]^+ \xrightarrow{-2e^-} [Nb_6Cl_{12}(C_6Me_6)_6]^{4+} \quad (64)$$

The paramagnetic purple trimer $[Nb_3(\mu\text{-}Cl)_6(C_6Me_6)_3]^{2+}$ could be stabilized when using the dimeric $(TCNQ)_2^{2-}$ as counterion.[655] It has *m* symmetry with two Nb—Nb distances of 3.327(2) and 3.344(3) Å. The overall structural arrangement is novel, and displays strongly bonded TCNQ dimers interacting with the C_6Me_6 ligands in a zigzag chain. The nearly isotropic ESR spectra at room temperature ($g = 1.996$) are consistent with the unpaired electron residing on the niobium cluster. The polycrystalline conductivity data indicate a semiconductor behavior.

All the hexamethylbenzene metal halide cluster cations exhibit characteristic electronic spectra which can be used for their identification in their green or brown solutions. Various attempts to obtain derivatives of $[Nb_3(\mu\text{-}Cl)_6(C_6Me_6)_3]$ under the action of donors or organometallic reagents failed.

34.5.3 Tetranuclear Clusters

The formation of $[CsNb_4X_{11}]$ (X = Cl, Br) and $[RbNb_4Cl_{11}]$ (equation 65) instead of A_2MX_6, $A_2M_2X_9$ or $A_6M_6X_8$ illustrates the importance of the alkali metal cation in stabilizing a given cluster.[647] These brown-black compounds are soluble only with decomposition. Their X-ray structures show the presence of planar Nb_4 clusters (Figure 35), and they can be described as $(Nb_4X^i_6)$ units linked together into chains by four and six X^{o-o} bridges. The holes between the chains are occupied by the alkali metal cations in a $Cs^2_\infty[(Nb_4X^i_6)X^{o-o}_{4/2}X^{o-o}_{6/2}]$ configuration. The structure is closely related to that of Nb_3X_8.

$$5Nb_3X_8 + Nb + 4AX \longrightarrow 4ANbX_{11} \qquad (65)$$
$$A = Cs, Rb; \quad X = Cl, Br, I$$

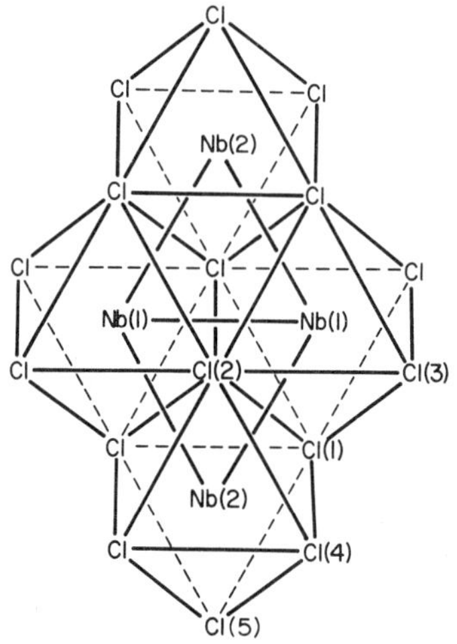

Figure 35 Drawing of the structure of CsM_4Cl_{11} showing the five different chlorine environments (Nb(1)—Nb(1): 2.84, Nb(1)—Nb(2): 2.95 Å; (reproduced from ref. 647 with permission)

In these clusters the metal has a formal oxidation state of 2.5, with 10 electrons available for M—M bonding in the Nb_4 group; the compounds are diamagnetic with a slight temperature independent paramagnetism. It is noticeable that Nb(1) and Nb(2) have different arrangements, which led to the proposal of different formal oxidation states: +II for Nb(1) and +III for Nb(2).

34.5.4 Octahedral Clusters

The lowest halides, obtained under strongly reducing conditions or thermal disproportionation of the higher halides, are Nb_6F_{15} and Ta_6X_{15} (X = Cl, Br, I), Nb_6X_{14} (X = Cl, Br), Ta_6X_{14} (X = Br, I) and Nb_6I_{11}.[1,5,643] Their oxidation numbers follow the expected trends: Ta > Nb and F > Cl > Br > I, and Nb_6I_{11} is the lowest niobium subhalide reported so far. All are based on octahedral $(M_6X_{12})^{q+}$ (X = Cl, Br; $q = 2, 3, 4$) or $(M_6X_8)^{q+}$ ($q = 2, 3$) units (Figure 36) as for Nb_6I_{11} (Section 34.5.4.2). Although their chemistry has long been investigated—mainly in aqueous solutions—their formulation and structure, and those of their derivatives, have been established only recently, or, in many cases, remain uncertain. The first evidence for the existence of the $[M_6X_{12}]^{2+}$ group is due to Vaughan *et al.*, who in 1950 studied ethanolic solutions of some of these complexes by X-ray scattering.[656]

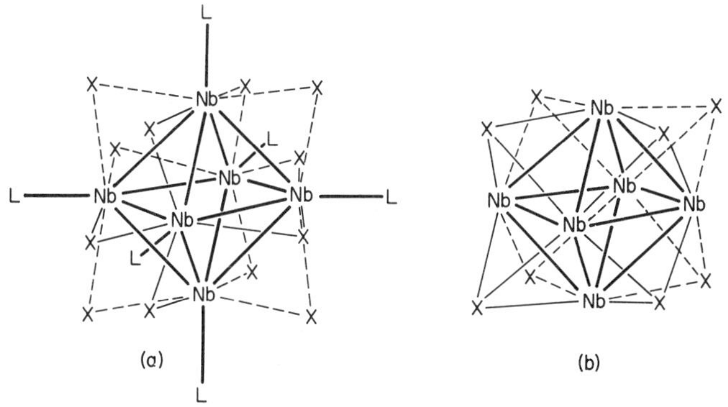

Figure 36 Octahedral clusters: (a) the $(M_6X_{12})^{q+}$ core with additional centrifugal ligands L; (b) the $(M_6X_8)^{q+}$ core

Conflicting data, due to studies performed on mixtures of compounds in which different oxidation states were present (oxido–reduction processes occur easily, especially in the presence of air[657–659]) are reported.

The metals of the octahedral clusters can readily add a variety of ligands in 'centrifugal' positions, resulting in compounds of type $[(M_6X_{12})Y_nL_{6-n}]^{(q-n)}$ (X = Cl, Br; Y = F, Cl, Br, I, OH; L = neutral donor; n = oxidations state). In the description of such species, the central core containing the metal atoms will be represented in parentheses.

34.5.4.1 Clusters based on the $(M_6X_{12})^{q+}$ core

(i) Neutral binary halides

The first metal-cluster fluoride known, $[(Nb_6F_{12})F_3]$, was prepared by disproportionation of NbF_4 or by reduction of NbF_5 by Nb. The compounds of composition M_6X_{14} have been obtained by reduction of the higher halides by various metals, and Nb_6Cl_{14} also by heating Nb_3Cl_8. Phase diagram studies established that the only tantalum cluster iodide formed was Ta_6I_{14}, whereas in the case of bromine both Ta_6Br_{14} and Ta_6Br_{15} formed. The compounds are dark, air stable solids; Ta_6Cl_{14} and Nb_6Br_{14} dissolve in water and ethanol, and Nb_6F_{15} and Ta_6Cl_{15} are unaffected by dilute mineral acids.

Regular octahedra of metal atoms were found with rather short metal–metal distances (Table 44) for Nb_6F_{15} and Ta_6Cl_{15}. Nb_6Cl_{14} and Ta_6I_{14} are isotypic and contain flattened octahedra, hence two types of centrifugal ligands. The strong equatorial Ta—Ta bonds of Ta_6I_{14} are accompanied by a greater distortion of the Ta_6 octahedron as compared to that of Nb_6 in Nb_6Cl_{14}. The regular O_h symmetry usually given for isolated clusters is obviously an oversimplified view here. For the M_6X_{14} class of compounds, the three-dimensional network can be described as ${}^3_\infty[(Nb_6Cl_{10}Cl_2^{i-o})Cl_{2/2}^{o-i}Cl_{4/2}^{o-i}]$.[646] For Nb_6F_{15} and Ta_6Cl_{15} additional intercluster bridging halogen atoms are bonded to each type of metal atom.

(ii) Ionic halides $[(M_6X_{12})X_6]^{4-}$ and hydrates with an $(M_6X_{12})^{2+}$ core

(*a*) *Synthesis and structure.* The hydrates $M_6X_{14}{\cdot}8H_2O$ (X = Cl, Br), being more soluble than the anhydrous lower halides, are useful starting materials for preparing subsequent derivatives.[658]

The preparation of $[M_6X_{14}]{\cdot}8H_2O$ generally involves high temperature reactions, the $(M_6X_{12})^{2+}$ cation-containing products being extracted with boiling water and crystallized from aqueous HX. When the extraction was achieved with warm water, $[M_6Cl_{14}]{\cdot}9H_2O$ was obtained.[660] The first of these hydrates, $[Ta_6Cl_{14}]{\cdot}8H_2O$, was prepared as early as 1907, but at that time it was formulated as $[TaCl_2]{\cdot}2H_2O$.

Harned's method, which has long been used to obtain the ternary halides required for the

Table 44 Structural Data for Various Niobium and Tantalum Clusters of Type $(M_6X_{12})^{q+}$

Compound	Symmetry	Bond lengths (Å) M—M	M—X bridging	M—X centrifugal	Observations	Ref.
$[(Nb_6F_{12})F_3]$	Cubic	2.80	2.05	2.11	Regular octahedron	1
$[(Nb_6Cl_{12})Cl_2]$	Orthorhombic	2.96(basal) 2.89(apex)	2.40 (2.36–2.39)	2.58 3.04	Tetragonal distortion	2
$[(Nb_6Cl_{12})Cl_6]^{4-}$	Monoclinic	2.92	2.48	2.61	Regular octahedron	3
$[(Nb_6Cl_{12})Cl_6]^{3-}$	Trigonal rhombohedral	2.97	2.43	2.52	Regular octahedron	4
$[(Nb_6Cl_{12})Cl_6]^{2-}$	Trigonal	3.02	2.42	2,46	Regular octahedron	5
$[(Ta_6I_{12})I_2]$	Orthorhombic	2.80(basal) 3.08(apex)	2.75–2.83		Tetragonal distortion	6

1. H. Schäfer, H. G. Von Schnering, K. J. Niehues and H. G. Niedervarenholz, *J. Less-Common Met.*, 1966, **9,** 95.
2. A. Simon, H. G. Von Schnering, H. Wohrle and H. Schäfer, *Z. Anorg. Allg. Chem.*, 1965, **339,** 155.
3. A. Simon, H. G. Von Schnering and H. Schäfer, *Z. Anorg. Allg. Chem.*, 1968, **361,** 235.
4. F. W. Koknat and R. E. McCarley, *Inorg. Chem.*, 1974, **13,** 295.
5. F. W. Koknat and R. E. McCarley, *Inorg. Chem.*, 1972, **11,** 812.
6. D. Bauer, H. G. Von Schnering and H. Schäfer, *J. Less-Common Met.*, 1965, **8,** 388.

preparation of the hydrates, involves the reduction of the pentahalide by cadmium and formation of $[Cd_2(M_6X_{12})X_6]$ (<20%).[661] The high temperature disproportionation of Nb_3X_8 with an alkali halide (equation 66) or its reduction to $A_4[(Nb_6X_{12})X_6]$ in the presence of niobium have also been proposed. A disadvantage of these methods is that they require Nb_3X_8 as a precursor. More efficient is the fast and quantitative comproportionation of niobium or tantalum whith their pentahalides in the presence of an alkali halide (equation 67).[648] This synthesis is facilitated by the stabilization of the cluster through complexation into an $[M_6X_{18}]^{4-}$ anion, and is particularly valuable for $[Nb_6Br_{14}]\cdot 8H_2O$. The selection of the appropriate alkali metal halide is essential for obtaining the $[M_6X_{18}]^{4-}$ cluster, since the relative stabilities of $[A_2MX_6]$, $[A_3M_2X_9]$ (Section 34.4.2.1) and $[AM_4X_{11}]$ (Section 34.5.3), which are competing with it, depend on the cation. For all compounds competing with $[A_4Nb_6Cl_{18}]$, the sodium derivatives have markedly lower stabilities than those of the heavier alkali metals; NaCl was therefore generally selected as complexing agent when the $NbCl_{18}$ derivatives were sought. For the less stable bromine derivatives, Na^+ was replaced by K^+.

$$5Nb_3X_8 + 14AX \longrightarrow 2A_4[(Nb_6X_{12})X_6] + 3[A_2NbX_6] \quad (66)$$
$$A = Na, K; \quad X = Cl, Br$$

$$14MX_5 + 16M + 20AX \xrightarrow[1\,d]{700\,°C} 5A_4[(M_6X_{12})X_6] \quad (67)$$
$$A = Na, K; \quad X = Cl, Br$$

An undistorted octahedral symmetry has been established for the $(Nb_6Cl_{12})^{2+}$ core in $[K_4Nb_6Cl_{18}]$. The structure of $[Ta_6Cl_{14}]\cdot 7H_2O$ shows that the compounds should be written as $[(Ta_6Cl_{12})Cl_2(H_2O)_4]\cdot 3H_2O$.[662] The short centrifugal Ta—Cl distance of 2.35 Å (2.50 for the Ta—Cl bridges) is accompanied by an elongation of the Ta_6 octahedron along this axis (Ta—Ta: 3.15 and 2.78 Å).

(*b*) *Reactivity of the hydrates* (*Table* 45). .The preparation of pure $(M_6X_{12})^{2+}$ (X = Cl, Br) complexes from aqueous solutions is not easy, as oxidation to $(M_6X_{12})^{3+}$ readily occurs. The latter may be detected by the emergence of a band near 8000 cm^{-1}. Addition of reductants such as $SnCl_2$ or Cd to oxidized solutions caused a return of the familiar green color of $(M_6X_{12})^{2+}$. The bromides are soluble in water and methanol, while the chlorides dissolve well only in methanol. Titration with NaOH precipitates $[(M_6X_{12})(OH)_2]\cdot 8H_2O$, which redissolves as $[(M_6X_{12})(OH)_4]^{2-}$ by further addition of NaOH.[659]

The $(M_6X_{12})^{2+}$ clusters act as Lewis acids and a number of $[(M_6Cl_{12})Cl_2L_4]$ complexes have been obtained, mainly with oxygen donors, including DMSO, DMF, $OAsPh_3$, $OPPh_3$ and pyO. The more recently obtained phosphine complexes are the only examples of Nb or Ta hexameric clusters with neutral ligands without oxygen donors. Whereas bidentate ligands have been widely used to form derivatives of the molybdenum and tungsten $(M'_6X_8)^{q+}$ clusters, only $[(Nb_6Cl_{12})Cl_2(bipyO_2)_2]$ has been reported.[663] Models showed that $bipyO_2$ is able to occupy two *cis* centrifugal positions, and the electronic spectrum of the complex was comparable to that of the adducts formed with monodentate ligands.

Table 45 Niobium and Tantalum Cluster Adducts Containing the $[M_6X_{12}]^{2+}$ Core

Compound	*Comments*	*Ref.*
$[(Nb_6X_{12})X_2(H_2O)_4]\cdot nH_2O$ ($X = F, Cl, Br$)	$n = 4$ extraction with H_2O at 100 °C; ESCA for X = Cl; $n = 5$ and X = Cl, Br extraction with H_2O at 70 °C; X-ray powder data	1, 2, 3, 4
$[(Ta_6X_{12})X_2(H_2O)_4]\cdot nH_2O$ ($X = Cl$, $n = 3, 4$; $X = Br$, $n = 4$)	X-Ray structure for X = Cl, $n = 3$: tetragonal distortion; X-ray powder data	2, 3, 5
$[(Nb_6Cl_{12})Y_2(H_2O)_4]\cdot nH_2O$ ($Y = BF_4$, NO_3, $1/2SO_4$)		6
$[(Nb_6Cl_{12})Cl_2L_4]$ (L = DMSO, DMF, Ph_3AsO, Ph_3PO, pyNO, 4-ZC_5H_4NO)	L = DMSO, ESCA, no analysis L = 4-ZC_5H_4NO, Z = Me_2NO, MeO, Me, Cl, NO_2, Br, UV study	8 7, 9
$[(Nb_6Cl_{12})Br_2(DMSO)_4]$	No analysis, only IR characterization	9, 10
$[(Nb_6Cl_{12})(ClO_4)_2(DMSO)_4]\cdot 2DMSO$	No analysis	10
$[(Nb_6Cl_{12})Cl_2(PR_3)_4]$ ($R_3 = Et_3$, Pr^n_3, Et_2Ph)	ESCA for R = Pr^n, cyclic voltammetry	11
$[(M_6Cl_{12})Cl_2[P(OR)_3]_4]$ (R = Me, Et)	Less soluble than the PR_3 derivatives	11
$[(Ta_6Cl_{12})Cl_2L_4]$ (L = DMSO, PPr^n_3)		11
$[(M_6X_{12})(OH)_2]\cdot 8H_2O$ (X = Cl, Br)		12
$[(M_6Cl_{12})(OH)_2]\cdot 6H_2O\cdot 2MeOH$	Obtained by titration in H_2O–MeOH; less stable than the octahydrate X-ray powder data	13
$(Et_4N)_4[(Nb_6Cl_{12})X_6]$ (X = F, Cl, Br)	Polarography for X = Cl; X = Cl, Br, isolation of adducts with 2 EtOH	1, 9, 14
$A_4[(Nb_6Cl_{12})Cl_6]$ (A = Li, Na, K, Cs, Me_4N)	X-Ray structure for A = K	2, 9, 15
$Ba_2[(Nb_6Cl_{12})(Cl_6)]$	X-Ray powder data	16
$(picH)_4[(Nb_6Cl_{12})I_6]$		15
$A_4[(M_6X_{12})X_6]$ (M = Nb, Ta, X = Br, A = K; M = Ta, X = Cl, A = Na)		2
$(Ph_4As)_4[(Nb_6Cl_{12})Cl_5(OH)]$		17
$H_4[(Ta_6Cl_{12})(CN)_6]\cdot 12H_2O$	X-Ray structure; regular octahedron of Ta_6	18

1. B. G. Hughes, J. L. Meyer, P. B. Fleming and R. E. McCarley, *Inorg. Chem.*, 1970, **9**, 1343.
2. F. W. Koknat, J. A. Parsons and A. Vongvusharintra, *Inorg. Chem.*, 1974, **13**, 1699.
3. B. Spreckelmeyer, *Z. Anorg. Allg. Chem.*, 1968, **358**, 148.
4. J. G. Converse and R. E. Carley, *Inorg. Chem.*, 1970, **9**, 1361.
5. R. D. Burbank, *Inorg. Chem.*, 1966, **5**, 1491.
6. P. M. Boorman and B. P. Straughan, *J. Chem. Soc. (A)*, 1966, 1515.
7. R. A. Field and D. L. Kepert, *J. Less-Common Met.*, 1967, **13**, 378.
8. R. A. Field, D. L. Kepert and D. Taylor, *Inorg. Chim. Acta*, 1970, **4**, 113.
9. S. A. Best and R. A. Walton, *Inorg. Chem.*, 1979, **18**, 484.
10. P. B. Fleming, J. L. Meyer, W. K. Grindstaff and R. E. McCarley, *Inorg. Chem.*, 1970, **9**, 1769.
11. D. D. Klenworth and R. A. Walton, *Inorg. Chem.*, 1981, **20**, 1151.
12. N. Brnicevic and H. Schäfer, *Z. Anorg. Allg. Chem.*, 1978, **441**, 219.
13. N. Brnicevic and B. Kojic-Prodic, *Z. Anorg. Allg. Chem.*, 1982, **489**, 235.
14. R. A. Mackay and R. F. Schneider, *Inorg. Chem.*, 1967, **6**, 549.
15. A. Simon, H. G. Von Schnering and H. Schäfer, *Z. Anorg. Allg. Chem.*, 1968, **361**, 235.
16. A. Broll and H. Schäfer, *J. Less-Common Met.*, 1970, **22**, 367.
17. P. B. Fleming, T. A. Dougherty and R. E. McCarley, *J. Am. Chem. Soc.*, 1967, **89**, 159.
18. S. S. Basson and J. G. Leipoldt, *Transition Met. Chem.*, 1982, **7**, 207.

The inner halogens of the $(M_6X_{12})^{q+}$ core are relatively inert; no exchange was detected (within a ±20% margin of error) between $(Ta_6Cl_{12})^{2+}$ and $*Cl^-$ over 13 hours at room temperature.[2] No solvolysis reactions have been reported. The higher lability of the outer halogen atoms allowed the production of mixed halide clusters, for instance $[(Nb_6Cl_{12})X'_2(H_2O)_4]\cdot 4H_2O$ (X′ = Br, F).

(*iii*) *Oxidized clusters*

(*a*) *Synthesis.* In the solid state, $[Ta_6Cl_{14}]\cdot 8H_2O$ was found to undergo reversible oxidation by topotactic electron/proton transfer (equation 68).[664]

$$[(Ta_6Cl_{12})Cl_2(H_2O)_4]\cdot 4H_2O \rightleftharpoons [(Ta_6Cl_{12})(OH)_x(H_2O)_{4-x}]\cdot 4H_2O + xe^- + xH^+ \quad (68)$$
$$0 \leq x \leq 1.5$$

In aqueous solution the green diamagnetic $(M_6X_{12})^{2+}$ is oxidized by air, I_2, or Hg^{II} to the paramagnetic yellow $(M_6X_{12})^{3+}$, and by two or more moles of strong oxidizing agents such as BrO^{3-}, H_2O_2, Cl_2, Fe^{III}, Ce^{IV} or VO_2^+ ions to the red-brown diamagnetic $(M_6X_{12})^{4+}$ (X = Cl, Br).[665–668] Oxidation is always faster for the chlorine than for the bromine derivatives. No redox chemistry is known so far for $(Nb_2F_{12})^{3+}$ or $(Ta_6I_{12})^{2+}$, the latter being unstable in aqueous solution. Standard potentials have been measured (Table 46).[657,659] The instability of the niobium bromine clusters with respect to hydrolysis hindered the obtaining of $E°$ values. Only the $(Ta_6Cl_{12})^{3+}/(Ta_6Cl_{12})^{2+}$ couple has found analytical applications.

Table 46 Standard Potentials of Niobium and Tantalum $(M_6X_{12})^{q+}$ in Neutral Aqueous Solutions

M	*X*	$E^{\circ a}$ (V)	$E^{\circ b}$ (V)
Nb	Cl	0.83	1.12
Ta	Cl	0.49	0.83
Ta	Br	0.59	0.89

[a] $(M_6X_{12})^{3+}/(M_6X_{12})^{2+}$ [b] $(M_6X_{12})^{4+}/(M_6X_{12})^{3+}$

Oxidation of $(M_6X_{12})^{2+}$ by dioxygen is slow in neutral solutions, but occurs rapidly in acidic or basic media (equations 69 and 70). Cyclic voltammetry data are available on the $(Ta_6Br_{12})^{q+}$ system in aqueous solutions.[669] Controlled oxidation of $[M_6X_{14}]\cdot 7H_2O$ led to the hydrates $[M_6X_{15}]\cdot 7H_2O$ or $[M_6H_{16}]\cdot 7H_2O$. Spontaneous reduction was observed for all the $(M_6X_{12})^{4+}$ clusters except $(Ta_6Br_{12})^{4+}$ upon redissolution. The $(M_6X_{12})^{q+}$ ($q = 3, 4$) ions are also reduced to $(M_6X_{12})^{2+}$ species by Cr^{II} or V^{II}.[668] Examples of reduction by ligands seem to be limited to that of $[Nb_6Cl_5]\cdot 7H_2O$ by phosphines to give $[(Nb_6Cl_{12})Cl_2(PR_3)_4]$ (R = Et, Pr^n).[670]

$$2(M_6X_{12})^{2+} + 0.5O_2 + 2H^+ \longrightarrow 2(M_6X_{12})^{3+} + H_2O \quad (69)$$

$$2(M_6X_{12})^{2+} + 0.5O_2 + 10OH^- \longrightarrow 2[(M_6X_{12})(OH)_6]^{3-} \quad (70)$$

$$X = Cl, Br$$

Reversible redox processes were also observed in non-aqueous solutions. The behavior of $[(M_6Cl_{12})Cl_2(PR_3)_4]$ (R = Et, Pr^n) in CH_2Cl_2 or MeCN was investigated by cyclic voltammetry, revealing several one-electron processes (equation 71). Chemical oxidation by $NOPF_6$ yielded $[(M_6Cl_{12})Cl_2(PPr^n_3)_4](PF_6)_n$ ($n = 1, 2$). The electron transfer is proposed to occur *via* M—Cl · · · + NO bridges. The monoanions $[(M_6Cl_{12})Cl_2(PR_3)_4]^-$ are formally derived from the previously unknown $(M_6Cl_{12})^+$ core. The stabilizing influence of the phosphine led to a redox chemistry which is more extentive than that in aqueous solutions. Attempts to obtain the monoanions by electrochemical or chemical reductions (Li, Mg, Na/Hg) failed. The average oxidation state of the metal in the monoanions remains higher than for $(Nb_6I_8)^{2+}$.

$$[(M_6Cl_{12})Cl_2(PR_3)_4]^- \underset{}{\overset{-e^-}{\rightleftharpoons}} [(M_6Cl_{12})Cl_2(PR_3)_4] \overset{-e^-}{\rightleftharpoons} [(M_6Cl_{12})Cl_2(PR_3)_4]^+ \overset{-e^-}{\rightleftharpoons} [(M_6Cl_{12})Cl_2(PR_3)_4]^{2+} \quad (71)$$

Within the $(M_6X_{12})^{q+}$ (X = Cl, Br; q = 2-4) cluster halides, the only series that cover all three oxidation states are the tantalum halide hydrates and the haloanions $[(M_6X_{12})X_6]^{(q-6)}$. The isolation of $[(Ta_6Cl_{12})Cl_2(PR_3)_4](PF_6)_n$ ($n = 1, 2$) led to the first complete series of group VA cluster stabilized by an organic ligand.

Polarographic measurements have also been performed on $[(Nb_6Cl_{12})Cl_6]^{(q-6)}$ ($q = 2, 3$) in DMSO. While in aqueous solutions iodine causes a two-electron oxidation, only a one-electron oxidation was observed in DMSO.

(*b*) *Reactivity*. The oxidation of the hydrates $[M_6X_{14}]\cdot 8H_2O$ in ethanolic or aqueous HX, in the presence of a suitable counterion, was used to prepare anions of type $[(M_6X_{12})X'_6]^{(q-6)}$ ($q = 2, 3$; equation 72).[658,672] The species that have been isolated include $(pyH)_2[(M_6Cl_{12})Cl_6]$

and $(A)_n[(M_6X_{12})X_6]$ (X = Cl, Br; A = R_4N, Ph_4As, *etc.*; Table 47). To obtain pure compounds containing the $(Ta_6X_{12})^{3+}$ unit, air must be rigorously excluded.

$$(Ta_6Br_{12})^{2+} + X'_2 + 2Et_4NX' \longrightarrow (Et_4N_2)[(Ta_6Br_{12})X'_6] \qquad (72)$$
$$X' = Cl, Br$$

The stability of the $(M_6X_{12})^{q+}$ compounds towards hydrolysis decreases in the order Ta > Nb, Cl > Br, and with increasing oxidation states. Evidence has been reported for partial replacement of the chlorines by hydroxo groups in $[Nb_6Cl_{15}]\cdot 7H_2O$.[672]

Irradiation of $(Ta_6Br_{12})^{2+}$ in deaerated HCl solution led to $(Ta_6Br_{12})^{3+}$ and evolution of H_2. It was assumed that the key step involves a two-electron transfer from $(Ta_6Br_{12})^{2+}$ to a water molecule in the solvent cage, yielding $(Ta_6Br_{12})^{4+}$ and H^-, which is scavenged by H^+, generating H_2.[673]

The mixed metal octahedral clusters $(Ta_5MoCl_{12})^{3+}$ and $(Ta_4Mo_2Cl_{12})^{4+}$ were obtained by reduction of $TaCl_5$–$MoCl_5$ mixtures with aluminum in $NaAlCl_4$–$AlCl_3$ melts.[674] Further reduction of $(Ta_5MoCl_{12})^{3+}$ by zinc offered the +2 cluster. Spectroscopic and magnetic data indicate that $[(Ta_5MoCl_{12})Cl_6]^{3+}$ and $[(Ta_4Mo_2Cl_{12})Cl_6]^{2-}$ are analogous to $[(Ta_6Cl_{12})Cl_6]^{4-}$ (16 M—M bonding electrons), and $[(Ta_5MoCl_{12})Cl_6]^{2-}$ to $[(Ta_6Cl_{12})Cl_6]^{3-}$ (15 bonding electrons).

(*c*) *Structure*. The influence of the oxidation state on the structure of $(M_6X_{12})^{q+}$ or $(M_6X_{18})^{(q-6)}$ can be evaluated from the comparison of X-ray data available on $K_4[(Nb_6Cl_{12})Cl_6]$, $(Me_4N)_3[(Nb_6Cl_{12})Cl_6]$ and $(pyH)_2[(Nb_6Cl_{12})Cl_6]$[675], and, in the case of

Table 47 Niobium and Tantalum Cluster Adducts Containing the $(M_6X_{12})^{q+}$ (q = 3, 4) Core

Compound	*Comments*	*Ref.*
$(M_6X_{12})^{3+}$		
$[(M_6X_{12})X_3(H_2O)_4]\cdot nH_2O$ (X = Cl, Br; n = 3, 4)	X-Ray powder data	1, 2, 3
$[(M_6Cl_{12})Cl_3(DMSO)_3]\cdot n$DMSO	M = Nb, n = 3, analysis; M = Ta, n = 0, IR characterization only	4
$(R_4N)_3[(M_6Cl_{12})Cl_6]$ (R = Me, Et)	M = Nb, R = Me_4N, X-ray structure and ESCA; M = Nb, R = Et_4N, polarography and ESR	5, 6, 7, 8
$(Bu_4N)_3[(Nb_6Cl_{12})I_6]$		2
$(PrNH_3)_3[(Nb_6Br_{12})Cl_6]$		2
$(Et_4N)_2[(Nb_6Cl_{12})Cl_5(DMSO)]$DMSO	Analysis	1
$(Ph_4As)_2[(NbCl_{12})Cl_5(H_2O)$		5
$(Ph_4As)_2[(Nb_6Cl_{12})Cl_4(OH)(H_2O)]$	μ = 1.46 BM	5
$(M_6X_{12})^{4+}$		
$[(Nb_6Cl_{12})X_4(EtOH)_2]$ (X = Cl, Br)		9
$[(Ta_6X_{12})X_4(H_2O)_2]\cdot nH_2O$ (X = Cl n = 7; X = Br n = 3)	$[(Ta_6Cl_{12})Cl_2(OH)_2]$ is obtained by room temperature vacuum dehydration	2, 4
$A_2[(M_6X_{12})X'_6]$ (M = Nb, X = X', A = $PrNH_3$; M = Ta X = X' = Cl A = Ph_4As; X = Br, X' = Cl, A = Et_4N)		2
$A_2[(M_6Cl_{12})X_6]$ (X = Cl, A = Et_4N, pyH; X = Br A = Bu_tN)	X-Ray structure for M = Nb and A = pyH NQR for M = Nb and A = Et_4N	10, 11 12, 6
$H_2[(Ta_6Cl_{12})Cl_6]\cdot 6H_2O$	X-Ray structure; regular Ta_6 octahedron	13

1. B. Spreckelmeyer, *Z. Anorg. Allg. Chem.*, 1968, **358**, 148.
2. B. G. Hughes, J. L. Meyer, P. B. Fleming and R. E. McCarley, *Inorg. Chem.*, 1970, **9**, 1343.
3. B. Spreckelmeyer and H. Schäfer, *J. Less-Common Met.*, 1967, **19**, 122.
4. P. B. Fleming, J. L. Meyer, W. K. Grindstaff and R. E. McCarley, *Inorg. Chem.*, 1970, **9**, 1769.
5. P. B. Fleming, T. A. Dougherty and R. E. McCarley, *J. Am. Chem. Soc.*, 1967, **89**, 159.
6. S. A. Best and R. A. Walton, *Inorg. Chem.*, 1979, **18**, 484.
7. F. W. Koknat and R. E. McCarley, *Inorg. Chem.*, 1974, **13**, 295.
8. R. A. Mackay and R. F. Schneider, *Inorg. Chem.*, 1967, **6**, 549.
9. N. Brnicevic and B. Kojic-Prodic, *Z. Anorg. Allg. Chem.*, 1982, **489**, 235.
10. B. Spreckelmeyer, *Z. Anorg. Allg. Chem.*, 1969, **368**, 18.
11. B. Spreckelmeyer and H. G. Von Schnering, *Z. Anorg. Allg. Chem.*, 1971, **386**, 27.
12. P. A. Edwards, R. E. McCarley and D. R. Torgeson, *Inorg. Chem.*, 1972, **11**, 1185.
13. C. B. Thaxton and R. A. Jacobson, *Inorg. Chem.*, 1971, **10**, 1460.

tantalum, on $H_2[(Ta_6Cl_{12})Cl_6]\cdot 6H_2O$[676] and $H_4[(Ta_6Cl_{12})(CN)_6]\cdot 12H_2O$.[677] In an M_6Cl_{18} unit each metal is surrounded by an approximately square planar set of four bridging chlorine atoms Cl_b. The metal is located slightly below this plane. One- and two-electron oxidations of, for instance, $[Nb_6Cl_{18}]^{4-}$ result in an increase of the Nb—Nb distances in two equal steps, from 2.92 to 2.97 and to 3.02 Å with retention of the octahedral symmetry (Table 44), which has been interpreted as meaning that the electrons are removed from bonding orbitals centered mainly on the M_6 cluster. Since $(Nb_6Cl_{12})^{3+}$ and $(Nb_6Cl_{12})^{4+}$ are diamagnetic, and since $(Nb_6Cl_{12})^{3+}$ shows a paramagnetism corresponding to one electron, these electrons appear to be removed from the same orbital.

As anticipated, a significant shortening of the Nb—Cl bond occurs after oxidation. However, the terminal Nb—Cl_t distances remain longer than that of Nb—Cl_b or the covalent bond (2.35–2.40 Å); even the strongest Nb—Cl_t bonds therefore appear relatively weak, which was attributed to the geometry of the Nb_6Cl_{18} unit: upon oxidation, the niobium atom moves up below the plane of its four Cl_b atoms, and becomes more accessible to Cl_t. Apparently only the Nb—Cl_t bonds benefit from the lowering in Nb—Nb bonding strength upon oxidation.

$[(Ta_6Cl_{12})(CN)_4]^{4-}$ consists of a regular Ta_6 octahedron with terminal cyanide ligands. The Ta—C and C—N distances are comparable to those found in normal cyanide complexes. Each nitrogen atom is hydrogen-bonded to two oxygen atoms.

(*d*) *Other characterizations*. The $(M_6X_{12})^{q+}$ clusters have been the subject of far-IR studies and normal coordinate analysis.[671,678–680] A uniform high shift (*ca.* 10 cm^{-1}) per unit increase in the charge of the cluster has been observed for the strong bands assigned to the M—X stretching modes. The M—M stretching vibration has been located around 140 cm^{-1} or in the 59 to 109 cm^{-1} range, *i.e.* at much lower frequencies than for $(Mo_6Cl_8)^{q+}$ (230 cm^{-1}). The force constant for the M—M bonds in the niobium or tantalum cluster halides was estimated to be less than 3×10^{-9} J $Å^{-1}$.

ESCA spectra have been recorded on several ionic or neutral niobium and tantalum clusters.[681] The $C12p_{1/2,3/2}$ spectra generally exhibit three doublets, which were assigned to intracluster and intercluster bridging Cl_b and to terminally bonded Cl_t halogens. The binding energy order was found to be $Cl_b > Cl_t$ with $\Delta E(Cl_b - Cl_t)$ between 1.4 and 2.1 eV, the lower values being characteristic of the hydrates. No discernible variations in the metal core electron-bonding energies of $[(M_6X_{12})X_6]^{(q-6)}$ clusters were observed upon oxidation. By contrast, an increase in the oxidation state in $[(M_6Cl_{12})Cl_2(PPr^n_3)_4](PF_6)_n$ ($n = 0$–2) is paralleled by an increase in the metal core electron-bonding energies $Nb3d_{5/2}$ or $Ta4f_{1/2}$ These results in conjunction with the ESR spectra and the fact that the cyclic voltammetry data are quite unaffected by changes of the phosphines[671] point to the HOMO of $(M_6Cl_{12})^{q+}$ ions' being almost exclusively metal-based in character.

NQR studies on $(Me_4N)_2[(Nb_6Cl_{12})Cl_6]$ revealed only the Cl_b resonance. It was suggested that the Cl_t resonance may have occurred outside the observed range (3 to 33 MHz) as a result of a high ionic character of the Nb—Cl_t bonds.

(*iv*) *Bonding, electronic spectra and magnetism*

A bonding scheme for the $(M_6X_{12})^{q+}$ clusters should account for the magnetic properties and observed electronic transitions. Several semiempirical MO calculations, for instance those of Cotton and Haas[682] based on the M_6 group, or those of Robin and Kuebler[683] on the $(M_6X_{12})^{2+}$ group, are available. A valence bond description has also been proposed.[2]

The effects of both X and L on the bonding have also been considered;[684] the energy level scheme closely resembles that of Cotton *et al.* Another model including both metal–metal and metal–ligand interactions, with spin–orbit coupling introduced as a final perturbation, is available.[685]

The observation that both the $(M_6X_{12})^{2+}$ (16 electrons) and $(M_6X_{12})^{4+}$ (14 electrons) clusters are diamagnetic, and the magnetic moments found for the $(M_6X_{12})^{3+}$ (15 electrons) cluster (1.62 BM), show that the fifteenth and sixteenth electrons occupy a non-degenerate MO. The ESR spectra obtained on $(Nb_6Cl_{12})^{3+}$ ($g = 1.95$) display a complex splitting, with 49 detectable components. The isotropic hyperfine structure accounts for a single electron uniformly delocalized over six equivalent Nb atoms (55 components expected) and suggests that the HOMO is an M–M orbital, of type A symmetry.

The absorption spectra of the $(M_6X_{12})^{q+}$ (X = Cl, Br; $q = 2$–4) clusters, as well as those of $(Nb_6F_{12})^{3+}$ and $(Ta_6I_{12})^{2+}$, have been measured over the range 4–50 kK;[663,678,683] they are fairly complex. The correspondence between the spectra of the niobium and tantalum complexes ($\varepsilon(Nb)/\varepsilon(Ta) = 0.5$–$0.8$) supports the absence of notable spin–orbit coupling influence. A compilation of detailed visible spectra on the solid state or in solution is available.[686] Several tentative assignments have been presented.[674,684]

Magnetic circular dichroism spectra[685] and magnetic susceptibilities[687,688] have been measured. The $(M_6X_{12})^{3+}$ ions are paramagnetic, and $[(M_6X_{12})X_6]^{3-}$ anions generally display simple Curie law behavior. The magnetic moments at room temperature are significantly larger than expected for one unpaired electron. Deviations from Curie law behavior were observed for M_6X_{15} (M = Nb, X = F; M = Ta, X = Cl, Br) and for $[(Ta_6Cl_{12})Cl_3(H_2O)_3]\cdot 3H_2O$. The existence for the anhydrous halides of bridging halogens between cluster units could cause Curie–Weiss behavior *via* superexchange interactions. The origin of the same behavior for the hydrate remains unexplained. Large temperature independent paramagnetic susceptibilities were found for the diamagnetic clusters. The higher values for niobium than for the analogous tantalum compounds indicate a decrease in energy separation between the excited and the ground states for niobium.

34.5.5 Clusters Based on the $(M_6X_8)^{q+}$ ($q = 2, 3$) Core

The group VA clusters containing the $(M_6X_8)^{q+}$ unit are limited to the niobium subiodides. Nb_6I_{11} is formed by thermal decomposition of Nb_3I_8 or by its reduction with niobium.[1,643] Nb_3Br_8 is not reduced under similar conditions. In contrast to the $(M'_6X_8)^{q+}$ (M′ = Mo, W) clusters, which possess a formal oxidation state of 2 and a closed shell electronic configuration, $(Nb_6I_8)^{3+}$ has a non-integral oxidation state and an open shell electronic configuration.

A salient feature of the structure of Nb_6I_{11} [689,690] is the pronounced orthorhombic distortion of the Nb_6 core with Nb—Nb distances ranging from 2.72 to 2.94 Å, comparable in average to those found in the metal (2.86 Å). In contrast to the $(Nb_6X_{12})^{q+}$ clusters, there are eight bridging halogen atoms, one above each face of the distorted octahedron (Figure 36b). The $(Nb_6I_8)^{3+}$ clusters are joined together by sharing centrifugal iodine atoms, forming $[(Nb_6I_8)I_{6/2}]$.

Nineteen valence electrons are available for the Nb_6 framework. The magnetic susceptibility of Nb_6I_{11}, studied over the 5–863 K temperature range, suggests the presence of one and three unpaired electrons per cluster at low and high temperatures, respectively.

Heating Nb_6I_{11} with H_2 at 300 °C under normal pressure yielded the remarkable hydride $[Nb_6HI_{11}]$. Neutron diffraction investigations on $[Nb_6HI_{11}]$ and $[Nb_6DI_{11}]$ permitted the location of the hydrogen atom at the center of the cluster. The compound is diamagnetic below 200 K, but approaches the magnetic susceptibility of Nb_6I_{11} between 400 and 800 K. A vibration was found at 1120 cm^{-1} for the hydride in the cluster.

A more reduced, moderately air stable cluster, $[CsNb_6I_{11}]$, containing the $(Nb_6I_8)^{2+}$ unit, was obtained by reaction of Nb_6I_{11} or of its precursor Nb_3I_8 with CsI in the presence of the metal (equations 73 and 74).[691] It has D_{3d} symmetry and Nb—Nb distances of 2.835 Å, comparable to those in Nb_6I_{11}. The distortions of the $(Nb_6I_8)^{q+}$ ($q = 2, 3$) octahedra were attributed to different modes of packing and to strains at the bridging iodine atoms. Hydrogenation of $[CsNb_6I_{11}]$ at 400 °C (1 atm) gave the isostructural hydride $[CsNb_6HI_{11}]$. No hydrogenation was observed in comparable conditions for the $(M_6X_{12})^{q+}$ clusters.

$$10Nb_6I_{11} + 6Nb + 11CsI \longrightarrow 11CsNb_6I_{11} \quad (73)$$

$$5Nb_3I_8 + 9Nb + 4CsI \longrightarrow 4CsNb_6I_{11} \quad (74)$$

34.6 OXIDATION STATE +II

Molecular compounds in which niobium or tantalum has a d^3 configuration are still very rare, and are mainly limited to adducts of monomeric dihalides with phosphorus donors $[MCl_2(dmpe)_2]$ and $[MCl_2(PMe_3)_4]$. Niobium aryloxo analogs have recently been described.[731a]

34.6.1 Dihalide Adducts

34.6.1.1 Synthesis and structure

The first member of this series, $[TaCl_2(dmpe)_2]$,[596] was reported in 1977; all the other compounds have been reported, since 1983, by Sattelberger.[692,588] $TaCl_5$ is reduced smoothly by sodium amalgam in ether in the presence of PMe_3 to give the brown, air sensitive $[TaCl_2(PMe_3)_4]$ (**49**) in a typical yield of 60% (equation 75). Excess phosphine is necessary to counter the tendency towards dimerization of the intermediate *mer*-$[TaCl_3(PMe_3)_3]$. Once (**31**) is formed, it is no longer reduced. The reduction of $NbCl_5$ $[NbCl_4(PMe_3)_2]$ in similar conditions gave poor, but reproducible, yields of $[NbCl_2(PMe_3)_4]$, the Nb^{III} dimer $[Nb_2Cl_4(PMe_3)_4(\mu\text{-}Cl)_2]$ being the major product.

$$TaCl_5 + 3Na/Hg + PMe_3\ (excess) \xrightarrow[1\ h]{Et_2O} \underset{(\mathbf{49})}{[TaCl_2(PMe_3)_4]} + 3NaCl \qquad (75)$$

The dmpe adducts have been obtained in high yields as brown sublimable thermally stable solids, soluble in hexane and other non-polar solvents (equation 76). Reduction of $TaCl_5$ by one equivalent of sodium amalgam in the presence of dmpe in THF and benzene solvent gave $[TaCl_4(dmpe)_2]$, together with $[TaCl_2(dmpe)_2]$ when the THF:C_6H_6 ratio was higher than 15%.[596] The Ta^{II} and Nb^{II} dmpe complexes were also obtained quantitatively from $[MCl_2(PMe_3)_4]$; there was no evidence for dinuclear products. $[TaBr_2(dmpe)_2]$ was identified by mass spectrometry as the major product of the reaction between $[TaCl_4(dmpe)_2]$ and an excess of LiMe containing LiBr.

$$\underset{(\mathbf{22})}{[MCl_4(dmpe)_2]} + 2Na/Hg \xrightarrow{THF} \underset{(\mathbf{50})}{[MCl_2(dmpe)_2]} + 2NaCl \qquad (76)$$

The products are paramagnetic ($\mu = 1.45$ BM, for instance, for **49**). Solutions of (**49**) in toluene are ESR silent at room temperature, and ^1H NMR data suggest a *trans* octahedral geometry. This was confirmed by X-ray crystallography (D_{2d} symmetry; Ta—Cl: 2.464(3) Å, Ta—P: 2.543(2) Å).[692]

Poorly characterized hexamethylbenzene adducts $[Nb_2Cl_4(C_6Me_6)_2]$ and $[TaBr_2(C_6Me_6)]_n$ formed in 9% and 51% yields, respectively, on application of the reducing Friedel–Crafts procedure to the pentahalides.[654] Short-lived M^{II} species have also been generated by electrochemical reduction of $[Nb_2Cl_6(PhPMe_2)_4]$[604] or electrochemical oxidation of seven-coordinated M^I complexes of types $[MX(CO)_2(dmpe)_2]$ (M = Ta, X = Cl, Br, I, Me, H; M = Nb, X = Cl) or $[TaCl(\eta^4\text{-}C_4H_8)(dmpe)_2]$.[693]

34.6.1.2 Reactivity of the dihalo adducts

The paramagnetic 15-electron M^{II} species react readily with H_2 or D_2 (equations 77 and 78) to give paramagnetic, thermally stable M^{IV} 17-electron hydrides, except for $[NbCl_2(PMe_3)_4]$.[588,692] These inorganic radicals are the first isolated M^{IV} monomeric hydrides, as $[TaH_4(dmpe)_2]$ and $[NbH_2Cp_2]$ were characterized only by low temperature ESR.[575] Ether solutions of $[NbCl_2(PMe_3)_4]$ react with H_2, but $[Nb_2Cl_4(PMe_3)_4(\mu\text{-}H)_4]$ was the only product isolated, even in the presence of excess ligand, and may arise from the rapid decomposition of transient $[NbH_2Cl_2(PMe_3)_4]$. Preliminary results show the potential of (**51**) as a starting material for preparing tantalum compounds in low oxidation states (Scheme 9), especially $[TaHCl_2(PMe_3)_4]$.

$$\underset{(\mathbf{50})}{[MCl_2(dmpe)_2]} + X_2 \xrightarrow[2.7\ atm,\ 25\ ^\circ C,\ 2\ h]{hexane} \underset{90\%}{[MX_2Cl_2(dmpe)_2]} \qquad (77)$$

$$\underset{(\mathbf{49})}{[TaCl_2(PMe_3)_4]} + X_2 \xrightarrow[2.7\ atm,\ 25\ ^\circ C,\ 4\ h]{OEt_2} \underset{(\mathbf{51})\ 75\%}{[TaX_2Cl_2(PMe_3)_4]} \qquad (78)$$

$$X_2 = H_2, D_2; \quad X = H$$

Strong terminal M—H stretching modes are found in these adducts. The magnetic moments are indicative of d^1 complexes with orbitally non-degenerate ground states. ESR spectra are observed at room temperature ($[NbH_2Cl_2(dmpe)_2]$: $\langle g \rangle = 1.96$, $\langle A \rangle_{Nb} = 109$ G, $\langle a \rangle_P = 25.5$ G, $\langle a \rangle_H = 11.1$ G).

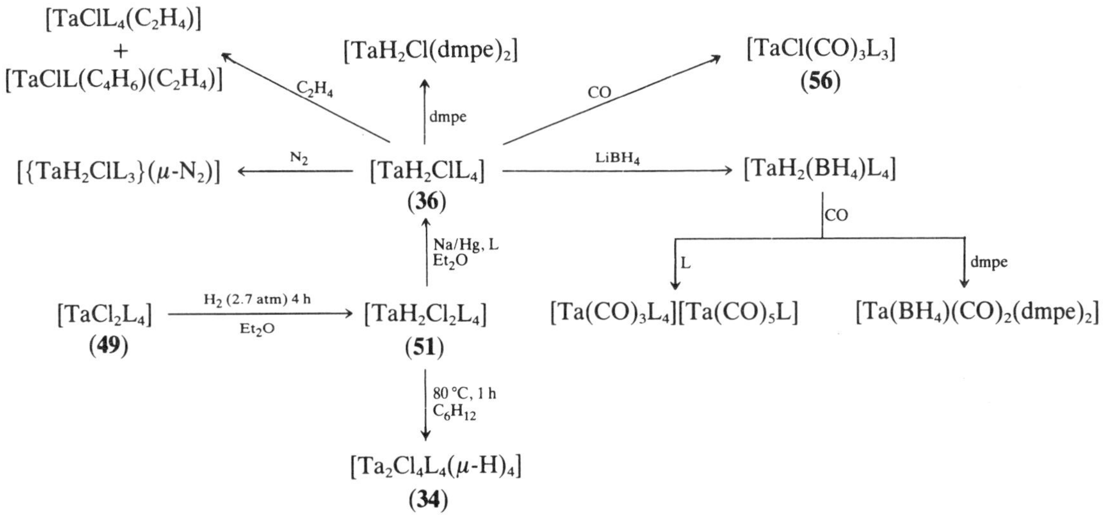

Scheme 9 (L = PMe_3)

The tantalum(IV) hydrides $[TaH_2Cl_2(PMe_3)_4]$ (**51**) and $[TaH_2Cl_2(dmpe)_2]$ (**50**) were characterized by low temperature X-ray crystallography.[588,692] Complex (**51**) adopts a distorted dodecahedral geometry in the solid state, while (**50**) is better described as a distorted square antiprismatic complex. The hydrogen atoms have been located.

Reductive carbonylation of $[NbCl_2(PMe_3)_4]$ or reduction of $[TaCl_2(dmpe)_2]$ (**50**) by vitride offered $[NbCl(CO)_3(PMe_3)_3]$ and a Ta^I aluminohydride adduct, respectively (Section 34.7).

No divalent product derived from solvolysis of the dihalide adducts is yet known. The reactions of (**50**) with Cp^- and PPh_2^- show that replacement of chloride by ligands having stronger fields leads to species resembling organic radicals in reactivity, which rapidly disproportionate or abstract hydrogen atoms from the solvent. The Ta^{III} derivatives $[Ta(dmpe)(Cp)_2]Cl$, $[Ta(dmpe)_2Cl(Cp)]Cl$ and $[TaH(PPh_2)_2(dmpe)_2]$ (Section 34.4.4) were obtained. No evidence for the $[Ta(PPh_2)_2(dmpe)_2]$ intermediate was found by ESR.

34.6.2 Miscellaneous

Air stable materials of formula MPc (Pc = phthalocyanine) have been obtained from MCl_5, phthalonitrile and quinoline at 220 °C.[694] Nb^{II} alanates $[LiNb_2(AlH_4)_5]$ and $[LiNb(AlH_4)_3]$ were obtained from the low temperature reaction of $NbCl_5$ with excess $LiAlH_4$.[642]

Reduction of (**44**) yielded a singly bonded Nb—Nb dimer (equation 79).[568]

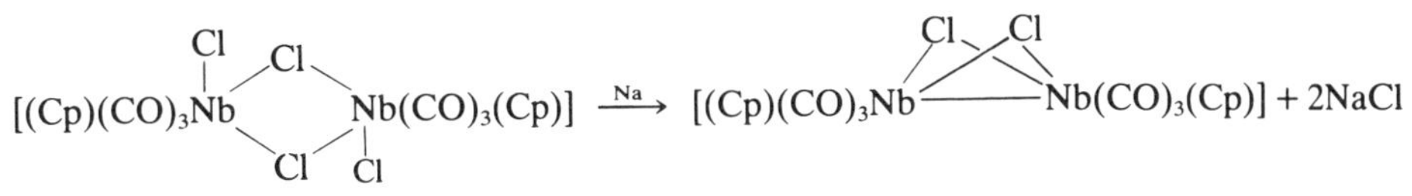

(79)

(**44**)

34.7 OXIDATION STATE +I

The number of authenticated niobium and tantalum coordination compounds in oxidation state +I is very limited. They were generally obtained through reduction of $[MCl_4(dmpe)_2]$, $[TaCl_2(dmpe)_2]$ or $Nb(OC_6H_3Me_2\text{-}3,5)_2(dmpe)_2$.[731a]

34.7.1 Phosphine Adducts

Treatment of $[TaCl_2(dmpe)_2]$ with two equivalents of vitride offered $[TaCl(CO)_2(dmpe)_2]$ (**53**) under CO, while a coordinatively unsaturated complex, $[Ta\{H_2Al(OC_2H_4OMe)_2\}(dmpe)_2]$

(54), was obtained under an argon or ethylene atmosphere. It is the only structurally characterized non-carbonyl M^I compound.[695]

Its molecular structure (Figure 37) consists of a centrosymmetric dimer with a bridging $H_2Al(OR)(\mu\text{-}OR)_2Al(OR)H_2$ entity. The Ta atoms are approximately square pyramidal, with the four phosphorus atoms forming the basal plane (Ta lies 0.64 Å out of it). The relatively short Ta—Al distances are comparable to those found in other transition metal aluminum complexes (Ta—Al: 2.79–3.13 Å). The hydrogen atoms have not been located, but were evidenced by chemical and spectroscopic techniques (IR: 1605, 1540 cm^{-1}; δ ^1H NMR: 16.30 p.p.m.). The Ta—(μ-H_2)Al unit is relatively stable, and **(54)** is inert to carbon monoxide or trimethylamine. It is a poor catalyst in the isomerization of 1-pentene. Formation of complexes analogous to **(54)** may explain the low yields often obtained from alkoxoaluminohydrides and metal halides.

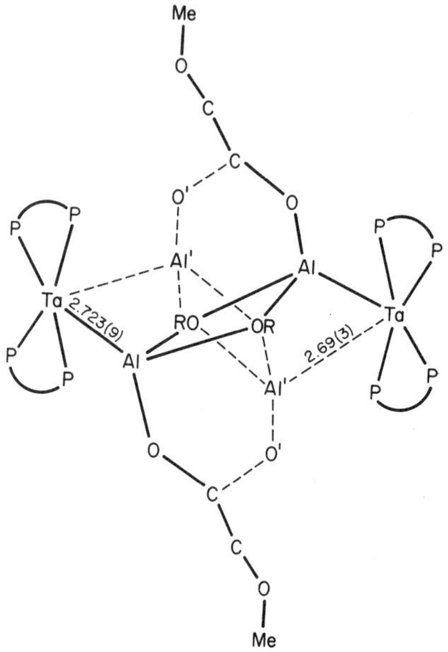

Figure 37 Molecular structure of $[Ta\{H_2Al(OC_2H_4OMe)_2\}(dmpe)_2]$ **(54)** (the dotted lines indicate the less abundant (20%) structural form resulting from a different orientation of the Al_2O_2 bridge) (reproduced from ref. 695 with permission)

$[\{NbCl(dmpe)_2\}(\mu\text{-}N_2)]$ was formed through reduction of $[NbCl_4(dmpe)_2]$ (equation 28). Electrochemical reduction of $[Nb_2Cl_6(PhPMe_2)_4]$ yielded a dimeric Nb^I derivative. Compounds of empirical formula $[Ta(py)](ClO_4)$ and $[Ta(Et_2NH)](ClO_4)$, generated by anodic dissolution, were also reported.[696]

34.7.2 Carbonyl Derivatives

34.7.2.1 Non-cyclopentadienyl carbonyls

The monomeric carbonyl complexes of d^4 M^I isolated so far are mainly the seven-coordinated $[MX(CO)_2(dmpe)_2]$ compounds (M = Nb, X = Cl, Br, N_3;[697] M = Ta, X = H, Cl, Br, CN, Me, Et, Pr^i); **(53)** has been obtained by reduction of $[TaCl_2(dmpe)_2]$ with one equivalent of sodium naphthalenide under CO; the other tantalum derivatives were generally prepared by oxidative additions to $[Ta(CO)_2(dmpe)_2]^-$ formed *in situ* by reduction of (**53**; Scheme 10). The niobium analogs were formed by reductive carbonylation of $[NbCl_4(dmpe)_2]$.[698]

The hydride **(55)**, also obtained through carbonylation of $[TaH_5(dmpe)_2]$, has a distorted capped octahedral structure (A) established by X-ray diffraction (Figure 38).[699] Spectroscopic data show that bulkier ligands ($X \neq H$) all favor a capped trigonal prismatic geometry (B). Stable seven-coordinated monomeric carbonyls $[MCl(CO)_3(PMe_3)_3]$ **(56)** could also be synthesized.[700] The smooth reaction of **(36)** with CO offered $[TaCl(CO)_3(PMe_3)_3]$ (equation 80). The niobium analogue of **(36)** is still unknown and $[NbCl(CO)_3(PMe_3)_3]$ was obtained by

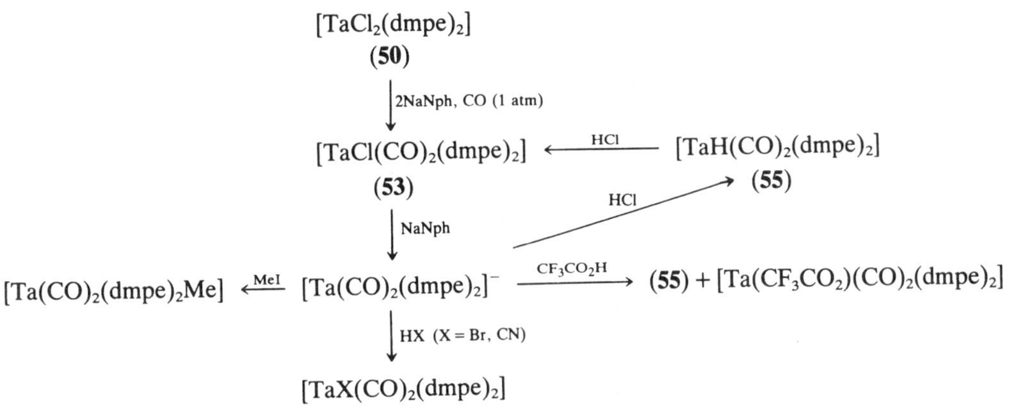

Scheme 10 (all reactions in THF, Nph = $C_{10}H_8$)

reductive carbonylation of $[NbCl_2(PMe_3)_4]$ (equation 81). Attempts to prepare (**56**) directly from $NbCl_5/PMe_3$ mixtures by reductive carbonylation have been unsuccessful. These new carbonyls (sharp CO stretching modes at 1962, 1862 and 1845 cm^{-1} for M = Ta) are diamagnetic, 18-electron monomers. They are stereochemically rigid in solution, as are also the $[MX(CO)_2(dmpe)_2]$ adducts. Complex (**56**) (M = Ta) has an overall C_1 symmetry and is best described as a capped trigonal prism with the chlorine atom in the apical position, three phosphines and one carbonyl ligand constituting the capped quadrilateral face.

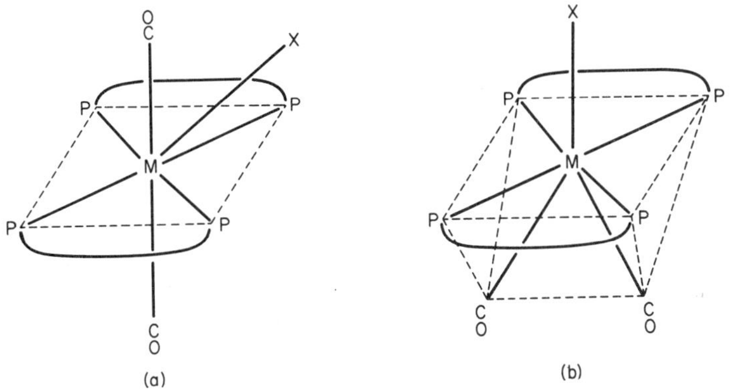

Figure 38 Coordination polyhedron found for the heptacoordinated species of type $[MX(CO)_2(dmpe)_2]$: X = H (structure a); X ≠ H (structure b)

$$\underset{(\mathbf{36})}{[TaH_2Cl(PMe_3)_4]} + 3CO \xrightarrow[2.7\ atm,\ 25\ ^\circ C]{Et_2O} \underset{(\mathbf{56})\ 80\%}{[TaCl(CO)_3(PMe_3)_3]} + PMe_3 + H_2 \qquad (80)$$

$$[NbCl_2(PMe_3)_4] + 3CO + Na/Hg \xrightarrow[2.7\ atm,\ 25\ ^\circ C]{THF} \underset{(\mathbf{56})\ 40\%}{[NbCl(CO)_3(PMe_3)_3]} + PMe_3 + NaCl \qquad (81)$$

The reaction of $[TaH_2(BH_4)(PMe_3)_4]$ in the presence of CO and PMe_3 led quantitatively to a salt $[Ta(CO)_3(PMe_3)_4][Ta(CO)_5(PMe_3)]$. The cation has a capped octahedral geometry, one PMe_3 capping a trigonal $Ta(CO)_3$ face.[698]

Dinuclear carbonyl anions $[M_2(CO)_8X_3]^-$ were obtained through a two-electron transfer to protons by treatment of $[M(CO)_6]^-$ with HX (X = Cl, OAc, OMe) (equation 82).[526] The anion has C_{2v} symmetry. The niobium atoms are heptacoordinated with four terminal carbonyl groups each and three bridging chlorides located at the vertices of an approximately equatorial triangle perpendicular to the metal–metal axis (non-bonded Nb - - - Nb distance: 3.63 Å). The carbonyl stretching vibrations are in agreement with the local C_{4v} symmetry of the $M(CO)_4$ moiety.

$$2[M(CO)_6]^- + 4HCl \xrightarrow{T_0} [M_2Cl_3(CO)_8]^- + 2H_2 + 2Cl^- + 4CO \qquad (82)$$

The $[M_2(CO)_8Cl_3]^-$ anions are the first halocarbonyl complexes (without other ligands in the coordination sphere) of group VA metals to be reported. They probably originate from the unstable $HM(CO)_6$ carbonyls.

34.7.2.2 Cyclopentadienyl carbonyls

The recent development of the synthesis of $[Nb(CO)_4(Cp)]$ by reductive carbonylation of $[NbCl_4(Cp)]$ (**57**) (i) by Zn in THF under CO (200 atm)[630] with 27% yield; or (ii) by Na sand, in the presence of a Cu/Al alloy as a halogen acceptor, in benzene under CO (135 °C, 330 atm, 90% yield)[701] has spawned a renaissance in its chemistry. Extension of the latter procedure to $[TaCl_4(Cp)]$ gave $[Ta(CO)_4(Cp)]$ in 7% yield, the main products being chlorocarbonyl derivatives, isolated as η^4-oxabutadiene Ta^{IV} compounds (equation 83).

$$[TaCl_4(Cp)] \xrightarrow[CO]{Na} [TaCl_2(CO)_y(Cp)] \xrightarrow{Me_2C{=}CHCOMe} (Cp)Cl_2Ta\text{—} \quad (83)$$

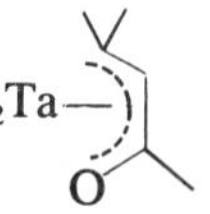

Photolysis of (**57**) in the presence of phosphines yielded $[Nb(CO)_2(LL)(Cp)]$ with LL = $2PEt_3$, $Ph_2P(CH_2)_nPPh_2$ ($n = 1$–5), $Ph_2PCH{=}CHPPh_2$, $Ph_2AsCH_2CH_2PPh_2$ (arphos), and $C_6H_4(AsPh_2)_2PPh_2$.[702] The air sensitive tetrahaptocycloheptatriene and cyclooctatetraene complexes $[(Cp)Nb(CO)_2(\eta^4\text{-}C_7H_8)]$ and $[(Cp)Nb(CO)_2(\eta^4\text{-}C_8H_8)]$ have also been obtained. The C_8H_8 ring is fluxional.

Photolysis of (**57**) in the absence of additional ligands provided an unusual cluster, $[Nb_3(CO)_7(Cp)_3]$, the sole carbonyl cluster of niobium reported to date (equation 84).[649] Its quantitative declusterification to the parent carbonyl is also uncommon. X-Ray diffraction techniques revealed that one of the CO ligands acts as an $\eta^2(\mu_3\text{-C},\mu_2\text{-O})$ bridge symmetrically spanning a nearly equilateral triangle, and was proposed to act as a six-electron donor towards the three metal atoms. Its length (1.303(14) Å) as compared to that of the terminal carbonyls (1.08–1.17 Å) is also reflected by a long-wave shift of its ν(CO) absorption in the IR ($1330\ cm^{-1}$). This $\eta^2\text{-}(\mu_3\text{-C},\mu_2\text{-O})$ ligand is mobile; a migratory process as well as its conversion from a six-electron to a four- or two-electron donor bridging Nb═Nb double bonds were observed.[703]

$$\underset{(\mathbf{57})}{[Nb(CO)_4(Cp)]} \underset{CO\ (20\ atm),\ THF}{\overset{h\nu,\ hexane\ (-5CO)}{\rightleftharpoons}} \underset{(\mathbf{47})}{[Nb_3(CO)_7(Cp)_3]} \quad (84)$$

Photolysis of (**57**) in the presence of H_2S or MeSH produced $[\{Nb(CO)_2(Cp)\}_2(\mu\text{-}S)_2]$ (**58**), $[\{Nb(CO)_2(Cp)\}_2(\mu\text{-}S)_3]$ and $[\{Nb(CO)_2(Cp)\}_2(\mu\text{-}SMe_2)_2]$ (**59**). Complexes (**58**) and (**59**) are isostructural with comparable Nb—Nb bond lengths (3.143(1) and 3.164(9) Å) and NbSNb angles (≈75°), but the introduction of a third sulfur atom leads to the opening of the bridge angle (up to 85°) and loss of the metal–metal bonding (3.555(1) Å).[704] Complex (**57**) afforded $[Nb(CO)_3I_2(Cp)]$ by oxidative addition of iodine. It may therefore provide a useful entry to Nb^{III} chemistry.[630] The reaction between (**57**) and $[\{Cr(SBu^t)(Cp)\}_2S]$ provided the tetranuclear heterometallic cluster $[NbCr_3(\mu\text{-}S)_4(Cp)_4]$.[705].

34.7.3 Alkene and Diene Adducts

Ta^I adducts with ethylene have been obtained as highly air sensitive solids by reduction of the corresponding Ta^{III} compounds under argon (equation 85),[292] or by reductive elimination of H_2 from $[TaH_2ClL_4]$ (Scheme 9). A similar procedure, but under dinitrogen, gave Ta^V nitrenes (Section 34.2.3.6). The same Ta^{III} precursor (**60**) provided σ alkyl derivatives (equation 86). Complex (**63**) catalyzes the selective dimerization of ethylene to 1-butene.

$$trans,mer\text{-}[TaCl_3L_2(C_2H_4)] + 2Na/Hg \xrightarrow[THF]{L} [TaClL_4(C_2H_4)] + 2NaCl \quad (85)$$

L = PMe_3, 0.5 dmpe

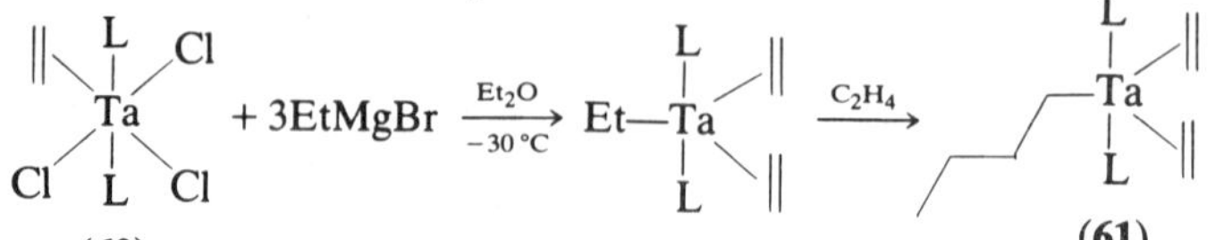

(86)

(**60**) (**61**)

L = PMe_3

Butadiene d^4 complexes were obtained (i) from $[NbCl_4(dmpe)_2]$ and magnesium butadiene (equation 87);[706] (ii) by dimerization of ethylene using alkylidenes (equation 88);[707] or (iii) by metal vapor techniques (equation 89), which yielded sublimable methylallyl derivatives.[708] Compound (62) could not be prepared by Na/Hg reduction of (22) in the presence of butadiene. Compound (63) is also accessible from $[TaH_2ClL_4]$ (Scheme 9).

$$[NbCl_4(dmpe)_2] + [Mg(THF)_2(C_4H_6)] \xrightarrow{\text{THF}} [NbCl(dmpe)_2(C_4H_6)] + MgCl_2 \qquad (87)$$

$$(22) \qquad\qquad\qquad\qquad\qquad (62)\ \ 24\%$$

(88)

L = PMe_3

(63)

$$M\ \text{atoms} + \quad \xrightarrow[\text{77 K}]{\text{THF}} \quad (89)$$

Treatment of $[TaCl_4(dmpe)_2]$ with sodium naphthalenide afforded $[TaCl(dmpe)_2(\eta^4\text{-}C_{10}H_8)]$ (64), which was reduced by further addition of sodium naphthalenide to unisolated $Na[Ta(dmpe)_2(\eta^4\text{-}C_{10}H_8)]$. Protonation or methylation of this anion afforded the Ta^I adduct $[TaX(dmpe)_2(\eta^4\text{-}C_{10}H_8)]$ (X = H, Me). Related procedures gave the analogous 1,3-cyclohexadiene adduct $[TaCl(dmpe)_2(\eta^4\text{-}C_6H_8)]$ in poor yields.[709]

The molecular structure of (64) may be considered as a pentagonal bipyramid with the naphthalene, the chlorine and two phosphorus atoms in the pentagonal plane. It is structurally related to (53) by a 45° rotation of the naphthalene unit about the Ta—Cl vector. The temperature dependent ^1H and ^{31}P NMR spectra indicate that such a process is operative in solution. The structural parameters suggest that the π-accepting interaction is substantial; this is consistent with the inertness of the Ta–naphthalene unit toward substitution.

An allyltetrakistrifluorophosphine Ta^I compound, unstable at room temperature, has also been obtained (equation 90). Attempts to substitute all the allyl groups under higher pressure led to metallic tantalum and propane.[710]

$$2[Ta(\eta^3\text{-allyl})_4] + 10PF_3 \xrightarrow{200\ \text{atm}} 2[Ta(PF_3)_5(\eta^3\text{-allyl})] + 3C_6H_{10} \qquad (90)$$

34.8 OXIDATION STATE 0

Examples of derivatives of zerovalent niobium and tantalum are very rare. Cocondensation of vapors of the metals, generated by an electron gun furnace at 3000 °C, operating at a positive potential with dmpe, gave excellent yields ($>60\%$) of stable crystalline $[M(dmpe)_3]$, which displays a distorted octahedral geometry.[711] Attempts to obtain the corresponding niobium compound by reduction of $NbCl_5$ with sodium naphthalenide at room temperature in the presence of an excess of dmpe failed and led to metal deposition.[712] By contrast, reduction of MX_5 by sodium amalgam in the presence of bipy or o-phen (LL) afforded the paramagnetic $M(LL)_3$ adducts in *ca.* 80% yield.[713]

Bis(arene) $[M(\eta^6\text{-}C_6H_3R_3)_2]$ niobium (R = H$_3$, H$_2$Me or 1,3,5-Me$_3$), and less stable tantalum derivatives (R = H$_3$) were prepared in a similar way.[714] They are extremely sensitive to oxygen, but relatively stable to water. The photoelectron spectra of the volatile bis(η^6-arene) niobium display low values for the first ionization potentials (5.18–5.57 eV), indicating that the compounds are electron rich. The IR spectra of $[Nb(C_6H_3R_3)_2]$ (R = H, Me) in an argon matrix at 80 K are in agreement with a symmetrical sandwich structure.

The ESR spectra of all zerovalent derivatives are consistent with the paramagnetism expected for 17-electron d^5 compounds with one unpaired electron ($[Nb(\eta^6\text{-}C_6H_5Me)_2]$: $\langle g \rangle = 1.992$, $\langle A \rangle = 45.9$ G). Cocondensation of Nb atoms with N$_2$ in an Ar matrix gave

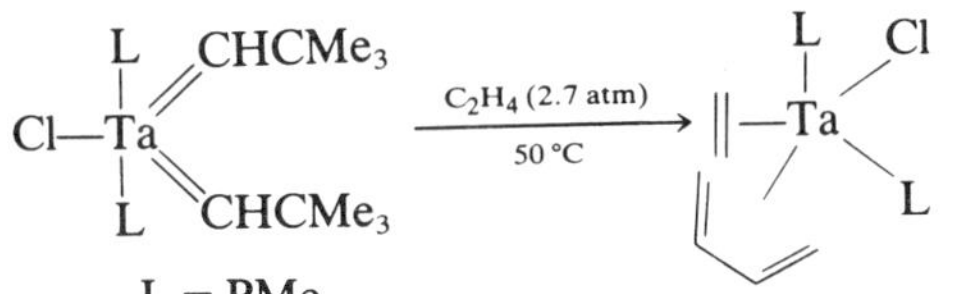

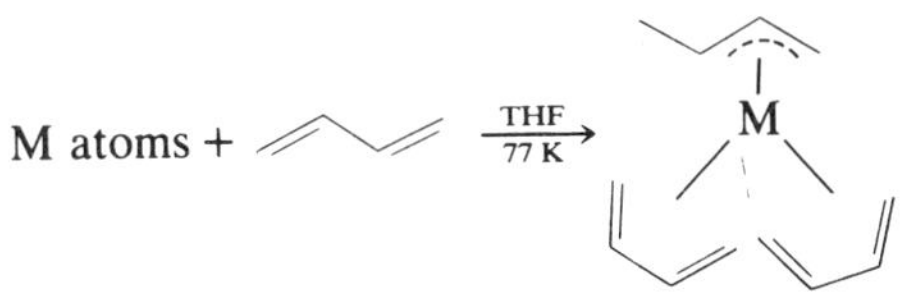

unusually numerous absorptions, some of which were assigned to $Nb(N_2)$ and to an $Nb(N_2)_4$ complex of D_{4h} symmetry. $[Nb(CO)_x(N_2)_y]$ species were similarly produced.[715]

34.9 OXIDATION STATES LOWER THAN 0

Compounds containing niobium or tantalum in negative formal oxidation states −I and −III are mainly metal carbonyl anions. Although these are organometallic derivatives, the report of efficient procedures for the synthesis of $[M(CO)_6]^-$ since the review of Labinger[8] merits mention, as it can be anticipated that these highly reduced and reactive species will be important precursors of a large variety of new coordination compounds and metal clusters.

34.9.1 Oxidation State −I

34.9.1.1 $[M(CO)_6]^-$

Until recently, the synthesis of the yellow $[M(CO)_6]^-$ required the reduction of Nb_2Cl_{10} by Na or Na–K alloy in diglyme as solvent under a high pressure of CO (>360 atm), and only poor and non-reproducible yields (<14%) of the anions were obtained.[716] Several mild and efficient carbonylation procedures have now become available. The reductive carbonylation of MCl_5 using the Mg/Zn couple as reductant and pyridine as reactive solvent occurs at room temperature under one atmosphere of CO (yields ≈ 48% and 35% for M = Nb and Ta respectively). The reduction is largely due to magnesium.[717]

Treatment of MX_5 with alkali metal naphthalenide in DME at −60 °C provides thermally unstable brown intermediates, assumed to be $[M(C_{10}H_8)_2]^-$ or $[M(1\text{-}MeC_{10}H_7)_2]$, which react with CO (1 atm) to give $[M(CO)_6]^-$ in 30–54% yields.[718] Carbonylation, using sodium in the presence of cyclooctatetraene in THF under CO (1 atm, 40 °C), has also been reported, although its efficiency is low in the case of Ta (12%).[719] Cyclooctatetraene complexes are thought to form as intermediates, but the mild carbonylation conditions compared to those of $[Nb(COT)_3]^-$ (50 atm, 100 °C) suggest that a compound of different stoichiometry may be involved.[720]

The $[M(CO)_6]^-$ anions were isolated with solvated sodium or organic cations as counterions. $[Ni(phen)_3][Nb(CO)_6]_2$ has also been obtained.[721] The unsolvated $Na[Ta(CO)_6]$ is pyrophoric, and the Nb analog is light- and air-sensitive.

The crystal structures of both $(PPN)[M(CO)_6]$ derivatives have been determined.[717] The coordination polyhedron is octahedral (Nb—C: 2.098(5) Å; CNbC: 89.2(2)°). The PPN moiety is constrained to be centrosymmetric, and thus linear. These compounds correspond to the lowest oxidation state of niobium and tantalum for which structural data are available. A single $\nu(CO)$ is found in the IR (1854 and 1852 cm^{-1} for Nb and Ta respectively). Comparable spectra are observed for $Na[M(CO)_6]$ in pyridine, but in solvents of lower dielectric constants such as tetrahydrofuran, additional bands attributed to distortion of the anion by the countercation are observed. $[Nb(CO)_6]^-$ appears to be the most labile carbonyl of the group VA analogs.

34.9.1.2 Other derivatives

$[Nb(PF_3)_6]^-$ has been prepared by photolysis of $[Nb(CO)_6]^-$ under PF_3 and studied by multinuclear NMR. The ^{93}Nb chemical shifts suggest weaker Nb–PF_3 than Nb–CO interactions.[722]

A diamagnetic compound whose analysis corresponds to $Li[Nb(bipy)_3]\cdot 3.5THF$ was reported to form by reduction of $NbCl_5$ with Li_2bipy. As bipy can behave as $bipy^{\cdot -}$, and in the absence of any spectroscopic data, the metal's oxidation state may still be questioned.[723,724] $Na[Ta(dmpe)_2(CO)_2]$ was characterized by its derivatives only (Scheme 10).

34.9.2 Oxidation State −III

34.9.2.1 $[M(CO)_5]^{3-}$ and $[HM(CO)_5]^{2-}$

The pentacarbonylmetallate trianions of Nb and Ta are the first compounds to contain these elements in a formal oxidation state of −III; they appear to be the first transition metal rianions isolated as analytically pure substances (equation 91).[725] The unsolvated $Na_3[M(CO)_5]$ are thermally unstable, but isolation was achieved as $Cs_3[M(CO)_5]$ in 40–50% yields. These salts are shock sensitive, and their reactivity was mainly explored in solution. Addition of Ph_3SnCl and NH_4Cl provided $[M(CO)_5(Ph_3Sn)]^{2-}$ and $[M(CO)_5(NH_3)]^-$, which were isolated as their Et_4N and Ph_4As salts, respectively. The amine complexes are very labile and react readily with a variety of π acceptor ligands L, such as PR_3, $P(OR)_3$ and RNC, to provide $[M(CO)_5L]^-$ in 50–80% yields. Addition of Et_4NBH_4 to $M(CO)_5^{3-}$ provided $(Et_4N)_2[HM(CO)_5]$.[525b]

$$Na[M(CO)_6] + 3Na \xrightarrow[-78\,°C]{NH_3\,liq} Na_3[M(CO)_5] + 0.5Na_2C_2O_2 \qquad (91)$$

34.9.2.2 $[M(CO)_3(Cp)]^{2-}$

Treatment of $[M(CO)_4(Cp)]$ with sodium in liquid ammonia at $-78\,°C$ generates the corresponding pyrophoric $Na_2[M(CO)_3(Cp)]$ salts.[726] Exchange reactions offered the thermally very stable $Cs_2[M(CO)_3(Cp)]$ (dec. $>250\,°C$). These dianions are the only established examples of transition metal complexes of formula $[M(CO)_x(Cp)]^{2-}$. Reactions of $[M(CO)_3(Cp)]^{2-}$ with weak Brönsted acids such as MeCN, Et_4N^+ gave $[MH(CO)_3(Cp)]^-$, for which ^{93}Nb NMR data are available.[727] The $[M(CO)_3(Cp)]^{2-}$ dianions were also characterized by their triphenylstannyl derivatives $(Et_4N)[M(CO)_3(SnPh_3)(Cp)]$.

34.10 MIXED VALENCE COMPOUNDS

The only well-established, structurally characterized mixed valence compounds are the niobium selenide derivatives $Nb_3Se_5Cl_7^{V/IV}$ and $[(MeCN)_2Br_2Nb(\mu\text{-}Se)_2(\mu\text{-}Se)NbBr_2(MeCN)_2]^{IV/III}$ (Section 34.3.4).

The niobyl phosphoniobate $[Nb^VONb^{III}(PO_4)_2]\cdot6H_2O$ has been described in aqueous solution (Section 34.4.6.1).

The reaction between $NbCl_5$ and $LiAlH_4$ in ether offered highly reactive, generally pyrophoric, lower valent niobium alanates $[Nb(AlH_4)_n]$ ($n < 5$).[642] The mean oxidation state n decreases with increasing temperatures (equation 92). In the presence of an excess of $LiAlH_4$, ionic derivatives were isolated: $[LiNb_2(AlH_4)_7]$ at $-70\,°C$, $[LiNb_2(AlH_4)_5]$ at $+25\,°C$ and $[LiNb(AlH_4)_3]$ at $+25\,°C$. No structural data are however available on these compounds, of formal oxidation states 3.5 or 2.5, and their re-examination might prove rewarding. Unstable mixed valence M^{II}–M^{III} compounds have also been formed by pulsed radiolysis of dinuclear M^{III} adducts.[728]

$$NbCl_5 + 5LiAlH_4 \xrightarrow{Et_2O} [Nb(AlH_4)_n] + (5-n)AlH_3 + \left(\frac{5-n}{2}\right)H_2 + 5LiCl \qquad (92)$$

Dicyclopentadienyl derivatives in a formal oxidation state of 3.5 appear to be relatively common. A purple, paramagnetic compound with one unpaired electron (1.32 BM) for two Ta atoms was isolated and formulated as a dimer (**65**; equation 93).[729] A similar reaction did not yield the niobium analog, but violet or red-brown compounds were observed as intermediates in the reduction of $[MX_2(Cp)_2]$ to $[MX(Cp)_2]$ (X = Cl, Br). They could be isolated in good yields ($\sim60\%$, Scheme 11).[730]

$$TaCl_5 + NaCp + LiPPh_2 \xrightarrow[THF]{} [Ta_2Cl_3(Cp)_4] + P_2Ph_4 + \cdots \qquad (93)$$
$$\text{(65)} \quad 20\%$$

Structural studies are as yet limited by the lack of stability of these M^{IV}–M^{III} compounds, especially in polar solvents. Their formation was considered a consequence of the acidic character of the $MX(Cp)_2$ entity, which achieves 18 electrons, and an alternative structure with

$$[MX_2(Cp)_2] \xrightarrow{Na/Hg} [MX(Cp)_2]$$

$$\begin{array}{ccc} & 0.5\,Na/Hg & THF \\ & \text{or} & \\ & 0.5\,NaNph & \searrow \quad \swarrow \, [MX_2(Cp)_2] \end{array}$$

$$[M_2X_3(Cp)_4] \qquad \begin{array}{l} M = Nb;\ X = Cl,\ Br \\ M = Ta;\ X = Cl \end{array}$$
$$\text{(65)}$$

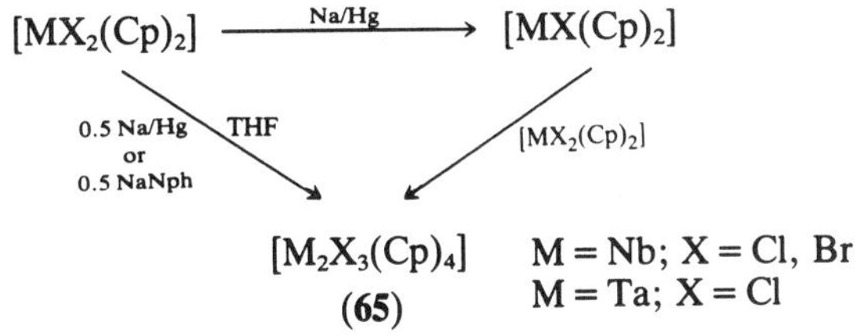

Scheme 11

only one bridging chlorine has been proposed. Reaction with $AgClO_4$ led to cationic complexes.

Note: Since the chapter was completed, relevant papers concerning the chemistry of niobium and tantalum in various oxidation states with bulky anionic oxygen donors [2,6-dialkylaryloxo[731] or silox (Bu^t_3SiO)[732]] have appeared. Reviews concerning the analysis and classification of X-ray data of niobium[733] and tantalum[734] are also now available.

34.11 REFERENCES

1. F. Fairbrother, 'The Chemistry of Niobium and Tantalum', Elsevier, New York, 1967.
2. D. L. Kepert, 'The Early Transition Metals', Academic, London, 1972.
3. R. A. Walton, in 'Progress in Inorganic Chemistry', ed. S. J. Lippard, Wiley, New York, 1972, vol. 16.
4. 'Gmelin Handbuch der Anorganische Chemie', Vanadium — Niobium — Tantalum, Springer Verlag, Berlin, 1973.
5. D. Brown, in 'Comprehensive Inorganic Chemistry', ed. J. C. Bailar, Jr., H. J. Emeléus, R. Nyholm and A. F. Trotman-Dickenson, Pergamon, Oxford, 1973, vol. 3.
6. R. C. Mehrotra, A. K. Rai, P. N. Kapoor and R. Bohra, *Inorg. Chim. Acta,* 1976, **16,** 237.
7. D. A. Miller and R. D. Bereman, *Coord. Chem. Rev.*, 1972–73, **9,** 107.
8. J. A. Labinger, in 'Comprehensive Organometallic Chemistry', ed. G. Wilkinson, F. G. A. Stone and E. W. Abel, Pergamon, Oxford, 1982, vol. 3.
9. T. A. O'Donnell and T. E. Peel, *Inorg. Nucl. Chem. Lett.*, H. H. Hyman Memorial Volume, 1976.
10. A. J. Edwards, *J. Chem. Soc.*, 1964, 3714.
11. V. P. Seleznev, E. G. Rakov and V. V. Mikulenok, *Zh. Fiz. Khim.*, 1971, **45,** 2941.
12. L. E. Alexander, I. R. Beattie and P. J. Jones, *J. Chem. Soc., Dalton Trans.*, 1972, 210.
13. J. Fawcett, A. J. Hewitt, J. H. Holloway and M. A. Stephen, *J. Chem. Soc., Dalton Trans.*, 1976, 2422.
14. J. Fawcett, J. H. Holloway, R. D. Peacock and D. K. Russell, *J. Fluorine Chem.*, 1982, **20,** 9.
15. J. Brunvoll, A. A. Ischenko, I. N. Maikshin, G. V. Romonov, V. B. Sokolov, V. P. Spirodonov and T. G. Strand, *Acta Chem. Scand., Ser. A,* 1979, **33,** 775.
16. S. J. Cyvin, H. Hovdan and W. Brockner, *J. Inorg. Nucl. Chem.*, 1975, **37,** 1905.
17. R. Huglen, F. W. Poulsen, G. Mamanton and G. M. Begun, *Inorg. Chem.*, 1979, **18,** 2551.
18. P. A. Edwards and R. E. McCarley, *Inorg. Chem.*, 1973, **12,** 900.
19. A. Leffler and P. Penque, *Inorg. Synth.*, 1970, **12,** 187.
20. U. Muller and P. Klingelhöfer, *Z. Naturforsch., Teil B,* 1983, **38,** 559.
21. U. Muller, *Acta Crystallogr., Sect. B,* 1979, **35,** 3502.
22. B. Krebs and D. Sinram, *Z. Naturforsch., Teil B,* 1980, **35,** 12.
23. L. Kolditz, C. Kürschner and U. Calov, *Z. Anorg. Allg. Chem.*, 1964, **329,** 172.
24. L. Kolditz and U. Calov, *Z. Anorg. Allg. Chem.*, 1970, **376,** 1.
25. L. Kolditz, U. Calov and A. R. Grimmer, *Z. Chem.*, 1970, **10,** 35.
26. B. Krebs, H. Janssen, N. J. Bjerrum, R. W. Berg and G. N. Papatheodorou, *Inorg. Chem.*, 1984, **23,** 164.
27. W. Bues, F. Demiray and W. Brockner, *Spectrochim. Acta, Part A,* 1976, **32,** 1623.
28. W. Brockner and S. J. Cyvin, *Monatsh. Chem.*, 1976, **107,** 1369.
29. N. D. Chikanov, *Zh. Neorg. Khim.*, 1981, **26,** 752.
30. F. Fairbrother and W. C. Frith, *J. Chem. Soc.*, 1951, 3051.
31. M. Siskin and J. Porcelli, *J. Am. Chem. Soc.*, 1974, **96,** 3640.
32. M. Siskin, *J. Am. Chem. Soc.*, 1976, **98,** 5413.
33. M. Siskin, *J. Am. Chem. Soc.*, 1978, **100,** 1838.
34. M. Siskin, *J. Am. Chem. Soc.*, 1974, **96,** 3641.
35. J. Wristers, *J. Am. Chem. Soc.*, 1975, **97,** 4312.
36. J. Wristers, *J. Am. Chem. Soc.*, 1977, **99,** 5051.
37. J. A. K. Du Plessis, *S. Afr. J. Chem.*, 1981, **34,** 87.
38. T. Masuda, T. Takahashi and T. Higashimura, *J. Chem. Soc., Chem. Commun.*, 1982, 1297.
39. T. Masuda, Y. X. Deng and T. Higashimura, *Bull. Chem. Soc. Jpn.*, 1983, **56,** 2798.
40. T. Masuda, E. Isobe, T. Higashimura and K. Takada, *J. Am. Chem. Soc.*, 1983, **105,** 7473.
41. J. E. Land and C. V. Osborne, *J. Less-Common Met.*, 1972, **29,** 147.
42. E. W. Baumann, *J. Inorg. Nucl. Chem.*, 1972, **34,** 687.
43. G. Neumann, *Ark. Kemi,* 1970, **32,** 229.
44. J. A. S. Howell and K. C. Moss, *J. Chem. Soc.* (*A*), 1971, 2481.
45. N. A. Matwiyoff, L. B. Asprey and W. E. Wageman, *Inorg. Chem.*, 1970, **9,** 2014.
46. R. Gut, *Inorg. Nucl. Chem. Lett.*, 1976, **12,** 149.
47. I. M. Kutyrev, E. S. Stoyanov, V. V. Bagreev and Yu. A. Zolotov, *Zh. Neorg. Khim.*, 1977, **22,** 1043.
48. R. L. Davidovich, T. F. Levchishina, T. A. Kaidalova and V. I. Sergienko, *J. Less-Common Met.*, 1972, **27,** 35.
49. G. M. Brown and L. A. Walker, *Acta Crystallogr.*, 1966, **20,** 220.
50. J. L. Hoard, *J. Am. Chem. Soc.*, 1939, **61,** 1252.
51. J. L. Hoard. W. J. Martin, M. E. Smith and J. F. Whitney, *J. Am. Chem. Soc.*, 1954, **76,** 3820.
52. L. K. Marinina, E. G. Rakov, B. V. Gromov and O. V. Markina, *Zh. Fiz. Khim.*, 1971, **45,** 1592.
53. G. A. Burkhalova, L. A. Kamenskaya, V. I. Konstantinov and A. M. Matveev, *Zh. Neorg. Khim.*, 1981, **26,** 1406.
54. A. I. Agulyanskii, *Zh. Neorg. Khim.*, 1980, **25,** 2998.
55. G. A. Yagodin, E. G. Rakov and N. A. Veleshko, *Zh. Neorg. Khim.*, 1980, **25,** 560.
56. A. J. Edwards and G. R. Jones, *J. Chem. Soc.* (*A*), 1970, 1891.
57. A. J. Edwards and G. R. Jones, *J. Chem. Soc.* (*A*), 1970, 1491.

58. A. J. Edwards, *J. Chem. Soc., Dalton Trans.*, 1972, 2325.
59. S. Brownstein, *Can. J. Chem.*, 1973, **51,** 2530.
60. S. Brownstein, *Inorg. Chem.*, 1973, **12,** 584.
61. J. E. Griffiths, W. A. Sunder and W. E. Falconer, *Spectrochim. Acta, Part A*, 1975, **31,** 1207.
62. R. D. W. Kemmitt, M. Murray, W. M. McRae, R. D. Peacock, M. C. R. Symons and T. A. O'Donnell, *J. Chem. Soc. (A)*, 1968, 862.
63. A. A. Artyukhov and S. S. Khoroshev, *Koord. Khim.*, 1977, **3,** 1478.
64. E. Stumpp and G. Piltz, *Z. Anorg. Allg. Chem.*, 1974, **409,** 53.
65. H. C. Gaebell, G. Meyer and R. Hoppe, *Z. Anorg. Allg. Chem.*, 1982, **493,** 65.
66. D. R. Sadoway and S. N. Flengas, *Can. J. Chem.*, 1978, **56,** 2013.
67. D. Brown, G. W. A. Fowles and R. A. Walton, *Inorg. Synth.*, 1970, **12,** 225.
68. Yu. A. Buslaev, V. D. Kopanev, S. M. Sinitsyna and V. G. Khlebodarov, *Russ. J. Inorg. Chem. (Engl. Transl.)*, 1973, **18,** 1362.
69. D. E. Davey, R. W. Catrall, T. J. Cardwell and R. J. Magee, *J. Inorg. Nucl. Chem.*, 1978, **40,** 1135.
70. D. E. Davey, R. W. Catrall, T. J. Cardwell and R. J. Magee, *J. Inorg. Nucl. Chem.*, 1978, **40,** 1141.
71. D. C. Bradley, M. B. Hursthouse, J. Newton and N. P. C. Walker, *J. Chem. Soc., Chem. Commun.*, 1984, 188.
72. J. I. Bullock, F. W. Parrett and N. J. Taylor, *J. Chem. Soc., Dalton Trans.*, 1973, 522.
73. R. Dübgen, U. Müller, F. Weller and K. Dehnicke, *Z. Anorg. Allg. Chem.*, 1980, **471,** 89.
74. U. Müller, R. Dübgen and K. Dehnicke, *Z. Anorg. Allg. Chem.*, 1981, **473,** 115.
75. H. Preiss, *Z. Anorg. Allg. Chem.*, 1971, **380,** 56.
76. G. Okon, *Z. Anorg. Allg. Chem.*, 1980, **469,** 68.
77. E. M. Smirnova and V. M. Tsintsius, *Zh. Neorg. Khim.*, 1971, **16,** 566.
78. M. Valloton and A. E. Merbach, *Helv. Chim. Acta*, 1974, **57,** 2345.
79. D. R. Sadoway and S. M. Flengas, *Can. J. Chem.*, 1978, **56,** 2538.
80. M. G. B. Drew, A. P. Wolters and J. D. Wilkins, *Acta Crystallogr., Sect. B*, 1975, **31,** 324.
81. Yu. A. Buslaev, E. G. Il'in, S. N. Bainova and M. N. Krutkina, *Dokl. Akad. Nauk SSSR*, 1971, **196,** 374.
82. Yu. A. Buslaev, E. G. Il'in and M. N. Krutkina, *Dokl. Akad. Nauk SSSR*, 1971, **200,** 1345.
83. Yu. A. Buslaev, V. D. Kopanev and V. P. Tarasov, *Chem. Commun.*, 1971, 1175.
84. Yu. A. Buslaev, E. G. Il'in and M. N. Krutkina, *Dokl. Akad. Nauk SSSR*, 1971, **201,** 99.
85. Yu. A. Buslaev and E. G. Il'in, *J. Fluorine Chem.*, 1974, **4,** 271.
86. A. Merbach and J. C. Bünzli, *Helv. Chim. Acta*, 1972, **55,** 580.
87. G. W. A. Fowles, D. A. Rice and J. D. Wilkins, *J. Chem. Soc., Dalton Trans.*, 1972, 2313.
88. G. W. A. Fowles, D. A. Rice and J. D. Wilkins, *J. Chem. Soc., Dalton Trans.*, 1974, 1080.
89. E. G. Il'in, M. E. Ignatov and Yu. A. Buslaev, *Koord. Khim.*, 1979, **5,** 949.
90. A. Merbach and J. C. Bünzli, *Helv. Chem. Acta*, 1971, **54,** 2543.
91. R. Good and A. E. Merbach, *Helv. Chim. Acta*, 1974, **57,** 1192.
92. J. C. Bünzli and A. E. Merbach, *Helv. Chim. Acta*, 1972, **55,** 2867.
93. C. M. P. Favez, H. Rollier and A. E. Merbach, *Helv. Chim. Acta*, 1976, **59,** 2383.
94. A. Merbach and J. C. Bünzli, *Helv. Chim. Acta*, 1972, **55,** 1903.
95. C. M. P. Favez and A. E. Merbach, *Helv. Chim. Acta*, 1977, **60,** 2695.
96. R. Good and A. E. Merbach, *J. Chem. Soc., Chem. Commun.*, 1974, 163.
97. R. Good and A. E. Merbach, *Inorg. Chem.*, 1975, **14,** 1030.
98. A. Kuz'min, S. I. Kuznetsov, S. M. Chizhikova, G. M. Denisova, G. N. Zviadadze and B. E. Dzevitskii, *Zh. Fiz. Khim.*, 1979, **53,** 155.
99. M. Valloton and A. E. Merbach, *Helv. Chim. Acta*, 1975, **58,** 2272.
100. P. R. Hammond and R. R. Lake, *J. Chem. Soc. (A)*, 1971, 3800.
101. L. V. Surpina, Yu. V. Kolodyazhnyi and O. A. Osipov, *Zh. Neorg. Khim.*, 1973, **18,** 2853.
102. J. C. Fuggle, D. W. A. Sharp and J. M. Winfield, *J. Fluorine Chem.*, 1972, **1,** 427.
103. K. C. Moss, *J. Chem. Soc. (A)*, 1970, 1224.
104. L. G. Hubert-Pfalzgraf and M. Tsunoda, *Inorg. Chim. Acta*, 1980, **38,** 43.
105. M. S. Gill, H. S. Ahuja and G. S. Rao., *J. Less-Common Met.*, 1970, **21,** 447.
106. M. S. Gill, H. S. Ahuja and G. S. Rao, *J. Indian Chem. Soc.*, 1978, **55,** 875.
107. L. V. Surpina, Yu. V. Kolodyazhnyi and A. O. Osipov, *Zh. Obshch. Khim.*, 1973, **43,** 1165.
108. J. D. Wilkins, *J. Organomet. Chem.*, 1974, **80,** 357.
109. L. V. Surpina, Yu. V. Kolodyazhnyi and O. A. Osipov, *Zh. Obshch. Khim.*, 1971, **41,** 1420.
110. J. R. Masaguer and M. R. Bermejo, *Quim. Nova*, 1973, **69,** 1099.
111. A. O. Baghlaf, K. Behzadi and A. Thompson, *J. Less-Common Met.*, 1978, **61,** 31.
112. E. G. Il'in, M. E. Ignatov and Yu. A. Buslaev, *Koord. Khim.*, 1977, **3,** 46.
113. E. G. Il'in, M. E. Ignatov, L. S. Butorina, T. A. Mastryukova and Yu. A. Buslaev, *Dokl. Akad. Nauk SSSR*, 1977, **236,** 609.
114. E. G. Il'in, M. E. Ignatov, A. A. Shuets and Yu. A. Buslaev, *Dokl. Akad. Nauk SSSR*, 1978, **243,** 1182.
115. E. G. Il'in, M. E. Ignatov, L. S. Butorina, T. A. Mastryukova and Yu. A. Buslaev, *Dokl. Akad. Nauk SSSR*, 1979, **248,** 626.
116. E. G. Il'in, M. E. Ignatov and Yu. A. Buslaev, *Dokl. Akad. Nauk SSSR*, 1979, **247,** 113.
117. R. J. Dorschner, *J. Inorg. Nucl. Chem.*, 1972, **34,** 2665.
118. J. R. Masaguer and J. Sordo, *Acta Cient. Compostelana*, 1971, **8,** 115.
119. J. R. Masaguer and J. Sordo, *An. Quim.*, 1973, **69,** 1263.
120. D. Brown, J. F. Easey and J. G. H. Du Preez, *J. Chem. Soc. (A)*, 1966, 258.
121. Y. Gushiken, O. L. Alues, Y. Hase and Y. Kawano, *J. Coord. Chem.*, 1977, **6,** 179.
122. A. V. Suvorov, A. M. German and N. P. Bondareva, *Zh. Neorg. Khim.*, 1971, **16,** 2413.
123. I. L. Nadelyaev, A. V. Suvorov and Yu. V. Kondrat'ev, *Zh. Obshch. Khim.*, 1976, **46,** 1647.
124. D. Budd, R. Churchman, D. G. Holah, A. N. Hughes and B. C. Hui, *Can. J. Chem.*, 1972, **50,** 1008.
125. D. Brown, J. Hill and C. E. F. Rickard, *J. Less-Common Met.*, 1970, **20,** 57.

126. S. G. Murray and F. R. Hartley, *Chem. Rev.*, 1981, **81**, 365.
127. S. R. Wade and G. R. Willey, *Inorg. Chim. Acta*, 1983, **72**, 201.
128. R. E. De Simone and T. M. Tighe, *J. Inorg. Nucl. Chem.*, 1976, **38**, 1623.
129. R. E. De Simone and M. D. Glick, *J. Am. Chem. Soc.*, 1975, **97**, 942.
130. R. E. De Simone and M. D. Glick, *J. Coord. Chem.*, 1976, **5**, 181.
131. H. Böhland and F. M. Schneider, *Z. Chem.*, 1972, **12**, 63.
132. H. Böhland and F. M. Schneider, *Z. Chem.*, 1972, **12**, 28.
133. J. D. Wilkins, *J. Organomet. Chem.*, 1974, **65**, 383.
134. C. Djordjevic and V. Katovic, *J. Chem. Soc. (A)*, 1970, 3382.
135. M. G. B. Drew and J. D. Wilkins, *J. Chem. Soc., Dalton Trans.*, 1973, 1830.
136. B. Begolli, V. Valjak, V. Allegretti and V. Katovic, *J. Inorg. Nucl. Chem.*, 1981, **43**, 2785.
137. J. A. S. Howell and K. C. Moss, *J. Chem. Soc. (A)*, 1971, 2483.
138. G. A. Ozin and R. Walton, *J. Chem. Soc. (A)*, 1970, 2236.
139. C. Chavant, G. Constant, Y. Jeannin and R. Morancho, *Acta Crystallogr., Sect. B*, 1975, **31**, 1823.
140. G. W. A. Fowles, D. A. Rice and J. D. Wilkins, *J. Chem. Soc., Dalton Trans.*, 1973, 961.
141. M. S. Gill, H. S. Ahuja and G. S. Rao, *Inorg. Chim. Acta*, 1973, **7**, 359.
142. M. S. Gill, H. S. Ahuja and G. S. Rao, *J. Inorg. Nucl. Chem.*, 1974, **36**, 3731.
143. M. J. Frazer, B. G. Gillespie, M. Goldstein and L. I. B. Haines, *J. Chem. Soc. (A)*, 1970, 703.
144. A. V. Leshchenko and O. A. Osipov, *Zh. Obshch. Khim.*, 1977, **47**, 2581.
145. L. V. Surpina, A. D. Garnovskii, Yu. A. Kolodyazhnyi and O. A. Osipov, *Zh. Obshch. Khim.*, 1971, **41**, 2279.
146. N. S. Biradar, T. R. Goudar and V. H. Kulkarni, *J. Inorg. Nucl. Chem.*, 1974, **36**, 1181.
147. N. S. Biradar and T. R. Goudar, *J. Inorg. Nucl. Chem.*, 1977, **39**, 358.
148. P. J. Ashley and E. G. Torrible, *Can. J. Chem.*, 1969, **47**, 2587.
149. U. Thewalt and G. Albrecht, *Z. Naturforsch, Teil B*, 1982, **37**, 1098.
150. K. P. Srivastava, G. P. Srivastava and S. K. Arya, *J. Electrochem. Soc. India*, 1979, **28**, 21 (*Chem. Abstr.*, 1981, **94**, 113 649).
151. R. J. H. Clark, D. L. Kepert and R. S. Nyholm, *J Chem. Soc.*, 1965, 2877.
152. J. C. Dewan, D. L. Kepert, C. L. Raston and A. H. White, *J. Chem. Soc., Dalton Trans*, 1975, 2031.
153. D. L. Kepert and K. R. Trigwell, *Aust. J. Chem.*, 1976, **29**, 433.
154. D. C. Bradley, B. N. Chakravarty and W. Wardlaw, *J. Chem. Soc.*, 1956, 2381.
155. D. C. Bradley, W. Wardlaw and A. Whitley, *J. Chem. Soc.*, 1955, 726.
156. P. N. Kapoor and R. C. Mehrotra, *Chem. Ind. (London)*, 1966, 1034.
157. I. M. Thomas, *Can. J. Chem.*, 1961, **39**, 1386.
158. D. C. Bradley and I. M. Thomas, *Chem. Ind. (London)*, 1958, 1231.
159. S. C. Goel and R. C. Mehrotra, *Z. Anorg. Allg. Chem.*, 1978, **440**, 281.
160. S. C. Goel, V. K. Singh and R. C. Mehrotra, *Z. Anorg. Allg. Chem.*, 1978, **447**, 253.
161. P. N. Kapoor, S. K. Mehrotra, R. C. Mehrotra, R. B. King and K. C. Naiman, *Inorg. Chim. Acta*, 1975, **12**, 273.
162. S. C. Goel, V. K. Singh and R. C. Mehrotra, *Synth. React. Inorg. Metal-Org. Chem.*, 1979, **9**, 459.
163. S. C. Goel and R. C. Mehrotra, *Synth. React. Inorg. Metal.-Org. Chem.*, 1981, **11**, 35.
164. S. C. Goel, S. K. Mehrotra and R. C. Mehrotra, *Synth. React. Inorg. Metal.-Org. Chem.*, 1977, **7**, 519.
165. S. C. Goel, *Synth. React. Inorg. Metal.-Org. Chem.*, 1983, **13**, 725.
166. R. C. Mehrotra and P. N. Kapoor, *J. Less-Common Met.*, 1966, **10**, 3540.
167. R. C. Mehrotra and P. N. Kapoor, *J. Less-Common Met.*, 1964, **7**, 98.
168. R. C. Mehrotra and P. N. Kapppr, *J. Indian Chem. Soc.*, 1967, **44**, 345.
169. S. Govil, P. N. Kapoor and R. C. Mehrotra, *J. Inorg. Nucl. Chem.*, 1976, **38**, 172.
170. S. Govil, P. N. Kapoor and R. C. Mehrotra, *Inorg. Chim. Acta*, 1975, **15**, 43.
171. L. G. Hubert-Pfalzgraf and J. G. Riess, *Inorg. Chem.*, 1975, **14**, 2854.
172. A. A. Pinkerton, D. Schwarzenbach, L. G. Hubert-Pfalzgraf and J. G. Riess, *Inorg. Chem.*, 1976, **15**, 1196.
173. V. I. Telynov, I. B. Rabinovich, B. I. Kozyrkin, B. A. Salamatin and K. V. Kir'yanov, *Dokl. Akad. Nauk SSSR*, 1972, **205**, 364.
174. J. G. Riess and L. G. Pfalzgraf, *Bull. Soc. Chim. Fr.*, 1968, 2401.
175. L. G. Pfalzgraf and J. G. Riess, *Bull. Soc. Chim. Fr.*, 1968, 4348.
176. J. G. Riess and L. G. Hubert-Pfalzgraf, *J. Chim. Phys. Phys. Chim.*, 1973, 646.
177. C. E. Holloway, *J. Coord. Chem.*, 1972, **1**, 253.
178. D. C. Bradley and C. E. Holloway, *J. Chem. Soc. (A)*, 1968, 219.
179. L. G. Hubert-Pfalzgraf and J. G. Riess, *J. Chem. Soc., Dalton Trans*, 1974, 585.
180. L. G. Hubert-Pfalzgraf and J. G. Riess, *Bull. Soc. Chim. Fr.*, 1973, 1201.
181. L. G. Hubert-Pfalzgraf, J. Guion and J. G. Riess, *Bull. Soc. Chim. Fr.*, 1971, 3855.
182. L. G. Hubert-Pfalzgraf, *Inorg. Chim. Acta*, 1975, **12**, 229.
183. K. C. Malhotra and R. C. Martin, *J. Organomet. Chem.*, 1982, **239**, 159.
184. M. Schönherr, D. Hass and K. Banfeld, *Z. Chem.*, 1975, **15**, 66.
185. S. K. Mehrotra, R. N. Kapoor, J. Uttamchandani and A. M. Bhandari, *Inorg. Chim. Acta*, 1977, **22**, 15.
186. R. C. Mehrotra and P. N. Kapoor, *J. Less-Common Met.*, 1966, **10**, 237.
187. R. C. Mehrotra and P. N. Kapoor, *J. Less-Common Met.*, 1965, **8**, 419.
188. A. K. Narula, B. Singh, P. N. Kapoor and R. N. Kapoor, *Transition Met. Chem.*, 1983, **8**, 195.
189. S. Prakash and R. N. Kapoor, *Inorg. Chim. Acta*, 1971, **5**, 372.
190. A. K. Narula, B. Singh, P. N. Kapoor and R. N. Kapoor, *J. Indian Chem. Soc.*, 1982, **59**, 195.
191. R. Gut, H. Buser and E. Schmid, *Helv. Chim. Acta*, 1965, **48**, 878.
192. H. D. Gillman, *J. Inorg. Nucl. Chem.*, 1975, **37**, 1909.
193. R. C. Mehrotra and P. N. Kapoor, *Indian J. Chem.*, 1967, **5**, 505.
194. R. C. Mehrotra and P. N. Kapoor, *J. Indian Chem. Soc.*, 1967, **44**, 467.
195. R. C. Mehrotra, A. K. Rai and R. Bohra, *Z. Anorg. Allg. Chem.*, 1973, **399**, 338.
196. R. Bohra, A. K. Rai and R. C. Mehrotra, *Indian J. Chem.*, 1974, **12**, 855.
197. R. Bohra, A. K. Rai and R. C. Mehrotra, *Indian J. Chem.*, 1976, **14**, 558.

198. A. Singh and R. C. Mehrotra, *Indian J. Chem.*, 1975, **13,** 1197.
199. P. Prashar and J. P. Tandon, *Z. Naturforsch., Teil B*, 1971, **36,** 11.
200. M. N. Moorkerjee, R. V. Singh and J. P. Tandon, *Gazz. Chim. Ital.*, 1981, **111,** 109.
201. J. Uttamchandani, S. K. Mehrotra, A. M. Bhandari and R. N. Kapoor, *Transition Met. Chem.*, 1976, **1,** 249.
202. M. N. Mookerjee, R. V. Singh and J. P. Tandon, *Ann. Soc. Sci. Bruxelles, Ser. 1*, 1980, **94,** 207 (*Chem. Abstr.*, 1981, **95,** 72 333).
203. M. N. Mookerjee, R. V. Singh and J. P. Tandon, *Indian J. Chem.*, 1981, **20,** 246.
204. R. C. Mehrotra, A. K. Rai and R. Bohra, *J. Inorg. Nucl. Chem.*, 1974, **36,** 1887.
205. R. Bohra, A. K. Rai and R. C. Mehrotra, *Inorg. Chim. Acta*, 1977, **25,** L147.
206. R. Dorschner, *Bull. Soc. Chim. Fr.*, 1968, 2787.
207. R. C. Mehrotra and P. N. Kapoor, *J. Less-Common Met.*, 1966, **10,** 348.
208. M. Schönherr, *Z. Chem.*, 1980, **20,** 155.
209. A. A. Jones and J. D. Wilkins, *J. Inorg. Nucl. Chem.*, 1976, **38,** 95.
210. H. Funk and K. Niederlander, *Ber. Bunsenges. Phys. Chem.*, 1928, **61,** 249.
211. K. C. Malhotra, G. Mathur, K. C. Mahajan and S. C. Chaudry, *J. Indian Chem. Soc.*, 1982, **59,** 922.
212. M. Schönherr and J. Koellner, *Z. Chem.*, 1978, **18,** 36.
213. L. Chamberlain, J. Keddington, I. P. Rothwell and J. C. Huffman, *Organometallics*, 1982, **1,** 1538; I. P. Rothwell, *Polyhedron*, 1985, **4,** 177.
214. G. W. A. Fowles, D. A. Rice and R. J. Shanton, *J. Chem. Soc., Dalton Trans.*, 1978, 1658.
215. J. H. Wengrovius and R. R. Schrock, *Organometallics*, 1982, **1,** 148.
216. D. Brown and C. E. F. Rickard, *J. Chem. Soc. (A)*, 1971, 81.
217. H. Kodama and M. Goto, *Z. Anorg. Allg. Chem.*, 1976, **421,** 71.
218. Yu. V. Kokunov, V. D. Kopanev, M. P. Gustyakova and Yu. A. Buslaev, *Koord. Khim.*, 1977, **3,** 1835.
219. Yu. V. Kokunov, V. D. Kopanev, M. P. Gustyakova and Yu. A. Buslaev, *Koord. Khim.*, 1976, **2,** 605.
220. Yu. A. Buslaev, Yu. V. Kokunov, V. D. Kopanev and M. P. Gustyakova, *J. Inorg. Nucl. Chem.*, 1974, **36,** 1569.
221. H. Koeppel, M. Schoenherr and L. Kolditz, *Z. Chem.*, 1971, **11,** 28.
222. M. G. B. Drew and J. D. Wilkins, *Inorg. Nucl. Chem. Lett.*, 1974, **10,** 549.
223. K. C. Malhotra, U. K. Banerjee and S. C. Chaudry, *Natl. Acad. Sci. Lett. (India)*, 1979, **2,** 175 (*Chem. Abstr.*, 1980, **92,** 68 793).
224. K. Behzadi and A. Thompson, *J. Less-Common Met.*, 1977, **56,** 9.
225. L. G. Hubert-Pfalzgraf, A. A. Pinkerton and J. G. Riess, *Inorg. Chem.*, 1978, **17,** 663.
226. K. Yamanouchi and S. Yamada, *Inorg. Chim. Acta*, 1976, **18,** 201.
227. K. C. Malhotra, U. K. Banerjee and S. C. Chaudry, *J. Indian Chem. Soc.*, 1979, **56,** 754.
228. N. S. Biradar and V. H. Kulkarni, *J. Less-Common Met.*, 1972, **26,** 355.
229. D. Stefanovic, J. Stambolija and V. Katovic, *Org. Mass Spectrom.*, 1973, **7,** 1357.
230. G. C. Shivahare and M. Pushpa-Sharda, *J. Indian Chem. Soc.*, 1982, **59,** 242.
231. A. K. Narula, B. Singh, P. N. Kapoor and R. N. Kapoor, *Indian J. Chem.*, 1983, **22,** 166.
232. K. C. Malhotra, U. K. Banerjee and S. C. Chaudry, *Indian J. Chem., Sect. A*, 1978, **16,** 987.
233. K. C. Malhotra and S. C. Chaudry, *Indian J. Chem., Sect. A*, 1976, **14,** 706.
234. J. W. Moncrief, D. C. Pantaleo and N. E. Smith, *Inorg. Nucl. Chem. Lett.*, 1971, **7,** 255.
235. B. Viard, A. Laarif, F. Theobald and J. Amaudrut, *J. Chem. Res. (M)*, 1983, 2301.
236. M. K. Rastoji and S. N. Nigam, *J. Inst. Chem. (India)*, 1979, **51,** 62 (*Chem. Abstr.*, 1980, **92,** 76 624).
237. S. Brownstein, *Can. J. Chem.*, 1978, **56,** 343.
238. D. M. Puri and S. Singh, *J. Indian. Chem. Soc.*, 1981, **58,** 327.
239. D. C. Pantaleo and R. C. Johnson, *Inorg. Chem.*, 1971, **10,** 1298.
240. S. N. Nigam, M. K. Rastogi and R. K. Multani, *Indian J. Chem., Sect. A*, 1978, **16,** 361.
241. M. J. Bennett, M. Cowie, J. L. Martin and T. Takats, *J. Am. Chem. Soc.*, 1973, **95,** 7504.
242. J. L. Martin and J. Takats, *Inorg. Chem.*, 1975, **14,** 1358.
243. (a) M. Cowie and M. J. Bennett, *Inorg. Chem.*, 1976, **15,** 1589; (b) K. Tatsumi, Y. Sekiguchi, A. Nakamura, R. E. Cramer and J. J. Rupp, *Angew. Chem., Int. Ed. Engl.*, 1986, **25,** 86.
244. J. C. Daran, K. Prout, A. De Cian, M. L. H. Green and N. J. Siganporia, *J. Organomet. Chem.*, 1977, **136,** C4.
245. J. C. Daran, B. Meunier and K. Prout, *Acta Crystallogr., Sect. B*, 1979, **35,** 1709.
246. D. C. Bradley and M. H. Gilitz, *Chem. Commun.*, 1965, 289.
247. J. N. Smith and T. M. Brown, *Inorg. Nucl. Chem. Lett.*, 1970, **6,** 44.
248. P. R. Heckley and D. G. Holah, *Inorg. Nucl. Chem. Lett.*, 1970, **6,** 865.
249. P. R. Heckley, D. G. Holah and D. Brown, *Can. J. Chem.*, 1971, **49,** 1151.
250. D. J. Machin and J. F. Sullivan, *J. Less-Common Met.*, 1969, **19,** 413.
251. K. A. Uvarova, Yu. I. Usatenko, N. V. Melynikova and Zh. G. Klopova, *Zh. Neorg. Khim.*, 1971, **16,** 2137.
252. J. Sala-Pala, J. L. Migot, J. E. Guerchais, L. Legall and F. Grosjean, *J. Organomet. Chem.*, 1983, **248,** 299.
253. R. C. Fay, D. F. Lewis and J. R. Weir, *J. Am. Chem. Soc.*, 1975, **97,** 7179.
254. D. F. Lewis and R. C. Fay, *Inorg. Chem.*, 1976, **15,** 2219; J. R. Weir and R. C. Fay, *Inorg. Chem.*, 1986, **25,** 2969.
255. M. F. Lappert, P. P. Power, A. R. Sanger and R. C. Srivastava, 'Metal and Metalloid Amides', Wiley, New York, 1980.
256. D. C. Bradley and M. H. Chisholm, *Acc. Chem. Res.*, 1976, **9,** 273.
257. D. C. Bradley and M. H. Chisholm, *Adv. Inorg. Chem. Radiochem.*, 1972, **15,** 259.
258. K. Dehnicke and J. Strähle, *Angew. Chem., Int. Ed. Engl.*, 1981, **20,** 413.
259. K. Dehnicke and J. Strähle, *Adv. Inorg. Chem. Radiochem.*, 1983, **26,** 169.
260. G. W. A. Fowles and F. H. Pollard, *J. Chem. Soc.*, 1952, 4938.
261. G. W. A. Fowles and C. M. Pleass, *J. Chem. Soc., Dalton Trans.*, 1957, 2078.
262. P. J. H. Carnell and G. W. A. Fowles, *J. Less-Common Met.*, 1962, **4,** 40; *J. Chem. Soc. (A)*, 1959, 4113.
263. D. C. Bradley and I. M. Thomas, *Can. J. Chem.*, 1962, **40,** 449, 1335.
264. D. C. Bradley and M. H. Gilitz, *J. Chem. Soc. (A)*, 1969, 980.
265. D. C. Bradley and M. H. Chisholm, *J. Chem. Soc.*, 1971, 1511.
266. J. S. Wood, *Prog. Inorg. Chem.*, 1972, **16,** 227.

267. C. E. Heath and M. B. Hursthouse, *J. Chem. Soc., Chem. Commun.*, 1971, 143.
268. M. H. Chisholm, A. H. Cowley and M. Lattman, *J. Am. Chem. Soc.*, 1980, **102**, 42.
269. J. C. Fuggle, D. W. A. Sharp and J. M. Winfield, *J. Chem. Soc., Dalton Trans.*, 1972, 1766.
270. M. H. Chisholm, J. C. Huffman and L. S. Tan, *Inorg. Chem.*, 1981, **20**, 1859.
271. T. G. Appleton, H. C. Clark and L. E. Manzer, *Coord. Chem.*, 1973, **10**, 335.
272. D. C. Bradley, M. B. Hursthouse, K. M. Abdulmalik and G. B. Vuru, *Inorg. Chim. Acta*, 1980, **44**, L5.
273. R. A. Andersen, *Inorg. Chem.*, 1979, **18**, 3622.
274. J. L. Morançais, Thesis, Nice, 1983.
275. J. L. Martin and J. Takats, *Inorg. Chem.*, 1975, **14**, 73.
276. M. H. Chisholm and M. W. Extine, *J. Am. Chem. Soc.*, 1977, **99**, 792.
277. C. Santini-Scampucci and G. Wilkinson, *J. Chem. Soc., Dalton Trans.*, 1976, 807.
278. (a) M. H. Chisholm and M. Extine, *J. Chem. Soc., Chem. Commun.*, 1975, 438; (b) *J. Am. Chem. Soc.*, 1977, **99**, 782; (c) *J. Am. Chem. Soc.*, 1975, **97**, 1623.
279. D. C. Bradley and M. H. Gitlitz, *J. Chem. Soc. (A)*, 1969, 1152.
280. M. H. Chisholm, F. A. Cotton and M. W. Extine, *Inorg. Chem.*, 1978, **17**, 2000.
281. W. A. Nugent, D. W. Ovenall and S. J. Holmes, *Organometallics*, 1983, **2**, 161.
282. Y. Takahashi, N. Onoyama, Y. Ishikawa, S. Motojima and K. Sugiyama, *Chem. Lett.*, 1978, 525.
283. C. Airoli, D. C. Bradley and G. Vuru, *Transition Met. Chem.*, 1979, **4**, 64.
284. J. M. Mayer, C. J. Curtis and J. E. Bercaw, *J. Am. Chem. Soc.*, 1983, **105**, 2651.
285. W. A. Nugent and B. L. Haymore, *Coord. Chem. Rev.*, 1980, **31**, 123.
286. S. Cenini and G. La Monica, *Inorg. Chim. Acta*, 1976, **18**, 279.
287. A. K. Rappe and W. A. Goddard, *J. Am. Chem. Soc.*, 1982, **104**, 448.
288. D. C. Bradley and I. M. Thomas, *Proc. Chem. Soc.*, 1959, 225.
289. W. A. Nugent and R. Harlow, *J. Chem. Soc., Chem. Commun.*, 1978, 579; W. A. Nugent, *Inorg. Chem.*, 1983, **22**, 965.
290. T. C. Jones, A. J. Nielson and C. E. F. Rickard, *J. Chem. Soc., Chem. Commun.*, 1984, 205.
291. L. S. Tan, G. V. Goeden and B. L. Haymore, *Inorg. Chem.*, 1983, **22**, 1744.
292. S. M. Rocklage and R. R. Schrock, *J. Am. Chem. Soc.*, 1982, **104**, 3077.
293. R. R. Schrock and J. D. Fellmann, *J. Am. Chem. Soc.*, 1978, **100**, 3359.
294. L. G. Hubert-Pfalzgraf and G. Aharonian, *Inorg. Chim. Acta*, 1985, **100**, L22.
295. P. A. Finn, M. S. King, P. A. Kilty and R. E. McCarley, *J. Am. Chem. Soc.*, 1975, **97**, 220.
296. F. A. Cotton and W. T. Hall, *J. Am. Chem. Soc.*, 1979, **101**, 5094.
297. F. A. Cotton and W. T. Hall, *Inorg. Chem.*, 1978, **17**, 3525.
298. M. R. Churchill and H. J. Wasserman, *Inorg. Chem.*, 1982, **21**, 223.
299. S. M. Rocklage and R. R. Schrock, *J. Am. Chem. Soc.*, 1980, **102**, 7808.
300. W. A. Nugent, R. J. McKinney, R. V. Kasowski and F. A. Van Catledje, *Inorg. Chim. Acta*, 1982, **65**, L91.
301. S. M. Rocklage, J. D. Fellmann, G. A. Rupprecht, L. N. Messerle and R. R. Schrock, *J. Am. Chem. Soc.*, 1981, **103**, 1440.
302. D. A. Lemenovskii, T. V. Baukova, E. G. Perevalova and A. N. Nesmeyanov, *Izv. Akad. Nauk SSSR, Ser. Khim.*, 1976, **10**, 2404.
303. R. J. Burt, G. J. Leigh and D. L. Hughes, *J. Chem. Soc., Dalton Trans.*, 1981, 793.
304. J. L. Morançais and L. G. Hubert-Pfalzgraf, *Transition Met. Chem.*, 1984, **9**, 130.
305. J. Chatt, J. R. Dilworth, G. J. Leigh and R. L. Richards, *Chem. Commun.*, 1970, 955.
306. S. M. Rocklage, H. W. Turner, J. D. Fellmann and R. R. Schrock, *Organometallics*, 1982, **1**, 703.
307. M. R. Churchill and H. J. Wasserman, *Inorg. Chem.*, 1981, **20**, 2899.
308. H. W. Turner, J. D. Fellman, S. M. Rocklage and R. R. Schrock, *J. Am. Chem. Soc.*, 1980, **102**, 7809.
309. M. R. Churchill and H. J. Wasserman, *Inorg. Chem.*, 1982, **21**, 218.
310. C. K. Kruger and Y. H. Tsay, *Angew. Chem., Int. Ed. Engl.*, 1973, **12**, 998.
311. (a) R. D. Sanner, J. M. Manriquez, R. E. Marsh and J. E. Bercaw, *J. Am. Chem. Soc.*, 1976, **98**, 8351; (b) R. D. Sanner, D. Duggan, T. C. McKenzie, R. E. Marsh and J. E. Bercaw, *J. Am. Chem. Soc.*, 1976, **98**, 8358.
312. J. Strähle, *Z. Anorg. Allg. Chem.*, 1974, **405**, 139.
313. H. Belzer and J. Strähle, *Z. Naturforsch., Teil B*, 1979, **34**, 1199.
314. H. Belzer and J. Strähle, *Z. Naturforsch., Teil B*, 1983, **38**, 317.
315. E. M. Shustorowich, M. A. Poraikoshits and Yu. A. Buslaev, *Coord. Chem. Rev.*, 1975, **17**, 1.
316. K. P. Frank, J. Strähle and J. Weidlein, *Z. Naturforsch, Teil B*, 1980, **35**, 300.
317. S. M. Sintitsyna, *Russ. J. Inorg. Chem. (Engl. Transl.)*, 1977, **22**, 402.
318. W. Weishaupt and J. Strähle, *Z. Anorg. Allg. Chem.*, 1977, **429**, 261.
319. M. A. Glushkova, M. M. Ershova and Yu. A. Buslaev, *Russ. J. Inorg. Chem. (Engl. Transl.)*, 1965, **10**, 1290; *Izv. Akad. Nauk Neorg. Mater.*, 1967, **3**, 1971.
320. E. M. Shustorowich, *J. Struct. Chem. (Engl. Transl)*, 1962, **3**, 204.
321. J. W. Buchler and K. Rohbock, *Inorg. Nucl. Chem. Lett.*, 1972, **8**, 1073.
322. M. Gouterman, L. K. Hanson, G. E. Khalil, J. W. Bucher, K. Rohbock and D. Dolphin, *J. Am. Chem. Soc.*, 1975, **97**, 3142.
323. P. Richard and R. Guilard, *J. Chem. Soc., Chem. Commun.*, 1983, 1454.
324. M. L. H. Green and J. J. E. Moreau, *Inorg. Chim. Acta*, 1978, **31**, L461.
325. B. D. Berezin and T. N. Lomova, *Zh. Neorg. Khim.*, 1981, **26**, 379.
326. R. Guilard, B. Flinniaux, B. Maume and P. Fournari, *C.R. Hebd. Seances Acad. Sci., Ser. C*, 1975, **281**, 461.
327. C. Lecomte, J. Protas and R. Guilard, *C.R. Hebd. Seances Acad. Sci., Ser. C*, 1976, **283**, 397.
328. J. F. Johnson and W. R. Scheidt, *J. Am. Chem. Soc.*, 1977, **99**, 294.
329. J. F. Johnson and W. R. Scheidt, *Inorg. Chem.*, 1978, **17**, 1280.
330. C. Lecomte, J. Protas, R. Guilard, B. Fliniaux and P. Fournari, *J. Chem. Soc., Dalton Trans.*, 1979, 1306.
331. K. Tatsumi and R. Hoffman, *J. Am. Chem. Soc.*, 1981, **103**, 3328.
332. C. Lecomte, J. Protas, P. Richard, J. M. Barbe and R. Guilard, *J. Chem. Soc., Dalton Trans.*, 1982, 247.

333. C. Lecomte, J. Protas, R. Guilard, B. Fliniaux and P. Fournari, *J. Chem. Soc., Chem. Commun.*, 1976, 434.
334. Y. Matsuda, S. Yamada, T. Goto and Y. Murakami, *Bull. Chem. Soc. Jpn.*, 1981, **54,** 452.
335. K. Varmuza, G. Maresch and A. Meller, *Monatsh. Chem.*, 1974, **105,** 327.
336. G. P. Shaposhnikov, V. F. Borodkin and M. I. Fedorov, *Izv. Vyssh. Uchebn. Zaved., Khim. Khim. Tekhnol.*, 1981, **24,** 1485 (*Chem. Abstr.*, 1982, **97,** 30 632).
337. D. L. Reger, C. A. Swift and L. Lebioda, *J. Am. Chem. Soc.*, 1983, **105,** 5343; *Inorg. Chem.*, 1984, **23,** 349.
338. L. G. Hubert-Pfalzgraf and M. Tsunoda, *Polyhedron,* 1983, **2,** 203.
339. J. D. Wilkins, *J. Organomet. Chem.*, 1974, **80,** 349.
340. J. D. Wilkins, *J. Organomet. Chem.*, 1974, **67,** 269; **65,** 383.
341. M. G. B. Drew and J. D. Wilkins, *J. Chem. Soc., Dalton Trans.*, 1974, 1973.
342. J. D. Wilkins and M. G. B. Drew, *J. Organomet. Chem.*, 1974, **69,** 111.
343. A. R. Middleton and G. Wilkinson, *J. Chem. Soc., Dalton Trans.*, 1980, 1888.
344. G. Brauer and H. Walz, *Z. Anorg. Allg. Chem.*, 1962, **319,** 236.
345. B. Crociani and R. L. Richards, *J. Chem. Soc., Chem. Commun.*, 1973, 127.
346. B. Crociani, M. Nicolini and R. L. Richards, *J. Organomet. Chem.*, 1975, **101,** C1.
347. M. Behnam-Dahkordy, B. Crociani and R. L. Richards, *J. Chem. Soc., Dalton Trans.* 1977, 2015.
348. J. Dehand, R. Dorschner and R. Rohmer, *C.R. Hebd. Seances Acad. Sci., Ser. C,* 1966, **262,** 1882.
349. H. A. Droll and A. Fregoso, *Inorg. Chim. Acta,* 1977, **24,** L79.
350. E. G. Il'yn V. P. Tarsov, M. M. Ershova, V. A. Ermakov, M. A. Glushkova and Yu. A. Buslaev, *Koord. Khim.*, 1978, **4,** 1370.
351. R. Kergoat, M.-C. Senechal-Tocquer, J. E. Guerchais and F. Dahan, *Bull. Soc. Chim. Fr.*, 1976, 1203.
352. N. Vuletic and C. Djordjevic, *J. Chem. Soc., Dalton Trans.*, 1972, 2322.
353. H. Bohland and E. Harke, *Z. Anorg. Allg. Chem.*, 1975, **413,** 102.
354. J. D. Wilkins, *J. Organomet. Chem.*, 1974, **65,** 383.
355. F. Dahan, R. Kergoat, M.-C. Tocquer and J. E. Guerchais, *Acta Crystallogr., Sect. B,* 1976, **32,** 1038.
356. F. Dahan, R. Kergoat, M.-C. Senechal-Torquer and J. E. Guerchais, *J. Chem. Soc., Dalton Trans.*, 1976, 2202.
357. R. G. Kidd and H. G. Spinney, *J. Am. Chem. Soc.*, 1981, **103,** 4759.
358. E. G. Rakov, K. N. Marushkin and A. S. Alikhanyan, *Zh. Neorg. Khim.*, 1983, **28,** 328.
359. V. I. Yampolyskii, E. G. Rakov, V. A. Davydov, V. V. Mikulenok and A. A. Mal'tsev, *Zh. Neorg. Khim.*, 1974, **19,** 2299.
360. L. V. Frevell and H. W. Rinn, *Acta Crystallogr.*, 1956, **9,** 626.
361. K. S. Vorres and J. Donohue, *Acta Crystallogr.*, 1955, **8,** 25.
362. K. A. Wilhelmi, L. Jahnberg and S. Andersson, *Acta Chem. Scand.*, 1970, **24,** 1472.
363. L. A. Nisel'son, K. V. Tret'yakova, S. M. Kireev and D. B. Drobot, *Zh. Neorg. Khim.*, 1982, **27,** 887.
364. V. Katovic and C. Djordjevic, *Inorg. Chem.*, 1970, **9,** 1720.
365. Yu. A. Buslaev, S. M. Sinitsyna, V. I. Sinyagin and M. A. Polikarpova, *Zh. Neorg. Khim.*, 1970, **15,** 2324.
366. J. Sola, Y. Do, J. M. Berg and R. H. Holm, *J. Am. Chem. Soc.*, 1983, **105,** 7784.
367. C. Crevecoeur, *Acta Crystallogr.*, 1964, **17,** 757.
368. F. Hulliger, *Helv. Phys. Acta,* 1961, **34,** 303.
369. M. Muller, M. J. F. Leroy and R. Rohmer, *C.R. Hebd. Seances Acad. Sci., Ser. C,* 1970, **270,** 1458.
370. G. W. A. Fowles, R. J. Hobson, D. A. Rice and K. J. Shanton, *J. Chem. Soc., Chem. Commun.*, 1976, 552.
371. A. O. Baghlaf and A. Thompson, *J. Less-Common Met.*, 1977, **53,** 291.
372. H. Schaeffer and W. Beckmann, *Z. Anorg. Allg. Chem.*, 1966, **347,** 225.
373. M. G. B. Drew, D. A. Rice and D. M. Williams, *J. Chem. Soc., Dalton Trans.*, 1983, 2251.
374. M. G. B. Drew, G. W. A. Fowles, R. J. Hobson and D. A. Rice, *Inorg. Chim. Acta,* 1976, **20,** L35.
375. M. G. B. Drew and R. J. Hobson, *Inorg. Chim. Acta,* 1983, **72,** 233.
376. H. Böhland and F. M. Schneider, *Z. Anorg. Allg. Chem.*, 1972, **390,** 53.
377. E. J. Peterson, R. B. Von Dreele and T. M. Brown, *Inorg. Chem.*, 1978, **17,** 1410.
378. V. I. Pakhomov, T. A. Kaidolova and R. L. Davidovich, *Zh. Neorg. Khim.*, 1974, **19,** 1823.
379. R. Stomberg, *Acta Chem. Scand., Ser. A.*, 1983, **37,** 453.
380. Yu. E. Gorbunova, U. I. Pakhamov, V. G. Kuznetsov and E. S. Kovaleva, *Zh. Strukt. Khim.*, 1972, **13,** 165.
381. J. Sala-Pala, J. Y. Calves, J. E. Guerchais, S. Brownstein, J. C. Dewan and A. J. Edwards, *Can. J. Chem.*, 1978, **56,** 1545.
382. J. C. Dewan, A. J. Edwards, J. Y. Calves and J. E. Guerchais, *J. Chem. Soc., Dalton Trans,* 1977, 978.
383. V. I. Pakhomov, R. L. Davidovich, T. A. Kaidalova and T. F. Levchishina, *Zh. Neorg. Khim.*, 1973, **18,** 1240.
384. J. Sala-Pala, J. E. Guerchais and A. J. Edwards, *Angew. Chem., Int. Ed. Engl.*, 1982, **21,** 870.
385. Yu. A. Buslaev, E. G. Il'in, V. D. Kopanev and V. P. Tarasov, *Zh. Struckt. Khim.*, 1972, **13,** 930.
386. Yu. A. Buslaev and E. G. Il'in, *Koord. Khim.*, 1975, **1,** 451.
387. Yu. A. Buslaev, E. G. Il'in, V. D. Kopanev and O. G. Gavrish, *Izv. Akad. Nauk SSSR, Ser. Khim.*, 1971, 1139.
388. Yu. A. Buslaev, V. D. Kopanev, S. M. Sinitsyna and V. G. Khlebodarov, *Zh. Neorg. Khim.*, 1973, **18,** 2567.
389. R. F. Weinland and L. Storz, *Z. Anorg. Allg. Chem.*, 1907, **54,** 223.
390. D. Brown, *J. Chem. Soc.*, 1964, 4944.
391. (a) J. Guerchais, B. Spinner and R. Rohmer, *Bull. Soc. Chim. Fr.*, 1965, 55; (b) E. Wendling and R. Rohmer, *Bull. Soc. Chim. Fr.*, 1967, 8.
392. L. V. Surpina, A. V. Leschenko and O. A. Osipov, *Zh. Obshch. Khim.*, 1970, **40,** 1915.
393. T. I. Beresneva, G. P. Gosteva and A. N. Ketov, *Zh. Neorg. Khim.*, 1970, **15,** 3274.
394. N. A. Ivanov, I. P. Alimarin, I. M. Gibalo and G. F. Bebikh, *Izv. Akad. Nauk SSSR, Ser. Khim.*, 1970, 2664.
395. F. I. Lobanov, V. S. Voskov, I. M. Gibalo and V. F. Kramenef, *Russ. J. Inorg. Chem.* (*Engl. Transl.*), 1972, **17,** 1405.
396. I. M. Gibalo, V. S. Voskov and F. I. Lobanov, *Zh. Anal. Khim.*, 1970, **25,** 1918.
397. U. Müller and I. Lorenz, *Z. Anorg. Allg. Chem.*, 1980, **463,** 110.
398. F. A. Cotton and R. C. Najjar, *Inorg. Chem.*, 1981, **20,** 1866.
399. K. Dehnicke and H. Prinz, *Z. Anorg. Allg. Chem.*, 1982, **490,** 171.

400. J. C. Green, C. P. Overton, K. Prout and J. M. Martin, *J. Organomet. Chem.*, 1982, **241,** C21.
401. E. Hey, F. Weller and F. Dehnicke, *Z. Anorg. Allg. Chem.*, 1983, **502,** 45.
402. A. I. Morozov and I. I. Leonova, *Zh. Neorg. Khim.*, 1972, **17,** 2128.
403. M. Schönherr and L. Kolditz, *Z. Chem.*, 1970, **10,** 72.
404. I. S. Morozov and N. Lipatova, *Zh. Neorg. Khim.*, 1968, **13,** 2128.
405. M. Y. Mirza and S. Ahmed, *J. Inorg. Nucl. Chem.*, 1981, **43,** 1935.
406. A. I. Morozov and I. I. Leonova, *Russ. J. Inorg. Chem.* (*Engl. Transl.*), 1972, **17,** 1107.
407. A. M. Karayannis, C. M. Mikulski, M. J. Stocko, L. L. Pytlewski and M. M. Labes, *J. Less-Common Met.*, 1970, **21,** 195.
408. C. Santini-Scampucci and J. G. Riess, *J. Chem. Soc., Dalton Trans.*, 1974, 1433.
409. (a) S. M. Sinitsyna, T. M. Gorlova and V. F. Chistyakov, *Zh. Neorg. Khim.*, 1973, **18,** 2114; (b) S. M. Sinitsyna, V. G. Khlebodarov and M. A. Bukhterva, *Zh. Neorg. Khim.*, 1974, **19,** 1532.
410. S. M. Sinitsyna, V. I. Siniyagin and Yu. A. Buslaev, *Izv. Akad. Nauk SSSR, Neorg. Mater.*, 1969, **5,** 605.
411. A. V. Leshchenko, B. T. Panyushkin, A. D. Garnovski and O. A. Osipov, *Zh. Neorg. Khim.*, 1966, **11,** 2156.
412. C. Djordjevic and V. Katovic, *Chem. Commun.*, 1966, 224.
413. N. Brnicevic and C. Djordjevic, *J. Less-Common Met.*, 1967, **13,** 470.
414. A. P. Bell and A. G. Wedd, *J. Organomet. Chem.*, 1979, **181,** 81.
415. C. Chavant, J. C. Daran, Y. Jeannin, G. Constant and R. Morancho, *Acta Crystallogr., Sect. B*, 1975, **31,** 1828.
416. L. G. Hubert-Pfalzgraf and A. A. Pinkerton, *Inorg. Chem.*, 1977, **16,** 1895.
417. (a) R. Broussier, J. D. Olivier and B. Gautheron, *J. Organomet. Chem.*, 1983, **251,** 307; (b) R. Broussier, H. Normand and B. Gautheron, *J. Organomet. Chem.*, 1976, **120,** C28; (c) *J. Organomet. Chem.*, 1978, **155,** 347; (d) *Tetrahedron Lett.*, 1976, 4077.
418. P. M. Treichel and G. W. Weber, *J. Am. Chem. Soc.*, 1968, **90,** 1753.
419. D. A. Lemenovskii, T. V. Baukova, V. A. Knizhnikov, E. V. Perevalova and A. N. Nesmeyanov, *Dokl. Akad. Nauk SSSR*, 1976, **226,** 585.
420. D. A. Lemenovskii, T. V. Baukova and V. D. Fedin, *J. Organomet. Chem.*, 1977, **132,** C14.
421. P. Köpf-Maier, M. Leitner and H. Köpf, *J. Inorg. Nucl. Chem.*, 1980, **42,** 1789.
422. W. E. Douglas and M. L. H. Green, *J. Chem. Soc., Dalton Trans.*, 1972, 1796.
423. K. Prout, T. S. Cameron, R. A. Forder, S. R. Critchley, B. Denton and G. V. Rees, *Acta Crystallogr., Sect. B*, 1974, **30,** 2290.
424. M. M. Kappes and R. H. Staley, *J. Am. Chem. Soc.*, 1981, **103,** 1286.
425. E. L. Muetterties and C. M. Wright, *J. Am. Chem. Soc.*, 1965, **87,** 4706.
426. A. K. Majumdar and A. K. Mukherjee, *Anal. Chim. Acta*, 1958, **19,** 23.
427. R. C. Johnson and A. Syamal, *J. Inorg. Nucl. Chem.*, 1971, **33,** 2547.
428. F. Fairbrother, N. Ahmed, K. Edgar and A. Thompson, *J. Less-Common. Met.*, 1962, **4,** 466.
429. C. Djordjevic and V. Katovic, *J. Inorg. Nucl. Chem.*, 1963, **25,** 1099.
430. R. Kergoat and J. E. Guerchais, *J. Less-Common Met.*, 1975, **39,** 91.
431. J. Y. Calves, J. E. Guerchais, R. Kergoat and N. Kheddar, *Inorg. Chim. Acta*, 1979, **33,** 95.
432. J. C. Daran, Y. Jeannin, J. E. Guerchais and R. Kergoat, *Inorg. Chim. Acta*, 1979, **33,** 81.
433. H. Funk and K. Niederländer, *Ber. Bunsenges. Phys. Chem.*, 1929, **62,** 1688.
434. K. Andrianov, M. Filimonova and V. Sidorov, *J. Organomet. Chem.*, 1977, **142,** 31.
435. B. Viard, M. Poulain, D. Grandjean and J. Amaudrut, *J. Chem. Res.* (*S*), 1983, 84.
436. R. N. Gupta and B. K. Sen, *J. Inorg. Nucl. Chem.*, 1975, **37,** 1548.
437. N. Brnicevic and C. Djordjevic, *J. Less-Common Met.*, 1971, **23,** 61.
438. M. Muller and J. Dehand, *Bull. Soc. Chim. Fr.*, 1971, 2837.
439. M. Muller and J. Dehand, *Bull. Soc. Chim. Fr.*, 1971, 2843.
440. N. Kheddar and B. Spinner, *Bull. Soc. Chim. Fr.*, 1972, 502.
441. G. Mathern, R. Weiss and R. Rohmer, *Chem. Commun.*, 1969, 70.
442. G. Mathern and R. Weiss, *Acta Crystallogr., Sect. B*, 1971, **27,** 1610.
443. M. Galesic, B. Markovic, M. Herceg and M. Sljukic, *J. Less-Common Met.*, 1971, **25,** 234.
444. B. Kojic-Prodic, R. Liminga and S. Scavnicar, *Acta Crystallogr., Sect. B*, 1973, **29,** 864.
445. N. Brnicevic and C. Djordjevic, *J. Chem. Soc., Dalton Trans.*, 1974, 165.
446. N. Brnicevic, *J. Inorg. Nucl. Chem.*, 1975, **37,** 719.
447. F. I. Lobanov, G. K. Nurtaeva and I. M. Gibalo, *Zh. Neorg. Khim.*, 1978, **23,** 982.
448. M. Siroki and C. Djordjevic, *J. Less-Common Met.*, 1971, **23,** 228.
449. A. K. Sen Gupta and N. G. Mandal, *J. Inorg. Nucl. Chem.*, 1977, **39,** 821.
450. J. F. Dietsch, M. Muller and J. Dehand, *C.R. Hebd. Seances Acad. Sci., Ser. C*, 1972, **272,** 471.
451. R. Kergoat and J. E. Guerchais, *Z. Anorg. Allg. Chem.*, 1975, **416,** 174.
452. N. Brnicevic and C. Djordjevic, *J. Less-Common Met.*, 1970, **21,** 469.
453. A. T. Casey, D. J. Mackey, R. L. Martin and A. H. White, *Aust. J. Chem.*, 1972, **25,** 477.
454. B. Ilmaier and R. C. Johnson, *J. Less-Common Met.*, 1971, **25,** 323.
455. S. Kumar and N. K. Kauskik, *Gazz. Chim. Ital.*, 1981, **111,** 57.
456. J. C. Dewan, D. L. Kepert, C. L. Raston, D. Taylor, E. N. Maslen and A. H. White, *J. Chem. Soc., Dalton Trans.*, 1973, 2082.
457. B. Kamenar and C. K. Prout, *J. Chem. Soc.* (*A*), 1970, 2379.
458. L. G. Hubert-Pfalzgraf, G, Le Borgne and M. Tsunoda, to be published.
459. (a) C. Djordjevic and B. Tamhina, *Anal. Chem.*, 1968, **40,** 1512; (b) B. Tamina and C. Djordjevic, *Croat. Chem. Acta*, 1975, **47,** 79 (*Chem. Abstr.*, 1976, **84,** 68 884).
460. G. B. Seifer, Yu. Ya. Kharitonov and Z. A. Tarasova, *Izv. Akad. Nauk SSSR, Neorg. Mater.*, 1968, **4,** 518.
461. D. Driss, A. Tazaïrt and B. Spinner, *C.R. Hebd. Seances Acad. Sci., Ser. C*, 1973, **277,** 945.
462. J. E. Guerchais and R. Rohmer, *C.R. Hebd. Seances Acad. Sci.*, 1964, **259,** 1135.
463. J. E. Guerchais and R. Rohmer, *C.R. Hebd. Seances Acad. Sci.*, 1964, **259,** 2647.
464. J. E. Guerchais, J. Dehand, M. P. Aufranc and R. Rohmer, *C.R. Hebd. Seances Acad. Sci.*, 1965, **261,** 2907.
465. R. N. Shchelokov, E. N. Traggeim, M. A. Michnik and S. V. Morozova, *Zh. Neorg. Khim.*, 1973, **18,** 2104.

466. G. Mathern and R. Weiss, *Acta Crystallogr., Sect. B,* 1971, **27,** 1598.
467. J. Fuchs and D. Lubkoll, *Z. Naturforsch., Teil B,* 1973, **28,** 590.
468. G. V. Jere, L. Surendra and M. K. Gupta, *Thermochim. Acta,* 1983, **63,** 229.
469. G. N. Latysh, A. S. Chernyak and T. E. Serebrennikova, *Zh. Neorg. Khim.,* 1973, **18,** 1014.
470. V. P. Vasilyev and G. A. Zaitseva, *Zh. Neorg. Khim.,* 1973, **18,** 139.
471. V. P. Vasilyev, G. A. Zaitseva and V. G. Malakhova, *Russ. J. Inorg. Chem., (Engl. Transl.),* 1971, **16,** 360.
472. V. P. Vasilyev, G. A. Zaitseva and V. G. Malakhova, *Russ. J. Inorg. Chem. (Engl. Transl.),* 1971, **16,** 1144.
473. G. A. Bogdanov, G. K. Yurchenko and O. V. Popov, *Zh. Neorg. Khim.,* 1982, **27,** 2143.
474. A. A. Merkulov, V. M. Berdnikov, Yu. V. Gatilov and M. M. Bazhin, *Zh. Neorg. Khim.,* 1972, **17,** 3253.
475. A. A. Merkulov and V. M. Berdnikov, *Zh. Neorg. Khim.,* 1978, **23,** 3250.
476. R. N. Shchelokov, E. N. Traggeim, M. A. Michnik and K. I. Petrov, *Zh. Neorg., Khim.,* 1972, **17,** 2439.
477. R. N. Shchelokov, E. N. Traggeim and M. A. Michnik, *Chem. Abstr.,* 1976, **84,** 25 283.
478. G. Mathern, R. Weiss and R. Rohmer, *Chem. Commun.,* 1970, 153.
479. J. Amaudrut, J. E. Guerchais and J. Sala-Pala, *J. Organomet. Chem.,* 1978, **157,** C10.
480. J. L. Migot, J. Sala-Pala and J. E. Guerchais, *J. Organomet. Chem.,* 1983, **243,** 427.
481. B. Viard, J. Sala-Pala, J. Amaudrut, J. E. Guerchais, C. Sanchez and J. Livage, *Inorg. Chim. Acta,* 1980, **39,** 99.
482. R. M. Roder, *Diss. Abstr. Int. B,* 1974, **34,** 4866.
483. R. Mercier, J. Douglade, J. Amaudrut, J. Sala-Pala and J. E. Guerchais, *Acta Crystallogr., Sect. B,* 1980, **36,** 2986.
484. A. Bali and K. C. Malhotra, *Aust. J. Chem.,* 1975, **28,** 481.
485. A. S. Chernyak and G. N. Latysh, *Russ. J. Inorg. Chem. (Engl. Transl.),* 1975, **20,** 698.
486. M. I. Andreeva, R. A. Popova and S. A. Kobycheva, *Zh. Neorg. Khim.,* 1971, **16,** 1036.
487. A. S. Chernyak, T. N. Yas'ko, V. K. Karnaukhova, M. L. Shepot'ko and E. A. Rozhkova, *Zh. Neorg. Khim.,* 1982, **27,** 2503.
488. A. S. Chernyak, T. N. Yas'ko and T. F. Tikhonova, *Zh. Neorg. Khim.,* 1981, **26,** 1568.
489. J. M. Longo and P. Kierkegaard, *Acta Chem. Scand.,* 1966, **20,** 72.
490. R. S. Roth, A. D. Wadsley and S. Andersson, *Acta Crystallogr.,* 1965, **18,** 643.
491. N. G. Chernorukov, N. P. Egorov, E. V. Shitova and Yu. Chigirinskii, *Zh. Neorg. Khim.,* 1981, **26,** 2714.
492. N. G. Chernorukov, N. P. Egorov and I. R. Mochalova, *Zh. Neorg. Khim.,* 1978, **23,** 2931.
493. G. I. Deulin, R. B. Dushin and V. N. Krylov, *Zh. Neorg. Khim.,* 1979, **24,** 2327.
494. N. G. Chernorukov, N. P. Egorov and V. F. Kutsepin, *Zh. Neorg. Khim.,* 1979, **24,** 1782.
495. R. G. Cavell and R. A. Sanger, *Inorg. Chem.,* 1972, **11,** 2016.
496. E. Niecke, R. Kröher and S. Pohl, *Angew Chem., Int. Ed. Engl.,* 1977, **16,** 864.
497. S. Jagner, E. Ljungstroem and A. Tullberg, *Acta Crystallogr., Sect. B,* 1980, **36,** 2213.
498. Yu. A. Ol'dekop and V. A. Knizhnikov, *Zh. Obshch. Khim.,* 1981, **51,** 1723.
499. (a) P. H. M. Budzelaar, K. H. Den Haan, J. Boersma, G. J. M. Van Der Kerk and A. L. Spek, *Organometallics,* 1984, **3,** 156; (b) P. H. M. Budzelaar, A. A. H. Van Der Zeijden, J. Boersma, G. J. M. Van Der Kerk, A. L. Spek and A. J. M. Duisenberg, *Organometallics,* 1984, **3,** 159.
500. R. Kergoat, M. M. Kubicki, J. E. Guerchais, N. C. Norman and A. G. Orpen, *J. Chem. Soc., Dalton Trans.,* 1982, 633.
501. P. W. Seabaugh and J. D. Corbett, *Inorg. Chem.,* 1965, **4,** 176.
502. R. E. McCarley and J. C. Boatman, *Inorg. Chem.,* 1963, **2,** 547.
503. D. R. Taylor, J. C. Calabrese and E. M. Larsen, *Inorg. Chem.,* 1977, **16,** 721.
504. L. F. Dahl and D. L. Wampler, *Acta Crystallogr.,* 1958, **11,** 615.
505. M. H. Whangbo and M. J. Foshee, *Inorg. Chem.,* 1981, **20,** 113.
506. S. S. Shaik and R. Bar, *Inorg. Chem.,* 1983, **22,** 735.
507. H. Kawamura, I. Shiratani and K. Tachikawa, *Phys. Lett. A,* 1978, **65,** 335.
508. G. W. A. Fowles, T. D. Tidmarsh and R. A. Walton, *Inorg. Chem.,* 1969, **8,** 631.
509. D. Lal and A. D. Westland, *J. Chem. Soc., Dalton Trans.,* 1974, 2505.
510. L. O. Gilpatrick and L. M. Toth, *Inorg. Chem.,* 1974, **13,** 2242.
511. F. E. Dickson, R. A. Hayden and W. G. Fateley, *Spectrochim. Acta, Sect. A,* 1969, **25,** 1875.
512. R. Gut and W. Perron, *J. Less-Common Met.,* 1972, **26,** 369.
513. L. G. Hubert-Pfalzgraf and M. Tsunoda, *Inorg. Chim. Acta,* 1980, **38,** 43.
514. J. H. Clark, D. L. Kepert, J. Lewis and R. S. Nyholm, *J. Chem. Soc. (A),* 1965, 2865.
515. R. E. McCarley and B. A. Torp, *Inorg. Chem.,* 1963, **2,** 540; M. Albutt, K. Feeman and G. W. Fowles, *J. Less-Common Met.,* 1964, **6,** 299.
516. G. W. A. Fowles and K. F. Gadd, *J. Chem. Soc. (A),* 1970, 2233.
517. R. L. Kepert and D. L. Deutscher, *Inorg. Chim. Acta,* 1970, **4,** 645.
518. K. Kirskey and J. B. Hamilton, *Inorg. Chem.,* 1972, **11,** 1945.
519. L. E. Manzer, *Inorg. Chem.,* 1977, **16,** 525.
520. F. A. Cotton, S. A. Duraj and W. J. Roth, *Inorg. Chem.,* 1984, **23,** 4046.
521. T. A. Dougherty, *Diss. Abstr. B,* 1967, **28,** 83.
522. A. J. Benton, M. G. B. Drew, R. J. Hobson and D. A. Rice, *J. Chem. Soc., Dalton Trans.,* 1981, 1304.
523. J. B. Hamilton and R. E. McCarley, *Inorg. Chem.,* 1970, **9,** 1333.
524. F. A. Cotton and W. J. Roth, *Inorg. Chem.,* 1984, **23,** 945.
525. A. Broll, H. G. Von Schnering and H. Schafer, *J. Less-Common Met.,* 1970, **22,** 243.
526. F. Calderazzo, M. Castellani, G. Pampaloni and P. F. Zanazzi, *J. Chem. Soc., Dalton Trans.,* 1985, 1989.
527. J. B. Hamilton and R. E. McCarley, *Inorg. Chem.,* 1970, **91,** 1339.
528. R. L. Deutscher and D. L. Kepert, *Inorg. Chem.,* 1970, **9,** 2305.
529. D. L. Kepert, B. W. Skelton and A. H. White, *J. Chem. Soc., Dalton Trans.,* 1981, 652.
530. F. A. Cotton, L. R. Falvello and R. C. Najjar, *Inorg. Chem.,* 1983, **22,** 770.
531. R. Gut, *Helv. Chim. Acta,* 1960, **43,** 830.
532. T. M. Brown and G. S. Newton, *Inorg. Chem.,* 1966, **5,** 1117.

533. D. J. Machin and J. F. Sullivan, *J. Less-Common Met.*, 1969, **19,** 405.
534. T. M. Brown and D. L. Bush, *J. Less-Common Met.*, 1971, **25,** 397.
535. E. J. Peterson, R. B. Von Dreele and T. M. Brown, *Inorg. Chem.*, 1976, **15,** 309.
536. P. M. Kierman and W. P. Griffith, *J. Chem. Soc., Dalton Trans.*, 1975, 2489.
537. P. M. Kierman, J. F. Gibson and W. P. Griffith, *J. Chem. Soc., Chem. Commun.*, 1973, 816.
538. I. F. Gainullin, N. S. Garifyanov and B. M. Kozyrev, *Dokl. Akad. Nauk SSSR,* 1968, **180,** 858.
539. Yu. A. Buslaev, M. A. Glushkova and N. A. Ovchinnikova, *Koord. Khim.*, 1975, **1,** 319.
540. D. A. Rice, *Coord. Chem. Rev.*, 1978, **25,** 199.
541. M. J. Atherton and J. H. Holloway, *Adv. Inorg. Chem. Radiochem.*, 1979, **22,** 171.
542. J. Fenner, A. Rabenau and G. Trageser, *Adv. Inorg. Chem. Radiochem.*, 1980, **23,** 329.
543. S. M. Sinitsyna, V. G. Khlebodarov and N. A. Bukhtera, *Russ. J. Inorg. Chem.*, (*Engl. Trans.*), 1975, **20,** 1267.
544. H. G. Von Schnering and W. Beckman, *Z. Anorg. Allg. Chem.*, 1966, **347,** 231.
545. J. Rijnsdorp, G. J. De Lange and G. A. Wiegers, *J. Solid State Chem.*, 1979, **30,** 365.
546. C. Perrin-Billot, A. Perrin and J. Prigent, *Chem. Commun.*, 1970, 676; C. Perrin, A. Perrin and P. Caillet, *J. Chim. Phys.*, 1973, 105.
547. H. F. Franzen, W. Hönle and H. G. Von Schnering, *Z. Anorg. Allg. Chem.*, 1983, **497,** 13.
548. J. Rijnsdorp and F. Jellinek, *J. Solid State Chem.*, 1979, **28,** 149.
549. I. S. Morozov and N. P. Dergacheva, *Otkritiya, Izobret. Prom. Obraztsy Tonvarnye Znaki,* 1974, **51,** 65 (*Chem. Abstr.*, 1975, **82,** 61 374).
550. A. J. Benton, M. G. B. Drew and D. A. Rice, *J. Chem. Soc., Chem. Commun.*, 1981, 1241.
551. M. Bochmann, G. Wilkinson, G. B. Young, M. B. Hursthouse and K. M. A. Malik, *J. Chem. Soc., Dalton Trans.*, 1980, 901.
552. R. L. Deutscher and D. Kepert, *Chem. Commun.*, 1969, 121.
553. T. J. Pinnavaia, B. L. Barnett and G. Podolski, *J. Am. Chem. Soc.*, 1975, **97,** 2712.
554. J. L. Morançais, L. G. Hubert-Pfalzgraf and P. Laurent, *Inorg. Chim. Acta,* 1983, **71,** 119.
555. A. A. Pasynskii, Yu. U. Skripkin and V. T. Kalinnikov, *J. Organomet. Chem.*, 1978, **150,** 51.
556. V. T. Kallinnikov, A. A. Passynskii, G. M. Larin, V. M. Novotortsev, Yu. T. Struchkov, A. I. Gusev and N. I. Kirilova, *J. Organomet. Chem.*, 1974, **74,** 91.
557. D. Attanasio, C. Belluto and A. Flamini, *Inorg. Chem.*, 1980, **19,** 3419.
558. J. Stach, R. Kirmse, W. Dietzsch, I. N. Marov and V. K. Belxaeva, *Z. Anorg. Allg. Chem.*, 1980, **466,** 36.
559. M. J. Bunker, A. De Cian, M. L. H. Green, J. J. E. Moreau and N. Siganporia, *J. Chem. Soc., Dalton Trans.*, 1980, 2155.
560. M. Sato and T. Yoshida, *J. Organomet. Chem.*, 1975, **87,** 217.
561. V. U. Tkachev, S. Shchepinov and T. O. Atoumyan, *J. Struct. Chem.* (*Engl. Transl.*), 1977, **18,** 823. R. N. McGinnis and J. B. Hamilton, *Inorg. Nucl. Chem. Lett.*, 1972, **8,** 245.
562. L. E. Manzer, *US Pat.* 4 042 610 (1975) (*Chem. Abstr.*, 1977, **87,** 184 686).
563. D. J. Machin and J. F. Sullivan, *J. Less-Common Met.*, 1969, **19,** 413.
564. N. Hovnanian and L. G. Hubert-Pfalzgraf, to be published.
565. R. T. Baker, P. J. Krusic, T. H. Tulip, J. C. Calabrese and S. S. Wreford, *J. Am. Chem. Soc.*, 1983, **105,** 6763.
566. R. Guilard, P. Richard, M. El Borai and E. Laviron, *J. Chem. Soc., Chem. Commun.*, 1980, 516.
567. K. Ukei, *Acta Crystallogr., Sect. B,* 1982, **38,** 1288.
568. M. D. Curtis and J. Real, *Organometallics,* 1985, **4,** 940.
569. M. Behnam-Dehkordy, B. Crociani, M. Nicolini and R. L. Richards, *J. Organomet. Chem.*, 1979, **181,** 69.
570. J. Rijnsdorp and J. F. Jellinek, *J. Solid State Chem.*, 1978, **25,** 325.
571. J. L. Hodeau, M. Marezio, C. Roucau, C. Ayroles, A. Meerschaut, J. Rouxel and P. Monceau, *J. Phys. Chim.*, 1978, **11,** 4117.
572. A. Meerschaut, L. Guemas and J. Rouxel, *C.R. Hebd. Seances Acad. Sci.*, 1980, **290,** 215.
573. R. Schoellhorn and A. Weiss, *Z. Naturforsch., Teil B,* 1973, **28,** 172, 716; B. Bach and J. M. Thomas, *J. Chem. Soc., Chem. Commun.*, 1972, 301; A. H. Thompson, *Nature* (*London*), 1974, **251,** 1492.
574. B. Di Pietro, M. Patriarca and B. Scrosati, *Synth. Met.*, 1982, **5,** 1.
575. I. H. Elson, J. K. Kochi, U. Klabunde, L. E. Manzer, G. W. Parshall and F. N. Tebbe, *J. Am. Chem. Soc.*, 1974, **96,** 7374.
576. P. A. Belmonte, F. G. N. Cloke and R. R. Schrock, *J. Am. Chem. Soc.*, 1983, **105,** 2643.
577. M. R. Churchill and H. J. Wasserman, *Inorg. Chem.*, 1982, **21,** 226.
578. M. R. Churchill and W. J. Youngs, *Inorg. Chem.*, 1981, **20,** 382.
579. M. R. Churchill, H. J. Wasserman, P. A. Belmonte and R. R. Schrock, *Organometallics,* 1982, **1,** 559; M. R. Churchill and H. J. Wasserman, *J. Organomet. Chem.*, 1983, **247,** 175.
580. R. E. McCarley and E. T. Maas, *Inorg. Chem.*, 1973, **12,** 1096.
581. J. L. Templeton, W. C. Dorman, J. C. Clardy and R. E. McCarley, *Inorg. Chem.*, 1978, **17,** 1263.
582. M. Pouchard, M. R. Torki, G. Demazeau and P. Hagenmüller, *C.R. Hebd. Seances Acad. Sci., Ser. C,* 1971, **273,** 1093.
583. A. D. Allen and S. Naito, *Can. J. Chem.*, 1976, **54,** 2948.
584. L. G. Hubert-Pfalzgraf, M. Tsunoda and J. G. Riess, *Inorg. Chim. Acta,* 1980, **41,** 283.
585. F. A. Cotton and R. C. Najjar, *Inorg. Chem.*, 1981, **20,** 1866.
586. L. G. Hubert-Pfalzgraf and J. G. Riess, *Inorg. Chim. Acta,* 1978, **29,** L251; L. G. Hubert-Pfalzgraf, M. Tsunoda and J. G. Riess, *Inorg. Chim. Acta,* 1981, **52,** 231.
587. A. P. Sattelberger, R. B. Wilson and J. C. Huffman, *J. Am. Chem. Soc.*, 1980, **102,** 7111; *Inorg. Chem.*, 1982, **21,** 2392.
588. M. L. Luetkens, W. L. Elcesser, J. C. Huffman and A. P. Sattelberger, *Inorg. Chem.*, 1984, **23,** 1718.
589. P. Sobota and B. Jezowska-Trzebiatowska, *Coord. Chem. Rev.*, 1978, **26,** 71; P. Sobota, T. Pluzinski, B. Jezowska-Trzebiatowska and S. Rummel, *J. Organomet. Chem.*, 1980, **185,** 69.
590. H. W. Turner, R. R. Schrock, J. D. Fellman and S. J. Holmes, *J. Am. Chem. Soc.*, 1983, **105,** 4942.
591. F. A. Cotton, M. P. Diebold and W. J. Roth, *Inorg. Chim. Acta,* 1985, **105,** 41.

592. E. Hey, F. Weller and K. Dehnicke, *Nature (London)*, 1983, **70,** 41; *Z. Anorg. Allg. Chem.*, 1984, **514,** 25.
593. M. E. Clay and M. T. Brown, *Inorg. Chim. Acta,* 1982, **58,** 1.
594. M. E. Clay and M. T. Brown, *Inorg. Chim. Acta,* 1983, **72,** 75.
595. F. A. Cotton, L. R. Falvello and R. C. Najjar, *Inorg. Chem.*, 1983, **22,** 375.
596. S. Datta and S. S. Wreford, *Inorg. Chem.*, 1977, **16,** 1134.
597. F. A. Cotton, L. R. Falvello and R. C. Najjar, *Inorg. Chim. Acta,* 1982, **63,** 107.
598. L. G. Hubert-Pfalzgraf, unpublished results.
599. L. G. Hubert-Pfalzgraf, *Inorg. Chim. Acta,* 1983, **76,** L233.
600. J. L. Templeton and R. E. McCarley, *Inorg. Chem.*, 1978, **17,** 2293.
601. S. Shaik, R. Hoffman, C. R. Fisell and R. H. Summerville, *J. Am. Chem. Soc.*, 1980, **102,** 4555.
602. F. A. Cotton and W. J. Roth, *Inorg. Chim. Acta,* 1983, **71,** 175.
603. G. Aharonian and L.G. Hubert-Pfalzgraf, to be published.
604. L. G. Hubert-Pfalzgraf, A. Chaloyard and N. E. Murr, *Inorg. Chim. Acta,* 1982, **65,** L173.
605. A. P. Sattelberger, R. B. Wilson and J. C. Huffman, *Inorg. Chem.*, 1982, **21,** 4179.
606. J. R. Wilson, A. P. Sattelberger and J. C. Huffman, *J. Am. Chem. Soc.*, 1982, **104,** 858.
607. A. P. Sattelberger, in 'Towards the 21st Century', ed. M. H. Chisholm, ACS Symposium Series, American Chemical Society, Washington, DC, 1983.
608. F. A. Cotton and W. T. Hall, *Inorg. Chem.*, 1980, **19,** 2352.
609. E. A. Robinson, *J. Chem. Soc., Dalton Trans.*, **1981,** 2373.
610. F. A. Cotton and W. T. Hall, *Inorg. Chem.*, 1981, **20,** 1285.
611. F. A. Cotton and W. T. Hall, *Inorg. Chem.*, 1980, **19,** 2354.
612. F. A. Cotton and K. J. Karol, *Macromolecules,* 1981, **14,** 233.
613. F. A. Cotton, L. R. Falvello and R. C. Najjar, *Organometallics,* 1982, **1,** 1640.
614. V. C. Gibson, P. D. Grebenik and M. L. H. Green, *J. Chem. Soc., Chem. Commun.*, 1983, 1101.
615. F. A. Cotton and W. J. Roth, *J. Am. Chem. Soc.*, 1983, **105,** 3734; F. A. Cotton, S. A. Duraj and W. J. Roth, *J. Am. Chem. Soc.*, 1984, **106,** 6989; F. A. Cotton and W. J. Roth, *Inorg. Chim. Acta,* 1987, **126,** 161.
616. (a) P. Laurent and L. G. Hubert-Pfalzgraf, unpublished results; (b) F. A. Cotton, M. P. Diebold and W. J. Roth, *Inorg. Chem.*, 1985, **24,** 3509.
617. A. V. Lobanov, G. A. Abakumov and G. A. Razuvaev, *Dokl. Akad. Nauk SSSR,* 1977, **235,** 824 (*Chem. Abstr.*, 1977, **87,** 126 509).
618. F. A. Cotton and W. J. Roth, *Inorg. Chem.*, 1983, **22,** 868.
619. L. G. Hubert-Pfalzgraf, M. Tsunoda and D. Katoch, *Inorg. Chim. Acta,* 1981, **51,** 81.
620. D. G. Blight, R. L. Deutscher and D. L. Kepert, *J. Chem. Soc., Dalton Trans.*, 1972, 87.
621. F. A. Cotton, S. A. Duraj and W. J. Roth, *J. Am. Chem. Soc.*, 1984, **106,** 3527; F. A. Cotton, M. P. Diebold, M. Matusz and W. J. Roth, *Inorg. Chim. Acta,* 1986, **112,** 147.
622. R. Bosselarr, B. G. Van Der Heyden and R. Mieras, *Inorg. Nucl. Chem. Lett.*, 1971, **7,** 1199.
623. I. L. Eremenko, Yu. V. Skripkin, A. A. Pasynskii, V. T. Kalinnikov, Yu. T. Struchkov and G. G. Aleksandrov, *J. Organomet. Chem.*, 1981, **220,** 159.
624. H. Funk and M. Hesselbarth, *Z. Chem.*, 1966, **6,** 227; K. Andra, *Z. Anorg. Allg. Chem.*, 1970, **373,** 209.
625. H. Malissa and H. Kolbe-Rohde, *Talanta,* 1961, **8,** 841.
626. P. J. Domaille, B. M. Foxman, T. J. McNeese and S. S. Wreford, *J. Am. Chem. Soc.*, 1980, **102,** 4114.
627. N. Hovnanian and L. G. Hubert-Pfalzgraf, *J. Organomet. Chem.*, 1986, **299,** C29; N. Hovnanian, Ph.D. Thesis, Universty of Nice, 1986.
628. B. M. Foxman, T. M. McNeese and S. S. Wreford, *Inorg. Chem.*, 1978, **17,** 2311.
629. A. M. Cardoso, R. J. H. Clarke and S. Moorhouse, *J. Organomet. Chem.*, 1980, **186,** 237.
630. R. R. King and C. D. Hoff, *J. Organomet. Chem.*, 1982, **225,** 245.
631. G. S. Bristow, P. B. Hitchcock and M. F. Lappert, *J. Chem. Soc., Chem. Commun.*, 1981, 1145.
632. R. Serrano and P. Royo, *J. Organomet. Chem.*, 1983, **247,** 33.
633. L. Acedo, A. Otera and P. Royo, *J. Organomet. Chem.*, 1983, **258,** 181.
634. G. Smith, R. R. Schrock, M. R. Churchill and W. J. Youngs, *Inorg. Chem.*, 1981, **20,** 387.
635. M. B. Hursthouse, A. M. Galas, A. M. Soares and W. P. Griffith, *J. Chem. Soc., Chem. Commun.*, 1980, 1167.
636. R. N. Gupta and B. K. Sen, *Z. Anorg. Allg. Chem.*, 1973, **398,** 312.
637. D. F. Hadjipanayoti-Stabaki and D. Katakis, *J. Inorg. Nucl. Chem.*, 1977, **39,** 995.
638. R. N. Gupta and B. K. Sen, *J. Inorg. Nucl. Chem.*, 1975, **37,** 1044.
639. B. K. Sen, R. Maity, R. N. Gupta and P. Bandyopadhyky, *Anal. Chim. Acta,* 1976, **81,** 173.
640. A. V. Saha, R. K. Maiti, R. N. Gupta and B. K. Sen, *Indian J. Chem., Sect. A,* 1977, **15,** 43.
641. Y. E. Federov, V. K. Eustafev, S. D. Kirik and A. V. Mishchenko, *Russ. J. Inorg. Chem. (Engl. Transl.)*, 1981, **26,** 1447.
642. E. Wiberg and H. Neumaier, *Z. Anorg. Allg. Chem.*, 1965, **340,** 189.
643. D. L. Kepert and K. Vrieze, in 'Halogen Chemistry', ed. V. Gutman, 1967, vol. 3.
644. A. Simon, *Angew. Chem., Int. Ed. Engl.*, 1981, **20,** 1.
645. A. Muller, R. Jostes and F. A. Cotton, *Angew. Chem., Int. Ed. Engl.*, 1980, **19,** 875.
646. A. Simon and H. G. Von Schnering, *Z. Anorg. Allg. Chem.*, 1965, **339,** 155.
647. A. Broll, A. Simon, H. G. Von Schnering and H. Schafer, *Z. Anorg. Allg. Chem.*, 1969, **367,** 1.
648. K. W. Koknat, J. A. Parsons and A. Vongvusharintra, *Inorg. Chem.*, 1974, **13,** 1699.
649. W. A. Herrmann, H. Biersack, M. L. Ziegler, K. Weidenhammer, R. Siegel and D. Rehder, *J. Am. Chem. Soc.*, 1981, **103,** 1692.
650. G. W. A. Fowles, T. D. Tidmarsh and R. A. Walton, *J. Inorg. Nucl. Chem.*, 1969, **31,** 2373.
651. (a) A. Bino, *J. Am. Chem. Soc.*, 1980, **102,** 7990; (b) F. A. Cotton, M. P. Diebold, R. Llusar and W. J. Roth, *J. Chem. Soc., Chem. Commun.*, 1986, 1276.
652. F. O. Fischer and F. Röhrscheid, *J. Organomet. Chem.*, 1966, **6,** 53.
653. M. R. Churchill and S. W. Y. Chang, *J. Chem. Soc., Chem. Commun.*, 1974, 248.
654. R. B. King, D. M. Braitsch and P. N. Kapoor, *J. Am. Chem. Soc.*, 1975, **97,** 60.

655. S. Z. Goldberg, B. Spivack, G. Stanley, R. Eisenberg, D. M. Braitsch, J. S. Miller and M. Abkowitz, *J. Am. Chem. Soc.*, 1977, **99**, 110.
656. P. A. Vaughlan, J. H. Sturdivant and L. Pauling, *J. Am. Chem. Soc.*, 1950, **72**, 5477.
657. H. Schäfer, H. Plautz and H. Baumann, *Z. Anorg. Allg. Chem.*, 1973, **401**, 63.
658. B. G. Hughes, J. L. Meyer, P. B. Fleming and R. E. McCarley, *Inorg. Chem.*, 1970, **9**, 1343.
659. N. Brnicevic and H. Schafer, *Z. Anorg. Allg. Chem.*, 1978, **441**, 219.
660. B. Spreckelmeyer, *Z. Anorg. Allg. Chem.*, 1968, **358**, 148.
661. H. S. Harned, C. Pauling and R. B. Corey, *J. Am. Chem. Soc.*, 1960, **82**, 4815.
662. R. B. Burbank, *Inorg. Chem.*, 1966, **5**, 1491.
663. R. A. Field, D. L. Kepert and D. Taylor, *Inorg. Chim. Acta*, 1970, **4**, 113.
664. R. Schöllhorn, K. Wagner and H. Jonke, *Angew. Chem., Int. Ed. Engl.*, 1981, **20**, 109.
665. R. J. Epenson, *Inorg. Chem.*, 1968, **7**, 636; R. J. Epenson and R. E. McCarley, *Inorg. Chem.*, 1968, **7**, 631.
666. B. Spreckelmeyer and H. Schäfer, *J. Less-Common Met.*, 1967, **13**, 127.
667. R. E. McCarley, B. G. Hughes, F. A. Cotton and R. Zimmerman, *Inorg. Chem.*, 1965, **4**, 1491.
668. J. P. Kuhn and R. E. McCarley, *Inorg. Chem.*, 1965, **4**, 1482.
669. N. E. Cooke, T. Kuwana and J. Espenson, *Inorg. Chem.*, 1971, **101**, 1081.
670. D. D. Klendworth and R. A. Walton, *Inorg. Chem.*, 1981, **20**, 1151.
671. R. A. McKay and R. F. Schneider, *Inorg. Chem.*, 1967, **6**, 549; 1968, **7**, 455.
672. R. J. Allen and J. C. Sheldon, *Aust. J. Chem.*, 1965, **18**, 277.
673. A. Vogler and H. Kunkely, *Inorg. Chem.*, 1984, **23**, 1360.
674. J. L. Meyer and R. E. McCarley, *Inorg. Chem.*, 1978, **17**, 1867.
675. B. Spreckelmeyer and H. G. Von Schnering, *Z. Anorg. Allg. Chem.*, 1971, **386**, 27.
676. C. B. Thaxton and R. A. Jacobson, *Inorg. Chem.*, 1971, **10**, 1460.
677. S. S. Basson and J. G. Leipoldt, *Transition Met. Chem.*, 1982, **7**, 207.
678. P. B. Fleming, J. L. Meyer, W. K. Grindstaff and R. E. McCarley, *Inorg. Chem.*, 1970, **9**, 1769.
679. P. M. Boorman and B. P. Straughan, *J. Chem. Soc. (A)*, 1966, 1515.
680. R. Mattes, *Z. Anorg. Allg. Chem.*, 1969, **346**, 279.
681. S. A. Best and R. A. Walton, *Inorg. Chem.*, 1979, **18**, 484.
682. F. A. Cotton and T. E. Haas, *Inorg. Chem.*, 1964, **3**, 10.
683. M. B. Robin and N. A. Kuebler, *Inorg. Chem.*, 1965, **4**, 978.
684. R. F. Schneider and R. A. McKay, *J. Chem. Phys.*, 1968, **42**, 843.
685. D. J. Robbins and A. J. Thomson, *J. Chem. Soc., Dalton Trans.*, 1972, 2350.
686. B. Speckelmeyer and H. Schäfer, *Z. Anorg. Allg. Chem.*, 1969, **365**, 225.
687. J. G. Converse, J. B. Hamilton and R. E. McCarley, *Inorg. Chem.*, 1970, **9**, 1366.
688. A. Simon, H. G. Von Schnering and H. Schäfer, *Z. Anorg. Allg. Chem.*, 1967, **355**, 295.
689. L. R. Bateman, J. F. Blount and L. F. Dahl, *J. Am. Chem. Soc.*, 1966, **88**, 1982.
690. A. Simon, *Z. Anorg. Allg. Chem.*, 1967, **355**, 311.
691. H. Imoto and J. D. Corbett, *Inorg. Chem.*, 1980, **19**, 1241.
692. M. L. Luetkens, J. C. Huffman and A. P. Sattelberger, *J. Am. Chem. Soc.*, 1983, **105**, 4474; M. L. Luetkens, W. L. Elcesser, J. C. Huffman and A. P. Sattelberger, *J. Chem. Soc., Chem. Commun.*, 1983, 1072.
693. A. M. Bond, J. W. Bixler, E. Mocellin, S. Datta, E. J. James and S. S. Wreford, *Inorg. Chem.*, 1980, **19**, 1760.
694. Yu. A. Buslaev, A. A. Kuznetsova and L. F. Goryachova, *Inorg. Mater. (Engl. Transl.)* 1967, **3**, 1488.
695. T. J. McNeese, S. S. Wreford and B. M. Foxman, *J. Chem. Soc., Chem. Commun.*, 1978, 500.
696. H. Schmidt and J. Noack, *Angew. Chem.*, 1957, **69**, 638.
697. R. J. Burt and G. J. Leigh, *J. Organomet. Chem.*, 1978, **148**, C19.
698. M. L. Luetkens, J. C. Huffman and A. P. Sattelberger, *J. Am. Chem. Soc.*, 1985, **107**, 3361.
699. F. N. Tebbe, *J. Am. Chem. Soc.*, 1973, **95**, 5823.
700. M. L. Luetkens, D. J. Santure, J. C. Huffman and A. P. Sattelberger, *J. Chem. Soc., Chem. Commun.*, 1985, 552.
701. E. Guggolz, M. L. Ziegler, H. Biersack and W. A. Herrmann, *J. Organomet. Chem.*, 1980, **194**, 317; W. A. Herrmann and H. Biersack, *Chem. Ber.*, 1979, **112**, 3942.
702. H. C. Bechtold and D. Rehder, *J. Organomet. Chem.*, 1981, **206**, 305.
703. L. N. Lewis and K. G. Caulton, *Inorg. Chem.*, 1980, **19**, 3201.
704. W. A. Herrmann, H. Biersack, M. L. Ziegler and B. Balbach, *J. Organomet. Chem.*, 1981, **206**, C33.
705. A. A. Pasynskii, J. L. Eremenko, B. Orazsakhatov, V. T. Kalinnikov, G. G. Aleksandrov and Yu. T. Syruchkov, *J. Organomet. Chem.*, 1981, **216**, 211.
706. S. S. Wreford and J. F. Whitney, *Inorg. Chem.*, 1981, **20**, 3918.
707. J. D. Fellmann, G. A. Rupprecht and R. R. Schrock, *J. Am. Chem. Soc.*, 1979, **101**, 5099.
708. P. R. Brown, F. G. N. Cloke and M. L. H. Green, *J. Chem. Soc., Chem. Commun.*, 1980, 1126.
709. J. O. Albright, S. Datta, B. Dezube, J. K. Kouba, D. S. Marynick, S. S. Wreford and B. M. Foxman, *J. Am. Chem. Soc.*, 1979, **101**, 611.
710. T. Kruck and H. U. Hempel, *Angew. Chem.*, 1971, **83**, 437.
711. F. G. N. Cloke, P. J. Fyne, V. C. Gibson, M. L. H. Green, M. J. Ledoux, R. N. Perutz, A. Dix, A. Gourdon and C. K. Prout, *J. Organomet. Chem.*, 1984, **277**, 61.
712. J. Chatt and H. R. Watson, *J. Chem. Soc. (A)*, 1962, 2545.
713. J. Quirk and G. Wilkinson, *Polyhedron*, 1982, **1**, 209.
714. F. G. N. Cloke and M. L. H. Green, *J. Chem. Soc., Dalton Trans.*, 1981, 1938.
715. D. W. Green, R. V. Hodges and D. M. Gruen, *Inorg. Chem.*, 1976, **15**, 970.
716. J. E. Ellis and A. Davidson, *Inorg. Synth.*, 1976, **16**, 68.
717. F. Calderazzo, U. Englert, G. Pampaloni, G. Pelizzi and R. Zamboni, *Inorg. Chem.*, 1983, **22**, 1865.
718. C. G. Dewey, J. E. Ellis, K. L. Fjare, K. M. Pfahl and G. F. P. Warnock, *Organometallics*, 1983, **2**, 388.
719. F. Calderazzo and G. Pampaloni, *J. Organomet. Chem.*, 1983, **250**, C33.
720. L. J. Guggenberger and R. R. Schrock, *J. Am. Chem. Soc.*, 1975, **97**, 6693.

721. F. Calderazzo, G. Pampaloni and G. Pelizzi, *J. Organomet. Chem.*, 1982, **233,** C41.
722. D. Rehder, H. C. Bechthold and K. Paulsen, *J. Magn. Reson.*, 1980, **40,** 305.
723. S. Herzog and E. Wulf, *Z. Chem.*, 1966, **11,** 434.
724. M. H. Chisholm, J. C. Huffman, I. P. Rothwell, P. G. Bradley, N. Kress and W. H. Woodruff, *J. Am. Chem. Soc.*, 1981, **103,** 4945.
725. (a) G. F. P. Warnock, J. Sprague, K. L. Fjare and J. E. Ellis, *J. Am. Chem. Soc.*, 1983, **105,** 672; (b) G. P. F. Warnock and J. E. Ellis, *J. Am. Chem. Soc.*, 1984, **106,** 5016.
726. K. M. Pfahl and J. E. Ellis, *Organometallics,* 1984, **3,** 230.
727. F. Näumann, K. Rehder and V. Pank, *J. Organomet. Chem.*, 1982, **240,** 363.
728. A. M. Koulkes-Pujo, B. Motais and L. G. Hubert-Pfalzgraf, *J. Chem. Soc., Dalton Trans.*, 1986, 1741.
729. J. Reed, *Inorg. Chim. Acta,* 1977, **21,** L36.
730. A. Antinolo, M. Fajardo, A. Otero and P. Royo, *J. Organomet. Chem.*, 1982, **234,** 309; 1983, **246,** 269.
731. (a) T. W. Coffindaffer, I. P. Rothwell, K. Folting, J. C. Huffman and W. E. Streib, *J. Chem. Soc., Chem. Commun.*, 1985, 1519; (b) L. R. Chamberlain, I. P. Rothwell and J. C. Huffman, *J. Am. Chem. Soc.*, 1986, **108,** 1502; (c) L. Chamberlain, I. P. Rothwell and J. C. Huffman, *J. Chem. Soc., Chem. Commun.*, 1986, 1203.
732. (a) R. E. La Pointe, P. T. Wolczanski and G. D. Van Duyne, *Organometallics,* 1985, **4,** 1810; (b) R. E. La Pointe and P. T. Wolczanski, *J. Am. Chem. Soc.*, 1986, **108,** 3535; (c) R. E. La Pointe, P. T. Wolczanski and J. F. Mitchell, *J. Am.. Chem. Soc.*, 1986, **108,** 6382.
733. C. E. Holloway and M. Melnik, *J. Organomet. Chem.*, 1986, **303,** 1.
734. C. E. Holloway and M. Melnik, *J. Organomet. Chem.*, 1986, **303,** 38.

35

Chromium

LESLIE F. LARKWORTHY and KEVIN B. NOLAN
University of Surrey, Guildford, UK

and

PAUL O'BRIEN
Queen Mary College, London, UK

35.1 INTRODUCTION TO CHROMIUM COMPLEXES

Chromium, molybdenum and tungsten are the transition metal members of group VI of the periodic table. Chromium has the outer electronic configuration $3d^54s^1$ and forms compounds in oxidation states −II to +VI (Table 1). It differs in its chemistry from molybdenum and tungsten, which are alike because of their similar atomic and ionic radii. Most resemblances are in the 0 and I oxidation states, which are stabilized by π-acceptor C,N heterocyclic and P donor ligands. The extensive organometallic chemistry of Cr, Mo and W is described in the companion series 'Comprehensive Organometallic Chemistry'.[1]

Chromium(II) has extensive coordination chemistry provided the strongly reducing complexes are protected against aerial oxidation. Coordination numbers from three to seven are exhibited by chromium(II), although it is commonly six-coordinate, forming complexes which are high-spin ($t_{2g}^3e_g^1$ configuration) or, with strong-field ligands, low-spin (t_{2g}^4 configuration). Marked Jahn–Teller distortion of the octahedron (4 + 2 coordination), arising from the uneven filling of the d-orbitals, has been found for the d^9 (Cr^{2+}) and high-spin d^4 (Cr^{2+}) ions, and complexes of these ions are often isomorphous. The distortion leads to weak ferromagnetism in the halide-bridged structures of $[NRH_3]_2[CrX_4]$. Examples of planar chromium(II) complexes (the limit of 4 + 2 coordination) are becoming more common, but none containing tetrahedral chromium(II) has been firmly established except the atypical complex $[Cr(NO)\{N(SiMe_3)_2\}_3]$. There are similarities to molybdenum(II) in dinuclear complexes like the carboxylates, but the ionic nature of CrX_2 is very different from the covalent clusters of MoX_2.

Complexes of chromium(III), the most stable and important oxidation state, continue to be discovered as the complexing ability of new ligands with the first transition series is investigated. There are now a few examples of coordination number greater than six. The coordination chemistry of chromium(III) far surpasses that of molybdenum(III) and tungsten(III). Octahedral chromium(III) complexes are kinetically inert and this, combined with ease of synthesis, is why so many have been isolated. Their magnetic moments are close to the spin-only value (t_{2g}^3 configuration, 3.87 BM), except in the numerous hydroxo-bridged and other polynuclear species where antiferromagnetism produces lower, temperature-variable values. Chromium(III) complexes have well-characterized electronic spectra and extensive photochemistry.

Chromium(IV) and chromium(V), previously encountered as the oxides and halides and as unstable intermediates in solution, are now represented by complexes of ligands stabilized by heavy substitution against oxidation by the metal ion, and oxo and nitrido derivatives. These remain unimportant oxidation states compared with molybdenum and tungsten.

Chromium(VI) is powerfully oxidizing and is found in comparatively few but important

Table 1 Oxidation States and Stereochemistry of Chromium[a]

Oxidation state	*Coordination number*	*Stereochemistry*	*Examples*
Cr^{0}, d^6	6	Octahedral	$Cr(bipy)_3$, $Cr(PF_3)_6$, $Cr(CNBu^t)_6$
Cr^{I}, d^5	6	Octahedral	$[Cr(CNPh)_6]^+$, $[Cr(bipy)_3]^+$
Cr^{II}, d^4	3	Distorted 'T'	$Cr(OCBu^t_3)_2 \cdot LiCl(THF)_2$
	4	Planar	$Cr\{N(SiMe_3)_2\}_2(THF)_2$, Cr(acacen), $[NEt_4]_2[Cr(S_2C_2H_4)_2]$
		Distorted tetrahedral	$Cr(NO)\{N(SiMe_3)_2\}_3$, $CrCl_2(MeCN)_2$?
	5	Trigonal bipyramidal	$[CrBr\{N(CH_2CH_2NMe_2)_3\}]^+$, $Cr_2(S_2CNEt_2)_4$?
		Square pyramidal	$[Cr(NH_3)_4H_2O]^{2+}$
	6	High-spin (distorted octahedral)	$[Cr(en)_3]^{2+}$, $[CrCl_4]_n^{2n-}$
		Low-spin (octahedral)	$[Cr(phen)_3]^{2+}$, $[Cr(CN)_6]^{4-}$
		$(Cr{\equiv}Cr)^{4+}$ units	$[Cr_2(O_2CMe)_4(H_2O)_2]$, $[Cr_2(2\text{-}(O), 6\text{-}Mepy)_4]$
	7	Tetragonal base: trigonal cap (4:3)	$[Cr(CNBu^t)_7]^{2+}$
		Pentagonal bipyramidal	$CrH_2\{P(OMe)_3\}_5$
Cr^{III}, d^3	3	Trigonal planar	$Cr(NPr_2)_3$
	4	Tetrahedral	$[CrCl_4]^-$?
	5	Trigonal bipyramidal	$CrCl_3(NMe_3)_2$
	6	Octahedral	$[Cr(NH_3)_6]^{3+}$, $Cr(acac)_3$, $[Cr(CNPh)_6]^{3+}$
	7	Pentagonal bipyramidal	$[CrC_5H_3N(CMeNNHCONH_2)_2(H_2O)_2]^{3+}$
Cr^{IV}, d^2	4	Tetrahedral	$Cr(OBu^t)_4$, CrO_4^{4-}, $CrCl_4$
	5	Square pyramidal	CrO(TPP)
	6	Octahedral	$[CrF_6]^{2-}$
	7	Pentagonal bipyramidal	$Cr(NH_3)_3(O_2)_2$
	8	Dodecahedral	$CrH_4(PMe_2CH_2CH_2PMe_2)_2$
Cr^{V}, d^1	4	Tetrahedral	CrO_4^{3-}
	5	Square pyramidal	$[CrOCl_4]^-$, CrN(salen), $[CrO(O_2COCMeEt)_2]^-$
		Trigonal bipyramidal	CrF_5
	6	Octahedral	$[CrOCl_5]^{2-}$
	8	Dodecahedral	$[Cr(O_2)_4]^{3-}$
Cr^{VI}, d^0	4	Tetrahedral	CrO_4^{2-}, CrO_2Cl_2, $Cr(NBu^t)_2(OSiMe_3)_2$
	5	Pentagonal pyramidal	$CrO(O_2)_2py$
	6	Octahedral	CrF_6, $[CrO_2F_4]^{2-}$
	7	Pentagonal bipyramidal	$CrO(O_2)_2(bipy)$

[a] There are many examples of organometallic compounds of chromium in oxidation states −II to III.[1]

compounds. These are usually oxo or oxo halide species. Besides chromates and dichromates, tri- and tetra-chromates are known, but compounds analogous to the isopoly- and heteropoly-molybdates and -tungstates, and the molybdenum and tungsten bronzes, are not formed by chromium. There are weak similarities to sulfur(VI).

The industrial and biochemical importance of coordination compounds of chromium is described in Volume 6, and aspects of theoretical, mechanistic and solution chemistry are discussed in Volume 1 of this series.

Earlier work on the coordination and general chemistry of chromium is described in several treatises:[2–4] there are reviews for the years 1979–1983,[5,6] and a list of crystal structures to mid-1976 is available.[7] Summaries of chromium chemistry have been provided in two recent textbooks.[8,9]

35.2 CHROMIUM(0) AND CHROMIUM(I)

The extensive organometallic chemistry of chromium, *i.e.* the hexacarbonyl and its derivatives, organochromium compounds without carbonyl ligands, cyanide and isocyanide complexes, alkene, allyl, diene, cyclopentadiene and arene derivatives, and complexes of σ-donor carbon ligands, has been recorded in Chapters 26.1 and 26.2 of Volume 3 of 'Comprehensive Organometallic Chemistry'.[1] In the present section, chromium complexes mostly in oxidation states 0 and I are considered, although where the ligands are such that redox reactions lead to easy interconversion between oxidation states, complexes of other oxidation states have been included.

35.2.1 Group IV Ligands

35.2.1.1 Cyanides

Homoleptic cyanochromium complexes are known in oxidation states 0 to III, examples being $K_6[Cr(CN)_6]$, $K_3[Cr(CN)_4]$, $K_4[Cr(CN)_6]$ and $K_3[Cr(CN)_6]$. In addition, there are the cyanonitrosyls $K_4[Cr(CN)_5NO]$ and $K_3[Cr(CN)_5NO]$, the chromium(IV) peroxo derivative $K_3[Cr(O_2)_2(CN)_3]$ and mixed cyanoamine derivatives such as *cis*-$[Cr(CN)_2(en)_2]I$. All except the last, for which see Section 35.4.2.3, have been discussed in the companion series (Volume 3, Section 26.1.3.2).[1] More recent research on chromium(III) cyano and cyanonitrosyl complexes is covered in Sections 35.4.1.1 and 35.4.2.6 and the present section is restricted to some recent work on chromium(II) and chromium(0) complexes.

Attempts to isolate pure cyanochromium(II) complexes from the deep red solutions containing $[Cr(CN)_6]^{4-}$ ions formed when chromium(II) acetate is added to aqueous potassium cyanide in excess, and in other ways, have not been very successful. Now, through careful control of concentration and of pH, the known but poorly characterized green crystalline hexacyanochromates(II) $M_4[Cr(CN)_6]{\cdot}2H_2O$ (where M = Na or K), the new dark green simple cyanide $Cr(CN)_2{\cdot}2H_2O$, and their anhydrous forms have been prepared pure (Scheme 1 and Table 2).[10]

$$K_4(Na)[Cr(CN)_6]{\cdot}2H_2O \xleftarrow[\text{add EtOH}]{\text{KCN(NaCN)}} Cr^{2+}(aq)\ (pH \approx 7) \xrightarrow{\text{KCN}} Cr(CN)_2{\cdot}2H_2O\downarrow \xrightarrow[\text{vac.}]{100\,°C} Cr(CN)_2$$

$$K_4(Na)[Cr(CN)_6]{\cdot}2H_2O \xrightarrow{100\,°C,\ \text{vac.}} K_4(Na)[Cr^{II}(CN)_6] \xrightarrow[NH_3(l)]{K(Na)} K_6(Na)[Cr^{0}(CN)_6] \xrightarrow[NH_3(l)]{NH_4CN} K_3[Cr^{III}(CN)_6]$$

Scheme 1

Table 2 Magnetic and Spectroscopic Data for Cyanochromium(II) and Cyanochromium(0) Complexes

Complex	μ_{eff}^{293K} (BM)	$\nu(CN)$ (cm^{-1})	$\delta(MCN)$ (cm^{-1})	$\nu(MC)$ (cm^{-1})	*Reflectance spectra* (10^{-3} cm^{-1})		
$Cr(CN)_2{\cdot}2H_2O$	2.73	2182	520	—	37.4	8.5	
Dark green		2062					
$Cr(CN)_2$	2.79	2181	528	439	37.6	8.7	
Pale brown							
$K_4[Cr(CN)_6]{\cdot}2H_2O$	3.14	2122	451	345	38.0	25.0	14.5
Green		2020					
$K_4[Cr(CN)_6]$	3.42	2122	451	345	48.4	37.0	26.0
Green		2022			14.5	9.1	7.4
$Na_4[Cr(CN)_6]{\cdot}2H_2O$	3.30	2132	458	358	46.0	37.0	29.6
Blue		2020			14.6		7.7
$Na_4[Cr(CN)_6]$	3.48	2132	458	357	47.6	36.5	26.0
Blue		2020				9.1	7.8
$K_6[Cr(CN)_6]$	diam.	2018, 1920	605	450			
Dark green		1800	535				
$Na_6[Cr(CN)_6]$	diam.	2020	535	455			
Dark green		1920					

The low-spin t_{2g}^4 configuration of these complexes gives rise to a $^3T_{1g}$ ground term. Spin-allowed transitions to the ^3E_g, $^3T_{2g}$, $^3A_{1g}$ and $^3A_{2g}$ terms are possible, but no assignments have been made because of the breadth of the reflectance bands. The CN stretching frequencies of $Cr(CN)_2$ and $Cr(CN)_2{\cdot}2H_2O$ suggest that the former contains bridging and the latter bridging and terminal cyano groups. The decahydrate $Na_4[Cr(CN)_6]{\cdot}10H_2O$ has been prepared by reduction of $Na_3[Cr(CN)_6]$ with amalgamated aluminum sheet and its crystal structure determined.[11] Apparently, the decahydrate crystallizes but the dihydrate is the stable phase.

In liquid ammonia, hexacyanochromates(II) can be reduced to the diamagnetic chromium(0) complexes $M_6[Cr(CN)_6]$ by the appropriate alkali metal. The K salt was earlier obtained by reduction of $K_3[Cr(CN)_6]$. Greater metal–ligand π bonding is presumably responsible for the lower CN stretching frequencies of $M_6[Cr(CN)_6]$. The potassium salt is oxidized to $K_3[Cr(CN)_6]$ by ammonium cyanide in liquid ammonia (equation 1).[12]

$$2K_6[Cr(CN)_6] + 6NH_4CN \longrightarrow 2K_3[Cr(CN)_6] + 6KCN + 6NH_3 + 3H_2 \quad (1)$$

35.2.1.2 Aryl and alkyl isocyanides

Only more recent investigations of organic isocyanide complexes of chromium are in this section, because general synthetic methods for metal isocyanide complexes, including mixed carbonyl isocyanides, and the nature of the metal–isocyanide bond are discussed in Sections 27.1.3.2.2 and 26.1.2.5.2.ii of Volume 3 of 'Comprehensive Organometallic Chemistry'.[1] Organic isocyanides are π-bonding ligands like CO but are stronger σ donors and weaker π acceptors. Chromium(0) complexes such as $Cr(CNPh)_6$ are formally isoelectronic with $Cr(CO)_6$, but differ markedly in being sequentially oxidizable to Cr^{I}, Cr^{II} and Cr^{III} derivatives. These oxidation states were first observed in electrochemical studies (equations 2 to 4)[13,14] but solids of all four oxidation states are known, so, unlike carbon monoxide, isocyanides, owing to their greater σ-donor ability, form homoleptic complexes in normal as well as low oxidation states (Table 3 and Schemes 2–4). The earlier assignment of the $E_{1/2}$ value of equation (2) (*ca.* −0.3 V) to the half reaction $Cr(CNPh)_6^0/Cr(CNPh)_6^-$ has been revised[15] since attempts to reduce $Cr(CNPh)_6$ were unsuccessful, and $Cr(CNR)_6^-$ would be one electron in excess of the 18-electron configuration.

$$Cr(CNAr)_6 \rightleftharpoons Cr(CNAr)_6^+ + e^-, E_{1/2} \sim -0.3\text{ V } (\textit{versus}\text{ SCE}) \quad (2)$$

$$Cr(CNAr)_6^+ \rightleftharpoons Cr(CNAr)_6^{2+} + e^-, E_{1/2} \sim 0.2\text{ V} \quad (3)$$

$$Cr(CNAr)_6^{2+} \rightleftharpoons Cr(CNAr)_6^{3+} + e^-, E_{1/2} \sim 1\text{ V} \quad (4)$$

Table 3 Complexes of Alkyl and Aryl Isocyanides

Complex	*Comments*	*Ref.*
Chromium(0)		
Alkyl isocyanides		
$Cr(CNR)_6$ Red-brown	R = Bu^t, diamagnetic, ^{1}H NMR, δ 1.40 p.p.m., ν(CN) = 2100w, 1960s, 1870s cm^{-1}; R = Bu^n	1–3
$Cr(CNC_6H_{11})_6$ Yellow-orange	Diamagnetic, ν(CN) = 1871 cm^{-1}	1, 3
Aryl isocyanides		
$Cr(CNC_6H_4R)_6$ Red or orange	R = H, 4-Me, 4-OMe, 4-Cl, 4-Br, 3-Cl, 3-Cl-2Me, 4-Cl-2-Me, 2,4-Me_2, 2,5-Cl_2, R = 2-Me, 3-Me, 4-NMe_2, 4-F, 3-OMe, 3-CF_3; from $Cr_2(O_2CMe)_4(H_2O)_2$/EtOH or MeOH, and ArNC; diamagnetic; ν(CN) = 1935–1993 cm^{-1} (most intense peak); stable except R = 4-NMe_2; R = H, octahedral, Cr—C, 1.938 Å; C—N, 1.176 Å; Cr—C—N, 173.7°	4 5 6
Chromium(I)		
Aryl isocyanides		
$[Cr(CNC_6H_4R)_6]PF_6$ Yellow to red	R = H, 2-Me, 4-Me, 4-OMe, 4-Cl; from $Cr(CNC_6H_4R)_6$ + $AgPF_6$; R = 4-Me also from air + $Cr(CNR)_6$ and NH_4PF_6; and $[Cr(CNPh)_6]BPh_4$ from $Cr(CNPh)_6$, I_2 and $NaBPh_4$, μ_{eff}^{295K} = 2.04–2.18 BM, ν(CN) = 2060 cm^{-1} (most intense peak), R = 4-Me also from Cr^{2+}/EtOH, PhNC and KPF_6	7 8
$[Cr(CNC_6H_3Me_2\text{-}2,6)_6]PF_6$ Yellow	Prepared as above; ν(CN) = 2051 cm^{-1}	9
$[Cr(CNPh)_6]CF_3SO_3$	From $Cr(CNPh)_6$ + $AgCF_3SO_3$, octahedral, Cr—C (av), 1.975 Å; C—N, 1.159 Å, Cr—C—N, 178.7°	10

Table 3 *(continued)*

Complex	*Comments*	*Ref.*
Chromium(II)		
Alkyl isocyanides		
$[Cr(CNR)_6](PF_6)_2$ Pale yellow	R = Bu^t: ν(CN) = 2180 cm^{-1}; $\mu_{eff}^{r.t.}$ = 2.9 (Gouy), 2.87 BM (CH_2Cl_2), R = C_6H_{11}: ν(CN) = 2185 cm^{-1}; $\mu_{eff}^{r.t.}$ = 2.75 BM (CH_2Cl_2); R = Pr^i	11 12
$[Cr(CNR)_7](PF_6)_2$ Yellow	R = Bu^t: ν(CN) = 2160 cm^{-1}; diamagnetic (Gouy); ^{1}H NMR, δ 1.64–4.1 p.p.m. (dec., $CDCl_3$); Cr—C, 1.966–2.016, C—N, 1.142–1.155 Å R = C_6H_{11}: ν(CN) = 2150 cm^{-1}; diamagnetic, ^{1}H NMR, δ 1.50, 4.35→1.4, 3.25, 1.4 p.p.m. (dec., $CDCl_3$)	11 13 11
Aryl isocyanides		
$[Cr(CNC_6H_4R)_6](PF_6)_2$ Yellow	R = H, 2-Me, 4-Me, 4-OMe (maroon), 4-Cl; from $Cr^0(CNC_6H_4R)_6$ + 2$AgPF_6$; ν(CN) ≈ 2150 cm^{-1}; $\mu_{eff}^{295 K}$ = 3.05–3.14 BM, R = H, octahedral, Cr—C(av), 2.014; C—N(av), 1.158 Å; angular distortion of octahedron	7 10
'$[Cr(CNC_6H_3Me_2$-2,6)$_6](BF_4)_2 \cdot 8H_2O \cdot HBF_4$' Orange	From $[Cr(CNC_6H_3Me_2$-2,6)$_6](BF_4)_3$ and air; ν(CN) = 2147 cm^{-1}	14
Mixed-ligand complexes (alkyl isocyanides)		
trans-$[Cr^{II}(CNR)_4(PR'_3)_2](PF_6)_2$	R = Bu^t, C_6H_{11}; R′ = Et, Pr, Bu^n: μ_{eff} = 2.6–2.9 BM (CH_2Cl_2); ν(CN) = 2128–2150 cm^{-1} (one band)	11
cis-$[Cr^{II}(CNR)_4(dppe)](PF_6)_2$	R = Bu^t: μ_{eff} = 2.69 BM, ν(CN) = 2190, 2154 cm^{-1}; R = C_6H_{11}: μ_{eff} = 2.78 BM, ν(CN) = 2197, 2162 cm^{-1}	11
cis-$[Cr^{II}(CNR)_5(dppe)](PF_6)_2$	R = Bu^t; ν(CN) = 2175, 2130, 2105, 2045sh cm^{-1}; R = C_6H_{11}: ν(CN) = 2186, 2131 cm^{-1}	11
Mixed-ligand complexes (aryl isocyanides)		
$[CrCl(CNPh)_5]PF_6$ Yellow green	Dec. to $[Cr(CNPh)_6](PF_6)_2$ on recrystallization; ν(CN), 2129s, 2178w, 2194w cm^{-1}; μ_{eff} = 2.68 BM (CH_2Cl_2)	8
trans-$[CrCl(CNPh)_4(PR_3)]PF_6$ Orange to yellow	R = Et, Pr, Bu^n; R_3 = Et_2Ph; ν(CN) ≈ 2112 cm^{-1}; μ_{eff} = 2.73–2.98 BM	8
$[CrCl(CNPh)_3(dppe)]PF_6$ Orange red	ν(CN) = 2144m, 2097s cm^{-1}; μ_{eff} = 2.59 BM (CH_2Cl_2)	8
$[Cr(CNPh)_5(dppm)](PF_6)_2$ Orange	ν(CN) = 2170m, 2121s, 2080s cm^{-1}; diamagnetic, stable in solution	8
Chromium(III)		
Aryl isocyanides		
$[Cr(CNPh)_6](SbCl_6)_3$ Purple	From $Cr^0(CNPh)_6$ and $SbCl_5$; ν(CN) = 2208 cm^{-1} Cr^{3+}/Cr^{2+} = 0.73 V[a] CH_2Cl_2 solvate, octahedral, Cr—C (av), 2.068; C—N, 1.141 Å	14 9 10
$[Cr(CNC_6H_3Pr^i_2$-2,6)$_6](SbCl_6)_3$ Blue	From $Cr(CNC_6H_3Pr^i_2$-2,6)$_6$ and $SbCl_5$; ν(CN) = 2180 cm^{-1} Cr^{3+}/Cr^{2+} = 1.27 V[a]	14 9
$[Cr(CNC_6H_3Me_2$-2,6)$_6]X_3$	X = $SbCl_6$ (deep purple), prep. as above; ν(CN) = 2188 cm^{-1} X = BF_4 (maroon); from $[Cr^I(CNC_6H_3Me_2$-2,6)$_6]BF_4$ and $NOBF_4$, ν(CN) = 2214 cm^{-1}; $\mu_{eff}^{297 K}$ = 3.89 BM, Cr^{3+}/Cr^{2+} = 0.82 V[a]	14 9

[a] Formal potentials *versus* corrected AgCl/Ag in CH_2Cl_2 with 0.1 M NBu_4PF_6.

1. E. P. Kündig and P. L. Timms, *J. Chem. Soc., Dalton Trans.*, 1980, 991.
2. J. Müller and W. Holzinger, *Z. Naturforsch., Teil B*, 1978, **33,** 1309.
3. K. W. Chiu, C. G. Howard, G. Wilkinson, A. M. R. Galas and M. B. Hursthouse, *Polyhedron*, 1982, **1,** 803.
4. L. Malatesta, *Prog. Inorg. Chem.*, 1959, **1,** 283.
5. G. J. Essenmacher and P. M. Treichel, *Inorg. Chem.*, 1977, **16,** 800.
6. E. Ljungström, *Acta Chem. Scand., Ser. A*, 1978, **32,** 47.
7. P. M. Treichel and G. J. Essenmacher, *Inorg. Chem.*, 1976, **15,** 146.
8. F. R. Lemke, D. E. Wigley and R. A. Walton, *J. Organomet. Chem.*, 1983, **248,** 321.
9. D. A. Bohling, J. F. Evans and K. R. Mann, *Inorg. Chem.*, 1982, **21,** 3546.
10. D. A. Bohling and K. R. Mann, *Inorg. Chem.*, 1984, **23,** 1426.
11. W. S. Mialki, D. E. Wigley, T. E. Wood and R. A. Walton, *Inorg. Chem.*, 1982, **21,** 480.
12. D. E. Wigley and R. A. Walton, *Organometallics*, 1982, **1,** 1322.
13. J. C. Dewan, W. S. Mialki, R. A. Walton and S. J. Lippard, *J. Am. Chem. Soc.*, 1982, **104,** 133.
14. D. A. Bohling and K. R. Mann, *Inorg. Chem.*, 1983, **22,** 1561.

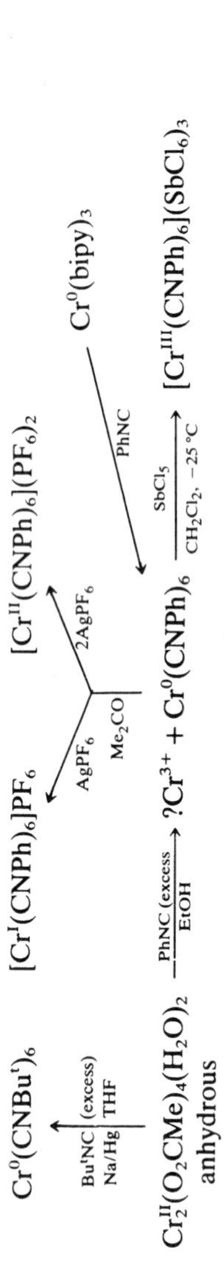

Scheme 2

Scheme 3

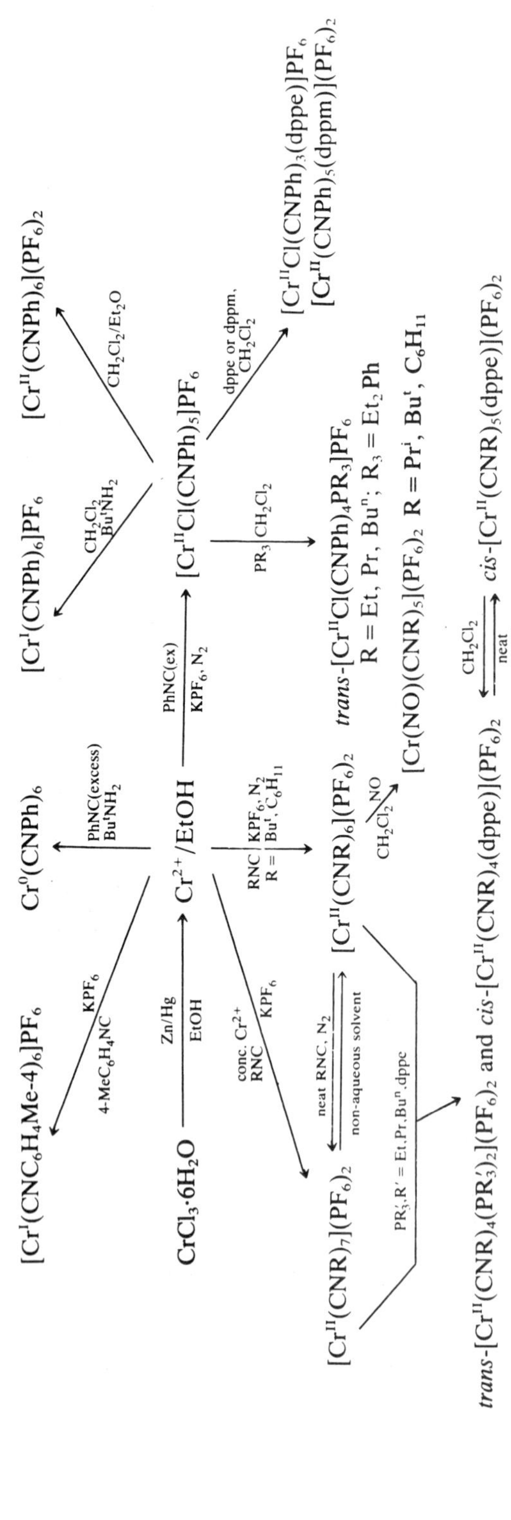

Scheme 4

Some recent advances are the isolation of aryl isocyanide complexes of chromium(III) $[Cr(CNAr)_6]X_3$, and six- and seven-coordinate alkyl isocyanide complexes of chromium(II) $[Cr(CNR)_{6,7}]X_2$. All but the chromium(0) and the seven-coordinate chromium(II) complexes have less than 18 electrons in the valency shell.

(i) Syntheses: oxidation state 0

Hexakis(aryl isocyanide)chromium(0) complexes have been known since 1952 when they were described by Malatesta and co-workers.[16] Alkyl isocyanide chromium(0) complexes were thought to be less stable than their aryl analogues, but this appears to be due mainly to a lack of suitable preparative methods because the complexes $Cr(CNR)_6$ (R = Bu, Bu^t or C_6H_{11}) have now been obtained by the action of RNC in excess on $Cr(C_{10}H_8)_2$, $Cr_2(C_8H_8)_3$,[17] the chromium(IV) compound $CrPr^i_4$ [18] or, more conveniently, anhydrous chromium(II) acetate in THF in the presence of Na/Hg.[19] The cleavage of the quadruple metal–metal bond in $[Cr_2(O_2CMe)_4]$ or the dihydrate by an excess of aryl isocyanide has long been standard[13] for the synthesis of the aryl derivatives $Cr(CNAr)_6$. This reaction (equation 5) involves disproportionation, although no Cr^{3+} complexes have been isolated from the reaction mixture. It may be that the use of Na/Hg as for the alkyl isocyanide complexes would improve the overall yield by reducing the Cr^{3+} species. The same reducing agent produces *trans*-$[Cr(CNBu^t)_2(dmpe)_2]$ from $[CrCl_2(dmpe)_2]$ (Scheme 8). Many zerovalent aryl isocyanide complexes are now known (Table 3).

$$3[Cr^{II}_2(O_2CMe)_4(H_2O)_2] + 12CNAr \longrightarrow 2Cr^0(CNAr)_6 + 4Cr^{3+} + 12MeCO_2^- + 6H_2O \qquad (5)$$

(ii) Syntheses: oxidation states I, II and III

Air oxidation of chromium(0) aryl isocyanides will produce chromium(I) cations $[Cr(CNAr)_6]^+$, but oxidation by stoichiometric amounts of $AgPF_6$ to give first the chromium(I) complexes $[Cr(CNAr)_6]PF_6$ and then the chromium(II) complexes $[Cr(CNAr)_6](PF_6)_2$ (Scheme 2 and Table 3) is far more efficient.[13,20] Chromium(I) alkyl isocyanide complexes have not been reported except for a comment[17] that $Cr(CNBu^t)_6$ oxidizes in air, presumably to $[Cr(CNBu^t)_6]^+$. In addition to the series $[Cr(CNAr)_6]PF_6$, $[Cr(CNPh)_6]BPh_4$ has been isolated[20] so that the species thought to be $[Cr(CNPh)_5]BPh_4$ was probably mischaracterized. Solid complexes in oxidation states I and II are often stable, although decomposition occurs in solution. Irradiation of $Cr(CNAr)_6$ in well-degassed $CHCl_3$ at 436 nm causes oxidation to $[Cr(CNAr)_6]^+$ (Ar = 2,6-$Pr^i_2C_6H_3$), whereas on irradiation in pyridine with Ar = Ph photosubstitution produces $Cr(CNPh)_5py$.[21]

Alkyl isocyanide chromium(II) complexes could not be prepared from $[Cr_2(O_2CMe)_4(H_2O)_2]$ or $[Cr_2(mhp)_4]$ (mhpH = 2-hydroxy-6-methylpyridine), but by the addition of alkyl isocyanides and KPF_6 to Cr^{2+} solutions from the reduction of $CrCl_3·6H_2O$ in ethanol with Zn/Hg, the homoleptic complexes $[Cr(CNR)_6](PF_6)_2$ (R = Bu^t or C_6H_{11}) are obtained (Scheme 4).[22] However, with phenyl isocyanide, the mixed-ligand complex $[CrCl(CNPh)_5]PF_6$ forms, apparently through its kinetic stability in the Cl^--containing medium since it decomposes to $[Cr(CNPh)_6](PF_6)_2$ in CH_2Cl_2. All are paramagnetic, low-spin complexes (t^4_{2g} configuration). It seems that the reaction product is substituent- and medium-dependent because the same procedure with the more reducing 4-methylphenyl isocyanide yields the chromium(I) salt $[Cr(CNC_6H_4Me\text{-}4)_6]PF_6$, and in the presence of Bu^tNH_2 phenyl isocyanide produces the chromium(0) complex $Cr(CNPh)_6$.[23] Also, the use of CrX_2 (X = Cl, Br) instead of the acetate in equation (5) produces crystalline orange-red $[CrCl_2(CNPh)_4]$ or olive brown $[CrBr_2(CNPh)_4]$. These low-spin chromium(II) complexes are stable in air but structures have not been determined.[24]

Partial replacement of isocyanide in $[Cr(CNR)_6](PF_6)_2$ or $[CrCl(CNPh)_5]PF_6$ by phosphines PR_3, bis(diphenylphosphino)ethane (dppe) or bis(diphenylphosphino)methane (dppm) provides two series of mixed-ligand complexes (Scheme 4).[22,23] The complexes $[Cr(CNR)_4(dppe)](PF_6)_2$, which must be *cis*, have more complicated spectra in the ν(CN) region than the complexes $[Cr(CNR)_4(PR'_3)_2](PF_6)_2$, which are therefore *trans* (Table 3). Nitric oxide reacts with $[Cr(CNR)_6](PF_6)_2$ to give $[Cr(NO)(CNR)_5](PF_6)_2$ from which many mixed nitrosyl isocyanide complexes of Cr^I and Cr^0 can be derived (Section 35.4.2.6).

The low-spin, diamagnetic complexes $[Cr(CNR)_7](PF_6)_2$, which contain seven-coordinate cations (**1**; see also Table 95), result when neat RNC is added to $[Cr(CNR)_6](PF_6)_2$. The reducing ability of the ligands and perhaps steric effects prevent the formation of seven-coordinate cations from aryl isocyanides.

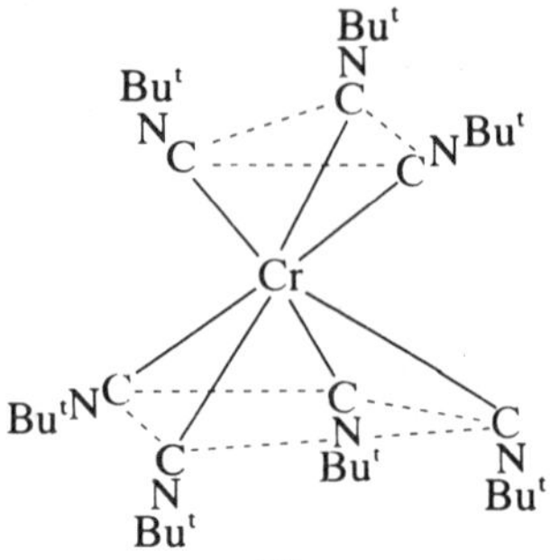

(**1**)

Chromium(III) complexes, of aryl isocyanides only, *e.g.* $[Cr(CNPh)_6](SbCl_6)_3$, have recently been prepared in good yield by oxidation of Cr^0 or Cr^I isocyanides with an excess of $SbCl_5$ or $NOBF_4$ (Schemes 2 and 3). Strong oxidizing agents are necessary because of the highly positive formal reduction potentials (equation 4 and Table 3). The Cr^{III} complexes are stable in the absence of water. On exposure to the atmosphere solid $[Cr(CNC_6H_3Me_2\text{-}2,6)_6](BF_4)_3$ is slowly reduced, presumably by water vapour, to an orange Cr^{II} complex (Table 3). Unlike typical Cr^{III} complexes, the isocyanide ligands are labile in polar non-aqueous solvents.[25]

(*iii*) *Spectroscopic and crystallographic results*

The general increase in the values of ν(CN) as the metal's oxidation state increases from 0 to III (Table 3) reflects the decrease in number of *d* electrons and the diminution in π donation to the ligand; the chromium(III) complexes have higher CN stretching frequencies even than the unbonded isocyanides (*ca.* 2130 cm^{-1}), suggesting a balance of σ-donor and π-acceptor behaviour by the ligands. There is also crystallographic evidence[26–28] for $d_\pi - p_\pi$ back-bonding, in that the Cr—C bonds shorten with decreasing oxidation state in the series $[Cr(CNPh)_6]^{n+}$ ($n = 0, 1, 2, 3$) and in $Cr(CNPh)_6^0$, compared with $[Cr^{II}(CNBu^t)_7](PF_6)_2$, where the difference in formal charge alone should have the opposite effect. There are parallel increases in C≡N bond lengths (Table 4).

Table 4 Bond Length Variation with Oxidation State in Hexaorganoisocyanide and Hexacyano Complexes

Complex	*Configuration*	*M—C (av) (Å)*	*C≡N (av) (Å)*	*N—C (Å)*	*Ref.*
$Cr(CNPh)_6$	LS d^6	1.938	1.176	1.388	1
$Cr(CNPh)_6^+$	LS d^5	1.975	1.159	1.397	1
$Cr(CNPh)_6^{2+}$	LS d^4	2.014	1.158	1.414	1
$Cr(CNPh)_6^{3+}$	d^3	2.068	1.141	1.396	1
$Mn(CNPh)_6^+$	LS d^6	1.901	1.162	1.411	1
$Cr(CN)_6^{4-}$	LS d^4	2.053	1.156	—	1
$Cr(CN)_6^{3-}$	d^3	2.071	1.156	—	1
$Cr(CNBu^t)_7^{2+}$	LS d^4	1.995	1.149	1.457	2

1. D. A. Bohling and K. R. Mann, *Inorg. Chem.*, 1984, **23**, 1426.
2. J. C. Dewan, W. S. Mialki, R. A. Walton and S. J. Lippard, *J. Am. Chem. Soc.*, 1982, **104**, 133.

The Cr $2p_{3/2}$ binding energies (XPES) increase from $Cr(CNPh)_6$ to $[Cr(CNPh)_6]^+$ to $[Cr(CNR)_6]^{2+}$, the values being 574.5, 575.3 and 576.7 eV respectively.[22,29] The shake-up satellite structure associated with the N 1*s* and C 1*s* binding energies in these spectra most probably arises from M $(d) \rightarrow \pi^*$ {CNAr(R)} excitations accompanying the primary photoemission.

(iv) Electrochemistry

Electrochemical studies have established the three reversible one-electron transfers in equations (2) to (4).[13–15,23] The electrochemical behaviour of a series of complexes $[Cr(CNAr)_6](PF_6)$, in which Ar is Ph, $4\text{-}Me_2NC_6H_4$, $4\text{-}MeOC_6H_4$, $4\text{-}MeC_6H_4$, $4\text{-}ClC_6H_4$, $2\text{-}MeC_6H_4$, $2,6\text{-}Me_2C_6H_3$ or $2,6\text{-}Pr^i_2C_6H_3$, has been studied by cyclic voltammetry in CH_2Cl_2 and MeCN.[14] With *para* substituents, and some *meta* substituents,[13] there is a linear relation between the formal potentials of equations (2) to (4) and the Hammett σ_p parameter, as expected from the usual operation of electronic effects. With *ortho* substituents this was true for the $0 \leftrightarrow 1+$ couple (equation 2) and the σ_o parameter, but not for the $1+ \leftrightarrow 2+$ and $2+ \leftrightarrow 3+$ couples (equations 3 and 4), particularly the latter. There were significant anodic potential shifts in the formal potentials beyond those expected from the σ_o values indicating destabilization of the higher oxidation state of a couple, particularly Cr^{3+} in the $2+ \leftrightarrow 3+$ couple. The anodic shifts also increased with the number of *ortho* substituents. These effects are attributed to weakening of the Cr^{III}—C bonds due to interligand repulsion among the substituents in the *ortho*-substituted complexes, a weakening also manifested through the appearance of new waves in certain media arising from substitution of *ortho*-substituted ligands by MeCN or ClO_4^-.[14] The photochemical behaviour of the *ortho*- and *para*-substituted complexes also differs markedly.[21]

Comparison between the half-wave potentials (equations 2 to 4) of $[Cr(CNR)_6](PF_6)_2$, *e.g.* for R = Bu^t, −1.04, −0.28 and 0.84 V (*versus* SCE),[22] with those for $[Cr(CNPh)_6](PF_6)_2$, *i.e.* −0.35, 0.25 and 1.00 V,[20] shows that alkyl and aryl isocyanides favour respectively the higher and the lower oxidation states as expected from the greater σ-donor and weaker π-acceptor capabilities of the alkyl over the aryl isocyanides. Similarly, the phosphines in the mixed ligand complexes (Table 3),[22,23] relative to isocyanide ligands, stabilize the Cr^{III} oxidation state. The great difference in the relative stabilities of Cr—C bonds in the cyano and phenyl isocyanide complexes is indicated by the magnitude of the shift (*ca.* 2.0 V) between the $Cr(CN)_6^{3-}/Cr(CN)_6^{4-}$ (−1.130 V) and the $Cr(CNPh)_6^{3+}/Cr(CNPh)_6^{2+}$ reduction potentials.[28]

(v) Tricyanomethide

The pale blue tricyanomethide complex $Cr[C(CN)_3]_2$ has a magnetic moment of *ca.* 4.7 BM and is thought to be polymeric.[30]

35.2.1.3 Si-, Ge-, Sn- and Pb-containing ligands

The yellow complex $[NPr_4]_6[Cr(GeCl_3)_6]$ is formed in poor yield if $Cr(C_6H_6)_2$ is refluxed in acetone with $[NPr_4][GeCl_3]$.[31]

35.2.2 Nitrogen Ligands

35.2.2.1 Complexes of bi- or tri-dentate N heterocyclic ligands

Since complexes of 2,2′-bipyridyl and 1,10-phenanthroline with chromium in oxidation states I and 0 can be obtained by reduction (Scheme 64) of the chromium(II) complexes, these oxidation states will be considered together. Oxidation, as shown in Schemes 65 and 68 of Section 35.4.2.5, gives chromium(III) complexes, which are often best prepared in this way. Earlier work has been extensively reviewed, and few complexes of 2,2′:6′,2″-terpyridyl are known.[32] A chromium(I) phthalocyanine derivative is mentioned in Section 35.4.9.3.

(i) Syntheses

Various methods to prepare tris(α-diimine) complexes of many metals in low oxidation states have been devised, mainly by Herzog and co-workers.[32] Those used for chromium complexes are illustrated in Schemes 5 and 6. Chromium(II) salts, anhydrous or hydrated, are the commonest starting materials. To these are added stoichiometric amounts of the diimine, and the Cr^{II} complex is then reduced. When bipy is added to the acetate suspended in water, disproportionation into $[Cr(bipy)_3]^0$ and $[Cr(bipy)_3]^{3+}$ occurs. The use of the anhydrous acetate in THF with bipy or phen and Na/Hg seems to provide a more efficient route to the zerovalent complexes.[33] The complexes $LiCr(bipy)_3 \cdot 4THF$, $Na_2Cr(bipy)_3 \cdot 7THF$, $Na_3Cr(bipy)_3 \cdot 7THF$ and $Ca_3Cr(bipy)_3 \cdot 7NH_3$, which are prepared from $Cr(bipy)_3$ and as appropriate Li_2bipy, Li or Na metal in THF, or Ca in liquid ammonia,[34] are not included in Schemes 5 and 6.

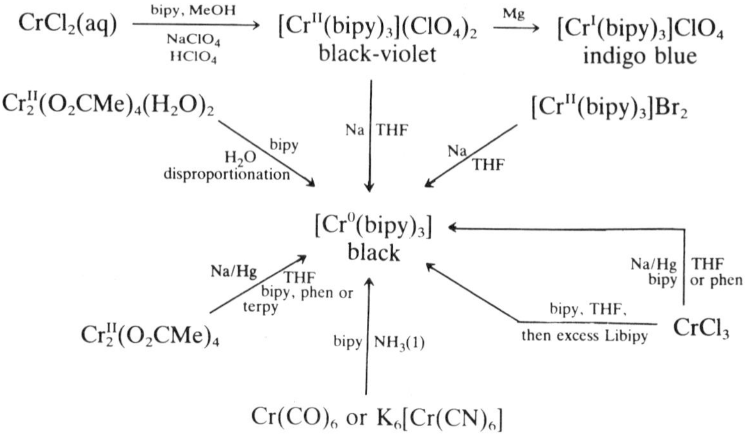

$$CrCl_2(aq) \xrightarrow[\substack{NaClO_4 \\ HClO_4}]{bipy, MeOH} [Cr^{II}(bipy)_3](ClO_4)_2 \xrightarrow{Mg} [Cr^I(bipy)_3]ClO_4$$

black-violet \qquad indigo blue

$Cr_2^{II}(O_2CMe)_4(H_2O)_2$ \qquad\qquad $[Cr^{II}(bipy)_3]Br_2$

Scheme 5

$$Cr(bipy)X_2 \xleftarrow[EtOH]{bipy \text{ or } phen} CrX_2 \text{hydrate} \xrightarrow[0\,°C]{3 \text{ bipy or phen}} [Cr(bipy)_3]X_2 \text{ or}$$

or $Cr(phen)X_2$ \qquad $X = Cl, Br, I$ \qquad $[Cr(phen)_3]X_2 \cdot 2H_2O$

$X = Cl, Br$

$[CrI_2(bipy)_2]$

$2 \text{ bipy}/EtOH$ \quad $\xrightarrow[H_2O]{CrBr_2 \cdot 2NH_4SCN}$ \quad $[Cr(NCS)_2(bipy)_2]$ or $[Cr(NCS)_2(phen)_2] \cdot H_2O$
(or phen) \qquad\qquad\qquad\qquad brown

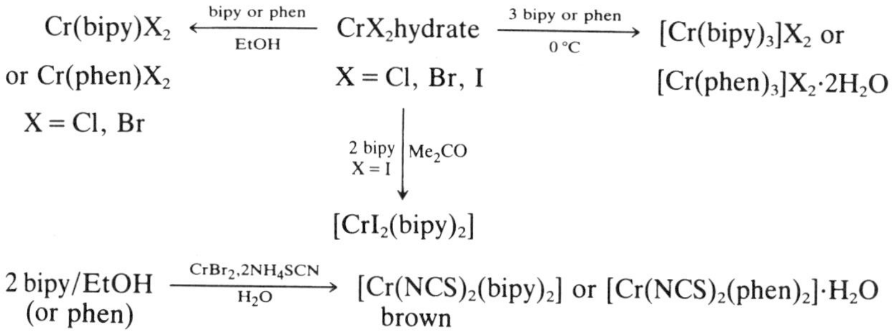

Scheme 6

(ii) Magnetic properties

The mono(amine)chromium(II) complexes, $CrCl_2(bipy)$ *etc.*, are high-spin ($t_{2g}^3 e_g^1$), but show antiferromagnetic interaction presumably through halide-bridged structures.[35] The bis- and the tris-(amine) chromium(II) complexes are low-spin (t_{2g}^4), as are all the terpy complexes and those of chromium in oxidation states below II (Table 5).[35,36] That spin pairing takes place in chromium(II) on the coordination of a second bipy molecule is in agreement with the observation that the second stepwise stability constant is greater than the first: $\log K_1 \approx 4$, $\log K_2 \approx 6.4$, $\log K_3 \approx 3.5$ (H_2O, ionic strength 0.3 M) (Table 39).[37]

The ground term of low-spin Cr^{II} is $^3T_{1g}$ and incomplete quenching of the orbital contribution is expected to produce magnetic moments in excess of the spin-only value (2.83 BM) which are temperature-dependent. This is not generally found with the low-spin bipy, phen or terpy complexes (Table 5), and attempts to account for the small or zero temperature dependence theoretically in terms of spin–orbit coupling, distortion from octahedral symmetry, and electron delocalization have not been very successful especially at low temperatures (~ 50 K) where the experimental values do not drop as the theory predicts. This may be because there is greater deviation from cubic symmetry than the theory can accommodate.[36] From further magnetochemical studies[38] and NMR measurements[39] it has been concluded that there is an appreciable trigonal component in the ligand field and relatively little delocalization of the unpaired electrons. The low-spin complexes $[CrX_2(diars)_2]$ also have essentially temperature-independent moments.[40]

Table 5 Complexes of bipy, phen and terpy with Chromium(0), (I) and (II) (see Section 35.4.2.5 for Chromium(III) Complexes)

Complex	*Comments*	*Ref.*
Chromium(0) complexes[a]		
$Cr(bipy)_3$	Diamagnetic	1–4
Black	Cr—N, 2.08 Å, N—Cr—N, 74.7°	5
$Cr(phen)_3$	Diamagnetic	1, 3, 4
Black		
$Cr(terpy)_3$	Diamagnetic	1, 3, 4
Chromium(I) complexes		
$[Cr(bipy)_3]X$	$X = I, ClO_4$, $\mu_{eff} = 2.07$ BM	1, 2
Indigo blue		
$[Cr(phen)_3]X$	$X = I, ClO_4$	1, 2
Chromium(II) complexes		
$[Cr(bipy)_3]X_2$	$X = Cl, Br, I, Br{\cdot}4H_2O$ (and below), $\mu_{eff} = 2.9$–3.4 BM with	2, 6
Deep violet	little temperature variation (300–90 K), similarly	
	$X = Br{\cdot}2H_2O$, ClO_4 (300–20 K), $X = Cl{\cdot}H_2O$, $Br{\cdot}H_2O$:	7
	$\mu_{eff} = 2.94$ and 2.73 BM (MeOH)	8
$[Cr(phen)_3]X_2{\cdot}2H_2O$	$X = Cl, Br, I, ClO_4$, μ_{eff} as above;	6–8
Olive green	$X = Cl$, $\mu_{eff} = 2.62$ BM (MeOH)	
$[CrI_2(bipy)_2]$	$\mu_{eff}^{295K} = 3.35$, $\mu_{eff}^{90K} = 3.32$ BM	9
$[CrX_2(bipy)_2]$	$X = OAc$, $\mu_{eff} = 3.01$ BM; $X = Cl$, $\mu_{eff} = 2.95$ BM; also phen compound	1
$[Cr(NCS)_2(bipy)_2]$	$\mu_{eff}^{295K} = 2.91$, $\mu_{eff}^{90K} = 2.86$ BM, $\nu(CN) = 2082, 2070$;	9
Brown	$\nu(M{-}NCS) = 362, 347\ cm^{-1}$	
$[Cr(NCS)_2(phen)_2]{\cdot}H_2O$	$\mu_{eff}^{295K} = 3.07$, $\mu_{eff}^{90K} = 3.06$ BM, $\nu(CN) = 2078, 2062$:	9
Brown	$\nu(M{-}NCS) = 363, 335\ cm^{-1}$	
$[Cr(R_2bipy)_3]Cl_2{\cdot}H_2O$	$R_2 = 4,4'\text{-}Me_2$, $\mu_{eff} = 2.76$ BM (MeOH); $R_2 = 5,5'\text{-}Me_2$	8, 10
$[Cr(R_2phen)_3]Cl_2{\cdot}xH_2O$	$R_2 = 4,7\text{-}Me$, $x = 2$, $\mu_{eff} = 2.90$ BM (MeOH)	8
	$R_2 = 5,6\text{-}Me_2$, $x = 1$, $\mu_{eff} = 2.87$ BM (MeOH)	
	$R_2 = H,4\text{-}Me$; $H,5\text{-}Me$; $3,4\text{-}Me_2$; $4,5\text{-}Me_2$; or $4,6\text{-}Me_2$; all $x = 1$	11
$CrX_2(bipy)$	$X = Cl$, $\mu_{eff}^{295K} = 4.19$, $\mu_{eff}^{90K} = 3.60$ BM, $\theta \approx 45°$	9, 12
	$X = Br$, $\mu_{eff}^{295K} = 4.37$, $\mu_{eff}^{90K} = 3.75$ BM, $\theta \approx 45°$	
$[CrX_2(phen)]{\cdot}H_2O$	$X = Cl$, $\mu_{eff}^{295K} = 4.57$, $\mu_{eff}^{90K} = 3.90$ BM, $\theta \approx 60°$	9
	$X = Br$, $\mu_{eff}^{295K} = 4.54$, $\mu_{eff}^{90K} = 3.90$ BM, $\theta \approx 60°$	
$[Cr(OAc)_2(Pr^iNH_2)_2(bipy)]$	$\mu_{eff} = 2.83$ BM, also phen compound	1
$[Cr(terpy)_2]X_2$	$X = I{\cdot}H_2O$, ClO_4, $\mu_{eff} \approx 2.9$ BM, almost temperature-invariant	7, 13

[a] Some complexes of formal oxidation states lower than 0 have been reported: $LiCr(bipy)_3{\cdot}4THF$, $Na_2Cr(bipy)_3{\cdot}7THF$, $Na_3Cr(bipy)_3{\cdot}7THF$ and $Ca_3Cr(bipy)_3{\cdot}7NH_3$; their magnetic moments are respectively *ca.* 1.83, 2.85, 3.85 and 2.46 BM.[1]

1. W. R. McWhinnie and J. D. Miller, *Adv. Inorg. Chem. Radiochem.*, 1969, **12**, 135.
2. F. Hein and S. Herzog, in 'Handbook of Preparative Inorganic Chemistry', ed. G. Brauer, Academic, New York, 2nd edn., 1965, vol. 2, sect. 24.
3. J. Quirk and G. Wilkinson, *Polyhedron*, 1982, **1**, 209.
4. H. Behrens and A. Müller, *Z. Anorg. Allg. Chem.*, 1965, **341**, 124.
5. G. Albrecht, *Z. Chem.*, 1963, **3**, 182.
6. A. Earnshaw, L. F. Larkworthy, K. C. Patel, K. S. Patel, R. L. Carlin and E. G. Terezakis, *J. Chem. Soc. (A)*, 1966, 511.
7. P. M. Lutz, G. J. Long and W. A. Baker, Jr., *Inorg. Chem.*, 1969, **8**, 2529.
8. G. N. La Mar and G. R. Van Hecke, *J. Am. Chem. Soc.*, 1969, **91**, 3442.
9. A. Earnshaw, L. F. Larkworthy, K. C. Patel and B. J. Tucker, *J. Chem. Soc., Dalton Trans.*, 1977, 2209.
10. I. Fujita, T. Yazaki, Y. Torii and H. Kobayashi, *Bull. Chem. Soc. Jpn.*, 1972, **45**, 2156.
11. G. N. La Mar and G. R. Van Hecke, *Inorg. Chem.*, 1970, **9**, 1546.
12. H. Lux, L. Eberle and D. Sarre, *Chem. Ber.*, 1964, **97**, 503.
13. S. Herzog and H. Aul, *Z. Chem.*, 1966, **6**, 382.

(iii) *Proton NMR studies of tris(bipy) and tris(phen) chelates*

Proton NMR spectroscopy has been used to investigate paramagnetic complexes of low-valent transition metal ions because the contact shifts provide information on the degree of covalency of the metal–ligand bond. The low-spin complexes $[Cr(phen)_3]Cl_2$ and $[Cr(bipy)_3]Cl_2$, and their methyl-substituted derivatives (Table 5), have been extensively investigated.[39,41,42] Narrow, well-resolved ^1H NMR spectra have been obtained from $[Cr(R_2phen)_3]^{2+}$ ($R = H$; $R_2 = 4,7$- and 5,6-dimethyl) and $[Cr(R_2bipy)_3]^{2+}$ ($R = H$; $R_2 = 4,4'$-dimethyl) and the shifts result from both $L \rightarrow M$ σ and $M \rightarrow L$ π charge transfer. The π bonding is not obvious in the tris chelates with symmetric ligands owing to near cancellation of two contributions of opposite sign.[41] The relative π-acceptor abilities of the differently substituted ligands, and the orbital ground states of the complexes were elucidated only after some mixed-ligand chelates (obtained in solution only) had been investigated. The M→L delocalized π spin density is centred predominantly at the 4,7 positions of phen or 4,4′ positions in bipy. This facilitates the reducing ability of these chelates and is believed[42] to be consistent with the stereoselective outer-sphere reduction of $[Co(phen)_3]^{3+}$ by $[Cr(phen)_3]^{2+}$: it is found that (+)-$[Co(phen)_3]^{3+}$ yields (−)-$[Cr(phen)_3]^{3+}$ as the main product on reduction with racemic $[Cr(phen)_3]^{2+}$.[43] Anomalously broadened methyl resonances are ascribed to solvent (water) complex interactions at the 4,7 positions which involve hydrogen bonding to pockets of electron density on the ligand assisted by the π bonding. From spectral changes it was suggested[35] that bipy separated from $[Cr(bipy)_3]X_2$ (X = Cl, Br, I) in ethanol, but there are no NMR signals of the free ligand from solutions of these salts in d_4-MeOH unless some is added.

(iv) *Electronic, ESR and IR spectra*

The electronic spectrum of $Cr(bipy)_3$ resembles that of $bipy^-$ in the lower wavenumber region, but formulation of the complex in terms of integral oxidation states is unrealistic.[44,45] According to an X-ray study, the molecule is trigonally distorted, the N—Cr—N angle being 74.7 ± 1.5°. The Cr—N distance is 2.08 Å.[46]

The ion $[Cr(bipy)_3]^+$ in solution exhibits an ESR spectrum from which $g = 1.9971$ and the hyperfine splitting constant of ^{53}Cr, A^{Cr}, is 20.4×10^{-4} cm^{-1}. There is further splitting of the ^{53}Cr hyperfine lines due to the ^{14}N nuclei of the ligands, which indicates strong σ bonding.[47] Its electronic spectrum consists of charge-transfer and intraligand bands.[45]

Ligand-field theory gives a reasonable account of the energies but not the intensities of the bands between 12 000 and 32 000 cm^{-1} in the spectrum of $[Cr(bipy)_3]^{2+}$, which are due to spin-allowed transitions of the low-spin d^4 configuration. There are also many charge-transfer or internal-ligand bands which are weakly perturbed by methyl substitution in bipy.[48] The assignment of formal oxidation states (II and III respectively) to the metal in $[Cr(bipy)_3]^{2+}$ and $[Cr(bipy)_3]^{3+}$ (Section 35.4.2.5) is valid but not in complexes of lower overall charge.[45] From spectral changes in aqueous solution the mixed complex $[Cr(bipy)_2(4,4'\text{-bipyH})(H_2O)]^{3+}$ has been identified in rapid equilibrium with $[Cr(bipy)_3]^{2+}$ and the free ligands.[49]

In the series $[Cr(bipy)_3]X_3$, $[Cr(bipy)_3]X_2$, $[Cr(bipy)_3]X$, $[Cr(bipy)_3]$ the Cr—N stretching frequency varies little with oxidation state from III to 0. Since this implies small changes in bond strength with decreasing oxidation state and greater occupation of the t_{2g} orbitals, it appears that increasing M→L π donation is offset by lower L→M σ donation as the oxidation state is lowered. Some complexes were labelled with ^{50}Cr or ^{53}Cr to aid in the assignment of the two Cr—N stretching bands (Table 6).[50] There are discrepancies with assignments for unlabelled complexes.[51] The splitting of the ν(CN) band in the IR spectra of the complexes $[Cr(NCS)_2(bipy)_2]$ and $[Cr(NCS)_2(phen)_2]$ indicates *cis* configurations but this has not been confirmed crystallographically.[35]

Table 6 Far IR Spectra of the $[Cr(bipy)_3](ClO_4)_n$ Series (cm^{-1})

$[Cr(bipy)_3]^{3+}$	$[Cr(bipy)_3]^{2+}$	$[Cr(bipy)_3]^{+}$	$[Cr(bipy)_3]^{0}$	*Assignment*
385m	351s	371m	382m	ν(Cr—N)
(Hidden)	359m	352m	357vw	Ligand
349m	343m	343m	308s	ν(Cr—N)
226w	225m	228w	227w	Ligand
193vw	205w			
	189w	195w	203s	Ligand
177w	175vw	179w	178w	Ligand

The absorption spectrum and first formation constant of $[Cr(bipy)]^{2+}$ have been determined spectrophotometrically in hexamethylphosphoramide (HMPA) (see also Table 39). Since $\log K_1$ (4.61) is slightly larger in HMPA than in water, it is said that $[Cr(bipy)(HMPA)_2]^{2+}$ is tetrahedral, although from the general chemistry of chromium(II) the ion would be expected to be in planar or distorted octahedral coordination.[52]

(*v*) *Electrochemistry*

Voltammetric investigations of chromium complexes of bipy, phen and terpy in MeCN and other solvents, as well as magnetic and other data above, show that the metal forms series of tris-(bipy) and -(phen) and bis(terpy) complexes related by reversible one-electron transfers in which the metal has formal oxidation states III, II, I, 0, −I (see also Section 35.4.2.5). Some half-wave potentials are given in Table 7.[53] In aqueous solution complications occur because $[Cr(bipy)_3]^{3+}$ undergoes catalyzed ligand exchange at the dropping mercury electrode to form $[Cr(bipy)_2(H_2O)_2]^{3+}$, which reduces more cathodically than the tris complex, to give $[Cr(bipy)_2(H_2O)_2]^{2+}$. The phen complexes and $[CrCl_2(bipy)_2]^+$ behave similarly [54,55]

Table 7 Half-wave Potentials

Complex couple	*DC polarography* (V)[a]	*AC polarography* (V)[a]
$[Cr(terpy)_2]$		
III→II	−0.110	−0.120
II→I	−0.516	−0.525
I→0	−1.030	−1.035
0→−I	−1.95	−1.95
$[Cr(bipy)_3]$		
III→II	−0.209	−0.200
II→I	−0.715	−0.710
I→0	−1.28	−1.28
0→−I	−1.91	−1.89
$[Cr(phen)_3]$		
III→II	−0.237	−0.228
II→I	−0.732	−0.838
I→0	−1.29	−1.28
0→−I	−1.79	−1.79

[a] *Versus* Ag/AgCl saturated NaCl electrode.

35.2.2.2 Nitrosyls and dinitrogen complexes

Nitrosyls are included in Section 35.4.2.6 and dinitrogen complexes with the phosphorus donors immediately below.

35.2.3 Phosphorus Ligands

35.2.3.1 Mixed complexes of dinitrogen and tertiary phosphines

There are few examples of complexes of chromium with molecular nitrogen, although molybdenum and tungsten form many such complexes.[1] Unstable species containing Cr—N_2 bonds produced by matrix isolation techniques have not been included.[56]

The chromium(0) complex $[Cr(N_2)_2(PMe_3)_4]$ is not well established because it decomposes at room temperature with liberation of PMe_3 and N_2. Since there are two IR absorption bands at 1990 and 1918 cm^{-1} [$\nu(N{\equiv}N)$] it has been assigned a *cis* octahedral structure. The dark brown complex can be prepared from anhydrous $CrCl_3$ by Scheme 7 or directly from the reactants. There is no spectroscopic or structural information on the intermediates $[Cr^{III}Cl_3(PMe_3)_3]$ and $[Cr^{II}Cl_2(PMe_3)_3]$.[57]

The systems $CrCl_2$–Mg–THF and $CrCl_2$–Mg–dppe–THF absorb nitrogen and some NH_3 and N_2H_4 are produced on hydrolysis. The paramagnetic solids $Cr_2N_2Mg_4Cl_4(THF)_5$ (black) and $[Cr(dppe)_2]_2N_2$ (brown), which are difficult to obtain reproducibly, have been isolated.[58]

The chromium(0) complex *trans*-$[Cr(N_2)_2(dmpe)_2]$ (**2**) has been prepared together with some incompletely characterized *cis* isomer from $[CrCl_2(dmpe)_2]$ according to Scheme 8, which also shows that its reactions all involve displacement of dinitrogen. The *trans* dinitrogen ligands are linear, and the N—N bond distance (1.122 Å) is normal. The IR spectrum of the *trans* isomer

$$CrCl_3 \xrightarrow[THF]{PMe_3} [Cr^{III}Cl_3(PMe_3)_3] \xrightarrow[PMe_3 (excess)]{Mg} [Cr^{II}Cl_2(PMe_3)_3]$$

$$\qquad\qquad\quad \text{brown violet} \qquad\qquad\qquad \text{from deep blue solution}$$
$$\qquad\qquad\quad \text{m.p. 144–147 °C} \qquad\qquad\qquad \text{m.p. 51 °C}$$

$$\left. \begin{array}{c} PMe_3 \\ Mg \end{array} \right| \begin{array}{c} N_2 \\ THF \end{array}$$

$$[Cr(CO)_3(PMe_3)_3] \xleftarrow[-PMe_3]{CO} [Cr(CO)_2(PMe_3)_4] \xleftarrow[-2N_2]{CO} [Cr(N_2)_2(PMe_3)_4]$$

$$\text{dark brown}$$
$$\text{dec. at r.t.}$$

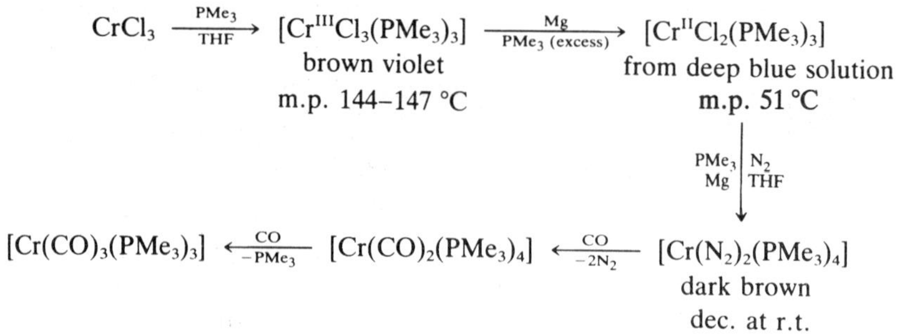

Scheme 7

has one band (N≡N stretch) at 1932 cm⁻¹ and the *cis* isomer bands at 1920 and 1895 cm⁻¹. Since the Cr—P distance (2.296 Å) is shorter than in [CrMe₂(dmpe)₂] (2.345 Å, Section 35.3.4.1) there is significant metal–ligand π back-bonding.[59]

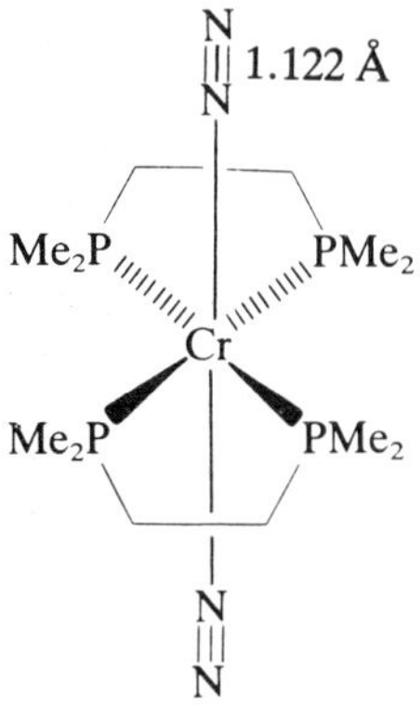

(2)

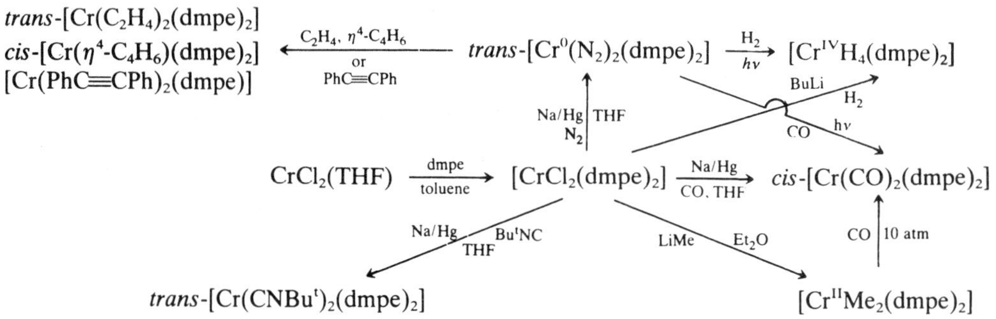

Scheme 8

The cocondensation of chromium vapour and the phosphorus ligand is used to prepare [Cr(dmpe)₃] and the trimethyl phosphite complex [Cr{P(OMe)₃}₆] (Table 8). The former has a distorted octahedral structure, and the latter forms a seven-coordinate hydride CrH₂{P(OMe)₃}₅ which, with other chromium(II) complexes of P donor ligands, is considered in Section 35.3.4.

35.2.3.2 Fluorophosphine complexes

The reaction of tris(π-allyl)chromium with the π-acceptor ligand PF₃ produces stable crystals of [Cr(PF₃)₆] *via* [(C₃H₅)₃Cr(PF₃)] (Scheme 9). At high pressure, only [Cr(PF₃)₆] is isolated;[60] it can also be obtained by metal vapour synthesis,[61] and its He(I) PE spectrum has been assigned.[62]

Complexes of chelating aminodifluorophosphines Cr{(PF₂)₂NR}₃ (R = Ph (m.p. 233 °C)[63] or Me (m.p. 182 °C)[64]) can be prepared by UV irradiation of Cr(CO)₆ with an excess of the ligand

Table 8 Tertiary Phosphine and Arsine Complexes of Chromium(0) Including Mixed Complexes with Dinitrogen

Complex	*Comments*	*Ref.*
trans-$[Cr(N_2)_2(dmpe)_2]$ Red-orange	M.p. > 350 °C, stable in solution to 90 °C, diamagnetic NMR: ^1H, δ 1.29 (PMe_2 and PCH_2, s); $^{31}P[^1H]$, δ 69.3 (PhH-d_6, 25 °C), $\nu(N_2)$, 1932 cm^{-1} (hexane), 2001 cm^{-1} (Raman) Cr—N, 1.874, N—N, 1.122, Cr—P (av), 2.296 Å, Cr—N—N, 178.2°	1
$[Cr(dmpe)_3]$ Yellow	$[CrCl_3(THF)_3]$, dmpe in THF + $NaC_{10}H_8$ or $LiAlH_4$; cocondensation of Cr and dmpe vapours, air-sensitive; distorted octahedron, Cr—P, 2.317 Å, P—Cr—P (chelate ring), 76.5°	2 3
$[Cr\{P(OMe)_3\}_6]$ Yellow	M.p. > 100 °C (dec.), cocondensation of Cr and $P(OMe)_3$ vapours	4
$Cr(L^2)_3$	L^2 = dppm, dppe or $Ph_2As(CH_2)_2AsPh_2$; from $K_6[Cr(CN)_6]$ and L^2 in NH_3(l) at −60 °C	5

1. J. E. Salt, G. S. Girolami, G. Wilkinson, M. Motevalli, M. Thornton-Pett and M. B. Hursthouse, *J. Chem. Soc., Dalton Trans.*, 1985, 685.
2. J. Chatt and H. R. Watson, *J. Chem. Soc.*, 1962, 2545.
3. F. G. N. Cloke, P. J. Fyne, M. L. H. Green, M. J. Ledoux, A. Gourdon and C. K. Prout, *J. Organomet. Chem.*, 1980, **198**, C69.
4. S. D. Ittel, F. A. Van-Catledge and C. A. Tolman, *Inorg. Chem.*, 1985, **24**, 62.
5. H. Behrens and A. Müller, *Z. Anorg. Allg. Chem.*, 1965, **341**, 124.

$$Cr(C_3H_5)_3 \xrightarrow[Et_2O]{\text{20–30 atm } PF_3} [(C_3H_5)_3Cr(PF_3)] \xrightarrow{PF_3} [(Cr(PF_3)_6] + 1.5C_6H_{10}$$

dark green; colourless, m.p. 193 °C

Scheme 9

in diethyl ether. Cocondensation of chromium vapour and $MeN(PF_2)_2$ also affords $[Cr\{(PF_2)_2NMe\}_3]$.[65,66] Although the ligands have a small bite and must form four-membered chelate rings, the colourless complexes (**3**) are volatile and stable. In accordance with the formulation as zerovalent tris(bidentate) chelates, the ^1H and ^{13}C NMR spectra each exhibit a single resonance from the three equivalent methyl groups, and there are intense molecular ions in the mass spectra. The strong P—F stretching absorptions are at 831 cm^{-1} (R = Me) and 865 cm^{-1} (R = Ph).

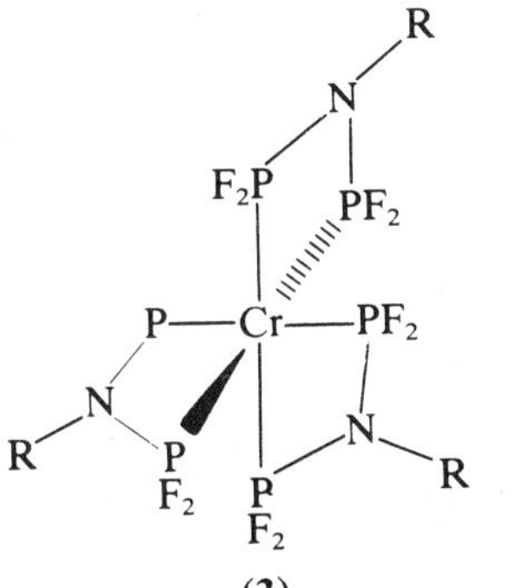

(**3**)

The cocondensation of chromium vapour with a 4:1 mixture of Me_2NPF_2 and $MeN(PF_2)_2$ forms the homoleptic chromium(0) derivative $[Cr(PF_2NMe_2)_6]$ and the mixed ligand derivative $[Cr(PF_2NMe_2)_4\{(PF_2)_2NMe\}]$. The former slowly decomposes when left in air, especially if in solution. This instability, compared with $[Cr(CO)_6]$ or $[Cr(PF_3)_6]$, is due to the steric hindrance of the six bulky monodentate ligands. The latter, with only four monodentate ligands and one bidentate ligand, can be sublimed and is considerably more stable. When prepared, it was contaminated with a little $[Cr(PF_2NMe_2)_2\{(PF_2)_2NMe\}_2]$. An X-ray determination of the structure of $[Cr(PF_2NMe_2)_4\{(PF_2)_2NMe\}]$ shows the expected octahedral configuration distorted by the small bite of bidentate $(PF_2)_2NMe$ (P—Cr—P = 66°), which relieves steric strain compared with $Cr(PF_2NMe_2)_6$; no bond distances were supplied.[65]

The stable zerovalent complexes $[Cr(PF_2OPr)_6]$ and $[Cr\{(PF(OMe)_2\}_6]$ can be obtained by UV irradiation of $[Cr(CO)_6]$ in pentane in the presence of a slight excess of the ligand.[67]

35.2.4 Miscellaneous Chromium(0) Complexes

Several chromium nitrosyls, and complexes of quinones which are members of redox series, can be considered to contain chromium(0) or chromium(I) (see Sections 35.4.2.6 and 35.4.4.8).

35.3 CHROMIUM(II)

Chromium(II) complexes, especially in solution, are very rapidly oxidized by air, and this restricted synthetic and other investigations until comparatively recently when efficient inert atmosphere boxes had been developed, and the use of Schlenk and similar techniques had become general.[68,69] Solids show varying behaviour towards aerial oxidation; some are pyrophoric but a few are stable for days or longer. The need to use oxygen-free conditions should be assumed throughout the Cr^{II} section. Few chromium(II) compounds, possibly the acetate and some anhydrous halides, are commercially available. Consequently, those wishing to investigate the coordination chemistry of chromium(II) must first prepare their starting materials.

Chromium(II) has been extensively used in investigations of inner-sphere electron transfer because inert chromium(III) captures the ligands attacked by chromium(II).[70]

35.3.1 General Synthetic Methods

Methods for the preparation of chromium(II) solutions and the isolation of solids for synthetic work are set out below.

35.3.1.1 Chromium(II) solutions

(i) Dissolution of chromium in acids

The impure metal dissolves easily in mineral acids and in fluoroboric, sulfamic and trifluoromethylsulfonic acids to give Cr^{2+} solutions, but oxidation of Cr^{2+} by hydrogen ions (equation 6), $E°(Cr^{3+}, Cr^{2+}) = -0.41$ V) even in an inert atmosphere is catalyzed by the impurities and various ions.[71] Indefinitely stable chromium(II) solutions can be obtained from the pure (electrolytic) metal (99.5% or better), although the reaction with acid may need to be initiated by heat and the inclusion of some metal previously attacked by acid. The use of an excess of metal, which can be filtered off, ensures that little acid remains. In near neutral solution the hydrogen potential is lowered and the Cr^{2+} ion is stable. In alkaline conditions brown $Cr(OH)_2$, which slowly reduces water, precipitates.[72,73]

$$Cr^{2+} + H^+ \longrightarrow Cr^{3+} + \tfrac{1}{2}H_2 \qquad (6)$$

(ii) Reduction of chromium(III) solutions

The reduction of aqueous chromium(III) solutions can be carried out electrolytically or chemically with zinc amalgam, zinc and acid or a Jones reductor.[2,24] Electrolytic procedures can be cumbersome, and with chemical reductants contamination with other products can occur. Chromium metal and acid can be used to reduce chromium(III) salts, and this requires less of the metal than in the method described in Section 35.3.1.1.i.

(iii) Non-aqueous solutions

The hydrated chloride, bromide and iodide (Table 9) are soluble in ethanol, butanol and other organic solvents, but in many systems traces of water cause oxidation, hydrolysis or failure to complex with weak donor ligands. Water can be avoided by dissolving the metal in THF, ethanol or diethyl ether through which hydrogen chloride is bubbled.[24,74,75] It is also possible to dissolve or suspend in organic solvents the anhydrous acetate or the halides CrX_2 (Table 9), and dehydration of the hydrated halides with 2,3-dimethoxypropane in ethanol, followed by vacuum removal of the liquid, produces mixed alcoholates suitable for use in water-free conditions.[76] Triethyl orthoformate may be used similarly.

Table 9 Preparations of Some Chromium(II) Compounds Used as Starting Materials

Compound	*Preparation*	*Ref.*
$[Cr_2(O_2CMe)_4(H_2O)_2]$ Dark red	Cr^{2+}(aq) + NaO_2CMe; brown $[Cr_2(O_2CMe)_4]$ on pumping at 100 °C	1, 2
$Cr_2(O_2CMe)_4 \cdot 2MeCO_2H$	Cr powder, anhydrous $MeCO_2H$ with some MeCOCl	3
$CrCl_2 \cdot 4H_2O$ Blue	Cr + HCl, conc. solution, add Me_2CO; [a]$CrCl_2$ by thermal dehydration at 140 °C	4–6
$CrBr_2 \cdot 6H_2O$ Blue	Cr + HBr, conc. solution, add Me_2CO; [a]$CrBr_2$ by thermal dehydration at 120 °C	4–7
$CrI_2 \cdot 6H_2O$ Blue	Cr + HI, conc. solution; soluble org. solvents, 'CrI_2' on pumping at r.t.[a]	4–6
$Cr(ClO_4)_2 \cdot 6H_2O$ Blue	Cr + $HClO_4$, conc. solution; soluble org. solvents, solid decomposes under N_2, internal oxidation/reduction	4–6
	$[Cr(bipy)_3](ClO_4)_2$ known but probably dangerous to prepare ClO_4 salts of Cr^{II} cations generally	2, 6
$CrSO_4 \cdot 5H_2O$ Blue	Cr + H_2SO_4, add acetone; forms $CrSO_4 \cdot H_2O$ and then $CrSO_4$ on heating, stable in air when dry	2, 4–6
$[Cr(acac)_2]$ Light yellow	$Cr_2(O_2CMe)_4(H_2O)_2$ + acacH, H_2O, reacts slowly with H_2O	1
$[CrCl_2(MeCN)_2]$ Pale blue	Anhydrous[a] $CrCl_2$ (above) in EtOH + MeCN	6
$[CrCl_2(THF)_2]$ Pale green	$CrCl_2$ extracted with THF	8
$[CrCl_2(THF)]$ White	Cr + HCl(g) in THF	9
$[CrBr_2(THF)_2]$ Pale blue-green	Crystallized from $CrBr_2$ in THF	10

[a] The anhydrous halides are also prepared by the methods in Section 35.3.7.

1. L. R. Ocone and B. P. Block, *Inorg. Synth.*, 1966, **8**, 125, 130.
2. D. N. Hume and H. W. Stone, *J. Am. Chem. Soc.*, 1941, **63**, 1197.
3. H.-D. Hardt and G. Streit, *Z. Anorg. Allg. Chem.*, 1970, **373**, 97.
4. A. Earnshaw, L. F. Larkworthy and K. S. Patel, *J. Chem. Soc.*, 1965, 3267.
5. H. Lux and G. Illmann, *Chem. Ber.*, 1958, **91**, 2143.
6. D. G. Holah and J. P. Fackler, Jr., *Inorg. Synth.*, 1967, **10**, 26.
7. M. A. Babar, L. F. Larkworthy and A. Yavari, *J. Chem. Soc., Dalton Trans.*, 1981, 27.
8. R. J. Kern, *J. Inorg. Nucl. Chem.*, 1962, **24**, 1105.
9. L. F. Larkworthy and M. H. O. Nelson-Richardson, *Chem. Ind. (London)*, 1974, 164.
10. D. E. Scaife, *Aust. J. Chem.*, 1967, **20**, 845.

A method which gives a chromium(II)-substituted alkoxide ($[Cr(OCBu^t_3)_2 \cdot LiCl(THF)_2]$, Section 35.3.5.2) directly from $CrCl_3$ is to add a solution of the alcohol previously treated with BuLi to a slurry of the halide in THF.[77] How the reduction to chromium(II) takes place has not been clarified.

35.3.1.2 Chromium(II) salts

The poorly soluble acetate $[Cr_2(O_2CMe)_4(H_2O)_2]$ can be crystallized from aqueous Cr^{2+} solutions by the addition of sodium acetate. When dry it reacts slowly with oxygen and can be handled briefly in air, but the anhydrous acetate chars on exposure.

The hydrates $CrSO_4 \cdot 5H_2O$, $CrCl_2 \cdot 4H_2O$, $CrBr_2 \cdot 6H_2O$ and $CrI_2 \cdot 6H_2O$, and the anhydrous halides, all most easily isolated as in Table 9, are useful starting materials. Thermal dehydration of halides can lead to hydrolysis, but this is not significant for preparative use of the anhydrous chromium(II) compounds. The iodide and perchlorate are difficult to obtain hydrated to a definite degree. On pumping $CrI_2·6H_2O$ loses its blue colour as water is removed, but the solid remains a chromium(II) species. The perchlorate should be used with extreme caution, if at all, because the juxtaposition of Cr^{2+} as part of a complex cation containing an organic ligand and ClO_4^- is intrinsically dangerous and not only impact but the accidental admission of air to the solid could cause an explosion. The use of BPh_4, BF_4, NH_2SO_3, PF_6 and CF_3SO_3 salts has not yet been much exploited.

Aqueous chromium(II) solutions have long been available as standard reducing agents in volumetric analysis.[78] More recently, the use of chromium(II) salts as reducing agents in organic chemistry has been reviewed. Chromium(II) chloride, sulfate and perchlorate have been classified as powerful reducing agents whereas the acetate, a relatively insoluble salt, acts as a mild reducing agent in near neutral conditions. Chromium(II) ethylenediamine complexes have been used in dimethylformamide,[79] and transition metal chelates have been reduced by chromium(II) acetylacetonate in toluene and THF.[80] Chromium(II) compounds also readily reduce various iron oxides.[81] These uses of chromium(II) compounds are likely to increase.

Metals can be used as sacrificial anodes in the electrochemical synthesis of anhydrous salts and complexes. This method frequently produces chromium(III) complexes, but electrochemical oxidation of chromium in a mixture of concentrated aqueous HBF_4 and acetonitrile yields $[Cr(MeCN)_6](BF_4)_2$.[82]

35.3.2 Group IV Ligands

Cyanide and organic isocyanide complexes of chromium(II) have been included with the lower oxidation states in Sections 35.2.1.1 and 35.2.1.2.

35.3.3 Nitrogen Ligands

35.3.3.1 Ammines

(*i*) *Syntheses*

A deep blue colour appears when concentrated solutions of ammonia and aqueous chromium(II) are mixed, but brown chromium(II) hydroxide separates before any ammine can be isolated.[83] Gaseous ammonia and solid $CrCl_2$ form $CrCl_2·nNH_3$ ($n = 2$, 3, 5 and 6) and when ammonia is bubbled through ethanolic solutions of the hydrated halides pale blue, violet and then greyish precipitates successively appear corresponding to the formation of diammines, pentaammines and hexaammines (Table 10).[84] The hexaammines lose ammonia so readily (and reversibly) to form pentaammines that they must be sealed in ammonia gas. The easy loss of ammonia appears to be a 'chemical' effect of the Jahn–Teller distortion (see later). The tetraammine $[Cr(NH_3)_4(H_2O)]SO_4$ is obtained by bubbling ammonia through an ethanolic suspension of $CrSO_4·5H_2O$[85] and the diammines by heating the tetraammine or the pentaammines.

(*ii*) *Magnetic and spectroscopic properties and structures*

Chromium(II) complexes may be high- or low-spin (Figure 1), and the hexaammines and pentaammines are typical, high-spin Cr^{2+} compounds with effective magnetic moments close to the spin-only value ($t_{2g}^3e_g^1$ configuration, $\mu_{so} = 4.90$ BM). With these magnetically dilute complexes the experimental moments are temperature-independent (Table 10) so the Curie law is obeyed ($\theta = 0$ °C). The diammines show weak antiferromagnetic interaction since the moments decrease as the temperature is lowered and the Curie–Weiss law is followed ($\theta \approx 35$ °C).

The σ antibonding effect, arising from the uneven occupancy of the d orbitals in high-spin chromium(II) complexes, has a marked effect (Jahn–Teller distortion) on the electronic spectra and structures. Chromium(II) has a ^5D free ion ground term and this is split in an octahedral (O_h) ligand field as shown in Figure 2 to give a ^5E_g ground term. The Jahn–Teller theorem

Table 10 Magnetic and Diffuse Reflectance Data of Ammines

Complex	μ_{eff} (BM) *295 K*	*90 K*	θ (°)	*Reflectance spectra*[a] (cm^{-1}) ν_2		ν_1
$[CrCl_2(NH_3)_2]$	4.57	4.05	40	17 000sh	14 000svb	11 300sh
Pale blue				17 000sh	14 100svb	11 000sh
$[Cr(NH_3)_5]Cl_2$	4.77	4.74	0	17 800svb		
Violet				18 200s	14 500sh	11 400sh
$[Cr(NH_3)_6]Cl_2$	4.85	4.78	0	16 000svb		8500m
Grey				17 300svb		8500m
$[CrBr_2(NH_3)_2]$	4.33	4.09	23	17 500sh	13 200s	9500sh
Pale blue				17 400sh	13 700s	9500sh
$[Cr(NH_3)_5]Br_2$	4.76	4.80	0	17 500svb		
Violet				18 200s	14 500sh	11 300sh
$[Cr(NH_3)_6]Br_2$	4.87	4.90	0	15 400sb		7500m
Greenish-grey				15 700sb		7700m
$[Cr(NH_3)_5]I_2$	4.87	4.87	0	17 600s		12 800sh
Violet				18 100s	14 000sh	11 200sh
$[Cr(NH_3)_6]I_2$	4.88	4.91	0	15 200s		7600m
Greenish-grey				15 400s		7800m
$[Cr(NH_3)_2(SO_4)]$	4.69	4.26	32	14 000vb		
Pale blue				14 000s	11 500sh	9800sh
$[Cr(NH_3)_4(H_2O)]SO_4$	4.89	4.74	8	18 000vb		
Violet				18 400vb		15 000sh

[a] Spectra at room and liquid nitrogen (second line) temperatures for each complex.

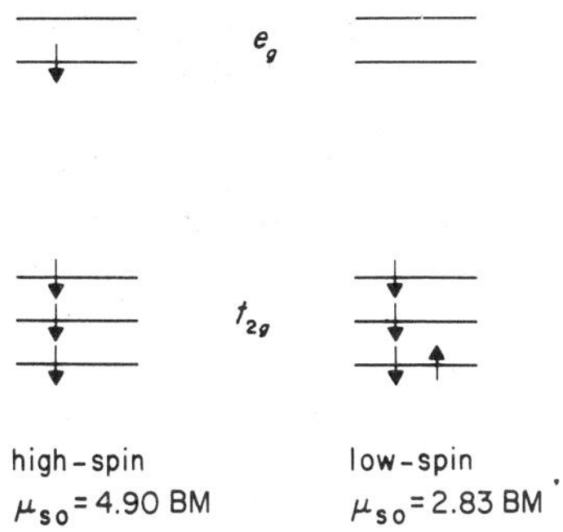

Figure 1 Occupancy of *d* orbitals in octahedral chromium(II) complexes

requires that in an orbitally degenerate, non-linear system, distortion will occur to remove the degeneracy, and this can take place in high-spin d^4 (and d^9) systems by elongation or compression along an octahedral four-fold axis (Figure 2, tetragonal distortion, D_{4h}). In solids, the Jahn–Teller distortion may be local to the coordination spheres of the individual Cr^{2+} ions or there may be interactions between neighbouring ions affecting the whole lattice, and this cooperative Jahn–Teller effect has been reviewed (Section 35.3.7).[86]

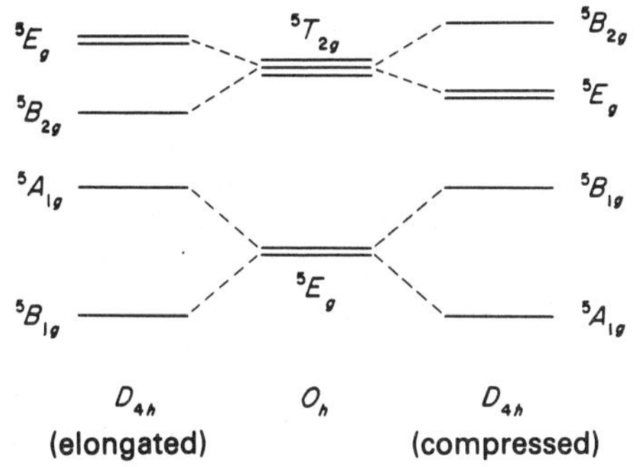

Figure 2 Energy levels of high-spin chromium(II) in ligand fields of O_h and D_{4h} symmetry

From Figure 2, one spin-allowed *d–d* transition is expected for a regular octahedral complex, and three such transitions for a tetragonally distorted complex. Experimentally, most high-spin chromium(II) complexes exhibit one absorption band (ν_2, the main band) with a weaker band or shoulder (ν_1, the distortion band) to lower wavenumber; ν_1 is usually assigned to the $^5B_{1g} \rightarrow {}^5A_{1g}$ transition and ν_2 to superimposed $^5B_{1g} \rightarrow {}^5B_{2g}$, ^5E_g transitions. There are also examples where there is one very broad band due to the overlap of all three transitions and others where they are resolved. There are few single crystal spectroscopic studies of chromium(II) complexes.

The spectra of the ammines follow the general patterns outlined above. From the magnetic and spectroscopic data, and X-ray powder comparisons with copper(II) complexes, the hexaammines and pentaammines have been assigned tetragonal octahedral and square pyramidal structures respectively, and the diammines anion-bridged six-coordinate structures (**4**). The tetraammine sulfate is square pyramidal, with coordinated water; other five-coordinate chromium(II) complexes are listed in Table 42. The acetates $[Cr_2(O_2CMe)_4(NH_3)_2]$ and $Cr(NH_3)_4(O_2CMe)_2$ are discussed in Section 35.3.5.5.

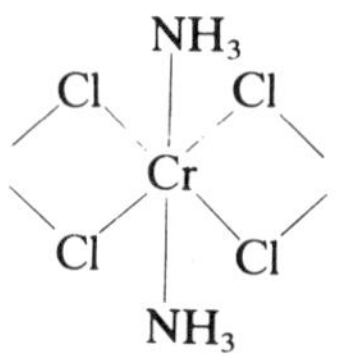

(**4**)

35.3.3.2 *Bidentate saturated amines, ethylenediamine, propylenediamine,* etc.

Although earlier attempts to isolate chromium(II) complexes of various bidentate amines from aqueous solutions produced chromium(III) complexes and hydrogen, the predominantly non-aqueous methods outlined in Scheme 10 provide complexes of ethylenediamine (en), 1,2-diaminopropane (pn), 1,3-diaminopropane (tmd), 1,2-diamino-2-methylpropane (dmp), *N,N*-dimethylethylenediamine (NNdmn) and *N,N'*-dimethylethylenediamine (NN'dmn) (Table 11). In general, ethanol is a suitable solvent but with some amines it is necessary to dehydrate the halide with 2,2-dimethoxypropane (DMP) and dry the ethanol carefully to prevent hydrolysis and oxidation.

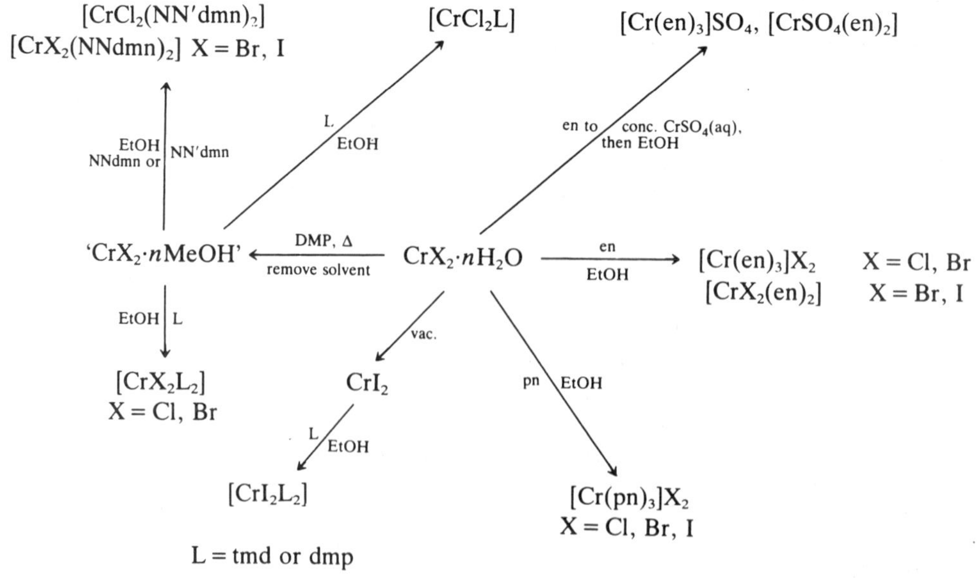

Scheme 10

Table 11 Bidentate Saturated Amines

Complex		μ_{eff} (BM) 300 K	90 K	θ (°)	Reflectance spectra (cm^{-1}) ν_2	ν_1
$[Cr(en)_3]Cl_2{\cdot}H_2O$	Blue	4.84	4.85	0	16 000m	8300m
$[Cr(en)_3]Br_2$	Blue	4.75	4.75	0	15 700m	8300m
$[Cr(en)_3]SO_4$	Light blue	4.87	4.85	0	16 800m	7800m
$[CrBr_2(en)_2]$	Violet	4.74	4.71	0	17 900m	13 200w
$[CrI_2(en)_2]$	Violet	4.73	4.73	0	18 200m	14 500sh
$[CrSO_4(en)_2]$	Violet	4.80	4.56	10	17 700m	11 600w
$[Cr(pn)_3]Cl_2{\cdot}2H_2O$	Light blue	4.80	4.75	0	16 000s	7700s
$[Cr(pn)_3]Br_2{\cdot}H_2O$	Blue	4.90	4.83	0	15 800s	7600s
$[Cr(pn)_3]I_2$	Blue	4.83	4.78	0	16 000s	7600s
$[CrCl_2(tmd)_2]$	Violet	4.80	4.75	0	17 600s	12 600sh
$[CrBr_2(tmd)_2]$	Violet	4.87	4.89	0	18 000s	14 200sh
$[CrI_2(tmd)_2]$	Mauve	4.80	4.83	0	18 300s	15 300sh
$[CrCl_2(tmd)]$	Light blue	3.87	2.14	—	18 600sh	14 100s
$[CrCl_2(dmp)_2]$	Violet	4.75	4.55	10	17 500s	
$[CrBr_2(dmp)_2]$	Purple	4.78	4.52	12	17 400s	
$[CrI_2(dmp)_2]$	Mauve	4.79	4.45	19	18 200s	
$[CrCl_2(dmp)]$	Light blue	4.02	2.67	—	18 600sh	13 400s
$[CrCl_2(NN'dmn)_2]$	Mauve	4.81	4.80	0	16 900s	11 400sh
$[CrBr_2(NNdmn)_2]$	Violet	4.82	4.81	0	17 700s	
$[CrI_2(NNdmn)_2]$	Purple	4.93	4.77	8	16 500s	

The $[Cr(en)_3]^{2+}$ and $[Cr(pn)_3]^{2+}$ salts have reflectance spectra (Table 11) resembling those of the hexaammines, and the six N donor atoms are assumed to complete tetragonally distorted octahedra around the metal. Stability constant measurements (Table 39) have shown that the ions $[Cr(en)(aq)]^{2+}$ ($\bar{\nu}_{max} = 18\,300\ cm^{-1}$, $\varepsilon = 25\ dm^3\ mol^{-1}\ cm^{-1}$) and $[Cr(en)_2(aq)]^{2+}$ ($\bar{\nu}_{max} = 17\,500\ cm^{-1}$, $\varepsilon = 17\ dm^3\ mol^{-1}\ cm^{-1}$) exist in aqueous solution, but that, as in the copper(II) system, the third ethylenediamine molecule is only weakly bound, and care is needed to prevent loss of en from tris(amine) complexes in the preparations. Several bis(amine) complexes, *e.g.* $[CrBr_2(en)_2]$, have been isolated, and these are assigned *trans* structures because of IR spectral resemblances to the corresponding copper(II) complexes. Since the spectrum of $[Cr(SO_4)(en)_2]$ also shows the presence of bidentate sulfate, this is assigned a *trans* octahedral structure with bridging anions.

The pale blue tris(amine) complexes $[Cr(pn)_3]X_3$ separate readily from solutions containing an excess of the amine. Violet solutions form when 2:1 ratios of pn and metal halide are used, but pure bis(amine) complexes have not been obtained. On the other hand, the other bidentate amines give only the bis(amine) complexes $[CrX_2(tmd)_2]$, $[CrX_2(dmp)_2]$, $[CrX_2(NNdmn)_2]$ and $[CrCl_2(NN'dmn)_2]$ (Table 11), even from solutions containing an excess of amine. These are also given *trans* structures. The antiferromagnetic monoamine complexes $[CrCl_2(tmd)]$ and $[CrCl_2(dmp)]$ are believed to be chloride-bridged polymers. No crystallographic evidence is available.

The bands ν_2 and ν_1, particularly the latter, move to higher wavenumber with weaker field axial ligands: $I > Br > SO_4 > NH_2$. This behaviour is characteristic of distorted systems.[87,88]

35.3.3.3 Polydentate amines, diethylenetriamine, triethylenetetramine and facultative ligands

(i) Diethylenetriamine

The ions $[Cr(dien)_2]^{2+}$ and $[Cr(dien)]^{2+}$ have been identified in aqueous solution by stability constant (Table 39) and other measurements: $\mu_{eff}^{293\,K} \approx 4.9$ BM, $\bar{\nu}_{max} = 17\,000\ cm^{-1}$, $\varepsilon = 39\ dm^3\ mol^{-1}\ cm^{-1}$ $[Cr(dien)_2]^{2+}$; $\bar{\nu}_{max} = 16\,300\ cm^{-1}$, $\varepsilon = 32\ dm^3\ mol^{-1}\ cm^{-1}$ $[Cr(dien)(aq)]^{2+}$, but the solutions decompose on standing, and the solids $[Cr(dien)_2]X_2$ and $[CrX_2(dien)]$ have been crystallized from alcoholic media. The bis(dien) complexes have similar magnetic moments and electronic spectra to the tris(en) complexes and so contain tetragonally distorted six-coordinate Cr^{2+}. Unlike $[Cr(en)_3]^2$ ions, which lose en, the $[Cr(dien)_2]^{2+}$ ions maintain their integrity in solution in organic solvents. Halide-bridged binuclear structures have been suggested for the antiferromagnetic mono(dien) complexes (Table 12).

Table 12 Diethylenetriamine Complexes

Complex[a]	μ_{eff} (BM) 300 K	μ_{eff} (BM) 90 K	θ (°)	Reflectance spectra (cm^{-1}) ν_2	Reflectance spectra (cm^{-1}) ν_1
$[Cr(dien)_2]Cl_2 \cdot H_2O$	4.82	4.80	0	16 100	b
$[Cr(dien)_2]Br_2$	4.88	4.82	2	16 000	8600
$[Cr(dien)_2]I_2$	4.84	4.84	0	16 100	b
				16 800 ($\varepsilon = 19$)[c]	8800 ($\varepsilon = 10$)[c]
$[CrCl_2(dien)]$	4.38	3.39	d	14 500vb	
$[CrBr_2(dien)]$	4.26	3.30	d	14 500vb	
$[CrI_2(dien)]$	4.28	3.74	e	14 900vb	

[a] All complexes are blue.
[b] Not measured below 10 000 cm^{-1}.
[c] DMF, ε in $dm^3\ mol^{-1}\ cm^{-1}$.
[d] *J ca.* 10 cm^{-1}, $g = 2$.
[e] *J ca.* 4 cm^{-1}, $g = 2$.

(ii) Triethylenetetramine

From stability constant measurements (Table 39) the $[Cr(trien)]^{2+}$ ion ($\mu_{eff} \approx 4.9$ BM, $\bar{\nu}_{max} = 16\,900\ cm^{-1}$, $\varepsilon = 46\ dm^3\ mol^{-1}\ cm^{-1}$) exists in aqueous solution, but only incompletely characterized solids have been isolated.[89]

(iii) Facultative ligands

Chromium(II) forms the trigonal bipyramidal complex $[CrBr(Me_6tren)]Br$ with the 'tripod' ligand tris(2-dimethylaminoethyl)amine (Me_6tren) (Section 35.3.4.3), and pyrazolyl-substituted ligands also form five-coordinate complex cations (Section 35.3.3.4.v; see also Table 42).

35.3.3.4 N heterocyclic ligands

(i) Pyridine and substituted pyridines

Like other bivalent metals of the first transition series, chromium(II) forms many complexes with pyridine and substituted pyridines. These are mostly of the types $[CrX_2L_4]$, in which X = Cl, Br and I, and $[CrX_2L_2]$, where X = Cl, Br and occasionally I, but a few of the general formulae $[CrX_2(H_2O)_2L_2]$ and $[Cr(py)_6]X_2$ are known (Table 13). The complexes $[CrX_2L_2]$ invariably obey the Curie–Weiss law, with large θ values, and this antiferromagnetic interaction is evidence for the halide-bridged structure (**5**), which is confirmed by the isomorphism of $[CrX_2(py)_2]$ and $[CuX_2(py)_2]$ (X = Cl, Br).

Table 13 Complexes of Pyridine and Substituted Pyridines

Complex	μ_{eff} (BM)	θ (°)	ν(Cr—X) and ν(Cr—N) (cm^{-1})	Reflectance spectra[a] (cm^{-1}) ν_2	Reflectance spectra[a] (cm^{-1}) ν_1	Ref.
Pyridine						
$[CrCl_2(py)_2]$	4.84 (r.t.)	—	322 ν(Cr—Cl)	15 050	10 800	1–4
Light green	4.62 (298 K)	40	281 ν(Cr—N)			
$[CrBr_2(py)_2]$	4.81 (r.t.)	—	272 } ν(Cr—Br) ν(Cr—N)	14 300	10 300	2, 4
Light green						
$[Cr(py)_6]Br_2$	4.94 (r.t.)	—	—	17 000	11 600	2
Dark green						
$[CrI_2(py)_4]$	5.03 (r.t.)			17 000	12 750	2
Brown	4.98 (298 K)	0				3
$[Cr(py)_6]I_2$	5.03 (r.t.)	—	—	CT	12 100	2
Dark green						
$[CrCl_2(H_2O)_2(py)_2]$	4.79 (292 K)	—		—		3
$[CrBr_2(H_2O)_2(py)_2]$	5.03 (298 K)	0	—	—		3
Green						

Table 13 *(continued)*

Complex	μ_{eff} (BM)		θ (°)	$\nu(Cr—X)$ and $\nu(Cr—N)$ (cm^{-1})	Reflectance spectra[a] (cm^{-1}) ν_2	ν_1	Ref.
$[CrI_2(H_2O)_2(py)_2]$ Green	5.01 (298 K)		0	—	—		3
Methylpyridines							
	300 K	86 K					
$[CrCl_2(3\text{-Mepy})_4]$ Brown	4.79	4.74	2	295, 280 ν(Cr—N)	—	9800	4
$[CrBr_2(3\text{-Mepy})_4]$ Golden-brown	4.90	4.79	5	286 ν(Cr—N)	17 000sh	10 800	4
$[CrI_2(3\text{-Mepy})_4]$ Olive brown	4.86	4.87	0	292 ν(Cr—N)	17 500sh	11 800	4
$[CrCl_2(4\text{-Mepy})_4]$ Yellow-brown	4.79	4.89	−4	298 ν(Cr—N)	17 000sh	9700	4
$[CrBr_2(4\text{-Mepy})_4]$ Olive-yellow	4.93	4.86	3	296 ν(Cr—N)	17 000	10 800	4
$[CrI_2(4\text{-Mepy})_4]$ Olive-yellow	4.90	4.80	5	298 ν(Cr—N)	17 500sh	12 000	4
$[CrI_2(H_2O)_2(3\text{-Mepy})_2]$ Green	4.88	4.80	4	283sh ν(Cr—N) 278 ν(Cr—N)	16 600	11 400	4
$[CrI_2(H_2O)_2(4\text{-Mepy})_2]$ Green	4.87	4.79	5	275 ν(Cr—N)	16 500	11 800	4
$[CrCl_2(3\text{-Mepy})_2]$ Light green	4.62	4.01	46	320 ν(Cr—Cl) 282 ν(Cr—N)	14 500	11 300sh	4
$[CrBr_2(3\text{-Mepy})_2]$ Light green	4.76	4.16	42	282 ν(Cr—N) 272 ν(Cr—Br)	14 200	11 000m	4
$[CrCl_2(4\text{-Mepy})_2]$ Blue-green	4.70	4.09	45	332 ν(Cr—Cl) 280 ν(Cr—N)	14 600	11 700sh	4
$[CrBr_2(4\text{-Mepy})_2]$ Light green	4.60	4.03	41	296 ν(Cr—Br) 271 ν(Cr—N)	14 200	11 500sh	4
Halopyridines							
	295 K	90 K					
$[CrCl_2(3\text{-Clpy})_2]$ Light green	4.08	3.65	37	310 ν(Cr—Cl) 272 ν(Cr—N)	14 300	10 700	5
$[CrBr_2(3\text{-Clpy})_2]$ Green	4.51	3.93	48	274 ν(Cr—N) 260 ν(Cr—Br)	13 800	10 100	5
$[CrI_2(H_2O)_2(3\text{-Clpy})_2]$ Olive green	4.80	4.85	−3	270 ν(Cr—N)	16 500 12 000	10 700sh	5
$[CrCl_2(3\text{-Brpy})_2]$ Light green	4.59	4.10	37	324 ν(Cr—Cl) 268 ν(Cr—N)	14 200 12 150sh	10 400sh	5
$[CrBr_2(3\text{-Brpy})_2]$ Dark green	4.59	4.11	37	272 ν(Cr—N) 252 ν(Cr—Br)	13 900	10 400	5
$[CrI_2(H_2O)_2(3\text{-Brpy})_2]$ Olive green	4.83	4.78	2	265 ν(Cr—N)	16 300 12 000	10 700sh	5
$[CrI_2(3\text{-Brpy})_2]$ Green	4.61	4.13	36	264 ν(Cr—N) 238 ν(Cr—I)	13 500	10 600	5
$[CrCl_2(3\text{-Ipy})_2]$ Light green	4.64	4.30	22	324 ν(Cr—Cl) 260 ν(Cr—N)	14 500 12 600sh	10 900	5
$[CrBr_2(3\text{-Ipy})_2]$ Bluish green	4.69	4.30	27	282 ν(Cr—Br) 258 ν(Cr—N)	14 100 12 000sh	10 400sh	5
$[CrI_2(3\text{-Ipy})_2]$ Green	4.60	4.27	22	272 ν(Cr—N) 234 ν(Cr—I)	13 600 12 100sh	11 000sh	5
$[CrCl_2(3,5\text{-Cl}_2py)_2]$ Light yellow	4.71	4.06	53	324 ν(Cr—Cl) 272 ν(Cr—N)	13 750	9600	5
$[CrBr_2(3,5\text{-Cl}_2py)_2]$ Yellow	4.67	4.00	57	272 ν(Cr—N), ν(Cr—Br)	13 600	9300	5
$[CrI_2(3,5\text{-Cl}_2py)_2]$ Dark yellow	4.69	4.33	24	262 ν(Cr—N) 250 ν(Cr—I)	13 200	8800	5
$[CrI_2(3,5\text{-Cl}_2py)_4]$ Dark red	4.83	4.63	11	270 ν(Cr—N)	15 200s,sh	7000	5

[a] Reflectance spectra at liquid nitrogen temperature.

1. H. Lux, L. Eberle and D. Sarre, *Chem. Ber.*, 1964, **97,** 503.
2. D. G. Holah and J. P. Fackler, Jr., *Inorg. Chem.*, 1965, **4,** 1112.
3. A. Earnshaw, L. F. Larkworthy and K. S. Patel, *Chem. Ind.* (*London*), 1965, 1521.
4. M. M. Khamar, L. F. Larkworthy and M. H. O. Nelson-Richardson, *Inorg. Chim. Acta*, 1978, **28,** 245.
5. M. M. Khamar, L. F. Larkworthy and D. J. Phillips, *Inorg. Chim. Acta*, 1979, **36,** 223.

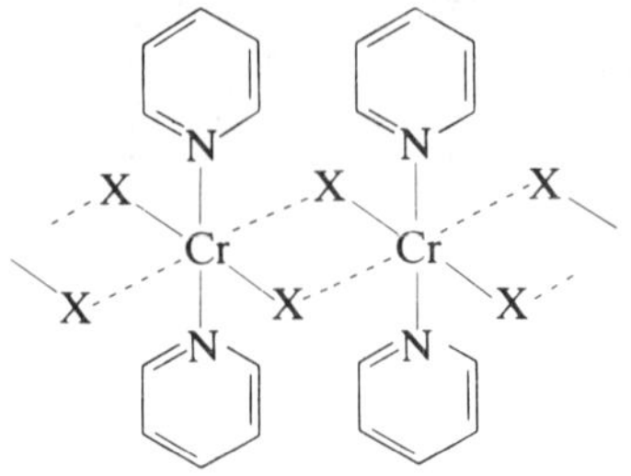

(5)

The complexes $[CrX_2L_4]$ and $[CrX_2(H_2O)_2L_2]$ are high-spin and magnetically dilute. The former are believed to have *trans* octahedral structures and there is some evidence that the latter have too.[85,90–93]

Some pyridine complexes, *e.g.* $[CrCl_2(py)_2]$, have been prepared from the hydrated chromium(II) halide in aqueous solution, but it is more common to use solvents such as acetone, which is particularly useful for iodo complexes, and ethanol. In some cases it is necessary to recrystallize from dimethylformamide to remove water arising from the use of the hydrated halide. It is also sometimes necessary to control the amount of pyridine fairly closely, use cold solutions and wash with solvents containing small amounts of pyridine if the desired complex is to be obtained. Some complexes, *e.g.* $[Cr(py)_6]Br_2$ and $[Cr(py)_6]I_2$, have been prepared by dissolving the bis- or tetrakis-amine complex in an excess of pyridine. These two complexes, which are isomorphous[91] with $[Cu(py)_6]Br_2$, are the only known hexakispyridine complexes of chromium(II), although more could probably be obtained with $[BPh_4]^-$ and similar anions. The iodo complex $[CrI_2(3,5\text{-}Clpy)_2]$ was prepared by warming the tetrakis(amine) in acetone so that it lost two chloropyridine molecules, and $[CrI_2(3\text{-}Brpy)_2]$ was obtained by the use of 2,2-dimethoxypropane in acetone.

Attempts to isolate complex halides containing 2-methylpyridine produced blue substances, which could not be characterized.[93] Bridging iodide is less common than bridging chloride or bromide and the few chromium(II) iodo complexes, $[CrI_2(3\text{-}Brpy)_2]$, $[CrI_2(3\text{-}Ipy)_2]$ and $[CrI_2(3,5\text{-}Clpy)_2]$, might be expected to be tetrahedral. However, because of their antiferromagnetic behaviour, they are apparently linear polymers; tetrahedral monomers would be expected to have magnetic moments greater than 4.90 BM. The reflectance spectra of the iodo complexes are also similar to those of the corresponding bromides and chlorides and the molar absorbances are low ($\varepsilon \approx 30$ dm^3 mol^{-1} cm^{-1} in ethanol).[92]

Bands in the far IR spectrum of $[CrCl_2(py)_2]$ at 328 and 303 cm^{-1} were assigned[94] to ν(Cr—Cl), and a band at 219 cm^{-1} to ν(Cr—N), but in a reinvestigation[93] rather different frequencies (322vs, 281s and 215s cm^{-1}) were found. The band at 322 cm^{-1} has been assigned to ν(Cr—Cl) since it was absent from the spectrum of $[CrBr_2(py)_2]$. Instead, the bromide has an intense, broad and asymmetric band at 272 cm^{-1}, and strong absorption rising to a maximum at 200 cm^{-1}. It has therefore been suggested that ν(Cr—N) is at 281 cm^{-1} in the chloride, the band at 272 cm^{-1} for the bromide contains superimposed ν(Cr—N) and ν(Cr—Br) absorptions, and the band near 200 cm^{-1} corresponds to δ(N—Cr—N). These spectra were recorded only to 200 cm^{-1}, but the assignments are confirmed by investigations of $[CrX_2(Mepy)_2]$ (X = Cl, Br) and $[CrX_2(\text{halo-py})_2]$ (X = Cl, Br, I), which show that ν(Cr—Cl) is near 320 cm^{-1}; ν(Cr—Br) and ν(Cr—N) sometimes overlap in the range 250–300 cm^{-1}, and, from a few examples, ν(Cr—I) is near 240 cm^{-1}. The ν(Cr—N) vibrations are generally at lower frequencies in the halopyridine complexes because of the greater mass of the amines (Table 13).[92,93]

The far IR spectra of $[CrX_2(Mepy)_4]$ (X = Cl, Br, I) do not differ significantly over the range 400–70 cm^{-1} except that the chlorides have two bands in the region of 136 cm^{-1} which may be due to 'long-bonded' ν(Cr—Cl) vibrations, *i.e.* in these complexes the halide ions lie along the elongated tetragonal axis.

The reflectance spectra show the usual two band pattern of distorted six-coordinate, high-spin chromium(II).

The complexes $[CrCl_2(py)_2]$ and $[Cr(CF_3SO_3)_2(py)_4]$ have been used in the preparation of macrocyclic complexes of Cr^{II} and Cr(NO) (Section 35.3.10.2 and Table 64).

The chromium(II) ion is in an uncommon *trans* planar environment (**6**) in bis(2,6-dimethylpyridine)bis(trifluoroacetato)chromium(II), which is prepared by the reaction sequence in

Scheme 11. Each ligand is monodentate and the methyl groups of the 2,6-Mepy molecules above and below the CrO_2N_2 plane block further coordination.[95]

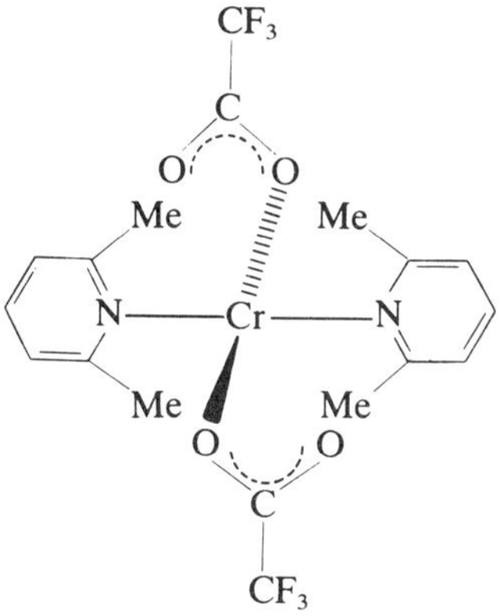

(6)

$$CrCl_3 \xrightarrow{NaBH_4} [(THF)Cr(BH_4)_2] \xrightarrow{CF_3CO_2H} [(THF)Cr(O_2CCF_3)_2]_2 \xrightarrow{2,6\text{-Mepy}} [Cr(O_2CCF_3)_2(2,6\text{-Mepy})_2]$$

Scheme 11

Thermal decomposition of $[CrBr_2(H_2O)_2py_2]$ takes place[96] according to reactions (7) to (9).

$$[CrBr_2(H_2O)_2(py)_2] \text{ (cryst.)} \longrightarrow [CrBr_2(py)_2] \text{ (cryst.)} + 2H_2O \text{ (g)} \qquad (7)$$

$$[CrBr_2(py)_2] \text{ (cryst.)} \longrightarrow [CrBr_2(py)] \text{ (cryst.)} + py \text{ (g)} \qquad (8)$$

$$[CrBr_2(py)] \text{ (cryst.)} \longrightarrow CrBr_2 \text{ (cryst.)} + py \text{ (g)} \qquad (9)$$

$[CrI_2(py)_4]$ was found to lose two pyridine molecules to give $[CrI_2(py)_2]$, with the loss of further pyridine not beginning until 237 °C, but it was not possible to distinguish the stepwise loss of the remaining two ligands.

The heat of reaction for the removal of $2H_2O$ from $[CrBr_2(H_2O)_2(py)_2]$ ($\Delta H = 117.5\,kJ\,mol^{-1}$) is close to that for the removal of 2py from $[CrI_2(py)_4]$, suggesting H_2O and pyridine form bonds of similar strength.

Values of ΔH for the loss of two pyridine molecules from $[CrX_2(py)_2]$ are 148.4, 132.1 and 103.7 $kJ\,mol^{-1}$, for X = Cl, Br and I respectively. The value for $[CrCl_2(py)_2]$ is greater than those for $[MCl_2(py)_2]$ where M = Mn through Cu. The chloride decomposes according to equations (8), (10) and (11).

$$[CrCl_2(py)] \text{ (cryst.)} \longrightarrow [CrCl_2(py)_{2/3}] \text{ (cryst.)} + \tfrac{1}{3}py \text{ (g)} \qquad (10)$$

$$[CrCl_2(py)_{2/3}] \text{ (cryst.)} \longrightarrow CrCl_2 \text{ (cryst.)} + \tfrac{2}{3}py \text{ (g)} \qquad (11)$$

(ii) Nicotinic acid and saccharine

Nicotinic acid (nic) forms a series of complexes $[M^{II}(nic)_2(H_2O)_4]$ in which it is N-bonded. The yellow, crystalline chromium(II) complex is prepared by running an aqueous Cr^{2+} solution into a nicotinic acid solution under hydrogen. It is air-stable and the analyses were consistent with $[Cr^{III}(nic)_2(H_2O)_3OH]$ or $[Cr^{II}(nic)_2(H_2O)_4]$, but the molecular structure (**7**) agrees with the chromium(II) formulation. The complex has a magnetic moment (3.1 BM) corresponding to low-spin chromium(II), which is unexpected since the complexes $[CrX_2(py)_4]$ are high-spin (Section 35.3.3.4.i above), and two bipy or phen molecules are necessary to cause spin pairing (Section 35.2.2.1).* This complex was prepared in an investigation of chromium(III) nicotinic acid complexes related to the glucose tolerance factor (Section 35.4.8.3). The X-ray powder patterns and IR spectra of the complexes $[M^{II}(nic)_2(H_2O)_4]$ in general are closely similar to those of the Cr^{II} complex.[97]

Saccharine forms a similar series of complexes $[M^{II}(C_7H_4NO_3S)_2(H_2O)_4]\cdot 2H_2O$. The chromium(II) complex is yellow and has a *trans* structure (**8**) analogous to that of

---

* Found later to be contaminated with isomorphous Zn^{II} complex; Cr—N = 2.128, Cr—O, two at 2.039, two at 2.471 Å (W. E. Broderick, M. R. Pressprich, U. Geiser, R. D. Willet and J. I. Legg, *Inorg. Chem.*, 1986, **25**, 3372).

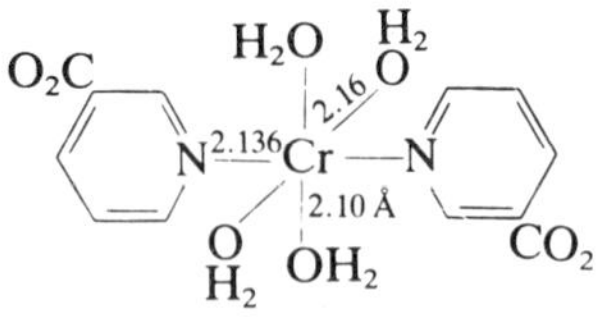

(7)

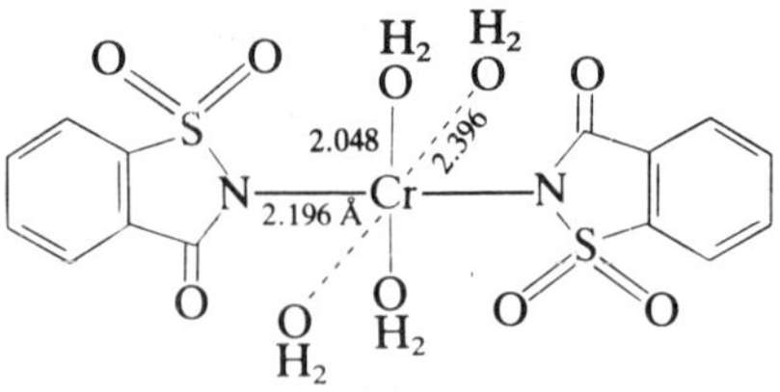

(8)

$[Cr(nic)_2(H_2O)_4]$ (**7**), but with the effect of Jahn–Teller distortion apparent in the Cr—O distances (Table 34).[98]

(*iii*) *Pyrazoles and imidazoles*

Complexes of imidazole (iz),[99,100] *N*-methylimidazole (Nmiz), pyrazole (pz), 3,5-dimethylpyrazole (dmpz),[99] 3(5)-methylpyrazole (5Mepz), 1-methylpyrazole (NMepz), 2-methylimidazole (2Meiz) and 1-ethylimidazole (NEtiz)[101] (**9**) have been prepared from ethanol or butanol by methods resembling those in Scheme 10. As with the analogous pyridine complexes, those of the general formula $[CrX_2L_4]$ are high-spin, tetragonally distorted, six-coordinate complexes, and those of the formula $[CrX_2L_2]$ are also six-coordinate, but are antiferromagnetic halide-bridged polymers; 2-methylimidazole, however, forms cations which are probably square pyramidal: $[CrX(2Meiz)_4]X$ (X = Cl, Br), $[Cr(2Meiz)_5](BPh_4)_2$ and $[Cr(NCS)_2(2Meiz)_2]$ (Table 14). In the last complex five-coordination (Table 42) is believed to be achieved through NCS bridges. The reduction of chromium(III)–imidazole complexes at the dropping mercury electrode has been studied (Section 35.4.2.5.iii).

(**9**) X = N, Y = CH, imidazole
X = CH, Y = N, pyrazole

A pale blue-green complex $[CrCl_2(PMT)_2]$ ($\mu_{eff}^{297\,K} = 4.55$ BM) of pentamethylenetetrazole (PMT) has been obtained by the addition of solid PMT in excess to a solution of hydrated chromium(II) chloride in a 50% DMP/MeOH mixture.[102] Pyrazines and Cr^{2+} give intensely coloured 'pyrazine green' products (Section 35.4.2.5.iv).

(*iv*) *Bidentate ligands: bipyridyls, phenanthrolines, 2-aminomethylpyridine, 8-aminoquinoline and adenine*

Chromium(II) complexes of bipyridyls, terpyridyl and the phenanthrolines have been discussed in Section 35.2.2.1. Complexes of the ligands 2-aminomethylpyridine (pic, 2-picolylamine) and 8-aminoquinoline (amq), which have one heterocyclic and one amino nitrogen donor atom, have been prepared by methods similar to those in Scheme 10. The bis(amine) complexes are typical high-spin, distorted octahedral complexes, and the mono(amine) complexes, from their antiferromagnetic behaviour and reflectance spectra, are six-coordinate, halide-bridged polymers (Table 15).[103] No tris(amine) complexes could be prepared so the attempt to find spin isomeric systems in octahedral chromium(II) systems was unsuccessful ($[Cr(en)_3]X_2$ are high-spin and $[Cr(bipy)_3]X_3$ and $[CrX_2(bipy)_2]$ low-spin).

Complexes of adenine (**10**; Ade) have been isolated by mixing the ligand in 2-methoxyethanol with chromium(II) halides in butanol. The low magnetic moments (Table 16)[104] indicate dinuclear species $[Cr_2X_2(Ade)_4^{2+}]$ (Section 35.3.5.5), or polynuclear structures with bidentate adenine.

(*v*) *Bi-, tri- and tetra-dentate pyrazolyl-substituted ligands*

Chromium(II) complexes of the ligands 1,1′-methylenedipyrazole (**11**; H_2Cpz_2), 1,1′-methylenebis(3,5-dimethylpyrazole) (**12**; H_2Cdmpz_2), tris(1-pyrazolyl)methane (**13**; $HCpz_3$),[105] tris(1-pyrazolylethyl)amine (**14**; R = H, TPyEA)[106,107] and tris(3,5-dimethyl-1-pyrazolylethyl)amine

Table 14 Complexes of Imidazoles and Pyrazoles

Complex	μ_{eff} (BM) 295 K	90 K	*Reflectance spectra* (cm^{-1}) ν_2	ν_1	*Ref.*
$[CrCl_2(iz)_4]$	4.87	4.86	17 500	13 300sh	1
			14 000 ($\varepsilon = 40$)[a]		
Violet-purple	4.87	4.86	17 500	13 000sh	2
$[CrBr_2(iz)_4]$	4.85	—	19 200	16 600sh	1
			17 400 ($\varepsilon = 30$)[a]		
Pale purple	4.84	4.87	19 300vb	16 400sh	2
$[CrI_2(iz)_4]$	4.78	4.77	19 000vb		2
Violet					
$[Cr(SO_4)(iz)_4]$	4.77	4.75	17 200	12 400sh	2
Blue-violet					
$[CrCl_2(pz)_4]$	4.93	4.90	18 200	12 500sh	1
			12 500 ($\varepsilon = 49$)[a]		
$[CrBr_2(pz)_4]$	4.93	—	18 500	13 300sh	1
			13 300 ($\varepsilon = 20$)[a]		
$[CrI_2(pz)_4]$	4.93	—	18 300	15 400sh	1
$[CrCl_2(Nmiz)_4]$	4.90	—	17 700	13 300sh	1
$[CrBr_2(Nmiz)_4]$	4.90	4.88	17 700	14 000sh	1
$[CrI_2(Nmiz)_4]$	4.90	—	17 700	15 500sh	1
$[CrCl_2(Nmiz)_2]$	4.69	4.30	13 950		1
$[CrCl_2(dmpz)_2]$	4.48	3.65	15 000	8350sh	1
$[CrBr_2(dmpz)_3]$	4.95	4.95	15 000		1
$[CrCl_2(pz)_2]$	4.62	4.06	14 700	11 100	1
$[CrCl_2(5Mepz)_4]$	4.81	—	17 400	12 100	3
$[CrBr_2(5Mepz)_4]$	4.82	—	17 550	12 900sh	3
$[CrI_2(5Mepz)_4]$	4.84	—	17 850	15 050	3
$[CrCl_2(2Meiz)_4]$	4.80	—	16 950		3
			16 000 ($\varepsilon = 40$)[a]		
$[CrBr_2(2Meiz)_4]$	4.80	—	17 850		3
			17 400 ($\varepsilon = 50$)[a]		
$[CrI_2(2Meiz)_4]$	4.84	—	17 850		3
			17 550 ($\varepsilon = 40$)[a]		
$[Cr(NCS)_2(2Meiz)_2]$	4.48	4.06 (84 K)	18 870		3
$[Cr(2Meiz)_5](BPh_4)_2$	4.78	—	18 200		3
			18 000 ($\varepsilon = 50$)[a]		
$[CrCl_2(NEtiz)_3]$	4.22	3.84	16 700	12 100	3
$[CrBr_2(NEtiz)_4]$	4.80	—	17 400	12 750sh	3
$[CrCl_2(NMepz)_2]$	4.74	4.46	13 400		3
$[CrBr_2(NMepz)_3]\cdot H_2O$	4.90	4.90	12 900		3

[a] $dm^3\ mol^{-1}\ cm^{-1}$ in DMF.

1. F. Mani and G. Scapacci, *Inorg. Chim. Acta*, 1976, **16**, 163.
2. L. F. Larkworthy and J. M. Tabatabai, *Inorg. Chim. Acta*, 1977, **21**, 265.
3. P. Dapporto and F. Mani, *J. Chem. Res. (S)*, 1979, 374.

Table 15 Magnetic and Reflectance Data for Complexes of 2-Aminomethylpyridine and 8-Aminoquinoline

Complex		μ_{eff} (BM) 300 K	90 K	θ (°)	*Reflectance spectra* (cm^{-1}) ν_2		ν_1
$[CrCl_2(pic)_2]\cdot H_2O$	Orange	4.86	4.82	2	22 400		13 000sh
$[CrBr_2(pic)_2]$	Orange	4.90	4.82	4	22 000		13 600
$[CrI_2(pic)_2]$	Orange	4.90	4.85	2	21 800		13 800
$[CrCl_2(pic)]$	Light green	4.45	3.70	$J = 6\ cm^{-1}$		14 000vb	
$[CrBr_2(pic)]$	Green	4.40	3.46	$J = 9\ cm^{-1}$		13 600vb	
$[CrCl_2(amq)_2]\cdot H_2O$	Dark red	4.86	4.82	2	CT		13 500sh
$[CrBr_2(amq)_2]\cdot 2H_2O$	Red	4.91	4.87	2	CT		13 700sh
$[CrI_2(amq)_2]$	Red	4.85	4.83	2	CT		13 800sh
$[CrCl_2(amq)]\cdot H_2O$	Greenish yellow	4.28	3.59	$J = 7\ cm^{-1}$	CT		13 500sh

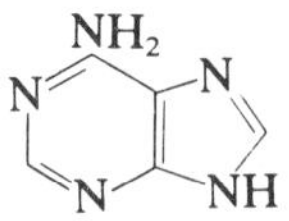
(10)

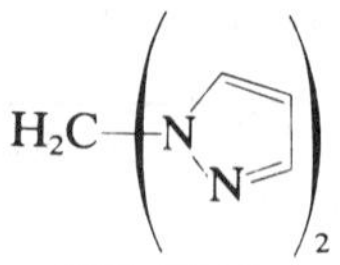
(11) H_2Cpz_2

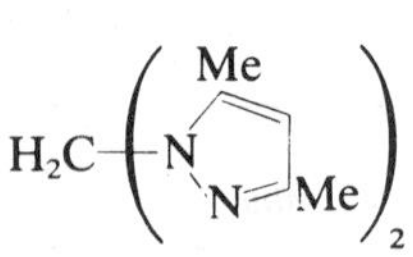
(12) H_2Cdmpz_2

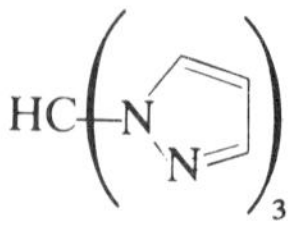
(13) $HCpz_3$

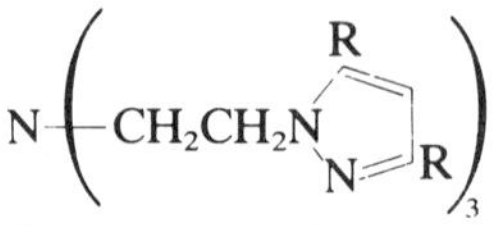
(14) R = H, TPyEA
R = Me, MeTPyEA

Table 16 Complexes of Adenine

Complex	μ_{eff} (BM)		*Reflectance spectra* (cm^{-1})			
$CrCl_2(Ade)_2 \cdot C_3H_8O_2$	1.63(295 K)	1.11(103 K)	19 800	14 300sh	10 500sh	8000
$CrBr_2(Ade)_2 \cdot C_3H_8O_2$	1.49(286 K)	0.75(85 K)	19 050	14 800sh	10 000sh	7800
$CrI_2(Ade)_2 \cdot 0.5C_3H_8O_2$	1.44(296 K)	0.64(85 K)	19 200	14 300sh	9500sh	

(**14**; R = Me, MeTPyEA) have been isolated from non-aqueous solvents (Table 17).[105–108] From their electronic spectra, simple paramagnetic behaviour and stoichiometry $[CrCl(H_2Cdmpz_2)_2]BF_4$, $[CrX(H_2Cdmpz_2)_2]BPh_4$ and $[CrX(H_2Cdmpz_2)_2]X$ (X = Br, I) are thought to be five-coordinate. The spectra differ from those of distorted six-coordinate or square pyramidal chromium(II) complexes and, except for intensity differences, resemble those of other trigonal bipyramidal chromium(II) complexes. Conductance and solution spectra suggest that the trigonal bipyramidal cations are present in solution too. From similar evidence $[CrX(MeTPyEA)]BPh_4$ (X = Br, NCS) contain square pyramidal cations.[108] The blue complex $[CrNCS(TPyEA)]BPh_4$, dissolved in acetone or acetonitrile, is oxidized very rapidly by air to give a green solution from which the μ-oxo chromium(III) complex $[\{Cr(NCS)(TPyEA)\}_2O](BPh_4)_2$ crystallizes (Section 35.4.2.5.iv).[107] The remaining complexes are mostly high-spin six-coordinate complexes with temperature-independent magnetic moments, but in a few, *e.g.* $[CrCl_2(H_2Cpz_2)]$ and $[CrBr(TPyEA)]BPh_4$, the halide bridges needed to produce six-coordination lead to weak antiferromagnetism.

Table 17 Complexes of Pyrazolyl-substituted Ligands

Complex		μ_{eff} (BM) (*298 K*)	(*84 K*)	*Reflectance spectra* (cm^{-1}) ν_2	ν_1
$[Cr(HCpz_3)_2](BPh_4)_2$	Greenish grey	4.82		18 900	10 200w
$[CrCl_2(H_2Cpz_2)]$	Greenish blue	4.45	3.60	13 200	10 500sh
$[CrBr_2(H_2Cpz_2)_2]$	Green	4.80		14 100	8200w
$[CrI_2(H_2Cpz_2)_2]$	Lilac-grey	4.80		17 400	13 200sh
$[CrCl(H_2Cpz_2)_2]BPh_4$	Green	4.42	3.80	14 700	8600w
$[CrBr(H_2Cpz_2)_2]BPh_4$	Greenish grey	4.14	3.15	17 200	11 600sh
$[CrBr(H_2Cdmpz_2)_2]Br$	Light blue	4.93		15 400	11 300
$[CrI(H_2Cdmpz_2)_2]I$	Light blue	4.83		15 300	11 100
$[CrCl(H_2Cdmpz_2)_2]BF_4$	Blue	4.80	4.85	17 000	12 500
$[CrBr(H_2Cdmpz_2)_2]BPh_4$	Blue	4.85	4.85	16 000	12 500
$[CrI(H_2Cdmpz_2)_2]BPh_4$	Blue	4.88	4.80	14 700	11 100
$[CrBr(MeTPyEA)]BPh_4 \cdot EtOH$		4.83	4.80	14 900	10 400[a]
$[Cr(NCS)(MeTPyEA)]BPh_4$		4.88	4.92	16 400	
$[CrBr(TPyEA)]BPh_4$		4.77	4.54	16 100	10 150
$[Cr(NCS)(TPyEA)]BPh_4$		4.74	4.42	$\nu(CN) = 2085\ cm^{-1}$	

[a] In acetone 14 300 ($\varepsilon = 40\ dm^{-3}\ mol^{-1}\ cm^{-1}$), 11 400sh cm^{-1}.

35.3.3.5 Nitrosyls and hydrazine

Nitrosyls are discussed in Section 35.4.2.6, and hydrazine is the only ligand of its type from which chromium(II) complexes have been prepared.

(i) Hydrazine

The pale blue complexes $[CrX_2(N_2H_4)_2]$ (X = Cl, Br or I), which are reasonably stable to dry air, separate rapidly when an excess of hydrazine is added to an aqueous solution of the appropriate halide.[109] The chloride has also been prepared by the addition of hydrazine to a solution of $[CrCl_2(H_2O)_4]Cl$ reduced with Na/Hg.[110] The complexes have been assigned the polymeric structure (**15**) because the IR spectra are almost identical to those of analogous complexes with crystallographically determined structures. The reflectance spectra (absorption bands at *ca.* 17 500s and 11 000m cm^{-1}) are independent of the halide ion suggesting that these lie along the distortion axis, and the magnetic moments show little temperature variation, consistent with weak antiferromagnetic interaction arising from hydrazine rather than halide bridges. The reaction of CrF_2 with N_2H_4 gives unstable and impure $[CrF_2(N_2H_4)_2]$.[111]

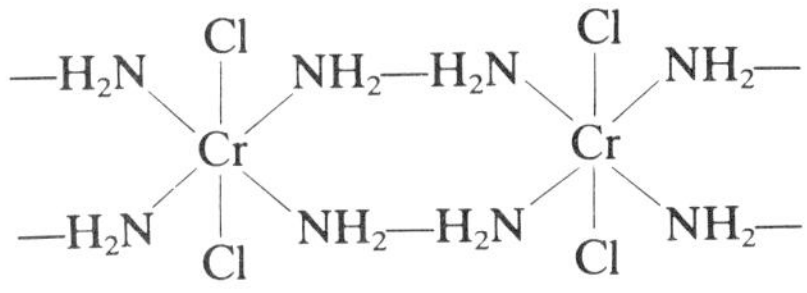

(**15**)

35.3.3.6 Dialkylamides and disilylamides

The disproportionation of chromium(III) dialkylamides, *e.g.* $[Cr(NEt_2)_3]$ into $[Cr^{IV}(NEt_2)_4]$ and $[Cr^{II}(NEt_2)_2]$, has been used to prepare several chromium(IV) dialkylamides (Section 35.5.4.3) but pure chromium(II) dialkylamides have not been obtained by this procedure. In fact, detailed work is restricted to silylamide derivatives. The first, $[Cr\{N(SiMe_3)_2\}_2(THF)_2]$, is *trans* planar (**16**), an uncommon stereochemistry (Table 41) for chromium(II); and it has a magnetic moment of 4.93 BM at 298 K.[112] Other presumably planar complexes in which one or more THF molecules are replaced by other bases are exemplified in Scheme 12. Some solids, *e.g.* $[Cr\{N(SiMe_3)_2\}_2(Et_2O)_2]$, decompose even in an inert atmosphere, but solutions in organic solvents are stable for several weeks so dietherate solutions especially were used in other reactions. The compounds were characterized by analyses and reflectance spectra, and it was necessary to use carefully prepared anhydrous $CrCl_2$ in the preparations. The reflectance spectra were not assigned and no magnetic measurements were reported.[113]

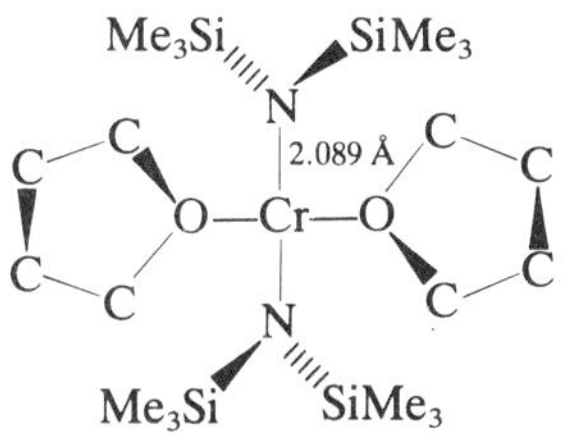

(**16**)

The nitrosyl $[Cr(NO)\{N(SiMe_3)_2\}_3]$ is diamagnetic and has a pseudotetrahedral structure (**135**). It can be formally considered to contain Cr^{II} and NO^+ ($\nu(NO) = 1698\ cm^{-1}$) and is the only example of a crystallographically confirmed tetrahedral Cr^{II} compound (Section 35.4.2.6).[112]

35.3.3.7 Thiocyanates

The deep blue solutions produced on the addition of thiocyanates to aqueous Cr^{2+} decompose on standing or concentration. Nevertheless, thiocyanato complexes can be

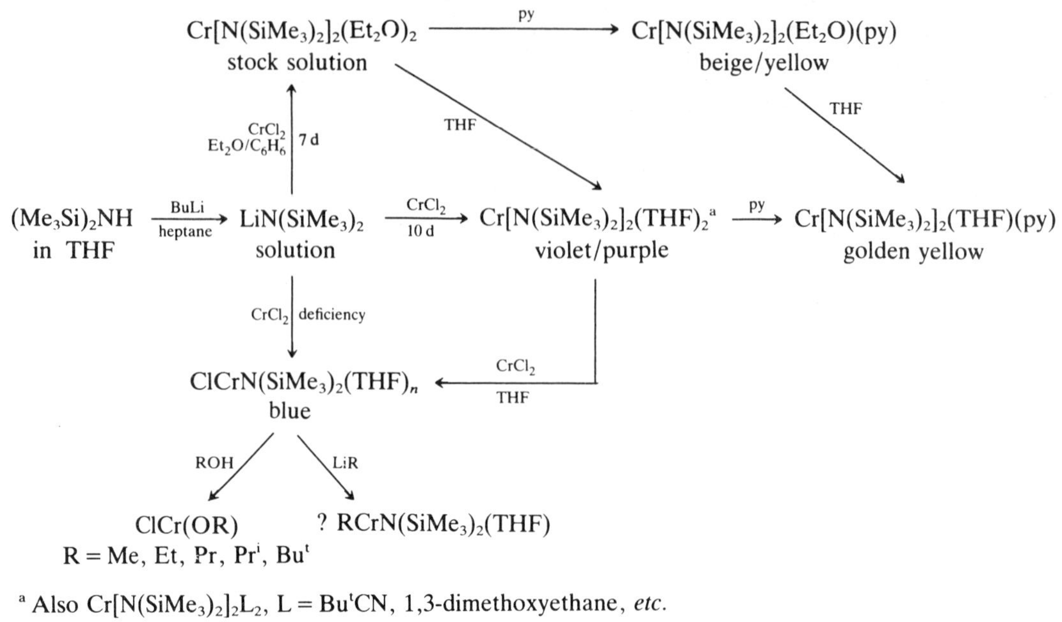

Scheme 12

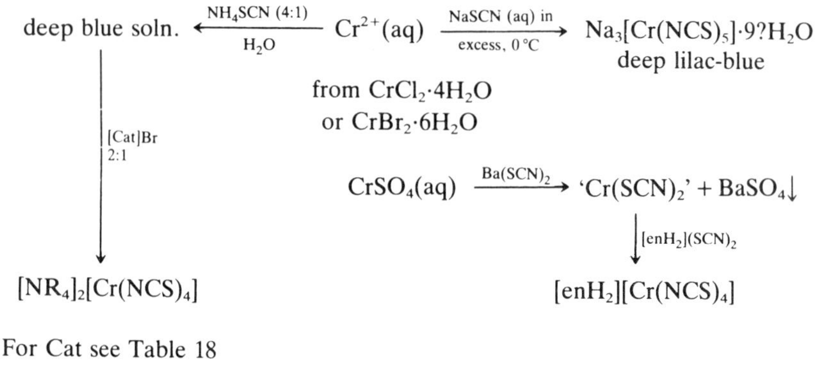

Scheme 13

crystallized if the cations are Na^+ (but not NH_4^+, K^+, Cs^+ or Ba^{2+}) or various substituted ammonium ions (Scheme 13).

The pentathiocyanato complex $Na_3[Cr(NCS)_5]\cdot 9–11H_2O$ is thought from its magnetically dilute behaviour and IR and electronic spectra to contain N-bonded NCS groups in a square pyramid around the metal ion with a water molecule possibly in the sixth position. With organic cations, tetrathiocyanatochromium(II) salts crystallize, and several were isolated in two forms dependent upon the experimental procedure: NBu_4^+, brown or blue; and with or without ethanol $hexH^+$, NEt_4^+ and NPr_4^+. One of each pair of complexes, and all the complexes which were obtained in one form only, are antiferromagnetic since their magnetic moments are below the spin-only value and are temperature-dependent. Thiocyanato-bridged structures have been proposed and this is compatible with the complexity of the ν(CN) absorptions and reflectance spectra. Dissolved in acetone, both NBu_4^+ salts and the NMe_4^+ salt give high-spin magnetic moments showing that the low moments of the solids arise from intermolecular antiferromagnetism. Tetrahedral anions are excluded by the low extinction coefficients. Those which are magnetically dilute *i.e.* $[hexH]_2[Cr(NCS)_4]$, $[NEt_4]_2[Cr(NCS)_4]\cdot EtOH$, $[NPr_4]_2[Cr(NCS)_4]\cdot EtOH$ and brown $[NBu_4]_2[Cr(NCS)_4]$, probably have planar $[Cr(NCS)_4]^{2-}$ anions with bridging prevented by the cations or coordinated EtOH molecules (Table 18).[114]

The tetragonally distorted ion $[Cr(NCS)_6]^{4-}$ has been identified in molten KSCN,[114] and a few mixed ligand complexes containing thiocyanate are known (Sections 35.2.2.1, 35.3.3.4 and 35.3.10.2). Thiocyanate will displace H_2O from $[Cr_2(O_2CMe)_4(H_2O)_2]$ (p. 746).

Table 18 Thiocyanatochromates(II)

Complex	μ_{eff} (BM) 295 K	μ_{eff} (BM) 90 K	θ (°)	IR spectra (cm^{-1}) $\nu(CN)$	IR spectra (cm^{-1}) $\nu(CS)$	Reflectance spectra (cm^{-1}) ν_2	Reflectance spectra (cm^{-1})	Reflectance spectra (cm^{-1}) ν_1
$Na_3[Cr(NCS)_5]\cdot 9–11H_2O$	4.5	4.5	0	2085	810w		16 800	
Deep lilac-blue						[a]17 200		13 500sh
$Na_3[Cr(NCS)_5]\cdot Me_2CO\cdot 6H_2O$	4.31	3.55	77	2118sh	810		—	
Pale blue				2085	724			
$[enH_2][Cr(NCS)_4]$	4.24	3.36	80	2118			17 700	
Grey-blue				2070sh				
$[hexH]_2[Cr(NCS)_4]$[b]	4.75	4.75	0	2090		16 800		12 000
Blue				2050sh				
$[hexH]_2[Cr(NCS)_4]\cdot EtOH$[b]	4.74	3.92	70	2080			17 000	
Blue				2050sh		[a]17 400		16 600sh
$[LH_2][Cr(NCS)_4]$[c]	4.29	3.27	90	2130sh, 2115			15 000	
Blue				2105, 2075				
$[pyH]_2[Cr(NCS)_4]$	3.78	3.36	35	2120sh	810		17 400	
Green-brown				2050				
$[NMe_4]_2[Cr(NCS)_4]$	3.81	2.98	90	2095	810		16 700	
Pale blue					793			
$[NEt_4]_2[Cr(NCS)_4]$	4.38	3.91	38	2080			18 000	
Mauve								
$[NEt_4]_2[Cr(NCS)_4]\cdot EtOH$	4.63	4.63	0	2080			17 000	
Blue								
$[NPr_4]_2[Cr(NCS)_4]$	4.47	3.47	82	2115	808		15 800	
Bright blue				2075	795sh			
$[NPr_4]_2[Cr(NCS)_4]\cdot EtOH$	4.69	4.66	0	2060	826		19 200	
Mauve								
$[NBu_4]_2[Cr(NCS)_4]$ Brown	4.76	4.75	0	2075		18 000		15 000sh
Blue	4.30	3.36	90	2126	812		14 500	
				2090sh	790sh			
				2070				

[a] Liquid nitrogen temperature.
[b] hex = hexamine (hexamethylenetetramine).
[c] L = 5,7,7,12,14,14-hexamethyl-1,4,8,11-tetraazacyclotetradeca-4,11-diene.

35.3.3.8 Polypyrazolylborates and carboranes

On the addition of anhydrous $CrBr_2$ dissolved in warm ethanol to a warm solution of the $[NEt_4]^+$ salt of the appropriate polypyrazolylborate **(17)** in the same solvent, crystals of the complexes $[Cr\{H_nB(pz)_{4-n}\}_2]$ ($n = 0$, 1 or 2 appear) (Table 19).[115] These are moderately air-stable, and the ethyl derivative $[Cr\{Et_2B(pz)_2\}_2]_2$, which was obtained from $[Cr_2(O_2CMe_2)_4]$ and $Na\{Et_2B(pz)_2\}$ in THF on which hexane was layered to induce crystallization, is stable to air.[116] An indazole (a 4,5-substituted pyrazole) derivative $[Cr(H_2B\{4,5\text{-}(5\text{-}NO_2benzo)pz\}_2)_2]$ is also known (Table 19). The isolation of $[Cr\{HB(pz)_3\}_2]$ and $[Cr\{B(pz)_4\}_2]$ from aqueous solution[117] could not be repeated.[115]

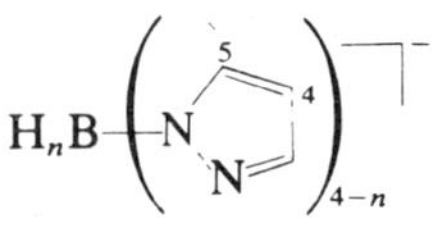

(17) $B(pz)_4^-$ ($n = 0$), $HB(pz)_3^-$ ($n = 1$), $H_2B(pz)_2^-$ ($n = 2$)

The complexes $[Cr\{HB(pz)_3\}_2]$ and $[Cr\{B(pz)_4\}_2]$ are high-spin, with similar reflectance and solution spectra indicative of distorted octahedral structures, which implies that $B(pz)_4^-$ is acting as a tridentate ligand. Both $[Cr\{H_2B(pz)_2\}_2]$[115] and $[Cr\{Et_2B(pz)_2\}_2]$ **(18)**[116] are planar, the latter with the Et groups positioned protectively over the Cr^{II} atoms. This stereochemistry is unusual for Cr^{II} (Table 41), and the extreme distortion is manifest in the reflectance spectrum of $[Cr\{H_2B(pz)_2\}_2]$, which has a band at 20 500 cm^{-1}, much further to high wavenumber than its less distorted six-coordinate analogues, and an ill-defined absorption at 11 000 cm^{-1}.

Table 19 Polypyrazolylborates

Complex	μ_{eff} (BM) 293 K	Reflectance spectra (cm^{-1})		Comment	Ref.
$[Cr\{B(pz)_4\}_2]$ Yellow	4.90	17 500sh 18 500 (33)[a]	11 000 12 000sh	Distorted octahedral	1
$[Cr\{HB(pz)_3\}_2]$ Light green	4.75	17 700sh 18 100(24)[a]	10 700 10 500	Distorted octahedral	1
$[Cr\{H_2B(pz)_2\}_2]$ Orange-red	4.80	20 500	11 000w,br	Square planar Cr—N, 2.055, 2.069 Å	1
$[Cr\{Et_2B(pz)_2\}_2]$ Bright orange	—	—	—	Square planar Cr—N, 2.058, 2.061 Å	2
$[Cr(H_2B\{4,5\text{-}(5\text{-}NO_2benzo)pz\}_2)_2]$	4.85	13 900	11 700w,br	Distorted octahedral	3

[a] ε ($dm^3\ mol^{-1}\ cm^{-1}$, in dichloroethane).

1. P. Dapporto, F. Mani and C. Mealli, *Inorg. Chem.*, 1978, **17**, 1323.
2. F. A. Cotton and G. N. Mott, *Inorg. Chem.*, 1983, **22**, 1136.
3. Z. A. Siddiqi, S. Khan and S. A. A. Zaidi, *Synth. React. Inorg. Metal-Org. Chem.*, 1982, **12**, 433.

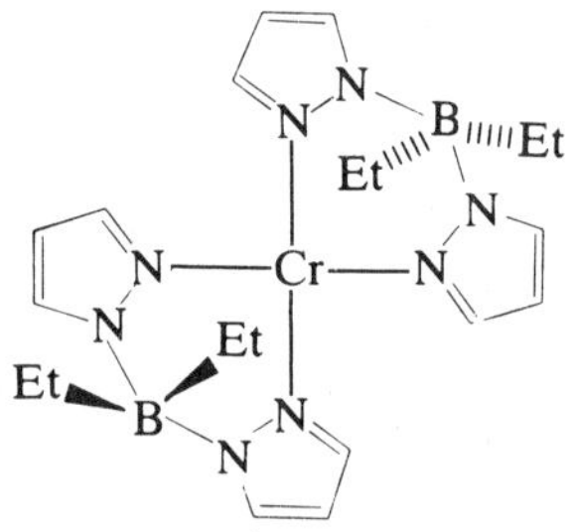

(**18**)

Magnetic data for the low-spin, air-stable chromium(II) metallocarborane sandwich[118] compound $[NEt_4]_2[Cr(C_2B_{10}H_{12})_2]$ have been compared with those for chromocene.

35.3.3.9 Nitriles

The acetonitrile complexes $[CrX_2(MeCN)_2]$ (X = Cl, Br or I) may be planar or tetrahedral monomers or halide-bridged polymers, and there is no clear evidence. The chloride and bromide are high-spin at room temperature, but variable temperature studies have not been carried out; the reflectance spectra are very similar, and typical of Cr^{II} in six-coordination presumably achieved through bridging halide. The much higher wavenumber of the main band in the spectrum of $[CrI_2(MeCN)_2]$ must be due to greater distortion suggestive of a planar structure (Table 20).[119] The addition of tetraalkylammonium halides to solutions of the chloride and bromide in acetonitrile and ethanol causes shoulders at *ca.* 10 000 cm^{-1} to be replaced by moderately intense bands ($\varepsilon \approx 45\ dm^3\ mol^{-1}\ cm^{-1}$), which were thought to be due to the formation of distorted tetrahedral anions $[CrX_4]^{2-}$. In view of the results in Section 35.3.7.3 it is likely that the changes are due to the formation of *trans*-$[CrX_4(solvent)_2]^{2-}$ species.

The complexes $[Cr(MeCN)_4](BF_4)_2$ and $[Cr(MeCN)_6](BF_4)_2$ have been prepared electrolytically (Table 20).[82]

35.3.4 Phosphorus and Arsenic Ligands

35.3.4.1 Mono- and bi-dentate P donor ligands

Strongly basic aliphatic tertiary phosphines form blue solutions with CrX_2 (X = Cl or Br) in benzene or xylene and it is believed that the equlibrium (12) is set up.

$$\underset{\text{blue}}{[CrCl_2(PR_3)_2]} \rightleftharpoons \underset{\text{white}}{[CrCl_2(PR_3)]} + PR_3\ (R = Et, Pr, Bu) \quad (12)$$

Table 20 Complexes of Acetonitrile and Chromium(II) Halides

Complex	*Comments*
$[CrCl_2(MeCN)_2]$ Pale blue	$CrCl_2$ in EtOH + MeCN; $\mu_{eff}^{295 K} = 4.81$ BM Reflectance: 13 300, 9800sh (cm^{-1}) MeCN soln: 13 200 (35),[a] 10 000sh
$[CrBr_2(MeCN)_2]$ Pale blue green	$CrBr_2$ extracted with hot MeCN; $\mu_{eff}^{299 K} = 4.85$ BM Reflectance: 13 200, 9600sh (cm^{-1}) MeCN soln: 11 800 (80),[a] 10 000sh
$[CrI_2(MeCN)_2]$ Pale blue	CrI_3 in MeCN reduced with Zn/Hg Reflectance: 16 200, 10 000sh MeCN soln: 15 400 (4),[a] 10 000 (1)
$[Cr(MeCN)_4](BF_4)_2$ Pink	Electrolysis with (sacrificial) Cr anode in MeCN–HBF_4(aq); bridging BF_4^-?
$[Cr(MeCN)_6](BF_4)_2$ Navy blue	Electrolysis with (sacrificial) Cr anode in MeCN–HBF_4–Et_2O

[a] ε ($dm^3 mol^{-1} cm^{-1}$).

The only solid isolated was white $[CrCl_2(PEt_3)]_n$, and the system is difficult to handle.[120] A related complex $[CrCl_2(PMe_3)_3]$ has been suggested as an intermediate (Section 35.2.3.1). Greater success has been achieved with PMe_2Ph in hexane since the complexes $[CrX_2(PMe_2Ph)_2]$ have been prepared with good analyses and magnetic susceptibilities approximating to high-spin chromium(II).[121] Polymers containing dppe and tetramethyldiphosphine have been prepared from $[CrCl_2(MeCN)_2]$ or $CrCl_2$ (Scheme 14); these phosphines also combine with coordinatively unsaturated Cr^{II} in a silica surface.[122] Chromium(II) surface compounds have catalytic activity and can be used for gas purification *etc.*[123] The complex $[CrCl_2\{o\text{-}C_6H_4(PMe_2)_2\}_2]$ has been formed electrochemically in MeCN but not isolated, and the Cr^{3+}/Cr^{2+} potential determined.[124]

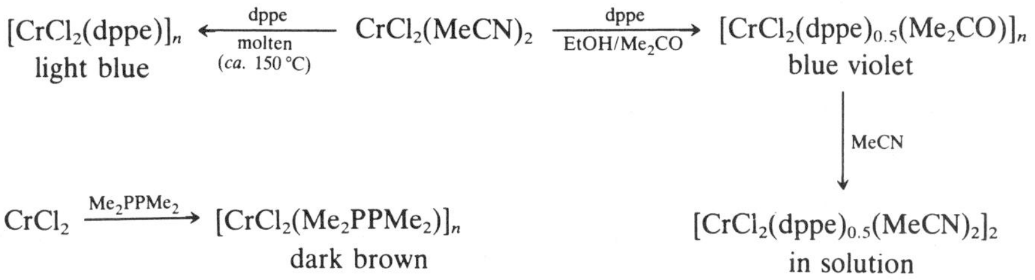

Scheme 14

The reactions of $[CrCl_2(dmpe)_2]$, from which $[Cr(N_2)_2(dmpe)_2]$ is prepared, are given in Scheme 8, Section 35.2.3.1. Included in Scheme 8 are the dimethyl compound, *trans*-$[CrMe_2(dmpe)_2]$ (**19**), which is an unusual Cr^{II} alkyl (others are the diamagnetic, quadruply bonded species in Section 35.3.5.5), and the hydride $[CrH_4(dmpe)_2]$, which is the first example of eight-coordinate Cr^{IV} (Section 35.5). The tetrahydride is diamagnetic, and the coordinated hydrogen appears as a binomial quintet in the ^{1}H NMR spectrum, in agreement with a dodecahedral structure.[59]

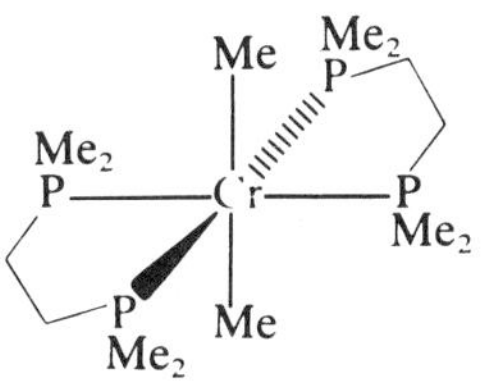

(19)

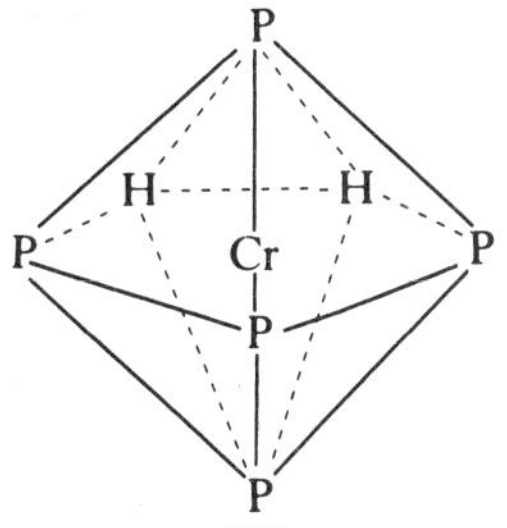

(20)

The chromium(II) hydride $[CrH_2\{P(OMe)_3\}_5]$ is fluxional and the seven-coordinate structure (**20**) is likely from an analysis of the ^{31}P and ^{1}H NMR spectra (Table 21).[125]
Some mixed ligand complexes with organic isocyanides are given in Table 3.

Table 21 Complexes of Chromium(II) and (IV) with P Donor Ligands[a]

Complex	*Comments*	*Ref.*
Chromium(II)		
$[CrX_2(PMe_2Ph)_2]$ Blue green	From CrX_2 (X = Cl, Br) and PMe_2Ph in hexane, high-spin	1
$[CrMe_2(dmpe)_2]$ Red-orange	M.p. 195 °C (dec.), $\mu_{eff}^{r.t.} = 2.7$ BM, ν_{CH}(CrMe), 2780 cm^{-1}, NMR, ^1H: δ 0.9 (PCH_2, s, $W_{1/2} = 260$ Hz), −28.6 (PMe_2, s, $W_{1/2} = 310$ Hz) (PhH-d_6, 25 °C), Cr—C, 2.168; Cr—P, 2.342 and 2.349 Å, isomorphous with $[Cr(N_2)_2(dmpe)_2]$	2
$[CrCl_2(dmpe)_2]$ Yellow-green	M.p. 270 °C (dec.), $\mu_{eff}^{r.t.} = 2.76$ BM (soln.) NMR, ^1H: δ −13.1 (PCH_2, s, $W_{1/2} = 240$ Hz), −33.5 (PMe_2, s, $W_{1/2} = 310$ Hz) (PhH-d_6, 25 °C)	2
$[CrCl_2(dppe)]$	Also $[CrCl_2(dppe)_{0.5}(Me_2CO)]$, $[CrCl_2(dppe)_{0.5}(MeCN)_2]$ and $[CrCl_2(Me_2PPMe_2)]$, Scheme 14	3
$[CrH_2\{P(OMe)_3\}_5]$ Off-white	$[Cr\{P(OMe)_3\}_6]$ in C_5H_{12} or toluene and H_2 at 1 atm, seven-coordinate, (**20**)	4
Chromium(IV)		
$[CrH_4(dmpe)_2]$ Yellow	M.p. 130 °C (dec.), diamagnetic, NMR, ^1H: δ 1.23 (PMe_2 and PCH_2, s), −6.91 (Cr—H quintet, $J_{P-H} = 56.1$ Hz); $^{31}P\{^1H\}$: δ 78.8 (PhH-d_6, 25 °C), ν(Cr—H), 1757m, 1725s, 1701m cm^{-1}, D_{2d} dodecahedral geometry: Cr—P (av), 2.255, Cr—H (av), 1.57 Å	2

[a] For chromium(III) complexes see Section 35.4.3.

1. W. Seidel and P. Schol, *Z. Chem.*, 1978, **18,** 106.
2. G. S. Girolami, J. E. Salt, G. Wilkinson, M. Thornton-Pett and M. B. Hursthouse, *J. Am. Chem. Soc.*, 1983, **105,** 5954; J. E. Salt, G. S. Girolami, G. Wilkinson, M. Motevalli, M. Thornton-Pett and M. B. Hursthouse, *J. Chem. Soc., Dalton Trans.*, 1985, 685; G. S. Girolani, G. Wilkinson, A. M. R. Galas, M. Thornton-Pett and M. B. Hursthouse, *J. Chem. Soc., Dalton Trans.*, 1985, 1339.
3. J. Ellermann, K. Hagen and H. L. Krauss, *Z. Anorg. Allg. Chem.*, 1982, **487,** 130.
4. S. D. Ittel, F. A. Van-Catledge and C. A. Tolman, *Inorg. Chem.*, 1985, **24,** 62.

35.3.4.2 o-Phenylenebis(dimethylarsine)

Diarsine forms the low-spin, *trans* octahedral chromium(II) complexes $[CrX_2(diars)_2]$ (X = Cl, Br, I); the magnetic moments are in the range 2.85 to 2.99 BM and independent of temperature (84–298 K). Reflectance spectra have been recorded, but no assignments made.[40]

35.3.4.3 Tetradentate 'tripod' ligands

A few five-coordinate complex cations, apparently of trigonal bipyramidal stereochemistry, are known. These contain coordinated halide and a tripod-like quadridentate ligand, which can be a P, mixed P,N or N donor: $[CrX\{P(CH_2CH_2PPh_2)_3\}]^+$, $[CrX\{P(2\text{-}C_6H_4PPh_2)_3\}]^+$, $[CrX(N\{CH_2CH_2P(C_6H_{11})_2\}_3)]^+$, $[CrX\{N(CH_2CH_2PPh_2)_3\}]^+$, $[CrBr\{N(CH_2CH_2PPh_2)(CH_2CH_2NEt_2)_2\}]^+$ and $[CrBr\{N(CH_2CH_2NMe_2)_3\}]^+$.[40,126–128] To prepare the complexes, the anhydrous chromium(II) halide in butanol or ethanol is added to the ligand in the same solvent or in 1,2-dichloromethane, and the cations isolated as BPh_4 or PF_6 salts (Table 22). Attempts to prepare complexes of $N(CH_2CH_2AsPh_2)_3$[127] and the linear quadridentate ligand $(—CH_2PPhCH_2CH_2PPh_2)_2$[40] have not succeeded.

The complexes are high-spin. As $[CrBr\{N(CH_2CH_2NMe_2)_3\}]Br$ is isomorphous with its trigonal bipyramidal cobalt(II) analogue, this stereochemistry is suggested for the chromium(II) complex too.[128] In the cobalt(II) complex the bromide lies along the three-fold axis and the symmetry is C_{3v} rather than D_{3h}.[129] All complexes are 1:1 electrolytes, and there are close resemblances between the solution ($1,2\text{-}C_6H_4Cl_2$) and reflectance spectra, which consist of a broad intense band with a shoulder or less intense band to higher frequency; thus the five-coordinate structures are retained in solution. The electronic bands are assigned to the $^5A_1' \rightarrow {}^5E'$ and $^5A_1' \rightarrow {}^5E''$ transitions. The bands are to higher wavenumber in the spectra of the phosphine complexes, which also exhibit extinction coefficients many times greater than those

Table 22 Complexes of Tetradentate P, PN and N Donor Ligands

Complex	*X*	μ_{eff}^{298K} (BM)	*Electronic spectra* ($10^{-3}\bar{\nu}$, cm^{-1})[a] $^5A_1' \rightarrow {}^5E'$	$^5A_1' \rightarrow {}^5E''$	*Ref.*
$[CrX\{P(CH_2CH_2PPh_2)_3\}]BPh_4$ Blue	Cl	4.49	14.7	19.0sh	1
			15.1 (560)		
	Br	4.70	14.3	18.1sh	
			14.8 (585)	18.1sh	
	I	4.70	14.3	18.7	
			14.4 (440)	18.7sh	
$[CrX\{P(2\text{-}C_6H_4PPh_2)_3\}]BPh_4$ Green	Cl	4.85	13.4		2
			14.1 (762)		
	Br	4.82	13.6		
			13.7 (360)		
	I	4.89	13.3		
			13.4 (830)		
$[CrX\{N(CH_2CH_2P(C_6H_{11})_2)_3\}]PF_6$ Blue	Cl	4.84	13.3	16.4	2
			13.3 (546)	16.4 (370)	
	Br	4.88	13.2	15.6	
			13.2 (415)	15.8 (340)	
$[CrX\{N(CH_2CH_2PPh_2)_3\}]BPh_4$	Cl	4.62	13.5	16.7	3
			13.8 (510)	17.0 (470)	
	Br	4.55[b]	13.0	16.1	
			13.3 (530)	16.0 (415)	
	I	4.95	12.7	15.4	
			13.15 (520)	15.6 (380)	
$[CrBr\{L\}]BPh_4$ ($L = N(CH_2CH_2PPh_2)(CH_2CH_2NEt_2)_2$)	—	4.97	11.6	14.1	3
$[CrBr\{N(CH_2CH_2NMe_2)_3\}]Br$	—	4.85	≈11.0	≈13.0	4
			0.8 (84)	14.0 (32)	5

[a] Reflectance spectra and solution spectra (ε, $dm^3\ mol^{-1}\ cm^{-1}$, in parentheses).
[b] Independent of temperature (77–293 K).

1. F. Mani, P. Stoppioni and L. Sacconi, *J. Chem. Soc., Dalton Trans.*, 1975, 461.
2. F. Mani and P. Stoppioni, *Inorg. Chim. Acta*, 1976, **16**, 177.
3. F. Mani and L. Sacconi, *Inorg. Chim. Acta*, 1970, **4**, 365.
4. M. Ciampolini, *Chem. Commun.*, 1966, 47.
5. M. Ciampolini and N. Nardi, *Inorg. Chem.*, 1966, **5**, 1150.

of $[CrBr\{N(CH_2CH_2NMe_2)_3\}]Br$, presumably because of the greater covalent character of the Cr—P bonds. The electronic spectrum of this last compound is quite different from those containing CrN_6 or CrX_2N_4 chromophores (Section 35.3.3.2), and square pyramidal structures would be expected to produce electronic spectra resembling those of the CrX_2N_4 chromophore; thus the trigonal bipyramidal structures are well established, although not yet confirmed in detail crystallographically. Other five-coordinate species are given in Table 42.

35.3.5 Oxygen Ligands

35.3.5.1 Aqua complexes

(*i*) *Hydrated chromium(II) salts*

Hydrated chromium(II) salts can be prepared by the methods in Section 35.3.1.2. They are high-spin and magnetically dilute except for the monohydrogen phosphate, which is weakly antiferromagnetic (Table 23).[130,131] The fluoride $CrF_2{\cdot}2H_2O$ has been little studied.[85] The reflectance spectra are typical of Cr^{II} in distorted six-coordination. Only one crystal structure determination has been reported, that of $CrCl_2{\cdot}4H_2O$; there are four normal length Cr—O bonds to water molecules in a square plane, and two long *trans* Cr—Cl bonds (Table 34).[132] Chromium(II) sulfate pentahydrate is isomorphous with $CuSO_4{\cdot}5H_2O$, and the parameters $D = 2.24\ cm^{-1}$, $E = 0.10\ cm^{-1}$, $g_x = g_y = 2.00$, $g_z = 1.96$ gave the best fit to its ESR spectrum.[133]

Aqueous Cr^{2+} solutions have a broad asymmetric absorption at 14 000 cm^{-1} and a weak shoulder at 9500 cm^{-1} due to tetragonally distorted $[Cr(H_2O)_6]^{2+}$ ions.[131]

Table 23 Magnetic and Spectroscopic Properties of Some Chromium(II) Salts

Complex	*Reflectance spectra* (cm^{-1}) ν_2	ν_1	μ_{eff} (BM) 95 K	295 K	*Ref.*
$CrSO_4\cdot 5H_2O$			4.98	4.92	1
Blue	14 300	10 300sh		4.92	2
$CrCl_2\cdot 4H_2O$			4.98	4.94	1
Blue	14 700	10 000sh		4.94	2
$CrBr_2\cdot 6H_2O$			4.94	4.98	1
Blue	14 900	10 500		4.90	2
$CrI_2\cdot 6H_2O$			4.99	4.96	1
Blue	≈14 300vb			4.98	2
$Cr(ClO_4)_2\cdot 6H_2O$			4.98	5.02	1
Blue	14 100	10 000sh			2
$Cr(HPO_4)\cdot 4H_2O$	—	—	4.13	4.91	1
Blue			($\theta = 66°$)		
$CrXOH\cdot H_3BO_3$ (X = Cl, Br, I)[a]			*ca.* 4.5		3

[a] Said to be air-stable, but chromium(II) borate is pyrophoric.

1. A. Earnshaw, L. F. Larkworthy and K. S. Patel, *J. Chem. Soc.*, 1965, 3267.
2. J. P. Fackler, Jr. and D. G. Holah, *Inorg. Chem.*, 1965, **4**, 954.
3. A. J. Deyrup, *Inorg. Chem.*, 1964, **3**, 1645.

(ii) *Hydrated double sulfates*

Chromium(II) double sulfates can be crystallized by the addition of ethanol to concentrated aqueous solutions containing equimolar quantities of the components. The hexahydrates $A_2SO_4\cdot CrSO_4\cdot 6H_2O$ (A = NH_4, Rb or Cs) are high-spin and have reflectance spectra compatible with the presence of tetragonally distorted $[Cr(H_2O)_6]^{2+}$ ions (Table 24).[134] The potassium and sodium salts crystallize as pale blue dihydrates with similar properties, and from the splittings of the SO_4^{2-} absorption bands in their IR spectra it seems that coordinated sulfate anions are present.

Table 24 Double Sulfates

Compound		μ_{eff} (BM) 300 K	90 K	θ (°)	*Reflectance spectra* (cm^{-1}) ν_2	ν_1
$(NH_4)_2SO_4\cdot CrSO_4\cdot 6H_2O$	Blue	4.88	4.85	0	14 000	8500
$Rb_2SO_4\cdot CrSO_4\cdot 6H_2O$	Blue	4.95	4.94	0	14 500	8800
$Cs_2SO_4\cdot CrSO_4\cdot 6H_2O$	Blue	4.93	4.90	0	14 100	9100
$Na_2SO_4\cdot CrSO_4\cdot 2H_2O$	Blue	4.89	4.86	4	14 300	10 500sh
$K_2SO_4\cdot CrSO_4\cdot 2H_2O$	Blue	4.77	4.69	4	14 900	11 500sh
$Cs_2SO_4\cdot CrSO_4\cdot 2H_2O$	Violet	0.88	0.48	—	18 000	12 400sh
$Cs_2SO_4\cdot CrSO_4$	Dark blue	4.50	4.20	16	13 200	
$(N_2H_5)_2SO_4\cdot CrSO_4\cdot H_2O$	Blue	4.89		5		

The cesium salt is unusual; it loses water readily at room temperature in vacuum or over P_4O_{10} to give a violet dihydrate. This dihydrate has a very low magnetic moment and its IR spectrum contains bands exactly as expected for bidentate sulfato groups. Consequently, it is proposed that the violet dihydrate has a sulfato-bridged dinuclear structure analogous to that of $[Cr_2(O_2CMe)_4(H_2O)_2]$ (Section 35.3.5.5). Complete dehydration affords dark blue $Cs_2SO_4\cdot CrSO_4$.

Hydrated and anhydrous chromium(II) hydrazinium sulfates $(N_2H_5)_2Cr(SO_4)_2$ are known. They are air-stable, and in the latter (from isomorphism with the Zn^{II} complex) each metal ion is coordinated to the nitrogen atoms of two hydrazinium ions and four O atoms of bridging sulfato anions.[135]

35.3.5.2 Alcohols, alkoxides and aryloxides

(i) Alcohols

Anhydrous and hydrated chromium(II) halides are soluble in alcohols, and these are useful reaction media. The hydrated halides can be dehydrated by heating with 2,2-dimethoxypropane or triethyl orthoformate and alcoholates remain on removal of the solvent, but none have been characterized. The spectra of $CrCl_2$ in EtOH and MeOH have been recorded.[136]

(ii) Alkoxides and aryloxides

Many alkoxides have been prepared by alcoholysis of $[Cr\{N(SiMe_3)_2\}_2(THF)_2]$ and similar compounds (Section 35.3.3.6). The general reaction is represented by equation (13), in which L and L′ are Et_2O, THF, py or Bu^tCN or various combinations of them, and ROH is an alcohol or a phenol [R = Me, Et, Pr, Pr^i, Bu^n, Bu^s, Bu^t, C_5H_{11}, CHMePr, $CHEt_2$, 1-adamantyl, CEt_3, CPh_3, Ph, 2,6-$Ph(Bu^t)_2$, 2,4,6-$Ph(Bu^t)_3$ and 2,4,6-$Ph(Ph_3)$]. The solvent can be a hydrocarbon, Et_2O, THF or a mixture of these. Some complexes of the type $[Cr(OR)_2(ROH)]$ were obtained from the phenols and the alcohols with bulky aliphatic groups. It is necessary to dry the alkoxides carefully to obtain materials free from solvent. Alkoxides containing branched alkyl chains and aromatic groups show reasonable solubility in organic solvents, but the alkoxides of primary alcohols are insoluble.[137]

$$Cr\{N(SiMe_3)_2\}_2(L)(L') + 2ROH \xrightarrow[\text{solvent}]{25\,°C} Cr(OR)_2 + 2HN(SiMe_3)_2 + L + L' \quad (13)$$

Many alkoxides have been prepared from chromocene in THF or hydrocarbons (equation 14). Only partial replacement of cyclopentadienyl by R occurs with Me_3COH and silanols (equation (15)).

$$[Cr(C_5H_5)_2] + 2ROH \longrightarrow [Cr(OR)_2]_x + 2C_5H_6 \quad (14)$$

R = Me, Et, Pr, Pr^i, Bu^n, Bu^i;[138] R = Me, Et, Pr^i, Me_3CCH_2 [139]

$$[Cr(C_5H_5)_2] + ROH \longrightarrow \tfrac{1}{2}[Cr(C_5H_5)(OR)]_2 + C_5H_6 \quad (15)$$

R = Bu^t, Me_3Si, Ph_3Si[139]

The product from Bu^tOH is an antiferromagnetic, butoxy-bridged dimer (**21**) with a non-planar central $(Cr—\mu\text{-}O)_2$ moiety and a Cr—Cr separation of 2.65 Å. The magnetic properties ($\mu_{eff} = 1.88$ BM at 333.5 K) suggest that there is one unpaired electron per metal atom. It is thought that the compound may contain a Cr—Cr single bond. The dimer structure is preserved by reaction with CO, NO and $CF_3C{\equiv}CCF_3$ but with CO_2 in THF, $[Cr(C_5H_5)_2]$ and $[Cr_2(O_2CBu^t)_4(THF)_2]$ (Cr≡Cr-bonded, Section 35.3.5.5) are formed. Antiferromagnetic interaction has been found in $Cr(OMe)_2$ (prepared from $CrCl_2$ and LiOMe in MeOH as well as from chromocene) and other alkoxides, and polymeric O-bridged structures have been proposed.[140]

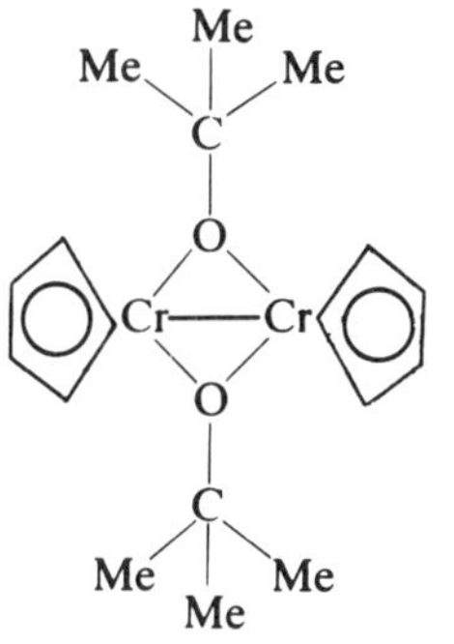

(**21**)

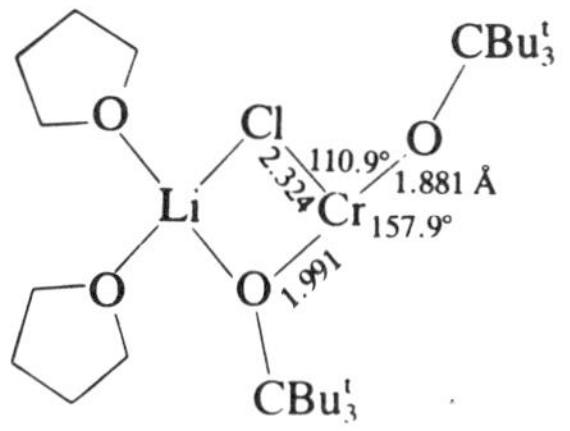

(**22**)

When the more sterically hindered alkoxides dissolve in donor solvents such as THF, adducts $[Cr(OR)_2(THF)_2]$ form and these are probably *trans* planar like $[Cr\{N(SiMe_3)_2\}_2(THF)_2]$. There are considerable colour variations among the alkoxides and their adducts, so there must be considerable structural differences imposed by the different R groups. The magnetic and spectroscopic properties have not been investigated in any detail.[137]

The complex [Cr(OCBu$_3^t$)$_2$·LiCl(THF)$_2$], which was obtained according to Scheme 15,[77] has a unique structure. The metal ion is three-coordinate, and the donor atoms outline a distorted T (22). It is not obvious how the reduction occurs because the BuLi is added in stoichiometric quantity to the alcohol and this solution then added to the CrCl$_3$ suspension. Since LiCl is eliminated on the addition of hexane to give [Cr(OCBu$_3^t$)$_2$(THF)$_2$], Scheme 15 may provide a simpler route to chromium(II) alkoxides than that from the very air- and moisture-sensitive complex [Cr{N(SiMe$_3$)$_2$}$_2$(THF)$_2$].

$$Bu_3^tCOH + BuLi \xrightarrow[18\,h,\ conc.]{CrCl_3,\ THF} Cr(OCBu_3^t)_2 \cdot LiCl(THF)_2$$

$$\text{(Et}_2\text{O)} \quad \text{(hexane)} \quad\quad\quad \text{emerald green}$$

$$\Big\downarrow \text{hexane}$$

$$[Cr(OCBu_3^t)_2(THF)_2] + LiCl$$
$$\text{m.p. } 196\text{--}200\,°C,\ \text{blue green}$$

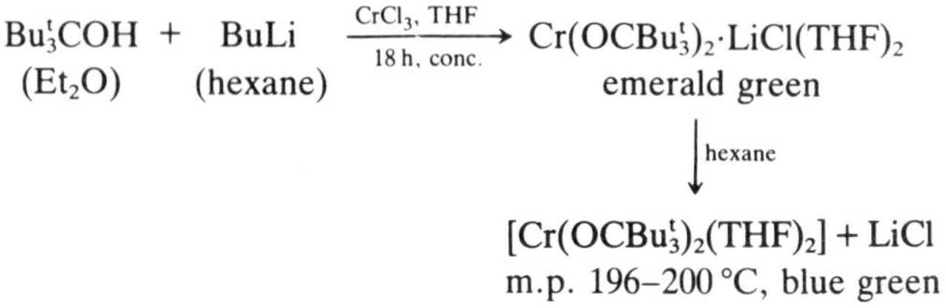

Scheme 15

On addition to halogenated solvents the alkoxides give intensely coloured solutions, stable to oxygen and believed to contain Cr^{III}—C species.[137]

The μ-oxoalkoxide [Cr{OAl(OPr^i)$_2$}$_2$] dissolves in benzene and cyclohexane to give blue solutions in which it is associated.[141]

35.3.5.3 β-Ketoenolates, tetrahydrofuran, ureas and biuret

(i) β-Ketoenolates

Complexes of acetylacetone (acacH), benzoylacetone (bzacH) and dipivaloylmethane (dpmH) have been reported. The acetylacetonate [Cr(acac)$_2$] has been prepared from chromium(II) acetate and acetylacetone.[142,143] It can also be obtained by the addition of aqueous sodium acetylacetonate to an aqueous solution of chromium(II) chloride, but in any preparation the yellow solid must be filtered off and dried as rapidly as possible, otherwise the chromium(III) compound is obtained. Its magnetic moment is 4.99 BM at room temperature consistent with a high-spin d^4 configuration.[142] The powerful reducing ability of [Cr(acac)$_2$] has been used to prepare iron(II) and chromium(II) complexes[80] of porphyrins and related ligands.

On slow sublimation[144] of [Cr(acac)$_2$] a mixture of separate purple and light brown crystals was obtained. The cell constants of the former showed them to be [Cr(acac)$_3$], and the latter were [Cr(acac)$_2$]. In the chromium(II) complex the metal atom lies on an inversion centre and has rigorously planar coordination to four oxygen atoms at a mean Cr—O distance of 1.98 Å (Table 41). Methine carbon atoms of adjacent molecules lie at 3.05 Å above and below the CrO$_4$ plane so that there is severely tetragonally distorted octahedral coordination of the chromium atom (23). At this distance there cannot be much Cr—C σ-bonding. Chromium(II), copper(II) and palladium(II) acetylacetonates are isomorphous. It is presumably the weak axial interactions that lead to the poor solubility of the essentially planar [Cr(acac)$_2$], but the structure is very different from that found in other metal(II) acetylacetonates, which are trimers (Ni, Zn) or tetramers (Co) with bridging anions coordinated by oxygen atoms.

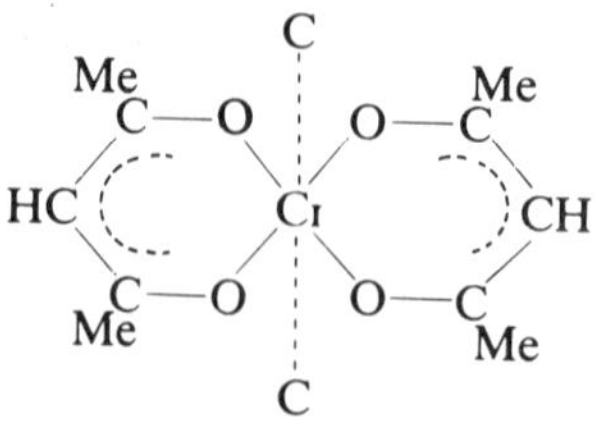

(23)

The benzoylacetonato complex is known only as the black bis(pyridine) adduct [Cr(bzac)$_2$(py)$_2$], which crystallizes when a mixture of the ligands in acetone is added to aqueous chromium(II) acetate.[145] It is a low-spin $3d^4$ compound: $\mu_{eff} \approx 3.2$ BM independent of temperature over the range 295 to 90 K. The IR spectra of the Cr^{II}, Ni^{II} and Zn^{II} complexes are almost identical and they are considered to have *trans* octahedral structures.

[Cr(dpm)$_2$] is prepared from anhydrous chromium(II) acetate in strictly air- and water-free conditions according to Scheme 16.[146] Attempts to prepare the monothio analogue were unsuccessful. There is no absorption band below 23 000 cm^{-1} in the mull spectrum of [Cr(dpm)$_2$], although the spectrum of a toluene solution has a weak shoulder at 16 500 cm^{-1} ($\varepsilon \approx 12$ dm^3 mol^{-1} cm^{-1}) as well as a band at 23 000 cm^{-1} superimposed on a more intense band in the UV region. It is high-spin ($\mu_{eff}^{292\,K} = 4.84$ BM in toluene), but the spectrum is quite different from high-spin complexes such as [CrX$_2$(en)$_2$] (Section 35.3.3.2) in which there is moderate tetragonal distortion. This is because [Cr(dpm)$_2$] is planar, it is isomorphous with planar [Ni(dpm)$_2$], and the extreme distortion has caused the *d–d* transitions to move to high frequency where metal-to-ligand charge transfer transitions occur. There appears to be no theoretical study of the spectra to be expected for high-spin, planar chromium(II).

$$dpmH + BuLi \longrightarrow Lidpm \xrightarrow[-20\,°C]{[Cr_2(O_2CMe)_4]} [Cr(dpm)_2] + LiO_2CMe$$
(THF) (hexane)

i, warm to r.t.
ii, remove solvent
iii, extract pentane

[Cr(dpm)$_2$], m.p. 193 °C
golden yellow

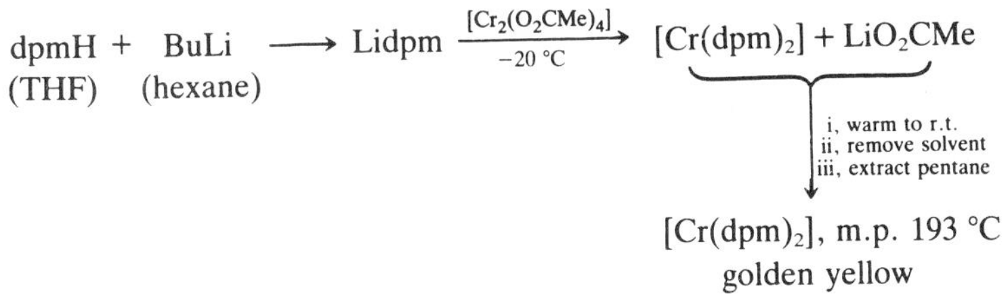

Scheme 16

(ii) Tetrahydrofuran

Adducts of chromium(II) halides with tetrahydrofuran [CrX$_2$(THF)$_2$] can be prepared by extraction of CrCl$_2$ with the hot solvent,[147] and by crystallization from solutions of CrBr$_2$ in THF[148] (Table 25). The latter method did not give solvates of reproducible composition with CrI$_2$ or CrCl$_2$. When dry hydrogen chloride is bubbled through THF over an excess of chromium, the metal dissolves with evolution of hydrogen to give a mauve solution (Cr^{III} spectrum). When the passage of HCl is stopped and the mixture heated to 65–70 °C, evolution of hydrogen from residual metal and dissolved HCl continues and the solution becomes bright blue (Cr^{II} spectrum). The light blue crystals which separate after filtration and the addition of more THF could not be characterized because on drying in vacuum white [CrCl$_2$(THF)], suitable as a source of Cr^{II} for non-aqueous preparations, is formed.[59,74]

Table 25 Tetrahydrofuran and Substituted Ureas

Complex		μ_{eff} (BM) 295 K	90 K	θ (°)	Reflectance spectra (cm^{-1}) ν_2	ν_1
[CrCl$_2$(THF)$_2$]	Pale green	4.71	—	—		
[CrCl$_2$(THF)]	White	—	—	—	11 600	8400
[CrBr$_2$(THF)$_2$]	Pale blue-green	4.86	—	15	12 700	10 000sh
[CrCl$_2$(ur)$_2$]	Pale blue	4.39	3.50	—	13 300	10 800sh
[CrCl$_2$(ur)$_4$]	Blue	4.78	4.83	−3	13 000	9400
[CrBr$_2$(ur)$_4$]	Pale blue	4.85	4.90	−2	14 400	12 800
[CrCl$_2$(mur)$_4$]	Pale blue	4.79	4.79	0	13 000	
[CrBr$_2$(mur)$_4$]	Pale blue	4.87	4.90	−2	13 500	
[CrCl$_2$(bi)$_2$]	Pale blue	4.85	4.86	0	15 000	11 000sh

(iii) Ureas and biuret

The complexes of urea (ur) and biuret (bi) in Table 25 crystallized when a solution of the ligand in ethanol was added to a solution of the hydrated chromium(II) halide in the same solvent (an ethanol/2,2-dimethoxypropane mixture in the case of [CrCl$_2$(ur)$_4$]). Complexes of methylurea would not crystallize from ethanol; they were obtained from the halide dissolved in a mixture of acetone and 2,2-dimethoxypropane, and methylurea in acetone. The reflectance spectra indicate *trans* octahedral complexes, achieved through chloride bridges in [CrCl$_2$(ur)$_2$], which is antiferromagnetic. The positions of the reflectance bands indicate O coordination as do the IR spectra. X-Ray powder data show that [CrCl$_2$(bi)$_2$] is isomorphous with [CuCl$_2$(bi)$_2$], which is known to have a *trans* structure with 'long' Cu—Cl bonds and O donor biuret. Copper(II) forms complexes in which anionic biuret is coordinated *via* nitrogen, but when biuret in alkaline solution was added to aqueous chromium(II) chloride, immediate oxidation occurred.[76]

35.3.5.4 Oxo anions

(i) Sulfato complexes

Sulfato complexes of chromium(II) are known, and the insolubility of the monohydrogen phosphate $Cr(HPO_4)\cdot 4H_2O$ suggests anion coordination (Section 35.3.5.1).

The high-spin phosphites $CrHPO_3\cdot H_2O$ ($\mu_{eff}^{298\,K} = 4.82$ BM, $\theta = 10°$) and $CrH_4P_2O_6\cdot H_2O$ ($\mu_{eff}^{298\,K} = 4.91$ BM) have reflectance bands at *ca.* 13 500 cm^{-1} typical of hexacoordinate Cr^{II}, and their Raman and IR spectra have been assigned. The loss of water on heating under nitrogen is followed by internal oxidation/reduction.[149]

The chlorosulfate $Cr(SO_3Cl)_2$ has been prepared from the acid and $[Cr_2(O_2CMe)_4]$. It forms adducts $Cr(SO_3Cl)_2L_4$ (L = MeCN, py, pyNO, acridine, L_2 = bipy).[150]

The formation of chromium(II) complexes of many oxo anions, *e.g.* NO_3^-, is precluded by the reducing ability of the metal ion.

(ii) Phosphinato complexes

A few chromium(II) phosphinates $[Cr(OPR_2O)_2]$ (R = Me, Ph, C_8H_{17}; R_2 = Me, Ph) have been prepared from chromium(II) acetate[151] or aqueous $CrCl_2$[152,153] and the potassium phosphinate. They have been little investigated because their preparation was incidental to that of the green, phosphinato-bridged and linear chromium(III) polymers to which they are easily oxidized by air. The *n*-octyl derivative $[Cr\{OP(C_8H_{17})_2O\}_2]$ is blue and anhydrous. Since there is a typical Cr^{II} band at 15 400 cm^{-1} and the symmetric and antisymmetric PO_2 stretching bands at 1055 and 1133 cm^{-1} are broad, it is believed that the Cr^{II} is six coordinate through different types of phosphinate bridge. When heated above 185 °C, there is a colour change from blue to pink, which is reversed on cooling. The diphenyl derivative is pink and a monohydrate, and insoluble in common solvents.[153] The colour suggests a chromium(II) acetate type structure, but the magnetic moment has not been reported.

35.3.5.5 Carboxylato and other complexes containing Cr—Cr quadruple bonds

Chromium(II) acetate monohydrate, discovered by Peligot in 1844, is the best known chromium(II) compound because it can be precipitated readily from aqueous Cr^{II} solutions by sodium acetate, is stable in air for short periods, and is used as a reducing agent and a source of other chromium(II) compounds (Table 9). Nonetheless, in its deep red colour, feeble paramagnetism and acetate-bridged, binuclear structure (**24**) (determined in 1953 and refined in 1971),[154,155] which are still not completely understood, it differs sharply from other chromium(II) salts (Section 35.3.5.1).

Extensive investigations by Cotton and his co-workers have shown that structure (**24**) is a prototype for a great variety of complexes of chromium(II) (and Mo^{II} and W^{II}) in which there are quadruple metal–metal bonds.[156] Besides carboxylates (**24**) to (**35**), the anionic bridging ligands in these complexes are zwitterionic glycinate (**36**), diethylcarbamate (**37**), carbonate (**38**), various 2-oxopyridines (**39**) to (**41**), 2,4-dimethyl-6-oxopyrimidine (**42**) and 2-methyl-6-iminopyridine (**43**), substituted acetanilido groups (**44**) to (**51**), carbanilido (**52**), 1,3-diphenyltriazinate (**53**) and *N,N'*-dimethylbenzamidinate (**54**). In addition, there are organochromium dimers containing 2-substituted phenyl (**55**) to (**58**), $Me_2P(CH_2)_2$ (**59**) and Me_3SiCH_2 (**60**) bridges, the methyl (**61**) and allyl (**62**) complexes which are binuclear but do not contain bridging ligands, and the solvated lithium dimers (**63**) and (**64**). There are also a few examples with mixed bridging ligands (**65**) and (**66**), complexes containing Cr_2^{4+} and Cr^{2+} units in the same lattice (**32**), and the mixed metal complex (**67**). In general only one bridge is illustrated in the structural formulae, and as indicated by the Cr–Cr and Cr–axial ligand distances, these structures have been determined by single crystal methods. Furthermore, there are the tetramers (**68**) and (**69**) which contain Cr_2^{4+} units.

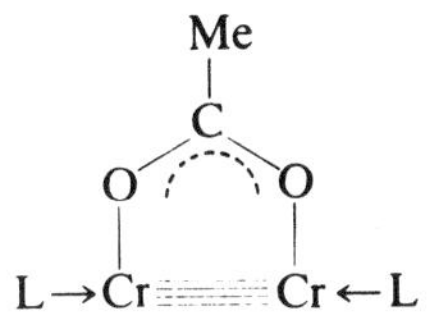

L = H_2O[155,185]
Cr—Cr = 2.362 Å (2.3532, 90 K)
Cr—OH_2 = 2.272 (2.2598, 90 K)

L = $MeCO_2H$[158]
Cr—Cr = 2.300
Cr—OC(OH)Me = 2.306

L = *pyridine*[160]
Cr—Cr = 2.369
Cr—N = 2.335

L = 0.5(*pyrazine*)[160]
Cr—Cr = 2.295
Cr—N = 2.314, pyrazine bridges Cr_2^{4+} units

L = *piperidine*[158]
Cr—Cr = 2.342
Cr—N = 2.338

L = 4-*cyanopyridine*[156]
Cr—Cr = 2.317
Cr—N(py) = 2.322

(**25**) $[Cr_2(O_2CMe)_4L_2]$

see also ref. 167b

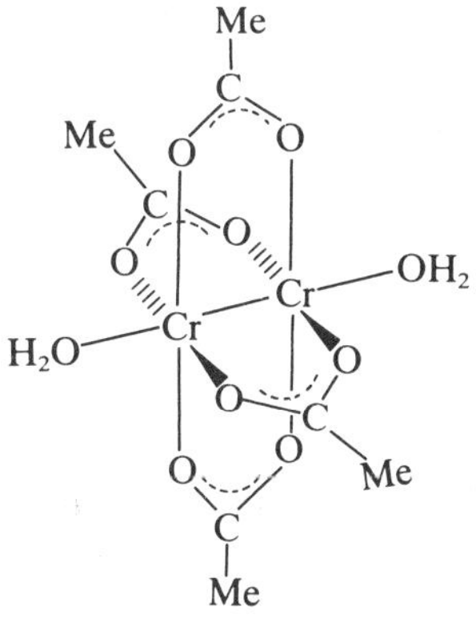

(**24**)

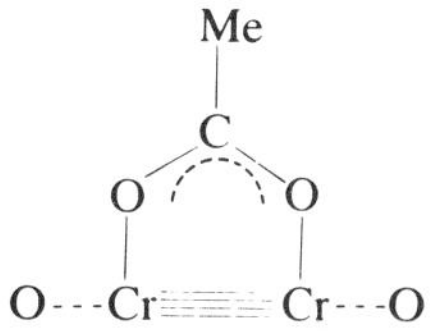

Cr—Cr = 2.288 Å
Cr- - -O = 2.327 (intermolecular)

(**26**) $[Cr_2(O_2CMe)_4]$[169]

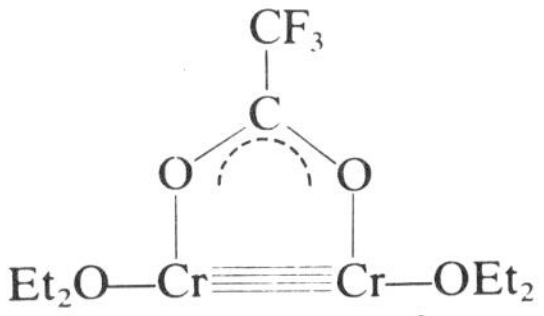

Cr—Cr = 2.541 Å
Cr—OEt_2 = 2.244

(**27**) $[Cr_2(O_2CCF_3)_4(Et_2O)_2]$[164]

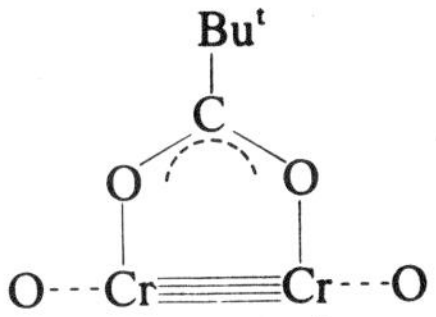

Cr—Cr = 2.388 Å
Cr—O = 2.44 and 2.47 (intermolecular)

(**28**) $[Cr_2(O_2CBu^t)_4]$[164]

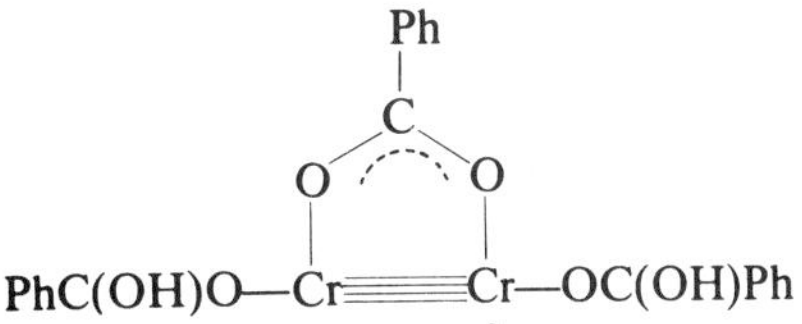

Cr—Cr = 2.352 Å
Cr—O = 2.295 (benzoic)

(**29**) $[Cr_2(O_2CPh)_4(PhCO_2H)_2]$[164]

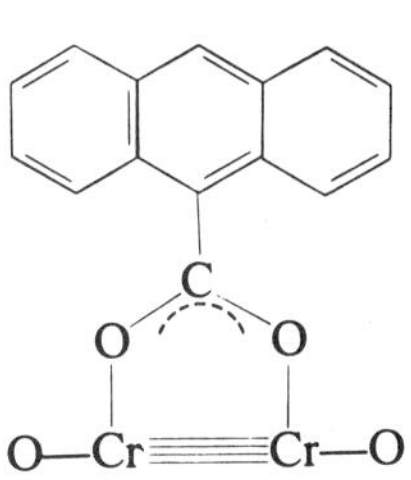

Cr—Cr = 2.283 Å
Cr—O = 2.283 $[CH_2(OMe)CH_2OMe]$

(**30**) $[Cr_2$(9-anthracenecarboyxlate)$_4$(1,2-DME)$]$[164]

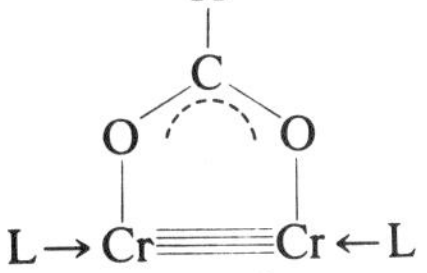

L = H_2O
Cr—Cr = 2.373 Å, Cr—OH_2 = 2.268
Cr—Cr = 2.360, Cr—OH_2 = 2.210
Two independent dimers
$[Cr_2(O_2CH)_4(H_2O)_2]_3 \cdot 10H_2O$[165]

L = *pyridine*[164]
Cr—Cr = 2.408, Cr—N = 2.308

(**31**) $[Cr_2(O_2CH)_4(L)_2]$

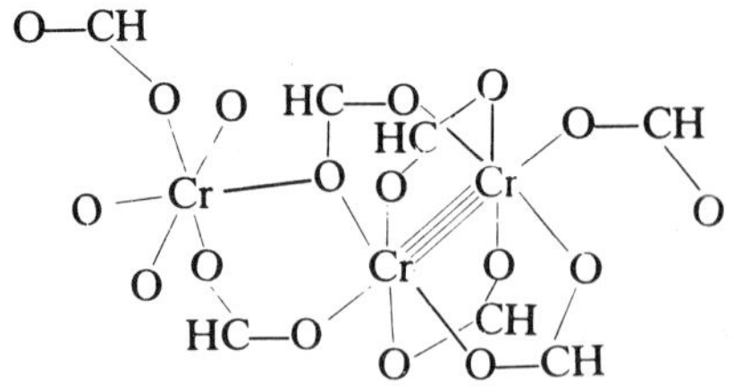

Cr—Cr = 2.451 Å
Cr—O = 2.224

(32) $[Cr_3(O_2CH)_6(H_2O)_2]$[164]

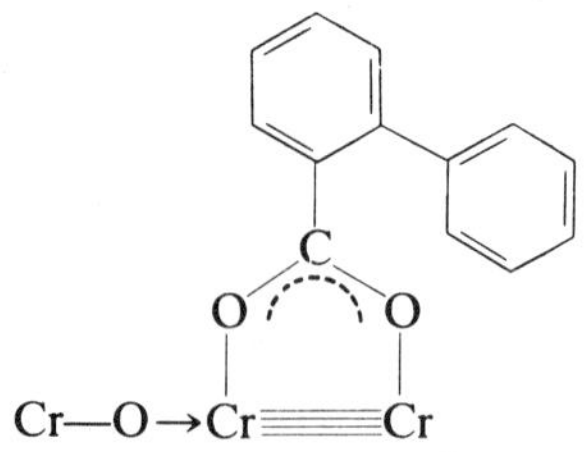

Cr—Cr = 2.348 Å
O→Cr = 2.309
(dimer of dimers)[181]

(33) $[Cr_2(2\text{-phenylbenzoato})_4]\cdot 2PhMe$

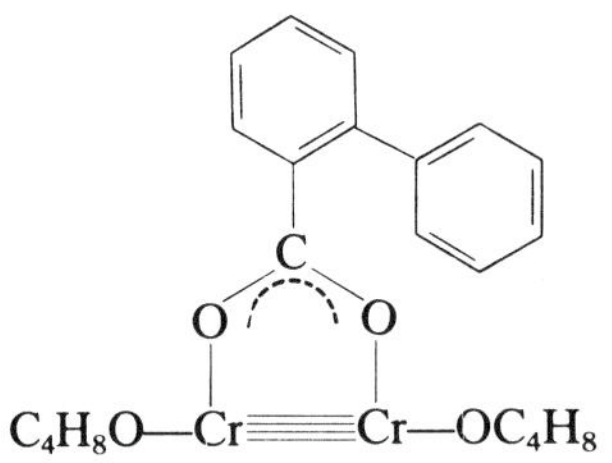

Cr—Cr = 2.316 Å
Cr—O(THF) = 2.275[181]

(34) $[Cr_2(2\text{-phenylbenzoate})_4(THF)_2]$

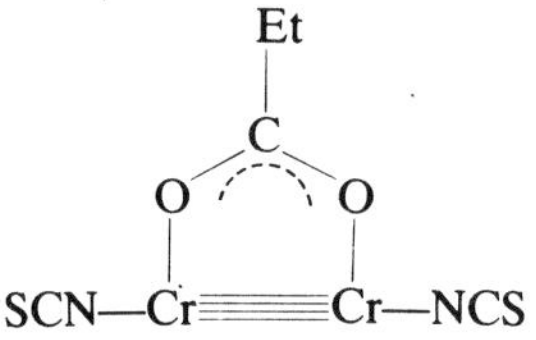

Cr—Cr = 2.467 Å
Cr—N = 2.249

(35) $[NEt_4]_2[Cr_2(O_2CEt)_4(NCS)_2]$[162]

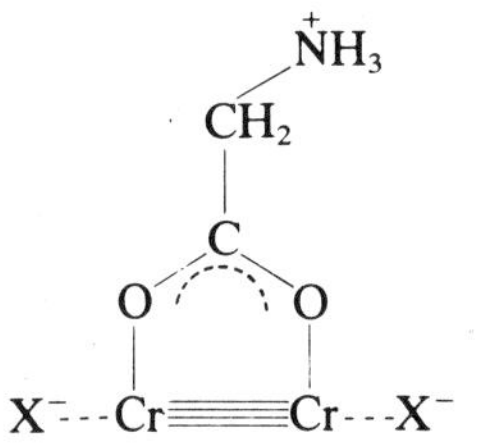

X = Cl, Cr—Cr = 2.524 Å
X = Br, Cr—Cr = 2.513
Cr—Cl = 2.581
Cr—Br = 2.736

(36) $[Cr_2(O_2CCH_2NH_3)_4X_4]\cdot xH_2O$[168]

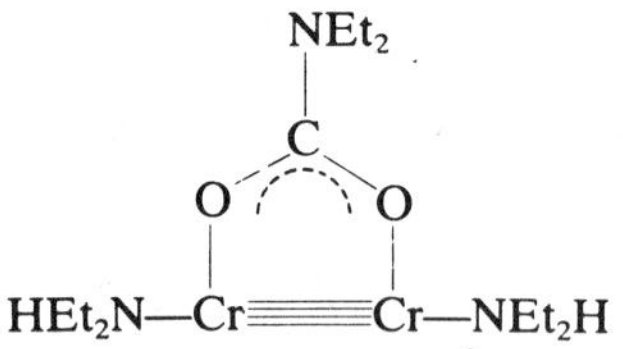

Cr—Cr = 2.384 Å
Cr—N = 2.452[163]

(37) $[Cr_2(\text{diethylcarbamato})_4(NEt_2H)_2]$

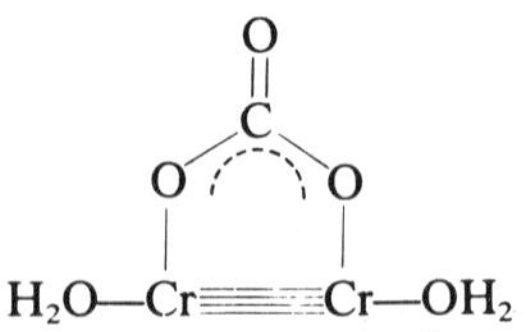

Cr—Cr = 2.214 Å
Cr—O = 2.300[158]

(38) $(NH_4)_4[Cr_2(CO_3)_4(H_2O)_2]\cdot 1\text{–}2H_2O$

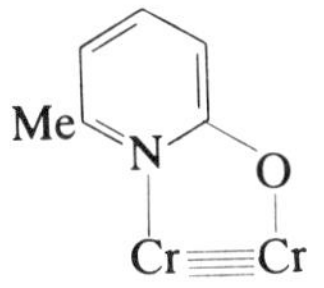

Cr—Cr = 1.889 Å (CH_2Cl_2 solvate)[170]
Cr—Cr = 1.879 (unsolvated, 74 K)[187]
6-methyl-2-oxopyridine
(**39**) $[Cr_2(mhp)_4]$

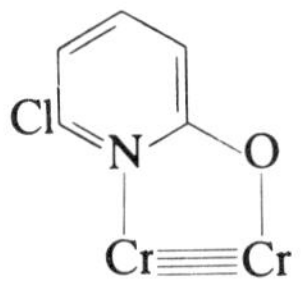

Cr—Cr = 1.955 Å
6-chloro-2-oxopyridine
(**40**) $[Cr_2(chp)_4]$[182]

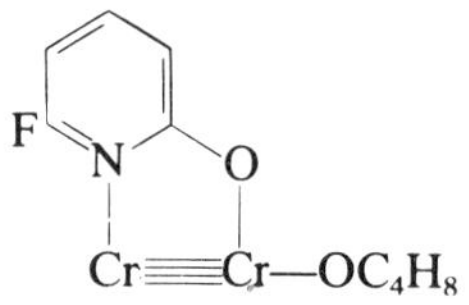

Cr – Cr′ = 2.150 Å
Cr—O(THF) = 2.266
6-fluoro-2-oxopyridine
(**41**) $[Cr_2(fhp)_4(THF)]\cdot THF$[186]

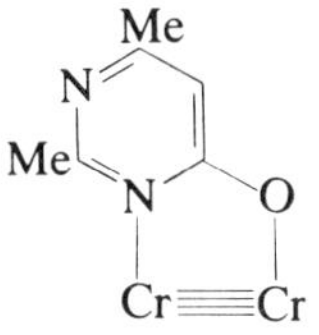

Cr—Cr = 1.898 Å, form I
Cr—Cr = 1.907, form II
2,4-dimethyl-6-oxopyrimidine
(**42**) $[Cr_2(dmhp)_4]$[173]

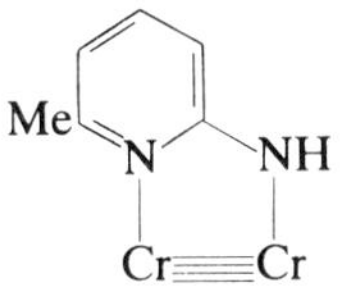

Cr—Cr = 1.870 Å
2-methyl-6-iminopyridine
(**43**) $[Cr_2(map)_4]$[176]

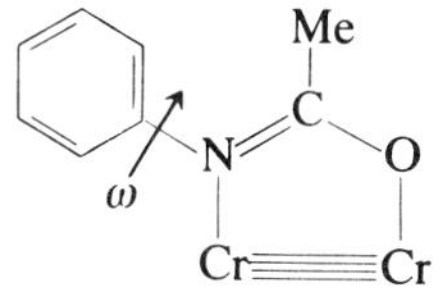

Cr—Cr = 1.873 Å, $\omega = 48°$
acetanilido
(**44**) $[Cr_2\{C_6H_5NC(O)Me\}_4]$[183]

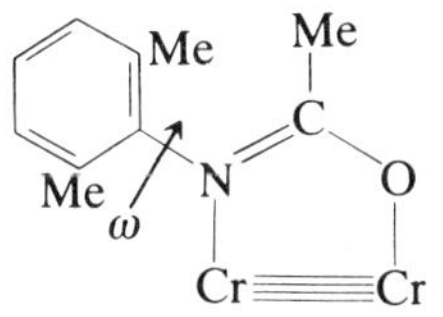

Cr—Cr = 1.937 Å, $\omega = 89.7°$
2,6-dimethylacetanilido[184]
(**45**) $[Cr_2\{2,6\text{-}Me_2C_6H_3NC(O)Me\}_4]\cdot 1.5C_6H_5Me$

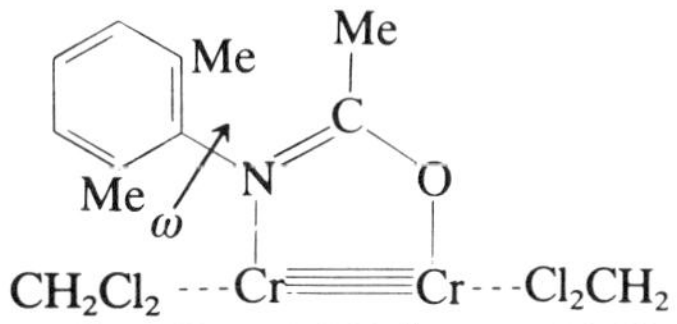

Cr—Cr = 1.949 Å, $\omega = 93.3°$
Cr—Cl = 3.354 and 3.58[184]
(**46**) $[Cr_2\{2,6\text{-}Me_2C_6H_3NC(O)Me\}_4]\cdot 2CH_2Cl_2$

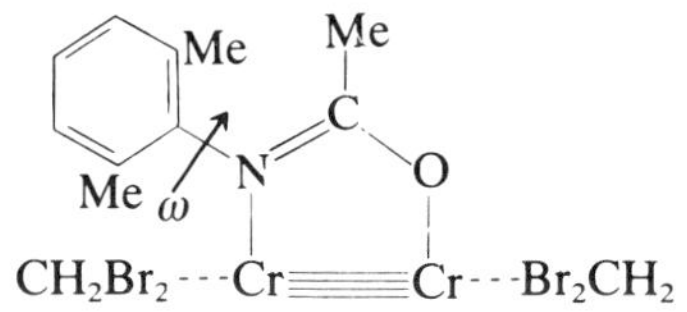

Cr—Cr = 1.961 Å
Cr—Br = 3.554 and 3.335[184]
(**47**) $[Cr_2\{2,6\text{-}Me_2C_6H_3NC(O)Me\}_4]\cdot 2CH_2Br_2$

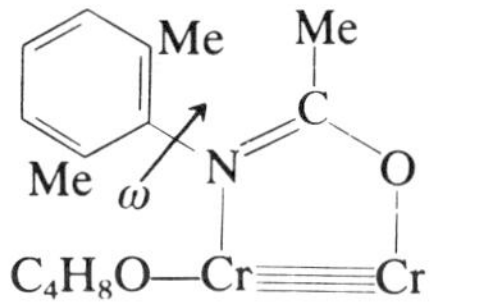

Cr—Cr = 2.023 Å, $\omega = 88.3°$
Cr—O(THF) = 2.315[183]
(**48**) $[Cr_2\{2,6\text{-}Me_2C_6H_3NC(O)Me\}_4](THF)]\cdot PhMe$

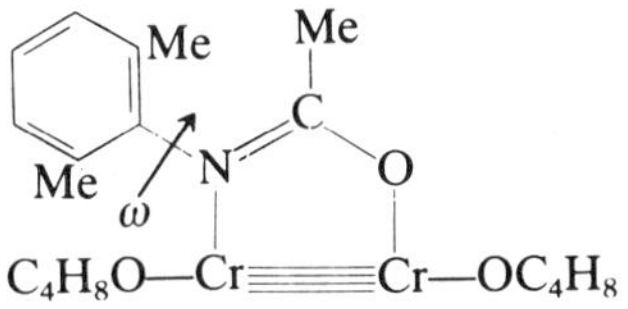

Cr—Cr = 2.221 Å, $\omega = 88°$
Cr—O(THF) = 2.32[183]
(**49**) $[Cr_2\{2,6\text{-}Me_2C_6H_3NC(O)Me\}_4(THF)_2]\cdot THF$

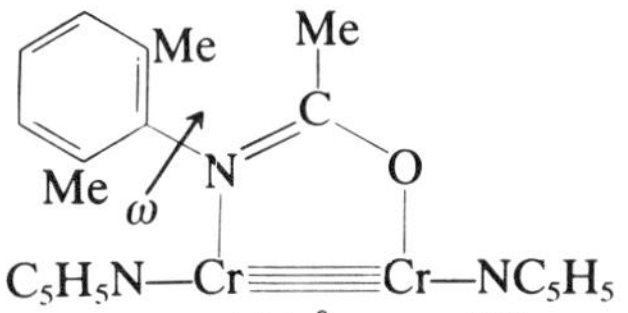

Cr—Cr = 2.354 Å, $\omega = 88°$
Cr—N(py) = 2.31 and 2.40[183]
(**50**) $[Cr_2\{2,6\text{-}Me_2C_6H_3NC(O)Me\}_4(py)_2]\cdot py$

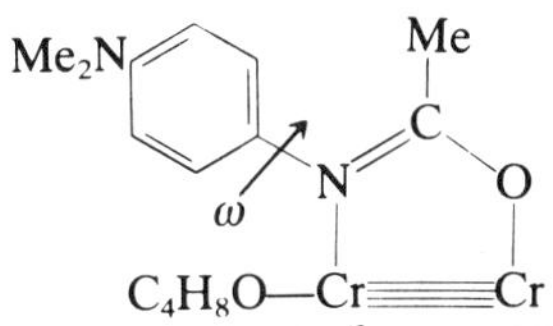

Cr—Cr = 2.006 Å, $\omega = 81.7°$
Cr—O(THF) = 2.350[183]
(**51**) $[Cr_2\{4\text{-}NMe_2C_6H_4NC(O)Me\}_4(THF)]$

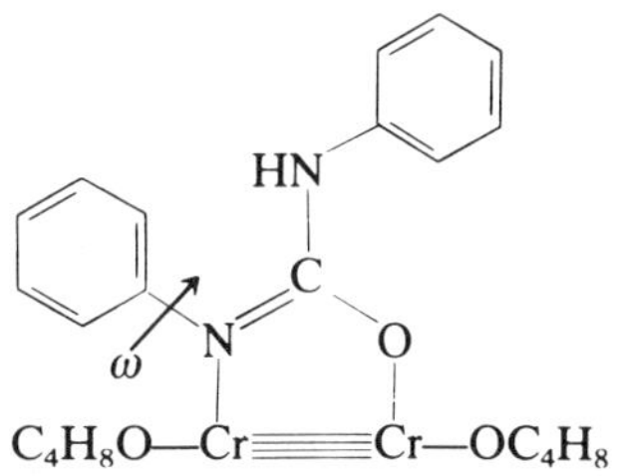

Cr—Cr = 2.246 Å, $\omega = 76.2°$
Cr—O(THF) = 2.350[183]
carbanilido
(**52**) $[Cr_2\{PhNC(O)NHPh\}_4(THF)_2]\cdot C_6H_{14}$

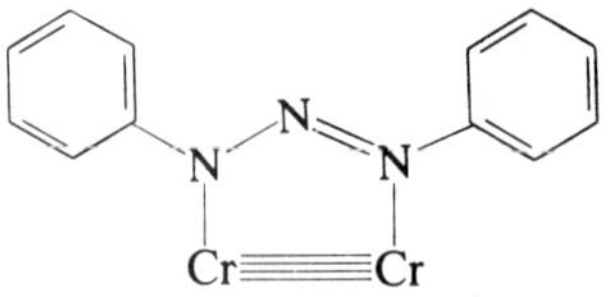

Cr—Cr = 1.858 Å
1,3-diphenyltriazinato
(**53**) $[Cr_2(PhNNNPh)_4]$[178]

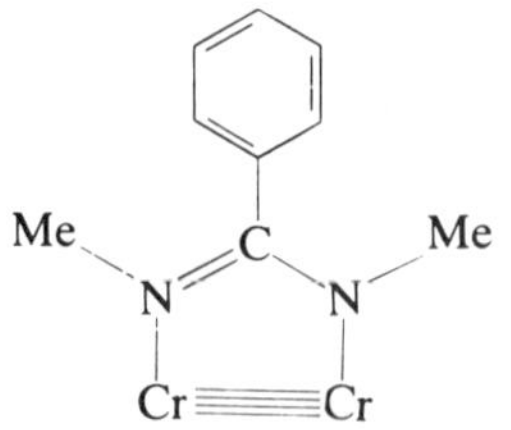

Cr—Cr = 1.843 Å
***N,N'*-dimethylbenzamidinato**
(**54**) $[Cr_2(MeNCPhNMe)_4]$[183]

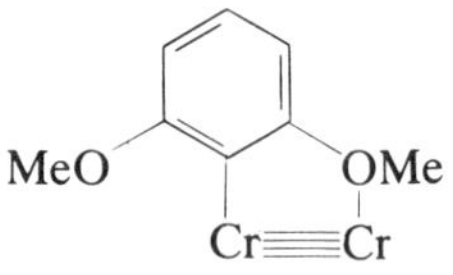

Cr—Cr = 1.847 Å
2,6-dimethoxyphenyl
(**55**) $[Cr_2(DMP)_4]$[1]

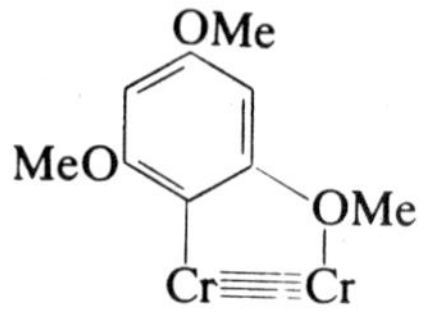

Cr—Cr = 1.849 Å
2,4,6-trimethoxyphenyl
(**56**) $[Cr_2(TMP)_4]$[1]

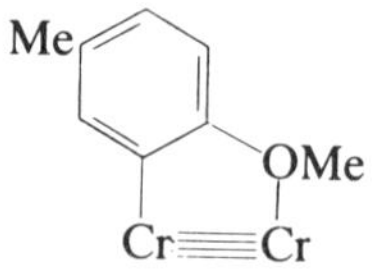

Cr—Cr = 1.828 Å
2-methoxy-5-methylphenyl
(**57**) $[Cr_2(2\text{-}MeO\text{-}5\text{-}MeC_6H_3)_4]$[1]

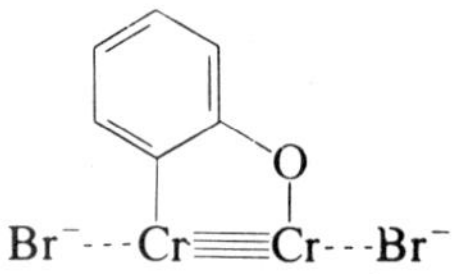

Cr—Cr = 1.830 Å
Cr---Br = 3.266
2-oxophenyl
(**58**) $Li_6[Cr_2(2\text{-}OC_6H_4)_4]Br_2\cdot 6Et_2O$[1]

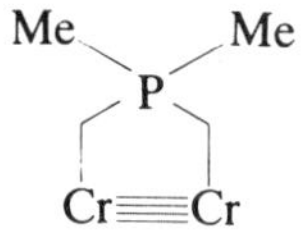

Cr—Cr = 1.895 Å
dimethylphosphoniumdimethylido
(**59**) $[Cr_2\{(CH_2)_2P(CH_3)_2\}_4]$[1]

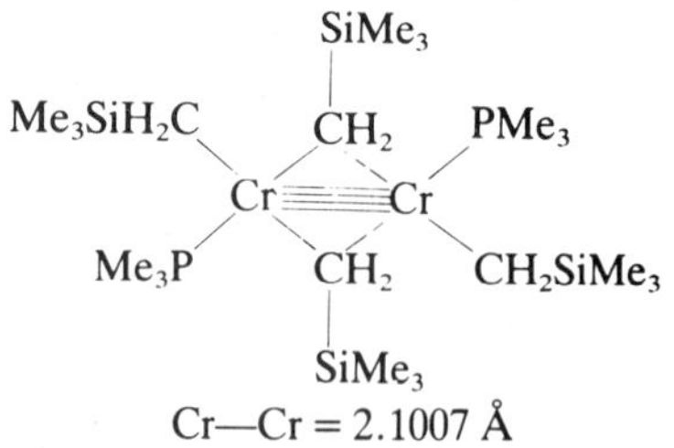

Cr—Cr = 2.1007 Å
(**60**) $[Cr_2(CH_2SiMe_3)_4(PMe_3)_2]$[1]

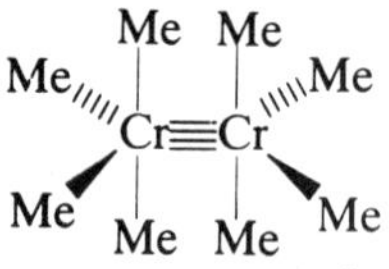

Cr—Cr = 1.980 Å
(**61**) $Li_4[Cr_2Me_8]$[1]

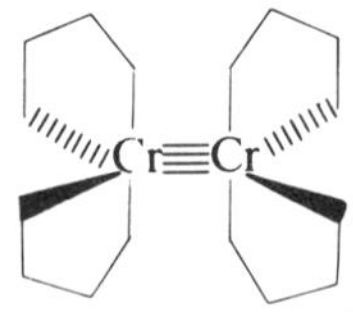

Cr—Cr = 1.975 Å
(**62**) $Li_4[Cr_2(C_4H_8)_4]$[1]

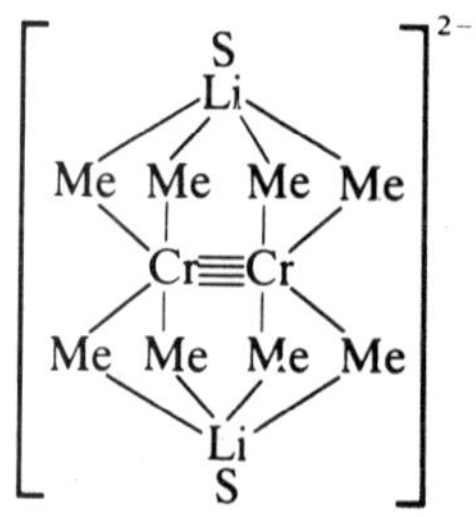

Cr—Cr = 1.980 Å
(**63**) $[Li_2CrMe_4(THF)_2]_2$[1]

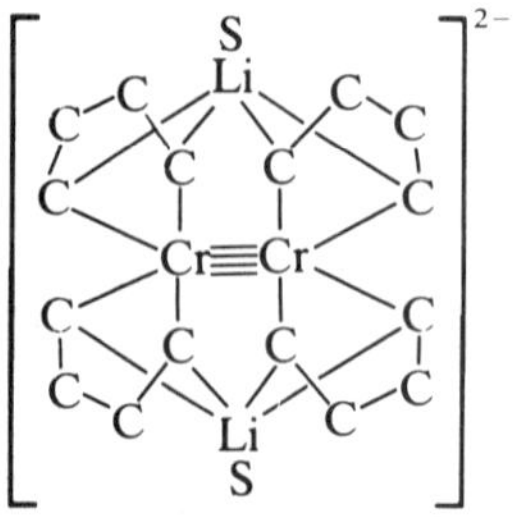

Cr—Cr = 1.975 Å
(**64**) $[Li_2Cr(C_4H_8)_2(Et_2O)_2]_2$[1]

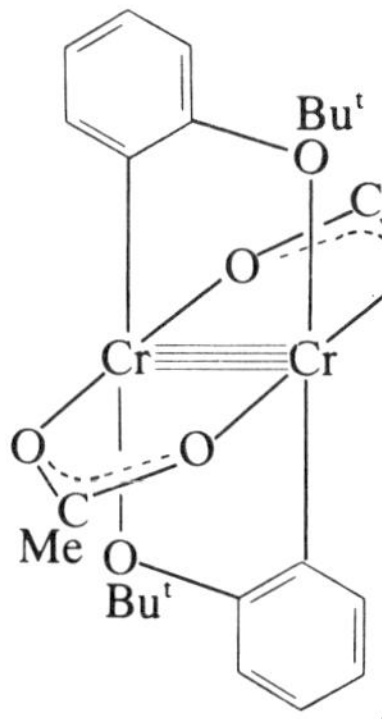

Cr—Cr = 1.862 Å
bis(2-oxophenyl)di(acetate)
(**65**) $[Cr_2(O_2CMe)_2(2\text{-}Bu^tOC_6H_4)_2]$[1]

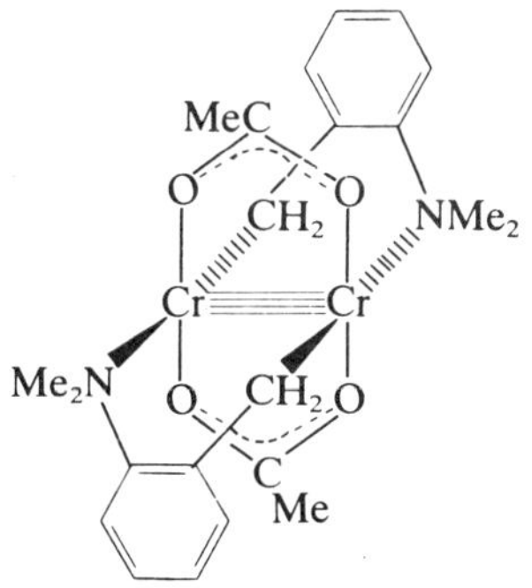

Cr—Cr = 1.870 Å
bis{(2-dimethylamino)benzyl}di(acetate)
(**66**) $[Cr_2(O_2CMe)_2(2\text{-}NMe_2C_6H_4CH_2)_2]$[180]

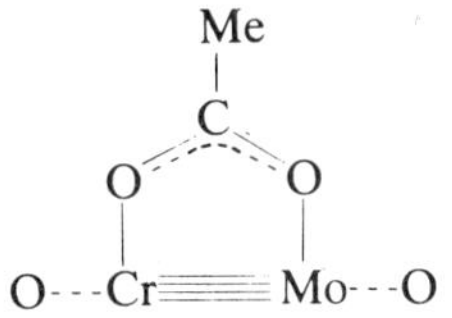

Cr—Mo = 2.050 Å
(CrMo)—O = 2.548 Å
intermolecular infinite chain structure
(**67**) $[CrMo(O_2CMe)_4]$[175]

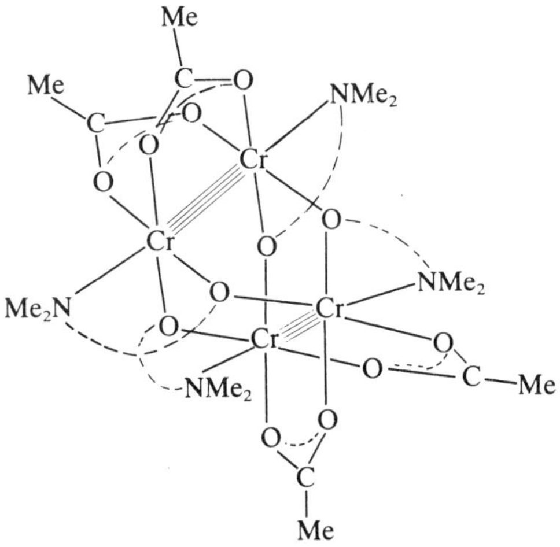

Cr—Cr = 2.531 Å
Cr—N = 2.444
(**68**) $[Cr_4(O_2CMe)_4(OCH_2CH_2NMe_2)_4]$[116]

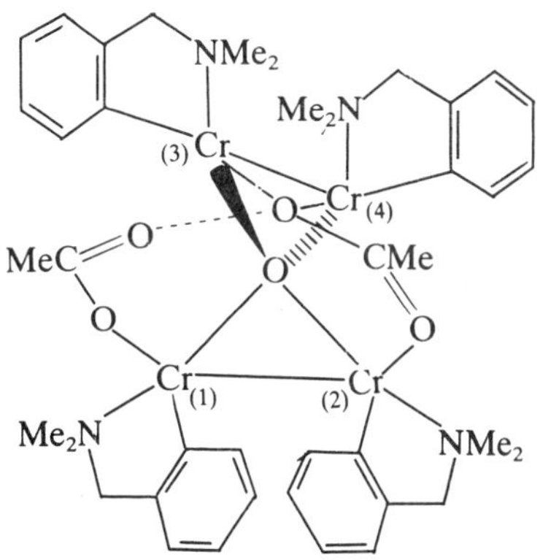

Cr(1)—Cr(2) = 2.544 Å
Cr(3)—Cr(4) = 2.554
Cr—O(centre) = 2.005 (av)
acetate and oxo bridges
(**69**) $[Cr_4(2\text{-}NMe_2CH_2C_6H_4)_4(O_2CMe)_2O]$[179]

(*i*) *Syntheses*

The method used by Peligot can be extended to carboxylates generally (equation 16). Since the complexes with $n > 1$ were dried in vacuum at 100 °C before analysis, their degree of hydration has not been established. All have low magnetic moments ($\mu_{eff} \approx 0.7$ BM) so it is likely that they have the acetate structure.[157]

$$2Cr^{2+} + 4RCO_2^- + 2H_2O \longrightarrow [Cr_2(O_2CR)_4(H_2O)_2]\downarrow \qquad (16)$$
$$R = C_nH_{2n+1},\ n = 1\text{–}7,\ 11,\ 17$$

Water can be replaced by ligands L through recrystallization from L if a liquid, or from an organic or aqueous solvent containing L (L = $MeCO_2H$, piperidine,[158] pyridine, pyrazine[160]). Adducts $[Cr_2(O_2CR)_4L_2]$ (R = CMe_3, L = pyridine, 2- and 4-methylpyridine, quinoline and 2-methylquinoline, R = Ph, C_4H_3S, L = THF; R = C_4H_3O, L = THF, pyridine, 2-methylpyridine, 2-methylquinoline) have been prepared[159] directly in THF by Scheme 11 in which RCO_2H and L replace CF_3CO_2H and $Me_2C_5H_3N$.[95]

When ammonia gas is bubbled through a suspension of the acetate in ethanol, ammonia molecules replace the water, but in liquid ammonia the dinuclear structure is destroyed to give the tetraammine (Scheme 17). Unless kept in an atmosphere of ammonia the tetraammine reverts to the dinuclear ammine adduct.[161]

$$\underset{\text{red-purple}}{[Cr_2(O_2CMe)_4(NH_3)_2]} \xleftarrow[\text{EtOH}]{NH_3(g)} [Cr_2(O_2CMe)_4(H_2O)_2] \xrightarrow{NH_3(l)} \underset{\text{violet}}{Cr(O_2CMe)_2(NH_3)_4}$$

Scheme 17

The water molecules can be displaced by thiocyanate to give $[NR'_4]_2[Cr_2(O_2CR)_4(NCS)_2]$ (R′ = Me, Et; R = Me, Et[162]) and a complicated reaction between $[Cr(NEt_2)_4]$ and CO_2 leads to a diethylcarbamato-bridged complex $[Cr_2(O_2CNEt_2)_4(NEt_2H)_2]$ (Section 35.5.4.3, Scheme 119).[163]

Formate gives a complicated series of hydrates according to conditions (Scheme 18),[157,164,165] and adducts of $Cr(O_2CH)_2$ with urea, THF *etc.* are known, but their structures have not been determined.[157]

$$CrCl_3{\cdot}6H_2O \xrightarrow[\text{reductor}]{\text{Jones}} Cr^{2+}(aq) \xrightarrow{NaO_2CH/HCO_2H} \underset{\text{red}}{[Cr_2(O_2CH)_4(H_2O)_2]_3{\cdot}10H_2O}$$

$$Cr^{2+}(aq) \xrightarrow{NH_4O_2CH} \text{purple gel}$$

$$\text{purple gel} \xrightarrow{\Delta} \underset{(Cr_2^{4+}\text{ and }Cr^{2+})}{\underset{\text{red}}{[Cr_3(O_2CH)_6(H_2O)_2]}} \xrightarrow{py} [Cr_2(O_2CH)_4(py)_2]$$

Scheme 18

Dimeric carbonato-bridged complexes $A_4[Cr_2(CO_3)_4(H_2O)_2]$ (A = NH_4, Li, Na, K, Rb, Cs; A_2 = Mg) can be precipitated[166] from aqueous suspensions of the acetate, and the ammonium salt (or $CrCO_3$)[157] suspended in ether will react with the stronger acids CF_3CO_2H[164] and CF_2HCO_2H[167] to give dinuclear carboxylates (equation 17).

$$[Cr_2(CO_3)_4(H_2O)_2]^{4-} + 4CF_3CO_2H \underset{\Delta}{\overset{Et_2O}{\rightleftharpoons}} [Cr_2(O_2CCF_3)_4(Et_2O)_2] + 4CO_2 + 4H_2O \qquad (17)$$

Pyridine and 4-cyanopyridine react with $[Cr_2(O_2CCF_2H)_4(Et_2O)_2]$ to give the mixed oxidation state, oxo-centred complexes $[Cr^{II}Cr_2^{III}O(O_2CCF_2H)_6(py)_3(Et_2O)]$ and $[Cr^{II}Cr_2^{III}O(O_2CCF_2H)_6(4\text{-}CNpy)_3]{\cdot}PhMe$ (see also p. 869).[167]

Dichromium(II) complexes with zwitterionic glycine bridges can be crystallized[168] from aqueous solutions (equation 18).

$$2CrX_2 + 4H_3\overset{+}{N}CH_2CO_2^- \xrightarrow{H_2O} \underset{\text{purple (X = Cl, Br)}}{[Cr_2(O_2CCH_2NH_3)_4X_4]} \qquad (18)$$

Crystals of the anhydrous acetate[169] and $[Cr_2(O_2CBu^t)_4]$[164] for X-ray investigation were grown by vacuum sublimation. The anhydrous acetate (Table 9) is the source of Cr_2^{4+} for the preparation of many water-sensitive dinuclear complexes. In general, the lithium salt (not usually isolated) of the anionic ligand which is to replace the acetato bridges is first prepared from BuLi and the acid. The solid anhydrous acetate is then added to the lithium salt solution, the solution filtered after reaction to remove the lithium acetate, and the $[Cr_2Y_4]$ crystallized from the filtrate, or, if poorly soluble, extracted from the LiO_2CMe (Scheme 19).

$$\text{BuLi/hexane} + \text{YH/THF or } Et_2O \xrightarrow{0\,°C} LiY \xrightarrow{[Cr_2(O_2CMe)_4]} [Cr_2Y_4] + LiO_2CMe$$

Scheme 19

These reactions were used to prepare complexes of 2-oxopyridines (**39**) to (**41**), 2-substituted phenyl derivatives (**55–57**) and others. One of the first oxopyridine complexes was prepared slightly differently (Scheme 20).[170]

$$NaOMe/EtOH + \underset{Hmhp}{\underset{Me}{\overset{}{\bigcirc}}} \xrightarrow{[Cr_2(O_2CMe)_4(H_2O)_2]} [Cr_2(mhp)_4]\cdot H_2O \xrightarrow[\text{vac.}]{100\,°C} [Cr_2(mhp)_4]$$

$$\downarrow CH_2Cl_2$$

$$[Cr_2(mhp)_4]\cdot CH_2Cl_2$$

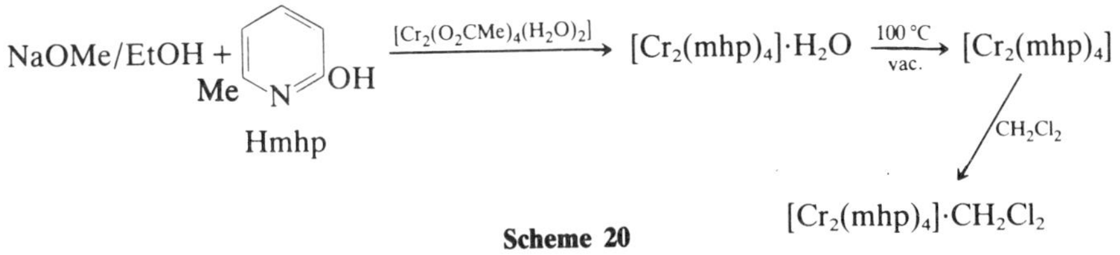

Scheme 20

In another variation the dilithium salt was obtained from 2-bromophenol and 'contamination' with LiBr followed by Soxhlet extraction to solubilize the binuclear anionic salt and obtain crystals of $Li_6Cr_2(2\text{-}OC_6H_4)_4Br_2\cdot6Et_2O$ (Scheme 21).[171]

$$\underset{Br}{\bigcirc OH} \xrightarrow{BuLi/hexane} \underset{Li}{\bigcirc OLi} \xrightarrow[\text{LiBr, Et}_2O]{[Cr_2(O_2CMe)_4]} Li_6Cr_2(2\text{-}OC_6H_4)_4Br_2\cdot6Et_2O$$

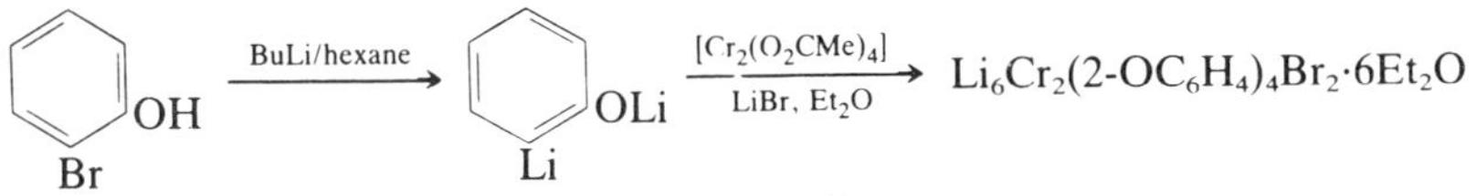

Scheme 21

The 2-methoxyphenyl complex, $[Cr_2(2\text{-}MeOC_6H_4)_4]$, has been prepared from $CrBr_2(THF)_2$ (equation 19).[172] The yellow, poorly soluble and pyrophoric product is diamagnetic and is likely to be a Cr_2^{4+} dimer (equation 19).

$$[CrBr_2(THF)_2] + 2\text{-}MeOC_6H_4MgBr \xrightarrow{THF} [Cr_2(2\text{-}MeOC_6H_4)_4] \tag{19}$$

Oxidation of the metal carbonyl is used in the preparations illustrated by equations (20) to (22),[173–175] in the third of which a heteronuclear dimer is formed. No reaction occurred between $Cr(CO)_6$ and 2-amino-6-methylpyridine.[176] The binuclear nature of red $[Cr\{PhC(NPh)_2\}_2]$ has not been established, and an impure sample of $[CrMo(mhp)_4]$ has been prepared from $[CrMo(O_2CMe)_4]$.[177]

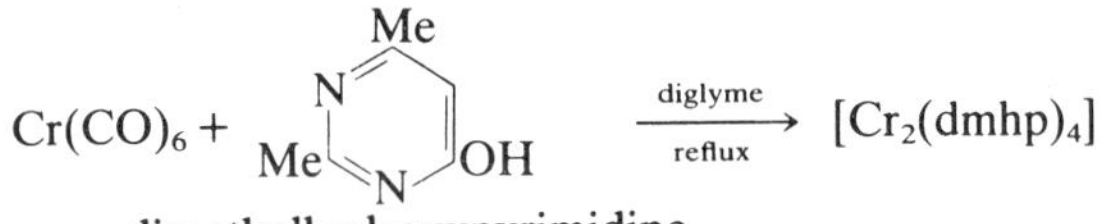

$$Cr(CO)_6 + \underset{\text{dimethylhydroxypyrimidine}}{\underset{(dmhpH)}{\underset{Me}{\overset{Me}{\bigcirc OH}}}} \xrightarrow[\text{reflux}]{\text{diglyme}} [Cr_2(dmhp)_4] \tag{20}$$

$$Cr(CO)_6 + \underset{N,N'\text{-diphenylbenzamidine}}{PhC(NPh)(NHPh)} \xrightarrow[\text{reflux}]{\text{pet. ether}} \underset{\text{red}}{[Cr\{PhC(NPh)_2\}_2]} \tag{21}$$

$$Mo(CO)_6 + \underset{\text{in excess}}{[Cr_2(O_2CMe)_4]} \xrightarrow[\substack{CH_2Cl_2 \\ MeCO_2H}]{(MeCO)_2O} \underset{\text{pale yellow}}{[CrMo(O_2CMe)_4]} \tag{22}$$

In the preparation[178] of the 1,3-diphenyltriazinato-bridged complex an organochromium dimer was the source of Cr_2^{4+} (equation 23).

$$[Li(THF)]_4[Cr_2Me_8] + 4PhNHNNPh \xrightarrow{THF} [Cr_2(PhN_3Ph)_4] + 4CH_4 + 4LiMe \tag{23}$$

Chromocene can also be used as a starting material (Scheme 22).[164]
Tetranuclear complexes, composed of two Cr_2^{4+} subunits are made by[116,179] reactions (24) and (25) in which the acetate bridges were not all displaced.

$$[Cr_2(O_2CMe)_4] + NaOCH_2CH_2NMe_2 \xrightarrow{THF} \underset{\text{red}}{[Cr_4(O_2CMe)_4(OCH_2CH_2NMe_2)_4]} \tag{24}$$

$$[Cr_2(O_2CMe)_4] + 2Li(2\text{-}NMe_2CH_2C_6H_4) \xrightarrow{THF} \underset{\text{orange}}{[Cr_4(2\text{-}NMe_2CH_2C_6H_4)_4(O_2CMe)_2O]} \tag{25}$$

$$[\text{Cr}_2(\text{O}_2\text{CBu}^\text{t})_4] \xleftarrow[\text{benzene}]{\text{Bu}^\text{t}\text{CO}_2\text{H}} [\text{Cr}(\eta^5\text{-C}_5\text{H}_5)_2] \xrightarrow[\text{toluene}]{\text{PhCO}_2\text{H}} [\text{Cr}_2(\text{O}_2\text{CPh})_4(\text{PhCO}_2)_2]$$

red   red

MeOCH$_2$CH$_2$OMe (1,2-DME)   toluene

[Cr$_2$(9-anthracenecarboxylate)$_4$(DME)]
red

[Cr$_2$(2-phenylbenzoate)$_4$]·2PhMe
red

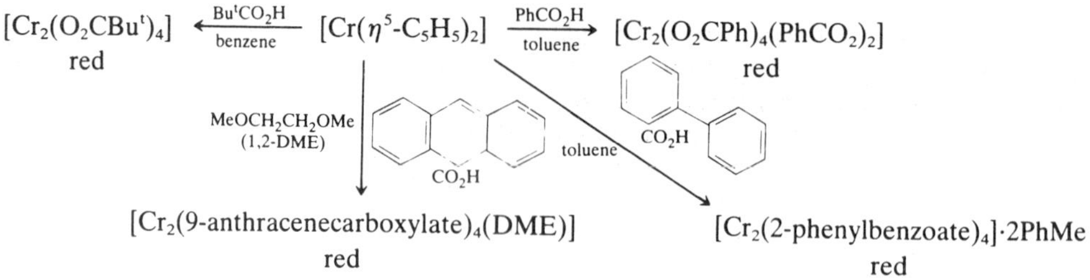

Scheme 22

Several organochromium(II) dimers containing short Cr≡Cr bonds (**55–64**) have been included for completeness. For their preparations the companion series[1] should be consulted. Complexes (**65**) and (**66**)[180] contain mixed carboxylato and organo bridges.

(ii) Structures

Crystallographic investigations demonstrate that the length of the Cr—Cr quadruple bond is primarily dependent upon whether axial ligands are present and, if so, upon their nature and number. This is apparent from a comparison of the Cr—Cr separations found for complexes (**25**) to (**38**), (**41**), (**46**) to (**52**) and (**67**), and, without axial ligands, (**39**), (**40**), (**42**) to (**45**) and (**53**) to (**66**). Those without axial ligands and $r(\text{Cr—Cr}) < 1.9$ Å are considered[156] to contain 'supershort' bonds compared with those in the carboxylates (2.28–2.55 Å). The Cr—Cr separations in the tetramers (**68**) and (**69**) are equal to the greatest found in the dinuclear carboxylates.

There is considerable variation of $r(\text{Cr—Cr})$ within the tetracarboxylate series mainly dependent upon the donor ability of the axial ligands. In [Cr$_2$(O$_2$CR)$_4$(R'py)$_2$] (R = Bu$^\text{t}$, Me, H, CClH$_2$ or CHF$_2$; R' = H, 4-NH$_2$, 4-CN *etc.*) the Cr—Cr and Cr—N distances are inversely related,[167b] and axial coordination of oxindole in [Cr$_2$(oxindolate)$_4$(oxindole)$_2$] produces a long Cr≡Cr bond (2.495 Å).[167c] No carboxylates without axial ligands are known: in anhydrous compounds oxygen atoms of neighbouring dimers bond axially (**26**) to give polymers; and substitution in the bridge is too distant to be effective (**30**) or is nullified by the final configuration (**33**). In (**33**) the four 2-phenyl groups are directed towards the same end of a Cr≡Cr unit and block association [though not coordination of THF (**34**)], but dimers of dimers are formed through association at the other end. Carboxylates producing steric hindrance at both ends of the Cr$_2$ units have not been devised.

Careful choice of bridging ligand is necessary to prevent axial donation or association and produce very short Cr—Cr bonds. In the acetanilido complex (**44**) this is achieved: the unlike ligand atoms on each chromium atom are *trans* and with a small torsion angle ω about the NPh bond the axial positions are blocked. Dichromium compounds (**46**) to (**50**), in which ω has been increased to approximately 90° to open up the axial positions, have been obtained by introducing 2,6-dimethyl groups. Steric effects then force the rings to be perpendicular to the O—Cr$_2$—N plane. Electron donation by a 4-NMe$_2$ substituent acts similarly. Access by axial ligands then lengthens the Cr—Cr bond in the series of complexes (**48**) to (**51**) compared with (**44**) without change in bridging ligand or overall geometry. Complexes (**44**) and (**45**) have no axial ligands and the smallest Cr—Cr separations, and the axial CH$_2$X$_2$ molecules in (**46**) and (**47**) interact at the most very weakly with the metal. Axial donation modifies the Cr—Cr distance as in the carboxylates, but in the acetanilido compounds the bond separation without axial ligands is known. Metal–metal separations in dimolybdenum complexes are not sensitive to axial donation.

There are no axial ligands in the 2-oxopyridine, 2-iminopyridine and 2-phenyl derivatives which have the shortest Cr—Cr bonds. The bromide ions in (**58**) are off-axis and beyond bonding distance.

The nature of the bridging ligand is also important; the quadruple bonds (the ranges are given) shorten with type of bridging bidentate ligand in the order:[183]

O—C—O(ax) > N—C—O(ax) > N—C—O > N—C—N ≈ N—N—N > C—C—O

2.55–2.21 Å   2.35–2.01 Å   1.96–1.87 Å   1.87–1.84 Å   1.86 Å   1.85–1.83 Å

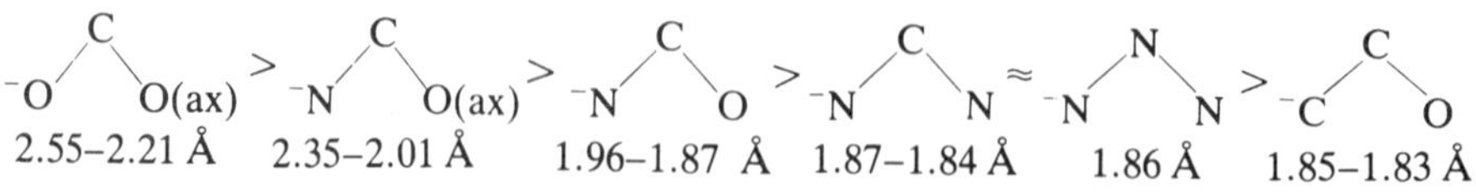

The presence of axial ligands is signified by (ax), and the formally negatively charged atom is shown, but with N—C—O bridges the formal charge is on N or O according to whether the anion is from an amide or an hydroxypyridine. In addition to the 2-phenyl derivatives (**55**) to (**58**) with CCO bridges, there are organodichromium species (**59**) to (**64**) with Cr—Cr separations in the range 2.1 to 1.89 Å, two of which do not contain bridging ligands. In the mixed ligand complexes (**65**) and (**66**) (CCCN bridge) the Cr—Cr distances are very short presumably because the C donor dominates. The general trend is that the Cr—Cr bonds shorten as the donor strength (basicity) of the anions increases: $O^- < N^- < C^-$.

In keeping with the so far perceived trends, the longest bond (2.541 Å) is in $[Cr_2(O_2CCF_3)_4(OEt_2)_2]$ (**27**) where there is axial ligation and the bridging ligand the least basic, and the shortest (1.828 Å) in $[Cr_2(2\text{-MeO-5-MeC}_6H_3)_4]$ (**57**). The ranges of metal–metal distances in the O—C—O(ax) and N—C—O(ax) series overlap so that dichromium compounds no longer fall into two distinct classes, *i.e.* carboxylates and others with much shorter Cr—Cr bonds.

In order to compare Cr—Cr bond lengths with other bond lengths a formal shortness ratio (FSR) for a bond A—B has been defined:

$$\text{FSR} = \frac{D_{\text{A—B}}}{R_{\text{A}} + R_{\text{B}}} = \frac{r(\text{Cr—Cr})}{2 \times 1.186(R_{\text{Cr}})}$$

The FSR values for complexes (**57**) and (**27**) are 0.771 and 1.07, compared with 0.786 for N≡N.[156]

Chromium resembles molybdenum and tungsten in its quadruply bonded compounds but differs in not forming dimetal compounds of lower bond order. Although salts of the stable binuclear anions $[Mo_2Cl_8]^{4-}$ and $[Mo_2(SO_4)_4]^{4-}$ are well known, chromium(II) forms very different chloro complexes (Section 35.3.7.3), and the violet, nearly diamagnetic sulfato-bridged dichromium complex $Cs_4[Cr_2(SO_4)_4(H_2O)_2]\cdot 2H_2O$, obtained only by heating $Cs_2SO_4\cdot Cr_2SO_4\cdot 6H_2O$, re-forms $[Cr(H_2O)_6]^{2+}$ with water.[162]

(*iii*) *Theoretical description of Cr≡Cr bonds*

The Cr—Cr bond lengths vary from supershort bonds of about 1.85 Å to long bonds approaching 2.6 Å. This is a wide range for what is formally considered a quadruple bond. Molecular orbital calculations of electron density in the formate $[Cr_2(O_2CH)_4]$ as a long-bonded example and in '$Cr_2\{H_2P(CH_2)_2\}_4$' as a model for the supershort Cr—Cr complexes $[Cr_2\{(CH_2)_2P(CH_3)_2\}_4]$ and $[Cr_2(mhp)_4]$ show agreement with the experimental electron densities obtained by high resolution low temperature single crystal X-ray investigations on $[Cr_2(O_2CMe)_4(H_2O)_2]$[185] and $[Cr_2(mhp)_4]$.[187] The contraction of the Cr—Cr bond in (**25**; L = H_2O) by 0.009 Å on cooling to 90 K agrees with a shallow Cr—Cr potential and the sensitivity of the bond to axial donors. From the X-ray experiment, there is an electron-deficient area along the Cr—Cr bond, and a broad region of excess electron density off it, but the computed map for $[Cr_2(O_2CH)_4]$ has a broad maximum around the centre of symmetry. The difference is tentatively related to weakening of the σ bond by interaction of the axial water molecules present in the acetate.

In the supershort cases, the electron density is at a maximum at the bond midpoint.[187] The pure quadruple bond configuration $\sigma^2\pi^2\delta^2$ (Figure 3) makes only a small contribution as a leading term in the calculations for the carboxylates, but is much more important in short Cr—Cr species (as for Mo_2^{4+});[185,187–190] in each case excited configurations must be included in the calculations. From comparisons of the results of calculations on the formate and the hypothetical tetrakis(formamidato) species $[Cr_2(NH(O)CH)_4]$, and the formate and '$[Cr_2\{H_2P(CH_2)_2\}_4]$', it has been suggested[188–190] that greater charge on the ligand donor atoms leads to better overlap of the *d* orbitals and a shorter quadruple bond. The importance of the inductive effects of the bridging ligands does not seem to have been clearly established experimentally: on the other hand the importance of the axial ligands on Cr—Cr bond lengths is clear from X-ray work, but not quantitatively explained theoretically.

Other calculations of the Cr—Cr interaction are based on a model in which there is weak antiferromagnetic interaction between high-spin Cr^{2+} atoms.[191] This is not believed to be realistic, at least for the supershort bonds.[187] A computational method which supposes multiple bonds to be single bonds intensified by screening of the internuclear repulsion is said not to support the antiferromagnetic interaction mode.[192]

Figure 3 Qualitative order of energies of orbitals formed by metal $d-d$ overlaps

The standard enthalpies of formation of crystalline (**25**; $L=H_2O$), (**26**), (**67**), (**39**), (**55**), $[Mo_2(O_2CMe)_4]$ and $[Mo_2(O_2CMe)_2(acac)_2]$ have been derived from calorimetric measurements of oxidative hydrolysis in solution. It has not been possible to distinguish between various theoretical treatments from the thermochemical data.[193,194]

(iv) Photoelectron spectroscopy

Assignments by different workers of the photoelectron spectra of $[Cr_2(mhp)_4]$ agree that the lowest energy peak A (6.7–6.8 eV) is due to ionization from the δ metal–metal bonding level. The next feature is complex with an intense peak B at 7.75 eV and a shoulder C at 8.15 eV. It was considered that B and C are the π and σ ionizations, but it seems that B is due to ionization from π levels of the ligand and C is the π metal–metal ionization.[177,190,195] The σ ionization has been assigned to the second of three components at 9.8, 10.2 and 11.1 eV of a complex peak, and the other two components to ligand π ionizations.[177] (When metal–ligand interactions are included, the molecular orbital diagram is of course more complicated than Figure 3.)

Ionizations from the δ and π levels of dimolybdenum tetracarboxylates have been resolved, but as shown in Table 26,[190] not those from $[Cr_2(O_2CMe)_4]$. The location of the ionization from the σ metal–metal bonding levels remains uncertain.[190,196] Table 26 also contains core electron ionization energies.

Table 26 Experimental Core[a] and Valence[b] Electron Ionization Energies (eV)

Orbital	$Cr_2(mhp)_4$	$Cr_2(O_2CMe)_4$
Cr $2p_{1/2}$	590.2	591.1
Cr $2p_{3/2}$	580.6	581.5
M—M (δ)	6.70	} 8.6–9.1
M—M (π)	8.15	
O 1*s*	536.4	537.8
N 1*s*	404.2	—
C 1*s*	291.2sh	294.3
	290.2	291.1
	$Mo_2(mhp)_4$	$Mo_2(O_2CMe)_4$[c]
Mo $3d_{3/2}$	236.9	237.7
Mo $3d_{5/2}$	233.8	234.7
M—M (δ)	5.84	6.92
M—M (π)	8.02	8.77
O 1*s*	536.5	537.9
N 1*s*	404.2	—
C 1*s*	290.0	295.7, 293.4
		291.1

[a] Gas-phase X-ray PES.
[b] He^{I} and He^{II} PES.
[c] Formate data also in ref. 190.

The oxygen core IE values of $[M_2(O_2CMe)_4]$ (M = Cr, Mo) are the same within experimental error, as are the oxygen and nitrogen IE values for $[M_2(mhp)_4]$. Also, the decrease in the binding energy of the Cr 2*p* levels from $[Cr_2(O_2CMe)_4]$ to $[Cr_2(mhp)_4]$ matches that of the Mo 3*d* levels. Thus, the charge distribution is the same within $[Cr_2(O_2CMe)_4]$ and $[Mo_2(O_2CMe)_4]$, and within $[Cr_2(mhp)_4]$ and $[Mo_2(mhp)_4]$. Consequently, the significant lengthening (0.40 Å) in going from $[Cr_2(mhp)_4]$ to $[Cr_2(O_2CMe)_4]$ as compared with the Mo pair (0.03 Å) is not due to differences in the bonding of the ligands, but to the different nature of the metal–metal interactions, notably the shallower Cr—Cr potential well, which makes the Cr—Cr separation more sensitive to changes in the electron density distribution.

(*v*) *Electronic spectra*

The absorption spectra of $[M_2(mhp)_4]$ (M = Cr, Mo, W) in THF are similar. The lowest frequency band (22 500 cm^{-1}, M = Cr, Table 27) has been assigned to the metal-localized $\delta \rightarrow \delta^*$ transition ($^1A_1 \rightarrow {}^1B_2$, D_{2d} symmetry). The vibronic structure on this band in the low temperature spectra of the solids, with progressions of 320 (M = Cr), 400 (M = Mo) and 320 cm^{-1} (M = W), is consistent with the assignment.

Table 27 Electronic Absorption Spectra of Some Dichromium Species

Complex	*Electronic spectra* (cm^{-1})					*r(Cr—Cr)* (Å)	*Ref.*
$[Cr_2(mhp)_4]$[a] (THF)	43 700 (39 000)	38 000 (22 000)	33 900 (31 000)	29 600sh (≈3400)	22 600 (480)	1.889	1
$[Li(Et_2O)_4][Cr_2(CH_3)_8]$[a] ($Et_2O$)					22 000 (<700)	1.980	2
			II		I		
$[Cr_2(O_2CMe)_4(H_2O)_2]$[b]			≈30 000		≈21 000	2.362	3
$[Cr_2(O_2CMe)_4(NH_3)_2]$[c]		37 000b	29 000		19 500vb	—	4
$[Cr_2(stearate)_4(H_2O)_2]$[c]	40 000vb		30 000b	28 400sh	21 900vb	—	4
$[NMe_4]_2[Cr_2(O_2CMe)_4(NCS)_2]$[c]		34 800b 33 900sh	28 600b	24 700w	19 000vb	—	4
$[NEt_4]_2[Cr_2(O_2CMe)_4(NCS)_2]$[c]		35 300b 33 800sh	28 700b	24 900sh	18 700vb	—	4
$[NEt_4]_2[Cr_2(O_2CEt)_4(NCS)_2]$[c]	39 100sh	36 000b 33 000sh	29 200b	24 900sh	18 500vb	2.467	4
$Cs_4[Cr_2(SO_4)_4(OH_2)_2]\cdot 2H_2O$	43 100	35 600	30 100 29 400sh	26 300sh	18 100vb	—	4

[a] Molar extinction coefficients in parentheses ($dm^3\ mol^{-1}\ cm^{-1}$).
[b] Polarized single crystal spectra at 6 K; exact wavenumbers not tabulated.
[c] Powder reflectance spectra.

1. M. C. Manning and W. C. Trogler, *J. Am. Chem. Soc.*, 1983, **105**, 5311.
2. A. P. Sattelberger and J. P. Fackler, *J. Am. Chem. Soc.*, 1977, **99**, 1258.
3. S. F. Rice, R. B. Wilson and E. I. Solomon, *Inorg. Chem.*, 1980, **19**, 3425.
4. P. D. Ford, L. F. Larkworthy, D. C. Povey and A. J. Roberts, *Polyhedron*, 1983, **2**, 1317.

Two Raman lines, either of which may be the Cr—Cr mode, have been observed at 340 and 400 cm^{-1} for $[Cr_2(mhp)_4]$. The former is preferred[197] since the vibronic progression is 320 cm^{-1}, and this is lower than earlier proposed.[169] Others believe that this is a Cr—O or Cr—N totally symmetric stretch.[198] The 22 500 cm^{-1} band and the shoulder at 29 600 cm^{-1} are metal-dependent since they are intensified, and move together and to lower frequency from Cr to Mo to W. The shoulder at 29 600 cm^{-1} is assigned to the $\delta \rightarrow \pi^*$ ($^1A_1 \rightarrow {}^1E$) transition.[197]

The low temperature single crystal spectrum of the acetate contains two polarization dependent bands near 21 000 cm^{-1} and 30 000 cm^{-1} (I and II, Table 27). Different assignments (I, $\delta \rightarrow \pi^*$; and II, $np\pi(O) \rightarrow \pi^*$) to the above have been made,[199] and further work seems necessary. Bands I and II move to lower wavenumber with increasing Cr—Cr separation, and powder reflectance spectra of several dinuclear Cr_2^{4+} species show the same general features as the detailed acetate spectrum.[162] The ground state Cr—Cr stretching frequency is estimated at 150 to 200 cm^{-1}. The Cr—Mo stretch of (**67**) is reported as 394 cm^{-1}. An accurate spectrum of $[Cr_2\{(CH_2)_2P(CH_3)_2\}_4]$ could not be obtained because of decomposition.[200]

(vi) *Magnetic properties*

The red chromium(II) carboxylates $[Cr(O_2CR)_4L_2]$ are weakly paramagnetic (μ_{eff} = 0.4–0.9 BM at room temperature)[157,161] and it is generally believed that this is due to paramagnetic impurities (chromium(III) species, from ESR measurements)[156] arising from inadvertent aerial oxidation. However, the temperature variation of susceptibility of several carefully isolated samples of carboyxlates (including **35**) can be explained by assuming strong antiferromagnetic interactions in the dinuclear structures ($2J \approx 700\ cm^{-1}$), and the presence of *ca.* 1% or less chromium(III) impurity.[161] It seems that at least in the carboxylates with the longest Cr—Cr separations the paramagnetism is genuine. Carboxylates can crystallize with a proportion of mononuclear chromium(II), *e.g.* in (**32**),[164] or form blue paramagnetic species on dehydration.[201] The carbonato complexes also have low magnetic moments.[202]

The yellow colour and temperature-dependent moment of the oxalate $CrC_2O_4{\cdot}H_2O$ ($\mu_{eff}^{340\ K}$ = 4.5, $\mu_{eff}^{100\ K}$ = 4.0 BM) suggest that it is polymeric.[201,203] The malonato complexes $Cr(C_3H_2CO_2)_2{\cdot}2H_2O$ (red) and $Na_2Cr(C_3H_2O_4)_2{\cdot}4H_2O$ (blue) have normal magnetic moments at room temperature ($\mu_{eff} \approx 4.8$ BM),[204] which is inconsistent with the colour of the former.

Red or blue, variously hydrated chromium(II) complexes of thiocarboxylic acids, *e.g.* $S[(CH_2)_2SCH_2CO_2H]_2$, have been isolated, and the red complexes are antiferromagnetic.[205] Some antiferromagnetic complexes of adenine may have binuclear structures (Table 16).

(vii) *Miscellaneous investigations of carboxylates and related complexes*

The molecular ions have been identified in the mass spectra of $[(Cr_2(O_2CR)_4]$ (R = Me, Et, Pr^i, CMe$_3$ and $C_{11}H_{23}$)[206] and two reversible one-electron oxidations have been found for $[Cr_2(map)_2]$ (**43**) (−0.53 and −0.09 V *vs.* Ag/AgNO$_3$).[207]

In the molecule $[Cr_3(\eta^5\text{-}C_5H_5)_2(O_2CCF_3)_6]$ the central metal atom is in oxidation state II and forms four short (2.01 Å) and two long (2.58 Å) Cr—O bonds (**70**).[208]

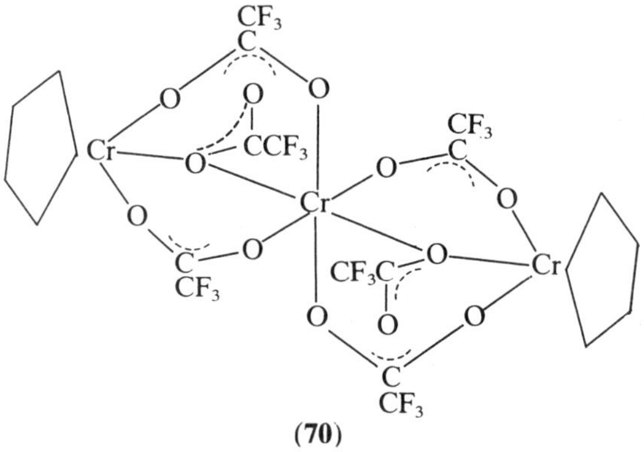

(**70**)

The pentaacetatodiborate $Cr[B_2O(OCOMe)_5]_2$ has been prepared ($\mu_{eff}^{293\ K}$ = 4.80 BM).[209]

Chromium(II) acetate is predominantly dimeric in water in which it is not very soluble. The equilibria are shown below.

$$Cr_2(O_2CMe)_4 \underset{}{\overset{K_{DO}}{\rightleftharpoons}} 2Cr(O_2CMe)_2 \qquad \log K_{DO} = -4.4\ (25\ ^\circ C,\ I = 3.0\ M)$$

$$Cr(O_2CMe)_2 \overset{K_2^{-1}}{\rightleftharpoons} Cr(O_2CMe)^+ + MeCO_2^- \qquad \log K_2 = 0.9$$

$$Cr(O_2CMe)^+ \overset{K_1^{-1}}{\rightleftharpoons} Cr^{2+} + MeCO_2^- \qquad \log K_1 = 0.8$$

Reaction of edta with the aqueous dimer causes rapid decomposition to the monomer which appears to be the effective reductant in redox reactions of the acetate. The difference between the Cr≡Cr and Cu—Cu bond energies has been estimated from the dissociation data for the acetates to be only about 45 kJ mol^{-1} even though the Cu—Cu bond is formally a single bond.[210] Oxidation of the dimer by I_2 has been investigated kinetically.

From a spectrophotometric study of the equilibria between various chelating agents and $[Cr_2(O_2CMe)_4]$ some stability constants of mononuclear Cr^{II} complexes have been determined (malonate, $\log \beta_2 = 6.0 \pm 0.2$; *N*-methyliminodiacetate, $\log \beta_2 = 12.3 \pm 0.5$; ethylenediamine-*N,N'*diacetate, $\log K_1 = 9.1 \pm 0.2$; at $T = 25\,°C$, $I = 1.0\,M\ NaClO_4$) and the values agree with prediction.[211]

The binuclear formate is incompletely formed over the intermediate formate concentration range. At the highest formate concentrations the binuclear complex dissociates to mononuclear complexes with $HCO_2^-/Cr > 3$.[212]

$$2Cr(O_2CH)^+ + HCO_2^- \overset{K}{\rightleftharpoons} Cr_2(O_2CH)_3^+ \quad K = 2.1 \pm 0.2$$

Stability constants have also been determined for oxalate and malonate (Table 39). The succinate $[Cr(C_4H_4O_4)H_2O]_2 \cdot 2H_2O$ is isostructural with the copper(II) compound.[213a]

35.3.5.6 Aliphatic and aromatic hydroxy acids

The glycolate $Cr(C_2H_3O_3)_2 \cdot H_2O$,[213] salicylate $Cr(C_6H_4(O)CO_2) \cdot 3H_2O$ and tartrate $Cr(CO_2CHOHCHOHCO_2)$[3] have been prepared. Stability constants have been determined for complexation of Cr^{II} with salicylate and 5-sulfosalicylate (Table 39).

35.3.5.7 Dimethyl sulfoxide, phosphine oxides and arsine oxides

Complexes of dimethyl sulfoxide[214] and some phosphine and arsine oxides[148,215,216] have been prepared by the methods in Scheme 23. The phosphine oxide complexes are decomposed by polar solvents, and were obtained by heating the reactants in sealed tubes, or in xylene or bromobenzene. The use of halogenated solvents can lead to Cr^{III}—C bonds but apparently not in this case. No structures have been clearly established. The complexes are all high-spin, and have reflectance spectra suggestive of distorted six-coordinate chromium(II). Thus those of the formula $[CrX_2L_4]$ have monomeric *trans* configurations and those with the formula $[CrX_2L_2]$, which are known to be weakly antiferromagnetic ($\theta \approx 0°$), have polymeric structures. However, the chloro complexes $[CrCl_2(R_3PO)_2]$ are non-electrolytes, and a dipole moment of 4.9 Debye for $[CrCl_2(Et_3PO)_2]$ in C_6H_6 excludes a *trans* planar structure. The complex $[CrBr_2(Ph_3PO)_2]$ has been obtained as a green polymer and as a yellow form which, like $[CrI_2(Ph_3PO)_2]$, is thought to have a flattened tetrahedral structure. Different reflectance spectra provide the main structural evidence, the yellow form having a broad band at lower wavenumber (Table 28, $10\,000\ cm^{-1}$). However, complexes such as $[NEtH_3]_2[CrBr_4]$, in which the Cr^{II} is six-coordinate through bromide bridges between planar $[CrBr_4]^{2-}$ units, absorb at approximately $10\,500\ cm^{-1}$ (Section 35.3.7.3); thus tetrahedral structures are unlikely. Possibly different ligands lie along the distortion axis in the two forms.

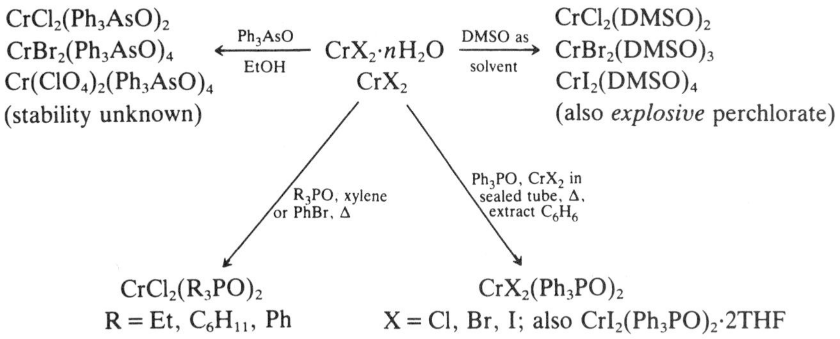

Scheme 23

The IR spectra of the DMSO complexes show that the ligand is O-bonded; and the lack of splitting of the perchlorate stretching absorptions of $[Cr(ClO_4)_2(Ph_3AsO)_4]$ shows that ClO_4^- interacts only weakly with the metal in the probably planar $Cr(Ph_3AsO)_4$ unit.

Table 28 Complexes of Dimethyl Sulfoxide and Phosphine and Arsine Oxides

Complex		μ_{eff} (293 K) (BM)	θ (°)	*Reflectance spectra* (cm^{-1})			*Ref.*
$[CrCl_2(DMSO)_2]$	Pale blue	4.91	—		12 000		1
$[CrBr_2(DMSO)_3]$	Pale blue	4.96	—		12 200		1
$[CrI_2(DMSO)_4]$	Pale blue	4.95	—		14 100		1
$[CrCl_2(Ph_3PO)_2]$[a]	Blue	4.86[b]	15	18 100	15 380sh	11 760	2, 4
					12 500sh		
$[CrBr_2(Ph_3PO)_2]$	Green	4.9[b,c]	8		14 600	11 100sh	2
	Yellow	4.80[b]	23		10 000		
$[CrI_2(Ph_3PO)_2]$	Dark brown	4.79[b]	34		10 800		2
$CrI_2(Ph_3PO)_2(THF)_2$	Blue-green	4.88[b]	15		14 100	10 000sh	2
$[CrCl_2(Ph_3AsO)_2]$	Violet-pink	4.6	0	18 000	15 500	13 000sh	3
$[CrBr_2(Ph_3AsO)_4]$	Pink	4.89	0	17 700	15 700sh		3
$[Cr(ClO_4)_2(Ph_3AsO)_4]$	Pale blue	4.89	0		16 400		3

[a] The blue complexes $[CrCl_2(R_3PO)_2]$ (R = Et, C_6H_{11}, Ph) are non-electrolytes with dipole moments in C_6H_6 of 4.9 D (ref. 4).
[b] These magnetic moments were calculated from $\mu_{eff} = 2.828\{\chi_A(T + \theta)\}^{1/2}$.
[c] Could not be obtained free from yellow form.

1. D. G. Holah and J. P. Fackler, Jr., *Inorg. Chem.*, 1965, **4**, 1721.
2. D. E. Scaife, *Aust. J. Chem.*, 1967, **20**, 845.
3. L. F. Larkworthy and D. J. Phillips, *J. Inorg. Nucl. Chem.*, 1967, **29**, 2101.
4. K. Issleib, A. Tzschach and H. O. Frohlich, *Z. Anorg. Allg. Chem.*, 1959, **298**, 164.

35.3.6 Sulfur Ligands

35.3.6.1 Dithiocarbamates

A few very air-sensitive dialkyldithiocarbamato complexes $[Cr(S_2CNR_2)_2]$ (R = Me, Et or $R_2 = C_4H_8$) have been prepared from aqueous or aqueous ethanolic solutions of $CrCl_2$ and the sodium dithiocarbamate (Table 29).[217,218] The diethyl-substituted derivative $[Cr(S_2CNEt_2)_2]$ is isomorphous with other metal(II) diethyldithiocarbamates[217] known to be S-bridged dimers with distorted trigonal bipyramidal arrangements of S atoms around the metal, *e.g.* $[Fe(S_2CNEt_2)_2]$.[219] The reflectance spectrum resembles the spectra of other square pyramidal rather than trigonal bipyramidal chromium(II) complexes (Table 42). The complexes are antiferromagnetic. (For R = Et, $J = 26$ cm^{-1}, $g = 2.19$; binuclear expression.)

Table 29 Complexes of Dialkyldithiocarbamates and Thiourea

Complex	μ_{eff} (BM) 295 K	90 K	*Reflectance spectra* (cm^{-1})		$\nu(Cr—Cl)$ (cm^{-1})
Dithiocarbamates					
$[Cr(S_2CNMe_2)_2]$ Red	4.45	4.07	—		—
$[Cr(S_2CNEt_2)_2]$ Yellow	3.98	2.21[a]	14 000	11 000sh	—
$[Cr(S_2CNC_4H_8)_2]$ Yellow	4.38	3.09	—		—
Thioureas					
$CrCl_2(tu)_2$ Yellow	4.49	3.94	11 600	8000sh	320
$CrCl_2(tu)_2 \cdot Me_2CO$ Yellow green	4.41	3.38	11 900	8400sh	—
$CrCl_2(etu)_2$ Light blue	4.33	3.16	11 400		320
$CrCl_2(dctu)$ Yellow	3.83	2.45	11 000	6700	315

[a] Néel point at *ca.* 100 K.

35.3.6.2 *Ethane-1,2-dithiol*

Chromium(II) is in planar coordination (**71**) in $[NEt_4]_2[Cr(S_2C_2H_4)_2]$ ($\mu_{eff}^{297\,K} = 4.95$ BM in MeCN), which separates as a violet solid when ether is added, after stirring for several days, to a mixture of $CrCl_2$, sodium ethane-1,2-dithiolate and $[NMe_4]Cl$.[220]

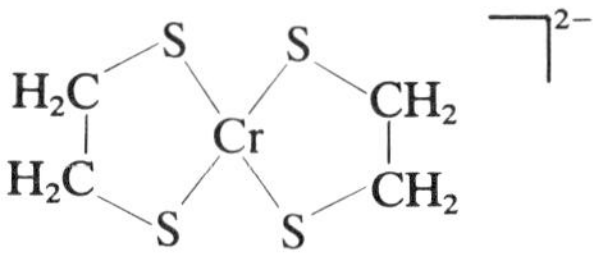

(**71**)

35.3.6.3 *Thioureas*

Complexes of chromium(II) halides with thiourea (tu), *N,N'*-ethylenethiourea (etu) and *N,N'*-dicyclohexylthiourea (dctu) have been isolated from solutions of the halides which had been dehydrated with an excess of 2,2-dimethoxypropane. Attempts to isolate complexes from more polar solvents such as water, DMF and the lower alcohols were unsuccessful. The bromides and iodides $CrBr_2{\cdot}4tu$, $CrI_2{\cdot}6tu$, $CrBr_2{\cdot}5etu$, $CrI_2{\cdot}4etu$, $CrBr_2{\cdot}5dctu$ and $CrI_2{\cdot}6dctu$ are high-spin, with temperature-invariant moments in the range 4.74–4.97 BM. There is a band in the range 11 000 to 14 000 cm^{-1} (ν_2) and a weaker band or shoulder to lower wavenumbers (ν_1) in the reflectance spectra. This is the usual pattern for distorted six-coordinate high-spin chromium(II) complexes. Since ν_2 is well below the values found for comparable complexes containing N donor ligands, *e.g.* for $[CrX_2(mepy)_4]$, $\nu_2 \approx 17\,000$ cm^{-1}, the thioureas are acting as S donors. Square pyramidal structures were not excluded for the cations in $CrBr_2{\cdot}5etu$ (or 5 dctu). The chloride $CrCl_2{\cdot}2tu$, its acetone adduct, $CrCl_2{\cdot}2etu$ and $CrCl_2{\cdot}dctu$ are antiferromagnetic and chloride- or perhaps thiourea-bridged polymers (Table 29).[221] No X-ray data are available.

35.3.7 Halogens as Ligands

35.3.7.1 *Anhydrous halides*

Only more recent work on the anhydrous halides CrX_2 (X = F, Cl, Br or I) will be considered as their preparations, properties and structures have already been ably summarized.[222] Generally, they are prepared by a variety of solid state reactions at high temperature, *e.g.* from the metal and the hydrogen halide or by reduction of a higher halide with hydrogen or the metal. Thermal dehydration of the hydrated chloride, bromide and iodide provides adequately pure anhydrous halides for preparative use (Table 9). It has been found that in the preparation of $CrCl_2$ by reduction of $CrCl_3$ with Al, Cr or H_2 in a temperature gradient of 450 to 350 °C, the $CrCl_2$ is transported as gaseous $CrAl_2Cl_8$ from the 450 °C zone, where reduction occurs, to the 350 °C zone.[223] Besides the solid state procedures,[222] CrF_2 can be obtained by thermal dehydration of $CrF_2{\cdot}2H_2O$[85] and by volatilization of NH_4F from NH_4CrF_3.[224] In oxygen at 300 °C, CrF_2 forms CrO_2F_2, and with sulfur Cr_2S_3 is formed. It liberates hydrogen from water, and NO at 600 °C produces CrF_3, Cr_2F_5 and Cr_2O_3.[225] It is also not inert to N_2 at high temperatures because at 650 °C over 100 h a dark product of unknown structure Cr_2NF_3 is obtained.[226] With an excess of XeF_6 at 120 °C, CrF_2 forms $XeF_6{\cdot}Cr^{IV}F_6$.[227]

In the halides bridging anions lead to distorted CrX_6 octahedra sharing corners in which there are four short approximately equal Cr—X bonds and two long *trans* Cr—X bonds. There is also angular distortion of the octahedra. The fluoride has the tetragonal (distorted) rutile structure and the other halides chain-type structures. The mixed crystal $(Mn, Cr)I_2$ is isotypic with monoclinic CrI_2 (Table 34).

The anhydrous difluoride is antiferromagnetic; the magnetic moment is 4.3 BM at room temperature and the Néel temperature 53 K (neutron diffraction). The magnetic axis is parallel to the direction of the long Cr—F bonds. The chloride also is antiferromagnetic ($T_N = 20$ K).[222]

In the single crystal absorption spectra of CrF_2 and $CrCl_2$ the pattern of spin-allowed (quintet–quintet) bands for the ammines and hydrates is repeated. Gaussian analysis of the higher wavenumber bands has allowed assignments of all three transitions and calculation of the ligand field parameters. There is some controversy over the ordering and extent of splitting of the D_{4h} levels but those in Table 30 and Figure 2 seem generally accepted. The many observed spin-forbidden transitions (mainly quintet–triplet transitions) in the single crystal spectra have also been assigned, and there is good agreement between observed and calculated energies. The decrease in intensity of these bands which takes place between 70 and 6 K is believed to be due to the onset of magnetic ordering; neutron diffraction studies show that below the Néel temperature the spins in $CrCl_2$ align parallel within the ferromagnetic (011) planes, but adjacent planes have opposite spins.[228]

Table 30 Spin-allowed Transitions and Ligand Field Parameters for CrF_2 and $CrCl_2$

Assignment	Theoretical formula	Experimental (6 K) CrF_2	$CrCl_2$
${}^5B_{1g} \rightarrow {}^5A_{1g}$	$4Ds + 5Dt$	10 495	9100
${}^5B_{1g} \rightarrow {}^5B_{2g}$	$10Dq$	11 623	11 050[a]
${}^5B_{1g} \rightarrow {}^5E_g$	$10Dq + 3Ds - 5Dt$	14 669	12 250[a]
Dq (cm^{-1})		1162	1100
Ds (cm^{-1})		1934	1470
Dt (cm^{-1})		551	640

[a] Maxima of Gaussian components of band at 11 750 cm^{-1}.

The gaseous dihalides are believed to be linear monomers, and the PE spectrum of $CrCl_2$ has been assigned on this basis with the assumption of a high spin molecule.[229] Anomalies in the heat capacity data of CrF_2 (10–300 K) have been associated with the magnetic ordering.[230] From the temperature dependence of the ^{129}I Mössbauer spectra of CrI_2 a Néel temperature of 45 ± 5 K has been deduced.[231]

The treatment of Cr metal with anhydrous HF acidified with the Lewis acids BF_3, AsF_5 and SbF_5 gives solutions believed to contain the ions $[Cr(FH)_6]^{2+}$ because there is a single broad absorption band (9900 cm^{-1}),[232] as is observed in aqueous solution (14 100 cm^{-1}), and in molten $AlCl_3$ (8400 cm^{-1}) and fluoride melts (14 000 cm^{-1}),[233]

The green adduct $CrF_2{\cdot}AsF_5$ ($\mu_{eff} = 4.68$ BM) has been prepared by the reaction of CrF_2 with AsF_5 in anhydrous HF.[234]

35.3.7.2 Complex fluorides

Much information on the correlation between magnetic and other properties and structure has come from investigations of the complex fluorides of the transition metals,[235] and chromium(II) provides several examples (Table 31) of such complexes.

(i) Preparations

Trifluorochromates, $ACrF_3$ (A = NH_4, K or Rb), can be prepared in aqueous conditions from chromium(II) chloride and AF or chromium(II) acetate and AHF_2.[224,236] The potassium salt has also been prepared by fusing CrF_2 with KF under argon.[237]

Sodium tetrafluorochromate has been prepared in aqueous conditions as above, and from solutions of chromium(II) sulfamate.[238] Some workers[236,238] have isolated Na_2CrF_4 only in spite of varying the reaction conditions, but others have obtained $NaCrF_3$[239,240] although Li and Cs salts could not be obtained from aqueous or dry conditions. The salt Na_2CrF_4 is the only one of its stoichiometry among the chromium(II) complex fluorides. Attempts to prepare tetrafluorochromates in the solid state from $MCrF_3$ (M = K or Rb) and MF lead[224] to disproportionation:

$$3MCr^{II}F_3 + MF = 2M_3Cr^{III}F_6 + Cr^0 \qquad (26)$$

The complexes $MCrF_4$ (M = Ca, Sr or Ba), Ba_2CrF_6, $CrZr(Hf)F_6$ and $CrMF_5$ (M = Al, Ti, V), and the mixed oxidation state species Cr_2F_5, Rb_xCrF_3 and $Cs_4Cr_5F_{18.24}$ have been obtained by solid state reactions (see Table 31). Cr_2F_5 is an example of a general class of chromium(III) fluoro complex $M^{II}Cr^{III}F_5$.[235]

Table 31 Complex Fluorides

Complex	*Comments*	*Ref.*
$KCrF_3$[a]	$\mu_{eff} = 4.9$ BM, $\theta = 0°$ (85–350 K)	1–3
Light blue	$\mu_{eff} = 4.6$ BM, $\theta = 5°$, $T_N = 40$ K (34–290 K)	
NH_4CrF_3	$\mu_{eff}^{330\,K} = 4.90$; $\mu_{eff}^{95\,K} = 4.40$ BM, $\theta = 32°$	1
Light blue		
$RbCrF_3$	$\mu_{eff} = 4.9$ BM, $\theta = 0°$ (60–570 K), antiferromagnetic <50 K	2
Na_2CrF_4	$\mu_{eff}^{300\,K} = 4.8$ BM, $\theta = 0°$	1, 4
Dark blue		
$MCrF_4$	M = Ca, Sr, Ba; planar from space group and optical spectra; CrF_2, SrF_2 at	5
	700 °C, M = Sr, Cr—F = 1.980 Å, tetragonal $KBrF_4$ type	6
Ba_2CrF_6	CrF_2, BaF_2 at 700 °C	5, 7
CrZr(Hf)F_6	$CrF_2 + ZrF_4 \xrightarrow[\text{Pt tube}]{830\,°C,\ 50\,h} CrZrF_6$ (trace of Cr^{3+})	8
Greenish grey	${}^5B_{1g} \rightarrow {}^5A_{1g}$, ${}^5B_{2g}$, ${}^5E_g$ transitions at 7000, 10 000sh, 11 700 cm^{-1}	
	$\mu_{eff} = 4.9$ BM above 200 K; weak interaction <200 K	
$CrMF_5$	M = Al (paramagnetic), M = Ti, V (ferromagnetic, $T_c = 26$ K and 40 K)	10
Cr_2F_5	Mixed Cr^{2+}, Cr^{3+}; prep. from CrF_2, CrF_3 at 800–900 °C; Cr^{II}—F, 2 at	9
	1.955 Å, 2 at 2.010 Å, 2 at 2.572 Å, Cr^{III}—F (av), 1.89 Å; antiferromag-	10
	netic, $T_N = 40$ K	
Rb_xCrF_3	Mixed Cr^{2+} ($x = 0.18$–0.30), Cr^{3+}; prep. from RbF, CrF_2, CrF_3 at 850 °C,	11
	28 d, amber, hexagonal bronze phase, $x = 0.18$–0.29, $T_N = 25$–35 K. Also	
	K, Cs analogues	
$Cs_4Cr^{II}Cr_4^{III}F_{18.24}$	Formed with $Cs_4CoCr_4^{III}F_{18}$ from $CoCl_2$, CsCl, CsF, CoF_2, CrF_3 at 850 °C	12

[a] Also $K_2[Cr_3Cl_2F_6]$. J. C. Dewan, A. J. Edwards and J. J. Guy, *J. Chem. Soc., Dalton Trans.*, 1986, 2623.

1. A. Earnshaw, L. F. Larkworthy and K. S. Patel, *J. Chem. Soc. (A)*, 1966, 363.
2. A. de Kozak, *Rev. Chim. Min.*, 1971, **8**, 301.
3. S. Yoneyama and K. Hirakawa, *J. Phys. Soc. Jpn.*, 1966, **21**, 183.
4. A. J. Deyrup, *Inorg. Chem.*, 1964, **3**, 1645.
5. D. Dumora, C. Fouassier, R. von der Muhll, J. Ravez and P. Hagenmuller, *C. R. Hebd. Seances Acad. Sci.*, 1971, **273**, 247.
6. H. G. von Schnering, B. Kolloch and A. Kolodziejczyk, *Angew. Chem. Int. Ed. Engl.*, 1971, **10**, 413.
7. R. von der Muhll, D. Dumora, J. Ravez and P. Hagenmuller, *J. Solid State Chem.*, 1970, **2**, 262.
8. C. Friebel, J. Pebler, F. Steffens, M. Weber and D. Reinen, *J. Solid State Chem.*, 1983, **46**, 253.
9. R. Colton and J. H. Canterford, 'Halides of the First Row Transition Metals', Wiley-Interscience, London, 1969, chap. 4.
10. A. Tressaud, J. M. Dance, J. Ravez, J. Portier, P. Hagenmuller and J. B. Goodenough, *Mat. Res. Bull.*, 1973, **8**, 1467.
11. Y. S. Hong, K. N. Baker, R. F. Williamson and W. O. J. Boo, *Inorg. Chem.*, 1984, **23**, 2787.
12. G. Courbion, R. de Pape, G. Knoke and D. Babel, *J. Solid State Chem.*, 1983, **49**, 353.

(*ii*) *Properties*

Down to liquid nitrogen temperatures, $KCrF_3$ and Na_2CrF_4 have essentially temperature-independent magnetic moments, but NH_4CrF_3 is antiferromagnetic.[236] Other measurements have shown that antiferromagnetism develops in $KCrF_3$ at lower temperatures ($T_N = 40$ K, $\theta = 5°$). Neutron diffraction experiments at 4.2 K have established that ferromagnetic (001) sheets are coupled antiferromagnetically along the (001) direction. The exchange constants in and between the layers are of similar magnitude, $J = 1.4°$ and $-2.2°$ respectively.[237]

Solid state electronic spectra show the expected pattern for high-spin chromium(II) complexes, and for $KCrF_3$,[236,241] $SrCrF_4$[242] and $CrZrF_6$[243] all three spin-allowed *d–d* transitions have been resolved. A theoretical analysis of the sharp spin-forbidden (quintet–triplet) electronic transitions in the solid state spectra of $KCrF_3$ has been made.[240]

The salts $KCrF_3$ and $RbCrF_3$ have distorted perovskite structures, and from analysis of powder data the former contains tetragonally compressed CrF_6 octahedra in which two Cr—F bond distances are 2.00 Å and four 2.14 Å.[244] The ionic radius of Cr^{2+} in fluoride perovskites has been estimated at 0.73 Å as a weighted average from lattice constants.[245]

The complex fluoride $CrZrF_6$ (Table 31) is an example of a class of compound $A^{II}M^{IV}F_6$ which undergoes phase transitions as the temperature changes. In the chromium(II) and copper(II) compounds additional modifications occur because of local and cooperative Jahn–Teller distortions. From the three spin-allowed *d–d* transitions for high-spin Cr^{2+} in strongly distorted CrF_6 octahedra (Table 31) an octahedral Δ value of 7600 cm^{-1} is calculated. Transitions between the monoclinic II′, pseudotetragonal I′ and cubic I phases have been

$$CrZrF_6\text{: II}' \xrightarrow{\approx 150\,K} \text{I}' \xrightarrow{\approx 400\,K} \text{I}$$

established by X-ray and neutron diffraction methods. Additionally, ^{57}Fe doped into $CrZrF_6$ has a Mössbauer spectrum with an isomer shift typical of high-spin iron(II), and a large quadrupole splitting below 400 K ($\Delta E \approx 1.4$ mm s^{-1}), which decreases almost to zero over a few degrees as the temperature rises above 400 K. The large ΔE arises from the presence of Fe^{2+} in lattices (II′ and I′) where there are statically distorted CrF_6 octahedra, but above *ca.* 400 K, *i.e.* in the cubic phase I, a dynamic Jahn–Teller effect takes over; the time for interchange of the tetragonal axis among the three four-fold axes of the FeF_6 octahedron being short on the Mössbauer timescale. The Mössbauer data also show that the FeF_6 octahedra in II′ and I′ are elongated,[243] and the phase transitions have been monitored by neutron powder diffraction studies.[246]

35.3.7.3 Complex chlorides, bromides and iodides

(i) Solvated tetrachloro- and tetrabromo-chromates(II)

The dihydrates $A_2[CrX_4(H_2O)_2]$ (X = Cl or Br) have been crystallized from solutions of the constituents in the appropriate halogen acid except for the pyridinium salts which were isolated from ethanol (Table 32 and Scheme 24). All are magnetically dilute, high-spin compounds with reflectance spectra as expected for tetragonally-distorted octahedral anions $[CrX_4(H_2O)_2]^{2-}$. The NH_4, Rb and Cs salts are isomorphous with the corresponding copper(II) complex halides, which contain *trans* anions $[CuBr_4(H_2O)_2]^{2-}$ with two short and two long CuX bonds. Similar structures are therefore likely for the chromium(II) anions.[247–249]

Table 32 Properties of Halochromate(II) Dihydrates and a Diacetic Acid Adduct

Compound	*Comments*	*Reflectance spectra* (cm^{-1}) ν_2	ν_1	*Ref.*
$A_2[CrCl_4(H_2O)_2]$ Light blue	A = NH_4, K (decomposes), Rb, Cs, $NPhH_3$, $\frac{1}{2}pipzH_2$, pyH (light green), $\mu_{eff} \approx 4.8$ BM, indep. of *T* (90–295 K), ν(Cr—Cl), *ca.* 325 cm^{-1}	*ca.* 13 000	*ca.* 10 000sh	1, 2
$A_2[CrBr_4(H_2O)_2]$ Light blue	A = NH_4, Rb, Cs, pyH (light green); $[pipzH_2]_2CrBr_6{\cdot}2H_2O$, $\mu_{eff} \approx 4.8$ BM, indep. of *T* (90–295 K), ν(Cr—Br), *ca.* 270 cm^{-1}	*ca.* 12 500vbr		3, 4
$Cs[CrCl_3(H_2O)_2]$ Off-white	$\mu_{eff}^{295 K} = 4.61$, $\mu_{eff}^{90 K} = 4.25$ BM, $\theta = 24°$; antiferromagnetic	12 300s	6600m	1
$[C(NH_2)_3][CrBr_4(MeCO_2H)_2]$ Light green	$\mu \approx 4.85$ BM, indep. of *T* (90–295 K) CO_2H absorptions 1615 and 1290 cm^{-1}	11 600s	7000m	5

1. L. F. Larkworthy, J. K. Trigg and A. Yavari, *J. Chem. Soc., Dalton Trans.*, 1975, 1879.
2. M. A. Babar, L. F. Larkworthy and S. S. Tandon, *J. Chem. Soc., Dalton Trans.*, 1983, 1081.
3. M. A. Babar, L. F. Larkworthy and A. Yavari, *J. Chem. Soc., Dalton Trans.*, 1981, 27.
4. M. F. C. Ladd, L. F. Larkworthy, G. A. Leonard, D. C. Povey and S. S. Tandon, *J. Chem. Soc., Dalton Trans.*, 1984, 2351.
5. M. A. Babar, M. F. C. Ladd, L. F. Larkworthy, G. A. Leonard and D. C. Povey, *J. Crystallogr. Spectrosc. Res.*, 1983, **13**, 325.

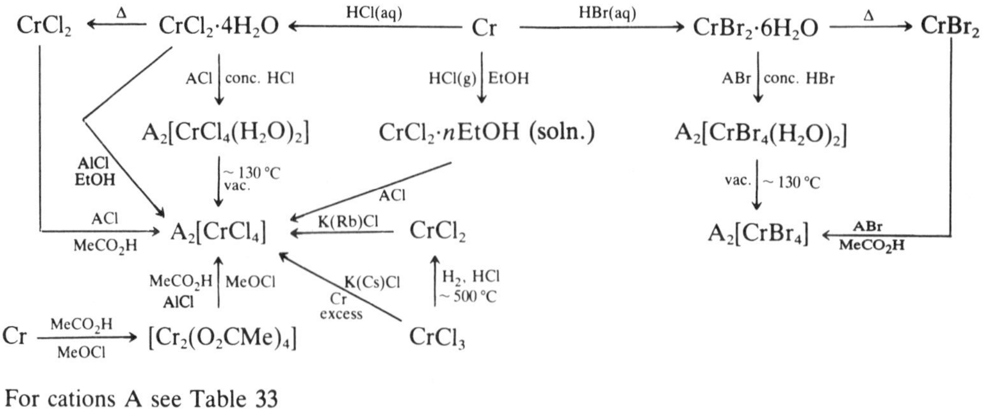

Scheme 24

From the diffuse reflectance and IR spectra and the absence of magnetic interaction, it was deduced that there are analogous *trans* anions containing coordinated acetic acid molecules in the guanidinium salt $[C(NH_2)_3][CrBr_4(MeCO_2H)_2]$; and this has been confirmed by a single crystal investigation which also shows that there are two distinctly different Cr—Br bond distances (Table 34 and **72**).[250] However, in $[pyH]_2[CrBr_4(H_2O)_2]$ there are discrete *trans* planar $CrBr_2(H_2O)_2$ units and bromide ions instead of the expected *trans* octahedral anions (Table 41). This is an example of Cr^{2+} in an uncommon planar environment.[251] The chloride $[pyH]_2[CrCl_4(H_2O)_2]$ may have the same structure since it is green unlike $[NH_4]_2[CrCl_4(H_2O)_2]$ which is blue.

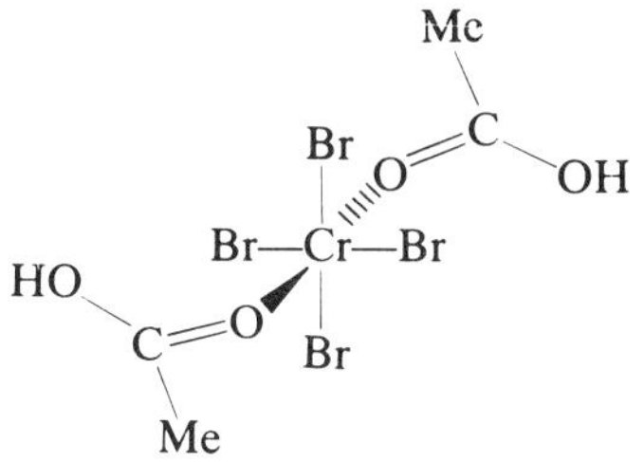

(72)

(*ii*) *Tetrachlorochromates(II): syntheses*

Complex salts of the general formula $A_2[CrCl_4]$, where A represents NH_4, K, Rb, Cs or an organic cation (usually unipositive), have been intensively investigated during the last decade particularly because many of them are weakly ferromagnetic. Ferromagnetism is well known for certain metals and alloys, but not for magnetically concentrated transition metal complexes, the great majority of which are antiferromagnets. Many tetrabromochromates(II) behave similarly and with the tetrachlorochromates(II) provide the most numerous class of ionic ferromagnets.[252]

The alkali metal tetrachlorochromates(II) can be crystallized from the melt (A = K, Rb, Cs)[253,254] or glacial acetic acid containing acetyl chloride (A = Rb, Cs, NH_4),[255] or obtained by thermal dehydration *in vacuo* of the dihydrates (A = NH_4, K, Rb, Cs)[256] (Scheme 24 and Table 33). Many salts of organic cations have been crystallized from concentrated hydrochloric acid or various non-aqueous solvents.[257–260] The complexes $[dienH_3][CrCl_4]Cl$ and $[trienH_4][CrCl_4]Cl_2$ are also tetrachlorochromates because, as indicated, the additional chlorides are ionic.[257] There is obviously scope for the preparation of salts of many more cations, and whereas the monoalkylammonium salts (and the few known monoarylammonium salts) are ferromagnetic, the more heavily substituted cations, *e.g.* NMe_2H_2 and pyH, produce different structures (see below) and antiferromagnetic tetrachlorochromates.

Large crystals of $Rb_2[CrCl_4]$ and $[N(CH_2Ph)H_3]_2[CrCl_4]$ have been grown; the former from the melt by the Bridgman and Czochralski methods, and the latter by careful temperature control from solutions in sealed tubes.[252,261]

(*iii*) *Ferromagnetic tetrachlorochromates(II): magnetic behaviour, structure and optical properties*

The ferromagnetic behaviour of these complexes was detected by magnetic moments which are considerably higher at room temperature ($\mu_{eff} \approx 5.7$ BM) than the spin-only value, and increase rapidly to *ca.* 9 BM at 90 K as the temperature is lowered.[262] At high temperatures, the reciprocal magnetic susceptibility follows the Curie–Weiss law with large positive intercepts on the absolute temperature axis, and it is apparent from the measurements that the Curie temperatures T_C lie a little below liquid nitrogen temperature. The susceptibility data become field dependent below the T_C values, which have been estimated for several complexes from plots of reduced magnetization against temperature. The persistence of short range order above T_C reduces the accuracy, and more accurate values have been obtained from neutron diffraction and optical spectroscopy (see below), which are zero field methods dependent on spontaneous magnetization.

Table 33 Tetrachlorochromates(II)

Complex	*Comments*	*Ref.*
$A_2[CrCl_4]$ Yellow to brown	A = NH_4 (colourless), K, Rb, Cs; ferromagnetic, $\theta = -65$ to $-85°$[a] $J = 8.5\ cm^{-1}$ (Rb),[b] $T_c = 60$–65 K (ref. 2), 52 K (ref. 4)	1–3
$A_2[CrCl_4]$[c] Silver-grey to green	A = NRH_3: R = Me, Et, Pr, n-C_5H_{11}, n-C_8H_{17}, n-$C_{10}H_{21}$, n-$C_{12}H_{25}$, Ph, o-MeC_6H_4, CH_2Ph; ferromagnetic, $\theta = -50$ to $-75°$; Me: $J = 9\ cm^{-1}$, $g = 2$, $T_C = 58$ K (ref. 5), $T_C = 41$ K (ref. 4)	3–9
	A = NMe_2H_2, pyH, NEt_3H, $\frac{1}{2}$pipzH; antiferromagnetic, $\theta = 29$ to $84°$	
$[enH_2][CrCl_4]$ Grey-green	From $CrCl_2{\cdot}4H_2O$ and $[enH_2]Cl_2$ in conc. HCl; ferromagnetic, $\theta = -64°$, $J = 4.9\ cm^{-1}$, $g = 1.94$	6
$[H_3N(CH_2)_3NH_3][CrCl_4]$ Yellow-green	From $CrCl_2{\cdot}4H_2O$ and $[tmdH_2]Cl_2$ in conc. HCl; ferromagnetic, $\theta = -67°$	7
$[dienH_3][CrCl_4]Cl$	From $CrCl_2{\cdot}4H_2O$ and $[dienH_3]Cl_3$ in conc. HCl; ferromagnetic, $\theta = -63°$, $J = 6.2\ cm^{-1}$, $g = 1.94$, $T_C \simeq 40$ K	6 10
$[trienH_4][CrCl_4]Cl_2$	From $CrCl_2{\cdot}4H_2O$ and $[trienH_4]Cl_4$ in conc. HCl; ferromagnetic, $\theta = -52°$, $J = 5.4\ cm^{-1}$, $g = 1.94$	6
$[N_2H_6]_2[CrCl_6]$ Green	From $CrCl_2$ and $[N_2H_6]Cl_2$ in $MeCO_2H$; antiferromagnetic, $\theta = 74°$	7
$Rb_2Cr_{1-x}Mn_xCl_4$	$0 < x < 0.5$, solid solutions from melts of mixed halides, Cr rich $0 < x < 0.41$, ferromagnetic; Mn rich $0.59 < x < 1$, antiferromagnetic	11
$Rb_2Cr_{1-x}Cd_xCl_4$	$x = 0$ (Rb_2CrCl_4) to *ca.* 0.5 ferromagnetic, $T_C = 53$ K (Rb_2CrCl_4), decreases with x	12

[a] Curie–Weiss law taken as $\chi_A^{-1} \alpha (T + \theta)$.
[b] $Ng^2\mu_B^2/\chi = \frac{1}{2}kT + J(-4 + 9y - 9.07y^2 + 55.73y^3 - 160.7y^4 + 116.6y^5 + \dots)$ where $y = J/kT$ ($H = -\Sigma JS_iS_j$) (ref. 8).
[c] ν(Cr—Cl) $\approx 300\ cm^{-1}$.

1. D. H. Leech and D. J. Machin, *J. Chem. Soc., Dalton Trans.*, 1975, 1609.
2. A. K. Gregson, P. Day, D. H. Leech, M. J. Fair and W. E. Gardner, *J. Chem. Soc., Dalton Trans.*, 1975, 1306.
3. L. F. Larkworthy, J. K. Trigg and A. Yavari, *J. Chem. Soc., Dalton Trans.*, 1975, 1879.
4. C. Bellitto, T. E. Wood and P. Day, *Inorg. Chem.*, 1985, **24**, 558.
5. C. Bellitto and P. Day, *J. Chem. Soc., Dalton Trans.*, 1978, 1207.
6. L. F. Larkworthy and A. Yavari, *J. Chem. Soc., Dalton Trans.*, 1978, 1236.
7. M. A. Babar, L. F. Larkworthy and S. S. Tandon, *J. Chem. Soc., Dalton Trans.*, 1983, 1081.
8. M. J. Stead and P. Day, *J. Chem. Soc., Dalton Trans.*, 1982, 1081.
9. C. Bellitto and P. Day, *J. Cryst. Growth*, 1982, **58**, 641.
10. M. A. Babar, L. F. Larkworthy, B. Sandell, P. Thornton and C. Vuik, *Inorg. Chim. Acta*, 1981, **54**, L261.
11. G. Münninghoff, W. Treutmann, E. Hellner, G. Heger and D. Reinen, *J. Solid State Chem.*, 1980, **34**, 289.
12. W. Kurtz, *Solid State Commun.*, 1982, **42**, 875.

The high temperature susceptibility data have been fitted to a series expansion formula suitable for a layer ferromagnetic (see below for structures) with $S = 2$, to give estimates of the nearest neighbour exchange integrals J, and some representative results for J and examples of T_C are in Table 33.

Values of J of 7 to $9\ cm^{-1}$ with g *ca.* 1.98 give reasonable fits for the chlorides. For the bromides J is generally larger (Table 35), indicating greater coupling through bridging bromide. J, however, varies little with increasing interlayer separation; thus there is strong magnetic coupling *via* the bridging halide between the cations in the layers and little interaction between the cations in adjacent layers so it is reasonable to assume that the magnetic properties are essentially two-dimensional.[259] Similarly, the Curie temperatures, where known, are not very sensitive to the interplanar spacing. The T_C values themselves are dependent on the method used to measure them. For example, T_C for $[NMeH_3]_2[CrCl_4]$ has been reported as 41 K from very low field magnetization experiments,[263] whereas higher field magnetization experiments and optical studies have given values of 58 K. There is no sign of a phase transition from heat capacity measurements[264] from 10 to 302 K although by comparison with data from the analogous Mn^{II} complex the corrected curve has a maximum at approximately 45 K, where the three-dimensional order breaks down as the temperature rises.

The alkali metal salts have the distorted K_2NiF_4 structure (*cf.* K_2CuF_4) in which there are sheets of chloride-bridged $(CrCl_4)^{2-}$ units separated by double layers of ACl from which chloride ions complete tetragonally elongated $(CrCl_6)$ octahedra. The principal axes of the octahedra lie in the sheets and are directed alternately at right angles. Related structures have been described from single crystal studies for $[NMeH_3]_2[CrCl_4]$, $[dienH_3][CrCl_4]Cl$ and $[tmdH_2][CrCl_4]$, in which the sheets $[CrCl_4]_n$ are separated by the organic cations (**73**) (and chloride also in the $dienH_3$ salt) (Table 34). In addition, the salts $A_2[CrCl_4]$ (A = $NMeH_3$,

$NEtH_3$ and $NPhH_3$) are isomorphous with the corresponding copper(II) complexes, which are known to have chloride-bridged sheets intercalated with the organic cations.

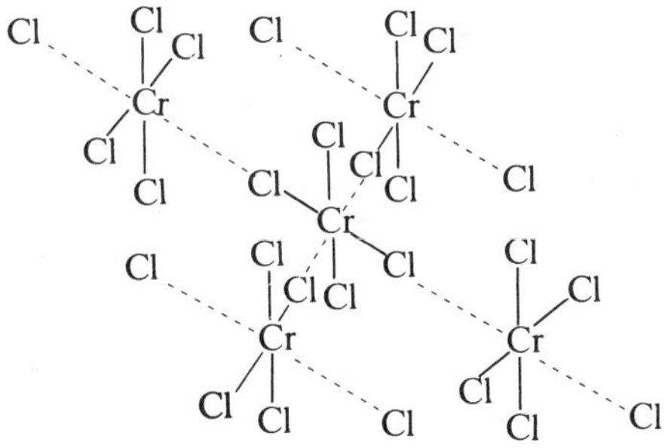

(73)

The ferromagnetic chloro complexes have unusual electronic spectra. Besides the usual broad, spin-allowed *d–d* transitions of Jahn–Teller distorted chromium(II), ${}^5B_{1g} \rightarrow {}^5A_{1g}(\nu_1)$(*ca.* 8000 cm^{-1}) and ${}^5B_{1g} \rightarrow {}^5B_{2g}$, ${}^5E_g(\nu_2)$ (*ca.* 11 250 cm^{-1}), the solid state spectra contain relatively intense spin-forbidden bands near 18 700 and 15 900 cm^{-1}. These bands all but vanish on cooling to 4.2 K. Owing to the density of triplet levels (see Tanabe–Sugano diagram) precise assignments are difficult, but the bands have been assigned to transitions to the ${}^3E_g({}^3H)$ and ${}^3A_{1g}({}^3G)$ levels. Transitions to these levels should be independent of ligand field strength, and in confirmation, ferromagnetic bromides (Table 35) have bands at 18 400 and 15 600 cm^{-1}. The greater nephelauxetic effect of bromide would account for the small decrease in wavenumber. The unusual intensity arises from the 'hot band' mechanism (exciton–magnon combination) by which, coincidently with the $\Delta S = -1$ change accompanying the optical excitation of one chromium(II) ion, there occurs a $\Delta S = +1$ change at an adjacent magnetically coupled ion, which drops from a thermally populated magnetic energy level to a lower one. Effectively, $\Delta S = 0$ and the spin-forbidden absorptions gain considerable intensity. At low enough temperatures, when all the metal ions are in the ferromagnetic ground state (all spins parallel), the intensity of the optical transition is zero because no compensating $\Delta S = +1$ changes can occur. Almost up to T_C, which has been estimated from the data, the intensity follows a T^2 law, and this has been explained theoretically.[252,263]

The transparency of the tetrachlorochromates(II) in the visible region could be of technological importance because other ferromagnets are opaque, but their relatively low Curie temperatures and air-sensitivity are severe limitations.

There are many pathways for magnetic interaction in 180° high-spin d^4 systems, some leading to ferromagnetic and others to antiferromagnetic coupling. The pathways have been described for the situation where two planar $[CrX_4]^{2-}$ units share a corner, and it has been calculated that the net result is likely to be antiferromagnetic coupling.[265] No complex with this structure has been isolated.

Parameters such as metal–metal separation, *d*-electron configuration, size and polarizability of bridging atom, bridging angle and planarity of the bridging system are important in determining whether the magnetic interaction will be ferromagnetic or antiferromagnetic. The local environments are very similar in the few ferromagnetic chromium(II) complexes for which structural data are available (Table 34). They all have layer structures, the bridges are approximately linear, the direct Cr—Cr separations are in the range 5.05–5.25 Å and the elongated distortion axes alternate at 90° in the plane of the metal ions. This cooperative Jahn–Teller distortion permits superexchange *via* bridging chloride anions between a half-filled d_{z^2} orbital on one metal ion and the empty $d_{x^2-y^2}$ orbital of a neighbour throughout the lattice. Departure of the bridging angles from 180° will affect the degree of interaction but too few structures have been determined for any pattern to emerge.

(*iv*) *Antiferromagnetic tetrachlorochromates*(*II*)

Several complexes: $A_2[CrCl_4]$ (A = pyH, NMe_2H_2 or NEt_3H) and $[N_2H_6]_2[CrCl_6]$ are antiferromagnets. This behaviour and the reflectance spectra of the complexes suggest that they contain distorted six-coordinate Cr^{II}. One structure has been determined; that of

Table 34 Anhydrous and Complex Halides and Other Chromium(II) Complexes Showing 4 + 2 Coordination

Complex[a,b]	*Coordination sphere* (Å)	*Ref.*
CrF_2	Cr—F, 4 at 1.98, 2 at 2.43	1
$CrCl_2$	Cr—Cl, 4 at 2.39, 2 at 2.92	1
$CrBr_2$	Cr—Br, 4 at 2.54, 2 at 3.00	1
CrI_2	Cr—I, 4 at 2.740, 2 at 3.237[c]	1, 2
$KCrF_3$	Cr—F, 2 at 2.00, 4 at 2.14[d]	3
$SrCrF_4$	Cr—F, 4 at 1.980, planar	4
$Cr^{II}Cr^{III}F_5$	Cr^{II}—F, 2 at 1.955, 2 at 2.572 2 at 2.010	5
$CrCl_2{\cdot}4H_2O$	Cr—O, 2.078 (av), Cr—Cl, 2.758	6
$[C(NH_2)_3][CrBr_4(MeCO_2H)_2]$	Cr—Br, 2 at 2.637, 2 at 2.839 Cr—O, 2 at 2.06	7
Cs_2CrCl_4	Cr—Cl, 4 at 2.609, 2 at 2.399[e]	8
Rb_2CrCl_4	Cr—Cl, 2 at 2.35, 2 at 2.70 2 at 2.38	9, 10
$[NMeH_3]_2[CrCl_4]$	Cr—Cl, 4 at 2.41, 2 at 2.82	10
$[NH_3(CH_2)_3NH_3][CrCl_4]$	Cr—Cl, 2 at 2.39, 2 at 2.87 2 at 2.41	11
$[NH_3(CH_2)_3NH_2(CH_2)_3NH_3][CrCl_5]$	Cr—Cl, 2 at 2.39, 2 at 2.79 2 at 2.41	12
$[NMe_2H_2]_2[CrCl_4]$[f] (**74**)	$[Cr_3Cl_{12}]^{6-}$ units Cr(central)—Cl, 4 at 2.62, 2 at 2.40 Cr(terminal)—Cl, 2 at 2.37, 1 at 2.70 2 at 2.44, 1 at 3.22	11
$[Cr_3(O_2CH)_6(H_2O)_2]$ (**32**)	Cr_2^{4+} units,[g] and Cr^{2+} with Cr—O, 2 at 2.022, 2 at 2.499 2 at 2.063	13
$[Cr_3(\eta^5\text{-}C_5H_5)_2(O_2CCF_3)_6]$ (**70**)	Cr—O, 4 at 2.01, 2 at 2.58 (central Cr^{2+})	14
$[Cr(C_7H_4NO_3S)_2(H_2O)_4]$ (**8**)	Cr—O, 2 at 2.048, 2 at 2.396 Cr—N, 2 at 2.196	15
CrS	Cr—S, 4 at 2.45, 2 at 2.88	16

[a] Other Cr^{II} complexes for which single crystal X-ray structural information is available are $[Cr(NO)\{N(SiMe_3)_2\}_3]$ (**135**), $[CrMe_2(dmpe)_2]$ (**19**), $[Cr(\eta^5\text{-}C_5H_5)_2(OBu^t)_2]$ (**21**), $Cr(OCBu^t_3)_2{\cdot}LiCl(THF)_2$ (**22**) and the complexes in Tables 41 and 95

[b] For the low-spin ($S = 1$) complexes $Na_4[Cr(CN)_6]{\cdot}10H_2O$, $[Cr(CNPh)_6](PF_6)_2$ and $[Cr(TPP)(py)_2]$ see Tables 3, 4 and 102 and for $[Cr(nic)_2(H_2O)_4]$ (**6**) Section 35.4.3.4.

[c] Monoclinic form, isotypic with (Mn, Cr)I_2; similar Cr—I distances in orthorhombic phase.

[d] From powder data.

[e] Later work[9,10] suggests that the tetragonally compressed structure is unlikely.

[f] For $ACrX_3$ and A_2CrI_6 see Table 36.

[g] For structures of other Cr_2^{4+} species see Section 35.3.5.5.

1. F. Besrest and S. Jaulmes, *Acta Crystallogr., Sect. B*, 1973, **29**, 1560.
2. L. Guen and N.-H.-Dung, *Acta Crystallogr., Sect. B*, 1976, **32**, 311.
3. A. J. Edwards and R. D. Peacock, *J. Chem. Soc.*, 1959, 4126.
4. H. G. von Schnering, B. Kolloch and A. Kolodziejczyk, *Angew. Chem., Int. Ed. Engl.*, 1971, **10**, 413.
5. R. Colton and J. H. Canterford, 'Halides of the First Row Transition Metals', Wiley-Interscience, London, 1969, chap. 4.
6. H. G. von Schnering and B.-H. Brand, *Z. Anorg. Allg. Chem.*, 1973, **402**, 159.
7. M. A. Babar, M. F. C. Ladd, L. F. Larkworthy, G. A. Leonard and D. C. Povey, *J. Crystallogr. Spectrosc. Res.*, 1983, **13**, 325.
8. M. T. Hutchings, A. K. Gregson, P. Day and D. H. Leech, *Solid State Chem.*, 1974, **15**, 313.
9. P. Day, M. T. Hutchings, E. Janke and P. J. Walker, *J. Chem. Soc., Chem. Commun.*, 1979, 711.
10. C. Bellitto, T. E. Wood and P. Day, *Inorg. Chem.*, 1985, **24**, 558.
11. M. A. Babar, M. F. C. Ladd, L. F. Larkworthy, D. C. Povey, K. J. Proctor and L. J. Summers, *J. Chem. Soc., Chem. Commun.*, 1981, 1046.
12. M. A. Babar, M. F. C. Ladd, L. F. Larkworthy, G. P. Newell and D. C. Povey, *J. Cryst. Mol. Struct.*, 1979, **8**, 43.
13. F. A. Cotton, M. W. Extine and G. W. Rice, *Inorg. Chem.*, 1978, **17**, 176.
14. F. A. Cotton and G. W. Rice, *Inorg. Chim. Acta*, 1978, **27**, 75.
15. F. A. Cotton, G. E. Lewis, C. A. Murillo, W. Schwotzer and G. Valle, *Inorg. Chem.*, 1984, **23**, 4038.
16. A. F. Wells, 'Structural Inorganic Chemistry', 5th edn., Clarendon, Oxford, 1984, p. 770.

$[NMe_2H_2]_2[CrCl_4]$.[266] It does not have a layer structure, but contains isolated trimeric units $[Cr_3Cl_{12}]^{6-}$ in which the metal ions are linearly arranged and antiferromagnetically coupled (**74**). The central chromium atom can be considered to be surrounded by a tetragonally flattened octahedron of chlorine atoms, which bridge to the terminal metal atoms. These form four short and two long Cr—Cl bonds. There is considerable distortion of the angles at the metal atoms from 90°. Now that the structure is known, it is apparent that the successful fitting

Table 35 Bromochromates(II)

Complex	*Comments*
A_2CrBr_4 Dark green or brown	A = Cs, ferromagnetic, $\mu_{eff}^{295\,K} = 5.0$ BM, $\theta = -44°$ A = Rb, NH_4, antiferromagnetic, $\mu_{eff}^{295\,K} \approx 4.2$ BM, $\theta = 82°$ and 94°
A_2CrBr_4 Greenish yellow	[a]A = NRH_3: R = Me, Et, Pr, n-C_5H_{11}, n-C_8H_{17}, n-$C_{12}H_{25}$; ferromagnetic, $\mu_{eff}^{295\,K} \approx 5.8$ BM, $\theta = -87$ to $-100°$, $J \approx 10$ cm^{-1}, $g = 1.94$, $T_C \approx 58$ K A = $NPhH_3$, pyH, NMe_2H_2, NEt_4; antiferromagnetic, $\mu_{eff}^{295\,K} \approx 4.0$ BM, $\theta = 35$ to 90°
$[pipzH_2]_2[CrBr_6]$ Light green	Thermal dehydration of dihydrate, $\mu_{eff}^{295\,K} = 4.75$ BM, $\theta = 4°$

[a] ν(Cr—Br) = 220–270 cm^{-1}.

of the magnetic data to the equation for a sheet antiferromagnet was fortuitous[257] and illustrates the difficulty often experienced in distinguishing between possible structures magnetochemically.

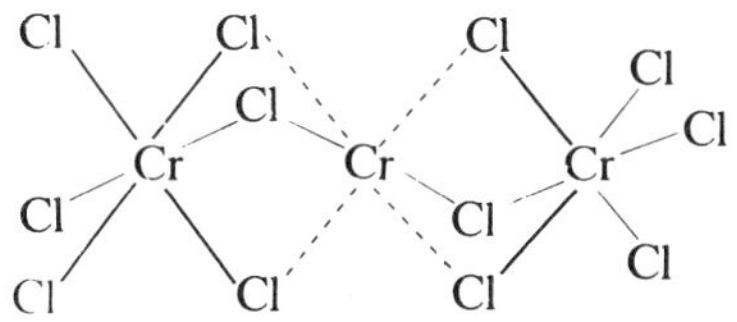

(**74**) $[Cr_3Cl_{12}]^{6-}$

There is evidence for the formation of a trimeric anion $[Cr_3Cl_{10}]^{4-}$ ($\mu_{eff} = 4.78$ BM) in solutions of $CrCl_2$ and $[NBu_4]Cl$ in CH_2Cl_2.[267] The antiferromagnetism of $[N_2H_6]_2[CrCl_6]$ suggests that its lattice does not contain $[CrCl_6]^{4-}$ units, since this would lead to magnetically dilute behaviour, but contains ionic chloride and halide-bridged complex anions. The powder photograph of $[NEt_3H]_2[CrCl_4]$ bears no resemblance to that of $[NEt_3H]_2[CuCl_4]$.[260] The latter contains discrete distorted tetrahedral $[CuCl_4]^{2-}$ anions for which no Cr^{II} analogues are known.

(*v*) *Tetrabromochromates(II)*

Like the chlorides, some salts A_2CrBr_4 are ferromagnetic and others antiferromagnetic. They can be prepared by thermal dehydration, but non-aqueous procedures are preferable (Scheme 24). Salts of enH_2^{2+}, $dienH_3^{3+}$ and p-$MeC_6H_4NH_3^+$ could not be obtained. Compared with the chlorides, few investigations have been carried out. The ferromagnetic complexes generally have slightly higher moments and larger J values than the chlorides suggesting that the magnetic interaction is greater. The Curie temperatures are known approximately for $[NRH_3]_2[CrBr_4]$ (R = Me, Pr or n-$C_{12}H_{25}$) and are similar to those of the chlorides[268,269] (Table 35). No structural data have been reported but there seems little doubt that the ferromagnetic complexes will be found to have sheet structures. Since J and T_C do not vary significantly with R, the interplanar coupling between the metal ions is much weaker than the intraplanar coupling as in the chlorides.

Many of the tetrabromochromates(II), *i.e.* the Rb, NH_4, $NPhH_3$, pyH, NEt_4 and NMe_2H_2 salts, are antiferromagnets. Since the chloro analogues of the first three are ferromagnets, structural comparisons would be informative.

Whether ferromagnetic or antiferromagnetic, the salts A_2CrBr_4 have a very broad band near 10 500 cm^{-1}, asymmetric or with a shoulder to lower wavenumber, in their reflectance spectra. These typical spectra of six-coordinate Cr^{II}, with the absorptions at lower wavenumber than in the chlorides (*ca.* 11 000 cm^{-1}), confirm that the $[CrBr_4]^{2-}$ units are polymerized. There are many spin-forbidden bands to higher wavenumber, including two bands near 18 400 and 15 600 cm^{-1}, which are unusually intense in the spectra of the ferromagnetic complexes. Their intensity presumably arises from the same mechanism as for the chlorides.

(vi) Trihalochromates(II): syntheses and structures

Complexes of the general formula AMX_3, where A is a unipositive cation, M a bivalent metal ion and X a halide ion (Cl^-, Br^- or I^-) crystallize in several different lattice types, but have structures comprising infinite parallel chains $(MX_3)_n^{n-}$ of face-sharing octahedra separated by the cations A (**75**). Since the metal–metal separation in the chains is 3.1 to 3.5 Å, increasing from Cl to Br to I, and the distance between chains is much greater at approximately 8 Å, there is strong intrachain interaction, but weak interchain interaction when M is a paramagnetic transition metal ion.[270] In addition, the M—Cl—M bridge angles are approximately 90° so that the structures of the trihalochromates(II) contrast sharply with those of the layer tetrahalochromates(II).

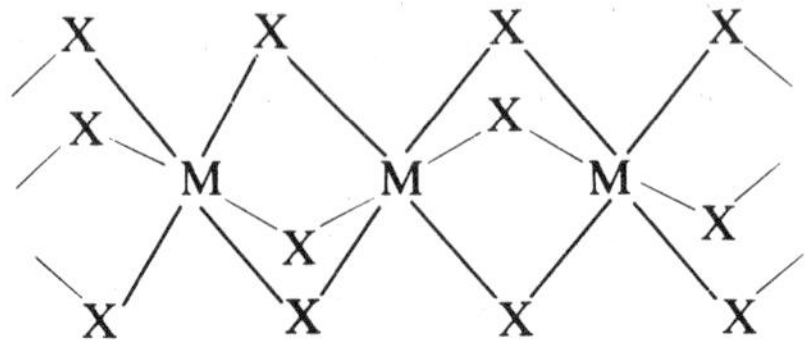

(**75**) $(MX_3)_n^{n-}$

The alkali metal trihalochromates, $ACrX_3$ (X = Cl, Br) can be crystallized from a melt of the constituent anhydrous halides[271–273] or from non-aqueous solvents, especially acetic acid. The latter method is particularly useful in the preparation of salts of non-metallic cations $ACrCl_3$ (A = NH_4, NMe_4, pyH and NMe_2H_2).[255] Thermal dehydration in vacuum has also been used to produce $pyHCrCl_3$ and $CsCrCl_3$[256] from $[pyH]_2[CrCl_4(H_2O)_2]$ and $CsCrCl_3(H_2O)_2$. The triiodides $ACrI_3$ (A = Rb, Cs[274]) and the hexaiodides $(Tl\ or\ In)_4CrI_6$[275] have been crystallized from the melt. There is no report of attempts to crystallize iodides from non-aqueous solvents.

There is agreement over the gross structural features of these linear chain chromium(II) complexes (and the isomorphous copper(II) analogues), but different models have been used for structural refinement. Phase transitions associated with the Jahn–Teller distortion (α to β to γ as the temperature is lowered) complicate the structural studies.

In the first full structural investigation, that of $CsCrCl_3$ (295 K, hexagonal, α phase), it was concluded that the space group was $P6_3mc$ and the $[CrCl_6]$ octahedra were of C_{3v} symmetry with three short and three long bonds (Table 36), *i.e.* there is no evidence of a static Jahn–Teller distortion, unlike in $CsCuX_3$ (X = Cl, Br).[271] Single crystal X-ray data for $CsCrBr_3$ were analyzed on the same model.[272] If the site symmetry of the Cr^{2+} ion is C_{3v}, the degeneracy of the E_g ground level is not removed and this modifies the assignments of the electronic spectra (see below). However, the space group of the α phases of $ACrX_3$ cannot be uniquely determined from the data,[271,273] and the choice of a disordered $P6_3/mmc$ space group rather than $P6_3mc$ is more logical in relation to the expected operation of the Jahn–Teller effect. With the assumption of pseudo D_{4h} symmetry at each Cr^{2+}, and a random distribution of elongated octahedra, better refinement results for α-$CsCrCl_3$ and α-$CsCrI_3$. The large anisotropic thermal motion of anions parallel to the *c* axis is now attributed to the disorder. The same disordered Jahn–Teller distorted structure is expected for α-$CsCrBr_3$ and α-$RbCrX_3$ (X = Cl, Br or I).[273]

In β-$RbCrCl_3$ the deformation around Cr-1 is elongation along a particular axis, but elongations along other axes average to an apparent compression about Cr-2.[276,277] In γ-$RbCrCl_3$ the elongated axes alternate along different directions.[276]

Whereas the Mössbauer spectra at 77 K of $CsM^{139}I_3$ (M = V, Mn) can be fitted assuming quadrupole splitting at a single metal site, two sites ($a/b = 2/1$), with coupling constants of opposite sign, must be assumed when M = Cr. In the *a* and *b* sites the ^{129}I atoms are in asymmetric and symmetric bridging positions respectively, consistent with axially distorted octahedra. The α to β phase transition (Table 36) is confirmed from X-ray powder diffraction, but at 165 ± 5 K.[278]

(vii) Trihalochromates(II): electronic spectra and magnetic behaviour

The single crystal electronic absorption spectra of $CsCrX_3$ (X = Cl or Br) have been examined in some detail. There are intense broad bands at 22 000 and 11 500 cm^{-1} with a

Table 36 Trihalochromates(II) and Hexaiodochromates(II)

Complex	*Comments*	*Ref.*
$ACrCl_3$	$A = K$, $\mu_{eff}^{290\,K} = 4.56$ BM, $\theta = 45°$, $J = -2.8$ to $-3.5\ cm^{-1}$	1–5
$A = K, Rb, Cs,$	$A = Cs$, $\mu_{eff}^{290\,K} = 3.41$ BM, $T_N \approx 140$ K, $J = -17.4$ to $-18.4\ cm^{-1}$	
(brownish green) NH_4,	$A = NMe_4$, $\mu_{eff}^{295\,K} = 3.81$ BM, $T_N \approx 104$ K, $J = -12.5\ cm^{-1}$	
NMe_4, NMe_2H_2, pyH	$A = pyH$, $\mu_{eff}^{295\,K} = 4.44$ BM, $\mu_{eff}^{90\,K} = 3.05$ BM	
Yellow	$A = NMe_2H_2$, $\mu_{eff}^{295\,K} = 3.80$ BM, $\mu_{eff}^{90\,K} = 2.46$ BM, $J \approx -11\ cm^{-1}$	
	Phase changes: $\alpha \xrightarrow[170\ K,\ A=Cs]{470\ K,\ A=Rb} \beta \xrightarrow{201\ K,\ A=Rb} \gamma$ hexagonal monoclinic monoclinic	6
	C_{3v} symmetry, Cr—Cl: 2.419 (3×) and 2.618 (3×), Cr—Cr = 3.112 Å ($CsCrCl_3$)	5
	D_{4h} symmetry, Cr—Cl: *ca.* 2.41 (4×) and *ca.* 2.72 (2×) Å	6
$ACrBr_3$	C_{3v} symmetry, Cr—Br: 2.574 (3×) and 2.780 (3×), Cr—Cr = 3.253 Å ($CsCrBr_3$)	5
$A = K, Rb, Cs, NMe_4$		2, 10
Yellow or brown		
$ACrI_3$	β-$CsCrI_3$, $T_C \approx 27$ K, $J/k = -14$ K	7
$A = Rb, Cs$	$RbCrI_3$, $J/k = -11$ K	
	Phase changes: $\alpha \xrightarrow{293\ K,\ A=Rb} \beta \xrightarrow{1.2\ K,\ A=Rb} \gamma$ hexagonal monoclinic monoclinic $\alpha \xrightarrow{150\ K,\ A=Cs} \beta$ hexagonal orthorhombic	
	D_{4h} symmetry, Cr—I, 2.79 (4×), 3.07 (2×) Å	6, 7
	C_{3v} symmetry, Cr—I, 2.83 (3×), 2.94 (3×) Å	8
A_2CrI_6	Phase changes: α-$Tl_4CrI_6 \xrightarrow{<77\ K} \beta$-form tetragonal orthorhombic	9
$A = Tl, In$		
	$A = Tl$, $\mu_{eff} = 4.5$ BM, $\theta = 7°$, $T_N = 2.7$ K	
	$A = In$, $\mu_{eff} = 4.6$ BM, $\theta = 8°$, $T_N = 2.9$ K	

1. D. H. Leech and D. J. Machin, *J. Chem. Soc., Dalton Trans.*, 1975, 1609.
2. H.-D. Hardt and G. Streit, *Z. Anorg. Allg. Chem.*, 1970, **373,** 97.
3. L. F. Larkworthy, J. K. Trigg and A. Yavari, *J. Chem. Soc., Dalton Trans.*, 1975, 1879.
4. L. F. Larkworthy and A. Yavari, *J. Chem. Soc., Dalton Trans.*, 1978, 1236.
5. T.-I. Li and G. D. Stucky, *Acta Crystallogr., Sect. B*, 1973, **29,** 1529.
6. W. J. Crama and H. W. Zandbergen, *Acta Crystallogr., Sect. B*, 1981, **37,** 1027.
7. H. W. Zandbergen and D. J. W. Ijdo, *J. Solid State Chemistry*, 1981, **38,** 199.
8. L. Guen, R. Marchand, N. Jouini and A. Verbaere, *Acta Crystallogr., Sect. B*, 1979, **35,** 1554.
9. H. W. Zandbergen, *J. Solid State Chem.*, 1981, **38,** 239.
10. C. Bellitto, D. Fiorani and S. Viticoli, *Inorg. Chem.*, 1985, **24,** 1939.

weaker broad band at 6500 cm^{-1}. The last two bands have been assigned[273] to the usual $^5B_{1g} \rightarrow {}^5B_{2g}$, ^5E_g and $^5B_{1g} \rightarrow {}^5A_{1g}$ transitions (Figure 2, D_{4h} symmetry), but because the E_g and T_{2g} levels of O_h symmetry remain unsplit in C_{3v} symmetry, an alternative assignment of the bands at 11 500 and 6500 cm^{-1} is to the $^5E \rightarrow {}^5T_2$ transition and the spin-forbidden $^5E \rightarrow {}^3T_1$ transition respectively.[279] The alternative seems less likely in view of the recent crystallographic work, which indicates pseudo D_{4h} symmetry for the $[CrCl_6]$ units in all phases.

The band at 22 000 cm^{-1} (anomalous because other Cr^{II} complexes of Cl^- and Br^- do not absorb strongly here) has been assigned to simultaneous excitation of a pair of antiferromagnetically coupled Cr^{II} ions so that the absorption is at *ca.* $2 \times 11\,500\ cm^{-1}$.

A number of weak sharp bands in the region 15 000 to 25 000 cm^{-1} have been assigned to spin-forbidden transitions and charge transfer is found beyond 25 000 cm^{-1}.[279] The iodide $CsCrI_3$ absorbs strongly throughout the visible–near IR region; a shoulder at 10 000 cm^{-1} may be a *d–d* transition.[270]

The trihalochromates(II) behave antiferromagnetically and the temperature-dependence of susceptibility has been analyzed in terms of linear chains of interacting spins (high spin Cr^{II}, $S = 2$, Table 36). Direct exchange should increase as the Cr—Cr separation decreases (I → Br → Cl), while superexchange *via* the bridging anions should increase with increasing covalency (Cr → Br → I). The second effect predominates in $CsNiX_3$,[270] but there have not been enough investigations for correlations to appear in the Cr^{II} case.

The only known hydrated trichlorochromate is $Cs[CrCl_3(H_2O)_2]$ (Table 32), the antiferromagnetic behaviour of which can be reproduced by substitution of $J = -3.0\ cm^{-1}$ and $g = 1.93$ in the expression for a binuclear compound with $S = 2$. This suggests[256] that it has the structure found in copper(II) compounds of the same stoichiometry in which binuclear $Cu_2Cl_6(H_2O)_2$ units, formed by symmetrical chloride bridges, are weakly linked in chains by further chloride bridges.

(*viii*) *Pentahalochromates(II)*

When sodium chloride is fused with anhydrous chromium(II) chloride the product is $NaCrCl_5$ regardless of the proportions of chlorides used.[222] Other cations give tetra- or tri-chlorochromates(II) as in Sections 35.3.7.3.ii and 35.3.7.3.vi above. The solvate $[pyH]_3[CrBr_5]\cdot 2MeCO_2H$ has been isolated from the metal acetate and pyridine in a mixture of acetyl bromide and acetic acid, and there is considerable splitting of the spin-allowed *d–d* band in the reflectance spectrum of this complex,[255] but no detailed investigations of pentahalochromates have been reported.

(*ix*) *Hexaiodochromates(II)*

The complexes A_2CrI_6 (A = Tl or In) contain isolated $[CrI_6]^{4-}$ units; in the α phase there is a random distribution of Jahn–Teller distorted octahedra, which are elongated perpendicular to the *c* axis, and below 77 K the β-phase forms in which the directions of elongation are ordered (Table 36). Refinement leading to axially compressed octahedra in the $[CrI_6]^{4-}$ units [Cr—I, 3.046 Å (4 ×), 2.738 Å (2 ×)][280] is said[275] to be due to an inadequate model.

35.3.8 Hydrogen and Hydrides as Ligands

The borohydride $Cr(BH_4)_2\cdot 2THF$ has been prepared from $[Cr(OBu^t)_4]$ and B_2H_6 (Section 35.3.4.2) and from $CrCl_3$ (p. 725). CrH_2 has a cubic close packed structure and in hexagonal close-packed CrH the CrH distance is 1.91 Å.[281] The hydride $[CrH_2\{P(OMe)_3\}_5]$ is discussed in Section 35.3.4.1.

35.3.9 Mixed Donor Atom Ligands

Several types of mixed donor ligand (N—O, C—O) bridge the $(Cr—Cr)^{4+}$ unit in dinuclear species (Section 35.3.5.5).

35.3.9.1 Complexes of β-ketoamines and Schiff's bases

(*i*) *β-Ketoamines*

Chromium(II) complexes (**76**) to (**79**) have been made[282,283] by the procedure outlined in Scheme 16 for the preparation of $[Cr(dpm)_2]$. Complexes (**76**; R = Me, Pr^i) are thought to be planar because they are high-spin (Table 37), with solution and solid state spectra[282] like those of planar high-spin $[Cr(dpm)_2]$. One member (R = Bu^n) of the related series (**78**) and [Cracacen] (**79**), have been shown to be planar by single crystal X-ray studies; resemblances between the electronic spectra suggest that series (**78**) are all planar. Unlike the nickel(II) series analogous to (**78**), bulky R groups do not induce the formation of tetrahedral chromium(II) presumably because Cr^{2+} is larger than Ni^{2+}

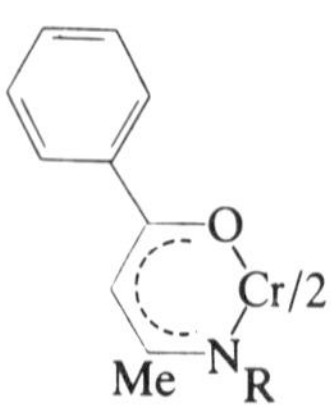

(**76**) R = Me, Pr^i

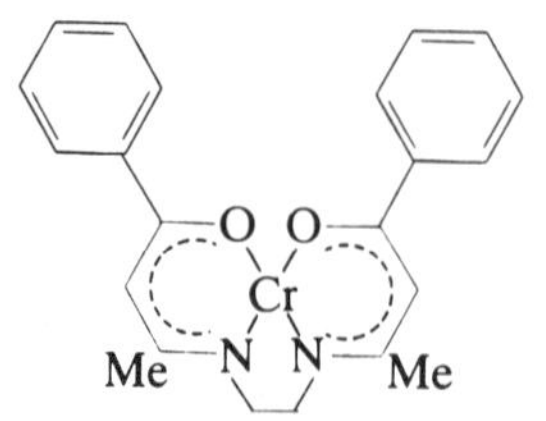

(**77**) Cr(bzacen)

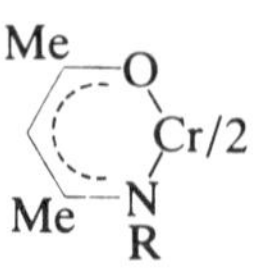

(**78**) R = Me, Et, Pr^n, Pr^i, Bu^n

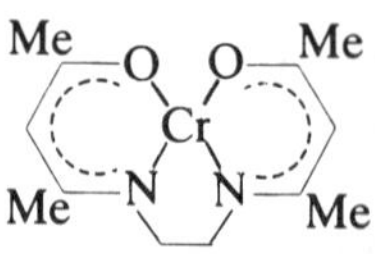

(**79**) Cr(acacen)

Table 37 Complexes of β-Ketoamines

Complex	Comments			Ref.
(**76**)				
R = Me	$\mu_{eff}^{297\,K} = 4.74$ BM; 4.62 BM in toluene; $\bar{\nu} = 18\,700$ cm^{-1},			1
Red-brown	$\varepsilon = 200$, C_6H_6			
R = Pr^i	$\mu_{eff}^{292\,K} = 4.92$ BM, 4.85 BM in toluene; $\bar{\nu} \approx 19\,500$sh cm^{-1},			1
Red-brown	$\varepsilon \approx 260$, toluene			
(**77**)				
[Cr(bzacen)]	$\mu_{eff}^{292\,K} = 2.22$ BM (solid); $\bar{\nu} = 18\,500$ cm^{-1},			1
Red-brown	$\varepsilon \approx 300$, toluene			
(**78**)	$\mu_{eff}^{293\,K}$ (BM)	$\mu_{eff}^{90\,K}$ (BM)	θ(°)	2
R = Me	4.26	2.99	126	
R = Et	4.49	3.16	147	
R = Pr	4.50	3.83	42	
R = Pr^i	4.69	4.31	24	
R = Bu^n	4.31	3.99	23	
Planar CrO_2N_2:	Cr—O (av) = 1.961, Cr—N (av) = 2.086 Å			
(**79**) [Cr(acacen)]	4.66	3.88	50	2
Planar CrO_2N_2:	Cr—O (av) = 1.972, Cr—N (av) = 2.012 Å			

1. D. H. Gerlach and R. H. Holm, *Inorg. Chem.*, 1970, **9**, 588.
2. L. F. Larkworthy, D. C. Povey and B. Sandell, *Inorg. Chim. Acta*, 1984, **83**, L29.

Series (**78**) and [Cracacen] have magnetic moments at room temperature which are below the value of 4.90 BM expected for high-spin chromium(II) and decrease still further at lower temperatures. Such magnetic behaviour is usually associated with antiferromagnetic interaction, but the crystal structures do not indicate any obvious pathways for the interaction. It is possible that the magnetic data can be accounted for by the progressive formation of low-spin forms ($S = 1$ and/or 0) as the temperature is lowered.[283] The magnetic moment of 2.22 BM found[282] for (**77**) [CrBzacen] at room temperature may imply strong antiferromagnetic interaction, or that the molecule is predominantly in a low-spin state. Further structural and magnetic work is needed. Complexes (**76**) react with chlorinated solvents to form chromium(III) species which have not been characterized (Table 37).

(ii) *Schiff's base complexes*

Chromium(II) chelates of 3-(N-2-furfuralideneimino)propionic acid (HFP), o-(N-α-furfuralidene)benzoic acid (HFB), and o-(N-α-furfuralideneimino)ethanesulfonic acid (HFE)[284–286] have been found to have magnetic moments of approximately 4.8 BM at room temperature. A band near 14 500 cm^{-1} ($CHCl_3$ solution) has been assigned to the $^5E_g \rightarrow {}^5T_{2g}$ transition and formation constants have been measured. However, there is no indication of difficulties arising from the air sensitivity of chromium(II) solutions.

A low-spin chromium(II) complex of pyruvic acid thiosemicarbazide has been reported (Table 96).[287]

(iii) *Complexes of polymeric Schiff's bases*

Chromium(II) complexes of the polymeric Schiff's bases (**80**) are said[288,289] to be produced when chromium(II) acetate dissolved in DMF is added under nitrogen to a hot solution of the ligand in which sodium acetate is suspended. The yellow-green complexes are insoluble in all common solvents, are surprisingly stable to air, and oxidation to the Cr^{3+} complex with hydrogen peroxide occurs only on heating. The magnetic moments are approximately 3.0 BM at room temperature, indicating a low-spin d^4 configuration. In view of the difficulties in the preparation of complexes of β-ketoamines, it would seem possible that the polymeric complexes are of chromium(III), and cross-linked oxo or hydroxo structures could, through antiferromagnetic interaction, lead to moments below the 3.87 BM expected for chromium(III) complexes. It is true that direct preparation of the chromium(III) complex of (**80**; R = —$(CH_2)_3NMe(CH_2)_3$—) gives a product with an essentially normal moment of 3.6 BM but in this case coordinated chloride is present. Irrespective of their precise nature, the chromium Schiff's base complexes have molecular sieve properties, and as gas chromatographic stationary phases can be used[290] for the separation of noble gases and oxygen at room temperature.

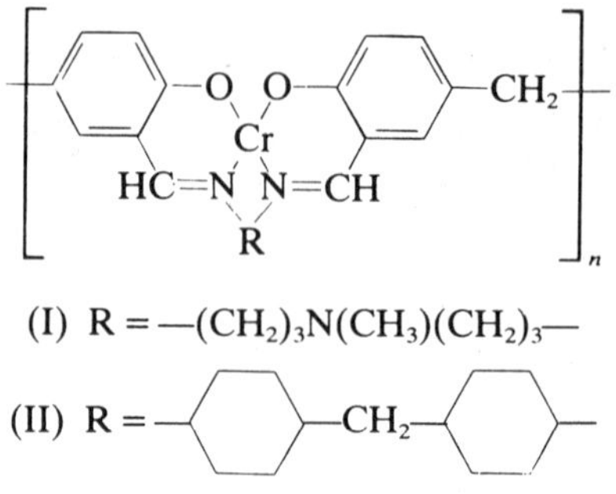

(80)

35.3.9.2 *Amino acids*

The glycinato complex $[Cr(gly\text{-}o)_2]\cdot H_2O$ has been prepared by mixing aqueous solutions of sodium glycinate and chromium(II) sulfate[85] or aqueous lithium glycinate with ethanolic $CrCl_2\cdot 4H_2O$. The second method reduces the chance of hydrolysis to brown $Cr(OH)_2$ and has been used with minor modifications in the preparations of the methionato, *S*-methyl-L-cysteinato and histidinato complexes: $[Cr(met\text{-}o)_2]$, $[Cr(mcys\text{-}o)_2]$ and $[CrCl(his\text{-}o)]\cdot H_2O$.[100] The general similarity in the spectra (Table 38) of the bis(amino acidato) complexes suggests that the metal environment is the same in each, *i.e.* NH_2 and CO_2^- coordination. The weak antiferromagnetic interaction apparent from the small variation of magnetic moment is likely to be due to extended structures arising through carboxylate bridges. The histidinato complex $[CrCl(his\text{-}o)]\cdot H_2O$ shows stronger interaction and may be chloro bridged.

Table 38 Amino Acid Complexes

Complex	μ_{eff} (BM) 295 K	90 K	θ (°)	Reflectance spectra (cm^{-1}) ν_2	ν_1
$[Cr(gly\text{-}o)_2]\cdot H_2O$ Pale violet	4.61	4.52	6	16 300	12 000sh
$[Cr(met\text{-}o)_2]$ Pale violet	4.76	4.61	11	17 000	12 100sh
$[Cr(mcys\text{-}o)_2]$ Pale violet	4.67	4.54	8	16 800	13 100sh
$[CrCl(his\text{-}o)]\cdot H_2O$ Blue-violet	3.70	3.38	24	18 000	12 700sh

A green, air-stable homocystine complex, $[Cr\{(\text{—}SCH_2CH_2CHNH_2CO_2)_2\}_2]$ ($\mu_{eff} = 4.83$ BM, $\bar{\nu}_{max} = 40\,800$, 39 200 and 14 300 cm^{-1}), has been prepared from aqueous $CrCl_2$ and homocystine (pH 8, $NaHCO_3$).[291]

Table 39 contains some stability constant data for chromium(II) and amino acids in aqueous solution.

35.3.9.3 *Complexones*

The chromium(II)–edta system is powerfully reducing (the half-wave potential is −1.48 V at pH 12 *vs.* SCE) and has been used in the reduction of iron–sulfur clusters.[292] No solid complex has been isolated because of its instability to oxidation, but Cr^{II}–edta is high-spin in aqueous solution ($\mu_{eff} = 5.12$ BM) and its stability constant has been determined. The edta is believed to be pentadentate with H_2O in the sixth position.[293]

Stability constants, usually determined pH-metrically, for complexation between Cr^{2+}(aq) and

Table 39 Stability Constants of Chromium(II) Complexes in Aqueous Solution

Ligand	*log β (CrL)*	*log β (CrL₂)*	*log β (CrLH)*	*Ref.*
Glycine	4.21	7.27		1
	7.72	15.26		2
β-Alanine	3.89			1
	7.53			2
Oxalate[a]	3.85	6.81		2
Malonate	3.57	5.49	6.45	1
	3.92	7.13		2
Acetylacetone	5.96	11.70		3
Ethylenediamine	5.48	9.63		1
	5.15	9.19		4
Diethylenetriamine	6.71	9.40		5
Triethylenetetramine	7.71			5
Iminodiacetate	5.01	8.18		1
Nitrilotriacetate[b]	6.52	9.66		1
Ethylenediaminetetraacetate	12.7		16.18	1
	13.6		16.6	6
Thiodiacetate	3.00	5.39		7
Ethanedithioacetate	1.99			8
Diethylenetrithiodiacetate	2.33			9
Salicylate	8.41	15.36		10
5-Sulfosalicylate	9.89			2
	7.14	12.88		10
8-Hydroxyquinoline	9.05			2
Bipy	4	6.4 (CrL_3, 3.5)		11
	4.61 (in HMPA)			12
OH^-		−5.3 [log β (CrH_{-1}) form]		1

[a] Values for acetate and formate are in Section 35.3.5.5.
[b] Some dimerization, log β (Cr_2L) = 8.54.

1. K. Micskei, F. Debreczeni and I. Nagypal, *J. Chem. Soc., Dalton Trans.*, 1983, 1335.
2. Y. Fukuda, E. Kyuno and R. Tsuchiya, *Bull. Chem. Soc. Jpn.*, 1970, **43**, 745.
3. W. P. Schaefer and M. E. Mathisen, *Inorg. Chem.*, 1965, **4**, 431.
4. R. L. Pecsok and J. Bjerrum, *Acta Chem. Scand.*, 1957, **11**, 1419.
5. R. L. Pecsok, R. A. Garber and L. D. Shields, *Inorg. Chem.*, 1965, **4**, 447.
6. R. L. Pecsok, L. D. Shields and W. P. Schaefer, *Inorg. Chem.*, 1964, **3**, 114.
7. J. Podlaha and J. Podlahova, *Inorg. Chim. Acta.*, 1970, **4**, 521.
8. J. Podlaha and J. Podlahova, *Inorg. Chim. Acta*, 1971, **5**, 413.
9. J. Podlaha and J. Podlahova, *Inorg. Chim. Acta*, 1971, **5**, 420.
10. R. L. Pecsok and W. P. Schaefer, *J. Am. Chem. Soc.*, 1961, **83**, 62.
11. J. M. Crabtree, D. W. Marsh, J. C. Tomkinson, R. J. P. Williams and W. C. Fernelius, *Proc. Chem. Soc.*, 1961, 336.
12. Y. Abe and G. Wada, *Bull. Chem. Soc. Jpn.*, 1981, **54**, 3334.

various ligands including edta and other complexones are given in Table 39. There are discrepancies between recent and earlier work, but there is agreement with 'preparative' observations. No complex formation could be detected in the Cr^{2+}–ammonia system even in 2 mol dm^{-3} NH_4Cl before hydrolytic precipitation of Cr^{2+}, and it was necessary to take various measures, *e.g.* use a large excess of ligand, to avoid hydrolytic precipitation in other systems. The greater ionic radius of Cr^{2+} (0.82 Å) compared with Cu^{2+} (0.69 Å) modifies the log K_1/K_2 ratios as well as leading to smaller overall constants.[294] The rates of formation and dissociation of $[Cr(gly\text{-}o)]^{2+}$ and $[Cr(gly\text{-}o)_2]^{2+}$ have been measured.[295]

35.3.9.4 Bidentate mixed donor atom ligands

(i) N—O ligands

Unidentified orange–yellow precipitates have been obtained from aqueous Cr^{2+} solutions and 8-hydroxyquinoline or an alkali metal 2-aminobenzoate.[83]

(ii) N—S ligands: 2-aminobenzenethiol

When aqueous ethanolic solutions of 2-aminobenzenethiol (abt) and chromium(II) chloride are mixed, the green suspension first formed turns light blue within one hour. The blue product

$[Cr(abt)_2]$ is high-spin ($\mu_{eff}^{322\,K} = 4.71$ BM, $\theta = 1°$), with a broad reflectance band at 16 900 cm^{-1}. Axial donation from adjacent $[Cr(abt)_2]$ units seems necessary to produce six-coordinate Cr^{II} so it is surprising that the magnetic moment is not temperature-dependent. Possibly the molecule is essentially planar.[296]

35.3.10 Multidentate Macrocyclic Ligands

35.3.10.1 Planar (closed) macrocycles; porphyrins, corrins and phthalocyanines

Chromium(II) porphyrins and phthalocyanines are dealt with in Sections 35.4.9.1 and 35.4.9.3 respectively.

35.3.10.2 Other polyazamacrocycles

The chromium(II) complex of 1,4,8,12-tetraazacyclopentadecane ([**15**]aneN_4; **81**) has been prepared in aqueous solution and used as an intermediate in the production of chromium(III) complexes (Scheme 25), including some which contain Cr—C σ bonds.[297] Its powerful reducing ability and the chemical inertness of the Cr^{III} products (it is preferable to Cr^{2+}aq) have been exploited[298] in kinetic studies of the reduction of ferredoxins.

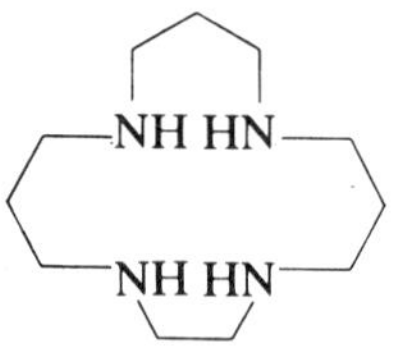

(**81**) [15]aneN_4

$CrCl_2 \cdot 4H_2O \xrightarrow[H_2O]{[15]aneN_4} [Cr^{II}([15]aneN_4)(H_2O)_2]^{2+} \xrightarrow{air} [Cr^{III}([15]aneN_4(H_2O)_2]^{3+}$

royal purple ($E^0 = -0.58$ V) — orange

$\downarrow$ RX

$[RCr^{III}([15]aneN_4)(H_2O)]^{2+} + [XCr^{III}([15]aneN_4)(H_2O)]^{2+}$

R = Me, Et, Pr, Pr^i, Bu^n, n-C_5H_{11}, Cy, 1-adamantyl, benzyl

Scheme 25

The reaction of chromium(II) with cyclam, 1,4,8,11-tetraazacyclotetradecane ([**14**]aneN_4; **82**) results in oxidation to Cr^{III} derivatives, but several complexes of 1,4,8,11-tetramethyl-1,4,8,11-tetraazacyclotetradecane (Me_4cyclam; **83**);[299] *meso*-5,12-dimethyl-1,4,8,11-tetraazacyclotetradecane (Me_2[**14**]aneN_4; **84**, R = H) and *meso*-5,7,7,12,14,14-hexamethyl-1,4,8,11-tetraazacyclotetradecane (Me_6[**14**]aneN_4; **84**, R = Me)[300] have been prepared according to Scheme 26. The use of CH_2Cl_2 as solvent and even as the source of chloride in the formation of $[CrCl(Me_4cyclam)]BPh_4 \cdot 0.5CH_2Cl_2$ would seem inadvisable in view of the formation of Cr^{III}—C species from organic halides referred to above. However, this complex and the others have magnetic moments consistent with high-spin chromium(II). The complexes $[CrX_2(Me_4cyclam)]$(X = Br, I) are non-electrolytes in CH_2Cl_2 and have closely similar solution and reflectance spectra. Since the reflectance spectra are not typical of *trans*-CrX_2N_4 species the complexes appear to be *cis* with Me_4cyclam in a folded configuration. In DMF five-coordinate cations $[CrBr(Me_4cyclam)]^+$ and $[Cr(DMF)(Me_4cyclam)]^{2+}$ form because the spectra resemble those of the BPh_4 salts of these cations (Table 40) and the conductance behaviour agrees. The cations in $[CrX(Me_4cyclam)]BPh_4$ and $[Cr(solv)(Me_4cyclam)](BPh_4)_2$ have respectively been assigned distorted trigonal bipyramidal and square pyramidal structures. There is no crystallographic evidence.

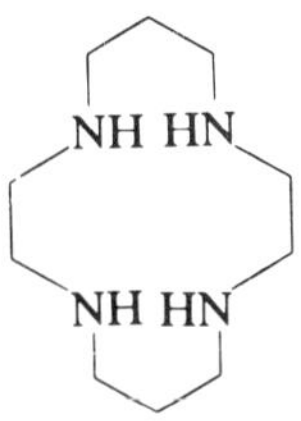

(82) [14]aneN$_4$ (cyclam)

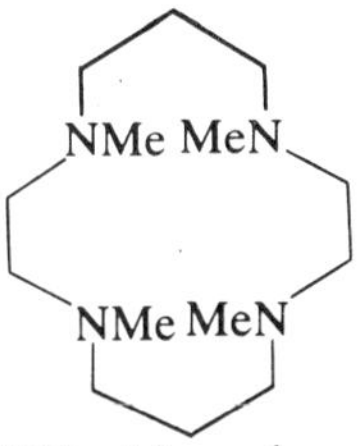

(83) Me$_4$cyclam

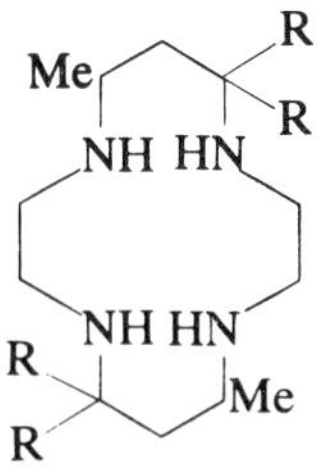

(84) R = H, Me$_2$[14]aneN$_4$
R = Me, Me$_6$[14]aneN$_4$

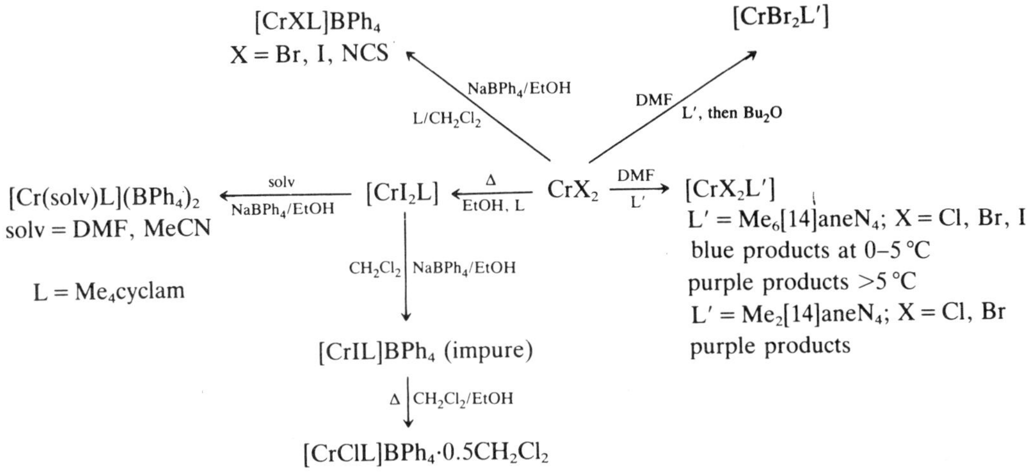

Scheme 26

$$[CrXL]BPh_4 \quad X = Br,\ I,\ NCS \xleftarrow{\ NaBPh_4/EtOH,\ L/CH_2Cl_2\ }$$

$$[CrBr_2L'] \xleftarrow{\ DMF,\ L',\ then\ Bu_2O\ }$$

$$[Cr(solv)L](BPh_4)_2 \quad solv = DMF,\ MeCN \xleftarrow[NaBPh_4/EtOH]{solv} [CrI_2L] \xleftarrow[EtOH,\ L]{\Delta} CrX_2 \xrightarrow[L']{DMF} [CrX_2L']$$

L = Me$_4$cyclam

$L' = Me_6[14]aneN_4$; X = Cl, Br, I
blue products at 0–5 °C
purple products >5 °C
$L' = Me_2[14]aneN_4$; X = Cl, Br
purple products

$$[CrI_2L] \xrightarrow[NaBPh_4/EtOH]{CH_2Cl_2} [CrIL]BPh_4 \text{ (impure)} \xrightarrow[CH_2Cl_2/EtOH]{\Delta} [CrClL]BPh_4 \cdot 0.5CH_2Cl_2$$

Table 40 Complexes of Me$_6$[14]aneN$_4$, Me$_2$[14]aneN$_4$ and Me$_4$cyclam

Complex	Colour	μ_{eff} (BM) 295 K	88 K	Reflectance spectra (cm^{-1})[a]		Ref.
[CrCl$_2$(Me$_6$[14]aneN$_4$)]	Blue	4.82	—		14 100	1
	Purple	4.80	—	19 400	14 700sh	
[CrBr$_2$(Me$_6$[14]aneN$_4$)]	Blue	4.77	—	18 200sh	15 000	1
	Purple	4.82		20 400	15 000sh	
[CrI$_2$(Me$_6$[14]aneN$_4$)]	Blue	4.82	—	20 300	15 400	1
	Purple	4.92		20 000		
[CrCl$_2$(Me$_2$[14]aneN$_4$)]	Purple	4.87	—	19 700	14 300	1
[CrBr$_2$(Me$_2$[14]aneN$_4$)]	Purple	4.90	—	19 600	15 400	1
[CrBr$_2$(Me$_4$cyclam)]		4.74	—	14 900 15 300 (122, CH$_2$Cl$_2$) 15 200 (111, DMF)		2
[CrI$_2$(Me$_4$cyclam)]	Blue	4.74	—	15 900 15 600 (127, CH$_2$Cl$_2$) 16 700 (66, DMF)		2
[CrCl(Me$_4$cyclam)]BPh$_4$·0.5CH$_2$Cl$_2$		4.72	4.66	18 200sh 14 800 (110, CH$_2$Cl$_2$) 14 900 (84, DMF)	15 400	2
[CrBr(Me$_4$cyclam)]BPh$_4$		4.66	4.68	16 950sh 14 700 (120, CH$_2$Cl$_2$) 15 200 (98, DMF)	15 200	2
[CrI(Me$_4$cyclam)]BPh$_4$		4.65	4.60	15 600 16 700 (70, DMF)		2
[CrNCS(Me$_4$cyclam)]BPh$_4$		4.70	—	15 900		2
[Cr(MeCN)(Me$_4$cyclam)](BPh$_4$)$_2$		4.75	—	18 000 16 700 (68, DMF)		2
[Cr(DMF)(Me$_4$cyclam)](BPh$_4$)$_2$		4.74	—	17 860 16 700 (81, DMF)	16 400sh	2

[a] ε (dm^3 mol^{-1} cm^{-1}) and solvent in parentheses.

1. A. Dei and F. Mani, *Inorg. Chem.*, 1976, **15**, 2574.
2. F. Mani, *Inorg. Chim. Acta*, 1982, **60**, 181.

The complexes $[CrX_2(Me_2[\mathbf{14}]aneN_4)]$ and the purple forms of $[CrX_2(Me_6[\mathbf{14}]aneN_4)]$ are high-spin with electronic spectra of the usual pattern for *trans* octahedral chromium(II). The blue complexes $[CrX_2(Me_6[\mathbf{14}]aneN_4)]$ are also high-spin but have a different spectral pattern. It is not clear whether they contain the folded macrocyclic ligand in *cis* octahedral or distorted trigonal bipyramidal cations. The brown complex $[Cr(Me_6[\mathbf{14}]4,11\text{-diene}N_4)(py)](PF_6)_2$ reacts with sodium nitrite to give $[Cr(Me_6[\mathbf{14}]\text{-}4,11\text{-diene}N_4)(NO_2)(NO)]PF_6$ (Section 35.4.2.6).

35.3.11 Stereochemistry of Bivalent Chromium

Excluding the dinuclear and related complexes (Section 35.3.5.5), the structures of comparatively few chromium(II) complexes have been determined by single crystal X-ray or other diffraction methods. From the structures reported it is apparent that high-spin chromium(II) has as rich a stereochemistry as copper(II). In most complexes (Table 34) Cr^{2+} forms four coplanar bonds with two longer bonds completing a distorted octahedron (4 + 2 coordination), but the number of planar complexes (the limit of 4 + 2 coordination) is increasing rapidly (Table 41), and there are many for which trigonal bipyramidal or square pyramidal structures (Table 42) have been inferred from the general properties of the complexes. Seven-coordinate chromium(II) complexes are also known (Table 95).

Table 41 Planar Chromium(II) Complexes

Complex	*Coordination sphere* (Å)	*Ref.*
$[Cr(O_2CCF_3)_2(2,6\text{-Mepy})_2]$	*trans*, Cr—N, 2.111, Cr—O, 2.028	1
$[Cr\{N(SiMe_3)_2\}_2(THF)_2]$[a]	*trans*, Cr—N, 2.089, Cr—O, 2.089	2
$[pyH]_2[CrBr_2(H_2O)_2]Br_2$	*trans*, Cr—O, 2.038, Cr—Br, 2.579	3
$[Cr(acac)_2]$[a]	Cr—O, 1.98 (av), intermolecular contact Cr—C, 2 at 3.05	4
$[Cr\{H_2B(pz)_2\}_2]$	Cr—N, 2.062 (av)	5
$[Cr\{Et_2B(pz)_2\}_2]$	Cr—N, 2.060 (av)	6
$[NEt_4]_2[Cr(S_2C_2H_4)_2]$	Cr—S, 2.388 (av)	7
$[Cr\{MeC(O){=}CHCMe{=}NBu\}_2]$	Cr—N, 2.086 (av), Cr—O, 1.963 (av)	8
[Cr(acacen)]	Cr—N, 2.012, Cr—O, 1.972 (av)	8
[Cr(TPP)]·2PhMe	Cr—N, 2.033 (av), CrN_4 plane to PhMe, 3.37	9

[a] The square species $[Cr\{N(SiMe_3)_2\}_2(PMe_3)_2]$ is referred to in G. S. Girolami, G. Wilkinson, A. M. R. Galas, M. Thornton-Pett and M. B. Hursthouse, *J. Chem. Soc., Dalton Trans.*, 1985, 1339.
[b] $[Cr(dpm)_2]$ is isomorphous with planar $[Ni(dpm)_2]$ (Section 35.3.5.3).

1. D. C. Bradley, M. B. Hursthouse, C. W. Newing and A. J. Welch, *J. Chem. Soc., Chem. Commun.*, 1972, 567.
2. I. L. Eremenko, A. A. Pasynskii, V. T. Kalinnikov, G. G. Aleksandrov and Yu. T. Struchkov, *Inorg. Chim. Acta*, 1981, **54**, L85.
3. M. F. C. Ladd, L. F. Larkworthy, G. A. Leonard, D. C. Povey and S. S. Tandon, *J. Chem. Soc., Dalton Trans.*, 1984, 2351.
4. F. A. Cotton, C. E. Rice and G. W. Rice, *Inorg. Chim. Acta*, 1977, **24**, 231.
5. P. Dapporto, F. Mani and C. Mealli, *Inorg. Chem.*, 1978, **17**, 1323.
6. F. A. Cotton and G. N. Mott, *Inorg. Chem.*, 1983, **22**, 1136.
7. J. R. Dorfman, C. P. Rao and R. H. Holm, *Inorg. Chem.*, 1985, **24**, 453.
8. L. F. Larkworthy, D. C. Povey and B. Sandell, *Inorg. Chim. Acta*, 1984, **83**, L29.
9. W. R. Schiedt and C. A. Reed, *Inorg. Chem.*, 1978, **17**, 710.

35.4 CHROMIUM(III)

The coordination chemistry of chromium(III) was first seriously investigated by Pfeiffer at the turn of the century; in many ways his studies parallel the work of Werner on cobalt(III). The complexes of chromium(III) are almost exclusively six-coordinate with an octahedral disposition of the ligands. Many are monomeric ($\mu_{eff} \approx 3.6$ BM), although hydroxy-bridged and other polynuclear complexes are known in which spins on neighbouring chromiums are coupled.

The electronic[301] and magnetic properties of mononuclear chromium(III) complexes are quite well understood; however there is a distinct tendency for octahedral symmetry to be invoked in cases where the true symmetry is much lower. Chromium(III) is a hard Lewis acid and many stable complexes are formed with oxygen donors. In particular hydroxide complexes are readily formed in aqueous solution, and this may be a problem in synthesis. Substitution at chromium(III) centres is slow[302,303] and may well have some associative character in many cases. The kinetic inertness of chromium(III) has led to the resolution of many optically active complexes; this work has been extensively reviewed.[304]

Table 42 Five-coordinate Chromium(II)

Complex[a]	*Comments*	*Ref.*
$[Cr(NH_3)_5]X_2$ X = Cl, Br, I	Square pyramidal from electronic spectra, magnetic and powder data	1
$[Cr(NH_3)_4(OH_2)]SO_4$	As above	1
$[CrX(2Meiz)_4]X$ X = Cl, Br, I	As above, also 1:1 electrolytes; similarly $[Cr(2Meiz)_5](BPh_4)_2$ (2:1 electrolyte)	2
$[Cr(NCS)_2(2Meiz)_2]$	Antiferromagnetic, bridging NCS^-	2
$[CrX(H_2Cdmpz_2)_2]Y$ X = Cl, Y = BF_4; Y = X, BPh_4	Trigonal bipyramidal from electronic spectra and magnetic data, 1:1 electrolytes	3
$[CrX(MeTPyEA)]BPh_4$ X = Br, NCS	Square pyramidal from data as above	4
$[Cr(S_2CNEt_2)_2]_2$	Antiferromagnetic, distorted trigonal bipyramidal? (Section 35.3.6.1)	—
$[CrX(Me_4cyclam)]BPh_4$ X = Cl, Br, I, NCS	Distorted trigonal bipyramidal, electronic spectra, 1:1 electrolytes	5
$[Cr(solv)Me_4cyclam](BPh_4)_2$ solv = MeCN, DMF	Distorted square pyramidal, electronic spectra, 1:1 electrolytes	5

[a] Trigonal bipyramidal Cr^{II} complexes containing tetradentate 'tripod' ligands are in Section 35.3.4.3.

1. L. F. Larkworthy and J. M. Tabatabai, *J. Chem. Soc., Dalton Trans.*, 1976, 814.
2. P. Dapporto and F. Mani, *J. Chem. Res. (S)*, 1979, 374.
3. F. Mani and R. Morassi, *Inorg. Chim. Acta*, 1979, **36,** 63.
4. M. Di Vaira and F. Mani, *Inorg. Chem.*, 1984, **23,** 409.
5. F. Mani, *Inorg. Chim. Acta*, 1982, **60,** 181.

The organometallic chemistry of chromium(III) has been the subject of two recent reviews.[1,305]

35.4.1 Group IV Ligands

The reported complexes of chromium(III) with group IV ligands are limited mainly to those containing CN^- and to a lesser extent those containing other C donor ligands.

35.4.1.1 Cyanides

(i) Hexacyanochromate(III) and its aquation products

By far the most widely studied cyanochromium(III) complex is the pale yellow, water-soluble $K_3[Cr(CN)_6]$.[306,307] The most convenient synthesis reported for this complex involves the addition of $Cr(OAc)_3$, prepared by H_2O_2 reduction of CrO_3 in acetic acid, to a solution of KCN.[308] Results of a number of structure determinations carried out on salts containing $[Cr(CN)_6]^{3-}$ are summarized in Table 43.

Table 43 Bond Distances in $[Cr(CN)_6]^{3-}$ Salts

Complex	*Cr—C* (pm)	*Cr—N* (pm)	*Comments*	*Ref.*
$Cs_2K[Cr(CN)_6]$	206.8–207.5	113.7–113.8	295 K	1
$K_3[Cr(CN)_6]$	203.8–209.0	113.6–119.8	4.2 K, neutron	2
$Cs_2Li[Cr(CN)_6]$	205.8	114	381 K	3
$Cs_2Li[Cr(CN)_6]$	199–222	110–116	295 K, neutron	4
$Cd_3[Cr(CN)_6]\cdot xH_2O$	204.7	113.7	295 K	5
$Mn_3[Cr(CN)_6]_2\cdot xH_2O$	206.3	112.3	295 K	6

1. B. N. Figgis, E. S. Kucharski, P. A. Reynolds and A. H. White, *Acta Crystallogr., Sect. C*, 1983, **39,** 1587.
2. B. N. Figgis, P. A. Reynolds and G. A. Williams, *Acta Crystallogr., Sect. B*, 1981, **37,** 504.
3. R. R. Ryan and B. I. Swanson, *Inorg. Chem.*, 1974, **13,** 1681.
4. M. R. Chowdhury, F. A. Wedgwood, B. M. Chadwick and H. J. Wilde, *Acta Crystallogr., Sect. B*, **33,** 46.
5. H. U. Güdel, *Acta Chem. Scand.*, 1972, **26,** 2169.
6. H. U. Güdel, H. Stucki and A. Ludi, *Inorg. Chim. Acta*, 1973, **7,** 121.

The Raman spectrum of $K_3[Cr(CN)_6]$ has recently been assigned and earlier work on its vibrational spectrum has been summarized.[309] Raman spectra of single crystals show $\nu(CN)$ at 2131, 2132 and 2135 cm^{-1} while $\nu(CN)$ for aqueous solutions occurs at 2132 cm^{-1}. The electronic spectrum of $[Cr(CN)_6]^{3-}$ in aqueous solution has *d–d* bands at 377 ($\varepsilon = 90\ dm^3\ mol^{-1}\ cm^{-1}$) and 309 nm ($\varepsilon = 65\ dm^3\ mol^{-1}\ cm^{-1}$) and a charge transfer band at 263 nm ($\varepsilon = 6240\ dm^3\ mol^{-1}\ cm^{-1}$).[310] Detailed *ab initio* SCF calculations have recently been carried out on the ligand field bands of this anion.[311] The X-ray photoelectron spectrum of $K_3[Cr(CN)_6]$ in the regions 24 and 9 eV is similar to that of KCN, indicating that the 3σ and 4σ molecular orbitals of CN^- are invariant to complex formation.[312] However, the broad lines due to photoemission from the 5σ and 1π orbitals in KCN are completely smeared in the spectrum of the complex suggesting that these ligand orbitals participate strongly with the metal in bonding. Discrete variational X_α-methods have been applied to the calculation of orbital energies and results are in good agreement with the experimentally determined XPS values.[313] The difference between the phosphorescent spectra of $Ag_3[Cr(CN)_6]$ and $K_3[Cr(CN)_6]$ has been attributed to an interaction between the nitrogen end of the cyano ligand and the Ag^+ ion in the former complex.[307]

Some representative reactions of $[Cr(CN)_6]^{3-}$ are depicted in Scheme 27.[306,308,314–318] The properties of $[Cr(CN)_5NO]^{3-}$ are discussed in Section 35.4.2.6.

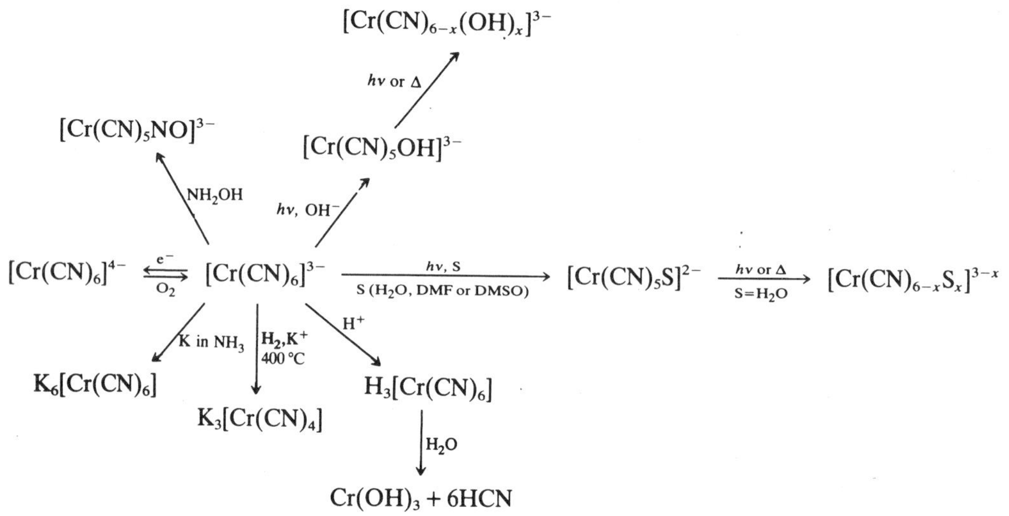

Scheme 27

The slow addition of $K_3[Cr(CN)_6]$ to a solution of $FeSO_4{\cdot}7H_2O$ at 0 °C gives a polymer of approximate composition $Fe_3[Cr(CN)_6]_2$, which on the basis of IR (ν_{CN} at 2170 cm^{-1} and therefore indicative of a bridged chromicyanide, $\nu_{MC} = 488\ cm^{-1}$) and Mössbauer (all the iron present as high-spin Fe^{2+}) spectroscopy seems to contain Cr^{3+}—C≡N—Fe^{2+} linkages.[307,319] On heating at 50 °C in an inert atmosphere half the Cr^{3+} ions are displaced by Fe^{2+} ions from their carbon 'sites'. In the presence of air further rearrangement and partial oxidation occurs to give a compound which on reduction with hydrazine affords an isomer of the starting material containing Cr^{3+}—N≡C—Fe^{2+} linkages ($\nu_{CN} = 2100\ cm^{-1}$, $\nu_{MC} = 525\ cm^{-1}$).

In aqueous acid solution $[Cr(CN)_6]^{3-}$ aquates *via* a series of stepwise stereospecific reactions to give $[Cr(H_2O)_6]^{3+}$ as the final product.[320] Some of the intermediate cyanoaqua complexes in this sequence have been isolated and details of their electronic spectra are presented in Table 44. These complexes aquate by both acid-independent and acid-dependent pathways, the latter involving protonation of the cyano ligands followed by aquation of the singly protonated species. Some of these reactions are found to be catalyzed by Cr^{2+} and kinetic data for the aquation of $[CrCN(H_2O)_5]^{2+}$ are consistent with the transition state structure (**85**).[307] In the presence of Hg^{2+} and Ag^+ cyanoaquachromium(III) complexes undergo strong association followed by 'flipping' of one or more of the cyano ligands, depending on the heavy metal to complex concentration ratio and the time elapsed since mixing the reactants (Scheme 28).[321] Addition of Cr^{2+} to solutions of cyanocobalt(III) complexes produces the metastable intermediate $[CrNC(H_2O)_5]^{2+}$ (at $[H^+] = 0.9$ M, $\lambda_{max} = 535$ nm, $\varepsilon = 19.6\ dm^3\ mol^{-1}\ cm^{-1}$; $\lambda_{max} = 396$ nm, $\varepsilon = 23\ dm^3\ mol^{-1}\ cm^{-1}$).[322] This isomerizes to $[CrCN(H_2O)_5]^{2+}$ ($\lambda_{max} = 525$ nm, $\varepsilon =$

25.2 $dm^3 mol^{-1} cm^{-1}$; $\lambda_{max} = 393$ nm, $\varepsilon = 20.0$ $dm^3 mol^{-1} cm^{-1}$) in a Cr^{2+}-catalyzed reaction which occurs by a ligand-bridge electron-exchange mechanism. In aqueous solution containing CN^- and OH^-, $[Cr(CN)_6]^{3-}$ exists in equilibrium with $[Cr(CN)_5OH]^{3-}$ when $[CN^-]$ and $[OH^-]$ are both greater than 0.1 M and when $[OH^-]/[CN^-] < 2$.[323] The Cr^{2+}-catalyzed aquation of $[Cr(CN)_6]^{3-}$ under these conditions seems to proceed by an outer-sphere redox mechanism represented by equation (27) although the formation of a short-lived bridged intermediate is not excluded by the kinetic data. The complex $K_3[Cr(CN)_5OH]\cdot H_2O$ may be obtained as orange-yellow crystals by reaction of $[CrCl(NH_3)_5]Cl_2$ and KCN in boiling water, followed by purification on a Sephadex column.[324]

$$[Cr(CN)_6]^{3-} + [Cr(CN)_5OH]^{4-} \rightleftharpoons [Cr(CN)_6]^{4-} + [Cr(CN)_5OH]^{3-} \qquad (27)$$

Table 44 Details of the Electronic[a] and IR[b] Spectra of Cyanoaqua and Cyanoammine/Aminechromium(III) Complexes

Complex	λ_{max} (ε) (nm, $dm^3 mol^{-1} cm^{-1}$)	λ_{min} (ε)	$\nu(CN)$ (cm^{-1})	*Ref.*
$[CrCN(H_2O)_5]^{2+}$	525 (26.0), 393 (20.5)[c]			1
cis-$[Cr(CN)_2(H_2O)_4]^+$	535sh (30.2), 474 (45.3), 464 (45.5), 378 (25.8)[d]		—	1
fac-$[Cr(CN)_3(H_2O)_3]$	467 (115), 360 (25.2)[e]		—	1
$[Cr(CN)_5H_2O]^{2-}$	427 (~110), 332 (~50)[f]			2, 3
$[Cr(CN)_5OH]^{3-}$	433 (~120), 365 (~75)[g]			3
$[CrCN(NH_3)_5]^{2+}$	451 (42.6), 347 (37.7)	391 (10.0)	2140w	4
trans-$[Cr(CN)_2(NH_3)_4]^+$	440 (42.6), 344 (41.5)	385 (12.5)	~2130w	5
cis-$[Cr(CN)_2(NH_3)_4]^+$	436 (49.0), 342 (37.6)	379 (14.0)	~2130w	5
trans-$[CrCN(NH_3)_4H_2O]^{2+}$	468 (48), 354 (32)	400 (10)[h]	—	5
cis-$[CrCN(NH_3)_4H_2O]^{2+}$	468 (34), 355 (32)	400 (11)[h]	—	5
trans-$[CrCN(OH)(NH_3)_4]^+$	494 (55), 381 (39)	426 (21)[i]	—	5
cis-$[CrCN(OH)(NH_3)_4]^+$	491 (46), 387 (44)	432 (25)[i]	—	5
cis-$[CrCN(NH_3)_4DMSO]^{2+}$	483 (44), 362 (36)	406 (13)	2139w	5
trans-$[Cr(CN)_2en_2]^+$	433 (50.1), 337 (42.7)		—	6
cis-$[Cr(CN)_2en_2]^+$	434 (70.0), 339 (62.3)		2139w	7
	434 (69.5), 339 (62.2)			6
cis-$[CrCN(en)_2H_2O]^{2+}$	467 (53.4), 356 (49.1)[j]		—	7
cis-$[Cr(CN)OH(en)_2]^+$	487 (59.4), 386 (54.5), 345sh (35.9)[k]		—	7
$[Cr(CN)_2(NH_3)_3H_2O]^+$				
1,6-CN-2-H_2O	449 (39), 351 (41)	395 (19)[l]	—	8
1,2-CN-6-H_2O	452 (53), 350 (34)	391 (17)[l]		8
$[Cr(CN)_2edtd(H_2O)]^{3-}$	512 (70.4), 386 (83.0)	450 (45.0)	2160w	9

[a] All in aqueous solution. [b] As Nujol mulls or KBr discs. [c] In 0.6 mol dm^{-3} $NaClO_4$. [d] pH 2.0–4.0. [e] pH 3.0–4.5. [f] pH 7. [g] pH 11. [h] pH 3. [i] pH 10. [j] pH 2.9, $I = 2.0$. [k] pH 10.4, $I = 2.0$. [l] 5×10^{-4} mol dm^{-3} $HClO_4$.

1. D. K. Wakefield and W. B. Schaap, *Inorg. Chem.*, 1969, **8**, 512.
2. L. Jeftić and S. W. Feldberg, *J. Am. Chem. Soc.*, 1970, **92**, 5272.
3. L. Jeftić and S. W. Feldberg, *J. Phys. Chem.*, 1971, **75**, 2381.
4. P. Riccieri and E. Zinato, *Inorg. Chem.*, 1980, **19**, 853.
5. P. Riccieri and E. Zinato, *Inorg. Chem.*, 1981, **20**, 3722.
6. A. D. Kirk and G. B. Porter, *Inorg. Chem.*, 1980, **19**, 445.
7. A. P. Sattelberger, D. D. Darsow and W. P. Schaap, *Inorg. Chem.*, 1976, **15**, 1412.
8. E. Zinato, P. Riccieri and M. Prelati, *Inorg. Chem.*, 1981, **20**, 1432.
9. Z. Chen, M. Cimilino and A. A. Adamson, *Inorg. Chem.*, 1983, **22**, 3035.

$$[(H_2O)_4\overset{\overset{\displaystyle CN}{|}}{Cr}—OH—Cr(H_2O)_5]^{3+}$$

(85)

$$fac\text{-}(H_2O)_3Cr(CN)_3 + M^{n+} \rightleftharpoons [(H_2O)_3(CN)_2Cr—C{\equiv}N—M]^{n+} \longrightarrow [(H_2O)_3(CN)_2Cr—N{\equiv}C—M]^{n+}$$

Scheme 28

(*ii*) *Cyanoammine- and cyanoamine-chromium*(*III*) *complexes*

Although the range of reported acidoammine/amine chromium(III) complexes is quite extensive, only a few such cyano species have so far been prepared, due mainly to the difficulties associated with the CN^- anation of the appropriate aqua complexes. These difficulties stem largely from the ionization of coordinated water in the presence of CN^-. This not only generates a leaving group of greater nucleophilicity but also makes complex decomposition competitive with anation.[325] Anation reactions of non-aqueous solvates circumvent these problems.[325] Hence the reaction between $[Cr(NH_3)_5DMSO](ClO_4)_3$ and excess NaCN in DMSO at 70 °C affords the monocyano complex $[Cr(CN)(NH_3)_5]^{2+}$, which was isolated as a perchlorate in 40% yield.[326] Details of the electronic and IR spectra of the complex are presented in Table 44. This complex phosphoresces in aqueous solution and its photoreactivity on LF band irradiation consists exclusively of NH_3 aquation. Thermal aquation of the complex on the other hand involves loss of CN^- and follows both acid-independent and acid-dependent pathways, the latter involving N protonation of the CN^- ligand.

The isomers *cis*- and *trans*-$[Cr(CN)_2(NH_3)_4]^+$ as well as *cis*-$[Cr(CN)(NH_3)_4DMSO](ClO_4)_2$ have also been synthesized by the anation reactions shown in Scheme 29.[325] While anation in DMSO is accompanied by stereochemical change, the reaction in H_2O is stereoretentive. The dicyano complexes undergo H^+-assisted thermal aquation, involving the successive loss of the CN^- ligands. The *trans* complex is about ten-fold more reactive in the first step than the *cis*, an observation attributed to the *trans* labilizing effect of CN^-.

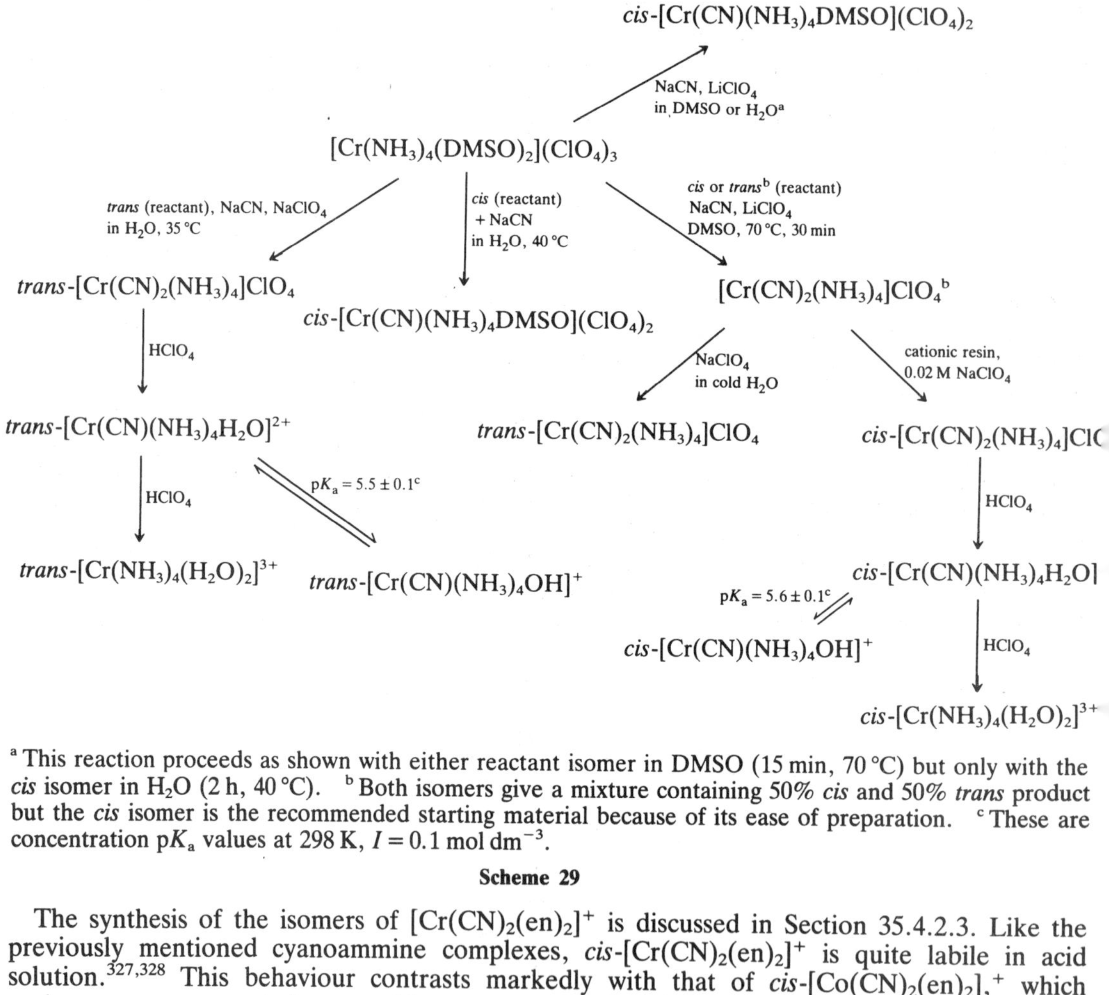

[a] This reaction proceeds as shown with either reactant isomer in DMSO (15 min, 70 °C) but only with the *cis* isomer in H_2O (2 h, 40 °C). [b] Both isomers give a mixture containing 50% *cis* and 50% *trans* product but the *cis* isomer is the recommended starting material because of its ease of preparation. [c] These are concentration pK_a values at 298 K, $I = 0.1$ mol dm^{-3}.

Scheme 29

The synthesis of the isomers of $[Cr(CN)_2(en)_2]^+$ is discussed in Section 35.4.2.3. Like the previously mentioned cyanoammine complexes, *cis*-$[Cr(CN)_2(en)_2]^+$ is quite labile in acid solution.[327,328] This behaviour contrasts markedly with that of *cis*-$[Co(CN)_2(en)_2]^+$, which undergoes no apparent decomposition over a period of five days at 80 °C in aqueous acid.[327] The addition of Ag^+ to an aqueous solution of *cis*-$[Cr(CN)_2(en)_2]^+$ results in ligand isomerization, equation (28), and formation of adduct **(86)** which has an electronic spectrum

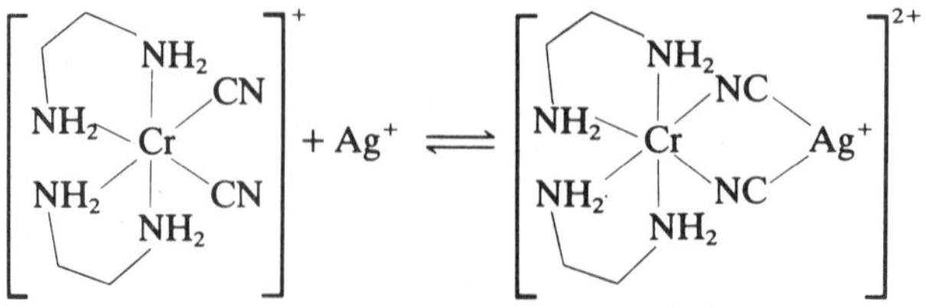

(28)

(86)

characteristic of a CrN_6 chromophore.[328] The IR spectrum of the adduct, which has been isolated as its perchlorate salt, has a CN stretching band ($2128\ cm^{-1}$) in a similar position to that of the parent complex ($2130\ cm^{-1}$) but of much greater intensity, consistent with extensive chromium cyanide π bonding. Addition of NaI to a solution of the adduct causes the precipitation of AgI and the regeneration of the original complex, a reaction which emphasizes the critical role played by Ag^+ in stabilizing the isocyano isomer. The preparation of *trans*-$[Cr(CN)_2(cyclam)]ClO_4$ is described in Section 35.4.9.2.

The photochemical behaviour of cyanoammine/amine chromium(III) complexes differs from that of other acido chromium(III) species. The higher ligand field strength of CN^- relative to ammine/amine ligands causes the $^4B_{2g}$ component of the first spin-allowed quartet excited state to lie at lower energy than the ^4E_g component for D_{4h} complexes.[329] Since this state is associated with selective population of the $d_{x^2-y^2}$ orbital, LF band irradiation causes photolabilization of the in-plane amine ligands. Hence the complexes *trans*-$[Cr(CN)_2(NH_3)_4]^+$ and *trans*-$[Cr(CN)_2(en)_2]^+$ behave 'antithermally' on photolysis, the former losing an ammine ligand (on 440 nm excitation, $\phi_{NH_3} = 0.24$, $\phi_{CN^-} = 0.005$)[329] and the latter undergoing a reaction involving release and protonation of one end of an en ligand.[330] In the case of *trans*-$[Cr(CN)_2(cyclam)]^+$, however, the presence of the macrocyclic ligand militates against cleavage of a chromium–amine bond and this complex shows no discernible photochemical reactivity.[331] The complex $Na_3[Cr(CN)_2edta(H_2O)]\cdot 4H_2O$ which contains tridentate (2N,O) edta, is referred to in Table 44.

(*iii*) *Bi- and multi-nuclear cyanochromium*(*III*) *complexes*

A number of binuclear complexes containing Cr^{III}—NC—M bridges have been reported. The methods of preparation and spectroscopic properties of these complexes are presented in Table 45. The IR spectrum of each complex displays two CN stretching bands, the higher wavenumber one attributed to bridging CN and the other due to terminal CN ligands.[332] Magnetic susceptibilities down to liquid helium temperatures have been determined for *cis*-$K[(N—N)_2FCr—NC—Cr(CN)_5]$ (N—N = en, tn) and *cis*-$K[(en)_2FCr—NC—CrNO(CN)_4]$.[333] Whereas the first two complexes show weak antiferromagnetic interaction between the chromium ions, the nitrosyl complex behaves as a Curie paramagnet. The absence of antiferromagnetic superexchange in this complex has been suggested to indicate that the NO ligand is *trans* to the cyano bridge. Reaction of $Na_2Cr(CO)_5$ with $Br_2CNNCBr_2$ in THF affords the tetranuclear complex $(THF)_3Cr^{III}[NCCr(CO)_5]_3$ (**87**).[334] A pair of binuclear linkage isomers involving the CN^- ligand has been prepared in solution according to Scheme 30.[335]

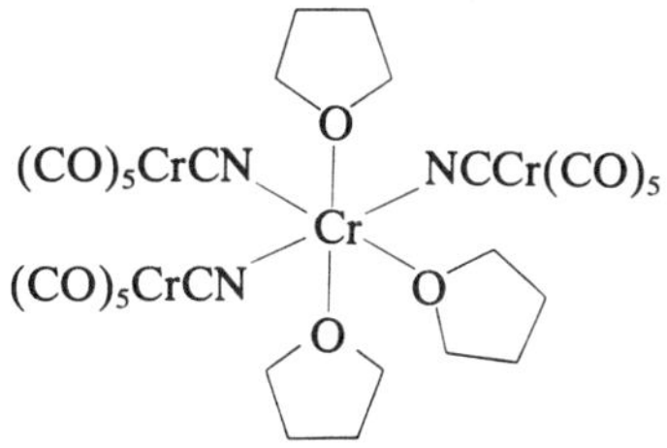

$\nu(CN) = 2102\ cm^{-1}$, $\nu(CO) = 2055, 1990, 1946\ cm^{-1}$

(87)

$$[Cr(H_2O)_6]^{3+} + [Co(CN)_6]^{3-} \longrightarrow (H_2O)_5Cr—NC—Co(CN)_5$$

$$\lambda_{max} = 560\ (\varepsilon = 16.2),\ 401\ (\varepsilon = 20.1),\ 310\ (\varepsilon = 218)$$

$$[Co(CN)_5N_3]^{3-} \xrightarrow{HNO_2} [Co(CN)_5]^{2-} \xrightarrow{[Cr(H_2O)_5CN]^{2+}} (H_2O)_5Cr—CN—Co(CN)_5$$

$$\lambda_{max} = 552\ (\varepsilon = 18.3),\ 347\ (\varepsilon = 231)$$

λ_{max} in nm, ε in $dm^3\ mol^{-1}\ cm^{-1}$

Scheme 30

Table 45 Preparation and Spectroscopic Properties of Binuclear Cyano-bridged Complexes[b]

Complex	Preparation	IR (cm^{-1})[a] $\nu(CN)_{brdg}$	IR (cm^{-1})[a] $\nu(CN)_{ter}$	Electronic spectrum[b] λ_{max}(nm) $^4A_{2g} \rightarrow {}^4T_{2g}$		$^4A_{2g} \rightarrow {}^4T_{1g}$	Other bands	Ref.
[H$_2$O(NH$_3$)$_4$Cr—NC—Fe(CN)$_5$]	[Cr(NH$_3$)$_4$(H$_2$O)$_2$](ClO$_4$)$_3$ + Na$_3$[Fe(CN)$_6$]	2137	2115	515			420, 405, 320	1
[(NH$_3$)$_5$Cr—NC—Fe(CN)$_5$]	[Cr(NH$_3$)$_5$H$_2$O](NO$_3$)$_3$ + K$_3$[Fe(CN)$_6$]	2144	2124	515			420, 405, 320	1
cis-[(en)$_2$FCr—NC—Ni(CN)$_3$]·2H$_2$O	Heat *trans*-[CrF(en)$_2$H$_2$O][Ni(CN)$_4$]	2160	2120	492 (84)			312 (890)[c]	2, 3
cis-[(en)$_2$ClCr—NC—Ni(CN)$_3$]·2H$_2$O	*cis*-[CrCl(en)$_2$H$_2$O]Br$_2$ + (NH$_4$)$_2$[Ni(CN)$_4$]	2160	2120	500			315	2
cis-[(en)$_2$ClCr—NC—Pd(CN)$_3$]	*cis*-[CrCl(en)$_2$H$_2$O]Br$_2$ + K$_2$Pd(CN)$_4$ or *cis*-[CrCl(en)$_2$H$_2$O][Pd(CN)$_4$] over P$_2$O$_5$	2175	2130	501		378		4
cis-[(en)$_2$ClCr—NC—Pt(CN)$_3$]	*cis*-[CrCl(en)$_2$H$_2$O][Pt(CN)$_4$] over P$_2$O$_5$	2180	2140	502			303	4
cis-[(en)$_2$BrCr—NC—Ni(CN)$_3$]·2H$_2$O	*cis*-[CrBr(en)$_2$H$_2$O]Br$_2$ + (NH$_4$)$_2$[Ni(CN)$_4$]	2160	2120	507			312	2
cis-[(en)$_2$BrCr—NC—Pd(CN)$_3$]	*cis*-[CrBr(en)$_2$H$_2$O]Br$_2$ + K$_2$[Pd(CN)$_4$], heat or *cis*-[CrBr(en)$_2$H$_2$O][Pd(CN)$_4$] over silica	2180	2135	510		383		4
cis-[(en)$_2$BrCr—NC—Pt(CN)$_3$]	*cis*-[CrBr(en)$_2$H$_2$O][Pt(CN)$_4$]·H$_2$O over silica	2165	2130	508			303	4
cis-[(en)$_2$FCr—NC—Pd(CN)$_3$]	Heat *trans*-[CrF(en)$_2$H$_2$O][Pd(CN)$_4$]	2175	2130	492 (84)		363 (63)[d]		3
cis-[(en)$_2$FCr—NC—Pt(CN)$_3$]	Heat *trans*-[CrF(en)$_2$H$_2$O][Pt(CN)$_4$]	2175	2130	492 (84)			305[d]	3
cis-K[(en)$_2$FCr—NC—Co(CN)$_5$]·3H$_2$O	Heat *trans*-K[CrF(en)$_2$H$_2$O][Co(CN)$_6$]	2170	2120	493			310	5
cis-K[(en)$_2$FCr—NC—Cr(CN)$_5$]·2H$_2$O	*trans*-[CrF(en)$_2$H$_2$O](ClO$_4$)$_2$ + K$_3$[Cr(CN)$_6$]	2160	2120	495		375–380		5
cis-M[(en)$_2$FCr—NC—Cr(CN)$_5$](M = Na, K, NH$_4$)	Heat *trans*-M[CrF(en)$_2$H$_2$O][Cr(CN)$_6$]·H$_2$O	2160	2128	496 (36.1)		377 (108)[c]		6
cis-M[(en)$_2$FCr—NC—CrNO(CN)$_4$](M = Na, K, NH$_4$)	Heat *trans*-M[CrF(en)$_2$H$_2$O][CrNO(CN)$_5$]·H$_2$O	2140	2100		458 (155)[c,f]			6
cis-[(pn)$_2$FCr—NC—Ni(CN)$_3$]·3H$_2$O	*trans*-[CrF(pn)$_2$H$_2$O](ClO$_4$)$_2$ + (NH$_4$)$_2$[Ni(CN)$_4$]	2170	2120	492–494				7
cis-[(pn)$_2$FCr—NC—Pd(CN)$_3$]·H$_2$O	*trans*-[CrF(pn)$_2$H$_2$O](ClO$_4$)$_2$ + H$_2$[Pd(CN)$_4$]	2175	2125	492–494		365		7
cis-[(pn)$_2$FCr—NC—Pt(CN)$_3$]	*trans*-[CrF(pn)$_2$H$_2$O](ClO$_4$)$_2$ + H$_2$[Pt(CN)$_4$]	2170	2125	492–494				7
cis-K[(pn)$_2$FCr—NC—Co(CN)$_5$]·H$_2$O	*trans*-[CrF(pn)$_2$H$_2$O](ClO$_4$)$_2$ + K$_3$[Co(CN)$_6$]	2165	2120	492–494		365		7
cis-K[(pn)$_2$FCr—NC—Cr(CN)$_5$]·H$_2$O	*trans*-[CrF(pn)$_2$H$_2$O](ClO$_4$)$_2$ + K$_3$[Cr(CN)$_6$]	2165	2130	492–494				7
cis-K[(pn)$_2$FCr—NC—CrNO(CN)$_4$]·H$_2$O	*trans*-[CrF(pn)$_2$H$_2$O](ClO$_4$)$_2$ + K$_3$[CrNO(CN)$_5$]	2140	2105	492–494				7
trans-[(tn)$_2$FCr—NC—Ni(CN)$_3$]	Heat *trans*-[CrF(tn)$_2$H$_2$O][Ni(CN)$_4$]	2160	2120	520–530[e] 465–470				3
trans-[(tn)$_2$FCr—NC—Pd(CN)$_3$]	Heat *trans*-[CrF(tn)$_2$H$_2$O][Pd(CN)$_4$]	2170	2135	520–530[e] 465–470				3
trans-[(tn)$_2$FCr—NC—Pt(CN)$_3$]	Heat *trans*-[CrF(tn)$_2$H$_2$O][Pt(CN)$_4$]	2170	2135	520–530[e] 465–470				3
cis-M[(tn)$_2$FCr—NC—Cr(CN)$_5$](M = Na, K, NH$_4$)	Heat *trans*-M[CrF(tn)$_2$H$_2$O][Cr(CN)$_6$]·H$_2$O	2160	2128	502 (40.6)		372 (107)[c]		6
cis-M[(tn)$_2$FCr—NC—CrNO(CN)$_4$](M = Na, K, NH$_4$)	Heat *trans*-M[CrF(tn)$_2$H$_2$O][CrNO(CN)$_5$]·H$_2$O	2140	2100		450 (199)[c,f]			6

[a] KBr disc or Nujol mull. [b] Reflectance spectra unless otherwise stated. [c] In aqueous solution; ε in dm^3 mol^{-1} cm^{-1}. [d] In HClO$_4$ solution. [e] The $^4A_{2g} \rightarrow {}^4T_{2g}$ band in these complexes is split, indicative of *trans* configurations. [f] Metal to NO charge transfer band.

1. J. Casabo, J. Ribas and S. Alvarez, *Inorg. Chim. Acta*, 1976, **16,** L15.
2. J. Ribas, J. Casabo, M. Serra and J. M. Coronas, *Inorg. Chim. Acta*, 1979, **36,** 41.
3. J. Ribas, M. Serra, A. Escuer, J. M. Coronas and J. Casabo, *J. Inorg. Nucl. Chem.*, 1981, **43,** 3113.
4. J. Ribas, M. Serra and A. Escuer, *Transition Met. Chem.*, 1984, **9,** 287.
5. J. Ribas, J. Casabo, M. Monfort, M. L. Alvarez and J. M. Coronas, *J. Inorg. Nucl. Chem.*, 1980, **42,** 707.
6. J. Ribas, M. Monfort and J. Casabo, *Transition Met. Chem.*, 1984, **9,** 407.
7. J. Ribas, M. L. Martinez, M. Serra, M. Monfort, A. Escuer and N. Navarro, *Transition Met. Chem.*, 1983, **8,** 87.

The thermal decomposition of a number of double complexes containing the $[Cr(CN)_6]^{3-}$ anion has been investigated and the results obtained under quasi-isothermal and -isobaric conditions are summarized in Scheme 31.[336] The 'CN-flipping' reactions involved in the scheme resemble similar reactions of Prussian blue analogues such as $Co_3[Cr(CN)_6]_2$[337] and $Fe_3[Cr(CN)_6]_2$.[338] The unusually low magnetic moment of $Cr(CN)_6Cr$ indicates considerable metal–metal interaction *via* the CN bridges.

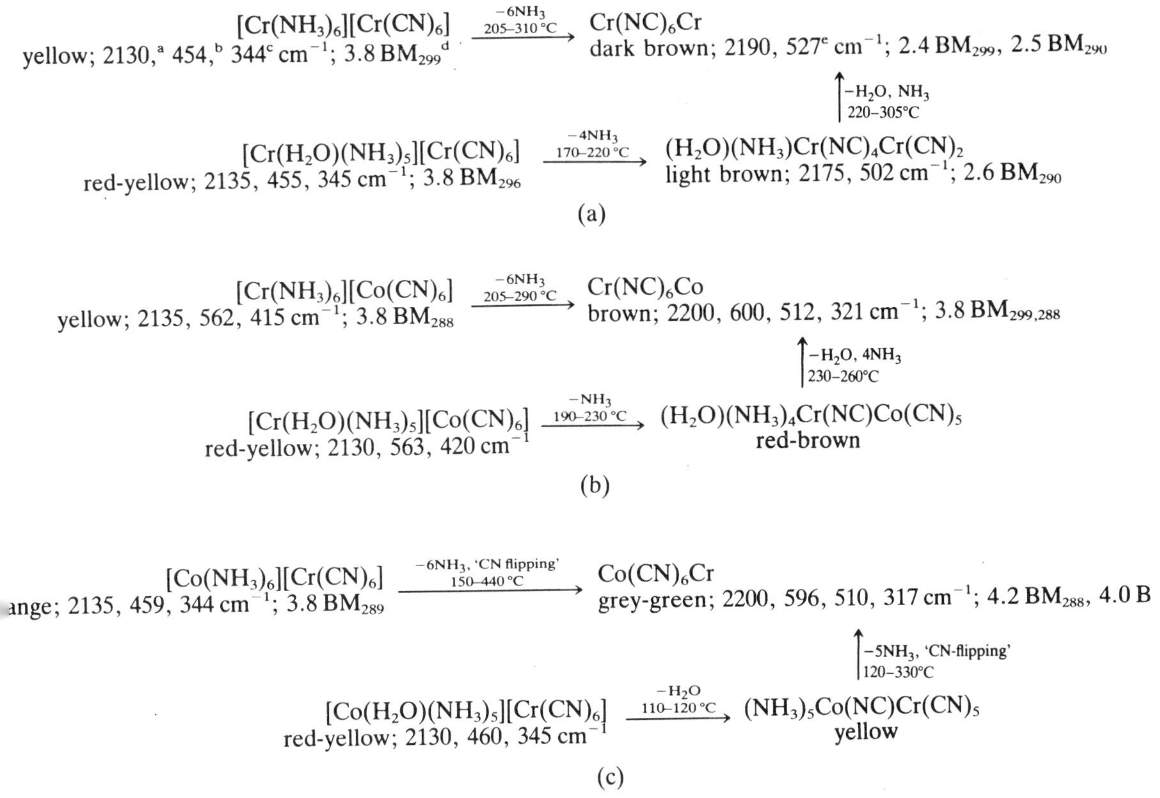

$$[Cr(NH_3)_6][Cr(CN)_6] \xrightarrow[205-310\,°C]{-6NH_3} Cr(NC)_6Cr$$

yellow; 2130,[a] 454,[b] 344[c] cm^{-1}; 3.8 BM$_{299}$[d]   dark brown; 2190, 527[e] cm^{-1}; 2.4 BM$_{299}$, 2.5 BM$_{290}$

$$\Big\uparrow {-H_2O,\ NH_3 \atop 220-305°C}$$

$$[Cr(H_2O)(NH_3)_5][Cr(CN)_6] \xrightarrow[170-220\,°C]{-4NH_3} (H_2O)(NH_3)Cr(NC)_4Cr(CN)_2$$

red-yellow; 2135, 455, 345 cm^{-1}; 3.8 BM$_{296}$ light brown; 2175, 502 cm^{-1}; 2.6 BM$_{290}$

(a)

$$[Cr(NH_3)_6][Co(CN)_6] \xrightarrow[205-290\,°C]{-6NH_3} Cr(NC)_6Co$$

yellow; 2135, 562, 415 cm^{-1}; 3.8 BM$_{288}$ brown; 2200, 600, 512, 321 cm^{-1}; 3.8 BM$_{299,288}$

$$\Big\uparrow {-H_2O,\ 4NH_3 \atop 230-260°C}$$

$$[Cr(H_2O)(NH_3)_5][Co(CN)_6] \xrightarrow[190-230\,°C]{-NH_3} (H_2O)(NH_3)_4Cr(NC)Co(CN)_5$$

red-yellow; 2130, 563, 420 cm^{-1} red-brown

(b)

$$[Co(NH_3)_6][Cr(CN)_6] \xrightarrow[150-440\,°C]{-6NH_3,\ 'CN\ flipping'} Co(CN)_6Cr$$

ange; 2135, 459, 344 cm^{-1}; 3.8 BM$_{289}$ grey-green; 2200, 596, 510, 317 cm^{-1}; 4.2 BM$_{288}$, 4.0 B

$$\Big\uparrow {-5NH_3,\ 'CN-flipping' \atop 120-330°C}$$

$$[Co(H_2O)(NH_3)_5][Cr(CN)_6] \xrightarrow[110-120\,°C]{-H_2O} (NH_3)_5Co(NC)Cr(CN)_5$$

red-yellow; 2130, 460, 345 cm^{-1} yellow

(c)

[a] $v(CN)$.   [b] $v(MCN)$.   [c] $\delta(CMC)$.
[d] Magnetic moment at the temperature (K) subscripted.
[e] $v(MC)$ or $\delta(CMC)$ not adequately assignable in the bridged complexes.

Scheme 31

35.4.1.2 Isocyanides

Some aryl isocyanide complexes of the type $[Cr(CNR)_6]X_3$ ($X = SbCl_6^-$, BF_4^-) have recently been reported. Details of the preparations and properties of these complexes are given in Table 3.

35.4.1.3 Alkyls and aryls

The chemistry of alkyl and aryl chromium(III) complexes is covered in the companion series[306] and has recently been the subject of a thorough review.[305] Some of these complexes are also referred to in Sections 35.4.2.5 and 35.4.8.1.iv.

35.4.2 Nitrogen Ligands

Complexes of chromium(III) with ammonia and amines were among the first to be recognized by the early coordination chemists. Consequently, there is a vast literature, and this section is restricted to work since the comprehensive review by Garner and House[302] in which new solids have been characterized. In many cases the compounds are tabulated without comment.

35.4.2.1 Ammonia

Some cyanoammines are discussed in Section 35.4.1.1; other ammines which are new or have been prepared by improved methods are listed in Table 46 or mentioned below.

Table 46 Ammines

Complex	*Ref.*
$CrXN_5$	
$[CrNCS(NH_3)_5]X_2$ ($X = Br, NO_3, ClO_4$)	6, 15
$[CrCN(NH_3)_5](ClO_4)_2$	26
$[CrX(NH_3)_5]^{(3-n)+}$ ($X^{n-} = HC_2O_4^-, NH_3^+CH_2CO_2^-, SO_4^{2-}, IO_3^-$)	5
$[CrX(NH_3)_5](ClO_4)_2$ ($X = MeCO_2, CHCl_2CO_2, CCl_3CO_2, CF_3CO_2, HCO_2, C_4H_3O_4, CH_2ClCO_2$)	15, 16, 21, 22
$[CrSeO_4(NH_3)_5]X$ ($X = HSeO_4, ClO_4, NO_3, Cl, Br$)	7
$[Cr(H_2AsO_4)(NH_3)_5](ClO_4)_2$	7
$[Cr(H_2PO_2)(NH_3)_5]X_2$, $[Cr(H_2PO_3)(NH_3)_5]X_2$ ($X = Br, ClO_4$)	9
$[CrSO_4(NH_3)_5]HSO_4$, $[CrSO_4(NH_3)_5]_2S_2O_6$	11
$[Cr(CrO_4)(NH_3)_5]X$ ($X = NO_3, I, \frac{1}{2}S_2O_6$)	23
$CrPO_4(NH_3)_5$, $[CrH_2PO_4(NH_3)_5]X_2$ ($X = Br, ClO_4$; $X_2 = S_2O_6$; also dithionato- and μ-oxalato-pentammines)	12
$[CrOP(OEt)_3(NH_3)_5](ClO_4)_3$	25
$[CrN_3(NH_3)_5](ClO_4)_2$	25
$[Cr(NH_3)_5(sol)](ClO_4)_3$ (sol = DMF, DMSO)	8, 14, 16, 20
$[CrX(NH_3)_5](S_2O_6)_{3/2}$ ($X = (NH_2)_2CO, (NHMe)_2CO$)	20, 28
$[Cr(NH_3)_5OCHNH_2](ClO_4)_{5/2}(CF_3SO_3)_{1/2}$, $[Cr(OSO_2CF_3)(NH_3)_5](CF_3SO_3)_2$	20, 28
$[Cr(NH_3)_5(H_2O)](ClO_4)_3$, $[Cr(NH_3)_5(CH_3CN)](CF_3SO_3)_3$	28
$[Cr(nic\text{-}O)(NH_3)_5](ClO_4)_2$	30
$[Cr(NCO)(NH_3)_5](NO_3)_2$	31
$[Cr(NCSe)(NH_3)_5]X_2$ ($X = Br, ClO_4, NO_3$)	32
CrX_2N_4	
trans-$[CrF_2(NH_3)_4]X$ ($X = I, ClO_4$)	1, 3, 6, 13
cis-$[CrF_2(NH_3)_4]X$ ($X = I, ClO_4$)	1, 13
cis-$[CrClH_2O(NH_3)_4]X_2$ ($X = Cl, ClO_4$)	1, 2
cis-$[Cr(H_2O)_2(NH_3)_4](ClO_4)_3$	2
trans-$[CrFH_2O(NH_3)_4](ClO_4)_2$	1
trans-$[CrBr_2(NH_3)_4]Br$	1, 13
cis-$[CrBr_2(NH_3)_4]Br$	13
trans-$[CrBrH_2O(NH_3)_4]Br_2$	1
trans-$[CrCl_2(NH_3)_4]Cl$	1, 13
cis-$[CrCl_2(NH_3)_4]X$ ($X = Cl, ClO_4$)	13
trans-$[CrClH_2O(NH_3)_4]X_2$ ($X = Cl, ClO_4$)	1, 2
cis-$[CrClH_2O(NH_3)_4]Cl_2$	29
trans-$[Cr(H_2O)(OH)(NH_3)_4](ClO_4)_2$	1
trans-$[Cr(H_2O)_2(NH_3)_4](ClO_4)_3$	1, 2
trans-$[CrXF(NH_3)_4]ClO_4$ ($X = Cl, Br$)	1, 3
trans-$[CrBrCl(NH_3)_4]X$ ($X = Cl, Br$)	27
trans-$[CrFNCS(NH_3)_4]ClO_4$	6
trans-$[CrClNCS(NH_3)_4]X$ ($X = SCN, ClO_4$)	19
trans-$[Cr(NCS)(H_2O)(NH_3)_4](ClO_4)_2$	19
trans-$[Cr(NCS)_2(NH_3)_4]ClO_4$	6
trans-$[CrIH_2O(NH_3)_4]X_2$ ($X = I, ClO_4$)	4
cis-$[CrIH_2O(NH_3)_4]X_2$ ($X = I, ClO_4$)	4
trans-$[CrClI(NH_3)_4]ClO_4$	4
cis- and *trans*-$[Cr(ONO)_2(NH_3)_4]ClO_4$	17
trans-$[CrCl(ONO)(NH_3)_4]ClO_4$	17
$[Cr(NH_3)_4(ONO)(O_2CCH_2NH_3)](ClO_4)_2$	5
$[Cr(NH_3)_4(ONO)(O_2CCH_2NH_2)](ClO_4)$	5
cis- and *trans*-$[CrH_2O(ONO)(NH_3)_4]ClO_4$	17
cis-$[CrNCS(ONO)(NH_3)_4]ClO_4$	17
trans-$[Cr(NH_3)_4(DMSO)_2](ClO_4)_3$	8, 14
cis-$[Cr(NH_3)_4(DMF)_2](ClO_4)_3$	8
trans-$[Cr(NH_3)_4(DMF)_2](ClO_4)_3$	14
cis-$[Cr(NH_3)_4(DMF)(DMSO)](ClO_4)_3$	8
cis-$[CrCl(sol)(NH_3)_4](ClO_4)_2$ (sol = DMF, DMSO, DMA)	10, 14

Table 46 *(continued)*

Complex	Ref.
trans-$[CrClDMSO(NH_3)_4](ClO_4)_2$	10, 14
cis-$[Cr(NH_3)_4(DMSO)_2](ClO_4)_3$	10, 14
trans-$[CrCl(DMF)(NH_3)_4](ClO_4)_2$	14
$[CrSO_4(H_2O)(NH_3)_4][Cr(SO_4)_2(NH_3)_4]$, $[CrSO_4(H_2O)(NH_3)_4]_2S_2O_6$	11
$[CrCrO_4(NH_3)_4]X$ ($X = I$, ClO_4, $\frac{1}{2}S_2O_6$)	23
trans-$[Cr(CN)_2(NH_3)_4]ClO_4$	24
cis-$[Cr(CN)_2(NH_3)_4]ClO_4$	24
cis-$[Cr(CN)DMSO(NH_3)_4]ClO_4$	24
cis- and *trans*-$[Cr(nic\text{-}O)_2(NH_3)_4]ClO_4$	30
trans-$[Cr(Hnic\text{-}O)_2(NH_3)_4](ClO_4)_3$[a]	30
$K[Cr(NCS)_4(NH_3)_2]$	33

[a] X-Ray single structure determination; structural determinations also on $[Cr(NH_3)_6]X_3$ ($X = [HgCl_5]$,[34] $[MnCl_5{\cdot}H_2O]$[35] and $[Cr(ONO)(NH_3)_5]Cl_2$.[36]

1. J. Glerup and C. E. Schäffer, *Inorg. Chem.*, 1976, **15,** 1408.
2. D. W. Hoppenjans and J. B. Hunt, *Inorg. Chem.*, 1969, **8,** 505.
3. G. Wirth, C. Bifano, R. T. Walters and R. G. Linck, *Inorg. Chem.*, 1973, **12,** 1955.
4. D. W. Hoppenjans, G. Gordon and J. B. Hunt, *Inorg. Chem.*, 1971, **10,** 754.
5. T. Ramasami, R. K. Wharton and A. G. Sykes, *Inorg. Chem.*, 1975, **14,** 359.
6. A. D. Kirk and C. F. C. Wong, *Inorg. Chim. Acta,* 1978, **27,** 265.
7. A. Creix and M. Ferrer, *Inorg. Chim. Acta,* 1982, **59,** 177.
8. J. Casabó, J. Ribas, V. Cubas, G. Rodríguez and F. J. Fernández, *Inorg. Chim. Acta,* 1979, **36,** 183.
9. J. M. Coronas, J. Casabó and M. Ferrer, *Inorg. Chim. Acta,* 1977, **25,** 109.
10. W. G. Jackson, P. D. Vowles and W. W. Fee, *Inorg. Chim. Acta,* 1977, **22,** 111.
11. J. M. Coronas, J. Ribas, J. Casabó and N. Cabayol, *Inorg. Chim. Acta,* 1977, **21,** 51.
12. J. Casabó, J. M. Coronas and M. Ferrer, *Inorg. Chim. Acta,* 1976, **16,** 47.
13. W. G. Jackson, P. D. Vowles and W. W. Fee, *Inorg. Chim. Acta,* 1976, **19,** 221.
14. P. Riccieri and E. Zinato, *J. Inorg. Nucl. Chem.*, 1981, **43,** 739.
15. E. Zinato, R. Lindholm and A. W. Adamson, *J. Inorg. Nucl. Chem.*, 1969, **31,** 449.
16. N. Al-Shalti, T. Ramasami and A. G. Sykes, *J. Chem. Soc., Dalton Trans.*, 1977, 94.
17. T. C. Matts and P. Moore, *J. Chem. Soc. (A)*, 1971, 1632.
18. T. C. Matts, P. Moore, D. M. W. Ogilvie and N. Winterton, *J. Chem. Soc., Dalton Trans.*, 1973, 992.
19. P. Riccieri and E. Zinato, *J. Am. Chem. Soc.*, 1975, **97,** 6071.
20. N. J. Curtis, G. A. Lawrance and A. M. Sargeson, *Aust. J. Chem.*, 1983, **36,** 1495.
21. R. Davies and R. B. Jordan, *Inorg. Chem.*, 1971, **10,** 2432.
22. E. Zinato, C. Furlani, G. Lanna and P. Riccieri, *Inorg. Chem.*, 1972, **11,** 1746.
23. J. Casabó, J. Ribas and J. M. Coronas, *J. Inorg. Nucl. Chem.*, 1976, **38,** 886.
24. P. Riccieri and E. Zinato, *Inorg. Chem.*, 1981, **20,** 3722.
25. E. A. Hosegood and J. L. Burmeister, *Synth. Inorg. Metal.-Org. Chem.*, 1971, **1,** 21.
26. P. Riccieri and E. Zinato, *Inorg. Chem.*, 1980, **19,** 853.
27. H. Ueno, A. Uehara and R. Tsuchija, *Bull. Chem. Soc. Jpn.*, 1981, **54,** 1821.
28. N. E. Dixon, G. A. Lawrance, P. A. Lay and A. M. Sargeson, *Inorg. Chem.*, 1984, **23,** 2940.
29. P. Andersen, T. Damhus, E. Pedersen and A. Petersen, *Acta Chem. Scand., Ser. A*, 1984, **38,** 359.
30. J. C. Chang, L. E. Gerdom, N. C. Baenziger and H. M. Goff, *Inorg. Chem.*, 1983, **22,** 1739.
31. T. Schönherr and H. H. Schmidtke, *Inorg. Chem.*, 1979, **18,** 2726.
32. N. V. Duffy and F. G. Kosel, *Inorg. Nucl. Chem. Lett.*, 1969, **5,** 519.
33. A. H. Norbury, *Adv. Inorg. Chem. Radiochem.*, 1975, **17,** 231.
34. W. Clegg, *J. Chem. Soc., Dalton Trans.*, 1982, 593.
35. W. Clegg, *Acta Crystallogr., Sect. B*, 1978, **34,** 3328.
36. E. Nordin, *Acta Crystallogr., Sect. B*, 1978, **34,** 2285.

(i) *Diammines and triammines*

Of the five possible isomers of the diammine $[CrCl_2(NH_3)_2(H_2O)_2]Cl$ the one with all-*trans* ligand pairs,[339] and that which is *cis* with respect to NH_3 and *trans* with respect to Cl[340] have been isolated; $[CrBr_2(NH_3)_2(H_2O)_2]Br$ is also known.[341] The triammines *mer*-$[CrX_3(NH_3)_3]$ ($X = Cl$, Br), *mer*-$[CrBrCl_2(NH_3)_3]$ and *mer*-$[CrBr_2Cl(NH_3)_3]$ have been obtained by thermal decomposition of the hexaammines *etc.* in a quasi-closed system which produces much purer intermediate compounds than the more usual open dynamic conditions;[342] *fac*-$[CrCl_3(NH_3)_3]$ and *fac*-$[Cr(NH_3)_3(H_2O)_3](ClO_4)_3$ have been isolated as in Scheme 34.[343]

(ii) *Tetraammines*

Schemes 32 and 33 illustrate the preparations of various members of the numerous *trans*-tetraammine Cr^{III} series[344] starting from *trans*-$[CrF_2(py)_4]I$. Analogous *cis* complexes

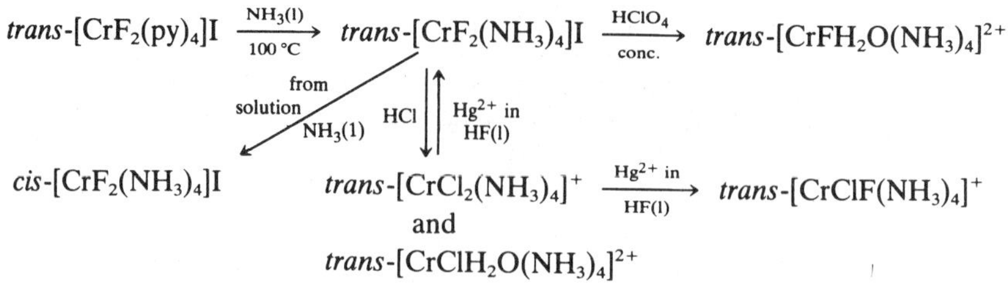

Scheme 32

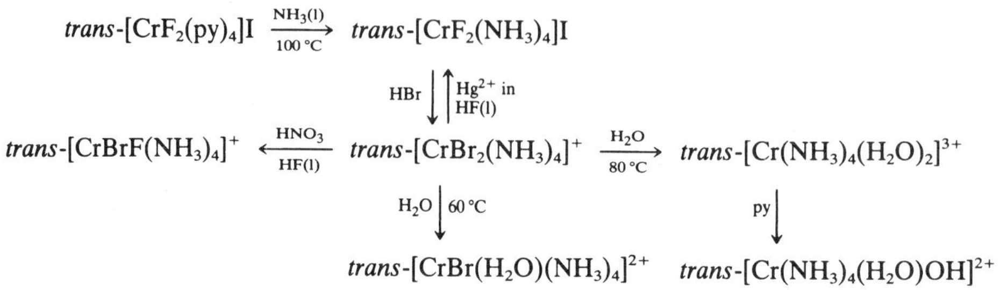

Scheme 33

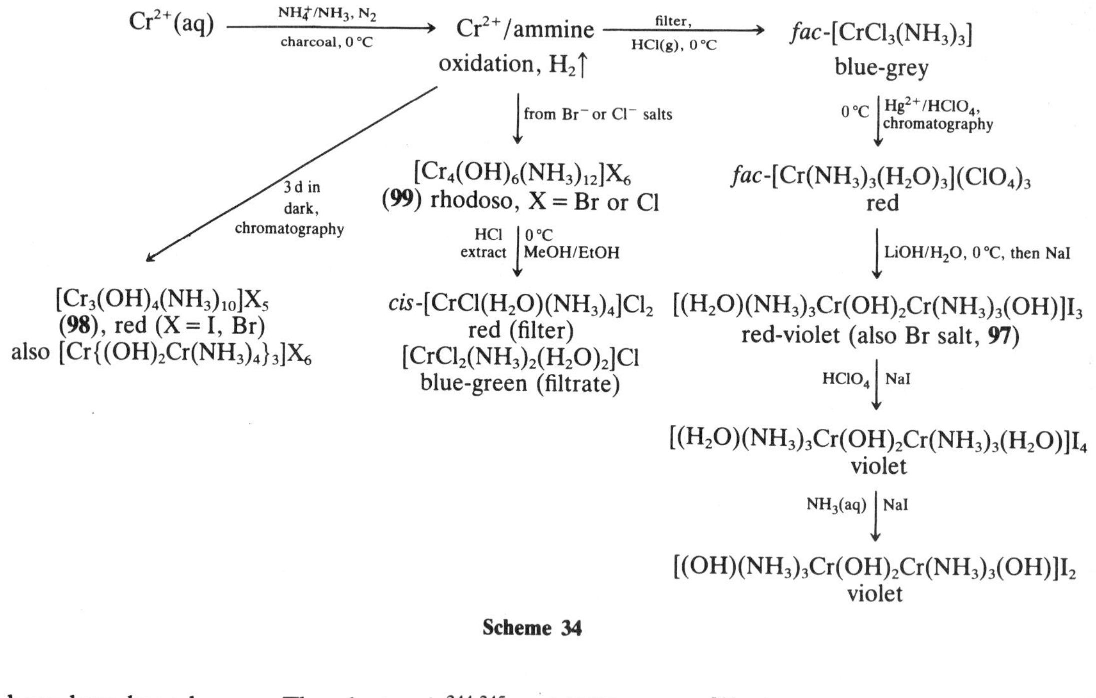

Scheme 34

have long been known. The electronic[344,345] and ESR spectra[346] of various *trans* ammines and $[CrXY(py)_4]^{n+}$ have been determined and ligand field parameters derived.

Cyanide anation in DMSO has proved useful in the preparation of cyano complexes (Scheme 29).[325] Cyanide displaces coordinated DMSO cleanly but deprotonates rather than displaces coordinated water. The preparation of *trans*- and *cis*-$[Cr(CN)_2(NH_3)_4]ClO_4$ starts from $[Cr(DMSO)_2(NH_3)_4]^{3+}$ or $[CrCl(DMSO)(NH_3)_4]^{2+}$. The insolubility of *trans*-$[Cr(CN)_2(NH_3)_4]ClO_4$ permitted its isolation from solution, but it was necessary to separate the *cis* isomer by ion exchange. Rapid *trans* to *cis* isomerization of *trans*-$[Cr(DMSO)_2(NH_3)_4]^{3+}$ occurs on the addition of CN^- before anation produces the dicyano complexes stepwise, both apparently *via* *cis*-$[CrCN(DMSO)(NH_3)_4]^{2+}$, which in an uncommon rearrangement gives a mixture of *trans*- and *cis*-dicyano complexes. Substitution of DMSO by CN^- in water is stereoretentive.

(iii) Pentaammines

Many new pentaammines have been isolated, mainly for photochemical and kinetic investigations. The absorption and luminescence spectra of single crystals and powders of $[Cr(OH)(NH_3)_5](ClO_4)_2$ have been investigated[347] from 7 K to room temperature. The OH^- group produces a large tetragonal component, which causes a large splitting of the second but not the first spin-allowed band. From the fine structure in the luminescence spectrum the totally symmetric Cr—O stretch is at $570\,cm^{-1}$. The far-IR, Raman and vibronic spectra of $[Cr(NCO)(NH_3)_5](NO_3)_2$ have also been studied.[348] There is some dispute over the ability of crystal field calculations to account for the detailed splittings. The ESR spectra of $[CrX(NH_3)_5]^{n+}$ and other tetragonal Cr^{III} complexes have been analyzed.[349] Little magnetic ordering occurs at very low temperatures in complexes such as $[Cr(H_2O)(NH_3)_5][Cr(CN)_6]$, $\theta = 0.34$ K.[350]

(iv) Hexaammines

As might be expected with the well-known hexaammines, there has been little additional work. Scheme 31 describes the thermal decomposition of some hexaammines and pentaammines; the borohydride $[Cr(NH_3)_6][BH_4]_3$ has been prepared by a modified method and its thermal stability studied,[351,352] and $[Cr(NH_3)_6][Co(N_3)_6]$ has been isolated.[353] The skeletal vibrational frequencies of $[Cr(NH_3)_6](NO_3)_3$ have been assigned,[354] and from low temperature single crystal investigations of $[Cr(NH_3)_6](ClO_4)_2Cl\cdot KCl$ it has been deduced[355] that the Cr—NH_3 bonds are lengthened by 0.12 Å along two axes and shortened by 0.02 Å along a third in the excited $^4T_{2g}$ state compared with the $^4A_{2g}$ ground state. The zero field splitting in $[Cr(NH_3)_6](ClO_4)_2Br\cdot CsBr$ has been determined[356] by single crystal magnetic susceptibility studies from 0.040 to 4.2 K, and the value is in agreement with ESR results. Even allowing for zero field splitting in $[Cr(NH_3)_6][CdCl_5]$ it appears that there is weak antiferromagnetic interaction between the cations ($\theta = 1.4$ K) although the Cr—Cr separation is 7.9 Å.[357]

The study of equilibria between Cr^{III} and ligands in aqueous solution is complicated by the robustness of Cr^{III} complexes. Equilibria can be established quickly in Co^{III} systems if charcoal is added because this reduces a little Co^{III} to Co^{II}, but in Cr^{III} systems Cr^{II} must be added as well as charcoal. Oxidation by the medium could remove all the Cr^{II} before equilibrium is reached among the Cr^{III} species, but it has been found[358] that a constant concentration of Cr^{II} can be maintained by electrolytic reduction until equilibrium is reached. Then the remaining Cr^{II} can be allowed to oxidize and the equilibrium concentrations of the different Cr^{III} species determined. In this way log β_6 for complexation with ammonia has been found to be 13 (24 °C, 4.5 M NH_4Cl) and log β_3 (en) = 19.5 (24 °C, 1 M NaCl). The method works also with pn and edta, but fails with glycine and non-amine ligands like ox and SCN^- where a steady state concentration of Cr^{II} cannot be built up.

The photoinduced reactions of the cations $[Cr(NH_3)_{6-n}(H_2O)_n]^{3+}$ have been investigated,[359] and their visible spectra recorded during a study[360] of the formation of $[Cr(NH_3)_6]^{3+}$ by the acid-catalyzed hydrolysis of $[Cr(NCO)(NH_3)_5]^{2+}$.

(v) Polynuclear complexes

Table 47 lists the polynuclear ammines which have been extensively investigated since the review by Garner and House.[302]

The interconversions between monobridged dinuclear Cr^{III} ammines with their traditional names are illustrated in Scheme 35.[302,361] Inter-relations exist between the mono-μ-hydroxo and di-μ-hydroxo species in acid solution, and these are outlined in Scheme 36. Acid hydrolysis of the mono-bridged dinuclear cations to mononuclear species is slow and salts of (**94**) and (**95**) as well as of the long-established μ-diols are known. In ammonia buffer $[(OH)(NH_3)_4Cr(OH)Cr(NH_3)_4(OH)]^{3+}$ forms.[362]

The basic rhodo complexes $[(NH_3)_5CrOCr(NH_3)_5]X_4$ (**88**) cannot be obtained free from the basic *erythro* salts $[(NH_3)_5Cr(OH)Cr(NH_3)_4OH]X_4$ (**90**) unless cold solutions protected from light and containing concentrated ammonia are used.[361] The pure samples exhibit reproducible antiferromagnetic behaviour; it seems that the linear Cr—O—Cr bridge with its short Cr—O distances (≈1.80 Å)[363,364] permits the transmission of strong exchange interactions in which π

Table 47 Oxo- and Hydroxo-bridged Ammines

Complex	*Comments*	*Ref.*
Dinuclear (monobridged)		
$[(NH_3)_5CrOCr(NH_3)_5]X_4$ Dark blue **(88)**	X = Cl, Br, ClO_4; X = Cl, Br, singlet–triplet sepn. 450 cm^{-1}, linear Cr—O—Cr, Cr—O, *ca.* 1.80, Cr—N (av), 2.12 Å	1
$[(NH_3)_5Cr(OH)Cr(NH_3)_5]Cl_5$ Red **(89)**	1 H_2O: singlet–triplet sepn. 31.5 cm^{-1}, Cr—O—Cr, 165.6°, Cr—O, 1.94, Cr—N, 2.07–2.15 Å 2 H_2O: Cr—O—Cr, 158.4°, Cr—O, 1.974, Cr—N, 2.08–2.09 Å	2–4
$[(NH_3)_5Cr(OH)Cr(NH_3)_4(OH)]X_4$ (*cis*) **(90)**	X = Cl, Br, ClO_4, $0.5S_2O_6$; X = Cl, Br, $0.5S_2O_6$, singlet–triplet sepn. 23 cm^{-1}; $X_4 = (S_2O_6)_2{\cdot}3H_2O$, singlet–triplet sepn. 21 cm^{-1}, Cr—O—Cr, 142.8°, Cr—O (br), 1.962, 1.989, Cr—O (term), 1.915, Cr—N, 2.05–2.11 Å	1, 2 5
$[(NH_3)_5Cr(OH)Cr(NH_3)_4(H_2O)]X_5$ (*trans*) **(91)**	X = Cl, NO_3; $X_5 = Cl_5{\cdot}3H_2O$, singlet–triplet sepn. 34.6 cm^{-1}, Cr—O—Cr, 155.1°, Cr—O (br), 1.983, 2.084, Cr—N (av), 2.070, disorder in *trans* ligands, Cr—O(N) (av), 2.071 Å	2, 6
$[(NH_3)_5Cr(OH)(NH_3)_4Cl]X_4$	X = Cl, ClO_4	2
$[(H_2O)(NH_3)_4Cr(OH)Cr(NH_3)_4(OH)][ZnCl_3OH](ClO_4)_2$ (*cis, cis*) Red	$(\varepsilon, \lambda)_{max}$: (97.7, 519.5), (77.4, 385.5)	7
$[(H_2O)(NH_3)_4Cr(OH)Cr(NH_3)_4(H_2O)][ZnCl_4]_2Cl$ (*cis, cis*) Red	$(\varepsilon, \lambda)_{max}$: (98.0, 514.0), (62.7, 380.0)	7
$[(OH)(NH_3)_4Cr(OH)Cr(NH_3)_4(OH)]^{2+}$ (*cis, cis*)	Ammonia buffer $(\varepsilon, \lambda)_{max}$: (100.0, 534.0), (91.0, 402.0)	7
Trinuclear		
$[Cr_3(OH)_4(NH_3)_{10}]Br_5$ Red	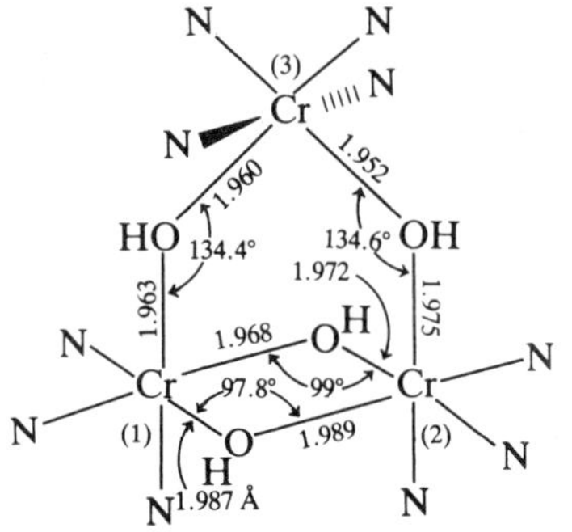 **(98)**	13
Tetranuclear		
$[Cr\{(OH)_2Cr(NH_3)_4\}_3]Br_6$ Blue violet	Analogue of Werner's brown salt (Co^{III})	15
$[Cr_4(OH)_6(NH_3)_{12}]Cl_6$	Rhodoso complex 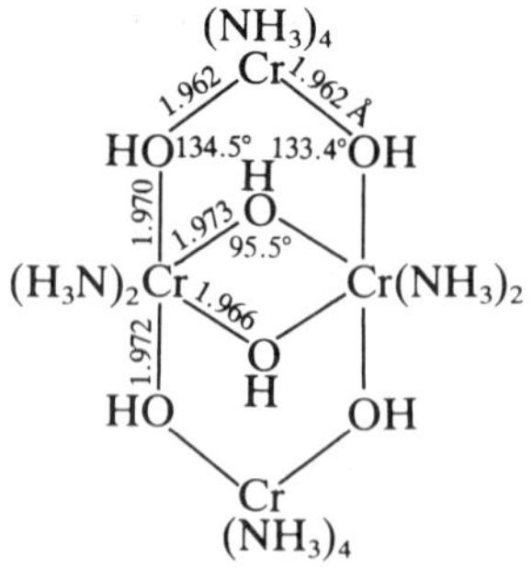 **(99)**	13, 14, 15
Dinuclear (dibridged)		
$[(NH_3)_4Cr(OH)_2Cr(NH_3)_4]X_4$ Red	X = Cl, Br, ClO_4, $0.5S_2O_6$ $(\varepsilon, \lambda)_{max}$: (124.5, 536.0), (67.3, 386.5)	8–10 7
$X_4 = Cl_4{\cdot}4H_2O$:	Triclinic form, singlet–triplet sepn. 1.43 cm^{-1}, Cr—O—Cr, 100.83°, Cr—O (av), 1.973, Cr—N, 2.068–2.081, Cr—Cr, 3.041 Å, θ = 50°	11
	Monoclinic form, singlet–triplet sepn. 4.1 cm^{-1}, Cr—O—Cr, 99.92°, Cr—O (av), 1.974, Cr—N, 2.074–2.089, Cr—Cr, 3.023 Å, θ = 41°	9
$X_4 = (S_2O_6)_2{\cdot}4H_2O$:	Singlet–triplet sepn. 5.8 cm^{-1}, Cr—O—Cr, 101.54°, Cr—O (av), 1.965, Cr—N, 2.078–2.082, Cr—Cr, 3.045 Å, θ = 24°	9

Table 47 *(continued)*

Complex	Comments	Ref.
$[(H_2O)(NH_3)_3Cr(OH)_2Cr(NH_3)_3(OH)]Br_3{\cdot}2H_2O$ *trans, facial,* red–violet	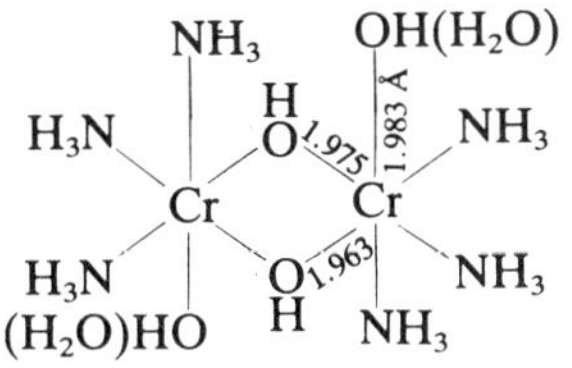 **(97)**	12
$[(H_2O)(NH_3)_3Cr(OH)_2Cr(NH_3)_3(H_2O)]I_4{\cdot}4H_2O$ Violet	Protonation of **(97)**	12
$[HO(NH_3)_3Cr(OH)_2Cr(NH_3)_3(OH)]I_2{\cdot}H_2O$ Violet	Deprotonation of **(97)**	12

1. E. Pedersen, *Acta Chem. Scand.*, 1972, **26**, 333.
2. D. W. Hoppenjans and J. B. Hunt, *Inorg. Chem.*, 1969, **8**, 505.
3. J. T. Veal, D. Y. Jeter, J. C. Hempel, R. P. Eckberg, W. E. Hatfield and D. J. Hodgson, *Inorg. Chem.*, 1973, **12**, 2928.
4. P. Engel and H. U. Güdel, *Inorg. Chem.*, 1977, **16**, 1589.
5. D. J. Hodgson and E. Pedersen, *Inorg. Chem.*, 1980, **19**, 3116.
6. S. J. Cline, J. Glerup, D. J. Hodgson, G. S. Jensen and E. Pedersen, *Inorg. Chem.*, 1981, **20**, 2229.
7. F. Christensson and J. Springborg, *Acta Chem. Scand., Ser. A*, 1982, **36**, 21.
8. J. Springborg and C. E. Schäffer, *Inorg. Synth.*, 1978, **18**, 75.
9. S. J. Cline, D. J. Hodgson, S. Kallesøe, S. Larsen and E. Pedersen, *Inorg. Chem.*, 1983, **22**, 637.
10. S. Decurtins and H. U. Güdel, *Inorg. Chem.*, 1982, **21**, 3598.
11. D. J. Hodgson and E. Pedersen, *Inorg. Chem.*, 1984, **23**, 2363.
12. P. Andersen, K. M. Nielsen and A. Petersen, *Acta Chem. Scand., Ser. A*, 1984, **38**, 593.
13. P. Andersen, T. Damhus, E. Pedersen and A. Petersen, *Acta Chem. Scand., Ser. A*, 1984, **38**, 359.
14. H. U. Güdel, U. Hauser and A. Furrer, *Inorg. Chem.*, 1979, **18**, 2730.
15. E. Bang, *Acta Chem. Scand., Ser. A*, 1984, **38**, 419.

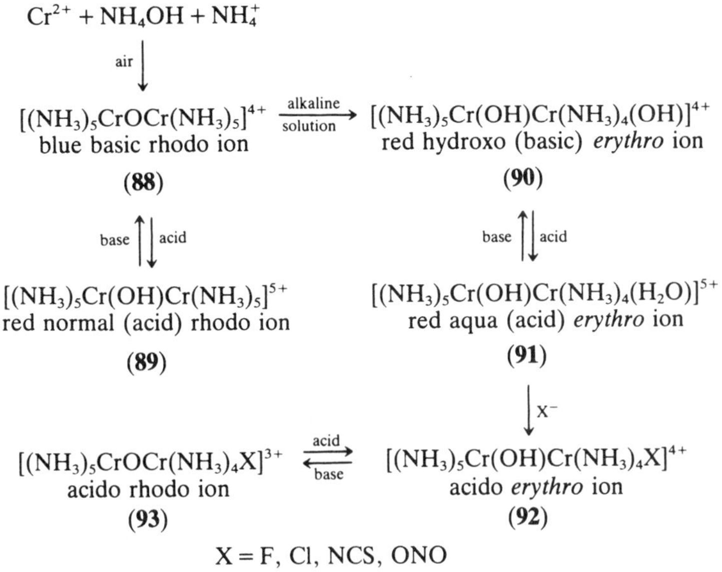

X = F, Cl, NCS, ONO

Scheme 35

bonding is important. An almost linear bridge is also present in $[\{Cr(NCS)(TPyEA)\}_2O](BPh_4)_2$.[107]

The normal (acid) rhodo complex $[(NH_3)_5Cr(OH)Cr(NH_3)_5]Cl_5{\cdot}H_2O$ (**89**) has a bent bridge (Cr—O—Cr angle = 165.6°) and shows much weaker antiferromagnetic interaction (Table 47).[365] The Cr—N bonds *trans* to the bridge are a little longer (0.07 Å) than the other Cr—N bonds. This may relate to the formation of the basic *erythro* salts by the substitution of a terminal NH_3 by OH. However, no bond lengthening is apparent in the structure of the dihydrate $[(NH_3)_5Cr(OH)Cr(NH_3)_5]Cl_5{\cdot}2H_2O$, the polarized electronic spectrum of which has been investigated in order to gain information on the exchange-induced intensity-gaining mechanisms for spin-forbidden transitions.[366]

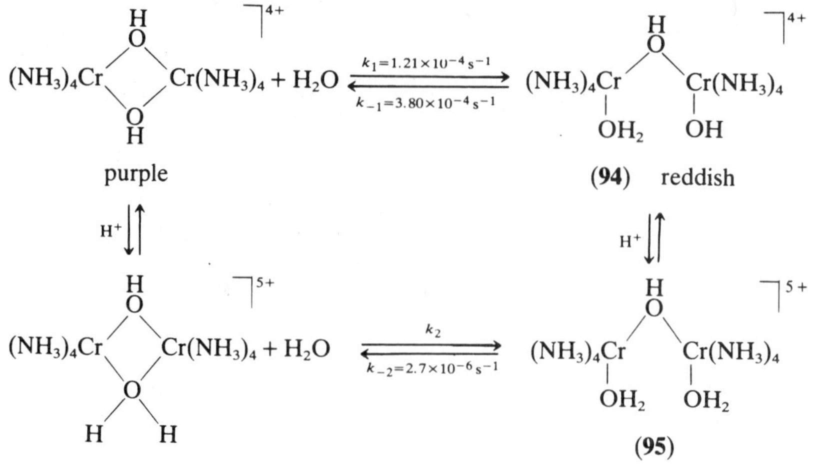

Scheme 36

Structural, spectroscopic and magnetic investigations have been carried out on the basic *erythro* complex *cis*-$[(NH_3)_5Cr(OH)Cr(NH_3)_4(OH)](S_2O_6)_2\cdot 3H_2O$[367] (**90**), the acid *erythro* complex *trans*-$[(NH_3)_5Cr(OH)Cr(NH_3)_4(H_2O)]Cl_5\cdot 3H_2O$[368] (**91**), for which the bridging angles are respectively 142.8° and 155.1°, and the di-μ-hydroxo complexes $[(NH_3)_4Cr(OH)_2Cr(NH_3)_4]X_4\cdot 4H_2O$ (X = Cl, Br or $0.5S_2O_6$).[369–373]

The dibridged ammines are examples of the general dinuclear complex (**96**), with L = NH_3 and R = H. Much research has been aimed at relating antiferromagnetic behaviour and spectroscopic properties to structure. There does not seem to be a simple correlation between the magnitude of the exchange parameter J (singlet–triplet splitting) and bridge geometry, *i.e.* the Cr—O—Cr angle ϕ, the Cr—O bond length r, or the Cr—Cr separation. Evidence is accumulating that there is a correlation between the magnitude of J and the angular displacement of the R (or H) atom out of the Cr_2O_2 plane (dihedral angle θ). The implications are that superexchange through a $p\pi$ orbital on O is dominant and this is at a maximum only if the H atom lies in the plane. Recently, a model has been proposed which relates the magnitude of the magnetic interaction to all three parameters r, ϕ and θ.[372,374]

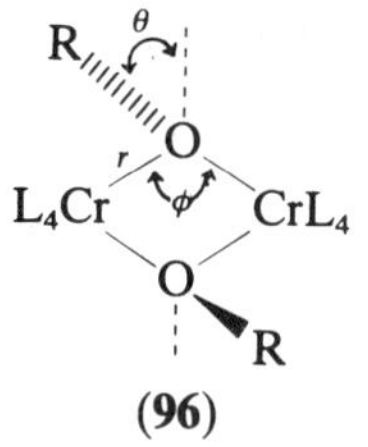

(**96**)

The synthetic work has been extended to di-μ-hydroxo-*fac*-triammine derivatives (**97**) (Table 47 and Scheme 34). The crystals of (**97**) contain two different centrosymmetric cations, the *trans* diaqua and *trans* dihydroxo species, linked alternately in infinite chains by short (2.45 Å) hydrogen bonds. The bond distances of the diaqua cation differ slightly from those of the dihydroxo cation.[375]

The preparations of the trinuclear complex $[Cr_3(OH)_4(NH_3)_{10}]Br_5$ (**98**) and the tetranuclear complexes $[Cr\{(OH)_2Cr(NH_3)_4\}_3]Br_6$ and $[Cr_4(OH)_6(NH_3)_{12}]Cl_6$ (**99**) (Table 47 and Scheme 34) start with aqueous Cr^{II}—NH_4^+/NH_3 buffer solutions and charcoal.[376] The antiferromagnetic behaviour of (**98**) has been analyzed in terms of its triangular structure [J_{12} (7.90 or 18.7 cm^{-1}), $J_{13} = J_{23}$ (18.7 or 4.3 cm^{-1})], which resembles that of (**99**) (Table 47). The complex $[Cr\{(OH)_2Cr(NH_3)_4\}_3]Br_6$ also behaves antiferromagnetically, and is structurally an analogue of Werner's brown salt $[Co\{(OH)_2Co(NH_3)_4\}_3]Cl_6$.

The tetranuclear 'rhodoso' complex $[Cr_4(OH)_6(NH_3)_{12}]Cl_6\cdot 4H_2O$ has the structure (**99**).[376,377] Its antiferromagnetic behaviour can be reproduced in terms of dominant exchange parameters along the side and shorter diagonal of the planar rhombus formed by the four chromium atoms. Some differences in parameters may be due to the use of slightly different crystal modifications.[377,378] Neutron inelastic scattering has been used to observe the exchange splittings of the ground electronic state and provide refined and additional exchange parameters. There is a sharp drop in magnetic moment below 10 K which requires that the lowest level of the ground state be a spin singlet with a low-lying triplet. The triplet from the temperature dependence of the neutron inelastic scattering spectra lies less than 4 cm^{-1} above the singlet. The exchange splittings of the ground state have also been observed through the temperature dependent intensities (<11 K) of the transitions to the ^2E and $^2T_1$ levels.[378]

35.4.2.2 Monodentate amines

A number of chromium(III) complexes containing monodentate amine ligands are known and details of their electronic spectra are given in Table 48.

Table 48 Electronic Spectra of Monodentate Amine Chromium(III) Complexes

Complex	λ_{max} (ε)(nm, $dm^3\ mol^{-1}\ cm^{-1}$)		*Medium*	*Ref.*
$[Cr(NH_3)_6]^{3+}$	462 (44.0)	346 (37.2)	DMSO/glycerol	1
$[Cr(MeNH_2)_6]^{3+}$	477 (71.8)	360 (62.5)	DMSO/glycerol	1
$[Cr(EtNH_2)_6]^{3+}$	478 (71.5)	363 (61.5)	DMSO/glycerol	1
$[CrCl(MeNH_2)_5]^{2+}$	526 (—)	—	Glycerol/H_2O	2
$[CrBr(MeNH_2)_5]^{2+}$	535 (46.3)	390 (54.7)	HBr/H_2O	3
$[Cr(OSO_2CF_3)(MeNH_2)_5]^{2+}$	506 (49.1)	374 (45.8)	CF_3SO_3H	4
$[Cr(MeNH_2)_5H_2O]^{3+}$	490 (55.6)	368 (49.2)	H_2O, pH 4	5
$[Cr(MeNH_2)_5NH_3]^{3+}$	474 (52.5)	359 (45.5)	HBr/H_2O	3
	474 (61.7)	360 (52.9)	DMSO/glycerol	1
$[Cr(EtNH_2)_5NH_3]^{3+}$	476 (65.2)	362 (56.0)	DMSO/glycerol	1
$[Cr(MeNH_2)_2(en)_2]^{3+a}$	465 (84.1)	354 (71.6)	DMSO/glycerol	1
trans-$[CrF_2(MeNH_2)_4]^+$	494 (22.4)	408 (14.5) 356 (16.3)	0.1 M HCl	6
trans-$[CrF_2(Pr^nNH_2)_4]^+$	502 (26.1)	413 (16.1) 356 (17.5)	0.1 M HCl	6

[a] Prepared from *trans*-$[CrBr_2(en)_2]Br$ and liquid $MeNH_2$ but product of unknown isomer composition. The pn and tn complexes were prepared similarly.
1. K. Kühn, F. Wasgestian and H. Kupka, *J. Phys. Chem.*, 1981, **85,** 665.
2. H. L. Schläefer, H. Gausmann and H. Witzke, *Z. Phys. Chem.* (*Frankfurt am Main*), 1967, **56,** 55.
3. K. Kühn and F. Wasgestian, *Inorg. Nucl. Chem. Letts.*, 1976, **12,** 803.
4. N. E. Dixon, G. A. Lawrance, P. A. Lay and A. M. Sargeson, *Inorg. Chem.*, 1984, **23,** 2940.
5. Yu. N. Shevchenko, V. V. Sachok and N. K. Davidenko, *Russ. J. Inorg. Chem.* (*Engl. Transl.*), 1979, **24,** 30.
6. J. Glerup, J. Josephsen, K. Michelsen, E. Pedersen and C. E. Schäffer, *Acta Chem. Scand.*, 1970, **24,** 247.

(*i*) *Aliphatic hexamines*

The reaction between $CrBr_3$ and liquid alkylamines RNH_2 (R = Me, Et) at reflux temperatures affords the yellow-coloured hexamines $[Cr(RNH_2)_6]^{3+}$, which have been isolated as nitrate (R = Me),[379,380] bromide (R = Me)[380] and perchlorate (R = Me,[380] Et[381]) salts. Hydrolysis of $[Cr(MeNH_2)_6]^{3+}$ in $HClO_4$ solution gives $[Cr(MeNH_2)_5H_2O]^{3+}$ but in solutions of other inorganic acids (HX) gives the acidopentamine complexes $[CrX(MeNH_2)_5]^{2+}$.[380,382] Reaction between $[Cr(MeNH_2)_6]Br_3$ and $(NH_4)_2[B_{10}H_{10}]$ in aqueous solution leads to crystalline $[Cr(MeNH_2)_6]Br(B_{10}H_{10})\cdot 2H_2O$ which undergoes thermal decomposition to a monodentate decaborane complex (Scheme 37).[383] The isomers *cis*- and *trans*-$[Cr(MeNH_2)_2(en)_2]Br(B_{10}H_{10})$ are stable up to 250 °C but above this temperature both lose $MeNH_2$ and give the common product *trans*-$[Cr(B_{10}H_{10})Br(en)_2]$.[383]

$$[Cr(MeNH_2)_6]Br(B_{10}H_{10})\cdot 2H_2O \xrightarrow{70\text{–}80\,^\circ C} [Cr(MeNH_2)_6]Br(B_{10}H_{10})$$

$$\downarrow 100\,^\circ C$$

$$Cr(B_{10}H_{10})Br(MeNH_2)_4 \xleftarrow{120\text{–}150\,^\circ C} [CrBr(MeNH_2)_5]B_{10}H_{10}$$

Scheme 37

(*ii*) *Aliphatic pentamines*

In contrast to $CrBr_3$ the reaction of $CrCl_3$ with neat alkylamines stops at the pentamine stage and gives the products $[CrCl(RNH_2)_5]Cl_2$ (R = Me, Et, Pr^n, Bu^n, Bu^i, C_3H_5).[384,385] The light-red-coloured complexes $[CrBr(RNH_2)_5]Br_2$ (R = Me, Et) may however be obtained by controlled addition of liquid amine to $CrBr_3$.[381,386] These complexes readily react with liquid NH_3 to give $[Cr(NH_3)(RNH_2)_5]Br_3$.[381] Owing to the general lability of the triflate ligand, the complex $[Cr(OSO_2CF_3)(MeNH_2)_5](CF_3SO_3)_2$, obtained from a solution of $[CrCl(MeNH_2)_5]Cl_2$ in CF_3SO_3H, is a potential starting material for the synthesis of $[CrX(MeNH_2)_5]^{n+}$ complexes.[387]

An X-ray crystal structure determination of $[CrCl(MeNH_2)_5]Cl_2$ shows that the cation is severely distorted having Cr—N—C and N—Cr—N bond angles which deviate by up to 15 and 6.6° from normal tetrahedral and octahedral angles respectively.[388] In accordance with the sterically crowded nature of its ground state, this complex undergoes base hydrolysis 225 times faster than $[CrCl(NH_3)_5]^{2+}$ at 298 K which is in the expected direction for a D_{cb} mechanism.[388] Aquation of $[CrCl(MeNH_2)_5]^{2+}$, however, is 38 times slower than that of the ammine complex at 298 K, an order which is reversed in the analogous cobalt(III) complexes.[388-390] This reversal in reactivity is attributed to a change in mechanism from I_d in cobalt(III) to I_a in chromium(III).[390] The orange-coloured complex $[Cr(MeNH_2)_5H_2O](NO_3)_3$ may be prepared by passing a solution of $[CrBr(MeNH_2)_5]^{2+}$ through an ion exchange resin in the OH^- form followed by acidification of the effluent with concentrated HNO_3.[391]

(iii) *Aliphatic tetramines, triamines and diamines*

Reaction of *trans*-$[CrF_2(py)_4]^+$ with amines (RNH_2) at 100 °C provides a convenient route to *trans*-$[CrF_2(RNH_2)_4]^+$ complexes, a number of which have been isolated and characterized (R = Me, Et, Pr^n, C_3H_5).[392] Thermal decomposition of $[CrCl(EtNH_2)_5]Cl_2$ in the solid state at 250 °C gives the green-coloured complex $CrCl_3(EtNH_2)_3$, while its reaction with KSCN in aqueous solution leads to the isomers 1,2,6- (red) and 1,2,3-$Cr(NCS)_3(EtNH_2)_3$ (lilac).[302]

The reaction between $CrCl_3$ and Me_3N in the presence of zinc dust catalyst gives the five-coordinate purple-blue solid $CrCl_3(Me_3N)_2$.[393] This complex has a distorted *trans* trigonal bipyramidal structure and a magnetic moment of 3.88 BM (298 K), which is virtually independent of temperature.[394] In benzene solution this complex undergoes either ligand substitution or decomposition as outlined in Scheme 38.[395,396] The hexacoordinated chromium(III) dimer $Cr_2Cl_6(Me_3N)_3$ has magnetic (μ_{eff} at 297 K = 3.68 BM; obeys the Curie–Weiss law over the temperature range 103–297 K) and spectroscopic ($^4A_{2g} \rightarrow {}^4T_{2g}$ at 13 300 cm^{-1}; $^4A_{2g} \rightarrow {}^4T_{1g}$ at 18 520 cm^{-1}) properties like those of $[Cr_2Cl_9]^{3-}$, suggesting a similar confacial bioctahedral structure with no significant metal–metal interaction. A mechanism involving halogen bridging with synchronous Me_3N expulsion and ligand reorientation is suggested for dimer formation. No reaction was observed between $CrCl_3(Me_3N)_2$ and the potential ligands CO_2, CS_2 and $(Pr^i)_2O$. In the reaction products, $CrCl_3$(tren) and $[Cr(tren)_2]Cl_3$, tren behaves as a tridentate ligand, bonded to the metal through its primary amino groups in both cases.

$$CrCl_3L_3 \xleftarrow[C_6H_6]{L (=py, THF)} CrCl_3(Me_3N)_2 \xrightarrow[C_6H_6]{tren (1:1 \text{ or } 2:1)} CrCl_3(tren) \text{ or } [Cr(tren)_2]Cl_3$$

$$\xleftarrow{py/C_6H_6}$$

$$\downarrow C_6H_6$$

$$Cr_2Cl_6(Me_3N)_3 + Me_3N$$

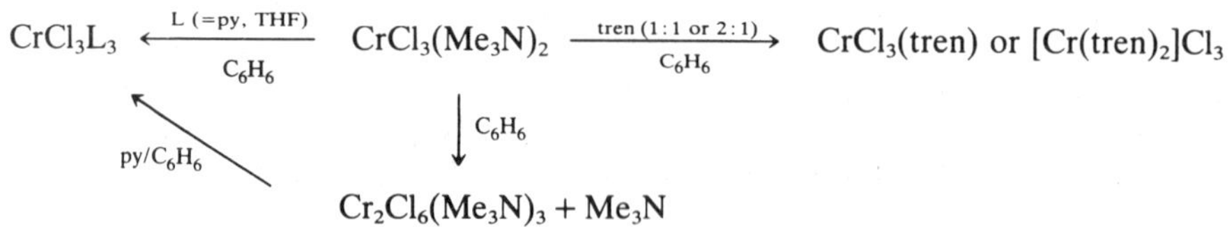

Scheme 38

(iv) *Anilines*

A number of chromium(III) complexes containing aniline $PhNH_2$ and substituted aniline ligands are known. These include $[CrCl_2(PhNH_2)_4]^+$ [302] and $CrCl_3(PhNH_2)_3$ [397] although the configuration is not stated in either case. The latter complex was prepared by refluxing a solution of $CrCl_3$ in aniline (the pyridine complex $CrCl_3(py)_3$, prepared analogously, was shown to have a *mer* configuration). Reaction of $CrCl_3(PhNH_2)_3$ and Br_2 in glacial acetic acid proceeds at a similar rate and with identical substitution orientation (*i.e.* 2,4,6-Br_3PhNH_2 product) as that of uncoordinated $PhNH_2$.[397] Coordination to chromium(III) therefore has a contrasting effect to protonation, which greatly deactivates aniline with respect to electrophilic substitution and causes it to occur at the *meta* position.

Some Reinecke-type salts containing aniline ligands have been prepared.[302] The complexes $Bi[Cr(NCS)_4L_2]_3$ (L = $PhNH_2$, morpholine, *p*-$MePhNH_2$, *p*-anisidine, 0.5bipy), $BiX_4[Cr(NCS)_4L_2]_3$ (X = thiourea, L = $PhNH_2$, morpholine), $CdX_2[Cr(NCS)_4(PhNH_2)_2]_2$ (X = en, piperazine) and $Cd[Cr(NCS)_4(PhNH_2)_2]_2$ may be used in the analysis of Bi and Cd by titrimetric, spectrophotometric or gravimetric methods.[398–400] A number of double complex salts of the type $A[Cr(NCS)_4(PhNH_2)_2]_n$ (A represents a range of cobalt(III) and silver(I) complexes) have been prepared and in the case of the Ag salts magnetic moments and positions of IR bands have been reported.[401] Reaction of *trans*-$[Cr(C_2O_4)_2(H_2O)_2]^-$ and aniline gives $[Cr(C_2O_4)_2(PhNH_2)_2]^-$, which can exist in a rose-coloured or in an unstable green-coloured modification.[302]

35.4.2.3 Ethylenediamine and related bidentate amines

The chemistry of amine complexes of chromium(III) up to 1969 has been extensively reviewed by Garner and House;[302] consequently earlier work will not be covered here. Fluorodiamine complexes of chromium(III) have also been carefully reviewed by Vaughn.[402,403]

(*i*) *General preparative methods*

The easy formation of hydroxo- or oxo-bridged Cr^{III} polymers in basic aqueous solution, the comparative lability of the Cr—N bond, and the precautions needed to obtain chromium(II) complexes compared with cobalt(II) complexes have meant that the preparative chemistry of chromium(III) is more difficult than that of cobalt(III). A greater variety of non-aqueous solvents is now in use, and there is greater knowledge of chromium(II) chemistry to be exploited in the preparation of chromium(III) complexes generally, but few new methods of preparation of amine complexes have been devised since the early work.

The general synthetic methods are: (a) reaction of anhydrous $CrCl_3$, $CrBr_3$ or the sulfate with anhydrous amines alone or in an organic solvent, sometimes with the addition of zinc powder to provide a catalytic pathway *via* chromium(II) complexes—the anhydrous halides can be prepared from the hydrated salts by reaction with 2,2-dimethoxypropane, triethylorthoformate or thionyl chloride, or by distillation of water from solutions of the hydrates in DMF or DMSO; (b) reaction of hydrated chromium(III) salts such as $CrCl_3{\cdot}6H_2O$, $[Cr(H_2O)_4Br_2]Br$ and $Cr(NO_3)_3{\cdot}6H_2O$ in solutions buffered with an amine/amine-salt mixture to prevent the formation of oxo and hydroxo polymers followed by acidification—carbon, added as a catalyst, can change the course of the reaction; (c) reaction of $[CrCl_3X_3]$ (X = py, THF, DMF, DMSO) with amines—the preliminary preparation of these compounds, even if only *in situ*, gives essentially anhydrous starting materials soluble in organic solvents, although aqueous solutions can be used in some cases; (d) reaction of salts such as $K_3[Cr(NCS)_6]$, $K_3[Cr(ox)_3]$ and $[Cr(urea)_6]Cl_3$ with amines; (e) reaction of diperoxochromium(IV) amines with acids and the reduction of chromium(VI) compounds—these methods have not been applied recently; (f) oxidation of chromium(II) complexes; and (g) substitution of aqua or acido ligands in amine complexes by ammonia (from liquid ammonia) or other amines to produce mixed ligand complexes.

Complexes containing ethylenediamine and related bidentate N donor ligands which are new since 1969, or have been prepared by new methods, are listed in Tables 49 and 50. Water of hydration has not been specified. In general, the complexes have been prepared for kinetic and photochemical investigations. The list is restricted to solids which have been characterized at least by analyses. Cations obtained only in solution have not been included. Most complexes have also been characterized by conductance measurements, and IR, UV and visible spectroscopy. Spectroscopic methods have been used to distinguish between *cis* and *trans* isomers, and where *cis* isomers have been resolved this is indicated in the tables.

(*a*) *Recent syntheses.* The most numerous complexes are those with the donor set X_2N_4 (X is anionic or neutral), especially with $N_4 = (en)_2$. The nitrosation of Cr—OH_2 groups has been used to synthesize many *trans* compounds containing Cr—ONO groups (NO^+ addition reactions), *e.g.* *trans*-$[CrBr(ONO)(en)_2]ClO_4$, but *cis* compounds could not be obtained because the ONO group labilizes *cis* ligands.[404] Destruction of Cr—ONO groups with acids

Table 49 Complexes of Ethylenediamine and Related Diamines (see also Table 50)

Complex	*Ref.*
CrN_6	
$[Cr(en)_3]X_3$ (X = F; X_3 = $Br(BH_4)_2$, $Cl_2(BH_4)$, $Cl(B_{10}H_{10})$)	1, 2
$[Cr(en)_3]Cl_3$	3, 4, 5, 6
$[Cr(en)_3](NCS)_3$ racemic and $(+)_{589}$ salts[a]	7
$[Cr(tmd)_3]Cl_3$	5
$[Cr(pn)_3]Cl_3$	3
$[Cr(en)_{3-n}(tmd)_n]Br_3$ (*n* = 1, 2, 3 resolved)	8
$[CrL_3]Cl_3$	
L = *trans*-1,2-cyclohexanediamine	5, 9, 10
L = *cis*-1,2-cyclohexanediamine, *mer*- and *fac*-Δ and -Λ isomers	11
L = *trans*-1,2-cyclopentanediamine	12
cis- and *trans*-$[Cr(en)_2(NH_2CH_3)_2]Br(B_{10}H_{10})$	2
$[Cr(NH_3)_4(en)]ClO_4$	13
For other CrN_6 species see $[Cr(NH_3)_2(en)_2]^{3+}$ and $[Cr(NCS)_2(en)_2]^+$ salts below and Table 50	
CrX_2N_4 (X is anionic or neutral)[b]	
trans-$[CrF_2(en)_2]X$ (X = Br, ClO_4)[a]	14, 15, 16
trans-$[CrF(H_2O)(en)_2]X_2$ (X = Br, ClO_4)	14
cis-$[CrF_2(en)_2][CrF_4(en)]$	17
cis-$[CrF_2(en)_2]F$, I, ClO_4 {$(-)_{546}$ isomer has Δ configuration}	1, 17, 19
cis-$[CrF(H_2O)(en)_2]X_2$ (X = I, ClO_4, BPh_4) {$(-)_{546}$ isomer has Δ configuration}	17, 18, 19
cis-$[CrClF(en)_2]X_2$ (X = I, ClO_4)	17, 18
cis-$[CrBrF(en)_2]I$	18
cis-$[CrF(NCS)(en)_2]X$ (X = I, SCN, ClO_4) {$(+)_{546}$ isomer has Λ configuration}	17, 19
trans-$[CrFX(en)_2]ClO_4$ (X = Cl, Br, SCN)	14, 20, 21
cis-$[CrF(NH_3)(en)_2]X_2$ (X = I, ClO_4 {$(-)_{546}$-isomer}, X_2 = Cl, I)	22, 23
trans-$[CrF(NH_3)(en)_2](ClO_4)_2$	22
cis- and *trans*-$[Cr(NH_3)_2(en)_2](ClO_4)_3$	13
cis-$[CrCl(NH_3)(en)_2]Br_2$	22
cis-$[CrBr(NH_3)(en)_2]Br_2$	22
trans-$[CrCl(NH_3)(en)_2]X_2$ (X = Br, ClO_4)	22
trans-$[Cr(NH_3)(H_2O)(en)_2]X_3$ (X = Cl, Br; X_3 = Br_2ClO_4)	22
cis- and *trans*-$[Cr(NH_2Me)_2(en)_2](ClO_4)_3$; *cis*-$[CrBr(NH_2Me)(en)_2]Br_2$ *cis*-$[Cr(H_2O)(NH_2Me)(en)_2](ClO_4)_3$	24
cis- and *trans*-$[Cr(NCS)(NH_3)(en)_2](SCN)_2$	25
trans-$[Cr(NCS)_2(en)_2]X$ (X = Cl, ClO_4)	26, 27
cis-$[Cr(NCS)_2(en)_2]SCN$	28, 29
cis-$[CrCl_2(en)_2]Cl$	5
cis-$[CrBr_2(en)_2]X$ (X = Br, ClO_4)	30
cis-$[CrClH_2O(en)_2]X_2$ [X = Br, I, $M(CN)_4$, (M = Pd, Pt)]	6, 31
cis-$[CrBrH_2O(en)_2]X_2$ [X = Br, I, $M(CN)_4$, (M = Ni, Pd, Pt)]	6, 31
cis-$[CrCl(DMSO)(en)_2]X_2$ (X = Cl, ClO_4)	5
trans-$[Cr(DMSO)(DMF)(en)_2](ClO_4)_3$	32
cis-$[Cr(DMSO)(DMF)(en)_2](ClO_4)_2(NO_3)$	33
trans-$[Cr(DMSO)(H_2O)(en)_2](ClO_4)_3$	33
cis-$[Cr(DMSO)_2(en)_2](ClO_4)_3$	33, 34
cis-$[CrBr(DMSO)(en)_2](ClO_4)_2$	33, 34
trans-$[CrBr(DMSO)(en)_2](ClO_4)_2$	33, 34
trans-$[CrX(ONO)(en)_2]ClO_4$ (X = Cl, Br)	6
cis-$[CrCl(ONO)(en)_2]ClO_4$	6
trans-$[CrBr(H_2O)(en)_2]X_2$ (X = Br, ClO_4)	6
trans-$[CrCl(H_2O)(en)_2]X_2$ (X = Br, ClO_4)	6
trans-$[Cr(H_2O)(DMF)(en)_2](ClO_4)_3$	6
trans-$[CrX(NO_3)(en)_2]I$ (X = Cl, Br)	6
trans-$[CrBrI(en)_2]X$ (X = I, ClO_4)	6
trans-$[CrClI(en)_2]I$	6
cis-$[CrXI(en)_2]I$ (X = Cl, Br; not isomerically pure)	6
cis-$[CrClBr(en)_2]X$ (X = Br, ClO_4)	6
trans-$[CrClBr(en)_2]X$ (X = Br, ClO_4)	6
trans-$[CrX(NCS)(en)_2]ClO_4$ (X = Cl, Br)	6
trans-$[Cr(ONO)(H_2O)(en)_2](ClO_4)_2$	6
trans-$[Cr(OH)(ONO)(en)_2]ClO_4$	6
trans-$[Cr(ONO)(DMF)(en)_2](ClO_4)_2$	6
trans-$[Cr(ONO)(DMSO)(en)_2](ClO_4)_2$	6
cis-$[Cr(ONO)_2(en)_2]ClO_4$, $(+)_{480}$ isomer isolated	35
cis- and *trans*-$[CrBr(DMF)(en)_2](ClO_4)_2$	6

Table 49 (*continued*)

Complex	*Ref.*
cis- and *trans*-$[CrCl(DMF)(en)_2](ClO_4)_2$	33
cis- and *trans*-$[CrBr(DMA)(en)_2](ClO_4)_2$	6, 33
cis-$[CrCl(DMA)(en)_2](ClO_4)_2$	33
trans-$[CrCl(DMSO)(en)_2](ClO_4)_2$	6
cis-$[CrCl(DMSO)(en)_2](ClO_4)(NO_3)$	33
trans-$[CrCl(DMF)(en)_2](ClO_4)_2$	6
cis-$[Cr(DMF)_2(en)_2](ClO_4)_3$ and $(ClO_4)_2NO_3$	6
trans-$[Cr(DMF)_2(en)_2](ClO_4)_3$	6
cis- and *trans*-$[Cr(DMA)_2(en)_2](ClO_4)_3$	6, 33
trans-$[Cr(DMSO)_2(en)_2](ClO_4)_3$	6
cis-$[Cr(DMA)(DMSO)(en)_2](ClO_4)_2NO_3$	33
trans-$[CrCl_2(en)_2]X$ ($X = Cl, ClO_4$)	6
trans-$[CrBr_2(en)_2]Br$	6
trans-$[CrI_2(en)_2]ClO_4$	6
trans-$[Cr(OH)(H_2O)(en)_2]X_2$ ($X = Br, ClO_4$)	33
cis-$[Cr(OH)(H_2O)(en)_2]S_2O_6$	33
trans-$[Cr(H_2O)_2(en)_2]Br_3$	33
$[CrI(H_2O)(en)_2]I_2$	17
$[CrI_2(en)_2]I$	17
cis-$[Cr(CN)_2(en)_2]X$ ($X = Cl$, I (resolved), ClO_4)	36, 37
trans-$[Cr(CN)_2(en)_2]ClO_4$	37, 38
cis-$[Cr(N_3)_2(en)_2]NO_3$, $(+)_{589}$-*cis*-$[Cr(N_3)_2(en)_2][(+)SbOT]$ and $[ClO_4]$, Δ conf.	29, 39
$[Cr(SR)(en)_2]ClO_4$ ($R = 2\text{-}C_6H_4CO_2^-$, $CH_2CH_2CO_2^-$)	40, 41
$[Cr(SCH_2CO_2)(en)_2]ClO_4$[a]	42
$[Cr(SCH_2CH_2NH_2)(en)_2]X_2$ ($X = ClO_4$, I)	43
$[Cr(SCMe_2CO_2)(en)_2]ClO_4$	41, 44
$[Cr(SBzCH_2NH_2)(en)_2](ClO_4)_3$ not fully characterized	45
trans-$[A][CrF(H_2O)(en)_2][X]$ ($A = K, Na, NH_4$; $X = [Cr(CN)_6], [CrNO(CN)_5]$)	46
cis-$[Cr(nic\text{-}O)_2(en)_2]Br$	47
$[Cr(CN)(NCS)(en)_2]SCN$	48
CrX_4N_2 (X is anionic or neutral)	
$[Cr(DMSO)_4(en)](ClO_4)_3$	49
$[Cr(DMF)_4(en)](ClO_4)_3$	49
$[A][CrF_4(en)]$ ($A = Na$, *cis*-$[CrF_2(en)_2]$)	17, 50
$[enH_2][CrF_4(en)]Cl$[a]	51
$[CrF_2(H_2O)_2(en)]X$ ($X = Cl$ (*trans*-F_2),[a] Br)	51, 52, 53, 54
$[Cr(SCH_2CH_2NH_2)_2(en)]ClO_4$	43
$[CrCl_2(H_2O)_2en]Cl$, (*trans*-Cl_2)[a]	55
$[CrBrF_2(H_2O)(en)]$	51
$K[Cr(CN)(NCS)_3(en)]$	48
cis-$[Cr(OSO_2CF_3)_2(en)_2]CF_3SO_3$	56

X-Ray single crystal structure determinations; see also refs. 57 and 58 for other $[Cr(en)_3]^{3+}$ salts.
Cyanato and selenocyanato complexes are dealt with in Section 35.4.2.8.

1. Yu. N. Shevchenko and N. K. Davidenko, *Russ. J. Inorg. Chem. (Engl. Transl.)*, 1979, **24**, 1193.
2. Yu. N. Shevchenko, N. I. Yashina, K. B. Yatsimirskii, R. A. Svitsyn and N. V. Egorova, *Russ. J. Inorg. Chem. (Engl. Transl.)*, 1983, **28**, 216.
3. R. D. Gillard and P. R. Mitchell, *Inorg. Synth.*, 1972, **13**, 184.
4. F. Galsbøl, *Inorg. Synth.*, 1970, **12**, 269.
5. E. Pedersen, *Acta Chem. Scand.*, 1970, **24**, 3362.
6. W. W. Fee, J. N. MacB. Harrowfield and W. G. Jackson, *J. Chem. Soc. (A)*, 1970, 2612.
7. K. Akabori and Y. Kushi, *J. Inorg. Nucl. Chem.*, 1978, **40**, 1317.
8. H. P. Jensen, *Acta Chem. Scand., Ser. A*, 1981, **35**, 127.
9. S. E. Harnung and T. Laier, *Acta Chem. Scand., Ser. A*, 1978, **32**, 41.
10. P. Andersen, F. Galsbøl, S. E. Harnung and T. Laier, *Acta Chem. Scand.*, 1973, **27**, 3973.
11. H. Toftlund and T. Laier, *Acta Chem. Scand., Ser. A*, 1977, **31**, 651.
12. H. Toftlund and E. Pedersen, *Acta Chem. Scand.*, 1972, **26**, 4019.
13. C. F. C. Wong and A. D. Kirk, *Inorg. Chem.*, 1978, **17**, 1672.
14. J. W. Vaughn, J. M. DeJovine and G. J. Seiler, *Inorg. Chem.*, 1970, **9**, 684.
15. J. Glerup, J. Josephsen, K. Michelsen, E. Pedersen and C. E. Schäffer, *Acta Chem. Scand.*, 1970, **24**, 247.
16. J. V. Brenčič and I. Leban, *Z. Anorg. Allg. Chem.*, 1981, **480**, 213.
17. J. W. Vaughn and A. M. Yeoman, *Inorg. Chem.*, 1976, **15**, 2320.
18. J. W. Vaughn and A. M. Yeoman, *Synth. React. Inorg. Metal-Org. Chem.*, 1977, **7**, 165.
19. J. W. Vaughn and G. J. Seiler, *Inorg. Chem.*, 1979, **18**, 1509.
20. G. Wirth, C. Bifano, R. T. Walters and R. G. Linck, *Inorg. Chem.*, 1973, **12**, 1955.
21. C. F. C. Wong and A. D. Kirk, *Can. J. Chem.*, 1976, **54**, 3794.
22. C. F. C. Wong and A. D. Kirk, *Can. J. Chem.*, 1975, **53**, 3388.
23. J. W. Vaughn, *Inorg. Chem.*, 1983, **22**, 844.
24. Yu. N. Shevchenko and V. V. Sachok, *Russ. J. Inorg. Chem. (Engl. Transl.)*, 1983, **28**, 1118.
25. A. D. Kirk and T. L. Kelly, *Inorg. Chem.*, 1974, **13**, 1613.

Table 49 (*footnotes continued*)

26. J. C. Hempel, L. O. Morgan and W. B. Lewis, *Inorg. Chem.*, 1970, **9**, 2064.
27. C. Bifano and R. G. Linck, *Inorg. Chem.*, 1974, **13**, 609.
28. D. A. House, *J. Inorg. Nucl. Chem.*, 1973, **35**, 3103.
29. M. Nakano and S. Kawaguchi, *Bull. Chem. Soc. Jpn.*, 1979, **52**, 3563.
30. W. G. Jackson and W. W. Fee, *Inorg. Chem.*, 1975, **14**, 1154.
31. J. Ribas, M. Serra and A. Escuer, *Transition Met. Chem.*, 1984, **9**, 287.
32. W. G. Jackson and W. W. Fee, *Inorg. Chem.*, 1975, **14**, 1174.
33. W. G. Jackson, P. D. Vowles and W. W. Fee, *Inorg. Chim. Acta*, 1977, **22**, 111.
34. D. A. Palmer and D. W. Watts, *Inorg. Chim. Acta*, 1972, **6**, 197.
35. S. Kaizaki, *Bull. Chem. Soc. Jpn.*, 1983, **56**, 3625.
36. A. P. Sattelberger, D. D. Darsow and W. B. Schaap, *Inorg. Chem.*, 1976, **15**, 1412.
37. S. Kaizaki, J. Hidaka and Y. Shimura, *Bull. Chem. Soc. Jpn.*, 1975, **48**, 902.
38. A. D. Kirk and G. B. Porter, *Inorg. Chem.*, 1980, **19**, 445.
39. I. J. Kindred and D. A. House, *J. Inorg. Nucl. Chem.*, 1975, **37**, 1359.
40. C. J. Weschler, J. C. Sullivan and E. Deutsch, *Inorg. Chem.*, 1974, **13**, 2360.
41. I. K. Adzamli and E. Deutsch, *Inorg. Chem.*, 1980, **19**, 1366.
42. R. C. Elder, L. R. Florian, R. E. Lake and A. M. Yacynych, *Inorg. Chem.*, 1973, **12**, 2690.
43. C. J. Weschler and E. Deutsch, *Inorg. Chem.*, 1973, **12**, 2682.
44. J. C. Sullivan, E. Deutsch, G. E. Adams, S. Gordon, W. A. Mulac and K. H. Schmidt, *Inorg. Chem.*, 1976, **15**, 2864.
45. G. J. Kennard and E. Deutsch, *Inorg. Chem.*, 1978, **17**, 2225.
46. J. Ribas, M. Monfort and J. Casabo, *Transition Met. Chem.*, 1984, **9**, 407.
47. J. C. Chang, L. E. Gerdom, N. C. Baenziger and H. M. Goff, *Inorg. Chem.*, 1983, **22**, 1739.
48. A. Botar, E. Blasius and H. Augustin, *Z. Anorg. Allg. Chem.*, 1976, **422**, 54.
49. W. G. Jackson and W. W. Fee, *Inorg. Chem.*, 1975, **14**, 1161.
50. J. W. Vaughn and J. Marzowski, *Inorg. Chem.*, 1973, **12**, 2346.
51. C. Diáz, A. Seguí, J. Ribas, X. Solans, M. Font-Altaba, A. Solans and J. Casabó, *Transition Met. Chem.*, 1984, **9**, 469.
52. J. W. Vaughn and G. J. Seiler, *Synth. React. Inorg. Metal-Org. Chem.*, 1979, **9**, 1.
53. J. W. Vaughn, *J. Cryst. Spectrosc. Res.*, 1983, **13**, 231.
54. S. Kaizaki, M. Akaho, Y. Morita and K. Uno, *Bull. Chem. Soc. Jpn.*, 1981, **54**, 3191.
55. R. Stomberg and I. Larking, *Acta Chem. Scand.*, 1969, **23**, 343.
56. N. E. Dixon, G. A. Lawrance, P. A. Lay and A. M. Sargeson, *Inorg. Chem.*, 1984, **23**, 2940.
57. C. Brouty, P. Spinat and A. Whuler, *Acta Crystallogr., Sect. B*, 1980, **36**, 2037.
58. P. Spinat, A. Whuler and C. Brouty, *Acta Crystallogr., Sect. B*, 1979, **35**, 2914.

Table 50 Complexes of Diamines Excluding Ethylenediamine Except for Mixed Diamine Complexes and Some Oxalato, Malonato and Acetylacetonato Complexes (see also Table 49)

Complex	*Ref.*
CrN_6	
$[Cr(en)_2(pn)]X_3$ (X = Cl, Br, SCN)	1, 2, 3
$[Cr(en)(pn)_2]X_3$ (X = Cl, SCN, ClO_4)	2, 4
$[Cr(en)_2(tmd)]X_3$ (X = Cl, Br, SCN, $(+)_D$-Br Λ conf.)	2, 3, 5
$[Cr(en)(tmd)_2]X_3$ (X = Cl, Br, I, SCN, $(+)_D$-Br Λ conf.)	2, 3, 5
$[Cr(pn)(tmd)_2]X_3$ (X = Cl, SCN)	2
$[Cr(pn)_2(tmd)]X_3$ (X = Cl, SCN)	2
$[Cr(en)(pn)(tmd)]X_3$ (X = Cl, SCN)	2
$[Cr(en)_2(ibn)]Br_3$	3
$[Cr(en)_2(chxn)]Br_3$	3
$[Cr(tmd)_2(ibn)]Br_3$	3
$[Cr(pn)_2(ibn)]Br_3$	3
$[Cr(pn)_2(chxn)]Br_3$	3
CrX_2N_4	
trans-$[CrF_2(tmd)_2]X$, (X = Cl, I, ClO_4)	6, 7
trans-$[CrBrF(tmd)_2]ClO_4$	8
trans-$[CrF(NH_3)(tmd)_2]X_2$ (X = ClO_4,[a] X_2 = Br, ClO_4)	8
trans-$[Cr(NH_3)(H_2O)(tmd)_2]Br_2ClO_4$	8
trans-$[CrCl_2(tmd)_2]X$ (X = Cl, Br, ClO_4, NO_3)	9, 10
trans-$[CrBr_2(tmd)_2]X$ (X = Br, ClO_4)	10
trans- and *cis*-$[Cr(NCS)_2(tmd)_2]SCN$, $(+)_{589}$- and $(-)_{589}$-isomers, Δ and Λ conf.	2, 11, 12
cis- and *trans*-$[Cr(H_2O)_2(tmd)_2](NO_3)_2$	11
cis- and *trans*-$[Cr(ONO)_2(tmd)_2]ClO_4$, $(+)_{480}$-ClO_4	11, 13
trans-$[Cr(OH)(H_2O)(tmd)_2]X_2$ (X = ClO_4, NO_3)	10, 11
cis-$[Cr(OH)(H_2O)(tmd)_2]X_2$ (X = NO_3, X_2 = S_2O_6)	11, 14
cis-$[CrCl_2(tmd)_2]X$ (X = Cl, ClO_4 and $(+)_D$-ClO_4)	9
trans- and *cis*-$[Cr(N_3)_2(tmd)_2]ClO_4$, $(+)_{589}$- and $(-)_{589}$-isomers, former Δ conf.	11
trans-$[Cr(CH_3CO_2)_2(tmd)_2]X$ (X = Cl, I, ClO_4)	7
trans-$[Cr(nic\text{-}O)_2(tmd)_2]Cl$ (nic-O = O-bonded nicotinate)	15
trans-[A]$[CrF(H_2O)(tmd)_2]$[X] [A = K, Na, NH_4; X = $Cr(CN)_6$, $CrNO(CN)_5$]	16
trans-$[CrX_2(en)(tmd)]X$ [X = Cl, Br, SCN]	2, 17
cis-$[CrCl_2(en)(tmd)]Cl$	2, 17
trans-$[CrCl_2(en)(tmd)]ClO_4$	18
trans-$[Cr(OH)(H_2O)(en)(tmd)]ClO_4$	18

Table 50 *(continued)*

Complex	*Ref.*
trans-$Cr[F_2(pn)_2]X$ (X = Br, ClO_4, $[CrF_4(pn)]$; also (*R*)-pn)	1, 6, 19, 20
cis-$Cr[F_2(pn)_2]X$ (X = ClO_4, BPh_4)	19, 21
trans-$[CrF(H_2O)(pn)_2]X_2$ (X = Br, I, ClO_4)	22
cis-$[CrFX(pn)_2]X$ (X = Cl, Br, I)	22
cis-$[CrX_2(pn)_2]X$ (X = Cl, Br)	23
trans-$[CrBr_2(pn)_2]Br$, also *R*-pn	24, 29
trans-$[CrCl_2(pn)_2]X$ (X = Cl, ClO_4; also *R*-pn)	25, 29
trans-$[Cr(NCS)_2(pn)_2]SCN$	2
cis- and *trans*-$[Cr(NCO)_2(pn)_2]NCO$	2, 30
trans-$[Cr(ONO)_2(R\text{-pn})_2]ClO_4$	13
cis- and *trans*-$[CrF_2(en)(pn)]Br$, also *R*-pn	1, 20, 21
trans-$[CrX_2(en)(pn)]X$ (X = Cl, Br, SCN)	2, 17
trans-$[CrF(H_2O)(en)(pn)]X_2$ (X = Cl, Br, I, ClO_4)	1, 21
trans-$[CrBr_2(en)(pn)]ClO_4$	21
trans-$[CrFCl(en)(pn)]ClO_4$	21
trans-$[Cr(ONO)_2(en)(R\text{-pn})]ClO_4$	13
cis-$[CrFCl(en)(pn)]Cl$	21
cis-$[CrFBr(en)(pn)]Br$	21
cis-$[CrX_2(en)(pn)]X$ (X = Cl, Br)	2, 17
trans-$[CrF_2(pn)(tmd)]Br$, also *R*-pn	1, 20
trans-$[CrF(H_2O)(pn)(tmd)]X_2$ (X = Cl, Br, I, NO_3, ClO_4)	1, 21
trans-$[Cr(ONO)_2(R\text{-pn})(tmd)]ClO_4$	13
trans-$[CrX_2(pn)(tmd)]X$ (X = Cl, Br, SCN)	2, 17
cis-$[CrX_2(pn)(tmd)]X$ (X = Cl, Br)	2, 17
cis-$[CrFCl(pn)(tmd)]Cl$	21
cis-$[CrFBr(pn)(tmd)]Br$	21
trans-$[CrF_2(bn)_2]ClO_4$	26
trans-$[CrX_2(bn)_2]X$ (X = Cl, Br)	26
cis-$[CrX_2(bn)_2]X$ (X = Cl, Br)	26
cis-$[CrCl_2(AA)_2]ClO_4$ (AA = ibn, *meso*- and (±)-bn)	27
trans-$[CrF_2(ptn)_2]ClO_4$, also *R,R*- and *S,S*-ptn	20, 26, 28, 29
trans-$[CrCl_2(ptn)_2]X$ (X = Cl, ClO_4, also *S,S*-ptn)	26, 29
trans-$[CrBr_2(S,S\text{-ptn})_2]Br$	29
cis-$[CrCl_2(ptn)_2]Cl$	26
trans-$[CrF_2(en)(R,R\text{-ptn})]Br$	20
trans-$[Cr(ONO)_2(en)(R,R\text{-ptn})]ClO_4$	13
trans-$[Cr(ONO)_2(R,R\text{-ptn})_2]ClO_4$	13
trans-$[CrF_2(R\text{-pn})(R,R\text{-ptn})]Br$, also *S*-pn or *R,S*-ptn	20
trans-$[Cr(ONO)_2(R\text{-pn})(R,R\text{-ptn})]ClO_4$, also *R,S*- and *S,S*-ptn	13
trans-$[CrF_2(chxn)_2]ClO_4$, *trans*-$[CrF_2(R,R\text{-chxn})_2]Cl$	20, 26
trans-$[CrF(ONO)(R,R\text{-chxn})_2]ClO_4$	13
trans-$[Cr(ONO)_2(R,R\text{-chxn})_2]ClO_4$	13
trans-$[CrCl_2(chxn)_2]Cl$	26
cis-$[CrCl_2(chxn)_2]Cl$	9
trans-$[CrF_2(R,R\text{-chxn})(en)]Br$	20
cis-$[CrF_2(chxn)(pn)]X$ (X = Br, ClO_4)	1
trans-$[CrF_2(R,R\text{-chxn})(R\text{-pn})]Br$, also *S*-pn	20
cis- and *trans*-$[Cr(NCO)_2(chxn)_2]NCO$	30
trans-$[CrF_2(S,S\text{-stien})_2]ClO_4$	29
trans-$[CrX_2(S,S\text{-stien})_2]X$ (X = Cl, Br)	29
$[CrCl_2(opd)_2]Cl$, $[CrCl_2(opd)_4]Cl$ (monodentate amine)	31
CrX_4N_2	
$[CrF_2(H_2O)_2(pn)]$ (X = Cl, Br, I)	1, 21, 32
$[pnH_2][CrF_4(pn)]Cl$	32
$[CrF_2(pn)_2][CrF_4(pn)]$, also Na salt	1
$[NBu_4][Cr(NCO)_4(pn)]$	30
$A[Cr(NCS)_4(tmd)]$ (A = K, Ag)	12
$[NBu_4][Cr(NCO)_4(chxn)]$	30
$[chxnH_2][CrF_4(chxn)]Cl$	32
$[CrF_2(H_2O)_2(chxn)]Br$	32
$[CrBrF_2(H_2O)(aa)]$ (aa = pn, chxn)	32
Oxalato-, malonato- and acetylacetonato-amine complexes	
$[Cr(ox)(L)_2]ClO_4$ (L = en, pn, tmd; $(-)_{589}$-cations, Δ conf.)	35
$[Cr(ox)(en)_2][Cr(ox)_2(en)]$[a]	36
$[Cr(mal)(en)_2]X$ (X = ClO_4, $[Cr(mal)_2(en)]$)	37
$[Cr(ox)(H_2O)_2(en)]Br$	37
$K[Cr(ox)_2(en)]\cdot KI\cdot 2H_2O$[a]	38

Table 50 *(continued)*

Complex	*Ref.*
$K[Cr(mal)_2(en)]$, $[\{Cr(OH)(mal)(en)\}_2]$	37
Na[*cis*-$\{CrF_2(ox)(en)\}$]	39
Li[*trans*-$\{CrF_2(mal)(en)\}$]	39
$[Cr(L)(en)_2]Cl_2$ (L = acac, 3-Cl- and 3-Br-acac)	40
$[Cr(L)_2(en)]Cl$ (L = acac, 3-Cl- and 3-Br-acac)	40
$[Cr(L)(L')(en)]Cl$ (L = acac, 3-Cl- and 3-Br-acac)	40
$[Cr(acac)(tmd)_2]I_2$	41
$[Cr(acac)_2(tmd)]I$, $(-)_{589}$-cation, Λ conf.[a]	41, 42
$[Cr(ox)(NH_3)_4]ClO_4$	35

[a] Crystal structure determinations; also for *trans*-$[CrF(H_2O)(tmd)_2](ClO_4)_2$[33] and $(-)_{589}$-$[Cr(NCS)_2(tmd)_2]\cdot 0.5[Sb_2(+)\text{-}tart_2]$[34]

1. J. W. Vaughn and J. Marzowski, *Inorg. Chem.*, 1973, **12**, 2346.
2. A. Uehara, Y. Nishiyama and R. Tsuchiya, *Inorg. Chem.*, 1982, **21**, 2422.
3. J. A. McLean, Jr. and A. Bhattacharyya, *Inorg. Chim. Acta*, 1981, **53**, L43.
4. K. Kühn, F. Wasgestian and H. Kupka, *J. Phys. Chem.*, 1981, **85**, 665.
5. M. Rancke-Madsen and F. Woldbye, *Acta Chem. Scand.*, 1972, **26**, 3405.
6. J. Glerup, J. Josephsen, K. Michelsen, E. Pedersen and C. E. Schäffer, *Acta Chem. Scand.*, 1970, **24**, 247.
7. J. W. Vaughn, G. J. Seiler, M. W. Johnson and G. L. Traister, *Inorg. Chem.*, 1970, **9**, 2786.
8. J. W. Vaughn, *Inorg. Chem.*, 1981, **20**, 2397.
9. E. Pedersen, *Acta Chem. Scand.*, 1970, **24**, 3362.
10. M. C. Couldwell and D. A. House, *Inorg. Chem.*, 1972, **11**, 2024.
11. M. Nakano and S. Kawaguchi, *Bull. Chem. Soc. Jpn.*, 1979, **52**, 3563.
12. E. Blasius and E. Mernke, *Z. Anorg. Allg. Chem.*, 1984, **509**, 167.
13. S. Kaizaki, *Bull. Chem. Soc. Jpn.*, 1983, **56**, 3625.
14. T. Laier, B. Nielsen and J. Springborg, *Acta Chem. Scand., Ser. A*, 1982, **36**, 91.
15. C. A. Green, R. J. Bianchini and J. I. Legg, *Inorg. Chem.*, 1984, **23**, 2713.
16. J. Ribas, M. Montfort and J. Casabó, *Transition Met. Chem.*, 1984, **9**, 407.
17. S. Mitra, T. Yoshikuni, A. Uehara and R. Tsuchiya, *Bull. Chem. Soc. Jpn.*, 1979, **52**, 2569.
18. M. C. Couldwell, D. A. House and H. K. J. Powell, *Inorg. Chem.*, 1973, **12**, 627.
19. J. Casabó, A. Solans and J. Real, *Synth. React. Inorg. Metal-Org. Chem.*, 1983, **13**, 835.
20. S. Kaizaki, *Bull. Chem. Soc. Jpn.*, 1983, **56**, 3620.
21. J. W. Vaughn and G. J. Seiler, *Inorg. Chem.*, 1974, **13**, 598.
22. J. Ribas, M. L. Martinez, M. Serra, M. Montfort, A. Escuer and N. Navarro, *Transition Met. Chem.*, 1983, **8**, 87.
23. J. A. McLean, Jr. and N. A. Maes, *Inorg. Nucl. Chem. Lett.*, 1972, **8**, 147.
24. J. A. McLean, Jr. and R. I. Goorman, *Inorg. Nucl. Chem. Lett.*, 1971, **7**, 9.
25. J. A. McLean, Jr. and R. Barona, *Inorg. Nucl. Chem. Lett.*, 1969, **5**, 385.
26. R. Tsuchiya, A. Uehara and T. Yoshikuni, *Inorg. Chem.*, 1982, **21**, 590.
27. C. P. Madhusudham and J. A. McLean, Jr., *Inorg. Chem.*, 1975, **14**, 82.
28. H. Toftlund and E. Pedersen, *Acta Chem. Scand.*, 1972, **26**, 4019.
29. S. Kaizaki and Y. Shimura, *Bull. Chem. Soc. Jpn.*, 1975, **48**, 3611.
30. E. Blasius and G. Klemm, *Z. Anorg. Allg. Chem.*, 1978, **443**, 265.
31. J. C. Chang, *J. Inorg. Nucl. Chem.*, 1975, **37**, 855.
32. C. Diáz, A. Seguí, J. Ribas, X. Solans, M. Font-Altaba, A. Solans and J. Casabó, *Transition Met. Chem.*, 1984, **9**, 469.
33. X. Solans, M. Font-Altaba, M. Montfort and J. Ribas, *Acta Crystallogr., Sect. B*, 1982, **38**, 2899.
34. K. Matsumoto, S. Ooi, H. Kawaguchi, M. Nakano and S. Kawaguchi, *Bull. Chem. Soc. Jpn.*, 1982, **55**, 1840.
35. D. A. House, *Inorg. Chim. Acta*, 1982, **60**, 145.
36. J. W. Lethbridge, L. S. Dent Glasser and H. F. W. Taylor, *J. Chem. Soc. (A)*, 1970, 1862.
37. J. W. Lethbridge, *J. Chem. Soc., Dalton Trans.*, 1980, 2039.
38. L. S. Dent Glasser and J. W. Lethbridge, *J. Chem. Soc., Dalton Trans.*, 1976, 2065.
39. S. Kaizaki, M. Akaho, Y. Morita and K. Uno, *Bull. Chem. Soc. Jpn.*, 1981, **54**, 3191.
40. S. Kaizaki, J. Hidaka and Y. Shimura, *Inorg. Chem.*, 1973, **12**, 135.
41. M. Nakano, S. Kawaguchi and H. Kawoguchi, *Bull. Chem. Soc. Jpn.*, 1979, **52**, 2897.
42. K. Matsumoto and S. Ooi, *Bull. Chem. Soc. Jpn.*, 1979, **52**, 3307.

provides aqua derivatives, *e.g.* *trans*-$[CrBr(H_2O)(en)_2]Br_2$, with rentention of configuration since the Cr—O bond does not break. Anation in dipolar aprotic solvents such as DMF, DMSO and DMA, and solvolysis induced by the addition of $AgClO_4$, HgF_2 or $Hg(MeCO_2)_2$ are also useful preparative procedures.[404,405] The inversion in relative solubility of perchlorates and halides in passing from water to these solvents has facilitated the isolation of certain water-soluble complexes.

Examples of the preparations of *cis*-$[Cr(CN)_2(en)_2]^{2+}$ salts, mono- and di-ammine- and fluoro-bis(ethylenediamine) salts are given in Schemes 39 to 46.

(*b*) *Cyano series.* The preparation of a series of *cis*-dicyanobis(ethylenediamine)-chromium(III) salts (Scheme 39) starts with the reaction of green *trans*-$[CrCl_2(H_2O)_4]Cl\cdot 2H_2O$ and liquid hydrogen cyanide, but no *trans*-dicyano complex was obtained, presumably because of the lability of systems with CN *trans* to CN.[327] The purity of *cis*-$[Cr(CN)_2(en)_2]^+$ was established by ion exchange chromatography, and its reactions and visible and IR spectra, and particularly the resolution of the iodide, support the *cis* assignment. The $\nu(C{\equiv}N)$ absorption of 2139 cm^{-1} was unexpectedly unsplit, but there were four ν(Cr—N, en) bands at

550, 536, 491 and 425 cm^{-1} and *trans* complexes show fewer bands in this region. Unusually, spin-forbidden bands were observed in the solution spectrum. From the CD spectra the $(+)_{589}$ isomer was assigned the Δ configuration (see also Section 35.4.1.1.ii).

$$HCN\ (0°C) + CrCl_3 \cdot 6H_2O \xrightarrow[\Delta]{en} cis\text{-}[Cr(CN)_2(en)_2]Cl$$
green

$$\downarrow NaI\ or\ NaClO_4$$

$$cis\text{-}[Cr(CN)(H_2O)(en)_2]^{2+} + HCN \xleftarrow[\text{then pH 4}]{\text{pH 1}} cis\text{-}[Cr(CN)_2(en)_2]Cl \xleftarrow[DMF]{LiCl} cis\text{-}[Cr(CN)_2(en)_2]X$$
lemon yellow

$$\downarrow (NH_4)(+)_D\text{-}\alpha\text{-}[C_{10}H_{14}SO_4Br] \qquad\qquad \downarrow HCl\ or\ HBr$$

$$cis\text{-}[CrX_2(en)_2]^+$$

$$(+)_{589}\text{-}[Cr(CN)_2(en)_2]I \xleftarrow[EtOH]{HI} [Cr(CN)_2(en)_2][C_{10}H_{14}SO_4Br] + filtrate \xrightarrow[0°C]{NaI} (-)_{589}\text{-}[Cr(CN)_2(en)_2]I$$

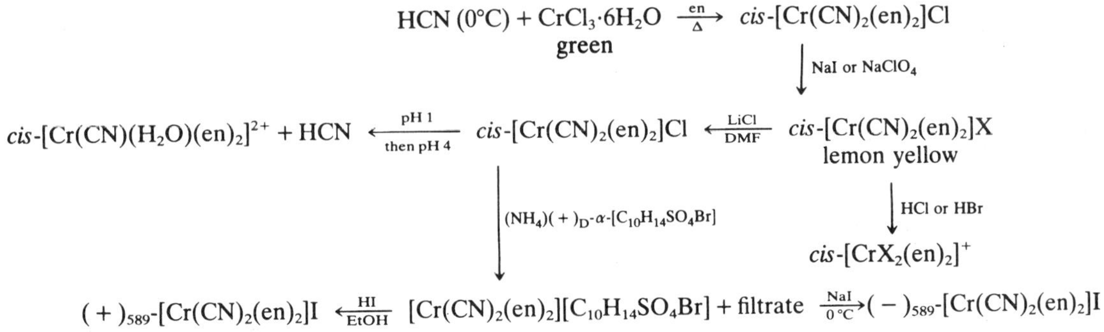

Scheme 39

A safer procedure, starting from aqueous [Cr(en)$_3$]Cl$_3$ and NaCN in the presence of charcoal, has been reported[406] to yield both *trans*- and *cis*-[Cr(CN)$_2$(en)$_2$]Cl, but others[330] have not been successful with this method.

(c) *Reactions with liquid ammonia.* Ammonolysis of [CrXY(en)$_2$]$^{n+}$ salts sealed in Carius tubes with liquid ammonia has afforded many mono- or di-ammine derivatives (Table 49 and Schemes 40–45).[407–409] The preparation of the *cis* isomers is straightforward, but thermal substitution with *trans → cis* rearrangement is common. Thus, *cis*-[CrCl$_2$(en)$_2$]$^+$ reacts smoothly to yield *cis*-[Cr(NH$_3$)$_2$(en)$_2$]$^{3+}$ with a few per cent of the *trans* compound, whereas *trans*-[CrCl$_2$(or Br$_2$)(en)$_2$]$^+$ gives 95% *cis*-[Cr(NH$_3$)$_2$(en)$_2$]$^{3+}$; *trans*-[Cr(NH$_3$)$_2$(en)$_2$]$^{3+}$ has been isolated[408] in 30% yield from the less accessible complex *trans*-[CrCl(NH$_3$)(en)$_2$](ClO$_4$)$_2$[407] (Scheme 44) by ammonolysis followed by fractional crystallization. For a particular pair of *cis* and *trans* isomers the geometrical assignment was based on IR spectral data and on the overall molar absorptivities, the *cis* having the more intense bands, because the electronic spectra are otherwise very similar. An easier route to *cis*-[CrF(NH$_3$)(en)$_2$]$^{2+}$ salts has been devised (Scheme 45), and the isolation of one chiral form has confirmed the *cis* geometry (Δ configuration).[409] HPLC was useful in determining the *cis/trans* ratio (4:1) in the crude ammonolysis product.

$$trans\text{-}[CrBrF(en)_2]ClO_4 \xrightarrow[\text{trace } NH_4ClO_4,\ 0°C]{NH_3(l)} 60/40\ cis/trans\text{-}[CrF(NH_3)(en)_2]BrClO_4$$

$$\xrightarrow[\text{with } HClO_4]{\text{HCl, then fract. cryst.}} trans\text{- and } cis\text{-}[CrF(NH_3)(en)_2](ClO_4)_2$$

Scheme 40

$$trans\text{-}[CrBr(NO_2)(en)_2]ClO_4 \xrightarrow[\text{trace } NH_4ClO_4,\ 0°C]{NH_3(l)} cis/trans\text{-}[Cr(NO_2)(NH_3)(en)_2]BrClO_4$$

$$\xrightarrow[\text{ii, HBr}]{\text{i, HCl } (NO_2\uparrow)} cis\text{-}[CrXBr(en)_2]Br_2$$
X = Cl, Br

Scheme 41

$$cis\text{-}[CrCl_2(en)_2]ClO_4 \xrightarrow{NH_3(l)} cis\text{-}[Cr(NH_3)_2(en)_2](ClO_4)_3$$

Scheme 42

$$trans\text{-}[CrCl(H_2O)(en)_2]Br_2 \xrightarrow[NaOH,\ 0°C]{NaClO_4} trans\text{-}[Cl(en)_2CrOHCr(en)_2Cl](ClO_4)_3$$

$$\xrightarrow[\text{then HBr}]{NH_3(l),\ LiNH_2,\ 20\,°C} trans\text{-}[Cr(NH_3)(H_2O)(en)_2]Br_3$$

Scheme 43

$$trans\text{-}[CrCl(NH_3)(en)_2](ClO_4)_2 \xrightarrow{NH_3(l)} 2:1\ cis/trans\text{-}[Cr(NH_3)_2(en)_2]Cl(ClO_4)_2$$

$$\xrightarrow[HCl,\ HClO_4]{\text{fract. cryst.}} trans\text{-}[Cr(NH_3)_2(en)_2](ClO_4)_3$$

Scheme 44

$$cis\text{-}[CrF_2(en)_2][CrF_4(en)] \longrightarrow cis\text{-}[CrF(H_2O)(en)_2]I_2 \longrightarrow cis\text{-}[CrClF(en)_2]I \xrightarrow[20\,°C]{NH_3(l)}$$

$$cis\text{-}[CrF(NH_3)(en)_2]ICl \xrightarrow{HX} (\pm)\text{-}cis\text{-}[CrF(NH_3)(en)_2]X_2 \xrightarrow[salt]{via \atop (+)_{589}\text{-}C_{10}H_{14}O_4BrS} (-)_{546}\text{-}cis\text{-}[CrF(NH_3)(en)_2](ClO_4)$$

$$(>70\%\ cis) \qquad\qquad X = I,\ ClO_4$$

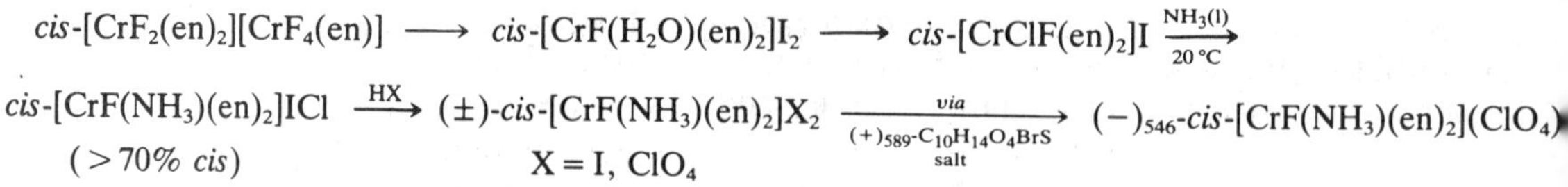

Scheme 45

(d) *Fluoro complexes.* The fluoro complex $[Cr(en)_2F_2][Cr(en)F_4]$ (Schemes 45 and 46) is easily prepared by reaction of anhydrous ethylenediamine with $CrF_3\cdot3.5H_2O$. Thus equation (29) is a general reaction of bidentate amines AA and many fluoro-containing complexes have been derived from the anion or the cation. The small, highly basic fluoride ion is strongly bound to Cr^{III} so it can be carried through various syntheses and removed when necessary in acid. Some procedures are in Schemes 45 and 46.[402,409] The fluoride $CrF_3\cdot3.5H_2O$, $CrCl_3\cdot6H_2O$ in dilute HF, *trans*-$[CrF_2(py)_4]^+$ in boiling 2-methoxyethanol and CrF_2 suspended in dry ethereal HF can be used to form fluoro-containing complexes in reactions with various amines. Other methods include anation of *cis*- or *trans*-$[CrF(H_2O)(AA)_2]^{2+}$ in methanol, thermal dehydration of *trans*-$[CrF_2(H_2O)(AA)_2]X_2$, chloride abstraction from *cis*-$[CrCl_2(AA)_2]^+$ with HgF_2 in liquid HF and thermal deamination of $[Cr(AA)_3]X_3$ in an NH_4F matrix.

$$3AA + 2CrF_3\cdot3.5H_2O \longrightarrow [CrF_2(AA)_2][CrF_4(AA)] + 7H_2O \qquad (29)$$

$$AA = en,\ pn,\ tmd\ or\ dach$$

cis-$[CrF(H_2O)(en)_2]^{2+}$ has been resolved and the absolute configuration of its chiral forms established by ORD spectral studies.[410] The cation in $(-)_{546}$-*cis*-$[CrF(H_2O)(en)_2](BPh_4)_2$ has the Δ configuration. Anation of the optical isomers with SCN^- or F^- in MeOH takes place with retention of configuration[410] and this provides an expeditious route to *cis*-$[CrFX(en)_2]X$ complexes.[411]

$$[CrF_2(en)_2][CrF_4(en)]$$
$$\downarrow {\scriptstyle HI}$$

$$cis\text{-}[CrBr(H_2O)(en)_2]Br_2 \xleftarrow{HBr} cis\text{-}[CrF_2(en)_2]I \xrightarrow{HCl} cis\text{-}[CrCl_2(en)_2]I$$

$$\downarrow {\scriptstyle HI}$$

$$cis\text{-}[CrFX(en)_2]I \xleftarrow{X^-} (\pm)\text{-}cis\text{-}[CrF(H_2O)(en)_2]I_2$$
$$X = Cl,\ Br,\ SCN \qquad\qquad via\ \big|\ (+)\text{-}NH_4C_{10}H_{14}O_4BrS$$

anation, MeOH | anation, MeOH

$$\Delta\text{-}(-)_{546}\text{-}cis\text{-}[CrF(NCS)(en)_2]SCN \quad \Delta\text{-}(-)_{546}\text{-}cis\text{-}[CrF(H_2O)(en)_2](BPh_4)_2 \quad \Delta\text{-}(-)_{546}\text{-}cis\text{-}[CrF_2(en)_2]I$$

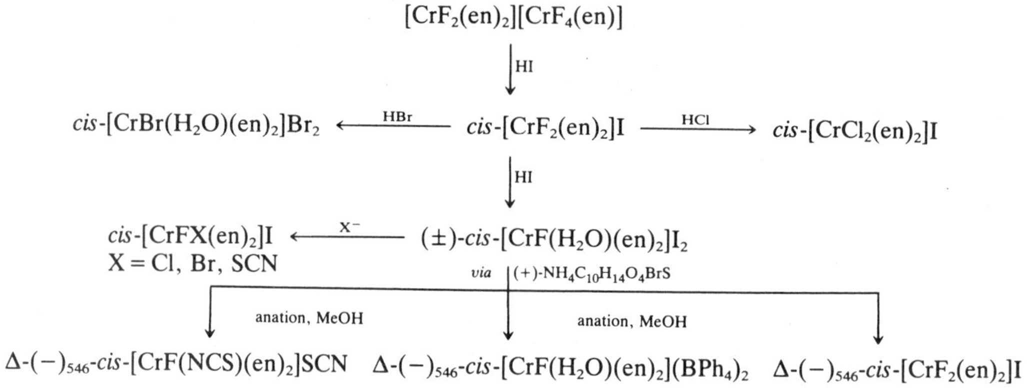

Scheme 46

(e) *Thiolato complexes.* During kinetic investigations of the efficiency of thiolato sulfur as an electron transfer bridge a number of thiolatobis(ethylenediamine)chromium(III) salts have been prepared by reduction of the appropriate disulfide with chromium(II) ethylenediamine perchlorate[412] (Table 49). The structure of the mercaptoacetato derivative $[Cr(en)_2(SCH_2CO_2)]ClO_4$ has been determined[413] and it contains enantiomeric forms ($\Delta\lambda\delta$ and $\Lambda\delta\lambda$) of the pseudooctahedral cation; the mercaptoacetato ligand is bidentate, and there is some lengthening of the Cr—N bond (2.112 Å) *trans* to the S atom. The mean Cr—N (*cis*) distance is 2.090 Å, and the Cr—O and Cr—S distances are 1.966 and 2.337 Å respectively.

(f) *General investigations.* The electronic spectra of amine complexes are routinely recorded to aid in the assignment of structure. For details the original literature and reviews[301,302,402,414,415] should be consulted. Measurements of the single crystal polarized electronic spectra accompanied by appropriate calculations provide considerable information concerning ligand parameters, site symmetry, the ordering of energy levels and selection rules.

Many tris(diamine)- and *cis*-diacidobis(diamine)chromium(III) complexes have been resolved into their enantiomers, thus providing proof of structure. Absolute configurations are frequently inferred from ORD and CD measurements, which have become of great importance because they provide quick structural information. The spectra–structure correlations are empirical, but when X-ray single crystal investigations have been carried out, they have confirmed the assignments.[304]

The four possible isomers of $[Cr(cis\text{-}1,2\text{-chxn})_3]^{3+}$ (*mer*-Λ, *mer*-Δ, *fac*-Λ and *fac*-Δ) have been separated by cation exchange using disodium hydrogen phosphate and disodium tartrate, and isolated as the perchlorates. They were identified from their electronic, CD and ESR behaviour and comparisons with the Co^{III} analogues. The ligand has two dissymmetric centres of opposite chirality (*R,S*) and *fac* and *mer* refer to the spatial arrangement of the dissymmetric centres.[416] Some correlations of X-ray powder photographs and absolute configurations of $[Cr(trans\text{-}1,2\text{-chxn})_3]Cl_3$ have been made.[417]

Developments in instrumentation have allowed measurements of the single-crystal CD spectra of $[(+)_D\text{-}Cr(en)_3]^{3+}$ doped in $[Ir(en)_3]Cl_3$ between 7 and 293 K. Transitions to the excited states $^4T_2$, $^4T_1$, ^2E, $^2T_1$ and $^2T_2$ were observed.[418] The discussion refers to much earlier work. The polarized electronic spectra of single crystals of several *trans*-$Cr[XY(en)_2]$ complexes have been assigned and ligand field parameters evaluated.[419] CD, absorption and circularly polarized emission spectra have been reported for $[Cr(en)_3]^{3+}$ in the region of the $^4A_{2g} \rightleftarrows {}^2E_g {}^2T_{1g}$ transitions.[420]

Thermal deamination of tris(ethylenediamine)chromium(III) complexes is a standard preparative method for *cis*- and *trans*-diacidobis(ethylenediamine) complexes[421,422] and the thermal behaviour of the starting materials has been related to their crystal structures.[423] The cyano complex *cis*-$[Cr(CN)_2(en)_2]ClO_4$ in DMSO undergoes stepwise reduction III→II→I at the DME.[424] The standard redox potential for the Cr^{III}/Cr^{II} couple is −1.586 V (*versus* SCE).

From Raman investigations the strong metal sensitive band at 450–600 cm^{-1} shown by $[Cr(en)_3]^{3+}$ compounds has been assigned to the totally symmetric M—N stretching vibration and the band at 280–320 cm^{-1} to a totally symmetric N—M—N bending mode. Studies of electronic spectra suggest that the vibration responsible for the latter band has at least as much M—N stretching character as the former.[425] It appears that mixed ring conformations can be identified from the IR spectra of tris(ethylenediamine) complexes.[426]

The XPES of a number of ethylenediamine and cyano complexes, including $[Cr(en)_3]Cl_3$ and $K_3Cr(CN)_6$, have been measured in order to obtain the difference between the N1*s* and C1*s* binding energies. For the en complexes the ΔE values are related to σ donation, and for the cyano complexes are mainly controlled by the nitrogen charge, which depends on the π back-donation.[427]

(*ii*) *Bidentate diamines excluding ethylenediamine*

Synthetic methods with bidentate amines generally follow those developed for complexes of ethylenediamine, but there can be differences in detail, *e.g.* thermal dehydration of *trans*-$[CrF(H_2O)(en)_2]X_2$ (X = Cl, Br, I, SCN) produces *cis*-FX isomers, whereas the major products from *trans*-$[CrF(H_2O)(tmd)_2]X_2$ are the *trans*-FX isomers (Table 50). Similarly, the reaction of liquid ammonia with *trans*-$[CrFBr(en)_2]ClO_4$ provides *cis*- and *trans*-$[CrF(NH_3)(en)_2]^{2+}$ in 60:40 ratio; with *trans*-$[CrFBr(tmd)_2]ClO_4$ the only product is *trans*-$[CrF(NH_3)(tmd)_2]^{2+}$; tmd complexes appear more resistant to stereochemical change than en complexes. Reactions of diamines with $CrCl_3{\cdot}6H_2O$ in aqueous HF, or *trans*-$[CrF_2(py)_4]ClO_4$ in 2-methoxyethanol under reflux are general synthetic methods for *trans*-$[CrF_2(AA)_2]^+$ complexes. However, the less accessible isomers, *cis*-$[CrF_2(AA)_2]^+$ (AA = en and pn) have now been isolated from the mother liquors left by the less soluble *trans* complexes or by using longer reaction times to allow *trans*→*cis* isomerization to occur.[402,428]

The structure of *trans*-$[CrF(NH_3)(tmd)_2](ClO_4)_2$ has been confirmed crystallographically.[429] In the absence of crystallographic data, assignments of geometry in $[CrX_2(diamine)_2]$ complexes can be aided by IR spectroscopy. Commonly, *trans* isomers show essentially three bands in the 390 to 550 cm^{-1} region, while *cis* isomers show four or more (Table 51),[430] although these criteria did not work well with a series of *cis* and *trans* complexes $[CrX_2(tmd)_2]^{n+}$ (X = NCS, ONO, N_3, H_2O; X_2 = OH, H_2O).[431] Complexes containing Cr—F groups give strong bands in the 350–370 cm^{-1} range due to Cr—F stretching vibrations coupled with other metal–ligand modes. The bands are low by comparison with other chromium(III) fluoride complexes (Table 51) and other chromium–halogen modes, *i.e.* Cr—Cl, 320–360 cm^{-1}, Cr—Br, 250–280 cm^{-1}.[430]

Table 51 IR Spectra (600–200 cm^{-1}) of Complexes $[CrFX(AA)_2]^+$ and $[CrF(H_2O)(AA)_2]^{2+}$

Complex	*Chelate ring deformation*	*ν(Cr—N)*	*ν(Cr—F)*	*ν(Cr—X) (X = Cl, Br, I)*
trans-$[CrF_2(en)_2]Cl$	552s, 525sh	492m, 452m	370s	
trans-$[CrF_2(en)_2]ClO_4$	560s, 520sh	488w, 442m	370s	
trans-$[CrCl_2(en)_2]ClO_4$	558s, 535sh	490w, 445m		350vs
trans-$[CrBr_2(en)_2]ClO_4$	552s, 531sh	488mw, 445m		285vs
trans-$[CrClF(tmd)_2]ClO_4$	550s	490m, 435m	370s	325m
trans-$[CrBrF(tmd)_2]ClO_4$	550s	490m, 435m	370s	250s
trans-$[CrF_2(pn)_2]Cl$	545sh, 525m	462w, 442m	370s	
trans-$[CrCl_2(NH_3)_4]ClO_4$		465m, 450m, 435m		330s 350sh
trans-$[CrF(OH_2)(en)_2](ClO_4)_2$	552s	475w, 445m	365ms	
trans-$[CrF(OH_2)(en)_2]Br_2$	565w, 550w	485w, 445m	365s	
trans-$[CrF(OH_2)(pn)_2](ClO_4)_2$	570sh, 523s	495sh, 450m	365s	
trans-$[CrF(NH_3)(tmd)_2](ClO_4)_2$	538	492, 435	358	
cis-$[CrClF(tmd)_2]Cl$	545vs	490mw, 445m, 420w	360vs	325m
cis-$[CrBrF(tmd)_2]Br$	545m, 520w	488w, 445w, 440w, 420w	365s	240s
cis-$[CrClF(en)_2]Cl$	575sh, 550w, 525w	490w, 445m, 420w	355m	320w
cis-$[CrBrF(en)_2]Br$	570sh, 550w, 525w	485w, 445m, 420w	355m	240m
cis-$[CrFI(en)_2]I$	575sh, 550m, 525w	480w, 445m, 420w	355m	225m
$NH_4[CrF_4en]$			489	
$K_3[CrF_6]$			535 522	

(iii) *Mixed bidentate amines including ethylenediamine*

At the time of the Garner and House review[302] there were only five known complexes of Cr^{III} with two different amines; this is no longer the situation (Table 50) and there seem to be no particular difficulties in obtaining mixed amine complexes.[432–434]

Thermal deamination in the solid state of the diamine and mixed diamine complexes $[Cr(AA)_3]X_3$, $[Cr(AA)_2(BB)]X_3$ and $[Cr(AA)(BB)(CC)]X_3$ (AA, BB and CC represent the amines en, pn and tmd and X = Cl or SCN) takes place with the loss of one diamine molecule (the lowest boiling amine with the mixed complexes). The thiocyanates give *trans*-bis(diamine)bis(isothiocyanato) complexes but the chlorides form *cis* complexes although isomerization to *trans* complexes occurs on further heating, especially when *cis*-$[CrCl_2(tmd)_2]Cl$ is the initial product.[432] The *cis* ↔ *trans* thermal isomerization has also been investigated in *cis*- and *trans*-$[CrX_2(AA)_2]X$[435] (AA = bn, chxn or ptn; X = Cl, Br). The presence of two six-membered chelate rings or large cations seems to encourage the formation of *trans* products.[432,435] Since intermediates such as *mer*-$[CrCl_3(bn)_2]\cdot H_2O$ have been isolated, isomerization goes through bond rupture.

A number of complexes with two identical chiral diamines or two mixed diamines of the types *trans*-$[CrF_2(N_4)]^+$ and *trans*-$[Cr(ONO)_2(N_4)]^+$ have been prepared. The diamines are en, tmd, (*R*)- and (*S*)-propylenediamine, (1*R*, 2*R*)-1,2-cyclohexanediamine (chxn) and (2*R*,4*S*)-2,4-pentanediamine (ptn) (Table 50). The *trans* configurations follow from the spectral resemblances to *trans*-$[CrX_2(en)_2]^{2+}$ complexes. Contributions to the CD of complexes with chiral diamines can come from the conformation of the puckered rings and the vicinal effect of asymmetric atoms on the ligands. With *trans*-$[CrF_2(N_4)]^+$ complexes the CD contributions are separable and additive, except for bis(*R*-pn), bis(*R*,*R*-chxn) complexes and mono(*R*-pn) complexes with en and tmd. The overall CD arises mainly from the effect of the λ ring conformations.[436] By studying the CD spectra of the complexes *trans*-$[Cr(ONO)_2(N_4)]^+$ with the same amines in the region of intraligand $n \rightarrow \pi^*$ nitrito absorption bands, the existence of chiral rotational isomers of the nitrito ligands has been deduced. Marked solvent variations of the intraligand CD spectra are attributed to the effects of solvent bulk and donor number on rotamer equilibria.[436]

From the crystal structure of *cis*-$[Cr(NCS)_2(tmd)_2][Sb_2(tart)_2]_2\cdot 2H_2O$, the cation has the Λ configuration and this agrees with the assignment from the CD spectra.[431,437]

The Pfeiffer effect (the shift in a chiral equilibrium on the addition of an optical isomer of a different compound) of racemic $[Cr(ox)_3]^{3-}$ has been examined using for the first time optically stable metal complexes *cis*-$[MXY(AA)_2]^{n+}$ (where M = Cr^{3+} or Co^{3+}, AA = en or tmd and X and Y = anionic monodentate ligand). It was found that the chiral equilibrium of $[Cr(ox)_3]^{3-}$ was always displaced in favour of its Δ enantiomer in the presence of Λ enantiomers of the *cis* complexes, and it is proposed that the absolute configurations of *cis* complexes could be inferred from the equilibrium shift induced in $[Cr(ox)_3]^{3-}$.[438] Laser irradiation of an aqueous solution of racemic $[Cr(ox)_2(phen)]^-$ or $[Cr(ox)(phen)_2]^+$ in the presence of (+)- or (−)-cinchonine hydrochloride rapidly shifts the chiral equilibrium in a direction opposite to that induced by the usual Pfeiffer effect in the dark.[439]

(*iv*) *Ethylenediamine and related diamines — polynuclear complexes*

Many analogues of the dinuclear ammines (Table 47) are known in which ammonia molecules may be considered replaced by ethylenediamine and other bidentate ligands (Table 52). The inter-relations among the complexes are similar to those in Schemes 35 and 36. There are additional examples in the review by Garner and House[302] and other bridged complexes are covered in Sections 35.4.1.1, 35.4.2.3.vi, 35.4.2.4, 35.4.2.5, 35.4.4.2, 35.4.4.10, 35.4.7.4 and 35.4.8.3. Cyano-bridged complexes are listed in Table 45.

In a recent investigation[440] the racemic binuclear cations Δ,Δ-Λ,Λ-$[(en)_2Cr(OH)_2Cr(en)_2]^{4+}$ and Δ,Δ-Λ,Λ-$[(H_2O)(en)_2Cr(OH)Cr(en)_2(OH)]^{4+}$ have been isolated from the *meso* isomer Δ,Λ-$[(en)_2Cr(OH)_2Cr(en)_2]^{4+}$ as various salts (Table 52). The equilibria between the cations have been studied. The binuclear cations could not be separated into their enantiomers Δ,Δ and Λ,Λ because they react with classical resolving agents to form μ-carboxylato compounds which have been isolated,[441] *e.g.* Δ,Λ-$[(en)_2Cr(OH)(MeCO_2)Cr(en)_2](ClO_4)_4$. They contain symmetrical carboxylate bridges and deprotonate in strong base to give μ-carboxylato-μ-oxo complexes. Deprotonation of $[(en)_2Cr(OH)_2Cr(en)_2]^{4+}$ to $[(en)_2Cr(O)(OH)Cr(en)_2]^{3+}$ takes place similarly[442] and sulfato-bridged systems are also known.[443]

As with the dinuclear ammines, most research has been aimed at relating the antiferromagnetic behaviour and spectroscopic properties of the dinuclear complexes to their structures. Great variations in J have been observed for the same cation $[(en)_2Cr(OH)_2Cr(en)_2]^{4+}$ (R = H in (**96**))[370] in different environments[371] and between these cations and $[(NH_3)_4Cr(OH)_2Cr(NH_3)_4]^{4+}$ (table in ref. 374). Thus external factors such as hydrogen bonding must have an appreciable effect on the exchange in these complexes, probably through altering θ.

It has been possible to derive J for several salts of $[(en)_2Cr(OH)_2Cr(en)_2]^{4+}$ from low-temperature single-crystal absorption and powder luminescence spectra in the region of the $^4A_2{}^4A_2 \rightarrow {}^4A_2{}^2E$ transition of the paired system.[444] The difference of up to 10% from the values derived from powder magnetic susceptibility data indicates a slight temperature dependence of J.

The complex $[(en)_2Cr(OH)_2Co(en)_2](S_2O_6)_2$ has been prepared by heating the cocrystallized optically active dithionates (equation 30).[445] It is an internal active racemate because the chiral centres are different, and it contains Cr^{III} in a dinuclear system not subject to magnetic interaction so that the ESR spectrum is simple.

$$\Lambda\text{-}cis\text{-}[Cr(OH)(H_2O(en)_2]S_2O_6 + \Delta\text{-}cis\text{-}[Co(OH)(H_2O)(en)_2]S_2O_6 \longrightarrow (-)_{589}\text{-}\Lambda,\Delta\text{-}[(en)_2Cr(OH)_2Co(en)_2](S_2O_6)_2 \quad (30)$$

A crystal structure determination has confirmed that Δ,Λ-$[HO(en)_2CrOHCr(en)_2OH]$-$(ClO_4)_3{\cdot}H_2O$ formed in the reversible ring opening of Δ,Λ-$[(en)_2Cr(OH)_2Cr(en)_2]^{4+}$ has the structure inferred from its chemical properties.[446] Analogous complexes are apparently formed by tmd.[447]

The linear trinuclear cation (**100**) and the tetranuclear cations (**101**) to (**103**)[448–450] (Table 52) exist with mononuclear complexes in aqueous ethylenediamine buffer solutions in which equilibrium has been established by the catalytic effect of Cr^{2+} and charcoal under well-specified conditions.[358] The cations were separated by ion exchange chromatography and fractional crystallization, and isolated as various salts (Table 52). The tetranuclear (rhodoso) compound (**104**)[451] has been reprepared[448] by a modification of Pfeiffer's method.

The antiferromagnetic behaviour of (**101**) and (**104**) has been correlated with their structures. For (**101**) all models indicate that $J_{12} = J_{34} > J_{23}$ and that a spin singlet lies lowest with a triplet

Table 52 Bridged Complexes of Bidentate Saturated Amines

Complex	*Comments*					*Ref.*
Dinuclear (monobridged)						
Δ,Δ-Λ,Λ[$(H_2O)(en)_2Cr(OH)Cr(en)_2(OH)$]$X_4$	X = Br(Δ,Λ-isomer), ClO_4					1, 2
Δ,Λ-[$(OH)(en)_2Cr(OH)Cr(en)_2(OH)$]$(ClO_4)_3$		Cr—O—Cr 135.4°	Cr—O 1.987 (br) 1.900 (term)	Cr—N 2.079 to 2.125	Cr—Cr (Å) 3.677	1, 3
Dinuclear (dibridged)						
Δ,Λ-[$(en)_2Cr(OH)_2Cr(en)_2$]X_4	X_4					1, 4
Red violet, also X = I	$(S_2O_6)_2$	100.0°	1.978 1.980	2.081	3.032	5
	$(ClO_4)_2Cl_2$	103.4°	1.946 1.952	2.081 to 2.097	3.059	6
	Cl_4	102.4°	1.940 1.949	2.078 to 2.096	3.030	1, 4
	Br_4	102.6°	1.946 1.946	2.068 to 2.089	3.038	4
Δ,Δ-Λ,Λ-[$(en)_2Cr(OH)_2Cr(en)_2$]X_4	X = Br, ClO_4					2
Δ,Λ-[$(en)_2CrO(OH)Cr(en)_2$]$(ClO_4)_3$	Green crystals from NaOH on diol (blue solution)					1
Δ,Λ-[$(en)_2Cr(OH)(SO_4)Cr(en)_2$]$X_3$	X	Cr—O—Cr	Cr—O	Cr—N	Cr—Cr (Å)	7, 8
Red, X = ClO_4, $0.5S_2O_6$, $X_3 = ZnCl_4{\cdot}Cl$	$0.5S_2O_6$	137.4°	1.989 (OH) 1.973 (SO_4)	2.070 to 2.085	3.706	
Δ,Λ-[$(en)_2CrO(SO_4)Cr(en)_2$]$(ClO_4)_2$ Bluish green	NaOH on above, then mono-ol forms					7
Δ,Λ-[$(en)(mal)Cr(OH)_2Cr(mal)(en)$]		99°	1.98 (OH) 1.95 (mal)	2.06 (av)		9
Δ,Λ-[$(en)_2Cr(OH)(RCO_2)Cr(en)_2$]$X_4$ Red	R = H, X = ClO_4, $0.5ZnCl_4$; R = Me, X = ClO_4, $0.5ZnCl_4$; R = $CH_2NH_3^+$, $X_5 = (ZnCl_4)_2Cl$; R = CH_2NH_2, X = Br; also similar salts of racemic (Δ,Δ-Λ,Λ) cation; in base blue μ-oxo-μ-carboxylate forms initially and then mono-ol					10
[$(tmd)_2Cr(OH)_2Cr(tmd)_2$]X_4	X = Br, ClO_4, $X_2 = S_2O_6$; red-purple; blue with OH^-, μ-OH-μ-O?					11
[$(tmd)F_2Cr(OMe)_2CrF_2(tmd)$]	Blue solid, extracted from *trans*-[$CrF_2(tmd)_2$]I with MeOH					12
Dinuclear (tribridged)						
[$Cr_2(OH)_3(en)_3(SO_3NH_2)_3$]	Pink crystals from ageing of aqueous $Cr(en)_3(SO_3NH_2)_3$; $\mu_{eff}^{300\,K} = 4.05$, $\mu_{eff}^{78\,K} = 2.21$ BM; also $Cr_2(OH)_3(SO_3NH_2)_3{\cdot}3H_2O$; no crystallographic data					13
Trinuclear (linear)						
[$Cr_3(OH)_4(en)_5$]X_5 (X = I, NO_3) **(100)**			X = I, NO_3; 70% $HClO_4$ gives 2:1, *cis*-[$Cr(H_2O)_2(en)_2$]$^{3+}$/[$Cr(H_2O)_4(en)$]$^{3+}$			14

Tetranuclear (linear)			
$[Cr_4(OH)_6(en)_6]X_6$ Violet-blue **(101)**	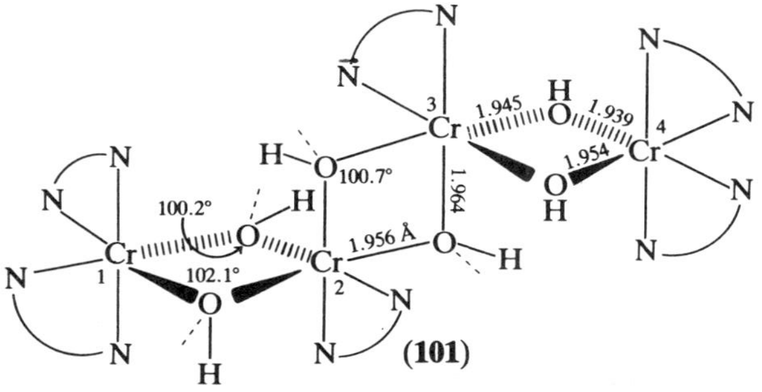 **(101)**	X = Br, I, NO_3 ΔΔΛΛ config., $J_{12} = J_{34} \approx 19\ cm^{-1}$, $J_{23} \approx 14\ cm^{-1}$	14, 15
Tetranuclear $[Cr\{(OH)_2Cr(en)_2\}_3](S_2O_6)_3$ **(102)**	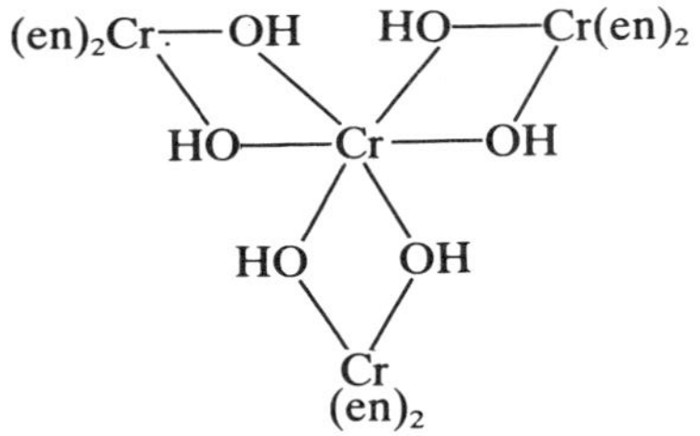 **(102)**	Trigonal planar Cr atoms, isomorphous cobalt(III) complex (Werner's brown salt), racemic Δ{ΔΛΛ}/Λ{ΛΔΔ}	14, 16
$[Cr_4(OH)_6(en)_6](NO_3)_6$ **(103)**		Red-violet crystals in low yield from same reaction mixture as (**100**) to (**102**), different properties from these and (**104**) below, centrosymmetric	14

Table 52 *(continued)*

Complex	*Comments*	*Ref.*
$[Cr_4(OH)_6(en)_6]X_6$ ($X = N_3$, BPh_4, NO_3, ClO_4, $0.5SO_4$) **(104)**	Pfeiffer's (rhodoso) compound, planar rhomboid of Cr atoms	14, 17

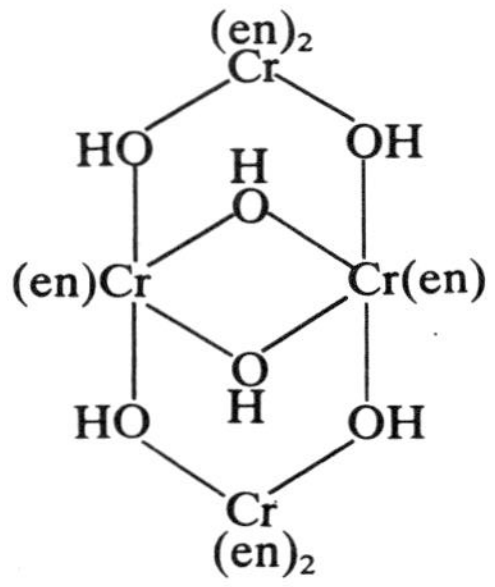

(104)

1. J. Springborg and H. Toftlund, *Acta Chem. Scand., Ser. A*, 1976, **30**, 171.
2. F. Christensson, J. Springborg and H. Toftlund, *Acta Chem. Scand., Ser. A*, 1980, **34**, 317.
3. K. Kaas, *Acta Crystallogr., Sect. B*, 1979, **35**, 1603.
4. A. Beutler, H. U. Güdel, T. R. Snellgrove, G. Chapuis and K. J. Schenk, *J. Chem. Soc., Dalton Trans.*, 1979, 983.
5. S. J. Cline, R. P. Scaringe, W. E. Hatfield and D. J. Hodgson, *J. Chem. Soc., Dalton Trans.*, 1977, 1662.
6. K. Kaas, *Acta Crystallogr., Sect. B*, 1976, **32**, 2021.
7. J. Springborg, *Acta Chem. Scand., Ser. A*, 1978, **32**, 231.
8. K. Kaas, *Acta Crystallogr., Sect. B*, 1979, **35**, 596.
9. J. W. Lethbridge, *J. Chem. Soc., Dalton Trans.*, 1980, 2039.
10. J. Springborg and H. Toftlund, *Acta Chem. Scand., Ser. A*, 1979, **33**, 31.
11. T. Laier, B. Nielsen and J. Springborg, *Acta Chem. Scand., Ser. A*, 1982, **36**, 91.
12. J. W. Vaughn, G. J. Seiler and D. J. Wierschke, *Inorg. Chem.*, 1977, **16**, 2423.
13. L. A. Baikova, K. A. Kotyaeva, E. A. Nikonenko and O. A. Antropova, *Russ. J. Inorg. Chem. (Engl. Transl.)*, 1982, **27**, 3089.
14. P. Andersen and T. Berg, *Acta Chem. Scand., Ser. A*, 1978, **32**, 989.
15. T. Damuus and E. Pedersen, *Inorg. Chem.*, 1984, **23**, 695.
16. H. U. Güdel and U. Hauser, *Inorg. Chem.*, 1980, **19**, 1325.
17. M. T. Flood, R. E. Marsh and H. B. Gray, *J. Am. Chem. Soc.*, 1969, **91**, 193.

and a quintet at about 10 cm^{-1} and 32 cm^{-1} respectively.[449] In (**102**) the dominant exchange is between the central and each of the three peripheral chromium ions. An $S = 3$ level lies lowest and there is a complicated temperature variation of μ_{eff}.[450]

(*v*) *Bidentate ligands with primary amine and N heterocyclic donor groups: (2-pyridyl)methylamine and 1-(2-pyridyl)ethylamine*

The geometrical and optical isomerism exhibited by tris(amine) complexes of the bidentate ligands (**105**) and (**106**) has been investigated. The *fac* and *mer* isomers of $[Cr(pic)_3]^{3+}$ (**107**) and (**108**) have been prepared by utilizing slightly different starting materials and solvents (Table 53). The *fac* complexes are yellow and less soluble than the orange *mer* complexes. The latter have more intense, less symmetrical electronic absorption bands to lower energy. The assignments are confirmed by the isomorphism with corresponding cobalt(III) complexes for which the ^{1}H NMR spectra of the methylene groups clearly indicate the isomeric types.[452]

Table 53 Complexes of (2-Pyridyl)methylamine (2-Picolylamine)

Complex	*Comments*	*Ref.*
fac-$[Cr(pic)_3]X_3$	Yellow, X = Cl, Br, I, *trans*-$[CrBr_2(py)_4]$I in 2-MeO$(CH_2)_2$OH/EtOH	1
	$[Cr(pic)_3]^{3+}$, log $K_1 = 5.65$, log $K_2 = 3.16$, log $K_3 = 2.21$, 25 °C	2
mer-$[Cr(pic)_3]X_3$	Orange, X = Br, I, *trans*-$[CrBr_2(py)_4]$Cl in py/EtOH	1
α-cis-$[CrX_2(pic)_2]$Y	X = F, Br, Y = Br, I; X = Cl, Y = Cl, I, X = H_2O, Y = Cl; X_2 = OH, H_2O,	3
	Y = Cl, $0.5S_2O_6$; in last Cr—N(py) (*trans*), 2.064; Cr—NH_2, 2.054, 2.089 Cr—OH, 1.926; Cr—OH_2, 1.998 Å	4
	$(-)_D$-*α-cis*-$[CrCl_2(pic)_2]ClO_4$, Δ conf.	5
β-cis-$[CrX_2(pic)_2]$Y	X = Cl, Y = Cl, I, $0.5S_2O_6$; X = Br, Y = I	3
trans-$[CrX_2(pic)_2]$Y	X = H_2O; Y = $(NO_3)_3$, X = Y = Cl or Br	6

1. K. Michelsen, *Acta Chem. Scand.*, 1970, **24**, 2003.
2. J. R. Siefker and R. D. Shah, *Talanta*, 1979, **26**, 505.
3. K. Michelsen, *Acta Chem. Scand.*, 1972, **26**, 1517.
4. S. Larsen, K. B. Nielsen and I. Trabjerg, *Acta Chem. Scand., Ser. A*, 1983, **37**, 833.
5. K. Michelsen, *Acta Chem. Scand., Ser. A*, 1976, **30**, 521.
6. K. Michelsen, *Acta Chem. Scand.*, 1973, **27**, 1823.

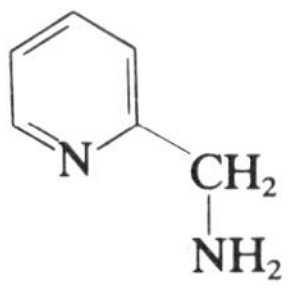

(**105**) (2-Pyridyl)methylamine (pic)

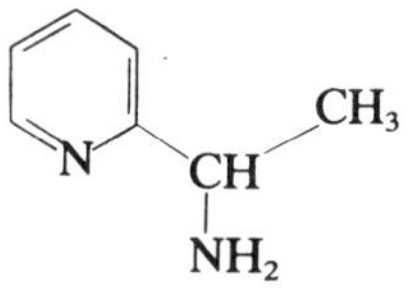

(**106**) 1-(2-Pyridyl)ethylamine (mepic)

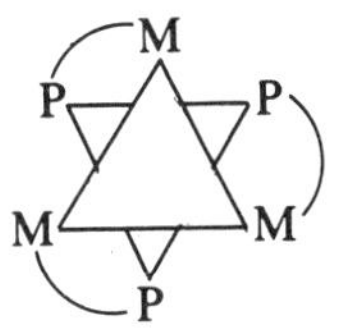

facial

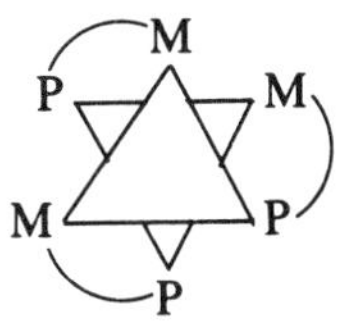

meridional

P = pyridine N, M = methylamine N

(**107**) (**108**)

The possible geometrical isomers of diacidobis(diamine) complexes containing two unsymmetrical amines such as pic are shown in (**109**) to (**113**). The *α* series are differentiated from the *β* series by an approximate C_2 axis, and from each other by the relative positions of the pyridine P and methylamine M nitrogen atoms. The preparative routes for *cis-α* and *cis-β* series are outlined in Schemes 47 and 48. Although *trans*-$[CrX_2(py)_4]^+$ salts, where X = F, Cl or Br, were used as starting materials, only *cis* products were produced. In many cases lithium salts LiX (X = Cl, Br or I) were added to aid crystallization. The *β* series were prepared from the chromium(II) halide by oxidation with I_2 in pyridine in the presence of pic, or from the *trans*-[CrX$_2$(py)$_4$]$^+$ salts in hot 2-methoxyethanol. The α-*cis* and β-*cis* assignments were made from isomorphous relationships between α-*cis*-[CrCl$_2$(pic)$_2$]Cl·H$_2$O and β-*cis*-[CrCl$_2$(pic)$_2$]$_2$S$_2$O$_6$, and the corresponding cobalt(III) complexes (^{1}H NMR). A decision between the *cis trans cis* or *cis cis trans* arrangements could not be made[453] but from the formation of members of the α series on acid treatment of related diols of crystallographically determined structures (see Section 35.4.2.3.vi below) the *cis trans cis* arrangement was proposed and confirmed by a structure determination on [Cr(H$_2$O)(OH)(pic)$_2$]S$_2$O$_6$, which showed that the cation belongs to the α-*cis* series with the pyridine nitrogen atoms *trans*.[454] The Cr—OH bond is considerably shorter than the Cr—OH$_2$ bond (Table 53).

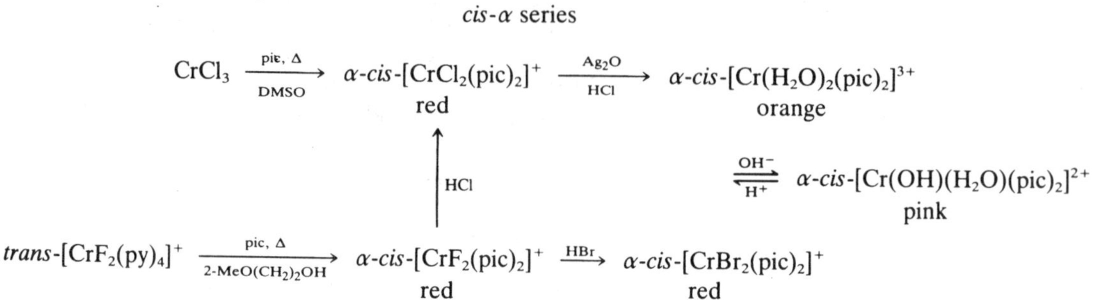

cis-α series

$$CrCl_3 \xrightarrow[\text{DMSO}]{\text{pic}, \Delta} \alpha\text{-}cis\text{-}[CrCl_2(pic)_2]^+ \xrightarrow[\text{HCl}]{Ag_2O} \alpha\text{-}cis\text{-}[Cr(H_2O)_2(pic)_2]^{3+}$$

red   orange

$$\rightleftharpoons[\text{H}^+]{\text{OH}^-} \alpha\text{-}cis\text{-}[Cr(OH)(H_2O)(pic)_2]^{2+}$$

pink

↑ HCl

$$trans\text{-}[CrF_2(py)_4]^+ \xrightarrow[\text{2-MeO(CH}_2)_2\text{OH}]{\text{pic}, \Delta} \alpha\text{-}cis\text{-}[CrF_2(pic)_2]^+ \xrightarrow{HBr} \alpha\text{-}cis\text{-}[CrBr_2(pic)_2]^+$$

red   red

Scheme 47

cis-β series

$$trans\text{-}[CrX_2(py)_4]^+ \xrightarrow[\text{2-MeO(CH}_2)_2\text{OH}]{\text{pic}, \Delta} \beta\text{-}cis\text{-}[CrX_2(pic)_2]^+ \xleftarrow[I_2, \Delta]{\text{pic, py}} CrX_2$$

violet red

X = Cl or Br

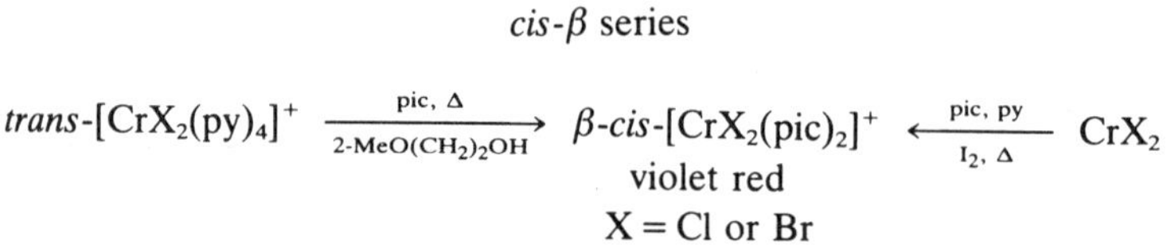

Scheme 48

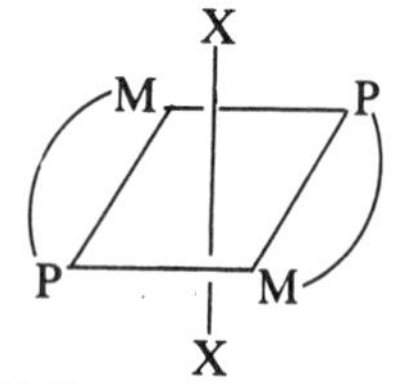

(109) *trans trans trans*   **(110)** *trans cis cis*   **(111)** *cis trans cis*   **(112)** *cis cis trans*

α series

β series   **(113)** *cis cis cis*

In the *trans* series (Scheme 49) only one geometrical isomer was obtained, presumably the *trans trans trans* because steric hindrance would prevent the formation of the *trans cis cis* isomer. The assignment was based on comparisons of colours and spectra with those of the bis(en) analogues.[455]

Complexes of the related amine 2-(2-pyridyl)ethylamine (or 2-(2-aminoethyl)pyridine, AEP) [CrX$_2$(AEP)$_2$]Y (where X = Y = Cl or Br) have been prepared by methods similar to those in Scheme 48, but *cis* or *trans* assignments were not made.[456]

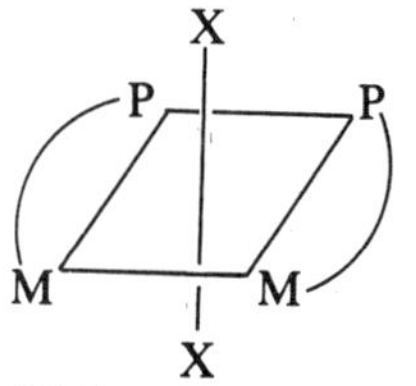

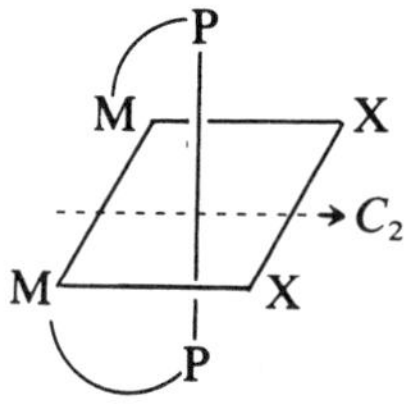

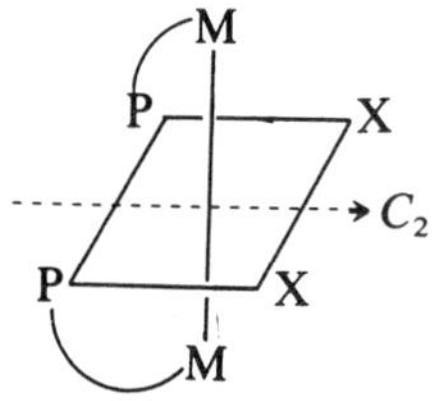

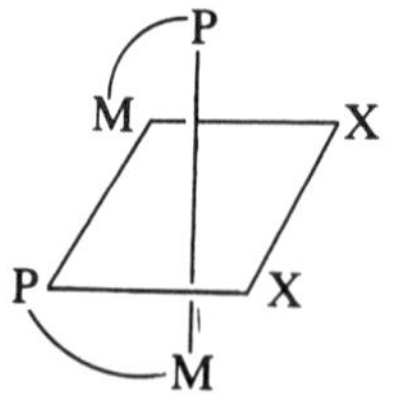

trans series

$$\alpha\text{-}cis\text{-}[CrCl_2(pic)_2]^+ \xrightarrow[\text{moist}]{Ag_2O} trans\text{- and }\alpha\text{-}cis\text{-}[Cr(OH)(H_2O)(pic)_2]^{2+}$$

$$\text{filter,}\ \Big|\ \begin{matrix}HNO_3\\EtOH\end{matrix}$$

$$trans\text{-}[CrX_2(pic)_2]Cl \xleftarrow[X=Cl,\ Br]{HX} trans\text{-}[Cr(H_2O)_2(pic)_2](NO_3)_3$$
$$\text{green} \qquad\qquad\qquad \text{yellow-orange}$$

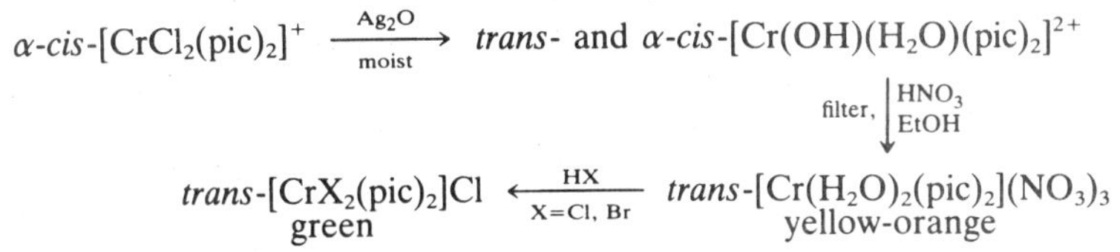

Scheme 49

(vi) Di-μ-hydroxo complexes containing (2-pyridyl)methylamine and 1-(2-pyridyl)ethylamine

In the preparation of β-*cis*-[CrBr₂(pic)₂]I a by-product assumed to contain the diol cation [(pic)₂Cr(OH)₂Cr(pic)₂]⁴⁺ was obtained.[453] Improved preparative procedures gave the optically active cations (−)_D- and (+)_D-di-μ-hydroxobis{bis(pic)chromium(III)} (Table 54).[457] Comparisons of CD spectral data suggest that the absolute configurations are ΔΔ and ΛΛ respectively. The binuclear cation can in principle exist in several geometrical and configurational isomeric forms, but from models all except (114) and its antipode are grossly sterically hindered and this explains why only one pair of isomers is obtained. Acid treatment of the diols gives (−)_D- and (+)_D-α-*cis*-[CrCl₂(pic)₂]⁺, identified by their absorption spectra; consequently they also have Δ and Λ configurations, assuming that cleavage occurs with retention of configuration. It is also believed from these results that the α-*cis* series have the *cis trans cis* configuration.

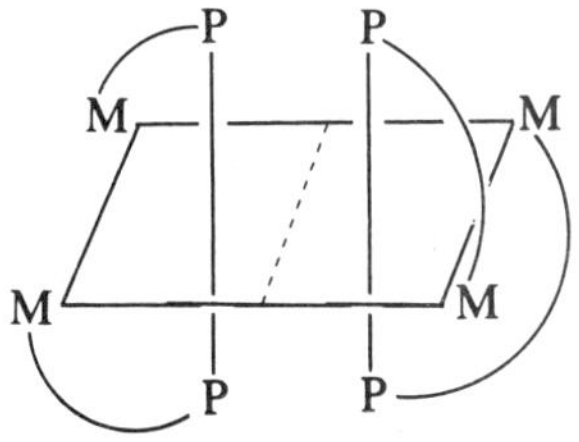

(114) ΛΛ isomer

The diols prepared from racemic mepic (Table 54), purified by column chromatography and resolved with sodium antimonyl (+)_D-tartrate, turned out to be identical with the pair prepared directly from (R)(+)_D-mepic and (S)(−)_D-mepic.[458] There are more potential geometrical isomers than in the pic series because of the introduction of the methyl group, but again almost all are ruled out because of steric hindrance. From models it appeared that the most stable diol would have the pyridine nitrogen atoms *trans*, and from the absorption and CD spectra it was concluded that the diol resulting from the preparation with (S)(−)_D-mepic has the ΛΛ absolute configuration. This last assignment was based on empirical rules for mononuclear complexes since all previously reported structures of chromium(III) diols have been either *meso* forms or racemates. An X-ray analysis of (+)_D-[{(S)(−)_D-mepic}₂Cr(OH)]₂(S₂O₆)₂·2H₂O supports the assignment (Table 54).[459] The cation has almost C₂ symmetry with the pyridine N atoms *trans*. The interplanar angle between the pyridine rings is 50° indicating considerable strain within the binuclear complex because in the related mononuclear complex *cis*-aquahydroxobis(pic)chromium(III) dithionate the pyridine planes are almost perpendicular. The hydrogen atoms of the bridge lie in the coordination plane.

The action of sodium hydroxide on the diol gives the μ-oxo-μ-hydroxo- and di-μ-oxo complexes [L₂Cr(O)(OH)CrL₂]³⁺ and [L₂Cr(O)₂CrL₂]²⁺ (L = mepic; Table 54).[460] The visible and CD spectral parameters have been obtained in media of appropriate pH. The structure of [(mepic)₂CrO(OH)Cr(mepic)₂]Br₃·5H₂O has been determined by X-ray methods. Equal numbers of enantiomorphic cations (ΔΔ and ΛΛ) are present and, as was the case with the diol, only the isomer with the pyridine nitrogen atoms *trans* has been obtained. The Cr—Cr separation of 2.883 Å is substantially less than in any dihydroxo bridged dimer so far reported (2.95–3.06 Å). The Cr—N distances are in the range 2.057 to 2.130 Å, and the oxo ligand may exert a *trans* effect since it is *trans* to the long Cr—N bond. The Cr₂O₂ bridging unit is significantly non-planar. The dinuclear complexes generally are antiferromagnetic and values of

Table 54 Dinuclear Complexes of (2-Pyridyl)methylamine, 1-(2-Pyridyl)ethylamine and Related Open Chain Tetradentate Amines

Complex	*Comments*	*Ref.*
(2-Pyridyl)methylamine (pic)		
rac-$[(pic)_2Cr(OH)_2Cr(pic)_2]X_4$ Red-violet, also $(-)_D$ and $(+)_D$ chlorides	X = Cl, Br: from $[CrX_2(H_2O)_4]X{\cdot}2H_2O$ with some $CrCl_2$ in 2-$MeO(CH_2)_2OH$; Zn gives X = $0.5ZnCl_4$; resolved *via* (+)-$[SbOC_4H_4O_6]$ salts; also X = I, ClO_4	1
1-(2-Pyridyl)ethylamine (mepic)		
$[(mepic)_2Cr(OH)_2Cr(mepic)_2]Br_4$ Red	From *racemic* mepic, $[CrBr_2(H_2O)_4]Br{\cdot}2H_2O$ in 2-$MeO(CH_2)_2OH$ and Zn, ion exchange to remove Zn salts, also I and ClO_4 salts	2
$(-)_D$-$[\{(+)_D\text{-mepic}\}_2Cr(OH)]_2X_4$ $(+)_D$-$[\{(-)_D\text{-mepic}\}_2Cr(OH)]_2X_4$	X = Br, ClO_4, from racemic complex above *via* $(+)_D$-$[SbOC_4H_4O_6]$ salts; cations also from $(+)_D$- and $(-)_D$-mepic respectively (below)	
$(+)_D$-$[\{(-)_D\text{-mepic}\}_2Cr(OH)]_2Br_4$ Red violet	From $(-)_D$-mepic as above; also ClO_4 salt and crystal structure of S_2O_6 salt: Cr—O—Cr, 101.9°; Cr—O, 1.946 (av); Cr—N (py and NH_2), 2.059 (av), Cr—Cr, 3.021 Å, (*S*)-mepic present, ΛΛ configuration, singlet–triplet splitting ≈33 cm^{-1}	2 3
$(-)_D$-$[\{(+)_D\text{-mepic}\}_2Cr(OH)]_2(ClO_4)_4$	From $(+)_D$-mepic as above	
$[(mepic)_2CrO(OH)Cr(mepic)_2]Br_3$ Olive green	From racemic diol above and NaOH (2 M); antiferro., singlet–triplet sepn. 46 cm^{-1}, Cr—O—Cr, 100.6°, Cr—OH—Cr, 95.0°, Cr—O, 1.874 (av), Cr—OH, 1.955 (av), Cr—N, 2.057 to 2.130, Cr—Cr, 2.883 Å; pyridine Ns *trans*, ΔΔ*RRRR* and ΛΛ*SSSS* cations in equal nos. $(-)_D$-isomer from $(-)_D$-diol and NaOH (2 M)	4
$[(mepic)_2Cr(O)_2Cr(mepic)_2]Cl_2$ Brown yellow	From diol chloride and NaOH (4 M); antiferro., singlet–triplet sepn. 83 cm^{-1}, Cr—O *ca.* 1.89 Å	4
Open chain tetramines (see also Section 35.4.2.4)		
$[(bispicen)Cr(OH)]_2(ClO_4)_4$ Blue violet	Antiferro., singlet–triplet sepn. 24 cm^{-1}, Cr—O—Cr, 103.0° (av), Cr—O, 1.91–1.98, Cr—NH, 2.075 (av), Cr—N (py), 2.072 (av), Cr—Cr, 3.046 Å, racemic mixture, ΔΔ or ΛΛ in cation, one *cis*-α, one *cis*-β ligand	5
$[(bispicpn)Cr(OH)]_2(ClO_4)_4$ Red	Antiferro., singlet–triplet sepn. 44 to 50 cm^{-1} for various isomers from optically active and racemic bispicpn	6, 7
$[(bispictmd)Cr(OH)]_2(ClO_4)_4$ Violet	Antiferro., singlet–triplet sepn. 32 cm^{-1}, Cr—O—Cr, 101.9° (av), Cr—O, 1.967 (av), Cr—NH, 2.080 (av), Cr—N (py), 2.085 (av), Cr—Cr, 3.054 Å, racemic mixture, ΔΔ or ΛΛ in cation, *cis*-β ligand; also red isomer and $[Cr\{Cr(bispictmd)(OH)_2\}_3](ClO_4)_6$	8
$[(bispicbn)Cr(OH)]_2(ClO_4)_4$	From *S,S*-picbn, ΛΛ conf., also *S,S*-picchxn complex	7

1. K. Michelsen, *Acta Chem. Scand., Ser. A*, 1976, **30**, 521.
2. K. Michelsen and E. Pedersen, *Acta Chem. Scand., Ser. A*, 1978, **32**, 847.
3. S. Larsen and B. Hansen, *Acta Chem. Scand., Ser. A*, 1981, **35**, 105.
4. K. Michelsen, E. Pedersen, S. R. Wilson and D. J. Hodgson, *Inorg. Chim. Acta*, 1982, **63**, 141.
5. M. A. Heinrichs, D. J. Hodgson, K. Michelsen and E. Pedersen, *Inorg. Chem.*, 1984, **23**, 3174.
6. K. Michelsen and E. Pedersen, *Acta Chem. Scand., Ser. A*, 1983, **37**, 141.
7. Y. Yamamoto and Y. Shimura, *Bull. Chem. Soc. Jpn.*, 1981, **54**, 2924.
8. H. R. Fischer, D. J. Hodgson, K. Michelsen and E. Pedersen, *Inorg. Chim. Acta*, 1984, **88**, 143.

the singlet–triplet separations are given in Table 54. The chemistry of the diols in Table 54 which contain open chain tetramines is discussed in Section 35.4.2.4.v.

35.4.2.4 Open-chain polyamines

(i) Diethylenetriamine (dien) and related ligands

The complex $[Cr(dien)_2]Cl_3$ may be prepared by a number of methods. These include heating $CrCl_3$ and dien directly,[461] or heating a solution of $CrCl_3$ and dien in DMSO,[462] or heating a solution of $CrCl_3{\cdot}6H_2O$ and dien in absolute alcohol.[463] The perchlorate $[Cr(dien)_2](ClO_4)_3$ has been obtained in crystalline form from a solution of the chloride in dilute $HClO_4$.[463] Although three geometric isomers, two of which should be optically active, are possible for this complex, no resolutions or structural assignments have so far been made. The kinetics of aquation[463] and the phosphorescence behaviour[464] have been reported.

A number of complexes of the type $[CrCl(AA)dien]^{2+}$ (AA = en, pn or tn) have been prepared by the reaction in equation (31).[465] Only the *a*(Cl)-*b*,*c*(AA)-*d*,*f*,*e*(dien) isomer was isolated in each case. These complexes undergo spontaneous, Hg^{2+}- and base-catalyzed hydrolysis at similar rates and all seemingly by an associative interchange mechanism.[466] Details of the solution electronic spectra of the complexes and their hydrolysis products are in Table 55.

$$CrCl_3 \cdot 6H_2O \xrightarrow{\text{i, ii, iii, iv}} [CrCl(AA)dien]ZnCl_4 \qquad (31)$$

i, DMSO, heat; ii, dien (1:1), heat; iii, en (1:1), heat; iv, $[ZnCl_4]^{2-}$

Table 55 Electronic Solution Spectra of $[CrX(AA)dien]^{n+}$ Complexes[a,1]

Complex	λ_{max} (ε)	λ_{min} (ε) (nm; $dm^3\,mol^{-1}\,cm^{-1}$)	λ_{max} (ε)
$[ClCl(en)dien]^{2+}$	514 (79.3)	428 (24.5)	373 (91.4)
$[CrCl(pn)dien]^{2+}$	514 (74.5)	417 (22.9)	372 (85.7)
$[CrCl(tn)dien]^{2+}$	526 (71.5)	432 (21.3)	375 (87.6)
$[Cr(en)(dien)H_2O]^{3+}$	482 (69.4)	410 (21.7)	365 (55.7)
$[Cr(pn)(dien)H_2O]^{3+}$	482 (65.3)	410 (19.5)	362 (52.9)
$[Cr(tn)(dien)H_2O]^{3+}$	490 (58.9)	414 (19.4)	362 (54.9)

[a] In $0.1\,M\,H^+$ solution.
1. D. A. House, *Inorg. Nucl. Chem. Lett.*, 1976, **12**, 259.

Dissolution of the chromium(IV) diperoxo compound, $[Cr(O_2)_2dien] \cdot H_2O$, in concentrated HCl affords the mauve crystalline product *mer*-$[CrCl_3(dien)]$.[467] The *fac* isomer may be obtained as green crystals by the addition of concentrated HCl to the oily residue obtained after evaporating a solution containing $[Cr(urea)_6]Cl_3$ and dien in EtOH.[468] The configurations of the complexes were initially assigned on the basis of spectral[469] and kinetic data[470] and later confirmed by X-ray crystallography, which showed the green material to be the *fac* isomer.[471] The bond lengths (pm) in this complex are shown in (**115**). Hydrolysis of these complexes proceeds as shown in Scheme 50.[468,470] The fact that the partly unwrapped dien complex $[Cr(H_2O)_4dienH]^{4+}$ may be obtained by hydrolysis of either 1,2,3- or 1,2,6-$[Cr(dien)(H_2O)_3]^{3+}$ implies that protonation occurs on a primary amino group. Protonation of the secondary amino group followed by isomerization is an unlikely alternative since such reactions are generally very slow.[468]

1,2,3-$CrCl_3$dien $\xrightarrow[\text{50 °C, 1–2 min}]{\text{0.5 M HClO}_4}$ 1,2,3-$[CrCl_2(dien)H_2O]^{+\ a}$
green   blue

$\big\downarrow$ 0.4 M $Hg(NO_3)_2$, 1 M $HClO_4$; 20–25 °C, 1 h $\big\downarrow$ 0.05 M $HClO_4$; 55–60 °C, 5 min

1,2,3-$[Cr(dien)(H_2O)_3]^{3+}$ 1,2,3-$[CrCl(dien)(H_2O)_2]^{2+}$
red   magenta

$\big\downarrow$ H^+, 60 °C

$[Cr(dienH)(H_2O)_4]^{4+}$ $\xrightarrow[\text{dark, 3 h}]{\text{2 M HClO}_4,\ 60\,°C,}$ $[Cr(dienH_2)(H_2O)_5]^{5+}$ $\xrightarrow[\text{dark, 3 h}]{\text{2 M HClO}_4,\ 60\,°C}$ $[Cr(H_2O)_6]$
pink   purple

$\big\uparrow$ 0.1–1 M $HClO_4$; 20–25 °C, dark, 5 h

1,2,6-$[Cr(dien)(H_2O)_3]^{3+}$ $\xleftarrow{H^+}$ 1,2,6-$Cr(OH)_3$dien $\xrightarrow{\text{20–25 °C, slow}}$ $Cr(OH)_3$ + dien
orange   blue

$\big\uparrow$ 0.1 M NaOH; 10 min, 20–25 °C

1,2,6-$[CrCl_2(dien)H_2O]^{+}$ $\xleftarrow[\text{50 °C, 5 min}]{\text{0.05 M HClO}_4}$ 1,2,6-$CrCl_3(dien)$
pink   purple

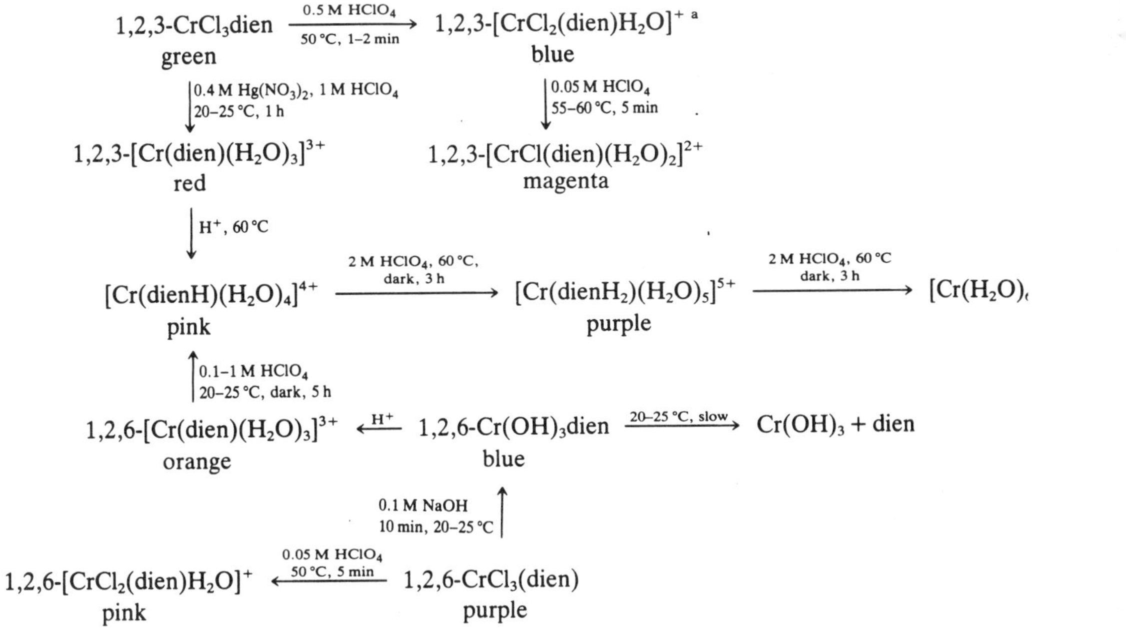

[a] This is thought to be the 5,6-dichloro isomer.

Scheme 50

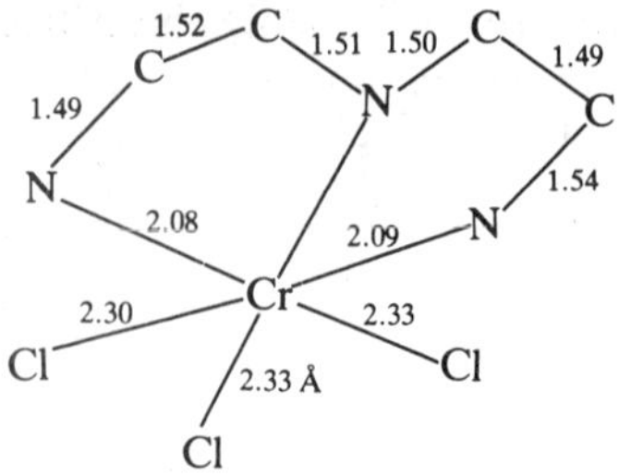

(115)

The mononuclear complex $[CrCl_2(tame)DMSO]I \cdot H_2O$ and the binuclear complexes $[Cr(\mu\text{-}OH)OH(tame)]_2I_2 \cdot 2H_2O$ and $[H_2O(tame)Cr(\mu\text{-}OH)_2CrOH(tame)]I_3 \cdot H_2O$ containing the tridentate ligand tris(aminomethyl)ethane (tame) have been reported.[472]

(ii) Triethylenetetramine (trien)

The isomers $cis\text{-}\alpha\text{-}[CrCl_2(trien)]Cl$ **(116)** and $cis\text{-}\beta\text{-}[CrCl_2(trien)]Cl$ **(117)** are prepared in 70–80% yields according to Scheme 51.[473] The use of $SOCl_2$ as solvent in the preparation of $cis\text{-}\beta\text{-}[CrCl_2(trien)]Cl$ offers the following advantages: (a) it dissolves the starting complex but not the product, (b) its reaction with H_2O provides a homogeneous source of HCl and (c) it provides an anhydrous medium and thus prevents the formation of chloroaquachromium(III) products. Although the isomers of $cis\text{-}[CrCl_2(trien)]^+$ are stereorigid under anhydrous conditions, the presence of water causes rapid conversion of the $cis\text{-}\beta$ to the $cis\text{-}\alpha$ form, probably via the stereomobile aquation product $[CrCl(trien)H_2O]^{2+}$ (Scheme 52). Indeed the first step in this reaction, the aquation of $cis\text{-}\beta\text{-}[CrCl_2(trien)]^+$, occurs 10^3-fold more rapidly than that of the α isomer[473] and is one of the most rapid aquations reported for chromium(III) complexes. Additional support for this mechanism is provided by the fact that isomerizations of analogous cobalt(III) complexes proceed much more readily after aquation. The lability of $cis\text{-}\beta\text{-}[CrCl_2(trien)]^+$ relative to its α isomer is consistent with an I_a mechanism in which H_2O attacks the metal ion at a position *trans* to the leaving group. In the $cis\text{-}\alpha$ isomer this site is blocked and rendered hydrophobic by methylene groups in the trien chain. In the $cis\text{-}\beta$ isomer, however, this site is open and is also hydrophilic because of the proximity of amino groups.[473]

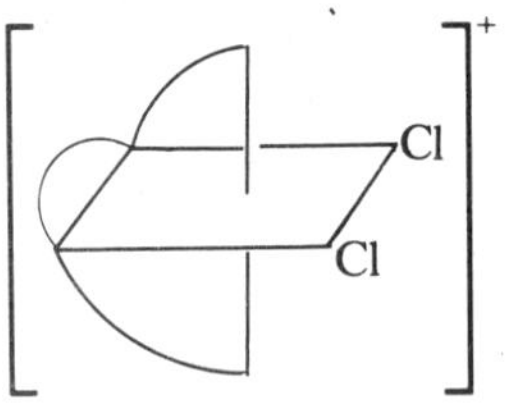

$$\left[\begin{array}{c} \text{Cl} \\ \text{Cl} \end{array}\right]^+ \qquad \left[\begin{array}{c} \text{Cl} \\ \text{Cl} \end{array}\right]^+$$

(116) *cis-α*   **(117)** *cis-β*

$$\alpha\text{-}[Cr(C_2O_4)trien]X \quad X = Br \text{ or } ClO_4$$

$$\downarrow SOCl_2/H_2O \ (50:1, \text{v/v})$$

$$cis\text{-}\beta\text{-}[CrCl_2(trien)]ClO_4 \xleftarrow{\text{LiClO}_4/\text{MeOH}} cis\text{-}\beta\text{-}[CrCl_2(trien)]^+ \xrightarrow{\text{NaClO}_4(\text{aq})} cis\text{-}\alpha\text{-}[CrCl_2(trien)]ClO_4$$

$$\swarrow{\substack{\text{dry} \\ \text{MeOH/HCl}}} \qquad \searrow{\text{HCl(aq)/MeOH}}$$

$$cis\text{-}\beta\text{-}[CrCl_2(trien)]Cl \qquad cis\text{-}\alpha\text{-}[CrCl_2(trien)]Cl \cdot H_2O$$

Scheme 51

The $cis\text{-}\alpha\text{-}$ and $\text{-}\beta\text{-}[CrCl_2(trien)]^+$ isomers may be distinguished by spectroscopic methods. The electronic spectrum of the $cis\text{-}\alpha$ isomer has a long wavelength shoulder on the first ligand field band similar to that observed in the spectra of other $cis\text{-}\alpha\text{-}[MCl_2(trien)]^+$ (M = Co, Rh) complexes. No such shoulder is observed in the spectra of the $cis\text{-}\beta$ isomers. The differences in

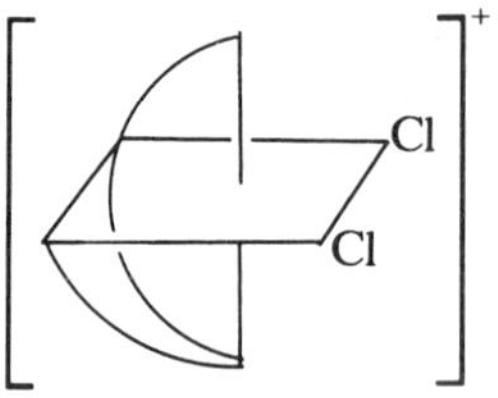

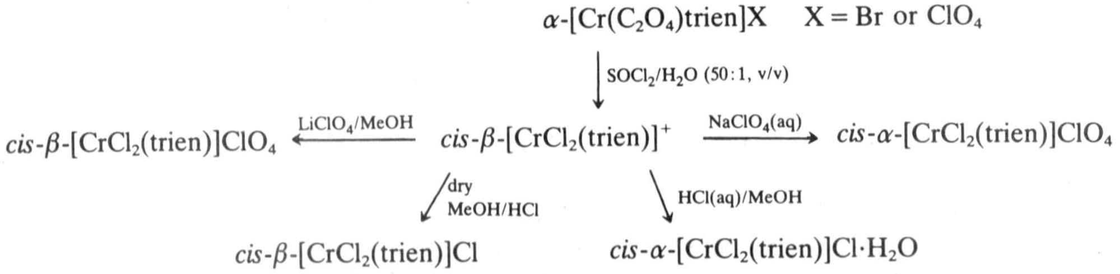

$$cis\text{-}\beta\text{-}[CrCl_2(trien)]^+ \xrightarrow{H_2O} cis\text{-}\beta\text{-}[CrCl(trien)H_2O]^{2+} + Cl^-$$

$$\rightleftharpoons$$

$$cis\text{-}\alpha\text{-}[CrCl_2(trien)]^+ \xleftarrow{Cl^-} cis\text{-}\alpha\text{-}[CrCl(trien)H_2O]^{2+}$$

Scheme 52

the IR spectra of the complexes, shown in Figure 4, provide support for the assignment of configurations.[473] Details of the electronic solution spectra of these and other chromium(III)–trien complexes are given in Table 56. While the $[Cr(C_2O_4)trien]^+$ starting material used in the above reactions was, on the basis of IR spectroscopy, originally thought to have the *cis-α* configuration, recent evidence based on chiral interconversions indicates that it is in fact the *cis-β* isomer.[474] The primary step in the aquation of this complex involves Cr—N bond rupture and leads to the formation of $[Cr(C_2O_4)(H_2O)trienH]^{2+}$.[475]

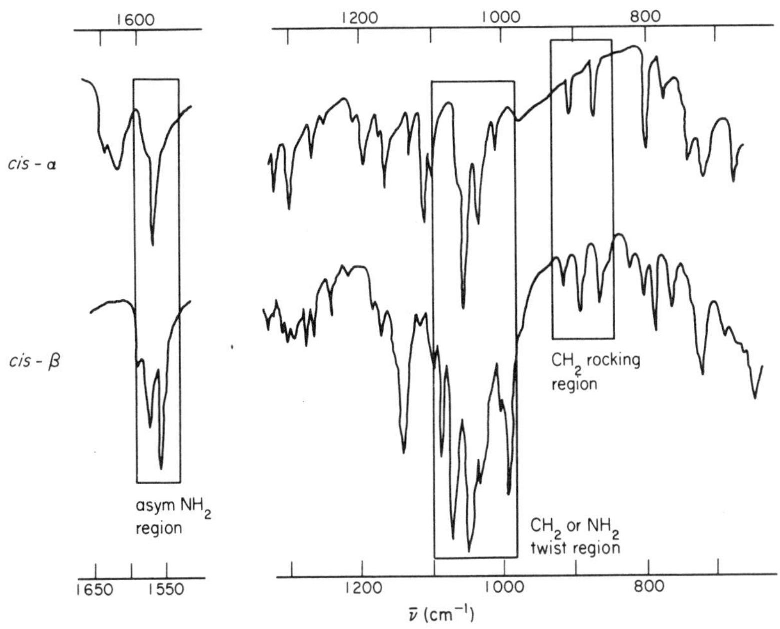

Figure 4 Differences in IR spectra (Nujol mulls) between *cis-α*- and *cis-β*-$[Cr(trien)Cl_2]Cl$ (adapted from *Inorg. Chem.*, 1977, **16,** 1154)

Reaction of $CrCl_3{\cdot}6H_2O$ with trien gives a blue-violet mixture from which some chromium(III)–trien complexes (electronic spectra given in Table 56) may be isolated as indicated in Scheme 53.[476] While the conversion of *cis-α*-$[Cr(OH)(trien)H_2O]Cl_2$ to the *cis-β* isomer in the temperature range 160–225 °C is reversed on cooling, the process is irreversible if the reaction is carried out above 225 °C.

A number of chromium(III) complexes containing 'partially unwrapped' trien ligands have been characterized. The complexes $[Cr(trienH_x)(H_2O)_{x+2}]^{x+3}$ are intermediates in the stepwise aquation of *cis-α*-$[CrCl_2(trien)]^+$ (Scheme 54) and can be isolated by cation exchange chromatography.[477] The aquation of *cis-α*-$[Cr(N_3)_2trien]^+$ similarly affords a series of $[Cr(N_3)(trienH_x)(H_2O)_{x+1}]^{x+2}$ ions as outlined in Scheme 55.[478] The colours and electronic spectra of these trien complexes are also included in Table 56.

Empirical[479] and theoretical[480] models have been developed to predict the photochemical behaviour of chromium(III) complexes. These rules predict that photolysis of *cis*-$[CrN_4Cl_2]^+$ complexes should proceed with aquation of the amino ligands *trans* to the weak field Cl^-. While the photoreactivity of *cis*-$[CrCl_2(en)_2]^+$ is, as expected, dominated by loss of amine, the behaviour of *cis-α*-$[CrCl_2(trien)]^+$ offers a crucial test to these models since photolabilization of the secondary amino groups should occur but because of the geometry of the complex these are constrained from leaving the primary coordination sphere of the metal ion. Indeed photolysis

Table 56 Electronic Spectra of Chromium(III)–trien Complexes

Complex	*Colour*	$\lambda_{max}(\varepsilon)$/nm($dm^3\ mol^{-1}\ cm^{-1}$)		$\lambda_{min}(\varepsilon)$	*Ref.*
cis-α-$[CrCl_2(trien)]^+$	Purple	535 (97.2)	396 (90.5)	455 (28.8)[a]	1
cis-β-$[CrCl_2(trien)]^+$	Purple	542 (100)	403 (85)	461 (35)[b]	1
cis-$[CrCl(trien)H_2O]^{2+}$ [c]	Pink	518 (105)	391 (70)	442 (38)[a]	1
cis-α-$[Cr(OH)(trien)H_2O]^{2+}$	Pink	521 (91.2)	386 (70.8)[d]		2
cis-α-$[CrOH(trien)]_2^{4+}$	Blue-violet	543 (129)	397 (123)[e]		2
cis-α-$[Cr(NCS)_2(trien)]^+$	Orange	488 (157)	372 (86.7)	420 (36.9)[a]	3
cis-α-$[Cr(N_3)_2(trien)]^+$	Brick red	518 (185)	402 (126)	452 (70.0)[a]	3
$[CrC_2O_4(trien)]^+$	Pink	500 (139)	372 (96.5)	425 (28.0)[a]	3
1,2,3-$[Cr(H_2O)_3trienH]^{4+}$	Red	513 (72.0)	375 (35.6)	430 (13.8)[f]	4
1,2,3-$[Cr(N_3)(H_2O)_2trienH]^{3+}$	Magenta	532 (158)	403 (102)	453 (51.8)[g]	5
1,2,6-$[Cr(H_2O)_3trienH]^{4+}$	Orange	493 (82.9)	368 (45.5)	418 (20.4)[f]	4
1,2,6-$[Cr(N_3)(H_2O)_2trienH]^{3+}$	Deep red	502 (189)	390 (120)	437 (59.7)[g]	5
$[Cr(H_2O)_4trienH_2]^{5+}$	Magenta	522 (48.7)	389 (27.5)	440 (10.1)[f]	4
$[Cr(N_3)(H_2O)_3trienH_2]^{4+}$	Purple	543 (117)	409 (82.1)	467 (33.7)[g]	5
$[Cr(H_2O)_5trienH_3]^{6+}$	Purple	552 (22.3)	398 (22.0)	465 (5.1)[g]	4
$[Cr(N_3)(H_2O)_4trienH_3]^{5+}$	Pink	560 (75.6)	423 (73.5)	488 (26.3)[h]	5

[a] In 0.1 M HCl. [b] In MeOH. [c] This is the product 5 min after dissolving *cis*-β-$[CrCl_2(trien)]^+$ in 0.1 M HCl at room temp. [d] In H_2O. [e] In DMSO. [f] In 2 M $HClO_4$. [g] In 3 M $HClO_4$. [h] In 4 M $HClO_4$.

1. W. A. Fordyce, P. S. Sheridan, E. Zinato, P. Riccieri and A. W. Adamson, *Inorg. Chem.*, 1977, **16**, 1154.
2. R. Tsuchiya, A. Uehara, K. Noji and H. Yamamura, *Bull. Chem. Soc. Jpn.*, 1978, **51**, 2942.
3. D. A. House and C. S. Garner, *J. Am. Chem. Soc.*, 1966, **88**, 2156.
4. R. L. Wilder, D. A. Kamp and C. S. Garner, *Inorg. Chem.*, 1971, **10**, 1393.
5. S. C. Tang, R. L. Wilder, R. K. Kurimoto and C. S. Garner, *Synth. React. Inorg. Metal-Org. Chem.*, 1971, **1**, 207.

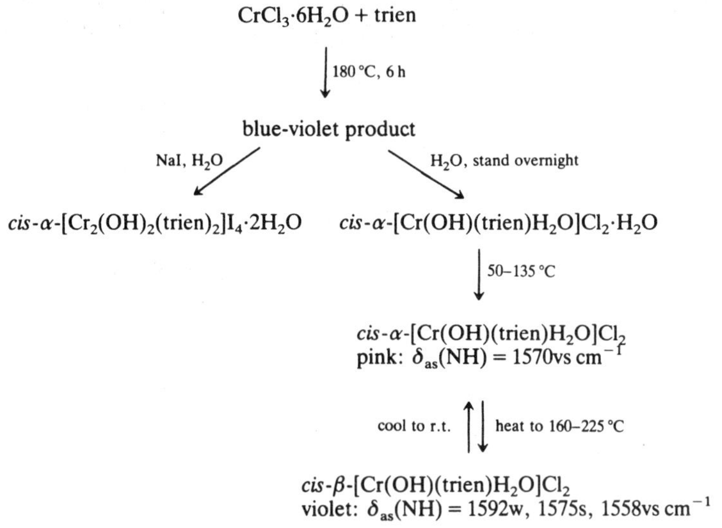

Scheme 53

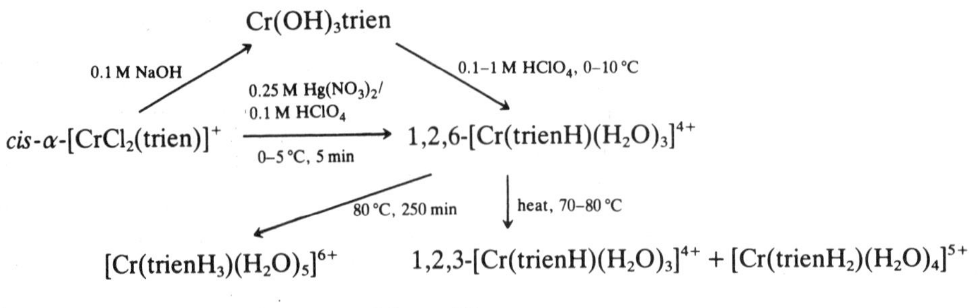

Scheme 54

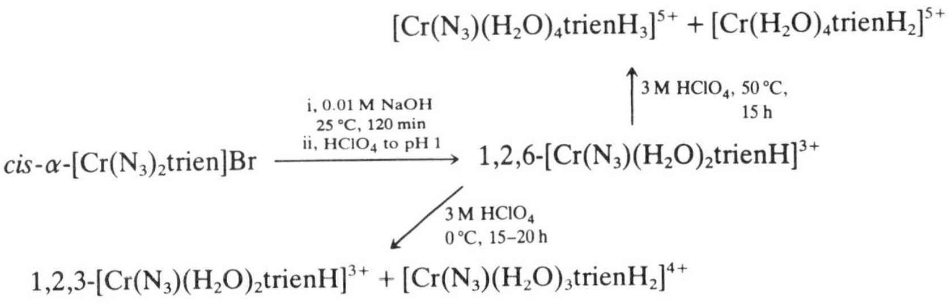

Scheme 55

of *cis*-α-$[CrCl_2(trien)]^+$ was observed to involve negligible amine loss, with Cl^- aquation being the dominant process. Since the axial coordination sites in this complex contain primary amino groups which are as unconstrained as the photolabile amines in *cis*-$[CrCl_2(en)_2]^+$, the fact that amine loss is not observed confirms that ligand labilization is confined to the *xy* plane. The photolysis proceeds with retention of configuration.

(*iii*) 2,2′,2″-*Triaminotriethylamine* (*tren*)

A number of chromium(III) complexes of tren, a tripodal isomer of trien, have been prepared from $[CrCl_2(tren)]Cl$ according to Scheme 56.[481] This complex is conveniently prepared by the addition of tren to a solution of $[CrCl_2(H_2O)_4]Cl{\cdot}H_2O$ in DMSO which had previously been heated to boiling. In acid solution $[CrCl_2(tren)]^+$ undergoes a two-step aquation, the first involving loss of Cl^- and the second involving Cr—N bond rupture to give the partly unwrapped tren complex $[CrCl(H_2O)trenH]^{3+}$. Solvolysis of $[Cr(Cl)_2tren]^+$ in formamide, *N*-methylformamide, DMF and DMSO appears to occur *via* dissociative mechanisms. A number of these solvolysis products have been isolated and characterized.[482] Reaction of *trans*-$[CrF_2(py)_4]ClO_4$ with tren in 2-methoxyethanol gives *cis*-$[CrF_2(tren)]ClO_4$ as product.[392] From this the tren complexes $[CrF(X)tren](ClO_4)_x$ ($X = H_2O$, NCS, Br, N_3, $MeCO_2$; $x = 0$, 1 or 2) have been prepared.[487] In acid solution *cis*-$[CrF_2(tren)]ClO_4$ is converted to *cis*-α-$[CrF(tren)H_2O]^{2+}$ (H_2O *trans* to the tertiary N) both by acid-catalyzed thermal aquation and by ligand field photolysis.[483] A second but pH-independent thermal aquation process involving rupture of a Cr—N bond has also been observed. The electronic solution spectra of chromium(III)–tren complexes display *d–d* bands of greater intensity than those of analogous Cr—N_4 complexes having higher symmetry.[481] Dissolution of $[Cr(C_2O_4)tren]^+$ in dilute $HClO_4$ at 15 °C leads to the rapid formation of $[Cr(C_2O_4)H_2O(trenH)]^{2+}$.[484] Heating the product solution to 60 °C causes stepwise red-shifts in the *d–d* spectral maxima until the formation of $[Cr(C_2O_4)(H_2O)_4]^+$ is complete. In concentrated HCl solution, however, $[Cr(C_2O_4)tren]^+$ reacts to give $[Cr(Cl)_2tren]^+$ in a two-step process.

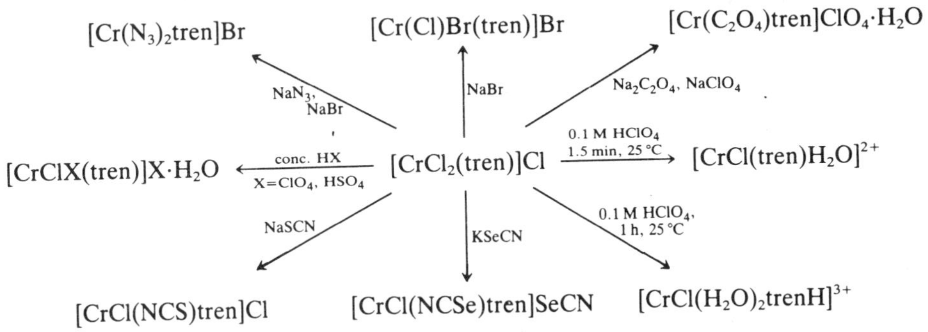

Scheme 56

Reaction of $CrCl_3(Me_3N)_2$ with tren in benzene (Scheme 57) leads to products in which tren behaves as a tridentate ligand coordinated to the metal through the primary amino groups[485] (Section 35.4.2.2).

$(Me_3N)_2CrCl_3$ —tren (1:1)→ $CrCl_3(tren)$
blue-grey, air-sensitive, non-electrolyte in CH_2Cl_2,
$\mu_{eff} = 3.83$ BM (295 K), $\nu(Cr—Cl) = 338, 300\ cm^{-1}$,
$\lambda_{max} = 591, 408$ nm (reflectance)

$(Me_3N)_2CrCl_3$ —tren (2:1)→ $[Cr(tren)_2]Cl_3$
pink, hygroscopic, $\mu_{eff} = 3.86$ BM (294 K),
3:1 electrolyte in H_2O, $\lambda_{max} = 498, 380$ nm (reflectance)

Scheme 57

(*iv*) *Linear tetramines other than trien*

The linear tetramine ligand, 1,9-diamino-3,7-diazanonane (2,3,2-tet), a homologue of trien, forms octahedral complexes which can adopt several isomeric configurations.[486] The complex *cis-β-RR,SS*-$[Cr(C_2O_4)2,3,2\text{-tet}]^+$ has been synthesized and resolved and then used to prepare racemic and chiral *cis-α*-$[Cr(X)(Y)(2,3,2\text{-tet})]^{n+}$ ($X = Y = Cl$, NCS, CF_3CO_2, H_2O; $X = Cl$, $Y = H_2O$) complexes.[486] The *meso* complexes *trans*-$[CrX_2(2,3,2\text{-tet})]^+$ ($X = Cl$, F, NCS) have also been prepared and the secondary NH isomeric configurations for two of these ($X = F$, NCS) have been confirmed by X-ray crystallography.[486–489] The complexes *cis-β-RR,SS*-$[CrX_2(3,2,3\text{-tet})]^+$ ($X_2 = C_2O_4^{2-}$ or $2Cl^-$) and *trans-RR,SS*-$[CrY_2(3,2,3\text{-tet})]^+$ ($Y = Cl$, NCS^-) which contain the ligand 1,10-diamino-4,7-diazadecane (3,2,3-tet) have been synthesized and resolved.[490] Details of the visible absorption spectra of the above complexes have been reported as well as CD spectra and aquation kinetic parameters in some cases.[486,490–493] The complex *trans*-$[CrF_2(3,3,3\text{-tet})]ClO_4$ (3,3,3-tet = 1,11-diamino-4,8-diazaundecane) has been prepared and its electronic spectrum compared with those of other $Cr^{III}X_4Z_2$ complexes (X = a tetramine or two diamines; Z = F, Cl, Br, I, DMF or CO_2^-) using crystal field and angular overlap models[494] (note that the tetramine ligands are incorrectly named in this reference). A spectrochemical series in *Cp*, the second-order crystal-field parameter, shows trends in chromium–ligand interactions which are not apparent from the conventional series in *Dq*.

(*v*) 1,6-*Bis*(2′-*pyridyl*)-2,5-*diazahexane* (*bispicen*) *and related ligands*

Mononuclear and binuclear chromium(III) complexes containing bispicen and related tetradentate ligands have been prepared. The isomers *cis-α*- (**118**) and -*β*- (**119**) $[CrCl_2(bispicen)]^+$ are obtained by the methods outlined in Scheme 58[495,496] but complexes with the *trans* geometry are unknown.

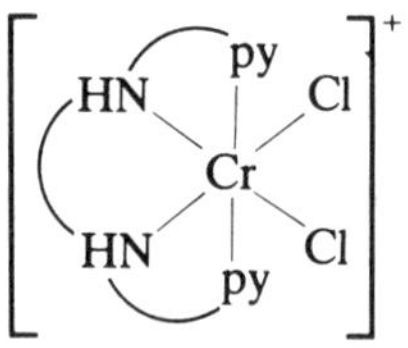

(**118**) *cis-α*

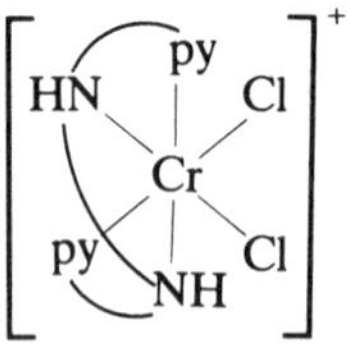

(**119**) *cis-β*

Addition of bispicen to a solution of $[CrBr_2(H_2O)_4]Br{\cdot}2H_2O$ in 2-methoxyethanol and in the presence of zinc powder gives the binuclear di-μ-hydroxo complex $[CrOH(bispicen)]_2^{4+}$ (see Table 54 for details of the structure).[495] In principle, 13 possible isomers of this complex exist but the use of molecular models shows that 11 of these are grossly sterically hindered. The remaining two isomers, (**120**) and (**121**), were isolated by the methods outlined in Scheme 59. Treatment of ΛΛ-(+)$_D$-$[CrOH(bispicen)]_2(ClO_4)_4{\cdot}4H_2O$ with concentrated HCl affords (+)$_D$-*cis-α*-$[CrCl_2(bispicen)]ClO_4$, which has therefore been assigned the Λ configuration.[495] The stereochemistries of the bispicen complexes were established using electronic and CD spectroscopy and by comparison of their X-ray powder photographs with those of analogous cobalt(III) complexes with structures known from ^1H NMR spectroscopy.

A number of complexes of the type *cis-α*-$[CrX_2(S\text{-bispicpn})]Y$ ($X = Y = Cl$, Δ and Λ;

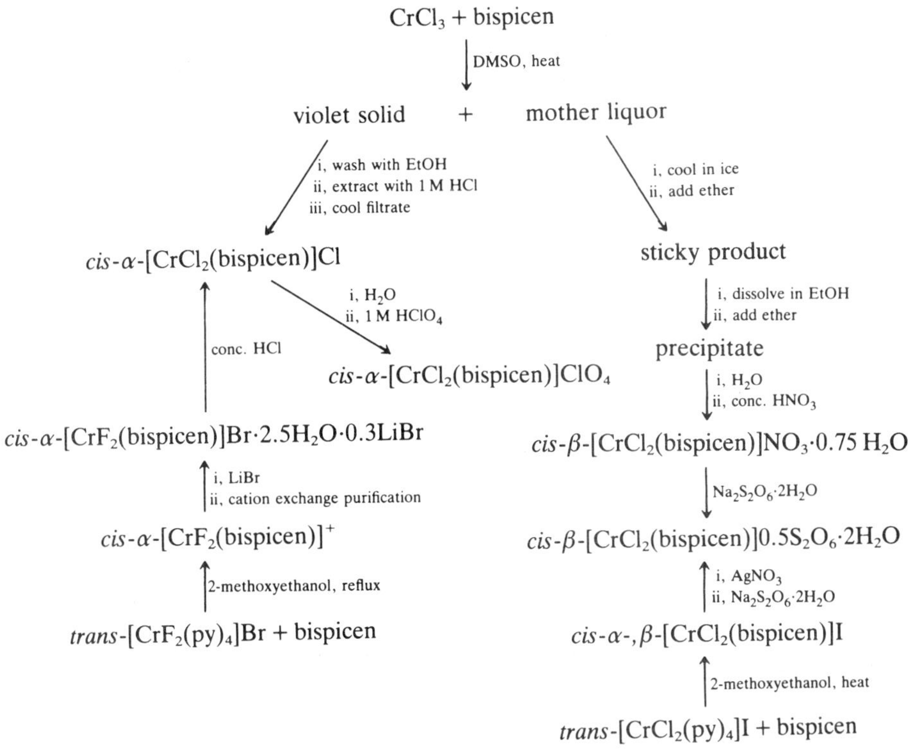

$$CrCl_3 + bispicen$$

$\downarrow$ DMSO, heat

violet solid + mother liquor

violet solid:
i, wash with EtOH
ii, extract with 1 M HCl
iii, cool filtrate

mother liquor:
i, cool in ice
ii, add ether

cis-α-[CrCl_2(bispicen)]Cl

sticky product

cis-α-[CrCl_2(bispicen)]Cl → (i, H_2O; ii, 1 M HClO_4) → *cis*-α-[CrCl_2(bispicen)]ClO_4

sticky product:
i, dissolve in EtOH
ii, add ether

precipitate

conc. HCl ↑

cis-α-[CrF_2(bispicen)]Br·2.5H_2O·0.3LiBr

precipitate:
i, H_2O
ii, conc. HNO_3

cis-β-[CrCl_2(bispicen)]NO_3·0.75 H_2O

i, LiBr
ii, cation exchange purification ↑

cis-α-[CrF_2(bispicen)]⁺

Na_2S_2O_6·2H_2O ↓

cis-β-[CrCl_2(bispicen)]0.5S_2O_6·2H_2O

2-methoxyethanol, reflux ↑

trans-[CrF_2(py)_4]Br + bispicen

i, AgNO_3
ii, Na_2S_2O_6·2H_2O ↑

cis-α-,β-[CrCl_2(bispicen)]I

2-methoxyethanol, heat ↑

trans-[CrCl_2(py)_4]I + bispicen

Scheme 58

[CrOH(bispicen)]_2I_4·3H_2O + (+)_D-NaSbOC_4H_4O_6

H_2O ↓

(−)_D-[CrOH(bispicen)]_2{(+)_D-SbOC_4H_4O_6}_2I_2·11H_2O + mother liquor

i, 1 M NaOH
ii, EtOH
iii, 4 M HCl
iv, NaI ↓

EtOH ↓

(+)_D-[CrOH(bispicen)]_2{(+)_D-SbOC_4H_4O_6}_4·14H_2O

ΔΔ-(−)_D-[CrOH(bispicen)]_2I_4·3H_2O

i, hot H_2O
ii, NaClO_4 ↓

ΛΛ-(+)_D-[CrOH(bispicen)]_2(ClO_4)_4·4H_2O

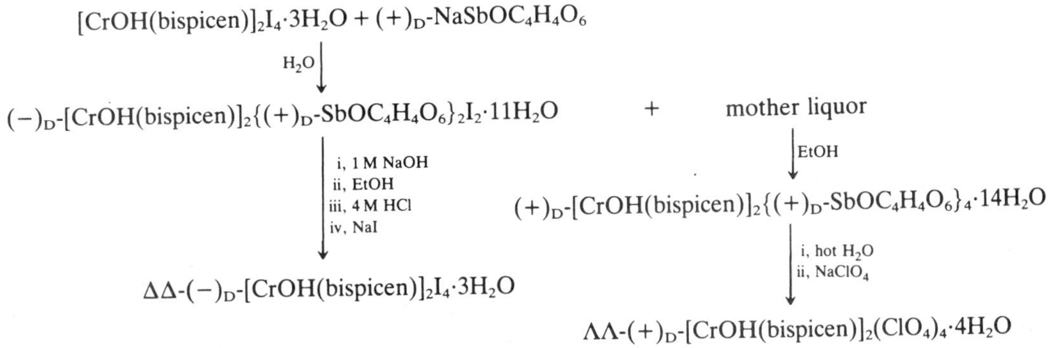

Scheme 59

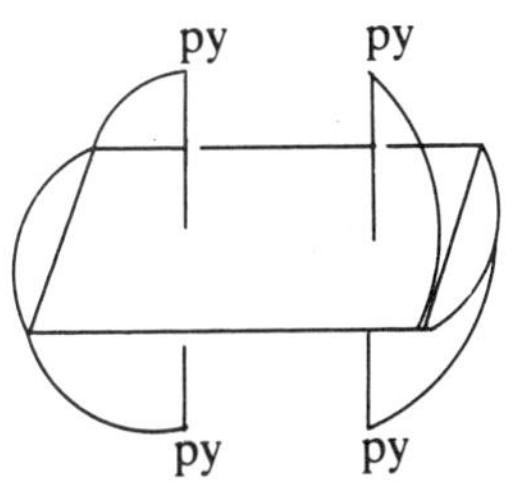

(120) ΛΛ

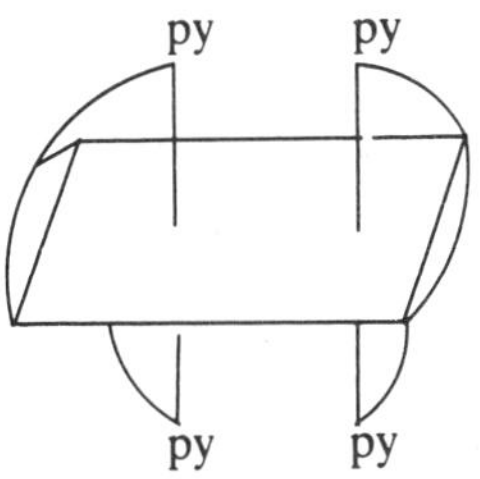

(121) ΔΔ

X = Cl, Y = I, Δ; X = Br, Y = I, Δ; X = Cl, Y = ClO_4, Δ; X = F, Y = Br, Λ; X = Y = Br, Λ; X = Y = NCS, Λ) containing the unsymmetrically substituted and optically active ligand 1,6-bis(2′-pyridyl)-3-methyl-2,5-diazahexane (bispicpn) have been prepared by methods similar to those in Scheme 58 and characterized by electronic and CD spectra.[497] An X-ray crystal structure determination of Δ-*cis*-α-[CrCl_2(S-bispicpn)]Cl confirmed the assignment of a Δ configuration, initially based on CD spectroscopy.[498] A number of analogous *cis*-α-[CrCl_2(R-bispicpn)]⁺ complexes and also *cis*-β-[CrCl_2(R-bispicpn)]I have been prepared and characterized.[499] The complexes $[CrCl_2(S,S\text{-bispicbn})]ClO_4$ (bispicbn = 1,6-bis(2′-pyridyl)-3,4-dimethyl-2,5-diazahexane) and $[CrCl_2(R,R\text{-bispicchxn})]ClO_4$ (bispicchxn = N,N'-bis(2-pyridylmethyl)-1,2-cyclohexanediamine) have been assigned Λ-*cis*-α structures on the basis of CD spectral comparisons.[500] Details of the electronic spectra of a number of bispicdiaminechromium(III) complexes are presented in Table 57.

Table 57 Electronic Spectra of Bispicen and Related Chromium(III) Complexes[a]

Complex[b]	λ_{max} (ε) (nm, $dm^3\ mol^{-1}\ cm^{-1}$)	*Ref.*
cis-α-$[CrCl_2(\text{bispicen})]^+$	545 (104), 407 (99)	1
cis-β-$[CrCl_2(\text{bispicen})]^+$	538 (131), 407 (100)	1
Δ-*cis*-α-$[CrCl_2(S\text{-bispicpn})]^+$	540 (102), 406 (102)	2
Λ-*cis*-α-$[CrCl_2(S\text{-bispicpn})]^+$	550 (100), 408 (93)	2
Λ-*cis*-α-$[CrCl_2(R\text{-bispicpn})]^+$	543 (103), 407 (101)	3
Δ-*cis*-α-$[CrCl_2(R\text{-bispicpn})]^+$	554 (95.4), 409 (90)	3
Λ-*cis*-β-$[CrCl_2(R\text{-bispicpn})]^+$	538 (129), 406 (104)	3
Δ-*cis*-α-$[CrBr_2(S\text{-bispicpn})]^+$	558 (117), 418 (145)	2
Λ-*cis*-α-$[CrBr_2(S\text{-bispicpn})]^+$	565 (105), 420 (107)	2
Λ-*cis*-α-$[CrF_2(S\text{-bispicpn})]^+$	523 (81), 378 (49)[c]	2
Λ-*cis*-α-$[Cr(NCS)_2(S\text{-bispicpn})]^+$	501 (117), 319 (3890)[c]	2
Λ-*cis*-α-$[CrCl_2(S,S\text{-bispicbn})]^+$	547 (91.2), 407 (83.2)	4
Λ-*cis* α-$[CrCl_2(S,S\text{-bispicchxn})]^+$	553 (93.3), 410 (87.1)	4
$[CrOH(\text{bispicen})]_2^{4+}$	534 (197), 385 (150)	1
ΛΛ-$[CrOH(S\text{-bispicpn})]_2^{4+}$	530 (174), 384 (162)[c]	4
ΛΛ-$[CrOH(R\text{-bispicpn})]_2^{4+}$	533 (220), 384 (148)	5
ΔΔ-$[CrOH(R\text{-bispicpn})]_2^{4+}$	529 (173), 386 (147)	5
ΛΛ-$[CrOH(S,S\text{-bispicbn})]_2^{4+}$	533 (186), 386 (148)[c]	4
ΛΛ-$[CrOH(S,S\text{-bispicchxn})]_2^{4+}$	527 (174), 386 (148)[c]	4

[a] In 0.1 M HCl unless otherwise indicated. [b] Complexes are either red, blue or violet coloured. [c] In H_2O.

1. K. Michelsen, *Acta Chem. Scand., Ser. A*, 1977, **31**, 429.
2. Y. Yamamoto and Y. Shimura, *Bull. Chem. Soc. Jpn.*, 1980, **53**, 395.
3. K. Michelsen and K. M. Nielsen, *Acta Chem. Scand., Ser. A.*, 1980, **34**, 755.
4. Y. Yamamoto and Y. Shimura, *Bull. Chem. Soc. Jpn.*, 1981, **54**, 2924.
5. K. Michelsen and E. Pedersen, *Acta Chem. Scand., Ser. A*, 1983, **37**, 141.

The di-μ-hydroxo complexes $[CrOH(\text{bispicpn})]_2(ClO_4)_4$,[500,501] $[Cr(OH)(S,S\text{-bispicbn})]_2(ClO_4)_4$, $[Cr(OH)(S,S\text{-bispicchxn})]_2(ClO_4)_4$[500] and $[CrOH(\text{bispictmd})]_2(ClO_4)_4$ (bispictmd is 1,7-bis(2′-pyridyl)-2,6-diazaheptane)[502,508] have been prepared by methods similar to those in Scheme 59. Once again, steric repulsion appears to be responsible for the limited number of the potential isomers isolated. From X-ray crystallography (Table 54) the violet form of $[CrOH(\text{bispictmd})]_2(ClO_4)_4$ is shown to be a racemic mixture of ΔΔ and ΛΛ isomers, and the singlet–triplet separation in this antiferromagnetic species is 33 cm^{-1}.[502] For various salts of $[Cr(OH)(\text{bispicpn})]_2^{4+}$ the values are in the range 44–50 cm^{-1}.[501]

(vi) *Tetraethylenepentamine (tetren)*

The complex $[Cr(\text{tetren})(H_2O)]^{3+}$, prepared by hydrolysis of α,β,*R*-$[CrCl(\text{tetren})]Cl_2$, undergoes aquation with successive rupture of Cr—N bonds in acid solution (Scheme 60).[503,504] The *N*-thiocyanato complex $[CrNCS(\text{tetren})](SCN)_2$ may be obtained by heating the chloro complex with KSCN in aqueous solution.[505]

$$[Cr(\text{tetren})H_2O]^{3+} \xrightarrow[k=5.5\times10^{-3}\,s^{-1}]{4\text{ M HClO}_4,\ 60\,^{\circ}C} [Cr(\text{tetrenH})(H_2O)_2]^{4+}$$

$$\xrightarrow[k=2.1\times10^{-2}\,s^{-1\,a}]{4\text{ M HClO}_4,\ 60\,^{\circ}C} 1,2,3\text{-}[Cr(\text{tetrenH}_2)(H_2O)_3]^{5+} \xrightarrow[k=2.4\times10^{-4}\,s^{-1}]{4\text{ M HClO}_4,\ 60\,^{\circ}C}$$

$$[Cr(\text{tetrenH}_3)(H_2O)_4]^{6+} \xrightarrow[k=1\times10^{-5}\,s^{-1}]{4\text{ M HClO}_4,\ 60\,^{\circ}C} [Cr(\text{tetrenH}_4)(H_2O)_5]^{7+} \xrightarrow[k=1.7\times10^{-6}\,s^{-1}]{4\text{ M HClO}_4,\ 60\,^{\circ}C} [Cr(H_2O)_6]^{3+} + \text{tetrenH}_5^{5+}$$

[a] Rate constant extrapolated from 10–35 °C values.

Scheme 60

35.4.2.5 *N heterocyclic ligands*

(i) *Pyridines*

A large number of chromium(III) complexes containing pyridine (py) ligands are known.

The addition of py to a solution containing high concentrations of Cr^{3+} and F^- gives *trans*-$[CrF_2(py)_4]^+$, which, owing to the high stability of the Cr^{III}—F bond in the absence of acid, is a convenient starting material for a variety of $[CrF_2N_4]^+$ complexes (Schemes 32, 33, 47, 58 and 61).[392,496] These reactions are usually carried out in boiling 2-methoxyethanol. Under acid conditions, the F^- ligands can be substituted to give complexes of the type $[Cr(X)(Y)N_4]^+$. A detailed analysis of the electronic spectra of the complexes *trans*-$[Cr(X)(Y)(py)_4]^+$ (X, Y = F, Cl, Br) using an angular overlap model to describe the Cr—py bond has led to the conclusion that py in these cases acts as a π acceptor.[506] However assumptions leading to this conclusion have been disputed,[507,508] since it has been pointed out that such behaviour is inconsistent with that of py in other complexes, *e.g.* $M(X)(Y)(py)_4$ (M = Fe, Co, Ni), where it acts as a π donor.[507] Frozen solution X-band ESR spectra have been recorded for a series of complexes of the type *trans*-$[Cr(X)(Y)(py)_4]^{n+}$ (X = Y = F, Cl, Br, I, H_2O, OH; X = F, Y = Cl, Br, H_2O, OH; X = Cl, Y = H_2O, OH; X = Br, Y = H_2O).[509] The dihalogeno complexes show tetragonal symmetry with $1.97 \leqslant g \leqslant 1.99$ and zero field splittings $0.30 \leqslant 2D \leqslant 3.5\ cm^{-1}$. The aqua and hydroxy complexes have rhombic symmetry.

trans-$[CrF_2(RNH_2)_4]X$

$\uparrow$ RNH_2 (R = alkyl, allyl)
X = I

cis-$[CrF_2N_4]X$ (N_4 = trien, tren)
or
trans-$[CrF_2N_4]X$ (N_4 = 2,3,2-tet or 3,2,3-tet)

$\xleftarrow{\begin{array}{c} N_4, \\ X = ClO_4 \end{array}}$ *trans*-$[CrF_2(py)_4]X$

$\downarrow$ N—N,
X = ClO_4

cis-$[CrF_2(N—N)_2]X$ (N—N = phen, bipy)
or
trans-$[CrF_2(N—N)_2]X$ (N—N = en, pn, tn, chxn)

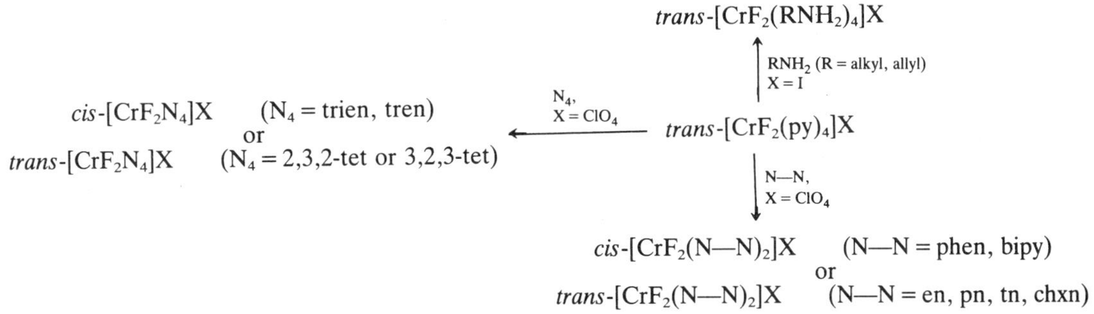

Scheme 61

Reaction of $CrCl_3$ and pyridine at reflux temperatures and under anhydrous conditions affords the green-coloured product $CrCl_3(py)_3$.[397,510] The far-IR spectrum of this complex shows three well-separated $\nu(Cr—Cl)$ bands indicative of a *mer* configuration.[511] Subsequent X-ray studies established that both this and $CrBr_3py_3$ are isostructural with the known *mer*-$MoCl_3(py)_3$.[512] The complexes $CrX_3(4-Mepy)_3$ (X = Cl, Br) obtained from CrX_3 and 4-Mepy were also shown to have *mer* configurations.[513] Reaction of $CrCl_3 \cdot 6H_2O$ with a ten-fold excess of each of the cyanopyridine isomers (CNpy) in either 1- or 2-butanol at 80 °C affords the complexes $CrCl_3(CNpy)_3$, although configurations have not been assigned.[514] Thermal decomposition of the complexes *mer*-$[CrCl_3(R-py)_3]$ (R = H, 3-Me, 4-Me) and $pyH[CrCl_4(py)_2]$ gives the dimers $Cr_2Cl_6(R-py)_4$.[515]

The complexes *mer*-$Cr(CF_3CO_2)_3py_3$ and $CpCr(CF_3CO_2)_2py$ have been prepared by the methods outlined in Scheme 62.[516] An X-ray analysis of the former complex showed it to have a slightly distorted octahedral structure with almost identical Cr—O (mean 195 pm) and Cr—N (mean 210 pm) bond lengths. Reaction of $Cr(MeSO_3)_3$, a CrO_6 complex containing bridging $MeSO_3$ ligands, with py and the appropiate amount of bipy gives the green complexes $Cr(MeSO_3)_3py_3$, $Cr(MeSO_3)_3bipy$ and $Cr(MeSO_3)_3(bipy)_2$ all of which appear to contain octahedrally-coordinated chromium(III).[517]

Reduction of $RpyCr(O)(O_2)_2$ (R = 3-Me, 3-Cl and 3-CN) with Fe^{2+} under acid conditions gives the species $[Cr(Rpy)(H_2O)_5]^{3+}$ in solution.[518] For the aquation of these and the unsubstituted py complex (equation 32) a linear relationship exists between $-\log k$ and the pK_a values of the py ligands.

$$[Cr(Rpy)(H_2O)_5]^{3+} + H_2O \xrightarrow{k} [Cr(H_2O)_6]^{3+} + Rpy \qquad (32)$$

Some multinuclear chromium(III) complexes which contain pyridine ligands have been described. The amide (**122** = H_4L) reacts with $CrCl_3 \cdot 6H_2O$ in pyridine solution containing Na_2CO_3 to give the dimer $[Cr_2(HL)_2(py)_4] \cdot 2py$, the structure of which is described in Section 35.4.2.11.[519]

$$Cp_2Cr \xrightarrow[\text{C}_6\text{H}_6]{\text{CF}_3\text{CO}_2\text{H in}} [Cp_2Cr]^+[CpCr(CF_3CO_2)_3]^-$$

↓ THF, boiling

$$CpCr(CF_3CO_2)_2py \xleftarrow[\text{C}_6\text{H}_6/\text{THF}(3:1)]{\text{py in}} CpCr(CF_3CO_2)_2 \cdot THF$$

violet: $\mu_{eff} = 3.44$ BM
(temp. independent)

↓ i, CF$_3$CO$_2$H in THF, reflux
ii, py

$$(CpCrSCMe_3)_2S \xrightarrow[\text{ii, py}]{\text{i, CF}_3\text{CO}_2\text{H in THF, heat}} mer\text{-}Cr(CF_3CO_2)_3(py)_3$$

violet: $\mu_{eff} = 3.96$ BM (77–295 K): $\nu_{CO_2} = 1770, 1410$ cm^{-1}

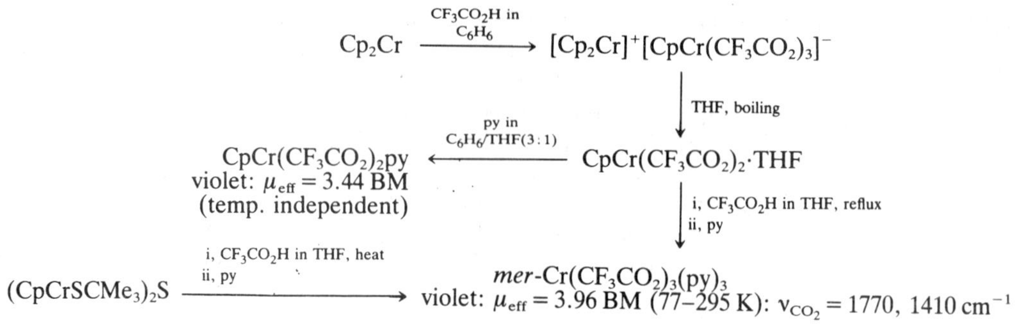

Scheme 62

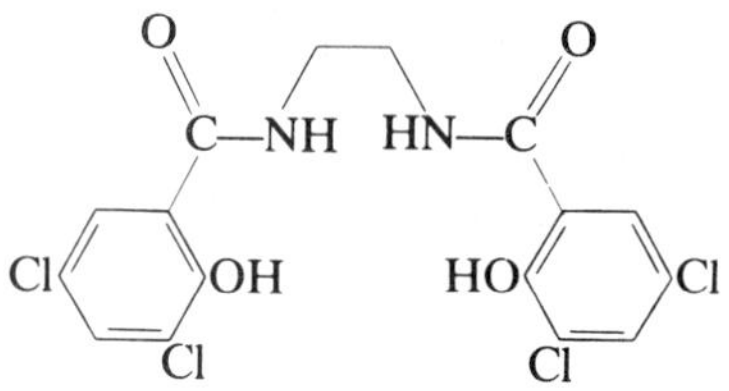

(122)

Reaction of chromium(II) acetate with various other divalent metal acetates in py gives a series of oxo-centred trinuclear complexes which contain bridging acetate ligands (Scheme 63).[520] The reaction between various poly-4-vinylpyridines and *cis*-[CrCl$_2$(en)$_2$]Cl in water at 70 °C gives complexed polymers which, despite having degrees of coordination between 0.36 and 0.71, show no magnetic interactions between the chromium(III) ions along the polymer chains.[521] Addition of py, 3-Mepy or py-O to benzene solutions of the chromium(III) alkoxides Cr(Cl)$_2$OR(ROH)$_2$ (R = Et, Pr^i, Bu^n) gives green or brown solids Cr(Cl)$_2$OR(py)$_2$, Cr(Cl)$_2$OR(3-Mepy)$_2$ and Cr(Cl)$_2$OR(py-O)$_2$, all of which appear to be dimeric with alkoxide bridging ligands.[522]

$$Cr_2^{III}M^{II}O(OAc)_6L_3$$

M = Mg, Co, Ni; L = py
M = Mn, Zn; L$_3$ = py + 2H$_2$O

↗ M(OAc)$_2$ in py
M = Mg, Co, Ni

$$Cr_2(OAc)_4$$

↘ Fe$_2^{III}$M^{II}O(OAc)$_6$L$_3$
in py

$$Cr^{III}Fe^{III}M^{II}O(OAc)_6L_3$$
M = Mn, Fe, Co, Ni; L = py

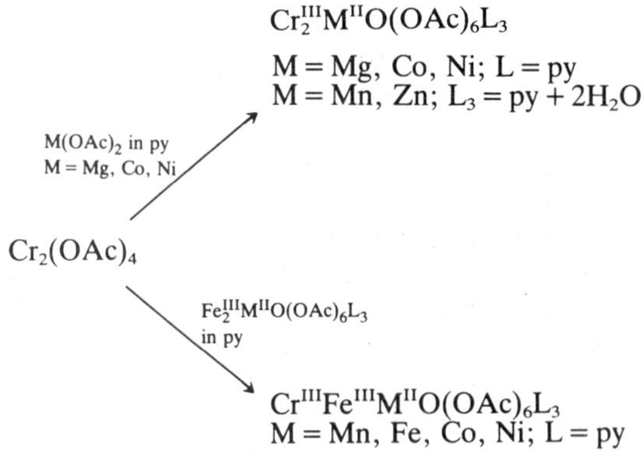

Scheme 63

Other reported chromium(III) complexes containing py and related ligands are listed in Table 58. Complexes of 2-pyridylmethylamine are dealt with in Section 35.4.2.3 and of pyridine carboxylic acids in Section 35.4.8.

(ii) Bipyridyls, phenanthrolines and terpyridyls

A number of chromium(III) complexes with bipy, phen, terpy and related ligands have been prepared and their photochemical and photophysical properties investigated, mainly because of their potential applicability as photosensitizers for solar energy conversion and storage. Much of this work has recently been thoroughly and critically reviewed[523,524] and a later article

Table 58 Chromium(III)–Pyridine and Related Complexes

Complex	*Ref.*	*Complex*	*Ref.*
$[Cr(C_2O_4)_2(py)H_2O]^-$	1	$[CrCl_2(Rpy)_2(tu^a)_2]Cl$	7
$[CrCl_2(C_2O_4)(py)H_2O]^-$	1	(R = 2-Me, 3-Me, 4-Me,	
$[CrCl_4(py)H_2O]^-$	1	$2,4\text{-}Me_2, 2,6\text{-}Me_2$)	
$[Cr(py)(H_2O)_5]^{3+}$	2	$[CrCl_2(quin)_2(tu)_2]Cl$	7
$(Et_4N)_2[Cr(NCS)_5py]$	3	$[CrCl_2(quin')_2(tu)_2]Cl$	7
$[Cr(py)_2(H_2O)_4]^{3+}$	1	$CrNO(CN)_2(py)_2$	8
$[Cr(OH)(py)_2(H_2O)_3]^{2+}$	1	$CrNO(CN)_2(quin)_2$	8
$[Cr(OH)_2(py)_2(H_2O)_2]^+$	1	CrF_3py_3	1
$[CrCl_2(py)_2(H_2O)_2]^+$	1	$Cr(N_3)_3py_3$	1
$[CrBr_2(py)_2(H_2O)_2]^+$	1	$Cr(OAc)_3py_3$	1
$Cr(OH)_3(py)_2(H_2O)$	1	$Cr(NCS)_3py_3$	9
$CrF_3(py)_2H_2O$	1	$Cr(NCS)_3(4\text{-}Mepy)_3$	10
$CrCl_3(py)_2H_2O$	1	$Cr(NCS)_3(quin)_3$	10
$[Cr(C_2O_4)_2(py)_2]^-$	1	$[Cr_3(OAc)_6(py)_3(H_2O)_2]^{3+}$	1
cis-, *trans*-$[Cr(py)_2(H_2O)_4]^{3+}$	4	$[Cr_3(OAc)_6(OH)(py)_3H_2O]^{2+}$	1
$[Cr(NCS)_4(3\text{-}NH_2py)_2]^-$	5	$[Cr_3(OAc)_6(OH)_2(py)_3]^+$	1
$[Cr(NCS)_4(4\text{-}NH_2py)_2]^-$	5	$[Cr_3(C_2H_5CO_2)_6(OH)_2(py)_3]^+$	1
$Cr(NCS)_2(4\text{-}NH_2py)_2(\beta\text{-}dik)$	6	$[Cr_3O(OAc)_6(py)_3]^+$	1
(β-dik = benzoylacetone,			
dibenzoylmethane)			

[a] tu = thiourea.

1. C. S. Garner and D. A. House, *Transition Met. Chem. (N.Y.)*, 1970, **6,** 59.
2. A. Bakac, R. Marcec and M. Orhanovic, *Inorg. Chem.*, 1974, **13,** 57.
3. S. T. D. Lo and D. W. Watts, *Aust. J. Chem.*, 1975, **28,** 1907.
4. A. Bakac and M. Orhanovic, *Z. Naturforsch., Teil B*, 1974, **29,** 134.
5. L. K. Mishra, H. Bhushan and N. K. Jha, *Indian J. Chem., Sect. A*, 1982, **21,** 639.
6. K. M. Purohit and D. V. Ramana Rao, *J. Indian Chem. Soc.*, 1983, **60,** 1009.
7. M. M. Khan, *J. Inorg. Nucl. Chem.*, 1975, **37,** 1621.
8. K. S. Nagaraja and M. R. Udupa, *Indian J. Chem., Sect. A*, 1983, **22,** 531.
9. S. J. Patel, *Bol. Soc. Chil. Quim.*, 1970, **16,** 18; *Chem. Abs.*, 1971, **75,** 14 464q.
10. S. J. Patel, *Bol. Soc. Chil. Quim.*, 1971, **17,** 61; *Chem. Abs.*, 1972, **77,** 83 056m.

describing a pulsed laser photochemical study of $[Cr(bipy)_3]^{3+}$ includes some additional references.[525] The ability of these heterocyclic ligands to stabilize a wide range of oxidation states of coordinated metals also renders their complexes suitable for participation in series of one-electron reversible charge transfers. Such electron transfer chains containing four reversible redox couples have been observed for chromium complexes containing bipy (Scheme 64), phen, terpy and related ligands in acetonitrile solution.[53,526] The first three couples for each complex involve redox orbitals which are primarily metal-centred while the last couple in each case involves a molecular orbital with mainly ligand character.[53] All five members of the bipy series, as well as the further reduced complexes $[Cr(bipy)_3]^{2-}$ and $[Cr(bipy)_3]^{3-}$, have been isolated and their magnetic moments and electronic spectra (the latter with the exception of the anionic complexes) reported.[45]

$$[Cr(bipy)_3]^{3+} \rightleftharpoons [Cr(bipy)_3]^{2+} \rightleftharpoons [Cr(bipy)_3]^{+} \rightleftharpoons Cr(bipy)_3 \rightleftharpoons [Cr(bipy)_3]^{-}$$

Scheme 64

A convenient synthesis of $[Cr(bipy)_3]^{3+}$ and its aquation behaviour under various conditions are outlined in Scheme 65.[527–529] The complex $[Cr(phen)_3](ClO_4)_3$ can be prepared in a similar manner[527] and its thermal- and photo-aquation reactions parallel those of the bipy complex under similar conditions.[530] Both complexes have been successfully resolved using $(+)_D$-tris{(−)-cysteinesulfinato}cobaltate(III) as the anionic resolving agent.[527] Similar tris chelates containing the ligands 5-Clphen, 4,7-Me_2phen, 4,7-Ph_2phen, 3,4,7,8-Me_4phen, 4,4′-Me_2bipy and 4,4′-Ph_2bipy as well as the bis complex $[Cr(terpy)_2](ClO_4)_3$, which was shown by X-ray crystallography to have a distorted *mer* configuration,[531] were obtained by chlorine oxidation of the corresponding chromium(II) complexes.[532] The influence of various factors, *e.g.* nature of oxidant and its rate of addition, on the oxidation of $[Cr(bipy)_3]^{2+}$ and $[Cr(phen)_3]^{2+}$ have been summarized in a recent article.[533] Details of the visible absorption spectra of these complexes are presented in Table 59.

$$CrCl_3 \cdot 6H_2O \xrightarrow{\text{i}} Cr_{aq}^{2+} \xrightarrow{\text{ii}} [Cr(bipy)_3](ClO_4)_2$$

$$\downarrow \text{vi} \qquad\qquad\qquad \downarrow \text{iii}$$

$$cis\text{-}[Cr(bipy)_2(H_2O)_2](ClO_4)_3 \xleftarrow{\text{vii}} cis\text{-}[Cr(OH)_2(bipy)_2]ClO_4 \xleftarrow{\text{iv or v}} [Cr(bipy)_3](ClO_4)_3$$

i, Zn amalgam; ii, bipy, $NaClO_4$, $HClO_4$; iii, air; iv, $h\nu$, pH 9; v, H_2O, OH^-; vi, bipy, O_2, pH 4; vii, $HClO_4$.

Scheme 65

Table 59 Visible Absorption Spectra of $[Cr(N-N)_3]^{3+}$ and $[Cr(N-N-N)_2]^{3+}$ Complexes

Complex ion	λ_{max} (ε) (nm, $dm^3 mol^{-1} cm^{-1}$)	Ref.
$[Cr(bipy)_3]^{3+}$	458 (269), 428 (676), 402 (933)[a]	1
$[Cr(4,4'\text{-}Me_2bipy)_3]^{3+}$	446 (269), 418 (631), 394 (933)[a]	1
$[Cr(4,4'\text{-}Ph_2bipy)_3]^{3+}$	445 (1740), 422 (3630), 404 (5370)[a]	1
$[Cr(phen)_3]^{3+}$	454 (324), 435 (603), 405 (871)[a]	1
$[Cr(5\text{-}Clphen)_3]^{3+}$	466 (389), 436 (708)[a]	1
$[Cr(4,7\text{-}Me_2phen)_3]^{3+}$	450 (501), 424 (891), 402 (1047)[a]	1
$[Cr(4,7\text{-}Ph_2phen)_3]^{3+}$	484 (1230), 445 (2040)[a]	1
$[Cr(3,4,7,8\text{-}Me_4phen)_3]^{3+}$	456 (741), 428 (1150), 400 (1350)[a]	1
$[Cr(terpy)_2]^{3+}$	473 (1410), 443 (2239), 422 (2040)[a]	1
	472 (1470), 442 (2390), 418 (2220)[b]	2
$[Cr(4'\text{-}Phterpy)_2]^{3+}$	467sh (1540), 438sh (2860), 397sh (4750)[b]	2
$[Cr(4,4',4''\text{-}Ph_3terpy)_2]^{3+}$	465sh (2540), 428sh (5320), 399sh (11 200)[b]	3

[a] In 1 M HCl. [b] In MeCN solution.

1. N. Serpone, M. A. Jamieson, M. S. Henry, M. Z. Hoffman, F. Bolletta and M. Maestri, *J. Am. Chem. Soc.*, 1979, **101**, 2907.
2. J. M. Rao, M. C. Hughes and D. J. Macero, *Inorg. Chim. Acta*, 1976, **18**, 127.
3. M. C. Hughes, D. J. Macero and J. M. Rao, *Inorg. Chim. Acta*, 1981, **49**, 241.

Despite a few earlier reports to the contrary it now appears that complexes of the type $[Cr(N-N)_2(X)(Y)]^{n+}$, $(N-N = bipy$ or phen; X, Y = monodentate ligands) invariably have *cis* configurations.[534,535] The dichloro complexes $(X = Y = Cl^-)$ are conveniently prepared by heating $CrCl_3$ and the amine in boiling ethanol in the presence of zinc dust catalyst.[534] The preparation, resolution, reactions and properties of a number of such complexes have been thoroughly investigated and some of the reported chemical interconversions are outlined in Scheme 66.[535] Visible spectral data for these species and for other heterochelates containing bipy or phen ligands are presented in Table 60 and pK_a values of the aqua complexes are listed in Table 61. Details of the crystal and molecular structure of $cis\text{-}[CrCl(bipy)_2H_2O](ClO_4)_2$ have been reported.[533] The binuclear di-μ-hydroxo complexes $[CrOH(N-N)_2]_2^{4+}$, some reactions of which are included in Scheme 66 and the pK_a values of which are given in Table 61, can be obtained by the gradual neutralization of a refluxing solution containing $Cr(NO_3)_3 \cdot 9H_2O$ and a two-fold excess of the amine in 1 M $HClO_4$.[536] Because of the acidic properties (pK_a values in Table 61) and robustness of these complexes the corresponding μ-oxo-μ-hydroxo (**124**) and di-μ-oxo species (**125**) can easily be isolated after the addition of base to (**123**; Scheme 67).[536] Magnetic susceptibilities of salts containing these complex ions over the temperature range 5–300 K are indicative of antiferromagnetic exchange interactions, which lead to singlet–triplet separations of between 40 cm^{-1} in (**123**) and 110 cm^{-1} in (**125**).[537] The complexes $[CrOH(bipy)_2]_2Cl_4 \cdot 9H_2O$ and $[CrOH(phen)_2]_2(NO_3)_4 \cdot 7H_2O$ have both been resolved.[538] The crystal and molecular structures of $[CrOH(phen)_2]_2I_4 \cdot 4H_2O$ and $[CrOH(phen)_2]_2Cl_4 \cdot 6H_2O$ have been reported.[539] Addition of phen·HNO_3 to an aqueous solution of $CrCl_3(bipy)DMF$ gives, after the addition of base, the complexes $[CrOH(bipy)phen]_2(NO_3)_4{\cdot}6H_2O$, which has been resolved, and $[CrOH(bipy)(phen)H_2O](NO_3)_2{\cdot}1.5H_2O$.[538] The addition of one equivalent of base and then KI to an aqueous solution of *cis*-$[Cr(bipy)_2(H_2O)_2]^{3+}$ gives a crystalline product which was shown by X-ray analysis to be the $H_3O_2^-$-bridged dimer $[Cr(H_3O_2)_2(bipy)_2]_2I_4{\cdot}2H_2O$.[540] The existence of this dimer in concentrated aqueous solutions has been demonstrated by three-phase vapour tensiometry.[541] The tartrate-bridged complexes $Cr_2(tart_{-H})_2(N{-}N)_2$ and $Na[Cr_2(tart_{-2H})(tart_{-H})(N{-}N)_2]$ (tart = tartate = $C_4H_4O_6^{2-}$; $tart_{-H} = C_4H_3O_6^{3-}$; $tart_{-2H} = C_4H_2O_6^{4-}$) containing either two optically active or two *meso* bridging ligands have been prepared and characterized.[542]

The complexes $[Cr(C_2O_4)(N{-}N)_2]Cl$, $K[Cr(C_2O_4)_2N{-}N]$ (N—N = bipy or phen) and $[Cr(C_2O_4)(bipy)phen]Cl$ have been resolved.[538,543–545] The kinetics of racemization of a number of chelates described in this section have been studied and evidence for both intermolecular

Table 60 Visible Absorption Spectra of *cis*-$[Cr(N{-}N)_2(X)(Y)]^{n+}$ Complexes

Complex ion	$\lambda_{max}{}^a(\varepsilon^b)$	$\lambda_{min}{}^a(\varepsilon^b)$	$\lambda_{sh}{}^a(\varepsilon^b)$	*Medium*	*Ref.*
$[Cr(bipy)_2(H_2O)_2]^{3+}$	492 (44.8)	468 (39.2)		0.1 M $HClO_4$	1
	448 (94.2)	442 (89.2)			
$[Cr(phen)_2(H_2O)_2]^{3+}$	496 (43.7)	458 (29.4)		0.1 M HNO_3	1
$[Cr(OH)_2(bipy)_2]^+$	517 (42.0)	465 (35.1)	440 (69.7)	0.1 M NaOH	1
			424 (160)		
$[Cr(OH)_2(phen)_2]^+$	519 (41.5)	468 (34.8)		0.1 M NaOH	1
$[Cr(F)_2(bipy)_2]^+$	519 (48.6)	461 (22.0)		H_2O	1
	443 (57.3)				
	415 (144)				
$[Cr(F)_2(phen)_2]^+$	522 (46.5)	455 (16.6)	420 (69.2)	H_2O	1
$[Cr(Cl)_2(bipy)_2]^+$	553 (43.7)	474 (18.9)	445 (90.0)	0.1 M HCl	1
	572 (47.3)	448 (24.0)		DMF	1
$[Cr(Cl)_2(phen)_2]^+$	558 (40.1)	478 (17.5)		0.1 M HCl	1
	578 (42.5)	492 (22.8)		DMF	1
$[Cr(Br)_2(bipy)_2]^+$	574 (52.7)	526 (43.9)	445 (132)	H_2O	1
	518 (45.6)	484 (29.0)			
$[Cr(Br)_2(phen)_2]^+$	598 (49.8)	538 (36.5)	421 (227)	DMF	1
	522 (42.8)	495 (32.0)			
$[Cr(Cl)(bipy)_2(H_2O)]^{2+}$	520 (56)			0.01 M HCl	2
$[Cr(Cl)(phen)_2(H_2O)]^{2+}$	522 (41.2)	461 (17.0)		0.1 M HNO_3	1
$[Cr(Cl)OH(bipy)_2]^+$	550 (65)			0.01 M NaOH	1
$[Cr(C_2O_4)(bipy)_2]^+$	497 (66.5)	464 (48.7)	448 (90.4)	H_2O	1[c]
			417 (220)		
$[Cr(C_2O_4)(phen)_2]^+$	501 (62.6)	454 (33.0)		H_2O	1[c]
$[Cr(C_2O_4)_2(phen)]^-$	543 (77)			H_2O	3
	397 (133)				
$[Cr(L\text{-tart}^e)(phen)_2]^+$	505 (72.4)		420 (178)		4
$[Cr(L\text{-tart})(bipy)_2]^+$	496 (61.7)		444 (126)		
$[Cr(C_2O_4)_2(bipy)]^-$	535 (86)			H_2O	3
	392 (178)				
$[CrOH(bipy)_2]_2^{4+}$	537 (106)	467 (45.6)		10^{-3} M HCl	1[d]
	447 (158)				
$[CrOH(phen)_2]_2^{4+}$	540 (114)	462 (42.9)	422 (303)	0.1 M HNO_3	1[d]
$[Cr_2(OH)(O)(bipy)_2]^{3+}$	537 (108[f])	467 (46[f])		0.1 M HCl	5
	447 (160[f])				
$[Cr_2(OH)(O)(phen)_2]^{3+}$	540(—)	462 (—)		0.1 M HCl	
$[CrO(bipy)_2]_2^{2+}$	537 (108[f])	467 (46[f])		0.1 M HCl	5
	447 (158[f])				
$[CrO(phen)_2]_2^{2+}$	540(—)	462(—)		0.1 M HCl	5

[a] In nm. [b] In $dm^3\,mol^{-1}\,cm^{-1}$.
[c] Spectra of these complexes are also reported in ref. 3 and are in moderately good agreement.
[d] Spectra of these complexes are also reported in ref. 5 and are in excellent agreement. [e] tart = tartrate.
[f] The absorption coefficients quoted in ref. 5 are per chromium(III) and have been adjusted here to molar absorption coefficients.

1. M. P. Hancock, J. Josephsen and C. E. Schäffer, *Acta Chem. Scand.*, 1976, **30**, 79.
2. W. A. Wickramasinghe, P. H. Bird, M. A. Jamieson, N. Serpone and M. Maestri, *Inorg. Chim. Acta*, 1982, **64**, L85.
3. J. A. Broomhead, M. Dwyer and N. Kane-Maguire, *Inorg. Chem.*, 1968, **7**, 1388.
4. A. Tatehata, *Inorg. Chem.*, 1977, **16(5)**, 1247.
5. J. Josephsen and C. E. Schaffer, *Acta Chem. Scand.*, 1970, **24**, 2929.

Table 61 pK_a Values for Ionization of Aqua and μ-Hydroxo Ligands in 1,10-Phenanthroline and 2,2′-Bipyridyl Complexes of Chromium(III)

Complex ion	pK_a	*Medium*	*Ref.*
cis-$[Cr(bipy)_2(H_2O)_2]^{3+}$	$pK_1 = 3.5$; $pK_2 = 6.1^a$	0.1 M $NaNO_3$	1
	$pK_1 = 3.0 \pm 0.2$; $pK_2 = 6.0 \pm 0.2^{a,b}$	1.0 M $NaNO_3$	2
cis-$[Cr(phen)_2(H_2O)_2]^{3+}$	$pK_1 = 3.4$; $pK_2 = 6.0^a$	1 M $NaNO_3$	1
cis-$[CrCl(bipy)_2(H_2O)]^{2+}$	4.6 ± 0.2^a	1.0 M $NaNO_3$	2
$[CrOH(bipy)_2]_2^{4+}$	$pK_1 = 7.6$; $pK_2 = 11.9^c$	1 M KCl	3
$[CrOH(phen)_2]_2^{4+}$	$pK_1 = 7.4$; $pK_2 = 11.8$	1 M KCl	3

[a] Temperature not stated but assumed to be 298 K.
[b] These values were erroneously attributed to $[Cr(phen)_2(H_2O)_2]^{3+}$ in ref. 2; clearly on the last line, column 1, L86, $[Cr(phen)_2(H_2O)_2]^{3+}$ was mistakenly written for $[Cr(bipy)_2(H_2O)_2]^{3+}$.
[c] $T = 297–300$ K.

1. R. G. Inskeep and J. Bjerrum, *Acta Chem. Scand.*, 1961, **15**, 62.
2. W. A. Wickramasinghe, P. H. Bird, M. A. Jamieson, N. Serpone and M. Maestri, *Inorg. Chim. Acta*, 1982, **64**, L85.
3. J. Josephsen and C. E. Schäffer, *Acta Chem. Scand.*, 1970, **24**, 2929.

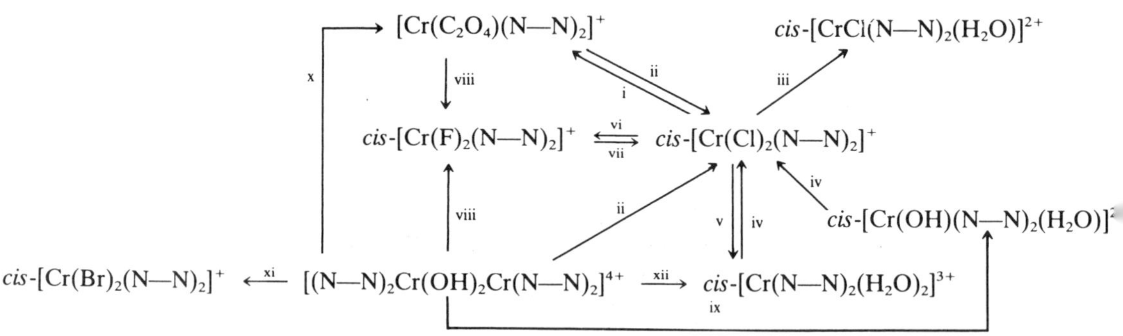

i, $K_2C_2O_4$ in H_2O, 100 °C, 4 min; ii, conc. HCl/HCl(g), 50 °C overnight in sealed tube; iii, 0.3 M HCl, 100 °C, 3 h, (N—N = phen); iv, conc. HCl, 70 °C, overnight in sealed tube; v, Ag_2O in H_2O, 25 °C, overnight; vi, $Hg(OAc)_2$ in HF, 25 °C, overnight; vii, conc. HCl, 100 °C, 5 h in sealed tube; viii, HF, 25 °C overnight; ix, conc. $HClO_4$, 60 °C, 24 h, adjust pH to 4; x, $Na_2C_2O_4/H_2C_2O_4$ in H_2O, 80 °C, 2 h; xi, conc. HBr/HBr(g), 50 °C, 24 h in sealed flask; xii, conc. $HClO_4$, 60 °C, 24 h.

Scheme 66

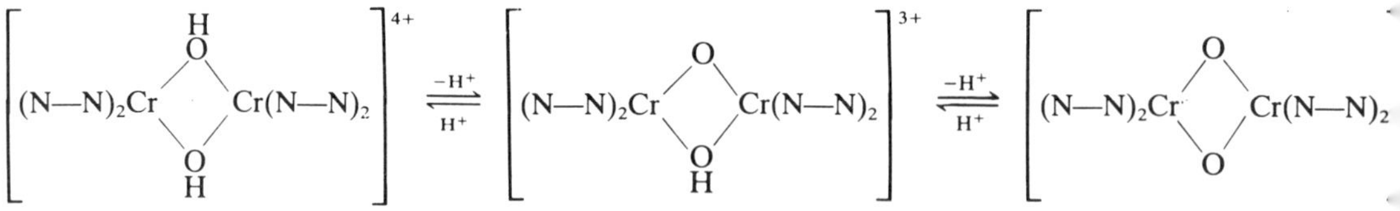

(123) **(124)** **(125)**

Scheme 67

(involving loss of one chelating ligand) and intramolecular (involving either one-ended dissociation of a bidentate ligand from the metal ion or a twist without bond rupture) mechanisms have been obtained for such reactions. Some of the most significant work in this area has been summarized in a recent article.[543]

The reaction between $CrCl_3$ and phen in DMF solution containing a little zinc dust catalyst results in precipitation of a dark green complex $[CrCl_3(phen)(DMF)]$ which according to X-ray crystallographic evidence contains the Cl^- ligands disposed in a *mer* configuration about the metal ion and the DMF coordinated through its O atom ($\nu_{CO} = 1635\ cm^{-1}$).[546,547] Heating a suspension of this complex in DMF causes it to isomerize to a pale green, DMF-soluble species, which presumably is the *fac* isomer. Both complexes lose DMF at 240–250 °C to give

$CrCl_3$(phen), a complex which on the basis of IR and mass spectroscopic evidence appears to contain five-coordinate chromium(III). The preparation of the related complex $CrCl_3$(terpy) is also reported.[547] A chromium(III) complex with the potentially pentadentate planar ligand 6,6″-bis(methylhydrazino)-4′-phenyl, 2,2′:6′2″-terpyridine (**126**) and formulated as [Cr(**126**)]-Cl_3·HCl was obtained by heating the ligand and $CrCl_3·6H_2O$ in refluxing methanol.[548] However this report is preliminary, dealing mainly with ligand synthesis and does not include information pertaining to the structure of the complex. Ligand (**126**) reacts with glyoxal in the presence of $CrCl_3·6H_2O$ to give the macrocyclic ligand (**127**), which after the addition of HCl and NH_4PF_6, can be isolated as [H_2(**127**)]$(PF_6)_2$.[549] Appropriately substituted bis(hydrazines) of bipy and phen similarly react with 2,6-pyridinedicarbaldehyde in the presence of $CrCl_3·6H_2O$ to give metal-free macrocyclic ligands such as (**128**).[550] In these reactions the true template is thought to be the proton and not the metal.

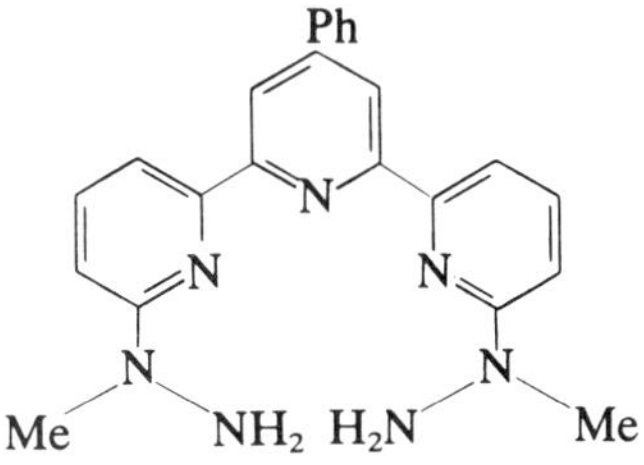

(**126**)

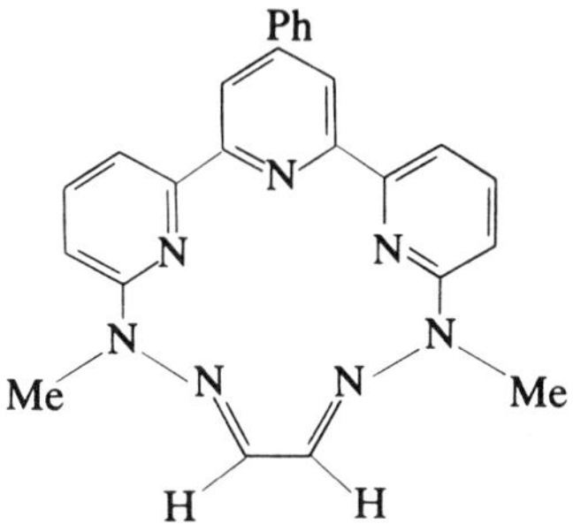

(**127**)

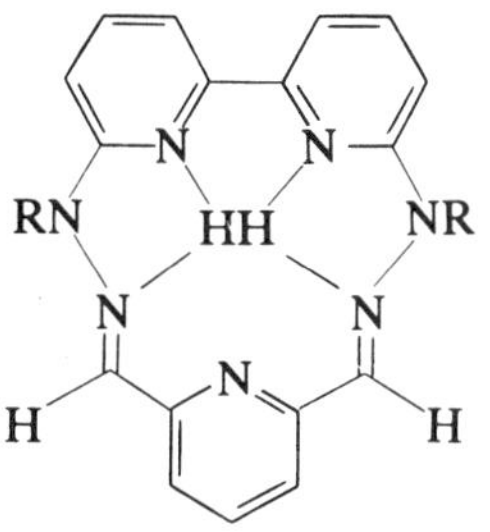

$R = CH_2CH_2OH$

(**128**)

Some air- and water-stable σ-bonded organochromium(III) complexes have been prepared according to Scheme 68 and the structures of three of these (R = Ph, 2-MeOPh and Me_3SiCH_2) have been determined by X-ray methods.[551–553] All three compounds have *cis* octahedral configurations with almost identical $Cr—C_R$ (2.087–2.107 Å); $Cr—N_{trans\ C}$ (2.147–2.156 Å) and $Cr—N_{trans\ N}$ (2.071–2.103 Å) bond distances. The Cr—C bond distances are appreciably longer than that found in the unstable complex *p*-MePhCr$(Cl)_2(THF)_3$.[554] These facts suggest that the $Cr—C_R$ bond distance is virtually insensitive to the nature of R (*e.g.* whether it is alkyl or aryl, or the type of substituents in the aryl ring) in a given chromium environment but is influenced strongly by changes in this environment.

$$2RMgBr + CrBr_2 + bipy \xrightarrow{THF} \underset{\text{deep blue solid}}{R_2Cr(bipy)_2} \xrightarrow[\text{ii, KI}]{\text{i, air}} cis\text{-}[R_2Cr(bipy)_2]I$$

$R = Ph$, 2-, 3-, or 4-MeOPh, 4-MePh, Me_3SiCH_2

Scheme 68

(*iii*) *Imidazoles and related ligands*

A number of chromium(III) complexes with imidazole (Him) and related ligands have been synthesized. Hence the reaction between $CrCl_3$ and imidazole, 1-methylimidazole (1-Meim) or 2-methylimidazole (2-Meim) in absolute MeOH and in the presence of zinc dust catalyst affords the complexes $[Cr(Cl)_{6-n}(Him)_n]Cl_{n-3}$ ($n = 4$–6), $[Cr(Cl)_{6-n}(1\text{-Meim})_n]Cl_{n-3}$ ($n = 4, 5$) and $[Cr(Cl)_2(2\text{-Meim})_4]Cl$, the product obtained with the first two ligands being determined largely by the ligand to metal mole ratio used in the reaction mixture. These complexes have been investigated by IR and ESR spectroscopy and their reduction to the corresponding chromium(II) complexes at the dropping mercury electrode in various electrolyte solutions has also been studied.[555,556] The mixed ligand complexes $Cr(NCS)_3(1\text{-Rim})_3$, $Cr(NCS)(OH)_2(1\text{-Rim})_3$, $Cr(NCS)_2(2\text{-R'im})_3$ and $Cr(NCS)_3(2\text{-R'im})_2(H_2O)$ (R = H, Me, Et, vinyl; R′ = Me, Et, Pr^i) have been prepared from $K_3[Cr(NCS)_6]$ and the appropriate imidazole in refluxing ethanol.[557] By comparing the positions of the $^4A_{2g} \rightarrow {}^4T_{2g}$ bands in the electronic spectra of the $Cr(NCS)_3(1\text{-Rim})_3$ and $Cr(NCS)_3(2\text{-R'im})_3$ complexes the spectrochemical order of imidazole ligands (all of which lie between NH_3 and NCS in the spectrochemical series) was found to be:

Him > 2-Meim > 1-Meim > 2-Etim ~ 1-Vinylim > 2-Pr^{i}im ~ 1-Etim. The complexes have magnetic moments characteristic of octahedral chromium(III) complexes and the thiocyanate groups are N-bonded. While benzimidazole (bzim) and 2-methylbenzimidazole react with $[Cr(NCS)_6]^{3-}$ to give Reinecke-type anions such as $[Cr(NCS)_4(bzim)_2]^-$, [558] bidentate ligands such as 2-guanidino-benzimidazole and 2-(2′-pyridyl)imidazole can under appropriate conditions cause substitution of 2-, 4- or 6-*N*-thiocyanato ligands.[559,560] Treatment of $(2\text{-MeOPh})_2Cr$ with a solution of 2-(2′-pyridyl)imidazole in THF followed by air oxidation gives the orange-red solid tris[2-(2′-pyridyl)imidazolato]chromium(III) dihydrate.[553] A number of 5-substituted tetrazolate (5-Rtet$^-$, R = *p*-$ClC_6H_4CH_2$, *o*-ClC_6H_4, *p*-ClC_6H_4, C_6H_5) complexes $[Cr(5\text{-Rtet})_2OH]\cdot xH_2O$ have been prepared and shown on the basis of spectroscopic and magnetic moment data to contain chromium(III) in octahedral ligand environments.[561]

(*iv*) *Pyrazines and pyrazoles*

The reaction between Cr^{2+} and pyrazines (pyz) in oxygen-free solutions produces intensely coloured 'pyrazine green' products. Data obtained from kinetic, thermodynamic and spectroscopic investigations indicate that formation of the green complex involves Cr^{2+} and H^+ coordination at the ligand N atoms and that the product $[Cr(pyzH)(H_2O)_5]^{3+}$ is best formulated as a chromium(III)–pyrazine radical anion complex.[562] The electrochemical behaviour of a number of these pyrazine greens has been investigated.[563] While the reaction between chromium(II) halides CrX_2 and a 4:1 excess of various pyrazoles (R-pyrH) in deoxygenated alcohol solutions generally leads to the expected $CrX_2(\text{R-pyrH})_4$ products, the addition of 5-methylpyrazole (**9**; 5-MepyrH) to a deoxygenated solution of $Cr(BF_4)_2$ in ethanol results instead in oxidation of the metal, F^- abstraction from the anion and formation of the crystalline complex *trans*-$[CrF_2(5\text{-MepyrH})_4]BF_4$ (**129**).[101] Step-wise and overall formation constants of various $[Cr(\text{R-pyrH})_6]^{3+}$ complexes increase with changing ligand in the order: pyrH < 5-MepyrH < 3,5-Me_2pyrH < 3,4,5-Me_3pyrH.[564] The dimeric chromium(III) complex $[\{Cr(NCS)(TPyEA)\}_2O](BPh_4)_2$ (**130**) containing the tetradentate tris(2-pyrazole-1-ylethyl)amine is prepared according to Scheme 69.[107] The values of μ_{eff} as a function of temperature (1.63 BM at 300 K; 0.98 BM at 86 K) indicate strong antiferromagnetic coupling through the oxygen bridge.

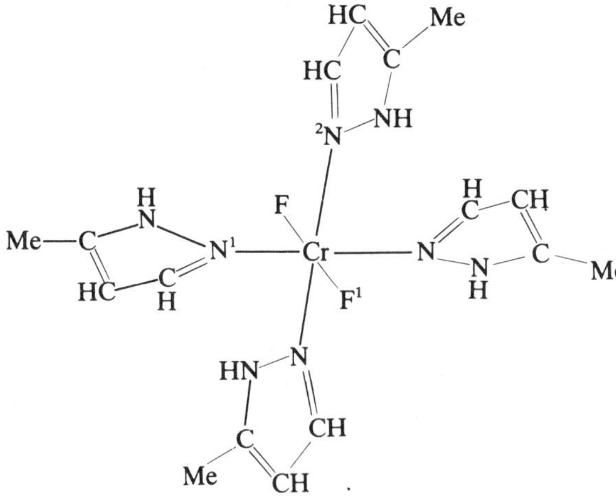

Cr—F¹ = 1.960 Å
Cr—N¹ = 2.027 Å
Cr—N² = 2.048 Å
N¹—Cr—N² = 86.9°
F¹—Cr—N¹ = 91.4°
F¹—Cr—N² = 92.4°

(129)

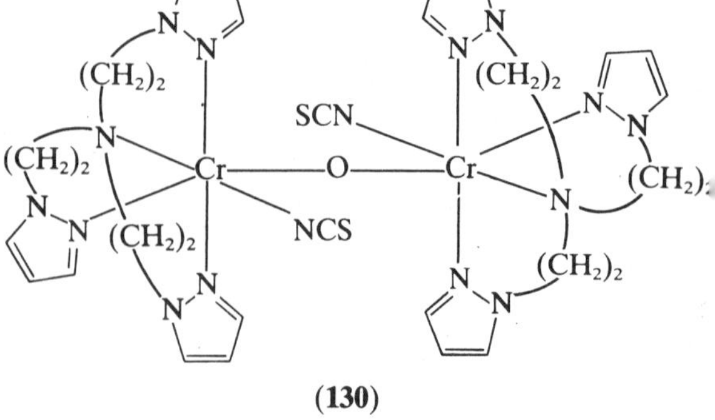

(130)

$$CrBr_2(EtOH)_n + KCNS \xrightarrow{EtOH} \text{'}Cr(NCS)_2\text{'}$$

$$\text{'}Cr(NCS)_2\text{'} \xrightarrow{\text{TPyEA, in EtOH, } NaBPh_4} [Cr(NCS)(TPyEA)]BPh_4$$

$$[Cr(NCS)(TPyEA)]BPh_4 \xrightarrow[Me_2CO \text{ or } MeCN]{O_2} [\{Cr(NCS)(TPyEA)\}_2O](BPh_4)_2$$

[{Cr(NCS)(TPyEA)}$_2$O](BPh$_4$)$_2$: green, $\nu(CN) = 2060\ cm^{-1}$; $\nu(Cr—^{16}O—Cr) = 830\ cm^{-1}$; $\nu(Cr—^{18}O—Cr) = 790\ cm^{-1}$

[Cr(NCS)(TPyEA)]BPh$_4$: blue, $\nu(CN) = 2085\ cm^{-1}$

Scheme 69

35.4.2.6 Nitrosyl, thionitrosyl, azo, hydrazine, hydroxylamine and related ligands

(*i*) *Nitrosyls*

Although the rough correlation between core electron binding energies, as determined by X-ray photoelectron spectroscopy, and atomic charges has in some instances permitted the assignment of oxidation states to metals in complexes,[565] for the majority of metal nitrosyls such assignments still remain ambiguous. Hence complexes containing the $CrNO^{2+}$ group, for example, are in some cases best formulated as Cr^{I}—(NO^+) and in others as Cr^{III}—(NO^-). Therefore for convenience this section deals with nitrosyl complexes of chromium in a number of different oxidation states.

(*a*) *[CrNO(CN)$_5$]$^{3-}$*. Reduction of CrO_3 by NH_2OH in basic solution containing CN^- affords the bright green solid $K_3[CrNO(CN)_5]\cdot H_2O$,[566] some properties of which are listed in Table 62. A crystal structure determination of $[Co(en)_3][CrNO(CN)_5]\cdot 2H_2O$ by X-ray diffraction showed that the Cr—NO group is almost linear.[567] The value of the chromium 3*p* electron-binding energy in this anion is almost identical to the values for known chromium(III) complexes and on this basis it may be formulated as $[Cr^{III}(NO^-)(CN)_5]^{3-}$ [565] contrary to previous work, which regards the chromium oxidation state as +1 and the ligand as NO^+.

In aqueous solution, $[Cr(NO(CN)_5]^{3-}$ undergoes hydrolysis as outlined in Scheme 70. Electronic absorption spectra and voltammetric properties of this and its hydrolysis products have been reported.[568] The formation, isomerization (Scheme 71) and reactions of bi- ($n = 1$) and tri- ($n = 2$) nuclear cyano-bridged adducts between $[CrNO(CN)(H_2O)_4]^+$ and Ag^+ and Hg^{2+} have been investigated.[569] That the isomerization step in Scheme 71 ($n = 1$) is some 2000 times faster than that of $[CrNO(CN)(H_2O)_4]^+$ alone seems rather surprising in view of the fact that a similar isomerization in $[Cr(CN)(H_2O)_5]^{2+}$ is enhanced only 100-fold on adduct formation with Hg^{2+}. The origin of this effect, caused by the presence of NO, remains unresolved. The binuclear complexes *cis*-M[(N—N)$_2$FCr—NC—CrNO(CN)$_4$] (M = Na, K or NH$_4$; N—N = en or tn) have been obtained by heating the double complex salts *trans*-M[CrF(N—N)$_2$H$_2$O][CrNO(CN)$_5$] at temperatures of 90–120 °C for one to three hours (Table 45). Reaction of $K_3[CrNO(CN)_5]\cdot H_2O$ with various nitrogen donors in aqueous acetic acid solutions gives the complexes CrNO(CN)$_2$L—L (L—L = (py)$_2$, (quin)$_2$, bipy or phen), for which tetrameric cyano-bridged structures are proposed.[570] Polarographic reduction of $[CrNO(CN)_5]^{3-}$, or $[Cr(CN)_6]^{3-}$, the latter in the presence of nitrite, leads to the same reduced complex $[CrNO(CN)_5]^{4-}$.[571,572]

(*b*) *Other cyano chromium nitrosyls,* N-*thiocyanato- and azido-chromium nitrosyls.* It has recently been shown that the reductive nitrosylation of tetraoxometallates using NH_2OH as the source of NO requires the presence of another reducing anion (*e.g.* CN^-, SCN^-) in the reaction medium and generally proceeds much more efficiently when a vast excess of nitrosylating agent over tetraoxometallate is used.[573] Hence very high yields of nitrosyl complexes (Table 62) may be obtained under the reaction conditions outlined in Scheme 72.[573–576]

(*c*) *Pyridine-, pyridinecarboxylate- and hydroxylamine-chromium nitrosyls.* The mononitrosylchromium complex $[CrNO(ONO)_2(py)_3]$ (**131**) was obtained by the addition of pyridine to a solution of $Cr(CO)_6$ in THF which had been purged with NO and then UV irradiated.[577] The source of the NO_2 ligands in the green coloured product is thought to be NO_2 impurity in the NO gas. The complex contains a linear CrNO group ($\nu_{NO} = 1708\ cm^{-1}$).

Table 62 Preparation and Properties of $[CrNOL_5]^{n+}$ and Related Complexes

Complex	*Preparation*	*Colour*	$\nu(NO)$ (cm^{-1} [a])	*Other IR bands* (cm^{-1} [a])	μ_{eff} (BM)[c]	*Ref.*
$K_3[CrNO(CN)_5]\cdot H_2O$	i	Green	1645	2137, 2095, 2170, 2125 (CN)	1.87	1, 2
$[CrNO(NH_3)_5]Cl_2$	ii	Yellow	1670	1609, 1290, 1273, 1240 (NH_3)	2.3	2, 3
$[CrNO(H_2O)_5]SO_4$	iii, iv	Red-brown	1747	1630 (H_2O)	2.2[d]	2, 4, 5
$[CrNO(CH_3OH)_5]Cl_2$	v	—	1719[b]		—	2
$(Ph_4P)_3[CrNO(N_3)_5]$	vi	Dark brown	1660	2080, 2040 (N_3)	2.0	6
$(Ph_4P)_3[CrNO(NCS)_5]$	vi	Yellow-green	1694	—	2.23	7, 8
$[CrNO(NCS)_2(bipy)]$	vi	Yellow-brown	1705	2080 (CN)	2.0	7
$[CrNO(NCS)_2(phen)]$	vi	Yellow-brown	1705	2080 (CN)	2.0	7
$[CrNO(N_3)_2(bipy)]$	vi	Dark brown	1665	2040 (N_3)	1.9	6
$[CrNO(N_3)_2(phen)]$	vi	Dark brown	1650	2030 (N_3)	2.0	6
$CrNO(dipic)(H_2O)_2$	vi	Red	1725	1630, 1350 (CO_2^-)	1.70	9
$CrNO(dipic)(NH_2OH)_2$	vi	Yellow-green	1710	1630, 1350 (CO_2^-)	1.71	9

[a] Measured as Nujol mulls unless otherwise stated. [b] Measured in methanol solution. [c] Measured by the Gouy method on solid samples at 293–298 K unless otherwise stated. [d] Measured in aqueous solution by the NMR method at 295 K.

i, CrO_3, KOH, KCN, $NH_2OH\cdot HCl$, heat.
ii, $CrCl_2$ in liquid NH_3 + NO or Cr^{2+} in H_2O + NO, NH_3, Cl^-.
iii, $Cr(ClO_4)_2$ + NO in H_2O under N_2. Elution from cation exchange resin with H_2SO_4 gave sulfate salt.
iv, Cr^{2+} + excess NO in $HClO_4$ under argon gives complex in solution.
v, $CrCl_2$ in CH_3OH + NO gave complex in solution.
vi, Described in text.

1. W. P. Griffith, J. Lewis and G. Wilkinson, *J. Chem. Soc.*, 1959, 872.
2. W. P. Griffith, *J. Chem. Soc.*, 1963, 3286.
3. H. B. Karak, S. Sen, P. Bandyopadhyay and B. K. Sen, *Indian J. Chem., Sect. A*, 1984, **23,** 657.
3. H. B. Karak, S. Sen, P. Bandyopadhyay and B. K. Sen, *Indian J. Chem. Sect. A*, 1984, **23,** 657.
4. M. Ardon and J. I. Herman, *J. Chem. Soc.*, 1962, 507.
5. J. N. Armor and M. Buchbinder, *Inorg. Chem.*, 1973, **12,** 1086.
6. R. G. Bhattacharyya and G. P. Bhattacharjee, *Polyhedron*, 1983, **2,** 1221.
7. R. G. Bhattacharyya, G. P. Bhattacharjee and P. S. Roy, *Inorg. Chim. Acta*, 1981, **54,** L263.
8. S. Sarkar and A. Müller, *Z. Naturforsch, Teil B*, 1978, **33,** 1053.
9. K. Wieghardt, U. Quilitzsch and J. Weiss, *Inorg. Chim. Acta*, 1984, **89,** L43.

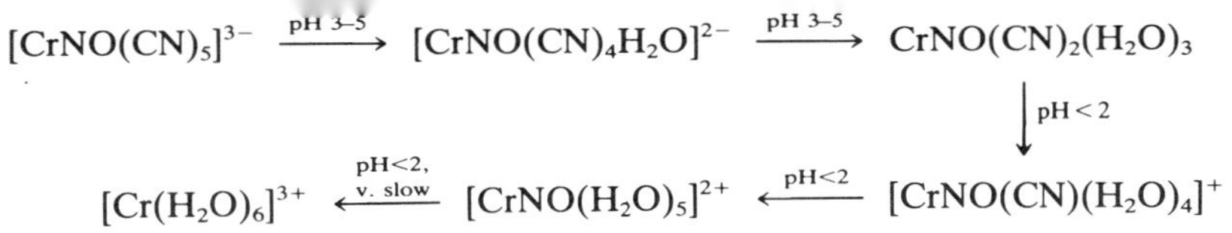

Scheme 70

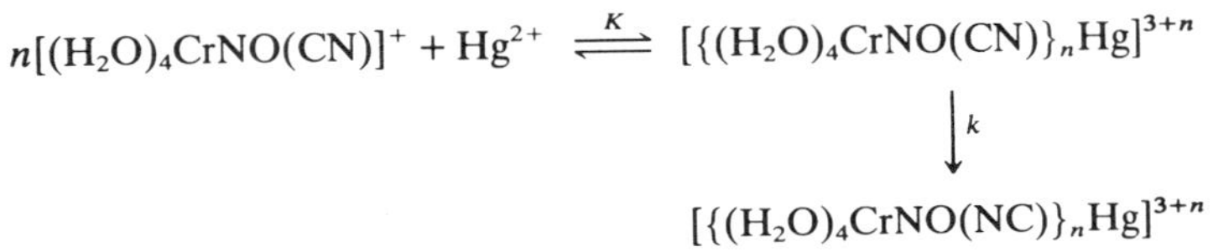

$[CrNO(CN)_5]^{3-} \xrightarrow{pH\ 3-5} [CrNO(CN)_4H_2O]^{2-} \xrightarrow{pH\ 3-5} [CrNO(CN)_2(H_2O)_3]$

$[Cr(H_2O)_6]^{3+} \xrightarrow[v.\ slow]{pH<2,} [CrNO(H_2O)_5]^{2+} \xrightarrow{pH<2} [CrNO(CN)(H_2O)_4]^+$

$\downarrow pH<2$

Scheme 71

$n[(H_2O)_4CrNO(CN)]^+ + Hg^{2+} \xrightleftharpoons{K} [\{(H_2O)_4CrNO(CN)\}_nHg]^{3+n}$

$\downarrow k$

$[\{(H_2O)_4CrNO(NC)\}_nHg]^{3+n}$

Scheme 72

$K_2CrO_4 (\approx 2.5\ \text{mmol}) + NH_2OH \cdot HCl\ (\approx 38\ \text{mmol})$

KCN (38.6 mmol) in
KOH solution (3 g/20 cm^3), 80 °C

Solution A → Me$_4$N[CrNO(CN)$_3$H$_2$O]

Solution A:
i, pH to 7.5
ii, Me$_4$NCl (6.4 mmol)
→ (Me$_4$N)$_2$[CrNO(CN)$_4$]·H$_2$O

i, pH to 5
ii, Me$_4$NCl (3.6 mmol)
→ Me$_4$N[CrNO(CN)$_3$H$_2$O]

i, pH to 5
ii, Ph$_4$PCl (6.4 mmol)
→ (Ph$_4$P)$_2$[CrNO(CN)$_4$]·H$_2$O

i, pH to 5
ii, N—N (bipy or phen, 3.8 mmol)
→ CrNO(CN)$_2$N—N

NH$_4$SCN (25 mmol)
reaction in H$_2$O at 80 °C
→ **Solution B**

Solution B:
NaEt$_2$dtc·3H$_2$O^a (6.6 mmol)
→ CrNO(dtc)$_2$

Ph$_4$PCl (7.5 mmol)
→ (Ph$_4$P)$_3$[CrNO(NCS)$_5$]

N—N (bipy or phen; 3.8 mmol)
→ CrNO(NCS)$_2$N—N

py in acetone
→ CrNO(NCS)$_2$(N—N)py

i, NaN$_3$ (26 mmol) reaction at pH 11, 60–70 °C
ii, acidification to pH 6.5–7
→ **Solution C**

Solution C:
Ph$_4$PBr (9 mmol)
→ (Ph$_4$P)$_3$[CrNO(N$_3$)$_5$]

N—N (bipy or phen; 3.8 mmol)
→ CrNO(N$_3$)$_2$N—N

^{a} Et$_2$dtc = N,N'-diethyldithiocarbamate.

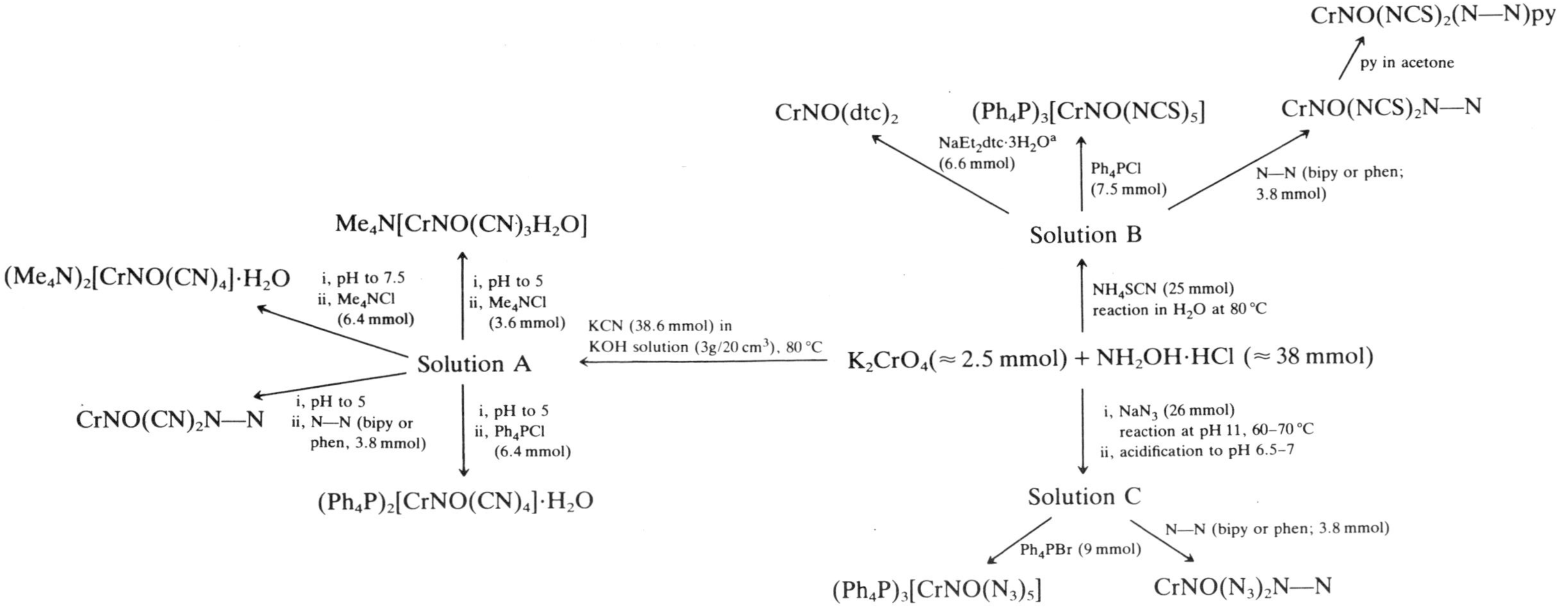

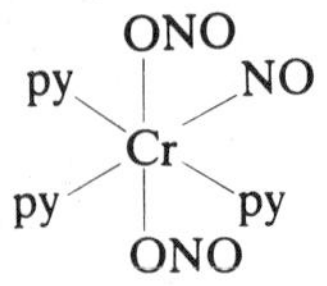

(**131**)

Addition of excess $(NH_3OH)Cl$ to an aqueous solution containing CrO_4^{2-} and pyridine-2,6-dicarboxylic acid (H_2dipic) leads to three products: the nitrosyls CrNO(dipic)$(H_2O)_2$ and CrNO(dipic)$(NH_2OH)_2$ (Table 62), as well as the complex $Cr(NH_2O)$(dipic)$(NH_2OH)_2$.[578] A crystal structure determination of the first of these shows that NO is bonded to the metal *trans* to the pyridine nitrogen of the tridentate dipic ligand. The Cr^{2+}-reduction of NO in the complex $[CrNO(H_2O)_5]^{2+}$ leads initially to coordinated hydroxylamine, but the product complex readily decomposes to the dimer $[Cr(OH)(H_2O)_4]_2^{4+}$, $[Cr(H_2O)_6]^{3+}$ and NH_3OH^+.[579] Reaction of the pentaammine instead of the pentaaqua complex follows a similar stoichiometry and rate law but produces two additional polymeric materials suggested to be $[(NH_3)_5Cr—X—Cr(H_2O)_4Y]^{n+}$ ($X = OH^-$ or NH_2OH, $Y = H_2O$ or NH_2OH).

(*d*) *Isocyanide-, nitrile-, chalcogeno-, phosphine- and arsine-chromium nitrosyls.* The oxidative nitrosylation of $(\eta^6\text{-}C_6H_6)Cr(CO)_3$ by NOX ($X = Cl^-$, PF_6^-) leads to the solvent (S) stabilized species $[CrNO(Cl)(CO)_2]\cdot x$S or the complex $[Cr(NO)(CO)_2S_3]PF_6$, both of which are useful precursors for the synthesis of a wide variety of $Cr(NO)^{2+}$ and $Cr(NO)_2^{2+}$-containing complexes.[580] Two of these complexes, the polymeric chloro-bridged species, $[Cr(NO)_2(Cl)_2]$ (**132**) and $[Cr(NO)S_5](PF_6)_2$ can also be prepared from $Cr(CO)_6$ by photonitrosylation and reaction with $NOPF_6$ respectively.[581] The above reactions as well as the colours and positions of NO stretching bands (cm^{-1}) in the intermediates and products are shown in Scheme 73.

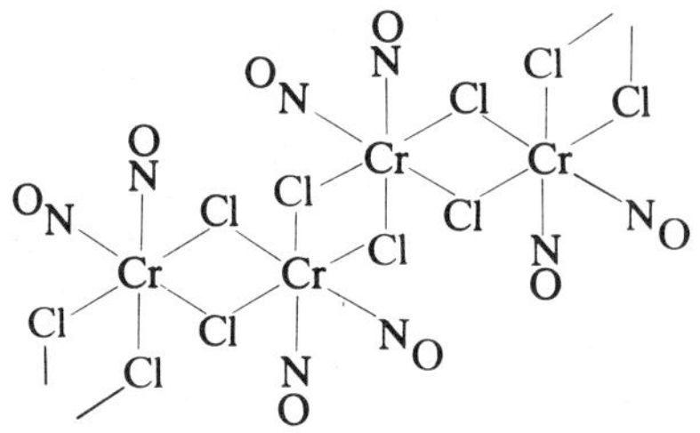

(**132**)

The octahedral dinitrosyl complexes *cis*-$Cr(NO)_2$(S—S)$_2$, *cis*-$Cr(NO)_2$(S—O) and *cis*-$Cr(NO)_2$(Se—O)$_2$ have been prepared by the reaction of $[Cr(NO)_2(Cl)_2]\cdot y$S (Scheme 73) with the dithio (S—S), thiooxo (S—O) and selenooxo (Se—O) four-membered chelating anions $Me_2AsS_2^-$, $Me_2PS_2^-$, $(MeO)_2PS_2^-$, $Me_2NCS_2^-$, $MeCS_2^-$, $MeSCS_2^-$, $MeOCS_2^-$, Me_2NCOS^-, $MeCOS^-$ and Me_2NCOSe^-.[581] The IR spectra of all the complexes show two NO stretching bands, which with the exception of the dithiocarbamate complex *cis*-$Cr(NO)_2(S_2CNMe_2)_2$ ($\nu_{NO} = 1805$, $1684\ cm^{-1}$) lie in the regions 1820–1840 and 1697–1718 cm^{-1}. The nitrile ligands in $[CrNO(MeCN)_5]^{2+}$ are readily displaced by neutral and anionic ligands as outlined in Schemes 73 (reaction xxiv) and 74.[582] An X-ray crystallographic investigation of $[CrNO(S_2CNEt_2)_3]$ (**133**) showed it to be one of the few known examples of seven-coordinate chromium complexes. The geometry of the complex is distorted pentagonal bipyramidal, the NO ligand and an S atom occupying the axial positions, with the remaining five S atoms forming the equatorial girdle. The complex in solution, as shown by its reaction in toluene under reflux, is thermally less stable than its Mo and W analogues, both of which are stable up to at least 130 °C. The F^- ligand in *trans*-$[CrNO(F)(dppe)_2]BF_4$ (Scheme 47) is provided by abstraction from the BF_4^- anion in the reactant. A number of nitrosylchromium complexes of the type represented in (**134**) have been formed in solution by reaction of $[CrNO(S—S)(H_2O)_3]$ (S—S = $EtOCS_2^-$, $Et_2NS_2^-$) with a number of arsine, phosphine and amine Lewis bases (L), and have been studied by ESR spectroscopy.[583]

The purple, diamagnetic isocyanide complexes $[CrNO(RNC)_5]PF_6$ (R = Me, Bu^t, *p*-ClC_6H_4) may be oxidized chemically (Ag^+ in CH_2Cl_2[584] or NO^+ in MeCN[580]) or voltammetrically[584] to the yellow paramagnetic species $[CrNO(RNC)_5]^{2+}$. The reaction of NO with the hexakis(isocyanide) complexes $[Cr(RNC)_6](PF_6)_2$ also gives $[CrNO(RNC)_5](PF_6)_2$ (R = Bu^t, Pr^i

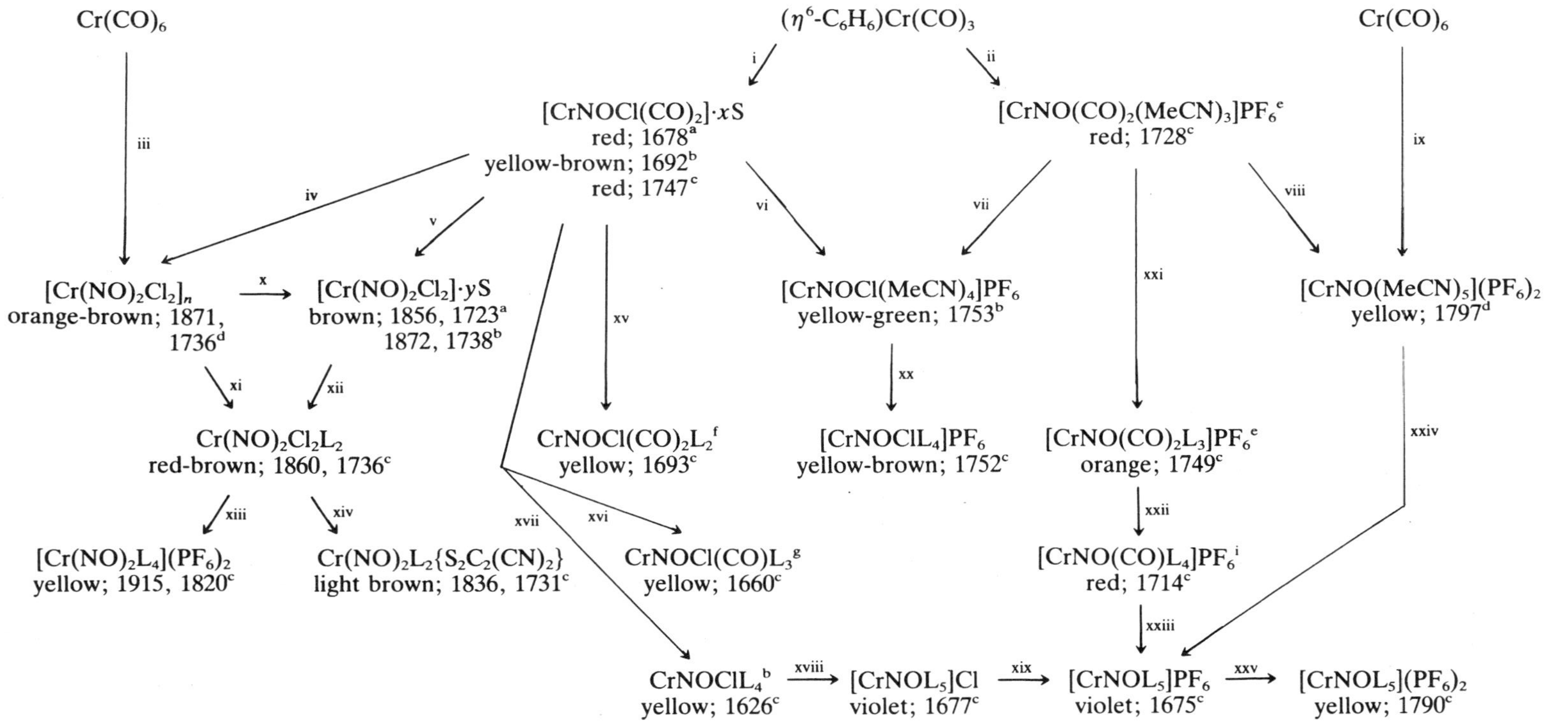

L = $Me_3CN{\equiv}C$

[a] In THF. [b] In MeCN. [c] In CH_2Cl_2. [d] KBr disc. [e] *fac*. [f] *trans*-(NO)(Cl), *cis*-$(CO)_2$, *cis*-L_2. [g] *trans*-(NO)(Cl), *mer*-L_3. [h] *trans*. [i] *cis*.

i, NOCl in CH_2Cl_2 or THF at −79 °C or MeCN at −30 °C; ii, $NOPF_6$ in MeCN at −30 °C; iii, NO, $h\nu$ in CCl_4; iv, NOCl in CH_2Cl_2 or hexane; v, NOCl in THF or MeCN at room temperature; vi, $NOPF_6$ in MeCN; vii, NOCl in MeCN; viii, $NOPF_6$ in MeCN; ix, $2NOPF_6$ in MeCN; x, THF or MeCN; xi, $2Me_3CN{\equiv}C$ in CH_2Cl_2; xii, $2Me_3CN{\equiv}C$; xiii, $2Me_3CN{\equiv}C$, $2NH_4PF_6$ in THF; xiv, $K_2S_2C_2(CN)_2$ in acetone; xv, $2Me_3CN{\equiv}C$ in CH_2Cl_2; xvi, Me_3CNC, 2:1, in THF at −78 °C, then 25 h at room temp; xvii, $Me_3CN{\equiv}C$, 4.5:1, in THF at −78 °C; xviii, $Me_3CN{\equiv}C$ in MeCN; xix, PF_6^- in H_2O; xx, $4Me_3CN{\equiv}C$ in MeCN; xxi, $3MeCN{\equiv}C$ in CH_2Cl_2; xxii, $Me_3CN{\equiv}C$ in CH_2Cl_2; xxiii, $Me_3CN{\equiv}C$ in CH_2Cl_2; xxiv, $Me_3CN{\equiv}C$ in MeCN; xxv, $NOPF_6$ in MeCN or $MeNO_2$.

Scheme 73

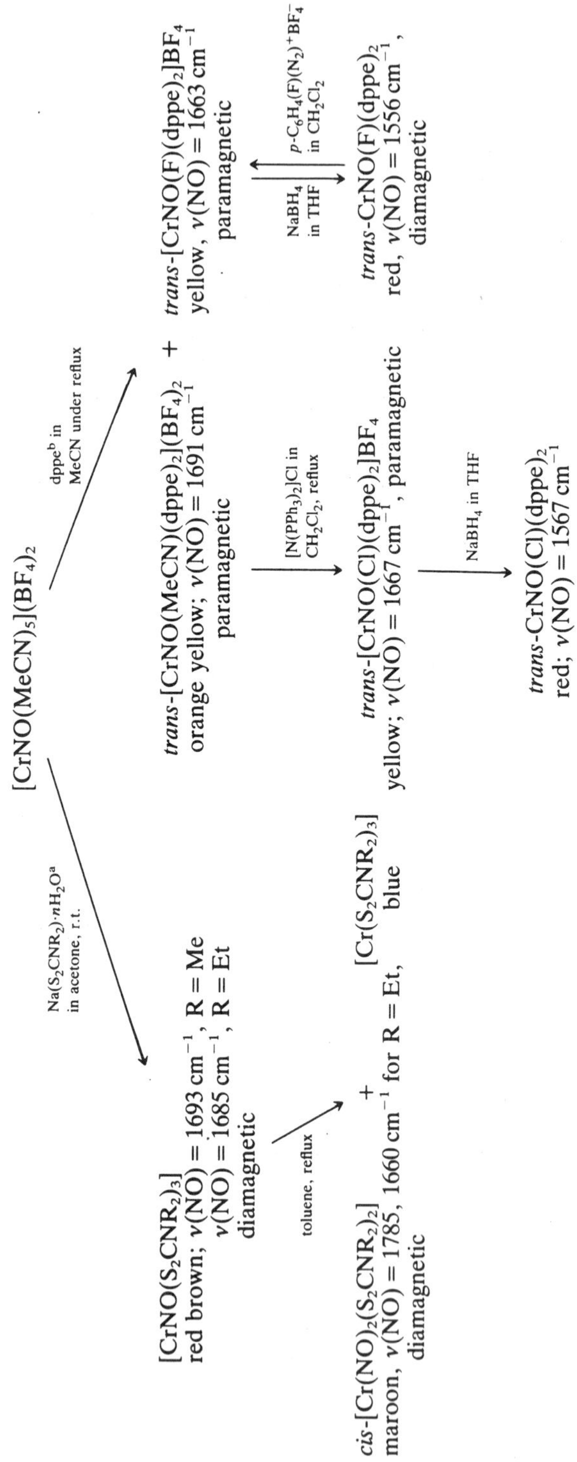

$[CrNO(MeCN)_5](BF_4)_2$
$Na(S_2CNR_2)\cdot nH_2O^a$ in acetone, r.t.
$dppe^b$ in MeCN under reflux
$[CrNO(S_2CNR_2)_3]$
red brown; $\nu(NO) = 1693\ cm^{-1}$, R = Me
$\nu(NO) = 1685\ cm^{-1}$, R = Et
diamagnetic
toluene, reflux
cis-$[Cr(NO)_2(S_2CNR_2)_2]$
maroon, $\nu(NO) = 1785, 1660\ cm^{-1}$ for R = Et,
diamagnetic
+
$[Cr(S_2CNR_2)_3]$
blue
trans-$[CrNO(MeCN)(dppe)_2](BF_4)_2$
orange yellow; $\nu(NO) = 1691\ cm^{-1}$
paramagnetic
+
trans-$[CrNO(F)(dppe)_2]BF_4$
yellow, $\nu(NO) = 1663\ cm^{-1}$
paramagnetic
$[N(PPh_3)_2]Cl$ in CH_2Cl_2, reflux
trans-$[CrNO(Cl)(dppe)_2]BF_4$
yellow; $\nu(NO) = 1667\ cm^{-1}$, paramagnetic
$NaBH_4$ in THF
trans-$CrNO(Cl)(dppe)_2$
red; $\nu(NO) = 1567\ cm^{-1}$
$NaBH_4$ in THF
p-$C_6H_4(F)(N_2)^+BF_4^-$ in CH_2Cl_2
trans-$CrNO(F)(dppe)_2$
red, $\nu(NO) = 1556\ cm^{-1}$,
diamagnetic

[a] R = Me, $n = 2$; R = Et, $n = 3$. [b] dppe = 1,2-bis(diphenylphosphino)ethane.

Scheme 74

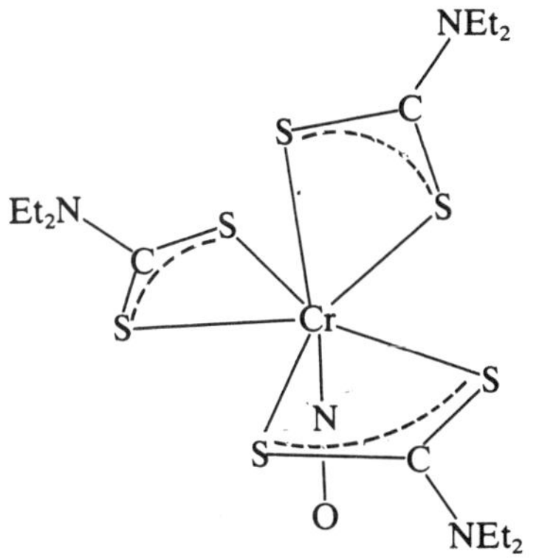

(133)

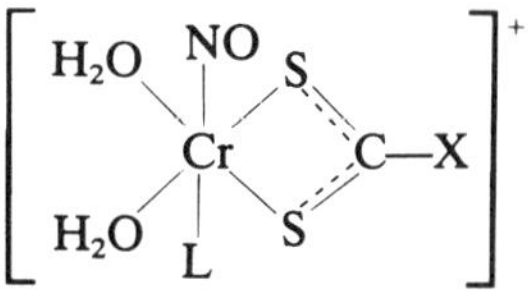

(134) X = OEt, NEt$_2$
L = NH$_3$, NH$_2$OH, amines,
phosphines, arsines

or C$_6$H$_{11}$), which undergoes substitution and/or reduction as outlined in Scheme 75.[585] A number of monodentate phosphine complexes were synthesized by substitution of isocyanide in [CrNO(RNC)$_5$]X and then oxidized according to Scheme 75.[586] If the oxidized forms are regarded as Cr^I—NO$^+$ species and the reduced forms as Cr^0—NO$^+$ species then the complexes [CrNO(RNC)$_5$]X$_n$ and *trans*-[CrNO(RNC)$_4$(PR$_3'$)]X$_n$ (n = 1 or 2) constitute a series of 17 or 18 electron complexes, which are of current interest in organometallic chemistry.[586] The reaction of [CrNO(RNC)$_5$](PF$_6$)$_2$ with the bidentate phosphine dppe, affords the reduction product *fac*-[CrNO(RNC)$_3$(dppe)]PF$_6$ and *trans*-[CrNO(F)(dppe)$_2$]PF$_6$, the F ligand of which is the result of abstraction from the PF$_6$ anion.

$$[Cr(RNC)_6]X_2$$

$$\downarrow \text{NO}^a$$

trans-[CrNO(RNC)$_4$(am)]X $\xleftarrow{\text{am in EtOH}^b}$ [CrNO(RNC)$_5$]X$_2$ $\xrightarrow{\text{Cl}^- \text{ in CH}_2\text{Cl}_2{}^c}$ [CrNO(Cl)(RNC)$_4$]X

dppe / in EtOH
65 °C, 4 h^g

one-electron reduction^d

fac-[CrNO(RNC)$_3$(dppe)]X + *trans*-[CrNO(F)(dppe)$_2$]X   [CrNO(RNC)$_5$]X

dppe in EtOH
65 °C, 5 h^h

PR$_3'$
in Pr^{n}OH, T °C^e

trans-[CrNO(RNC)$_4$PR$_3'$]X

fac-[CrNO(RNC)$_3$(dppe)]X

NOPF$_6$, acetone^f

trans-[CrNO(RNC)$_4$PR$_3'$]X$_2$

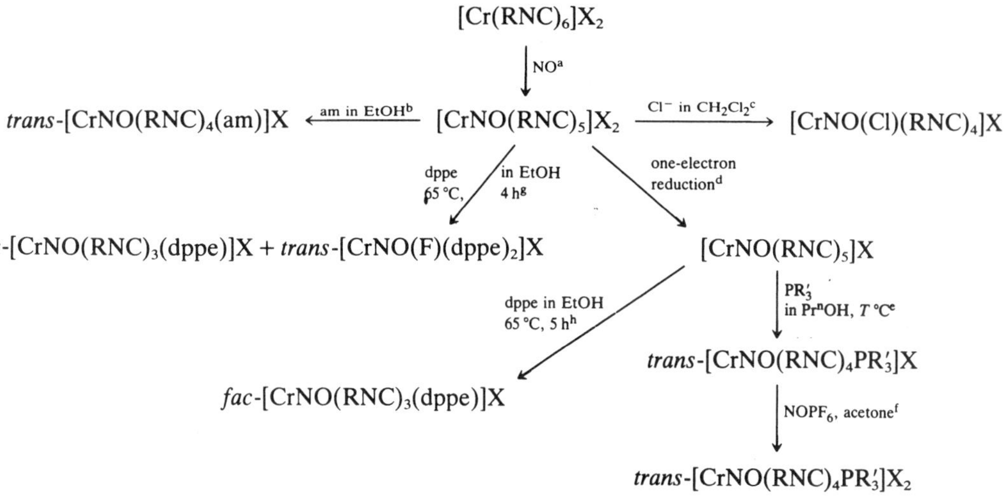

^{a} R = Bu^t, Pr^i, C$_6$H$_{11}$; X = PF$_6$. ^{b} am = Bu^{t}NH$_2$, Pr^{i}NH$_2$, piperidine; R = Bu^t, Pr^i; X = PF$_6$. ^{c} R = Bu^t, Pr^i; X = PF$_6$. ^{d} R = Bu^t, Pr^i, C$_6$H$_{11}$; X = PF$_6$. ^{e} R = Bu^t, Pr^i; R' = Et, Pr^n, Bu^n, Me$_2$Ph, Et$_2$Ph; X = PF$_6$; T = 65 °C: R = Me; R' = Et, Bu^n, Me$_2$Ph; T = reflux temp.: R = Me; R' = Et; X = BF$_4^-$; T = reflux temp. ^{f} R = Bu^t; R' = Bu^n, Me$_2$Ph; X = PF$_6$; R = Pr^i; R' = Et$_2$Ph; X = PF$_6^-$; R = Me, R' = Et, X = BF$_4^-$. ^{g} R = Bu^t; X = BF$_4$. ^{h} R = Bu^t; X = PF$_6$.

Scheme 75

The products of oxidative nitrosylation of the complexes Cr(CO)$_4$P—P (P—P = dcpe, dmpe) by NOPF$_6$ in MeCN depend on the ratio of nitrosylating agent to complex used in the reaction mixture as illustrated in Scheme 76.[587] The observed decrease in NO stretching frequency on converting *fac*-[CrNO(CO)$_3$P—P]$^+$ to *fac*-[CrNO(MeCN)$_3$P—P]$^{2+}$ shows that the effect of replacing three strong π-acid ligands by weaker ones, a substitution which normally results in a lowering of ν_{NO}, outweighs the effect of increased metal oxidation state which normally affects ν_{NO} in the opposite direction. The above reactions of the dcpe and dmpe complexes contrast with that of Cr(CO)$_4$dppe. In MeOH/toluene, reaction of this complex with NOPF$_6$ affords *cis*-[Cr(CO)$_4$dppe]PF$_6$ but in MeCN the product, although initially thought to be *trans*-[Cr(NO)$_2$(MeCN)$_4$](PF$_6$)$_2$,[588,589] is in fact [CrNO(MeCN)$_5$](PF$_6$)$_2$.[582]

$$\textit{fac-}[Cr(NO)(MeCN)_3P\text{—}P](PF_6)_2{}^a \xrightarrow{Me_2CO} \textit{fac-}[CrNO(Me_2CO)_3P\text{—}P](PF_6)_2{}^b$$

orange; $v(NO) = 1722\ cm^{-1}$ 　　　　　　　　　　　　　red

NOPF$_6$ (2:1)
in MeCN

$\textit{cis-}Cr(CO)_4P\text{—}P$　　　　　　　　　　MeCN　　　　Me$_2$CO

NOPF$_6$ (1:1)
in MeCN

$\textit{fac-}[CrNO(CO)_3P\text{—}P]PF_6$ 　　　+　　　　　$[Cr(CO)_4P\text{—}P]PF_6$
$v(NO) = 1754\ cm^{-1}$ in CH_2Cl_2(P—P = dmpe)

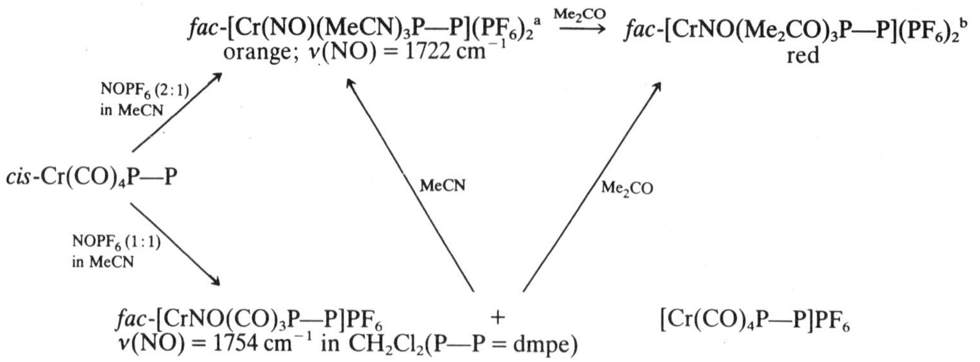

[a] $\lambda_{max}\ (\varepsilon)/nm\ (dm^3\ mol^{-1}\ cm^{-1})$ in MeCN; 393 (1320), 460 (423) for P—P = dmpe; 412 (1000), 470 (200) for P—P = dcpe: $v(CN)$ in Nujol = 2315, 2290 cm^{-1} for P—P = dmpe; 2310, 2282 cm^{-1} for P—P = dcpe: ESR in MeCN; 1:2:1 triplet, $g_{(av)} = 1.993$ for P—P = dmpe; 1.991 for P—P = dcpe. [b] $\lambda_{max} = 354$, 534 nm in Me$_2$CO: ESR in Me$_2$CO; 1:2:1 triplet, $g_{(av)} = 1.987$

Scheme 76

A nitrosylchromium complex stabilized with *o*-phenylenebis(dimethylarsine) (diars) ligands has been prepared and reduced according to reaction Scheme 77.[590] The product [CrNO(Cl)(diars)$_2$]ClO$_4$ has an X-ray powder pattern similar to that of *trans*-[MNO(Cl)(diars)$_2$]ClO$_4$ (M = Fe, Co) and therefore must have a six-coordinate *trans* configuration. Its ESR spectrum in CH$_2$Cl$_2$ consists of thirteen lines due to interaction of the unpaired electron with four ^{75}As ($I = 3/2$) nuclei. Each line is further split into a triplet due to interaction with the N ($I = 1$) nucleus.

$$CrCl_2 \xrightarrow{i} \textit{trans-}[CrNO(Cl)(diars)_2]Cl$$
$$\Big\downarrow ii$$

CrNO(Cl)(diars)$_2$　　　　　$\xleftarrow{iii}$　　　　*trans*-[CrNO(Cl)(diars)$_2$]ClO$_4$
orange, diamagnetic, $v(NO) = 1590\ cm^{-1}$　　　yellow, $\mu_{eff} = 1.70$ BM at 298 K, $v(NO) = 1690\ cm^{-1}$

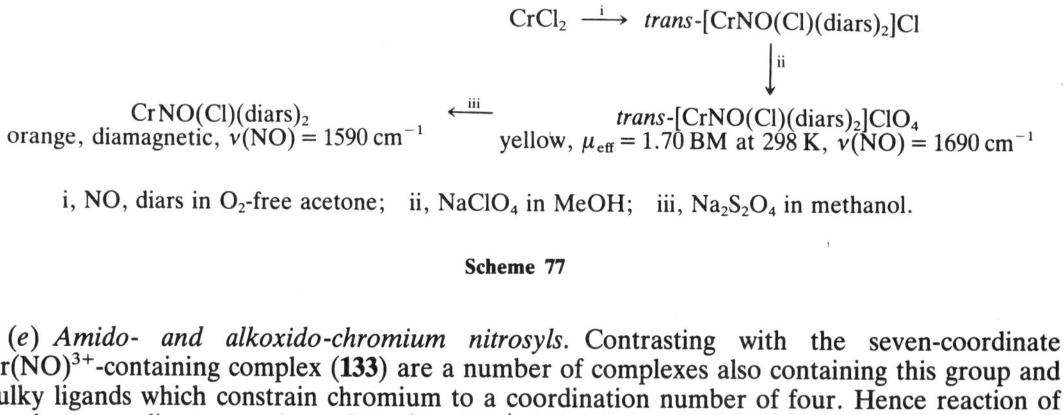

i, NO, diars in O$_2$-free acetone; ii, NaClO$_4$ in MeOH; iii, Na$_2$S$_2$O$_4$ in methanol.

Scheme 77

(e) Amido- and alkoxido-chromium nitrosyls. Contrasting with the seven-coordinate Cr(NO)$^{3+}$-containing complex (**133**) are a number of complexes also containing this group and bulky ligands which constrain chromium to a coordination number of four. Hence reaction of the three-coordinate complexes CrL$_3$ (L = NPr$^i_2$, N(SiMe$_3$)$_2$ or 2,6-dimethylpiperidide) with NO in pentane solution gave the diamagnetic, chemically stable mononitrosyls CrNOL$_3$.[591,592] The crystal structure of the bis(trimethylsilylamide) complex (**135**) shows that the molecule has space-group-imposed C_3 symmetry with a linear Cr—N—O group. The chemical stability of this group is demonstrated by the reactions of CrNO(NPr$^i_2$)$_3$ with bulky alcohols (Scheme 78), which occur without loss of the NO ligand.[592,593] The dimer Cr$_2$(NO)$_2$(OPr^i)$_6$, represented schematically in (**136**) has virtual C_{2h} symmetry with the two chromium atoms, the two nitrosyl ligands and the bridging alkoxy oxygen atoms in the σ_h plane.[593] Above 80 °C in toluene solution, exchange of bridging and terminal isopropoxy ligands, for which an intramolecular mechanism is implicated, occurs rapidly on the NMR time-scale. The addition of Mo$_2$(NO)$_2$(OPr^i)$_6$ or ligands L (L = NH$_3$, pyridine or 2,4-lutidine) to Cr$_2$(NO)$_2$(OPr^i)$_6$ in toluene leads to the equilibria represented by equations (33) and (34). The existence of the heteronuclear dimer CrMo(NO)$_2$(OPr^i)$_6$ was confirmed by ^{1}H NMR, IR and mass spectroscopy. The NO stretching frequencies in the IR spectra of Cr(NO)L$_3$ complexes (Table 63) are remarkably sensitive to the nature of the ligand in the trigonal plane.

$$Cr_2(NO)_2(OPr^i)_6 + Mo_2(NO)_2(OPr^i)_6 \rightleftharpoons CrMo(NO)_2(OPr^i)_6 \qquad (33)$$

$$Cr_2(NO)_2(OPr^i)_6 + 2L \rightleftharpoons 2Cr(NO)(OPr^i)_3L \qquad (34)$$

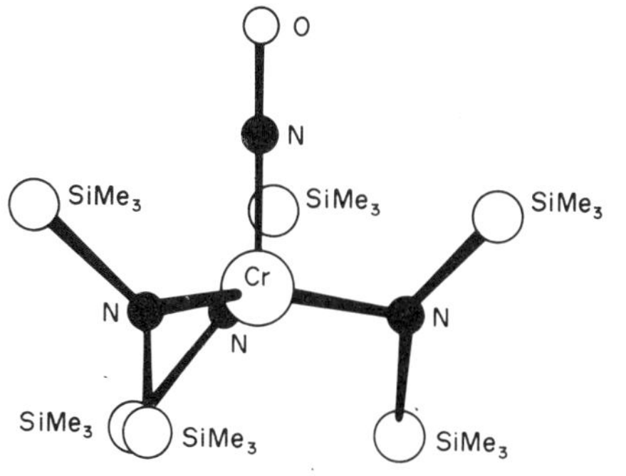

(135)

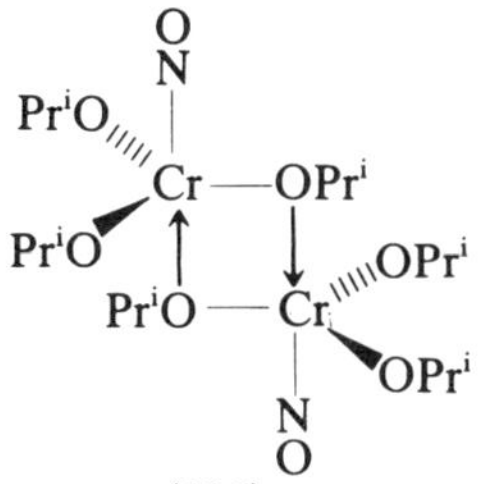

(136)

Table 63 Colours and NO Stretching Frequencies for Nitrosylchromium Complexes Containing Bulky Ligands[1]

Complex	*Colour*	$\nu(N{-}O)$ (cm^{-1})^a
CrNO(NPr$^i_2$)$_3$	Orange	1641
CrNO[N(SiMe$_3$)$_2$]$_3$	Deep purple	1698
CrNO(Me$_2$pip)$_3$	Brown	1673
CrNO(OBu^t)$_3$	Orange-red	1707
CrNO(NPr$^i_2$)(OBu^t)$_2$	Orange	1683
Cr$_2$(NO)$_2$(OPr^i)$_6$	Brick-red	1720

^{a} As Nujol mulls.
1. D. C. Bradley, C. W. Newing, M. H. Chisholm, R. L. Kelly, D. A. Haitko, D. Little, F. A. Cotton and P. E. Fanwick, *Inorg. Chem.*, 1980, **19**, 3010.

$$\text{CrNO(OBu}^t)_3 + \text{Cr(NO)(NPr}^i_2)(\text{OBu}^t)_2$$

$$\overset{\text{Bu}^t\text{OH}}{\nearrow}$$

$$\text{CrNO(NPr}^i_2)_3$$

$$\underset{\text{Pr}^i\text{OH}}{\searrow}$$

$$\text{Cr}_2(\text{NO})_2(\text{OPr}^i)_6$$

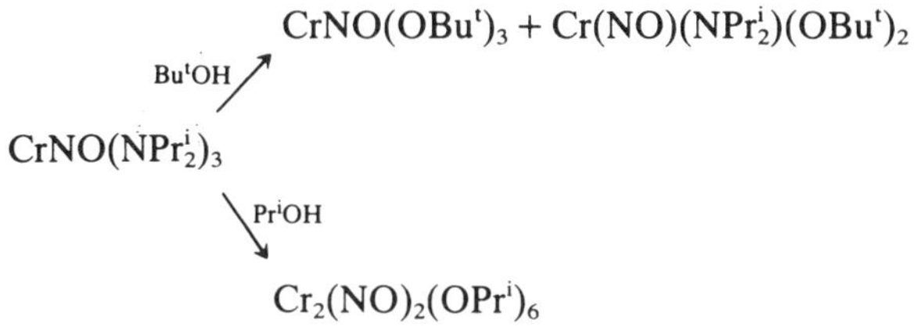

Scheme 78

The electrochemical reductions of CrNO[N(SiMe$_3$)$_2$]$_3$, CrNO(NPr$^i_2$)$_3$ and CrNO(OBu^t)$_2$(NPr$^i_2$) in acetonitrile have been studied and the results interpreted in terms of quasi-reversible or irreversible one-electron reductions with the possibility of side reactions.[594]

(*f*) *Macrocyclic chromium nitrosyls.* A number of NO adducts with chromium macrocyclic complexes have been synthesized. The addition of NO to a solution of Cr(TPP) (TPP = tetraphenylporphyrin) in toluene leads to the red-coloured CrNO(TPP) ($\nu_{NO} = 1700$ cm^{-1}, the visible spectrum in toluene solution has λ_{max} at 600, 555, 520sh and 435 nm).[595] This complex is also obtained by reductive nitrosylation of Cr(TPP)(OMe) in chloroform–methanol solution, a reaction which probably proceeds by a mechanism involving nucleophilic attack by MeOH on coordinated NO (Scheme 79). The complex has an $S = \frac{1}{2}$ ground state and its ESR spectrum, which in toluene solution shows hyperfine splitting from both porphyrin and nitric oxide ^{14}N, is diagnostic of a configuration which is predominantly metal $(d_{xz}d_{yz})^4 \, d^1_{xy}$ in character. Its formulation as a Cr^I(NO$^+$)(TPP) complex contrasts with that of [CrNO(CN)$_5$]$^{3-}$, shown by X-ray photoelectron spectroscopy to be [Cr^{III}(NO$^-$)(CN)$_5$]$^{3-}$,[565] despite striking similarities in their ESR spectra. Recent work on ligand addition and electron transfer reactions of CrNO(TPP) is referred to in Section 35.4.9.1. The β-polymorph (but not the α) of phthalocyaninatochromium(II), β-PcCr, absorbs NO at room temperature to give the adduct CrNO(Pc) ($\nu_{NO} = 1690$ cm^{-1}).[596,597] Treatment of this complex with anhydrous, degassed pyridine affords the stable product CrNO(Pc)py ($\nu_{NO} = 1680$ cm^{-1}).

The nitrosylchromium complexes (**137**) and (**138**) are produced in the reactions of Cr(Me$_2$[14]tetraenatoN$_4$)(py)$_2$]PF$_6$ (Me$_2$[14]tetraenatoN$_4$ = dianion of *trans*-5,14-dimethyl-1,4,8,11-tetraazacyclotetradeca-4,6,11,13-tetraene) and [Cr(Me$_6$[14]-4,11-dieneN$_4$)py](PF$_6$)$_2$

$$Cr(TPP)(OMe) + NO \longrightarrow CrNO(TPP)(OMe)$$
$$CrNO(TPP)(OMe) + MeOH \longrightarrow Cr(TPP) + MeONO + MeOH$$
$$Cr(TPP) + NO \longrightarrow CrNO(TPP)$$

Scheme 79

(Me$_6$[14]-4,11-dieneN$_4$ = 5,7,7,12,14,14-hexamethyl-1,4,8,11-tetraazacyclotetradeca-4,11-diene) respectively with alcoholic $NaNO_2$.[598] The crystal and molecular structure of (**138**) shows that the MNO group is linear and that the NO and NO_2^- ligands are mutually *trans*. A combination of IR, ESR and magnetic data has been interpreted to imply the presence in the complexes of a $Cr^I(NO^+)$ group having, as in the case of CrNO(TPP), a $(d_{xz}, d_{yz})^4\ d_{xy}^1$ configuration. Some properties of these complexes are presented in Table 64. The electronic spectra of complexes (**137**) and (**138**) are similar in band positions despite differences in their macrocyclic ligands and coordination numbers. This tends to suggest that the NO ligand dominates the electronic configurations of such complexes.

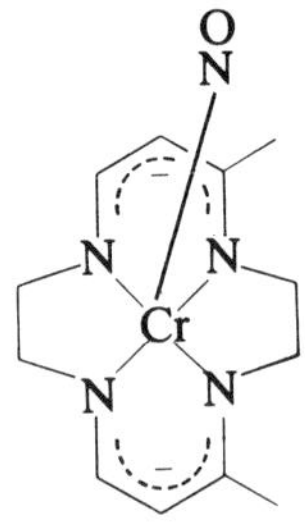

(**137**)

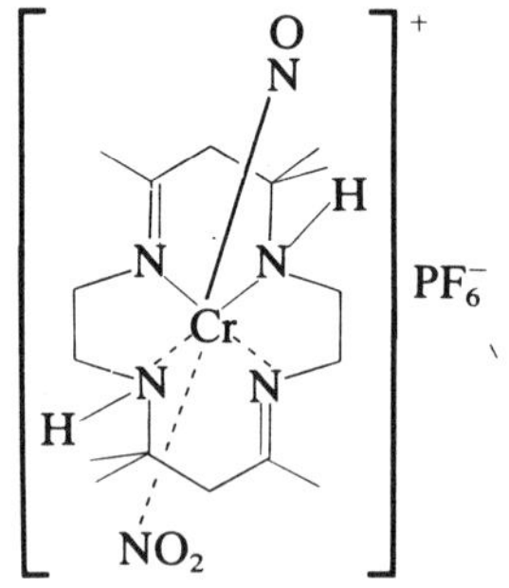

(**138**)

Table 64 Some Properties of Macrocyclic-chromium Nitrosyl Complexes[1]

Complex	*Colour*	$\nu(NO)$ (cm^{-1})[a]	μ_{eff} (BM)[b]	$\lambda_{max}(\varepsilon)$ (nm, dm^3 mol^{-1} cm^{-1})
(**137**)	Green	1620	1.81	654 (1330), 450sh (2100), 385 (7100)[c]
(**138**)	Green	1640	1.72	649 (45), 450 (99), 361 (340)[d]

[a] As Nujol mulls. [b] Room temperature. [c] In THF solution. [d] In MeCN solution.
1. D. Wester, R. C. Edwards and D. H. Busch, *Inorg. Chem.*, 1977, **16**, 1055.

(*g*) *Cr(NO)*$_4$. The binary nitrosyl $Cr(NO)_4$, a dark red volatile solid which is both isoelectronic and isostructural with $Ni(CO)_4$, may be obtained by photolyzing a solution of $Cr(CO)_6$ in pentane while purging with NO.[599] The evidence from IR and Raman spectroscopy suggests that the molecule has T_d symmetry both in solution and in the solid state. Photolysis of $Cr(CO)_6$ and NO in frozen argon and methane matrices also leads to $Cr(NO)_4$ *via* the intermediate $Cr(CO)_3(NO)_2$.[600]

The chemistry of a number of carbonyl and cyclopentadienylchromium–nitrosyl complexes has been discussed in the companion series.[1]

(*ii*) *Thionitrosyls*

The relative scarcity of transition metal complexes containing NS ligands compared to the number containing NO has been attributed to the lower stability of NS and to the lack of reagents for introducing it into complexes.[601] The preparation and reactions of the first reported organometallic thionitrosyl complex, $CpCr(CO)_2NS$, are shown in Scheme 80.[601] Gas-phase core-electron binding energies of this compound have been determined.[602] Addition of a nitromethane solution of $NS^+PF_6^-$, obtained *in situ* from $N_3S_3Cl_3$ and $AgPF_6$, to $C_6H_6Cr(CO)_3$ in ethanenitrile gives the thionitrosyl complex $[CrNS(MeCN)_5](PF_6)_2$.[603] Reactions of this complex and some properties of it and its reaction products are outlined in Scheme 81. The $\nu(NS)$ band in the IR spectrum of $[CrNS(Bu^tNC)_5]PF_6$ occurs at lower wavenumbers than the corresponding band in the spectrum of its oxidation product consistent with greater electron density on the metal and increased back-bonding into a π^* orbital on NS.

$$Na[CpCr(CO)_3] + N_3S_3Cl_3 \longrightarrow CpCr(CO)_2NS + CO + NaCl$$

$$CpCr(CO)_2NS \xrightarrow{NOPF_6\ \text{in}\ MeCN/CH_2Cl_2} [CpCrCO(NO)NS]PF_6$$

$$CpCr(CO)_2NS \xrightarrow{NOCl} CpCr(NO)_2Cl$$

Scheme 80

$[CrNS(MeCN)_5](PF_6)_2$
blue; $\nu(NS) = 1245\ cm^{-1}$; $\nu(CN) = 2325, 2305\ cm^{-1}$

$\downarrow$ Bu^tNC, Zn in MeCN

$[CrNS(Bu^tNC)_5]PF_6$
green; $\nu(NS) = 1135\ cm^{-1}$; $\nu(CN) = 2203, 2135, 2070\ cm^{-1}$

$\downarrow$ $NOPF_6$ in MeCN

$[CrNS(Bu^tNC)_5](PF_6)_2$
blue; $\nu(NS) = 1220\ cm^{-1}$; $\nu(CN) = 2225\ cm^{-1}$

Scheme 81

(*iii*) *Azo ligands*

Complexes of chromium(III) with multidentate ligands containing azo linkages as well as other donor groups are of interest because of their applications in spectrophotometric analysis and because of their uses as commercial dyestuffs.[604] Some ligands used in the analysis of chromium(III) are listed in Table 65.

Table 65 Azo Ligands Used for the Analysis of Chromium(III)

Ligand	*Application*	*Ref.*
$C_5H_2NBr_2$—N=N—$C_{10}H_{15}NO$	Cr^{III} and Cr^{VI} at pH 3.3–4.0	1
$C_6H_6AsO_3$—N=N—$C_{10}H_6O_8S_2$—N=N—$C_6H_2RR^1R^2$ ($R, R^1, R^2 = H, CO_2H, PO_3H_2, AsO_3H_2, NO_2, Cl$)	Cr^{III} at pH 2	2, 3
$C_7H_5O_2$—N=N—$C_{17}H_{11}O_8S_2Na_2$	Cr^{III} at pH 3–4.5	4
Ph—N=N—$C_6H_5O_2$	Cr^{III} at pH 3–6	5
$C_{10}H_7$—N=N—$C_6H_5O_2$	Cr^{III} at pH 3–6	5
C_5H_4N—N=N—$C_6H_5O_2$	Cr^{III} at pH 5–5.5	6
C_3H_2NS—N=N—$C_6H_5O_2$	Cr^{III} at pH 5	7

1. F. Zhang and G. Chen, *Chem. Abstr.*, 1985, **102,** 67121h.
2. S. Fusheng and L. Xu, *Chem. Abstr.*, 1984, **100,** 150240a.
3. V. K. Guseinov and O. A. Tataev, *Chem. Abstr.*, 1977, **87,** 177042n.
4. L. Xu and F. Li, *Chem. Abstr.*, 1983, **98,** 118679e.
5. I. K. Guseinov, N. Kh. Rustamor, Ya. A. Asimov and M. M. Agamalieva, *Chem. Abstr.*, 1978, **89,** 35914r.
6. K. L. Cheng, K. Ueno and T. Imamura, 'Handbook of Organic Analytical Reagents', CRC, Florida, 1982, p. 197.
7. K. L. Cheng, K. Ueno and T. Imamura, 'Handbook of Organic Analytical Reagents', CRC, Florida, 1982, p. 208.

The structures of a number of chromium(III)–azo complexes have been determined by X-ray diffraction methods. The ligands 2,2′-dihydroxyazobenzene and 5-nitro-2-(2-hydroxynaphthylazo)phenol form 2:1 complex anions with chromium(III) which have *mer* configurations.[605] In the 2:1 complex anion formed between 3-methyl-2-[3-methyl-1-(*p*-bromophenyl)-5-oxo-2-pyrazolin-4-ylazo]benzoic acid and chromium(III) (**139**) the azo, carboxylate and enolate groups of one ligand are arranged *trans* to the corresponding groups in the other ligand.[606]

A number of isomeric 2:1 chromium(III) complexes containing ligands derived from

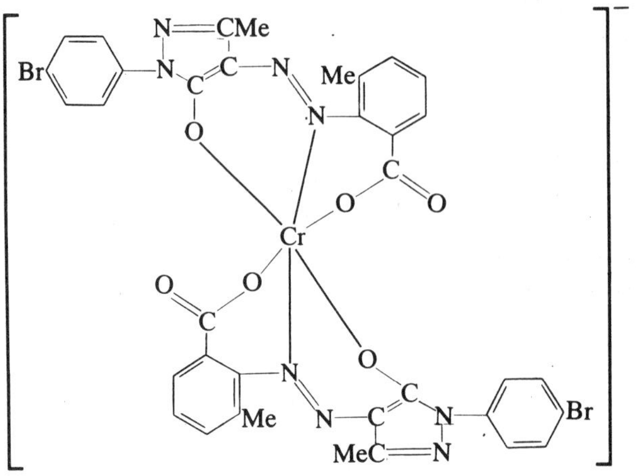

(**139**)

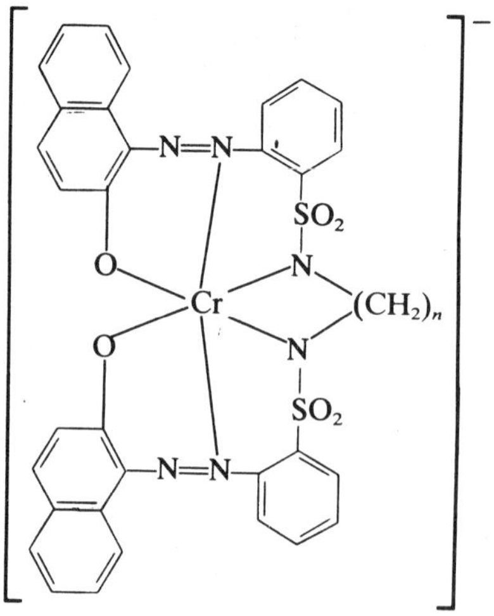

(**140**) $n = 2$, 3 or 6

2,2′-dihydroxyazobenzene, *o*-(2-hydroxyl-1-naphthylazo)phenol, 2,2′-dihydroxyazonaphthalene and *o*-(2-hydroxy-1-naphthylazo)pyridine have been prepared.[607] Chromium(III) complexes containing the tridentate ligands 4-methylsulfonyl-2-(4,5-diphenylimidazol-2-ylazo)phenol and 4-chloro-2-(4,5-diphenylimidazol-2-ylazo)benzoic acid have also been prepared.[608] The effects of central ring size on the electronic spectra and stabilities of the pentacyclic chromium(III) complexes (**140**) have been investigated.[609] The preparation and properties of a number of 1:1 chromium(III) complexes containing substituted naphthol-azopyrazolone ligands (L) have been reported.[604] These complexes appear to be of the type *fac*-$[CrL(H_2O)_3]^+$ and *fac*-$CrL(Cl)(H_2O)_2$ and react with various bidentate ligands such as acac, salicylaldehyde, amino- and hydroxy-acids (Y) to give the ternary complexes, $Cr(L)(Y)H_2O$. The effects of substituents on the spectra of the dyes and their complexes are discussed.

(*iv*) *Hydrazine, hydroxylamine and related ligands*

Addition of alcoholic $N_2H_4{\cdot}H_2O$ to an aqueous solution of $Cr(ClO_4)_3$ affords a violet crystalline compound of composition $Cr(N_2H_4)_2(ClO_4)_3$.[610] The IR spectrum of the product shows split absorption bands at 1100 and 630 cm^{-1} characteristic of chelating ClO_4^- and a band at 950 cm^{-1} characteristic of a monodentate hydrazine ligand. On this basis structure (**141**) is proposed for this complex.

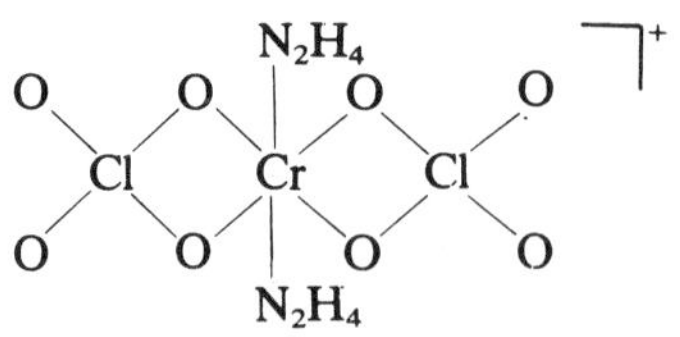

(**141**)

Addition of $Cr_2(C_2O_4)_3{\cdot}6H_2O$ to a cold alcoholic solution of N_2H_4 gives a violet powder of composition $Cr_2(C_2O_4)_3{\cdot}4N_2H_4{\cdot}4H_2O$.[611] This product is also obtained when the starting complex is stored in a desiccator over a 60% alcoholic solution of N_2H_4. Exposure to a higher partial pressure of N_2H_4 leads instead to the product $Cr_2(C_2O_4)_3{\cdot}7N_2H_4{\cdot}H_2O$. Evidence based on spectroscopic[611,612] and magnetic[613] studies indicates that the former product is polymeric containing an infinite —N_2H_4—Cr—N_2H_4— chain in the equatorial plane, while the latter complex contains $C_2O_4^{2-}$ as well as N_2H_4 bridging ligands.

Acetylhydrazide ($AcNHNH_2$) and benzoylhydrazide ($BzNHNH_2$) form complexes with $CrCl_3{\cdot}6H_2O$ having compositions $CrCl_3(AcNHNH_2)_3{\cdot}3H_2O$ and $CrCl_3(BzNHNH_2)_3$ respectively.[614] IR and electronic spectral data suggest that each ligand is coordinated to the metal through the unsaturated nitrogen atom of its imidol form (**142**).

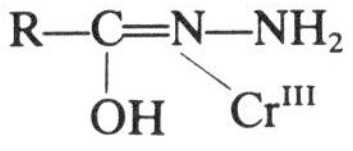

(142)

Chromium(III) forms a number of complexes with NH_2OH depending on the reaction conditions as outlined in Scheme 82.[615] Reaction of $KCr(SO_4)_2 \cdot 12H_2O$ with $NH_2OH \cdot HCl$ (12-fold excess) and KOH (6-fold excess) in aqueous solution affords the lilac-violet complex $[CrCl(SO_4)(NH_2OH)_4] \cdot 2H_2O$ (for which $\delta(NOH) = 1275\ cm^{-1}$). All of these complexes contain N-coordinated hydroxylamine as shown by the characteristic shift to higher wavenumbers in the $\delta(NOH)$ band relative to that in $NH_2OH \cdot HCl$. The electronic spectrum of $[CrOH(NH_2OH)_5]Cl_2$, the only one of these complexes to dissolve in water, shows maxima at 552 and 407 nm; the position of the first implies that the NH_2OH ligand lies between NH_3 and H_2O in the spectrochemical series. Reductive nitrosylation of CrO_4^{2-} by NH_2OH at pH 4–5 in the presence of pyridine-2,6-dicarboxylic acid (H_2dipic) gives the products $CrNO(dipic)(H_2O)_2$ (red, $\mu_{eff} = 1.70$ BM at 298 K), $[CrNO(dipic)(NH_2OH)_2] \cdot 3H_2O$ (yellow-green, $\mu_{eff} = 1.71$ BM at 298 K) and $Crdipic(NH_2O)(NH_2OH)_2$ (dark green, $\mu_{eff} = 3.6$ BM at 298 K, Table 62).[578] The modes of coordination of hydroxylamine and the hydroxylamine anion in these complexes have not been established.

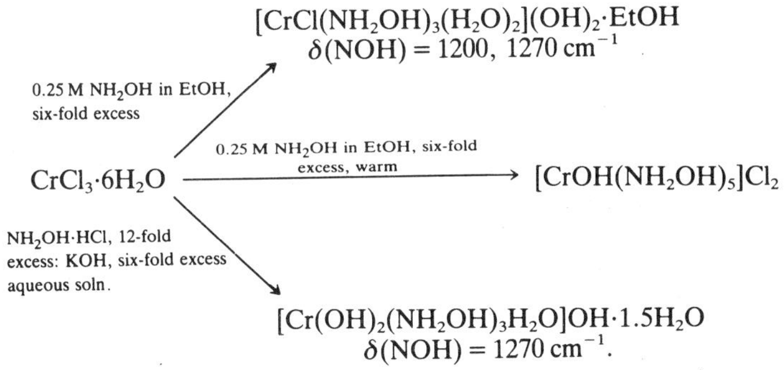

Scheme 82

Hydroxylaminechromium(III) complexes have been observed amongst the products of reduction of $[CrNO(NH_3)_5]^{2+}$ by chromium(II) in acid solution.[579] Ion exchange experiments show that at least 50% of the NH_2OH produced (by the two-electron reduction of NO^-) in this reaction is initially coordinated in highly charged complexes such as $[(NH_3)_5CrXCr(H_2O)_4Y]^{n+}$ ($X = OH^-$ or NH_2OH; $Y = H_2O$ or NH_2OH).

In the complex tris(*N*-benzoyl-*N*-phenylhydroxylamine)chromium(III) the ligand acts as a bidentate two O donor.[616]

35.4.2.7 Amido ligands

This section is dominated by dialkyl- and disilyl-amide ligands, which, because of steric factors, constrain chromium(III) to the unusual coordination numbers of three or four.

The reaction between $LiNPr^i_2$ and $CrCl_3$ in THF solution leads to the air- and water-sensitive red-brown solid $Cr(NPr^i_2)_3$.[592] The complex is monomeric in solution and in the solid state ($\mu_{eff} = 3.80$ BM at 298–123 K). An X-ray crystal structure analysis shows that the molecule contains a planar CrN_3 and three planar $CrNC_2$ units with 68–70° dihedral angles between the former and each of the latter planes (**143**).[617] The planarity of the nitrogen atoms suggests involvement of their lone pairs in π bonding to the metal, an idea which is further supported by the relatively short Cr—N bond lengths. A number of other chromium(III) dialkylamide complexes, all of which are highly sensitive to air and water, have been prepared and their coordination properties clearly depend on the steric requirements of the specific ligand. Hence in solution both $Cr(NMe_2)_3$ and $Cr(NEt_2)_3$ appear to exist as dialkylamido-bridged dimers in which each chromium is surrounded by a tetrahedral ligand field (**144**).[592] Under vacuum and at mild temperatures the dimers disproportionate according to equation (35). This reaction has been used to synthesize a number of $Cr(NR_2)_4$ compounds ($R = NEt_2$, NPr^n_2, NBu^n_2, piperidide)

(Section 35.5.4.3).[618] The inability of chromium(III) to achieve its preferred hexacoordination because of steric hindrance accounts in part for this unusual disproportionation. In the tetrahedral chromium(IV) compounds the d^2 electrons occupy low energy orbitals. In the dimeric chromium(III) complexes, however, one of the d electrons on each chromium must occupy a high energy orbital. Transfer of an electron from one chromium to another results in formation of the volatile $Cr(NR_2)_4$ and the polymeric residue $[Cr(NR_2)_2]_n$.[618] The monomeric complex $Cr(NPr^i_2)_3$ is stable thermodynamically with respect to disproportionation.[592] While many metal–dialkylamide complexes react with CS_2 to give dithiocarbamate complexes, this insertion reaction is not successful with $Cr(NPr^i_2)_3$. It does, however, react with CO_2, NO (see Section 35.4.2.6) and O_2 (Scheme 83). The species $Cr(O_2)(NPr^i_2)_3$ may be a peroxo chromium(V) compound. At low temperatures the reaction with O_2 produces an explosive compound of composition $CrO_3(NPr^i_2)_3$.[592]

$$[Cr_2(NR_2)_6] \longrightarrow Cr(NR_2)_4 + [Cr(NR_2)_2]_n \qquad (35)$$

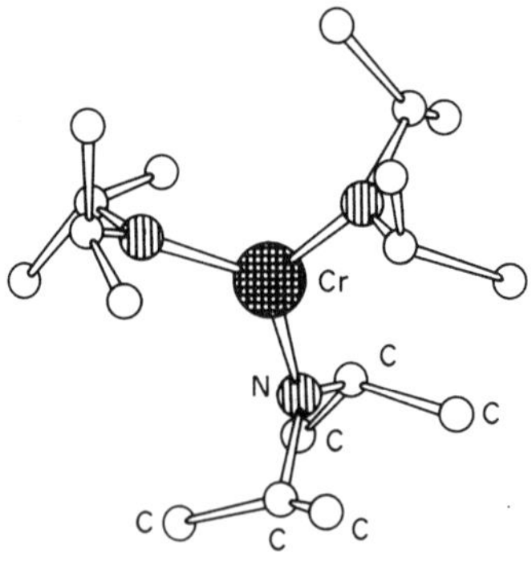

(143)

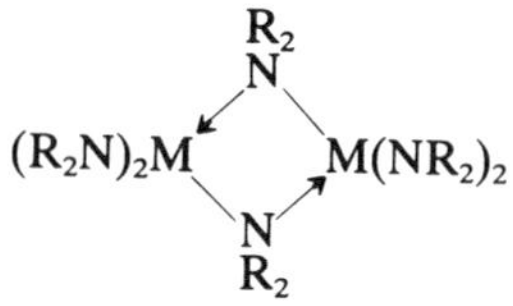

$$(R_2N)_2M \underset{\underset{R_2}{N}}{\overset{\overset{R_2}{N}}{\diamond}} M(NR_2)_2$$

(144) R = Me or Et

$$Cr(NPr^i_2)_2(O_2CNPr^i_2) \qquad\qquad CrNO(NPr^i_2)_3$$

$$\overset{CO_2}{\nwarrow} \qquad \overset{NO}{\nearrow}$$

$$Cr(NPr^i_2)_3$$

$$\downarrow{O_2}$$

$$CrO_2(NPr^i_2)_3$$

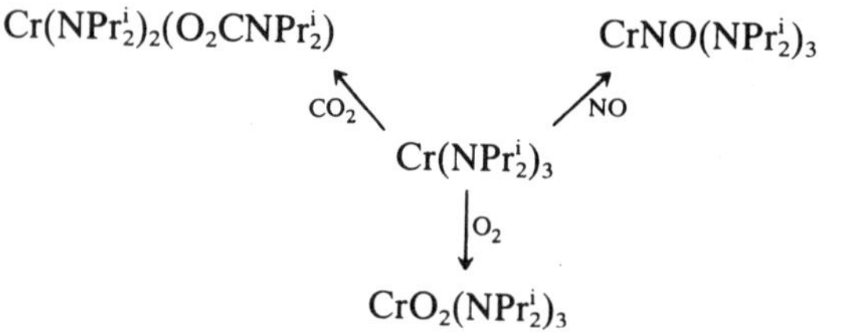

Scheme 83

The reaction of $LiN(SiMe_3)_2$ with $CrCl_3$ in THF affords the bright green complex $Cr[N(SiMe_3)_2]_3$.[592] A single crystal X-ray analysis showed that the complex is similar structurally to $Cr(Pr^i_2)_3$ with a trigonal planar CrN_2 and planar $CrNSi_2$ units. Like the dialkylamide complexes, $Cr[N(SiMe_3)_2]_3$ is rapidly hydrolyzed although it is less susceptible to insertion reactions as shown by its inability to react with either CS_2 or CO_2.

Since $Cr(NPr^i_2)_3$ and $Cr[N(SiMe_3)_2]_3$ are both volatile and can be sublimed without decomposition, they are amenable to study by photoelectron spectroscopy. The gas phase He-I photoelectron spectra of these species show Cr—N σ bond ionization energies in similar regions (9.9 and 10 eV respectively).[619] However the nitrogen lone pair ionization and the d (^3A, ^3E) bands occur at significantly lower energies in the spectrum of the dialkylamido complex. This confirms that the nitrogen lone pair is stabilized by delocalization on to the silicon atoms in $N(SiMe_3)_2$, thus making the ligand unable to act as a π donor to the metal. No such delocalization is possible in the dialkylamido ligand, and so in the complex $Cr(NPr^i_2)_3$ π donation from ligand to metal can occur, with a consequent lowering of metal d electron ionization energies.

A study of the electrochemical reductions of $Cr(NPr^i_2)_3$ and $Cr[N(SiMe_3)_2]_3$ in acetonitrile solution shows that the reductions are either quasi-reversible or irreversible and that the dialkylamido complex has a slightly lower reduction potential.[594]

Magnetic susceptibility measurements confirm that $Cr(NPr^i_2)_3$ and $Cr[N(SiMe_3)_2]_3$ behave as magnetically dilute species with μ_{eff} values of 3.80 and 3.73 BM respectively over the temperature range 123–298 K.[620] Neither complex gives an ESR signal in solution at room temperature. However, frozen solutions of the former in toluene at 130 K and polycrystalline samples of the latter show g anisotropy with principal values $g_{\parallel} = 2.0$, $g_{\perp} = 4.0$ corresponding to an axially symmetric system with a large zero-field splitting of the chromium ion.[620]

35.4.2.8 N-*Thiocyanato,* N-*cyanato,* N-*selenocyanato, azido and related ligands*

The vast majority of known chromium(III) complexes with the ambidentate chalcogenocyanate ligands NCO^-, NCS^-, $NCTe^-$ are those containing *N*-bonded thiocyanate.

(*i*) N-*Thiocyanates*

(*a*) $[Cr(NCS)_6]^{3-}$. A variety of salts containing the $[Cr(NCS)_6]^{3-}$ anion have been prepared.[621,622] Prior to its structure determination by X-ray diffraction methods, criteria based mainly on IR spectroscopy were used to establish N coordination of the ligand to Cr^{III} in this anion. Hence the IR spectrum of $K_3[Cr(NCS)_6]$ (Table 66) has NCS bending and CS stretching vibrations at 483 and 820 cm^{-1} respectively, positions which are characteristic of N-bonded thiocyanate complexes.[621] In addition a band at 364 cm^{-1} in the IR spectrum of $(Et_4N)_3[Cr(NCS)_6]$ has been assigned to the triply degenerate (T_{1u}) Cr—N stretch of the $Cr(NCS)_6$ group. Some properties of $[Cr(NCS)_6]^{3-}$ are listed in Table 66.

Table 66 Some Properties of $[Cr(NCS)_6]^{3-}$

Cation	*Property*		*Ref.*
$[Ho(C_6H_4NCO_2H)(H_2O)_2]$	Bond lengths (pm):	Cr—N, 200.2; N—C, 114.4; C—S, 161.9	1
	Bond angles (°):	NCS, 176.6; CrNC, 164.3	
$(Et_4N)_3$	IR spectrum (cm^{-1}):	ν(CN), 2078; δ(NCS), 483; ν(CrN), 364 (Nujol mull)	2
K_3	IR spectrum (cm^{-1}):	ν(CN) = 2058s, 2098s, 2118s; δ(NCS) = 474s; ν(CS) = 820w, (KBr disc)	3
K_3	Electronic spectrum [nm ($dm^3\ mol^{-1}\ cm^{-1}$)]:	550 (102), 410sh (95), 380sh (79), 310 (17 380) in 96% MeOH	4
K_3	Magnetic moment (BM):	3.79 at 295 K, 3.74 at 82 K	5

1. A. H. Norbury, *Adv. Inorg. Chem. Radiochem.*, 1975, **17,** 231.
2. A. Sabatini and I. Bertini, *Inorg. Chem.*, 1965, **4,** 959.
3. M. A. Bennett, R. J. H. Clark and A. D. J. Goodwin, *Inorg. Chem.*, 1967, **6,** 1625.
4. R. Ripan, I. Ganescu and C. Varhelyi, *Z. Anorg. Allg. Chem.*, 1968, **357,** 140.
5. J. J. Salzmann and H. H. Schmidtke, *Inorg. Chim. Acta*, 1969, **3,** 207.

(*b*) *Mixed ligand* N-*thiocyanates*. A large number of mixed ligand *N*-thiocyanatochromium(III) complexes, many of which are of the type $[Cr(NCS)_4L_2]^{n+}$ and similar to the original Reinecke's salt ($L = NH_3$), are known.[621,623] Generally they are prepared by refluxing a solution of $[Cr(NCS)_6]^{3-}$ and the appropriate ligand L in alcohol and give *trans* isomers except when L = py, $PhEt_2P$ or of course when L_2 is a bidentate ligand. Some representative reactions of $[Cr(NCS)_6]^{3-}$ are shown in Scheme 84.[558–560,623–630] The electronic spectra of the products have been used in many cases to deduce the positions of ligands L in the spectrochemical (the first spin-allowed ligand-field band $^4A_{2g} \rightarrow {}^4T_{2g}$ is a direct measure of $10Dq$) and nephelauxetic series. The reaction of $[Cr(NCS)_6]^{3-}$ with secondary and tertiary phosphines ($R_2R'P$) gives, after acidification, one of the products $R_2R'PH[Cr(NCS)_4(R_2R'P)_2]$ (A), $(R_2R'PH)_2[Cr(NCS)_5R_2R'P]$ (B) or $(R_2R'PH)_3[Cr(NCS)_6]$ (C) depending on the steric requirements of R and R′.[628] Hence straight chain alkylphosphines give products of type A, $EtPh_2P$ gives B, while branching in the alkyl group increases the likelihood of obtaining product C. Neither PPh_3 nor $PHPh_2$ undergoes any of these reactions. The replacement of ligands in $[Cr(NCS)_6]^{3-}$ by imidazole and its derivatives in 95% EtOH has also been investigated.[558,626] While benzimidazole (bzim) and 2-methylbenzimidazole (2-Mebzim) give Reinecke type salts such as $[Cr(NCS)_4(bzim)_2]^-$, imidazole (Him) and its 1- and 2-substituted derivatives give instead complexes of the type $[Cr(NCS)_3(Him)_3]\cdot H_2O$ or $[Cr(NCS)_3(2\text{-}Etim)_2(H_2O)]$. Details of the IR and electronic spectra of some mixed ligand *N*-thiocyanatochromium(III) complexes are presented in Table 67.

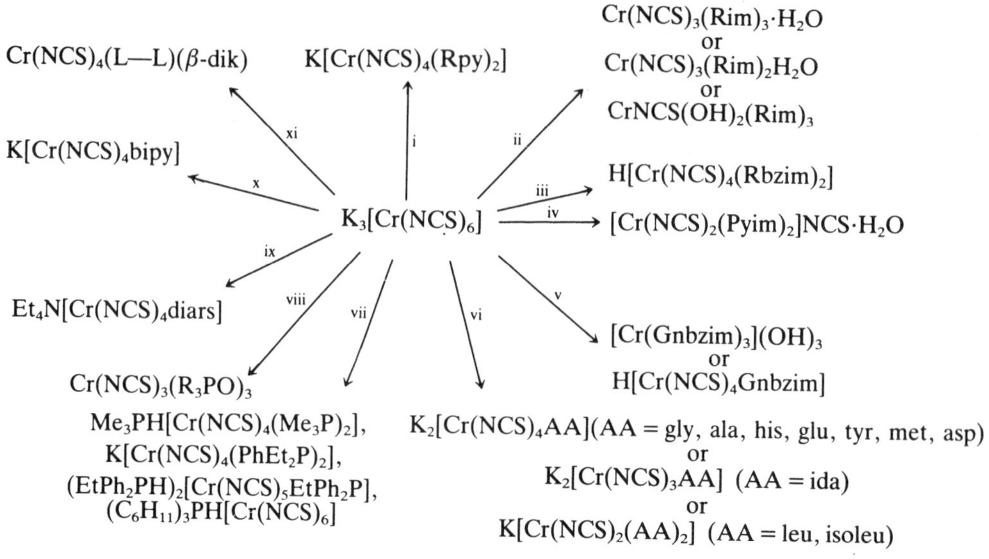

i, R = H; heat with py at 110 °C: R = 3-NH_2, 4-NH_2; reflux in EtOH; ii, Rim = imidazole or substituted imidazole (R = H, 1-Me, 1-Et, 1-Vi, 2-Me, 2-Et, 2-Pr^i); reflux in 95% EtOH; iii, Rbzim = benzimidazole or substituted benzimidazole (R = H, 2-Me); reflux in 95% EtOH; iv, Pyim = 2,2′-pyridylimidazole; reflux in EtOH; v, Gnbzim = 2-guanidinobenzimidazole; reflux in EtOH or Pr^nOH; vi, AA = amino acid; 1:1 or 2:1 mole ratio as appropriate, in H_2O; vii, phosphine in Pr^iOH, reflux, H^+; viii, R_3PO (R = Et, Bu^n, C_6H_{11}) in Pr^iOH, reflux; ix, diars in EtOH, reflux, Et_4NCl; x, bipy in EtOH, reflux; xi, β-dik = β-diketone (benzoylacetone, dibenzoylmethane); L—L = bipy, phen, $(4\text{-}NH_2py)_2$; reflux in EtOH.

Scheme 84

A successful separation of the species $[Cr(NCS)_n(H_2O)_{6-n}]^{(3-n)+}$ ($n = 0$–6) including the *cis, trans* isomers of $[Cr(NCS)_2(H_2O)_4]^+$ and the *mer, fac* isomers of $Cr(NCS)_3(H_2O)_3$ has been accomplished by adsorption chromatography on Sephadex gel.[631] The complete series of anions $[Cr(CN)_{6-n}(NCS)_n]^{3-}$ ($n = 0$–6) has been prepared and separated by gel electrophoresis and by chromatography on alumina.[632] Ligand substitution in $[Cr(NCS)_6]^{3-}$ by CN^- in acetonitrile affords the *trans* ($n = 2$) and *mer* ($n = 3$) mixed ligand intermediates, while substitution in $[Cr(CN)_6]^{3-}$ by NCS^- in the same solvent leads to the *cis* and *fac* isomers (Scheme 85).[633] The position of the low energy ${}^4A_{2g}(F) \rightarrow {}^4T_{2g}(F)$ band undergoes a step-wise blue shift of about 30 nm on decreasing n (Figure 5) and log ε for the charge transfer band at ~310 nm increases almost linearly with n, the number of thiocyanate ligands. These observations suggest that no change in the mode of NCS^- coordination occurs throughout this series. The reaction of $[Cr(CN)(NCS)_5]^{3-}$ with en in acetonitrile affords the complexes *mer*-$[Cr(CN)(NCS)_3en]^-$ and *cis*-$[Cr(CN)(NCS)(en)_2]^+$.[634]

The solid state deamination of $[Cr(en)_3](SCN)_3$ on its own and in the presence of trace NH_4SCN catalyst, which functions by protonating and breaking loose one end of the coordinated amine in the rate-determining step, gives in a two-step process *trans*-$[Cr(NCS)_2(en)_2]SCN$.[635] At high catalyst levels the *cis* product is obtained. Other *N*-thiocyanatochromium(III) complexes prepared by solid state deaminations include *cis*-$[Cr(NCS)_2(tn)_2]SCN$ and *cis*-$[Cr(NCS)_2(pn)_2]SCN$,[636] while the *trans* isomer of the former was obtained by deamination of a suspension of $[Cr(tn)_3](SCN)_3$ in acetonitrile.[637] A number of *N*-thiocyanatoammine and amine chromium(III) complexes are referred to in Tables 46, 49 and 50. The complex $[\{Cr(NCS)(TPyEA)\}_2O](BPh_4)_2$ contains N-bonded thiocyanate (**130**).

(c) *Bridged NCS^- complexes.* Complexes containing both bridging and non-bridging NCS^- ligands have been prepared as outlined in Scheme 86.[638] Evidence for formation of bridged complexes is based mainly on the IR spectra which show CN stretching bands at ~2135 cm^{-1} and at ~2100 cm^{-1} corresponding to bridging and terminal *N*-thiocyanate ligands respectively. The homonuclear dimer is both photochemically and thermally the more stable of the two complexes. Evidence from IR and electronic spectroscopy points to the existence of bridging thiocyanate ligands in the salts $M_3[Cr(NCS)_6]$ ($M_3 = Cu_3, Ag_3, Cd_{3/2}, Hg_{3/2}, Tl_3, Pb_{3/2}$) and also in adducts formed between Reinecke's anion and a number of M^{2+} ions of the first row *d*-block

Table 67 Details of the IR and Electronic Spectra of Mixed Ligand *N*-Thiocyanatochromium(III) Complexes

Complex	IR (cm^{-1}) $\nu(CN)$	$\nu(CS)$	$\delta(NCS)$	$\nu(Cr{-}NCS)$	Medium	Electronic spectra: $\lambda_{max}(\varepsilon)$ (nm, $dm^3\ mol^{-1}\ cm^{-1}$) ${}^4A_{2g} \rightarrow {}^4T_{2g}$	${}^4A_{2g} \rightarrow {}^4T_{1g}$	CT or ligand bands	Medium	Ref.
trans-$NH_4[Cr(NCS)_4(NH_3)_2]$	2057vs	827m	467s	342s	[a]	522 (105)	396 (83)	310 (34 000)	EtOH	1
	2127s									
cis-$K[Cr(NCS)_4(py)_2]$	2074vs	840vw	484s	382s	[a]	542 (150)	426sh	328 (13 000)	EtOH	1
				362s						
				335s						
trans-$Me_3PH[Cr(NCS)_4(Me_3P)_2]$	2067vs	846m	480s	368s	[a]	501 (180)	—	385 (14 000)	EtOH	1
								315 (16 000)		
								238 (10 000)		
trans-$PhEt_2PH[Cr(NCS)_4(PhEt_2P)_2]$	2073vs	855[b]	484s	365s	[a]	525 (255)	—	392 (10 000)	MeOH	1
	2055vs							327 (20 000)		
trans-$K[Cr(NCS)_4(urea)_2]$	2092vs	823vw	—	—	KBr disc	560 (88)	415 (82)	220 (8000)	H_2O	2
trans-$K[Cr(NCS)_4(thiourea)_2]$	2105vs	702w	—	—	KBr disc	566 (80)	421 (77)	237 (17 000)	H_2O	2
	2060sh									
trans-$H[Cr(NCS)_4(bzim^c)_2]\cdot EtOH$	2080s	785w			KBr disc	540 (—)	406 (—)	—	95% EtOH	3
trans-$H[Cr(NCS)_4(2\text{-}Mebzim^d)_2]\cdot 2H_2O$	2080s	775w	—	—	KBr disc	565 (—)	419 (—)		95% EtOH	3
$K[Cr(NCS)_4bipy]$	2110vs	847vw	478s	382s	[a]	538 (137)	—	341 (9000)	MeOH	1
	2092vs			360s				331, 318, 306 (12 000)		
	2080vs									
$Ag[Cr(NCS)_4tn^e]$	2090vs	—	480m	355m	KBr disc	~540 (88)	~405 (77)	~320 (15 000)	EtOH or MeOH	4
$Et_4N[Cr(NCS)_4diars]$	2051vs	—	486s	358s	[a]	556 (—)	—	—	MeOH	1
	2060vs									
	2090vs									
$K_2[Cr(NCS)_4gly]$	2089vs	748m	—	—	KBr disc	545 (118)	410 (92)	—	H_2O	2
	2113sh									

Table 67 *(continued)*

Complex	IR (cm^{-1}) $\nu(CN)$	$\nu(CS)$	$\delta(NCS)$	$\nu(Cr—NCS)$	Medium	Electronic spectra: $\lambda_{max}(\varepsilon)$ (nm, $dm^3\ mol^{-1}\ cm^{-1}$) $^4A_{2g}\rightarrow{}^4T_{2g}$	$^4A_{2g}\rightarrow{}^4T_{1g}$	CT or ligand bands	Medium	Ref.
mer-$[Cr(NCS)_3py_3]$	2050vs	—	476s	374s	KBr disc	546 (121)	399 (93)	—	Acetone	5
fac-$[Cr(NCS)_3py_3]$	2082vs	—	478s	384s	KBr disc	545 (133)	398 (97)	—	Acetone	5
$[Cr(NCS)_3(Him)_3]\cdot H_2O$	2105vs	840–850s	490		KBr disc	520 (—)	388 (—)	—	—	6
$[Cr(NCS)_3(1\text{-}Meim^g)_3]_2\cdot H_2O$	2105vs	800vw	491s		KBr disc	532 (—)	396 (—)	—	—	6
$[Cr(NCS)_3(2\text{-}Meim^g)_3]\cdot H_2O$	2120vs	845–860s	487m	—	KBr disc	527 (—)	398 (—)	—	—	6
cis-α-$[Cr(NCS)_2trien]Cl$	2065vs	810w	—		KBr disc	488 (157)	372 (86.7)		0.1 M HCl	7
trans-$[Cr(NCS)_2(en)_2]SCN$	2080vs	—	—	—	Unstated	483 (93)	364 (67)	—	MeOH/H_2O(50:50)	8
cis-$[Cr(NCS)_2(en)_2]SCN$						485 (—)	373 (—)	—	—	8
trans-$[Cr(NCS)_2(tn)_2]SCN$	2050vs		480m	355m	KBr disc	~500 (92)	~375 (72)	~325 (12 000)	EtOH or MeOH	4
$[Cr(NCS)_2(bigH^h)_2]NCS$	2095s	770m	490m	350m	KBr disc	490 (—)	375 (—)	—	Nujol	9
cis-$[Cr(NCS)Cl(en)_2]Br\cdot H_2O$	2045				Nujol mull	506 (92.6)	385 (70.0)	—	0.8 M HCl	10
cis-$[Cr(NCS)F(en)_2]NCS$	—	—	—	—	—	500 (85)	375 (52)	—	H_2O	11
cis-$[Cr(NCS)(en)_2H_2O]^{2+}$						488 (104)	380 (59)	—	1.5 M $HClO_4$	12
trans-$[Cr(NCS)(en)_2H_2O]^{2+}$	—	—	—	—	—	527 (73)	380 (45)	—	1 M $HClO_4$	13

[a] NaCl, KBr or polythene mulls. [b] Assignment uncertain because of ligand absorption in this region. [c] bzim = benzimidazole. [d] 2-Mebzim = 2-methylbenzimidazole. [e] tn = 1,3-diaminopropane. [f] Him = imidazole. [g] Meim = methylimidazole. [h] bigH = biguanide.

1. M. A. Bennett, R. J. H. Clark and A. D. J. Goodwin, *Inorg. Chem.*, 1967, **6,** 1625.
2. G. Contreras and R. Schmidt, *J. Inorg. Nucl. Chem.*, 1970, **32,** 127.
3. S. P. Ghosh and A. Mishra, *J. Inorg. Nucl. Chem.*, 1971, **33,** 4199.
4. E. Blasius and E. Mernke, *Z. Anorg. Allg. Chem.*, 1984, **509,** 167.
5. S. DasSarma and B. DasSarma, *Inorg. Chem.*, 1979, **18,** 3618.
6. A. Mishra and J. P. Singh, *J. Inorg. Nucl. Chem.*, 1979, **41,** 905.
7. D. A. House and C. S. Garner, *J. Am. Chem. Soc.*, 1966, **88,** 2156.
8. W. A. Baker, Jr. and M. G. Phillips, *Inorg. Chem.*, 1966, **5,** 1042.
9. A. Syamal, *Z. Naturforsch., Teil B*, 1974, **29,** 492.
10. D. A. House and C. S. Garner, *J. Inorg. Nucl. Chem.*, 1966, **28,** 904.
11. J. W. Vaughn, O. J. Stvan, Jr. and V. E. Magnuson, *Inorg. Chem.*, 1968, **7,** 736.
12. C. Bifano and R. G. Linck, *Inorg. Chem.*, 1974, **13,** 609.
13. J. M. Veigel and C. S. Garner, *Inorg. Chem.*, 1965, **4,** 1569.

$$[Cr(NCS)_6]^{3-} \xrightarrow{CN^-} [CrCN(NCS)_5]^{3-} \xrightarrow{CN^-} trans\text{-}[Cr(CN)_2(NCS)_4]^{3-} \xrightarrow{CN^-} mer\text{-}[Cr(CN)_3(NCS)_3]^{3-} \xrightarrow{CN^-}$$

$$trans\text{-}[Cr(CN)_4(NCS)_2]^{3-} \xrightarrow{CN^-} [Cr(CN)_5NCS]^{3-} \xrightarrow{CN^-} [Cr(CN)_6]^{3-} \xrightarrow{SCN^-} fac\text{-}[Cr(CN)_3(NCS)_3]^{3-}$$

$$\xrightarrow{SCN^-} [Cr(CN)_5NCS]^{3-} \xrightarrow{SCN^-} cis\text{-}[Cr(CN)_4(NCS)_2]^{3-}$$

Scheme 85

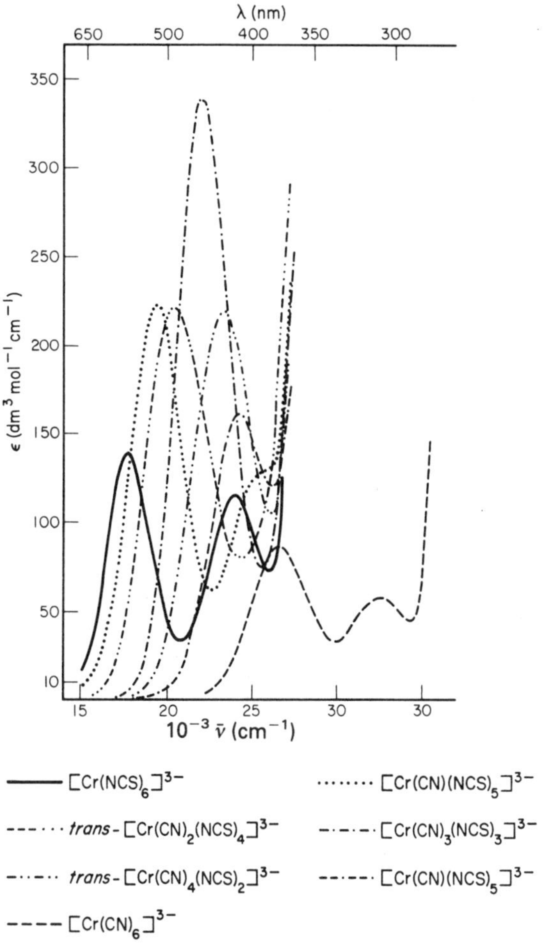

Figure 5 Electronic spectral changes (in aqueous solution) accompanying stepwise substitution in $[Cr(NCS)_6]^{3-}$ by CN^- (adapted from *Z. Anorg. Allg. Chem.*, 1975, **417,** 55)

elements.[621] The measured formation constant of $[(NH_3)_5Cr—NCS—Ag]^{3+}$ ($\log K = 5.11$, $I = 0.03$ M, $T = 298$ K) is comparable to the formation constant of AgSCN ($\log K = 4.8$) under the same conditions.[621]

(*d*) *NCS^- linkage isomers.* Although the thiocyanate ion is a potentially ambidentate ligand, examples of linkage isomerism in thiocyanatochromium(III) complexes are rare. This can be attributed to the class *a* nature of the metal ion and its almost exclusive preference for N over S donor atoms. One of the few known *S*-thiocyanatochromium(III) complexes, $[Cr(SCN)(H_2O)_5]^{2+}$, is obtained by the inner-sphere reductions of *trans*-$[CoX(NCS)(en)_2]^{n+}$ ($X = H_2O$, NH_3, Cl^-, SCN^-), $[Fe(NCS)(H_2O)_5]^{2+}$ (remote attack) or $[Co(SCN)(NH_3)_5]^{2+}$ (adjacent attack) by aqueous Cr^{2+}.[639,640] In aqueous solution $[Cr(SCN)(H_2O)_5]^{2+}$ undergoes (as well as aquation) spontaneous and metal-ion-catalyzed isomerization. Catalysis by Cr^{2+} occurs by a redox mechanism involving initial metal ion attack at the N atom of the S-coordinated ligand while catalysis by Hg^{2+} occurs by the non-redox pathway shown in Scheme 87.[639]

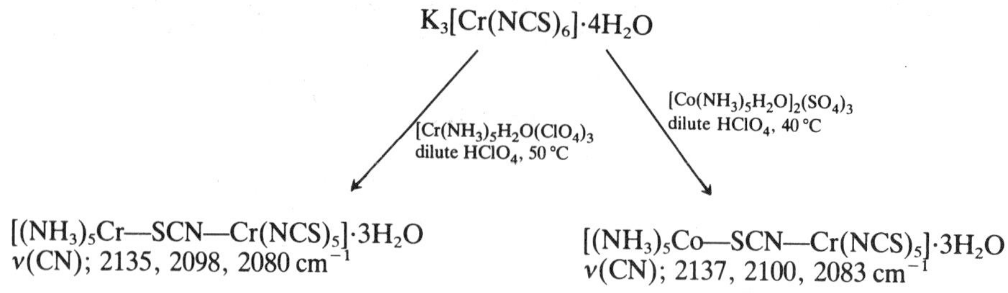

$$K_3[Cr(NCS)_6]\cdot 4H_2O$$

[Cr(NH₃)₅H₂O(ClO₄)₃ dilute HClO₄, 50 °C]   [Co(NH₃)₅H₂O]₂(SO₄)₃ dilute HClO₄, 40 °C

$[(NH_3)_5Cr\!-\!SCN\!-\!Cr(NCS)_5]\cdot 3H_2O$
$\nu(CN)$; 2135, 2098, 2080 cm^{-1}

$[(NH_3)_5Co\!-\!SCN\!-\!Cr(NCS)_5]\cdot 3H_2O$
$\nu(CN)$; 2137, 2100, 2083 cm^{-1}

Scheme 86

$$[Cr(SCN)(H_2O)_5]^{2+} + Hg^{2+} \rightleftharpoons [(H_2O)_5CrSCN]^{4+}$$ (with Hg bonded)

$$[HgCl]^+ + [Cr(NCS)(H_2O)_5]^{2+} \underset{Hg^{2+}}{\overset{Cl^-}{\rightleftharpoons}} [(H_2O)_5Cr\!-\!NCS\!-\!Hg]^{4+}$$

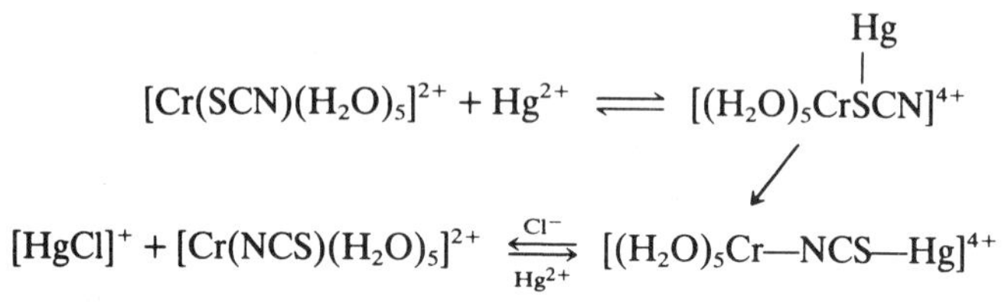

Scheme 87

The products of reduction of $Fe(NCS)^{2+}$ by Cr^{2+} in aqueous solution containing thiocyanate ions (equation 36), include the isomers of $[Cr(NCS)(SCN)(H_2O)_4]^+$, which contains both N- and S-bonded thiocyanate ligands.[641] These isomers undergo spontaneous decomposition by parallel aquation (loss of the S-thiocyanato ligand) and isomerization ($Cr\!-\!SCN \rightarrow Cr\!-\!NCS$) reactions. Details of the solution visible spectra of these isomers as well as those of similar chromium(III) complexes for comparison are listed in Table 68. From the compiled data it is apparent that the S-bonded thiocyanate ligand lies very close to Br^- in the spectrochemical series while the N-bonded form lies between N_3^- and NO.

$$Fe(NCS)^{2+} + Cr^{2+} \xrightarrow{H^+} Cr^{3+} + Cr(NCS)^{2+} + Cr(SCN)^{2+} + cis, trans\text{-}Cr(NCS)_2^+$$

$$+ cis, trans\text{-}[Cr(NCS)SCN]^+ + Fe^{2+} \quad (36)$$

Table 68 Visible Spectra of Isomeric Thiocyanato and other Chromium(III) Complexes[a]

Complex	λ_{max} (nm)(ε in dm³ mol^{-1} cm^{-1})			*Ref.*
$[Cr(NCS)(H_2O)_5]^{2+}$	570 (31.4)	410 (33.5)		1
$[Cr(SCN)(H_2O)_5]^{2+}$	620 (26)	435 (20)	262 (8000)	2[b]
$[CrI(H_2O)_5]^{2+}$	650 (36.1)	474 (32.6)		1
$[CrBr(H_2O)_5]^{2+}$	622 (19.9)	432 (22.4)		1
$[CrCl(H_2O)_5]^{2+}$	609 (16.4)	428 (20.8)		1
$[CrF(H_2O)_5]^{2+}$	595 (12.2)	417 (11.8)		1
$[CrN_3(H_2O)_5]^{2+}$	585 (67.5)	434 (66.4)		1
$[Cr(NC)(H_2O)_5]^{2+}$	560–545[a]			1
$[CrNO(H_2O)_5]^{2+}$	559 (28)	449 (120)		1
$[Cr(H_2O)_5NH_3]^{2+}$	545 (22.1)	397 (21.8)		1
cis-$[Cr(NCS)_2(H_2O)_4]^+$	562 (~50)	420 (~54)		3
$trans$-$[Cr(NCS)_2(H_2O)_4]^+$	575 (~47)	425 (~47)		3
$[Cr(NCS)(SCN)(H_2O)_4]^{+ c}$	605 (55)	440 (43)[a]		4

[a] In dilute HClO₄. [b] Note that values of ε for this complex in ref. 1 have been corrected in ref. 2. [c] An undetermined composition of *cis*, *trans* isomers.

1. A. Haim and N. Sutin, *J. Am. Chem. Soc.*, 1966, **88**, 434.
2. M. Orhanovic and N. Sutin, *J. Am. Chem. Soc.*, 1968, **90**, 4286.
3. J. T. Hougen, K. Schug and E. L. King, *J. Am. Chem. Soc.*, 1957, **79**, 519.
4. L. D. Brown and D. E. Pennington, *Inorg. Chem.*, 1971, **10**, 2117.

(ii) N-cyanates and N-selenocyanates

Compared with *N*-thiocyanatochromium(III) complexes, those containing *N*-cyanato and *N*-selenocyanato ligands are relatively few in number. Some crystalline salts $M_3[Cr(NCSe)_6]\cdot nS$ (M = Na, K) have been prepared by heating acetone or dioxane (S) solution mixtures containing MSeCN and $CrCl_3$, followed by slow evaporation of solvent.[642] The purple-coloured complex $(Me_4N)_3[Cr(NCSe)_6]$ was obtained by addition of Me_4NCl to a cooled solution containing $CrCl_3$ and a ten-fold excess of KSeCN which had been refluxed for four hours.[642] The green-coloured complex $(Ph_4As)_3[Cr(NCO)_6]$ was obtained by the addition of AgCNO to a solution of $CrCl_3$ and Ph_4AsCl in acetone containing a trace of zinc dust.[643] Details of the IR and electronic spectra of these complexes together with those of $[Cr(NCS)_6]^{3-}$ for comparative purposes are listed in Table 69. The suggested presence of *N*-selenocyanato ligands in the anion $[Cr(NCSe)_6]^{3-}$ is based on these data. Hence the appearance of ν(C—Se) in the region 600–700 cm^{-1} indicates N-coordination of this ligand to the metal. In selenium-bonded selenocyanato complexes the ν(CN) and ν(C—Se) bands are generally observed above 2080 cm^{-1} and in the region 500–600 cm^{-1} respectively. The almost identical values to 10 *Dq* for $[Cr(NCS)_6]^{3-}$ (18 080 cm^{-1}), which is known to contain N-bonded ligands, and $[Cr(NCSe)_6]^{3-}$ (17 920 cm^{-1}) further suggest N-bonding by the ligand in the latter species. The situation regarding $[Cr(NCO)_6]^{3-}$ however is less clear. While the ν(CO) band at 1336 cm^{-1} in the IR spectrum of the complex is in the range expected for *N*-cyanato ligands[644] and the strong band at 345 cm^{-1} has been assigned to a Cr—N stretching vibration, the very low value of 10 *Dq* (16 390 cm^{-1}) is more typical of a $Cr^{III}O_6$ rather than a $Cr^{III}N_6$ chromophore. Spectroscopic evidence for the existence of *N*-cyanato ligands in the complex *mer*-$Cr(NCO)_3(py)_3$ is somewhat more convincing.[645] While the ν(CO) and ν(CrN) bands in the IR spectrum of this complex occur at positions close to those reported for $[Cr(NCO)_6]^{3-}$, its visible spectrum is quite similar to that of *mer*-$[Cr(NCS)_3(py)_3]$, from which it is prepared by oxidation with BrO_3^- (equation 37).

$$mer\text{-}[Cr(NCS)_3(py)_3] + 4BrO_3^- \xrightarrow{py} mer\text{-}[Cr(NCO)_3(py)_3] + 4Br^- + 3SO_3 \qquad (37)$$

$$\begin{array}{lll} \lambda_{max}: & 546\ nm\ (\varepsilon = 121\ dm^3\ mol^{-1}\ cm^{-1}) & 556\ nm\ (110) \\ & 399\ nm\ (\varepsilon = 93\ dm^3\ mol^{-1}\ cm^{-1}) & 393\ nm\ (83) \end{array}$$

Table 69 Main Bands in the IR and Electronic Spectra of $[Cr(NCX)_6]^{3-}$ Complexes

Complex	IR $\nu(CN)$	$\delta(NCX)$	$\nu(Cr—N)$ (cm^{-1})	$\nu(CX)$	Electronic $^4A_{2g}\rightarrow{}^4T_{2g}$	$^4A_{2g}\rightarrow{}^4T_{1g}(F)$ (nm)	CT	Ref.
$[Cr(NCS)_6]^{3-}$	2078s	483m	364s	820w[a]	553	409	309[b]	1–3
$[Cr(NCSe)_6]^{3-}$	2067s	420m	—	663m[c]	556[d]	—	342[e]	4, 5
$[Cr(NCO)_6]^{3-}$	2205s	619m 601m	345s	1335m[f]	610	436[g]	—	6

[a] Et_4N^+ salt as KBr disc or Nujol mull. [b] In MeOH. [c] Me_4N^+ salt in Nujol. [d] In acetone. [e] In DMSO. [f] Ph_4As^+ salt in KBr disc. [g] Reflectance.

1. M. A. Bennett, R. J. H. Clark and A. D. J. Goodwin, *Inorg. Chem.*, 1967, **6**, 1625.
2. A. Sabatini and I. Bertini, *Inorg. Chem.*, 1965, **4**, 959.
3. V. Alexander, J. L. Hoppé and M. A. Malati, *Polyhedron*, 1982, **1**, 191.
4. A. I. Brusilovets, V. V. Skopenko and G. V. Tsintsadze, *Russ. J. Inorg. Chem. (Engl. Transl.)*, 1969, **14**, 239.
5. K. Michelsen, *Acta Chem. Scand.*, 1963, **17**, 1811.
6. R. A. Bailey and T. W. Michelson, *J. Inorg. Nucl. Chem.*, 1972, **34**, 2935.

The reaction of $K_3[Cr(NCX)_6]$ or $(Bu^n_4N)_3[Cr(NCX)_6]$ (X = O, Se) with chelating diamine ligands in acetonitrile solution followed by column chromatography on alumina gives salts of the complexes $[Cr(NCX)_4AA]^-$ and *cis,trans*-$[Cr(NCX)_2(AA)_2]^+$ (AA = en, pn, chxn) all of which have been characterized by electronic and IR spectroscopy.[646] Details of the preparations and spectra of $[Cr(NCX)(NH_3)_5]^{2+}$ (X = O, S, Se) complexes are presented in Table 70.

(iii) Azides

The reaction of $[Cr(NH_3)_5H_2O](ClO_4)_3$ and NaN_3 in dilute acetic acid solution at 60 °C affords the monoazido complex $[CrN_3(NH_3)_5](ClO_4)_2$.[647] The addition of less than one equivalent of KN_3 to a solution of chromium(III) in molten KCNS at 185 °C produces a bathochromic shift in the visible spectrum consistent with the formation of the thermally stable complex $[Cr(NCS)_5N_3]^{3-}$.[648] The neutral triazido complexes $Cr(N_3)_3(NH_3)_3$ and $Cr(N_3)_3(py)_3$

Table 70 Preparations and Spectra of $[Cr(NCX)(NH_3)_5]^{2+}$ (X = O, S, Se) Complexes

Complex	*Preparation*	*Electronic spectrum*[a] λ_{max} (nm) (ε in $dm^3\,mol^{-1}\,cm^{-1}$)		*IR spectrum*[b]	*Ref.*
$[Cr(NCO)(NH_3)_5](NO_3)_2$	$Cr(NO_3)_3{\cdot}10H_2O$ + urea at 170 °C	493 (63)	366 (38)	$\nu(CN) = 2240\ cm^{-1}$	1
$[Cr(NCS)(NH_3)_5](ClO_4)_2$	$[Cr(NH_3)_5H_2O](NO_3)_2{\cdot}NH_4NO_3$ + NaSCN in H_2O at 60–65 °C; + $NaClO_4$	491 (83.1)	370sh (52.5)	$\nu(CN) = 2084\ cm^{-1}$	2, 3
$[Cr(NCSe)(NH_3)_5](ClO_4)_2$	$[Cr(NH_3)_5H_2O](ClO_4)_3$ + NaSeCN in DMSO at 80 °C	476 (67.6)	—	$\nu(CN) = 2075\ cm^{-1}$	4

[a] In H_2O. [b] KBr disc. sh = shoulder.

1. H. H. Schmidtke and T. Schönherr, *Z. Anorg. Allg. Chem.*, 1978, **443**, 225.
2. E. Zinato, R. Lindholm and A. W. Adamson, *J. Inorg. Nucl. Chem.*, 1969, **31**, 449.
3. C. S. Garner and D. A. House, *Transition Met. Chem. (N.Y.)*, 1970, **6**, 59.
4. N. V. Duffy and F. G. Kosel, *Inorg. Nucl. Chem. Lett.*, 1969, **5**, 519.

have been prepared, the first by refluxing an aqueous solution containing $[Cr(NH_3)_6](ClO_4)_3$ and NaN_3 and the second by the addition of pyridine to a hot ethanolic solution containing $Cr(NO_3)_3{\cdot}xH_2O$ and NaN_3.[302] The green crystals of the latter complex are only slightly water soluble and should be treated with caution because of their potential explosive nature. When an aqueous solution mixture of $[Cr(en)_3]Cl_3$ and NaN_3 is heated the colour changes to rose-red and the salts *cis*-$[Cr(N_3)_2(en)_2]Cl$ and *cis*-$[Cr(N_3)_2(en)_2]N_3$ are coprecipitated from solution on cooling.[302] The brick-red complex *cis*-α-$[Cr(N_3)_2(trien)]Br$ has been prepared by heating an aqueous solution of *cis*-α-$[Cr(Cl)_2(trien)]Cl{\cdot}H_2O$ in the presence of excess NaN_3 and then adding KBr.[649] The configuration of the complex was established on the basis of its IR spectrum. Hence in the asymmetric NH_2 bending region (1560–1660 cm^{-1}) it shows a split absorption, whereas in the CH_2 rocking region (860–940 cm^{-1}) it shows two bands of medium intensity, features which distinguish the *cis*-α from the *cis*-β and *trans* configurations.[649]

The violet complex $(Bu^n_4N)_3[Cr(N_3)]_6$ (μ_{eff} at 295 K = 3.76 BM[650]), in which a large counterion is employed to stabilize the $[Cr(N_3)_6]^{3-}$ anion, was obtained by the addition of Bu^n_4NCl to the filtrate from a reaction suspension containing $CrCl_3{\cdot}6H_2O$ and NaN_3 which had been heated at 50–60 °C for one hour.[651] Evidence from IR spectroscopy points to the fact that $[Cr(N_3)_6]^{3-}$ contains bent M—N—N groups. Comparison of band positions in the electronic spectrum of the complex with those for other hexacoordinated chromium(III) complexes indicates that the azide ligand occupies a position slightly lower than F^- but higher than $(Et_2O)PS_2^-$ in the spectrochemical series and roughly the same position as I^- in the nephelauxetic series.[651]

Electronic spectral data corresponding to the ${}^4A_{2g} \rightarrow {}^4T_{2g}(F)$ and ${}^4A_{2g} \rightarrow {}^4T_{1g}(F)$ electronic transitions in a number of azido–chromium(III) complexes are compiled in Table 71.

Table 71 Electronic and IR Spectroscopic Data for Some Azidochromium(III) Complexes

Complex	$\nu_{as}(N_3^-)$ (cm^{-1})	λ_{max} (nm) (ε in $dm^3\,mol^{-1}\,cm^{-1}$)	λ_{min} (ε)	*Medium*	*Ref.*
$[CrN_3(NH_3)_5]^{2+}$	2094	500 (154), 382 (93.4)		H_2O	1, 2
cis-$[CrN_3(en)_2(H_2O)]^{2+}$		502 (166), 366 (99)		2 M $HClO_4$	3
		515 (186), 399 (121)	446	0.01 M $HClO_4$	3
$[CrN_3(trienH)(H_2O)_2]^{3+}$		509 (154), 392 (93)	438 (60)	1.5 M $HClO_4$	2
$[CrN_3(edta)]^{2-}$		551 (213), 407 (128)		1 M $NaClO_4$	4
$[CrN_3(H_2O)_5]^{2+}$		585 (67.5), 434 (66.4)		1 M $HClO_4$	5
cis-$[Cr(N_3)_2(NH_3)_4]^+$		514 (195), 397 (129)		0.01 M $HClO_4$	2
		515 (224), 398 (148)	447 (68.5)	H_2O	2
cis-$[Cr(N_3)_2(en)_2]^+$		514 (195), 397 (129)		0.01 M $HClO_4$	2
		515 (224), 398 (148)	447 (68.5)	H_2O	2
cis-α-$[Cr(N_3)_2(trien)]^+$	2090, 2055	518 (185), 402 (126)	452 (70.0)	0.1 M HCl	2
		520 (220), 400 (135)	450 (70)	1 M $NaClO_4$	2
cis-$[Cr(N_3)_2(H_2O)_4]^+$		603 (87.5), 443 (87.2)		1 M $HClO_4$	5
$[Cr(N_3)_6]^{3-}$	2102, 2048	667 (204), 502 (170)		CH_2Cl_2	6

1. R. Davies and R. B. Jordan, *Inorg. Chem.*, 1971, **10**, 1102.
2. C. S. Garner and D. A. House, in 'Transition Metal Chemistry', ed. R. L. Carlin, Dekker, New York, 1970, vol. 6, p. 59.
3. I. J. Kindred and D. A. House, *J. Inorg. Nucl. Chem.*, 1975, **37**, 1320.
4. H. Ogino, T. Watanabe and N. Tanaka, *Inorg. Chem.*, 1975, **14**, 2093.
5. T. W. Swaddle and E. L. King, *Inorg. Chem.*, 1964, **3**, 234.
6. W. Beck, W. P. Fehlhammer, P. Pöllmann, E. Schuierer and K. Feldl, *Chem. Ber.*, 1967, **100**, 2335.

The photochemical conversion of azidochromium(III) to nitridochromium(V) complexes has in recent years attracted considerable interest.[652] The complexes $[CrN_3(edta)]^{2-}$, $[CrN_3(nta)(H_2O)]^-$, $[Cr(N_3)_2(nta)]^{2-}$, $[CrN_3(VDA)(H_2O)]^-$ and $[Cr(N_3)_2(VDA)]^{2-}$ (VDA = *N,N'*-valinediacetate) when exposed to UV radiation liberated N_2 and gave products having ESR spectra (narrow line, no fine structure) characteristic of species with one unpaired electron and visible spectra similar to those of oxochromium(V) complexes.[652] It was therefore concluded that the products were nitridochromium(V) species. UV irradiation of $CrN_3(salen)(H_2O)$[653] and $Cr(TPP)N_3{\cdot}nH_2O$[654] (Scheme 88) also give the corresponding nitridochromium(V) complexes. The evidence for formation of an intermediate nitrene complex $[CrN(NH_3)_5]^{2+}$ in the 313 nm photolysis of $[CrN_3(NH_3)_5]^{2+}$ has been disputed and the results reinterpreted to indicate formation of $[Cr^V(N^{3-})(NH_3)_5]^{2+}$, which undergoes rapid aquation to the nitridopentaaquachromium(V) cation.[653]

$$Cr(TPP)OH{\cdot}nH_2O \xrightarrow{HN_3} Cr(TPP)N_3{\cdot}nH_2O \xrightarrow{h\nu} [Cr^V(TPP)(N^{3-})]$$

Scheme 88

The kinetics of aquation of a number of azidochromium(III) complexes have been investigated.[303,655] Compared with other acidochromium(III) complexes, the chromium–azide bonds in these species seem remarkably stable to thermal substitution. Hence in the base hydrolysis of $[CrN_3(NH_3)_5]^{2+}$ a pathway involving initial loss of NH_3 concurs with the usual base hydrolysis pathway involving loss of N_3^-. The aquation of azidochromium(III) complexes is H^+-assisted with protonation of the azido ligand accounting for the enhanced reactivity.

While the action of oxidants on coordinated azide is generally quite complicated, the peroxymonosulfate ion HSO_5^- reacts cleanly with $[CrN_3(NH_3)_5]^{2+}$ in aqueous solution according to equation (38).[656] Tracer experiments using HSO_5^- enriched with ^{18}O at either the terminal peroxide position or at all other positions indicate that the oxygen transferred from oxidant to reductant in reaction (38) is the terminal peroxide of HSO_5^-. Peroxymonosulfate also reacts cleanly with uncoordinated HN_3 and N_3^- (to give N_2O and N_2) and the observed reactivity follows the order $HN_3 \ll [CrN_3(NH_3)_5]^{2+} < N_3^-$. That HSO_5^- is a more effective oxidant than SO_5^{2-} is consistent with a reaction mechanism which involves nucleophilic attack on the peroxide oxygen.

$$[CrN_3(NH_3)_5]^{2+} + HSO_5^- \longrightarrow [CrNO(NH_3)_5]^{2+} + HSO_4^- + N_2 \quad (38)$$

(iv) Triazenides

The distorted octahedral complex $Cr(PhN_3Ph)_3$ (**146**) which contains bidentate triazenido ligands is a major product of the reaction between $[Li(THF)]_4[Cr_2Me_8]$ and diphenyltriazine in THF under argon, equation (39).[178] The formation of this product under the stated reaction conditions seems surprising in view of the expected weak oxidizing capability of the triazene ligand and the severe angle strain in the four-membered chelate rings of the product. In contrast the minor product (**145**) which contains bridging triazenido ligands is virtually strain free.

$$[Li(THF)]_4[Cr_2Me_8] \xrightarrow[THF]{PhN_3HPh} Cr_2(PhN_3Ph)_4 + \quad (39)$$

(**145**)

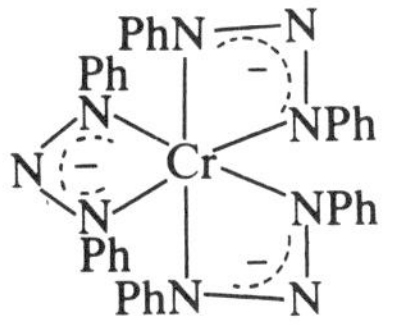

(**146**)

35.4.2.9 Polypyrazolylborates

Despite the vast interest shown by coordination chemists in polypyrazolylborate ligands (**17**) since their discovery in 1966 few complexes of chromium(III) with these ligands have been mentioned in the literature. The complex $[\{B(pz)_4\}_2Cr]PF_6$ (**147**) may be prepared by either of the routes outlined in Scheme 89.[657]

$$Et_4N[B(pz)_4CrCO_3] \xrightarrow[\text{in MeCN}]{C_3H_5Br} B(pz)_4Cr(CO)_2(\pi\text{-}C_3H_5)$$

i, Cr(CO)$_6$ in DMF, 100 °C,
ii, 1 M Et$_4$NCl

heat in C$_6$H$_6$

KB(pz)$_4$

yellow-green crystals

i, CrCl$_3$, Zn dust in DMF
ii, NH$_4$PF$_6$

i, heat in CHCl$_3$
ii, NH$_4$PF$_6$

$$\left[\left\{\left(\underset{N-N}{\bigcirc}-B\left(\underset{N-N}{\bigcirc}\right)_3\right)\right\}_2Cr\right]PF_6$$

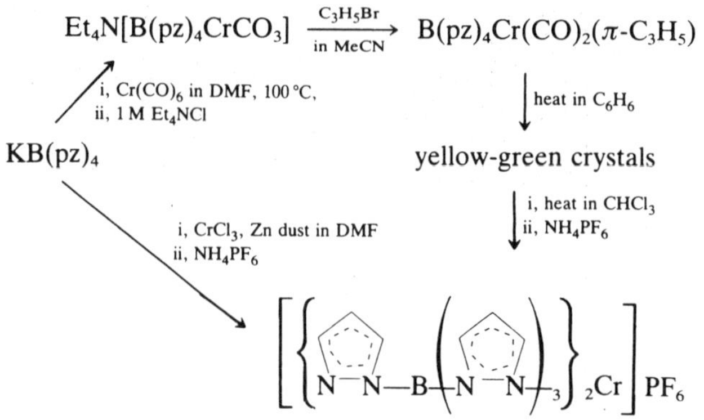

yellow needles: m.p. 358–359 °C (dec.)

(**147**)

Scheme 89

Although some polypyrazolylborate complexes of chromium(II) have been reported to undergo oxidation on exposure to air,[115] the resulting chromium(III) complexes have not been investigated.

35.4.2.10 Nitriles

Chromium metal may be oxidized by Cl_2, Br_2 or I_2 in ethanenitrile solution to give the octahedral complexes $[CrCl_3(MeCN)_3]\cdot MeCN$, $CrBr_3(MeCN)_3$ and $[CrI_2(MeCN)_4]I$ although the nitrile content of the bromo complex is somewhat variable.[658] The chloro complex loses the uncoordinated nitrile when heated at 80 °C under vacuum. Some properties of these complexes are listed in Table 72. Reaction of $CrCl_3$ with propanenitrile and acrylonitrile affords the dark purple complexes $CrCl_3(nitrile)_3$.[659]

Table 72 Properties of $CrX_3(MeCN)_y$ Complexes[1]

Complex	Colour	μ_{eff} (BM)[a]	IR spectrum (cm^{-1}) $\nu(CN_{ligand})$	$\nu(CN_{free})$	Reflectance spectrum (cm^{-1})
$[CrCl_3(MeCN)_3]\cdot MeCN$	Grey	3.90	2285 2305	2255[b]	15 400 20 200 22 200(sh)[b]
$[CrBr_3(MeCN)_3]$	Green	—	2285 2305	—	16 600
$[CrI_2(MeCN)_4]I$	Red-brown	3.78	2285 2300	—	16 000

[a] At room temperature. [b] Disappears on heating the complex at 80 °C under vacuum.
1. B. J. Hathaway and D. G. Holah, *J. Chem. Soc.*, 1965, 537.

The complex $[Cr(MeCN)(H_2O)_5]^{3+}$ may be prepared in solution either by the reaction of iodoethanenitrile and chromium(II) in aqueous acid (equation 40) or by the reduction of $[Co(NH_3)_4(H_2O)_2]^{3+}$ with chromium(II) in 50% aqueous ethanenitrile.[660] The product, which is purified by cation exchange chromatography, undergoes aquation with loss of the nitrile ligand at a rate which is independent of $[H^+]$ and which is some 2×10^3 faster than aquation of the corresponding cobalt(III) complex. Details of the electronic spectrum of $[Cr(MeCN)(H_2O)_5]^{3+}$ are given in Table 73. The reduction of $[Co(NH_3)_5(4\text{-}CNpy)]^{3+}$ by chromium(II) in acid solution occurs by an inner-sphere mechanism, giving the chromium(III) complex (**148**), which undergoes Cr^{2+}-catalyzed linkage isomerization to (**149**; Scheme 90).[661]

$$2[Cr(H_2O)_6]^{2+} + ICH_2CN \longrightarrow [CrI(H_2O)_5]^{2+} + [CrCH_2CN(H_2O)_5]^{2+} + [Cr(MeCN)(H_2O)_5]^{3+} \quad (40)$$

100%   25%   75%

Table 73 Properties of Ethanenitrilechromium(III) Complexes

Complex	Colour	IR $\nu(CN)$ (cm^{-1})	IR *Others* (cm^{-1})	*Electronic spectra* $^4A_{2g} \rightarrow {}^4T_{2g}$ λ_{max} (nm) (ε in $dm^3\ mol^{-1}\ cm^{-1}$)	$^4A_{2g} \rightarrow {}^4T_{1g}$ (*F*)	*CT*		*Ref.*
$[Cr(MeCN)(H_2O)_5]^{3+}$		—	—	566 (15.7)	397 (21.2)	—	($HClO_4/NaClO_4$ solution)	1
$CrCl_3$(bipy)MeCN	Green	2285	2312[a] 365, 353, 337[b]	652 (28) 633	444 (86) 450sh, 420, 398	—	(MeCN solution) (Reflectance)	2
(Hquin)[c]$[CrCl_4(MeCN)_2]\cdot$MeCN	Dark purple	2292 2247	2322[a] 340[b]	690	495	317, 310, 279, 224	(Reflectance)	3
$NEt_4[CrCl_4(MeCN)_2]$	Grey	2302	2334[a] 342, 336[b]	685	480	310, 278, 222	(Reflectance)	3
$Hpy_2[CrCl_5(MeCN)]$	Purple	2294	2320[a] 331, 322, 272[b]	741	521	317, 290, 263, 230	(Reflectance)	3
$[CrNO(MeCN)_5](PF_6)_2$[d]	Yellow	2327 2305	1796[e]		—			4
$[CrNS(MeCN)_5](PF_6)_2$[d]	Blue	2325 2305	1245[f]		—			5

[a] $\delta(CH_3) + \nu(C—C)$. [b] $\nu(Cr—Cl)\cdot$. [c] Hquin = quinolinium. [d] Oxidation state of metal ambiguous. [e] $\nu(NO)$. [f] $\nu(NS)$.

1. D. Jordan, W. C. Kupferschmidt and R. B. Jordan, *Inorg. Chem.*, 1984, **23,** 1986.
2. D. A. Edwards and S. C. Jennison, *Transition Met. Chem.*, 1981, **6,** 235.
3. K. R. Seddon and V. H. Thomas, *J. Chem. Soc., Dalton Trans.*, 1977, 2195.
4. S. Clamp, N. G. Connelly, G. E. Taylor and T. S. Louttit, *J. Chem. Soc., Dalton Trans.*, 1980, 2162.
5. M. Herberhold and L. Haumaier, *Z. Naturforsch., Teil B*, 1980, **35,** 1277.

$$[(NH_3)_5Co-N\bigcirc-CN]^{3+} \xrightarrow{Cr^{2+}} [(H_2O)_5CrNC-\bigcirc N]^{3+} \underset{}{\overset{H^+}{\rightleftharpoons}} [(H_2O)_5CrNC-\bigcirc NH]^{4+}$$

(148)

$$[(H_2O)_5Cr-N\bigcirc-CN]^{3+}$$

(149)

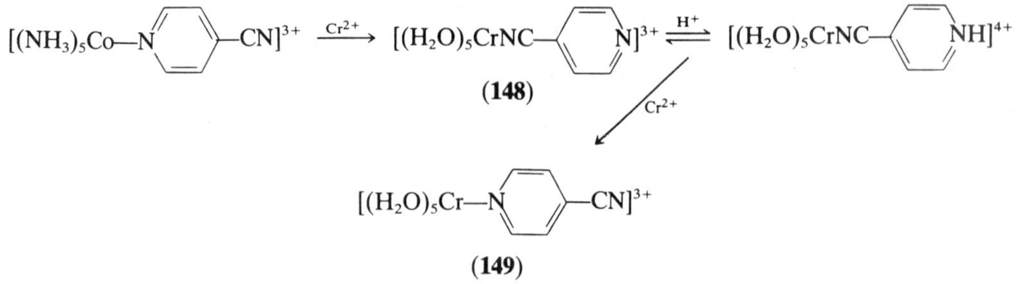

Scheme 90

The ability of various organic solvents and reagents to reduce tetrachlorooxochromate(V) salts is well known. Hence when a solution of (bipyH$_2$)[CrOCl$_5$] (Table 109) in ethanenitrile is refluxed, the colour changes from red-brown to green and crystalline CrCl$_3$(bipy)MeCN ($\mu_{eff} = 3.83$ BM at 293 K) can be isolated.[662] Some reactions of this complex, the stereochemistry of which could not be established with the available spectroscopic data, are outlined in Scheme 91.

$$
\begin{array}{c}
\text{CrCl}_3\text{(bipy)PPh}_3\text{O} \\
\nearrow \qquad \uparrow \qquad \text{CrOCl}_3\text{(bipy)} \\
\text{Ph}_3\text{PO in} \qquad \text{PPh}_3 \text{ in CHCl}_3, \quad \text{MeCN} \nearrow \\
\text{MeCN, boil} \qquad \text{heat} \\
\text{(bipyH}_2)[\text{CrOCl}_5] \xrightarrow{\text{MeCN}} \text{CrCl}_3\text{(bipy)MeCN} \\
\downarrow\text{MeNO}_2, \quad \text{heat} \quad \downarrow \text{py in CHCl}_3 \qquad \text{pyO in CHCl}_3 \\
\text{warm} \quad \text{in vacuum,} \qquad \searrow \\
270\,^{\circ}\text{C} \\
[\text{CrCl}_3\text{(bipy)}]_n \xleftarrow{200\,^{\circ}\text{C}} \text{CrCl}_3\text{(bipy)py} \qquad \text{CrCl}_3\text{(bipy)pyO}
\end{array}
$$

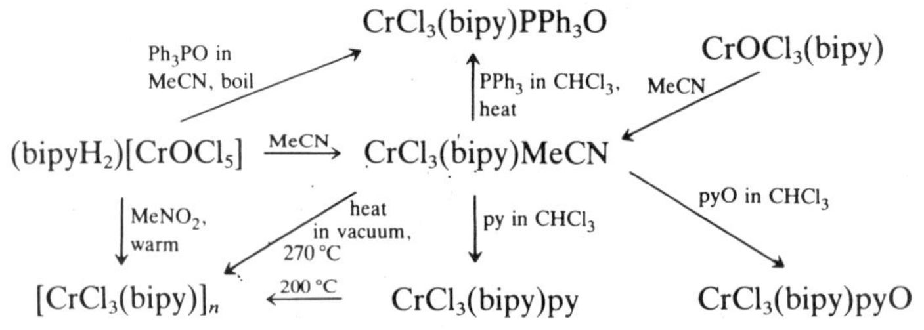

Scheme 91

The complexes *cis*-A[CrCl$_4$(MeCN)$_2$] and A$_2$[CrCl$_5$(MeCN)] (A = quinolinium, NEt$_4^+$ or pyridinium) have been prepared by heating the corresponding ethanoic acid complexes (rare examples of coordinated MeCO$_2$H) in ethanenitrile (Scheme 92).[663]

$$
\begin{array}{c}
\text{CrO}_3 + \text{ACl} \\
\Big| \text{MeCO}_2\text{H} \\
\text{A = quinolinium} \qquad\qquad \text{A = pyridinium} \\
\text{or NEt}_4^+ \qquad\qquad\qquad\qquad \\
\swarrow \qquad\qquad\qquad\qquad \searrow \\
\textit{cis-}\text{A[CrCl}_4(\text{MeCO}_2\text{H})_2] \qquad\qquad \text{A}_2[\text{CrCl}_5(\text{MeCO}_2\text{H})] \\
\downarrow \text{MeCN, reflux} \qquad\qquad \downarrow \text{MeCN, reflux} \\
\textit{cis-}\text{A[CrCl}_4(\text{MeCN})_2] \qquad\qquad \text{A}_2[\text{CrCl}_5(\text{MeCN})]
\end{array}
$$

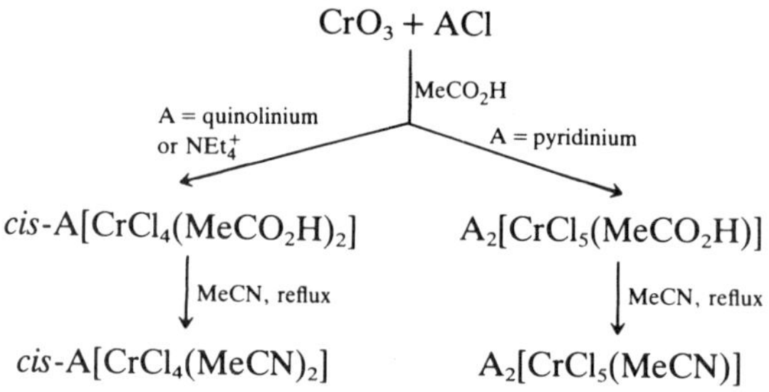

Scheme 92

Reaction of tricarbonyltoluenechromium with tetracyanoquinodimethane (TCNQ; **150**) in ethanenitrile gives the dark purple complex Cr(TCNQ)$_2$(MeCN)$_2$, which contains chromium(III), the radical anion TCNQ$^{\bar{\cdot}}$ and TCNQ^{2-}.[664] The effective magnetic moment of 4.08 BM at 298 K is close to the predicted value of 4.24 BM for a molecule which contains two magnetic centres, one having three unpaired electrons (metal ion) and the other having one unpaired electron (radical anion).[665] The magnetic moment decreases on lowering the temperature, indicative of antiferromagnetic coupling.

$$
\begin{array}{c}
\text{NC} \qquad\qquad\qquad \text{CN} \\
\diagdown \qquad\qquad\qquad \diagup \\
\bigcirc \\
\diagup \qquad\qquad\qquad \diagdown \\
\text{NC} \qquad\qquad\qquad \text{CN}
\end{array}
$$

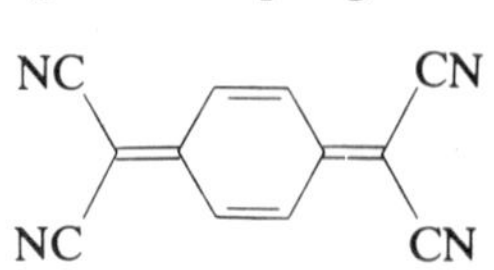

(150)

The complexes [CrNO(MeCN)$_5$]X$_2$ (X = PF$_6$, BF$_4$) and [CrNS(MeCN)$_5$](PF$_6$)$_2$, formed by oxidative nitrosylation of $Cr(CO)_6$ and thionitrosylation of $C_6H_6Cr(CO)_3$ respectively in ethanenitrile solution are, because of the lability of the nitrile ligands, useful starting materials in the synthesis of other $CrNO^{n+}$-containing complexes.[582,603] Reactions of these complexes are dealt with in Section 35.4.2.6.

The colours and spectroscopic properties of a number of ethanenitrilechromium(III) complexes dealt with in this section are listed in Table 73.

35.4.2.11 Oximes, biguanides, guanidines, N-coordinated ureas and amides

(i) Oximes

Pyridine-2-aldoxime (Hpox) forms complexes $[Cr(Hpox)_3]Cl_3 \cdot 3H_2O$ (**151**), $[Cr(Hpox)_2X_2]ClO_4$ (X = F, Cl, Br) and $[Cr(Hpox)_2(H_2O)_2]Br_3 \cdot H_2O$ with chromium(III) in which it acts as a bidentate 2-N donor ligand.[666] The first of these may be obtained by simply heating a solution of $[Cr(urea)_6]Cl_3 \cdot 3H_2O$ with an excess of Hpox in ethanol, while the others are conveniently prepared as in Scheme 93. The dihalogeno and diaqua complexes have X-ray powder patterns almost identical to *trans*-$Fe(py)_4(NCS)_2$ and by analogy may be assigned *trans* octahedral configurations. Chromium(III) complexes of 6-methylpyridine-2-aldoxime have also been prepared as in Scheme 93.[667] The positions of bands in the reflectance spectra of some of these aldoxime complexes are given in Table 74. While the spectrum of $[Cr(Hpox)_3]Cl_3 \cdot 3H_2O$ shows bands at 22 300 and 28 820 cm^{-1} corresponding to the ${}^4A_{2g} \rightarrow {}^4T_{2g}$ and ${}^4A_{2g} \rightarrow {}^4T_{1g}(F)$ transitions respectively, in the dihalogeno and diaqua complexes each of these bands is resolved into two components because of splitting of the ${}^4T_{2g}$ and ${}^4T_{1g}$ terms in the fields of lower symmetry. The IR spectra of the complexes show bands at *ca.* 3250, 1660 and 1074 cm^{-1} which are assigned to ν_{OH}, ν_{CN}(acyclic) and ν_{NO}, respectively. The effective magnetic moments of the complexes agree well with the theoretical spin-only value of 3.87 BM and are temperature invariant.

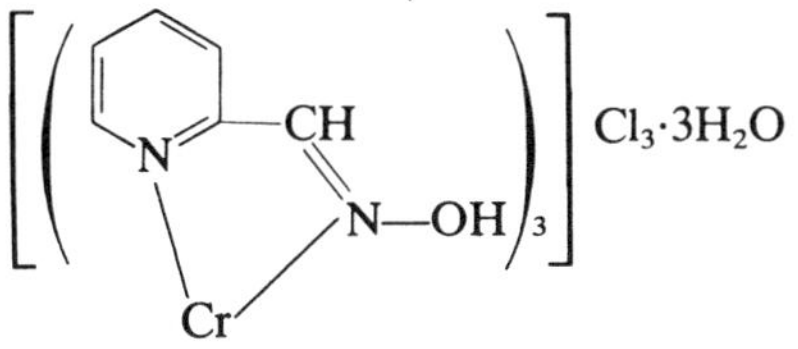

(**151**)

$$CrX_3 + Hpox + NaClO_4 \xrightarrow[heat]{EtOH} [Cr(Hpox)_2X_2]ClO_4 \xrightarrow[]{X = Br,\ HBr/H_2O,\ heat} [Cr(Hpox)_2(H_2O)_2]Br_3 \cdot H_2O$$

Scheme 93

Table 74 Reflectance Spectra (cm^{-1}) of Pyridine-2-aldoxime (Hpox) Chromium(III) Complexes

Complex	${}^4B_{1g} \rightarrow {}^4E_g$	${}^4B_{1g} \rightarrow {}^4B_{2g}$	${}^4B_{1g} \rightarrow {}^4A_{2g}$	${}^4B_{1g} \rightarrow {}^4E_g$	*Ref.*
$[Cr(Hpox)_3]Cl_3 \cdot 3H_2O$	22 300[a]			28 820[b]	1
trans-$[Cr(Hpox)_2F_2]ClO_4$	18 940	22 150	29 750	25 750	1
trans-$[Cr(Hpox)_2Cl_2]ClO_4$	17 750	22 450	25 750	26 950	1
trans-$[Cr(Hpox)_2Br_2]ClO_4$	17 405	23 150	24 655	26 418	1
trans-$[Cr(Hpox)_2(H_2O)_2]Br_3 \cdot H_2O$	20 128	22 920	30 117	27 920	1
trans-$[Cr(6\text{-}MeHpox)_2F_2]ClO_4$	18 930	22 125	29 725	25 725	2
trans-$[Cr(6\text{-}MeHpox)_2Cl_2]ClO_4$	17 730	22 430	25 730	26 930	2
trans-$[Cr(6\text{-}MeHpox)_2Br_2]ClO_4$	17 390	23 140	24 645	24 420	2
trans-$[Cr(6\text{-}MeHpox)_2(H_2O)_2]Br_3 \cdot H_2O$	20 120	22 910	30 110	27 910	2

[a] ${}^4A_{2g} \rightarrow {}^4T_{2g}$. [b] ${}^4A_{2g} \rightarrow {}^4T_{1g}$.

1. M. Mohan, S. G. Mittal, H. C. Khera and A. K. Sirivastava, *Monatsh. Chem.*, 1980, **111**, 63.
2. M. Mohan, H. C. Khera, S. G. Mittal and A. K. Sirivastava, *Gazz. Chim. Ital.*, 1979, **109**, 65.

The addition of β-furfuraldoxime (Hfox) to a solution of $CrCl_3{\cdot}6H_2O$ in absolute ethanol affords, after refluxing and removal of solvent, a green crystalline product $Cr(Hfox)_3Cl_3$ (μ_{eff} at 297 K = 3.94 BM) which contains monodentate *O*(furan)-bonded Hfox ligands.[668] Stability constants as a function of ionic strength and temperature have been determined for the 1:1, 2:1 and 3:1 complexes between β-furfuraldoxime and chromium(III) in a dioxane–water (75:25) solvent system.[669] The ligand 4-benzoyloxime-3-methyl-1-phenyl-2-pyrazolin-5-one, HL, forms a complex CrL_3 in which it appears, on the basis of evidence from IR spectroscopy, to be coordinated to the chromium through the oxime nitrogen and the oxygen of its enolate tautomer (**152**).[670] Phenanthroquinone monoxime (Hpqox) reacts with $CrCl_3{\cdot}6H_2O$ in aqueous ethanol to give a red-brown complex of composition $Cr(pqox)_2(OH)H_2O$.[671]

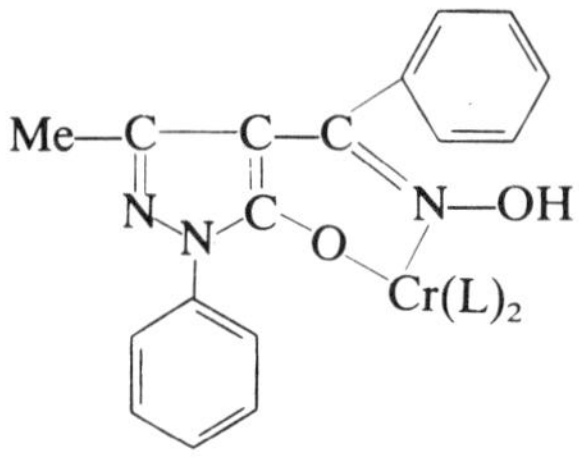

(**152**)

(ii) Biguanide and tetramethylguanidine

Biguanide $NH_2C({=}NH)NHC({=}NH)NH_2$ (bgH) and its derivatives (R—bgH) form two types of complexes with chromium(III): those which contain deprotonated ligands and are uncharged, *e.g.* $Cr(bg)_3$, $Cr(R{-}bg)_3$ and those which contain the parent ligands and give charged species such as $[Cr(bgH)_3]^{3+}$ and $[Cr(bgH)(H_2O)(OH)]^{2+}$. The complex $Cr(bg)_3{\cdot}H_2O$ (**153**) prepared from $KCr(SO_4)_2{\cdot}12H_2O$ and biguanide under alkaline conditions, contains an octahedral arrangement of bidentate biguanidato ligands around the metal.[672] Each of the deprotonated ligands contains a delocalized π electron system and the rings are planar. The structure of $(-)_D$-$[Cr(bgH)_3]^{3+}$ as its (+)-10-camphorsulfonate salt has also been determined by X-ray analysis and although the molecular geometry has not been reported it appears that the complex conforms closely to D_3 symmetry and that the chelate rings are planar.[673]

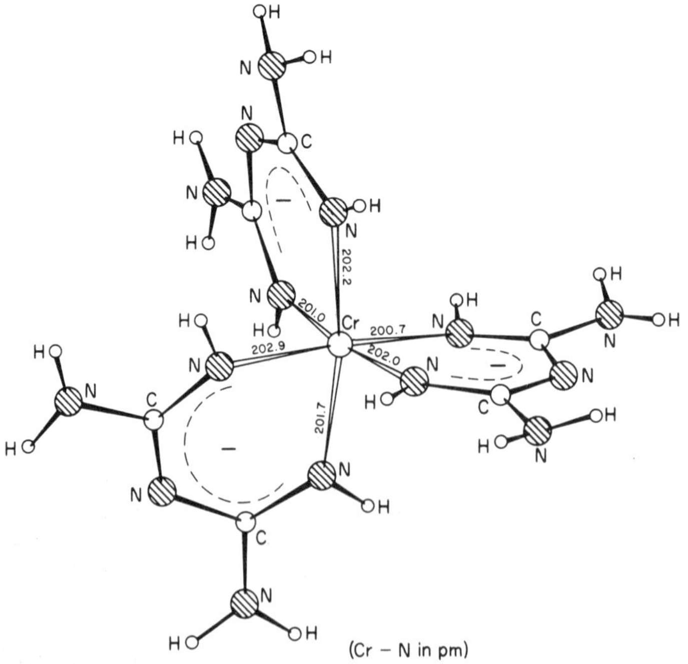

(**153**)

Other reported biguanide complexes of chromium(III) include $[Cr(RbgH)_3]X_3$ (R = *N'*-$C_6H^n_{13}$, X = OH; R = *N'*-Ph, X = OH, Cl, Br, I, NO_3, $0.5SO_4$, $0.5CrO_4$; R = *N'*-$PhCH_2$, X = Cl, $0.5SO_4$) and $Cr(R'bg)_3$ (R' = *N'*-Ph; *N'*-$PhCH_2$).[674] Generally the neutral complexes are red whereas the salts are orange or yellow. The kinetics of hydrolysis of $[Cr(bgH)_3](ClO_4)_3$, $[Cr(bgH)_2(H_2O)_2](ClO_4)_3$ and the corresponding substituted biguanide (RbgH) complexes have been investigated.[675] Under acid conditions a first-order dependence of rate on $[H^+]$ was observed and rationalized in terms of rapid preequilibrium formation of a conjugate acid of the reacting complex, with the central nitrogen atom acting as the basic site, followed by its dissociation in the rate-determining step. Consistent with this mechanism is the observation that the reactivity order towards acid hydrolysis correlates with the order of ligand basicity, *i.e.* bgH > *N'*-$C_6H^n_{13}$bgH > *N'*-PhbgH. Replacement of the aqua ligands in *cis*-$[Cr(bgH)_2(H_2O)_2]^{3+}$ by phen, bipy[676] and $C_2O_4^{2-}$ [677] in aqueous solution proceeds by a pathway which is first order in ligand concentration but with values of $\Delta H^{\neq}$ which are similar for all three incoming groups. A mechanism involving outer-sphere association of reactants followed by transformation of outer-sphere complexes into products by a dissociative pathway has been proposed for these reactions. The products have been isolated as the perchlorate salts $[Cr(bgH)_2L](ClO_4)_n$ (L = phen, bipy, $n = 3$; L = C_2O_4, $n = 1$) and characterized. Some chromium(III) complexes with dibiguanide ligands which contain ethylene, hexamethylene or phenylene bridges linking two biguanide residues have also been reported.[674]

When an aqueous paste containing $Cr(bg)_3 \cdot H_2O$ and NH_4SCN is heated at 80 °C until evolution of ammonia ceases, the red crystalline complex $[Cr(bgH)_2(NCS)_2]NCS$ is obtained.[678] While evidence from IR spectroscopy ($\nu_{CN} = 2095\ cm^{-1}$, $\nu_{CS} = 770\ cm^{-1}$, $\delta_{NCS} = 490\ cm^{-1}$) supports the presence of *N*-thiocyanato ligands, no information regarding the geometrical configuration of the complex can be deduced from its IR and electronic spectra. By comparing the positions of the $^4A_{2g} \rightarrow {}^4T_{2g}$ bands in the electronic spectra of appropriate complexes (Table 75) it may be inferred that biguanide is slightly higher in the spectrochemical series than either the *N*-thiocyanato or biguanidato ligands.

Table 75 Electronic Spectra and Magnetic Moments of Chromium(III) Biguanide Complexes

Complex	$^4A_{2g} \rightarrow {}^4T_{2g}$ (nm)	$^4A_{2g} \rightarrow {}^4T_{1g}(F)$ (nm)	μ_{eff} (277–278 K) (BM)	*Ref.*
$[Cr(bgH)_3]Cl_3$	483	369[a]	3.88	1
$[Cr(bgH)_2(NCS)_2]NCS$	489	375[a]	3.87	1
$Cr(bg)_3$	500	384[b]	—	2
$[Cr(bgH)_2C_2O_4]^+$	490 (87)	385 (92)[c]	—	3
$[Cr(bgH)_2phen]^{3+}$	465 (86)[c]	—	—	4
$[Cr(bgH)_2bipy]^{3+}$	465 (87)[c]	—	—	4
cis-$[Cr(bgH)_2(H_2O)_2]^{3+}$	490 (74)	378 (71)[c]	—	3–5

[a] Nujol mull supported on filter paper. [b] Reflectance spectra. [c] In aqueous solution.
1. A. Syamal, *Z. Naturforsch., Teil B*, 1974, **29**, 492.
2. R. H. Skabo and P. W. Smith, *Aust. J. Chem.*, 1969, **22**, 659.
3. D. Banerjea and S. Sengupta, *Z. Anorg. Allg. Chem.*, 1973, **397**, 215.
4. S. Sengupta and D. Banerjea, *Z. Anorg. Allg. Chem.*, 1976, **424**, 93.
5. D. Banerjea and C. Chakravarty, *J. Inorg. Nucl. Chem.*, 1964, **26**, 1233.

Complexes of chromium(III) with tetramethylguanidine, $HN{=}C(NMe_2)_2$, have been prepared according to Scheme 94 and although they have not been fully characterized because of the presence of free ligand impurities, the appearance of C=N stretching bands at lower wavenumbers in the IR spectra of the complexes suggest that the ligand is coordinated to the metal through the imine nitrogen.[679]

$$rCl_3 + HN{=}C(NMe_2)_2 \xrightarrow[\text{acetone}]{\text{Zn dust}} CrCl_3 \cdot 4HN{=}C(NMe_2)_2 \xrightarrow[]{\text{excess } HN{=}C(NMe_2)_2,\ 120\,^\circ C} CrCl_3 \cdot 6HN{=}C(NMe_2)_2$$

$\nu(C{=}N) = 1609\ cm^{-1}$ green, $\nu(C{=}N) = 1563\ cm^{-1}$ grey-green, $\nu(C{=}N) = 1562\ cm^{-1}$

Scheme 94

(iii) N-*coordinated ureas and amides*

In most of the known urea complexes of chromium(III) the ligand is O-bonded to the metal ion.[679] However the complexes $[Cr(NH_3)_5\{OC(NH_2)_2\}]^{3+}$ and $[Cr(NH_3)_5\{OC(NHMe)_2\}]^{3+}$, both of which can be prepared in high yield by appropriate substitution of the labile triflate ligand in $[Cr(NH_3)_5(OSO_2CF_3)]^{2+}$, undergo isomerization without competitive hydrolysis in basic solution to give deprotonated N-bonded urea complexes (Scheme 95).[680] Reacidification of the product solutions regenerates the O-bonded isomers rapidly and quantitatively. The IR spectrum of the O-bonded urea complex possesses characteristic absorptions at 1570 and 1640 cm^{-1} both of which are absent from the spectrum of the deprotonated N-bonded form. A similar change from O-bonded to deprotonated N-bonded formamide occurs when a solution of $[Cr(NH_3)_5(OCHNH_2)]^{3+}$ is made basic.

$$[(NH_3)_5Cr-OC(NH_2)_2]^{3+} \;\underset{}{\overset{OH^-}{\rightleftharpoons}}\; [(NH_3)_4(NH_2)Cr-OC(NH_2)_2]^{2+} \;\underset{k_b}{\overset{k_f}{\rightleftharpoons}}\; [(NH_3)_4(NH_2)Cr-NH_2CONH_2]$$

λ_{max} 504 ($\varepsilon = 58$), 370 ($\varepsilon = 38$)

$\Big\Updownarrow$ pK_a = 13.5, 298 K, I = 0.1 M NaClO$_4$ $\qquad\qquad\qquad\qquad\qquad\qquad\qquad\qquad$ $\Big\downarrow$ fast

$[(NH_3)_5Cr-OC(NH_2)NH]^{2+} + H^+$ $\qquad\qquad\qquad\qquad\qquad\qquad$ $[(NH_3)_5Cr-NHCONH_2]^{2+}$

$\qquad\qquad\qquad\qquad\qquad\qquad\qquad\qquad\qquad\qquad\qquad\qquad\qquad\qquad$ λ_{max} 490 ($\varepsilon = 48$), 372 ($\varepsilon = 43$)

λ_{max} in nm, ε in dm^3 mol^{-1} cm^{-1}

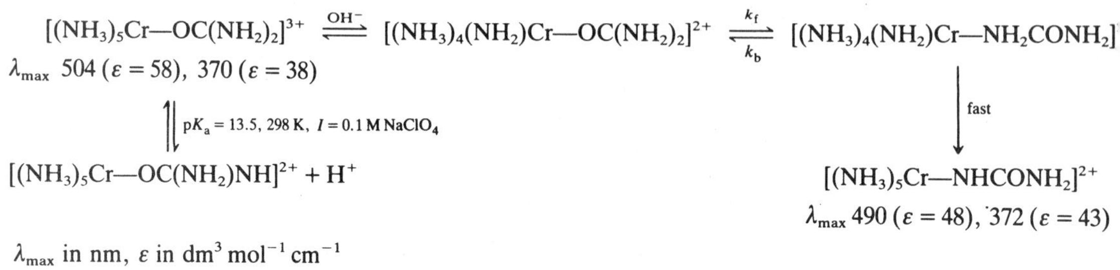

Scheme 95

Addition of amide (**122**, H$_4$L) and Na$_2$CO$_3$ to a boiling solution of CrCl$_3$·6H$_2$O in pyridine results in formation of the complex [Cr(HL)py$_2$]$_2$·2py (**154**), which contains chromium(III) coordinated to the amide oxygen and deprotonated amide nitrogen of different molecules.[519] This is the first reported example of chromium(III) coordinated to a nitrogen of a deprotonated amide. The structure consists of discrete dimers and disordered solvent molecules of pyridine.

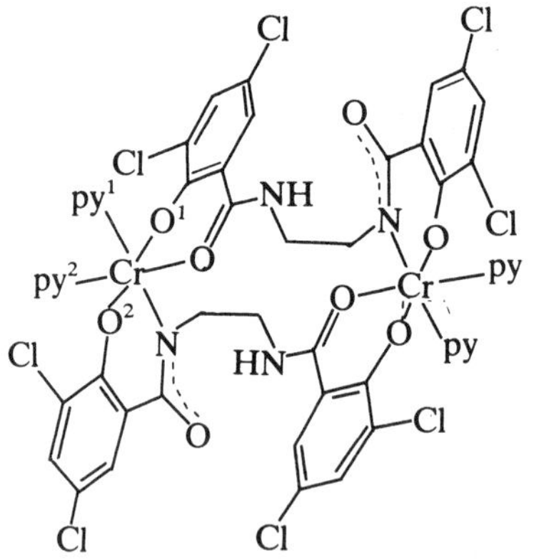

$Cr-N_{py1} = 2.145$ Å
$Cr-N_{py2} = 2.097$ Å
$Cr-O^1 = 1.931$ Å
$Cr-O^2 = 1.915$ Å
$Cr-O_{amide} = 1.976$ Å
$Cr-N_{amide} = 2.030$ Å

(**154**)

35.4.3 Phosphorus, Arsenic and Antimony Ligands

Complexes of hard Cr^{III} with ligands containing soft P and As donor atoms have not been easy to prepare. It has usually been necessary to use weak donor solvents in the absence of oxygen and moisture, although some hydrates have been obtained by admitting traces of moisture. Structures have generally been inferred from conductance and molecular weight measurements, and IR and electronic spectra. The Cr^{III} is invariably six-coordinate with magnetic moments close to 3.9 BM.

Complexes of a few monodentate tertiary phosphines are known. There are dimers $[CrX_3(PR_3)_2]_2$ (X = Cl, R = Et, Bu^n; X = Br, R = Et),[681,682] polymers $[CrCl_3(PR_3)]_n$ (R = Et, Bu^n, Ph) and anionic complexes $[PPh_4][CrCl_4(PR_3)_2]$[682] (Table 76). The tris(phosphine) complexes CrCl$_3$(PHEt$_2$)$_3$ and CrCl$_3$(PH$_2$Ph)$_3$ have also been reported,[683] but a number of organophosphines would not react with chromium(III) halides.[681–683] From $K_3[Cr(SCN)_6]$, several salts $[R_3PH][Cr(SCN)_4(PR_3)_2]$ have been obtained (see also Section 35.4.2.8.i.b).[684] Monodentate phosphines with $[Cr(THF)_6]^{3+}$ solutions (see below) produce green oils.[685]

Earlier work with ditertiary arsines and phosphines produced the compounds: $CrX_3[o\text{-}C_6H_4(AsMe_2)_2]_{1.5}$, $[CrX_2\{o\text{-}C_6H_4(AsMe_2)_2\}_2]ClO_4$ (X = Cl, Br, I), $[CrX_3(H_2O)\{o\text{-}C_6H_4(AsMe_2)_2\}]$ (X = Cl, Br),[686] $[CrX_3\{o\text{-}C_6H_4(AsMe_2)_2\}]_2$ (X = Br, I), $[CrI_2\{o\text{-}C_6H_4(AsMe_2)_2\}_2]I_3$,[687] $CrX_3(Ph_2PCH_2CH_2PPh_2)_{1.5}\cdot nH_2O$ (X = Cl, Br, $n = 2$; X = I, $n = 4$; X = NCS, $n = 0$) and $[CrX_2\{o\text{-}C_6H_4)(PMe_2)_2\}_2]$ ClO_4 (X = Cl, Br).[689] Complexes of multidentate tertiary phosphines and arsines were also obtained: $[CrX_3\{P(o\text{-}C_6H_4PPh_2)_3\}]$ (X = Cl, Br), $[CrI_2\{P(o\text{-}C_6H_4PPh_2)_3\}]I_3$,[690] $CrX_3\{PPh(o\text{-}C_6H_4PPh_2)_2\}$ (X = Cl, Br, NCS),[691] $CrX_3\{AsMe(o\text{-}C_6H_4AsMe_2)_2\}$ (X = Cl, Br, I)[692,693] and $CrCl_3\{CMe(CH_2AsMe_2)_3\}$.[693]

Table 76 Complexes of Phosphorus, Arsenic and Antimony Ligands

Complex	*ν(Cr—X)* (cm^{-1})	*Electronic spectra*[a] (cm^{-1})	*Ref.*
Monodentate phosphines			
$[PPh_4][CrCl_4(PEt_3)_2]$	329s, 325s	12 800sh, 15 800, 20 350	1
$[PPh_4][CrCl_4(PBu^n_3)_2]$	334sh, 328s	12 800, 13 500, 15 900, 20 450	1
$[AsPh_4][CrCl_4(PEt_3)_2]$	333sh, 327s	—	1
$[CrCl_3(PEt_3)_2]_2$ Green	370s, 345s, br	13 000sh, 15 600, 20 500	1
$[CrCl_3(PBu^n_3)_2]_2$ Green	355s, 341s	13 500sh, 15 150, 20 900	1
$[CrBr_3(PEt_3)_2]_2$ Green	220w, 306w	15 800	1
$[CrCl_3(PEt_3)]_n$ Violet	345s, 319s	12 800, 17 400	1
$[CrCl_3(PBu^n_3)]_n$ Violet	349s, 321s	13 150, 17 550	1
$[CrCl_3(PPh_3)]_n$ Violet	364sh, 356sh	13 300, 18 450	1
Multidentate ligands			
$[NPr^n_4][CrCl_4\{o\text{-}C_6H_4(PMe_2)_2\}]$ Purple	355m, 330s, 305sh	15 580, 19 840	2
$[NPr^n_4][CrCl_4\{o\text{-}C_6H_4(AsMe_2)_2\}]$ Blue	358sh, 333m, 325s, 312m	15 000 (502), 19 300 (328)	2
$[PPh_3(CH_2Ph)][CrBr_4\{o\text{-}C_6H_4(AsMe_2)_2\}]$ Blue	312m, 285s, 258s	14 800 (993), 18 000 (601)	2
$[NBu^n_4][CrI_4\{o\text{-}C_6H_4(AsMe_2)_2\}]$ Yellow-green	—	13 440, 16 450	2
$[NBu^n_4][CrCl_4\{o\text{-}C_6H_4(PMe_2)(SbMe_2)\}]$[b] Blue	354, 326s, 302s	15 530, 20 800	2
$[NPr^n_4][CrCl_4(Ph_2PCHCHPPh_2)]$[c] Purple	340s, 315s	14 660 (378), 19 500 (170)	2
$[NBu^n_4][CrBr_4(Ph_2PCHCHPPh_2)]$ Blue	284s, 259s, 250sh	14 800 (408), 19 000 (150)	2
$[NPr^n_4][CrCl_4(Ph_2PCH_2CH_2PPh_2)]$ Purple	366sh, 340m, 315sh, 300s	14 700 (543), 19 700 (343)	2
$[NBu^n_4][CrBr_4(Ph_2PCH_2CH_2PPh_2)]$ Blue	284s, 255s	14 700 (402), 18 800 (337)	2
$[NBu^n_4][CrI_4(Ph_2PCH_2CH_2PPh_2)]$ Yellow	—	14 490, 17 540	2
$[NPr^n_4][CrCl_4(Ph_2AsCHCHAsPh_2)]$ Purple	346m, 335s, 302s	14 000, 18 200	2
$[NPr^n_4][CrCl_4\{Me_2As(CH_2)_3AsMe_2\}]\cdot H_2O$ Blue	326s, 319s	15 250 (367), 19 600 (273)	2
$[CrCl_3(Ph_2PCH_2CH_2PPh_2)_2]$ Green	362, 340w, 316	15 660 (253), 21 660 (173)	2
$[CrBr_3(Ph_2PCH_2CH_2PPh_2)_2]$ Green	290, 250	14 790, 19 460	2
$[CrI_3(Ph_2PCH_2CH_2PPh_2)_2]$ Yellow	—	14 970, 18 240	2
$[CrCl_3(Ph_2PCHCHPPh_2)_2]$ Blue	366, 336, 316	15 660 (352), 21 000 (163)	2
$[CrBr_3(Ph_2PCHCHPPh_2)_2]$ Green	298, 252	15 150 (206), 20 400 (110)	2

Table 76 (*continued*)

Complex	$\nu(Cr—X)$ (cm^{-1})	*Electronic spectra*[a] (cm^{-1})	*Ref.*
$[CrCl_3(Ph_2AsCH_2CH_2AsPh_2)_2]$ Green	375, 324	15 800, 21 300	2
$[CrCl_3(H_2O)(Ph_2PCHCHPPh_2)]$ Blue	349, 336, 317	16 130, 20 830	2
$[CrBr_3(H_2O)(Ph_2PCHCHPPh_2)]$ Blue	269, 250	15 820, 20 000	2
$[CrCl_3(H_2O)(Ph_2PCH_2CH_2PPh_2)]$ Green	337br	—	2
$[CrCl_2\{o\text{-}C_6H_4(PMe_2)_2\}_2]Cl$ Purple-brown	335m	16 100sh, 18 300sh, 20 000, 22 530	2
$[CrCl_2\{o\text{-}C_6H_4(PMe_2)_2\}_2]PF_6$ Pink-brown	355m	16 200sh, 18 400sh, 19 960, 22 700	2
$[CrBr_2\{o\text{-}C_6H_4(PMe_2)_2\}_2]PF_6$ Green	282m	16 610sh, 18 150, 21 300sh, 22 830	2
$[CrBr_2\{o\text{-}C_6H_4(PMe_2)_2\}_2]Br$ Green	280m	16 720sh, 18 180, 21 400sh, 22 700	2
$[CrCl_2\{o\text{-}C_6H_4(AsMe_2)_2\}_2]ClO_4$ Bright green	380m	15 970sh, 17 010, 21 500sh, 24 000	2, 3, 4
$[CrBr_2\{o\text{-}C_6H_4(AsMe_2)_2\}_2]ClO_4$ Bright green	320m	15 300sh, 16 550, 20 000sh, 22 600	2, 3, 4
$CrBr_3[o\text{-}C_6H_4(PMe_2)_2]_{1.5}$ Blue	290sh, 275s	—	2
$CrCl_3[o\text{-}C_6H_4(PMe_2)_2]_{1.5}$ Pink	355m, 352sh, 338m, 306sh	—	2
$CrCl_3[o\text{-}C_6H_4(AsMe_2)_2]_{1.5}$ Green	380m, 333sh, 321s, 307sh	—	2, 3
$CrBr_3[o\text{-}C_6H_4(AsMe_2)_2]_{1.5}$ Green	314, 285s, 258sh	—	2, 3
$CrI_3[o\text{-}C_6H_4(AsMe_2)_2]_{1.5}$ Yellow-brown	—	—	2, 3
$CrCl_3[o\text{-}C_6H_4(AsMe_2)(PMe_2)]_{1.5}$ Purple	375sh, 352s, 338, 320sh, 308sh	—	2
$CrBr_3[o\text{-}C_6H_4(AsMe_2)(PMe_2)]_{1.5}$ Blue-purple	306, 295s, 275s	—	2
$CrCl_3[o\text{-}C_6H_4(PMe_2)(SbMe_2)]_{1.5}$[b] Grey-green	375w, 345s, 333sh, 308m	—	2
$CrBr_3[o\text{-}C_6H_4(PMe_2)(SbMe_2)]_{1.5}$[b] Yellow-brown	320br, 280s	—	2
$CrCl_3[o\text{-}C_6H_4(NMe_2)(PMe_2)]_{1.5}$ Blue	370sh, 345s, 306s	—	2
$CrCl_3[o\text{-}C_6H_4(AsMe_2)(NMe_2)]_{1.5}$ Green	375sh, 345sh, 340sh, 315m	—	2
$CrCl_3[Me_2As(CH_2)_3AsMe_2]_{1.5}$ Blue	350s, 338s	14 500, 17 000, 21 000	2
$CrBr_3[Me_2As(CH_2)_3AsMe_2]_{1.5}$ Purple	290br	14 120, 17 000, 20 000	2
[CrCl₃L]		*Isomer*	
$PPh(CH_2CH_2PPh_2)_2$ Blue	362, 342, 320	*mer*	5
$[—CH_2P(Ph)CH_2CH_2PPh_2]_2$ Blue	354, 334, 320sh	*mer*	5
$P(CH_2CH_2PPh_2)_3$[d] Purple	360sh, 340, 318	*mer*	5
$CMe(CH_2PPh_2)_3$ Blue-green	350, 334	*fac*	5
$AsMe(CH_2CH_2CH_2AsMe_2)_2$ Purple	364br	*fac*	5
$As(CH_2CH_2CH_2AsMe_2)_3$ Purple	346, 340sh	*fac*	5
$CMe(CH_2AsMe_2)_3$ Blue-purple	333sh, 326	*fac*	6
[CrBr₃L]			
$PPh(CH_2CH_2PPh_2)_2$ Blue	298, 265, 230	*mer*	5
$[—CH_2P(Ph)CH_2CH_2PPh_2]_2$ Blue	305, 282	*mer*	5

Table 76 *(continued)*

Complex		$\nu(Cr—X)$ (cm^{-1})	*Electronic spectra*[a] (cm^{-1})	*Ref.*
$P(CH_2CH_2PPh_2)_3$ Blue		290sh, 263, 245sh	*mer*	5
$CMe(CH_2PPh_2)_3$ Green		290, 256	*fac*	5
$AsMe(CH_2CH_2CH_2AsMe_2)_2$ Purple		288br	*fac*	5
$As(CH_2CH_2CH_2AsMe_2)_3$ Purple		286br	*fac*	5
$CMe(CH_2AsMe_2)_3$ Blue		298, 278	*fac*	5
[CrF_3L]				
$PPh(CH_2CH_2PPh_2)_2$ Blue		544, 517, 505	*mer*	5
$P(CH_2CH_2PPh_2)_3$ Purple		542, 519, 507	*mer*	5
$CMe(CH_2AsMe_2)_3$ Blue		538, 492	*fac*	5
[CrX_2L]BF_4	*X*			
$P(CH_2CH_2PPh_2)_3$ Dark blue	Cl	350, 310	*cis*	5
$P(CH_2CH_2PPh_2)_3$ Dark blue	Br	296, 255	*cis*	5
$P(CH_2CH_2PPh_2)_3$ Turquoise	I		*cis*	5
[—$CH_2P(Ph)CH_2CH_2PPh_2$]$_2$ Dark blue	Cl	347, 318	*cis*	5
$As(CH_2CH_2CH_2AsMe_2)_3$ Purple	Cl	346, 326	*cis*	5

[a] ε ($dm^3\ mol^{-1}\ cm^{-1}$) in parentheses. [b] First examples of Cr—Sb bonds.
[c] $[NPr^n_4][CrCl_4(cis\text{-}Ph_2PCHCHPPh_2)]$, Cr—P, 2.485, 2.511, Cr—Cl, 2.331, 2.319, 2.318 Å.
[d] *mer*-$[CrCl_3\{P(CH_2CH_2PPh_2)_3\}]$, Cr—$PPh_2$, 2.466, 2.489, Cr—$P(CH_2)_3$, 2.399, Cr—Cl, 2.306, 2.292 and 2.320 Å.

1. M. A. Bennett, R. J. H. Clark and A. D. J. Goodwin, *J. Chem. Soc. (A)*, 1970, 541.
2. L. R. Gray, A. L. Hale, W. Levason, F. P. McCullough and M. Webster, *J. Chem. Soc., Dalton Trans.*, 1983, 2573.
3. R. S. Nyholm and G. J. Sutton, *J. Chem. Soc.*, 1958, 560.
4. R. D. Feltham and W. Silverthorn, *Inorg. Chem.*, 1968, **7**, 1154.
5. L. R. Gray, A. L. Hale, W. Levason, F. P. McCullough and M. Webster, *J. Chem. Soc., Dalton Trans.*, 1984, 47.
6. R. J. H. Clark, M. L. Greenfield and R. S. Nyholm, *J. Chem. Soc. (A)*, 1966, 1254.

The above complexes were generally prepared from poorly soluble chromium(III) chloride, or $[CrCl_3(THF)_3]$ *in situ* in THF, although a few were obtained by oxidation of phosphine– or arsine–chromium carbonyls[687,690,692] with halogens. By using $[CrX_3(THF)_3]$ (X = Cl, Br or I) in dry CH_2Cl_2 [694] instead of THF, competition for the hard Cr^{III} by an O donor solvent has been avoided, and extensive series[685,695,696] of complexes of multidentate P, As and S (see p. 886) donor ligands have been obtained (Table 76). The pseudooctahedral structure of $[CrCl_4(cis\text{-}Ph_2PCHCHPPh_2)]^-$ has been confirmed crystallographically.[695] The diphosphine does not exert any *trans* influence, consistent with weak bonding to the hard Cr^{III} ion, which is also evident from the electronic spectral parameters of the complexes generally. It has been confirmed that the complexes typified by $CrCl_3[o\text{-}C_6H_4(AsMe_2)_2]_{1.5}$ have the *trans*-$[CrX_2(L^2)_2]$*cis*-$[CrX_4(L^2)]$ structure favoured earlier,[686] but for $CrX_3(Me_2As(CH_2)_3AsMe_2)_{1.5}$ a ligand-bridged structure $X_3(L^2)Cr(L^2)Cr(L^2)X_3$ has been put forward because the ν(Cr—Cl) absorptions are different from those in $[NPr^n_4][CrCl_4\{Me_2As(CH_2)_3AsMe_2\}]$ (Table 76). Repeat preparations of $CrX_3(Ph_2P(CH_2)_2PPh_2)_{1.5}\cdot nH_2O$, for which similar diphosphine-bridged structures were proposed,[688] failed. The weaker donors $Ph_2PCH_2CH_2PPh_2$, *cis*-$Ph_2PCHCHPPh_2$ or $Ph_2AsCH_2CH_2AsPh_2$ produce complexes $[CrX_3(L^2)_2]$, which are believed to be meridional, containing one uni- and one bi-dentate diphosphine and diarsine, and give $[CrX_3(H_2O)(L^2)]$ in the presence of moisture.

It remains necessary to generate *in situ* solutions[685] presumed to contain $[CrF_3(THF)_3]$ and $[Cr(THF)_6][BF_4]_3$ for the preparation[696] of the fluoro complexes CrF_3L (Table 76), and the salts $[Cr(L^2)_3][BF_4]_3$ [695] ($L^2 = Ph_2PCH_2CH_2PPh_2$, $Me_2PCH_2CH_2PMe_2$, o-$C_6H_4(PMe_2)_2$, o-$C_6H_4(AsMe_2)_2$ or *cis*-$Ph_2AsCH{=}CHAsPh_2$) and $[Cr(L^3)_2][BF_4]_3$ ($L^3 = PhP(CH_2CH_2PPh_2)_2$, $MeAs(CH_2CH_2CH_2AsMe_2)_2$, $MeC(CH_2PPh_2)_3$ or $MeC(CH_2AsMe_2)_3$) because attempts to isolate solid starting materials caused decomposition. Of these octahedral complexes,[695] only $[Cr(Me_2PCH_2CH_2PMe_2)_3][BF_4]_3$ undergoes reversible electrochemical one-electron reduction in DMSO (as does $[CrX_2\{o\text{-}C_6H_4(PMe_2)_2\}_2]ClO_4$ in MeCN[689]) and reactions with LiX lead to partial displacement of the neutral ligands to give, for example $[CrCl_3\{PhP(CH_2CH_2PPh_2)_2\}_2]$. An attempt[686] to prepare $[Cr\{o\text{-}C_6H_4(AsMe_2)_2\}_3][ClO_4]_3$ from $[Cr(H_2O)_6][ClO_4]_3$ failed.

In the neutral complexes $CrX_3(L^3$ or $L^4)$ (X = Cl or Br) those P or As ligands which are potentially quadridentate act as tridentates.[696] An X-ray study has established that $[CrCl_3\{P(CH_2CH_2PPh_2)_3\}]$ has a *mer* arrangement of the P_3Cl_3 donor set with one PPh_2 group not bonded. This phosphine is usually quadridentate in complexes of other metal ions. Structures have been assigned to the other complexes from the number of ν(Cr—Cl) vibrations. Tripodal tridentates, *e.g.* $CMe(CH_2PPh_2)_3$, constrain monomeric complexes to form *fac* isomers and this provides a further check on the IR data. No 2:1, L^3 or L^4 to Cr complexes were obtained when X = Cl or Br.

The reaction of $[CrI_3(THF)_3]$ with multidentate ligands in a 1:1 ratio gave[696] the complexes CrI_3L [$L = PPh(CH_2CH_2PPh_2)_2$, $P(CH_2CH_2PPh_2)_3$, $CMe(CH_2AsMe_2)_3$, $AsMe(CH_2CH_2CH_2AsMe_2)_2$ or $As(CH_2CH_2CH_2AsMe_2)_3$]. Few physical data are available, but CrI_3L are considered analogous to the corresponding chlorides and bromides. Reaction of $[CrI_3(THF)_3]$ in a 1:2 molar ratio gave two complexes of the type CrI_2L_3: with $L = PPh(CH_2CH_2PPh_2)_2$ the complex is formulated as $[Cr\{PPh(CH_2CH_2PPh_2)_3\}]I_3$ and with $L = CMe(CH_2AsMe_2)_3$ as $[CrI_2\{CMe(CH_2AsMe_2)_3\}_2]I$, the arsine being bidentate.

Treatment of the corresponding halide in CH_2Cl_2 with $AgBF_4$ gave the complexes $[CrX_2L][BF_4]$ (Table 76). The two ν(Cr—X) stretches are consistent with *cis* cations and tetradentate L.

The Cr—P bonds in $[NPr^n_4][CrCl_4(cis\text{-}Ph_2PCHCHPPh_2)]$ and *mer*-$[CrCl_3\{P(CH_2CH_2PPh_2)_3\}]$[696] are long (Table 76) and this is attributed to steric repulsions and the weak binding of the neutral phosphine ligands to the hard Cr^{III} ion. The Cr—Cl bonds are normal.[695] Analysis of the electronic spectra[682,685,688,689,693,695–697] is consistent with octahedral Cr^{III}, and the derived Racah parameters are always much lowered from the free ion value; this also is attributed to weak binding of the soft ligands. An alternative assignment[697] has been proposed[695] for $[CrX_2\{o\text{-}C_6H_4(AsMe_2)_2\}_2]^+$ which places the diarsine lower in the spectrochemical series.

35.4.4 Oxygen Ligands

35.4.4.1 Aqua complexes

The hexaaquachromium(III) ion (**155**) is a regular octahedron and is found in aqueous solution and in many hydrated salts.[698,699,700,701] Typical examples are the violet hydrate $[Cr(H_2O)_6]Cl_3$ and the chrome alums, $MCr(SO_4)_2{\cdot}12H_2O$. The acidity of the aqua complex is marked; the first hydrolysis has $pK_a \approx 4.0$ (25 °C). Further hydrolytic equilibria are complicated by polymerization and are dealt with in the following section. The most recent detailed reports on $[Cr(H_2O)_6]^{3+}$ in the solid state are IR[702] and crystallographic studies[699] of cesium alum. In the vibrational study, metal–ligand vibrations were assigned at 555 cm^{-1} (ν_{as} stretch) and 329 cm^{-1} (ν_{as} bend). In the structural paper, the classification of alums is discussed; the Cr—O bond length was found to be 1.959 Å. The hexaaqua ion is also found in crystals of $Cr_4H_2(SO_4)_7{\cdot}24H_2O$; the Cr—O bond distances are in the range 1.915–1.991 Å.[703]

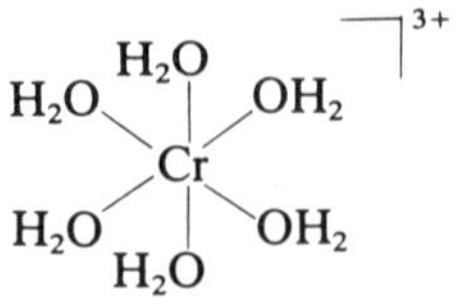

(**155**)

The rate of exchange of H_2O at $[Cr(H_2O)_6]^{3+}$ is given as rate = $k[Cr(H_2O)_6]^{3+}$; this reaction has been extensively studied.[704] The most reliable results are summarized in Table 77. The substantial and negative value of $\Delta V^{\ddagger}$ together with $\Delta S^{\ddagger}$ near to zero suggests an I_a mechanism for exchange.

Table 77 Kinetic Data for Water Exchange at Chromium(III)

Compound	$k \times 10^7$ (s^{-1})	$\Delta H^{\ddagger}$ (kJ mol^{-1})	$\Delta S^{\ddagger}$ (J K^{-1} mol^{-1})	$\Delta V^{\ddagger}$ (cm^3 mol^{-1})	*Ref.*
$[Cr(H_2O)_6]^{3+}$	4.5	109.6	1.3	−9.3	1, 2
$[Cr(NH_3)_5(H_2O)]^{3+}$	601	97.1	0.0	−5.8	3, 4

1. J. P. Hunt and R. A. Plane, *J. Am. Chem. Soc.*, 1954, **76**, 5960.
2. D. R. Stranks and T. W. Swaddle, *J. Am. Chem. Soc.*, 1971, **93**, 2783.
3. N. V. Duffy and J. E. Earley, *J. Am. Chem. Soc.*, 1967, **89**, 272.
4. T. W. Swaddle and D. R. Stranks, *J. Am. Chem. Soc.*, 1972, **94**, 8357.

35.4.4.2 Polymeric hydroxy complexes

The existence of such complexes was first postulated by Bjerrum in 1908.[705] There have been a number of studies in recent years. Several early members of the series of hydrolytic polymers formed on the addition of base to aqueous chromium(III) solutions have been isolated;[706] purification was achieved by ion-exchange chromatography on Sephadex-SP C25 resin. The structures suggested for these complexes are illustrated (**156–162**). The structural unit (**162**) was held to be singularly important, both as a constituent of higher polymers and in the mechanism of dimerization. Equilibrium data for these complexes are summarized in Table 78.

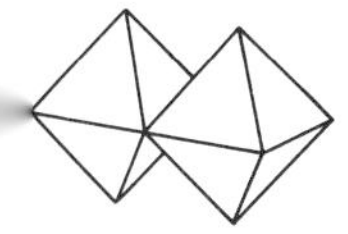

(156)

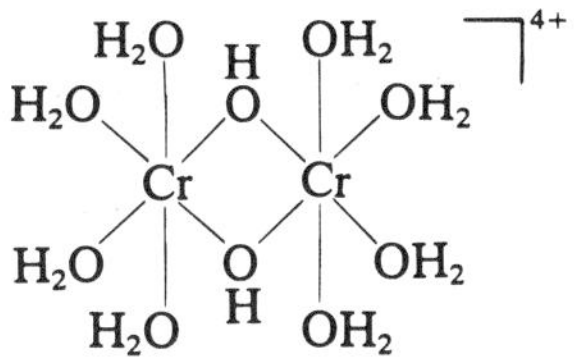

(156)

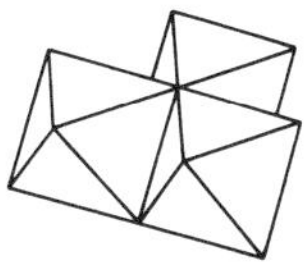

(157)

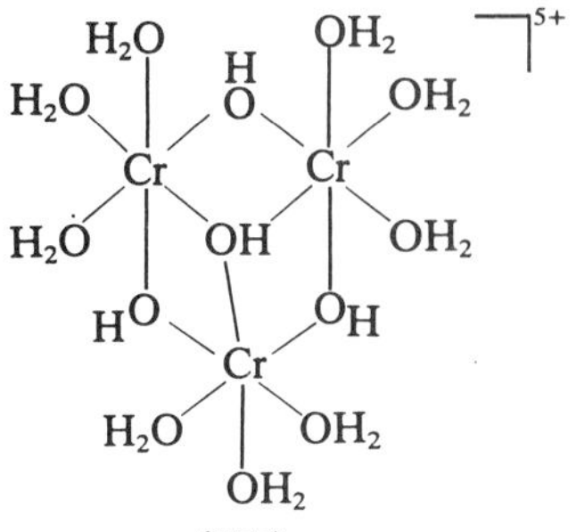

(157)

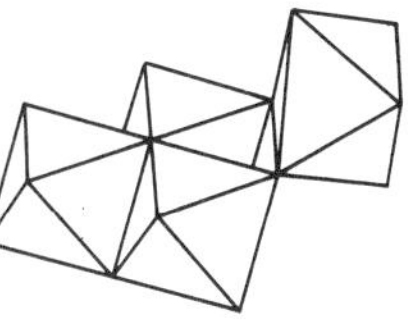

(158)

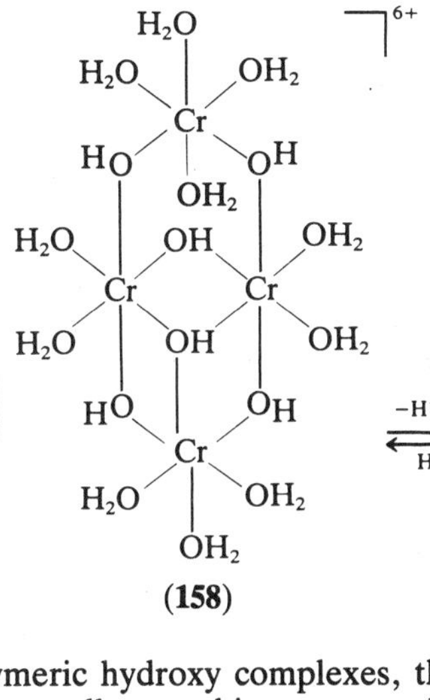

(158)

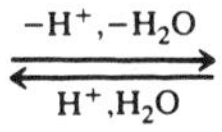

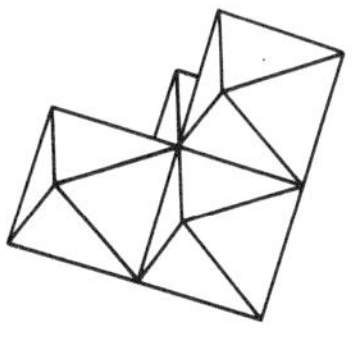

(159)

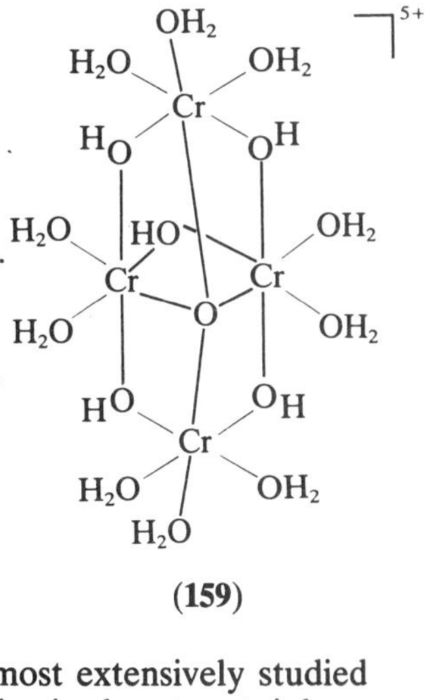

(159)

Among the polymeric hydroxy complexes, the dimer has been the most extensively studied species. There is no totally unambiguous crystal structure but the complex is almost certainly as illustrated (**156**). In addition to this doubly bridged dimer a singly bridged $(H_2O)_5CrOHCr(H_2O)_5]^{5+}$ (**163**) species has been reported.[707] The magnetic properties of both (**156**) and (**163**) have been studied. Coupling in the doubly bridged dimer (**156**) was

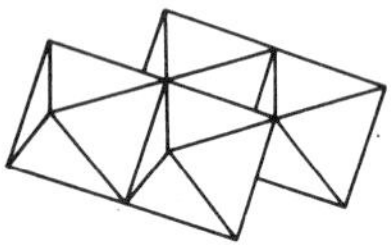

(160)   (160)   (161)   (161)

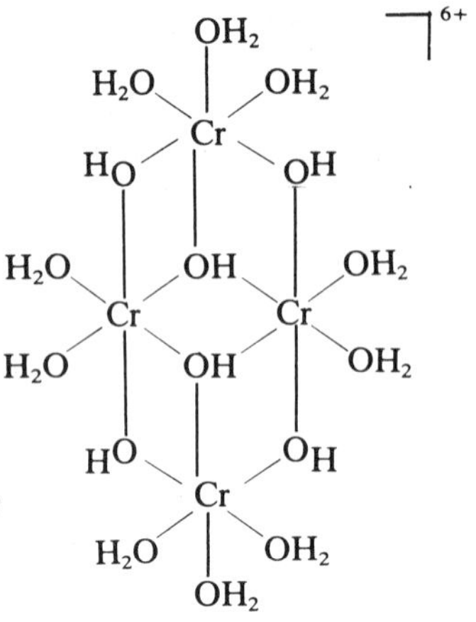

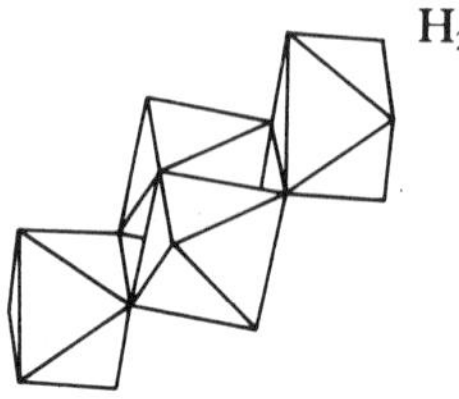

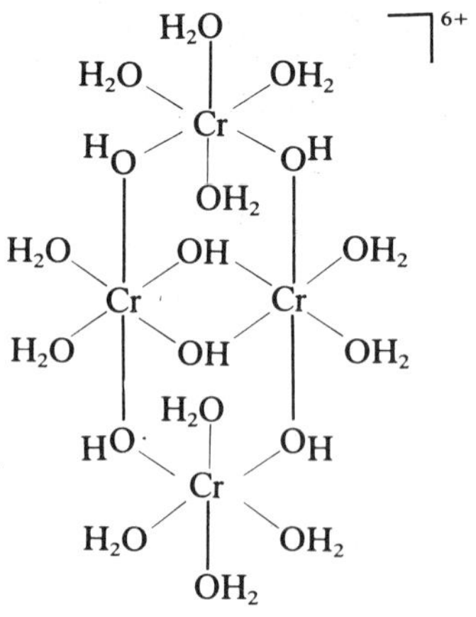

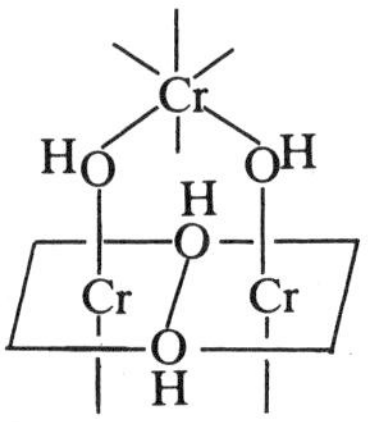

(162)

Table 78 pK_a Values for Coordinated Water in some Chromium Complexes

Complex	Cr^{3+}	$Cr_2(OH)_2$	$Cr_3(OH)_4$	$Cr_4(OH)_6$
pK_a values	4.29	3.68	4.35	2.55
	6.10	6.04	5.63	5.08
			6.01	

[a] After Marty (ref. 706); charges on complexes omitted, $I = 1.0\,M\,NaClO_4$, 25 °C, structures are given in (155)–(159).

characterized by $-J/k = 7.5°$ and $g = 1.96$, and for the singly bridged species (163) by $-J/k = 16°$ and $g = 1.96$. A more favourable bond angle was suggested to explain the stronger coupling in the latter case; direct Cr/Cr interactions were not believed to be important. The dimer (156) has also been the subject of an extensive NMR study, in solution, by the inert probe method.[708] Changes in structure at high proton activities were investigated; p-dioxane was used as the additive. Magnetic moments were measured, in solution, by the Evans method; values of 3.94 BM and 3.89 BM were obtained for the monomer and dimer respectively. There is obviously no strong spin interaction at ambient temperatures. At high concentrations of perchlorate (>8 M), the complex turns from blue/green to green ($t_{0.5}$ 15 min). Leffler suggested[708] that perchlorate complexation (164) causes this change (equation 41) based on the fact that monobridged dimer (163) and dimer (156) should be distinguishable in NMR line broadening experiments.

(163)

$$[(H_2O)_4Cr(OH)_2Cr(H_2O)_4]^{4+} + ClO_4^- \rightleftharpoons [(H_2O)_4Cr(OH)_2Cr(H_2O)_3(ClO_4)]^{3+} + H_2O \quad (41)$$

(156)   (164)

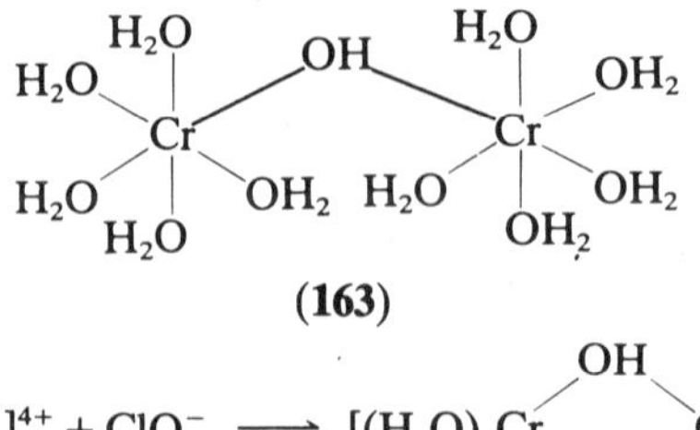

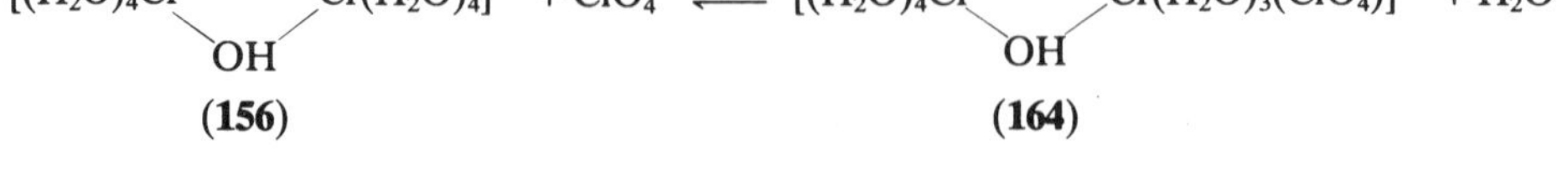

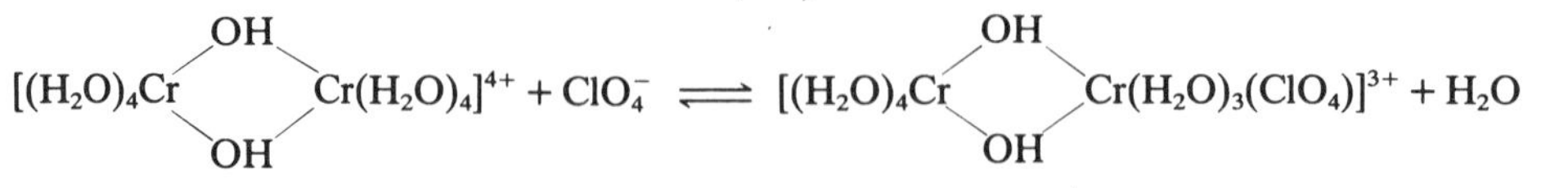

A trimeric complex has also been the subject of a detailed study.[709] The mole ratio [Cr]:[OH] was found to be 4:3; the degree of polymerization was assessed, by freezing point depression, to be three. A triangular structure, as shown in (**157**), was suggested on the basis of magnetic studies.

The closely related oxo-bridged complex $[(H_2O)_5CrOCr(H_2O)_5]^{4+}$ is obtained in low yields on the reduction of 1,4-benzoquinone by aqueous solutions of chromium(II).[710] The sole product of the hydrolysis of this dimer[711] ($HClO_4/LiClO_4$ $I = 1.0$) is hexaaquachromium(III).

35.4.4.3 Oxides and oxide hydroxides

The oxide systems deriving from chromium(III) are numerous. Chromium(III) oxide, Cr_2O_3, is the final product of the calcination of many chromium(III) complexes. Chromium(III) oxide has the Al_2O_3 structure (D_5) with hexagonal lattice parameters $a = 4.95$, $c = 13.66$ Å and $c/a = 2.761$.[712]

The oxide is also the product of the spectacular ignition of $(NH_4)_2Cr_2O_7$. The hydrated form of the oxide $Cr_2O_3{\cdot}2H_2O$ is an important pigment,[713] Guignet's green. Chromites formally contain the $Cr_2O_4^{3-}$ ion and are generally prepared by fusing Cr_2O_3 with a binary oxide.[714] They are typical spinels; chromium(III) occupies octahedral sites. The parent spinel, chromite—$FeCr_2O_4$—is the only commercially important chromium mineral. Calcium chromite is not a spinel and exists in two modifications: the β form, which is isomorphous with calcium ferrite, space group P_{nam},[715] and the α form, which is a high-temperature stable modification,[716] prepared by fusing $CaCO_3$ and Cr_2O_3 (1350 °C, 24 h). The spinels formed by chromium are all of the normal type; lattice energy calculations for a large number of such species have recently been reported.[717]

Another important class of oxide-containing solids are the compounds of general formulation CrO_2H, referred to as oxide hydroxides or, occasionally, as chromous acids. The system is polymorphic; a rhombohedral form is obtained by the decomposition of an aqueous solution of Cr_2O_3 (Mallinkrodt analytical reagent) in a high pressure vessel at 300 °C.[718] The material is red/brown and has a layer structure; oxygen atoms are coordinated to chromium(III) in a distorted octahedron. Each octahedron shares six edges with six surrounding coplanar octahedra to form a continuous sheet of close-packed oxygen atoms. The sheets are superimposed, oxygen atoms in one sheet falling directly above those of the sheet below. The structure as a whole is not close-packed; the layers are held together by short H bonds.[719] A second form of CrO_2H has been prepared by the hydrothermal treatment of CrO_2 (derived from CrO_3, 72 h, 450 °C, 40 000 psi); this form is olive green. The compound is orthorhombic and isomorphous with InO_2H, *i.e.* a deformed rutile structure.[720,721] The coprecipitation of γ-(Al, Cr)O_2H and H(Cr, Al)O_2 has been investigated.[722]

35.4.4.4 Other oxide systems

The novel material $Cr_2Te_4O_{11}$ contains the binuclear ion $[Cr_2O_{10}]^{14-}$, as isolated units, formed by two CrO_6 octahedra sharing an edge.[723] The exchange interaction between the chromium centres has been modelled by the Heisenberg–Dirac–Van Vleck phenomenological model, with the exchange parameter $J_{SASB} = -6\ (\pm 0.5)\ cm^{-1}$. The substitution into the $M^{III}TeO_6$ system of VO^{2+} has been investigated for a number of metals including chromium(III).[724] The transition between monoclinic and rutile forms can be modelled on the basis of the size of the M^{III} ion. The substitution of chromium(III) into yttrium, iron and gallium garnets has been investigated.[725] In the system $[Y_3][Cr_xFe_{2-x}][Fe_3]O_2$ single-phase specimens were obtained with values of $x = 0.5$. For the gallium system, $x < 0.75$ gave a single phase. Perovskites $A^{III}B^{III}O_3$ involving indium or thallium and chromium(III) have been synthesized.[726] The coupling of the chromium(III) centres in the trirutile compound MCr_2O_6 (M = Te or W) has been investigated.[727] The coupling is dependent on the second metal ion and has been correlated with slight structural variations.

35.4.4.5 Peroxides

Two green peroxychromium(III) species have been reported.[728] The products of the reaction of H_2O_2 and Cr^{VI} in 2–6 M perchloric acid were purified by ion-exchange chromatography

(Dowex 50W-X2). The compounds were tentatively identified as CrO_2Cr^{4+} and $CrO_2CrO_2Cr^{5+}$ species; their decompositions were not simple processes. Attempts to synthesize these or similar species by the oxidation of chromium(II) with oxygen were unsuccessful. The compound $[(H_2O)_5CrO_2Cr(H_2O)_5]^{4+}$ was also reported earlier.[729]

35.4.4.6 Alcohol and alkoxide complexes

(i) Alcohol complexes

Alcoholate complexes of the general formula $CrCl_3 \cdot nROH$ (R = Me, Et, Pr^i, Bu^n and *n*-hexyl) have been reported.[730] These purple insoluble complexes were prepared by refluxing $CrCl_3 \cdot 3THF$ in the anhydrous alcohol and are typical chromium(III) species. The structure of the related complex *trans*-$[Cr(MeOH)_4Cl_2]Cl$ has been determined.[731] It is similar to that of the related aqua complex and is of approximately C_{4v} symmetry. A detailed study of the complexes formed by ethylene glycol and chromium(III) has been reported;[732] both monodentate and bidentate coordination of chromium by the glycol was postulated.

(ii) Alkoxide complexes

Alkoxide complexes including those of chromium(III) have been extensively and excellently reviewed.[733,734] Methoxy complexes, $[Cr(OMe)Cl_2]_n \cdot 2nMeOH$ and $[Cr(OMe)Cl_2]_n \cdot nMeOH$, have been thoroughly studied.[735] The former is prepared by the reaction of anhydrous $CrCl_3$ with methanol, the latter by heating the former at 100 °C for several hours. Adducts of the dimethanoates can be prepared with acetone, acrylonitrile and dioxane. These compounds are unusual among alkoxides in showing marked solubility: the acetone adduct appears to be a dimer in both solution and the solid state. A tetrameric structure was suggested for the monomethanoate (**165**).

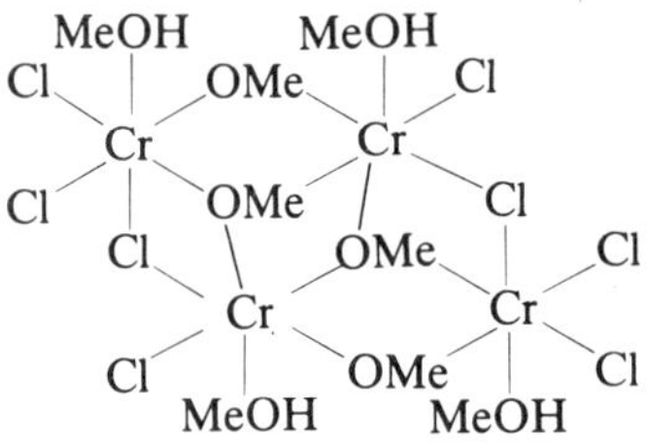

(**165**)

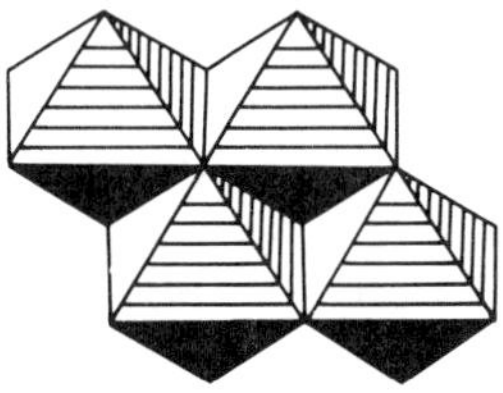

(**165**)

The physical properties of samples of $Cr(OMe)_3$ prepared by a variety of methods have been investigated.[736] Good crystals could not be obtained but limited crystallography suggested a layered lattice with chromium occupying octahedral sites. A useful summary of earlier preparations and attempted preparations of chromium methoxides can be found in this paper.

The insertion of various isocyanates into chromium(III) alkoxide M—O bonds has been reported.[737] The complexes are prepared by refluxing the isocyanates with a suspension of the alkoxide in benzene. No structural data were given for the products. Unusual bimetallic alkoxides have recently been prepared[738] by the reaction of $Cr[Al(OPr^i)_4]_3$ with alcohols and acetylacetone (**166**). A wide range of spectroscopic methods were used to study them. In general, the results were in accord with a monomeric formulation similar to (**166**) below; $Cr[Al(OMe)_4]_3$ was grossly insoluble; the small size of the methyl groups may permit extensive polymerization.

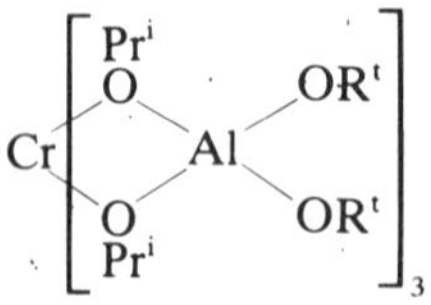

(**166**)

35.4.4.7 *β-Ketoenolates and related ligands*

(i) Synthesis

The tris and bis complexes of acetylacetone (2,4-pentanedione) (**167**) with chromium(III) have been known for many years (**168, 169**).[739] The tris compound is generally prepared by the reaction of an aqueous suspension of anhydrous chromium(III) chloride with acetylacetone, in the presence of urea.[740] Recently a novel, efficient synthesis of tris(acetylacetonato)chromium-(III) from CrO_3 in acetylacetone has been reported.[741] The crystal structure of the tris complex has been determined.[742] A large anisotropic motion was observed for one of the chelate rings, attributed to thermal motion, rather than a slight disorder in the molecular packing.

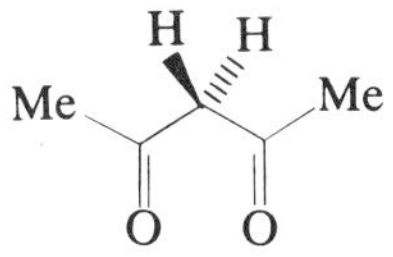
(**167**)

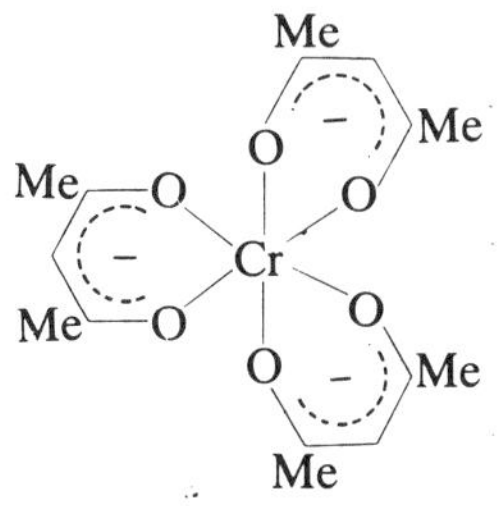
(**168**)

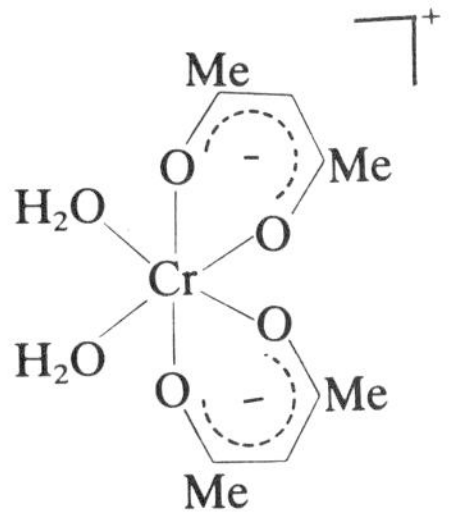
(**169**)

Derivatives of, or ligands closely related to, acetylacetonate have been extensively studied. Compounds of analogous structure are, in general, obtained. Ligands studied include 1-phenylbutane-1,3-dione, 1,1,1-trifluoropentane-2,4-dione, 1,1,1,5,5,5-hexafluoropentane-2,4-dione, 2,2,6,6-tetramethylheptane-3,5-dione and 1-phenyl-5,5-dimethylhexane-2,4-dione.[743] Preparations include the original one detailed above[740] and methods from chromium(III) acetate and metallic chromium.[744] The earlier work and advantages of the various methods are summarized by Dilli and Robards,[743] who noted, particularly with derivatives of acetylacetone, that in the classic method,[740] reaction of the free ligand with urea may markedly reduce the yield.

The closely related tris complex of malonaldehyde (**170**) has been synthesized by the reaction of 1,1,3,3-tetramethoxypropane with anhydrous $CrCl_3$ in an ether slurry.[745] The complex was purified by sublimation and contains the simplest structural unit capable of enolizing to form a β-ketoenolate chelate.

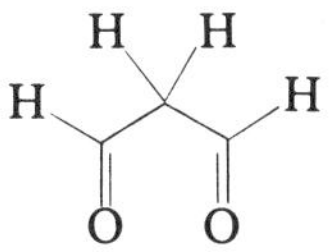
(**170**)

Several unusual complexes containing neutral acetylacetonate have been reported.[746] The complex $[CrBr_2(acac)(acacH)]$ was obtained from the reaction of HBr with $[Cr(acac)_3]$; $[CrCl_2(acac)(acacH)]$ was obtained from the reaction of $CrCl_3(THF)_3$ with acetylacetonate. Other unusual β-diketonates include the tris(rhenaacetonates) (**171**).[747,748] Two different forms of the tris complex of 2-nitroacetophenone with chromium(III) have been reported;[749] these are probably geometric isomers.

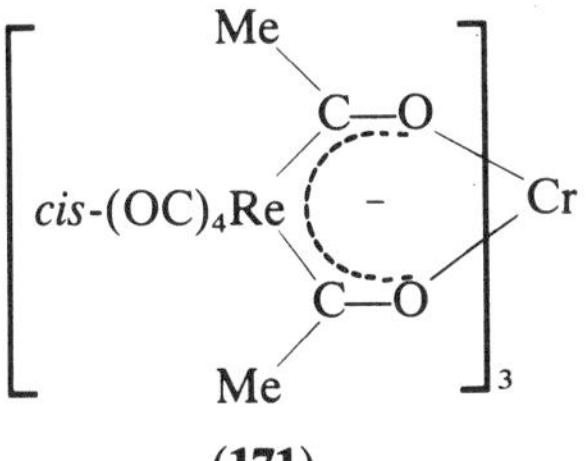
(**171**)

Complexes containing two different ligands of the β-diketonate type have also been prepared[750] including the series $[Cr(CF_3COCHCOMe)_n(MeCOCHCOMe)_{3-n}]$ ($n = 1$ or 2); *cis* and *trans* isomers were isolated. The linkage isomerization and disproportionation of the unsymmetrically substituted acetylacetonate complex (**172**) have been studied[751] and were said to be intermolecular.

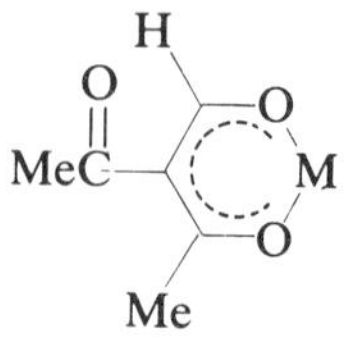

(**172**)

Camphorate complexes of chromium(III) have been studied. The four possible isomers of the tris complex of (+)-3-acetylcamphorate (**173**) were isolated,[752,753] and absolute configurations were tentatively assigned. The photoisomerization of these complexes has been investigated;[754] quantum yields of the order of 10^{-3} were obtained with visible or ultraviolet radiation at temperatures around 100 °C. Bond-breaking processes were held to be important in the reactivity of *cis* isomers.

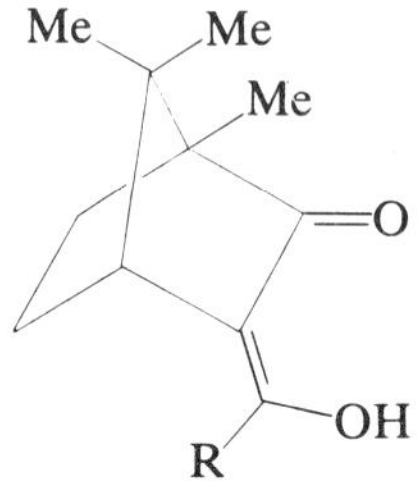

(**173**) R = H or Me

Dimeric complexes of bis(acetylacetonates) have also been reported: di-μ-diphenylphosphinatoacetylacetonatochromium(III) prepared by the reaction of diphenylphosphinic acid with an excess of $[Cr(acac)_3]$ has been investigated.[755] It crystallizes in the triclinic space group $P1$, the bridge $(CrOPOCr)_2$ forming a puckered eight-membered ring. The crystal structure of the alkoxy-bridged complex bis(μ-methoxy)bis[bis(2,4-pentanedionato)chromium(III)] has been determined; it crystallizes in a racemic rather than *meso* form.[756] Other alkoxy-bridged complexes have also been synthesized.[757]

(*ii*) *Physical studies*

Acetylacetonates have occupied a central position in the study of complexes by physical methods. Some of this work is summarized in Table 79.

Table 79 Some Physical Properties of $[Cr(acac)_3]$

Property	*Value*	*Comment*	*Ref.*
Cr—O bond energy	256 kJ mol^{-1}	Combustion calorimetry	1
10*Dq*	217 kJ mol^{-1}	Combustion calorimetry	2
10*Dq*	17.8×10^3 cm^{-1}	Electronic spectra	3
	17.5×10^3 cm^{-1}	Also substituted acacs	4
		All $17.5(\pm 3) \times 10^3$ cm^{-1}	
β_{35}	0.55 cm^{-1}	Nephelauxetic parameters	5
β_{55}	0.79 cm^{-1}	Also for substituted acacs	
$E_{0.5}$	−1.73 V	*vs.* SCE in DMSO	6
		Also 17 substituted acacs	

1. M. M. Jones, B. J. Yow and W. R. May, *Inorg. Chem.*, 1962, **1**, 166.
2. J. L. Wood and M. M. Jones, *Inorg. Chem.*, 1964, **3**, 1553.
3. R. L. Lintvedt and L. K. Kernitsky, *Inorg. Chem.*, 1970, **9**, 491.
4. C. J. Ballhausen, 'Introduction to Ligand Field Theory', McGraw-Hill, New York, 1962, p. 239.
5. A. M. Fatta and R. L. Lintvedt, *Inorg. Chem.*, 1971, **10**, 478.
6. R. F. Handy and R. L. Lintvedt, *Inorg. Chem.*, 1974, **13**, 893.

(*a*) *Vapour phase chemistry*. A remarkable feature of the chemistry of the complexes of the tris(acetylacetonate) type is their extreme stability in the vapour phase: there are hence many mass spectral and gas chromatographic studies of them. Much of this emanates from the group of Sievers. The complexes of hexafluoroacetylacetonate are considerably more volatile than those of the parent acetylacetonate. The partial resolution of tris(1,1,1,5,5,5-hexafluoro-2,4-pentanedionato)chromium(III) by gas chromatography on pulverised d-quartz has been reported.[758,759] The *cis* and *trans* isomers of the tris(trifluoroacetylacetonate) complexes of chromium(III) can be separated efficiently by gas chromatography (5% silicone grease on Chromasorb-W, 115 °C).[759] To provide better understanding of such studies, vapour pressure measurements have been made on a number of metal β-diketonates.[760] Earlier related measurements are reviewed in this paper. In typical gas chromatographic phases (Apiezon-L, QF1) the results of the above study indicated that the solution behaviour of these tris octahedral complexes is determined solely by non-specific van der Waals type forces; there was no evidence for a direct interaction such as H bonding.

A rate and equilibrium study of the *cis/trans* isomerization of tris(1,1,1-trifluoro-2,4-pentanedionato)chromium(III) in the vapour phase has been reported.[761] The equilibrium constant 3.56 was independent of temperature in the range 118–144.8 °C, the same value being measured by both static and kinetic methods. The activation energies for isomerization were both less than 115 kJ mol^{-1}; the Cr—O bond energy is 210 kJ mol^{-1}. On these grounds, a bond-rupture mechanism was rejected. A similar mechanism may operate for the isomerization in solution.

The mass spectra of various β-diketonates have been extensively studied.[762,763] The availability of ionization energies, measured by mass spectra, has led to a fairly detailed debate as to the nature of the orbital from which the first electron ionizes.[764] Gas phase rearrangements,[765] often involving fluorine migration to the metal, have been observed. The SIMS spectrum of $[Cr(acac)_3]$ is particularly simple,[766] yielding the base peak $[Cr(acac)_2]^+$.

The structure of tris(1,1,1,5,5,5-hexafluoro-2,4-pentanedionato)chromium(III) has been determined by gas-phase electron diffraction.[767] A most interesting feature of this study was that the normalized ligand bite angle [(O—O/M—O) 30.1°] was very close to the value found in the solid state by X-ray diffraction (30.8°) but quite different to the value predicted by Kepert's model.[768] The results of this and related work on vapour phase structures are summarized in Table 80.

Table 80 Summary of Electron Diffraction Data[a]

Compound	*C—O*	*C—C (ring)*	*Method*	*Ref.*
$[Cr(hfa)_3]$	1.270	1.409	ED	1
$[Cr(hfa)_2]$	1.276	1.392	ED	2
$[Cr(acac)_3]$	1.263	1.388	X-Ray	3
hfa	1.259	1.407	ED	4
acac	1.287	1.405	ED	5

[a] Abbreviations: hfa = 1,1,1,5,5,5-hexafluoro-2,4-pentanedione, acac = 2,4-pentanedione, ED = electron diffraction. All bond lengths in Å.
1. Ref. 767.
2. B. G. Thomas, M. L. Morris and R. L. Hilderbrandt, *J. Mol. Struct.*, 1976, **35**, 241.
4. A. L. Andressen, D. Zebelman and S. H. Bauer, *J. Am. Chem. Soc.*, 1971, **93**, 1148.
5. A. H. Lowery, C. George, P. D'Antonio and J. Karle, *J. Am. Chem. Soc.*, 1971, **93**, 6399.

(*b*) *Optical activity*. As tris β-diketonates are neutral D_3 complexes, resolution by the more-popular, less-soluble diastereoisomer and ion-exchange methods is not possible. Various methods have been reported. Potentially most useful are: a stereospecific synthesis from an unisolated (+)-tartrate complex of chromium(III),[769] chiral chromatography using (2*R*,3*R*)-(−)-dibenzoyltartaric acid[770] and chromatography on an L-nickel(II)tris(1,10-phenanthroline)/montmorillonite column.[771] Other methods of resolving $[Cr(acac)_3]$ are extensively reviewed in these papers.

The crystal structure and absolute configuration of L-(−)-$[Cr(acac)_3]$ has been reported.[772] The circular dichroism of D-$[Cr(acac)_3]$ has been studied in the solid state and in solution.[773] The rotational strengths of the *d–d* transitions are extensively discussed; competing rather than reinforcing effects lead to the smaller rotational strengths observed for tris(acetylacetonates) as compared to tris-(ethylenediamine) or -oxalate complexes.

The Cotton effects in mixed amino acidate/acetylacetonate complexes [Cr(acac)$_2$L] (L = L-alanine, L-valine or L-phenylalanine) have been studied:[774] absolute configurations were assigned by reference to the parent tris(acetylacetonate) complexes. Synthesis was achieved by the photolysis of mixtures of the amino acid and [Cr(acac)$_3$]. The partial photoresolution of both *cis*- and *trans*-(1,1,1-trifluoro-2,4-pentanedionato)chromium has been accomplished by irradiation with circularly polarized light (5461 Å) in chlorobenzene solution.[775] The results indicated that both bond rupture and twist mechanisms were important. A number of other β-diketonates have also been investigated.[776]

(*c*) *NMR*. The complex [Cr(acac)$_3$] has found use as a spin relaxant in ^{13}C FTNMR spectroscopy.[777] The solution dynamics of [Cr(acac)$_3$] in methanol, acetonitrile, benzene and DMSO have been studied in some detail.[778] Different models were proposed for solvation by the various solvents. In methanol, the hydroxyl proton is apparently closer to the chromium than the methyl group. An outer-sphere dipolar mechanism of exchange was suggested in DMSO, the oxygen atom of DMSO pointing toward the octahedral face of the [Cr(acac)$_3$]. In benzene a random orientation of the flat and cleft positions, with respect to the solvent, best fits the data. In acetonitrile the results, surprisingly, suggest that the methyl group points toward the complex; weak H bonding may account for this observation. A summary of other spin relaxation studies can be found in this paper.[778]

A detailed understanding of the many factors involved when [Cr(acac)$_3$] is used as a spin relaxant is required. It has been shown that line widths are affected by the reagent.[779] If temperature dependent features are to be studied, it was concluded that the ratio of relaxant to substrate must be kept as low as possible. Caution in the use of [Cr(acac)$_3$] as a spin relaxant was also emphasized in a study of cholesterol chloride.[780] Despite the problems with the use of this complex as a spin relaxant, it remains widely used.[779,780]

(*iii*) *Chemical reactivity*

The bromination of tris(acetylacetonato)chromium(III) was first reported by Reihlen.[781] There have been many studies of electrophilic substitution at complexes of both acetylacetonate and its derivatives; this work has been extensively reviewed.[782,783] Some typical reactions are outlined below (equation 42). In this section, we shall briefly mention some more recent work; the interested reader is recommended to study the extensive, although somewhat dated, review by Collman,[782] and Mehrotra's book.[783]

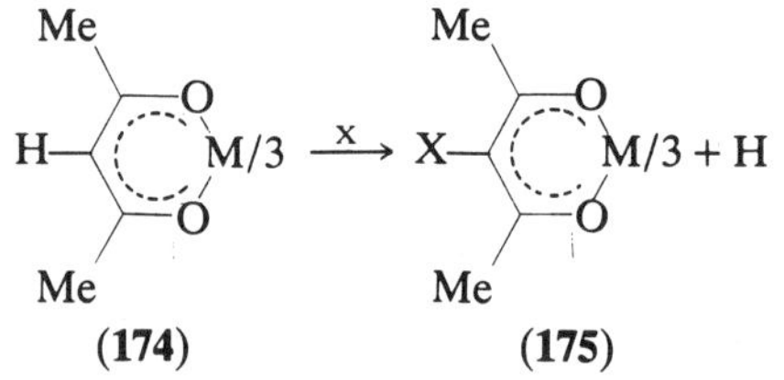

(42)

(174) (175)

X = I, Br, Cl, SCN, SAr, SCl, NO_2, CH_2Cl, CH_2NMe, COR, CHO

The electrophilic substitutions of acetylacetonate complexes have been taken as suggesting 'aromatic character' in the chelate ring. Results with seventeen different 1,3-diketonatochromium(III) complexes were recently held to support this suggestion (**176–178**; equation 43).[784] The bromination of tris(1,1,1-trifluoro-2,4-pentanedionato)chromium(III), previously claimed to be unreactive,[785] has been reported.[786]

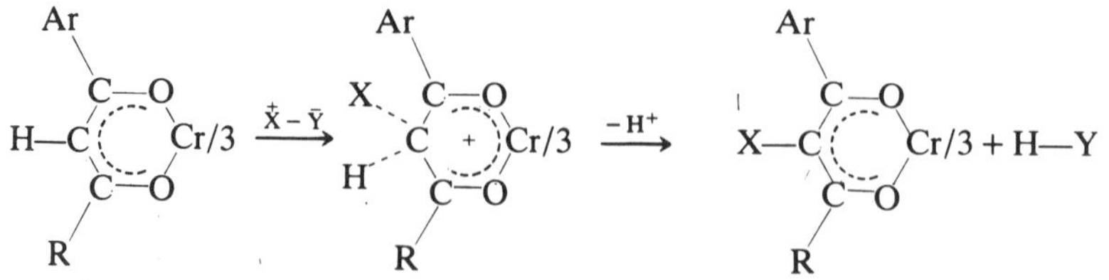

(43)

(176) (177) (178)

$X^+ = Cl^+$, Br^+, or NO_2^+
Ar = 4-FC_6H_4 or a derivative
R = Me, Et, Pr^n, CF_3 or Ph

An isolated study of the tris tropolonate complex (**179**) of chromium(III) has appeared;[787] mass spectra were recorded and fragmentation patterns discussed.

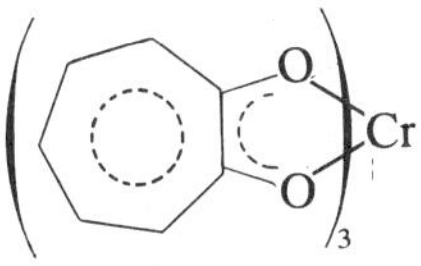

(**179**)

35.4.4.8 Catecholates, quinones, tetrahydrofuran and O-bonded urea

The chemistry of the transition metals including chromium(III) with these ligands has been the subject of a recent and extensive review,[788] with references to the early literature. The close relationship between the catechol (**180**), semiquinone (**181**) and quinone (**182**) complexes may be appreciated by considering the redox equation below (equation 44).[789] The formal reduction potentials for the chromium(III) complexes (**183–186**; equation 45) are +0.03, −0.47 and −0.89 V (*vs.* SCE in acetonitrile) respectively.

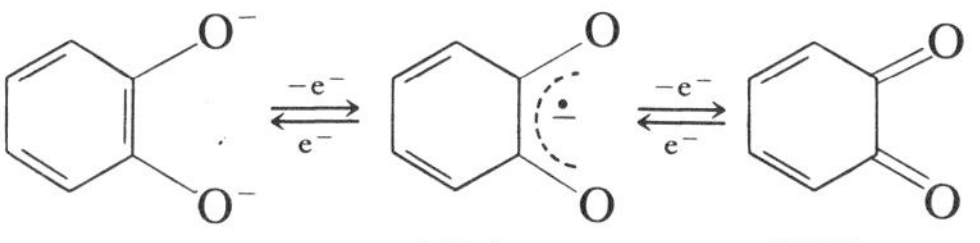

(44)

(**180**)cat (**181**) (**182**)

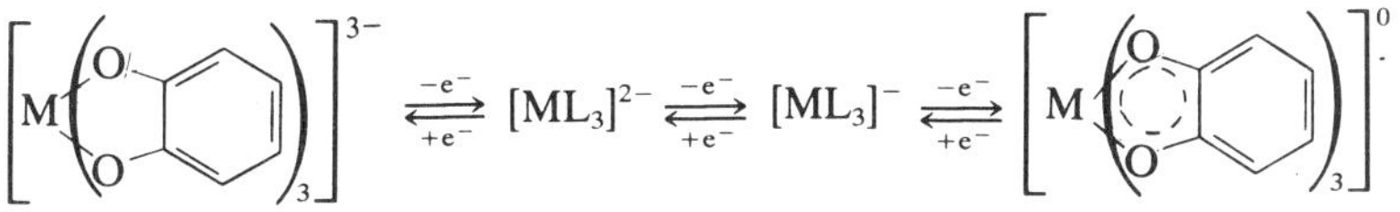

(45)

(**183**) (**184**) (**185**) (**186**)

(i) o-Semiquinones

The preparation of tris(*o*-semiquinone) complexes of chromium(III) has been reported;[790] complexes of formulae $Cr(o\text{-}Cl_4SQ)_3\cdot 4Ph$ and $Cr(phenSQ)_3\cdot$anisole were prepared ($o\text{-}Cl_4SQ$ = tetrachloro-1,2-benzoquinone; **189**). The complexes exhibit temperature-dependent magnetic properties, μ_{eff} 1.1 BM at 286 K dropping to 0.35 BM at 4.2 K. This is the result of intramolecular spin-coupling between chromium(III) and the semiquinone. The complexes were prepared by the reaction of chromium hexacarbonyl with the quinone (equation 46) but are correctly formulated,[790] by comparison with authentic samples, as semiquinones of chromium(III). The above reaction has proved difficult to repeat in the absence of photolysis,[791] and it was hence suggested to be a photochemical, oxidative substitution. In related work, the reaction of $Crbipy(CO)_4$ with a number of quinones was studied.[792] The resulting redox series: $[Cr(SQ)_2(bipy)]^+/[Cr(SQ)(Cat)(bipy)]/[Cr(Cat)_2(bipy)]^-$ was studied; the oxidation state of the chromium remains at three during these reactions.

$$\underset{(\mathbf{187})}{Cr(CO)_6} + \text{quinone} \longrightarrow \underset{(\mathbf{188})}{Cr(\text{quinone})_3} \qquad (46)$$

e.g. quinone =

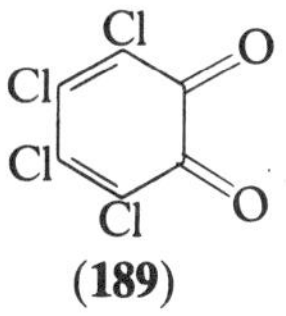

(**189**)

(*ii*) *Catecholates*

The resolution of tris(catecholato)chromate(III) has been achieved by crystallization with L-$[Co(en)_3]^{3+}$; the diastereomeric salt isolated contained the L-$[Cr(cat)_3]^{3-}$ ion.[793] Comparison of the properties of this anion with the chromium(III) enterobactin complex suggested that the natural product stereospecifically forms the L-*cis* complex with chromium(III) (**190**). The tris(catecholate) complex $K_3[Cr(Cat)_3]\cdot 5H_2O$ crystallizes in space group *C*2/*c* with $a = 20.796$, $b = 15.847$ and $c = 12.273$ Å and $\beta = 91.84°$; the chelate rings are planar.[794] Electrochemical and spectroscopic studies of this complex have also been undertaken.[795] Recent molecular orbital calculations[796] on quinone complexes are consistent with the ligand-centred redox chemistry generally proposed for these systems.[788]

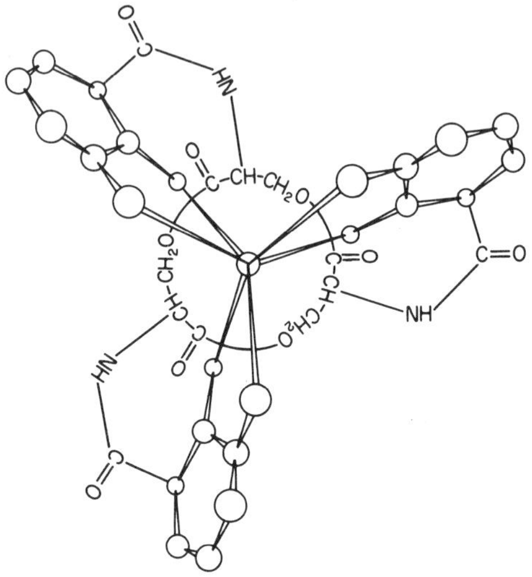

(**190**) A diagram of the Δ-*cis* chromium(III) enterobactin trianion

(*iii*) *Hydroxamates*

Chromium(III) complexes of a number of polyhydroxamic acids, microbial iron sequestering and transport agents (siderochromes) have been reported.[797,798] The kinetic inertness of the chromium(III) complexes allows the facile separation of isomers; for the model complex tris(*N*-methyl-(−)-methoxyacetylhydroxamato)chromium(III), D-*cis*, L-*cis* and the L/D-*trans* isomers have been separated.[798] The chromium complexes of desferrioxamine B (**191**) have been investigated; the possible isomers are illustrated below (**192–196**). The *cis* isomer was isolated in relatively pure form.[799] Thiohydroxamate[800] and dihydroxamate (rhodotorulic acid) complexes have also been studied.[801]

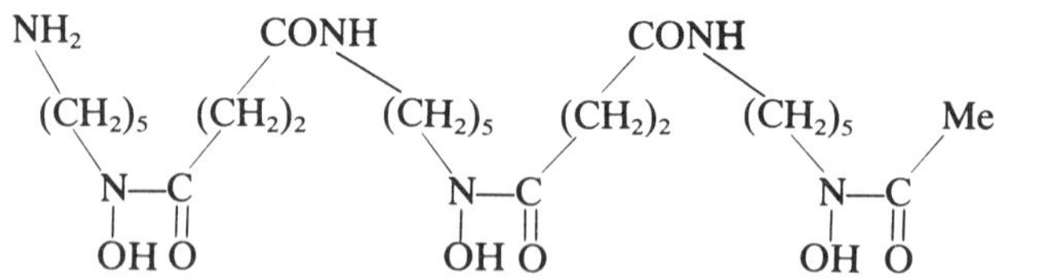

(**191**)

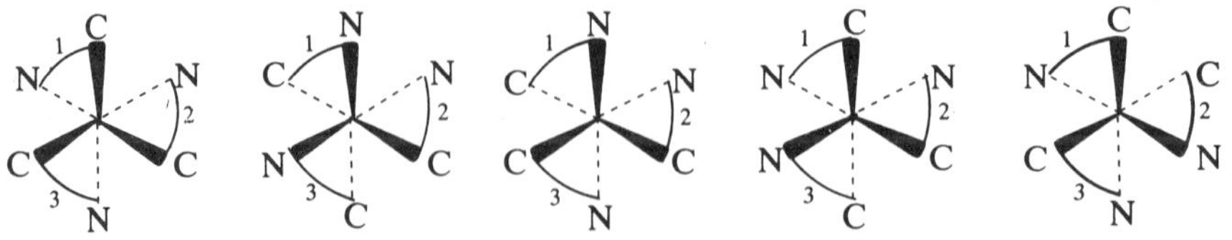

Λ-N-*cis*,*cis* (**192**) Λ-C-*trans*,*cis* (**193**) Λ-C-*trans*,*trans* (**194**) Λ-N-*cis*,*trans* (**195**) Λ-N-*trans*,*cis* (**196**)

The five enantiomeric geometrical isomers of ferrioxamine B. The oxygen donor atoms of each hydroxamate group have been omitted for clarity. The Λ optical isomer is shown in each case.

(iv) THF

The complex $[Cr(THF)_3Cl_3]$ was first prepared in 1958.[802] However pure bromide, iodide and thiocyanate complexes were reported only in 1983.[803] The preparation of $[Cr(THF)_6]^{3+}$ by the reaction of $[Cr(THF)_3Cl_3]$ with $AgBF_4$ in THF was attempted; only an impure brown/pink oil was obtained. The tris complexes are useful starting materials in both organometallic and coordination chemistry.

(v) Urea (O-bonded)

Chromium is unusual in that it forms a stable hexakis O-bonded urea complex. The complex was first prepared as the chloride salt by Pfeiffer[804] and a crystal structure of the complex salt $[Cr\{OC(NH_2)_2\}_6][Cr(CN)_6]\cdot 2DMSO\cdot 2EtOH$ has recently been reported.[805] Coordination at chromium(III) is octahedral; r(Cr—O) is in the range 1.96–1.98 Å. The reduction of the perchlorate salt of this complex to a chromium(II) species has been studied polarographically.[806] Detailed studies of the luminescence spectra of several salts of the chromium urea complex have been reported.[807,808]

35.4.4.9 Non-C oxo anions

(i) Nitrogen-containing

Anhydrous chromium(III) nitrate may be prepared by the reaction of dinitrogen pentoxide with chromium carbonyl in dry carbon tetrachloride.[809] The product has rather low thermal stability, is involatile and decomposition begins at 60 °C.

The complex 1,2,3-trinitrotriamminechromium(III) has been prepared and is a nitrito isomer.[810] The nitrito complexes of *cis*- and *trans*-bis(ethylenediamine)chromium(III) have been prepared,[811] *cis*-$[Cr(en)_2(ONO)_2]ClO_4$ and *trans*-$[Cr(en)_2(ONO)_2]X$ ($X = ClO_4^-$ or NO_3^-). The nitrito complexes *cis*- and *trans*-$[Cr(en)_2(ONO)(X)]^+$ ($X = ONO^-$, OH^-, F^-, Cl^-, Br^-) have also been studied.[812]

(ii) Sulfur-containing

There are old reports of tris sulfate complexes[813,814] and a number of chromium(III) sulfate hydrates are known. In recent years monosulfatopentaaquachromium(III) complexes have been prepared[815] and characterized by IR spectroscopy;[816] sulfate probably behaves as a monodentate ligand. The basic chromium(III) sulfate $Cr(OH)SO_4\cdot H_2O$ crystallizes in the non-centrosymmetric, monoclinic, space group *Cc*.[817] Each chromium atom is coordinated octahedrally by two hydroxo groups, three sulfate O atoms and a water molecule. The metal atoms are joined by single hydroxo bridges forming an infinite chain.

In acidic aqueous solutions the reaction between hexaaquachromium(III) and sulfate has been studied in detail; ion-pairing can be detected in this system.[818] Polymeric species involving μ-hydroxy and μ-sulfato groups have been isolated.[819,820]

Trifluoromethylsulfate forms weak complexes with chromium(III);[821] this anion may provide a viable alternative to perchlorate as a 'non-coordinating anion' (for $Cr(O_3SCF_3)_3$ see Section 35.7.2). Tris(O,O'-sulfinate) complexes of chromium(III) have been prepared (RSO_2^{2-}, R = Me, C_6H_5, p-$CH_3C_6H_4$); typical monomeric complexes are obtained.[822]

(iii) Phosphorus-containing

(a) Complexes of H_3PO_3 and HPO_2. Ebert prepared a complex which he formulated as $H_3[Cr(HPO_3)_3]^{3-}$ (trisphosphitochromic acid) by refluxing chromium(III) hydroxide with H_3PO_3.[823] The complex was said to be tris-chelated (**197**) and a resolution as the strychninium salt reported.[823,824] It may provide one of the very few examples of molecular optical activity in a non-carbon-containing molecule. Its IR spectrum has been studied in detail.[825] Good analyses, particularly for phosphorus, have not been obtained;[824,825] until such data become available there must be some doubt as to the exact composition of these species. A number of other complexes involving this ligand, including a series of bisphosphites, have been reported.[826,827]

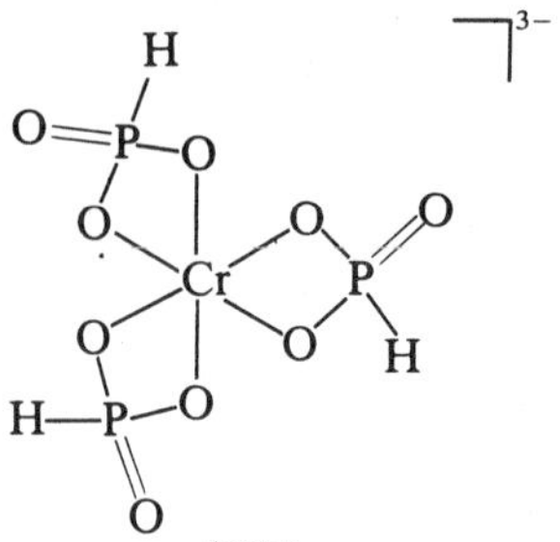

(197)

A monophosphite complex has been prepared by the reaction of either chromium(III) perchlorate[828] or chromate[829] with the ligand. The related monohypophosphite complex $[Cr(H_2O)_4H_2PO_2]^{2+}$ has also been reported.[830]

(*b*) *Phosphates.* Although chromium phosphates have been known for years,[831] there are few that are well characterized. A crystalline hexahydrate is the best-characterized species and exists in at least three forms.[832] Simple complexes of phosphate and chromium(III) may be important in the absorption of chromium(III) by biological systems.[833,834] Phosphate derivatives have been studied and simple complexes have been prepared from chromium(III) in combination with: tri-*p*-tolylphosphate,[835] alkoxyalkylphosphate,[836] isopropylphosphate,[837] triethylphosphate[838] and dimethylmethylphosphonate (for $Cr(O_2PF_2)_3$ see Section 35.7.2).[839]

Chromium(III) forms stable complexes with adenosine-5′-triphosphate.[840,841,842] These are kinetically inert analogues of magnesium ATP complexes and may be used to study enzyme systems. The complexes prepared are chiral and may be distinguished in terms of chirality at the metal centre (**198, 199**).[843] The related complex of chromium(III) with adenosine-5′-(1-thiodiphosphate) has been prepared; the diastereoisomers were separated.[844] The stereospecific synthesis of chromium(III) complexes of thiophosphates has been reported[845] by the method outlined in equation (47), enabling the configuration of the thiophosphoryl centre to be determined. The availability of optically pure substrates will enable the stereospecificity of various enzyme systems to be investigated.[845]

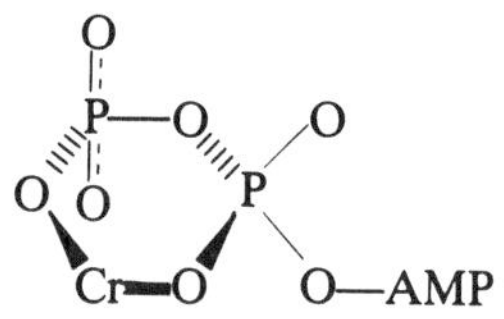

(**198**) Δ-$Cr(H_2O)_4$ATP

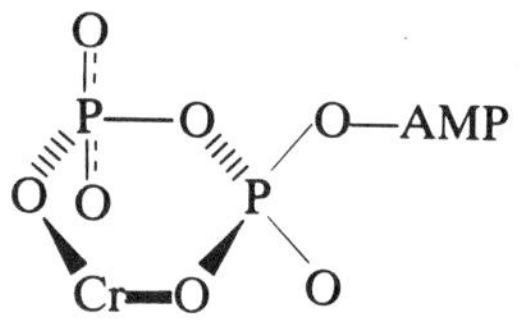

(**199**) Λ-$Cr(H_2O)_4$ATP

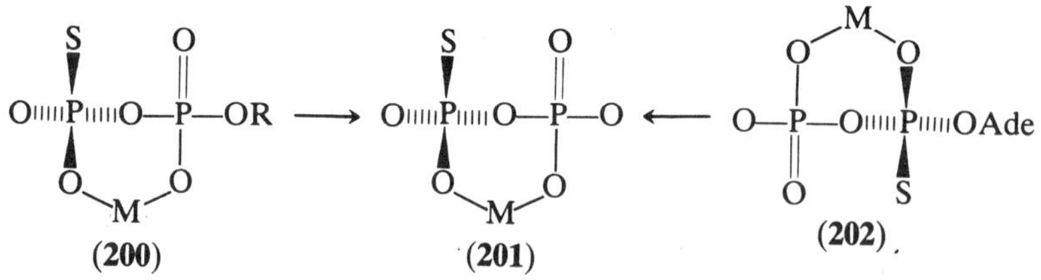

(**200**) (**201**) (**202**) (47)

(*c*) *Phosphinates.* Typical chromium(III) phosphinates are polymeric materials soluble in non-polar solvents (**203**). Poly{di-μ-(diphenylphosphinato)}aquahydroxychromium(III) may be prepared by the aerial oxidation of the chromium(II) precursor.[846] This method leads to a molecular weight of 6000 (number average). Polymers of identical composition may also be prepared from chromium(III) in 1:1 water/THF mixtures.[847,848] Polymers of the general formula $[Cr(H_2O)OH(OPRR'O)_2]n$ ($R = R' = C_6H_5$, $R = Me$, $R' = Ph$ and $R = R' = C_8H_{17}$) were also prepared; on heating to constant weight *in vacuo* these materials are converted to the corresponding hydroxooxo polymers.

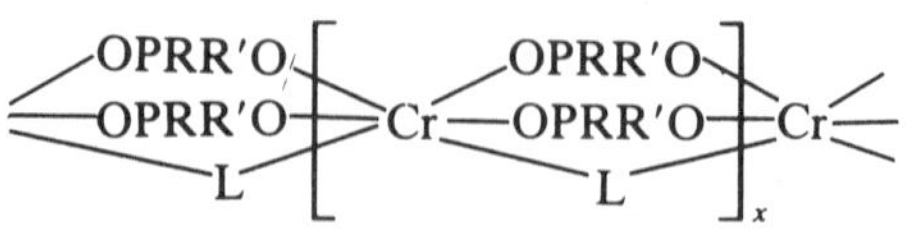

(**203**)

Many polymeric chromium(III) complexes in this general class exist. For the general formula $[Cr(L)(OPRR'O)_2]_n$, compounds have been prepared involving hydroxide, perfluorocarboxylate, alkoxide, aryloxide and carboxylates.[849,850]

(*d*) *Miscellaneous.* A perchlorate complex of chromium(III) has been reported; IR evidence was used to suggest bidentate coordination of the anion.[851] Chromium(III) borates have been prepared by solid-state methods;[852] Cr_2O_3 was reacted with molten B_2O_3 at 1100 °C. The solid isolated had the calcite structure. A monomeric complex $[Cr(O_2SeMe)_3]$ has been prepared; the seleninate was a stronger ligand than DMSO but weaker than urea.[853]

35.4.4.10 Carbon-containing oxo anions

(*i*) Carboxylates

Simple carboxylates of chromium(III) find industrial application as catalysts for the polymerization of ω-alkenes[854,855] and in the preparation of chrome-tanning solutions.[856,857] There seems to be no simple carbonate of chromium(III). Compounds formed on the surface of Cr_2O_3, sometimes formulated $Cr_2(CO_3)_3 \cdot nH_2O$, are best viewed as carbon dioxide adsorbed on the oxide.[858] Carboxylate complexes of chromium(III) will now be considered in terms of the various ligand types.

(*ii*) Acetates and simple carboxylates

Early investigations[859] showed the product of the reaction between acetic anhydride and chromium(III) chloride to have the composition $[Cr_3(MeCO_2)_6(OH)_2(H_2O)]^+$. Orgel[860] suggested a structure of the kind illustrated below (**204**) for the trimer and crystallography[861] has confirmed this. The three carboxylate-bridged chromium(III) atoms form an equilateral triangle around a shared oxygen atom. The vertex of each chromium(III) octahedron *trans* to the central oxygen atoms is occupied (in the trihydrate) by a water molecule. This is the only ligand in this complex which will readily undergo substitution.

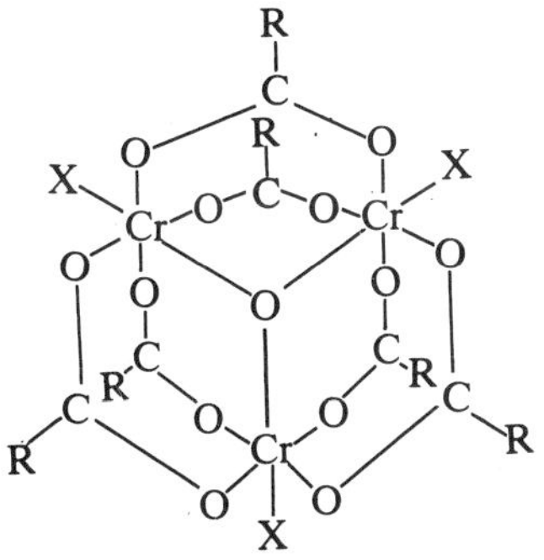

(**204**)

The physical properties of a large number of trimeric acetates of this type have now been investigated. Magnetic coupling between the chromium centres within the trimer occurs, *e.g.* in $[Cr_3O(MeCO_2)_6(H_2O)_3]Cl \cdot H_2O$, $J_0 = 10.5\ cm^{-1}$ and $J_1 = 2.5\ cm^{-1}$.[862] How such results are best interpreted is controversial. Brown and co-workers suggested that inter-trimer coupling is important.[863,864] In contrast, Tsukerblat *et al.* find the majority of results[865] to be interpretable in terms of models which do not invoke inter-trimer interactions. The gross features of the electronic spectrum of trinuclear chromium(III) acetates may be explained[866] in terms of tetragonally distorted chromium(III) centres; fine structure in the 650–750 nm region is the result of exchange coupling. The electronic spectra of a series of acetates $[Cr_3O(MeCO_2)_6R_3]^+$ ($R = H_2O$, NH_3, py, DMSO) have also been successfully explained in terms of ligand field theory.[867] Related complexes of chloroacetic acids have been studied.[868,869]. Mixed valence[870] and mixed metal[871,520] acetates of this type are also known (see also Sections 35.4.2.5.i and 35.7.2). Other simple derivatives of acetate (RCO_2H; $R = Me$, Et, Pr, CH_2Cl, $CHCl_2$, CCl_3) give trimeric complexes of similar structure.[872]

Simple monomeric acetate complexes have been reported to form in solution from the reaction of aqua[873] and ammine[874] complexes of chromium(III) with acetate. Chloro carboxylates $CrCl_2(O_2CR)\cdot THF$ and $CrCl(O_2CR)_2$ ($R = C_{17}H_{15}–C_{21}H_{43}$) have been prepared in THF. These are dimeric and trimeric respectively, in benzene.[875] Basic trimeric formate complexes of chromium(III) are formed when anhydrous chromium(III) chloride is reacted with HCO_2H.[876]

(*iii*) *Oxalates*

Oxalate complexes of chromium(III) were first characterized at the turn of the century by Rosenheim and Cohen.[877] The most extensively studied are the tris species and the *cis*- and *trans*-bisoxalates (**205–207**): these formulations were first suggested by Werner.[878] All may be made by the reduction of chromate with oxalate. Reliable preparations have been reported for tris by Kauffman and Faoro[879] and for *cis*- and *trans*-diaquabisoxalatochromate(III) by Bailar.[880]

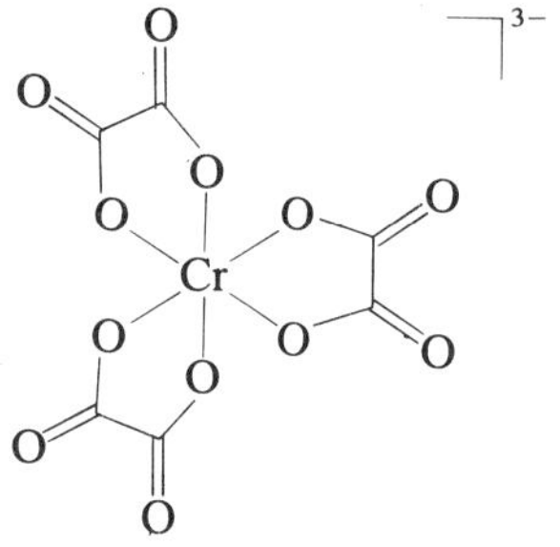

(205)

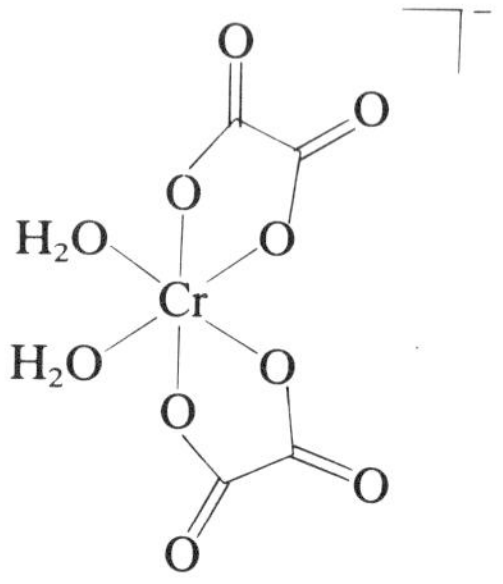

(206)

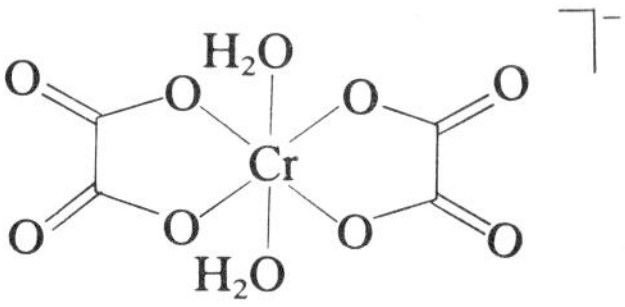

(207)

In aqueous solution, the equilibrium between the *cis*- and *trans*-diaqua complexes lies almost completely toward the *cis* isomer[881] ($K \approx 0.17$, pH 3–4). The sparingly soluble potassium salt of the *trans* isomer may, however, be prepared by the slow evaporation of a saturated solution at room temperature,[878,880] and the *cis* isomer by cooling a hot solution or by allowing potassium dichromate and oxalic acid to react in the presence of a minimal quantity of water.[878,879,882] The tris complex was resolved by Werner in 1912,[883] providing the first example of an anionic optically active coordination complex.

(*iv*) *Tris(oxalato)chromate(III)*

Structural and other studies of the salts of this ion are numerous. The crystal structure of potassium tris(oxalato)chromate(III) has recently been reinvestigated.[884] EPR studies of $[Cr(ox)_3]^{3-}$ doped in $K_3[Al(ox)_3]\cdot 3H_2O$ had revealed four magnetically distinguishable sites for substitution,[885] a result in conflict with the reported crystal structure[886] of $K_3[Cr(ox)_3]\cdot 3H_2O$ with which $K_3[Al(ox)_3]\cdot 3H_2O$ is isomorphous. The earlier crystal structure[886] was found to be unsatisfactory. Crystals of $K_3[Cr(ox)_3]\cdot 3H_2O$ are monoclinic ($a = 7.714$, $b = 19.687$, $c = 10.316$ Å and $\beta = 108.06°$). The structure consists of discrete $[M(ox)_3]^{3-}$ ions, K^+ ions and water molecules. One potassium and one water are cooperatively disordered over two possible pair sites (termed α and β) with an occupancy ratio of 3:1. The result is in excellent agreement with the EPR study.[885] The structure of $(NH_4)_3[Cr(ox)_3]\cdot 2H_2O$ has also been reported.[887]

The electronic spectrum of $[Cr(ox)_3]^{3-}$ has been studied extensively both in solution and in the solid state.[888,889] The assignment of the spin-allowed transitions from solution electronic spectroscopy is a popular undergraduate exercise.[890] The low energy spin-forbidden ${}^4A_2 \rightarrow {}^2E$ transition is readily observed for this complex.[889,890] Table 81 summarizes the most significant features of the electronic spectrum.

Werner resolved the tris oxalate using strychninium ion in 1912.[891] Other effective resolving agents for the tris complex include: (+)-tris(1,10-phenanthroline)nickel(II),[892] (+)- or (−)-tris(ethylenediamine)cobalt(III)[893] and (−)-tris(ethylenediamine)rhodium(III).[896] The (−)-$[Co(ox)_3]^{3-}$ ion has been studied by single-crystal methods[894] and has the D configuration. It has been related to (+)-$[Cr(ox)_3]^{3-}$ by X-ray powder photography,[895] so its absolute configuration is also established as D by the more exact form of the rule of least-soluble diastereoisomers.[897]

Table 81 The Electronic Spectrum of Potassium Tris(oxalato)chromate(III) (after ref. 889)

λ (nm)	$\varepsilon_{corr.}$ ($dm^3\ mol^{-1}\ cm^{-1}$)	*Assignment*	*Assignment 'O_h'*
711	0.13	O–O + 543 } 267	$^4A_2 \rightarrow {}^2E_2, T_1$
697	2.65	O–O + 810 } 267	
676	0.31	O–O + 310 } 95 } 215	
672	0.23	O–O + 405 } 95 } 215	
663	0.41	O–O + (2 × 130) or O–O + 405 + 225 } 215	
570	75.8	Spin-allowed	$^2A_2 \rightarrow {}^4T_2$
488	0.13	O–O + 225 } 318	
481	0.09	O–O + 543 } 318	
420	96.4	Spin-allowed	$^4A_2 \rightarrow {}^4T_1$

The 'Pfeiffer effect' is a term used to describe changes in the optical activity of solutions containing a chiral compound (the 'environmental substance') on the addition of a racemic dissymmetric complex. The effect is generally attributed to a shift in the position of the equilibrium between D and L isomers for the racemic complex. The exact mechanism involved in mediating the chiral interaction is unknown. Perhaps surprisingly, both 'environmental substance' and complex may simultaneously be cations. Studies of the 'Pfeiffer effect' usually involve a moderately labile racemic complex; $[Cr(ox)_3]^{3-}$ is a popular choice for such studies, summarized in Table 82. Other studies of the optical activity of tris oxalates include work on photoinduced optical activity,[898] photoracemization[899] and the solid-state racemization of $K_3[Cr(ox)_3]$.[900,901]

Table 82 Pfeiffer Effect Studies on $[Cr(ox)_3]$

Environmental compound	*Solvent*	*% Resolution*	*Ref.*
(+)-Cinchonine	Water	2–6	1, 2
(+)-$[Co(phen)_3]^{3+}$	Water/dioxane	6.0	3
(+)-Cinchonine	Water/THF	—	4
	Water/DMSO	—	4
Chiral α-hydroxy esters	Water/ester	70 (max)	5

1. S. Kirschner, *Rec. Chem. Prog.*, 1971, **32,** 29.
2. K. T. Kan and D. G. Brewer, *Can. J. Chem.*, 1971, **49,** 2161.
3. K. Miyoshi, K. Wada and H. Yoneda, *Inorg. Chem.*, 1978, **17,** 751.
4. K. Miyoshi, Y. Kuroda, J. Takeda, H. Yoneda and H. Takagi, *Inorg. Chem.*, 1979, **18,** 1425.
5. A. F. Drake, J. R. Levey, S. F. Mason and T. Prosperi, *Inorg. Chim. Acta*, 1982, **57,** 151.

The thermal decomposition of $K_3[Cr(ox)_3]\cdot 3H_2O$ proceeds, after dehydration, *via* the loss of CO and CO_2 to Cr_2O_3.[902] When $[Cr(H_2O)_6][Cr(ox)_3]$ was heated at 303 K in the solid state, along with decomposition, some exchange of the chromium(III) centres could be detected by ^{51}Cr labelling.[903]

(*v*) *Bis oxalate complexes*

Crystallographic studies of the bis oxalates of chromium(III) are not abundant. However, the structure of both *trans*[904] and *cis*[905] isomers has been confirmed crystallographically. Potassium *trans*-bis(oxalato)diaquachromate(III) is monoclinic (space group *P2/c*); the oxalates are strictly coplanar. The crystal structure of the complex salt $[Cr(en)_2(ox)][Cr(en)(ox)_2]$ has been determined:[905] this red salt is obtained as an intermediate in the preparation of salts of mixed ethylenediamine/oxalate chromium(III) complexes. The structure consists of discrete complex ions linked by H bonding to water molecules and neighbouring ions.

The bis oxalato complexes are readily assigned to the *cis* or *trans* geometry on the basis of their electronic spectrum.[906,907] The EPR spectra of a large series of *trans* isomers have been studied (axial ligands studied included H_2O, OH, py and NH_3).[908] The spectra are quite insensitive to the nature of the axial ligand, but ESR spectra are useful in distinguishing *cis* and *trans* isomers. Potassium *cis*-bis(oxalato)diaquachromate(III) has been resolved using strychnine.[909] The rate of *cis/trans* isomerization paralleled that of racemization, suggesting a single mechanism. Many reports concern the kinetics of isomerization and/or aquation of bis oxalato complexes. A number of the more recent and/or significant papers are collated in Table 83.

Table 83 Studies of Bis Oxalato Complexes

Ligand	*Nature of work*	*Ref.*
H_2O	Reaction with oxalate, *cis/trans* synthesis and kinetics of anation with oxalate, nitrate catalysis	1
H_2O	Aquation and isomerization of *cis* and *trans* complexes, isomerization. (Also for malonate). Fourteen references to studies of the isomerization.	2
H_2O	Solid state isomerization	3
SCN^-	Monothiocyanato, monoaqua, synthesis and kinetic study	4
N_3^-	Mono and bis aquation kinetics	5
$CH_3CO_2^-$	Synthesis, *cis/trans,* hydrolysis kinetics, isomerization kinetics	6
NH_3	Synthesis	7
DMF	*Cis,* synthesis and a study of aquation kinetics	8
DMSO	*Cis,* preparation and aquation kinetics	9
CO_3^{2-}	Monocarbonato, *cis* and aquation kinetics	10
en	Synthesis	11
1,10-phen/2,2′-bipy	Photoresolution/preparation	12
	Preparation/resolution	13
	Racemization kinetics	14
Gly^-	Preparation and acid hydrolysis	15

1. T. W. Kallen and E. J. Senko, *Inorg. Chem.*, 1983, **22,** 2924.
2. P. L. Kendall, G. A. Lawrance and D. R. Stranks, *Inorg. Chem.*, 1978, **17,** 1166.
3. M. Malati, M. McEvoy and M. W. Raphael, *Inorg. Chim. Acta,* 1976, **19,** L5.
4. K. R. Ashley and S. Kulprathipanja, *Inorg. Chem.*, 1972, **11,** 444.
5. K. R. Ashley and R. S. Lamba, *Inorg. Chem.*, 1974, **13,** 2117.
6. M. Casula, G. Illuminati and G. Ortaggi, *Inorg. Chem.*, 1972, **11,** 1062.
7. E. Kyuno, M. Kamada and N. Tanaka, *Bull. Chem. Soc. Jpn.*, 1967, **40,** 1848.
8. K. R. Ashley, D. K. Pal and V. D. Ghanekar, *Inorg. Chem.*, 1979, **18,** 1517.
9. K. R. Ashley and R. E. Hamm, *Inorg. Chem.*, 1966, **5,** 1645.
10. D. A. Palmer, T. P. Dasgupta and H. Kelm, *Inorg. Chem.*, 1978, **17,** 1173.
11. A. Mead, *Trans. Faraday Soc.*, 1934, **30,** 1052.
12. K. Miyoshi, Y. Matsumoto and H. Yoneda, *Inorg. Chem.*, 1982, **21,** 790.
13. J. A. Broomhead, *Aust. J. Chem.*, 1962, **15,** 228.
14. J. A. Broomhead, N. Kane-Maguire and I. Lauder, *Inorg. Chem.*, 1970, **5,** 1243.
15. T. W. Kallen and R. E. Hamm, *Inorg. Chem.*, 1977, **16,** 1147.

Bridged hydroxy complexes (**208**) of the bisoxalate type have been known for many years.[910,911] Both a trihydrate[910,911,912] and an anhydrous[913] form are known. The trihydrate is isomorphous[912] with the corresponding cobalt(III) complex (Durrant's salt) and the hydrous complex[913] is slowly converted to the hydrate on exposure to moisture.[912] Only a preliminary crystallographic characterization has appeared;[914] it crystallizes as a *meso* rather than racemic complex. IR[915] and magnetic studies[912] are in accord with the hydroxy-bridged structure. Despite the crystallographic *meso* structure, the resolution of the complexes has been reported either photochemically[915] or with *cis*-(−)bis(ethylenediamine)dinitrocobalt(III)[916] and leads to optically active complexes with spectra similar to those of the corresponding tris oxalate.

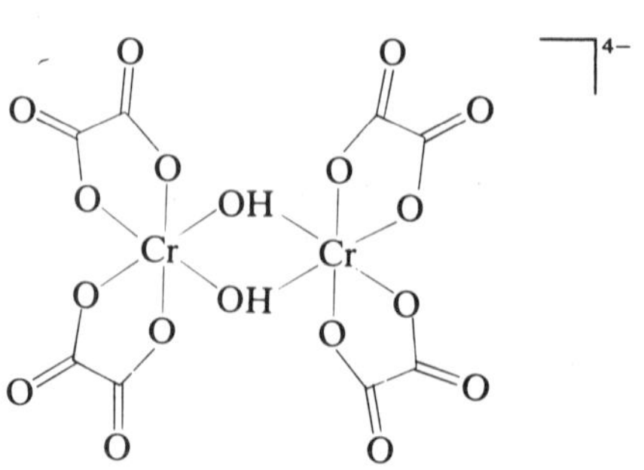

(**208**)

(vi) Malonates

Malonic acid $CH_2(CO_2H)_2$ (H_2mal) (**209**) has a coordination chemistry with chromium(III) closely resembling that of oxalate. Malonic acid is a slightly weaker acid than oxalic acid and slightly more labile complexes are formed. The tris complex is the most extensively studied, prepared by the reduction of chromate solutions or the reaction of chromium(III) hydroxide with malonate.[917,918,919] The *cis* and *trans* diaqua complexes may be prepared by the reduction of chromate with malonate; the isomers are separated by fractional crystallization. The electronic spectrum of the tris complex is similar to that of the tris oxalate and a detailed analysis of these spectra has appeared.[889]

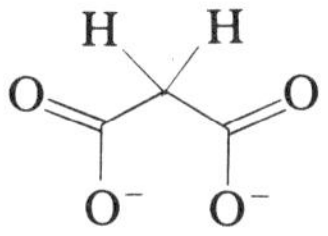

(**209**)

The resolution of $[Cr(mal)_3]^{3-}$ was achieved with L-[Co{(−)-1,2-diaminopropane}$_3$]Br$_3$;[920] the only diastereoisomer to precipitate contained D-$[Cr(mal)_3]^{3-}$. A crystal structure of the diastereoisomer was reported, permitting the correction of earlier work which had suggested the $(+)_{589}$-$[Cr(mal)_3]^{3-}$ ion to have the L-configuration.[921,922] Polarized crystal spectra of $[Cr(mal)_3]^{3-}$ doped into $[NH_4]_3[Fe(mal)_3]$ are extremely similar to those of $[Cr(ox)_3]^{3-}$ under similar conditions.[923]

Solution studies of malonate complexes include the unusual observation that $[(H_2O)_5Cr(Hmal)]^{2+}$ has considerable kinetic stability.[924] The mechanism of ring closure has been the subject of detailed study.[925] The methylenic protons of $[Cr(mal)_3]^{3-}$ undergo bromination when a suspension of the potassium salt is treated with bromine in ether (2 h).[926] Legg and co-workers[927] have recently used a number of malonate complexes to demonstrate that the ^{2}H NMR spectra are both readily observed and useful for solving structural problems. The resolution that may be obtained by this method may be appreciated by the fact that in the complex $[Cr(mal\text{-}d^2)_2(bipy)]^{2-}$ the geminal protons of the malonate are clearly resolved.

(vii) Squarates

Complexes of chromium(III) with the squarate ligand (**210**) have been prepared.[928] Subsequently the complex $Cr(C_4O_4)(OH)\cdot 3H_2O$ was investigated crystallographically and shown to contain pairs of chromium atoms joined by two μ-hydroxy bridges and two bridging $C_4O_4^{2-}$ ions.[929] In the same study, a compound of formulation $Cr(C_4O_4)_{3/2}\cdot 7H_2O$ was isolated, a salt of the hexaaquachromium(III) ion.

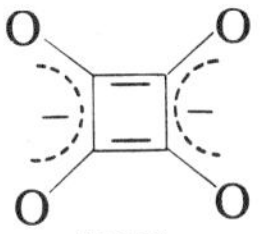

(**210**)

35.4.4.11 Hydroxy acids

(i) Tartrates and aliphatic hydroxy acids

While undertaking a study of alkaline solutions of chromium(III) and (+)-tartrate in 1896, Cotton[930] first observed the anomalous rotatory dispersion of light in the region of an absorption band—'The Cotton Effect'. The solutions used probably contained a tris tartrato species. Mason concluded that complexation in such solutions was stereospecific;[931] the complex formed in the presence of an excess of doubly deprotonated tartrate is probably *fac*-D-$[Cr(R,R\text{-tart})_3]^{3-}$.[932] This diastereoisomer may be markedly stabilized by hydrogen bonding when the ligand is doubly deprotonated, a suggestion supported by the fact that the addition of a single proton to the complex greatly decreases the intensity of the CD spectrum (**211**, **212**). Alkaline solutions of chromium(III) in the presence of racemic tartrate become optically active when irradiated with circularly polarized light.[933] Photo inversions at the chromium(III) centre are postulated to account for this observation. In the dark, the optical activity slowly decreases to zero.

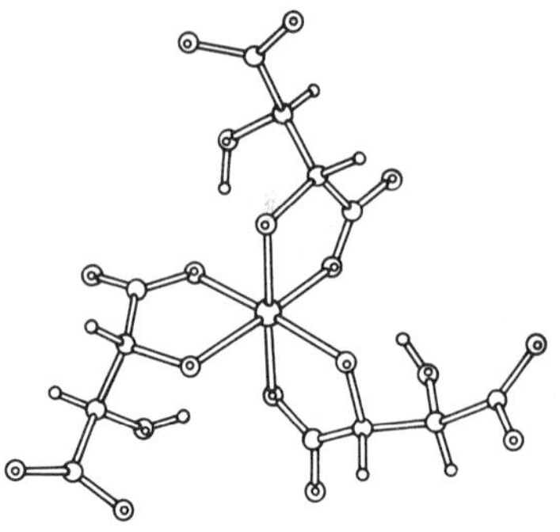

(211) Δ-*fac*

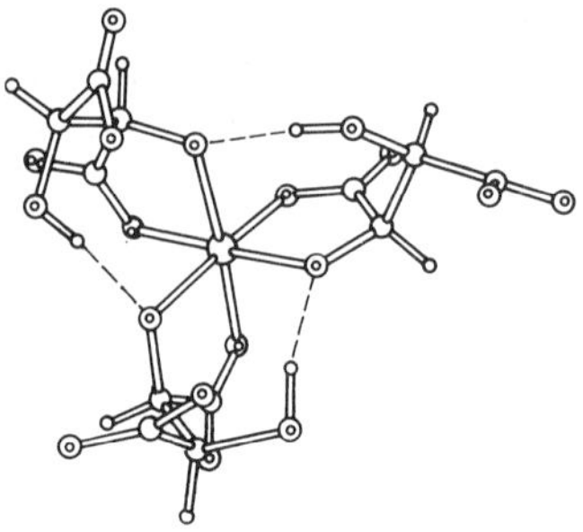

(212) Λ-*fac*

Mixed dinuclear complexes of tartrates and 2,2′-bipyridyl or 1,10-phenanthroline and chromium(III) have been prepared;[934,935] the typical structure is illustrated below **(213)**. Stereochemical correlations have been carried out by oxidatively cleaving the tartrate bridge **(214, 215)**.[936] The crystal structure of sodium hydrogen bis(μ-*meso*-tartrato)bis(2,2′-bipyridyl)dichromate(III) heptahydrate has been reported.[937]

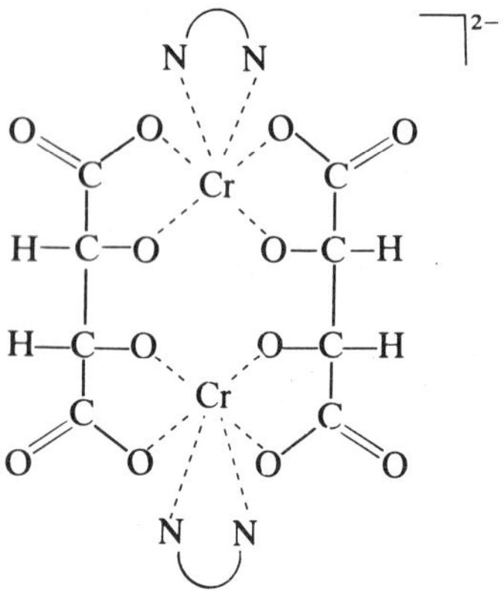

(213)

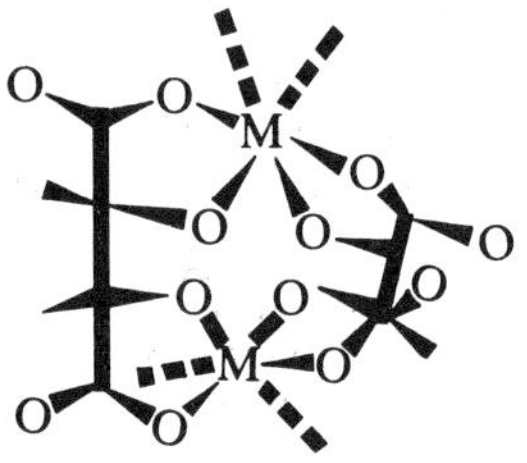

(214) ΔΔ(*dd*)

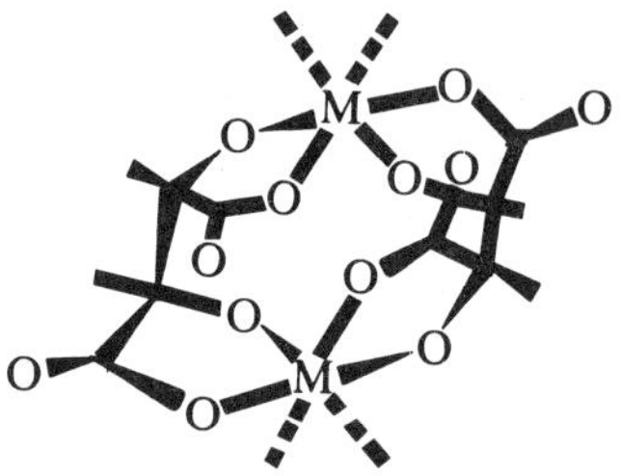

(215) β-ΔΔ(*meso-meso*)

An isolated report of the synthesis and characterization of a number of citrate complexes of chromium(III) has appeared.[938]

(*ii*) *Aromatic hydroxy acids*

Complexes of salicylate with chromium(III) have not been reported but the tris complexes of salicylaldehyde and chromium(III) may be prepared by refluxing $[Cr(THF)_3Cl_3]$ with salicylaldehyde and sodium acetate in ethanol.[939] The acid hydrolysis of this complex was studied in detail, but the isomerism obviously possible for this complex was not apparently considered. Khan and Tyagi[940] studied the formation of phthalate complexes of chromium(III).

35.4.4.12 Sulfoxides, N-*oxides and* P-*oxides*

(*i*) *DMSO*

The complex $[Cr(DMSO)_6][ClO_4]_3$ was first prepared by Cotton in 1959[941] as green needles by reacting concentrated $Cr(ClO_4)_3$ solutions with DMSO. Subsequently Drago established the spectrochemical series for chromium(III) to be $H_2O > DMSO > TMSO > C_6H_5NO$.[942] A detailed IR study of the hexakis DMSO ion as its perchlorate, and the corresponding d_6-DMSO complex, suggested that the point group of the cation was D_{3d} or S_6.[943] The preparation of $[Cr(DMSO)_6]^{3+}$ from anhydrous $CrBr_3$ has been reported.[944]

The solvolysis of hexaaquachromium(III) in DMSO proceeds *via* the series of complexes $[Cr(DMSO)_n(H_2O)_{6-n}]^{3+}$ ($n = 1$ to 6).[945,946] The anation and solvolysis of several chromium(III) complexes in DMSO have been studied;[947,948] in general the reactions proceed by I_d mechanisms. A direct electrochemical synthesis of $[Cr(DMSO)_6][BF_4]_3$ has been reported.[949] A comprehensive review of metal ion complexation by DMSO contains an extensive section on chromium(III).[950]

(*ii*) *DMF*

The complex $[Cr(DMF)_6][ClO_4]_3$ has been prepared from the reaction of chromium(III) perchlorate with DMF (90 °C, 10 h).[951] Solvent exchange and substitution proceed by I_a mechanisms.[947,952] The complex $[Cr(DMF)_6]Br_3{\cdot}DMF$ was prepared by the reaction of chromium metal with bromine in DMF;[953] the electronic spectrum of this complex was very similar to that of the perchlorate salt.

(*iii*) N-*Oxides*

An extensive, if somewhat dated, review of the coordination chemistry of aromatic *N*-oxides is available.[954] Most studies of the *N*-oxide complexes of chromium(III) involve the preparation and characterization (usually by magnetic and spectroscopic methods) of monomeric or hydroxy-bridged complexes and are summarized in Table 84.

Table 84 N-Oxide Complexes

Ligand	*Complex*	*Ref.*
Trimethylamine *N*-oxide	$[Cr(L)_6][ClO_4]_3$	1
Pyridine *N*-oxide	$[Cr(L)_6][ClO_4]_3$	2
4-Substituted pyridine *N*-oxide	$[Cr(L)_6][ClO_4]_3$	3
2,6-Lutidine *N*-oxide	$[Cr(L)_6][BF_4]_3$	4, 5
	$[Cr(L)_2Cl_2(OH_2)_2]Cl{\cdot}4H_2O$	
Picolinic acid *N*-oxide	$[Cr(L)_2(LH)(OH_2)]ClO_4$	6
Pyridine carboxamide *N*-oxide	$[Cr(L)_3]Cl_3{\cdot}2H_2O$	7
1,10-Phenanthroline *N*-oxide	$[Cr(L)_3][ClO_4]_3$	8
2,2′-Bipyridyl *N*-oxide	$[Cr(L)_3][ClO_4]_3$	8
Quinoxaline *N*-oxide	$[Cr(L)_4(OH_2)(OClO_3)][ClO_4]_2{\cdot}5H_2O$	9
3-Methylisoquinoline *N*-oxide	$[Cr(L)_4(OClO_3)(OH_2)][ClO_4]_2$	10
Quinoxaline 1,4-dioxide	$[CrCl_3L(OH_2)_2]{\cdot}H_2O$	11, 12
	Polymeric	
Phenazine 5,10-dioxide	$Cr(ClO_4)_3L_2{\cdot}H_2O$	13, 14
	$[Cr(ClO_4)_3]_2L{\cdot}12H_2O$	
3,4,5-Trimethyl-1-hydroxypyrazole 2-oxide	$[CrL_3(H_2O)]$	15
Purine *N*(1)-oxide	$[Cr(LH)_2(EtOH)_2(OClO_3)_2]ClO_4$	16
Adenine *N*(1)-oxide	$[Cr(LH)_2(OClO_3)_2(EtOH)_2]ClO_4$	17
N,N,N′,N′-Tetramethylethylenediamine *N,N′*-dioxide	$[Cr(L)_3][ClO_4]_3$	18

1. R. S. Drago, J. T. Donoghue and D. W. Herloker, *Inorg. Chem.*, 1965, **4**, 836.
2. C. Valdemoro, *C.R. Hebd. Seances Acad. Sci.*, 1961, **253**, 277.
3. L. C. Nathan and R. O. Ragsdale, *Inorg. Chim. Acta*, 1974, **10**, 177.
4. C. M. Mikulski, L. S. Gelfand, E. S. C. Schwartz, L. L. Pytlewski and N. M. Karayannis, *Inorg. Chim. Acta*, 1980, **39**, 143.
5. C. M. Mikulski, L. S. Gelfand, L. L. Pytlewski, N. S. Skryantz and N. M. Karayannis, *Inorg. Chim. Acta*, 1977, **21**, 9.
6. F. J. Iaconianni, L. S. Gelfand, L. L. Pytlewski, C. M. Mikulski, A. N. Speca and N. M. Karayannis, *Inorg. Chim. Acta*, 1979, **36**, 97.
7. A. E. Landers and D. J. Phillips, *Inorg. Chim. Acta*, 1982, **59**, 125.
8. A. N. Speca, N. M. Karayannis and L. L. Pytlewski, *Inorg. Chim. Acta*, 1974, **9**, 87.
9. D. E. Chasan, L. L. Pytlewski, C. Owens and N. M. Karayannis, *J. Inorg. Nucl. Chem.*, 1977, **39**, 585.
10. A. N. Speca, L. S. Gelfand, F. J. Iaconianni, L. L. Pytlewski, C. M. Mikulski and N. M. Karayannis, *Inorg. Chim. Acta*, 1979, **33**, 195.
11. D. E. Chasan, L. L. Pytlewski, C. Owens and N. M. Karayannis, *Inorg. Chim. Acta*, 1977, **24**, 219.
12. D. E. Chasan, L. L. Pytlewski, C. Owens and N. M. Karayannis, *J. Inorg. Nucl. Chem.*, 1978, **40**, 1019.
13. D. E. Chasan, L. L. Pytlewski, C. Owens and N. M. Karayannis, *Transition Met. Chem.* (*Weinheim, Ger.*), 1976, **1**, 269.
14. D. E. Chasan, L. L. Pytlewski, C. Owens and N. M. Karayannis, *J. Inorg. Nucl. Chem.*, 1979, **41**, 13.
15. D. X. West and M. A. Vanek, *J. Inorg. Nucl. Chem.*, 1978, **40**, 1027.
16. C. M. Mikulski, R. DePrince, T. BaTran, F. J. Iaconianni and N. M. Karayannis, *Inorg. Chim. Acta*, 1981, **56**, 27.
17. C. M. Mikulski, R. DePrince, T. BaTran, F. J. Iaconianni, L. L. Pytlewski, A. N. Speca and N. M. Karayannis, *Inorg. Chim. Acta*, 1981, **56**, 163.
18. M. J. Bigley, K. J. Radigan and L. C. Nathan, *Inorg. Chim. Acta*, 1976, **16**, 209.

Interesting complexes have been prepared from the chiral ligands derived from 3,3′-dimethyl-2,2′-bipyridyl *N,N′*-dioxide (**216, 217**).[955,956] A number of diastereomeric tris complexes were isolated. These could be interconverted photochemically.

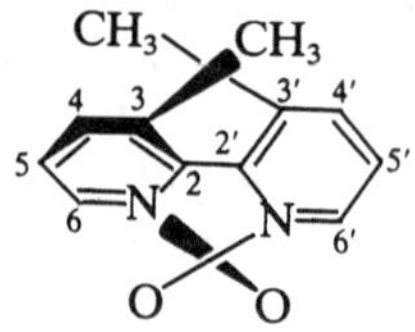

(**216**) *R* (λ)

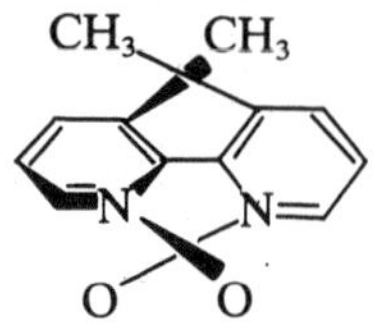

(**217**) *S* (δ)

(iv) P-Oxides

The complex of tri-*n*-butylphosphine oxide (tbo), $[Cr(tbo)_4(ClO_4)_2]ClO_4$, has been prepared.[957] Crystalline samples were obtained, and, on the basis of IR and electronic spectra, monodentate coordination of the perchlorate was suggested. The geometric isomerism possible for this complex was not considered. The hexakis complex of tris(hydroxymethyl)phosphine oxide and chromium(III), $[Cr(L)_6][ClO_4]_3$ has been prepared;[958] it was concluded that bonding in these complexes was weaker than in the corresponding complexes of trimethylphosphine oxide.[959] Tris(dimethylamino)phosphine oxide forms complexes CrL_3Cl_3, for which a facial arrangement of the ligands has been suggested.[960] The reaction of $Cr(CO)_6$ with diisopropylmethylphosphonate results in a series of polymeric complexes.[961] A complex of triphenylphosphine oxide has been said to contain a trigonally bipyramidal chromium(III) ion.[962]

35.4.5 Sulfur Ligands

35.4.5.1 Thiolate, disulfido and thioether ligands

The chemistry of chromium(III) complexes containing sulfur donor ligands, although until recently unexplored, is now an area of considerable research activity. Mechanisms for reactions of these complexes are described in a recent review.[70]

(i) $[Cr(SH)(H_2O)_5]^{2+}$

The tendency of chromium to capture whatever ligands are in its first coordination sphere on being oxidized from the +II to the +III state[963] has been exploited in the preparation of $[Cr(SH)(H_2O)_5]^{2+}$ in aqueous solution (equation 48).[964,965] Hence oxidation of $[Cr(H_2O)_6]^{2+}$ by PbS, Ag_2S or $S_2O_3^{2-}$ under N_2 gives the desired product, although in low yield. However the use of polysulfide as oxidant increases the yield of green-coloured product to 10–20%. The scope of this reaction can be extended to include the synthesis of a range of alkyl-, aryl- and chelating-thiolatochromium(III) complexes by choice of an appropriate disulfide oxidant. However, complexes containing simple alkylthiolato ligands cannot be synthesized by this method due to the inability of chromium(II) to reduce the corresponding disulfides.[966]

$$[Cr(H_2O)_6]^{2+} + \text{various S-containing oxidants} \longrightarrow [Cr(SH)(H_2O)_5]^{2+} \quad (48)$$

Some reactions of $[Cr(SH)(H_2O)_5]^{2+}$ are outlined in Scheme 96.[964–968] The ion exchange behaviour of this complex is characteristic of a 2+ species, thus confirming the presence of HS^- rather than H_2S as the sulfur ligand. No spectroscopic evidence for protonation of this ligand was observed even at pH = 0. This, however, is not too surprising, since by analogy with $[Cr(H_2O)_6]^{3+}$, which is more acidic by a factor of 5×10^{11} than free H_2O, the pK_a of H_2S in the complex $[Cr(H_2O)_5H_2S]^{3+}$ should be considerably less than 0. However, kinetic evidence has been obtained for the existence of this species. While the complex $[Cr(SH)(H_2O)_5]^{2+}$ is remarkably stable under anaerobic conditions in aqueous solution at room temperature,[964] it undergoes aquation by both acid-independent and acid-dependent pathways, the latter attributed to a rapid protonation equilibrium preceding the rate-determining step.[965] Aquation rate constants of $3.1 \times 10^{-5}\ s^{-1}$ and $\gg 3.8 \times 10^{-5}\ s^{-1}$ have been estimated for $[Cr(SH)(H_2O)_5]^{2+}$ and its conjugate acid respectively, both at 313 K, $I = 1.0$.

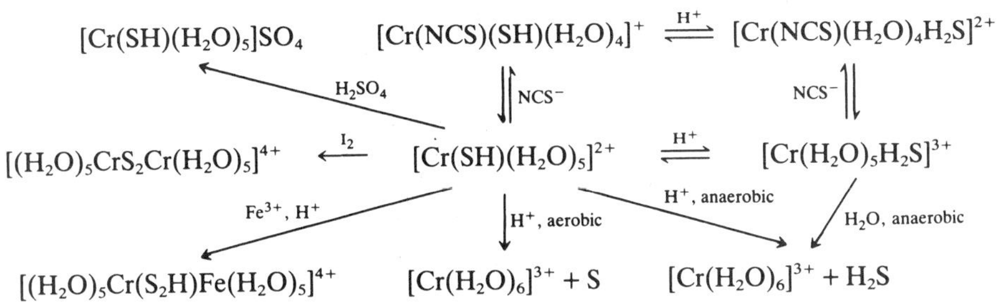

Scheme 96

The UV–visible spectrum of $[Cr(SH)(H_2O)_5]^{2+}$ in aqueous solution (Table 85) is typical of thiolatochromium(III) complexes. The bands at 575 and 435 nm are identified with the $^4A_{2g} \rightarrow {}^4T_{2g}$ and $^4A_{2g} \rightarrow {}^4T_{1g}$ transitions respectively, while the very intense band at 258 nm is due to an S ligand to metal charge transfer transition.[966] The aquation of $[Cr(SH)(H_2O)_5]^{2+}$ is characterized by the disappearance of the charge transfer band from the spectrum. Addition of H_2SO_4 to a concentrated solution of $[Cr(SH)(H_2O)_5]^{2+}$ leads to the isolation of the brown-green solid $[Cr(SH)(H_2O)_5]SO_4$.[965] The IR spectrum of this complex has bands at 2560 (ν_{SH}) and 340 cm^{-1} (ν_{MS}), while the Raman spectra of aqueous solutions show a band at 337 cm^{-1} (ν_{MS}).

(ii) $[Cr(SAr)(H_2O)_5]^{3+}$ (*Ar* = *anilinium*)

The aromatic thiolatochromium(III) complexes $[Cr(SAr)(H_2O)_5]^{3+}$ (Ar = 4-$PhNH_3^+$, 4-$PhNMe_3^+$) have been prepared in solution under N_2 (equation 49) and purified by cation exchange chromatography.[969] Initial elution with 1 M $LiClO_4$ solution separated species of charge +3 from more highly charged polymeric products, while subsequent elution with 2 M $LiClO_4$ separated the green thiolato complexes from the blue and slightly faster moving $[Cr(H_2O)_6]^{3+}$. Despite repeated chromatography, however, the complex [Cr(4-$SPhNH_3^+$)$(H_2O)_5]^{3+}$ could not be obtained free from the hexaaqua species, in any greater than about 95% purity. The solution spectrum of this complex (Table 85) shows an internal ligand band at 230 nm, a very intense charge-transfer band at 295 nm which masks the $^4A_{2g} \rightarrow {}^4T_{1g}$ *d–d* band and an unsymmetrical band at 595 nm with a shoulder at 640 nm. The shape of this band is due to splitting of the $^4T_{2g}$ excited state and the absorptions at 595 and 640 nm may be attributed to the $^4B_1(^4A_{2g}) \rightarrow {}^4B_2, {}^4E_a(^4T_{2g})$ transitions respectively.

$$[Cr(H_2O)_6]^{2+} + ArSSAr \xrightarrow{HClO_4,\ N_2} [Cr(SAr)(H_2O)_5]^{3+} \quad (49)$$

As in the case of $[Cr(SH)(H_2O)_5]^{2+}$, the complexes $[Cr(SAr)(H_2O)_5]^{3+}$ undergo anaerobic aquation with loss of the S ligands by both acid-independent and acid-dependent pathways, the kinetically active species in the latter resulting from protonation on the S atom. Aerobic aquation of the thiolatoanilinium complex gives $[Cr(H_2O)_6]^{3+}$ and *p*-aminophenyl disulfide as products. This reaction is autocatalytic (*p*-aminothiophenol acting as catalyst) and is also catalyzed by H_2O_2 and Fe^{3+}, all seemingly by a mechanism which involves oxidation of the thiolate ligand to a labile coordinated free radical.

(iii) $[Cr(SR)(H_2O)_5]^{2+}$ (*R* = *aliphatic*)

Because of the thermodynamic inability of chromium(II) to reduce simple alkyl disulfides, the synthetic method used for the previous complexes is unsuccessful for chromium(III) complexes containing simple alkylthiolato ligands. Hence, whereas complexes containing chelating thiolate and 4-$^-SPhNH_3^+$ ligands are obtained rapidly by chromium(II) reduction of the corresponding disulfides, the complex containing $^-SCH_2NMe_3^+$ is produced only slowly and those containing the ligands MeS^-, EtS^- and $^-S(CH_2)_2NH_3^+$ are unobtainable by this

Table 85 Colours and Electronic Solution Spectra of Thiolatochromium(III) Complexes

Complex	*Colour*	λ_{max} (nm) (ε in $dm^3\ mol^{-1}\ cm^{-1}$) $^4A_{2g} \rightarrow {}^4T_{2g}$	$^4A_{2g} \rightarrow {}^4T_{1g}$	*CT* ($L \rightarrow M$)	*Medium*	*Ref.*
$[Cr(SH)(H_2O)_5]^{2+}$	Green	575 (27.5)	435 (43.1)	258 (6520)	0.1 M $HClO_4$	1
$[Cr(SCH_2CH_2NH_3)(H_2O)_5]^{3+}$	Green	578 (26)	438 (46)	274 (7000)	0.001 M $HClO_4$(d–d) 0.1 M $HClO_4$ (CT)	2
$[Cr(4\text{-}SPhNH_3)(H_2O)_5]^{3+}$	Green	640sh (42), 595 (52)	—	295 (8250)	0.001 M $HClO_4$	3
$[Cr(SCH_2CO_2)(H_2O)_4]^+$	—	548 (68.3)	437 (53.4)	264 (5070)	0.1 M $HClO_4$	4
$[Cr\{SCH(CH_3)CO_2\}(H_2O)_4]^+$	—	545 (71.2)	440 (52.2)	265 (5050)	—	4
$[Cr(SCH_2CH_2NH_2)(en)_2]^{2+}$	Brown	560sh (39), 468 (99)	—	256 (7600)	Aq. ClO_4^-, pH 3	5
$[Cr(SCH_2CO_2)(en)_2]^+$	—	500 (99)	385 (91)	250 (7440)	Aq. ClO_4^-, pH 3	5
$[Cr(SCH_2CH_2CO_2)(en)_2]^+$	Deep purple	510 (88)	395 (65)	258 (6050)	Dilute $HClO_4$	6
$[Cr\{SC(CH_3)_2CO_2\}(en)_2]^+$	—	495 (94)	385 (87)	251 (7420)	Dilute $HClO_4$	6
$[Cr(SCH_2CH_2NH_2)_2en]^+$	Red	528 (157)	424 (117)	282 (15 600)	Aq. ClO_4^-, pH 3	5
$[Cr(SCH_2CO_2)_3]^{3-}$	—	632 (202)	476 (202)	—	—	6

1. T. Ramasami and A. G. Sykes, *Inorg. Chem.*, 1976, **15**, 1010.
2. L. E. Asher and E. Deutsch, *Inorg. Chem.*, 1973, **12**, 1774.
3. L. E. Asher and E. Deutsch, *Inorg. Chem.*, 1972, **11**, 2927.
4. R. H. Lane, F. A. Sedor, M. J. Gilroy and L. E. Bennett, *Inorg. Chem.*, 1977, **16**, 102.
5. C. J. Weschler and E. Deutsch, *Inorg. Chem.*, 1973, **12**, 2682.
6. C. K. Jorgensen, *Inorg. Chim. Acta Rev.*, 1968, **2**, 65.

method.[966] However some of these complexes, *e.g.* $[Cr(SCH_2CH_2NH_3)(H_2O)_5]^{3+}$, can be prepared by chromium(II) reduction of the alkylthiolatocobalt(III) complexes, a reaction which occurs by an inner sphere mechanism and which results in ligand transfer (Scheme 97).[966] The product in solution can be partly purified by cation exchange chromatography but cannot be obtained free from $[Cr(H_2O)_6]^{3+}$, which is present to an extent of 5% in the final effluent. The solution electronic spectrum of this complex (Table 85) contrasts with that of the thiolato-anilinium complex in that the charge transfer band is at a shorter wavelength (274 nm) and does not obscure the ${}^4A_{2g}\rightarrow{}^4T_{1g}$ band, which is clearly observed at 438 nm. Using this and a series of related complexes $[CrX(H_2O)_5]^{2+}$ (X = F, N_3, OAc, CN) evidence is presented that in the case of ligands derived from weak acids (HX) the acid-independent term in the aquation rate law, like the acid-dependent one, depends on the ability of the coordinated ligand to be protonated and corresponds to the removal of HX from $[Cr(OH)HX(H_2O)_4]^{2+}$ rather than the removal of X from $[CrX(H_2O)_5]^{2+}$.[70,966] Potential ligands have been found to enhance the rate of Cr—S bond cleavage in $[Cr(SR)(H_2O)_5]^{2+}$ and $[Cr(SAr)(H_2O)_5]^{2+}$ complexes and to be incorporated into the coordination sphere of the metal at greatly enhanced rates, indicative of a high *trans* effect associated with thiolate ligands (equation 50).[970]

$$[Cr(SR)(H_2O)_5]^{2+} + L^{n-} + H^+ \longrightarrow [CrL(H_2O)_5]^{(3-n)+} + HSR \qquad (50)$$

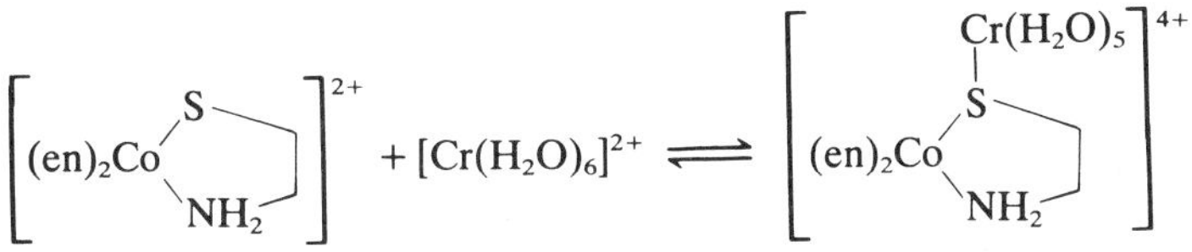

$$\xrightarrow{H_3O^+} [Cr(SCH_2CH_2NH_3)(H_2O)_5]^{3+} + [Co(H_2O)_6]^{2+} + 2enH_2^{2+}$$

Scheme 97

(iv) Complexes containing S, N and S, O chelating ligands

Complexes containing these ligands are also dealt with in Sections 35.4.2.3 and 35.4.8.

The thiolatoamine complexes $[Cr(SCH_2CO_2)(en)_2]ClO_4$ and $[Cr(SCH_2CH_2NH_2)(en)_2](ClO_4)_2$ have been prepared by chromium(II) reduction of the disulfides in the presence of en (Scheme 98).[971] In the preparation of the aminoethanethiol complex a second product, $[Cr(SCH_2CH_2NH_2)_2en]ClO_4$, was also produced. These complexes could be separated by either fractional crystallization of the perchlorate salts from dilute $NaClO_4$ solution or by cation exchange chromatography. The proportion of unipositively charged species in the product mixture was found to increase by using excess disulfide in the preparation. The electronic solution spectra of these complexes are presented in Table 85. The aquation reactions of the complexes in dilute $HClO_4$ solutions proceed predominantly through H^+-catalyzed Cr—S bond fission, although acid-independent processes involving Cr—S and Cr—N bond breaking are also competitive.[971] The aquation of $[Cr(SCH_2CO_2)(en)_2]^+$ in $HClO_4$ proceeds *via* several steps the first of which gives *cis*-$[Cr(O_2CCH_2SH)(en)_2H_2O]^{2+}$. Subsequent aquation steps involving Cr—N and at a later stage Cr—carboxylate bond fission (analogous with aquation reactions of oxalatopolyammine[475,972] and acetatopentaamminechromium(III)[973] complexes) result in $[Cr(en)(H_2O)_4]^{3+}$, the predominant product following extensive hydrolysis.[971] The aquation of $[Cr(SCH_2CH_2NH_2)(en)_2]^{2+}$ in 4 M $HClO_4$ gives $[Cr(en)_2(NH_2CH_2CH_2SH)H_2O]^{3+}$ as the primary aquation product and is much slower than aquation of $[Cr(SCH_2CO_2)(en)_2]^+$. Both complexes however are some 10^4-fold more reactive than their cobalt(III) analogues. Since X-ray crystallographic evidence[413] indicates that (after allowance is made for the different ionic radii of the metal ions) the Cr—S bond is actually stronger than the Co—S bond in the $[M(SCH_2CO_2)(en)_2]^+$ complexes,[970] and since the larger crystal field activation energy of cobalt(III) would account for only a 10^1–10^2 decrease in rate constant,[974] it appears that the greater lability of the Cr—S bond may be due to the greater basicity and ease of protonation of sulfur bound to Cr. The metal–ligand bond distances in the $[M(SCH_2CO_2)(en)_2]^+$ (M = Co, Cr) cations are compared in Table 86.

Oxidation of some of the afore-mentioned complexes as well as $[Cr\{S(CH_2)_2CO_2\}(en)_2]ClO_4$ and $[Cr\{SC(Me_2)CO_2\}(en)_2]ClO_4$ (electronic spectra of these complexes are included in Table

$$Cr \xrightarrow{i} [Cr(H_2O)_6]^{2+} \xrightarrow{ii-iv} [Cr(S\text{—ligand})(en)_2](ClO_4)_x{}^{a}$$

i, excess 20% $HClO_4$, anaerobic conditions; ii, 10% deaerated en solution, inert atmosphere; iii, deaerated solution of disulfide, inert atmosphere, so that Cr:en:disulfide = 2:6:1; iv, $HClO_4$ after removal of $Cr(OH)_3$ by filtration.

[a] S ligand = $^-SCH_2CO_2^-$, $^-SCH_2CH_2NH_2$, $x = 1, 2$, respectively.

Scheme 98

Table 86 Bond Distances (pm) in $[M(SCH_2CO_2)(en)_2]^+$ (M = Co, Cr) Complexes[1]

M	*M—S*	*M—O*	*M—N*[1 a]	*M—N*[2 b]	*M—N*[3 c]	*M—N*[4 d]
Cr	233.7[e]	196.6	209.6	209.9	211.2	207.6
Co	224.3[f]	191.8	197.1	195.8	200.5	196.5

[a] N^1 = nitrogen *trans* to oxygen.
[b] N^2 = other nitrogen in same en ring as N^1.
[c] N^3 = nitrogen *trans* to sulfur.
[d] N^4 = other nitrogen in same en ring as N^3.
[e] In the complex $Cr(S_2COEt)_3$ the Cr—S distances are 238.7 and 239.9 pm.[2]
[f] In $Co(S_2COEt)_3$ the Co—S distances are 227.6 and 227.7 pm.[3]

1. R. C. Elder, L. R. Florian, R. E. Lake and A. M. Yacynych, *Inorg. Chem.*, 1973, **12**, 2690.
2. S. Merlino and F. Sartori, *Acta Crystallogr., Sect. B*, 1972, **28**, 972.
3. S. Merlino, *Acta Crystallogr., Sect. B*, 1969, **25**, 2270.

85) by H_2O_2 has been investigated.[412] Reaction of $[Cr(SCH_2CO_2)(en)_2]^+$ produces a complex product mixture, the composition of which has proved difficult to elucidate. However, the thiolate ligand is oxidized and it appears that the oxidation state of sulfur changes from −II to +II (RSO_2^-, sulfinate) and possibly +IV (RSO_3^-, sulfonate), with the likelihood of a transient 0 (RSO^-, sulfenate) state. The mechanism of oxidation by H_2O_2 involves nucleophilic attack by coordinated sulfur on the O—O bond. The reactions follow second-order kinetics and while reactivity is relatively insensitive to the nature of the thiolato complex, steric crowding at the sulfur atom causes decreased reactivity for the complexes $[Cr(SCMe_2CO_2)(en)_2]^+$ and $[Cr(4\text{-}SPhNH_3)(en)_2]^{3+}$.

The nucleophilicity of coordinated sulfur in the complex $[Cr(SCH_2CH_2NH_2)(en)_2]^{2+}$ towards MeI (equation 51) has been measured in DMF/H_2O and compared with sulfur nucleophilicities in thiolatocobalt(III) systems.[975] As in the case of the H_2O_2 oxidations, the nucleophilicity of thiolate coordinated to chromium(III) is only slightly less than when it is coordinated to cobalt(III), implying that nucleophilic attack by coordinated sulfur does not involve any appreciable distortion in the first coordination sphere of the metal.

$$[Cr(SCH_2CH_2NH_2)(en)_2]^{2+} + MeI \longrightarrow [Cr\{S(Me)CH_2CH_2NH_2\}(en)_2]^{3+} + I^- \quad (51)$$

(v) $Cr(SR)_3$, $CrX_3(SR_2)_3$, $CrCl_3(THT)_3$ (X = Cl, Br; R = Me, Et; THT = tetrahydrothiophene)

The preparation and properties of a number of these thiolato and thioether complexes are presented in Table 87. X-Ray powder photographs of $Cr(SMe)_3$ show that the compound has a layer-type lattice with chromium(III) occupying octahedral $Cr(SMe)_6$ sites.[976,977] The chemical behaviour of the mixed halogenothioetherchromium(III) complexes is illustrated using $CrCl_3(SMe_2)_3$ as an example in Scheme 99.[978] It was not possible to isolate solid samples of the dimer $[CrCl_3(SEt_2)_2]_2$ or the monomers $CrBr_3(SR_2)_3$ (R = Me, Et). The complexes $CrCl_3(SR_2)_3$ (R = Me, Et) have been assigned *mer* geometric configurations on the basis of their far IR spectra. Hence in the Cr—Cl stretching region these complexes show three absorption bands, which for the *mer* isomer (C_{2v} symmetry) is expected, since the three Cr—Cl stretching modes ($2a_1 + b_1$) are IR active. For the *fac* isomer, however, only two Cr—Cl stretching modes ($a_1 + e$ of C_{3v}) are active.[978] These criteria have also been applied to distinguish between the isomers of $CrCl_3(THT)_3$ (THT = tetrahydrothiophene).[979]

Table 87 Preparation and Properties of Thiolato- and Thioether-chromium(III) Complexes

Complex	Colour	Method of preparation	IR spectra[a] (cm^{-1})	Electronic spectra[b] $^4A_{2g} \to {}^4T_{2g}$	Electronic spectra[b] $^4A_{2g} \to {}^4T_{1g}$	CT	Comments	Ref.
$Cr(SMe)_3$	Green	Irradiate $C_6H_6Cr(CO)_3$ or $Cr(CO)_6$ in $(MeS)_2$ or reflux $CrCl_3$ in $(MeS)_2$ or reflux $CrCl_3 + Na_2S$ in $(MeS)_2$	[c]	600	400[d]		$\mu_{eff} = 3.49$ BM, $\mu_{corr} = 3.88$ at room temp. Insoluble in common organic solvents. Unreactive towards cold or boiling water. Layer structure with Cr^{III} in octahedral $[Cr(SMe)_6]^{3-}$ sites	1
$Cr(SEt)_3$	Green	Irradiate $C_6H_6Cr(CO)_3$ in $(EtS)_2$	[e]	610	385[d]		$\mu_{eff} = 3.56$ BM, room temp. Stable to hydrolysis.	1
$Cr(SPh)_3$	Green	Irradiate $C_6H_6Cr(CO)_3$ in $(PhS)_2$	[e]	620	400[d]		$\mu_{eff} = 3.60$ BM, room temp. Stable to hydrolysis.	1
$Cr(SCH_2Ph)_3$	Green	Irradiate $C_6H_6Cr(CO)_3$ in $PhCH_2SPh$	[e]	600	370[d]		$\mu_{eff} = 3.50$ BM, room temp. Stable to hydrolysis.	1
fac-$CrCl_3(THT^f)_3$	Blue-purple	$CrCl_3(NMe_3)_2$ + THT in C_6H_6, room temp. or $CrCl_3$ + zinc dust in THT, room temp.	819, 685: ν(CSC) 368, 340: ν(Cr—Cl)	685	480	350, 254sh[d]	Air, moisture-sensitive	2
mer-$CrCl_3(THT)_3$	Lilac	Isomerize *fac*-$CrCl_3(THT)_3$ by stirring in C_6H_6; room temp.	819, 685: ν(CSC) 380, 351, 278: ν(Cr—Cl)	699	497	380, 260sh[d]	Air, moisture-sensitive	2
mer-$CrCl_3(SMe_2)_3$	Blue-violet	$CrCl_3$ + zinc dust + Me_2S, room temp.[g]	256 vw, 234w, 210w: ν(Cr—S)[h] 375vs, 346vs, 317s: ν(Cr—Cl)	690 690 (112)	569, 486 574 (56), 480 (68)[i]	370[d]		3
mer-$CrBr_3(SMe_2)_3$	—	Dissolve $[CrBr_3(SMe_2)_2]_2$ in Me_2S	—	720 (118)	588 (81), 518 (86)[i]		Cannot be isolated in solid state. Reverts to dimer on removal of solvent	3
mer-$CrCl_3(SEt_2)_3$	Blue-violet	$CrCl_3$ + zinc dust + Et_2S, room temp.[g]	255m, 250sh: ν(Cr—S)[h] 368vs, 345m, 307s: ν(Cr—Cl)	700 706 (102)	570 484 580 (58), 486 (78)[j]	373[d]		3
mer-$CrBr_3(SEt_2)_3$	—	Dissolve $[CrBr_3(SEt_2)_2]_2$ in Et_2S	—	733 (110)	600 (76), 525 (84)[j]		Cannot be isolated in solid state. Reverts to dimer on removal of solvent	3
$[CrCl_3(SMe_2)_2]_2$	Red-violet	Recrystallize $CrCl_3(SMe_2)_3$ from C_6H_6	257w, 235w: ν(Cr—S)[h] 381s, 350s, 316m: ν(Cr—Cl)	699 699 (103)	526[d] 534 (91)[k]		Contains six-coordinate chromium ions linked by a Cl^- bridging ligand	3
$[CrBr_3(SMe_2)_2]_2$	Deep green	$CrBr_3$ + zinc dust + Me_2S, room temp.[g]	244w: ν(Cr—S)[h] 316s, 290s: ν(Cr—Cl)	744 (120)	570 (145)[k]		Contains six-coordinate chromium ions linked by a Br^- bridging ligand	3
$[CrBr_3(SEt_2)_2]_2$	Deep green	$CrBr_3$ + zinc dust + Et_2S, room temp.[g]	245m: ν(Cr—S)[h] 311s, 287s: ν(Cr—Cl)	750 760 (113)	570[d] 574 (126)[k]		Contains six-coordinate chromium ions linked by a Br^- bridging ligand	3

[a] Nujol or fluorolube mulls.
[b] Wavelengths in nm, ε values in $dm^3\ mol^{-1}\ cm^{-1}$ in parentheses.
[c] Full details of IR spectrum and band assignments given in ref. 1.
[d] Diffuse reflectance spectrum.
[e] Full details of IR spectrum in ref. 1.
[f] THT = tetrahydrothiophene.
[g] Chromium(III) halide should be treated with appropriate thionyl halide before reaction.
[h] Because of weakness of bands assignment of ν(Cr—S) is tentative.
[i] In Me_2S solution.
[j] In Et_2S solution.
[k] In benzene solution.

1. D. A. Brown, W. K. Glass and B. Kumar, *J. Chem. Soc. (A)*, 1969, 1510.
2. J. Hughes and G. R. Willey, *Inorg. Chim. Acta*, 1974, **11**, L25.
3. R. J. H. Clark and G. Natile, *Inorg. Chim. Acta*, 1970, **4**, 533.

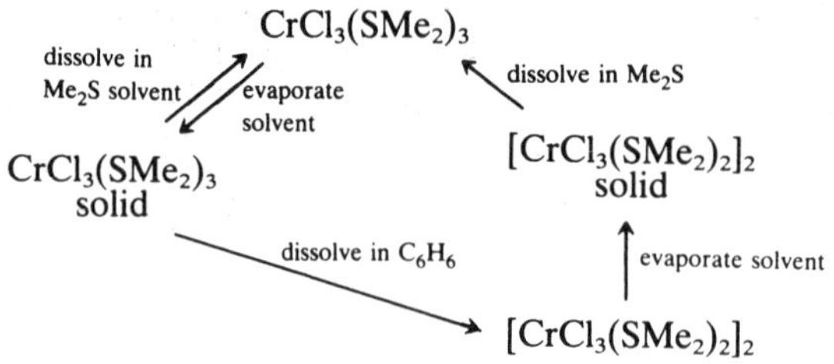

Scheme 99

(*vi*) *Complexes containing* S_2^{2-} *and* HS_2^- *ligands*

The oxidation of $[Cr(SH)(H_2O)_5]^{2+}$ by I_2 or Fe^{3+} under aerobic conditions in acid solutions gives the disulfido-bridged complexes $[(H_2O)_5CrS_2Cr(H_2O)_5]^{4+}$ and $[(H_2O)_5Cr(S_2H)Fe(H_2O)_5]^{4+}$ respectively (Scheme 100).[967,968] The latter complex can also be obtained by substitution of chromium(III) in the former complex by iron(II) under acid conditions. The product distribution in the iron(III) oxidation of $[Cr(SH)(H_2O)_5]^{2+}$ is pH dependent and at 298 K, pH = 1 the heteronuclear dimer $[(H_2O)_5Cr(S_2H)Fe(H_2O)_5]^{4+}$ constitutes over 80% of the product mixture. The rate of this reaction shows a $[H^+]^{-1}$ dependence, an observation consistent with $[CrS(H_2O)_5]^+$ being the kinetically active species.

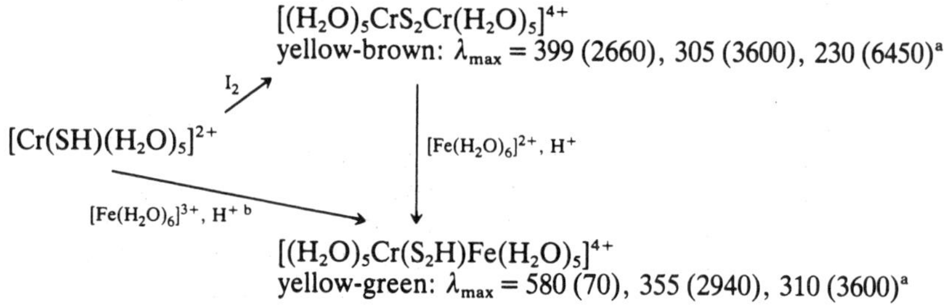

[a] λ_{max} in nm (ε in $dm^3\,mol^{-1}\,cm^{-1}$). [b] This reaction also gives $[(H_2O)_5CrS_2Cr(H_2O)_5]^{4+}$ as a minor (pH-dependent) product.

Scheme 100

(*vii*) *Binary and ternary sulfides*

A number of binary compounds in the chromium–sulfur system have been identified and their structures established. These include CrS, Cr_2S_3, Cr_3S_4, Cr_5S_6 and Cr_5S_8.[980,981] The origin of ferromagnetism in Cr_2S_3 and Cr_5S_6 lies in the ordered arrangement of cation vacancies with resulting uncompensated spins.[982] Since the magnetic and electrical properties of Cr_5S_8 indicate that the *d* electrons of chromium are localized, the formula $Cr_4^{III}Cr^{IV}S_8$ has been suggested for this compound with chromium(III) occupying the filled cation layers and chromium(IV) occupying the partly filled layers.[981]

The ternary sulfide $NaCrS_2$, which can be prepared by fusion of either Na_2CrO_4, Na_2CO_3 and sulfur[983] or K_2CrO_4, $NaKCO_3$ and sulfur,[984] contains octahedral $Cr^{III}S_6$ groups in which the Cr—S distances are 244 pm.[985] On lowering the temperature from 300 to 20 K a shoulder at 13 350 cm^{-1} on the ${}^4A_{2g} \rightarrow {}^4T_{2g}$ band (14 600 cm^{-1}) in the reflectance spectrum becomes prominent.[985] Although this shoulder was originally attributed to splitting of the ${}^4T_{2g}$ state in a trigonally distorted octahedral field it now appears that it is instead due to a spin-forbidden transition to the ${}^2E_g$ state.[986] The sulfides $CuCrS_2$ and $AgCrS_2$ have been prepared by heating the mixed binary chalcogenides.[987] The crystal structures of these compounds show that the sulfur atoms are arranged in a deformed cubic close pack with chromium occupying octahedral sites and copper or silver in tetrahedral sites. Like $NaCrS_2$, these sulfides are antiferromagnetic and Néel temperatures for the three compounds fall in the range 19–40 K.[987] Ferromagnetic interactions occur in the chromium layers of $NaCrS_2$. The structure of $LiCrS_2$ is of the NiAs type with lithium and chromium atoms occupying octahedral sites.[988] In $CrUS_3$ the chromium atoms have distorted octahedral coordination, while each uranium is located at the centre of a bicapped trigonal prism of sulfur atoms.[989,990]

The semiconducting thiospinels MCr_2S_4 (M = Mn, Fe, Co and Zn), which can be prepared by the high temperature combination of the metals and sulfur, contain $M^{II}S_4$ and $Cr^{III}S_6$ groups.[991] However, $CuCr_2S_4$ is metallic and appears from crystallographic and magnetic evidence to contain copper(I) and not copper(II). The chromium atoms occupy octahedral $Cr^{III}S_6$ sites. The metallic properties of this compound have been attributed to a conduction band arising from the absence of a fraction of the valence electrons of sulfur.[986] The structure of $CrPS_4$, determined by X-ray crystallography, consists of puckered hexagonally close-packed sulfur layers stacked parallel to (100) with chromium in octahedral and phosphorus in tetrahedral interstices.[992] This is the first reported metal thiophosphate having the metal ion in octahedral sulfur coordination.

35.4.5.2 Bidentate S donor ligands: dithiocarbamates, dithiophosphates, 1,1- and 1,2-dithiolates, etc.

The chemistry of the complexes formed by the dithio ligands (**218**) to (**225**) with transition metals including chromium(III) has been well described in several reviews.[583,986,993–996]

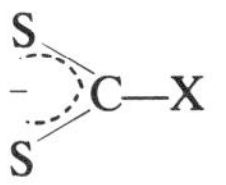

(**218**) X = NR_2, dithiocarbamate
X = OR, xanthate
X = SR, thioxanthate
X = R, dithiocarboxylate

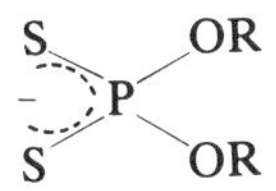

(**219**) Dithiophosphate

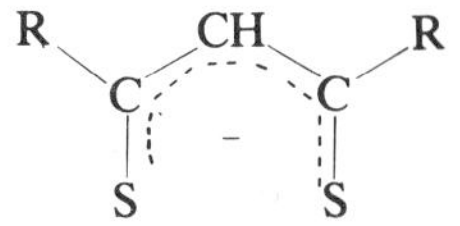

(**220**) 1,3-Dithiodiketone

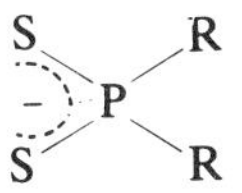

(**221**) Dithiophosphinate

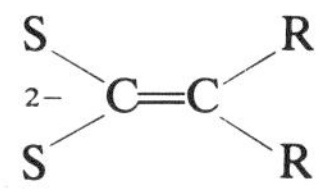

(**222**) 1,1-Dithiolate

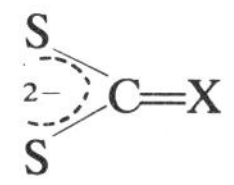

(**223**) X = O, dithiocarbonate
X = S, trithiocarbonate
X = NCN, *N*-cyanodithiocarbimate

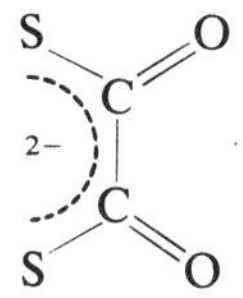

(**224**) Dithiooxalate

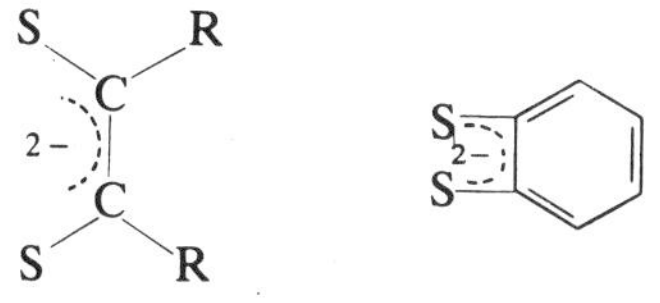

(**225**) 1,2-Dithiolates (dithiolenes)

In general, the complexes are prepared by mixing aqueous or alcoholic solutions of the sodium, ammonium or potassium salt of the ligand and $CrCl_3{\cdot}6H_2O$, sometimes with Zn dust present.[997,998] However, in some cases anhydrous conditions are necessary to prevent the formation of polymeric hydrolysis products,[986] and dithiophosphate is much more resistant to temperature and extremes of pH than dithiocarbamate or xanthate. In the preparation of CrL_3, where L is 4-aminophenazonedithiocarbamate, aqueous $Cr(NO_3)_3$ was added to the reaction mixture from the preparation of the ligand.[999]

Another general method is illustrated by the preparation of $[Cr(S_2CNR_2)_3]$ (where R = Me, Et, Pr, Bu^n, Bu^i, CH_2Bu; $R_2 = (CH_2)_4O$, $(CH_2)_4$, $(CH_2)_5$, MePh or EtPh; equation 52). In this a chromium(II) salt is added to the ligand and the mixture allowed to oxidize.[1000] An alternative is to reduce a precursor, *e.g.* the 3,5-dimethyl-1,2-dithiolium cation, with a chromium(II) salt to the ligand which then coordinates to the Cr^{III} produced (equation 53).[1001]

$$[Cr(O_2CMe)_2H_2O] + NaS_2CNR_2 \xrightarrow[HCl(aq)]{air} [Cr(S_2CNR_2)_3] \qquad (52)$$

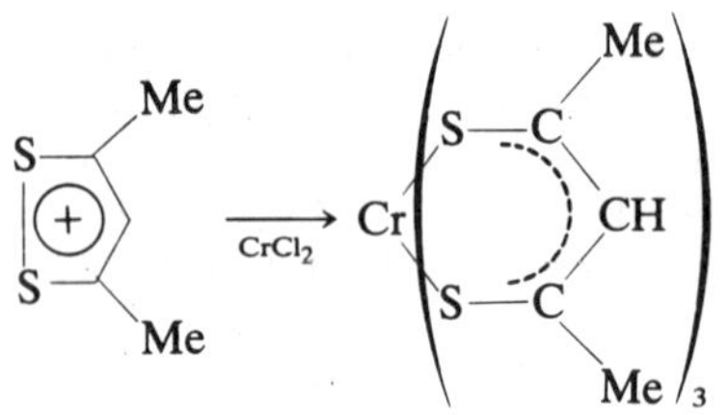

(53)

Chromium(III) complexes of some strong acids, *e.g.* HS_2PF_2, have been obtained directly from the metal and the acid.[1002,1003] The complex $[Cr(S_2PF_2)_3]$ can also be made by reduction of CrO_2Cl_2 by the acid.[1004] A specific method is the insertion of CS_2 into the Cr—N bond accompanied by reduction of Cr^{IV} to Cr^{III} (equation 54).[618] An account of the redox behaviour of many complexes containing the CrS_6 core also provides many references to the preparation of the ligands and their chromium(III) complexes.[1005]

$$[Cr(NEt_2)_4] + 4CS_2 \longrightarrow [Cr(S_2CNEt_2)_3] + \tfrac{1}{2}(S_2CNEt_2)_2 \qquad (54)$$

A new method[1006] is the oxidative decarbonylation of $Cr(CO)_6$ with $[Hg(S_2CNEt_2)_2]$ in 2:3 molar ratio to give the chromium(II) complex followed by aerial oxidation of the reaction mixture (Scheme 101). Although the unoxidized intermediate solution has the electronic spectrum of the chromium(II) complex (Section 35.3.6.1) the reaction pathway is not completely clear.

$$Cr(CO)_6 \xrightarrow[N_2,\ toluene]{[Hg(S_2CNEt_2)_2]} [Cr(S_2CNEt_2)_2] + Hg + CO$$

green solution

↓ air

$[Cr(S_2CNEt_2)_3]$

dark blue

Scheme 101

The oxidation of $Na(NH_4)S_2CNR_2$ (R = Me, Et; $R_2 = (CH_2)_5$) with $K_2Cr_2O_7$ produces a mixture of $[Cr(S_2CNR_2)_3]$ and the dithioperoxycarbamato complex $[Cr(S_2CNR_2)_2(OS_2CNR_2)]$, which has been separated chromatographically. The new dithioperoxy derivatives are in the second fraction and have additional bands in their IR spectra at 1009 and 488 cm^{-1}, assigned to the ν(S—O) and ν(Cr—O) vibrations respectively.[1007–1009] A redetermination of the crystal structure of $[Cr(S_2CNEt_2)_2(OS_2CNEt_2)]$ taking into account the disorder of the *O,S*-chelate has confirmed the gross structure, improved the refinement and provided distances comparable with those expected from other chromium(III) complexes.[1010] The X-ray results make unlikely the assignment of a dinuclear structure $[Cr_2O(S_2CNR_2)_2(R_2NCSS_2CSNR_2)]$ to the second fraction.[1011] The tris(dithiocarbamates) with $R_2 = (CH_2)_4S$ and $(CH_2)_4NMe$ have also been prepared by the oxidative method.[1012]

Some data from X-ray investigations are given in Table 88. The CrS_6 octahedra show small trigonal distortions but these are less than in related MS_6 complexes, and the Cr—S bond distances fall within a small range.

Perfluoroalkyl derivatives $[R_FCr(S_2CNR_2)_2py]$ (R = Et, $R_F = C_2F_5$, C_3F_7, C_4F_9; R = Me, Pr^i, Bz, $R_F = C_3F_7$) and $[C_3F_7Cr(S_2CNEt_2)_2(cyclohexylamine)]$ have been prepared in the same way as $[C_2F_5Cr(salen)py]$ (Section 35.4.8.1).[1013] The complexes are light sensitive and produce the cations $[R_FCr(H_2O)_5]^{2+}$ in water, but are otherwise stable σ-bonded chromium(III) species; no stable alkylchromium(III) complexes were obtained. The doublet found for the methylene protons in the ^1H NMR spectrum of $[C_3F_7Cr(S_2CNEt_2)_2py]$ suggested that enantiomers with a high barrier to interconversion were present in solution. A *cis* configuration was found in $[C_3F_7Cr(S_2CNMe_2)_2py]$ and the bond *trans* to the C_3F_7 group is considerably longer than the other Cr—S linkages.

The crystal structure of $[Cr(S_2COEt)_3]$ is composed of enantiomorphous molecules with trigonal symmetry. The short S_2C—O bond distance (1.297 Å) is taken to mean that the resonance structure S_2^{2-}—C=$\overset{+}{O}$—R contributes considerably.[1014]

The magnetic moments of the complexes are usually near 3.8 BM at room temperature. The moment of $[Cr(S_2CNEt_2)_3]$ is essentially independent of temperature down to *ca.* 40 K. Then it falls slowly to approximately 3.6 BM at 4 K and more rapidly to 3.5 BM at 2.5 K.[1015] It was not

Table 88 Structural Data for Complexes of Bidentate S Donor Ligands

Complex	*Cr—S* (Å)	*Ref.*
$[Cr(S_2CNC_4H_8O)_3]\cdot CH_2Cl_2$	2.40	1
$[Cr(S_2CNC_4H_8O)_3]\cdot 2C_6H_6$	2.396	2
$[Cr(S_2CNC_4H_8)_3]\cdot 0.5C_6H_6$	2.404	3
$[Cr(S_2CNEt_2)_3]$	2.39	4
$[Cr\{S_2CN(CH_2)_5\}_3]\cdot 2CHCl_3$	2.397 (av)	5
$[Cr(S_2CNEt_2)_2(OS_2CNEt_2)]$	2.405 (S_2O ligand) 2.405, 2.403 2.379, 2.370 Cr—O, 1.988	6
$[C_3F_7Cr(S_2CNMe_2)_2py]$	*trans*, 2.457 others, 2.392	7
$[Cr(S_2COEt)_3]$	2.393	8
$[Cr(S_2P(OEt)_2)_3]$	2.421–2.430	9

1. R. J. Butcher and E. Sinn, *J. Chem. Soc., Dalton Trans.*, 1975, 2517.
2. R. J. Butcher and E. Sinn, *J. Am. Chem. Soc.*, 1976, **98,** 2440.
3. E. Sinn, *Inorg. Chem.*, 1976, **15,** 369.
4. C. L. Raston and A. H. White, *Aust. J. Chem.*, 1977, **30,** 2091.
5. V. Kettman, J. Garaj and J. Majer, *Collect. Czec. Chem. Commun.*, 1981, **46,** 6.
6. R. L. Martin, J. M. Patrick, B. W. Skelton, D. Taylor and A. H. White, *Aust. J. Chem.*, 1982, **35,** 2551.
7. A. M. Van Den Bergen, K. S. Murray, R. M. Sheahan and B. O. West, *J. Organomet. Chem.*, 1975, **90,** 299.
8. S. Merlino and F. Sartori, *Acta Crystallogr., Sect. B*, 1972, **28,** 972.
9. H. V. F. Schousboe-Jensen and R. G. Hazell, *Acta Chem. Scand.*, 1972, **26,** 1375.

possible to decide whether the low moments at low temperature are due to small zero-field splittings or very weak exchange interactions, or both.[1015] The complex [Cr(SacSac)$_3$] (**220**; R = Me) has a moment which decreases steadily as the temperature is lowered.[1001] The polymers $[Cr_2(S_2CNHC_2H_4CNHCS_2)_3]$ and $[Cr_2(S_2CNHC_6H_{12}NHCS_2)_3]$ show weak antiferromagnetic interaction, with θ values (negative intercepts on T axis) of 32° and 8° respectively. They are amorphous, and insoluble in water and common solvents.[1016]

The temperature dependence of the ^{1}H NMR spectra of many $[Cr(S_2CNR_2)_3]$ complexes has been investigated.[1000,1017] There were broad double resonances from the N—CH_2 protons indicating that inversion between the Δ and Λ isomers is slow on the ^{1}H NMR time-scale. No exchange broadening was observed up to 84 °C when $[Cr(S_2CNEt_2)_3]$ decomposed. Dithiocarbamato complexes of other tervalent metal ions do not exhibit this stereochemical rigidity.

Assignments of spin-forbidden bands in the spectra of $[Cr(S_2CNR_2)_3]$ and $[Cr(S_2COR)_3]$ (R = Me or Et) have been made and the emission spectra studied.[1018]

MCD spectroscopy has been useful in the location of the spin-forbidden transitions.[1019] That the Cr—S bond is comparatively strong has been adduced from the presence of relatively intense molecular ion peaks in the mass spectra of several Cr^{III} dialkyldithiocarbamates.[1020]

The complexes $[Cr(S_2CNMe_2)_3]$, $[Cr\{S_2P(OEt)_2\}_3]$ and $[Cr(S_2COEt)_3]$ have Pfeiffer CD activity but this develops only when there is a large ratio of optically active environment compound to the racemate.[1021] A normal coordinate analysis of the IR spectra of $[Cr(S_2CNR_2)_3]$ (where R = Me, Et, Pr or Bu^n) has shown that the alkyl substituent affects the thioureide band at *ca.* 1550 cm^{-1} through kinematic as well as electronic effects.[1022]

Complexes $[Cr(L)_2Cl_2]Cl$ in which L is a tetraalkylthiuram sulfide or disulfide have been characterized.[1023]

Other typical chromium(III) complexes of bidentate S donor ligands are the xanthates $[Cr(S_2COR)_3]$ (R = Me, Et, Bz, Ph), thioxanthates $[Cr(S_2CSR)_3]$ (R = Me, Et, Pr, Bu^t, Bz), dithiocarboxylates $[Cr(S_2CR)_3]$ (R = Ph, C_6H_4Me, C_6H_4OMe, $C_6H_4NMe_2$, $C_6H_4NEt_2$, C_4H_3O, $C_5H_3NH_2$, Bz),[1005,1024–1029] dithiooxalate $[Cr(C_2S_2O_2)_3]^{3-}$, trithiocarbonate $[Ph_4P]_3$ or $[Ph_4As]_3[Cr(CS_3)_3]$,[1030–1031] dialkyldithiophosphates $[Cr\{S_2P(OR)_2\}_3]$ (R = Me, Et, Pr^n, Pr^i, Bu^s, Bu^i),[1005,1032,1033] dithiophosphinates $[Cr(S_2PR_2)_3]$ (R = F, CF_3, Me, Ph, OEt)[1002,1005] and, for example, $[Cr\{S_2PMe(C{\equiv}CMe)\}_3]$,[1034] dithio-$\beta$-diketonates $[Cr(SCRCHCR'S)_3]$ (R = R' = Me, Ph; R = Me, R' = OMe, OEt)[1001,1005] and dithiocarbazates such as $[Cr(S_2CNHNH_2)_3]$ and $[Cr(S_2CNHNPh_2)_3]$.[1035]

Complexes of Se donor ligands, $[Cr(Se_2CNR_2)_3]$ (R = Et; R_2 = $(CH_2)_5$, $(CH_2)_4O$ or $(CH_2)_4S$) and $[Cr\{Se_2P(OEt)_2\}_3]$ are similar to the corresponding S donor complexes, but their electronic spectra exhibit a pronounced nephelauxetic effect.[986,1036]

A number of chromium(III) complexes of 1,1-dithiolato dianions (**222**) are known: $[Cr(S_2C{=}X)_3]^{3-}$ (where $X = C(CN)_2$, $C(CO_2Me)_2$, NCN, $C(CN)CO_2Et$, $C(CO_2Et)_2$, $C(CN)CO_2Me$ and $C(CN)CONH_2$)[583,993,1005] and the thioselenyl analogues $[Cr\{SSeC{=}C(CN)_2\}_3]^{3-}$ and $[Cr\{SSeCO(CH_2)_2OMe\}_3]$ have been obtained.[1037]

The complex $[Cr(S_2PF_2)_3]$ is volatile and monomeric in the gas phase.[1002] Its He^{I} photoelectron spectrum has been reported.[1038] The absence of interfering ionization bands of substituent groups outside the coordination sphere permitted detailed assignments. The UVPES spectrum of $[Cr\{S_2P(OEt)_2\}_3]$ has been obtained,[1039] and the symmetric ν(Cr—S) vibration at 282 cm^{-1} has been identified in the RR spectrum of $[Cr(S_2PPh_2)_3]$.[1040]

Complexes of 1,2-dithiolate dianions (**225**) are of interest, although because they undergo successive one-electron transfer processes which involve orbitals of ligand character so unlike those of the dithiocarbamate type of ligand, it is unprofitable to attempt to assign oxidation states.[995,1005] The chromium complexes are shown in Table 89.

Table 89 1,2-Dithiolato Complexes

Complex	*Comments*	*Ref.*
$[A]_3[Cr\{S_2C_2(CN)_2\}_3]$ Brown	$A = Ph_4As$, $\mu_{eff} = 3.90$ BM	1, 2
	$A = Ph_4P$, $\mu_{eff} = 3.88$ BM (95 K), $\theta = 9°$	3
	$A = Ph_3BzP$	4
$[A]_2[Cr\{S_2C_2(CN)_2\}_3]$ Yellow-brown	$A = Ph_4As$, $\mu_{eff} = 2.89$ BM	1, 2
	$A = Ph_4P$, $\mu_{eff} = 2.76$ BM (88 K), $\theta = 8°$	3
$[Ph_4As]_2[Cr\{S_2C_2(CF_3)_2\}_3]$ Olive green	$\mu_{eff} = 2.95$ BM	1
$[Ph_4As][Cr\{S_2C_2(CF_3)_2\}_3]$ Dark green	$\mu_{eff} = 1.89$ BM	1
$[Cr\{S_2C_2(CF_3)_2\}_3]$ Red-purple	Diamagnetic	1
$[Cr(S_2C_2Ph_2)_3]$ Dark red	Diamagnetic	1, 5
$[Ph_4P][Cr(S_2C_2Ph_2)_3]$	$g = 1.996$	4, 5
$[A]_2[Cr(S_2C_6Cl_4)_3]$ Green	$A = Et_4N$ or Bu^n_4N, $\mu = 2.88$ BM	6
$[Bu_4N][Cr(S_2C_6F_4)_2]$	Unusual bis(chelate) complex	7
$[Me_4N]_n[Cr_n(S_2C_2H_4)_{2n}]$ Gold	$\mu_{eff} = 2.50$ BM, polymer?	8
	$\mu_{eff} = 3.85$ BM (DMSO)	

1. A. Davison, N. Edelstein, R. H. Holm and A. H. Maki, *J. Am. Chem. Soc.*, 1964, **86**, 2799.
2. E. I. Stiefel, L. E. Bennett, Z. Dori, T. H. Crawford, C. Simo and H. B. Gray, *Inorg. Chem.*, 1970, **9**, 281.
3. J. A. McCleverty, J. Locke, E. J. Wharton and M. Gerloch, *J. Chem. Soc. (A)*, 1968, 816.
4. P. Vella and J. Zubieta, *J. Inorg. Nucl. Chem.*, 1978, **40**, 613.
5. G. N. Schrauzer and V. P. Mayweg, *J. Am. Chem. Soc.*, 1966, **88**, 3235.
6. E. J. Wharton and J. A. McCleverty, *J. Chem. Soc. (A)*, 1969, 2258.
7. A. Callaghan, A. J. Layton and R. S. Nyholm, *Chem. Commun.*, 1969, 399.
8. J. R. Dorfman, Ch.P. Rao and R. H. Holm, *Inorg. Chem.*, 1985, **24**, 453.

The general electrochemical behaviour of the tris(chelate) complexes of Cr^{III} with the ligands (**218**) to (**225**) has been extensively investigated.[996,1008,1027,1041] New data have been correlated with early work.[1005] In general, the Cr^{III} complexes are resistant to oxidation and reduction consistent with the stable t_{2g}^3 subshell. Comparisons have been made between complexes of different ligand types and substituent effects analyzed for a particular ligand type in terms of the Hammett σ^+ constant. The dithiocarbamate ligand stabilizes high and low oxidation states and the redox processes are modified compared with the tris(dithiocarbamates) when one ligand is OS_2CNR_2 and by the presence of bipy in solution, which forms mixed ligand complexes in the lower oxidation states. Substituent effects are prominent in the 1,1-ethyleneditholate series. With these ligand systems the redox processes are metal based whereas in complexes of the 1,2-dithiolates (dithiolenes) the redox processes are primarily ligand based.

Published data on the volatility of complexes of the transition metals including chromium with bidentate sulfur and sulfur–oxygen donor ligands have been summarized.[1042]

The octahedral complexes $[Cr(cis\text{-}MeSCH{=}CHSMe)_3](BF_4)_3$, $[Cr\{S(CH_2CH_2CH_2SMe)_2\}_2](BF_4)_3$,[685] *mer*-$[Cr\{S(CH_2CH_2CH_2SMe)_2\}X_3]$, (X = Cl, Br or I), *fac*-$[Cr\{CMe(CH_2SMe)_3\}X_3]$ (X = Cl or Br) and *fac*-$[Cr(CH_2SCH_2CH_2SMe)_2Cl_3]$[696] have been prepared and characterized by methods used for similar P and As donor ligands.

Table 90 Colours and Main IR Bands for Thioureachromium(III) Complexes[1,2,a]

Complex	Colour	IR bands (cm^{-1}) $\nu(NH_2)$	$\delta(NH_2)$	$\nu_{as}(NCN)$	$\nu_{sym}(NCN) + \nu(CS) + \delta(NH_2)^2$	$\nu_{sym}(NCN) + \nu(CS) + \rho(NH_2)^2$	$\nu(CS)$	$\nu(Cr—S)$
tu[b]		3300, 3175, 3075	1617	1476	1417	1086	733	—
$CrCl_3(tu)_3$	Dark red	3400, 3310, 3220	1617	1490	1420	1100	705	282
$[Cr(phen)_2(tu)_2]Cl_3$	Grey	3225, 3030	1600	1515, 1469	1390	1105	722	—
$[Cr(bipy)_2(tu)_2]Cl_3$	Brown	3226, 3125, 3160	1610	1493, 1471	1390	1075	730	—
$[CrCl_2(py)_2(tu)_2]Cl$	Light green	3320, 3240, 3160	1610	1480	1400	1090	740	—
$[CrCl_2(2\text{-Mepy})_2(tu)_2]Cl$	Green	3330, 3250, 3180	1590	1460	1390	1050	760, 720	—
$[CrCl_2(quin)_2(tu)_2]Cl$	Green	3360, 3260, 3160	1600	1450	1390	1075, 1050	725	—
$[Cr(PhAsO_3H)_2(tu)_2]Cl$	Light green	3300, 3125	1610	1460	1400	1090	735	—
N-Phtu[b]		—		1528, 1515		1074	705	—
$CrCl_3(N\text{-Phtu})_3$	Red-brown	—		1537, 1516		1071	696	304
N-3-MePhtu[b]		—		1508		1079	741	
$CrCl_3(N\text{-3-MePhtu})_3$	Red-brown	—		1555		1103	708	273
N-3-ClPhtu[b]		—		1515		1074	706	
$CrCl_3(N\text{-3-ClPhtu})_3$	Red-brown	—		1525		1072	696	296, 274

[a] IR spectra were recorded of KBr discs (4000–650 cm^{-1}) and Nujol mulls (650–200 cm^{-1}). [b] Ligand IR bands included for comparison.

1. M. M. Khan, *J. Inorg. Nucl. Chem.*, 1975, **37,** 1621.
2. M. H. Sonar and A. S. R. Murty, *J. Inorg. Nucl. Chem.*, 1977, **39,** 2155.

35.4.5.3 *Thiourea*

Complexes of thiourea (tu) with chromium(III) have not been studied in detail and the lac[] of structural investigations in particular has led to uncertainties regarding the ligand donor site[] and geometrical configurations of some of the reported species.

The complex $CrCl_3(tu)_3$ is conveniently prepared by evaporating a methanolic solutio[] containing $CrCl_3 \cdot 6H_2O$ and tu (1:4) to dryness, and then washing the residue with ho[] acetone.[1043] Evidence based on the IR spectrum of this complex (Table 90) indicates that th[] ligand is bonded through sulfur to the metal ion. Hence the $\nu(NH)$ bands in the spectrum o[] the complex occur at higher wavenumbers than in the spectrum of free tu, indicating reduction in H bonding and therefore S coordination of the ligand to the metal. Similarly th[] $\nu(CS)$ band at 733 cm^{-1} in the free ligand spectrum is shifted to a lower wavenumber in th[] spectrum of the complex consistent with a reduction in carbon–sulfur bond order an[] coordination through the sulfur atom.[1043,1044] Furthermore, the expected increase in th[] carbon–nitrogen bond order due to S coordination is confirmed by the shift to highe[] wavenumber of the B_1 NCN band in the complex (1490 cm^{-1}) relative to free tu (1476 cm^{-1}). [] strong band at 282 cm^{-1} assigned to $\nu(MS)$ lends additional support to the presence [] S-thiourea ligands. The reflectance spectra of $CrCl_3(N,N'\text{-}Et_2tu)_3$ and $CrCl_3(N,N'\text{-}Bu^ntu)_3$ ar[] strikingly similar to that of $CrCl_3(tu)_3$ ($\lambda_{max} = \sim 680, 510$ nm) implying similar ligand fields in a[] three cases.[1045] No mention is made of the geometrical disposition of the ligands in any of thes[] complexes.

$CrCl_3(tu)_3$ reacts with various Lewis bases in hot methanol solution as outlined in Schem[] 102.[1043] Salient features of the IR spectra of a selection of these complexes are presented i[] Table 90. Application of the afore-mentioned criteria to the IR spectra of these products lead[] to the conclusion that in some cases the complexing site of tu changes from sulfur to nitroge[] when heterocyclic amines or phenylarsonate ligands are introduced into the complex. In othe[] cases, however, *e.g.* $[CrCl_2(py)_2(tu)_2]Cl$, for which both $\nu_{asym}(NCN)$ and $\nu(CS)$ occur at highe[] wavenumbers than the corresponding IR bands for free tu, no definite conclusions can b[] reached on the basis of IR spectra regarding the mode of ligand coordination. Clearly the I[] data on their own are insufficient to establish structures with certainty and X-ray crysta[] lographic studies are needed to clarify both the ligand donor sites and the geometrica[] configurations of these complexes. The preparation and spectra of complexes of genera[] formula $M[Cr(NCS)_4(tu)_2]$ (M = K or various Ni^{II}–amine complexes) have bee[] reported.[1046,1047]

$$[CrCl_2(Rpy)_2(tu)_2]Cl$$

$$\uparrow Rpy^a$$

$$[Cr(L-L)_2(tu)_2]Cl_3 \xleftarrow{L-L^c} CrCl_3(tu)_3 \xrightarrow{Rquin^b} [CrCl_2(Rquin)_2(tu)_2]Cl$$

$$\downarrow PhAsO_3H_2$$

$$[Cr(PhAsO_3H)_2(tu)_2]Cl$$

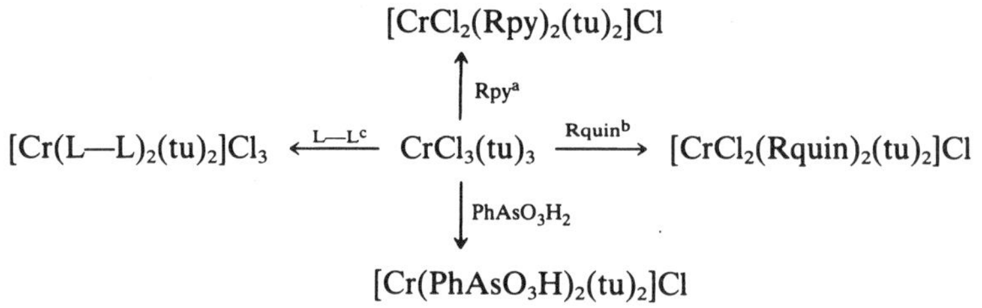

^{a} Rpy = pyridine; 2-, 3- and 4-methylpyridine; 2,4- and 2,6-dimethylpyridine. ^{b} Rquin = quinolin[] isoquinoline and acridine. ^{c} L—L = phen or bipy.

Scheme 102

A number of substituted tu complexes of chromium(III) have been prepared an[] characterized. These include $[Cr(N\text{-}Metu)_6](ClO_4)_3$, $[Cr(N,N'\text{-}Me_2tu)_6](ClO_4)_3$, $CrCl_3(N[]$ Metu)$_3$, $CrX_3(N,N'\text{-}Me_2tu)_3$ (X = Cl, Br),[1048] and the N-phenylthiourea complexes $CrCl_3(N[]$ Phtu)$_3$ and $CrCl_3(N\text{-}RPhtu)_3$ (R = 2- or 3-Me; 2-MeO; 2-, 3- or 4-Cl; 3-Br; 4-I; 2-OH).[104] Evidence from IR (Table 90) and electronic spectroscopy strongly suggests that in all thes[] complexes the substituted tu ligands are coordinated through sulfur to the metal ion. Details o[] the electronic spectra of some of these complexes are given in Table 91. The ligan[] 3-(4-pyridyl)triazoline-5-thione (L) reacts with $CrCl_3 \cdot 6H_2O$ in refluxing EtOH to give th[] product *trans*-$[Cr(H_2O)_4L_2]Cl_3$ in which it is S-bonded to the metal ion.[1050] The mixed ligan[] chromium(III) complexes $[Cr(H_2O)_5Etu](OAc)_3$ and $[Cr(OAc)OH(T\text{-}5\text{-}T)] \cdot 2H_2O$, containin[] ethylenethiourea (Etu, an S donor) and 1-substituted tetrazoline-5-thiones (T-5-T; N,S donors[] have been prepared and characterized.[1051]

Dissolution of thioureachromium(III) complexes in coordinating solvents such as MeOH[] DMF, DMSO, MeCN generally results in solvolysis with replacement of the tu ligands.

Table 91 Reflectance Spectra of Thioureachromium(III) Complexes

Complex	$^4A_{2g}\rightarrow{}^4T_{2g}$ λ (nm)	$^4A_{2g}\rightarrow{}^4T_{1g}$ λ (nm)	Ref.
$[Cr(N\text{-Metu})_6](ClO_4)_3$	690	526	1
$[Cr(N,N'\text{-}Me_2tu)_6](ClO_4)_3$	714	549	1
$CrCl_3(tu)_3$	680	510	2
$CrCl_3(N\text{-Metu})_3$	699	518	1
$CrCl_3(N,N'\text{-}Me_2tu)_3$	725	521	1
$CrCl_3(N\text{-Phtu})_3$	700	510	3
$CrCl_3(N\text{-3-MePhtu})_3$	682	510	3
$CrCl_3(N\text{-3-ClPhtu})_3$	675	510	3
$[CrCl_6]^{3-}$ [a]	730	529	4

[a] Included for comparison.
1. P. Askalani and R. A. Bailey, *Can. J. Chem.*, 1969, **47**, 2275.
2. E. Cervone, P. Cancellieri and C. Furlani, *J. Inorg. Nucl. Chem.*, 1968, **30**, 2431.
3. M. H. Sonar and A. S. R. Murty, *J. Inorg. Nucl. Chem.*, 1977, **39**, 2155.
4. C. K. Jørgensen, 'Absorption Spectra and Chemical Bonding in Complexes', Pergamon, Oxford, 1962, p. 290.

35.4.6 Selenium and Tellurium Ligands

A number of binary and ternary compounds in the chromium–selenium and the chromium–tellurium systems have been investigated. These include CrSe, Cr_2Se_3, Cr_3Se_4, Cr_7Se_8, Cr_5Se_8, CrTe, Cr_2Te_3, Cr_3Te_4, Cr_5Te_6, Cr_5Te_8, Cr_7Te_8, $Cr_5S_{4.8}Se_{3.2}$,[981] $NaCrSe_2$, $CuCrSe_2$ and $AgCrSe_2$.[986] Some Se donor ligands are included in Section 35.4.5.2.

35.4.7 Halogens as Ligands

35.4.7.1 *Simple halides CrX_3*

These have been known for many years.[1052–1054] Chromium(III) is approximately octahedral (μ_{eff} = 3.69–4.1 BM); the compounds have a layer structure. In the chloride, *r*(Cr—Cl) is 5.76 Å between layers and 3.46 Å within layers. The iodide is isomorphous with the chloride and the bromide has a similar but distinct structure. All may be prepared by the direct halogenation of the metal. Other methods are available, *e.g.* $CrCl_3$ may be prepared by heating $Cr_2O_3\cdot xH_2O$ in CCl_4 vapour at 650 °C.[1055] The anhydrous halides are insoluble in water, however reducing agents such as zinc catalyze dissolution. The trichloride reacts with liquid ammonia to form ammine complexes.

35.4.7.2 *$[CrX_4]^-$ and related ions*

There has been some controversy[1056,1057] concerning the correct formulation of $[CrCl_4]^-$ ions in salts such as $[PCl_4][CrCl_4]$. In general the evidence supports polymeric six-coordinate anions.[1057] This is supported by a crystal structure of $K[CrF_4]$,[1058] which contains columnar $[CrF_4]^-$ ions. Related compounds such as $NaCrF_4$[1059] and NH_4CrF_4[1060] also involve six-coordinate chromium(III); these compounds are weak antiferromagnets. A related antiferromagnetic species is $MnCrF_5$, where a neutron diffraction study shows chromium(III) to be octahedral with a single bridging fluoride;[1061] this is an example of a general class of chromium(III) fluoro complexes $M^{II}Cr^{III}F_5$.[235] The anions $[CrCl_4(MeCO_2H)_2]^-$ and $[CrCl_5(MeCO_2H)]^{2-}$ are known (Scheme 92).

35.4.7.3 *$[CrX_6]^{3-}$ anions*

Complexes containing anions of the above formulation have attracted a large number of studies because of their alleged simplicity. This is illustrated by the central position such complexes have played in the evolution of crystal field, ligand field and molecular orbital models of bonding in transition metal complexes.

In a study of the heats of formation of first row transition metal fluorides, ΔH_f for K_3CrF_6 was found to be 2977 kJ mol^{-1}.[1062] The usual double-humped distribution of ΔH_f *vs.* d^n was observed. Semi-empirical[1063] and self-consistent-field molecular-orbital calculations[1064] give similar results for the $[CrF_6]^{3-}$ ion and are in accord with the observed electronic spectrum.[1065] These studies are summarized in Table 92.

Table 92 Electronic Spectra of Hexafluorochromate(III) (all energies are $cm^{-1} \times 10^3$, after ref. 1063)

One electron $2t_{2g}$–$3e_g$ Uncorrected (correction ref. 1)	$10Dq$	$3t_{1u} \rightarrow 2t_{2g}$ $3t_{1u} \rightarrow 3e_g$	$t_{2g} \rightarrow 2t_{2g}$ $t_{2u} \rightarrow 3e_g$	$\pi_u \rightarrow t_{2g}$	Ref.
	16.54				2
	6.88 (12.7)				3
12.68 (16.4)	16.44	27.8	36.76		7
		40.50	49.44		7
		16.10		7.50	4, 6
		15.20			1

1. Spin corrections after A. Dutta-Ahmed and E. A. Boudreaux, *Inorg. Chem.*, 1973, **12**, 1590.
2. R. F. Fenske, G. Caulton, D. D. Radtke and C. C. Sweeney, *Inorg. Chem.*, 1966, **5**, 960.
3. P. O. Offenhartz, *J. Chem. Phys.*, 1967, **47**, 2951.
4. D. L. Wood, J. Ferguson, K. Knox and J. F. Dillon, Jr., *J. Chem. Phys.*, 1963, **39**, 890.
5. C. K. Jørgenson, 'Absorption Spectra and Chemical Bonding in Complexes', Pergamon, London, 1962.
6. Ref. 1065, and G. C. Allen and K. D. Warren, *Struct. Bonding (Berlin)*, 1971, **9**, 49.
7. Ref. 1063.

Luminescence spectrocopy is potentially a powerful technique for studying chromium(III) complexes and a series of fluoro and aqua complexes have been studied.[1066] The luminescence correlates well with ligand field strength and, at liquid air temperatures, the lifetime of the doublet state from which phosphorescence originates is $2 \times 10^{-7}\,s^{-1}$.

A detailed study of the spin density and bonding of the $[CrF_6]^{3-}$ ion in $K_3[CrF_6]$ has appeared.[1067] The results of polarized neutron diffraction experiments[1068] were reinterpreted. A chemically based model of the $[CrF_6]^{3-}$ ion has been fitted to the 92 observed magnetic structure factors by a least squares procedure. The spin density in the chromium(III) orbitals has t_{2g} symmetry ($t_{2g}^{2.66(5)}$, $e_g^{-0.06(5)}$). There is also a region of spin density centred on the chromium atom, radially more diffuse than the theoretical $3d$ orbital and containing 0.4(1) spins parallel to the spin density of t_{2g} symmetry. There are small parallel spin densities in $2p_\pi$ of fluoride and antiparallel spin populations in fluoride $2p_\sigma$ and along the Cr—F vector. There is no significant spin population in fluoride $2s$. The results of this modelling are in agreement with X_α,[1069] semi-empirical[1070] and *ab initio* molecular orbital calculations.[1071] The magnetic circular dichroism of chromium(III) in dicesium yttrium hexachloride has been studied[1072] down to 6 K. Transitions to all three quartet terms, $^4T_{2g}$ and $^4T_{1g}$ (I and II), were observed. The parameters $D = 1280\,cm^{-1}$ and $a = 78.2\,cm^{-1}$ were determined. At 6 K rich vibronic fine structure was observed, which could satisfactorily be explained only in terms of O_h geometry.

35.4.7.4 Dimeric structures

Many complexes contain the $[Cr_2X_9]^{3-}$ ion.[1073–1076] This unit consists of two distorted octahedra which share a face; there are three bridging and three terminal halides (**226**). The distortion of the octahedron is such that the two metal ions are displaced from each other. The magnetic properties of these complexes have been the subject of extensive study summarized in Table 93.

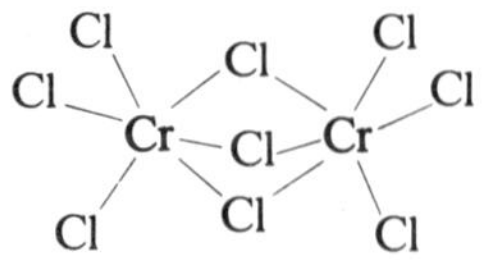

(**226**) Typical structure of a $[Cr_2X_9]^{3-}$ ion

The complexes undergo weak antiferromagnetic interactions with exchange proceeding *via* the halide bridges. The counterion may have a marked effect on the structure; for $[Cr_2Cl_9]^{3-}$, change of the counterion from potassium to tetraethylammonium is accompanied by a decrease

Table 93 Magnetic Properties of $[Cr_2X_9]^{3-}$ Ions

M—M (Å)	μ_{eff} (BM)	$-J$ (cm^{-1})	*Complex*	*Ref.*
3.05	3.77–4.00	16	$K_3[Cr_2Cl_9]$	1, 2
3.07	3.77–3.88	13.3	$Rb_3[Cr_2Cl_9]$	1, 2
3.12	3.76–3.83	9.3	$Cs_3[Cr_2Cl_9]$	1, 2, 3
—	3.56	—	$[Et_2NH_2]_3[Cr_2Cl_9]$	4
3.89	3.91–3.96	5	$[Et_4N]_3[Cr_2Cl_9]$	2, 5
—	3.91	4	$[Bu_4N]_3[Cr_2Cl_9]$	1
3.26, 3.32	3.8--3.87	8.8	$Cs_3[Cr_2Br_9]$	2, 6
4.00	3.89	0	$[Et_4N]_3[Cr_2Br_9]$	2
—	3.88	—	$Cs_3[Cr_2I_9]$	1

1. R. Saillant and R. A. D. Wentworth, *Inorg. Chem.*, 1968, **7,** 1606.
2. G. J. Wessel and D. J. W. Ijdo, *Acta Crystallogr.*, 1957, **10,** 466.
3. I. E. Grey and P. W. Smith, *Aust. J. Chem.*, 1971, **24,** 73.
4. P. C. Crouch, G. W. A. Fowles and R. A. Walton, *J. Chem. Soc.*, 1969, 972.
5. A. Earnshaw and J. Lewis, *J. Chem. Soc.*, 1961, 396.
6. R. Saillant, R. B. Jackson, W. E. Streib, K. Folting and R. A. D. Wentworth, *Inorg. Chem.*, 1971, **10,** 1453.

in *J* from 16 to 5 cm^{-1}. An extreme example of similar behaviour is $[Et_4N]_3[Cr_2Br_9]$ which obeys the Curie–Weiss law.

The vibrational spectra of $[Cr_2X_9]^{3-}$ have been variously studied.[1077,1078] The agreement between observed and calculated spectra was excellent for D_{3h} symmetry; no evidence was found for any Cr—Cr interaction.[1077] In related work, the doping of chromium(III) ions into $CsMgCl_3$ and related materials has been investigated.[1079,1080] ESR spectroscopy has shown that there is a distinct tendency for chromium ions to cluster in pairs [for single crystals with 1 part in 1000 chromium(III)]. $CsMgCl_3$ adopts the $CsNiCl_3$ structure (**227**); the structure proposed for the impurity centre is illustrated. The forces controlling the distribution of ions were held to be electrostatic; other trivalent metal ions should behave in a similar way to chromium(III).

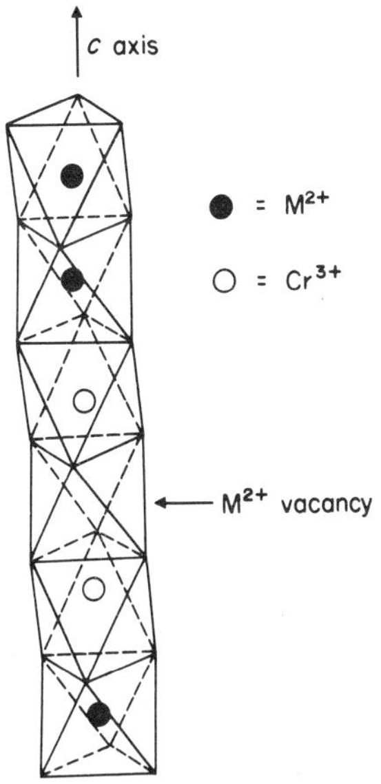

(**227**)

A perspective view of an $[MX_3^-]_n$ chain showing the proposed structure of the Cr^{III}—Cr^{III} pair. The corners of the octahedra are occupied by halide ions.[1080]

35.4.7.5 Solution chemistry

Werner's early work on coordination chemistry involved the preparation of many of the isomers in the $CrX_3{\cdot}nH_2O$ system.[1081,1082] Crystal structures are available for several of these complexes. The green complex *trans*-$[Cr(H_2O)_4Cl_2]Cl\cdot 2H_2O$, the normal compound supplied as chromium(III) chloride, has been studied on a number of occasions.[1083,1084] Recently a low-temperature structure of the complex dicesium *trans*-dichlorotetraaquachromium(III) trichloride[1085] was determined; the electronic spectrum was also studied. The mechanism of the geometric isomerization of the dichloro complex has been investigated.[1086] The pure *cis* complex may be prepared by the aquation of trichloro complexes followed by purification by ion-exchange chromatography.[1086]

The formations of chloro,[1087,1088] bromo[1089] and iodo[1090] complexes have been studied. Recently, electrochemical methods have been used to measure equilibrium constants for the formation of chloro (0.086), bromo (1.1×10^{-3}) and iodo (1.1×10^{-5}) complexes of chromium(III).[1091]

35.4.8 Mixed Donor Atom Ligands

35.4.8.1 Schiff's bases, β-ketoamines and related ligands

Compared with other transition elements,[1092] few complexes of Schiff's bases or β-keto amines with chromium(III) are known. Most work has been done with ligands (**228**) to (**231**) and their complexes and the interrelations between the more important of them are set out in Table 94 and Schemes 103 and 104.

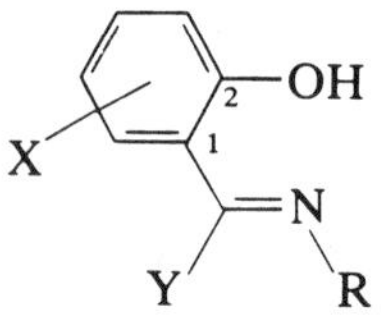

(**228**)

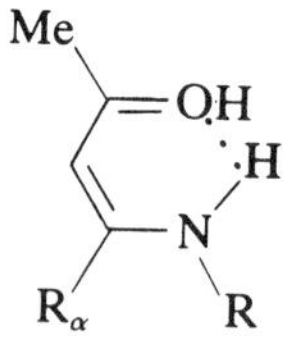

(**229**)

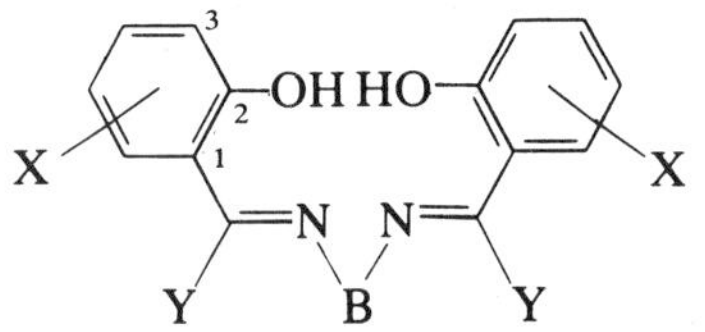

(**230**) X = Y = H, B = $(CH_2)_2$: salenH_2
X = Y = H, B = *o*-C_6H_4: salphenH_2
X = Y = Me, B = $(CH_2)_2$: MesalenH_2
X = 3-CO_2H, Y = H, B = $(CH_2)_2$: 3-CO_2salenH_2

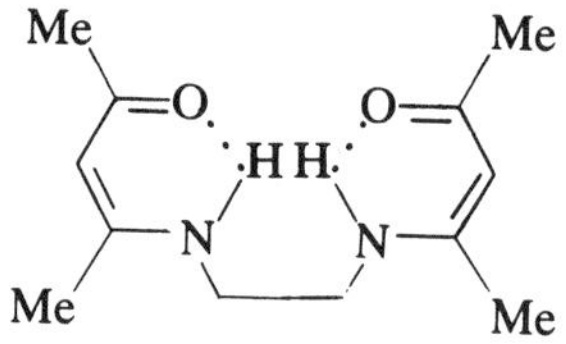

(**231**) acacenH_2

(i) Syntheses

Many preparative methods have been used, with variable success: oxidative decarbonylation of $Cr(CO)_6$ by the ligand,[1093,1094] reaction of the appropriate amine with [Cr(sal)$_3$] (salH is salicylaldehyde), reaction of the ligand, either produced in the reaction mixture or preformed,[1095] with hydrated or anhydrous $CrCl_3$, hydrated Cr^{III} acetate,[1096] or [$CrCl_3(THF)_3$],[1097–1099] and aerial oxidation of the chromium(II) complex produced *in situ*.[1099–1101] No well-established chromium(II) complexes, except of a few β-keto amines (Section 35.3.9.1), have been isolated.

The source of Cr^{2+} can be [$Cr_2(MeCO_2)_4(H_2O)_2$],[1100] aqueous Cr^{2+} produced by the reduction of Cr^{3+} electrolytically[1101] or by Zn/Hg in hydrochloric acid,[1099] or anhydrous or hydrated $CrCl_3$ reduced by zinc dust in suspension in the reaction mixture.[1100,1102] Presumably the ligands can displace solvent molecules more readily from the labile Cr^{2+} species.

Several hours reflux in organic solvents is usually required for preparations directly from chromium(III) starting materials. No method is of general applicability; for example, [Cr(salen)$(H_2O)_2$]Cl (Scheme 103) is readily prepared from [$Cr_2(MeCO_2)_4(H_2O)_2$] but this method affords [Cr(acac)$_3$] with acacenH_2.[1100]

Table 94 Preparations of Some Schiff Base and β-Ketoamine Complexes

Complex	*Synthesis*	*Ref.*
Tris(chelates)		
*Ligand (**228**), bidentate, X = Y = H*		
R = Me	$Cr(CO)_6 + C_6H_4(CHNMe)OH$, toluene, reflux	1
R = Pr^n, Bu^i,	$Cr(sal)_3$ + amine in benzene, or	2
p-C_6H_4Me, p-C_6H_4Br	$CrCl_3{\cdot}6H_2O$ + Schiff base, Na_2CO_3	
R = Me, Et, Pr^n, Pr^i,	Hydrated Cr^{III} acetate, suspended in ethylene glycol, salH, amine, 2/3 theoretical Na_2CO_3	3
*Ligand (**228**), bidentate, X = 5,6-benzo, Y = H*		
R = Me, Et, Pr^n, Pr^i, Ph	As above	3
*Ligand (**228**), bidentate, X = 5-NO_2, Y = H*		
R = (*R* or *S*)-α-CH(Bz)Me	As above	4
R = (*R* or *S*)-α-CH(Ph)Me		5
*Ligand (**230**), bidentate*		
R_α = Me, R = CH_2Ph, Ph, *o*-, *p*-tolyl, *o*-, *m*-, *p*-C_6H_4Cl, *p*-anisyl, *p*-C_6H_5-C_6H_4, β-naphthyl	$KOBu^t$ in Bu^tOH, ligand, Zn, $CrCl_3(THF)_3$	6
R_α = H, R = *p*-tolyl		
Bis(chelates)		
*Ligand (**228**), tridentate, X = Y = H*		
R = $CH_2CH_2NH_2$	$Cr(sal)_3$ + en, MeOH, add	2
$[Cr(en\text{-}sal)_2]X$	NaX, X = Br, I or ClO_4	
R = *o*-$C_6H_4CO_2H$,[a]	$[Cr_2(CH_3CO_2)_4{\cdot}2H_2O]$, salH, anthranilic acid,	7
o-C_6H_4OH, CH_2CO_2H	*o*-aminophenol or glycine	
R = CH_2CH_2OH	As above, add $NaClO_4$	7
$[Cr\{C_6H_4(CHNCH_2CH_2OH)O\}_3]ClO_4$		
*Ligand (**228**), tridentate, R = CH_2CH_2-2-pyridyl*		
Y = H, X = H, 5-Cl, 5-NO_2, 5,6-benzo	2-(2′-Aminoethylpyridine), salH in EtOH +	8
X = H, Y = Me, Et, Pr	$CrCl_3{\cdot}6H_2O$, NaOAc; formulae $[CrL_2]Cl$	

[a] Complexes formulated $H[Cr\{C_6H_4(CHNR)O\}_2]$ as proton from one R group must remain to balance charge.

1. F. Calderazzo, C. Floriani, R. Henzi and F. L'Epplattenier, *J. Chem. Soc. (A)*, 1969, 1378.
2. R. Lancashire and T. D. Smith, *J. Chem. Soc., Dalton Trans.*, 1982, 693.
3. S. Yamada and K. Iwasaki, *Bull. Chem. Soc. Jpn.*, 1968, **41,** 1972.
4. J. E. Gray and G. W. Everett, Jr., *Inorg. Chem.*, 1971, **10,** 2087.
5. K. S. Finney and G. W. Everett, Jr., *Inorg. Chim. Acta*, 1974, **11,** 185.
6. J. P. Collman and E. T. Kittleman, *Inorg. Chem.*, 1962, **1,** 499.
7. K. Dey and K. C. Ray, *Inorg. Chim. Acta*, 1974, **10,** 135.
8. D. K. Rastogi and P. C. Pachauri, *Indian J. Chem., Sect. A*, 1977, **15,** 748.

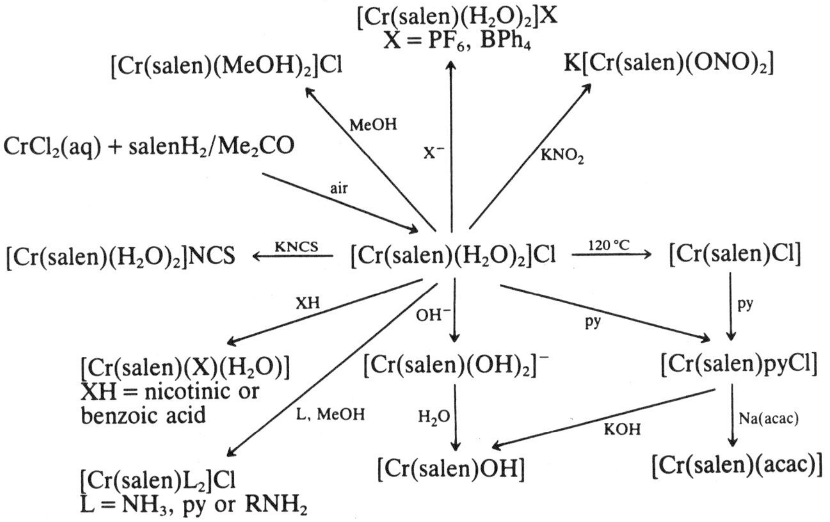

Scheme 103

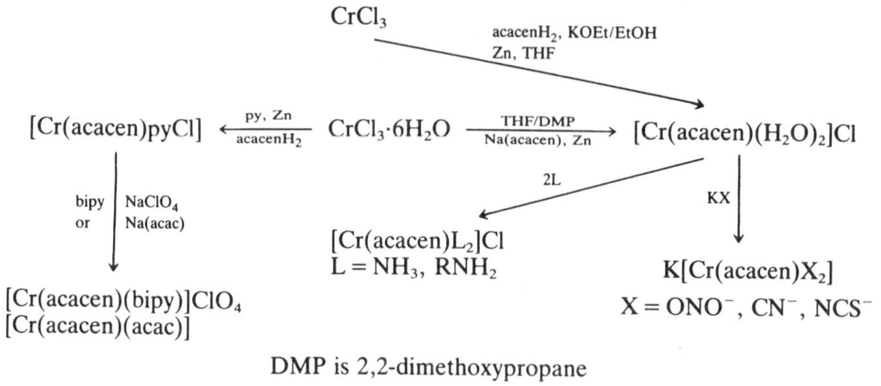

Scheme 104

Since oxidation of [Fe^{II} salen] gives {Fe(salen)}$_2$O, the formation of [Cr(salen)(H_2O)$_2$]Cl from Cr^{II} solutions is unexpected, but it is consistent with the observation[1103] that aerial oxidation of aqueous chromium(II) leads to the formation of hydroxo or aqua species. The elusive {Cr(salen)}$_2$O was eventually prepared by oxidative decarbonylation[1094] of Cr(CO)$_6$ by [Hg(salen)], a method which could have more general application. The reaction of salenH$_2$ with Cr(CO)$_6$ gives a product similar to [Cr(salen)(OH)]·0.5H_2O.[1099] Dichromium and mixed metal μ-oxo complexes such as (TPP)CrOFe(TPP) are now known (Section 35.4.9.1).

Complexes obtained by methods similar to those in Scheme 103 include [Cr(salen)(H_2O)$_2$]ClO_4,[1101] [Cr(salen)(X)(H_2O)] (XH = nicotinic acid or benzoic acid),[1104] [Cr(salen)X(H_2O)] (X = NCS^- or N_3^-),[1105] [Cr(salphen)(H_2O)$_2$]X (X = Cl^- or OAc^-),[1106] [Cr(salen)(AA)]BPh_4, [Cr(salphen)(AA)]ClO_4 (AA = bipy, phen or en).[1107] Complexes related to those in Scheme 104 are [Cr(acacen)A$_2$]BPh_4, [Cr(acacen)ClA], [Cr(acacen)(NCS)A], [Cr(acacen)A$_2$]ClO_4, [Cr(acacen)(H_2O)$_2$]SCN (A = py or 4-methylpyridine)[1108] and [CrL(H_2O)$_2$]Cl, in which L is formed from acac and benzidine.[1109] The salts [Cr(L)(NH_3)$_2$]SCN (L = salen, salphen or Mesalen) have been prepared[1100] from NH_4[Cr(NCS)$_4$(NH_3)$_2$].

(*ii*) *Properties*

Photolysis of the azido complex CrN_3(salen)·2H_2O produces[1110] the nitrido chromium(V) complex CrN(salen)·H_2O and the reaction[1111] of [Cr(salen)(H_2O)$_2$]$^+$ with iodosylbenzene gives the oxo chromium(V) derivative [CrO(salen)]$^{2+}$; chromium(III) porphyrin complexes behave similarly (Section 35.4.9.1).

As expected for chromium(III), the complexes generally possess magnetic moments near 3.8 BM at room temperature. A few, (H[Cr{C_6H_4(CHNR)O}$_2$] (R = *o*-C_6H_4OH or CH_2CH_2OH)[1100] and [Cr(salen)(OH)]·0.5H_2O) have low moments (≈3.5 BM) ascribed to antiferromagnetic interactions. The magnetic susceptibility of [Cr(salen)(OH)]·0.5H_2O deviates markedly from the Curie law, but as the results were sample-dependent, no detailed interpretation was attempted.[1099] The room temperature magnetic moment of {Cr(salen)}$_2$O is 2.2 BM, consistent with an oxo-bridged structure. The ESR spectrum confirms that the chromium is trivalent,[1094] so further magnetic and structural investigations are desirable.

The complexes have yellow to red colours and the electronic spectra are dominated by intense ligand absorptions. A weak band ususally found in the range 18 000 to 21 000 cm^{-1} has been assigned to the $^4A_{2g} \rightarrow {}^4T_{2g}$ transition and provides an approximate value for Δ.

The *trans* structure of the cation in [Cr(salen)(H_2O)$_2$]Cl[1099] and the *mer* arrangement of the tridentate ligands in [Cr(en-sal)$_2$]I [en-sal is *N*-(2-aminoethyl)salicylaldimine)] have been confirmed by single-crystal X-ray investigations (Cr—O$_{(av)}$ = 1.93, Cr—N$_{(av)}$ = 2.08 Å).[1112] In [Cr(salen)(H_2O)$_2$]Cl, the Cr—N bond lengths are 1.977 and 2.005 Å, but the Cr—O distances vary considerably (1.916 and 1.951 Å (salen); 1.923 and 2.085 Å to H_2O) and the angles at the chromium atom are distorted from right angles. The ground-state distortion has been related to the lability of the axial ligands.[1113] There must be considerable distortion of salen and acacen from their usual essentially planar geometry in complexes such as [Cr(acacen)(acac)].

The complexes [Cr(en-sal)$_2$]X were prepared (Table 94) in conditions which could have given [Cr(salen)X], but, as indicated by other data and confirmed by the crystal structure of the iodide, en and salH condensed to give the tridentate ligand en-sal. The solubility of the nitrate has been exploited[1114] in the resolution of $[Cr(en\text{-}sal)_2]^+$ as the hydrogen (*R,R*)-*O,O'*-dibenzoyltartrate. The diastereoisomers were converted into the iodides which formed twinned crystals unsuitable for X-ray investigation. When (*R*)-1,2-diaminopropane (*R*-pn) is used instead of ethylenediamine in the reaction with $Cr(sal)_3$, the salicylaldehyde may condense with either NH_2 group. The product was (**232**), with the ligands meridional, as expected, and the —CH_2NH_2 groups condensed.[1115]

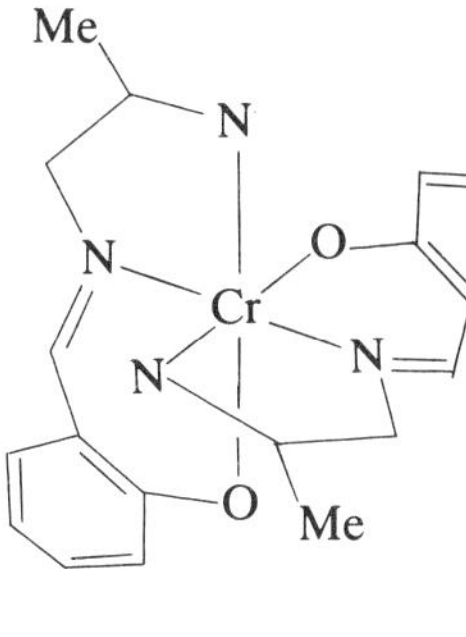

(**232**)

Tris[*N*-(*R* or *S*)-α-benzylethyl-5-nitrosalicylaldiminato]- and tris[*N*-(*R* or *S*)-α-phenylethyl-5-nitrosalicylaldiminato]-chromium(III) have been prepared[1116,1117] from the (*R*)- or (*S*)-amine as appropriate. For ligands of a given chirality, four diastereoisomers are possible (*fac*, Δ and Λ; *mer*, Δ and Λ). Chromatographic separation of the complex with R = (*S*)-α-CH(Bz)Me gave two products A and B (the more rapidly eluted), and with R = (*R*)-α-CH(Bz)Me, A′ and B′ were obtained. A and A′ have identical absorption spectra but from their CD spectra have opposite configurations about the metal. Since they differ in ligand chirality also, they are enantiomers. Steric effects would favour the *mer*(*trans*) configuration, and the corresponding cobalt(III) complexes, which chromatographically separate similarly, have been assigned *mer* configurations from their ^{1}H NMR spectra. Analogous relationships were found among the pair of diastereoisomers obtained with R = (*R*)- or (*S*)-α-CH(Ph)Me. The more rapidly eluted isomers are thought to have the Λ-*mer* configuration from analogy with the cobalt(III) case, and because studies with the chiral β-diketonate (+)-3-acetylcamphor have shown that diastereoisomers with the same absolute configuration have the same relative rates of elution. The isomers did not interconvert under reflux in the presence or absence of charcoal.

The complexes CrL_3, where LH is formed by condensation of salicylaldehyde with 2-furfurylamine or 2-thenylamine, contain the ligand in isomeric form (**233**).[1118]

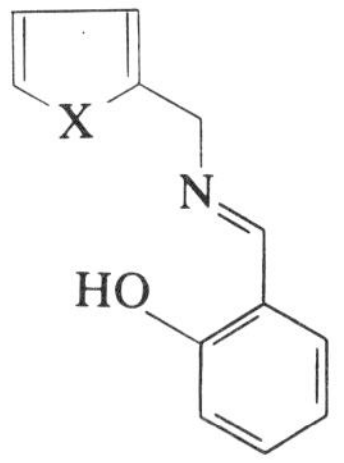

(**233**) X = O or S

The addition of en or 1,3-diaminopropane to a mixture of chromium(III) acetate hexahydrate and diacetyl under reflux, followed by addition of $NaClO_4$, has given the complexes $[Cr(diacen)(H_2O)_2]ClO_4$ and $[Cr(diacpn)(H_2O)_2]ClO_4$, which contain the tetradentate Schiff bases (**234**).[1119] The thiocyanates $[CrL(H_2O)(NCS)](NCS)_2$ were also isolated. From their IR spectra, and because the coordinated Schiff base would not react with another molecule of diacetyl to form a macrocycle, they have been assigned *cis* configurations.

In the chromium(III) complex of the potentially heptadentate ligand (**235**, X = 5-Cl) the apical nitrogen atom is not bonded, the Cr—N distance being 3.229 Å. The Cr—O (1.979 Å) and Cr—N (2.137 Å) bonds are longer than in $[Cr(salen)(H_2O)_2]Cl$ and $[Cr(en\text{-}sal)_2]I$ (see above) and the ligand seems to fix a minimum sized 'hole' into which the metal ion fits.[1120] The

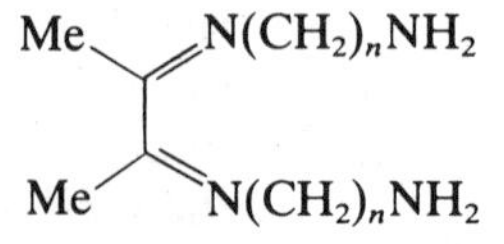

(**234**) diacen, $n = 2$
diapn, $n = 3$

Mn^{III} and Fe^{III} complexes are isomorphous with the Cr^{III} complex. The chromium(III) complexes with X = H, 3-OMe, 5-Cl, 5-Br, 5-Me[1121] or 3,5-Cl_2 [1122] have been prepared by methods similar to those for the salen derivatives.

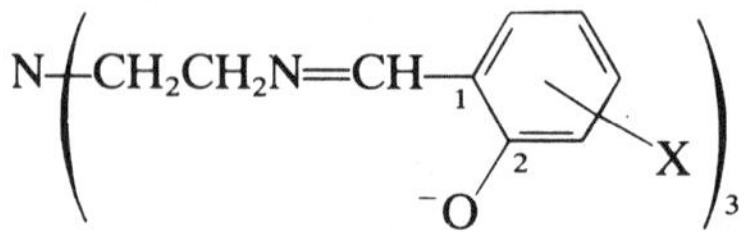

(**235**)

Chromium(III) does become seven-coordinate in two complexes in which the organic ligands are formed by condensation respectively of 2,6-diacetylpyridine and semicarbazide,[1123] and 2,6-pyridinedialdehyde and 6,6′-bis(α-2-hydroxyethylhydrazino)-2,2′-bipyridyl.[550] Seven-coordinate complexes are formed by chromium in several oxidation states (Table 95).

Table 95 Seven-coordinate Chromium[a]

Complex	*Structure*	*Ref.*
$[Cr^{II}(CNBu^t)_7](PF_6)_2$	4:3 Tetragonal base-trigonal cap (**1**) (or 4:3 piano stool, C_s)	1, 2
$[Cr^{II}(CNR)_5(dppe)](PF_6)_2$	R = Bu^t, C_6H_{11}	2
$Cr^{II}[P(OMe)_3]_5H_2$	Pentagonal bipyramid	3
$CrNO(S_2CNEt_2)_3$	Pentagonal bipyramid, axial NO (**133**)	4
$[Cr^{III}(dapsc)(H_2O)_2]OH(NO_3)_2$[b]	Pentagonal bipyramid (**238**)	5
$[Cr^{II}(CNPh)_5(dppm)](PF_6)_2$	Unknown structure	6
$[CrL(H_2O)_2]Cl_3$[c]	Pentagonal bipyramid, see (**128**)	7

[a] Several mixed ligand complexes containing carbon monoxide[8] or peroxide (Section 35.7.7) are seven-coordinate.
[b] dapsc is 2,6-diacetylpyridine bis-semicarbazone.
[c] L is the planar pentadentate macrocycle formed by condensation of 2,6-pyridinedialdehyde and 6,6′-bis(α-2-hydroxyethylhydrazino)-2,2′-bipyridine.

1. J. C. Dewan, W. S. Mialki, R. A. Walton and S. J. Lippard, *J. Am. Chem. Soc.*, 1982, **104**, 133.
2. W. S. Mialki, D. E. Wigley, T. E. Wood and R. A. Walton, *Inorg. Chem.*, 1982, **21**, 480.
3. F. A. Van-Catledge, S. D. Ittel, C. A. Tolman and J. P. Jesson, *J. Chem. Soc., Chem. Commun.*, 1980, 254.
4. S. Clamp, N. G. Connelly, G. E. Taylor and T. S. Louttit, *J. Chem. Soc., Dalton Trans.*, 1980, 2162.
5. G. J. Palenik, D. W. Wester, U. Rychlewska and R. C. Palenik, *Inorg. Chem.*, 1976, **15**, 1814.
6. F. R. Lemke, D. E. Wigley and R. A. Walton, *J. Organomet. Chem.*, 1983, **248**, 321.
7. L.-Y. Chung, E. C. Constable, M. S. Khan, J. Lewis, P. R. Raithby and M. D. Vargas, *J. Chem. Soc., Chem. Commun.*, 1984, 1425.
8. M. G. B. Drew, *Prog. Inorg. Chem.*, 1977, **23**, 67.

Chromium(III) is also in an unusual situation in the heterobinuclear complex (**236**) synthesized by the addition of $CrCl_3·6H_2O$ to Cu(3CO_2-salenH_2).[1124] There is ferromagnetic interaction between the Cu^{II} and Cr^{III} ions because of the orthogonality of the magnetic orbitals (quintet–triplet separation = $120\ cm^{-1}$).

Chromium(III) complexes of Schiff bases derived from pyridoxal and glycylglycine have been prepared in solution in attempts to find complexes that show good intestinal absorption[1125] (see also Section 35.4.8.3).

(iii) *Miscellaneous Schiff's base and related complexes*

Often, in studies of complexation between metal ions and the great variety of Schiff's bases and related ligands, chromium(III) complexes are included amongst many from the first

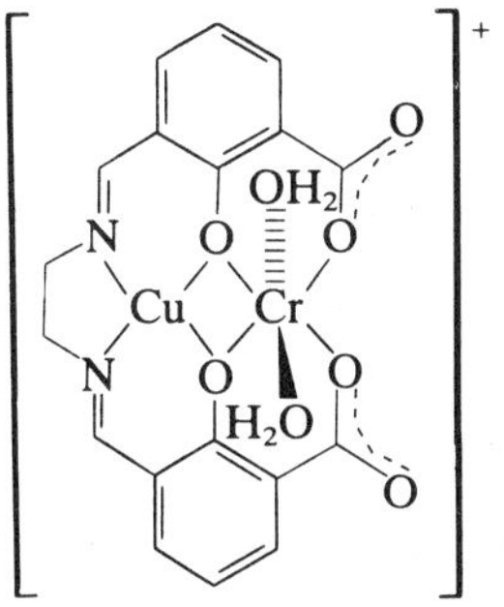

(236) $[CuCr(3\text{-}CO_2salen)(H_2O)_2]Cl\cdot 3H_2O$

transition series generally. These Cr^{III} complexes are typical of the oxidation state with no unusual properties. Consequently, only examples are collected in Table 96.

(*iv*) *Organometallic derivatives*

Although tetradentate Schiff's bases stabilize Co^{III}–alkyl σ bonds this behaviour is less common with Cr^{III}. Several perfluoroalkyl derivatives have been obtained by Scheme 105 and the same method yields complexes of bidentate ligands too: $R_FCr(sal\text{-}NR)_2py$, $R_F = C_2F_5$, C_3F_7, $R = 4\text{-}MeC_6H_4$; $R_F = C_3F_7$, $R = Bu^n$; and $C_3F_7CrL_2py$ where LH = salH, acacH or bzacH. The intermediates '$R_FCrCl_2(MeCN)_3$' have not been isolated, but pyridine, bipy, phen and terpy analogues have (Scheme 105). An incompletely characterized mixed chloro–iodo–chromium pyridinate separates first in the reaction with pyridine and is removed before crystallization of $[C_3F_7CrCl_2(py)_3]$. The perfluoroalkyl–chromium(III) derivatives are stable, their electronic and ESR spectra resemble those of other Cr^{III}–salen complexes but the magnetic moments are higher than usual.[1126,1127]

Alkyl halides would not react with $[CrCl_2(MeCN)_2]$, but the alkyl derivatives $[RCr(salen)(H_2O)]$ and $[RCr(salphen)(H_2O)]$ (R = Me, Ph) are said to form on reaction of the organic hydrazines with the Cr^{III}–Schiff's base complexes in MeCN under nitrogen followed by oxidation with oxygen and hydrolysis.[1128]

35.4.8.2 Miscellaneous mixed donor atom ligands

(*i*) *Complexes of N—O chelating ligands*

The majority of papers concerned with such complexes are synthetic studies of six-coordinate species with octahedral ligation. In general, such complexes have magnetic moments in the range of 4.0 BM and electronic spectra which have usually been assigned under pseudo-octahedral symmetry. Such studies are summarized in Table 97. A number of ligands have received considerably more attention; these are now considered individually.

There have been a number of studies of ethanolamine (**237**) complexes of chromium(III). A tris complex with deprotonated alcohol groups $[Cr_2(Eta)_3]\cdot 3H_2O$ and a more complicated complex $[Cr_2(Eta)_3(HEta)_3][ClO_4]_3$ have been prepared.[1129] The tris complex was prepared by reacting anhydrous $CrCl_3$ with 2-aminoethanol for 24 h, followed by recrystallization. The complex $[Cr_2(Eta)_3(HEta)_3][ClO_4]_3$ was prepared by hydrolysis of the tris complex in aqueous ammonium perchlorate. Electronic spectra of $[Cr(Eta)_3]$ were consistent with chelated amino alcohol ligands, though isomerism was not considered. The solid complex $[Cr_2(Eta)_3(EtaH)_3](ClO_4)_3$ was nearly isomorphous with the corresponding cobalt(III) complex. The similarities would suggest a structure involving a hydrogen-bonded dimer with three protons holding two facial $[Cr(Eta)_3]$ molecules together. Other studies of chromium(III) 2-aminoethanol complexes have led to the isolation of tris complexes and dinuclear complexes of various compositions.[1130,1131,1132,1134,1135]

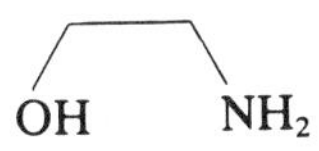

(237)

Table 96 Mixed Donor Atom Ligands: Semicarbazides, Hydrazones, Thiosemicarbazides, *etc.*

Ligand	*Complexes*	*Ref.*
(**228**), X = Y = H; R = $NHCONH_2(ssaH_2)$; tridentate, $-2H^+$ in base	$[Cr(ssaH)_2]Z$, Z = NO_3, Cl; $[Cr(ssa)(ssaH)]$, $NH_4(K)[Cr(ssa)_2]$	1
X = 5-Cl, 5-Br, Y = H, R = $NHCONH_2$	Analogous 5-Cl and 5-Br complexes	2
(**228**), X = H, Y = H, Me, Et, Pr, X = 5, 6-benzo Y = H; R = NHCOPh; $-2H^+$ in base	$[CrL_2]Cl$ octahedral; $[CrLX]_2$ in base, X = Cl, Br, NO_3, five-coordinate? μ_{eff} ~3.0 BM	3
(**228**), X = H, Y = Me, R = NHCO-2-py (tetradentate?) or NHCO-4-py (tridentate?)	$[CrLCl_2]$(pH 2–3), μ_{eff} ~3.8 BM; $[CrLCl]_2$(pH 5) μ_{eff} ~3.3 BM, all non-electrolytes	4
(**228**), X = Y = H, R = CHMeC(OH) : NOH, tridentate	$[CrL(NO_3)(H_2O)_2]\cdot 2H_2O$, $\mu_{eff}^{304\,K}$ = 3.6 BM	5
(**228**), X = Y = H, R = $NHCSNH_2(stscH_2)$; tridentate, $-2H^+$ in base	$[Cr(stscH)_2Cl]$, $[Cr(stsc)(stscH)]$, $NH_4[Cr(stsa)_2]$	6
Mixed complexes of stsc and LH, LH = oxine, picolinic acid or glycine	$[Cr(stsc)L(H_2O)]$, non-electrolytes, μ = 3.8 BM	7
(**228**), Y = Me, X = 3-Me, 4-Me, 5-Me, R = $NHCSNH_2$	CrL_3	8
(**228**), LH_2: X = Y = H, R = N : C(SMe)NH_2	$[Cr(LH)_2]Z$, Z = Cl, NO_3, $[Cr(LH)L]$	9
(**228**), X = Y = H, R = $NHCSeNH_2$, ($sesaH_2$)	$[Cr(sesa)(sesaH)]\cdot H_2O$	10
acacH + picolinic acid hydrazide(acac2ph)	$[Cr(acac2ph)X_2]$, X = Cl, Br, NO_3, NCS	11
acacH + isonicotinic acid hydrazide, (acac2iH)	$[Cr(acac2ih)X_2]$, X = Cl, Br, NO_3, NCS	
acacH + 2,6-diaminopyridine	[CrLX], L = macrocycle (tetradentate?), X = Cl, Br, NO_3, NCS	12
$4\text{-}NMe_2C_6H_4CHO + 2\text{-}HOC_6H_4CONHNH_2$	$[CrLCl_2]Cl$, L neutral keto form	13
$4\text{-}NMe_2C_6H_4CHO + C_6H_5CONHNH_2$	$[CrLCl_2]Cl$, L neutral keto form	
salH + $NH_2NHCO(CH_2)_nCONHNH_2$, n = 0, 1, 2	[CrL(OAc)]	14
5,5-Methylenebis(salicyladehyde) + $PhNH_2$	$[CrL(H_2O)_2]Cl$	15
$5\text{-}XC_6H_4CHO + NH_2NHCSNH_2$	$[CrL_3]$, X = Cl, NO_2, MeO, NH_2	16
Diacetyl monoxime + hydrazides: $RCONHNH_2$, R = Bu, Ph, 3-ClPh, 2-NO_2Ph:	[Cr(LH)L], $[CrL(H_2O)_3]Cl_2$, $[CrCl_3(LH_2)]$	17, 18, 19
dihydrazides, R = $CONHNH_2$, $(CH_2)_4CONHNH_2$:	$[Cr(LH_2)L]$, $[Cr_2L(H_2O)_6]Cl_4$, $[Cr_2Cl_6(LH_4)]$	
Diacetyl monoxime + semicarbazide, thiosemicarbazide or selenosemicarbazide	[Cr(LH)L]; $[Cr(LH)(H_2O)_3]Cl_2$ (semicarbazone only)	20
Pyruvic acid + $NH_2CSNHNH_2$ ($thpuH_2$)	[Cr(thpuH)(thpu)], $NH_4[Cr(thpu)_2]$, μ_{eff} ~3.8 BM; $[Cr(thpuH)_2]$, μ_{eff} ~3.1 BM, low spin Cr^{II}	21
Phenylpyruvic acid + $NH_2CXNHNH_2$ 4-Mephenylpyruvic acid + $NH_2CXNHNH_2$	X = S, O X = S, O } $[Cr(LH)_2]X$, X = Cl, Br, I, NO_3	22
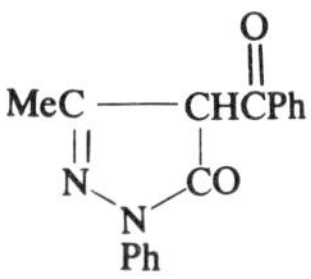 + B	B = $NH_2NHCSNH_2$: $[CrL(NO_3)_2(H_2O)_2]$	23
	B = $RC_6H_4NH_2$, R = H, 2-Me, 3-Me, 4-Me: CrL_3	24
	Further derivatives	25
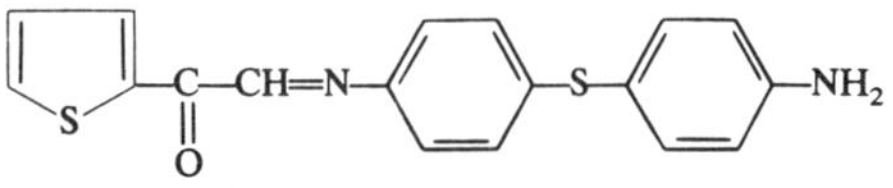	$[CrL_2]Cl_3$	26

1. N. M. Samus and V. G. Chebanu, *Russ. J. Inorg. Chem. (Engl. Transl.)*, 1969, **14**, 1097.
2. V. G. Chebanu and N. M. Samus, *Russ. J. Inorg. Chem. (Engl. Transl.)*, 1976, **21**, 1807.
3. D. K. Rastogi, S. K. Dua and S. K. Sahni, *J. Inorg. Nucl. Chem.*, 1980, **42**, 323.
4. V. B. Rana, J. N. Gurtu and M. P. Teotia, *J. Inorg. Nucl. Chem.*, 1980, **42**, 331.
5. M. K. Ghosh, B. Sur and A. K. Chakraburtty, *J. Indian Chem. Soc.*, 1984, **61**, 282.
6. A. V. Ablov and N. V. Gerbeleu, *Russ. J. Inorg. Chem. (Engl. Transl.)*, 1965, **10**, 33.
7. K. M. Purohit and D. V. Ramana Rao, *Indian J. Chem., Sect. A*, 1983, **22**, 520.
8. M. S. Patil and J. R. Shah, *J. Indian Chem. Soc.*, 1981, **58**, 944.
9. V. M. Leovac, N. V. Gerbeleu and V. D. Canic, *Russ. J. Inorg. Chem. (Engl. Transl.)*, 1982, **27**, 514.
10. A. V. Ablov, N. V. Gerbeleu and A. M. Romanov, *Russ. J. Inorg. Chem. (Engl. Transl.)*, 1968, **13**, 1558.
11. M. P. Teotia, I. Singh, P. Singh and V. B. Rana, *Synth. React. Inorg. Metal-Org. Chem.*, 1984, **14**, 603.
12. V. B. Rana, P. Singh, D. P. Singh and M. P. Teotia, *Transition Met. Chem.*, 1982, **7**, 174.
13. Y. M. Temerk, S. A. Ibrahim and M. M. Kamal, *Z. Naturforsch., Teil B*, 1984, **39**, 812.
14. K. K. Narang and U. S. Yadav, *Curr. Sci.*, 1980, **49**, 852.
15. A. M. Karampurwala, R. P. Patel and J. R. Shah, *Angew. Makromol. Chem.*, 1980, **89**, 57.
16. U. N. Pandy, *J. Indian Chem. Soc.*, 1978, **55**, 645.
17. V. Yu. Plotkin, M. G. Felin and N. A. Subbotina, *Russ. J. Inorg. Chem.*, 1983, **28**, 668.
18. V. I. Spitsyn, M. G. Felin, V. Yu. Plotkin, A. I. Zhirov and N. A. Subbotina, *Russ. J. Inorg. Chem. (Engl. Transl.)*, 1980, **25**, 1654.
19. V. I. Spitsyn, M. G. Felin, V. Yu. Plotkin, N. A. Subbotina and A. I. Zhirov, *Russ. J. Inorg. Chem. (Engl. Transl.)*, 1982, **27**, 1278.
20. V. Yu. Plotkin, M. G. Felin, N. A. Subbotina and V. V. Zelentsov, *Russ. J. Inorg. Chem. (Engl. Transl.)*, 1983, **28**, 825.
21. A. V. Ablov and N. V. Gerbeleu, *Russ. J. Inorg. Chem. (Engl. Transl.)*, 1970, **15**, 952.
22. S. Chanda and K. B. Pandeya, *Gazz. Chim. Ital.*, 1981, **111**, 419.
23. A. K. Rana and J. R. Shah, *Indian J. Chem., Sect. A*, 1982, 929.
24. A. K. Rana and J. R. Shah, *Indian J. Chem., Sect. A*, 1982, **21**, 177.
25. N. R. Shah and J. R. Shah, *Indian J. Chem., Sect. A*, 1982, **21**, 312.
26. H. S. Verma, A. Pal, R. C. Saxena and J. L. Vats, *J. Indian Chem. Soc.*, 1982, **59**, 1184.

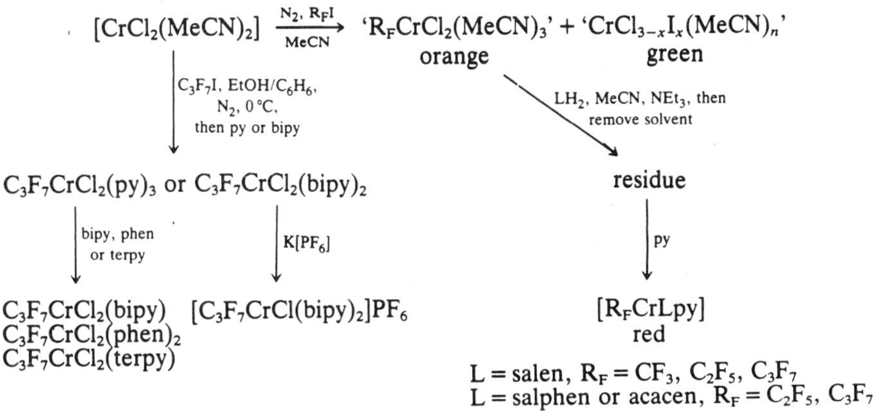

Scheme 105

Table 97 Complexes with N,O Chelate Donors

Ligand	*Form*	*Complex*	*Ref.*
Violurate	H_3L	$[Cr(H_2L)_3]\cdot 5H_2O$	1
2-Acetylpyrolate	HL	$[Cr(L)_3]$	2
Acetyl hydrazide	L	$[Cr(L)_3Cl_3]$	3
Oxamic acid	H_2L	$K_3[Cr(L)_3]$	4
Dipicolinic acid	H_2L	$[Cr(L)(L')]$ L' = amino acid, *etc.*	5
D-Cycloserine	H_2L	$[Cr(H_2L)X_3]$ $X = Cl^-, Br^-, I^-$	6
Quinolinic acid	H_2L	$[Cr(HL)_3]$	7

1. E. Garcia-Espana, J. Moratal and J. Faus, *J. Coord. Chem.*, 1982, **12,** 41.
2. J. J. Habeeb, D. G. Tuck and F. H. Walters, *J. Coord. Chem.*, 1978, **8,** 27.
3. P. K. Biswa, M. K. Dasgupta, S. Mitra and N. R. Chaudhuri, *J. Coord. Chem.*, 1982, **11,** 225.
4. A. G. Gallinos, J. M. Tsangaris and J. K. Kouinis, *Z. Naturforsch., Teil B,* 1977, **32,** 645.
5. S. K. Sengupta, S. K. Sahni and R. N. Kapoor, *Synth. React. Inorg. Metal-Org. Chem.*, 1983, **13,** 117.
6. C. Preti and G. Tosi, *Aust. J. Chem.*, 1980, **33,** 57.
7. B. Chatterjee, *J. Inorg. Nucl. Chem.*, 1981, **43,** 2553.

Picolinic acid (pyridine-2-carboxylic acid) complexes of chromium(III) have been the subject of a number of studies. Complexation by picolinic acid in water/ethanol (30% v/v) follows an ion-pairing, Eigen–Wilkins type mechanism.[1136] Activation parameters suggest an associative character for the reaction of the aqua complex. Chelated complexes of chromium(III) and picolinic acid are the products of the rapid, inner-sphere reduction of $[Co^{III}(pico)(NH_3)_5]^{2+}$ with chromium(II).[1137] The reaction of the related 4-carboxylic acid complex of cobalt(III) with chromium(II) is also rapid; in contrast, pyridine-3-carboxylic acid (nicotinic acid) complexes undergo slower reactions. A μ-hydroxy-bridged dimeric complex $[Cr_2(pico)_4(OH)_2]$ has also been prepared. A study of magnetic properties in the temperature range 16–300 K leads to $J = -6\ cm^{-1}$ and $g = 2$, typical for such complexes.[1138]

Complexes of 8-hydroxyquinoline and chromium(III) have been known for many years.[1139] Complexes have been mercurated and the deuterated quinoline isolated.[1139] The bromination of chelated 8-hydroxyquinoline proceeds about thirty-five times faster than that of the free ligand.[1140] Tris(8-hydroxyquinolinato)chromium(III) absorbs large amounts of hydrochloric, hydrobromic and hydrofluoric acids. Chemical reaction with the complex was considered a more likely explanation than solid solution or clathrate formation, even though more than one mole of acid was absorbed per mole of complex.[1141]

The complex diaqua[2,6-diacetylpyridinebis(semicarbazone)]chromium(III) hydroxide dinitrate hydrate (**238**) has a most unusual structure.[1123] Chromium(III) is coordinated in an approximate pentagonal bipyramid (PBP), with the ligand forming the pentagonal plane and two water molecules in the axial positions. The solution magnetic moment is 4.05 BM. An interesting feature of such a complex is that a regular PBP would lead to a single unpaired electron in the degenerate $d_{x^2-y^2}$ or d_{xy} orbitals. The distortion of the ligand plane removes this degeneracy and provides the first examples of the Jahn–Teller effect in a pentagonal bipyramidal complex (Table 95).

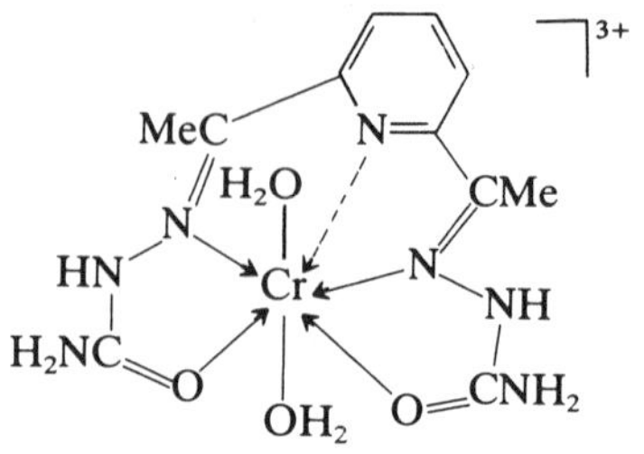

(238)

The coordination of chromium(III) is also unusual in the dimeric complex $[Cr\{H(chba\text{-}Et)\}(py)_2]_2\cdot 2py$ (**239**)[519] (H_4(chba-Et) = 1,2-bis(3,5-dichloro-2-hydroxybenzamido)ethane). The structure of this complex is described in Section 35.4.2.11.

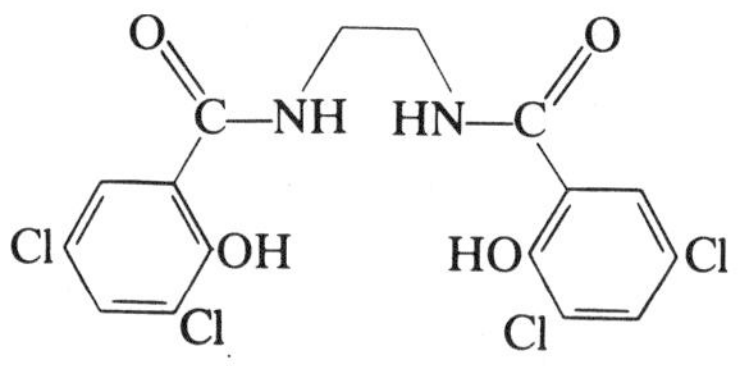

(239)

(ii) Chelating ligands containing sulfur and related donor atoms

There are a number of papers reporting the synthesis and characterization of monomeric complexes of this type; the more significant and/or recent are summarized in Table 98.

Table 98 Complexes with S,O and N,S Chelate Donors

Ligand	*Form*	*Complex*	*Ref.*
Dithiocarbazic acids	H_2L	$[Cr(L)_3]^{3-}$	1
Thiosemicarbazide	HL	$[Cr(HL)_3]Cl_3$ $[Cr(L)_3]$	2
Cyclohexanone thiosemicarbazone	L	$[Cr(L)_2]Cl_3$	3
Monothiobenzoates	HL	$[Cr(L)_3]$	4
Thiosemicarbazones	L	$[Cr(L)_2Cl_2]Cl$	5
Mercaptoacetate	H_2L	$[Cr(L)(H_2O)_4]^+$	6

1. R. A. Haines and J. W. Louch, *Inorg. Chim. Acta*, 1983, **71,** 1.
2. W. Shibutani, K. Shinra and C. Matsumoto, *J. Inorg. Nucl. Chem.*, 1981, **43,** 1395.
3. S. Chandra, K. B. Padeya and R. R. Singh, *J. Inorg. Nucl. Chem.*, 1980, **42,** 1075.
4. V. V. Savant, J. Gopalakrishnan and C. C. Patel, *Inorg. Chem.*, 1970, **9,** 748.
5. S. Chandra and K. K. Sharma, *Synth. React. Inorg. Metal-Org. Chem.*, 1982, **12,** 647.
6. R. J. Balahura and N. A. Lewis, *Inorg. Chem.*, 1977, **16,** 2213.

The complex bis(2-mercaptoethylamine)ethylenediaminechromium(III) perchlorate has been prepared by the reaction of cystamine with aqueous solutions of chromium(II) in the presence of ethylenediamine.[1142] The complex crystallizes from aqueous solution to form opaque violet rectangular bipyramids, the crystal structure of which has been determined. The reaction was said to go *via* an intermediate (**240**) to the chromium(III) complex (**241**) which underwent rapid ring closure. The stereochemistry of (**241**) is dictated by the strong Jahn–Teller effect in the chromium(II) precursor (**240**). A series of thiolato complexes, *e.g.* $[Cr(H_2O)_4SCH_2CO_2]^+$, has been oxidized with H_2O_2.[976] The chromium(III) complexes thus formed, in contrast to the cobalt(III) analogues, decompose, presumably *via* an unstable sulfanato–chromium(III) intermediate.

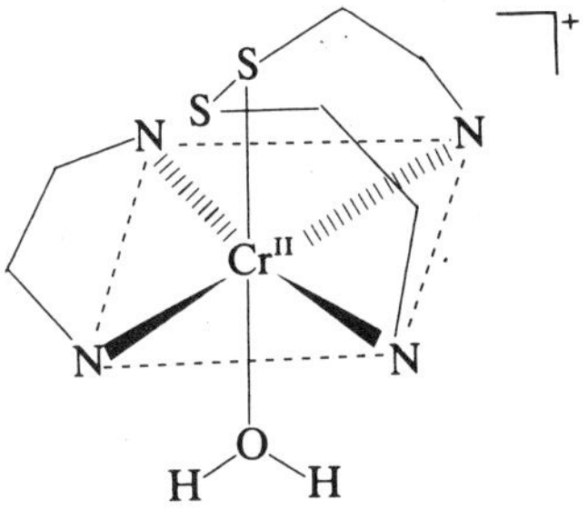

(**240**)

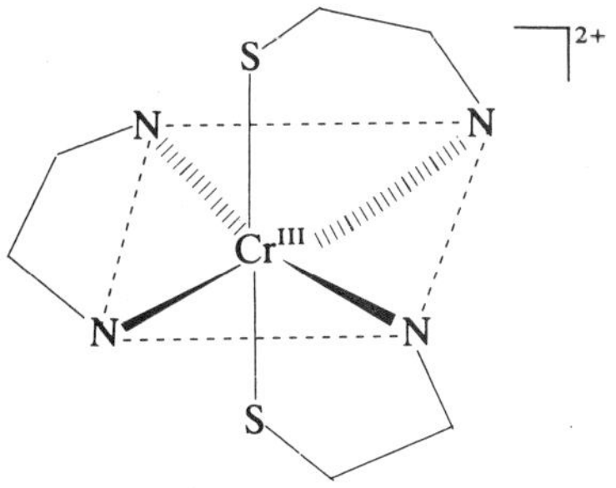

(**241**)

The complexes of the monothio-β-diketonate RC(SH)=CHCOR′ (R = R′ = phenyl; R = phenyl, 2-thienyl, β-naphthyl; R′ = CF_3), which are acetylacetonate analogues have been synthesized (**242**).[1143] Dipole moment measurements are consistent with facial structures (**243**). Mass spectra indicate no metal-containing peaks for the complex $[Cr(PhC(S)=CHCOPh)_3)]$: however, for the fluorinated monothio-β-diketonates, various metal-containing peaks, *e.g.* M—2L=F, *were* observed. Such ions involve fluoride migration. Monothiooxalate complexes of chromium(III) have been prepared; fairly unusual complexes, exemplified by $[Cr\{(C_2SO_3)Cu(Ph_3)_2\}_3]$, were reported. On refluxing under chloroform (7 h) reactions of the kind illustrated were alleged to occur (equation 55).[1144]

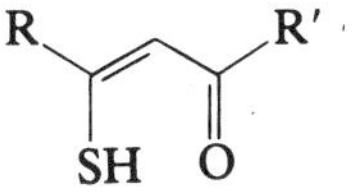

(**242**)

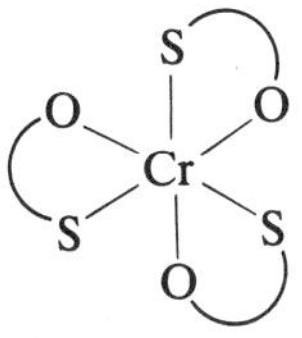

(**243**)

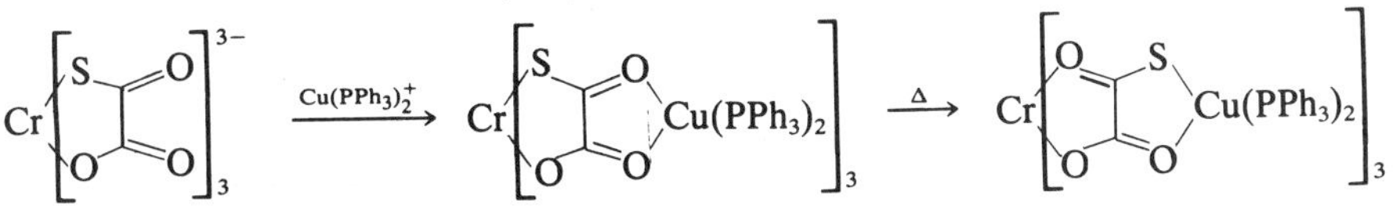

(55)

(**244**) (**245**) (**246**)

The complexes formed between thioglycolic acid (mercaptoacetic acid) and chromium(III) have been investigated.[1145] In alkaline solutions above 60 °C, a blue green complex $[Cr(SCH_2CO_2)_3]^{3-}$ is formed. In acidic solutions, two red complexes $[Cr(H_2O)_4(SCH_2CO_2H)]^{2+}$ and $[Cr(H_2O)_2(SCH_2CO_2H)_2]^+$ are predominant with a species $[Cr(H_2O)_2(SCH_2CO_2)_2]^-$ detectable at neutral values of pH. The rates of ring opening and closure for the chromium(III) mercaptoacetate complex $[Cr(H_2O)_4SCH_2CO_2]^+$ have been measured.[1146]

A large number of complexes with arsenic-containing ligands have been reported (**247**). Three types of complex of formula $CrO(o\text{-}R_2AsC_6H_4CO_2)\cdot nH_2O$, $Cr(Ph_2AsC_6H_4CO_2)_2(OH)\cdot 2.5H_2O$ and $Cr(o\text{-}R_2AsC_6H_4CO_2)(OH)_2\cdot nH_2O$ were prepared. The authors concluded that arsenic was probably not coordinated to the chromium in any of these complexes.[1147]

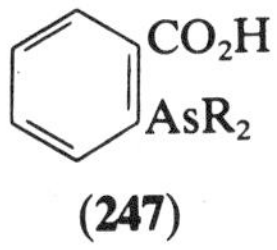

(**247**)

35.4.8.3 Amino acids

(*i*) *Potentially bidentate amino acids*

Complexes of simple amino acids with chromium(III) were first prepared by Ley.[1148] The isomers possible for tris chelated complexes of this type are illustrated below (**248–251**). The consequences of such isomerism were first seriously considered by Gillard.[1149] Red complexes of the formulae [Cr(gly)$_3$] and [Cr(L-ala)$_3$] were prepared by neutralizing refluxed solutions of hexaaquachromium(III) and the amino acid in ratios between 1:5 and 1:10. These complexes were shown to be isomorphous with β-[Co(gly)$_3$] and D-β-[Co(L-ala)$_3$] respectively. The crystal structure of red β-[Cr(gly)$_3$] has also been reported.[1150]

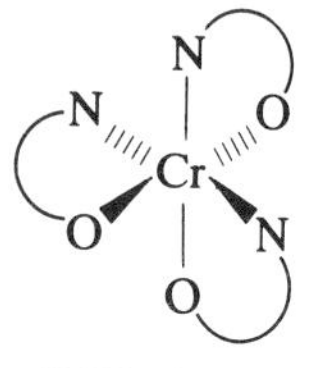
(**248**) Δ-α

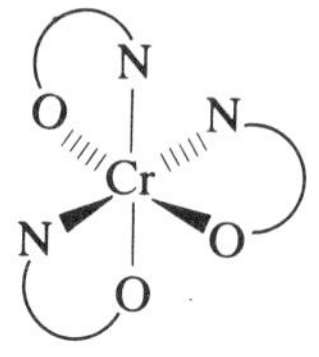
(**249**) Λ-α

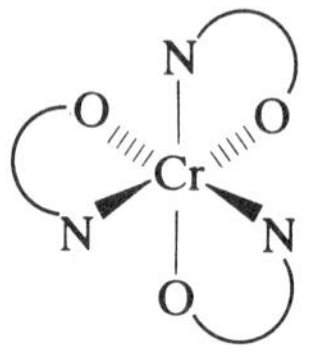
(**250**) Δ-β

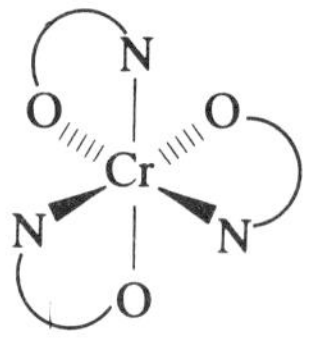
(**251**) Λ-β

There is some controversy concerning the existence of the α isomers of these tris complexes. Israily[1151] reported a purple complex to be α-[Cr(gly)$_3$]; however, subsequent workers have shown that this substance most probably was the dihydroxy dimer [Cr$_2$(gly)$_4$(OH)$_2$].[1149,1152,1153] Careful chromatography, on potato starch, of solutions from chromium(III)/glycine reactions yielded red and purple fractions,[1153] the electronic spectra of which were consistent with β and α isomers respectively. Solutions of the 'α complex' were unstable even in the dark and cold. Hoggard has recently claimed[1154] the preparation of the α isomer of the glycine complex by a fractional crystallization. The complex was anhydrous, unlike its cobalt(III) analogue. X-Ray powder methods could hence not be used to confirm the identity of the complex; the luminescence spectra were held to be consistent with meridional coordination. There have been a number of studies of the physical properties of β-[Cr(gly)$_3$], summarized in Table 99.

Table 99 Physical Properties of Tris Amino Acidates

	D-β-[Cr(L-ala)$_3$]	β-[Cr(gly)$_3$]	Ref.
Crystal structure	—	$P2_1/c$, $z=4$	1
ΔH_f° (kJ mol^{-1})	2380	2264	2
μ_{eff} (BM)	3.83	3.82	3
	—	3.89	4
Electronic spectrum			
70% HClO$_4$	534 (46.6)	539 (44.6)	
λ_{max} (nm), (ε_m)	396 (39.5)	400 (37.8)	
CD, 70% HClO$_4$	512 (+1.00)	—	3
$\lambda_{max/min}$, ($\Delta\varepsilon_m$)	452 (−0.25)	—	
	418 (+0.08)	—	
	382 (−0.19)	—	

1. Ref. 1150.
2. C. E. Skinner and M. M. Jones, *Inorg. Nucl. Chem. Lett.*, 1967, **3**, 185.
3. Ref. 1153.
4. T. Morishita, K. Hori, E. Kyuno and R. Tsuchiya, *Bull. Chem. Soc. Jpn.*, 1965, **88**, 1276.

Many other amino acids form complexes which are red β isomers with 'glycine-like' coordination at chromium(III). The complexes are most frequently prepared by refluxing the amino acid with chromium(III), followed by neutralization. The problem with this kind of method is that significant quantities of the corresponding bridged hydroxy complex are formed; indeed the solid tris complexes are converted to such complexes on standing under water. Repetition of the literature preparations of these complexes is often difficult and the ease of hydrolysis/kinetic inertness of chromium(III) probably leads to some of the discrepancies in the literature. An approach which potentially avoids the problems of hydrolysis is solid-state synthesis. Amino acid complexes have been prepared by heating (135 °C) solid hexaamminechromium(III) nitrate with the amino acid followed by recrystallization. Complexes of glycine, (±)-alanine, (±)-isoleucine, (±)-leucine, (±)-aminobutyric acid,

(±)-norvaline and (±)-valine[1155] have been prepared by this method as have the corresponding L-amino-acid complexes.[1156] In general, *β-fac* isomers were obtained. In related work, complexes of various basic amino acids were prepared.[1157] The tris complex of (±)-methionine has been prepared by conventional methods.[1158] A number of simple mixed complexes of amino acids with chromium(III) and aminoethanol or 2,2′-iminodiethanol have been reported.[1159]

(*ii*) *Hydroxy-bridged complexes*

These are extremely common and can be a problem when the synthesis of the related tris complexes is being attempted, particularly as they are of rather limited solubility. They have been known for many years.[1148,1160,1161] The isomers possible for these species are profuse.[1149] For the general formula $[(aa)_2Cr\text{-}(\mu\text{-}OH)_2Cr(aa)_2]$ there are three possible geometric configurations about each metal ion: *trans* N *cis* O, *cis* N *trans* O and *cis* N *cis* O. In addition, each metal ion is a dissymmetric centre bringing the total number of geometric and optical isomers to 21.

A crystal structure of the all-*trans* isomer of $[Cr_2(gly)_4(OH)_2]$ has been reported.[1162] Its low temperature magnetic susceptibility has been fitted to both the Van Vleck and modified Van Vleck models. The uncorrected model leads to $\mu_{eff} = 3.80$ BM with $2J = 8.4\ cm^{-1}$;[1162,1163,1164] with the inclusion of quadratic exchange $2J = 7.4\ cm^{-1}$.[1162] Related studies of alanine,[1153,1165] valine, phenylalanine, leucine,[1163,1149] proline[1166,1167] and histidine[1168] complexes have appeared. The proline complex is unusual in that it is soluble in methanol and DMSO. Circular dichroism spectra have been measured; the X-ray structure shows the complex to be the L-*trans*(N), L-*trans*(O) isomer.[1166] More complicated dimers of unusual stoichiometry have been reported.[1169]

(*iii*) *Solution studies*

Stability constant determinations are few;[1170] they are summarized in Table 100. Complexation by acidic amino acids is obviously of relevance to the tanning of leather; the stability constants for L-glutamic and aspartic acid[1171] complexes are much greater than those for glycine or L-alanine.[1172,1173,1174] This is probably because the acidic amino acids form tridentate complexes. In contrast, cysteine[1173] appears to form glycine-like complexes in moderately acidic solution; however, in the solid state L-cysteine is known to be tridentate (*vide infra*).

Table 100 Typical Equilibrium Constants of Amino Acid Complexes[a]

Ligand	K_1	K_2	*Conditions*	*Ref.*
L-Aspartic acid	12.15	8.98	50 °C	1
	10.1	9.5		2
L-Glutamic acid	11.39	7.57		1
α-Alanine	8.53	7.44	25 °C	3
Glycine	8.62	7.65		3
	8.4	6.4		4
L-Cysteine	8.05	7.45	25 °C	5
DL-Methionine	7.45	6.45		5
Glycine	7.60	—	40 °C 0.4 M $NaClO_4$	6

[a] $\log_{10}$ values in 0.1 M $NaClO_4$, except ref. 1174.
1. Ref. 1171.
2. M. Mizuochi, S. Shirakata, E. Kyuno and R. Tsuchiya, *Bull. Chem. Soc. Jpn.*, 1970, **43**, 397.
3. Ref. 1172.
4. A. A. Khan and W. U. Malik, *J. Indian Chem. Soc.*, 1963, **40**, 565.
5. Ref. 1173.
6. Ref. 1174.

Kinetic studies have concentrated on the glycine system.[1174,1175,1176] In the pH range 3.0–3.8 (40 °C), the reaction proceeds by acid-dependent and -independent paths; reaction *via* the hydroxy complex is probably I_d, whereas the aqua complex appears to undergo I_a substitution.

In more acidic solutions, monodentate O-bonded complexes are formed.[1177,1175,1178] Ring closure is rapid if the pH is raised; γ, δ and ε amino acid complexes do not undergo ring closure.[1179]

(*iv*) *Potentially tridentate amino acids*

For tridentate amino acids with three non-equivalent donor atoms such as L-aspartic acid or L-cysteine, the isomers possible are illustrated below (**252–254**). There have been a number of reports of the preparation of L-aspartic acid complexes.[1180,1181,1182]. In the earlier work the isomers were not identified, however in the later study, the complexes were tentatively identified by comparison of their spectroscopic properties with those of the corresponding cobalt(III) complexes.[1183] The order of elution of the complexes on HPLC was also similar to that observed for the corresponding cobalt(III) complexes. Mixed complexes containing L- or D-aspartate and L-histidine were also prepared.[1182] A crystal structure of one salt obtained from this kind of system, bis(L-histidinato-*O,N,N′*)chromium(III) nitrate, has been determined.[1184]

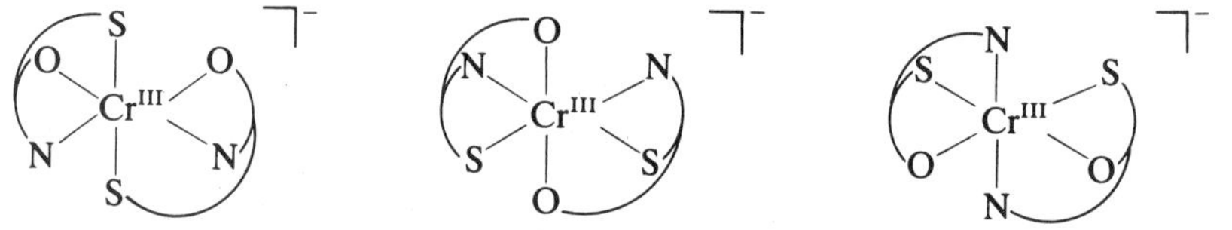

(**252**) (**253**) (**254**)

Sodium bis(L-cysteinato)chromate(III) dihydrate has been prepared by refluxing L-cysteine with chromium(III) nitrate and neutralizing the solution.[1185]. The product is a dark blue solid in which cysteine is coordinated with the sulfur atoms *trans* (see **252–254**). A number of chloro and other complexes of cysteine and related amino acids have been studied.[1186] A related complex L-histidinato-D-penicillaminatochromium(III) has been prepared[1187] and its crystal structure reported.

The reaction of chromium(III) with D-pencillamine has been investigated.[1188] Chromium(III) complexes are also rapidly obtained as the products of the reaction of chromium(VI) with D-penicillamine. Several complexes were prepared including a monomeric (S,N,O chelate); equilibrium constants were measured. These studies may be relevant to the treatment of chromium(VI) toxicity with D-penicillamine (see biological chromium(VI) chemistry p. 947). Unusual S-bonded complexes of chromium(III) have been prepared by reacting aqueous solutions of chromium(II) with $[Co(en)_2(\text{L-cys})]^{2+}$ or the corresponding methionine complex.[1189] Under the fairly acidic conditions used, S,O [excess of chromium(II)] or S,N [limiting chromium(II)] complexes were obtained; these both eventually rearranged to give O,N species.

(*v*) *Chromium(III) and glucose tolerance*

Chromium was recognized as an essential trace element in 1955.[1190] Rats fed a chromium-deficient diet developed an impaired tolerance for intravenous glucose, which could be reversed by an insulin-potentiating factor present in brewer's yeast, meat and various other foods. The insulin-potentiating factor was found to be a complex of chromium(III)[1191] and such substances have been termed Glucose Tolerance Factor(s) (GTFs). Chromium was demonstrated to be essential for humans in 1975.[1192] There are several reviews of the chemistry of chromium(III) and its relationship to glucose tolerance.[1193–1196]

A detailed procedure for the extraction of GTF from *Saccharomyces carlsbergenis* has been reported,[1197] involving ethanolic extraction, hydrolysis and ion exchange chromatography. Nicotinic acid (**255**) was detected in the complex by its characteristic UV and mass spectra. Glycine, glutamic acid and cysteine together with trace amounts of other amino acids were found in the purified sample of GTF. This is probably the most refined GTF yet isolated. Although IR spectra were recorded, no detailed spectroscopic studies were undertaken, which is unfortunate as studies of the ligand field spectra might help to suggest likely structures.

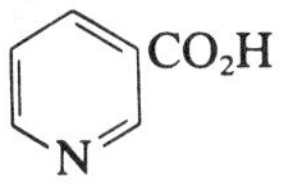

(**255**)

Nicotinic acid is present in the more active GTF preparations isolated from yeasts. The coordination chemistry of this ligand is particularly relevant to glucose tolerance and the presence of this substance is apparently essential for the maximal activity of complexes in tests *in vitro*. The instability of highly purified GTF fractions has frequently been noted; this may arise because the substance *in vivo* is stabilized by a protein.

The synthesis and characterization of materials showing biological activity similar to that of GTF isolated from yeast is a logical objective. As already mentioned, Mertz found aqua and similar complexes of chromium to be more active than chelates with strong ligands. An exception to this was an unspecified cysteine complex,[1193] prepared by C. L. Rollinson, which showed marked, but erratic, behaviour in GTF tests. Further investigation of this observation would be interesting, particularly as the crystal structure of a cysteine complex is now known.[1185]

A complex with considerable activity in tests *in vitro* has been synthesized[1197] by reacting chromium(III) acetate in 80% alcohol with two equivalents of nicotinic acid followed by the addition of one equivalent, in turn, of glycine, glutamic acid and cysteine. A complex containing nicotinic acid was isolated from the reaction mixture by ion exchange chromatography. The physical properties and biological activity of this complex were very similar to those of GTF isolated from yeast.[1197] Although this material is probably a mixture, it should be further investigated as it is one of the most potent synthetic materials yet obtained.

The amino acids believed to be involved in GTF are the constituents of the naturally occurring tripeptide glutathione (**256**). Anderson *et al.* have reported a complex of glutathione/nicotinic acid and chromium(III) to be particularly active in their *in vitro* assay.[1198] The complex was described as being purified by HPLC, but the details of the preparation are not available.

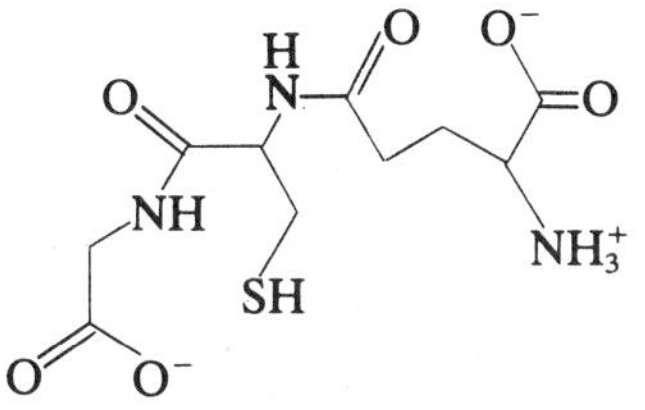

(**256**) Glutathione (GSH)
(shown as $[H_4L]^-$)

Recently, three different kinds of compounds containing chromium(III) and glutathione have been synthesized:[1199] a bisglutathione complex $K_2[Cr(H_3L)(H_2L)]$, mixed complexes with the amino acids L-cysteine, L-glutamic acid and L-aspartic acid $K_2[Cr(H_2L)(A)]$ (A is the dianion of the amino acid), and a mixed complex with glycine $K_2[Cr(H_2L)(gly\text{-}O)(OH)]$. All exhibit an intense UV charge transfer band characteristic of a Cr—S bond. The sulfhydryl to chromium linkage undergoes an acid-catalyzed hydrolysis. The complexes have been characterized by elemental and thermogravimetric analysis, electronic and IR spectroscopy and circular dichroism. Comparison of these properties with those of known chromium(III) complexes leads to the conclusion that glutathione is bound to chromium(III) by the O-terminal glycine group {N,O} and the deprotonated sulfur of cysteine. The glutamic acid residue does not apparently interact with the chromium centre; the suggested mode of coordination of glutathione is illustrated in (**257**).

The lack of structural information means that any suggested structure for GTF must be highly speculative. Mertz has suggested[1200] that two *trans* nicotinic acids are coordinated to chromium as illustrated in (**258**). The first coordination sphere is completed by the various amino acids involved. There is little evidence to support this and it seems unlikely that the hard chromium(III) centre would bind preferentially to the nitrogen group of nicotinic acid; the idea of N coordination has attracted support and influenced other workers. One product of the reaction of nicotinic acid and chromium(III) (mole ratio 3:1, pH 3.2) has been said to be

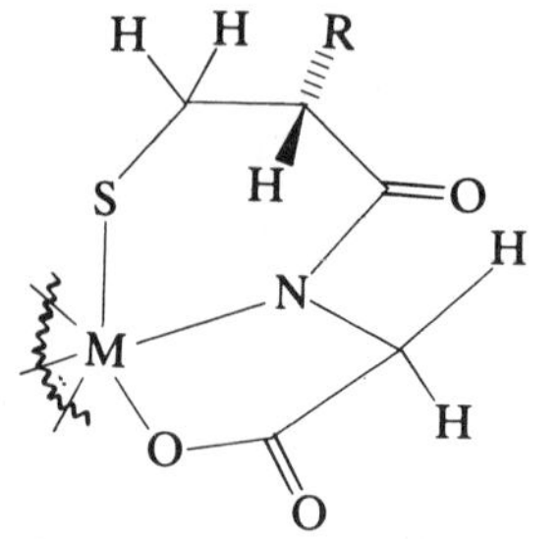

(**257**) Suggested coordination mode of glutathione
$R = NHCOCH_2CH_2CHNH_2CO_2H$

tetraaqua-*trans*-bis(nicotinate).[1200] It showed some biological activity but no spectral or other characterization was reported. It would be surprising if chloro ligands were not coordinated in this complex and nicotinic acid is most likely to be O-bonded. Indeed, Legg has recently shown,[1201] using deuterium NMR, that carboxyl-bonded nicotinic acid complexes are readily formed in aqueous solution. A crystal structure of a pentaammine nicotinic acid chromium(III) complex shows O-bonded nicotinate,[1202] which again supports the dominance of O-bonded nicotinic acid in chromium(III) chemistry.

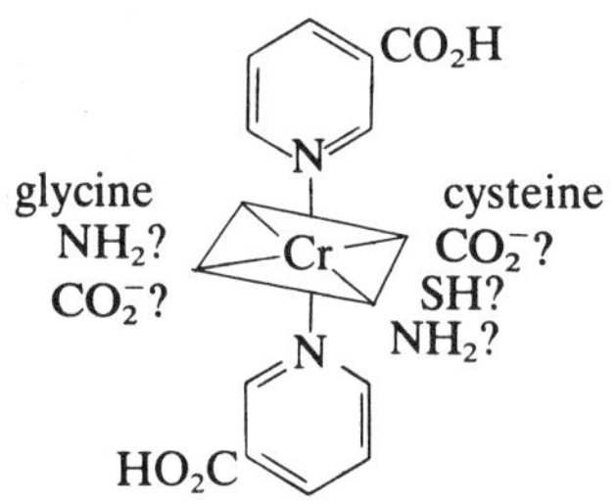

(**258**)

The view that N-bonded *trans* nicotinic acids are involved in GTF has also been challenged[1203,1204] from a more biological viewpoint; an O-bonded nicotinic acid complex has been shown to have some biological activity. The biological significance of chromium(III) was suggested to relate to the ability of the relatively hard chromium(III) centre to form stable complexes with the carboxylate group of nicotinic acid. The crucial grouping recognized by the biological system would hence involve the nitrogen of the O-coordinated nicotinic acid (**259–262**). The surface topology of the complex may be the most important factor and the effect of 1,4-diguanidinobutane (**262**) on yeast metabolism, which is similar to that of chromium(III) complexes, may be explained on this basis.[1204]

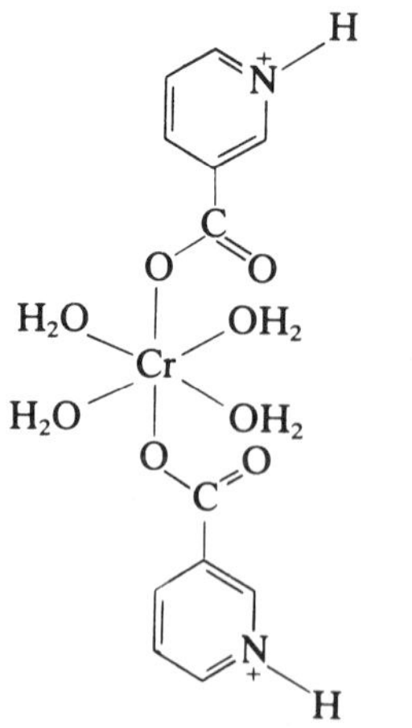

(**259**) $[Cr(nic)_2(H_2O)_4]^{3+}$

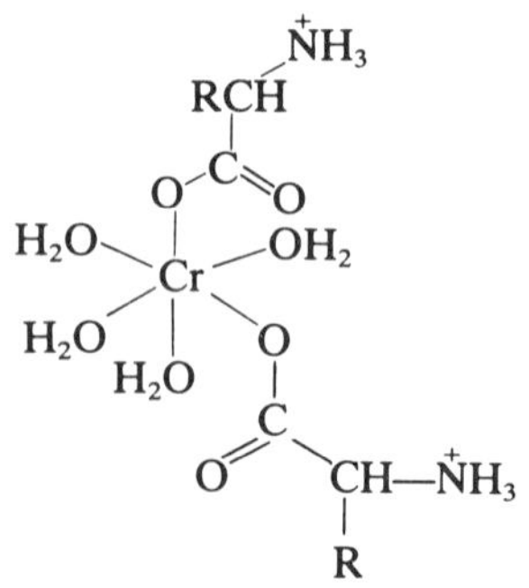

(**260**) $[Cr(gly)_2(H_2O)_4]^{3+}$

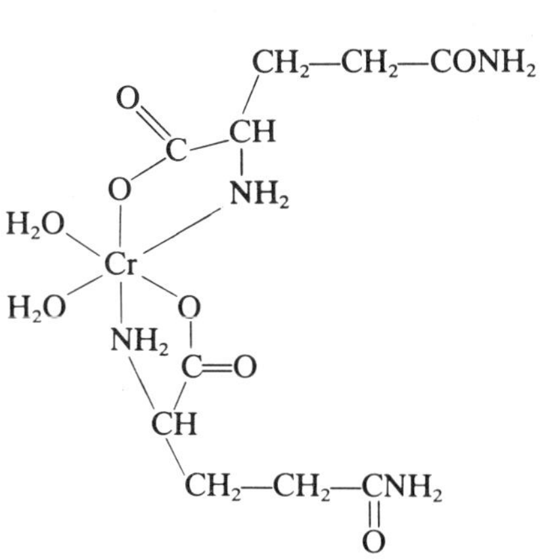

(261) $[Cr(gln)_2(H_2O)_2]^+$

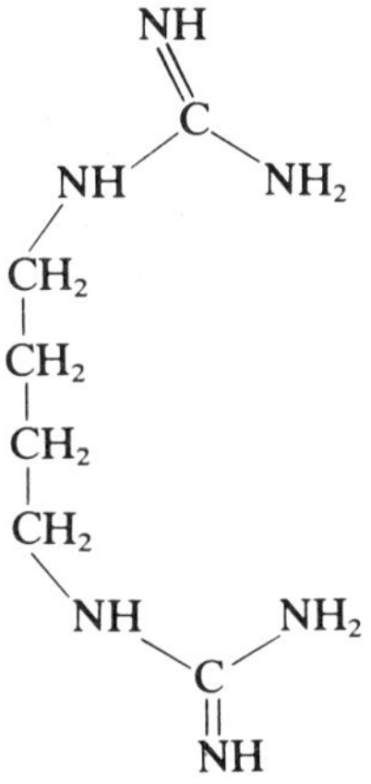

(262) 1,4-Diguanidinobutane

Suggested structures for biological activity

(vi) Chromium and tanning

Knapp discovered in 1858 that chromium chloride converted raw skins into leather, but he failed to realize the commercial significance of his discovery. Others developed his ideas. A two-bath process due to Schultz (1894) and an adaptation of Knapp's original one-bath process by Dennis (1893), meant that by the early twentieth century chromium tannage was commercially important. At present it accounts for the vast majority of leather production.

In the tanning process hides are first washed or soaked, hair and keratinous debris are removed, bated (enzymes are used to break down non-collagenous components, which are washed out) and the hide is acid-pickled to prepare for the addition of the chromium salt. Contemporary processes are exclusively based on one-bath procedures and utilize chromium(III). The older two-bath process is now obsolete, mainly because it involved the *in situ* reduction of chromate, a major environmental and toxicological hazard (*cf.* chromate toxicity p. 947) to chromium(III) on the hide. A useful review of the history of chromium tannage processes is available.[1205]

The most popular tanning solution in current use is termed 33% basic chromium(III) sulfate and corresponds to the empirical formula $CrOHSO_4$. In older procedures, chromate was often reduced at the plant using glucose/sulfuric acid mixtures. At present, 33% basic chromium sulfate is provided as either a solution (chrome liquor) or as a commercially available powder,[1206] of constant, but at present slightly uncertain, chemical composition.

The composition of such tanning liquors has been investigated.[1207] As many as 12 components have been identified, including monomeric di- and oligo-nuclear complexes involving hydroxide and/or sulfate bridges. Chrome tanning involves the careful control of the hydrolysis of chromium(III) (*cf.* O donors p. 857). The tanning mixture is introduced to the pelt at fairly acidic values of pH (2.5–3.0). The pH is then increased and the chromium reacts with and is fixed to the collagen. The low pH of introduction minimizes the processes of olation and oxolation (see O donors), which could easily lead to oligomeric and even insoluble material unable to penetrate the hide. At low pH the collagen is protonated and this also favours the penetration of the chromium complex into the hide. The hydroxy complexes, formed on raising the pH, are more reactive than aqua species[1199] due to the *trans* effect, a point not generally mentioned in the literature of tanning. The unfavourable effects of hydroxide are thus balanced against the requirement for a reasonably rapid reaction. Olation may further be limited by the use of a complexing ('masking') agent. However, the reaction of chromium with collagen will be less favourable if the metal is strongly complexed.[1208,1209] Carboxylates of various kinds have been used as masking agents in tanning,[1208] but the present trend seems to be to use 'simple basic sulfate solutions' or relatively low concentrations of carboxylates such as formate under carefully controlled conditions.

An exception to this is the preparation of special leathers, where a suitable chromium-conplexing agent, *e.g.* a hydrophobic ligand, may be used to impart a required property, *e.g.* shower resistance or lubrication, to the finished hide[1208] and similarly the use of fluorocarboxylates to produce soil-resistant leathers.

Gustavson first recognized[1210] that chrome tanning was due to the reaction of chromium with the carboxylic acid groups of collagen. Collagen, the principal component of hide, is an unusual protein consisting of a thin rod *ca.* 3000 Å long and 14 Å thick, composed of three peptide chains of equal length each containing about 1000 amino acid residues. Glycine accounts for about 30% of the amino acid residues and there are large quantities of proline and hydroxyproline. Metal ion binding to collagen has recently been reviewed.[1211]

Collagen fibrils tanned with 33% basic sulfate show a distinct cross striation (electron microscopy, 50 000:1), presumably owing to an increased ordering of the collagen monomers.[1212,1213] The exact way in which chromium is incorporated into collagen is not yet fully understood. In the tanning process, chromium may be envisaged as reacting with the free carboxylates of the protein to give three distinct kinds of coordinate linkage: (1) with a single carboxylate (**263**), (2) with cross-linking between strands *via* a single chromium (**264**), or (3) with cross-linking between strands by a di- or oligo-nuclear chromium species polymerized either *via* ol, oxo or sulfato bridges, or a combination of them (**265**). The relative importance of the different kinds of interaction has not been fully established. Bridging by di- and oligo-nuclear complexes, as illustrated in (**265**), is believed to be the most important interaction for the tanning of leather, although as much as 90% of the bound chromium may be bound only to a single collagen strand as in (**263**).

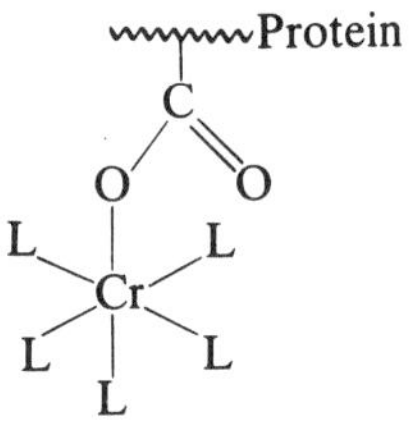

(**263**)

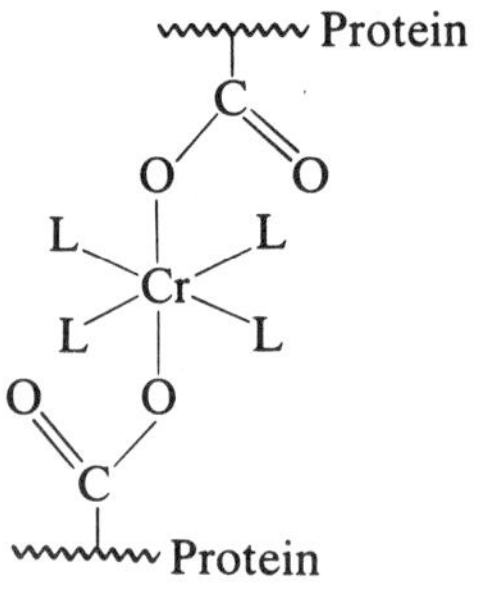

(**264**)

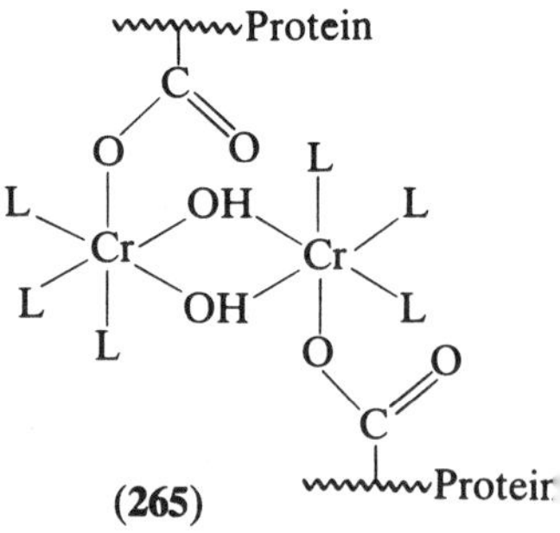

(**265**)

Schematic representations of possible collagen interactions. L may be a masking ligand, water or hydroxide

In (**265**) the bridging ligands, shown as ol, could be oxo or sulfato. There are obvious possibilities for isomerism in this interaction

The chemistry of tanning is very complicated. Although reliable procedures have been developed and the general principles of the chromium–collagen interaction are well understood, there is the possibility for much new work in this area.

35.4.8.4 Complexones: edta, pdta, nta and related ligands

(*i*) *Ethylediaminetetraacetic acid* (*edta*)

A violet complex [Cr(Hedta)H_2O] was first prepared in 1943 by Brintzinger, Thiele and Muller.[1214] It was transformed on the addition of base to a blue complex, then formulated as [Cr(Hedta)OH]$^-$. Thus, even from the earliest studies, edta in its chromium(III) complex was thought to be pentadentate; this has now been confirmed crystallographically (**266**).[1215] Coordination at chromium(III) is a distorted octahedron; the average deviation from 90° is 5.1°. The currently accepted protonation schemes for chromium(III)–edta complexes are illustrated in equations (56)–(61).[1216] Deuterium NMR spectroscopy has been used to confirm these solution equilibria.[1217] As well as the quinquedentate ones,[1214,1215] complexes in which edta is of lower denticity have been prepared.[1218] [Cr(Hedta)H_2O] reacted with hydrochloric and hydrobromic acids to give complexes of formulae [Cr(H_3edta)X_2(H_2O)]·$3H_2O$. The most likely structure was considered to have one of the nitrogens and two of the carboxylates of an edta ligand uncoordinated.

The equilibrium constant (log K) for the formation of the quinquedentate complex is 23.40.[1219] Perhaps the most remarkable feature of chromium(III)–edta chemistry is the kinetic

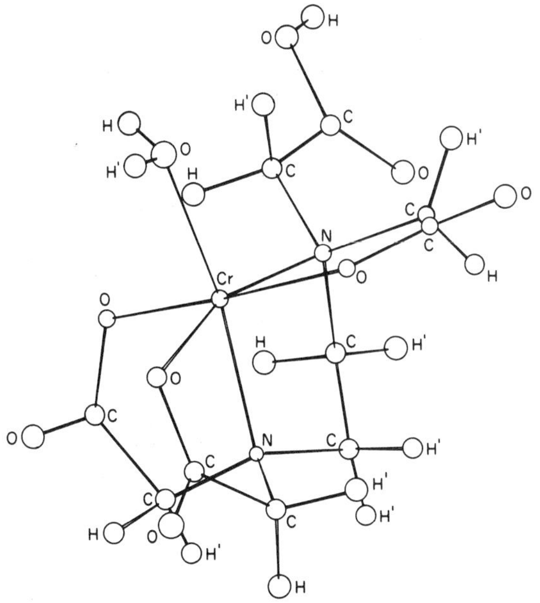

(266) Molecular structure of $[Cr(H_2O)Hedta]$

Protonation Scheme for Chromium(III) edta Complexes (after ref. 1216)

	pK_a	
$[Cr(Hedta)H_2O] \longrightarrow [Cr(edta)H_2O]^- + H^+$	3.1	(56)
$[Cr(edta)H_2O]^- \longrightarrow [Cr(edta)(OH)]^{2-} + H^+$	7.5	(57)
$[Cr(H_2edta)(H_2O)_2]^+ \longrightarrow [Cr(Hedta)(H_2O)_2] + H^+$	2.49	(58)
$[Cr(Hedta)(H_2O)_2] \longrightarrow [Cr(edta)(H_2O)_2]^- + H^+$	2.92	(59)
$[Cr(edta)(H_2O)_2]^- \longrightarrow [Cr(edta)(OH)(H_2O)]^{2-} + H^+$	6.16	(60)
$[Cr(edta)(OH)(H_2O)]^{2-} \longrightarrow [Cr(edta)(OH)_2]^{3-} + H^+$	8.0	(61)

lability of the coordinated water of $[Cr(Hedta)H_2O]$. The exchange of this water occurs on the stopped-flow time-scale, and has hence attracted a large number of kinetic studies, summarized in Table 101.

Chromium(III)–edta complexes are typically prepared[1220] by refluxing the ligand with a suitable chromium(III) salt for about half an hour followed by recrystallization. The formation of $[Cr^{III}(Hedta)H_2O]$ from acetato, formamato and trifluoroacetato chromium(III) complexes

ble 101 Rate Constants of the Aquation Reactions [k_b, (s^{-1})] of $[CrX(H_2O)_5]^{2+}$, $[CrX(NH_3)_5]^{2+}$ and $[CrXY]^{(n-2)}$ Complexes at 25 °C and $I = 1$ M (after ref. 1224)

Complex	X^- N_3^-	OAc^-	ONO^-	NCS^-	Cl^-	Br^-
$:X(H_2O)_5]^{2+}$	4.1×10^{-8} [a]	4.1×10^{-7} [b]		9.2×10^{-9} [c]	2.8×10^{-7} [a]	4.3×10^{-6} [a]
$:X(NH_3)_5]^{2+}$	3.6×10^{-8} [a,d]			8.7×10^{-8} [a,e]	7.4×10^{-6} [a]	8.1×10^{-5} [f]
:X(edtra)]$^-$		$(1.63 \pm 0.13) \times 10^{-3}$ [g]		$(2.40 \pm 0.17) \times 10^{-6}$ [h]		
·X(medtra)]$^-$		$(5.89 \pm 0.28) \times 10^{-5}$ [g]	$(8 \pm 3) \times 10^{-4}$ [h]			
:X(hedtra)]$^-$	0.189 ± 0.012 [h,i]	0.450 ± 0.030 [g]	6.24 ± 0.32 [h]	0.244 ± 0.019 [h]	0.697 ± 0.044 [j,k]	2.0 [j]
·X(aeedtra)]$^-$		0.38 ± 0.06 [l]				
:X(Hedta)]$^-$				$(3.17 \pm 0.33) \times 10^{-2}$ [h]		
·X(edta)]$^{2-}$	13.4 [m]	5.4 ± 0.6 [g]		26.8 ± 1.9 [h]		

ef. 303. [b] E. Deutsch and H. Taube, *Inorg. Chem.*, 1968, **7,** 1532. [c] C. Postmus and E. L. King, *J. Phys. Chem.*, 1955, **59** 6. [d] $I = 0.2$ M. [e] $I = 0.07$ M. [f] T. Ramasami and A. G. Sykes, *Inorg. Chem.*, 1976, **15,** 2885. [g] Ref. 1223. [h] Ref. 10. [i] $\Delta H^{\neq} = 60.4 \pm$ kJ mol^{-1}, $\Delta S^{\neq} = -55 \pm 4$ J mol^{-1} K^{-1}. [j] H. Ogino, M. Shimura and N. Tanaka, *Bull. Chem. Soc. Jpn.*, 1978, **51,** 1380. [k] $\Delta H^{\neq} =$ ± 3.3 kJ mol^{-1}, $\Delta S^{\neq} = 68 \pm 11$ J mol^{-1} K^{-1}. [l] H. Ogino, M. Shimura, A. Masuku and N. Tanaka, *Chem. Lett.*, 1979, 71. [m] Y. S. Sulfab, R aylor and A. G. Sykes, *Inorg. Chem.*, 1976, **15,** 2388; edtra = ethylenediamine-*N,N′,N′*-triacetate, medtra = *N*-methylenediamine-*N,N′,N′*-tri ate, hedtra = *N*-(hydroxyethyl)ethylenediamine-*N,N′,N′*-triacetate, aeedtra = *N*-(acetoxyethyl)ethylenediamine-*N,N′,N′*-triacetate.

has been studied;[1221] the acetato complex reacted 103 times faster than hexaaqua chromium(III).[1220] The thermal decomposition of $[Cr(NH_3)_6]Naedta\cdot 3H_2O$ provides[1222] an interesting method of synthesizing $[Cr(Hedta)H_2O]$.

(*ii*) *Other complexones derived from ethylenediamine*

Ligand exchanges with a variety of ligands closely related to edta (**267**, **268** and **269**) are rapid.[1223,1224] Typical results are summarized in Table 101. Chromium(III) ethylenediamine-*N*,*N′*-diacetato complexes $[Cr(edda)(H_2O)_2]^+$ have been synthesized.[1225] $CrCl_3\cdot 6H_2O$ was reacted with H_2edda and neutralized; separation of the isomers (**270**, **271** and **272**) by ion-exchange chromatography was attempted. On the basis of electronic spectra and comparison with the cobalt(III) system, the first complex to elute was assigned as *α-cis* (94%), the second band as *β-cis* (6%). The equilibrium constant for this system is $\log K = 12.42$ (room temperature),[1226] presumably an overall value for both isomers.

(**267**) $hedtra^{3-}$

(**268**) $edtra^{3-}$

(**269**) $medtra^{3-}$

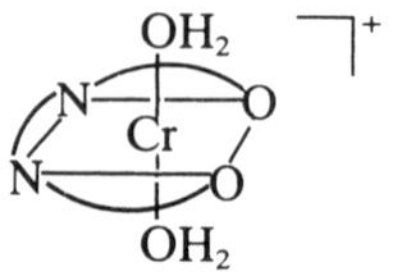

(**270**) *trans*

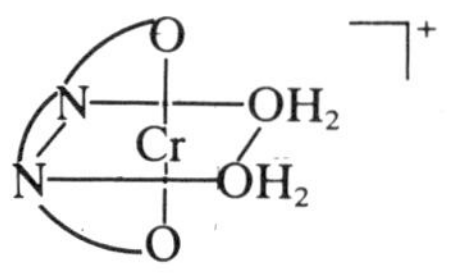

(**271**) *α-cis*

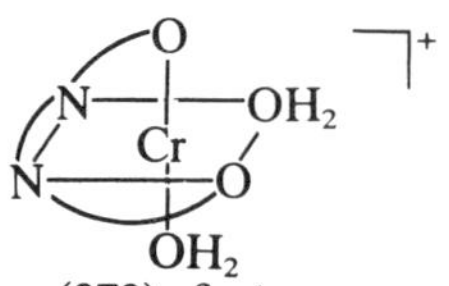

(**272**) *β-cis*

(*iii*) *Other complexones derived from polyamines*

In contrast to edta, the complexes of 1,3-propylenediaminetetraacetic acid are sexidentate.[1227] This may well be because ring strain is reduced (**273**) by the additional —CH_2 group in each chelate ring. The complex (2*S*,4*S*)-2,4-pentanediaminetetraacetatochromate(III) monohydrate has been prepared[1228] and forms stereospecifically. By comparison with Co^{III} analogues, absolute configurations have been assigned. Complexes with ethylenediamine-*N′*,*N′*-diacetic-*N*,*N′*-dipropionic acid (H_4edda) and *S*,*S*-2,2″-(ethylenediimino)disuccinic acid (H_4S,S-edds) were also made.[1228] An extensive study of triethylenetetraaminehexaacetic acid complexes of chromium demonstrated both mononuclear, dinuclear and heteronuclear chelates with copper(II).[1229]

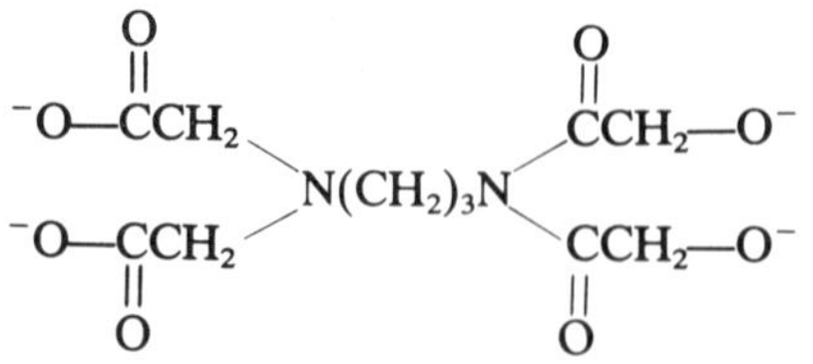

(**273**) pdta

(*iv*) *Imino acetic acids*

Several complexes with iminodiacetic acid (H_2ida) and methyliminodiacetic acid (H_2mida) have been prepared.[1225] The bis mida complex exists only as the *trans* isomer, ida forms bis *cis* N complexes; this is also the arrangement adopted in the mixed mida/ida complex.

Spectroscopic measurements were used to support these assignments. Related studies of *trans*-bis(*N*-isopropyliminodiacetato)chromium(III) dihydrate[1230,1231] support the earlier suggestions[1225] of specificity in the coordination of mida.

The oxygen-18 exchange of *trans*-(*fac*)-bis(*N*-methyliminodiacetato)chromate(III) has been studied in detail.[1232]. Acid-dependent and -independent paths are involved; one-ended ligand dissociation was held to be important. The equilibrium constants for complexation of ida are ($I = 0$, 25 °C) $\log K_1 = 9.30$ and $\log K_2 = 7.14$.[1233] The chromium complex of 2-hydroxyethyliminodiacetic acid (H_2OH-A), K[Cr(HO-A)$_2$],[1234] has been made. It is approximately octahedral and the hydroxy groups are not deprotonated.

(*v*) *Nitrilotriacetic acid*

Complexation of chromium(III) by nitrilotriacetic acid (H_2nta) has been investigated. Values of equilibrium constants ($I = 0$, 25 °C) are $\log K_1 = 10.66$ and $\log K_2 = 8.73$.[1234] A large number of mixed complexes with nta/chromium(III) have been prepared, including mixed aminoacidates,[1235] acetylacetonates, catecholates, oxalates and complexes with 1,10-phenanthroline and 2,2′-bipyridyl.[1236–1238]

(*vi*) *Sulfur-containing ligands*

Two different isomers (**274** and **275**) have been obtained from the reaction of thiobis(ethylenenitrilo)tetraacetic acid (tedta) with $Cr(ClO_4)_3$, depending on the pH of the reaction. The *cis* isomer has pentadentate (*cf.* edta) ligation, the *trans* isomer is sexadentate; the equilibrium is pH-dependent.[1239] The related complexes of thiobis(ethylenenitrilo)-tetraacetic acid have also been investigated.[1240] The above studies were directed to finding ligands which would bind to the surface of electrodes by spontaneous absorption. Although the chemistry of these systems is complicated, this objective was achieved in both studies.

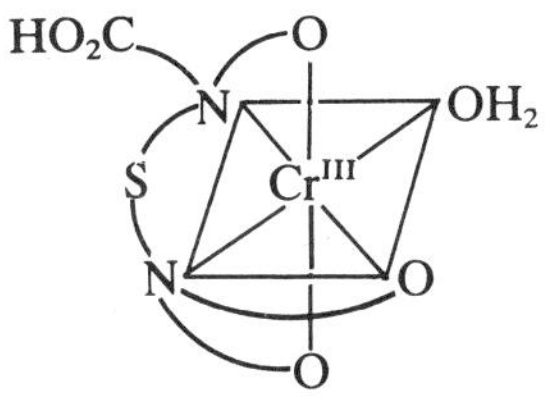

(274)

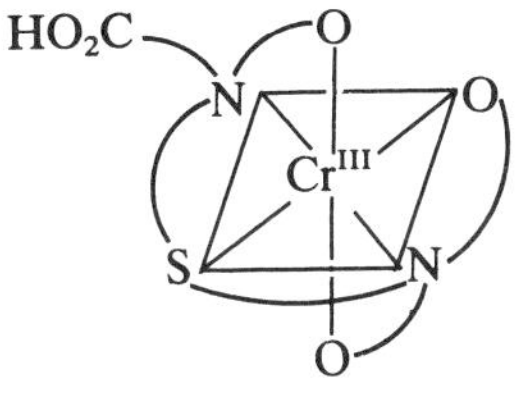

(275)

35.4.9 Multidentate Macrocyclic Ligands

35.4.9.1 Porphyrins and corrins: oxidation states II to V

Chromium porphyrins constitute a comparatively new and important class of compound. In order to avoid an artificial separation, complexes of various oxidation states starting with chromium(II) are dealt with in this section and some chromium(V) salen derivatives are included. Electronic, ESR and IR spectra are routinely recorded to characterize chromium porphyrin complexes and chromatography is extensively used in their purification; for details the original references should be consulted. The chemistry of the porphyrins is covered in a recent series.[1241]

(*i*) *Chromium(II) porphyrin complexes*

The chromium(II) complex of mesoporphyrin IX dimethyl ester (**276**) can be prepared by metal insertion with an excess of chromium hexacarbonyl under nitrogen.[1242] After removal of solvent and extraction with toluene, Cr^{II}(MPDME) is crystallized by the addition of *n*-pentane. Presumably the weakly acidic NH protons of the porphyrin oxidize Cr^0 to Cr^{II} and form H_2.

The oxidation of Cr(MPDME) in solution by O_2 or H_2O_2 can be reversed by sodium dithionite. Since $\mu_{eff} = 2.84$ BM in the solid and 5.19 BM in $CHCl_3$, there are axial interactions in the solid state which are removed on dissolution.

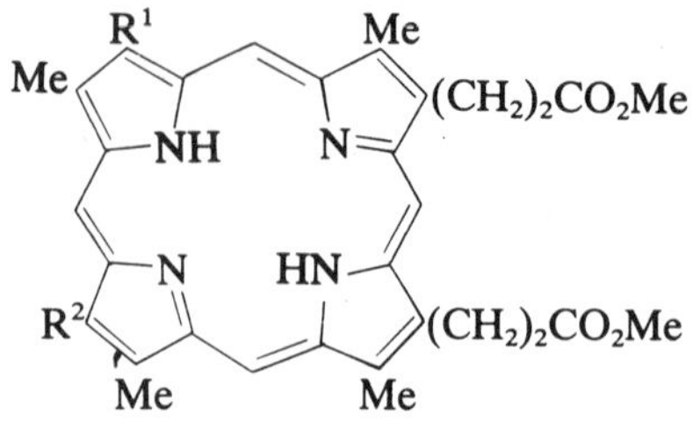

(**276**) $R^1 = R^2 =$ Et, mesoporphyrin dimethyl ester, H_2MPDME
$R^1 = R^2 =$ H, deuteroporphyrin dimethyl ester, H_2DPDME
$R^1 = R^2 = CH{=}CH_2$, protoporphyrin IX dimethyl ester, H_2PPDME
$R^1 = R^2 =$ CH(OH)Me, hematoporphyrin dimethyl ester, H_2HPDME

The *meso*-tetraphenylporphyrinate, Cr(TPP), first reported as a tetrahydrate[1243] with $\mu_{eff} = 4.90$ BM, was prepared by metal insertion into H_2TPP (**277**) with $Cr(CO)_6$ in *n*-butyl ether (Scheme 106). The reaction was carried out under nitrogen containing small quantities of air, the need for which was not explained, and the formulation is in doubt because products from other procedures[80] have different visible spectra. It can be prepared by reduction of CrCl(TPP) with $NaBH_4$ in EtOH/$CHCl_3$,[595] activated zinc in THF/py/H_2O,[80] or [Cr(acac)$_2$] in toluene or THF. The last method is of general application to M^{III}Cl(porphyrin) complexes (equation 62) because [Cr(acac)$_2$] is a powerful one-electron reducing agent, soluble in toluene, which is oxidized to a soluble Cr^{III} species, presumably [CrCl(acac)$_2$], which firmly binds the Cl^-. The solvate Cr(TPP)·2PhMe is obtained and Cr(OEP) (**278**) can be prepared similarly. A disadvantage is that Cr(acac)$_2$ is very air-sensitive and an excess is required because it does not react stoichiometrically.

$$M^{III}Cl(porphyrin) + Cr(acac)_2 \longrightarrow M^{II}(porphyrin) + CrCl(acac)_2 \quad (62)$$

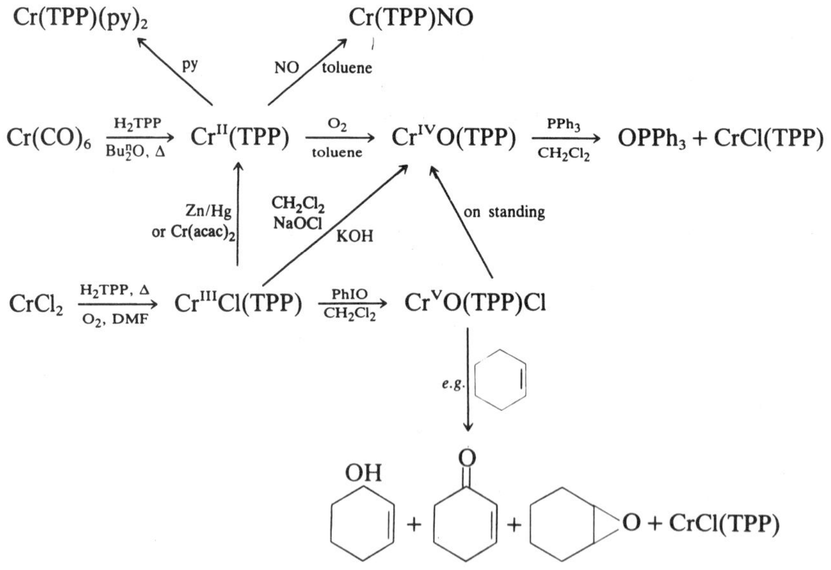

Scheme 106

From its magnetic moment (4.9 BM), Cr(TPP)·2PhMe is high spin. A crystal structure determination[1244] shows that the molecule is centrosymmetric and the Cr atom essentially four-coordinate; each toluene interacts very weakly with the metal and a pyrrole ring.

Planar Cr(TPP) will add two molecules of bases such as pyridine axially to form intermediate

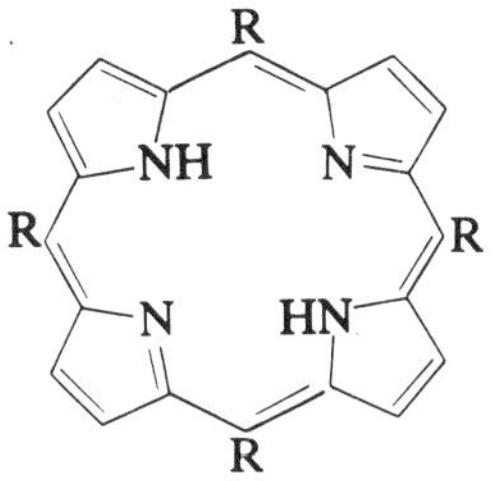

(277) R = Ph, tetraphenylporphyrin, H_2TPP
R = *p*-tolyl, tetra-*p*-tolylporphyrin, H_2TTP
R = mesityl, tetramesitylporphyrin, H_2TMP
R = *p*-$C_6H_4SO_3^-$, tetra-*p*-sulfonatophenylporphyrin, H_2TPPS

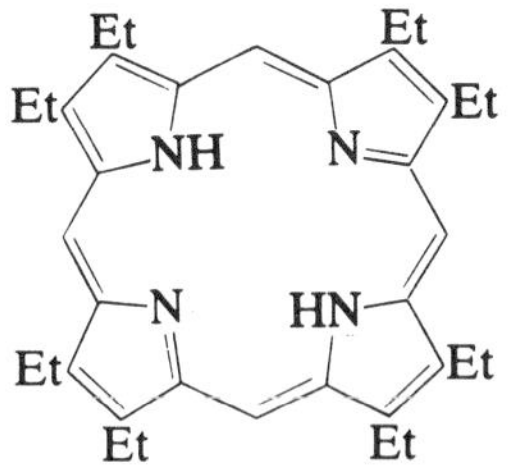

(278) Octaethylporphyrin, H_2OEP

spin complexes (Table 102). The spin-pairing has little effect on the average Cr—N (porphyrin) distances since they are 2.033 Å in Cr(TPP)·2PhMe ($S = 2$)[1244] and 2.027 Å in $Cr(TPP)(py)_2$ ($S = 1$).[1245] However, the Cr—N distances to the axial pyridine molecules (*ca.* 2.13 Å) are considerably less than in isoelectronic high-spin manganese(III) complexes (*ca.* 2.4 Å). This is attributed to the depopulation of the $3d_{z^2}$ orbital on spin-pairing in the Cr^{II} complex. The alternative formulation of the Cr^{II} porphyrin complexes as Cr^{III} porphyrin radical anion complexes is excluded by their visible spectra and oxidation–reduction behaviour.[80,1246]

The rate of oxidation of Cr(TPP) in solution can be lowered by cooling and by having high axial ligand concentrations, but no evidence has been found for reversible oxygenation. Powdered $Cr(TPP)(py)_2$ (the five-coordinate structure earlier proposed[1246] was in error[1244]) irreversibly forms an O_2 adduct which was thought from its magnetic moment (2.7 BM) and IR spectrum to be a Cr^{III} superoxide complex $Cr(O_2)(TPP)(py)$.[1246] In view of recent results (see below) which show that $Cr^{IV}O(TPP)$ and $(Cr^{III}TPP)_2O$ exist, it is possible that the adduct is a mixture. Carbon monoxide does not react with Cr(TPP),[1246] but Cr(TPP)NO can be obtained.[595,1247]

(*ii*) *Chromium(III) porphyrin complexes*

Evidence is growing that chromium is an essential trace element (Section 35.4.8.3). If mammalian diets are supplemented with simple ^{51}Cr-labelled salts it is found that less than 1% of the dose is absorbed because at intestinal pHs (>6) insoluble Cr^{III} hydrolysis products are produced. Complexes which remain intact in the acid conditions of the stomach as well as in more alkaline conditions are required. As Cr^{III} porphyrin complexes are stable in these respects, methods for synthesis of the complexes and incorporation of ^{51}Cr have been surveyed.[1248]

Chromium(III) porphyrin complexes have been synthesized *via* aerial oxidation of the chromium(II) complex prepared from $Cr(CO)_6$,[1242] from $CrCl_2$ and the porphyrin in refluxing DMF,[1249] and directly from $CrCl_3$ or $[Cr(acac)_3]$ in DMF[1250] and other high boiling solvents such as benzonitrile.[1251] Since the ^{51}Cr radiolabel is supplied as $CrCl_3$ in 0.1 M HCl, the carbonyl method and methods needing an excess of metal[1251] are unsuitable. To incorporate the label the $^{51}Cr^{3+}$ was reduced essentially completely by a 20-fold excess of $CrCl_2{\cdot}4H_2O$ in a 1:1 $EtOH/H_2O$ mixture according to the rapidly established equilibrium (63) for which the equilibrium constant is 1. The dilution of the ^{51}Cr was acceptable. DMF and the porphyrin were then added in an adaptation[1252] of the earlier method.[1249] The labelled complexes, shown in Table 102, were purified chromatographically. The method does not work well for porphyrins with free carboxylate side chains.

$$^{*}Cr^{III}Cl^{2+} + Cr^{II} \rightleftharpoons {}^{*}Cr^{II} + Cr^{III}Cl^{2+} \qquad (63)$$

Kinetic studies have shown that axial ligands in the complexes $Na_3[Cr(TPPS)(H_2O)_2]$, CrCl(TPP)L (L = monodentate base) and $[Cr(Schiff's\ base)(H_2O)_2]^+$ are unusually labile,[1113,1253,1254] and equilibrium constants for proton dissociation from coordinated water in the first complex[1252] have been determined.

The chloro complex CrCl(TPP) is a non-electrolyte in DMSO and neutral N, O and S donor ligands will bind to give six-coordinate species.[1255] There is little difference in the stability

Table 102 Some Synthetic Details and Properties of Porphyrin and Cr^{V}–Salen Complexes

Complex	*Comments*	*Ref.*
Oxidation state II		
Cr(MPDME) Violet	$Cr(CO)_6$, H_2MPDME in *n*-decane (170 °C) or decalin (205 °C), $\mu_{eff} = 2.84$ (solid), 5.19 BM ($CHCl_3$)	1
Cr(OEP) Purple	[$Cr(acac)_2$] reduction of CrCl(OEP), $\mu_{eff} = 4.8$ BM	2
Cr(TPP)·2PhMe Purple	[$Cr(acac)_2$] reduction of CrCl(TPP), $\mu_{eff} = 4.9$ BM	2
	Cr—N (av), 2.033 Å; planar CrN_4	3
$Cr(TPP)(py)_2$·PhMe Green-purple	Cr(TPP)·2PhMe and py in toluene or CrCl(TPP), Zn, THF/py/H_2O, $\mu_{eff} = 2.93$ BM indep. of *T*; Cr—N(TPP) (av),	2
	2.033, Cr—N(py) (av), 2.13 Å	4
$Cr(TPP)(L)_2$·PhMe Green-purple	Cr(TPP)·2PhMe and L in toluene, L = 3-pic, 4-pic, 1-MeIm; $\mu_{eff} = 2.9$ BM	2
Oxidation state III		
CrCl(MPDME)	$CrCl_2·4H_2O$, H_2MPDME, reflux DMF, expose to air as for CrCl(TPP) below, ^{51}Cr incorporated	5
CrCl(DPDME)	As above	5
CrCl(PPDME)	As above	5
CrCl(HPDME)	As above	5
CrCl(TPP) Deep blue	$CrCl_2$ or $CrCl_2·4H_2O$, H_2TPP, reflux DMF, expose to air. ^{51}Cr incorporated (see text)	6, 7, 5
CrBr(TPP)	From CrCl(TPP), excess [NBu_4]Br in $CHCl_3$	7
Na_3[Cr(TPPS)]	$CrCl_2·4H_2O$, Na_4[TPPS], reflux DMF, expose to air; also from $Cr(CO)_6$; sometimes as $(H_2O)_2$, $\mu_{eff} = 3.87$ BM	5, 8, 9
Cr(OMe)(TPP)·2MeOH	$Cr(CO)_6$, H_2TPP, reflux decalin, expose to air, crystallized from MeOH, $\mu_{eff} = 3.63$ (294 K), 3.60 BM (79 K)	10
Cr(OEt)(TPP)·2EtOH	As above, crystallized from EtOH, $\mu_{eff} = 3.60$ (294 K), 3.59 BM (79 K)	10
Cr(OH)(TPP)·$2H_2O$	As above, reaction mixture left one month, no alcohols in work-up	10
$Cr(ClO_4)$(TPP) Dark green	CrCl(TPP) and $AgClO_4$ in THF; monodentate ClO_4	11
$Cr(O_2CMe)$(TPP)	$Cr(O_2CMe)_3$, H_2TPP, reflux PhCN	12
Cr(TPP)NO Red	$\nu(NO) = 1700\ cm^{-1}$, $S = \frac{1}{2}$ (ESR)	11, 13
$[Cr(TPP)]_2O$ Purple	Monohydrate, CrCl(TPP) in CH_2Cl_2, NaOH; CrO(TPP) + Cr(TPP); $\mu_{eff}^{300\ K} = 1.61$ BM	14, 15, 16
(TPP)CrOFe(TPP) Purple	Per dimer, $\mu = 3.12$ (300 K), 1.9 BM (4.2 K), $J = -153\ cm^{-1}$, $g = 2.04$; intermolecular; $J = -1.45\ cm^{-1}$, $g = 2.14$; Cr—O—Fe, 842 cm^{-1}. Also mixed porphyrins and (TPP)CrOFe(PC); Cr—O—Fe linear	7, 18
CrCl(TTP) Green	$CrCl_2$, H_2TTP, reflux DMF, NaCl aq, HCl, Poor analyses, occludes CH_2Cl_2	14
CrCl(TMP) Green	As above, poor analyses, decomp. to ?CrOH(TMP)	14
CrOH(TTP)$H_2O·H_2O$ Violet	$CrCl_3$, H_2TTP, PhCN, reflux in stream of N_2 to remove HCl, chromat., add NaOH	19
CrN_3(TTP) Violet	—	20
CrOH(OEP)·$0.5H_2O$ Violet	$CrCl_3$, H_2OEP, PhCN, reflux stream of N_2 to remove HCl, chromat., add NaOH	19
CrBr(OEP)py Dark brown	[$Cr(acac)_3$], H_2OEP, reflux diethylene glycol	21
Cr(OPh)(OEP)PhOH Violet	[$Cr(acac)_3$], H_2OEP, PhOH, 250 °C, $\mu_{eff} = 3.2$ BM	22
CrCl(OEP)	Prepared as CrCl(TPP)	7
Oxidation state(IV)		
CrO(OEP) Red	CrOH(OEP)·$0.5H_2O$, NaOCl and NaOH, ν(Cr═O), 1015 cm^{-1}, diam.	23
CrO(TPP) Red	Cr^{II}(TPP), O_2 in toluene, ν(Cr═O), 1020 cm^{-1}, diam., Cr—O, 1.62, Cr—N, 2.036, Cr *ca.* 0.5 Å above N_4 plane; also CrO(TFP), CrO(TXP)	15, 16
	CrCl(TPP), PhIO in CH_2Cl_2, KOH; similarly from NaOCl, Bu^tO_2H, or *m*-chloroperoxybenzoic acid, ν(Cr═O), 1025 cm^{-1}, $\nu(Cr═^{18}O)$, 981 cm^{-1}, diam.	14
CrO(TMP) Purple	CrCl(TMP), PhIO in CH_2Cl_2, similarly from Bu^tO_2H and KOH, ν(Cr═O), 1023 cm^{-1}, diam.	14

Table 102 (*continued*)

Complex	*Comments*	*Ref.*
CrO(TTP)	CrCl(TTP), PhIO in C_6H_6, ν(Cr=O), 1020 cm^{-1}, diam., Cr—O,	14
Lavender	1.572, Cr—N(av), 2.032, Cr 0.469 Å above N_4 plane	
Red	Cr(OH)(TTP)·$2H_2O$, PhIO in CH_2Cl_2 or NaOCl and NaOH, ν(Cr=O), 1020 cm^{-1}, diam.	23
Oxidation state (V)		
CrO(TPP)Cl	CrCl(TPP), PhIO in CH_2Cl_2, μ_{eff} = 2.05 BM, ν(Cr=O), 972 cm^{-1}	14, 24, 25
$[CrO(TPP)]^+$	generated electrochemically	30
Red solution		
CrO(TETMC)	$CrCl_2$, H_3TETMC, NaOAc, DMF, exposure to air, μ_{eff} =	26
Dark red brown	1.71 BM, ν(Cr=O), 964 cm^{-1}, g = 1.987	
CrN(OEP)	CrOH(OEP)·$0.5H_2O$ as below, g = 1.9825	19
Red		
CrN(TTP)	CrOH(TTP)·$2H_2O$, CH_2Cl_2/EtOH, aq. NH_3, NaOCl, g = 1.9825	19
Red-purple	Photolysis CrN_3(TTP), g = 1.985, ν($Cr\equiv{}^{14}N$), 1017 cm^{-1},	20
	ν($Cr\equiv{}^{15}N$), 991 cm^{-1}, Cr≡N, 1.565, Cr—Na(TTP)(av), 2.042,	
	Cr 0.42 Å above N_4 plane (C_6H_6 adduct), RR spectrum, CrN(TPP) and CrN(TMP) also known	27
Salen complexes		
$[CrO(salen)]PF_6$	$[Cr(salen)(H_2O)_2]PF_6$, PhIO in MeCN, g = 1.977, ν(Cr=O),	28
Dark brown	997 cm^{-1}; also CrO(salen)X, X = F, Cl, Br	
CrN(salen)	CrN_3(salen)·$2H_2O$, photolysis, ν(Cr≡N), 1012 cm^{-1}, g = 1.974	29
Red brown		

1. M. Tsutsui, R. A. Velapoldi, K. Suzuki, F. Vohwinkel, M. Ichikawa and T. Koyano, *J. Am. Chem. Soc.*, 1969, **91**, 6262.
2. C. A. Reed, J. K. Kouba, C. J. Grimes and S. K. Cheung, *Inorg. Chem.*, 1978, **17**, 2666.
3. W. R. Scheidt and C. A. Reed, *Inorg. Chem.*, 1978, **17**, 710.
4. W. R. Scheidt, A. C. Brinegar, J. F. Kirner and C. A. Reed, *Inorg. Chem.*, 1979, **18**, 3610.
5. D. P. Riley, *Synth. React. Inorg. Metal-Org. Chem.*, 1980, **10**, 147.
6. A. D. Adler, F. R. Longo, F. Kampas and J. Kim, *J. Inorg. Nucl. Chem.*, 1970, **32**, 2443.
7. D. A. Summerville, R. D. Jones, B. M. Hoffman and F. Basolo, *J. Am. Chem. Soc.*, 1977, **99**, 8195.
8. J. G. Leipoldt, R. van Eldik and H. Kelm, *Inorg. Chem.*, 1983, **22**, 4146.
9. E. B. Fleischer and M. Krishnamurthy, *J. Coord. Chem.*, 1972, **2**, 89.
10. E. B. Fleischer and T. S. Srivastava, *Inorg. Chim. Acta*, 1971, **5**, 151.
11. S. L. Kelly and K. M. Kadish, *Inorg. Chem.*, 1984, **23**, 679.
12. T. N. Lomova, B. D. Berezin, L. V. Oparin and V. V. Zvezdina, *Russ. J. Inorg. Chem.*, 1982, **27**, 383.
13. B. B. Wayland, L. W. Olson and Z. U. Siddiqui, *J. Am. Chem. Soc.*, 1976, **98**, 94.
14. J. T. Groves, W. J. Kruper, Jr., R. C. Haushalter and W. M. Butler, *Inorg. Chem.*, 1982, **21**, 1363.
15. J. R. Budge, B. M. K. Gatehouse, M. C. Nesbit and B. O. West, *J. Chem. Soc., Chem. Commun.*, 1981, 370.
16. D. J. Liston and B. O. West, *Inorg. Chem.*, 1985, **24**, 1568.
17. D. J. Liston, K. S. Murray and B. O. West, *J. Chem. Soc., Chem. Commun.*, 1982, 1109.
18. D. J. Liston, B. J. Kennedy, K. S. Murray and B. O. West, *Inorg. Chem.*, 1985, **24**, 1561.
19. J. W. Buchler, C. Dreher, K.-L. Lay, A. Raap and K. Gersonde, *Inorg. Chem.*, 1983, **22**, 879.
20. J. T. Groves, T. Takahashi and W. M. Butler, *Inorg. Chem.*, 1983, **22**, 884.
21. J. W. Buchler, G. Eikelmann, L. Puppe, K. Rohbock, H. H. Schneehage and D. Weck, *Liebigs Ann. Chem.*, 1971, **745**, 135.
22. J. W. Buchler, L. Puppe, K. Rohbock and H. H. Schneehage, *Chem. Ber.*, 1973, **106**, 2710.
23. J. W. Buchler, K.-L. Lay, L. Castle and V. Ullrich, *Inorg. Chem.*, 1982, **21**, 842.
24. J. T. Groves and W. J. Kruper, Jr., *J. Am. Chem. Soc.*, 1979, **101**, 7613.
25. J. T. Groves, W. J. Kruper, Jr., T. E. Nemo and R. S. Myers, *J. Mol. Catal.*, 1980, **7**, 169.
26. Y. Matsuda, S. Yamada and Y. Murakami, *Inorg. Chim. Acta*, 1980, **44**, L309.
27. C. Campochiaro, J. A. Hofmann, Jr. and D. F. Bocian, *Inorg. Chem.*, 1985, **24**, 449.
28. T. L. Siddall, N. Miyaura, J. C. Huffman and J. K. Kochi, *J. Chem. Soc., Chem. Commun.*, 1983, 1185.
29. S. I. Arshankow and A. L. Poznjak, *Z. Anorg. Allg. Chem.*, 1981, **481**, 201.
30. S. E. Creager and R. W. Murray, *Inorg. Chem.*, 1985, **24**, 3824.

constants for binding of s-$BuNH_2$ ($\log K = 4.28$) and π-acceptor ligands such as pyridine ($\log K = 4.28$) to CrCl(TPP) in acetone (equation 64).[1256] It therefore seems that metal to ligand π bonding is unimportant. However, 1-methylimidazole is anomalously strongly bound ($\log K = 6.71$), consistent with some ligand to chromium π bonding. N donors will displace O donors and S donors but not the coordinated chloride.

$$\text{CrCl(TPP)(acetone)} + \text{L} \underset{}{\overset{K}{\rightleftharpoons}} \text{CrCl(TPP)L} + \text{acetone} \qquad (64)$$

Cyclic voltammetry shows that the electrochemical reduction of CrCl(TPP) in non-aqueous media is a function of the solvent.[1257] In non-coordinating solvents, *e.g.* CH_2Cl_2, the Cr^{III}/Cr^{II} reduction (equation 65) is irreversible and the Cr^{II}/Cr^{II}-anion radical reduction (equation 66) is reversible. In weakly coordinating solvents, *e.g.* Me_2CO, the solvent L modifies process (65) through weak coordination to give CrCl(TPP)L, and in coordinating media such as DMF is involved in both reductions and renders equation (65) reversible. The effect of N donors was studied, and in $C_2H_4Cl_2$ and DMF the Cr^{II} and Cr^{II}-anion radical species axially coordinate two substituted pyridine molecules. From the reduction potentials, this stabilized the Cr^{III} oxidation state over the Cr^{II} state, and the Cr^{II} state over the Cr^{II}-anion-radical state. There was no indication that coordinating ligands would displace Cl^- from CrCl(TPP)L. In $Cr(ClO_4)(TPP)$ the axial interactions between the metal centre and the counterion are much less than in CrCl(TPP).[1247] Consequently it is easier to reduce and less easy to oxidize and species $Cr(TPP)(L)_2^+$ (where L = DMF, DMSO and py) form in solution.

$$Cr^{III}Cl(TPP) \longrightarrow Cr^{II}(TPP) + Cl^- \quad (65)$$

$$Cr^{II}(TPP) \longrightarrow [Cr(TPP)]^- \quad (66)$$

The reaction of NO with Cr^{II}(TPP) affords Cr(TPP)NO which is a red solid with $\nu(NO) = 1700\ cm^{-1}$ and an ESR spectrum typical of an $S = \frac{1}{2}$ species.[595] When solutions of Cr(OMe)(TPP) in $CHCl_3$/MeOH are exposed to NO, spectroscopic changes indicative of the formation of an NO complex occur. On degassing, the original spectrum reappears. Since the electronic and ESR spectra of the NO complex are similar to those of Cr(TPP)NO, it is believed that reductive nitrosylation has taken place according to equations (67) to (69). It is not stated whether Cr(TPP)NO as the solid or redissolved loses NO. More recent work[1247] shows that DMF, DMSO and py coordinate axially to Cr(TPP)NO in CH_2Cl_2 and formation constants have been determined; at high concentrations some displacement of NO occurs.

$$Cr(OMe)(TPP) + NO \rightleftharpoons Cr(OMe)(TPP)NO \quad (67)$$

$$Cr(OMe)(TPP)NO + MeOH \longrightarrow Cr^{II}(TPP) + MeONO + MeOH \quad (68)$$

$$Cr^{II}(TPP) + NO \longrightarrow Cr(TPP)NO \quad (69)$$

(*iii*) *Chromium*(*IV*) *and* (*V*) *porphyrin complexes*

Chlorotetraphenylporphyrinatochromium(III) dissolved in CH_2Cl_2 is oxidized by iodosylbenzene or *m*-chloroperbenzoic acid to Cr^VO(TPP)Cl, which is stable in solution for several hours (Scheme 106).[1258] The addition of alkenes rapidly regenerates CrCl(TPP) and gives alcohols and epoxides; similarly cyclohexanol gives cyclohexanone. The same products are obtained with catalytic amounts of CrCl(TPP), and FeCl(TPP) behaves in a like manner.[1259] The red complex CrO(TPP)Cl has been well characterized in solution by IR, magnetic (Table 102) and ESR investigations. As it forms in CH_2Cl_2, the typical ESR spectrum of CrCl(TPP) is replaced by a sharp signal near $g = 2$ with the expected hyperfine splittings for Cr^V.[1260] Details of the ESR spectra show that the unpaired electron is primarily on the metal so that CrO(TPP)Cl is an oxochromium(V) species and not a chromium(IV) porphyrin π cation radical. Coupling effects in the ESR spectra consequent upon the formation of $Cr^{17}O$(TPP)Cl (from $PhI^{17}O$) and substitution of Cl by $BuNH_2$, OH^- or F^- confirm the formulation. On standing at room temperature CrO(TPP)Cl decomposes in CH_2Cl_2 to give a bright red solution containing a diamagnetic oxochromium(IV) complex. This happens so readily that the spectrum of Cr^VO(TPP)Cl first published[1258] is, in fact, that of red $Cr^{IV}O$(TPP);[1261] CrO(TPP)Cl is also red but the spectrum is different.[1250,1260] $Cr^{IV}O$(TPP) can be directly obtained by oxidation of CrCl(TPP) by iodosylbenzene, Bu^tO_2H, NaOCl or *m*-chloroperbenzoic acid, followed by treatment with base (Scheme 106); CrO(TMP), CrO(TTP) and CrO(OEP)[1250,1261] and others have been prepared similarly (Table 102).

In the crystalline state, the oxochromium(IV) complexes are stable for months. Unlike oxochromium(V) species they do not oxidize hydrocarbons, even in the presence of PhIO, although CrO(TPP) will oxidize $PhCH_2OH$ to PhCHO.[1261] Another route to CrO(TPP) is from Cr^{II}(TPP) by reaction with oxygen.[1262] The structures of CrO(TPP)[1262] and CrO(TTP)[1250] are similar to those of other five-coordinate oxometalloporphyrins. In both complexes the metal atom is ~0.5 Å out of the porphyrin plane towards the oxygen atom, and the Cr—O distances are similar: 1.572 Å [CrO(TTP)] and 1.62 Å [CrO(TPP)]. The average Cr—N distances are almost identical at *ca.* 2.03 Å. The Cr^{II}(TPP) was prepared by zinc amalgam or $Cr(acac)_2$ reduction of CrCl(TPP) and its oxidation to CrO(TPP) was monitored spectrophotometrically. The reaction (equation 70) takes place in two distinct stages believed to involve a μ-oxo intermediate,[1262] although others consider that autoxidation of the solvent toluene is ıvolved.[1250] $Cr^{IV}O(TPP)$ oxidizes $Fe^{II}(TPP)(pip)_2$ (pip = piperidine) to the mixed-metal μ-oxo ɔmplex $(TPP)Cr^{III}OFe^{III}(TPP)$ which contains a strongly antiferromagnetically coupled $S = \frac{3}{2}, S = \frac{5}{2}$) pair (Table 102). Similar complexes can be obtained with two different ɜtraphenylporphyrins, and $Cr^{IV}O(TPP)$ can also be prepared by stirring CrCl(TPP), $CHCl_3$ nd aqueous HCl in air.[1263]

$$4Cr^{II}(TPP) \xrightarrow{O_2} 2\{Cr^{III}(TPP)\}_2O \xrightarrow{O_2} 4Cr^{IV}O(TPP) \tag{70}$$

Oxidation of CrCl(TPP) in CH_2Cl_2 by the radical cation from phenoxathiin hexachloroanmonate produces a species with $\mu_{eff} = 2.8$ BM. The electronic spectrum is consistent with $Cr^{III}(TPP^{\cdot +})$ or $Cr^{IV}(TPP)$ and the former is favoured.[1264] An IR band in the 1270 cm^{-1} region is ıdicative of a ring-oxidized metalloporphyrin, and there is a strong band at 1285 cm^{-1} in the pectrum of $[CrCl(TPP)][SbCl_6]$.[1265]

One oxochromium(V) complex, CrO(TETMC), containing the trinegative anion of a corrole **279**), has been characterized as the solid.[1266] It is prepared (Table 102) simply by exposure to ir of a solution presumably containing a Cr^{II} complex. Aerial oxidation of $Cr^{II}(TPP)$ produces he oxochromium(IV) complex CrO(TPP) so the corrole ligand apparently facilitates autoxidaion. The redox behaviour of CrO(TETMC) has been examined by cyclic voltametry.[1267]

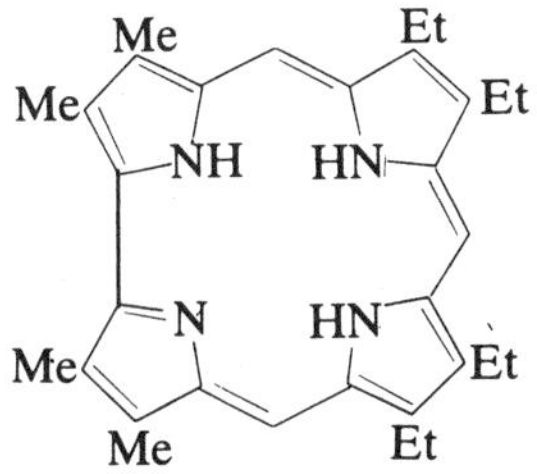

(**279**) 7,8,12,13-Tetraethyl-2,3,17,18-tetramethylcorrole, H_3TETMC

Nitridochromium(V) porphyrins, stable to air, moisture and mild reducing agents unlike xochromium-(IV) and -(V), have been prepared by two methods (equations 71[1268] and 2).[654,1269] The Cr^{III} porphyrins probably take up NH_3 as one axial ligand and oxidative lehydrogenation produces the nitride (equation 73).

$$CrOH(TTP) + NaOCl + NH_3 \longrightarrow CrN(TTP) + NaCl + 2H_2O \tag{71}$$

$$CrN_3(TTP) \xrightarrow[h\nu]{CH_2Cl_2 \text{ or } C_6H_6} CrN(TTP) + N_2 \tag{72}$$

$$\underset{Cr^{III},\, d^3}{HO\!-\!:\!Cr\!\cdot\!-\!NH_3} \xrightarrow[-H_2O]{-2H} \underset{Cr^{V},\, d^1}{\cdot Cr\!\equiv\!N:} \tag{73}$$

The nitrido complexes have been thoroughly characterized. The oxidation state has been confirmed by ESR and ENDOR spectroscopy. The $Cr\equiv N^{2+}$ group can be considered weakly electron donating; only the d_{xy} orbital is not involved in σ or π bonding, and it accommodates the single d electron. The stability to reduction compared with $Cr=O^{2+}$ species is ascribed to the high energy and low electron affinity of the empty d orbitals.[1268] The expected square pyramidal structure has been confirmed by X-ray methods for $CrN(TTP)\cdot C_6H_6$.[1269] It is isostructural with CrO(TPP). Coordination shell bond distances are given in Table 102, and the $Cr\equiv N$ distance is short, consistent with a triple bond.

(*iv*) *Chromium*(*V*)–*salen derivatives*

Chromium(III) salen will form chromium(V) derivatives in much the same way as porphyrin complexes do. By reaction of iodosylbenzene with chromium(III)salen oxochromium(V) complexes are obtained (equation 74) which will effect oxygen atom transfer to phosphines and alkenes such as norbornene (equation 75) in stoichiometric and catalytic systems.[1270] The halides [CrO(salen)]X (X = F, Cl, Br) can be prepared from $[CrO(salen)]PF_6$ by metathesis with $[NR_4]X$.

$$[Cr^{III}(salen)(H_2O)_2]PF_6 + PhIO \xrightarrow[MeCN]{-H_2O} [Cr^{V}O(salen)]PF_6 + PhI \quad (74)$$

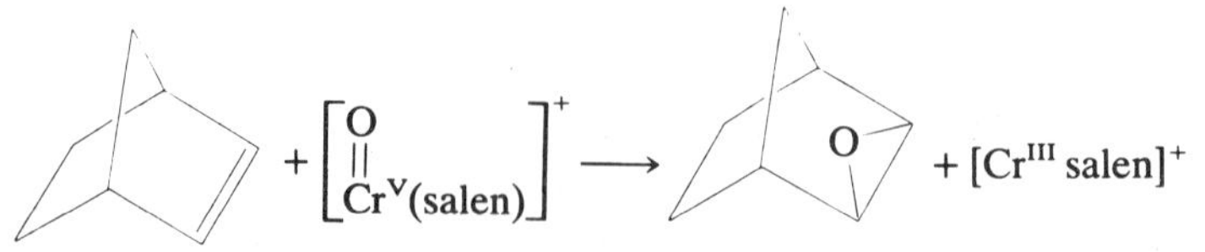

(75)

There is disorder in the crystal structure of $[CrO(salen)]PF_6$, but the average out-of-plane displacement (which takes place on oxygen transfer to the metal) of the Cr atom is 0.52 Å, and the average Cr=O distance 1.56 Å.

The nitrido complex CrN(salen) can be obtained from $[CrN_3(salen)]\cdot 2H_2O$ by photolysis (equation 72).[653] The ν(Cr≡N) frequency is 1012 cm^{-1}.

35.4.9.2 Polyazamacrocycles: cyclam, tet a, *tet* b *and related ligands*

In comparison with some other metal ions, limited synthetic work has been carried out on chromium(III) complexes with macrocyclic ligands. The vast majority of reported complexes involve tetraazamacrocycles.

(*i*) *Cyclam*

A number of complexes of the type $[Cr(cyclam)X_2]^+$ (cyclam = 1,4,8,11-tetraazacyclotetradecane, **82**) have been prepared as outlined in Scheme 107.[1271–1273] The difficulty in synthesizing *trans*-$[CrX_2(cyclam)]^+$ complexes directly from the free cyclic ligand is in marked contrast to the behaviour of other ions of the first transition series, *e.g.* cobalt(III), which forms stable *trans* complexes. The colours and spectroscopic properties of these cyclamchromium(III) complexes are presented in Table 103. The use of IR spectroscopy to differentiate between *cis* and *trans* cyclam complexes has been widely explored and the region 790–910 cm^{-1} was found to provide confirmatory evidence of structure, independent of metal or counteranions present.[1272] Hence *trans* complexes generally show two groups of bands in this region, a doublet near 890 cm^{-1} associated with the secondary amine vibration and a singlet near 810 cm^{-1} due to the methylene vibration. The spectra of *cis* isomers on the other hand are more complex with at least three bands between 840 and 890 cm^{-1} and two between 790 and 810 cm^{-1} (Table 104).

Table 103 Colours and Electronic Spectra of Cyclamchromium(III) Complexes

Complex	*Colour*	λ_{max}[a] (ε)[b]	λ_{min}[a] (λ)[b]	*Ref.*
cis-$[CrCl_2(cyclam)]^+$	Dark red	529 (111) 404 (106)	468 (26)	1
cis-$[CrBr_2(cyclam)]^+$	Grey-violet	527 (94) 408 (72)	455 (48) 356 (38)	1
cis-$[Cr(OH)_2(cyclam)]^+$		547 (84) 370 (66)		1
cis-$[Cr(cyclam)(H_2O)_2]^{3+}$ [c]	Orange	483 (126) 370 (38)	417 (26)	1
cis-$[Cr(NCS)_2(cyclam)]^+$	Orange-red	486 (189) 368 (101)	418 (35)	1
cis-$[Cr(ONO)_2(cyclam)]^+$ [d]	Red-brown	481 (134) 355 (211)	415 (29)	1
cis-$[Cr(N_3)_2(cyclam)]^+$	Red-violet	517 (276) 400 (144)	445 (60)	1
cis-$[CrCl_2(Me_2cyclam)]^+$		559 (123) 412 (97)		2
cis-$[Cr(Me_2cyclam)(H_2O)_2]^{3+}$		506 (75) 380 (53)		2
trans-$[CrCl_2(cyclam)]^+$	Purple	567 (20) 404sh (30), 364 (33.5)		3
trans-$[Cr(NCS)_2(cyclam)]^+$	Orange	483 (139) 364sh (117)		3
trans-$[Cr(cyclam)(H_2O)_2]^{3+}$		510 (24) 405 (39) 350 (53)		5
trans-$[CrCl_2(Me_2cyclam)]^+$		571 (20) 386 (31)		2
trans-$[Cr(CN)_2(cyclam)]^+$	Yellow	414 (62.5) 328 (62.5)	365 (22)	4

[a] λ in nm of aqueous solutions. [b] ε in $dm^3\ mol^{-1}\ cm^{-1}$. [c] $pK_1 = 3.8$, $pK_2 = 7.0$, 20 °C, $I = 0.1$ M. [d] Dinitrito complex, since spectrum is similar to that of *cis*-$[Cr(en)_2(ONO)_2]^+$ which is known to contain O-bonded NO_2^- ligands.

1. J. Ferguson and M. L. Tobe, *Inorg. Chim. Acta*, 1970, **4**, 109.
2. D. A. House, R. W. Hay and M. A. Ali, *Inorg. Chim. Acta*, 1983, **72**, 239.
3. C. K. Poon and K. C. Pun, *Inorg. Chem.*, 1980, **19**, 568.
4. N. A. P. Kane-Maguire, J. A. Bennett and P. K. Miller, *Inorg. Chim. Acta*, 1983, **76**, L123.
5. R. G. Swisher, G. A. Brown, R. C. Smierciak and E. L. Blinn, *Inorg. Chem.*, 1981, **20**, 3947.

Table 104 IR Spectra of $[Cr(cyclam)X_2]^+$ Isomers Between 790–910 cm^{-1} [1]

Complex	NH vibration	CH_2 vibration
trans-$[CrCl_2(cyclam)]Cl$	890s, 882s	804s
cis-$[CrCl_2(cyclam)]Cl$	872m, 862m, sh, 854m	815w, 805m
trans-$[Cr(NCS)_2(cyclam)]SCN$	885s, 878m	802m
cis-$[Cr(NCS)_2(cyclam)]SCN$	870m, 860m, 850m, sh	818m, 810m

1. C. K. Poon and K. C. Pun, *Inorg. Chem.*, 1980, **19**, 568.

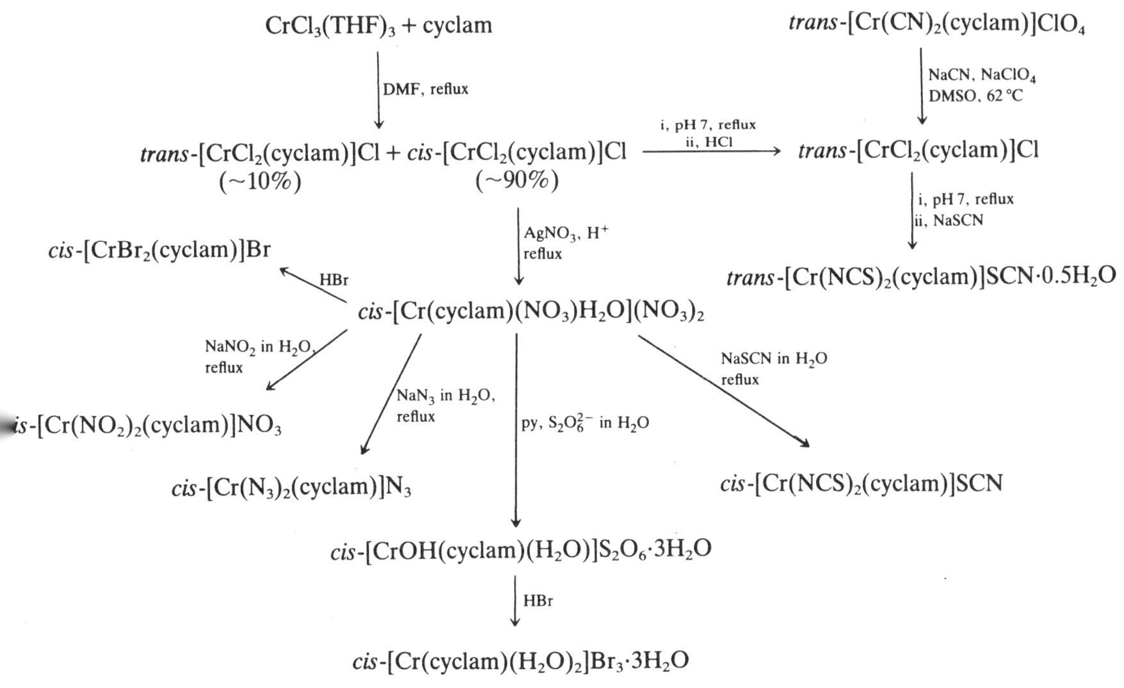

Scheme 107

In the attempted preparation of *trans*-$[Cr(cyclam)(NH_3)_2]^{3+}$ from the dichloro complex in liquid NH_3 only a crude product contaminated by an orange material could be isolated.[1274] This impurity was shown to contain the *O*-carbamato complex *trans*-$[Cr(OCONH_2)_2(cyclam)]^+$. Suitable crystals of *trans*-$[Cr(OCONH_2)_2(cyclam)]ClO_4\cdot1.5H_2O$ were grown and the structure determined by X-ray diffraction methods. The structure contains two independent centrosymmetrical complex ions, which are almost identical except for the orientation of the carbamate ligands. The macrocyclic ligand has a conformation in which the five-membered ring has a *gauche* and the six-membered ring a chair conformation. The average C—C (151.3 pm), C—N (148.0 pm) and Ī—N (205.9 pm) distances in the complex are similar to those in the doubly protonated ligand (150.9, 148.2, 204.5 pm respectively). The average Cr—N bond length (205.9 pm) is close to the ideal value (207 pm) calculated for *trans*-cyclam complexes[1275] and also similar to those in chromium(III) complexes with other amines.[1274] The relative substitution inertness of *trans*-cyclamchromium(III) complexes in solution[1276] may be due to steric hindrance by the two axial C—H groups of the six-membered rings towards the entering group in an associative interchange mechanism.[1274]

The photochemistry of *cis*- and *trans*-$[CrCl_2(cyclam)]^+$ on irradiation of the *d*–*d* bands in dilute acid solution involves monosubstitution by solvent to give the corresponding isomeric $[CrCl(cyclam)(H_2O)]^{2+}$ products.[1277] However, the photobehaviour of $[Cr(CN)_2(cyclam)]^+$ in solution at room temperature is in marked contrast to that normally displayed by chromium(III) complexes.[1273,1278] For this complex, like *trans*-$[Cr(CN)_2(NH_3)_4]^+$,[329] the $^4B_{2g}$ component of the first spin-allowed quartet excited state lies at lower energy than the ^4E_g component, an ordering which in the tetraammine complex restricts photolabilization to the in-plane NH_3 positions. However, in the cyclam complex the nature of the in-plane ligand prevents the occurrence of such a reaction with the result that even after extensive ligand field photolysis no substitution photochemistry is detected. Furthermore the complex exhibits an intense long-lived $^2E_g \rightarrow {}^4B_{1g}$ phosphorescence which shows equal enhancements in intensity and lifetime on deuteration of the NH groups. These properties of its ^2E_g state make *trans*-[Cr(CN)$_2$(cyclam)]$^+$ an attractive compound to study in terms of energy- and electron-transfer with various substrate molecules.

(ii) Tet a and tet b

A number of chromium(III) complexes of the diastereoisomeric C-*meso*- and C-*racemo*-5,7,7,12,14,14-hexamethyl-1,4,8,11-tetraazacyclotetradecane (tet *a* and tet *b* respectively; **280** and **281**) have been prepared.[1279,1280] While the tet *b* isomer like cyclam can readily fold to give *cis*-[CrX$_2$(tet *b*)]$^+$ complexes, tet *a* only folds with difficulty because of unfavourable interactions between the methyl substituents and axially coordinated ligands. Hence these isomers react with CrCl$_3$(DMF)$_3$, prepared *in situ* by dehydration of CrCl$_3$·6H$_2$O in DMF, according to Scheme 108. Replacement of Cl$^-$ by other anions provides a convenient route to a range of *trans*-[CrX$_2$(tet *a*)]X and *cis*-[CrX$_2$(tet *b*)]X complexes. Details of the electronic spectra of these products are presented in Table 105.

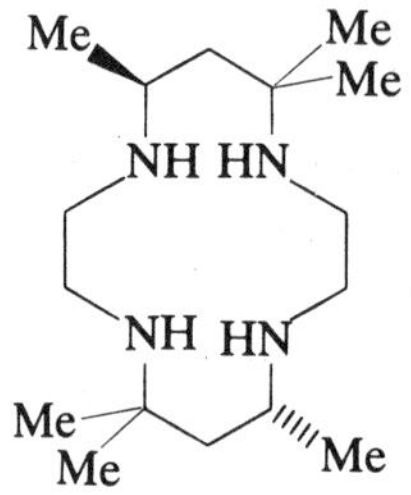

(280) teta *a* (C-*meso*)   **(281)** tet *b* (C-racemic)

trans-[CrCl$_2$(tet *a*)]Cl·HCl·2H$_2$O
green; $\lambda_{max}(\varepsilon) = 574$ (25), 440sh (27), 387 (47) nm (dm^3 mol^{-1} cm^{-1})

CrCl$_3$·6H$_2$O $\xrightarrow{\text{DMF, boil}}$ CrCl$_3$(DMF)$_3$

tet *a* in DMF 100 °C

tet *b* in DMF, boil

cis-[CrCl$_2$(tet *b*)]Cl
sea-green; $\lambda_{max} = 598$, 430 nm^b

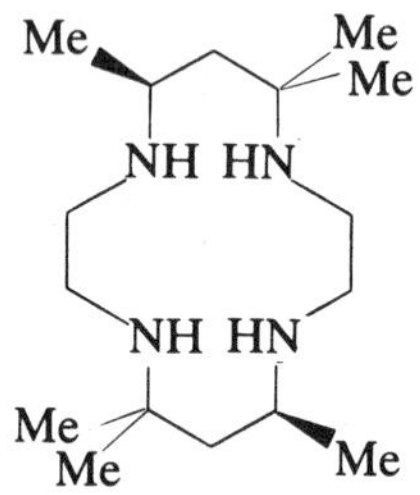

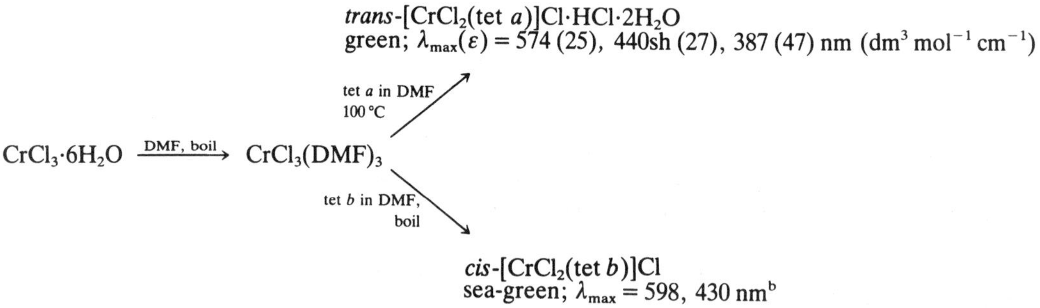

^{a} In DMF. ^{b} Nujol mull spectrum.

Scheme 108

The first step in the base hydrolysis of *trans*-[CrCl$_2$(tet *a*)]$^+$ (Scheme 109; $k_1 = 145\,\text{dm}^3\,\text{mol}^{-1}\,\text{s}^{-1}$ at 298 K, ionic strength 0.1 M; $\Delta H^{\ddagger} = 114\,\text{kJ}\,\text{mol}^{-1}$, $\Delta S^{\ddagger} = 179\,\text{J}\,\text{K}^{-1}\,\text{mol}^{-1}$)[1281] proceeds some 112-fold faster than the corresponding reaction of *trans*-[CrCl$_2$(cyclam)]$^+$ ($k_1 = 1.3\,\text{dm}^3\,\text{mol}^{-1}\,\text{s}^{-1}$ at 298 K).[1276] On the basis of the steric acceleration due to C-methyl substitution, the first-order dependence of rate on [OH$^-$] and the large positive entropy of activation an S_N1CB mechanism is suggested for this reaction.[1279] Acid hydrolysis of the tet *a* complex is also considerably faster than its cyclam analogue.[1279] The crystal structure of one of the possible isomers of *trans*-[CrCl(tet *a*)H$_2$O](NO$_3$)$_2$ has been reported.[1282] Addition of base to an aqueous solution of *cis*-[Cr(mal)(tet *b*)]$^+$ (mal = malonate) gives rise to a bathochromic shift in the longer wavelength band of its visible spectrum.[1280] This spectral change has been attributed to ionization of the CH$_2$ group in the malonate ligand (equation 76), although such an explanation must be treated with at least some scepticism since it implies an abnormally large increase in the acidity of this group due to coordination of the neighbouring CO$_2^-$ groups.

Table 105 Colours and Electronic Spectra of Chromium(III)–tet Complexes[1, 2]

Complex	*Colour*	λ_{max} (nm) (ε in $dm^3\ mol^{-1}\ cm^{-1}$)	*Medium*
trans-$[CrCl_2(\text{tet } a)]^+$	Green	574 (25) 440sh (27) 387 (47)	DMF
trans-$[Cr(NCS)_2(\text{tet } a)]^+$	Orange	500 420 322	Nujol[a]
trans-$[Cr(\text{tet } a)(H_2O)_2]^{3+}$	Pink	530 408 340	Nujol[a]
trans-$[Cr(\text{tet } a)(H_2O)_2]^{3+}$	Pink	535 (65) 423 (96)	MeOH
trans-$[CrBr_2(\text{tet } a)]^+$	Green	602 (30) 410sh 374 (47)	DMF
trans-$[CrBr_2(\text{tet } a)]^+$	Green	600 (33) 410sh (38) 382 (43)	MeOH
cis-$[CrCl_2(\text{tet } b)]^+$	Sea green	598 430	Nujol[a]
cis-$[Cr(NCS)_2(\text{tet } b)]^+$	Maroon	526 (210) 390 (99)	MeCN
cis-$[CrBr_2(\text{tet } b)]^+$	Magenta	595 438	Nujol[a]
cis-$[Cr(NO_3)_2(\text{tet } b)]^+$	Pink	525 392	Nujol[a]
cis-$[Cr(NO_3)_2(\text{tet } b)]^+$	Pink	524 (202) 390 (116)	DMF
cis-$[Cr(N_3)_2(\text{tet } b)]^+$	Blue-violet	570 420	Nujol[a]
cis-$[Crox(\text{tet } b)]^+$	Pink	522 390	Nujol[a]
cis-$[Crox(\text{tet } b)]^+$	Pink	516 (157) 384 (102)	H_2O or 1 M NaOH
cis-$[Cracac(\text{tet } b)]^{2+}$	Red	512 385 325	Nujol[a]
cis-$[Cracac(\text{tet } b)]^{2+}$	Red	535 (161) 390 (256)	1 M HCl or NaOH
cis-$[Cr(OH)_2(\text{tet } b)]^+$	Dark blue	609 (111) 380 (73)	1 M NaOH
cis-$[Cr(OH)(\text{tet } b)(H_2O)]^{2+}$	—	572 (130) 407 (53)	1 M NaCl
cis-$[Cr(\text{tet } b)(H_2O)_2]^{3+}$		529 (169) 388 (82)	$HClO_4/NaClO_4$
cis-$[CrCl(\text{tet } b)(H_2O)]^{2+}$	Violet	554 (173) 405 (105)	1 M HCl
cis-$[CrF_2(\text{tet } b)]^+$	Violet	541 (145) 379 (58)	1 M HCl or NaOH
cis-$[Crmal(\text{tet } b)]^+$	Violet	527 (167) 387 (82)	1 M HCl

[a] Because of insolubility in the common solvents, spectra of Nujol mulls on filter paper were recorded.
1. D. A. House, R. W. Hay and M. A. Ali, *Inorg. Chim. Acta*, 1983, **72**, 239.
2. J. Eriksen and O. Mønsted, *Acta Chem. Scand., Ser. A*, 1983, **37**, 579.

$$\textit{trans}\text{-}[CrCl_2(\text{tet } a)]^+ + OH^- \xrightarrow{k_1} \textit{trans}\text{-}[CrCl(OH)(\text{tet } a)]^+ + Cl^-$$

$$\downarrow OH^-, k_2$$

$$\textit{trans}\text{-}[Cr(OH)_2(\text{tet } a)]^+ + Cl^-$$

Scheme 109

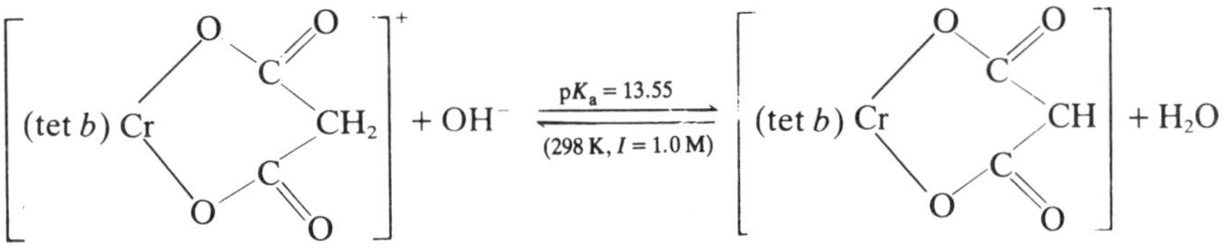

$\lambda_{max} = 527$ nm $\qquad$ $\lambda_{max} = 570$ nm

(76)

(iii) $[12]aneN_4$*, unsymmetrical* $[14]aneN_4$*,* $[15]aneN_4$

The preparation and properties of some chromium(III) complexes of 1,4,7,10-tetraazacyclododecane ($[12]aneN_4$) are outlined in Scheme 110.[1283] The reaction between $CrCl_3\cdot 3THF$ and the unsymmetrical 14-membered macrocycle 1,4,7,11-tetraazacyclotetradecane ($1,4,7,11[14]aneN_4$) in DMF leads to equal amounts of *cis*- and *trans*-$[CrCl_2(1,4,7,11[14]aneN_4)]Cl$ and these can be separated on the basis of the insolubility of the former and the solubility of the latter in methanol (Scheme 111). The reaction between $CrCl_3\cdot 3THF$ and the 15-membered macrocycle $[15]aneN_4$ (**81**) in DMF produces only the *trans* isomer of $[CrCl_2([15]aneN_4)]Cl\cdot 2H_2O$ (green; $\lambda_{max} = 586$ nm, $\varepsilon = 24\ dm^3\ mol^{-1}\ cm^{-1}$; 476 nm, $\varepsilon = 48\ dm^3\ mol^{-1}\ cm^{-1}$; 410 nm, $\varepsilon = 63\ dm^3\ mol^{-1}\ cm^{-1}$), which undergoes a slow one-stage hydrolysis in aqueous solution.[1283]

The reaction between the strongly reducing complex $[Cr([15]aneN_4)(H_2O)_2]^{2+}$ and organic halides (RX) produces the monoalkylchromium(III) complexes *trans*-$[RCr([15]aneN_4)H_2O]^{2+}$, equation (77).[305,1284] The reaction follows second-order kinetics and the reactivity of alkyl halides follows the pattern tertiary > secondary > primary and RI > RBr > RCl (Table 106). This and other information suggests a mechanism which involves rate-limiting halogen atom

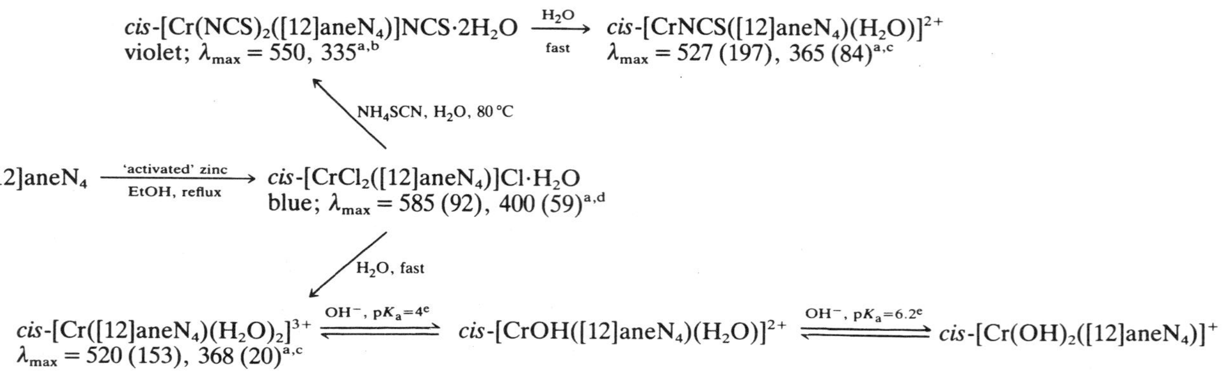

[a] λ_{max} in nm (ε in $dm^3\,mol^{-1}\,cm^{-1}$). [b] Reflectance spectrum. [c] In 0.1 M HNO_3 solution. [d] In MeOH solution. [e] $I = 0.5$ M, temperature not stated.

Scheme 110

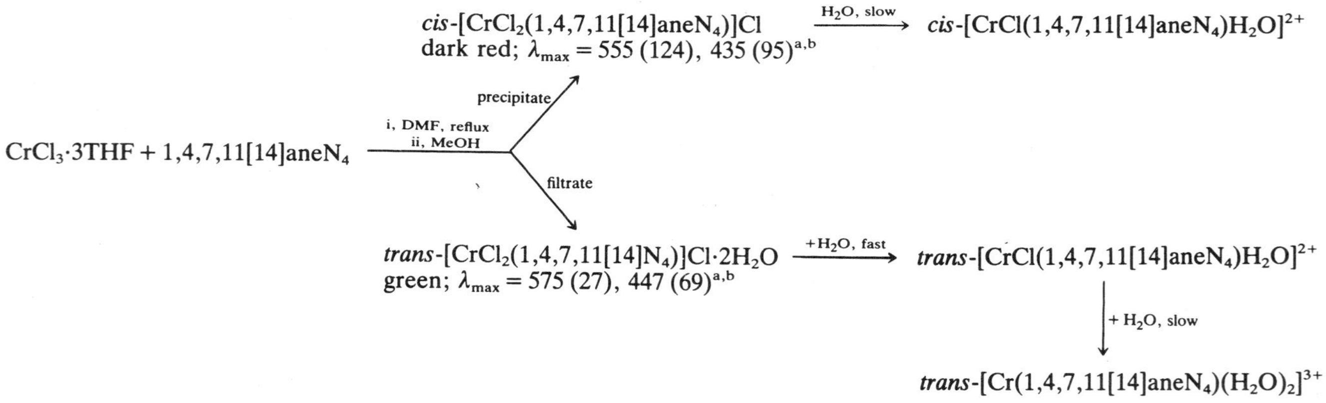

[a] λ_{max} in nm (ε in $dm^3\,mol^{-1}\,cm^{-1}$). [b] In MeOH.

Scheme 111

abstraction from RX by the chromium(II) complex to generate a carbon-centred free radical which rapidly couples with a second chromium(II) complex (Scheme 112). In the presence of electrophiles such as Hg^{2+} or $MeHg^+$, the alkylchromium(III) complexes undergo electrophilic dealkylation as in Scheme 113.[297] These reactions also follow a second-order rate law and the rate constants fall sharply with increasing bulk of R and by two to three orders of magnitude as the electrophile is changed from Hg^{2+} to $MeHg^+$ (Table 106). On the basis of such observations a bimolecular electrophilic substitution mechanism, S_E2 is proposed for these reactions. Furthermore these complexes are somewhat less reactive towards Hg^{2+} than the corresponding $[RCr(H_2O)_5]^{2+}$ species[1285] owing to the considerable steric bulk added to the site of reactivity by the macrocyclic ligand. Details of the electronic spectra of *trans*-$[RCr([15]aneN_4)H_2O]^{2+}$ and related complexes are listed in Table 107.

$$2[Cr([15]aneN_4)(H_2O)_2]^{2+} + RX \longrightarrow trans\text{-}[RCr([15]aneN_4)H_2O]^{2+} + trans\text{-}[XCr([15]aneN_4)H_2O]^{2+} \quad (77)$$

Table 106 Second-order Rate Constants for Alkylation of $[Cr([15]aneN_4)(H_2O)_2]^{2+}$ by RX (k_{RX}) and for Electrophilic Dealkylation of the Chromium(III) Products by Hg^{2+}(k_{Hg}) and $MeHg^+$(k_{MeHg}).

RX	k_{RX} [a,b]	k_{Hg} [a,c]	k_{MeHg} [a,c]
MeI	0.046	3.1×10^6	1.63×10^3
EtBr	0.164	2.53×10^3	9.9
EtI	0.413	—	—
Pr^nBr	0.169	8.21×10^1	—
Bu^nBr	0.138	4.88×10^1	—
$PhCH_2Cl$	3.23×10^2	1.14×10^3	5.2
$PhCH_2Br$	1.91×10^4		
Pr^iBr	1.91	4.3×10^{-3}	—
CyBr	0.836	1.6×10^{-3}	
1-AdamantylBr	0.98	3.1×10^{-3}	

[a] In $dm^3\,mol^{-1}\,s^{-1}$. [b] All at 298 K in 1:1 Bu^tOH/H_2O, $I = 0.2$ M. [c] All at 298 K, $I = 0.5$ M, $[H^+] = 0.25$ M.
1. G. J. Samuels and J. H. Espenson, *Inorg. Chem.*, 1979, **18**, 2587.
2. G. J. Samuels and J. H. Espenson, *Inorg. Chem.*, 1980, **19**, 233.

Table 107 Electronic Spectra[a] of *trans*-$[LCr([15]aneN_4)H_2O]^{2+}$ (L = alkyl, halide or aqua) Complexes[1]

L	λ_{max} (nm) (ε in $dm^3\,mol^{-1}\,cm^{-1}$)		
Me	468 (69)	375 (227)	258sh (3300)
Et	467 (66)	383 (387)	264 (3100)
Pr^n	468 (71)	383 (465)	265 (3440)
Bu^n	468 (69)	383 (459)	268 (3300)
Pr^i	510sh (69)	396 (550)	287 (3280)
Cy	500sh (67)	400 (422)	298 (3070)
1-Adamantyl	463 (88)	383 (347)	268 (3070)
$PhCH_2$	353 (2170)	297 (7470)	273 (7920)
Cl	564 (22)	462 (62)	393 (76)
Br	560 (23)	460 (68)	393 (78)
I[b]	562 (26)	462 (75)	394 (90)
H_2O[c]	540sh (28)	454 (87)	377 (88)

[a] In aqueous solution pH 3. [b] Complex thermally unstable. [c] 3+ complex.
1. G. J. Samuels and J. H. Espenson, *Inorg. Chem.*, 1979, **18**, 2587.

$$[Cr([15]aneN_4)(H_2O)_2]^{2+} + RX \xrightarrow{\text{slow}} trans\text{-}[XCr([15]aneN_4)H_2O]^{2+} + R\cdot$$

$$[Cr([15]aneN_4)(H_2O)_2]^{2+} + R\cdot \xrightarrow{\text{fast}} trans\text{-}[RCr([15]aneN_4)H_2O]^{2+}$$

Scheme 112

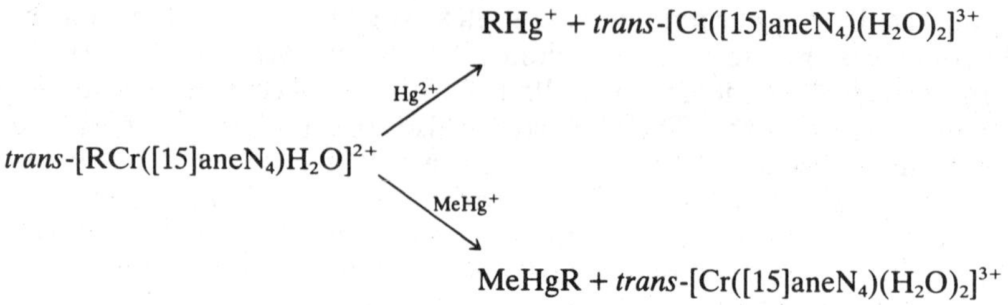

Scheme 113

(iv) Other macrocycles

A green chromium(III) complex (**282**) containing the dianionic unsaturated macrocycli ligand 5,14-dimethyl-1,4,8,11-tetraazacyclotetradeca-4,6,11,13-tetraenate, Me_2[14]tetraenateN_4 has been prepared by the method in Scheme 114.[598] This complex reacts with $NaNO_2$ i refluxing alcohol to give the mononitrosyl Cr(Me_2[14]tetraenatoN_4)NO (**137**), which on th basis of experimental evidence appears to contain chromium(I) and the cationic NO^+ ligand Section 35.4.2.6. Some chromium(III) complexes with macrocyclic ligands derived from N heterocycles are mentioned in Section 35.4.2.5. A number of chromium(III) complexes o the ligand 1,4,7-triazacyclononane, [9]aneN_3, and its *N,N,N*-trimethylated derivative Me_3[9]aneN_3, have been investigated.[472,1286] From $[Cr([9]aneN_3)(H_2O)_3]^{3+}$ the binuclea complex $[Cr(\mu\text{-OH})([9]aneN_3)H_2O]_2^{4+}$ and the trinuclear complex $[Cr_3(\mu$-$OH)_4OH([9]aneN_3)_3]^{4+}$ have been obtained and the crystal structure of the latter as the iodid salt (+$5H_2O$) has been determined by X-ray diffraction methods. This cation (**283**) contain two different pairs of μ-OH bridges with Cr—O—Cr bond angles of 98° and 126° and a ver short H bond which links the terminal and one of the bridging OH ligands. The crysta structures of the binuclear complexes $[Cr_2(\mu\text{-OH})_2(\mu\text{-CO}_3)([9]aneN_3)_2]I_2{\cdot}H_2O$ and $[Cr_2(\mu$-$OH)_3(Me_3[9]aneN_3)_2]I_3{\cdot}3H_2O$, the latter representing the first genuine tris(μ-hydroxo) comple of chromium, have also been reported.

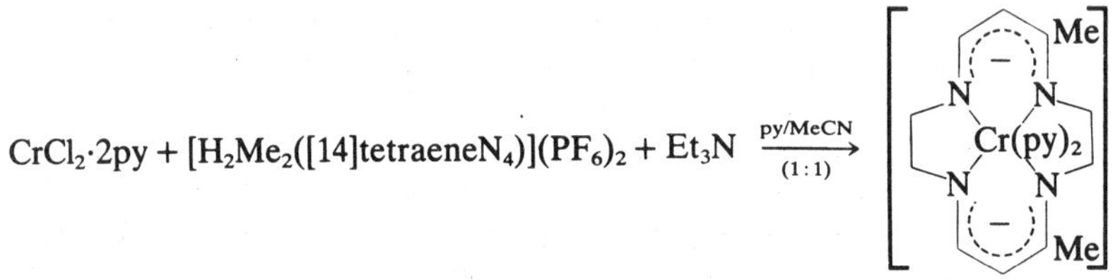

(**282**)

Scheme 114

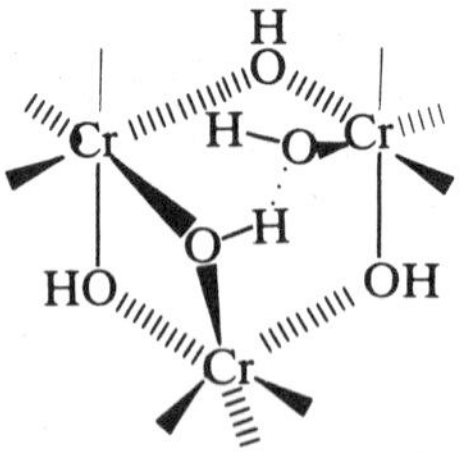

(**283**)

35.4.9.3 Phthalocyanines

Relatively few chromium complexes with phthalocyanine ligands have been reported and, i view of the chemical relationship between the few known compounds, all of them, irrespectiv of the metal oxidation state, are dealt with in this section. In the following discussion Pc represents the dianionic phthalocyanine ligand.

(i) CrPc and $[CrPc(\mu\text{-}O)]_2$

The chromium(II) complex CrPc can be prepared by either of the routes in Scheme 115.[1287–1290] While the $Cr(CO)_6$ method conveniently gives the β polymorph of CrPc (40% yield) and this on sublimation at 300–350 °C can be converted to α-CrPc,[1288] the chromium(III) acetate method gives, after washing the product mixture with organic solvents, both CrPc and an oxidized species which was long thought to be the chromium(III) complex CrPc(OH).[1287,1290] Repeated sublimation of this mixture at 400 °C/10^{-6} mmHg was reported to give the pure hydroxide. In contrast to α-CrPc, which is air stable, the β-polymorph is air sensitive and under anhydrous conditions picks up O_2 (0.5 O_2 per chromium) to give a deep blue product, which on the basis of available experimental evidence seems to be the di-μ-oxochromium(IV) dimer (**284**) rather than CrPc(OH).[1288] Hence the product: (i) lacks characteristic OH vibrations in its IR spectrum, (ii) shows no evidence for $CrPc(OH)^+$ or the characteristic fragments of this ion in its mass spectrum, and (iii) is formed in the absence of water. Furthermore ^{18}O isotopic substitution effects on the IR spectra support a structure containing a Cr=O group and eliminate possible structures such as those with peroxo bridges or those with polymeric —Cr—O—Cr—O— groups. The electronic spectra of the oxidation product and CrPc are quite similar in band positions and shapes, thus eliminating the phthalocyanine ring as the possible site of oxidation.

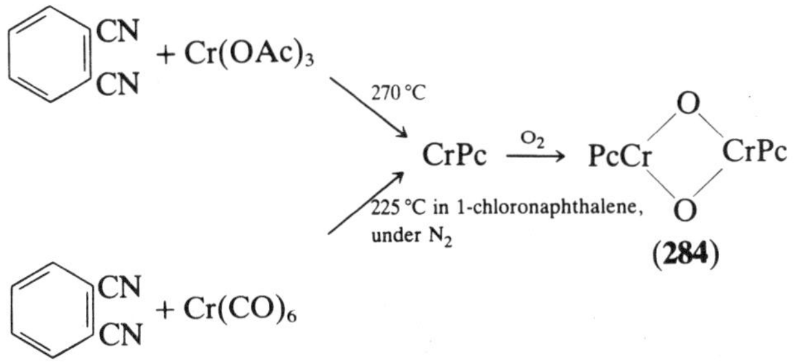

(**284**)

Scheme 115

The magnetic moment of the oxidation product is 1.9 BM at room temperature ($\mu_{\text{spin only}}$ for a d^2 ion = 2.9 BM) and is temperature-dependent, pointing to the existence of antiferromagnetic coupling between the chromium ions.[1288] The reflectance spectrum of the product has bands at 755sh, 720, 660sh, 490 and 380 nm. Its Raman spectrum has a band at 1041 cm^{-1}, which shifts by 45 cm^{-1} on ^{18}O isotopic substitution and which seems to be due to a Cr=O stretching vibration. The fragmentation pattern in the mass spectrum is similar to that of PcCr except for a group of peaks between 578 and 582 thought to originate from $CrPcO^+$. When ^{18}O is used as oxidant these peaks are observed in the range 580–594. Some of the many reported reactions of the oxidation product are illustrated in Scheme 116.[1287,1291] The magnetic moment of the high-spin CrPC (3.49 BM at 298 K) falls sharply with decreasing temperature ($\theta = 306$ K) to below the spin-only value for a low spin complex.[1290] This behaviour has been attributed to antiferromagnetic interactions ($J = -38.2°$, $g = 2.0$) in stacks of planar molecules. By contrast the relatively air-stable bis(pyridine) adduct, $CrPc(py)_2$, is low spin (μ_{eff} at 298 K = 3.16 BM) and shows a much lower temperature dependence ($\theta = 35$ K).

(ii) CrPc(NO)

Solid β-CrPc reacts reversibly with gaseous NO at ambient temperature to give a 1:1 adduct, which reacts with both py and O_2 (Scheme 117).[596] No structural information is available on the oxidation product of the nitrosyl complex. In contrast to the β form, solid α-CrPc is unreactive towards either O_2 or NO, a difference ascribed to the solid-state structures of the two polymorphs.[597] Although the distance between the phthalocyanine planes is similar (340 pm), in both cases the metal–metal distance increases from 340 to 480 pm on going from the α to the β form. These distances are sufficient to permit access of O_2 or NO in the β form, while excluding them from the α form. The addition of py to CrPc(NO) causes the NO stretching band in the IR spectrum to intensify and to sharpen considerably. The diffuse nature of this band in CrPc(NO) indicates the presence of a perturbed NO ligand but the introduction of py causes lattice expansion and eliminates the perturbation.

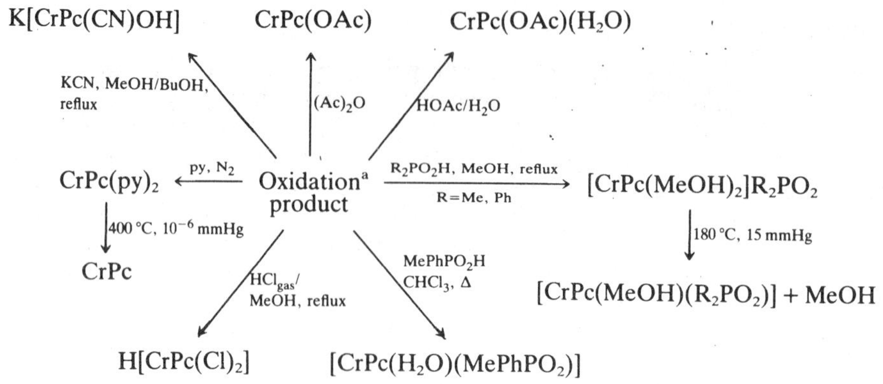

[a] In the references to this work the oxidation product was described as CrPc(OH) but was later described as the dimer $[CrPc(\mu\text{-}O)]_2$.

Scheme 116

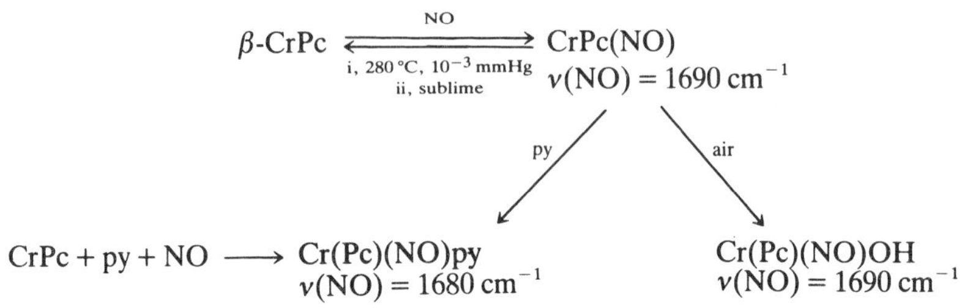

Scheme 117

(iii) *Oxidation and reduction of CrPc*

Reduction of CrPc with Na gives different products in THF and HMPA.[1292,1293] In THF the product obtained is $[Cr^{I}Pc]^-$, which has an ESR spectrum consistent with the low-spin chromium(I) configuration $e_g^4b_{2g}^1$. The product in HMPA on the other hand has an ESR spectrum consistent with the low-spin configuration $b_{2g}^2e_g^3$. The inversion of orbital energies is due to axial binding of a π donor solvent molecule in HMPA but not in THF. Two-electron reduction of CrPc or $CrCl(Pc)H_2O$ by dilithium benzophenone in THF gives the green crystalline solid $Li_2[CrPc]\cdot 6THF$.[1293,1294] The magnetic moment of the complex indicates that it has two unpaired electrons and this points to the formulation $[Cr^{I}(Pc^{\dot{-}})]^{2-}$, which has the electronic configuration $e_g^4b_{2g}^1\ e_g(\pi^*)^1$ (as Pc represents the doubly negatively charged phthalocyanine ligand, $Pc^{\dot{-}}$ represents the radical anion obtained by one-electron addition to Pc). Oxidation of CrPc by $SOCl_2$ affords the complex $CrCl_2Pc$, which is two oxidation equivalents above the starting complex.[665] The magnetic moment of the complex (4.40 BM at 298 K) is consistent with the formulation $Cr^{III}Cl_2(Pc^{\dot{+}})$, which contains two magnetic centres, one (Cr^{III}) having three unpaired electrons and the other ($Pc^{\dot{+}}$) one unpaired electron. It also eliminates the alternative formulation $Cr^{IV}Cl_2Pc$ for the oxidation product. Furthermore, the Cr—Cl stretching bands in the far IR spectrum of the product occur at positions close to those in the spectra of known Cr^{III}Cl-containing complexes. Since the reactivity of thionyl chloride towards MPc complexes parallels that of NO, it appears that an initial stage of the oxidation involves coordination of thionyl chloride to the metal.

(iv) *Other complexes*

The binuclear complexes $H_2O(Pc)CrOCr(Pc)H_2O$ and $NH_3(Pc)CrOHCr(Pc)OH$ form when $[Cr_2O(NH_3)_{10}]Cl_4\cdot H_2O$ and $[Cr_2(OH)(NH_3)_{10}]Cl_5$, respectively, are allowed to react with phthalonitrile at 543 °C.[1295] The 4′,4″,4‴,4⁗-tetrasulfonated phthalocyanine (TSPc) complex, OH(TSPc)CrOHCr(TSPc)H_2O was obtained by fusing $[Cr_2OH(NH_3)_{10}]Cl_5$ with ammonium-4-sulfophthalate and urea.[1295]

The complex $Na_3[Cr(TSPc)]\cdot 9H_2O$ has been prepared from cobalt(II) acetate, sodium 4-sulfophthalic acid, ammonium chloride and ammonium molybdate in nitrobenzene.[1296] Electrochemical reduction of this complex at a platinum electrode at −0.7 V *vs.* SCE gave the highly air-sensitive $[Cr(TSPc)]^{4-}$ in DMF solution. The action of $SOCl_2$ on $Na_3[Cr(TSPc)]$ affords the corresponding tetrasulfonyl chloride which on refluxing with octylamine gives the tetraoctylsulfonamide (TOPc) derivative Cr(TOPc)OH.[1297] The ability of this and related complexes to act as photosensitizers in the reduction of methyl viologen in THF/H_2O (15:85) solutions containing triethanolamine has been investigated. Irradiation of the system containing the chromium(III) complex did not result in formation of reduced methyl viologen. However the corresponding chromium(II) complex, prepared *in situ* by irradiation of the chromium(III) complex in the presence of cysteine, was found to be a more effective photocatalyst than any of the other metallophthalocyanines studied, but the quantum yield even in this case was low ($\phi = 2.6 \times 10^{-3}$).

35.5 CHROMIUM(IV)

Chromium(IV) does not have any aqueous solution chemistry except for the formation of intermediates in the reduction of Cr^{VI} to Cr^{III}. Chromium(IV) compounds tend to disproportionate into Cr^{III} and Cr^{VI} species (equation 78) and the metal ion in this oxidation state is powerfully oxidizing towards organic compounds. An eight-coordinate complex $[CrH_4(dmpe)_4]$ is known (Section 35.3.4.1).

$$3Cr^{IV} \longrightarrow 2Cr^{III} + Cr^{VI} \tag{78}$$

35.5.1 Anhydrous Halides and Complex Fluorides

Chromium tetrafluoride can be prepared by fluorination of the metal, CrF_3 or $CrCl_3$ at high temperature.[1364] It is an amorphous green solid and turns brown with traces of moisture, but is unexpectedly inert towards some reagents at room temperature. It is of intermediate volatility and is antiferromagnetic ($\mu = 3.02$ BM, $\theta = 78°$); it is therefore likely to consist of CrF_6 units sharing edges. The preparations of several CrF_4 adducts such as $XeF_6\cdot CrF_4$ are shown in Scheme 120 (Section 35.6.1). $CrOF_2$ ($\mu_{eff} = 1.1$ BM) is produced on thermal decomposition of CrO_2F_2, and a complex $(NO_2)_2CrO_2F_2$ is known (Section 35.7.1).

At high temperature in the gas phase the chloride and bromide exist in equilibrium with the halogen and solid CrX_3[222] (equation 79) and from their Raman spectra $CrCl_4$ and $CrBr_4$ are tetrahedral.[1298]

$$2CrX_3(s) + X_2(g) \xrightleftharpoons{800\text{–}1300\ K} 2CrX_4(g) \tag{79}$$

The hexafluorochromates(IV), $M_2[CrF_6]$ (M = Li, K, Rb, Cs) and $M[CrF_6]$ (M = Ba, Sr, Ca, Mg, Zn, Cd, Hg, Ni) are prepared by fluorinating appropriate mixtures of MCl and $CrCl_3$.[1299] The nitrosonium salt $(NO)_2[CrF_6]$ has been obtained[1300] from CrF_5 and NOF and characterized by its Raman spectrum. The symmetric stretch of the $[CrF_6]^{2-}$ unit was found at 610 cm^{-1}. A band at 20 200 cm^{-1} in the reflectance spectra of $A_2[CrF_6]$ (A = K, Rb, Cs) has been assigned to the first spin-allowed *d–d* transition $^3T_{1g} \rightarrow {}^3T_{2g}$. The second transition ($^3T_{1g} \rightarrow {}^3T_{1g}$) may be superimposed on a charge transfer band at ≈30 000 cm^{-1}.[1301] Pentafluorochromates(IV) $M[CrF_5]$ (M = K, Rb, Cs[222]), considered to contain condensed CrF_6 units, and the pink heptafluorochromates(IV) Rb_3CrF_7 and Cs_3CrF_7, which have the $(NH_4)_3SiF_7$ structure,[1302] are also known.

35.5.2 Chloro and Cyano Complexes

Alcoholic reduction of $K[CrO_3X]$ in the presence of pyridine, the pyridinium halide and a little water is reported[1303] to give the stable dimeric complexes $[Cr(OH)_2X_2py]_2$ (X = Cl, Br, I) ($\mu_{eff}^{297\ K} = 2.8$ BM); and hygroscopic $K_2[CrO(CN)_4py]$ ($\mu_{eff} = 3.0$ BM) is obtained on addition of KCN to $[pyH][CrO_3Cl]$ in pyridine to which ethanol has been added.[1304]

35.5.3 Chromates(IV) and CrO_2

There are several types of mixed oxide or chromate(IV). These are $MCrO_3$, M_2CrO_4, M_3CrO_5 and M_4CrO_6 (where M = Sr, Ba) and Na_4CrO_4;[1305] equations 80–82 illustrate typical preparations.[2,24,1306] The reaction of CrO_3 with liquid sodium does not yield Na_2CrO_3 as previously thought.[1307] The compounds M_2CrO_4 are air-stable, coloured substances, which contain tetrahedral CrO_4^{4-} groups and have magnetic moments of 2.8 BM at room temperature.

$$Cr_2O_3 + SrCrO_4 + 5Sr(OH)_2 \xrightarrow[\text{Ar}]{900\,^{\circ}\text{C}} 3Sr_2CrO_4 + 5H_2O \tag{80}$$

$$NaCrO_2 + 2Na_2O \xrightarrow[\text{vac.}]{410\,^{\circ}\text{C}} Na_4CrO_4 + Na\uparrow \tag{81}$$

$$Ba_2CrO_4 + CrO_2 \xrightarrow[\text{60–65 kbar}]{1000\,^{\circ}\text{C}} 2BaCrO_3 \tag{82}$$

High pressure and temperature are required to produce the compounds $MCrO_3$, *e.g.* $BaCrO_3$, which exists as several polytypes.[1308–1310] The Cr^{IV} is octahedrally coordinated in these compounds,[1306] the structure consisting of pairs of face-sharing octahedra linked to other pairs by sharing corners. The Cr—Cr distances are *ca.* 2.6 Å.

Chromium(IV) oxide can be prepared in various ways, for example by thermal decomposition of CrO_3 and CrO_2Cl_2.[2,1306] Two recent methods are the thermal decomposition of $Cr(NO_3)_3\cdot 9H_2O$ and the oxidation of Cr_2O_3 with NH_4ClO_4 under pressure in a heated sealed gold tube.[1311–1313] Its ferromagnetic behaviour ($T_C \approx 120\,^{\circ}C$) makes it suitable for magnetic tapes. The formula $Cr_4^{III}Cr^{IV}S_8$ has been suggested for Cr_5S_8 (Section 35.4.5.1.vii). For peroxo complexes see Section 35.7.7.

35.5.4 Alkyls, Alkoxides and Amides

35.5.4.1 Alkyls

The chromium(IV) tetraalkyls CrR_4 are air and light sensitive and thermally unstable to different degrees dependent upon R, but if kinetic paths are blocked through the absence of β-hydrogen atoms, or the presence of bulky organic groups, many can be isolated. The chemistry of these tetrahedral monomers has been covered in the companion series.[1] A new tetraalkyl is $Cr(1\text{-adamantylmethyl})_4$[1314] and there is fuller information on the alkenyl compound $Cr(CPh{=}CMe_2)_4$.[1315]

35.5.4.2 Alkoxides

The stability problems with the tetraalkyls also apply to the tetraalkoxides; in addition, the high oxidative power of Cr^{IV} means that alcoholysis with primary alcohols and most secondary alcohols leads to oxidation to the aldehyde or ketone and the formation of a chromium(III) alkoxide. Relatively stable chromium(IV) alkoxides are obtained only from tertiary alcohols and some heavily substituted secondary alcohols.

(*i*) *Syntheses and properties*

The tertiary butoxide $Cr(OBu^t)_4$ can be prepared by the following methods: (1) the action of di-*t*-butyl peroxide on $Cr(C_6H_6)_2$, (2) alcoholysis of the alkylamide, *e.g.* $Cr(NEt_2)_4$, (3) oxidation of $Cr(OBu^t)_3$ (prepared *in situ* from $Cr(NEt_2)_3$ and Bu^tOH in excess) by various reagents, *e.g.* $CrO_2(OBu^t)_2$, Br_2, $Bu^t_2O_2$, $Pb(OAc)_4$ or O_2, (4) oxidation of a mixture of $NaOBu^t$ and $[CrCl_3(THF)_3]$ with CuCl and (5) the addition of CuCl to a suspension of $Li[Cr(OBu^t)_4]$ in refluxing THF.

These methods, particularly (2), have been used in the preparation[1316] of other chromium(IV) tertiary alkoxides (Table 108). These are blue liquids or low melting solids which can be distilled in vacuum and are monomeric in cyclohexane. Their magnetic moments are close to 2.8 BM and that of $Cr(OBu^t)_4$ is almost independent of temperature as expected for tetrahedral Cr^{IV} (d^2, $\mu_{so} = 2.83$ BM). The electronic absorption spectrum of blue $Cr(OBu^t)_4$

contains bands centred at 9100, 15 200 and 25 000 cm^{-1} assigned to the $^3A_2(F) \rightarrow {}^3T_2(F)$, $^3T_1(F)$ and $^3T_1(P)$ transitions in T_d symmetry and a band at 41 000 cm^{-1} assigned to charge transfer. No ESR signal was detected for $Cr(OBu^t)_4$ in toluene solution down to 98 K due to fast relaxation in the zero-field split levels of the $^3A_2$ ground term; a small zero-field splitting was confirmed by the detection of ESR signals from frozen toluene samples at much lower temperature (10 K). From thermochemical measurements, $Cr(OBu^t)_4$ has considerable thermodynamic stability ($\Delta H_f^\circ = -1275$ kJ mol^{-1}).

Table 108 Chromium(IV) Alkoxides and Amides; Preparation and Properties

Compound	*Comments*	*Ref.*
$Cr(OR)_4$	R = Bu^t, CMe_2Et, $CMeEt_2$, CEt_3, $SiEt_3$; $Cr(OBu^t)_2(OCMe_2Et)_2$	1
Blue	ν(CrO) *ca.* 600 cm^{-1}; from $Cr(NEt_2)_4$ and alcohol	
	R = CHMeCMe$_3$, from $Cr(OBu^t)_4$ and $CMe_3CHMeOH$	2
	R = 1-adamantoxo, see Scheme 118	3
	R = $CHBu^t_2$, see Scheme 118, tetrahedral monomer:	4
	Cr—O (av), 1.773 Å, O—Cr—O, 108.8–112.4°, ν(CrO) = 718 cm^{-1}	
$Cr(NR_2)_4$	R = Et, Pr, R_2 = MeBu, C_5H_{10}; liquids, except	5
Green	$R_2 = C_5H_{10}$, m.p. *ca.* 60 °C	
	μ_{eff} *ca.* 2.8 BM, $\theta = -10°$ (R = Et), $-3°$ ($R_2 = C_5H_{10}$)	

1. E. C. Alyea, J. S. Basi, D. C. Bradley and M. H. Chisholm, *J. Chem. Soc. (A)*, 1971, 772.
2. G. Dyrkacz and J. Rocek, *J. Am. Chem. Soc.*, 1973, **95,** 4756.
3. M. Bochmann, G. Wilkinson, G. B. Young, M. B. Hursthouse and K. M. A. Malik, *J. Chem. Soc., Dalton Trans.*, 1980, 901.
4. M. Bochmann, G. Wilkinson and G. B. Young, *J. Chem. Soc., Dalton Trans.*, 1980, 1879.
5. J. S. Basi, D. C. Bradley and M. H. Chisholm, *J. Chem. Soc. (A)*, 1971, 1433.

Chromium(II) borohydride is formed[1317] by reduction of $Cr(OBu^t)_4$ with B_2H_6 (equation 83).

$$Cr(OBu^t)_4 \xrightarrow[THF]{B_2H_6} Cr(BH_4)_2\cdot 2THF \qquad (83)$$

Although the alcoholysis of $Cr(OBu^t)_4$ with several primary and secondary alcohols leads to oxidation by Cr^{IV} to the aldehyde or ketone,[1316] alcoholysis by the secondary alcohol 3,3-dimethyl-2-butanol affords a Cr^{IV} alkoxide (equation 84)[1318] which is sensitive to oxygen and moisture, but otherwise stable. Presumably the bulky *t*-butyl groups prevent the molecule from achieving the conformation required for hydrogen transfer in the oxidation step.

$$Cr(OBu^t)_4 + 4HOCHMeCMe_3 \xrightarrow[36\ h]{70\ °C} Cr(OCHMeCMe_3)_4 + 4Bu^tOH \qquad (84)$$

The methods for the preparation of chromium(IV) alkoxides are well illustrated by Scheme 118, which shows the preparations of adamantoxo (1-adamantanol = adoH, **285**) and bis(*t*-butylmethoxo) complexes of Cr^{III} and Cr^{IV}.[1319,1320] The blue chromium(IV) complex Cr(1-ado)$_4$ is air-stable, presumably because the small covalent radius of Cr^{IV} allows close packing of the bulky adamantyl groups which protect against encroaching ligands. Owing to its stability and easy preparation from $Cr(NEt_2)_4$, and because it does not undergo redox or disproportionation reactions, Cr(1-ado)$_4$ is preferable to $Cr(OBu^t)_4$ as a precursor for organic derivatives of chromium(IV), *e.g.* $Cr(CH_2SiMe_3)_4$ (Scheme 118). The room temperature magnetic moment of 2.68 BM is consistent with the Cr^{IV} formulation; the IR spectrum is consistent with a monomer, and a molecular ion has been observed in the mass spectrum. As with $Cr(OBu^t)_4$,[1316] no ESR signal has been detected down to 98 K.

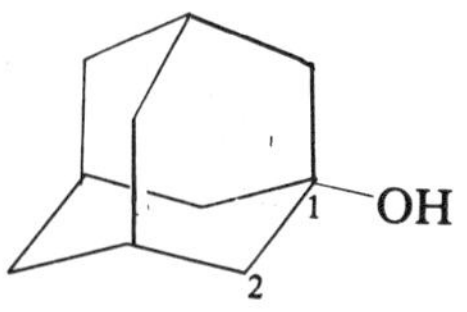

(285)

The sterically demanding secondary alcohol bis(*t*-butyl)methanol (2,2,4,4-tetramethylpentan-3-ol) does not react with $Cr\{N(SiMe_3)_2\}_3$ or $Cr(NEt_2)_4$, but the chromium(III) complexes $Cr(OCHBu^t_2)_3\cdot THF$ and $LiCr(OCHBu^t_2)_4$, and the chromium(IV) complex $Cr(OCHBu^t_2)_4$ have

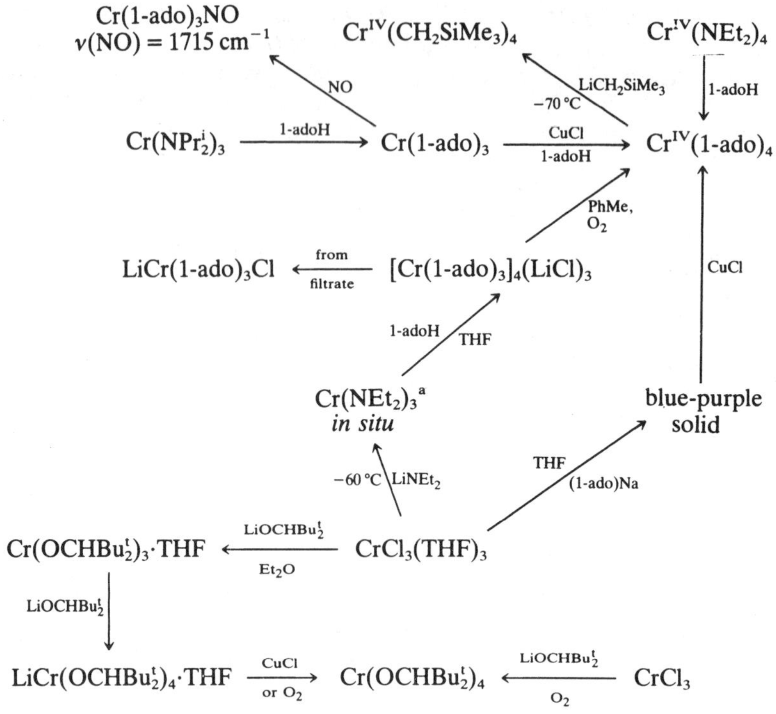

a Low yield of $Cr(1\text{-ado})_4$ by this route.

Scheme 118

been prepared as in the lower part of Scheme 118.[1320] All are soluble in non-polar solvents and $Cr(OCHBu^t_2)_3$ is a monomer in benzene, but the solid has a low moment ($\mu_{eff} = 3.2$ BM at 296 K) and the broad IR band, ν(Cr—O), at 555 cm^{-1} also suggests that it is a dimer. The absence of strong bands between 450 and 750 cm^{-1} suggests that $Cr(OCHBu^t_2)_4$ is a monomer, and this has been confirmed by an X-ray investigation. The O—Cr—O angles are close to tetrahedral and the crowding in the molecule prevents coordination oligomerization (**286**). The short Cr—O bond lengths (1.773 Å) indicate O→Cr $p\pi$–$d\pi$ bonding, which may account for the rather large Cr—O—C bond angles (140.5, 141.1°).

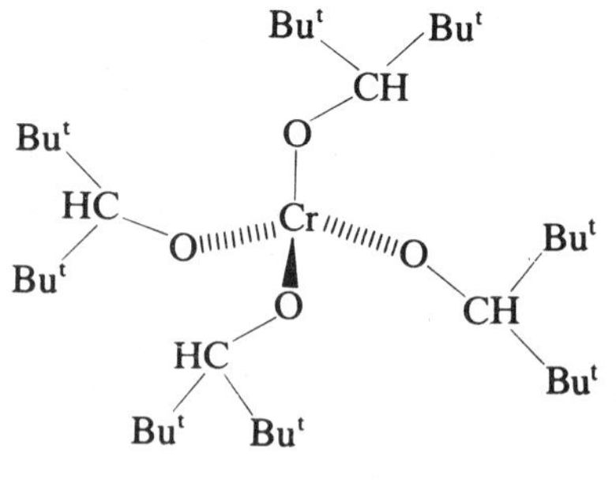

(**286**)

35.5.4.3 *Amides*

Equations (85) and (86) express the general method[592,618] for the preparation of chromium(IV) dialkylamides. The chromium(III) dialkylamide (Section 35.4.2.7) is extracted into pentane from the residue after removal of the solvent; careful evaporation of the pentane gives $Cr(NR_2)_3$ which disproportionates on further heating in vacuum because of the volatility of $Cr^{IV}(NR_2)_4$. The method has been successful with R = Et, Pr and R_2 = MeBu, C_5H_{10}, but failed with higher homologues because of the low volatility of the Cr^{IV} species, and with

$R = Me$ because the initial reaction is then more complicated than expressed by equation (85). In dimeric $Cr_2(NR_2)_6$, steric hindrance prevents chromium(III) from obtaining its preferred octahedral coordination; this, the ligand field stabilization in $Cr(NR_2)_4$ (tetrahedral, d^2), the covalency of the $Cr^{IV}—NR_2$ bond, and the polymeric structure of involatile $Cr^{II}(NEt_2)_2$ all contribute to the driving force for reaction (86). The structural changes are represented for $R = Et$ in equation (87).

$$\underset{\text{anhydrous}}{CrCl_3} + 3LiNR_2 \xrightarrow[0\,°C]{THF} Cr^{III}(NR_2)_3 + 3LiCl \qquad (85)$$

$$2Cr^{III}(NR_2)_3 \xrightarrow[10^{-3}\,mm\,Hg]{40\text{–}100\,°C} Cr^{IV}(NR_2)_4\uparrow + Cr^{II}(NR_2)_2 \qquad (86)$$

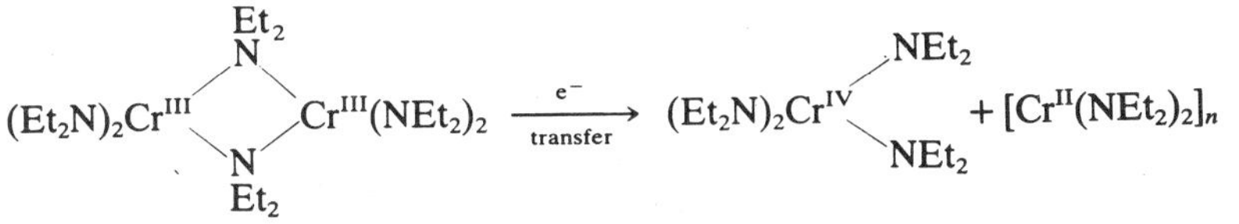

(87)

The magnetic moments (*ca.* 2.8 BM, Table 108) vary little with temperature as expected for tetrahedral Cr^{IV}, and molecular ions were found in the mass spectra. Like $Cr(OBu^t)_4$, no ESR signal was detected over the temperature range 98–298 K. A band in the electronic absorption spectra (for $R = Et$, $\bar{\nu} = 13\,700\ cm^{-1}$, $\varepsilon = 1200$) could not be assigned with confidence to simple d–d transitions because of the covalent nature of the Cr—N bonds in which there is likely to be $N \rightarrow Cr$ σ and p_π–d_π donor character. The He^I UVPES of $Cr(NEt_2)_4$ has been interpreted; it remains unclear why $Cr(NEt_2)_4$ is paramagnetic and $Mo(NEt_2)_4$ diamagnetic.[1321]

An unstable blue complex, which may be $Cr(O_2)(NPr^i_2)_3$ or the chromium(IV) nitroxide complex $CrO(NPr^i_2)_2(ONPr^i_2)$, is formed in a rapid reaction between oxygen and $Cr(NPr^i_2)_3$ in pentane.[592,1322]

Reactions of $Cr(NEt_2)_4$, some examples of which have been discussed elsewhere, are presented in Scheme 119. The unusual reactions with CO_2, which produce the chromium(III) and chromium(II) carbamato complexes (**287**) and (**288**), are believed[163] to proceed by CO_2 insertion into a Cr—N bond, which promotes β-hydrogen elimination from a coordinated diethylamido ligand, and then reductive elimination of Et_2NH produces a reactive chromium(II) species $Cr^{II}(O_2CNEt_2)(NEt_2)$. The subsequent reaction is dependent upon the relative concentration of CO_2.

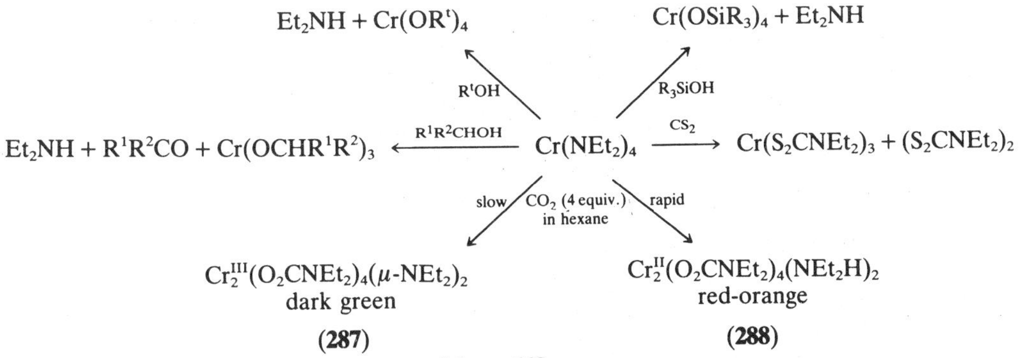

Scheme 119

35.5.5 Porphyrins

Chromyl(IV) porphyrins are included in Section 35.4.9.1.

35.6 CHROMIUM(V)

Chromium(V) chemistry is not extensive. Species of this oxidation state are frequently postulated as reactive intermediates during the reduction of chromium(VI). However, compounds of this oxidation state with reasonable stability are being discovered; there are several classes of well-defined complex: the pentafluoride, oxyhalides and various oxohalochromates; oxo and peroxo complexes; water-soluble chromyl(V) derivatives of α-hydroxy carboxylates; and chromyl(V) complexes of porphyrins and other multidentate ligands (Section 35.4.9.1). There is a recent review.[1323]

35.6.1 Halogen Compounds

Earlier work on CrF_5, $CrOF_3$, $CrOCl_3$ and the oxohalochromates(V), *e.g.* $K[CrOF_4]$ and $K_2[CrOCl_5]$, is described in refs. 222 and 1306. More recent information is outlined in Schemes 120–122 and Table 109.

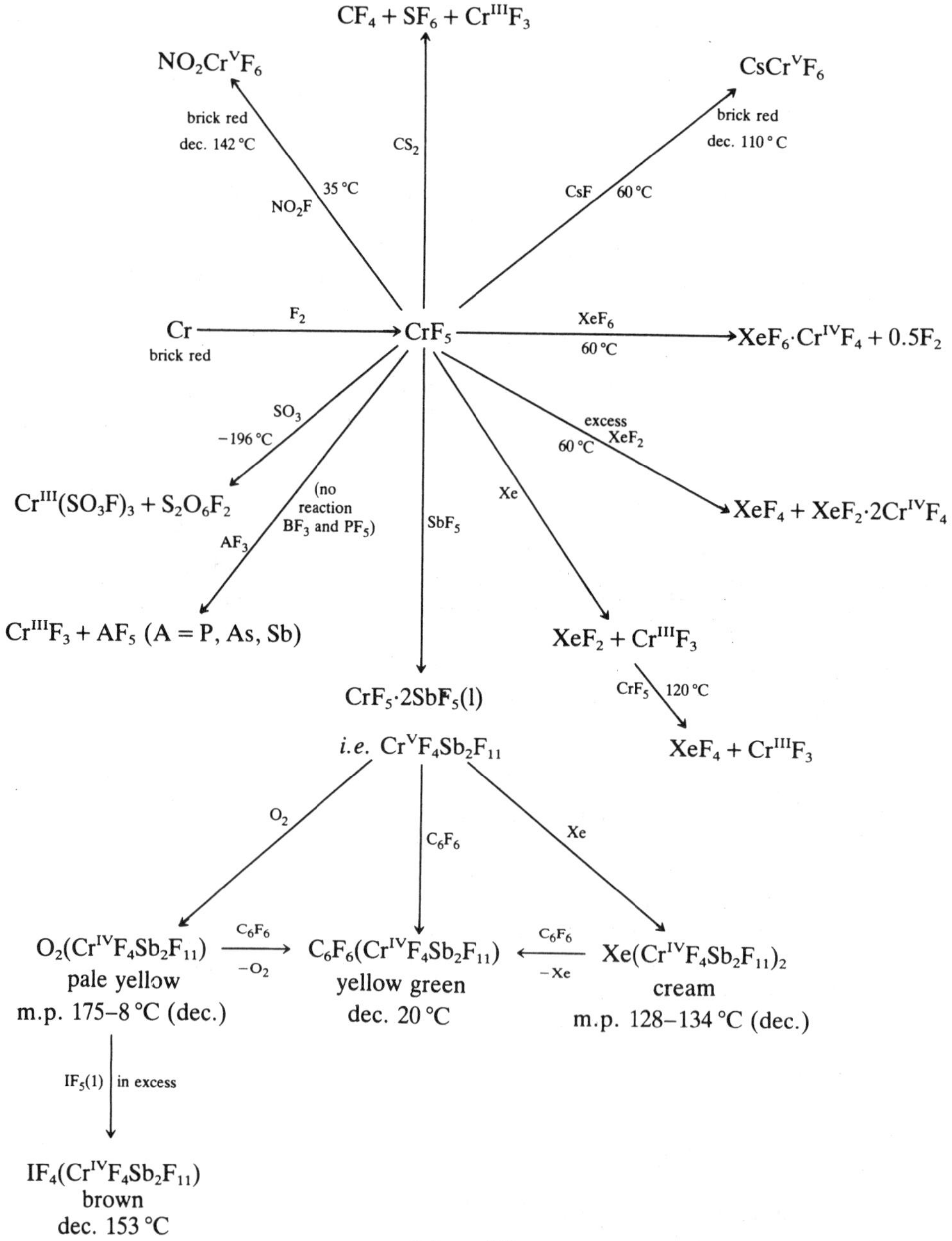

Scheme 120

35.6.1.1 Chromium(V) fluoride

The pentafluoride is a red solid (m.p. 30 °C) which can be prepared[1324,1325] by the action of fluorine on chromium powder and probably has a distorted trigonal bipyramidal structure in

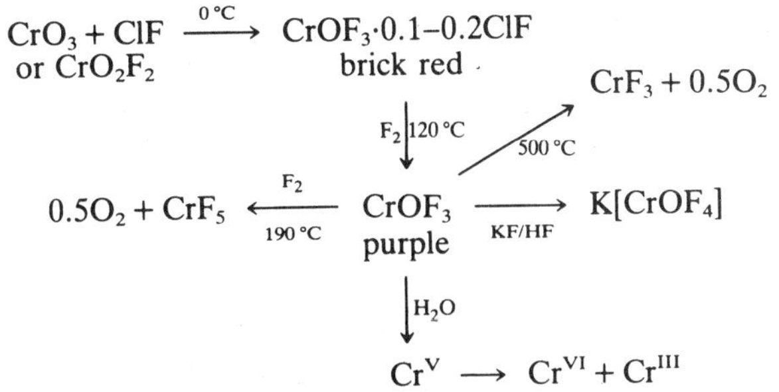

$$CrO_3 + ClF \xrightarrow{0\,°C} CrOF_3 \cdot 0.1\text{--}0.2ClF$$
$$\text{or } CrO_2F_2 \qquad\qquad \text{brick red}$$

$$\searrow F_2 \big| 120\,°C \quad \nearrow CrF_3 + 0.5O_2$$

Scheme 121:

$$0.5O_2 + CrF_5 \xleftarrow[190\,°C]{F_2} \quad CrOF_3 \xrightarrow{KF/HF} K[CrOF_4]$$
$$\text{purple}$$

$$\downarrow H_2O$$

$$Cr^{V} \longrightarrow Cr^{VI} + Cr^{III}$$

Scheme 121

$$CrO_2Cl_2 + BCl_3 \xrightarrow{-20\,°C} CrOCl_3 \xrightarrow{r.t.} CrO_2Cl_2 + ?$$
$$\text{red-black}$$

$$[pyH][CrOCl_4]$$

$$CrOCl_3(bipy)$$

$$[phenH][CrO_2Cl_2]$$

$$[AsPh_4][CrOCl_4] \xleftarrow{1:1\ [AsPh_4]Cl}$$

$$CrCl_3(acacH)(H_2O)$$

$$[AsPh_4][CrO_3Cl] + Cr^{III} \xleftarrow[\ [AsPh_4]Cl\ ]{excess} \qquad \xrightarrow{PPh_3} Cr^{III}$$

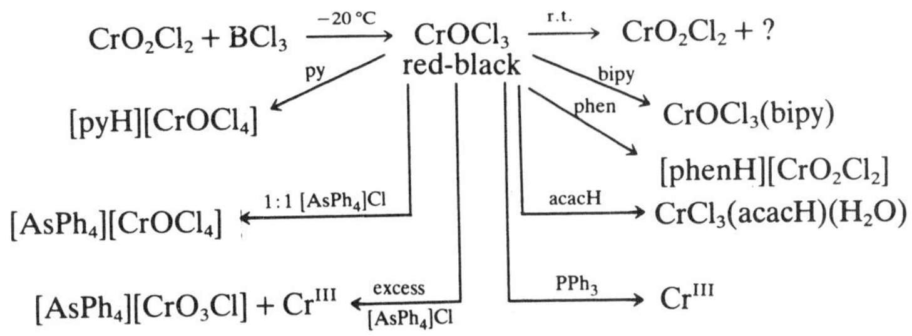

Scheme 122

the vapour phase.[1325] The solid and the liquid are fluorine-bridged polymers.[1364] CrF_5 is much more reactive than MoF_5 and WF_5; it is readily hydrolyzed and attacks glass, and is a powerful oxidant and fluorinating agent: for example, it will oxidize Xe to XeF_2 and XeF_4;[1326] and in an unusual reaction with XeF_6, fluorine is evolved and the brick red solid $XeF_6 \cdot Cr^{IV}F_4$, which decomposes rapidly in air, obtained. Curie law behaviour and a magnetic moment of 2.76 BM support the Cr^{IV} formulation. The action of XeF_6 in excess on CrF_2 also gives $XeF_6 \cdot CrF_4$.[227] Reaction of SO_3 and CrF_5 produces peroxydisulfuryl difluoride and $Cr(SO_3F)_3$.[1327] The viscous deep brown liquid $CrF_4Sb_2F_{11}$, which can be obtained from SbF_5 and CrF_5 and distils unchanged in vacuum, is another vigorous oxidizing and fluorinating agent comparable to PtF_6 in its ability to oxidize dioxygen and xenon (Scheme 120). It does not react with krypton or molecular nitrogen. There is one strong absorption band at 453 nm in its visible spectrum consistent with a d^1 system. The IR and Raman spectra support the presence of $Sb_2F_{11}^-$ but not CrF_4^+; thus the preferred structure is $CrF_4Sb_2F_{11}$ with fluoride bridges and partial charge separation.

A strong absorption at 1860 cm^{-1} in the Raman spectrum of $O_2(CrF_4Sb_2F_{11})$ is assignable to the dioxygenyl cation, and various bands at lower frequency in the IR and Raman spectra of this and related complexes (Scheme 120) to the $(CrF_4Sb_2F_{11})^-$ group. The magnetic moment of 3.08 BM at 295 K is low for three unpaired electrons (3.87 BM), suggesting that there is some interaction between the unpaired electrons of O_2^+ and $(Cr^{IV}F_4Sb_2F_{11})^-$. Bands characteristic of the cations are found in the IR and Raman spectra of $C_6F_6(CrF_4Sb_2F_{11})$ and $IF_4(CrF_4Sb_2F_{11})$.

The two T_{1u} modes of the octahedral CrF_6^- ion are found at 600 cm^{-1} and near 280 cm^{-1} in the spectra of NO_2CrF_6 and $CsCrF_6$, and the bands of the former at 2315 and 1360 cm^{-1} are assigned to the symmetric and asymmetric stretches of the $[O{=}N{=}O]^+$ group.[1328]

35.6.1.2 Chromyl(V) halides

It is difficult to obtain pure $CrOF_3$ by the reaction between CrO_3 or CrO_2F_2 and ClF because to remove the residual ClF from $CrOF_3 \cdot 0.1\text{--}0.2ClF$ requires repeated treatments with F_2 (Scheme 121). The purple $CrOF_3$ is slightly hygroscopic; it attacks certain organic solvents and glass above 230 °C.[1329] An apparently purer red-purple product, which hydrolyzes readily but is unreactive towards organic solvents, has been prepared by treatment of the adduct $CrOF_3 \cdot 0.3BrF_3$ with fluorine under mild conditions in which reaction to form CrF_5 (Scheme 121) does not occur.[1330]

Chromyl(V) chloride (Scheme 122) slowly decomposes at room temperature and in organic solvents.[1331,1332] From its IR spectrum $CrOCl_3$ in an argon matrix is a molecular species of C_{3v} symmetry (Cl—Cr—Cl angle $\approx 105°$), but an additional ν(Cr—Cl) band, and the shift of other

Table 109 Chromyl(V) Compounds

Complex	μ_{eff} (BM)	ν(*Cr=O*)(cm^{-1})	ν(*Cr—X*) (cm^{-1})	*Ref.*
Chromyl(V) halides				
$CrOF_3$	1.82	1000	740–600 (Cr—F)	1
Purple	1.87	~990	~670 (Cr—F)	2
			~565 (Cr—F—Cr)	1
			620–480 (Cr—F—Cr)	2
$CrOCl_3$		1022	—	3
Red-black	1.8	1024	435sh, 408, 333	4
		condensed on CsI, 250 K		
		1018	462, 410w	4
		Ar matrix, 12 K		
Oxotetrahalochromates(V)[a]				
$K[CrOF_4]$	1.76	1020	640 Cr—F	1,5
			500 Cr—F—Cr?	
$Cs[CrOF_4]$		1005	650 Cr—F	2
			505 Cr—F—Cr	
$Ag[CrOF_4]$	—	—	—	6
$[pyH][CrOCl_4]$[b]	1.78	1019	398, 343sh	7
Dark red				
$[quinH][CrOCl_4]$	1.75	1016	397, 347sh	7
Dark red				
$[iquinH][CrOCl_4]$	1.84	1016	398, 348sh	7
Dark red				
$[NMe_4][CrOCl_4]$	1.80	1016	408sh, 380br	7
Brown red	1.79	1020	400	8
$[NEt_4][CrOCl_4]$	1.78	1022	395, 348sh	7
Dark red	1.78 (1.64)[c]	1035, 960	395	8
$[NPr_4][CrOCl_4]$	1.74 (1.77)[c]	1005, 965	400	8
Gold				
$[PBu_4][CrOCl_4]$	1.74[c]	1025, 950	385	8
Red				
$[PPh_3Bz][CrOCl_4]$	1.73[c]	1020, 940	415	8
Dark brown				
$[AsPh_4][CrOCl_4]$	1.62	1019	398	7
Yellow brown	1.87 (1.96)[c]	1020, 940	405	8
$[PCl_4][CrOCl_4]$	1.91	1017	401, 348w	9
Red-brown				
Oxopentahalochromates(V)				
'$[NEt_4]_2[CrOF_5]$'	1.52[b]	960, 890	480, 450	2, 8
Yellow			{ν(Cr—F)}	
$[bipyH_2][CrOCl_5]$[b]	1.77	—	366	7
Brown	1.90	955		10
$[4,4'\text{-}bipyH_2][CrOCl_5]$	1.78	—	354	7
Dark brown				
$[phenH_2][CrOCl_5]$	1.97	950	—	10
$Cs_2[CrOCl_5]$		927	334, 312sh	7
Dark red	1.85	945	340	8
		933		11
$Rb_2[CrOCl_5]$	1.92	945	320	8
Dark red		950		11
$K_2[CrOCl_5]$		956		11
Amine adducts of $CrOCl_3$				
$CrOCl_3$(bipy)	1.85	954	—	12
Brown, prep. by thermal				
decomp.		898		3
$CrOCl_3$(bipy)	—	964	360	3
Direct reaction (Scheme 122)				
from $CrOCl_3$	1.76	945		10
$CrOCl_3$(phen)	1.87	957	—	12
	1.86	945		10
$CrOCl_3$(2-CNpy)	1.78	930		10
Red				
Periodato complex				
$Cr(IO_6)(phen)_2$	1.94	—	—	13

[a] $NO\cdot CrO_2F_2$ and $SO_2\cdot 2CrO_2F_2$ (Section 35.7.1) may contain Cr^V.
[b] Ref. 14 contains the preparation of additional salts of $[CrOCl_4]^-$ and $[CrOCl_5]^{2-}$.
[c] Solution measurement (NMR).

Table 109 *(continued)*

1. P. J. Green, B. M. Johnson, T. M. Loehr and G. L. Gard, *Inorg. Chem.*, 1982, **21**, 3562.
2. E. G. Hope, P. J. Jones, W. Levason, J. S. Ogden, M. Tajik and J. W. Turff, *J. Chem. Soc., Dalton Trans.*, 1984, 2445.
3. E. A. Seddon, K. R. Seddon and V. H. Thomas, *Transition Met. Chem.*, 1978, **3**, 318.
4. W. Levason, J. S. Ogden and A. J. Rest, *J. Chem. Soc., Dalton Trans.*, 1980, 419.
5. C. Rosenblum and S. L. Holt, *Transition Met. Chem.*, 1972, **7**, 87.
6. R. Colton and J. H. Canterford, 'Halides of the First Row Transition Metals', Wiley-Interscience, London, 1969, chap. 4.
7. K. R. Seddon and V. H. Thomas, *J. Chem. Soc., Dalton Trans.*, 1977, 2195.
8. O. V. Ziebarth and J. Selbin, *J. Inorg. Nucl. Chem.*, 1970, **32**, 849.
9. K. R. Seddon and V. H. Thomas, *Inorg. Chem.*, 1978, **17**, 749.
10. H. K. Saha and S. K. Ghosh, *J. Indian Chem. Soc.*, 1983, **60**, 599.
11. J. E. Ferguson, A. M. Greenaway and B. R. Penfold, *Inorg. Chim. Acta*, 1983, **71**, 29.
12. S. Sarkar and J. P. Singh, *J. Chem. Soc., Chem. Commun.*, 1974, 509.
13. Y. Sulfab, N. I. Al-Shatti and M. A. Hussein, *Inorg. Chim. Acta*, 1984, **86**, L59.
14. H. L. Krauss, M. Leder and G. Münster, *Chem. Ber.*, 1963, **96**, 3008.

bands to lower frequency, in the spectrum of the poorly volatile solid suggest polymerization.[1332] The IR spectrum of $CrOF_3$ also indicates halide bridging (Table 109).

35.6.1.3 Oxohalochromates(V) and other derivatives of chromyl(V) halides

$CrOCl_3$(bipy) exists in two forms because the product from direct reaction (Scheme 122) exhibits a Cr—O stretching frequency at 964 cm^{-1}, but the product from thermal decomposition of [bipyH_2][$CrOCl_5$] has a broad band at 898 cm^{-1} suggestive of polymerization through Cr=O···Cr interactions. The protons needed to form [phenH][CrO_2Cl_2] and [pyH][$CrOCl_4$] (Scheme 122) are presumed to have come from oxidation of the amine. The magnetic moments, ESR and electronic spectra of the oxochlorochromate derivatives are characteristic of Cr^V complexes, and they all disproportionate in water.[1331] The ionization energy (10 eV) of the *d* electron in $CrOCl_3$ has been derived from He^I and He^{II} PES.[1333]

In a typical preparation of a tetrachlorooxochromate(V) salt, the chloride ACl in glacial acetic acid is added to CrO_3 in the same solvent saturated with hydrogen chloride (reaction 88). With some cations pentachlorooxochromates(V) can be obtained but the relation between cation size and coordination number is unclear.[1334,1335] A pentafluorooxochromate(V), $[NEt_4]_2[CrOF_5]$ was prepared by the reaction of AgF/HF with $[NEt_4][CrOCl_4]$ in dichloromethane,[1335] but its composition is in doubt.[1330] Care is needed to prevent contamination by Cr^{VI} or Cr^{III} impurities,[1334,1335] *e.g.* $K_2[CrOCl_5]$ is usually contaminated with $K_2[CrCl_5(H_2O)]$, although pure samples have been obtained,[1336] and attempts to prepare bromooxochromates(V) from CrO_3 produced perbromide (Br_3^-), although the preparation of [bipyH_2][$CrOBr_5$] has been claimed.[1337] The ν(Cr=O) and ν(Cr—Cl) vibrations (Table 109) of $[CrOCl_4]^-$ and $[CrOCl_5]^-$ are in the ranges expected for isolated anions, and decrease with increase in coordination number from five to six as commonly found in other systems. An X-ray study of $[AsPh_4][CrOCl_4]$ has confirmed the presence of square pyramidal anions (Cr—O = 1.519, Cr—Cl = 2.240 Å).[1338] The short Cr—O distance indicates considerable $p_\pi \rightarrow d_\pi$ bonding. From a single crystal study[1339] the electronic absorption bands of $[CrOCl_4]^-$ at *ca.* 13 000 cm^{-1} and 18 000 cm^{-1} have been assigned to $d_{xy} \rightarrow d_{xz,yz}$ and Cr—O(π)→ Cr—O(σ^*) transitions, and bands at higher frequency to charge transfer, but other calculations[1340] assign the band at 18 000 cm^{-1} to halogen-to-metal charge transfer. The ESR spectra have been interpreted. The fact that the addition of $[NEt_4]Cl$ to a solution of $[AsPh_4][CrOCl_4]$ changed $\bar{g}$ from 1.988 to 1.970 indicates that compounds earlier thought to contain $[CrOCl_5]^{2-}$ contained $[CrOCl_4]^-$.[1341]

$$CrO_3 + ACl \xrightarrow[HCl(g)]{MeCO_2H} [A][CrOCl_4] \qquad (88)$$

The product of reaction between CrO_2Cl_2 and PCl_5 (equation 89) at high temperature or in the presence of a reducing agent has been identified as $[PCl_4][CrCl_4]$ (Section 35.4.7.2). The product at room temperature has been shown to be the reactive chromyl(V) salt $[PCl_4][CrOCl_4]$ (Table 109) and not the chromium(VI) adduct $CrO_2Cl_2{\cdot}PCl_5$.[1342] The pyridine adducts, $CrO_2Cl_2(py)_2$ and CrO_2Cl_2(py), are believed to be mixtures of chromyl(V) species, and from reaction in acetic acid [pyH][$CrOCl_4$], [pyH][CrO_2Cl_2] and $[Cr_3O(MeCO_2)_6(py)_2(MeCO_2H)]Cl$ have been isolated; similarly [bipyH][CrO_2Cl_2] has been obtained rather than CrO_2Cl_2(bipy), and Ph_4AsCl and CsCl with CrO_2Cl_2 in acetic acid yield mixtures of $[AsPh_4][CrO_3Cl]$ and $[AsPh_4][CrOCl_4]$, and $Cs[CrO_3Cl]$ and $Cs_2[CrOCl_5]$, respectively.[1343]

$$CrO_2Cl_2 + 2PCl_5 \xrightarrow{20\,°C} [PCl_4][CrOCl_4] + POCl_3 + 0.5Cl_2 \quad (89)$$

Solutions containing $[CrOCl_4]^-$ and $[CrOCl_5]^{2-}$ anions slowly generate Cr^{III}. This has been exploited in the preparation[1334] of the Cr^{III} anions $[CrCl_4(MeCO_2H)_2]^-$ and $[CrCl_5MeCO_2H]^{2-}$, and various bipy complexes, *e.g.* $[CrCl_3(bipy)(NCMe)]$, from $[bipyH_2][CrOCl_5]$.[1344]

Oxidation of *cis*-$[Cr(H_2O)_2(phen)_2](NO_3)_3$ with sodium periodate at $pH \approx 4$ produces the periodato complex $[Cr(IO_6)(phen)_2]$ (Table 109).

35.6.2 Complexes of Oxide and Peroxide

There are some well-defined chromium(V) complexes[1306] in which oxide is the ligand: $M^I_3CrO_4$ (M = Li, Na,[1305] K, Rb, Cs),[1345] $M^{II}_3[(CrO_4)_2]$ (M = Ca, Sr, Ba),[1306] $M^{II}_5(CrO_4)_3OH$ (M = Ca, Sr, Ba) and $M_5(CrO_4)_3X$ (M = Ca, Sr, Ba, X = F or Cl). In these the Cr occupies a tetrahedral site. There are also the peroxides[1346] $M^I_3[Cr(O_2)_4]$ (M = NH_4, Na, K), which are obtained by the action of 30% H_2O_2 on alkaline chromate solutions. They are red-brown, paramagnetic salts (μ_{eff} = 2.94 BM) of some stability, and in $K_3[CrO_8]$ the anion has a distorted dodecahedral arrangement (**289**) with the oxygen coordination occurring in pairs, and the O—O distance close to that for peroxo bonds (O_2^{2-} = 1.49 Å).[1306,1346,1347] Not much coordination chemistry of these complexes has been published since the reviews by Rosenblum and Holt[1306] and Connor and Ebsworth.[1346] Some investigations of the ESR[1348] and electronic spectra[1349] of CrO_4^{3-} doped in chloroapatites [$(PO_4)_3Cl^-$ salts] have been reported. The belief that $Sr_5(PO_4)_3Cl$ is isotypic with apatite has been disproved by a single crystal study.[1350] The three equilibrium phases in the ternary system $CaO—Cr^{VI}O_3—Cr_2^{III}O_3$ (the ternary compound, thought to have a chromate chromite formulation such as $Ca_9(Cr_4^{VI}Cr_2^{III})O_{24}$, calcium chromate $CaCr^{VI}O_4$ and monocalcium chromite $CaCr_2^{III}O_4$) have been further studied. Comparisons of the solid state IR, diffuse reflectance and ESR spectra of the three phases suggest that Cr^{III} is not present in the ternary phase, but $Cr^VO_4^{3-}$ anions are. Hence, the solid ternary phase is best described as $Ca_3(Cr^VO_4)_2$ rather than $Ca_9(Cr_4^{VI}Cr_2^{III})O_{24}$. The latter represents what happens when the ternary phase is dissolved in acid, where Cr^V disproportionates into Cr^{VI} and Cr^{III}.[1351] Disproportionation of Li_3CrO_4 (equation 90) takes place in a KCl/LiCl melt.[1352] The thermal decomposition of some rare earth chromates(V) has been related to their structures,[1353] and magnetic measurements on $KSrCrO_4$ and $KBaCrO_4$ confirm the V oxidation state.[1354]

$$3Li_3Cr^VO_4 \longrightarrow 2Li_2Cr^{VI}O_4 + LiCr^{III}O_2 + 2Li_2O \quad (90)$$

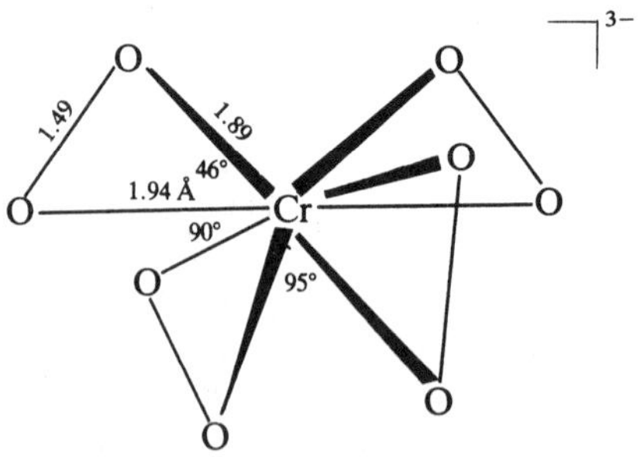

(**289**)

35.6.3 Complexes of Tertiary α-Hydroxy Carboxylates

The formation and decomposition of Cr^V in aqueous and non-aqueous media during the oxidation of organic substrates such as oxalic acid and ethylene glycol by potassium dichromate has been recognized for some time. No study resulted in the isolation of a stable, well-characterized chromium(V) complex until 1978 when potassium bis(2-hydroxy-2-methylbutyrato)oxochromate(V) monohydrate was prepared from chromium trioxide and the tertiary α-hydroxy acid in dilute perchloric acid according to equation (91). The Cr^V, which is produced in equimolar quantity with Cr^{III}, undergoes further reduction, first slowly and then very rapidly. Good yields of the potassium salt are obtained after neutralization with $KHCO_3$, provided the separation procedures are carried out expeditiously, at low temperature as far as possible, and with careful control of pH.[1355] A simpler, generally applicable route with good yields to this type of complex is the reaction of anhydrous $Na_2Cr_2O_7$ with the tertiary α-hydroxy acids RR'C(OH)CO_2H (R = R' = Me, Et, Bu; R = Me, R' = Et, Ph, R,R' = $(CH_2)_n$, $n = 4, 5$) in acetone (equation 92).[1356] The complexes are insoluble in non-polar solvents, soluble in water and acetone and heat stable. Whilst higher alkyl groups stabilize the complex, phenyl groups have a destabilizing effect. Many other hydroxy acids form unstable Cr^V complexes in solution.

$$2HCrO_4^- + 4EtMeC(OH)CO_2H \xrightarrow[2\text{–}3\ d]{4H^+,\ -16\,^\circ C} [OCr(O_2COCMeEt)_2]^- + Cr^{3+} + 2MeCOEt + 2CO_2 + 5H_2O \quad (91)$$

$$Na_2Cr_2O_7 + 5RR'C(OH)CO_2H \longrightarrow 2Na[OCr(O_2COCRR')_2] + RR'CO + CO_2 + 5H_2O \quad (92)$$

The salt $Na[OCr(O_2COCEt_2)_2]$ has been used in recent mechanistic studies of the oxidation of NH_3OH^+ and $N_2H_5^+$ by Cr^V; the former yields Cr—NO derivatives (see also Section 35.4.2.6) and the latter N_2 and Cr^{III}.[1357]

The structure of $K[OCr(O_2COCMeEt)_2]$ (**290**), obtained with difficulty from a poor crystal, contains pentacoordinate Cr^V and *trans* bidentate ligands of the same chirality, although enantiomers of the complete anions are present in the crystal. The short apical Cr—O bond distance (Table 110) is compatible with a Cr^V=O group. The other Cr—O distances are greater and differ from each other, and all O—Cr—O bond angles depart considerably from 90° or 180° so that the coordination polyhedron is much distorted from a square pyramid. In the solid the ethyl groups close off the position *trans* to the chromyl oxygen. The oxidation state of (**290**), and of the sodium salts,[1356] has been established by analytical methods, including a new iodometric method which determines Cr^V separately from its ability to oxidize iodide rapidly at much lower acidities than Cr^{VI} does. The electronic spectrum, magnetic moment and ESR data (Table 110) confirm the Cr^V assignment, and the Cr=O stretching frequency is consistent with a terminal Cr—O multiple bond. The stability of these Cr^V complexes to oxidative decomposition has not yet been explained.

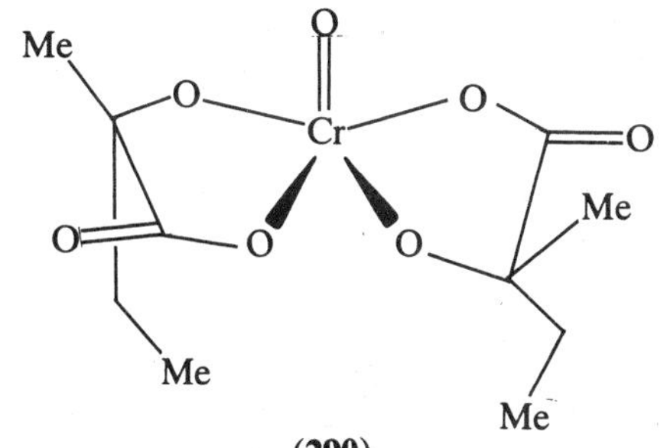

(**290**)

Table 110 Complexes of Tertiary α-Hydroxy Carboxylic Acid Dianions

Complex	μ_{eff} (BM)	ν(Cr=O) (cm^{-1})	$\bar{g}$	*Ref.*
$K[OCr(O_2COCMeEt)_2]\cdot H_2O$[a]	2.05 (H_2O)	994 (K salt)	1.9780	1
Dark brown		1005 (Na salt)		2
$M[OCr(pfp)_2]$	1.79 (K)	—	1.9794	3, 4
[M = K, Cs (purple),	1.71 (Cs)	—		
NEt_4 (blue)]	2.33 (NEt_4)		—	

[a] Bond distances: Cr=O, 1.554; Cr—O (carboxyl), 1.911 (av), Cr—O (OH-derived), 1.781 Å (av).

1. M. Krumpolc, B. G. DeBoer and J. Roček, *J. Am. Chem. Soc.*, 1978, **100**, 145.
2. M. Krumpolc and J. Roček, *J. Am. Chem. Soc.*, 1979, **101**, 3206.
3. N. Rajasekar, R. Subramaniam and E. S. Gould, *Inorg. Chem.*, 1983, **22**, 971.
4. C. J. Willis, *J. Chem. Soc., Chem. Commun.*, 1972, 944.

Other chromium(V) complexes stable to reduction and hydrolysis can be prepared[1358,1359] by reaction of chromate with perfluoropinacol $(CF_3)_2C(OH)C(OH)(CF_3)_2$ (H_2pfp) (equation 93). Hydroxy acids slowly displace two chloride ligands of $Cs_2[CrOCl_5]$ in hot MeCN to give $CrO(AB)Cl_3$ in which AB is tartrate, hydroxymalonate or malate. ESR data indicate that these are Cr^V complexes but the magnetic moments are high.[1335] IR bands near 940 cm^{-1} and 340 cm^{-1} have been assigned to the $\nu(CrO)$ and $\nu(CrCl)$ vibrations respectively. A dibenzoato complex $Cr(O_2CC_6H_5)_2Cl_3$ has also been described.[1360]

$$K_2CrO_4 + H_2pfp \xrightarrow[H_2SO_4,\ boil]{EtOH/H_2O} K[OCr(pfp)_2] \quad (93)$$

35.6.4 Porphyrins and Schiff's Bases

Chromium(V) complexes of these ligands are included in Section 35.4.9.1.

35.7 CHROMIUM(VI)

Although there are several very important chromium(VI) compounds, the powerful oxidizing ability of this oxidation state leads to the oxidation of many potential ligands, and its coordination chemistry is limited. However, the oxidizing ability is exploited in many organic syntheses,[2] and a new oxidizing system is $CrO_3/HBF_4/MeCN$.[1361] A safer method of preparing the reagent *t*-butyl chromate has been reported.[1362] Aqueous solutions of CrO_4^{2-} react with excess NH_2OH in the presence of various coligands to generate complexes containing the $[Cr(NO)]^{2+}$ moiety (Section 35.4.2.6).

References to the many extensive accounts of the preparation, properties and reactions of chromium(VI) compounds, and their industrial and analytical importance have been collected.[2,1363]

35.7.1 Hexafluoride and Chromyl(VI) Halides

Chromium hexafluoride (Table 111), the only hexafluoride of the first transition series, was first prepared in small yields together with CrF_5 from the metal powder and fluorine at 400 °C and 350 atm. It decomposes rapidly to CrF_5 and F_2. Recently, it has been prepared from CrO_3 and F_2 under less extreme conditions, *i.e.* 170 °C and 25 atm. The lemon yellow CrF_6 sublimes at room temperature from the CrF_5 which is also formed. IR spectra (N_2 matrix) showed Cr—F stretching vibrations centred at 759 cm^{-1}. Weaker bands due to some CrO_2F_2, $CrOF_4$, CrF_5 and CrF_4 were also identified in the spectra.[1364] It has been suggested that the fluorination of CrO_3 by F_2 at 4 atm is stepwise, because as the reaction temperature increases the main products are CrO_2F_2 (150 °C), $CrOF_4$ (220 °C) and a mixture of CrF_5 and CrF_4 (250 °C) believed to come from the decomposition of CrF_6.[1365] This is confirmed by the new preparation of CrF_6.

The oxotetrafluoride $CrOF_4$ is an easily hydrolyzed solid obtained as a by-product when fluorine is passed over chromium,[1365] but it is better prepared from CrO_3 as just outlined. It is a powerful fluorinating and oxidizing agent, is fluoride bridged in the solid state, molecular (C_{4v}) in a low temperature matrix, and forms the salt $Cs[CrOF_5]$ (Table 111).[1366]

The dioxofluoride CrO_2F_2 can be obtained in many ways. The methods starting from CrO_3 outlined in equations (94) to (97) are generally more convenient[1367] than earlier ones which involve F_2[1365], HF, SF_4, SeF_4, CoF_3 or IF_5. In reactions (96) and (97) purification is simplified because CrO_2F_2 is volatile and WOF_4 and $MoOF_4$ are not. Treatment of CrO_2Cl_2 with F_2 or ClF also produces CrO_2F_2. The reaction between CrO_2F_2 and CrO_2Cl_2 gives CrO_2ClF, which from its IR and mass spectra is a true compound.[1368] Decomposition of CrO_2F_2 to $CrOF_2$ occurs at high temperature (500–525 °C).[1369] It is a powerful oxidizing agent and reacts vigorously with NO, NO_2 and SO_2 as equations (98) to (100) show.

$$CrO_3 + \text{excess } ClF \xrightarrow{0\,°C} CrO_2F_2 + Cl_2 + O_2 + ClO_2F \quad (94)$$

$$CrO_3 + COF_2 \xrightarrow{185\,°C} CrO_2F_2 + CO_2 \quad (95)$$

Table 111 Chromium(VI) Fluoride, the Chromyl(VI) Halides and $Cs[CrOF_5]$

Compound	*Comments*	*Ref.*
CrF_6 Yellow	Unstable above −100 °C, dec. to CrF_5 and F_2, ν(Cr—F), 759 cm^{-1}	1
$CrOF_4$ Dark red	M.p. 55 °C, separated by fractional sublimation, readily hydrolyzed, reacts with Pyrex, ν(Cr=O), 1020, Cr—F, 760–690, Cr—F—Cr, ~500 cm^{-1}	2 3
CrO_2F_2 Dark violet-red	M.p. 31.6 °C, reactive, attacks glass, sublimes 29.6 °C, stable in dark; ν(Cr—O), 1006, 1016 cm^{-1}, ν(Cr—F), 727, 789 cm^{-1}	2,4 5
CrO_2Cl_2 Deep red (l)	B.p. 117 °C, fumes in moist air, heat- and light-sensitive, ignites some organic compounds; Cr=O, 1.581 Å, Cr—Cl, 2.126 Å, OCrO, 108.5°, ClCrCl, 113.3°, OCrCl, 108.7°	6, 7
CrO_2Br_2	Unstable even below room temp.	6
$Cs[CrOF_5]$ Orange-brown	CsF, $CrOF_4$, 100 °C, N_2; ν(Cr=O), 955, ν(Cr—F), ~690 cm^{-1}	3

1. E. G. Hope, P. J. Jones, W. Levason, J. S. Ogden and M. Tajik, *J. Chem. Soc., Chem. Commun.*, 1984, 1355.
2. A. J. Edwards, W. E. Falconer and W. A. Sunder, *J. Chem. Soc., Dalton Trans.*, 1974, 541.
3. E. G. Hope, P. J. Jones, W. Levason, J. S. Ogden, M. Tajik and J. W. Turff, *J. Chem. Soc., Dalton Trans.*, 1985, 529.
4. I. R. Beattie, C. J. Marsden and J. S. Ogden, *J. Chem. Soc., Dalton Trans.*, 1980, 535.
5. P. J. Green and G. L. Gard, *Inorg. Chem.*, 1977, **16**, 1243.
6. R. Colton and J. H. Canterford, 'Halides of the First Row Transition Metals', Wiley-Interscience, London, 1969, vol. 1, chap. 4.
7. C. J. Marsden, L. Hedberg and K. Hedberg, *Inorg. Chem.*, 1982, **21**, 1115.

$$CrO_3 + WF_6 \xrightarrow{125\,°C} CrO_2F_2 + WOF_4 \quad (96)$$

$$CrO_3 + MoF_6 \xrightarrow{125\,°C} CrO_2F_2 + MoOF_4 \quad (97)$$

$$NO + CrO_2F_2 = NO{\cdot}CrO_2F_2 \quad (98)$$

$$SO_2 + 2CrO_2F_2 = SO_2{\cdot}2CrO_2F_2 \quad (99)$$

$$2NO_2 + CrO_2F_2 = (NO_2)_2CrO_2F_2 \quad (100)$$

Magnetic moment measurements suggest that the solids produced in equations (98) and (99) contain Cr^V and that produced in equation (100) contains Cr^{IV}.[1370] Other reactions of CrO_2F_2 are shown in Scheme 123.

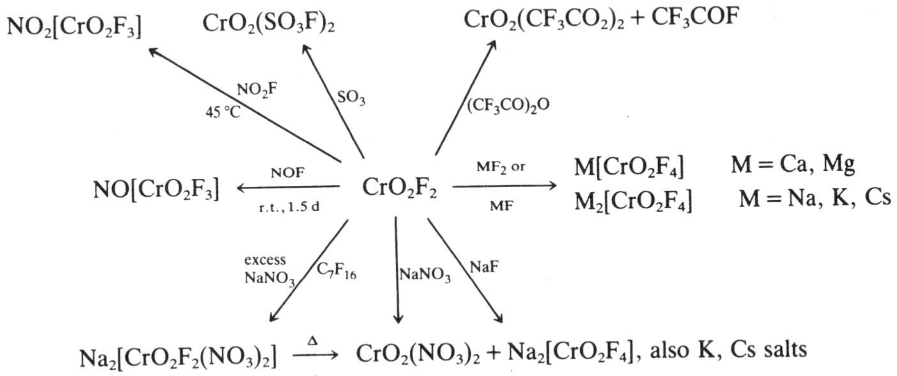

Scheme 123

Chromyl chloride CrO_2Cl_2 can be prepared by distillation from a mixture of an alkali metal chloride and concentrated sulfuric acid with an alkali metal dichromate or CrO_3. It can also be prepared from dichromate and HCl, CrO_3 and an acid chloride such as PCl_5 or HSO_3Cl,[222] or CrO_3 and $AlCl_3$ in molten LiCl/NaCl/KCl.[1371]

Chromyl bromide CrO_2Br_2 is thermally unstable and little is known of its properties.[222] It can be prepared from CrO_2Cl_2 and HBr in excess at low temperature, and by reaction of CrO_3 with CF_3COBr and HBr in the presence of P_4O_{10}. The existence of the iodide has been claimed but as mentioned in a review[1372] of chromyl(VI) compounds this is unlikely.

Since the metal in the chromyl(VI) halides CrO_2F_2 and CrO_2Cl_2 is formally in the d^0 oxidation state, the structures of these molecules would be expected to follow the valence shell electron pair repulsion theory, *i.e.* $O{=}\widehat{Cr}{=}O > X{-}\widehat{Cr}{-}X$ (X = F or Cl). From the gas phase Raman and matrix IR spectra[1373] of CrO_2F_2 the angles are $O\widehat{Cr}O$ 102.5° and $F\widehat{Cr}F$ 124°, in agreement with the earlier values from electron diffraction,[1374] but not with those from microwave spectroscopy. Later electron diffraction values[1375] are $O\widehat{Cr}O$ 107.8° and $F\widehat{Cr}F$ 111.9°. Irrespective of the discrepancies, it is clear that molecular CrO_2F_2, in common with many d^0 complexes of the transition metals, does not follow the VESPR theory. The angles in CrO_2Cl_2 follow[1376] the same pattern: $O\widehat{Cr}O = 108.5°$, $Cl\widehat{Cr}Cl = 113.3°$ (electron diffraction). The IR spectrum of CrO_2Cl_2 in an argon matrix has been reported.[1377]

The Raman spectrum of liquid CrO_2F_2 can be assigned in C_{2v} symmetry, and a band at 708 cm^{-1} arising from the symmetric Cr—F stretch is replaced in the solid-state spectrum by one at 540 cm^{-1}.[1378] The lowered frequency suggests polymerization through fluoride bridges in the solid, and this is consistent with the similar unit cell parameters of CrO_2F_2[1379] and VOF_3, which is known to be polymeric.

35.7.2 Other Chromyl(VI) Compounds

Chromyl derivatives CrO_2X_2 (X = nitrate, sulfate, perchlorate, fluorosulfate, borate, acetate, trichloroacetate, azide and benzoate) have long been known.[2] The perchlorate $CrO_2(ClO_4)_2$ has been obtained as a minor product in the reaction of Cl_2O_6 with $CrCl_3$. From its IR spectrum $CrO_2(ClO_4)_2$ contains monodentate perchlorate groups so the chromium has a tetrahedral environment like CrO_2Cl_2.[851]

Chromyl nitrate is a red, easily hydrolyzed liquid that can be prepared from N_2O_5 and CrO_3, CrO_2Cl_2 or $KCrO_3Cl$. A method which avoids the use of N_2O_5 is the reaction of CrO_2F_2 with sodium nitrate (Scheme 123).[1380] Besides the chromyl derivatives in Scheme 123, CrO_2F_2 forms the adduct $CrO_2F_2{\cdot}TaF_5$, $CrO_2F_2{\cdot}SbF_5$ and $CrO_2F_2{\cdot}2SbF_5$ which may be $[CrO_2F]^+$ salts.[1381]

The fluoro- and chloro-sulfates can be isolated as green solids, stable if dry, by reactions (101) and (102). Brown crystals can be sublimed from green $CrO_2(SO_3F)_2$,[1382] and if the reaction mixture in equation (101) is heated, dark brown crystals separate.[1383] In each case the brown form and the green form have substantially the same analyses, IR spectra and powder photographs $\{CrO_2(SO_3F)_2\}$ and are diamagnetic; the different colours may be due to different degrees of polymerization. Besides CrO_2Cl_2, CrO_3, K_2CrO_4 and $K_2Cr_2O_7$ are solvolyzed to $CrO_2(SO_3Cl)_2$ by HSO_3Cl.[1383,1384] In disulfuric acid $CrO_2(HSO_4)_2$ forms, identified by conductance and freezing point data and electronic spectroscopy.[1384,1385]

$$CrO_2Cl_2 + S_2O_6F_2 \longrightarrow CrO_2(SO_3F)_2 + Cl_2 \qquad (101)$$

$$CrO_2Cl_2 + 2HSO_3Cl \longrightarrow CrO_2(SO_3Cl)_2 + 2HCl \qquad (102)$$

The reactions (103) to (110) afford further chromyl derivatives: $CrO_2(O_2CR)_2$ (R = CF_3, $CClF_2$ or C_3F_7),[1386] $CrO_2(O_3SCF_3)_2$, $CrO_2(NO_3)(O_2CCF_3)$,[1387] $CrO_2(O_2PF_2)_2$,[1388] $CrO_2(OTeF_5)_2$,[1389] $CrO_2(OSeF_5)_2$,[1390] CrO_2ClN_3,[1391] $CrO_2Cl_2(ICN)_2$[1392] and $[CrO_2Cl(NCS)_3]_2$. The reaction of CrO_2Cl_2 in CS_2 or CCl_4 with various donor molecules (acetone, DMSO, THF, Ph_3P, *etc.*) does not produce stable adducts of higher coordination number but oxidation of the base and ill-defined Cr^{IV} products.[1393]

$$CrO_3 + (RCO)_2O \longrightarrow CrO_2(O_2CR)_2 \qquad (103)$$

$$CrO_2(O_2CCF_3)_2 + CF_3SO_3H \longrightarrow CrO_2(O_3SCF_3)_2 + CF_3CO_2H \qquad (104)$$

$$CrO_2(O_2CCF_3)_2 + HNO_3 \longrightarrow CrO_2(NO_3)(O_2CCF_3) + CrO_2(NO_3)_2 + CF_3CO_2H \qquad (105)$$

$$CrO_3 + P_2O_3F_4 \longrightarrow CrO_2(O_2PF_2)_2 \qquad (106)$$

$$CrF_5 + B(OTeF_5)_3 \longrightarrow CrO_2(OTeF_5)_2 + BF_3,\ TeF_6,\ CrF_4? \qquad (107)$$

$$CrO_2Cl_2 + Hg(OSeF_5)_2 \longrightarrow CrO_2(OSeF_5)_2 \quad (108)$$

$$CrO_2Cl_2 + ICN \longrightarrow CrO_2Cl_2(ICN)_2 \quad (109)$$

$$CrO_2Cl_2 + (SCN)_2 \longrightarrow [CrO_2Cl_2(NCS)_3]_2 \quad (110)$$

The chromyl derivatives generally hydrolyze readily to chromate and the acid corresponding to the coordinated anion. Thermal decomposition of yellow $CrO_2(O_3SCF_3)_2$ and $CrO_2(O_2CCF_3)_2$ produces oxygen and the green anhydrous chromium(III) salts $Cr(O_3SCF_3)_3$[1387] and $Cr(O_2CCF_3)_3$:[1386] $CrO_2(O_2PF_2)_2$ is much less stable and readily yields $Cr(O_2PF_2)_3$; its solutions in CCl_4 are moderately stable, like those of $CrO_2(SCN)_2$ and $CrO_2(NCO)_2$, and have electronic spectra similar to those of other chromyl complexes.[1388] Chromyl chloride reacts with monocarboxylic acids to form trinuclear cations $[Cr_3O(O_2CR)_6(RCO_2H)_3]^+$ (R = H, Me); the low moments of the formate and chloride salts and of derivatives such as $[Cr_3O(O_2CMe)_6L_3]Cl$ (L = py, NH_3) suggest similar structures to other trinuclear complexes[1394] (Section 35.4.4.10). The anionic difluorophosphato complex $K_2[CrO_2(O_2PF_2)_4]$ is prepared by reaction of K_2CrO_4 with $P_2O_3F_4$.[1388] From their IR and Raman spectra, it is concluded that the adduct $CrO_2Cl_2(ICN)_2$ has a *cis* structure with N-bonded ICN and $[CrO_2Cl_2(NCS)_3]_2$ has an —NCS—SCN-bridged structure.[1392]

35.7.3 Chromium(VI) Oxy Compounds

35.7.3.1 Chromium(VI) oxide

Most chromium(VI) compounds may be considered to be derived from the orange-red trioxide CrO_3, the anhydride of chromic acid, which can be crystallized from aqueous solutions of chromates and dichromates by the addition of sulfuric acid. When heated above its melting point, CrO_3 loses oxygen to form Cr_3O_8, Cr_2O_5, CrO_2 and finally Cr_2O_3. It consists of infinite chains of distorted CrO_4 tetrahedra sharing two corners; the bridging Cr—O distances are 1.748 Å and the terminal Cr—O distances 1.599 Å. The angle at the bridging oxygen is 143°.[1306] Electron diffraction data for gaseous CrO_3 can be best interpreted by assuming that it consists of trimers and tetramers.[1395] Depolymerization occurs on dissolution in water (to form H_2CrO_4) and in the few organic solvents in which it is soluble. It is a powerful oxidizing agent. Adducts, CrO_3py and CrO_3py_2, used as organic oxidants in CH_2Cl_2, are formed by pyridine bases, and with alcohols chromate esters are obtained.[2,1396]

35.7.3.2 Chromates, dichromates and polychromates

When CrO_3 is dissolved in water the equilibria (111) to (115) are set up. Above pH 8 the main species is yellow chromate CrO_4^{2-}; in the pH range 2–6 $HCrO_4^-$ and orange-red dichromate $Cr_2O_7^{2-}$ are in equilibrium; and below pH 1 H_2CrO_4 predominates. Chromates and dichromates of ammonium and the common metal ions are known: the ammonium, magnesium and alkali metal chromates are soluble in water, but those of silver, lead and the alkaline earth metals are only slightly soluble. Whether chromates or dichromates crystallize depends on the pH and the cation present. At high Cr^{VI} concentrations trichromates and tetrachromates form and some have been isolated as solids, *e.g.* $K_2Cr_3O_{10}$ and $K_2Cr_4O_{13}$, but, unlike molybdenum and tungsten, chromium does not exhibit an extensive iso- or hetero-polyanion chemistry. Many double chromates and basic chromates are also known.[2]

$$H_2CrO_4 \rightleftharpoons HCrO_4^- + H^+ \quad (111)$$

$$HCrO_4^- \rightleftharpoons CrO_4^{2-} + H^+ \quad (112)$$

$$Cr_2O_7^{2-} + H_2O \rightleftharpoons 2HCrO_4^- \quad (113)$$

$$Cr_2O_7^{2-} + OH^- \rightleftharpoons HCrO_4^- + CrO_4^{2-} \quad (114)$$

$$HCrO_4^- + OH^- \rightleftharpoons CrO_4^{2-} + H_2O \quad (115)$$

Chromates and polychromates of organic cations can be prepared by careful hydrolysis of the appropriate ester with the organic base in organic solvents. Examples are $(NBu^nH_3)_2CrO_4$, $(pyH)_2Cr_3O_{10}$ and $(NBu^n_4)_2Cr_4O_{13}$.[1397]

Whereas the ester bis(trimethylsilyl) chromate is explosive and readily hydrolyzed,[1398] the crystal structure[1399] of the triphenyl derivative has been obtained (Table 112). The germanium, tin and lead analogues have been prepared by reactions (116) to (118). These compounds are better regarded as chromate esters than chromyl(VI) compounds.[1400] Electrochemical reduction of $(Ph_3XO)_2CrO_2$ (X = Si, Ge, Sn or Pb) is generally irreversible and of little use in the characterization of chromium(VI) complexes.[1401] Chromates of the type Me_3SbCrO_4 and $[Me_3Sn]_2[CrO_4]$ are also known.[1402]

$$2Ph_3GeBr + Ag_2CrO_4 \longrightarrow (Ph_3GeO)_2CrO_2 + 2AgBr \quad (116)$$

$$2Ph_3SnCl + Ag_2CrO_4 \longrightarrow (Ph_3SnO)_2CrO_2 + 2AgCl \quad (117)$$

$$2Ph_3PbCl + Na_2CrO_4 \longrightarrow (Ph_3PbO)_2CrO_2 + 2NaCl \quad (118)$$

Table 112 Bond Distances in CrO_3 and Representative Chromates and Polychromates

Compound	*Bond distances* (Å)	*Ref.*
CrO_3	Cr—O (av), 1.599, Cr—O (br), 1.748 Å, Cr—O—Cr, 143°	1
$(NH_4)_2CrO_4$	Cr—O (av), 1.658 Å, O—Cr—O (av), 109.7°	2
$Ce(CrO_4)_2·2H_2O$	Cr—O, 1.583 Å Cr—O·····Ce^{4+}, 1.665, 1.685, 1.636 Å	3
$CaCrO_4·H_2O$	Cr—O (av), 1.644 Å	4
Ag_2CrO_4	Cr—O, 1.69, 1.67, 1.63 Å Cr—O····Ag^+, 111.6, 110.1, 109.3, 106.1 Å angular distortion through interaction Ag^+	5
$(NH_4)_2[Zn(NH_3)_2(CrO_4)_2]$	Cr—O, 1.65, Cr—O····Zn^{2+}, 1.67 Å (bidentate CrO_4^{2-})	6
$[Co(NH_3)_6](CrO_4)Cl·3H_2O$	Cr—O (av), 1.64 Å; uncommon coexistence of CrO_4^{2-} and Cl^-	7
$Hf_4(OH)_8(CrO_4)_4·H_2O$	Cr—O (av), 1.65 Å, chains of edge-sharing pentagonal bipyramids of HfO_8 interconnected by CrO_4 tetrahedra; Zr compound isomorphous	8
$M^IM_3^{III}(OH)_6(CrO_4)_2$	M^I = K, M^{III} = Fe, Cr—O (av), 1.63 Å M^I = Na, M^{III} = Al, Cr—O (av), 1.69 Å	9
$(Ph_3SiO)_2CrO_2$	CrO_4^{2-} linked to Si by O bridges Cr—O, 1.54, Cr—O(Si), 1.74 Å Si—O—Cr, 133.1° and 162.7°	10
$KBi(CrO_4)(Cr_2O_7)·H_2O$	$[CrO_4]$, Cr—O (av), 1.653 Å $[Cr_2O_7]$, Cr—O (av), 1.623, Cr—O (br), 1.782 Å, Cr—O—Cr, 118.2°	11
α-$Al_2(CrO_4)_2(Cr_2O_7)·4H_2O$	Cr—O (av), 1.65 Å Cr—O (br), 1.76 Å	12
$Rb_2Cr_2O_7$	$P2_1/n$ form, Cr—O (av), 1.596 Å Cr—O (br), 1.796 Å, Cr—O—Cr, 122.9°	13
$(NH_4)_2Cr_2O_7$	Cr—O, 1.634, Cr—O (br), 1.781 Å Cr—O—Cr, 121.0°	14
$CaCr_2O_7·2[(CH_2)_6N_4]·7H_2O$	Cr—O (av), 1.595, Cr—O (br), 1.764 Å Cr—O—Cr, 129.0°	15
$[C(NH_2)_3]_2Cr_3O_{10}$	Cr—O (av). 165. Cr—O—Cr (av), 1.75 Å Cr—O—Cr, 132.7°	16
$RbCr_4O_{13}$	Cr—O, 1.576–1.621, Cr—O (br), 1.691–1.846 Å Cr—O—Cr, 147.2°, 139.3°, 120.5°	17
$BaH[O_3CrOPO_2OCrO_3]·H_2O$	Cr(1)—O, 1.585, 1.630, 1.600, Cr(1)—O···P, 1.826 Å Cr(2)—O, 1.616, 1.611, 1.602, Cr(2)—O···P, 1.828 Å	18

1. J. S. Stephens and D. W. J. Cruikshank, *Acta Crystallogr., Sect. B*, 1970, **26**, 222.
2. J. S. Stephens and D. W. J. Cruikshank, *Acta Crystallogr., Sect. B*, 1970, **26**, 437.
3. O. Lindgren, *Acta Chem. Scand., Ser. A*, 1977, **31**, 167.
4. O. Bars, J. Y. Le Marouille and D. Grandjean, *Acta Crystallogr., Sect. B*, 1977, **33**, 3751.
5. M. L. Hackert and R. A. Jacobson, *J. Solid State Chem.*, 1971, **3**, 364.
6. M. Harel, C. Knobler and J. D. McCullough, *Inorg. Chem.*, 1969, **8**, 11.
7. B. N. Figgis, B. W. Skelton and A. H. White, *Aust. J. Chem.*, 1979, **32**, 417.
8. M. Hansson and W. Mark, *Acta Chem. Scand.*, 1973, **27**, 3467.
9. Y. Cudennec, A. Riou, A. Bonnin and P. Caillet, *Rev. Chim. Miner.*, 1980, **17**, 158.
10. B. Stensland and P. Kierkegaard, *Acta Chem. Scand.*, 1970, **24**, 211.
11. A. Riou, Y. Gerault and Y. Cudennec, *Acta Crystallogr., Sect. B*, 1982, **38**, 1693.
12. Y. Cudennec, *J. Inorg. Nucl. Chem.*, 1977, **39**, 1711.
13. P. Löfgren, *Acta Chem. Scand.*, 1971, **25**, 44.
14. G. A. P. Dalgaard, A. C. Hazell and R. G. Hazell, *Acta Chem. Scand., Ser. A*, 1974, **28**, 541.
15. F. Dahan, *Acta Crystallogr., Sect. B*, 1975, **31**, 423.
16. A. Stepien and M. J. Grabowski, *Acta Crystallogr., Sect. B*, 1977, **33**, 2924.
17. P. Löfgren, *Acta Crystallogr., Sect. B*, 1973, **29**, 2141.
18. M. T. Averbuch-Pouchot, A. Durif and J. C. Guitel, *Acta Crystallogr., Sect. B*, 1977, **33**, 1431.

In recent investigations (see also Section 35.6.2) the double chromates $KLn(CrO_4)_2$ (Ln = La–Sm), $RbLn(CrO_4)_2$ (Ln = La–Nd) and $CsLa(CrO_4)_2$ have been found to be isostructural with croconite ($PbCrO_4$),[1403] the chromates $A_4Pb_4(CrO_4)_6$ (A = K, NH_4) have been obtained,[1404] and the reactions of chromate, dichromate or CrO_3 in alkali carbonate[1405] and bisulfate[1406] eutectics and fused sodium nitrite[1407] have been investigated. Interaction between Ag^+ and CrO_4^- in $(K, Na)NO_3$ melts leads to the formation of the weak complexes $[AgCrO_4]^-$, $[Ag_2CrO_4]$ and $[Ag(CrO_4)_2]^{3-}$.[1408]

(*i*) *Spectroscopic investigations*

There have been a number of investigations of the electronic and IR spectra of chromates, and attempts to account for the electronic spectrum by molecular orbital theory.[1306,1409]

From the splittings of the IR absorption bands of CrO_4^{2-} it can be determined whether the anion is mono- or bi-dentate, *e.g.* it is monodentate in $[Cr(CrO_4)(NH_3)_5]I$ and bidentate in $[Cr(CrO_4)(NH_3)_4]I$.[1410] Complete vibrational assignments have been made for K_2CrO_4;[1411] and molecular K_2CrO_4 has been isolated in an argon matrix; the IR spectrum shows that its structure is D_{2d} and the O—Cr—O bond angles are $96 \pm 5°$.[1412] RR spectroscopy has been used to investigate the unusual red colour of Ag_2CrO_4.[1413] Crystallographic and IR studies on the hydroxychromates $M^IM_3^{III}(OH)_6(CrO_4)_2$ (M^I = Na, K, NH_4; M^{III} = Fe, Al) have confirmed that they are isomorphous.[1414]

The NMR shifts of the ^{17}O spectra of CrO_4^{2-}, $Cr_2O_7^{2-}$ and CrO_2Cl_2 have been related to the π character of the Cr—O linkages, and from them it has been deduced that the Cr—O—Cr linkage in dichromate is angular rather than linear in solution as well as in the solid.[1415]

(*ii*) *Structures of chromates and polychromates*

The tetrahedral structure of the anion in simple chromates such as K_2CrO_4 has been confirmed by many single crystal X-ray studies. Within the CrO_4 groups (**291**) the O—Cr—O angles are close to the tetrahedral value and the Cr—O bond distances are essentially equal and close to 1.6 Å unless there is interaction of some oxygen atoms with other cations, *e.g.* Ce^{4+} in $Ce(CrO_4)_2 \cdot 2H_2O$ (Table 112). Chromates are often isomorphous with corresponding sulfates.

The dichromates consist of two distorted CrO_4 tetrahedra with a corner in common. Rubidium dichromate, for example, exists in several crystal forms which contain anions shaped like (**292**); the bridging Cr—O bonds are considerably longer (~1.8 Å) than the terminal Cr—O bonds, which are similar to those in CrO_4^{2-}, and the Cr—O—Cr angles are 123°. In $SrCr_2O_7$ there are distinct $Cr_2O_7^{2-}$ units of types (**292**) and (**293**) with Cr—O—Cr angles of 133° and 140° respectively.[1306] A more accurate determination of the structure of $(NH_4)_2Cr_2O_7$ has shown that $Cr—O_{br}$ is 1.78 Å (not 1.98 Å as earlier reported).

The structures of the tri- and tetra-chromates, *e.g.* $Rb_2Cr_3O_{10}$ (**294**) and $Rb_2Cr_4O_{13}$ (**295**), and of polymeric CrO_3 (**296**) follow the same general pattern, although the detailed structure can vary with the cation.

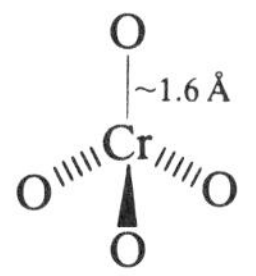

(**291**) CrO_4^{2-}

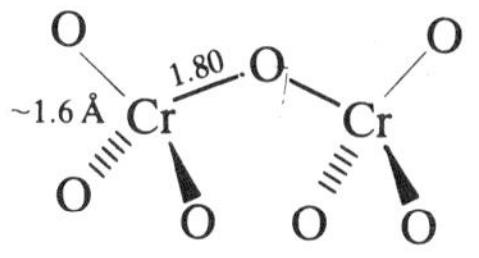

(**292**) $Cr_2O_7^{2-}$

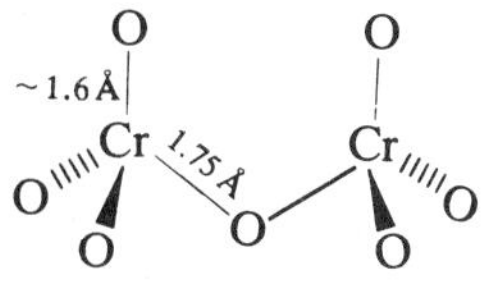

(**293**) $Cr_2O_7^{2-}$

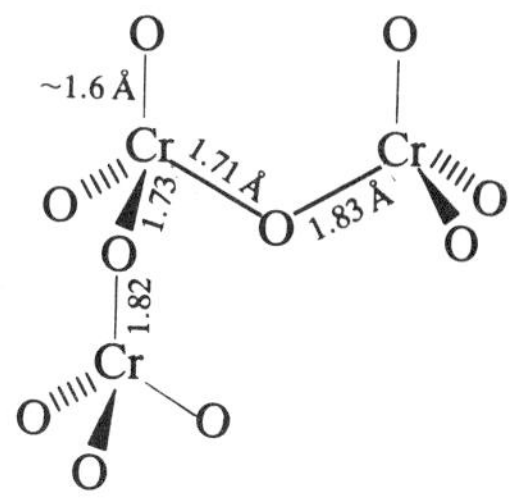

(**294**) $Cr_3O_{10}^{2-}$

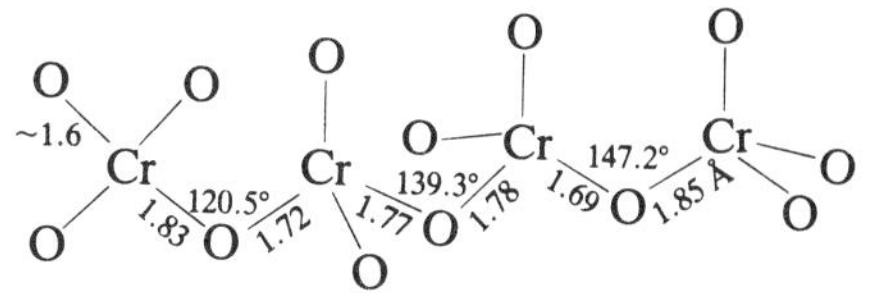

(**295**) $Cr_4O_{13}^{2-}$

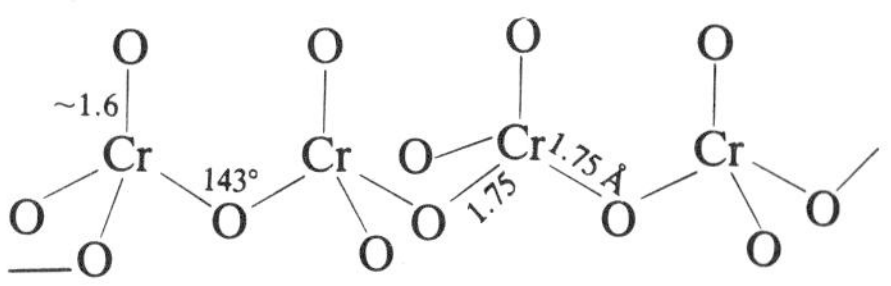

(**296**) CrO_3

The Cr—O—Cr angle varies between 115 and 148° and the Cr—O_{br} distance between 1.7 and 1.9 Å in the condensed chromates. Brief structural details for examples of the different chromates are in Table 112. There are earlier results in Ref. 1306.

Chromates(VI) of the divalent metals (V, Ni, Co, Cu, Zn and Cd) are isomorphous, and each Cr atom is surrounded by six oxygen atoms in distorted, edge-sharing octahedra.[1306]

35.7.4 Anionic Oxo Halides and Other Substituted Chromates

The equilibria (111) to (115) exist only in HNO_3 and $HClO_4$ and are modified when other anions are present because of the formation of substituted chromates, *e.g.* chlorochromate and sulfatochromate, as in equations (119) and (120). Dichromate is slow to oxidize, otherwise no halochromates could be isolated.[2,222]

$$HCr_2O_7^- + HCl \rightleftharpoons CrO_3Cl^- + H_2O \quad (119)$$

$$HCr_2O_7^- + HSO_4^- \rightleftharpoons [CrO_3SO_4]^{2-} + H_2O \quad (120)$$

Salts of the trioxohalochromates(VI) $[CrO_3X]^-$ (X = F, Cl, Table 113) can be crystallized from solutions of the appropriate dichromate and HX and have uses as oxidants of organic compounds. The fluorochromates can also be obtained from CrO_3, HF and the alkali metal carbonate, and the chlorochromates from CrO_3, MCl and HCl. Potassium trioxochlorochromate may also be prepared from K_2CrO_4 and CrO_2Cl_2. Salts of $[CrO_3Br]^-$ and $[CrO_3I]^-$ have been reported but they decompose through oxidation of the halide.

Table 113 Fluoro- and Chloro-chromates(VI)

Compound	*Comments*	*Ref.*
$M[CrO_3F]$ Red	M = K, Rb, Cs, NH_4, easily hydrolyzed; Cr—O, 1.60 and 1.62, Cr—F, 1.68 Å (NH_4) 2 Cr—O, 1.60, 2 Cr—(O, F), 1.67 Å, disorder (Rb)	1, 2
$M[CrO_3Cl]$ Orange	M = Li, K, Rb, Cs, NH_4, Ph_3MeAs, hydrolyzed in water, stable in acid, Cr—O (av), 1.53 (K salt), 1.52 (NH_4), 1.606 (Rb), 1.612 Å (Cs), Cr—Cl (av), 2.16 (K salt), 2.15 (NH_4), 2.194 (Rb), 2.197 Å (Cs)	1, 3 4 5

1. R. Colton and J. H. Canterford, 'Halides of the First Row Transition Metals', Wiley-Interscience, London, 1969, vol. 1, chap. 4.
2. W. Granier, S. Vilminot, J. D. Vidal and L. Cot, *J. Fluorine Chem.*, 1981, **19**, 123.
3. J. J. Foster and A. N. Hambly, *Aust. J. Chem.*, 1977, **30**, 251.
4. R. Armstrong and N. A. Gibson, *Aust. J. Chem.*, 1968, **21**, 897.
5. J. J. Foster and M. Sterns, *J. Cryst. Mol. Struct.*, 1974, **4**, 149.

There are three major processes in the thermal decomposition of the alkali metal chlorochromates $M[CrO_3Cl]$ (M = Li, K, Rb or Cs); these are the formation of CrO_2Cl_2, MCr_3O_8 and Cr_2O_3; in each case there is simultaneous formation of $M_2Cr_2O_7$ and MCl.[1416] The rubidium salt is isostructural with the potassium and ammonium salts, but not the cesium salt. The anions are approximately tetrahedral. The average values for the Cr—O bond distances in Cs(Rb)$[CrO_3Cl]$ are rather longer than determined earlier for K(NH_4)$[CrO_3Cl]$ (Table 113).[1417] Some disorder was found in the tetrahedral $[CrO_3F]^-$ anions of $KCrO_3F$ and the average Cr—(O, F) distance is 1.65 Å.[1418]

Salts $M[CrO_3X]$ (where X = iodate, sulfate, NH_2, *etc.*) have earlier been described.[2] Adducts of $KCrO_3X$ (X = F, Cl, Br, IO_3) with primary and secondary aliphatic amines have been characterized in solution by their UV and IR spectra.[1419]

The fluorinated anhydride $(CF_3CO)_2O$ removes chloride from $K[CrO_3Cl]$ and adds across one Cr=O bond according to equation (121).[1420] The same anionic complex is obtained from $K_2Cr_2O_7$ and the anhydride; but $K_2[CrO_2(O_2CCF_3)_4]$ is formed from K_2CrO_4. Other chromates and dichromates, and other fluorinated anhydrides, react similarly. The complexes are red-brown solids, easily hydrolyzed and unstable to heat and light.

$$K[CrO_3Cl] + 2(CF_3CO)_2O \longrightarrow K[CrO_2(O_2CCF_3)_3] + CF_3COCl \quad (121)$$

Boron trifluoride adducts approximating to $K_2[CrO_4(BF_3)_2]$ and $K_2[CrO_4(BF_3)_3]$ have been reported.[1421]

The salts $A[CrO_2F_3]$, ($A = NO^+$ or NO_2^+, Scheme 123) and the other dioxo compounds below are believed to contain *cis*-dioxo groups and fluorine bridges because there are IR bands in the 900–1000 cm^{-1} region, which are assignable to the $\nu_s(CrO_2)$ and $\nu_{as}(CrO_2)$ vibrations, and bands in the regions of 550–750 cm^{-1} and 350 cm^{-1} assignable to terminal and bridging fluoride respectively.[1422] Salts of the anions $[CrO_4F]^{3-}$ and $[CrO_2F_4]^{2-}$ have been described;[2] and $M_2[CrO_2F_4]$ (M = Na, K, Cs), $M[CrO_2F_4]$ (M = Ca, Mg)[1370] and $Na_2[CrO_2F_2(NO_3)_2]$ can be derived from CrO_2F_2 (Scheme 123).[1380]

35.7.5 Organoimidochromium(VI) Complexes

Among the *t*-alkylimido complexes of the d^0 transition metals of groups V to VII is deep red $(Me_3SiO)_2Cr(NBu^t)_2$, which can be prepared from CrO_3 or CrO_2Cl_2 and *t*-butyl(trimethylsilyl)amine and is monomeric from preliminary X-ray data. After removal of the hexane (equation 122), the residue is dissolved in hexamethyldisiloxane and the imide separates at −40 °C.[1423] The preparation from CrO_2Cl_2 (equation 123) has some advantages.[1424] The diimide forms the oxo derivative $(Me_3SiO)_2CrO(NBu^t)$ on reaction with benzaldehyde.

$$CrO_3 + NHBu^t(SiMe_3) \xrightarrow[\text{reflux}]{\text{hexane}} (Me_3SiO)_2Cr(NBu^t)_2 \qquad (122)$$

$$\underset{\text{in excess}}{CrO_2Cl_2 + NHBu^t(SiMe_3)} \xrightarrow[\text{reflux}]{\text{hexane}} (Me_3SiO)_2Cr(NBu^t)_2 + 2[NH_2Bu^t(SiMe_3)]Cl \qquad (123)$$

The reaction of bis(trimethylsilyl)diazene with chromocene or $CpCrCl_2$ affords the dark violet crystalline imide with structure (**297**). The imide shows distinct trimethylsilyl signals in the NMR spectrum, and in methanol forms $[CpCr(NH)(NSiMe_3)]_2$.[1425]

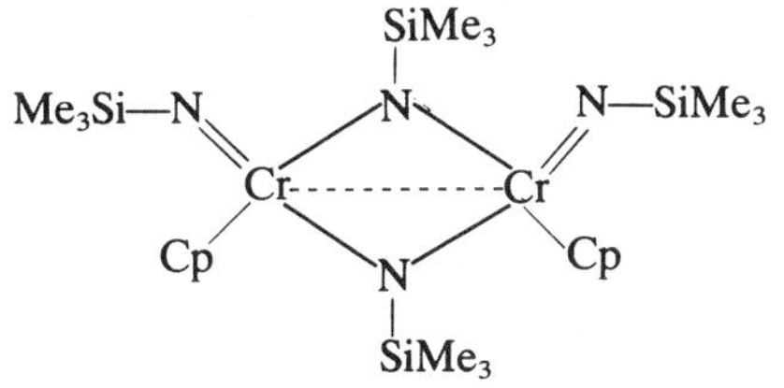

(**297**)

35.7.6 Mixed Oxidation State (III and VI) Oxo Compounds

The compounds MCr_3O_8 have long been known. In their structures $Cr^{III}O_6$ octahedra and $Cr^{VI}O_4$ tetrahedra are linked together in the ratio 1:2 by sharing corners and edges.[1306] Several new methods of preparation have been described (equations 124 to 127). Trichromates ($M_2Cr_3O_{10}$, M = K, Cs) give the same products as tetrachromates (equation 127).[1426]

$$KCrO_3Cl \xrightarrow[12\text{ h}]{275\,^\circ C} K_2Cr_2O_7 + KCr_3O_8 + 2KCl + 1.5Cl_2 \qquad (124)$$

$$KCrO_3Cl + 2CrO_3 \xrightarrow{350\text{–}400\,^\circ C} KCr_3O_8 + 0.5Cl_2 + 0.5O_2 \qquad (125)$$

$$3MX + 5CrO_3 \xrightarrow{250\,^\circ C} M_2Cr_2O_7 + MCr_3O_8 + 1.5X_2 \qquad (126)$$

$$2K_2Cr_4O_{13} \xrightarrow{\Delta} K_2Cr_2O_7 + 2KCr_3O_8 + 1.5O_2 \qquad (127)$$

From the XPES, Cr_2O_5 can be described as a mixed valence compound containing Cr^{3+} and Cr^{6+} in the ratio 1:2 as in KCr_3O_8.[1427] The oxide Cr_5O_{12} is formulated $Cr_2(CrO_4)_3$.[1306]

35.7.7 Peroxo Complexes of Chromium(IV), (V) and (VI)

The action of H_2O_2 on acidic dichromate solutions gives an unstable blue species which can be extracted into ether (equation 128).[2,518,1428] On the addition of pyridine the blue compound $Cr^{VI}O(O_2)_2py$ crystallizes. The molecules are distorted pentagonal pyramids with sideways-bonded peroxo groups (**298**);[1429] some workers with less refined data have reported rather different bond distances.[1430] In water the blue species is considered to be $CrO(O_2)_2(OH_2)$, in ether it is $CrO(O_2)_2OEt_2$, and amines such as aniline, bipy and phen can replace the pyridine. With bidentate bipy the coordination sphere becomes essentially a pentagonal bipyramid (**299**), and the phen compound is similar; it also has one Cr—N bond (2.23 Å) much longer than the other (2.11 Å).[1431] By extraction into tri-*n*-butylphosphate (TBP) containing Ph_3AsO, the adduct $CrO(O_2)_2OAsPh_3$ can be obtained.[1432]

$$HCrO_4^- + 2H_2O_2 + H^+ \longrightarrow \underset{\text{blue}}{CrO_5{\cdot}H_2O} + 2H_2O \qquad (128)$$

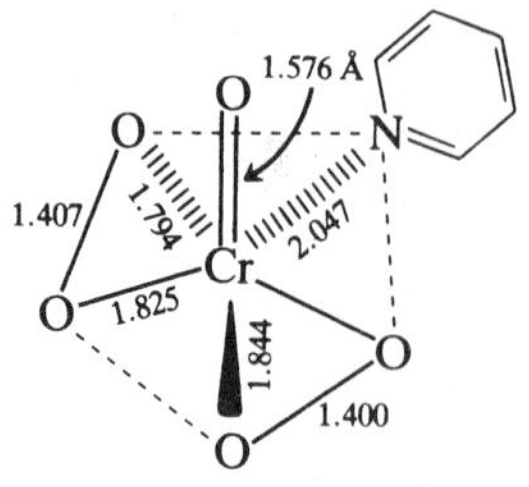

(**298**)

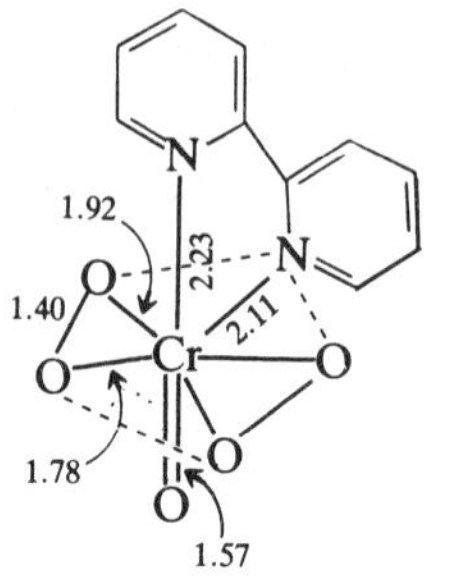

(**299**)

In neutral or weakly acidic solutions of dichromates and H_2O_2 the violet species $CrO_5(OH)^-$ forms (equation 129) and explosive violet salts $M[Cr^{VI}O(O_2)_2OH]$ (M = NH_4, K, Tl) can be isolated.[1433] The salt $[Ph_3MeAs][CrO(O_2)_2OH]$ is more stable, and from the lack of an OH band in the IR spectrum it is believed that the anion contains a strong hydrogen bond; related blue peroxohalochromates $[Ph_3MeAs][CrO(O_2)_2X]$ (where X is Cl or Br) are also known (Scheme 124).[1432] Kinetic studies show that the blue and the violet anions are formed by similar mechanisms.[1434]

$$HCrO_4^- + 2H_2O_2 \longrightarrow CrO_5(OH)^- + 2H_2O \qquad (129)$$

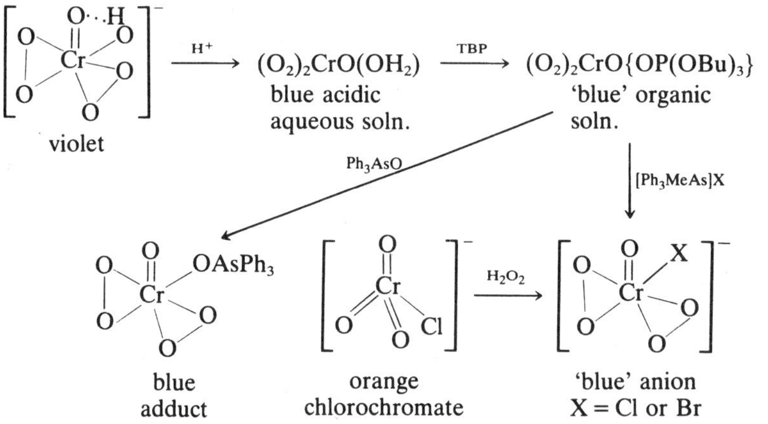

Scheme 124

The unstable, violet black compound $(NH_4)_2[Cr_2^{VI}O_2(O_2)_5]{\cdot}2H_2O$ is produced[1435] by reaction of H_2O_2 with CrO_3 in the presence of NH_4Cl and HCl, and $Cr(O_2)(NPr^i_2)_3$ may be a peroxo chromium(V) compound (p. 138).

From the RR spectrum of $CrO(O_2)_2{\cdot}H_2O$, its instability has been related to the relatively low (952 cm^{-1}) Cr—O stretching vibration and the relatively high (1012 cm^{-1}) O—O stretching vibration.[1436] The peroxidic stretch was earlier assigned to a band in the 865 cm^{-1} region.[1433]

The chromium(V) complexes $M_3^I[Cr(O_2)_4]$, obtained from alkaline solutions, contain dodecahedral anions (see Section 35.6.2).

Brown crystals of paramagnetic, reasonably stable $Cr^{IV}(NH_3)_3(O_2)_2$ can be prepared[1437] according to equation (130). Some reactions of $Cr(NH_3)_3(O_2)_2$ are given in Scheme 125, and its structure (**300**) has been determined.[1438] The anion in $K_3[Cr(CN)_3(O_2)_2]$ is structurally similar,[1439] and the O—O bond distance is 1.45 Å. In $[Cr(en)H_2O(O_2)_2]{\cdot}H_2O$ this distance is 1.46 Å.[1440]

$$CrO_3 + 3NH_3 + 3H_2O_2 \xrightarrow{<-6\,^\circ C} Cr(NH_3)_3(O_2)_2 + 3H_2O + O_2 \qquad (130)$$

The action of H_2O_2 on aqueous solutions of CrO_3 containing amines gives the chromium(IV) complexes $[CrNH_3en(O_2)_2]{\cdot}H_2O$, $[Crdien(O_2)_2]{\cdot}H_2O$, $[CrenH_2O(O_2)_2]{\cdot}H_2O$, $[CrpnH_2O(O_2)_2]{\cdot}2H_2O$ and $[CribnH_2O(O_2)_2]{\cdot}H_2O$. These and $Cr(NH_3)_3(O_2)_2$ are useful sources of Cr^{III}–

$$[Cr^{III}(NH_3)_3H_2OCl_2]Cl \xleftarrow{\text{HCl}} Cr^{IV}(NH_3)_3(O_2)_2 \xrightarrow{\text{acid}} O_2 + Cr^{III}$$

$$Cr^{IV}(C_6H_{12}N_4)(O_2)_2 \qquad\qquad [Cr^{IV}enH_2O(O_2)_2]\cdot H_2O \qquad\qquad K_3[Cr^{IV}(CN)_3(O_2)_2]$$

hexamine / en / KCN

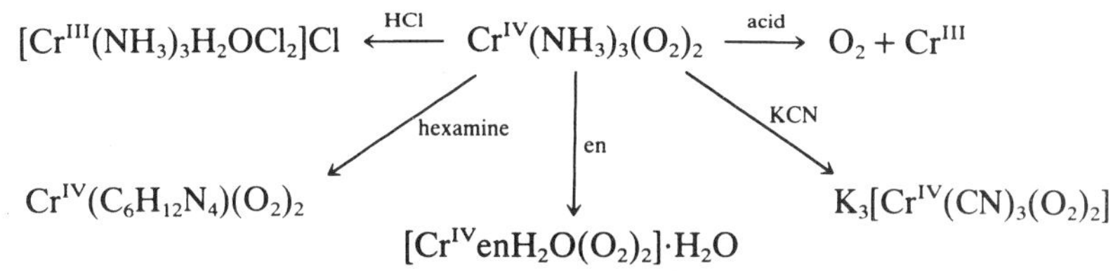

Scheme 125

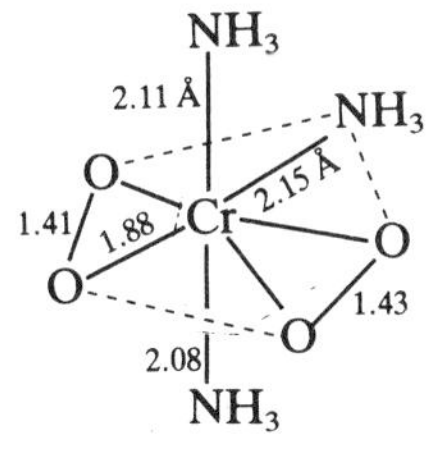

(300)

amine complexes,[302] but their explosive nature is a disadvantage and rather few amines form this type of complex so the synthetic use is limited.

Further information on peroxo chromium complexes is given in Pascal.[3]

35.7.8 Biological Effects of Chromate(VI)

The biological effects of chromate are apparent in three main areas: mutagenicity and carcinogenicity, an adverse immune response and the use of chromate in clinical chemistry to tag erythrocytes. The clinical use of chromate has recently been reviewed[1441] and although reviews of the immune response to chromate are available[1442] little is known of the chemistry involved; this is surprising as chromate present in cement is a major cause of dermatitis.

The carcinogenicity and mutagenicity of chromium(VI) are well established.[1443] The toxicity is usually considered in terms of the uptake/reduction model[1443] since chromium(VI) readily passes into the cell, *via* anion channels, and once within the cell it is eventually reduced by cellular components to chromium(III) species. Figure 6 illustrates the likely fate of chromate within a mammalian cell. As can be appreciated from Figure 6 it is complexes trapped within the cell that are the agents responsible for the toxic effects of chromate. The systems which reduce the chromium(VI) are as yet unknown, as are the final products of the reaction. However, microsomes[1444] are capable of reducing chromium(VI) as are various nucleotides[1445] and even fulvic acids.[1446] In these cases chromium(V) species of considerable stability have been observed using EPR spectroscopy. Within the cell reduction by a sulfide is the most probable reaction.

Glutathione is an intracellular peptide important in the maintenance of redox status;[1446,1447] it is found at millimolar concentrations in typical mammalian cells and is implicated, in general,

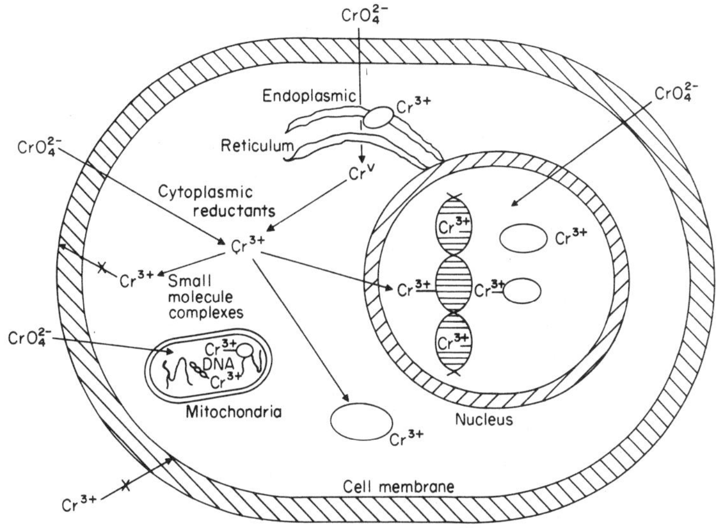

Figure 6 The uptake reduction model for chromate carcinogenicity (after Wetterhahn). Possible sites for reduction of chromate include the cytoplasm, endoplasmic reticulum, mitochondria or the nucleus[1443]

in the defence of cells challenged with oxidizing agents. However, in the case of chromate toxicity, cells with elevated levels of GSH have been shown to have an increased susceptibility to damage,[1449,1450] which perhaps indicates that a chromium complex of glutathione is an active intracellular toxin. It is most unusual for a toxic metabolite to be generated by a reaction with GSH.

The reaction of GSH with chromium(VI) has been studied in acidic solutions[1451,1452] under which the complete reduction of chromate to chromium(III) is rapid. Mechanisms involving thiolate esters of chromium(VI), which react with either protons or GSH, were suggested to be important. A detailed study of the reduction of chromate by various biological reducing agents (in neutral, buffered solutions, at low GSH concentrations) indicates that chromium(V) species are not generated[1453] and also suggests the importance of thiolate esters.

However at neutral pH, solutions of chromium(VI) containing an excess of glutathione (1:10) rapidly develop a green colour,[1454] which slowly decays to a purple colour typical of chromium(III) complexes. Concurrent with the green colour the typical EPR spectrum of a chromium(V) species is observed. Chromium(V) has also been observed in the EPR spectrum of frozen glasses obtained from reaction mixtures of chromium(VI) and glutathione.[1455]

A related reaction, that of L-cysteine with Cr^{VI} at neutral pH, has been studied in detail;[1456] the product was suggested to be the bis-*trans-S* complex of cysteine characterized crystallographically by DeMeester *et al.*,[1187] but no evidence for Cr^{V} species could be obtained, and the rate-determining step was suggested to involve two-electron transfer. It has been reported[1457] that initial one-electron reduction of chromium(VI) leads to chromium(III) complexes in which the oxidized form of the substrate is coordinated. The presence of chromium(V) in GSH/chromate reaction mixtures suggests that GSSG may be captured in the reaction of GSH with chromium(VI). This is in marked contrast to the results of Pennington[1456] with L-cysteine, but Wetterhahn[1443] has some evidence for Cr^{III}/GSSG complexes from such reactions.

The way in which the dominant reduction mechanism for chromate changes with the reaction conditions and how this is related to the toxicity of chromate is not as yet clear. As outlined above, the products of the reaction may depend on the mechanism of reduction and these, as yet, unidentified chromium complexes are probably the agents responsible for the mutagenicity of chromate. The substantial stability of the chromium(V) complexes and thiolate esters generated in the reaction of GSH with chromate suggests that if similar complexes were formed *in vivo* they would have time to reach many intracellular compartments and could hence be the crucial active intermediates in the toxicity of chromate.

The complexes responsible for the toxicity have not yet been identified. The possibilities include: the thiolate ester, a relatively stable chromium(V) species, or a chromium(III) complex. More work is needed in this area.

35.8 REFERENCES

1. G. Wilkinson, F. G. A. Stone and E. W. Abel (eds.), 'Comprehensive Organometallic Chemistry', Pergamon, Oxford, 1982, vol. 3.
2. C. L. Rollinson, in 'Comprehensive Inorganic Chemistry', ed. J. C. Bailar, Jr., H. J. Emeléus, R. S. Nyholm and A. F. Trotman-Dickenson, Pergamon, Oxford, 1973, vol. 3.
3. P. Pascal (ed.), 'Nouveau Traité de Chimie Minerale', Masson, Paris, 1959, vol. 14.
4. 'Gmelins Handbuch der Anorganische Chemie, Teil 52C, Chromium', Springer-Verlag, Berlin, 1965.
5. L. F. Larkworthy, *Coord. Chem. Rev.*, 1981, **37,** 91; 1982, **45,** 105; 1984, **57,** 189.
6. R. Colton, *Coord. Chem. Rev.*, 1984, **58,** 245; 1985, **62,** 85.
7. I. D. Brown, *Coord. Chem. Rev.*, 1978, **26,** 161.
8. F. A. Cotton and G. Wilkinson, 'Advanced Inorganic Chemistry', 4th edn., Wiley-Interscience, New York, 1980.
9. N. N. Greenwood and A. Earnshaw, 'Chemistry of the Elements', Pergamon, Oxford, 1984.
10. J. P. Eaton and D. Nicholls, *Transition Met. Chem.*, 1981, **6,** 203.
11. E. Ljungström, *Acta Chem. Scand., Ser. A.*, 1977, **31,** 104.
12. E. S. Dodsworth, J. P. Eaton, M. P. Ellerby and D. Nicholls, *Inorg. Chim. Acta,* 1984, **89,** 143.
13. G. J. Essenmacher and P. M. Treichel, *Inorg. Chem.*, 1977, **16,** 800.
14. D. A. Bohling, J. F. Evans and K. R. Mann, *Inorg. Chem.*, 1982, **21,** 3546.
15. P. M. Treichel, D. W. Firsich and G. P. Essenmacher, *Inorg. Chem.*, 1979, **18,** 2405.
16. Y. Yamamoto, *Coord. Chem. Rev.*, 1980, **32,** 193.
17. E. P. Kündig and P. L. Timms, *J. Chem. Soc., Dalton Trans.*, 1980, 991.
18. J. Müller and W. Holzinger, *Z. Naturforsch., Teil B,* 1978, **33,** 1309.
19. K. W. Chiu, C. G. Howard, G. Wilkinson, A. M. R. Galas and M. B. Hursthouse, *Polyhedron,* 1982, **1,** 803.
20. P. M. Treichel and G. J. Essenmacher, *Inorg. Chem.*, 1976, **15,** 146.

21. K. R. Mann, H. B. Gray and G. S. Hammond, *J. Am. Chem. Soc.*, 1977, **99**, 306.
22. W. S. Mialki, D. E. Wigley, T. E. Wood and R. A. Walton, *Inorg. Chem.*, 1982, **21**, 480.
23. F. R. Lemke, D. E. Wigley and R. A. Walton, *J. Organomet. Chem.*, 1983, **248**, 321.
24. F. Hein and S. Herzog, in 'Handbook of Preparative Inorganic Chemistry', 2nd edn., ed. G. Brauer, Academic, New York, 1965, vol. 2, p. 1361.
25. D. A. Bohling and K. R. Mann, *Inorg. Chem.*, 1983, **22**, 1561.
26. J. C. Dewan, W. S. Mialki, R. A. Walton and S. J. Lippard, *J. Am. Chem. Soc.*, 1982, **104**, 133.
27. E. Ljungström, *Acta Chem. Scand., Ser. A*, 1978, **32**, 47.
28. D. A. Bohling and K. R. Mann, *Inorg. Chem.*, 1984, **23**, 1426.
29. W. S. Mialki and R. A. Walton, *Inorg. Chem.*, 1982, **21**, 2791.
30. G. L. Silver, *Inorg. Nucl. Chem. Lett.*, 1968, **4**, 533.
31. T. Kruck and W. Molls, *Z. Naturforsch., Teil B*, 1974, **29**, 198.
32. W. R. McWhinnie and J. D. Miller, *Adv. Inorg. Chem. Radiochem.*, 1969, **12**, 135.
33. J. Quirk and G. Wilkinson, *Polyhedron*, 1982, **1**, 209.
34. S. Herzog, U. Grimm and W. Waicenbauer, *Z. Chem.*, 1967, **7**, 355.
35. A. Earnshaw, L. F. Larkworthy, K. C. Patel and B. J. Tucker, *J. Chem. Soc., Dalton Trans.*, 1977, 2209.
36. P. M. Lutz, G. J. Long and W. A. Baker, Jr., *Inorg. Chem.*, 1969, **8**, 2529.
37. J. M. Crabtree, D. W. Marsh, J. C. Tomkinson, R. J. P. Williams and W. C. Fernelius, *Proc. Chem. Soc.*, 1961, 336.
38. Y. M. Udachin and M. E. Dyatkina, *Zh. Strukt. Khim.*, 1967, **8**, 368 (*Chem. Abstr.*, 1967, **67**, 58 546g).
39. G. N. La Mar and G. R. Van Hecke, *J. Am. Chem. Soc.*, 1969, **91**, 3442.
40. F. Mani, P. Stoppioni and L. Sacconi, *J. Chem. Soc., Dalton Trans.*, 1975, 461.
41. G. N. La Mar, *J. Am. Chem. Soc.*, 1972, **94**, 9055.
42. G. N. La Mar and G. R. Van Hecke, *Inorg. Chem.*, 1973, **12**, 1767.
43. J. H. Sutter and J. B. Hunt, *J. Am. Chem. Soc.*, 1969, **91**, 3107.
44. R. Pappalardo, *Inorg. Chim. Acta*, 1968, **2**, 209.
45. E. König and S. Herzog, *J. Inorg. Nucl. Chem.*, 1970, **32**, 585.
46. G. Albrecht, *Z. Chem.*, 1963, **3**, 182.
47. E. König, *Z. Naturforsch., Teil A*, 1964, **19**, 1139.
48. I. Fujita, T. Yazaki, Y. Torii and H. Kobayashi, *Bull. Chem. Soc. Jpn.*, 1972, **45**, 2156.
49. J. Ulstrup, *Acta Chem. Scand.*, 1971, **25**, 3397.
50. Y. Saito, J. Takemoto, B. Hutchinson and K. Nakamoto, *Inorg. Chem.*, 1972, **11**, 2003.
51. G. C. Percy and D. A. Thornton, *J. Mol. Struct.*, 1971, **10**, 39.
52. Y. Abe and G. Wada, *Bull. Chem. Soc. Jpn.*, 1981, **54**, 3334.
53. M. C. Hughes and D. J. Macero, *Inorg. Chem.*, 1976, **15**, 2040.
54. D. M. Soignet and L. G. Hargis, *Inorg. Chem.*, 1972, **11**, 2349, 2921.
55. D. M. Soignet and L. G. Hargis, *Inorg. Chem.*, 1973, **12**, 877.
56. J. J. Turner, M. B. Simpson, M. Poliakoff, W. B. Maier, II and M. A. Graham, *Inorg. Chem.*, 1983, **22**, 911.
57. H. H. Karsch, *Angew. Chem., Int. Ed. Engl.*, 1977, **16**, 56.
58. P. Sobota and B. Jezowska-Trzebiatowska, *J. Organomet. Chem.*, 1977, **131**, 341.
59. J. E. Salt, G. S. Girolami, G. Wilkinson, M. Motevalli, M. Thornton-Pett and M. B. Hursthouse, *J. Chem. Soc., Dalton Trans.*, 1985, 685.
60. T. Kruck, H. L. Diedershagen and A. Engelmann, *Z. Anorg. Allg. Chem.*, 1973, **397**, 31.
61. P. L. Timms, *Chem. Commun.*, 1969, 1033.
62. R. A. Head, J. F. Nixon, G. J. Sharp and R. J. Clark, *J. Chem. Soc., Dalton Trans.*, 1975, 2054.
63. R. B. King and S. Goel, *Synth. React. Inorg. Metal-Org. Chem.*, 1979, **9**, 139.
64. R. B. King and J. Gimeno, *Inorg. Chem.*, 1978, **17**, 2390.
65. M. Chang, M. G. Newton and R. B. King, *Inorg. Chim. Acta*, 1978, **30**, L341.
66. R. B. King and M. Chang, *Inorg. Chem.*, 1979, **18**, 364.
67. R. Mathieu and R. Poilblanc, *Inorg. Chem.*, 1972, **11**, 1858.
68. D. F. Shriver, 'The Manipulation of Air-Sensitive Compounds', McGraw-Hill, New York, 1969.
69. L. F. Larkworthy, *J. Chem. Soc.*, 1961, 4025.
70. E. Deutsch, M. J. Root and D. L. Nasco, *Adv. Inorg. Bioinorg. Mech.*, 1982, **1**, 269.
71. M. Kranz and W. Duczmal, *Rocz. Chem.*, 1973, **47**, 1823.
72. H. Lux and G. Illmann, *Chem. Ber.*, 1958, **91**, 2143.
73. D. N. Hume and H. W. Stone, *J. Am. Chem. Soc.*, 1941, **63**, 1197.
74. L. F. Larkworthy and M. H. O. Nelson-Richardson, *Chem. Ind.* (*London*), 1974, 164.
75. C. Bellitto and P. Day, *J. Chem. Soc., Dalton Trans.*, 1978, 1207.
76. L. F. Larkworthy and M. H. O. Nelson-Richardson, *Inorg. Chim. Acta*, 1978, **28**, 251.
77. J. Hvoslef, H. Hope, B. D. Murray and P. P. Power, *J. Chem. Soc., Chem. Commun.*, 1983, 1438.
78. A. I. Vogel, 'A Textbook of Quantitative Inorganic Analysis', 3rd edn., Longmans, London, 1961.
79. T. L. Ho, *Synthesis*, 1979, 1.
80. C. A. Reed, J. K. Kouba, C. J. Grimes and S. K. Cheung, *Inorg. Chem.*, 1978, **17**, 2666.
81. M. G. Segal and R. M. Sellers, *J. Chem. Soc., Chem. Commun.*, 1980, 991.
82. J. J. Habeeb, F. F. Said and D. G. Tuck, *J. Chem. Soc., Dalton Trans.*, 1981, 118.
83. D. N. Hume and H. W. Stone, *J. Am. Chem. Soc.*, 1941, **63**, 1200.
84. L. F. Larkworthy and J. M. Tabatabai, *J. Chem. Soc., Dalton Trans.*, 1976, 814.
85. H. Lux, L. Eberle and D. Sarre, *Chem. Ber.*, 1964, **97**, 503.
86. D. Reinen and C. Friebel, *Struct. Bonding* (*Berlin*), 1979, **37**, 1.
87. A. Earnshaw, L. F. Larkworthy and K. S. Patel, *J. Chem. Soc.* (*A*), 1969, 1339.
88. L. F. Larkworthy, K. C. Patel and J. K. Trigg, *J. Chem. Soc.* (*A*), 1971, 2766.
89. A. Earnshaw, L. F. Larkworthy and K. C. Patel, *J. Chem. Soc.* (*A*), 1969, 2276.
90. A. Earnshaw, L. F. Larkworthy and K. S. Patel, *Chem. Ind.* (*London*), 1965, 1521.

91. D. G. Holah and J. P. Fackler, Jr., *Inorg. Chem.*, 1965, **4,** 1112.
92. M. M. Khamar, L. F. Larkworthy and D. J. Phillips, *Inorg. Chim. Acta,* 1979, **36,** 223.
93. M. M. Khamar, L. F. Larkworthy and M. H. O. Nelson-Richardson, *Inorg. Chim. Acta,* 1978, **28,** 245.
94. R. J. H. Clark and C. S. Williams, *Inorg. Chem.*, 1965, **4,** 350.
95. I. L. Eremenko, A. A. Pasynskii, V. T. Kalinnikov, G. G. Aleksandrov and Yu. T. Struchkov, *Inorg. Chim. Acta,* 1981, **54,** L85.
96. G. Beech, C. T. Mortimer and E. G. Tyler, *J. Chem. Soc. (A),* 1969, 512.
97. J. A. Cooper, B. F. Anderson, P. D. Buckley and L. F. Blackwell, *Inorg. Chim. Acta,* 1984, **91,** 1.
98. F. A. Cotton, G. E. Lewis, C. A. Murillo, W. Schwotzer and G. Valle, *Inorg. Chem.*, 1984, **23,** 4038.
99. F. Mani and G. Scapacci, *Inorg. Chim. Acta,* 1976, **16,** 163.
100. L. F. Larkworthy and J. M. Tabatabai, *Inorg. Chim. Acta,* 1977, **21,** 265.
101. P. Dapporto and F. Mani, *J. Chem. Res. (S),* 1979, 374.
102. D. M. Bowers and A. I. Popov, *Inorg. Chem.*, 1968, **7,** 1594.
103. A. Earnshaw, L. F. Larkworthy and K. C. Patel, *J. Chem. Soc. (A),* 1970, 1840.
104. A. Dei and F. Mani, *J. Chem. Res. (S),* 1981, 358.
105. F. Mani and R. Morassi, *Inorg. Chim. Acta,* 1979, **36,** 63.
106. F. Mani, *Inorg. Chim. Acta,* 1982, **65,** L197.
107. M. Di Vaira and F. Mani, *Inorg. Chem.*, 1984, **23,** 409.
108. F. Mani, *Inorg. Nucl. Chem. Lett.*, 1981, **17,** 45.
109. A. Earnshaw, L. F. Larkworthy and K. S. Patel, *Z. Anorg. Allg. Chem.*, 1964, **334,** 163; R. Ya. Aliev, *Zh. Obshch. Khim.*, 1972, **42,** 2370.
110. G. B. Kauffman and N. Sugisaka, *Z. Anorg. Allg. Chem.*, 1966, **344,** 92.
111. P. Glavic, J. Slivnik and A. Bole, *Vestn. Slov. Kem. Drus.*, 1982, **29,** 227.
112. D. C. Bradley, M. B. Hursthouse, C. W. Newing and A. J. Welch, *J. Chem. Soc., Chem. Commun.*, 1972, 567.
113. B. Horvath, J. Strutz and E. G. Horvath, *Z. Anorg. Allg. Chem.*, 1979, **457,** 38.
114. L. F. Larkworthy, A. J. Roberts, B. J. Tucker and A. Yavari, *J. Chem. Soc., Dalton Trans.*, 1980, 262.
115. P. Dapporto, F. Mani and C. Mealli, *Inorg. Chem.*, 1978, **17,** 1323.
116. F. A. Cotton and G. N. Mott, *Inorg. Chem.*, 1983, **22,** 1136.
117. P. Burchill and M. G. H. Wallbridge, *Inorg. Nucl. Chem. Lett.*, 1976, **12,** 93.
118. V. H. Crawford, W. E. Hatfield, C. G. Salentine, K. P. Callahan and M. F. Hawthorne, *Inorg. Chem.*, 1979, **18,** 1600.
119. D. G. Holah and J. P. Fackler, Jr., *Inorg. Chem.*, 1966, **5,** 479.
120. K. Issleib and H. O. Fröhlich, *Z. Anorg. Allg. Chem.*, 1959, **298,** 84.
121. W. Seidel and P. Scholz, *Z. Chem.*, 1978, **18,** 149.
122. J. Ellermann, K. Hagen and H. L. Krauss, *Z. Anorg. Allg. Chem.*, 1982, **487,** 130.
123. B. Rebenstorf, B. Jonson and R. Larsson, *Acta Chem. Scand., Ser. A,* 1982, **36,** 695.
124. L. F. Warren and M. A. Bennett, *Inorg. Chem.*, 1976, **15,** 3126.
125. S. D. Ittel, F. A. Van-Catledge and C. A. Tolman, *Inorg. Chem.*, 1985, **24,** 62; F. A. Van Catledge, S. D. Ittel and J. P. Jesson, *Organometallics,* 1985, **4,** 18.
126. F. Mani and P. Stoppioni, *Inorg. Chim. Acta,* 1976, **16,** 177.
127. F. Mani and L. Sacconi, *Inorg. Chim. Acta,* 1970, **4,** 365.
128. M. Ciampolini, *Chem. Commun.*, 1966, 47.
129. M. Ciampolini and N. Nardi, *Inorg. Chem.*, 1966, **5,** 1150.
130. A. Earnshaw, L. F. Larkworthy and K. S. Patel, *J. Chem. Soc.*, 1965, 3267.
131. J. P. Fackler, Jr. and D. G. Holah, *Inorg. Chem.*, 1965, **4,** 954.
132. H. G. von Schnering and B.-H. Brand, *Z. Anorg. Allg. Chem.*, 1973, **402,** 159.
133. K. Ono, *J. Phys. Soc. Jpn.*, 1957, **12,** 1231.
134. A. Earnshaw, L. F. Larkworthy, K. C. Patel and G. Beech, *J. Chem. Soc. (A),* 1969, 1334.
135. D. W. Hand and C. K. Prout, *J. Chem. Soc. (A),* 1966, 168.
136. H. L. Schläfer and H. Skoludek, *Z. Phys. Chem.*, 1957, **11,** 277.
137. B. Horvath and E. G. Horvath, *Z. Anorg. Allg. Chem.*, 1979, **457,** 51.
138. J. Votinsky, J. Kalousova, M. Nadvornik, J. Klikorka and K. Komarek, *Collect. Czech. Chem. Commun.*, 1979, **44,** 80.
139. M. H. Chisholm, F. A. Cotton, M. W. Extine and D. C. Rideout, *Inorg. Chem.*, 1979, **18,** 120.
140. R. W. Adams, E. Bishop, R. L. Martin and G. Winter, *Aust. J. Chem.*, 1966, **19,** 207.
141. T. Ouhadi, J. P. Bioul, C. Stevens, R. Warin, L. Hocks and P. Teyssié, *Inorg. Chim. Acta,* 1976, **19,** 203.
142. G. Costa and A. Puxeddu, *J. Inorg. Nucl. Chem.*, 1958, **8,** 104.
143. L. R. Ocone and B. P. Block, *Inorg. Synth.*, 1966, **8,** 125.
144. F. A. Cotton, C. E. Rice and G. W. Rice, *Inorg. Chim. Acta,* 1977, **24,** 231.
145. R. Nast and H. Rückemann, *Chem. Ber.*, 1960, **93,** 2329.
146. D. H. Gerlach and R. H. Holm, *Inorg. Chem.*, 1969, **8,** 2292.
147. R. J. Kern, *J. Inorg. Nucl. Chem.*, 1962, **24,** 1105.
148. D. E. Scaife, *Aust. J. Chem.*, 1967, **20,** 845.
149. L. Kavan and M. Ebert, *Collect. Czech. Chem. Commun.*, 1979, **44,** 1353.
150. S. A. A. Zaidi, M. Shakir, M. Aslam, S. R. A. Zaidi and Z. A. Siddiqi, *Indian J. Chem., Sect. A,* 1983, **22,** 1078.
151. A. J. Saraceno and B. P. Block, *Inorg. Chem.*, 1964, **3,** 1699.
152. K. D. Maguire, *Inorg. Synth.*, 1970, **12,** 258.
153. H. D. Gillman, *Inorg. Chem.*, 1974, **13,** 1921.
154. J. N. Van Niekerk, F. R. L. Schoening and J. F. de Wet, *Acta Crystallogr.*, 1953, **6,** 501.
155. F. A. Cotton, B. G. DeBoer, M. D. LaPrade, J. R. Pipal and D. A. Ucko, *Acta Crystallogr., Sect. B,* 1971, **27,** 1664.
156. F. A. Cotton and R. A. Walton, 'Multiple Bonds Between Metal Atoms', Wiley, New York, 1982.
157. S. Herzog and W. Kalies, *Z. Anorg. Allg. Chem.*, 1967, **351,** 237.

158. F. A. Cotton and G. W. Rice, *Inorg. Chem.*, 1978, **17**, 2004.
159. A. A. Pasynskii, I. L. Eremenko, T. Ch. Idrisov and V. T. Kalinnikov, *Koord. Khim.*, 1977, **3**, 1205.
160. F. A. Cotton and T. R. Felthouse, *Inorg. Chem.*, 1980, **19**, 328.
161. L. F. Larkworthy and J. M. Tabatabai, *Inorg. Nucl. Chem. Lett.*, 1980, **16**, 427.
162. P. D. Ford, L. F. Larkworthy, D. C. Povey and A. J. Roberts, *Polyhedron*, 1983, **2**, 1317.
163. M. H. Chisholm, F. A. Cotton, M. W. Extine and D. C. Rideout, *Inorg. Chem.*, 1978, **17**, 3536.
164. F. A. Cotton, M. W. Extine and G. W. Rice, *Inorg. Chem.*, 1978, **17**, 176.
165. F. A. Cotton and G. W. Rice, *Inorg. Chem.*, 1978, **17**, 688.
166. (a) R. Ouahes, Y. Maouche, M.-C. Perucaud and P. Herpin, *C. R. Hebd. Seances Acad. Sci.*, 1973, **276**, 281; (b) *Nouv. J. Chim.*, 1984, **8**, 831; (c) *Polyhedron*, 1985, **4**, 1735.
167. F. A. Cotton and W. Wang, *Inorg. Chem.*, 1982, **21**, 2675.
168. M. Ardon, A. Bino, S. Cohen and T. R. Felthouse, *Inorg. Chem.*, 1984, **23**, 3450.
169. F. A. Cotton, C. E. Rice and G. W. Rice, *J. Am. Chem. Soc.*, 1977, **99**, 4704.
170. F. A. Cotton, P. E. Fanwick, R. H. Niswander and J. C. Sekutowski, *J. Am. Chem. Soc.*, 1978, **100**, 4725.
171. F. A. Cotton and S. Koch, *Inorg. Chem.*, 1978, **17**, 2021.
172. R. P. A. Sneeden and H. H. Zeiss, *J. Organomet. Chem.*, 1973, **47**, 125.
173. F. A. Cotton, R. H. Niswander and J. C. Sekutowski, *Inorg. Chem.*, 1979, **18**, 1152.
174. F. A. Cotton, T. Inglis, M. Kilner and T. R. Webb, *Inorg. Chem.*, 1975, **14**, 2023.
175. C. D. Garner, R. G. Senior and T. J. King, *J. Am. Chem. Soc.*, 1976, **98**, 3526.
176. F. A. Cotton, R. H. Niswander and J. C. Sekutowski, *Inorg. Chem.*, 1978, **17**, 3541.
177. B. E. Bursten, F. A. Cotton, A. H. Cowley, B. E. Hanson, M. Lattman and G. G. Stanley, *J. Am. Chem. Soc.*, 1979, **101**, 6244.
178. F. A. Cotton, G. W. Rice and J. C. Sekutowski, *Inorg. Chem.*, 1979, **18**, 1143.
179. F. A. Cotton and G. N. Mott, *Organometallics*, 1982, **1**, 38.
180. F. A. Cotton and G. N. Mott, *Organometallics*, 1982, **1**, 302.
181. F. A. Cotton and J. L. Thompson, *Inorg. Chem.*, 1981, **20**, 1292.
182. F. A. Cotton, W. H. Ilsley and W. Kaim, *Inorg. Chem.*, 1980, **19**, 1453.
183. F. A. Cotton, W. H. Ilsley and W. Kaim, *J. Am. Chem. Soc.*, 1980, **102**, 3464.
184. S. Baral, F. A. Cotton and W. H. Ilsley, *Inorg. Chem.*, 1981, **20**, 2696.
185. M. Benard, P. Coppens, M. L. DeLucia and E. D. Stevens, *Inorg. Chem.*, 1980, **19**, 1924.
186. F. A. Cotton, L. R. Falvello, S. Han and W. Wang, *Inorg. Chem.*, 1983, **22**, 4106.
187. A. Mitschler, B. Rees, R. Wiest and M. Benard, *J. Am. Chem. Soc.*, 1982, **104**, 7501.
188. R. A. Kok and M. B. Hall, *J. Am. Chem. Soc.*, 1983, **105**, 676.
189. R. A. Kok and M. B. Hall, *Inorg. Chem.*, 1985, **24**, 1542.
190. P. M. Atha, J. C. Campbell, C. D. Garner, I. H. Hillier and A. A. MacDowell, *J. Chem. Soc., Dalton Trans.*, 1983, 1085.
191. P. C. de Mello, W. D. Edwards and M. C. Zerner, *J. Am. Chem. Soc.*, 1982, **104**, 1440.
192. J. C. A. Boeyens and D. J. Ledwidge, *Inorg. Chem.*, 1983, **22**, 3587.
193. K. J. Cavell, C. D. Garner, G. Pilcher and S. Parkes, *J. Chem. Soc., Dalton Trans.*, 1979, 1714.
194. K. J. Cavell, C. D. Garner, J. A. Martinho-Simoes, G. Pilcher, H. Al-Samman, H. A. Skinner, G. Al-Tekhin, I. B. Walton and M. T. Zafarani-Moattar, *J. Chem. Soc., Faraday Trans. 1*, 1981, **77**, 2927.
195. C. D. Garner, I. H. Hillier, M. J. Knight, A. A. MacDowell and I. B. Walton, *J. Chem. Soc., Faraday Trans. 2*, 1980, **76**, 885.
196. A. W. Coleman, J. C. Green, A. J. Hayes, E. A. Seddon, D. R. Lloyd and Y. Niwa, *J. Chem. Soc., Dalton Trans.*, 1979, 1057.
197. M. C. Manning and W. C. Trogler, *J. Am. Chem. Soc.*, 1983, **105**, 5311.
198. P. E. Fanwick, B. E. Bursten and G. B. Kaufmann, *Inorg. Chem.*, 1985, **24**, 1165.
199. S. F. Rice, R. B. Wilson and E. I. Solomon, *Inorg. Chem.*, 1980, **19**, 3425.
200. F. A. Cotton and P. E. Fanwick, *J. Am. Chem. Soc.*, 1979, **101**, 5252.
201. A. Earnshaw, L. F. Larkworthy and K. S. Patel, *Proc. Chem. Soc.*, 1963, 281.
202. H. Suquet and C. Malard, *C. R. Hebd. Seances Acad. Sci., Ser. C*, 1968, **267**, 770.
203. T. G. Aminov, V. M. Allenov, V. V. Zelentsov and V. B. Evdokimov, *Russ. J. Phys. Chem. (Engl. Transl.)*, 1965, **39**, 368.
204. R. W. Asmussen, 'Magnetochemiske Undersögelser over Uorganiske Kompleks forbindelser', Gjellerups Forlag, Copenhagen, 1944, p. 123.
205. J. Podlaha and J. Podlahová, *Inorg. Chim. Acta*, 1971, **5**, 420.
206. N. V. Gerbeleu, G. A. Popovich, K. M. Indrichan and G. A. Timko, *Russ. J. Inorg. Chem. (Engl. Transl.)*, 1983, **28**, 1720.
207. J. Watanabe, T. Saji and S. Aoyagui, *J. Electroanal. Chem. Interfacial Electrochem.*, 1981, **129**, 369.
208. F. A. Cotton and G. W. Rice, *Inorg. Chim. Acta*, 1978, **27**, 75.
209. H. U. Kibbel, *Z. Anorg. Allg. Chem.*, 1968, **359**, 272.
210. (a) R. D. Cannon, *Inorg. Chem.*, 1981, **20**, 2341; (b) L. M. Wilson and R. D. Cannon, *Inorg. Chem.*, 1985, **24**, 4366.
211. R. D. Cannon and M. J. Gholami, *Bull. Chem. Soc. Jpn.*, 1982, **55**, 594.
212. E. H. Abbott and J. M. Mayer, *J. Coord. Chem.*, 1977, **6**, 135.
213. (a) P. Sharrock, T. Théophanides and F. Brisse, *Can. J. Chem.*, 1973, **51**, 2963; (b) W. Traube and A. Goodson, *Chem. Ber.*, 1916, **49**, 1679.
214. D. G. Holah and J. P. Fackler, Jr., *Inorg. Chem.*, 1965, **4**, 1721.
215. K. Issleib, A. Tzschach and H. O. Fröhlich, *Z. Anorg. Allg. Chem.*, 1959, **298**, 164.
216. L. F. Larkworthy and D. J. Phillips, *J. Inorg. Nucl. Chem.*, 1967, **29**, 2101.
217. J. P. Fackler, Jr. and D. G. Holah, *Inorg. Nucl. Chem. Lett.*, 1966, **2**, 251.
218. L. F. Larkworthy and R. R. Patel, *Inorg. Nucl. Chem. Lett.*, 1972, **8**, 139.
219. O. A. Ileperuma and R. D. Feltham, *Inorg. Chem.*, 1975, **14**, 3042.
220. J. R. Dorfman, C. P. Rao and R. H. Holm, *Inorg. Chem.*, 1985, **24**, 453.

221. L. F. Larkworthy and M. H. O. Nelson-Richardson, *Inorg. Chim. Acta*, 1980, **40**, 217.
222. R. Colton and J. H. Canterford, 'Halides of the First Row Transition Metals', Wiley-Interscience, London, 1969, chap. 4.
223. H. Schäfer and R. Laumans, *Z. Anorg. Allg. Chem.*, 1977, **436**, 47.
224. A. de Kozak, *Rev. Chim. Miner.*, 1971, **8**, 301.
225. E. G. Ippolitov, T. A. Tripol'skaya and B. M. Zhigarnovskii, *Russ. J. Inorg. Chem.* (*Engl. Transl.*), 1979, **24**, 299.
226. A. D. Westland, *J. Inorg. Nucl. Chem.*, 1973, **35**, 451.
227. B. Zemva, J. Zupan and J. Slivnik, *J. Inorg. Nucl. Chem.*, 1973, **35**, 3941.
228. D. R. Rosseinsky and I. A. Dorrity, *J. Phys. Chem.*, 1977, **81**, 2672.
229. E. P. F. Lee, A. W. Potts, M. Doran, I. H. Hillier, J. J. Delaney and R. W. Hawksworth, *J. Chem. Soc., Faraday Trans. 2*, 1980, **76**, 506.
230. W. O. J. Boo and J. W. Stout, *J. Chem. Phys.*, 1979, **71**, 9.
231. J. M. Friedt, J. P. Sanchez and G. K. Shenoy, *J. Chem. Phys.*, 1976, **65**, 5093.
232. C. G. Barraclough, R. W. Cockman, T. A. O'Donnell and W. S. J. Schofield, *Inorg. Chem.*, 1982, **21**, 2519.
233. J. P. Young, *Inorg. Chem.*, 1969, **8**, 825.
234. B. Frlec, D. Gantar and J. H. Holloway, *J. Fluorine Chem.*, 1982, **20**, 385.
235. A. Tressaud and J.-M. Dance, *Struct. Bonding* (*Berlin*), 1982, **52**, 87.
236. A. Earnshaw, L. F. Larkworthy and K. S. Patel, *J. Chem. Soc.* (*A*), 1966, 363.
237. S. Yoneyama and K. Hirakawa, *J. Phys. Soc. Jpn.*, 1966, **21**, 183.
238. A. J. Deyrup, *Inorg. Chem.*, 1964, **3**, 1645.
239. D. Dumora, J. Ravez and P. Hagenmuller, *J. Solid State Chem.*, 1972, **5**, 35.
240. L. Seijo, L. Pueyo and F. G. Beltran, *J. Solid State Chem.*, 1982, **42**, 28.
241. D. Oelkrug, *Struct. Bonding* (*Berlin*), 1971, **9**, 1.
242. D. Dumora, C. Fouassier, R. von der Mühll, J. Ravez and P. Hagenmuller, *C. R. Hebd. Seances Acad. Sci.*, 1971, **273**, 247.
243. C. Friebel, J. Pebler, F. Steffens, M. Weber and D. Reinen, *J. Solid State Chem.*, 1983, **46**, 253.
244. A. J. Edwards and R. D. Peacock, *J. Chem. Soc.*, 1959, 4126.
245. K. Knox, *Acta Crystallogr.*, 1961, **14**, 583.
246. H. W. Mayer, D. Reinen and G. Heger, *J. Solid State Chem.*, 1983, **50**, 213.
247. L. F. Larkworthy, J. K. Trigg and A. Yavari, *J. Chem. Soc., Dalton Trans.*, 1975, 1879.
248. M. A. Babar, L. F. Larkworthy and S. S. Tandon, *J. Chem. Soc., Dalton Trans.*, 1983, 1081.
249. M. A. Babar, L. F. Larkworthy and A. Yavari, *J. Chem. Soc., Dalton Trans.*, 1981, 27.
250. M. A. Babar, M. F. C. Ladd, L. F. Larkworthy, G. A. Leonard and D. C. Povey, *J. Crystallogr. Spectrosc. Res.*, 1983, **13**, 325.
251. M. F. C. Ladd, L. F. Larkworthy, G. A. Leonard, D. C. Povey and S. S. Tandon, *J. Chem. Soc., Dalton Trans.*, 1984, 2351.
252. P. Day, *Acc. Chem. Res.*, 1979, **12**, 236.
253. H. J. Seifert and K. Klatyk, *Z. Anorg. Allg. Chem.*, 1964, **334**, 113.
254. D. H. Leech and D. J. Machin, *J. Chem. Soc., Dalton Trans.*, 1975, 1609.
255. H. D. Hardt and G. Streit, *Z. Anorg. Allg. Chem.*, 1970, **373**, 97.
256. L. F. Larkworthy, J. K. Trigg and A. Yavari, *J. Chem. Soc., Dalton Trans.*, 1975, 1879.
257. L. F. Larkworthy and A. Yavari, *J. Chem. Soc., Dalton Trans.*, 1978, 1236.
258. C. Bellitto and P. Day, *J. Chem. Soc., Dalton Trans.*, 1978, 1207.
259. M. J. Stead and P. Day, *J. Chem. Soc., Dalton Trans.*, 1982, 1081.
260. M. A. Babar, L. F. Larkworthy and S. S. Tandon, *J. Chem. Soc., Dalton Trans.*, 1983, 1081.
261. C. Bellitto and P. Day, *J. Cryst. Growth*, 1982, **58**, 641.
262. L. F. Larkworthy and J. K. Trigg, *Chem. Commun.*, 1970, 1221.
263. C. Bellitto, T. E. Wood and P. Day, *Inorg. Chem.*, 1985, **24**, 558.
264. A. Rahman, L. A. K. Staveley, C. Bellitto and P. Day, *J. Chem. Soc., Faraday Trans. 2*, 1982, **78**, 1895.
265. A. P. Ginsberg, *Inorg. Chim. Acta Rev.*, 1971, **5**, 45.
266. M. A. Babar, M. F. C. Ladd, L. F. Larkworthy, D. C. Povey, K. J. Proctor and L. J. Summers, *J. Chem. Soc., Chem. Commun.*, 1981, 1046.
267. M. S. Matson and R. A. D. Wentworth, *Inorg. Chem.*, 1976, **15**, 2139.
268. M. A. Babar, L. F. Larkworthy and A. Yavari, *J. Chem. Soc., Dalton Trans.*, 1981, 27.
269. M. A. Babar, L. F. Larkworthy, B. Sandell, P. Thornton and C. Vuik, *Inorg. Chim. Acta*, 1981, **54**, L261.
270. L. V. Interrante (ed.), 'Extended Interactions between Metal Ions in Transition Metal Complexes', ACS Symposium Series No. 5, American Chemical Society, Washington, DC, 1974, chaps. 11–13.
271. G. L. McPherson, T. J. Kistenmacher, J. B. Folkers and G. D. Stucky, *J. Chem. Phys.*, 1972, **57**, 3771.
272. T.-I. Li and G. D. Stucky, *Acta Crystallogr., Sect. B*, 1973, **29**, 1529.
273. W. J. Crama and H. W. Zandbergen, *Acta Crystallogr., Sect. B*, 1981, **37**, 1027.
274. H. W. Zandbergen and D. J. W. Ijdo, *J. Solid State Chem.*, 1981, **38**, 199.
275. H. W. Zandbergen, *J. Solid State Chem.*, 1981, **38**, 239.
276. W. J. Crama, M. Bakker, G. C. Verschoor and W. J. A. Maaskant, *Acta Crystallogr., Sect. B*, 1979, **35**, 1875.
277. W. J. Crama, W. J. A. Maaskant and G. C. Verschoor, *Acta Crystallogr., Sect. B*, 1978, **34**, 1973.
278. J. P. Sanchez, J. M. Friedt, B. Djermouni and G. Jehanno, *J. Phys. Chem. Solids*, 1979, **40**, 585.
279. N. W. Alcock, C. F. Putnik and S. L. Holt, *Inorg. Chem.*, 1976, **15**, 3175.
280. N. Jouini, L. Guen, R. Marchand and M. Tournoux, *Ann. Chim. Fr.*, 1980, **5**, 493.
281. A. F. Wells, 'Structural Inorganic Chemistry', 5th edn., Clarendon Press, Oxford, 1984.
282. D. H. Gerlach and R. H. Holm, *Inorg. Chem.*, 1970, **9**, 588.
283. L. F. Larkworthy, D. C. Povey and B. Sandell, *Inorg. Chim. Acta*, 1984, **83**, L29.
284. N. K. Sankhla, D. C. Sehgal and R. K. Mehta, *J. Indian Chem. Soc.*, 1979, **56**, 1030.
285. D. C. Sehgal, C. P. Gupta and R. K. Mehta, *Indian J. Chem., Sect. A*, 1978, **16**, 910.

286. N. K. Sankhla, C. P. Gupta and R. K. Mehta, *Z. Phys. Chem. (Leipzig)*, 1979, **260**, 1188.
287. A. V. Ablov and N. V. Gerbeleu, *Russ. J. Inorg. Chem. (Engl. Transl.)*, 1970, **15**, 952.
288. W. Sawodny and M. Riederer, *Angew. Chem., Int. Ed. Engl.*, 1977, **16**, 859.
289. W. Sawodny, M. Riederer and E. Urban, *Inorg. Chim. Acta*, 1978, **29**, 63.
290. M. Riederer and W. Sawodny, *Angew. Chem., Int. Ed. Engl.*, 1978, **17**, 610.
291. K. S. Siddiqi, M. R. H. Siddiqi, P. Khan, S. Khan and S. A. A. Zaidi, *Synth. React. Inorg. Metal-Org. Chem.*, 1982, **12**, 521.
292. R. A. Henderson and A. G. Sykes, *Inorg. Chem.*, 1980, **19**, 3103.
293. R. L. Pecsok, L. D. Shields and W. P. Schaefer, *Inorg. Chem.*, 1964, **3**, 114.
294. K. Micskei, F. Debreczeni and I. Nagypál, *J. Chem. Soc., Dalton Trans.*, 1983, 1335.
295. I. Nagypãl, K. Micskei and F. Debreczeni, *Inorg. Chim. Acta*, 1983, **77**, L161.
296. L. F. Larkworthy, J. M. Murphy and D. J. Phillips, *Inorg. Chem.*, 1968, **7**, 1436.
297. G. J. Samuels and J. H. Espenson, *Inorg. Chem.*, 1980, **19**, 233.
298. I. K. Adzamli, R. A. Henderson, J. D. Sinclair-Day and A. G. Sykes, *Inorg. Chem.*, 1984, **23**, 3069.
299. F. Mani, *Inorg. Chim. Acta*, 1982, **60**, 181.
300. A. Dei and F. Mani, *Inorg. Chem.*, 1976, **15**, 2574.
301. L. S. Forster, *Transition Met. Chem.*, 1969, **5**, 1.
302. C. S. Garner and D. A. House, *Transition Met. Chem.*, 1970, **6**, 59.
303. J. O. Edwards, F. Monacelli and G. Ortaggi, *Inorg. Chim. Acta*, 1974, **11**, 47.
304. H. P. Jensen and F. Woldbye, *Coord. Chem. Rev.*, 1979, **29**, 213.
305. J. H. Espenson, *Adv. Inorg. Bioinorg. Mech.*, 1982, **1**, 1.
306. S. W. Kirtley, in 'Comprehensive Organometallic Chemistry', ed. G. Wilkinson, F. G. A. Stone and E. W. Abel, Pergamon, Oxford, 1982, vol. 3, chap. 26.1.
307. A. G. Sharpe, 'The Chemistry of Cyano Complexes of the Transition Metals', Academic, London, 1976, pp. 46–52.
308. A. M. Qureshi, *J. Inorg. Nucl. Chem.*, 1969, **31**, 2281.
309. P. Waage Jensen, *J. Mol. Struct.*, 1973, **17**, 377.
310. J. J. Alexander and H. B. Gray, *J. Am. Chem. Soc.*, 1968, **90**, 4260.
311. L. G. Vanquickenborne, L. Haspeslagh, M. Hendrickx and J. Verhulst, *Inorg. Chem.*, 1984, **23**, 1677.
312. A. Calabrese and R. G. Hayes, *J. Am. Chem. Soc.*, 1974, **96**, 5054.
313. M. Sano, H. Adachi and H. Yamatera, *Bull. Chem. Soc. Jpn.*, 1981, **54**, 2898.
314. A. Marchaj and Z. Stasicka, *Polyhedron*, 1983, **2**, 485.
315. H. F. Wasgestian, *J. Phys. Chem.*, 1972, **76**, 1947.
316. N. Sabbatini, M. A. Scandola and V. Carassiti, *J. Phys. Chem.*, 1973, **77**, 1307.
317. D. C. McCain, *Inorg. Nucl. Chem. Lett.*, 1969, **5**, 873.
318. D. F. Banks and J. Kleinberg, *Inorg. Chem.*, 1967, **6**, 1849.
319. D. B. Brown, D. F. Shriver and L. H. Schwartz, *Inorg. Chem.*, 1968, **7**, 77.
320. D. K. Wakefield and W. B. Schaap, *Inorg. Chem.*, 1969, **8**, 512.
321. S. N. Frank and F. C. Anson, *Inorg. Chem.*, 1972, **11**, 2938.
322. J. P. Birk and J. H. Espenson, *J. Am. Chem. Soc.*, 1968, **90**, 1153.
323. L. Jeftić and S. W. Feldberg, *J. Phys. Chem.*, 1971, **75**, 2381.
324. Y. Sakabe and Y. Matsumoto, *Bull. Chem. Soc. Jpn.*, 1981, **54**, 1253.
325. P. Riccieri and E. Zinato, *Inorg. Chem.*, 1981, **20**, 3722.
326. P. Riccieri and E. Zinato, *Inorg. Chem.*, 1980, **19**, 853.
327. A. P. Sattelberger, D. D. Darsow and W. B. Schaap, *Inorg. Chem.*, 1976, **15**, 1412.
328. A. Heatherington, S. Min Oon, R. Vargas and N. A. P. Kane-Maguire, *Inorg. Chim. Acta*, 1980, **44**, L279.
329. E. Zinato, P. Riccieri and M. Prelati, *Inorg. Chem.*, 1981, **20**, 1432.
330. A. D. Kirk and G. B. Porter, *Inorg. Chem.*, 1980, **19**, 445.
331. N. A. P. Kane-Maguire, W. S. Crippen and P. K. Miller, *Inorg. Chem.*, 1983, **22**, 696.
332. J. Casabo, J. Ribas and S. Alvarez, *Inorg. Chim. Acta*, 1976, **16**, L15.
333. J. Ribas, M. Monfort and J. Casabo, *Transition Met. Chem.*, 1984, **9**, 287.
334. F. Edelmann and U. Behrens, *J. Organomet. Chem.*, 1977, **131**, 65.
335. D. Gaswick and A. Haim, *J. Inorg. Nucl. Chem.*, 1978, **40**, 437.
336. A. Uehara, S. Terabe and R. Tsuchiya, *Inorg. Chem.*, 1983, **22**, 2864.
337. A. Ludi and H. U. Güdel, *Struct. Bonding (Berlin)*, 1973, **14**, 1.
338. J. E. House, Jr. and J. C. Bailar, Jr., *Inorg. Chem.*, 1969, **8**, 672.
339. M. C. Couldwell, D. A. Pickering and D. A. House, *J. Inorg. Nucl. Chem.*, 1971, **33**, 3455.
340. P. Andersen, T. Damhus, E. Pedersen and A. Petersen, *Acta Chem. Scand., Ser. A*, 1984, **38**, 359.
341. D. A. House, *Aust. J. Chem.*, 1969, **22**, 647.
342. H. Ueno, A. Uehara and R. Tsuchiya, *Bull. Chem. Soc. Jpn.*, 1981, **54**, 1821.
343. P. Andersen, K. M. Nielsen and A. Petersen, *Acta Chem. Scand., Ser. A*, 1984, **38**, 593.
344. J. Glerup and C. E. Schäffer, *Inorg. Chem.*, 1976, **15**, 1408.
345. J. Glerup, O. Mønsted and C. E. Schäffer, *Inorg. Chem.*, 1976, **15**, 1399.
346. E. Pedersen and H. Toftlund, *Inorg. Chem.*, 1974, **13**, 1603.
347. S. Decurtins, H. U. Güdel and K. Neuenschwander, *Inorg. Chem.*, 1977, **16**, 796.
348. T. Schoenherr and H.-H. Schmidtke, *Inorg. Chem.*, 1979, **18**, 2726.
349. E. Pedersen and S. Kallesøe, *Inorg. Chem.*, 1975, **14**, 85.
350. R. L. Carlin, R. Burriel, J. Pons and J. Casabó, *Inorg. Chem.*, 1983, **22**, 2832.
351. K. N. Semenenko, S. E. Kravchenko and O. V. Kravchenko, *Russ. J. Inorg. Chem. (Engl. Transl.)*, 1976, **21**, 1000.
352. Yu. N. Shevchenko and N. K. Davidenko, *Russ. J. Inorg. Chem. (Engl. Transl.)*, 1979, **24**, 1193.
353. H. Siebert and R. Macht, *Z. Anorg. Allg. Chem.*, 1982, **489**, 77.
354. K. H. Schmidt and A. Müller, *Inorg. Chem.*, 1975, **14**, 2183.

355. R. B. Wilson and E. I. Solomon, *Inorg. Chem.*, 1978, **17,** 1729.
356. R. D. Chirico and R. L. Carlin, *Inorg. Chem.*, 1980, **19,** 3031.
357. W. E. Estes, D. Y. Jeter, J. C. Hempel and W. E. Hatfield, *Inorg. Chem.*, 1971, **10,** 2074.
358. P. Andersen, T. Berg and J. Jacobsen, *Acta Chem. Scand., Ser. A,* 1975, **29,** 381, 599.
359. L. Mønsted and O. Mønsted, *Acta Chem. Scand., Ser. A,* 1984, **38,** 679.
360. D. Yang and D. A. House, *Polyhedron,* 1983, **2,** 1267.
361. E. Pedersen, *Acta Chem. Scand.*, 1972, **26,** 333.
362. F. Christensson and J. Springborg, *Acta Chem. Scand., Ser. A,* 1982, **36,** 21.
363. A. Urushiyama, *Bull. Chem. Soc. Jpn.*, 1972, **45,** 2406.
364. M. Yevitz and J. A. Stanko, *J. Am. Chem. Soc.*, 1971, **93,** 1512.
365. J. T. Veal, D. Y. Jeter, J. C. Hempel, R. P. Eckberg, W. E. Hatfield and D. J. Hodgson, *Inorg. Chem.*, 1973, **12,** 2928.
366. P. Engel and H. U. Güdel, *Inorg. Chem.*, 1977, **16,** 1589.
367. D. J. Hodgson and E. Pedersen, *Inorg. Chem.*, 1980, **19,** 3116.
368. S. J. Cline, J. Glerup, D. J. Hodgson, G. S. Jensen and E. Pedersen, *Inorg. Chem.*, 1981, **20,** 2229.
369. S. J. Cline, D. J. Hodgson, S. Kallesøe, S. Larsen and E. Pedersen, *Inorg. Chem.*, 1983, **22,** 637.
370. S. Decurtins, H. U. Güdel and A. Pfeuti, *Inorg. Chem.*, 1982, **21,** 1101.
371. S. Decurtins and H. U. Güdel, *Inorg. Chem.*, 1982, **21,** 3598.
372. D. J. Hodgson and E. Pedersen, *Inorg. Chem.*, 1984, **23,** 2363.
373. M. Morita, T. Schönherr, R. Linder and H.-H. Schmidtke, *Spectrochim. Acta, Part A,* 1982, **38,** 1221.
374. J. Glerup, D. J. Hodgson and E. Pedersen, *Acta Chem. Scand., Ser. A,* 1983, **37,** 161.
375. P. Andersen, K. M. Nielsen and A. Petersen, *Acta Chem. Scand., Ser. A,* 1984, **38,** 593.
376. P. Andersen, T. Damhus, E. Pedersen and A. Petersen, *Acta Chem. Scand., Ser. A,* 1984, **38,** 359.
377. E. Bang, *Acta Chem. Scand., Ser. A,* 1984, **38,** 419.
378. H. U. Güdel, U. Hauser and A. Furrer, *Inorg. Chem.*, 1979, **18,** 2730.
379. M. Parris and N. F. Feiner, *Inorg. Nucl. Chem. Lett.*, 1967, **3,** 337.
380. Yu. N. Shevchenko, V. V. Sachok and N. K. Davidenko, *Russ. J. Inorg. Chem.* (*Engl. Transl.*), 1979, **24,** 1475.
381. K. Kühn, F. Wasgestian and H. Kupka, *J. Phys. Chem.*, 1981, **85,** 665.
382. Yu. N. Shevchenko, V. V. Sachok and N. K. Davidenko, *Russ. J. Inorg. Chem.* (*Engl. Transl.*), 1979, **24,** 30.
383. Yu. N. Shevchenko, N. I. Yashina, K. B. Yatsimirskii, R. A. Svitsyn and N. V. Egorova, *Russ. J. Inorg. Chem.* (*Engl. Transl.*), 1983, **28,** 216.
384. H. L. Schläfer, H. Gausmann and H. Witzke, *Z. Phys. Chem.* (*Frankfurt am Main*), 1967, **56,** 55.
385. A. Rogers and P. J. Staples, *J. Chem. Soc.*, 1965, 6834.
386. K. Kühn and F. Wasgestian, *Inorg. Nucl. Chem. Lett.*, 1976, **12,** 803.
387. N. E. Dixon, G. A. Lawrance, P. A. Ley and A. M. Sargeson, *Inorg. Chem.*, 1984, **23,** 2940.
388. B. M. Foxman, *Inorg. Chem.*, 1978, **17,** 1932.
389. M. Parris and W. J. Wallace, *Can. J. Chem.*, 1969, **47,** 2257.
390. T. W. Swaddle, *Coord. Chem. Rev.*, 1974, **14,** 217.
391. Yu. N. Schevchenko, V. V. Sachok and N. K. Davidenko, *Russ. J. Inorg. Chem.* (*Engl. Transl.*), 1979, **24,** 30.
392. J. Glerup, J. Josephsen, K. Michelsen, E. Pedersen and C. E. Schäffer, *Acta Chem. Scand.*, 1970, **24,** 247.
393. M. W. Duckworth, G. W. A. Fowles and P. T. Greene, *J. Chem. Soc.* (*A*), 1967, 1592.
394. G. W. A. Fowles, P. T. Greene and J. S. Wood, *Chem. Commun.*, 1967, 971.
395. J. Hughes and G. R. Willey, *Inorg. Chim. Acta,* 1975, **12,** 143.
396. J. Hughes and G. R. Willey, *J. Coord. Chem.*, 1974, **4,** 33.
397. J. C. Taft and M. M. Jones, *J. Am. Chem. Soc.*, 1960, **82,** 4196.
398. I. Ganescu, C. Varhelyi, L. Boboc and G. Brinzan, *Chem. Anal.* (*Warsaw*), 1977, **22,** 413 (*Chem. Abstr.*, 1978, **88,** 114 669x).
399. I. Ganescu, C. Varhelyi and A. Popescu, *Chem. Abstr.*, 1970, **73,** 126 617g.
400. I. Ganescu, C. Varhelyi and G. Briznan, *Mikrochim. Acta,* 1977, **1,** 139.
401. P. K. Mathur and L. N. Srivastava, *J. Inorg. Nucl. Chem.*, 1973, **35,** 2112.
402. J. W. Vaughn, *Coord. Chem. Rev.*, 1981, **39,** 265.
403. J. W. Vaughn, *Synth. React. Inorg. Metal-Org. Chem.*, 1979, **9,** 585.
404. W. W. Fee, J. N. MacB. Harrowfield and W. G. Jackson, *J. Chem. Soc,* (*A*), 1970, 2612.
405. W. G. Jackson, P. D. Vowles and W. W. Fee, *Inorg. Chim. Acta,* 1977, **22,** 111.
406. S. Kaizaki, J. Hidaka and Y. Shimura, *Bull. Chem. Soc. Jpn.*, 1975, **48,** 902.
407. C. F. C. Wong and A. D. Kirk, *Can. J. Chem.*, 1975, **53,** 3388.
408. C. F. C. Wong and A. D. Kirk, *Inorg. Chem.*, 1978, **17,** 1672.
409. J. W. Vaughn, *Inorg. Chem.*, 1983, **22,** 844.
410. J. W. Vaughn and G. J. Seiler, *Inorg. Chem.*, 1979, **18,** 1509.
411. J. W. Vaughn and A. M. Yeoman, *Synth. React. Inorg. Metal-Org. Chem.*, 1977, **7,** 165.
412. I. K. Adzamli and E. Deutsch, *Inorg. Chem.*, 1980, **19,** 1366.
413. R. C. Elder, L. R. Florian, R. E. Lake and A. M. Yacynych, *Inorg. Chem.*, 1973, **12,** 2690.
414. J. R. Perumareddi, *Coord. Chem. Rev.*, 1969, **4,** 73.
415. L. S. Forster, *Transition Met. Chem.*, 1969, **5,** 1.
416. H. Toftlund and T. Laier, *Acta Chem. Scand., Ser. A,* 1977, **31,** 651.
417. P. Andersen, F. Galsbøl, S. E. Harnung and T. Laier, *Acta Chem. Scand., Ser. A,* 1973, **27,** 3973.
418. U. Geiser and H. U. Güdel, *Inorg. Chem.*, 1981, **20,** 3013.
419. R. L. Klein, Jr., N. C. Miller and J. R. Perumareddi, *Inorg. Chim. Acta,* 1973, **7,** 685.
420. G. L. Hilmes, H. G. Brittain and F. S. Richardson, *Inorg. Chem.*, 1977, **16,** 528.
421. K. Akabori, *J. Inorg. Nucl. Chem.*, 1981, **43,** 677.
422. L. E. Metcalf and J. E. House, Jr., *J. Inorg. Nucl. Chem.*, 1980, **42,** 961.
423. K. Akabori and Y. Kushi, *J. Inorg. Nucl. Chem.*, 1978, **40,** 625.
424. N. Maki, *J. Inorg. Nucl. Chem.*, 1975, **37,** 1207.
425. C. D. Flint and A. P. Matthews, *Inorg. Chem.*, 1975, **14,** 1219.

426. R. E. Cramer and J. T. Huneke, *Inorg. Chem.*, 1975, **14**, 2565.
427. M. Sano and H. Yamatera, *Bull. Chem. Soc. Jpn.*, 1981, **54**, 2023.
428. J. Casabó, A. Solans and J. Real, *Synth. React. Inorg. Metal-Org. Chem.*, 1983, **13**, 835.
429. J. W. Vaughn, *Inorg. Chem.*, 1981, **20**, 2397.
430. D. H. Brown, D. R. MacSween and D. W. A. Sharp, *Inorg. Nucl. Chem. Lett.*, 1969, **5**, 991.
431. M. Nakano and S. Kawaguchi, *Bull. Chem. Soc. Jpn.*, 1979, **52**, 3563.
432. A. Uehara, Y. Nishiyama and R. Tsuchiya, *Inorg. Chem.*, 1982, **21**, 2422.
433. J. A. McLean, Jr. and A. Bhattacharyya, *Inorg. Chim. Acta*, 1981, **53**, L43.
434. H. P. Jensen, *Acta Chem. Scand., Ser. A*, 1981, **35**, 127.
435. R. Tsuchiya, A. Uehara and T. Yoshikuni, *Inorg. Chem.*, 1982, **21**, 590.
436. S. Kaizaki, *Bull. Chem. Soc. Jpn.*, 1983, **56**, 3620, 3625.
437. K. Matsumoto, S. Ooi, H. Kawaguchi, M. Nakano and S. Kawaguchi, *Bull. Chem. Soc. Jpn.*, 1982, **55**, 1840.
438. K. Miyoshi, Y. Matsumoto and H. Yoneda, *Inorg. Chem.*, 1981, **20**, 1057.
439. K. Miyoshi, Y. Matsumoto and H. Yoneda, *Inorg. Chem.*, 1982, **21**, 790.
440. F. Christensson, J. Springborg and H. Toftlund, *Acta Chem. Scand., Ser. A*, 1980, **34**, 317.
441. J. Springborg and H. Toftlund, *Acta Chem. Scand., Ser. A*, 1979, **33**, 31.
442. J. Springborg and H. Toftlund, *Acta Chem. Scand., Ser. A*, 1976, **30**, 171.
443. J. Springborg, *Acta Chem. Scand., Ser. A*, 1978, **32**, 231.
444. A. Beutler, H. U. Güdel, T. R. Snellgrove, G. Chapuis and K. J. Schenk, *J. Chem. Soc., Dalton Trans.*, 1979, 983.
445. J. Springborg and C. E. Schäffer, *Acta Chem. Scand., Ser. A*, 1976, **30**, 787.
446. K. Kaas, *Acta Crystallogr., Sect. B*, 1979, **35**, 1603.
447. T. Laier, B. Nielsen and J. Springborg, *Acta Chem. Scand., Ser. A*, 1982, **36**, 91.
448. P. Andersen and T. Berg, *Acta Chem. Scand., Ser. A*, 1978, **32**, 989.
449. T. Damhus and E. Pedersen, *Inorg. Chem.*, 1984, **23**, 695.
450. H. U. Güdel and U. Hauser, *Inorg. Chem.*, 1980, **19**, 1325.
451. M. T. Flood, R. E. Marsh and H. B. Gray, *J. Am. Chem. Soc.*, 1969, **91**, 193.
452. K. Michelsen, *Acta Chem. Scand.*, 1970, **24**, 2003.
453. K. Michelsen, *Acta Chem. Scand.*, 1972, **26**, 1517.
454. S. Larsen, K. B. Nielsen and I. Trabjerg, *Acta Chem. Scand., Ser. A*, 1983, **37**, 833.
455. K. Michelsen, *Acta Chem. Scand.*, 1973, **27**, 1823.
456. D. K. Rastogi, K. C. Sharma, S. K. Dua and M. P. Teotia, *J. Inorg. Nucl. Chem.*, 1975, **37**, 685.
457. K. Michelsen, *Acta Chem. Scand., Ser. A*, 1976, **30**, 521.
458. K. Michelsen and E. Pedersen, *Acta Chem. Scand., Ser. A*, 1978, **32**, 847.
459. S. Larsen and B. Hansen, *Acta Chem. Scand., Ser. A*, 1981, **35**, 105.
460. K. Michelsen, E. Pedersen, S. R. Wilson and D. J. Hodgson, *Inorg. Chim. Acta*, 1982, **63**, 141.
461. O. Kling and H. L. Schläfer, *Z. Anorg. Allg. Chem.*, 1961, **313**, 187.
462. R. Lansberg, P. Janietz and M. Prügel, *Monatsh. Chem.*, 1979, **110**, 831.
463. R. E. Hamm and F. C. Fushimi, *J. Indian Chem. Soc.*, 1977, **54**, 33.
464. K. Kühn, F. Wasgestian and H. Kupka, *J. Phys. Chem.*, 1981, **85**, 665.
465. D. A. House, *Inorg. Nucl. Chem. Lett.*, 1976, **12**, 259.
466. B. S. Dawson and D. A. House, *Inorg. Chem.*, 1977, **16**, 1354.
467. D. A. House and C. S. Garner, *Inorg. Chem.*, 1966, **5**, 840.
468. S. H. Caldwell and D. A. House, *J. Inorg. Nucl. Chem.*, 1969, **31**, 811.
469. H. H. Schmidtke and D. Garthoff, *Inorg. Chim. Acta*, 1968, **2**, 357.
470. D. K. Lin and C. S. Garner, *J. Am. Chem. Soc.*, 1969, **91**, 6637.
471. A. D. Fowlie, D. A. House, W. T. Robinson and S. S. Rumball, *J. Chem. Soc. (A)*, 1970, 803.
472. K. Wieghardt, W. Schmidt, H. Endres and C. R. Wolfe, *Chem. Ber.*, 1979, **112**, 2837.
473. W. A. Fordyce, P. S. Sheridan, E. Zinato, P. Riccieri and A. W. Adamson, *Inorg. Chem.*, 1977, **16**, 1154.
474. D. A. House, *Aust. J. Chem.*, 1982, **35**, 659.
475. J. M. Veigel, *Inorg. Chem.*, 1968, **7**, 69.
476. R. Tsuchiya, A. Uehara, K. Noji and H. Yamamura, *Bull. Chem. Soc. Jpn.*, 1978, **51**, 2942.
477. R. L. Wilder, D. A. Kamp and C. S. Garner, *Inorg. Chem.*, 1971, **10**, 1393.
478. S. C. Tang, R. L. Wilder, R. K. Kurimoto and C. S. Garner, *Synth. React. Inorg. Metal-Org. Chem.*, 1971, **1**, 207.
479. A. W. Adamson, *J. Phys. Chem.*, 1967, **71**, 798.
480. L. G. Vanquickenborne and A. Ceulemans, *J. Am. Chem. Soc.*, 1977, **99**, 2208.
481. S. G. Zipp and S. K. Madan, *Inorg. Chem.*, 1976, **15**, 587.
482. M. J. Saliby, D. West and S. K. Madan, *Inorg. Chem.*, 1981, **20**, 723.
483. M. J. Saliby, P. S. Sheridan and S. K. Madan, *Inorg. Chem.*, 1980, **19**, 1291.
484. M. J. Saliby and S. K. Madan, *Inorg. Chem.*, 1981, **20**, 289.
485. J. Hughes and G. R. Willey, *J. Coord. Chem.*, 1974, **4**, 33.
486. D. A. House and D. Yang, *Inorg. Chim. Acta*, 1983, **74**, 179.
487. B. Bosnich, R. D. Gillard, E. D. McKenzie and G. A. Webb, *J. Chem. Soc. (A)*, 1966, 1331.
488. E. Bang and E. Pedersen, *Acta Chem. Scand., Ser. A*, 1978, **32**, 833.
489. H. R. Mäcke, B. F. Mentzen, J. P. Puaux and A. W. Adamson, *Inorg. Chem.*, 1982, **21**, 3080.
490. D. Yang and D. A. House, *Inorg. Chem.*, 1982, **21**, 2999.
491. B. Bosnich, J. MacB. Harrowfield and H. Boucher, *Inorg. Chem.*, 1975, **14**, 815.
492. B. Bosnich and J. MacB. Harrowfield, *Inorg. Chem.*, 1975, **14**, 828.
493. D. A. House and O. Nor, *Inorg. Chim. Acta*, 1983, **70**, 13.
494. T. J. Barton and R. C. Slade, *J. Chem. Soc., Dalton Trans.*, 1975, 650.
495. K. Michelsen, *Acta Chem. Scand., Ser. A*, 1977, **31**, 429.
496. Y. Yamamoto and Y. Shimura, *Bull. Chem. Soc. Jpn.*, 1981, **54**, 3351.
497. Y. Yamamoto and Y. Shimura, *Bull. Chem. Soc. Jpn.*, 1980, **53**, 395.

498. Y. Hata, Y. Yamamoto and Y. Shimura, *Bull. Chem. Soc. Jpn.*, 1981, **54,** 1255.
499. K. Michelsen and K. M. Nielsen, *Acta Chem. Scand., Ser.* 1980, **34,** 755.
500. Y. Yamamoto and Y. Shimura, *Bull. Chem. Soc. Jpn.*, 1981, **54,** 2924.
501. K. Michelsen and E. Pedersen, *Acta Chem. Scand., Ser. A,* 1983, **37,** 141.
502. H. R. Fischer, D. J. Hodgson, K. Michelsen and E. Pedersen, *Inorg. Chim. Acta,* 1984, **88,** 143.
503. S. J. Ranney and C. S. Garner, *Synth. React. Inorg. Metal-Org. Chem.*, 1971, **1,** 179.
504. S. J. Ranney and C. S. Garner, *Inorg. Chem.*, 1971, **10,** 2437.
505. D. A. House and C. S. Garner, *Inorg. Chem.*, 1967, **6,** 272.
506. J. Glerup, O. Mønsted and C. E. Schäffer, *Inorg. Chem.*, 1976, **15,** 1399.
507. D. W. Smith, *Inorg. Chem.*, 1978, **17,** 3153.
508. J. Glerup, O. Mønsted and C. E. Schäffer, *Inorg. Chem.*, 1980, **19,** 2855.
509. E. Pedersen and H. Toftlund, *Inorg. Chem.*, 1974, **13,** 1603.
510. J. C. Taft and M. M. Jones, *Inorg. Synth.*, 1963, **7,** 132.
511. R. J. H. Clark and C. S. Williams, *Inorg. Chem.*, 1965, **4,** 350.
512. J. V. Brencic, *Z. Anorg. Allg. Chem.*, 1974, **403,** 218.
513. J. V. Brencic and I. Leban, *Z. Anorg. Allg. Chem.*, 1980, **465,** 173.
514. J. C. Chang, M. A. Haile and G. R. Keith, *J. Inorg. Nucl. Chem.*, 1972, **34,** 360.
515. D. H. Brown and R. T. Richardson, *J. Inorg. Nucl. Chem.*, 1973, **35,** 755.
516. I. L. Eremenko, A. A. Pasynskii, O. G. Volkov, O. G. Ellert and V. T. Kalinnikov, *J. Organomet. Chem.*, 1981, **222,** 235.
517. S. K. Sharma, R. K. Mahajan, B. Kapila and V. P. Kapila, *Polyhedron,* 1983, **2,** 973.
518. A. Bakac, R. Marcec and M. Orhanovic, *Inorg. Chem.*, 1974, **13,** 57.
519. T. J. Collins, B. D. Santarsiero and G. H. Spies, *J. Chem. Soc., Chem. Commun.*, 1983, 681.
520. A. B. Blake and A. Yavari, *J. Chem. Soc., Chem. Commun.*, 1982, 1247.
521. H. Nishide, S. Hata and E. Tsuchida, *Makromol. Chem.*, 1978, **179,** 1445.
522. R. C. Mehrotra and K. N. Mahendra, *Inorg. Chim. Acta,* 1984, **81,** 163.
523. M. A. Jamieson, N. Serpone and M. Z. Hoffman, *Coord. Chem. Rev.*, 1981, **39,** 121.
524. A. D. Kirk, *Coord. Chem. Rev.*, 1981, **39,** 225.
525. J. Lilie and W. L. Waltz, *Inorg. Chem.*, 1983, **22,** 1473.
526. M. C. Hughes, D. J. Macero and J. M. Rao, *Inorg. Chim. Acta,* 1981, **49,** 241.
527. N. A. P. Kane-Maguire and J. S. Hallock, *Inorg. Chim. Acta,* 1979, **35,** L309.
528. M. Maestri, F. Bolletta, N. Serpone, L. Moggi and V. Balzani, *Inorg. Chem.*, 1976, **15,** 2048.
529. B. R. Baker and B. D. Mehta, *Inorg. Chem.*, 1965, **4,** 848.
530. F. Bolletta, M. Maestri, L. Moggi, M. A. Jamieson, N. Serpone, M. S. Henry and M. Z. Hoffman, *Inorg. Chem.*, 1983, **22,** 2502.
531. W. A. Wickramasinghe, P. H. Bird and N. Serpone, *Inorg. Chem.*, 1982, **21,** 2694.
532. N. Serpone, M. A. Jamieson, M. S. Henry, M. Z. Hoffman, F. Bolletta and M. Maestri, *J. Am. Chem. Soc.*, 1979, **101,** 2907.
533. W. A. Wickramasinghe, P. H. Bird, M. A. Jamieson, N. Serpone and M. Maestri, *Inorg. Chim. Acta,* 1982, **64,** L85.
534. L. H. Berka, R. R. Gagne, G. E. Philippon and C. E. Wheeler, *Inorg. Chem.*, 1970, **9,** 2705.
535. M. P. Hancock, J. Josephsen and C. E. Schäffer, *Acta Chem. Scand., Ser. A,* 1976, **30,** 79.
536. J. Josephsen and C. E. Schäffer, *Acta Chem. Scand.*, 1970, **24,** 2929.
537. J. Josephsen and E. Pedersen, *Inorg. Chem.*, 1977, **16,** 2534.
538. S. Kaizaki, J. Hidaka and Y. Shimura, *Inorg. Chem.*, 1973, **12,** 135.
539. R. P. Scaringe, P. Singh, R. P. Eckberg, W. E. Hatfield and D. J. Hodgson, *Inorg. Chem.*, 1975, **14,** 1127.
540. M. Ardon and A. Bino, *J. Am. Chem. Soc.*, 1983, **105,** 7747.
541. M. Ardon and B. Magyar, *J. Am. Chem. Soc.*, 1984, **106,** 3359.
542. G. L. Robbins and R. E. Tapscott, *Inorg. Chem.*, 1976, **15,** 154.
543. G. A. Lawrance and D. R. Stranks, *Inorg. Chem.*, 1977, **16,** 929.
544. J. A. Broomhead, *Aust. J. Chem.*, 1962, **15,** 228.
545. J. A. Broomhead, M. Dwyer and N. Kane-Maguire, *Inorg. Chem.*, 1968, **7,** 1388.
546. J. A. Broomhead and F. P. Dwyer, *Aust. J. Chem.*, 1961, **14,** 250.
547. J. A. Broomhead, J. Evans, W. D. Grumley and M. Sterns, *J. Chem. Soc., Dalton Trans.*, 1977, 173.
548. E. C. Constable and J. Lewis, *Polyhedron,* 1982, **1,** 303.
549. E. C. Constable, J. Lewis, M. C. Liptrot and P. R. Raithby, *J. Chem. Soc., Dalton Trans.*, 1984, 2177.
550. L.-Y. Chung, E. C. Constable, M. S. Khan, J. Lewis, P. R. Raithby and M. D. Vargas, *J. Chem. Soc., Chem. Commun.*, 1984, 1425.
551. J. J. Daly and F. Sanz, *J. Chem. Soc., Dalton Trans.*, 1972, 2584.
552. J. J. Daly, F. Sanz, R. P. A. Sneeden and H. H. Zeiss, *J. Chem. Soc., Dalton Trans.*, 1973, 1497.
553. R. P. A. Sneeden and H. H. Zeiss, *J. Organomet. Chem.*, 1973, **47,** 125.
554. J. J. Daly, R. P. A. Sneeden and H. H. Zeiss, *J. Am. Chem. Soc.*, 1966, **88,** 4287.
555. J. Losada and M. Gayoso, *An. Quim.*, 1978, **74,** 576.
556. M. J. Gonzalez Garmendia, J. Losada and M. Moran, *J. Inorg. Nucl. Chem.*, 1981, **43,** 2269.
557. A. Mishra and J. P. Singh, *J. Inorg. Nucl. Chem.*, 1979, **41,** 905.
558. S. P. Ghosh and A. Mishra, *J. Inorg. Nucl. Chem.*, 1971, **33,** 4199.
559. H. Bhushan, L. K. Mishra and N. K. Jha, *Indian J. Chem., Sect A,* 1982, **21,** 1135.
560. A. Mishra, *Indian J. Chem., Sect. A,* 1977, **15,** 919.
561. P. Labine and C. H. Brubaker, Jr., *J. Inorg. Nucl. Chem.*, 1971, **33,** 3383.
562. T. G. Dunne and J. K. Hurst, *Inorg. Chem.*, 1980, **19,** 1152.
563. J. Swartz and F. C. Anson, *Inorg. Chem.*, 1981, **20,** 2250.
564. S. N. Poddar, A. K. Das and P. Bhattacharyya, *Indian J. Chem., Sect. A,* 1983, **22,** 544.
565. D. N. Hendrickson, J. M. Hollander and W. L. Jolly, *Inorg. Chem.*, 1970, **9,** 612.

556. W. P. Griffith, J. Lewis and G. Wilkinson, *J. Chem. Soc.*, 1959, 872.
557. J. H. Enemark, M. S. Quinby, L. L. Reed, M. J. Steuck and K. K. Walthers, *Inorg. Chem.*, 1970, **9**, 2397.
558. J. Mocák, D. Bustin and M. Ziahová, *Inorg. Chim. Acta*, 1977, **22**, 185.
559. D. I. Bustin, M. Rievaj and J. Mocák, *Inorg. Chim. Acta*, 1978, **27**, 59.
560. K. S. Nagaraja and M. R. Udupa, *Indian J. Chem., Sect. A*, 1983, **22**, 531.
561. W. P. Griffith, *J. Chem. Soc.*, 1963, 3286.
562. J. G. Mason and J. M. Vander Meer, *Inorg. Chem.*, 1974, **13**, 2443.
563. R. G. Bhattacharyya, G. P. Bhattacharjee and P. S. Roy, *Inorg. Chim. Acta*, 1981, **54**, L263.
564. R. G. Bhattacharyya, G. P. Bhattacharjee and N. Ghosh, *Polyhedron*, 1983, **2**, 543.
565. R. G. Bhattacharyya and G. P. Bhattacharjee, *Polyhedron*, 1983, **2**, 1221.
566. R. C. Maurya and R. K. Shukla, *J. Indian Chem. Soc.*, 1982, **59**, 340.
567. C. M. Lukehart and J. M. Troup, *Inorg. Chim. Acta*, 1977, **22**, 81.
568. K. Wieghardt, U. Quilitzsch and J. Weiss, *Inorg. Chim. Acta*, 1984, **89**, L43.
569. J. N. Armor, M. Buchbinder and R. Cheney, *Inorg. Chem.*, 1974, **13**, 2990.
570. M. Herberhold and L. Haumaier, *Chem. Ber.*, 1982, **115**, 1399.
571. M. Herberhold and L. Haumaier, *Chem. Ber.*, 1983, **116**, 2896.
572. S. Clamp, N. G. Connelly, G. E. Taylor and T. S. Louttit, *J. Chem. Soc., Dalton Trans.*, 1980, 2162.
573. R. P. Burns, F. P. McCullough and C. A. McAuliffe, *Adv. Inorg. Chem. Radiochem.*, 1980, **23**, 211.
574. M. K. Lloyd and J. A. McCleverty, *J. Organomet. Chem.*, 1973, **61**, 261.
575. D. E. Wigley and R. A. Walton, *Organometallics*, 1982, **1**, 1322.
576. D. E. Wigley and R. A. Walton, *Inorg. Chem.*, 1983, **22**, 3138.
577. J. A. Connor, P. I. Riley and C. J. Rix, *J. Chem. Soc., Dalton Trans.*, 1977, 1317.
578. N. G. Connelly, *J. Chem. Soc., Dalton Trans.*, 1973, 2183.
579. R. Davis and L. A. P. Kane-Maguire, in 'Comprehensive Organometallic Chemistry', ed. G. Wilkinson, F. G. A. Stone and E. W. Abel, Pergamon, Oxford, 1982, vol. 3, p. 1037.
580. R. D. Feltham, W. Silverthorn and G. McPherson, *Inorg. Chem.*, 1969, **8**, 344.
581. D. C. Bradley and C. W. Newing, *Chem. Commun.*, 1970, 219.
582. D. C. Bradley and M. H. Chisholm, *Acc. Chem. Res.*, 1976, **9**, 273.
583. D. C. Bradley, C. W. Newing, M. H. Chisholm, R. L. Kelly, D. A. Haitko, D. Little, F. A. Cotton and P. E. Fanwick, *Inorg. Chem.*, 1980, **19**, 3010.
584. D. C. Bradley and M. Ahmed, *Polyhedron*, 1983, **2**, 87.
585. R. B. Wayland, L. W. Olsen and Z. U. Siddiqui, *J. Am. Chem. Soc.*, 1976, **98**, 94.
586. C. Ercolani and C. Neri, *J. Chem. Soc. (A)*, 1967, 1715.
587. C. Ercolani, C. Neri and G. Sartori, *J. Chem. Soc. (A)*, 1968, 2123.
588. D. Wester, R. C. Edwards and D. H. Busch, *Inorg. Chem.*, 1977, **16**, 1055.
589. B. I. Swanson and S. K. Satija, *J. Chem. Soc., Chem. Commun.*, 1973, 40.
590. S. K. Satija, B. I. Swanson, O. Crichton and A. J. Rest, *Inorg. Chem.*, 1978, **17**, 1737.
591. H. W. Roesky and K. K. Pandey, *Adv. Inorg. Chem. Radiochem.*, 1983, **26**, 337.
592. H. W. Chen, W. L. Jolly, S. F. Xiang, I. S. Butler and J. Sedman, *J. Electron Spectrosc. Relat. Phenom.*, 1981, **24**, 121.
593. M. Herberhold and L. Haumaier, *Z. Naturforsch, Teil. B*, 1980, **35**, 1277.
594. M. Idelson, I. R. Karady, B. H. Mark, D. O. Rickter and V. H. Hooper, *Inorg. Chem.*, 1967, **6**, 450.
595. R. Grieb and A. Niggli, *Helv. Chim. Acta*, 1965, **48**, 317.
596. H. Jaggi, *Helv. Chim. Acta*, 1968, **51**, 580.
597. G. Shetty, *Helv. Chim. Acta*, 1970, **53**, 1437.
598. G. Shetty, *Helv. Chim. Acta*, 1968, **51**, 505.
599. G. Shetty, *Helv. Chim. Acta*, 1967, **50**, 2212.
600. K. C. Patil, K. C. Nesamani and V. R. P. Verneker, *Chem. Ind. (London)*, 1979, 901.
601. M. G. Lyapilina, E. I. Krylov and V. A. Sharov, *Russ. J. Inorg. Chem. (Engl. Transl.)*, 1975, **20**, 216.
602. M. G. Lyapilina, E. I. Krylova and V. A. Sharov, *Russ. J. Inorg. Chem. (Engl. Transl.)*, 1975, **20**, 458.
603. V. A. Sharov, N. V. Povarova, V. A. Perelyaev, E. I. Krylov and G. M. Adamova, *Koord. Khim. (Engl. Transl.)*, 1979, **5**, 415 (*Chem. Abstr.*, 1979, **91**, 67 434s).
604. A. D. Ahmed and N. R. Chandhuri, *J. Inorg. Nucl. Chem.*, 1971, **33**, 189.
605. E. M. Ivashkovich and L. M. Zhelekhorskaya, *Russ. J. Inorg. Chem. (Engl. Transl.)*, 1979, **24**, 33.
606. J. C. Machado, M. M. Braga, A. M. P. R. Da Luz and G. Duplatre, *J. Chem. Soc., Faraday Trans. 1*, 1980, **76**, 152.
607. D. C. Bradley, M. B. Hursthouse and C. W. Newing, *Chem. Commun.*, 1971, 411.
608. J. C. Basi, D. C. Bradley and M. H. Chisholm, *J. Chem. Soc. (A)*, 1971, 1433.
609. J. C. Green, M. Payne, E. A. Seddon and R. A. Andersen, *J. Chem. Soc., Dalton Trans.*, 1982, 887.
610. P. G. Eller, D. C. Bradley, M. B. Hursthouse and D. W. Meek, *Coord. Chem. Rev.*, 1977, **24**, 1.
611. A. H. Norbury, *Adv. Inorg. Chem. Radiochem.*, 1975, **17**, 231.
612. J. E. House, Jr. and G. L. Jepsen, *J. Inorg. Nucl. Chem.*, 1978, **40**, 697.
613. M. A. Bennett, R. J. H. Clark and A. D. J. Goodwin, *Inorg. Chem.*, 1967, **6**, 1625.
614. G. Brauer, 'Handbook of Preparative Inorganic Chemistry', 2nd edn., Academic, London, 1962, vol. 2, p. 1379.
615. L. K. Mishra, H. Bhushan and N. K. Jha, *Indian J. Chem., Sect. A*, 1982, **21**, 639.
616. A. Mishra and J. P. Singh, *J. Inorg. Nucl. Chem.*, 1979, **41**, 905.
617. C. L. Sharma and V. P. Mishra, *J. Indian Chem. Soc.*, 1983, **60**, 809.
618. G. Thomas, *Z. Anorg. Allg. Chem.*, 1968, **360**, 15.
619. K. Issleib, A. Tzschach and H. O. Fröhlich, *Z. Anorg. Allg. Chem.*, 1959, **298**, 164.
620. K. M. Purohit and D. V. Ramana Rao, *J. Indian Chem. Soc.*, 1983, **60**, 1009.
621. K. Yoshimura and T. Tarutani, *J. Chromatogr.*, 1982, **237**, 89.
622. E. Blasius and H. Augustin, *Z. Anorg. Allg. Chem.*, 1975, **417**, 47.
623. E. Blasius and H. Augustin, *Z. Anorg. Allg. Chem.*, 1975, **417**, 55.

634. A. Botar, E. Blasius and H. Augustin, *Z. Anorg. Allg. Chem.*, 1976, **422,** 54.
635. J. E. House, Jr., G. L. Jepsen and J. C. Bailar, Jr., *Inorg. Chem.*, 1979, **18,** 1397.
636. W. W. Wendlandt and L. K. Sveum, *J. Inorg. Nucl. Chem.*, 1967, **29,** 975.
637. E. Blasius and E. Mernke, *Z. Anorg. Allg. Chem.*, 1984, **509,** 167.
638. R. C. Buckley and J. G. Wardeska, *Inorg. Chem.*, 1972, **11,** 1723.
639. M. Orhanovic and N. Sutin, *J. Am. Chem. Soc.*, 1968, **90,** 4286.
640. C. Shea and A. Haim, *J. Am. Chem. Soc.*, 1971, **93,** 3055.
641. L. D. Brown and D. E. Pennington, *Inorg. Chem.*, 1971, **10,** 2117.
642. A. I. Brusilovets, V. V. Skopenko and G. V. Tsintsadze, *Russ. J. Inorg. Chem. (Engl. Transl.)*, 1969, **14,** 239.
643. R. A. Bailey and T. W. Michelsen, *J. Inorg. Nucl. Chem.*, 1972, **34,** 2935.
644. S. J. Anderson and A. H. Norbury, *J. Chem. Soc., Chem. Commun.*, 1974, 37.
645. S. DasSarma and B. DasSarma, *Inorg. Chem.*, 1979, **18,** 3618.
646. E. Blasius and G. Klemm, *Z. Anorg. Allg. Chem.*, 1978, **443,** 265.
647. M. Linhard and W. Berthold, *Z. Anorg. Allg. Chem.*, 1955, **279,** 173.
648. H. C. Egghart, *J. Inorg. Nucl. Chem.*, 1981, **43,** 1390.
649. D. A. House and C. S. Garner, *J. Am. Chem. Soc.*, 1966, **88,** 2156.
650. J. J. Salzmann and H. H. Schmidtke, *Inorg. Chim. Acta*, 1969, **3,** 207.
651. W. Beck, W. P. Fehlhammer, P. Pöllmann, E. Schuierer and K. Feldl, *Chem. Ber.*, 1967, **100,** 2335.
652. S. I. Arzhankov and A. L. Poznyak, *Russ. J. Inorg. Chem. (Engl. Transl.)*, 1981, **26,** 850.
653. S. I. Arzhankov and A. L. Poznyak, *Z. Anorg. Allg. Chem.*, 1981, **481,** 201.
654. J. W. Buchler and C. Dreher, *Z. Naturforsch., Teil B*, 1984, **39,** 222.
655. D. A. House, *Coord. Chem. Rev.*, 1977, **23,** 223.
656. R. C. Thompson, P. Wieland and E. H. Appelman, *Inorg. Chem.*, 1979, **18,** 1974.
657. S. Trofimenko, *J. Am. Chem. Soc.*, 1969, **91,** 588.
658. B. J. Hathaway and D. G. Holah, *J. Chem. Soc.*, 1965, 537.
659. R. J. Kern, *J. Inorg. Nucl. Chem.*, 1963, **25,** 5.
660. D. Jordan, W. C. Kupferschmidt and R. B. Jordan, *Inorg. Chem.*, 1984, **23,** 1986.
661. R. J. Balahura, *J. Am. Chem. Soc.*, 1976, **98,** 1487.
662. D. A. Edwards and S. C. Jennison, *Transition Met. Chem.*, 1981, **6,** 235.
663. K. R. Seddon and V. H. Thomas, *J. Chem. Soc., Dalton Trans.*, 1977, 2195.
664. A. R. Siedle, G. A. Candela and T. F. Finnegan, *Inorg. Chim. Acta*, 1979, **35,** 125.
665. J. F. Myers, G. W. Rayner-Canham and A. B. P. Lever, *Inorg. Chem.*, 1975, **14,** 461.
666. M. Mohan, G. Mittal, H. C. Khera and A. K. Sirivastava, *Monatsh. Chem.*, 1980, **111,** 63.
667. M. Mohan, H. C. Khera, S. G. Mittal and A. K. Sirivastava, *Gazz. Chim. Ital.*, 1979, **109,** 65.
668. B. Sen and M. E. Pickerell, *J. Inorg. Nucl. Chem.*, 1973, **35,** 2573.
669. K. M. J. Al-Komser and B. Sen, *Inorg. Chem.*, 1963, **2,** 1219.
670. A. K. Rana and J. R. Shah, *Indian J. Chem., Sect. A*, 1981, **20,** 142.
671. A. Wasey, R. K. Bansal, B. K. Puri, F. Kamil and S. Chandra, *Transition Met. Chem.*, 1983, **8,** 341.
672. L. Coghi, M. Nardelli and G. Pelizzi, *Acta Crystallogr., Sect. B*, 1976, **32,** 842.
673. G. R. Brubaker and L. E. Webb, *J. Am. Chem. Soc.*, 1969, **91,** 7199.
674. P. Ray, *Chem. Rev.*, 1961, **61,** 313.
675. D. Banerjea and B. Chakravarty, *J. Inorg. Nucl. Chem.*, 1964, **26,** 1233.
676. S. Sen Gupta and D. Banerjea, *Z. Anorg. Allg. Chem.*, 1976, **424,** 93.
677. D. Banerjea and S. Sen Gupta, *Z. Anorg. Allg. Chem.*, 1973, **397,** 215.
678. A. Syamal, *Z. Naturforsch., Teil B*, 1974, **29,** 492.
679. R. Longhi and R. S. Drago, *Inorg. Chem.*, 1965, **4,** 11.
680. N. J. Curtis, G. A. Lawrance and A. M. Sargeson, *Aust. J. Chem.*, 1983, **36,** 1495.
681. K. Issleib and H. O. Fröhlich, *Z. Anorg. Allg. Chem.*, 1959, **298,** 84.
682. M. A. Bennett, R. J. H. Clark and A. D. J. Goodwin, *J. Chem. Soc. (A)*, 1970, 541.
683. K. Issleib and G. Döll, *Z. Anorg. Allg. Chem.*, 1960, **305,** 1.
684. K. Issleib and A. Tzschach, *Z. Anorg. Allg. Chem.*, 1958, **297,** 121.
685. A. L. Hale and W. Levason, *J. Chem. Soc., Dalton Trans.*, 1983, 2569.
686. R. S. Nyholm and G. J. Sutton, *J. Chem. Soc.*, 1958, 560.
687. J. Lewis, R. S. Nyholm, C. S. Pande, S. S. Sandhu and M. H. B. Stiddard, *J. Chem. Soc.*, 1964, 3009.
688. W. A. Baker and P. M. Lutz, *Inorg. Chim. Acta*, 1976, **16,** 5.
689. L. F. Warren and M. A. Bennett, *Inorg. Chem.*, 1976, **15,** 3126.
690. I. V. Howell and L. M. Venanzi, *J. Chem. Soc. (A)*, 1967, 1007.
691. I. V. Howell, L. M. Venanzi and D. C. Goodall, *J. Chem. Soc. (A)*, 1967, 395.
692. C. D. Cook, R. S. Nyholm and M. L. Tobe, *J. Chem. Soc.*, 1965, 4194.
693. R. J. H. Clark, M. L. Greenfield and R. S. Nyholm, *J. Chem. Soc. (A)*, 1966, 1254.
694. P. J. Jones, A. L. Hale, W. Levason and F. P. McCullough, Jr., *Inorg. Chem.*, 1983, **22,** 2642.
695. L. R. Gray, A. L. Hale, W. Levason, F. P. McCullough, Jr. and M. Webster, *J. Chem. Soc., Dalton Trans.* 1983, 2573.
696. L. R. Gray, A. L. Hale, W. Levason, F. P. McCullough, Jr. and M. Webster, *J. Chem. Soc., Dalton Trans.* 1984, 47.
697. R. D. Feltham and W. Silverthorn, *Inorg. Chem.*, 1968, **7,** 1154.
698. M. Magini, *J. Chem. Phys.*, 1980, **73,** 2499.
699. J. K. Beattie, S. P. Best, B. W. Skelton and A. H. White, *J. Chem. Soc., Dalton Trans.*, 1981, 2105.
700. M. Epple and W. Massa, *Z. Anorg. Allg. Chem.*, 1978, **444,** 47.
701. W. Massa, *Z. Anorg. Allg. Chem.*, 1977, **436,** 29.
702. S. P. Best, R. S. Armstrong and J. K. Beattie, *Inorg. Chem.*, 1980, **19,** 1958.
703. T. Gustafsson, J. O. Lundgren and I. Olovsson, *Acta Crystallogr., Sect. B*, 1980, **36,** 1323.
704. H. Gamsjager and R. K. Murmann, *Adv. Inorg. Bioinorg. Mech.*, 1982, **2,** 317.

705. N. Bjerrum, Ph.D. Thesis, Copenhagen, 1908.
706. H. Stünzi and W. Marty, *Inorg. Chem.*, 1983, **22,** 2145.
707. M. Thompson and R. E. Connick, *Inorg. Chem.*, 1981, **20,** 2279.
708. I. Ostrich and A. J. Leffler, *Inorg. Chem.*, 1983, **22,** 921.
709. J. E. Finholt, M. E. Thompson and R. E. Connick, *Inorg. Chem.*, 1981, **20,** 4151.
710. R. A. Holwerda and J. S. Petersen, *Inorg. Chem.*, 1980, **19,** 1775.
711. R. F. Johnston and R. A. Holwerda, *Inorg. Chem.*, 1983, **22,** 2942.
712. M. Hansen, 'Constituents of Binary Alloys', McGraw-Hill, New York, 1958, p. 546.
713. W. H. Hartford and R. L. Copson, Chromium Compounds, in 'Kirk-Othmer Encyclopedia of Chemical Technology,' 2nd edn., Interscience, New York, 1962, p. 507.
714. J. Amiel, 'Nouveau Traité de Chimie Minerale', Masson et Cie, Paris, 1951, vol. 14, p. 170.
715. P. M. Hill, H. S. Peiser and J. R. Rait, *Acta Crystallogr.*, 1956, **9,** 981.
716. F. P. Glasser and L. S. Dent-Glasser, *Inorg. Chem.*, 1962, **1,** 429.
717. C. Glidewell, *Inorg. Chim. Acta,* 1976, **19,** L45.
718. B. J. Thamer, R. M. Douglass and E. Staritzky, *J. Am. Chem. Soc.*, 1957, **79,** 547.
719. W. C. Hamilton and J. A. Ibers, *Acta Crystallogr.*, 1963, **16,** 1209.
720. N. C. Tombs, W. J. Croft, J. R. Carter and J. F. Fitzgerald, *Inorg. Chem.*, 1964, **3,** 1791.
721. N. Christensen, *Inorg. Chem.*, 1966, **5,** 1452.
722. J. A. Laswick, *Inorg. Chem.*, 1962, **1,** 775.
723. O. Khan, B. Briat and J. Galy, *J. Chem. Soc., Dalton Trans.*, 1977, 1453.
724. R. R. Neurgaonkar and R. Roy, *Inorg. Chem.*, 1976, **15,** 2809.
725. G. P. Espinosa, *Inorg. Chem.*, 1964, **3,** 848.
726. R. D. Shannon, *Inorg. Chem.*, 1967, **6,** 1474.
727. M. Drillon, L. Padel and J. C. Bernier, *J. Chem. Soc., Faraday Trans. 2,* 1980, **76,** 1224.
728. A. C. Adams, J. R. Crook, F. Bockhoff and E. L. King, *J. Am. Chem. Soc.*, 1968, **90,** 5761.
729. M. Ardon and B. Bleicher, *J. Am. Chem. Soc.*, 1966, **88,** 858.
730. R. C. Mehrotra, P. C. Bharara and K. N. Mahendra, *Transition Met. Chem.* (*Weinheim, Ger.*), 1977, **2,** 161.
731. K. I. Hardcastle, D. O. Skovlin and A.-H. Eidawad, *J. Chem. Soc., Chem. Commun.*, 1975, 190.
732. H. B. Klonis and E. L. King, *Inorg. Chem.*, 1972, **11,** 2933.
733. D. C. Bradley, R. C. Mehrotra and D. P. Gaur, 'Metal Alkoxides', Academic, London, 1978.
734. R. C. Mehrotra, *Adv. Inorg. Chem. Radiochem.*, 1983, **26,** 269.
735. L. Dubicki, G. A. Kakos and G. Winter, *Aust. J. Chem.*, 1968, **21,** 1461.
736. D. A. Brown, D. Cunningham and W. K. Glass, *J. Chem. Soc.* (*A*), 1968, 1563.
737. K. N. Mahendra, P. C. Bharara and R. C. Mehrotra, *Inorg. Chim. Acta,* 1977, **25,** L5.
738. R. C. Mehrotra and J. Singh, *Inorg. Chem.*, 1984, **23,** 1046.
739. H. Dieh, *Chem. Rev.*, 1937, **21,** 39.
740. W. C. Fernelius and J. E. Blanch, *Inorg. Synth.*, 1957, **5,** 130.
741. M. K. Chaudhuri, and N. Roy, *Synth. React. Inorg. Metal-Org. Chem.*, 1982, **12,** 879.
742. B. Morosin, *Acta Crystallogr.*, 1965, **19,** 131.
743. S. Dilli and K. Robards, *Aust. J. Chem.*, 1978, **31,** 1833.
744. R. G. Charles, *Inorg. Synth.*, 1966, **8,** 135.
745. E. W. Berg, N. M. Herrera and C. L. Modenbach, *Inorg. Chem.*, 1983, **22,** 1991.
746. Y. Nakamura, K. Isobe, H. Morita, S. Yamazaki and S. Kawaguchi, *Inorg. Chem.*, 1972, **11,** 1573.
747. C. M. Lukehart and G. P. Torrence, *Inorg. Chem.*, 1979, **18,** 3150.
748. B. D. Beaver, L. C. Hall, C. M. Lukehart and L. D. Preston, *Inorg. Chim. Acta,* 1981, **47,** 25.
749. R. Astolfi, I. Collamati and C. Ercolani, *J. Chem. Soc., Dalton Trans.*, 1973, 2238.
750. R. A. Palmer, R. C. Fay and T. S. Piper, *Inorg. Chem.*, 1964, **3,** 875.
751. J. P. Collman and J.-Y. Sun, *Inorg. Chem.*, 1965, **4,** 1273.
752. R. M. King and G. W. Everett, Jr., *Inorg. Chem.*, 1971, **10,** 1237.
753. L. L. Borer and R. L. Lintvedt, *Inorg. Chem.*, 1971, **10,** 2113.
754. S. S. Minor and G. W. Everett, Jr., *Inorg. Chem.*, 1976, **15,** 1526.
755. C. E. Wilkes and R. A. Jacobson, *Inorg. Chem.*, 1965, **4,** 99.
756. H. R. Fischer, J. Glerup, D. J. Hodgson and E. Pedersen, *Inorg. Chem.*, 1982, **21,** 3063.
757. K. N. Mahendra, G. K. Parashar and R. C. Mehrotra, *Synth. React. Inorg. Metal-Org. Chem.*, 1979, **9,** 213.
758. R. E. Sievers, R. W. Moshier and M. L. Morris, *Inorg. Chem.*, 1962, **1,** 966.
759. R. E. Sievers, B. W. Ponder, M. L. Morris and R. W. Moshier, *Inorg. Chem.*, 1963, **2,** 693.
760. W. R. Wolf, R. E. Sievers and G. H. Brown, *Inorg. Chem.*, 1972, **11,** 1995.
761. C. Kutal and R. E. Sievers, *Inorg. Chem.*, 1974, **13,** 897.
762. G. M. Bancroft, C. Reichert and J. B. Westmore, *Inorg. Chem.*, 1968, **7,** 870.
763. G. M. Bancroft, C. Reichert, J. B. Westmore and H. D. Gesser, *Inorg. Chem.*, 1969, **8,** 474.
764. M. M. Bursey and P. F. Rogerson, *Inorg. Chem.*, 1970, **9,** 676.
765. M. L. Morris and R. D. Koob, *Inorg. Chem.*, 1981, **20,** 2737.
766. J. L. Pierce, K. L. Busch, R. G. Cooks and R. A. Walton, *Inorg. Chem.*, 1982, **21,** 2597.
767. B. G. Thomas, M. L. Morris and R. L. Hilderbrandt, *Inorg. Chem.*, 1978, **17,** 2901.
768. D. L. Kepert, *Prog. Inorg. Chem.*, 1977, **23,** 1 and refs. therein.
769. S. F. Mason, P. D. Peacock and T. Prosperi, *J. Chem. Soc., Dalton Trans.*, 1977, 702.
770. A. F. Drake, J. M. Gould, S. F. Mason, C. Rosini and F. J. Woodley, *Polyhedron,* 1983, **2,** 537.
771. A. Yamagishi, *Inorg. Chem.*, 1982, **21,** 3393.
772. R. Kuroda and S. F. Mason, *J. Chem. Soc., Dalton Trans.*, 1979, 273.
773. R. D. Peacock, *J. Chem. Soc., Dalton Trans.*, 1983, 291.
774. S. S. Minor, G. Witte and G. W. Everett, Jr., *Inorg. Chem.*, 1976, **15,** 2052.
775. K. L. Stevenson and T. P. vanden Driesche, *J. Am. Chem. Soc.*, 1974, **96,** 7964.
776. K. L. Stevenson and R. L. Baker, *Inorg. Chem.*, 1976, **15,** 1086.

777. O. A. Gansow, A. R. Burke and G. N. LaMar, *J. Chem. Soc., Chem. Commun.*, 1972, 456.
778. G. S. Vigee and C. L. Watkins, *Inorg. Chem.*, 1977, **16,** 709.
779. F. A. Cotton, D. L. Hunter and A. J. White, *Inorg. Chem.*, 1975, **14,** 703.
780. G. C. Levy and U. Edlund, *J. Am. Chem. Soc.*, 1975, **97,** 4482.
781. H. Reihlen, R. Illig and R. Wittig, *Chem. Ber.*, 1925, **58B,** 12.
782. J. P. Collman, *Angew. Chem., Int. Ed. Engl.*, 1965, **4,** 132.
783. R. C. Mehrotra, R. Bohra and D. P. Gaur (eds.), 'Metal β-Diketonates and Allied Derivatives', Academic, London, 1978, p. 31.
784. K. C. Joshi and V. N. Pathak, *J. Chem. Soc., Perkin Trans. 1,* 1973, 57.
785. J. P. Collman, R. L. Marshall, W. L. Young, III and S. D. Goldby, *Inorg. Chem.*, 1962, **1,** 704.
786. T. J. Cardwell and T. Lorman, *Inorg. Chim. Acta,* 1981, **53,** L103.
787. J. Charalambous, *Inorg. Chim. Acta,* 1976, **18,** 241.
788. C. G. Pierpont and R. M. Buchanan, *Coord. Chem. Rev.*, 1981, **38,** 45.
789. H. H. Downs, R. M. Buchanan and C. G. Pierpont, *Inorg. Chem.*, 1979, **18,** 1736.
790. R. M. Buchanan, S. L. Kessel, H. H. Downs, C. G. Pierpont and D. N. Hendrickson, *J. Am. Chem. Soc.*, 1978, **100,** 7894.
791. I. Smidova, A. Vlček, Jr. and A. A. Vlček, *Inorg. Chim. Acta,* 1982, **64,** L63.
792. R. M. Buchanan, J. Claflin and C. G. Pierpont, *Inorg. Chem.*, 1983, **22,** 2552.
793. S. S. Isied, G. Kuo and K. N. Raymond, *J. Am. Chem. Soc.*, 1976, **98,** 1763.
794. K. N. Raymond, S. S. Isied, L. D. Brown, F. R. Fronczek and J. Hunter Nibert, *J. Am. Chem. Soc.*, 1976, **98,** 1767.
795. S. R. Sofen, D. C. Ware, S. R. Cooper and K. N. Raymond, *Inorg. Chem.*, 1979, **18,** 234.
796. D. J. Gordon and R. F. Fenske, *Inorg. Chem.*, 1982, **21,** 2907.
797. T. Emery, 'Metal Ions in Biological Systems', ed. H. Sigel, Dekker, New York, 1978, vol. 7, p. 77.
798. J. Leong and K. N. Raymond, *J. Am. Chem. Soc.*, 1974, **96,** 1757.
799. J. Leong and K. N. Raymond, *J. Am. Chem. Soc.*, 1975, **97,** 293.
800. K. Abu-dari and K. N. Raymond, *Inorg. Chem.*, 1977, **16,** 807.
801. C. J. Carrano and K. N. Raymond, *J. Am. Chem. Soc.*, 1978, **100,** 5371.
802. W. Herwig and H. H. Zeiss, *J. Org. Chem.*, 1958, **23,** 1404.
803. P. J. Jones, A. L. Hale, W. Levason and F. P. McCullough, Jr., *Inorg. Chem.*, 1983, **22,** 2642.
804. P. Pfeiffer, *Chem. Ber.*, 1903, **36,** 1926.
805. B. N. Figgis, E. S. Kucharski, J. M. Patrick and A. H. White, *Aust. J. Chem.*, 1984, **37,** 265.
806. A. Yamada, Y. Nakabayashi, T. Yoshikuni and N. Tanaka, *Bull. Chem. Soc. Jpn.*, 1980, **53,** 3596.
807. C. D. Flint and D. J. D. Palacio, *J. Chem. Soc., Faraday Trans. 2,* 1980, **76,** 82.
808. C. D. Flint and D. J. D. Palacio, *J. Chem. Soc., Faraday Trans. 2,* 1979, **75,** 1159.
809. C. C. Addison and D. J. Chapman, *J. Chem. Soc.*, 1964, 539.
810. M. Nakahara, *Bull. Chem. Soc. Jpn.*, 1962, **35,** 782.
811. W. W. Fee, C. S. Garner and J. N. MacB. Harrowfield, *Inorg. Chem.*, 1967, **6,** 87.
812. W. W. Fee, J. N. MacB. Harrowfield and C. S. Garner, *Inorg. Chem.*, 1971, **10,** 290.
813. A. Recoura, *C. R. Hebd. Seances Acad. Sci.*, 1926, **183,** 719.
814. See refs. in ref. 823.
815. N. Fogel, J. Tai and J. Yarborough, *J. Am. Chem. Soc.*, 1962, **84,** 1145.
816. J. E. Finholt, R. W. Anderson, J. A. Fyfe and K. G. Caulton, *Inorg. Chem.*, 1965, **4,** 43.
817. A. Riou and A. Bonnin, *Acta Crystallogr., Sect. B,* 1981, **37,** 1031.
818. J. E. Finholt and S. E. Deming, *Inorg. Chem.*, 1967, **6,** 1533.
819. N. P. Slabert, *J. Inorg. Nucl. Chem.*, 1977, **39,** 883.
820. J. E. Finholt, K. Caulton, K. Kimball and E. Uhlenhopp, *Inorg. Chem.*, 1968, **7,** 611.
821. A. Scott and H. Taube, *Inorg. Chem.*, 1971, **10,** 62.
822. E. Konig, E. Lindner, I. P. Lorenz, G. Ritter and H. Gausmann, *J. Inorg. Nucl. Chem.*, 1971, **33,** 3305.
823. J. Podlaha and M. Ebert, *Nature (London)*, 1960, **188,** 658.
824. J. Podlaha and M. Ebert, *Russ. J. Inorg. Chem. (Engl. Transl.)*, 1962, **7,** 1130.
825. R. N. Puri and R. O. Asplund, *Inorg. Chim. Acta,* 1982, **57,** 57.
826. M. Ebert and L. Kavan, *Collect. Czech. Chem. Commun.*, 1979, **44,** 2737.
827. M. Ebert and L. Kavan, *Collect. Czech. Chem. Commun.*, 1976, **41,** 23.
828. L. S. Brown and J. N. Cooper, *Inorg. Chem.*, 1972, **11,** 1154.
829. J. N. Cooper, *J. Phys. Chem.*, 1970, **74,** 955.
830. J. H. Espenson and D. E. Binau, *Inorg. Chem.*, 1966, **5,** 1365.
831. A. T. Ness, R. E. Smith and R. L. Evans, *J. Am. Chem. Soc.*, 1952, **74,** 4685 and refs. therein.
832. E. Rodek, W. Sterzel and N. Theile, *Z. Anorg. Allg. Chem.*, 1980, **462,** 42.
833. C. L. Rollinson and E. W. Rosenbloom, in 'Coordination Chemistry', ed. S. Kirschener, Plenum, New York, 1969, p. 108.
834. E. Gonzalez-Vergara, D. Held and H. M. Goff, Abstracts of the 23rd International Conference on Coordination Chemistry, Boulder, Colorado, p. 435.
835. C. M. Mikulski, L. L. Pytlewski and N. M. Karayannis, *Inorg. Chim. Acta,* 1979, **32,** 263.
836. C. M. Mikulski, N. M. Karayannis, M. J. Strocko, L. L. Pytlewski and M. M. Labes, *Inorg. Chem.*, 1970, **9,** 2053.
837. C. M. Mikulski, T. Tran, L. L. Pytlewski and N. M. Karayannis, *J. Inorg. Nucl. Chem.*, 1979, **41,** 1671.
838. N. M. Karayannis, E. E. Bradshaw, L. L. Pytlewski and M. M. Labes, *J. Inorg. Nucl. Chem.*, 1970, **32,** 1079.
839. N. M. Karayannis, C. Owens, L. L. Pytlewski and M. M. Labes, *J. Inorg. Nucl. Chem.*, 1969, **31,** 2767.
840. M. L. De Pamphillis and W. W. Cleland, *Biochemistry,* 1973, **12,** 3714.
841. D. Dunaway-Mariano and W. W. Cleland, *Biochemistry,* 1980, **19,** 1496.
842. M. J. Bossard, G. S. Samuelson and S. M. Schuster, *J. Inorg. Biochem.*, 1982, **17,** 61.
843. S. H. McClaugherty and C. M. Grisham, *Inorg. Chem.*, 1982, **21,** 4133.

844. I. Lin, A. Hsueh and D. Dunaway-Mariano, *Inorg. Chem.*, 1984, **23,** 1692.
845. I. Lin and D. Dunaway-Mariano, *J. Am. Chem. Soc.*, 1984, **106,** 6074.
846. K. D. Maguire, *Inorg. Synth.*, 1970, **12,** 258.
847. P. Nannelli, H. D. Gillman and B. P. Block, *J. Polym. Sci.* (*A1*), 1971, **9,** 3027.
848. H. D. Gillman, P. Nannelli and B. P. Block, *Inorg. Synth.*, 1976, **16,** 89.
849. H. D. Gillman and P. Nannelli, *Inorg. Chim. Acta,* 1977, **23,** 259.
850. C. M. Mikulski, J. Unruh, R. Rabin, F. J. Iaconianni and L. L. Pytlewski, *Inorg. Chim. Acta,* 1980, **44,** L77.
851. M. Chaabouni, T. Chausse, J. L. Pascal and J. Potier, *J. Chem. Res.* (*S*), 1980, 72.
852. N. C. Tombs, W. J. Croft and H. C. Mattraw, *Inorg. Chem.*, 1963, **2,** 872.
853. I. P. Lorenz, *Inorg. Nucl. Chem. Lett.*, 1979, **15,** 127.
854. C. Oldham, *Prog. Inorg. Chem.*, 1968, **10,** 223.
855. R. C. Mehrotra and R. Bohra, 'Metal Carboxylates', Academic, London, 1983.
856. See section 35.2.3.8.iii.f, p. 907 on tanning.
857. D. B. Sowerby and L. F. Audrieth, *J. Chem. Educ.*, 1960, **37,** 2.
858. D. L. Perry, L. Tsao and J. A. Taylor, *Inorg. Chim. Acta,* 1984, **85,** L57.
859. R. F. Weinland and H. Reihlen, *Z. Anorg. Allg. Chem.*, 1913, **82,** 426.
860. L. E. Orgel, *Nature* (*London*), 1960, **187,** 504.
861. B. N. Figgis and G. B. Robertson, *Nature* (*London*), 1965, **205,** 694.
862. J. T. Schriempf and S. A. Friedberg, *J. Chem. Phys.*, 1964, **40,** 296.
863. J. T. Wrobleski, C. T. Dziobkowski and D. B. Brown, *Inorg. Chem.*, 1981, **20,** 684.
864. C. T. Dziobkowski, J. T. Wrobleski and D. B. Brown, *Inorg. Chem.*, 1981, **20,** 671.
865. B. S. Tsukerblat, M. I. Belinski and B. Ya. Kuyavskaya, *Inorg. Chem.*, 1983, **22,** 995 and refs. therein.
866. L. Dubicki and P. Day, *Inorg. Chem.*, 1972, **11,** 1868.
867. L. Dubicki and R. L. Martin, *Aust. J. Chem.*, 1969, **22,** 701.
868. M. Makles-Grotowska and W. Wojciechowski, *Bull. Acad. Pol. Sci., Ser. Sci. Chim.*, 1979, **27,** 59.
869. Y. V. Rakitin, T. A. Zhemchuzhnikova and V. V. Zelentsov, *Inorg. Chim. Acta,* 1977, **23,** 145.
870. F. A. Cotton and W. X. Wang, *Inorg. Chem.*, 1982, **21,** 2675.
871. M. K. Johnson, R. D. Cannon and D. B. Powell, *Spectrochim. Acta, Part A,* 1982, **38,** 307.
872. R. Kapoor and R. Sharma, *Z. Naturforsch., Teil B,* 1979, **34,** 1369.
873. E. J. Serfass, C. D. Wilson and E. R. Theis, *J. Am. Leather Chem. Assoc.*, 1949, **44,** 647.
874. R. Davies, G. B. Evans and R. B. Jordon, *Inorg. Chem.*, 1969, **8,** 2025.
875. G. K. Parashar and A. K. Rai, *Synth. React. Inorg. Metal-Org. Chem.*, 1978, **8,** 427.
876. R. C. Paul, P. Kapoor, O. B. Baidya and R. Kapoor, *Z. Naturforsch., Teil B,* 1979, **34,** 160.
877. A. Rosenheim and R. Cohn, *Z. Anorg. Allg. Chem.*, 1901, **28,** 337.
878. A. Werner and H. Surber, *Liebigs Ann. Chem.*, 1914, **406,** 298, 304.
879. G. B. Kauffman and D. Faoro, *Inorg. Synth.*, 1977, **17,** 147.
880. J. C. Bailar, Jr. and E. M. Jones, *Inorg. Synth.*, 1939, **1,** 35.
881. M. W. Rophael and M. A. Malati, *J. Chem. Soc.* (*A*), 1971, 1903.
882. G. C. Pass and H. H. Sutcliffe, 'Practical Inorganic Chemistry', Barnes and Noble, New York, 1968, p. 91.
883. G. B. Kauffman, *Coord. Chem. Rev.*, 1974, **12,** 105.
884. D. Taylor, *Aust. J. Chem.*, 1978, **31,** 1455.
885. D. C. Doetschman and B. J. McCool, *J. Chem. Phys.*, 1975, **8,** 1.
886. J. N. van Niekerk and F. R. L. Schoening, *Acta Crystallogr.*, 1952, **5,** 196.
887. J. N. van Niekerk and F. R. L. Schoening, *Acta Crystallogr.*, 1952, **5,** 499.
888. C. A. Bunton, J. H. Carter, D. R. Llewellyn, A. L. Odell and S. Y. Yih, *J. Chem. Soc.*, 1964, 4615 and subsequent papers.
889. R. W. Olliff, D. B. Rands and D. Shooter, *Aust. J. Chem.*, 1974, **27,** 2057 and refs. therein.
890. S. S. Eaton, T. D. Yager and G. R. Eaton, *J. Chem. Educ.*, 1979, **56,** 635.
891. A. Werner, *Chem. Ber.*, 1912, **45,** 3061.
892. F. P. Dwyer and A. M. Sargeson, *J. Phys. Chem.*, 1956, **60,** 1331.
893. J. W. Vaughn, V. E. Magnuson and G. J. Seiler, *Inorg. Chem.*, 1969, **8,** 1201.
894. K. R. Butler and M. R. Snow, *J. Chem. Soc.* (*A*), 1971, 565.
895. K. R. Butler and M. R. Snow, *Chem. Commun.*, 1971, 550.
896. K. Garbett and R. D. Gillard, *J. Chem. Soc.* (*A*), 1966, 802.
897. R. D. Gillard, D. J. Shepherd and D. A. Tarr, *J. Chem. Soc., Dalton Trans.*, 1976, 594.
898. V. S. Sastri, *Inorg. Chim. Acta,* 1973, **7,** 381.
899. S. T. Spees and A. W. Adamson, *Inorg. Chem.*, 1962, **1,** 531.
900. D. M. Chowdhury and G. M. Harris, *J. Phys. Chem.*, 1969, **73,** 3366.
901. P. O'Brien, *Polyhedron,* 1983, **2,** 233.
902. D. Dollimore, D. L. Griffiths and D. Nicholson, *J. Chem. Soc.*, 1963, 2617.
903. A. M. Passaglia Schuch and A. G. Maddock, *Inorg. Chim. Acta,* 1982, **63,** 27.
904. J. N. van Niekerk and F. R. L. Schoening, *Acta Crystallogr.*, 1951, **4,** 35.
905. J. W. Lethbridge, L. S. Dent-Glasser and H. F. W. Taylor, *J. Chem. Soc.* (*A*), 1970, 1862.
906. J. R. Perumareddi, *Coord. Chem. Rev.*, 1969, **4,** 73.
907. J. M. Veigel and C. S. Garner, *Inorg. Chem.*, 1965, **4,** 1569.
908. W. T. M. Andriessen, *Inorg. Chem.*, 1975, **14,** 792.
909. G. L. Welch and R. E. Hamm, *Inorg. Chem.*, 1963, **2,** 295.
910. H. Croft, *Philos. Mag.*, 1842, **21,** 197.
911. A. Werner, *Ann.*, 1914, **406,** 261.
912. D. C. Price, Ph.D. Thesis, University of Kent at Canterbury, 1970, chap. 5, p. 194.
913. D. M. Grant and R. E. Hamm, *J. Am. Chem. Soc.*, 1956, **78,** 3006.
914. American Crystallographers, Spring Meeting, 1976, **4,** 11.
915. J. R. Ferraro, R. Driver, W. R. Walker and W. Wozniak, *Inorg. Chem.*, 1967, **6,** 1586.

916. K. L. Stevenson and J. F. Verdieck, *J. Am. Chem. Soc.*, 1968, **90**, 2974.
917. F. M. Jaeger, *Recl. Trav. Chim. Pays-Bas,* 1919, **38**, 171.
918. J. C. Chang, *Inorg. Synth.*, 1976, **16**, 80.
919. H. T. S. Britton and M. E. Jarrat, *J. Chem. Soc.*, 1935, 1728.
920. K. R. Butler and M. R. Snow, *J. Chem. Soc., Dalton Trans.*, 1976, 251.
921. A. J. McCaffery, S. F. Mason and R. E. Ballard, *J. Chem. Soc.*, 1965, 2883.
922. B. E. Douglas, R. A. Haines and J. G. Brushmiller, *Inorg. Chem.*, 1963, **2**, 1194.
923. W. E. Hatfield, *Inorg. Chem.*, 1964, **3**, 605.
924. D. H. Huchital and H. Taube, *Inorg. Chem.*, 1965, **4**, 1660.
925. M. V. Olson and C. E. Behnke, *Inorg. Chem.*, 1974, **13**, 1329.
926. A. A. Abdel Khalek, M. S. Al-Obadie and A. A. K. Al-Mahdi, *J. Inorg. Nucl. Chem.*, 1980, **42**, 1407.
927. W. Wheeler, S. Kaizaki and J. I. Legg, *Inorg. Chem.*, 1981, **21**, 3248.
928. R. West and H. Y. Niu, *J. Am. Chem. Soc.*, 1963, **85**, 2589.
929. J. P. Chesick and F. Doany, *Acta Crystallogr., Sect. B,* 1981, **37**, 1076.
930. A. Cotton, *Ann. Chim. Phys.*, 1896, **8**, 347.
931. A. J. McCaffery and S. F. Mason, *Trans. Faraday Soc.*, 1963, **59**, 1.
932. L. Johansson and B. Norden, *Inorg. Chim. Acta,* 1978, **29**, 189.
933. B. Norden, *Inorg. Nucl. Chem. Lett.*, 1977, **13**, 355.
934. S. Kaizaki, J. Hidaka and Y. Shimura, *Bull. Chem. Soc. Jpn.*, 1969, **42**, 988.
935. G. L. Robbins and R. E. Tapscott, *Inorg. Chem.*, 1976, **15**, 154.
936. G. L. Robbins and R. E. Tapscott, *Inorg. Chem.*, 1981, **20**, 2343.
937. R. B. Ortega, R. E. Tapscott and C. F. Campana, *Inorg. Chem.*, 1982, **21**, 2517.
938. K. K. Tripathy and R. K. Patnaik, *J. Indian Chem. Soc.*, 1966, **43**, 772.
939. K. Bauer, H. Elias, R. Gaubatz and G. Lang, *Inorg. Chim. Acta,* 1979, **36**, 55.
940. S. C. Tyagi and A. A. Khan, *J. Chem. Soc., Dalton Trans.*, 1979, 420.
941. F. A. Cotton and R. Francis, *J. Am. Chem. Soc.*, 1960, **82**, 2986.
942. D. W. Meek, R. S. Drago and T. S. Piper, *Inorg. Chem.*, 1962, **1**, 285.
943. C. V. Berney and J. H. Weber, *Inorg. Chem.*, 1968, **7**, 283.
944. J. J. Habeeb and D. G. Tuck, *Inorg. Synth.*, 1979, **19**, 126.
945. K. R. Ashley, R. E. Hamm and R. H. Magnuson, *Inorg. Chem.*, 1967, **6**, 413.
946. L. P. Scott, T. J. Weeks, Jr., D. E. Bracken and E. L. King, *J. Am. Chem. Soc.*, 1969, **91**, 5219.
947. S. T. D. Lo and D. W. Watts, *Aust. J. Chem.*, 1975, **28**, 491, 501.
948. S. T. D. Lo and D. W. Watts, *Aust. J. Chem.*, 1975, **28**, 1907.
949. J. J. Habeeb, F. F. Said and D. G. Tuck, *J. Chem. Soc., Dalton Trans.*, 1981, 118.
950. J. A. Davies, *Adv. Inorg. Chem. Radiochem.*, 1981, **24**, 115.
951. S. T. D. Lo and T. W. Swaddle, *Inorg. Chem.*, 1975, **14**, 1878.
952. S. T. D. Lo and T. W. Swaddle, *Inorg. Chem.*, 1976, **15**, 1881.
953. R. W. Windolph and A. J. Leffler, *Inorg. Chem.*, 1972, **11**, 594.
954. R. G. Garvey, J. H. Nelson and R. O. Ragsdale, *Coord. Chem. Rev.*, 1968, **3**, 375.
955. H. Kanno, K. Kashiwabara and J. Fujita, *Bull. Chem. Soc. Jpn.*, 1979, **52**, 1408.
956. H. Kanno, K. Kashiwabara and J. Fujita, *Bull. Chem. Soc. Jpn.*, 1980, **53**, 2881.
957. N. M. Karayannis, C. M. Mikulski, L. L. Pytlewski and M. M. Labes, *Inorg. Chem.*, 1970, **9**, 582.
958. C. M. Mikulski, J. S. Skryantz, N. M. Karayannis, L. L. Pytlewski and L. S. Gelfand, *Inorg. Chim. Acta,* 1978, **27**, 69.
959. N. M. Karayannis, C. M. Mikulski and L. L. Pytlewski, *Inorg. Chim. Acta, Rev.*, 1971, **5**, 69.
960. S. R. Wade and G. R. Willey, *J. Inorg. Nucl. Chem.*, 1980, **42**, 1133.
961. C. M. Mikulski, N. Harris, F. J. Iaconianni, L. L. Pytlewski and N. M. Karayannis, *Inorg. Nucl. Chem. Lett.*, 1980, **16**, 79.
962. N. M. Karayannis, C. M. Mikulski, M. J. Strocko, L. L. Pytlewski and M. M. Labes, *J. Inorg. Nucl. Chem.*, 1970, **32**, 2629.
963. H. Taube and E. S. Gould, *Acc. Chem. Res.*, 1969, **2**, 321.
964. M. Ardon and H. Taube, *J. Am. Chem. Soc.*, 1967, **89**, 3661.
965. T. Ramasami and A. G. Sykes, *Inorg. Chem.*, 1976, **15**, 1010.
966. L. E. Asher and E. Deutsch, *Inorg. Chem.*, 1973, **12**, 1774.
967. T. Ramasami, R. S. Taylor and A. G. Sykes, *J. Chem. Soc., Chem. Commun.*, 1976, 383.
968. T. Ramasami, R. S. Taylor and A. G. Sykes, *Inorg. Chem.*, 1977, **16**, 1931.
969. L. E. Asher and E. Deutsch, *Inorg. Chem.*, 1972, **11**, 2927.
970. L. E. Asher and E. Deutsch, *Inorg. Chem.*, 1976, **15**, 1531.
971. C. J. Weschler and E. Deutsch, *Inorg. Chem.*, 1973, **12**, 2682.
972. A. D. Kirk, K. C. Moss and J. G. Valentin, *Can. J. Chem.*, 1971, **49**, 375.
973. E. Zinato, C. Furlani, G. Lanna and P. Riccieri, *Inorg. Chem.*, 1972, **11**, 1746.
974. D. A. House, *Coord. Chem. Rev.*, 1977, **23**, 223.
975. M. J. Root and E. Deutsch, *Inorg. Chem.*, 1981, **20**, 4376.
976. D. A. Brown, W. K. Glass and B. Kumar, *Chem. Commun.*, 1967, 736.
977. D. A. Brown, W. K. Glass and B. Kumar, *J. Chem. Soc. (A)*, 1969, 1510.
978. R. J. H. Clark and G. Natile, *Inorg. Chim. Acta,* 1970, **4**, 533.
979. J. Hughes and G. R. Willey, *Inorg. Chim. Acta,* 1974, **11**, L25.
980. F. Jellinek, *Acta Crystallogr.*, 1957, **10**, 620.
981. A. W. Sleight and T. A. Bither, *Inorg. Chem.*, 1969, **8**, 566.
982. M. Yuzuri and Y. Nakamura, *J. Phys. Soc. Jpn.*, 1964, **19**, 1350.
983. A. L. Companion and M. Mackin, *J. Chem. Phys.*, 1965, **42**, 4219.
984. G. Brauer, 'Handbook of Preparative Inorganic Chemistry', Academic, New York, 1965, vol. 2, p. 1394.
985. S. L. Holt and A. Wold, *Inorg. Chem.*, 1967, **6**, 1594.

986. C. K. Jørgensen, *Inorg. Chim. Acta, Rev.*, 1968, **2,** 65.
987. P. F. Bongers, C. F. Van Bruggen, J. Koopstra, W. P. F. A. M. Omloo, G. A. Wiegers and F. Jellinek, *J. Phys. Chem. Solids,* 1968, **29,** 977.
988. J. G. White and H. L. Pinch, *Inorg. Chem.*, 1970, **9,** 2581.
989. P. Wolfers, G. Fillion, M. Bachman and H. Noel, *J. Phys.* (*Orsay, Fr.*), 1976, **37,** 57.
990. F. L. Carter, P. Wolfers and G. Fillion, *J. Chem. Soc., Dalton Trans.*, 1978, 1686.
991. R. J. Bouchard, P. A. Russo and A. Wold, *Inorg. Chem.*, 1965, **4,** 685.
992. R. Diehl and C. D. Carpentier, *Acta Crystallogr., Sect. B,* 1977, **33,** 1399.
993. D. Coucouvanis, *Prog. Inorg. Chem.*, 1979, **26,** 301.
994. T. N. Lockyer and R. L. Martin, *Prog. Inorg. Chem.*, 1980, **27,** 223.
995. R. P. Burns and C. A. McAuliffe, *Adv. Inorg. Chem. Radiochem.*, 1979, **22,** 303.
996. A. M. Bond and R. L. Martin, *Coord. Chem. Rev.*, 1984, **54,** 23.
997. F. Galsbøl and C. E. Schäffer, *Inorg. Synth.*, 1967, **10,** 42.
998. K. B. Pandeya, T. S. Waraich, R. C. Gaur and R. P. Singh, *Transition Met. Chem.*, 1982, **7,** 146.
999. H. B. Singh, S. Maheshwari, S. Srivastava and V. Rani, *Synth. React. Inorg. Metal-Org. Chem.*, 1982, **12,** 659.
1000. R. M. Golding, P. C. Healy, P. Colombera and A. H. White, *Aust. J. Chem.*, 1974, **27,** 2089.
1001. G. A. Heath, R. L. Martin and A. F. Masters, *Aust. J. Chem.*, 1972, **25,** 2547.
1002. R. G. Cavell, W. Byers and E. D. Day, *Inorg. Chem.*, 1971, **10,** 2710.
1003. F. N. Tebbe and E. L. Muetterties, *Inorg. Chem.*, 1970, **9,** 629.
1004. R. G. Cavell and A. R. Sanger, *Inorg. Chem.*, 1972, **11,** 2011.
1005. P. Vella and J. Zubieta, *J. Inorg. Nucl. Chem.*, 1978, **40,** 613.
1006. R. Lancashire and T. D. Smith, *J. Chem. Soc., Dalton Trans.*, 1982, 845.
1007. J. M. Hope, R. L. Martin, D. Taylor and A. H. White, *J. Chem. Soc., Chem. Commun.*, 1977, 99.
1008. A. M. Bond and G. G. Wallace, *Inorg. Chem.*, 1984, **23,** 1858.
1009. V. K. Sinha, M. N. Srivastava and H. L. Nigam, *Indian J. Chem., Sect. A,* 1982, **21,** 732.
1010. R. L. Martin, J. M. Patrick, B. W. Skelton, D. Taylor and A. H. White, *Aust. J. Chem.*, 1982, **35,** 2551.
1011. G. Soundararajan and M. Subbaiyan, *Indian J. Chem., Sect. A,* 1983, **22,** 1058.
1012. C. Preti, G. Tosi and P. Zannini, *J. Mol. Struct.*, 1980, **65,** 283.
1013. A. L. Marchese, M. Scudder, A. M. Van Den Bergen and B. O. West, *J. Organomet. Chem.*, 1976, **121,** 63.
1014. S. Merlino and F. Sartori, *Acta Crystallogr., Sect. B,* 1972, **28,** 972.
1015. B. N. Figgis and G. E. Toogood, *J. Chem. Soc., Dalton Trans.*, 1972, 2177.
1016. W. Kwoka, R. O. Moyer, Jr. and R. Lindsay, *J. Inorg. Nucl. Chem.*, 1975, **37,** 1889.
1017. L. Que, Jr. and L. H. Pignolet, *Inorg. Chem.*, 1974, **13,** 351.
1018. K. DeArmond and W. J. Mitchell, *Inorg. Chem.*, 1972, **11,** 181.
1019. P. J. Hauser, A. F. Schreiner and R. S. Evans, *Inorg. Chem.*, 1974, **13,** 1925.
1020. M.-L. Riekkola, *Acta Chem. Scand., Ser. A,* 1983, **37,** 691.
1021. J. D. Gunter and A. F. Schreiner, *Inorg. Chim. Acta,* 1975, **15,** 117.
1022. D. A. Brown, W. K. Glass and M. A. Burke, *Spectrochim. Acta, Part A,* 1976, **32,** 137.
1023. G. Contreras and H. Cortes, *J. Inorg. Nucl. Chem.*, 1971, **33,** 1337.
1024. G. C. Pelizzi and C. Pelizzi, *Inorg. Chim. Acta,* 1970, **4,** 618.
1025. M. Maltese, *J. Chem. Soc., Dalton Trans.*, 1972, 2664.
1026. C. Furlani and M. L. Luciani, *Inorg. Chem.*, 1968, **7,** 1586.
1027. G. Cauquis and A. Deronzier, *J. Inorg. Nucl. Chem.*, 1980, **42,** 1447.
1028. K. Nag and D. S. Joardar, *Inorg. Chim. Acta,* 1976, **17,** 111.
1029. C. A. Tsipis, M. P. Sigalas and C. C. Hadjikostas, *Z. Anorg. Allg. Chem.*, 1983, **505,** 53.
1030. J. Hidaka and B. E. Douglas, *Inorg. Chem.*, 1964, **3,** 1724.
1031. A. Müller, P. Christophliemk, I. Tossidis and C. K. Jørgensen, *Z. Anorg. Allg. Chem.*, 1973, **401,** 274.
1032. J. R. Wasson, S. J. Wasson and G. M. Woltermann, *Inorg. Chem.*, 1970, **9,** 1576.
1033. J. D. Lebedda and R. A. Palmer, *Inorg. Chem.*, 1971, **10,** 2704.
1034. W. Kuchen, R. Uppenkamp and K. Diemert, *Z. Naturforsch., Teil B,* 1979, **34,** 1398.
1035. C. Battistoni, G. Mattogno, A. Monaci and F. Tarli, *Inorg. Nucl. Chem. Lett.*, 1971, **7,** 1081.
1036. D. De Filippo, P. Deplano, A. Diaz and E. F. Trogu, *Inorg. Chim. Acta,* 1976, **17,** 139.
1037. W. Dietzsch, J. Kaiser, R. Richter, L. Golic, J. Siftar and R. Heber, *Z. Anorg. Allg. Chem.*, 1981, **477,** 71.
1038. M. V. Andreocci, P. Dragoni, A. Flamini and C. Furlani, *Inorg. Chem.*, 1978, **17,** 291.
1039. E. Ciliberto, L. L. Costanzo, I. Fragala and G. Granozzi, *Inorg. Chim. Acta,* 1980, **44,** L25.
1040. S. Sunder, L. Hanlan and H. J. Bernstein, *Inorg. Chem.*, 1975, **14,** 2012.
1041. G. K. Budnikov, A. V. Il'yasov, V. I. Morozov and N. A. Ulakhovich, *Russ. J. Inorg. Chem.* (*Engl. Transl.*), 1976, **21,** 255.
1042. S. V. Larionov, *Russ. J. Inorg. Chem.* (*Engl. Transl.*), 1979, **24,** 802.
1043. M. M. Khan, *J. Inorg. Nucl. Chem.*, 1975, **37,** 1621.
1044. A. Yamaguchi, R. B. Penland, S. Mizushima, T. J. Lane, C. Curran and J. V. Quagliano, *J. Am. Chem. Soc.*, 1958, **80,** 527.
1045. E. Cervone, P. Cancellieri and C. Furlani, *J. Inorg. Nucl. Chem.*, 1968, **30,** 2431.
1046. G. Contreras and R. Schmidt, *J. Inorg. Nucl. Chem.*, 1970, **32,** 127.
1047. M. Gaur, P. K. Mathur and S. N. Kapoor, *Indian J. Chem., Sect. A,* 1984, **23,** 774.
1048. P. Askalani and R. A. Bailey, *Can. J. Chem.*, 1969, **47,** 2275.
1049. M. H. Sonar and A. S. R. Murty, *J. Inorg. Nucl. Chem.*, 1977, **39,** 2155.
1050. H. N. Tiwari, H. H. Khan, R. N. Pandey and B. Singh, *Indian J. Chem., Sect. A,* 1982, **21,** 315.
1051. B. Singh and R. D. Singh, *Indian J. Chem., Sect. A,* 1982, **21,** 648.
1052. J. E. Fergusson, in 'Halogen Chemistry 3', Academic, New York, p. 227.
1053. H. von Wartenburg, *Z. Anorg. Allg. Chem.*, 1942, **249,** 100.
1054. Ref. 2, p. 2664.
1055. G. B. Heisig, B. Fawkes and R. Hedin, *Inorg. Synth.*, 1946, **2,** 193.

1056. K. R. Seddon and V. H. Thomas, *Inorg. Chem.*, 1978, **17,** 749.
1057. D. J. Machin, D. F. C. Morris and E. L. Short, *J. Chem. Soc.*, 1964, 4658.
1058. J. C. Dewan and A. J. Edwards, *J. Chem. Soc., Chem. Commun.*, 1977, 533.
1059. G. Knoke, W. Verscharen and D. Babel, *J. Chem. Res. (S)*, 1979, 213.
1060. G. Knoke, D. Babel and T. Hinrichsen, *Z. Naturforsch., Teil B*, 1979, **34,** 934.
1061. P. G. Ferey, R. DePape and B. Boucher, *Acta Crystallogr., Sect. B*, 1978, **34,** 1084.
1062. P. G. Nelson and R. V. Pearse, *J. Chem. Soc., Dalton Trans.*, 1983, 1977.
1063. A. Dutta-Ahmed and E. A. Boudreaux, *Inorg. Chem.*, 1973, **12,** 1597.
1064. See Table 92 and refs. 1005–1068.
1065. G. C. Allen, G. A. M. El-Sharkaway and K. D. Warren, *Inorg. Chem.*, 1971, **10,** 2538.
1066. H. L. Schlafer, H. Gausmann and H.-U. Zander, *Inorg. Chem.*, 1967, **6,** 1528.
1067. B. N. Figgis, P. A. Reynolds and G. A. Williams, *J. Chem. Soc., Dalton Trans.*, 1980, 2348.
1068. F. A. Wedgewood, *Proc. R. Soc. London, Ser. A*, 1976, **359,** 447.
1069. S. Larsson and J. W. D. Connolly, *J. Chem. Phys.*, 1974, **60,** 1514.
1070. R. D. Brown and P. G. Burton, *Theor. Chim. Acta,* 1970, **18,** 309.
1071. Ref. 23 in ref. 1067; G. S. Chandler and R. A. Phillips, personal communication.
1072. R. W. Schwartz, *Inorg. Chem.*, 1976, **15,** 2817.
1073. D. J. Hodgson, *Prog. Inorg. Chem.*, 1975, **19,** 173.
1074. R. Saillant and R. A. D. Wentworth, *Inorg. Chem.*, 1968, **7,** 1606.
1075. G. J. Wessel and D. J. W. Ijdo, *Acta Crystallogr.*, 1957, **10,** 466.
1076. I. E. Grey and P. W. Smith, *Aust. J. Chem.*, 1971, **24,** 73.
1077. R. J. Ziegler and W. M. Risen, Jr., *Inorg. Chem.*, 1972, **11,** 2796.
1078. J. D. Black, J. T. Dunsmuir, I. W. Forest and A. P. Lane, *Inorg. Chem.*, 1975, **14,** 1257.
1079. G. L. McPherson, J. A. Varga and M. H. Nodine, *Inorg. Chem.*, 1979, **18,** 2189.
1080. G. L. McPherson, W.-M. Heung and J. J. Barraza, *J. Am. Chem. Soc.*, 1978, **100,** 469.
1081. A. Werner and A. Gruber, *Chem. Ber.*, 1901, **34,** 1579.
1082. A full discussion of this chemistry can be found in G. B. Kauffman, *Coord. Chem. Rev.*, 1973, **11,** 161.
1083. I. G. Dance and H. C. Freeman, *Inorg. Chem.*, 1965, **4,** 1555.
1084. B. Morosin, *Acta Crystallogr.*, 1966, **21,** 280.
1085. P. J. McCarthy, J. C. Lauffenburger, P. M. Skonezy and D. C. Rohrer, *Inorg. Chem.*, 1981, **20,** 1566.
1086. J. D. Salzman and E. L. King, *Inorg. Chem.*, 1967, **6,** 426.
1087. R. J. Baltisberger and E. L. King, *J. Am. Chem. Soc.*, 1964, **86,** 795.
1088. C. F. Hale and E. L. King, *J. Phys. Chem.*, 1967, **71,** 1779.
1089. L. O. Spreer and E. L. King, *Inorg. Chem.*, 1971, **10,** 916.
1090. T. W. Swaddle and G. Guastalla, *Inorg. Chem.*, 1968, **7,** 1915.
1091. P. K. Wrona, *Inorg. Chem.*, 1984, **23,** 1558.
1092. R. H. Holm, G. W. Everett, Jr. and A. Chakravorty, *Prog. Inorg. Chem.*, 1966, **7,** 83.
1093. F. Calderazzo, C. Floriani, R. Henzi and F. L'Eplattenier, *J. Chem. Soc. (A)*, 1969, 1378.
1094. R. Lancashire and T. D. Smith, *J. Chem. Soc., Dalton Trans.*, 1982, 693.
1095. M. J. O'Connor and B. O. West, *Aust. J. Chem.*, 1968, **21,** 369.
1096. S. Yamada and K. Iwasaki, *Bull. Chem. Soc. Jpn.*, 1968, **41,** 1972.
1097. S. Yamada and K. Iwasaki, *Inorg. Chim. Acta*, 1971, **5,** 3.
1098. J. P. Collman and E. T. Kittleman, *Inorg. Chem.*, 1962, **1,** 499.
1099. P. Coggon, A. T. McPhail, F. E. Mabbs, A. Richards and A. S. Thornley, *J. Chem. Soc. (A)*, 1970, 3296.
1100. K. Dey and K. C. Ray, *Inorg. Chim. Acta,* 1974, **10,** 139.
1101. D. R. Prasad, T. Ramasami, D. Ramaswamy and M. Santappa, *Synth. React. Inorg. Metal-Org. Chem.*, 1981, **11,** 431.
1102. K. Yamanouchi and S. Yamada, *Bull. Chem. Soc. Jpn.*, 1972, **45,** 2140.
1103. R. W. Kolaczkowski and R. A. Plane, *Inorg. Chem.*, 1964, **3,** 322.
1104. L. E. Gerdom and H. M. Goff, *Inorg. Chem.*, 1982, **21,** 3847.
1105. D. R. Prasad, T. Ramasami, D. Ramaswamy and M. Santappa, *Inorg. Chem.*, 1980, **19,** 3181.
1106. K. Dey, R. L. De and K. C. Ray, *Indian J. Chem.*, 1972, **10,** 864.
1107. F. Brezina and Z. Sindelar, *Z. Chem.*, 1979, **19,** 72.
1108. M. Macicek and F. Brezina, *Acta Univ. Palacki. Olomuc., Fac. Rerum Nat.*, 1978, **57,** 23.
1109. A. M. Sayeed, K. Iftikhar and N. Ahmad, *Synth. React. Inorg. Metal-Org. Chem.*, 1982, **12,** 55.
1110. S. I. Arshankow and A. L. Poznjak, *Z. Anorg. Allg. Chem.*, 1981, **481,** 201.
1111. T. L. Siddal, N. Miyaura, J. C. Huffman and J. K. Kochi, *J. Chem. Soc., Chem. Commun.*, 1983, 1185.
1112. A. P. Gardner, B. M. Gatehouse and J. C. B. White, *Acta Crystallogr., Sect. B*, 1971, **27,** 1505.
1113. D. R. Prasad, T. Ramasami, D. Ramaswamy and M. Santappa, *Inorg. Chem.*, 1982, **21,** 850.
1114. T. H. Benson, M. S. Bilton and N. S. Gill, *Aust. J. Chem.*, 1977, **30,** 261.
1115. T. H. Benson, M. S. Bilton, N. S. Gill and M. Sterns, *J. Chem. Soc., Chem. Commun.*, 1976, 936.
1116. J. E. Gray and G. W. Everett, Jr., *Inorg. Chem.*, 1971, **10,** 2087.
1117. K. S. Finney and G. W. Everett, Jr., *Inorg. Chim. Acta,* 1974, **11,** 185.
1118. U. A. Bhagwat, V. A. Mukhedkar and A. J. Mukhedkar, *J. Chem. Soc., Dalton Trans.*, 1980, 2319; 1982, 1899.
1119. B. U. Nair, T. Ramasami and D. Ramaswamy, *Polyhedron,* 1983, **2,** 103.
1120. N. W. Alcock, D. F. Cook, E. D. McKenzie and J. M. Worthington, *Inorg. Chim. Acta,* 1980, **38,** 107.
1121. D. F. Cook, D. Cummins and E. D. McKenzie, *J. Chem. Soc., Dalton Trans.*, 1976, 1369.
1122. A. Malek, G. C. Dey, A. Nasreen, T. A. Chowdhury and E. C. Alyea, *Synth. React. Inorg. Metal-Org. Chem.*, 1979, **9,** 145.
1123. G. J. Palenik, D. W. Wester, U. Rychlewska and R. C. Palenik, *Inorg. Chem.*, 1976, **15,** 1814.
1124. Y. Journaux, O. Kahn, J. Zarembowitch, J. Galy and J. Jaud, *J. Am. Chem. Soc.*, 1983, **105,** 7585.
1125. E. Gonzalez-Vergara, B. C. de Gonzalez, J. Hegenauer and P. Saltman, *Isr. J. Chem.*, 1981, **21,** 18.
1126. A. M. Van Den Bergen, K. S. Murray, R. M. Sheahan and B. O. West, *J. Organomet. Chem.*, 1975, **90,** 299.

1127. A. L. Marchese and B. O. West, *J. Organomet. Chem.*, 1984, **266**, 61.
1128. K. Dey and R. L. De, *J. Inorg. Nucl. Chem.*, 1977, **39**, 153.
1129. L. Kolosci and K. Schug, *Inorg. Chem.*, 1983, **22**, 3053.
1130. A. Kuntzel and H. Trabizsch, *Z. Anorg. Allg. Chem.*, 1959, **299**, 188.
1131. A. Kuntzel and H. Trabizsch, *Z. Anorg. Allg. Chem.*, 1959, **302**, 10.
1132. V. V. Udovenko and O. N. Stepanenko, *Russ. J. Inorg. Chem.* (*Engl. Transl.*), 1967, **12**, 1109.
1133. V. V. Udovenko and O. N. Stepanenko, *Russ. J. Inorg. Chem.* (*Engl. Transl.*), 1969, **14**, 91.
1134. E. G. Timofeeva and Ya. V. Tkacheva, *Russ. J. Inorg. Chem.* (*Engl. Transl.*), 1972, **17**, 1003.
1135. J. A. Bertrand, P. G. Eller, E. Fujita, M. D. Lively and D. G. Van Deveer, *Inorg. Chem.*, 1979, **18**, 2419.
1136. M. Bhattacharya and G. S. De, *Indian J. Chem., Sect. A*, 1981, **20**, 780.
1137. E. S. Gould and H. Taube, *J. Am. Chem. Soc.*, 1964, **86**, 1318.
1138. H. J. Schugar, G. R. Rossman and H. B. Gray, *J. Am. Chem. Soc.*, 1969, **91**, 4564.
1139. R. J. Kline and J. G. Wardeska, *Inorg. Chem.*, 1969, **8**, 2153.
1140. J. E. Hix, Jr. and M. M. Jones, *J. Inorg. Nucl. Chem.*, 1964, **26**, 781.
1141. M. M. Jones, K. V. Dandh and G. T. Fisher, *J. Inorg. Nucl. Chem.*, 1964, **26**, 773.
1142. C. Stein, S. Bouma, J. Carlson, C. Cornelius, J. Maeda, C. Weschler, E. Deutsch and K. O. Hodgson, *Inorg. Chem.*, 1976, **15**, 1183.
1143. S. E. Livingstone and J. E. Oluka, *Aust. J. Chem.*, 1976, **29**, 1913.
1144. M. Leitheiser and D. Coucouvanis, *J. Inorg. Nucl. Chem.*, 1977, **39**, 811.
1145. E. Jacobsen and W. Lund, *Acta Chem. Scand.*, 1965, **19**, 2379.
1146. R. H. Lane and L. E. Bennett, *Chem. Commun.*, 1971, 491.
1147. S. S. Parmar and H. Kaur, *Transition Met. Chem.* (*Weinheim, Ger.*), 1982, **7**, 79.
1148. H. Ley and K. Ficken, *Chem. Ber.*, 1912, **45**, 377.
1149. R. D. Gillard, S. H. Laurie, D. C. Price, D. A. Phipps and C. F. Weick, *J. Chem. Soc., Dalton Trans.*, 1974, 1385.
1150. R. F. Bryan, P. T. Green, P. T. Stokely and E. W. Wilson, Jr., *Inorg. Chem.*, 1971, **10**, 1468.
1151. N. Israily, *C. R. Hebd. Seances Acad. Sci., Ser. C*, 1966, **262**, 1426.
1152. R. Trujillo, A. Baca, C. Castillo and W. F. Coleman, *J. Inorg. Nucl. Chem.*, 1981, **43**, 631.
1153. D. C. Price, Ph.D. Thesis, University of Kent at Canterbury, 1970, chap. 4, p. 135.
1154. W. M. Wallace and P. E. Hoggard, *Inorg. Chim. Acta*, 1982, **65**, L3.
1155. H. Oki and K. Otsuka, *Bull. Chem. Soc. Jpn.*, 1976, **49**, 1841.
1156. H. Oki, *Bull. Chem. Soc. Jpn.*, 1977, **50**, 680.
1157. H. Oki and Y. Takahashi, *Bull. Chem. Soc. Jpn.*, 1977, **50**, 2288.
1158. C. A. McAuliffe, J. V. Quagliano and L. M. Vallarino, *Inorg. Chem.*, 1966, **5**, 1996.
1159. V. Spitsyn, S. V. Mozgin, M. G. Felin, N. A. Subbotina and A. I. Zhirov, *Russ. J. Inorg. Chem.* (*Engl. Transl.*), 1982, **27**, 386.
1160. L. M. Volshtein, *Dokl. Akad. Nauk SSSR*, 1945, **48**, 105 (*Chem. Abstr.*, 1946, **40**, 4614).
1161. L. M. Volshtein, *Dokl. Akad. Nauk Arm. SSR*, 1945, **48**, 111.
1162. J. T. Veal, W. E. Hatfield, D. V. Jeter, J. C. Hempel and D. J. Hodgson, *Inorg. Chem.*, 1973, **12**, 342.
1163. A. Earnshaw and J. Lewis, *J. Chem. Soc.*, 1961, 396.
1164. D. J. Hodgson, J. T. Veal, W. E. Hatfield, D. Y. Jeter and J. C. Hempel, *J. Coord. Chem.*, 1972, **2**, 1.
1165. T. Morishita, K. Hori, E. Kyuno and R. Tsuchiya, *Bull. Chem. Soc. Jpn.*, 1965, **38**, 1276.
1166. H. Oki and H. Yoneda, *Inorg. Chem.*, 1981, **20**, 3875.
1167. S. Kellesoe and E. Pederson, *Acta Chem. Scand., Ser. A*, 1982, **36**, 859.
1168. P. E. Hoggard, *Inorg. Chem.*, 1981, **20**, 415.
1169. V. I. Spitsyn, N. A. Subbotina, M. G. Felin, M. I. Aizenberg and A. I. Zhirov, *Russ. J. Inorg. Chem.* (*Engl. Transl.*), 1980, **25**, 713.
1170. (a) L. G. Sillen and A. E. Martell, Chemical Society Special Publication No. 17, 1964; (b) L. G. Sillen and A. E. Martell, Chemical Society Special Publication No. 25, 1971; (c) D. D. Perrin, 'Stability Constants of Metal–Ion Complexes Part B. Organic Ligands', Pergamon, Oxford, 1982.
1171. K. Venkatachalapathi, M. S. Nair, D. Ramaswamy and M. Santappa, *J. Chem. Soc., Dalton Trans.*, 1982, 291.
1172. H. Matsukawa, M. Ohta, S. Takata and R. Tsuchiya, *Bull. Chem. Soc. Jpn.*, 1965, **38**, 1235.
1173. J. Maslowska and L. Chruscinski, *J. Inorg. Nucl. Chem.*, 1981, **43**, 3398.
1174. M. Abdullah, J. Barrett and P. O'Brien, *J. Chem. Soc., Dalton Trans.*, 1984, 1647.
1175. D. Banerjea and S. D. Chauduri, *J. Inorg. Nucl. Chem.*, 1968, **30**, 871.
1176. I. A. Khan and K. U-Din, *J. Inorg. Nucl. Chem.*, 1981, **43**, 1082.
1177. S. G. Shuttleworth and R. L. Sykes, *J. Am. Leather Chem. Assoc.*, 1959, **54**, 259; 1960, **55**, 154.
1176. R. W. Green and K. P. Ang, *J. Am. Chem. Soc.*, 1955, **77**, 5482.
1177. L. A. Tschugaev and E. Serbin, *C. R. Hebd. Seances Acad. Sci.*, 1910, **151**, 1361 (*J. Chem. Soc. Abstr.*, 1911, **i**, 115).
1178. M. Volshtein, G. G. Motyagina and L. S. Anakhova, *Zh. Neorg. Khim.*, 1956, **1**, 2378 (*Chem. Abstr.*, 1957, **52**, 5618e).
1179. G. Grouhi-Witte and E. Weiss, *Z. Naturforsch., Teil B*, 1970, **31**, 1190.
1180. M. Watabe, H. Yano, Y. Odaka and H. Kobayshi, *Inorg. Chem.*, 1981, **20**, 3623.
1181. S. Yamada, J. Hidaka and B. E. Douglas, *Inorg. Chem.*, 1971, **10**, 2187.
1182. W. T. Pennington, A. W. Cordes, D. Kyle and E. W. Wilson, Jr., *Acta Crystallogr., Sect. C*, 1984, **40**, 1322.
1185. P. de Meester, D. J. Hodgson, H. C. Freeman and C. J. Moore, *Inorg. Chem.*, 1977, **16**, 1494.
1186. M. Freni, A. Gervasini, P. Romiti, T. Beringhelli and F. Morazzoni, *Spectrochim. Acta, Part A*, 1983, **39**, 85.
1187. P. de Meester and D. J. Hodgson, *J. Chem. Soc., Dalton Trans.*, 1977, 1604.
1188. Y. Hoho, Y. Sugiura and H. Tanaka, *J. Inorg. Nucl. Chem.*, 1977, **39**, 1859.
1189. R. J. Balahura and N. A. Lewis, *Inorg. Chem.*, 1977, **16**, 2213.
1190. W. Mertz and K. Schwarz, *Arch. Biochem. Biophys.*, 1955, **58**, 504.
1191. W. Mertz and K. Schwarz, *Am. J. Physiol.*, 1959, **196**, 614.

1192. K. N. Jeejeebhoy, R. C. Chu, E. B. Marliss, G. R. Greenberg and A. Bruce-Robinson, *Am. J. Clin. Nutr.*, 1977, **30,** 531.
1193. W. Mertz, *Physiol. Rev.*, 1969, **49,** 163.
1194. R. A. Anderson, *Sci. Total Environ.*, 1981, **17,** 13.
1195. R. A. Anderson and W. Mertz, *TIBS*, 1977, **2,** 277.
1196. J. Barrett, P. O'Brien and J. Pedrosa de Jesus, *Polyhedron Report No. 9*, 1985, **4,** 1.
1197. E. W. Toepfer, W. Mertz, M. M. Polansky, E. E. Roginsky and W. R. Wolf, *J. Agric. Food. Chem.*, 1977, **25,** 162.
1198. R. A. Anderson, J. H. Branter and M. M. Polansky, *J. Agric. Food Chem.*, 1978, **26,** 1219.
1199. M. Abdullah, J. Barrett and P. O'Brien, *J. Chem. Soc., Dalton Trans.*, 1985, 2085.
1200. W. Mertz, E. W. Toepfer, E. E. Roginsky and M. M. Polansky, *Fed. Proc.*, 1974, **33,** 2275.
1201. C. A. Green, R. J. Bianchini and J. I. Legg, *Inorg. Chem.*, 1984, **23,** 2713.
1202. J. C. Chang, L. E. Gerdom, N. C. Baeziger and H. M. Goff, *Inorg. Chem.*, 1983, **22,** 1739.
1203. J. A. Cooper, B. F. Anderson, P. A. Buckley and L. F. Blackwell, *Inorg. Chim. Acta*, 1984, **91,** 1.
1204. J. A. Cooper, L. F. Blackwell and P. A. Buckley, *Inorg. Chim. Acta*, 1984, **92,** 23.
1205. R. M. Lollar, 'Chromium in the Tanning Industry', p. 283 in 'Chromium', ed. M. J. Udy, Reinhold, 1956.
1206. See refs. 1208 and 1209. Technical specifications are available from manufacturers, *e.g.* Yorkshire Chemicals (Selby, UK), 'Chrome Tanning Products and Auxilliaries', Technical Bulletin, 1984.
1207. S. Indubala and D. Ramasvamy, *J. Inorg. Nucl. Chem.*, 1973, **35,** 2055.
1208. M. Santappa, T. Ramasami and K. J. Kedlaya, *J. Sci. Ind. Res.*, 1982, 616.
1209. K. Bienkiewski, 'The Physical Chemistry of Leather Making', 1983, Krieger, Florida, 1983.
1210. K. H. Gustavson, *J. Am. Leather Chem. Assoc.*, 1924, **19,** 446.
1211. H. Hormann, in 'Metal Ions in Biological Systems', ed. H. Sigel, Dekker, New York, 1974, vol. 3, p. 89.
1212. A. F. Woodlock and B. S. Harrap, *Int. J. Pept. Protein Res.*, 1969, **1,** 93.
1213. K. Kuhn and E. Gebhardt, *Z. Naturforsch., Teil B*, 1960, **15,** 23.
1214. H. Brintzinger, H. Thiele and U. Müller, *Z. Anorg. Allg. Chem.*, 1943, **251,** 285.
1215. L. E. Gerdom, N. A. Baenziger and H. M. Goff, *Inorg. Chem.*, 1981, **20,** 1606.
1216. Z. Chen, M. Cimolino and A. W. Adamson, *Inorg. Chem.*, 1983, **22,** 3035.
1217. W. D. Wheeler and J. I. Legg, *Inorg. Chem.*, 1984, **23,** 3798.
1218. R. N. F. Thorneley and A. G. Sykes, *J. Chem. Soc. (A)*, 1969, 742.
1219. L. Pescok, L. D. Shields and W. P. Schaefer, *Inorg. Chem.*, 1964, **3,** 114.
1220. R. E. Hamm, *J. Am. Chem. Soc.*, 1953, **75,** 5670.
1221. M. V. Olson, *Inorg. Chem.*, 1973, **12,** 1416.
1222. R. Tsuchiya, A. Uehara and K. Kobayashi, *Bull. Chem. Soc. Jpn.*, 1980, **53,** 921.
1223. H. Ogino, T. Wanatabe and N. Tanaka, *Inorg. Chem.*, 1975, **14,** 2093.
1224. H. Ogino, M. Shimura and N. Tanaka, *Inorg. Chem.*, 1979, **18,** 2497.
1225. J. A. Weyh and R. L. Price, *Inorg. Chem.*, 1971, **10,** 858.
1226. D. S. Veselinovic, D. J. Radanovic and S. A. Grujic, *Inorg. Nucl. Chem. Lett.*, 1980, **16,** 211.
1227. J. A. Weyh and R. E. Hamm, *Inorg. Chem.*, 1968, **7,** 2431.
1228. S. Kaizaki and H. Mori, *Bull. Chem. Soc. Jpn.*, 1981, **54,** 3562.
1229. R. J. Lancashire and T. D. Smith, *Aust. J. Chem.*, 1975, **28,** 2137.
1230. D. Mootz and H. Wunderlich, *Acta Crystallogr., Sect. B*, 1980, **36,** 721.
1231. R. Wernicke, H. H. Schmidtke and P. E. Hoggard, *Inorg. Chim. Acta*, 1977, **24,** 145.
1232. S. Dutta-Chauduri, C. J. O'Connor and A. L. Odell, *J. Chem. Soc., Dalton Trans.*, 1975, 1921.
1233. S. N. Dubey, A. Singh and D. M. Puri, *J. Inorg. Nucl. Chem.*, 1981, **43,** 407.
1234. R. Krause, *Inorg. Chem.*, 1963, **2,** 297.
1235. C. L. Sharma, P. K. Jain and T. K. De, *J. Inorg. Nucl. Chem.*, 1980, **42,** 1681.
1236. Y. Fujii, E. Kyuno and R. Tsuchiya, *Bull. Chem. Soc. Jpn.*, 1969, **42,** 1301.
1237. A. Uehara, E. Kyuno and R. Tsuchiya, *Bull. Chem. Soc. Jpn.*, 1967, **40,** 2317; 1970, **43,** 1397.
1238. A. Uehara, E. Kyuno and R. Tsuchiya, *Bull. Chem. Soc. Jpn.*, 1967, **40,** 2322.
1239. P. J. Peerce, H. B. Gray and F. C. Anson, *Inorg. Chem.*, 1979, **18,** 2593.
1240. P. J. Peerce and F. C. Anson, *J. Electroanal. Chem. Interfacial Electrochem.*, 1979, **105,** 317.
1241. D. Dolphin (ed.), 'The Porphyrins', Academic, New York, 1978, vols. 1–7.
1242. M. Tsutsui, R. A. Velapoldi, K. Suzuki, F. Vohwinkel, M. Ichikawa and T. Koyano, *J. Am. Chem. Soc.*, 1969, **91,** 6262.
1243. N. J. Gogan and Z. U. Siddiqui, *Can. J. Chem.*, 1972, **50,** 720.
1244. W. R. Scheidt and C. A. Reed, *Inorg. Chem.*, 1978, **17,** 710.
1245. W. R. Scheidt, A. C. Brinegar, J. F. Kirner and C. A. Reed, *Inorg. Chem.*, 1979, **18,** 3610.
1246. S. K. Cheung, C. J. Grimes, J. Wong and C. A. Reed, *J. Am. Chem. Soc.*, 1976, **98,** 5028.
1247. S. L. Kelly, D. Lancon and K. M. Kadish, *Inorg. Chem.*, 1984, **23,** 1451.
1248. D. P. Riley, *Synth. React. Inorg. Metal-Org. Chem.*, 1980, **10,** 147.
1249. A. D. Adler, F. R. Longo, F. Kampas and J. Kim, *J. Inorg. Nucl. Chem.*, 1970, **32,** 2443.
1250. J. T. Groves, W. J. Kruper, Jr., R. C. Haushalter and W. M. Butler, *Inorg. Chem.*, 1982, **21,** 1363.
1251. J. W. Buchler, L. Puppe, K. Rohbock and H. H. Schneehage, *Chem. Ber.*, 1973, **106,** 2710.
1252. G. McLendon and M. Baily, *Inorg. Chem.*, 1979, **18,** 2120.
1253. J. G. Leipoldt, R. van Eldik and H. Kelm, *Inorg. Chem.*, 1983, **22,** 4146.
1254. P. O'Brien and D. A. Sweigart, *Inorg. Chem.*, 1982, **21,** 2094.
1255. D. A. Summerville, R. D. Jones, B. M. Hoffman and F. Basolo, *J. Am. Chem. Soc.*, 1977, **99,** 8195.
1256. F. Basolo, R. D. Jones and D. A. Summerville, *Acta Chem. Scand., Ser A*, 1978, **32,** 771.
1257. L. A. Bottomley and K. M. Kadish, *Inorg. Chem.*, 1983, **22,** 342.
1258. J. T. Groves and W. J. Kruper, Jr., *J. Am. Chem. Soc.*, 1979, **101,** 7613.
1259. J. T. Groves, W. J. Kruper, Jr., T. E. Nemo and R. S. Myers, *J. Mol. Catal.*, 1980, **7,** 169.
1260. J. T. Groves and R. C. Haushalter, *J. Chem. Soc., Chem. Commun.*, 1981, 1165.

1261. J. W. Buchler, K. L. Lay, L. Castle and V. Ullrich, *Inorg. Chem.*, 1982, **21,** 842.
1262. J. R. Budge, B. M. K. Gatehouse, M. C. Nesbit and B. O. West, *J. Chem. Soc., Chem. Commun.*, 1981, 370.
1263. D. J. Liston, K. S. Murray and B. O. West, *J. Chem. Soc., Chem. Commun.*, 1982, 1109.
1264. N. Carnieri and A. Harriman, *Inorg. Chim. Acta,* 1982, **62,** 103.
1265. E. T. Shimomura, M. A. Phillippi, H. M. Goff, W. F. Scholz and C. A. Reed, *J. Am. Chem. Soc.*, 1981, **103,** 6778.
1266. Y. Matsuda, S. Yamada and Y. Murakami, *Inorg. Chim. Acta,* 1980, **44,** L309.
1267. Y. Murakami, Y. Matsuda and S. Yamada, *J. Chem. Soc., Dalton Trans.*, 1981, 855.
1268. J. W. Buchler, C. Dreher, K.-L. Lay, A. Raap and K. Gersonde, *Inorg. Chem.*, 1983, **22,** 879.
1269. J. T. Groves, T. Takahashi and W. M. Butler, *Inorg. Chem.*, 1983, **22,** 884.
1270. T. L. Siddall, N. Miyaura, J. C. Huffman and J. K. Kochi, *J. Chem. Soc., Chem. Commun.*, 1983, 1185.
1271. J. Ferguson and M. L. Tobe, *Inorg. Chim. Acta,* 1970, **4,** 109.
1272. C. K. Poon and K. C. Pun, *Inorg. Chem.*, 1980, **19,** 568.
1273. N. A. P. Kane-Maguire, J. A. Bennett and P. K. Miller, *Inorg. Chim. Acta,* 1983, **76,** L123.
1274. E. Bang and O. Mønsted, *Acta Chem. Scand., Ser. A,* 1982, **36,** 353.
1275. L. Y. Martin, L. J. DeHayes, L. J. Zompa and D. H. Busch, *J. Am. Chem. Soc.*, 1974, **96,** 4046.
1276. E. Campi, J. Ferguson and M. L. Tobe, *Inorg. Chem.*, 1970, **9,** 1781.
1277. C. Kutal and A. W. Adamson, *Inorg. Chem.*, 1973, **12,** 1990.
1278. N. A. P. Kane-Maguire, W. S. Crippen and P. K. Miller, *Inorg. Chem.*, 1983, **22,** 696.
1279. D. A. House, R. W. Hay and M. Akbar Ali, *Inorg. Chim. Acta,* 1983, **72,** 239.
1280. J. Eriksen and O. Mønsted, *Acta Chem. Scand., Ser. A,* 1983, **37,** 579.
1281. D. A. House and R. W. Hay, *Inorg. Chim. Acta,* 1981, **54,** L145.
1282. R. Temple, D. A. House and W. T. Robinson, *Acta Crystallogr., Sect. C,* 1984, **40,** 1789.
1283. R. G. Swisher, G. A. Brown, R. C. Smierciak and E. L. Blinn, *Inorg. Chem.*, 1981, **20,** 3947.
1284. G. J. Samuels and J. H. Espenson, *Inorg. Chem.*, 1979, **18,** 2587.
1285. J. P. Leslie, II and J. H. Espenson, *J. Am. Chem. Soc.*, 1976, **98,** 4839.
1286. K. Wieghardt, P. Chandhuri, B. Nuber and J. Weiss, *Inorg. Chem.*, 1982, **21,** 3086.
1287. J. A. Elvidge and A. B. P. Lever, *J. Chem. Soc.*, 1961, 1257.
1288. K. H. Nill, F. Wasgestian and A. Pfeil, *Inorg. Chem.*, 1979, **18,** 564.
1289. F. H. Moser and A. L. Thomas, 'The Phthalocyanins', CRC Press, Boca Raton, FL, 1983, vol. 2, p. 6.
1290. A. B. P. Lever, *Adv. Inorg. Chem. Radiochem.*, 1965, **7,** 52.
1291. E. G. Melini, L. R. Ocone and B. P. Block, *Inorg. Chem.*, 1967, **6,** 424.
1292. C. M. Guzy, J. B. Raynor, L. P. Stodulski and M. C. R. Symons, *J. Chem. Soc. (A)*, 1969, 997.
1293. L. J. Boucher, in 'Metal Complexes of Phthalocyanines', ed. G. A. Melson, Plenum, New York, 1979, p. 477.
1294. R. Taube and K. L. Lunkenheimer, *Z. Naturforsch., Teil B,* 1964, **19,** 653.
1295. H. Przywarska-Boniecka and W. Wojciechowski, *Mater. Sci.*, 1975, **1,** 35.
1296. A. B. P. Lever, S. R. Pickens, P. C. Minor, S. Licoccia, B. S. Ramaswamy and K. Magnell, *J. Am. Chem. Soc.*, 1981, **103,** 6800.
1297. A. B. P. Lever, S. Licoccia, B. S. Ramaswamy, S. A. Kandil and D. V. Stynes, *Inorg. Chim. Acta,* 1981, **51,** 169.
1298. B. Cuoni, F. P. Emmenegger, C. Rohrbasser, C. W. Schläpfer and P. Studer, *Spectrochim. Acta, Part A,* 1978, **34,** 247.
1299. G. Siebert and R. Hoppe, *Z. Anorg. Allg. Chem.*, 1972, **391,** 113, 126.
1300. W. A. Sunder, A. L. Wayda, D. Distefano, W. E. Falconer and J. E. Griffiths, *J. Fluorine Chem.*, 1979, **14,** 299.
1301. G. C. Allen and G. A. M. El-Sharkawy, *Inorg. Nucl. Chem. Lett.*, 1970, **6,** 493.
1302. B. Hofmann and R. Hoppe, *Z. Anorg. Allg. Chem.*, 1979, **458,** 151.
1303. S. Sarkar, R. C. Maurya, S. C. Chaurasia and S. K. Srivastava, *Indian J. Chem., Sect. A,* 1977, **15,** 744.
1304. S. Sarkar, B. Sharma, R. C. Maurya, S. C. Chaurasia and S. K. Srivastava, *Indian J. Chem., Sect. A,* 1977, **15,** 747.
1305. M. G. Barker and A. J. Hooper, *J. Chem. Soc., Dalton Trans.*, 1975, 2487.
1306. C. Rosenblum and S. L. Holt, *Transition Met. Chem.*, 1972, **7,** 87.
1307. M. G. Barker and A. J. Hooper, *J. Chem. Soc., Dalton Trans.*, 1976, 1093.
1308. B. L. Chamberland, *J. Solid State Chem.*, 1982, **43,** 309.
1309. B. L. Chamberland, *J. Solid State Chem.*, 1983, **48,** 318.
1310. B. L. Chamberland and L. Katz, *Acta Crystallogr., Sect. B,* 1982, **38,** 54.
1311. W. D. Hill, Jr., *Inorg. Chim. Acta,* 1982, **65,** L100.
1312. G. Demazeau, P. Maestro, T. Plante, M. Pouchard and P. Hagenmuller, *Mater. Res. Bull.*, 1979, **14,** 121.
1313. A. Lerm, G. Elbinger and P. Rosemann, *Z. Anorg. Allg. Chem.*, 1978, **442,** 5.
1314. M. Bochmann, G. Wilkinson and G. B. Young, *J. Chem. Soc., Dalton Trans.*, 1980, 1879.
1315. C. J. Cardin, D. J. Cardin, J. M. Kelly, R. J. Norton, A. Roy, B. J. Hathaway and T. J. King, *J. Chem. Soc., Dalton Trans.*, 1983, 671.
1316. E. C. Alyea, J. S. Basi, D. C. Bradley and M. H. Chisholm, *J. Chem. Soc. (A)*, 1971, 772.
1317. H. Nöth and M. Seitz, *J. Chem. Soc., Chem. Commun.*, 1976, 1004.
1318. G. Dyrkacz and J. Rocek, *J. Am. Chem. Soc.*, 1973, **95,** 4756.
1319. M. Bochmann, G. Wilkinson, G. B. Young, M. B. Hursthouse and K. M. A. Malik, *J. Chem. Soc., Dalton Trans.*, 1980, 901.
1320. M. Bochmann, G. Wilkinson, G. B. Young, M. B. Hursthouse and K. M. A. Malik, *J. Chem. Soc., Dalton Trans.*, 1980, 1863.
1321. M. H. Chisholm, A. H. Cowley and M. Lattman, *J. Am. Chem. Soc.*, 1980, **102,** 46.
1322. J. C. W. Chien, W. Kruse, D. C. Bradley and C. W. Newing, *Chem. Commun.*, 1970, 1177.
1323. M. Mitewa and P. R. Bontchev, *Coord. Chem. Rev.*, 1985, **61,** 241.
1324. T. A. O'Donnell and D. F. Stewart, *Inorg. Chem.*, 1966, **5,** 1434.

1325. W. E. Falconer, G. R. Jones, W. A. Sunder, M. J. Vasile, A. A. Muenter, T. R. Dyke and W. Klemperer, *J. Fluorine Chem.*, 1974, **4,** 213.
1326. J. Slivnik and B. Zemva, *Z. Anorg. Allg. Chem.*, 1971, **385,** 137.
1327. S. D. Brown and G. L. Gard, *Inorg. Nucl. Chem. Lett.*, 1975, **11,** 19.
1328. S. D. Brown, T. M. Loehr and G. L. Gard, *J. Fluorine Chem.*, 1976, **7,** 19.
1329. P. J. Green, B. M. Johnson, T. M. Loehr and G. L. Gard, *Inorg. Chem.*, 1982, **21,** 3562.
1330. E. G. Hope, P. J. Jones, W. Levason, J. S. Ogden, M. Tajik and J. W. Turff, *J. Chem. Soc., Dalton Trans.*, 1984, 2445.
1331. E. A. Seddon, K. R. Seddon and V. H. Thomas, *Transition Met. Chem.*, 1978, **3,** 318.
1332. W. Levason, J. S. Ogden and A. J. Rest, *J. Chem. Soc., Dalton Trans.*, 1980, 419.
1333. M. Doran, R. W. Hawksworth and I. H. Hillier, *J. Chem. Soc., Faraday Trans. 2,* 1980, **76,** 164.
1334. K. R. Seddon and V. H. Thomas, *J. Chem. Soc., Dalton Trans.*, 1977, 2195.
1335. O. V. Ziebarth and J. Selbin, *J. Inorg. Nucl. Chem.*, 1970, **32,** 849.
1336. J. E. Fergusson, A. M. Greenaway and B. R. Penfold, *Inorg. Chim. Acta,* 1983, **71,** 29.
1337. A. K. Banarjee and N. Banerjee, *Inorg. Chem.*, 1976, **15,** 488.
1338. B. Gahan, C. D. Garner, L. H. Hill, F. E. Mabbs, K. D. Hargrave and A. T. McPhail, *J. Chem. Soc., Dalton Trans.*, 1977, 1726.
1339. C. D. Garner, J. Kendrick, P. Lambert, F. E. Mabbs and I. H. Hillier, *Inorg. Chem.*, 1976, **15,** 1287.
1340. J. Weber and C. D. Garner, *Inorg. Chem.*, 1980, **19,** 2206.
1341. C. D. Garner, I. H. Hillier, F. E. Mabbs, C. Taylor and M. F. Guest, *J. Chem. Soc., Dalton Trans.*, 1976, 2258.
1342. K. R. Seddon and V. H. Thomas, *Inorg. Chem.*, 1978, **17,** 749.
1343. E. K. Mooney and K. R. Seddon, *Transition Met. Chem.*, 1977, **2,** 215.
1344. D. A. Edwards and S. C. Jennison, *Transition Met. Chem.*, 1981, **6,** 235.
1345. R. Scholder, F. Schwochow and H. Schwarz, *Z. Anorg. Allg. Chem.*, 1968, **363,** 10.
1346. J. A. Connor and E. A. V. Ebsworth, *Adv. Inorg. Chem. Radiochem.*, 1964, **6,** 279.
1347. M. Roch, J. Weber and A. F. Williams, *Inorg. Chem.*, 1984, **23,** 4571.
1348. K. Forster, M. Greenblatt and J. H. Pifer, *J. Solid State Chem.*, 1979, **30,** 121.
1349. R. Borromei, P. Day and L. Oleari, *J. Chem. Soc., Faraday Trans. 2,* 1979, **75,** 401.
1350. H. Müller-Buschbaum and K. Sander, *Z. Naturforsch., Teil B,* 1978, **33,** 708.
1351. E. A. El-Rafei, *Inorg. Chem.*, 1981, **20,** 222.
1352. K. Niki and H. A. Laitinen, *J. Inorg. Nucl. Chem.*, 1975, **37,** 91.
1353. A. Roy, M. Chaudhury and K. Nag, *Bull. Chem. Soc. Jpn.*, 1978, **51,** 1243.
1354. R. Olazcuaga, J.-M. Reau and G. Le Flem, *C. R. Hebd. Seances Acad. Sci., Ser. C,* 1972, **275,** 135.
1355. M. Krumpolc, B. G. DeBoer and J. Rocek, *J. Am. Chem. Soc.*, 1978, **100,** 145.
1356. M. Krumpolc and J. Rocek, *J. Am. Chem. Soc.*, 1979, **101,** 3206.
1357. N. Rajasekar, R. Subramaniam and E. S. Gould, *Inorg. Chem.*, 1983, **22,** 971.
1358. C. J. Willis, *J. Chem. Soc., Chem. Commun.*, 1972, 944.
1359. P. F. Bramman, T. Lund, J. B. Raynor and C. J. Willis, *J. Chem. Soc., Dalton Trans.*, 1975, 45.
1360. H. L. Krauss, M. Leder and G. Münster, *Chem. Ber.*, 1963, **96,** 3008.
1361. T. Takeya, E. Kotani and S. Tobinaga, *J. Chem. Soc., Chem. Commun.*, 1983, 98.
1362. W. M. B. Koenst and J. T. M. F. Maessen, *Synth. Commun.*, 1980, **10,** 905.
1363. 'Kirk-Othmer Encyclopedia of Chemical Technology', 3rd edn., Wiley-Interscience, New York, 1979, vol. 6.
1364. E. G. Hope, P. J. Jones, W. Levason, J. S. Ogden, M. Tajik and J. W. Turff, *J. Chem. Soc., Dalton Trans.*, 1985, 1443.
1365. A. J. Edwards, W. E. Falconer and W. A. Sunder, *J. Chem. Soc., Dalton Trans.*, 1974, 541.
1366. E. G. Hope, P. J. Jones, W. Levason, J. S. Ogden, M. Tajik and J. W. Turff, *J. Chem. Soc., Dalton Trans.*, 1985, 529.
1367. P. J. Green and G. L. Gard, *Inorg. Chem.*, 1977, **16,** 1243.
1368. G. D. Flesch and H. J. Svec, *J. Am. Chem. Soc.*, 1959, **81,** 1787.
1369. W. V. Rochat, J. N. Gerlach and G. L. Gard, *Inorg. Chem.*, 1970, **9,** 998.
1370. S. D. Brown, P. J. Green and G. L. Gard, *J. Fluorine Chem.*, 1975, **5,** 203.
1371. R. S. Drago and K. W. Whitten, *Inorg. Chem.*, 1966, **5,** 677.
1372. W. H. Hartford and M. Darrin, *Chem. Rev.*, 1958, **58,** 1.
1373. I. R. Beattie, C. J. Marsden and J. S. Ogden, *J. Chem. Soc., Dalton Trans.*, 1980, 535.
1374. C. D. Garner, R. Mather and M. F. A. Dove, *J. Chem. Soc., Chem. Commun.*, 1973, 633.
1375. R. J. French, L. Hedberg, K. Hedberg, G. L. Gard and B. M. Johnson, *Inorg. Chem.*, 1983, **22,** 892.
1376. C. J. Marsden, L. Hedberg and K. Hedberg, *Inorg. Chem.*, 1982, **21,** 1115.
1377. E. L. Varetti and A. Mueller, *Spectrochim. Acta, Part A,* 1978, **34,** 895.
1378. S. D. Brown, G. L. Gard and T. M. Loehr, *J. Chem. Phys.*, 1976, **64,** 1219.
1379. A. J. Edwards and P. Taylor, *Chem. Commun.*, 1970, 1474.
1380. S. D. Brown and G. L. Gard, *Inorg. Chem.*, 1973, **12,** 483.
1381. S. D. Brown, P. J. Green and G. L. Gard, *J. Fluorine Chem.*, 1975, **5,** 203.
1382. W. V. Rochat and G. L. Gard, *Inorg. Chem.*, 1969, **8,** 158.
1383. Z. A. Siddiqi, Lutfullah, N. A. Ansari and S. A. A. Zaidi, *J. Inorg. Nucl. Chem.*, 1981, **43,** 397.
1384. J. K. Puri and J. M. Miller, *Inorg. Chim. Acta,* 1983, **75,** 215.
1385. R. C. Paul, J. K. Puri and K. C. Malhotra, *J. Inorg. Nucl. Chem.*, 1971, **33,** 4191.
1386. J. N. Gerlach and G. L. Gard, *Inorg. Chem.*, 1970, **9,** 1565.
1387. S. D. Brown and G. L. Gard, *Inorg. Chem.*, 1975, **14,** 2273.
1388. S. D. Brown, L. M. Emme and G. L. Gard, *J. Inorg. Nucl. Chem.*, 1975, **37,** 2557.
1389. P. Huppmann, H. Labischinski, D. Lentz, H. Pritzkow and K. Seppelt, *Z. Anorg. Allg. Chem.*, 1982, **487,** 7.
1390. K. Seppelt, *Chem. Ber.*, 1975, **108,** 1823.
1391. K. Dehnicke and J. Strähle, *Z. Anorg. Allg. Chem.*, 1965, **339,** 171.
1392. V. Fernandez and J. Rujas, *Inorg. Nucl. Chem. Lett.*, 1979, **15,** 285.

1393. R. C. Makhija and R. A. Stairs, *Can. J. Chem.*, 1969, **47**, 2293.
1394. R. C. Paul, O. B. Baidya, S. K. Gupta and R. Kapoor, *Z. Naturforsch., Teil B*, 1977, **32**, 148.
1395. A. A. Ivanov, A. V. Demidov, N. I. Popenko, E. Z. Zasorin, V. P. Spiridonov and I. Hargittai, *J. Mol. Struct.*, 1980, **63**, 121.
1396. E. Mappus and C.-Y. Cuilleron, *J. Chem. Res. (S)*, 1979, 42.
1397. W. J. Behr and J. Fuchs, *Z. Naturforsch., Teil B*, 1975, **30**, 299.
1398. M. Schmidt and H. Schmidbaur, *Angew. Chem.*, 1958, **70**, 704.
1399. B. Stensland and P. Kierkegaard, *Acta Chem. Scand.*, 1970, **24**, 211.
1400. K. Handlir, J. Holecek, M. Nadvornik and J. Klikorka, *Z. Chem.*, 1980, **20**, 31.
1401. M. Otto, R. Wagener and H. Hennig, *Inorg. Chim. Acta*, 1981, **64**, L11.
1402. H. C. Clark and R. G. Goel, *Inorg. Chem.*, 1966, **5**, 998.
1403. T. I. Kuzina, G. A. Sidorenko, I. V. Shakhno and T. I. Bel'skaya, *Russ. J. Inorg. Chem. (Engl. Transl.)*, 1980, **25**, 200.
1404. T. Fukasawa and M. Iwatsuki, *Bull. Chem. Soc. Jpn.*, 1979, **52**, 3697.
1405. D. A. Habboush, D. H. Kerridge and S. A. Tariq, *Thermochim. Acta*, 1979, **28**, 143.
1406. B. J. Meehan and S. A. Tariq, *Aust. J. Chem.*, 1980, **33**, 647.
1407. R. N. Kust and R. W. Fletcher, *Inorg. Chem.*, 1969, **8**, 687.
1408. B. Holmberg and G. Thomé, *Inorg. Chem.*, 1980, **19**, 2247.
1409. S. Felps, S. I. Foster and S. P. McGlynn, *Inorg. Chem.*, 1973, **12**,. 1389.
1410. J. Casabo, J. Ribas and J. M. Coronas, *J. Inorg. Nucl. Chem.*, 1976, **38**, 886.
1411. D. M. Adams and M. M. Hargreave, *J. Chem. Soc., Dalton Trans.*, 1973, 1426.
1412. I. R. Beattie, J. S. Ogden and D. D. Price, *J. Chem. Soc., Dalton Trans.*, 1982, 505.
1413. R. J. H. Clark and T. J. Dines, *Inorg. Chem.*, 1982, **21**, 3585.
1414. Y. Cudennec, A. Riou, A. Bonnin and P. Caillet, *Rev. Chim. Miner.*, 1980, **17**, 158.
1415. R. G. Kidd, *Can. J. Chem.*, 1967, **45**, 605.
1416. J. J. Foster and A. N. Hambly, *Aust. J. Chem.*, 1977, **30**, 251.
1417. J. J. Foster and M. Sterns, *J. Cryst. Mol. Struct.*, 1974, **4**, 149.
1418. W. Granier, S. Vilminot, J. D. Vidal and L. Cot, *J. Fluorine Chem.*, 1981, **19**, 123.
1419. R. Mitzner and R. Sommer, *Z. Chem.*, 1979, **19**, 76.
1420. J. N. Gerlach and G. L. Gard, *Inorg. Chem.*, 1971, **10**, 1541.
1421. V. Gutmann, U. Mayer and R. Krist, *Synth. React. Inorg. Metal-Org. Chem.*, 1974, **4**, 523.
1422. P. J. Green and G. L. Gard, *Inorg. Nucl. Chem. Lett.*, 1978, **14**, 179.
1423. W. A. Nugent, *Inorg. Chem.*, 1983, **22**, 965.
1424. W. A. Nugent and R. L. Harlow, *Inorg. Chem.*, 1980, **19**, 777.
1425. N. Wiberg, H. W. Häring and U. Schubert, *Z. Naturforsch., Teil B*, 1978, **33**, 1365.
1426. J. J. Foster and A. N. Hambly, *Aust. J. Chem.*, 1976, **29**, 2137.
1427. T. Tsutsumi, I. Ikemoto, T. Namikawa and H. Kuroda, *Bull. Chem. Soc. Jpn.*, 1981, **54**, 913.
1428. J. A. Connor and E. A. V. Ebsworth, *Adv. Inorg. Chem. Radiochem.*, 1964, **6**, 279.
1429. R. Stomberg, *Ark. Kemi*, 1963, **22**, 29.
1430. B. F. Pedersen and B. Pedersen, *Acta Chem. Scand.*, 1963, **17**, 557.
1431. R. Stomberg and I.-B. Ainalem, *Acta Chem. Scand.*, 1968, **22**, 1439.
1432. R. Armstrong and N. A. Gibson, *Aust. J. Chem.*, 1968, **21**, 897.
1433. W. P. Griffith, *J. Chem. Soc.*, 1962, 3948.
1434. S. N. Witt and D. M. Hayes, *Inorg. Chem.*, 1982, **21**, 4014.
1435. F. Hein and S. Herzog, in 'Handbook of Preparative Inorganic Chemistry', 2nd edn., ed. G. Brauer, Academic, New York, 1965, vol. 2, p. 1392.
1436. R. E. Hester and E. M. Nour, *J. Raman Spectrosc.*, 1981, **11**, 39.
1437. G. B. Kauffman and G. Acero, *Inorg. Synth.*, 1966, **8**, 132.
1438. R. Stomberg, *Ark. Kemi*, 1963, **22**, 49.
1439. R. Stomberg, *Ark. Kemi*, 1965, **23**, 401.
1440. R. Stomberg, *Ark. Kemi*, 1965, **24**, 47.
1441. C. J. Sanderson, in 'Biological and Environmental Aspects of Chromium', ed. S. Langard, Elsevier, Amsterdam, p. 102.
1442. D. Burrows, 'Chromium Metabolism and Toxicity', CRC Press, Boca Raton, FL, 1984.
1443. P. H. Connett and K. E. Wetterhahn, *Struct. Bonding (Berlin)*, 1983, **54**, 93.
1444. K. W. Jennette, *J. Am. Chem. Soc.*, 1982, **104**, 874.
1445. D. M. L. Goodgame, P. B. Hayman and D. E. Hathway, *Polyhedron*, 1982, **1**, 497.
1446. D. M. L. Goodgame, P. B. Hayman and D. E. Hathway, *Inorg. Chim. Acta*, 1984, **91**, 113.
1447. D. L. Rabenstein, G. Guevremont and G. Evans, in 'Metal Ions in Biological Systems', ed. H. Sigel, Dekker, New York, 1979, vol. 9.
1448. A. Meister, *Science*, 1983, **22**, 472.
1449. K. E. Wetterhahn and D. Y. Cupo, Abstracts of the 23rd International Conference on Coordination Chemistry, Boulder, USA, 1984, p. 474.
1450. K. E. Wetterhahn and D. Y. Cupo, *Carcinogenesis*, 1984, **5**, 1705.
1451. A. McAuley and M. A. Olatunji, *Can. J. Chem.*, 1977, **55**, 3335.
1452. A. McAuley and M. A. Olatunji, *Can. J. Chem.*, 1977, **55**, 3328.
1453. P. H. Connett and K. E. Wetterhahn, *J. Am. Chem. Soc.*, 1985, **107**, 4282; 1986, **108**, 1842.
1454. P. O'Brien, J. Barrett and F. Swanson, *Inorg. Chim. Acta*, 1985, **108**, L19.
1455. K. E. Wetterhahn, D. Y. Cupo and P. H. Connett, 'Trace Substances in Environmental Health – XVIII', Proceedings of a Conference at the University of Missouri, ed. D. D. Hemphill, June, 1984, p. 154.
1456. D. W. J. Kwong and D. E. Pennington, *Inorg. Chem.*, 1984, **23**, 2528.
1457. J. N. Cooper, G. E. Staudt, M. L. Smalser and G. P. Haight, *Inorg. Chem.*, 1973, **12**, 2075.

36

Molybdenum

A. GEOFFREY SYKES
University of Newcastle upon Tyne, UK

G. JEFFERY LEIGH and RAYMOND L. RICHARDS
AFRC Unit of Nitrogen Fixation, University of Sussex, Brighton, UK

C. DAVID GARNER and JOHN M. CHARNOCK
University of Manchester, UK

and

EDWARD I. STIEFEL
Exxon Research and Engineering Company, Annandale, NJ, USA

Because of factors beyond the editors' control, the submission of manuscripts for this chapter was delayed. In order to minimize any delay in publishing 'Comprehensive Coordination Chemistry' as a whole, the coverage of molybdenum appears at the end of this volume, commencing on page 1229.

37

Tungsten

ZVI DORI
Technion—Israel Institute of Technology, Haifa, Israel

37.1 INTRODUCTION

The chemistry of tungsten is varied and complex not only because it covers nine oxidation states (−2 to +6), but also because of its ability to form complexes with different coordination numbers and geometries, and because of its tendency to form clusters and polynuclear complexes with a variety of metal atoms.

The chemistry of this element has been studied since the characterization of tungstic acid in

1781,[1] and the oxychlorides, the hexachloride and the hexabromide in 1857. Until the 1930s most of the development centered around the higher oxidation states of the element. The development of the lower oxidation states had to await the discovery of $W(CO)_6$. The successful synthesis of cyclopentadienyl compounds has led to the development of its organometallic chemistry. This was followed in the late 1960s and the 1970s by the synthesis and development of the chemistry of complexes containing metal–metal bonds.[2]

This chapter, which is based on the literature that had appeared through the end of 1983, surveys the coordination chemistry of the element with special emphasis on structure, bonding, synthesis and chemical behavior. The discussion is limited to well-characterized complexes and well-understood systems.

37.2 THE ELEMENT

Tungsten was discovered in 1781 in a mineral now known as scheelite,[1] whose principal constituent is $CaWO_4$. Tungsten steel was invented in 1855[3] and the first alloys containing tungsten were manufactured in 1868. These 'high speed steels' have been used ever since as cutting tools because of their hardness even at high temperatures. Because of its high melting point, it has been used as a filament for incandescent lamps[4] since 1904, and for a long time this was its principal application. More recently, there has been increased interest in alloys containing tungsten, as materials for high temperature applications. Tungsten carbide, when cemented, is almost as hard as diamond and is therefore very useful for dies, tools and other applications requiring extreme hardness and wear resistance.

The element is obtained primarily from the two commercially important ores, scheelite ($CaWO_4$) and wolframite [(Fe,Mn)WO_4]. WO_3 is produced from these ores and then reduced to the metal with hydrogen.[5]

37.3 THE CHEMISTRY OF W^{VI}

In addition to complexes containing the WO, WO_2 and WO_3 structural units, the element shows quite a varied chemistry in this oxidation state, undoubtedly because the hexahalides are reasonably stable and give rise to a variety of substitution products of the type $[WX_{6-n}L_n]$ ($n = 1$–6). In addition, coordination numbers higher than six can also be obtained. W^{VI} also shows considerable tendency to form bonds of order higher than one with good π donor atoms such as S, Se, N, NR, CHR and CR.

37.3.1 W^{VI} Halides and Derivatives

The three hexahalides WF_6, WCl_6 and WBr_6 are known and synthetic routes to their preparation are summarized in Table 1. WF_6 is a colorless gas while WCl_6 and WBr_6 are blue-black moisture sensitive solids. Under well-defined conditions, the octahedral hexahalides undergo substitution to give products ot the type $[WX_{6-n}L_n]$ ($n = 1$–6). Mixed fluoro–chloro compounds have been reported to result from reaction of WCl_6 with F_2 or from WF_6 with Me_3SiCl.

Table 1 Synthetic Routes to W^{VI} Halides

Compound	*Preparation*	*Ref.*
WF_6	W with F_2	124
	W with ClF, ClF_3 or BrF_3	125
	WO_3 with HF, BrF_3 or SF_4	75
WCl_6	W with Cl_2 or S_2Cl_2	126
	WO_3 with Cl_2, S_2Cl_2, HCl, CCl_4, PCl_5	75
	WO_3 with hexachloropropene	127
WBr_6	W or $W(CO)_6$ with Br_2	75

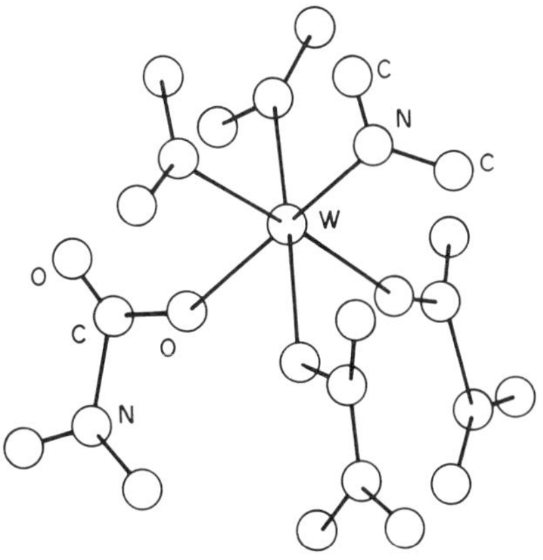

Figure 1 The structure of $[W(NMe_2)_3(O_2CNMe_2)_3]$[14]

^{1}H and ^{19}F NMR data have established the existence of *cis*- and *trans*-WCl_2F_4, *cis*- and *trans*-WF_4Cl_2, *mer*-WCl_3F_3 and *fac*-WCl_3F_3. WF_6 also reacts with $B(OTeF_5)_3$ at 120 °C to give $[WF_n(OTeF_5)_{6-n}]$. $[WF_5(OTeF_5)]$ and *cis*-$[WF_4(OTeF_5)_2]$ have been isolated and characterized.[7]

Halide substitution by ligands containing nitrogen, oxygen and sulfur donor atoms is also known. The reaction of WCl_6 with NCS^- leads to $[W(NCS)_6]$.[8] WF_6 reacts with Me_3SiNEt_2 to give $[WF_5NEt_3]$ and *cis*-$[WF_4(NEt_2)_2]$.[9] With Me_3SiN_3 one obtains $[WF_5N_3]$.[10]

An interesting six-coordinate complex results from the reaction of WCl_6 with $LiNMe_2$.[11] Crystal structure analysis[12] of $[W(NMe_2)_6]$ shows that the complex is octahedral with an average W—N distance of 2.025 Å. It reacts with ROH (R = Me, Et, Pr^n, Pr^i)[13] to give the known $W(OR)_6$ compounds. With MeOH, under certain conditions, *fac*-$[W(NMe_2)_3(OMe)_3]$ is obtained.[13] $[W(NMe_2)_6]$ reacts with CO_2 to give *fac*-$[W(NMe_2)_3(O_2CNMe_2)_3]$ whose structure is shown in Figure 1.[14] The W—N distance of 1.922 Å is significantly shorter than the one observed in $[W(NMe_2)_6]$ indicating significant $N \rightarrow W\ \pi$ donation. This is believed to be an important factor in limiting the insertion to only three CO_2 molecules. It is not surprising that with CS_2 insertion is coupled with reduction to the known W^{IV} compound $[W(S_2CNMe_2)_4]$.[13]

Both WCl_6 and WF_6 react with Me_3SiOR (R = Me, Ph, C_6F_5) to give $[WX_{6-n}(OR)_n]$ ($n = 1$–4).[9] For $n = 2$ and 4 both *cis* and *trans* isomers are known[15,16] and for $n = 3$ both *mer* and *fac* isomers have been reported. Further substitution to give $[WX(OMe)_5]$, $[W(OMe)_6]$ and $[W(OPh)_6]$ is also known. Crystal structure determination of *trans*-$[WCl_2(OPh)_4]$[17] and *cis*-$[WCl_2(OMe)_4]$[18] has revealed relatively short W—O distances of 1.82 and 1.84 Å respectively. These distances are just 0.1–0.15 Å longer than the ones found for W—O_t indicating considerable $O \rightarrow W\ \pi$ donation. Another example of a short W—O distance (1.908 Å) is furnished by the structure of $[W(O_2C_2H_4)_3]$.[19] With Me_3SiMe, WCl_6 reacts to give $[WCl_5SMe]$.[20]

Methyl derivatives of W^{VI} have also been prepared. Dimethyl mercury reacts with WCl_6 to give $[WMeCl_5]$,[21] while with $AlMe_3$, $[W(Me)_6]$ is obtained.[27] This red air-sensitive material reacts with several reagents[23] to give known compounds such as $[W(OMe)_6]$,[24] $[W(NMe_2)_6]$,[11] $[W(OPh)_6]$[25] and $[W(tdt)_3]$.[26] With NO, the compound $[WMe_4(O_2N_2Me)_2]$ is obtained. Crystal structure analysis has shown that the W atom is eight-coordinate with a geometry between square antiprismatic and dodecahedral (Figure 2).[27] In solution, this complex is stereochemically non-rigid down to −50 °C.

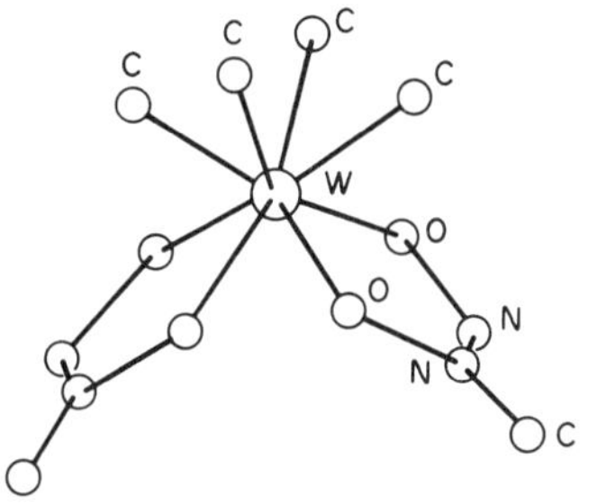

Figure 2 The structure of $[WMe_4(O_2N_2Me)_2]$[27]

The stability of complexes of the type $[W(OR)_6]$ is quite remarkable. It has been shown[28] that short exposure of $[W_2(OPr^i)_6py_2]$ to approximately two molar equivalents of CO gives two products, one of which is (**1**). This structure can best be viewed as a substituted tungsten carbonyl in which two *cis* CO groups have been replaced by donor atoms of the bidentate neutral ligand $[W(OPr^i)_6]$. This complex is unusual in that it contains two atoms of the same element in oxidation states that differ by six units.

$$[W(OPr^i)_4(\mu\text{-}OPr^i)_2W(CO)_4]$$
(1)

Several other W^{VI} complexes with coordination number greater than six have been reported. $[W(Me)_6]$ reacts with tertiary phosphines to give the seven-coordinate adducts $[W(Me)_6L]$ ($L = PMePh_2$, PMe_2PH, $PEtPh_2$, PMe_3).[29,23] The best characterized of these is the PMe_3 adduct, which was shown by ^1H and ^{13}C NMR to be fluxional at room temperature, although a rigid structure such as a capped biprism exists at low temperature. With excess MeLi, $[W(Me)_6]$ gives the bright yellow eight-coordinate compound $Li_2[W(Me)_8]$ which is reasonably stable as the dioxane solvate.[27] $[WMeCl_5]$ was reported to react with octamethylphosphoramide and diphos to give the adducts $[WMeCl_5{\cdot}L]$ which are believed to be eight-coordinate.[21] WF_6 is known to react with TIF and CuF_2 to yield[30] WF_7^{-1} which has physical properties similar to those reported for $(NO)[WF_7]$.[31] WF_7^- can also be prepared from LiN_3 and two equivalents of WF_6.[32] The existence of WF_8^{2-} has also been claimed.[31,33] WCl_6 reacts with Me_2S, 2,5-dithiahexane (dth) and tetrahydrothiophene (tht) to give the seven- and eight-coordinate adducts $[WCl_6L]$ and $[WCl_6L_2]$.[34] $[WF_6N_3]^-$ has been prepared from WF_6 and NH_4N_3.[32]

37.3.2 Complexes Containing the Oxo Ligand

The oxo ligand forms a strong bond with W^{VI} by utilizing both σ and π donation resulting in a short W—O bond length. Complexes containing the WO, *cis*-WO_2 and *cis*-WO_3 structural units are known.

37.3.2.1 Complexes with one oxo ligand

Synthetic routes to the oxohalo complexes of W^{VI} are summarized in Table 2. The oxohalo complexes with one terminal oxygen atom are of the general type $[WOX_4]$ (X = F, Cl, Br). The structures of $[WOCl_4]$ and $[WOBr_4]$ consist of chains of slightly distorted octahedra connected by oxygen bridges with W—O distances of 1.8 and 2.1 Å.[35] In contrast, WOF_4 is tetrameric in the solid state with fluoro bridges[36,37] as depicted in Figure 3. Chloride substitution by NCS^- to givc $[WO(NCS)_4]$ has been reported.[8]

The coordinatively unsaturated $[WOX_4]$ complexes react with ligands such as alkyl cyanides, THF or py to give the six-coordinate $[WOX_4L]$ complexes.[38,39] $[WOX_5]^-$ ($X = F^-, Cl^-, Br^-$) are also known.[40,41] Interesting adducts result from the reaction of WOF_4 with XeF_2 in HF. White stable crystalline materials of composition $[WOF_4{\cdot}XeF_2]$ and $[(WOF_4)_2{\cdot}XeF_2]$ have been

Table 2 Synthetic Routes to W^{VI} Oxyhalides

Compound	*Preparation*	*Ref.*
WOF_4	$WOCl_4$ with HF	75
	WO_3 with either F_2, IF_5, SeF_4	75
	W with F_2 and O_2	75
$WOCl_4$	WO_3 with $SOCl_2$, CCl_4, O_2 octachlorocyclopentene	128, 129, 130
	WCl_6 with SO_2 or WO_3	40, 131
$WOBr_4$	WO_3 with CBr_4	75
	WBr_6 with WO_3	131
WO_2Cl_2	WCl_6 with WO_3	131
	WO_3, WS_3, WO_4^{2-} with CCl_4, Cl_2 and O_2, CCl_4 and O_2	75
	W with Cl_2 in air	75
WO_2Br_2	WBr_6 with WO_3	131
	W with Br_2 in air	75
	WO_3 with Br_2 or CBr_4	75
WO_2I_2	W with NO_3 and I_2 in sealed tube	32

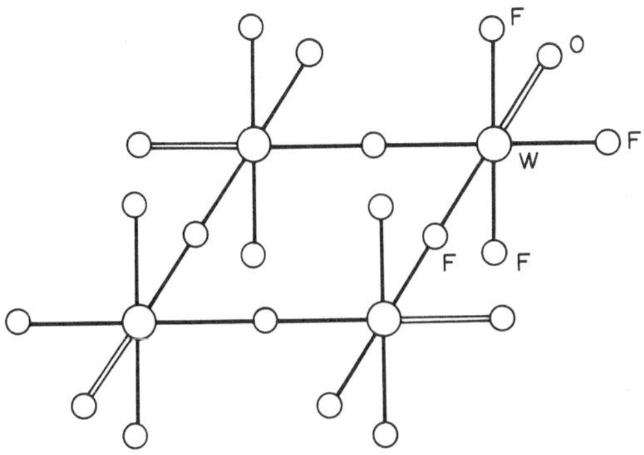

Figure 3 Proposed structure for $[WOF_4]_4$[37]

obtained.[42,43] On the basis of Raman measurements and ^{19}F NMR spectra it has been suggested that the XeF_2 molecule is bonded to the sixth coordination position of the W atom through the fluoro atom, forming a bent W—F—Xe bond. Adduct formation between $[WOF_4]$ and SbF_5 is also known.[44]

$WOCl_4$ reacts with bidentate Schiff bases to give complexes of the type $[WOCl_3$(Schiff base)].[45] These compounds show one $\nu(W—O_t)$ at around 970 cm^{-1}. With tridentate Schiff bases, $[WOCl_2$(Schiff base)] is obtained. Complexes of the type $[WOF_3L]^-$ (L = bidentate hydroxy acids) have also been reported.[46]

The structure of the azoxybenzene adduct of $WOCl_4$ has been reported.[47] The azoxybenzene ligand is coordinated through the oxygen atom *trans* to the oxo ligand with a bond length of 2.276 Å. The $W—O_t$ distance is 1.669 Å. The structure reported for the diars adduct of $WOCl_4$[48] appears to be in error. In this structure, the W—Cl bond *trans* to the oxo ligand is shorter than the *cis* W—Cl bond being 2.26 and 2.40 Å respectively. This result is of course contrary to our knowledge of the *trans* influence of the oxo ligand.[49]

With hydrogen peroxide, W^{VI}, like molybdenum, forms peroxy compounds in which the ratio of peroxy groups to metal is 4:1, 3:1, 2:1, and 1:1, but most are not well characterized.[50] A 2:1 peroxy complex, obtained from acidic solution containing alkali metal tungstate and a high concentration of H_2O_2 was shown by X-ray crystallography to be an oxygen-bridged dimer[51] with formula $[(O_2)_2W(O)OW(O)(O_2)_2(H_2O)_2]^{2-}$. The coordination geometry is best viewed as a pentagonal bipyramid with two pentagonal planes rotated by 90°, one with respect to the other. The structure of the 1:1 complex ion $[WO(O_2)F_4]^{2-}$ has been determined[52] and is best described as a distorted octahedron in which the center of the O—O bond occupies a corner. A 2:1 and a 1:1 peroxy complex can also be obtained in the presence of the ligands pyridine carboxylate and pyridine 2,6-dicarboxylate respectively.[53] Phosphine and arsine oxides have been shown to stabilize peroxo complexes of W^{VI}.[59] Thus, complexes of the type $[W(O)(O_2)_2L_2]$ (L = OPR_3; R = Pr^n, Bu^n, Ph; $OAsR_3$, R = Pr^n, Bu^n, Ph) have been prepared and characterized. These compounds show the expected $\nu(W—O_t)$ around 950 cm^{-1}, and $\nu(O—O)$ around 850 cm^{-1}.

37.3.2.2 *Complexes with two oxo ligands*

The oxohalo complexes containing two terminally bonded oxygen atoms are of the type WO_2X_2 (X = F, Cl, Br, I). WO_2Cl_2 was shown to be polymeric in the solid state[55] with oxygen bridges and W—O bond lengths ranging from 1.63 to 2.34 Å. WO_2F_2, WO_2Br_2 and WO_2I_2 are believed to be polymeric as well.[56] WS_2Cl_2 is also known.[57]

In spite of its polymeric structure and insolubility in common solvents, it has been shown[58] that complexes of the type $WO_2Cl_2L_2$ and WO_2Cl_2(L–L) can be prepared with mono- and bi-dentate ligands respectively. $WO_2Cl_2L_2$ can also be prepared with Schiff bases.[45]

$[WO_2(S_2CNR_2)_2]$ (R = Me, Et, Pr^n) have been prepared[59] by an oxo transfer utilizing a dinuclear Mo^V complex as depicted in equation (1). Because of their moisture sensitivity, these

$$[W(CO)_2(PPh_3)(S_2CNR_2)] + 2[Mo_2O_3(S_2P(OEt)_2)_4] \longrightarrow$$
$$[WO_2(S_2CNR_2)_2] + 4[MoO\{S_2P(OEt)_2\}_2] + 2CO + PPh_3 \quad (1)$$

Table 3 Complexes Containing the WO_2 Structural Unit.

Compound	ν(*cis*-WO_2) (cm^{-1})	*Ref.*
$[WO_2Cl_2(DMF)_2]$	913, 965	58
$[WO_2Cl_2(MeCN)_2]$	922, 972	58
$[WO_2Cl_2(OPPh_3)_2]$	915, 960	58
$[WO_2Cl_2(bipy)]$	915, 953	58
$[WO_2(S_2CNEt_2)_2]$	890, 933	60
$(Et_4N)[WO_2F_3(OPMe_3)]$	905, 950	133
$(Me_4N)_2[WO_2Cl_4]$	895, 943	45
$(Me_4N)[WO_2Cl_2(acac)]$	895, 940	45
$(Et_4N)[WO_2F_3(H_2O)]$	900, 997	133
$[WO_2Br_2(bipy)]$	907, 954	134
$[WO_2(ox)_2]$[a]	918, 958	134
$[WO_2(tox)_2]$[b]	910, 950	134
$[WO_2(acac)_2]$	912, 960	134

[a] OX = 8-hydroxyquinoline. [b] tox = 8-mercaptoquinoline

complexes cannot be prepared from WO_4^{2-} in a manner analogous to the preparation of $[MoO_2(S_2CNR_2)_2]$.[60]

All complexes containing the WO_2 structural unit have *cis* configuration and exhibit two strong IR absorption bands around 950 and 900 cm^{-1}. The structure of dichloro-dioxo-pentane-2,4-dionato W^{VI} has been determined and it contains the expected *cis* arrangement of the oxygen atoms with a W—O_t distance of 1.730 Å.[61] Table 3 lists well-characterized compounds containing the WO_2 structural unit.

37.3.2.3 Complexes with three oxo ligands

Complexes containing the WO_3 structural unit are the least studied. Of the oxyhalides, $WO_3F_3^{3-}$, $WO_3F_2^{2-}$, WO_3F^- and WO_3Cl^- have been prepared. The isolation of $[WO_3(dien)]$[62] has been reported. The complex is unstable in solution and hydrolyzes to WO_4^{2-} and protonated dien. Formation constants for WO_4^{2-} with aspartic acid and glutamic acid have also been measured,[63] but because of the competitive formation of isopolytungstate in acidic solution and the stability of WO_4^{2-} in basic solution, the pH range over which formation constants of such complexes can be measured is very small. The reaction of WO_3 with oxalate has been reported[64] to give $[WO_3(ox)]^{2-}$ and $[W_2O_5(ox)_3]^{4-}$.

Slow air oxidation of an acidic W^V solution results in the formation of the mixed valence W^{VI}–W^V complex[65] $[W_4O_8Cl_8(H_2O)_4]^{2-}$. It is believed that this ion is obtained by an equilibrated reaction between $[WO_2Cl_4]^{2-}$ and $[WOCl_5]^{2-}$. Its structure consists of four $WO_3Cl_2(H_2O)$ octahedra sharing corners as depicted in Figure 4. The four W atom form a

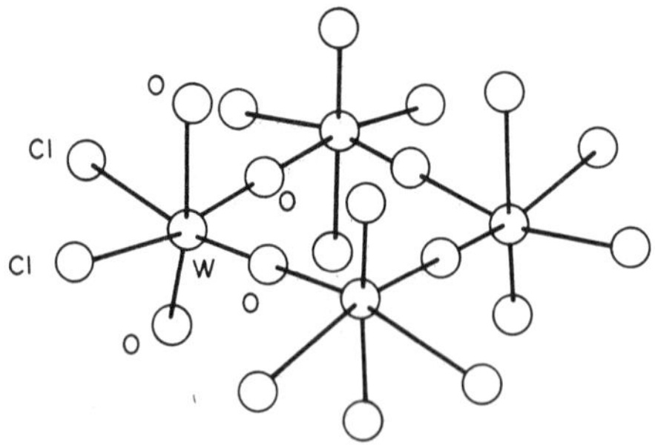

Figure 4 The structure of $[W_4O_8Cl(H_2O)_4]^{2-}$ [65]

nearly planar square and are joined through linear oxygen bridges. The W—O_b distance of 1.86 Å, although short, is significantly longer than the W—O_t bond length of 1.711 Å, as expected.

Substitutions without disruption of the $W_4O_8^{6+}$ core led to the isolation[66] of (**2**), (**3**) and (**4**).

$[W_4O_8Cl_6(DMF)_6]$	$[W_4O_8(NCS)_{12}]^{6-}$	$[W_4O_8(NCS)_4(C_2O_4)_4]^{6-}$
(**2**)	(**3**)	(**4**)

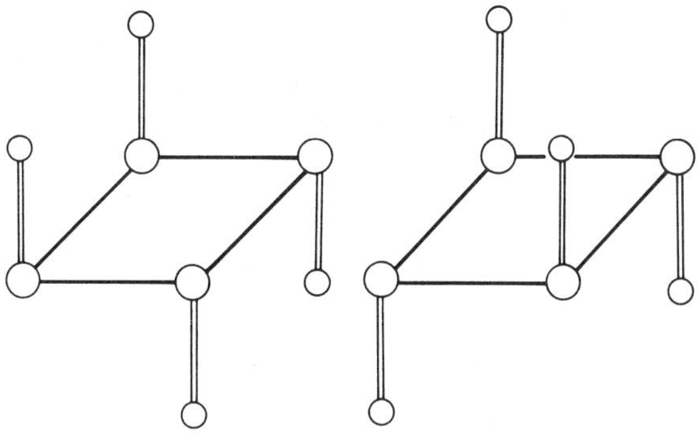

Figure 5 The dispositions of O_t around the W_4O_8 square[66]

The structure of (**3**) is quite similar to that of $[W_4O_8Cl_8(H_2O)_4]^{2-}$ with the exception of the spatial disposition of the terminally bonded oxygen atoms as depicted in Figure 5. As expected from the *trans* effect of the O_t ligand, the W—N bond lengths *trans* to O_t are on the average 0.2 Å longer than the W—N bond lengths *cis* to O_t. Both structures reveal no significant differences between the four tungsten atoms, which, together with theoretical consideration,[67] suggest that these mixed valence complexes are best described as delocalized systems.

37.3.3 Complexes Containing Multiple Bonds to Sulfur, Selenium, Nitrogen and Carbon Atoms

Several good π donor ligands, in addition to the oxo ligand, have been shown to form multiple bonds with W^{VI}. This tendency undoubtedly results from the d^0 electronic configuration of W^{VI} and its high formal positive charge. Thus, the reaction of WX_6 (X = Cl, Br) with Sb_2S_3 and Sb_2Se_3 leads to complexes of the type WSX_4 and $WSeX_4$.[68,69] These compounds are moisture sensitive and evolve H_2S and H_2Se on exposure to moist air. Structure analysis of WSX_4[70] shows that they are dimeric in the solid state with two WSX_4 square pyramidal units linked through unsymmetrical halogen bridges as depicted in Figure 6. The short $W—S_t$ distance of 2.090 Å clearly indicates a W—S bond order higher than one. The IR bands at 569 and 555 cm^{-1} for $[WSCl_4]$ and $[WSBr_4]$ respectively have been assigned to $\nu(W—S_t)$. It is believed that $[WSeX_4]$ has similar structures and an IR band at 396 cm^{-1} has been assigned to $\nu(W—Se_t)$.[69] The slight paramagnetism of these compounds (~0.4 BM at 25 °C) is believed to arise from W^V impurities, or, at least in part, from TIP.[38]

WF_6 also reacts with Sb_2S_3 to give WSF_4.[71] The reaction of $[WSCl_4]$ with XeF_2 leads to the mixed fluoro–chloro complexes. These have not been isolated, but ^{19}F NMR has established[71] the presence in solution of $[WSF_2Cl_2]$, $[WSFCl_3]$ and $[WSF_3Cl]$ in addition to the ions $[WSF_5]^-$, $[WSF_4Cl]^-$, $[WSFCl_4]^-$ and $[WSF_3Cl_2]^-$. The reaction of $[WYCl_4]$ (Y = O, S, Se) with Sb_2Y_3 in CS_2 leads to complexes of the type WS_2Cl_2, $WSSeCl_2$ and $WOSCl_2$.[72]

Compounds of the type $[WYX_4]$ are coordinatively unsaturated and thus react with a variety of ligands or with donor solvents to give simple six-coordinate adducts.[73a] An example of such

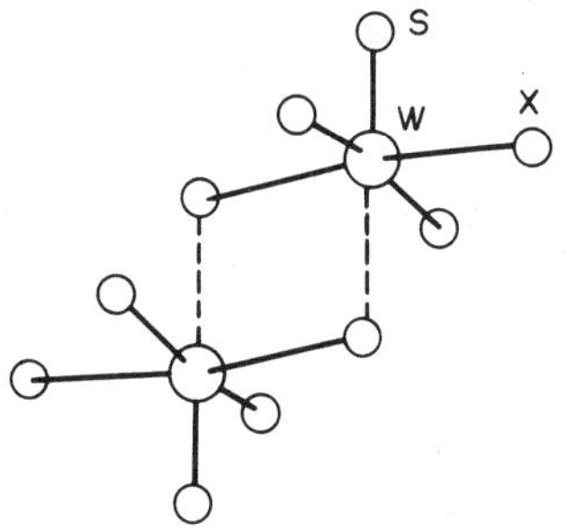

Figure 6 The structure of $[WSCl_4]$[70]

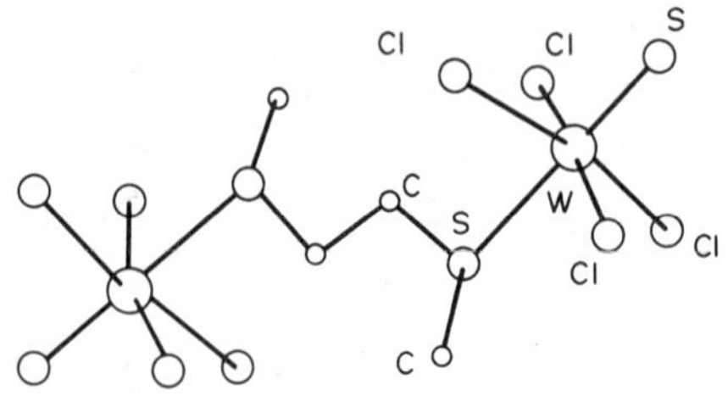

Figure 7 The structure of $[WSCl_4(\mu\text{-dth})WSCl_4]$[73b]

an adduct is shown in Figure 7. Important structural parameters of this and other structurally characterized complexes of W^{VI} are given in Table 4.

Table 4 Structural Parameters of some W^{VI} Complexes

Compound	*W—X$_t$*[a] (Å)	*W—A$_{trans}$*[b] (Å)	*W—A$_{cis}$*[b] (Å)	*Ref.*
$(qH)_2[WO(O_2)F_4]$	1.74 (O)	2.07 (F)	1.92 (F)	52
$[WOCl_4$(azoxybenzene)]	1.67 (O)	2.28 (O)	2.29 (Cl)	47
$(Me_3NH)_2[W_4O_8Cl_8(H_2O)_4]$	1.71 (O)	2.29 (O)	2.40 (Cl)	65
$(Cs_5NH_4)[W_4O_8(NCS)_{12}]\cdot 6H_2O$	1.68 (O)	2.28 (N)	2.08 (N)	66
$K_2[W_2O_3(O_2)_2]\cdot 4H_2O$	1.68 (O)	2.36 (O)	1.93 (O)	51
$(C_{10}H_{13}O_2)[WO_2C_{12}$(acac)]	1.73 (O)	2.16 (O)	2.39 (Cl)	61
$[WCl_4(NC_2Cl_5)]\cdot NCCCl_3$	1.70 (N)	2.37 (N)	2.30 (Cl)	86a
$(Ph_4As)[WCl_5(NC_2Cl_5)]$	1.68 (N)	2.42 (Cl)	2.33 (Cl)	89
$(Ph_4As)_2[W_2NCl_{10}]$	1.71 (N)	2.43 (Cl)	2.30 (Cl)	80
$[WSCl_4]$	2.10 (S)	3.05 (Cl_b)	2.37 (Cl_b)	70
$[WSBr_4]$	2.08 (S)	3.03 (Br_b)	2.55 (Br_b)	70
$(Ph_4As)[WSCl_5]$	2.17 (S)	2.35 (Cl)	2.33 (Cl)	
$[WSCl_4][WOSCl_2(DME)]$	2.10 (S)	2.26 (O_b)	2.32 (Cl)	74
$[WSCl_4]_2$(dth)	2.07 (S)	2.79 (*S*-thioether)	2.30 (Cl)	73a
$[WSeCl_3(CH_2OC_2H_4O)]$	2.26 (Se)	2.46 (*O*-ether)	1.93 (O)	76b

[a] Terminally bonded atom. [b] Atom *trans* or *cis* to X.

In some cases, the interaction of $[WYX_4]$ with donor solvents leads to oxygen abstraction or cleavage of an ether linkage. Thus, $[WSCl_4]$ abstracts an oxygen atom from 1,2-dimethoxyethane (DME) to form a complex that was shown by X-ray crystallography[74] to consist of the $[WOSCl_2DME]$ moiety bonded to $[WSCl_4]$ through an oxygen bridge. This oxygen abstraction is characteristic of the high oxidation state halides of the early transition metals,[75] and in some cases can be used as a simple route to oxyhalide complexes. For example, $[WOF_4\cdot Et_2O]$ can be prepared from WF_6 and diethylether.[76] The reaction of $[WSeCl_4]$ with the same solvent leads to cleavage of the ether linkage and the formation of a W–alkoxide bond as shown in Figure 8. A similar reaction was reported for $[WOCl_4]$.

Several nitrido complexes of W^{VI} are known. WCl_6 reacts with ClN_3[77] to give $[WNCl_3]$ whose structure is believed to be tetrameric similar to that of its molybdenum analogue.[78] $[WNCl_3]$ can also be prepared from WCl_6 and IN_3.[79] Impure $[WNBr_3]$ has been prepared from WBr_6 and IN_3.[79] The preparation of $[WNCl_4]^-$ has been reported and $[W_2NCl_{10}]^{2-}$ was isolated from this

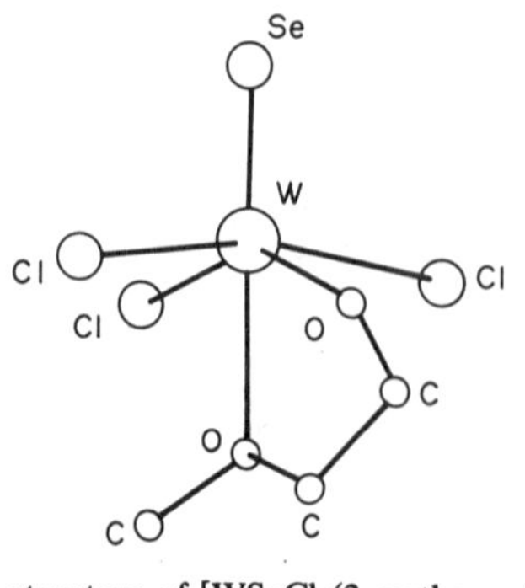

Figure 8 The structure of $[WSeCl_3$(2-methoxyethoxide)][76b]

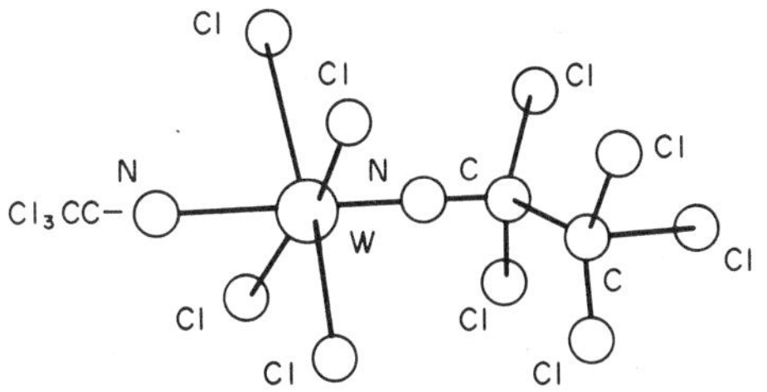

Figure 9 The structure of $[WCl_4(NC_2Cl_5)]$ CCl_3CN[86a]

preparation.[80] The structure of this dinuclear complex consists of W^{VI} and W^{V} linked by a linear asymmetric nitride bridge.[80]

With py, the adduct $[WNCl_3 \cdot 3py]$, of unknown structure, is obtained.[81] Other characterized complexes derived from $[WNCl_3]$ are $[WNCl_3 \cdot PPh_3]$, $[WNCl_5]^-$ and $[WNCl_3(bipy)]$.[82] The reaction of $[WNCl_3]$ with $POCl_3$ gives the tetranuclear compound $[WNCl_3(OPCl_3)]_4$. This structure contains a planar square W_4N_4 eight-membered ring with alternating W—N bond lengths. The oxygen atom of the $POCl_3$ ligand is coordinated *trans* to the shorter W—N bond.[83] The presence of the nitride ligand is easily detected in these complexes by a strong IR absorption band around 1050 cm^{-1} assigned to $\nu(W—N_t)$.[84]

Complexes containing the WNR group are also known. The reaction of WCl_6 with aliphatic and aromatic nitriles[85] leads to the formation of imido W^{VI} complexes. Strong IR absorption around 1280 cm^{-1} is characteristic of the WNR group. These complexes are thought to be intermediates in the alkylcyanide reduction of WCl_6.

The structures of several of these complexes have been determined crystallographically. The structure[86a] of $[WCl_4(NC_2Cl_5)] \cdot NCCCl_3$, which is depicted in Figure 9, reveals a short W—N bond length of 1.70 Å, as expected for a bond order of two or higher. When CCl_3CN reacts with WCl_5, a dichloro-bridged W^{VI} imido complex is obtained (Figure 10). Complexes of the type $[W(NPh)(OCMe_3)_4]$ and $[W(NPh)Cl_5]^-$ have also been prepared.[87] In all known structures containing WNR groups, the *trans* influence of the imido group is clearly seen (Table 4).

The coordinatively unsaturated complex $[WCl_4(NC_2Cl_5)]$ reacts with monodentate ligands to form six-coordinate adducts[88,89] with structures similar to the one depicted in Figure 9. When WCl_6 is reacted with cyanogen in $POCl_3$, a dinuclear complex in which NC_2Cl_4N bridges two $[WCl_4(OPCl_3)]$ moieties is obtained.[90] With bidentate ligands it is suggested that seven-coordinate complexes are formed.[88]

Oxygen abstractions are also observed for the imido W^{VI} complexes. Thus, in addition to the 1:1 adduct of $OPPh_3$ with $[WCl_4(NC_2Cl_5)]$ the complex $[WCl_2O_2(OPPh_3)_2]$ is formed.[91] Also, $W(NMe_2)_6$ reacts with Bu^tOH to give $[WO(OBu^t)_4]$.[13]

It has been shown recently that alkylidene and alkylidyne complexes of W^{VI} also exist, and that some of these are important catalysts for the metathesis of terminal and internal alkenes.[87,92] Crystal structure analysis of the alkylidene complex $[W(O)(CHCMe_3)Cl_2PEt_3]$ shows that the geometry around the metal atom is distorted trigonal bipyramidal with the $W—O_t$ and $W═CHCMe_3$ units in the equatorial plane. The short W—C distance of 1.882 Å is in accord with a W═C double bond.[92] $[W(NPh)(CHCMe_3)Cl_2L_2]$ has also been prepared. Here the tungsten atom is six-coordinate with the W═NPh and $W═CHCMe_3$ units in *cis* position, as expected, similar to complexes containing the WO_2 structural unit.[87,93,94]

The alkylidyne complexes are best prepared by cleavage of a W≡W triple bond[92] as depicted

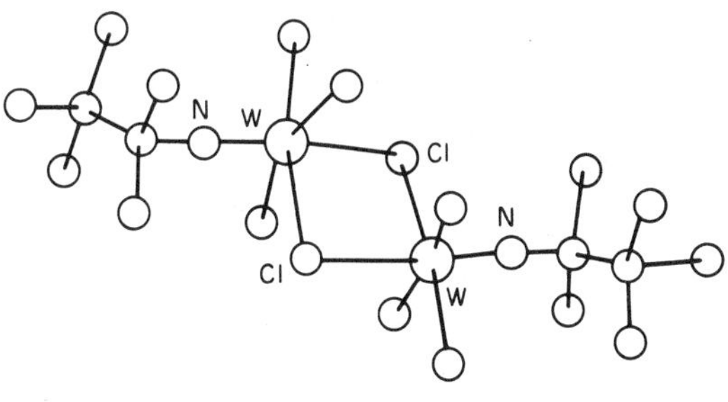

Figure 10 The structure of $[W_2Cl_6(\mu\text{-}Cl)_2(NC_2Cl_5)_2]$[86b]

schematically in equation (2). (These reactions will be discussed in more detail in Section 37.6.2.2.) Thus, the reaction of $[W_2(OBu^t)_6$ with RC≡N and RC≡CR leads to cleavage of the

$$\begin{aligned} W{\equiv}W + R{-}C{\equiv}N &\longrightarrow W{\equiv}N + W{\equiv}CR \\ W{\equiv}W + R{-}C{\equiv}C{-}R &\longrightarrow 2W{\equiv}CR \end{aligned} \tag{2}$$

W≡W triple bond and formation of both nitrido and alkylidyne complexes of W^{VI}. Crystal structure determination[94] of $[(OBu^t)_3W{\equiv}N]_n$ and $[(OBu^t)_3W{\equiv}CMe]_2$ has shown that the nitrido complex is polymeric with unsymmetrical W—N—W bridges. The alkylidyne complex is dimeric with a short W—C bond length of 1.76 Å. As expected, this distance is significantly shorter than the one observed for the W—C double bond.

37.3.4 WS_4^{2-}, WOS_3^{2-} and $WO_2S_2^{2-}$ as Ligands

The thiotungstate ion WS_4^{2-} has been known since 1826[95] and its exact chemical composition was determined in 1886.[96] Its structure[97] showed the expected tetrahedral geometry with a relatively short W—S distance of 2.165 Å. The mixed oxothio anions are also known.[98] WS_4^{2-} and to a lesser extent WOS_3^{2-} and $WO_2S_2^{2-}$ have been shown in recent years to act as ligands for several metal ions. Complexes of the type $[M(WS_4)_2]^{2-}$ ($M = Ni^{2+}$, Co^{2+}, Zn^{2+}, Pd^{2+}, Pt^{2+}), where the thiotungstate ion acts as a bidentate ligand, have been prepared.[99,100,101,102] Similar compounds with WOS_3^{2-} and $WO_2S_2^{2-}$ are also known,[103,104,104] and in all cases the coordination is through sulfur. It is interesting to note that with trivalent ions such as Cr^{3+}, Eu^{3+} and Dy^{3+}, complexes such as $[M(WO_2S_2)_3]^{3-}$ are not obtained, but rather $WO_2S_2^{2-}$ hydrolyzes to the tungstate ion.[106]

Of particular interest are those complexes which serve as model compounds for biological systems such as the active center of the MoFe protein in nitrogenase.[107] These complexes will be discussed in detail in the relevant chapters. It should be noted here, however, that WS_4^{2-} is a non-innocent ligand, and thus complexes such as $[Co(WS_4)_2]^{n-}$ ($n = 2, 3$)[108,109] and $[Fe(WS_4)_2]^{n-}$ ($n = 2, 3$) can easily be prepared.[110,111,112] In addition, some unusual complexes have been synthesized with WS_4^{2-} as ligands. One such example is $[Fe_3W_3S_{12}]^{4-}$ which contains the $Fe_3(\mu_3\text{-}S)_2$ center as depicted in Figure 11.

Complexes of the type $[Au_2(WS_4)_2]^{2-}$ and $[(Ph_2MeP)_2Au]_2[WS_4]$ have also been prepared and structurally analyzed.[113,114] In the first case, two WS_4^{2-} ions are bridged by gold atoms as shown in Figure 12 (identical structure is obtained with WOS_3^{2-}) and in the second, WS_4^{2-} acts as a bridging group for two $[(Ph_2MeP)_2Au]$ moieties. WS_4^{2-} and WOS_3^{2-} have also been shown to act as tridentate ligands.[115,116,117] One such complex is shown in Figure 13.[117]

Crystal structure analysis of complexes containing the WS_4^{2-}, WOS_3^{2-} and $WO_2S_2^{2-}$ ligands have shown that these ions are tetrahedral with the $W—S_b$ bond length slightly but significantly longer than the $W—S_t$ bond length. Finally it should be mentioned that both WSe_4^{2-} and $WO_2Se_2^{2-}$ are known[118] and $[Zn(WSe_4)]^{2-}$ has been prepared.[119]

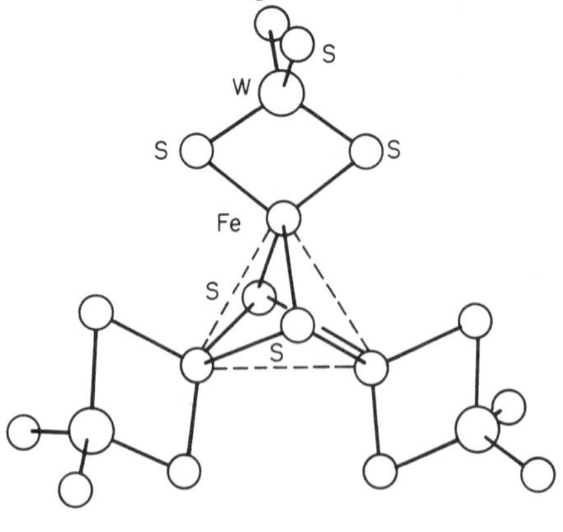

Figure 11 The structure of $[Fe_3W_3S_{12}]^{4-}$ [112]

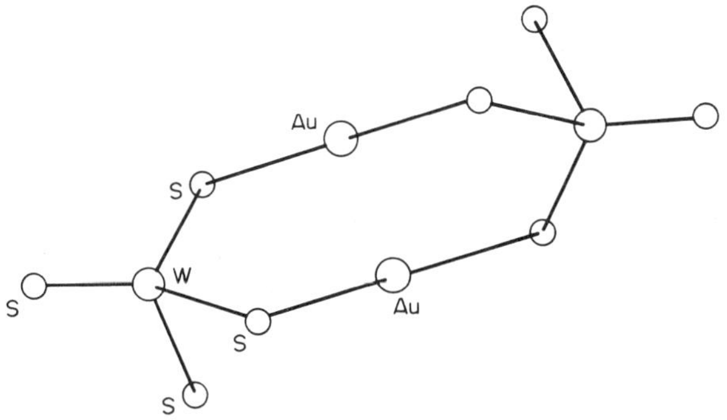

Figure 12 The structure of $[Au_2(WS_4)_2]^{2-}$ [113]

37.3.5 Dithiolene Complexes

Dithiolene complexes attracted considerable interest during the 1960s and early 1970s because of their unusual electronic and structural properties.

Tungsten complexes of the type $W(L\text{–}L)_3$ are known for L–L = sdt, bdt, tdt, $S_2C_2Me_2$, $S_2C_2H_2$, $S_2C_2(CF_3)_2$ and $Se_2C_2(CF_3)_2$. These highly colored complexes undergo two reversible one-electron reductions leading to mono- and di-anionic complexes.[26,120] Structurally, the neutral complexes represent examples of a trigonal prismatic coordination in six-coordinate complexes similar to the analogous Re and Mo complexes.[121] The interesting problems associated with this class of compounds, such as their electronic structure and the relation between this and geometry, have been thoroughly reviewed.[121,120]

Of the reduced complexes several have been isolated. Hydrazine reduction of $[W(sdt)_3]$ gives the dianion which reacts with two equivalents of MeI. It is interesting that the methylation reaction occurs only on one of the three ligands. The methylated product reacts with diphos to give $[W(sdt)_2(diphos)]$. The mixed ligand complex $[W(sdt)_2(S_2CNEt_2)]^-$ was also reported. Crystal structure analysis[122] of $(Ph_4As)_2[W(mnt)_3]$ shows that the symmetry of the WS_6 framework is close to D_3 and that the geometry is midway between octahedral and trigonal prism.

The complex $[W\{Se_2C_2(CF_3)_2\}_3]$ has also been prepared[123] and like its sulfur analogue, it undergoes two reversible one-electron reductions. The dianionic complex was isolated as the tetrabutylammonium salt.

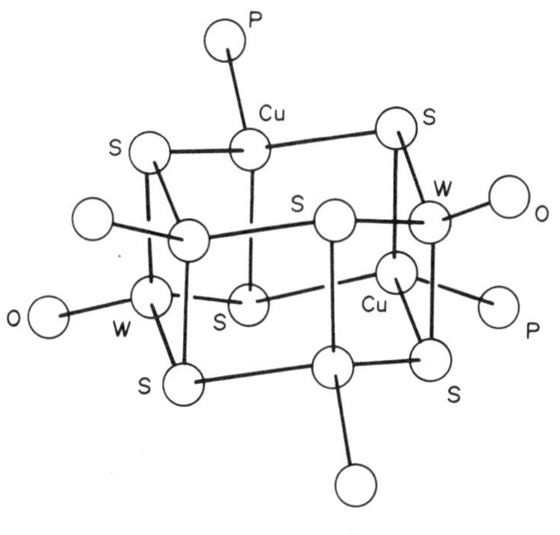

Figure 13 The structure of $[Cu_4(WOS_3)_2(PPh_3)_4]$; phenyl rings are omitted[117]

37.4 THE CHEMISTRY OF W^V

W^V, unlike Mo^V, is not dominated by dinuclear species. Most well-characterized complexes of W^V are monomeric and there are only a few reported examples of dinuclear compounds. These, with the exception of $[W_2Cl_{10}]$, all contain an M—M bond of order one.

37.4.1 W^V Halides and Their Derivatives

The three halides WF_5, WCl_5 and WBr_5 are known. WF_5 can be prepared from WF_6[135] and it was shown to be isostructural with MoF_5, NbF_5 and TaF_5,[136] so therefore probably has the same tetrameric structure with fluoride bridges. The dark green WCl_5 can be prepared by reduction of WCl_6[137] or by disproportionation[138] of WCl_4. It can also be prepared by reduction of WCl_6 with tetrachloroethylene in the presence of strong light.[139] The black-green WBr_5 is prepared by reduction of WCl_6 in HBr, or by thermal decomposition of WBr_6.[140] WCl_5 is dimeric in the solid state[141] and contains bridging Cl atoms. The long W—W distance of 3.814 Å, and the magnetic moment of ~1.0 BM per W[142] at room temperature, are consistent with the absence of a W—W bond.

The octahedral WX_6^- anions (X = F, Cl, Br) are best prepared by reaction of the higher halides in a suitable solvent.[41,143,144] In some cases they can be prepared by the reaction of the pentahalides with halide ions.[142,145,146] Crystal structure determination of $C_s[WCl_6]$ and $(Et_4N)[WCl_6]$[147] has shown that all the W—Cl bond lengths are practically the same, with an average value of 2.33 Å. When dry, they are quite stable, but easily hydrolyzed. The WX_6^- ions have low magnetic moments compared with the spin-only value for a d^1 electronic configuration. This is believed to result from large spin–orbit coupling. The reaction of $W(CO)_6$ with IF_5 in the presence of KI[148] gives the WF_8^{3-} ion which is believed to be eight-coordinate W^V.

Several adducts of WX_5 (X = Cl, Br) have been reported.[149,150] The best characterized is $[WCl_5(diars)]$[149] which is obtained by reduction of WCl_6 with excess diars. This paramagnetic complex (1.19 BM) is isomorphous with the seven-coordinate $[NbCl_5(diars)]$ and $[TaCl_5(diars)]$.[151]

It is well known that nitrogen-containing ligands can reduce W^{VI} and W^V halides to complexes of W^{IV}.[152,153,154] However, several substitution products of WX_5 (X = Cl, Br) have been prepared, and they are best formulated as *cis*-$[WX_4L_2]^+$ (L = Me_3py, py, PhCN, bipy, diphos).[138,155] These adducts are all paramagnetic with magnetic moments in the range of 0.63–1.45 BM. Although magnetic moments below the spin-only value are expected for W^V complexes, some of the reported moments seem to be unreasonably low, perhaps as a result of diamagnetic W^{IV} impurities, or magnetic interactions of adjacent d^1 centers.

Monomeric alkoxo complexes of W^V are also known. The dinuclear alkoxo complexes will be discussed in Section 37.4.4. The blue paramagnetic complex $[WCl_3(OR)_2]$ (R = Me, Et) were first isolated from the reaction of WCl_6 with the corresponding alcohols.[156,157] When ethanol is used, the reaction liberates chlorine which oxidizes the ethanol to acetaldehyde[157] as depicted in equations (3) and (4). Chloroalkoxo complexes of W^V can also be prepared directly from WCl_5 and alcohols at −70 °C. Thus, $M[WCl_4(OR)_2]$ and $M[WCl_5(OR)]$ (M = tetraalkyl ammonium; R = Me, Et, Pr^n) as well as the seven-coordinate $M[WCl_6(OEt)]$ have been isolated[158,159] and characterized. Similarly, $[WBr_4(OMe)_2]^-$ has been prepared from WBr_5 in methanol. The monomeric chloroalkoxo complexes are paramagnetic and exhibit electronic absorption spectra similar to the W^V oxyhalides.

$$WCl_6 + 2EtOH \longrightarrow \tfrac{1}{2}Cl_2 + [WCl_3(OEt)_2] + 2HCl \quad (3)$$

$$Cl_2 + EtOH \longrightarrow MeCHO + 2HCl \quad (4)$$

It has been suggested previously (Section 37.3.1) that the relatively short W^{VI}—OR bond length is indicative of bond order greater than one. If this holds true for the W^V—OR bonds then for complexes of the type $[WX_4(OR)_2]^-$ the π donor ligands OR^- will prefer *trans* stereochemistry since the *d* electron can be accommodated in the d_{xy} atomic orbital which is orthogonal to the orbitals involved in the bonding of the π donor atom. Indeed, ESR spectra of the dialkoxo complexes are consistent with axial symmetry.[160] This idea has also been put forth to explain the preference of the dioxo complexes with the d^2 electronic configuration for the *trans* stereochemistry.

An interesting solid state reaction has been reported for $(R_4N)[WCl_5(OR)]$[157] as depicted in equation (5). For a given R_4N^+ cation, the rate of alkyl chloride elimination follows the order Me > EtCl > PrCl. The rate of evolution of ethyl chloride decreased with increase in the size of the cation.

$$(R_4N)[WCl_5(OR)] \xrightarrow{-RCl} (R_4N)[WOCl_4] + \text{unknown products} \quad (5)$$

37.4.2 The Oxyhalides and Their Derivatives

Of the oxyhalides of W^V, both $[WOCl_3]$ and $[WOBr_3]$ are known. These can be prepared by reduction of $[WOX_4]$[161] or by the reaction of WX_5 with Sb_2O_3[162] as depicted in equation (6).

$$3WX_5 + Sb_2O_3 \longrightarrow 3WOX_3 + 2SbX_3 \quad (X = Cl, Br) \quad (6)$$

The structures of WOX_3 are believed to be similar to that of $[NbOCl_3]$[163] which reveals the presence of infinite chains of Nb—O—Nb bonds. The absence of a strong ν(W—O_t) around 950 cm^{-1} and the low magnetic moment of ~0.5 BM are in agreement with this structure. In addition, WOX_3 are antiferromagnetic, which is suggested to occur by superexchange mechanism through the W—O—W bridges.[161]

The monomeric $[WOX_5]^{2-}$ (X = Cl, Br) anions have been isolated with a variety of +1 cations. They are easily prepared from WCl_5 or $K_3[WO_2(C_2O_4)_2]$ in concentrated solutions of HCl or HBr in the presence of the desired cation.[164] $[WOX_4]^-$ are also known.[164] These are all paramagnetic with magnetic moments close to the spin-only value, presumably because tetragonal distortion destroys the spin–orbit coupling.[143] Their electronic spectra have been interpreted.[164,165]

Complexes of the type $[WOX_3L_2]$ and $[WOX_3(L–L)]$ are known for a variety of mono- and bi-dentate ligands.[166,167,168] They are best prepared by reacting $[WOX_3(THF)_2]$ or $[WOX_3(MeCN)_2]$ with the desired ligand.[166,167] With some nitrogen containing ligands, complexes of this type can be prepared by reduction of WOX_4. Thus, the reaction of WOX_4 with py or bipy, gives $[WOCl_3py_2]$ and $[WOCl_3(bipy)]$ respectively.[38] All the oxytrihalide adducts are paramagnetic and exhibit the characteristic ν(W—O_t) around 950 cm^{-1} (Table 5).

Three different isomers are possible, in principle, for the $[WYX_3L_2]$ complexes (Y = O, S, Se, NR) (Figure 14). ESR spectra of $[WOCl_3L_2]$ (L = Ph_3P, Me_2PhP) was reported to be consistent with the *cis-mer* isomer[166] in agreement with the solid state structure of $[WOCl_3(OPPh_3)_2]$. However, the complexes $[WOCl_3(PEt_3)_2]$ and $[W(NPh)Cl_3L_2]$ (L = $OPMe_3$, PMe_3, PMe_2Ph) have been shown to have *trans–mer* stereochemistry in the solid state.[169,170,171]

With bidentate ligands, (Figure 14) it has been suggested that the *fac* isomer is preferred.[167] However, structure determination of $[WSCl_3(mte)]$ (mte = 1,2-bis(methylthio)ethane) has shown that it has the *mer* stereochemistry in the solid state.[172] Whether this is indeed the preferred stereochemistry for complexes of the type $[WYX_3(L–L)]$ will have to await further studies.

The electronic spectra in the visible region of the $[WOX_3L_2]$ and $[WOX_3(L–L)]$ complexes are similar to those reported for WOX_5^- and similar assignments can be made.[166]

Several porphyrin complexes containing the WO unit are known.[173,174] The paramagnetic

Table 5 Properties of Some W^V Oxo Complexes

Compound	μ_{eff} (BM)	(*W—O_t*) (cm^{-1})	*Ref.*
$[WOCl_3(MeCN)_2]$	1.51	992	166
$[WOCl_3(py)_2]$	1.48	970	166
$[WOCl_3(bipy)]$	1.52	980	166
$[WOCl_3(C_4H_8O_2)_2]$	1.44	998	166
$[WOBr_3(MeCN)_2]$	1.44	990	166
$[WOCl_3(diphos)]$	1.52	955	167
$[WOCl_3(dmpe)]$	1.22	960	167
$[WOCl_3(Pcy_3)(THF)]$	1.48	970	167
$[WOCl_3(DMPA)]$[a]	1.20	950	167
$(Et_4N)[WO(SPh)_4]$	1.34	949	189
$(Et_4N)[WO(SePh)_4]$	1.30	943	189

[a] DMPA = $Me_2PC_2H_4AsMe_2$.

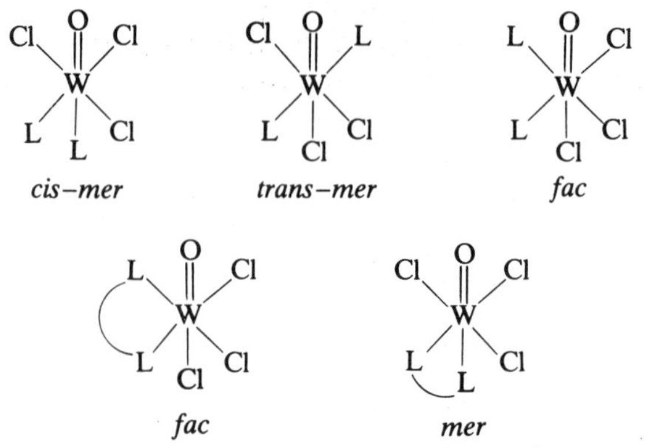

Figure 14 The possible isomers of $[WOCl_3L_2]$ and $[WOCl_3(L–L)]$

complexes [WO(OAP)L] (OAP = octaethylporphyrin; L = OPh, OMe) are obtained by reduction of H_2WO_4 in the presence of PhOH or by oxidation of $K_3W_2Cl_9$ in methanol. For L = OPh, $\mu_{eff} = 1.4$ BM and $\nu(W—O_t) = 946\ cm^{-1}$ and for L = OMe, $\mu_{eff} = 1.7$ BM and $\nu(W—O_t) = 910\ cm^{-1}$. The reaction of WF_5 with $OAPH_2$ leads to the oxo-bridged dimer $[WO(OAP)]_2O$. [WO(TPP)OMe] and $[WO(TPPS)H_2O]$ are also known. Base treatment of these leads to the oxo-bridged dimers.[175] The TPP complex undergoes electrochemical reduction to the W^{IV} compound (−0.85 V *vs.* SCE independent of solvent).[176]

37.4.3 Complexes with Terminal Sulfur and Selenium Atoms

The pentahalides WX_5 (X = Cl, Br) react with Sb_2S_3 and Sb_2Se_3[177] to give $WSCl_3$, $WSeCl_3$ and $WSeBr_3$. These complexes are insoluble polymeric materials and much less reactive than the corresponding WSX_4 and $WSeX_4$ complexes (Section 37.3.3). They show no IR bands which are expected for $\nu(W—S_t)$ and $\nu(W—Se_t)$. Whether their structures are similar to those of WOX_3 is not known. The only adducts that have been isolated and characterized are $[WSCl_3(mte)]$ and $[WSCl_3(bipy)]$. Both are paramagnetic and exhibit the expected IR absorption for $\nu(W—S_t)$: for $[WSCl_3(mte)]$, $\mu_{eff} = 1.52$ BM and $\nu(W—S_t) = 535\ cm^{-1}$; for $[WSCl_3(bipy)]$, $\mu_{eff} = 1.45$ BM and $\nu(W—S_t) = 531\ cm^{-1}$. On heating, both $WSCl_3$ and $WSeBr_3$ disproportionate to give the W^{IV} compounds $WSCl_2$ and $WSeBr_2$.[177] Attempts to prepare $WSBr_3$ were unsuccessful since the reaction of WBr_5 with Sb_2S_3 leads to the W^{IV} compound $WSBr_2$.[177]

Several complexes containing the W_2S_3 and W_2O_3 structural units have been prepared with thioxanthates and dithiocarbamates.[178,179,180] Whether these have structures similar to the analogous molybdenum complexes is yet to be determined.

37.4.4 Complexes with Metal–Metal Bonds

A dominant structural feature of complexes of Mo^V is the $Mo(\mu\text{-}X)_2Mo$ bridge which can be either planar or bent. It is believed that in these complexes there exists a Mo—Mo bond of order one which arises from the overlap of the d_{xy} atomic orbitals (Figure 15) (each with one electron) leading to a bonding orbital that accommodates the two electrons, in agreement with their diamagnetism. The same type of structures have only recently been shown to exist for W^V. Structural determination of the barium salt[181,182] of $[W_2O_4(edta)]^{2-}$ shows the presence of a bent $W(\mu\text{-}O)_2W$ bridge. The W—W separation is 2.542 Å, the dihedral angle between the

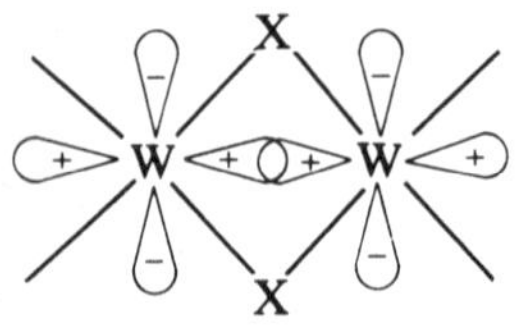

Figure 15 Overlap of W d_{xy} orbitals across the $[W(\mu\text{-}X)_2W]$ bridge; z axis is perpendicular to the plane of the paper

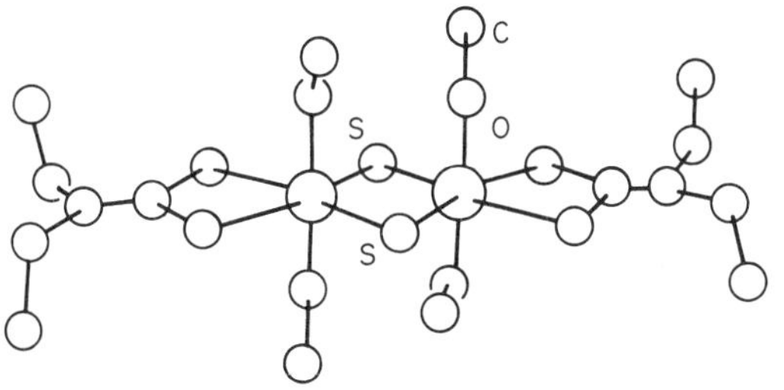

Figure 16 The structure of $[W_2(\mu\text{-}S)_2(S_2CNEt_4)_2(OMe)_4]$[186]

two WO_2 planes is 147°, the W atoms are displaced by 0.34 Å from the plane defined by the six oxygen atoms towards O_t, and the W—W—O_t angles are obtuse. These structural parameters are consistent with the presence of a W—W bond.

Electrochemical or tin reduction of WO_4^{2-} in the presence of oxalic acid and potassium oxalate leads to the isolation of $K_3[W_2O_4(C_2O_4)_{2.5}]\cdot 5H_2O$[183] which contains the $W_2O_4^{2+}$ moiety. This oxalate complex reacts with KF in HF to give $K_3H[W_2O_4F_6]$.[184] In this complex, the $W_2O_4^{2-}$ unit is almost planar with a W—W separation of 2.62 Å. It has been suggested[185] that electrolytic reduction of WO_4^{2-} leads to the formation of compound (**5**) which decomposes to compound (**6**).

W^V complexes containing the $W(\mu\text{-}S)_2W$ bridge are also known. Compound (**7**), prepared by

$[W_4O_8(C_2O_4)_5(H_2O)_4]^{6-}$	$[W_2O_4(C_2O_4)_2(H_2O)_2]^{2-}$	$[W_2S_2(S_2CNEt_2)_2(OMe)_4]$
(**5**)	(**6**)	(**7**)

reacting $[W(CO)_3(MeCN)_3]$ with tetraethyldithiuramdisulfide in methanol,[186] contains a planar $W(\mu\text{-}S)_2W$ bridge (Figure 16). The W—W distance of 2.791 Å, coupled with the other structural parameters, is indicative of a W—W bond. This conclusion is particularly clear when this structure is compared with that reported for $[W_2Cl_{10}]$[141] which has no W—W bond. In $[W_2Cl_{10}]$, the W—W bond length is 3.814 Å, the angles at Cl_b are obtuse and the angles Cl_b—W—Cl_b are acute. These features are expected for a non-bonded structure, since in the absence of an attractive force, there must be a net repulsion that causes the bridge to stretch along the M · · · M direction.

The synthesis of the mixed valence W^V, W^{VI} complex $[W_4S_{12}]^{2-}$ has been reported. Its structure analysis[187] reveals the presence of the $W_2S_4^{2+}$ structural unit with a planar $W(\mu\text{-}S)_2W$ bridge, and a W—W distance of 2.950 Å. The coordination sphere around each W atom is completed by a terminal sulfur atom (W—S_t = 2.10 Å; ν(W—S_t) = 495 cm^{-1}) and the bidentate ligand WS_4^{2-}. The W—W distance of 2.950 Å is somewhat longer than expected for a W—W bond of order one which is believed to result from the interaction of the W^V and W^{VI} centers. The structure of $[Cl_2W(O)(\mu\text{-}S)_2W(O)Cl_2]^{2-}$ has also been reported.[188] This complex contains a bent $W(\mu\text{-}S)_2W$ bridge (dihedral angle of 149°) with a W—W separation of 2.844 Å.

An interesting dinuclear complex of W^V has been obtained from the reaction of $[WSeCl_3]$ with Ph_4AsCl in a 1:1 mole ratio. The diamagnetic $[W_2Cl_8Se_3]^{2-}$ contains both an Se^{2-} and Se_2^{2-} bridging units with a W—W distance of 2.862 Å. The presence of a W—W bond in this compound is further supported by the acute angle of 73.3° subtended at the bridging selenium atom.[190b]

Complexes containing the bridge $W(\mu\text{-}SR)_2(\mu\text{-}X)W$ (X = Cl, OMe)[189,190] have also been prepared. The structure of compound (**8**) which is depicted in Figure 17[190] reveals the presence of a W—W bond of order one with a W—W separation of 2.854 Å. Complexes of this type have been shown to have interesting electrochemistry.[189]

$[W_2O_2(\mu\text{-}SBu^i)_2(\mu\text{-}Cl)Cl_4]^-$
(**8**)

Reaction of WCl_6 with ethanol gives the blue paramagnetic $[WCl_3(OEt)_2]$ compound (see Section 37.4.1). Treatment of this with ethanol affords the red diamagnetic dimer of composition $[W_2Cl_4(OEt)_6]$.[157] Crystal structure analysis shows[191] that the compound contains a planar $W(\mu\text{-}OEt)_2W$ bridge (Figure 18) with a W—W bond of order one (W—W separation of 2.715 Å). It is now known that the best method of preparing the ditungsten species $[W_2Cl_4(OR)_6]$ is by $AgNO_3$ oxidation of the W^{IV} complexes (**10**)[192] (see Section 37.5.4.1).

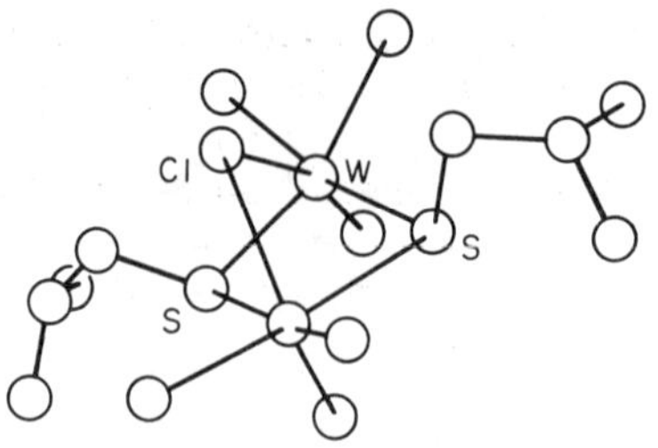

Figure 17 The structure of $[W_2O_2(\mu\text{-}SBu^i)_2(\mu\text{-}Cl)Cl_4]^-$ [190a]

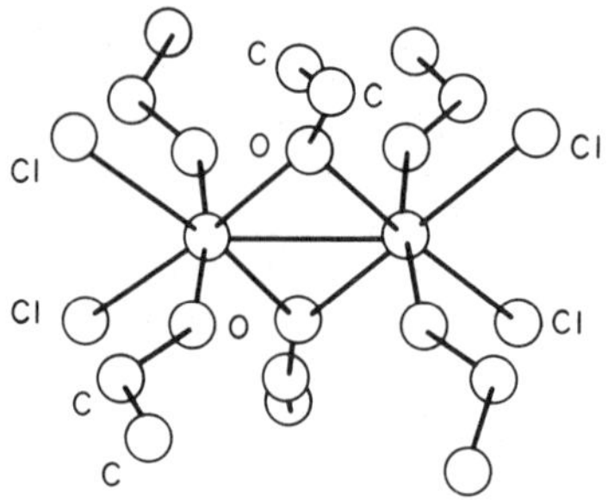

Figure 18 The structure of $[W_2Cl_4(OEt)_6]$[191]

37.4.5 Thiocyanato Complexes

Reduction of WO_4^{2-} with thiocyanic acid leads to colored species which have been used for tungsten analysis.[193] The color of these species are pH dependent and the structures of some of them are still uncertain. The following sequence of equilibria have been suggested on the basis of several isolated amine salts (equations 7–11).[88,194] Of these species, it is believed that the

$$[WO(OH)_2(NCS)_2]^- + NCS^- \rightleftharpoons [WO_2(NCS)_3]^{2-} + H_2O \quad (7)$$

$$[WO_2(NCS)_3]^{2-} + H^+ + NCS^- \rightleftharpoons [WO(OH)(NCS)_4]^{2-} \quad (8)$$

$$[WO(OH)NCS)_4]^{2-} + H^+ \rightleftharpoons [WO(NCS)_4]^{2-} + H_2O \quad (9)$$

$$2[WO(OH)(NCS)_4]^{2-} \rightleftharpoons [W_2O_3(NCS)_8]^{4-} + H_2O \quad (10)$$

$$[W_2O_3(NCS)_8]^{4-} + 2H^+ + 2NCS^- \rightleftharpoons 2[WO(NCS)_5]^{2-} \quad (11)$$

orange diamagnetic $[W_2O_3(NCS)_8]^{4-}$ and the green paramagnetic $[WO(NCS)_5]^{2-}$ have structures which are similar to the analogous molybdenum complexes. The complex $[W(NCS)_6]^-$ can be prepared[195] by direct reaction of WCl_5 and NCS^-.

37.5 THE CHEMISTRY OF W^{IV}

Monomeric complexes of W^{IV} such as the halides, the oxyhalides and their derivatives have been known for quite some time. In most cases these are six-coordinate, but higher coordination numbers are also known. During the last two decades, new structural types have been reported. Of these, the most interesting are the dinuclear complexes containing W═W double bonds and the trinuclear clusters containing W—W single bonds.

37.5.1 W^{IV} Halides and Their Derivatives

All four tetrahalides, WF_4, WCl_4, WBr_4 and WI_4, are known. WF_4 can be prepared by the reaction of WF_6 with benzene in a bomb at 110 °C.[196] Its structure is not known, but is is certainly polymeric with W—F—W bridges. WCl_4 and WBr_4 are best prepared as black needles by aluminum reduction of the higher halides in a controlled temperature gradient.[154] These are diamagnetic and isomorphous with their Nb and Ta analogues which consist of infinite linear chains of octahedra sharing a common edge. The alternating short and long metal distances imply the existence of direct metal–metal bonds,[197] in agreement with the diamagnetism of these compounds.

WI_4 is prepared by the reaction of WO_2 with AlI_3 or by the reduction of WCl_6 with HI. The tetraiodide is unstable and decomposes even at room temperature to WI_2 and I_2.

The hexahalide complexes WX_6^{2-} (X = Cl, Br) are easily prepared[144,198] by reduction of the higher halides with alkali iodide. These octahedral complexes have low magnetic moments and high θ values, suggesting antiferromagnetism. WCl_6^{2-} has been reported to react with vicinal diols[199] to give alkenes and species containing the WO_2^{2+} unit. The hexaisothiocyanate complex $[W(NCS)_6]^{2-}$ is also known.[195]

The six-coordinate adducts WX_4L_2 (X = Cl, Br) can be prepared by reduction of the higher halides, by direct addition to WX_4, or by oxidation of lower valent complexes. Certain nitrogen ligands have long been known to reduce WX_6 and WX_5 to complexes of the type WX_4L_2 (L = py, MeCN, EtCN, PrCN).[164] The acetonitrile complex, which is a useful starting material for the preparation of other adducts, can also be prepared by the reaction of WCl_5 with $W(CO)_6$ or by the oxidation of $W(CO)_6$ with Br_2 in acetonitrile[200] as depicted in equations (12) and (13).

$$4WCl_5 + W(CO)_6 \xrightarrow{MeCN} 5[WCl_4(MeCN)_2] + 6CO \quad (12)$$

$$W(CO)_6 + Br_2 \xrightarrow{MeCN} [WBr_4(MeCN)_2] + 6CO \quad (13)$$

Phosphine and arsine adducts of WCl_4 are conveniently prepared by reduction of WCl_6 with amalgamated zinc in the presence of the desired ligand.[201] Some of these complexes can also be prepared by Br_2 or Cl_2 oxidation of *cis*-$[W(CO)_4L_2]$ (L = tertiary phosphine).[202]

The stereochemistry of the WX_4L_2 complexes depends on the nature of L. Crystal structure analysis of $[WX_4(py)_2]$,[203] $[WCl_4(PMe_2Ph)_2]$[204] and $[WCl_4(PPh_3)_2]$[205] (prepared by reduction of $WYCl_4$ (Y = O, S, Se) with excess phosphine) have established *trans* stereochemistry. Far IR measurements suggest the same stereochemistry[200] for L = PrCN and C_4H_2S. However, with L = MeCN, EtCN and Et_2S, *cis*-$[WX_4L_2]$ is formed. The six-coordinate $[WX_4L_2]$ complexes, with the d^2 electronic configuration, are paramagnetic with magnetic moments in the range of 1.7–2.1 BM (Table 6).

The orange complex *trans*-$[WCl_4(PMe_2Ph)_2]$ reacts reversibly with PMe_2Ph to give the red paramagnetic (μ_{eff} = 2.68 BM) seven-coordinate complex $[WCl_4(PMe_2Ph)_3]$.[201,202] Other seven-coordinate W^{IV} complexes have been reported.[144,201,206] With an excess of PMe_3, $[WCl_4(PMe_3)_3]$ is obtained directly from WCl_6.[207] No structural information is available for the seven-coordinate W^{IV} complexes, but it appears that this coordination number is accessible for tungsten in this oxidation state, and that in this respect, the six-coordinate W^{IV} complexes need not be considered coordinatively saturated.

Several β-diketonate complexes of the type [W(L–L)X_2] (X = Cl, Br; L = acac, 6-methyl-2,4-heptanedione, dibenzoylmethane) have been prepared from reaction of the β-diketone ligands with WX_5. The suggested *cis* stereochemistry, based on IR measurements, must be considered tentative.

Table 6 Properties of Some W^{IV} Complexes

Compound	*Color*	μ_{eff} (BM)	*ν(W—O)* (cm^{-1})	*Ref.*
$[WCl_4(PMe_2Ph)_2]$	Orange	2.05	—	202
$[WCl_4(PMe_2Ph)_3]$	Red	2.68	—	202
$[WCl_4(AsMe_2Ph)_2]$	Orange	1.88	—	201
$[WO(NCO)_2(PMe_2Ph)_3]$	Blue	Diamagnetic	945	201
$[WO(NCS)_2(PMe_2Ph)_3]$	Blue	Diamagnetic	952	201
$[WCl_4(dth)\cdot MeCN]$	Brown	1.67	—	206
$[WOCl(diphos)_2]BPh_4$	Pink	Diamagnetic	955	167
$[WCl_4(Me_3N)_3]$	Green-brown	1.43	—	149
$[WOCl_2(PMe_2Ph)_3]$	Purple	Diamagnetic	960	201

37.5.2 The Oxyhalides and Their Derivatives

The oxyhalides WOF_2, $WOCl_2$ and $WOBr_2$ are known. WOF_2 is an inert black non-volatile powder. It is prepared by the reaction of WO_2 with HF at 500 °C. $WOCl_2$ was reported to result from the thermal decomposition of $WOCl_3$[162] as a black-purple solid. Golden-brown crystals of $WOCl_2$ have been obtained by $SnCl_2$ reduction of $WOCl_4$ followed by sublimation,[209] and by the reaction of a mixture of W, WO_3 and WCl_6 followed by chemical transport.[210] It is likely that $WOCl_2$ exhibits polymorphism, and that the different preparative routes simply lead to different polymorphs. Nonetheless, its structure, and that of $WOBr_2$, is polymeric, probably involving both oxygen and halogen bridges.

The oxohalo derivatives of W^{IV} can be prepared by reduction of the oxo complexes of W^{V} and W^{VI},[167,201] by reduction of WCl_6 in wet ethanol, or by the reaction of WCl_6^{2-} with the appropriate ligand in wet solvents.[167] Thus the complexes $[WOX_2L_3]$ (L = PMe_2Ph, PEt_2Ph and $PMePh_2$) are prepared by reacting $(Et_4N)[WCl_6]$, $WOCl_4$ or WCl_6 with the desired ligand in wet ethanol. These complexes exhibit the expected strong IR band around 950 cm^{-1} which is associated with $\nu(W—O_t)$. The $[WOX_2L_3]$ complexes can undergo halide substitution and complexes such as $[WOX_2(PMe_2Ph)_3]$ (X = NCO, NCS) result from the reaction of the halide derivatives with NaNCO and KNCS.[167,207]

It has been shown[207] that the seven-coordinate complex $[WCl_4(PMe_3)_3]$ can abstract oxygen to give $[WOCl_2(PMe_3)_3]$. By this method, $[WOCl_2L_3]$ (L = PMe_2Ph, $PMePh_2$, $P(OMe)_3$) have been prepared.

The $[WOX_2L_3]$ complexes are diamagnetic. Their proton and phosphorus NMR and far IR spectra suggest the *cis–mer* stereochemistry (see Section 37.4.2). This suggestion is strongly supported by the crystal structures of blue $[MoOCl_2(PMe_2Ph)_3]$ and green $[MoOCl_2(PEt_2Ph)_3]$ which show that the halide ions are indeed *cis* to one another.[211]

With certain bidentate ligands, it has been shown[167] that W^{V} complexes of the type $[WOCl_3(L–L)]$ (L–L = diphos, *cis*-$Ph_2PCH{=}CHPPh_2$, 1,2-bis(diphenylphosphino)benzene) can be reduced with excess ligand to $[WOCl(L–L)_2]^+$. These diamagnetic pink complexes, isolated as the tetraphenylborate salts, are 1:1 electrolytes in nitromethane and exhibit the expected $\nu(W—O_t)$ around 950 cm^{-1}. ^{31}P NMR data clearly favor the structure in which the two bidentate ligands occupy the equatorial plane and the Cl^- ligand is *trans* to O_t.

37.5.3 The $W_2OCl_{10}^{4-}$ Ion

The incomplete reduction by tin of K_2WO_4 in concentrated HCl leads to a very deep purple solution which was suggested to result from a compound of apparent composition $K_2[W(OH)Cl_5]$.[212] The high intensity of the electronic absorption band of $[W(OH)Cl_5]^{2-}$ at 19 100 cm^{-1} (ε = 10 000) is inconsistent with a mononuclear formulation (especially with ligands such as OH^- and Cl^-), and indeed, the suggestion has been made[213] that this ion should be formulated as $[W_2OCl_{10}]^{4-}$, an oxo-bridged W^{III}–W^{V} mixed valence dimer.

The dinuclear structure of $[W_2OCl_{10}]^{4-}$ was established by crystallography.[214] The W—O—W bridge is linear and the W—O separation is quite short (1.871 Å) indicating a bond order higher than one. However, the mixed valence formulation has been questioned. Careful spectral and magnetic measurements suggest that the two tungsten atoms have equivalent d^2 electronic configurations and that they are antiferromagnetically coupled with an exchange energy of ~75 K.[215] The intense absorption at 19 100 cm^{-1} is believed to be associated with the W—O—W chromophore.[215]

37.5.4 Complexes with Metal–Metal Bonds

As mentioned previously, one of the more interesting developments in the chemistry of W^{IV} has been the synthesis and the characterization of dinuclear complexes containing W—W double bonds and trinuclear clusters containing W—W single bonds.

In discussing complexes containing two interacting metal atoms, which are considered to have a metal–metal double bond, and those that contain three interacting metal atoms and have metal–metal single bonds, it should be pointed out that the assignment of bond orders is problematic since these complexes, almost invariably, also contain bridging ligands. For W^{V} (d^1 electronic configuration), the W—W bond order cannot exceed unity, but for W^{IV} with a d^2

electronic configuration, a metal–metal double bond may be at least postulated. However, arguments for deciding between a double bond on the one hand, and a single bond together with indirect pairing of the spins of the remaining two electrons through the bridging atoms on the other hand, are rarely, if ever, conclusive, because the only data on which to base them are structural. For an isolated case, these structural data can be used to show that there is a metal–metal bond, but they do not generally allow one to conclude that the bond order must be two rather than one. However, there are today a few reported cases in which a W—W double bond can be assigned almost unambiguously on the basis of comparison of structural parameters of two related complexes, which differ from one another only by two electrons, *i.e.* dinuclear complexes of W^{IV} and W^{V}.

37.5.4.1 Complexes with metal–metal double bonds

The synthesis and structural characterization of a W^{IV} complex containing a planar $W(\mu\text{-}S)_2W$ bridge[186] has provided a rare opportunity to develop an unambiguous argument for the existence of a double bond between two W atoms when bridging atoms are also present. Compound (**9**) (Figure 19) was prepared by allowing $[W(CO)_3(MeCN)_3]$ to react with tetraethyldithiuramdisulfide in acetone. The $W(\mu\text{-}S)_2W$ bridge is strictly planar (there is a crystallographic inversion center between the two metal atoms), with a W—W separation of 2.530 Å, indicating a direct metal–metal bond. The presence of such a bond is further supported by the very small $W—S_b—W$ angle of 65.5° and the obtuse $S_b—W—S_b$ angle of 114.5°. The chemical environment and geometry of the metal atoms in this complex are almost identical to those found for the W^V compound (**7**) (see Section 37.4.4) where the W—W separation was found to be 2.791 Å. The shortening of the W—W separation by 0.26 Å clearly indicates that a considerably stronger force exists between the metal atoms. This conclusion is further supported by the contraction of the $W—S_b—W$ angle and expansion of the $S_b—W—S_b$ angle (for the W^V compound, these angles are 73.2° and 106.8° respectively), whereas the $W—S_b$ distances in both are practically the same. Thus it seems clear that in this case we are dealing with a double bond between the two tungsten atoms.

Other examples which strongly support the existence of a metal–metal double bond are furnished by the dinuclear alkoxo-bridged dimers (**10**).[216] Several synthetic procedures have been reported for the preparation of these compounds. They include the oxidation of quadruply bonded tungsten compounds[217] in alcoholic HCl solutions, the oxidation of $[W_2Cl_9]^{3-}$ in alcohols,[218] the electrolytic reduction of alcoholic solutions of WCl_6[219] and the alcoholysis of WCl_4.[220] Other alkoxy derivatives can be prepared by a simple alcohol exchange reaction.[221]

$[W_2(\mu\text{-}S)_2(\mu\text{-}S_2CNEt_2)_2(S_2CNEt_2)_2]$ (**9**) $[W_2Cl_4(\mu\text{-}OR)_2(OR)_2(ROH)_2]$ (R = Me, Et) (**10**)

Crystal structures of (**10**)[216] consist of two edge-sharing distorted octahedra. The metal–metal vector bisects the edge formed by the bridging oxygen atoms and the complete $Cl_2W(\mu\text{-}O)_2WCl_2$ unit is planar (Figure 20). Above and below this plane, there is one OR on one W atom and HOR on the other, with a strong hydrogen bond between the two oxygen atoms (O···O = 2.48 Å), and therefore the W—O bonds are perpendicular to the plane

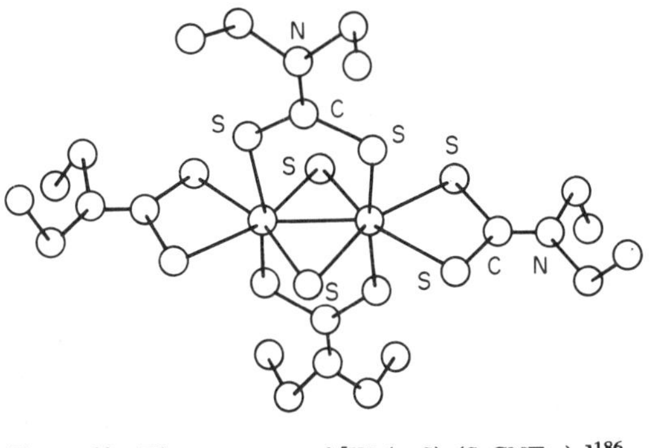

Figure 19 The structure of $[W_2(\mu\text{-}S)_2(S_2CNEt_2)_4]$[186]

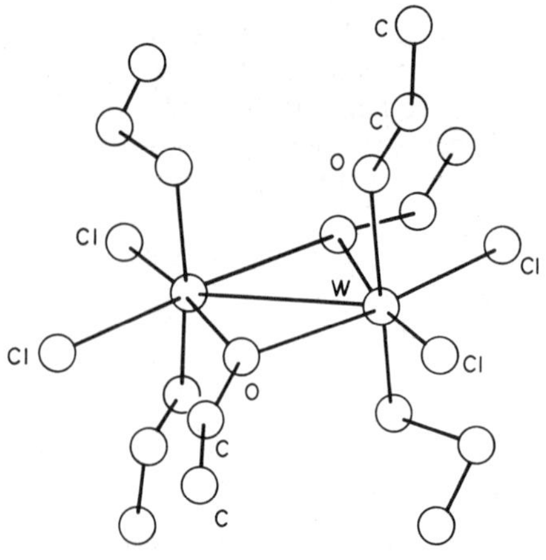

Figure 20 The structure of $[W_2Cl_4(OEt)_4(EtOH)_2]$[216]

instead of bending away from each other because of O · · · O repulsion as is observed in the W^V structure (O · · · O distance is 3.04 Å and the W—W—OR angle is 95.7°) (see Section 37.4.4).

The structures of the alkoxo-bridged dimers of W^{IV} and W^V provide an excellent opportunity to make a detailed analysis of M—M single and double bonding. The number of compositional and structural changes on going from the W^{IV} to the W^V compound are minimal. In composition, only two hydrogen atoms are removed, and structurally, other than the bending away of the two OR groups, all changes are those caused by the increase in the M—M distance from 2.482 Å in the W^{IV} complex to 2.715 Å in the W^V complex. These changes are associated with the W—O_b—W and O_b—W—O_b angles. Thus by going from the d^1–d^1 to d^2–d^2 electronic configuration the bond order increases from one to two. Moreover, qualitative considerations suggest that the double bond consists of the σ component responsible for the d^1–d^1 single bond and a π bond. Fenske–Hall calculations support this interpretation.[216]

As mentioned above, a variety of complexes of type (**10**) can easily be prepared by alcohol exchange and, in this way, compounds with R = Pr^n, Bu^n, Pen^n, Oct^n have been isolated and characterized.[221] However, when secondary alcohols have been used for exchange with (**10**), mixed alkoxy species were obtained. ^1H NMR showed that the bridging ethoxide ligands had not been replaced, a result that was confirmed by the crystal structure determination of the mixed ethoxide–isopropoxide and ethoxide–*sec*-pentoxide derivatives.

The reductive coupling of acetone by (**10**) to give a W^V compound in which the ditungsten center (W—W = 2.701 Å) is bridged by the resulting organic molecule (Figure 21) is another manifestation of the relative ease by which compounds (**10**) can be converted to dinuclear alkoxo-bridged W^V compounds.[222]

In addition to compounds of type (**10**) there is another class of W^{IV} alkoxides with a different structure. The oxidation of $[W_2(NMe_2)_6]$ by methanol and ethanol in hydrocarbon solvent leads to the tetrameric $[W_4(OR)_{16}]$ compounds.[223] These molecules do not possess the localized W—W double bonding as found in (**10**) but, rather, have a structure similar to that of

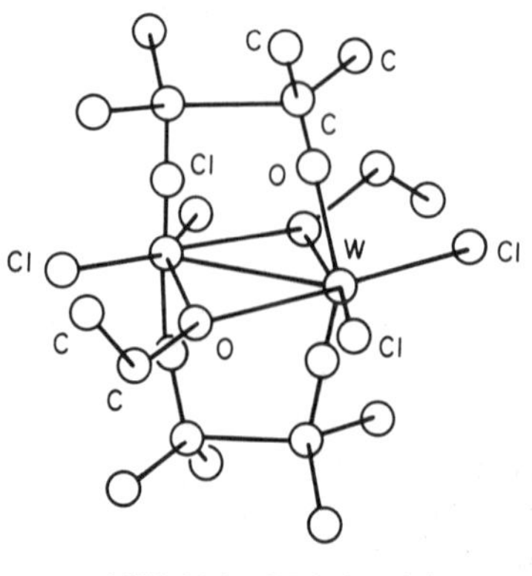

Figure 21 The structure of $[W_2Cl_4(\mu\text{-OEt})_2(\mu\text{-L})_2]$ (L = $Me_2COCOMe_2$)[222]

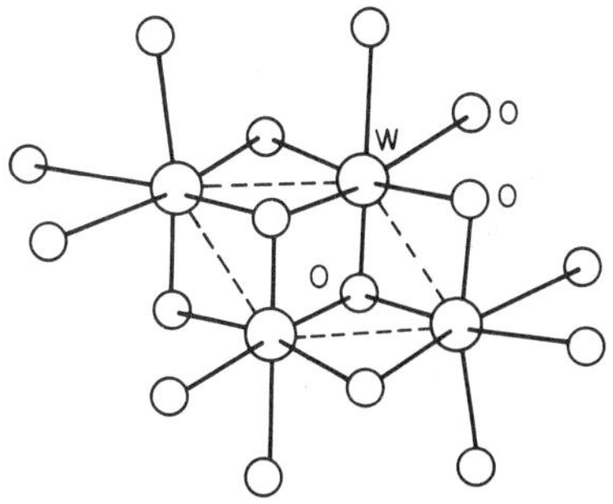

Figure 22 The structure of $[W_4(OR)_{16}]$; carbon atoms are omitted[223]

$[Ti_4(OEt)_{16}]$[224] with distortions due to M—M bonding. Crystal structure analysis of $[W_4(OEt)_{16}]$ (Figure 22) reveals the presence of two short W—W bonds (2.645 and 2.76 Å) and two long bonds of 2.93 Å. In this structure, there is a total of five possible W—W interactions, and thus, to form five bonds of order one, ten electrons are required. Only eight metal electrons are available for metal–metal bonding, which leads to the observed distortions. It has been suggested that this distortion results from a novel second order Jahn–Teller effect.[225]

The reaction of $[W_2(NMe_2)_6]$ with excess Pr^iOH leads to yet another type of tetranuclear W^{IV} alkoxy compound $[W_4(\mu\text{-H})_2(OPr^i)_{14}]$ as depicted in Figure 23.[226] The molecule is centrosymmetric with alternating short (2.446 Å) and long (3.407 Å) W—W distances, consistent with the presence of a W—W double bond and a non-bonding distance, respectively.

Finally, WO_2, which has a modified rutile structure,[227] exhibits alternate W—W distances of 2.475 and 3.096 Å across shared edges of WO_6 octahedra. Thus, the short distance of 2.475 Å which indicates a W—W double bond makes a recognizable dinuclear unit in the infinitely extended array of the metal oxide.

Dinuclear complexes with a W—W double bond containing three bridging groups are also known. The reaction of $[WCl_4L_2]$ (L = THF, Me_2S) with Me_3SiSR (R = Me, Et, Bu^i, CH_2Ph, Ph) has been shown to give compounds of type (**11**).[228,229,230,231] Crystal structure analysis of

$[LCl_2W(\mu\text{-S})(\mu\text{-SR})_2WCl_2L]$ (L = THF, Me_2S, Cl; R = Me, Et, Bu^i, CH_2Ph)
(11)

several of these have revealed a distorted confacial bi-octahedral structure with W—W separation of close to 2.5 Å. The acute bond angles subtended at the bridging atoms agree with a strong metal–metal interaction.

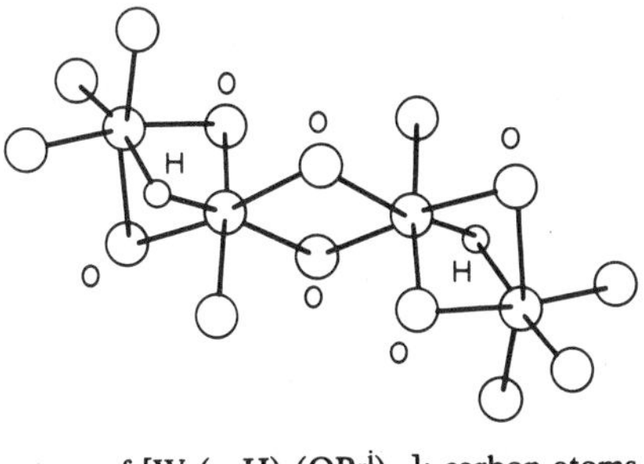

Figure 23 The structure of $[W_4(\mu\text{-H})_2(OPr^i)_{14}]$; carbon atoms are not shown[226]

37.5.4.2 *Trinuclear clusters*

During the past several years it has been shown that both molybdenum and tungsten in oxidation states of +4 or thereabouts, have marked predilection to form trinuclear cluster species. For the tungsten cluster species, there are three different structural types with respect to the ligand arrangements (Figure 24 a, b, c) which are based on an equilateral triangle of M—M bonded atoms. Structure (a) contains the $[W_3(\mu_3\text{-X})(\mu\text{-Y})_3]$ nucleus, where X = O, Cl,

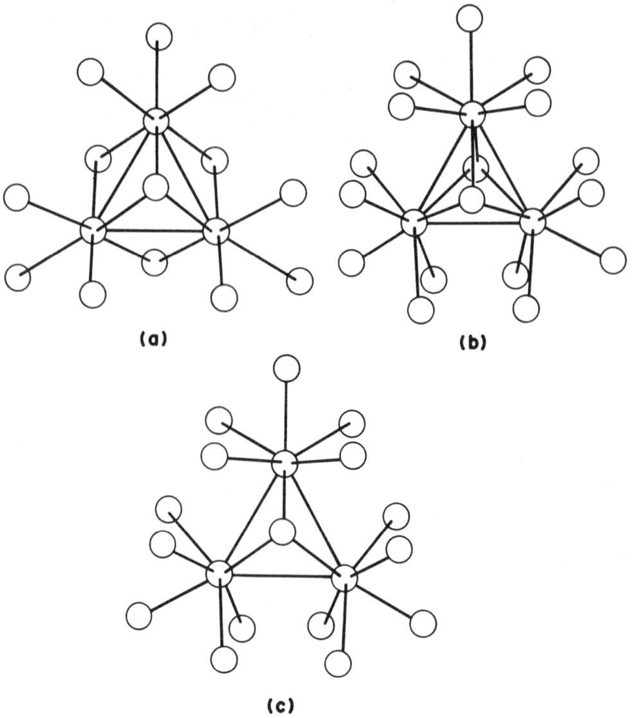

Figure 24 Structural types of trinuclear tungsten clusters

OR and Y = O. This structure is basically the same as the one observed several decades ago for the mixed metal oxide system $Zn_2Mo_3O_8$,[232] and is exemplified by the complexes (**12**), (**13**) and (**14**).

$[W_3O_4F_9]^{5-}$ (**12**) $[W_3O_3Cl_5(O_2CMe)(PBu^n_3)_3]$ (**13**) $[W_3OCH_2CMe_3O_3Cr_3(O_2CCMe_3)_{12}]$ (**14**)

Complex (**12**) has been prepared by treatment of $[W_2O_4(C_2O_4)_{2.5}]^{3-}$ with 40% HF at 70 °C. Crystal structure analysis has established the presence of the $W_3(\mu_3\text{-}X)(\mu\text{-}Y)_3$ nucleus (X = Y = O) with an average W—W distance of 2.515 Å. Without counting the M—M bonds, the coordination number of each metal atom is six and the geometry is distorted octahedral.[233]

Treatment of $[W_2Cl_4(PBu^n_3)_4]$ with acetic acid in diglyme at 160 °C leads to the isolation of (**13**) (Figure 25).[234] In this structure X = Cl, Y = O, and the average W—W distance is 2.609 Å. In both these complexes the metal is in the +4 oxidation state, and therefore each metal atom

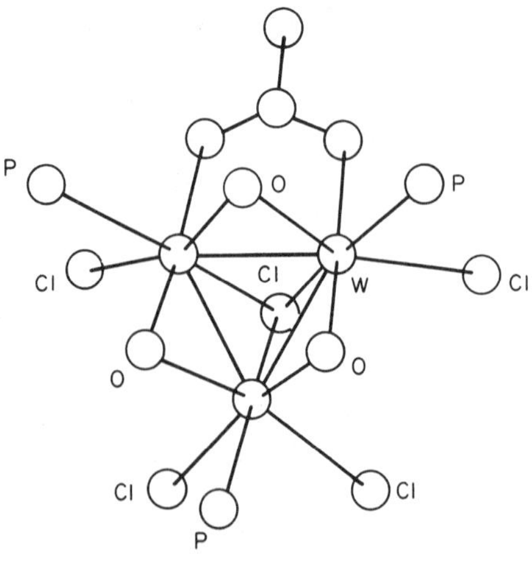

Figure 25 The structure of $[W_3O_3Cl_5(O_2CMe)(PBu^n_3)_3]$; the butyl groups are omitted[234]

has only two *d* electrons to use for M—M bonds. Thus, the two bonds formed by each metal atom cannot exceed a single bond.

Compound (**14**) has been prepared by refluxing 1:1:2 mole ratio of $W(CO)_6$, $Cr(CO)_6$ and pivalic acid in *o*-dichlorobenzene.[235] In this structure $X = OCH_2C(Me)_3$ and Y = O, and the average W—W separation is 2.610 Å. The average oxidation state of the tungsten atoms is $+3\frac{1}{3}$ or formally W^{III}, W^{III}, W^{IV}. In this eight electron system, the extra pair of electrons occupies a non-bonding or slightly antibonding molecular orbital. It is therefore reasonable to suggest that two electron oxidation to a six electron configuration may be possible without disrupting the cluster nucleus.

A comparison between the three W—W distances shows that with (μ_3-Cl), the W—W distance is approximately 0.1 Å longer than that with (μ_3-O), as expected when a larger capping atom is placed above the metal triangle. It is tempting to suggest that the longer bond length observed in the eight electron cluster is due to the additional pair of electrons. However, it is important to realize that the W—W bond length is also affected by the nature of the bridging atoms.

Trinuclear clusters of type (b) contain two capping groups, one above and one below the triangular plane. In addition, each pair of tungsten atoms is bridged by two carboxylates, and the coordination sphere is completed by a radial ligand which is bonded to each metal atom. With capping oxygen atoms, the $[W_3O_2(O_2CR)_6]$ unit can be the basis for a varied family of compounds. First, the R group of the carboxylate is variable. Among the four compounds isolated to date are those having R = Me, Et and Bu^t. Second, the radial coordination site may be occupied by neutral ligands, anionic ligands or a mixture of both, to give cationic clusters such as (**15**), neutral ones such as (**16**) and anionic ones such as (**17**).[236]

$[W_3O_2(O_2CR)_6(H_2O)_3]^{2+}$
(**15**)

$[W_3O_2(O_2CCMe_3)_6(O_2CCMe_3)_2(H_2O)]$
(**16**)

$[W_3O_2(O_2CMe)_6(O_2CMe)_3]^-$
(**17**)

These clusters are prepared by refluxing $W(CO)_6$ in the appropriate acid with or without its anhydride. Cationic clusters are best purified by ion exchange chromatography. The structure of (**15**) is depicted in Figure 26. Structure analysis has shown that the W_3O_2 nucleus is essentially invariant to the carboxylate ligands and that the W—W distance which averages 2.75 Å is only slightly affected by the radial ligands. In these clusters, the oxidation state of the metals is +4 (d^2 electronic configuration) and, therefore, the W—W bond order cannot exceed unity.

The stereochemistry of the tungsten atoms in these species is of particular interest. Each tungsten atom has a coordination number of nine, counting the neighboring metal atoms as well as the coordinated oxygen atoms. The geometry can be regarded as either distorted capped square antiprism or a distorted tricapped trigonal prism. The metal atoms are coordinatively saturated which accounts for the fact that the radical ligands are substitutionally inert.

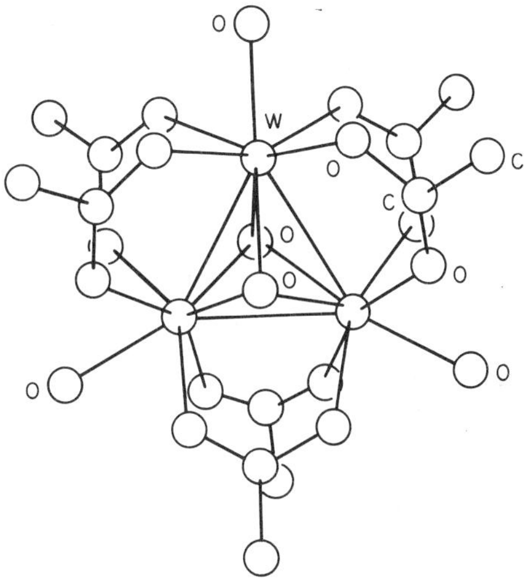

Figure 26 The structure of $[W_3O_2(O_2CMe)_6(H_2O)_3]^{2+}$ [236]

Recently, clusters (**18**) with one capping oxygen atom and one capping ethylidyne moiety have also been prepared.[237] The +1 cluster is prepared by refluxing $W(CO)_6$ with a mixture of acetic acid and its anhydride in MeCN. A one electron oxidation of the +1 ion leads to the +2 ion, whose crystal structure analysis reveals a W—W distance of 2.81 Å. This distance is significantly longer than the one observed for (**15**), in agreement with the fact that the +2 ion is a five electron system and therefore the W—W bond order is less than unity.

$[W_3O(CMe)(O_2CMe)_6(H_2O)_3]^{n+}$ ($n = 1, 2$)
(**18**)

$[W_3O(O_2CR)_6(H_2O)_3]^{2+}$ (R = Me, Et)
(**19**)

Clusters of type (c) (**19**) have been prepared by refluxing $W(CO)_6$ in a 1:1 mixture of the acid and its anhydride in the presence of small amounts of triethylamine.[238] These clusters might be considered a variant of (b) in which one capping group is absent. The structure of (**19**) (Figure 27) is virtually identical with the previously reported structures of tritungsten clusters with two capping oxygen atoms except for the absence of one capping oxygen. Of the quantitative differences between the hemicapped and the bicapped structures, the largest and most unambiguous by far is in the W—W distances. In the four bicapped structures,[236] the W—W distance averages 2.75(1) Å, while in the present coumpound, the average distance is 2.710 Å. This is clearly a significant difference in both the statistical and chemical senses.

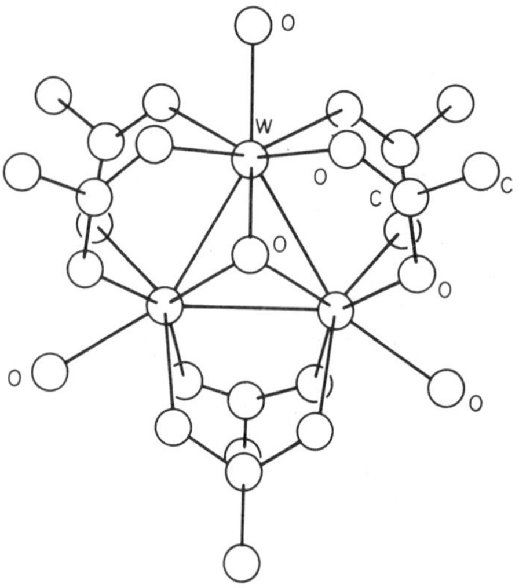

Figure 27 The structure of $[W_3O(O_2CMe)_6(H_2O)_2]^{2+}$ [238]

The hemicapped cluster is an eight electron system and a Fenske–Hall type molecular orbital calculation[238] shows that in addition to the six M—M bonding electrons comparable to those in the bicapped species, an additional electron pair occupies an orbital which is weakly M—M bonding, in agreement with the observed shortening of the W—W bond length.

Finally, although the difference in the W—W distances between structures (a) and (b) are approximately 0.2 Å, they both represent W—W single bonds. On the basis of molecular orbital calculations it has been suggested that this is due in part to the difference in the distribution of charge density in these molecules.[239] However, there is no doubt that changing the bridging ligands will have a considerable effect on the M—M distance, especially when dealing with metal–metal bonds of low order.

37.5.5 Eight-coordinate Complexes

Eight-coordinate dodecahedral complexes of the general type $W(L\text{–}L)_4$ are known for several bidentate ligands such as dithiocarbamate, 8-quinolinol, substituted quinolinols, picoline and substituted picolines.

The reaction of $[W(CO)_3(MeCN)_3]$ with tetraethyldithiuram disulfide in refluxing chloroform gives $[W(S_2CNEt_2)_4]$ which can also be prepared from $[WCl_4(MeCN)_2]$ and the dithiocarbamate ligand.[240] It was also reported that the addition of CS_2 to $[W(NMe_2)_6]$ leads to

$[W(S_2CNMe_2)_4]$.[241] $[W(S_2CNEt_2)_4]$ can be oxidized with Br_2 to the eight-coordinate W^V complex $[W(S_2CNEt_2)_4]^+$. Crystal structure analysis of this compound reveals the expected dodecahedral geometry[242] with W—S(A) distance of 2.529 Å and W—S(B) distance of 2.494 Å (A and B refer to the Hoard–Silverton notation).[243] Recently, $[W(S_2CSR)_4]$ (R = Et, Pr^i, Bu^t) have also been prepared and I_2 oxidation leads to the corresponding W^V complexes $[W(S_2CSR)_4]I_5$.[178]

With 8-quinolinol or its derivatives, $W(L–L)_4$ complexes have been prepared by a sealed-tube melt reaction with either $K_3[W_2Cl_9]$ or $W(CO)_6$.[244] Crystal structure analysis of $W(L–L)_4$ (L–L = 5-bromoquinolinol) has established the dodecahedral geometry for this and related compounds.[245] The four quinolinol ligands span the Hoard–Silverton *m* edges, with the oxygen atoms occupying A positions and the nitrogen atoms the B positions.

Another series of eight-coordinate complexes has been prepared with picolinates. All eight-coordinate complexes of the type $W(L–L)_4$ to date are diamagnetic, with the two electrons of the d^2 configuration paired in the low-lying $d_{x^2-y^2}$ level of the D_{2d} dodecahedron.[245] The IR electronic spectra consist of intense low energy bands ($\varepsilon > 10^4$), attributed to $d \rightarrow \pi^*$ charge transfer transitions.

$[W(picolinate)_4]$ and several quinolinate derivatives can be oxidized to the corresponding W^V complexes.[246] These are paramagnetic with moments of about 1.7 BM, close to the spin-only value. The two low-intensity near IR transitions are reasonably assigned to *d–d* transitions from the $d_{x^2-y^2}$ level. Liquid nitrogen ESR measurements suggest lower symmetry for these W^V complexes.[246] They are reasonably stable in the solid, but disproportionate rapidly in solution to the W^{IV} compounds and tungstate.[247]

37.5.6 Complexes with Multiple Bonds to Sulfur, Nitrogen and Carbon

The tendency of W^{IV} to form multiple bonds to good π donor atoms is much less pronounced than for W^V and W^{VI}; however, several such complexes are known.

The reaction of aqueous WS_4^{2-} with dilute HCl gives the $[W_3OS_8]^{2-}$ ion.[248] Its structure consists of the WO unit bonded to two WS_4^{2-} ligands with a short W—O_t distance of 1.68(2) Å. $[W_3S_9]^{2-}$ has also been prepared.[248,249] Its structure is very similar to that of $[W_3OS_8]^{2-}$ with a W—S_t bond length of 2.07(1) Å. In this complex, the geometry around the central W^{IV} ion is that of a distorted square pyramid.

The multiple bond between W^{IV} and a carbon atom is exemplified by $[W(CR)L_4Cl]$ (R = H, CMe_3; L = PMe_3)[250] and by $[W(CPh)(CO)_2(py)_2Br]$.[251] Crystal structure analysis of this last compound revealed the expected short W—C bond length of 1.84 Å.

Complexes containing multiple bonded nitrogen ligands are usually obtained from dinitrogen complexes. The complexes $[W(NH)X(diphos)]^+$ (X = F, Cl, Br) are known[252] and in basic methanol yield ammonia *via trans*-$[W(NH)(OMe)(diphos)_2]$.[253] The nitrido group is readily attacked by electrophiles to give imido, dialkylimido or thionitrosyl complexes. Thus, $[W(N)(N_3)(diphos)_2]$ reacts with EtI to give $[W(NEt)(N_3)(diphos)_2]I$.

Na/Hg reduction of $[W(NR)Cl_4]$ in the presence of the desired ligands leads to the complexes $[W(NR)Cl_2L_3]$ (R = Ph, Et; L = PMe_3, PMe_2Ph, $CNBu^t$).[171] Crystal structure analysis of $[W(NPh)Cl_2(PMe_3)_3]$ has revealed a short W—N distance of 1.755 Å in agreement with ν(W—NR) around 1080 cm^{-1}. The phosphine ligands occupy *mer* positions *cis* to the W(NR) unit. This stereochemistry is similar to that suggested for the analogous complex $[WOCl_2(PMe_2Ph)_3]$.[201]

37.5.7 Cyanide Complexes

The octacyanotungstate ion $[W(CN)_8]^{4-}$ has been known since 1914.[254] It is easily obtained by reacting a variety of starting materials containing W^{III}, W^{IV} and W^V with aqueous cyanide.[255] The trivalent metal is usually oxidized by either air or solvent, while the pentavalent starting materials disproportionate to $[W(CN)_8]^{4-}$ and WO_4^{2-}. The $[W(CN)_8]^{4-}$ ion can be easily oxidized to $[W(CN)_8]^{3-}$, but it does not appear possible to obtain it directly from W^V compounds. $[W(CN)_8]^{4-}$ is diamagnetic while $[W(CN)_8]^{3-}$ is paramagnetic with μ_{eff} = 1.76 BM.[255]

Both $[W(CN)_8]^{4-}$ and $[W(CN)_8]^{3-}$ are stable in solution in the absence of light and no exchange of $^{14}CN^-$ is detected. This exchange is strongly catalyzed by light and the rate is dependent on the light intensity. Exposing the yellow solution of $[W(CN)_8]^{4-}$ to light leads to a rapid change to red-brown, and then to purple. On the basis of crystallographic studies[256] the red-brown color is due to *trans*-$[WO_2(CN)_4]^{4-}$ and the purple color to *trans*-$[W(O)(OH)(CN)_4]^{3-}$. Recently, the $[W(O)(OH_2)(CN)_4]^{2-}$ ion has been isolated and characterized as its cadmium salt,[257] and protonation of $[WO_2(CN)_4]^{4-}$ was reported to lead[258] to $[W_2O_3(CN)_8]^{5-}$.

The photolysis of $[W(CN)_8]^{3-}$ has been investigated[259] in the pH range 1–13. The photolysis consists of a two-step sequence involving a primary intermolecular redox process and a consecutive thermal process both leading to $[W(CN)_8]^{4-}$. The rate of the oxidation of $[W(CN)_8]^{4-}$ with several oxidizing agents has been studied. With IO_4^- for example,[260] the reaction is first order with respect to both reagents and is catalyzed by H^+. Inorganic cations have an accelerating effect on the oxidation in the order, $Mg \gg Cs > Rb > K > Na > Li$.

The reaction of $[W(CN)_8]^{4-}$ with H_2 at 400 °C was reported to give a green-black product of composition $K_4[W(CN)_6]$.[255] Recently the same reaction has been shown to give $[W(CN)_7H]^{4-}$, and in the presence of base the following equilibrium is established,[261] as depicted in equation (14). $[W(CN)_7]^{5-}$ can also be prepared directly from $[W_2Cl_9]^{3-}$ with an excess of cyanide. The sharpness of the ^{13}C NMR lines suggests that on the NMR time scale there is no exchange of coordinated cyanide and that both structures are fluxional in solution.

$$[W(CN)_7H]^{4-} + OH^- \rightleftharpoons [W(CN)_7]^{5-} + H_2O \qquad (14)$$

Salts of $[W(CN)_8]^{4-}$ are dodecahedral both in solution and in the solid state with the exception of the cadmium salt which is believed to have the square antiprismatic structure in both phases.[255] ESR and vibrational spectra suggest that the W^V ion $[W(CN)_8]^{3-}$ has D_{4d} symmetry in solution.[262] However, evidence for both the dodecahedral and square antiprismatic ground state for this ion has also been reported.[263]

Mixed cyanoisonitrile complexes have been reported to result from the reaction of $Ag_4[W(CN)_8]$ with RNC (R = Me, Et, Pr^n, Pr^i, Bu^t and CPh_3).[264] The resulting $[W(CN)_4(CNR)_4]$ complexes are dodecahedral, with the more strongly π accepting isonitrile ligands occupying the B position. These complexes can also be obtained by the reaction of the silver salt with alkyl iodide.[255]

37.6 THE CHEMISTRY OF W^{III}

With a few exceptions, the chemistry of W^{III} is almost entirely that of dinuclear species containing a metal–metal bond of order three. There are only a few well-characterized examples of mononuclear complexes. This undoubtedly results from the tendency of W^{III} to form metal–metal bonds and the susceptibility of mononuclear complexes to oxidation. Of the halides of W^{III}, only WBr_3 and WI_3 are known.[269] The thermally unstable WBr_3 is prepared by the reaction of WBr_2 with excess Br_2 in a sealed tube. It is inert to H_2O and concentrated HCl. WI_3 is prepared from $W(CO)_6$ and I_2 at 120 °C.

37.6.1 Mononuclear Complexes of W^{III}

Complexes of the type WX_3py_3 (X = Cl, Br) have been prepared[265] by treatment of $[W(CO)_4X_2]$ with py at 140 °C. These are paramagnetic with $\mu_{eff} = 3.39$ and 3.40 BM for the chloro and bromo derivatives respectively. $[WX_4L_2]^-$ (X = Cl, Br; L = Py, 4-Mepy) have also been prepared by pyridine reduction at 350 K of the W^{IV} compounds $[WX_4L_2]$. Crystal structure analysis of $[WCl_4py_2]^-$ and $[WBr_4(Mepy)_2]^-$ shows that both have the *trans* stereochemistry.[266,267] Recently, it has been shown that $[LW(CO)_3]^-$ (L = polypyrazolylborate) react with thionyl chloride to give $[LWCl_3]$ which can be reduced[268] to the W^{III} complex $[LWCl_3]^-$.

37.6.2 Dinuclear Complexes of W^{III}

The dinuclear complexes of W^{III} can be divided into two groups. In one, the metal–metal bond is supported by bridging ligands as exemplified by the $[W_3Cl_9]^{3-}$ ion, while in the second, the dimeric species are of the W_2X_6 type, recognized by the $\sigma^2\pi^4$ electronic configuration, where the two metal atoms are held together by metal–metal bonds only. Structurally, what set these two groups of complexes apart are the M—M bond length, the coordination number and the geometry around the metal atom, and chemically, there is very little that the two groups have in common.

37.6.2.1 The $[W_2X_9]^{3-}$ ion and related compounds

The $[W_2X_9]^{3-}$ (X = Cl, Br) ions were first prepared years ago by electrochemical or chemical reduction of tungstates in concentrated HCl or HBr solutions.[269] More recently, it has been shown that $[W_2X_9]^{3-}$ is readily prepared by passage of HX gas through a solution containing CsX and $[W_2(mhp)_4]$ (mhp = anion of 2-hydroxy-6-methylpridine). It is interesting that when CsX is excluded, compound (**10**) is formed.[270] The bromide derivative can also be prepared by halogen exchange (equation 15). Reaction (15) reaches equilibrium after approximately three hours and $[W_2ClBr_8]^{3-}$ is the major tungsten-containing species.[276] A continuous flow of HBr throughout the course of the reaction leads to complete exchange within 24 hours. Of the several possible mechanisms for this halogen exchange, the most likely seems to involve an intermediate with a $W(\mu\text{-Cl})_2W$ bridge, and four terminal chlorides bonded to each metal atom. This is a viable intermediate in view of the structure of $[W_2Cl_6py_4]$ (Figure 28).

$$[W_2Cl_9]^{3-} + 9Br^- \rightleftharpoons [W_2Br_9]^{3-} + 9Cl^- \quad (15)$$

$[W_2Cl_9]^{3-}$ has the confacial bioctahedral structure with a W—W bond length of 2.418 Å. (This structure was first solved in 1958[272] and then re-refined in 1983).[273] This experimental result coupled with the electronic configuration of the metals is consistent with a W—W bond order of three.

$[W_2Cl_9]^{3-}$ can be oxidized with oxidants such as Cl_2, Br_2 and I_2.[274] The violet mixed valence complex $[W_2Cl_9]^{2-}$ is paramagnetic with $\mu_{eff} = 1.87$ BM per formula unit. The dianion can also be prepared in low yield by the reaction of $[W(CO)_5Cl]^-$ with WCl_6 in methylene chloride,[275] or in 40–50% yield by the reduction of WCl_4 with Na/Hg in THF[273] in the presence of $[(Ph_3PNPPh_3)Cl]$. Structurally $[W_2Cl_9]^{2-}$ is very similar to $[W_2Cl_9]^{3-}$ with a W—W bond length of 2.540 Å in agreement with a bond order of 2.5. The increase of 0.122 Å in the W—W distance in going from $[W_2Cl_9]^{3-}$ to $[W_2Cl_9]^{2-}$ is believed to result not only from the decrease in the bond order, but also because of the increase in the effective charge of the metal atoms.[273] The electronic absorption spectra of $[W_2X_9]^{n-}$ (X = Cl, Br; $n = 2, 3$) have been interpreted recently.[276] $[W_2Br_9]^{2-}$ is also known, and is obtained in reasonable yield according to equation (16). Its structure is similar to that of $[W_2Cl_9]^{2-}$ with a W—W bond length of 2.601 Å and a magnetic moment of 1.72 BM per formula unit.[277]

$$4[W(CO)_5Br]^- + 7C_2H_4Br_2 \longrightarrow 2[W_2Br_9]^{2-} + 7C_2H_4 + 2CO_2 \quad (16)$$

The reaction of $[Mo_2(O_2CMe)_4]$ with a warm concentrated solution of HCl or HBr affords the $[Mo_2X_8H]^{3-}$ anions. These anions contain the $Mo(\mu\text{-H})(\mu\text{-Cl})_2Mo$ bridging unit,[278] and possess structures similar to that of $[W_2Cl_9]^{3-}$. This type of oxidative addition is also possible with the quadruply bonded dimer $[MoW(O_2CCMe_3)_4]$. This molecule when treated in the same way[279] affords $[MoWCl_8H]^{3-}$ with a Mo—W distance of 2.495 Å.

$[W_2Cl_9]^{3-}$ reacts with heterocyclic organic bases[280] to give the diamagnetic compounds

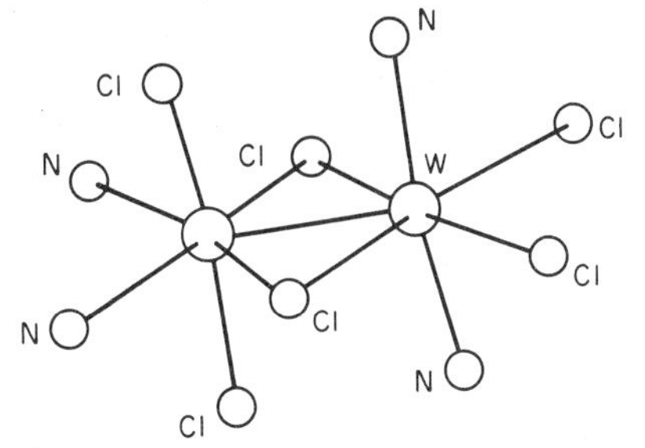

Figure 28 The structure of $[W_2Cl_6py_4]$; pyridine rings are not shown[281]

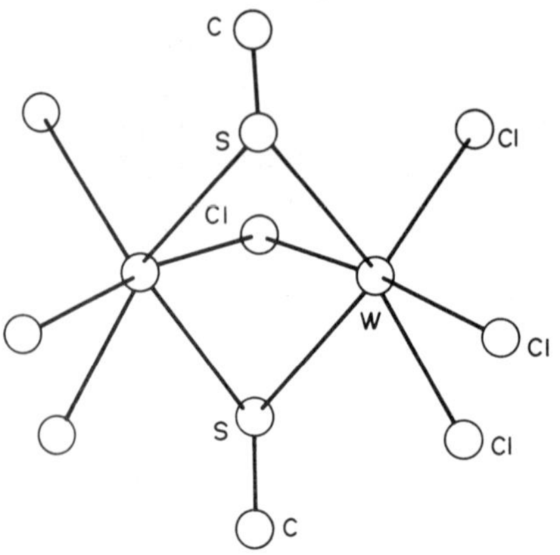

Figure 29 The structure of $[Cl_3(\mu\text{-}Cl)(\mu\text{-}SPh)_2WCl_3]^{2-}$; phenyl carbon atoms are omitted[231]

$[W_2Cl_6L_4]$ (L = py, pic and 4-isopropylpyridine). Structure analysis of $[W_2Cl_6py_4]$ has established[281] the presence of a planar $W(\mu\text{-}Cl)_2W$ bridge with a W—W distance of 2.737 Å (Figure 28). This distance is considerably larger than that expected for a W^{III}–W^{III} (d^3–d^3) dimer of this type with a triple bond. In fact, this distance is consistent with a W—W single bond. It can be argued that the situation here is similar to the one described for the d^4–d^4 $[Re_2Cl_{10}]^{4-}$ ion.[282] The complexes $[W_2Cl_6(PMe_3)_4]$ and $[W_2Cl_6(THF)_4]$ which are prepared by Na/Hg reduction of WCl_4 in THF with or without PMe_3,[283] may possess structures which are related to that of $[W_2Cl_6py_4]$.

Isoelectronic with $[W_2Cl_9]^{3-}$ are the paramagnetic thiolato-bridged dimers (**20**)[284] and (**21**).[231] They possess the confacial bioctahedral structure (Figure 29) with W—W distances of 2.505 and 2.519 Å respectively. These distances are not significantly different than those found for compounds (**11**) (see Section 37.5.4.1) which contain a W—W bond order of two. This result lent strong support to the suggestion that the difference in bond length of 0.122 Å observed in going from $[W_2Cl_9]^{3-}$ to $[W_2Cl_9]^{2-}$ is not caused only by the decrease in the bond order.[273]

The reaction of $[WCl_4(MeS)_2]$ with Et_2SiH gives compound (**22**) from which the anion (**23**) can be prepared.[285] The structure of (**23**) possesses the $W(\mu\text{-}H)(\mu\text{-}SMe_2)_2W$ bridge with a W—W distance of 2.410 Å, consistent with a W—W bond of order three.

$[(Me_2S)Cl_2W(\mu\text{-}SEt)_3WCl_2(SMe_2)]$
(**20**)

$[Cl_3W(\mu\text{-}Cl)(\mu\text{-}SPh)_2WCl_3]^{2-}$
(**21**)

$[Cl_3W(\mu\text{-}H)(\mu\text{-}SMe_2)_2WCl_2(SMe_2)]$
(**22**)

$[Cl_3W(\mu\text{-}H)(\mu\text{-}SMe_2)_2WCl_3]^-$
(**23**)

37.6.2.2 Complexes of the type W_2X_6

The metal–metal triple bond is formed by overlap of the metals' d_{z^2} orbitals to give a σ bond, and the d_{xz} and d_{yz} orbitals to give the π bonds. In a d^3–d^3 situation, the six metal electrons are accommodated in the three bonding molecular orbitals resulting in the $\sigma^2\pi^4$ electronic configuration. In the absence of the δ component, the triple bond allows free rotation about the M—M axis, and thus the adopted staggered configuration is determined by the steric requirement of the ligands.

The first W^{III} complex of the W_2X_6 type[286] was prepared by the reaction of WCl_6 with $LiCH_2SiMe_3$. The orange-brown organometallic compound $[W_2(CH_2SiMe_3)_6]$ can be sublimed in vacuum at 120 °C. It is stable in air for short periods, but oxidizes very rapidly in solution. The proton NMR spectra indicate that the six Me_3SiCH_2 groups are equivalent. $[W_2(CH_2SiMe_3)_6]$ crystallizes with two independent but virtually identical molecules in the asymmetric unit[287] with a mean W—W bond length of 2.255 Å. The stereochemistry in the solid is staggered as expected (Figure 30).

The compound $[W_2(NMe_2)_6]$ was first obtained as a minor product from the reaction of WCl_6 with $LiNMe_2$.[88] Attempts to prepare this material by reacting $LiNMe_2$ with lower valent starting materials such as $K_3[W_2Cl_9]$ or W_6Cl_{12} were not successful.[289] It is now known that the best preparative route to the compounds $[W_2X_6]$ (X = NMe_2, NMeEt, NEt_2) is by reacting X with a form of WCl_4 which is prepared by the reaction of $W(CO)_6$ with two equivalents of

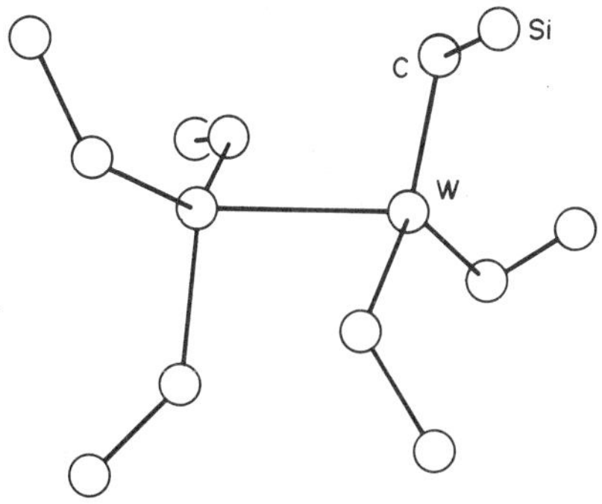

Figure 30 The structure of $[W_2(CH_2SiMe_3)_6]$; methyl groups are omitted[287]

WCl_6 in refluxing chlorobenzene.[284] As can be seen from Figure 31, $[W_2(NMe_2)_6]$ has the expected staggered, ethane-like stereochemistry with a W—W bond length of 2.294 Å. In addition, the methyl groups are not equivalent. One set of methyl groups lie approximately over the W—W bond (for example C-1; these are called the proximal methyl groups), while the other set, called the distal methyl groups (for example C-2), lie away from the W—W bond.

Compounds of the type $[W_2(OR)_6]$ are also known,[290] but they are much less stable than their molybdenum analogues. For example, $[Mo_2(OBu^t)_6]$ can easily be purified by vacuum sublimation at 100 °C, while the corresponding tungsten compound is thermally unstable and is very sensitive to both oxygen and moisture. In addition to $[W_2(OBu^t)_6]$, $[W_2(OSiMe_3)_6(NHMe_2)_2]$ and $[W_2(OPr^i)_6py_2]$[290] have been prepared by the reaction of $[W_2(NMe_2)_6]$ with Me_3SiOH in hydrocarbon or with Pr^iOH in pyridine. These compounds are also thermally unstable. The reaction of $[W_2(NMe_2)_6]$ with MeOH or EtOH gives the tetranuclear cluster compound $[W_4(OR)_{16}]$[223] while the reaction with Pr^iOH in the absence of pyridine gives $[W_4(\mu\text{-}H)_2(OPr^i)_{14}]$[226] (see Section 37.5.4.1).

Compounds of the type $[W_2Cl_2(NR_2)_4]$ (R = Me, Et) are known, and are best prepared by the reaction depicted in equation (17).[291] The lability of the W—Cl bond has led to the preparation of $[WX_2(NEt_2)_4]$ (X = Br, I),[292] and a series of dialkyl derivatives $[W_2R_2(NR'_2)_4]$ (R = Me, Et, Bu^n, Pr^i, CH_2CMe_3, CMe_3, CH_2SiMe_3; R′ = Me, Et).[293] The dialkyl derivatives are red-orange, thermally stable, and in most cases can be sublimed in vacuum around 100 °C. The structures of these molecules are very similar; they have the *anti* rotational conformation with bond length close to 2.3 Å (Table 7).

$$[W_2(NR_2)_6] + 2Me_3SiCl \longrightarrow [W_2Cl_2(NR_2)_4] + 2Me_3SiNR_2 \qquad (17)$$

The reaction of *anti*-$[W_2Cl_2(NEt_2)_4]$ with two equivalents of $LiCH_2SiMe_3$ is stereospecific[294] giving exclusively *anti*-$[W_2(CH_2SiMe_3)_2(NEt_2)_4]$ which then slowly isomerizes to an equilibrium mixture of *anti* and *gauche* rotamers. The stereospecificity clearly requires that the W≡W bond remains intact during the reaction, and that the Me_3SiCH_2 ligand replaces Cl with retention of configuration at the metal atom. NMR studies of this reaction suggest that there is a *trans* labilizing effect which is transmitted through the triple bond. $[W_2Me_2(NEt_2)_4]$[295] was also

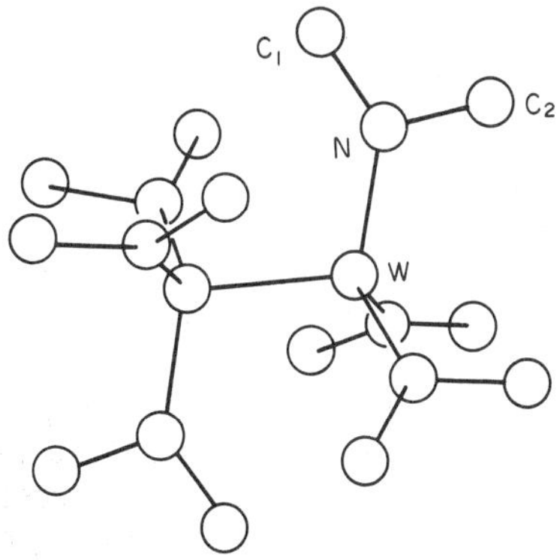

Figure 31 The structure of $[W_2(NMe_2)_6]$[12]

Table 7 Triple Bond Distances Between W Atoms

Compound	*W—W* (Å)	*Ref.*	*Compound*	*W—W* (Å)	*Ref.*
$[W_2(CH_2SiMe_3)_6]$	2.255(2)	289	$[W_2I_2(NEt_2)_4]$	2.300(4)	292
$[W_2(NMe_2)_6]$	2.294(2)	12	$[W_2Me_2(O_2CNEt_2)_4]$	2.272(1)	297
$[W_2Cl_2(NMe_2)_4]$	2.285(2)	291	$[W_2(O_2CNMe_2)_6]$	2.279(1)	297
$[W_2Cl_2(NEt_2)_4]$	2.301(1)	289	$[W_2Me_2(NEt_2)_4]$	2.291(1)	295
$[W_2Br_2(NEt_2)_4]$	2.301(2)	292	$[W_2(OPr)_6py_2]$	2.332(1)	290
$[W_2(OEt)_6(Me(H)NC_2H_4N(H)Me)]$	2.296(2)	299			

shown to exist in solution in an approximately 3:2 equilibrium mixture of *anti* and *gauche* rotamers. The *gauche* rotamer predominates as the bulk of the R group increases.[293] It is interesting that alcoholysis of the $[W_2R_2(NR_2)_4]$ compounds with bulky alcohols leads to $[W_2R_2(OR)_4]$ with retention of the W—R bonds.[296]

Insertion of CO_2 into M—NR_2 and M—OR bonds occurs quite readily in certain cases, leading to dialkylcarbamato and alkylcarbonato groups respectively. Thus, $[W_2(NMe_2)_6]$ reacts completely with excess CO_2 to give $[W_2(O_2CNMe_2)_6]$, while the reaction of $[W_2Me_2(NEt_2)_4]$ with CO_2 leads to $[W_2Me_2(O_2CNEt_2)_4]$. The structures of these molecules (Figures 32 and 33)[297] are quite similar in that both have two bridging and two chelating carbamato ligands similarly disposed around the metal atoms, and in the one case, the methyl groups are *trans* to the chelating ligand while in the other, an oxygen of a unidentate carbamato group is *trans* to the chelating ligand. In both, the W—W triple bond is retained (Table 7). These compounds are fluxional in solution and below −60 °C, the NMR spectrum of $[W_2(O_2CNEt_2)_6]$ is consistent with the solid state structure. The NMR spectra of $[W_2Me_2(O_2CNEt_2)_4]$ has not been completely interpreted.[289]

Until recently, the only two examples of complexes of the type $X_4W{\equiv}WX_4$ were $[W_2(OPr^i)_6py_2]$[290] and $[W_2(NMe_2)_4(PhNNNPh)_2$.[298] In these complexes, the coordination number of each metal atom is four, and the W—W bond lengths are only slightly longer than those found for the W_2X_6 compounds. It is now known that when $[W_2(NMe_2)_2]$ is allowed to react with EtOH in the presence of $Me(H)NCH_2CH_2N(H)Me$, the adduct (**24**) is formed. In this structure (Figure 34) the W—W triple bond is retained (2.296 Å), and the two ends of the molecule are almost perfectly staggered. $[W_2(NMe_2)_6]$ was also shown to react with alcohols in the presence of PMe_3 to give the phosphine adducts $[W_2(OR)_6(PMe_3)_2]$ ($R = Pr^i$, CH_2Bu^t and Et). The structure of $[W_2(OCH_2Bu^t)_6(PMe_3)_2]$ reveals two square planar WO_3P units joined by a W—W triple bond of distance 2.362 Å.

$[W_2(OEt)_6(Me(H)NC_2H_4N(H)Me)]$
(24)

1,2-$[W_2\{Si(SiMe_3)_3\}_2(NMe_2)_4]$
(25)

1,2-$[W_2\{CpFe(CO)_2\}_2(NMe_2)_4]$
(26)

As mentioned previously, $[WCl_2(NR_2)_4]$ reacts with LiR to give the 1,2-dialkyl-substituted derivatives. Recently, this reaction has been extended to other anionic ligands.[300] Thus, the reaction of $[W_2Cl_2(NMe_2)_4]$ with two equivalents of $[(Me_3Si)Si]^-$ or $[CpFe(CO)_2]^-$ results in the

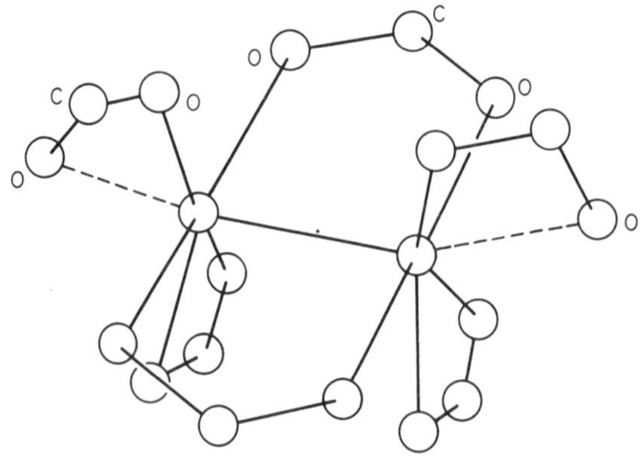

Figure 32 The structure of $[W_2(O_2CNMe_2)_6]$[297]

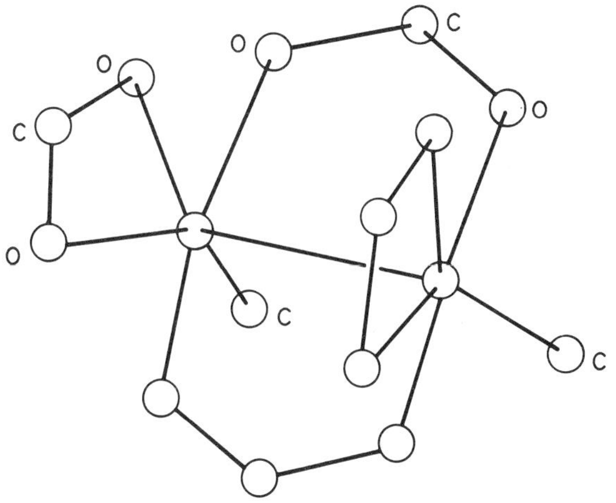

Figure 33 The structure of $[W_2Me_2(O_2CNEt_2)_4]$[297]

formation of compounds (**25**) and (**26**) respectively. These are the first examples of a W—W triple bond supporting a metal–metal single bond between two different metal atoms.

Of considerable interest are reactions in which the metal–metal bond order changes. So far, there are no known examples in which a W—W triple bond is converted to a W—W quadruple bond, a reaction which is known for molybdenum.[289] However, several reactions in which the W—W triple bond is oxidized are known. The preparation of alkoxo clusters of W^{IV} have been mentioned (Section 37.5.4.1) and the reaction in which W≡W reacts with RC≡CR or RC≡N to give mononuclear complexes containing the W≡CR and W≡N units have been discussed (Section 37.3.3).

Recently, several new and important reactions, in which the dinuclear unit remains intact, have been reported. The reaction $[W_2(OBu^t)_6]$ in the presence of pyridine or $[W_2(OPr^i)_6py_2]$ with ethyne gives (**27**).[301] The structure of the isopropoxide derivative contains the $W(\mu\text{-}OPr^i)_2(\mu\text{-}C_2H_2)W$ bridge with a W—W distance of 2.567 Å. In solution, this compound is fluxional showing rapid terminal bridge OR exchange. At low temperature, the NMR spectra is consistent with the solid state structure. The addition of C_2Me_2 to $[W_2(OCH_2Bu^t)_6py_2]$ in hexane yields the blue compound (**28**). In this structure there is only one bridging alkoxy group

$[W_2(O_2CH_2Bu^t)_6py_2(\mu\text{-}C_2H_2)]$
(27)

$[W_2(O_2CH_2Bu^t)_6py(\mu\text{-}C_2Me_2)]$
(28)

$[W_2(OPr^i)_6(\mu\text{-}C_4Me_4)(C_2Me_2)]$
(29)

$[W_2(O_2CH_2Bu^t)_6py\{\mu\text{-}N(CMe_4)N\}]$
(30)

(W—W = 2.602 Å), and the C_2Me_2 unit is closer to one of the W atoms (Figure 35). It has also been shown that changing the alkoxy groups and/or the alkyl substituents greatly influences the course of this reaction. Thus, reacting $[W_2(OPr^i)_6py_2]$ with an excess of C_2Me_2 leads to compound (**29**) in which two alkynes are coupled and a third is present as a terminal ligand. In

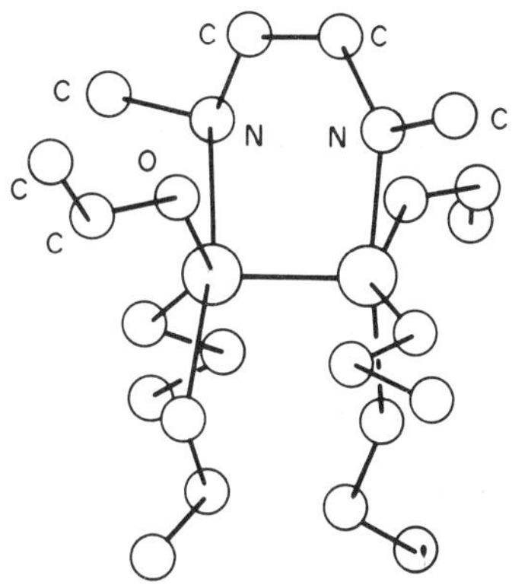

Figure 34 The structure of $[W_2(OEt)_6(Me(H)NC_2H_4N(H)Me)]$[299]

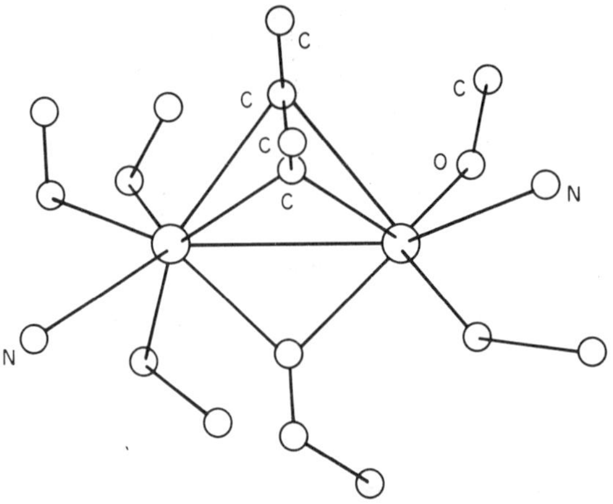

Figure 35 The structure of $[W_2(O_2CH_2Bu^t)_6py(\mu\text{-}C_2Me_2)]$; the tertiary butyl groups and py ring are omitted[301]

this molecule (W—W distance = 2.85 Å), as with the others, W—W interaction is retained. When compound (**28**) is reacted with MeCN, compound (**30**) is obtained.[302a] Its structure (Figure 36) reveals the presence of the deprotonated bridging ligand (μ-N(CMe)$_4$N) which results from coupling of the (μ-C_2Me_2) moiety with two equivalents of MeCN. The W—W bond length of 2.617 Å is consistent with a bond of order one. It has also been shown that $[W_2(OR)_6]$ reacts with α-diketones to give compounds in which the reduced organic moiety is bonded to a dinuclear complex with a W—W bond of order one.[302b]

The reaction of $[W_2(OBu^t)_6]$ with NO in pyridine leads to cleavage of the W—W triple bond,[303] and the formation of $[W(OBu^t)_3(NO)py]$. In this structure, the geometry around the metal atom is trigonal bipyramidal, with the py ligand *trans* to NO.

The reaction of $[W_2(OR)_6]$ ($R = Pr^i$, CH_2CMe_3) with CO in the presence of excess pyridine leads to $[W_2(OR)_6py_2(\mu\text{-}CO)]$.[304] The structure of the isopropoxide derivative is that of a confacial bioctahedron, with the CO ligand and a pair of OPr^i groups making up the bridging face. The W—W distance of 2.499 Å is consistent with a bond of order two. However, when $[W_2(OPr^i)_6py_2]$ is reacted with CO without additional pyridine,[305] two different products are obtained depending on the amount of CO added. With four equivalents of CO compound (**1**) is formed (Section 37.3.1), while with two equivalents of CO the tetranuclear compound $[W_2(OPr^i)_6py(CO)]_2$ is obtained. The structure of this molecule consists of two $[W_2(OPr^i)_6py(CO)]$ units which are bridged through the oxygen atoms of the CO group (Figure 37), an unprecedented bonding arrangement for the CO ligand. The W—W bond length of 2.654 Å is significantly longer than that found for $[W_2(OPr^i)_6py_2(CO)]$ and falls within the range typical of single rather than double W—W bonds.

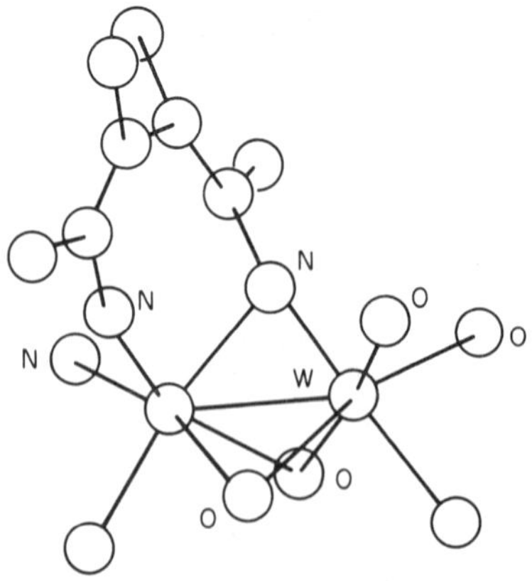

Figure 36 The structure of $[W_2(O_2CH_2Bu^t)_6py(\mu\text{-}N(CMe)_4N)]$; the tertiary butyl groups and py ring are omitted[302a]

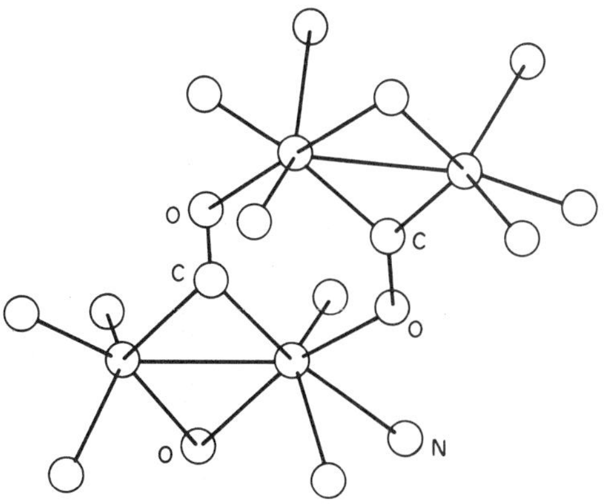

Figure 37 The structure of $[W_2(OPr^i)_6py(CO)]_2$; propyl groups are omitted[305]

37.7 THE CHEMISTRY OF W^{II}

Until several years ago, most of the known compounds of W^{II} were monomeric complexes, which exhibit mainly coordination number seven, and cluster compounds containing the octahedral W_6 nucleus with W—W single bonds. The most important advance in the chemistry of W^{II} has undoubtedly been the preparation and characterization of dinuclear compounds containing W—W quadruple bonds.

37.7.1 Monomeric Complexes

There are two general procedures for the preparation of monomeric W^{II} complexes.[269] In one, substituted tungsten carbonyls are oxidized by halogens under controlled conditions as exemplified by equations (18) and (19). In the second, $W(CO)_6$ is first oxidized by Cl_2 or Br_2 at low temperature, followed by reaction of the oxidized product with the appropriate ligands. This preparative procedure is exemplified by equation (20).

$$[W(CO)_4(bipy)] + Br_2 \longrightarrow [W(CO)_3(bipy)Br_2] + CO \quad (18)$$

$$[W(CO)_2(diars)_2] + Br_2 \longrightarrow [W(CO)_2(diars)_2Br]Br \quad (19)$$

$$[W(CO)_4Cl_2] + 2PPh_3 \longrightarrow [W(CO)_3(PPh_3)_2Cl_2] + CO \quad (20)$$

There are numerous examples of this type of complex. The majority are seven-coordinate; some representatives are given in Table 8. With bidentate and tridentate ligands, when the

Table 8 Properties of Some Monomeric W^{II} Complexes

Compound	*Carbonyl stretches* (cm^{-1})	*Visible spectra* (nm) (*ε or color*)	*Ref.*
$(Et_4N)[W(CO)_4Br_3]$	2080, 2041, 2000, 1925	Yellow	269
$[W(CO)_4Br_2]$	2110, 2025, 1980	Yellow	309
$[W(CO)_3(PPh_3)_2Br_2]$	2015, 1940, 1900	Yellow	309
$[W(CO)_2(PPh_3)_2Br_2]$	1960, 1875	Blue	309
$[W(CO)_2(diphos)_2]I_3$	1853	Red	307
$[W(CO)_3(diars)Br_2]$	2030, 1942, 1905	Yellow	269
$[W(CO)_3(dipy)Br_2]$	2037, 1959, 1908	Yellow	269
$[W(CO)_2(PPh_3)(S_2CNEt_2)_2]$	1930, 1835	685 (223); 490 (1060)	59
$[W(CO)(C_2H_2)(S_2CNEt_2)_2]$	1960	689 (90); 360 (3440)	269
$[W(CO)_2(PPh_3)(dcq)_2]$	1916, 1824	506 (38); 419sh	269
$[W(Bu^tNC)_6Br]Br$	—	405 (1960)	316
$[W(Bu^tNC)_7](PF_6)_2$	—	417 (1814)	316
$[W(CO)_3(diphos)I_2]$	2039, 1963, 1909	Orange	269
$[W(CO)_2(CS)(diphos)I_2]$	2036, 1972, 1247 (CS)	—	322

stoichiometry suggests a coordination number higher than seven, it is invariably found that not all the donor atoms are coordinated to the metal.

The halocarbonyl anions $[W(CO)_4X_3]^-$ (X = Cl, Br, I) are prepared by the addition of the desired halide to $[W(CO)_4X_2]$ or by oxidation of $[W(CO)_5X]^-$. Mixed halogen derivatives are prepared similarly.[306] For example, oxidation of $[W(CO)_5I]^-$ with Br_2 leads to $[W(CO)_4IBr_2]^-$.

Oxidation of *cis*-$[W(CO)_2(diphos)_2]$ with I_2 leads to the unexpected one electron oxidation product $[W(CO)_2(diphos)_2]^+$. This paramagnetic complex exhibits only one carbonyl stretching frequency, indicating a *trans* stereochemistry.[307] Reaction of $[W(CO)_4diars]$ with excess bromine also results in the seven-coordinate complex $[W(CO)_3(diars)Br_2]^+$, which appears to be a W^{III} derivative.[308] It is unusual that three carbonyls are retained by the metal in this oxidation state.

Refluxing of $[W(CO)_3(PPh_3)_2X_2]$ (X = Cl, Br) in CH_2Cl_2 gives the deep blue crystalline derivatives $[W(CO)_2(PPh_3)_2X_2]$.[309] The structure of this six-coordinate complex is probably similar to that of $[Mo(CO)_2(PPh_3)_2Br_2]$ described as having an unusual nonoctahedral six-coordinate geometry.[310] The complex reacts reversibly with CO as depicted in equation (21).

$$[W(CO)_2(PPh_3)_2X_2] + CO \rightleftharpoons [W(CO)_3(PPh_3)_2X_2] \quad (21)$$

A series of seven-coordinate complexes of the type $[W(CO)_2(PPh_3)(L\text{–}L)_2]$ (L–L = S_2CNR_2, S_2COMe, $S_2P(OEt)_2$) have been prepared by the reaction of $[W(CO)(PPh_3)_3Cl_2]$ with excess ligand.[60] With excess $HS_2P(Pr^i)_2$ or with an equimolar amount of $HS_2P(OEt)_2$, only one Cl is displaced and $[W(CO)_2(PPh_3)(L\text{–}L)Cl]$ is obtained. The visible absorption spectra of these complexes are solvent dependent. This, together with conductance measurements, strongly suggests the presence of the cationic species $[W(CO)_2(PPh_3)(L\text{–}L)]^+$ in polar solvents such as methanol. $[W(CO)_3(PPh_3)(S_2CNEt_2)]$ has been shown to react with acetylene to afford $[W(CO)(C_2H_2)(S_2CNEt_2)_2]$.[311] Crystal structure analysis reveals a distorted pentagonal bipyramidal geometry around the metal with sulfur atoms S-2 and S-3 (Figure 38) occupying the apices of the pyramid. The CO and C—C bonds are nearly coplanar, enabling the acetylene ligand to act as a four-electron donor. More recently it has been shown that $[W(CO)_3(S_2CNR_2)_2]$ also reacts with alkynes to give $[W(CO)(R^1C{\equiv}CR^2)(S_2CNR_2)_2]$ where $R^1 = R^2$ = H, Me, Et, Ph and R^1 = H, R^2 = Ph.[312]

The seven-coordinate complexes $[W(CO)_3(YPPh_3)X_2]$ (Y = P, As, Sb; X = Cl, Br) can undergo a two-electron reduction in CO atmosphere as depicted in equation (22). No intermediates are observed in this reaction.[313]

$$[W(CO)_3(YPh_3)_2X_2] + 2CO + 2e^- \longrightarrow [W(CO)_5X]^- + 2YPh_3 + X^- \quad (22)$$

Several examples of W^{II} complexes with no carbonyl ligands are also known. The reaction of WI_3 with excess diars at 160 °C gives the paramagnetic complex $[W(diars)_2I_2]$ with μ_{eff} = 2.70 BM.[314] A related complex $[WCl_2(PMe_2Ph)_4]$ can be prepared by Na/Hg reduction of $[WCl_4(PMe_2Ph)_2]$ under argon.[315] This paramagnetic complex (μ_{eff} = 2.30 BM) exhibits only one ν(W—Cl) band at 280 cm^{-1} suggesting the *trans* configuration.

A general procedure has been outlined for the preparation of isocyanide complexes of W^{II} (equations 23, 24).[316] $[W(CNR)_7]^{2+}$ (R = CMe_3, C_6H_{11}) can also be prepared by reaction of the

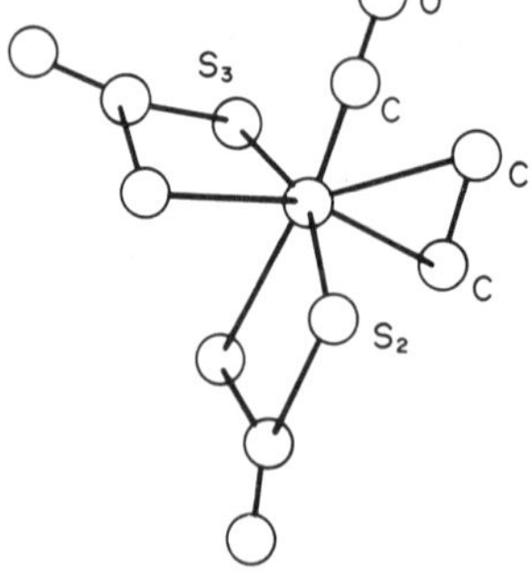

Figure 38 The structure of $[W(CO)(C_2H_2)(S_2CNEt_2)_2]$[311]

Table 9 Structures of Seven-coordinate W^{II} Complexes

Compound	*Suggested structure*	*Capping group*	*Ref.*
$(Et_4N)[W(CO)_4Br_3]$	CO	CO	269
$[W(CO)_3(dpam)_2Br_2]$	CO	CO	269
$[W(CO)_3(PMe_2Ph)_3I]BPh_4$	CO/CTP	I	269
$[W(CO)_3(dmpe)I_2]$	CO	CO	269
$[W(CO)_3(difas)I_2]$	CO	CO	269
$[W(CO)_3(dpam)I_2]$	CO	CO	269
$[W(CO)_3(bipy)(GeBr_3)Br]$	CO/CTP	Br	269
$[W(CO)_4(diars)I]I$	CTP	I	269
$[W(CO)_3(dth)(SnCl_3)$	CO	Sn	269
$[W(CO)_2(dmpe)_2I]I$	CTP	I	269
$[W(CO)(C_2H_2)(S_2CNEt_2)_2]$	Distorted PBP	—	311
$[W(CO)_2(Bu^tNC)_3I_2]$	4:3 piano stool	—	322
$[W(CO)_3(S_2CNMe_2)_2]$	4:3 piano stool	—	323

RNC ligand with $[W_2(mhp)_4]$.[318] NMR measurements show that the isocyanide complexes are fluxional down to −30 °C.[316]

$$W(CO)_6 + X_2 \xrightarrow{RNC} [W(CNR)_6X]X \quad (X = Cl, Br) \tag{23}$$

$$[W(CNR)_6X] + 2I^- \longrightarrow [W(CNR)_6I]I + 2X^-$$

$$[W(CNR)_6X]X \xrightarrow[PF_6^-]{RNC} [W(CNR)_7](PF_6)_2 \tag{24}$$

The structures of several of the seven-coordinate complexes of W^{II} have been determined and these are listed in Table 9. There are three recognized geometries for coordination number seven,[319,320] the pentagonal bipyramid (PBP), capped octahedron (CO) and capped trigonal prism (CTP). The 4:3 piano stool arrangement, which consists of a plane of four donor atoms parallel to a plane of three donor atoms, is also recognized.[321] Two extreme cases for this geometry, both with C_s symmetry, are considered, and these are related to one another by a rotation of the triangular face. It has been suggested however, that the 4:3 geometry is not sufficiently different from the CO and CTP to warrant its inclusion as a separate geometry. This point was recently questioned, and it is argued that when the dihedral angle between the two planes defining the 4:3 arrangement is close to 0°, the 4:3 geometry is a better representation of the molecular geometry.[322] A case in point is the structure of $[W(CO)_2(CNBu^t)_3I_2]$ (Figures 39 and 40). The 4:3 geometry is also considered as the best description for the structure of $[W(CO)_3(S_2CNMe_2)_2]$.[323] This molecule was also shown to be fluxional, and dynamic ^{13}C NMR measurements suggest two distinct intramolecular rearrangement processes.

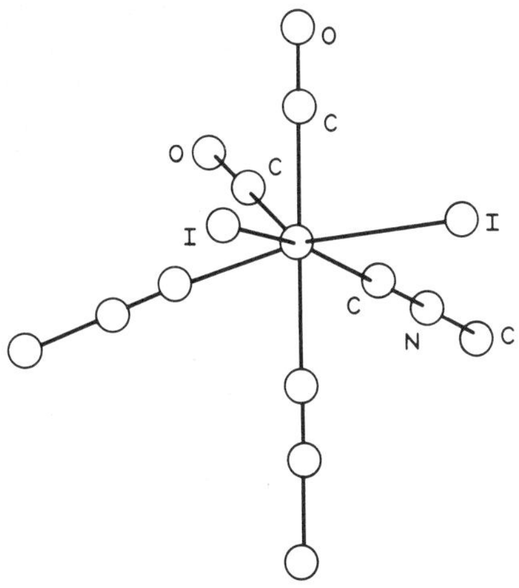

Figure 39 The structure of $[W(CO)_2(CNBu^t)_3I_2]$; the tertiary butyl groups are omitted[322]

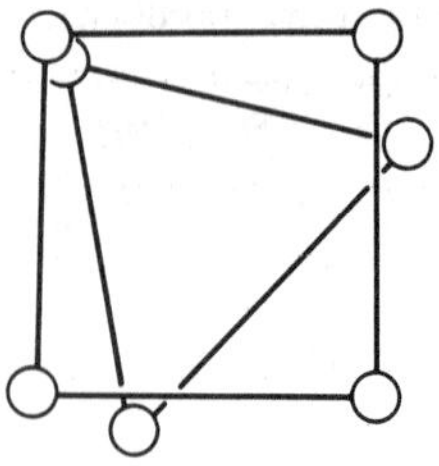

Figure 40 The two planes which define the 4:3 geometry in $[W(CO)_2(CNBu^t)_3I_2]$[322]

The structures of $[W(CO)_3Cl(dcq)(PPh_3)]$ and $[W(CO)_2(dcq)_2(PPh_3)]$ (dcq = 5,7-dichloro-8-quinolinato) have recently been determined.[324] These are best described as having the CTP geometry. The positions of all the donor atoms in the coordination sphere of these complexes are in agreement with electronic arguments.[325]

The first example of a PBP geometry in which a bidentate ligand spans one of the pentagonal edges is furnished by the structure of $[W(CO)_3I_2(Ph_2PCH_2PPh_2)]$.[326]

37.7.2 Complexes with Quadruple Bonds

The existence of quadruple bonds was first recognized in 1964[289] and since then a great many compounds of rhenium, technetium, chromium and molybdenum have been shown to contain them. Both bridged and unbridged compounds are known with a large variety of ligands. The bonding has been described in detail.[289]

The first W—W quadruply bonded complexes characterized were $[W_2Me_nCl_{8-n}]^{4-}$ prepared by treatment of WCl_4 with methyllithium. Crystal structure analysis has established the short W—W distances of 2.264 and 2.263 Å for the methyl and the mixed chloromethyl derivatives respectively. In addition, these compounds have the expected eclipsed stereochemistry.[327,328]

The reaction of WCl_4 with $K_2C_8H_8$ in THF gives a reasonably air-stable material that was shown to be $[W_2(C_8H_8)_3]$. The W—W bond length of 2.375 Å, although long, is consistent with the presence of a quadruple bond.[329] However, it is recognized that this compound (and the related chromium and molybdenum) is electronically different from other quadruply bonded species.

Treatment of $Mo(CO)_6$ with 6-methylpyridone (Hmhp) and 2-amino-6-methylpyridine (Hmap) in refluxing diglyme has led to the isolation of $[Mo_2(map)_4]$ and $[Mo_2(mhp)_4]$ which contain very short Mo—Mo distances.[289] $[W_2(mhp)_4]$ was also prepared by the same way. It reacts with Li(map) in THF to give $[W_2(map)_4]$.[330,331] Both compounds have been structurally characterized and have been shown to contain short W—W distances of 2.164 and 2.161 Å for the map and mhp compounds, respectively. These are the shortest known internuclear distances between any two atoms in the third transition series. $[W_2(mhp)_4]$ was the first and remains the only quadruply bonded tungsten complex to be obtained by oxidation of $W(CO)_6$. The structure of $[W_2(map)_4]$ is shown in Figure 41. The arrangement of the ligands around the W_2 unit is such that the idealized symmetry of the complex is D_{2d}.

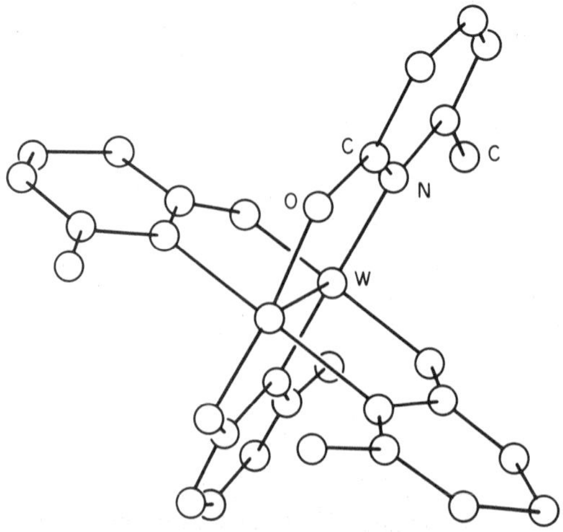

Figure 41 The structure of $[W_2(map)_4]$[331]

A special case in this family of complexes is represented by the tetracarboxylates of the form $[M_2(O_2CR)_4]$. Following the successful preparation of $[Mo_2(O_2CMe)_4]$ from $Mo(CO)_6$ in refluxing acetic acid–acetic anhydride mixture,[289] several attempts to prepare the analogous tungsten derivative by the same procedure have failed. This reaction leads invariably to the trinuclear clusters of type (**15**) (see Section 37.5.4.2). However, the continuous efforts to prepare complexes of the type $[W_2(O_2CR)_4]$ have recently been rewarded by the synthesis of $[W_2(TFA)_4]$. This was achieved by Na/Hg reduction of $[W_2Cl_6(THF)_4]$ in THF followed by addition of NaTFA, or by Na/Hg reduction of $[WCl_4]_x$ in the presence of NaTFA at 0 °C.[332,333]

$[W_2(TFA)]_4$ easily adds two ligands along the W—W bond and the structure of both $[W_2(TFA)_4 \cdot \frac{2}{3}diglyme]$ and $[W_2(TFA)_4 \cdot 2PPh_3]$ have been determined. The compounds have the expected geometry with short W—W bond lengths of 2.209 and 2.242 Å respectively. The increase of about 0.03 Å in the W—W bond length most likely indicates a stronger M—L σ bonding and a weaker M—M bonding in the phosphine adduct. On the basis of the Raman spectra of $[W_2(TFA)_4]$, $[W_2(TFA)_4 \cdot 2PPh_3]$ and $[Mo_2(TFA)_4 \cdot 2py]$ it is predicted that the W—W bond length in pure $[W_2(TFA)_4]$ should be in the range of 2.20–2.21 Å,[333] some 0.1 Å longer than that found in $[Mo_2(TFA)_4]$, indicating a weaker M—M bonding in the tungsten dimers. This is in agreement with the chemical behavior of these two compounds.

In contrast to the unsuccessful early attempts to produce $[W_2(O_2CR)_4]$, the heteronuclear compound $[MoW(O_2CCMe_3)_4]$ was obtained by reacting a 3:1 mixture of $W(CO)_6$ and $Mo(CO)_6$ in refluxing dichlorobenzene.[334] The heteronuclear complex was freed from $[Mo_2(O_2CCMe_3)_4]$ by careful oxidation with I_2. Structure analysis of $[MoW(O_2CCMe_3)_4]I \cdot MeCN$ shows the expected idealized D_{4h} symmetry with a short W—Mo separation of 2.194 Å. The iodide ion is coordinated to the W atom and the MeCN molecule is coordinated to the molybdenum atom.

$[MoW(O_2CCMe_3)_4]^+$ is reduced with zinc dust to $[MoW(O_2W(O_2CCMe_3)_4]$ whose structure reveals a W—Mo distance of 2.080 Å.[335] The longer distance observed for the iodo derivative is consistent with the fact that in $[MoW(O_2CCMe_3)_4]^+$, the bond order is formally 3.5.

$[MoW(mhp)_4]$ is also known. It is prepared by the reaction of a 1.5:1 mole ratio of $W(CO)_6$ and $Mo(CO)_6$ with Hmhp in refluxing diglyme. $[MoW(mhp)_4]$ is purified from the other products by I_2 oxidation followed by reduction with zinc amalgam.[336] The potentials (*vs.* SCE) of the one electron reversible oxidation of $[Mo_2(mhp)_4]$, $[MoW(mhp)_4]$ and $[W_2(mhp)_4]$ of 0.2, −0.16 and −0.35 V respectively, indicates that in a mixture of these, the heteronuclear compound can be selectively oxidized and thus purified.

The Mo—W distance of 2.091 Å[336] is shorter than the average M—M bond lengths of 2.065 and 2.161 Å found in $[Mo_2(mhp)_4]$ and $[W_2(mhp)_4]$ respectively,[330] suggesting a stronger metal–metal bonding in the heteronuclear species. Raman spectra and force constant calculations are in agreement with the crystallographic results. Also in agreement with this conclusion are the results of photoelectron spectra, which have been measured for a series of M—M quadruply bonded complexes.[337]

Attempts to prepare the octahalo anion $[W_2Cl_8]^{2-}$ were not successful until 1980, when a general procedure for the preparation of mixed halo–phosphine dimers was reported.[338] In this procedure WCl_4 is reduced with Na/Hg in the presence of a phosphine ligand, as depicted in equation (25). With PMe_3, if only one equivalent of reductant is used, the red crystalline $[W_2Cl_6(PMe_3)_4]$ is obtained. Treatment of this compound with an additional equivalent of reducing agent leads to $[W_2Cl_4(PMe_3)_4]$.[339] With the bidentate phosphines, dmpe and diphos, reaction (25) fails to give the desired products $[W_2Cl_4(dmpe)_2]$ and $[W_2Cl_4(diphos)_2]$. However, these can be obtained by treating $[W_2Cl_4(PBu^n)_4]$ with the bidentate ligands in toluene at 80 °C. With diphos, a green α isomer and a brown β isomer are formed in an analogous way to molybdenum. $[W_2Cl_4(dmpe)_2]$ and the green isomer $[W_2Cl_4(diphos)_2]$ possess a centrosymmetric eclipsed structure as depicted in Figure 42. The β isomer is structurally similar to $[Mo_2Cl_4(diphos)_2]$[289] and contains intramolecular diphos bridges. The tungsten compounds adopt a staggered conformation and thus can be considered to have a bond order of three.[340] In contrast, the molybdenum analog has a torsional angle of 30°. This difference undoubtedly reflects a weaker δ component in the W—W quadruple bond and therefore steric requirements are able to enforce a fully staggered conformation.

$$WCl_4 + 2Na/Hg + 2L \xrightarrow{THF} \tfrac{1}{2}[W_2Cl_4L_4] \quad L = PMe_3, PMe_2Ph, PMePh_2, PBu^n_3 \tag{25}$$

When the reduction of $[W_2Cl_6(THF)_4]$ is carried out at low temperature without the phosphine ligand, $[W_2Cl_8]^{4-}$ is formed.[339] Its structure reveals the expected eclipsed stereochemistry with a W—W bond length of 2.259 Å.[341]

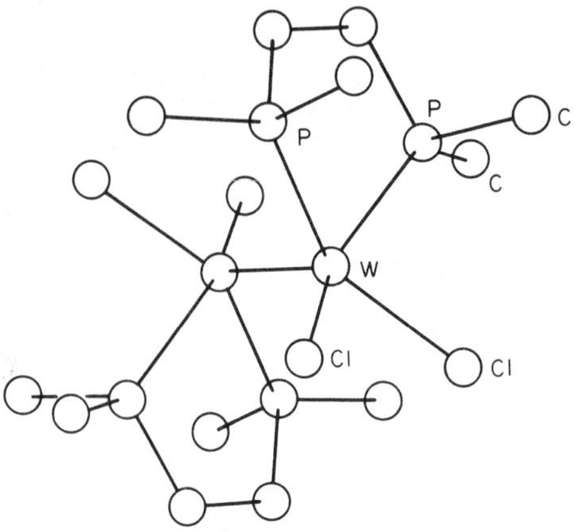

Figure 42 The structure of $[W_2Cl_4(dmpe)_2]$[340]

The complexes $[W_2Cl_4L_4]$ undergo a reversible one-electron oxidation. A comparison of the oxidation potential of analogous molybdenum and tungsten compounds clearly shows that the tungsten compounds are considerably easier to oxidize. This ease of oxidation is believed to be responsible for the fact that it is difficult, if not impossible, to obtain quadruply bonded tungsten compounds by synthetic procedures which are known for molybdenum.[339]

Chemical oxidation was also noted for $[W_2Cl_4(PBu^n_3)_4]$. Treatment of $[W_2Cl_2(PBu^n_3)_4]$ with $[Ag(MeCN)_4]^+$ leads to the paramagnetic $[W_2Cl_4(PBu^n_3)_4]^+$. Reaction of the butyl phosphine derivatives with acetic acid at 160 °C in diglyme affords the trinuclear cluster **(13)** (see Section 37.5.4.2), and with benzoic acid to give compound **(31)** having a W—W triple bond with a W—W distance of 2.423 Å.[342]

$$[W_2(\mu\text{-}H)(\mu\text{-}Cl)(PBu_3)_2Cl_2(O_2CPh)_2]$$
(31)

The ease of oxidation of the quadruply bonded tungsten dimers is also exemplified by the reaction of $[W_2(mhp)_4]$[343] or $[W_2(TFA)_4]$[333] with HCl to give $[W_2Cl_9]^{3-}$, which is also readily obtained from $[W_2Cl_8]^{4-}$ even at −78 °C. It is believed that these reactions proceed *via* the hydride intermediate $[W_2Cl_8H]^{4-}$.

Recently, it has been shown that the tetranuclear compound $[W_4Cl_8(PBu^n_3)_4]$[344] can be prepared by Na/Hg reduction of $[W_2Cl_6(THF)_2(PBu^n_3)_2]$. The structure of this molecule is similar to that of $[Mo_4Cl_8(PEt_3)_4]$[345] with W—W distances of 2.309 and 2.840 Å along the edges of the rectangular cluster.

The structures of several of the quadruply bonded tungsten dimers have been determined (Table 10) and in all cases the W—W bond length is significantly longer than those found in analogous molybdenum compounds. The crystallographic results coupled with theoretical,[346] spectroscopic[346,347] and electrochemical[339] studies clearly indicate that the W—W quadruple bond is indeed weaker than the Mo—Mo quadruple bond as was intuitively suggested originally by consideration of the core size of these two elements.

Table 10 Quadruple Bond Distances Between W Atoms

Compound	*W—W (Å)*	*Ref.*	*Compound*	*W—W (Å)*	*Ref.*
$Li_4[W_2Me_8]\cdot 4Et_2O$	2.264(1)	327	$W_2Cl_4(PMe_3)_4$	2.262(1)	340
$Li_4[W_2Me_{8-x}Cl_x]\cdot 4THF$	2.263(2)	327	$[W_2Cl_4(dmpe)_2]\cdot$toluene	2.287(1)	340
$[W_2(C_8H_8)_3]$	2.375(1)	329	$[W_2Cl_4(dppe)_2]\cdot\frac{1}{2}H_2O$	2.280(1)	340
$[W_2(mhp)_4]\cdot CH_2Cl_2$	2.161(1)	330			
$[W_2(Chp)_4]$	2.177(1)	289			
$[W_2(map)_4]\cdot THF$	2.164(1)	331	$Na_4(TMEDA)_4[W_2Cl_8]$	2.259(2)	341
$[W_2(dmhp)_2(Ph_2NNNPh_2)_2]$	2.169(1)	289	$[W_2(TFA)_4]\cdot\frac{2}{3}$dighgm	2.209	333
			$[W_2(TFA)_4]\cdot 2PPh_3$	2.242	333
			$[W_2(O_2CC_6H_3)_4\cdot 2THF$	2.196	357

37.7.3 The Hexanuclear Clusters $[W_6X_8]^{4+}$

The dihalides of W^{II}, WCl_2, WBr_2 and WI_2 are in reality the hexanuclear compounds W_2X_{12} which contain the $[W_6X_8]^{4+}$ cluster (Figure 43). The metal atoms occupy the corners of an octahedron with a bridging halogen atom above each triangular face. The entire unit has full O_h symmetry and it is connected to other units by halogen atoms (four per unit). There are two non-bridging halogen atoms per cluster.

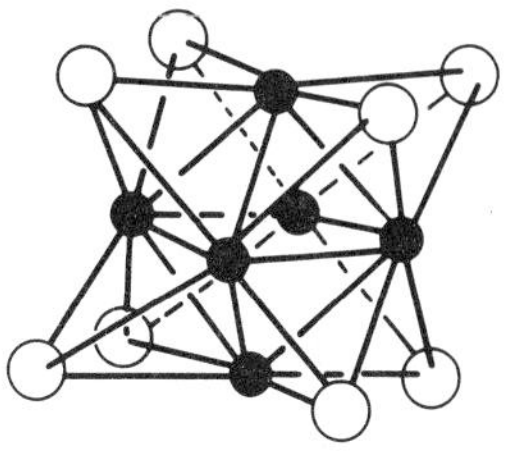

Figure 43 The $[W_6X_8]^{4+}$ core

W_6Cl_{12} and W_6Br_{12} are best prepared by disproportionation of the tetrachloride at about 500 °C or by reduction of WCl_5 and WBr_5 with aluminum in a temperature gradient.[154] W_6I_{12} is obtained by fusion of W_6Cl_{12} with a large excess of a KI–LiI mixture at 550 °C. As might be expected from the structures, the external halogens are readily replaceable by other halogens. Thus, $[W_6Cl_8]Br_4$, $[W_6Cl_8]I_4$, $[W_6Br_8]Cl_4$ and $[W_6Br_8]F_4$ have been prepared.[348,349] The $[W_6X_8]^{4+}$ clusters also readily form complex anions of the type $[W_6X_{14}]^{2-}$ (X = Cl, Br, I) and $[W_6X_8]Y_6^{2-}$ (X = Y = Cl, Br, I).[349]

These clusters are fairly good reducing agents in aqueous solutions and are oxidized even by water.[350] They can also be oxidized by halogens at elevated temperatures. With W_6Br_{12} the products are W_6Br_{14}, W_6Br_{16} and W_6Br_{18}.[351,352] In all these products, the W_6Br_8 unit is oxidized by two electrons. Thus, W_6Br_{14} is best formulated as $[W_6Br_8]Br_6$ while the others contain bridging Br_4^{2-} ions. The reaction of Cl_2 with W_6Cl_{12} at 100 °C leads to a product of stoichiometry WCl_3 which has been shown to contain the $[W_6Cl_{12}]^{6+}$ unit[353] isostructural with the Nb and Ta clusters.

The $[W_6X_8]^{4+}$ clusters are diamagnetic and with the 24 available metal electrons there is a metal–metal bond of order one between each pair of tungsten atoms, in agreement with theoretical treatment of the electronic structure of these clusters.[354,355]

Recently, it has been shown that $[W_6Cl_{14}]^{2-}$ is luminescent and that it undergoes a facile electrochemical oxidation at 1.14 V *vs.* SCE. The resulting monoanion is a powerful oxidizing agent.[356]

37.8 COMPLEXES OF DINITROGEN

Dinitrogen complexes of tungsten have been studied in great detail in relation to the function of nitrogenase. They show considerable versatility both with respect to reduction of the coordinated N_2 and in reactions leading to carbon–nitrogen bond formation.

Dinitrogen complexes of tungsten were first reported in 1970.[358] Na/Hg reduction of $[WCl_2(PMe_2Ph)_4]$ and $[WCl_2(diphos)_2]$ in the presence of excess phosphine under dinitrogen led to the isolation of *trans*-$[W(N_2)_2(diphos)_2]$ and *cis*-$[W(N_2)_2(PMe_2Ph)_4]$. The IR spectra of the *trans* compound show one band at 1953 cm^{-1} assigned to $\nu(N_2)$. The *cis* complex exhibits two such bands at 1931 and 1998 cm^{-1}. Magnesium reduction of *trans*-$[WCl_4(PMePh_2)_2]$ under N_2 in the presence of excess ligand leads to *trans*-$[W(N_2)_2(PMePh_2)_4]$.[359] The complex reacts with nitrogen donor ligands to give $[W(N_2)_2(PMePh_2)_3L]$ (L = py, 3-Mepy, 4-Mepy and 4-(CO_2H)py). It can also be oxidized to the monocation with $FeCl_3$. Reaction of *trans*-$[W(N_2)_2(PMePh_2)_4]$ with diphos gives the mixed phosphine complex $[W(N_2)_2(PMePh_2)_2(diphos)]$. The py-substituted complex exhibits only one $\nu(N_2)$ band in the IR indicating a *trans* configuration and a *mer* arrangement of the phosphine ligands. The reduction of $[WCl_4(PMe_3)_3]$ with dispersed sodium leads to *cis*-$[W(N_2)_2(PMe_3)_4]$ which reacts with excess PMe_3 to give $[W(N_2)(PMe_3)_5]$.[360] When the reduction is performed under ethylene, *trans*-$[W(C_2H_4)_2(PMe_3)_4]$ is obtained. The compound *trans*-$[WCl(N_2)(PMe_3)_4]$ is also known.[361]

Anionic dinitrogen complexes *trans*-$[WX(N_2)(diphos)_2]^-$ were prepared for X = NCS, CN

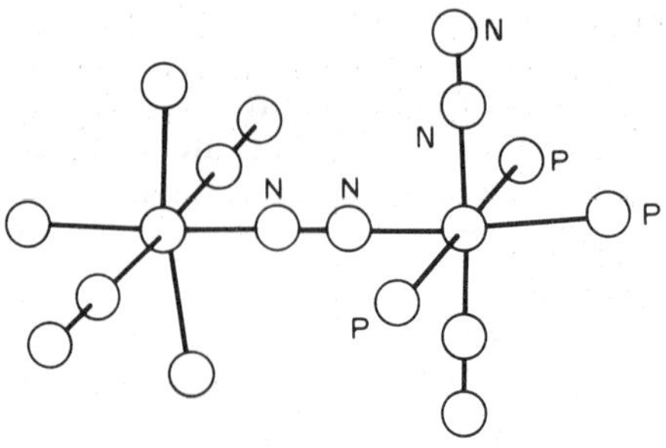

Figure 44 The structure of $[\{W(N_2)_2(PEt_2Ph)_3\}_2(\mu\text{-}N_2)]$; ethyl groups are omitted[363]

and N_3. The effect of ligands L (uncharged) or X^- on the redox properties of *trans*-[W(L or X^-)N_2(diphos)] as well as the correlation of the redox properties with $\nu(N_2)$ have been determined. These properties were also correlated with the susceptibility of the complexes towards attack by electrophilic reagents.[362]

Recently it has been shown that magnesium reduction of *trans*-$[WCl_4(PEt_2Ph)_2]$ under N_2 in THF in the presence of excess ligand leads in low yield to complex **(32)** which contains both terminal and bridging dinitrogen ligands.[363] Its structure is depicted in Figure 44. It has also been shown that *cis*-$[W(N_2)_2(PMe_2Ph)_4]$ reacts with $AlCl_3$ in py to give **(33)** in which two units of $[WClpy(PMe_2Ph)_3N_2]$ are bridged by two $AlCl_2$ moieties through the end nitrogen atoms.[364]

$[\{W(N_2)_2(PEt_2Ph)_3\}_2(\mu\text{-}N_2)]$
(32)

$[WXpy(PMe_2Ph)_3(\mu_3\text{-}N_2)]_2(AlCl_2)_2$ (X = Cl, Br)
(33)

There are two types of interesting and important reactions involving the coordinated dinitrogen, its reduction to hydrazine and ammonia, and the reactions leading to carbon–nitrogen bond formation.

37.8.1 Reduction of Coordinated Dinitrogen

The protonation of coordinated dinitrogen in a tungsten complex was first observed in the reaction of *trans*-$[W(N_2)_2(diphos)_2]$ with HCl as depicted in equation (26).[365] In the resulting complex, one chloride ligand is labile, and in the presence of BPh_4^- $[WCl(diphos)_2N_2H_2]BPh_4$ is obtained. The N_2H_2 group can coordinate to the metal as a diazene (W—NH═NH) or as a hydrazido(2−) (W═N—NH_2) group. Spectroscopic data suggest that in the parent complex the N_2H_2 ligand is coordinated as a diazene, whereas in the cationic complex it coordinates as a hydrazido(2−) group. The stabilization of diazene by coordination to a tungsten carbonyl complex has been noted.[366,367]

$$[W(N_2)_2(diphos)] \xrightarrow{HCl} [WCl_2(diphos)(N_2H_2)] + N_2 \quad (26)$$

Crystal structure analysis of $[WCl(diphos)_2(N_2H_2)]BPh_4$[368] clearly shows that the N_2H_2 ligand is indeed coordinated as a hydrazido(2−) group. The W—N—N linkage is linear with W—N and N—N distances of 1.73 and 1.37 Å respectively. The short W—N distance clearly suggests a considerable multiple bond character to the W—N bond.

Treatment of $[W(N_2)_2(diphos)_2]$ with two equivalents of HCl gives the hydrido complex $[WH(N_2)_2(diphos)_2](HCl_2)$, which in refluxing methanol yields the dihydride of W^{IV} $[WH_2Cl_2(diphos)_2]$.[369] Treatment of $[WX_2(diphos)_2(N_2H_2)]$ (X = Cl, Br) with $NaBH_4$ in ethanol, or irradiation of the dinitrogen complex under H_2 atmosphere leads to the tetrahydrido complex $[WH_4(diphos)_2]$.[370]

Deprotonation of $[WX_2(diphos)_2(N_2H_2)]$ and $[WF(diphos)_2(N_2H_2)]^+$ with Et_3N or aqueous K_2CO_3 under N_2 yields $[WX(diphos)_2(N_2H)]$.[371] The resulting diazenido complexes (W═N═N—H) are characterized by a strong IR band at 1890 cm^{-1} assigned to ν(N—N). This reaction can be reversed by an acid. The diazenido group is formally analogous to NO and thus the complex reacts smoothly with NO to give $[WX(NO)(diphos)_2]$ (X = F, Cl, Br).[371]

Ammonia is produced in good yield from dinitrogen complexes containing monodentate phosphine ligands.[372] Thus, treatment of *cis*-[$W(N_2)_2(PMe_2Ph)_4$] or *trans*-[$W(N_2)_2(PMePh_2)_4$] with H_2SO_4 in methanol gives approximately 1.9 moles of NH_3 per gram-atom of tungsten. The fact that hydrazine is a minor product in these reactions indicates that ammonia is formed by splitting of the N—N bond.[372] The amount of hydrazine can be increased if, prior to the addition of H_2SO_4, *trans*-[$W(N_2)_2(PMePh_2)_4$] is treated with anhydrous HCl to produce the hydrazido(1−) complex [$WCl_3(PMePh_2)_2(NHNH_2)$]. Thus it appears that hydrazine production depends on the formation of the W—$NHNH_2$ moiety.[373]

The reaction of [$WBr_2(PMe_2Ph)_3(NNH_2)$] with one equivalent of HCl gas in DME affords [$WClBr_2(PMe_2Ph)_3(N_2H_3)$], which on acid treatment produces hydrazine in preference to ammonia. The complex has one labile halide, and its correct formulation as [$WClBr(PMe_2Ph)_3(N_2H_3)$]Br has been established by crystal structure analysis. The linear W—N—N linkage indicates that the N_2H_3 group is bonded to the metal as W—N—NH_3 rather than as W—NH—NH_2.[374] It has also been shown that *cis*-[$W(N_2)_2(PMe_2Ph)_4$] reacts with excess HCl to give hydrazine in moderate yield.[375] [$WBr_2(PMe_2Ph)_3(NNH_2)$] reacts with one equivalent of HCl to give the hydrido, hydrazido(2−) complex (**34**). It reacts with $NaBPh_4$ to give a novel diazenido complex (**35**) with an N—B bond.[375]

[$WHClBr(PMe_2Ph)_3(NNH_2)$]Br
(**34**)

[$WHClBr(PMe_2Ph)_3(NN(H)BPh_3)$]
(**35**)

37.8.2 Reactions Leading to Carbon–Nitrogen Bond Formation

The reactions of *trans*-[$W(N_2)_2(diphos)_2$] with acid chloride in the presence of HCl were the first reported in which a stable dinitrogen complex was converted to an organonitrogen compound.[376] The products, [$WCl_2(diphos)_2(N_2HCOR)$] (R = Me, Et, Ph, *p*-$MeOC_6H_4$), are pink diamagnetic complexes that have no $\nu(N_2)$ stretching bands in their IR spectra. In the presence of a base such as Et_3N these compounds lose HCl to give the diamagnetic complexes [$WCl(diphos)_2(N_2COR)$] in which the azo ligand is also coordinated through the carbonyl oxygen. The reaction can be reversed by addition of HCl. [$WCl_2(diphos)_2(N_2H_2)$][369] and [$WX(diphos)_2(N_2H)$] react in an analogous way with MeCOCX to give [$WX_2(diphos)_2(N_2HCOMe)$] (X = Cl, Br).[371]

Visible light irradiation of a benzene solution containing alkyl bromides and *trans*-[$W(N_2)_2(diphos)$] produces the alkylated complexes [$WBr(diphos)_2(N_2R)$] (R = Me, Et, Bu^t).[377,378] These can be protonated to give complexes containing the (W═N—NHR) unit. [$WBr(diphos)_2(N_2HMe)$] can also be obtained from the reaction of [$WBr(diphos)_2(N_2H)$] with MeBr.[377] When the irradiation is carried out in THF rather than in benzene[379] the ω-diazobutanol compound (**36**) is obtained. Crystal structure analysis is consistent with this hydrazido(2−) formulation. The W—N—N linkage is linear and the W—N and N—N distances are 1.778 and 1.306 Å respectively. In the PF_6^- salt these distances are 1.772 and 1.32 Å.[380] Several structures of complexes containing the W═N—NHR or W═N—N═CR_2 groups have been determined. They are all similar in that the W—N—N linkage is linear with a short W—N distance suggesting a considerable multiple bond character to the W—N bond. It should be noted that although there are small differences in the W—N and N—N bond lengths in these structures, the trend expected is usually that observed, namely, the lengthening of the W—N bond is accompanied by the shortening of the N—N bond.[269,380]

[$WBr(diphos)_2(N—N═CH(CH_2)_3OH)$]Br
(**36**)

The reactions of [$W(N_2)_2(diphos)_2$] with RBr are clearly catalyzed by visible light. The homolytic fission of the R—Br bond that takes place at the metal center[381] is preceded by the loss of one N_2 molecule. The resulting C—N bond is formed by an alkyl radical attack on the remaining coordinated dinitrogen. Product distribution in these photocatalyzed reactions depends on the solvent and the stability of the free radical.[381,382] This mechanism is strongly supported by flash photolysis experiments.[370] When *gem*-dibromides are used in the photocatalyzed reaction, diazoalkanes are produced. With CH_2Br_2 for example, the diazomethane complex [$WBr(diphos)_2(N_2CH_2)$]Br is obtained.[383] More recently it has been shown that some of the diazoalkanes do not react with protonic acids, but that the unique carbon atom is attacked by nucleophiles such as MeLi to yield diazenido complexes.[384]

The reaction of [$WBr(diphos)_2(N_2H_2)$]$^+$ and (Ph_2I)Cl in the two phase system $CHCl_3$–aqueous K_2CO_3 leads to [$WBr(diphos)_2(NN═CCl_2)$]$^+$, which reacts with a variety of nucleophiles to give novel organonitrogen ligands. It reacts with $MeNH_2$, for example, to give $[WBr(diphos)_2\{N{-}N{=}C(NHMe)\}_2]^+$ and with F^- to give $[WBr(diphos)_2(N{=}N{-}CF_3)]$.[385] $[WBr(diphos)_2(N_2H_2)]^+$ also reacts with cyanoalkenes to give vinyldiazenido complexes or, after loss of the hydrazido(2−) ligand, to form nitrile-derived methylene amino complexes.[386]

Hydrazido(2−) and hydrazido(1−) complexes have also been shown to condense with aldehydes and ketones to give diazoalkane complexes containing the $W{=}N{-}N{=}CR_1R_2$ unit.[387,388] Treatment of these complexes with $LiAlH_4$ gives secondary amines and ammonia, whereas treatment with acid produces hydrazine, keto azines and N_2-free tungsten compounds. Amines can also be produced from organohydrazido(2−) complexes.[389,390]

37.9 NITROSYL COMPLEXES

Nitrosyl complexes of tungsten were first prepared in 1964 by reaction of $W(CO)_6$ with nitrosyl chloride in an inert atmosphere. $[W(NO)_2Cl_2]_n$ is a dark green hygroscopic, air-sensitive material, soluble only in coordinating solvents.[391] The analogous $[W(NO)_2Br_2]_n$ can be prepared by using NOBr[392] or by the reaction of NOBr with $[W_2(CO)_8Cl_4]$.[393]

$[W(NO)_2Cl_2]_n$ reacts with a variety of ligands to form octahedral complexes of the type $[W(NO)_2Cl_2L_2]$ ($L = PPh_3$, $AsPh_3$, py, $MeC_6H_4NH_2$, $C_6H_{11}NH_2$, Ph_3PO).[391,393] The IR spectra of these derivatives consist of two ν(NO) bands around 1750 and 1650 cm^{-1}, indicating a *cis* dinitrosyl arrangement. The positions of these bands suggests the NO^+ formulation and thus formally can be considered as octahedral complexes of W^0. With the bidentate dimethyldithiocarbamate ligand, $[W(NO)_2Br_2]_n$ gives *cis*-$[W(NO)_2(S_2CNMe_2)_2]$.[394]

Several nitrosyl complexes have been prepared by the reaction of nitrosyl chloride with anionic carbonyls. Thus, $[W(CO)_5Cl]^-$ reacts with NOCl[395] to give $[W(CO)_4(NO)Cl]$ while $[W(NO)Cl_2L]$ and $[W(NO)_2ClL]$ ($L = RB(P_z)_3$) were prepared from $[LW(CO)_3]^-$ with variable amounts of NOCl.[396] $[W_2(CO)_8Cl_4]$ also reacts with NOCl to give in addition to $[W(NO)_2Cl_2]_n$ the mononitrosyl $[W(NO)Cl_3]$[393] which exhibits only one ν(NO) bond at 1590 cm^{-1}.

WCl_6 or WCl_5 reacts with NO in benzene to give red air-sensitive material which reacts with phosphine ligands to give $[WCl_3(NO)(OPR_3)_2]$ ($PR_3 = PPh_3$, $PMePh_2$, $PEtPh_2$).[397] $[WCl_3(NO)(OPPh_3)_2]$ can also be obtained from $[W(CO)_4(PPh_3)_2]$ and NOCl.[393] Careful choice of reaction conditions also gives the $[WCl_3(NO)L_2]$ compounds with $L = PMePh_2$, $PEtPh_2$ and py.[397] The mononitrosyl complexes $[WX(NO)(diphos)_2]$ (X = Cl, Br) are also known.[371]

Cationic nitrosyl complexes have also been prepared. The reaction of $[W(CO)_3(MeCN)_3]$ with $NOPF_6$ in acetonitrile affords *cis*-$[W(NO)_2(CO)(MeCN)_3]$ $(PF_6)_2$ which exhibit two ν(NO) IR bands at 1867 and 1775 cm^{-1}. The proton NMR spectra consists of two bands of relative intensities of 2:1 consistent with both *mer* and *fac* arrangement of the MeCN ligands.[24] *cis*-$[W(NO)_2(CO)(MeCN)_3]^{2+}$ reacts with NaS_2CNEt_2 to give the known *cis*-$[W(NO)_2(S_2CNEt_2)_2]$ and with acac in the presence of base to give *cis*-$[W(NO)_2(acac)_2]$.[394]

The reaction of $[W(CO)_4(diphos)]$ with $NOPF_6$ gives $[W(CO)_3(NO)(diphos)]PF_6$.[398] It reacts with halides and dithiocarbamates to yield $[W(CO)_2(NO)(diphos)X]$ (X = Cl, Br, I) and $[W(CO)(NO)(diphos)(S_2CNR_2]$ (R = Me, Et). Spectroscopic data suggest that in the last compound the CO and NO ligands are *cis* to one another. Similar complexes can also be obtained with other bidentate ligands.[398]

Treatment of $[W(NO)_2Cl_2]_n$ with $AgBF_4$ in MeCN leads to *cis*-$[W(NO)_2(MeCN)_4](BF_4)_2$. NMR measurements have shown that this complex undergoes stereospecific exchange of coordinated MeCN *via* a dissociative pathway. The two labile MeCN groups are those *trans* to the nitrosyl ligands.[399]

Anionic nitrosyl complexes were prepared by the reaction of $[W(NO)_2Cl_2]_n$ with Ph_4AsCl,[39] and Ph_4PBr. $[W(NO)_2Cl_4]^{2-}$ and $[W(NO)_2Cl_2Br_2]^{2-}$ are green materials with *cis* dinitrosyl arrangement. They react with dithiolene type ligands to give *cis*-$[W(NO)_2(L{-}L)_2]^{2-}$ (L–L = mnt, i-mnt, tbd).[400] The two IR bands associated with ν(NO) decrease in the order dtc > mnt > i-mnt > tbd. This trend has been rationalized in terms of the effects of charge and π acceptor ability of the various dithiolene ligands.

37.10 HYDRIDO COMPLEXES

Transition metal hydrides are clearly an important class of metal complexes, and a number of books and review articles have appeared during the last 15 years. Hydrido complexes of tungsten contain in addition to the hydride such ligands as C_5H_5, phosphine, NO and CO.

With phosphine ligands, complexes with a different number of hydrides are known. $[WH_4L_4]$ (L = PMe_2Ph, $PMePh_2$) were first prepared by the reaction of $[WCl_4L_4]$ with $NaBH_4$ in alcoholic solutions in the presence of excess phosphine.[401] The crystal structures of $[WH_4(PEtPh_2)_4]$ and $[WH_4\{P(OPr^i)_3\}_4]$[402] have been determined. These are all yellow air-stable materials, and exhibit a complex IR spectrum in the 1700–1850 cm^{-1} region associated with ν(WH). The proton NMR spectrum of $[WH_4(PMePh_2)_4]$ at room temperature is consistent with a rigid structure. A fluxional behavior develops above 50 °C. $[WH_4(PMePh)_4]$ behaves similarly.[401] With other phosphines the same fluxional behavior is observed in solution.[403] An intramolecular rearrangement pathway has been suggested to explain this behavior. With the diphos ligand, $[WH_4(diphos)_2]$ is obtained by the reaction of $[WX_2(diphos)_2(N_2H_2)]$ (X = Cl, Br) with $NaBH_4$.[368]

Complexes containing the WH_5 and WH_6 moieties are also known.[404] *trans*-$[WX_4(PMe_2Ph)_2]$ (X = Cl, Br) complexes react with $NaBH_4$ in ethanol to give $[WH_6(PMe_2Ph)_3]$. NMR measurements show that the six hydrides are coupled equally to all three phosphorus nuclei. A reasonable structure for this complex is a trigonal prism of hydrogen atoms with phosphorus donor atoms capping the square faces of the prism.[403]

Hydrogenolysis of $[WMe_6]$ in the presence of phosphine ligands leads to different hydrido complexes depending on the size of the phosphine ligands. Thus, while PMe_2Ph gives primarily $[WH_4L_4]$, with the more bulky PPr^i_2Ph and PPr^i_3 ligands $[WH_6L_3]$ is obtained. $[WH_6(PPr^i_2Ph)_3]$ has been shown by ^{31}P NMR and X-ray diffraction to have approximately C_{2v} symmetry in the solid and in solution.[405]

The pentahydrido complex $[WH_5(PMePh_2)_4]^+$ is obtained[406] from the reaction of $[WH_4(PMePh_2)_4]$ with acids such as HBF_4, HPF_6 and TFA. The isolated salt with TFA, loses hydrogen on melting to give the dihydride $[WH_2(TFA)_2(PMePh_2)_3]$. The dihydride $[WH_2Cl_2(diphos)_2]$ is also known.[369]

The photolysis of a hexane solution of $[WL_6]$ (L = $P(OMe)_3$) in an atmosphere of H_2 yields a mixture of $[WH_2L_5]$ and $[WH_4L_4]$.[407]

Several monohydrido complexes have also been prepared. Treatment of *cis*-$[W(CO)_2(dpm)_2]$ with $O_2/HClO_4$ yields *trans*-$[WH(CO)_2dpm)_2]^+$ which is believed to be seven-coordinate with the capped octahedral geometry.[408] Similar complexes with other bidentate phosphine ligands are also known.[409,410] The preparation of $[WH(N_2)_2(diphos)_2]^+$ and $[WH(CNR)_2(diphos)_2]^+$ have been reported.[369,411] Although $[WH(N_2)_2(diphos)_2]^+$ has a pentagonal bipyramid geometry in the solid state, with the hydride ligand in the pentagonal plane, NMR data are compatible with both the pentagonal bipyramid and capped octahedron structures in solution.[411]

Recently it has been shown that Na/Hg reduction of $[WCl_4(PMe_3)_3]$ under H_2 gives the dihydride $[WH_2Cl_2(PMe_3)_4]$ and that the reaction of $[WCl_2(PMe_3)_4]$ with methanol in THF leads to the same product.[412]

Finally, it should be mentioned that several μ-hydrido-bridged tungsten carbonyl and mixed carbonyl nitrosyl complexes are known. When only one bridging hydride is present the W—H—W bond is always bent.[413] Bridging hydrides are also found together with other bridging groups in several tungsten complexes. For example, the reduction of $[W_2Cl_4(PMe_3)_4]$ in THF with Na/Hg leads to a species which is best formulated as $[W_2H_4(\mu\text{-}H)(\mu\text{-}PMe_2)(PMe_3)_5]$. This complex is described as a W^{II}–W^{IV} dimer (W—W distance = 2.588 Å) with four terminal hydrido ligands and one bridging hydride.[414]

37.11 REFERENCES

1. M. E. Weeks and H. M. Leicester, *J. Chem. Educ.*, 1968, 241.
2. F. A. Cotton, *Acc. Chem. Res.*, 1978, **11**, 225.
3. H. R. Schubert, 'The Steel Industry', p. 64.
4. R. Chadwick, 'New Extraction Processes for Metals', in 'A History of Technology', Oxford University Press, London, 1958, vol. 5, p. 98.
5. G. D. Rieck, 'Tungsten and Its Compounds', Pergamon, Oxford, 1967, p. 4.
6. G. W. Fraser, C. J. W. Gibbs and R. D. Peacock, *J. Chem. Soc. (A)*, 1970, 1708.
7. O. Leitzke and F. Sladky, *Z. Anorg. Allg. Chem.*, 1981, **480,** 7.
8. H. Funk and H. Böhland, *Z. Anorg. Allg. Chem.*, 1963, **324,** 168.
9. A. Majid, D. W. A. Sharp, J. M. Winfield and I. Hanley, *J. Chem. Soc., Dalton Trans.*, 1973, 1876.
10. J. Fawcett, R. D. Peacock and D. R. Russel, *J. Chem. Soc., Dalton Trans.*, 1980, 2294.
11. D. C. Bradley, M. H. Chisholm and M. W. Extine, *Inorg. Chem.*, 1977, **16,** 1791.
12. M. H. Chisholm, F. A. Cotton, M. W. Extine and B. R. Stults, *J. Am. Chem. Soc.*, 1976, **98,** 4477.

13. D. C. Bradley, M. H. Chisholm, M. W. Extine and M. E. Stager, *Inorg. Chem.*, 1977, **16**, 1794.
14. M. H. Chisholm and M. W. Extine, *J. Am. Chem. Soc.*, 1974, **96**, 6214.
15. L. B. Handy and F. E. Brinckman, *Chem. Commun.*, 1970, 214.
16. A. M. Noble and J. M. Winfield, *J. Chem. Soc. (A)*, 1970, 2574.
17. L. B. Handy and C. K. Fair, *Inorg. Nucl. Chem. Lett.*, 1975, **11**, 497.
18. L. B. Handy, *Acta Crystallogr., Sect. B*, 1975, **31**, 300.
19. J. Schearle and F. A. Schröder, *Acta Crystallogr., Sect. B*, 1974, **30**, 2772.
20. P. M. Boorman, T. Chivers, K. N. Mahadev and B. D. O'Dell, *Inorg. Chim. Acta*, 1976, **19**, L35.
21. C. Santini-Scampucci and J. G. Riess, *J. Chem. Soc., Dalton Trans.*, 1976, 195.
22. A. L. Galyer and G. Wilkinson, *J. Chem. Soc., Dalton Trans.*, 1976, 2235.
23. A. J. Shortland and G. Wilkinson, *J. Chem. Soc., Dalton Trans.*, 1973, 872.
24. M. Green and S. H. Taylor, *J. Chem. Soc., Dalton Trans.*, 1972, 2629.
25. P. I. Mortimer and M. I. Strong, *Aust. J. Chem.*, 1965, **18**, 1579.
26. E. I. Stiefel, L. E. Bennett, Z. Dori, T. H. Crawford, C. Simo and H. B. Gray, *Inorg. Chem.*, 1970, **9**, 281.
27. S. R. Fletcher, A. Shortland, A. C. Skapski and G. Wilkinson, *J. Chem. Soc., Chem. Commun.*, 1972, 922.
28. F. A. Cotton and W. Schwotzer, *J. Am. Chem. Soc.*, 1983, **105**, 5639.
29. R. A. Jones, G. Wilkinson, A. M. R. Galas and M. B. Hursthouse, *J. Chem. Soc., Chem. Commun.*, 1979, 926.
30. A. Prescott, D. W. A. Sharp and J. M. Winfield, *J, Chem. Soc., Dalton Trans.*, 1975, 934.
31. N. Bartlett, S. P. Beaton and N. K. Jha, *Chem. Commun.*, 1966, 168.
32. B. Glavincevsky and S. Brownstein, *Inorg. Chem.*, 1981, **20**, 3580.
33. B. Cox, D. W. A. Sharp and A. G. Sharpe, *J. Chem. Soc.*, 1956, 1242.
34. P. M. Boorman, M. Islip, M. M. Reimer and K. J. Reimer, *J. Chem. Soc., Dalton Trans.*, 1972, 890.
35. H. Hess and H. Hartung, *Z. Anorg. Allg. Chem.*, 1966, **344**, 157.
36. A. J. Edwards and G. R. Jones, *J. Chem. Soc. (A)*, 1968, 2074.
37. M. J. Bennett, T. E. Haas and J. T. Purdham, *Inorg. Chem.*, 1972, **11**, 207.
38. G. W. A. Fowles and J. L. Frost, *J. Chem. Soc. (A)*, 1967, 671.
39. H. Funk and G. Mohaupt, *Z. Anorg. Allg. Chem.*, 1962, **315**, 204.
40. G. W. A. Fowles and J. L. Frost, *J. Chem. Soc. (A)*, 1966, 1631.
41. D. M. Adams, J. Chatt, J. M. Davidson and J. Gerratt, *J. Chem. Soc.*, 1963, 2189.
42. J. H. Holloway, G. J. Schrobilgen and P. Taylor, *J. Chem. Soc., Chem. Commun.*, 1975, 40.
43. J. H. Holloway and G. J. Schrobilgen, *Inorg. Chem.*, 1980, **19**, 2632.
44. J. Fawcett, J. H. Holloway and D. R. Russell, *J. Chem. Soc., Dalton Trans.*, 1981, 1212.
45. K. Yamanouchi and S. Yamada, *Inorg. Chim. Acta*, 1974, **11**, 223.
46. Y. V. Kokunov, V. A. Bochkareva and Y. A. Guslaev, *Koord. Khim.*, 1981, **7**, 1853.
47. I. W. Bassi and R. Scordamaglia, *J. Organomet. Chem.*, 1975, **99**, 127.
48. M. G. B. Drew and R. Mandyczewsky, *Chem. Commun.*, 1970, 292.
49. B. Spivack and Z. Dori, *Coord. Chem. Rev.*, 1975, **17**, 99.
50. J. A. Connor and E. A. V. Ebsworth, *Adv. Inorg. Chem. Radiochem.*, 1964, **6**, 313.
51. F. W. B. Einstein and B. R. Penfold, *Acta Crystallogr. (Suppl.), Sect. A*, 1963, **16**, 35.
52. Z. Ruzic-Toros, B. Kojic-Prodii, F. Gabela and M. Sljukic, *Acta Crystallogr., Sect. B*, 1977, **33**, 692.
53. S. E. Jacobson, R. Tang and F. Mares, *Inorg. Chem.*, 1978, **17**, 3055.
54. A. D. Westland, F. Haque and J. M. Bouchard, *Inorg. Chem.*, 1980, **19**, 2255.
55. O. Jarchow, F. Schroder and H. Schultz, *Z. Anorg. Allg. Chem.*, 1968, **363**, 58.
56. C. G. Barraclough and J. Stals, *Aust. J. Chem.*, 1966, **19**, 741.
57. K. M. Sharma, S. K. Anand, R. K. Multani and B. D. Jain, *Chem. Ind. (London)*, 1969, 1556.
58. B. J. Brisdon, *Inorg. Chem.*, 1967, **6**, 1791.
59. G. J.-J. Chen, R. O. Yelton and J. W. McDonald, *Inorg. Chim. Acta*, 1977, **22**, 249.
60. G. J.-J. Chen, J. W. McDonald and W. E. Newton, *Inorg. Chim. Acta*, 1976, **19**, L67.
61. M. G. B. Drew, G. W. A. Fowles, D. A. Rice and K. J. Shanton, *J. Chem. Soc., Chem. Commun.*, 1974, 614.
62. R. S. Taylor, P. Gans, P. F. Knowles and A. G. Sykes, *J. Chem. Soc., Dalton Trans.*, 1972, 24.
63. D. L. Rabenstein, M. S. Greenberg and R. Saetre, *Inorg. Chem.*, 1977, **16**, 1241.
64. A. K. Sengupta and S. K. Nath, *Indian J. Chem., Sect. A*, 1981, **20**, 203.
65. Y. Jeannin, J.-P. Launay, J. Livage and A. Nel, *Inorg. Chem.*, 1978, **17**, 374.
66. J.-P. Launay, Y. Jeannin and A. Nel, *Inorg. Chem.*, 1983, **22**, 277.
67. J.-P. Launay and F. Babonneau, *Chem. Phys.*, 1982, **67**, 295.
68. D. Britnell, G. W. A. Fowles and R. Mandyczewsky, *Chem. Commun.*, 1970, 608.
69. D. Britnell, G. W. A. Fowles and D. A. Rice, *J. Chem. Soc., Dalton Trans.*, 1974, 2191.
70. M. G. B. Drew and R. Mandyczewsky, *J. Chem. Soc. (A)*, 1970, 2815.
71. M. J. Atherton and J. H. Holloway, *J. Chem. Soc., Chem. Commun.*, 1977, 424.
72. G. W. A. Fowles, R. J. Hobson, D. A. Rice and K. J. Shanton, *J. Chem. Soc., Chem. Commun.*, 1976, 552.
73. (a) D. Britnell, G. W. A. Fowles and D. A. Rice, *J. Chem. Soc., Dalton Trans.*, 1975, 213; (b) D. Britnell, M. G. B. Drew, G. W. A. Fowles and D. A. Rice, *Inorg. Nucl. Chem. Lett.*, 1973, **9**, 501.
74. D. Britnell, M. G. B. Drew, G. W. A. Fowles and D. A. Rice, *J. Chem. Soc., Chem. Commun.*, 1972, 462.
75. R. A. Walton, *Prog. Inorg. Chem.*, 1972, **16**, 135.
76. (a) A. M. Noble and J. M. Winfield, *Inorg. Nucl. Chem. Lett.*, 1968, **4**, 339; (b) D. Britnell, M. G. B. Drew, G. W. A. Fowles and D. A. Rice, *Inorg. Nucl. Chem. Lett.*, 1973, **9**, 415.
77. K. Dehnicke, *J. Inorg. Nucl. Chem.*, 1965, **27**, 809.
78. J. Strähle, *Z. Anorg. Allg. Chem.*, 1970, **375**, 238.
79. P. Ruschke and K. Dehnicke, *Z. Naturforsch., Teil B*, 1980, **35**, 1589.
80. F. Weller, W. Liebelt and K. Dehnicke, *Z. Anorg. Allg. Chem.*, 1982, **484**, 124.
81. K. Dehnicke and J. Strähle, *Z. Anorg. Allg. Chem.*, 1965, **339**, 171.
82. K. Dehnicke, H. Prinz, W. Kafitz and R. Kujanek, *Liebigs Ann. Chem.*, 1981, **1**, 20.

83. W. Musterle, J. Straehle, W. Liebelt and K. Dehnicke, *Z. Naturforsch., Teil B,* 1979, **34,** 942.
84. W. Kolitsch and K. Dehnicke, *Z. Naturforsch., Teil B,* 1970, **25,** 1080.
85. G. W. A. Fowles, D. A. Rice and K. J. Shanton, *J. Chem. Soc., Dalton Trans.,* 1977, 1212.
86. (a) M. G. B. Drew, K. C. Moss and N. Rolfe, *Inorg. Nucl. Chem. Lett.,* 1971, **7,** 1219; (b) M. G. B. Drew, G. W. A. Fowles, D. A. Rice and N. Rolfe, *J. Chem. Soc., Chem. Commun.,* 1971, 231.
87. S. F. Pedersen and R. R. Schrock, *J. Am. Chem. Soc.,* 1982, **104,** 7483.
88. G. W. A. Fowles, D. A. Rice and K. J. Shanton, *J. Chem. Soc., Dalton Trans.,* 1977, 2128.
89. U. Weiher and K. Dehnicke, *Z. Anorg. Allg. Chem.,* 1979, **457,** 105.
90. P. Ruschke and K. Dehnicke, *Z. Anorg. Allg. Chem.,* 1980, **466,** 157.
91. D. G. Blight and D. L. Kepert, *J. Chem. Soc. (A),* 1968, 534.
92. J. H. Wengrovius, R. R. Schrock, M. R. Churchill, J. R. Missert and W. J. Youngs, *J. Am. Chem. Soc.,* 1980, **102,** 4515.
93. M. H. Chisholm, J. C. Huffman and N. S. Marchant, *J. Am. Chem. Soc.,* 1983, **105,** 6162.
94. M. H. Chisholm, D. M. Hoffman and J. C. Huffman, *Inorg. Chem.,* 1983, **22,** 2903.
95. J. J. Berzelius, *Ann. Phys. (Leipzig),* 1826, **8,** 277.
96. A. Coleis, *Liebigs Ann. Chem.,* 1886, **232,** 268.
97. K. Sasvari, *Acta Crystallogr.,* 1963, **16,** 719.
98. A. Müller, H.-H. Heinsen and G. Vandrish, *Inorg. Chem.,* 1974, **13,** 1001.
99. K. P. Callahan and P. A. Piliero, *J. Chem. Soc., Chem. Commun.,* 1979, 13.
100. J. C. Huffman, R. S. Roth and A. R. Siedle, *J. Am. Chem. Soc.,* 1976, **98,** 4340.
101. A. Müller and E. Diemann, *J. Chem. Soc., Chem. Commun.,* 1971, 65.
102. A. Müller, M. C. Chakravorti and H. Dornfeld, *Z. Naturforsch., Teil B,* 1975, **30,** 162.
103. A. Müller and H. H. Heinsen, *Chem. Ber.,* 1972, **105,** 1730.
104. A. Müller, N. Wernstock, B. Krebs, B. Buss and A. Ferwanah, *Z. Naturforsch., Teil B,* 1971, **26,** 268.
105. A. Müller, E. Koeniger-Ahlborn, E. Krickemeyer and R. Jostes, *Z. Anorg. Allg. Chem.,* 1981, **483,** 69.
106. A. R. Siedle, T. Negas and J. Broussalian, *J. Inorg. Nucl. Chem.,* 1975, **37,** 2024.
107. D. Coucouvanis, *Acc. Chem. Res.,* 1981, **14,** 201.
108. A. Müller, W. Hellmann, U. Schimanski and W. E. Newton, *Z. Naturforsch.,* 1983, **38,** 528.
109. K. P. Callahan and P. A. Piliero, *Inorg. Chem.,* 1980, **19,** 2619.
110. J. W. McDonald, G. D. Friesen, W. E. Newton, A. Muller, W. Hellmann, U. Schimanski, A. Trautwein and U. Bender, *Inorg. Chim. Acta,* 1983, **76,** L297.
111. A. Müller and S. Sarkar, *Angew. Chem., Int. Ed. Engl.,* 1977, **16,** 705.
112. P. Stremple, N. C. Baenziger and D. Coucouvanis, *J. Am. Chem. Soc.,* 1981, **103,** 4601.
113. A. Müller, H. Dornfeld, G. Henkel, B. Krebs and M. P. A. Viegers, *Angew. Chem., Int. Ed. Engl.,* 1978, **17,** 52.
114. A. Müller, E. Diemann and H.-H. Heinsen, *Chem. Ber.,* 1971, **104,** 975.
115. A. Müller, T. K. Hwang and H. Bogge, *Angew. Chem., Int. Ed. Engl.,* 1979, **18,** 628.
116. A. Müller, I. Paulet-Böschen, B. Krebs and H. Dornfeld, *Angew. Chem., Int. Ed. Engl.,* 1976, **15,** 633.
117. A. Müller, H. Bogge and U. Schimanski, *Inorg. Chim. Acta,* 1983, **69,** 5.
118. V. Lenher and A. G. Fruchan, *J. Am. Chem. Soc.,* 1927, **49,** 3076.
119. A. Müller, E. Ahlborn and H. H. Heinsen, *Z. Anorg. Allg. Chem.,* 1971, **386,** 102.
120. J. A. McCleverty, *Prog. Inorg. Chem.,* 1968, **10,** 145.
121. R. Eisenberg, *Prog. Inorg. Chem.,* 1970, **12,** 295.
122. G. E. Brown and E. I. Stiefel, *Inorg. Chem.,* 1973, **12,** 2140.
123. A. Davison and E. T. Shawl, *Inorg. Chem.,* 1970, **9,** 1820.
124. H. F. Priest, *Inorg. Synth.,* 1950, **3,** 181.
125. J. J. Pitts and A. W. Jache, *Inorg. Chem.,* 1968, **7,** 1661.
126. M. H. Lietzke and M. L. Holt, *Inorg. Synth.,* 1950, **3,** 163.
127. W. W. Porterfield and S. Y. Tyree, *Inorg. Synth.,* 1967, **9,** 133.
128. R. Colton, I. B. Tomkins and P. W. Wilson, *Aust. J. Chem.,* 1964, **17,** 496.
129. J. E. Drake and G. W. A. Fowles, *J. Less-Common Met.,* 1960, **2,** 401.
130. S. E. Feil, S. Y. Tyree and F. N. Collier, Jr., *Inorg. Synth.,* 1967, **9,** 123.
131. P. C. Crouch, G. W. A. Fowles and R. A. Walton, *J. Inorg. Nucl. Chem.,* 1970, **32,** 329.
132. J. Tillack, P. Eckerlin and J. H. Dettingmeijer, *Angew Chem., Int. Ed. Engl.,* 1966, **5,** 421.
133. M. Postel, J. G. Riess, J. Y. Calves and J. Guerchais, *Inorg. Chim. Acta,* 1979, **32,** 175.
134. C. A. Rice, P. M. H. Kroneck and J. T. Spence, *Inorg. Chem.,* 1981, **20,** 1996.
135. W. E. Falconer, G. R. Jones, W. A. Sunder, I. Haigh and R. D. Peacock, *J. Inorg. Nucl. Chem.,* 1973, **35,** 751.
136. A. J. Edwards and G. R. Jones, *J. Chem. Soc. (A),* 1968, 2074.
137. F. Hein and S. Herzog, in 'Handbook of Preparative Inorganic Chemistry', 2nd edn., Academic, New York, 1965, vol. 2, p. 1419.
138. T. M. Brown and B. Ruble, *Inorg. Chem.,* 1967, **6,** 1335.
139. T. M. Brown and E. L. McCann, III, *Inorg. Chem.,* 1968, **7,** 1227.
140. J. E. Fergusson, in 'Halogen Chemistry', ed. V. Gutmann, Academic, New York, 1967, p. 265.
141. F. A. Cotton and C. E. Rice, *Acta Crystallogr., Sect. B,* 1978, **34,** 2833.
142. B. J. Brisdon, D. A. Edwards, D. J. Machin, K. S. Murray and R. A. Walton, *J. Chem. Soc. (A),* 1967, 1825.
143. K. W. Bagnall, D. Brown and J. G. H. du Pree, *J. Chem. Soc.,* 1964, 2603.
144. R. N. Dickinson, S. E. Feil, F. N. Collier, W. W. Horner, S. M. Horner and S. Y. Tyree, *Inorg. Chem.,* 1964, **3,** 1600.
145. B. J. Brisdon and R. A. Walton, *J. Inorg. Nucl. Chem.,* 1965, **27,** 1101.
146. B. J. Brisdon and R. A. Walton, *J. Chem. Soc.,* 1965, 2274.
147. W. Eichler and H.-J. Seifert, *Z. Anorg. Allg. Chem.,* 1977, **431,** 123.
148. G. B. Hargreaves and R. D. Peacock, *J. Chem. Soc.,* 1958, 2170.

149. D. G. Blight, D. L. Kepert, R. Mandyczewsky and K. R. Trigwell, *J. Chem. Soc., Dalton Trans.*, 1972, 313.
150. H. Funk and H. Schauer, *Z. Anorg. Allg. Chem.*, 1960, **306**, 203.
151. R. J. H. Clark, D. L. Kepert and R. S. Nyholm, *J. Chem. Soc.*, 1965, 2877.
152. E. A. Allen, B. J. Brisdon and G. W. A. Fowles, *J. Chem. Soc.*, 1964, 4531.
153. W. M. Carmichael, D. A. Edwards and R. A. Walton, *J. Chem. Soc. (A)*, 1966, 97.
154. R. E. McCarley and T. M. Brown, *Inorg. Chem.*, 1964, **3**, 1232.
155. P. M. Boorman, N. N. Greenwood, M. A. Hildon and R. V. Parish, *J. Chem. Soc. (A)*, 1968, 2002.
156. A. Fischer and L. Michielis, *Z. Anorg. Allg. Chem.*, 1913, **81**, 102.
157. O. J. Klejnot, *Inorg. Chem.*, 1965, **4**, 1668.
158. D. P. Rillema and C. H. Brubaker, Jr., *Inorg. Chem.*, 1969, **8**, 1645.
159. D. P. Rillema, W. J. Reagan and C. H. Brubaker, Jr., *Inorg. Chem.*, 1969, **8**, 587.
160. D. A. McClung, L. R. Dalton and C. H. Brubaker, Jr., *Inorg. Chem.*, 1966, **5**, 1985.
161. P. C. Crouch, G. W. A. Fowles, J. L. Frost, P. R. Marshall and R. A. Walton, *J. Chem. Soc. (A)*, 1968, 1061.
162. P. C. Crouch, G. W. A. Fowles, I. B. Tomkins and R. A. Walton, *J. Chem. Soc. (A)*, 1969, 2412.
163. D. E. Sands, A. Zalkin and R. E. Elson, *Acta Crystallogr.*, 1959, **12**, 21.
164. E. A. Allen, B. J. Brisdon, D. A. Edwards, G. W. A. Fowles and R. G. Williams, *J. Chem. Soc.*, 1963, 4649.
165. H. B. Gray and C. R. Hare, *Inorg. Chem.*, 1962, **1**, 363.
166. P. C. Crouch, G. W. A. Fowles, P. R. Marshall and R. A. Walton, *J. Chem. Soc. (A)*, 1968, 1634.
167. W. Levason, C. A. McAuliffe and F. P. McCullough, *Inorg. Chem.*, 1977, **16**, 2911.
168. W. Levason, C. A. McAuliffe and F. P. McCullough, *Inorg. Chim. Acta*, 1977, **24**, L13.
169. D. Britton, D. W. Macomber and P. G. Gassman, *Cryst. Struct. Commun.*, 1982, **11**, 1501.
170. A. J. Nielson and J. M. Waters, *Aust. J. Chem.*, 1983, **36**, 243.
171. D. C. Bradley, M. B. Hursthouse, K. M. Abdul Malik, A. J. Nielson and R. L. Short, *J. Chem. Soc., Dalton Trans.*, 1983, 2651.
172. M. G. B. Drew, G. F. Griffin and D. A. Rice, *Inorg. Chim. Acta*, 1979, **34**, L192.
173. J. W. Buchler and K. Rohbock, *Inorg. Nucl. Chem. Lett.*, 1972, **8**, 1073.
174. J. W. Buchler, L. Puppe, K. Rohbock and H. H. Schneehage, *Chem. Ber.*, 1973, **106**, 2710.
175. S. R. Fletcher, A. Shortland, A. C. Skapski and G. Wilkinson, *J. Chem. Soc., Chem. Commun.*, 1972, 922.
176. C. M. Newton and D. G. Davis, *J. Magn. Reson.*, 1975, **20**, 446.
177. D. Britnell, G. W. A. Fowles and D. A. Rice, *J. Chem. Soc., Dalton Trans.*, 1974, 2191.
178. K. Hanewald and G. Gattow, *Z. Anorg. Allg. Chem.*, 1980, **471**, 165.
179. R. Lozano, E. Alarcon, A. L. Doadrio and A. Doadrio, *Polyhedron*, 1983, **2**, 435.
180. R. Lozano, A. L. Doadrio, E. Alarcon, J. Roman and A. Doadrio, *Rev. Chim. Miner.*, 1983, **20**, 109.
181. S. Khalil, B. Sheldrick, A. B. Soares and A. G. Sykes, *Inorg. Chim. Acta*, 1977, **25**, L83.
182. S. Khalil and B. Sheldrick, *Acta Crystallogr., Sect. B*, 1978, **34**, 3751.
183. G. W. Fowles, 'Halides and Oxyhalides Complexes', in 'Preparative Inorganic Reaction', Interscience, New York, 1964, vol. 7, p. 137.
184. R. Mattes and K. Mennemann, *Z. Anorg. Allg. Chem.*, 1977, **437**, 175.
185. N. B. T. Yoshino, *J. Appl. Electrochem.*, 1982, **12**, 607.
186. A. Bino, F. A. Cotton, Z. Dori and J. C. Sekutowski, *Inorg. Chem.*, 1978, **17**, 2946.
187. F. Sécheresse, J. Lefebvre, J. C. Daran and V. Jeannin, *Inorg. Chim. Acta*, 1980, **45**, L45.
188. M. G. B. Drew, E. M. Page and D. A. Rice, *Inorg. Chim. Acta*, 1983, **76**, L33.
189. G. R. Hanson, A. A. Brunette, A. C. McDonell, K. S. Murray and A. G. Wedd, *J. Am. Chem. Soc.*, 1981, **103**, 1953.
190. (a) V. D. Patel, P. M. Boorman, K. A. Kerr and K. J. Moynihan, *Inorg. Chem.*, 1982, **21**, 1383; (b) M. G. B. Drew, G. W. A. Fowles, E. M. Page and D. A. Rice, *J. Am. Chem. Soc.*, 1979, **101**, 5827.
191. F. A. Cotton, D. DeMarco, B. W. S. Kolthammer and R. A. Walton, *Inorg. Chem.*, 1981, **20**, 3048.
192. F. A. Cotton, L. R. Falvello, M. F. Fredrich, D. DeMarco and R. A. Walton, *J. Am. Chem. Soc.*, 1983, **105**, 3088.
193. F. Feigl and P. Krumholz, *Angew. Chem.*, 1932, **45**, 674.
194. H. Böhland and E. Niemann, *Z. Anorg. Allg. Chem.*, 1965, **336**, 225.
195. C. J. Horn and T. M. Brown, *Inorg. Chem.*, 1972, **11**, 1970.
196. H. Paulssen-von Beck, *Z. Anorg. Allg. Chem.*, 1931, **196**, 85.
197. D. L. Kepert and R. Mandyczewsky, *Inorg. Chem.*, 1968, **7**, 2091.
198. C. D. Kennedy and R. D. Peacock, *J. Chem. Soc. (A)*, 1963, 3392.
199. K. B. Sharples and T. C. Flood, *J. Chem. Soc., Chem. Commun.*, 1972, 370.
200. M. A. S. King and R. E. McCarley, *Inorg. Chem.*, 1973, **12**, 1972.
201. A. V. Butcher, J. Chatt, G. J. Leigh and R. L. Richards, *J. Chem. Soc., Dalton Trans.*, 1972, 1064.
202. J. R. Moss and B. L. Shaw, *J. Chem. Soc. (A)*, 1970, 595.
203. J. W. Brencic, B. Oeh and P. Segedin, *Z. Anorg. Allg. Chem.*, 1979, **454**, 181.
204. L. Aslanov, R. Mason, A. G. Wheeler and P. O. Whimp, *J. Chem. Soc., Chem. Commun.*, 1970, 30.
205. M. G. B. Drew, E. M. Page and D. A. Rice, *J. Chem. Soc., Dalton Trans.*, 1983, 61.
206. W. Levason, C. A. McAuliffe, F. P. McCullough, S. G. Murray and C. A. Rice, *Inorg. Chim. Acta*, 1977, **22**, 227.
207. E. Carmona, L. Sanchez, M. L. Poveda, R. A. Jones and J. G. Hefner, *Polyhedron*, 1983, **2**, 797.
208. G. Doyle, *Inorg. Chem.*, 1971, **10**, 2348.
209. S. S. Eliseev, I. A. Glukhov and N. V. Gaidaenko, *Russ. J. Inorg. Chem. (Engl. Transl.)*, 1969, **14**, 328.
210. J. Tillack, R. Kaiser, G. Fisher and P. Eckerlin, *J. Less-Common Met.*, 1970, **20**, 171.
211. J. Chatt, L. Manojlovic-Muir and K. W. Muir, *Chem. Commun.*, 1971, 655.
212. O. Olsson, *Chem. Ber.*, 1913, **46**, 566.
213. E. Konig, *Inorg. Chem.*, 1969, **8**, 1278.
214. T. Glowiak, M. Sabat and B. Jezowska-Trzebiatowska, *Acta Crystallogr., Sect. B*, 1975, **31**, 1783.

215. J. SanFilippo, Jr., P. G. Fagan and F. J. Di Salvo, *Inorg. Chem.*, 1977, **16**, 1016.
216. L. B. Anderson, F. A. Cotton, D. DeMarco, A. Fang, W. H. Ilsley, B. W. S. Kolthammer and R. A. Walton, *J. Am. Chem. Soc.*, 1981, **103**, 5078.
217. D. DeMarco, T. Nimry and R. A. Walton, *Inorg. Chem.*, 1980, **19**, 575.
218. P. W. Clark and R. A. D. Wentworth, *Inorg. Chem.*, 1969, **8**, 1223.
219. H. J. Seifert, F. Petersen and H. Wöhrmann, *J. Inorg. Nucl. Chem.*, 1973, **35**, 2735.
220. W. J. Ragan and C. H. Brubaker, Jr., *Inorg. Chem.*, 1970, **9**, 827.
221. F. A. Cotton, L. R. Falvello, M. F. Fredrich, D. DeMarco and R. A. Walton *J. Am. Chem. Soc.*, 1983, **105**, 3088.
222. F. A. Cotton, D. DeMarco, L. R. Falvello and R. A. Walton, *J. Am. Chem. Soc.*, 1982, **104**, 7375.
223. M. H. Chisholm, J. C. Huffmann and J. Leonelli, *J. Chem. Soc., Chem. Commun.*, 1981, 270.
224. D. A. Wright and D. A. Williams, *Acta Crystallogr., Sect. B*, 1968, **24**, 1107.
225. F. A. Cotton and A. Fang, *J. Am. Chem. Soc.*, 1982, **104**, 113.
226. M. Akiyama, D. Little, M. H. Chisholm, D. A. Haitko, F. A. Cotton and M. W. Extine, *J. Am. Chem. Soc.*, 1979, **101**, 2504.
227. D. J. Palmer and P. G. Dickens, *Acta Crystallogr., Sect. B*, 1979, **35**, 2199.
228. V. D. Patel and P. M. Boorman, *Can. J. Chem.*, 1982, **60**, 1339.
229. P. M. Boorman, K. A. Kerr and V. D. Patel, *J. Chem. Soc., Dalton Trans.*, 1981, 506.
230. P. M. Boorman, P. W. Codding, K. A. Kerr, K. J. Moynihan and V. D. Patel, *Can. J. Chem.*, 1982, **60**, 1333.
231. J. M. Ball, P. M. Boorman, K. J. Moynihan, V. D. Patel, J. F. Richardson, D. Collison and F. E. Mabbs, *J. Chem. Soc., Dalton Trans.*, 1983, 2479.
232. W. H. McCarroll, L. Katz and R. Ward, *J. Am. Chem. Soc.*, 1957, **79**, 5410.
233. K. Mennemann and R. Mattes, *Angew. Chem., Int. Ed. Engl.*, 1976, **15**, 118.
234. F. A. Cotton, T. R. Felthouse and D. G. Lay, *Inorg. Chem.*, 1981, **20**, 2219.
235. V. Katovic, J. L. Templeton and R. E. McCarley, *J. Am. Chem. Soc.*, 1976, **98**, 5705.
236. A. Bino, F. A. Cotton, Z. Dori, S. Koch, H. Kuppers, M. Millar and J. C. Sekutowski, *Inorg. Chem.*, 1978, **17**, 3245.
237. F. A. Cotton, Z. Dori and M. Shaia, unpublished results.
238. M. Ardon, F. A. Cotton, Z. Dori, A. Fang, M. Kapon, G. M. Reisner and M. Shaia, *J. Am. Chem. Soc.*, 1982, **104**, 5394.
239. B. E. Bursten, F. A. Cotton, M. B. Hall and R. C. Najjar, *Inorg. Chem.*, 1982, **21**, 302.
240. J. N. Smith and T. M. Brown, *Inorg. Nucl. Chem. Lett.*, 1970, **6**, 441.
241. M. H. Chisholm and M. W. Extine, *J. Am. Chem. Soc.*, 1977, **99**, 782.
242. J. W. Wijenhoven, *Cryst. Struct. Commun.*, 1973, **2**, 637.
243. J. L. Hoard and J. V. Silverton, *Inorg. Chem.*, 1963, **2**, 235.
244. W. D. Bonds, Jr. and R. D. Archer, *Inorg. Chem.*, 1971, **10**, 2057.
245. W. D. Bonds, Jr., R. D. Archer and W. C. Hamilton, *Inorg. Chem.*, 1971, **10**, 1764.
246. C. J. Donahue and R. D. Archer, *Inorg. Chem.*, 1977, **16**, 2903.
247. R. D. Archer, W. D. Bonds, Jr. and R. A. Parbush, *Inorg. Chem.*, 1972, **11**, 1550.
248. A. Müller, H. Bögge, E. Krickemeyer, G. Henkel and B. Krebs, *Z. Naturforsch, Teil B*, 1982, **37**, 1014.
249. W.-H. Pan, M. E. Leonowicz and E. I. Stiefel, *Inorg. Chem.*, 1983, **22**, 672.
250. S. J. Holmes, D. N. Clark, H. W. Turner and R. R. Schrock, *J. Am. Chem. Soc.*, 1982, **104**, 6322.
251. F. A. Cotton and W. Schwotzer, *Inorg. Chem.*, 1983, **22**, 387.
252. R. A. Henderson, *J. Chem. Soc., Dalton Trans.*, 1983, 51.
253. P. C. Bevan, J. Chatt, J. R. Dilworth, R. A. Henderson and G. J. Leigh, *J. Chem. Soc., Dalton Trans.*, 1982, 821.
254. O. O. Olsson, *Z. Anorg. Allg. Chem.*, 1914, **88**, 49.
255. R. V. Parish, *Adv. Inorg. Chem. Radiochem.*, 1966, **9**, 315.
256. S. J. Lippard and B. J. Russ, *Inorg. Chem.*, 1967, **6**, 1943.
257. M. Dudek, A. Kanas and A. Samotus, *J. Inorg. Nucl. Chem.*, 1980, **42**, 1701.
258. A. Samotus, A. Kanas and M. Dudek, *J. Inorg. Nucl. Chem.*, 1979, **41**, 1129.
259. B. Sieklucka, A. Kanas and A. Samotus, *Transition Met. Chem. (Weinheim, Ger.)*, 1982, **7**, 131.
260. P. Guardado, A. Maestre and M. Balon, *J. Inorg. Nucl. Chem.*, 1981, **43**, 1391.
261. A. M. Soares, P. M. Kiernan, D. J. Cole-Hamilton and W. P. Griffith, *J. Chem. Soc., Chem. Commun.*, 1981, 84.
262. P. M. Kiernan and W. P. Griffith, *J. Chem. Soc., Dalton Trans.*, 1975, 2489.
263. R. A. Pribush and R. D. Archer, *Inorg. Chem.*, 1974, **13**, 2556.
264. R. V. Parish and P. G. Simms, *J. Chem. Soc., Dalton Trans.*, 1972, 2389.
265. A. D. Westland and N. Muriithi, *Inorg. Chem.*, 1973, **12**, 2356.
266. J. V. Brencic, B. Ceh and I. Leban, *Acta Crystallogr., Sect. B*, 1979, **35**, 3028.
267. J. V. Brencic, B. Ceh and P. Segedin, *J. Inorg. Nucl Chem.*, 1980, **42**, 1409.
268. M. Millar, S. Lincoln and S. A. Koch, *J. Am. Chem. Soc.*, 1982, **104**, 288.
269. Z. Dori, *Prog. Inorg. Chem.*, 1981, **28**, 239.
270. D. DeMarco, T. Nimry and R. A. Walton, *Inorg. Chem.*, 1980, **19**, 575.
271. J. L. Hayden and R. A. D. Wentworth, *J. Am. Chem. Soc.*, 1968, **90**, 5291.
272. W. H. Watson, Jr. and J. Waser, *Acta Crystallogr.*, 1958, **11**, 689.
273. F. A. Cotton, L. R. Falvello, G. N. Mott, R. R. Schrock and L. G. Sturgeoff, *Inorg. Chem.*, 1983, **22**, 2621.
274. R. Saillant and R. A. D. Wentworth, *J. Am. Chem. Soc.*, 1969, **91**, 2174.
275. W. H. Delphin and R. A. D. Wentworth, *Inorg. Chem.*, 1973, **12**, 1914.
276. W. C. Trogler, *Inorg. Chem.*, 1980, **19**, 697.
277. J. L. Templeton, R. A. Jacobson and R. E. McCarley, *Inorg. Chem.*, 1977, **16**, 3320.
278. F. A. Cotton and B. J. Kalbacher, *Inorg. Chem.*, 1976, **15**, 522.

279. V. Katovic and R. E. McCarley, *Inorg. Chem.*, 1978, **17,** 1268.
280. R. Saillant, J. L. Hayden and R. A. D. Wentworth, *Inorg. Chem.*, 1967, **6,** 1497.
281. R. B. Jackson and W. E. Streib, *Inorg. Chem.*, 1971, **10,** 1760.
282. S. Shaik and R. Hoffmann, *J. Am. Chem. Soc.*, 1980, **102,** 1194.
283. P. R. Sharp and R. R. Schrock, *J. Am. Chem. Soc.*, 1980, **102,** 1430.
284. P. M. Boorman, V. D. Patel, K. A. Kerr, P. W. Codding and P. Van Roey, *Inorg. Chem.*, 1980, **19,** 3508.
285. P. M. Boorman, K. J. Moynihan and K. A. Kerr, *J. Chem. Soc., Chem. Commun.*, 1981, 1286.
286. W. Mowat, A. Shortland, G. Yagupsky, N. J. Hill, M. Yagupsky and G. Wilkinson, *J. Chem. Soc., Dalton Trans.*, 1972, 533.
287. M. H. Chisholm, F. A. Cotton, M. Extine and B. R. Stults, *Inorg. Chem.*, 1976, **15,** 2252.
288. D. C. Bradley, M. H. Chisholm, C. E. Heath and M. B. Hursthouse, *Chem. Commun.*, 1969, 1261.
289. F. A. Cotton and R. A. Walton, 'Multiple Bonds Between Metal Atoms', Wiley, New York, 1982.
290. A. Akiyama, M. H. Chisholm, F. A. Cotton, M. W. Extine, D. A. Haitko, D. Little and P. E. Fanwick, *Inorg. Chem.*, 1979, **18,** 2266.
291. M. Akiyama, M. H. Chisholm, F. A. Cotton, M. W. Extine and C. A. Murillo, *Inorg. Chem.*, 1977, **16,** 2407.
292. M. H. Chisholm, F. A. Cotton, M. W. Extine, M. Millar and B. R. Stults, *Inorg. Chem.*, 1977, **16,** 320.
293. M. H. Chisholm and D. A. Haitko, *J. Am. Chem. Soc.*, 1979, **101,** 6784.
294. M. H. Chisholm, F. A. Cotton, M. W. Extine, M. Millar and B. R. Stults, *Inorg. Chem.*, 1976, **15,** 2244.
295. M. H. Chisholm, F. A. Cotton, M. W. Extine, M. Millar and B. R. Stults, *J. Am. Chem. Soc.*, 1976, **98,** 4486.
296. M. H. Chisholm and M. W. Extine, *J. Am. Chem. Soc.*, 1975, **97,** 5625.
297. M. H. Chisholm, F. A. Cotton, M. W. Extine and B. R. Stults, *Inorg. Chem.*, 1977, **16,** 603.
298. M. H. Chisholm, J. C. Huffman and P. L. Kelly, *Inorg. Chem.*, 1979, **18,** 3554.
299. M. J. Chetcuti, M. H. Chisholm, J. C. Huffman and J. Leonelli, *J. Am. Chem. Soc.*, 1983, **105,** 292.
300. M. J. Chetcuti, M. H. Chisholm, H. T. Chiu and J. C. Huffman, *J. Am. Chem. Soc.*, 1983, **105,** 1060.
301. M. H. Chisholm, K. Folting, D. M. Hoffman, J. C. Huffman and J. Leonelli, *J. Chem. Soc., Chem. Commun.*, 1983, 589.
302. (a) M. H. Chisholm, D. M. Hoffman and J. C. Huffman, *J. Chem. Soc. Chem. Commun.*, 1983, 967; (b) M. H. Chisholm, J. C. Huffman and A. L. Raterman, *Inorg. Chem.*, 1983, **22,** 4100.
303. M. H. Chisholm, F. A. Cotton, M. W. Extine and R. L. Kelly, *Inorg. Chem.*, 1979, **18,** 116.
304. M. H. Chisholm, J. C. Huffman, J. Leonelli and I. P. Rothwell, *J. Am. Chem. Soc.*, 1982, **104,** 7030.
305. F. A. Cotton and W. Schwotzer, *J. Am. Chem. Soc.*, 1983, **105,** 4955.
306. M. W. Anker, R. Cotton and I. B. Tomkins, *Rev. Pure Appl. Chem.*, 1968, **18,** 23.
307. J. Lewis and R. Whymann, *J. Chem. Soc.*, 1965, 5486.
308. J. Lewis, R. S. Nyholm, C. S. Pande and M. H. B. Stiddard, *J. Chem. Soc.*, 1963, 3600.
309. M. W. Anker, R. Colton and I. B. Tomkins, *Aust. J. Chem.*, 1967, **20,** 9.
310. M. G. B. Drew, I. B. Tomkins and R. Colton, *Aust. J. Chem.*, 1970, **23,** 2517.
311. L. Ricard, R. Weiss, W. E. Newton, G. J.-J. Chen and J. W. McDonald, *J. Am. Chem. Soc.*, 1978, **100,** 1318.
312. B. C. Ward and J. L. Templeton, *J. Am. Chem. Soc.*, 1980, **102,** 1532.
313. A. M. Bond, R. Colton and J. J. Jackowski, *Inorg. Chem.*, 1978, **17,** 105.
314. C. Djordjevic, R. S. Nyholm, C. S. Pande and M. H. B. Stiddard, *J. Chem. Soc. (A)*, 1966, 16.
315. B. Bell, J. Chatt and G. J. Leigh, *J. Chem. Soc., Dalton Trans.*, 1972, 2492.
316. D. L. Lewis and S. J. Lippard, *J. Am. Chem. Soc.*, 1975, **97,** 2697.
317. C. T. Lam, M. Novotny, D. L. Lewis and S. J. Lippard, *Inorg. Chem.*, 1978, **17,** 2127.
318. W. S. Mialki, R. E. Wild and R. A. Walton, *Inorg. Chem.*, 1981, **20,** 1380.
319. M. G. B. Drew, *Prog. Inorg. Chem.*, 1977, **23,** 67.
320. D. L. Kepert, *Prog. Inorg. Chem.*, 1979, **25,** 41.
321. E. L. Muetterties and C. M. Wright, *Q. Rev., Chem. Soc.*, 1967, **21,** 109.
322. E. B. Dreyer, C. T. Lam and S. J. Lippard, *Inorg. Chem.*, 1979, **18,** 1904.
323. J. L. Templeton and B. C. Ward, *Inorg. Chem.*, 1980, **19,** 1753.
324. R. O. Day, W. H. Batschelet and R. D. Archer, *Inorg. Chem.*, 1980, **19,** 2113.
325. R. Hoffmann, B. F. Beier, E. L. Muetterties and A. R. Rossi, *Inorg. Chem.*, 1977, **16,** 511.
326. R. M. Foy, D. L. Kepert, C. L. Raston and A. H. White, *J. Chem. Soc., Dalton Trans.*, 1980, 440.
327. D. M. Collins, F. A. Cotton, S. A. Koch, M. Millar and C. A. Murillo, *Inorg. Chem.*, 1978, **17,** 2017.
328. F. A. Cotton, S. Koch, K. Mertis, M. Millar and G. Wilkinson, *J. Am. Chem. Soc.*, 1977, **99,** 4989.
329. F. A. Cotton and S. A. Koch, *J. Am. Chem. Soc.*, 1977, **99,** 7371.
330. F. A. Cotton, P. E. Fanwick, R. H. Niswander and J. C. Sekutowski, *J. Am. Chem. Soc.*, 1978, **100,** 4725.
331. F. A. Cotton, R. H. Niswander and J. C. Sekutowski, *Inorg. Chem.*, 1978, **17,** 3541.
332. A. P. Sattelberger, K. W. McLaughlin and J. C. Huffman, *J. Am. Chem. Soc.*, 1981, **103,** 2880.
333. D. J. Santure, K. W. McLaughlin, J. C. Huffman and A. P. Sattelberger, *Inorg. Chem.*, 1983, **22,** 1877.
334. V. Katovic, J. L. Templeton, R. J. Hoxmeier and R. E. McCarley, *J. Am. Chem. Soc.*, 1975, **97,** 5300.
335. V. Katovic and R. E. McCarley, *J. Am. Chem. Soc.*, 1978, **100,** 5586.
336. F. A. Cotton and B. E. Hansen, *Inorg. Chem.*, 1978, **17,** 3237.
337. B. E. Bursten, F. A. Cotton, A. H. Cowley, B. E. Hansen, M. Lattman and G. G. Stanley, *J. Am. Chem. Soc.*, 1979, **101,** 6244.
338. P. R. Sharp and R. R. Schrock, *J. Am. Chem. Soc.*, 1980, **102,** 1430.
339. R. R. Schrock, L. G. Sturgeoff and P. R. Sharp, *Inorg. Chem.*, 1983, **22,** 2801.
340. F. A. Cotton, T. R. Felthouse and D. G. Lay, *J. Am. Chem. Soc.*, 1980, **102,** 1431.
341. F. A. Cotton, G. N. Mott, R. R. Schrock and L. G. Sturgeoff, *J. Am. Chem. Soc.*, 1982, **104,** 6781.
342. F. A. Cotton and G. N. Mott, *J. Am. Chem. Soc.*, 1982, **104,** 5978.
343. D. DeMarco, T. Nimry and R. A. Walton, *Inorg. Chem.*, 1980, **19,** 575.
344. R. E. McCarley, T. R. Ryan and C. C. Tovardi, in 'Reactivity of Metal–Metal Bonds', ed. M. H. Chisholm, ACS Symposium 155, p. 41.

345. R. N. McGinnis, T. R. Ryan and R. E. McCarley, *J. Am. Chem. Soc.*, 1978, **100,** 7900.
346. F. A. Cotton, J. L. Hubbard, D. L. Lichtenberger and I. Shim, *J. Am. Chem. Soc.*, 1982, **104,** 679.
347. G. M. Bancroft, E. Pellach, A. P. Sattelberger and K. W. McLaughlin, *J. Chem. Soc., Chem. Commun.*, 1982, 752.
348. F. A. Cotton, R. M. Wing and R. A. Zimmerman, *Inorg. Chem.*, 1967, **6,** 11.
349. R. D. Hogue and R. E. McCarley, *Inorg. Chem.*, 1970, **9,** 1354.
350. K. Lindner and A. Köhler, *Z. Anorg. Allg. Chem.*, 1924, **140,** 364.
351. H. Schäfer and R. Siepmann, *Z. Anorg. Allg. Chem.*, 1968, **357,** 273.
352. R. Siepmann and H. G. von Schnering, *Z. Anorg. Allg. Chem.*, 1968, **357,** 289.
353. H. Schaefer and J. Tillack, *J. Less-Common Met.*, 1964, **6,** 152.
354. F. A. Cotton and T. E. Haas, *Inorg. Chem.*, 1964, **3,** 10.
355. F. A. Cotton and G. G. Stanley, *Chem. Phys. Lett.*, 1978, **58,** 450.
356. A. W. Maverick, J. S. Najdzionek, D. MacKenzie, D. G. Nocera and H. B. Gray, *J. Am. Chem. Soc.*, 1983, **105,** 1878.
357. F. A. Cotton and W. Wang, *Inorg. Chem.*, 1982, **21,** 3859.
358. B. Bell, J. Chatt and G. J. Leigh, *Chem. Commun.*, 1970, 842.
359. J. Chatt, A. J. Pearman and R. L. Richards, *J. Chem. Soc., Dalton Trans.*, 1977, 2139.
360. E. Carmona, J. H. Marin, M. L. Poveda, R. D. Rogers and J. L. Atwood, *J. Organomet. Chem.*, 1982, **238,** C63.
361. E. Carmona, J. M. Marin, M. L. Poveda, J. L. Atwood and R. D. Rogers, *Polyhedron,* 1983, **2,** 185.
362. J. Chatt, G. J. Leigh, H. Neukomm, C. J. Pickett and D. R. Stanley, *J. Chem. Soc., Dalton Trans.*, 1980, 121.
363. S. N. Anderson, R. L. Richards and D. L. Hughes, *J. Chem. Soc., Chem. Commun.*, 1982, 1291.
364. T. Takahashi, T. Kodama, A. Watakabe, Y. Uchida and M. Hidai, *J. Am. Chem. Soc.*, 1983, **105,** 1680.
365. J. Chatt, G. A. Heath and R. L. Richards, *J. Chem. Soc., Chem. Commun.*, 1972, 1010.
366. D. Sellmann, A. Brandell and R. Endell, *Angew. Chem., Int. Ed. Engl.*, 1973, **12,** 1019.
367. M. N. Ackermann, D. J. Dobmeyer and L. C. Hardy, *J. Organomet. Chem.*, 1979, **182,** 561.
368. G. A. Heath, R. Mason and K. M. Thomas, *J. Am. Chem. Soc.*, 1974, **96,** 259.
369. J. Chatt, G. A. Heath and R. L. Richards, *J. Chem. Soc., Dalton Trans.*, 1974, 2074.
370. R. J. W. Thomas, G. S. Laurence and A. A. Diamantis, *Inorg. Chim. Acta,* 1978, **30,** L353.
371. J. Chatt, A. J. Pearman and R. L. Richards, *J. Chem. Soc., Dalton Trans.*, 1976, 1520.
372. J. Chatt, A. J. Pearman and R. L. Richards, *J. Chem. Soc., Dalton Trans.*, 1977, 1852.
373. J. Chatt, A. J. Pearman and R. L. Richards, *J. Chem. Soc., Dalton Trans.*, 1978, 1766.
374. T. Takahashi, Y. Mizobe, M. Sato, Y. Uchida and M. Hidai, *J. Am. Chem. Soc.*, 1979, **101,** 3405.
375. T. Takahashi, Y. Mizobe, M. Sato, Y. Uchida and M. Hidai, *J. Am. Chem. Soc.*, 1980, **102,** 7461.
376. J. Chatt, G. A. Heath and G. J. Leigh, *J. Chem. Soc., Chem. Commun.*, 1972, 444.
377. P. C. Bevan, J. Chatt, A. A. Diamantis, R. A. Head and G. J. Leigh, *J. Chem. Soc., Dalton Trans.*, 1977, 1711.
378. J. Chatt, A. A. Diamantis, G. J. Leigh and G. A. Heath, *J. Organomet. Chem.*, 1975, **84,** C11.
379. P. C. Bevan, J. Chatt, R. A. Head, P. B. Hitchcock and G. J. Leigh, *J. Chem. Soc., Chem. Commun.*, 1976, 509.
380. R. A. Head and P. B. Hitchcock, *J. Chem. Soc., Dalton Trans.*, 1980, 1150.
381. J. Chatt, R. A. Head, G. J. Leigh and C. J. Pickett, *J. Chem. Soc., Chem. Commun.*, 1977, 299.
382. J. Chatt, A. A. Diamantis, G. A. Heath, N. E. Hooper and G. J. Leigh, *J. Chem. Soc., Dalton Trans.*, 1977, 688.
383. R. Ben-Shoshan, J. Chatt, W. Hussain and G. J. Leigh, *J. Organomet. Chem.*, 1976, **112,** C9.
384. R. Ben-Shoshan, J. Chatt, G. J. Leigh and W. Hussain, *J. Chem. Soc., Dalton Trans.*, 1980, 771.
385. H. M. Colquhoun and T. J. King, *J. Chem. Soc., Chem. Commun.*, 1980, 879.
386. H. M. Colquhoun, A. E. Crease, S. A. Taylor and D. J. Williams, *J. Chem. Soc., Chem. Commun.*, 1982, 736.
387. P. C. Bevan, J. Chatt, M. Hidai and G. J. Leigh, *J. Organomet. Chem.*, 1978, **160,** 165.
388. M. Hidai, Y. Mizobe, T. Takahashi and Y. Uchida, *Chem. Lett.*, 1978, 1187.
389. P. C. Bevan, J. Chatt, G. J. Leigh and E. G. Leelamani, *J. Organomet. Chem.*, 1977, **130,** C59.
390. D. C. Busby and T. A. George, *Inorg. Chim. Acta,* 1978, **29,** L273.
391. F. A. Cotton and B. F. G. Johnson, *Inorg. Chem.*, 1964, **3,** 1609.
392. B. F. G. Johnson, *J. Chem. Soc. (A)*, 1967, 475.
393. R. Davis, B. F. G. Johnson and K. H. Al-Obaidi, *J. Chem. Soc., Dalton Trans.*, 1972, 508.
394. B. F. G. Johnson, K. H. Al-Obaidi and J. A. McCleverty, *J. Chem. Soc. (A)*, 1969, 1668.
395. P. Legzdins and J. T. Malito, *Inorg. Chem.*, 1975, **14,** 1875.
396. S. Trofimenko, *Inorg. Chem.*, 1969, **8,** 2675.
397. F. King and G. J. Leigh, *J. Chem. Soc., Dalton Trans.*, 1977, 429.
398. N. G. Connelly, *J. Chem. Soc., Dalton Trans.*, 1973, 2183.
399. B. F. G. Johnson, A. Khair, C. G. Savary and R. H. Walton, *J. Chem. Soc., Chem. Commun.*, 1974, 744.
400. N. G. Connelly, J. Locke, J. A. McCleverty, D. A. Phipps and B. Ratliff, *Inorg. Chem.*, 1970, **9,** 278.
401. B. Bell, J. Chatt, G. J. Leigh and T. Ito, *J. Chem. Soc., Chem. Commun.*, 1972, 34.
402. E. B. Lobkovskii, A. P. Borisov, V. D. Makhaev and K. N. Semenenko, *Zh. Strukt. Khim.*, 1980, **21,** 126.
403. P. Meakin, L. J. Guggenberger, W. G. Peet, E. L. Muetterties and J. P. Jesson, *J. Am. Chem. Soc.*, 1973, **95,** 1467.
404. J. R. Moss and B. L. Shaw, *J. Chem. Soc., Dalton Trans.*, 1972, 1910.
405. D. Gregson, J. A. K. Howard, J. N. Nicholls, J. L. Spencer and D. G. Turner, *J. Chem. Soc., Chem. Commun.*, 1980, 572.
406. E. Carmona-Guzman and G. Wilkinson, *J. Chem. Soc., Dalton Trans.*, 1977, 1716.
407. H. W. Choi, R. M. Gavin and E. L. Muetterties, *J. Chem. Soc., Chem. Commun.*, 1979, 1085.
408. A. M. Bond, R. Colton and J. J. Jackowski, *Inorg. Chem.*, 1975, **14,** 2526.
409. J. A. Connor, P. I. Riley and C. J. Rix, *J. Chem. Soc., Dalton Trans.*, 1977, 1317.
410. B. D. Dombek and R. J. Angelici, *Inorg. Chem.*, 1976, **15,** 2397.

411. J. Chatt, A. J. L. Pombeiro and R. L. Richards, *J. Chem. Soc., Dalton Trans.*, 1979, 1585.
412. K. W. Chiu, D. Lyons, G. Wilkinson, M. Thornton-Pett and M. B. Hursthouse, *Polyhedron*, 1983, **2**, 803.
413. R. A. Love, H. B. Chin, T. F. Koetzle, S. W. Kirtley, B. R. Wittlesey and R. Bau, *J. Am. Chem. Soc.*, 1976, **98**, 4491.
414. K. W. Chiu, R. A. Jones, G. Wilkinson, A. M. R. Galas and M. B. Hursthouse, *J. Chem. Soc., Dalton Trans.*, 1981, 1892.

38

Isopolyanions and Heteropolyanions

MICHAEL T. POPE
Georgetown University, Washington DC, US

38.1 INTRODUCTION

Hexavalent molybdenum and tungsten, pentavalent vanadium and, to a more limited extent, niobium and tantalum form a very large number of polyoxoanions ('heteropolyanions' such as $[PW_{12}O_{40}]^{3-}$ and 'isopolyanions' such as $[Mo_7O_{24}]^{6-}$).[1-3] Unlike the polyoxoanions of the post transition elements the heteropolyanions for the most part are discrete, compact species of high symmetry and low basicity. Salts and free acids of heteropolyanions are usually highly soluble in water and other solvents. Their extensive acid–base and oxidation–reduction chemistry, and their structures, which resemble fragments of close-packed metal oxide lattices, make heteropolyanions attractive as potential (and actual) catalysts.[4] Other demonstrated applications of heteropolyanions include their use as analytical reagents, electron microscope stains, and antiviral and antitumoral agents.

Why should the early transition metals form so many polyoxoanions? The answer lies in the size of the $M^{5/6+}$ cations and their π-acceptor properties.[1,5] The effective ionic radii of V^{5+} (0.68 Å), Mo^{6+} (0.77 Å) and W^{6+} (0.74 Å) are consistent with the observation that these cations adopt four-, five- and six-fold coordination by oxide ion. With very few exceptions V, Mo and W atoms in heteropolyanions are six-coordinate. On the other hand Cr^{6+} (0.58 Å) has a maximum coordination number of four in oxides and oxoanions. Few isopoly- and heteropoly-chromates are known and they are all based on groups of corner-shared CrO_4 tetrahedra: $[Cr_2O_7]^{2-}$, $[Cr_3O_{10}]^{2-}$, $[Cr_4O_{13}]^{2-}$, $[O_3SOCrO_3]^{2-}$, $[O_2IOCrO_3]^-$, $[HOAs(OCrO_3)_3]^{2-}$, *etc.* These species bear little similarity to the other heteropolyanions of Groups 5a and 6a, and, with the exception of the dichromate anion, have a limited or nonexistent solution chemistry. Several post transition elements have ionic radii similar to those of Mo^{6+} and W^{6+}: Ge^{4+} (0.67 Å), Sb^{5+} (0.74 Å), Te^{6+} (0.70 Å) and I^{7+} (0.67 Å). The polyoxoanions of these elements are quite unlike the molybdates and tungstates and are often polymeric chains and networks of XO_6 octahedra. The typical bond lengths found in oxoanions of Mo^{6+} and I^{7+}, shown in Figure 1, reveal the essential difference between transition and post transition cations. Note that Mo^{6+}, by virtue of its accessible vacant *d* orbitals, forms stronger (shorter) bonds with terminal oxygen atoms than does I^{7+}. The greater bond order of the Mo—O bonds reduces the basicity of the oxygen atoms that lie on the exterior of the heteropolyanion. The consequent low charge density on the polyanion surface minimizes protonation, cation binding (insolubility) and further polymerization. Some years ago Lipscomb[6] speculated that the structures of heteropolyanions might be limited to those in which each MO_6 octahedron has no more than two unshared vertices. At present, no structure has been found that contradicts Lipscomb's proposal, and it is a simple matter to rationalize this. In view of the large *trans* influence of the M—O(terminal) bonds, see Figure 1, an MO_6 octahedron with three terminal oxygens would be attached to the remainder of the polyanion structure by three long, and therefore weak, bonds. It seems probable that such a polyanion would be readily susceptible to the loss of the MO_6 octahedron either directly as a neutral MO_3 group or hydrolytically as $HOMO_3^-$.

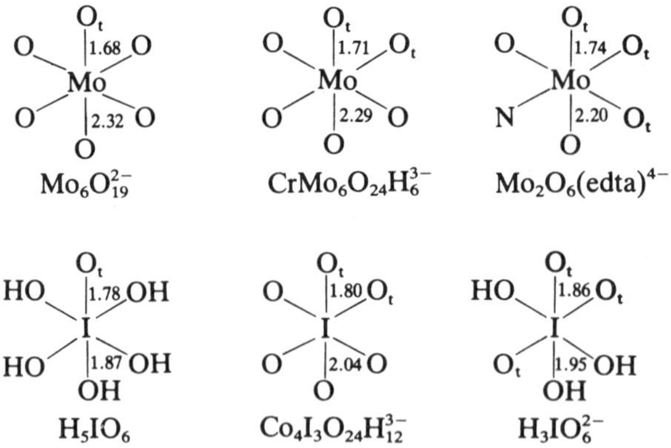

Figure 1 Comparison of the *trans* influence of terminal oxide ligands in octahedral complexes of Mo^{6+} and I^{7+}

Much synthetic and descriptive chemistry of heteropolyanions dates from 1860–1920 when the terms 'isopoly acid' and 'heteropoly acid' were introduced. Reviews of the early work are available[7] and they provide much valuable, if descriptive, information. In the following sections we shall discuss the chemistry of isopolyanions (general formula: $[M_mO_y]^{p-}$) as well as that of heteropolyanions ($[X_xM_mO_y]^{q-}$; $x \leq m$). Excluded from consideration here are substances with similar formula that are the result of high temperature solid state or melt reactions and which are polymeric mixed oxides with no defined solution chemistry.

38.2 ISOPOLYANIONS

38.2.1 General

The chemistry of isopolyanions has to a large extent developed from the study of hydrolytic equilibria in aqueous solution (equation 1).[8] The stoichiometry and equilibrium constants for such reactions are generally evaluated from precise EMF measurements (*i.e.* of the hydrogen ion concentration as a function of metal ion concentration and added acid or base). Measurements of this kind, pioneered by Sillén and his school,[9] have to be made with extraordinary precision, to permit distinction between the large number of species that are often present in polyanion solutions. Even when all experimental difficulties (attainment of equilibrium, liquid junction potentials, *etc.*) appear to have been overcome or allowed for, the EMF method has other limitations. An equilibrium such as equation (1) is defined only by the values of p and q; the number of oxygen atoms in the isopolyanion cannot be determined. For example, the species $[H_2V_3O_{10}]^{3-}$ and $[V_3O_9]^{3-}$ cannot be distinguished. Furthermore, it is becoming clear that factors such as the nature of the counterions and the ionic strength can have significant effects upon polyanion equilibria. In general, investigations have been made at one ionic strength, and the possibilities of specific counterion binding (as observed in some heteropolyanion systems) have not always been excluded.

$$pH^+ + q[MO_4]^{r-} \rightleftharpoons [H_xM_qO_y]^{n-} + (4q-y)H_2O \qquad (1)$$

Virtually every experimental chemical technique that can be applied to solutions has at one time or another been used for the investigation of isopolyanions. The often conflicting results that pervade the literature are caused by the investigators' failure (a) to appreciate the complexity of the equilibria and kinetics involved and (b) realistically to assess the experimental limitations of the technique used. Clearly, there is no *one* experimental method that can reveal the complete story, and the literature must therefore be read extremely critically.

Although aqueous hydrolytic equilibria such as equation (1) provide an overall framework for the discussion of isopolyanions, several species have been synthesized in nonaqueous media, and are unstable in water. Other isopolyanions that are crystallized from aqueous solution, may be incongruently soluble and have polymeric structures. It will, however, prove to be convenient to treat all isopolyanions as though they were derived by equation (1), and characterize each by its 'acidity', Z $(=p/q)$. Note that Z for a specific anion is not necessarily equal to the value of p/q for the solution from which the anion was crystallized.[10]

38.2.2 Isopolyvanadates

38.2.2.1 *Aqueous solutions*

Vanadium(V) oxide is amphoteric, dissolving in alkali to give salts of tetrahedral VO_4^{3-} anion, and in acid to give VO_2^+. The latter cation is almost certainly *cis*-$[VO_2(H_2O)_4]^+$, as in $[VO_2(ox)_2]^{3-}$ and $[VO_2(edta)]^{3-}$ (ox = oxalate dianion; edta = ethylenediamine tetraacetate tetraanion). All dioxo complexes of d^0 metals have *cis* configurations, an arrangement that maximizes $d\pi$–$p\pi$ interactions.

Salts of the orthovanadate anion are frequently isomorphous with the analogous phosphates and arsenates, *e.g.* $Na_3XO_4{\cdot}12H_2O$. In dilute solutions, *ca.* 10^{-5} M, the protonation equilibria parallel those of PO_4^{3-} but with pK values that are greater by 1–2 units (see Table 1). In solutions more concentrated than 10^{-4} M, a multitude of isopolyvanadates is formed. These may roughly be divided into colorless species that predominate above pH 6–7, and orange species in the range pH 2–6. Interconversion between colorless and orange polyanions is kinetically controlled, and equilibrium EMF measurements become unreliable in the region where both types of polyanion exist. The distribution of isopolyvanadate species according to the 'best' fit of potentiometric data for pH 6–14 is shown in Figure 2 for $[V]_{total} = 1.0$ and 10^{-3} M.[11] These conclusions are supported by high field ^{51}V and ^{17}O NMR spectroscopy. The predominant polyanions are listed in Table 2. The di- and noncyclic tri-vanadates are presumed to be analogous to the corresponding phosphates, and their protonation constants (Table 1) are consistent with this. NMR spectroscopy indicates the presence of traces of linear tetramer also. The divanadate anion has been structurally characterized in several salts, *e.g.* $Ca_2V_2O_7{\cdot}2H_2O$

Table 1 Comparison of Protonation Constants for Phosphates and Vanadates[a]

	$\log K_1$	$\log K_2$	$\log K_3$
$[VO_4]^{3-}$	13.95	7.8	3.8
$[PO_4]^{3-}$	12.0	6.5	1.7
$[V_2O_7]^{4-}$	9.73	8.78	
$[P_2O_7]^{4-}$	8.0	5.8	
$[V_3O_{10}]^{5-}$	9.39	7.4 (?)	
$[P_3O_{10}]^{5-}$	7.5	5.2	

[a] In 0.5 M Na (Cl); 25 °C (K.H. Tytko and J. Mehmke, *Z. Anorg. Allg. Chem.*, 1983, **503,** 67).

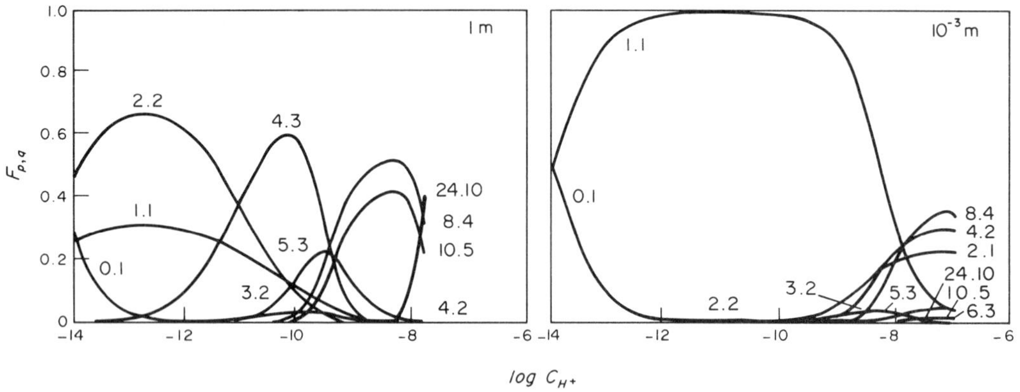

Figure 2 Distribution diagrams for vanadate(V) species in aqueous solution at 25 °C and $\mu = 0.5$ M Na (Cl). Curves are labelled with the values of p, q for each anion (see Table 2) (reprinted with permission from K. H. Tytko and J. Mehmke, *Z. Anorg. Allg. Chem.*, 1983, **503,** 67)

Table 2 Major Oxovanadium(V) Species in Aqueous Solution[a]

Formula	p, q	$\log \beta_{p,q}$ [b]	$\delta(^{51}V)$ (p.p.m.)[c]	$\delta(^{17}O)$ (p.p.m.)[c]
$[VO_4]^{3-}$	0, 1	—	−541.2	+565
$[HVO_4]^{2-}$	1, 1	13.95	−538.8	+573
$[H_2VO_4]^-$	2, 1	21.72	−560.4	
$[V_2O_7]^{4-}$	2, 2	28.46	−561.0	+405, +695
$[HV_2O_7]^{3-}$	3, 2	38.00	−563.5	
$[H_2V_2O_7]^{2-}$	4, 2	46.89	−572.7	
$[V_3O_{10}]^{5-}$	4, 3	53.96	−556.3, −590.4	+438, +721, +852
$[HV_3O_{10}]^{4-}$	5, 3	63.36	*ca.* −570	
$[V_4O_{12}]^{4-}$	8, 4	97.39	−577.6	+472, +928
$[V_5O_{15}]^{5-}$	10, 5	121.77	−586.0	+472, +928
$[V_{10}O_{28}]^{6-}$	24, 10	276.01	e	f
$[HV_{10}O_{28}]^{5-}$	25, 10	d	e	f
$[H_2V_{10}O_{28}]^{4-}$	26, 10	d	e	f
$[VO_2]^+$	4, 1	28.7	−545	

[a] Minor components provisionally identified by ^{51}V NMR include linear $[V_4O_{13}]^{6-}$, $[HV_4O_{13}]^{5-}$, $[H_2V_4O_{13}]^{4-}$, $[V_5O_{16}]^{7-}$ and cyclic $[V_6O_{18}]^{6-}$.
[b] Equilibrium constant for equation (1), $\mu = 0.5$ M Na (Cl) 25 °C (K. H. Tytko and J. Mehmke, *Z. Anorg. Allg. Chem.*, 1983, **503,** 67).
[c] E. Heath and O. W. Howarth, *J. Chem. Soc., Dalton Trans.*, 1981, 1105; S. E. O'Donnell and M. T. Pope, *J. Chem. Soc., Dalton Trans.*, 1976, 2290.
[d] Protonation constants for $[V_{10}O_{28}]^{6-}$ in 1.0 M $NaClO_4$, 25 °C; $\log K_1$ 5.8; $\log K_2$ 3.6 (F. J. C. Rossotti and H. S. Rossotti, *Acta Chem. Scand.*, 1956, **10,** 957).
[e] pH dependent δ, −421, −496, −512 (pH 6.5) (M. A. Habayeb and O. E. Hileman, Jr., *Can. J. Chem.*, 1980, **58,** 2255).
[f] pH dependent δ, +62, 406, 766, 780, 893, 1143 (pH 6.0) (W. G. Klemperer and W. Shum, *J. Am. Chem. Soc.*, 1977, **99,** 3544).

with V—O(terminal) 1.69 Å, V—O(bridging) 1.80 Å and ⟨VOV, 139°. No structures containing $[V_3O_{10}]^{5-}$ have yet been reported.

The cyclic tetramer and pentamer are the major constituents of so-called 'metavanadate' solutions (pH 6.5–8) but the only salts that have so far been crystallized from such solutions contain polymeric anions. Typical structures of metavanadates are (a) infinite chains of corner-shared VO_4 tetrahedra (KVO_3, α-$NaVO_3$, NH_4VO_3, $CsVO_3$), and (b) chains of edge-shared distorted VO_5 trigonal bipyramids ($KVO_3{\cdot}H_2O$, $NaVO_3{\cdot}1.89H_2O$, $Ca(VO_3)_2{\cdot}2H_2O$, $Sr(VO_3)_2{\cdot}4H_2O$).[1] The cyclic tetrameric anion has been structurally characterized in $(Bu^n_4N)_3HV_4O_{12}$, which was prepared by dissolving V_2O_5 in alcoholic Bu^n_4NOH.[13] The Raman spectrum of this salt is similar to that of an aqueous metavanadate solution. Earlier UV and potentiometric measurements had suggested the presence of a trimer in metavanadate solutions, either cyclic $[V_3O_9]^{3-}$ or linear $[H_2V_3O_{10}]^{3-}$. There is no compelling evidence for significant quantities of such a species according to later measurements. The ^{51}V NMR data indicate that traces of a cyclic hexamer may be present and that this could be a precursor to the orange decavanadates.

The decavanadates (Table 2) were first convincingly identified from potentiometric investigations of alkalinized VO_2^+ solutions, since the metavanadate–decavanadate equilibria are established very slowly and involve NMR-detectable intermediates. The structure of the decavanadate anion has been established by several X-ray investigations and is shown in Figure 3.[14] The anion consists of an arrangement of 10 edge-shared VO_6 octahedra with D_{2h} symmetry. The octahedra are distorted in order to maintain approximate valence balance at terminal and bridging oxygens and bond lengths range from 1.60 Å for V—O(terminal) to 2.32 Å for V—O(central). The ^{51}V NMR spectrum of the anion in solution has three lines in the ratio 1:2:2 and is consistent with the solid state structure. Both ^{51}V and ^{17}O NMR have been used to determine the protonation sites of $[V_{10}O_{28}]^{6-}$, with different conclusions. The crystal structure of tetrakis(4-ethylpyridinium)dihydrogen decavanadate reveals protonation at yet a third type of oxygen atom.[15] The kinetics of decomposition of decavanadate by acid and alkali have been investigated,[16] and ^{18}O exchange between water and the polyanion occurs with a half-life of *ca.* 15 h at 25 °C and pH 4–6.3.[17] A mechanism to account for the exchange and for the alkaline degradation has been proposed.

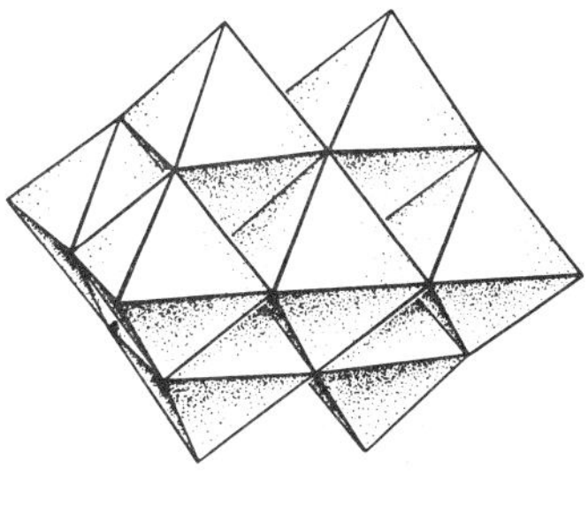

Figure 3 The structure of $[V_{10}O_{28}]^{6-}$ anion shown as an arrangement of VO_6 octahedra

38.2.2.2 Other isopolyvanadates

(i) Polymeric anions

The polymeric metavanadate anions have been described above. Upon warming or ageing solutions that contain the decavanadate ion, it is possible to precipitate salts of the dark red 'trivanadate' anion, MV_3O_8 (M = NH_4, K, Rb, Cs). The formula is sometimes written as $M_2V_6O_{16}$, and the salts called 'hexavanadates' in the literature. Potassium and cesium trivanadate contain infinite buckled layers of distorted VO_5 and VO_6 polyhedra.[18] Another layer structure is found in yellow $K_3V_5O_{14}$ which also is deposited from aqueous solution. The $[V_5O_{14}]_x^{3x-}$ anion is composed of layers of corner-shared VO_4 tetrahedra and VO_5 square pyramids.[19]

(*ii*) *Polyvanadates containing vanadium(IV)*

Vanadium(IV) oxide, like V_2O_5, is amphoteric and dissolves in alkali to yield brown, easily oxidizable solutions. Salts obtained from these solutions were originally formulated as $M_2V_4O_9{\cdot}nH_2O$ but were later shown to have the empirical formula $M_2V_3O_7{\cdot}nH_2O$. The structure of the potassium salt reveals the anion $[V_{18}O_{42}]^{12-}$ consisting of an almost spherical arrangement of edge- and corner-shared VO_5 square pyramids with axial V—O bonds directed outwards (see Figure 4).[20] The anion is stable in solution between pH 9 and 13, and is paramagnetic with about eight free spins at 22 °C. Dilute solutions of VO_2 in alkali appear to contain $[VO(OH)_3]^-$ according to ESR spectroscopy. In strongly basic solution (greater than 1 M OH^-) vanadate(IV) disproportionates into vanadium(III) and vanadium(V).

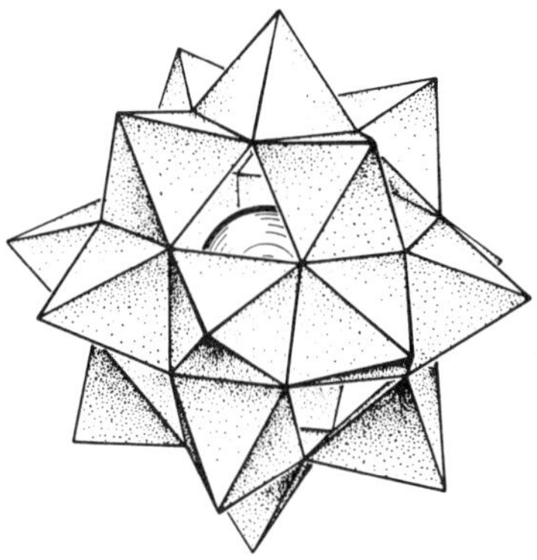

Figure 4 The structure of $[V_{18}O_{42}]^{12-}$ anion. The central cavity has a radius of 4.5 Å and is fractionally occupied by K^+ or H_2O in the crystal

Several mixed-valence polyvanadates(IV, V) are reported in the literature. Almost without exception the anions have been formulated as decavanadates, often with no real justification. Thus Ostrowetsky[21] reported the existence of six anions in the pH range 4–6.5. The anions were given formulas such as $[V_3^{IV}V_7^{V}O_{26}H]^{4-}$ and contained $V^{IV}:V^{V}$ ratios ranging from 2:8 to 7:3. More recently Johnson isolated, from similar solutions, salts of a blue-violet anion $[V_6^{IV}V_{13}^{V}O_{40}H]^{8-}$ and determined its structure.[20] The anion has a structure of virtual C_3 symmetry composed of 18 VO_n polyhedra (tetrahedra, octahedra and square pyramids) surrounding a central VO_4 tetrahedron. The valence stoichiometry of the anion varies from one preparation to another ($V^{IV}:V^{V} = 5:14–7:12$), and given that Johnson's formula is approximately twice those reported by Ostrowetsky, it seems likely that some of Ostrowetsky's complexes are similarly constituted.

Salts of the deep violet anion $[V_2^{IV}V_8^{V}O_{26}]^{4-}$ have been obtained adventitiously from nonaqueous media. The crystal structure of the tetramethylammonium salt shows an anion of virtual S_8 symmetry composed of a ring of eight corner-shared $V^{V}O_4$ tetrahedra capped on either side by $V^{IV}O_5$ square pyramids.[22] The polymeric anion in $K_2V_3O_8$, which is obtained from aqueous solution, has a structure comprising $V^{IV}O_5$ square pyramids linked to ditetrahedral V_2O_7 units. The salt $(NH_4)_2V_3O_8{\cdot}0.5H_2O$, prepared by homogeneous acidification of thiovanadate solutions,[23] probably has the same anion.

38.2.3 Isopolyniobates

Aqueous niobate solutions contain the hexaniobate anion, $[H_xNb_6O_{19}]^{(8-x)-}$, throughout most of the accessible pH range (above *ca.* pH 7). Normal and acid salts of $[Nb_6O_{19}]^{8-}$ may be crystallized from aqueous solutions that are generally prepared by dissolution of freshly precipitated Nb_2O_5 in alkali. The structure of the anion, in crystals of $Na_7HNb_6O_{19}{\cdot}15H_2O$, is shown in Figure 5. As in the decavanadate structure, the metal atoms are displaced towards the unshared oxygen atoms. The Nb—O bond lengths are 1.75–1.78 Å (terminal), 1.970–2.056 Å (bridging) and 2.371–2.386 Å (central).[24] That the same structure persists in aqueous solutions of sodium and potassium salts is strongly supported by ultracentrifugation, light scattering, Raman spectroscopy and ^{17}O NMR. Vibrational spectra of the crystalline potassium salt have been subjected to normal coordinate analysis. The successive protonation constants (log *K*) for $[Nb_8O_{19}]^{8-}$ in 1 M KCl are 12.3(1), 10.94(5) and 9.71(2).[25]

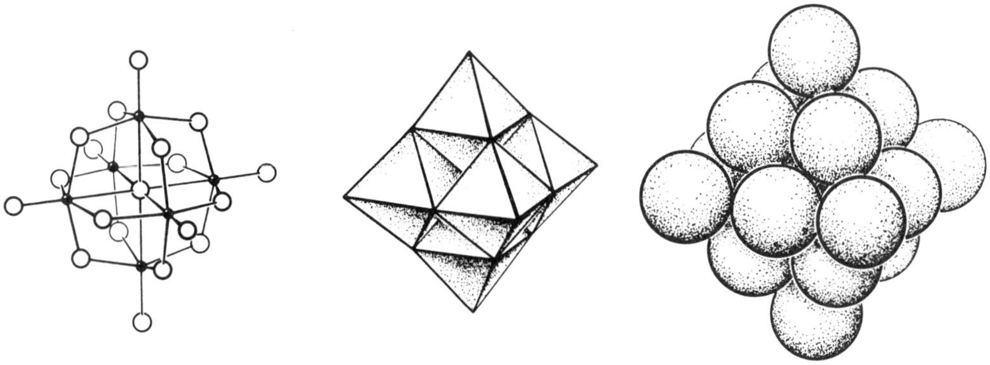

Figure 5 The structure of $[Nb_6O_{19}]^{8-}$ in bond, polyhedral and space-filling representations

On the basis of Raman and UV spectroscopy it has been suggested that $[Nb_6O_{19}]^{8-}$ is degraded into tetrameric and monomeric species in 1–12 M KOH.[26] Acidification of niobate solutions in the presence of NMe_4^+ or NEt_4^+ ion is said to lead to an Nb_9 species. The latter may well be a decaniobate since hydrolysis of a methanolic solution of $Nb(OEt)_5$ with aqueous NMe_4OH and NaOH yields $(NMe_4)_6Nb_{10}O_{28}{\cdot}6H_2O$ and $(NMe_4)_4Na_2Nb_{10}O_{28}{\cdot}8H_2O{\cdot}0.5MeOH$. The structures of both salts have been determined.[27] The $[Nb_{10}O_{28}]^{6-}$ anion is exactly analogous to $[V_{10}O_{28}]^{6-}$ (Figure 3).

38.2.4 Isopolytantalates

The only well-established isopolytantalate is $[Ta_6O_{19}]^{8-}$, isostructural with the corresponding niobate. The bond lengths in $K_4Na_2H_2Ta_6O_{19}{\cdot}2H_2O$ are Ta—O(terminal), 1.786–1.817; Ta—O(bridging), 1.976–2.012; Ta—O(central), 2.356–2.426 Å.[28] Hexatantalate solutions have been examined by ultracentrifugation, light scattering, Raman spectroscopy and ^{17}O NMR. Normal coordinate analysis of vibrational spectra reveal that the force constant for the Ta—O stretch is about 8% larger than for Nb—O. Oxygen exchange between water and $[Ta_6O_{19}]^{8-}$ is much slower than that for $[Nb_6O_{19}]^{8-}$ and occurs more rapidly at the terminal oxygen atoms, according to ^{17}O NMR.[29] Protonation constants (log *K*) of 12.68, 10.81 and 9.28 in 1.0 M KCl have been reported and are similar to those of $[Nb_6O_{19}]^{8-}$.

Solvolysis of $Ta(OEt)_5$ with 85% H_2O_2 in the presence of Bu^tNH_2 leads to a product formulated as a dodecatantalate derivative, $(BuNH_3)_5H_5Ta_{12}O_{17}(O_2)_{20}{\cdot}21H_2O$.[30] Ultracentrifugation of an aqueous solution of this material shows two molecular weight fractions, one corresponding approximately to a Ta_{12} species and the other to $[Ta_6O_{19}]^{8-}$. As the solution ages, the proportion of the heavier solute species decreases.

38.2.5 Isopolymolybdates

Salts of about a dozen isopolymolybdate anions have been isolated from aqueous or nonaqueous solutions. The formulas of these anions are listed in Table 3 and their structures (see below) are based predominantly on six-coordinate molybdenum in contrast to those of vanadates and chromates.

38.2.5.1 Aqueous solutions

In dilute solutions ($\sim 10^{-5}$ M) the equilibria (2) are governed by log *K* values of 3.87 and 3.70 respectively (3 M $NaClO_4$, 25 °C). While the MoO_4^{2-} anion is certainly tetrahedral, the anomalously small value of log K_1 has been taken to indicate an expansion of coordination number upon protonation. In view of the prevalence of octahedral *cis* MoO_2 units in molybdenum(VI) chemistry, formulas such as $[MoO_2(H_2O)(OH)_3]^-$ and $[MoO_2(H_2O)_2(OH)_2]$

Table 3 Isopolymolybdate Anions

Formula	*Acidity, Z*[a]	*Ref. for structure*
MoO_4^{2-}	0	b
$[Mo_2O_7]^{2-}$	1.00	c
$[Mo_2O_7]_x^{2x-}$	1.00	d
$[Mo_7O_{24}]^{6-}$	1.14	e
$[Mo_{10}O_{34}]^{8-}$	1.20	f
$[Mo_8O_{26}(OH)_2]^{6-}$	1.25	g
$[Mo_8O_{27}]_x^{6x-}$	1.25	h
$[Mo_3O_{10}]_x^{2x-}$	1.33	i
$[Mo_5O_{17}H]^{3-}$	1.40	—
α-$[Mo_8O_{26}]^{4-}$	1.50	j
β-$[Mo_8O_{26}]^{4-}$	1.50	k
$[Mo_6O_{19}]^{2-}$	1.67	l
$[Mo_{36}O_{112}(H_2O)_{16}]^{8-}$	1.78	m
$[Mo_5O_{16}H(H_2O)]_x^{x-}$	1.80	n

[a] $Z = p/q$ as defined in equation (1).
[b] O. Nagano and Y. Sasaki, *Acta Crystallogr., Sect. B*, 1979, **35**, 2387.
[c] V. W. Day, M. F. Fredrich, W. G. Klemperer and W. Shum, *J. Am. Chem. Soc.*, 1977, **99**, 6146.
[d] I. Knoepnadel, H. Hartl, W. D. Hunnius and J. Fuchs, *Angew. Chem., Int. Ed. Engl.*, 1974, **13**, 823.
[e] H. T. Evans, Jr., B. M. Gatehouse and P. Leverett, *J. Chem. Soc., Dalton Trans.*, 1975, 505.
[f] J. Fuchs, H. Hartl, W. D. Hunnius and S. Mahjour, *Angew. Chem., Int. Ed. Engl.*, 1975, **14**, 644.
[g] M. Isobe, F. Marumo, T. Yamase and T. Ikawa, *Acta Crystallogr., Sect. B*, 1978, **34**, 2728.
[h] I. Boeschen, B. Buss and B. Krebs, *Acta Crystallogr., Sect. B*, 1974, **30**, 48.
[i] H.-U. Kreusler, A. Forster and J. Fuchs, *Z. Naturforsch., Teil B*, 1980, **35**, 242.
[j] J. Fuchs and H. Hartl, *Angew. Chem., Int. Ed. Engl.*, 1976, **15**, 375.
[k] H. Vivier, J. Bernard and H. Djomaa, *Rev. Chim. Miner.*, 1977, **14**, 584.
[l] See, for example, W. Clegg, G. M. Sheldrick, C. D. Garner and I. B. Walton, *Acta Crystallogr., Sect. B*, 1982, **38**, 2906.
[m] B. Krebs and I. Paulat-Boeschen, *Acta Crystallogr., Sect. B*, 1982, **38**, 1710.
[n] B. Krebs and I. Paulat-Boeschen, *Acta Crystallogr., Sect. B*, 1976, **32**, 1697.

have been proposed. Molybdenum(VI) is amphoteric and in 0.2–3.0 M H^+ yields cationic species with empirical formulas $[HMoO_3]^+$, $[MoO_2]^{2+}$, $[Mo_2O_5]^{2+}$ and $[HMo_2O_5]^+$, but which also undoubtedly are based on six-coordinate Mo.

$$MoO_4^{2-} \underset{}{\overset{K_1}{\rightleftharpoons}} HMoO_4^- \underset{}{\overset{K_2}{\rightleftharpoons}} H_2MoO_4 \qquad (2)$$

Acidification of aqueous solutions of Na_2MoO_4 at concentrations greater than about 10^{-3} M leads successively to solutions containing the polyanions $[Mo_7O_{24}]^{6-}$ ('paramolybdate'), β-$[Mo_8O_{26}]^{4-}$ and $[Mo_{36}O_{112}(H_2O)_{16}]^{8-}$. The evidence for these species is based on potentiometric and spectroscopic investigations of solutions, and on X-ray structural determinations of salts isolated from these solutions. A reevaluation of the potentiometric data[31] has suggested the presence of additional species, $[H_2Mo_{12}O_{42}]^{10-}$ (analogue of 'paratungstate B', see Section 38.2.6.1) and $[Mo_{18}O_{54}(OH)_4(H_2O)_8]^{4-}$, but confirmatory evidence for these is lacking at present. No stable intermediates seem to exist between $[MoO_4]^{2-}$ and $[Mo_7O_{24}]^{6-}$ in aqueous solution, but $(NH_4)_2Mo_2O_7$ crystallizes from hot solutions after some hours. The anion in this salt is polymeric however with infinite chains of MoO_6 octahedra and MoO_4 tetrahedra.[32] A discrete ditetrahedral $[Mo_2O_7]^{2-}$ anion has been synthesized in acetonitrile solution and isolated as a tetrabutylammonium salt.[33] Although $[Mo_2O_7]^{2-}$ is unchanged in aqueous–acetonitrile solutions, addition of alkali metal counterions caused immediate precipitation of salts of $[Mo_7O_{24}]^{6-}$.

The heptamolybdate anion and its protonated forms, $[H_nMo_7O_{24}]^{(6-n)-}$, $n = 1$–3, predominate in solutions of pH 3–*ca.* 5.5. The structure of the anion, Figure 6, has been determined in sodium, potassium, ammonium and isopropylammonium salts, and has been confirmed in solution by low angle X-ray scattering.[34] The anion structure can be viewed as $\frac{7}{10}$

of the decavanadate structure and involves an arrangement of octahedra with *cis*-MoO_2 stereochemistry for each metal atom. IR, Raman and ^{17}O NMR spectra have been reported. Crystals of the isopropylammonium salt are photoreducible.[35]

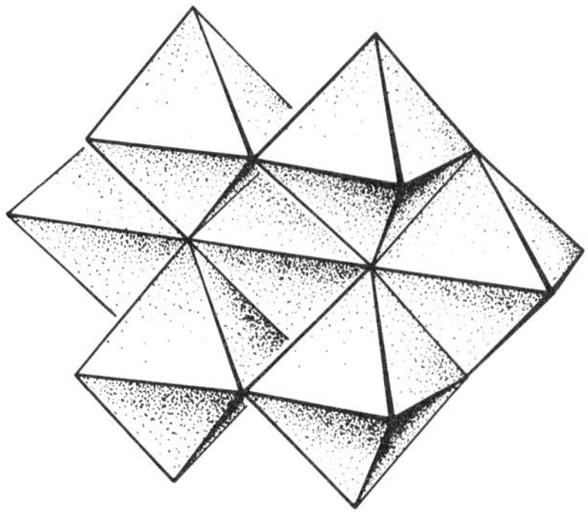

Figure 6 The structure of $[Mo_7O_{24}]^{6-}$ anion

Salts of β-$[Mo_8O_{26}]^{4-}$ may be crystallized from more acidic solutions, pH 2–3. The structure of the anion (Figure 7) is also formally derivable from that of $[V_{10}O_{28}]^{6-}$. The existence of significant concentrations of β-$[Mo_8O_{26}]^{4-}$ in aqueous solutions has been controversial, but Raman and X-ray scattering studies provide more positive evidence.[34]

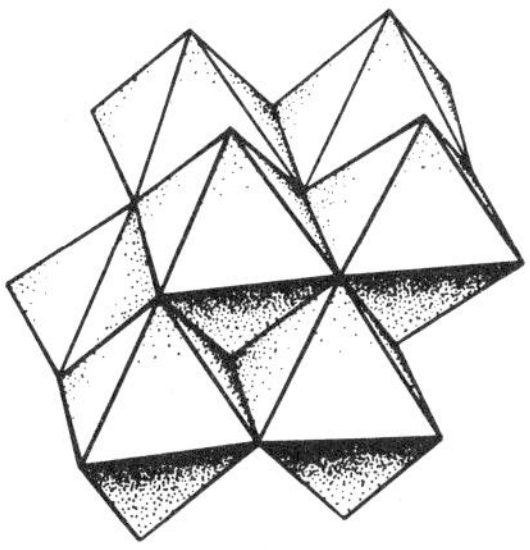

Figure 7 The structure of β-$[Mo_8O_{26}]^{4-}$ anion

The presence of a very large anion in solutions acidified to H^+/MoO_4^{2-} ~1.8 has been inferred from EMF and ultracentrifugation studies. Sodium, potassium, ammonium and barium salts of this anion were isolated from such solutions and the structure of $K_8[Mo_{36}O_{112}(H_2O)_{16}]\cdot38–40H_2O$ has been reported. The anion, Figure 8, consists of two Mo_{18} units related by inversion symmetry. Each unit contains an Mo_7O_{24} moiety surrounded by edge- and corner-linked MoO_6 octahedra containing terminal and bridging water molecules as shown. Two of the Mo atoms in the Mo_7 units become seven-coordinate (quasi pentagonal bipyramidal) and an identical feature is seen in the peroxo polymolybdate $[Mo_7O_{22}(O_2)_2]^{6-}$.[37]

38.2.5.2 Nonaqueous solutions

Four isopolymolybdates have been synthesized or stabilized in nonaqueous or mixed solvents: $[Mo_2O_7]^{2-}$, $[Mo_6O_{19}]^{2-}$, α-$[Mo_8O_{26}]^{4-}$ and $[Mo_5O_{17}H]^{3-}$. The first of these has already been discussed. Several salts of $[Mo_6O_{19}]^{2-}$ have been prepared by both rational and serendipitous routes, and five X-ray structural determinations reported.[38] The anion is isostructural with $[Nb_6O_{19}]^{8-}$ and has no significant distortions from O_h symmetry, except in one case, which is attributed to crystal packing. The structure has a rare feature for an oxomolybdate(VI), namely a C_{4v} monooxo site symmetry for the metal atoms. As a consequence (see Section 38.5) the anion can be electrolytically reduced to mixed valence $[Mo_6O_{19}]^{3-}$ and $[Mo_6O_{19}]^{4-}$ species. The oxidized anion is yellow (λ_{max} = 325 nm) whereas all other isopolymolybdates are colorless. The IR, Raman and ^{17}O and ^{95}Mo NMR spectra have been reported.

The α-$[Mo_8O_{26}]^{4-}$ anion is precipitated by Bu_4N^+ from an aqueous molybdate solution, and has a structure isotypic with $[(MeAsO_3)_2Mo_6O_{18}]^{4-}$ (see Section 38.6.2) and thus contains two

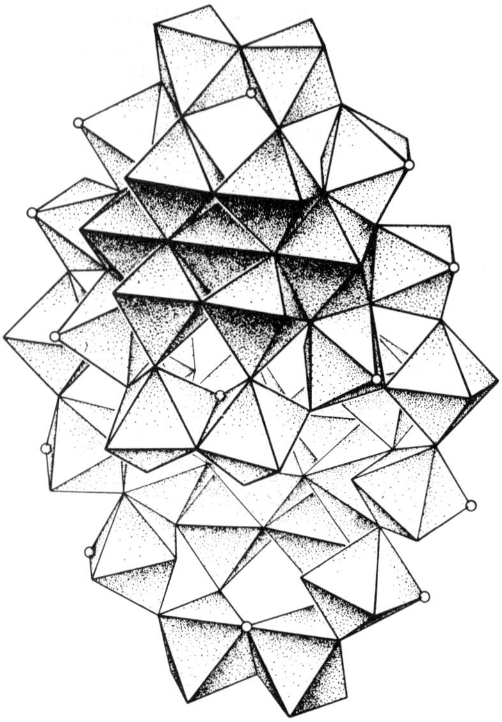

Figure 8 The structure of $[Mo_{36}O_{112}(H_2O)_{16}]^{8-}$ anion. A heptamolybdate unit in one of the two Mo_{18} moieties is emphasized. Coordinated water molecules are indicated by open circles

MoO_4 tetrahedra capping a planar ring of six edge-shared octahedra.[39] According to ^{17}O NMR the MoO_4 tetrahedra undergo intermolecular or intramolecular reorientation in acetonitrile solution. Addition of a small counterion such as K^+ to such solutions induces isomerization to β-$[Mo_8O_{26}]^{4-}$. A unimolecular mechanism for the isomerization has been proposed.[40]

A structure for the unstable $[Mo_5O_{17}H]^{3-}$ anion, isotypic with $[Me_2AsMo_4O_{15}H]^{2-}$ (Section 38.6.3), has been proposed from IR and ^1H NMR spectroscopy. The anion is isolated as an amorphous Bu_4N salt from equilibrium (3) in acetonitrile.[42]

$$7Mo_2O_7^{2-} + H_2O \rightleftharpoons 2[Mo_5O_{17}H]^{3-} + 4MoO_4^{2-} \tag{3}$$

38.2.5.3 Other isopolymolybdates

Several isopolymolybdates which have been isolated from solution are polymeric or appear to have structures that do not persist in solution. These include the trimolybdates $M_2Mo_3O_{10}{\cdot}xH_2O$, obtained as fibrous crystals from appropriately acidified solutions. The structures of $Rb_2Mo_3O_{10}{\cdot}H_2O$ and $RbNaMo_3O_{10}$ contain infinite double chains of edge-shared MoO_6 octahedra.[42] From more acidic solution polymeric pentamolybdates $[MMo_5O_{16}H(H_2O)]$ slowly crystallize. The structure of the potassium salt reveals a complex three-dimensional network of MoO_6 octahedra.[43]

Finally, the salts $(NH_4)_6[Mo_8O_{27}]{\cdot}4H_2O$, $(NH_4)_8[Mo_{10}O_{34}]$ and $(Pr^iNH_3)_6[Mo_8O_{26}(OH)_2]$ all contain a common Mo_8O_{28} unit of edge-shared MoO_6 octahedra[1] (see Section 38.6.4).

38.2.6 Isopolytungstates

Isopolytungstate anions isolated from solution in the form of crystalline salts are listed in Table 4. Several other species undoubtedly remain to be characterized, but analysis of aqueous polytungstate equilibria is complicated by the extreme range of rates involved. Many apparently stable isopolytungstates may be kinetic intermediates. For this reason, no equilibrium distribution diagrams will be given, nor would such diagrams provide a useful introduction to aqueous polytungstate chemistry. Scheme 1 is believed to represent the sequence of reactions in aqueous solution at moderate to high ionic strength.

Table 4 Isopolytungstate Anions

Formula	Acidity, Z^a	Ref. for structure
WO_4^{2-}	0	b
$[W_4O_{16}]^{8-}$	0	c
$[W_7O_{24}]^{6-}$	1.14	d
$[W_{12}O_{42}H_2]^{10-}$	1.17	e
$\alpha\text{-}[(H)W_{12}O_{40}]^{7-}$	1.48	—
$\alpha\text{-}[(H_2)W_{12}O_{40}]^{6-}$	1.50	f
$\beta\text{-}[(H_2)W_{12}O_{40}]^{6-}$	1.50	—
$[W_{10}O_{32}]^{4-}$	1.60	g
$[W_6O_{19}]^{2-}$	1.67	h

^{a} $Z = p/q$ as defined in equation (1).
^{b} A. Thiele and J. Fuchs, *Z. Naturforsch., Teil B*, 1979, **34**, 145.
^{c} A. Hüllen, *Angew. Chem.*, 1964, **76**, 588.
^{d} K. G. Burtseva, T. S. Chernaya and M. I. Sirota, *Sov. Phys.-Dokl. (Engl. Transl.)*, 1978, **23**, 784.
^{e} H. T. Evans, Jr. and E. Prince, *J. Am. Chem. Soc.*, 1983, **105**, 4838.
^{f} M. Asami, H. Ichida and Y. Sasaki, *Acta Crystallogr., Sect. C*, 1984, **40**, 35.
^{g} J. Fuchs, H. Hartl, W. Schiller and U. Gerlach, *Acta Crystallogr., Sect. B*, 1975, **32**, 740.
^{h} J. Fuchs, W. Freiwald and H. Hartl, *Acta Crystallogr., Sect. B*, 1978, **34**, 1764.

$$[WO_4]^{2-} \rightleftharpoons [W_7O_{24}]^{6-} \overset{\text{H}^+}{\underset{\text{OH}^-}{\rightleftharpoons}} \psi \longleftrightarrow [W_{10}O_{32}]^{4-}$$

$$\text{(A)}$$

$$\psi'$$

$$[W_{12}O_{42}H_2]^{10-} \qquad \beta\text{-}[(H_2)W_{12}O_{40}]^{6-}$$

$$\text{(B)}$$

$$\alpha\text{-}[(H_2)W_{12}O_{40}]^{6-}$$

paratungstates   metatungstates

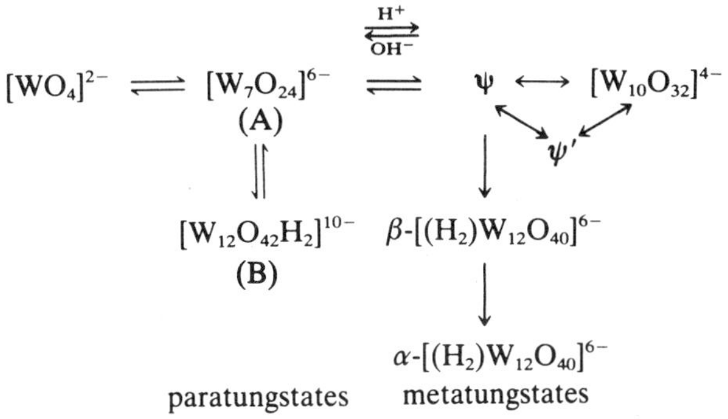

Scheme 1

38.2.6.1 Paratungstates

As for the molybdates, the protonation constants for WO_4^{2-} ($\log K_1 = 3.5$; $\log K_2 = 4.6$; 0.1 M NaCl, 20 °C) indicate an expansion of coordination number, *e.g.* to $[WO_2(H_2O)(OH)]^-$, *etc.* Potentiometric titration of sodium tungstate solutions ($> \sim 10^{-5}$ M) leads to pH inflections at H^+/WO_4^{2-} values of ~ 1.1–1.2 and ~ 1.5. The first corresponds to the formation of 'paratungstates' and the second to the formation of 'metatungstates'.

At least two paratungstate species are known. 'Paratungstate A' is formed rapidly upon acidification, and, after decades of formulation as '$[HW_6O_{21}]^{5-}$', now appears to be a heptatungstate, $[W_7O_{24}]^{6-}$, with a structure identical to that of the corresponding molybdate (Figure 6),[44] although the simultaneous presence of hexatungstate species in these solutions cannot be convincingly ruled out by potentiometric data. Crystals of $Na_6W_7O_{24}\cdot21H_2O$ are isolated from paratungstate solutions after prolonged boiling and are more soluble than those of 'paratungstate B', $Na_{10}[W_{12}O_{42}H_2]\cdot27H_2O$, the usual product of such solutions. Freshly prepared solutions of paratungstate B hydrolyze slowly to an 'equilibrium' mixture of A and B (equation 4). In dilute solutions (<0.02 M) the paratungstates appear to yield $[WO_4]^{2-}$ and $\alpha\text{-}[(H_2)W_{12}O_{40}]^{6-}$ (see below).

$$12[W_7O_{24}]^{6-} + 2H^+ + 6H_2O \rightleftharpoons 7[W_{12}O_{42}H_2]^{10-} \tag{4}$$

The structure of the paratungstate B anion has been established by X-ray and neutron diffraction. The somewhat open structure (Figure 9) contains two internal protons that probably help its stabilization. The neutron diffraction study confirms the conclusion from broadline NMR that the interproton separation is 2.24 Å.[45]

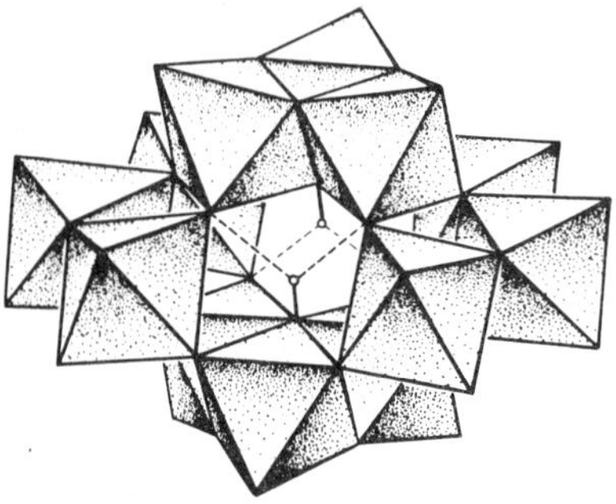

Figure 9 The structure of paratungstate-B anion, $[H_2W_{12}O_{42}]^{10-}$, showing location of two 'internal' protons

Although no intermediates between $[WO_4]^{2-}$ ($Z=0$) and $[W_7O_{24}]^{6-}$ ($Z=1.14$) have been detected in solution, $[W_4O_{16}]^{8-}$ ($Z=0$) is found in crystals of $7Li_2WO_4 \cdot 4H_2O$ (or $Li_{14}(WO_4)_3(W_4O_{16}) \cdot 4H_2O$) which is isolated from alkaline lithium tungstate solutions. The anion structure consists of a tetrahedral arrangement of four edge-shared WO_6 octahedra.[46] The incongruently soluble ditungstate, $Na_2W_2O_7 \cdot 5H_2O$, is probably polymeric.[47]

38.2.6.2 Metatungstates

The initially formed product in tungstate solutions acidified to $H^+/WO_4^{2-}=1.5$ is so-called pseudometatungstate anion (ψ) which has to date been isolated in noncrystalline form only, but which has a characteristic polarogram. Solutions of ψ are slowly converted successively to β and α isomers of the metatungstate anion $[(H_2)W_{12}O_{40}]^{6-}$. At 80 °C and pH 3 the half-lives of both processes are ~70 min.[48]

The ultimate product, α-$[(H_2)W_{12}O_{40}]^{6-}$, is a particularly stable anion with a structure and chemistry that parallel those of the heteropolyanions $[XW_{12}O_{40}]^{n-}$ (X = P, Si, *etc*). The structure ('Keggin'), deduced from early observations of isomorphism, and from analyses of $[Bu_3NH]^+$ and $[Me_4N]^+$ salts, is described in more detail in Section 38.3.1.1. Two protons occupy the central tetrahedral site, normally occupied by the heteroatom, X. The protons are not susceptible to rapid exchange with solvent water and are detectable by NMR of aqueous solutions.[49] As discussed in Section 38.5, Keggin anions have a rich redox chemistry. Highly reduced metatungstate anions spontaneously lose one of the internal protons and may be reoxidized to metastable $[(H)W_{12}O_{40}]^{7-}$ which has been isolated and characterized by NMR and polarography.[50] Fluoro derivatives of metatungstates have been prepared by indirect methods.[51]

The β isomer of $[(H_2)W_{12}O_{40}]^{6-}$ ('tungstate X') is difficult to isolate in pure form, but can be obtained indirectly from reduced ψ. There seems little doubt, based on comparisons with heteropolyanions, and from ^1H and ^{183}W NMR, that the anion structure corresponds to that of β-$[SiW_{12}O_{40}]^{4-}$.[52] The composition and structure of ψ are still unknown. Although W_6 and W_{24} formulas have been suggested, recent ultracentrifugation measurements indicate a dodecamer. Reduction of ψ by two equivalents/12W causes transformation to a new species ψ', which, upon reoxidation, slowly is converted to ψ. A potassium salt of oxidized ψ' has been isolated.

38.2.6.3 Decatungstate and hexatungstate

Strongly acidified solutions of sodium tungstate ($H^+/WO_4^{2-}>1.5$) contain a pale yellow species ($\lambda_{max}=320$ nm) that may be isolated as a metastable potassium salt, $K_4W_{10}O_{32}$(aq). Solutions of this species yield ψ, β-$[(H_2)W_{12}O_{40}]^{6-}$ or $[W_7O_{24}]^{6-}$ depending upon pH and ionic strength. The anion is more conveniently studied as a tetraalkylammonium salt in nonaqueous solutions. The structure of $[W_{10}O_{32}]^{4-}$ in the tri-n-butylammonium salt may be viewed as a dimer of D_{4h} symmetry derived from two W_6O_{19} units, each of which has lost a WO_6 octahedron.[53] The anion is electrochemically and photochemically[123] reducible (see Section 38.5).

The hexatungstate anion, $[W_6O_{19}]^{2-}$, first obtained, like $[Mo_6O_{19}]^{2-}$, from controlled hydrolysis of tungstate esters, is formed by acidification of methanolic solutions of sodium tungstate.[54] Addition of water to a solution of $[W_6O_{19}]^{2-}$ in methanol causes its conversion to $[W_{10}O_{32}]^{4-}$. In methanol solution, $[W_{10}O_{32}]^{4-}$ is slowly (days) converted to $[W_6O_{19}]^{2-}$.

38.3 HETEROPOLYANIONS

At present count, some 68 elements (other than Mo and W) have been observed as heteroatoms in polyanions (see Table 5). Since each element may form more than one heteropolyanion (*e.g.* at least 18 tungstophosphates) and there are polyanions with combinations of several heteroatoms, the total number of known heteropolyanions is extremely large. It is therefore out of the question to attempt a complete accounting of every type of heteropolyanion on the following pages. We shall restrict discussion to the main structural types. More complete reviews are available.[1–3]

Table 5 Confirmed Heteroatoms in Heteropolyanions

H																
Li	Be											B	C			
Na	Mg											Al	Si	P	S	
K	Ca		Ti	V	Cr	Mn	Fe	Co	Ni	Cu	Zn	Ga	Ge	As	Se	
Rb	Sr	Y	Zr	Nb	Mo		Ru	Rh		Ag		In	Sn	Sb	Te	I
Cs	Ba	La	Hf	Ta	W	Re	Os		Pt		Hg	Tl	Pb	Bi		
			Ce	Pr	Nd		Sm	Eu	Gd	Tb		Ho	Er		Yb	
			Th		U	Np	Pu	Am	Cm		Cf					

The heteroatoms may be divided into (i) those ('primary') that are essential for the maintenance of the polyanion structure and are therefore not susceptible to chemical exchange, *e.g.* Co^{2+} in $[CoW_{12}O_{40}]^{6-}$, and (ii) those ('secondary') that in principle and usually in practice may be removed from the polyanion structure without destroying it. Heteropolyanions containing secondary heteroatoms may be viewed as coordination complexes with polyanion ligands, *cf.* $[(H_3N)_5Cr(OH_2)]^{2+}$ with $[(SiW_{11}O_{39})Cr(OH_2)]^{5-}$. Although there will be occasions in which a clear distinction between primary and secondary heteroatoms may not be possible, in most cases the classification is unambiguous. Heteropolyanions containing secondary heteroatoms will be discussed in Section 38.4.

38.3.1 Four-coordinate Primary Heteroatoms

38.3.1.1 The Keggin structure and derivatives

The structure of $[PW_{12}O_{40}]^{3-}$ in the hexahydrate of the free acid was first reported by Keggin in 1933 on the basis of powder data, and has been confirmed by more recent single crystal X-ray and neutron measurements. The same structure has been found in several other heteropolytungstates and molybdates and is shown in Figure 10. Table 6 lists the currently known Keggin anions. The structure has idealized T_d point symmetry and comprises four trigonal groups of edge-shared MO_6 octahedra, each group sharing corners with its neighbors and with the enclosed central tetrahedron. When examined in noncubic space groups, the heteropolymolybdates are found to have lower (T) symmetry as a result of alternating bond lengths for the Mo—O—Mo bridges. In every case (Mo or W) the metal atoms are displaced outwards from the center of the anion toward the unshared oxygen atoms, the displacement conferring approximate C_{4v} local symmetry on each metal. Although the polyhedral representation may suggest otherwise, the Keggin structure, like that of most polyanions, is a close-packed arrangement of oxygen atoms.

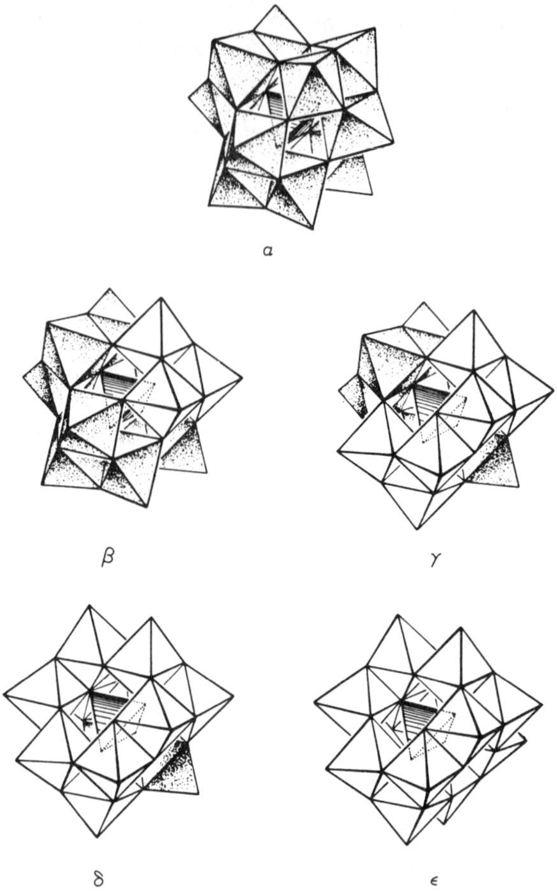

Figure 10 The structures of $[PW_{12}O_{40}]^{3-}$ ('Keggin', α) and four possible skeletal isomers (β, γ, δ, ε) formed by $\pi/3$ rotation of one, two, three and four W_3O_{13} groups, respectively. The rotated groups are shown unshaded

Also shown in Figure 10 are *skeletal* isomers of the Keggin structure in which one or more of the edge-shared groups of octahedra have been rotated by $\pi/3$. The β-isomer structure has been confirmed in $[SiW_{12}O_{40}]^{4-}$ and $[SiMoW_{11}O_{40}]^{4-}$ salts and is slowly converted to the α (Keggin) form in solution, see Section 38.3.1.1.i. The γ, δ and ε structures are expected to be less stable on two accounts, an increased number of coulombically unfavorable edge-shared contacts between MO_6 octahedra (M $\cdots$ M, 3.4 *vs.* 3.7 Å for corner sharing) and a decreased number of quasi-linear M—O—M bridges resulting in decreased $d\pi$–$p\pi$ interactions.[55] It is of interest that Al^{3+} which cannot form a $d\pi$–$p\pi$ bond and which has a smaller charge than W^{6+} adopts the ε structure for $[Al_{13}O_4(OH)_{24}(OH_2)_{12}]^{7+}$.

A second form of isomerism occurs with mixed metal polyanions, such as $[PW_8V_4O_{40}]^{7-}$, in which the relative positions of the W and V atoms may vary.[56] That such *positional* isomers do exist has been amply demonstrated by NMR and ESR spectroscopy.[57] The numbers of possible isomers increase with the lowering of skeletal symmetry, *e.g.* five for α-$[PW_{10}V_2O_{40}]^{5-}$, fourteen for β-$[SiMo_2W_{10}O_{40}]^{4-}$.

Several deficit or lacunary derivatives of the Keggin structure exist as individual species or as fragments of other heteropolyanion structures. Removal of one MO_6 octahedron (stoichiometric loss of MO^{n+}) from α- or β-Keggin anions leads to isomers of $[XM_{11}O_{39}]^{n-}$. Removal of three adjacent octahedra leads to A- and B-type XM_9O_{34} anions. The structures of these are shown in Figure 11.

(i) *Silicon and germanium heteroatoms*

Both α- and β-$[SiW_{12}O_{40}]^{4-}$ were first described over a century ago and the conversion of $\beta \rightarrow \alpha$ was observed in 1909. The free acids are stable crystalline materials and may be obtained by ion exchange or by extraction as 'etherates' from acidified aqueous solution. The

Table 6 Keggin Structure Heteropolyanions

Reported heteroatoms, X[a]

H												
	Be								B			
									Al	Si	P	
		V	Cr	Mn[b]	Fe	Co	Cu	Zn	Ga	Ge	As	(Se)[c]
											(Sb)[c]	(Te)[c]
											(Bi)[c]	

Anion	*Ref. to structure*
$[(H_2)W_{12}O_{40}]^{6-}$	d
$[BeW_{11}Co^{III}(OH_2)O_{39}]^{7-}$	e
$[BW_{11}Co^{III}(OH_2)O_{39}]^{6-}$	f
$[SiW_{12}O_{40}]^{4-}$	g
$[SiMo_{12}O_{40}]^{4-}$	h
$[PW_{12}O_{40}]^{3-}$	i
$[PMo_{12}O_{40}]^{3-}$	j
$[(V)W_9V_3O_{40}]^{6-}$	k
$[(V)Mo_{10}V_2O_{40}]^{6-}$	l
$[Co^{III}W_{12}O_{40}]^{5-}$	m
$[Co^{II}W_{11}Co^{II}(OH_2)O_{39}]^{8-}$	n
$[GeMo_{12}O_{40}]^{4-}$	o

[a] All tungstates; molybdates known with X = Si, Ge, P, As and V only.
[b] Unconfirmed.
[c] Heteroatoms in (N-2) oxidation state, no structural investigations.
[d] M. Asami, H. Ichida and Y. Sasaki, *Acta Crystallogr., Sect. C*, 1984, **40**, 35.
[e] T. J. R. Weakley, *J. Chem. Soc. Pak.*, 1982, **4**, 251.
[f] T. J. R. Weakley, *Acta Crystallogr., Sect. C*, 1984, **40**, 16.
[g] J. Fuchs, A. Thiele and R. Palm, *Z. Naturforsch., Teil B*, 1981, **36**, 161.
[h] H. Ichida, A. Kobayashi and Y. Sasaki, *Acta Crystallogr., Sect. B*, 1980, **36**, 1382.
[i] G. M. Brown, M. R. Noe-Spirlet, W. R. Busing and H. A. Levy, *Acta Crystallogr., Sect. B*, 1977, **33**, 1038.
[j] H. D'Amour and R. Allmann, *Z. Kristallogr., Kristallogeom., Kristallphys., Kristallchem.*, 1976, **143**, 1.
[k] K. Nishikawa, A. Kobayashi and Y. Sasaki, *Bull. Chem. Soc. Jpn.*, 1975, **48**, 3152.
[l] A. Rjörnberg and B. Hedman, *Acta Crystallogr., Sect. B*, 1980, **36**, 1018.
[m] N. F. Yannoni, *Diss. Abstr.*, 1961, **22**, 1032.
[n] A. S. Barrett, *Diss. Abstr. B*, 1972, **33**, 1475.
[o] R, Strandberg, *Acta Crystallogr., Sect. B*, 1977, **33**, 3090.

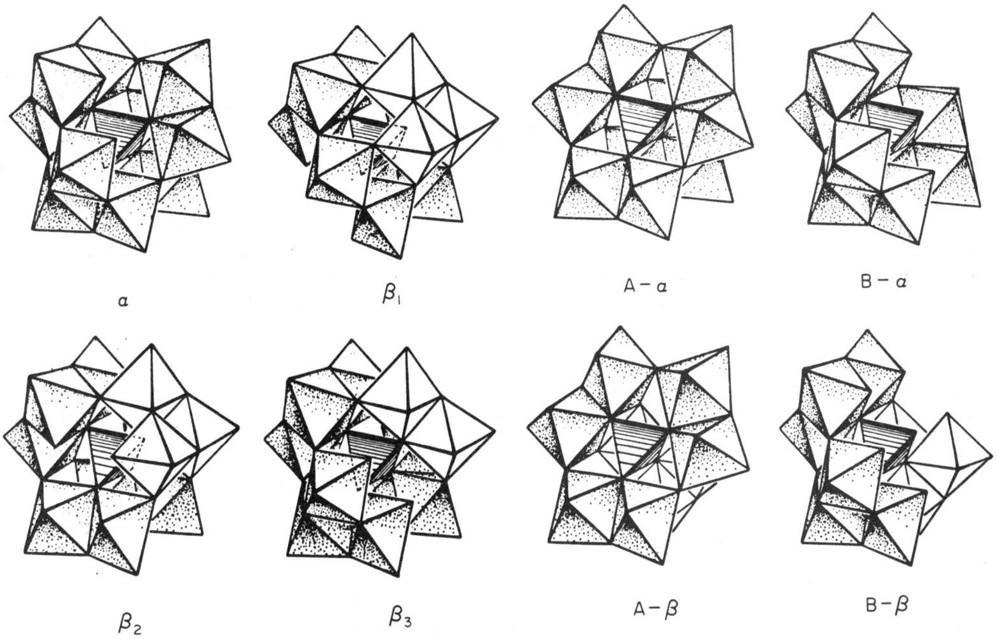

Figure 11 Lacunary structures, $[XM_{11}O_{39}]^{(n+4)-}$ and $[XM_9O_{34}]^{(n+6)-}$, derived from α- and β-$[XM_{12}O_{40}]^{n-}$ anions by removal of one or three adjacent MO_6 octahedra. No example of the B-β-XW_9 structure is currently known

latter behavior is common to all Keggin anions and certain other heteropolyanions.[58] In view of its ease of synthesis and its stability there have been innumerable investigations of α-$[SiW_{12}O_{40}]^{4-}$. In solution, the anion behaves hydrodynamically as an (unsolvated) sphere of radius *ca.* 5.5 Å, according to viscosity and diffusion data. The many spectroscopic methods applied[1] include IR and Raman (solution and solid state), electronic, ^{17}O and ^{183}W NMR, X-ray photoelectron spectroscopy and ^{182}W Mössbauer spectroscopy.[59]

When solutions of $[XW_{12}O_{40}]^{4-}$ (X = Si, Ge) are treated with alkali (pH > 5) a complex series of hydrolyses ensues. Based on polarographic studies the series of equilibria and reactions in aqueous solutions of tungstate and silicate (germanate) are summarized in Scheme 2. All species listed have been isolated as salts, although many are metastable in solution, and the underlined species have been structurally characterized by X-ray diffraction. The structure of β-$[SiW_9O_{34})]^{10-}$ is type A (Figure 11).[60] The anion $[SiW_{10}O_{36}]^{8-}$ has a C_{2v} structure based on the γ-Keggin isomer (Figure 10) with two edge-shared octahedra missing. The three β isomers of $[SiW_{11}O_{39}]^{8-}$ are distinguishable polarographically and the structure of $K_4[SiMoW_{11}O_{40}]\cdot9H_2O$ prepared from the β_2 isomer shows the Mo atom in the ring of six octahedra adjacent to the rotated W_3O_{13} group (Figure 10). It is concluded that the β_1 and β_3 isomers have vacant sites as shown in Figure 11. All the lacunary tungstosilicate structures conform to the principle enunciated by Lipscomb that no MO_6 octahedron in a polyanion has more than two unshared vertices (see Section 38.1).

$$\beta\text{-}[SiW_9O_{34}]^{10-} \quad \rightleftharpoons \quad \beta_1\text{-}[SiW_{11}O_{39}]^{8-}$$

$$\Updownarrow \qquad\qquad\qquad\qquad\qquad \downarrow$$

$$[SiO_3]^{2-}, [WO_4]^{2-} \longleftarrow [SiW_{10}O_{36}]^{8-} \longleftarrow \beta_2\text{-}[SiW_{11}O_{39}]^{8-} \rightleftharpoons \beta\text{-}[SiW_{12}O_{40}]^{4-}$$

$$\uparrow \qquad\qquad\qquad\qquad\qquad\qquad \downarrow$$

$$\qquad\qquad\qquad\qquad\qquad\qquad\qquad\qquad \beta_3\text{-}[SiW_{11}O_{39}]^{8-}$$

$$\qquad\qquad\qquad\qquad\qquad\qquad\qquad\qquad \downarrow$$

$$\alpha\text{-}[SiW_9O_{34}]^{10-} \quad \rightleftharpoons \quad\quad\quad \alpha\text{-}[SiW_{11}O_{39}]^{8-} \rightleftharpoons \alpha\text{-}[SiW_{12}O_{40}]^{4-}$$

$$\xrightleftharpoons[OH^-]{H^+, [WO_4]^{2-}}$$

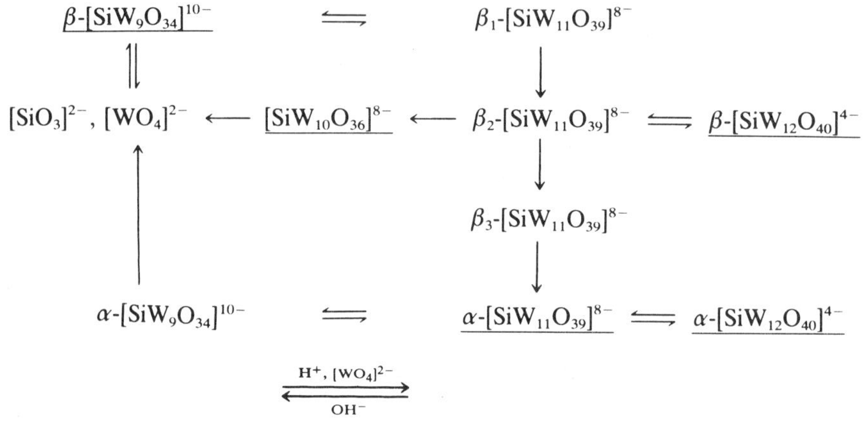

Scheme 2

The molybdo-silicates and -germanates seem to parallel the tungsto species, but are considerably more labile, and fewer structures have been determined. Like the corresponding tungstate, α-$[SiMo_{12}O_{40}]^{4-}$ has been intensively studied. The β isomer has recently been shown[124] to have the same structure as β-$[SiW_{12}O_{40}]^{4-}$. Isomerization of $\beta \rightarrow \alpha$ is rapid in aqueous solution, but may be arrested in nonaqueous solvents. The reverse isomerization ($\alpha \rightarrow \beta$) occurs when the polyanion is reduced to a heteropoly blue form (see Section 38.5.1.2). The lacunary anions $[XMo_{11}O_{39}]^{8-}$ and $(?\beta)$-$[XMo_9O_{34}]^{10-}$ (X = Si, Ge) are quite unstable and difficult to isolate. They react with divalent transition metal cations to yield $[XMo_{11}MO_{40}]^{n-}$ and other uncharacterized products.[61]

(ii) *Phosphorus and arsenic heteroatoms*

There are probably more heteropolyanions of phosphorus than of any other heteroatom (Table 7), and most of these polyanions have structures derived from the Keggin anion $[PW_{12}O_{40}]^{3-}$. Scheme 3 outlines some of the major known species. The pattern of reactions is similar to that of the tungstosilicates except that the β isomers, with the exception of β-$[XW_9O_{34}]^{10-}$, are much less stable and have never been unambiguously identified. The Keggin anion $[PW_{13}O_{40}]^{3-}$ has been the subject of innumerable papers and has been thoroughly characterized by X-ray and neutron diffraction, vibrational, electronic and NMR (^{31}P, ^{183}W and ^{17}O) spectroscopy. The free acid is very soluble in water (*ca.* 85% by weight) and in numerous polar solvents. The acid is levelled in water and crystals of $H_3PW_{12}O_{40}\cdot6H_2O$ contain $[PW_{12}O_{40}]^{3-}$ and $[H_5O_2]^+$ ions only.[62] The trimethyl ester has recently been

Table 7 Heteropolyanions with Phosphorus(V) Heteroatoms

	Formula[a]	$\delta(^{31}P)$[b]	*Ref.*
1	α-$[PW_{12}O_{40}]^{3-}$	−14.9	c
2	α-$[PW_{11}O_{39}]^{7-}$	−10.4	c, d
3	$[P_2W_{21}O_{71}(H_2O)_2]^{6-}$	−13.3	c, e
4	$[PW_{10}O_{36}]^{7-}$	−10.5	f
5	$[KP_2W_{20}O_{72}]^{13-}$	—	g
6, 7	$\alpha(\beta)$-$[PW_9O_{34}]^{9-}$	—	c
8, 9	$\alpha(\beta)$-$[P_2W_{18}O_{62}]^{6-}$	−12.7 (−11.0, −11.6)	c
10	α_1-$[P_2W_{17}O_{61}]^{10-}$	−9.0, −13.1	c, h
11	α_2-$[P_2W_{17}O_{61}]^{10-}$	−7.1, −13.6	c, h
12	β-$[P_2W_{17}O_{61}]^{10-}$	—	h
13	$[P_2W_{15}O_{56}]^{12-}$	—	i
14	$[H_2P_2W_{12}O_{48}]^{12-}$	−7.8	c, h
15	$[NaP_5W_{30}O_{110}]^{14-}$	−9.9	c, j
16	$[P_8W_{48}O_{184}]^{40-}$	−6.6	k
17	$[PW_3O_{13}]_x^{3x-}$	—	l
18	$[P_2W_5O_{23}]^{6-}$	−2.4	f
19	$[P_4W_8O_{40}]^{12-}$	—	m
20, 21	$\alpha(\beta)$-$[PMo_{12}O_{40}]^{3-}$	−4 (−5.2)	n
22	α-$[PMo_9O_{31}(OH_2)_3]^{3-}$		n
23, 24	$\alpha(\beta)$-$[P_2Mo_{18}O_{62}]^{6-}$	−3.4 (—)	n
25	$[P_2Mo_5O_{23}]^{6-}$	+1.8	o
26	$[PV_{14}O_{42}]^{8-}$	+1	p

[a] Arsenate(V) analogues known for all species except 4, 5, 13–19, 25.
[b] Chemical shift (p.p.m. from 85% H_3PO_4).
[c] R. Massart, R. Contant, J. M. Fruchart, J. P. Ciabrini and M. Fournier, *Inorg. Chem.*, 1977, **16**, 2916.
[d] J. Fuchs, A. Thiele and R. Palm, *Z. Naturforsch., Teil B*, 1981, **36**, 544.
[e] T. J. R. Weakley, 'Synthesis, Structures and Reactivity of Polyoxoanions', NSF/CNRS Workshop, St. Lambert-des-Bois, 1983.
[f] W. H. Knoth and R. L. Harlow, *J. Am. Chem. Soc.*, 1981, **103**, 1865.
[g] J. Fuchs and R. Palm, *Z. Naturforsch., Teil B*, 1984, **39**, 757.
[h] R. Contant and J. P. Ciabrini, *J. Chem. Res. (S)*, 1982, 50.
[i] R. Contant and J. P. Ciabrini, *J. Inorg. Nucl. Chem.*, 1981, **43**, 1525.
[j] M. Alizadeh, S. P. Harmalker, Y. Jeannin, J. Martin-Frère and M. T. Pope, *J. Am. Chem. Soc.*, 1985, **107**, 2662.
[k] R. Contant and A. Tézé, *Inorg. Chem.*, 1985, **24**, 4610.
[l] S. Dubois and P. Souchay, *Ann. Chim.*, 1948, **3**, 105.
[m] B. M. Gatehouse and A. J. Jozsa, *Acta Crystallogr., Sect. C*, 1983, **39**, 658.
[n] L. P. Kazansky, I. V. Potapova and V. I. Spitsyn, *Chem. Uses Molybdenum, Proc. Int. Conf., 3rd.*, 1979, 67.
[o] D. E. Katsoulis, A. N. Lambrianidou and M. T. Pope, *Inorg. Chim. Acta*, 1980, **46**, L55.
[p] R. Kato, A. Kobayashi and Y. Sasaki, *Inorg. Chem.*, 1982, **21**, 240.

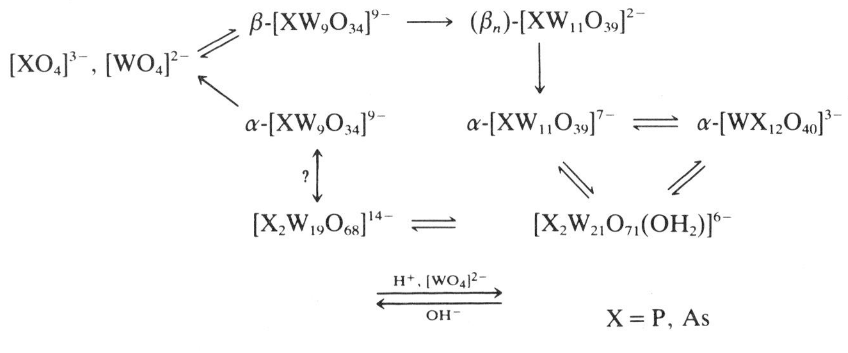

Scheme 3

reported.[125] Solutions of the 12-acid are rapidly converted to $[PW_{11}O_{39}]^{7-}$ at pH > *ca.* 1.5; the latter anion appears to be stable between pH 2 and 6. The corresponding AsW_{12} and AsW_{11} anions are much less stable and must be synthesized and studied in aqueous organic solvents.

The $[X_2W_{19}O_{68}]^{14-}$ and $[X_2W_{21}O_{71}(H_2O)_2]^{6-}$ species have no tungstosilicate analogues. Salts of the 2:21 anion were first described in 1892, but its structure has only recently been determined.

The anion has almost D_{3h} symmetry with two A-α-PW$_9$O$_{34}$ units (see Figure 11) that sandwich two WO$_5$(H$_2$O) octahedra and a WO$_5$ square pyramid, *i.e.* the equatorial plane contains two radial H$_2$O—W=O groups and one W=O group. The structure of [P$_2$W$_{19}$O$_{68}$]$^{14-}$ is currently unknown, but it appears to be a lacunary derivative of P$_2$W$_{21}$ for it reacts with several divalent metal ions to give P$_2$W$_{19}$M$_2$ species (see Section 38.4.2.2).

When solutions of sodium tungstate are boiled with a large excess of phosphoric acid (P:W ~ 4:1) other tungstophosphates are produced. The main species present in such solutions appear to be α and β isomers of [P$_2$W$_{18}$O$_{62}$]$^{6-}$ and the cryptate [NaP$_5$W$_{30}$O$_{110}$]$^{14-}$ (originally formulated as a P$_3$W$_{18}$ species). The structure of α-[P$_2$W$_{18}$O$_{62}$]$^{6-}$ (the 'Dawson' anion) is shown in Figure 12 and is seen to be a fused dimer of A-α-PW$_9$O$_{34}$ moieties. Rotation of one edge-shared W$_3$O$_{13}$ 'cap' of this structure by $\pi/3$ leads to the β structure.[64] Other possible isomers, not yet known for the phosphates, but identified *via* ^{183}W NMR for [As$_2$W$_{18}$O$_{62}$]$^{6-}$, are γ (both caps rotated) and δ (D_{3d}-symmetrized fusion of PW$_9$ moieties). The γ and β isomers spontaneously isomerize to the α form in aqueous solution.

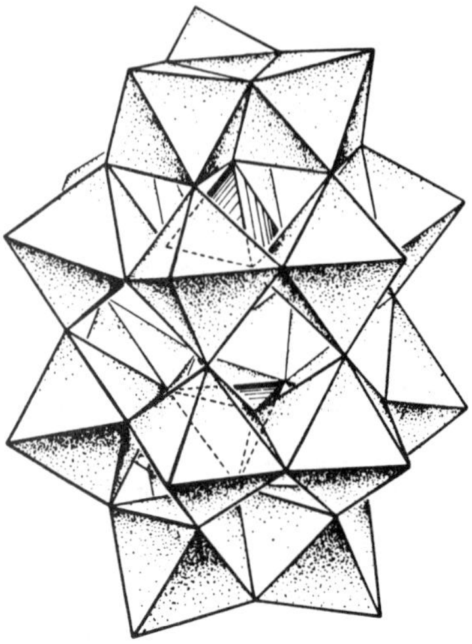

Figure 12 The structure of α-[P$_2$W$_{18}$O$_{62}$]$^{6-}$ ('Dawson') anion

Several lacunary derivatives of the Dawson structure can be synthesized by controlled alkalinization of aqueous solutions, as shown in Scheme 4.

$$\alpha\text{-[P}_2\text{W}_{18}\text{O}_{62}]^{6-} \rightleftharpoons \alpha_2\text{-[P}_2\text{W}_{17}\text{O}_{61}]^{10-} \rightleftharpoons \alpha\text{-[P}_2\text{W}_{15}\text{O}_{56}]^{12-}$$

$$\alpha_1\text{-[P}_2\text{W}_{17}\text{O}_{61}]^{10-} \rightleftharpoons \alpha\text{-[P}_2\text{W}_{12}\text{O}_{48}]^{14-} \longrightarrow \text{[PO}_4]^{3-}, \text{[WO}_4]^{2-}$$

$$\beta\text{-[P}_2\text{W}_{18}\text{O}_{62}]^{6-} \rightleftharpoons \beta\text{-[P}_2\text{W}_{17}\text{O}_{61}]^{10-} \rightleftharpoons \beta\text{-[P}_2\text{W}_{15}\text{O}_{56}]^{12-}$$

$$\xrightleftharpoons[\text{H}^+, \text{[WO}_4]^{2-}]{\text{OH}^-}$$

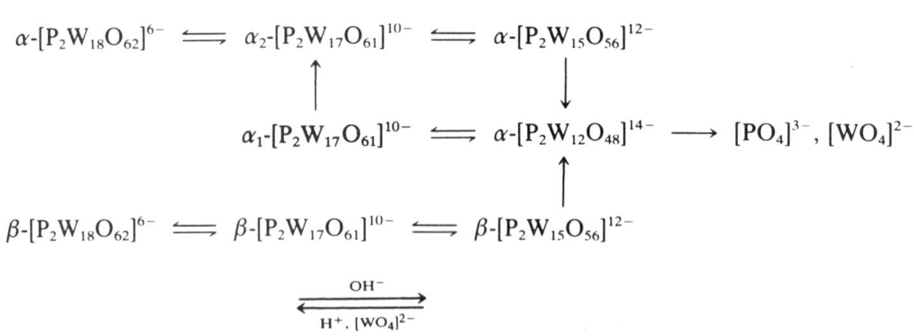

Scheme 4

The α_1- and α_2-[P$_2$W$_{17}$O$_{61}$]$^{10-}$ isomers have been shown by ^{183}W NMR and IR spectroscopy to contain tungsten vacancies in the equatorial ring and terminal cap sites respectively. Numerous metal-ion-substituted derivatives of these species are known (Section 38.4.2). The structures of the P$_2$W$_{15}$ and P$_2$W$_{12}$ anions are not known with certainty, although from NMR the two phosphorus atoms are equivalent in P$_2$W$_{12}$. A likely structure for the latter corresponds to loss of six WO$_6$ octahedra from one long side of the Dawson anion (a 'peeled banana'); the anion $[P_8W_{48}O_{184}]^{40-}$ which can be isolated from solutions of P_2W_{12} consists of a cyclic tetramer of just such units.[65] The P_8W_{48} anion is one of an increasing number of polyanions shown to act as cryptands for alkali and alkaline earth cations (Section 38.4.2.3).

Although formally a cryptate, the sodium ion in $[NaP_5W_{30}O_{110}]^{14-}$ is very tightly bound and is not removed when the free acid $H_{14}[NaP_5W_{30}O_{110}]$ is prepared by ion exchange. The heteropolyanion is one of those rare molecules with full five-fold symmetry, and may be viewed as an assembly of 10 corner-shared W_3O_{13} units forming the faces of a pentagonal bipyramid (*cf.* the same units in the Keggin and Dawson structures that form tetrahedral and octahedral assemblies respectively). The presence of the sodium ion in a cavity on the C_5 axis of the polyanion reduces the overall symmetry from D_{5h} to C_{5v}. The anion has been characterized by vibrational, electronic and NMR spectroscopy, and is stable in solution to pH 11 (*cf.* pH 1.5 for α-PW_{12} and pH *ca.* 6 for α-P_2W_{18}).[66]

Although molybdophosphates and molybdoarsenates corresponding to $[PW_{12}O_{40}]^{3-}$ and $[P_2W_{18}O_{62}]^{6-}$ are well known ($[PMo_{12}O_{40}]^{3-}$ was probably prepared by Berzelius in 1826) these are usually not the most predominant species in aqueous solutions (see Section 38.3.1.2). The Keggin anions $[XMo_{12}O_{40}]^{3-}$ are best studied in aqueous organic solvents to retard hydrolytic degradation. Both α and β isomers are known, the latter being stabilized in the reduced state (Section 38.5.1.2). The Dawson anions are more stable in aqueous solution and are formed *via* XMo_9 species (equation 5). Salts of $[PMo_9O_{31}(OH_2)_3]^{3-}$ and the corresponding arsenate have been isolated. The anion structure is a distorted form of A-α-XW_9O_{34} (Figure 11) in which three of the six (nominally equivalent) terminal oxygen atoms have been converted into weakly bound water molecules (Mo—OH_2, 2.23 Å). The locations of the water molecules and the resulting Mo—O—Mo bond alternation confers a pronounced chirality upon the overall structure, and the chirality is retained in the resulting $[X_2Mo_{18}O_{62}]^{6-}$ anions (D_3 *vs.* D_{3h} for P_2W_{18}).[67] Numerous examples of 'mixed metal' heteropolyanions (Mo + W, V + W, V + Mo) exist with apparent Keggin or Dawson structures, although crystallographic investigation has been limited to salts containing α-$[V_5W_8O_{40}]^{7-}$, α-$[V_3Mo_{10}O_{40}]^{6-}$ and β_1-$[SiW_{11}MoO_{40}]^{4-}$. In each case the high symmetry of the α or β Keggin framework results in crystallographic disorder. Further complications arise with the vanadium-containing anions which are mixtures of isomers (Section 38.3.1.1).

$$2[XMo_9O_{31}(OH_2)_3]^{3-} \longrightarrow [X_2Mo_{18}O_{62}]^{6-} + 6H_2O \quad (5)$$

A modification of the Keggin structure is seen in the heteropolyvanadates, $[XV_{14}O_{42}]^{9-}$ (X = P, As). In these anions two extra VO^{3+} groups are attached to opposite sides of the Keggin core forming a cluster of reduced charge (see Figure 13).[68]

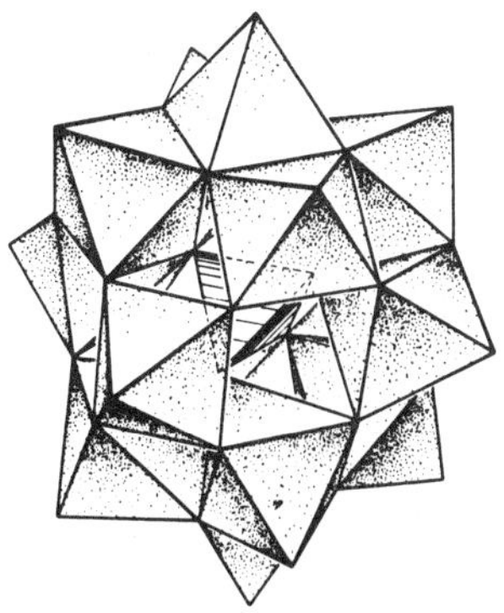

Figure 13 The structure of $[PV_{14}O_{42}]^{9-}$ anion

(iii) Other Keggin species

Several other heteropolytungstates are known or presumed to have the Keggin structure (Table 6). Two 12-tungstoborates have been known for over a century, but these prove not to be simple α or β isomers. The anion α-$[BW_{12}O_{40}]^{5-}$ seems to be an undoubted Keggin species based on isomorphism, vibrational spectrum and its behavior upon reduction. The second isomer ('h'), crystallizing as the acid in hexagonal crystals, clearly is not a β isomer for its polarographic and spectroscopic properties[69] are quite unlike those of all other β-XW_{12} species. A crystal structure determination shows h to be $H_{21}B_3W_{39}O_{132}$.[126] No molybdoborates are known.

Of the remaining anions listed in Table 6 the transition-metal-centered species are noteworthy. The cobalt(III) anion was the first example of a tetrahedral (high spin) d^6 complex.[70] Its reduction to $[Co^{II}W_{12}O_{40}]^{6-}$ (E = +1.00 V *vs.* nhe) renders it an attractive outer sphere oxidant.[71] The corresponding Fe^{III} anion does not undergo reduction to an Fe^{II} species but directly to a heteropoly blue ($Fe^{III}W^{V,VI}$, see Section 38.5), but $[CuW_{12}O_{40}]^{6-}$ is reducible to the brown Cu^I analogue (E = +0.09 V). The Mn- and Cr-centered anions have been little studied and are difficult to prepare.

The 12-tungstates of Se^{IV}, Te^{IV}, Sb^{III} and Bi^{III} are unlikely to have an undistorted Keggin structure in view of the unshared pair of electrons on the heteroatoms. No detailed structural studies have been made. There are however several characterized heteropolytungstates of As^{III} and Sb^{III} in which the lone pair is stereochemically active. These are summarized in Table 8. Most of these contain B-α-XW_9O_{33} units (see Figure 11), linked by extra WO_6 or WO_5 polyhedra. See Figure 14 for an example.

Table 8 Heteropolyanions Containing Trivalent Arsenic and Antimony

Anion	*Ref.*
$[AsW_9O_{33}]^{9-}$	a
$[SbW_9O_{33}]^{9-}$	a
$[As_2W_{19}O_{67}]^{14-}$	a, b
$[As_2W_{20}O_{68}]^{10-}$	b
$[As_2W_{21}O_{69}(H_2O)]^{6-}$	b, c
$[As(H_2)W_{18}O_{60}]^{7-}$	d
$[KAs_4W_{40}O_{140}]^{27-}$	e
$[(Hg_2)_2As_2W_{19}O_{67}(H_2O)]^{10-}$	f
$[NaSb_9W_{21}O_{86}]^{18-}$	g, h
$[NaSb_6As_3W_{21}O_{86}]^{18-}$	h

[a] C. Tourné, A. Revel, G. Tourné and M. Vendrell, *C.R. Hebd. Seances Acad. Sci., Ser. C.*, 1973, **277**, 643.
[b] M. Leyrie, J. Martin-Frère and G. Hervé, *C.R. Hebd. Seances Acad. Sci., Ser. C.*, 1974, **279**, 895.
[c] Y. Jeannin and J. Martin-Frère, *J. Am. Chem. Soc.*, 1981, **103**, 1664.
[d] Y. Jeannin and J. Martin-Frère, *Inorg. Chem.*, 1979, **18**, 3010.
[e] F. Robert, M. Leyrie, G. Hervé, A. Tézé and Y. Jeannin, *Inorg. Chem.*, 1980, **19**, 1746.
[f] J. Martin-Frère and Y. Jeannin, *Inorg. Chem.*, 1984, **23**, 3394.
[g] J. Fischer, L. Ricard and R. Weiss, *J. Am. Chem. Soc.*, 1976, **98**, 3050.
[h] M. Michelon, G. Hervé and M. Leyrie, *J. Inorg. Nucl. Chem.*, 1980, **42**, 1583.

38.3.1.2 Non-Keggin metallophosphates and metalloarsenates

Potentiometric and ^{31}P NMR investigations show that in aqueous solutions of molybdate and phosphate the major polyanion solute species are pale yellow PMo_9 species and colorless $[H_nP_2Mo_5O_{23}]^{(6-n)-}$ anions.[71] The structure of the latter anion (Figure 15) has been determined in several crystalline salts and in solution and has a C_2 structure that leaves a vertex of each PO_4 tetrahedron unshared.[72] The anion is quite labile and possibly fluxional (see Section 38.6.1). Protonation (pK values 5.10, 3.65 in 3.0 M $NaClO_4$, 25 °C) occurs on the terminal oxygens of the phosphate groups. Analogous anions are formed by phosphonates (RPO_3^{2-}),[73] phosphate monoesters,[74] sulfites[75] and selenites. The corresponding tungstophosphate $[P_2W_5O_{23}]^{6-}$ has been reported but it appears not to be a major solute species in aqueous solution.

Aqueous solutions of molybdate and arsenate, pH 3–5, contain $[AsMo_9O_{31}(H_2O)_3]^{3-}$ and an As_2Mo_6 anion.[71] The exact structure of the latter ion in solution is in some doubt, although the tetramethylammonium sodium salt contains the anion shown in Figure 15. Low angle X-ray scattering measurements of aqueous molybdoarsenate solutions are *not* compatible with this structure and the anion may be hydrated in a fashion similar to that observed for the analogous

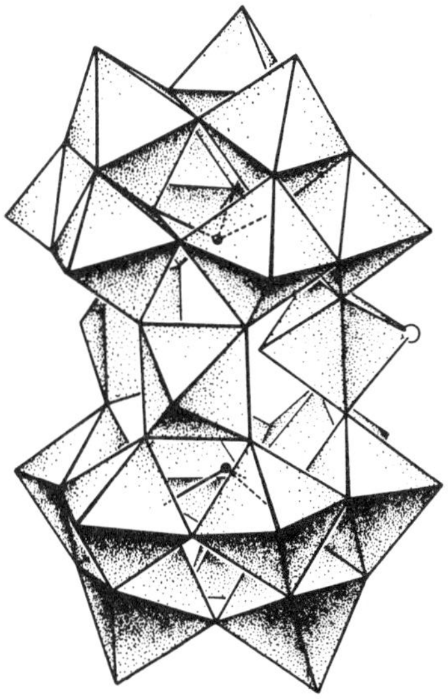

Figure 14 The structure of $[As_2W_{21}O_{69}(H_2O)]^{6-}$ anion. The water molecule is indicated by the open circle

molybdoarsonates (see Section 38.6.2). The D_{3h} structure for $[As_2Mo_6O_{26}]^{6-}$ shown in Figure 15 is also found for α-$[Mo_8O_{26}]^{4-}$ (see Section 38.2.5.2), $[(MeAs)_2Mo_6O_{24}]^{4-}$ (Section 38.6.2) and $[V_2Mo_6O_{26}]^{6-}$. Several other molybdoarsenates are formed in acidic solutions. One long-known species is $[H_4As_4Mo_{12}O_{50}]^{4-}$ which has an arrangement of MoO_6 octahedra like an inverted Keggin anion, a central cavity and external $HOAsO_3$ tripod groups.[76] Analogous organoarsonate derivatives are described in Section 38.6.2.

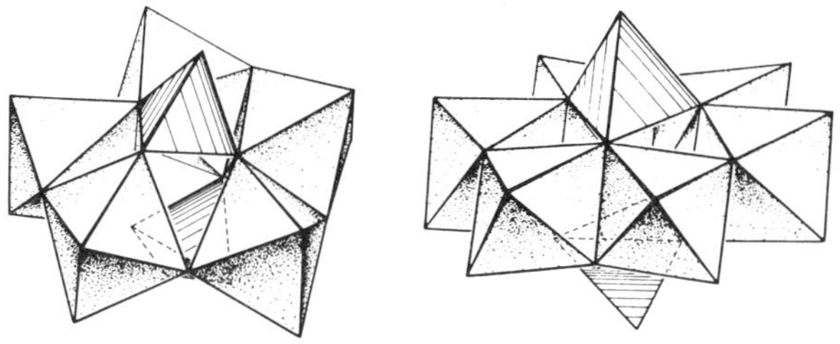

Figure 15 The structures of (a) $[P_2Mo_5O_{23}]^{6-}$ and (b) $[As_2Mo_6O_{26}]^{6-}$ anions

38.3.2 Six-coordinate Primary Heteroatoms

38.3.2.1 The Anderson and related structures

In 1936 Anderson proposed that several 1:6 heteropolyanions and the isopolyanion $[Mo_7O_{24}]^{6-}$ had the planar (D_{3d}) structure shown in Figure 16. The structure was verified for $[TeMo_6O_{24}]^{6-}$ in 1948, but heptamolybdate was shown to have the isomeric (C_{2v}) form shown in Figure 6. Both D_{3d} and C_{2v} structures confer local C_{2v} symmetry on the molybdenum or tungsten atoms; each MO_6 octahedron has two (*cis*) terminal oxygens. The currently known anions are listed in Table 9 and include a periodato complex in which the ring of external octahedra comprises alternate IO_6 and $CoO_4(OH_2)_2$ units. Protonation of the anions (see Cr^{III}, Ni^{II} and Pt^{IV} complexes in Table 9) is believed to take place on the oxygen atoms directly attached to the heteroatom. In most cases, attempts to neutralize the protons result in hydrolytic decomposition of the heteropolyanion. Isomerism has been observed for crystalline protonated $Pt^{IV}Mo_6$ anions (see Table 9).

Some of the anions have been shown to be labile by isotope exchange. Thus Mo exchange between $[Mo_7O_{24}]^{6-}$ and $[H_6XMo_6O_{24}]^{3-}$ at pH 2.5 and 0 °C has $t_{\frac{1}{2}} = 35$ min for X = Cr^{III} and $t_{\frac{1}{2}} \ll 1$ min for X = Fe^{III}. Chromium exchange between the polyanion and $[Cr(H_2O)_6]^{3+}$ is fast

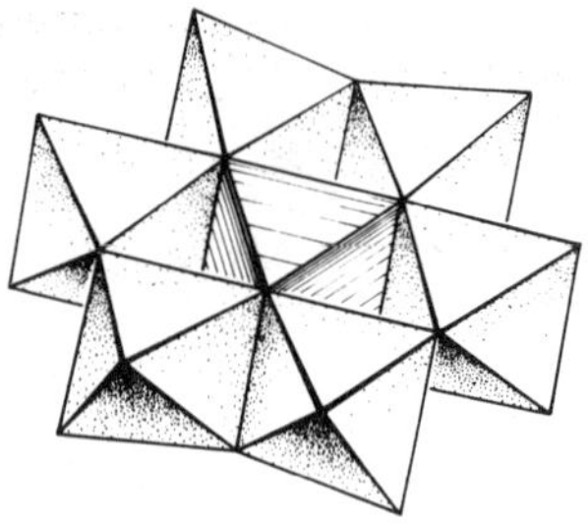

Figure 16 The structure of $[TeMo_6O_{24}]^{6-}$ ('Anderson') anion

Table 9 Anderson and Related Heteropolyanions

Anion	*Ref.*
D_{3d} Structure (Figure 16)	
$[H_6Cr^{III}Mo_6O_{24}]^{3-}$	a, b
$[Mn^{IV}W_6O_{24}]^{8-}$	c
$[H_6Ni^{II}W_6O_{24}]^{4-}$	d
$[Ni^{IV}W_6O_{24}]^{8-}$	e
$[Pt^{IV}W_6O_{24}]^{8-}$	f
$[H_3Pt^{IV}W_6O_{24}]^{5-}$	g
α-$[H_{4.5}PtMo_6O_{24}]^{3.5-}$	h
$[Te^{VI}Mo_6O_{24}]^{6-}$	i
$[I^{VII}Mo_6O_{24}]^{5-}$	j
$[Co^{III}I_3^{VII}Co_3^{III}O_{18}(OH_2)_6]^{3-}$	k
C_{2v} Structure (Figure 6)	
$[Mo_7O_{24}]^{6-}$	l
$[W_7O_{24}]^{6-}$	m
β-$[H_4PtMo_6O_{24}]^{4-}$	h
$[H_2Sb^VMo_6O_{24}]^{5-}$	n

[a] A. Perloff, *Inorg. Chem.*, 1970, **9,** 2228.
[b] Analogous molybdates of divalent Mn, Fe, Co, Ni, Cu and Zn and of trivalent Al, Ga, Fe, Co and Rh are considered to have the same structure.
[c] V. S. Sergienko, V. N. Molchanov, M. A. Porai-Koshits and E. A. Torchenkova, *Sov. J. Coord. Chem. (Engl. Transl.)*, 1979, **5,** 740.
[d] U. C. Agarwala, *Diss. Abstr.*, 1960, **21,** 749.
[e] H. H. K. Hau, *Diss. Abstr. B*, 1970, **31,** 2600.
[f] U. Lee, H. Ichida, A. Kobayashi and Y. Sasaki, *Acta Crystallogr., Sect. C*, 1984, **40,** 5.
[g] U. Lee, A. Kobayashi and Y. Sasaki, *Acta Crystallogr., Sect. C*, 1983, **39,** 817.
[h] U. Lee and Y. Sasaki, *Chem. Lett.*, 1984, 1297.
[i] H. T. Evans, Jr., *Acta Crystallogr., Sect. B*, 1974, **30,** 2095.
[j] H. Kondo, A. Kobayashi and Y. Sasaki, *Acta Crystallogr., Sect. B*, 1980, **36,** 661.
[k] L. Lebioda, M. Ciechanowicz-Rutkowska, L. C. W. Baker and J. Grochowski, *Acta Crystallogr., Sect. B*, 1980, **36,** 2530.
[l] H. T. Evans, Jr., B. M. Gatehouse and P. Leverett, *J. Chem. Soc., Dalton Trans.*, 1975, 505.
[m] K. G. Burtseva, T. S. Chernaya and M. I. Sirota, *Sov. Phys.-Dokl. (Engl. Transl.)*, 1978, **23,** 784.
[n] Y. Sasaki and A. Ogawa, in 'Some Recent Developments in the Chemistry of Chromium, Molybdenum and Tungsten', ed. J. R. Dilworth and M. F. Lappert, Royal Society of Chemistry, London, 1982, p. 59.

(4–45 min at 29.5 °C) and involves transfer of Mo as MoO_4^{2-} or $HMoO_4^-$ rather than breaking of Cr—O bonds.[77]

When solutions of $[H_6CoMo_6O_{24}]^{3-}$ are treated with H_2O_2 in the presence of activated charcoal the anion is converted quantitatively into $[H_4Co_2M_{10}O_{38}]^{4-}$, an olive green species that may be resolved into stable enantiomers.[78] The new polyanion has a structure of D_2 symmetry that may be 'constructed' as follows. Removal of an MoO_6 octahedron from the Anderson anion leaves a chiral $CoMo_5$ fragment of C_2 symmetry. Two (+)- (or two (−)-) $CoMo_5$ groups are slotted together so that the two CoO_6 octahedra share an edge and produce

(+ +)-(− −)-$[H_4Co_2Mo_{10}O_{38}]^{4-}$.[79] Crystals of the ammonium salt are pleochroic and transmit red, blue-green and yellow light along the three molecular axes.

38.3.2.2 Other structures incorporating octahedral heteroatoms

The Anderson structure discussed above can be viewed as a 'close-packed' layer of edge-sharing MO_6 octahedra. Addition of three octahedra above and three octahedra below this layer produces a cluster, $XM_{12}O_{38}$, of octahedral symmetry in which the central XO_6 octahedron shares each of its edges with an MO_6 octahedron.[80] The mineral sherwoodite contains the only naturally occurring heteropolyanion, $[AlV_2^{IV}V_{12}^{V}O_{40}]^{9-}$, and has a structure consisting of the $XM_{12}O_{38}$ cluster with two additional VO_6 octahedra placed in *trans* positions on a four-fold axis (Figure 17).[81] The synthetic heteropolyvanadates, $[XV_{13}O_{38}]^{7-}$ ($X = Mn^{IV}$, Ni^{IV}) have the sherwoodite structure less one of the equatorial VO_6 octahedra.[82] The heteropolymolybdates $[XMo_9O_{32}]^{6-}$ ($X = Mn^{IV}$, Ni^{IV}) also have a structure that can be derived from the $XM_{12}O_{38}$ cluster by removal of three alternate MO_6 octahedra from the original Anderson fragment.[83] The resulting structure has D_3 symmetry. Attempts to resolve the 1:9 complexes have been unsuccessful, although the circular dichroism spectrum of the Mn species has been recorded on crystals of the ammonium salt. The ESR spectrum of the Mn complex has been analyzed and the ^{61}Ni Mössbauer spectrum has been reported for $[NiMo_9O_{32}]^{6-}$.

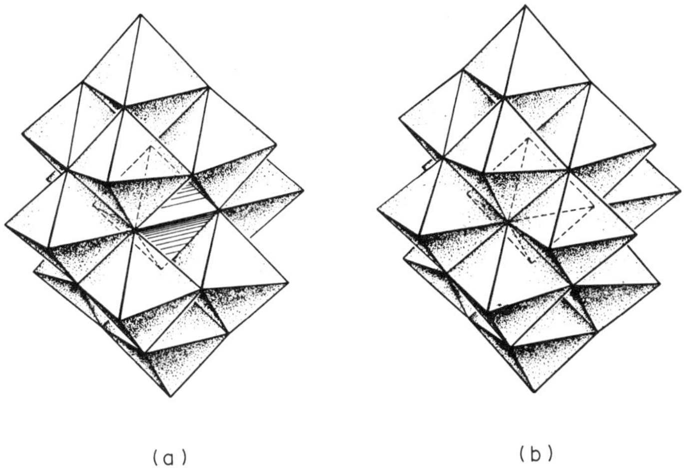

Figure 17 The structures of (a) $[MnV_{13}O_{38}]^{7-}$ and (b) $[AlV_{14}O_{40}]^{9-}$ anions

38.3.3 Eight- and Twelve-coordinate Primary Heteroatoms

38.3.3.1 Decatungstometalates, $[XW_{10}O_{36}]^{n-}$

The heteroatoms listed in Table 10 form complexes of stoichiometry $[XW_{10}O_{36}]^{8-}$. The structure, based on determinations of the Ce^{IV} and U^{IV} anions, can be regarded as the attachment of two quadridentate ligands, derived from W_6O_{19} through the loss of one WO_6 octahedron, to the heteroatom (Figure 18). The site symmetry of the latter is approximately D_{4d}. The XW_{10} anions are of only moderate stability in aqueous solution (pH 5.5–8.5). Electronic, vibrational and ^{17}O NMR spectra have been recorded. The emission spectra and luminescence properties of several of the lanthanide anions have been discussed.[84]

38.3.3.2 Dodecamolybdometalates, $[XMo_{12}O_{42}]^{n-}$

The structure shown in Figure 19 was first described for the $[CeMo_{12}O_{42}]^{8-}$ anion.[85] It contained two unprecedented features, an icosahedrally coordinated heteroatom, and face-shared pairs of MoO_6 octahedra. Other examples of octahedral face-sharing have since been observed, but this feature remains uncommon owing to the high coulombic repulsion between

Table 10 Decatungstometalate Heteropolyanions, $[XW_{10}O_{26}]^{8,9-}$

Heteroatom	Ref.
X^{III}: Y, La, Ce, Pr, Nd, Sm, Eu, Gd, Ho, Er, Yb	a, b
Am	c
X^{IV}: Ti (?), Hf (?)	d
Zr	e
Ce	a, f
Th, U	b, g

[a] R. D. Peacock and T. J. R. Weakley, *J. Chem. Soc. (A)*, 1971, 1836.
[b] L. P. Kazanskii, A. M. Golubev, I. I. Baburina, E. A. Torchenkova and V. I. Spitsyn, *Bull. Acad. Sci. USSR, Div. Chem. Sci.*, 1978, **27**, 1956.
[c] A. S. Saprykin, V. P. Shilov, V. I. Spitsyn and N. N. Krot, *Dokl. Chem. (Engl. Transl.)*, 1976, **226**, 114.
[d] G. Marcu, I. Todorut and A. Botar, *Rev. Roum. Chim.*, 1971, **16**, 1335.
[f] J. Iball, J. N. Low and T. J. R. Weakley, *J. Chem. Soc., Dalton Trans.*, 1974, 2021.
[g] A. M. Golubev, L. A. Muradyan, L. P. Kazanskii, E. A. Torchenkova, V. I. Simonov and V. I. Spitsyn, *Sov. J. Coord. Chem. (Engl. Transl.)*, 1977, **3**, 715.

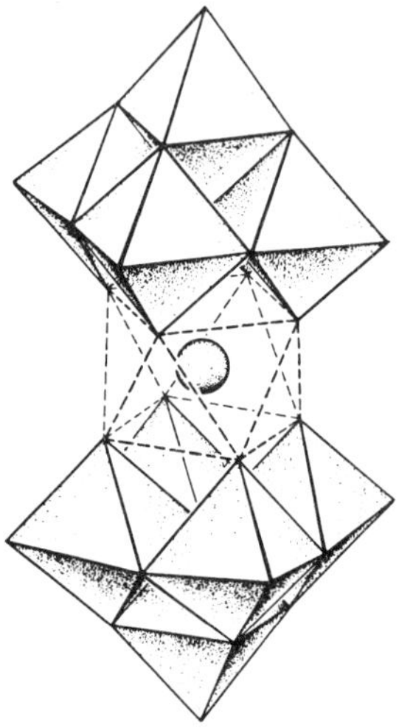

Figure 18 The structure of $[CeW_{10}O_{36}]^{8-}$ anion

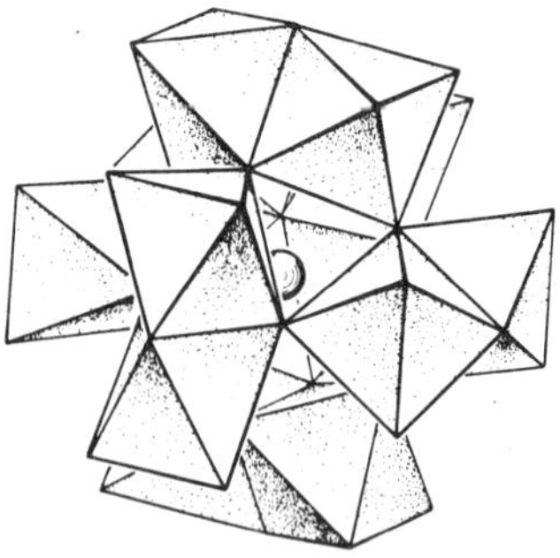

Figure 19 The structure of $[CeMo_{12}O_{42}]^{8-}$ anion

the metal atoms (Pauling's rules). An alternative description of the structure is as a Ce^{4+} cation surrounded by six bidentate ditetrahedral $[Mo_2O_7]^{2-}$ ligands. The overall point symmetry is T_h.

The structure has been confirmed for Ce^{IV} and U^{IV} heteroatoms, and analogous complexes exist with Th^{IV}, Np^{IV} and perhaps Sn^{IV}. All are molybdates. The Ce^{IV} and U^{IV} anions undergo reversible reduction and oxidation respectively to Ce^{III} (unstable) and U^{V} analogues.[86]

38.4 HETEROPOLYANIONS AS LIGANDS: SECONDARY HETEROATOMS

The complexes formed by the attachment of metal ions or organometallic moieties to the surface oxygen atoms of complete or lacunary (deficit) heteropolyanions comprise a rapidly growing area of polyanion chemistry. Among other reasons, such chemistry is perceived as being particularly pertinent to mechanisms of catalysis on metal oxide surfaces, or indeed to the development of new catalysts.

38.4.1 Complete Polyanion Ligands

The first complexes incorporating an intact polyanion were $[X(Nb_6O_{19})_2]^{12-}$ ($X = Mn^{IV}$, Ni^{IV}). In these complexes the Mn and Ni atoms are bound to three bridging oxygens on the surface of the $[Nb_6O_{19}]^{8-}$ anions, which therefore act as tridentate ligands.[87] The salts $Na_5X(en)(Nb_6O_{19})\cdot nH_2O$ ($X = Cr^{3+}$, Co^{3+}) probably contain a similar ligand (*i.e.* $[(H_2O)(en)X(Nb_6O_{19})]^{5-}$). More recently, analogous derivatives of *cis*-$[Nb_2W_4O_{19}]^{4-}$ and $[MW_5O_{19}]^{3-}$ (M = Nb, Ta) have been synthesized, *e.g.* $[(OC)_3Re(Nb_2W_4O_{19})]^{3-}$, $[Me_5CpRh(Nb_2W_4O_{19})]^{2-}$ and $[Cp_3X(MW_5O_{19})_2]^{5-}$ (X = U, Th).[88] With the 16-electron d^8 species $[(C_7H_8)RhCl]_2$ (C_7H_8 = norbornadiene), a bizarre bridged complex results, $[\{(C_7H_8)Rh\}_5(Nb_2W_4O_{19})_2]^{3-}$.[89]

Similar complexes should be formed by Keggin anions, for each edge-shared triplet of WO_6 octahedra in the Keggin structure is analogous to one half of the M_6O_{19} structure. Two Keggin derivatives have been reported: $[Me_5CpRh(SiW_9Nb_3O_{40})]^{5-}$ and $[(cod)Rh(SiW_9Nb_3O_{40})]^{6-}$ (cod = cyclooctadiene).[90] All of the above organometallic derivatives have been structurally characterized by single crystal X-ray diffraction and/or ^{17}O, ^{183}W NMR spectroscopy.

The anions $[XMo_{12}O_{42}]^{8-}$ (Figure 19), form weak complexes in aqueous solution with divalent Mn, Fe, Co, Ni, Zn, Cu, Cd, trivalent Y, Er, Yb and tetravalent Th.[91] Two crystal structures have been determined and contain the anions $[(UMo_{12}O_{42})\{Er(OH_2)_5\}_2]^{2-}$ and $[(UMo_{12}O_{42})Th(OH_2)_3]_x^{4x-}$.[1,2] In both cases the external metal atoms are attached to opposite sides of the heteropolyanion *via* three terminal oxygen atoms, one from each of three different Mo_2O_7 units.

38.4.2 Lacunary Polyanion Ligands

Heteropolyanion structures in which the local site symmetry of each metal atom is approximately C_{4v} (*i.e.* each metal has one terminal oxygen) frequently undergo partial hydrolysis to lose an MO^{n+} group and generate a so-called lacunary structure, see for example the $[XM_{11}O_{39}]^{n-}$ (Figure 11) and $[X_2M_{17}O_{61}]^{n-}$ anions. The lacunary anions function as tetra- or penta-dentate ligands and react with an astonishing variety of metal ions or organometallic groups to form 1:1 or 2:1 complexes. In some cases such complexes are stable even although the lacunary 'ligand' itself may have no independent existence. The $[XW_{10}O_{36}]^{6-}$ anions described in Section 38.3.3.1 may be viewed in this light as 2:1 complexes of the hypothetical lacunary anion $[W_5O_{18}]^{6-}$. The organometallic derivatives $[CpTiM_5O_{18}]^{3-}$ (M = Mo, W) are analogous '1:1' complexes. In the last two examples the $[TiCp]^{3+}$ group plays the same role as WO^{4+} does in $[W_6O_{19}]^{2-}$. Further organometallic groups may be attached to the surface oxygens of the substituted polyanion in the manner described in Section 38.4.1, *e.g.* $[Me_5CpRh(Cp)TiMo_5O_{18}]^{3-}$.[92]

38.4.2.1 *Derivatives of $[XM_{11}O_{39}]^{n-}$ and $[X_2M_{17}O_{61}]^{n-}$*

The number of complexes of this type is legion. The general formula is $[XM_{11}Z(L)O_{39}]^{n-}$ (or $[X_2M_{17}Z(L)O_{61}]^{n-}$) where Z is a secondary heteroatom and L is a terminal ligand on Z. In some cases L is another lacunary polyanion, an organic or organometallic moiety, or is absent. Scheme 5 illustrates some of the possibilities.

When Z is a simple aquacation, two types of complex are formed depending upon the ionic radius of Z. For alkali, alkaline earth and most transition metal cations the product contains Z^{n+} in quasi-octahedral coordination. Equilibrium constants for reaction (6) have been determined for Li^+, Na^+, K^+, Mg^{2+}, Ca^{2+}, Sr^{2+}, Ba^{2+}, Mn^{2+}, Fe^{2+}, Co^{2+}, Ni^{2+}, Cu^{2+} and Zn^{2+}.[93] For the transition metals, log K lies between 3 and 9, and is sensitive both to Z and to the lacunary polyanion involved. Larger cations, Sr^{2+}, Ba^{2+}, and tri- and tetra-valent lanthanides and actinides are also able to bind two lacunary ligands in a manner similar to that illustrated in Figure 18. Although the stepwise formation of 1:1 and 2:1 complexes of the

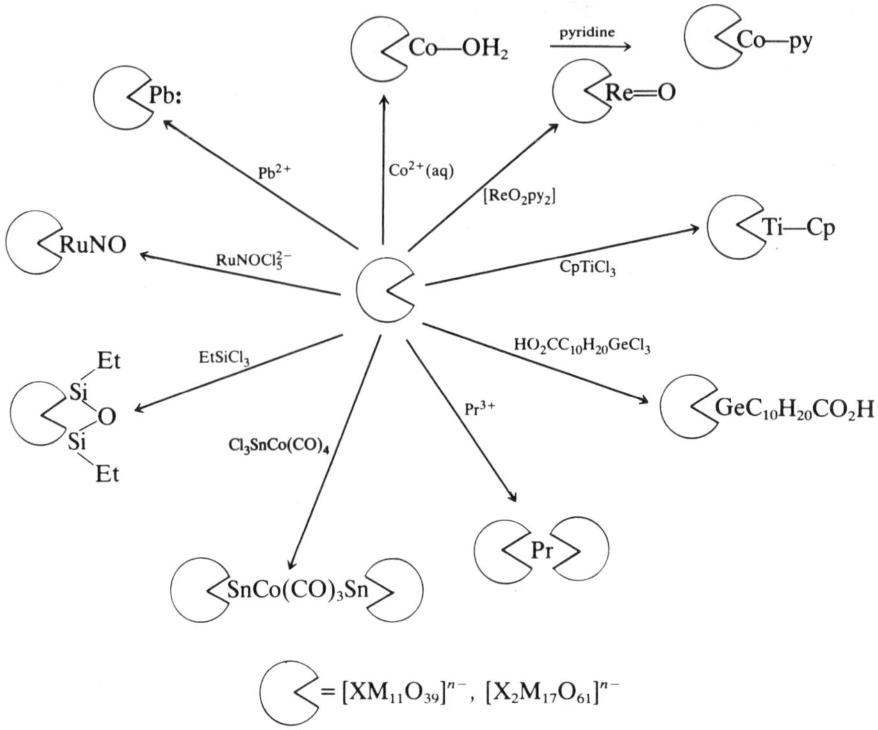

Scheme 5

lanthanide cations can be demonstrated spectroscopically, only the 2:1 complexes have been isolated. The local symmetry around the cation is approximately square antiprismatic, as confirmed in structures of $Cs_{12}[U(\alpha\text{-}GeW_{11}O_{39})_2]\cdot aq$ and $K_{16}[Ce(\alpha_2\text{-}P_2W_{17}O_{61})_2]\cdot aq$.[94] These anions generally have a high charge, and favor the stabilization of unusual oxidation states, *e.g.* tetravalent Pr, Tb, Am, Cm and Cf and pentavalent U.

$$[XM_{11}O_{39}]^{n-} + Z^{p+}(aq) \rightleftharpoons [XM_{11}Z(OH_2)O_{39}]^{(n-p)-} \tag{6}$$

Although lacunary derivatives of Keggin anions with divalent heteroatoms have never been detected, derivatives of these species are readily prepared by other methods. Examples are $[(H_2)W_{11}Co(OH_2)O_{39}]^{7-}$, $[CoW_{11}Fe(OH_2)O_{39}]^{7-}$ and $[BeW_{11}Mn(OH_2)O_{39}]^{7-}$. The ligand field exerted by the lacunary anion seems to be weaker than that of the corresponding aqua complex, *e.g.* for $[SiW_{11}Ni(OH_2)O_{39}]^{6-}$ $\Delta = 8100\ cm^{-1}$ *vs.* $8600\ cm^{-1}$ for $[Ni(H_2O)_6]^{2+}$. In the Anderson anions $[H_6NiM_6O_{24}]^{4-}$ (Section 38.3.2.1) Δ is significantly larger: $9520\ cm^{-1}$ (Mo), $9800\ cm^{-1}$ (W).

The terminal water molecule may be replaced by other ligands, *e.g.* NH_3, pyridine, Cl^-.[95] With pyrazine or 4,4′-bipyridyl two polyanions can be linked together. In aqueous solution the equilibrium constants for water replacement are not very large, although only a handful have been measured. As an example, $K = 250\ M^{-1}$ for the substitution of NO_2^- for water in $[BW_{11}Co^{III}(OH_2)O_{39}]^{6-}$. Typically, the aqua derivatives are stable at $4 < pH < 8$, although several exceptions are known. It has been possible to remove the coordinated water molecules of such anions in nonpolar solvents.[131] Kinetic inertness of Z^{n+} frequently extends the effective pH range over which the complexes may be studied. For example the free acid of $[SiMo_{11}Cr^{III}(OH_2)O_{39}]^{5-}$ can be isolated, and the anion also can be deprotonated to $[SiMo_{11}Cr(OH)O_{39}]^{6-}$ at pH 9.[96]

Within the last few years there has been considerable activity directed towards the synthesis of organic and organometallic derivatives of heteropolyanions. Reactions of lacunary polyanions with $RMCl_3$, where R is an organic or organometallic radical, provide a simple route to such derivatives.[97] Some examples of these reactions are presented in Scheme 5. For the most part the reactions proceed smoothly and the products are remarkably stable towards hydrolysis. Apart from their structural characterization *via* crystallography and NMR spectroscopy the chemistry of these complexes has not been completely described.

38.4.2.2 Multisubstituted heteropolyanions

Heteropolyanion structures in which more than one W or Mo atom has been replaced range from the mixed metal species such as $[PW_8V_4O_{40}]^{7-}$ to anions like $[P_2W_{18}(SnC_6H_5)_2O_{60}]^{8-}$ and α-$[SiW_9(CoOH_2)_3O_{37}]^{10-}$. In some cases there are systematic methods of synthesis from an isolated lacunary species, *e.g.* α-$[SiW_9O_{34}]^{10-}$; in others, 'brute force' methods lead to complex mixtures of isomeric polyanions. As more work is done in this area, patterns are beginning to emerge. Thus several anions have been shown to contain B-α-XW_9 units (see Figure 11): $[(P^VW_9O_{34})_2Co_4(H_2O)_2]^{10-}$,[98] $[(As^{III}W_9O_{33})_2Cu_3(H_2O)_2]^{12-}$,[99] $[(Se^{IV}W_9O_{33})_2$-$Mn_2^{III}WO(H_2O)_n]^{8-}$ [100] and $[(AsW_9O_{33})_2Co(WO)_2(H_2O)_n]^{8-}$. On the other hand the A-type anions $[SiW_9O_{34}]^{10-}$ and $[PW_9O_{34}]^{9-}$ react to give complete Keggin-like structures such as $[SiW_9Fe_3(OH_2)_3O_{37}]^{7-}$ [127] and $[PW_9V_3O_{40}]^{6-}$.[128]

38.4.2.3 Heteropoly cryptands

In Section 38.3.1.1.ii we noted that the large tungstophosphate anions $[P_5W_{30}O_{110}]^{15-}$ and $[P_8W_{48}O_{184}]^{40-}$ act as cryptands towards alkali metal cations. Two other leviathans, $[(NH_4)As_4W_{40}O_{140}(CoOH_2)_2]^{23-}$ (a)[101] and $[NaSb_9W_{21}O_{86}]^{18-}$ (b)[102] are illustrated in Figure 20. Anion (a) is constructed of four B-α-AsW_9 units linked by four WO_6 octahedra; it has a central crypt (optimum size for K^+ and Ba^{2+}) occupied by NH_4^+, and has four 'lacunae', two of which are occupied by Co^{2+}. The anion shows novel allosteric behavior; when it binds two small cations like Co^{2+}, the remaining two coordination sites expand and are incapable of binding further Co^{2+}, although two Ag^+ ions can be complexed.

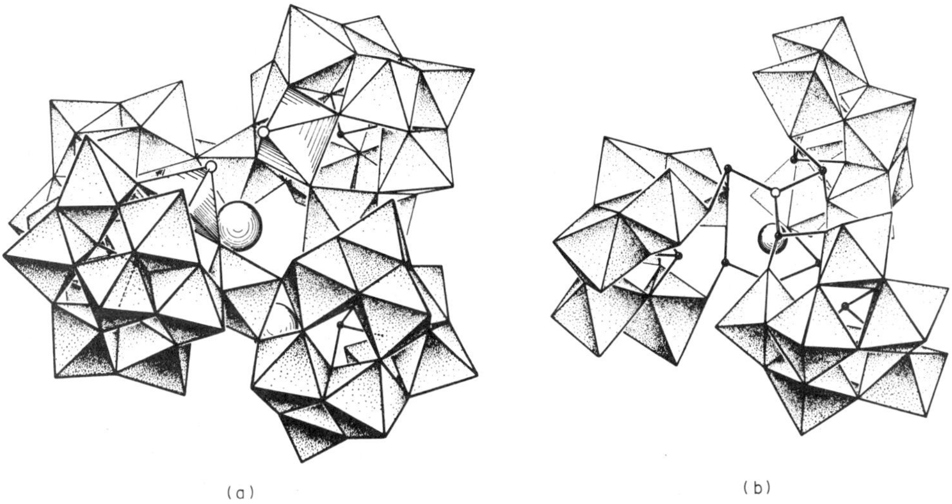

Figure 20 (a) $[(NH_4)As_4W_{40}O_{140}(CoOH_2)_2]^{23-}$ with aquacobalt(II) ions represented in the square pyramids with open circles. The large circle represents the ammonium ion in the central crypt. Two other ammonium ions (one partly visible) are weakly coordinated in positions corresponding to the cobalt(II) sites on the other side of the anion. (b) $[NaSb_9W_{21}O_{86}]^{18-}$. The β-SbW_7 units are shown in polyhedral fashion. The Sb atoms are represented as small dark spheres and the Na^+ ion as the large sphere

Anion (b) contains three B-β-SbW_7 units linked to an Sb_6O_2 'axis'. The central cavity is optimum size for Na^+. Both (a) and (b) have been found to exhibit antiviral and antitumoral properties at non-cytotoxic doses *in vitro* and *in vivo,* and are potent inhibitors of cellular, bacterial and viral DNA and RNA polymerases. Anion (b) has recently been used to treat patients with Acquired Immune Deficiency Syndrome.[129]

38.5 HETEROPOLY BLUES AND OTHER REDUCED SPECIES

The formation of intensely colored species ('heteropoly blues') upon partial reduction of certain heteropolyanions forms the basis of widely used analytical procedures for the heteroelements in such anions (P, Si, *etc.*), or for the reductant (*e.g.* uric acid). Photochemical reduction, which can, for example, lead to hydrogen evolution from aqueous solution is of particular current interest.[103] Although a pioneering study of the reduced molybdo- and tungsto-phosphates was made in 1920,[104] most of our present knowledge has developed from more recent electrochemical and spectroscopic studies.[1,105] The reduction of a heteropolyanion to a heteropoly blue involves the reversible addition of one or more electrons. In most cases there are no significant structural changes during the first stages of reduction. The added electrons must therefore enter orbitals that are mainly nonbonding in character. Such orbitals are present only in those anions in which each metal atom has a single terminal oxygen atom (approximate local symmetry, C_{4v}), *e.g.* the Keggin (Figure 10), Dawson (Figure 12) and M_6O_{19} (Figure 5) structures. Such structures are referred to as Type I and contain a nonbonding d orbital on each metal atom (in C_{4v} the orbital (d_{xy}) has B_2 symmetry). Other polyanion structures (Type II) have metal atoms with two (*cis*) terminal oxygens. In this symmetry all of the metal's d orbitals are involved in σ and π bonding. Consequently Type II polyanions (*e.g.* the Anderson anions, Figure 16) do not undergo reversible reduction to heteropoly blue species.

38.5.1 Electrochemistry

38.5.1.1 Polytungstates

Careful polarographic and voltammetric studies of several Keggin anions have revealed an elaborate redox behavior. The most complete study has been made on the metatungstate anion, α-$[(H_2)W_{12}O_{40}]^{6-}$.[50] In Scheme 6 and the subsequent discussion, the oxidized anion will be designated as **0** and each reduced species by a roman numeral representing the number of added electrons. In the scheme the horizontal double-headed arrows represent polarographically reversible electron transfer processes. Slower disproportionations or reductions are represented by the vertical arrows. The unprimed species are blue, and are mixed valence $W^{V,VI}$ anions, *e.g.* $[(H_2)W_2^VW_{10}^{VI}O_{40}]^{8-}$ (**II**). The primed species are red-brown, are less air sensitive and appear to contain W^{IV}, *e.g.*, **VI′** is $[(H_2)W_3^{IV}W_9^{VI}O_4H_6]^{6-}$ with a metal–metal bonded group of three edge-shared WO_6 octahedra. As the anion is reduced further, the remaining W_3 groups become metal–metal bonded. Electrons added to **XXIV′** enter delocalized nonbonding orbitals of which there is one for each set of three W^{IV} atoms. It is remarkable that the Keggin framework does not appear to be destroyed in this process; **XXIV′**, which is amphoteric, an orange cation (pH < 3) or a violet anion (pH > 12), can be reoxidized quantitatively to **0** under acidic conditions. As mentioned in Section 38.2.6.2, **XXIV′** slowly loses one of the central protons to yield the corresponding reduced version of $[(H)W_{12}O_{40}]^{7-}$.

$$\mathbf{0} \longleftrightarrow \mathbf{I} \longleftrightarrow \mathbf{II} \longleftrightarrow \mathbf{III} \longleftrightarrow \mathbf{IV} \longleftrightarrow \mathbf{(VI)} \xrightarrow{\text{fast}} \mathbf{VI'}$$

$$\mathbf{VI'} \longleftrightarrow \mathbf{VII'} \longleftrightarrow \mathbf{VIII'}$$

$$\mathbf{XII'} \longleftrightarrow \mathbf{XIV'}$$

$$\mathbf{XVIII'}$$

$$\mathbf{XXIV'} \longleftrightarrow \mathbf{XXVIII'} \longleftrightarrow \mathbf{XXXII'}$$

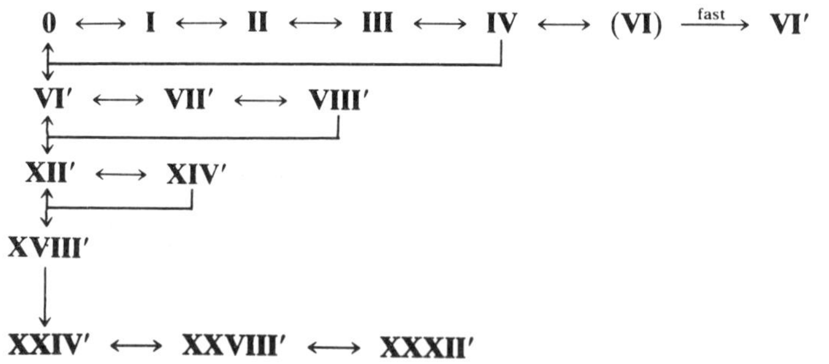

Scheme 6

The other Keggin anions undergo similar redox processes, although there are individual variations. The initial reduction potentials, **0/I**, of $[XW_{12}O_{40}]^{n-}$ vary approximately linearly with n, from *ca.* 0 V (*vs.* SCE) for $[PW_{12}O_{40}]^{3-}$ to -0.7 V for $[Cu^IW_{12}O_{40}]^{7-}$. The maximum reduction states attainable for $[PW_{12}O_{40}]^{3-}$, $[SiW_{12}O_{40}]^{4-}$ and $[BW_{12}O_{40}]^{5-}$ are **VI**, **XX′** and **XVIII′** respectively. This behavior is attributed to coulombic repulsions between the heteroatoms and the W^{IV} atoms which must move towards the center of the anion in **VI′** and more reduced species. Reduction of other XW_{12} anions, $[X_2W_{18}O_{62}]^{6-}$, $[W_6O_{19}]^{2-}$, $[W_{10}O_{32}]^{4-}$ and lacunary species has not been explored much beyond stages **I** and **II**.

38.5.1.2 *Polymolybdates*

The electrochemistry of heteropolymolybdates parallels that of the tungstates but with the following differences: the reduction potentials are more positive and the primed species (metal–metal bonded?) are much less stable. Scheme 7 applies for α-$[SiMo_{12}O_{40}]^{4-}$. Species in parentheses are detectable only by rapid scan cyclic voltammetry, and **XVIII'** decomposes rapidly at $0\,°C$. The reduced anions such as **II** and **IV** are easily obtained by controlled potential electrolysis or by careful chemical reduction, *e.g.* with ascorbate. The use of metal ion reductants generally leads to other reactions, (equation 7). The reduced anions slowly isomerize (equation 8). The isomerization can be followed polarographically (all β potentials are more positive) or by NMR spectroscopy. By this means β isomers of most Keggin and Dawson molybdates have been prepared.

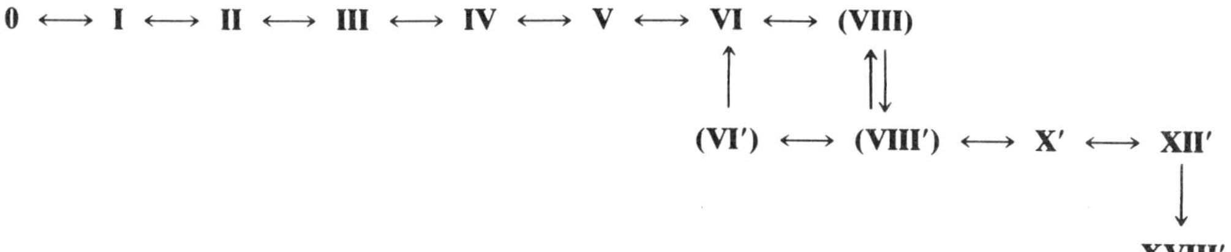

$$\mathbf{0} \longleftrightarrow \mathbf{I} \longleftrightarrow \mathbf{II} \longleftrightarrow \mathbf{III} \longleftrightarrow \mathbf{IV} \longleftrightarrow \mathbf{V} \longleftrightarrow \mathbf{VI} \longleftrightarrow (\mathbf{VIII})$$

$$(\mathbf{VI'}) \longleftrightarrow (\mathbf{VIII'}) \longleftrightarrow \mathbf{X'} \longleftrightarrow \mathbf{XII'}$$

$$\mathbf{XVIII'}$$

Scheme 7

$$\mathbf{0}\,(\alpha\text{-}SiMo_{12}) \xrightarrow{Sn^{II}} \mathbf{IV}\,(SiMo_{10}Sn_2) \tag{7}$$

$$\mathbf{IV}\,(\alpha\text{ series}) \longrightarrow \mathbf{IV}\,(\beta\text{ series}) \tag{8}$$

38.5.2 Spectroscopy and Electronic Structures

The electronic spectra of heteropoly blues show broad bands in the visible and near IR regions that are attributable to intervalence charge transfer (*e.g.* $W^V \rightarrow W^{VI}$) and to d–d transitions. The intensities of these bands vary considerably with the structure of the polyanion. The reduced species, **I**, which have received the greatest attention so far, are currently assumed to be Class II mixed valence complexes in the Robin–Day classification.[106] This implies that the ground state may be described in terms of an electron that is weakly trapped on one of the W (or Mo) atoms. Partial delocalization of the electron on to the neighboring W^{VI} atom is defined by an electron-transfer integral (electron-tunnelling matrix element), J. Valence interchange between the metal centers can take place by optical excitation or by adiabatic electron transfer ('hopping'). In the high temperature limit, $kT \gg h\nu$ (where ν is the frequency of the vibrational mode(s) coupled to the electron transfer), $E_{th} = E_{op}/4$, where E_{th} is the activation energy for electron hopping and E_{op} is the energy of the intervalence transition. When allowance is made for partial electron delocalization, equation (9) applies. In favorable cases, unambiguous evidence for temperature dependent electron hopping is provided by ESR spectroscopy. The high and low temperature ESR spectra of $[P_2V^{IV}V_2^VW_{15}O_{62}]^{10-}$ (Figure 21) show that the electron interacts equally with the three $I = \frac{1}{2}$ ^{51}V nuclei at $350\,K$ and is trapped on a single vanadium atom at $5\,K$. The high temperature activation energy (E_{th}) for this compound is $29\,kJ\,mol^{-1}$ and J is $13\,kJ\,mol^{-1}$. Values for J and E_{th} for some polymolybdates and polytungstates have been estimated from ESR line broadening and intervalence band parameters, and are given in Table 11.

$$E_{th} = \tfrac{1}{4}E_{op} - J + \frac{J^2}{E_{op}} \tag{9}$$

The electronic spectra of heteropoly blue anions (see Figure 22 for examples) show other intense broad bands in addition to the intervalence transitions. These are believed to result from intensity-enhanced d–d transitions.[108] For a given polyanion structure, the intensity of the spectra increase approximately linearly with the degree of reduction until state **VI** is reached. This behavior implies that successive electrons occupy orbitals of different metal atoms. The intensities of the intervalence bands are structure dependent: $M_6O_{19} < \alpha\text{-}XM_{12}O_{40} < \beta\text{-}XM_{12}O_{40} < \alpha,\beta\text{-}X_2M_{18}O_{62} < W_{10}O_{32}$. The molar extinction coefficient per W^V increases from 280 for $[W_6O_{19}]^{3-}$ to 7500 for $[W_{10}O_{32}]^{5-}$. The variation has been attributed to the change in M^V—O—M^{VI} bond angles ($100°$ to $175°$ over the series). A greater bond angle leads to more

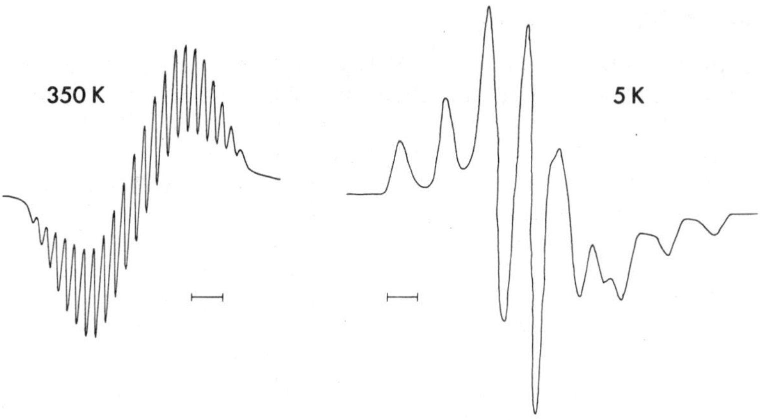

Figure 21 ESR spectra of α-$[P_2W_{15}V^{IV}V_2^VO_{62}]^{10-}$ at 350 K and 5 K. Scale bars represent 100 G

Table 11 Intervalence Parameters for some Heteropoly Blues

Anion	*Intervalence band frequency* (cm^{-1})	E_{th} (kJ mol^{-1})	J (kJ mol^{-1})
$[Mo_6O_{19}]^{3-}$	9000	15	14
$[PMo_{12}O_{40}]^{4-}$	6400	3.4	33
$[GeMo_{12}O_{40}]^{5-}$	7300	4.3	24
$[W_6O_{19}]^{3-}$	10 200	5.3	37
$[SiW_{12}O_{40}]^{5-}$	8000	3.4	30
$[(H_2)W_{12}O_{40}]^{7-}$	8400	3.4	32
$[As_2W_{18}O_{62}]^{7-}$	11 000	3.9	43

efficient electron delocalization *via* π bonding, and this is also reflected in the values of J (Table 11).

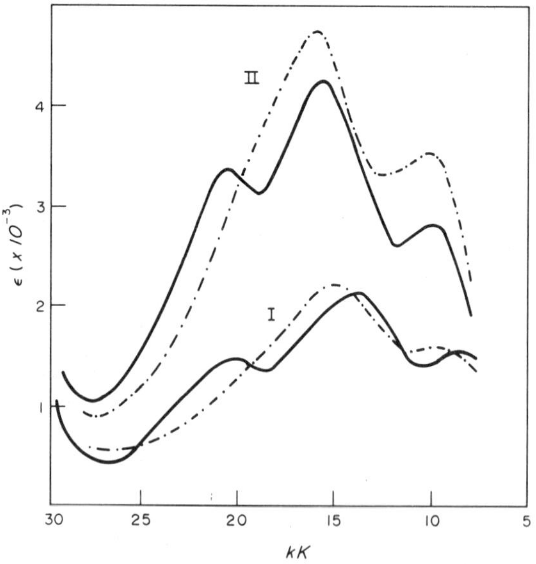

Figure 22 Electronic absorption spectra of reduced species **I** and **II** of $[PW_{12}O_{40}]^{3-}$ (—) and $[CoW_{12}O_{40}]^{6-}$ (—·—·—)

38.6 ORGANIC DERIVATIVES OF POLYANIONS

The introduction of organic and organometallic moieties into lacunary polyanion structures has been described in Sections 38.4.2.1 and 38.4.2.2. Other organic derivatives of polyanions are discussed in the following sections.

38.6.1 Organophosphonate Complexes

Pentamolybdobis(organophosphonates), $[(RPO_3)_2Mo_5O_{15}]^{4-}$, are readily formed in weakly acidic solution (pH 2–5).[73] The following derivatives have been isolated: R = H, Me, Et, $C_2H_4NH_3^+$, Ph, p-$C_6H_4NH_3^+$ and n-$C_{22}H_{45}$. The last-mentioned complex has been shown to be an effective electron-dense stain for electron microscopy of phospholipid vesicles.[109] X-Ray investigations of the methyl and β-aminoethyl derivatives show the anions to be isostructural with $[P_2Mo_5O_{23}]^{6-}$ (Figure 15) with R groups occupying the unshared vertices of the PO_4 tetrahedra.[110] A 'neutral' zwitterionic analog, prepared from *cis*-$[Co(en)_2(H_2O)OPO_3H)]^+$, has been isolated, as have several derivatives of phosphate monoesters.[74]

The corresponding pentatungsto complexes are less common; only those with R = Ph and $C_2H_4NH_3^+$ have been reported. According to ^{31}P and ^1H NMR the phenyl derivative is partially dissociated in aqueous solution, in contrast to the molybdates. In organic solvents, however, the tungstate appears to be undissociated, but ^{183}W and ^{17}O NMR spectra of these solutions reveal the anion to be fluxional at temperatures above −38 °C. The dynamic process is described as a pseudorotation of the nonplanar ring of five tungsten atoms, coupled to an oscillation of the $PhPO_3$ groups.[111] There is little doubt that the corresponding molybdates behave analogously.

38.6.2 Organoarsonate Complexes

Three polyanion structures incorporating $RAsO_3^{2-}$ groups have been reported. Two of these are analogous to $[As_2Mo_6O_{26}]^{6-}$ and $[H_4As_4Mo_{12}O_{50}]^{4-}$ described in Section 38.3.1.2. The As_2Mo_6 complexes are formed under similar conditions to the P_2Mo_5 species and the structure of $(Me_4N)_2Na_2[(MeAsO_3)_2Mo_6O_{18}]\cdot 6H_2O$ corresponds to that shown in Figure 15 for $[As_2Mo_6O_{26}]^{6-}$.[112] However, a hydrated version of the anion was observed in crystals of $(CN_3H_6)_4[(PhAsO_3)_2Mo_6O_{18}(OH_2)]\cdot 4H_2O$ and is shown in Figure 23. In aqueous solution the hydrated anion is the predominant species (IR and Raman spectra), but the R groups are made NMR equivalent by rapid exchange of the bridging water molecule.[113] The corresponding tungstate is isolated as a deprotonated species, $[(PhAsO_3)_2W_6O_{18}(OH)]^{5-}$, and the NMR spectrum of this complex at pH 3.6 shows two nonequivalent phenyl groups. As the pH of the solution is lowered, the phenyl resonances coalesce. This behavior is attributed to increased protonation of the anion to $[(PhAsO_3)_2W_6O_{18}(OH_2)]^{4-}$ and consequent water exchange as for the molybdate.[114]

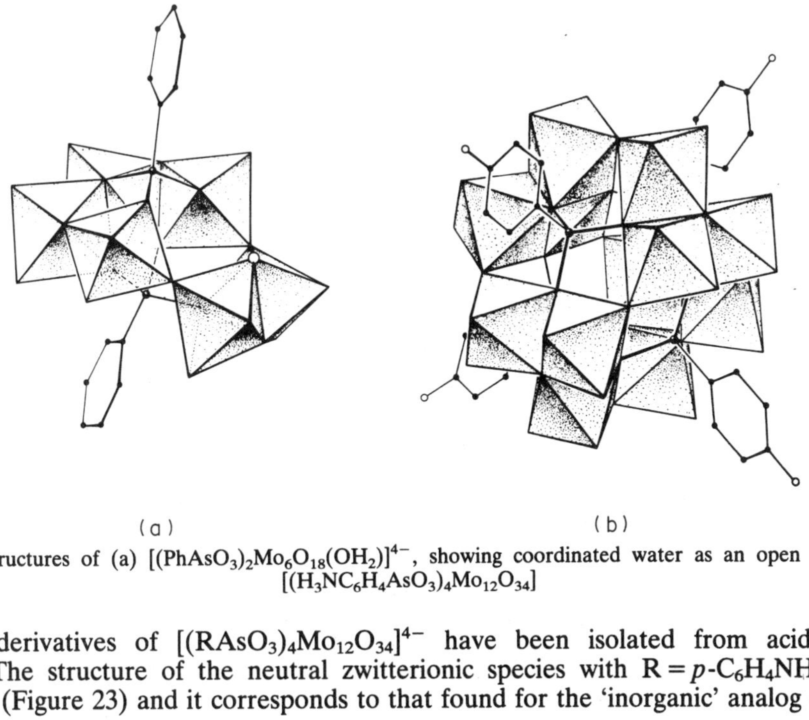

Figure 23 Structures of (a) $[(PhAsO_3)_2Mo_6O_{18}(OH_2)]^{4-}$, showing coordinated water as an open circle, and (b) $[(H_3NC_6H_4AsO_3)_4Mo_{12}O_{34}]$

Several derivatives of $[(RAsO_3)_4Mo_{12}O_{34}]^{4-}$ have been isolated from acidic solutions (pH < 1). The structure of the neutral zwitterionic species with R = p-$C_6H_4NH_3^+$ has been reported[115] (Figure 23) and it corresponds to that found for the 'inorganic' analog (R = OH).

The anion $[(MeAsO_3)Mo_6O_{18}(H_2O)_6]^{2-}$ has been reinvestigated.[116] Its structure contains a ring of $MoO_5(H_2O)$ octahedra and can be viewed as the structure of A-α-$[AsMo_9O_{31}(H_2O)_3]^{3-}$ (Figure 11) with the Mo_3O_{13} 'cap' removed.

38.6.3 Diorganoarsinate and Related Complexes

The anion shown in Figure 24, $[(Me_2AsO_2)Mo_4O_{12}(OH)]^{2-}$, is readily formed in aqueous solution at pH 3–5. Among its unusual features is the presence of a hydroxide group that bridges four molybdenum atoms.[117] Anions of this structure have proved to be remarkably effective inhibitors of certain acid phosphatases.[118] The structure may be considered to have been assembled from $R_2AsO_2^-$, OH^- and cyclic Mo_4O_{12}. Similar complexes based on $R_2C(OH)_2$ (acetal) units have been prepared by reaction of $[Mo_2O_7]^{2-}$ with aldehydes in nonaqueous solution. The structure of $(Bu_4N)_3[(H_2CO_2)Mo_4O_{12}(OH)]$ has been reported.[119] When the glyoxal derivative, $[(OHCCHO_2)Mo_4O_{12}(OH)]^{3-}$, is treated with HX ($X = F^-$, HCO_2^-), the second carbonyl group inserts into an Mo—O bond, and the four-fold bridging OH^- group is replaced by doubly bridging X^- (see Figure 25).[120]

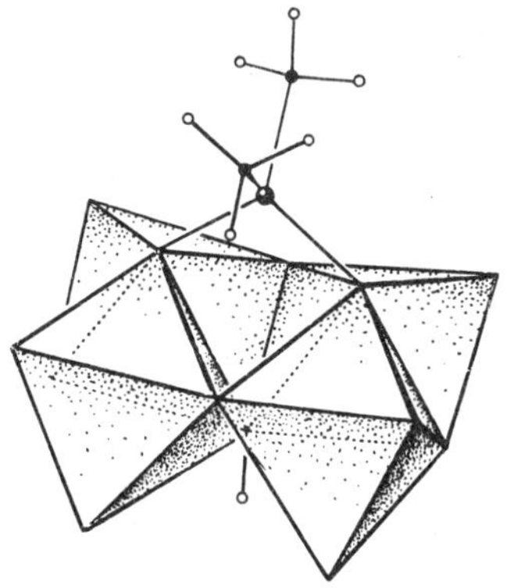

Figure 24 The structure of $[(Me_2AsO_2)Mo_4O_{12}(OH)]^{2-}$ anion

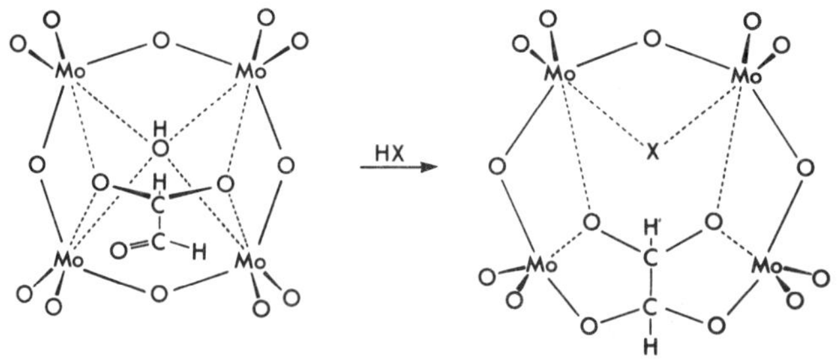

Figure 25 Structural scheme showing the insertion, upon treatment with HX, of the free carbonyl group of $[(OHCCHO_2)Mo_4O_{12}(OH)]^{3-}$ into an Mo—O bond to yield $[HCCHMo_4O_{15}X]^{3-}$. Broken lines denote Mo—O bonds of length 2.00–2.40 Å

38.6.4 Octamolybdate Derivatives

Addition of formic acid to a solution of $(NH_4)_6Mo_7O_{24}$ yields crystals of $(NH_4)_6[(HCO)_2Mo_8O_{26}]\cdot 2H_2O$.[120] When $MoO_3\cdot 2H_2O$ is refluxed with absolute MeOH over a molecular sieve, crystals of $Na_4[Mo_8O_{24}(OMe)_4]\cdot 8MeOH$ are isolated from the mother liquor.[122] The anions of both of these compounds contain the octamolybdate moiety shown in Figure 26. Terminal formyl and methoxy groups are attached to the MoO_6 octahedra with three unshared vertices; two methoxy groups occupy bridging sites. UV irradiation or thermal decomposition of the methoxy derivative yields formaldehyde, and this behavior illustrates some of the potential value of organopolyanion chemistry in modelling heterogeneous catalysis by metal oxides.

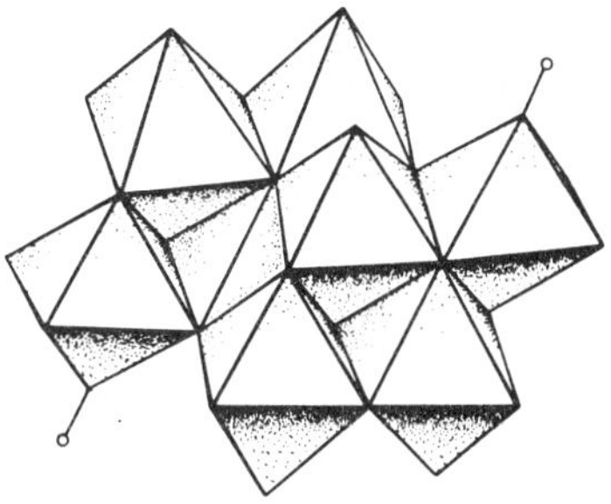

Figure 26 The Mo_8O_{26} unit found in $[(HCO)_2Mo_8O_{26}]^{6-}$ and $[Mo_8O_{24}(OMe)_4]^{4-}$ anions. The small circles indicate points of attachment of the formyl groups

ADDENDUM

Research activity continues to be high with groups in Europe, Japan, Canada and the USA. Considerable interest is focused on catalytic and photocatalytic applications.[130] In this context, transferral of heteropolyanions into nonpolar solvents opens up new possibilities such as oxygen-carrying species.[131] Recent work on organometallic derivatives of polyoxoanions from groups at the Universities of Illinois and Nebraska has been reviewed.[132] High pressure liquid chromatography has been used to separate several heteropolymolybdates, including α- and β-$[SiMo_{12}O_{40}]^{4-}$.[134] The structure of the heteropoly blue species (four-electron-reduced β-$[PMo_{12}O_{40}]^{3-}$) formed in the 'molybdenum blue' determination of phosphorus has been reported[135] and a review of molybdate heteropoly blues has appeared.[136] A 'stability index' of polyanion structures has been discussed.[137] The presence of two PMo_9 anions in aqueous solutions of molybdate and phosphate has been demonstrated by ^{31}P NMR.[138] Solid state NMR (nonspinning and MAS techniques) has been used for characterization of heteropolyanions.[139] Molybdenum-95 NMR spectra of some polymolybdates have been reported.[140]

Acknowledgement: It is a pleasure to acknowledge the contribution of my colleague Dr Howard T. Evans, Jr., US Geological Survey, in the preparation of the structural figures for this chapter.

38.7 REFERENCES

1. M. T. Pope, 'Heteropoly and Isopoly Oxometalates', Inorganic Chemistry Concepts 8, Springer-Verlag, Berlin, 1983.
2. V. I. Spitsyn, L. P. Kazansky and E. A. Torchenkova, *Sov. Sci. Rev., Sect. B, Chem. Rev.*, 1981, **3,** 111.
3. G. A. Tsigdinos, *Top. Curr. Chem.*, 1978, **76,** 1.
4. I. V. Koshevnikov and K. I. Matveev, *Russ. Chem. Rev. (Engl. Transl.)*, 1982, **51,** 1075.
5. M. A. Porai-Koshits and L. O. Atovmyan, *Russ. J. Inorg. Chem. (Engl. Transl.)*, 1981, **26,** 1697.
6. W. N. Lipscomb, *Inorg. Chem.*, 1965, **4,** 132.
7. A. Rosenheim and H. Jaenicke, *Z. Anorg. Allg. Chem.*, 1917, **100,** 304; A. Rosenheim, in 'Handbuch der Anorganischen Chemie', ed. R. Abegg and F. Auerbach, Hirzel Verlag, Leipzig, 1921, vol. 4, part 1, ii, pp. 977–1064; 'Gmelins Handbuch der Anorganischen Chemie', Verlag Chemie, Berlin, System-Nr. 53 (Molybdan), 1935, pp. 312–393; System-Nr. 54 (Wolfram), 1933, pp. 324–397.
8. C. F. Baes, Jr. and R. E. Mesmer, 'The Hydrolysis of Cations', Wiley-Interscience, New York, 1976.
9. L. G. Sillén, *Q. Rev., Chem. Soc.*, 1959, **13,** 146.
10. K. H. Tytko and O. Glemser, *Adv. Inorg. Chem. Radiochem.*, 1976, **19,** 239.
11. K. H. Tytko and J. Mehmke, *Z. Anorg. Allg. Chem.*, 1983, **503,** 67.
12. E. Heath and O. W. Howarth, *J. Chem. Soc., Dalton Trans.*, 1981, 1105.
13. J. Fuchs, S. Mahjour and J. Pickardt, *Angew. Chem., Int. Ed. Engl.*, 1976, **15,** 374.
14. See, for example, H. T. Evans, Jr., *Inorg. Chem.*, 1966, **5,** 967.
15. H. T. Evans, Jr. and M. T. Pope, *Inorg. Chem.*, 1984, **23,** 501.
16. D. M. Druskovich and D. L. Kepert, *J. Chem. Soc., Dalton Trans.*, 1975, 947; F. Corigliano and S. DiPasquale, *J. Chem. Soc., Dalton Trans.*, 1978, 1329; J. B. Goddard and A. M. Gonas, *Inorg. Chem.*, 1973, **12,** 574; B. Viossat, *Bull. Soc. Chim. Fr.*, 1975, 1111.
17. R. K. Murmann and K. C. Giese, *Inorg. Chem.*, 1978, **17,** 1160.
18. H. T. Evans, Jr. and S. Block, *Inorg. Chem.*, 1966, **5,** 1808.
19. A. M. Bystrom and H. T. Evans, Jr., *Acta Chem. Scand.*, 1959, **13,** 377.
20. G. K. Johnson, Ph.D. Thesis, University of Missouri-Columbia, 1977; *Diss. Abstr. B,* 1978, **38,** 4801; G. K. Johnson and E. O. Schlemper, *J. Am. Chem. Soc.*, 1978, **100,** 3645.
21. S. Ostrowetsky, *Bull. Soc. Chim. Fr.*, 1964, 1003, 1012, 1018.
22. A. Bino, S. Cohen and C. Heitner-Wirguin, *Inorg. Chem.*, 1982, **21,** 429.

23. E. Hayek and U. Pallasser, *Monatsh. Chem.*, 1968, **99,** 2494.
24. A. Goiffon, E. Philippot and M. Maurin, *Rev. Chim. Minér.*, 1980, **17,** 466.
25. A. Goiffon, R. Granger, C. Bockel and B. Spinner, *Rev. Chim. Minér.*, 1973, **10,** 487.
26. A. Goiffon and B. Spinner, *Rev. Chim. Miner.*, 1974, **11,** 262; *Bull. Soc. Chim. Fr.*, 1975, 2435.
27. E. J. Graeber and B. Morosin, *Acta Crystallogr., Sect. B.* 1977, **33,** 2136.
28. A. Thiele and J. Fuchs, private communication, 1979.
29. M. Filowitz, R. K. C. Ho, W. G. Klemperer and W. Shum, *Inorg. Chem.*, 1979, **18,** 93.
30. J. Fuchs and D. Lubkoll, *Z. Naturforsch., Teil B*, 1973, **28,** 590.
31. K. H. Tytko, G. Baethe, E. R. Hirshfeld, K. Mehmke and D. Stellhorn, *Z. Anorg. Allg. Chem.*, 1983, **503,** 43.
32. I. Knoepnadel, H. Hartl, W. D. Hunnius and J. Fuchs, *Angew. Chem., Int. Ed. Engl.*, 1974, **13,** 823.
33. V. W. Day, M. F. Fredrich, W. G. Klemperer and W. Shum, *J. Am. Chem. Soc.*, 1977, **99,** 6146.
34. G. Johansson, L. Pettersson and N. Ingri, *Acta Chem. Scand., Ser. A*, 1977, **33,** 305.
35. T. Yamase, *J. Chem. Soc., Dalton Trans.*, 1982, 1987.
36. B. Krebs and I. Paulat-Boeschen, *Acta Crystallogr., Sect. B*, 1982, **38,** 1710.
37. I. Larking and R. Stomberg, *Acta Chem. Scand.*, 1972, **26,** 3708.
38. See, for example, W. Clegg, G. M. Sheldrick, C. D. Garner and I. B. Walton, *Acta Crystallogr., Sect. B*, 1982, **38,** 2906.
39. J. Fuchs and H. Hartl, *Angew. Chem., Int. Ed. Engl.*, 1976, **15,** 375.
40. W. G. Klemperer and W. Shum, *J. Am. Chem. Soc.*, 1976, **98,** 8291.
41. M. Filowitz, W. G. Klemperer and W. Shum, *J. Am. Chem. Soc.*, 1978, **100,** 2580.
42. H. R. Kreusler, A. Forster and J. Fuchs, *Z. Naturforsch., Teil B*, 1980, **35,** 242.
43. B. Krebs and I. Paulat-Boeschen, *Acta Crystallogr., Sect. B*, 1976, **32,** 1697.
44. K. G. Burtseva, T. S. Chernaya and M. I. Sirota, *Sov. Phys.-Dokl.* (*Engl. Transl.*), 1978, **23,** 784.
45. H. T. Evans, Jr. and E. Prince, *J. Am. Chem. Soc.*, 1983, **105,** 4838.
46. A. Hüllen, *Angew. Chem.*, 1964, **76,** 588.
47. E. L. Simons, *Inorg. Chem.*, 1964, **3,** 1079.
48. P. Souchay, M. Boyer and F. Chauveau, *K. Tek. Hoegsk. Handl.* No. 259, 1972 (Contributions to Coordination Chemistry in Solution, Stockholm, 1972), p. 159.
49. M. T. Pope and G. M. Varga, Jr., *Chem. Commun.*, 1966, 653.
50. J. P. Launay, *J. Inorg. Nucl. Chem.*, 1976, **38,** 807.
51. F. Chauveau, P. Doppelt and J. Lefebvre, *J. Chem. Res.* (*S*), 1978, 130; C. Sanchez, J. Livage, P. Doppelt, F. Chauveau and J. Lefebvre, *J. Chem. Soc., Dalton Trans.*, 1982, 2439.
52. J. Lefebvre, F. Chauveau, P. Doppelt and C. Brévard, *J. Am. Chem. Soc.*, 1981, **103,** 4589.
53. J. Fuchs, H. Hartl, W. Schiller and U. Gerlach, *Acta Crystallogr., Sect. B*, 1975, **32,** 740.
54. M. Boyer and B. LeMeur, *C.R. Hebd. Seances Acad. Sci., Ser. C*, 1975, **281,** 59.
55. D. L. Kepert, 'The Early Transition Metals', Academic, New York, 1972, pp. 46–60; M. T. Pope, *Inorg. Chem.*, 1976, **15,** 2008.
56. M. T. Pope and T. F. Scully, *Inorg. Chem.*, 1975, **14,** 953.
57. M. T. Pope, S. E. O'Donnell and R. A. Prados, *J. Chem. Soc., Chem. Commun.*, 1975, 22; P. J. Domaille and W. H. Knoth, *Inorg. Chem.*, 1983, **22,** 818.
58. E. Schill, in 'Handbuch der Praparativen Anorganischen Chemie', ed. G. Brauer, Enke Verlag, Stuttgart, vol. 3, 1981, 1774–98.
59. A. G. Maddock, R. H. Platt, A. F. Williams and R. Gancedo, *J. Chem. Soc., Dalton Trans.*, 1974, 1314.
60. F. Robert and A. Tézé, *Acta Crystallogr., Sect. B*, 1981, **37,** 318.
61. M. Leyrie, M. Fournier and R. Massart, *C.R. Hebd. Seances Acad. Sci., Ser. C*, 1971, **273,** 1569.
62. G. M. Brown, M. R. Noe-Spirlet, W. G. Busing and H. A. Levy, *Acta Crystallogr., Sect. B*, 1977, **33,** 1038.
63. Y. Jeannin and J. Martin-Frère, *J. Am. Chem. Soc.*, 1981, **103,** 1664.
64. H. D'Amour, *Acta Crystallogr., Sect. B*, 1976, **32,** 729; R. Acerete, S. Harmalker, C. F. Hammer, M. T. Pope and L. C. W. Baker, *J. Chem. Soc., Chem. Commun.*, 1979, 777.
65. R. Contant and A. Tézé, *Inorg. Chem.*, 1985, **24,** 4610.
66. M. Alizadeh, Y. Jeannin, J. Martin-Frère and M. T. Pope, *J. Am. Chem. Soc.*, 1985, **107,** 2662.
67. J. F. Garvey and M. T. Pope, *Inorg. Chem.*, 1978, **17,** 1115.
68. R. Kato, A. Kobayashi and Y. Sasaki, *Inorg. Chem.*, 1982, **21,** 240.
69. G. Hervé and A. Tézé, *C.R. Hebd. Seances Acad. Sci., Ser. C*, 1974, **278,** 1417.
70. L. C. W. Baker and V. E. Simmons, *J. Am. Chem. Soc.*, 1959, **81,** 4744; L. Eberson, *J. Am. Chem. Soc.*, 1983, **105,** 3192.
71. R. Contant, *Bull. Soc. Chim. Fr.*, 1973, 3277; P. Souchay and R. Contant, *Bull. Soc. Chim. Fr.*, 1973, 3287; L. Lyhamn and L. Pettersson, *Chem. Scr.*, 1980, **16,** 52.
72. See, for example, B. Hedman, *Acta Chem. Scand.*, 1977, **27,** 3335.
73. W. Kwak, M. T. Pope and T. F. Scully, *J. Am. Chem. Soc.*, 1975, **97,** 5735.
74. D. E. Katsoulis, A. N. Lambrianidou and M. T. Pope, *Inorg. Chim. Acta*, 1980, **46,** L55.
75. K. Y. Matsumoto, M. Kato and Y. Sasaki, *Bull. Chem. Soc. Jpn.*, 1976, **49,** 106.
76. K. Nishikawa and Y. Sasaki, *Chem. Lett.*, 1975, 1185.
77. G. A. Tsigdinos, Ph.D. Thesis, Boston University, 1961; *Diss. Abstr. B*, 1961, **22,** 732; K. H. Lee, Ph.D. Thesis, Georgetown University, 1970, *Diss. Abstr. B*, 1971, **31,** 5240.
78. T. Ama, J. Hidaka and Y. Shimura, *Bull. Chem. Soc. Jpn.*, 1970, **43,** 2654.
79. H. T. Evans, Jr. and J. S. Showell, *J. Am. Chem. Soc.*, 1969, **91,** 6881.
80. D. L. Kepert, *Inorg. Chem.*, 1969, **8,** 1556.
81. H. T. Evans, Jr. and J. A. Konnert, *Am. Miner.*, 1978, **63,** 863.
82. A. Kobayashi and Y. Sasaki, *Chem. Lett.*, 1975, 1123.
83. See for example, T. J. R. Weakley, *J. Less-Common Met.*, 1977, **54,** 129.
84. G. Blasse and F. Zonnevijlle, *Recl. Trav. Chim. Pays-Bas*, 1982, **101,** 434; R. Ballardini, Q. G. Mulazzani, M. Venturi, F. Bolleta and V. Balzani, *Inorg. Chem.*, 1984, **23,** 300.

85. D. D. Dexter and J. V. Silverton, *J. Am. Chem. Soc.*, 1968, **90,** 3589.
86. S. C. Termes and M. T. Pope, *Transition Met. Chem.*, 1978, **3,** 103.
87. C. M. Flynn, Jr. and G. D. Stucky, *Inorg. Chem.*, 1969, **8,** 332.
88. C. J. Besecker and W. G. Klemperer, *J. Am. Chem. Soc.*, 1980, **102,** 7598; C. J. Besecker, V. W. Day, W. G. Klemperer and M. R. Thompson, *Inorg. Chem.*, 1985, **24,** 44; V. W. Day, W. G. Klemperer and D. J. Maltbie, *Organometallics,* 1985, **4,** 104; R. G. Finke and M. W. Droege, *J. Am. Chem. Soc.*, 1984, **106,** 7274.
89. C. J. Besecker, W. G. Klemperer and V. W. Day, *J. Am. Chem. Soc.*, 1982, **104,** 6158.
90. R. G. Finke, M. W. Droege, B. Rapko and R. Saxton, 'Synthesis, Structures and Reactivity of Polyoxoanions', NSF/CNRS Workshop, St. Lambert-des-Bois, 1983.
91. V. I. Spitsyn, E. A. Torchenkova, L. P. Kazanskii and P. Bajdala, *Z. Chem.*, 1974, **14,** 1.
92. V. W. Day, M. F. Fredrich and M. R. Thompson, 'Synthesis, Structures and Reactivity of Polyoxoanions', NSF/CNRS Workshop, St. Lambert-des-Bois, 1983.
93. R. Contant and J. P. Ciabrini, *J. Chem. Res.* (*S*), 1982, 50.
94. V. N. Molchanov, L. P. Kazanskii, E. A. Torchenkova and V. I. Simonov, *Sov. Phys.-Crystallogr.* (*Engl. Transl.*), 1979, **24,** 96; C. M. Tourné, G. F. Tourné and M. C. Brianso, *Acta Crystallogr., Sect. B,* 1980, **36,** 2012.
95. L. C. W. Baker and J. S. Figgis, *J. Am. Chem. Soc.*, 1970, **92,** 3794; T. J. R. Weakley, *J. Chem. Soc., Dalton Trans.*, 1973, 341; W. H. Knoth, P. J. Domaille and D. C. Roe, *Inorg. Chem.*, 1983, **22,** 198.
96. M. Fournier and R. Massart, *C.R. Hebd. Seances Acad. Sci., Ser. C.*, 1973, **276,** 1517.
97. R. K. C. Ho and W. G. Klemperer, *J. Am. Chem. Soc.*, 1978, **100,** 6772; W. H. Knoth, *J. Am. Chem. Soc.*, 1979, **101,** 759, 2211; F. Zonnevijlle and M. T. Pope, *J. Am. Chem. Soc.*, 1979, **101,** 2731.
98. T. J. R. Weakley, H. T. Evans, Jr., J. S. Showell, G. F. Tourné and C. M. Tourné, *J. Chem. Soc., Chem. Commun.*, 1973, 139; R. G. Finke, M. Droege, J. R. Hutchinson and O. Gansow, *J. Am. Chem. Soc.*, 1981, **103,** 1587; R. G. Finke and M. W. Droege, *Inorg. Chem.*, 1983, **22,** 1006.
99. F. Robert, M. Leyrie and G. Hervé, *Acta Crystallogr., Sect. B,* 1982, **38,** 358; A. R. Siedle, F. Padula, J. Baranowski, C. Goldstein, M. DeAngelo, G. F. Kokoszka, L. Azevedo and E. L. Venturini, *J. Am. Chem. Soc.*, 1983, **105,** 7447.
100. T. J. R. Weakley, 'Synthesis, Structures and Reactivity of Polyoxoanions', NSF/CNRS Workshop, St. Lambert-des-Bois, 1983.
101. F. Robert, M. Leyrie, G. Hervé, A. Tézé and Y. Jeannin, *Inorg. Chem.*, 1980, **19,** 1746.
102. J. Fischer, L. Ricard and R. Weiss, *J. Am. Chem. Soc.*, 1976, **98,** 3050; M. Michelon, G. Hervé and M. Leyrie, *J. Inorg. Nucl. Chem.*, 1980, **42,** 1583.
103. J. R. Darwent, *J. Chem. Soc., Chem. Commun.*, 1982, 2205; E. Papaconstantinou and A. Ioannidis, *Inorg. Chim. Acta,* 1983, **75,** 235; T. Yamase and T. Kurozumi, *J. Chem. Soc., Dalton Trans.*, 1983, 2205; Chem. Uses Molybdenum, Proc. Int. Conf., 4th, 1982, 208; C. L. Hill and D. A. Bouchard, *J. Am. Chem. Soc.*, 1985, **107,** 5148. E. N. Savinov, S. S. Saidkhanov, V. M. Nekipelov and V. N. Parmon, *Khim. Fiz.*, 1983, 1215.
104. H. Wu, *J. Biol. Chem.*, 1920, **43,** 189.
105. M. T. Pope, in 'Mixed Valence Compounds', ed. D. B. Brown, Reidel, Dordrecht, 1980, p. 365.
106. M. B. Robin and P. Day, *Adv. Inorg. Chem. Radiochem.*, 1967, **10,** 247; P. Day, *Int. Rev. Phys. Chem.*, 1981, **1,** 149.
107. S. P. Harmalker, M. A. Leparulo and M. T. Pope, *J. Am. Chem. Soc.*, 1983, **105,** 4286.
108. C. Sanchez, J. Livage, J. P. Launay, M. Fournier and Y. Jeannin, *J. Am. Chem. Soc.*, 1982, **104,** 3194.
109. S. Mann, R. J. P. Williams, P. R. Sethuraman and M. T. Pope, *J. Chem. Soc., Chem. Commun.*, 1981, 1083.
110. J. K. Stalick and C. O. Quicksall, *Inorg. Chem.*, 1976, **15,** 1577.
111. P. R. Sethuraman, M. A. Leparulo, M. T. Pope, F. Zonnevijlle, C. Brévard and J. Lemerle, *J. Am. Chem. Soc.*, 1981, **103,** 7665.
112. W. Kwak, L. M. Rajković, J. K. Stalick, M. T. Pope and C. O. Quicksall, *Inorg. Chem.*, 1976, **15,** 2778.
113. W. Kwak, L. M. Rajković, M. T. Pope, C. O. Quicksall, K. Y. Matsumoto and Y. Sasaki, *J. Am. Chem. Soc.*, 1977, **99,** 6463.
114. S. H. Wasfi, W. Kwak, M. T. Pope, K. M. Barkigia, R. J. Butcher and C. O. Quicksall, *J. Am. Chem. Soc.*, 1978, **100,** 7786.
115. K. M. Barkigia, L. M. Rajković-Blazer, M. T. Pope and C. O. Quicksall, *Inorg. Chem.*, 1981, **20,** 3318.
116. K. Y. Matsumoto, *Bull. Chem. Soc. Jpn.*, 1979, **52,** 3284.
117. K. M. Barkigia, L. M. Rajković-Blazer, M. T. Pope, E. Prince and C. O. Quicksall, *Inorg. Chem.*, 1980, **19,** 2531.
118. R. H. Glew, M. S. Czuczman, W. F. Diven, R. L. Berens, M. T. Pope and D. E. Katsoulis, *Comp. Biochem. Physiol. B,* 1982, **72,** 581.
119. V. W. Day, M. F. Fredrich, W. G. Klemperer and R. S. Liu, *J. Am. Chem. Soc.*, 1979, **101,** 491.
120. V. W. Day, M. R. Thompson, C. S. Day, W. G. Klemperer and R. S. Liu, *J. Am. Chem. Soc.*, 1980, **102,** 5971.
121. R. D. Adams, W. G. Klemperer and R. S. Liu, *J. Chem. Soc., Chem. Commun.*, 1979, 256.
122. E. M. McCarron, III and R. L. Harlow, *J. Am. Chem. Soc.*, 1983, **105,** 6179.
123. T. Yamase, N. Takabayashi and M. Kaji, *J. Chem. Soc., Dalton Trans.*, 1984, 793.
124. Y. Sasaki, personal communication, 1984.
125. W. H. Knoth and R. D. Farlee, *Inorg. Chem.*, 1984, **23,** 4765.
126. G. Hervé, reported at US-Japan Seminar on the Catalytic Activity of Polyoxoanions, Shimoda, 1985.
127. F. Ortéga, Ph.D. Thesis, Georgetown University, 1983; M. Leyrie and A. Tézé, personal communication.
128. P. Domaille, *J. Am. Chem. Soc.*, 1984, **106,** 7677.
129. W. Rozenbaum, D. Dormont, B. Spire, E. Vilmer, M. Gentilini, C. Griscelli, L. Montaigner, F. Barre-Sinoussi and J. C. Chermann, *Lancet,* 1985, 450.
130. See, for example, I. V. Kozhevnikov and K. I. Matveev, *Appl. Catal.*, 1983, **5,** 135; J. B. Moffat, *J. Mol. Catal.*, 1984, **26,** 385; N. Mizuno, T. Watanabe and M. Misono, *J. Phys. Chem.*, 1985, **89,** 80; A. Ioannidis and E. Papaconstantinou, *Inorg. Chem.*, 1985, **24,** 439; T. Yamase and T. Kurozumi, *Inorg. Chim. Acta,* 1984, **83,** L25.
131. D. E. Katsoulis and M. T. Pope, *J. Am. Chem. Soc.*, 1984, **106,** 2737.
132. V. W. Day and W. G. Klemperer, *Science,* 1985, **228,** 533.

133. R. G. Finke, M. W. Droege, J. C. Cook and K. S. Suslik, *J. Am. Chem. Soc.*, 1984, **106,** 5750.
134. M. Braungart and H. Russel, *Chromatographia,* 1985, **19,** 185.
135. J. N. Barrows, G. B. Jameson and M. T. Pope, *J. Am. Chem. Soc.*, 1985, **107,** 1771.
136. R. I. Buckley and R. J. H. Clark, *Coord. Chem. Rev.*, 1985, **65,** 167.
137. K. Nomiya and M. Miwa, *Polyhedron,* 1984, **3,** 341; 1985, **4,** 89.
138. L. Pettersson, I. Andersson and L. O. Oehman, *Acta Chem. Scand., Ser. A,* 1985, **39,** 53.
139. J. B. Black, N. J. Clayden, L. Griffiths and J. D. Scott, *J. Chem. Soc., Dalton Trans.*, 1984, 2765; W. H. Knoth, P. J. Domaille and R. D. Farlee, *Organometallics,* 1985, **4,** 62.
140. S. F. Gheller, M. Sidney, A. F. Masters, R. T. C. Brownlee, M. J. O'Connor and A. G. Wedd, *Aust. J. Chem.*, 1984, **37,** 1825.

39

Scandium, Yttrium and the Lanthanides

F. ALAN HART
Queen Mary College, London, UK

39.1 SCANDIUM

39.1.1 Introduction

Scandium can truly be described as a little-studied element. In their thorough review[1] 'The Coordination Chemistry of Scandium', Melson and Stotz showed that, at that date, though many of the fundamental topics relating to this element had been to some extent investigated, much careful work needed still to be done to bring our knowledge of scandium chemistry to the level of that of its neighbours such as titanium or vanadium. By the present date, a number of advances have indeed been made but the chemistry of scandium is still backward compared with other elements. Of course, the element is very expensive but it is hard to see this as a major deterrent to study. With essentially only one oxidation state, the chemistry may lack variety to some extent, but still has much of interest to offer to the researcher.

With an ionic radius of 0.81 Å (Pauling), the ion Sc^{3+} lies on the borderline between six-coordination and higher coordination numbers. In those crystal structures which are known, Sc^{3+} is predominantly six-coordinated, but examples of seven-, eight- and nine-coordination do occur, together with a limited number of examples of lower coordination numbers than six where ligands are very bulky.

Regarding ligand preferences, Sc^{3+} is a typical Chatt–Ahrland Class A metal ion. Thus we see ready coordination with ligands based on oxygen, nitrogen or halides, but virtually no complexes with sulfur- or phosphorus-based ligands. While there is a small but interesting organoscandium chemistry in which considerable development has recently taken place, that is not dealt with in these pages.

39.1.2 Nitrogen Donor Ligands

39.1.2.1 Ammonia

There is little doubt that ammines are formed. Thus scandium hydroxide is incompletely precipitated by ammonium hydroxide, perhaps owing to the formation of ammine complexes.[2] Also the oxalato–ammine $\{Sc(NH_3)_6\}_2(C_2O_4)_3$ and its 12-hydrate have been obtained,[3] though some of the ammonia molecules in these species might be hydrogen bonded to the carboxylate anions rather than being coordinated. A further amino–carboxylate $Sc_2ac_6(NH_3)_3$ is also known.[4] A series of ammines of ScX_3 (X = Cl, Br or I) have been prepared by direct interaction of the anhydrous salt and ammonia.[5–7] The initial products are $ScX_3{\cdot}nNH_3$ ($n = 7$–8) but gentle heat (~100–115 °C) gives $ScCl_3(NH_3)_4$ or $ScBr_3(NH_3)_5$. IR studies[8] of deuterated halo–ammines have been interpreted in terms of structures such as $[ScCl(NH_3)_5]Cl_2$, and dissociation pressure measurements[9] gave heats of formation ($ScCl_3(NH_3)_5$, 72 kJ mol^{-1}; $ScCl_3(NH_3)_4$, 76 kJ mol^{-1}). Localized molecular orbitals have been calculated for $Sc(NH_3)_5X^{3+}$ ($X = H^-$, NH_3, F^-, Cl^-, H_2O). d-Orbital participation and π bonding were discussed and electron density contours given.[10] A hydrazine adduct $ScF_3{\cdot}N_2H_4$ has been described,[11] but its constitution is uncertain.

39.1.2.2 Amines

Both bis and tris ethylenediamine complexes have been described;[12] the latter have been formulated $[Sc(en)_3]X_3$ (X = Cl or Br), this being supported by IR data. Heating in vacuum leads to formation of *cis*-$[ScX_2(en)_2]X$. A number of complexes of $ScCl_3$ with primary and secondary amines, mainly aliphatic, have been prepared and their thermochemistry and IR spectroscopy studied.[13–15] The compounds are $ScCl_3{\cdot}n(RR'NH)$, where $n = 1$–4 and R = R' = Me, Et, Pr, Bu or Pe; or R = H and R^1 = Et, Pr, Bu, Pe, phenyl, benzyl or 2-naphthyl. Further investigation would be necessary to establish the structural nature of these complexes.

Several investigations have explored monodentate and bidentate aromatic amine complexes. Treatment of $ScCl_3$ or $ScBr_3$ with hot neat pyridine gave $ScX_3 \cdot 4py$ (X = Cl or Br) where the IR spectrum suggested the presence of coordinated and uncoordinated pyridine; under vacuum these compounds were converted into $ScX_3 \cdot 3py$.[16] IR studies suggested that $\nu_{Sc—N} = 260–280\ cm^{-1}$, $\nu_{Sc—Cl} = 300–350\ cm^{-1}$ and $\nu_{Sc—Br} = 266–268\ cm^{-1}$. A series of pyridine thiocyanato complexes has also been prepared. The stoichiometries are $Sc(NCS)_3(py)_n(EtOH)_x$, where $n = 1–4$ and $x = 0$ or 1, the most stable complex being $Sc(NCS)_3(py)_3$. The anions of course dissociate in water, in which medium the presence of $[Sc(py)_n(H_2O)_y]^{3+}$ cations was suggested.[17]

The bidentate ligands 1,10-phenanthroline and 2,2′-bipyridyl form complexes with scandium chloride and thiocyanate. From anhydrous THF solutions $ScCl_3(bipy)_2$ and $ScCl_3(phen)_2$ were isolated. They dissociate to some extent as 1:1 electrolytes in acetonitrile or in nitromethane. Far IR evidence is that the structures are *cis*-$[ScCl_2L_2]Cl$ (L = phen or bipy). Somewhat similar thiocyanato complexes were also obtained.[18,19] However, in this case the tris complexes $Sc(NCS)_3L_3$ were both isolated and were assigned the formulation $[ScL_3](NCS)_3$ from IR and conductivity data. In the thiocyanate series, the bis complexes $[Sc(NCS)_2L_2]NCS$ were each obtained in *cis* and *trans* forms, again as determined mainly from IR spectra. Here $\nu(Sc—NCS) = 300–325\ cm^{-1}$. It would be interesting to have single crystal X-ray data on these compounds.

A comparatively large amount of work has been done on polydentate amine complexes, which show considerable stability and force high coordination numbers on the metal ion. Thus $Sc(NO_3)_3 \cdot 6H_2O$ and 2,2′:6′,2″-terpyridyl (terpy) gave, from acetonitrile, $[Sc(NO_3)_3(terpy)]$ as golden needles. A single crystal X-ray determination showed nine-coordination, the nitrate ions being bidentate, and the organic ligand essentially coplanar, with Sc—N = 2.272–2.322 Å and Sc—O = 2.234–2.322 Å but with one longer Sc—O distance at 2.482 Å. This seems to be a case where one of the nitrate oxygen atoms is almost being expelled from the coordination sphere by what must be a very considerable interligand repulsion. The terpyridyl ligand dissociates in water but the complex retains its integrity in nitromethane. The ligand tripyridyltriazine (tpta), which is similar in coordination geometry to terpyridyl if one of its pyridine groups remains uncoordinated, forms a complex $ScCl_3(tpta)$ where IR evidence suggests the presence of an uncoordinated pyridine, and the formulation as six-coordinated $[ScCl_3(tpta)]$ is suggested. Interconversion of the coordinated and uncoordinated pyridine groups takes place in methanol solution.[20]

A variety of tetradentate amine ligands have been used to form scandium complexes. Thus 1,2-bis(pyridine-α-carbaldimino)ethane (bpce) forms a dimeric complex $[Sc_2(NO_3)_2(\mu\text{-}OH)_2(bpce)_2](NO_3)_2 \cdot 3MeCN$ whose structure has been determined by single crystal X-ray.[20] Each metal ion is eight-coordinated and each bpce ligand is essentially coplanar, while the two hydroxy and two nitrate oxygen atoms which are coordinated to one particular scandium ion form an approximately coplanar system which is perpendicular to the plane of the tetraimine, producing a distorted dodecahedral coordination. The four Sc—N distances are 2.341–2.403 Å, and Sc—O(nitrate) are 2.333 and 2.295 Å. The Sc—OH distances are notably short at 2.053 and 2.089 Å and an $sd_{z^2}–sp^2$ covalent interaction is suggested. A complex $Sc(NO_3)_3(bpce)(H_2O)_2$ was also obtained and the formulation $[Sc(NO_3)_3(bpce)] \cdot 2H_2O$ was suggested. The bpce ligand, despite being quadridentate, dissociates in water and in DMSO.[20]

The cyclic hexaimine (L) formed by cocondensation of two molecules each of hydrazine and 2,6-diacetylpyridine under template conditions gives a complex formulated as $[ScL(H_2O)_2](ClO_4)_3 \cdot 4H_2O$. Only four nitrogen atoms of the six are geometrically able to coordinate.[21] The best-known tetradentate macrocyclic amine complexes are those of the porphyrins, however, and a good deal of work has been done on these, usually in conjunction with studies of similar complexes of other metal ions. Much of this attention is due to the relevance to photosynthetic chlorophylls, where a dimer similar to a dimer formed by scandium is proposed to participate in the photosynthetic reaction centres. Octaethylporphyrin (H_2OEP) gives complexes $Sc(OEP)(O_2CMe)$[22] and a dimer (OEP)Sc—O—Sc(OEP). The latter was first prepared by the action of H_2OEP on $Sc(acac)_3$ in imidazole, followed by chromatography on Al_2O_3 and was characterized by NMR, IR and mass spectrometry.[23] A preferred method uses H_2OEP and $ScCl_3$ reacting in hot pyridine, followed by treatment with water and extraction into, and recrystallization from, CH_2Cl_2 giving purple plates. The corresponding tetraphenylporphyrin complex was prepared similarly. The dimer readily dissociates to monomers in alcohols. It shows a two band absorption at ~410 nm and ~550 nm and very strong fluorescence with moderate phosphorescence. The phosphorescence lifetime of

$Sc(OEP)O_2CMe$ is 0.4 s, which is the longest for any porphyrin, while the fluorescent yield of 0.2 is very high. The radiative properties are explained in terms of covalent interactions between the metal and the ring as modified by the probable location of the metal ion above the porphyrin plane.[24] Scandium OEP complexes are reduced to the α,γ-dihydro derivatives on reduction with sodium anthracenide and methanol.[25] The redox potentials of Sc(OEP)OH have been determined by cyclic voltammetry to be: ligand oxidation in PrCN, 1.03 and 0.70; ligand reduction in DMSO, −1.54 ($E_{1/2}$ values in V *vs.* SCE); no metal redox wave was observed.[26]

The dimer (OEP)Sc—O—Sc(OEP) has been examined by optically detected magnetic resonance with the conclusion that the rings are approximately coplanar.[27] The NMR spectra of the dimer and the monomer show that the methylene protons are anisochronous, this probably not being caused by hindered rotation but by the inherent asymmetry of the ligands.[28]

Scandium phthalocyanine (H_2pc) complexes have been obtained and quantum mechanical calculations carried out on these systems. Thus Sc(pc)Cl is obtained from reaction of $ScCl_3$ with either phthalonitrile or with $Li_2(pc)$ and gives a molecular ion in the mass spectrum.[29] The molecular orbitals of scandium (and other) phthalocyanines have been calculated and the electron distribution over the ring system studied in relation to the effect of the scandium ion.[30]

The uncharged tetradentate macrocyclic tetramines Me_4Bz_4[14]-tetraene-N_4 (L) and Me_4[14]-tetraene-N_4 (L′) (**1** and **2**) form complexes $Sc(NCS)_3L$ and $Sc(NCS)_3L'\cdot 0.5(THF)$ when the ligands react with ethereal $Sc(NCS)_3$. IR evidence, namely the appearance of non-hydrogen-bonded NH around 3200 cm^{-1} and the lowering of ν(CN) by 10–40 cm^{-1}, tends to indicate coordination of the tetramines. It is suggested that the complex of the more rigid L is five-coordinated square pyramidal, while the more flexible L′ may permit another anion to coordinate giving six-coordination.[31]

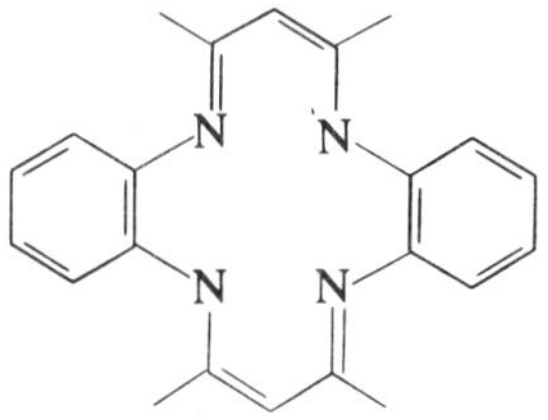

(1)

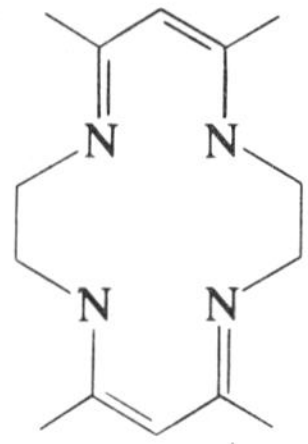

(2)

39.1.2.3 Amido complexes

Scandium (and the lanthanides) forms a very interesting tris-silylamide $[Sc\{N(SiMe_3)_2\}_3]$ where the eighteen methyl groups are effective in reducing the scandium coordination to its lowest known value of three. Surprisingly, the coordination is pyramidal, with N—Sc—N = 115.5°, possibly owing to van der Waals attractive forces between the ligands. The Sc—N distance is 2.049 Å.[32] This configuration contrasts with the trigonal planar configuration of the tris-aryloxide (*vide infra*). The compound is volatile and very air sensitive. It is prepared by the reaction of $LiN(SiMe_3)_2$ with $ScCl_3$.[33]

39.1.3 Oxygen Donor Ligands

39.1.3.1 β-Diketonates

Scandium tris-β-diketonates are monomers, in contrast to the lanthanide analogues which form oxygen-bridged dimers except in the cases of bulky β-diketonates combined with smaller-radius lanthanides. $Sc(acac)_3$ itself is a compound of some antiquity[34] and has been studied by a number of authors.[35–38] Other β-diketonates are also known: $Sc(tfaa)_3\cdot 3H_2O$,[36] $Sc(hfaa)_3$,[36] $Sc(tta)_3$,[39] $Sc(baa)_3$,[35] $Sc(bza)_3$,[40] $Sc(dbm)_3$[39] and $Sc(dpm)_3$[36,41] (Htfaa = 1,1,1-trifluoroacetylacetone, Hhfaa = 1,1,1,5,5,5-hexafluoroacetylacetone, Htta = thienoyltrifluoroacetone, Hbaa = 3-bromoacetylacetone, Hbza = benzoylacetone, Hdbm = dibenzoylmethane and Hdpm = dipivaloylmethane). The heat of sublimation of $Sc(acac)_3$ is 99.6 kJ mol^{-1} which is comparable with those of the iron and vanadium analogues.[42] The first stability constant K_1 for Sc^{3+}–acetylacetonate is two to three orders of magnitude larger than for the lanthanides.[43] The

^{1}H NMR spectrum of $Sc(acac)_3$ has been measured,[35,41] and a ^{13}C NMR study in $CDCl_3$ has also been reported.[44]

Two single crystal structural determinations of tris-β-diketonates have been reported. $[Sc(acac)_3]$ shows nearly D_3 symmetry[45] with monomeric six-coordination. The Sc—O interatomic distance is 2.070 Å which would give an ionic radius for Sc^{3+} of 0.67 Å for six-fold coordination to oxygen. The other example[46] is $[Sc(trop)_3]$ which shows crystallographic D_3 symmetry with Sc—O = 2.102 Å.

The mass spectrometry of scandium β-diketonates, which are fairly volatile, has been the subject of several studies. In an earlier study, dimeric species were detected in the mass spectra of the vapour.[47] The compounds $M(RCOCHCOCF_3)_3$ where M = Sc or Y and R = Me, Ph, 2-thienyl or 2-furyl showed fluorine migration in their mass spectra, ions containing CF_2 groups being observed.[48] The mass spectra of scandium (and Y and lanthanide) acetylacetonates and dipivaloylmethanates and some adducts with bipyridyl, phenanthroline or Ph_3PO have been examined and detailed fragmentation schemes proposed. Appearance potentials of the mono and bis β-diketonates were measured, but no adducts of the dipivaloylmethanates were detected in the spectra.[49,50] Russian authors have also examined the vapour phase chromatography of $Sc(acac)_3$.[51]

The tris-β-diketonates fairly readily form the tetrakis-diketonate anions. Thus the compounds $M[Sc(diket)_4]$, where M = K, Rb, Cs and diket = $CF_3COCHCOMe$, $CF_3COCHCOMe$ or $CF_3COCHCOCF_3$, have been synthesized by the reaction of $Sc(diket)_3$ with M(diket) in equimolar proportions in ethanol or ethanol/CCl_4. X-Ray powder and thermogravimetric studies were made. The hexafluoro compounds $M[Sc(CF_3COCHCOCF_3)_4]$ were particularly stable, subliming without decomposition at 0.1 mmHg, though $Cs[Sc(CF_3COCHCOMe)_4]$ decomposes into $[Sc(diket)_3]$ on attempted sublimation.[52,53] The tetrakis-tropolonate adducts $HSc(trop)_4$ and $NaSc(trop)_4$ have been prepared[54] and the structure of the former has been determined by single crystal X-ray.[55] Scandium is eight-coordinated to four bidentate ligands: three tropolonate anions and one uncharged tropolone molecule. The latter forms a hydrogen bond with a neighbouring molecule giving a dimeric structure, each dimer unit having two hydrogen bonds with O···O = 2.484 Å. The coordination geometry is intermediate between bicapped trigonal prismatic and dodecahedral, with Sc—O in the range 2.161–2.310 Å.

39.1.3.2 Carboxylates and hydrated carboxylates

Scandium complexes readily with carboxylic acid anions, and the resulting solid compounds often have a polymeric structure with a resulting tendency towards insolubility. The formate, prepared from the hydroxide and the acid,[56] is a three-dimensional network where scandium ions are connected by bridging HCO_2^- ions. The coordination around Sc^{3+} is nearly octahedral with the average Sc—O distance equalling 2.03 Å.[57] The hydrated oxalate $Sc_2(C_2O_4)_3(H_2O)_6$ has a polymeric structure in which the Sc^{3+} ion is approximately dodecahedrally coordinated with two water molecules and three bidentate $C_2O_4^{2-}$ anions, the latter bridging to other Sc^{3+} ions by means of their remaining carboxylate oxygen atoms. The Sc—O distances lie in the range 2.18–2.26 Å.[58]

The hydrated basic malonate $Sc(OH)(C_3H_2O_4)H_2O{\cdot}H_2O$ is approximately octahedrally coordinated, each Sc^{3+} being linked to three carboxylate oxygen atoms, two hydroxyl ions and one water molecule, with Sc—O in the range 2.05–2.12 Å. In this structure a linear hydroxo- and malonato-bridged polymer is three-dimensionally cross-linked by hydrogen bonds.[59] The hydrated hydroxyacetate $Sc(HOCH_2CO_2)_3(H_2O)_2$ is found to have the constitution $[Sc(HOCH_2CO_2)_2(H_2O)_4][Sc(HOCH_2CO_2)_4]$ with each type of Sc^{3+} being approximately dodecahedrally coordinated and the hydroxyacetate anions each coordinating by the hydroxo oxygen and one carboxylate oxygen giving a five-membered ring chelate.[60] It will be noticed that in these carboxylate structures, scandium is very much on the borderline between six- and eight-coordination with Sc—O at about 2.05 Å for the former and about 2.20 Å for the latter. The sterically inefficient seven-coordination is not observed.

The complexation constants K_1 for interaction of Sc^{3+} ions with substituted and unsubstituted carboxylates in aqueous solution have been determined potentiometrically at 25 °C and 0.1 M KNO_3 ionic background. They are listed in Table 1, the ΔH values being derived with the aid of additional measurements at 40 °C.[61] These values for scandium are greater than those for the lanthanides by roughly one order of magnitude.[62]

Table 1 Complexation Between Aqueous Carboxylic Acids and Scandium[61]

Acid	*log K*	$-\Delta G$ (kJ mol^{-1})	$-\Delta H$ (kJ mol^{-1})
α-Hydroxyisobutyric	4.84	27.6	115
Isobutyric	4.47	25.5	90.1
Glycolic	4.40	25.5	121
Acetic	3.48	19.6	1.2
Propionic	3.77	21.3	8.8
Mandelic	2.91	16.7	−5.4
D-Gluconic	4.21	23.8	−52.3
Acetylacetone	7.77	44.3	388

39.1.3.3 Hydrates and complexes with oxyanions

There is rather little definitive information on the hydrated Sc^{3+} ion, which is generally agreed to be $[Sc(H_2O)_6]^{3+}$. This ion, together with its interaction with anions in aqueous solution, has been studied by ^{45}Sc NMR. Thus nitrate, perchlorate and chloride were all found to complex with the hydrated Sc^{3+} ion,[63] and in aqueous solution containing HCl at various molarities, the following species were believed to be present: $[ScCl(H_2O)_5]^{2+}$ (4 M), $[ScCl_2(H_2O)_4]^+$ (7 M), and $[ScCl_4(H_2O)_2]^-$ (saturated).[64] The IR and Raman spectra of the hydrated Sc^{3+} ion have been reported.[65]

A variety of hydroxoscandates are known. On hydrolysis of Sc^{3+}(aq) in 0.1 M KNO_3 at 25 °C, potentiometric studies indicated the formation of $[Sc(OH)(H_2O)_x]^{2+}$, $[Sc_2(OH)_2(H_2O)_x]^{4+}$ and $[Sc_3(OH)_5(H_2O)_x]^{4+}$, and formation constants were evaluated.[66] A number of solid hydroxoscandates have been isolated but their structures are not well established. Most reported species are tetra- or hexa-hydroxoscandates but $K_2Sc(OH)_5(H_2O)_4$ has been obtained from a solution of scandium and potassium hydroxides.[67] The hexahydroxo compound $Na_3Sc(OH)_6(H_2O)_2$ has been isolated from an aqueous solution of scandium and sodium hydroxides[68] and decomposes into the tetrahydroxo compound and NaOH at temperatures above 120 °C.[69] The anhydrous compound $Na_3Sc(OH)_6$ may be obtained from reaction in methanol[70] or by gently heating the hydrate. The Group II analogues $M_3Sc_2(OH)_{12}$ (M = Ca or Ba) and $MSc_2(OH)_8(H_2O)_2$ (M = Ca, or Sr) have been obtained in crystalline form.[71]

A number of well-defined nitrato complexes have been established. As mentioned above, the nitrate ion complexes with Sc^{3+} even in water, and the ion $[Sc(NO_3)_5]^{2-}$ is stable in concentrated aqueous nitric acid. Salts of this anion of the type $M_2[Sc(NO_3)_5]$, where M = K, Rb or Cs, have been obtained.[72] From $ScCl_3$ and N_2O_4 in ethyl acetate was obtained $(NO)_2[Sc(NO_3)_5]$ from which a single crystal X-ray analysis showed scandium with the very high, for this metal, coordination number of nine, four NO_3^- being bidentate and one monodentate, with Sc—O being 2.283 Å and 2.176 Å respectively. The coordination polyhedron is irregular.[73] Some double sulfates $MSc(SO_4)_2$, where M = K, Rb or Cs, have been studied by means of high temperature X-ray powder photography and their phase diagrams determined,[74] but their structures are uncertain.

39.1.3.4 Other uncharged oxygen donor ligands

Scandium coordinates fairly readily with several other types of oxygen donors, especially those with a little negative charge on the oxygen atom, as in $Ph_3P \rightarrow O$. It usually takes a coordination number of six, judging from the stoichiometry of the compounds formed and from structures in the limited number of cases where there are single crystal X-ray data available.

Tetrahydrofuran complexes are among the most studied. The chloride $[ScCl_3(THF)_3]$ has a meridional octahedral structure with Sc—O = 2.18 Å and Sc—Cl = 2.41 Å (mean values).[75] This complex may be simply prepared from a suspension of $ScCl_3 \cdot 6H_2O$ in THF by addition of $SOCl_2$ and reflux.[76] Alternatively, scandium metal may be added to a refluxing solution of $HgCl_2$ in THF; the material obtained thus was characterized by IR spectrum.[77] Other THF complexes include the tetrahydroborate $[Sc(BH_4)_3(THF)_2] \cdot 2(THF)$. The cation is trigonally bipyramidally coordinated with the two THF molecules in the axial positions.[78] Tetrahydrofuran often stabilizes organometallic compounds, and the neopentyls $Sc(Me_3MCH_2)_3(THF)_2$, where M = C or Si, can be isolated from reaction of the appropriate lithium reagent with $ScCl_3$ in THF–hexane–ether at 0 °C. NMR data are consistent with a trigonal bipyramidal structure with axially coordinated THF.[79]

A number of compounds of scandium salts with cyclic crown polyethers have been described.[80] They comprise $Sc(NCS)_3$(b-15-c-5)·1.5(THF)·$2H_2O$, $Sc(NCS)_3$(db-18-c-6)$_2$·3(THF), $(ScCl_3)_3$(b-15-c-5)$_2$·H_2O, $ScCl_3$(db-18-c-6)·1.5(THF) and $ScCl_3$(db-18-c-6)·1.5MeCN where (db-18-c-6) is dibenzo-18-crown-6. The characteristic lowering of ν(CO) in the IR spectrum when a crown ether coordinates was minimal in most cases and though evidence for complexation was adduced from the UV spectra, coordination of the crown ligands cannot be regarded as established. In the compound $[Sc(NO_3)_3(H_2O)_2]$·(b-15-c-5), the crown ether is uncoordinated, the Sc^{3+} being eight-coordinated to two H_2O at 2.131 and 2.145 Å and to six O (NO_3) at 2.166–2.256 Å.[81]

There have been a number of studies of urea complexes. Thus the Sc^{3+}–acetate–water–urea (or thiourea) system at 30 °C shows the presence of a phase $Sc(MeCO_2)_3$·(urea)·H_2O but no thiourea complex was observed.[82] The complexes ScX_3 (urea)$_x$ (X = Cl, x = 6 or 4; X = Br, ClO_4 or NO_3, x = 6) have all been isolated[83,84] but structural data other than X-ray powder photographs are lacking, though coordination by the urea is fairly certain as ν(CO) was lowered by some 110 cm^{-1}. However, single crystal data are available for $[Sc(NO_3)_2(urea)_4]NO_3$, where there is coordination to oxygen atoms of four urea molecules and two nitrate groups.[85]

Amine *N*-oxides, phosphine oxides and other phosphoryl compounds and analogous arsenic derivatives, and sulfoxides all give well-defined complexes with scandium. In many cases, authors deal with several of these classes of ligand in the same paper, so it will be most convenient to consider them together here.

Complexes of scandium perchlorates, chlorides and thiocyanates with pyridine oxide, 2-, 3- and 4-picoline oxides, lutidine oxide, bipyridyl dioxide, hexamethylphosphoramide, bis(diphenylphosphino)ethane dioxide $Ph_2P(O)C_2H_4(O)PPh_2$ (dpeo) and the corresponding arsenic compound (daeo), tributylphosphate, trimorpholinophosphine oxide (tmpo) and thioxane oxide (toxo) have been examined. They are mostly obtained from ethanolic solution. The stoichiometry ScX_3L_3 (X = Cl, L = hmpa, tbp or tmpo; X = NCS, L = DMSO, tppo, tpao, pyo, 2-, 3-, 4-pico or luto) was most prevalent for the fairly strongly coordinating chloride or thiocyanate anions, but for the perchlorates, $Sc(ClO_4)_3L_6$ (L = DMSO, pyo, 2-pico, hmpa or toxo) or $Sc(ClO_4)_3L_3$ (L = bipo, dpeo and daeo) were preferred. While no single crystal data are available, it is highly likely that six-coordination is present in most, if not all, of these compounds. The lowering of the S=O or P=O IR frequency on coordination is quite variable, 95 cm^{-1} for $Sc(ClO_4)_3(toxo)_6$, 10 cm^{-1} for $Sc(ClO_4)_3(hmpa)_6$ and almost nil for $Sc(ClO_4)_3(daeo)_3$. IR data indicate the SCN^- ions to be N-coordinated.[86–90]

Structural data are available for $[Sc(NO_3)_3(OPPh_3)_2]$. This complex was obtained by reaction of hydrated scandium nitrate with the phosphine oxide in hot, very dilute, benzene solution. The P—O stretching mode was lowered from 1195 cm^{-1} in the free ligand to 1170, 1140 cm^{-1} in the complex. The scandium ion is eight-coordinated to both phosphine oxides and the three bidentate nitrates in a very irregular dodecahedral arrangement. Leading bond lengths (Å) and angles are Sc—O(P) = 2.036, 2.076; Sc—O(NO_3) = 2.207–2.338; Sc—O—P = 161.2°, 174.8°.[91]

The kinetics of ligand exchange of some phosphoryl-coordinated complexes of scandium are of a convenient rate to be studied by NMR lineshapes. $[Sc\{OP(OMe)_3\}_6]^{3+}$ exchanges with free ligand in CD_3CN, *sym*-$C_2D_4Cl_2$ or CD_3NO_2. In the first two solvents, rate is independent of the concentration of added ligand (CD_3CN: $\Delta H^{\ddagger} = 29.8$ kJ mol^{-1}; $\Delta S^{\ddagger} = -111$ J K^{-1} mol^{-1}) but in CD_3NO_2, exchange rates are first order in each reactant.[92] Similar studies have been made on analogous $(MeO)_2P(O)Me$ systems.[93]

39.1.3.5 Alkoxides

Some well-defined scandium alkoxides have been synthesized and studied (together with lanthanide analogues). The compounds are $[Sc(OR)_3]$, $Sc(OR)_3(THF)$, $Sc(OR)_3(tppo)$ and $(RO)_2Sc(\mu\text{-}OR)_2Na(THF)$, where R = 2,6-di-*t*-butyl-4-methylphenate. $[Sc(OR)_3]$ was prepared by reaction of $Sc\{N(SiMe_3)_2\}_3$ with ROH in hexane at 20 °C followed by reflux, or by reaction of NaOR and $ScCl_3$ in boiling THF, or by sublimation of $Sc(OR)_3(THF)$. The latter and

$Sc(OR)_3(tppo)$ are prepared by direct addition of ligand to $Sc(OR)_3$ in benzene at 20 °C. The bridged compound is obtained together with $Sc(OR)_3(THF)$ by reaction of NaOR·THF on $ScCl_3$ followed by appropriate work-up. These compounds are soluble in hydrocarbons and of course very easily hydrolyzed. Single crystal X-ray data are available for $Sc(OR)_3$; the coordination is trigonal planar (O—Sc—O = 123.3° + 120.1° + 115.1° = 358.5°) with Sc—O = 1.854–1.889 Å. Sc—O—C is not quite linear, averaging 168.4°.[94]

39.1.3.6 Amides

A few complexes of organic amides such as dimethylformamide and dimethylacetamide are known, *e.g.* $Sc(NCS)_3(DMF)_3$, and $Sc(ClO_4)_3L_3$, where L = *N,N,N′,N′*-tetramethylmalonamide. In the latter complex, IR, conductimetric (1:3 electrolyte) and X-ray measurements indicate a six-coordinated complex $[ScL_3](ClO_4)_3$.[95] NMR rate measurements on ligand exchange of $[ScL_6]^{3+}$, where L = *N*-mono (or di) alkyl formamides or acetamides, showed that the acetamides exchanged considerably faster than the formamides. It was concluded that steric interactions due to the introduction of the carbonyl-attached alkyl group were of importance, rather than inductive effects.[96]

39.1.4 Ligands Containing Both Nitrogen and Oxygen Donors

Scandium complexes readily with 8-hydroxyquinolinate ion. Many years ago, $Sc(hq)_3$·Hhq, which was obtained from aqueous solution, was described,[97] and this compound does not lose Hhq cleanly on heating but does so in benzene solution.[98] However, it has been stated that the solid deposited at about pH 7 is in fact $Sc(hq)_3$.[99] More recently, stability constant values have been determined for the Sc^{3+}/hq^-/50% aqueous dioxane system and the corresponding 2-methyl-8-hydroxyquinoline system.[100] Chinese workers, who are showing some interest in spectrophotometric analysis, have detected Sc^{3+} on the microgram scale by fluorometry using 8-hydroxyquinoline-5-sulfonate and cetyltrimethylammonium bromide as an extractant system.[101]

There have been rather few investigations describing complexes of scandium with simple amino acids. There are Russian papers describing the isolation from aqueous solutions of several such complexes but it is not clear whether both the amino and the carboxylate groups are coordinated; in some cases IR criteria were used as evidence for coordination of the carboxylate group. Among compounds isolated were $Sc(glut)(Hglut)·3H_2O$, $Sc_2(glut)(Hglut)_4·5H_2O$, and several compounds of the type $Sc_2(SO_4)_3·A_x·(H_2O)_y$ where A is an uncharged amino acid.[102] Various concentrations of $ScCl_3$, acid and potassium salt of acid at pH 2–7 gave $Sc(OH)(anth)_2·H_2O$ and $Sc(anth)_3$ (anth = anthranilate anion) and the corresponding *p*-aminobenzoates.[103]

With amino acids of the polydentate chelating type, there are a number of studies. Thus $Sc_2(ida)_3·H_2O$, $ScCl(ida)·2H_2O$ and $KSc(ida)_2·3H_2O$ have been isolated and studied by thermal and IR methods.[104] Association constant values have been obtained for edta ($\log K = 23.1 \pm 0.15$) at 20 °C and $\mu = 0.1$ by polarography,[105] and for diethylenetriaminepentaacetic acid, where $\log K = 26.28 \pm 0.37$.[106] These values are rather greater than corresponding ones for the lanthanides. The rate of exchange of KSc(edta) with added ligand has been determined by NMR and was found to be slower than for the lanthanum analogue.[107] The exchange of $^{46}Sc^{3+}$ with the scandium complex of *N*-(2-hydroxyethyl)ethylenediamine-*N,N′,N′*-triacetate has been studied, also by NMR. The exchange proceeds by way of dissociation of a protonated complex. Activation parameters were obtained.[108] The interaction has been studied by potentiometry between edta or nitrilotriacetic acid complexes of scandium and a variety of added phenols in anionic form. Mixed ligand complexes were observed and *K* values obtained.[109]

There has been some interest in complexes of a class of ligands formed by aldimine-type condensations, either free or scandium-templated, giving complexes where polydentate ligands, which may be coplanar or non-coplanar, complex the scandium ion. Thus β,β',β''-triaminotriethylamine condenses with three molecules of 3,5-dichlorosalicylaldehyde to yield a potentially heptadentate ligand H_3saltren, $N(CH_2CH_2N{=}C{-}R)_3$, where R = 3,5-dichloro-6-hydroxybenzene, which yields the complex [Sc(saltren)]. However, on steric grounds it is unlikely that the tertiary amine nitrogen coordinates with scandium, though it probably coordinates in the analogous lanthanide complexes. This is because of steric limitations of the ligand, not because of the incapacity of Sc^{3+} to accept seven-coordination.[110] There is a case, however, where scandium is forced into the inefficient pentagonal bipyramidal seven-coordination by a coplanar quinquedentate ligand. This is formed by condensation of 2,6-diacetylpyridine with two molecules of semicarbazide giving a three-nitrogen, two-oxygen donor, dapsc (**3**). Four of the nitrogen atoms are unable to coordinate for steric reasons. The complex $[Sc(dapsc)(H_2O)_2](OH)(NO_3)_2$ was obtained from aqueous alcohol. The ligand is very nearly coplanar with an average deviation of only 0.023 Å, and the two water molecules are in the axial positions. Internuclear distances are Sc—O = 2.110 (av) and Sc—N = 2.283 (av) Å.[111]

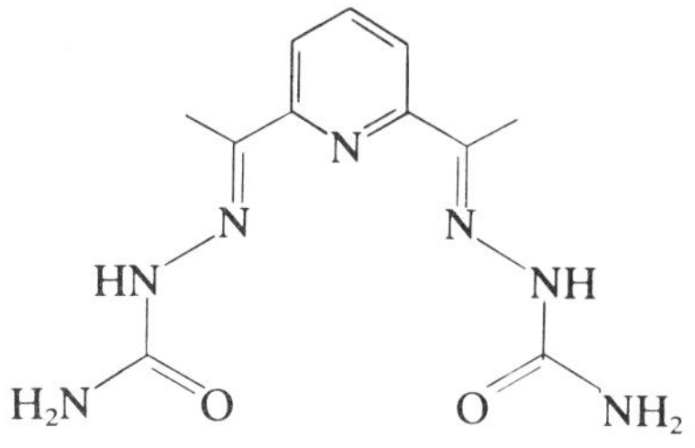

(**3**)

39.1.5 Complex Halides

Association constants between Sc^{3+} and fluoride ion in aqueous solution have been obtained.[112] They are as follows ($\Delta H^{\ominus}_{298}$ values in kJ mol^{-1} in parenthesis): $K_1 = 1.2 \times 10^7$ (1.3), $K_2 = 6.4 \times 10^5$ (−6.3), $K_3 = 3.0 \times 10^4$ (−3.3), $K_4 = 7 \times 10^2$. Although only ScF_4^- has thus been observed in solution, salts of ScF_6^{3-} have been isolated in the solid state. Thus M_3ScF_6 (M = NH_4, Na or K) is obtained by action of aqueous scandium ion.[113] Both $(NH_4)_3ScF_6$ and Na_3ScF_6 exist in high temperature (α) and low temperature (β) forms; the former salt[114] is cubic (α) and tetragonal (β) while the latter[115] has a monoclinic β form. The Raman spectrum of $(NH_4)_3ScF_6$ has been reported in terms of O_h symmetry for the ScF_6^{3-} ion, as A_{1g}, 504; E_g, 370; T_{2g}, 240 cm^{-1}.[116] Other fluoro species have been obtained from aqueous media, namely $NaScF_4{\cdot}H_2O$ and $MScF_4$ (M = Na or K).[115,117] The thermal decomposition of $(NH_4)_3ScF_6$ yields NH_4ScF_4 at around 200 °C.[118]

Scandium complexes with chloride ion in aqueous solution, and there is ion-exchange evidence for anionic species, presumably $ScCl_4^-$ aq, in concentrated hydrochloric acid,[119] while values of $K_1 = 90$ and $K_2 = 37$ for the first two association constants have been reported.[120] Solid salts of the anions $ScCl_6^{4-}$, $Sc_2Cl_9^{3-}$ and $ScCl_5^{2-}$ have been isolated with alkali metal cations.[121,122] It seems likely that both the first two species are octahedral and bridged-octahedral respectively, in line with $Cs_2NaScCl_6$ which has an X-ray powder pattern in accord with a face-centred cubic structure.[123]

The first two association constants between Sc^{3+} and Br^- in aqueous solution are $K_1 = 120$ and $K_2 = 10$ respectively.[120] Despite the greater radius of the bromide ion, Sc^{3+} should be amply large for six-coordination by this ligand and the complexes M_3ScBr_6 and $M_3Sc_2Br_9$ (M = Na, K, Rb or Cs) which are formed in the solid state[124,125] are almost certainly six-coordinated like the chloride which they resemble in stoichiometry. The formation of $[ScX_6]^{3-}$ (X = Cl or Br) in aqueous solution is proposed as an intermediate in the preparation of ScX_3 from aqueous media;[126] ScI_3 cannot be obtained in this way and indeed, the formation of any substantial quantity of iodo complex would not be expected in water. As noted above (Section 39.1.3.3) however, ^{45}Sc NMR detected only $[ScCl_4(H_2O)_2]^{2+}$ in aqueous solution,[64] no $ScCl_6^{3-}$ species being recorded. The nature of scandium and lanthanide chloro complexes has been examined by ultrasonic absorption in 0.2 M solutions of aqueous methanol. Scandium behaved differently from the lanthanides but no firm conclusions were possible.[127]

There are a number of solid phases of the types $MScCl_x$ and $ScCl_x$ where the formal oxidation state of scandium is less than three. They are usually made by direct combination at elevated temperatures of MCl, $ScCl_3$ and metallic scandium. Their structures often show evidence of Sc—Sc bonds. Thus $CsScCl_3$ is made by action of Sc on $Cs_3Sc_2Cl_9$ at 700 °C. The shiny blue product has the hexagonal perovskite $CsNiCl_3$ structure. This is similar to the $Cs_3Sc_2Cl_9$ structure but with all Sc positions filled. Non-stoichiometric phases exist between the two end structures.[128] When scandium is heated with $ScCl_3$ at 940–960 °C in a sealed Ta container for several weeks, Sc_5Cl_8 is obtained. Its structure contains two types of chain which are interlinked by bridging chlorine atoms. Both types consist of linear arrays of octahedra, each linked to the next octahedron at two adjacent vertices. One chain consists of chlorine atoms with Sc at the centre of each octahedron, while the other type consists of scandium atoms, some of which are coordinated to Cl. The bonded Sc—Sc distance is 3.021 Å (Sc—Sc is 3.26 Å in the metal).[129] At 890–900 °C, Sc_7Cl_{12} is obtained. This contains Sc^{3+} ions and $Sc_6Cl_{12}^{3-}$ clusters, in which Sc—Sc distances of 3.204 and 3.234 Å are observed.[130]

39.2 YTTRIUM AND THE LANTHANIDE ELEMENTS

39.2.1 Introduction

The study of coordination compounds of the lanthanides dates in any practical sense from around 1950, the period when ion-exchange methods were successfully applied to the problem of the separation of the individual lanthanides,[131–133] a problem which had existed since 1794 when J. Gadolin prepared mixed rare earths from gadolinite, a lanthanide iron beryllium silicate. Until 1950, separation of the pure lanthanides had depended on tedious and inefficient multiple crystallizations or precipitations, which effectively prevented research on the chemical properties of the individual elements through lack of availability. However, well before 1950, many principal features of lanthanide chemistry were clearly recognized, such as the predominant trivalent state with some examples of divalency and tetravalency, ready formation of hydrated ions and their oxy salts, formation of complex halides,[134] and the line-like nature of lanthanide spectra.[135]

Since 1950 and at an ever-increasing pace during the last 20 years, the picture of lanthanide coordination chemistry has become much broader and clearer. The use of anhydrous conditions and a sophisticated variety of ligands has led to an extensive nitrogen ligand chemistry, a great extension of oxygen ligand chemistry, the discovery of reactive organolanthanide complexes, and the beginnings of phosphorus and sulfur donor chemistry. There have been other important advances also, including syntheses of macrocyclic complexes and cryptates, investigations of electronic spectra in terms of crystal field effects, and the introduction of lanthanide NMR shift reagents. So extensive has lanthanide chemistry now become, that coverage and selection of material for inclusion in this review poses, in some areas of activity, considerable problems, and an apology is offered to those authors whose fine work has through pressure of space, or perhaps even through inadvertence, been omitted.

The most stable state of the lanthanides in their complexes is in all cases the M^{3+} ion, in which all outer electrons reside in the 4*f* shell, in contrast to the uncharged atom M where, in addition to the 4*f* and 6*s* electrons, the 5*d* shell may be partly occupied. The M^{4+} ions also have all outer electrons in the 4*f* shell, but there is partial 5*d* occupation in some M^{2+} ions.

Ionic radii of the M^{3+} ions range from 1.06 to 0.85 Å and a graphical plot shows a shallow cusp shape whose origin has been discussed,[136–138] and which arises only partially in an analogous manner to the corresponding effect in the 3*d* transition metals (being much less pronounced because the ligand field–4*f* electron interaction is much weaker than in the 3*d* series) but is probably mainly due to 4*f*–4*f* interactions. A strictly comparable set of values for ionic radii is difficult to obtain, as structures and coordination numbers usually change along the series.

The large ionic radii, combined with the generally electrostatic nature of the metal–ligand bonding in complexes of these metals, lead to high coordination numbers.[139] Where the ligands are not sterically hindered, but are not chelate or polydentate with some of the ligand–ligand repulsion subsumed within the ligand, coordination numbers of nine to six are usually shown, depending on the relative cation/anion radii. Examples are $[La(H_2O)_9]^{3+}$ and $[LaBr_6]^{3-}$. Where the ligands, such as NO_3^-, themselves contain close pairs of donor atoms, coordination numbers up to 12 may be obtained, as in $[La(NO_3)_6]^{3-}$. Coordination polyhedra are not governed by considerations arising from covalency or strong crystal fields, as in the *d* transition metals, with their emphasis on regular geometries such as square planar or octahedral. Lanthanide coordination geometries are governed purely by minimization of repulsive terms. These terms usually arise from ligated atom–ligated atom repulsions but in the case of bulky ligands non-ligated interactions are important also. The steric requirements of polydentate ligands, in the sense that for example a terpyridyl complex must of course contain a roughly coplanar LnN_3 system, have also to be taken into account.

In a useful series of papers,[140–143] Kepert has considered high coordination number stereochemistries, including lanthanide examples, on the basis of a point-charge repulsion model, where the point charges lie on the surface of a sphere and are constrained only by intraligand factors.

For bulky ligands, the cone-angle approach, developed mainly for actinide complexes but equally applicable to lanthanides, is available.[144] A cone of sufficient solid angle to enclose the whole ligand is drawn, using suitable van der Waals radii for the non-ligated atoms. The ligand cone angles are then summed and for stable complexes the sum is found empirically to lie near to $0.80 \times 4\pi$. This enables coordination numbers to be rationalized or predicted. Coordination numbers lower than six are sometimes obtained with bulky ligands, particularly in the organo- and alkylamido-lanthanide area. In the case of all the ligands of a complex having very similar steric requirements, regular or semiregular polyhedra may be obtained, particularly in the cases of six-coordination (octahedron), eight-coordination (dodecahedron) and nine-coordination (face-centred trigonal prism). Other regular or semiregular polyhedra, namely the tetrahedron, trigonal prism, square antiprism and distorted icosahedron (point group *O*) are sometimes observed. Where ligands are unequal in their steric requirements then irregular polyhedra occur. There is a considerable tendency on the part of some authors to rationalize these irregular polyhedra of high coordination number in terms of semiregular polyhedra; thus a nine-coordinated complex might well be described as of distorted capped square antiprismatic shape. These descriptions tend to have little significance and such a complex could probably be described equally well as distorted face-centred trigonal prismatic.

39.2.2 Nitrogen Donor Ligands

39.2.2.1 Bipyridyls, o-phenanthrolines and 1,8-naphthyridines

These ligands are now well represented in complexes with the lanthanides, but were not investigated until 1962, when a study[145] in aqueous solution using a Job's plot method, *e.g.* at 451 nm for Ho^{3+}, showed that a complex $Ho(phen)_2^{3+}$ was formed. Pr, Nd and Er tripositive ions were also studied, but no solid products were characterized. Indeed, stability constants in water are possibly too low for a solid complex to be isolated. A little later, there were several reports[146–151] of the isolation of bipyridyl and 1,10-phenanthroline complexes from ethanolic solution. It is perhaps of interest that at this time, the lanthanides were believed not to form stable amine complexes and the role of higher coordination numbers in lanthanide complexes was not fully appreciated. Complexes isolated included the stoichiometries $M(NO_3)_3(bipy)_2$, where M = La–Lu;[146,148,151] $M(NCS)_3(bipy)_3$, where M = La, Ce, Dy;[151] $M(MeCO_2)_3(bipy)$, where M = Pr, Nd, Yb;[151] $MCl_3(bipy)_2H_2O$, where M = La–Nd;[151] $MCl_3(bipy)_2$, where M = Eu, Gd, Ho–Lu;[151] $M(NO_3)_3(phen)_2$, where M = La–Lu;[147,150] $M(NCS)_3(phen)_3$, where M = La–Lu;[147,149] $M(NCS)_3(phen)_2$, where M = Yb;[147–149] and $La(ClO_4)_3(phen)_4$.[152] The crystal structure of $[La(NO_3)_3(bipy)_2]$ has been determined. The La^{3+} ion is irregularly ten-coordinated, with La—N = 2.59 Å average.[153]

These complexes, which either are salts, or at least contain coordinated anionic ligands in which case a large Ln–ligand bond moment should result from the rather electrostatic nature of the bond, tend to be insoluble in solvents other than polar solvents in which they dissociate. Hence physicochemical studies in solution are limited. However, solid state fluorescence spectra of phenanthroline and bipyridyl complexes of Eu have been obtained[154] and correlated with the molecular site symmetry (for a discussion of this type of fluorescence, see Section 39.2.7.2). Magnetic susceptibility values have been measured[150] for the series of complexes $M(NO_3)_3(phen)_2$ at 293 K and are (BM) as follows: La, 0; Ce, 2.46; Pr, 3.48; Nd, 3.44; Sm, 1.64; Eu, 3.36; Gd, 7.97; Tb, 9.81; Dy, 10.6; Ho, 10.7; Er, 9.46; Tm, 7.51; Yb, 4.47; Lu, 0. These values correspond closely with the calculated values.[155] The complexes $Ln(NO_3)_3$(4,4-di-*n*-butyl-2,2′-bipyridyl)$_2$ and $Ln(NO_3)_3$(5,5′-di-*n*-butyl-2,2′-bipyridyl)$_2$ are soluble in organic solvents and give NMR spectra which show paramagnetic shifts.[156] These were at first believed to be of a covalently induced, or contact, nature but it later[157] became evident that they incorporate a substantial dipolar, or pseudocontact, component also. Electronic spectra of these complexes were also obtained in solution.[158] They showed only a small nephelauxetic effect with $\beta \simeq 0.96$, which indicated very little covalent character in the Ln—N bond.

The solubility in chloroform of the 4,4′-di-*n*-butyl-2,2′-bipyridyl (L) complex $Eu(NO_3)_3L_2$ enabled an NMR study of the kinetics of the ligand exchange shown in equation (1) to be carried out.[157] Temperature–line width plots showed that the reaction was first order in $Eu(NO_3)_3L_2$ but zero order in L and therefore presumably proceeded by a dissociative mechanism, with $\Delta G^{\ominus}_{273} = 72$ kJ mol^{-1}.

$$Eu(NO_3)_3L_2 + L^* \longrightarrow Eu(NO_3)_3LL^* + L \qquad (1)$$

The ligand 1,8-naphthyridine (**4**; naph) is enabled by its very short N—N distance to form complexes with lanthanide ions which have very high coordination numbers. Thus when anhydrous solutions of lanthanide perchlorates in ethyl acetate were treated with this ligand, complexes $[Ln(naph)_6](ClO_4)_3$, where Ln = La–Pr, or $[Ln(naph)_5](ClO_4)_3$, where Ln = Nd–Eu, were formed.[159] These formulations as 12- or 10-coordinated complexes respectively were fairly well established by IR, conductivity, and particularly by their paramagnetic NMR shifts which were independent of dilution, showing that all 1,8-naphthyridines were coordinated in nitromethane solution. An X-ray structure of $[Pr(naph)_6](ClO_4)_3$ showed[160] icosahedral coordination distorted by the difference between intraligand N—N distances (2.257 Å) and interligand distances (2.890–3.195 Å). (Figure 1). Pr–N distances are comparatively great at 2.735–2.768 Å, and this undoubtedly reflects the considerable interligand repulsions. Other complexes of 1,8-naphthyridines have been described,[161] namely $M(NO_3)_3L_2$ (M = Y, La–Yb, L = naph; M = Y, Sm–Yb, L = 2,7-dimethyl-1,8-naphthyridine), where the coordination number is probably 10.

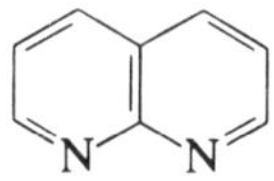

(**4**)

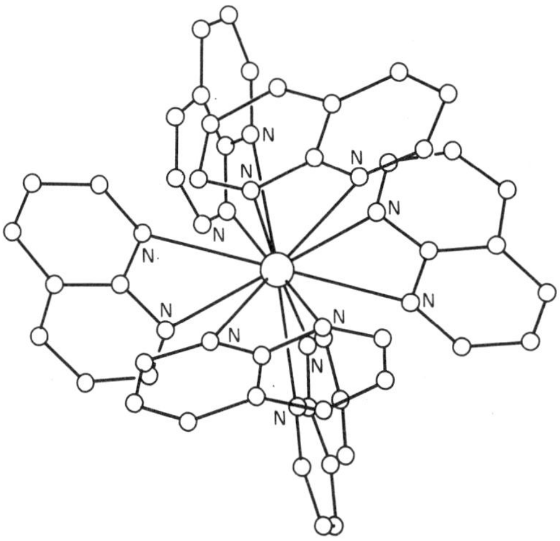

Figure 1 Structure of the $[Pr(naphthyridine)_6]^{3+}$ ion (reproduced with permission from ref. 160. Copyright, American Chemical Society, 1977)

39.2.2.2 2,2′:6′,2″-Terpyridyl

Despite being tridentate, this ligand is inferior to water in competition for coordination sites on lanthanide ions and non-aqueous solvents are used for preparative reactions. Using the poorly coordinating perchlorate as counterion, a complex $[M(terpy)_3](ClO_4)_3$, where M = La, Eu or Lu, is obtained.[162] The fluorescence spectrum of the Eu complex supported a nine-coordinated D_3 configuration with three terpyridyl ligands each bound by all three nitrogen atoms, a structure confirmed by X-ray.[163] Under different conditions, the 1:1 complex hydrated chlorides $MCl_3(terpy)(H_2O)_n$, where M = Ce–Gd may be obtained.[164] They have variable amounts of water of hydration, and the X-ray structure[165] of $[PrCl(terpy)(H_2O)_5]Cl_2{\cdot}3H_2O$ showed nine-coordination also (Figure 2). In these compounds,

Eu—N distances were 2.57–2.62 Å, while the Pr—N distances were 2.63–2.64 Å, with Pr—Cl at 2.88 Å and Pr—O at 2.48–2.54 Å. The visible spectra of the monoterpyridyl complexes have been used to investigate the nephelauxetic effect arising from the M—N bonds.[148]

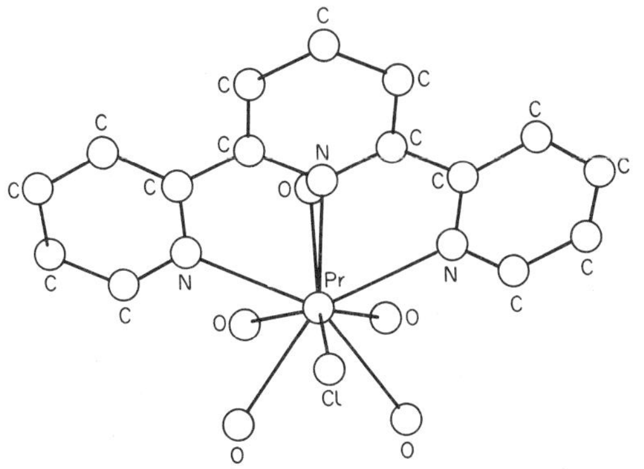

Figure 2 Structure of the cation of $[PrCl(terpyridyl)(H_2O)_5]Cl_2 \cdot 3H_2O$. The structure of this complex would have been impossible to establish by other means than diffraction, and the same is true for many other lanthanide complexes, owing to the lanthanides' great variety of coordination patterns (Reproduced with permission from ref. 165. Copyright, American Chemical Society, 1971)

Complexes of 2,4,6-tri-α-pyridyl-1,3,5-triazine (**5**) have also been prepared.[166] This readily available ligand is rather similar to terpyridyl in its stereochemistry and coordinating properties, and 3:1, 2:1 and 1:1 complexes were obtained. The rates of ligand-exchange reactions of the tris-terpyridyl complexes are quite slow, doubtless because of the considerable negative entropy change associated with dissociation of the tridentate ligand. Reaction (2) was investigated[167] in nitromethane solution.

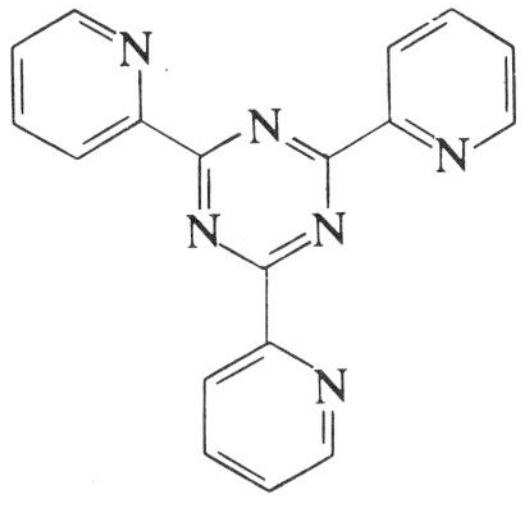

(**5**)

$$[Eu(terpy)_3]^{3+} + Tb^{3+} \longrightarrow [Eu(terpy)_2]^{3+} + [Tb(terpy)]^{3+} \quad (2)$$

When irradiated at 360 nm, $[Eu(terpy)_3]^{3+}$ fluoresces at 595 nm while $[Tb(terpy)]^{3+}$ fluoresces at 540 nm; the other species do not fluoresce and hence the reaction kinetics could be followed by measurement of fluorescence intensity giving $k = 1.9 \times 10^{-3}\ s^{-1}$. The reaction was first order in $[Eu(terpy)_3]^{3+}$ and zero order in Tb^{3+}, proceeding as shown in equations (3) and (4). The rapid nature of the second step was confirmed directly in separate experiments by similar means. Dye-laser excited emission and excitation spectra of $[Eu(terpy)_3](ClO_4)_3$ in the solid state and in acetonitrile solution have been measured[168] and evidence has been obtained which was interpreted as pointing towards the presence of some mono- or bi-dentate terpyridyl ligands in addition to those which are tridentate as observed in the X-ray structural determination.

$$[Eu(terpy)_3]^{3+} \xrightarrow{\text{rate determining}} [Eu(terpy)_2]^{3+} + terpy \quad (3)$$

$$terpy + Tb^{3+} \xrightarrow{\text{fast}} [Tb(terpy)]^{3+} \quad (4)$$

39.2.2.3 Dialkylamido complexes

These compounds are not numerous but have some important and unique features. The disubstituted amido group, R_2N^-, forms many complexes with *d* transition metals which have been extensively investigated, and the case R = $SiMe_3$ has been particularly closely examined.[169] In these compounds, interest attaches both to the effect of the very large cone angle of the $N(SiMe_3)_2$ group and to the nature of the bonding, which may involve metal *d*–N*p*–Si*d* interactions. The lanthanides form the tris(amides) $M(NR_2)_3$ and some adducts and derivatives of these, and the considerable bulk of the ligand is effective in producing a coordination number of three, the lowest known for lanthanides, in some of these compounds.

The compounds $[M\{N(SiMe_3)_2\}_3]$ where M = Y, La–Eu, Gd, Ho, Yb and Lu, have been synthesized by reaction of MCl_3 with $LiN(SiMe_3)_2$.[170,171] They are slightly volatile and very soluble in non-polar solvents. The M—N bond is avidly attacked by traces of water. Single crystal X-ray structural determinations of the europium[172] and neodymium[173] compounds of this isostructural series show an unusual situation. The coordination about the metal ion is pyramidal with Eu—N = 2.259 Å and N—Eu—N = 116.6°. If the metal ion is ignored the symmetry of the ligands is D_3, with the europium ion placed randomly above or below the N_3 triangle to give the slightly pyramidal configuration observed; the pyramidal situation of the metal ion does not reduce the D_3 symmetry of the group of three ligands considered alone. The dipole moment of these compounds is close to zero and hence they must have coplanar MN_3 groups in solution. It is possible that attractive van der Waals forces among the ligands are responsible for a 'squeezing' of the metal ion upwards into a pyramidal position in the crystalline state.

Reaction of $OPPh_3$ with the tris(silylamides) gave two different products.[174,175] With about one molecular proportion of Ph_3PO, the expected adducts $[M\{N(SiMe_3)_2\}_3OPPh_3]$ were obtained, and the crystal structure of the lanthanum compound showed a nearly tetrahedral configuration with La—N = 2.38–2.41 Å, O—La—N = 104.8–107.8° and N—La—N = 109.2–116.4°. The P—O—N system was nearly linear at 174.6°. Attempts to prepare bis(phosphine oxide)-substituted compounds using 2–5 molar proportions of $OPPh_3$ surprisingly yielded $[(Ph_3PO)\{N(SiMe_3)_2\}_2M(\mu\text{-}O_2)M\{N(SiMe_3)_2\}_2(OPPh_3)]$. The origin of the peroxy group is obscure; it could not have been introduced by air or traces of peroxides. The same compounds could be obtained by reaction of 0.5 mol $(Ph_3PO)_2{\cdot}H_2O_2$ with the tris(silylamides). An X-ray structural determination of the lanthanum complex showed a five-coordinated structure with each metal ion coordinated to two silylamide N, one phosphine oxide O and the two peroxide O atoms. The $La{\cdot}O_2{\cdot}La$ system is coplanar, with La—O = 2.33 and 2.34 Å, and O—O = 1.65 Å. Again, La—O—P is nearly linear at 172.6°, with La—O = 2.42 Å; La—N is similar to the tris(silylamide) and the monophosphine oxide adduct, at 2.37 and 2.49 Å. There was evidence from the UV absorption spectra that the amide ligand was a particularly effective σ donor relative to donors such as H_2O or Cl^-, the nephelauxetic parameter β taking values of 0.993 (Pr) and 0.978 (Nd) for the tris(amides).

Compounds of the less severely hindered di-2-propylamido group have also been reported.[176] These are $Nd\{N(CHMe_2)_2\}_3(THF)$ and $M\{N(CHMe_2)_2\}_3$ where M = Y or Yb. It is significant that the Nd compound was isolated with coordinated THF which was not removed under vacuum, while the compounds of the smaller Y and Yb ions were presumably too sterically crowded to allow adduct formation by this weak ligand. These compounds were less stable than the tris (trimethylsilylamido) analogues, decomposing at 80 °C and 10^{-4} mmHg.

39.2.2.4 Other nitrogen donors

Although ammonia undoubtedly forms complexes with lanthanide ions, these materials are not fully characterized.[177] The aliphatic polyamines, however, form a fairly extensive series of complexes, but must be prepared in completely anhydrous conditions as they are avidly hydrolyzed to hydroxy species which form insoluble precipitates. With ethylenediamine, tetrakis complexes $MX_3(en)_4$ were obtained for M = La–Yb, X = NO_3; or for M = La–Lu, X = Cl; or for M = La, Nd, Gd, X = Br. These series showed IR evidence of discontinuities of structure, the nitrates being formulated $[M(en)_4(NO_3)](NO_3)_2$ for M = La–Sm, but $[M(en)_4](NO_3)_3$ for the remainder of the series. The tris complexes $[M(en)_3(NO_3)_2]NO_3$, where M = Gd–Ho, and $MX_3(en)$ where M = Gd, Er and X = Cl; or M = Gd and X = Br, were also obtained, as were tetrakis(perchlorates) $[M(en)_4](ClO_4)_3$ where M = La–Nd.[178,179] Bis complexes of triethylenetetramine[180,181] and tris and bis complexes of diethylenetriamine[182] have also been described. They are exemplified by $[M(dien)_3](NO_3)_3$ where M = La–Gd, and $[M(trien)_2](ClO_4)_3$ where M = La–Ho. The exchange of ligand between $[Nd(trien)_2]^{3+}$ and trien in acetonitrile has been studied and is believed to proceed by way of an intermediate when trien is not fully coordinated.[183]

Turning now to complex thiocyanates and azides, where the pseudohalide is a nitrogen donor, the series $[NBu^t_4][M(NCS)_6]$, where M = Pr–Yb or Y, has been obtained and the erbium compound has been submitted to X-ray ($R = 12\%$).[184,185] The latter has octahedral coordination by six nitrogen atoms at Er—N = 2.34 Å (average) and Er—N—C = 174°. This compound shows SCN^- IR frequencies of 2090 sh, 2045 and 480 cm^{-1}.

Association between lanthanide ions and azide or thiocyanate ions has been studied in solution by electronic, Raman and NMR spectroscopy.[186,187] The complexation constant in water between Nd^{3+} and N_3^- is approximately 2.5. Longitudinal relaxation time studies for Gd–Dy indicate that the M—N—NN angle is bent (135° approximately).

The solution NMR[188] and structure[189] of the tripyrazolylborohydride $Yb\{(C_3H_3N_2)_3BH\}_3$ have been correlated in terms of a dipolar-shifted spectrum, which corresponds with the X-ray solid structure determination, where bicapped trigonal prismatic eight-coordination was observed from one bidentate and two tridentate anions. Surprisingly, the NMR spectrum indicated slow interconversion of the two types of ligand in solution. The range of NMR shifts was 200 p.p.m.

39.2.3 Phosphorus Donor Ligands

Under most normal circumstances, lanthanide–phosphine complexes are not formed, but phosphines have been shown to be capable of coordinating when the lanthanide ion is in the form of an organolanthanide. Thus $[Yb(C_5Me_5)_2Cl(Me_2PCH_2PMe_2)]$ is obtained from the reaction of $[Yb(C_5Me_5)_2(Me_2PCH_2PMe_2)]$ with $YbCl_3$ in toluene (see also Section 39.2.11.4 for the Yb^{2+}–diphosphine complexes). Here the Yb is bonded to two η^5-C_5Me_5 with Yb—C = 2.612–2.718 Å, one Cl at 2.532 Å and one P at 2.941 Å (the diphosphine having only one phosphorus atom coordinated).[190] It seems clear from the large Yb—P internuclear distance that the bond cannot be a strong one.

39.2.4 Oxygen Donor ligands

39.2.4.1 Hydrated ions and salt hydrates

(*i*) *Thermodynamic parameters of hydrated lanthanide ions*

Values of free energies of formation, heats of formation and entropies for the series La–Yb, including Pm, have been given,[191–193] together with standard reduction potentials. The $\Delta H_f^\ominus$ values were obtained from $\Delta H_f^\ominus$ for MCl_3,[194] together with heats of solution for MCl_3.[195] $S^\ominus$ values were obtained from the entropies of the hydrated ions[196] and those of the metals. The $\Delta G_f^\ominus$ and $E^\ominus$ values were obtained by calculation from the foregoing data. These data are shown in Table 2.

Table 2 Thermodynamic Parameters of Hydrated Lanthanide Ions[191,192]

	$-\Delta H_f^\ominus(M^{3+}, aq)$ (kJ mol^{-1})	$-S^\ominus(M^{3+}, aq)$ (J K^{-1} mol^{-1})	$-\Delta G_f^\ominus(M^{3+}, aq)$ (kJ mol^{-1})	$-E^\ominus[M^{3+}/M]$ (V)
La	709.4 ± 1.6	218 ± 13	685 ± 5	2.37 ± 0.02
Ce	700.4 ± 2.1	205 ± 17	672 ± 8	2.32 ± 0.03
Pr	706.2 ± 1.6	205 ± 13	677 ± 8	2.34 ± 0.03
Nd	696.6 ± 1.7	205 ± 13	673 ± 5	2.32 ± 0.02
Pm	695 ± 17	213 ± 29	669 ± 21	2.31 ± 0.07
Sm	691.1 ± 1.7	218 ± 17	661 ± 13	2.28 ± 0.04
Eu	605.6 ± 2.3	222 ± 17	573 ± 13	1.98 ± 0.04
Gd	687.0 ± 2.1	222 ± 13	657 ± 13	2.27 ± 0.04
Tb	698 ± 6	230 ± 17	657 ± 13	2.27 ± 0.04
Dy	695.9 ± 2.8	230 ± 13	669 ± 17	2.32 ± 0.06
Ho	707 ± 8	226 ± 13	686 ± 13	2.37 ± 0.04
Er	705 ± 10	243 ± 13	673 ± 8	2.33 ± 0.03
Tm	705.2 ± 3.0	243 ± 29	665 ± 17	2.30 ± 0.06
Yb	674.5 ± 3.0	238 ± 17	644 ± 13	2.23 ± 0.04
Lu	702.6 ± 2.6	—	—	—

Alternative values for heats of formation of some hydrated tripositive ions have been obtained[197] from a reanalysis of the first half-wave amalgamation potentials, $(E_{1/2})_1$. These values, together with standard reduction potentials,[198] are given in Table 3.

Table 3 Thermodynamic and Polarographic Parameters of Hydrated Lanthanide Ions[197,198]

	$-(E_{\frac{1}{2}})$ (V)	$-E^{\ominus}[M^{3+}/M]$ (V)	$-\Delta H_f^{\ominus}(M^{3+}, aq)$ (kJ mol^{-1})
La	1.75	2.36	706.3
Ce	1.83	2.32	695.0
Sm	1.71	2.28	686.6
Eu	1.70	1.98	605.0
Gd	1.78	2.26	682.0
Tb	1.73	2.27	687.8
Ho	1.77	2.34	708.8
Yb	1.81	2.23	674.5
Lu	1.68	2.28	697.5

(ii) Hydrated lanthanide ions—structures and properties

X-Ray structural determinations of fully hydrated lanthanide ions in salts with weakly coordinating anions show the predominating coordination number is likely to be nine water molecules. Thus after some early structural determinations[199–202] of $[M(H_2O)_9]X_3$, where M = Pr, Nd, Sm or Er and X = BrO_3 or $EtOSO_3$, the structures of the hydrated Pr and Yb bromates and ethyl sulfates have been more accurately determined.[203,204] The hydrated bromates and ethylsulfates each form an isostructural series for all lanthanide ions, the bromates *P*6$_3$/*mmc* and the ethylsulfates in space group *P*6$_3$/*m*. Tricapped trigonal prisms of water oxygen atoms with symmetry D_{3h} in the bromates and C_{3h} in the ethylsulfates surround the metal ions. M—O distances are (a) for the bromates: Pr—O (prism) = 2.49(1), Pr—O (equatorial) = 2.52(1), Yb—O (prism) = 2.32(1), Yb—O (equatorial) = 2.43(1), and (b) for the ethyl sulfates: Pr—O (prism) = 2.470(2), Pr—O (equatorial) = 2.592(3), Yb—O (prism) = 2.321(3), Yb—O (equatorial) = 2.518 Å. The decrease in the prismatic M—O distances is what is expected from the lanthanide contraction, 0.155 Å, but van der Waals repulsions between prismatic and equatorial oxygen atoms make the decrease in the equatorial M—O bonds about half this value. The anions are hydrogen bonded to the complex ions. Accurate values of unit cell dimensions were obtained for both these series.

In the light of these results, it is very surprising therefore that $M(ClO_4)_3\cdot 6H_2O$ (M = La, Tb or Er) when studied by X-ray single crystal methods[205] shows the lanthanide ion octahedrally coordinated to six water molecules in the *Fm*3*m* crystal, with M—O = 2.48 (La), 2.35 (Tb) and 2.25 Å (Er). The perchlorate anions show rotational disorder and are held by weak M—O—H···O (perchlorate) hydrogen bonding at O(water)—O(perchlorate) distances of 3.23 (La), 3.20 (Tb) and 3.27 Å (Er).

The problem of lanthanide coordination numbers in aqueous solutions of the tripositive lanthanide ions was for long unsolved. Thus Choppin stated in a 1971 review:[206] 'At this time we know little about the coordination numbers of trivalent lanthanide and actinide ions in complexes in aqueous solution.' This remark referred particularly to the role of the water molecules in complexing the cation. Some of the lines of evidence then used to estimate *n* in $[Ln(H_2O)_n]^{3+}$(aq) were the following. X-Ray diffraction measurements[207] on concentrated solutions of $ErCl_3$ and ErI_3 seemed to show octahedral cations $[ErCl_2(H_2O)_4]^+$ or $[ErI_2(H_2O)_4]^+$. However, electronic spectra of $Nd(BrO_3)_3\cdot 9H_2O$ in aqueous solution were very similar to the solid state spectra of the same compound,[208] thus pointing towards $[Nd(H_2O)_9]^{3+}$ in solution. There was considerable evidence from discontinuities in various properties near Gd^{3+} or Tb^{3+} in the lanthanide series that *n* in $[M(H_2O)_n]^{3+}\cdot$aq took a different value for the later members of the series as opposed to the earlier. Thus values of ΔG and *K*, ΔH and ΔS for complexation of tripositive lanthanide ions in aqueous solution with edta, nta and several related ligands showed these discontinuities;[206] but significantly, no discontinuity (apart from a small discontinuity of about 10 J mol^{-1} at Eu^{3+}, the f^7 member, doubtless related to an ionic radius or crystal field discontinuity) was observed for reaction (5).[209]

$$[M(H_2O)_9]X_3(\text{solid}) + 3L^{2-}(aq) \longrightarrow [ML_3]^{3-}(aq) + 3X^-(aq) \tag{5}$$

$$X = BrO_3^- \quad \text{or} \quad EtOSO_3^-; \quad H_2L = O(CH_2CO_2H)_2 \quad \text{or} \quad 2,6\text{-}NC_5H_3(CO_2H)_2$$

Other properties of aqueous solutions where discontinuities were observed were molar volumes, heat capacities and viscosities.[210–212] From all these pieces of evidence, the general opinion was that n in $[M(H_2O)_n]^{3+}(aq)$ was nine for the lighter lanthanide ions, but eight for the heavier,[213] though some workers were of the opinion that n did not change along the series.[214] Solution X-ray studies have provided evidence for an average coordination number of 8.9 for $Nd^{3+}(aq)$,[215] while neutron diffraction supports a value of 8.5 for the same ion.[216]

However, considerable weight must attach to the following luminescence lifetime studies, in which it has been established from measurements made in dilute aqueous solution[217,218] that the hydrated europium ion is $[Eu(H_2O)9.6 \pm 0.5]^{3+}$ and the corresponding terbium ion is $[Tb(H_2O)9.0 \pm 0.5]^{3+}$. The error limits should certainly be no larger and are probably smaller than the 0.5 molecule stated. In these experiments, the lanthanide ions are excited by a tunable pulsed dye laser, and there are then two competing deexcitation routes (Figure 3). The first is by direct emission of a quantum of light in the visible region of the spectrum, *e.g.* at around 545 nm for the $^5D_4 \rightarrow {}^7F_5$ deexcitation of Tb^{3+}. The second route is a deexcitation in which the

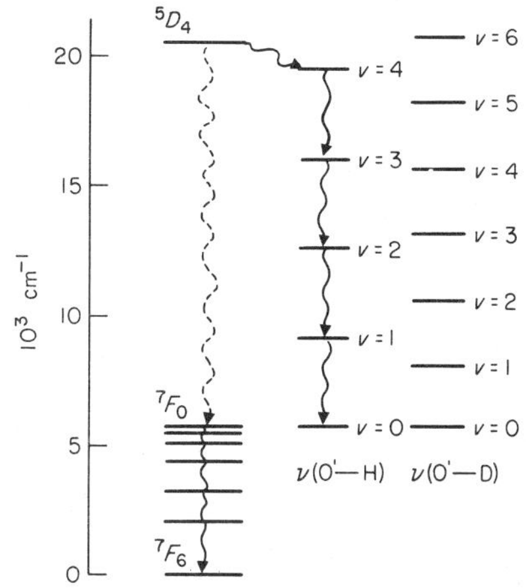

Figure 3 Alternative deactivation pathways for a Tb^{3+} ion coordinated to H_2O or to D_2O (adapted from ref. 218)

energy of the excited ion is transferred to O—H vibrational energy in water molecules in the first coordination sphere by excitation of perhaps the third vibrational overtone at around 19 000 cm^{-1}. The result of the second mechanism is that the observed luminescence lifetime is reduced according to equation (6), where K_{nat} is the natural rate constant for photon emission, K_{nonrad} is the rate constant for the transfer of energy to the O—H vibrations of coordinated water and K_{obs} is the observed exponential decay constant. The contribution of K_{H_2O} can be quantified by a similar experiment using D_2O where this mechanism is virtually unavailable owing to the closer spacing of the O—D vibrational levels, and is found to be proportional to the number of coordinated water molecules in crystalline solids where these numbers are known. The principle can be extended to solutions and then gives the hydration numbers referred to above for the plain hydrated ions. Also, very reasonable values for hydration numbers of complexes in solution have been obtained, such as $[Eu(nta)(H_2O)_{6.0}]$ and $[Tb(nta)(H_2O)_{5.0}]$.

$$K_{obs} = K_{nat} + K_{nonrad} + K_{H_2O} \tag{6}$$

(*iii*) *Salt hydrates*

In this section are discussed hydrates of lanthanide salts where the water molecules form a majority of the coordination sphere and where there is evidence concerning the number or geometry of the coordinated water molecules. Using laser excitation as described above, it has been shown[219] that the 1:1 association product between Eu^{3+} ions and NO_3^- in aqueous solution is $[Eu(NO_3)(H_2O)_{6.8\pm0.4}]^{2+}$, the nitrate ion presumably being bidentate. However, luminescence lifetime measurements have shown that there is little complexation by Cl^- in water, though SO_4^{2-} and NO_3^- do associate.[220] Enthalpy and free energy changes on crystallization of hydrated chlorides, nitrates and perchlorates have been calculated from literature activity data, and the results were interpreted in terms of anion crowding at higher lanthanide atomic numbers.[221]

X-Ray single crystal data show that in the solid state, coordination numbers of around nine predominate in this type of complex. Thus $La_2(SO_4)_3{\cdot}9H_2O$ contains La^{3+} coordinated to six H_2O in a trigonal prism and three SO_4^{2-} in the equatorial face-capping positions,[222] while $[M(NCS)_3(H_2O)_6]$, where M = Dy or Pr, both contain distorted nine-coordinated polyhedra.[223] However, with the slightly more bulky chloride ion present, $[MCl_2(H_2O)_6]Cl$, where M = Eu[224] or Gd,[225] has distorted square antiprismatic eight-coordination with one Cl^- in each square in *cis* positions. Interatomic distances are Gd—O = 2.39–2.42 and Gd—Cl = 2.768 Å. The complex cations are linked together in two ways, directly by $Gd—OH_2\cdots Cl—Gd$ hydrogen bonds and indirectly by the system $Gd—OH_2\cdots Cl\cdots H_2O—Gd$. The O···Cl distances of the first sort are 3.17–3.24 Å, and the second type are 3.16–3.18 Å.

Russian workers have continued to study hydrated salts of lanthanide ions. Some of these studies are of a rather detailed nature and a summary of some of the more recent work will be given here to indicate its general nature. There is considerable emphasis on phase studies, X-ray powder data, thermogravimetric and differential thermal analysis, and IR data.

$RbM(SO_4)_2{\cdot}H_2O$ (M = Gd—Lu) have been obtained from solution and belong to space group $P2_1/c$.[226] The deposition from aqueous solution of $Ce_2Sr_3(SO_4)_6{\cdot}3H_2O$ has been studied, and the product characterized by DTA, TGA, powder pattern and IR.[227] The thermal properties of $M_2SO_4{\cdot}Gd_2(SO_4)_3{\cdot}2H_2O$ (M = Na, K, Rb, Cs, NH_4) have been studied by DTA, TGA and powder pattern.[228,229] $M_2SO_4{\cdot}Gd_2(SO_3)_3{\cdot}8H_2O$ (M = Rb, Cs, NH_4) have been obtained from aqueous solution and characterized by analysis and powder pattern.[230]

Solid phases have been isolated from the systems $M(NO_3)_3$–quinolinium nitrate–H_2O (M = Ce, Sm, Tb, Ho).[231,232] The composition of many solid phases, *e.g.* $Gd(NO_3)_3{\cdot}4H_2O$, has been determined in the system $M(NO_3)_3$–HNO_3–H_2O (M = Y, Gd, Lu).[233] $M_2Nd(NO_3)_5{\cdot}H_2O$ (M = Li, Na, K) has been prepared and characterized by TGS, DTA and IR.[234] The system piperidinium (Hpip) nitrate–$Sm(NO_3)_3$–H_2O yielded $(Hpip)Sm(NO_3)_4{\cdot}4H_2O$ and $(Hpip)_3Sm(NO_3)_6{\cdot}2H_2O$ which were studied by DTA and TGA.[235] The system Me_4NNO_3–$Tb(NO_3)_3$–H_2O yielded $Me_4NTb(NO_3)_4{\cdot}2H_2O$ which was characterized by analysis, IR and powder pattern.[236] A solubility study identified phases in the system $Ca(NO_3)_2$–$M(NO_3)_3$–H_2O (M = Ce, Pr, Nd) at 25 °C.[237] The thermal decomposition of $M(NO_3)_3{\cdot}6H_2O$ (M = Sm–Gd) was studied by DTA, TGA and IR.[238]

The single crystal X-ray structure of hydrated neodymium hypophosphate, $NdHP_2O_6{\cdot}4H_2O$, has been determined.[239] The Nd ion is eight-coordinated to four H_2O and two bidentate hypophosphate anions. The details of interatomic distances and bond angles were compared with other members of this isostructural series, which extends at least from Nd to Er.[240]

A number of hydrated carboxylates have been studied. Isothermal solubility studies of $Y(HCO_2)_3$–$KHCO_2$–H_2O gave the compounds $KY(HCO_2)_4{\cdot}H_2O$ (of which a single-crystal X-ray structure was obtained) and $K_5Y(HCO_2)_8$.[241] The enthalpies of dissolution of $M(Me_3CO_2)_3{\cdot}2H_2O$ (M = La–Lu) were measured and standard heats of formation calculated.[242] $K_4Pr_2(C_2O_4)_5{\cdot}nH_2O$ (n = 7–9) have been characterized by analysis, powder pattern and IR.[243] M_2 (dimethylmalonate)$_3{\cdot}nH_2O$ (n = 5–8, M = La–Lu) have been synthesized and characterized by analysis, IR, powder pattern, DTA and TGA.[244]

The thermal behaviour of $MCl_3{\cdot}xH_2O$ (x = 6 or 7, M = Y, La–Lu) was investigated by DTA, TGA and X-ray powder photography.[245] Equilibrium diagrams for the MCl_3–H_2O systems (M = Y, Gd, Tb, Dy, Er) were obtained by DTA and evidence obtained for the hydrates $MCl_3{\cdot}nH_2O$ (n = 6, 8, 9 and 15).[246]

The system $Nd(IO_3)_3$–MIO_3–H_2O (M = K, Rb) were studied by the isothermal saturation method, and $3Nd(IO_3)_3{\cdot}2KIO_3{\cdot}2H_2O$ was described and studied by TGA and IR.[247]

Perrhenates and pertechnetates $M(M'O_4)_3{\cdot}4H_2O$ (M = Y, Ho–Lu; M′ = Tc or Re) are shown to be isostructural by powder pattern[248] with the single crystal X-ray structure which had been obtained for M = Yb, M′ = Re.[249] IR studies were also made.[250]

Behaviour in aqueous alkali has been investigated. Thus $M(OH)_3$ (M = Yb or Lu) when heated at 170 °C for 12 h in aqueous NaOH gave $Na_4M(OH)_7{\cdot}2{\cdot}5H_2O$ and a single crystal

X-ray structure showed octahedral YbO_6 groups to be present.[251] Ion exchange, dialysis and spectrophotometry using Arsenazo-III have been used to study the hydrolysis of the hydrated Y^{3+} ion at 10^{-5} to 10^{-2} M and pH 1 to 9. Mono- and poly-nuclear complexes were present, and K_1 for Y(OH)·aq was 2.52×10^6.[252]

39.2.4.2 β-Diketonates and their adducts

Under this section are considered lanthanide complexes of β-diketones and their adducts, but work specifically in the area of lanthanide shift reagents is dealt with in Section 39.2.9. Of course, the majority of lanthanide shift reagents are lanthanide β-diketonates, and when they function as shift reagents they do so by forming adducts in solution. Furthermore, interest in shift reagents has directly stimulated a considerable amount of fundamental research on lanthanide β-diketonates and their adducts. Much of this fundamental work was, therefore, carried out in the early 1970s. All this means that the division between this section and Section 39.2.9 is not entirely clear cut.

The general position regarding lanthanide complexes of the simple parent β-diketone, acetylacetone, is that its anion readily complexes, but that the resulting uncharged compounds $M(acac)_3$ are coordinatively unsaturated, and accordingly form hydrates in hydrous conditions, or polymerize in order to attain a higher coordination number, the process being driven by the consequent gain in bond energy. There have been several investigations[253–8] of anhydrous and hydrated lanthanide acetylacetonates, of which the study by Richardson *et al.* will be discussed here. The initial product prepared from aqueous solution is $M(acac)_3·3H_2O$, except for M = La when it is the dihydrate. Complexes may be converted into the anhydrous compound, or into the monohydrate or dihydrate, or remain as the trihydrate, by recrystallization from appropriately hydrous solvents under various conditions, or by vacuum drying at room temperature. These compounds are very easily partially hydrolyzed to $M(acac)_2(OAc)·nH_2O$ or $M(acac)_2(OH)·nH_2O$, hence some early reports may need reinterpretation. The freshly prepared $M(acac)_3$ readily rehydrate, but after storage an insoluble material is obtained which is probably polymeric. None of the compounds is appreciably volatile. Several X-ray structures of the hydrates have appeared, thus $[M(acac)_3(H_2O)_2]$[259,260] (M = La, Nd) are eight-coordinated, as is $[Eu(acac)_3(H_2O)_2]·H_2O$[261] and $[Ho(acac)_3(H_2O)_2]·2H_2O$,[262] while $[Yb(acac)_3(H_2O)]$[263,264] shows seven-coordination of the smaller Yb^{3+} ion. However, $Gd(acac)_3(H_2O)_4$ is dimeric, with Gd^{3+} coordinated to six oxygen atoms from the diketonate anions, two from water molecules and one oxygen atom from a diketonate anion which is bidentate towards the other Gd^{3+} ion. The relationship between the repulsion energy and coordination polyhedron of this M(bidentate)$_3$(monodentate)$_3$ coordination was explored.[265]

Solid state studies of adducts where the β-diketone is not very bulky include the monohydrate $[Ho(PhCOCHCOPh)_3H_2O]$, where the X-ray structure[266] shows a capped octahedron, resembling a three-bladed propeller with the H_2O molecule located on the trigonal axis. Here Ho—O(water) = 2.39 Å, with Ho—O(diketone) = 2.28 and 2.31 Å. The corresponding complexes having M = La–Ho are isostructural[267] and the oscillator strength of the $^4I_{9/2} \rightarrow {}^4G_{5/2}$ transition of the Ho compound at 17 000 cm^{-1} has been related to the Judd–Ofelt equation.

Lanthanide complexes of β-diketones $RCOCH_2COR$, where R is not methyl, have steric demands which are greater than those of acetylacetone, either slightly as when R = CF_3 or Ph, or considerably as when R = CMe_3. Many investigations have been reported into the state of aggregation of the complexes M(diketonate)$_3$ and their capability of forming adducts [M(diketonate)$_3L_x$], and the relationship of these properties to the lanthanide radius and the nature of the group R.

The dipivaloylmethanates $M(Me_3CCOCHCOCMe_3)_3$ have been well studied. This highly hindered ligand is sufficiently space filling that the monomeric tris complex is reasonably stable in the case of the smaller lanthanides and yttrium, though a mono adduct $M(dpm)_3L$ (dpm = $Me_3CCOCHCOCMe_3$, dipivaloylmethanate) is readily formed, where L = H_2O, ROH, R_3PO *etc.* The larger lanthanide ions do not form monomers but bridged dimers $M_2(Me_3CCOCHCOCMe_3)_6$. The complete series $M(dpm)_3$, where M = Y, La–Lu, was first prepared by Eisentraut and Sievers,[268] using a standard method (β-diketone, metal salt and ammonia in aqueous alcohol) but under reduced pressure. These complexes are remarkably volatile and can be separated by gas chromatography.[267] The solids are dimeric in the cases of La–Dy as exemplified by the X-ray crystal structure determination[269] of $[Pr_2(dpm)_6]$, and monomeric for Y, Tb–Lu, as exemplified by the X-ray structure[270] of $[Er(dpm)_3]$. The terbium and dysprosium complexes can apparently have either structure.[271] Bond lengths are in the region 2.39–2.44 Å, with one bond distance at 2.59 Å, for $[Pr_2(tmd)_6]$ and 2.156–2.262 Å for $[Er(dpm)_3]$. The coordination polyhedron of the praseodymium complex is irregular, while the erbium complex shows trigonal prismatic coordination, probably a consequence of the comparatively large ionic radius of Er^{3+} compared with the intraligand O—O distance, giving the angle O—Er—O = 74.4°. The complexes are monomeric in non-polar solvents but avidly form hydrated adducts with traces of water.[272] An X-ray crystal structure of the hydrate $[Dy(dpm)_3H_2O]$ shows seven-coordination with a side-capped trigonal prismatic coordination, with Dy—OH_2 = 2.36 and Dy—O(dpm) = 2.25, 2.36 Å. The molecules are arranged in dimeric pairs, but the interconnection is by means of hydrogen bonding, and not by Dy—O—Dy interaction.

The vapour pressures of $M(dpm)_3$, where M = La–Lu, have been determined[273] over a range of temperature within 410–530 °C and the heats of sublimation, and heats of vapourization of the ligands, were calculated. The heats of sublimation of the lighter lanthanides (*e.g.* Pr, 165.4 kJ mol^{-1}) are considerably greater than those of the heavier (*e.g.* Yb, 133.3 kJ mol^{-1}) and doubtless this increase reflects the enthalpy of cleavage of the dimeric $[M_2(dpm)_6]$.[274]

The complexes $M(fod)_3$, where fod = $CF_3CF_2CF_2COCHCOCMe_3$, have been synthesized[275] and studied because they are probably the most effective lanthanide shift reagents known for general applications. They readily form hydrates, including the interesting $[Pr_2(fod)_6H_2O]\cdot H_2O$ where the X-ray determination[276] shows a similar structure to that of $[Pr_2(dpm)_6]$ but with an additional water molecule bridging the two Pr^{3+} ions and H-bonded to the fluorocarbon chains, giving eight-coordination. For one Pr^{3+} ion, the coordination is distorted dodecahedral; the other has irregular coordination. The smaller Lu^{3+} ion gives[227] seven-coordinated $[Lu(fod)_3H_2O]$ with Lu—O(diketone) = 2.17 and Lu—O(water) = 2.29 Å. In an influential early paper,[275] Sievers and co-workers prepared $M(fod)_3\cdot H_2O$, where M = Y, La–Lu, using standard methods (β-diketone, $M(NO_3)_3\cdot aq$ and NaOH in aqueous methanol). The hydrated complexes were converted into $M(fod)_3$ by P_4O_{10}/vacuum, but are avidly rehydrated. They are very soluble in benzene and very volatile, volatilizing completely by 260 °C (La) or 180 °C (Lu). They may easily be separated by gas chromatography, a plot of retention times against ionic radii giving a smooth curve. It is of interest that in these two last volatility-dependent properties, $Y(fod)_3$ has properties intermediate between those of the Ho and Er complexes; in broader terms, all types of complexes of Y resemble the corresponding complexes of the Dy, Ho, Er group (except, of course, in those properties which are directly related to electronic configuration, such as electronic spectra).

$M(fod)_3$, where M = Pr or Eu, have been shown[278] to undergo association in carbon tetrachloride solution, doubtless of a similar nature to that observed[269] in the case of $[Pr_2(dpm)_6]$. Vapour pressure osmometry on carefully dried solutions showed association into dimers and trimers, with association quotients as follows (mol^{-1}): $Pr_2(fod)_6$, 140 ± 8; $Pr_3(fod)_9$, 45 ± 5; $Eu_2(fod)_6$, 367 ± 22; $Eu_3(fod)_9$, 12 ± 2; all at 37 °C. Controlled addition of traces of water was observed to reduce these values; presumably it coordinates and splits the β-diketonate bridge. This association in solution is in contrast to the more highly hindered $M(dpm)_3$ complexes which are monomeric in carbon tetrachloride.[279]

The lanthanide β-diketonates readily form stable, isolable adducts with amines, phosphine oxides and other ligands with a fairly strong affinity for lanthanide ions. They do not generally form stable, isolable complexes with alcohols or ethers, although their interaction in solution with the species is evident from their effectiveness as NMR shift reagents for alcohols and, in some cases, ethers. Studies[280] in carbon tetrachloride of $[Ho(dpm)_3]$ and its adducts with triphenylphosphine oxide, borneol, cedrol and camphor by means of Job's plots of the electronic absorption spectra showed that only 1:1 association took place with stability constants ($\log K_1$) as follows: triphenylphosphine oxide, 4.48 ± 0.3; borneol, 2.66 ± 0.1; cedrol, 2.66 ± 0.1; campor, 1.75 ± 0.1. The large difference between the coordinating abilities of the phosphoryl group and those of the alcohols and ketone is noteworthy. However, with $Eu(fod)_3$ and methyldimethylcarbamate in carbon tetrachloride, both 1:1 and 1:2 association were found with $\log K_1 = 3.20 \pm 0.14$ and $\log K_2 = 2.03 \pm 0.04$.[278]

Evidence for a cooperative type of association in forming 1:2 adducts, in which the second association constant K_2 is greater than K_1, was found by emission titration studies.[281] These involve study of the dependence of intensity of the ${}^5D_o \rightarrow {}^7F_n$ emission of the Eu^{3+} ion upon the nature of the emitting complex. An example of the results is that $\log K_1 = 2.34$ and $\log K_2 = 5.00$ for association between $Eu(dpm)_3$ and n-propylamine. However, some evidence was obtained from circularly polarized luminescence (CPL) and IR studies that Schiff bases are formed by interaction between $Eu(dbm)_3$ or $Eu(benzoylacetonate)_3$ and aminoalcohols such as D-α-phenylglycinol.[282]

The thermodynamic parameters for association between pyridine or dimethyl sulfoxide and lanthanide β-diketonates have been determined by titration calorimetry. In a valuable paper,[283] complexation was investigated between $M(dpm)_3$, where M = Eu or Yb, or $M(fod)_3$, where M = Eu or Pr, and pyridine, 2-, 3- or 4-picoline, 2,4,6-trimethylpyridine, or Me_2SO. Typical results are shown in Table 4. Various points are exemplified here. The more crowded $Yb(dpm)_3$ complex gives only a mono adduct, while the two fod complexes coordinate with two molecules of ligand. The inductive effect of the 4-methyl group is just discernible, but the steric effects of the 2-methyl or 2,6-dimethyl groups are plainly evident. K_2 is, as is normal, not greater than K_1, thus contrasting with the results[281] with *n*-propylamine. Finally, the *K* values for DMSO are roughly comparable with those of the unhindered pyridines.

Table 4 Interaction Between Lanthanide β-Diketonates and Ligands in Benzene[283]

		log K^a		$-\Delta H^\circ$ (kJ mol^{-1})		$-\Delta G^\circ$ (kJ mol^{-1})	
$Yb(dpm)_3$	py	3.14		22.7		18.2	
	4-pic	3.27		22.8		18.9	
	2-pic	2.16		18.0		12.5	
		log K_1	log K_2	$-\Delta H^\circ_1$	$-\Delta H^\circ_2$	$-\Delta G^\circ_1$	$-\Delta G^\circ_2$
$Pr(fod)_3$	py	3.3	3.0	24	9	19.1	17.4
	4-pic	4.3	3.0	24	13	24.9	17.4
	2-pic	3.0	2.0	11	20	17.4	11.6
		log K_1	log K_2				
$Eu(fod)_3$	py	4.0	2.7				
	DMSO	3.3	3.3				
	2,4,6-Me_3py	1.0	0.5				

^{a} *K* in dm^3 mol^{-1}.

Ten-coordination has ben claimed for $M(fod)_3(bipy)(H_2O)_2$, where M = Y, La, Pr, Eu or Lu, but the evidence (thermogravimetric analysis) is slender and a coordination number of eight as $[M(fod)_3bipy]\cdot 2H_2O$ seems more likely, in line with the complexes $M(fod)_3$ phen where M = Y, La, Pr, Yb, or Lu, prepared at the same time.[284] The electronic spectra were also studied and related to covalency parameters and the NMR spectra were discussed, but as might be expected, the complexes show no ability to act as shift reagents as they are already too crowded.

As in the case of most classes of lanthanide complexes, an X-ray structure determination is necessary to give convincing proof of structure and configuration, and many lanthanide β-diketonate adducts with amines and other ligands have been subjects of such reports. In general, the method of preparation of the adducts is straightforward, namely an addition reaction in an inert solvent or in excess ligand, followed by removal of solvent and recrystallization. Coordination numbers are usually seven or eight, depending on ionic radius, bulk of β-diketone and bulk of added ligand, but can rise to nine. Thus for $[Eu(acac)_3phen]$ where the β-diketone is unhindered, the lanthanide is reasonably large, and the phenanthroline is fairly undemanding sterically, eight-coordination is achieved,[285] with a rather distorted square antiprismatic configuration having Eu—O at 2.356–2.409 and Eu–N = 2.643 Å (average). The molecule has approximate C_s symmetry, which is consistent with the number of lines observed in the fluorescence spectrum. Eight-coordination can still be achieved even with the bulky dpm ligand. Thus both $[Eu(dpm)_2py_2]$,[286,287] and $[Ho(dpm)_3(4\text{-picoline})_2]$[288] have very similar structures. The former has near C_2 symmetry with a distorted square antiprismatic, nearly dodecahedral, structure with Eu—O = 2.312–2.372 and Eu—N = 2.649 Å (average). With the very bulky ligand, quinuclidine (**6**), only seven-coordination is achieved[289] in $[Eu(dpm)_3(quin)]$. Here the symmetry interestingly is almost C_3 with a capped octahedral configuration in which the three quinuclidine strands bisect the positions of the set of three oxygen atoms closest to the amine. The interatomic distances are Eu—O = 2.33 (average) and Eu—N = 2.60 Å (Figure 4). The molecular configuration is thus similar to that of

[Ho(PhCOCHCOPh)$_3$H$_2$O] mentioned above. In the case of the smallest lanthanides, coordination may be restricted to one ligand even when it is quite small. Thus [Lu(dpm)$_3$(3-picoline)] has a seven-coordinated, distorted, side-face-capped trigonal prismatic configuration.[290] An oxygen is the capping atom. Interatomic distances are Lu—O = 2.24 (average) and Lu—N = 2.49 Å.

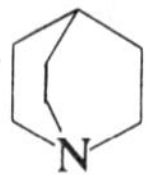

(6)

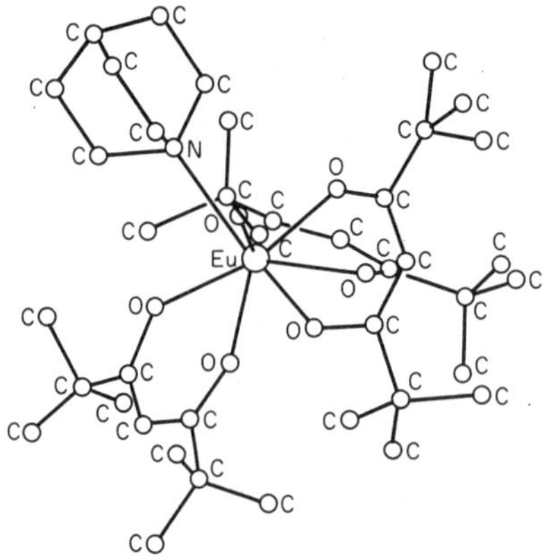

Figure 4 Structure of [Eu(Me$_3$CCOCHCOCMe$_3$)$_3$ (quinuclidine)] (reproduced with permission from ref. 289)

Among oxygen donors, the groups C=O, P=O and S=O are effective. The adduct [Pr$_2$(facam)$_6$(dmf)$_3$] provides an interesting example of nine-coordination.[291] Here facam is 3-trifluoroacetyl-(+)-camphor (**7**) a bulky ligand but one in which the main bulk is located at some distance from the metal ion. The three nitrogen atoms from the dimethylformamide ligands act as symmetrically bridging donors to both Pr^{3+} ions, giving distorted capped square antiprismatic coordination. Interatomic distances are Pr—O(diketone) = 2.46 Å; Pr—O(dmf) = 2.60 Å. The bridging angle Pr—O—Pr is 103.6°.

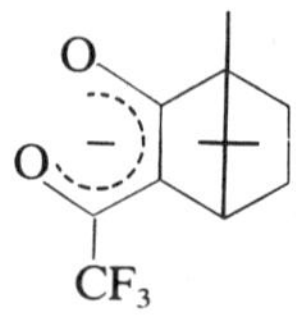

(7)

The phosphine oxide adduct [Pr(dpm)$_3$(OPPh$_3$)] is of interest as although Pr^{3+} is a large ion, and Ph$_3$PO is not perhaps too sterically demanding except at a distance from the ion, only a seven-coordinated mono adduct is obtained. The X-ray structure[292] shows Pr—O(diketone) = 2.30–2.39 and Pr—O(phosphoryl) = 2.35 Å. The comparatively short Pr–phosphoryl link is to be noted (compare Section 39.2.4.4) and would increase repulsion between the ligating atoms. The seven-coordination is in agreement with the solution studies mentioned above.[283]

The X-ray crystal structure[293] of a sulfoxide adduct, [Eu(dpm)$_3$L], where L is the cyclic sulfoxide (**8**), shows irregular seven-coordination with Eu—O(diketone) = 2.33 (average) and Eu—O(sulfoxide) = 2.40 Å; the sulfoxide group, like the phosphine oxide group just previously mentioned, forms a short, and presumably strong, bond.

(8)

In the case of β-diketones which are not too sterically demanding, the $[M(\beta\text{-diketonate})_4]^-$ anion may be obtained. An example is $NH_4[M(CF_3COCHCOCF_3)_4]$, where M = Y, Nd, Gd or Er. Preparation is by treatment of a benzene solution of $[M(CF_3COCHCOCF_3)_3]$ with $CF_3COCH_2COCF_3$ (1 mol) and gaseous ammonia.[294] The (rhodamine 6G)$^+$ salts of $[EuL_4]^-$, where L is benzoylacetonate or thenoyltrifluoroacetonate, have also been prepared and their fluorescence spectra studied.[295]

Europium, and to a lesser extent terbium, complexes of β-diketones have been studied in solution and in the solid state by means of their fluorescence (luminescence) spectra. As explained further in Section 39.2.10, it is possible to relate the splitting of the $^5D_o \rightarrow {}^7F_n$ transitions of Eu^{3+} to the symmetry of the emitting complex, and studies of circularly polarized luminescence (CPL) spectra can give related information. Thus a study[296] of $EuCl_3$ and complexes of Eu^{3+} with hexafluoroacetylacetone and four other β-diketones in methanol or DMF showed that while $EuCl_3$ itself had axial symmetry in solution, the complexes had orthorhombic symmetry. The emission spectra of solutions of adducts of $Eu(dpm)_3$ with Ph_3PO or borneol have been studied at low temperatures where conformal lability is reduced; the Ph_3PO adduct has uniaxial symmetry but the bulky, less symmetrical borneol molecule confers lower symmetry on its adducts.[297]

Russian workers have been active recently in this field. Thus [Eu(thenoyltrifluoroacetate)$_3$(phen)] has been shown to retain its structure in polystyrene or PVC;[298] the second order crystal field parameters B_2^0 and B_2^2 have been obtained for a series of complexes $[EuL_3(phen)]$ where L was a series of eight β-diketonates,[299] and the temperature broadening has been investigated, for similar adducts and for $EuL_2(NO_3)(Ph_3PO)_2$, of the Stark components of the ^7F_J levels.[300]

39.2.4.3 Amine oxides

These compounds have been known for some time, having been described in 1964 by Melby *et al.*[301] who investigated the fluorescence spectrum of $Eu(tfa)_3$(4-picoline *N*-oxide)$_2$. Pyridine oxides bond firmly to lanthanides, the strength of the interaction being doubtless associated with the partial charge on the oxygen atom, and a considerable number of papers have appeared in which the preparation and general properties of pyridine oxide or substituted pyridine oxide complexes with lanthanide salts, often with poorly coordinating anions such as perchlorate, methylsulfate or hexafluorophosphate, but also with halides or nitrates, have been studied.

Before describing these, it is of interest to survey X-ray crystallographic data on this type of complex. The compounds $M(ClO_4)_3(pyo)_8$, where M = La or Nd and pyo = pyridine *N*-oxide, are both dimorphic, existing in a $P2_1/a$ and a $C2/c$ modification.[302] The first modification consisted of twinned crystals only, but in the case of the Nd compound the crystal structure was obtained despite this. The $C2/c$ version of the lanthanum compound was also submitted to structure determination. Both compounds are formulated $[M(pyo)_8](ClO_4)_3$ with all *N*-oxide atoms coordinated, the leading dimensions being: La—O = 2.493–2.500; Nd—O = 2.356–2.447 Å; La—O—N = 127.7–136.3°; Nd—O—N = 129.0–133.3°. The Nd compound has only C_1 symmetry but very nearly S_8, with the coordination polyhedron very nearly a square antiprism. The La compound had C_2 symmetry, with near D_4 symmetry, the polyhedron being distorted from a square antiprism towards a cube by rotation of the four upper pyridine oxide molecules around the 'D_4' axis some way towards a cube configuration. The La—O—N linkage seems sterically unconstrained and there is little doubt that this is a natural bonding angle of about 132° reflecting localized lanthanide–lone pair interaction.

In the complex $[La(bpyo)_4](ClO_4)_3$, where bpyo = 2,2'-bipyridyl *N,N'*-dioxide, the configuration is cubic, with one bidentate ligand spanning each vertical edge. The complex ion has symmetry C_2, but is very near to D_4. The two rings in each ligand are twisted relative to each other at a dihedral angle of 61.4° average. The cube is slightly compressed along the 'D_4' axis, the angle between this axis and the La—O vectors being 56.1° average (compare 54.7° required for the cube). Leading dimensions are La—O = 2.495–2.512 Å; La—O—N = 121.7–124.3°.[303]

The lanthanide trifluoromethylsulfonates $M(CF_3SO_3)_3$ have been prepared and their complexes with pyo and bpyo have been prepared in ethanol or ethanol/triethylformate and examined. Complexes $[M(pyo)_8](CF_3SO_3)_3$ and $[M(bpyo)_4](CF_3SO_3)_3$ (M = La, Nd, Eu or Ho in each case) were obtained and results indicated that the anions were not coordinated.[304,305] A number of complexes of pyo, bpyo and 2,2',6',2''-terpyridyl-*N,N',N''*-trioxide (tpyo) with lanthanide trifluoromethylsulfonates, perchlorates, nitrates and thiocyanates have been isolated from ethanolic solution and their constitution and emission spectra studied. The complexes are as follows: $M(NO_3)_3(pyo)_8$ (M = La–Ho); $M(NCS)_3(pyo)_4(H_2O)_2$ (M = La–Tb); $M(NCS)_3(pyo)_3(H_2O)_2$ (M = Ho–Tm); $M(NO_3)_3(bpyo)_2$ (M = La–Tm); $M(NCS)_3(bpyo)_2(H_2O)_n$ (M = La–Eu, $n = 1$; M = Gd–Er, $n = 2$); $M(ClO_4)_3(tpyo)_3(H_2O)_2$ (M = La–Tb); $M(NO_3)_3(tpyo)(H_2O)$ (M = La–Ho); and $M(CF_3SO_3)_3(tpyo)_3(H_2O)_2$ (M = La–Tb). Conductivity studies in MeCN established the constitution of some of these, *e.g.* $[M(pyo)_8](NO_3)_3$, or $[M(NO_3)_3(bpyo)_2]$ (earlier lanthanides) and $[M(NO_3)_2(bpyo)_2](NO_3)$ (later lanthanides), but where water was present its status was uncertain. The bpyo and tpyo ligands remained coordinated in aqueous solution. The intensity of the hypersensitive absorption bands is very considerably increased by these ligands; thus the $^5I_8 \rightarrow {}^5G_6$ Ho^{3+} band is increased four-fold by bpyo in aqueous solution. The magnitude of this effect increases in the series pyo < tpyo < bpyo. The emission spectra of the Tb and Eu complexes were studied and it was concluded that the octa pyo complexes have high symmetry.[306–308]

Other complexes of pyo with chloride or bromide have been studied.[309,310] Products were examined by IR, emission spectra, NMR and conductance. Compounds isolated include $[MCl_3(pyo)_8]$ (M = Y, La, Pr, Nd, Sm or Er); $[MBr_2(pyo)_8]Br$ (M = La, Pr, Nd, Sm, Tb or Er); and $[MBr_2(pyo)_6(H_2O)_2]Br$ (M = Y, La, Pr, Nd, Sm, Gd, Dy or Yb).

Complexes of 2-, 3- and 4-picoline oxides have been made in large numbers. These include $[M(2po)_7](PF_6)_3$ (M = La–Sm) and $[M(2po)_6](PF_6)_3$ (M = Sm–Lu or Y; 2po = 2-picoline oxide).[311] This seems a good example of reduction of coordination number with ionic radius of metal ion. Another series showing this effect is $[M(NCS)_3(2po)_x]$ (M = La–Pr, $x = 5$; M = Nd–Eu, $x = 4$; M = Eu–Lu, Y; $x = 3$). Powder X-ray photographs show three series and all are non-conductors. In the IR, $\nu(NO)$ is reduced from 1240 cm^{-1} to 1220–1240 cm^{-1} and $\delta(NO)$ increased from 840 cm^{-1} to 845–850 cm^{-1}.[312] Other series are $M(MeSO_3)_3(2po)_x$ (M = Ce–Sm, $x = 2$; M = La–Lu or Y, $x = 4$);[313,314] $[M(NO_3)_3(2po)_3]$ (M = Nd, Sm, Tb, Dy, Yb);[315,316] $[MCl_3(2po)_x]$ (M = La, Pr, $x = 4$; M = Nd, Sm, Tb, Er or Y; $x = 5$;[316] $M(ClO_4)_3(2po)_7(H_2O)_x$ (M = La–Lu, Y; $x = 3$ or 2);[317] and $MCl_3(2po)_3(H_2O)_2$ (M = La–Ho), where IR and conductivity (MeOH) suggested a formulation $[MCl_2(2po)_3(H_2O)_2]Cl$.[318] On the whole, these coordination numbers tend to be lower than for the unsubstituted pyridine *N*-oxide. In the case of the oxides 3-picoline and 4-picoline, steric effects should however be less pronounced. Series isolated with the 3-picoline oxide ligand include $[M(3po)_8](PF_6)_3$ (M = La–Lu, Y);[319] $[M(3po)_8](ClO_4)_3$ (M = La–Yb (some) or Y)[320] and $MI_3(3po)_8 \cdot (H_2O)_x$ (M = La–Yb (some), Y; $x = 0$, 1 or 2). In the last series, NMR results indicated restricted rotation of the pyridine ring in the Yb compound.[321] This has also been observed in $[MI(2po)_5]I_2$.[322] Series with 4-picoline oxide as ligand include $MBr_3(4po)_8$ (M = La–Er (some), Y)[323] and $MI_3(4po)_8$ {M = La–Yb (some)}.[322]

39.2.4.4 Phosphine and arsine oxides and related ligands

Lanthanide ions complex more readily with phosphine or arsine oxides and other ligands containing the P=O or As=O groups than with most other uncharged monodentate ligands. Doubtless much of this affinity can be ascribed to the negative charge on the oxygen atom produced by the canonical form $P^+–O^-$, which confers an electrostatic component on the cation ligand bond energy. The fact that trivalent lanthanide cations formed stable isolable complexes with phosphine or arsine oxides was fairly early recognized. Thus the complex $EuCl_3(OPPh_3)_3$ was reported[324] in 1964 and its fluorescence spectrum examined, while a wide variety of phosphine and arsine oxide complexes were reported shortly afterwards.[325] These were prepared in ethanol or acetone solution and included $M(NO_3)_3L_2(EtOH)$, where M = Y or La–Lu and L = $OPPh_3$ or $OAsPh_3$; $M(NO_3)_3L_3$, where M = Y or La—Lu and L = $OPPh_3$ or where M = Y or La–Yb and L = $OAsPh_3$. Other complete series included $M(NO_3)_3(OPPh_3)_3(Me_2CO)_2$ and $M(NO_3)_3(OAsPh_3)_4$, and other complexes prepared included $M(NO_3)_3(OPPh_3)_4(EtOH)_x$ where $x = 0$ and 1, M = Y and Tb–Lu. The thiocyanates $M(NCS)_3L_4$, where L = $OPPh_3$ or $OAsPh_3$, $M(NCS)_3(OPPh_3)_3$, and the chlorides $MCl_3(OPPh_3)_x$ where M = Nd or Sm and $x = 3$ or 4 have also been prepared and characterized.[326–328] Some of these were soluble and monomeric in organic solvents, *e.g.* $SmCl_3(OPPh_3)_4$ which should thus be seven-coordinate, but others contained ionic nitrate, *e.g.*

$[Eu(NO_3)_2(OAsPh_3)_4]NO_3$, as shown by conductivity and IR data. There is a characteristic lowering of the PO IR stretching frequency on coordination, from 1195 cm^{-1} to about 1160 cm^{-1}, while the AsO frequency is raised on coordination, from 878 cm^{-1} in the free ligand to about 910 cm^{-1}. As coordination is expected to strengthen the PO and AsO σ bonds inductively and weaken the corresponding π bonds inductively, this result is in accord with the expected greater importance of the π bond in PO as compared with AsO.[327] For both ligands the stretching frequency increases along the series La→Lu. Many of the complexes of different stoichiometries could be interconverted by heat or action of solvents or addition of further ligand. Some apparently rather similar complexes have been reported in the Russian literature.[329] These comprised $M(NCS)_3\{OP(i\text{-}C_5H_{11})_3\}_4$, where M = La–Sm; $M(NCS)_3\{OP(i\text{-}C_5H_{11})_3\}_3$, where M = Y, Sm–Yb; and $M(NO_3)_3\{OP(i\text{-}C_5H_{11})_3\}_3$, where M = Y, La, Nd, Sm, Gd, Er, Yb, Lu. Other work on simple complexes of phosphine oxides with lanthanide salts is not extensive but includes the preparation by a solvent extraction method of MX_3L_3, where M = La–Dy and Er, X = NO_3, Cl or NCS, and L = $OP(n\text{-}C_4H_9)_3$ or $OP(n\text{-}C_8H_{17})_3$.[330] Complexes of Y and lanthanide perchlorates of the stoichiometry $M(ClO_4)_3\{OP(n\text{-}C_4H_9)_3\}_4$ have also been reported,[331] as have the complexes $Nd(NO_3)_3(OPR_3)$, where $R_3 = Ph_3$, Ph_2(benzyl), Ph(benzyl)$_2$, (benzyl)$_3$, Et_3 or Me_3.[332]

Phosphine oxides are effective ligands in forming adducts with preexisting lanthanide complexes, particularly β-diketonates. The crystal structure of the 2:1 adduct $[Nd(ttf)_3(OPPh_3)_2]$, where Httf = thenoyltrifluoroacetone, shows an eight-coordinated dodecahedral monomer. The Nd—O(P) distances are 2.399 and 2.423 Å and the Nd—O(C) distances lie in the range 2.40–2.50 Å. The coordination polyhedron is very closely dodecahedral.[333] In contrast to this 2:1 adduct, $M(dbm)_3$, where Hbdm = dibenzoylmethane, give 1:1 adducts $M(dbm)_3L$, where L = OPR_3 and R = Ph, $n\text{-}C_8H_{17}$ or OBu. These were prepared in alcohol and isolated, and there was evidence that the stability of the adducts increased in the sequence $R = C_8H_{17} < Ph < OBu$.[334] Other papers have also described the preparation of other adducts of phosphine oxides with lanthanide β-diketonates.[335,336]

Phosphine oxides are found to exert a synergic effect on the extraction of lanthanides from aqueous into organic phases by means of β-diketones or other ligands; doubtless the replacement of coordinated water in $M(diket)_3(H_2O)_x$ by a trialkylphosphine oxide would have a favourable effect on solubility in the non-aqueous phase. In an investigation of the effect of $OP(n\text{-}C_8H_{17})_3$ on extraction of Tm^{3+} ions from an aqueous phase using thenoyltrifluoroacetone at a variable temperature, values of ΔH and ΔS for the equilibria involved were obtained and mechanisms were discussed.[337] The synergic effect of $OP(C_8H_{17})_3$ on extraction of lanthanide ions by means of 1-phenyl-3-methyl-4-acyl-5-pyrazolones has been also studied and extraction constants and separation factors determined.[338]

Adducts of triphenylphosphine oxide with complexes of the chelate sulfur-coordinating ligand $S_2P(OEt)_2^-$ have been investigated and X-ray structures obtained. Reaction of Ph_3PO with $[M\{S_2P(OEt)_2\}_4]^-$ gave, for M = La–Pr, $[M\{S_2P(OEt)_2\}_3(OPPh_3)_2]$, and where M = Y or Nd–Lu, $[M\{S_2P(OEt)_2\}_2(OPPh_3)_3]^+$ was produced. The lanthanum complex is eight-coordinated and approximately square antiprismatic, with La—O = 2.456 or 2.422 Å, O—La—O = 152.8°, and La—O—P = 168.2° or 165.7°. The longer La—S interatomic distances lie within the range 2.982–3.092 Å. In the case of the Sm complex, an example of the group of smaller cations where two chelate dithiophosphate ligands are displaced by three monodentate phosphine oxides, a very closely pentagonal bipyramidal configuration is adopted with the Ph_3PO occupying both axial positions and one equatorial. Here Sm—O = 2.28–2.32 Å and Sm—O—P = 171.2–175.5°.[339]

Another class of adducts where there are structural data is formed by adducts of complexes of the bulky amido anion $N(SiMe_3)_2^-$. The action of Ph_3PO (1.5 molar proportions) on $[M\{N(SiMe_3)_2\}_3]$, where M = La, Eu or Lu, gave $[M\{N(SiMe_3)_2\}_3OPPh_3]$. The X-ray structure of the La complex showed an approximately tetrahedral geometry, with N—La—N and O—La—N in the range 104.8° to 116.4°. La—O at 2.40 Å was approximately equal to La—N at 2.38–2.41 Å, while La—O—P was not quite colinear, at 174.6°. Curiously, treatment of $[M\{N(SiMe_3)_2\}_3]$, where M = La, Pr, Sm or Eu, with $OPPh_3$ (2–5 molar proportions) gave peroxo complexes $[M_2\{N(SiMe_3)_2\}_4(O_2)(OPPh_3)_2]$, despite rigorous exclusion of oxygen and preparation of the phosphine oxide from Ph_3PBr_2 rather than treatment of Ph_3P with H_2O_2. In the La complex, two pyramidal $La\{N(SiMe_3)_2\}_2(OPPh_3)$ entities are bridged by a peroxo group, such that O—O is at right angles to La—La, while La_2 and O_2 are coplanar.[174,175] Interatomic distances and angles were given in Section 39.2.2.3.

(*i*) *Other ligands containing the phosphoryl group*

Hexamethylphosphoramide $OP(NMe_2)_3$ forms two series of complexes with lanthanide perchlorates. Thus complexes $[M\{OP(NMe_2)_3\}_6](ClO_4)_3$, where M = La, Nd, Gd, Ho or Yb, were obtained from acetonitrile solution, IR spectra showing that the perchlorate anions remained uncoordinated.[340] These complexes had also been studied previously.[341,342] From ethanolic solution, complexes $M(ClO_4)_3\{OP(NMe_2)_3\}_4$, where M = La, Nd, Gd, Ho or Yb, were obtained.[340] The PO IR stretching frequency was reduced by about 17 cm^{-1} compared with the free ligand. The cation in $[Nd\{OP(NMe_2)_3\}_6](ClO_4)_3$ has been shown by X-ray crystallography to be octahedrally coordinated, with Nd—O at 2.37 Å.[343]

The hexamethylphosphoramide perchlorates have been studied in acetonitrile solution by means of ^{139}La NMR spectrometry.[344] This nucleus has an electric quadrupole and therefore the line width of the ^{139}La resonance is much broadened by an unsymmetrical coordination geometry (which produces an electric field gradient at the nucleus). Accordingly, the coordination of added $OP(NMe_2)_3$ can be observed to alter the ^{139}La line-width from a broadened resonance characteristic of unsymmetrical cationic species $[La\{OP(NMe_2)_3\}_x(MeCN)_y]^{3+}$ where $x < 6$, to a narrow resonance characteristic of $x = 6$, $y = 0$. This latter species exchanges ligand only slowly on the NMR time scale with the other species present in solution, which suggests that the La—O bond may be quite strong.

Structures of an interesting pair of hexamethylphosphoramide complexes[345] where the coordinated anion is nitrosodicyanomethanide [X, O—N=C(CN)₂] have been obtained. Thus $[NdX_3\{OP(NMe_2)_3\}_4]$ is a seven-coordinated distorted capped octahedron, where X is monodentate *via* oxygen, while $[YbX_2\{OP(NMe_2)_3\}_4]^+$ is nearly *trans* octahedral with Yb—O (hmpa) = 2.197 Å average.

Some complex chlorides $[MCl_3\{OP(NMe_2)_3\}_3]$ have been described[346] and the X-ray crystal structure of the Pr complex has been reported.[347] It is very closely octahedral, with O—P—O, Cl—P—Cl and Cl—P—O in the range 82.6–96.6°. Interatomic distances include Pr—Cl at 2.706–2.733 Å, Pr—O at 2.351–2.356 Å and P—O at 1.483–1.493 Å; the coordination is meridional. Some bromides $MBr_3\{OP(NMe_2)_3\}_4$ have been described,[348] as have a number of nitrates.[349,350] Two series of trifluoromethanesulfonates have been obtained;[351] for La—Eu, the hexakis complexes $[M\{OP(NMe_2)_3\}_6](CF_3SO_3)_3$, and for Y and Nd–Lu, the tetrakis complexes $[M(CF_3SO_3)\{OP(NMe_2)_3\}_4](CF_3SO_3)_2$. The formulations shown here are based on conductivity values in acetonitrile appropriate to 3:1 and 2:1 electrolytes respectively. Fluorescence spectra of europium hexamethylphosphoramide complexes have been discussed in this paper and another[352] and their relationship to the symmetry of the emitting species discussed.

The anion $(EtO)_2PO_2^-$ forms the complex $Nd\{EtO)_2PO_2\}_3$ on treatment of anhydrous $NdCl_3$ with $OP(OEt)_3$ in non-aqueous solution. X-Ray analysis shows a linear chain of slightly distorted octahedrally coordinated Nd^{3+} ions, with Nd—O = 2.40 Å average. Each pair of adjacent Nd^{3+} ions is bridged by three anions, and each anion bridges thus: Nd—O—P(OEt)₂—O—Nd.[353]

Complexes of some bidentate or terdentate phosphoryl ligands have been studied. The tetraphenylimidodiphosphate anion $\{(PhO)_2P(O)\}_2N^-$ (= L) acts as a bidentate ligand towards lanthanide ions, giving complexes ML_3.[354] The Yb complex, studied by X-ray crystallography shows coplanar Yb—O—P—N—P—O chelate rings, the three ligands giving almost octahedral coordination with nearly D_3 symmetry. The coordinated atoms are well separated (interligand O—O = 3.07–3.39 Å, intraligand O—O = 2.883 Å). The angle Yb—O—P, at 135.4–137.7°, is constrained to be smaller than in monodentate complexes, while Yb—O = 2.199–2.219 and P—O = 1.490–1.496 Å.[355] Diisopropyl-*N,N*-diethylcarbamylmethylenephosphonate (L) $Pr^i_2P(O)CH_2C(O)NEt_2$, which is an extractant agent for lanthanide ions from aqueous acid into hydrocarbons,[356] gives complexes $M(NO_3)_3L_2$, where M = La–Gd, and $M(NO_3)_3L_2(H_2O)$, where M = Tb–Lu.[357] The structures of these complexes have been investigated by X-ray diffraction, which shows a 10-coordinated monomeric structure $[Sm(NO_3)_3L_2]$ with L being coordinated by the phosphoryl (Sm—O = 2.418 Å) and carbonyl (Sm—O = 2.433 Å) oxygen atoms, and the nitrate groups are bidentate. The other complexes, exemplified by Er, are nine-coordinated $[Er(NO_3)_3L_2(H_2O)]$. Here, L is monodentate only, being coordinated by the phosphoryl oxygen (Er—O = 2.275 Å). The carbonyl oxygen is hydrogen bonded to the coordinated water (Er—OH₂ = 2.302 Å). The coordination geometry is nearly a face-centred trigonal prism.[358] The extraction chemistry of the dihexyl analogue of L has been studied and monodentate tris coordination in the organic phase was suggested.[359]

The bidentate ligand nonamethylimidodiphosphoramide $(Me_2N)_2P(O)N(Me)P(O)(NMe_2)_2$

(= L) forms a series of complexes $[ML_3](ClO_4)_3$, where M = La–Lu, which are X-ray isomorphous and in which the PO stretching frequency is reduced from 1205 cm^{-1} in the free ligand to 1170 cm^{-1}, while the PO bend increases in frequency by 25 cm^{-1}. The IR spectra and the conductivity in nitromethane support the formulation as a six-coordinated tris-chelate cation.[360] Other ligands, rather similar to that just mentioned, whose lanthanide complexes have been investigated are $(Me_2N)_2P(O)OP(O)(NMe_2)_2$, octamethylpyrophosphoramide;[361] $(Pr^iO)_2P(O)CH_2P(O)(Pr^iO)_2$, tetraisopropylmethylenediphosphonate;[362] and $(Pr^iO)_2P(O)$-$C_2H_4P(O)(Pr^iO)$, tetraisopropylethylenediphosphonate.[363]

Complexes of the interesting terdentate ligand $(Me_2N)_2P(O)N(Me)(P(O)(NMe_2)$-$N(Me)P(O)(NMe_2)_2$, bis(methylimido)triphosphoric acid (pentakisdimethylamide) (L) have been prepared, and investigated by NMR. The complexes are $[ML_3](ClO_4)_3$, where M = Y, or La–Lu, this presumably nine-coordinated structure being supported by the IR spectra. The NMR spectra show that there is (on the NMR time scale) no intermolecular ligand exchange, which is hardly surprising since we are dealing with a presumably terdentate ligand bonding *via* its powerful phosphoryl groups. However, there is evidence for intramolecular conformational rearrangements of some complexity. The very small ^1H paramagnetic shifts, up to 0.35 p.p.m., suggest that a very symmetrical coordination polyhedron is adopted; however, the ^{31}P shifts are quite large, up to 111 p.p.m., and it is suggested that a contact mechanism is responsible for these.[364]

Mono and bis complexes of Pr^{3+} and Nd^{3+} in aqueous solutions of iminodiaceticmonomethylenephosphonic acid have been studied over the pH range 1–12 by visible spectroscopy and association constant values determined.[365]

A short review of lanthanide phosphoryl complexes has appeared, which deals particularly with recent Russian work.[366]

39.2.4.5 Other oxygen donors

The lanthanides complex readily with the carboxylate anion and a variety of anhydrous and hydrated carboxylates have been described; most are hydrated. Preparation of the solid hydrated compounds is usually straightforward from aqueous solution, but long chain aliphatic anhydrous carboxylates may be prepared from organic solvents, either directly or by ligand exchange with the acetate.[367] A number of X-ray single crystal determinations are available, and these often show a polymeric structure. Thus the structures of hydrated and anhydrous oxalates show chelation of oxalate ion forming five-membered rings, not four-membered rings, with the lanthanide ions in nine-coordinated polymeric $\{M(ox)_3H_2O\}\cdot 2H_2O$, where M = Nd–Tb,[368] and in $K_8M_2(ox)_7\cdot 14H_2O$, where M = Y, Tb, Dy, Er or Yb.[369] In both these compounds M—(oxalate)—M bridges are present, where one anion forms two five-membered rings with two lanthanide ions. The bridging capability of the carboxylate ligands extends into solutions. Thus citric acid (H_4L) forms heterocationic complexes $GdM(H_2A)_4^{2-}$, $GdM(H_2A)(HA)_2^{2-}$ and $GdM(HA)_3^{3-}$, where M = Tb–Tm. These complexes were established by Job-type plots of relaxation times, which showed $1/T_1$ to maximize at a 1:1 Gd:M ratio.[370]

In the lanthanide tris-glycolates $M(O_2CCH_2OH)_3$, the glycolate anion chelates *via* one carboxylate and one hydroxyl oxygen, and also forms carboxylate bridges to neighbouring cations.[371–373] The diglycolates form anions, as in the salts $Na_3[M(O_2CCH_2OCH_2OCH_2CO_2)_3]$-$2NaClO_4\cdot 6H_2O$, where the cations are nine-coordinated in an approximately D_3 configuration, each diglycolate anion being tridentate.[374] The ether–lanthanide bond distances are slightly longer than those of the carboxylates, thus Nd—O = 2.523 and 2.428 Å. The chloroanilate anion (*ca*, **9**) coordinates to two separate lanthanide ions, each in a bidentate manner leading to a three-dimensional network. In $Pr_2(ca)_3\cdot 8EtOH$ each Pr^{3+} ion is nine-coordinated with three ca and three EtOH (a relatively rare example of an ethanol complex of known structure) with Pr—O(ca) = 2.483 (average) and Pr—O(EtOH) = 2.553 Å (average).[375]

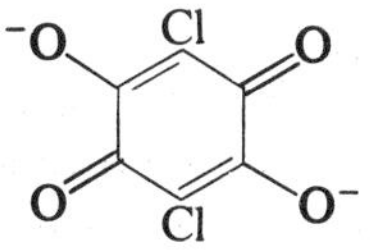

(9)

There are many complexes of lanthanides with organic ligands which bond by means of a carbonyl oxygen atom. Representative examples are $M(Me_2NCOMe)_n(ClO_4)_3$, where M = La–Nd, $n = 8$ and M = Sm–Er or Y, $n = 7$ and M = Tm–Lu, $n = 6$ (this is a progression doubtless due to decreasing ionic radius);[376] $M\{CO(NMe_2)_2\}_6(ClO_4)_3$, where M = La–Lu or Y;[377] $M(NO_3)_3\{CO(NMe_2)_2\}_3$, where M = La–Lu or Y;[378] and $M(N,N'\text{-ethyleneurea})_n(NO_3)_3$, where M = La or Ce, $n = 8$ and M = Ho or Tm, $n = 6$.[379] An extensive solution study has been made[380] of the equilibrium between $[M(DMF)_n](ClO_4)_3$, where $n = 8$ or $n = 9$ in DMF solution. Variable temperature and pressure studies of absorption spectra and NMR showed that $n = 8$ is the major species for Ce–Nd and the only species for Tb—Lu. For Nd, equilibrium parameters are $\Delta H^\ominus = -14.9 \pm 1.3\ \text{kJ mol}^{-1}$; $\Delta S^\ominus = -69.1 \pm 4.2\ \text{J k}^{-1}\ \text{mol}^{-1}$, and $\Delta V^\ominus = -9.8 \pm 1.1\ \text{cm}^3\ \text{mol}^{-1}$. The kinetic parameters for DMF exchange in $[Tb(DMF)_8]^{3+}$ were $\Delta H^\ddagger = 14.1 \pm 0.4\ \text{kJ mol}^{-1}$, $\Delta S^\ddagger = -58 \pm 2\ \text{J mol}^{-1}\ \text{K}^{-1}$ and $\Delta V^\ddagger = 5.2 \pm 0.2\ \text{cm}^3\ \text{mol}^{-1}$. In solid $[M(DMF)_8](ClO_4)_3$ there is eight-coordination for M = La–Lu and NMR integration showed the species $[Tm(DMF)7.7 \pm 0.2]^{3+}$ in $Tm(ClO_4)_3$ dissolved in DMF/CD_2Cl_2.[381]

The oxyanions OH^-, SO_4^{2-} and NO_3^- coordinate strongly. The two first named have a considerable tendency to give polymeric structures. A rare single crystal X-ray determination is that of $Na_4Yb(OH)_7 \cdot 2.5H_2O$, mentioned in Section 39.2.4.1. It shows octahedral coordination of the Yb^{3+} ion.[251]

Sulfato complexes have been included in a recent review.[382] To exemplify sulfate structures, the hydrated sulfate $Ce_2(SO_4)_3 \cdot 9H_2O$ contains, in addition to nine-coordinated $Ce(H_2O)_6(SO_4)_3$ groups, Ce^{3+} ions twelve-coordinated to sulfate groups.[383] A number of other sulfates have been structurally investigated, and coordination by monodentate and bidentate sulfate groups is observed. Thus $Nd_2(SO_4)_3 \cdot 5H_2O$ shows Nd^{3+} in pairs, coordinated in a very distorted tricapped trigonal prismatic manner to two H_2O, and six or seven SO_4^{2-}. Two SO_4^{2-} are bridging groups, monodentate or bidentate to each Nd^{3+} ion in the pair, and the pairs are further linked in a three-dimensional network.[384]

Nitrate complexes, including hydrated nitrates, have been well studied. This ion is almost always bidentate towards a lanthanide ion, and discrete complexes or complex anions are usually formed, in contrast to the sulfate complexes mentioned above. The complex anions $[M(NO_3)_5]^{2-}$, where M = Ce, Eu, Ho or Er, have had their structures reported and are pseudotrigonal bipyramidal. The M—O distances are Ce—O = 2.553–2.591 Å and Er—O = 2.392–2.493 Å.[385–8] The $(Hpy)NO_3\text{–}Yb(NO_3)_3$ aqueous system has been studied, and $(Hpy)_2[Yb(NO_3)_5]$ isolated from it and characterized by X-ray powder pattern.[389]

The nine-coordinated uncharged aqua complex $[M(NO_3)_3(H_2O)_3]$ is found in $[Ho(NO_3)_3(H_2O)_3] \cdot 2(4\text{-bipy})$, in which Ho—O(nitrate) = 2.42–2.63 Å and Ho—O(water) = 2.30–2.35 Å,[390] and in $[Gd(NO_3)_3(H_2O)_3]$, in which Gd—O(nitrate) = 2.411–2.533 Å and Gd—O(water) = 2.308–2.459 Å.[391] The slightly larger ion Nd^{3+} will, however, accept either ten-coordination in $[Nd_2(NO_3)_8(H_2O)_4(4\text{-bipy})]^{2-}$ where each Nd^{3+} ion is coordinated to three bidentate and one monodentate NO_3^-, two H_2O and one nitrogen atom of the bipyridyl;[392] or eleven-coordination where all four nitrate ions are bidentate. Another hydrated nitrate is the irregularly ten-coordinated $[Nd(NO_3)_3(H_2O)_4] \cdot 2H_2O$ where Nd—O(nitrate) = 2.547–2.718 Å and Nd—O(water) = 2.429–2.462 Å,[393] while the rare eleven-coordination is observed in $[M(NO_3)_3(H_2O)_5]$ where M = La or Ce.[394,395]

39.2.5 Sulfur Donor Ligands

Sulfur is a comparatively weak donor towards lanthanide ions, but in favourable circumstances coordination does occur. The sulfur should preferably be part of a bidentate or polydentate system, and should preferably carry a negative charge, to judge from the nature of the few compounds of this class which have been isolated.

The bidentate anions $R_2PS_2^-$ give a number of complexes which are fairly stable, soluble in organic solvents, and have been studied by single crystal X-ray and by NMR. Thus treatment of hydrated lanthanide chlorides with salts of $S_2PR_2^-$ (R = Me, C_6H_{11} or OEt) gives complexes $[M(S_2P(C_6H_{11})_2)_3]$ or $M'[M(S_2PR_2)_4]$ (M′ = Na, Ph_4As or Ph_4P; M = La–Lu and Y). The electronic spectra of these complexes show oscillator strengths of three to nine times those of the hydrated ions, and this may be caused by a transfer of intensity from charge-transfer bands.[396,397]

The dithiophosphate complexes react with Ph_3PO to give adducts, whose crystal structures have been determined, of the types $[M\{S_2P(OEt)_2\}_3(Ph_3PO)_2]$ (M = La–Pr) or $[M\{S_2P(OEt)_2\}_2(Ph_3PO)_3]^+$ (M = Nd–Lu and Y) (see Section 39.2.4.4).

The X-ray crystal structures of $[M(S_2P(C_6H_{11})_2)_3]$, where M = Pr or Sm, have been determined. Interatomic distances are Pr—S = 2.804–2.877 Å, and Sm—S = 2.781–2.796 Å. The Pr complex is isostructural with the Nd analogue and the Sm complex with the Eu–Lu and Y analogues.[398] Both these complexes are trigonal prismatic, slightly distorted by a small rotation of all three ligands about the M—P vector. This geometry, as opposed to trigonally distorted octahedral, is necessitated by the small ratio of ligand bite to metal ionic radius. This last point was explored by crystal structure determinations of the Dy and Lu analogues. The angle of twist of the ligands about the M—P vector increases from 18.5–17.3° (Sm) to 19.1–17.8° (Dy) to 20.0–18.6° (Lu) as the bite:M—S distance ratio increases in the sequence 1.203, 1.219, 1.234. In the Lu complex, coordination geometry is best described as a trigonally distorted octahedron.[399]

The tetrakis-dithiophosphinate complex $[PPh_4][Pr(S_2PMe_2)_4]$, whose crystal structure has also been determined, has a distorted tetragonal antiprismatic geometry with Pr—S = 2.888–3.015°.[400] The complex ions $[M\{S_2P(OEt)_2\}_4]^-$, have been the subject of an NMR paramagnetic shift study (see Section 39.2.9.4).

Early among the lanthanide–sulfur complexes were the dithiocarbamates, of which two types $[M(Et_2NCS_2)_3]$ and $[NEt_4][M(Et_2NCS_2)_4]$ were first reported.[401] Later the series $Na[M(Et_2NCS_2)_4]$, where M = La–Yb, was prepared[402] and the X-ray structure of the La example showed distorted dodecahedral coordination with La—S = 2.97 Å average. Charge-transfer bands obtruded considerably into the visible, the Eu compound being brick-red and the Yb bright yellow. The f–f bands showed definite red shifts, and the $f \rightarrow d$ transition of the Ce complex was brought into the visible. All these effects are as usually observed with conjugated or polarizable ligands.

Uncharged thioester sulfur atoms are forced to coordinate with lanthanides in a series of thia-18-crown-6 macrocyclic ligands. The ligands are 1,10-dithia-18-crown-6 (L^2), 1,4,10,13-tetrathia-18-crown-6 (L^4) and hexathia-18-crown-6 (L^6). Complexes obtained were $M(ClO_4)_3(L^2)\cdot H_2O$ (M = La–Pr), $Nd(ClO_4)_3(L^2)$, $Eu(ClO_4)_3(L^4)\cdot H_2O$ and $M(ClO_4)_3(L^6)\cdot H_2O$ (M = Sm, Eu or Yb), together with some similar acetonitrile adducts. There is no doubt that all three ligands are coordinated at least with europium merely from the colours: the Eu complexes are respectively yellow (L^2), yellow-orange (L^4) and orange (L^6). This is confirmed by a study of the charge-transfer bands in the near UV, which are due to $n \rightarrow f$ transitions, and of course by the crystal structure of $[La(ClO_4)_2(H_2O)(L^2)]ClO_4$ which shows a slightly folded six-coordinated macrocycle with a bidentate perchlorate on one side of it, and a monodentate perchlorate and a water molecule coordinated on the other side. Interatomic distances are La—S = 3.030, 2.045; La—O(L^2) = 2.591–2.623; La—O(water) = 2.539; La—O(perchlorate) = 2.494 (monodentate) and 2.683, 2.738 Å (bidentate). The definitely bonded La—S distance of over 3 Å is noteworthy.[403]

The Bu^tS^- group acts as a bridging ligand in $[Gd_2(\mu\text{-SBu})_2\{N(SiMe_3)_2\}_4]$. This was made from the rather similar chloro-bridged complex $[Gd_2(\mu\text{-Cl})_2\{N(SiMe_3)_2\}_4(THF)_2]$ where each Gd^{3+} ion is five-coordinated. X-Ray structures of both these complexes were obtained. In the sulfide complex, which can also be made by reaction of $[Gd\{N(SiMe_3)_2\}_3]$ with Bu^tSH, Gd—S = 2.760 and 2.823 Å and Gd—N = 2.230 and 2.244 Å, while the coordination is distorted tetrahedral with angles ranging from 135° to 71°.[404]

39.2.6 Ligands Containing Nitrogen and Oxygen Donors

Of the compounds which fall into this category, complexes of edta and its derivatives form an important group and will be treated separately. Among complexes of other ligands, there has been a fair amount of work directed at synthesis or at obtaining stability constants for multifunctional organic ligands where the structure of the resulting complexes must be in substantial doubt in the absence of specific structural data. In this section, emphasis is placed upon those studies where, for one reason or another, the experimental results being discussed can be viewed in terms of an at least reasonably probable structure.

39.2.6.1 Edta and related ligands

Complexation between lanthanide ions and edta or related ligands was invesigated in a number of early studies, and has been the subject of a short review[405] which is still useful (though there have been changing views on coordination numbers of the hydrated lanthanide ions). There are two main points. First, complexation by these ligands is often entropy driven, values of ΔH being only moderately negative or even slightly positive, while ΔS values are strongly positive. This seems to be in very reasonable agreement with the chemical process involved, namely the displacement of a number of water molecules from a strongly hydrated tripositive ion by a polydentate ligand whose N or O donors are forming bonds with the lanthanide ion, probably of a similar enthalpy to those formed by water.

The second salient feature is the decidedly non-monotonic trend usually observed when ΔH or ΔS values are plotted against atomic number or any simple function of ionic radius. This is usually interpreted in terms of a discontinuity of hydration number at some point in the lanthanide series, both in the hydrated ion $[M(H_2O)]^{3+}$ and in the complex $[M(ligand^{n-})(H_2O)_y]^{(3-n)+}$. Certainly the fact that the value of y in complexes of this type is finely balanced is shown by the two structures adopted by La–edta complexes, depending on ligand charge, in the solid state. These are $K[La(edta)(H_2O)_3]\cdot5H_2O$ which is nine-coordinated[406] with average interatomic distances of 2.755 (La—N), 2.507 (La—O(edta)) and 2.580 Å (La—O(water)); and $[La(Hedta)(H_2O)_4]$ which is ten-coordinated[407] with corresponding distances of 2.865, 2.555 and 2.592 Å respectively. These two structure determinations, as a point of historical interest, first made high (greater than six) coordination numbers for lanthanide complexes widely accepted, though there had been previous evidence in their favour.[408,409]

The complex anions $[M(edta)(H_2O)_3]^-$ where M = Pr, Sm, Gd,[410] Tb[411] and Dy[412] all have similar coordination to the lanthanum complex, but for Er and Yb, $[M(edta)(H_2O)_2]^-$, which are eight-coordinated, are obtained.[412,413] The change in coordination number occurs at some point in the Ho area. Of course, it is by no means certain that a change, if there is one, occurs at exactly the same position in the series in solution.

In many cases, the inflexions present in the plots of ΔH and ΔS *vs.* atomic number compensate, leading to much smoother curves for corresponding plots of ΔG or K. Examples of this type of behaviour are shown in reference 405, where values of ΔG, ΔH and ΔS for lanthanide complexes with edta and related ligands are plotted. Extensive lists of association constants for edta-type and other carboxylate ligands have been given.[414,415]

NMR has been used to study the mode of complexation of La^{3+} and Lu^{3+} with 2-hydroxyethylenediaminetriacetic acid (H_4L) at 0–95 °C and various pH values. Both six-coordinated and five-coordinated (*i.e.* without the hydroxyl oxygen coordinated) complexes were observed, depending on the conditions.[416]

Edta-related lanthanide complexes are of course coordinatively unsaturated in the absence of other ligands. In water, they are hydrated and the water of hydration may be displaced by other species. The kinetics of such a process, the reaction of europium 1,2-diaminocyclohexanetetraacetate, $[Eu(dcta)aq]^-$ with iminodiacetate ion, ida^{2-}, to give $[Eu(dcta)(ida)]^{3-}$, have been studied by the elegant method of selective tunable laser excitation of the $^7F_0 \rightarrow {}^5D_0$ transition in either species. If the fluorescent lifetimes of the excited states are comparable with the rate of reaction, a study of these in the reacting system will give kinetic parameters. Those obtained in this instance at 23.5 °C were k (forward) $= 1.6 \times 10^7\ M^{-1}\ s^{-1}$; k (reverse) $= 2.6 \times 10^3\ s^{-1}$; and $\log K = 3.80$. The lifetimes of the species were: $[Eu(dcta)aq]^-$, 0.320 ms; $[Eu(dcta)(ida)]^{3-}$, 0.805 ms.[417] A similar system, but using edta instead of dcta, has been investigated by NMR[418] and pH titration methods[419] and gave greater $\log K$ values of 4.28 and 4.23 respectively. It is possible that the greater bulk and rigidity of the dcta ligand is related to this difference, but the ionic strengths were different in the various studies.

In addition to these kinetic studies, there has been a considerable amount of work on the thermodynamic parameters associated with this type of reaction. Thus the interaction of $[M(edta)aq]^-$ with anions of 8-hydroxyquinoline-5-sulfonic acid (oxs^-), ida^{2-} and nitrilotriacetic acid (nta^{3-}) was thoroughly studied for M = Y, La–Lu (except Pm) by calorimetry and pH titration,[419] and ΔG, ΔH, ΔS and K found. Some of these values are given in Table 5. It was, however, concluded from the variation of these parameters with atomic number that all M^{3+} in aqueous solution have the same coordination number (nine) but that the coordination number of the $[M(edta)aq]^-$ species changes between Sm^{3+} and Tb^{3+}.

It will be seen from Table 5 that the variation of parameters with atomic number is by no means always monotonic. The explanation of this must again be connected with variation of coordination number along the series of lanthanides. If the strongly coordinating organic ligands are reasonably assumed to exert their full ligancy, then any variation must be in the extent of inner-sphere hydration of the various species, as it is rather unlikely that there is a

Table 5 Thermodynamic Parameters[419] for the Reaction

$$M(edta)^- + L^{n-} \longrightarrow M(edta)L^{(n+1)-} \quad (M = Y, La, Gd, Lu; \quad L = oxs, ida, nta)$$

	oxs^-				ida^{2-}				nta^{3-}			
	log K	$-\Delta G$	$-\Delta H$	ΔS	*log K*	$-\Delta G$	$-\Delta H$	ΔS	*log K*	$-\Delta G$	$-\Delta H$	ΔS
Y	4.31	24.2	21.1	10.5	3.24	18.2	27.3	−31.0	3.73	21.0	29.3	−28.5
La	3.36	18.9	12.6	21.3	2.80	15.7	6.5	31.8	4.79	26.9	25.4	5.02
Gd	4.36	24.5	25.5	−3.8	4.30	24.1	27.7	−12.1	4.86	27.3	33.1	−19.7
Lu	4.82	27.0	26.4	2.1	2.51	14.1	16.7	−9.2	2.81	15.8	15.1	2.5

ΔG, ΔH in kJ mol^{-1}, ΔS in J mol^{-1} k^{-1}.

dramatic alteration in outer-sphere hydration at some point in the series. The detailed interpretation of these changes is of course not easy.

Though edta forms very stable 1:1 complexes with lanthanides, its 2:1 complexes are much less stable. In the case of praseodymium, these have been investigated by ^{1}H NMR using shift and broadening data. The 1:1 complex shows resonances which are consistent with the two types of proton present. The 2:1 complex shows, however, two distinguishable molecules of edta, one similar in line widths to the 1:1 complex, and another more broadened. The second molecule exchanges rapidly on the NMR time scale with an excess of edta and it is postulated that the first molecule is fully coordinated while the second is coordinated by only one iminodiacetate residue, there being rapid intra- and inter-molecular rearrangement involving this molecule only. These studies were at pH 5–12, edta:Pr^{3+} ratios of 1–50 and temperatures of 8–98 °C. At the higher temperature, inner-sphere exchange between the two coordinated edta molecules began to take place also.[420]

The kinetics of exchange of cerium with other lanthanide ions in lanthanide edta complexes has been investigated in some detail by isotope exchange and spectrophotometry.[421–423] Exchange takes place by two concurrent routes: by the proton-catalysed dissociation of the complexes, or by direct attack of the exchanging metal on the complex. In the interval $3 < pH < 5$ the dissociative path predominates. By spectrophotometric methods using the Ce^{3+} absorption at 280 nm, the very fast rates of the lanthanide–edta complex formation can be obtained.[424] They are shown in Table 6. It seems probable that the rate-controlling step in the reaction between M^{3+} and $Hedta^-$ is the ring closure, which follows the initial formation of a metal–carboxylate bond.

Table 6 Rate Constants for Formation of Lanthanide–edta Complexes[424]

k (M^{-1} s^{-1})	Nd^{3+}	Gd^{3+}	Er^{3+}	Y^{3+}
10^{-8} k(assoc, Hedta)	2.7	3.0	1.8	0.9
10^{-6} k(assoc, H_2edta)	1.1	1.1	0.54	0.18
k(exch)	0.35	0.28	0.22	0.14

The reactions referred to are as follows:

$$Hedta^{3-} + M^{3+} \xrightarrow{k(\text{assoc, Hedta})} M(edta)^- + H^+$$

$$H_2edta^{2-} + M^{3+} \xrightarrow{k(\text{assoc, H}_2\text{edta})} M(edta)^- + 2H^+$$

$$Ce(edta)^- + M^{3+} \xrightarrow{k(\text{exch})} M(edta)^- + Ce^{3+}$$

In another investigation,[425] the exchange between $[Ce(edta)aq]^-$ and hydrated Pb^{2+}, Ni^{2+} or Co^{2+} ions again show reaction by dissociation of protonated [Ce(Hedta)aq] as well as by the direct attack of metal ions on $[Ce(edta)aq]^-$ or [Ce(Hedta)aq]. The kinetic parameters for the Ni^{2+} or Co^{2+} ions could be related to the relatively slow ($k = 2.6 \times 10^6$ s^{-1} for Co^{2+} and 3.4×10^4 s^{-1} for Ni^{2+}) water exchange reactions of these ions. The direct attack was interpreted in terms of an intermediate in which one of the carboxylate groups was coordinated to the incoming ion rather than to Ce^{3+}. These reactions were followed by spectrophotometry at 280 nm, where the absorbance of Ce^{3+}aq is much lower than the edta complex.

39.2.6.2 Other nitrogen and oxygen donors

The tris(dipicolinate) complexes of the lanthanides have been submitted to X-ray crystallographic examination. These are rather important compounds, being very stable, water-soluble and having nine-coordinated D_3 symmetry, and several examples have been examined. All these show nearly side-face-centred trigonal prismatic coordination, with the triangular faces slightly rotated relative to one another to accommodate the bite of the tridentate ligands. Complexes examined were $Na_3[Nd(dipic)_3]\cdot 15H_2O$ (Ce–Dy are isostructural) with Nd—O = 2.37–2.61 Å and Nd—N = 2.57–2.59 Å;[426] $Na_3[Yb(dipic)_3]\cdot 13H_2O$ (Ho—Tm are isostructural) with Yb—O = 2.34–2.43 Å and Yb—N = 2.50–2.53 Å;[427] $Na_3[Yb(dipic)_3]\cdot NaClO_4\cdot 10H_2O$ (Ho—Lu isostructural) with Yb—O = 2.38 Å and Yb—N = 2.43 Å;[428] and $Na_3[Yb(dipic)_3]\cdot 14H_2O$ with Yb—O = 2.35 Å and Yb—N = 2.35 Å av.[429] The $[Eu(dipic)_3]^{3-}$ ion has been examined by magnetic field-induced circularly polarized emission spectroscopy at 0–4.2 T using a c.w. Ar ion laser. The $^7F_0 \rightarrow {}^5D_2$ emission indicated that D_3 symmetry is maintained in solution.[430]

There are difficulties associated with the use of ordinary electronic absorption spectra of lanthanide complexes in solution to provide detailed information regarding coordination number and geometry. However, difference spectra *versus* $NdCl_3$ are reported for Nd^{3+}–ligand (L) solutions for the $^4I_{9/2} \rightarrow {}^4G_{5/2}$, $^4G_{7/2}$ transitions (L = dipicolinate, oxydiacetate, iminodiacetate, malate, methyliminodiacetate and *N,N'*-ethylenebis{*N*-(*o*-hydroxyphenyl)glycinate}). Hypersensitive behaviour was examined and transition dipole strengths were discussed in terms of the nature of the complex species present.[431]

A somewhat similar investigation[432] of aqueous complexes of Er^{3+} with the ligands dipicolinate, oxydiacetate, iminodiacetate, (methylimino)diacetate and *N,N'*-ethylenebis{*N*-(*O*-hydroxyphenyl)glycine} has been carried out over a range of pH. The oscillator strengths of the hypersensitive transitions $^4I_{15/2} \rightarrow {}^2H_{11/2}$, $^4G_{11/2}$ were rationalized in terms of ligand structure and coordination properties, and ligand field geometry. Intensity calculations, based on a theoretical model for $4f \rightarrow 4f$ electric dipole strengths,[433] were carried out for some of the structures believed to be those adopted, and the spectra–structure correlations discussed. There was remarkably good agreement between calculated and observed oscillator strengths, but though encouraging, this should not be taken as either confirmation of the validity of the intensity model or as proof of the solution structures.

Complexes of the ligand acetylhydrazine, *N*H$_2$NHC*O*Me (ah) which coordinates by the italicized atoms, have been prepared and examined by single crystal X-ray. The complexes are $[Er(ah)_4(H_2O)](NO_3)_3$ and an isostructural pair $[M(ah)_3(H_2O)_3]Cl_3$, where M = Ho or Dy. The polyhedron of the erbium complex approximates to a capped square antiprism, with Er—N = 2.54 (average), Er—O(ah) = 2.337 (average) and Er—O(water) = 2.363 Å. The chloride complexes form a face-capped trigonal prism, with Dy—N = 2.571, Dy—O(ah) = 2.360 and Dy—O(water) = 2.425 Å.[434] However, in the related ligand H$_2$*N*NHC*O*NH*N*H$_2$, carbohydrazide (ch) a series of complexes $M(ch)_3Cl_3$, where M = La–Eu or Ho, has been obtained from anhydrous ethanol, where IR spectra indicate that a six-membered chelate ring is formed by coordination of the italicized nitrogen atoms; this needs to be confirmed.[435] Another hydrazine-related ligand, thiosemicarbazidediacetic acid (H_2tsda) H$_2$*N*CS*N*H*N*(CH$_2$COOH)$_2$, gives hydrated complexes of the types M(tsda)Cl, M(tsda)(OH) and NaM(tsda)$_2$, where M = Nd, Ho or Er, in which coordination by 3N and 2O as italicized was postulated, and the visible spectra were investigated.[436]

Still another hydrazide ligand, 4-pyridyl-C(*O*)NH*N*H$_2$ (L) gives complexes $[ML_3(H_2O)_3](NO_3)_3$, where M = La, Nd, Sm, Gd, Ho and Yb. The X-ray structure of the Sm example shows irregular nine coordination involving bidentate L, using the amide oxygen and the NH_2 nitrogen; the pyridine nitrogen is not coordinated.[437]

Yet another hydrazine-based ligand, 2,6-diacetylpyridine bis(benzoic acid hydrazone) (dpbh; **10**) forms an interesting complex with $La(NO_3)_3$ in which the coplanar penannular pentadentate dpbh coordinates together with three bidentate nitrate ions to give an 11-coordinated complex. One nitrate ion is disposed near the gap in the dpbh chain with the other two above and below the dpbh plane. Interatomic distances are: La—N(ligand) = 2.769–2.790, La—O(ligand) = 2.507 and 2.560, and La—O(nitrate) = 2.563–2.648 Å.[438]

A structurally very interesting complex is formed by the related ligand 2,6-diacetylpyridine bis(semicarbazone) (dapsc; **3**). This ligand and $Ce(ClO_4)_3$ in aqueous alcohol gave $[Ce(dapsc)_2](ClO_4)_3\cdot 3H_2O$ in which the ligands are each pentadentate and are essentially arranged each in one of two mutually perpendicular planes, though each ligand is slightly

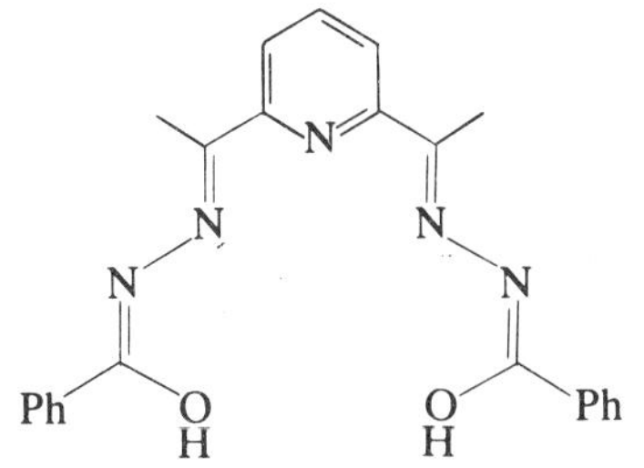

(10)

twisted out of planarity. The IR CO band shifted from 1690 cm^{-1} in the free ligand to 1650 cm^{-1}. Interatomic distances are Ce—N = 2.642–2.724 while Ce—O = 2.483–2.597 Å.[439]

A series of complexes between lanthanide ions and the transition metal complexes *N,N'*-1,3-propylene-bis(salicylideniminato)nickel(II), Ni(psal), or the corresponding copper complex, has been obtained (**11**).[440] In these compounds, the two phenate oxygen atoms of the copper or nickel complex are able to form a chelate group which apparently coordinates with lanthanide ions, giving the stoichiometries $MM'(psal)X_3 \cdot nH_2O$ where M' = Ni, M = La–Sm, X = NO_3, $n = 0$; M' = Ni, M = Eu–Lu, X = NO_3, $n = 2$; M' = Ni or Cu, M = La–Dy, X = NCS, $n = 3$; M' = Ni or Cu, M = Ho–Lu, X = NCS, $n = 4$; M' = Ni or Cu, M = La–Lu, X = ClO_4, $n = 4$; M' = Ni or Cu, M = La–Sm, X = Cl, $n = 4$; and M' = Ni or Cu, M = Eu–Lu, X = Cl, $n = 2$.

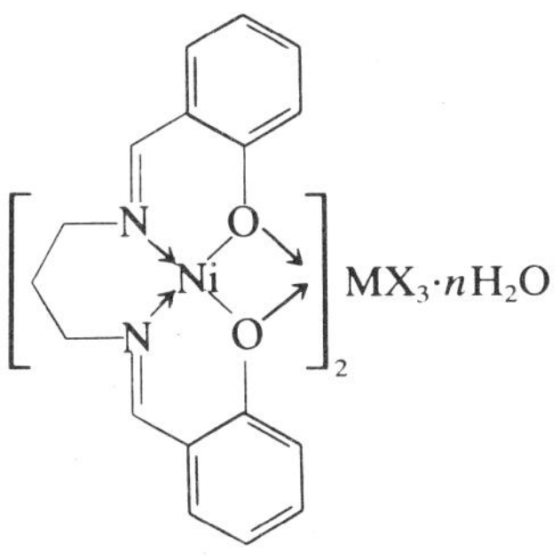

(11)

39.2.7 Complexes with Macrocyclic Ligands

39.2.7.1 Introduction

Since Pedersen's remarkable paper[441] appeared as long ago as 1967, in which he described a wide variety of cyclic polyethers (crown ethers) and their complexes with Group I and Group II metal ions, a considerable interval of about 10 years occurred before the principle was extended to the lanthanide ions. This is surprising, because the ionic radii of the tripositive lanthanides together with the non-directional electrostatic nature of their bonding should indicate that they might form strong complexes with such ligands. The factor that may have prevented early extension of this work into the lanthanide area is the large hydration energy of the tripositive lanthanide ions which in fact leads to dissociation of most lanthanide–crown ether complexes in aqueous solution. However, the use of non-aqueous solvents allows the ready formation of these interesting complexes. Later work, detailed below, has shown that crown thioethers will also complex, though not strongly, while nitrogen analogues of crown ethers will form extremely stable lanthanide complexes. Crown nomenclature is illustrated by 12-crown-4 (**12**), cyclohexyl-18-crown-6 (**13**) and dibenzo-18-crown-6 (**14**).

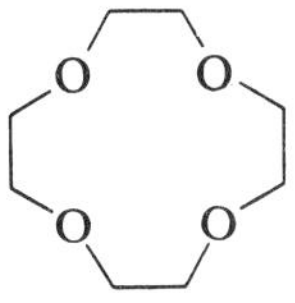

(12)

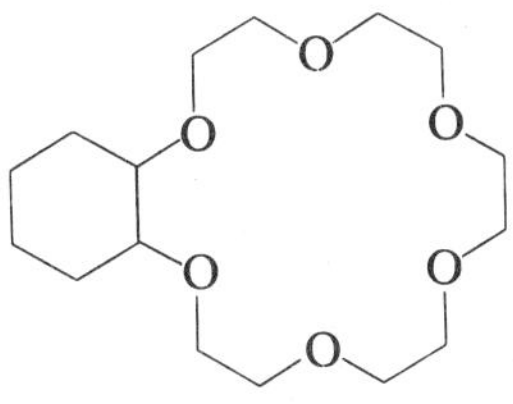

(13)

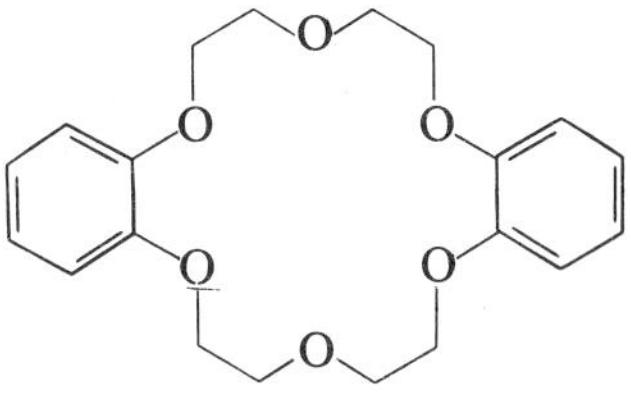

(14)

39.2.7.2 The crown ethers $(CH_2CH_2O)n$ and their derivatives

Molecular models indicate that 15-crown-5 and its substituted derivatives should on the basis of ionic radii give stable complexes with lanthanides,[442] and an early report[443] of lanthanide–crown coordination in fact described the formation in acetonitrile solution of complexes $M(NO_3)_3$(benzo-15-crown-5) and $M(NCS)_3$(benzo-15-crown-5)·H_2O, where M = La–Lu. Other contemporary work[444] described a large number of lanthanide complexes of this same ligand, $M(NO_3)_3$(benzo-15-crown-15), where M = La–Sm, and $M(NO_3)_3$(benzo-15-crown-5)·$3H_2O$·Me_2CO, where M = Sm–Lu, together with $M(NO_3)_3$(dibenzo-18-crown-6), where M = La–Nd. Conductance measurements showed that the nitrate ions were coordinated in solution. Coordination of the macrocycle was indicated by UV or IR absorption; thus Pedersen[441] had observed that the absorption band of dibenzo-18-crown-6 at 275 nm was modified by addition of sodium hydroxide, while the C—O stretching frequency of the polyethers at about 1100 cm^{-1} is moved by about 30 cm^{-1} to lower wavenumber by coordination. The latter effect seems to be a reliable indicator of coordination to a lanthanide ion.

A number of X-ray crystal structure determinations, mainly involving six-oxygen crowns and nitrate or perchlorate ions, have made the structural principles of lanthanide–crown complexes clearer. It appears that lanthanides, when coordinating with six-oxygen crowns (in which case only the lighter lanthanides form stable complexes) will form total coordination numbers of twelve, eleven or ten, depending on the nature of the anionic ligands and the size of the lanthanide cation (Table 7). Thus in $[La(NO_3)_3$(18-crown-6)] where the lanthanide radius is the largest possible and the steric demands of the closely bidentate nitrate anion are quite small, twelve-coordination results.[445] However, the smaller Nd^{3+} ion, besides giving an analogous twelve-coordinated complex,[446,447] also gives $[Nd(NO_3)_2$(18-crown-6)$]_3[Nd(NO_3)_6]$ where the complex cations are ten-coordinated only,[448] and with the more bulky chloride anion, nine-coordinated $[GdCl_2$(EtOH)(18-crown-6)]Cl is the resulting complex.[449] In $[Y(NCS)_3$(18-crown-6)][450] where the N-coordinated anions are not sterically demanding, but the Y^{3+} cation is smaller than in the preceding examples, nine-coordination again occurs, and the crown ligand shows some steric strain. Interestingly, in $[Sm(ClO_4)_3$(dibenzo-18-crown-6)],[451] the rather wide bite and poorer donor properties of the perchlorate ligand compared with nitrate lead to ten-coordination only, with one bidentate and two monodentate perchlorate ions.

Table 7 Lanthanide–Ether Distances in Crown Ether Complexes

Complex	*Coordination number*	*M—O$_{ether}$(Å)*	*Ref.*
$[La(NO_3)_3$(18-crown-6)]	12	2.644–2.680	445
$[La(ClO_4)_2$(dithia-18-crown-6)$(H_2O)]^+$	10	2.59–2.62	461
$[Nd(NO_3)_3$(18-crown-6)]	12	2.580–2.615	446, 447
$[Nd(NO_3)_2$(18-crown-6)$]^+$	10	2.56–2.71	448
$[Eu(NO_3)_3$(15-crown-5)]	11	2.55–2.69	453
$[Eu(NO_3)_3$(12-crown-4)]	10	2.477–2.570	452

X-Ray structure determinations of complexes of smaller crown ligands are much fewer than those of the six-oxygen crowns, but Bünzli and co-workers have reported ten-coordinated $[Eu(NO_3)_3$(12-crown-4)][452] and eleven-coordinated $[Eu(NO_3)_3$(15-crown-5)],[453] both of which show pyramidal coordination of the small crowns (Figure 5).

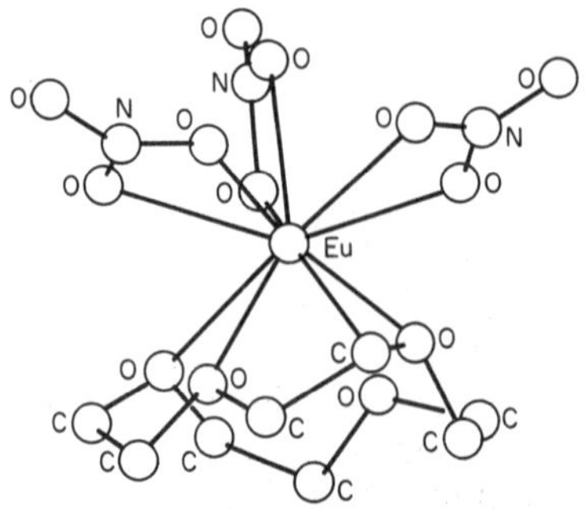

Figure 5 Structure of $[Eu(NO_3)_3$(15-crown-5)] (reproduced with permission from ref. 453)

Where the lanthanide ionic radius and the macrocyclic cavity are incompatible, though in hydrous conditions the crown ether is likely to be displaced by water ligands, the crown may still be present in the structure of the crystal as a hydrogen-bonded adduct. This behaviour is seen in $[Gd(NO_3)_3(H_2O)_3]\cdot$(18-crown-6).[445] This type of compound is quite well known in the case of *s* block metals also, *e.g.* $[Mg(H_2O)_6]Cl_2\cdot$(12-crown-4)[454] and, a more subtle case, [Ca(nitrobenzoate)$_2$(benzo-15-crown-5)]$\cdot 3H_2O\cdot$(benzo-15-crown-5)[455] in which an apparent 2:1 complex has only half its crown ligand coordinated to Ca^{2+}.

The modified crown ligand (**15**), which is sterically unsuitable for coordination of all six ether oxygen atoms, forms a complex $[Sm(NO_3)_3(H_2O)(L)]$ where L is the 2-methoxy-1,3-xylyl crown (**15**).[456] This is an irregular bicapped square antiprism in which three consecutive oxygen atoms of the crown ligand (the third being the OMe) are coordinated.

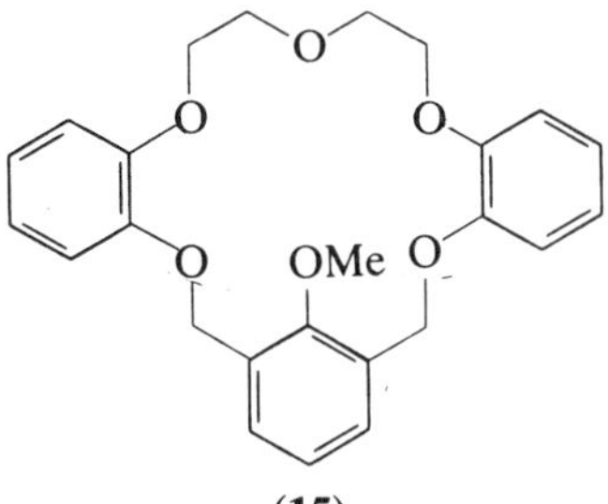

(**15**)

Although they are not macrocyclic ligands, polyethyleneglycols behave somewhat similarly to the crown ethers regarding lanthanide complexation, as established by a number of crystal structure determinations. Thus a series of neodymium complexes shows consistent 10-coordination, namely $[Nd(NO_3)_3$(tri-eg)],[458] where tri-eg is triethyleneglycol, $[Nd(NO_3)_2$(penta-eg)]NO_3,[458] and $[Nd(NO_3)_2(NO_3')$(tetra-eg)],[459] where NO_3' is monodentate. The larger La^{3+} ion shows 11-coordination in $[La(NO_3)_3$(tetra-eg)].[460]

Lanthanides form far fewer complexes with sulfur-based ligands than with oxygen-based ligands, and this appears to be a real effect based on chemical properties and not simply on lack of exploration. It is very interesting, therefore, that the relatively favourable entropy change upon complexation of the crown ligands is sufficient to force sulfur into coordination.[461] Thus dithia-18-crown-6 (**16**) forms, from acetonitrile solution, a series of complexes of the type $M(ClO_4)_3$(dithia-18-crown-6)$(H_2O)_x(MeCN)_y$ where M = La–Eu, Ho or Yb; $x = 0$ or 1; $y = 0$, 1.5 or 2. The crystal structure of $[La(ClO_4)_2(H_2O)$(dithia-18-crown-6)]ClO_4 shows ten-coordination with one bidentate and one monodentate perchlorate ion, but with both sulfur atoms coordinated. The related ligands (**17**) and (**18**) also gave 1:1 complexes $M(ClO_4)_3$(ligand)(H_2O) where IR evidence indicated coordination by sulfur, but no X-ray data are available. See also Section 39.2.5.

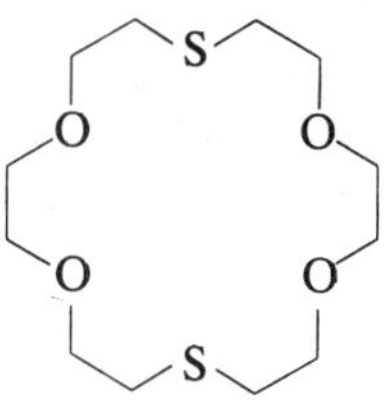

(**16**)

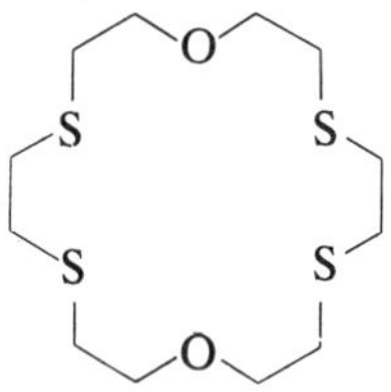

(**17**)

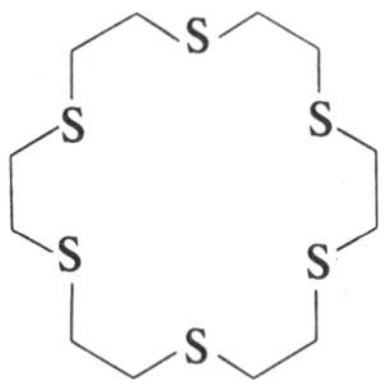

(**18**)

Although the lanthanide crown ether complexes tend towards insolubility in non-polar solvents and dissociate in water, they may be studied by physicochemical methods in solvents such as acetonitrile. Thus a ^{13}C and ^{1}H study[462] of $[Pr(ClO_4)_3$(*cis*-syn-*cis*-dicyclohexyl-18-crown-6)] in $CD_3CN/CDCl_3$ was carried out and the results of an analysis of paramagnetically shifted spectra indicated that the conformation of the coordinated crown determined[463] by X-ray (for the lanthanum nitrate analogue) was adopted in solution by the $Pr(ClO_4)_3$ complex. The ^{1}H paramagnetic shifts were largely dipolar in origin, but the ^{13}C shifts had a considerable contact contribution. In acetonitrile, ^{1}H NMR and conductivity measurements[445] provided evidence for an equilibrium with a stability constant $\log K = 2.82$ and an enthalpy of dissociation $\Delta H_{dis} = 0 \pm 2\ \mathrm{kJ\ mol^{-1}}$ (equation 7).

$$Yb(NO_3)_3(CD_3CN)_n + \text{18-crown-6} \longrightarrow [Yb(NO_3)_2(\text{18-crown-6})]^+ + NO_3^- + nCD_3CN \quad (7)$$

Similar measurements[463] led to the following values of $\log K$ for $M(NO_3)_3$ with 18-crown-6: La, 4.4; Pr, 3.7; Nd, 3.5; Eu, 2.6; Yb. 2.3. The increased stability of the complexes of the larger lanthanides is very noticeable here. The low-temperature spectrum[463] of $[Pr(NO_3)_3(\text{18-crown-6})]$ in CD_3CN shows splitting into two equal peaks, both paramagnetically shifted, whereas only one (in addition to a little dissociated ligand) is seen at room temperature. It must be probable that the X-ray structure, with two nitrate groups on one side of the crown and one on the other side, is being observed at low temperature whereas rapid nitrate exchange occurs at the higher temperatures, while rotation of crown relative to nitrate occurs at all temperatures.

In a potentiometric study in propylene carbonate, using Pb^{II} or Tl^{I} as auxiliary ions, stability constants have been determined for a variety of crown ethers. Some results[464] are shown in Table 8. They show that the 'wrap-around' ligand dibenzo-30-crown-10 is relatively quite effective, while the 2:1 complexes, presumably of the sandwich type, are favoured for larger lanthanides and smaller crowns.

Table 8 Log Association Constant Values for Lanthanide–Crown Ether Complexes in Propylene Carbonate[464]

Crown	*I*		*II*		*III*	*IV*	*V*
	$\log K_1$	$\log K_2$	$\log K_1$	$\log K_2$	$\log K_1$	$\log K_1$	$\log K_1$
La	5.0	2.0	6.5	3.7	5.2	3.2	4.3
Yb	5.0	—	5.9	2.0	2.6	2.8	4.8

I = 12-crown-4; II = 15-crown-5; III = *tert*-butylbenzo-15-crown-5; IV = di(*tert*-butylbenzo)18-crown-6; V = dibenzo-30-crown-10.

The excitation and fluorescence spectra of some crown ether complexes have been interpreted in terms of the molecular point group symmetry and vibronic coupling.[452,453] This approach tends to be limited to europium(III) complexes.

Non-cyclic long-chain polyethers also form stable complexes with lanthanide ions, as exemplified by $[La(NO_3)_2(pe)]_3([La(NO_3)_6]$, where pe = $MeO(C_2H_4O)_5Me$.[465] Here the cation contains 10-coordinated La^{3+} in a structure similar to that of $[Nd(NO_3)_2(\text{18-crown-6})]^+$.

39.2.7.3 Nitrogen donor macrocycles

As the amines as a class form much stronger complexes with lanthanides than do ethers, so it would be expected that nitrogen-based macrocycles should complex more strongly than do the crown ethers. That this is in fact the case is confirmed by the preparation and properties of a small but representative number of lanthanide complexes of nitrogen donor macrocycles. These may be divided into two groups: complexes of porphyrins and phthalocyanines, which have four nitrogen donors, and complexes of ligands usually with six nitrogen donors which are analogues of crown ethers. Much the earliest compounds to be reported were those of phthalocyanine[466,467] (H_2pc; **19**), and this area has since been extensively developed by Russian workers.

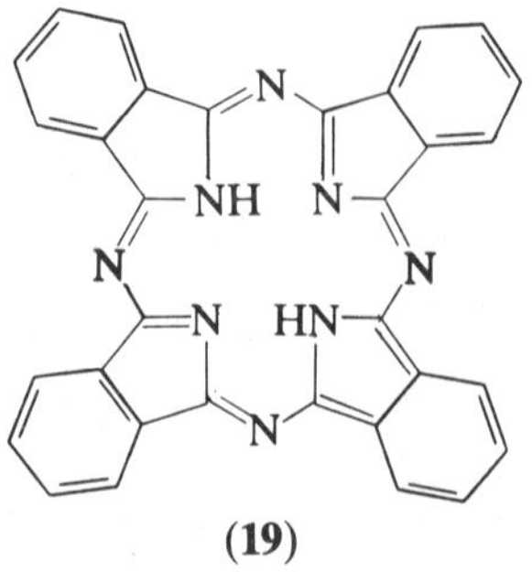

(19)

Lanthanide phthalocyanines may be prepared by heating the metal chlorides[468] or formates[469] with 1,2-cyanobenzamide, 1:1 complexes, for example Er(pc)Cl, being the products. An alternative and better route used phthalonitrile (equation 8).[470,471]

$$C_6H_4(CN)_2 + MCl_3 \xrightarrow{280\,^{\circ}C} M(pc)Cl.nH_2O \qquad (8)$$
$$(M = Y,\ n = 4;\ M = Er,\ n = 2)$$

It was later established that with neodymium, 2:1 and 3:2 complexes of constitution Nd(pc)(Hpc) and $Nd_2(pc)_3$, respectively, could be prepared,[472–474] and 2:1 complexes of Y, La, Ce, Nd, Eu, Er and Yb have been reported.[475–477] The lighter lanthanides tend to give 2:1 complexes, while the heavier ones give 1:1 preferentially.[478,9] The X-ray crystal structure of [Nd(pc)Hpc)][480] shows a sandwich structure with one coplanar macrocycle and one concave towards the Nd^{3+} ion. The two ligands are staggered, and Nd—N is in the range 2.39–2.49 Å. The acidic H atom was not located. The 2:1 complexes M(pc)(Hpc) can be anodically oxidized to $M(pc)_2$ (M = Nd, Eu, Ho or Lu).[481] These products $M(pc)_2$ give adducts with molecular iodine with the I_2:complex ratio increasing along the series and reaching 2:1 for Er (Pr, 0.85:1). The addition is reversible on heating, and the resistivity of the brown iodine adducts is less by a factor of ten than the green $M(pc)_2$ complexes.[482]

A variety of substituted phthalocyanines has been used to confer solubility in various solvents upon these complexes. Thus water-soluble sulfonated complexes, stable even in strong ammonia, were obtained[483,4] by sulfonation of M(pc)(Hpc) (M = Y, Eu, Gd, Tm, Yb or Lu). The *t*-butyl-substituted complexes M(pc′)(Hpc′), where pc′ is tetra-4-butylphthalocyanine, are quite soluble in ethanol in contrast to the unsubstituted compounds and their solution absorption spectra, dominated by the ligand, have been studied[485] alone and in the presence of NaOH. These spectral changes, which must correspond with the deprotonation of the Hpc′ ligand, are discussed in terms of the magnitude of the splitting of the excited states and the relative orientation of the moments of the electronic transitions of the interacting ligands. The pronounced electrochromism of the lanthanide phthalocyanines has been investigated.[486]

As might be expected in the light of the phthalocyanine complexes described above, the lanthanides also form complexes with porphyrins. Thus the reaction of $Eu(acac)_3$ with *meso*-tetraphenylporphyrin (**20**; R = Ph, R′ = H; phporph) gave Eu(acac)(phporph).[487] This reaction was extended[428] to other lanthanides, β-diketonates and porphyrins, giving M(acac)(tolporph) and M(dpm)(tolporph), where M = Pr through to Lu, and tolporph = (**20**; R = *p*-tolyl, R′ = H). Soluble in organic solvents, these complexes functioned as upfield ring-current diamagnetic shift reagents. A further extension to *p*-sulfonatophenylporphyrin gave compounds $M(sulporph)(OH)(imidazol)_x$, where M = Gd–Lu, sulporph = (**20**; R = *p*-sulfonatophenyl, R′ = H) and $x \leqslant 2$. These complexes were stable in neutral or highly basic aqueous solution, and induced paramagnetic shifts in cations, anions and neutral species. These shifts received detailed study.[489] An ESCA study[490] of complexes of octaethylporphyrin (**20**; R = H, R^1 = Et, etporph) and phporph of the type M(etporph)(OH), where M = Eu, Gd, Yb or Lu, and M(phporph)(acac), where M = Eu, Gd, Yb or Lu, has been carried out. The results implied the presence of covalent interaction of lanthanide 4*f* and outer 6*s*, 6*p* and 5*d* orbitals, together or as alternatives, with nitrogen valence orbitals.

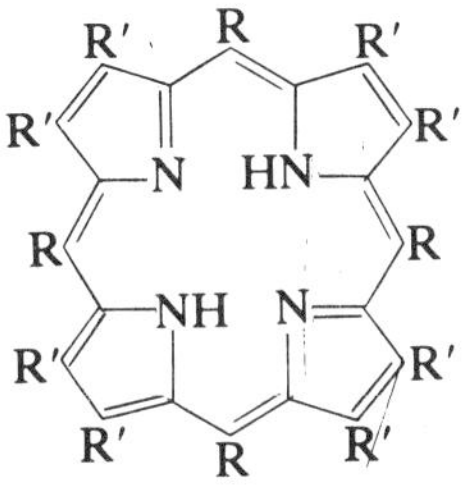

(**20**)

There has been recent interest in lanthanide complexes of ligands of the type (**21**). These are in some sense nitrogen analogues of the crown ethers. As Schiff bases, their lanthanide complexes may be synthesized by a template condensation, for example from 2,6-diacetylpyridine, 1,2-diaminoethane and lanthanum nitrate. These components react in alcohols to give $[La(NO_3)_3L]$, where the aza-crown L = (**21**; R = Me). This compound has a structure (Figure 6), quite analogous to that of $[La(NO_3)_3(18\text{-crown-6})]$, in which La—N

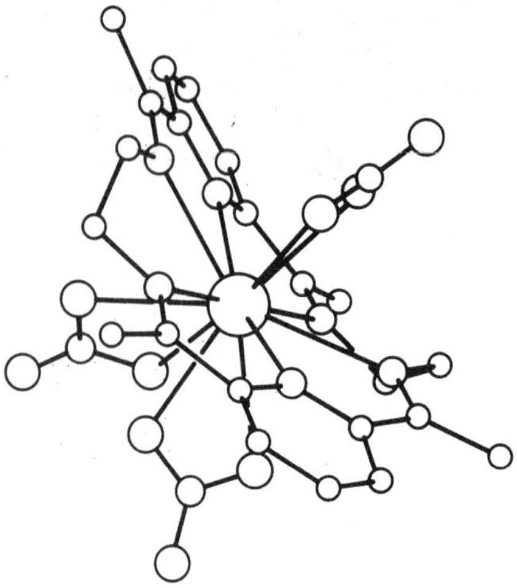

Figure 6 Structure of the aza-crown complex $[La(NO_3)_3(C_{22}H_{26}N_6)]$ ($C_{22}H_{26}N_6$ is **21**; R = Me) (reproduced with permission from ref. 491)

interatomic distances lie in the range 2.672–2.764 Å. Unlike 18-crown-6, however, the aza-crown remains coordinated in aqueous solution and even in an alkaline or fluoride-containing aqueous medium.[491] The use of lanthanum perchlorate in the condensation reaction gives $La(ClO_4)_3L(H_2O)_2$.[492] A number of complexes of the ligand (**21**; R = H) and related ligands formed by condensations using $H_2N(CH_2)_3NH_2$ or $H_2NCH(Me)CH_2NH_2$, rather than $H_2NCH_2CH_2NH_2$, have also been reported.[493,494] In these, the imine double bond has a tendency to hydrate and the original product of the template reaction for the series Nd, Sm, Gd–Lu shows hydration of one of the imine double bonds to give the partial structure —NH—CH(OH)— while the larger ions La—Pr and Eu yield complexes of the unhydrated macrocycle. The Sm complex, which is in the hydrated form, may be converted into a complex of (**21**; R = H; L′) by recrystallization from water, and X-ray analysis of the recrystallized complex shows a ten coordinated Sm^{3+} ion in $[Sm(NO_3)(OH)(H_2O)L']NO_3 \cdot 2MeOH$. The Sm—OH internuclear distance (2.46 Å) is shorter than the Sm—OH_2 distance (2.52 Å).[493] It is also possible to prepare this type of complex by a metal transfer reaction, the complex $[Ba(ClO_4)_2L)]$ reacting in aqueous solution with lanthanide nitrates to give $[M(NO_3)_2L(H_2O)]NO_3 \cdot H_2O$ (M = La–Pr) or $[M(NO_3)L(H_2O)_2]_2(NO_3)(ClO_4)_3 \cdot 4H_2O$ (M = Nd–Er). The structures of the eleven-coordinated Ce and ten-coordinated Nd compounds were established by X-ray.[495] Lanthanide complexes of the macrocycle (**22**) which, though it contains six nitrogen atoms, is sterically only capable of four-coordination, have been also obtained.[496]

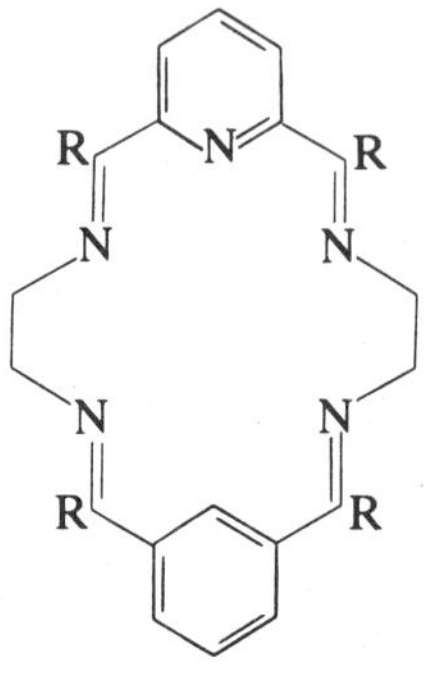

(21)

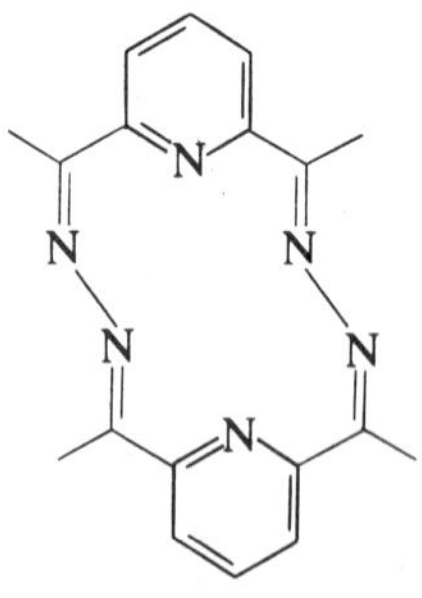

(22)

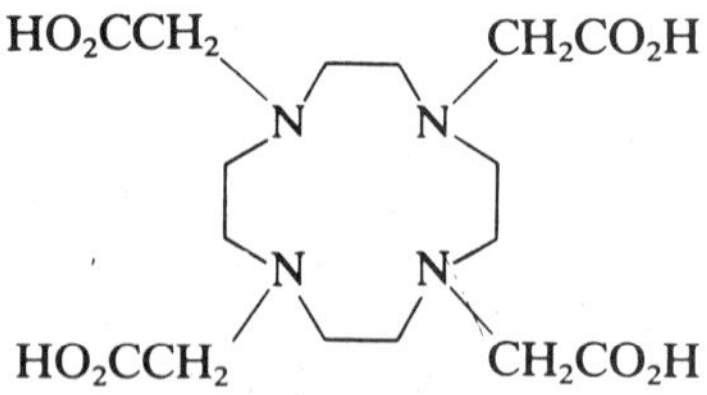

(23)

A most interesting functionalized four-nitrogen crown ligand (**23**; H_4dota) has been synthesized.[497] The X-ray structure of the complex Na[Eu(dota)(H_2O)] (Figure 7) shows that the nine-coordinated geometry is approximately capped square antiprismatic with the four N forming a basal square, four carboxylate O forming an upper square, and the H_2O capping the latter.[498] This geometry leads to very large paramagnetic NMR shifts, and the complex acts as an effective shift reagent for H_2O and Cl^-.[499]

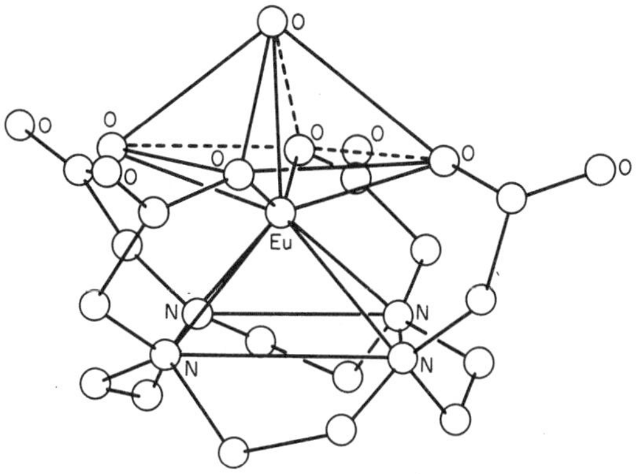

Figure 7 Structure of the anion of Na[Eu(dota)] (H_4dota is **23**) (reproduced with permission from ref. 498. Copyright, American Chemical Society, 1984)

39.2.7.4 Mixed nitrogen and oxygen macrocycles

There have been one or two reports of lanthanide complexes of this type of ligand. Thus three furan Schiff bases (**24**; R = $(CH_2)_2$, $(CH_2)_3$ or CH_2CHMe) have been obtained as their complexes with the series La–Nd. These slowly dissociate in water.[500] The functionalized crown ligand 1,10-diaza-4,7,13,16-tetraoxacyclooctadecane-*N,N'*-diacetic acid (dacda) is 18-crown-6 with NCH_2CO_2H units replacing two opposite oxygen atoms. It is characterized by very high association constants for lanthanide ions, determined by potentiometric titration. Thus log *K* values are: La, 12.21; Gd, 11.93; Lu, 10.84. Values for Group I and II ions are much lower.[501]

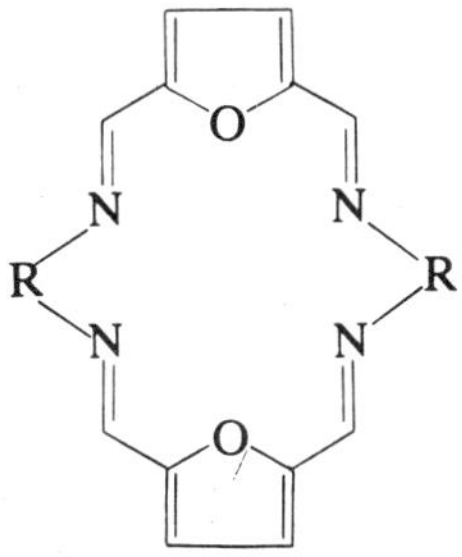

(**24**)

39.2.7.5 Cryptates

Lanthanides form stable complexes with cryptands typified by $N(CH_2CH_2OCH_2CH_2OCH_2CH_2)_3N$ (this molecule is 2,2,2-cryptate and similar ligands are named analogously). In such complexes the lanthanide ion lies completely inside the cryptate ligand which, in the example given, fills eight positions on the lanthanide coordination sphere. Depending on the relative size of the lanthanide ion and the cryptate ligand, other donors may also coordinate in positions between two —$CH_2CH_2OCH_2CH_2OCH_2CH_2$— strands of the cryptate.

The complexes MCl_3(2,2,1-cryptate), where M = La, Pr, Eu, Gd or Yb, may be prepared from the ligand and MCl_3 in anhydrous organic solvents, and together with $La(NO_3)_3$(2,2,2-cryptate) were the subject of the first report of this type of lanthanide complex.[502] Cyclic voltammetry showed that the redox potentials of the europium and ytterbium complexes of 2,2,1-cryptate and 2,2,2-cryptate were significantly affected in water by complexation. Thus the redox potential decreased from 626 mV for the aqueous Eu^{3+} ion to 205 mV for the Eu^{3+} complex with 2,2,2-cryptate.[502] Using polarography, one-electron reduction kinetics for [Eu(2,2,1-cryptate)]$^{3+}$ and the corresponding 2,2,2 complex with Eu^{2+} or V^{2+} were investigated, together with the oxidation kinetics of [Eu(2,2,1-cryptate)]$^{2+}$ when oxidized by $[Co(NH_3)_6]^{3+}$. The rate constant for electron exchange was estimated to increase over the $Eu^{2+}_{aq}/Eu^{3+}_{aq}$ rate by a factor of 10^7 upon encapsulation by 2,2,1-cryptate. The results indicate that the low reaction rates observed for the $Eu^{2+}_{aq}/Eu^{3+}_{aq}$ exchange are not due to adiabaticity, but probably due to large Franck–Condon barriers.[503]

These complexes, unlike the crown ether complexes but similar to the aza-crown and phthalocyanine complexes, are fairly stable in water. Their dissociation kinetics have been studied and not surprisingly they showed marked acid catalysis.[504] Association constant values for lanthanide cryptates have been determined.[505,506] A study in dimethyl sulfoxide solution by visible spectroscopy using murexide as a lanthanide indicator showed that there was little lanthanide specificity (but surprisingly the K values for Yb are higher than those of the other lanthanides). The values are set out in Table 9.[507]

Table 9 Log Association Constant Values for Lanthanide Cryptates in DMSO[507]

Cryptate	2,2,2	2,2,1	2,1,1
Pr	3.22	3.47	3.86
Nd	3.26	3.01	3.97
Gd	3.45	3.26	3.87
Ho	3.47	3.11	3.80
Yb	4.11	4.00	4.43

Other methods of preparation of lanthanide cryptates have been reported,[508] including the functionalized cryptate (**25**) whose La, Pr and Eu nitrate complexes were obtained. Paramagnetic shifts over a range of about 40 p.p.m. showed complexation with the lanthanide enclosed in the ligand cavity. Preparation was in dry acetonitrile to which triethylorthoformate was added, permitting the use of hydrated lanthanide nitrates.[509] The preparation of an organo derivative $LaCl_2(CPh_3)$(2,2,2-cryptate) has also been reported.[510] The difficulty in straightforward preparation of lanthanide cryptates probably arises from an initial tendency to form insoluble 'external' (*i.e.* not enclosed) complexes and from the formation of hydroxyl ion if any water is present. Cryptates of Eu^{2+} may be prepared in aqueous solution, however, and then converted into the corresponding trivalent complex by electrochemical oxidation.[503]

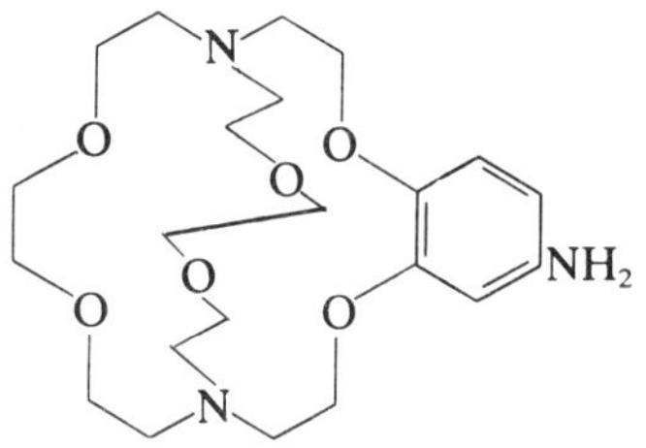

(**25**)

A number of X-ray crystal determinations have made the principles of lanthanide cryptate structural chemistry fairly clear. In $[La(NO_3)_2$(2,2,2-cryptate)$][La(NO_3)_6]$ (Figure 8), the La^{3+} ion is 12-coordinated with two bidentate nitrate ions coordinating in two of the three spaces between the cryptate chains; the third space is thus too compressed to be occupied also.[508] $[Sm(NO_3)$(2,2,2-cryptate)$][Sm(NO_3)_5(H_2O)]$ shows only one such space occupied[511] and the structure of $[Eu(ClO_4)$2,2,2-cryptate$](ClO_4)_2\cdot MeCN$ is similar to the samarium cryptate.[512,513] Internuclear distances in these complexes are shown in Table 10.

Interesting conformation information has been given by a ^1H NMR study of the series of complexes MX_3(2,2,1-cryptate), where X = Cl, NO_3 or ClO_4 and M = La–Yb. A plot of paramagnetic shift against Bleaney's anisotropy constant gives a good straight line, which is evidence that all these lanthanide cryptate species are isostructural in D_2O. Furthermore, a single conformation is either dynamically favoured or stable, because the same AMXY pattern for all four $NCH_2CH_2OCH_2CH_2OCH_2CH_2N$ protons is observed.[514]

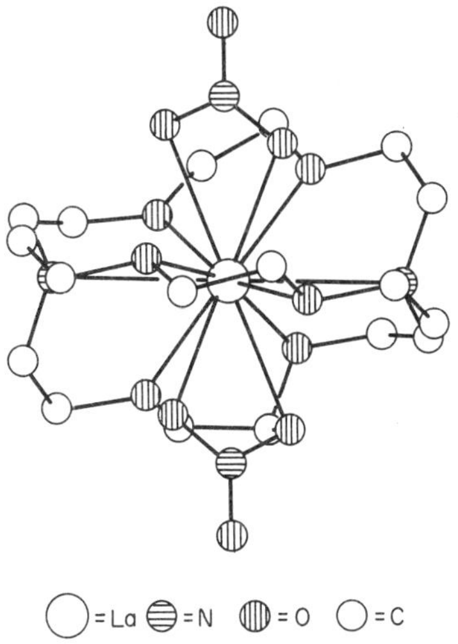

Figure 8 Structure of the cation in $[La(NO_3)_2(2,2,2\text{-cryptate})][La(NO_3)_6]$ (reproduced with permission from ref. 508)

Table 10 Lanthanide–Ligand Distances in Cryptate Complexes

Complex	*Coordination Number*	*M—O$_{ether}$(Å)*	*M—N*	*M—O$_{anion}$*	*Ref.*
$[La(NO_3)_2(2,2,2\text{-cryptate})]^+$	12	2.64–2.74	2.81–2.85	2.63–2.69	508
$[Sm(NO_3)(2,2,2\text{-cryptate})]^{2+}$	10	2.441–2.566	2.748–2.779	2.484, 2.402	511
$[Eu(ClO_4)(2,2,2\text{-cryptate})]^{2+}$	10	2.44–2.52	2.64–2.70	2.67, 2.71	512, 513

39.2.8 Complex Halides

Phase diagrams have been given for the NaF–MF_3 systems, where M = La–Lu.[515] In all these systems, hexagonal phases $NaMF_4$ exist below 700 °C. Above this temperature, for M = Pr–Lu, these phases become converted into disordered fluorite-like cubic phases of variable composition. Particularly for the heavier lanthanides, orthohombic phases $Na_5M_9F_{32}$ also exist at lower temperatures, $Na_5Lu_9F_{32}$ being stable from 778 °C to room temperature; other $Na_5M_9F_{32}$ are not stable below 450 °C. The system NaF–YbF_3 has also been investigated[516] by DTA and X-ray powder methods; $NaYbF_4$ was again found, together with a fluorite phase $Na_{0.38}Yb_{0.62}F_{2.24}$ and another phase $Na_{0.32}Yb_{0.68}F_{2.36}$ which is somewhat different from the $Na_5M_9F_{32}$ previously reported.[515] The structure of $NaMF_4$ (M = La–Tm or Y) in the lower-temperature hexagonal phase contains two types of M^{3+}, both nine-coordinated to tricapped trigonal prisms, with Nd–F = 2.377–2.503 Å in the Nd^{3+} compound.[517] The high temperature fluorite-like phase of course shows cubic eight-coordination.[518] The 1:1 stoichiometry $M'MF_4$ has also been observed for M′ = Li, M = Eu–Lu or Y and these compounds have the $CaWO_4$ structure.[519]

In the case of the larger alkali metal cations, the 1:1 complexes $M'MF_4$ (where M′ = K, M = La or Ce; M′ = Rb, M = La–Pr) can be obtained, but the predominant stoichiometry is M'_3MF_6, examples of which are K_3YF_6, $RbMF_6$ (M = La–Pr, Sm, Tb, Er or Y), and Cs_3MF_6 (M = La–Pr, Sm, Gd, Dy–Er or Y), the X-ray parameters of which have been reported.[520,521]

The chloride and bromide complexes show four structure types, $M'M_2X_7$, $M'_3M_2X_9$, M'_2MX_5 and M'_3MX_6 (M′ = alkali metal, M = lanthanide or Y).[522] They may be conveniently prepared by evaporation of aqueous acidic solutions of the appropriate metal ratios, followed by heating in a stream of HCl or HBr gas. The most stable phases are generally $M'MX_6$, but are considerably subject to phase transitions at fairly low temperatures so that single crystals for X-ray study are difficult to obtain and structures are dependent on powder data. Thus K_3LuCl_6 crystallizes in the K_3MoCl_6 structure[523] containing individual $LuCl_6^{3-}$ octahedra. The compounds $(NH_4)_3MCl_6$ (M = Sm–Lu or Y) and $(NH_4)_2MCl_5$ (M = La–Tb) have been obtained as two isostructural series.[524,525] The compounds K_3MCl_6 (M = Nd–Lu and Y) and K_2MCl_5 (M = Nd–Tb) have also been characterized by powder diffraction[526] following phase studies.[527]

Synthesis was by heating above 450 °C in evacuated quartz vessels of the stoichiometric components. The isostructural K_2MCl_6 series underwent a phase transition in the range 350–410 °C, and became monohydrated in air. The K_2MCl_5 series was also isostructural.

The complexes M'_3MX_6, where M′ = Ph_3PH or C_5H_5NH and X = Cl with M = Pr–Sm or Dy–Tm, or X = Br with M = Pr–Eu, have been prepared from acetonitrile solution and their electronic spectra have been studied. This is of importance because there are limited numbers of lanthanide complexes where the metal ion is octahedrally coordinated to identical ligands in separated octahedra. As expected, in the absence of unsymmetrical components of the crystal field, these complexes show much weaker oscillator strengths than other lanthanide complexes. The hexaiodo complexes $[Ph_3PH][MI_6]$, where M = La–Yb, and the corresponding pyridinium salts, have been prepared, working rapidly on the small scale, by action of liquid HI on the solid $[MCl_6]^{3-}$ compound. These compounds are moisture sensitive and dissociate even in non-aqueous solvents in the presence of excess iodide ion. The series of halo complexes have been used to obtain values of optical electronegativities from the charge-transfer spectra. The values of $d\beta$, the nephelauxetic constant, from the $f \rightarrow f$ transitions, are similar for all three halides.[528,529]

The structure of Cs_2DyCl_5 has been determined from single crystal data.[530] It contains $DyCl_6^{3-}$ octahedra connected by *cis* corners to form chains, the connecting Dy—Cl—Dy angle being 180°. The heavier lanthanides and yttrium are isostructural.

All the compounds of the type $M'_3M_2Cl_9$ which have been obtained (M′ = K, M = Ce–Nd; M′ = Cs, M = Ho–Lu or Y) contain paired octahedra $M_2Cl_9^{3-}$ joined at one face; they have one or other of the $Cs_3Cr_2Cl_9$ or $Cs_3Tl_2Cl_9$ structures.[531] These structures are also adopted by a number of bromides or iodides, namely $Cs_3Y_2I_9$ and $Cs_3M_2Br_9$ (M = Sm–Lu or Y). Single crystal data are available for $Cs_3Y_2I_9$ which has the dimensions Y—I—Y = 80.8°, Y—I(terminal) = 2.90 Å and Y—I(bridging) = 3.13 Å.[532]

In the case of the stoichiometries $M'M_2X_7$, structures tend not towards more highly condensed MX_6 octahedra but towards seven-coordinated metal ions, as in KM_2Cl_7 (M = Gd–Tm or Y), where monocapped trigonal prisms are interconnected to form layers, and in $M'M_2Cl_7$ (M′ = Rb, M = Sm–Yb or Y; M′ = Cs, M = La or Pr) where the lanthanide ions are seven- and eight-coordinated.[522]

39.2.9 NMR Spectra of Paramagnetic Lanthanide Complexes

39.2.9.1 Introductory summary

In 1969, Hinckley reported[533] the extraordinary effect produced by $[Eu(dpm)_3(py)_2]$ upon the NMR spectrum of cholesterol in non-polar organic solvents. The spectrum was greatly shifted to low field, and in general the resonances of those protons nearer to the hydroxyl group were shifted more. This meant that firstly, accidental degeneracies in spectra of large organic molecules may often be removed because of differential shifts. Secondly, assignments of peaks may be confirmed, where there might otherwise be doubt, on the basis of their geometric position relative to the lanthanide ion. Thirdly and perhaps most important, structural information concerning the organic molecule may be obtained by means of a knowledge of the relative positions of the protons (or other nuclei) obtained through their differential shifts. It soon became clear that the β-diketonates themselves rather than their pyridine adducts should be used[534] and that the process depended on the formation of an adduct $M(\beta\text{-diketonate})_3L_x$, where L is the molecule under investigation, usually termed the substrate, which is often an alcohol, amine or ketone. Furthermore, the use of different lanthanides produced shifts in either direction and of various magnitudes.[535] Another variable is of course the nature of the β-diketone. This must be sufficiently bulky to prevent polymerization by intermolecular coordination, but not so very bulky that an adduct with the substrate cannot be formed.

The NMR shift is produced by an interaction between the resonating nucleus and the magnetic field produced by the unpaired f electrons of the lanthanide. This interaction may occur in a through-space dipole-dipole manner, or by delocalization of the f-electron density towards the resonating nucleus through the bonds of the substrate. Usually the dipole–dipole effect predominates. Much of the research on the mode of action of lanthanide shift reagents (LSRs) has been directed towards study of the following: the stoicheometry and geometry of the LSR–substrate adduct; the stability constant(s) relevant to this association; and the quantification of the two shift mechanisms in terms of molecular properties and the establishment of the relative importance of the two mechanisms. There is of course much literature concerned with the application of LSRs to the solution of individual problems in organic chemistry, but this will not be dealt with here in detail.

There are a number of reviews on the principles and applications of LSRs[536–540] and on the important uses of lanthanides as NMR probes in biological systems.[541–543] It should be noted that adducts of LSRs of the tris(β-diketonate) type constitute only a special case of paramagnetic shifts of NMR resonances of lanthanide complexes. All the nuclei of any lanthanide complex (except La, Gd, Lu) will in principle show a paramagnetic shift unless the complex is cubic, and some of the most interesting work has in fact been carried out on widely differing types of lanthanide complexes, particularly complexes of uncertain structure formed in solution, including aqueous solution, by interaction of lanthanide salts with various ligands.[544]

39.2.9.2 The magnetic interaction between lanthanide ions and coordinated ligands

The most important magnetic interaction between lanthanide ions and nuclei which are components of coordinated molecules is probably the through-space dipole–dipole interaction, usually termed the dipolar, or in earlier work the pseudocontact, interaction. The reason for this statement is that the dipolar shift is related to the geometry of the adduct, obeying equation (9).[545] Here, r, θ and ϕ are the polar coordinates of the particular nucleus under study, relative to the principal magnetic axes of the molecule (whose origin of coordinates is the lanthanide ion). A is a constant which is different for each lanthanide and depends on the lanthanide electronic configuration; and F_n^m are crystal field coefficients which depend on the nature and geometry of the ligands.

$$\Delta H_d/H = \{K_1(3\cos^2\theta - 1) + K_2\sin^2\theta\cos^2\phi\}/r^3, \quad \text{where} \quad K_1,\ K_2 \text{ are functions of } A\langle F_n^m\rangle \qquad (9)$$

The values of A have been tabulated.[545] In the cases of Eu^{3+} and Sm^{3+}, there are substantial magnetic contributions from excited states. Eu^{3+} with a ground state of $^7F_0$, which is diamagnetic, depends totally upon the $^7F_{1,2,3}$ excited states, which are thermally accessible, for its magnetic properties. From the equation it follows that for one complex of any individual lanthanide, the relative dipolar shifts of all the nuclei are determined by their geometric relationship in terms of θ, ϕ and r. Also for an isostructural series of complexes of different lanthanides, the shifts of analogous nuclei are related only by the values of A, assuming the crystal field parameters are not appreciably different.

Since each nucleus has three parameters, θ, ϕ and r, but yields only one experimental datum, ΔH, it is not of course possible to determine the overall structure of the complex *ab initio*. However, uncertainty between alternative structures of substrates is capable of being resolved, as is uncertainty in alternative assignments of the NMR resonances, or uncertainty in the orientation or position of coordination of substrates of known structure. A computer optimization of agreement between observed shifts and coordination orientation (or whatever other uncertain features are to be investigated) is the normal procedure, as first demonstrated for the [Pr(dpm)$_3$(borneol)] adduct.[546]

The mechanism of this type of paramagnetic shift can be understood in a purely qualitative way by considering a lanthanide complex tumbling in the magnetic field of an NMR spectrometer. This field causes the magnetic dipole of the lanthanide ion on average to provide a component opposed to the field. The magnetic field due to the lanthanide ion, as experienced by a hydrogen nucleus which is part of the complex, will be in opposite directions depending on whether the nucleus, during molecular tumbling, is on the axis of, or at right angles to, the magnetic dipole of the lanthanide ion. For isotropic complexes, it can be shown that tumbling over all orientations averages the resultant shift to zero (except for the Evans shift[547] on all nuclei). However, for anisotropic complexes the magnitudes of the lanthanide magnetic dipole will be different in the two situations mentioned and the resultant time-averaged field is non-zero and will provide a paramagnetic shift.

It has been stated above that for a series of isostructural lanthanide complexes the paramagnetic shifts of analogous nuclei should be proportional to the parameter A, if a purely dipolar mechanism operates. The paramagnetic shift is usually taken as the difference between the chemical shift of the nucleus in the paramagnetic complex under investigation, and the corresponding shift of the diamagnetic lanthanum complex (or the mean value of the lanthanum and lutetium shifts). In many series of complexes it is found that this simple relationship does not apply, certainly to all nuclei, even if the series is isostructural. Furthermore, it is almost always found that the geometrical relationship of the Bleaney equation is not valid for all nuclei when high levels of accuracy are concerned and in many cases it is not even approximately valid, especially for nuclei other than hydrogen or those connected to the lanthanide ion through a small number of bonds. This situation is caused by the presence of contact shifts arising from the through-bond delocalization of unpaired electron density associated with the lanthanide ion, which then directly influences the nucleus concerned. Equation (10) expresses the contact shift,[548] where a is the hyperfine coupling constant, while S_z is the spin expectation value of the particular lanthanide. Equation (11) then expresses the total paramagnetic shift.

$$\Delta H_c/H = a\langle S_z\rangle \qquad (10)$$

$$\Delta H_{\text{total}} = \Delta H_d + \Delta H_c \qquad (11)$$

It is clearly necessary to estimate the relative contributions of the contact and dipole mechanisms to the overall shift in order to make use of the structural information inherent in the dipolar shifts. This may be done in different ways.[549–551] Firstly, the shift of any gadolinium complex is entirely contact, as the $^8S_{7/2}$ ground state of Gd^{3+} is necessarily isotropic. Hence a, the hyperfine coupling constant, may be found and if it is assumed constant for the series of lanthanides, then the contact component for each lanthanide complex may be found from the appropriate value of S_z. Unfortunately, the very long relaxation time for Gd^{3+} confers considerable broadening on the NMR spectra of its complexes which usually prevents resonances from being observed. An alternative method compares the observed shifts of analogous nuclei in two complexes, in which case a is assumed constant, A and S_z are known for each ion, and hence a can be calculated and ΔH_c and ΔH_d found. If a is constant over a series of several lanthanides, then computer optimization of such a series of complexes would be preferable. If this type of treatment, of which there are a number of variations,[552] gives good quantitative agreement with the observed data for a series of lanthanide complexes, then they may normally be assumed isostructural. Furthermore, the values of θ, ϕ and r from which follow the value of ΔH_d for the various different nuclei can be computer optimized for consistency of the shifts with hypothetical molecular models which are structurally acceptable.

The dipolar paramagnetic shifts depend, as has been stated, on unsymmetrical terms in the crystal field expansion. An unsymmetrical crystal field can be caused by unsymmetrically disposed ligands of similar type, or by the presence in the same complex of ligands which exert dissimilar crystal fields although the coordination polyhedron may be fairly symmetrical. The practical effect of the unsymmetrical crystal field is to confer anisotropy on the magnetic susceptibility tensor by perturbing the f electron shell. The intensity of magnetization of an individual lanthanide ion thus varies with the orientation of the complex, of which it forms part, relative to the prevailing magnetic field. Bleaney[545] has given the relationship between the magnetic anisotropy and the resulting paramagnetic shift. It is as shown in equation (12), where $\chi_{x,y,z}$ are the principal components of the molecular magnetic susceptibility tensor and $\bar{\chi}$ is their mean.

$$\Delta H_d/H = \{(3\cos^2\theta - 1)(\chi_z - \bar{\chi}) + \sin^2\theta\cos 2\phi(\chi_x - \chi_y)\}/2Nr^3 \qquad (12)$$

Hence, if the magnetic anisotropy of the crystalline complex is measured experimentally, usually by single crystal measurements using the torsion fibre method, and if its crystal structure is known, then the molecular anisotropy usually can be calculated. Provided that the complex has the same conformation in solution as in the solid state, the solution dipolar shifts can be calculated and compared with the experimental data. This procedure and variations of it may be used to investigate the relative crystal fields of different ligands, and to test experimentally the validity of the theoretical approach employed.[553,554] A direct NMR measurement of anisotropic magnetic susceptibilities can be obtained from the ^2H spectra of lanthanide complexes of known structure. The method depends on quadrupole coupling of the ^2H nuclei, which splits absorption lines at high field owing to partial ordering of the molecules in the field.[555]

The question of the effect of molecular symmetry on dipolar shifts may be briefly discussed next. If a complex has a principal axis of symmetry of order three or more, that is, it is uniaxial, then χ_x is equal to χ_y and a simplified geometric factor $(3\sin^2\theta - 1)r^{-3}$ is valid. Although some lanthanide complexes have such symmetry in the solid state, most do not and are said to be biaxial. However, if an individual ligand undergoes NMR-rapid rotation about a single bond relative to the remainder of the complex, then an averaging effect is naturally observed for the nuclei of that ligand even though the complex instantaneously may have even C_1 symmetry.[556,557] This situation is common in solutions of lanthanide complexes and the simplified geometric factor is often capable of fitting the experimental data.

There is some evidence that while mono adducts of the β-diketonate shift reagents may show time-averaged uniaxial symmetry, bis adducts often do not. Thus the spectra of $[M(dpm)_3L_2]$, where M = Pr, Sm, Dy, Ho, Er or Yb and L = 3-picoline or 3,5-lutidine, have been interpreted in terms of biaxial magnetic anisotropy. One magnetic axis is defined by the molecular C_2 axis present in the X-ray crystal structure and the other two axes lie in similar orientations in all the adducts. A variable-temperature study of this system shows restricted rotation of the aromatic rings with a very solvent-dependent conformational equilibrium resulting from this.[558–560]

39.2.9.3 Nuclear relaxation rates

A further, most important, feature which does not produce a paramagnetic shift but which falls naturally into place for discussion at this point is the lanthanide contribution to the nuclear relaxation. This is a dipole–dipole interaction and hence follows an r^{-6} law. The longitudinal relaxation rate of a nucleus of a diamagnetic atom near a paramagnetic lanthanide ion (except Gd^{3+}) is given by equation (13), where $1/T_1$ = longitudinal relaxation rate of the nucleus, γ_1 = magnetogyric ratio, τ_e = electron spin relaxation time, μ = magnetic moment, β = Bohr magneton and r = ion–nucleus distance.[561]

$$\frac{1}{T_1} = \frac{4}{3}\left(\frac{\mu_0}{4\pi}\right)^2 \gamma_1^2 \mu^2 \beta^2 \tau_e r^{-6} \tag{13}$$

The effect of this equation is that the ratios of the paramagnetic contributions to the relaxation rates of a set of, for example, ^{1}H nuclei belonging to the same complex are determined by the lanthanide ^{1}H distance only. Hence a similar type of computer optimization as for the paramagnetic shifts may be used to test compatibility with hypothetical structures which incorporate distance-dependent parameters such as the orientation or conformation of an organic ligand.

A further application of relaxation rate measurements is that similar $1/T_1$ ratios in a series of lanthanide complexes may be taken to indicate an isostructural series. However, this approach has the limitation that if only part of the complex is studied, perhaps an organic ligand, its T_1 ratios would be independent of changes, for example changes in the extent of hydration in the remainder of the complex, provided that the conformation of the ligand relative to the lanthanide ion were preserved. An excellent example of the use of T_1 data in a quite different way is its use to determine hydration numbers of lanthanide dipicolinate complexes.[562]

39.2.9.4 Examples of applications of shift or relaxation measurements

A good example of an investigation into the conformation of coordination of an organic ligand is the work of Singh, Reynolds and Sherry[563] on lanthanide L-proline complexation. The ^{1}H and ^{13}C paramagnetic shifts and relaxation rates were measured for all H and C atoms of aqueous proline–lanthanide ion solutions at pH 3 for ten lanthanides. It was concluded that while the relaxation data were consistent with an isostructural series, there were inconsistencies in the dipolar shift data, after separation from contact shifts, which could be resolved either by division into two, rather than one, isostructural series, or by invoking biaxial rather than uniaxial symmetry.

Adenosine-2′-monophosphate and adenosine-3′-monophosphate have been investigated in aqueous solution and were indicated by shift and relaxation data to be in conformational equilibrium when bound to lanthanide ions. The lanthanides bind to bidentate phosphate groups.[564]

Adenosine-5′-monophosphate has also been examined similarly. It was studied in $D_2O/(CD_3)_2SO$ in the presence of lanthanide ions under various conditions of solvent composition and temperature. Shifts and relaxation times for the series Pr–Yb gave results more nearly consistent with axial symmetry for the first half of the lanthanide series than for the second.[565] The extraction of structural information from NMR data has been described for aqueous systems where lanthanide edta and related complexes interact with cytidine-5′-monophosphate, (–)-alanine[566] and *endo-cis*-bicyclo[2.2.1]hept-5-ene-2,3-dicarboxylic acid.[567] The shift reagents $[M(fod)_3]$ where M = Pr, Eu, Gd, Ho, Er or Yb have been used to elucidate the solution structure of nicotine *N*-methiodide, ^{13}C and ^{1}H spectra being studied. The experimental data indicated effective axial symmetry, good agreement with theory being obtained. Calculation of the relative magnitudes of ΔH_d and ΔH_c showed that the dipolar mechanism predominated.[568] Of course, many other examples could be given of studies of adducts of LSRs and organic substrates but these fall within the purview of organic chemistry rather than the present discussion.

There has been some work on interactions between LSRs and transition metal complexes, both NMR shifts and relaxation rates being studied. Presumably the mode of interaction is by means of bridges formed by the donor atoms of one complex which bond in a labile manner to the metal ion of the second complex. Interactions examined include those between $[Eu(fod\text{-}d_9)_3]$ and azido{*N,N*′-ethylenebis(acetylacetoneiminato)}pyridinecobalt(III), where a 1:1 adduct with a lifetime of 8.4×10^{-4} s was formed in $CDCl_3$;[569] $[Gd(fod)_3]$ and *N,N*′-ethylenebis(acetylacetoniminato)nickel(II)[570,571] or *N,N*′-vinylenebis(acetylacetoniminato)nickel(II); $[Pr(fod)_3]$ and $[Pt(acac)_2]$;[572] and $[M(fod)_3]$ and cobalt(III) tris-β-diketonates (M = La–Eu).[573]

Other more specialized interactions of β-diketonate lanthanide shift reagents are those with alkenes, which require the mediation of a silver complex.[574] Thus $[M(fod)_3]$, where M = Pr or Yb, together with Ag(fod) will shift the resonances of alkenes, aromatic molecules or phosphines, which are ligands which will not coordinate directly to lanthanide β-diketonates.[575,576] The shifted spectrum resulting from the interaction between a platinum alkene complex and $[Eu(hfc)_3]$ has also been studied (Hhfc = 3-heptafluoropropylhydroxymethylene-(+)-camphor). In this case the alkene complex is prochiral and the shift reagent is of course optically active.[577] Optically active shift reagents such as $Eu(hfc)_3$ have often been used to demonstrate quantitatively the presence of chiral isomers of organic substrates; two sets of resonances are obtained.[578,579] Chiral interactions in aqueous solution are also possible,[580] and various types of such interaction, leading to spectral resolution of substrates or to assignment of chirality, have been discussed.[581]

Paramagnetic shifts may be induced in cations which associate with suitable lanthanide complexes. Thus in aqueous solution, $[M(hpda)_3]^{6-}$, where M = Dy or Tm and H_3hpda = 4-hydroxypyridine-2,6-dicarboxylic acid, associates with $^{25}Mg^{2+}$, $^{39}K^+$, $^{23}Na^+$, $^{87}Rb^+$ or $^{14}NH_4$ and produces shifts of up to 25 p.p.m. which are most marked above pH 7.[582] A similar effect is shown for cations (and anions and uncharged molecules) by lanthanide tetra-*p*-sulfonatophenylporphyrins in aqueous solutions.[583] This is discussed in Section 39.2.6.3, as are the large shifts produced in aqueous solution by $Na[Eu(dota)(H_2O)]$ (H_4 dota = **23**).[584]

Most examples of the application of paramagnetic lanthanide NMR spectra quoted here have so far been based on transient association in solution between a preexisting lanthanide species and a substrate, often organic. However, the shifted NMR spectra of stable isolable lanthanide complexes have been studied since it was established[585] that in contrast to many paramagnetic complexes of the *d* transition metals, lanthanide complexes in general gave fairly sharp spectra. The kinetics of ligand exchange processes were early studied by these means,[586] and in more recent work, the conformations of stable isolable lanthanide complexes and their interconformational or dissociative equilibria have often been usefully investigated. Thus the paramagnetically shifted spectra of $[M(dipic)_3]^{3-}$, $[M(dipic)_2(H_2O)_3]^-$ and $[M(dipic)(H_2O)_6]^+$, where M = Ce–Yb and H_2dipic = pyridine-2,6-dicarboxylic acid, have been shown to be consistent with an approximately tricapped trigonal prismatic configuration of all three types of complex,[587] similar to the X-ray crystal structure of the Yb tris complex.[588] The lanthanide diethyldithiophosphonates $[M\{S_2P(OEt)_2\}_4]^-$ which are dodecahedral in the solid state, divide into two series in CD_2Cl_2 solution, the heavier lanthanides having square antiprismatic geometry.[589] Other examples, drawn from the NMR spectra of macrocyclic complexes are given in Section 39.2.7.

39.2.9.5 Studies of complexation with lanthanide shift reagents

Owing to their special utility in structure elucidation of organic molecules in non-hydroxylic solvents, the lanthanide tris(β-diketonate) complexes have undergone much investigation into their complexation with lone pair donors, mainly with the objectives of determination of association constant values and determining whether 1:1 or 2:1 adducts are formed. Many of these papers naturally rely on NMR measurements though other methods have been used. The complexes mainly investigated are $M(RCOCHCOR')_3$ where M = Eu, Pr or Yb. These ions are used because they give fairly large shifts but without too much broadening. The β-diketones are usually R = R' = CMe_3 (dpm) or R = CMe_3, R' = n-C_3H_7 (fod). This topic is discussed in more detail in Section 39.2.4.2.

39.2.10 Electronic Spectra of Lanthanide Complexes

This area has been the subject of recent extensive reviews,[590-592] which deal particularly with the physical and theoretical aspects. In this section, a more chemical and descriptive approach will be adopted.

The tripositive lanthanide ions have the general outer configuration $4f^n5s^2p^6$ and follow the Russell–Saunders L,S,J coupling scheme but with a certain amount of configuration interaction. The 5s and 5p shells have a radial dispersion which is greater than the 4f electrons[593] and hence shield the latter from the effect of coordinated ligands to a very large extent, but not completely. Thus the electronic spectra of tripositive lanthanide compounds can be considered as derived from the spectra of the gaseous ions by a fairly small perturbation. This perturbation may be of two general types. The first is a general nephelauxetic effect owing to a drift of ligand or 5s, 5p electron density into the lanthanide ion, slightly expanding the 4f shell and reducing the 4f–4f interactions. The second is a crystal field splitting effect, which removes the degeneracy of the free-ion levels, and which can be represented with fair success by a point negative charge model having the appropriate molecular symmetry.

The experimental energy level schemes of the gaseous free tripositive ions are obtainable with some difficulty but are available in some cases, *e.g.* praseodymium[594] or gadolinium.[595] Values for Pr^{3+} (gaseous) *versus* $PrCl_3$ (solid) are shown in Table 11, and differ by about 2–5%, those for $PrCl_3$ being the smaller, as expected. Most of the basic experimental data regarding the energy levels of lanthanides must therefore be derived from the solid state, in

Table 11 Positions (cm^{-1}) of Energy Levels of Pr^{3+} (Gaseous, Obtained from Arc Spectra) Compared with $PrCl_3$ (Crystalline)[594]

	$^3H_4$	$^3H_5$	$^3H_6$	$^3F_2$	$^3F_3$	$^3F_4$	$^1G_4$	$^1D_2$	$^3P_0$	$^3P_1$	$^1I_6$	$^3P_2$
E(crystal)	0	2117.4	4306.3	4846.6	6232.3	6681.7	9697.6	16 639.3	20 383.4	20 984.9	21 324.5	22 139.1
E(vapour)	0	2152.2	4389.1	4996.7	6415.4	6854.9	9921.4	17 334.5	21 390.1	22 007.6	22 211.6	23 160.9

which the ion under study is often present in dilute solid solution in a lanthanum salt matrix. Of course, except for the ground state multiplet, identifiable from Hund's rules, the observed levels cannot be definitely assigned to particular $^{(2S+1)}L_J$ terms unless a theoretically derived set of energy levels is available for comparison. Considerable effort has been devoted to theoretical prediction of the energy levels of f^n configurations. The total energy H of a system consisting of a point nucleus surrounded by N electrons can be represented by the Hamiltonian in equation (14), where H_0 is the kinetic energy and nuclear interaction term, H_E is the electronic repulsion term, and H_{SO} is the spin–orbit interaction term.

$$H = H_0 + H_E + H_{SO} \tag{14}$$

Model Hartree–Fock calculations which include only the electrostatic interaction in terms of the Slater integrals F_0, F_2, F_4 and F_6, and the spin–orbit interaction ζ, result in differences between calculated and experimentally observed levels[596] which can be more than 500 cm^{-1} even for the f^2 ion Pr^{3+}. However, inclusion of configuration interaction terms, either two-particle or three-particle, considerably improves the correlations.[597,598] In this way, an ion such as Nd^{3+} can be described in terms of 18 parameters (including crystal field parameters) which reproduced 101 of the energy levels observed for Nd^{3+} in $LaCl_3$ with an rms deviation from experimental values of $8\,cm^{-1}$.[599] Although the parameters can be evaluated theoretically, it is usual to obtain them from a best fit of the observed spectra. The energy levels for all the tripositive lanthanide ions in $LaCl_3$ solid solution have been given.[600]

Using modern computers it is possible to diagonalize a complete free-ion and crystal-field Hamiltonian. However, for the lanthanides, if not for the actinides, the effect of the crystal field can usually be treated as a perturbation on the unsplit $^{2S+1}L_J$ levels. The levels are split into $2J+1$ components in a crystal field of C_1 symmetry, or into fewer components in higher symmetries, depending on the degeneracy of the J vector representation in the crystal field point group. Typical values of the crystal field parameters $B_g^{(k)}$ are 200–2000 cm^{-1}. Owing to the small effect of the ligands on the $4f$ levels, the $4f \rightarrow 4f$ spectra of lanthanide complexes tend to have narrow bands that are also weak owing to the transitions being Laporte forbidden. The intensities of the transitions have been the subject of much study and are usually quantified by the Judd–Ofelt theory. Judd[601] and Ofelt[602] used the odd parity terms of the ligand field to mix electronic states of opposite parity, including states derived from d configurations, into the f^n configurations, thus obtaining non-zero matrix elements for the electric dipole operator. The Judd–Ofelt theory also accommodates the existence of the hypersensitive transitions. These are transitions that are dramatically increased in intensity in complexes, particularly those of low symmetry and with polarizable ligands, compared with for example the hydrated ions. These transitions correspond to those having large values of Judd's second rank tensor operator $U^{(2)}$, and mostly have $\Delta J = \pm 2$. These selection rules correspond with those for a radiative electric quadrupolar process. It is, however, fairly certain that these hypersensitive transitions are electric dipolar in the main. Their mechanism can be described in two physically related ways, the inhomogeneous dielectric theory and the ligand polarization model.[603,607] The dipolar components of the radiation field induce a set of transient electric dipoles in the ligand environment which can couple with the $4f$ electrons by way of $4f$ electrostatic quadrupolar–ligand dipolar interactions. In a non-centrosymmetric complex, or in a centrosymmetric complex subject to an ungerade vibrational mode, large amplifications of $4f \rightarrow 4f$ transitions may be obtained from these quadrupole–induced dipolar interactions. These transitions are thus often referred to as pseudoquadrupolar. For a short readable discussion of this still controversial area, see Judd.[608]

In a recent paper,[609] the hypersensitive transitions $^4I_{15/2} \rightarrow {}^2H_{11/2}$ or $^4G_{11/2}$ of Er^{3+} have been examined for complexes with five amino- or oxo-carboxylate anions. They showed four- to six-fold increase in oscillator strengths over the hydrated ion Er^{3+} aq. Using likely models for the structures of the complexes and calculating crystal field wave functions and energy levels for the $4f^{11}$ configuration of Er^{3+}, electric and magnetic dipole transition strengths were calculated. The structural models were based on X-ray solid state determinations and charge and polarizability parameters were assigned to the ligand molecules. The calculated magnetic dipole intensitities were at least two orders of magnitude less than the electric dipolar; the latter were very successful in accounting quantitatively for the observed intensitities. In a related paper, the oscillator strengths of Nd, Ho and Er acetates and propionates in aqueous solution have been related to solution structures and hypersensitivity mechanisms discussed.[610]

In practice, the hypersensitive transitions are often used for the determination of stability constants in aqueous solution. Lanthanide absorption bands in solution do not normally change in position on complexation to such an extent that bands due to the complexed and uncomplexed ion can be clearly observed independently, as is often the case for d transition metal ions, but the marked change of intensity of the hypersensitive bands is sufficient to allow determination of K values, for example as demonstrated for various adducts of $[Ho(dpm)_3]$.[611]

There has been a marked increase recently in the interest shown in fluorescence, phosphorescence or emission spectra. There is little useful distinction between these terms when applied to lanthanide ions, and the term emission will be preferred here. The two ions mainly studied are Tb^{3+} and, especially, Eu^{3+}. The energy level schemes for Eu^{3+} and Tb^{3+} are given in Figure 9. In both these ions there is a large energy gap between the ^7F and ^5D levels, which corresponds with emission in the red and the green regions respectively of the visible spectrum. The main emitting state in each case is the lowest component of the excited multiplet, as higher components rapidly descend by a cascade process to $^5D_0$ or $^5D_4$ respectively. In the case of Eu^{3+}, emission corresponding to $^5D_0 \rightarrow {}^7F_{0,1,2,3}$ can be observed with $^5D_0 \rightarrow {}^7F_1$ or $^7F_2$ the strongest.

This emission is useful in two respects. Firstly, for Eu^{3+}, the crystal field splitting of the $^7F_1$ and $^7F_2$ levels can be observed and can be related to the symmetry of the complex if that is

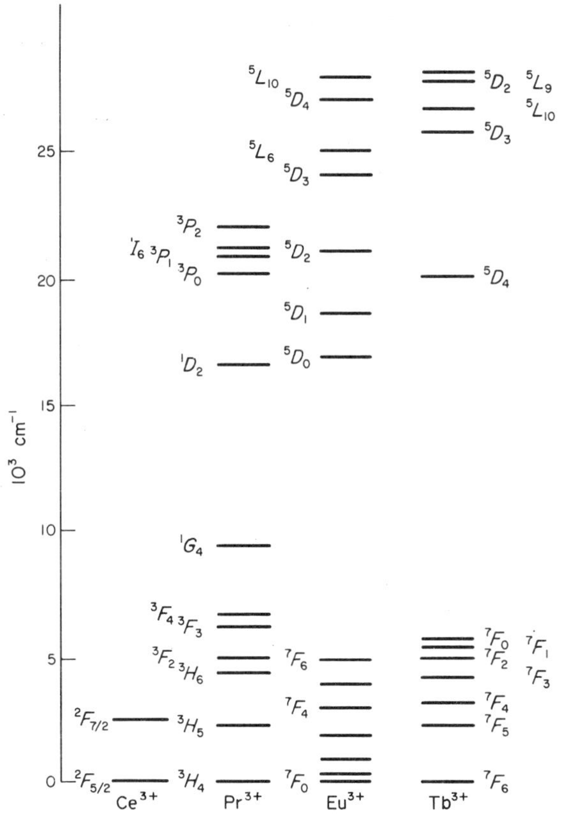

Figure 9 Energy levels of some representative lanthanide tripositive ions. Ce and Pr, full set; Eu and Tb, partial set (energy values taken from S. Hüfner, 'Optical Spectra of Transparent Rare Earth Compounds', Academic Press, 1978)

unknown.[612,613] The *J* values for both the excited and ground states of Tb^{3+} are too large for this effect to be so useful for that ion, as it is difficult to assign the multiply split transitions. Most of these transitions could in principle be observed in absorption but are too weak to be easily studied.

Secondly, the emission of a particular complex species is characteristic of that species and can be used to identify the species present. Particularly is this so if excited state lifetimes are measured, as these vary dramatically depending on the number of OH groups coordinated to the Eu^{3+} or Tb^{3+} ion. This is because multiple excitation of the OH stretching mode provides an alternative deexcitation route. Measurement of lifetimes thus can be used to determine the number of coordinated water molecules.[218]

Excitation of the Eu^{3+} or Tb^{3+} ions has traditionally been indirect, by broad-band UV excitation of a conjugated organic ligand which is followed by intramolecular energy transfer to the lanthanide ion *f* system, followed in turn by $f \rightarrow f$ emission.[614] However, more recently, following the advent of tunable dye lasers, direct excitation of an excited f^n level is in many cases preferable. By scanning this frequency, an excitation spectrum can be obtained whose energy values are independent of the resolution of a monochromator and not subject to spectral interferences.

Excitation spectra have been of considerable use recently in studying both hydration numbers (by lifetime measurements) and inner-sphere complexation by anions (by observing appearance of the characteristic frequencies for *e.g.* the Eu^{3+} $^5D_0 \rightarrow {}^7F_0$ transition for the different possible species). Thus using a pulsed dye laser source, it was possible to demonstrate the occurrence of inner sphere complexes of Eu^{3+} with SCN^-, Cl^- or NO_3^- in aqueous solution, the *K* values being 5.96 ± 2, 0.13 ± 0.01 and 1.41 ± 0.2 respectively. The ClO_4^- ion did not coordinate. Excited state lifetimes suggest the nitrate species is $[Eu(NO_3)(H_2O)_{6.8\pm0.4}]^{2+}$; the technique here is to compare the lifetimes of the H_2O and the corresponding D_2O species, where the vibrational deactivation pathway is virtually inoperative.[219] The reduction in lifetime is proportional to the number of water molecules complexed.[217,218]

A study of hydration of the complex $[EuL_2(H_2O)_x](BF_4)_3$ where $L = \{(C_3H_7)_2N{\cdot}CO{\cdot}CH_2{-}OCH(Me){-}\}_2$ shows by lifetime studies of a similar sort that in aqueous acetone, $x = 0.9 \pm 0.4$ which gives a coordination number of nine. This complex was prepared from $Eu(MeCN)_x(BF_3)_3$, itself prepared from the action of $NOBF_4$ on europium metal in acetonitrile.[615]

The emission spectra of the complexes $R_2[Eu(NO_3)_5]$, where $R = Ph_3EtP^+$ or Ph_4As^+, have been investigated and it was found that the spectrum of the former could be interpreted in terms of D_{3h} pseudosymmetry (each NO_3 ion considered as occupying one ligand position in the coordination sphere) but the position as regards the Ph_4As^+ salt was less clear.[616] In a similar study, using lifetime measurements, the species $Tb(NO_3)_3L_3$, where $L = DMSO$ or H_2O, were identified in MeCN solution. Also, efficiencies were determined of energy transfer from excited Tb^{3+} ions in $Tb(ClO_4)_3$ or $Tb(NO_3)_3$ to Nd^{3+} ions in MeCN and other more polar organic solvents.[617]

The ${}^4F_{3/2} \rightarrow {}^4F_{5/2}$ emission of Nd^{3+} at 885 nm can also be monitored in a rather similar way, and a study of emission lifetimes together with absorption spectra for the hypersensitive Nd^{3+} transitions, *e.g.* ${}^4I_{9/2} \rightarrow {}^4G_{5/2}$, has been made in order to investigate the solvation of neodymium nitrate and perchlorate in MeCN, Me_2CO, DMSO or DMF. Excited lifetimes varied from 300 ns in Me_2CO to 2300 ns in CD_3CN. It was found that species $\{Nd(NO_3)_n\}^{(3-n)+}$ were formed, where $n = 1$–5, on addition of NO_3^- ions to perchlorate solutions, but that addition of DMSO to $Nd(ClO_4)_3$ in MeCN gave $[Nd(DMSO)_n]^{3+}$, where $n = 9.7 \pm 0.8$.[618]

A further variation on the theme of emission is circularly polarized emission, where chiral interactions, for example between a lanthanide complex and a chiral ligand in solution, can be studied. Selection rules have been given[619] based on *S*, *L* and *J* values for 4*f* states perturbed by spin–orbit coupling and 4*f* electron–crystal field interactions, and four types of transition were predicted to be highly active chiroptically. These are given in Table 12.

Table 12 Active Chiroptical Transitions in Lanthanide Complexes

Type	$\lvert\Delta S\rvert$	$\lvert\Delta L\rvert$*	$\lvert\Delta J\rvert$
a	0	0	$0, 1 (J \neq 0 \neq J')$
b	0	0	$1 (J \text{ or } J' = 0)$
c	>0	⩾0	$0, 1 (J \neq 0 \neq J')$
d	>0	⩾0	$1 (J \text{ or } J' = 0)$

(From ref. 619; a = type 1, b = type 2, c = type 9, d = type 10).
* $\Delta L \not> 6$.

The Pfeiffer effect, the outer-sphere interaction of a chiral substrate with a rapidly interconverting racemic solution of a chiral lanthanide complex, can be investigated by measurement of the luminescence dissymmetry factor (the ratio of circularly polarized luminescence to total luminescence) for Eu^{3+} or Tb^{3+} complexes. Thus the racemic D_3 chiral complexes $[M(dpa)_3]^{3+}$, where M = Eu or Tb, interact in an outer-sphere manner with the following optically active species: cationic chiral transition metal complexes,[620] ascorbic acid,[621] aminocarboxylates,[622,623] tartrates,[624] amines[625] and phenols.[626] Association constants can be obtained from limiting values of the dissymmetry factors. In some cases, inner-sphere complexation[626] can be demonstrated, as judged by changes in the general nature of the circularly polarized luminescence spectrum and pH irreversibility of the complexation.

The interaction of aspartic acid and other ligands with complexes of Tb^{3+} with edta and related ligands has also been studied and association constants determined.[627] The complex formation between Tb^{3+} or Eu^{3+} and (R)-(−)-1,2-propanediaminetetraacetic acid or (R,R)-*trans*-1,2-cyclohexanediaminetetraacetic acid has been similarly investigated. The pH dependence of the circularly polarized and total luminescence shows a drastic configurational change of the chelate system at pH 10.5–11, corresponding, it is believed, to formation of hydroxide complexes.[628] The technique of magnetic-field-induced circularly polarized emission has been introduced for lanthanide ions;[629] the mechanisms of lanthanide transition intensities are also discussed in the paper.

A useful short general review[630] of the utility of Eu^{3+} and Tb^{3+} emission spectra in investigations of biological systems gives also an account of the relevant basic spectroscopy of the lanthanide ions. The importance of these techniques for exploring proteins and nucleotides is that Eu^{3+} is easily bound by such molecules, and that it readily substitutes Ca^{2+} naturally present. The effects of coordinated ligands upon the fluorescence spectrum can then be used for structural exploration. An example is the investigation of complexes of Eu^{3+} or Tb^{3+} with the ionophores lasalocid A and A23187 in MeOH. These are polyfunctional compounds containing keto, ether, carboxylate or amine groups. Mono and bis species were observed, and the high K values showed multidentate complexing. Suggestions as to the nature of this could be made.[631]

The bulk magnetic susceptibilities of lanthanide complexes are accurately predicted by theory, as the crystal field splitting of the ground state has little effect on these. The effective magnetic moment μ(eff) is thus given by equation (15) where g_J is the Landé factor. Values of these parameters together with a typical set of room temperature experimental values, for $[M(NO_3)_3(phen)_2]$,[150] are given in Table 13. In the case of Sm^{3+} and particularly Eu^{3+}, the excited states $^6H_{7/2}$ (Sm) and $^7F_{1,2}$ (Eu) are thermally populated at room temperature and lead to higher observed values of μ(eff) than those calculated from the ground term only.

$$\mu_{eff} = g_J\sqrt{J(J+1)}, \quad \text{where} \quad g_J = \{S(S+1) - L(L+1) + 3J(J+1)\}/2J(J+1) \qquad (15)$$

Table 13 Ground Terms and Magnetic Moments of the Lanthanide Ions

M^{3+}	La	Ce	Pr	Nd	Pm	Sm	Eu	Gd	Tb	Dy	Ho	Er	Tm	Yb	Lu
Ground term	$^1S_0$	$^2F_{5/2}$	$^3H_4$	$^4I_{9/2}$	$^5I_4$	$^6H_{5/2}$	$^7F_0$	$^8S_{7/2}$	$^7F_6$	$^6H_{15/2}$	$^5I_8$	$^4I_{15/2}$	$^3H_6$	$^2F_{7/2}$	$^1S_0$
g_J	0	6/7	4/5	8/11	3/5	2/7	—	2	3/2	4/3	5/4	6/5	7/6	8/7	0
$g_J\sqrt{J(J+1)}$	0	2.54	3.58	3.62	2.83	0.84[a]	0[a]	7.94	9.7	10.6	10.6	9.6	7.6	4.5	0
μ_{eff} (exp)[b]	0	2.46	3.48	3.44	—	1.64	3.36	7.97	9.81	10.6	10.7	9.46	7.51	4.47	0

[a] Non-ground state terms contribute to the observed magnetic moments.
[b] Values are those found for $M(NO_3)_3(phen)_2$ at 20 °C.[150]

The crystal field does, however, have a dramatic effect on the magnetic anisotropy of lanthanide complexes. For complexes of less than cubic symmetry the three principal values of the susceptibility tensor are unequal. For uniaxial symmetry, $\chi_x = \chi_y \neq \chi_z$ and for biaxial symmetry $\chi_x \neq \chi_y \neq \chi_z$. Very extensive studies[632–640] have been carried out on the single crystal susceptibilities of the D_{3d} lanthanide hexakis(antipyrene) triiodides over the temperature range 80–300 K, and crystal field parameters were obtained. This crystal field-induced anisotropy is responsible for the effectiveness of lanthanide complexes as NMR shift reagents, and single crystal anisotropies of lanthanide complexes have been determined in this connection also.[553]

39.2.11 Lanthanides in the Dipositive Oxidation State

Although all the lanthanides are stable in the solid state as M^{2+} ions doped into CaF_2 crystals, only in the cases of europium, ytterbium and samarium is there any real coordination chemistry, and that is very limited. There is a small but developing organometallic chemistry of the lower oxidation states,[641] but that is not within the scope of the present review. Much of the chemistry of the dipositive state depends on solvated species[642] and it is convenient to begin with these.

39.2.11.1 Hydrated species

The hydrated ions of europium, ytterbium or samarium can be obtained by electrolytic reduction or by dissolution of MCl_2 in water, and are reasonably stable. Colourless or pale yellow-green Eu^{2+}(aq) is stable for more than ten days in the absence of oxidizing agents or of platinum catalysts.[643] Pale green Yb^{2+}(aq) decays in water with $k = 2.4 \times 10^{-5}\,s^{-1}$, and red Sm^{2+}(aq) has $k = 0.6 \times 10^{-4}\,s^{-1}$, both reactions being first order.[646] These rates of decomposition follow the sequence of standard electrode potentials which are listed together with other parameters of the M^{2+}(aq) ions in Table 14. Of course, as not all the M^{2+}(aq) ions exist, many of these parameters cannot be determined directly. Pulse radiolysis has been used to demonstrate the short-lived existence of red Tm^{2+}(aq), Er^{2+}(aq) and surprisingly Gd^{2+}(aq).[648] Electrochemical reduction has been used to investigate stabilities and reduction potentials of

M^{2+} ions.[649,650] Thus polarography and cyclic voltammetry of 10^{-4}–10^{-3} M Sm^{3+} or Gd^{3+} in 0.001 M H_2SO_4 showed, for the cyclic voltammetry, a two step reduction in each case, which was reversible for samarium with fast scan rates (50 V s^{-1}) but which showed no anodic waves for gadolinium. The reduction for the latter was interpreted as being similar to dysprosium[651] where there is a two-stage reduction of water *via* lanthanide hydroxy species, no M^{2+} species being involved (equations 16 and 17).

$$2M^{3+} + 2H_2O + 2e^- \longrightarrow 2M(OH)^{2+} + H_2 \qquad (16)$$

$$M(OH)^{2+} + 2H_2O + 2e^- \longrightarrow M(OH)_3 + H_2 \qquad (17)$$

Table 14 Thermodynamic and Electrochemical Parameters for Hydrated Dipositive Lanthanide Ions

	La	*Ce*	*Pr*	*Nd*	*Pm*	*Sm*	*Eu*	*Gd*	*Tb*	*Dy*	*Ho*	*Er*	*Tm*	*Yb*
I_3(eV)[a,645]	19.2	20.1	21.6	22.1	—	23.7	24.9	20.8	21.9	22.9	22.8	22.7	23.8	25.0
$-\Delta G_f^\circ(M^{2+}aq)$[b,646]	121	268	385	406	427	515	540	184	322	418	406	381	460	540
$-\Delta G_{hyd}^\circ(M^{2+})$[c,647]	—	1397	1309	1317	1346	1344	1361	1550	1463	1437	1433	1452	1451	1461
$-E^\circ$ (M^{3+}(aq)/M^{2+}(aq))[d,646]	3.8	3.5	3.0	2.8	2.5	1.5	0.35	3.6	3.5	2.6	2.9	3.0	2.1	1.1

[a] Third ionization potential.
[b] Standard free energy of formation of the hydrated dipositive ion of $4f^n$ configuration (La, Ce and Gd, whose hydrated dipositive ions would be $4f^{n-1}5d^1$, are uncorrected).
[c] Standard free energy of hydration of the gaseous M^{2+} ions.
[d] Standard electrode potential relative to NHE.

There has been some exploration of the mechanism of reduction of *d* transition metal complexes by M^{2+}(aq) (M = Eu, Yb, Sm). Both inner- and outer-sphere mechanisms are believed to operate. Thus the ready reduction of $[Co(en)_3]^{3+}$ by Eu^{2+}(aq) is necessarily outer-sphere.[652] However, the strong rate dependence on the nature of X when $[Co(NH_3)_5X]^{2+}$ or $[Cr(H_2O)_5X]^{2+}$ (X = F, Cl, Br or I) are reduced by Eu^{2+}(aq) possibly suggests an inner-sphere mechanism.[653] The more vigorous reducing agent Yb^{2+} reacts with $[Co(NH_3)_6]^{3+}$ and $[Co(en)_3]^{3+}$ by an outer-sphere route but with $[Cr(H_2O)_5X]^{2+}$ (X = halide) by the inner-sphere mechanism.[654] Outer-sphere redox reactions are catalyzed by electron-transfer catalysts such as derivatives of isonicotinic acid, one of the most efficient of which is *N*-phenyl-methylisonicotinate, as the free radical intermediate does not suffer attenuation through disproportionation. Using this catalyst, the outer-sphere reaction between Eu^{2+}(aq) and $[Co(py)(NH_3)_5]^{3+}$ proceeds as in reactions (18) and (19). Values found were $k_1 = 5.8 \times 10^2\ M^{-1}\ s^{-1}$ and $k_2/k_1 = 16$.[655]

$$Eu^{2+} + cat \underset{k_{-1}}{\overset{k_1}{\rightleftharpoons}} cat^- + Eu^{3+} \qquad (18)$$

$$cat^- + Co^{3+} \xrightarrow{k_2} cat + Co^{2+} \qquad (19)$$

The crown ether 18-crown-6 has been shown by polarography to stabilize aqueous solutions of dipositive europium and samarium relative to the tripositive state.[656]

39.2.11.2 Other solvated species

Solutions of dipositive lanthanide cations have been obtained in liquid ammonia, ethanol, THF, acetonitrile and hexamethylphosphoramide. Ions stabilized, or for which there is evidence for stabilization, include Nd^{2+}, Dy^{2+} and Tm^{2+} as well as Eu^{2+}, Yb^{2+} and Sm^{2+}. The non-hydroxylic solvents are best at stabilizing M^{2+} ions. Thus $NdCl_2(THF)_2$ has been reported from the reduction of $NdCl_3$ in THF by Na(naphthalene),[657] and corresponding reductions of MCl_3 (M = Eu, Yb or Sm) have also been achieved.[658] Solutions of solvated MI_2 (M = Sm or Yb) in THF may easily be made by the quantitative reaction of the metal with 1,2-diiodoethane, producing ethane. The solid THF adducts may be isolated.[659]

In hexamethylphosphoramide, sodium reduction of lanthanide trihalides gives solutions of Eu^{2+} (pale yellow), Yb^{2+} (yellow), Sm^{2+} (red-violet) and Tm^{2+} (red-brown). The latter has a half-life at 4.5 hours. The solution of $SmCl_2$ tends to be more stable (half-life, 60 hours) than those of $SmBr_2$ (30 hours) or SmI_2 (25 hours).[660,661] In liquid ammonia, Eu^{2+} or Yb^{2+} may be obtained by dissolution of the metal, but sodium reduction of $Sm(ClO_4)_3$ must be used to prepare solutions of Sm^{2+}.[662]

39.2.11.3 Complexes with nitrogen donors

Some interesting examples of this class of complexes have recently emerged. Most of them are based on the bis(trimethylsilyl)amido ligand. Silylamides of two types have been prepared by the action of EuI_2 or YbI_2 on $NaN(SiMe_3)_2$ (NaL) in diethyl ether or glyme (reactions 20 and 21). Using YbI_2, reaction (21) gave $YbL_2(Et_2O)_2$ when the product was recrystallized from ether, but $NaYbL_3$ upon recrystallization from toluene. A single crystal X-ray of $NaML_3$ showed a planar three-coordinated configuration (**26**) with Eu—N = 2.448 Å, Yb—N = 2.38 Å, Eu—N′ = 2.554 Å, Yb—N′ = 2.45 Å, Eu—N″ = 2.539 Å, Yb—N″ = 2.47 Å; a = 94.3° (Eu) or 97.4° (Yb). The metal–carbon distances are rather short at Eu—C = 3.21 Å average and may indicate some interaction.[663]

$$MI_2 + 2NaL \xrightarrow[M=Eu,\ Yb]{MeOCH_2CH_2OMe} ML_2(glyme)_2 + 2NaI \quad (20)$$

$$EuI_2 + 3NaL \xrightarrow{Et_2O} NaEuL_3 + 2NaI \quad (21)$$

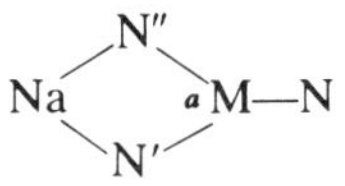

(**26**)

The X-ray structure of $[EuL_2(glyme)_2]$ has also been determined.[664] The six-coordination polyhedron has only C_2 symmetry, doubtless relating to the presence of two very bulky ligands and two bidentate ones with rather short 'bites', giving N—Eu—N = 134.5°. Bond lengths are: Eu—N = 2.530, Eu—O = 2.638, 2.756 Å. Magnetic susceptibility measurements show the Curie law obeyed, with μ = 8.43 BM. Other compounds isolated[664] were $EuL_2(bipy)_2$ and $EuL_2(THF)_2$. These were yellow or orange and soluble in hydrocarbons; doubtless they are six- and four-coordinated monomers.

Complexes of $EuCl_2$ with 1,10-phenanthroline and 2,2′:6′,2″-terpyridyl have been obtained from acetonitrile. They are purple $EuCl_2(phen)_2$ and dark blue $EuCl_2(terpy)$, both air-sensitive solids. In the NMR spectrum of the latter in Me_2SO solution, ligand peaks were broadened but not shifted, as expected for the f^7 configuration. In Me_2SO solution, only 1:1 association occurred with the phenanthroline complex, giving an association constant value $\log K = 3.9$. Chloride-bridged associated structures are likely for these complexes in the solid state.[665]

39.2.11.4 Complexes with phosphorus donors

The compound $[Yb\{N(SiMe_3)_2\}_2(Me_2PCH_2CH_2PMe_2)]$ is a well-established member of this rare group. It is made by the action of diphosphine on $YbL_2(Et_2O)_2$ and the structure has been determined. It is monomeric and four-coordinated with N—Yb—N = 123.6°, P—Yb—P = 68.4° and N—Yb—P = 101.2°. Bond lengths are Yb—P = 3.012 Å, Yb—N = 2.331 Å, and Yb—C (silylamide) shows a probable weak interaction at 3.04 Å. Also obtained similarly were $ML_2\{PBu^n_3\}_2$ where M = Eu (orange) or Yb (brown-red).[666] Phosphines also complex with divalent lanthanide organo compounds.[667]

39.2.11.5 Complexes with oxygen donors

A number of complexes in which ethers and silylamido ligands shared the coordination sphere were mentioned in the preceding two sections, and there are also several examples of ethers participating in the coordination spheres of dipositive lanthanide organo compounds.[667] There seem to be no examples of well-defined complexes with coordination spheres composed completely of oxygen donors, although M^{2+} in solutions of water or $OP(NMe_2)_3$ must contain these. Also, though not strictly complexes, MCO_3, where M = Sm, Eu or Yb, adopt the aragonite structure with nine-oxygen coordination.[668] An interesting organo complex with oxygen coordination is $\{(C_5Me_5)_2Yb\}_2Fe_3(CO)_{11}$ where each $(C_5Me_5)_2Yb$ group is also coordinated by the oxygen atoms of two carbonyl groups (the carbon atoms of which bridge two iron atoms). The Yb—O distance is quite short at 2.243 Å, showing a strong interaction. The compound is made by treating $(C_5Me_5)_2Yb(Et_2O)_2$ with $Fe_3(CO)_{12}$.[669]

39.2.11.6 Electronic spectra of dipositive lanthanides

All the lanthanides can be obtained as M^{2+} ions in dilute solid solution in the CaF_2 lattice by means of either γ-irradiation or by reduction by metal vapour. These materials have been instrumental in providing good spectra of the dipositive lanthanide ions[670] in an electrostatic field provided by the cube of fluoride ions which of course surround the M^{2+} ions in this structure. Stabilization of dipositive lanthanides in SrF_2 or BaF_2 matrices is also possible. Perovskite phases such as yellow $CsEuF_3$, bright green $CsYbF_3$ and orange $RbYbF_3$ can also be made by reduction of MF_3 with Cs or Rb;[671] these ternary halides have been reviewed.[672]

The spectra of dipositive lanthanides differ from those of tripositive ions (except Ce^{3+}) in that the excited $4f^{n-1}5d^1$ configuration lies at a level up to about 40 000 cm^{-1} and dominates the spectrum with more intense absorption than is the case for the tripositive ions. The single d electron is subject to crystal field splitting of 10^4 cm^{-1} order of magnitude and will, in a cubic field, give d_ϵ and d_γ levels. This contrasts with the $4f^{n-1}$ configuration where the crystal field parameters are small compared with the spin–orbit coupling. There are therefore two states to be considered separately, the Russell–Saunders coupled $4f^{n-1}$ shell and the crystal field affected d^1 state. These in turn will interact and the details of the interaction will depend on the relative energies of the ^SL_J split levels of the $4f^{n-1}$ configuration. An approximate energy diagram for the Er^{2+} ion in CaF_2 is shown in Figure 10 and the experimental values for the parameters Δ, E_{ff} and E_{fd} are listed in Table 15. The La^{2+} and Gd^{2+} ions have d^1 and f^7d^1 ground states respectively. It will be noticed that the values of Δ decrease towards the centre of each half series La–Gd and Gd–Yb. This is because there is little or no f–d interaction for f^0, f^7 or f^{14} (S states) but in other cases increasing f–d coupling broadens the levels so that Δ as measured is no longer a simple crystal field parameter but also depends on the nature and extent of the f–d interaction.[673]

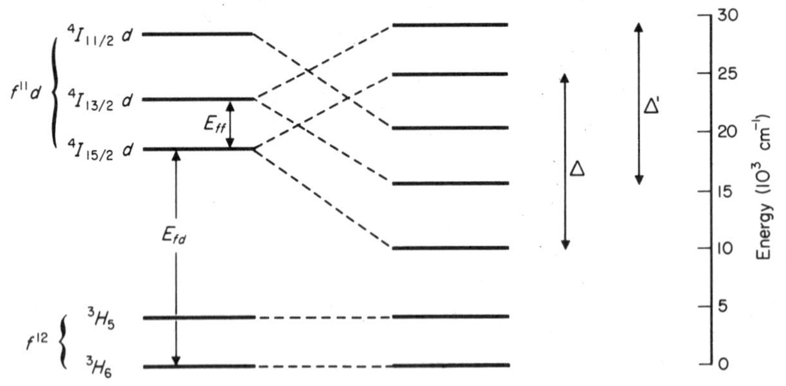

Figure 10 Energy levels of Er^{2+} in CaF_2 (adapted from ref. 673).

Table 15 Energies (10^3 cm^{-1}) Associated with Dipositive Lanthanide Ions in CaF_2

	La	Ce	Pr	Nd	Pm	Sm	Eu	Gd	Tb	Dy	Ho	Er	Tm	Yb
E_{fd} [a]	−3.4	25	10.9	13.6		25.6	37.7	−3.3	25.2	17.8	17.6	19.3	26.5	37.3
E_{ff} [a]			4.8	14.2		7.7			8.5	3.8	4.0	5.0	6.5	
Δ [a]	24	15	11.5	9.8		16.1	17	24.3	14.5	11.1	9.3	14.2	15	16.3

[a] As shown in Figure 10.

The spectra of a number of dipositive lanthanide species have been investigated, often in solution. The bands are fairly broad, owing to the effect of varying crystal fields on the 5*d* electron. Spectra have been measured in water, ethanol, acetonitrile, THF or HMPA and values of Δ and E_{fd} derived. In THF, the maximum observed value of Δ was 14 600 cm^{-1} for Tm^{2+}.[661,674] The absorption spectra of $[EuCl(phenanthroline)]^+$ and $[EuCl(terpyridyl)]^+$ in dimethyl sulfoxide have been reported.[665] The absorption spectrum of the Eu^{2+}–2,2,2-cryptate complex has been measured, together with its emission spectrum at 77 K.[675] Eu^{2+}–crown ether complexes when suitably excited give a strong blue emission at ~440 nm.[676] When excited at 365 nm, Sm^{2+} stabilized in ternary fluorides gives a deep red to orange fluorescence believed to be due to a $^5D \rightarrow {}^7F$ transition within the f^6 configuration.[677]

39.2.12 Complexes of Lanthanides in the Tetrapositive Oxidation State

39.2.12.1 Introduction

This is a somewhat restricted area in two respects. Firstly, only cerium has a tetrapositive oxidation state which is sufficiently stable to coexist with organic ligands. Although praseodymium and terbium give one or two binary salts and simple halide complexes, that is the limit of their achievement as tetravalent complex formers, and the remaining elements either form no tetrapositive compounds at all, or are restricted to fluoro complexes only. Secondly, the oxidizing nature of even cerium as Ce^{4+} is normally sufficiently great to limit its complexes to those with ligands which are not too easily oxidized. Some data relevant to the behaviour of tetrapositive lanthanide ions are given in Table 16. The fourth ionization potential (I_4) values are derived spectroscopically, and the standard reduction potential ($E^\ominus$) values are also derived spectroscopically, specifically from positions of charge-transfer and $f \rightarrow d$ bands particularly in hexahalo complexes $[MX_6]^{3-}$. It will be noticed that the correlation between these values and the stabilities of various classes of tetrapositive compounds is very good.

Table 16 Data Concerning the 4+ State in Lanthanides

anthanide	*La*	*Ce*	*Pr*	*Nd*	*Pm*	*Sm*	*Eu*	*Gd*	*Tb*	*Dy*	*Ho*	*Er*	*Tm*	*Yb*	*Lu*
nown compounds	*A*	*D*	*C*	*B*	*A*	*A*	*A*	*A*	*C*	*B*	*A*	*A*	*A*	*A*	*A*
	—	36.76	38.98	40.41	41.09	41.37	42.65	44.01	39.79	41.47	42.48	42.65	42.69	43.74	45.19
$E(M^{3+}(aq)/M^{4+}(aq))$	—	1.8	3.4	4.6	4.9	5.2	6.4	7.9	3.3	5.0	6.2	6.1	6.1	7.1	8.5
		(1.74)	(3.2)	(5.0)					(3.1)	(5.2)					

, no compounds known; *B*, complex fluorides only; *C*, complex fluorides, binary salts and alkaline-aqueous ions only; *D*, binary salts and a riety of complexes. I_4 values in eV. *E* values for $M^{3+}(aq)/M^{4+}(aq)$ are in volts relative to NHE. I_4 and *E* values[678,679] are derived from ectroscopic data except those in parentheses which are experimental values.

39.2.12.2 Hydrated ions

Although a single value is given in Table 16 for $E^\ominus$ ($Ce^{3+}(aq)/Ce^{4+}(aq)$), the oxidation potential of Ce^{3+} ion in aqueous media varies considerably with the nature of the counterion, doubtless due to stabilization of Ce^{4+} to different extents by complexation. Thus in 1 M H_2SO_4, $E^\ominus = 1.4435$ V,[680] but in perchloric acid solutions, $E^\ominus = 1.6400$–1.7310 V depending on concentation,[681] while a range of 1.6085–1.6104 V was observed in nitric acid solutions.[682] There is some evidence for the formation of polymeric species in aqueous nitric acid solutions, and this may explain some anomalies in electrode potential determinations. Values for association constants of 17 and 2 respectively for Ce^{4+}–Ce^{4+} and Ce^{4+}–Ce^{3+} interactions were obtained.[683]

The value of $E^\ominus$ ($Ce^{3+}(aq)/Ce^{4+}(aq)$) is rather greater than that for the oxidation of water, since $E^\ominus$ (H_2O/O_2) = 1.23 V, and Ce^{4+} ion is in fact capable of oxidizing water to oxygen in 0.5 M H_2SO_4 in the presence of RuO_2 catalyst. Yields are up to 73% theoretical.[684,685]

Tetrapositive Pr^{4+} and Tb^{4+} are quite stable in strongly alkaline carbonate solutions. Thus electrolytic oxidation for three hours of 2–5.5 mol dm^{-3} aqueous M_2CO_3 (M = K or Cs), 1 mol dm^{-3} in KOH, containing $CeCl_3$, $PrCl_3$ or $TbCl_3$ (0.1–0.01 mol dm^{-3}) gave stable solutions of yellow Pr^{4+} ($\lambda_{max} = 283$ nm), dark reddish-brown Tb^{4+} ($\lambda_{max} = 365$ nm), and of course Ce^{4+}. The alkaline medium provided considerable stabilization of the tetrapositive ions; $E_0^{\ominus}$ for Ce^{3+}/Ce^{4+} being only +0.05 V.[686] The ion Tb^{4+} is also stabilized in solutions of sodium tripolyphosphate $Na_5P_3O_{10}$. The solutions show λ_{max} at 375 nm ($\varepsilon = 1095$) and oxidize Ce^{3+} or Mn^{2+} at an acid pH, while slowly reducing ascorbic acid in neutral solution.[687] Oxidation of Tb^{3+}(aq) by the conducting glass electrode, which exhibits an unusually high overvoltage towards the oxidation of water, produces Tb^{4+} in a solid form, perhaps a hydroxycarbonate, as the aqueous medium was 2.5 K_2CO_3/0.5 M KOH.[688]

39.2.12.3 Nitrogen donor ligands

These complexes are rare, doubtless owing to the comparatively ready oxidation of amines. However, the template preparation of $Ce(NO_3)_4L{\cdot}3H_2O$ has been described; L is the eighteen-membered macrocycle formed by cocondensation of two molecules of ethylenediamine and two molecules of 2,6-diacetylpyridine. The structure of this compound is unknown and it is uncertain to what extent the macrocycle is bonded to the metal ion.[689]

39.2.12.4 Oxygen donor ligands

These compounds, together with the halo complexes described in the next section (39.2.7.5), form the bulk of the known complexes of tetrapositive lanthanides.

(i) Complexes of inorganic oxy anions

A number of double sulfates of Ce^{4+}, for example $(NH_4)_4Ce(SO_4)_4{\cdot}2H_2O$, may be obtained by concentration of acidic solutions of ceric sulfate and alkali sulfates. The ceric sulfate may be obtained by heating ceric oxide with fairly concentrated sulfuric acid (it dissolves rather slowly), cooling, filtration and washing with acetic acid. A series of hydrated double sulfates of caesium and cerium(IV) have been prepared and their IR spectra have been interpreted in terms of crystal symmetry and point group symmetry.[690] Cerium(IV) salts are solvolyzed in chlorosulfuric acid and the properties of the resulting solutions have been interpreted in terms of the formation of $Ce(SO_3Cl)_4$ which, however, polymerizes at higher concentrations.[691]

Ceric ammonium nitrate, the well-known volumetric standard oxidant, is formulated $(NH_4)_2[Ce(NO_3)_6]$ with 12-coordinated Ce^{4+} ions (each NO_3 ion being bidentate) in approximately T_d symmetry. The Ce—O distances are in the range 2.488–2.530 Å. The ligand 'bite' is only 50.6–50.9° and the other O—Ce—O angles are 65.0–69.9°.[692] The carbonato complexes $Na_6[Ce(CO_3)_5]{\cdot}12H_2O$ and $[C(NH_2)_3]_6[Ce(CO_3)_5]{\cdot}4H_2O$ each show cerium ten-coordinated to five bidentate carbonate ions with Ce—O = 2.379–2.504 Å in the first complex and 2.388–2.488 Å in the second. The coordination polyhedra are irregular.[693,694]

The preceding three structures show that the ionic radius of Ce^{4+} is about 1.10 Å for 12-coordination and about 1.04 Å for 10-coordination. Thus the ionic radius of Ce^{4+} is surprisingly similar to that of Ce^{3+}, but as noted below (Section 39.2.12.4.iii) can vary according to the nature of the ligand.

(ii) Complexes of organic oxy anions

Cerium tetrakis(acetylacetonate), a venerable compound,[695] exists as α and β modifications. Both these contain $[Ce(acac)_4]$ monomers of D_2 square antiprismatic coordination but differ in the crystal packing. In the α form,[696] Ce—O = 2.36–2.43 Å with the ring angle O—Ce—O = 72°, while in the β form[697] Ce—O = 2.32 Å and O—Ce—O = 71.3°. The tetrakis(acetylacetonates) of Zr, Hf, Th, U and Pu also show dimorphism. The tetrakis(dibenzoylmethanate) $[Ce(dbm)_4]$ adopts a similar square antiprismatic structure with Ce—O = 2.299–2.363 Å and O—Ce—O = 70.8° or 71.2°.[698] There is some tendency for Ce^{3+} β-diketonates to pass into the Ce^{4+} tetrakis(β-diketonate). A series of tri- and tetra-valent cerium β-diketonates has been examined from the point of view of the effect of additional ligands such as Ph_3PO on this process, and it was found that Ce^{3+} β-diketonates were stabilized by adduct formation, particularly by 1,10-phenanthroline.[699]

A very interesting species is the very stable catechol complex $Na_4[Ce(cat)_4]{\cdot}21H_2O$, where H_2 cat = 1,2-$C_6H_4(OH)_2$, which forms deep red crystals, the colour (λ_{max} = 517 nm) probably being due to a ligand-to-4*f* charge-transfer band. It has D_{2d} chelated dodecahedral symmetry with ring O—Ce—O angles of 68.3° and Ce—O distances of 2.357 and 2.362 Å. Cyclic voltammetry showed a quasireversible one-electron reduction at −692 mV *vs.* SCE. The complex is diamagnetic. It is rather surprising that this complex exists, when the highly reducing nature of the catecholate ligand is considered, but the $[Ce(cat)]^{4-}$ ion is unequivocally established.[700]

(iii) Ligands with uncharged oxygen donors

Tetravalent cerium forms complexes with amine oxides or phosphine oxides fairly readily, as do the trivalent lanthanides. Thus the following complexes have been described: $CeCl_4L_2$, where L = Ph_3PO,[701] bipyridyl dioxide or hexamethylphosphoramide,[702] $CeCl_4(DMSO)_3$[703] and $Ce(NO_3)_4(Ph_3PO)_2$.[704] A single crystal X-ray structure is available of the orange Ph_3PO complex, which was prepared by the action of the ligand on $(NH_4)_2Ce(NO_3)_6$ in acetone. The metal ion is ten-coordinated in a pseudo-*trans*-octahedral manner to two *trans* phosphoryl oxygen atoms and four bidentate nitrate ions, but the CeO_{10} polyhedron is irregular. Leading distances and angles are: Ce—O(P) = 2.21 and 2.23 Å, Ce—O(NO_3) = 2.41–2.54 Å, O(P)—Ce—O(P) = 174° and P—O—Ce = 168°. The fairly short Ce—O(P) distances, compared with the Ce—O(NO_3) distances, are noteworthy. The corresponding thorium complex is isostructural.

A number of complexes of Ce^{4+} may conveniently be prepared as follows. Hydrous ceric oxide is added to $SOCl_2$ in cold diglyme and yields stable $H_2CeCl_6{\cdot}3$(diglyme) which is soluble in organic solvents, to which solutions of suitable ligands may be added to form complexes such as $CeCl_4\{OP(NMe_2)_3\}_2$.[705]

39.2.12.5 Complex halides

(i) Complex fluorides

This is arguably the most prolific area of tetravalent lanthanide chemistry, and Ce, Pr, Nd, Tb and Dy are all represented.

A variety of cerium(IV) ammonium fluorides are known. Thus the CeF_4–NH_4F–H_2O system gives $(NH_4)_4CeF_8$ at concentrations of NH_4F greater than 28.9%. At lower concentrations, a monohydrate $(NH_4)_3CeF_7{\cdot}H_2O$ is obtained, which on dehydration is converted into $(NH_4)_4CeF_8$ and $(NH_4)_2CeF_6$. The structure of $(NH_4)_2CeF_6$ contains chains of distorted square antiprisms with Ce—F = 2.198–2.318 Å, while $(NH_4)_4[CeF_8]$ contains isolated distorted square antiprisms. $(NH_4)_3CeF_7{\cdot}H_2O$ contains dimeric ions $[Ce_2F_{14}]^{6-}$ in which the metal ions are eight-coordinated in a distorted dodecahedral manner with Ce—F = 2.133–2.375 Å. Other complex ammonium cerium(IV) fluorides reported are $(NH_4)_7Ce_6F_{31}$ and NH_4CeF_5.[706–708]

The alkali metal fluoro complexes M_2CeF_6 (M = Na, K, Rb, Cs) and M_3CeF_7 (M = Na, Rb, Cs) have been obtained as colourless diamagnetic crystals by fluorination of MCl/CeO_2 mixtures. They liberate iodine from aqueous iodide and decompose in water.[709] The analogous terbium compound Cs_3TbF_7, which is colourless and paramagnetic (μ = 7.4 BM), has been obtained similarly by fluorination at 390 °C, and its magnetic moment is much more consistent with the presence of Tb^{4+} (theory, 7.9 BM) than with Tb^{3+} (9.7 BM).[710]

Compounds of tetrapositive praseodymium can be made by the action of fluorine at 400–500 °C on MCl and Pr_6O_{11} mixtures. Thus obtained were M_3PrF_7 and M_2PrF_6 (M = Na, K, Rb or Cs). They are colourless and decompose in water. The Curie–Weiss law is obeyed with observed room-temperature moments of 2.10–2.45 BM (Pr^{4+}, 2.56 BM theory; Pr^{3+}, 3.61 BM).[711] The compounds $Na_7Pr_6F_{31}$ and Na_2PrF_6 have been obtained by fluorination of $PrCl_3/NaCl$ at 200–400 °C. These products were characterized by X-ray powder pattern and iodine equivalent.[712,713] More recently, a wide variety of compounds M_3LnF_7 (M = K, Rb or Cs; Ln = Ce, Pr, Nd, Tb or Dy) have been obtained, mainly by fluorination of M_3LnCl_6 at 50–150 bar and 400 °C for 2–3 hours. All the fluorides were isostructural with $(NH_4)_3ZrF_7$; examples are $KRbCsPrF_7$, $RbCs_2NdF_7$ (orange) and KCs_2DyF_7 (orange). Compounds $MPrF_6$

(M = Sr, Ba or 2Li) were also obtained starting from $MPrO_3$ (itself made from the action of O_2 on the mixed lower oxides).[714]

The series of fluoro complexes Cs_3MF_7 (M = Ce, Pr, Nd, Tb or Dy) have also been made by reaction of Cs_3MCl_6 with XeF_2 at temperatures of 100 °C for Ce, Pr or Tb, and 400 °C for Nd or Dy. Corresponding treatment of Cs_3MCl_6 (M = Er or Yb) yielded only Cs_3MF_6.[715]

It should be noted that definitive X-ray single crystal work is really needed on some of these compounds, as structures are often based on powder correlations with type structures of non-lanthanide complex fluorides, some of which are themselves quite early single crystal studies.

Good electronic spectra may be obtained from these fluoro complexes which thus provide a unique source of information about electronic energy levels in tetravalent lanthanides. The electronic spectra of Cs_3NdF_7 have been obtained in absorbance and fluorescence and an excellent correlation is observed with expected behaviour for the Nd^{4+} f^2 ion in intermediate-field coupling. Values of $F_2 = 352$, $F_4 = 52.3$, $F_6 = 5.56$, and spin–orbit coupling constant $\xi = 1109\ cm^{-1}$ explained the spectra, with the first nine excited states up to $^3P_1$ at 25 700 cm^{-1} being observed above the $^3H_4$ ground state.[716] Similar spectra for Cs_3DyF_7 have been obtained and agree with theoretical values calculated using only two parameters, $F_2 = 465.8\ cm^{-1}$ and $\xi = 2270\ cm^{-1}$. Only the ^7F ground state multiplet and the $^5D_3$ excited state at 21 600 cm^{-1} were observed.[717]

(ii) Complex chlorides

Complex chlorides of Ce^{4+} may be prepared by dissolution of cerium(IV) hydroxide in methanolic hydrochloric acid followed by addition of a base B, giving $(BH)_2[CeCl_6]$. $(NH_4)_2CeCl_6$ can be used as a selective precipitant for Cs^+ ions in presence of other alkali metals, with separation coefficients ranging from 5300 (Na) to 60 (Rb).[718]

The thermal decomposition of $Cs_2[CeCl_6]$ has been studied. In nitrogen at 600 K it decomposes into $CeCl_3$, Cl_2 and CsCl,[719] but in air CeO_2, Cl_2 and CsCl are formed.[720] The enthalpies of decomposition and formation were determined as: $\Delta H^{\ominus}_{f,298} = -1974.09 \pm 8.37\ kJ\ mol^{-1}$ and ΔH (decomp. 600 K) = $25.5 \pm 3.3\ kJ\ mol^{-1}$. Hexachlorocerates of organic cations vary in their thermal decomposition behaviour. Thus the pyridinium or some tetraalkylammonium compounds give chlorine, but the PPh_4^+ or $AsPh_4^+$ compounds do not.[721]

The IR and Raman spectra of $R_2[CeCl_6]$, where R = Cs or various substituted ammonium, has been studied and assignments to the vibrational modes of the octahedral $[CeCl_6]^{2-}$ ion have been made.[722,723] These are: A_{1g} stretch, 295 cm^{-1}; T_{1u} stretch, 265 cm^{-1}; T_u bend, 116 cm^{-1}, and T_{2g} bend, 123 cm^{-1}. The complexes are made from $(NH_4)_2CeCl_6$ and RCl.[724]

39.3 REFERENCES

1. G. A. Melson and R. W. Stotz, *Coord. Chem. Rev.*, 1971, **7**, 133.
2. R. C. Vickery, *J. Chem. Soc.*, 1955, 251.
3. R. C. Vickery, *J. Chem. Soc.*, 1955, 255.
4. A. I. Grigorev, E. G. Pogodilova and A. V. Novoselova, *Russ. J. Inorg. Chem. (Engl. Transl.)*, 1965, **10**, 416.
5. G. F. Huttig and W. Danschen, *Monatsh. Chem.*, 1951, **82**, 746.
6. I. V. Tananev and V. P. Orlovskii, *Russ. J. Inorg. Chem. (Engl. Transl.)*, 1962, **7**, 1192.
7. I. V. Tananev and V. P. Orlovskii, *Russ. J. Inorg. Chem. (Engl. Transl.)*, 1963, **8**, 572.
8. A. I. Grigorev, N. K. Evseeva and V. A. Sipachev, *J. Struct. Chem. (Engl. Transl.)*, 1969, **10**, 387.
9. V. P. Orlovskii and I. V. Tananaev, *Russ. J. Inorg. Chem. (Engl. Transl.)*, 1969, **14**, 786.
10. D. S. Marynick and C. M. Kirkpatrick, *J. Phys. Chem.*, 1983, **87**, 3273.
11. P. Glavic, *Polyhedron*, 1982, **1**, 735.
12. M. R. Wagner and G. A. Melson, *J. Inorg. Nucl. Chem.*, 1973, **35**, 869.
13. E. M. Kirmse, *Z. Chem.*, 1961, **1**, 334.
14. E. M. Kirmse and U. Schilbach, *Z. Chem.*, 1967, **7**, 238.
15. N. L. Firsova, Yu. V. Kolodyazhnyi and O. A. Osipov, *J. Gen. Chem. USSR (Engl. Transl.)*, 1969, **39**, 2101.
16. N. N. Greenwood and R. L. Tranter, *J. Chem. Soc. (A)*, 1969, 2878.
17. T. M. Sas, L. N. Komissarova and N. I. Anatskaya, *Russ. J. Inorg. Chem. (Engl. Transl.)*, 1971, **16**, 45.
18. N. P. Crawford and G. A. Melson, *Inorg. Nucl. Chem. Lett.*, 1968, **4**, 399.
19. N. P. Crawford and G. A. Melson, *J. Chem. Soc. (A)*, 1969, 427.
20. A. M. Arif, F. A. Hart, M. B. Hursthouse, M. Thornton-Pett and W. Zhu, *J. Chem. Soc., Dalton Trans.*, 1984 2449.
21. W. Radecka-Paryzek, *Inorg. Chim. Acta*, 1979, **35**, L349.

22. J. W. Buchler, G. Eikelmann, L. Puppe, K. Rohbock, H. H. Schneehage and D. Weck, *Liebigs Ann. Chem.*, 1971, **745**, 135.
23. J. W. Buchler and H. H. Schneehage, *Z. Naturforsch., Teil B*, 1973, **28**, 433.
24. M. Gouterman, L. K. Hanson, G. E. Khalil, J. W. Buchler, K. Rohbock and D. Dolphin, *J. Am. Chem. Soc.*, 1975, **97**, 3142.
25. J. W. Buchler and H. H. Schneehage, *Tetrahedron Lett.*, 1972, **36**, 3803.
26. J. H. Furhop, K. M. Kadish and D. G. Davis, *J. Am. Chem. Soc.*, 1973, **95**, 5140.
27. W. R. Leenstra, M. Gouterman and A. L. Kwiram, *J. Chem. Phys.*, 1979, **71**, 3535.
28. C. A. Busby and D. Dolphin, *J. Magn. Reson.*, 1976, **23**, 211.
29. K. Varmuza, G. Maresch and A. Meller, *Monatsh. Chem.*, 1974, **105**, 327.
30. S. C. Mathur and J. Singh, *Int. J. Quantum Chem.*, 1974, **8**, 79.
31. D. J. Olzanski and G. A. Melson, *Inorg. Chim. Acta*, 1977, **23**, L4.
32. J. S. Ghotra, M. B. Hursthouse and A. J. Welch, *J. Chem. Soc., Chem. Commun.*, 1973, 669.
33. E. C. Alyea, D. C. Bradley and R. G. Copperthwaite, *J. Chem. Soc., Dalton Trans.*, 1972, 1580.
34. G. T. Morgan and H. W. Moss, *J. Chem. Soc.*, 1914, **105**, 189.
35. R. H. Holm and F. A. Cotton, *J. Am. Chem. Soc.*, 1958, **80**, 5658.
36. E. W. Berg and J. J. C. Acosta, *Anal. Chim. Acta*, 1968, **40**, 101.
37. P. R. Singh and R. Sahai, *Inorg. Chim. Acta*, 1968, **2**, 102.
38. L. A. Gribov, Y. A. Zolotov and M. P. Noskova, *J. Struct. Chem. (Engl. Transl.)*, 1968, **9**, 378.
39. M. P. Noskova, L. A. Gribov and Y. A. Zolotov, *J. Struct. Chem. (Engl. Transl.)*, 1969, **10**, 391.
40. D. Purushottam and B. S. V. R. Rao, *Indian J. Chem.*, 1966, **4**, 109.
41. K. J. Eisentraut and R. E. Sievers, *J. Am. Chem. Soc.*, 1965, **87**, 5254.
42. T. P. Melia and R. Merrifield, *J. Inorg. Nucl. Chem.*, 1970, **32**, 2573.
43. T. Sekine, A. Koczcimi and M. Sakairi, *Bull. Chem. Soc. Jpn.*, 1966, **39**, 2681.
44. C. A. Wilkie and D. T. Haworth, *J. Inorg. Nucl. Chem.*, 1978, **40**, 195.
45. T. J. Anderson, M. A. Neuman and G. A. Melson, *Inorg. Chem.*, 1973, **12**, 927.
46. T. J. Anderson, M. A. Neuman and G. A. Melson, *Inorg. Chem.*, 1974, **13**, 158.
47. C. G. Macdonald and J. S. Shannon, *Aust. J. Chem.*, 1966, **19**, 1545.
48. M. Das, *Inorg. Chim. Acta*, 1984, **83**, L1.
49. E. M. Gavrischuk, N. G. Dzyubenko, L. I. Martynenko and P. E. Gaivoronskii, *Russ. J. Inorg. Chem. (Engl. Transl.)*, 1983, **28**, 493.
50. E. M. Gavrischuk, N. G. Dzyubenko and L. I. Martynenko, *Russ. J. Inorg. Chem. (Engl. Transl.)*, 1984, **29**, 399.
51. V. I. Spitsyn, I. A. Murav'eva, L. I. Martynenko, D. N. Sokolov and V. I. Golubtsova, *Russ. J. Inorg. Chem. (Engl. Transl.)*, 1982, **27**, 478.
52. M. Z. Gurevich, B. D. Stepin and V. V. Zelentsov, *Russ. J. Inorg. Chem. (Engl. Transl.)*, 1970, **15**, 454.
53. M. Z. Gurevich, B. D. Stepin, L. N. Komissarova, N. E. Lebedeva and T. M. Sas, *Russ. J. Inorg. Chem. (Engl. Transl.)*, 1971, **16**, 48.
54. E. L. Muetterties and C. M. Wright, *J. Am. Chem. Soc.*, 1965, **87**, 4706.
55. T. J. Anderson, M. A. Neuman and G. A. Melson, *Inorg. Chem.*, 1974, **13**, 1884.
56. E. L. Head and C. E. Holley, *J. Inorg. Nucl. Chem.*, 1964, **26**, 525.
57. M. K. Guseinova, A. S. Antsyshkina and M. A. Porai-Koshits, *J. Struct. Chem. (Engl. Transl.)*, 1968, **9**, 926.
58. E. Hansson, *Acta Chem. Scand.*, 1972, **26**, 1337.
59. E. Hansson, *Acta Chem. Scand.*, 1973, **27**, 2841.
60. L. M. Dikareva, A. S. Antsyshkina, M. A. Porai-Koshits, V. N. Ostrikova, I. V. Arkhangel'skii and A. Z. Zamanov, *Dokl. Akad. Nauk Az.SSR*, 1978, **34**, 41.
61. H. Itoh, N. Itoh and Y. Suzuki, *Bull. Chem. Soc. Jpn.*, 1984, **57**, 716.
62. Y. Suzuki, S. Yokoi, M. Katoh and N. Takizawa, 'The Rare Earths in Modern Science and Technology,' ed. G. J. McCarthy, H. B. Silber and J. J. Rhyne, Plenum, 1980, vol. 2, p. 121.
63. V. P. Yarasov, G. A. Kirakosyan, S. V. Trots, Yu. A. Buslaev and V. T. Panyushkin, *Koord. Khim.*, 1983, **9**, 205.
64. G. A. Kirakosyan and V. P. Tarasov, *Koord. Khim.*, 1982, **8**, 261.
65. V. A. Sipachev and A. I. Grigor'ev, *J. Struct. Chem. (Engl. Transl.)*, 1969, **10**, 710.
66. P. L. Brown, J. Ellis and R. L. Sylva, *J. Chem. Soc., Dalton Trans.*, 1983, 35.
67. B. N. Ivanov-Emin, L. D. Borzova, B. E. Zaitsev and M. M. Fisher, *Russ. J. Inorg. Chem. (Engl. Transl.)*, 1968, **13**, 1156.
68. B. N. Ivanov-Emin and E. A. Ostroumov, *Russ. J. Inorg. Chem. (Engl. Transl.)*, 1959, **4**, 27.
69. B. N. Ivanov-Emin and L. A. Niselson, *Russ. J. Inorg. Chem. (Engl. Transl.)*, 1959, **4**, 623.
70. B. N. Ivanov-Emin, L. D. Borzova, S. G. Malyugina and B. E. Zaitsev, *Russ. J. Inorg. Chem. (Engl. Transl.)*, 1969, **14**, 610.
71. B. N. Ivanov-Emin, L. D. Borzova, B. E. Zaitsev and R. T. Kirillova, *Russ. J. Inorg. Chem. (Engl. Transl.)*, 1969, **14**, 927.
72. L. N. Komissarova, G. Ya. Pushkina and V. I. Spitsyn, *Russ. J. Inorg. Chem. (Engl. Transl.)*, 1971, **16**, 1262.
73. C. C. Addison, A. J. Greenwood, M. J. Haley and N. Logan, *J. Chem. Soc., Chem. Commun.*, 1978, 580.
74. F. M. Korytnaya, S. N. Putilin and A. N. Pokrovskii, *Russ. J. Inorg. Chem.*, 1983, **28**, 968.
75. J. L. Atwood and K. D. Smith, *J. Chem. Soc., Dalton Trans.*, 1974, 921.
76. L. E. Manzer, *Inorg. Synth.*, 1982, **21**, 135.
77. O. V. Kravchenko, S. E. Kravchenko, V. D. Makhaev, V. B. Polyakova, G. V. Slobodenchuk and K. N. Semenenko, *Koord. Khim.*, 1982, **8**, 1356.
78. E. B. Lobkovskii, S. E. Kravchenko and N. N. Semenenko, *J. Struct. Chem. (Engl. Transl.)*, 1977, **18**, 389.
79. M. F. Lappert and R. Pearce, *J. Chem. Soc., Chem. Commun.*, 1973, 126.
80. D. J. Olzanski and G. A. Melson, *Inorg. Chim. Acta*, 1978, **26**, 263.
81. A. M. Arif, Ph.D. Thesis, University of London, 1984.
82. Z. M. Zholalieva, K. Sulaimankulov, B. Murzubraimov and N. Khudaibergenova, *Izv. Akad. Nauk Kirg. SSR*, 1983, 29 (*Chem. Abstr.*, **99**, 219 535).

83. N. L. Firsova, Yu. V. Kolodyazhnyi and O. A. Osipov, *J. Gen. Chem. USSR (Engl. Transl.)*, 1969, **39**, 2101.
84. F. Kutek and B. Dusek, *J. Inorg. Nucl. Chem.*, 1969, **31**, 1544.
85. V. I. Kuskov, E. N. Treushnikov and N. V. Belov, *Kristallografiya*, 1978, **23**, 1196.
86. N. P. Crawford and G. A. Melson, *J. Chem. Soc. (A)*, 1969, 427.
87. N. P. Crawford and G. A. Melson, *J. Chem. Soc. (A)*, 1969, 1049.
88. N. P. Crawford and G. A. Melson, *J. Chem. Soc. (A)*, 1970, 141.
89. J. O. Edwards, R. J. Goetsch and J. A. Stritar, *Inorg. Chim. Acta*, 1967, **1**, 360.
90. J. T. Donoghue, E. Fernandez, J. A. McMillan and D. A. Peters, *J. Inorg. Nucl. Chem.*, 1969, **31**, 1431.
91. A. M. Arif, Ph.D. Thesis, University of London, 1984.
92. D. L. Pisaniello, S. F. Lincoln and E. H. Williams, *J. Chem. Soc., Dalton Trans.*, 1979, 1473.
93. D. L. Pisaniello and S. F. Lincoln, *Inorg. Chim. Acta*, 1979, **36**, 85.
94. P. B. Hitchcock, M. F. Lappert and A. Singh, *J. Chem. Soc., Chem. Commun.*, 1983, 1499.
95. G. Vicentini and R. Isuyama, *An. Acad. Bras. Cienc.*, 1972, **44**, 423.
96. D. L. Pisaniello and S. F. Lincoln, *Inorg. Chem.*, 1981, **20**, 3689.
97. L. Pokras and P. M. Bernays, *J. Am. Chem. Soc.*, 1951, **73**, 7.
98. L. Pokras, M. Kilpatrick and P. M. Bernays, *J. Am. Chem. Soc.*, 1953, **75**, 1254.
99. J. N. Petronio and W. E. Ohnesorge, *Anal. Chem.*, 1967, **39**, 460.
100. M. Thompson, *Anal. Chim. Acta*, 1978, **98**, 357.
101. W. Cui and H. Shi, *Fen Hsi Hua Hsueh*, 1983, **11**, 778 (*Chem. Abstr.*, **100**, 150 230).
102. I. A. Fedorov, T. A. Balakaeva and A. N. Kuchumova, *Russ. J. Inorg. Chem. (Engl. Transl.)*, 1966, **11**, 906.
103. T. V. Shestakova, Z. N. Prozorovskaya and L. N. Komissarova, *Russ. J. Inorg. Chem. (Engl. Transl.)*, 1974, **19**, 1459.
104. N. M. Prutkova, A. I. Grigorev and N. D. Mitrofanova, *Russ. J. Inorg. Chem. (Engl. Transl.)*, 1967, **12**, 1766.
105. G. Schwarzenbach, R. Gut and G. Anderegg, *Helv. Chim. Acta*, 1954, **37**, 937.
106. V. T. Krumina, K. V. Astaklov and S. A. Barkov, *Russ. J. Phys. Chem. (Engl. Transl.)*, 1969, **43**, 334.
107. A. Merbach and F. Gnaegi, *Chimia*, 1969, **23**, 271.
108. W. D'Olieslager, M. Wevers and M. De Jonghe, *J. Inorg. Nucl. Chem.*, 1981, **43**, 423.
109. S. D. Makhijana and S. P. Sangel, *Chem. Era*, 1981, **17**, 154.
110. E. C. Alyea, A. Malek and A. E. Vougioukas, *Can. J. Chem.*, 1982, **60**, 667.
111. D. D. McRitchie, R. C. Palenik and G. J. Palenik, *Inorg. Chim. Acta*, 1976, **20**, L27.
112. J. Kury, A. D. Paul, L. G. Hepler and R. E. Connick, *J. Am. Chem. Soc.*, 1959, **81**, 4185.
113. R. J. Meyer, A. Wassjuchnow, N. Drapier and E. Bodlander, *Z. Anorg. Allg. Chem.*, 1914, **86**, 257.
114. H. Bode and E. Voss, *Z. Anorg. Allg. Chem.*, 1957, **290**, 1.
115. R. E. Thoma and R. H. Karraker, *Inorg. Chem.*, 1966, **5**, 1933.
116. K. Wieghardt and H. H. Eysel, *Z. Naturforsch., Teil B*, 1970, **25**, 105.
117. B. N. Ivanov-Emin, T. N. Susanina and L. A. Ogorodnikova, *Russ. J. Inorg. Chem. (Engl. Transl.)*, 1966, **11**, 274.
118. B. N. Ivanov-Emin, T. N. Susanina and A. I. Ezhov, *Russ. J. Inorg. Chem. (Engl. Transl.)*, 1967, **12**, 11.
119. K. A. Kraus, F. Nelson and G. W. Smith, *J. Phys. Chem.*, 1954, **58**, 11.
120. A. D. Paul, *J. Phys. Chem.*, 1962, **66**, 1248.
121. N. Y. Fedorov and E. S. Petrov, *Izv. Sib. Otd. Akad. Nauk SSSR, Ser. Khim. Nauk*, 1967, 57 (*Chem. Abstr.*, **67**, 68 084).
122. R. Gut and D. M. Gruen, *J. Inorg. Nucl. Chem.*, 1961, **21**, 259.
123. L. R. Morss, M. Siegal, L. Stenger and N. Edelstein, *Inorg. Chem.*, 1970, **9**, 1771.
124. N. Y. Fedorov and E. S. Petrov, *Izv. Sib. Otd. Akad. Nauk SSSR, Ser. Khim. Nauk*, 1969, 47 (*Chem. Abstr.*, **72**, 93 776).
125. N. Y. Fedorov and E. S. Petrov, *Izv. Sib. Otd. Akad. Nauk SSSR, Ser. Khim. Nauk*, 1969, 55 (*Chem. Abstr.*, **72**, 71 361).
126. R. W. Stotz and G. A. Melson, *Inorg. Chem.*, 1972, **11**, 1720.
127. H. B. Silber and T. Mioduski, *Inorg. Chem.*, 1984, **23**, 1577.
128. K. R. Poeppelmeir, J. D. Corbett, T. P. McMullen, D. R. Torgeson and R. G. Barnes, *Inorg. Chem.*, 1980, **19**, 129.
129. K. R. Poeppelmeir and J. D. Corbett, *J. Am. Chem. Soc.*, 1978, **100**, 5039.
130. J. D. Corbett, R. L. Daake, K. R. Poeppelmeir and D. H. Guthrie, *J. Am. Chem. Soc.*, 1978, **100**, 652.
131. F. H. Spedding, A. F. Voight, E. M. Gladrow and N. R. Sleight, *J. Am. Chem. Soc.*, 1947, **69**, 2777.
132. F. T. Fitch and D. S. Russell, *Can. J. Chem.*, 1951, **29**, 363.
133. R. C. Vickery, *J. Chem. Soc.*, 1952, 4357.
134. J. Koppel, *Z. Anorg. Allg. Chem.*, 1898, **18**, 305.
135. W. Crookes, *Proc. R. Soc., London*, 1883, **35**, 262.
136. V. I. Spitsyn, V. G. Vokhmin and G. V. Ionova, *Russ. J. Inorg. Chem. (Engl. Transl.)*, 1983, **28**, 465.
137. V. I. Spitsyn, V. G. Vokhmin and G. V. Ionova, *Russ. J. Inorg. Chem. (Engl. Transl.)*, 1983, **28**, 925.
138. B. F. Dzhurinskii, *Russ. J. Inorg. Chem. (Engl. Transl.)*, 1983, **28**, 617.
139. G. J. Palenik, 'Systematics and the Properties of the Lanthanides', ed. S. P. Sinha, Reidel, Dordrecht, 1983, 153.
140. D. L. Kepert, *Prog. Inorg. Chem.*, 1977, **23**, 1.
141. D. L. Kepert, *Prog. Inorg. Chem.*, 1978, **24**, 179.
142. D. L. Kepert, *Prog. Inorg. Chem.*, 1979, **25**, 41.
143. M. C. Favas and D. L. Kepert, *Prog. Inorg. Chem.*, 1981, **28**, 309.
144. K. W. Bagnall and X.-F. Li, *J. Chem. Soc., Dalton Trans.*, 1982, 1365.
145. L. I. Kononeko and N. S. Poluektov, *Russ. J. Inorg. Chem. (Engl. Transl.)*, 1962, **7**, 965.
146. N. I. Lobanov and V. A. Smirnova, *Zh. Neorg. Khim.*, 1963, **9**, 2208.
147. F. A. Hart and F. P. Laming, *Proc. Chem. Soc., London*, 1963, 107.
148. S. P. Sinha, *Spectrochim. Acta*, 1964, **20**, 879.
149. F. A. Hart and F. P. Laming, *J. Inorg. Nucl. Chem.*, 1964, **26**, 579.

150. F. A. Hart and F. P. Laming, *J. Inorg. Nucl. Chem.*, 1965, **27**, 1605.
151. F. A. Hart and F. P. Laming, *J. Inorg. Nucl. Chem.*, 1965, **27**, 1825.
152. S. S. Krishnamurthy and S. Soundararajan, *Z. Anorg. Allg. Chem.*, 1966, **348**, 309.
153. A. R. Al-Kharaghouli and J. S. Wood, *Inorg. Chem.*, 1972, **11**, 2293.
154. E. Butter, *Z. Anorg. Allg. Chem.*, 1968, **356**, 294.
155. J. H. Van Vleck and N. Frank, *Phys. Rev.*, 1929, **34**, 1494.
156. F. A. Hart, J. E. Newbery and D. Shaw, *Chem. Commun.*, 1967, 45.
157. F. A. Hart, J. E. Newbery and D. Shaw, *J. Inorg. Nucl. Chem.*, 1970, **32**, 3585.
158. F. A. Hart and J. E. Newbery, *J. Inorg. Nucl. Chem.*, 1969, **31**, 1725.
159. R. J. Foster, R. L. Bodner and D. G. Hendricker, *J. Inorg. Nucl. Chem.*, 1982, **34**, 3795.
160. A. Clearfield, R. Gopal and R. W. Olsen, *Inorg. Chem.*, 1977, **16**, 911.
161. D. G. Hendricker and R. J. Foster, *J. Inorg. Nucl. Chem.*, 1972, **34**, 1949.
162. D. A. Durham, G. H. Frost and F. A. Hart, *J. Inorg. Nucl. Chem.*, 1969, **31**, 833.
163. G. H. Frost, F. A. Hart, C. Heath and M. B. Hursthouse, *Chem. Commun.*, 1969, 1421.
164. S. P. Sinha, *Z. Naturforsch., Teil A*, 1965, **20**, 552.
165. L. J. Radonovich and M. D. Glick, *Inorg. Chem.*, 1971, **10**, 1463.
166. D. A. Durham, G. H. Frost and F. A. Hart, *J. Inorg. Nucl. Chem.*, 1969, **31**, 571.
167. G. H. Frost and F. A. Hart, *J. Chem. Soc., Chem. Commun.*, 1980, 836.
168. R. D. Chapman, R. T. Loda, J. P. Riehl and R. W. Schwartz, *Inorg. Chem.*, 1984, **23**, 1652.
169. M. F. Lappert, P. P. Power, A. R. Sanger and R. C. Srivastava, 'Metal and Metalloid Amides', Ellis Horwood, Chichester, 1980.
170. D. C. Bradley, J. S. Ghotra and F. A. Hart, *J. Chem. Soc., Chem. Commun.*, 1972, 349.
171. D. C. Bradley, J. S. Ghotra and F. A. Hart, *J. Chem. Soc., Dalton Trans.*, 1973, 1021.
172. J. S. Ghotra, M. B. Hursthouse and A. J. Welch, *J. Chem. Soc., Chem. Commun.*, 1973, 669.
173. R. A. Anderson, D. H. Templeton and A. Zalkin, *Inorg. Chem.*, 1978, **17**, 2317.
174. D. C. Bradley, J. S. Ghotra, F. A. Hart, M. B. Hursthouse and P. R. Raithby, *J. Chem. Soc., Chem. Commun.*, 1974, 40.
175. D. C. Bradley, J. S. Ghotra, F. A. Hart,. M. B. Hursthouse and P. R. Raithby, *J. Chem. Soc., Dalton Trans.*, 1977, 1166.
176. D. C. Bradley, J. S. Ghotra and F. A. Hart, *Inorg. Nucl. Chem. Lett.*, 1976, **12**, 735.
177. T. Moeller and G. Vicentini, *J. Inorg. Nucl. Chem.*, 1965, **27**, 1477.
178. J. H. Forsberg and T. Moeller, *J. Am. Chem. Soc.*, 1968, **90**, 1932.
179. J. H. Forsberg and T. Moeller, *Inorg. Chem.*, 1969, **8**, 883.
180. J. H. Forsberg, T. M. Kubik, T. Moeller and K. Gucwa, *Inorg. Chem.*, 1971, **10**, 2656.
181. M. F. Johnson and J. H. Forsberg, *Inorg. Chem.*, 1976, **15**, 734.
182. J. H. Forsberg and C. A. Wathen, *Inorg. Chem.*, 1971, **10**, 1379.
183. M. F. Johnson and J. H. Forsberg, *Inorg. Chem.*, 1972, **11**, 2683.
184. J. L. Burmeister and E. A. Deardorff, *Inorg. Chim. Acta*, 1970, **4**, 97.
185. J. L. Martin, L. C. Thompson, L. S. Radonovich and M. D. Glick, *J. Am. Chem. Soc.*, 1968, **90**, 4493.
186. C. Musikas, C. Cuillerdier, J. Livet, A. Forchioni and C. Chachaty, *Inorg. Chem.*, 1983, **22**, 2513.
187. C. Musikas, C. Cuillerdier and C. Chachaty, *Inorg. Chem.*, 1978, **17**, 3610.
188. M. V. R. Stainer and J. Takats, *J. Am. Chem. Soc.*, 1983, **22**, 410.
189. M. V. R. Stainer and J. Takats, *Inorg. Chem.*, 1982, **21**, 4044.
190. T. D. Tilley, R. A. Andersen and A. Zalkin, *Inorg. Chem.*, 1983, **22**, 856.
191. D. A. Johnson, *J. Chem. Soc., Dalton Trans.*, 1974, 1671.
192. L. R. Morss, *Chem. Rev.*, 1976, **76**, 827.
193. C. Hurtgen, D. Brown and J. Fuger, *J. Chem. Soc., Dalton Trans.*, 1980, 70.
194. D. A. Johnson, *J. Chem. Soc. (A)*, 1969, 2578.
195. L. R. Morss, *J. Phys. Chem.*, 1971, **75**, 392.
196. R. J. Hinchey and J. W. Cobble, *Inorg. Chem.*, 1970, **9**, 917.
197. L. J. Nugent, *J. Inorg. Nucl. Chem.*, 1975, **37**, 1767.
198. W. M. Latimer, 'Oxidation Reduction Potentials', 3rd edn. (rev. J. M. Cobble), Prentice Hall, New York, 1975.
199. L. Helmholtz, *J. Am. Chem. Soc.*, 1939, **61**, 1544.
200. S. K. Sikka, *Acta Crystallogr., Sect. A*, 1969, **25**, 621.
201. J. A. A. Ketelaar, *Physica (Amsterdam)*, 1937, 619.
202. D. R. Fitzwater and R. E. Rundle, *Z. Kristallogr.*, 1959, **112**, 362.
203. C. R. Hubbard, C. O. Quicksall and R. A. Jacobson, *Acta Crystallogr., Sect. B*, 1974, **30**, 2613.
204. J. Albertsson and I. Elding, *Acta Crystallogr., Sect. B*, 1977, **34**, 1460.
205. J. Glaser and G. Johansson, *Acta Chem. Scand., Ser. A*, 1981, **35**, 639.
206. G. R. Choppin, *Pure Appl. Chem.*, 1971, **27**, 23.
207. G. W. Brady, *J. Chem. Phys.*, 1960, **33**, 1079.
208. D. G. Karraker, *Inorg. Chem.*, 1968, **7**, 473.
209. L. A. K. Staveley, D. R. Markham and M. R. Jones, *J. Inorg. Nucl. Chem.*, 1968, **30**, 231.
210. F. H. Spedding, M. J. Pikal and B. O. Ayers, *J. Phys. Chem.*, 1966, **70**, 2440.
211. F. H. Spedding and K. C. Jones, *J. Phys. Chem.*, 1966, **70**, 2450.
212. F. H. Spedding and M. J. Pikal, *J. Phys. Chem.*, 1966, **70**, 2430.
213. M. L. Smith and D. L. Wertz, *Inorg. Chem.*, 1977, **16**, 1225.
214. G. Geir and V. Karlen, *Helv. Chim. Acta*, 1971, **54**, 135.
215. A. Habenschuss and F. H. Spedding, *J. Chem. Phys.*, 1979, **70**, 3758.
216. A. H. Marten and R. L. Hahn, *Science*, 1982, **217**, 1249.
217. W. De W. Horrocks and D. R. Sudnick, *J. Am. Chem. Soc.*, 1979, **101**, 334.
218. W. De W. Horrocks and D. R. Sudnick, *Acc. Chem. Res.*, 1981, **14**, 384.
219. P. J. Breen and W. De W. Horrocks, *Inorg. Chem.*, 1983, **22**, 536.

220. F. Tanaka and S. Yamashita, *Inorg. Chem.*, 1984, **23,** 2044.
221. E. I. Onstott, L. B. Brown and E. J. Peterson, *Inorg. Chem.*, 1984, **23,** 2430.
222. E. G. Sherry, *J. Solid State Chem.*, 1976, **19,** 271.
223. P. I. Lazarev, L. A. Aslanov and M. A. Porai-Koshits, *Koord. Khim.*, 1975, **1,** 979.
224. N. K. Belskii and Y. T. Struchkov, *Sov. Phys.-Crystallogr.* (*Engl. Transl.*), 1965, **10,** 15.
225. M. Marezio, H. A. Plettinger and W. H. Zachariasen, *Acta Crystallogr.*, 1961, **14,** 234.
226. N. L. Sarukhanyan, L. D. Iskhakova, E. P. Morochenets and V. K. Trunov, *Russ. J. Inorg. Chem.* (*Engl. Transl.*), 1982, **27,** 1109.
227. K. I. Tobelko, G. I. Tsizin, G. I. Malofeeva, Yu. A. Zolotov and V. S. Urusov, *Russ. J. Inorg. Chem.* (*Engl. Transl.*), 1983, **28,** 502.
228. D. A. Storozhenko, *Russ. J. Inorg. Chem.* (*Engl. Transl.*), 1983, **28,** 654.
229. D. A. Storozhenko, *Russ. J. Inorg. Chem.* (*Engl. Transl.*), 1983, **28,** 656.
230. D. A. Storozhenko, A. K. Molodkin, V. G. Sherchuk, V. M. Akimov and Yu. A. Grigor'ev, *Russ. J. Inorg. Chem.* (*Engl. Transl.*), 1983, **28,** 506.
231. D. A. Khisaeva, E. F. Zhuravlev and E. V. Semenova, *Russ. J. Inorg. Chem.* (*Engl. Transl.*), 1982, **27,** 1188.
232. D. A. Khisaeva, E. F. Zhuravlev and E. V. Semenova, *Russ. J. Inorg. Chem.* (*Engl. Transl.*), 1982, **27,** 1190.
233. Y. A. Afanasev, L. T. Azhipa and L. V. Salnik, *Russ. J. Inorg. Chem.* (*Engl. Transl.*), 1982, **27,** 431.
234. A. K. Molodkin, Z. K. Odinets, T. V. S'edina and T. N. Ivanova, *Russ. J. Inorg. Chem.* (*Engl. Transl.*), 1983, **28,** 56.
235. D. A. Khisaeva and E. F. Zhuravlev, *Russ. J. Inorg. Chem.* (*Engl. Trans.*), 1983, **28,** 131.
236. E. F. Zhuravlev, D. A. Khisaeva and M. K. Boeva, *Russ. J. Inorg. Chem.* (*Engl. Transl.*), 1983, **28,** 895.
237. V. G. Shevchuk, A. G. Dryuchko, N. V. Bunyakina and D. A. Storozhenko, *Russ. J. Inorg. Chem.* (*Engl. Transl.*), 1983, **28,** 1365.
238. Z. K. Odinets, O. Vargas Ponse, B. E. Zaitsev and A. K. Molodkin, *Russ. J. Inorg. Chem.* (*Engl. Transl.*), 1983, **28,** 1580.
239. K. K. Palkina, S. I. Maksimova and N. T. Chibiskova, *Russ. J. Inorg. Chem.* (*Engl. Transl.*), 1983, **28,** 501.
240. K. K. Palkina, S. I. Maksimova, V. S. Mironova, N. T. Chibiskova and I. V. Tananaev, *Russ. J. Inorg. Chem.* (*Engl. Transl.*), 1983, **28,** 315.
241. L. S. Itkina, E. V. Petrova and A. S. Antsyskhina, *Russ. J. Inorg. Chem.* (*Engl. Transl.*), 1981, **26,** 1663.
242. M. Batkibekova, E. D. Mikheeva and Dzh. U. Usubaliev, *Russ. J. Inorg. Chem.* (*Engl. Transl.*), 1983, **28,** 287.
243. V. G. Zubarev and N. N. Krot, *Russ. J. Inorg. Chem.* (*Engl. Transl.*), 1983, **28,** 1595.
244. S. B. Pirkes, T. V. Zakharova, G. A. Martynova and V. D. Porokh, *Russ. J. Inorg. Chem.* (*Engl. Transl.*), 1983, **28,** 1585.
245. L. G. Sokolova, A. V. Lapitskaya, A. F. Bolshakov, S. B. Pirkes and B. V. Abalduev, *Russ. J. Inorg. Chem.* (*Engl. Transl.*), 1981, **26,** 936.
246. N. P. Sokolova, *Russ. J. Inorg. Chem.* (*Engl. Transl.*), 1983, **28,** 443.
247. G. N. Tarasova, E. E. Vinogradov and L. B. Kudinov, *Russ. J. Inorg. Chem.* (*Engl. Transl.*), 1981, **26,** 1520.
248. E. D. Bakhareva, M. B. Varfolomeev, V. P. Mashonkin and V. V. Ilyukhin, *Koord Khim.*, 1976, **2,** 1135.
249. M. B. Varpolomeev, L. L. Zaitseva, A. A. Gruglov and S. V. Mikheikin, *Russ. J. Inorg. Chem.* (*Engl. Transl.*), 1983, **28,** 754.
250. E. G. Teterin and L. L. Zaitseva, *Russ. J. Inorg. Chem.* (*Engl. Transl.*), 1983, **28,** 1717.
251. N. N. Nevskii, B. N. Ivanov-Emin and N. A. Nevskaya, *Russ. J. Inorg. Chem.* (*Engl. Transl.*), 1983, **28,** 1074.
252. Yu. P. Davydov and N. I. Voronik, *Russ. J. Inorg. Chem.* (*Engl. Transl.*), 1983, **28,** 1270.
253. G. W. Pope, J. F. Steinbach and W. F. Wagner, *J. Inorg. Nucl. Chem.*, 1962, **20,** 304.
254. J. M. Koehler and W. G. Bos, *Inorg. Nucl. Chem. Lett.* 1967, **3,** 545.
255. C. Brecher, H. Samelson and A. Lempicki, *J. Chem. Phys.*, 1965, **42,** 1081.
256. R. C. Mehrotra, T. V. Misra and S. M. Misra, *Indian J. Chem.*, 1965, **3,** 525.
257. B. S. Sankhla and R. N. Kapoor, *Aust. J. Chem.*, 1967, **20,** 685.
258. M. F. Richardson, W. F. Wagner and D. E. Sands, *Inorg. Chem.*, 1968, **7,** 2495.
259. T. Phillips, D. E. Sands and W. F. Wagner, *Inorg. Chem.*, 1968, **7,** 2295.
260. L. A. Aslanov, M. A. Porai-Koshits and M. O. Dekaprilevich, *J. Struct. Chem.* (*Engl. Transl.*), 1971, **12,** 431.
261. A. L. Il'inskii, L. A. Aslanov, V. I. Ivanov, A. D. Khalilov and O. M. Petrukin, *J. Struct. Chem.* (*Engl. Transl.*), 1969, **10,** 263.
262. L. A. Aslanov, E. F. Korytnyi and M. A. Porai-Koshits, *J. Struct. Chem.* (*Engl. Transl.*), 1981, **12,** 600.
263. J. A. Cunningham, D. E. Sands, W. F. Wagner and M. R. Richardson, *Inorg. Chem.*, 1969, **8,** 22.
264. E. D. Watkins, J. A. Cunningham, T. Phillips, D. E. Sands and W. F. Wagner, *Inorg. Chem.*, 1969, **8,** 29.
265. M. C. Favas, D. L. Kepert, B. W. Skelton and A. H. White, *J. Chem. Soc., Dalton Trans.*, 1980, 454.
266. A. Zalkin, D. H. Templeton and D. G. Karraker, *Inorg. Chem.*, 1969, **8,** 2680.
267. A. F. Kirby and R. A. Palmer, *Inorg. Chem.*, 1981, **20,** 1030.
268. K. J. Eisentraut and R. E. Sievers, *J. Am. Chem. Soc.*, 1965, **87,** 5254.
269. C. S. Erasmus and J. C. A. Boeyens, *Acta Crystallogr., Sect. B*, 1970, **26,** 1843.
270. J. P. R. de Villiers and J. C. A. Boeyens, *Acta Crystallogr., Sect. B*, 1971, **27,** 2335.
271. V. A. Mode and G. S. Smith, *J. Inorg. Nucl. Chem.*, 1969, **31,** 1857.
272. J. S. Ghotra, F. A. Hart, G. P. Moss and M. L. Staniforth, *J. Chem. Soc., Chem. Commun.*, 1973, 113.
273. J. E. Sicre, J. T. Dubois, K. J. Eisentraut and R. E. Sievers, *J. Am. Chem. Soc.*, 1969, **91,** 3476.
274. J. C. A. Boeyens, *J. Chem. Phys.*, 1971, **54,** 75.
275. C. S. Springer, D. W. Meek and R. E. Sievers, *Inorg. Chem.*, 1967, **6,** 1105.
276. J. P. R. de Villiers and J. C. A. Boeyens, *Acta Crystallogr., Sect. B*, 1971, **27,** 692.
277. J. C. A. Boeyens and J. P. R. de Villiers, *J. Cryst. Mol. Struct.*, 1971, **1,** 297.
278. A. H. Brudet, S. R. Tanny, H. A. Rockefeller and C. S. Springer, *Inorg. Chem.*, 1974, **13,** 880.
279. J. S. Ghotra, F. A. Hart, G. P. Moss and M. L. Staniforth, *J. Chem. Soc., Chem. Commun.*, 1973, 113.
280. G. A. Catton, F. A. Hart and G. P. Moss, *J. Chem. Soc., Dalton Trans.*, 1976, 208.
281. H. G. Brittain, *J. Chem. Soc., Dalton Trans.*, 1979, 1187.

282. X. Yang and H. G. Brittain, *Inorg. Chim. Acta,* 1982, **59,** 261.
283. D. P. Graddon and L. Muir, *J. Chem. Soc., Dalton Trans.*, 1981, 2434.
284. K. Iftikhar, M. Sayeed and N. Ahmad, *Inorg. Chem.*, 1982, **21,** 80.
285. W. H. Watson, R. J. Williams and N. R. Stemple, *J. Inorg. Nucl. Chem.*, 1982, **34,** 501.
286. R. E. Cramer and K. Seff, *J. Chem. Soc., Chem. Commun.*, 1982, 400.
287. R. E. Cramer and K. Seff, *Acta Crystallogr., Sect. B,* 1982, **28,** 3281.
288. W. de Horrocks, J. P. Sipe and J. R. Luber, *J. Am. Chem. Soc.*, 1971, **93,** 5258.
289. E. Bye, *Acta Chem. Scand., Ser. A.*, 1984, **28,** 731.
290. S. J. S. Wasson, D. E. Sands and W. F. Wagner, *Inorg. Chem.*, 1973, **12,** 187.
291. J. A. Cunningham and R. E. Sievers, *J. Am. Chem. Soc.*, 1975, **97,** 1586.
292. L. A. Aslanov, V. M. Ionov, V. B. Rybakov, E. F. Korythyi and L. I. Martynenko, *Koord. Khim.*, 1978, **4,** 1427.
293. R. M. Wing, J. J. Uebel and K. K. Anderson, *J. Am. Chem. Soc.*, 1973, **95,** 6046.
294. M. F. Richardson, W. F. Wagner and D. E. Sands, *J. Inorg. Nucl. Chem.*, 1968, **30,** 1275.
295. V. E. Karasev and I. V. Shchukina, *Russ. J. Inorg. Chem.* (*Engl. Transl.*), 1983, **28,** 746.
296. D. R. Foster, F. S. Richardson, L. M. Vallarino and D. Shillady, *Inorg. Chem.*, 1983, **22,** 4002.
297. G. A. Catton, F. A. Hart and G. P. Moss, *J. Chem. Soc., Dalton Trans.*, 1975, 221.
298. V. E. Karasev, A. G. Mirochnik and R. N. Shchelokov, *Russ. J. Inorg. Chem.* (*Engl. Transl.*), 1983, **28,** 1280.
299. V. E. Karasev, A. G. Mirochnik and E. N. Murav'eva, *Russ. J. Inorg. Chem.* (*Engl. Transl.*), 1984, **29,** 148.
300. V. E. Karasev, A. G. Mirochnik and R. N. Shchalokov, *Russ. J. Inorg. Chem.* (*Engl. Transl.*), 1984, **29,** 395.
301. L. R. Melby, N. H. Rose, E. Abramson and J. C. Caris, *J. Am. Chem. Soc.*, 1964, **84,** 5117.
302. A. Al-Karaghouli, A. Razzack and J. S. Wood, *Inorg. Chem.*, 1979, **18,** 1177.
303. A. R. Al-Karaghouli, R. O. Day and J. S. Wood, *Inorg. Chem.*, 1978, **17,** 3702.
304. A. Seminara and E. Rizzarelli, *Cong. Naz. Chim. Inorg. Atti Trieste, 12th,* 1979, 131.
305. A. Seminara and E. Rizzarelli, *Inorg. Chim. Acta,* 1980, **40,** 249.
306. A. Musumeci, R. P. Bonomo, E. Rizzarelli and A. Seminara, *Cong. Naz. Chim. Inorg. Atti Camerino, 13th,* 1980, 131.
307. A. Musumeci, R. P. Bonomo, V. Cucinotta and A. Seminara, *Inorg. Chim. Acta,* 1982, **59,** 133.
308. A. Musumeci and A. Seminara, *Inorg. Chim. Acta,* 1981, **54,** L81.
309. P. V. Sivapullaiah and S. Soundararajan, *Bull. Soc. Chim. Belg.*, 1981, **90,** 147.
310. P. V. Sivapullaiah and S. Soundararajan, *Rev. Roum. Chim.*, 1977, **22,** 1493.
311. L. B. Zinner, G. Vicentini and A. M. P. Felicissimo, *Rev. Latinoam. Quim.*, 1979, **10,** 109.
312. A. M. P. Felicissimo, L. B. Zinner, G. Vicentini and K. Zinner, *J. Inorg. Nucl. Chem.*, 1978, **40,** 2067.
313. G. Vicentini, A. M. P. Felicissimo and L. B. Zinner, *An. Acad. Bras. Cienc.*, 1981, **53,** 323.
314. L. Ramakrishnan and S. Soundararajan, *Monatsh. Chem.*, 1976, **107,** 1095.
315. L. B. Zinner and A. M. P. Felicissimo, *An. Acad. Bras. Cienc.*, 1977, **49,** 221.
316. P. V. Sivapullaiah and S. Soundararajan, *Curr. Sci.*, 1976, **45,** 711.
317. L. B. Zinner and A. M. P. Felicissimo, *An. Acad. Bras. Cienc.*, 1977, **49,** 139.
318. A. M. P. Felicissimo and L. B. Zinner, *An. Acad. Bras. Cienc.*, 1977, **49,** 537.
319. G. Vicentini and W. F. de Giovani, *J. Inorg. Nucl. Chem.*, 1978, **40,** 1448.
320. D. K. Koppicar and S. Soundararajan, *J. Inorg. Nucl. Chem.*, 1976, **38,** 1875.
321. L. Ramakrishnan and S. Soundararajan, *Can. J. Chem.*, 1976, **54,** 3169.
322. L. Ramakrishnan and S. Soundararajan, *Proc. Indian Acad. Sci., Sect. A,* 1977, **86,** 59.
323. P. V. Sivapullaiah and S. Soundararajan, *Inorg. Nucl. Chem. Lett.*, 1977, **13,** 291.
324. L. R. Melby, N. J. Rose, E. Abramson and J. C. Caris, *J. Am. Chem. Soc.*, 1964, **86,** 5117.
325. D. R. Cousins and F. A. Hart, *Chem. Ind.* (*London*), 1965, 1259.
326. D. R. Cousins and F. A. Hart, *J. Inorg. Nucl. Chem.*, 1967, **29,** 1745.
327. D. R. Cousins and F. A. Hart, *J. Inorg. Nucl. Chem.*, 1967, **29,** 2965.
328. D. R. Cousins and F. A. Hart, *J. Inorg. Nucl. Chem.*, 1968, **30,** 3009.
329. M. Kh. Karapetyants, N. N. Guseva, A. I. Mikhailichenko and E. V. Sklenskaya, *Tezisy Dokl.-Vses. Chugaevskoe Soveshch. Khim. Kompleksn. Soedin., 12th,* 1975, **3,** 352 (*Chem. Abstr.*, **85,** 186 077).
330. V. K. Manchanda, K. Chander, N. P. Singh and G. M. Nair, *J. Inorg. Nucl. Chem.*, 1977, **39,** 1039.
331. C. M. Mikulski, M. N. Karayannis and L. L. Pythewski, *J. Less-Common Met.*, 1977, **51,** 201.
332. A. M. G. Massabni, M. L. R. Gibran and O. A. Serra, *Inorg. Nucl. Chem. Lett.*, 1978, **14,** 419.
333. J. G. Leipoldt, L. D. C. Bok, A. E. Laulscher and S. S. Basson, *J. Inorg. Nucl. Chem.*, 1975, **37,** 2477.
334. N. Denliev, I. A. Muraveva, L. I. Martynenko, V. I. Spitzyn, A. V. Davydov and B. F. Myasoedov, *Izv. Akad. Nauk SSSR., Ser. Khim.*, 1976, **7,** 1455.
335. B. T. Khan and S. F. M. Ali, *Indian J. Chem. Sect. A,* 1980, **19,** 701.
336. I. A. Muraveva, L. I. Martynenko and V. I. Spitsyn, *Zh. Neorg. Khim.*, 1977, **22,** 3009.
337. A. T. Kandil and K. Farah, *Radiochim. Acta,* 1981, **28,** 165.
338. Y. Sasaki and H. Freiser, *Inorg. Chem.*, 1983, **22,** 2289.
339. A. A. Pinkerton and D. Schwarzenbach, *J. Chem. Soc., Dalton Trans.*, 1976, 2466.
340. R. P. Scholer and A. E. Merbach, *Inorg. Chim. Acta,* 1975, **15,** 15.
341. E. Giesbrecht and L. B. Zinner, *Inorg. Nucl. Chem. Lett.*, 1969, **5,** 575.
342. M. T. Durney and R. S. Marisnelli, *Inorg. Nucl. Chem. Lett.*, 1970, **6,** 895.
343. L. A. Aslanov, V. M. Ionov, A. A. Stepanov, M. A. Porai-Koshits, V. G. Lebedev, B. N. Kulikovskii, O. N. Gilyalov and T. L. Novoderezhkina, *Izv. Akad. Nauk SSSR, Neorg. Mater.*, 1975, **11,** 1331.
344. D. F. Evans and P. H. Missen, *J. Chem. Soc., Dalton Trans.*, 1982, 1929.
345. G. G. Sadikov, Yu. L. Zub, V. V. Skopenko, V. P. Nicolaev and M. A. Porai-Koshits, *Koord. Khim.*, 1984, **10,** 1253.
346. J. T. Donoghue and D. A. Peters, *J. Inorg. Nucl. Chem.*, 1969, **31,** 467.
347. L. J. Radonovich and M. D. Glick, *J. Inorg. Nucl. Chem.*, 1973, **35,** 2745.

348. N. B. Mikheev, A. N. Kamenskaya, N. A. Konovalova, T. A. Bidakova and L. M. Mikheeva, *Zh. Neorg. Khim.*, 1977, **22,** 3243.
349. J. A. Sylvanovich and S. K. Madan, *J. Inorg. Nucl. Chem.*, 1972, **34,** 1675.
350. S. P. Sinha, *Z. Anorg. Allg. Chem.*, 1977, **434,** 227.
351. L. B. Zinner and G. Vincentini, *J. Inorg. Nucl. Chem.*, 1981, **43,** 193.
352. S. P. Sinha, *Inorg. Chim. Acta,* 1978, **28,** 145.
353. V. G. Lebedev, K. K. Palkina, S. I. Maksimova, E. N. Lebedova and O. V. Galaktionova, *Russ. J. Inorg. Chem.* (*Engl. Transl.*), 1982, **27,** 1689.
354. E. Herrman, Hoang Ba Nang and R. Dreyer, *Z. Chem.*, 1979, **19,** 187.
355. S. Kulpe, I. Seidel, K. Szulzewsky and G. Kretschmer, *Acta Crystallogr., Sect. B,* 1982, **38,** 2813.
356. T. H. Siddall, *J. Inorg. Nucl. Chem.*, 1964, **26,** 1991.
357. W. E. Stewart and T. H. Siddall, *J. Inorg. Nucl. Chem.*, 1970, **32,** 3599.
358. S. M. Bowen, E. N. Duesler and R. T. Paine, *Inorg. Chim. Acta,* 1982, **61,** 155.
359. E. P. Horwitz, A. C. Muscatello, D. G. Kalina and L. Kaplan, *Sep. Sci. Technol.*, 1981, **16,** 417.
360. C. Airoldi, Y. Gushikem and C. R. Puschel, *J. Coord. Chem.*, 1976, **6,** 17.
361. J. A. Sylvanovich and S. K. Madan, *J. Inorg. Nucl. Chem.*, 1972, **34,** 2569.
362. W. E. Stewart and T. H. Siddall, *J. Inorg. Nucl. Chem.*, 1968, **30,** 1513.
363. W. E. Stewart and T. H. Siddall, *J. Inorg. Nucl. Chem.*, 1971, **33,** 2965.
364. R. P. Scholer, *Inorg. Chim. Acta,* 1979, **35,** 79.
365. G. S. Kholodnaya, A. I. Kirillov, N. A. Kostromina and E. K. Kolova, *Russ. J. Inorg. Chem.* (*Engl. Transl.*), 1983, **28,** 809.
366. V. V. Skopenko, Yu. L. Zub and A. A. Kapshuk, *Russ. J. Inorg. Chem.* (*Engl. Transl.*), 1984, **29,** 337.
367. R. C. Paul, J. S. Ghotra and M. S. Bains, *J. Inorg. Nucl. Chem.*, 1965, **27,** 265.
368. I. A. Kahwa, F. R. Fronczek and J. Selbin, *Inorg. Chim. Acta,* 1984, **82,** 161.
369. I. A. Kahwa, F. R. Fronczek and J. Selbin, *Inorg. Chim. Acta,* 1984, **82,** 167.
370. Yu. I. Sal'nikov and F. V. Devyatov, *Russ. J. Inorg. Chem.* (*Engl. Transl.*), 1983, **28,** 339.
371. I. Grenthe, *Acta Chem. Scand.*, 1971, **25,** 3347.
372. I. Grenthe, *Acta Chem. Scand.*, 1971, **25,** 3721.
373. I. Grenthe, *Acta Chem. Scand.*, 1972, **26,** 1479.
374. J. Albertsson and I. Elding, *Acta Chem. Scand., Ser. A,* 1977, **31,** 21.
375. P. E. Riley, S. F. Haddad and K. N. Raymond, *Inorg. Chem.*, 1983, **22,** 3090.
376. T. Moeller and G. Vicentini, *J. Inorg. Nucl. Chem.*, 1965, **27,** 1477.
377. E. Giesbrecht and M. Kawashita, *J. Inorg. Nucl. Chem.*, 1970, **32,** 2461.
378. G. Vicentini and R. Najjar, *Inorg. Nucl. Chem. Lett.*, 1970, **6,** 571.
379. A. Seminara and G. Gondorelli, *Ann. Chim.* (*Rome*), 1969, **59,** 990.
380. D. L. Pisaniello, L. Helm, P. Meier and A. E. Merbach, *J. Am. Chem. Soc.*, 1983, **22,** 4528.
381. D. L. Pisaniello and A. E. Merbach, *Helv. Chim. Acta,* 1982, **65,** 573.
382. L. Niinisto, 'Systematics and the Properties of the Lanthanides', ed. S. P. Sinha, Reidel, Dordrecht, 1983, p. 125.
383. A. Dereigne and G. Pannetier, *Bull. Soc. Chim. France,* 1968, 174.
384. L. O. Larsson, S. Londerbrandt, L. Niinisto and U. Skoglund, *Suom. Kemistil. B,* 1973, **46,** 314.
385. A. R. Al-Karaghouli and J. S. Wood, *J. Chem. Soc., Dalton Trans.*, 1973, 2318.
386. J.-C. G. Bünzli, B. Klein, G. Chapins and K. J. Schenk, *J. Inorg. Nucl. Chem.*, 1980, **42,** 1307.
387. G. E. Toogood and C. Chieh, *Can. J. Chem.*, 1975, **53,** 831.
388. E. G. Sherry, *J. Inorg. Nucl. Chem.*, 1978, **40,** 257.
389. E. F. Zhuralov, D. A. Kisaeva and L. V. Ismailova, *Russ. J. Inorg. Chem.* (*Engl. Transl.*), 1983, **28,** 122.
390. T. J. R. Weakley, *Inorg. Chim. Acta,* 1984, **95,** 317.
391. J. D. J. Backer-Dirks, J. E. Cooke, A. M. R. Galas, J. S. Ghotra, C. J. Gray, F. A. Hart and M. B. Hursthouse, *J. Chem. Soc., Dalton Trans.*, 1980, 2191.
392. K. Al-Rasoul and T. J. R. Weakley, *Inorg. Chim. Acta,* 1982, **60,** 191.
393. D. J. Rogers, N. J. Taylor and G. E. Toogood, *Acta Crystallogr., Sect, C,* 1983, **39,** 939.
394. B. Eriksson, L. O. Larsson and L. Nüniskö, *Inorg. Chem.*, 1980, **19,** 1207.
395. M. Milinski, B. Ribar and M. Satarie, *Cryst. Struct. Commun.*, 1980, **9,** 473.
396. A. A. Pinkerton, Y. Meseri and C. Rieder, *J. Chem. Soc., Dalton Trans.*, 1978, 85.
397. A. A. Pinkerton and D. Schwartzenbach, *J. Chem. Soc., Dalton Trans.*, 1976, 2466.
398. Y. Meseri, A. A. Pinkerton and G. Chaperis, *J. Chem. Soc., Dalton Trans.*, 1977, 725.
399. A. A. Pinkerton and D. Schwartzenbach, *J. Chem. Soc., Dalton Trans.*, 1980, 1300.
400. A. A. Pinkerton and D. Schwartzenbach, *J. Chem. Soc., Dalton Trans.*, 1976, 2464.
401. D. Brown and D. G. Holah, *Chem. Commun.*, 1968, 1545.
402. M. Ciampolini, N. Nardi, P. Colamarino and P. Orioli, *J. Chem. Soc., Dalton Trans.*, 1977, 379.
403. M. Ciampolini, C. Mealli and N. Nardi, *J. Chem. Soc., Dalton Trans.*, 1980, 376.
404. H. C. Aspinall, D. C. Bradley, M. B. Hursthouse, K. D. Sales and N. P.C. Walker, *J. Chem. Soc., Chem. Commun.*, 1985, 1585.
405. G. R. Choppin, *Pure Appl. Chem.*, 1971, **27,** 23.
406. J. L. Hoard, B. Lee and M. D. Lind, *J. Am. Chem. Soc.*, 1965, **87,** 1612.
407. M. D. Lind, B. Lee and J. L. Hoard, *J. Am. Chem. Soc.*, 1965, **87,** 1611.
408. L. C. Thompson and J. A. Loraas, *Inorg. Chem.*, 1963, **2,** 89.
409. G. Anderegg, P. Nägeli, F. Müller and G. Schwartzenbach, *Helv. Chim. Acta,* 1959, **42,** 827.
410. L. K. Templeton, D. H. Templeton, A. Zalkin and H. W. Ruben, *Acta Crystallogr., Sect. B,* 1982, **38,** 2155.
411. B. Lee, *Diss. Abstr. B,* 1967, **28,** 84.
412. T. V. Philippova, T. N. Polynova, A. L. Il'Inskii, M. A. Porai-Koshits and L. I. Martynenko, *Zh. Strukt. Khim.*, 1977, **18,** 1127.
413. L. R. Massimbeni, M. R. W. Wright, J. C. Van Niekerk and P. A. McCallum, *Acta Crystallogr., Sect. B,* 1979, **35,** 1341.

414. T. Moeller, E. R. Birnbaum, J. H. Forsberg and R. B. Gayhart, *Prog. Sci. Tech. Rare Earths,* 1968, **2,** 61.
415. A. E. Martell and R. M. Smith, 'Critical Stability Constants', Plenum, New York, 1977, vol. 3, p. 24.
416. N. A. Kostromina and N. N. Tananaeva, *Koord. Khim.*, 1979, **5,** 1633.
417. W. de W. Horrocks, V. K. Arkle, F. J. Liotta and D. R. Sudnick, *J. Am. Chem. Soc.*, 1983, **105,** 3455.
418. R. V. Southwood-Jones and A. E. Merbach, *Inorg. Chim. Acta,* 1978, **30,** 135.
419. G. Geier and U. Karlen, *Helv. Chim. Acta,* 1971, **54,** 135.
420. L. A. Fedorov, L. V. Ruzaikina, N. I. Grebenshchikov, V. A. Ryabukhin and A. N. Ermakov, *Russ. J. Inorg. Chem.* (*Engl. Transl.*), 1983, **28,** 1167.
421. W. D'Olislager and G. R. Choppin, *J. Inorg. Nucl. Chem.*, 1971, **33,** 127.
422. T. Ryhl, *Acta Chem. Scand.*, 1973, **27,** 303.
423. E. Brücher and G. Laurenczy, *Inorg. Chem.*, 1983, **22,** 338.
424. G. Laurenczy and E. Brücher, *Inorg. Chim. Acta,* 1984, **95,** 5.
425. E. Brücher and G. Laurenczy, *Inorg. Chem.*, 1983, **22,** 338.
426. J. Albertsson, *Acta Chem. Scand.*, 1972, **26,** 1023.
427. J. Albertsson, *Acta Chem. Scand.*, 1972, **26,** 985.
428. J. Albertsson, *Acta Chem. Scand.*, 1972, **26,** 1005.
429. J. Albertsson, *Acta Chem. Scand.*, 1970, **24,** 1213.
430. D. R. Foster and F. S. Richardson, *Inorg. Chem.*, 1983, **22,** 3996.
431. E. M. Stephens, K. Schoene and F. S. Richardson, *Inorg. Chem.*, 1984, **23,** 1641.
432. S. A. Davis and F. S. Richardson, *Inorg. Chem.*, 1984, **23,** 184.
433. F. S. Richardson, J. D. Saxe, S. A. Davis and T. R. Faulkner, *Mol. Phys.*, 1981, **42,** 1401.
434. V. S. Pangani, V. M. Agre and V. K. Trunov, *Russ. J. Inorg. Chem.* (*Engl. Transl.*), 1983, **28,** 1213.
435. I. I. Kalinichenko, M. G. Ivanov and N. M. Titov, *Russ. J. Inorg. Chem.* (*Engl. Transl.*), 1984, **29,** 534.
436. N. V. Gerbeleu, M. A. Tishchenko, V. I. Lozan, O. A. Bolga, I. V. Bezlutskaya, Zh. Yu. Vaisbein and N. S. Poluetkov, *Russ. J. Inorg. Chem.* (*Engl. Transl.*), 1983, **28,** 183.
437. L. B. Zinner, D. E. Crotty, T. J. Anderson and M. D. Glick, *Inorg. Chem.*, 1979, **18,** 2045.
438. J. E. Thomas, R. C. Palenik and G. J. Palenik, *Inorg. Chim. Acta,* 1979, **37,** L459.
439. J. E. Thomas and G. J. Palenik, *Inorg. Chim. Acta,* 1980, **44,** L303.
440. A. Semanara, S. Giuffrida, A. Musumeci and I. Fragalia, *Inorg. Chim. Acta,* 1984, **95,** 201.
441. C. J. Pedersen, *J. Am. Chem. Soc.*, 1967, **89,** 7017.
442. C. J. Pedersen, *J. Am. Chem. Soc.*, 1970, **92,** 386.
443. A. Cassol, A. Seminara and G. de Paoli, *Inorg. Nucl. Chem. Lett.*, 1973, **9,** 1163.
444. R. B. King and P. R. Heckley, *J. Am. Chem. Soc.*, 1974, **96,** 3118.
445. J. D. J. Backer-Dirks, J. E. Cooke, A. M. R. Galas, J. S. Ghotra, C. J. Gray, F. A. Hart and M. B. Hursthouse, *J. Chem. Soc., Dalton Trans.*, 1980, 2191.
446. J.-C. G. Bünzli, B. Klein and D. Wessner, *Inorg. Chim. Acta,* 1980, **44,** L147.
447. G. Bombieri, G. de Paoli, F. Benetollo and A. Cassol, *J. Inorg. Nucl. Chem.*, 1980, **42,** 1417.
448. J.-C. G. Bünzli, B. Klein, D. Wessner, J. K. Schenk, G. Chapuis, G. Bombieri and G. de Paoli, *Inorg. Chim. Acta,* 1981, **54,** L43.
449. E. Forsellini, F. Benetollo, G. de Paoli and G. Bombieri, *Eur. Cryst. Meeting,* 1982, **7,** 211.
450. Z. B. Xu, Y. X. Gu, Q. T. Zheng, F. L. Shen, C. K. Yao, S. P. Yan and G. L. Wang, *Wuli Xuebao,* 1982, **31,** 956 (*Chem. Abstr.*, 1982, **97,** 172 794).
451. M. Ciampolini, N. Nardi, R. Cini, S. Mangari and P. Orioli, *J. Chem. Soc., Dalton Trans.*, 1979, 1983.
452. J.-C. G. Bünzli, B. Klein, D. Wessner and N. W. Alcock, *Inorg. Chim. Acta,* 1982, **59,** 269.
453. J.-C. G. Bünzli, B. Klein, G. Chapuis and K. J. Schenk, *Inorg. Chem.*, 1982, **21,** 808.
454. M. A. Neuman, E. C. Steiner, F. P. van Remoortere and F. P. Boer, *Inorg. Chem.*, 1975, **14,** 734.
455. P. D. Cradwick and N. S. Poonia, *Acta Crystallogr., Sect. B,* 1977, **33,** 197.
456. G. Tomat, G. Valle, A. Cassol and P. di Bernardo, *Inorg. Chim. Acta,* 1983, **76,** L13.
457. Y. Hirashima, T. Tsutsui and J. Shiokawa, *Chem. Lett.*, 1982, 1405.
458. Y. Hirashima, K. Kanetsuki, J. Shiokawa and N. Tanaka, *Bull. Chem. Soc. Jpn,* 1981, **54,** 1567.
459. Y. Hirashima, T. Tsutsui and J. Shiokawa, *Chem. Lett.*, 1981, 1501.
460. V. Casellato, G. Tomat, P. di Bernado and R. Graziani, *Inorg. Chim. Acta,* 1982, **61,** 181.
461. M. Ciampolini, C. Mealli and N. Nardi, *J. Chem. Soc., Dalton Trans.*, 1980, 376.
462. G. A. Catton, M. E. Harman, F. A. Hart, G. E. Hawkes and G. P. Moss, *J. Chem. Soc., Dalton Trans.*, 1978, 181.
463. J.-C. G. Bünzli, D. Wessner and B. Klein, 'The Rare Earths in Modern Science and Technology', ed. G. J. McCarthy, J. J. Rhyne and H. B. Silber, Plenum, New York, 1980, vol. 2.
464. J. F. Desreux and J. Massaux, 'The Rare Earths in Modern Science and Technology', ed. G. J. McCarthy, H. B. Silber and J. J. Rhyne, Plenum, New York, 1982, vol. 3.
465. J.-C. G. Bünzli, J.-M. Pfefferle and B. Ammann, *Helv. Chim. Acta,* 1984, **67,** 1121.
466. W. Herr, *Angew. Chem.*, 1953, **65,** 303.
467. W. Herr, *Z. Naturforsch., Teil A,* 1954, **11,** 1980.
468. M. G. Gurevich and K. N. Solv'ev, *Dokl. Akad. Nauk BSSR,* 1961, **5,** 291.
469. L. P. Shklover and V. E. Plyushchev, *Russ. J. Inorg. Chem.* (*Engl. Transl.*), 1964, **9,** 991.
470. V. E. Plyuskchev and L. P. Shklover, *Russ. J. Inorg. Chem.* (*Engl. Transl.*), 1964, **9,** 1086.
471. V. E. Plyushchev and L. P. Shklover, *Russ. J. Inorg. Chem.* (*Engl. Transl.*), 1964, **9,** 183.
472. I. S. Kirin, P. N. Moskalev and Y. A. Makashev, *Russ. J. Inorg. Chem.* (*Engl. Transl.*), 1965, **10,** 1065.
473. I. S. Kirin, P. N. Moskalev and Y. A. Makashev, *Russ. J. Inorg. Chem.* (*Engl. Transl.*), 1967, **12,** 369.
474. I. S. Kirin, P. N. Moskalev and Y. A. Makashev, *Russ. J. Inorg. Chem.* (*Engl. Transl.*), 1967, **12,** 497.
475. P. N. Moskalev and I. S. Kirin, *Russ. J. Inorg. Chem.* (*Engl. Transl.*), 1970, **15,** 7.
476. S. Misumi and K. Kasuga, *Nippon Kagaku Zasshi,* 1971, **92,** 335.
477. A. G. MacKay, J. F. Boas and G. J. Troup, *Aust. J. Chem.*, 1974, **27,** 995.
478. M. Yamana, *Nippon Kagaku Kaishi,* 1977, 144.

479. M. Yamana, *Res. Rep. Tokyo Denki Univ.*, 1977, **25**, 39.
480. K. Kasuga, M. Tsutsui, R. C. Petterson, K. Tatsumi, N. Van Opdenbosch, G. Pepe and E. F. Mayer, *J. Am. Chem. Soc.*, 1980, **102**, 4835.
481. P. N. Moskalev, G. N. Shapkin and A. N. Darovskikh, *Russ. J. Inorg. Chem.* (*Engl. Transl.*), 1979, **24**, 188.
482. P. N. Moskalev, G. N. Shapkin and N. I. Alimova, *Russ. J. Inorg. Chem.* (*Engl. Transl.*), 1982, **27**, 794.
483. P. N. Moskalev and I. S. Kirin, *Zh. Obshch. Khim.*, 1968, **38**, 1191.
484. P. N. Moskalev and I. S. Kirin, *Russ. J. Inorg. Chem.* (*Engl. Transl.*), 1971, **16**, 57.
485. E. I. Kapinus, L. I. Smykalova and I. I. Dilung, *Russ. J. Inorg. Chem.* (*Engl. Transl.*), 1980, **25**, 225.
486. G. A. Corker, B. Grant and N. J. Clecak, *J. Electrochem. Soc.*, 1979, **126**, 1339.
487. C.-P. Wong, R. F. Venteicher and W. de W. Horrocks, *J. Am. Chem. Soc.*, 1974, **96**, 7149.
488. W. de W. Horrocks and C.-P. Wong, *J. Am. Chem. Soc.*, 1976, **98**, 7157.
489. W. de W. Horrocks and G. E. Hove, *J. Am. Chem. Soc.*, 1978, **100**, 4386.
490. K. Tatsumi and M. Tsutsui, *J. Am. Chem. Soc.*, 1980, **102**, 882.
491. J. D. J. Backer-Dirks, C. J. Gray, F. A. Hart, M. B. Hursthouse and B. C. Schoop, *J. Chem. Soc., Chem. Commun.*, 1979, 774.
492. W. Radecka-Paryzek, *Inorg. Chim. Acta*, 1980, **45**, L147.
493. K. K. Abid, D. E. Fenton, U. Casellato, P. A. Vigato and R. Graziani, *J. Chem. Soc., Dalton Trans*, 1984, 351.
494. K. K. Abid and D. E. Fenton, *Inorg. Chim. Acta*, 1984, **95**, 119.
495. C. J. Gray, Ph.D. Thesis, University of London, 1982.
496. W. Radecka-Paryzek, *Inorg. Chim. Acta*, 1981, **52**, 261.
497. J. F. Desreux, *Inorg. Chem.*, 1980, **19**, 1319.
498. M.-R. Spirlet, J. Rebizant, J. F. Desreux and M.-F. Locin, *Inorg. Chem.*, 1984, **23**, 359.
499. C. C. Bryden, C. N. Reilley and J. F. Desreux, *Anal. Chem.*, 1981, **53**, 1418.
500. K. K. Abid and D. E. Fenton, *Inorg. Chim. Acta*, 1984, **82**, 223.
501. C. A. Chang and M. E. Rowland, *Inorg. Chem.*, 1983, **22**, 3866.
502. O. A. Gansow, A. R. Kauser, K. M. Triplett, M. J. Weaver and E. L. Yee, *J. Am. Chem. Soc.*, 1977, **99**, 7087.
503. E. L. Yee, J. T. Hupp and M. J. Weaver, *Inorg. Chem.*, 1983, **22**, 3465.
504. E. L. Yee, O. A. Gansow and M. J. Weaver, *J. Am. Chem. Soc.*, 1980, **102**, 2278.
505. J. H. Burns and C. F. Baes, *Inorg. Chem.*, 1981, **20**, 616.
506. G. Anderegg, *Helv. Chim. Acta*, 1981, **64**, 1790.
507. R. Pizer and R. Selzer, *Inorg. Chem.*, 1983, **22**, 1359.
508. F. A. Hart, M. B. Hursthouse, K. M. A. Malik and S. Moorhouse, *J. Chem. Soc., Chem. Commun.*, 1978, 549.
509. O. A. Gansow and A. R. Kausar, *Inorg. Chim. Acta*, 1983, **72**, 39.
510. G. Campari and F. A. Hart, *Inorg. Chim. Acta*, 1982, **65**, L217.
511. J. H. Burns, *Inorg. Chem.*, 1979, **18**, 3044.
512. M. Ciampolini, P. Dapporto and N. Nardi, *J. Chem. Soc., Chem. Commun.*, 1978, 788.
513. M. Ciampolini, P. Dapporto and N. Nardi, *J. Chem. Soc., Dalton Trans.*, 1979, 974.
514. O. A. Gansow, D. J. Pruett and K. B. Triplett, *J. Am. Chem. Soc.*, 1979, **101**, 4408.
515. R. E. Thoma, H. Insley and G. M. Herbert, *Inorg. Chem.*, 1966, **5**, 1222.
516. P. P. Fedorov, A. V. Rappo, F. M. Spiridonov and B. P. Sobolev, *Russ. J. Inorg. Chem.* (*Engl. Transl.*), 1983, **28**, 420.
517. J. H. Burns, *Inorg. Chem.*, 1965, **4**, 881.
518. F. Hund, *Z. Anorg. Allg. Chem.*, 1950, **261**, 106.
519. C. Keller and H. Schmutz, *J. Inorg. Nucl. Chem.*, 1965, **27**, 900.
520. H. Bode and E. Voss, *Z. Anorg. Allg. Chem.*, 1957, **290**, 1.
521. G. A. Bukhalov and E. P. Babaeva, *Russ. J. Inorg. Chem.* (*Engl. Transl.*), 1966, **11**, 337.
522. G. Meyer, 'The Rare Earths in Modern Science and Technology', ed. G. J. McCarthy, H. B. Silber and J. J. Rhyne, Plenum, New York, 1982, vol. 3, pp. 317.
523. Z. Amilius, B. van Laar and H. M. Rietveld, *Acta Crystallogr., Sect. B*, 1969, **25**, 400.
524. V. A. Abramets, M. B. Varfolomeev and L. G. Zhavoronkova, *Russ. J. Inorg. Chem.* (*Engl. Transl.*), 1980, **25**, 393.
525. V. A. Abramets, M. B. Varfolomeev, L. G. Zhavoronkova, L. Ya. Petrenko and N. B. Shamrai, *Russ. J. Inorg. Chem.* (*Engl. Transl.*), 1981, **26**, 825.
526. M. B. Varfolomeev, P. B. Shamrai, A. V. Markov, V. A. Abramets and D. V. Drobot, *Russ. J. Inorg. Chem.* (*Engl. Transl.*), 1984, **29**, 536.
527. D. V. Drobot, B. G. Korshunov and G. P. Borodulenko, *Russ. J. Inorg. Chem.* (*Engl. Transl.*), 1968, **13**, 855.
528. J. L. Ryan and C. K. Jørgensen, *J. Phys. Chem.*, 1966, **70**, 2845.
529. J. L. Ryan, *Inorg. Chem.*, 1969, **8**, 2053.
530. G. Meyer, *Z. Anorg. Allg. Chem.*, 1980, **469**, 149.
531. G. Meyer, *Z. Anorg. Allg. Chem.*, 1978, **445**, 140.
532. D. M. Guthrie, G. Meyer and J. D. Corbett, *Inorg. Chem.*, 1981, **20**, 1192.
533. C. C. Hinckley, *J. Am. Chem. Soc.*, 1969, **91**, 5160.
534. J. K. M. Sanders and D. H. Williams, *Chem. Commun.*, 1970, 422.
535. J. Briggs, G. H. Frost, F. A. Hart, G. P. Moss and M. L. Stainforth, *Chem. Commun.*, 1970, 749.
536. A. F. Cockerill, G. L. O. Davies, R. C. Harden and D. M. Rackham, *Chem. Rev.*, 1973, **73**, 553–588.
537. B. C. Mayo, *Chem. Soc. Rev.*, 1973, **2**, 49–74.
538. B. D. Flockhart, *CRC Crit. Rev. Anal. Chem.*, 1976, **6**, 69–130.
539. K. A. Kime and R. E. Sievers, *Aldrichim. Acta*, 1977, **10**, 54–62.
540. F. Inagaki and T. Miyazawa, *Prog. Nucl. Magn. Reson. Spectrosc.*, 1980, **14**, 67–111.
541. S. P. Sinha, in '9th International Summer School Coordination Chemistry', ed. B. Jezowska-Trzebiatowski, B. Legendziewica and M. F. Rudolf, Panst. Wydawn. Nauk, Warsaw, 1977, p. 235–287.
542. R. Reuben, in 'Handbook of the Physics and Chemistry of the Rare Earths', ed. K. A. Gschneidner and Le Roy Eyring, Elsevier, 1979, vol. 4, pp. 515–552.

543. R. J. P. Williams, *Struct. Bonding (Berlin)*, 1982, **50**, 79.
544. J. A. Balschi, V. P. Cirillo, W. J. Le Noble, M. M. Pike, E. C. Schreiber, S. R. Simon and C. S. Springer, in 'The Rare Earths in Modern Science and Technology', ed. G. J. McCarthy H. B. Silber and J. J. Rhyne, Plenum, New York, 1982, p. 15.
545. B. Bleaney, *J. Magn. Reson.*, 1972, **8**, 91.
546. J. Briggs, F. A. Hart and G. P. Moss, *Chem. Commun.*, 1970, 1506.
547. D. F. Evans, J. N. Tucker and G. C. de Villardi, *J. Chem. Soc., Chem. Commun.*, 1975, 205.
548. R. M. Golding and M. P. Halton, *Aust. J. Chem.*, 1972, **25**, 2577.
549. C. N. Reilley, B. W. Good and J. F. Desreux, *Anal. Chem.*, 1975, **47**, 2110.
550. C. N. Reilley, B. W. Good and R. D. Allendoerfer, *Anal. Chem.*, 1976, **48**, 1446.
551. J. Reuben and G. A. Elgavish, *J. Magn. Reson.*, 1980, **39**, 421.
552. J. Reuben, *J. Magn. Reson.*, 1982, **50**, 233.
553. W. de W. Horrocks and J. P. Sipe, *Science*, 1972, **177**, 994.
554. W. de W. Horrocks, H. P. Sipe and D. Sudnick, Nuclear Magnetic Shift Reagents, ed. R. E. Sievers, Academic, 1973, p. 53.
555. P. J. Domaille, *J. Am. Chem. Soc.*, 1980, **102**, 5392.
556. J. M. Briggs, G. P. Moss, E. W. Randall and K. D. Sales, *J. Chem. Soc., Chem. Commun.*, 1972, 1180.
557. W. de W. Horrocks, *J. Am. Chem. Soc.*, 1974, **96**, 3022.
558. R. E. Cramer, R. Dubois and K. Seff, *J. Am. Chem. Soc.*, 1974, **96**, 4125.
559. R. E. Cramer, R. Dubois and C. K. Furuike, *Inorg. Chem.*, 1975, **14**, 1005.
560. R. E. Cramer, R. B. Maynard and R. Dubois, *J. Chem. Soc., Dalton Trans.*, 1979, 1350.
561. R. A. Dwek, in 'NMR in Biochemistry', Clarendon, Oxford, 1973, p. 208.
562. B. M. Alsaadi, F. J. C. Rossotti and R. J. P. Williams, *J. Chem. Soc., Dalton Trans.*, 1980, 813.
563. M. Singh, J. J. Reynolds and A. D. Sherry, *J. Am. Chem. Soc.*, 1983, **105**, 4172.
564. C. F. G. C. Geraldes, *J. Magn. Reson.*, 1979, **36**, 89.
565. C. F. G. C. Geraldes and J. R. Ascenso, *J. Chem. Soc., Dalton Trans.*, 1984, 267.
566. A. D. Sherry, C. A. Stark, J. R. Ascenso and C. F. G. C. Geraldes, *J. Chem. Soc., Dalton Trans.*, 1981, 2078.
567. M. Delepierre, C. M. Dobson and S. L. Menear, *J. Chem. Soc., Dalton Trans.*, 1981, 678.
568. R. L. Bassfield, *J. Am. Chem. Soc.*, 1983, **105**, 4168.
569. L. F. Lindoy and H. W. Louie, *Inorg. Chem.*, 1981, **20**, 4186.
570. J. Y. Lee, D. A. Hanna and G. W. Everett, *Inorg. Chem.*, 1981, **20**, 2004.
571. G. W. Everett, D. A. Hanna and J. Y. Lee, *Inorg. Chim. Acta*, 1982, **64**, 215.
572. M. Hirayama and Y. Sasaki, *Chem. Lett.*, 1982, 195.
573. M. Kirayama, Y. Kawamata, Y. Fujii and Y. Nakono, *Bull. Chem. Soc. Jpn.*, 1982, **55**, 1788.
574. D. F. Evans, J. C. Tucker and G. C. de Villardi, *J. Chem. Soc., Chem. Commun.*, 1975, 205.
575. T. J. Wenzel, T. C. Bettes, J. E. Sadlowski and R. E. Sievers, *J. Am. Chem. Soc.*, 1980, **102**, 5903.
576. T. J. Wenzel and R. E. Sievers, *J. Am. Chem. Soc.*, 1982, **104**, 382.
577. A. de Renzi, G. Morelli, A. Panunzi and S. Wurzburger, *Inorg. Chim. Acta*, 1983, **76**, L285.
578. G. M. Whitesides and D. W. Lewis, *J. Am. Chem. Soc.*, 1970, **92**, 6979.
579. C. Kutal in 'NMR Shift Reagents', ed. R. E. Sievers, Academic, New York and London, 1973, p. 87.
580. J. Reuben, *J. Chem. Soc., Chem. Commun.*, 1979, 68.
581. J. Reuben, *J. Am. Chem. Soc.*, 1980, **102**, 2232.
582. M. M. Pike, D. M. Yarmush, J. A. Balschi, R. E. Lenkinsky and C. S. Springer, *Inorg. Chem.*, 1983, **22**, 2388.
583. W. de W. Horrocks and E. G. Hove, *J. Am. Chem. Soc.*, 1978, **100**, 4386.
584. C. C. Bryden, C. N. Reilley and J. F. Desreux, *Anal. Chem.*, 1981, **53**, 1418.
585. D. R. Eaton, *J. Am. Chem. Soc.*, 1965, **87**, 3097.
586. F. A. Hart, J. E. Newbery and D. Shaw, *Nature (London)*, 1967, **216**, 261.
587. B. M. Alasaadi, F. J. C. Rossotti and R. J. P. Williams, *J. Chem. Soc., Dalton Trans.*, 1980, 597.
588. J. Albertsson, *Acta Chem. Scand.*, 1972, **26**, 1005.
589. A. A. Pinkerton and D. Schwarzenback, *J. Chem. Soc., Dalton Trans.*, 1981, 1470.
590. S. Hüfner, 'Systematics and the Properties of the Lanthanides', ed. S. P. Sinha, Reidel, Dordrecht, 1983, p. 313.
591. W. T. Carnall, J. V. Beitz, H. Crosswhite, K. Rajnak and J. B. Mann, 'Systematics and the Properties of the Lanthanides', ed. S. P. Sinha, Reidel, Dordrecht, 1983, p. 389.
592. J. Rossat-Mignod, 'Systematics and the Properties of the Lanthanides', ed. S. P. Sinha, Reidel, Dordrecht, 1983, p. 255.
593. H. M. Crosswhite, H. Crosswhite, W. T. Carnall and A. P. Paszek, *J. Chem. Phys.*, 1980, **72**, 5103.
594. J. Sugar, *Phys. Rev. Lett.*, 1965, **14**, 731.
595. J. F. Kielkopf and H. M. Crosswhite, *J. Opt. Soc. Am.* 1970, **60**, 347.
596. F. H. Spedding, *Phys. Rev.*, 1940, **58**, 255.
597. J. C. Morrison, *Phys. Rev.*, 1972, **A6**, 643.
598. H. Crosswhite, H. M. Crosswhite and B. R. Judd, *Phys. Rev.*, 1968, **174**, 89.
599. H. M. Crosswhite, H. Crosswhite, F. W. Kaseta and R. Sarup, *J. Chem. Phys.*, 1976, **64**, 1981.
600. J. P. Hessler and W. T. Carnall, 'Lanthanide and Actinide Chemistry and Spectroscopy', ed. N. M. Edelstein, American Chemical Society, Washington, DC, 1980, p. 349.
601. B. R. Judd, *Phys. Rev.*, 1962, **127**, 750.
602. G. S. Ofelt, *J. Chem. Phys.*, 1962, **37**, 511.
603. C. K. Jørgensen and B. R. Judd, *Mol. Phys.*, 1964, **8**, 281.
604. S. F. Mason, R. D. Peacock and B. Stewart, *Mol. Phys.*, 1975, **30**, 1829.
605. R. D. Peacock, *Struct. Bonding (Berlin)*, 1975, **22**, 83.
606. B. R. Judd, *J. Chem. Phys.*, 1979, **70**, 4830.
607. S. F. Mason, *Struct. Bonding (Berlin)*, 1980, **39**, 43.
608. B. R. Judd, 'Lanthanide and Actinide Chemistry and Spectroscopy', ed. N. M. Edelstein, American Chemical Society, Washington, DC, 1980, p. 267.

609. S. A. Davis and F. S. Richardson, *Inorg. Chem.*, 1984, **23,** 184.
610. K. Bukietynska and A. Mondry, *Inorg. Chim. Acta,* 1983, **72,** 109.
611. G. A. Catton and F. A. Hart, *J. Chem. Soc., Dalton Trans.*, 1976, 208.
612. D. A. Durham, G. H. Frost and F. A. Hart, *J. Inorg. Nucl. Chem.*, 1969, **31,** 833.
613. J. J. Degnan, C. R. Hurt and N. Filipescu, *J. Chem. Soc., Dalton Trans.*, 1972, 1158.
614. F. Halverson, J. S. Brinen and J. R. Leto, *J. Chem. Phys.*, 1964, **41,** 157.
615. M. Albin, A. G. Goldstone, A. S. Withers and W. de W. Horrocks, *Inorg. Chem.*, 1983, **22,** 3182.
616. J.-C. G. Bünzli, B. Klein, G.-O. Pradervand and P. Porcher, *Inorg. Chem.*, 1983, **22,** 3763.
617. J.-C. G. Bünzli and M. M. Vuckovic, *Inorg. Chim. Acta,* 1983, **73,** 53.
618. J.-C. G. Bünzli and M. M. Vuckovic, *Inorg. Chim. Acta,* 1984, **95,** 105.
619. F. S. Richardson, *Inorg. Chem.*, 1980, **19,** 2806.
620. J. S. Madaras and H. G. Brittain, *Inorg. Chem.*, 1980, **19,** 3841.
621. J. S. Madaras and H. G. Brittain, *Inorg. Chim. Acta,* 1980, **42,** 109.
622. H. G. Brittain, *Inorg. Chem.*, 1981, **20,** 3007.
623. F. Yan, R. A. Copeland and H. G. Brittain, *Inorg. Chem.*, 1982, **21,** 1180.
624. F. Yan and H. G. Brittain, *Polyhedron,* 1982, **1,** 195.
625. H. G. Brittain, *Inorg. Chem.*, 1982, **21,** 2955.
626. F. Yan, R. Copeland and H. G. Brittain, *Inorg. Chim. Acta,* 1983, **72,** 211.
627. H. G. Brittain, *Inorg. Chim. Acta,* 1983, **70,** 91.
628. H. G. Brittain and K. H. Pearson, *Inorg. Chem.*, 1983, **22,** 78.
629. D. R. Foster and F. S. Richardson, *Inorg. Chem.*, 1983, **22,** 2996.
630. F. S. Richardson, *Chem. Rev.*, 1982, **82,** 541.
631. M. Albin, B. M. Cader and W. de W. Horrocks, *Inorg. Chem.*, 1984, **23,** 3045.
632. M. Gerlock and D. J. Mackey, *J. Chem. Soc. (A)*, 1970, 3030.
633. M. Gerlock and D. J. Mackey, *J. Chem. Soc. (A)*, 1970, 3040.
634. M. Gerlock and D. J. Mackey, *J. Chem. Soc. (A)*, 1971, 2605.
635. M. Gerlock and D. J. Mackey, *J. Chem. Soc. (A)*, 1971, 2612.
636. M. Gerlock and D. J. Mackey, *J. Chem. Soc. (A)*, 1971, 3372.
637. M. Gerlock and D. J. Mackey, *J. Chem. Soc. (A)*, 1972, 37.
638. M. Gerlock and D. J. Mackey, *J. Chem. Soc. (A)*, 1972, 42.
639. M. Gerlock and D. J. Mackey, *J. Chem. Soc. (A)*, 1972, 410.
640. M. Gerlock and D. J. Mackey, *J. Chem. Soc. (A)*, 1972, 415.
641. W. J. Evans, *J. Organomet. Chem.*, 1983, **250,** 217.
642. A. N. Kamenskaya, *Russ. J. Inorg. Chem. (Engl. Transl.)*, 1984, **29,** 439.
643. D. J. Meier and C. S. Garner, *J. Phys. Chem.*, 1952, **56,** 853.
644. A. N. Kamenskaya, N. B. Mikheev and A. I. Strekalov, *Russ. J. Inorg. Chem. (Engl. Transl.)*, 1983, **28,** 965.
645. D. A. Johnson, *Adv. Inorg. Chem. Radiochem.*, 1977, **20,** 1.
646. D. A. Johnson, *J. Chem. Soc., Dalton Trans.*, 1974, 1671.
647. N. B. Mikheev, *Russ. J. Inorg. Chem. (Engl. Transl.)*, 1984, **29,** 450.
648. S. Gordon, J. C. Sullivan and W. A. Mulac, *Proc. IV Sympos. Rad. Chem.*, Keszthely, 1976, 763.
649. C. Musikas, R. G. Haire and J. R. Peterson, *J. Inorg. Nucl. Chem.*, 1981, **43,** 2935.
650. L. O. de S. Bulhoes and T. Rabochai, *Electrochim. Acta,* 1982, **27,** 1071.
651. R. F. Large and A. Tunnick, *Anal. Chem.*, 1964, **36,** 1258.
652. J. Doyle and A. G. Sykes, *J. Chem. Soc. (A)*, 1968, 2836.
653. A. Adin and A. G. Sykes, *J. Chem. Soc. (A)*, 1968, 354.
654. R. J. Christensen, J. H. Espenson and A. B. Butcher, *Inorg. Chem.*, 1973, **12,** 564.
655. M. S. Ram and E. S. Gould, *Inorg. Chem.*, 1983, **22,** 2454.
656. H. Yamana, T. Mitsugashira, Y. Shiokawa and S. Suzuki, *Bull. Chem. Soc. Jpn.*, 1982, **55,** 2615.
657. K. Rossmanith, *Monatsh. Chem.*, 1979, **110,** 1019.
658. K. Rossmanith, *Monatsh. Chem.*, 1979, **110,** 109.
659. P. Girard, J. L. Namy and H. B. Kagan, *J. Am. Chem. Soc.*, 1980, **102,** 2693.
660. A. N. Kamenskaya, N. B. Mikheev and N. A. Konovalova, *Russ. J. Inorg. Chem. (Engl. Transl.)*, 1977, **22,** 1152.
661. A. N. Kamenskaya, K. Bukietynska, N. B. Mikheev, V. I. Spitsyn and B. Jesowska-Trzebiatowska, *Russ. J. Inorg. Chem. (Engl. Transl.)*, 1979, **24,** 633.
662. N. B. Mikheev, R. A. D'yachkova, L. I. Telegina and N. A. Rozenkevich, *Russ. J. Inorg. Chem. (Engl. Transl.)*, 1972, **17,** 1065.
663. T. D. Tilley, R. A. Andersen and A. Zalkin, *Inorg. Chem.*, 1984, **23,** 2271.
664. T. D. Tilley, A. Zalkin, R. A. Andersen and D. H. Templeton, *Inorg. Chem.*, 1981, **20,** 551.
665. F. A. Hart and W. Zhu, *Inorg. Chim. Acta,* 1981, **54,** L275.
666. T. D. Tilley, R. A. Andersen and A. Zalkin, *J. Am. Chem. Soc.*, 1982, **104,** 3725.
667. T. D. Tilley, R. A. Andersen and A. Zalkin, *Inorg. Chem.*, 1983, **22,** 856.
668. L. B. Asprey, F. H. Ellinger and E. Staritsky, 'Rare Earth Research', ed. K. S. Vorres, Gordon and Breach, New York, 1964, vol. 2, p. 11.
669. T. D. Tilley and R. A. Andersen, *J. Am. Chem. Soc.*, 1982, **104,** 1772.
670. D. S. McClure and Z. Kiss, *J. Chem. Phys.*, 1963, **39,** 3251.
671. G. Wu and R. Hoppe, *Z. Anorg. Allg. Chem.*, 1983, **504,** 55.
672. G. Meyer, *J. Less-Common Met.*, 1983, **93,** 371.
673. K. E. Johnson and J. N. Sandoe, *J. Chem. Soc. (A)*, 1969, 1694.
674. A. N. Kamenskaya, N. B. Mikheev and N. P. Kholmogorova, *Russ. J. Inorg. Chem. (Engl. Transl.)*, 1983, **28,** 1420.
675. N. Sabbatini, M. Ciano, S. Dellonte, A. Bonazzi and V. Balzani, *Chem. Phys. Lett.*, 1982, **90,** 265.
676. G.-Y. Adachi, K. Tomokiyo, K. Sorita and J. Shiokawa, *J. Chem. Soc., Chem. Commun.*, 1980, 914.

677. F. Gaume, A. Gros and J. C. Bourcet, 'The Rare Earths in Modern Science and Technology', ed. G. J. McCarthy, H. B. Silber and J. J. Rhyne, Plenum, New York, 1982, vol. 3, p. 143.
678. L. J. Nugent, R. D. Baybarz, J. L. Burnett and J. L. Ryan, *J. Phys. Chem.*, 1973, **77,** 1528.
679. J. Sugar and J. Reader, *J. Chem. Phys.*, 1973, **59,** 2083.
680. A. H. Kunz, *J. Am. Chem. Soc.*, 1931, **53,** 98.
681. M. S. Sherrill, C. B. King and R. C. Spooner, *J. Am. Chem. Soc.*, 1943, **65,** 170.
682. A. A. Noyes and C. S. Garner, *J. Am. Chem. Soc.*, 1936, **58,** 1265.
683. B. D. Blaustein and J. W. Gryder, *J. Am. Chem. Soc.*, 1957, **79,** 540.
684. A. Mills, *J. Chem. Soc., Dalton Trans.*, 1982, 1213.
685. J. Kiwi, M. Grätzel and G. Blondeel, *J. Chem. Soc., Dalton Trans.*, 1983, 2215.
686. D. E. Hobart, K. Samhoun, J. P. Young, V. E. Norvell, G. Mamantov and J. R. Peterson, *Inorg. Nucl. Chem. Lett.*, 1980, **16,** 321.
687. R. Yang and X. Sun, *Lanzhou Daxue Xuebao, Ziran Kescueban,* 1982, **18,** 162 (*Chem. Abstr.*, **97,** 207 083).
688. R. C. Propst, *J. Inorg. Nucl. Chem.*, 1974, 36, 1085.
689. G. Wang, R. Zhang and Z. Jiang, *Gaodeng Xuexiao Huaxue Xuebao,* 1983, **4,** 535 (*Chem. Abstr.*, **100,** 78 872).
690. S. D. Nikitina and S. A. Bondar, *Russ. J. Inorg. Chem. (Engl. Transl.),* 1983, **28,** 1275.
691. J. K. Puri, P. Singh and J. M. Miller, *Inorg. Chim. Acta,* 1983, **74,** 139.
692. T. A. Beineke and J. Delgaudio, *Inorg. Chem.*, 1968, **7,** 715.
693. S. Voliotis and A. Rimsky, *Acta Crystallogr., Sect. B,* 1975, **31,** 2620.
694. S. Voliotis, A. Rimsky and J. Faucherre, *Acta Crystallogr., Sect. B,* 1975, **31,** 2607.
695. A. Job and P. Goissedet, *C. R. Hebd. Seances. Acad. Sci.*, 1913, **50,** 157.
696. B. Matkovic and D. Grdenic, *Acta Crystallogr.*, 1963, **16,** 456.
697. H. Tutze, *Acta Chem. Scand.*, 1969, **23,** 399.
698. H. Barninghausen, *Z. Kristallogr.*, 1971, **134,** 449.
699. V. I. Spitsyn, L. I. Martynenko, N. I. Pechurova, N. I. Snezhko, I. A. Murav'eva and I. Anufrieva, *Izv. Akad. Nauk SSSR, Ser. Khim.*, 1982, **4,** 772.
700. S. R. Sofen, S. R. Cooper and K. N. Raymond, *Inorg. Chem.*, 1979, **18,** 1611.
701. F. Brezina, *Collect. Czech. Chem. Commun.*, 1971, **36,** 2889.
702. F. Brezina, *Collect. Czech. Chem. Commun.*, 1973, **39,** 2162.
703. J. G. H. Du Preez, H. E. Rohwer, J. F. de Wet and M. R. Caira, *Inorg. Chim. Acta,* 1978, **26,** L59.
704. Mazhar-ul-Haque, C. N. Caughlan, F. A. Hart and R. Van Nice, *Inorg. Chem.*, 1971, **10,** 115.
705. J. Barry, J. G. H. Du Preez, E. Els, H. E. Rohwer and P. J. Wright, *Inorg. Chim. Acta,* 1981, **53,** L17.
706. R. A. Penneman and A. Rosenzweig, *Inorg. Chem.*, 1969, **8,** 627.
707. R. R. Ryan, A. C. Larson and F. H. Kruse, *Inorg. Chem.*, 1969, **8,** 33.
708. R. R. Ryan and R. A. Penneman, *Acta Crystallogr., Sect. B,* 1971, **27,** 1939.
709. R. Hoppe and K.-M. Rödder, *Z. Anorg. Allg. Chem.*, 1961, **313,** 154.
710. R. Hoppe and K.-M. Rödder, *Z. Anorg. Allg. Chem.*, 1961, **312,** 277.
711. R. Hoppe and W. Liebe, *Z. Anorg. Allg. Chem.*, 1961, **313,** 221.
712. L. B. Asprey and T. K. Keenan, *J. Inorg. Nucl. Chem.*, 1961, **16,** 260.
713. L. B. Asprey, J. S. Coleman and M. J. Reisfeld, *Adv. Chem. Ser.*, 1967, **71,** 122.
714. R. Hoppe, 'The Rare Earths in Modern Science and Technology', ed. G. J. McCarthy, H. B. Silber and J. J. Rhyne, Plenum, New York, 1982, **3,** 315.
715. Yu. M. Kiselev, S. A. Goryachenkov and L. I. Martynenko, *Russ. J. Inorg. Chem. (Engl. Transl.),* 1984, **29,** 38.
716. L. P. Varga and L. B. Asprey, *J. Chem. Phys.*, 1968, **49,** 4674.
717. L. P. Varga and L. B. Asprey, *J. Chem. Phys.*, 1968, **48,** 139.
718. A. Brandt, Yu. M. Kiselev, L. I. Martynenko and V. I. Spitsyn, *Dokl. Akad. Nauk SSSR,* 1981, **257,** 1189.
719. A. Brandt, Yu. M. Kiselev and L. I. Martynenko, *Russ. J. Inorg. Chem. (Engl. Transl.),* 1981, **26,** 499.
720. A. P. Bayanov, Z. A. Temerdashev and Yu. A. Afanas'ev, *Russ. J. Inorg. Chem. (Engl. Transl.),* 1978, **23,** 1778.
721. Yu. M. Kiselev and A. I. Popov, *Russ. J. Inorg. Chem. (Engl. Transl.),* 1983, **28,** 190.
722. A. Brandt, Yu. M. Kiselev and L. I. Martynenko, *Z. Anorg. Allg. Chem.*, 1981, **474,** 233.
723. A. Brandt, Yu. M. Kiselev and L. I. Martynenko, *Russ. J. Inorg. Chem. (Engl. Transl.),* 1983, **28,** 1593.
724. Yu. M. Kiselev, A. Brandt, L. I. Martynenko and V. I. Spitsyn, *Dokl. Akad. Nauk SSSR,* 1979, **246,** 879.

40

The Actinides

KENNETH W. BAGNALL
University of Manchester, UK

40.1 GENERAL SURVEY

40.1.1 Introduction

Most of this chapter relates to two earlier members of the actinide group, thorium and uranium, both available in large quantities and relatively cheap. The radioactivity arising from the naturally occurring isotopic mixture of these two elements presents no serious handling problems, whereas ^{231}Pa, ^{237}Np, ^{239}Pu, *etc.*, are much more intensely radioactive, requiring very costly handling facilities. These isotopes are also very expensive. Consequently, basic research on these elements is restricted to government-supported research establishments, such as AERE, Harwell in the UK and the Argonne and Oak Ridge National Laboratories in the USA, where much of the effort is of an applied nature. Beyond californium ($Z = 98$), the elements can only be produced in very small quantities and all known isotopes of $Z = 99$ to 103 are short-lived and, consequently, of very high specific radioactivity. This makes it exceedingly difficult to prepare solid, characterizable compounds of any kind, and most of the scanty information that is available is based on tracer level studies.

40.1.2 Oxidation States

In contrast to the lanthanide 4*f* transition series, for which the normal oxidation state is +3 in aqueous solution and in solid compounds, the actinide elements up to, and including, americium exhibit oxidation states from +3 to +7 (Table 1), although the common oxidation state of americium and the following elements is +3, as in the lanthanides, apart from nobelium ($Z = 102$), for which the +2 state appears to be very stable with respect to oxidation in aqueous solution, presumably because of a high ionization potential for the $5f^{14}$ No^{2+} ion. Discussions of the thermodynamic factors responsible for the stability of the tripositive actinide ions with respect to oxidation or reduction are available.[1,2]

Table 1 Known Oxidation States of the Actinide Elements

Th	*Pa*	*U*	*Np*	*Pu*	*Am*	*Cm*	*Bk*	*Cf*	*Es*	*Fm*	*Md*	*No*	*Lw*
					2			2	2	2	2	2	
	3?	3	3	3	3	3	3	3	3	3	3	3	3
4	4	4	4	4	4	4	4	4					
	5	5	5	5	5								
		6	6	6	6								
			7	7	7?								

40.1.3 Steric Crowding

Ligand field interactions in 5*f* transition element complexes are very small,[3] so that there is little or no contribution to the kinetic stability of the complexes, all of which can be regarded as kinetically labile. Ligand exchange will therefore occur quite readily for there will be virtually no gain (or loss) of ligand field stabilization energy in an inter- or intra-molecular rearrangement. There will also be a marked degree of flexibility in the coordination geometries adopted by the complexes formed by these elements. Consequently, steric crowding about the central metal atom is the principal factor governing the coordination number and geometry adopted by actinide complexes.

One simple way of quantifying steric effects in actinide complexes [the cone angle factor (caf) approach] uses the sum of the solid angles subtended by the ligands to the centre of the metal atom in a given complex, divided by 4π (abbreviated to Σ caf) to represent the fraction of the total sphere surface enclosing the metal atom which is occupied by the ligands.[4,5] Obviously the value of Σ caf can never be unity, which would represent complete filling of the sphere surface about the metal atom, because of ligand–ligand repulsions, and for thorium(IV) and uranium(IV) complexes of known structure, values for Σ caf are 0.80 ± 0.04 and 0.80 ± 0.03 respectively. This approach has some predictive value; for example, seven-coordination would be expected for complexes of thorium tetrachloride with medium sized ligands, and both six- and seven-coordination for similar complexes of uranium tetrachloride.

However, prediction of the most likely geometry to be adopted in a given compound is best based on a consideration of the repulsions between the metal–ligand bonds in the complexes, and will depend on whether the ligands are uni-, bi- or poly-dentate. The actual combinations of the different ligands in individual compounds seem to be very important in determining which particular geometry and coordination number are adopted, as shown in the possible geometries for eight-coordination.[6]

40.1.4 Coordination Numbers and Geometries

The radii of the actinide cations at the beginning of the series are, for any given oxidation state, larger than *d* transition metal cations in the same oxidation state, so that high (>6) coordination numbers are common in actinide complexes. Because the ionic radii decrease with increasing atomic number (the actinide contraction), the observed coordination number also decreases along the series for a given oxidation state with the same ligand. Much published work concerned with the coordination chemistry of these elements has compared the behaviour of the early and later members of the series. A useful survey of structural information is available.[7]

Examples of the known coordination numbers and geometries are given in Table 2. Low (<6) coordination numbers are observed only with very bulky ligands, *e.g.* $[N(SiMe_3)_2]^-$, or high oxidation states, *e.g.* $[NpO_4]^-$, and the highest (12 to 14) are found only with relatively small multidentate ligands, such as bidentate NO_3 and bi- and ter-dentate BH_4 groups. Seven-coordination is very common in dioxoactinide(VI) complexes, in which the rigid O═M═O group provides the axis for pentagonal bipyramidal geometry, but several examples of this geometry have been recorded for complex cations (*e.g.* $[UCl_3(EtCONEt_2)_4]^+$),[8] the neutral complex $[U(NCS)_4(Me_2CHCONMe_2)_3]$[9] and the anion $[UF_7]^{3-}$ in $K_3[UF_7]$.[10] The analytical stoichiometry of a complex may well mislead, for $UCl_4{\cdot}3DMSO$ does not involve seven-coordinate uranium(IV), but is actually $[UCl_2(DMSO)_6]^{2+}[UCl_6]^{2-}$.[11]

40.1.5 Literature Sources

Exhaustive accounts of the chemistry of thorium, protactinium, uranium and the trans-uranium elements appear in the most recent supplementary volumes of the Gmelin series.[12] Unless otherwise indicated by the inclusion of references to specific papers, further information on the compounds discussed in this chapter will be found in the Gmelin volumes and in the work cited in reference 2.

40.2 THORIUM, PROTACTINIUM, URANIUM, NEPTUNIUM AND PLUTONIUM

40.2.1 The +3 Oxidation State

Thorium(III) and protactinium(III) complexes are unknown, and relatively few uranium(III), neptunium(III) and plutonium(III) compounds have been described. This is mainly because of the ease of oxidation to the +4 state in all three cases, accentuated for plutonium(III) by the oxidizing nature of the α-radiolysis products formed in solutions.

40.2.1.1 Nitrogen ligands

(i) Ammonia

The only reported complexes are those with uranium and plutonium trihalides. The products of composition $UCl_3{\cdot}(6.8\text{–}6.9)NH_3$ and $UCl_3{\cdot}(7.0\text{–}7.4)NH_3$ lose ammonia at 20 or 45 °C to

Table 2 Coordination Numbers and Geometries of Actinide Compounds

Coordination number	*Complex*	*Coordination geometry*	*Ref.*
3	$[U(N\{SiMe_3\}_2)_3]$	Probably pyramidal	a
4	$[UH(N\{SiMe_3\}_2)_3]$	Probably trigonal pyramidal	b
	$[U(NPh_2)_4]$	Highly distorted tetrahedron	c
5	$[U(NEt_2)_4]_2$	Distorted trigonal pyramidal	d
6	$[UCl_6]^{2-}$	Octahedral	e
	$[UCl_4\{(Me_2N)_3PO\}_2]$	*Trans* octahedral	f
	$[UCl_4(Ph_3PO)_2]$	*Cis* octahedral	g
7	$[UCl(Me_3PO)_6]^{3+}$	Distorted monocapped octahedron	h
	$[UF_7]^{3-}$ (in K_3UF_7)	Pentagonal bipyramidal	i
	$[UCl_3(EtCONEt_2)_4]^+$	Pentagonal bipyramidal	j
	$[UO_2(NCS)_5]^{3-}$	Pentagonal bipyramidal	k
	$[PuF_7]^{2-}$ (in Rb_2PuF_7)	Capped trigonal prism	l
8	$[U(NCS)_8]^{4-}$(NEt_4 salt)	Cube	m
	$[PaF_8]^{3-}$ (in Na_3PaF_8)	Cube	n
	$[U(NCS)_8]^{4-}$ (Cs salt)	Square antiprism	o
	$[UCl_2(DMSO)_6]^{2+}$	Distorted dodecahedron	p
	$PuBr_3$	Bicapped trigonal prism	q
	$[UO_2(S_2CNEt_2)_3]^-$	Hexagonal bipyramid	r
9	UCl_3	Tricapped trigonal prism	q
	$[UF_5]^-$ (in $LiUF_5$)	Tricapped trigonal prism	s
	$[Th(C_7H_5O_2)_4(DMF)]$ $C_7H_6O_2$ = tropolone	Monocapped square antiprism	t
10	$[Th(NO_3)_4(Ph_3PO)_2]$	Best described as *trans* octahedral with four bidentate NO_3 groups in the equatorial plane	u
	$[Th(NO_3)_3(Me_3PO)_4]^+$	1:5:4 geometry	v
	$[Th(C_2O_4)_4]^{4-}$ [in $K_4Th(C_2O_4)_4$]	Bicapped square antiprism	w
11	$[Th(NO_3)_4(H_2O)_3]\cdot 2H_2O$	Best described as a monocapped trigonal prism with four bidentate NO_3 groups occupying four apices	x
12	$[Th(NO_3)_6]^{2-}$	Icosahedron	v
	$[Th(NO_3)_5(Me_3PO)_2]^-$	Disordered	v
14	$[U(BH_4)_4]$	Bicapped hexagonal antiprism	y
	$[U(BH_4)_4(THF)_2]$	Bicapped hexagonal antiprism	z

[a] R. A. Andersen, *Inorg. Chem.*, 1979, **18**, 1507.
[b] R. A. Andersen, A. Zalkin and D. H. Templeton, *Inorg. Chem.*, 1981, **20**, 622.
[c] J. G. Reynolds, A. Zalkin, D. H. Templeton and N. M. Edelstein, *Inorg. Chem.*, 1977, **16**, 1090.
[d] J. G. Reynolds, A. Zalkin, D. H. Templeton, N. M. Edelstein and L. K. Templeton, *Inorg. Chem.*, 1976, **15**, 2498.
[e] W. H. Zachariasen, *Acta Crystallogr.*, 1948, **1**, 268.
[f] J. F. de Wet and S. F. Darlow, *Inorg. Nucl. Chem. Lett.*, 1971, **7**, 1041.
[g] G. Bombieri, D. Brown and R. Graziani, *J. Chem. Soc., Dalton Trans.*, 1975, 1873.
[h] G. Bombieri, E. Forsellini, D. Brown and B. Whittaker, *J. Chem. Soc., Dalton Trans.*, 1976, 735.
[i] W. H. Zachariasen, *Acta Crystallogr.*, 1954, **7**, 792.
[j] K. W. Bagnall, R. L. Beddoes, O. S. Mills and Li Xing-fu, *J. Chem. Soc., Dalton Trans.*, 1982, 1361.
[k] G. Bombieri, E. Forsellini, R. Graziani and G. C. Pappalardo, *Transition Met. Chem.*, 1979, **4**, 70.
[l] R. A. Penneman, R. R. Ryan and A. Rosenzweig, *Struct. Bonding (Berlin)*, 1973, **13**, 1.
[m] R. Countryman and W. S. McDonald, *J. Inorg. Nucl. Chem.*, 1971, **33**, 2213.
[n] D. Brown, J. F. Easey and C. E. F. Rickard, *J. Chem. Soc. (A)*, 1969, 1161.
[o] G. Bombieri, P. T. Moseley and D. Brown, *J. Chem. Soc., Dalton Trans.*, 1975, 1520.
[p] G. Bombieri and K. W. Bagnall, *J. Chem. Soc., Chem. Commun.*, 1975, 188.
[q] W. H. Zachariasen, *Acta Crystallogr.*, 1948, **1**, 265.
[r] K. Bowmann and Z. Dori, *Chem. Commun.*, 1968, 636.
[s] G. Brunton, *Acta Crystallogr.*, 1966, **21**, 814.
[t] V. W. Day and J. L. Hoard, *J. Am. Chem. Soc.*, 1970, **92**, 3626.
[u] Mazur-ul-Haque, C. N. Caughlin, F. A. Hart and R. van Nice, *Inorg. Chem.*, 1971, **10**, 115.
[v] N. W. Alcock, S. Esperãs, K. W. Bagnall and Wang Hsian-Yun, *J. Chem. Soc., Dalton Trans.*, 1978, 638.
[w] M. N. Akhtar and A. J. Smith, *Chem. Commun.*, 1969, 705.
[x] T. Ueki, A. Zalkin and D. H. Templeton, *Acta Crystallogr.*, 1966, **20**, 836.
[y] E. R. Bernstein, W. C. Hamilton, T. A. Keiderling, S. J. La Placa, S. J. Lippard and J. J. Mayerle, *Inorg. Chem.*, 1972, **11**, 3009.
[z] R. R. Rietz, A. Zalkin, D. H. Templeton, N. M. Edelstein and L. K. Templeton, *Inorg. Chem.*, 1978, **17**, 658.

form $UCl_3\cdot 3NH_3$, which decomposes to $UCl_3\cdot NH_3$ above 45 °C. $UBr_3\cdot 6NH_3$ is apparently obtained by treating the tribromide with gaseous ammonia, whereas $UBr_3\cdot(3–4)NH_3$ is said to form with liquid ammonia. $PuCl_3$ and PuI_3 yield products of composition $PuCl_3\cdot$*ca.* $8NH_3$ and $PuI_3\cdot$*ca.* $9NH_3$ with either gaseous or liquid ammonia;[13] these lose ammonia readily to yield the products $PuCl_3\cdot(4.1–4.5)NH_3$ and $PuI_3\cdot(5.5–6)NH_3$ respectively. No structural information is available for any of these products.

(ii) N-Heterocyclic ligands; pyridine

The existence of the only recorded complex, $[U(py)_6][Cr(NCS)_6]$, is doubtful because uranium(III) reduces the $[Cr(NCS)_6]^{3-}$ ion.[14]

(iii) Amides

The compound with the bulky silylamide,[15] $U\{N(SiMe_3)_2\}_3$, may be a monomer, in which case the molecular geometry is probably pyramidal.

(iv) Nitriles

The only recorded compounds are $UCl_3 \cdot MeCN$, which may be polymeric, and $NpCl_3 \cdot 4MeCN$, for which the ^{237}Np Mössbauer spectrum has been reported.[16a] Mössbauer studies indicate[16a] that products of composition $M^INpCl_4 \cdot xMeCN$ are mixtures of $NpCl_3 \cdot 4MeCN$ and $M^I_2NpCl_5$.

40.2.1.2 Phosphorus ligands

The bis diphosphine complex, $[U(BH_4)_3(dmpe)_2]$ (dmpe = $Me_2PCH_2CH_2PMe_2$), has been prepared by treating $U(BH_4)_3 \cdot THF$ with an excess of dmpe in THF. The uranium centre is formally 13-coordinate, with the three BH_4 groups tridentate.[16b]

40.2.1.3 Oxygen ligands

(i) Aqua species, hydroxides and oxides

A selection of the known hydrated compounds is given in Table 3. In many cases they are easily dehydrated and the water is presumably held in the lattice. Water does not appear to be coordinated to the metal atom in the hydrated trifluorides, such as $PuF_3 \cdot (0.4–0.75)H_2O$, whereas it is coordinated in $PuCl_3 \cdot 6H_2O$ and the bromides $MBr_3 \cdot 6H_2O$ (M = U, Np, Pu), which are isostructural with the lanthanide hydrates $LnCl_3 \cdot 6H_2O$ and with $AmCl_3 \cdot 6H_2O$, being of the form $[MX_2(H_2O)_6]^+X^-$. The hydrated plutonium(III) hexacyanoferrate(III), $Pu[Fe(CN)_6] \cdot$*ca.* $7H_2O$, is black, indicating a considerable degree of charge transfer.

Table 3 Some Hydrates of Actinide(III) Compounds

$PuF_3 \cdot (0.4 \text{ to } 0.75)H_2O$	
$PuCl_3 \cdot 6H_2O$	
$M^{III}Br_3 \cdot 6H_2O$	M^{III} = U, Np, Pu
$M^IUCl_4 \cdot 5H_2O$	M^I = Rb, NH_4
$M_2^{III}(SO_4)_3 \cdot xH_2O$	M^{III} = U, $x = 8$; Pu, $x = 5$ or 7
$NaNp(SO_4)_2 \cdot xH_2O$	
$M^IPu(SO_4)_2 \cdot xH_2O$	$x = 1, 2, 4$ or 5 variously with M^I = Na, K, Rb, Cs, Tl, NH_4
$(NH_4)_2U_2(SO_4)_4 \cdot 9H_2O$	
$Pu_2(SO_3)_3 \cdot xH_2O$	
$PuPO_4 \cdot 0.5H_2O$	
$HM^{III}[Fe^{II}(CN)_6] \cdot xH_2O$	M^{III} = U, $x = 9$ to 10; Pu, x uncertain
$Pu[Fe^{III}(CN)_6] \cdot$*ca.* $7H_2O$	
$M_2^{III}(C_2O_4)_3 \cdot xH_2O$	M^{III} = Np, x = *ca.* 11; Pu, $x = 1, 2, 3, 6, 9$ or 10

The only recorded hydroxide, $Pu(OH)_3 \cdot xH_2O$, rapidly oxidizes to a plutonium(IV) species, but two forms of Pu_2O_3 are known, one with the hexagonal La_2O_3 structure and the other cubic; in these the Pu^{3+} is seven- and six-coordinate respectively. Ternary oxides, such as $PuAlO_3$, are also known.

(ii) β-Ketoenolates and ethers

β-Ketoenolates have not been isolated for the tripositive elements; in the attempted preparation of $Pu(MeCOCHCOMe)_3$, oxidation to plutonium(IV) occurred very readily.[17] Complexes with ethers appear to be restricted to THF solvates; apart from $UCl_3 \cdot THF$, a useful starting material for uranium(III) preparative chemistry,[18] the only other examples are solvated organometallic compounds, such as $Cp_3Np \cdot 3THF$ and $Cp_3Pu \cdot THF$.

(iii) Oxoanions as ligands

Hydrated sulfato complexes of the type $M^IM^{III}(SO_4)_2 \cdot xH_2O$ (Table 4) are known for $M^{III} = U$ and Pu, and $(NH_4)_2U_2(SO_4)_4 \cdot 9H_2O$ may be of the same type. $KPu(SO_4)_2 \cdot H_2O$ and the corresponding dihydrate are isostructural with the neodymium(III) analogues and $NH_4Pu(SO_4)_2 \cdot 4H_2O$ is isomorphous with the corresponding cerium(III) compound. Salts of other complex anions, such as $K_5M^{III}(SO_4)_4$ ($M^{III} = Np$, Pu), are also known.

Table 4 Some Actinide(III) Sulfato complexes

$M^IM^{III}(SO_4)_2 \cdot xH_2O$	$x = 1$, $M^I = K$, $M^{III} = Pu$ $x = 2$, $M^I = Na$, $M^{III} = U$ and $M^I = K$, $M^{III} = Pu$ $x = 4$, $M^I = Rb$, $M^{III} = U$ and $M^I = Na$, NH_4, Rb, Cs or Tl, $M^{III} = Pu$
$M^I_2U_2(SO_4)_4 \cdot xH_2O$	$x = 9$, $M^I = NH_4$ and $x = 11$, $M^I = Cs$
$K_3U(SO_4)_3$	
$K_5M^{III}(SO_4)_4$	$M^{III} = Np$, Pu
$K_8Pu_2(SO_4)_7$	

(iv) Carboxylates and carbonates

Uranium(III) formate, $U(HCO_2)_3$, isostructural with $Np(HCO_2)_3$,[19] $Pu(HCO_2)_3$[20a] and $Nd(HCO_2)_3$, is precipitated when a solution of UCl_4 in 95% formic acid is treated with a large excess of zinc amalgam. The corresponding uranium(III) acetate and propionate could not be obtained by this route. $Np(HCO_2)_3$ has been prepared[19] in the same way as $U(HCO_2)_3$ and $Pu(HCO_2)_3$ by treating $Pu(OH)_3$ with 90% formic acid.[20a]

The hydrated oxalates, $Np_2(C_2O_4)_3 \cdot$*ca.* $11H_2O$ and $Pu_2(C_2O_4)_3 \cdot 10H_2O$, are precipitated on addition of oxalic acid to aqueous solutions containing the tripositive elements. The hydrated oxalato complexes, $M^IPu(C_2O_4)_2 \cdot xH_2O$ ($M^I = Li$, Na, K, Cs and NH_4, $x = 0.3–3.5$) have also been recorded.[20b] Very little else is known about the metal(III) carboxylates. A carbonate, $Pu_2(CO_3)_3$, has also been recorded.

(v) Aromatic hydroxyacids

Neptunium(III) and plutonium(III) are precipitated from aqueous solution by salicyclic acid; the neptunium salicylate, $Np(C_6H_4(OH)CO_2)_3 \cdot 1.5H_2O$, can be dehydrated at 50 to 150 °C.

(vi) N,N-Dimethylformamide and related ligands

The DMF and antipyrine (2,3-dimethyl-1-phenylpyrazol-5-one, ap) complexes $[UL_6][Cr(NCS)_6]$, and the pyrimidone (4-dimethylaminoantipyrine, pym) analogue $[U(pym)_3][Cr(NCS)_6]$, are probably uranium(IV) compounds.[14] However, $[U(ap)_6]Cl_3$ has been obtained by dissolving $RbUCl_4 \cdot 5H_2O$ in ethanol containing the stoichiometric quantity of the ligand, followed by evaporation at low temperature,[14] and the corresponding tetraphenylborates, $[U(ap)_6](BPh_4)_3$ and $[U(pym)_4](BPh_4)_3$, are precipitated when ethanolic $Na(BPh_4)$ is added to a solution of $NH_4UCl_4 \cdot 5H_2O$ and the ligand in the same solvent, a method also used to obtain the dicarboxylic acid amide complexes,[21] $[UL_4]X_3$ where $L = (CH_2)_n(CONR_2)_2$ ($n = 1–4$) and $Me_2C(CH_2CONR_2)_2$ with R = Me or Et, and $X = BPh_4$ or PF_6.

40.2.1.4 *Sulfur ligands*

(i) Sulfides

The sulfides M_2S_3 (M = Np, Pu) exist in three crystal modifications. Th_2S_3 and M_2S_3, however, may be metallic in character like the monosulfides MS (M = Th, U, Np, Pu). Plutonium disulfide, PuS_2, is a plutonium(III) polysulfide and the sulfides of composition M_3S_5 (M = U, Np, Pu) are mixed phases, $M_2S_3 \cdot MS_2$. Other known sulfides include Pu_3S_4 and Th_7S_{12}. Isostructural oxosulfides, Np_2O_2S and Pu_2O_2S, are also known. A detailed discussion of actinide sulfides is available.[22]

(ii) N,N-*Diethyldithiocarbamates (dtc)*

The complexes $M^{III}(dtc)_3$ (M^{III} = Np, Pu) are obtained by treating the metal tribromide with Na(dtc) in anhydrous ethanol. $Pu(dtc)_3$ is fairly stable to oxidation, but $Np(dtc)_3$ and the even less stable $U(dtc)_3$ are rapidly oxidized to $M(dtc)_4$, so that neither can be isolated. However, the anionic complexes, $(NEt_4)[M^{III}(dtc)_4]$ (M^{III} = Np, Pu), have been prepared and the geometry about the metal atom is a distorted dodecahedron, best regarded as a planar pentagon of five S atoms with one S atom above and two S atoms below the pentagon.[23] These salts of the $[M(dtc)_4]^-$ ion are isostructural with the analogous lanthanide complexes, whereas $Pu(dtc)_3$ is not isostructural with any of the lanthanide $Ln(dtc)_3$.

40.2.1.5 *Selenides and tellurides*

A variety of selenides and tellurides of composition MX (M = Th, U, Pu; X = Se or Te), M_2X_3 (M = Th, U, Pu, X = Se; M = Th, X = Te), M_3X_4 (M = U, Np, Pu, X = Se; M = U, Np, X = Te), U_3X_5 (X = Se, Te) and Th_7Se_{12} have been recorded. These are probably structurally similar to the corresponding sulfides.

40.2.1.6 *Halogens as ligands*

The known trifluorides, MF_3 (M = U, Np, Pu), possess the LaF_3-type structure in which the coordination geometry about the metal atom is a fully capped trigonal prism, an interesting example of 11-coordination.[24] The trichlorides, MCl_3 (M = U, Np, Pu), adopt the UCl_3-type structure in which the nine-coordinate metal atom lies at the centre of a tricapped trigonal prism, a structure also found for UBr_3 and α-$NpBr_3$. β-$NpBr_3$ and $PuBr_3$ have the eight-coordinate $PuBr_3$-type structure in which the coordination geometry is a bicapped trigonal prism and this is found also for the triiodides, MI_3 (M = U, Np, Pu). Amido chlorides, $U(NH_2)Cl_2$ and $U(NH_2)_2Cl$, obtained by heating UCl_3 in NH_3 at 450 to 500 °C, are also known; the other trihalides would probably react in a similar manner. Examples of the known halogeno complexes are given in Table 5. $NaPuF_4$ is isostructural with $NaNdF_4$; the geometry about the M^{III} centre is a tricapped trigonal prismatic arrangement of fluorine atoms.[25] The $[M^{III}Cl_6]^{3-}$ anions (M = U, Np, Pu) are octahedral and the geometry about the central metal atom in the complex anions of the salts $M^I_2M^{III}Cl_5$ is a monocapped trigonal prism. Crystallographic data for $M^I_2M^{III}Cl_5$ (M^I = K, Rb, Cs), $CsUCl_4$ and RbU_2Cl_7, obtained in M^ICl/UCl_3 phase studies, have been reported.[26a]

40.2.1.7 *Hydrides as ligands*

Hydrides of composition MH_2 (M = Th, Np, Pu; all fluorite structure) and MH_3 (M = Pa, U, Np, Pu) have been recorded, as well as phases of intermediate composition, such as $NpH_{2.7}$. The only recorded borohydride, $U(BH_4)_3$, obtained by thermal or photochemical decomposition of $U(BH_4)_4$, probably involves multiple U—H—B bonding.

Table 5 Some Actinide(III) Halogeno Complexes

$M^IM^{III}F_4$	M^{III} = U, Pu with M^I = Na, K
$M^I_2UF_5$	M^I = Na, K
$M^I_3UF_6$	M^I = K, Rb, Cs
KPu_2F_7	
$CsUCl_4$[a]	
$M^IUCl_4{\cdot}5H_2O$	M^I = Rb, NH_4
$M^I_2M^{III}Cl_5$	M^{III} = U, M^I = K, Rb, Cs, NH_4
	M^{III} = Np,[b] M^I = K, Rb, NH_4
	M^{III} = Pu, M^I = K, Rb
$M^IM^{III}_2Cl_7$	M^{III} = U,[c] M^I = K, Rb, NH_4, PPh_4, $AsPh_4$
	M^{III} = Np,[b] M^I = K, Rb, NH_4
$Cs_2Na[M^{III}Cl_6]$	M^{III} = U, Np, Pu
$(Ph_3PH)_3[PuX_6]$	X = Cl, Br
$M^I_3UBr_6$	M^I = Rb, Cs

[a] I. G. Suglobova and D. E. Chirkst, *Sov. J. Coord. Chem. (Engl. Transl.)*, 1981, **7**, 52.
[b] D. G. Karracker and J. A. Stone, *Inorg. Chem.*, 1980, **19**, 3545.
[c] J. Drozdzynski, *Inorg. Chim. Acta*, 1979, **32**, L83.

40.2.1.8 *Mixed donor atom ligands*

The neutral 2-(diphenylphosphino)pyridine complex, $[U(BH_4)_3(Ph_2Ppy)_2]{\cdot}0.5C_6H_6$, has been prepared by treating $U(BH_4)_3{\cdot}THF$ with the ligand. If each triply bridging BH_4 group is regarded as occupying one coordination site, the geometry of the formally 13-coordinate uranium atom is close to pentagonal bipyramidal.[26b]

The only other recorded compound for the +3 oxidation state is the 8-hydroxyquinoline complex, $Pu(C_9H_6NO)_3$, precipitated when an aqueous solution containing plutonium(III) is added dropwise, in a stream of nitrogen, to an aqueous solution of the ligand in the presence of reducing agents. It is rapidly oxidized in air.

40.2.1.9 *Multidentate macrocyclic ligands*

The only examples of complexes with this type of ligand are the crown ether compounds UCl_3L (L = 15-crown-5,[18] benzo-15-crown-5[18] and 18-crown-6[27]), which precipitate when the ligand is added to a solution of $UCl_3{\cdot}THF$ in THF, and $U(BH_4)_3{\cdot}$18-crown-6, prepared by treating the UCl_3 complex with $NaBH_4$. It appears that the 15-crown-5 ligands provide a better cavity match than 18-crown-6.

40.2.2 The +4 Oxidation State

40.2.2.1 *Group IV ligands*

(*i*) *Cyanides*

Actinide cyanide systems have scarcely been investigated. A product described as $Th[Th(CN)_8]$ $[Th(CN)_4?]$ is reported to be formed when a thorium(IV) hydroxide gel is treated with a 10–20% excess of a solution of stabilized HCN in an autoclave.[28] Uncharacterized cyanide-containing products are obtained when $ThCl_4$ or $ThBr_4$ is treated with an alkali metal cyanide in anhydrous liquid ammonia, a reaction which yields $UCl_3(CN){\cdot}4NH_3$ [ν(CN) = 2120 cm^{-1}] in the case of UCl_4.[29] UBr_4 and UI_4 appear to react in the same way as UCl_4. An impure product of approximate composition $(Et_4N)_2UCl_4(CN)_2$ has also been recorded. Alkali metal cyanides do not react with UCl_4 in anhydrous liquid HCN, the only product being a UCl_4–HCN adduct.

(*ii*) *Isocyanides, RNC*

A cyclohexyl isocyanide complex, probably $[ThI_4(C_6H_{11}NC)_4]$ [ν(NC) = 2194 cm^{-1}, $\Delta\nu$(NC) = +59 cm^{-1}] has been obtained by treating ThI_4 with $C_6H_{11}NC$ in *n*-hexane. The analogous uranium(IV) complex [$\nu(NC) = 2190\ cm^{-1}$, $\Delta\nu(NC) = +55\ cm^{-1}$] was obtained in the same way. Similar complexes appear to be formed with UCl_4 and UBr_4.[30]

40.2.2.2 *Nitrogen ligands*

(*i*) *Ammonia and amines*

(*a*) *Ammonia.* A selection of the known thorium(IV), uranium(IV) and plutonium(IV) ammines is given in Table 6. Most of them were obtained from the parent compound or ($PuCl_4 \cdot xNH_3$) from $Cs_2[PuCl_6]$ by treatment with gaseous or liquid ammonia. $ThBr_4 \cdot 4NH_3$ (reported[31a] as $(NH_4)_2ThBr_4(NH_2)_2)$) is obtained when $(Et_4N)_2[ThBr_6]$ is treated with liquid ammonia at low temperature. Its nature is uncertain. Although no structural data are available for any of these compounds, two of them, $UCl_4 \cdot 4NH_3$ and $UBr_4 \cdot 4NH_3$, are reported to be white solids, which suggests a centrosymmetric structure, possibly cubic. $UCl_3(CN) \cdot 4NH_3$, in contrast, is light green. Hydrated products, formulated earlier as $Th(NO_3)_4 \cdot 3NH_3 \cdot H_2O$ and $2Th(NO)_3)_4 \cdot 7NH_3 \cdot 3H_2O$, are doubtful; $Th(NO_3)_4 \cdot 4H_2O$ and $Th(SO_4)_2$ undergo ammonolysis in liquid ammonia, but $Th(C_2O_4)_2$ is soluble, which may indicate the formation of a complex.

Table 6 Some Ammonia Complexes of Actinide(IV) Compounds

$UF_4 \cdot xNH_3$	$x = 0.5$ (?), 1 or 4.2 to 4.6
$MCl_4 \cdot xNH_3$	M = Th, $x = 3$ (?), 6; M = U, $x = 1, 2, 4, 5, 7.3$ to 7.6, 8, 9 to 10 or 12; M = Pu, $x =$ *ca.* 5 or *ca.* 7
$UCl_3(CN) \cdot 4NH_3$	
$MBr_4 \cdot xNH_3$	M = Th, $x = 4$; M = U, $x = 4$, 5 to 6 or 10
$MI_4 \cdot xNH_3$	M = Th, $x = 7$ or 8; M = U, $x = 4$ to 5 or 10
$ThI_3(NH_2) \cdot 4NH_3$	
$ThI_2(NH_2)_2 \cdot 3NH_3$	
$U(OR)_4 \cdot xNH_3$	$x = 1$, R = Ph, *o*-ClC_6H_4, *o*-, *p*-MeC_6H_4, α-, β-$C_{10}H_7$ $x = 2$, R = *m*-, *p*-ClC_6H_4
$U(OR)_4 \cdot 2ROH \cdot NH_3$	R = Ph, *p*-ClC_6H_4, *o*-, *p*-MeC_6H_4
$U(RCO_2)_4 \cdot xNH_3$	R = H, $x = 2$; R = Me, $x = 1$
$U(C_2O_4)_2 \cdot 3NH_3 \cdot H_2O$	
$UO(OR)_2 \cdot 4ROH \cdot NH_3$	R = Ph, *p*-ClC_6H_4
$UO(RCO_2)_2 \cdot xNH_3$	R = H, $x = 1$ or 2; R = Me, $x = 1$

(*b*) *Aliphatic and aromatic amines.* Amine complexes are known only for thorium(IV) and uranium(IV) species (Table 7), prepared either by treating the parent compound with the amine, removing the excess by vacuum evaporation, or by precipitation from a solution of the parent compound in a non-aqueous solvent by addition of a solution of the amine in the same solvent.

The thorium atom in $[ThCl_4(Me_3N)_3]$ is seven-coordinate in a capped octahedral environment, with one Cl in the cap position; three N atoms occupy the capped face and three Cl atoms occupy the uncapped face.[31b]

The IR spectra of $Th(NO_3)_4 \cdot x\text{amine} \cdot yH_2O$ indicate that no ionic nitrate is present when $x = 1$ and $y = 3$, or $x = y = 2$, whereas with $x = 3$ and $y = 4$ both coordinated and ionic nitrate groups are present. When $x = y = 4$ nearly all the nitrate groups are ionic.[32] No other structural information is available for these compounds.

(*c*) *Hydrazines.* The few known complexes are listed in Table 8. $ThCl_4 \cdot 4PhNHNH_2$ behaves as a 1:4 electrolyte in DMF but this could be due to displacement of all chloride anions and hydrazine molecules by solvent.

(*d*) *Ethylenediamine and other diamines.* Complexes with these ligands (Table 9) are normally prepared from the parent compound and the ligand, either alone or in a non-aqueous solvent. For example, the diaminobenzene complexes, $ThCl_4 \cdot 2C_8H_8N_2$, precipitate when ethereal solutions of the chloride and the ligand are mixed. It is not known whether the water molecules in the hydrated complexes $ThBr_4 \cdot xL \cdot yH_2O$ (Table 9) are bonded to the thorium atom. The *N*,*N*,*N*′,*N*′-tetramethylethylenediamine (tmed) complex, $\{(CF_3)_2CHO\}_4U \cdot (\text{tmed})$, is obtained by treating $\{(CF_3)_2CHO\}_4U \cdot 2THF$ with tmed.[33] The majority of the known complexes probably involve eight-coordinate actinide(IV) ions, and it is possible that the 1,2-diaminobenzene complex of composition $2UCl_4 \cdot 5L$ is of the form $[UCl_3L_4]^+[UCl_5L]^-$,

Table 7 Some Amine Complexes of Thorium(IV) and Uranium(IV) Compounds

$ThCl_4{\cdot}4L$	$L = RNH_2$, with R = Me, Et, Pr^n, $PhCH_2$, *m*-, *p*-MeC_6H_4, *o*-, *p*-$MeOC_6H_4$, *p*-$EtOC_6H_4$, α-, β-$C_{10}H_7$ (naphthyl) $L = RR'NH$, with R = Ph and R′ = Me, Et, $PhCH_2$, Ph $L = R_2R'N$, with R = Me or Et and R′ = Ph
$ThBr_4{\cdot}4RNH_2$	R = Et, Ph
$[ThCl_4(Me_3N)_3]$	
$ThCl_4{\cdot}3MeC_6H_4NH_2$	(toluidine)
$ThCl_4{\cdot}2Et_3N$	
$ThCl_4{\cdot}\beta\text{-}C_{10}H_7NH_2$	
$Th(acac)_4{\cdot}PhNH_2$	
$Th(NO_3)_4{\cdot}xBu^nNH_2{\cdot}yH_2O$	$x = 1$, $y = 3$; $x = 2$, $y = 1$; $x = 3$ or 4, $y = 4$
$Th(NO_3)_4{\cdot}xMe_2NH{\cdot}yH_2O$	$x = 1$, $y = 2$ and $x = 4$, $y = 8$
$Th(NO_3)_4{\cdot}xEt_3N{\cdot}H_2O$	$x = 1$ or 2
$Th(C_2O_4)_2{\cdot}4Bu^nNH_2{\cdot}2H_2O$	
$UCl_4{\cdot}xRNH_2$	$x = 1$, R = Et,[a] Pr^n;[a] $x = 2$, R = Me,[a] Et,[a] Ph; $x = 3$, R = Bu^t; $x = 4$. R = Pr^n,[a] Bu^n
$UCl_4{\cdot}xR_2NH$	$x = 2$, R = Et; $x = 3$, R = Me, Pr^i, Bu^t
$UCl_4{\cdot}xR_3N$	$x = 1$, R = Et; $x = 2$, R = Me
$UBr_4{\cdot}2Et_2NH$	
$U(OPh)_4{\cdot}2PhOH{\cdot}Et_3N$	
$UO(OPh)_2{\cdot}4PhOH{\cdot}Et_3N$	

[a] Evidence for the formation of these species was obtained from X-ray and thermal desorption studies of the solid phases in the UCl_4–amine systems (I. Kalnins and G. Gibson, *J. Inorg. Nucl. Chem.*, 1958, **7**, 55).

Table 8 Some Complexes with Hydrazine, N_2H_4 and Phenylhydrazine, $PhNHNH_2$

$M^{IV}F_4{\cdot}xN_2H_4$	M^{IV} = Th, $x = 1$, 1.66; M^{IV} = U, $x = 1$, 1.5 or 2
$ThCl_4{\cdot}4PhNHNH_2$	
$UCl_4{\cdot}xN_2H_4$	$x = 6$ or 7
$Th(SO_4)_2{\cdot}xN_2H_4$	$x = 1.5$ or 2

Table 9 Some Complexes of Ethylenediamine and other Diamines with Actinide(IV) Compounds

Ethylenediamine, en	
$Th(C_9H_6NO)_4{\cdot}C_9H_7NO{\cdot}en$	C_9H_7NO = 8-hydroxyquinoline
$UCl_4{\cdot}4en$	
N,N,N′,N′-*Tetramethylethylenediamine, tmed*	
$U\{(CF_3)_2CHO\}_4{\cdot}tmed$	
Diaminoalkanes	
$ThBr_4{\cdot}xL{\cdot}yH_2O$	$x = 2$, $y = 5$, L = 1,2-diaminopropane and 1,4-diaminobutane $x = 2$, $y = 2$ and $x = 4$, $y = 6$, L = 1,4-diaminobutane $x = 4$, $y = 2$, L = 1,2-diaminopropane
Diaminoarenes	
$ThCl_4{\cdot}2L$	L = 1,2-, 1,3- or 1,4-diaminobenzene, 4,4′-diaminobiphenyl (benzidine), 4,4′-diamino-3,3′-dimethylbiphenyl (*o*-tolidine) or 4,4′-diamino-3,3′-dimethoxybiphenyl (*o*-dianisidine)
$UCl_4{\cdot}2L$	L = 1,8-diaminonaphthalene
$2UCl_4{\cdot}5L$	L = 1,2-diaminobenzene
$UBr_4{\cdot}4L$	L = 1,2-diaminobenzene
$[Th(NO_3)_2L_2](NO_3)_2$	L = 1,2-diaminobenzene

analogous to the known amide complexes of this stoichiometry. Thorium tetraacetate is soluble in ethylenediamine, but the complex formed has not been identified.

(*ii*) *N-Heterocyclic ligands*

(*a*) *Pyrrolidine, C_4H_9N, and β-pyridyl-α-N-methylpyrrolidine (nicotine), $C_{10}H_{14}N_2$.* The only recorded complexes appear to be $ThCl_4{\cdot}C_{10}H_{14}N_2$ and $UCl_4{\cdot}3C_4H_9N$. The latter may be an example of seven-coordinate uranium(IV).

(*b*) *Pyridine, C_5H_5N, py, and derivatives.* Thorium(IV) complexes with pyridines (Table 10) are commonly of the form ThX_4L_4, where X = Cl, Br or NCS and L is pyridine or a variety of substituted pyridines. They are precipitated when a solution of the parent compound in, for example, absolute ethanol, is treated with an excess of the ligand, and are non-electrolytes in nitrobenzene.[34] The thorium atom is evidently eight-coordinate, and the same applies to $ThI_4{\cdot}6L$, which behaves as $[ThI_2(L)_6]I_2$ in nitrobenzene, and complexes of the perchlorate, $[ThL_8](ClO_4)_4$, formed with pyridine and less sterically demanding substituted pyridines; with 2,4- and 2,6-dimethylpyridine, $[ThL_6](ClO_4)_4$ are obtained. The coordination numbers of the metal sites in the bis complexes $Th(NO_3)_4{\cdot}2L$ (L = py or a substituted pyridine) and $Th(Cl_3CCO_2)_4{\cdot}2py$ are uncertain. Thorium(IV) and uranium(IV) complexes of composition $[M(py)_8]_3[Cr(NCS)_6]_4$ have also been recorded, and bis pyridine complexes $UX_4{\cdot}2py$ (X = Cl, Br), which presumably involve six-coordinate uranium(IV), have been reported.

(*c*) *Piperidine, $C_5H_{11}N$.* The piperidine complexes $ThCl_4{\cdot}4C_5H_{11}N$ and $UX_4{\cdot}4C_5H_{11}N$ (X = Cl, Br) are probably examples of the eight-coordinate metals. The nature of the hydrated compound, $Th(NO_3)_4{\cdot}C_5H_{11}N{\cdot}6H_2O$, is uncertain.

Table 10 Some Complexes of N-Heterocyclic Ligands with Actinide(IV) Compounds

Pyridine, C_5H_5N (py) and substituted pyridines	
$ThCl_4{\cdot}2L$	L = 2-Me- or 2-H_2N-C_5H_4N
$UX_4{\cdot}2py$	X = Cl, Br
$ThX_4{\cdot}4L$	X = Cl, Br, NCS; L = py, 2-Me-, 2,4-Me_2- and 2,6-Me_2-pyridine
$UCl_4{\cdot}4L$	L = (2-H_2N, 3-HO)C_5H_3N
$ThI_4{\cdot}6L$	L = py, 2-Me-, 2,4-Me_2- and 2,6-Me_2-pyridine
$Th(NO_3)_4{\cdot}2L$	L = py, 2-Me-, 2-H_2N-, 2,4-Me_2- and 2,6-Me_2-pyridine
$Th(ClCCO_2)_4{\cdot}2py$	
$[ThL_6](ClO_4)_4$	L = 2-H_2N-, 2,4-Me_2- and 2,6-Me_2-pyridine
$[ThL_8](ClO_4)_4$	L = py, 2-Me-C_5H_4N
$[M(py)_8]_3[Cr(NCS)_6]_4$	M = Th, U
$UCl_2(C_7H_5O_2)_2{\cdot}2py$	$C_7H_6O_2$ = 2-hydroxybenzaldehyde
Quinoline and isoquinoline, C_9H_7N	
$ThX_4{\cdot}4C_9H_7N$	X = Cl, Br, NCS
$Th(NCS)_4{\cdot}4$-iso-C_9H_7N	
$ThI_4{\cdot}6C_9H_7N$	
$Th(NO_3)_4{\cdot}2L$	L = C_9H_7N or *iso*-C_9H_7N
$[ThL_8](ClO_4)_4$	L = C_9H_7N or *iso*-C_9H_7N

(*d*) *Quinoline and isoquinoline, C_9H_7N.* The quinoline complexes $ThX_4{\cdot}4C_9H_7N$ (X = Cl, Br, NCS) (Table 10) are non-electrolytes in nitrobenzene, as is the isoquinoline complex $Th(NCS)_4{\cdot}4isoC_9H_7N$, whereas the complex $ThI_4{\cdot}6C_9H_7N$ behaves as a 1:2 electrolyte, $[ThI_2(C_9H_7N)_6]I_2$. The complexes of the perchlorate with quinoline and isoquinoline behave as 1:4 electrolytes, $[ThL_8](ClO_4)_4$, and in all of these, the thorium(IV) ion is evidently eight-coordinate. The complexes of both ligands with thorium tetranitrate, $Th(NO_3)_4{\cdot}2L$, are non-electrolytes in nitrobenzene and may be examples of 10-coordinate thorium(IV). The only recorded uranium(IV) complex with quinoline, $UCl_4{\cdot}C_9H_7N$, is doubtful.

(*e*) *2,2′-Bipyridyl, bipy.* The complexes $ThX_4{\cdot}2bipy$ (Table 11) (X = Cl, Br,[34] NCS[35]) and $Th(NO_3)_4{\cdot}bipy$[35] are non-electrolytes in nitrobenzene, whereas $ThI_4{\cdot}3bipy$ behaves as a 1:2 electrolyte,[34] $[ThI_2(bipy)_3]I_2$, and $Th(ClO_4)_4{\cdot}4bipy$ as a 1:4 electrolyte.[35] The metal ion is evidently eight-coordinate in these compounds, and in the uranium(IV) complexes, $UX_4{\cdot}2bipy$ (X = Cl, Br), except for the nitrate in which it is presumably 10-coordinate.

The tetraethylammonium salt of the rather unusual complex anion $[U(NCS)_5(bipy)_2]^-$ is obtained by reaction of $(NEt_4)_4[U(NCS)_8]$ with bipy in MeCN. In the structure of this anion the uranium atom is bonded to nine nitrogen atoms in a highly distorted monocapped square antiprismatic arrangement in which one bipy molecule occupies a position bridging the cap and the near square of the antiprism, while the other bipy molecule spans the opposite edge of the far square of the antiprism (Figure 1). The structure could also be described in terms of a seven-coordinate pentagonal bipyramidal arrangement in which each bipy molecule occupies an axial position.[36a]

The nature of the other recorded bipy complexes, and of those with 4,4′-dimethyl-2,2′-bipyridyl, $ThX_4{\cdot}1.5C_{12}H_{12}N_2$ (X = Cl or NO_3) is unknown.

Table 11 2,2′-Bipyridyl (bipy) Complexes with Actinide(IV) Compounds

$M^{IV}X_4·2bipy$	M^{IV} = Th, X = Cl[a], Br[a], NCS[b] and M^{IV} = U, X = Cl, Br
$ThI_4·3bipy$[a]	
$Th(NO_3)_4·bipy$[b]	
$Th(ClO_4)_4·4bipy$[b]	
$ThCr_2O_7·0.5bipy$	
$Th(C_8H_4F_3O_2S)_4·bipy$	$C_8H_5F_3O_2S$ = 4,4,4-trifluoro-1-(2-thienyl) butane-1,3-dione
$UCl_3(OEt)·bipy$	
$UCl_2(acac)_2·bipy$	
$UCl_2L·bipy$	$H_2L = HOC_6H_4CH{=}NCH_2CH_2N{=}CHC_6H_4OH$
$(NEt_4)[U(NCS)_5(bipy)_2]$[c]	
$ThX_4·1.5L$	X = Cl, NO_3; L = 4,4′-dimethyl-2,2′-bipy

[a] A. K. Srivastava, R. K. Agarwal, M. Srivastava, V. Kapoor and T. N. Srivastava, *J. Inorg. Nucl. Chem.*, 1981, **43**, 1393.
[b] R. K. Agarwal, A. K. Srivastava, M. Srivastava, N. Bhakru and T. N. Srivastava, *J. Inorg. Nucl. Chem.*, 1980, **42**, 1775.
[c] R. O. Wiley, R. B. von Dreele and T. M. Brown, *Inorg. Chem.*, 1980, **19**, 3351.

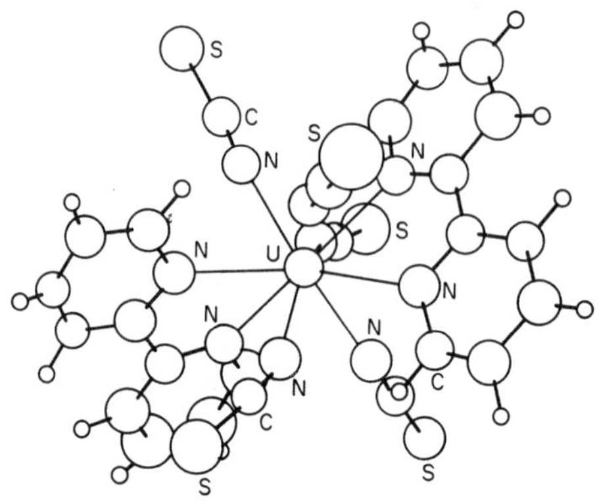

Figure 1 Perspective view of the structure of the $[U(NCS)_5(bipy)_2]^-$ anion[36a] (reproduced with permission from *Inorg. Chem.*, 1980, **19**, 3353, Copyright 1980, American Chemical Society)

(*f*) *Terpyridyl,* $C_{15}H_{11}N_3$. The only recorded complexes are $ThCl_4·2C_{15}H_{11}N_3·8H_2O$ and $Th(NO_3)_4·1.5C_{15}H_{11}N_3$.

(*g*) *Imidazole,* $C_3H_4N_2$. Complexes ThX_4L_2, some solvated (*e.g.* X = Cl), have been recorded for L = imidazole or substituted imidazole. Conductivity data for the 2-(2′-pyridyl)benzimidazole ($C_{12}H_9N_3$) complexes $ThX_4(C_{12}H_9N_3)_2$ indicate that the chloride complex is a non-electrolyte in MeOH, $MeNO_2$ or DMF, whereas the thiocyanate and nitrate are appreciably dissociated in solution. The molar conductivity of $Th(ClO_4)_4·4C_{12}H_9N_3$ is lower than that expected for a 1:4 electrolyte.

(*h*) *Dipyridylethanes and dipyridylamine.* Hydrated complexes with 1-(3-pyridyl)-2-(4-pyridyl)- and 1,2-di(4-pyridyl)-ethane, $Th(NO_3)_4·L·H_2O$, and di(2-pyridyl)amine, $ThCl_4·2L·2H_2O$ and $Th(NO_3)_4·2L$, have been reported. Very little is known about them.

(*i*) 1,10-*Phenanthroline, phen.* $[Th(NCS)_4(phen)_2]$ (Table 12) is a non-electrolyte in nitrobenzene, whereas $ThI_4·3phen$ behaves as a 1:2 electrolyte, $[ThI_2(phen)_3]I_2$. The perchlorate behaves as a 1:4 electrolyte in methyl cyanide, $[Th(phen)_4](ClO_4)_4$, so that the thorium ion is evidently eight-coordinate in these complexes. $Th(NO_3)_4·2phen$, and the 1:1 complexes with the β-diketonates, presumably involve 12- or 10-coordinate thorium(IV). A few complexes with 2,9-dimethyl-1,10-phenanthroline, ThX_4L_2 (X = Cl or NO_3), are also known.

(*j*) *Pyrazine,* $C_4H_4N_2$. The hydrated bis pyrazine complex, $Th(NO_3)_4·2C_4H_4N_2·6H_2O$, and the anhydrous tris complex, $Th(NO_3)_4·3C_4H_4N_2$, have been recorded. The IR spectrum of the latter suggests that the pyrazine is acting as a monodentate ligand and that the nitrate groups are bidentate, indicating 11-coordination. However, the tris complex behaves as a 1:4 electrolyte in DMF and as a 1:2 electrolyte in methyl cyanide or nitromethane. Complexes of UCl_4 with pyrazine or pyrazine-2-carboxamide, $UCl_4·2.5L$, and with pyrazine-2,3-dicarboxamide, $UCl_4·2L$, have also been reported.[36b]

Table 12 1,10-Phenanthroline (phen) Complexes with Actinide(IV) Compounds

$ThX_4 \cdot phen$	$X = Cl$, $C_8H_4F_3O_2S$ (tta) or $C_{10}H_6F_3O_2$ (4,4,4-trifluoro-1-phenylbutane-1,3-dionate)
$M^{IV}X_4 \cdot 2phen$	$M^{IV} = Th$ or U, $X = Cl$ or Br; $M^{IV} = Th$, $X = NO_3$ or NCS
$ThI_4 \cdot 3phen$	
$Th(NO_3)_4 \cdot 2phen \cdot MeCO_2Et$	
$[Th(phen)_4](ClO_4)_4$	
$M^{IV}Cl_3(OEt) \cdot 2phen$	$M^{IV} = Th$ or U
$UCl_2(acac)_2 \cdot phen$	
$ThCr_2O_7 \cdot 0.5phen$	

(*k*) *Piperazine, $C_4H_{10}N_2$, and its derivatives.* Several complexes with thorium(IV) compounds (Table 13) are known. They are probably polymeric, with the ligands coordinated through both nitrogen atoms.

Table 13 Complexes of Piperazine, $C_4H_{10}N_2$ and its Derivatives with Actinide(IV) Compounds

Piperazine, $C_4H_{10}N_2$	
$ThCl_4(C_4H_{10}H_2)_2$	
$Th(NO_3)_4(C_4H_{10}N_2)_x \cdot yS$	$x = 1$, $yS = MeOH$; $x = 2$, $yS = THF$ or 2EtOH
N-Methylpiperazine, $C_5H_{12}N_2$	
$ThCl_4(C_5H_{12}N_2)_2$	
$Th(NO_3)_4(C_5H_{12}N_2)_x \cdot yS$	$x = 1$, $yS = 4MeOH$; $x = 2$, $yS = 2EtOH$; $x = 3$, $yS = THF$
2-Methylpiperazine, $C_5H_{12}N_2$	
$ThCl_4(C_5H_{12}N_2)_2$	
$Th(NO_3)_4(C_5H_{12}N_2)_x \cdot yS$	$x = 1$, $yS = THF$ or 4MeOH $x = 2$, $yS = 2EtOH$
N,N'-Dimethylpiperazine, $C_6H_{14}N_2$	
$ThCl_4 \cdot 1.5C_6H_{14}N_2$	
$Th(NO_3)_4(C_6H_{14}N_2)_x \cdot yS$	$x = 1$, $yS = 4MeOH$, $x = 2$, $yS = THF$
N-Phenylpiperazine, $C_{10}H_{14}N_2$	
$ThCl_4 \cdot 1.5C_{10}H_{14}N_2$	
$Th(NO_3)_4 \cdot 2.5C_{10}H_{14}N_2 \cdot S$	$S = MeOH$, EtOH or THF
N,N'-Diaminopiperazine, $C_4H_{12}N_4$	
$Th(NCS)_4(C_4H_{12}N_4)_6 \cdot 4H_2O$	

(*iii*) *Dialkyl- and diaryl-amides*

Compounds of the type $M(NR_2)_4$ ($M = Th$ or U) are obtained by reaction of the metal tetrachloride with $LiNR_2$ in non-aqueous solvents. They are all highly reactive towards protonated species and they readily undergo CO_2, COS, CS_2 or CSe_2 insertion into the M—N bond to form carbamates.

When the group R is very bulky, as in $U(NPh_2)_4$,[37] $HU\{N(SiMe_3)_2\}_3$[38] and $ClTh\{N(SiMe_3)_2\}_3$,[39] the compounds are monomeric. The coordination arrangement in the diphenylamide is highly distorted from tetrahedral or other regular geometry,[37] whereas in $U(NEt_2)_4$ the compound is a dimer in the solid state and the geometry about each uranium atom is a distorted trigonal bipyramid.[40]

A partially hydrolyzed derivative of the diphenylamide, $[UO(Ph_2N)_3Li(OEt_2)]_2$, has been isolated as a by-product in the preparation of $U(NPh_2)_4$ and in this compound each five-coordinate uranium atom is bonded to two bridging oxygen atoms and three terminal nitrogen atoms with approximately trigonal bipyramidal geometry; one nitrogen atom and one oxygen atom occupy the axial positions.[37] In the preparation of the uranium(IV) alkylamides derived from *N,N'*-dimethyldiaminoethane, the main product is the trimer, $U_3(MeNCH_2CH_2NMe)_6$; the tetramer, $U_4(MeNCH_2CH_2NMe)_8$, is obtained as a by-product. The central uranium atom in the trimer is octahedrally coordinated to six bridging nitrogen atoms, and the other two uranium atoms are coordinated to three bridging and three terminal nitrogen atoms in a distorted trigonal prismatic array,[41] whereas in the tetramer the bridging leads to a closed unit in which each uranium atom is bonded to six nitrogen atoms in a highly distorted trigonal prismatic geometry.[42]

(iv) *Triazenes, thio- and seleno-cyanates*

(*a*) *Triazenes and related compounds.* A complex of thorium tetranitrate with 2,4,6-tris(dimethylamino)triazene ($C_9H_{18}N_6$, hexamethylmelamine), $Th(NO_3)_4{\cdot}C_9H_{18}N_6$, and the probably polymeric derivatives of purine and adenine, $Th(C_5H_3N_4)_2Cl_2$, have been recorded,[43] but little is known about them.

(*b*) *Thiocyanates.* Anhydrous $Th(NCS)_4$ has been reported, on the basis of thermogravimetric evidence, to be the product of the thermal decomposition of complexes of the type $Th(NCS)_4L_2$ (*e.g.* L = 2,6-lutidine *N*-oxide); it is said to decompose to $ThO(NCS)_2$ above 490 °C.[44] The hydrate, $[Th(NCS)_4(H_2O)_4]$, is better established; the uranium(IV) analogue is known only as the solvate of composition $[U(NCS)_4(H_2O)_4]{\cdot}(18\text{-crown-6})_{1.5}{\cdot}3H_2O{\cdot}MeCOBu^i$, in which the structure of the $[U(NCS)_4(H_2O)_4]$ molecule is a distorted square antiprism in which the upper square comprises one $O(H_2O)$ and three N(NCS) atoms and the lower square three $O(H_2O)$ and one N(NCS) atoms.[45]

Anionic complexes of the type $M^I_4[M^{IV}(NCS)_8]$ are well known (Table 14). The tetraethylammonium salts of this anion have been reported for M^{IV} = Th, Pa, U, Np and Pu and all have the same crystal structure. The coordination geometry of the anion in $(Et_4N)_4[U(NCS)_8]$ is a cube, which requires *f*-orbital participation in the bonding, but the $[U(NCS)_8]^{4-}$ anion in the cesium salt is an almost perfect square antiprism. The IR spectra of the dihydrated rubidium and cesium salts of the $[M(NCS)_8]^{4-}$ ion (M = Th, U) suggest that the coordination geometry of the anion is dodecahedral in the solid salts, but square antiprismatic in acetone solution.[46] (Crystalline $Cs_4[Th(NCS)_8]{\cdot}2H_2O$ is not isostructural with the uranium(IV) analogue.) The hydrated rubidium and cesium salts dehydrate at 60 to 100 °C. The other thiocyanato complexes listed in Table 14 require further investigation.

Table 14 Actinide(IV) Thiocyanates and Thiocyanato Complexes

$[Th(NCS)_4(H_2O)_4]$	
$[U(NCS)_4(H_2O)_4]{\cdot}(18\text{-crown-6})_{1.5}{\cdot}3H_2O{\cdot}MeCOBu^i$	
$Rb[Th(NCS)_5(H_2O)_3]$	
$Na_2[Th(NCS)_5(OH)(H_2O)_x]$	x = 2 to 3
$(NH_4)_3Th(NCS)_7{\cdot}5H_2O$	
$K_4[U(NCS)_8]{\cdot}xH_2O$	x = 0 or 2
$(NH_4)_4[M(NCS)_8]{\cdot}xH_2O$	M = Th or U, x = 0 and M = Th, x = 2
$Rb_4[M(NCS)_8]{\cdot}xH_2O$	M = Th, x = 0, 2 or 3 and M = U, x = 0 or 1
$Cs_4[M(NCS)_8]{\cdot}xH_2O$	M = Th, x = 0 or 2 and M = U, x = 0, 1 or 2
	M = Pu, x = 0
$(NEt_4)_4[M(NCS)_8]$	M = Th, Pa, U, Np and Pu

(*c*) *Selenocyanates.* The salts $(Et_4N)_4[M(NCSe)_8]$ (M = Pa, U) are isostructural with the corresponding thiocyanates. Compounds of composition $M^I_2Th(NCSe)_6{\cdot}x$DMF (M^I = Na, x = 3 and K, x = 4.5) and $K_4Th(NCSe)_8{\cdot}2$DMF have been recorded, but their nature is uncertain.

(v) *Polypyrazol-1-ylborates*

The known thorium(IV) and uranium(IV) complexes are listed in Table 15. The attempted preparation of $Th(H_2BPz_2)_4$ (Pz = $C_3H_3N_2$) yields the salt $KTh(H_2BPz_2)_5$. Although the ^1H NMR spectrum of this salt indicates dissociation to $Th(H_2BPz_2)_4$ and $K(H_2BPz_2)$ in $(CD_3)_2CO$, $Th(H_2BPz_2)_4$ could not be isolated.

Table 15 Actinide(IV) Polypyrazol-1-ylborates

$ThCl(HBPz_3)_3$	Pz = $C_3H_3N_2$
$M^{IV}X_2L_2$	M^{IV} = Th, L = $HBPz_3$ with X = Cl or Br; L = BPz_4 with X = Br M^{IV} = U, L = $HBPz_3$, $HB(3,5\text{-}Me_2Pz)_3$, Ph_2BPz_2 or BPz_4, with X = Cl
$U(H_2BPz_2)_4$	
$M^{IV}(HBPz_3)_4$	M^{IV} = Th, U
$U[H_2B(3,5\text{-}Me_2Pz)_2]_4$	
$KTh(H_2BPz_2)_5$	

The ^1H NMR spectra of the thorium(IV) complexes $ThX_2(HBPz_3)_2$ (X = Cl, Br) in CD_2Cl_2 or $(CD_3)_2CO$ (X = Cl) indicate that the $HBPz_3$ ligand is tridentate, in contrast to $UCl_2(HBPz_3)_2$ and $UCl_2(HB\{3,5\text{-}Me_2Pz\}_3)_2$, in which the ligand appears to be bidentate, a conclusion in agreement with the solid reflectance spectra of the two compounds which are consistent with six-coordinate uranium(IV). The ^1H NMR spectrum of $ThBr_2(BPz_4)_2$ in CD_2Cl_2 is more complicated and suggests that the proportion of $C_3H_3N_2$ rings that are bonded to thorium to those that are not bonded is 3:2, so that the complex may be of the form[47] $[(Pz_2BPz_2)_2Th(\mu\text{-}Pz_2BPz_2)Th(Pz_2BPz_2)_2]^{3+}$ $[Th(BPz_4)Br_6]^{3-}$. In contrast, the ^1H NMR spectrum of $UCl_2(BPz_4)_2$ is consistent with the presence of one bidentate and one tridentate BPz_4 ligand, suggesting seven-coordination. The ^{13}C NMR spectrum of $U(HBPz_3)_4$ indicates the presence of two bidentate and two tridentate $HPBz_3$ ligands, suggesting 10-coordination. Cyclopentadienylthorium(IV) and -uranium(IV) pyrazol-1-ylborates of the types $(\eta^5\text{-}C_5H_5)MCl_2(H_nBPz_{4-n})$ (M = Th, $n = 0$; M = U, $n = 1$ or 2) are also known.

(*vi*) *Nitriles*

The majority of the known complexes are with the tetrahalides (Table 16) and are obtained directly from the halide and the ligand. In most cases they have the composition $MX_4(RCN)_4$ and the UV–visible spectra of the uranium(IV) compounds are consistent with eight-coordination. However, with the more sterically demanding nitrile, Bu^tCN, tris complexes, $UX_4(Bu^tCN)_3$ (X = Cl, Br), are obtained in which the metal centre is presumably seven-coordinate. In the IR spectra of the nitrile complexes the $\nu(CN)$ feature is always shifted to higher frequency by 20 to 30 cm^{-1} on complexation, except for the UF_4 compounds for which the shifts are 44 to 54 cm^{-1}. The shifts in $\nu(CN)$ for the cyanogen chloride complexes, $MCl_4{\cdot}xNCCl$ (Table 16), are +23 to +24 cm^{-1}. $ThCl_4{\cdot}2MeCN$ and $ThOCl_2{\cdot}MeCN$ have been obtained respectively by thermal decomposition and by controlled hydrolysis of $ThCl_4{\cdot}4MeCN$.

Table 16 Nitrile Complexes with Actinide(IV) Compounds

$UF_4{\cdot}MeCN{\cdot}yHF$	$y = 0$ or 2
$ThCl_4{\cdot}2MeCN$	
$MCl_4{\cdot}4MeCN$	M = Th, Pa, U, Np
$MBr{\cdot}4MeCN$	M = Th, Pa, U, Np, Pu
$MI_4{\cdot}4MeCN$	M = Th, Pa
$ThCl_4{\cdot}4RCN$	R = Ph, $PhCH_2$
$UX_4{\cdot}4RCN$	X = Cl, Br; R = Et, Pr^n, Pr^i, Bu^n, Ph
$UX_4{\cdot}3Bu^tCN$	X = Cl, Br
$ThOCl_2{\cdot}MeCN$	
$MCl_4{\cdot}xNCCl$	M = Th, U, Np; $x = 1$ or 2

(*vii*) *Oximes and related ligands*

Very few complexes with oximes have been recorded. A pyridine-2-aldoxime $\{(C_5H_4N)CH{=}NOH\}$ complex, possibly $Th(C_6H_5N_2O)_4$, is obtained as a green solid by evaporating an absolute ethanol solution containing $Th(NO_3)_4$ and the ligand. Isatin β-oxime $(\overline{C_6H_4C({=}NOH)C(OH){=}N})$ and the isomeric α-oxime $(\overline{C_6H_4C(O)C({=}NOH)NH})$ yield yellow and red precipitates respectively with ammoniacal solutions of thorium(IV) salts, and phenyl pyruvic acid oxime $(PhCH_2C({=}NOH)CO_2H)$ precipitates thorium from aqueous solutions of the tetranitrate, but the composition of these products is unknown.

A large number of urea and substituted urea complexes with actinide(IV) compounds are known; in these the ligands are bonded to the metal *via* the carbonyl oxygen atom, and these complexes are therefore described in the section dealing with carboxylic acid amide complexes (p. 1164).

40.2.2.3 Phosphorus and arsenic ligands

Most of the early reports of phosphine or arsine complexes are erroneous, the products being phosphine oxide or arsine oxide compounds.

(i) Phosphines

A complex with trimethylphosphine, $UCl_4(PMe_3)_3$, has been reported, and complexes with $Me_2PCH_2CH_2PMe_2$(dmpe) are now well established. $ThI_4(dmpe)_2$ separates when a solution of ThI_4 in CH_2Cl_2 containing an excess of dmpe is cooled to −20 °C and $ThCl_4(dmpe)_2$ is obtained by heating $ThCl_4$ with dmpe at 80 °C.

$ThCl_4(dmpe)_2$ reacts with MeLi at 0 °C to form $ThMe_4(dmpe)_2$, and this product reacts readily with hydrogen chloride or phenol in toluene to form $ThCl_4(dmpe)_2$ and $[Th(OPh)_4(dmpe)_2]\cdot C_7H_8$, which is isostructural with $[U(OPh)_4(dmpe)_2]$. This last is obtained either from $UMe_4(dmpe)_2$ and phenol or by direct reaction of UCl_4 with LiOPh in the presence of dmpe. The uranium(IV) complexes $UX_4(dmpe)_2$ (X = Br, Me) have also been reported.

In the structure of $[U(OPh)_4(dmpe)_2]$, four phosphorus and four oxygen atoms occupy the A and B sites respectively of a D_{2d} dodecahedron.[48] Although the attempted preparation of complexes of $Ph_2P(CH_2)_2PPh_2$(dppe) with $ThCl_4$ and $Th(NO_3)_4$ led to the diphosphine dioxide complexes, UCl_4(dppe) has been obtained as a precipitate from a THF solution of the chloride and dppe.

(ii) Arsines

The complex $PaCl_4$(diars) [diars = *o*-phenylenebis(dimethylarsine)] is reported to be formed by reaction of the ligand with $PaCl_5$ in benzene.[49] Treatment of $UCl_5\cdot C_3Cl_4O$ (C_3Cl_4O = trichloroacryloyl chloride) or UCl_4 with diars yields the analogous UCl_4(diars) as a yellow (? oxidized) or blue-green solid respectively.

40.2.2.4 Oxygen ligands

(i) Aqua species, hydroxides and oxides

(*a*) *Aqua species.* Some of the reported hydrates of actinide(IV) compounds are listed in Table 17. A very large number of other hydrates have been recorded, but in many cases it is uncertain whether any or all of the water molecules are coordinated to the actinide metal atom. In the absence of a structure determination, the only evidence for bonded water is a high dehydration temperature. For example, the octahydrated sulfates, $M(SO_4)_2\cdot 8H_2O$ (M = Th, U, Pu) lose four molecules of water at about 70 °C and the remaining water can be removed only at temperatures exceeding 400 °C, indicating that four molecules of water are bonded to the metal atom, a conclusion confirmed by the determination of the structure of $U(SO_4)_2(H_2O)_4$. The uranium atoms are surrounded by a square antiprism of oxygen atoms, with each uranium atom bonded to four molecules of water and linked by bridging sulfate groups to other uranium atoms.[53] The coordination geometry is similar to that of $[U(NCS)_4(H_2O)_4]$ (p. 1142). Higher coordination numbers have been reported for hydrates of compounds which involve more compact bidentate ligands. The nine-coordinate arrangement in $[Th(CF_3COCHCOMe)_4(H_2O)]$ consists of a capped square antiprism with the bidentate ligands spanning the edges linking the two square faces[54] and in the hydrated tropolonate, $[Th(trop)_4(H_2O)]$, one trop ligand occupies a slant position on the cap of the square antiprism and the water molecule occupies another vertex of the cap[55] (Fig. 2). The coordination geometry in the hydrated γ-isopropyltropolonate, $[ThL_4(H_2O)]$, is also a singly capped square antiprism.[56a] The same geometry has been reported for the diaqua (oxodiacetato) sulfato thorium(IV) complex, $[Th\{O(CH_2CO_2)_2\}(SO_4)(H_2O)_2]\cdot H_2O$. In the polymeric network, each oxodiacetate group is chelated to one thorium atom *via* the ether oxygen atom and two carboxylate oxygen atoms, and also links two different thorium atoms through the two remaining carboxylate oxygen atoms. The sulfate groups are also bridging and the polyhedron is completed by the two bonded water molecules.[56b]

In the hydrated hydroxyacetate, $[U(HOCH_2CO_2)_4(H_2O)_2]$, the uranium atom is 10-coordinate in a bicapped square antiprismatic arrangement in which the eight carboxylate oxygen atoms occupy the corners of the square antiprism and the oxygen atoms from the two water molecules occupy the cap positions (Figure 3).[57] Bicapped square antiprismatic geometry has also been reported[58a] for the hydrated pyridine-2,6-dicarboxylate, $[Th\{C_5H_3N\text{-}2,6\text{-}(CO_2)_2\}_2(H_2O)_4]$, as has that of the oxodiacetate, $[Th\{O(CH_2CO_2)_2\}_2(H_2O)_4]\cdot 6H_2O$.[58b]

Table 17 Some Hydrates of Thorium(IV), Protactinium(IV), Uranium(IV), Neptunium(IV) and Plutonium(IV) Compounds

$MF_4{\cdot}2.5H_2O$	M = Th, U, Pu
$U[M^{II}(CN)_6]{\cdot}xH_2O$	M^{II} = Fe, x = 6; M^{II} = Ru, Os, x = 10
$M(NO_3)_4{\cdot}xH_2O$	M = Th, x = 2, 4 or 5; M = Np, x = 2; M = Pu, x = 5
$M(SO_4)_2{\cdot}xH_2O$	M = Th, U, Pu, x = 8; M = Th, Pa, U, Np, Pu, x = 4
$MF_2SO_4{\cdot}2H_2O$	M = Pa, U
$Th(ClO_4)_4{\cdot}xH_2O$	x = 2, 3, 4, 6 or 8
$Th(HIO_6){\cdot}5H_2O$	
$M(C_2O_4)_2{\cdot}6H_2O$	M = Th, Pa, U, Np, Pu
$[U(HOCH_2CO_2)_4(H_2O)_2]$	

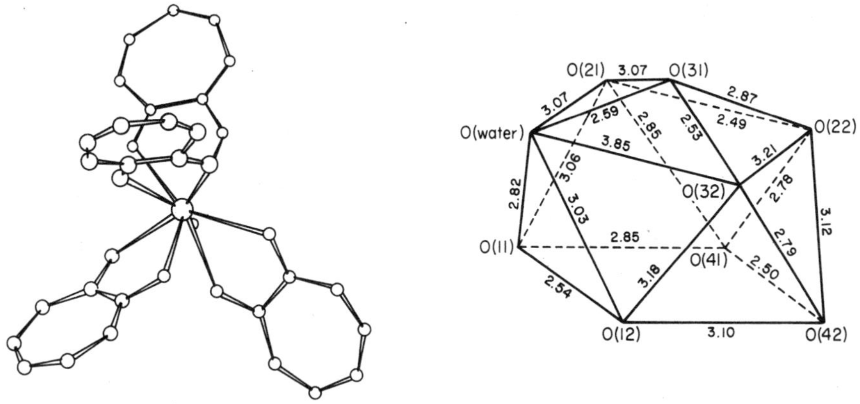

Bond distances (Å) in the coordination polyhedron

Th—O(water)	2.608(13)
Th—O(11)	2.491(14)
Th—O(12)	2.413(14)
Th—O(21)	2.395(17)
Th—O(22)	2.362(17)
Th—O(31)	2.496(14)
Th—O(32)	2.427(13)
Th—O(41)	2.417(14)
Th—O(42)	2.412(15)

Bond angles (°) subtended at thorium atoms

O(11)—Th—O(12)	62.2(0.5)
O(21)—Th—O(22)	63.1(0.7)
O(31)—Th—O(32)	61.8(0.6)
O(41)—Th—O(42)	62.3(0.7)

Figure 2 Perspective view of the $[Th(trop)_4(H_2O)]$ (trop = tropolonate) molecule and of the coordination polyhedron[55]

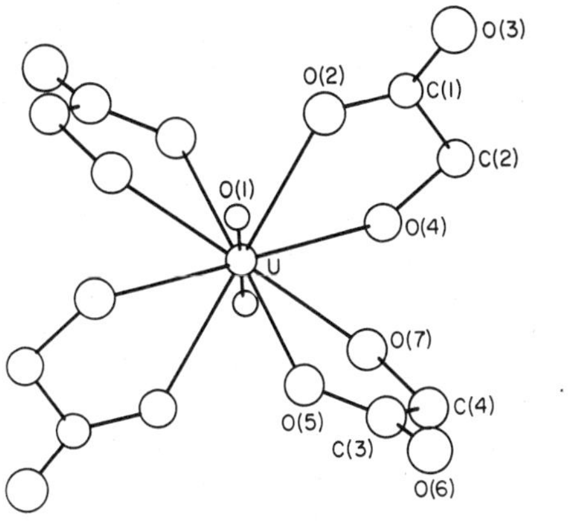

Figure 3 The structure of the uranium(IV) glycolate dihydrate molecule, $[U(HOCH_2CO_2)_4(H_2O)_2]$[57]

The coordination geometry about the 11-coordinate thorium atom in $Th(NO_3)_4{\cdot}5H_2O([Th(NO_3)_4(H_2O)_3]{\cdot}2H_2O)$ is basically a monocapped trigonal prism in which four of the prism apices are occupied by bidentate nitrate groups.[59] In the dimeric basic nitrate, $[Th_2(OH)_2(NO_3)_6(H_2O)_6]{\cdot}2H_2O$, the thorium atoms are bridged by two OH groups, and each thorium atom is also coordinated to three bidentate nitrate groups and three molecules of water. The geometry can be considered as a rather distorted dodecahedron in which the nitrate groups occupy three apices.[60]

(*b*) *Hydroxides.* The structures of the known hydroxides have not been recorded; a few structures have been reported for basic compounds. The structure of $Th(OH)_2CrO_4{\cdot}H_2O$ is built up of infinite chains, $[Th(OH)_2]_n^{2n+}$, containing two almost parallel rows of OH groups so that each thorium atom is in contact with four OH groups; the CrO_4 groups are so packed that each thorium atom is in contact with four oxygen atoms of four different CrO_4 groups, making up a square antiprismatic arrangement of oxygen atoms about each thorium atom.[61] The structure of $Th(OH)_2SO_4$ is the same.[62]

(*c*) *Oxides.* All the dioxides have the cubic fluorite lattice, with the eight-coordinated metal ion at the centre of a cube of oxygen atoms. A considerable number of ternary oxides, such as K_2ThO_3, $BaM^{IV}O_3$ (M^{IV} = Th, Pa, U, Np or Pu), Li_8PuO_6 and $BaCe$(or Ti)PuO_6, are also known.

(*ii*) *Peroxides*

Because of the ease of oxidation of protactinium(IV) and uranium(IV), peroxides and peroxo complexes are limited to their higher oxidation states. The compounds $MO_4{\cdot}xH_2O$ precipitated from dilute acid solutions of neptunium(IV) and plutonium(IV) by hydrogen peroxide appear to be actinide(IV) compounds. Soluble intermediates of the type $[Pu(\mu\text{-}O_2)_2Pu]^{4+}$ are formed at low hydrogen peroxide concentrations.

The hydrated thorium peroxide sulfate, $Th(O_2)SO_4{\cdot}3H_2O$, is very stable to heat, and thorium(IV) peroxo compounds of variable composition, approximating to $Th(O_2)_{1.6}(A^-)_{0.5}O^{2-}_{0.15}{\cdot}2.5H_2O$, with A = Cl or NO_3, are obtained from aqueous solutions of hydrogen peroxide and the appropriate thorium compound, and in the case of the nitrate the analytical results of one such product are compatible with a hexameric formulation,[63] $Th_6(O_2)_{10}(NO_3)_4{\cdot}10H_2O$. Peroxocarboxylato and phenoxo compounds are also known;[64] examples are $Th(RCO_2)_2(O_2)$ (R = C_5H_4N-2- and 2-$H_2NC_6H_4$), $Th(2\text{-}H_2NC_6H_4O)_2)_2(O_2)$ and $Th(2,6\text{-}(O_2C)_2C_5H_3N)(O_2){\cdot}H_2O$.

(*iii*) *Alcohols, phenols, alkoxides, aryloxides and silyloxides*

(*a*) *Alcohols and phenols.* Alcohol solvates of thorium and uranium tetrachlorides, $MCl_4{\cdot}4ROH$ (Table 18), are obtained by evaporating solutions of the tetrachloride in the alcohol (M = Th, U) or by treating the hydrated chloride with a mixture of the alcohol and benzene, removing the liberated water by azeotropic distillation (M = Th). $ThCl_4{\cdot}4Bu^nOH$ and $ThCl_4{\cdot}4Bu^iOH$ are, however, best prepared by heating $ThCl_4{\cdot}4Pr^iOH$ with a mixture of benzene and the butanol, removing the liberated Pr^iOH as the benzene azeotrope. The attempted preparation of the butanol (R = Bu^n, Bu^{sec} or Bu^t) adducts of UCl_4 led to solvolysis.

Table 18 Alcohol and Phenol Adducts of Actinide(IV) Compounds

$MCl_4{\cdot}4ROH$	M = Th, R = Me, Et, Pr^n, Pr^i, Bu^n, Bu^i M = U, R = Me, Et, Pr^n, Pr^i
$ThCl_4{\cdot}C_7H_7OH$	C_7H_7OH = *o*- or *m*-cresol
$Th(C_9H_6NO)_4{\cdot}EtOH$	
$Th(C_9H_6NO)_4{\cdot}2ROH$	R = 2,4-$(O_2N)_2C_6H_3$, 2,4,6-$(O_2N)_3C_6H_2$
$Th(C_9H_6NO)(OMe)(Cl_3CCO_2)_2{\cdot}MeOH$	
$U(CF_3COCHCOPh)_4{\cdot}Bu^nOH$	
$Np(OEt)_4{\cdot}EtOH$[a]	
$Pu(OPr^i)_4{\cdot}Pr^iOH$	

[a] A. K. Solanki and A. M. Bhandari, *Radiochem. Radioanal. Lett.*, 1980, **43,** 279.

The bis 2,4-dinitro- and 2,4,6-trinitro-phenol adducts of the 8-hydroxyquinolinate, $Th(C_9H_6NO)_4{\cdot}2ROH$, are obtained when $Th(C_9H_6NO)_4{\cdot}C_9H_7NO$ is heated with the phenol. The frequencies of the asymmetric and symmetric ν(NO) features due to the nitrophenols are almost identical in their IR spectra to those of the free phenols, which may therefore be hydrogen bonded to oxygen atoms of the C_9H_6NO groups.

(*b*) *Alkoxides.* The tetraalkoxides, $M(OR)_4$, are best prepared by reaction of the tetrachlorides or their alcohol solvates with the sodium or lithium alkoxide in the corresponding alcohol [$Th(OPr^i)_4$, $M(OR)_4$ with M = U or Np, R = Me or Et] or in dimethyl cellosolve [$U(OR)_4$; R = Pr^n or Pr^i]. $PuCl_4$ does not exist, but $Pu(OPr^i)_4$ has been obtained by treating a suspension of $(pyH)_2[PuCl_6]$ in benzene and Pr^iOH with ammonia; $Th(OPr^i)_4$ has been prepared from $(pyH)_2[ThCl_6]$ in the same way. An alternative, but expensive, preparative route is by reaction of the dialkylamide [*e.g.* $U(NR_2)_4$] with the alcohol in ether. The uranium *t*-butoxide seems only to have been obtained by reaction of UCl_4 and Bu^tOH with a solution of KNH_2 in anhydrous liquid ammonia and by dropwise addition of Bu^tOH in ether to a solution of the allyl, $U(\eta\text{-}C_3H_5)_4$, in the same solvent at −30 °C. Because of the relative ease of preparation of the isopropoxides, $M(OPr^i)_4$ (M = Th, Pu), other alkoxides are commonly prepared from them by reaction with an excess of the appropriate alcohol in boiling benzene. The only recorded fluoroalkoxides are the THF solvates, $U\{OC(CF_3)_3\}_4{\cdot}2THF$ and $U\{(OCH(CF_3)_2\}_4{\cdot}2THF$ (Table 19).

Table 19 Actinide(IV) Alkoxides

$M(OR)_4$	M = Th, R = Me, Et, Pr^n, Pr^i, Bu^n, Bu^i, Bu^t, $Pent^n$, $Pent^{neo}$, CMe_2Et, $CMeEt_2$, CMe_2Pr^n, CMe_2Pr^i, $CMeEtPr^n$, $CMeEtPr^i$, CEt_3
	M = U, R = Me, Et, Pr^n, Pr^i, Bu^t
	M = Np, R = Me, Et
	M = Pu, R = Pr^i, Bu^t, $CMeEt_2$
$UCl_2(OR)_2$	R = Me, Bu^n
$MTh_2(OPr^i)_9$	M = Li, Na
$Li_2U(OMe)_6$	
$Np_2O(OEt)_6$[a]	

[a] A. K. Solanki and A. M. Bhandari, *Radiochem. Radioanal. Lett.*, 1980, **43**, 279.

All the known tetraalkoxides are very easily hydrolyzed by water vapour and the uranium(IV) compounds oxidize rapidly in air, so their preparation must be carried out under nitrogen. Molecular weight determinations (M = Th, U) indicate a considerable degree of polymerization, approximately tetrameric in the case of $Th(OR)_4$ with R = Pr^i or MeEtCH, but the molecular complexity decreases to about 3.4 for R = Bu^t, and with R = CEt_3 and $CMeEtPr^i$ the alkoxides are monomers in boiling benzene.[65a] The plutonium compound $Pu(OCMeEt_2)_4$ is volatile at 150 °C/0.05 torr, suggesting a low molecular complexity.

The oxoalkoxides $U_3O(OBu)_{10}$[65b] and $U_3O(OCMe_3)_{10}$[65c] are structurally similar to the molybdenum analogue, $Mo_3O(OCH_2CMe_3)_{10}$, but the U—U bond distances, 3.576(1) Å and 3.574(1) Å, show that metal–metal bonding does not occur.

Anionic alkoxide complexes of the type $M^ITh_2(OPr^i)_9$ (M^I = Li, Na) have been mentioned in a review[66] and $Li_2U(OMe)_6$ precipitates when UCl_4 is added to the stoichiometric quantity of LiOMe in MeOH. Compounds of the type $U[Al(OPr^i)_4]_4$ and $U[Al(OPr^i)_4]_2Cl_2$ are also known. The dialkoxide dichlorides, $U(OR)_2Cl_2$ (R = Me, Bu^n), are obtained by reaction of UCl_4 with the stoichiometric quantity of the sodium alkoxide in ether (R = Me) or THF (R = Bu^n).

(*c*) *Aryloxides.* A few thorium(IV) aryloxides have been recorded (Table 20); the 4,6-dinitro-2-aminophenol derivative, $Th[OC_6H_2(NO_2)_2(NH_2)]_4{\cdot}2H_2O$, is precipitated when aqueous sodium picramate is added to an aqueous solution of thorium tetranitrate. Its IR spectrum suggests that the amino nitrogen atom may be bonded to the thorium atom.[67] Unsolvated uranium(IV) tetraaryloxides are not known; for solvates with ammonia, amines and phenols see Tables 6 (p. 1137), 7 (p. 1138) and 18 (p. 1146) respectively. However, $U(OPh)_2Cl_2$ is obtained in the same way as $U(OMe)_2Cl_2$ (see above).

(*d*) *Silyloxides.* The only actinide(IV) compound of this type appears to be $U(OSiMeEt_2)_2Cl_2$, which is precipitated when UCl_4 is treated with $Li(OSiMeEt_2)$ in a mixture of benzene and dimethyl cellosolve. It oxidizes readily in air.[68]

Table 20 Actinide(IV) Aryloxides

$Th(OR)_4 \cdot 2H_2O$	4,6-Dinitro-2-aminophenoxide, $R = \{4,6\text{-}(NO_2)_2\}(2\text{-}H_2N)C_6H_2$
$MCl_2(OR)_2$	M = Th, R = Ph, *o*- or *p*-$NO_2C_6H_4$, α- or β-$C_{10}H_7$
	M = U, R = Ph
$MCl_3(OR)$	M = Th, R = *o*- or *m*-MeC_6H_4, α- or β-$C_{10}H_7$

(*vi*) *β-Ketoenolates, Tropolonates, Catecholates, Quinones, Ethers, Ketones and Esters*

(*a*) *β-Ketoenolates.* The wide range of known β-diketone, ketoaldehyde and ketoester complexes is summarized in Table 21. They are commonly prepared by treating a dichloromethane, ethanol, aqueous or aqueous methanol solution of the actinide(IV) with the ligand (or its sodium or thallium salt) followed, when necessary, by addition of ammonia or aqueous alkali to precipitate the complexes ML_4. Alternative methods include reaction of the metal tetrachloride with the ligand in the presence of sodium (*i.e.* reaction with NaL) or reaction of a carboxylate [*e.g.* $U(EtCO_2)_4$] with the ligand in dibutyl ether or benzene. Solvent extraction is also a useful preparative method; examples are the extraction of $Np(MeCOCHCOMe)_4$ into benzene by shaking an aqueous solution of neptunium(IV), saturated with the ligand, with benzene at pH 4.5 to 5.0, and extraction of $Pu(MeCOCHCOPh)_4$ from an aqueous plutonium(IV) solution (at pH 8) into a chloroform solution of the ligand. The product is recovered from the extract by evaporation to dryness. The complexes are normally purified by recrystallization from benzene, mixtures of hydrocarbons or methanol and many of them can be sublimed under vacuum.

Analogous salicylaldehyde complexes, $M(OC_6H_4CHO)_4$ (M = Th, U) and $Th(OC_6H_4CHO)_3(NO_3) \cdot H_2O$, are also known.

A complex of composition $Pu(MeCOCHCOMe)_3Cl$ has been mentioned in the patent literature.[69] The reaction of $Th(MeCOCHCOMe)_4$ in benzene with carboxylic acids, RCO_2H ($R = CF_3$, CCl_3, $CHCl_2$, 2-C_5H_4N) leads to partial displacement of the β-diketone, yielding products of the type $Th(MeCOCHCOMe)_3(RCO_2) \cdot xH_2O$ ($x = 4$ for $R = CF_3$, CCl_3 and $CHCl_2$ and $x = 1$ for R = 2-C_5H_4N). The products of further displacement of the β-diketone appear to contain acetate (*e.g.* $Th(MeCOCHCOMe)(MeCO_2)_2(C_5H_4NCO_2) \cdot 2H_2O$) which presumably results from the decomposition of the acetylacetonate group.

The attempted preparation of lithium or thallium salts of the complex anion $[U(MeCOCHCOMe)_5]^-$ have been unsuccessful.

Structural information is available for several β-diketonates. $Th(MeCOCHCOMe)_4$ exists in two crystal forms, the coordination geometry in the α form being principally dodecahedral, with significant distortion to square antiprismatic geometry, while in the β form the geometry is principally square antiprismatic. It has also been suggested that the α form approximates most closely to a C_{2v} bicapped trigonal prism.[70] In the benzene solvate, $[Th(MeCOCHCOMe)_4] \cdot 0.5C_6H_6$, the geometry is square antiprismatic.[71] Uranium(IV) acetylacetonate also exists in the α and β forms, isostructural with the thorium analogues, but the protactinium(IV), neptunium(IV) and plutonium(IV) complexes have only been reported as the β form. The α and β forms of $Th(Bu^tCOCHCOBu^t)_4$ are isomorphous with the uranium(IV) analogues.[72]

$Th(PhCOCHCOPh)_4$ has been reported as being isomorphous with the corresponding protactinium(IV), uranium(IV) and cerium(IV) complexes; the coordination geometry in the last is a triangular faced dodecahedron, but a more recent publication[73] reports the coordination geometry of the uranium(IV) compound as square antiprismatic.

The coordination geometry in anhydrous $Th(CF_3COCHCOMe)_4$ is a 1111 (D_4-422) antiprism;[74] the structure of the monohydrate has been discussed earlier (p. 1144). $Th[CF_3COCHCO(2\text{-}C_4H_3S)]_4$ is isostructural with the cerium(IV), uranium(IV) and plutonium(IV) analogues. The coordination polyhedron is a distorted dodecahedron in which the four ligands span the two perpendicular trapezoids of the dodecahedron.[75] In the complexes $M(n\text{-}C_3F_7COCHCOBu^t)_4$, the thorium(IV), uranium(IV) and neptunium(IV) compounds are isomorphous, but the plutonium compound is not.

The coordination geometry about the thorium atom in the salicylaldehyde complex, $Th(OC_6H_4CHO)_4$, is dodecahedral;[76] the compound is isostructural with $U(OC_6H_4CHO)_4$.

(*b*) *Tropolonates.* The compounds $M(trop)_4$ ($trop = C_7H_5O_2$), have been reported for M = Th, Pa, U, Np and Pu. The thorium complex is precipitated when a slight excess of

Table 21 Actinide(IV) β-Ketoenolates, $M(R^1COCR^2COR^3)_4$

$R^1 = Me$,	$R^2 = H$,	$R^3 = H$, M = U
		$R^3 = Me$, M = Th, Pa,[a] U, Np, Pu
		$R^3 = CH_2OEt$, Et, Pr^n, Pr^i, Bu^n, M = U
		$R^3 = Bu^t$, M = Th, U
		$R^3 = CH_2CHMe_2$, M = Th
		$R^3 = n\text{-}C_5H_{11}$, $n\text{-}C_6H_{13}$, M = U
		$R^3 = n\text{-}C_{17}H_{35}$, M = Th
		$R^3 = Ph$, M = Th, U, Pu
		$R^3 = p\text{-}MeOC_6H_4$, M = Th
		$R^3 = C_4H_3O$ (2-furyl), M = U
		$R^3 = (\eta\text{-}C_5H_4)Fe(\eta\text{-}C_5H_5)$, M = Th
		$R^3 = OEt$, M = U
	$R^2 = CN$,	$R^3 = Me$, M = Th
	$R^2 = Cl$,	$R^3 = Me$, M = U
	$R^2 = C_6H_2Me_3$,	$R^3 = Me$, M = Th
$R^1 = Et$,	$R^2 = H$,	$R^3 = Et$, M = Th, U
$R^1 = Pr^n$,	$R^2 = H$,	$R^3 = Pr^n$, M = Th, U
		$R^3 = CH_2CHMe$, M = Th
$R^1 = Pr^i$,	$R^2 = H$,	$R^3 = Pr^i$, M = Th, U
$R^1 = Bu^t$,	$R^2 = H$,	$R^3 = Bu^t$, M = Th, U, Np,[b] Pu[b]
$R^1 = Ph$,	$R^2 = H$,	$R^3 = H$, Et, M = U
		$R^3 = Pr^n$, M = Th, U
		$R^3 = Pr^i$, M = U
		$R^3 = CH_2Ph$, M = Th
		$R^3 = Ph$, M = Th, Pa,[a] U, Pu
		$R^3 = m$- or $p\text{-}O_2NC_6H_4$, $p\text{-}MeOC_6H_4$, $m\text{-}Br(p\text{-}MeOC_6H_3)$, $m\text{-}O_2N(p\text{-}MeOC_6H_3)$, M = Th
		$R^3 = C_4H_3O$ (2-furyl), M = Th, U
		$R^3 = C_4H_4NO$ (3-methylisoxazol-5-yl), M = U
		$R^3 = C_5H_4N$ (3-pyridyl), M = U
$R^1 = PhC{\equiv}C$,	$R^2 = Ph$,	$R^3 = OEt$, M = Th
$R^1 = CF_3$,	$R^2 = H$,	$R^3 = Me$, M = Th, U
		$R^3 = Et$, M = U, Pu
		$R^3 = Pr^n$, M = Th, U
		$R^3 = Pr^i$, M = U
		$R^3 = Bu^n$, Bu^t, CH_2CHMe_2, $n\text{-}C_5H_{11}$, M = U
		$R^3 = Ph$, M = Th, U
		$R^3 = C_4H_3S$ (2-thienyl), M = Th, U, Np (?), Pu
		$R^3 = CF_3$, M = Th, U
		$R^3 = OMe$, OEt, OBu^t, M = U
$R^1 = n\text{-}C_3F_7$,	$R^2 = H$,	$R^3 = Bu^t$, M = Th, U, Np, Pu

[a] D. Brown, B. Whittaker and J. Tacon, *J. Chem. Soc., Dalton Trans.*, 1975, 34.
[b] E. M. Rubtsov, V. Y. Mischin and V. K. Isupov, *Sov. Radiochem.* (*Engl. Transl.*), 1981, **23**, 465.

tropolone in methanol is added to a solution of $Th(NO_3)_4{\cdot}4H_2O$ in 33% aqueous methanol, and by the reaction of $ThCl_4$ with an excess of tropolone in oxygen free dichloromethane. The hydrate, $[Th(trop)_4(H_2O)]$, is obtained from preparations in water or ethanol, and also by recrystallizing $Th(trop)_4$ from a mixture of ethanol, methyl cyanide and water. Its structure is described on p. 1144. $Pa(trop)_4$ is rather easily oxidized, but is obtained by reaction of $PaCl_4$ or $PaBr_4$ with Li(trop) in oxygen-free dichloromethane. The uranium(IV) compound precipitates when a slurry of $U(MeCO_2)_4$ in tropolone and methyl cyanide is heated under reflux. It is also obtained by reaction of UCl_4 with an excess of tropolone in dichloromethane, ethanol or water; $Np(trop)_4$ is prepared in the same way from ethanol solution. $Pu(trop)_4$ precipitates on mixing methanol solutions of tropolone and $Pu(NO_3)_4$; it can be recrystallized from DMF. $U(trop)_4$ is isomorphous with the cerium(IV), neptunium(IV) and plutonium(IV) analogues, but not with $Th(trop)_4$.

The complexes ML_4 (M = Th, U) with HL = α-, β- or γ-isopropyl tropolone are also known. Molecular weight determinations (benzene solution) for the thorium compounds indicate that the α compound is close to a dimer, whereas the β and γ derivatives are approximately trimeric. The corresponding uranium(IV) complexes are all monomers. All three thorium compounds readily form monohydrates and the structure of the hydrated γ-isopropyltropolonate is described on p. 1144.

Uranium(IV) complexes, UL_4, are known also for kojic acid and chlorokojic acid, both of which behave as the enolic forms of α-diketones.

Anionic pentakis tropolonato complexes, $M^I[M^{IV}(trop)_5]$, are known for $M^{IV} = Th$ ($M^I = Li$, Na or K), $M^{IV} = Pa$ or U ($M^I = Li$). The thorium compounds are prepared by heating $Th(trop)_4$ with H trop and M^IOH in a mixture (1:1:2 by volume) of ethanol, water and methyl cyanide. $Li[Pa(trop)_5]$ has been obtained from $Pa(trop)_4$ in the presence of Li(trop) in DMF, and $Li[U(trop)_5]$ has been prepared by both of the above methods. Sodium salts, $Na[ThL_5]$, are also known for HL = α- and γ-isopropyltropolone.

(*c*) *Catecholates*. Complexes formed by catechol and the related compounds resorcinol, phloroglucinol, orcinol and pyrogallol are listed in Table 22. Thorium dichloride catecholate and the corresponding resorcinolate, phloroglucinolate and orcinolate have been obtained by evaporating an ethereal solution of the components to dryness and heating the residue until the evolution of hydrogen chloride ceased;[77] when thorium tetrachloride is added to an excess of molten catechol, the product is $H_2[Th(C_6H_4O_2)_3]$.[78] The anionic catecholato complex salts, $Na_4[M(C_6H_4O_2)_4]{\cdot}21H_2O$ (M = Th, U), separate when a solution of catechol in aqueous sodium hydroxide is added to aqueous solutions of the tetrachlorides. The geometry of the anion is a trigonal faced dodecahedron and the oxygen atoms of the water molecules form a hydrogen-bonded network throughout the crystal.[79] The other compounds noted in Table 22 were also obtained from aqueous media. In addition to the listed compounds, thorium(IV) bis derivatives of 2,2′-dihydroxybiphenyl or -dinaphthyl and 1,8-dihydroxynaphthalene, ThL_2, are precipitated from methanolic solutions of the tetrachloride and the diol in the presence of di- or tri-ethylamine as a hydrogen chloride acceptor.[78]

Table 22 Actinide(IV) Catecholates and Related Compounds

Catechol, 1,2-*dihydroxybenzene*, $C_6H_6O_2$	
$ThCl_2(C_6H_4O_2)$	
$M^I_2[Th(C_6H_4O_2)_3]{\cdot}xH_2O$	$M^I = H$, $x = 0$; NH_4, $x = 5$
$2(NH_4)_2[U(C_6H_4O_2)_3]{\cdot}C_6H_6O_2{\cdot}8H_2O$	
$Na_4[M^{IV}(C_6H_4O_2)_4]{\cdot}21H_2O$	$M^{IV} = Th$, U
$(NH_4)_2H_2[Th(C_6H_4O_2)_4]$	[or $(NH_4)_2[Th(C_6H_4O_2)_3]{\cdot}C_6H_6O_2$]
$M^I_4H_2[U_2(C_6H_4O_2)_7]{\cdot}xH_2O$	$M^I = K$, $x = 3$; NH_4, $x = 6$; CN_3H_6 (guanidinium), $x = 14$
$M^I_2[Th_3(C_6H_4O_2)_7]{\cdot}20H_2O$	$M^I = Na$, K
$M^I[U(C_6H_4O_2)_2(OH)]{\cdot}xH_2O$	$M^I = pyH$, $x = 4$; $C_2N_4H_5$ (dicyandiamidinium), $x = 20$
$(NH_4)_2[Th_3(C_6H_4O_2)_6(OH)_2]{\cdot}10H_2O$	
Resorcinol, 1,3-*dihydroxybenzene, and hydroquinone*, 1,4-*dihydroxybenzene*, $C_6H_6O_2$	
$ThCl_2(C_6H_4O_2)$	
Orcinol, 2,5-*dihydroxytoluene*, $C_7H_8O_2$	
$ThCl_2(C_7H_6O_2)$	
Phloroglucinol, 1,3,5-*trihydroxybenzene*, $C_6H_6O_3$	
$ThCl_2(C_6H_4O_3)$	
Pyrogallol, 1,2,3-*trihydroxybenzene*, $C_6H_6O_3$	
$M^I_2[Th(C_6H_3O_3)_2]{\cdot}7H_2O$	$M^I = Na$, K

(*d*) *Quinones*. A few thorium(IV) complexes with hydroxyquinones have been recorded. The complex with 2,5-dihydroxybenzo-1,4-quinone ($C_6H_4O_4$), $Th(C_6H_2O_4)_2$, precipitates when an excess of aqueous $Th(NO_3)_3$ is added to an aqueous solution of the ammonium salt of the quinone at pH 3, 7 or 11. The corresponding 3,6-dichloroquinone ($C_6H_2Cl_2O_4$) and 3,6-diphenylquinone ($C_{18}H_{12}O_4$, polyporic acid) derivatives, $Th(C_6Cl_2O_4)_2$, $Th(C_6Cl_2O_4)F_2$ and $Th(C_{18}H_{10}O_4)_2$ (?), have also been reported. With 2,3,5,6-tetrahydroxybenzo-1,4-quinone ($C_6H_4O_6$) products formulated as $[Th_2(C_6H_2O_6)](NO_3)_6$ and $[Th_4(C_6O_6)](NO_3)_{12}$ separate when a mixture of $Th(NO_3)_4$ and the ligand is heated.[80a] A complex with 2-hydroxynaphtho-1,4-quinone ($C_{10}H_6O_3$), presumably $Th(C_{10}H_5O_3)_4$, precipitates from aqueous solutions of thorium(IV) on addition of the quinone and, as precipitation is quantitative, this has some value for the gravimetric determination of thorium in a procedure in which the precipitate is ignited and weighed as ThO_2. All of these compounds require further investigation.

(*e*) *Ethers*. In addition to the ether complexes with thorium(IV) and uranium(IV) compounds listed in Table 23, a considerable number of THF solvates of Schiff base complexes with these elements are known [see ref. 12, Thorium (vol. E) and Uranium (vol. E1)]. Diethyl ether solvates of imidazole and related complexes of the type $ThCl_4{\cdot}2L{\cdot}n Et_2O$, where L = imidazole, 1-benzylimidazole and 4,5-diphenylimidazole, with $n = 14$, 8 and 6 respectively, have been recorded but it is uncertain whether any of the ether molecules are coordinated to the metal atom. Otherwise, the only simple ether complexes which have been reported are those of dimethyl and diethyl ether with the tetrahalides, and the dimethoxyethane, THF and dioxane complexes. These compounds are usually obtained by treating the parent compounds with the appropriate ether. For example, $UBr_4 \cdot 2THF$ is prepared by adding a 10% solution of THF in carbon disulfide to a solution of uranium tetrabromide in the same solvent. In some cases the ether complexes result from the method of preparation of the parent compound. Thus treatment of a solution of uranium tetrachloride in THF with the stoichiometric quantity of thallium acetylacetonate, followed by addition of *n*-hexane to the supernatant, yields a precipitate of $UCl_2(acac)_2 \cdot 2THF$. In the dimethoxyethane (DME) complex, $[\{(Me_3Si)_2N\}_2UCl_2(DME)]$, obtained by reaction of UCl_4 with $NaN(SiMe_3)_2$ in DME, the uranium atom is coordinated to two nitrogen, two chlorine and two oxygen (from DME) atoms.[80b]

Table 23 Ether Complexes with Actinide(IV) Compounds

$UX_4 \cdot R_2O$	R = Me, X = Cl; R = Et, X = Cl, Br
$MCl_4 \cdot 2C_4H_{10}O_2$	M = Th, U; $C_4H_{10}O_2 = MeOCH_2CH_2OMe$
$UCl_2(acac)_2 \cdot C_4H_{10}O_2$	
$UX_4 \cdot y$THF	X = Cl, *y* = 1 or 3; X = Br, *y* = 2
$ThCl_3(HBPz_3)_3 \cdot L \cdot THF$	Pz = $C_3H_3N_2$, L = $MeCONMe_2$
$UBr_2(HBPz_3)_2 \cdot THF$	
$UCl_2\{HB(3,5\text{-}Me_2Pz)_3\}_2 \cdot THF$	
$UCl_2(acac)_2 \cdot 2THF$	
$U(H_2BPz_2)_4 \cdot THF$	
$M(HBPz_3)_4 \cdot THF$	M = Th, U
$U(OR)_4 \cdot 2THF$[a]	R = $CH(CF_3)_2$, $C(CF_3)_3$
$Na_4[Th(bipy))_4] \cdot 8THF$	
$UCl_4 \cdot L$	L = $MeOC_2H_4OC_2H_4OMe$, $MeOC_2H_4OC_2H_4OC_2H_4OMe$
$MX_4 \cdot yC_4H_8O_2$	$C_4H_8O_2$ = dioxane; M = Th, X = Cl, *y* = 3; M = U, X = Cl or Br, *y* = 2 or 3

[a] R. A. Andersen, *Inorg. Nucl. Chem. Lett.*, 1979, **15**, 57.

(*f*) *Ketones and aldehydes.* The only complexes with ketones and aldehydes appear to be those with thorium and uranium tetrahalides. The complexes $MCl_4 \cdot x MeCOMe$ (M = Th, $x = 2$ and M = U, $x = 3$), $ThCl_4 \cdot 4RCOPh$ (R = Me or Ph), $2ThBr_4 \cdot 3MeCOPh$, $ThCl_4 \cdot 2RCHO$ (R = Me, $C_6H_4(OH)$(salicyl) or PhCH=CH) and $ThBr_4 \cdot 4PhCHO$ are obtained from the components. The acetone solvated lactam complex, $ThCl_4 \cdot 4C_7H_{13}NO \cdot MeCOMe$ ($C_7H_{13}NO$ = 1-azacyclooctan-2-one), is precipitated when ethyl acetate is added to a heated solution made up from thorium tetrachloride in 1:1 acetone-nitromethane and an excess of the lactam in acetone.

(*g*) *Esters.* Ethyl acetate complexes of composition $UX_4 \cdot y MeCO_2Et$ ($y = 3$, X = Cl, Br; $y = 2$, X = Br) are reported to be formed from the components. However, both ethyl and *n*-propyl acetates react with uranium tetrachloride, yielding the complexes $UCl_3(MeCO_2) \cdot L$, where L is the ester. The bis isopropyl acetate complex, $ThCl_4 \cdot 2MeCO_2Pr^i$, is obtained when the alkoxide, $Th(OPr^i)_4$, is heated with the stoichiometric quantity (molar ratio, 1:4) of acetyl chloride in benzene under reflux. With 1:3 and 1:2 molar ratios of the reactants under similar conditions, the products are reported to be $ThCl_3(OPr^i) \cdot MeCO_2Pr^i$ and $ThCl_2(OPr^i)_2 \cdot 0.5MeCO_2Pr^i$ respectively.

The ethyl carbamate complexes, $MCl_4 \cdot x R_2NCO_2Et$ ($R_2 = Me_2$ or MeH, M = Th, $x = 3$ and M = U, $x = 2$) are obtained by treating a solution of the tetrachloride in THF with an excess of the ligand, followed by dropwise addition of *n*-hexane until the solution becomes turbid. $ThCl_4 \cdot 3(MeEt)NCO_2Et$ is obtained by adding an excess of the ligand to a suspension of the tetrachloride in benzene. ^1H NMR data are available for these compounds[81] and their IR spectra indicate that the ligands are bonded to the metal *via* the carbonyl oxygen atom.[81]

Hydrated *N*-pentylisonicotinate adducts, $ThX_4 \cdot yL \cdot zH_2O$ (X = Cl, $y = 2$, $z = 8$; X = Br, $y = 4$, $z = 9$; X = I, $y = 2$, $z = 9$) have been obtained by grinding the solid, hydrated tetrahalide with an excess of the ester, but their IR spectra indicate that the ester molecules are not bonded to the metal atom.

(*v*) *Oxoanions as ligands*

(*a*) *Nitrates.* Hydrated nitrates, $M(NO_3)_4 \cdot xH_2O$ (M = Th, Np and Pu), and the structure of $[Th(NO_3)_4(H_2O)_3] \cdot 2H_2O$, are included in Section 40.2.2.4.i.a (pp. 1145, 1146). Anhydrous

$Th(NO_3)_4$ is obtained by heating $(NO_2)_2[Th(NO_3)_6]$ at 150 °C or $(NO)_3[Th(NO_3)_6]$ at 90 °C under vacuum. Hexanitrato complexes (Table 24) are obtained from moderately concentrated (8 M to 14 M) nitric acid, in the presence of sulfamic acid to inhibit oxidation by nitrite in the case of uranium(IV). The nitrate groups in these compounds are bidentate and the structure of the anion in $[Mg(H_2O)_6][Th(NO_3)_6]\cdot 2H_2O$ is an irregular icosahedron.[82] The corresponding zinc, cobalt and nickel salts are isomorphous with the magnesium salt. The high symmetry of the complex anion is apparent in the almost white appearance of the salts $M^I_2[U(NO_3)_6]$. The product of composition $Np(NO_3)_4\cdot 1.2N_2O_5$, obtained by the reaction of $NpCl_4$ with N_2O_5, is presumably a nitrato complex.

Table 24 Hexanitrato Actinide(IV) Complexes

$M^I_2[M^{IV}(NO_3)_6]$	$M^I = NH_4$, $M^{IV} = Th, U, Pu^a$
	$M^I = K$, $M^{IV} = Pu$
	$M^I = Rb$, $M^{IV} = Th, U, Pu$
	$M^I = Cs$, $M^{IV} = Th, U, Pu$
	$M^I = Tl$, $M^{IV} = Pu$
	$M^I = Et_4N$, $M^{IV} = Th, U, Np, Pu$
	$M^I = Bu^n$, $M^{IV} = Pu$
	$M^I = Me_2(PhCH_2)_2N$, $M^{IV} = Th, U, Np, Pu$
	$M^I = Me_3(PhCH_2)N$, $M^{IV} = Th, U, Np, Pu$
	$M^I = C_5H_5NH$, $M^{IV} = U, Pu(+ca.\ 14H_2O)$
	$M^I = (PhCH_2)C_5H_4NH$, $M^{IV} = Th, U, Np, Pu$
	$M^I = (\alpha\text{-}H_2N)C_5H_4NH$, $M^{IV} = U$
	$M^I = C_9H_7NH$ (quinolinium), $M^{IV} = U, Pu$
	$M^I = NO, NO_2$, $M^{IV} = Th$
$(bipyH_2)[U(NO_3)_6]$	
$M^{II}[M^{IV}(NO_3)_6]\cdot 8H_2O$	$M^{II} = Mg$ or Zn, $M^{IV} = Th, U, Pu$
	$M^{II} = Co$ or Ni, $M^{IV} = Th, Pu$
	$M^{II} = Mn$, $M^{IV} = Th$
$M^ITh(NO_3)_5\cdot xH_2O$	$M^I = Na$, $x = 8.5$; K, $x = 6$
$K_3H_3M^{IV}(NO_3)_{10}\cdot xH_2O$	$M^{IV} = Th$, $x = 4$; U, $x = 3$

[a] A dihydrate has also been reported.

In the hydrated compounds (included in Table 24) the water molecules are almost certainly not bonded to the actinide(IV) ion, with the possible exception of the complexes formulated as $M^ITh(NO_3)_5\cdot xH_2O$ (M^I = Na or K) and $K_3H_3Th(NO_3)_{10}\cdot xH_2O$, the nature of which is uncertain.

(*b*) *Phosphates*. Metaphosphates, $M(PO_3)_4$ (M = Th, Pa, U, Pu), the hydrated compounds $Pu(HPO_4)_2\cdot xH_2O$, $Pu_2H(PO_4)_3\cdot xH_2O$ and $Pu_3(PO_4)_4\cdot xH_2O$, and the pyrophosphates $M(P_2O_7)$ (M = Th, Pa, U, Np, Pu) have been recorded. The structure of the triclinic form of $U(PO_3)_4$ consists of eight-coordinate square antiprisms, UO_8, connected by $(P_4O_{12})^{4-}$ rings.[83a]

Alkyl phosphates, $U\{O_2P(OR)_2\}_4$ (R = Me, Et or Bu^n) and $U\{O_2PH(OR)\}_4$ (R = Me, Et, Pr^i or Bu^n), and the phenyl derivative, $U(O_3PPh)_2$, have also been reported.[83b]

(*c*) *Sulfites*. The only simple sulfite appears to be $Th(SO_3)_2\cdot 4H_2O$. Salts of hydrated complexes of the type $M^I_{2n}M^{IV}(SO_3)_{n+2}\cdot xH_2O$ (Table 25) are known for thorium(IV) and uranium(IV), both of which form a series of hydrated salts of what appear to be sulfitooxalato complex anions, $Na_{2n}M^{IV}(SO_3)_n(C_2O_4)_2\cdot xH_2O$, obtained by dissolving $Th(C_2O_4)_2$ in concentrated aqueous sodium sulfite in varying molar proportions, and then pouring the resulting solution into ethanol to yield a syrup which becomes crystalline on treatment with ethanol.[84,85] The product of composition $Na_2U(SO_3)(C_2O_4)_2\cdot 5H_2O$ is formulated as $Na_2[U(SO_3)(C_2O_4)_2(H_2O)_2]\cdot 3.5H_2O$ in the original paper.[84] All these products require further investigation.

(*d*) *Sulfates*. Hydrated sulfates, including the structure of $[U(SO_4)_2(H_2O)_4]$, have been described in Section 40.2.2.4.i.a (pp. 1144, 1145); some anhydrous sulfates are also known. Fluorosulfates $U(SO_3F)_4$, $UO(SO_3F)_2$ and $M^{II}U(SO_3F)_6$ (M^{II} = Mg, Zn) have been obtained by treating $U(MeCO_2)_4$ or $M^{II}U(MeCO_2)_6$ with HSO_3F. The compound $U(SO_3F)_4$ appears to involve two mono- and two bi-dentate SO_3F groups.[86] Salts of sulfato complex anions of the type $M^I_{2n}M^{IV}(SO_4)_{2+n}\cdot xH_2O$ (n = 1 to 4) are listed in Table 26, together with sulfatooxalates and salts of complex anions derived from them.

The structure of the anion in $K_4Th(SO_4)_4\cdot 2H_2O$ consists of chains of thorium atoms linked by pairs of bridging sulfate groups, and the coordination geometry about the thorium atom is a

Table 25 Actinide(IV) Sulfites and Sulfito Complexes

$Th(SO_3)_2 \cdot 4H_2O$	
$M^I_2Th(SO_3)_3 \cdot xH_2O$	$M^I = Na$, $x = 5$; K, $x = 7.5$; NH_4, $x = 4$; CN_3H_6 (guanidinium), $x = 12$
$M^I_4Th(SO_3)_4 \cdot xH_2O$	$M^I = Na$, $x = 3$ or 6; NH_4, $x = 5$
$Na_{2n}U(SO_3)_{n+2} \cdot xH_2O$	$n = 3$, 4, 5 and 6; x, unspecified
$Na_{2n}M^{IV}(SO_3)_n(C_2O_4)_2 \cdot xH_2O$	$M^{IV} = Th$; $n = 3$, 4, 5 or 7, $x = 5$ to 6; $n = 6$, $x = 5$, 6 or 8; $n = 9$, $x = 6$; $M^{IV} = U$; $n = 1$, $x = 5.5$; $n = 2$, $x = 2$; $n = 3$, $x = 5$; $n = 4$, $x = 4$; $n = 5$, $x = 7.5$; $n = 6$, $x = 7$ to 8
$Ba_6Th(SO_3)_6(C_2O_4)_2 \cdot 7H_2O$	

Table 26 Actinide(IV) Sulfato Complexes

$M^I_2[M^{IV}(SO_4)_3] \cdot xH_2O$	$M^{IV} = Th$; $M^I = Na$, $x = 6$; K, $x = 4$; NH_4, $x = 0$ or 5; Rb, $x = 0$ or 2; Cs, $x = 2$; Tl, $x = 4$
	$M^{IV} = U$; $M^I = K$ or Cs, $x = 2$; NH_4 or Rb, $x = 0$
$M^I_4[M^{IV}(SO_4)_4] \cdot xH_2O$	$M^{IV} = Th$; $M^I = Na$, $x = 4$; K, $x = 2$; NH_4, $x = 0$ or 2; Cs, $x = 1$
	$M^{IV} = U$; $M^I = Na$, $x = 6$: K, $x = 2$; NH_4, $x = 0$ or 3; Rb, $x = 2$; enH, $x = 2$
	$M^{IV} = Np$; $M^I = K$, $x = 3$
	$M^{IV} = Pu$; $M^I = K$ or NH_4, $x = 2$; Rb, $x = 0$, 1 or 2; Cs, $x = 0$
$[Co(NH_3)_6]Na[Np(SO_4)_4] \cdot 8H_2O$[a]	
$M^I_6[M^{IV}(SO_4)_5] \cdot xH_2O$	$M^{IV} = Th$; $M^I = NH_4$ or Cs, $x = 3$
	$M^{IV} = U$; $M^I = NH_4$, $x = 4$
	$M^{IV} = Pu$; $M^I = Na$, $x = 1$; K, $x = 0$; NH_4 $x = 2$ to 4
$M^I_8[M^{IV}(SO_4)_6] \cdot xH_2O$	$M^{IV} = Th$; $M^I = NH_4$, $x = 2$
	$M^{IV} = U$; $M^I = NH_4$, $x = 3$
$Na_6[U_2(SO_4)_7] \cdot 4H_2O$	
$U(SO_4)(C_2O_4) \cdot xH_2O$	$x = 0$, 1, 2 or 3
$U_2(SO_4)(C_2O_4)_3 \cdot xH_2O$	$x = 0$, 2, 4, 8 or 12
$M^I_6U_2(SO_4)_4(C_2O_4)_3 \cdot xH_2O$	$M^I = NH_4$ or Rb, $x = 0$, 2 or 4
$Rb_4U_2(SO_4)_3(C_2O_4)_3 \cdot xH_2O$	$x = 0$, 4 or 6

[a] M. P. Mefod'eva, N. N. Krot and A. D. Gel'man, *Russ. J. Inorg. Chem.* (*Engl. Transl.*), 1972, **17**, 885.

tricapped trigonal prism.[87] It would be useful to have similar information for the other complex anions.

(*e*) *Selenites and selenates.* Hydrated selenites, $M(SeO_3)_2 \cdot xH_2O$ (M = Th, U, Pu), have been isolated from aqueous solution, and a hydrated thorium selenate, $Th(SeO_4)_2 \cdot 9H_2O$, has been recorded; very little is known about these compounds.

(*f*) *Tellurites and tellurates.* Actinide(IV) tellurites,[88] $M(TeO_3)_2$ (M = Th, U, Np, Pu), and a basic thorium(IV) tellurate, $ThO(TeO_4) \cdot xH_2O$ ($x = 4$ or 8), which may be $ThO(H_4TeO_6) \cdot yH_2O$ ($y = 2$ or 6), have been recorded. There is also evidence for the formation of complexes of composition $M^{II}_5Th(TeO_3)_7$ (M^{II} = Ba or Ca) from phase studies.[89]

(*vi*) *Carboxylates, carbonates and nitroalkanes*

(*a*) *Monocarboxylates.* In addition to the compounds listed in Table 27, which includes basic and mixed carboxylates, a large number of hydroxocarboxylato compounds, such as $[Th_3(HCO_2)_6(OH)_5X] \cdot yH_2O$, where X = NO_3, ClO_3, ClO_4 or NCS, with y in the range 7 to 16, and anionic derivatives formulated as $K_2[Th_3(HCO_2)_6(O)_2(OH)_2(NCS)_2]$, have been recorded, but these require further investigation.

The formates and acetates, $M(RCO_2)_4$, are prepared either by reaction of the tetrachloride with the acid or (M = U) by reduction of a dioxouranium(VI) compound in non-aqueous media. The product obtained by reaction of the tetrachloride with the carboxylic acid depends on the temperature of the reaction, lower temperatures giving only partial replacement of chloride, so that compounds of the type $M(RCO_2)_xCl_{4-x}$ are obtained (see Table 27). Other

Table 27 Actinide(IV) Monocarboxylates and Carboxylato Complexes

$M^{IV}(RCO_2)_4$	M^{IV} = Th (also +0.67 or $3H_2O$), Pa, U and Np, R = H M^{IV} = Th and U, R = Me, CH_2Cl, $CHCl_2$, CCl_3, CF_3, Et, Pr^n, Pr^i, Bu^n, Bu^i, Bu^t, n-C_5H_{11}, $PhCH_2$, Ph, Me_2N, Et_2N M^{IV} = Th, R = $PhOCH_2$, PhCH=CH, $2MeC_6H_4$, 3-$O_2NC_6H_4$, 3-IC_6H_4, (α-$C_{10}H_7$)CH_2 M^{IV} = U, R = n-C_6H_{13}, $MeOCH_2$ (also +$1H_2O$)
$Th(RCO_2)_3(MeCO_2)$	R = (C_4H_3S)-2-(thiophene)
$M^{IV}(RCO_2)_2(MeCO_2)_2$	M^{IV} = Th; R = 3- or 4-$H_2NC_6H_4$, 4-(MeO)C_6H_4 M^{IV} = U; R = (C_4H_3S)-2
$M^{IV}(RCO_2)_xCl_{4-x}$	M^{IV} = Th; R = H, x = 1 or 2; R = Et, x = 1, 2 or 3; R = $PhCH_2$, x = 2 or 3; R = 2-$H_2NC_6H_4$, x = 2 M^{IV} = U; R = Me, x = 2 or 3
$U(MeCO_2)_2Br_2$	
$M^{IV}O(RCO_2)_2$	M^{IV} = Th, U; R = PhCH=CH, 2-MeC_6H_4, 2-(MeO)C_6H_4 M^{IV} = U; R = H (also hydrates), Me (also +1, 1.5, 2 or 2.5 H_2O), CH_2Cl (also +1 or 2.5 H_2O), $CHCl_2$ (also +$2H_2O$), CCl_3 (+$1H_2O$), CH_2I (+$1H_2O$) Et (+$1H_2O$), (C_4H_3S)-2-, 4-(MeO)C_6H_4
$U_4O_2(Et_2NCO_2)_{12}$	
$M^{IV}(OH)(RCO_2)_3{\cdot}xH_2O$	M^{IV} = Th, R = Me, x = 0 or 1.5; $PhCH_2$, x = 2; CCl_3, x = 2 or 3; Et, x = 4; 2-HO(3,5-$I_2C_6H_2$), x = 0; 2-IC_6H_4, x = 0 M^{IV} = U, R = Me, x = 0
$M^{IV}(OH)_2(RCO_2)_2$	M^{IV} = Th, R = H (also +$1.3H_2O$), Me (also +1, 2 or $2.5H_2O$), CCl_3 (+$2H_2O$), $PhCH_2$, Ph_2CH, Et, Pr^n,[a] Bu^n, (+$4H_2O$), n-C_5H_{11} (+$4H_2O$), Ph, 2-ClC_6H_4, 3-IC_6H_4, 2,3,5-$I_3C_6H_2$, 4-$O_2NC_6H_4$ (+$1H_2O$) M^{IV} = U, R = H (also +$2H_2O$), CH_2Cl (+$1H_2O$), CCl_3 (+$3H_2O$), Et, 2-MeC_6H_4
$M^{IV}(OH)_3(RCO_2)$	M^{IV} = Th, Pu; R = Et, Pr^n [a] M^{IV} = Th; R = (α-$C_{10}H_7$)CH_2 M^{IV} = U; R = $PhOCH_2$
$M^IM^{IV}(RCO_2)_5$	M^{IV} = Th, M^I = pyH, K, Rb, Cs, R = H; M^{IV} = U, M^I = Na, R = CH_2Cl (+$1H_2O$)
$M^I_2Th(RCO_2)_6$	R = H, M^I = NH_4, Rb ($2H_2O$), Cs; R = Me, M^I = NH_4, CN_3H_6 (guanidinium; also +*ca.* $8H_2O$) R = CCl_3, M^I = Na, K
$M^{II}M^{IV}(RCO_2)_6{\cdot}xH_2O$	M^{IV} = Th, M^{II} = Sr, Ba, x = 0 and 2, R = H M^{IV} = U, M^{II} = Mg, Fe, Zn, x = 0, R = Me
$Cs_3Th(HCO_2)_7$	
$M^I_4Th(HCO_2)_8$	M^I = Na, Rb, Cs
$Zn_2U(HCO_2)_8$	
$KU(OH)(HCO_2)_4{\cdot}3H_2O$	
$Na_2U(OH)(MeCO_2)_5{\cdot}H_2O$	
$M^IUO(HCO_2)_3{\cdot}xH_2O$	M^I = NH_4, x = 0; M^I = Li, K, NH_4, x = 1
$NH_4(UO)_2(MeCO_2)_5$	
$Rb_2U_2(OH)_2(HCO_2)_5(NCS)_3{\cdot}xH_2O$	

[a] V. S. Shmidt and V. G. Andryushin, *Sov. Radiochem.* (*Engl. Transl.*), 1982, **24**, 498.

carboxylates are commonly prepared by heating the tetraacetate with the appropriate carboxylic acid, a method used also to obtain compounds of the type $M(RCO_2)_2(MeCO_2)_2$. In some cases [*e.g.* $UO(2\text{-}MeC_6H_4CO_2)_2$] a basic product precipitates and similar basic compounds are often obtained from aqueous media, although many of the known tetracarboxylates can be obtained anhydrous from aqueous solution. Some authors report the basic species as hydroxocarboxylato compounds, and others report them as hydrated monoxo compounds, and it is uncertain whether such compounds should be regarded as $MO(RCO_2)_2{\cdot}xH_2O$ or $M(OH)_2(RCO_2)_2{\cdot}(x-1)H_2O$.

Anhydrous thorium, protactinium and neptunium tetraformates are isostructural,[90] but $U(HCO_2)_4$ does not possess the same crystal structure and appears to exist in three crystal modifications. Hydrated thorium formate, $Th(HCO_2)_4{\cdot}3H_2O$, has a formato bridged structure in which each thorium atom is surrounded by eight oxygen atoms from eight HCO_2 groups in a bicapped trigonal prismatic array.[91] Thorium tetraacetate is isomorphous with $U(MeCO_2)_4$ in which the structure consists of infinite columns parallel to the c axis with acetate groups bridging the uranium atoms, giving rise to square antiprismatic geometry. There are two additional oxygen atoms in neighbouring polyhedra at 2.80 Å from each uranium atom.[92] Although salts of formato complex anions of composition $[Th(HCO_2)_{4+n}]^{n-}$ (n = 1–4, Table

27) have been recorded, no structural data are available for them. $Zn_2[U(HCO_2)_8]$ is obtained by reduction of UO_2Cl_2 in formic acid by zinc amalgam, but otherwise only basic uranium(IV) formato complexes have been reported. Acetato complexes are always $M^I_2[M^{IV}(MeCO_2)_6]$ or $M^{II}[M^{IV}(MeCO_2)_6]$. The uranates(IV) (M^{II} = Mg, Fe or Zn) are easily obtained by reduction of $UO_2(MeCO_2)_2 \cdot 2H_2O$ with the appropriate metal in a mixture of glacial acetic acid and acetic anhydride, but the zinc salt readily loses $Zn(MeCO_2)_2$ to yield the tetraacetate. A few trichloroacetato complex salts, $M^I[Th(Cl_3CCO_2)_6]$, are also known, and, somewhat surprisingly, a sodium salt formulated as $NaU(CH_2ClCO_2)_5 \cdot H_2O$ has been reported. Otherwise virtually nothing is known about anionic complexes with other carboxylic acids or with other actinides(IV).

(*b*) *Chelating carboxylates*. Pyridine-2- and -3-carboxylic acids appear to form simple carboxylates [*e.g.* $U(NC_5H_4\text{-}2\text{-}CO_2)_4$, Table 28], whereas pyridine-4-carboxylic acid can also act as a neutral ligand,[93] as in the complex $Th(NO_3)_4 \cdot (NC_5H_4\text{-}4\text{-}CO_2H) \cdot H_2O$. Quinoline-2-carboxylates are also known. It is not clear whether the heterocyclic nitrogen atom is bonded to the metal in these carboxylates.

Table 28 Actinide(IV) Compounds with Chelating Carboxylates

Pyridine and quinoline carboxylates	
$U(RCO_2)_4$	R = NC_5H_4-2-, ONC_5H_4-2-
$M(RCO_2)_2X_2$	M = Th; R = NC_5H_4-4-, X = OH; R = NC_5H_4-2-, X = Cl M = U; R = NC_5H_4-3-, X = $MeCO_2$; R = NC_9H_6-2-, X = Cl or $MeCO_2$
$U(NC_9H_6\text{-}2\text{-}CO_2)(MeCO_2)_3$	
$M(pdc)_2 \cdot xH_2O$	$H_2pdc = NC_5H_3\text{-}2,6\text{-}(CO_2H)_2$; M = Th ($x$ = 0 or 4), U (x = 0)
$U\{R(CO_2)_2\}(MeCO_2)_2$	R = NC_5H_3-2,3-
$U(Hpdc)(MeCO_2)_3$	
$(Ph_4As)_2[M(pdc)_3] \cdot 3H_2O$	M = Th, U
Imino- and oxo-diacetates	
$MO\{N(CH_2CO_2)_2\} \cdot 2H_2O$	M = Th,[a] U
$Th\{N(CH_2CO_2)_2\}_2 \cdot 4H_2O$[a]	
$[Th\{O(CH_2CO_2)_2\}_2]_n$[b]	
$U\{O(CH_2CO_2)_2\}_2 \cdot 2H_2O$	
$Na_2[Th\{O(CH_2CO_2)_2\}_3] \cdot xH_2O$[b]	x = 2 or 3
$Na_2[Th\{O(CH_2CO_2)_2\}_3] \cdot 2NaNO_3$[b]	
$Na_2[U\{O(CH_2CO_2)_2\}_3] \cdot 2\{O(CH_2CO_2H)_2\} \cdot H_2O$	
$Th_2\{CH_2(C_6H_4NHCH_2CO_2)_2\}_4 \cdot 8H_2O$[c]	

[a] G. A. Battiston, G. Sbrignadello, G. Bandoli, D. A. Clemente and G. Thomat, *J. Chem. Soc., Dalton Trans.*, 1979, 1965.
[b] G. Sbrignadello, G. Thomat, G. Battiston, G. de Paoli and L. Magon, *Inorg. Chim. Acta*, 1976, **18**, 195.
[c] C. G. Macarovici and E. Chis, *Rev. Roum. Chim.*, 1977, **22**, 657.

The simple hydrated pyridine-2,6-dicarboxylate (pdc), $[Th(pdc)_2(H_2O)_4]$, is precipitated from methanol solutions of $Th(NO_3)_4 \cdot 5H_2O$ by the acid; the thorium atom is 10-coordinate in a bicapped square antiprismatic arrangement in which the pyridine nitrogen atoms cap the two opposite faces of the antiprism, the corners of which are occupied by two oxygen atoms from each dicarboxylate group and the four oxygen atoms from the water molecules.[94] The tetrahydrate yields the anhydrous polymer, $[Th(pdc)_2]_n$, on heating. The corresponding uranium(IV) compound, $U(pdc)_2$, is prepared by heating $U(MeCO_2)_4$ with a solution of the acid in methanol under reflux, and mixed carboxylates, such as $U(Hpdc)(MeCO_2)_3$, are obtained in a similar manner using methanol or ethanol as the solvent.

The tetraphenylarsonium salts, $(Ph_4As)_3[M(pdc)_3] \cdot 3H_2O$, have been prepared by adding the metal tetrachloride (M = Th, U) or tetranitrate (M = Th) to an excess of the acid and Ph_4AsCl in water. In the structure of the anion of the uranium compound, the uranium atom is nine-coordinated with distorted tricapped trigonal prismatic geometry in which the three nitrogen atoms of the pdc groups cap the rectangular faces of the prism.[95]

The imino-, oxo- and 4,4′-diaminodiphenylmethane-*N*,*N*′-diacetates have been isolated from aqueous solution (Table 28). The central nitrogen or oxygen atoms of these diacetate groups are probably bonded to the metal as found for the pyridine-2,6-dicarboxylates.

(*c*) *Carbonates*. The simple carbonates (Table 29) are not well known, and their preparation usually requires autoclave techniques; for example, $Th(CO_3)_2 \cdot 0.5H_2O$ is obtained[96] from thorium(IV) hydroxide and carbon dioxide at 1800–3000 atm and 100–150 °C. The best established carbonato complexes are salts of the $[M^{IV}(CO_3)_4]^{4-}$ and $[M^{IV}(CO_3)_5]^{6-}$ anions, but tri-, hexa- and octa-carbonato complexes (Table 29) have also been reported. The two last require confirmation, for the preparation of the hexa- and octa-carbonato plutonates(IV) was by dissolution of $Pu(C_2O_4)_2$ in a solution of the appropriate alkali carbonate, after which the solution was poured into aqueous ethanol, yielding an oil which became crystalline on standing or on treatment with 99% ethanol. The same procedure can also be used to prepare the tetra- and penta-carbonato complexes.[97]

Table 29 Actinide(IV) Carbonates and Carbonato Complexes

$M^{IV}(CO_3)_2 \cdot xH_2O$	$M^{IV} = Th$, $x = 0.5$, 3–4; Pu, x unspecified
$M^{IV}O(CO_3) \cdot xH_2O$	$M^{IV} = Th$, $x = 2$, 8; U, $x = 0$; Pu, $x = 2$
$Th(OH)_2(CO_3) \cdot 2H_2O$	
$xM^{IV}O_2 \cdot M^{IV}O(CO_3) \cdot yH_2O$	$M^{IV} = Th$, $x = 1$, $y = 1.5$ or 4; $x = 3$, $y = 1$; $x = 6$, $y = 0$
	$M^{IV} = Pu$, $x = 1$, $y = 0$ or 3
$M^I_2[Th(CO_3)_3] \cdot xH_2O$	$M^I = NH_4$, $x = 6$; CN_3H_6(guanidinium), $x = 0$, 4
$(enH_2)[U(CO_3)_3(H_2O)_2] \cdot 2H_2O$	
$M^I_4[M^{IV}(CO_3)_4] \cdot xH_2O$	$M^{IV} = Th$, $M^I = CN_3H_6$, $x = 0$, 6; Na, $x = 0$, 7
	$M^{IV} = U$, $M^I = CN_3H_6$, $x = 4$, 5; $M^I_4 = (CN_3H_6)_3(NH_4)$, $x = 4$
	$M^{IV} = Np$[a], $M^I = Na$, x unspecified
	$M^{IV} = Pu$, $M^I = Na$, $x = 3$; NH_4, $x = 4$; K, x unspecified
$M^I_6[M^{IV}(CO_3)_5] \cdot xH_2O$	$M^{IV} = Th$, $M^I = Na$, $x = 0$, 3, 5, 10–11, 12, 20; K, $x = 10$, 13; NH_4, $x = 3$; Tl, $x = 1$, 2; CN_3H_6, $x = 0$, 4, 10
	$M^I_6 = Ca_3$, Ba_3, $x = 7$; $(CN_3H_6)_3(NH_4)_3$, $x = 3$; $[Co(NH_3)_6]_2$, $x = 4$, 6, 9–10
	$M^{IV} = U$, $M^I = Na$, $x = 10$, 12; K, $x = 6$; CN_3H_6, $x = 4$, 5
	$M^I_6 = (CN_3H_6)_4(NH_4)_2$, $x = 1$; $(CN_3H_6)_3(NH_4)_3$, $x = 2$; $[Co(NH_3)_6]_2$, $x = 4$, 5
	$M^{IV} = Np$, $M^I = K$,[a] x unspecified
	$M^{IV} = Pu$, $M^I = Na$, $x = 2$, 4; K, NH_4, x unspecified
	$M^I_6 = [Co(NH_3)_6]_2$, $x = 5$
$M^I_8[Pu(CO_3)_6] \cdot xH_2O$	$M^I = Na$, K, NH_4, x unspecified
$M^I_{12}[Pu(CO_3)_8] \cdot xH_2O$	$M^I = Na$, K, NH_4, x unspecified
$Na[Th(OH)(CO_3)_2(H_2O)_3] \cdot 3H_2O$	
$M^I_2[Th(OH)_2(CO_3)_2(H_2O)_2] \cdot xH_2O$	$M^I = Na$, $x = 8$; K, $x = 3$
$K_3[Th(OH)(CO_3)_3(H_2O)_2] \cdot 3H_2O$	
$Na_5[Th(OH)(CO_3)_4(H_2O)] \cdot 8H_2O$	
$(CN_3H_6)_5[M^{IV}(OH)_3(CO_3)_3] \cdot 5H_2O$	$M^{IV} = Th$, U
$(enH_2)[U(OH)_2(CO_3)_2] \cdot 3H_2O$	
$(enH_2)_2[U_2(OH)_2(CO_3)_5(H_2O)_4] \cdot 2H_2O$	
$(CN_3H_6)_5[Th(CO_3)_3F_3]$	
$(CN_3H_6)_4(NH_4)[Th(CO_3)_3F_3]$	
$M^I_2[U(HCO_3)_2F_4]$	$M^I = Na$, NH_4

[a] Yu. Ya. Kharitonov and A. I. Moskvin, *Sov. Radiochem. (Engl. Transl.)*, 1973, **15**, 240.

In the cases of the hydrated salts, some authors include a part of the water in the coordination sphere (*e.g.* $M^I_6[Th(CO_3)_5] \cdot xH_2O$ is sometimes written as $M^I_6[Th(CO_3)_5(H_2O)] \cdot (x-1)H_2O$ and $(NH_4)_2[Th(CO_3)_3] \cdot 6H_2O$ as $(NH_4)_2[Th(CO_3)_3(H_2O)_3] \cdot 3H_2O$), but there does not appear to be sufficient experimental evidence to support this. The same applies to the formulations of the basic carbonato complexes listed in Table 29.

The coordination polyhedron of the anion in $(CN_3H_6)_6[Th(CO_3)_5] \cdot 4H_2O$ is an irregular decahexahedron[98] and the geometry is a bicapped square antiprism. Replacement of CN_3H_6 by Na, in $Na_6[Th(CO_3)_5] \cdot 12H_2O$, leads only to a slight deformation of the polyhedron.[99] The water molecules are not bonded to the thorium atom in either salt. The thorium atom in the carbonatofluorothorate(IV), $(CN_3H_6)_5[Th(CO_3)_3F_3]$, is nine-coordinate, with three bidentate carbonate groups and three terminal fluorine atoms making up a monocapped square antiprism.[100]

(*d*) *Oxalates.* A large number of oxalato (Table 30) and mixed oxalato complexes (Table 31) have been recorded. The hydrated oxalates, $M(C_2O_4)_2 \cdot xH_2O$, are precipitated from aqueous media, the basis of a useful gravimetric method for the elements. The thorium compounds ($x = 0$, 1, 2 or 6) are isomorphous with their uranium(IV) analogues. The hydrated basic oxalate, $UO(C_2O_4) \cdot 6H_2O$, precipitates on photoreduction of $UO_2(HCO_2)_2$ in the presence of oxalic acid; other hydrates are known and some authors describe them as hydroxo compounds [*e.g.* $U(OH)_2(C_2O_4) \cdot 5H_2O$], but this requires confirmation.

Table 30 Actinide(IV) Oxalates and Oxalato Complexes

$M(C_2O_4)_2{\cdot}xH_2O$	M = Th, x = 0, 1, 2, 4, 6; U, x = 0, 1, 2, 3, 5, 6; Np, Pu, x = 2, 6
$UO(C_2O_4){\cdot}xH_2O$	x = 0, 4, 6
$M^I_2M^{IV}(C_2O_4)_3{\cdot}xH_2O$	M^{IV} = Th, M^I = CN_3H_6, x = 6, 8; NH_4, x unspecified $M^I_2 = [(PhCH_2)N(C_9H_7)]^+H^+$, M^{IV} = Th, U (NC_9H_7 = quinoline)
$H_2Ca[U_2(C_2O_4)_6]{\cdot}24H_2O$	
$M^I_4[M^{IV}(C_2O_4)_4]{\cdot}xH_2O$	M^{IV} = Th; M^I = Na, x = 0, 5.5, 6; K, x = 0, 4; NH_4, x = 0, 3, 4.7, 6.5, 7; Me_2NH_2, x = 0, 2, 9; $Bu^n_2NH_2$, x = 0, 4; CN_3H_6, x = 2 $M^I_4 = (CN_3H_6)_3(NH_4)$, x = 3 M^{IV} = U; M^I = K, x = 0, 1, 2, 4, 4.5, 5; NH_4, x = 0, 3, 5, 6, 7; Cs, x = 3; CN_3H_6, x = 0, 2 M^{IV} = Np; M^I = Na, x = 3; K, x = 4; NH_4, x unspecified M^{IV} = Pu; M^I = Na, x = 5; K, x = 4
$M^{II}_2[M^{IV}(C_2O_4)_4]{\cdot}xH_2O$	M^{IV} = Th, M^{II} = Ba, x = 11; enH_2, x = 2.5 M^{IV} = U, M^{II} = Ca, x = 0, 1, 4, 6, 10; Sr, x = 0, 4, 6; Ba, x = 0, 6, 6.5, 7, 8, 9; Cd, x = 0, 6, 7; Pb, x = 0, 6, 8; $[Pt(NH_3)_4]$, x = 3
$M^{III}_4[M^{IV}(C_2O_4)_4]_3{\cdot}xH_2O$	M^{IV} = Th, M^{III} = $[Co(en)_3]$, x = 22; $[Co(tn)_3]$, x = 3; tn = $H_2N(CH_2)_3NH_2$ M^{IV} = U, M^{III} = La, x = 22; $[Cr(urea)_6]$, x = 6 to 11 M^{IV} = Pu, M^{III} = $[Cr(urea)_6]$, x unspecified
$[Pt(NH_3)_6][U(C_2O_4)_4]{\cdot}3H_2O$	
$M^I_6[M^{IV}(C_2O_4)_5]{\cdot}xH_2O$	M^{IV} = Th, M^I = NH_4, x = 3, 7.5; M^{IV} = Pu, M^I = K, x = 4; NH_4, x unspecified
$M^{III}_2[M^{IV}(C_2O_4)_5]{\cdot}xH_2O$	M^{IV} = Th, M^{III} = $[Co(NH_3)_6]$, x = 3; $[Cr(NH_3)_6]$, x = 20; $[Cr(urea)_6]$, x = 0.5 M^{IV} = U, M^{III} = $[Cr(urea)_6]$, x = 0.5
$M^I_2[M^{IV}_2(C_2O_4)_5]{\cdot}xH_2O$	M^{IV} = Th, M^I = H, x = 9;[a] NH_4, x = 2, 7 M^{IV} = U, M^I = H, x = 0, 4, 8; Na, K, x = 8; NH_4, x = 0, 2, 4, 8; CN_3H_6, x = 0, 1, 4
$M^I_8[Th(C_2O_4)_6]{\cdot}xH_2O$	M^I = Et_3NH, x = 0, 3; $Bu^n_2NH_2$, x = 0
$[Co\ en_3]_8[Th(C_2O_4)_6]_3{\cdot}42H_2O$	
$H_2CaU_2(C_2O_4)_6{\cdot}24H_2O$	
$M^I_6Th_2(C_2O_4)_7{\cdot}xH_2O$	M^I = Et_2NH_2, x = 0, 6; $Pr^n_2NH_2$, x = 0, 8; CN_3H_6, x = 5, 8, 12.5 to 13.7

[a] Also reported as $(H_3O)_2[Th_2(C_2O_4)_5]{\cdot}5H_2O$.

A few salts of the tris oxalato actinide(IV) anions are known, such as the acid benzylquinolinium compounds (Table 30), but the more usual complexes are the tetraoxalato and pentaoxalato species. The coordination geometry of the 10-coordinate thorium atom in the anion of $K_4[Th(C_2O_4)_4]{\cdot}4H_2O$ is a slightly irregular bicapped square antiprism in an oxalate bridged structure cross-linked into a three-dimensional framework by hydrogen bonding (Figure 4).[101] The geometry in both crystal modifications of $K_4[U(C_2O_4)_4]{\cdot}4H_2O$ is the same as in the thorium compound; in one phase the three bidentate C_2O_4 groups and a tetradentate bridging C_2O_4 group link the metal atoms in a one dimensional polymeric array and the other phase is isostructural with the thorium compound.[102] It would be useful to have structural information for the $[Th(C_2O_4)_6]^{8-}$, $[U_2(C_2O_4)_6]^{4-}$ and $[Th_2(C_2O_4)_7]^{6-}$ anions, for it is not clear whether these are genuine complexes or mixtures involving other known oxalato anions.

Mixed oxalates and oxalato complexes (Table 31) also require further investigation. The sulfito and sulfato oxalates have been mentioned earlier (p. 1152) and an equally large number of carbonatooxalato species have been recorded,[103,104] some of which may well be mixtures. In addition to the compounds listed in Table 31, products of the rather unlikely compositions $K_7[U(OH)(C_2O_4)_2(CO_3)_3]{\cdot}6H_2O$ and $K_{16}[U_2(OH)_2(C_2O_4)_3(CO_3)_8]{\cdot}10H_2O$ have been reported.[105]

(*e*) *Other dicarboxylates*. Compounds with dicarboxylic acids other than oxalates or chelating carboxylates have scarcely been investigated (Table 32). In some cases the *bis* compounds, $M^{IV}L_2$, are precipitated from aqueous solution, often as hydrates, a preparative method which also gives rise to a variety of basic compounds reported either as $M^{IV}OL{\cdot}xH_2O$ or $M^{IV}(OH)_2L{\cdot}(x-1)H_2O$. These are presumably identical compounds, but no structural information is available. A better preparative route, used for example for the phthalates, is by heating the metal tetraacetate with the dicarboxylic acid in ethanol under reflux, a method which could usefully be applied to the preparation of a much wider range of dicarboxylates. The only recorded dicarboxylato complex anions are the malonato compounds, $M^I[M^{IV}\{CH_2(CO_2)_2\}_3]{\cdot}xH_2O$ and $Na_2[Th\{O(CH_2CO_2)_3\}_3]{\cdot}2NaNO_3$. In the structure of the

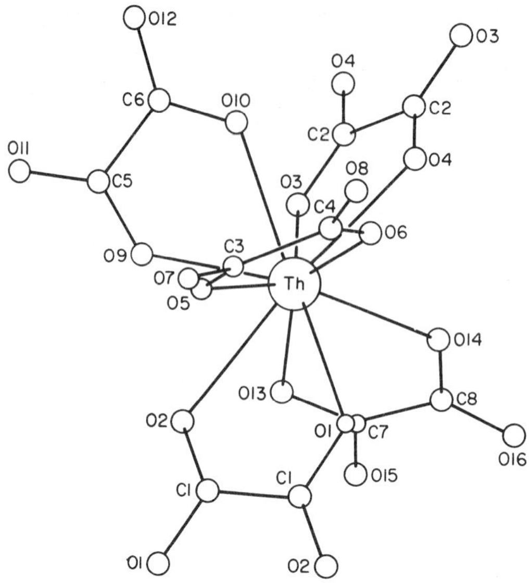

Figure 4 The five oxalate groups coordinated to one thorium atom in the structure of $K_4Th(C_2O_4)_4{\cdot}4H_2O$[101]

Table 31 Actinide(IV) Mixed Oxalates and Oxalato Complexes

$UF_2(C_2O_4){\cdot}1.5H_2O$	
$U_2X_2(C_2O_4)_3{\cdot}yH_2O$	$X = F$, $y = 0$; $X = Cl$, $y = 0$, 2, 4 or 12
$M^I_4M^{IV}F_4(C_2O_4)_2{\cdot}xH_2O$	$M^{IV} = Th$, $M^I = K$, $x = 0$
	$M^{IV} = U$, $M^I = NH_4$, $x = 4$
$K_2(Pu(C_2O_4)(CO_3)_2{\cdot}$*ca.* $1.5H_2O$	
$M^I_4M^{IV}(C_2O_4)_x(CO_3)_{4-x}{\cdot}yH_2O$	$M^{IV} = Th$, $M^I = K$, $x = 1$, $y = 4.6$
	$M^I = NH_4$, $x = 2$, $y = 0.5$
	$M^I_4 = (CN_3H_6)_3(NH_4)$, $x = 1$, $y = 1.5$, or 2–3.5, and $x = 2$, $y = 3$
	$M^{IV} = U$, $M^I_4 = (CN_3H_6)_3(NH_4)$, $x = 1$, $y = 2$
	$M^{IV} = Pu$, $M^I = Na$, K, $x = 1$, y unspecified
	$M^I = Na$, $x = 2$, $y = 3$
$(NH_4)_4Th_2(C_2O_4)(CO_3)_5{\cdot}10H_2O$	
$K_6M^{IV}(C_2O_4)_x(CO_3)_{5-x}{\cdot}yH_2O$	$M^{IV} = Th$, $x = 1$, $y = 6$–8 and $x = 2$, $y = 0$, 1 or 4
	$M^{IV} = Pu$, $x = 2$, y unspecified
$M^I_6Th_2(C_2O_4)_x(CO_3)_{7-x}{\cdot}yH_2O$	$M^I = K$, $x = 3$, $y = 6$
	$M^I = CN_3H_6$, $x = 2$, $y = 4$ or 8 and $x = 3$, $y = 14$
$Na_8Th(C_2O_4)_x(CO_3)_{6-x}{\cdot}yH_2O$	$x = 1$, $y = 10$ to 11 and $x = 2$, $y = 9$ to 10.5 or 11
$M^I_8Th_2(C_2O_4)_x(CO_3)_{8-x}{\cdot}yH_2O$	$M^I = K$, $x = 3$, $y = 13$ or 16
	$M^I = CN_3H_6$, $x = 1$, $y = 6$ and $x = 3$, $y = 0$
$Na_{10}Th(C_2O_4)_2(CO_3)_5{\cdot}xH_2O$	$x = 10$, 11, 11.5 or 16
$M^I_{10}M^{IV}_2(C_2O_4)_x(CO_3)_{9-x}{\cdot}yH_2O$	$M^{IV} = Th$, $M^I = K$, $x = 2$, $y = 8$, 12 or 14 and $x = 4$, $y = 5$ or 7
	$M^I = CN_3H_6$, $x = 1$, $y = 8$
	$M^{IV} = U$, $M^I_{10} = (CN_3H_6)_8(NH_4)_2$, $x = 1$, $y = 4$ or 8
	$M^I_{10} = [Cr(urea)_6]_3(NH_4)$, $x = 1$, $y = 6$
$Na_{12}Th(C_2O_4)_2(CO_3)_6{\cdot}13H_2O$	
$K_2Th_2(OH)_2(C_2O_4)(CO_3)_3{\cdot}xH_2O$	$x = 0$, 1 or 2
$K_5Th_2(OH)(C_2O_4)_2(CO_3)_4{\cdot}2H_2O$	
$Na_4[M^{IV}_2(OH)_{2x}(C_2O_4)(CO_3)_{5-x}]{\cdot}yH_2O$	$M^{IV} = Th$, $x = 1$, $y = 4$ and $x = 3$, $y = 2$
	$M^{IV} = U$, $x = 2$, $y = 4$
$Na_{10}Th(OH)_2(C_2O_4)_3(CO_3)_3{\cdot}xH_2O$	$x = 8$–9
$(NH_4)_4U_2(C_2O_4)_3(HCO_2)_3{\cdot}H_2O{\cdot}2HCO_2H$	
$K[U(C_2O_4)_2(NCS)(H_2O)_3]$	
$Cs[U(C_2O_4)(NCS)_2(H_2O)_x]$	$x = 0$ or 2
$K_4Th(C_2O_4)_2(HPO_4)_2{\cdot}6H_2O$	
$K_4[Th(C_2O_4)_2(C_4H_4O_6)_2]{\cdot}3H_2O$	$C_4H_6O_6$ = tartaric acid
$K_4[Th(C_2O_4)(C_6H_5O_7)_2]{\cdot}3H_2O$	$C_6H_8O_7$ = citric acid

Table 32 Actinide(IV) Dicarboxylates and Dicarboxylate Complexes

$M^{IV}\{R(CO_2)_2\}_2 \cdot xH_2O$	$M^{IV} = Th$; $R = CH_2$ (malonate), $x = 2$; $(CH_2)_4$ (adipate), $x = 0$; $(CH_2)_8$ (sebacate), $x = 0$; CH═CH (maleate), $x = 2$; C_6H_4-1,2-(phthalate), $x = 0, 2$; 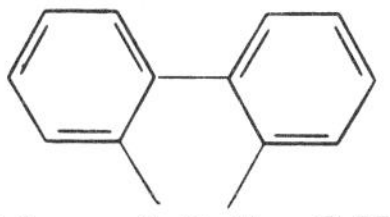 (2,2′-biphenyldicarboxylate), $x = 0, 2$; $R = C_6H_4$-2-$NHCH_2$, $x = 2$ $M^{IV} = U$; $R = (CH_2)_4$, $x = 0.5$; C_6H_4-1,2-, $x = 0$
$Th_2\{CH_2(C_6H_4$-4,4′-$NHCH_2CO_2)_2\}_4 \cdot 8H_2O$	
$M^I[M^{IV}\{CH_2(CO_2)_2\}_3] \cdot xH_2O$	$M^I = Na$, $M^{IV} = U$, $x = 2$; $M^I = K$, $M^{IV} = Th$, $x = 0$
$(M^{IV}OH)_2\{R(CO_2)_2\}_3 \cdot xH_2O$	$M^{IV} = Th$, $R = (CH_2)_2$ (succinate) or $(CH_2)_3$ (glutarate), $x = 8$; $(CH_2)_4$, $x = 4.5, 5$; $(CH_2)_7$ (azelaate), $x = 4$ $M^{IV} = U$, $R = CH_2$ or $(CH_2)_3$, $x = 2$; $(CH_2)_4$, $x = 4$; $(CH_2)_7$, $x = 4$
$M^IO\{R(CO_2)_2\} \cdot xH_2O$	$M^{IV} = Th$, $R = (CH_2)_4$, $x = 3$; C_8H_{14} (camphorate), $x = 0, 2$ $M^{IV} = U$, $R = CH_2$, $x = 6$; $(CH_2)_2$, $x = 2, 3$; C_6H_4-1,2-, $x = 3$
$Th(OH)_2\{R(CO_2)_2\} \cdot xH_2O$	$R = (CH_2)_3$, $x = 4$; CH═CH (fumarate), $x = 4$; C_6H_4-1,2-, $x = 0$; $C_9H_7NO_3$ (2′-carboxymaleanilate), $x = 2$
$Th(OH)_2\{O_2C(CH_2)_2CO_2H\}_2$	
$Th(OH)_3\{O_2C(CH_2)_4CO_2H\}$	
$U_2O\{CH_2(CO_2)_2\}_3$	

anion of the latter, the nine-coordinated thorium atom is at the centre of a tricapped trigonal prism, with three tridentate $O(CH_2CO_2)_2$ groups bonded to the metal.[58b]

(*f*) *Carbamates. N,N*-Dialkylcarbamates, $M^{IV}(R_2NCO_2)_4$ ($M^{IV} = Th, U$) are obtained by reaction of the appropriate dialkylamide, $M(NR_2)_4$, with carbon dioxide. A much simpler route is by reaction of UCl_4 with R_2NH and carbon dioxide in benzene ($R = Et$) or toluene ($R = Me$). The carbamates precipitate on addition of *n*-heptane after evaporation to small volume. $U(Et_2NCO_2)_4$ is a monomer in benzene, and the ^1H NMR spectra of the known compounds (Table 27) indicate that the alkyl groups are equivalent.

A by-product of the preparation of $U(Et_2NCO_2)_4$ from the tetrachloride and the amine is a product of composition $U_4O_2(Et_2NCO_2)_{12}$; this is a tetramer in which there are two non-equivalent uranium(IV) sites. In one of them the coordination geometry is a distorted tricapped trigonal prism, and the geometry of the other does not fit any type of regular polyhedron.[106]

(*g*) *Tetracarboxylates.* Thorium pyromellitate, $[Th(1,2,4,5\text{-}C_6H_2(CO_2)_4)]_n$, and the 2,3,6,7-naphthalenetetracarboxylate, $[Th(2,3,6,7\text{-}C_{10}H_4(CO_2)_4)]_n$, are both polymers as expected.

(*h*) *Nitroalkanes.* The only recorded complexes appear to be $UCl_4 \cdot xMeNO_2$; $UCl_4 \cdot MeNO_2$ is formed initially when $MeNO_2$ is condensed onto UCl_4 at low temperature and the mixture is allowed to warm to room temperature, and $UCl_4 \cdot 2.5MeNO_2$ is obtained from the mixture on standing.

(*vii*) *Aliphatic and aromatic hydroxyacids*

(*a*) *Aliphatic hydroxyacids.* The majority of the known hydroxycarboxylates (Table 33) are precipitated from aqueous solutions of the actinide(IV) ion on addition of the appropriate acid; many of these products are hydrated. Uranium(IV) glycollate, $[U(HOCH_2CO_2)_4(H_2O)_2]$, precipitates on photochemical reduction of $UO_2(NO_3)_2$ in aqueous solution in the presence of an excess of the acid[57] and the structure of this compound has been described earlier (p. 1144). The thorium analogue probably has the same molecular structure.

The alkali metal salts, such as those of the malato complex ions, have been isolated by dissolving the actinide carboxylate [*e.g.* $(U_2OCCH_2CH(OH)CO_2)_2 \cdot H_2O$] in a hot, aqueous solution containing the stoichiometric quantity of the alkali carboxylate, and then evaporating the resulting solution. The hydroxocarboxylato salts of the type $M^I_2M^{IV}(OH)_2X_n \cdot yH_2O$, where X_n represents the hydroxycarboxylate anions, have also been reported as $M^I_2M^{IV}OX_n \cdot zH_2O$ (Table 33); these are obviously identical compounds, but it is uncertain which formulation is

Table 33 Actinide(IV) Compounds with Aliphatic Hydroxyacids

Glycollates and related carboxylates	
$M^{IV}(HOCH_2CO_2)_4 \cdot 2H_2O$	$M^{IV} = Th, U$
$Th(OH)(HOCH_2CO_2)_3 \cdot H_2O$	
$ThO(HOCH_2CO_2)_2 \cdot 2H_2O$	
$Th(PhCH(OH)CO_2)_4 \cdot H_2O$	
$Th(OH)_2\{PhCH(OH)CO_2\}_2$	
$Th\{Me_2C(OH)CO_2\}_4 \cdot 2H_2O$	
$M^{IV}\{Ph_2C(OH)CO_2\}_4$	$M^{IV} = Th, U$
$Th\{Ph_2C(O)CO_2\}_2$	
Malates; $H_3L = HO_2CCH_2CH(OH)CO_2H$	
$M^{IV}(HL)_2 \cdot xH_2O$	$M^{IV} = Th, x = 1$ to 2; $U, x = 1$
$(ThOH)_2(HL)_3 \cdot 4H_2O$	
$Th_2Cl_2(HL)_3$	
$NaTh(OH)(HL)_2 \cdot 6H_2O$	
$Na_2Th(OH)_2(HL)_2 \cdot 4H_2O$	
$M^I_2ThO(HL)_2 \cdot xH_2O$	$M^I = Na, x = 6; K, NH_4, x = 4$
$M^I_2[U(HL)_3] \cdot H_2O$	$M^I = Na, K$
$M^I_2[U_2(HL)_5] \cdot xH_2O$	$M^I = Na, K, x = 5; NH_4, x = 2$
Tartrates; $H_4L = HO_2C(CHOH)_2CO_2H$	
$ThL \cdot 9H_2O$	
$Th_3(OH)_4(H_2L)_4 \cdot xH_2O$	$x = 0, 2, 5, 18, 22$ (the compound is sometimes written as $Th_3O_2(H_2L)_4 \cdot (x+2)(H_2O)$)
$U(H_2L)_2 \cdot 2H_2O$	
$U_2O(H_2L)_3 \cdot 5H_2O$	
$2U(H_2L)_2 \cdot UO_2 \cdot 12H_2O$	
$KTh(OH)L \cdot 10H_2O$	
$KTh(OH)(H_2L)_2 \cdot 7H_2O$	
$K_2Th(OH)_2L \cdot 4H_2O$	
$M^I_2ThOL \cdot 4H_2O$	$M^I = K, NH_4$
$K_2Th(OH)_2(H_2L)_2 \cdot 8H_2O$	
$M^I_2ThO(H_2L)_2 \cdot xH_2O$	$M^I = Na, K, x = 8; NH_4, x = 3$
$M^I_4[U(H_2L)_2(C_2O_4)_2] \cdot xH_2O$	$M^I = K, x = 3; M^I_4 = [Pt(NH_3)_4]_2, x = 0$
Trihydroxyglutarates; $H_5L = HO_2C(CHOH)_3CO_2H$	
$Th(H_3L)_2 \cdot 4H_2O$	
$Th(OH)(H_2L) \cdot 2H_2O$	
$(ThOH)_2(H_3L)_3 \cdot xH_2O$	$x = 0, 1$
$NaTh(OH)(H_3L)_2 \cdot H_2O$	
$NaTh(OH)_2(H_2L) \cdot H_2O$	
$Na_2Th(OH)_2(H_3L)_2$	
Mucates; $H_6L = HO_2C(CHOH)_4CO_2H$	
$(ThOH)_2(H_4L)_3 \cdot 6H_2O$	
$NaTh(OH)(H_2L) \cdot 2H_2O$	
$NaTh(OH)(H_4L)_2 \cdot 6H_2O$	
$Na_2Th(OH)_2(H_4L)_2 \cdot 10H_2O$	
Citrates; $H_4L = HOC(CH_2CO_2H)_2(CO_2H)$	
$U_3(HL)_4 \cdot xH_2O$	x unspecified
$UO(H_2L)$	
$[Co(NH_3)_6]_2[M^{IV}(HL)_2]_3 \cdot xH_2O$	$M^{IV} = Th,$[a] $x = 0$; $Pu,$[a] $x = 12$
$K_4[U(HL)_2(C_2O_4)] \cdot 3H_2O$	

[a] M. Hoshi and K. Ueno, *Radiochem. Radioanal. Lett.*, 1977, **30**, 145.

correct. The tartrato- and citrato-oxalato compounds are precipitated on dropwise addition of ethanol to a solution of $U(C_2O_4)_2$ in an aqueous mixture of the hydroxy acid and alkali in the stoichiometric proportions.

(*b*) *Aromatic hydroxyacids.* Most of the compounds with aromatic hydroxyacids (Table 34) are precipitated from aqueous solution; the composition of the product depends on the pH; for example, with 2-hydroxy-3-naphthoic acid, $Th(OC_{10}H_6CO_2)_2 \cdot 4H_2O$ is reported to precipitate at at pH 4 to 5, whereas $Th(HOC_{10}H_6CO_2)_4$ is said to precipitate at pH 6 to 7. In both cases the stoichiometric quantities of the reagents were used. It is quite possible that compounds such as $Th(OC_{10}H_6CO_2)_2 \cdot 4H_2O$ are really basic species of the type $ThO(HOC_{10}H_6CO_2)_2 \cdot 3H_2O$. In addition to the compounds listed in Table 34, 3- and 5-nitro-, 3,5-dinitro- and 5-bromo-salicylates of composition ThL_2 (all) or $ThL(HL)_2$ (3- and 5-nitro) have been recorded. The hydroxynaphthoate acetates, $M^{IV}(HL)_x(MeCO_2)_{4-x}$, have been obtained by heating the metal tetraacetate with the naphthoic acid in ethanol under reflux.

Table 34 Actinide(IV) Compounds with Aromatic Hydroxyacids

Salicylates; $H_2L = 2\text{-}HOC_6H_4CO_2H$

$ThL(HL)_2$	
$M^{IV}L_2 \cdot xH_2O$	$M^{IV} = Th, x = 0; U, x = 2, 4$
$M^{IV}O(HL)_2 \cdot xH_2O$	$M^{IV} = Th, x = 3; U, x = 0; Pu, x = 0$
$ThCl_x(HL)_{4-x}$	$x = 1$ or 2
Na_2ThOL_2	

3- and 4-Hydroxybenzoates; $H_2L = HOC_6H_4CO_2H$

$Th(OH)_2(HL)_2 \cdot xH_2O$	$x = 1$ (3-HO) and 3 (4-HO)

Salicylsulfonic acid, $HOC_6H_3(SO_3H)CO_2H$

$U(HOC_6H_3(SO_3)CO_2)_2$

Thiodisalicylic acid,

$$Th\left\{S\left(\underset{CO_2}{-\!\!\!\bigcirc\!\!\!-}OH\right)_2\right\}_2 \cdot 4H_2O \qquad S\left(\underset{CO_2H}{-\!\!\!\bigcirc\!\!\!-}OH\right)_2$$

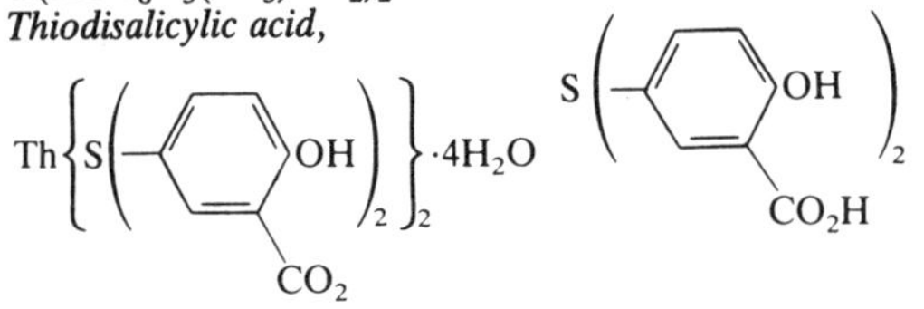

Hydroxynaphthoic acids; $H_2L = HOC_{10}H_6\text{-}2\text{-}CO_2H$

$ThL_2 \cdot xH_2O$	$x = 2,4$ (1-hydroxy)
$Th(HL)_4$	(1-hydroxy)
$Th(OH)_3(HL) \cdot 2H_2O$	(1-hydroxy)
$Th(OH)_2L \cdot 3H_2O$	(1-hydroxy-4-nitro)
$M^{IV}(HL)_2(MeCO_2)_2$	$M^{IV} = Th$ (1- and 3-hydroxy), U (1-hydroxy)
$U(HL)(MeCO_2)_3$	(3-hydroxy)

2-Hydroxy-3-naphthoic acid, H_2L

$ThL_2 \cdot 4H_2O$
$Th(HL)_4$
$ThOL \cdot 3H_2O$

(*viii*) *Amides and related compounds, carboxylic acid hydrazides, ureas, N-oxides, P-oxides, As-oxides and S-oxides*

The majority of the known complexes with neutral oxygen donor ligands have been prepared from the parent compound and the ligand in a non-aqueous solvent of relatively poor donor ability. When the parent compound does not exist (*e.g.* $PuCl_4$) or is difficult to obtain anhydrous (*e.g.* $M^{IV}(NCS)_4$), the alternative route is to react salts such as $Cs_2M^{IV}Cl_6$ with the ligand in a non-aqueous solvent, relying on the low solubility of CsCl in such solvents to shift the following equilibrium to the right:

$$Cs_2MCl_6 + xL \rightleftharpoons MCl_4L_x + 2CsCl(\text{solid})$$

In the thiocyanate case, the complexes of the metal tetrachlorides are simply treated with the stoichiometric quantity of KNCS in suitable solvents. In some instances the preparations can be carried out in one step from the metal tetrachloride, potassium thiocyanate and the ligand, but this procedure can make the purification of the final product quite difficult. A number of complexes of the type $(\eta^5\text{-}C_5H_5)MX_3L_2$ (M = Th, U, Np) have been recorded; these are described elsewhere.[107]

(*a*) *Amides and related compounds.* Complexes with formamide do not appear to have been recorded, but several acetamide complexes are known (Table 35). In the group $MCl_4 \cdot 6L$ (M = U, Np, Pu), the uranium compound is ionic and is probably of the form $[UCl_2(MeCONH_2)_6]^{2+}Cl_2$ rather than $[U(MeCONH_2)_6]^{4+}Cl_4$. The complex with the sulfate, $U(SO_4)_2 \cdot 4MeCONH_2$, has been prepared by heating the sulfate in molten acetamide, and the anionic complex, $(NH_4)_2[U(SO_4)_3(MeCONH_2)] \cdot xH_2O$, was obtained from aqueous solution.

Relatively few complexes with *N*-alkylacetamides have been reported. No structural information is available for any of them, but the thorium nitrate complexes,

$Th(NO_3)_4{\cdot}2.5MeCONHR$ (R = Me, Et), and the analogous DMF complex, are probably ionic $[Th(NO_3)_3L_4]^+[Th(NO_3)_5L]^-$ like the diethylpropionamide complex, $UCl_4{\cdot}2.5EtCONEt_2$ (p. 1164). In the case of the uranium(IV) perchlorate complexes with MeCONHMe (Table 35), it appears that the $[U(MeCONHMe)_8]^{4+}$ anion is unstable with respect to loss of ligand, and the corresponding seven-coordinate cation may be formed; this is surprising, for in cone-angle terms (p. 1130), the estimated value of Σ caf is well within the stable region.

A much wider range of complexes has been reported for *N,N*-dimethylformamide (DMF) and for *N,N*-dimethylacetamide (DMA). Structural information is available for four DMF complexes. The mixed oxidation state compound, $[U^{IV}Cl(DMF)_7]_2[U^{VI}O_2Cl_4]_3$, is formed by treating UCl_4 in acetone with DMF, and in this compound the coordination geometry of the cation is dodecahedral, and the anion is *trans* octahedral.[108] In the complex with the 8-hydroxyquinolinate, $[Th(C_9H_6NO)_4(DMF)]$, which separates from a solution of $Th(C_9H_6NO)_4{\cdot}C_9H_7NO$ in warm DMF, the coordination polyhedron is a slightly distorted tricapped trigonal prism.[109a] In the complex with the tropolonate, $[Th(trop)_4(DMF)]$, which is precipitated when $Th(trop)_4$ is heated in anhydrous DMF, the structure of the molecule is a monocapped square antiprism in which one tropolone ligand spans a slant edge of the cap, and the DMF molecule is bonded at another vertex of the cap.[110] The structure of the 1-oxo-2-thiopyridinato compound, $[ThL_4(DMF)]$, is very similar.[109b]

The DMF complex with thorium(IV) perchlorate behaves as a 1:4 electrolyte in methanol or nitromethane and its IR spectrum suggests that the perchlorate groups are ionic, indicating that the complex is $[Th(DMF)_8](ClO_4)_4$. The same complex cation is reported for the compounds $[M(DMF)_8]_3[Cr(NCS)_6]_4$ (M = Th, U). *N,N*-Diphenylformamide complexes of composition $UX_4{\cdot}4HCONPh_2$ (X = Cl, Br) have also been recorded.

No structural data are available for any of the DMA complexes (Table 35). However, the UV–visible spectrum of $2UCl_4{\cdot}5DMA$ is virtually identical to that of $2UCl_4{\cdot}5EtCONEt_2$, indicating the presence of both six- and seven-coordinate uranium(IV) centres, and it is therefore probable that the DMA complex, as well as the other complexes of composition $2MX_4{\cdot}5L$ (Table 35), is ionic and of the form $[MX_3L_4]^+[MX_5L]^-$, even though the conductance of $2UCl_4{\cdot}5DMA$ in nitromethane is low.

The thorium atom in $ThCl_4{\cdot}4DMA$ appears, on the basis of the cone-angle approach to steric crowding (p. 1130), to be very overcrowded, and it is probable that this complex should be written as $[ThCl_3(DMA)_4]^+Cl^-$, even though its conductivity in nitromethane is very low. Repeated dissolution of this complex in THF and reprecipitation with a mixture of *n*-pentane and toluene yields the tris complex, $[ThCl_4(DMA)_3]$.[111a]

The conductances of $2Th(NO_3)_4{\cdot}5DMA$ (in nitromethane) and $Th(NO_3)_4{\cdot}3DMA{\cdot}3H_2O$ (in

Table 35 Complexes of Actinide(IV) Compounds with Amides

Acetamide, MeCONH₂	
$M^{IV}Cl_4{\cdot}6MeCONH_2$	M^{IV} = U, Np, Pu
$U(SO_4)_2{\cdot}4MeCONH_2$	
$(NH_4)_2[U(SO_4)_3(MeCONH_2)_2]{\cdot}xH_2O$	x = 0, 4
N-*Alkyl acetamides, MeCONHR*	
$UX_4{\cdot}4MeCONHR$	X = Cl, R = Me; X = Cl and Br, R = Et, Pr^i
$M^{IV}(NO_3)_4{\cdot}xMeCONHR$	M^{IV} = Th, x = 3, R = Me, Et; M^{IV} = U, x = 2.5, R = Et; M^{IV} = Pu, x = 2, R = Et; HR = Pr_2
$U(ClO_4)_4{\cdot}xMeCONHMe$	x = 6, 7.4–7.68
N,N-*Dimethylformamide, DMF*	
$M^{IV}X_4{\cdot}yDMF$	M^{IV} = Th; X = NO_3, y = 2.5; X = Cl, NCSe, NO_2, y = 4 M^{IV} = U; X = Cl, y = 2.5, 3; X = I, y = 4
$[UCl(DMF)_7]_2[UO_2Cl_4]_3$	
$[Th(DMF)_8](ClO_4)_4$	
$[M^{IV}(DMF)_8]_3[Cr(NCS)_6]_4$	M^{IV} = Th, U
$[Th(C_9H_6NO)_4(DMF)]$	C_9H_7NO = 8-hydroxyquinoline
$[M^{IV}(trop)_4(DMF)]$	M^{IV} = Th, U; H trop = tropolone
$U(C_2O_4)_2{\cdot}xDMF{\cdot}yH_2O$	x = 1, y = 2; x = 1.5, y = 1
$M^I_2Th(NCSe)_6{\cdot}xDMF$	M^I = Na, x = 3; K, x = 4.5
$K_4Th(NCSe)_8{\cdot}2DMF$	
N,N-*Dimethylacetamide, DMA*	
$2M^{IV}X_4{\cdot}5DMA$	X = Cl, M^{IV} = Pa, U, Np, Pu; X = Br, M^{IV} = Pa, U; X = NO_3, M^{IV} = Th, U, Np

Table 35 *(continued)*

N,N-*Dimethylacetamide, DMA* (*continued*)	
$M^{IV}X_4 \cdot y$DMA	X = Cl; M^{IV} = Th, y = 1, 2, 3, 4; M^{IV} = Pa, y = 3 X = Br; M^{IV} = Th, y = 2, 4 (acetone solvate), 5; M^{IV} = Pa, y = 5; M^{IV} = U, y = 4 (acetone solvate), 5 X = I; M^{IV} = Th, y = 6; M^{IV} = U, y = 4 X = NCS; M^{IV} = Th, U, y = 4 X = NO_3; M^{IV} = Th, y = 3 (+3H_2O) X = ClO_4; M^{IV} = Th, y = 6 (also +3H_2O), 8; M^{IV} = U, y = 6 (also +3H_2O) X = Cl_3CCO_2, M^{IV} = Th, U, y = 3 X = F_3CCO_2, M^{IV} = Th, y = 3 X = Cl_2CHCO_2, M^{IV} = Th, y = 1 X = $ClCH_2CO_2$, M^{IV} = Th, y = 1; M^{IV} = U, y = 0.5
$UI_2Cl_2 \cdot x$DMA	x = 3, 5
$UI_3Cl \cdot x$DMA	x = 3, 5
$UCl_2(HBPz_3)_2 \cdot$DMA	Pz = $C_3H_3N_2$
$ThCl_3(HBPz_3) \cdot$DMA·THF	
$U(H_2BPz_2)_4 \cdot$DMA	
$Th(HBPz_3)_4 \cdot$2DMA	
$\{U(RCO_2)_3(DMA)\}_2O$	R = Cl_2CH, CF_3
Other N,N-*Dialkylamides*	
$M^{IV}X_4 \cdot 2RCONR'_2$	R = Me; R' = Pr^i, X = NO_3, M^{IV} = Pu[d] R' = Ph, X = Cl, M^{IV} = U;[a] X = NO_3, M^{IV} = Th R' = cyclo-C_6H_{11}, X = Cl, M^{IV} = Th, U[a] R = Et; R' = Me,[b] Et,[b] Pr^i,[a] X = Cl, M^{IV} = U R = Pr^i; R' = Me, X = Cl, Br, M^{IV} = U R' = Pr^i,[a] X = Cl, M^{IV} = Th, U; X = Br, M^{IV} = U R = Bu^t; R' = Me, Pr^i,[a] X = Cl, M^{IV} = Th, U; X = Br, M^{IV} = U R = Me_2CHCH_2; R' = Me, X = Cl, Br, M^{IV} = U
$M^{IV}X_4 \cdot 2.5RCONR'_2$	R = Me; R' = Et, X = NO_3, M^{IV} = Th, U R' = Pr^i, X = NO_3, M^{IV} = Th[d], U[d] R' = Ph, X = Cl, M^{IV} = U[a] R = Et; R' = Me, X = Cl, M^{IV} = U;[b] X = NO_3, M^{IV} = Th, U R' = Et, X = Cl, M^{IV} = U;[b] X = NO_3, M^{IV} = Th R = Pr^i; R' = Me, X = Cl, M^{IV} = Th R = Bu^t; R' = Me, X = NO_3, M^{IV} = Th R = Ph_2CH; R' = Me, X = Cl, Br, M^{IV} = U
$M^{IV}X_4 \cdot 3RCONR'_2$	R = Me; R' = cyclo-C_6H_{11}, Pr^i, X = Cl, M^{IV} = Th[a] R = Et; R' = Me, Et, Pr^i,[a] X = Cl, M^{IV} = Th R = Pr^i; R' = Me, X = Cl, M^{IV} = Th; X = NCS, M^{IV} = U[c] R' = Pr^i, X = Cl, M^{IV} = Th[a] R = Bu^t; R' = Me, X = Cl, Br, NCS, M^{IV} = Th; X = NCS (also MeCOMe solvate[c]), M^{IV} = U R = Ph; R' = Me, X = Cl, Br, M^{IV} = U
$M^{IV}(NCS)_4 \cdot 4RCONR'_2$	R = Et; R' = Me, M^{IV} = Th, U[c] R' = Et, M^{IV} = Th R = Pr^i; R' = Me, M^{IV} = Th, U[c] R = Bu^t; R' = Me, M^{IV} = Th
$[M^{IV}(RCO_2)_3L]_2O$	L = $MeCONPh_2$, R = $CHCl_2$, M^{IV} = Th, U L = Bu^tCONMe_2, R = F_3C, Cl_3C, Cl_2CH, M^{IV} = Th, U
$UCl_4 \cdot$3MeCON(Me)(Ph)	
$ThCl_4 \cdot 2MeCONPh_2 \cdot$THF[a]	
$UCl_4 \cdot$2PhCON(Me)(Ph)·MeCOMe	
$Th(Cl_2CHCO_2)_4 \cdot MeCONPh_2$	
$Th(CF_3COCHCOC_4H_3S)_4 \cdot MeCON(Bu^n)(Ph)$	

[a] A. G. M. Al-Daher and K. W. Bagnall, *J. Less-Common Met.*, 1984, **87**, 343.
[b] K. W. Bagnall, R. L. Beddoes, O. S. Mills and Li Xing-fu, *J. Chem. Soc., Dalton Trans.*, 1982, 1361.
[c] K. W. Bagnall, Li Xing-fu, G. Bombieri and F. Benetollo, *J. Chem. Soc., Dalton Trans.*, 1982, 19.
[d] K. W. Bagnall, O. Velasquez Lopez and D. Brown, *J. Inorg. Nucl. Chem.*, 1976, **38**, 1997.

DMA) are, however, consistent with 1:1 electrolyte behaviour and $ThI_4 \cdot$6DMA behaves as a 1:2 electrolyte in nitromethane, presumably ionizing as $[ThI_2(DMA)_6]^{2+}I_2$. The conductivity of $Th(ClO_4)_4 \cdot$6DMA in nitromethane is low for a 1:4 electrolyte, and the $[Th(DMA)_6]^{4+}$ ion would be very much undercrowded in cone-angle terms (p. 1130), so that one or more ClO_4 groups may be covalently bound to the thorium atom. The IR spectrum of $Th(ClO_4)_4 \cdot$8DMA, however, is consistent with presence of ionic perchlorate only, and the $[Th(DMA)_8]^{4+}$ cation would be well within the stable region insofar as steric crowding is concerned (Σ caf = 0.80, p. 1130).

The ^{1}H NMR spectrum of $ThCl_3(HBPz_3)\cdot DMA\cdot THF$ ($Pz = C_3H_3N_2$) indicates that the $HBPz_3$ ligand is bidentate, but it uncertain whether the THF molecule is bonded to the thorium atom or not.

A selection of a very wide variety of complexes with other *N,N*-dialkylamides is included in Table 35. The UV–visible spectra and magnetic susceptibility results for many of the uranium(IV) complexes of composition UX_4L_2 are consistent with octahedral geometry and all of the bis complexes MX_4L_2 probably adopt this geometry. The complex of composition $2UCl_4\cdot 5EtCONEt_2$ is ionic, $[UCl_3(EtCONEt_2)_4]^+[UCl_5(EtCONEt_2)]^-$, and the cation is pentagonal bipyramidal with the Cl—U—Cl axis normal to the pentagon.[8] The other known complexes of composition $2MX_4\cdot 5L$ are probably of the same structure. Pentagonal bipyramidal geometry has also been found for the neutral complex[9] $[U(NCS)_4(Me_2CHCONMe_2)_3]$ (Figure 5) and for $[ThCl_4(EtCONEt_2)_3]$.[111b] The other known tris complexes may have similar coordination geometries. In the complexes $M(NCS)_4\cdot 4RCONR'_2$ the metal centres are presumably eight-coordinate; the coordination geometry of $[Th(NCS)_4(EtCONMe_2)_4]$ is a distorted square antiprism.[111b]

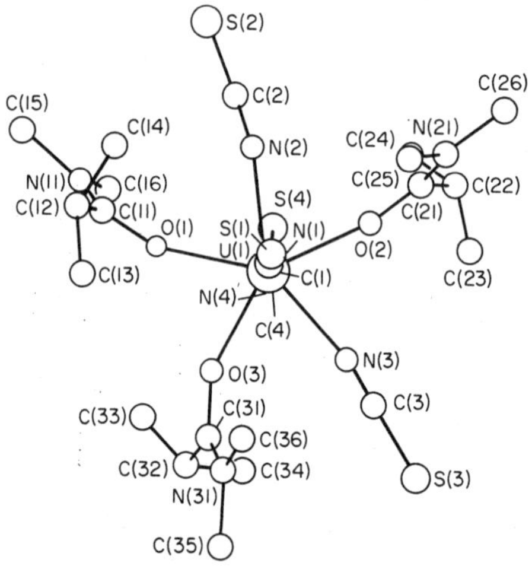

Figure 5 Perspective view of the molecular structure of $[U(NCS)_4(Me_2CHCONMe_2)_3]$ viewed down the SCN—U—NCS axis[9]

A number of complexes with ligands related to amides, lactams and antipyrine (2,3-dimethyl-1-phenylpyrazol-5-one), as well as complexes with dicarboxylic acid amides are also known (Table 36), but little structural information is available for them. The diantipyrylmethane complex, $UI_4\cdot 4L\cdot 2H_2O$, is probably ionic, $[UL_4]I_4\cdot 2H_2O$.

(*b*) *Carboxylic acid hydrazides.* Ligands of this type can be regarded as analogous to amides; the few known complexes are listed in Table 37.

(*c*) *Ureas.* A selection of the known complexes with urea and substituted ureas is given in Table 38. The hydrated compounds have been obtained by fusing together the stoichiometric quantities of the components, or by simply grinding them together and finally drying the product to constant mass in a desiccator; alternatively, some have been prepared from aqueous methanol solutions of the components. The anhydrous complexes have been prepared from solutions of the components in non-aqueous solvents, such as ethyl acetate. The complexes with $UICl_3$, UI_3Cl and $UIBr_3$ were obtained from mixtures of the component tetrahalides ($UI_4 + UCl_4$ or UBr_4) and urea in the same way as the complexes with the tetrahalides. The product of composition $ThBr_4\cdot 11.6$urea is probably a mixture of $ThBr_4\cdot 8$urea and free urea; this product, and the majority of the other urea complexes, require further investigation.

The IR spectra of all the complexes indicate that the ligands are bonded to the metal atom *via* the carbonyl oxygen atom. From the IR spectra of $Th(NO_3)_4\cdot 2$urea$\cdot$(1 or 6)H_2O it is clear that ionic nitrate is absent, but the spectra of most of the other urea complexes of thorium tetranitrate show that both covalent and ionic nitrate are present.

The systems with substituted ureas are less complicated than those with urea, and most of these compounds have been obtained from non-aqueous solutions of the components.

Table 36 Complexes of Actinide(IV) Compounds with Lactams, Antipyrine (2,3-Dimethyl-1-phenylpyrazol-5-one, ATP) and its Derivatives, and Dicarboxylic Acid Amides

Lactams $HN(CH_2)_xCO$	
$UCl_4 \cdot 4L$	$x = 4, 5, 6$
$UBr_4 \cdot 6L$	$x = 4, 5$
$[UBr_2L_6](ClO_4)_2$	$x = 4$
$M^{IV}I_4 \cdot 8L$	$M^{IV} = Th, U; x = 4, 5, 6$
$2UCl_4 \cdot 5L$	$L = MeN(CH_2)_5CO$
Antipyrine, ATP	
$ThX_4 \cdot y ATP$	$X = Cl, Br, y = 1, 3, 5, 6; X = I,$[a] $y = 6; X = NCS, y = 2, 4; X = NO_3, y = 2, 2.5; X = ClO_4, y = 6, 7$
$ThX_4 \cdot y(4\text{-}H_2NATP)$	$X = Cl, Br, y = 2, 3; X = I,$[a] $y = 3; X = NCS, y = 1, 2; X = NO_3, y = 2, 4; X = ClO_4, y = 4$
$[M^{IV}L_6]_3[Cr(NCS)_6]_4$	$M^{IV} = Th, U, L = ATP, 4\text{-}Me_2N(ATP)$ (pyrimidone)
$M^{IV}X_4 \cdot yL$	$L = 4,4'$-methylenediantipyrine (= diantipyrylmethane)
	$M^{IV} = Th, X = I, NCS, NO_3, y = 3$
	$M^{IV} = U, X = Cl, Br, y = 1; X = I, y = 4\ (+2H_2O)$
Dicarboxylic acid amides, $Me_2NCOXCONMe_2$	
$UCl_4 \cdot L$	$X = CMe_2$
$M^{IV}Cl_4 \cdot 1.5L$	$M^{IV} = Th, U, X = (CH_2)_3, CH_2CMe_2CH_2$
	$M^{IV} = Th, X = CMe_2; M^{IV} = U, X = CH_2$
$ThCl_4 \cdot 2L$	$X = CH_2$

[a] R. K. Agarwal, A. K. Srivastava and T. N. Srivastava, *Proc. Natl. Acad. Sci., India, Sect. A*, 1981, **51**, 79; TGA and DTA results included for these and other ATP and 4-H_2NATP complexes.

Table 37 Actinide(IV) Complexes with Carboxylic Acid Hydrazides, $RCONHN{=}CHR^1$

$ThCl_4 \cdot 2L$	$R = Ph, R^1 = CH{=}CHPh, (4\text{-}MeO)(3\text{-}HO)C_6H_3$
$Th(NO_3)_4 \cdot 2L$	$R = Ph, R^1 = 2\text{-}HOC_6H_4$
	$R = 2\text{-}HOC_6H_4, R^1 = 2\text{-}HOC_6H_4, (3\text{-}MeO)C_6H_4, (4\text{-}MeO)(3\text{-}HO)C_6H_3$

$ThBr_4 \cdot 3\{(Me_2N)_2CO\}$ behaves as a 1:1 electrolyte in methyl cyanide and the coordination geometry of $[Th(NCS)_4\{(Me_2N)_2CO\}_4]$ is a slightly distorted dodecahedron.[112] The bis complexes, $M^{IV}X_4 \cdot 2L$, are probably octahedral, but otherwise structural information is lacking for complexes with these ligands.

(*d*) N-*Oxides*. Most of the known complexes are thorium(IV) compounds (Table 39) because of the oxidizing nature of the ligand. Conductivity data are available for most of these complexes; the complexes with $Th(NCS)_4$ are non-electrolytes in nitrobenzene, as are $ThBr_4 \cdot 2(2,6\text{-}Me_2pyNO)$ (in methyl cyanide) and the 2,2′-bipyridyl *N,N*′-dioxide complexes $Th(NO_3)_4 \cdot L$ (in nitrobenzene or DMSO) and $ThX_4 \cdot 3L$ (X = Cl, Br; in DMSO). However, $ThI_4 \cdot 4(2,6\text{-}Me_2pyNO)$ and $ThI_4 \cdot 4(\text{bipy-}N,N'\text{-}O_2)$ behave as $[ThI_2L_4]I_2$ in methyl cyanide or nitrobenzene and DMSO or nitrobenzene respectively. $Th(NO_3)_4 \cdot 4(2,6\text{-}Me_2pyNO)$ behaves as $[Th(NO_3)_2L_4](NO_3)_2$ in methyl cyanide, consistent with its IR spectrum, which indicates the presence of both ionic and covalent nitrate. The conductivities of the complexes with thorium(IV) perchlorate are consistent with 1:4 electrolyte behaviour.

The IR spectrum of $Th(NO_3)_4 \cdot 8pyNO$ shows ionic nitrate only, and the complex is presumably $[Th(pyNO)_8](NO_3)_4$, whereas the spectrum of the quinoline *N*-oxide complex, $Th(NO_3)_4 \cdot 3L$, indicates bidentate nitrate only, and thorium is presumably 11-coordinate in this compound. The IR spectra of $Th(NO_3)_4 \cdot 2(\text{bipy-NO})$ and $Th(NO_3)_4 \cdot (\text{bipy-}N,N'\text{-}O_2)$ also indicate that only bidentate nitrate groups are present, so that in these two complexes thorium is evidently 10-coordinate.

The complex with nitrosyl chloride, $ThCl_4 \cdot 2NOCl$, is a nitrosonium salt, $(NO)_2[ThCl_6]$.

(*e*) P-*Oxides*. Because of the importance of phosphate esters in the separation of uranium from other actinides by solvent extraction, ligands of this type, as well as phosphine oxides and phosphinate esters (Table 40), have attracted a great deal of research interest and the structures of a number of complexes with these ligands have been reported.

In the structure of $[UCl_4\{(Me_2N)_3PO\}_2]$, the coordination is *trans* octahedral,[113] as also found in the structure[114] of $[UBr_4(Ph_3PO)_2]$, whereas in $[UCl_4(Ph_3PO)_2]$ the coordination

Table 38 Complexes of Actinide(IV) Compounds with Ureas, $L = R^1R^2NCONR^3R^4$

$R^1 = R^2 = R^3 = R^4 = H$	
$M^{IV}X_4 \cdot yL \cdot zH_2O$	$M^{IV} = Th$, $X = Cl$, $y = 4$ and $z = 3$ or 4; $y = 5$, $z = 6$; $y = 6$, $z = 2$; $y = 8$, $z = 0$ $X = Br$, $y = 2$, $z = 6$; $y = 4$, $z = 4$; $y = 6$, $z = 2$; $y = 8$, 11.6, $z = 0$ $X = I$, $y = 4$, $z = 4$; $y = 6$, $z = 2$; $y = 8$, $z = 0$ $X = NCS$, $y = 4$, $z = 4$ $X = NO_3$, $y = 2$, $z = 1$, 2 or 6; $y = 3$, $z = 1$; $y = 4$, $z = 4$; $y = 5$, $z = 3$; $y = 6$, $z = 0$, 2 or 4; $y = 7$, $z = 2.5$; $y = 8$, 10, $z = 0$; $y = 11$, $z = 2.5$ $X = HCO_2$, $y = 1.5$, $z = 0$ $M^{IV} = U$, $X = Cl$, $y = 3$, $z = 0$ or 1; $y = 4$, $z = 0$ or 3; $y = 7$, $z = 0$ or 3 $X = Br$, $y = 2$, $z = 0$, 7 or 9; $y = 3$, $z = 0$, 5 or 8; $y = 6$, $z = 0$ or 2; $y = 8$, $z = 0$ $X = I$, $y = 5$, $z = 2$; $y = 8$, $z = 0$
$UICl_3 \cdot 8L$	
$UI_3Cl \cdot 5L \cdot 2H_2O$	
$UIBr_3 \cdot 6L \cdot 3H_2O$	
$M^{IV}(SO_4)_2 \cdot xL \cdot yH_2O$	$M^{IV} = Th$, $x = 4$, $y = 0$ or 1; $x = 5$, $y = 3$; $x = 6$, 8, $y = 0$ $M^{IV} = U$, $x = 4$, $y = 0$ or 4
$M^{IV}(C_2O_4)_2 \cdot xL \cdot yH_2O$	$M^{IV} = Th$, $x = 2$, $y = 0$ or 2; $x = 4$, $y = 0$; $x = 5$, $y = 1$ $M^{IV} = U$, $x = 2$, $y = 0$; $x = 3$, $y = 0$ or 1
$ThX_2 \cdot 4L$	$X = C_5H_3N(CO_2)_2$, pyridine-2,6-dicarboxylate
$M^{II}Th(NO_3)_6 \cdot xL \cdot yH_2O$	$M^{II} = Mg$, $x = 6$, $y = 4$; $M^{II} = Ni$, $x = 5$, $y = 3$
$R^1 = R^3 = H$, $R^2 = R^4 = Me$	
$M^{IV}Cl_4 \cdot xL$	$M^{IV} = Th$, $x = 6$; $M^{IV} = U$, $x = 4$
$R^1 = H$, $R^2 = R^3 = R^4 = Me$	
$M^{IV}Cl_4 \cdot 4L$	$M^{IV} = Th$, U
$R^1 = R^2 = R^3 = R^4 = Me$	
$M^{IV}X_4 \cdot yL$	$M^{IV} = Th$, $X = Cl$, Br, $y = 3$; $X = NCS$, $y = 4$ $M^{IV} = U$, $X = Cl$, Br, $y = 2$
$ThCl_4 \cdot 2L \cdot MeNO_2$	
$R^1 = R^3 = Me$, $R^2 = R^4 = Ph$	
$M^{IV}X_4 \cdot yL$	$M^{IV} = Th$, $X = Cl$, $y = 3$ $M^{IV} = U$, $X = Cl$, Br, $y = 2$
$R^1 = R^2 = Me$, $R^3 = R^4 = Et$	
$UX_4 \cdot 2L$	$X = Cl$, Br
$R^1 = R^2 = R^3 = R^4 = Et$	
$UX_4 \cdot 2L$	$X = Cl$, Br

geometry is *cis* octahedral as a result of a ring–ring interaction between two Ph rings in the molecule, one from each of the two adjacent Ph_3PO molecules.[115] The thorium, protactinium, neptunium and plutonium complexes, $MCl_4 \cdot 2Ph_3PO$, are isostructural with this form of $[UCl_4(Ph_3PO)_2]$, and the complexes $MBr_4 \cdot 2Ph_3PO$ (M = Th, Pa, Np, Pu) are isostructural with the uranium(IV) bromide complex. A second crystal form of $[UCl_4(Ph_3PO)_2]$, isostructural with the bromide complex, has recently been reported.[114] The complex with trimethylphosphine oxide, $UCl_4 \cdot 6Me_3PO$, is ionic, $[UCl(Me_3PO)_6]^{3+}Cl_3$, and the coordination geometry of the seven-coordinate cation approximates to a singly capped octahedron.[116] The coordination geometry in $[U(NCS)_4(Me_3PO)_4]$,[117] $[U(NCS)_4\{(Me_2N)_3PO\}_4]$[118] and $[U(NCS)_4(O\{Me_2N)_2PO\}_2)_2]$[119] is square antiprismatic, whereas in $[ThCl_4(O\{(Me_2N)_2PO\}_2)_2]$[119] the geometry is dodecahedral.

$[Th(NO_3)_4(Ph_3PO)_2]$ is isostructural with the cerium(IV) analogue, in which the coordination geometry is best described as distorted *trans* octahedral, with the four bidentate nitrate groups each occupying one corner of the equatorial plane.[120] Several apparently non-stoichiometric compounds have been reported for complexes with the actinide(IV) nitrates. The complex of composition $Th(NO_3)_4 \cdot 2.67Me_3PO$ is ionic, $[Th(NO_3)_3(Me_3PO)_4]_2^+$ $[Th(NO_3)_6]^{2-}$, and the 10-coordinate geometry in the cation can be described as a 1:5:4 arrangement with a planar face (Figure 6).[121] The coordination geometry in the cation $[Th(NO_3)_3\{(Me_2N)_3PO\}_4]^+$ of the analogous complex $Th(NO_3)_4 \cdot 2.67(Me_2N)_3PO$ is described as a 1:6:3 arrangement.[122] $Th(NO_3)_4 \cdot 2.67Pr_3^nPO$ is probably similar in structure to the Me_3PO complex. The products of composition $Th(NO_3)_4 \cdot 2.33$ and $3.67Me_3PO$ appear to be mixtures. It has been suggested that $Th(NO_3)_4 \cdot 3Me_3PO$ may be ionic, $[Th(NO_3)_2(Me_3PO)_5]^{2+}[Th(NO_3)_5(Me_3PO)_2]_2^-$, but this requires confirmation. The complexes $Th(NO_3)_4 \cdot 4L$ ($L = Me_3PO$, Bu_3^nPO) behave as 1:1 electrolytes in nitromethane, and these may be of the form $[Th(NO_3)_3L_4]^+NO_3^-$. Their IR spectra confirm the presence of both ionic and covalent nitrate groups. Similarly,

Table 39 Complexes of Actinide(IV) Compounds with *N*-Oxides

L = pyridine *N*-oxides, $R^1R^2C_5H_3NO$	
$ThCl_4 \cdot 2L$	$R^1 = H$ and $R^2 = H$[a] (also $+2H_2O$), 2-, 3- or 4-Me;[a] R^1 = 2-Me, R^2 = 6-Me
$UCl_4 \cdot 2L$	$R^1 = R^2 = H$;[b] $R^1 = H$, R^2 = 2-, 3- or 4-CO_2H
$ThBr_4 \cdot 2L$	$R^1 = H$ and $R^2 = H$,[a] 2-, 3- or 4-Me[a]
$ThI_4 \cdot 4L$	$R^1 = H$ and $R^2 = H$,[a] 2-, 3- or 4-Me;[a] R^1 = 2-Me, R^2 = 6-Me (all $[ThI_2L_4]I_2$)[a]
$Th(NCS)_4 \cdot xL$	$R^1 = R^2 = H$, $x = 4$; $R^1 = H$, R^2 = 2-, 3- or 4-Me, $x = 4$; R^1 = 2-Me, R^2 = 6-Me, $x = 2$
$Th(NO_3)_4 \cdot xL$	$R^1 = R^2 = H$, $x = 2$ ($+MeCO_2Et$), 8; $R^1 = H$, R^2 = 2-Me, $x = 3$; R^1 = 2-Me, R^2 = 6-Me, x = 3, 4
$Th(ClO_4)_4 \cdot xL \cdot yH_2O$	$R^1 = H$; $R^2 = H$, x = 8, 9, $y = 0$; R^2 = 2-Me, 4-Me, $x = 8$, $y = 0$; R^2 = 3-Me, $x = 8$, y = 1 or 3; R^2 = 4-Cl, 4-NO_2, 4-MeO, $x = 8$, $y = 0$; R^1 = 2-Me, R^2 = 6-Me, x = 6 or 8, $y = 0$
$Th(Cl_3CCO_2)_4 \cdot 2L$	$R^1 = R^2 = H$
$Th\{(O_2CCH_2)_2O\}_2 \cdot 3L \cdot 3H_2O$	$R^1 = R^2 = H$
$Th(NO_3)_4 \cdot 3C_8H_{11}NO$	$C_8H_{11}NO$ = collidine
L = quinoline *N*-oxide	
$ThX_4 \cdot yL$	X = Cl, $y = 2$ ($+2H_2O$); X = NCS, $y = 4$; X = NO_3, $y = 3$; X = ClO_4, $y = 6$
$Th(OH)_2(NO_3)_2 \cdot 2L$	
L = isoquinoline *N*-oxide	
$Th(NCS)_4 \cdot 4L$	
L = 2,2′-bipyridyl *N*-oxide or 1,10-phenanthroline *N*-oxide	
$ThX_4 \cdot 2L$	X = Cl, Br, NCS, NO_3
$ThX_4 \cdot 3L$	X = I, ClO_4
L = 2,2′-bipyridyl *N,N*′-dioxide	
$ThX_4 \cdot L$	X = NCS, NO_3
$ThX_4 \cdot 3L$	X = Cl, Br
$ThX_4 \cdot 4L$	X = I, ClO_4
L = 1,10-phenanthroline *N,N*′-dioxide	
$ThX_4 \cdot 2L$	X = Cl, Br, NCS, NO_3, ClO_4
$ThI_4 \cdot 3L$	

[a] R. K. Agarwal and S. C. Rastogi, *Thermochim. Acta*, 1983, **63**, 363.
[b] A. D. Westland and M. T. H. Tarafdar, *Inorg. Chem.*, 1981, **20**, 3992.

Table 40 Complexes of Actinide(IV) Compounds with *P*-Oxides

$M^{IV}Cl_4 \cdot xR_3PO$	R = Me,[a] $x = 2$, M^{IV} = Th, U, Np; $x = 3$, M^{IV} = Th, U; $x = 6$, M^{IV} = Th, Pa, U, Np, Pu R = Et, $x = 2$, M^{IV} = Th, U R = Bu^n, x = 1.5($+6H_2O$), 2, 3.5, 4, 5, 8, M^{IV} = U R = Bu^nO, x = 2, 3, M^{IV} = U R = Ph, $x = 2$, M^{IV} = Th, Pa, U, Np, Pu; $x = 3$, M^{IV} = Th R = Me_2N, $x = 2$, M^{IV} = Th, Pa, U, Np, Pu $R_3 = (Me_2N)_2Ph$, $x = 2$, M^{IV} = U $R_3 = Et_2Ph$, $EtPh_2$, $x = 2$, M^{IV} = Th, U R = Cl, $x = 4$, M^{IV} = U
$ThCl_3(OH) \cdot 2Ph_3PO$	
$UCl_3(OH) \cdot 3Bu^n_3PO$	
$UCl_2(acac)_2 \cdot 2Ph_3PO$	
$M^{IV}Br_4 \cdot xR_3PO$	R = Me, $x = 2$, M^{IV} = U; $x = 6$, M^{IV} = Th, U R = Et, Bu^n, $x = 2$, M^{IV} = U R = Ph, $x = 2$, M^{IV} = Th, Pa, U, Np, Pu; $x = 3$, M^{IV} = Th R = Me_2N, $x = 2$, M^{IV} = Th, Pa, U, Np, Pu; $x = 3$, M^{IV} = Th $R_3 = (Me_2N)_2Ph$, $x = 2$, M^{IV} = U
$ThBr_3(OEt) \cdot xPh_3PO$	x = 2, 3
$M^{IV}(NCS)_4 \cdot xR_3PO$	R = Me, $x = 4$, M^{IV} = U, Np, Pu; $x = 6$, M^{IV} = Th R = Ph, $x = 4$, M^{IV} = Th, U, Np R = Me_2N, $x = 4$, M^{IV} = Th, U, Np, Pu
$U(NCSe)_4 \cdot 4R_3PO$[c]	R = Bu^n, Me_2N

Table 40 (*continued*)

$M^{IV}(NO_3)_4 \cdot xR_3PO$	R = Me, x = 2.33, 2.67,[b] M^{IV} = Th; x = 3, 4, 5, M^{IV} = Th,[b] U;
	x = 3, M^{IV} = Np[b]
	R = MeO, x = 3, 4, M^{IV} = Th
	R = Et, x = 2.67, M^{IV} = Th
	R = Pr^n, x = 2.67, M^{IV} = Th, U, Np[b]
	R = Bu^n, x = 2, M^{IV} = U; x = 4, M^{IV} = Th
	R = Bu^nO, x = 2, M^{IV} = U; x = 2, 2.33, M^{IV} = Th
	R = Bu^tO, x = 3, M^{IV} = Th
	R = i-C_5H_{11}, x = 2, M^{IV} = U
	R = n-C_8H_{17}, x = 2, 3, M^{IV} = Th
	R = Ph, x = 2, M^{IV} = Th, U, Np,[b] Pu
	R = Me_2N, x = 2, M^{IV} = Th,[b] U,[b] Np[b]
	x = 2.67, 3, M^{IV} = Th; x = 4, M^{IV} = Th, U
	R_3 = $(MeN)_2Ph$, x = 2, M^{IV} = U
	R_3 = $Bu^n(Bu^nO)_2P$, x = 2.21, 2.67, M^{IV} = Th
	R_3 = $(MeO)(PhO)_2$, x = 4, M^{IV} = Th
$[U(NO_3)_3\{(Me_2N)_3PO\}_4](BPh_4)$	
$[Th(NO_3)_2(Me_3PO)_5](BPh_4)_2$	
$ThO(NO_3)_2 \cdot 2(MeO)_3PO$	
$(Ph_4P)_2[Th(NO_3)_5(Me_3PO)_2]$	
$M^{IV}(ClO_4)_4 \cdot xR_3PO$	R = Me, x = 6, M^{IV} = U; R = Et, x = 4, M^{IV} = Th;
	R = Pr^n, x = 4–5, M^{IV} = U; R = Bu^n, x = 5 (+3H_2O), M^{IV} = Th;
	R = Ph, x = 4, 5, M^{IV} = Th; x = 5–6, M^{IV} = U
	R = Me_2N, x = 6, M^{IV} = Th, x = 5, 6, M^{IV} = U
$Th(CF_3COCHCOR)_4 \cdot L$	R = Me, L = Ph_3PO, $(Bu^nO)_3PO$, $(n\text{-}C_8H_{17})_3PO$
	R = CF_3, L = Ph_3PO, $(n\text{-}C_8H_{17})_3PO$, $(C_2H_2F_3O)_3PO$,
	$(C_3H_2F_5O)_3PO$, $(Bu^nO)_3PO$, $(PhO)_3PO$
	R = 2-C_4H_3S, L = Bu^n_3PO, $Bu^n(Bu^nO)_2PO$, $(n\text{-}C_8H_{17})_3PO$,
	Ph_3PO, $(Bu^nO)_3PO$
$U(CF_3COCHCOR)_4 \cdot 2L$	R = Me, CF_3, L = $(Bu^nO)_3PO$
$Th(Cl_3CCO_2)_4 \cdot 2Ph_3PO$	
L = $R_2P(O)(CH_2)_nP(O)R_2$	
$M^{IV}Cl_4 \cdot L$	R = Ph, n = 1, 2, M^{IV} = Th, U; n = 4, M^{IV} = U
$Th(NO_3)_4 \cdot xL$	R = Ph, n = 1, 2, x = 1.5 and n = 2, x = 2; R = $MeCH(Et)CH_2CH_2$,
	n = 1, x = 1.5
	R = Et, n = 2, x = 3
	R = *cyclo*-C_6H_{11}, n = 1, x = 1
L = $[(Me_2N)_2PO]_2O$	
$M^{IV}X_4 \cdot 1.5L$	X = Cl, M^{IV} = U, Np,[a] Pu;[a] X = NO_3, M^{IV} = Th, U, Np
$M^{IV}X_4 \cdot 2L$	X = Cl, M^{IV} = Th; X = NCS, M^{IV} = Th, U, Np
$Th(NO_3)_4 \cdot 2.5L$	
$M^{IV}(ClO_4)_4 \cdot 4L$	M^{IV} = Th, U
L = $[(Me_2N)_2PO]_2NMe$	
$Th(ClO_4)_4 \cdot 4L$	
L = 1,2,5-$Ph_3(C_4H_2PO)$, 1,2,5-triphenylphosphole oxide	
$UCl_4 \cdot 2L$	
L = $(EtO)_2P(O)CH_2C(O)NEt_2$	
$[Th(NO_3)_4L_2]$	

[a] Z. M. S. Al-Kazzaz, K. W. Bagnall and D. Brown, *J. Inorg. Nucl. Chem.*, 1973, **35**, 1493.
[b] K. W. Bagnall and M. W. Wakerley, *J. Chem. Soc., Dalton Trans.*, 1974, 889.
[c] V. V. Skopenko, Yu. L. Zub, V. N. Yankovich, R. N. Shchelokov, A. V. Rotov and G. T. Bolotova, *Ukr. Khim. Zh.* (*Russ. Ed.*), 1984, **50**, 1011 (*Chem. Abstr.*, 1985, **102**, 55 069).

$Th(NO_3)_4 \cdot 5Me_3PO$ may be $[Th(NO_3)_2(Me_3PO)_5]^{2+}(NO_3)_2$. The geometry of 12-coordinate thorium in the anion of $(Ph_4P)[Th(NO_3)_5(Me_3PO)_2]$ is distorted icosahedral[121] and the same geometry has been reported for the 12-coordinate thorium atom in the diethyl-*N,N'*-(diethylcarbamyl) methylene phosphonate complex, $[Th(NO_3)_4\{(EtO)_2P(O)CH_2C(O)NEt_2\}_2]$, in which two carbonyl, two phosphoryl and eight nitrate oxygen atoms are bonded to the thorium atom.[123]

The geometry of the nine-coordinate thorium atom in the thenoyltrifluoroacetonate complex, $[Th\{CF_3COCHCO\text{-}(2\text{-}C_4H_3S)\}_4\{(n\text{-}C_8H_{17})_3PO\}]$, is a 4,4,4-tricapped trigonal prism (Figure 7).[124]

The complexes $MBr_4 \cdot 2(Me_2N)_3PO$ (M = Th, Pa, U, Np, Pu) are isostructural, and the tris complex, $ThBr_4 \cdot 3(Me_2N)_3PO$, behaves as a 1:1 electrolyte in nitromethane.[125]

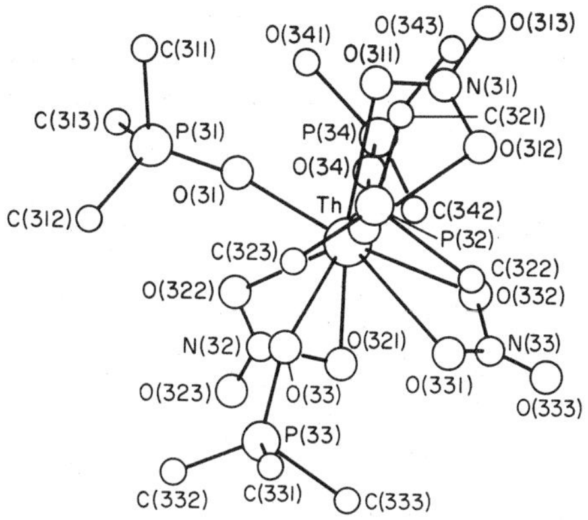

Bond lengths (Å)	
Th—O(31)	2.24(3)
Th—O(32)	2.38(3)
Th—O(33)	2.25(3)
Th—O(34)	2.25(3)
Th—O(311)	2.71(2)
Th—O(312)	2.70(3)
Th—O(321)	2.64(3)
Th—O(322)	2.64(4)
Th—O(331)	2.61(3)
Th—O(332)	2.67(5)

Bond angles (°)	
O(31)—Th—O(32)	79.9(11)
O(31)—Th—O(33)	83.9(11)
O(31)—Th—O(34)	95.6(10)
O(31)—Th—O(311)	68.2(9)
O(31)—Th—O(312)	113.3(10)
O(31)—Th—O(321)	119.5(10)
O(31)—Th—O(322)	72.1(12)
O(31)—Th—O(331)	145.0(10)
O(31)—Th—O(332)	161.9(13)
O(32)—Th—O(33)	72.6(12)
O(32)—Th—O(34)	140.7(11)
O(32)—Th—O(311)	70.6(9)
O(32)—Th—O(312)	77.7(11)
O(32)—Th—O(321)	136.2(10)
O(32)—Th—O(322)	136.5(13)
O(32)—Th—O(331)	71.4(11)
O(32)—Th—O(332)	114.3(14)
O(33)—Th—O(34)	146.3(12)

Bond angles (°)	
O(33)—Th—O(311)	136.8(10)
O(33)—Th—O(312)	142.3(11)
O(33)—Th—O(321)	71.4(10)
O(33)—Th—O(322)	72.0(13)
O(33)—Th—O(332)	110.4(13)
O(34)—Th—O(311)	71.6(9)
O(34)—Th—O(312)	68.3(10)
O(34)—Th—O(321)	79.9(10)
O(34)—Th—O(322)	75.9(12)
O(34)—Th—O(331)	119.3(11)
O(34)—Th—O(332)	66.3(13)
O(311)—Th—O(312)	45.2(8)
O(311)—Th—O(321)	151.2(8)
O(311)—Th—O(322)	124.7(10)
O(311)—Th—O(331)	118.0(9)
O(311)—Th—O(332)	104.9(12)
O(312)—Th—O(312)	120.0(9)

Figure 6 The structure of the $[Th(NO_3)_3(PMe_3O)_4]^+$ cation[121]

Some interatomic distances (Å)	
Th—O(11)	2.44(2)
Th—O(12)	2.44(2)
Th—O(9)	2.30(2)
Th—O(42)	2.46(2)
Th—O(31)	2.45(2)
Th—O(21)	2.49(2)
Th—O(22)	2.42(2)
Th—O(41)	2.37(2)
Th—O(32)	2.42(2)

Some bond angles (°)	
O(11)—Th—O(12)	69.5
O(21)—Th—O(22)	68.4
O(31)—Th—O(32)	67.7
O(41)—Th—O(42)	66.3
Th—O(11)—C(11)	137.1
Th—O(12)—C(13)	140.5
Th—O(21)—C(21)	132.8
Th—O(22)—C(23)	137.1
Th—O(31)—C(31)	135.2
Th—O(32)—C(32)	139.2
Th—O(41)—C(41)	138.8
Th—O(42)—C(43)	142.6
Th—O(9)—P	173.1

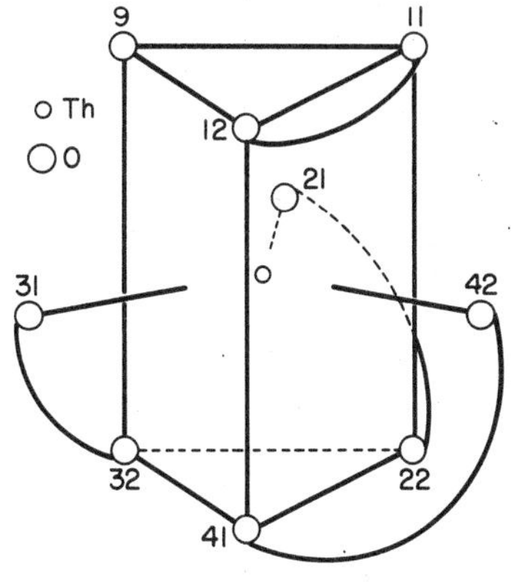

Figure 7 The coordination geometry of $[Th(CF_3COCHCO(2\text{-}C_4H_3S))_4\{(n\text{-}C_8H_{17})_3PO\}]$[124]

The complexes with thorium(IV) perchlorate, $Th(ClO_4)_4{\cdot}4L$ behave as 1:4 electrolytes in nitromethane when $L=[(Me_2N)_2PO]_2O$ or $[(Me_2N)_2PO]_2NMe$, but with $L=Ph_3PO$ the conductance is consistent with 1:3 electrolyte behaviour in nitromethane, perhaps as $[Th(Ph_3PO)_4(O_2ClO_2)]^{3+}(ClO_4)_3$. The IR spectrum of $Th(ClO_4)_4{\cdot}5Bu^n_3PO$ shows that both ionic and coordinated perchlorate groups are present, and this compound may have a similar ionic structure.

(*f*) As-*Oxides*. The few known complexes with arsine oxides (Table 41) appear to be very similar in composition and behaviour to those formed by the analogous phosphine oxides. The only reported structure is that of the complex [$UCl_4(Et_3AsO)_2$], in which the coordination geometry is *trans* octahedral.[126]

Table 41 Complexes of Actinide(IV) Compounds with Arsine Oxides

$M^{IV}Cl_4 \cdot xR_3AsO$	R = Me, $x = 2$, M^{IV} = U; $x = 6$, M^{IV} = Th, U
	R = Et, $x = 2, 4$, M^{IV} = U; $x = 6$, M^{IV} = Th
	R = Ph, $x = 2$, M^{IV} = U; $x = 4$, M^{IV} = Th
$UBr_4 \cdot xR_3AsO$	R = Me, Et, $x = 2, 6$
	R = Ph, $x = 2, 4$
$UI_4 \cdot 6R_3AsO$	R = Me, Et
$Th(ClO_4)_4 \cdot 5Ph_3AsO$	
$ThL_2 \cdot Ph_3AsO \cdot xH_2O$	$H_2L = C_5H_3N\text{-}2,6\text{-}(CO_2H)_2$; $x = 0, 3$
$Pu(NO_3)_4 \cdot 2Ph_3AsO$	
$L = Ph_2As(O)(CH_2)_nAs(O)Ph_2$	
$UCl_4 \cdot xL$	$x = 1, 2$ for $n = 2$; $x = 1$ for $n = 4$

(*g*) S-*Oxides*. The complexes with sulfoxides are usually prepared from the parent compound and the ligand in dry, non-aqueous solvents; the Ph_2SO complexes of $NpCl_4$ and $PuCl_4$ can, however, be obtained by treating a hot 6 M HCl solution of Cs_2MCl_6 with the ligand. The complexes with the nitrates have been prepared either by treating the salts $Cs_2M(NO_3)_6$ with the ligand or by treating the corresponding complex of the metal tetrachloride with silver nitrate in methyl cyanide (*e.g.* refs. 127, 128).

In addition to the large number of complexes with these ligands listed in Table 42, products of composition $ThX_4 \cdot yMe_2SO \cdot zH_2O$ (X = Cl, Br, I, with values of y ranging up to 12) have been recorded, prepared by treating the hydrated parent thorium compound either with the stoichiometric amount of Me_2SO or, in some instances, with an excess of the ligand. These products require further investigation.

$ThCl_4 \cdot 3Me_2SO$ is isostructural with $UCl_4 \cdot 3Me_2SO$, which is an ionic compound, [$UCl_2(Me_2SO)_6$][UCl_6]. The coordination geometry of the uranium atom in the cation is a distorted dodecahedron.[129] The complex of composition $ThCl_4 \cdot 5Me_2SO$ may also be ionic, possibly [$ThCl_3(DMSO)_5$]Cl, but this requires verification. The IR spectra of the complexes $M^{IV}Cl_4 \cdot 7Me_2SO$ (M^{IV} = Th, U, Np, Pu) indicate the presence of both bonded and free ligand, but the structures are not known. Three crystal modifications of $Th(NO_3)_4 \cdot 3Me_2SO$ have been reported, one of which (designated the γ form) is isostructural with the analogous complexes of uranium(IV), neptunium(IV) and plutonium(IV).

The coordination geometry in the Me_2SO complex of thorium 8-hydroxyquinolinate, $Th(C_9H_6NO)_4 \cdot 2Me_2SO$, is a monocapped square antiprism in which only one Me_2SO molecule is coordinated to the metal.[130]

In the structure of [$ThCl_4(Ph_2SO)_4$] the Cl atoms occupy the B sites and the sulfoxide O atoms the A sites in the dodecahedral complex.[131]

(*ix*) *Hydroxamates, cupferron and related ligands*

The known complexes, $M^{IV}L_4$ (Table 43), precipitate from aqueous solutions of actinide(IV) compounds on addition of the ligand. Somewhat surprisingly, one hydroxamic acid, (PhCO)NHOH, behaves as a neutral ligand, and the complex $Th(NO_3)_4 \cdot 2$(PhCO)NHOH precipitates from aqueous solution at pH 7. Structural information is available for [$Th\{(Bu^tCO)(Pr^i)NO\}_4$] and [$Th\{(t\text{-}C_5H_{11}CO)(Pr^i)NO\}_4$]; the coordination geometry in the former is a distorted cube, whereas in the latter the geometry is an *mmmm* dodecahedron.[132]

40.2.2.5 Sulfur ligands

(*i*) *Sulfides and thiols*

(*a*) *Sulfides*. Disulfides, MS_2, are known for M = Th, U and Pu, but the plutonium compound is a plutonium(III) polysulfide (p. 1135). The compounds M_2S_5 (M = Th, U, Np) and

Table 42 Complexes of Actinide(IV) Compounds with Sulfoxides

$M^{IV}Cl_4 \cdot xR_2SO$	R = Me; $x = 3$, M^{IV} = Th, Pa, U, Np, Pu; $x = 5$, M^{IV} = Th, Pa, U, Np; $x = 6$, M^{IV} = Th; $x = 7$, M^{IV} = Th, U, Np, Pu
	R = Et; $x = 2.5$, M^{IV} = Np, Pu; $x = 3$, M^{IV} = Th, U, Np; $x = 4$, M^{IV} = Th
	R = Pr^n; $x = 3$, M^{IV} = U
	R = Bu^n; $x = 2$, M^{IV} = Th; $x = 4$, M^{IV} = U
	R = Bu^i; $x = 2$, M^{IV} = U; $x = 3$, M^{IV} = Th
	R = Bu^t; $x = 2$, M^{IV} = U
	R = n-C_5H_{11}, n-C_6H_{13}, n-C_7H_{15}, n-C_8H_{17}, $x = 2$, M^{IV} = Th
	R = Ph, $x = 2$ (+2H_2O), M^{IV} = U; $x = 3$, M^{IV} = U, Np; $x = 4$, M^{IV} = Th, U, Np
	R = α-$C_{10}H_7$; $x = 3$, M^{IV} = Th, U, Np
$[UCl_3(Me_2SO)_5]ClO_4$	
$[UCl_2(Me_2SO)_6](ClO_4)_2$	
$M^{IV}Br_4 \cdot xR_2SO$	R = Me; $x = 6, 8$, M^{IV} = Th, U
	R = Et; $x = 5, 6$, M^{IV} = U
	R = Pr^n; $x = 7$, M^{IV} = U
	R = Bu^n; $x = 8$, M^{IV} = U
	R = Bu^i; $x = 4$, M^{IV} = U
	R = Bu^t; $x = 2$, M^{IV} = U
	R = Ph; $x = 4$, M^{IV} = Th, U
$ThI_4 \cdot 6Ph_2SO$	
$Th(NCS)_4 \cdot 4Ph_2SO$	
$M^{IV}(NO_3)_4 \cdot xR_2SO$	R = Me; $x = 3$, M^{IV} = Th, U, Np, Pu; $x = 6$, M^{IV} = Th, Np, Pu
	R = Et; $x = 3$, M^{IV} = Th, U, Np
	R = Bu^n, n-C_5H_{11}, n-C_6H_{13}, n-C_7H_{15}, n-C_8H_{17}, $x = 2$, M^{IV} = Th
	R = n-C_8H_{17}; $x = 3$, M^{IV} = Th
	R = $PhCH_2$; $x = 4$, M^{IV} = Th
	R = Ph; $x = 3, 4$, M^{IV} = Th, U, Np, Pu
	R = α-$C_{10}H_7$; $x = 3$, M^{IV} = Th, Np
$Th(ClO_4)_4 \cdot xR_2SO$	R = Me, $x = 6, 12$; R = Ph, $x = 6$
$Th(R^1COCHCOR^2)_4 \cdot Bu^n_2SO$	$R^1 = CF_3$, R^2 = Me, CF_3; $R^1 = R^2 = C_2F_5$
$ThX_4 \cdot yMe_2SO$	X = HCO_2, OSC_7H_5 (thiotroponate), $y = 1$; X = C_9H_6NO (8-hydroxyquinolinate), $y = 2$
$M^{IV}(trop)_4 \cdot Me_2SO$	Htrop = tropolone; M^{IV} = Th, U
$ThX_2 \cdot yMe_2SO$	X = SO_4, $y = 4$ with 3 or 9H_2O; X = C_2O_4, $y = 2$ with 1H_2O
L = thianthrene 5-oxide, $C_{12}H_8OS_2$	
$M^{IV}Cl_4 \cdot 4L$	M^{IV} = Th, U
L = tetrahydrothiophene *S*-oxide	
$ThX_4 \cdot yL$	X = Cl, Br, $y = 4$; X = I, $y = 8$; X = NCS, $y = 2$; X = NO_3, $y = 6$; X = ClO_4, $y = 10$

Table 43 Actinide(IV) Complexes with Hydroxamates, Cupferron and Related Ligands

Hydroxamates, HL = $(R^1CO)(R^2)NOH$	
$Th(NO_3)_4 \cdot 2HL$	R^1 = Ph, R^2 = H
ThL_4	$R^1 = Bu^t$, t-C_5H_{11}, $R^2 = Pr^i$; R^1 = Ph, 2-O_2NPh, R^2 = Ph; R^1 = 3-O_2NPh, R^2 = 3MeC_6H_4
UL_4	$R^1 = Bu^t$, $R^2 = Pr^i$
Cupferron, $NH_4L = NH_4[(ON)PhNO]$	
$M^{IV}L_4$	M^{IV} = Th, U, Pu
Neocupferron, $NH_4L = NH_4[(ON)(1\text{-}C_{10}H_7NO)]$	
$M^{IV}L_4$	M^{IV} = Th, U
3-Phenylcupferron, $NH_4L = NH_4[(ON)(C_6H_4Ph)NO]$	
ThL_4	
2-Phenyl-3-toluenesulfonylcupferron, $NH_4L = NH_4(C_{13}H_{11}N_2O_5S)$	
ThL_4	

MS_3 (M = U, Np) are actinide(IV) polysulfides; the geometry of 10-coordinate thorium in Th_2S_5, which is isomorphous with U_2S_5 and Np_2S_5, is derived from a square-based antiprism surmounted by a capped, approximately pentagonal array of S atoms.[133] Oxide sulfides, such as the compounds $M^{IV}OS$ (M^{IV} = Np, Pu), $Pu_2O_2S_3$ $[(Pu^{4+})_2(O^{2-})_2S^{2-}(S\text{—}S)^{2-}]$ and $M_4O_4S_3$ (M = Np, Pu; possibly $2MO_2 \cdot M_2S_3$) are also known.

(*b*) *Thiols*. The only recorded compounds appear to be $U(SR)_4$ with R = Et or Bu^n, prepared by reaction of $U(NEt_2)_4$ with the thiol in ether; both compounds are oxidized very easily and they inflame in air.

(ii) *Thioethers*

A complex with 1,2-dimethylthioethane, $UCl_4 \cdot 2(MeSCH_2CH_2SMe)$, has been recorded.

(iii) *Monothio- and dithio-carbamates, and thiohydroxamates*

(a) *Monothiocarbamates.* The thorium(IV) and uranium(IV) compounds, $M^{IV}(R_2NCOS)_4$ (M^{IV} = Th, R = Me, Et; M^{IV} = U, R = Et) have been obtained by insertion of COS into the M—N bond of the dialkylamides, $M(NR_2)_4$. The ^1H NMR spectrum of $Th(Et_2NCOS)_4$ indicates that rotation about the $(Et_2)N—C$ bond is slow on the NMR time scale, and the two ethyl groups are not equivalent.

(b) *Dithiocarbamates.* The compounds $M^{IV}(R_2NCS_2)_4$ (R = Me, M^{IV} = Th; R = Et, M^{IV} = Th, U, Np, Pu) have been prepared by reaction of $M^{IV}Cl_4$ or $M^I_2M^{IV}Cl_6$ with the alkali metal dithiocarbamate in non-aqueous solvents, and by CS_2 insertion into the M—N bond of the dialkylamides (M^{IV} = Th, U only). A much simpler route is by reaction of UCl_4 with the dialkylamine and CS_2 in toluene.

$U(Et_2NCS_2)_4$ is a monomer in benzene and is very sensitive to air oxidation in solution. It is isomorphous with $Th(Et_2NCS_2)_4$, in which the coordination geometry is close to an ideal dodecahedron.[134] The ^1H NMR spectra of $Th(R_2NCS_2)_4$ (R = Me, Et) indicate that the two alkyl groups are equivalent.

Mixed complexes with Schiff bases, $M^{IV}(Et_2NCS_2)_2(2\text{-}OC_6H_4CH{=}NCH_2CH_2N{=}CHC_6H_4O\text{-}2)$ (M^{IV} = Th, U), have also been recorded.

(c) *Thiohydroxamates.* The acids $R^1C(S)R^2NOH$ (R^1 = Ph or 4-MeC_6H_4, R^2 = Me) yield precipitates of composition $Th(R^1C(S)R^2NO)_4 \cdot H_2O$ when aqueous thorium tetrachloride is added to an aqueous solution of the sodium hydroxamate.[135]

(iv) *Dithiolates*

The maleonitriledithiolate complex, $(Ph_4As)_4[U(cis\text{-}S_2C_2(CN)_2)_4]$, precipitates when Ph_4AsCl is added to a solution containing UCl_4 and $Na_2\text{-}(cis\text{-}S_2C_2(CN)_2)$; it oxidizes on exposure to air.

Uranium tetraiodide reacts with toluene-3,4-dithiol in hexane under reflux to yield a product of composition $U_2I_4(C_7H_6S_2) \cdot C_7H_8S_2$, which may be a uranium(III) compound and is possibly a sulfur-bridged dimer. A product of somewhat similar composition, $U_2Cl_4(S_2C_4F_6)$, is obtained when UCl_4 is heated with bis(trifluoromethyl)-1,2-dithiete, whereas the reaction of UI_4 with $S_2C_4F_6$ in ethylcyclohexane yields a product of composition[136] $U_3F_3I(S_2C_4F_6)_2$. These last two products are probably dithiolates, but they require further investigation.

(v) *Thiocarboxamides*

The only recorded complex is the product of composition $UF_4 \cdot SC_3N_2H_2$ which results from the reaction of UF_6 with dithiooxamide.[137]

(vi) *Thioureas*

A few complexes of composition $ThCl_4 \cdot xL$ (L = $SC(NH_2)_2$, $x = 2, 8$; L = $SC(NH_2)(2\text{-}ClC_6H_4)NH$, $x = 4$ and L = $\overline{SCNHCH_2CH_2NH}$, $x = 8$) have been reported. They are 1:1 electrolytes in methanol and although there is no detectable shift in $\nu(CS)$ in their IR spectra as compared with the free ligand, the ligands do not appear to be bonded to the metal *via* the thiourea nitrogen atoms.

(vii) Dithiophosphinates

Thorium compounds of composition $Th(S_2PR_2)_4$ ($R = Me$, Et, Pr^i, Ph, C_6H_{11}, OEt and OPr^i) have been reported. In the structures of the compounds with $R = Me$ and $R = C_6H_{11}$, the thorium atoms are dodecahedrally surrounded by eight sulfur atoms.[138]

40.2.2.6 Selenium and tellurium ligands

(i) Selenides and tellurides

Diselenides, MSe_2, and ditellurides, MTe_2 ($M = Th, U, Pu$), are known but the oxidation state of the metal in these compounds is uncertain. Other selenides and tellurides [*e.g.* Th_2Se_5, USe_3, MTe_3 ($M = Th, U$) and UTe_5] are presumably analogous to the actinide(IV) polysulfides (p. 1135). A more detailed description of these systems is available.[139]

(ii) Diselenocarbamates

The thorium(IV) and uranium(IV) compounds, $M^{IV}(Et_2NCSe_2)_4$, have been prepared by insertion of CSe_2 into the M—N bond of the diethylamides, $M(NEt_2)_4$, and ($M = U$) by reaction of $Et_2NH_2(Et_2NCSe_2)$ with UCl_4 in methanol under nitrogen. The ^1H NMR spectrum of the thorium(IV) compound indicates that the two ethyl groups of the ligand are equivalent.

(iii) Phosphine selenides

The complex $ThCl_4{\cdot}2Ph_3PSe$ separates when $ThCl_4$ is heated with the stoichiometric quantity of Ph_3PSe in benzene under reflux. It is a non-electrolyte in DMF.

40.2.2.7 Halogens as ligands

The known halides, and a selection of the halogeno complexes derived from them, are listed in Table 44.

(i) Tetrahalides and oxohalides

These compounds are halogen bridged polymers in the solid state. All of the tetrafluorides, MF_4 ($M = Th–Pu$), have the same structure in which the M^{IV} centre is surrounded by eight F atoms in a slightly distorted square antiprismatic array. In the hydrated neptunium compound, $Np_3F_{12}{\cdot}H_2O$, the water molecule is not bonded to the metal atom, and three kinds of fluorine polyhedra surround the neptunium atoms. These are a tricapped trigonal prism with two vacant apices, a bicapped trigonal prism and a square antiprism.[140] The coordination geometry in solid $ThCl_4$ is dodecahedral[141] and the other tetrachlorides, MCl_4 ($M = Pa, U, Np$), have the same structure. Two of the tetrabromides, β-$ThBr_4$ and $PaBr_4$, are isostructural with the tetrachlorides, whereas the coordination geometry in ThI_4 is a distorted square antiprism.[142]

Several oxohalides, MOX_2, have been recorded (Table 44); the structure of $PaOCl_2$ consists of an infinite polymer in which the Pa atoms are seven-, eight- and nine-coordinate.[143]

(ii) Halogeno complexes

A detailed account of the very wide range of fluoro complexes is available.[144] The structures of the main types of fluoro complex are well established. In $LiUF_5$ the U atom is surrounded by nine F atoms in a tricapped trigonal prismatic array, with adjacent prisms sharing edges and corners.[145] The analogous Th, Pa, Np and Pu compounds adopt the same structure. The coordination geometry about the metal(IV) centres in the compounds $KM^{IV}_2F_9$ ($M^{IV} = Th–Pu$)[146] and in the structure of $NH_4U_3F_{13}$, which is isostructural with other $M^IM^{IV}_3F_{13}$ ($M^I = NH_4$, Rb; $M^{IV} = Th, U, Np$), is also a tricapped trigonal prism.[147]

Table 44 Actinide(IV) Halides and Halogeno Complexes

$M^{IV}X_4$	$X = F$, $M^{IV} = Th, Pa, U, Np, Pu$
	$X = Cl, Br$, $M^{IV} = Th, Pa, U, Np$
	$X = I$, $M^{IV} = Th, Pa, U$
$M^{IV}OX_2$	$X = F$, $M^{IV} = Th$
	$X = Cl, Br$, $M^{IV} = Th, Pa, U, Np$
	$X = I$, $M^{IV} = Th, Pa$
$M^IM^{IV}F_5$	$M^I = Li$, $M^{IV} = Th, Pa, U, Np, Pu$
	$M^I = Na, K, Rb$, $M^{IV} = U$
	$M^I = Cs$, $M^{IV} = Th, U, Pu$
	$M^I = NH_4$, $M^{IV} = Th, U, Np$,[a] Pu
$M^IM_2^{IV}F_9$	$M^I = Li$, $M^{IV} = Th$; $M^I = Na$, $M^{IV} = Th, U$
	$M^I = K$, $M^{IV} = Th, U, Np, Pu$; $M^I = Rb$, $M^{IV} = Th, U$;
	$M^I = Cs$, $M^{IV} = U$
$M^IM_3^{IV}F_{13}$	$M^I = Li$, $M^{IV} = Th, U, Np$,[b] Pu;[b]
	$M^I = K$, $M^{IV} = Th, U$; $M^I = Rb, Cs$, $M^{IV} = Th$;
	$M^I = NH_4$, $M^{IV} = U, Np$,[a] Pu
$M_2^IM^{IV}F_6$	$M^I = Na$, $M^{IV} = Th, U, Np, Pu$
	$M^I = K$, $M^{IV} = Th, U, Np$
	$M^I = Rb$, $M^{IV} = Th, Pa, U, Np, Pu$
	$M^I = Cs$, $M^{IV} = Th, U, Pu$
	$M^I = NH_4$, $M^{IV} = U, Np$,[a] Pu
	$M^I = Et_4N$, $M^{IV} = Pa, U$
$M^{II}M^{IV}F_6$	$M^{II} = Ca, Sr$, $M^{IV} = Th, U, Np, Pu$
	$M^{II} = Ba, Pb$, $M^{IV} = Th, U, Np$
	$M^{II} = Cd, Eu$, $M^{IV} = Th$
	$M^{II} = Co$, $M^{IV} = U, Np\ (+3H_2O)$
$M_3^IM^{IV}F_7$	$M^I = Li$, $M^{IV} = Th, U$; $M^I = Na, K$, $M^{IV} = Th, Pa, U$;
	$M^I = Rb, Cs$, $M^{IV} = Th, U$; $M^I = NH_4$, $M^{IV} = Th$
$M_4^IM^{IV}F_8$	$M^I = Li$, $M^{IV} = U, Np$,[b] Pu
	$M^I = NH_4$, $M^{IV} = Th, Pa, U, Np, Pu$
$M_7^IM_6^{IV}F_{31}$	$M^I = Na, K$, $M^{IV} = Th, Pa, U, Np, Pu$
	$M^I = Rb$, $M^{IV} = Th, Pa, U, Np, Pu$
	$M^I = NH_4$, $M^{IV} = U, Np, Pu$
$M_2^I[M^{IV}Cl_6]$	$M^{IV} = Th, U$; $M^I = Li–Cs, Me_4N, Et_4N$
	$M^{IV} = Pa, Np$; $M^I = Cs, Me_4N, Et_4N$
	$M^{IV} = Pu$; $M^I = Na–Cs, Me_4N, Et_4N$
$M^{II}[UCl_6]$	$M^{II} = Ca, Sr, Br$
$M_2^I[M^{IV}Br_6]$	$M^{IV} = Th, Pa$; $M^I = Me_4N, Et_4N$
	$M^{IV} = U$; $M^I = Na–Cs, Me_4N, Et_4N$
	$M^{IV} = Np$; $M^I = Cs, Et_4N$
	$M^{IV} = Pu$; $M^I = Et_4N$
$M_2^I[MI_6]$	$M^{IV} = Th$,[c] U;[c] $M^I = Et_4N, Me_3PhN$
	$M^{IV} = Pa$;[c] $M^I = Et_4N, Me_3PhN, Me_3PhAs$

[a] H. Abazli, J. Jové and M. Pagès, *C. R. Hebd. Seances Acad. Sci., Ser. C*, 1979, **288**, 157.
[b] J. Jové and A. Cousson, *Radiochim. Acta*, 1977, **24**, 73.
[c] D. Brown, P. Lidster, B. Whittaker and N. Edelstein, *Inorg. Chem.*, 1976, **15**, 511.

The structures of the compounds $Rb_2M^{IV}F_6$ ($M^{IV} = U, Np, Pu$) consist of chains of $M^{IV}F_8$ units with triangular faced dodecahedral geometry (K_2ZrF_6 type)[148] whereas in δ-Na_2UF_6 the coordination geometry in the UF_9 chains is a tricapped trigonal prism,[149] as in the structure of β_1-K_2UF_6.[150] The compounds $M^{II}M^{IV}F_6$ ($M^{II} = Ca, Sr, Ba, Pb, Cd, Eu$; Table 44) adopt the LaF_3 structure.[151] In contrast, the nine-coordination geometry about the Np atom in the $[NpF_8(H_2O)]$ chains in $CoNpF_6 \cdot 3H_2O$ consists of singly capped square antiprisms sharing two F atoms, with the water molecule occupying the cap position. The uranium compound is analogous.[152]

In the structures of compounds of the type $M_3^IUF_7$ the seven F atoms are statistically distributed over fluorite lattice sites.[153] The nine-coordinate thorium atom in $(NH_4)_4ThF_8$ is surrounded by a distorted tricapped trigonal prismatic array of fluorine atoms, with the prisms sharing edges to form chains, whereas the uranium(IV) compound contains discrete dodecahedrally coordinated $[UF_8]^{4-}$ ions. The protactinium(IV), neptunium(IV) and plutonium(IV) analogues are isostructural with the uranium compound.[154]

Compounds of the type $M_7^IM_6^{IV}F_{31}$ (Table 44) are isostructural with $Na_7Zr_6F_{31}$, in which the basic coordination geometry about the Zr atom is approximately square antiprismatic, and six antiprisms share corners to form an octahedral cavity which encloses the additional F atom.[155]

The complex anions formed with the other halogens are all of the form $[MX_6]^{2-}$, with octahedral geometry. For example, $Cs_2[NpBr_6]$ is isomorphous with $Cs_2[UBr_6]$ (face-centred cubic, $K_2[PtCl_6]$ type).[156] The IR and Raman spectra of the salts $M^I_2[M^{IV}Cl_6]$ (M^{IV} = Th, Pa, U, Np, Pu; M^I variously Cs, Et_4N, Bu^n_4N, Ph_4As), $(Me_4N)_2[PaBr_6]$ and $(Et_4N)_2[M^{IV}X_6]$ (M^{IV} = Th, Pa, U; X = Br, I) have been reported.[157] The far IR spectra of the compounds $M^I_2[NpCl_6]$ (M^I = Cs, Me_4N, Et_4N) have also been published.[158]

40.2.2.8 Hydrogen and hydrides as ligands

The only examples of compounds of this type are the borohydrides, $M^{IV}(BH_4)_4$ (M^{IV} = Th–Pu) and $M^{IV}(MeBH_3)_4$ (M^{IV} = Th, U, Np). These compounds are conveniently prepared by reaction, for example, of the metal tetrafluoride with $Al(BH_4)_3$ in a sealed tube, followed by vacuum sublimation of the product.[159] The neptunium and plutonium compounds are liquids at room temperature and are more volatile than the thorium, protactinium or uranium analogues.

In the structures of $Np(BH_4)_4$ and $Pu(BH_4)_4$ the four BH_4 groups are arranged tetrahedrally around the metal atom and three hydrogen atoms from each BH_4 group are bonded to the metal, making it 12-coordinate,[160] whereas in the polymeric uranium(IV) compound the uranium atoms are each surrounded by six BH_4 groups, two of which provide three hydrogen atoms each as above, and the other four provide two hydrogen atoms to each of two uranium atoms, making a 14-coordinate arrangement which could be regarded as a distorted bicapped hexagonal antiprism.[161] In the structures of the methylborohydrides, $M^{IV}(MeBH_3)_4$(M^{IV} = Th, U, Np) the $MeBH_3$ units are tetrahedrally coordinated to the metal *via* triple hydrogen bridges.[162]

40.2.2.9 Mixed donor atom ligands

(*i*) *Schiff base ligands*

There is a wide variety of complexes with neutral Schiff bases derived from benzaldehyde, as well as mono- and di-basic ligands derived from salicylaldehyde. A selection of these is given in Table 45, from which it can be seen that the mono- and di-basic derivatives can also act as neutral ligands. The range of complexes can be further extended by utilizing substituted salicylaldehydes, an example[163] being the compounds $M^{IV}L_2$ (M^{IV} = Th, U) formed with $(5\text{-}Bu^t)(2\text{-}HOC_6H_3)CH{=}NCH_2CH_2N{=}CH(C_6H_3OH\text{-}2)(5\text{-}Bu^t)$.

The coordination geometry in the complex $[Th(2\text{-}OC_6H_4CH{=}NC_6H_4N{=}CHC_6H_4O\text{-}2)_2]$ is a distorted square antiprism in which the corners of each square face are occupied by the two N and two O atoms of one ligand.[164] In the structure of $[Th\{(3\text{-}MeO)(2\text{-}O)C_6H_3CH{=}N(CH_2)_2N{=}CHC_6H_3(2\text{-}O)(3\text{-}OMe)\}_2]\cdot py$, the pyridine molecule is not bonded to the metal atom, and the coordination geometry is dodecahedral, with the two Schiff base ligands each occupying one of the trapezia which make up the polyhedron; the imine nitrogen atoms occupy the dodecahedral A sites and the phenolic oxygen atoms the B sites.[165]

(*ii*) *Amino acids*

The majority of the aminocarboxylates listed in Table 46 have been prepared by precipitation from aqueous solution, a procedure which frequently yields partially hydrolyzed products. No structural information is available for any of these compounds. The iminodiacetates, derived from $HN(CH_2CO_2H)_2$, are discussed on p. 1155.

(*iii*) *Complexones*

The simple thorium(IV) and uranium(IV) compounds of the type $[M^{IV}L]\cdot xH_2O$ (Table 47) separate from aqueous solutions of the actinide(IV) salts on addition of the acid, or on concentrating the resulting aqueous solution. The salts of the complex anions (*e.g.* $M^I_2H_2[U_2L_3]$) are obtained from concentrated solutions of UL in aqueous $M^I_2H_2L$. The pentaacetic acid compound, $(NH_4)_6[UL_2]\cdot 4H_2O$, is precipitated when ethanol is added to a solution of $U(C_2O_4)_2\cdot 6H_2O$ in the aqueous acid after neutralizing with ammonia and evaporation to small volume.

Table 45 Some Actinide(IV) Schiff Base Complexes

Neutral ligands: $L = C_6H_5CH{=}NR$	
$M^{IV}Cl_4 \cdot 2L$	$M^{IV} = Th, U, R = Ph; M^{IV} = Th, R = 4\text{-}MeC_6H_4$
$Th(NO_3)_4 \cdot 2L$	$R = 3\text{-}(Me_2N)C_6H_4$
Monobasic ligands: $HL = 2\text{-}HOC_6H_4CH{=}NR$	
$M^{IV}Cl_4 \cdot 2HL$	$M^{IV} = Th, R = 2\text{-}, 3\text{-}$ or $4\text{-}MeC_6H_4$; $M^{IV} = U, R = Ph, 4\text{-}ClC_6H_4, 4\text{-}O_2NC_6H_4$
$ThX_4 \cdot yHL$	$R = 4\text{-}MeC_6H_4$; $X = I, NO_3, y = 2$; $X = NCS, y = 3$; $R = Ph, X = NO_3, y = 2$
$ThCl_2L_2$	$R = Ph, 4\text{-}MeC_6H_4$
$M^{IV}L_4$	$M^{IV} = Th, R = Me, Et, Ph; M^{IV} = U, R = Me, Pr^n$
Dibasic ligands: $H_2L = 2\text{-}HOC_6H_4CH{=}NXN{=}CHC_6H_4\text{-}2\text{-}OH$	
$ThA_4 \cdot 2H_2L$	$A = Cl, Br, I, NCS, NO_3; X = CH_2CH_2$
$UCl_4 \cdot H_2L$	$X = CH_2CH_2, CH_2CH_2CH_2, 1,2\text{-}C_6H_4$
$M^{IV}LCl_2$	$M^{IV} = Th, U; X = CH_2CH_2$
$M^{IV}L_2$	$M^{IV} = Th, U; X = CH_2CH_2; M^{IV} = Th, X = 1,2\text{-}C_6H_4$; $M^{IV} = U, X = CH_2CH_2CH_2$
	$H_2L = 2\text{-}HOC_6H_4CH{=}NR$
$UCl_4 \cdot 2H_2L$	$R = 3\text{-}$ or $4\text{-}HOC_6H_4$
$ThLA_2$	$A = Cl, NO_3; R = 2\text{-}HOC_6H_4, CH(CH_2CHMe_2)CO_2H$
$Th(HL)_2A_2 \cdot yH_2O$[a]	$A = Cl, y = 4; A = I, NCS, ClO_4, y = 2; A = NO_3, y = 3$; $R = C_6H_4\text{-}2\text{-}CO_2H$
ThL_2	$R = 2\text{-}HOC_6H_4, CH(CH_2CHMe_2)CO_2H$
UL_2[b]	$R = 2\text{-}HOC_6H_4, C_6H_4CO_2H$

[a] H. Mohanta and K. C. Dash, *J. Indian Chem. Soc.*, 1980, **57**, 26; the water molecules are only held weakly and the compounds dehydrate at 120 °C.
[b] A. D. Westland and M. T. H. Tarafdar *Inorg. Chem.*, 1981, **20**, 3992.

Table 46 Actinide(IV) Compounds with Amino Acids

HL = aminoacetic acid (glycine), $H_2NCH_2CO_2H$
- $[ThL(OH)(H_2O)_5](NO_3)_2$
- $[UL_2Cl_2(H_2O)_2] \cdot 2H_2O$
- $UL_4 \cdot xH_2O$ — $x = 1.5$ or 2

HL = *N*-phenylglycine, $PhNHCH_2CO_2H$
- $UL_2Cl_2 \cdot 4H_2O$

H_2L = phenylglycine-4-carboxylic acid, $HO_2CC_6H_4NHCO_2H$
- $ThL_3(OH) \cdot 2H_2O$

HL = α- and β-aminopropionic acid (α- and β-alanine), $MeCH(NH_2)CO_2H$ and $H_2NCH_2CH_2CO_2H$
- $ThL_2(ClO_4)_2 \cdot 2H_2O$[a](?) inadequately characterized

H_2L = aminosuccinic (aspartic) acid, $HO_2CCH_2CH(NH_2)CO_2H$
- $[ThL(H_2O)_3](NO_3)_2$
- $[ThL(HL)(H_2O)](NO_3) \cdot 3H_2O$

H_2L = aminoglutaric (glutamic) acid, $HO_2C(CH_2)_2CH(NH_2)CO_2H$
- $[ThL(H_2O)_3](NO_3)_2$
- $[ThL(HL)(H_2O)](NO_3) \cdot H_2O$

HL = aminobenzoic acid, $H_2NC_6H_4CO_2H$
- $ThL_2(OH)_2$ — 2-amino
- $THL(OH)_3$
- UOL_2
- $UL_2(OH)_2 \cdot 2H_2O$
- UL_2Cl_2[b]
- $UL_3(MeCO_2)_2$ — 3-and 4-amino
- $ThL(O)(OH) \cdot 4H_2O$ — 4-amino

H_2L = β-aminosalicylate, $(H_2N)(HO)C_6H_3CO_2H$
- $Th(HL)(OH)_3 \cdot 3H_2O$

HL = 2-(*N*-phenylamino)benzoic acid, $2\text{-}(PhNH)C_6H_4CO_2H$
- ThL_4

HL = 5-iodo-2-aminobenzoic acid, (5-I), $(2\text{-}H_2N)C_6H_3CO_2H$
- $ThL_{4-n}(OH)_n$ — $n = 0$, 1 or 2

HL = 3-amino-2-naphthoic acid, $2\text{-}H_2NC_{10}H_6\text{-}3\text{-}CO_2H$
- ThL_4

[a] M. M. Mostafa, T. A. El-Awady and M. A. Girges, *J. Less-Common Met.*, 1980, **70**, 59.
[b] A. D. Westland and M. T. H. Tarafdar, *Inorg. Chem.*, 1981, **20**, 3992.

Table 47 Actinide(IV) Complexes with Complexones

Tricarboxylic acids	
Nitrilotriacetic acid, $H_3L = N(CH_2CO_2H)_3$	
$(NH_4)_2[ThL_2]\cdot 4H_2O$	
Tetracarboxylic acids	
Ethylenediaminetetraacetic acid, $H_4L = (HO_2CCH_2)_2NCH_2CH_2N(CH_2CO_2H)_2$	
$M^{IV}L\cdot xH_2O$	$M^{IV} = Th$, $x = 2, 3, 9$; $M^{IV} = U$, $x = 0, 2, 2.5–3.5$
$U_2L(OH)_4\cdot xH_2O$	
$M^I_2H_2[U_2L_3]\cdot xH_2O$	$M^I = K$, $x = 9–18$; NH_4, $x = 8–16$
$M^{II}_2[U_2L_3]\cdot xH_2O$	$M^{II} = Ca$, $x = 8, 12$; Sr, Ba, $x = 4, 12, 18$
$M^I[ThL(OH)]\cdot 4H_2O$	$M^I = Na, NH_4$
1,6-Diaminohexane-*N,N,N',N'*-tetraacetic acid, $H_4L = (HO_2CCH_2)_2N(CH_2)_6N(CH_2CO_2H)_2$	
$M^{IV}L\cdot xH_2O$	$M^{IV} = Th$, $x = 4.5$; $M^{IV} = U$, $x = 0$
1,2-Diaminocyclohexane-*N,N,N',N'*-tetraacetic acid, $H_4L = (HO_2CCH_2)_2N(C_6H_{10})N(CH_2CO_2H)_2$	
$ThL\cdot xH_2O$	$x = 0, 1$
$NH_4[ThL(OH)]\cdot 5H_2O$	
Pentacarboxylic acids	
Diethylenetriaminepentaacetic acid, $H_5L = (HO_2CCH_2)_2N(CH_2)_2N(CH_2CO_2H)(CH_2)_2N(CH_2CO_2H)_2$	
$H[ThL]\cdot H_2O$	
$Na[ThL]\cdot 4H_2O$	
$(NH_4)_6[UL_2]\cdot xH_2O$	$x = 0, 4$

In addition to the complexes listed in Table 47, the compounds *N,N'*-di(2-hydroxyphenyl)ethylenediamine-*N,N'*-diacetic acid and *N'*-(2-hydroxyethyl)ethylenediamine-*N,N,N'*-triacetic acid behave as tetrabasic ligands, forming $ThL\cdot xH_2O$ ($x = 6$ and 2 respectively).

(*iv*) *Ligands containing N,O donor sites*

(*a*) *Hexafluoroacetonylpyrazolides.* These ligands, derived from hexafluoroacetone and pyrazole or 3-methylpyrazole, form complexes of the type $M^{IV}[OC(CF_3)_2\overline{NCHCRCH{=}N}]_4$ ($R = H$, Me; $M^{IV} = Th$, U, Np, Pu) which are prepared by adding a methanol solution of $(CF_3)_2CO$ to the solid obtained when, for example, solid $(Et_4N)_2UCl_6$ is added to a toluene solution of potassium pyrazolide.

The four compounds with $R = H$ are isostructural, and in the thorium complex the eight-coordinate metal atom lies at the centre of a distorted square antiprismatic array of four N and four O atoms. In benzene solution, however, the molecular weights of the complexes with $R = H$ indicate that they are dimers, whereas those with $R = Me$ are monomers.[166]

(*b*) 8-*Hydroxyquinolines.* 8-Hydroxyquinoline can act as a neutral ligand, forming complexes of the type $ThCl_4\cdot xC_9H_7NO$ ($x = 2$, 3, 4 or 6), but the majority of the known compounds are of composition $M^{IV}L_4$ (HL = 8-hydroxyquinoline or the 5,7-dibromo ligand, $M^{IV} = Th$, U, Np, Pu; 5-chloro- and 5-chloro-7-iodo-8-hydroxyquinoline, $M^{IV} = Np$, Pu; 5,7-dichloro-8-hydroxyquinoline, $M^{IV} = Th$, Np, Pu; 2-methyl-8-hydroxyquinoline, $M^{IV} = Th$, Np). Many other substituted 8-hydroxyquinoline complexes of this type are known for $M^{IV} = Th$; most of them are precipitated on addition of the ligand to aqueous solutions of actinide(IV) compounds. In some cases (for example, HL = 8-hydroxyquinoline, $M^{IV} = Th$, U and the 5,7-dibromoligand, $M^{IV} = Th$) the product obtained from aqueous solution has the approximate composition $M^{IV}L_4\cdot HL$, but detailed studies have shown that the composition is variable, an example being $Th(C_9H_6NO)_4\cdot xC_9H_7NO$ with $x = 0.5$, 0.8 or close to unity. A hydrate, $Th(C_9H_6NO)_4\cdot 2H_2O$, has also been recorded.

Chloro and hydroxo compounds, $ThCl(C_9H_6NO)_3$ and $Th_2(C_9H_6NO)_7(OH)\cdot xH_2O$ ($x = 0$ or 4) are known, and a variety of mixed 8-hydroxyquinolinate/trichloroacetate species, obtained by heating a suspension of $Th(C_9H_6NO)_4$ with the appropriate quantity of trichloroacetic acid in benzene, have been recorded.

(*c*) 2,6-*Diacetylpyridine bis*(p-*methoxybenzoyl hydrazone*) ($= H_2L$). The coordination geometry in the complex of composition $[UL_2]\cdot 2C_6H_6\cdot 2H_2O$ is a pseudo bicapped square antiprism in which the two N_3O_2 donor ligands are chelated to the uranium atom.[167]

40.2.2.10 Multidentate macrocyclic ligands

(i) Phthalocyanine

The bis phthalocyanine complexes, $M^{IV}Pc_2$ (M^{IV} = Th, Pa, U, Np, Pu), can be prepared by heating a tetrahalide, usually the iodide (M = Th, Pa, U), or a triiodide (M = Np, Pu) with an excess of *o*-phthalodinitrile alone at 240–250 °C or, better, in 1-chloronaphthalene (see, *e.g.* ref. 168). The thorium compound has also been prepared[169] by heating the metal powder with the dinitrile at about 270 °C. An iodine adduct, $UPc_2{\cdot}I_2$, is also known.

In the structure of UPc_2, the nitrogen atoms of the pyrrole rings of the two Pc ligands form a distorted prism in which the Pc ring systems are rotated by about 37° from the prismatic configuration.[170] The structure of $ThPc_2$ has also been published.[169] The five complexes (M = Th, Pa, U, Np, Pu) are isostructural. The bonding in UPc_2 has been investigated by X-ray PE spectroscopy.[171]

(ii) Crown ethers and cryptands

The structures of some of the known crown ether compounds (Table 48) have been reported; the dicyclohexyl-18-crown-6 compound of composition $3UCl_4{\cdot}2L$ is ionic, $[UCl_3L]_2[UCl_6]$, with the uranium atom in the cation bonded to all six oxygen atoms of the crown ether and to three chlorine atoms.[172] The 18-crown-6 complex of the same composition is probably of this form, for on recrystallization from nitromethane the mixed oxidation state complex $[UCl_3L]_2[UO_2Cl_3(OH)(H_2O)]{\cdot}MeNO_2$ is obtained, in which the nine-coordinate uranium atoms in the cations are bonded to six oxygen atoms of the crown ether and to three chlorine atoms in an irregular polyhedron, while the anions adopt the pentagonal bipyramidal geometry commonly found for dioxouranium(VI) species.[173] However, in the uranium(IV) thiocyanate compound with 18-crown-6, the ether (L) is not bonded to the metal atom and the structure of this product, $[U(NCS)_4(H_2O)_4]{\cdot}1.5L{\cdot}3H_2O{\cdot}MeCOBu^i$, is described on p. 1142.

Table 48 Some Complexes of Actinide(IV) Compounds with Crown Ethers and Cryptands

L = 12-crown-4 = $C_8H_{16}O_4$	
$ThCl_4{\cdot}L$	
L = 15-crown-5 = $C_{10}H_{20}O_5$ and benzo-15-crown-5 = $C_{14}H_{20}O_5$	
$[Th(NO_3)_4(H_2O)_3]_3{\cdot}5L$	
L = 18-crown-6 = $C_{12}H_{24}O_6$	
$Th(NO_3)_4{\cdot}L{\cdot}xH_2O$	x = 1 or 3
$2Th(NO_3)_4{\cdot}3L{\cdot}2H_2O$	
$UX_4{\cdot}L$	X = Cl, CF_3CO_2
$3UCl_4{\cdot}2L$	
$[UCl_3]_2[UO_2Cl_3(OH)(H_2O)]{\cdot}MeNO_2$	
L = dibenzo-18-crown-6 = $C_{20}H_{24}O_6$	
$3Th(NO_3)_4{\cdot}5L{\cdot}3H_2O$	
L = dicyclohexyl-18-crown-6 = $C_{20}H_{36}O_6$	
$Th(NO_3)_4{\cdot}L{\cdot}3H_2O$	
$Th(NO_3)_4{\cdot}2L{\cdot}HNO_3$	
$3UCl_4{\cdot}2L = [UCl_3L]_2[UCl_6]$	
L = dibenzo-24-crown-8 = $C_{24}H_{32}O_8$	
$Th(NO_3)_4{\cdot}L{\cdot}3H_2O$	
L = dicyclohexyl-24-crown-8 = $C_{24}H_{44}O_8$	
$Th(NO_3)_4{\cdot}2L{\cdot}HNO_3$	
L = kryptofix ⟨2,2,2⟩, 4,7,13,16,21,24-hexaoxa-1,10-diazabicyclo[8.8.8]hexacosane = $C_{18}H_{36}N_2O_6$	
$Th(NO_3)_4{\cdot}L{\cdot}4H_2O$	
$UX_4{\cdot}L$	X = Cl, NCS

^1H NMR spectroscopy of the complexes of UCl_4 with 18-crown-6 and dicyclohexyl-18-crown-6, and of $U(CF_3CO_2)_4$ with these two ligands and with the cryptands [2.1.1], [2.2.1], [2.2.2] and [2.2.2.B] in nitromethane or methyl cyanide indicate that the metal atom is bonded to the oxygen atoms of the ligand.[172]

40.2.3 The +5 Oxidation State

40.2.3.1 Nitrogen ligands

(i) Ammonia and amines

(*a*) *Ammonia.* The only recorded ammine appears to be the alkoxide compound, $U(OCH_2CF_3)_5 \cdot (6–12)NH_3$, described as a green liquid. It is obtained by the reaction of UCl_5 with CF_3CH_2OH in the presence of an excess of ammonia.

(*b*) *Aliphatic amines.* Complexes of the trifluoroethoxide, $U(OCH_2CF_3)_5 \cdot 2RNH_2$ ($R = Pr^n, Pr^i$), $U(OCH_2CF_3)_5 \cdot xR_2NH$ ($x = 3$, R = Me and $x = 2$, $R = Pr^n$) and $U(OCH_2CF_3)_5 \cdot 2Me_3N$ are prepared by treating an ethereal solution of the alkoxide with an excess of the amine in ether, followed by vacuum evaporation of the mixture below 40 °C and, finally, vacuum distillation of the green liquid products. No other actinide(V) complexes with amines appear to have been reported.

(*c*) *Dihydroazirine (ethyleneimine, C_2H_5N).* The complex $U(OCH_2CF_3)_5 \cdot 3C_2H_5N$, a volatile green liquid, is obtained in the same way as the amine complexes (see above).

(ii) N-Heterocyclic ligands

(*a*) *Pyridine, C_5H_5N, py, and derivatives.* Complexes of uranium pentachloride, $UCl_5 \cdot xL$ ($x = 2$, L = py or $2\text{-}HSC_5H_4N$ and $x = 3$, L = py) and of the *t*-butoxide, $U(OBu^t)_5 \cdot py$, have been recorded. Very little is known about them.

(*b*) *Quinoline and isoquinoline, C_9H_7N.* Quinoline and isoquinoline complexes of uranium pentachloride, $UCl_5 \cdot xL$ ($x = 2$ and 3) have been recorded, but it is not known whether the uranium atoms are seven ($x = 2$) and eight-coordinate ($x = 3$), or whether the compounds are ionic (*e.g.* $[UCl_4L_2]^+Cl^-$).

(*c*) *2,2′-Bipyridyl, bipy.* The only known complex, $UCl_5 \cdot bipy$, behaves as a 1:1 electrolyte in nitromethane[174] and is evidently $[UCl_4(bipy)]^+Cl^-$.

(*d*) *1,10-Phenanthroline, phen.* An anionic complex of composition $(Et_4N)_2UOCl_5(phen)_2$ has been recorded.[175] It is probable that some of the chloride ions are displaced from the coordination sphere.

(*e*) *Pyrazole, substituted pyrazoles and 1,8-naphthyridine.* Complexes of composition $UCl_5 \cdot 1.5L$ (L = pyrazole or 3,5-dimethylpyrazole) and $UCl_5 \cdot 2L$ (L = 1-phenylpyrazole or 1,8-naphthyridine) have been obtained by treating a solution of UCl_5 in $SOCl_2$ with the appropriate ligand.[176]

(*f*) *Pyrazine, $C_4H_4N_2$, phthalazine, $C_8H_6N_2$, and phenazine, $C_{12}H_{18}N_2$.* Complexes of UCl_5 with these ligands ($UCl_5 \cdot 2C_4H_4N_2$, $UCl_5 \cdot 2C_8H_6N_2$ and $UCl_5 \cdot C_{12}H_8N_2$) are precipitated from benzene solutions of the trichloroacryloyl chloride compound, $UCl_5 \cdot C_3Cl_4O$, and the ligand. An anionic product of composition $(Et_4N)_2UOCl_5 \cdot C_{12}H_8N_2$ has also been reported, but the reaction of salts of the $[UOCl_5]^{2-}$ ion with phenazine in nitromethane appears to yield a salt of a free radical cation, $(C_{12}H_8N_2)H_2Cl$.[177]

(iii) Thiocyanates

A salt of a dioxoneptunium(V) thiocyanato complex, $Cs_4[NpO_2(NCS)_5]$, has been recorded; the coordination geometry of the complex anion is presumably pentagonal bipyramidal.

(iv) Nitriles

Reaction of UF_5 with Me_3SiCl or UCl_5 in methyl cyanide yields a complex with the mixed halide, $UCl_2F_3 \cdot MeCN$.[178] Complexes with the pentabromides, $M^VBr_5 \cdot xMeCN$ ($M^V = Pa$, $x = 3$; $M^V = U$, $x = 2$ to 3) and with an oxonitrate, $Pa_2O(NO_3)_8 \cdot 2MeCN$, are also known.

40.2.3.2 Phosphorus and arsenic ligands

The only recorded compound appears to be the complex with *o*-phenylenebis(dimethylarsine), $PaCl_5 \cdot xC_6H_4(AsMe_2)_2$ ($1 < x < 2$).[49]

40.2.3.3 Oxygen ligands

(i) Aqua species, hydroxides and oxides

(a) Aqua species. A selection of the reported hydrates is given in Table 49. The structure of one of them, $K_3NpO_2(CO_3)_2{\cdot}xH_2O$ ($x \leqslant 2$), has been reported to consist of layers of $[KNpO_2(CO_3)_2]^{2-}$ ions with K^+ ions in between the layers,[179a] but in a later paper[180] the M^VO_2 group in the compounds $M^I_3M^VO_2(CO_3)_2{\cdot}xH_2O$ ($M^V = Np$ or Pu with $M^I = Na$, K or Rb) is reported to be coordinated to six oxygen atoms from three bidentate carbonate groups to form a distorted hexagonal bipyramid. In most cases it is not known whether any or all of the water molecules are bonded to the actinide(V) metal ion in these compounds. In the hydrated carbonates, $KM^VO_2(CO_3){\cdot}xH_2O$ and $K_3M^VO_2(CO_3)_2{\cdot}xH_2O$ ($M^V = Np$, Pu) the water is considered to be zeolite type.[179a]

Table 49 Some Hydrates of Protactinium(V), Uranium(V), Neptunium(V) and Plutonium(V) Compounds

$PaF_5{\cdot}xH_2O$	$x = 1, 2$
$(Et_4N)_2(UOF_5){\cdot}2H_2O$	
$NpOF_3{\cdot}2H_2O$	
$NpO_2(ClO_4){\cdot}xH_2O$	$x = 3, 7$
$M^IM^VO_2(CO_3){\cdot}xH_2O$	$M^I = Na$, $M^V = Np$, $x = 0.5, 1, 2, 3, 3.5$ or 4; $M^V = Pu$, x unspecified
	$M^I = K$, Rb, $M^V = Np$, Pu
	$M^I = NH_4$, $M^V = Np$, Pu ($x = 3$)
$(NH_4)_2PuO_2(CO_3)(OH){\cdot}xH_2O$	
$K_3M^VO_2(CO_3)_2{\cdot}xH_2O$	$M^V = Np$, Pu ($x \leqslant 2$)
$Pa(C_2O_4)_2(OH){\cdot}6H_2O$	
$PaO(C_2O_4)(OH){\cdot}xH_2O$	$2 < x < 4$
$NpO_2(HC_2O_4){\cdot}2H_2O$	
$(NpO_2)_2C_2O_4{\cdot}H_2O$[a]	
$M^INpO_2(C_2O_4){\cdot}xH_2O$	$M^I = Na$, $x = 1, 3$; K, $x = 2$; NH_4, $x = 2$; Cs, $x = 2, 3$
$M^I_3NpO_2(C_2O_4)_2{\cdot}xH_2O$	$M^I = Na$, K, NH_4
$M^I_5NpO_2(C_2O_4)_3{\cdot}xH_2O$	$M^I = Na$, K, NH_4, Cs
$PaO(NO_3)_3{\cdot}xH_2O$	$1 < x < 4$
$NpO_2(NO_3){\cdot}xH_2O$	$x = 1, 5$
$RbNpO_2(NO_3)_2{\cdot}H_2O$	
$PaO(H_2PO_4)_3{\cdot}2H_2O$	
$NH_4PuO_2HPO_4{\cdot}4H_2O$	
$[Co(NH_3)_6]NpO_2(SO_4)_2{\cdot}3H_2O$	
$[Co(NH_3)_6]NpO_2(SO_4)_2{\cdot}M^I_2SO_4{\cdot}xH_2O$	$M^I = Na$, K
$[Co(NH_3)_6](NpO_2)_2(IO_3)_5{\cdot}4H_2O$	
$PaO(ReO_4)_3{\cdot}xH_2O$	

[a] V. G. Zubarev and N. N. Krot, *Sov. Radiochem.* (*Engl. Transl.*), 1981, **23**, 690.

The basic structure of the hydrated oxalato complex, $NaNpO_2C_2O_4{\cdot}3H_2O$, is made up of infinite chains of composition $[NpO_2C_2O_4(H_2O)]_n^{n-}$ connected by Na^+ ions and H_2O molecules in a three-dimensional network. The neptunium atom is at the centre of a pentagonal bipyramid, coordinated to four O atoms from two C_2O_4 groups and a fifth from the water molecule in the equatorial plane.[179b]

(b) Hydroxides and oxides. Protactinium(V), neptunium(V) and plutonium(V) hydroxides precipitate from alkaline aqueous solutions of the actinides(V); the last two appear to be of the form $M^VO_2(OH){\cdot}xH_2O$.

Oxides such as Pa_2O_5, Np_2O_5 and U_3O_8 (+5 and +6 oxidation state metal centres) are well established and a number of mixed oxides such as $Li_3M^VO_4$, $Li_7M^VO_6$ ($M^V = Pa$, U, Np, Pu) and $Ba_2U_2O_7$ have been recorded.

(ii) Alcohols, alkoxides, aryloxides and silyloxides

(a) Alcohols. The only known actinide(V) alcohol adducts appear to be the uranium(V) compounds $UOCl_3{\cdot}EtOH$, $UO(OR)_3{\cdot}ROH$ (R = Et, Bu^t) and $U(OBu^t)_5{\cdot}Bu^tOH$. No structural information is available for any of these compounds.

(*b*) *Alkoxides.* A very large number of uranium(V) alkoxides are known (Table 50), for these compounds are more stable with respect to disproportionation than any other uranium(V) compound. However, protactinium(V) and neptunium(V) alkoxides have scarcely been investigated, and no examples of plutonium(V) alkoxides have been recorded.

Table 50 Actinide(V) Alkoxides

$M^V(OR)_5$	M^V = Pa, R = Et
	M^V = U, R = Me, Et, Pr^n, Pr^i, Bu^n, Bu^s, Bu^i, Bu^t, n-C_5H_{11}, $CHMePr^n$, $CHEt_2$, CH_2Bu^s, CMe_2Et, $CHMePr^i$, CH_2Bu^i, CH_2Bu^t, CMe_2Pr^n, CMe_2Pr^i, $CMeEt_2$, $CMeEtPr^i$, CEt_3, $CH_2CH{=}CH_2$, $CH_2CH_2NH_2$, $CH_2CH_2NEt_2$, CH_2CF_3
$U^VO(OR)_3$	R = Et, Pr^i
$U^V(OR)_2F_3$	R = Me, Et
$U^V(OR)_{5-n}Cl_n$	R = Et, n = 1, 2, 3 or 4 ($MeCO_2Et$ adduct)
	R = Pr^n, n = 1, 2
	R = Pr^i, n = 1, 2, 3 or 4 ($MeCO_2Pr^i$ adduct)
	R = Bu^t, n = 1
$U(OR)_{5-n}Br_n$	R = Et, n = 1, 2, 3 or 4 ($MeCO_2Et$ adduct)
	R = n-C_5H_{11}, n = 1, 3 or 4 ($MeCO_2n$-C_5H_{11} adduct)
$Np(OEt)_4Br$	
$NpO(OEt)Br_2$	
$U(OR)_2Cl_2Br$	R = Me, Bu^n
$M^I[U(OR)_6]$	R = Me, M^I = Li, Na, K, Rb, Cs, Me_4N, Et_4N, Ph_4As
	R = Et or Bu^t, M^I = Na
$Ca[U(OEt)_6]_2$	
$Al[U(OEt)_6]_3$	
$M^I[U(OEt)_2X_4]$	X = Cl, M^I = Et_4N, Ph_4As
	X = Br, M^I = Et_4N

$U(OEt)_5$ is easily prepared by oxidation of $U(OEt)_4$ by bromine in ethanol, followed by addition of the calculated quantity of sodium ethoxide, a reaction which yields[181] only $Np(OEt)_4Br$ in the case of $Np(OEt)_4$. $U(OEt)_5$ is commonly used as the starting material for the preparation of other uranium(V) alkoxides by alcohol exchange reactions. A comprehensive review of metal alkoxides includes a useful discussion of the uranium(V) compounds.[182]

Molecular weight determinations indicate that in solution $U(OMe)_5$ is probably a tetramer in equilibrium with a dimer, whereas the other alkoxides are principally dimeric in non-aqueous media unless the alkyl group is bulky as in the compounds $U\{OCEt_3\}_5$ and $U\{OC(MeEtPr^i)\}_5$, which are monomers in solution.

The reaction of $U(OEt)_5$ with bidentate ligands, such as β-diketones, HL, has been reported to yield products of composition $U(OR)_4L$, $U(OR)_3L_2$ and $U(OR)_2L_3$, depending on the quantities of HL used, but the last two products are probably mixtures of $U(OR)_4L$, UL_4 and dioxouranium(VI) compounds.

The alkoxide fluorides, $U(OR)_2F_3$ (R = Me, Et) are obtained by treating UF_5 with the calculated quantities of Me_3SiOMe or $U(OEt)_5$. The other alkoxide halides are commonly prepared by reaction of $U(OR)_5$ with the calculated quantities of hydrogen chloride, acetyl chloride or bromide in ether or benzene.

Anionic compounds of the type $M^I[U(OR)_6]$ are also known (Table 50) and in these compounds the uranium atom is probably octahedrally coordinated.

(*c*) *Aryloxides.* The only known compounds of this type are the phenoxides $U(OPh)_4(OEt)$, $U(OPh)_2Cl_2Br$, $Na[U(OPh)_6]$ and a mixed uranium(V)/uranium(VI) compound of composition $U_4O_6(OPh)_{10}\cdot 4THF$. This last is obtained by atmospheric oxidation of the product of the reaction of $UCl_3\cdot x$THF with NaOPh in THF. The structure of the compound consists of two uranium(V) and two uranium(VI) centres which are connected by four bridging phenoxide groups and two triply bridging oxygen atoms, that is, $\{[U^V(OPh)_3THF]_2[U^{VI}O_2(THF)]_2(\mu\text{-}OPh)_4(\mu_3\text{-}O)_2\}$. The coordination polyhedron about each uranium atom is a distorted pentagonal bipyramid.[183]

(*d*) *Silyloxides.* The few known uranium(V) silyloxides are similar in properties to the corresponding alkoxides; $U(OSiMe_3)_5$ is mainly dimeric in non-aqueous media and the degree of dimerization of the compounds $U(OSiR_3)_5$ in solution decreases sharply in the order $U(OSiMe_3)_5 \gg U(OSiMe_2Et)_5 > U(OSiMeEt_2)_5 > U(OSiEt_3)_5$, the last being a monomer. Anionic species, $Na[U(OSiR_3)_6]$ (R = Me, Et), are also known.

(iii) β-Ketoenolates, tropolonates, ethers, ketones and esters

(a) β-Ketoenolates. Apart from the protactinium(V) compound of composition $PaCl_3(MeCOCHCOMe)_2$, all the reported compounds are uranium(V) products obtained by reaction of $U(OR)_5$ with the stoichiometric quantity of the β-diketone in benzene under reflux. These products are of composition $U(OR)_{5-n}(R^1COCHCOR^2)_n$ ($R = Et$ or Bu^t, $R^1 = R^2 = Me$ and $R^1 = R^2 = Ph$, both with $n = 1$, 2 or 3; $R = Et$ and $n = 1$, 2 or 3, or Bu^t and $n = 1$ or 3, $R^1 = Ph$, $R^2 = Me$). As mentioned earlier (p. 1181) only the products of composition $U(OR)_4(R^1COCHCOR^2)$ are likely to be genuine compounds. The same applies to the majority of the β-ketoester products of composition $U(OR)_{5-n}(R^1COCR^2CO_2R^3)_n$ ($R = Et$, $R^1 = Me$, $R^2 = H$ and $R^3 = Me$ or Et or $R^1 = R^2 = Me$, $R^3 = Et$; $R = Bu^t$, $R^1 = R^2 = Me$, $R^3 = Et$, all $n = 1$, 2 or 3), although the products of composition $U(OR)_3(CF_3COCHCO_2R^1)_2$ ($R = Me$, $R^1 = Et$ and $R = Et$, $R^1 = Me$, Et or Bu^n) may be genuine compounds.

(b) Tropolonates. The tropolonates $M^V(trop)_5$ ($M^V = Pa$, U) and $M^V(trop)_4X$ ($M^V = Pa$, $X = Cl$, Br, ClO_4; U, $X = Cl$) have been recorded. $U(trop)_5$ is precipitated when Htrop is added to a solution of $U(OEt)_5$ in ether and is very unstable with respect to decomposition to $U(trop)_4$. $U(trop)_4Cl$ is isostructural with $Pa(trop)_4Cl$. The ethoxide, $Pa(trop)_4(OEt)$, was subsequently shown to be $Pa(trop)_5$.

(c) Ethers. The only recorded compounds are $UCl_5{\cdot}R_2O$ ($R = Me, Et, Pr^i, Bu^n$, i-C_5H_{11}), $UCl_5{\cdot}THF$ and the dioxane complexes, $UCl_5{\cdot}xC_4H_8O_2$ ($x = 1$ or 3).

(d) Ketones. 1:1 Complexes, $UCl_5{\cdot}L$ [L = anthr-10-one ($C_{14}H_{10}O$), 9-methyleneanthr-10-one ($C_{15}H_{10}O$), 1,9-benzoanthr-10-one ($C_{17}H_{10}O$) and 9-benzylideneanthr-10-one ($C_{21}H_{14}O$)] and the bis complexes, $UCl_5{\cdot}2L$ with $L = C_{14}H_{10}O$ or $C_{17}H_{10}O$ have been reported. Benzil adducts, $UCl_5{\cdot}xPhCOCOPh$ ($x = 1$ or 2) are also known. These products presumably involve six- and seven-coordinate uranium(V).

(e) Esters. Ester complexes of composition $U(OR)X_4{\cdot}MeCO_2R$ ($R = Et$, $X = Cl$ or Br; $R = Pr^i$, $X = Cl$) have been obtained by reaction of $U(OR)_5$ with the acyl halide, MeCOX, in benzene under reflux.

(iv) Oxoanions as ligands

(a) Nitrates. The hydrated nitrates $PaO(NO_3)_3{\cdot}xH_2O$ ($1 < x < 4$) and $NpO_2NO_3{\cdot}xH_2O$ ($x = 1, 5$), as well as the hydrated nitrato complex, $RbNpO_2(NO_3)_2{\cdot}H_2O$, have been mentioned earlier (p. 1180). The hexanitratoprotactinates(V), $M^I[Pa(NO_3)_6]$ ($M^I = Cs$, Me_4N, Et_4N) have been prepared by treating the chloro complex salts, $M^I[PaCl_6]$ with liquid N_2O_5. The acid, $HPa(NO_3)_6$ is also known. In these compounds the protactinium(V) centre is presumably 12-coordinate.

(b) Phosphates. Some hydrated phosphates and phosphatocomplexes are included in Table 49, p. 1180. Hydrated protactinium(V) phosphate, $PaO(H_2PO_4)_3{\cdot}2H_2O$, dehydrates at 250 °C and decomposes to $PaO(PO_3)_3$ at 700 °C, then to $(PaO)_4(P_2O_7)_3$ at 1000 °C. The structures of these compounds are not known.

(c) Arsenates. A protactinium(V) phenylarsonate, $H_3PaO_2(PhAsO_3)_2$ and a number of hydrated salts of neptunium(V) arsenato complexes have been recorded, but no structural information is available for them.

(d) Sulfates. Protactinium(V) oxosulfate, $H_3PaO(SO_4)_3$, is obtained on evaporating HF/H_2SO_4 solutions of protactinium(V); it decomposes at 375 to 400 °C to $HPaO(SO_4)_2$, which yields $HPaO_2SO_4$ at 525 °C. $(NpO_2)_2SO_4$ has been reported and hydrated hexammine cobalt(III) salts of the $[NpO_2(SO_4)_2]^{3-}$ anion (Table 49, p. 1180) are also known.

(e) Selenates. The protactinium(V) compound, $H_3PaO(SeO_4)_3$, is obtained from HF/H_2SeO_4 solutions of protactinium(V). Unlike the sulfate, it decomposes directly to Pa_2O_5 on heating.

(f) Iodates. Hydrated neptunium(V) iodate and a salt of the complex anion $[(NpO_2)_2(IO_3)_5]^{3-}$ have been recorded (Table 49, p. 1180).

(v) Carboxylates and carbonates

(a) Monocarboxylates. Hydrated neptunium(V) formate, $NpO_2(HCO_2){\cdot}2H_2O$, and the formato complex, $Cs_2NpO_2(HCO_2)_3$, have been reported, the latter precipitated from aqueous

solution. The HCO_2 group appears to be bidentate in both compounds.[184] Salts of acetatodioxoactinates(V),[185] $Cs_2M^VO_2(MeCO_2)_3$ (M^V = Np, Pu) and $Na_4NpO_2(MeCO_2)_5$, have been reported. The two cesium salts are isostructural[186] and presumably have the same hexagonal bipyramidal coordination as the anion of the barium salt, $Ba[NpO_2(MeCO_2)_3]\cdot 2H_2O$, in which the oxygen atoms from the three bidentate acetate groups are in the equatorial plane.[187]

A number of mixed alkoxide/monocarboxylate compounds of the type $U(OR)_{5-n}(R^1CO_2)_n$ (R = Et, R^1 = Me, Et, Pr^n with n = 1, 2, 3 or 4; R = Et or Bu^t, R^1 = Ph with n = 1, 2 or 3) have been obtained by heating the alkoxide $U(OR)_5$ (R = Et) with the stoichiometric quantity of the acid in benzene under reflux or (R = Bu^t) by heating the appropriate ethoxide carboxylate with an excess of *t*-butanol in benzene.

(*b*) *Chelating carboxylates.* The anhydrous pyridine-2-carboxylate, $NpO_2(C_5H_4NCO_2)$, is obtained by heating the dihydrate at 125 °C; its IR spectrum indicates that the pyridine N atom is bonded to the Np atom. Similarly, in the *N*-oxide analogue, $NpO_2(C_5H_4N(O)CO_2)$, the amine oxide O atom is bonded to the metal.

The hydrated furan-2-carboxylate, $NpO_2(C_4H_3OCO_2)\cdot 2H_2O$, and the thiophene-2-carboxylate, $NpO_2(C_4H_3SCO_2)\cdot 2H_2O$, dehydrate at 90 °C and 100 °C respectively; it is possible that the ring O and S atoms in these two carboxylates are also bonded to the metal atom.

(*c*) *Carbonates.* Hydrated neptunium(V) carbonate, $NpO_2CO_3\cdot xH_2O$, and salts of hydrated carbonatocomplexes of the type $KM^VO_2CO_3\cdot xH_2O$ and $K_3M^VO_2(CO_3)_2\cdot xH_2O$ (M^V = Np, Pu) are included in Table 49 (p. 1180); the structure of $K_3NpO_2(CO_3)_2\cdot xH_2O$ is described on p. 1180. The basic structure of the salts $M^INpO_2CO_3\cdot xH_2O$ (M^I = Na, K) consists of $(NpO_2)CO_3$ layers with the M^+ cations distributed in between.[188] Anhydrous salts of carbonato complexes, $KM^VO_2CO_3$ (M^V = Np, Pu),[185] $NH_4PuO_2CO_3$ and $M^I_5NpO_2(CO_3)_3$ (M^I = K, Cs) have also been recorded.

(*d*) *Oxalates.* Hydrated oxalates and oxalato complex salts are included in Table 49, p. 1180. $(NpO_2)_2C_2O_4\cdot H_2O$ dehydrates at 180 °C and the anhydrous salt, $CsNpO_2C_2O_4$, has been recorded.

(*e*) *Polycarboxylates.* Salts of dioxoneptunium(V) mellitate and pyromellitate complex ions, $Na_4(NpO_2)_2(C_6(CO_2)_6)\cdot 8H_2O$ and $\{Na_3NpO_2(C_6H_2(CO_2)_4)\}_2\cdot 11H_2O$, have been reported; in the pyromellitate complex the NpO_2^+ ion is coordinated to four pyromellitate anions, the carboxylate oxygen atoms of which form a pentagon in the pentagonal bipyramidal coordination geometry.[189]

(*vi*) *Aliphatic and aromatic hydroxyacids*

(*a*) *Aliphatic hydroxyacids.* Uranium(V) 3-hydroxypropionates, formulated as $U(OEt)(OCH_2CH_2CO_2)_2$, $U(OEt)_3(OCH_2CH_2CO_2)$, $U(OCH_2CH_2CO_2)_2(HOCH_2CH_2CO_2)$ and $U_2(OCH_2CH_2CO_2)_5$, are obtained in the same way as the ethoxide monocarboxylates. Phenyl hydroxyacetates of composition $U(OR)_3(PhCH(O)CO_2)$, $U(OR)(PhCH(O)CO_2)_2$ (R = Et, Bu^t), $U(PhCH(O)CO_2)_2(PhCH(OH)CO_2)$ and $U_2(PhCH(O)CO_2)_5$ have been prepared in a similar manner.

(*b*) *Aromatic hydroxyacids.* The only recorded compounds appear to be the uranium(V) salicylates, $U(OEt)_3(OC_6H_4CO_2)$, $U(OEt)(OC_6H_4CO_2)_2$, $U_2(OC_6H_4CO_2)_5$, $HU(OC_6H_4CO_2)_3$ and the analogous *t*-butoxide salicylates, $U(OBu^t)_{5-2n}(OC_6H_4CO_2)_n$ (n = 1 or 2), obtained in the same way as the analogous alkoxide monocarboxylates.

(*vii*) *Amides,* P-*Oxides and* S-*Oxides*

(*a*) *Amides.* The only amide complex appears to be $U(OPh)_4Cl\cdot 2DMF$, obtained by reaction of $CsUCl_6$ with Na(OPh) in benzene under reflux, followed by extraction of the residue into DMF and evaporation of the extract to dryness. The reaction of UBr_5 with DMA in a mixture of acetone and dichloromethane results in disproportionation to the mixed oxidation state compound $[U^{VI}O_2(DMA)_5][U^{IV}Br_6]$.

(*b*) *Trichloroacryloyl chloride.* A product of composition $UCl_5\cdot Cl_2C{=}CClCOCl$ is obtained as an intermediate in the reaction of UO_3 with hexachloropropene (C_3Cl_6); its IR spectrum indicates that the acid chloride molecule is bonded to the metal atom *via* the carbonyl oxygen atom.

(*c*) P-*Oxides*. The known phosphine oxide complexes (Table 51) are nearly all derived from the actinide pentahalides. They are obtained from the pentahalide and the ligand in a non-aqueous solvent or by reaction of a chloro complex salt (*e.g.* $Cs[UCl_6]$) with the ligand in dichloromethane. $UBr_5{\cdot}(Me_2N)_3PO$ is obtained by bromine oxidation of a 1:1 mixture of UBr_4 and the ligand in MeCN, and the complexes $UF_5{\cdot}2Ph_3PO$ and $UCl_2F_3{\cdot}2Ph_3PO$ have been isolated by addition of Ph_3PO to UF_5 and to UCl_2F_3 respectively. The structure of $[UCl_5(Ph_3PO)]$ is a slightly distorted octahedron.[190]

Table 51 Complexes of Actinide(V) Compounds with *P*-Oxides

$MX_5{\cdot}R_3PO$	$R = n\text{-}C_8H_{17}$, M = U, X = Cl $R = Me_2N$, M = U, X = Br R = Ph, M = Pa, U, X = Cl, Br
$PaCl_5{\cdot}Ph_2(PhCH_2)PO$	
$MX_5{\cdot}2Ph_3PO$	M = Pa, X = F, Cl, Br; M = U, X = F, Cl
$UCl_2F_3{\cdot}2Ph_3PO$	
$PaBr_5{\cdot}(Ph_2PO)_2CH_2$	
$Pa(OEt)_2X_3{\cdot}Ph_3PO$	X = Cl, Br
$PaOX_3{\cdot}2Ph_3PO$	X = Cl, Br
$NpO_2NO_3{\cdot}3(n\text{-}C_8H_{17})_3PO$	
$[NpO_2\{(n\text{-}C_8H_{17})_3PO\}_4]ClO_4{\cdot}H_2O$	

(*d*) S-*Oxides*. The dimethyl sulfoxide complex, $Pa(trop)_4Cl{\cdot}DMSO$ appears to be the only known complex with this type of ligand. The thionyl chloride compound, $UCl_5{\cdot}SOCl_2$, obtained by heating UO_3 with $SOCl_2$ in a sealed tube for 24 hours at 150–200 °C, is probably ionic, possibly $SOCl^+UCl_6^-$.

40.2.3.4 Sulfur ligands

(*i*) *Thiols*

A compound of composition $[UCl_4(SPh)]_2$, which is presumably a thiophenolate, is precipitated when diphenyl disulfide is added to a benzene solution of the trichloroacryloyl chloride complex, $UCl_5{\cdot}Cl_2C{=}CClCOCl$.[174] The thioglycolate ester compounds $U(OEt)_{5-n}L_n$ (HL = $HSCH_2CO_2R$ with R = Me, Et or Bu^n and $n = 1, 2, 3, 4$ or 5) have been obtained by heating $U(OEt)_5$ with the appropriate quantity of the ester in benzene under reflux and the *t*-butoxide analogues $U(OBu^t)_{5-n}L_n$ ($n = 1, 2, 3$ or 4) have been prepared by treating the corresponding ethoxide compounds with Bu^tOH in benzene under reflux. An ethoxide-*p*-thiocresolate $U(OEt)(1\text{-}S\text{-}4\text{-}MeC_6H_4)_4$, thiosalicylate and thiolactate compounds of composition $U(OR)_{5-2n}L_n$ and $UL_2(HL)$ ($H_2L = 2\text{-}HSC_6H_4CO_2H$ or $C_2H_4(SH)CO_2H$, R = Et or Bu^t and $n = 1$ or 2) and the thiobenzoate $U_2(2\text{-}SC_6H_4CO_2)_5$ have been obtained in the same way as the ethoxide–thioester compounds. It is not clear whether the thiolactates[191] are α- or β-mercaptopropionates.

(*ii*) *Thiocarboxylates*

Alkoxide-thiobenzoates of composition $U(OR)_{5-n}(PhCOS)_n$ (R = Et or Bu^t and $n = 2, 3$ or 4) have been prepared in the same way as the thioglycollate ester analogues.

(*iii*) *Dithiocarbamates*

The only recorded complexes appear to be the protactinium(V) compounds, $Pa(Et_2NCS_2)_4X$ (X = Cl, Br), prepared by treating a suspension of the pentahalide in dichloromethane with an excess of $Na(Et_2NCS_2)$, followed by vacuum evaporation of the filtrate.

(*iv*) *Thiourea*

A product of composition $UF_5{\cdot}2SC(NH_2)_2$ has been obtained by treating UF_6 in methyl cyanide with thiourea; this may be a mixture of uranium(IV) and uranium(VI) species.

(v) Phosphine sulfides

The protactinium(V) complexes of composition PaX_5L (X = Cl, Br; R = Ph_3PS, $(Ph_2PS)_2CH_2$) have been recorded.

40.2.3.5 Selenium and tellurium ligands

(i) Diphenyl diselenide and ditelluride

The complexes $UCl_5{\cdot}Ph_2X_2$ (X = Se or Te) are precipitated when a solution of the diselenide or ditelluride in benzene is added to a solution of the trichloroacryloyl chloride complex, $UCl_5{\cdot}C_3Cl_4O$, in the same solvent.

(ii) Phosphine selenides

The protactinium(V) complexes PaX_5L (X = Cl, Br; L = Ph_3PSe, $(Ph_2PSe)_2CH_2$) have been recorded.

40.2.3.6 Halogens as ligands

(i) Pentahalides

Two modifications of UF_5 are known; in the high temperature form, α-UF_5, the structure consists of an infinite chain of UF_6 octahedra linked by opposite corners. The low temperature form, β-UF_5, with which PaF_5 and NpF_5 are isostructural, was originally thought to consist of an infinite chain of UF_7 pentagonal bipyramids[192], but is now known to be an eight-coordinate arrangement with the geometry intermediate between dodecahedral and square antiprismatic.[193] The structure of $PaCl_5$ is a chain of $PaCl_7$ pentagonal bipyramids[194] whereas UCl_5 is a dimer with UCl_6 octahedra sharing an edge.[195] β-$PaBr_5$ and UBr_5 are also dimers, like UCl_5. The only pentaiodide known is PaI_5.

Intermediate oxidation state fluorides, such as Pa_2F_9 and M_4F_{17} [M = Pa, U and Pu(?)] are also known.

(ii) Oxohalides

The main types of actinide(V) oxohalide are given in Table 52. The oxohalides can be prepared by heating the parent pentahalide with the appropriate quantity of oxygen; for example, Pa_2OCl_8 is obtained by this method at 400 °C. $NpOF_3$ has been prepared by heating the hydrate in dry HF at 100–150 °C.

Table 52 Actinide(V) Oxohalides

$M^V_2OX_8$	M^V = Pa, X = F, Cl; M^V = U, X = F
$NpOF_3$	
M^VOX_3	M^V = Pa, X = Br, I; M^V = U, X = Cl, Br
M^VO_2X	M^V = Pa, X = F, Cl, Br, I; M^V = U, X = Cl, Br; M^V = Np, X = F
$Pa_2O_3Cl_4$	
Pa_3O_7F	

The structure of $PaOBr_3$ consists of a polymer in which the protactinium atoms are surrounded by a slightly distorted pentagonal bipyramidal arrangement of three oxygen and four bromine atoms; two of the latter occupy the axial positions,[196] and the same geometry has been reported for UO_2Br, again with three oxygen and two bromine atoms in the equatorial plane.[197]

(iii) Halogeno complexes

Hexahalogeno complexes (Table 53) are well known; the cesium salts, $CsMF_6$, are readily obtained by fluorination of the tetrafluoride in the presence of cesium fluoride (M = Np, Pu) and others have been prepared by reaction of the pentafluoride with the cation fluoride; for example, reaction of CuF_2 with UF_5 in methyl cyanide yields $Cu[UF_6]_2$ which is reduced to $Cu[UF_6]$ by metallic copper.[198] In the compound $[(Ph_3P)_2N][UF_6]$ the anion is octahedral[199] as in the salts $Cs[M^VF_6]$ (M^V = U, Np, Pu), whereas the protactinium atom in $RbPaF_6$ is dodecahedrally coordinated.[200] In the protactinium(V) compounds of the type $M^I_2PaF_7$, the metal atom is surrounded by nine fluorine atoms in a slightly distorted tricapped trigonal prismatic array (K_2PaF_7 type),[201] while the salts $M^I_2UF_7$ and $Rb_2M^VF_7$ (M^V = Np, Pu) have the seven-coordinate (K_2TaF_7 type) structure. In the octafluoro complexes, $Na_3[M^VF_8]$ (M^V = Pa, U, Np), the coordination geometry is cubic.[202] The chloro complexes are obtained from the pentachloride and, in the case of uranium(V), from thionyl chloride solutions of $UCl_5{\cdot}SOCl_2$; examples are Cs_2UCl_7 and K_3UCl_8.[203] The anion in $[Ph_3(PhCH_2)P][UCl_6]$ is octahedral[204] and the structure of $PCl_4^+UCl_6^-$ has also been reported.[205]

Table 53 Actinide(V) Halogeno Complexes

$M^I[M^VF_6]$	M^V = Pa, M^I = Li, Na, K, Rb, Cs, NH_4, Ag
	M^V = U, M^I = Li, Na, K, Rb, Cs, NH_4, Ag, Tl, Cu, $(Ph_3P)_2N$, NO, NO_2
	M^V = Np or Pu, M^I = Cs
$Cu[UF_6]_2$	
$M^I_2M^VF_7$	M^V = Pa or U, M^I = K, Rb, Cs, NH_4
	M^V = Np or Pu, M^I = Rb
$M^I_3M^VF_8$	M^V = Pa, M^I = Li, Na, K, Rb, Cs
	M^V = U, M^I = Na, K, Rb, Cs, NH_4, Ag, Tl
	M^V = Np, M^I = Na
$M^I[M^VCl_6]$	M^V = Pa, M^I = Cs, Me_4N, Et_4N, Ph_4As
	M^V = U, M^I = Na,[a] K,[a] Rb,[a] Cs,[a] $Ph_3(PhCH_2)P$
Cs_2UCl_7	
$M^I_3M^VCl_8$	M^V = Pa, M^I = Me_4N, NO
	M^V = U, M^I = K
$M^I[M^VBr_6]$	M^V = Pa, M^I = Me_4N, Et_4N
	M^V = U, M^I = Na,[b] K,[b] Rb,[b] Cs,[b] Et_4N, Ph_4As
$Ph_3MeAs[PaI_6]$	

[a] V. L. Kudryashov, I. G. Suglobova and D. E. Chirkst, *Sov. Radiochem.* (*Engl. Transl.*), 1978, **20**, 314.
[b] I. G. Suglobova and D. E. Chirkst, *Koord. Khim.*, 1980, **6**, 409.

(iv) Oxohalocomplexes

A few salts of composition $(Et_4N)_2[M^VOX_5]$ (M^V = Pa, X = Cl, Br; M^V = U, X = F) are known; the uranium(V) compound is obtained from the dihydrate under vacuum, and the protactinium(V) chlorocompound is prepared by hydrolysis of the hexachloro complex salt in methyl cyanide containing 0.5% water. Salts of composition $(Et_4N)[Pa(OEt)_2X_4]$ (X = Cl, Br) are also known.

Difluorodioxoactinates(V), $M^IM^VO_2F_2$ (M^V = Np, M^I = K, Rb; M^V = Pu, M^I = Rb, NH_4) and the tetrachlorodioxoactinates(V), $Cs_3[M^VO_2Cl_4]$ (M^V = Np, Pu) are obtained from aqueous acid solutions of the actinide(V). $Cs_3[NpO_2Cl_4]$ and $Cs_3[PuO_2Cl_4]$ are isostructural and the $[M^VO_2Cl_4]^{3-}$ anion has a distorted octahedral geometry.[206]

A mixed oxidation state compound, $Cs_7(Np^VO_2)(Np^{VI}O_2)_2Cl_{12}$ is obtained by treating neptunium(V) hydroxide with ethanolic hydrochloric acid saturated with cesium chloride or from $Cs_3[Np^VO_2Cl_4]$ and hydrochloric acid.[207] The analogous neptunium(V)/uranium(VI) compound, $Cs_7(Np^VO_2)(U^{VI}O_2)_2Cl_{12}$, is isostructural with the Np^V/Np^V complex.[208] Products of the composition $M^I_2Np^VOCl_5$ (M^I = Cs, Ph_4As) obtained in a similar manner appear to be disproportionation mixtures.

40.2.3.7 *Mixed donor atom ligands*

(*i*) *Complexones*

The only compounds with this type of ligand appear to be the hydrated neptunium(V) complexes with ethylenediaminetetraacetic acid (H_4edta), $(NpO_2)_2H_2(edta)\cdot 5H_2O$ and $[Co(NH_3)_6]NpO_2(edta)\cdot 3H_2O$.

(*ii*) 8-*Hydroxyquinolines*

Complexes in which 8-hydroxyquinoline acts as a neutral donor ligand, $UCl_5\cdot xC_9H_7NO$ ($x = 2$ or 4), have been isolated from non-aqueous media. Hydrated dioxoneptunium(V) and dioxoplutonium(V) compounds of composition $Np^VO_2(C_9H_6NO)\cdot 2H_2O$, $M^VO_2L\cdot HL\cdot H_2O$ (M^V = Np, Pu; HL = 2-MeC_9H_6NO), $Pu^VO_2L\cdot 2HL\cdot xH_2O$ (HL = C_9H_7NO, x unspecified; 5-ClC_9H_6NO, $x = 1$; 5,7-$Cl_2C_9H_5NO$, $x = 2$) and $(Ph_4As)NpO_2(C_9H_6NO)_2\cdot H_2O$ have been obtained from aqueous solution.

40.2.4 The +6 Oxidation State

40.2.4.1 *Group IV ligands*

(*i*) *Cyanides*

The only recorded compound has the composition $K_2UO_2(CN)_2(NO_3)_2$; it was reported to be precipitated when a saturated solution of KCN in a 1:1 mixture of ethanol and ether was added to a solution of $UO_2(NO_3)_2\cdot 6H_2O$;[209] however, this was not confirmed in later work.[210]

(*ii*) *Isocyanides*

A cyclohexyl isocyanide complex with UO_2Cl_2 has been mentioned in the literature.[30]

40.2.4.2 *Nitrogen ligands*

The available IR data for the complexes with ureas indicate that the ligands are bonded to the metal *via* the carbonyl oxygen atom and these compounds are therefore discussed in the section on oxygen ligands.

(*i*) *Ammonia and amines*

(*a*) *Ammonia.* Although a number of ammonia adducts of dioxouranium(VI) compounds are known (Table 54), analogous neptunium(VI) and plutonium(VI) systems do not appear to have been investigated. The uranium(VI) complexes are obtained by treating the parent compounds with anhydrous liquid ammonia, with gaseous ammonia in a non-aqueous solvent or (the 7-methyl-8-hydroxyquinolate) by precipitation from aqueous solution in the presence of ammonia. The complexes with β-diketonates, originally reported as $UO_2(acac)_2\cdot NH_3$ and $UO_2(ba)_2$ (Hba = 1-phenylbutane-1,3-dione) are now known to be β-ketoimine adducts[211] but the analogous ammines $UO_2L_2\cdot NH_3$ (HL = $PhCOCH_2COPh$ and $CF_3COCH_2COCF_3$) are genuine compounds. In the molecule $[UO_2(CF_3COCHCOCF_3)_2(NH_3)]$ the coordination geometry is pentagonal bipyramidal;[212] this appears to be the only actinide ammine for which a crystallographic structure determination has been recorded. Basic compounds of composition $U_2O_5Cl_2\cdot xNH_3$ ($x = 2$ or 4), $(NH_4)_2UO_3Cl_2\cdot xNH_3$ ($x = 1$ or 2) and $(NH_4)_2U_2O_5Cl_3\cdot xNH_3$ ($x = 1$, 2, 4 or 7) have also been recorded.

(*b*) *Aliphatic and aromatic amines.* The known amine complexes, all with dioxouranium(VI) compounds (Table 55), are obtained in a similar manner to the actinide(IV) complexes (p. 1137).

Table 54 Ammonia Complexes of Dioxouranium(VI) Compounds

$UO_2F_2 \cdot xNH_3$	$x = 2$, 3 or 4
$UO_2Cl_2 \cdot xNH_3$	$x = 0.5$, 1, 2, 3, 4, 5 or 10
$UO_2Y_2 \cdot xNH_3$	Y = Br, I, $x = 2$, 3 or 4
$UO_2Y_2 \cdot xNH_3 \cdot Et_2O$	Y = Cl, Br, I, NO_3 and $x = 2$; Y = Cl, NO_3 and $x = 3$
$UO_2(NO_3)_2 \cdot xNH_3$	$x = 2$ or 4
$UO_2(NO_3)_2 \cdot 3NH_3 \cdot 2H_2O$	
$UO_2(NH) \cdot xNH_3$	$x = 1$ or 2
$UO_2Y_2 \cdot xNH_3$	Y = $MeCO_2$, $0.5SO_4$ and $x = 2$, 3 or 4; Y = HCO_2, $0.5C_2O_4$ and $x = 2$
$UO_2(OH)Y \cdot NH_3$	Y = HCO_2, $0.5C_2O_4$
$UO_2(C_{10}H_8NO)_2 \cdot NH_3$	$C_{10}H_9NO$ = 7-methyl-8-hydroxyquinoline
$UO_2(RCOCHCOR)_2 \cdot NH_3$	R = CF_3, Ph

Table 55 Amine Complexes of Dioxouranium(VI) Compounds

$UO_2Cl_2 \cdot 2RNH_2$	R = $PhNH_2$, $H_2NC_6H_4NO_2$, 2-, 3-, 4-$MeC_6H_4NH_2$, 2-, 3-$MeOC_6H_4NH_2$, 4-$EtOC_6H_4NH_2$, $H_2NC_6H_3$-3, 4-Me_2, 1-, 2-$H_2NC_{10}H_7$
$UO_2Cl_2 \cdot 2$ to $3RNH_2$	R = Me, Et, Pr[n], Bu[n]
$UO_2Cl_2 \cdot R_2NH$	R_2 = Ph_2, $(PhCH_2)_2$, (Ph)(Et), (Ph)($PhCH_2$)
$UO_2Cl_2 \cdot 2L$	L = Ph_3N, PhCH=NPh, $Et_2N(C_2H_4OH)$
$UO_2(NO_3)_2 \cdot 2L$	L = $MeNH_2$, $PhNH_2$, 2-$H_2NC_6H_4OH$, Et_2NH, Et_3N
$UO_2(HCO_2)_2 \cdot 2MeNH_2$[a]	
$UO_2(MeCO_2)_2 \cdot 2RNH_2$	R = Me,[a] Ph
$UO_2(EtCO_2)_2 \cdot 2MeNH_2$[a]	
$UO_2SO_4 \cdot 2PhNH_2 \cdot 3H_2O$	
$UO_2C_2O_4 \cdot 2PhNH_2 \cdot 2H_2O$	
$UO_2(C_7H_3NO_4) \cdot 2C_3H_7N$	C_3H_7N = allylamine; $C_7H_5NO_4$ = pyridine-2,6-dicarboxylic acid
$UO_2A_2 \cdot L$	HA = H(acac) with L = Me_3N and Et_3N; $PhCOCH_2COPh$, $C_7H_6O_2$ (tropolone), C_9H_7NO (8-hydroxyquinoline) with L = $PhNH_2$
$[UO_2(S_2CNEt_2)_2(Et_2NH)]$	

[a] IR data are given in G. N. Mazo, N. A. Santalova and K. M. Dunaeva, *Sov. Radiochem.* (*Engl. Transl.*), 1981, **23**, 547.

The structure of the complex with the *N,N*-diethyldithiocarbamate $[UO_2(S_2CNEt_2)_2(Et_2NH)]$ is pentagonal bipyramidal, with four sulfur atoms and the amine nitrogen atom in the equatorial plane.[213] No other structural information is available for the remaining amine complexes, but it is probable that the monoamine complexes of the β-diketonates, tropolonate and 8-hydroxyquinolinate, UO_2A_2L, have the same coordination geometry. The bis amine complexes of dioxouranium(VI) nitrate and the monocarboxylates, $UO_2A_2L_2$, may involve eight-coordinate uranium(VI) if the anions A^- are bidentate, in which case hexagonal bipyramidal geometry would be expected. In other *bis* amine complexes of the type $UO_2Cl_2L_2$ the geometry is presumably octahedral.

A number of aminoacid adducts, such as $UO_2(MeCO_2)_2 \cdot xH_2NCH_2CO_2H$ ($x = 1$, 2 or 3) have been recorded and in these the amino nitrogen is probably the only donor atom. Di- and tri-ethylamine adducts of several diolates are also known.

(*c*) *Hydrazines.* The only recorded complexes are those with dioxouranium(VI) compounds (Table 56). Very little is known about them.

Table 56 Complexes of Dioxouranium(VI) Compounds with Hydrazines

$UO_2X_2 \cdot 4L$[a]	X = Cl, Br, I, NO_3, NCS; L = N_2H_4, $PhNHNH_2$, Me_2NNH_2
$(UO_2Cl_2)_2 \cdot N_2H_4 \cdot 6H_2O$	
$UO_2X \cdot yN_2H_4 \cdot zH_2O$	X = SO_4, $y = 2$, $z = 0$; X = C_2O_4, $y = 1$, $z = 0.75$ and $y = 2$, $z = 1$
$UO_2SO_4 \cdot PhNHNH_2$	
$(UO_2)_2(OH)_2X_2 \cdot yN_2H_4 \cdot zH_2O$	X = Cl, $y = 1$, $z = 6$ and $y = 3$, $z = 0$; X_2 = SO_4, $y = 1$, $z = 0$

[a] A. K. Srivastava, R. K. Agarwal, V. Kapur, S. Sharma and P. C. Jain, *Transition. Met. Chem.* (*Weinheim, Ger.*), 1982, **7**, 41.

(*d*) *Ethylenediamine and other diamines.* In addition to the complexes listed in Table 57, a considerable number of 1:1 complexes of dioxouranium(VI) Schiff base compounds with en and Me_4en are also known. No structural information is available for these compounds. The phosphate compound, UO_2HPO_4·en, precipitated when en is added to a 95% ethanol solution of $UO_2(NO_3)_2·2H_2O$ containing phosphate ion, is probably an ethylenediammonium salt, $(enH)UO_2PO_4$. The hydrated 2,3-diaminotoluene complex, $UO_2(NO_3)_2·2L·2H_2O$, is reported to be maroon violet,[214] a colour which is unusual for dioxouranium(VI) compounds.

Table 57 Complexes of Ethylenediamine and other Polydentate Nitrogen Ligands with Dioxouranium(VI) Compounds

Ethylenediamine, en	
$UO_2(NO_3)_2·x$en	$x = 1$ or 2
$UO_2(HPO_4)$·en (?)	
L = 1,2-diaminopropane	
$UO_2(NO_3)_2·L$	
L = hexamethylenetetramine, $(CH_2)_6N_4$	
$UO_2X_2·L$	X = NCS, $MeCO_2$
Diaminoarenes	
$UO_2Cl_2·xL$	$x = 1$ or 2, L = 1,2-diaminobenzene
$UO_2Cl_2·2L$	L = 1,3- or 1,4-diaminobenzene
$UO_2(NO_3)_2·2L·H_2O$	L = 1,2-diaminobenzene
$UO_2(NO_3)_2·xL·yH_2O$	$x = 3$, $y = 0$, L = 3,4-diaminotoluene; $x = y = 2$, L = 2,3-diaminotoluene
$UO_2SO_4·L·THF$	L = 4,4′-diaminobiphenyl (benzidine)

(*ii*) *N-Heterocyclic Ligands*

(*a*) *Pyridine, C_5H_5N, and derivatives.* The central metal atom in the complexes $UO_2Cl_2·2L$ (Table 58), where L is pyridine or a monosubstituted pyridine, is probably six-coordinate, whereas in the complexes UO_2X_2L in which X is a β-diketone, *N,N*-dialkyldithiocarbamate or 8-hydroxyquinolinate, and in $[UO_2(py)_5](ClO_4)_2$, the coordination number is almost certainly seven, with pentagonal bipyramidal geometry, as in the diethylamine complex $[UO_2(S_2CNEt_2)_2(NHEt_2)]$ (p. 1188). The metal atom is probably eight-coordinate in the complexes $UO_2Cl_2·4py$ and $UO_2(NO_3)_2·2py$, but this requires confirmation.

Table 58 Some Complexes of N-Heterocyclic Ligands with Dioxouranium(VI) Compounds

Pyridine and substituted pyridines	
$UO_2X_2·yL$	X = Cl, $y = 1$ (+H_2O) or 2, L = py; $y = 2$, L = 2-Me-, 3-Me- and 2-H_2N-C_5H_4N; $y = 4$, L = py
	X = NO_3, $y = 1$ (+H_2O or Et_2O) or 2, L = py
	X = OPh, O(2- or 4-MeC_6H_4), O(4-ClC_6H_4), $y = 1$, L = py
	X = O(2-MeC_6H_4), $y = 3$, L = py
	X = MeCOCHCOMe, MeCOCHCOPh, PhCOCHCOPh, $y = 1$, L = py
	X = MeCOCHCOMe, $y = 1$, L = 3- or 4-H_2N-, 4-HO-, 3- or 4-MeCO-, 3- or 4-NC-, 3-Cl-, or 4-Me-C_5H_4N
	X = S_2CNEt_2, $y = 1$, L = py, 4-Bu^t-, 4-Ph-, 4-Ph$(CH_2)_3$-, or 2-$(CHEt_2)$-C_5H_4N
	X = C_9H_6NO, $y = 1$, L = py; C_9H_7NO = 8-hydroxyquinoline
$UO_2X·y$py	X = SO_4, $y = 2$ or 3; Cr_2O_7, $y = 2$; dibasic Schiff base anions, $y = 1$
$[UO_2(py)_5](ClO_4)_2$	
Piperidine, $C_5H_{11}CN$	
$UO_2L_2·C_5H_{11}N$	HL = $CF_3COCH_2CO(2\text{-}C_4H_3S)$ or *n*-$C_3F_7COCH_2COBu^t$
Quinolines	
$UO_2X_2·L$	X = NO_3, L = quinoline (+H_2O) or 2-methylquinoline (+H_2O)
	X = MeCOCHCOMe, L = quinoline
	X = C_9H_6NO (8-hydroxyquinolinate), L = quinoline
$UO_2SO_4·L$	L = 2-methylquinoline
$UO_2Cr_2O_7·2L$	L = isoquinoline

(*b*) *Piperidine*, $C_5H_{11}N$. 1:1 complexes with the β-diketonates $UO_2\{CF_3COCHCO(C_4H_3S)\}_2$ and $UO_2(n\text{-}C_3F_7COCH_2COBu^t)_2$ are known; in both cases the uranium atom is probably seven-coordinate.

(*c*) *Quinoline and isoquinoline*, C_9H_7N. The few known quinoline, isoquinoline and 2-methylquinoline complexes with dioxouranium(VI) compounds are included in Table 58. The uranium atom in the quinoline complexes with the acetylacetonate and 8-hydroxyquinolinate, $UO_2(acac)_2{\cdot}C_9H_7N$ and $UO_2(C_9H_6NO)_2{\cdot}C_9H_7N$, the uranium atom is presumably seven-coordinate.

(*d*) 2,2′-*Bipyridyl, bipy, and* 4,4′-*bipyridyl*, $C_{10}H_8N_2$. In addition to the complexes listed in Table 59, a number of compounds with dioxouranium(VI) Schiff base products are also known. The complex with the cyanocyanamide, $UO_2[N(CN)_2]_2{\cdot}2bipy$, is obtained by reaction of $UO_2(NO_3)_2$ with $K[N(CN)_2]$ and bipy.[215] The dioxoplutonium(VI) compound formulated as $PuO_2Cl_2(bipy{\cdot}HCl)_2$ should be regarded as the chloro complex salt, $(bipy{\cdot}H)_2[PuO_2Cl_4]$. $UO_2(MeCO_2)_2{\cdot}bipy$ and $UO_2(OH)(MeCO_2){\cdot}bipy$ behave as non-electrolytes in water or methanol whereas $UO_2Cl_2{\cdot}bipy{\cdot}H_2O$ and $UO_2(NO_3)_2{\cdot}bipy$ behave as 1:1 electrolytes, $[UO_2Cl(H_2O)(bipy)]Cl$ and $[UO_2(NO_3)(bipy)](NO_3)$.[216]

Table 59 Complexes of 2,2′-Bipyridyl, bipy, and 4,4′-Bipyridyl, $C_{10}H_8N_2$, with Dioxouranium(VI) Compounds

$UO_2X_2{\cdot}y$bipy	X = Cl, NO_3, $MeCO_2$, 2-, 3- or 4-$H_2NC_6H_4CO_2$, $C_8H_4F_3S$ (tta), $y = 1$
	X = NCS, NCSe, NO_3, $N(CN)_2$, $y = 2$
$UO_2X_2{\cdot}yC_{10}H_8N_2$	X = Cl, $y = 1$; X = NO_3, $y = 1.5$
$UO_2Cl_2{\cdot}bipy{\cdot}xH_2O$	$x = 1$ or 2
$UO_2(OH)X{\cdot}bipy$	X = NO_3, $MeCO_2$
$UO_2X{\cdot}y$bipy	X = SO_4, $y = 1$; X = Cr_2O_7, $y = 2$
$UO_2SO_4{\cdot}C_{10}H_8N_2$	
$UO_2X{\cdot}bipy{\cdot}yH_2O$	X = SO_4, $y = 4$; X = CrO_4, C_2O_4, $y = 1$

(*e*) *Imidazoles*, $C_3H_4N_2$. The only recorded complexes with this type of ligand appear to be those formed by 2-(2′-pyridyl)benzimidazole ($C_{12}H_9N_3$) with dioxouranium(VI) compounds, for which products of composition $UO_2X_2(C_{12}H_9N_3)$ (X = Cl, I, NO_3 or $0.5SO_4$), $UO_2(NCS)_2(C_{12}H_9N_3)_2$ and $UO_2(ClO_4)_2(C_{12}H_9N_3)_3$ have been reported.

(*f*) *Dipyridylalkanes*. Complexes of dioxouranium(VI) compounds with these ligands are listed in Table 60. Very little is known about them.

Table 60 Complexes of Dipyridylalkanes with Dioxouranium(VI) Compounds

1-(2-Pyridyl)-2-(3-pyridyl)ethane, $C_{12}H_{12}N_2$	
$UO_2(NCS)_2{\cdot}C_{12}H_{12}N_2{\cdot}H_2O$	
1-(3-Pyridyl)-2-(4-pyridyl)ethane, $C_{12}H_{12}N_2$	
$UO_2X_2{\cdot}yC_{12}H_{12}N_2$	X = NO_3, $MeCO_2$, $y = 1$
	X = NCS, ClO_4, $y = 2$
1,2-Di(4-pyridyl)ethane, $C_{12}H_{12}N_2$	
$UO_2X_2{\cdot}2C_{12}H_{12}N_2$	X = NCS, NO_3, ClO_4, $MeCO_2$
1,3-Di(4-pyridyl)propane, $C_{13}H_{14}N_2$	
$UO_2X_2{\cdot}C_{13}H_{14}N_2$	X = Cl, NO_3, $0.5SO_4$

(*g*) *Di(2-pyridyl)amine*, $C_{10}H_9N_3$. Dioxouranium(VI) compounds of composition $UO_2X_2{\cdot}yC_{10}H_9N_3$ (X = Cl, NCS, $y = 1$; X = NO_3, $0.5SO_4$, $y = 2$) have been reported. The imine nitrogen atom is probably not bonded to the uranium atom in these compounds.

(*h*) 1,10-*Phenanthroline, phen*. In addition to the compounds listed in Table 61, a few complexes with dioxouranium(VI) Schiff base compounds are also known. Conductivity measurements indicate that $UO_2Cl_2{\cdot}2phen$ is ionic in methanol, $[UO_2(phen)_2]Cl_2$, whereas $UO_2Cl_2{\cdot}phen$ is a non-electrolyte and $UO_2(NCS)_2{\cdot}2phen$ behaves as a 1:1 electrolyte, $[UO_2(NCS)(phen)_2]NCS$, in that solvent. Only one of the water molecules in the complex formulated as $(UO_2)_2(OH)_2SO_4(phen)_2{\cdot}2H_2O$ appears to be bonded to the metal atom. A hydroxo-bridged structure has been suggested for the complex with the basic acetate, $UO_2(OH)(MeCO_2){\cdot}phen$.

Table 61 Complexes of 1,10-Phenanthroline (phen) with Dioxouranium(VI) Compounds

UO_2X_2·phen	X = F,[a] Cl,[a] NO_3 (also +0.5H_2O),[a] $MeCO_2$ (also +2.5H_2O), Me_3CCO_2, 2-, 3- or 4-$H_2NC_6H_4CO_2$, $CF_3COCHCO$(2-C_4H_3S), *n*-$C_3F_7COCHCOBu^t$, C_9H_6NO (8-hydroxyquinolinate)
UO_2X_2·2phen	X = Cl, NCS, NCSe, $N(CN)_2$[c]
UO_2X·phen	X = SO_3 (+H_2O),[b] SO_4 (also +5H_2O), CrO_4, $C_5H_3N(O)$-2,6-$(CO_2)_2$,[d] C_2O_4 (+H_2O)
$UO_2Cr_2O_7$·2phen	
$UO_2(OH)X$·phen·yH_2O	y = 0, X = $MeCO_2$; y = 2, X = Cl, NO_3; y = 2.5, X = NO_3
$UO_2(NO_3)X$·phen·yH_2O	y = 0, X = 2-$H_2NC_6H_4CO_2$; y = 2, X = 4-$H_2NC_6H_4CO_2$
$(UO_2)_2(OH)_2(SO_4)$·2phen·yH_2O	y = 1 or 2

[a] The corresponding 2,9-dimethyl-1,10-phenanthroline complexes are also known.
[b] R. N. Shchelokov, G. T. Bolotova and N. A. Golubkova, *Sov. J. Coord. Chem. (Engl. Transl.)*, 1977, **3**, 810.
[c] E. L. Ivanova, V. V. Skopenko and Kh. Keller, *Ukr. Khim. Zh. (Russ. Ed.)*, 1981, **47**, 1024 (*Chem. Abstr.*, 1982, **96**, 45 232).
[d] Polymeric; S. Degetto, L. Baracco, M. Faggin and E. Celon, *J. Inorg. Nucl. Chem.*, 1981, **43**, 2413.

(*i*) *5-Methylpyrazole, $C_4H_6N_2$.* One complex only, $UO_2(S_2CNEt_2)_2·C_4H_6N_2$, has been recorded.

(*j*) *Pyrazine, C_4H_4N and derivatives.* Hydrated complexes of composition $UO_2Cl_2·L·H_2O$, where L = pyrazine, 2-methylpyrazine and 2,5- and 2,6-dimethylpyrazine, have been recorded.

(*iii*) *Triazenes, cyanates and thiocyanates*

(*a*) *Triazenes.* A few complexes with substituted 1,3,5-triazenes have been recorded, including those with the 2,4,6-tris(dimethylamino) derivative (hexamethylmelamine), $UO_2Cl_2·C_9H_{18}N_6$, and the hydrated complex with the tris(2-pyrimidyl) derivative, $UO_2(NO_3)_2·C_{15}H_9N_9·7H_2O$.

(*b*) *Cyanates.* The hydrated complex salt of composition $(Et_4N)_2[UO_2(NCO)_4(H_2O)]$ is precipitated when an acetone solution of $UO_2(NO_3)_2·6H_2O$ is added to an ethanol solution of $Et_4N(NCO)$[217] and a basic salt, $K_3[(U_2O_5)(NCO)_5]·H_2O$, has also been reported. There is IR evidence[218] for the formation of the cyanato complex anions $[UO_2(NCO)_4]^{2-}$ and $[UO_2Cl_3(NCO)]^{2-}$ in non-aqueous solutions containing $R_4N[UO_2Cl_3]$ and $R_4N[Ag(NCO)_2]$ (R = $C_{10}H_{21}$).

(*c*) *Thiocyanates.* The anhydrous and a number of hydrated forms of dioxouranium(VI) thiocyanate, $UO_2(NCS)_2$, are known, but no structural information is available for these. The complex thiocyanato anion is usually of the form $[UO_2(NCS)_5]^{3-}$ (Table 62), and in the cesium salt of this anion the coordination geometry is a distorted pentagonal bipyramid.[219] A few salts of tri-, tetra- and hexa-thiocyanato complex anions are also known, but no structural information is available for them. Mixed thiocyanato-acetato and -oxalato complexes have been recorded, as well as salts of complex anions of the type $[UO_2(NCS)_3L_2]^-$, where L is a neutral ligand, but very little is known about them.

Table 62 Dioxouranium(VI) Thiocyanate and Thiocyanato Complexes

$UO_2(NCS)_2·xH_2O$	x = 0, 1, 2 or 8
$M^I[UO_2(NCS)_3(H_2O)_x]$	x = 0, M^I = $N(C_{10}H_{21})_4$; x = 2, M^I = K, Rb and NH_4
$(NH_4)_2[UO_2(NCS)_4(H_2O)_x]$	x = 0 or 1
$M^I_3[UO_2(NCS)_5]·xH_2O$	x = 0, M^I = K, NH_4, Cs, Et_3NH, $N(C_{10}H_{21})_4$, bipyH x = 1, M^I = K, NH_4, Et_3NH, pyH, C_9H_7NH and $C_5H_{11}NH$ x = 2 or 6, M^I = K, NH_4
$[Co(en)_3]_2[UO_2(NCS)_6](NO_3)_2·5H_2O$	
$\{N(C_{10}H_{21})_4\}_2[(UO_2)_3(NCS)_8]$	

(*iv*) *Nitriles*

Methyl cyanide complexes of composition $UO_2X_2·y$MeCN (X = Cl, y = 1; X = NO_3, y = 2; X = ClO_4, y = 3), $\{(C_{10}H_{21})_4N\}NpO_2Cl_3$·MeCN and $UO_2SB·x$MeCN, where SB represents a Schiff base anion, have been recorded. In the IR spectra of the last, the frequency of ν(CN) is almost identical to that of the free ligand. A cyanogen chloride adduct, $UO_2Cl_2{\cdot}NCCl$, is also known.

(v) *Oximes*

A wide variety of dioxouranium(VI) aldoxime and ketoxime complexes of composition UO_2L_2 are known. Examples of the ligands used are 2-hydroxy-naphthaldehydeoxime, $HOC_{10}H_6CH{=}NOH$, salicylaldoxime, $HOC_6H_4CH{=}NOH$, isatin α- and β-oximes and substituted 2-hydroxyacetophenone oximes.[220] In the structure of the 1,2-naphthoquinone-2-oxime complex, $[UO_2L_2(H_2O)_2]{\cdot}2CHCl_3$, the coordination geometry is hexagonal bipyramidal.[221]

40.2.4.3 *Oxygen ligands*

(i) *Aqua Species, hydroxides and oxides*

(*a*) *Aqua species.* A short selection of the very wide range of aquated actinide(VI) compounds is given in Table 63. Although the hydrated thiocyanato complex salts have been reported as $M^I[UO_2(NCS)_3(H_2O)_2]$ (M^I = K, Rb, NH_4), the number of bonded water molecules is uncertain. The coordination geometry of the anion in $(enH_2)[UO_2F_4(H_2O)]$ is pentagonal bipyramidal[222] and the same coordination geometry has been found for the uranium atom in $UO_2(MeCO_2)_2{\cdot}2H_2O$, in which only one water molecule is bonded to the metal atom. Half of the acetate groups are chelated to individual uranium atoms, and the remainder link adjacent bipyramids to form chains.[223] In the hydrated oxalate, $UO_2C_2O_4{\cdot}3H_2O$, each uranium atom is again seven-coordinate in a pentagonal bipyramidal array with four oxygen atoms from two independent C_2O_4 groups and one from the water molecule in the equatorial plane; the C_2O_4 groups are tetradentate in this structure.[224]

Table 63 Some Hydrates of Dioxoactinide(VI) Compounds

$UO_2X_2{\cdot}yH_2O$	X = Cl, y = 1, 3; X = Br, y = 3; X = NCS, y = 1, 3;
	X = acac, y = 1
$M^{VI}O_2(NO_3)_2{\cdot}6H_2O$	M^{VI} = U, Np, Pu
$M^{VI}O_2(ClO_4)_2{\cdot}xH_2O$	M^{VI} = U, x = 3, 5, 7; M^{VI} = Pu, x = 6
$M^{VI}O_2(IO_3)_2{\cdot}xH_2O$	M^{VI} = U, x = 1, 2; M^{VI} = Np, x = 2
$UO_2(RCO_2)_2{\cdot}xH_2O$	R = H, x = 1; R = Me, Et, Pr^n, Pr^i, Bu^n, x = 2
$M^{VI}O_2(C_5H_4N\text{-}3\text{-}CO_2)_2{\cdot}2H_2O$	M^{VI} = U, Np, Pu
$M^{VI}O_2C_2O_4{\cdot}3H_2O$	M^{VI} = U, Np, Pu
$UO_2(H_2PO_4)_2{\cdot}3H_2O$	
$(UO_2)_3(PO_4)_2{\cdot}3H_2O$	
$(enH_2)[UO_2F_4(H_2O)]$	
$M^IPuO_2F_3{\cdot}H_2O$	M^I = Na, K, Rb, Cs, NH_4
$M^{II}UO_2F_4{\cdot}4H_2O$	M^{II} = Zn, Cd, Cu, Mn, Co, Ni
$Rb_2UO_2Cl_4{\cdot}2H_2O$	
$M^I_2U_2O_5Cl_4{\cdot}2H_2O$	M^I = Rb, Cs
$M^I[UO_2(NCS)_3(H_2O)_2]$	M^I = K, Rb, NH_4
$M^{II}[UO_2(MeCO_2)_3]_2{\cdot}xH_2O$	M^{II} = Mg, x = 6, 7, 8, 12; M^{II} = Ca, x = 6;
	M^{II} = Sr, x = 2, 6; M^{II} = Ba, x = 2, 3, 6, 10
$M^{II}_2[UO_2(CO_3)_3]{\cdot}xH_2O$	M^{II} = Mg, x = 16–18, 20; M^{II} = Ca, x = 4;
	M^{II} = Sr, x = 9; M^{II} = Ba, x = 5, 6
$M^IM^{VI}O_2PO_4{\cdot}xH_2O$	M^{VI} = U, M^I = H, Na, NH_4, x = 3; M^I = Li, Na, K, x = 4;
	M^{VI} = Np, M^I = H, Li, x = 4; M^I = Na, K, NH_4, x = 3
$M^{II}(M^{VI}O_2PO_4)_2{\cdot}xH_2O$	M^{VI} = U, M^{II} = Ca, Sr, x = 6; Cu, x = 8;
	M^{VI} = Np, M^{II} = Ca, Sr, Ba, x = 6
$M^IM^{VI}O_2AsO_4{\cdot}xH_2O$	M^{VI} = U, M^I = H, x = 4; M^I = NH_4, x = 3
	M^{VI} = Np, M^I = H, Li, x = 4; M^I = Na, x = 3.5;
	M^I = K, NH_4, x = 3
$M^{II}(M^{VI}O_2AsO_4)_2{\cdot}xH_2O$	M^{VI} = U, M^{II} = Mg, Zn,[a] Ni,[a] Co,[a] x = 8
	M^{VI} = Np, M^{II} = Mg, x = 8, 10; M^{II} = Ca, x = 6, 10;
	M^{II} = Sr, x = 8; M^{II} = Ba, x = 7
$Na_4NpO_2(O_2)_3{\cdot}9H_2O$	

[a] M. A. Nabar and V. J. Iyer, *Bull. Soc. Fr. Mineral. Crystallogr.*, 1977, **100**, 272.

Four crystal modifications of $[UO_2(acac)_2(H_2O)]$ have been reported; in the β form the molecular geometry is a pentagonal bipyramid with four oxygen atoms from the acac groups and one from the water molecule in the equatorial plane.[225]

The compounds of composition $M^I_2U_2O_5Cl_4 \cdot 2H_2O$ (M^I = Rb, Cs), obtained by the reaction of $UO_2(OH)_2$ with $M^IUO_2Cl_4 \cdot xH_2O$ (M^I = Rb, $x = 2$ and M^I = Cs, $x = 0$), are actually $M^I_4[(UO_2)_4O_2Cl_8(H_2O)_2] \cdot 2H_2O$, with the uranium atoms in the equatorial plane of a pentagonal bipyramidal arrangement of triply bridging oxygen and doubly bridging chlorine atoms.[226] In the analogous hydrated basic chloride, $K_2(UO_2)_4O_2(OH)_2Cl_4 \cdot 6H_2O$, the structure consists of discrete tetranuclear $[(UO_2)_4O_2(OH)_2Cl_4(H_2O)_4]^{2-}$ ions, with four almost parallel OUO groups linked through double oxo or mixed oxo-chloro or oxo-hydroxo bridges; there are two coordination sites, both pentagonal bipyramidal, one with three oxygen and two chlorine atoms, and the other with four oxgyen and one chlorine atoms in the equatorial plane.[227]

It has been reported[228] that $UO_2HPO_4 \cdot 3H_2O$ behaves as a solid ionic conductor because of H^+ mobility across the water network in the compound, and it appears that the two-dimensional network of hydrogen bonds serves as the conduction path.[229] In the IR spectra of the perchlorate hydrates, $UO_2(ClO_4)_2 \cdot xH_2O$, the ClO_4 groups appear to be ionic for $x = 5$ and 7, but covalent and monodentate for $x = 3$. Structural information on other hydrated dioxoactinide(VI) compounds is included in the following sections which deal with the parent compounds.

(*b*) *Hydroxides.* The trioxide monohydrates, $M^{VI}O_3 \cdot H_2O$ (M^{VI} = U, Np, Pu), can be regarded as hydroxides, $M^{VI}O_2(OH)_2$; however, the IR spectrum of $NpO_2(OH)_2 \cdot H_2O$ does not appear to exhibit any feature assignable to the O=Np=O group.[230] The neptunium(VI) and plutonium(VI) compounds are obtained by ozone oxidation of aqueous suspensions of neptunium(V) and plutonium(IV) hydrated oxides; they cannot be dehydrated and decompose on heating to lower oxides.

The basic compound of composition $Zn(UO_2)_2SO_4(OH)_4 \cdot 1.5H_2O$ involves tridentate bridging OH groups; the structure consists of chains of pentagonal bipyramidal $UO_2(OH)_3O_2$ groups.[231] Many other hydroxo complexes, such as $TlUO_2(OH)_2(MeCO_2)$ and $M^I_3[UO_2(OH)(C_2O_4)_2]$ (M^I = Na, K, Et_2NH_2, CN_3H_6), have been recorded, but structural information is lacking for them.

(*c*) *Oxides.* UO_3 is the only anhydrous trioxide known, but a number of actinide(VI) ternary oxides have been recorded. The principal types are $M^I_2M^{VI}O_4$ (M^{VI} = U, Np, Pu[232]; M^I = Li, Na, K, Rb, Cs), $M^{II}M^{VI}O_4$ (M^{VI} = U, M^{II} = Mg, Ca, Sr, Ba; M^{VI} = Np, M^{II} = Ca, Sr, Ba; M^{VI} = Pu, M^{II} = Ca, Sr), $Li_4M^{VI}O_5$ (M^{VI} = Np, Pu), $M^{II}_2UO_5$ (M^{II} = Ca, Sr), $M^{II}_3UO_6$ (M^{II} = Ca, Sr, Ba), $Li_2U_3O_{10}$, $Cs_2Np_3O_{10}$,[232] $Li_6M^{VI}O_6$ (M^{VI} = Np, Pu) and[232] $M^I_2Np_2O_7$ (M^I = K, Rb, Cs).

The structure of the anion in $Li_2U_3O_{10}$ (= $Li_2[(UO_2)_3O_4]$) consists of octahedral $(UO_2)O_4$ and pentagonal bipyramidal $(UO_2)O_5$ groups linked by oxygen bridges.[231]

(*d*) *Superoxides and peroxides.* The only simple actinide(VI) peroxide appears to be the hydrated compound, $UO_4 \cdot 2H_2O$, which is probably of the form $[UO_2(O_2)(H_2O)_2]$. Hydrated salts of peroxo complex anions, such as $Na_4M^{VI}O_2(O_2)_3 \cdot 9H_2O$ (M^{VI} = U, Np) have been mentioned earlier, and complexes of composition $[U(O)(O_2)_2L_2] \cdot H_2O$, L = Ph_3PO, Ph_3AsO, pyNO) have also been reported.[233]

$UO_4 \cdot 2H_2O$ reacts with hydrated dioxouranium(VI) oxalate, $UO_2(C_2O_4) \cdot H_2O$ in the presence of aqueous alkali oxalate to form salts of peroxooxalato complex anions, $M^I_6[(UO_2)_2(\mu\text{-}O_2)(C_2O_4)_4]$ (M^I = Na, K, Rb). Salts of the peroxocarbonato complex anion, $[(UO_2)_2(\mu\text{-}O_2)(CO_3)_4]^{6-}$, are obtained in a similar manner.[234] Other reported salts of peroxooxalato and peroxocarbonato complex anions include $M^I_2[UO_2(O_2)(C_2O_4)]$ (M^I = Na, K, NH_4), $M^I_4[(UO_2)_2(O_2)_2(C_2O_4)_2(H_2O)]_4 \cdot 2H_2O$ (M^I = Na, K), $Na_8[(UO_2)_2(O_2)_2(C_2O_4)_4] \cdot 7H_2O$, $K_2[UO_2(O_2)(CO_3)(H_2O)_2] \cdot xH_2O$, $K_4[UO_2(O_2)(CO_3)_2]$ and $K_8[(UO_2)_2(O_2)(CO_3)_5]$.

(*iii*) *Alcohols, alkoxides and aryloxides*

(*a*) *Alcohols.* A considerable number of alcohol adducts of dioxouranium(VI) compounds have been recorded, some of which are listed in Table 64. They are obtained either as the primary product in the preparation of the parent compound using an alcohol as the solvent, or by treating the parent compound with an alcohol. For example, $UO_2Cl_2 \cdot 2EtOH$ is obtained by dissolving the hydrated chloride in a mixture of ethanol and benzene, the displaced water being

Table 64 Some Alcohol Adducts of Dioxouranium(VI) Compounds

$UO_2X_2{\cdot}y$ROH	X = Cl, R = Et, $y = 2$
	X = NO_3, R = Me, Et, $y = 2$ and R = Bu^i, Bu^t, $y = 3$
	X = $MeCO_2$, R = Me, Et, Pr^i, $Me_2CHCH_2CH_2$, $y = 2$
	X = R'_2NCS, R = R' = Et, $y = 4$; R' = Pr^n, Bu^n, Bu^i, Bu^s, $CH_2{=}CHCH_2$, R = Et, $y = 6$
	X = OR, R = Me, Pr^i, Bu^i, $Me_2CHCH_2CH_2$, $y = 1$; R = Et, $y = 2$
	X = tropolonate, R = Me, Et, $y = 1$
	X = C_9H_6NO (8-hydroxyquinolinate), R = Me, $y = 1$
$[UO_2L_2(ROH)](ClO_4)_2$[a]	R = Me, Et and L = $MeN[(Me_2N)_2PO]_2$
$[UO_2L(ROH)]$	L = $OC_6H_4CH{=}NCH_2CH_2N{=}CHC_6H_4O$, R = Me
	L = $OC_6H_4CH{=}NC_6H_4N{=}CHC_6H_4O$, R = Et
$UO(OR)_4{\cdot}ROH$	R = Pr^i, Bu^s, Bu^t
$U_2O_5(OR)_2{\cdot}2ROH$	R = Pr^i, Bu^s, Bu^t

[a] L. Rodehuser, P. R. Rubini, K. Bokolo and J. J. Delpuech, *Inorg. Chem.*, 1982, **21**, 1061.

removed as the H_2O—EtOH—C_6H_6 azeotrope, followed by evaporation of the solution to dryness under reduced pressure.

The coordination geometry about the uranium atom in the ethanol complex of the tropolonate, $[UO_2(trop)_2(EtOH)]$, is a distorted pentagonal bipyramid[235] and the same geometry has been found for the alcohol complexes with the 2,2,6,6-tetramethylheptane-3,5-dionate (thd), $[UO_2(thd)_2(MeOH)]$,[236] and with the Schiff base compounds, $[UO_2(MeC(O){=}CHCOCH{=}C(Me)NHCH_2CH_2NHC(Me){=}CHCOCH{=}C(O)Me)(MeOH)]$,[237] $[UO_2(OC_6H_4CH{=}NCH_2CH_2N{=}CHC_6H_4O)(MeOH)]$[238] and $[UO_2(OC_6H_4CH{=}NC_6H_4N{=}CHC_6H_4O)(EtOH)]$.[239] In the last two compounds two oxygen and two nitrogen atoms from the Schiff base anion and the alcohol oxygen atom take up the equatorial plane positions, whereas in the first three compounds four oxygen atoms are provided by the anion or anions and one by the alcohol molecule.

(*b*) *Alkoxides.* The main types of uranium(VI) alkoxide are summarized in Table 65. The hexaalkoxides, $U(OR)_6$, are monomeric, slightly volatile, but thermally unstable compounds. $U(OEt)_6$ was originally prepared by oxidation of $Na[U(OEt)_6]$ with benzoyl peroxide, and this was then used as the starting material to prepare other hexaalkoxides by reaction with higher alcohols, but the most convenient preparative route to compounds of this type is *via* $U(OMe)_6$, which can be obtained by the reaction of UF_6 with $Si(OMe)_4$ in dichloromethane, followed by alcohol exchange as above. The compound $U(OMe)F_5$ has been prepared by reaction of UF_6 with the stoichiometric quantity of MeOH at 90 °C in $CFCl_3$ and other fluoride alkoxides, $U(OMe)_{6-n}F_n$ ($n = 2$–5), have been obtained by reaction of UF_6 with the appropriate quantities of $U(OMe)_6$ or $Me_3Si(OMe)$. $U(OEt)_5Cl$ has been prepared by reaction of $U(OEt)_6$ with the stoichiometric quantity of dry hydrogen chloride in ether.

Table 65 Uranium(VI) Alkoxides

$U(OR)_6$	R = Me, Et, Pr^n, Pr^i, Bu^n, Bu^s, Bu^t, CH_2CMe_3
$U(OBu^t)_5(OMe)$	
$U(OMe)_{6-n}F_n$	$n = 1$ to 5
$U(OEt)_5Cl$	
$UO(OR)_4{\cdot}ROH$	R = Pr^i, Bu^s, Bu^t
$U_2O_5(OR)_2{\cdot}2ROH$	R = Pr^i, Bu^s, Bu^t
$UO_2(OR)_2$	R = Me, Et, Pr^n, *i*-C_5H_{11}, *t*-C_5H_{11}

The monooxoalkoxides, $UO(OR)_4{\cdot}ROH$, result from the reaction of $UO_2(OMe)_2{\cdot}MeOH$ with the appropriate alcohol, and the isopropoxide of this type has also been obtained by reaction of UO_2Cl_2 in Pr^{i}OH with ammonia or NaOPr^i, apparently as a result of disproportionation:

$$3UO_2(OR)_2{\cdot}ROH \xrightarrow{ROH} UO(OR)_4{\cdot}ROH + U_2O_5(OR)_2{\cdot}2ROH(\text{solid})$$

The nature of these products is unknown, but they may be mixed oxidation state U^V/U^{VI} species. Dioxouranium(VI) alkoxides, $UO_2(OR)_2$, and some alcohol solvates are also known.

(*c*) *Aryloxides.* The only aryloxides known are the dioxouranium(VI) compounds

$UO_2(OR)_2$ with $R = 2\text{-}MeOC_6H_4$ or $1\text{-}NOC_{10}H_6\text{-}2$, and the products formed when $UO_2(MeCO_2)_2$ is dissolved in a carbon tetrachloride solution of a phenol in the presence of an amine; these may be salts of the form $M^I_2[UO_2(OR)_4]$ where $R = Ph$, 2- or $4\text{-}MeC_6H_4$ with $M^I = Et_2NH_2$ or Et_3NH, and $R = 2\text{-}$ or $4\text{-}ClC_6H_4$ with $M^I = Et_3NH$.

(*d*) *Silyloxides*. The only recorded compounds are $U(OSiR_3)_6$ ($R = Me, Et$), $U\{OSi(Me_2Et)\}_6$ and $U\{OSi(MeEt_2)\}_6$, prepared by reaction of $U(OEt)_6$ with a slight excess of the trialkylsilanol in benzene under reflux.

(*iv*) *β-Ketoenolates, tropolonates, catecholates, quinones, ethers, ketones, aldehydes and esters*

(*a*) *β-Ketoenolates*. In a few instances the β-ketoenol can act as a neutral ligand, as in $UO_2(NO_3)_2{\cdot}2PhCOCH_2CO(2\text{- or }4\text{-}C_5H_4N)$, but the common type of complex is of composition UO_2L_2, for which a wide range of compounds is known (Table 66). In the compound $UO_2(PhCOCHCOCO_2H)_2$, the carboxylate group is probably bonded to the metal atom. Several β-ketoamide complexes, such as $UO_2(MeCOCHCONHR)_2{\cdot}MeCOCH_2CONHR$ ($R = Ph$, $2\text{-}ClC_6H_4$, $2\text{-}MeOC_6H_4$, $2\text{-}MeC_6H_4$, $3,5\text{-}Me_2C_6H_3$) and $UO_2(MeCOCHCONHR)_2{\cdot}2H_2O$ ($R = Ph$, $3\text{-}O_2NC_6H_4$), as well as a considerable number of complexes derived from cyclic β-diketones, such as 2-hydroxy-1-naphthaldehyde, 2-hydroxybenzaldehyde, 1-hydroxy-anthracene-9,10-dione and pyrimidine-2,4,6-trione (barbituric acid), are also known.

Table 66 Some Dioxouranium(VI) β-Ketoenolates

$UO_2(R^1COCHCOR^2)_2$
$R^1 = Me$, $R^2 = H$, Me, Et, Pr^n, Pr^i, Bu^t, $n\text{-}C_5H_{11}$, $n\text{-}C_6H_{11}$, $n\text{-}C_{17}H_{35}$, Ph, $C_6H_4(p\text{-}NHCH_2CO_2H)$, $2\text{-}C_4H_3O$ (furyl), 2- or $3\text{-}C_5H_4N$ ($+H_2O$)
$R^1 = Pr^n$, $R^2 = H$, Pr^n
$R^1 = Bu^t$, $R^2 = H$, Bu^t ($+H_2O$)
$R^1 = CF_3$, $R^2 = Me$, CF_3 ($+H_2O$), Bu^t, Ph, OEt ($+H_2O$), $2\text{-}C_4H_3S$ (thienyl), 2- or $3\text{-}C_5H_4N$
$R^1 = n\text{-}C_3F_7$, $R^2 = Bu^t$
$R^1 = Ph$, $R^2 = H$, Ph ($+H_2O$), *m*- or $p\text{-}H_2NC_6H_4$, *o*-, *m*- ($+H_2O$) or $p\text{-}O_2NC_6H_4$ ($+H_2O$), *o*-, *m*- or $p\text{-}MeOC_6H_4$, $2\text{-}C_4H_3O$ (furyl), CO_2H, CO_2Me ($+H_2O$), CO_2Et
$R^1 = R^2 = 3,5\text{-}Me_2C_6H_5$
$UO_2(MeCOCHClCOMe)_2$
$UO_2\{RCOCHCO(\eta\text{-}C_5H_4)Fe(\eta\text{-}C_5H_5)\}_2$
$R = Me$, CF_3, Ph ($+H_2O$)
$UO_2(Fe\{PhCOCHCO(\eta\text{-}C_5H_4)\}_2{\cdot}H_2O$
$UO_2(MeCOCHCO\{\eta\text{-}C_6H_5Cr(CO)_3\})_2{\cdot}H_2O$

(*b*) *Tropolonates*. Tropolonates of composition $M^{VI}O_2(trop)_2$ ($M^{VI} = U$, Pu), $M^{VI}O_2(trop)_2{\cdot}Htrop$ ($M^{VI} = U$, Pu) and $UO_2(trop)_2{\cdot}H_2O$ have been recorded, as well as complexes of composition UO_2L_2 with ligands related to tropolone, such as 3-hydroxy-2-methyl-4*H*-thiopyran-4-one (thiomaltol) and 3-hydroxy-2-methyl-4*H*-pyran-4-one (maltol; trihydrate), and derivatives of more complex 1,2-diketones.

(*c*) *Catecholates*. Pyrocatecholates of composition $UO_2(1,2\text{-}C_6H_4O_2){\cdot}xH_2O$ ($x = 1$ or 3) and $(pyH)H[UO_2(1,2\text{-}C_6H_4O_2)(OH)]{\cdot}H_2O$, as well as the resorcinol compound, $UO_2(1,3\text{-}C_6H_4O_2)$, have been reported.

(*d*) *Quinones*. A complex of composition UO_2L_2, where HL = 2-hydroxy-1,4-naphthaquinone, is known.

(*e*) *Ethers*. Several complexes of dioxouranium(VI) compounds with ethers have been reported (Table 67). In the case of $[UO_2(NO_3)_2(THF)_2]$ the eight-coordination geometry is hexagonal bipyramidal, with four oxygen atoms from two bidentate nitrate groups and two oxygen atoms from the THF molecules in the equatorial plane.[240] The coordination geometry in the THF complexes with dioxouranium(VI) *N*-phenylbenzohydroxamate (L) chloride, $[UO_2Cl(L)(THF)_2]$,[241] and the volatile hexafluoroacetylacetonate, $[UO_2(CF_3COCHCOCF_3)_2(THF)]$,[242] are pentagonal bipyramidal. In the former the equatorial plane positions are occupied by two hydroxamate and two THF oxygen atoms, together with the chlorine atom.

(*f*) *Ketones*. A selection of complexes of dioxouranium(VI) compounds with ketones is given in Table 68. In the dibenzylidene acetone complex, $[UO_2Cl_2\{(PhCH{=}CH)_2CO\}_2]{\cdot}2MeCO_2H$, the acetic acid molecules, which were introduced when the complex was

Table 67 Some Ether Complexes of Dioxouranium(VI) Compounds

$UO_2Cl_2 \cdot xEt_2O \cdot yH_2O$	$x=1$, $y=0$ or 2; $x=2$, $y=1$
$UO_2(NO_3)_2 \cdot xL \cdot yH_2O$	$L=Et_2O$, $x=2$, $y=0$ or 2; $x=1$, $y=3$; $x=4$, $y=2$
	$L=Pr^n_2O$, $x=2$, $y=0$ or 2; $x=1$, $y=3$
	$L=Pr^i_2O$, $x=2$, $y=0$; $x=3$ or 4, $y=3$
	$L=Bu^n_2O$, $(ClC_2H_4)_2O$, THF or $C_4H_8O_2$ (dioxane), $x=2$, $y=0$
	$L=EtOCH_2CH_2OEt$ or $(Bu^nOC_2H_4)_2O$, $x=1$, $y=3$; $x=y=2$
$UO_2(NCS)_2 \cdot 3THF$	
$UO_2(NCX)_2 \cdot 3C_4H_8O_2$	X = S, Se
$UO_2(ClO_4)_2 \cdot xL \cdot yH_2O$	$L=Et_2O$, $x=3$, $y=0$; $x=4$, $y=1$
	$L=Bu^n_2O$, $x=2$, $y=0$
	$L=C_4H_8O_2$, $x=5$, $y=6$
$UO_2(acac)_2 \cdot xC_4H_8O_2$	$x=1$ or 1.25
$UO_2(OBu^t)_2 \cdot 2THF$	
$UO_2(Cl_3CCO_2)_2 \cdot xEt_2O \cdot yH_2O$	$x=1$, $y=0$, 1 or 2; $x=y=2$

Table 68 Some Ketone Complexes of Dioxouranium(VI) Compounds

L = ketones RCOR′	
$[UO_2Cl_2L_2] \cdot 2MeCO_2H$	R = R′ = PhCH=CH
$UO_2(NO_3)_2 \cdot xL \cdot yH_2O$	R = R′ = Me, $x=1$, $y=2$ or 3; $x=2$, $y=0$
	R = Me, R′ = Et, $x=1$, $y=3$; $x=2$, $y=0$
	R = Me, R′ = Bu^i, $x=1$, $y=2$; $x=2$, $y=0$
$UO_2X_2 \cdot L$	R = R′ = Me, X = NCO, acac, trop
$UO_2SO_4 \cdot L \cdot 2H_2O$	R = R′ = Me
L = cyclic ketones	
$UO_2(NO_3)_2 \cdot 2L$	L = cyclohexanone
$UO_2(trop)_2 \cdot L$	L = cyclopentanone, cyclohexanone

recrystallized from that acid, are not bonded to the uranium atom and the octahedral six-coordinate geometry is made up of two *trans* ketone oxygen atoms, two *trans* chlorine atoms and the two *trans* oxygen atoms of the UO_2 group.[243]

(*g*) *Aldehydes.* Very few complexes with aldehydes are known. The formaldehyde complex of dioxouranium(VI) phosphite, $UO_2(HPO_3) \cdot 2HCHO \cdot H_2O$, has been obtained from aqueous solution[244] and the 2-furaldehyde complex $UO_2(NO_3)_2 \cdot OC_4H_3CHO \cdot H_2O$, has also been reported.[245] The anhydrous complex, $UO_2(NO_3)_2 \cdot H_2NCH{=}CHCH{=}CHCHO$, is obtained by irradiation of $UO_2(NO_3)_2 \cdot py \cdot EtOH$ in ethanol at 254 nm; the IR spectrum of the compound indicates that the aldehyde molecule is bonded to the uranium atom *via* the carbonyl oxygen atom only.

(*h*) *Esters.* Complexes of composition $UO_2X_2 \cdot 2RCO_2R'$ (X = Cl, R = Me, R′ = Et; X = NO_3, R = R′ = Me and R = Me, H_2N, MeNH and Me_2N with R′ = Et) are obtained by vacuum evaporation of a solution of anhydrous UO_2Cl_2 in the ester (R = Me, R′ = Et), whereas the ethyl carbamate complexes are prepared by heating the parent nitrate with an excess of the molten ester at 70 to 100 °C for one to two hours. The coordination geometry about the uranium atom in $[UO_2(NO_3)_2(H_2NCO_2Et)_2]$ is an irregular hexagonal bipyramid, with the equatorial positions occupied by four oxygen atoms from the two bidentate nitrate groups and two carbonyl oxygen atoms from the ester molecules.[246] An ill-defined methyl formate complex, $UO_2(NO_3)_2 \cdot xHCO_2Me$, has been recorded, and ethyl carbamate complexes of the tropolonate, $UO_2(trop)_2 \cdot H_2NCO_2Et$, and the basic acetate, $[UO_2(MeCO_2)(OH)(H_2NCO_2Et)]_n$, are also known. Polymeric ethyl carbamate complexes, $[UO_2X(L)]_n$, have also been recorded for X = C_2O_4, C_6H_4-1,2-$(CO_2)_2$ and C_5H_3-2,6-$(CO_2)_2$.

(*V*) *Oxoanions as ligands*

(*a*) *Nitrates.* Hydrated dioxoactinide(VI) nitrates, $M^{VI}O_2(NO_3)_2 \cdot 6H_2O$ (M^{VI} = U, Np, Pu), are included in Table 63 (p. 1192). A product of composition $NpO_2(NO_3)_2 \cdot N_2O_5 \cdot H_2O$ is obtained by vacuum evaporation of a solution of $NpO_3 \cdot H_2O$ in N_2O_5; this may be $H_2NpO_2(NO_3)_4$, but the normal nitrato complex anions are $[M^{VI}O_2(NO_3)_3]^-$ (M^{VI} = U, Np,

Pu), for which alkali metal, ammonium and tetraalkylammonium salts are known. The salts $M^{I}[M^{VI}O_2(NO_3)_3]$ (M^{VI} = U, Np, Pu with M^{I} = K, Rb, Cs, NH_4) are isostructural.[247] The anion in $Rb[NpO_2(NO_3)_3]$ consists of a hexagonal bipyramidal arrangement of oxygen atoms about the neptunium atom, with six oxygen atoms from the bidentate nitrate groups in the equatorial plane.[248] The IR spectra of $Rb[M^{VI}O_2(NO_3)_3]$ (M^{VI} = U, Np, Pu) have been reported.[249]

(*b*) *Phosphites and phosphates.* A phosphite of composition $UO_2(H_2PO_2)_2$ is precipitated from aqueous solution and the salts $NaUO_2(H_2PO_2)_3{\cdot}xH_2O$ (x = 3.5 or 5) are also known. Treatment of $UO_2(NO_3)_2$ with KH_2PO_3 or K_2HPO_3 in aqueous solution yields products of composition $K_2(UO_2)_2(HPO_3)_3{\cdot}4H_2O$, and other hydrated salts of this type with Na, Rb, Cs or NH_4 in place of K, as well as $Ba(UO_2)_2(HPO_3)_3{\cdot}6H_2O$, have also been obtained. Other salts of composition $M^{I}UO_2(HPO_3)(H_2PO_3){\cdot}nH_2O$ (M^{I} = Na, K, Rb, Cs, NH_4), and $BaUO_2(HPO_3)(H_2PO_3)_2{\cdot}5H_2O$, are obtained by treating the corresponding $M^{I}_2UO_2(HPO_3)_2$ with a three- to five-fold excess of H_3PO_3.[250] Several hydrated phosphates and phosphato complexes are included in Table 63 (p. 1192).

In the structure of the phosphato complex, $K_4UO_2(PO_4)_2$, the coordination geometry about the uranium atom is described as a tetragonal bipyramid with four oxygen atoms from four tetrahedral PO_4^{3-} groups in the equatorial plane, making up a $[UO_2(PO_4)_2]_n^{4n-}$ layer structure.[251] The phosphate of composition $UO_2H(PO_3)_3$ is obtained[252] by heating $UO_2(NO_3)_2{\cdot}6H_2O$ with 85% H_3PO_4 at 340–380 °C and the salt $NaUO_2(PO_3)_3$ crystallizes[253] from 85% H_3PO_4 at 360 °C.

The pyrophosphates $M^{I}_2UO_2P_2O_7$ (M^{I} = Na, K, Rb, Cs) have been prepared by heating $UO_2(NO_3)_2{\cdot}6H_2O$ with $(NH_4)_2HPO_4$ in the presence of the alkali metal pyrophosphate or nitrate.[254] The rubidium and cesium compounds are isostructural and in the latter, layers of $UO_2P_2O_7^{2-}$ groups are held together by cesium ions. The UO_2^{2+} ion is coordinated by four oxgyen atoms from four P_2O_7 groups in the equatorial positions of a tetragonal bipyramid.[255]

(*c*) *Arsenates.* A number of hydrated arsenates are included in Table 63 (p. 1192). Products of composition $UO_2HAsO_4{\cdot}4H_2O$, $UO_2(H_2AsO_4)_2{\cdot}H_2O$, $(UO_2)_3(AsO_4)_2$, $(UO_2)_2As_2O_7$ and $UO_2(AsO_3)_2$ have also been reported.[256]

(*d*) *Sulfites.* Hydrated dioxouranium(VI) sulfite is precipitated when aqueous $UO_2(NO_3)_2$ is treated with sodium thiosulfate;[257] it is completely dehydrated at 130–150 °C. The sulfito complex, $[Co(NH_3)_6]_4[UO_2(SO_3)_3]_3{\cdot}22H_2O$, is precipitated from aqueous solution,[258] and a basic compound of composition $bipyH[(UO_2SO_3)_2(OH)(H_2O)_2]$ has also been reported.[259]

(*e*) *Sulfates.* Dioxouranium(VI) sulfate hydrates, $UO_2SO_4{\cdot}xH_2O$ (x = 1, 2, 2.5, 3 or 4), and the fluorosulfate, $UO_2(SO_3F)_2$, have been recorded; the last is obtained[86] by treating $UO_2(MeCO_2)_2$ with HSO_3F. Salts of sulfato complex anions, such $M^{I}_2UO_2(SO_4)_2{\cdot}2H_2O$ (M^{I} = K, NH_4), $(NH_4)_4UO_2(SO_4)_3{\cdot}4H_2O$ and $(NH_4)_2(M^{VI}O_2)_2(SO_4)_3{\cdot}5H_2O$ (M^{VI} = U, Pu), have been recorded. Other known salts include $M^{II}UO_2(SO_4)_2{\cdot}xH_2O$ (M^{II} = Ni, Zn, x = 4; M^{II} = Cu, x = 3), $M^{II}(UO_2)_3(SO_4)_5{\cdot}xH_2O$ (M^{II} = Mn, x = 5; M = Cd, x = 2; M = Hg, x = 3), $Co_3UO_2(SO_4)_4{\cdot}H_2O$, $Ce(UO_2)_2(SO_4)_4{\cdot}3H_2O$[261] and the two isomorphous hexammine cobalt(III) salts,[262] $\{[Co(NH_3)_6]HSO_4\}_2[M^{VI}O_2(SO_4)_3]{\cdot}xH_2O$ (M^{VI} = U, Np), all of which are obtained from aqueous solution.

The coordination geometry about the uranium atom in $(NH_4)_2UO_2(SO_4)_2{\cdot}2H_2O$ is pentagonal bipyramidal, with sulfate groups joining the polyhedra to make up a layered structure.[263] The structure of $K_4UO_2(SO_4)_3$ involves the anion $[(SO_4)_2UO_2(SO_4)_2UO_2(SO_4)_2]^{8-}$, in which each UO_2 group is bonded to five oxygen atoms from four non-equivalent sulfate groups in the equatorial plane of a pentagonal bipyramid.[264] The structure of $MgUO_2(SO_4)_2{\cdot}11H_2O$ consists of infinite layers of composition $[UO_2(SO_4)_2(H_2O)]_n^{2n-}$ in which the coordination geometry about the uranium atoms is also pentagonal bipyramidal,[265a] and the same coordination geometry has been found for the metal atom in $NH_4UO_2SO_4F$, in which the equatorial plane is made up from three O atoms from three tridentate sulfate groups together with two bridging fluorine atoms.[265b] Crystallographic data for the hydrated salts $M^{II}UO_2(SO_4){\cdot}5H_2O$ (M^{II} = Mg, Fe, Co, Ni, Cu, Zn, Mn, Cd) are also available.[266]

(*f*) *Selenates.* Hydrated dioxouranium(VI) selenate, $UO_2SeO_4{\cdot}4H_2O$, is not isostructural with $UO_2SO_4{\cdot}4H_2O$; its structure consists of an infinite strip, $[UO_2SeO_4(H_2O)_2]_\infty$, with the equatorial positions of the pentagonal bipyramidal arrangement of oxygen atoms surounding each uranium atom occupied by the oxygen atoms from three monodentate SeO_4 groups and two water molecules.[267] Hydrated salts of the selenato complex anion, $M^{II}UO_2(SeO_4)_2{\cdot}6H_2O$ (M^{II} = Mg, Co, Zn), are also known.[268]

(*g*) *Tellurites.* Compounds of composition $UTeO_5(UO_3{\cdot}TeO_2)$ and $UTe_3O_9(UO_3{\cdot}3TeO_2)$ are obtained by heating UO_3 with the appropriate quantity of TeO_2. The structure of $UTeO_5$

consists of infinite chains, $[UO_5]_n^{4n-5}$, in which pentagonal bipyramidal UO_7 groups share edges,[269] whereas UTe_3O_9 involves eight-coordinate uranium in which the UO_2^{2+} group is surrounded by six oxygen atoms in a flattened octahedral array.[270] The structure of the lead compound, $Pb_2UO_2(TeO_3)_3$, prepared by hydrothermal treatment of a stoichiometric mixture of PbO, $UO_2(MeCO_2)_2{\cdot}2H_2O$ and TeO_2 at 230 °C, consists of UO_7 pentagonal bipyramids in sheets, $[UO_2(TeO_3)_3]_n^{4n-}$, connected by Pb atoms.[271]

(*vi*) *Carboxylates, carbonates and nitroalkanes*

(*a*) *Monocarboxylates.* Most of the available information concerns dioxouranium(VI) compounds and very few neptunium(VI) or plutonium(VI) analogues have been reported (Table 69). The hydrated formate, $UO_2(HCO_2)_2{\cdot}H_2O$, precipitates when solid $UO_2(NO_3)_2{\cdot}xH_2O$ is added to anhydrous formic acid and when aqueous $UO_2(NO_3)_2$ is heated with an excess of the acid. The structure of the compound consists of an infinite chain of seven-coordinate (pentagonal bipyramidal) uranium units linked by bridging HCO_2 groups and by hydrogen bonds involving the water molecules.[272] The monohydrated neptunium(VI) and plutonium(VI) formates are isostructural with the uranium(VI) compound.[273] The anhydrous formate is obtained by washing the hydrate with dry methanol or by heating it at 105–170 °C. The basic formate, $UO_2(OH)(HCO_2){\cdot}H_2O$, is obtained by hydrolysis of $UO_2(HCO_2)_2{\cdot}H_2O$; the structure again consists of a pentagonal bipyramidal arrangement of oxygen atoms about the uranium atom with the uranium atoms doubly bridged by OH groups and linked to the two second nearest neighbouring uranium atoms *via* HCO_2 groups. The pentagonal plane is completed by the water oxygen atom.[274]

In the structure of $Na[UO_2(HCO_2)_3]{\cdot}H_2O$ the coordination arrangement is also pentagonal bipyramidal, with the bipyramids connected through two types of bridging bidentate HCO_2 groups and linked with the Na^+, H_2O layers by a third type of bridging bidentate HCO_2 group.[275] The tetraformato complex salt, $(NH_4)_2UO_2(HCO_2)_4$, is obtained when a solution of $UO_2(HCO_2)_2$ in aqueous HCO_2H and HCO_2NH_4 is evaporated and then cooled; the pentagonal bipyramidal coordination geometry is made up of two oxygen atoms from one bidentate HCO_2 group, and the remaining three HCO_2 groups bridge U and N atoms, making a layer structure.[276] The strontium salt, $SrUO_2(HCO_2)_4.1{\cdot}38H_2O$, obtained when a hot solution of $Sr(HCO_2)_2{\cdot}2H_2O$ and $UO_2(HCO_2)_2{\cdot}H_2O$ in aqueous formic acid is cooled slowly from 80 °C, shows a similar arrangement in which the HCO_2 groups bridge either two U atoms or one Sr and one U atom.[277]

The structure of the 2,6-dihydroxybenzoate, $[UO_2\{(HO)_2C_6H_3CO_2\}_2(H_2O)_2]{\cdot}8H_2O$ consists of an irregular hexagonal bipyramidal arrangement with two bidentate carboxylate groups and two H_2O molecules in *trans* positions in the equatorial plane.[278b]

Five crystal modifications of anhydrous $UO_2(MeCO_2)_2$ have been reported. The structure of the dihydrate consists of chains of pentagonal bipyramidal coordination polyhedra in which half of the $MeCO_2$ groups link adjacent bipyramids and the remainder are chelated to individual uranium atoms; one water molecule is bonded to each uranium atom.[278a]

The coordination polyhedron of the anion in $Na[UO_2(MeCO_2)_3]$ is a hexagonal bipyramid with three bidentate $MeCO_2$ groups in the equatorial plane[279] and this has also been found for the anion in $Na[NpO_2(MeCO_2)_3]$.[248] Hexagonal bipyramidal geometry has also been reported for the anthracene-9-(or 10)-carboxylate, $[UO_2(C_{14}H_9CO_2)_2(H_2O)_2]{\cdot}3C_4H_8O_2$; the dioxane molecules are not bonded to the uranium atom, and the six equatorial positions of the bipyramid are occupied by four oxygen atoms from two bidentate carboxylate groups and two oxygen atoms from the water molecules.[280] The IR spectra of $Na[M^{VI}O_2(MeCO_2)_3]$ (M^{VI} = U, Np or Pu) have also been reported.[249] The coordination geometry in the crotonate, $[UO_2(C_4H_5O_2)_2(H_2O_2)]$, is similar to that of the anthracene carboxylate, with the oxygen atoms of two bidentate carboxylate groups and two water molecules coordinated to the UO_2^{2+} ion.[281]

The precipitation of salts $M^IM^{II}[UO_2(RCO_2)_3]_3$ (Table 69) has been used for the gravimetric determination of alkali metals (particularly M^I = Li and Na). For this purpose an anhydrous product is desirable, but in many cases hydrates of varying composition are obtained.

Salts of basic species, such as $TlUO_2(MeCO_2)(OH)_2$, and of mixed thiocyanato/carboxylato anions, such as $Cs_3(UO_2)_3(MeCO_2)_7(NCS)_2$, have been reported, but require further investigation. A variety of halo- and mercapto-propionates, $UO_2(RCO_2)_2$, are also known.

(*b*) *Chelating carboxylates.* Furan-2- and thiophene-2-carboxylates are included in Table 70 because the heterocyclic oxygen or sulfur atoms may be bonded to the metal atom. The

Table 69 Dioxoactinide(VI) Monocarboxylates and Carboxylato Complexes

$UO_2(RCO_2)_2{\cdot}xH_2O$	R = H (x = 0, 1 or 2), Me, Et, Pr^n (x = 0 or 2), CF_3 (x = 0, 1 or 1.5), CCl_3 (x = 0, 1, 2 or 3), CH_2Cl (x = 0 or 1), $PhCH_2$ (x = 0), Pr^i, Bu^n, Me_2CHCH_2 (x = 2), Me_3C, n-$C_{11}H_{23}$, n-$C_{13}H_{27}$, n-$C_{15}H_{31}$, n-$C_{17}H_{35}$ (x = 0), PhCH=CH (x = 0 or 3), PhC≡C (x = 3), 1-$C_{10}H_7$ (x = 3), Ph (x = 0, 1 or 2), 2-$C_{10}H_7$ (x = 1)
$M^{VI}O_2(HCO_2)_2{\cdot}H_2O$	M^{VI} = Np, Pu
$[UO_2(C_{14}H_9CO_2)_2(H_2O)_2]{\cdot}3C_4H_8O$	$C_{14}H_9CO_2H$ = anthracene-9(or 10)-carboxylic acid; C_4H_8O = dioxane
$UO_2(OH)(RCO_2){\cdot}xH_2O$	R = H (x = 1); R = Me (x = 0 or 2)
$U_2O_5(HCO_2)_2$	
$M^I[M^{VI}O_2(RCO_2)_3]$	R = H, M^{VI} = U, M^I = Na or Tl (+H_2O), Cs
	R = Me, M^{VI} = U, M^I = H, Li (+2 or $5H_2O$), Na, K, Rb, Cs, Tl, Ag, NH_4, NEt_4
	M^{VI} = Np, Pu, M^I = Na, Cs
	R = CH_2Cl, M^{VI} = U, M^I = Na (+1 or $2H_2O$), Rb, Cs
	R = $CHCl_2$, M^{VI} = U, M^I = Na
	R = CCl_3, M^{VI} = U, M^I = Li, Na, K, Rb, Cs, Tl, NH_4
	R = $PhCH_2$, M^{VI} = U, M^I = Na
	R = Et, M^{VI} = U, M^I = Li (also +$2H_2O$), Na (also +H_2O), K, Rb, Cs, Tl, NH_4 (also +$2H_2O$)
	R = Pr^n, M^{VI} = U, M^I = Na, K, Rb, Cs, Tl, NH_4
	R = PhCH=CH, M^{VI} = U, M^I = NH_4
	R = Ph, M^{VI} = U, M^I = H, Na (also +$2H_2O$), NH_4, $AsPh_4$
$TlUO_2(HCO_2)(OH)_2$	
$M^{II}[UO_2(RCO_2)_3]_2{\cdot}xH_2O$	R = Me, M^{II} = Be, x = 0 or 2; Mg, x = 0, 6, 7, 8 or 12; Ca, x = 0 or 6; Sr, x = 0, 2 or 6; Ba, x = 0, 2, 3, 6 or 10; Zn, x = 0, 7 or 8; Mn, x = 0 or 9; Ni, x = 0 or 7
	R = CCl_3, M^{II} = Be, Mg, Ni, x = 0
	R = Et, M^{II} = Ca, x = 6; Sr, x = 4; Ba, x = 3; Zn, x = 0 or 6; Mn, Co, x = 0 or 7; Ni, x = 0 or 8
	R = Pr^n, M^{II} = Mg, Mn, Ni, Co, x = 0; Mg, Mn, x = 6; Ni, Co, x = 7
$M^{II}[NpO_2(MeCO_2)_3]_2$	M^{II} = Sr, Ba
$M^IM^{II}[UO_2(MeCO_2)_3]_3$	M^I = Li, M^{II} = Mg, Zn, Cd, Hg, Mn, Ni, Co, Fe, Cu
	M^I = Na, M^{II} = Be, Mg,[a] Ca,[a] Zn,[a] Cd,[a] Hg,[a] Mn,[a] Ni,[a] Co,[a] Fe, Cu[a]
	M^I = K, M^{II} = Be, Zn
	M^I = Rb, NH_4, M^{II} = Be
$M^IM^{II}[UO_2(EtCO_2)_3]_3{\cdot}xH_2O$	M^I = Na, M^{II} = Zn, x = 8; Mn, x = 0 or 6; Ni, Co, x = 7
	M^I = NH_4, M^{II} = Mg, Zn, x = 8; Ni, x = 7; Ca, Mn, Co, x = 6; Sr, x = 4; Ba, x = 3
$M^I_2[UO_2(RCO_2)_4]{\cdot}xH_2O$	R = H, M^I = NH_4
	R = Me, M^I = NMe_4, NEt_4
$M^{II}[UO_2(RCO_2)_4]{\cdot}xH_2O$	R = H, M^{II} = Sr, x = 1.38
	R = Me, M^{II} = Cd, Mn, x = 6
$(NH_4)_2[(UO_2)_2(MeCO_2)_5]$	
$M^{III}UO_2(RCO_2)_5{\cdot}xH_2O$	R = Me, M^{III} = La, x = 2 or 5; Nd, x = 3
	R = Et, M^{III} = La, Ce, Nd, x = 2; La, Nd, Eu, x = 3

[a] Hydrated.

pyridine-2-carboxylates, $M^{VI}O_2(NC_5H_4CO_2\text{-}2)_2$ (M^{VI} = U, Np, Pu), precipitate from aqueous solutions of the actinide(VI) in the presence of the acid, but at lower pH and higher $NC_5H_4CO_2H$-2 concentrations the acid salts $HM^{VI}O_2(NC_5H_4CO_2\text{-}2)_3$ are obtained. In the case of the *N*-oxopyridine-2-carboxylates, $M^{VI}O_2(ONC_5H_4CO_2\text{-}2)_2$, the primary products are hydrated, but are easily dehydrated at 120 °C. Similarly, the hydrated pyridine-3-carboxylates lose water at 160 °C (U) or 130 °C (Pu). In addition to the compounds listed in Table 70, 6-methylpyridine-2-carboxylates of composition UO_2L_2, HUO_2L_3 and $UO_2L(OH){\cdot}3H_2O$ (dehydrated at 95 °C) are known (L = 6-$MeC_5H_3CO_2$-2).

The pyridine-2,6-dicarboxylate groups in the polymeric compound $[UO_2(C_7H_3NO_4)(H_2O)]_n$ are terdentate to each uranium atom and each of the $UO_2(C_7H_3NO_4)$ units is linked to an adjacent unit through one of the carboxylate group oxygen atoms; the slightly distorted pentagonal bipyramidal coordination of the uranium atom is completed by the oxygen atom of the coordinated water molecule.[282] The coordination geometry of the anion in $(AsPh_4)_2[UO_2(C_7H_3NO_4)_2]{\cdot}6H_2O$ is an irregular hexagonal bipyramid with four oxygen and

Table 70 Dioxoactinide(VI) Compounds with Chelating Carboxylates

Furan-2- and thiophene-2-carboxylates	
$UO_2(RCO_2)_2 \cdot xH_2O$	$R = OC_4H_3$, SC_4H_3, $x = 0$ or 1
$(M^{VI}O_2)_2(RCO_2)_3(OH) \cdot xH_2O$	$M^{VI} = U$, $R = OC_4H_3$, SC_4H_3, $x = 0$
	$M^{VI} = U$, Np, Pu, $R = OC_4H_3$, SC_4H_3, $x = 4$
$M^I[M^{VI}O_2(RCO_2)_3] \cdot xH_2O$	$M^{VI} = U$, Np, Pu, $R = OC_4H_3$, SC_4H_3, $M^I = NH_4$ (x indefinite)
	$M^{VI} = Np$, Pu, $R = OC_4H_3$, SC_4H_3, $M^I = Na$ (x indefinite)
Pyridine-2-carboxylates	
$M^{VI}O_2(NC_5H_4\text{-}2\text{-}CO_2)_2$	$M^{VI} = U$ (also $+H_2O$), Np, Pu
$HM^{VI}O_2(NC_5H_4\text{-}2\text{-}CO_2)_3$	$M^{VI} = U$, Np, Pu
$M^IUO_2(NC_5H_4\text{-}2\text{-}CO_2)_3$	$M^I = Na$, $AsPh_4(+3H_2O)$
Pyridine-3-carboxylates	
$M^{VI}O_2(NC_5H_4\text{-}3\text{-}CO_2)_2 \cdot xH_2O$	$M^{VI} = U$, Np, Pu, $x = 0, 2$
N-*Oxopyridine-2-carboxylates*	
$M^{VI}O_2(ONC_5H_4\text{-}2\text{-}CO_2)_2 \cdot xH_2O$	$M^{VI} = U$, Np, Pu, $x = 0, 2$
Pyridine-2,6-dicarboxylates (acid = $C_7H_5NO_4$)	
$[UO_2(C_7H_3NO_4)]_n$	
$[UO_2(C_7H_3NO_4)(H_2O)]_n$	
$UO_2(C_7H_3NO_4) \cdot 2H_2O$	
$(AsPh_4)_2[UO_2(C_7H_3NO_4)_2] \cdot xH_2O$	$x = 0, 2, 5$ or 6
N-*Oxopyridine-2,6-dicarboxylates*[a]	
$[UO_2\{ONC_5H_3(CO_2)_2\}(H_2O)_2]_2 \cdot 2H_2O$	
$[UO_2\{ONC_5H_3(CO_2)_2\}]_n$	
$[UO_2\{ONC_5H_3(CO_2)_2\}(H_2O)]_n \cdot H_2O$	
Pyridine-2,3-dicarboxylates (quinolinates)	
$UO_2(C_7H_3NO_4) \cdot xH_2O$	$x = 0, 1$ or 2
Quinoline carboxylates (acid = $C_{10}H_7NO_2$)	
$UO_2(C_{10}H_6NO_2)_2 \cdot xH_2O$	$x = 0$ or 2
Oxydiacetates (acid = $O(CH_2CO_2H)_2$)	
$[UO_2\{O(CH_2CO_2)_2\}]_n$	
$Na_2[UO_2\{O(CH_2CO_2)_2\}_2 \cdot xH_2O$	$x = 2$ or 3
Iminodiacetates (acid = $HN(CH_2CO_2H)_2$)	
$[UO_2\{HN(CH_2CO_2)_2\}]_n$	
$UO_2\{HN(CH_2CO_2)_2H\}_2$	
$Na[UO_2\{HN(CH_2CO_2)_2\}_2H] \cdot 2H_2O$	

[a] S. Degetto, L. Baracco, M. Faggin and E. Celon, *J. Inorg. Nucl. Chem.*, 1981, **43**, 2413.

two nitrogen atoms of the two terdentate pyridine-2,6-dicarboxylate groups in the equatorial plane.[283]

In the structure of the dimeric *N*-oxopyridine-2,6-dicarboxylate, $[UO_2(ONC_5H_3(CO_2)_2)(H_2O)_2]_2 \cdot 2H_2O$, the two units of the dimer are bridged by carboxylate group oxygen atoms and the coordination geometry is pentagonal bipyramidal, with one oxygen atom each from a monodentate carboxylate group, a bridging group and the *N*-oxide, and two from the water molecules, all in the equatorial plane.[284]

Both carboxylate groups in the iminodiacetate, $UO_2\{HN(CH_2CO_2)_2H\}_2$, are ionized, with the extra proton linked to the imino nitrogen atom. The iminodiacetate groups bridge UO_2^{2+} ions in infinite chains, with one CO_2 group of each ligand monodentate and one bidentate to two adjacent uranium atoms, making up an irregular hexagonal bipyramid about each uranium atom.[285] In the polymeric iminodiacetate, $[UO_2\{HN(CH_2CO_2)_2\}]_n$, each UO_2^{2+} ion is bonded to two carboxylate oxygen atoms, the imino nitrogen atom of one $HN(CH_2CO_2)_2$ group and oxygen atoms from two other dicarboxylate ligands to form an irregular pentagonal bipyramid.[286] The polymeric oxydiacetate, $[UO_2\{O(CH_2CO_2)_2\}]_n$, has the same coordination geometry, with the ether oxygen atom in place of the imino nitrogen atom.[287] In the anion of the oxydiacetato complex, $Na_2[UO_2\{O(CH_2CO_2)_2\}_2] \cdot 2H_2O$, each $O(CH_2CO_2)_2$ group is terdentate, making up an irregular hexagonal bipyramid about the uranium atom.[288]

(*c*) *Carbonates.* The main types of carbonato complex are listed in Table 71; in addition, peroxocarbonato and fluorocarbonato complexes, such as $K_4[UO_2(O_2)(CO_3)_2]$ and $K_3[UO_2F_3(CO_3)]$, are also known. The coordination geometry about the uranium atom in UO_2CO_3 is approximately hexagonal bipyramidal, with two bidentate and two monodentate carbonate groups in the equatorial plane.[289] The tricarbonato complex anion in $(NH_4)_4[UO_2(CO_3)_3]$ is a distorted hexagonal bipyramid with three bidentate carbonate groups in the equatorial plane[290] and the same geometry has been reported for the anion in $K_4[UO_2(CO_3)_3]$.[291]

Table 71 Dioxoactinide(VI) Carbonates and Carbonato Complexes

$M^{VI}O_2CO_3{\cdot}xH_2O$	M^{VI} = U, Pu, $x = 0$; M^{VI} = U, $x = 1$, 2.2 or 2.5
$(UO_2)_6(CO_3)_5(OH)_2{\cdot}7H_2O$	The mineral sharpite
$M^I_2[UO_2(CO_3)_2]$	M^I = Li (hydrated), Na, K, Rb, Cs, Tl (hydrated), NH_4, CN_3H_6, $(CH_2)_6N_4H_2$
$M^{II}[UO_2(CO_3)_2]{\cdot}xH_2O$	M^{II} = Ca, $x = 10$; Ba, $x = 4$
$Ba[PuO_2(CO_3)_2(H_2O)_2]{\cdot}H_2O$ [a]	
$M^I_4[M^{VI}O_2(CO_3)_3]$	M^{VI} = U, M^I = Li (hydrated), K, Rb, Cs, Tl, NH_4
	M^{VI} = Np, M^I = K, NH_4,[b] Ag
	M^{VI} = Pu, M^I = Li,[c] Na,[c] K,[c] NH_4
$M^{II}_2[UO_2(CO_3)_3]{\cdot}xH_2O$	M^{II} = Mg, $x = 16$, 18 or 20; Ca, $x = 0$, 4, 8, 9 or 10; Sr, $x = 9$; Ba, $x = 5$ or 6; Pb, Mn, $x = 0$; Ni, $x = 6$; Co, $x = 2$; Cu, $x = 1$
$M^I_6[(M^{VI}O_2)_2(CO_3)_5]$	M^{VI} = U, M^I = Li (hydrated), Na, NH_4 (hydrated)
	M^{VI} = Pu, M^I = Rb,[c] Cs,[c] NH_4 (+2H_2O)
$Ba_3[(PuO_2)_2(CO_3)_5(H_2O)_2]$ [a]	
$NH_4M^{VI}O_2(CO_3)(OH){\cdot}3H_2O$	M^{VI} = U, Pu
$Ba[PuO_2(CO_3)(OH)(H_2O)_3]_2{\cdot}8H_2O$ [a]	
$Ba_3[(PuO_2)_2(CO_3)_3(OH)(H_2O)_5]_2{\cdot}6H_2O$ [a]	

[a] The assignment of water molecules within the coordination sphere requires confirmation.
[b] Yu. Ya. Kharitonov and A. I. Moskvin, *Sov. Radiochem.* (*Engl. Transl.*), 1973, **15**, 240.
[c] J. D. Navratil, US report RFP-1760, 1971 (*Chem. Abstr.*, 1972, **77**, 66 876).

(*d*) *Oxalates.* Only one water molecule is bonded to the metal in $UO_2C_2O_4{\cdot}3H_2O$; the pentagonal bipyramidal coordination geometry is made up of four oxygen atoms from two independent C_2O_4 groups and one from the water molecule in the equatorial plane. Each C_2O_4 group is tetradentate, bridging two UO_2^{2+} ions.[292]

In addition to the salts of the type $M^I[UO_2(C_2O_4)_2]$ listed in Table 72, a wide range of hydrates are also known; some have been reported in the form $M^I_2[UO_2(C_2O_4)_2(H_2O)_x]{\cdot}yH_2O$ without adequate justification. The coordination geometry in the polymeric anion $[UO_2(C_2O_4)_2]_n^{2n-}$ of the ammonium salt is approximately pentagonal bipyramidal, with one bidentate C_2O_4 group and one C_2O_4 group which is bidentate to one uranium atom and monodentate to a second uranium atom.[293] The oxalate groups in the anion of $(NH_4)_4[UO_2(C_2O_4)_3]$ are all bidentate, giving rise to distorted hexagonal bipyramidal coordination geometry,[294] whereas in the polymeric anion of $(NH_4)_2[(UO_2)_2(C_2O_4)_3]$ the coordination geometry is pentagonal bipyramidal, with one quadridentate C_2O_4 group coordinated to two uranium atoms and the other bidentate to one and unidentate to a second uranium atom, forming infinite double chains,[295] $[(C_2O_4)UO_2(C_2O_4)UO_2(C_2O_4)]_n^{2n-}$. In $K_6[(UO_2)_2(C_2O_4)_5]{\cdot}10H_2O$, the coordination geometry is again pentagonal bipyramidal, with two oxygen atoms each from two bidentate C_2O_4 groups and one from the bridging C_2O_4 group in the equatorial plane.[296]

A number of basic oxalates, such as $(UO_2)_2C_2O_4(OH)_2{\cdot}2H_2O$, and salts of basic oxalato complex ions of the types $M^I_3UO_2(C_2O_4)_2(OH)$ and $M^I_5(UO_2)_2(C_2O_4)_4(OH)$ have also been reported, as well as salts of a wide range of peroxo-, halogeno-, sulfito-, sulfato-, selenito-, selenato-, thiocyanato- and carbonato-oxalato complex anions: all need further investigation.

(*e*) *Other dicarboxylates.* A number of peroxomalonato compounds, such as $K_4[(UO_2)_2\{CH_2(CO_2)_2\}_3(O_2)(H_2O)_4]$, have been recorded in addition to the compounds listed in Table 73. The anion in $(NH_4)_2[UO_2\{CH_2(CO_2)_2\}_2]{\cdot}H_2O$ is polymeric, with one malonate group bidentate to one uranium atom and the second terdentate, with two oxygen atoms bonded to one uranium atom and one to an adjacent uranium atom, producing infinite chains in which the coordination geometry about the uranium atom is approximately pentagonal bipyramidal.[297] The coordination geometry in the polymeric anions of the salts

Table 72 Dioxoactinide(VI) Oxalates

$M^{VI}O_2C_2O_4{\cdot}3H_2O$	M^{VI} = U, Np,[b] Pu[b]
$UO_2C_2O_4{\cdot}xH_2O$	x = 0 or 1
$M^I_2[UO_2(C_2O_4)_2]$[a]	M^I = Li, Na, K, Rb, Cs, NH_4, Tl, CN_3H_6
$Cs_2[NpO_2(C_2O_4)_2]{\cdot}2H_2O$[b]	
$M^{II}[UO_2(C_2O_4)_2]$	M^{II} = Sr, Ba
$M^I_4[UO_2(C_2O_4)_3]$	M^I = H(hydrated), NH_4, Tl, NBu^n_4
$M^I_2[(UO_2)_2(C_2O_4)_3]$[a]	M^I = Li, Na, K, Rb, Cs, NH_4, Tl
$Cs_2[(NpO_2)_2(C_2O_4)_3]$[b]	
$M^I_6[(UO_2)_2(C_2O_4)_5]$[a]	M^I = Na, K, Rb, Cs, NH_4, CN_3H_6
$K_6[(NpO_2)_2(C_2O_4)_5]{\cdot}2\text{–}4H_2O$[b]	

[a] Hydrates are also known.
[b] M. P. Mefod'eva, M. S. Grigor'ev, T. V. Afonos'eva and E. B. Kryukov, *Sov. Radiochem. (Engl. Transl.)*, 1981, **23**, 565. The neptunium(VI) and plutonium(VI) compounds are isostructural with their uranium(VI) analogues.

$M^{II}[UO_2\{CH_2(CO_2)_2\}_2]{\cdot}3H_2O$ (M^{II} = Sr, Ba) is similar, except that the terdentate malonate group in these salts has a boat conformation whereas a chair conformation is adopted in the ammonium salt.[298] The coordination geometry in the succinate, $[UO_2\{C_2H_4(CO_2)_2\}(H_2O)]_n$ is an irregular pentagonal bipyramid, with four carboxylate oxygen atoms from four different succinate groups and one from the water molecule in the equatorial plane.[299]

Table 73 Dioxouranium(VI) Dicarboxylates

Malonates	$UO_2\{CH_2(CO_2)_2\}$
	$M^I_2[UO_2\{CH_2(CO_2)_2\}_2]$, M^I = Li, Na, K, Rb, Cs, NH_4
	$M^{II}[UO_2\{CH_2(CO_2)_2\}_2]{\cdot}3H_2O$, M^{II} = Sr, Ba
Succinates	$UO_2\{C_2H_4(CO_2)_2\}{\cdot}xH_2O$, x = 1, 2 or 3
	$Ca[UO_2\{C_2H_4(CO_2)_2\}_2]$
Fumarate and maleate	$UO_2\{$*cis*- or *trans*-$CH{=}CH(CO_2)_2\}$
	$(NH_4)_2[UO_2\{$*cis*-$CH{=}CH(CO_2)_2\}_2]$
Glutarate	$UO_2\{(CH_2)_3(CO_2)_2\}$
	$M^I[(UO_2)_2\{(CH_2)_3(CO_2)_2\}_3]$, M^I = Li, Na, K
	$UO_2\{(CH_2)_3(CO_2)_2\}{\cdot}LiH\{(CH_2)_3(CO_2)_2\}{\cdot}4H_2O$
Benzenedicarboxylates	$UO_2\{C_6H_4(CO_2)_2\}$ 1,2-, 1,3- and 1,4-dicarboxylates; hydrates are also known

In the glutarate of composition $UO_2(C_5H_6O_4){\cdot}Li(C_5H_7O_4){\cdot}4H_2O$, one glutarate group bridges the UO_2^{2+} ions in infinite chains and the other bridges UO_2^{2+} and Li^+ ions, the coordination geometry about the uranium atom being hexagonal bipyramidal;[300] the same coordination geometry has been reported for the hydrated fumarate, $UO_2(C_4H_2O_4)(H_2O)_2$, in which fumarate groups bridge the UO_2^{2+} ion in chains, and the water molecules are *trans* in the hexagonal ring.[301]

(*f*) *Carbamates.* The only recorded compound of this type appears to be $UO_2(Et_2NCO_2)_2$, which is precipitated when the uranium(IV) carbamate, $U(Et_2NCO_2)_4$, is oxidized in heptane.

(*g*) *Hexacarboxylates.* A mellitate (benzene-1,2,3,4,5,6-hexacarboxylate), $(UO_2)_3\{C_6(CO_2)_6\}{\cdot}12H_2O$, has been recorded.

(*h*) *Nitroalkanes and nitroarenes.* The nitromethane solvate, $UO_2(ClO_4)_2{\cdot}2MeNO_2$, is obtained by treating the anhydrous perchlorate with nitromethane, either alone or in solution in carbon tetrachloride, followed by evaporation of the resulting solution in the latter case. The ClO_4 groups may be covalent and bidentate in this product. Complexes of the corresponding nitrate, $UO_2(NO_3)_2{\cdot}2RNO_2$ (R = Me, Ph) crystallize from saturated solutions of $UO_2(NO_3)_2$ in RNO_2. The complex $UO_2(NO_3)_2{\cdot}MeNO_2$ has also been reported, and in this the $MeNO_2$ molecule may be bidentate.

(*vii*) *Aliphatic and aromatic hydroxyacids*

(*a*) *Aliphatic hydroxyacids.* A considerable number of dioxouranium(VI) compounds with aliphatic hydroxyacids are known (Table 74); alkali metal and barium salts of tartrato complexes, such as $Na_2[UO_2(C_4H_4O_6)_2]{\cdot}2H_2O$, have also been recorded. The glycolate,

$UO_2(HOCH_2CO_2)_2$, is obtained on cooling a warm aqueous solution of the acid which has been saturated with UO_3. The polymeric structure consists of infinite chains of pentagonal bipyramidal $UO_2(O)_5$ units with two sets of glycolate groups, one tridentate, which is bonded to one uranium atom and bridges two other uranium atoms, and the other bidentate, which bridges two uranium atoms.[302] Salts of the type $M^I[UO_2(HOCH_2CO_2)_3]$ are not known and very little information is available for the other compounds listed in Table 74.

Table 74 Some Dioxoactinide(VI) Compounds with Aliphatic Hydroxyacids

$UO_2(RCO_2)_2$	R = $HOCH_2$, PhCHOH (also $+2H_2O$), $Ph_2C(OH)(+2H_2O)$, MeCHOH (also +2 or $5H_2O$), MeC(OH)Ph ($+2H_2O$), $HOCH_2CHPh$, 2-$HOC_6H_4CH_2CH_2$
$PuO_2(C_6H_{13}CHOHCO_2)_2$	1-Hydroxycaprylate
$UO_2(MeCH(O)CO_2)$	
$UO_2(PhCH(O)CO_2)\cdot 2H_2O$	
$M^I_2[UO_2\{PhCH(O)CO_2\}_2]$	M^I = Na, K, NH_4, pyH
$M^I[(UO_2)_2(O)(HOCH_2CO_2)_2(OH)(H_2O)]$	M^I = Na, Cs
$Na_2[UO_2\{Ph_2C(O)CO_2\}_2]$	
$UO_2\{C(OH)_2(CO_2)_2\}$	Mesoxalate
$UO_2\{HOCHCH_2(CO_2)_2\}\cdot xH_2O$	Malate; $x = 0$, 1 or 3
$K[UO_2\{OCHCH_2(CO_2)_2\}\cdot H_2O]$	Malate
$[Co(NH_3)_6]_2[UO_2\{HOCHCH_2(CO_2)_2\}_2]_3\cdot xH_2O$	Malate; $x = 0$ or 2
$UO_2\{(CHOH)_2(CO_2)_2\}\cdot xH_2O$	Tartrate; $x = 1$ or 4
$K_2[UO_2\{(CHOH)_2(CO_2)_2\}_2]\cdot 2H_2O$	Tartrate
$[(UO_2)_2\{cis\text{-}O_2CC(O){=}C(O)CO_2\}(H_2O)_2]_2$	Dihydroxymaleate
$UO_2\{(CHOH)_3(CO_2)_2\}\cdot xH_2O$	Trihydroxyglutarate; $x = 1$ or 1.5
$UO_2(C_6H_6O_7)\cdot 4H_2O$	$C_6H_8O_7$ = citric acid, $HOC(CH_2CO_2H)_2CO_2H$
$(UO_2)_3(C_6H_5O_7)_2\cdot 2H_2O$	Citrate
$[Co(NH_3)_6]_2[UO_2(C_6H_6O_7)_2]_3\cdot 4H_2O$	Citrate

(*b*) *Aromatic hydroxyacids.* In addition to the hydroxybenzoates listed in Table 75, a number of compounds derived from substituted hydroxybenzoic acids have been recorded; a 2,3,4-trihydroxybenzoate (pyrogallate) and salts of complex anions derived from 3,4,5-trihydroxybenzoic acid (gallic acid) are also known.

Table 75 Some Dioxouranium(VI) Compounds with Aromatic Hydroxyacids

$UO_2(2\text{-}HOC_6H_4CO_2)_2\cdot xH_2O$	$x = 0$, 2, 3, 5 or 9
$[UO_2\{(2\text{-}HOC_6H_4CO_2)(NO_3)\}\{4\text{-}Me_2NC_5H_4N\}_2]_2$	
$M^I[UO_2(2\text{-}HOC_6H_4CO_2)_3]$	M^I = K, NH_4 ($+4H_2O$)
$M^I_2[UO_2(2\text{-}OC_6H_4CO_2)_2]\cdot xH_2O$	M^I = Na, $x = 9$; K, $x = 2$ or 10; NH_4, $x = 13$
$UO_2(3\text{-}HOC_6H_4CO_2)_2\cdot 6H_2O$	
$pyH[UO_2(3\text{-}HOC_6H_4CO_2)_3]\cdot 7H_2O$	
$UO_2(4\text{-}HOC_6H_4CO_2)_2\cdot 4H_2O$	
$pyH[UO_2(4\text{-}HOC_6H_4CO_2)_3]\cdot 4H_2O$	
$UO_2(2,4\text{-}(HO)_2C_6H_3CO_2)_2\cdot 10H_2O$	β-Resorcylate
$UO_2(2,5\text{-}(HO)_2C_6H_3CO_2)_2\cdot 3H_2O$	Gentisate
$UO_2(3,4\text{-}(HO)_2C_6H_3CO_2)_2\cdot 8.5H_2O$	Protocatechuate

The salicylato nitrato complex, $[\{UO_2(2\text{-}HOC_6H_4CO_2)(NO_3)\}\{(4\text{-}Me_2N)C_5H_4N\}_2]_2$, is obtained when the aminopyridine is added to a mixture of $UO_2(NO_3)_2\cdot 6H_2O$ and 2-$HOC_6H_4CO_2H$ in methanol. In this compound the pyridine molecules are not coordinated to the uranium atoms, but are hydrogen bonded to the phenolic oxygen atom of the salicylate group. The hexagonal bipyramidal arrangement of oxygen atoms about each uranium atom in the dimer is made up of one carboxylate oxygen from each of the two salicylate groups which bridge the two uranium atoms, and the second carboxylate oxygen atom of one salicylate group and the phenolic oxygen atom of the other, together with two oxygen atoms from the bidentate nitrate group.[303] In the structure of the *p*-aminosalicylato complex anion in $Na[UO_2(2\text{-}HO\text{-}4\text{-}H_2NC_6H_3CO_2)_3]\cdot 6H_2O$, the coordination geometry is again hexagonal bipyramidal, with the six carboxylate oxygen atoms from three nearly planar aminosalicylate groups in the equatorial plane.[304]

(viii) *Amides, carboxylic acid hydrazides, ureas,* N-*oxides,* P-*oxides,* As-*oxides and* S-*oxides*

(*a*) *Amides and related compounds.* The normal preparative route for complexes with these ligands (Table 76) is from the parent compound, usually hydrated, and an excess of the ligand in a non-aqueous solvent; in most cases the complex crystallizes on addition of a small amount of ether or a hydrocarbon to the solution. In some instances the complexes have been obtained from aqueous solution. The complexes with the peroxide chloride, formulated as $ClL_2UO_2(\mu_2\text{-}O_2)UO_2L_2Cl$ (L = DMF or DMA), in which the bridging O_2^{2-} group appears to act as a tetradentate ligand, are obtained by treating the complexes $UO_2Cl_2L_2$ with H_2O_2 in aqueous ethanol.[305]

Table 76 Amide Complexes of Dioxouranium(VI) Compounds

$[UO_2F_2(L)]_n$	L = $HCONMe_2$(DMF), $MeCONMe_2$(DMA), $MeCONH_2$
$UO_2Cl_2{\cdot}xL$	$X = 1$, L = $Me_2NCO(R)CONMe_2$ with R = CMe_2 or $CH_2C(Me)_2CH_2$
	$x = 1.5$, L = $Me_2NCO(CH_2)_nCONMe_2$ with $n = 1$ or 3
	$x = 2$, L = $MeCONH_2(+H_2O)$; MeCONHR with R = Pr^i, $p\text{-}H_2NC_6H_4$ or $p\text{-}EtOC_6H_4$; $HCONMe_2$,[a] $MeCONR_2$ with R = Me, Pr^n, Pr^i or $n\text{-}C_8H_{17}$; $RCONMe_2$ with R = Pr^i, Me_2CHCH_2 or Me_3C
	$x = 3$, L = $MeCONH(p\text{-}EtOC_6H_4)$, MeCONHEt, $HCONMe_2$
$UO_2Br_2{\cdot}xL$	$x = 2$, L = DMA
	$x = 3$, L = DMF
	$x = 4$, L = $MeCONH(p\text{-}EtOC_6H_4)$
$[UO_2(DMA)_5][UBr_6]$	
$UO_2I_2{\cdot}4DMF$	
$Cl(L)_2UO_2\{\mu_2\text{-}(O_2)\}UO_2(L)_2Cl$	L = DMF,[a] DMA[a]
$UO_2(NO_3)_2{\cdot}xL$	$x = 1$, L = $R^1_2NCOCH_2CONR^2_2$ with R^1 = Bu^n or Bu^i and R^2 = $Me(CH_2Ph)$
	$x = 2$, L = DMF, $HCON(Me)(CH_2Ph)$, $MeCONH_2$, $MeCONH(p\text{-}EtOC_6H_4)$, $MeCONR_2$ (with R = Et, Pr^i, $n\text{-}C_8H_{17}$, $n\text{-}C_{10}H_{21}$, $n\text{-}C_{12}H_{25}$ or Ph), $MeCONEt(MeC_6H_4)$, $RCONMe_2$ (with R = Pr^i, Bu^n, Me_2CHCH_2 or Me_3C), $Pr^nCONBu^n_2$, $Me_3CCONBu^n_2$
	$x = 3$, L = MeCONHPh
$[UO_2L_5](ClO_4)_2$	L = DMF, DMA or $MeCONEt_2$[d]
$UO_2(NCS)_2{\cdot}2L{\cdot}H_2O$	L = $MeCONH_2$
$UO_2(MeCO_2)_2{\cdot}xL$	L = DMF, $x = 1$ or 2; L = DMA,[b] $MeCONPr^i_2$, $x = 1$
$UO_2(Me_3CCO_2)_2{\cdot}L$	L = $MeCONPr^i_2$
$UO_2(C_2O_4){\cdot}xL$	$x = 1$, L = $MeCONH_2$
	$x = 2$, L = $MeCONH(p\text{-}EtOC_6H_4)$
	$x = 3$, L = DMF
$UO_2SO_3{\cdot}xL{\cdot}yH_2O$	L = $HCONH_2$, $x = 1$, $y = 2$; L = $MeCONH_2$, $x = 1$, $y = 1.5$ or $x = 1.5$, $y = 0.5$; L = DMF, $x = 1$, y unspecified
$UO_2SO_4{\cdot}xL{\cdot}yH_2O$	L = $MeCONH_2$, $x = 1$, $y = 2$; $x = 2$, $y = 0$ or $x = 3$, $y = 1$; L = DMF, $x = 2$, $y = 0$
$UO_2CrO_4{\cdot}2L$	L = $MeCONH_2$, DMF
$UO_2(HPO_3){\cdot}L$	L = $H_2NCOCH_2CONH_2$[c]
$UO_2(HPO_3){\cdot}2L$	L = $MeCONH_2$[c]
$[UO_2(HPO_3)(H_2O)L]$	L = $MeCONH_2$,[c] DMF[c]
$UO_2(Et_2NCS)_2{\cdot}xL$	$x = 1$, L = DMF; $x = 2$, L = $HCON(CH_2Ph)(Me)$
$UO_2L_2{\cdot}DMF$	L = $CF_3COCHCO(2\text{-}C_4H_3S)$
$UO_2L{\cdot}DMF$	L = $OC_6H_4CH{=}NCH_2CH_2N{=}CHC_6H_4O$

[a] R. N. Shchelokov, I. M. Orlova and A. V. Sergeev, *Dokl. Chem. (Engl. Transl.)*, 1980, **250/5**, 573.
[b] V. E. Mistryukov, Yu. N. Mikhailov, I. A. Yuranov, V. V. Kolesnik and K. M. Dunaeva, *Sov. J. Coord. Chem. (Engl. Transl.)*, 1983, **9**, 163.
[c] K. A. Avduevskaya, N. B. Ragulina, I. A. Rozanov, Yu. N. Mikhailov, A. S. Kanishcheva and T. G. Grevtseva, *Russ. J. Inorg. Chem. (Engl. Transl.)*, 1981, **26**, 546.
[d] A. M. Hounslow, S. F. Lincoln, P. A. Marshall and E. H. Williams, *Aust. J. Chem.*, 1981, **34**, 2543.

In the polymeric acetamide complex of the chromate, $[UO_2CrO_4(MeCONH_2)_2]_n$, the structure consists of chains of $(UO_2)O_5$ pentagonal bipyramidal coordination polyhedra in which the equatorial plane is made up from three oxygen atoms from $[CrO_4]^{2-}$ tetrahedra and two from the acetamide molecules.[306] Similarly, in the dimeric DMA complex with the acetate $[UO_2(MeCO_2)_2(DMA)]_2$, the structure consists of two pentagonal bipyramidal polyhedra bridged by two bidentate acetate groups, the other two acetate groups each chelating a uranium atom.[307]

Several complexes with ligands related to amides, such as lactams, lactones and antipyrine (atp) have also been recorded (Table 77); apart from preparative and IR spectroscopic information, little is known about them. The precipitation of $UO_2(NCS)_2 \cdot 3atp$ from aqueous acid in the presence of thiocyanate has been used as a method for the determination of uranium in minerals.

Table 77 Some Complexes of Lactams, Lactones and Antipyrine (atp) with Dioxouranium(VI) Compounds

Lactams and lactones	
$UO_2(Et_2NCS_2)_2 \cdot L$	$L = MeN(CH_2)_3CO$
$UO_2X_2 \cdot yL$	$L = O(CH_2)_3CO$ $X = F$, Cl, Br, y unspecified; $X = NO_3$, $y = 2$; $X = 0.5SO_4$, $y = 1$ $(+2H_2O)$
Antipyridine (2,3-*dimethyl*-1-*phenylpyrazol*-5-*one*)	
$UO_2F_2 \cdot atp \cdot 2H_2O$	Probably polymeric
$UO_2X_2 \cdot 3atp$	X = NCS, NCSe
$UO_2(ClO_4)_2 \cdot 5atp$	
$UO_2SO_3 \cdot x atp$[a,b]	$x = 1.5$ (also $+5H_2O$) or 2 $(+H_2O)$
$UO_2SO_4 \cdot 3atp \cdot xH_2O$	$x = 0$ or 3
$UO_2C_2O_4 \cdot 3atp$	

[a] A. Yu. Tsivadze, A. N. Smirnov, G. T. Bolotova, N. A. Golubkova and R. N. Shchelokov, *Russ. J. Inorg. Chem. (Engl. Transl.)*, 1979, **24**, 904.
[b] R. N. Shchelokov, G. T. Bolotova and N. A. Golubkova, *Sov. J. Coord. Chem. (Engl. Transl.)*, 1977, **3**, 810.

(*b*) *Carboxylic acid hydrazides.* The IR spectra of many of the known complexes with this type of ligand (Table 78) suggest that the hydrazides are bidentate, with bonding to the metal atom *via* a carbonyl oxygen and a hydrazine nitrogen atom in the case of the ligands $RCONHNH_2$.

Table 78 Some Complexes of Carboxylic Acid Hydrazides with Dioxouranium(VI) Compounds

$UO_2X_2 \cdot 2RCONHNH_2$	X = Cl, NO_3; R = p-$MeOC_6H_4$, o-, m- and p-$O_2NC_6H_4$ (all hydrated) X = NO_3, R = Ph (also $+2H_2O$), R = 2-HOC_6H_4 $(+2H_2O)$
$UO_2X_2 \cdot 3RCONHNH_2$	X = Cl, R = Ph; X = NO_3, R = Ph
$UO_2X_2 \cdot 3RCONHNH_2$	X = Cl, NO_3; R = Me, o-, m- and p-HOC_6H_4, o-, m- and p-$MeOC_6H_4$ (all hydrated)
$UO_2(NO_3)_2 \cdot 2RCONHNHCOMe$	R = Me $(+2H_2O)$, Ph

(*c*) *Ureas.* A very large number of urea complexes with dioxouranium(VI) compounds are known, a selection of which is given in Table 79. In all of them the urea, or substituted urea, molecules are bonded to the metal atom through the carbonyl oxygen atom. The complex $UO_2F_2 \cdot 3urea$ is obtained by treating $UO_2F_2 \cdot 2H_2O$ with urea in aqueous solution, and the dimeric complex, $[UO_2F_2(urea)_2]_2$, in which two pentagonal bipyramidal $[UO_2F_3(urea)_2]$ groups share a common F(1)—F(1′) edge, is obtained in the same way by slow crystallization from dilute aqueous solutions.[308] The coordination geometry in the complex $[UO_2(urea)_4(H_2O)](NO_3)_2$ is again pentagonal bipyramidal, with four urea and one water molecule oxygen atoms in the equatorial plane[309] and the structure of the dimeric basic iodide, $[UO_2(OH)(urea)_3]_2I_4$, is made up of two pentagonal bipyramidal UO_7 units bridged along one edge by two hydroxyl groups.[310] The same coordination geometry has been reported for the complex with the phosphite,[311] $[UO_2(HPO_3)(NH_2CONMe_2)(H_2O)]$ and for the cation in the complex $[UO_2(MeCO_2)(urea)_3]^+$ $[UO_2(MeCO_2)_3]^-$ $(= UO_2(MeCO_2)_2 \cdot 1.5urea)$, in which two oxygen atoms from the chelating acetate group and three from the urea molecules lie in the equatorial plane.[307]

The structures of three urea complexes of dioxouranium(VI) sulfate have also been reported. In the bis complex, $UO_2SO_4 \cdot 2urea$, the structure consists of infinite ribbons of OUO groups

Table 79 Some Urea Complexes of Dioxouranium(VI) Compounds

L = urea, $OC(NH_2)_2$	
$UO_2X_2 \cdot yL \cdot zH_2O$	X = F, $y = 2$, $z = 0$ (dimer) and $z = 1$; $y = 3$, $z = 0$
	X = Cl, $y = 2$ or 3, $z = 1$; $y = 4$, $z = 0$
	X = NCS, $y = 2$, $z = 0$ or 1; $y = 3$, $z = 0$
	X = NO_3, $y = z = 1$; $y = 2$ or 3, $z = 0$; $y = 4$, $z = 1$ or 2; $y = 5$, $z = 0$ or 1; $y = 6$, $z = 0$
	X = ClO_4, $y = 5$, $z = 0$
	X = $MeCO_2$, $y = 1$, 1.5, 2, 4 or 6, $z = 0$
$[UO_2XL_3][UO_2X_3]$[c]	X = $EtCO_2$
$[UO_2(OH)L_3]_2I_4$	
$UO_2X \cdot yL$	X = SO_3, $y = 2$
	X = SO_4, $y = 2$, 3 or 4
	X = C_2O_4, $y = 1$ or 3
L = MeHNCONHMe	
$UO_2X_2 \cdot yL$	X = Cl, $y = 2$, 4 or 5
	X = NCS, $y = 2$
	X = NO_3, $y = 4$, 5 or 6
	X = ClO_4, $y = 5$ or 6
$UO_2X \cdot yL$	X = SO_4, $y = 3$
	X = C_2O_4, $y = 1$
L = Me_2NCONH_2	
$UO_2(HPO_3) \cdot L \cdot H_2O$	
L = $Me_2NCONMe_2$	
$UO_2X_2 \cdot yL$	X = Cl, NCS, NO_3, $MeCO_2$,[a] $y = 2$
	X = ClO_4, $y = 5$
$UO_2X \cdot yL$	X = SO_4, CrO_4, $y = 2$
	X = C_2O_4, $y = 1$
$(UO_2)_2(O_2)Cl_2 \cdot 4L$[b]	
L = $H_2NCONHEt$	
$UO_2X_2 \cdot yL$	X = Cl, $y = 3$
	X = NO_3, $y = 2$, 3 or 5
$UO_2X \cdot yL$	X = SO_4, $y = 3$
	X = C_2O_4, $y = 1$ or 3
L = EtNHCONHEt	
$UO_2X_2 \cdot yL$	X = Cl, $y = 3$
	X = NO_3, $y = 2$
$UO_2X \cdot yL$	X = SO_4, $y = 2$ or 3
	X = C_2O_4, $y = 1$ or 3

[a] V. I. Spitsyn, K. M. Dunaeva, V. V. Kolesnik and I. A. Yuranov, *Russ. J. Inorg. Chem. (Engl. Transl.)*, 1982, **27**, 473.
[b] R. N. Shchelokov, I. M. Orlova and A. V. Sergeev, *Sov. J. Coord. Chem. (Engl. Transl.)*, 1981, **7**, 441.
[c] V. I. Spitsyn, V. V. Kolesnik, V. E. Mistryukov, Yu. N. Mikhailov and K. M. Dunaeva, *Dokl. Chem.*, 1983, **268/273**, 362.

coordinated to two urea oxygen atoms and to three oxygen atoms from tridentate sulfate groups, making an approximately pentagonal bipyramidal arrangement about each uranium atom,[312] and in $UO_2SO_4 \cdot 3$urea the polymeric structure is similar, with three urea oxygen atoms and one oxygen each from two bridging bidentate sulfate groups.[313] The same coordination geometry has been found for $[UO_2SO_4(urea)_4]$ in which the four oxygen atoms from the urea molecules and one from a monodentate sulfate group lie in the equatorial plane.[314]

(*d*) N-*Oxides*. A selection of the known Me_3NO and pyNO complexes with dioxouranium(VI) compounds is included in Table 80; in addition, a large number of pyridine *N*-oxide complexes with *β*-diketonates, $UO_2L_2 \cdot$pyNO, and many complexes with substituted pyridine *N*-oxides, quinoline and isoquinoline *N*-oxides, 2,2′-bipyridyl *N*-oxide and 2,2′-bipyridyl *N,N′*-dioxide have also been recorded. Several complexes with *p*-nitroso *N,N*-dimethyl and -diethylaniline are also known; they are precipitated when an ethanol solution of UO_2X_2 (X = Cl, Br, NO_3 or $MeCO_2$) or UO_2X (X = SO_4 or C_2O_4) is treated with the ligand in the same solvent; their IR spectra indicate that the ligand is bonded to the uranium atom *via* the nitroso oxygen atom only.

The uranium atom in $[UO_2(Et_2NCS_2)_2(Me_3NO)]$ is seven-coordinate, with four sulfur and one oxygen atom from the amine oxide in the equatorial plane of a pentagonal bipyramid[315]

Table 80 Some Complexes of *N*-Oxides with Dioxouranium(VI) Compounds

$UO_2X_2{\cdot}yL$	$L = Me_3NO$; $X = NO_3$, $y = 1$ or 4
	$X = ClO_4$, $y = 4$
	$X = Et_2NCS_2$ or Et_2NCSe_2, $y = 1$
	$L = C_5H_5NO$; $X = Cl$, $y = 2$ or 3
	$X = NCS$, $y = 3$
	$X = NO_3$, $y = 2$ or 3
	$X = ClO_4$, $y = 5$
	$X =$ acac, trop or 2-OC_6H_4CHO, $y = 1$
$[UO_2(Me_3NO)_4][BPh_4]_2$	
$UO_2X{\cdot}yL$	$L = C_5H_5NO$; $X = SO_4$, $y = 2$
	$X = NC_5H_3$-2,6-$(CO_2)_2$, $y = 2$
	$X = O(CH_2CO_2)_2$, $y = 1$ (polymer) or 2
	$L = 2$-MeC_5H_4NO; $X = ONC_5H_3$-2,6-$(CO_2)_2$, $y = 2$[a]

[a] S. Degetto, L. Baracco, M. Faggin and E. Celon, *J. Inorg. Nucl. Chem.*, 1981, **43**, 2413.

and the coordination geometry in $[UO\{O(CH_2CO_2)_2\}(pyNO)_2]$ is probably the same, with three oxygen atoms from the $O(CH_2CO_2)_2$ group, including the ether oxygen atom, and the two *N*-oxide oxygen atoms bonded to the UO_2 group.[288]

(*e*) P-*Oxides*. In addition to the complexes listed in Table 81, several 1:1 complexes of $UO_2(NO_3)_2$ with bis(dialkylphosphinyl)alkanes, $R_2P(O)(CH_2)_nP(O)R_2$, and many complexes of phosphine oxides, phosphate and phosphonate esters, $(RO)_3PO$ and $(RO)_2R'PO$, with dioxouranium(VI) β-diketonates and Schiff base compounds have been recorded.

In the complexes $[M^{VI}O_2Cl_2L_2]$, *trans* octahedral geometry has been reported for $M^{VI} = U$ with $L = (Me_2N)_3PO$[316] or Ph_3PO[317] and for $M^{VI} = Np$ with Ph_3PO;[318] in the complexes $[M^{VI}O_2(NO_3)_2(Ph_3PO)_2]$ ($M^{VI} = U$ or Np), the coordination geometry is hexagonal bipyramidal, with the bidentate nitrate groups *trans*,[318] as found also for the structure of $[UO_2(NO_3)_2\{(EtO)_3PO\}_2]$.[319] The cation $[UO_2\{Me_2N)_3PO\}_4]^{2+}$ in the perchlorate[320] and in the polyiodide, $[UO_2L_4](I_3)_2$, obtained by atmospheric oxidation of a mixture of UI_4 and $(Me_2N)_3PO$,[321] exhibits *trans* octahedral geometry.

In the dimeric compound $[UO_2(MeCO_2)_2(Ph_3PO)]_2$, which crystallizes when a cold acetone solution of $UO_2(MeCO_2)_2{\cdot}2H_2O$ containing the stoichiometric quantity of Ph_3PO is stirred for 20–30 minutes, the coordination geometry about the uranium atom is pentagonal bipyramidal with two bridging and two chelating $MeCO_2$ groups.[322] The same coordination geometry has been reported for monomeric $[UO_2(MeCSS)_2(Ph_3PO)]$, which is precipitated from methanol solutions of the dithioacetate and Ph_3PO,[323a] for the dithiophosphinate complex, $[UO_2(S_2PR_2)_2(Me_3PO)]$[323b], for the complex $[UO_2\{CF_3COCHCO(2\text{-}C_4H_3S)_2\}\{(n\text{-}C_8H_{17})_3PO\}]$[324] and for the α and β forms of the complex $[UO_2(CF_3COCHCOCF_3)_2\{(MeO)_3PO\}]$, which differ in that the β-diketonate groups in the α form are tilted by 22.5° to the plane of the pentagon in a boat conformation[325] whereas the pentagon is nearly planar in the β form.[326]

(*f*) As-*Oxides*. Complexes with these ligands (Table 82) are usually obtained by treating the parent compound with the ligand in a non-aqueous solvent. The complexes $[UO_2(MeCO_2)_2(Ph_3AsO)]_2$ and $[UO_2(MeCO_2)_2(Ph_3AsO)_2]$ are isomorphous with their Ph_3PO analogues. The coordination geometry in the dithio- and diseleno-carbamate compounds, $[UO_2(Et_2NCX_2)(Ph_3AsO)]$ is pentagonal bipyramidal, with four sulfur[327] or four selenium[328] and one oxygen atom in the equatorial plane. As expected, the coordination geometry in $[UO_2(NO_3)_2(Ph_3AsO)_2]$ is hexagonal bipyramidal, as in the phosphine oxide analogue; the U—O(As) bond length is shorter than the U—O(P) bond length in the latter.[329]

(*g*) S-*Oxides*. In addition to the compounds listed in Table 83, a large number of R_2SO ($R = Me$, Bu^n, *i*-C_5H_{11}, *n*-C_6H_{13}, $PhCH_2$ and Ph) complexes with dioxouranium(VI) β-diketonates and Me_2SO complexes with analogous Schiff base compounds are known and a few complexes of dioxouranium(VI) compounds with thioxane *S*-oxide, thianthrene 5-oxide, substituted sulfoxidothiophenes (RC_4H_3SO) and thiourea dioxide ($(H_2N)_2CSO_2$) have also been reported.

The structure of the polymeric complex $[UO_2F_2(Me_2SO)]_n$ consists of zigzag chains of uranium atoms linked by di-μ-fluoro bridges; the coordination geometry about each uranium

Table 81 Some Complexes of *P*-Oxides with Dioxoactinide(VI) Compounds; L = R_3PO

$UO_2X_2 \cdot yL$	R = Me; X = Cl, Br, NO_3, $y = 2$; X = ClO_4, $y = 4$
	R = Et; X = Cl, Br, NO_3, $y = 2$
	R = Pr^n; X = NO_3, $y = 2$; X = ClO_4, $y = 4$
	R = Bu^n; X = F, $y = 4/3$; X = Cl, Br, NO_3, ClO_4, $y = 2$; X = I, NCS, $y = 3$; X = I, ClO_4, $y = 4$; X = $MeCO_2$, $y = 1.5$; X = Et_2NCS_2, $(\eta\text{-}C_5H_5)Fe(\eta\text{-}C_5H_4CO_2)$, $(CO_3)Cr(\eta\text{-}PhCO_2)$, $y = 1$
	R = $n\text{-}C_8H_{17}$: X = F, $y = 1.5$ or 3; X = Cl, NO_3, $y = 2$; X = $MeCO_2$, $CF_3COCHCO\text{-}2\text{-}C_4H_3S$, $CF_3COCHCOPh$, $y = 1$; X = $CF_3COCHCO\text{-}2\text{-}C_4H_3S$, $y = 3$
	R = Me_2N; X = F, Cl, Br, NCS, N_3, NO_3, $N(CN)_2$,[b] $y = 2$; X = NCS, $y = 3$; X = ClO_4, $y = 4$ or 5; X = I_3, $y = 4$; X = $MeCO_2$ (dimer), Et_2NCS, Et_2NCSe_2, $y = 1$
	R = Ph; X = Cl, Br, I, NCS, NO_3, $MeCO_2$, $y = 2$; X = ClO_4, $y = 4$; X = $N(CN)_2$,[b] $y = 3$; X = $MeCO_2$ (dimer), MeCOS, $MeCS_2$, $PhCS_2$, Et_2NCS_2, Et_2NCSe_2, S_2PPr_2 (with R = Me, OMe, OEt, OPr^n, OPr^i, OBu^n, Ph), trop, X = R_2PS_2[a] (with R = OMe, OEt, OPr^n, OPr^i, OBu^n, Me or Ph), $y = 1$
$NpO_2X_2 \cdot 2L$	R = Ph; X = Cl, NO_3
$UO_2X \cdot yL$	R = Bu^n; X = SO_4, $y = 2.5$ or 3; X = C_2O_4, $y = 1$
	R = Me_2N; X = SO_3, $y = 1$;[e] X = SO_4, $y = 2$; X = $ONC_5H_3\text{-}2,6\text{-}(CO_2)_2$,[c] $y = 1.5$
$UO_2X_2 \cdot yL$	R = OMe; X = $CF_3COCHCOCF_3$, $y = 1$; X = ClO_4, $y = 5$
	R = OEt; X = NO_3, $y = 2$; X = ClO_4, $y = 5$
	R = OBu^n; X = Cl, Br, NCS, NO_3, ClO_4, $y = 2$; X = NCS, $y = 3$; X = $CF_3COCHCOR'$ with R' = Me, CF_3, Ph, $2\text{-}C_4H_3S$, $y = 1$
PuO_2X_2L	R = OBu^n; X = $CF_3COCHCO(2\text{-}C_4H_3S)$[d]

[a] I. Haiduc and M. Curtui, *Synth. React. Inorg. Metal-Org. Chem.*, 1976, **6,** 125.
[b] E. L. Ivanova, V. V. Skopenko and Kh. Keller, *Ukr. Khim. Zh.* (*Russ. Ed.*), 1981, **47,** 1024 (*Chem. Abstr.*, 1982, **96,** 45 232).
[c] S. Degetto, L. Baracco, M. Faggin and E. Celon, *J. Inorg. Nucl. Chem.*, 1981, **43,** 2413.
[d] J. P. Shukla, V. K. Manchanda and M. S. Subramanian, *J. Radioanal. Chem.*, 1976, **29,** 61.
[e] R. N. Shchelokov, G. T. Bolotova and N. A. Golubkova, *Sov. J. Coord. Chem.* (*Engl. Transl.*), 1977, **3,** 810.

Table 82 Some Complexes of *As*-Oxides with Dioxouranium(VI) Compounds

$UO_2X_2 \cdot R_3AsO$	R = Ph, X = $MeCO_2$ (dimer), MeCSS, PhCSS, Et_2NCSS, $Et_2NCSeSe$, trop ($C_7H_5O_2^-$, tropolonate), PhCOCHCSPh
$UO_2X_2 \cdot 2R_3AsO$	R = $n\text{-}C_8H_{17}$, X = Cl
	R = Ph, X = Cl, NCS, NO_3, $MeCO_2$
$UO_2X_2 \cdot 4R_3AsO$	R = Ph, X = ClO_4

atom is pentagonal bipyramidal, with four fluorine and one oxygen atom in the equatorial plane.[330] The same coordination geometry has been reported[331] for $[UO_2(Me_2SO)_5](ClO_4)_2$ and for the complex with the phosphite,[311] $[UO_2(HPO_3)(Me_2SO)(H_2O)]$.

(*ix*) *Hydroxamates, cupferron and related ligands*

A selection of the known dioxouranium(VI) complexes with this type of ligand is given in Table 84. These compounds are precipitated from aqueous solution, usually after the addition of alkali. The coordination geometry of the anion in the ammonium salt of the cupferron complex, $NH_4[UO_2\{(ON)N(Ph)O\}_3]$, is close to hexagonal bipyramidal.[332]

Table 83 Some Complexes of *S*-Oxides with Dioxouranium(VI) Compounds

$UO_2X_2 \cdot yR_2SO$	$y=1$; R = Me, X = F (polymer), $MeCO_2$, trop
	$y=2$; R = Me, Et, Bu^n, n-C_6H_{13}, X = NO_3
	R = n-C_8H_{17}, X = Cl
	R = Ph, X = Cl, Br, NCS, NO_3, $MeCO_2$
	$y=3$; R = Me, X = Cl, Br, NCS, NCSe
	R = Ph, X = Cl, Br
	$y=4$; R = Me, X = Br, NO_3, ClO_4
	R = $PhCH_2$, X = ClO_4
	R = Ph, X = I, ClO_4
	$y=4.5$; R = Me, X = Br
	$y=5$; R = Me, X = NO_3, ClO_4
	R = Ph, X = ClO_4
$UO_2X \cdot 2Me_2SO$	X = $SO_3(+0.5H_2O)$, SO_4, CrO_4, ONC_5H_3-2,6-$(CO_2)_2$ (polymer)[a]
$3UO_2(C_2O_4) \cdot 5Me_2SO$	
$(UO_2)_2(O_2)Cl_2 \cdot 4Me_2SO$[b]	

[a] S. Degetto, L. Baracco, M. Faggin and E. Celon, *J. Inorg. Nucl. Chem.*, 1981, **43**, 2413.
[b] R. N. Shchelokov, I. M. Orlova and A. V. Sergeev, *Sov. J. Coord. Chem.* (*Engl. Transl.*), 1981, **7**, 441.

Table 84 Some Dioxouranium(VI) Complexes with Hydroxamates, Cupferron and Related Ligands

Hydroxamates; *HL* = *(RCO)(R')NOH*	
UO_2L_2	R = Ph, R' = H (also $+1H_2O$), Ph (also $+2H_2O$), 2-MeC_6H_4 ($+1H_2O$ or HL)
	R = 2-HOC_6H_4, R' = H or Me ($+2H_2O$)
	R = 2-C_4H_3O, PhCH=CH ($+1H_2O$), 3- or 4-$O_2NC_6H_4$, R' = Ph
$H_2L = PhN(OH)C(O)(CH_2)_nC(O)N(OH)Ph$	
$UO_2L \cdot xH_2O$	$n=3$, $x=3$; $n=4$, $x=2$;
$UO_2(HL)_2 \cdot xH_2O$	$n=4$, $x=5$; $n=5$, $x=2$
$(UO_2)_2L(HL)_2$	$n=2$
Cupferron $\{(ON)N(Ph)O^-\}$	
$M^I[UO_2\{ONN(Ph)O\}_3]$	M^I = Na, K, Rb, Cs or NH_4
N-*Nitroso*-N-*cyclohexylhydroxylamine*, $C_6H_{12}N_2O_2$	
$NH_4[UO_2(C_6H_{11}N_2O_2)_3]$	

40.2.4.4 *Sulfur ligands*

(*i*) *Monothiocarbamates, dithiocarbamates and xanthates*

(*a*) *Monothiocarbamates.* The diethylammonium salt, $(Et_2NH_2)[UO_2(OEt)(Et_2NCOS)_2]$, precipitates when a saturated solution of $UO_2Cl_2 \cdot 3H_2O$ in ethanol is added to the solution obtained by passing COS through ethanolic diethylamine at 0 °C. The coordination geometry in the anion is pentagonal bipyramidal, with two sulfur and three oxygen atoms in the rather irregular pentagonal plane.[333] The anion structure is the same in the di-*n*-propylammonium salt of the $[UO_2(OEt)(Pr^n_2NCOS)_2]^-$ ion, prepared in a similar manner.[334] In the structure of an analogous compound, $(Pr^n_2NH_2)_2[UO_2(Pr^n_2NCOS)_2(S_2)]$, the uranium atom is at the centre of an irregular hexagonal bipyramid, with the equatorial plane positions occupied by the $(S—S)^{2-}$ group bonded sideways on, together with two oxygen and two sulfur atoms from the monothiocarbamate groups.[335]

(*b*) *Dithiocarbamates.* The existence of the compounds reported as $UO_2(R_2NCS_2)_2$ (R = Et, Pr^n, Bu^t) and $UO_2(RHNCS_2)_2$ (R = Et, Bu^i) is dubious and these are more likely to be salts of the anion $[UO_2(R_2NCS_2)_3]^-$. A number of salts of composition $M^I[UO_2(Et_2NCS_2)_3] \cdot xH_2O$ [M^I = Na ($x=2$, 3 or 6), K ($x=1$), Rb, Cs, NH_4 and NMe_4 (all $x=0$)] and $(R'R''NH_2)[UO_2(R'R''NCS_2)_3]$, with R' = Me and R'' = Me, Pr^i, CH_2CHMe_2, $CH(CHMe_2)_2$, $CH_2CH_2CHMe_2$, $CH(Me)CMe_3$ and $CHMeCH_2CMe_2OH$, or R' = R'' = CH_2CHMe_2, Bu^n or CH_2CH_2OH, have been reported. The coordination geometry in the anion of $(NMe_4)[UO_2(Et_2NCS_2)_3]$ is approximately hexagonal bipyramidal, with six sulfur atoms in the non-planar hexagon.[336]

(*c*) *Xanthates*. The salts $K[UO_2(ROCS_2)_3]$ (R = Et, Pr^i) are precipitated when an excess of the potassium alkyl xanthate is added to an aqueous solution of UO_2SO_4. Early reports of the formation of xanthates of the type $UO_2(ROCS_2)_2$ from aqueous solution appear to be doubtful.

(*ii*) *Dithiolates*

The *cis* maleonitriledithiolate salts $(NR_4)_2[UO_2(cis\text{-}C_4N_2S_2)_2]$ (R = Et, Pr^n) and the corresponding isomaleonitrile compounds $(NR_4)_2[UO_2\{(CN)_2C{=}CS_2\}_2]$ (R = Et, Pr^n) have been recorded.

(*iii*) *Thioureas*

A number of complexes of dioxouranium(VI) compounds with thioureas have been reported (Table 85) but little is known about them. In addition to those listed in the table, a number of complexes with dioxouranium(VI) nitrate formulated as $[UO_2L_x(NO_3)](NO_3)$, with $x = 4$ and L = *o*-chlorophenyl thiourea, $x = 3$ and L = *o*- or *p*-hydroxyphenyl thiourea, and $x = 2$ and L = *o*-methoxyphenylthiourea, as well as $[UO_2L(NO_3)_2]$ with L = benzylthiourea, have also been recorded.[337]

Table 85 Some Complexes of Thioureas with Dioxouranium(VI) Compounds

$L = SC(NH_2)_2$	
$UO_2X_2{\cdot}yL$	X = Cl, $y = 2$; $X = NO_3$, $y = 2$ or 4;[a] $X = MeCO_2$, $y = 1$, 2 or 4
$UO_2SO_4{\cdot}yL$	$y = 1$ or 2
$L = SC(NHMe)_2$	
$UO_2(NCS)_2{\cdot}3L$	
$L = SC(NH_2)(NHPh)$	
$UO_2(MeCO_2)_2{\cdot}2L$	
$L = SC(NMe_2)_2$	
$UO_2X_2{\cdot}yL$	X = Cl,[b] NO_3,[b] $y = 2$; X = NCS, $y = 3$
$UO_2(NCS)(NO_3){\cdot}2L$	

[a] A. S. R. Murty and M. B. Adi, *Proc. Nucl. Chem. Radiochem. Symp.*, 1980 (published 1981), p. 357, (*Chem. Abstr.*, 1982, **96**, 134 758).
[b] A. O. Baghlaf and K. W. Bagnall, *Bull. Fac. Sci. King Abdul Aziz Univ.*, 1980, **4**, 143.

(*iv*) *Dithiophosphinates*

Dithiophosphinates, $UO_2(R_2PS_2)_2$, have been obtained by ether[50] (R = Me) or benzene[51] (R = Ph) extraction of a mixture of concentrated aqueous solutions of $UO_2(NO_3)_2$ and $Na_2(R_2PS_2)$, followed by concentration of the extract, or (R = *p*-tolyl, $p\text{-}ClC_6H_4$) by heating a methanol solution of the ligand under reflux with the calculated quantity of[52] $UO_2(MeCO_2)_2$. These last, and the phenyl analogue, are monomers in benzene.[52] Alcohol adducts of the type $UO_2(S_2PR_2)_2{\cdot}R^1OH$ are obtained when UO_2Cl_2 in R^1OH (R^1 = Me, Et, Pr^i) is treated with a salt of the dithiophosphinic acid, R_2PS_2H (R = Me, Et, Pr^i, OMe, OEt, OPr^i, Ph or cyclo-C_6H_{11}). In the presence of an excess of chloride ion, the adducted R^1OH is displaced and salts of the type $Et_4N[UO_2(S_2PR_2)_2Cl]$ are formed. The coordination geometry about the uranium atom in $UO_2(S_2PR_2)_2{\cdot}R^1OH$ (R = Ph, cyclo-C_6H_{11}; R^1 = Et) and in the anion $[UO_2(S_2PR_2)_2Cl]^-$ (R = Me, Ph) is pentagonal bipyramidal with four S atoms from two bidentate S_2PR_2 ligands in the plane together with either one O atom from coordinated EtOH or one Cl atom, as in the structure of $[UO_2S_2PMe_2(Me_3PO)]$ (p. 1207).[323b]

40.2.4.5 Halogens as ligands

(*i*) *Hexahalides*

Octahedral hexafluorides, MF_6 (M = U, Np and Pu), are usually prepared by fluorination of lower fluorides, MF_4, at relatively high temperatures (*e.g.* M = Np, at *ca.* 500 °C). They are low melting, volatile solids. The only known hexachloride is UCl_6, prepared by chlorination of UCl_4, for example in $SbCl_5$. This is also octahedral.[338]

(*ii*) *Oxohalides*

The oxotetrafluorides, $M^{VI}OF_4$, are prepared by hydrolysis of the hexafluoride in HF with the stoichiometric quantity of water or, more elegantly, by using the calculated quantity of silica wool to provide the water by reaction with the solvent (M^{VI} = U,[339] Np[340] and Pu[341]). The coordination geometry about the metal atom in α-UOF_4 is close to pentagonal bipyramidal, with an oxygen and a fluorine atom in the axial positions.[339] $NpOF_4$[340] and $PuOF_4$[341] are isostructural with α-UOF_4. The coordination geometry in β-UOF_4 is also pentagonal bipyramidal, but with the two non-bridging axial fluorine atoms and a non-bridging equatorial oxygen atom together with four equatorial bridging fluorine atoms.[342] Complexes of the composition $UOF_4 \cdot xSbF_5$ ($x = 1$, 2 or 3) are also known. In these the structure consists of a fluorine-bridged network of UOF_4 and SbF_5 molecules in which each uranium atom is surrounded by a pentagonal bipyramidal array of one oxygen and six fluorine atoms.[343] The dioxodihalides, $M^{VI}O_2X_2$, are known for X = F (M^{VI} = U, Np, Pu), X = Cl [M^{VI} = U, Pu (+6H_2O)] and X = Br (M^{VI} = U, hydrated).

(*iii*) *Halo complexes*

Heptafluorouranates(VI), M^IUF_7 (M^I = Na, K, Rb, Cs or NF_4), have been prepared by reaction of M^IF with UF_6 in C_7F_{16}; the NF_4^+ compound is obtained from UF_6 and $NF_4 \cdot HF_2$ in HF.[344] Octafluorouranates(VI), $M_2^I[UF_8]$ are obtained by heating M^IUF_7 (M^I = Na, K, Rb or Cs) at 100 °C (M^I = Na) to 210 °C (M^I = Cs).[345] The U—F bond lengths in $Na_2[UF_8]$ are all equal.[346a] Oxidation of $(NH_4)_4U^{IV}F_8$ with XeF_2 at 55 °C yields a product of composition $(NH_4)_4UF_{10}$ which appears[346b] to be an example of uranium(VI) with coordination number greater than eight. The compounds $M_3^IUF_9$ (M^I = K or Rb) are prepared by fluorination of $M_3^IUF_7$ at 400 (K) to 500 °C (Rb).[347]

(*iv*) *Oxohalogeno complexes*

A few salts of complex anions derived from monoxoactinide(VI) halides are known; $(Ph_4P)[UOCl_5]$ is obtained when a solution of $(Ph_4P)_2[UO_2Cl_4]$ in thionyl chloride is heated under reflux for a few minutes and then cooled. The U—Cl bond length *trans* to the oxygen atom is shorter than the equatorial U—Cl bond lengths. The chloro complex reacts with HF at low temperature and with HBr in dichloromethane to form $(Ph_4P)[UOF_5]$ and $(Ph_4P)[UOClBr_4]$ respectively.[348]

The more common complexes are derived from dioxoactinide(VI) dihalides. The structure of the aquated anion in $(enH_2)[UO_2F_4(H_2O)]$ has been described earlier (p. 1192) and a number of hydrated salts of the type $M^IM^{VI}O_2F_3 \cdot xH_2O$ (M^{VI} = U, M^I = an organic base and $x = 0$–6; M^{VI} = Pu, M^I = Na, K, Rb, Cs, NH_4, C_9H_7NH (quinolinium), $x = 1$), $M_2^IPuO_2F_4$ (M^I = Na, K, Rb, NH_4), $M^{II}UO_2F_4 \cdot 4H_2O$ (M^{II} = Zn, Cd, Cu, Mn, Co, Ni), $M^I(M^{VI}O_2)_2F_5 \cdot xH_2O$ (M^{VI} = U, M^I = an organic base, $x = 0$–6; M^{VI} = Pu, M^I = Cs, $x = 3$) and $M^I(UO_2)_3F_7 \cdot xH_2O$ (M^I = an organic base) have been reported. The more usual fluoro complex anion is $[M^{VI}O_2F_5]^{3-}$ (M^{VI} = U, Np, Pu), for which several alkali metal salts are known. The potassium salts for the three elements are isostructural; in $K_3[UO_2F_5]$[349a] and $K_3[NpO_2F_5]$[349b] the coordination geometry of the anion is pentagonal bipyramidal. The IR spectra of $K_3[M^{VI}O_2F_5]$ (M^{VI} = U, Np, Pu) have also been reported.[249]

The compounds $Cs_2NH_4(UO_2)_2F_7$,[350] $Na_3(UO_2)_2F_7 \cdot 6H_2O$[351] and $Ni_3\{(UO_2)_2F_7\}_2 \cdot 18H_2O$[352] each involve infinite chains, $[(UO_2)_2F_7]_n^{3n-}$, formed by UO_2F_5 pentagonal bipyramids joined by a common edge and a common apex, with three bridging and two terminal fluorine atoms in the equatorial plane.

Chloro complexes are normally of composition $M_2^I[M^{VI}O_2Cl_4]$(M^{VI} = U, Np, Pu, with M^I = Cs, NEt_4; M^{VI} = U, with M^I = Li or Na (+6H_2O), K (+2H_2O), NH_4, 1,10-phenH, PPh_4, PBu_4^n; M^{VI} = Pu, M^I = NMe_4, NPr_4^n, Et_3NH), $M^{II}[UO_2Cl_4] \cdot xH_2O$ (M^{II} = Be, $x = 4$; Mg, Ca, Ba or VO, $x = 2$; Sr, $x = 1$). The coordination geometry of the anion in $(1,10\text{-phenH})_2[UO_2Cl_4]$ is a flattened square bipyramid.[353] The IR spectra of $Cs_2[M^{VI}O_2Cl_4]$ (M^{VI} = U, Np and Pu) have also been reported.[249] Products of composition $M^{III}[UO_2Cl_5]\cdot xH_2O$ (M^{III} = Y, $x = 3$ or La, $x = 2$), $M^{IV}[UO_2Cl_6]\cdot xH_2O$ (M^{IV} = Ce, $x = 5$ or Th, $x = 4$), $Mn_3[UO_2Cl_8]\cdot 6H_2O$, $Al_2[UO_2Cl_8]\cdot 6H_2O$ and a number of salts which appear to contain the anion $[(UO_2)_2Cl_7]^{3-}$ have been reported,[354] but these could be salts of the UO_2^{2+} ion. For example, $Ce[UO_2Cl_6]\cdot 5H_2O$ is more likely to be $[UO_2(H_2O)_5][CeCl_6]$.

Relatively few salts of bromo and iodo complex anions, $M^I_2[UO_2X_4]$, have been reported (X = Br, M^I = PPh_3Bu^n, $PPh_3(CH_2Ph)$, $HPPh_3$, $HPEt_3$, $HPPr^n_3$,[355] NBu^n_4; X = I, M^I = PPh_3Bu^n). The coordination geometry of the anion in $(NBu^n_4)_2[UO_2Br_4]$ is again a flattened square bipyramid.[356]

40.2.4.6 Mixed donor atom ligands

(i) Schiff bases

A very large number of dioxouranium(VI) compounds with this type of ligand have been recorded (see ref. 12, Uranium, vol. E2), examples of which are given in Table 86; these include some instances in which the mono- and di-basic ligands behave as neutral donors. Structure determinations indicate that the usual coordination geometry is approximately pentagonal bipyramidal; examples are $[UO_2(2\text{-}OC_6H_4CH{=}N(CH_2)_2NH(CH_2)_2NMe_2)(NO_3)]$, in which four equatorial positions are occupied by one oxygen and three nitrogen atoms of the tetradentate Schiff base and the fifth by an oxygen atom of the monodentate nitrate group;[357] $[UO_2(2\text{-}OC_6H_4CH{=}N(CH_2)_2NMe_2)_2]$, in which the equatorial sites are occupied by two oxygen and three nitrogen atoms from the two potentially tridentate ligands;[358] $[UO_2(2\text{-}OC_6H_4CH{=}N(CH_2)_2NH(CH_2)_2N{=}CHC_6H_4O\text{-}2)]$, with two oxygen and three nitrogen atoms of the ligand in the puckered equatorial plane;[359] and $[UO_2(2\text{-}OC_6H_4N{=}CHCH{=}NC_6H_4O\text{-}2)(H_2O)]$ with two oxygen and two nitrogen atoms from the ligand and one oxygen atom from the water molecule in the near planar pentagon.[360a]

Table 86 Some Dioxouranium(VI) Schiff Base Complexes

$UO_2(MeCO_2)_2\cdot LH$	LH^a = $2\text{-}HOC_6H_4CH{=}NR$ and $2\text{-}HOC_{10}H_6CH{=}NR$ with R = Et, Pr^n, Bu^n and Ph
$UO_2(MeCO_2)_2\cdot LH_2$	$LH_2{}^a$ = $2\text{-}HOC_6H_4CH{=}N(A)OH$ and $2\text{-}HOC_{10}H_6CH{=}N(A)OH$ with A = CH_2CHMe or $(CH_2)_3$
$UO_2X_2\cdot 2LH$	LH = $2\text{-}HOC_6H_4CH{=}NR$ with X = Cl or NO_3 and R = Et, Pr^n, Bu^n, $(CH_2)_2NHPh$
UO_2XL	LH = $2\text{-}HOC_6H_4CH{=}NR$ with R = $(CH_2)_2NH(CH_2)_2NMe_2$ and X = F, Cl, Br, NCS, NO_3, $MeCO_2$ or trop (tropolonate)
UO_2L_2	LH = 3-Me, $2\text{-}HOC_6H_3CH{=}NHCOPh$ and $2\text{-}HOC_6H_4CH{=}N(CH_2)_2NMe_2$
UO_2L	LH_2 = $2\text{-}HOC_6H_4CH{=}N(CH_2)_2NH(CH_2)_2N{=}CHC_6H_4\text{-}2\text{-}OH$
$[UO_2L(H_2O)]$	LH_2 = $2\text{-}HOC_6H_4N{=}CHCH{=}NC_6H_4\text{-}2\text{-}OH$

[a] R. G. Vijay and J. P. Tandon, *Monatsh. Chem.*, 1979, **110**, 889.

(ii) Amino acids

Dioxouranium(VI) compounds with aminoacetic acid (glycine), $UO_2(H_2NCH_2CO_2)_2$ and $UO_2(H_2NCH_2CO_2)(OH)\cdot 2H_2O$, and similar derivatives of α and β-aminopropionic acid (α- and β-alanine), $UO_2(MeCHNH_2CO_2)_2\cdot 4H_2O$, $UO_2(MeCHNH_2CO_2)(ClO_4)\cdot 2H_2O$ and $UO_2(H_2NCH_2CH_2CO_2)_2\cdot xH_2O$ ($x = 0$ or 4), have been recorded. In the glycine compound $[UO_2(O_2CCH_2NH_3)_4](NO_3)_2$ all of the glycine ligands are in the zwitterionic form and act as oxygen donor ligands; two are bidentate and two unidentate, making up the equatorial plane of a hexagonal bipyramidal arrangement of oxygen atoms around the uranium atom.[360b] L-Arginine $[H_2NC({=}NH)NH(CH_2)_3CHNH_2CO_2H$ (= HX)] compounds of the type $UO_2XY\cdot zH_2O$ (Y = NO_3, $z = 2$; Y = ClO_4 or $MeCO_2$, $z = 3$) are also known.[361] 2-, 3- and $4\text{-}H_2NC_6H_4CO_2H$ form compounds of composition UO_2L_2; an acid salt of the 2-amino acid (= HL), $UO_2L_2\cdot HL\cdot 2H_2O$, and a mixed oxalato salt, $Na_2[UO_2(C_2O_4)(L)_2]\cdot 2H_2O$, have been recorded.

(iii) Complexones

Nitrilotriacetic acid, $N(CH_2CO_2H)_3(=H_3L)$, forms the complexes $UO_2(HL)\cdot xH_2O$ ($x=0$, 2 or 5) and $(UO_2)_3(L)_2\cdot xH_2O$ ($x=3$ or 10), and the analogous phosphorus tripropionic acid, $P(CH_2CH_2CO_2H)_3(=H_3L)$, forms the salt of the complex anion, $Na[UO_2L]\cdot H_2O$. The hydrates are precipitated from aqueous solution. Similarly, ethylenediamine-*N,N,N′,N′*-tetraacetic acid, $(HO_2CCH_2)_2NCH_2CH_2N(CH_2CO_2H)_2$ ($=H_4$edta), yields compounds of composition $UO_2(H_2edta)\cdot xH_2O$ ($x=0$, 1 or 2), $(UO_2)_2(edta)\cdot xH_2O$ ($x=0$, 2, 3 or 4) and salts of the complex anion, such as $M^I_2[UO_2(edta)]\cdot xH_2O$ ($M^I=K$, x unspecified; $M^I=NH_4$, $x=0$, 2 or 3), $[Co(NH_3)_6]_2[UO_2(edta)]_3\cdot 22H_2O$ and $Na_2[UO_2(H_3edta)_4]$. Mixed peroxo-edta compounds of composition $[(UO_2)_2(H_2edta)(O_2)]\cdot 4H_2O$, $[(UO_2)_2(H_4edta)(O_2)_2]\cdot 9H_2O$, $[(UO_2)_2(H_4edta)_2(O_2)_2]\cdot 10H_2O$ and $Na_2[(UO_2)_2(edta)(O_2)]\cdot 7H_2O$ have also been recorded.

Hydrated compounds of the type $UO_2(H_2L)\cdot xH_2O$ have been reported for 1,2-diaminopropane-*N,N,N′,N′*-tetraacetic acid (x unspecified), *trans* diaminocyclohexane-*N,N,N′,N′*-tetraacetic acid ($x=0$ or 3; $(UO_2)_2L\cdot xH_2O$, with $x=0$ or 6 also known), 2,2′-diaminodiethyl ether-*N,N,N′,N′*-tetraacetic acid ($x=3$), and 2,2′-diaminodiethyl sulfide-*N,N,N′,N′*-tetraacetic acid ($x=0$ or 4); the hydrates are precipitated from aqueous media, but in the last instance a mixture of ethanol and acetone was added to induce precipitation. The analogous di(2-aminoethoxy)ethane-*N,N,N′,N′*-tetraacetic acid (H_4L) forms the compounds $[(UO_2)_2L(H_2O)_2]$ and $Na[UO_2(HL)(H_2O)]$.

(iv) Ligands containing N,O and S,O donor sites

(*a*) *8-Hydroxyquinolines.* Neptunium(VI) and plutonium(VI) are reduced to lower oxidation states by 8-hydroxyquinoline or its derivatives[362] and only dioxouranium(VI) compounds are known; these are of composition UO_2L_2, $UO_2L_2\cdot HL$ and $M^I[UO_2L_3]$ (Table 87). In addition, 1:1 adducts with the pyridine-2-carboxylate, $UO_2(NC_5H_4\text{-}2\text{-}CO_2)_2\cdot HL$, are formed by 8-hydroxyquinoline and the 2-methyl derivative. Further work on the wide range of substituted 8-hydroxyquinolates is desirable, for the existence of some of the reported solvates, $UO_2L_2\cdot HL$, is uncertain.

Table 87 Dioxouranium(VI) Complexes with 8-Hydroxyquinoline and Substituted 8-Hydroxyquinolines

UO_2L_2	HL = 8-hydroxyquinoline,[a] 2-, 5- or 7-methyl,[b] 5-acetyl, 5-chloro, 5-nitro-, 5-phenyl, 7-*t*-butyl-, 2,7-dimethyl-,[b] 5-chloro-7-iodo, 5,7-dibromo-[b] and 5,7-dichloro-8-hydroxyquinoline[b]
$UO_2L_2\cdot HL$	HL = 8-hydroxyquinoline,[c] 2- or 5-methyl-, 5-nitro-, 5-phenyl-, 5-acetyl-, 5-chloro-, 5-iodo-, 7-fluoro-, 7-phenyl-, 7-(α-anilinobenzyl)-, 7-[α](*m*- or *p*-nitroanilinobenzyl]-, 7-(*p*-carboxyphenylaminobenzyl)-, 5,7-dichloro-, 5,7-dibromo- and 5,7-diiodo-8-hydroxyquinoline
$M^U[UO_2L_3]$	HL = 8-hydroxyquinoline

[a] Mono-, tri- and tetra-hydrate also known.
[b] Monohydrate also known.
[c] Mono- and di-hydrate also known.

In the complex $[UO_2L_2(HL)]\cdot CHCl_3$ (HL = 8-hydroxyquinoline), the additional molecule of HL is coordinated to the uranium atom through the phenolic oxygen atom; the coordination geometry is approximately pentagonal bipyramidal.[363]

(*b*) *Ligands with S,O donor sites.* The complex with $S(cyclo\text{-}C_6H_{10}OH)_2$, $[UO_2Cl_2L]$, is obtained by adding an excess of the ligand in ethyl acetate to a solution of $UO_2Cl_2\cdot 3H_2O$ in the same solvent. The ligand is terdentate in this compound, with the two oxygen atoms from the hydroxyl groups and the sulfur atom, together with the two chlorine atoms, in the equatorial positions of a distorted pentagonal bipyramid.[364] Thiovanol ($HSCH_2CH(OH)CH_2OH=C_3H_8SO_2$) complexes, $[UO_2(C_3H_7SO_2)_2(H_2O)_2]$ and $[UO_2(C_3H_7SO_2)Cl(H_2O)_3]\cdot H_2O$, have been recorded,[365] and these may involve bonding to sulfur as well as oxygen, but the assignment of the water molecules to the coordination sphere requires confirmation. An 8-mercaptoquinolinate, $UO_2(C_9H_6NS)_2$, the sulfur analogue of the 8-hydroxyquinolinate, has been prepared by reaction of $UO_2(NO_3)_2\cdot 6H_2O$ with $Ca(C_9H_6NS)_2$ in benzene.[366]

40.2.4.7 Multidentate macrocyclic ligands

(i) Phthalocyanine and superphthalocyanine

The dioxouranium(VI) phthalocyanine complex, $UO_2(C_{32}H_{16}N_8){\cdot}C_6H_4(CN)_2$ ($C_{32}H_{18}N_8$ = phthalocyanine), has been reported to be formed by reaction of UO_2Cl_2 with *o*-phthalodinitrile, $C_6H_4(CN)_2$, but this reaction in DMF has been shown to yield the cyclopentakis(2-iminoisoindoline) ($C_{40}H_{22}N_{10}$) derivative, $[UO_2(C_{40}H_{20}N_{10})]$, in which the coordination geometry about the uranium atom can be described as a compressed pentagonal bipyramid, with five nitrogen atoms of the $C_{40}H_{20}N_{10}$ ligand in the irregularly distorted pentagonal girdle.[367] This compound reacts with di- and tri-valent metal halides with contraction of the ring and formation of the corresponding metal phthalocyanine complex.[368]

Complexes with 2,3,9,10,16,17,23,24,30,31-decaalkyl superphthalocyanines, $(4,5\text{-}R_2)_5PcUO_2$ (R = Me, Bu^n), and the analogous pentamethylphthalocyanine complex, $(4\text{-}Me)_5PcUO_2$, have been reported. In these the uranium atom is bonded to five nitrogen atoms of the superphthalocyaninate group.[369]

(ii) Crown ethers and kryptands

A number of crown ether and kryptand solvates of dioxouranium(VI) compounds have been recorded (Table 88). The IR and ^1H NMR spectra of many of these products suggest that the crown ether is not bonded to the metal atom, and an X-ray study has shown that the 18-crown-6 solvate, $UO_2(NO_3)_2{\cdot}L{\cdot}4H_2O$, consists of layers of hexagonal bipyramidal $[UO_2(NO_3)_2(H_2O)_2]$ units with the crown ether lying between the uranium atoms.[370] However, IR evidence suggests that the ether in the hydrated and anhydrous 15-crown-5, 18-crown-6 and dicyclo-hexyl-18-crown-6 solvates of UO_2Cl_2 may be bonded to the metal.[371] The constitution of the salt $Na_2UO_2Cl_4{\cdot}2$-benzo-15-crown-5 has been shown to be $[Na(\text{benzo-15-crown-5})]_2[UO_2Cl_4]$, with the anion octahedral.[372]

Table 88 Crown Ether and Kryptand Solvates of Dioxouranium(VI) Compounds

L = 12-crown-4; 1,4,7,10-tetraoxacyclododecane	
$UO_2X_2{\cdot}L{\cdot}yH_2O$	X = Cl, $y = 0$ or 2; X = NO_3, $y = 2$
L = 15-crown-5; 1,4,7,10,13-pentaoxacyclopentadecane	
$UO_2X_2{\cdot}L{\cdot}yH_2O$	X = Cl, $y = 3$; X = NO_3, $y = 2$
$(UO_2F_2)_2{\cdot}L{\cdot}xH_2O$	$x = 0$ or *ca.* 4
$(UO_2Cl_2)_2{\cdot}3L{\cdot}2HCl$	
L = benzo-15-crown-5; 2,3,5,6,8,9,11,12-octahydrobenzo[*b*]-1,4,7,10,13-pentaoxacyclopentadecin	
$UO_2X_2{\cdot}L{\cdot}yH_2O$	X = Cl, $y = 2$; X = NO_3, $y = 2$ or 5; X = ClO_4, $y = 7$
L = 18-crown-6; 1,4,7,10,13,16-hexaoxacyclooctadecane	
$UO_2X_2{\cdot}L{\cdot}yH_2O$	X = Cl, $y = 2$ or 3; X = NO_3, $y = 2$, 3, 4 or 5; X = $MeCO_2$, $x = 3$ or 4
$(UO_2Cl_2)_2{\cdot}L$	
L = dibenzo-18-crown-6; 2,3,5,6,11,12,14,15-octahydrodibenzo[*b*,*k*]-1,4,7,10,13,16-hexaoxacyclooctadecin	
$UO_2X_2{\cdot}L{\cdot}yH_2O$	X = Cl, NO_3, $y = 2$; X = ClO_4, $y = 6$
L = dicyclohexyl-18-crown-6; icosahydrodibenzo[*b*,*k*]-1,4,7,10,13,16-hexaoxacyclooctadecin	
$UO_2(NO_3)_2{\cdot}L{\cdot}2H_2O$	
$(UO_2Cl_2)_2{\cdot}L$	
L = kryptofix⟨2,2,2⟩; 4,7,13,16,21,24-hexaoxa-1,10-diazabicyclo[8.8.8]-hexacosane	
$UO_2X_2{\cdot}L{\cdot}yH_2O$	X = Cl, $y = 2$; X = NCS, $y = 1$; X = NO_3, $y = 4$; X = $MeCO_2$, $y = 0$ or 3

40.2.5 The +7 Oxidation State

Salts of oxoderivatives of the types $[MO_4]^-$, $[MO_5]^{3-}$ and $[MO_6]^{5-}$ have been reported for neptunium(VII) and plutonium(VII). Compounds of composition $M^I_3M^{VII}O_5$ (M^{VII} = Np, M^I = K, Rb or Cs; M^{VII} = Pu, M^I = Rb or Cs) have been prepared by heating the alkali metal peroxides with the actinide dioxide at 320 to 355 °C (Np) or 250 °C (Pu). Hydrated compounds, $M^{II}_3(NpO_5)_2{\cdot}xH_2O$ [M^{II} = Ca, Sr, Ba ($x = 3$)] have been prepared from aqueous alkaline solutions of neptunium(VII), as have the hydrated salts $M^{III}NpO_5{\cdot}xH_2O$ [$M^{III} = [Co(NH_3)_6]$, $[Co(en)_3]$ ($x = 5$) and $[Pt(NH_3)_5Cl]$ ($x = 1$)]. The IR spectra of $Ba_3(M^{VII}O_5)_2{\cdot}xH_2O$ ($M^{VII} = Np$, $x = 3$; Pu, $x = 3$), $[Co(NH_3)_6]M^{VII}O_5{\cdot}xH_2O$ ($M^{VII} = Np$, $x = 5$; Pu, $x = 3$),[373] $[Pt(NH_3)_5Cl]M^{VII}O_5{\cdot}xH_2O$ ($M^{VII} = Np$, $x = 1$; Pu[373], $x = 3$) suggest that the oxoanion structure consists of infinite chains of $[NpO_6]_n^{5n-1}$ units.[374] Structural investigations of compounds of the general composition $M^INpO_5{\cdot}xH_2O$ indicate that they are really $M^I_3[NpO_4(OH)_2]{\cdot}(x-1)H_2O$; the coordination geometry of the neptunium(VII) centre in the anion is described as tetragonal bipyramidal, with the two OH groups in the axial positions ($M^I = Na$,[375] $x = 5$; K,[376] $x = 3$). The anion preserves the same coordination geometry in $Na_3[NpO_4(OH)_2]{\cdot}xH_2O$ ($x = 0$,[377a] 2[377b] and 4[377c]); both hydrates lose all their water at 373 K. $Ba_3(NpO_5)_2$ has also been prepared by heating $NpO_3{\cdot}H_2O$ with barium peroxide in oxygen at 500 °C.

Compounds of composition M^INpO_4 ($M^I = K, Rb, Cs$) are prepared in the same way as the hydrated salts $M^{II}_3(NpO_5)_2{\cdot}xH_2O$ but with concentrated alkaline solutions of neptunium(VII).[378] The ammonium salt, $NH_4NpO_4{\cdot}nH_2O$ ($n = 1.5$ to 3) has been prepared by anodic oxidation of neptunium(VI) in $(NH_4)_2CO_3$ solution; it is isostructural with $M^INpO_4{\cdot}2.5H_2O$ ($M^I = Li$ or Na).[379]

The compounds $Li_5M^{VII}O_6$ ($M^{VII} = Np, Pu$) have been obtained by heating the actinide dioxide with lithium oxide in oxygen at *ca.* 400 °C(Np), and the analogous $Ba_2M^INpO_6$ ($M^I = Li$ or Na) by heating $NpO_3{\cdot}H_2O$ with the alkali metal peroxide in oxygen above 400 °C. The $[NpO_6]^{5-}$ anion appears to be octahedral.

The IR spectrum of $Li_3NpO_2(OH)_6$ has been reported; it was prepared by dissolving $NpO_2(OH)_3{\cdot}3H_2O$ in aqueous lithium hydroxide and then evaporating the solution.[380] In the structure of the compound $Li[Co(NH_3)_6][Np_2O_8(OH)_2]{\cdot}2H_2O$, precipitated from a lithium hydroxide solution of neptunium(VII) by allowing aqueous $[Co(NH_3)_6]Cl_3$ to diffuse into it, there are two independent neptunium centres, each coordinated by an octahedron of oxygen atoms; the octahedra share corners to form a chain.[381]

40.3 THE TRANSPLUTONIUM ELEMENTS

40.3.1 The +2 Oxidation State

The americium halides AmX_2 (X = Cl, Br[382] and I[383]) have been obtained by heating americium metal with the appropriate mercury(II) halide; they are isostructural with the corresponding europium dihalides. The californium(II) compounds CfX_2 [X = Cl,[384] Br[385] and I (dimorphic[386])], as well as the einsteinium analogues EsX_2 (X = Cl, Br, I),[387] have been prepared by hydrogen reduction of the trihalides at high temperatures. $CfBr_2$ has also been obtained[385] by heating Cf_2O_3 in HBr at 500–625 °C, and there is also tracer level evidence[388] for the formation of $FmCl_2$. The known monoxides and monosulfides, such as AmO and AmS, are probably semimetallic in character.

40.3.2 The +3 Oxidation State

Complexes involving group IV or group V ligands do not appear to have been recorded.

40.3.2.1 Oxygen ligands

(i) Aqua species, hydroxides and oxides

(a) Aqua species. A selection of the known hydrates is given in Table 89. The hydrated xenate(VIII), $Am_4(XeO_6)_3{\cdot}40H_2O$, has also been reported.[389] The hydrated trichlorides, $MCl_3{\cdot}6H_2O$ (M = Am or Bk), and tribromides, $MBr_3{\cdot}6H_2O$ (M = Am or Cf), are isostructural with the hydrated lanthanide chlorides, $LnCl_3{\cdot}6H_2O$, and involve aqua ions, such as[390] $[AmCl_2(OH_2)_6]^+Cl$. In the hydrated salicylate, $Am(C_7H_5O_3)_3{\cdot}H_2O$, each metal atom is linked to six different salicylate groups and is surrounded by nine oxygen atoms, eight from the salicylate groups and one from the water molecule; two salicylate groups are bidentate, one *via* its carboxylate group and the other *via* its carboxylate and phenolic groups, and the other four are monodentate *via* the carboxylate group.[391]

Table 89 Some Hydrates of Transplutonium Actinide(III) Compounds

$M^{III}X_3{\cdot}6H_2O$	X = Cl, M^{III} = Am, Bk; X = Br, M^{III} = Am, Cf
$M_2^{III}(SO_4)_3{\cdot}xH_2O$	M^{III} = Am, $x = 5$; M^{III} = Cf, $x = 12$
$M^IAm(SO_4)_2{\cdot}xH_2O$	M^I = Na, $x = 1$; K, $x = 2$; Rb, Cs, Tl, $x = 4$
$K_3Am(SO_4)_3{\cdot}H_2O$	
$Am_2(CO_3)_3{\cdot}xH_2O$	$x = 4$ or 5
$NaAm(CO_3)_2{\cdot}4H_2O$	
$Na_3Am(CO_3)_3{\cdot}3H_2O$	
$M_2^{III}(C_2O_4)_3{\cdot}xH_2O$	M^{III} = Am, $x = 0.5$, 1, 3, 4, 7, 9 or 11
	M^{III} = Cm, $x = 1$, 2, 3, 4, 5, 6, 7, 8, 9 or 10
$M^{III}PO_4{\cdot}0.5H_2O$	M^{III} = Am, Cm
$[Am(C_7H_5O_3)_3(H_2O)]$	$C_7H_6O_3$ = salicyclic acid

(*b*) *Hydroxides.* The hydroxides $M(OH)_3$ are known for M = Am, Cm, Bk and Cf; $Am(OH)_3$ is isostructural with $Nd(OH)_3$.[392]

(*c*) *Oxides.* Three crystal modifications have been reported for both Am_2O_3 and Cm_2O_3, two of which (the A and C types) have been found for Bk_2O_3 and the A and B types have been reported for Cf_2O_3. Ternary oxides, such as $LiAmO_2$, obtained by heating AmO_2 with Li_2O in hydrogen at 600 °C, $M^{II}(AmO_2)_2$ (M^{II} = Sr or Ba) and $M^{III}AlO_3$ (M^{III} = Am or Cm) have been reported.

(*ii*) *β-Ketoenolates*

The hydrated americium compounds, $AmL_3{\cdot}xH_2O$ (HL = $MeCOCH_2COMe$, $x = 1$; CF_3COCH_2COPh and $CF_3COCH_2CO(2\text{-}C_4H_3S)$, $x = 3$), are precipitated from aqueous media[393] and $Am(Bu^tCOCHCOBu^t)_3$ is precipitated from an aqueous ethanol solution of the components. It is isomorphous with the praseodymium and neodymium chelates, in which (Pr) the metal atoms are seven-coordinate.[394] The berkelium compounds, $Bk(CF_3COCHCOR)_3$ (R = CF_3, Bu^t and $2\text{-}C_4H_3S$), have been obtained by a solvent extraction method[395] and the sublimation behaviour of $M(Bu^tCOCHCOBu^t)_3$ (M = Am,[396,397] Cm[397] and Cf[396,397]) has also been reported. Compounds of composition $CsM(CF_3COCHCOCF_3)_4{\cdot}xH_2O$ have been isolated for M = Am ($x = 0$ or 1) and Cm ($x = 1$), and there is evidence for the formation of analogous compounds with M = Bk, Cf or Es. In the structure of the anhydrous americium compound, the metal atom is surrounded by a dodecahedral arrangement of oxygen atoms.[398]

(*iii*) *Oxoanions as ligands*

(*a*) *Phosphates.* Hydrated phosphates, $MPO_4{\cdot}0.5H_2O$ (M = Am, Cm), are obtained from aqueous solution and can be dehydrated at about 200 °C.

(*b*) *Sulfates.* Hydrated sulfates and sulfato complexes are included in Table 89. Anhydrous $Am_2(SO_4)_3$ is obtained by heating the pentahydrate at 500 °C; anhydrous sulfato complexes of composition $KAm(SO_4)_2$ and $M_8^IAm_2(SO_4)_7$ (M^I = K, Cs or Tl) are also known. The oxosulfates, $M_2^{III}O_2SO_4$ (M^{III} = Cm or Cf), have been reported; the curium compound is obtained by calcining the Cm^{3+} form of Dowex 50W-X8 resin in air at 900 °C. $Cm_2(SO_4)_3$ may be formed at 760 to 870 °C in this process[399] and the californium compound is obtained by heating $Cf_2(SO_4)_3{\cdot}12H_2O$ in air at 700 to 800 °C.[400]

(*iv*) *Carboxylates and carbonates*

(*a*) *Monocarboxylates.* The only recorded compound appears to be the formate, $Am(HCO_2)_3$, prepared by evaporating a solution of $Am(OH)_3$ in concentrated formic acid at 50 °C.

(*b*) *Carbonates.* Hydrated americium carbonate, $Am_2(CO_3)_3{\cdot}4H_2O$, is precipitated from aqueous solutions containing americium(III) by $NaHCO_3$. It decomposes to $Am_2O(CO_3)_2$, and then to $Am_2O_2(CO_3)$ on heating. The anhydrous carbonates, $M_2^{III}(CO_3)_3$ (M^{III} = Am or Cm) are formed in the radiolytic decomposition of the oxalate (Am) or by heating the anhydrous oxalate at 360 °C (Cm); $Cm_2O(CO_3)_2$ and $Cm_2O_2(CO_3)$ are obtained at 420 and 510 °C respectively.

(c) *Oxalates.* The hydrated oxalate, $Am_2(C_2O_4)_3{\cdot}xH_2O$ ($x = 7, 9, 10$ or 11), is precipitated from aqueous solutions containing americium(III) by oxalic acid and the anhydrous oxalate is obtained by heating the hydrate above 240 °C. The corresponding berkelium and californium compounds are precipitated from aqueous nitric acid solutions of Bk^{III} or Cf^{III} by oxalic acid.[401a] The oxalato complexes, $MAm(C_2O_4)_2{\cdot}xH_2O$ ($M = NH_4$, $x = 5$, $M = Na$, Cs), have been prepared from $Am_2(C_2O_4)_3{\cdot}10H_2O$ and $M_2C_2O_4$ in neutral solution.[401b] The potassium salts ($x = 2.5$, 3.5 and 5) have been obtained in the same way.[401c]

(v) *Aliphatic and aromatic hydroxyacids*

The only recorded compounds appear to be the citrates, $Am(C_6H_5O_7){\cdot}xH_2O$ and $[Co(NH_3)_6][Am(C_6H_5O_7)_2]{\cdot}xH_2O$,[402] and the hydrated salicylate, $Am(C_7H_5O_3)_3{\cdot}H_2O$, the structure of which is described on p. 1215.

(vi) *P-Oxides*

The only complexes known (Table 90) are those of β-ketoenolates with $(n\text{-}C_8H_{17})_3PO$ and $(Bu^nO)_3PO$, prepared by solvent extraction from aqueous solutions of the actinides(III) with a mixture of the β-ketoenol and the *P*-oxide. There is also solution chemistry evidence for the formation[403] of $Am[CF_3COCHCO(2\text{-}C_4H_3S)]_3{\cdot}2(Bu^nO)_3PO$. $Am[CF_3COCHCOCF_3)_3{\cdot}2(Bu^nO)_3PO$ is volatile at 175 °C[404] and a gas chromatography study of this complex and the curium analogue has been reported.[405]

Table 90 Complexes of Transplutonium Actinide(III) β-Ketoenolates with *P*-Oxides

$M^{III}L_3{\cdot}x(n\text{-}C_8H_{17})_3PO$	$HL = CF_3COCH_2COR$, with $R = Me$, CF_3 or Bu^t, M^{III} = Am or Cm, x unspecified[a]
$M^{III}L_3{\cdot}x(Bu^nO)_3PO$	$HL = CF_3COCH_2COR$, with $R = Me$ or Bu^t, M^{III} = Am or Cm, x unspecified[a]
	$HL = CF_3COCH_2COCF_3$, M^{III} = Am, $x = 2$;[a,b] Cm, x unspecified;[a] Bk, $x = 1$[c]

[a] A. V. Davydov, B. F. Myasoedov, S. S. Travnikov and E. V. Fedoseev, *Sov. Radiochem.* (*Engl. Transl.*), 1978, **20**, 217.
[b] A. V. Davydov, B. F. Myasoedov and S. S. Travnikov, *Dokl. Chem.* (*Engl. Transl.*), 1975, **220/5**, 672.
[c] E. V. Fedoseev, L. A. Ivanova, S. S. Travnikov, A. V. Davydov and B. F. Myasoedov, *Sov. Radiochem.* (*Engl. Transl.*), 1983, **25**, 343.

40.3.2.2 Sulfur ligands

Compounds with sulfur containing ligands do not appear to have been reported, but the sulfides $M_2^{III}S_3$ (M^{III} = Am, Cm, Bk or Cf) and $AmS_{1.9}$ [probably an americium(III) polysulfide] are known. Am_2S_3 is polymorphic, the α form having the α-Ce_2S_3 structure, and the γ form the Th_3P_4 structure which is also adopted by the other sesquisulfides. Oxysulfides, $M_2^{III}O_2S$ (M^{III} = Am, Cm or Cf), are also known; Cf_2O_2S is obtained by heating $Cf_2(SO_4)_3{\cdot}xH_2O$ ($x = 0$ or 12) in hydrogen or in a vacuum at 800 °C.

40.3.2.3 Selenium and tellurium ligands

The only recorded compounds appear to be americium selenide and telluride, Am_3X_4.

40.3.2.4 Halogens as ligands

(i) Trihalides

The trifluorides, $M^{III}F_3$ (Table 91), possess the 11-coordinate LaF_3 structure (M^{III} = Am, Cm and one form of BkF_3), and a second modification of BkF_3 has the eight-coordinate high temperature LaF_3-type structure, found also for CfF_3 and EsF_3.[406] The trichlorides, $M^{III}Cl_3$, have the nine-coordinate UCl_3-type structure (M^{III} = Am, Cm, Bk and one form of $CfCl_3$), and the second modification of $CfCl_3$ has the eight-coordinate $PuBr_3$ structure. $EsCl_3$ has the UCl_3-type structure at 425 °C. The structures of the tribromides, $M^{III}Br_3$ (M^{III} = Am, Cm and Bk), are $PuBr_3$-type, and α-AmI_3 has this structure also. However, β-AmI_3 and the other triiodides $M^{III}I_3$ (M^{III} = Cm, Bk and Cf) have the six-coordinate BiI_3 structure. The structure of the hydrated trichlorides and tribromides, $MX_3{\cdot}6H_2O$, has been discussed earlier (p. 1215).

Table 91 Transplutonium Actinide(III) Halides and Halogeno Complexes

$M^{III}X_3$	X = F, Cl, Br or I, M^{III} = Am, Cm, Bk, Cf and Es[a]
$M^{III}OX$	X = F, M^{III} = Cf; X = Cl, M^{III} = Am, Cm, Bk, Cf and Es; X = Br, M^{III} = Cm, Bk and Cf; X = I, M^{III} = Am,[b] Bk and Cf
M^IAmX_4	X = F, M^I = Na, K,[c] Rb;[c] X = Cl, M^I = Cs (also $+4H_2O$)
K_2AmF_5[c]	
K_3AmF_6[c]	
KAm_2F_7	
$Cs_2NaM^{III}Cl_6$	M^{III} = Am,[d,e] Cm[e] and Bk[d,e]
$(Et_4N)_2LiAmCl_6$	
$M^I_3AmX_6$	X = Cl, M^I = Cs, Et_4N and Ph_3PH; X = Br, M^I = Ph_3PH

[a] For EsF_3, see D. D. Ensor, J. R. Peterson, R. G. Haire and J. P. Young, *J. Inorg. Nucl. Chem.*, 1981, **43**, 2425.
[b] R. G. Haire, J. P. Young and J. R. Peterson, *J. Less-Common Met.*, 1983, **93**, 339.
[c] J. Jové and M. Pagès, *Inorg. Nucl. Chem. Lett.*, 1977, **13**, 329.
[d] L. R. Morss, M. Siegal, L. Stenger and N. Edelstein, *Inorg. Chem.*, 1970, **9**, 1771.
[e] M. E. Hendricks, E. R. Jones, Jr., J. A. Stone and D. G. Karracker, *J. Chem. Phys.*, 1974, **60**, 2095.

(ii) Halogeno complexes

The fluoro complexes M^IAmF_4 (M^I = K or Rb) are isostructural with the analogous plutonium compounds[407] and the anion in the salts $Cs_2Na[M^{III}Cl_6]$ is octahedral (*e.g.* M^{III} = Bk).[408]

40.3.2.5 Hydrides as ligands

The only recorded compounds are the hydrides AmH_3 and MH_{2+x} (M = Am, Cm or Bk, $x < 1$).

40.3.2.6 Mixed donor atom ligands

Compounds of composition AmL_3, where HL is 8-hydroxyquinoline, 5-chloro- or 5,7-dichloro-8-hydroxyquinoline, are precipitated when an aqueous solution containing americium(III) is added to an aqueous, or aqueous dioxane, solution of the ligand in the pH range 5.1–6.5.[409] There is solvent extraction evidence for the californium(III) 5,7-dichloro-8-hydroxyquinoline complex.[410]

40.3.2.7 Multidentate macrocyclic ligands

The bis phthalocyanine (Pc) complex, $AmPc_2$, has been obtained by heating AmI_3 with *o*-phthalodinitrile in 1-chloronaphthalene[411] or from americium(III) acetate and *o*-phthalodinitrile; it is probably a sandwich compound similar to those obtained with the tripositive lanthanides.[412]

40.3.3 The +4 Oxidation State

40.3.3.1 Oxygen ligands

(i) Oxides

All the reported dioxides, $M^{IV}O_2$ (M^{IV} = Am, Cm, Bk and Cf), possess the fluorite structure. Ternary oxides of the types $M^I_2Am^{IV}O_3$ (M^I = Li or Na), $M^{II}Am^{IV}O_3$ (M^{II} = Sr or Ba) and $Li_8M^{IV}O_6$ (M^{IV} = Am or Cm) have been recorded; they are obtained by heating the dioxides with the alkali or alkaline earth metal oxide at high temperatures under vacuum or in nitrogen.

(ii) β-Ketoenolates

The berkelium(IV) complexes BkL_4 [HL = $CF_3COCH_2COBu^t$[395] and $CF_3COCH_2CO(2\text{-}C_4H_3S)$[395,413]] are formed by solvent extraction of aqueous Bk^{IV} solutions with the β-ketoenol.

40.3.3.2 Halogens as ligands

(i) Tetrahalides

The tetrafluorides $M^{IV}F_4$ (M^{IV} = Am, Cm, Bk and Cf) are obtained by fluorination of lower oxidation state compounds; for example, CfF_4 has been prepared[414] by heating CfF_3, Cf_2O_3 or $CfCl_3\cdot xH_2O$ in fluorine at 400–450 °C under pressure (3 atm). All have the eight-coordinate UF_4 structure (p. 1173); for crystallographic data see refs. 415 and 416.

(ii) Halogeno complexes

The fluoro complexes $LiM^{IV}F_5$ (M^{IV} = Am or Cm) and $Rb_2M^{IV}F_6$ (M^{IV} = Am or Cm) have the $LiUF_5$ and Rb_2UF_6 structures (p. 1173) respectively. Compounds of composition $M^I_7M^{IV}_6F_{31}$ (M^I = Na, M^{IV} = Am, Cm, Bk or Cf and M^I = K, M^{IV} = Am, Cm or Bk) are also known (see p. 1174).

Cs_2BkCl_6 is not isomorphous with Cs_2PuCl_6 or Cs_2CeCl_6;[408] this compound would provide a useful starting material for the preparation of oxygen donor complexes of $BkCl_4$, which are at present unknown, by the methods used to prepare $PuCl_4$ complexes of this type (p. 1161).

40.3.4 The +5 Oxidation State

There is tracer scale evidence[455] for the formation of the CfO_2^+ ion during ozonization of ^{249}Bk and subsequent β decay to ^{249}Cf. However, the only compounds isolated in this oxidation state are all americium(V) species.

40.3.4.1 Oxygen ligands

(i) Aqua species, hydroxides and oxides

(a) Aqua species. Hydrated americium(V) carbonato and oxalato complexes are discussed below in the sections on carbonates and oxalates. The hydrated hydroxide, $AmO_2(OH)\cdot ca.2.3H_2O$, is precipitated from aqueous solution by alkali. It decomposes to AmO_2 on heating.[418]

(b) Oxides. Pentoxides, $M^V_2O_5$, are unknown, but ternary oxides, such as $Li_3Am^VO_4$, Li_7AmO_6 and Na_3AmO_4, are obtained by heating AmO_2 with the alkali metal oxide (*e.g.* Li_7AmO_6 in N_2/O_2 at 900 °C; this decomposes to Li_3AmO_4 at 1000 °C). Crystallographic data have been reported for these compounds.[419]

(ii) Carboxylates and carbonates

(a) Monocarboxylates. The acetato complex salt, $Cs_2AmO_2(MeCO_2)_3$, is isostructural with the analogous neptunium(V) and plutonium(V) compounds (p. 1183)[186] and its vibrational spectrum has also been reported.[185]

(b) Carbonates. The hydrated carbonato complexes $KAmO_2CO_3 \cdot xH_2O$ and $K_3AmO_2(CO_3)_2 \cdot xH_2O$ have been obtained by electrochemical reduction of americium(VI) solutions in the presence of carbonate. The carbonato complex salts $M^I_3AmO_2(CO_3)_2 \cdot xH_2O$ are isostructural with the analogous neptunium(V) compounds[180] (p. 1180). The vibrational spectrum of $CsAmO_2CO_3$ has also been reported.[185]

(c) Oxalates. The hydrated oxalato complex salts $M^IAmO_2(C_2O_4) \cdot xH_2O$ ($M^I = K$ or Cs) are precipitated from aqueous solution.[418]

40.3.4.2 Halogens as ligands

The chlorocomplex salt, $Cs_3[AmO_2Cl_4]$, is isostructural with the analogous neptunium(V) and plutonium(V) compounds.[206] Its vibrational spectrum has also been reported.[185]

40.3.5 The +6 Oxidation State

40.3.5.1 Oxygen ligands

(i) Aqua species and oxides

(a) Aqua species. Hydrated phosphato and arsenato complex salts, $M^IAmO_2XO_4 \cdot yH_2O$ ($M^I = K$, Rb, Cs or NH_4 and $X = P$ or As, with y in the range 0 to 3) are precipitated from americium(VI) solutions[421] at pH 3.5. $NH_4AmO_2PO_4 \cdot 3H_2O$ is isomorphous with the analogous dioxouranium(VI) and neptunium(VI) compounds[422] (Table 63, p. 1192). The hydrated sulfatocomplex salt, $([Co(NH_3)_6]HSO_4)_2AmO_2(SO_4)_3 \cdot xH_2O$, is isostructural[262] with the corresponding dioxouranium(VI) and neptunium(VI) compounds (p. 1197).

(b) Oxides. AmO_3 is unknown, but ternary oxides of the type $M^I_2AmO_4$ ($M^I = K$, Rb or Cs), $M^I_4AmO_5$ ($M^I =$ Li or Na), $M^I_6AmO_6$ ($M^I =$ Li or Na) and Ba_3AmO_6 have been recorded, prepared by heating AmO_2 with the metal hydroxide or oxide in oxygen (*e.g.* $M^I_2AmO_4$).[423]

(ii) Oxoanions as ligands

The nitratocomplexes $M^IAmO_2(NO_3)_3$ ($M^I =$ Rb or Cs[424]) are precipitated from nitric acid solutions of americium(VI). The IR spectrum of $RbAmO_2(NO_3)_3$ has been reported.[249] Hydrated phosphato, arsenato and sulfato complex salts are included under 'Aqua species' in the preceding section.

(iii) Carboxylates and carbonates

(a) Monocarboxylates. The IR spectrum of $NaAmO_2(MeCO_2)_3$ has been reported.[249]

(b) Chelating carboxylates. The pyridine- and *N*-oxopyridine-2-carboxylates, AmO_2L_2 and $HAmO_2L_3$, have been obtained from aqueous solutions of americium(VI) and the acid HL; the *N*-oxide 2-carboxylate is formed as the dihydrate which is dehydrated at 100 °C.[425]

(c) Carbonates. The carbonato complex salts, $M^I_4AmO_2(CO_3)_3$ ($M^I =$ Cs or NH_4), are obtained on addition of $M^I_2CO_3$ to a bicarbonate solution containing americium(VI).[424]

40.3.5.2 Halogens as ligands

AmO_2F_2 has been prepared by the reaction of $NaAmO_2(MeCO_2)_3$ with anhydrous HF and fluorine at −196 °C. It is isostructural with UO_2F_2, NpO_2F_2 and PuO_2F_2.[426] The IR spectrum of $Cs_2AmO_2Cl_4$ has been reporetd[249] and the corresponding rubidium salt has been mentioned in a report.[427] The cubic phase of $Cs_2AmO_2Cl_4$[428] is probably a mixed oxidation state compound, $Cs_7(Am^VO_2)(Am^{VI}O_2)_2Cl_{12}$.[207]

40.4 REFERENCES

1. K. W. Bagnall, *Essays Chem.*, 1972, **3**, 39.
2. K. W. Bagnall, in 'Comprehensive Inorganic Chemistry', ed. J. C. Bailar, Jr., H. J. Emeléus, R. Nyholm and A. F. Trotman-Dickenson, Pergamon, Oxford, 1973, vol. 5, p. 1.
3. K. W. Bagnall, in 'Organometallics of the f-Elements', ed. T. J. Marks and R. D. Fischer, Reidel, Dordrecht, 1979, p. 221.
4. K. W. Bagnall and Li Xing-fu, *J. Chem. Soc., Dalton Trans.*, 1982, 1365.
5. K. W. Bagnall, *Inorg. Chim. Acta,* 1984, **94,** 3.
6. D. L. Kepert, *J. Chem. Soc.*, 1965, 4736.
7. P. T. Moseley, in 'MTP International Review of Science, Inorganic Chemistry, Series 2', Butterworths, London, 1975, vol. 7, p. 65.
8. K. W. Bagnall, R. L. Beddoes, O. S. Mills and Li Xing-fu, *J. Chem. Soc., Dalton Trans.*, 1982, 1361.
9. K. W. Bagnall, Li Xing-fu, G. Bombieri and F. Benetollo, *J. Chem. Soc., Dalton Trans.*, 1982, 19.
10. W. H. Zachariasen, *Acta Crystallogr.*, 1954, **7,** 792.
11. G. Bombieri and K. W. Bagnall, *J. Chem. Soc., Chem. Commun.*, 1975, 188.
12. 'Gmelins Handbuch der Anorganischen Chemie'; Thorium, vol. E, 1985; Protactinium, vol. 2, 1977; Uranium, vols. E1, 1979; E2, 1980; C8, 1980; C9, 1979; C13, 1983; The Transuranium Elements, vol. C4, 1972.
13. J. M. Cleveland, G. H. Bryan and R. J. Sironen, *Inorg. Chim. Acta,* 1972, **6,** 54.
14. R. Barnard, J. I. Bullock, B. J. Gellatly and L. F. Larkworthy, *J. Chem. Soc., Dalton Trans.*, 1972, 1932.
15. R. A. Andersen, *Inorg. Chem.*, 1979, **18,** 1507.
16. (a) D. G. Karracker and J. A. Stone, *Inorg. Chem.*, 1980, **19,** 3545; (b) H. J. Wasserman, D. C. Moody and R. R. Ryan, *J. Chem. Soc., Chem. Commun.*, 1984, 532.
17. D. Brown, B. Whittaker and J. Tacon, *J. Chem. Soc., Dalton Trans.*, 1975, 34.
18. D. C. Moody, A. J. Zozulin and K. V. Salazar, *Inorg. Chem.*, 1982, **21,** 3856.
19. K. Schwochau and J. Drozdzynski, *Inorg. Nucl. Chem. Lett.*, 1980, **16,** 423.
20. (a) L. R. Crisler, *J. Inorg. Nucl. Chem.*, 1972, **34,** 3263; (b) V. G. Zubarev and N. N. Krot, *Radiokhimiya,* 1984, **26,** 176; *Sov. Radiochem. (Engl. Transl.),* 1984, **26,** 161.
21. J. I. Bullock, A. E. Storey and P. Thompson, *J. Chem. Soc., Dalton Trans.*, 1979, 1040.
22. R. M. Dell, in ref. 2, p. 319.
23. D. Brown, D. G. Holah and C. E. F. Rickard, *J. Chem. Soc. (A)*, 1970, 786.
24. J. C. Taylor, *Coord. Chem. Rev.*, 1976, **20,** 197.
25. J. H. Burns, *Inorg. Chem.*, 1965, **4,** 881.
26. (a) I. G. Suglobova and D. E. Chirkst, *Sov. J. Coord. Chem. (Engl. Transl.)*, 1981, **7,** 52; (b) H. J. Wasserman, D. C. Moody, R. T. Paine, R. R. Ryan and K. V. Salazar, *J. Chem. Soc., Chem. Commun.*, 1984, 533.
27. D. C. Moody, R. A. Penneman and K. V. Salazar, *Inorg. Chem.*, 1979, **18,** 208.
28. R. A. Paris, P. A. Amblard and A. C. Rousset, *Fr. Pat.* 2 157 203 (1973) (*Chem. Abstr.*, 1974, **80,** 49 880).
29. K. W. Bagnall and J. O. Baptista, *J. Inorg. Nucl. Chem.*, 1970, **32,** 2283.
30. F. Lux and U. E. Bufe, *Angew. Chem., Int. Ed. Engl.*, 1971, **10,** 274.
31. (a) G. W. Watt and D. W. Baugh, *Inorg. Nucl. Chem. Lett.*, 1974, **10,** 1025; (b) M. G. B. Drew and G. R. Willey, *J. Chem. Soc., Dalton Trans.*, 1984, 727.
32. A. K. Molodkin, Yu.Ya. Kharitonov, V. I. Nefedov and Z. V. Belyakova, *Russ. J. Inorg. Chem. (Engl. Transl.)*, 1976, **21,** 1355.
33. R. A. Andersen, *Inorg. Nucl. Chem. Lett.*, 1979, **15,** 57.
34. A. K. Srivastava, R. K. Agarwal, M. Srivastava, V. Kapoor and T. N. Srivastava, *J. Inorg. Nucl. Chem.*, 1981, **43,** 1393.
35. R. K. Agarwal, A. K. Srivastava, M. Srivastava, N. Bhakru and T. N. Srivastava, *J. Inorg. Nucl. Chem.*, 1980, **42,** 1775.
36. (a) R. O. Wiley, R. B. von Dreele and T. M. Brown, *Inorg. Chem.*, 1980, **19,** 3351; (b) S. C. Jain, M. S. Gill and G. S. Rao, *J. Indian Chem. Soc.*, 1981, **58,** 210.
37. J. G. Reynolds, A. Zalkin, D. H. Templeton and N. M. Edelstein, *Inorg. Chem.*, 1977, **16,** 1090.
38. R. A. Andersen, A. Zalkin and D. H. Templeton, *Inorg. Chem.*, 1981, **20,** 622.
39. D. C. Bradley, J. S. Ghotra and F. A. Hart, *Inorg. Nucl. Chem. Lett.*, 1974, **10,** 209.
40. J. G. Reynolds, A. Zalkin, D. H. Templeton, N. M. Edelstein and L. K. Templeton, *Inorg. Chem.*, 1976, **15,** 2498.
41. J. G. Reynolds, A. Zalkin, D. H. Templeton and N. M. Edelstein, *Inorg. Chem.*, 1977, **16,** 599.
42. J. G. Reynolds, A. Zalkin, D. H. Templeton and N. M. Edelstein, *Inorg. Chem.*, 1977, **16,** 1858.
43. C. M. Mikulski, S. Cocco, N. de Franco and N. M. Karayannis, *Inorg. Chim. Acta,* 1982, **67,** 61.
44. A. K. Srivastava and R. K. Agarwal, *Thermochim. Acta,* 1982, **56,** 247.
45. P. Charpin, R. M. Costes, G. Folcher, P. Plurien, A. Navaza and C. de Rango, *Inorg. Nucl. Chem. Lett.*, 1977, **13,** 341.
46. I. E. Grey and P. W. Smith, *Aust. J. Chem.*, 1969, **22,** 311.
47. K. W. Bagnall, A. Beheshti and F. Heatley, *J. Less-Common Met.*, 1978, **61,** 171.
48. P. G. Edwards, R. A. Andersen and A. Zalkin, *J. Am. Chem. Soc.*, 1981, **103,** 7792.
49. A. G. Maddock and A. Pires de Matos, *Radiochim. Acta,* 1973, **19,** 163.
50. P. N. M. Das, K. Diemert and W. Kuchen, *Indian J. Chem. Sect. A.*, 1977, **15,** 29.
51. R. N. Mukherjee, A. Y. Sonsale and J. Gupta, *Indian J. Chem.*, 1966, **4,** 500.
52. R. N. Mukherjee, S. V. Shambhag, M. S. Venkateshan and M. D. Zingde, *Indian J. Chem.*, 1973, **11,** 1066.
53. P. Kierkegaard, *Acta Chem. Scand.*, 1956, **10,** 599.
54. T. W. Hambley, D. L. Kepert, C. L. Raston and A. H. White, *Aust. J. Chem.*, 1978, **31,** 2635.
55. R. J. Hill and C. E. F. Rickard, *J. Inorg. Nucl. Chem.*, 1975, **37,** 2481.
56. (a) E. Huber-Buser, *Cryst. Struct. Commun.*, 1975, **4,** 731; (b) R. Graziani, G. A. Battiston, U. Casellato and G. Sbrignadello, *J. Chem. Soc., Dalton Trans.*, 1983, 1.

57. N. W. Alcock, T. J. Kemp, S. Sostero and O. Traverso, *J. Chem. Soc., Dalton Trans.*, 1980, 1182.
58. (a) S. Degetto, L. Baracco, R. Graziani and E. Celon, *Transition Met. Chem.*, 1978, **3,** 351; (b) F. Benetollo, G. Bombieri, G. Tomat, C. Bisi Castellani, A. Cassol and P. Di Bernardo, *Inorg. Chim. Acta*, 1984, **95,** 251.
59. T. Ueki, A. Zalkin and D. H. Templeton, *Acta Crystallogr.*, 1966, **20,** 836.
60. G. Johansson, *Acta Chem. Scand.*, 1968, **22,** 389.
61. G. Lundgren and L. G. Sillen, *Ark. Kemi,* 1949, **1,** 277.
62. G. Lundgren, *Ark. Kemi,* 1950, **2,** 535.
63. G. L. Johnson, M. J. Kelly and D. R. Cuneo, *J. Inorg. Nucl. Chem.*, 1965, **27,** 1787.
64. A. D. Westland and M. T. H. Tarafder, *Inorg. Chem.*, 1982, **21,** 3228.
65. (a) D. C. Bradley, M. A. Saad and W. Wardlaw, *J. Chem. Soc.*, 1954, 3488; (b) F. A. Cotton, D. O. Marler and W. Schwotzer, *Inorg. Chim. Acta*, 1984, **85,** L31; (c) F. A. Cotton, D. O. Marler and W. Schwotzer, *Inorg. Chim. Acta,* 1984, **95,** 207.
66. R. C. Mehrotra and A. Mehrotra, *Inorg. Chim. Acta Rev.*, 1971, **5,** 127.
67. R. S. Srivastava, S. P. Agarwal and H. N. Bhargava, *Propellants Explos.*, 1976, **1,** 101.
68. D. C. Bradley, R. N. Kapoor and B. C. Smith, *J. Inorg. Nucl. Chem.*, 1962, **24,** 863.
69. L. R. Crisler, *US Pat.* 3 919 273 and 3 919 274 (1975) (*Chem. Abstr.*, 1976, **84,** 90 309).
70. W. L. Steffen and R. C. Fay, *Inorg. Chem.*, 1978, **17,** 779.
71. M. Lenner, *Acta Crystallogr., Sect. B,* 1978, **34,** 3770.
72. E. M. Rubtsov, V. Ya. Mischin and V. K. Isupov, *Sov. Radiochem.* (*Engl. Transl.*), 1981, **23,** 465.
73. D. L. Kepert, J. M. Patrick and A. H. White, *J. Chem. Soc., Dalton Trans.*, 1983, 567.
74. G. F. S. Wessels, J. G. Leipoldt and L. D. C. Bok, *Z. Anorg. Allg. Chem.*, 1972, **393,** 284.
75. M. Lenner and O. Lindqvist, *Acta Crystallogr., Sect. B,* 1979, **35,** 600.
76. R. J. Hill and C. E. F. Rickard, *J. Inorg. Nucl. Chem.*, 1977, **39,** 1593.
77. S. Prasad and S. Kumar, *J. Indian Chem. Soc.*, 1963, **40,** 531.
78. K. Andrä, *Z. Anorg. Allg. Chem.*, 1968, **361,** 254.
79. S. R. Sofen, K. Abu-Dari, D. P. Freyberg and K. N. Raymond, *J. Am. Chem. Soc.*, 1978, **100,** 7882.
80. (a) M. R. Mahmoud, A. M. Hammam and S. A. Ibrahim, *J. Inorg. Nucl. Chem.*, 1980, **42,** 1199; (b) L. G. McCullough, H. W. Turner, R. A. Andersen, A. Zalkin and D. H. Templeton, *Inorg. Chem.*, 1981, **20,** 2869.
81. B. Zarli, G. Faraglia, L. Sindellari and G. Casotto, *Transition Met. Chem.*, 1981, **6,** 17.
82. S. Šćavničar and B. Prodić, *Acta Crystallogr.*, 1965, **18,** 698.
83. (a) S. A. Linde, Yu. E. Gorbunova and A. V. Lavrov, *Russ. J. Inorg. Chem.* (*Engl. Transl.*), 1983, **28,** 785; (b) C. Miyake, Y. Hinatsu and S. Imoto, *J. Inorg. Nucl. Chem.*, 1981, **43,** 2407.
84. A. D. Gel'man, L. N. Essen, F. A. Zakharova, D. P. Alekseeva and M. M. Orlova, *Dokl. Chem.* (*Engl. Transl.*), 1963, **148/153,** 317.
85. L. N. Essen, D. P. Alekseeva and A. D. Gel'man, *Russ. J. Inorg. Chem.* (*Engl. Transl.*), 1966, **11,** 853.
86. R. C. Paul, S. Singh and R. D. Verma, *J. Fluorine Chem.*, 1980, **16,** 153.
87. E. G. Arutyunyan, M. A. Porai-Koshits and A. K. Molodkin, *J. Struct. Chem.* (*Engl. Transl.*), 1966, **7,** 683.
88. J. Wroblewska, J. Dobrowolski, M. Pagès and W. Freundlich, *Radiochem. Radioanal. Lett.*, 1979, **39,** 241.
89. J. W. Wroblewska, A. Erb, J. Dobrowolski and W. Freundlich, *Ann. Chim.* (*Paris*), 1979, **4,** 353.
90. O. Greis, E. W. Bohres and K. Schwochau, *Z. Anorg. Allg. Chem.*, 1977, **433,** 111.
91. E. G. Arutyunyan, A. S. Antsyshkina and E. Ya. Balta, *J. Struct. Chem.* (*Engl. Transl.*), 1966, **7,** 448.
92. I. Jelenić, D. Grdenić and A. Bezjak, *Acta Crystallogr.*, 1964, **17,** 758.
93. A. K. Molodkin, T. A. Balakaeva, Ya. V. Salyn, O. M. Ivanova, Z. V. Belyakova and L. E. Kolesnikova, *Russ. J. Inorg. Chem.* (*Engl. Transl.*), 1980, **25,** 1054.
94. S. Degetto, L. Baracco, R. Graziani and E. Celon, *Transition Met. Chem.*, 1978, **3,** 351.
95. L. Baracco, G. Bombieri, S. Degetto, E. Forsellini, R. Graziani and G. Marangoni, *Inorg. Nucl. Chem. Lett.*, 1974, **10,** 1045.
96. H. Ehrhardt, H. Schweer and H. Seidel, *Z. Anorg. Allg. Chem.*, 1980, **462,** 185.
97. A. D. Gel'man and L. M. Zaitsev, *Russ. J. Inorg. Chem.* (*Engl. Transl.*), 1959, **4,** 1243.
98. S. Voliotis and A. Rimsky, *Acta Crystallogr., Sect. B,* 1975, **31,** 2612: S. Voliotis, F. Fromage, J. Faucherre and J. Dervin, *Rev. Chim. Miner.*, 1977, **14,** 441.
99. S. Voliotis and A. Rimsky, *Acta Crystallogr., Sect. B,* 1975, **31,** 2615.
100. S. Voliotis, *Acta Crystallogr., Sect. B,* 1979, **35,** 2899.
101. M. N. Akhtar and A. J. Smith, *Acta Crystallogr., Sect. B,* 1975, **31,** 1361.
102. M. C. Favas, D. L. Kepert, J. M. Patrick and A. H. White, *J. Chem. Soc., Dalton Trans.*, 1983, 571.
103. A. K. Molodkin and G. A. Skotnikova, *Russ. J. Inorg. Chem.* (*Engl. Transl.*), 1964, **9,** 308.
104. L. N. Essen, D. P. Alekseeva and A. D. Gel'man, *Russ. J. Inorg. Chem.* (*Engl. Transl.*), 1966, **11,** 853.
105. V. A. Golovnya and G. T. Bolotova, *Russ. J. Inorg. Chem.* (*Engl. Transl.*), 1961, **6,** 1259.
106. F. Calderazzo, G. Dell'Amico, M. Pasquali and G. Perego, *Inorg. Chem.*, 1978, **17,** 474.
107. T. J. Marks and R. D. Ernst, in 'Comprehensive Organometallic Chemistry', ed. G. Wilkinson, F. G. A. Stone and E. W. Abel, Pergamon, Oxford, 1982, vol. 3, pp. 221–223.
108. D. L. Kepert, J. M. Patrick and A. H. White, *J. Chem. Soc., Dalton Trans.*, 1983, 381.
109. (a) R. J. Barton, R. W. Dabeka, Hu Shengzhi, L. M. Mihichuk, M. Pizzey, B. E. Robertson and W. J. Wallace, *Acta Crystallogr., Sect. C,* 1983, **39,** 714; (b) U. Casellato, S. Sitran, S. Tamburini, P. A. Vigato and R. Graziani, *Inorg. Chim. Acta,* 1984, **95,** 37.
110. V. W. Day and J. L. Hoard, *J. Am. Chem. Soc.*, 1970, **92,** 3626.
111. (a) K. W. Bagnall, Li Xing-fu, Po-jung Pao and A. G. M. Al-Daher, *Can. J. Chem.*, 1983, **61,** 708; (b) G. Bombieri, F. Benetollo, U. Croatto, C. Bisi Castellani, K. W. Bagnall and Li Xing-fu, *Inorg. Chim. Acta,* 1984, **95,** 237.
112. C. E. F. Rickard and D. C. Woollard, *Aust. J. Chem.*, 1980, **33,** 1161.
113. J. F. de Wet and S. F. Darlow, *Inorg. Nucl. Chem. Lett.*, 1971, **7,** 1041.
114. G. Bombieri, F. Benetollo, K. W. Bagnall, M. J. Plews and D. Brown, *J. Chem. Soc., Dalton Trans.*, 1983, 343.
115. G. Bombieri, D. Brown and R. Graziani, *J. Chem. Soc., Dalton Trans.*, 1975, 1873.

116. G. Bombieri, E. Forsellini, D. Brown and B. Whittaker, *J. Chem. Soc., Dalton Trans.*, 1976, 735.
117. C. E. F. Rickard and D. C. Woollard, *Aust. J. Chem.*, 1979, **32,** 2182.
118. D. L. Kepert, J. M. Patrick and A. H. White, *J. Chem. Soc., Dalton Trans.*, 1983, 385.
119. D. L. Kepert, J. M. Patrick and A. H. White, *J. Chem. Soc., Dalton Trans.*, 1983, 559.
120. Mazhar-ul-Haque, C. N. Caughlan, F. A. Hart and R. van Nice, *Inorg. Chem.*, 1971, **10,** 115.
121. N. W. Alcock, S. Esperas, K. W. Bagnall and Wang Hsian-Yun, *J. Chem. Soc., Dalton Trans.*, 1978, 638.
122. R. P. English, J. G. H. du Preez, L. R. Nassimbeni and C. P. J. van Vuuren, *S. Afr. J. Chem.*, 1979, **32,** 119.
123. S. M. Bowen, E. N. Duesler and R. T. Paine, *Inorg. Chem.*, 1982, **21,** 261.
124. J. G. Leipoldt, G. F. S. Wessels and L. D. C. Bok, *J. Inorg. Nucl. Chem.*, 1975, **37,** 2487.
125. D. Brown, J. Hill and C. E. F. Rickard, *J. Chem. Soc. (A)*, 1970, 497.
126. P. Sommerville and M. Laing, *Acta Crystallogr., Sect. B*, 1976, **32,** 1551.
127. P. J. Alvey, K. W. Bagnall and D. Brown, *J. Chem. Soc., Dalton Trans.*, 1973, 2326.
128. P. J. Alvey, K. W. Bagnall, O. Velasquez Lopez and D. Brown, *J. Chem. Soc., Dalton Trans.*, 1975, 1277.
129. G. Bombieri and K. W. Bagnall, *J. Chem. Soc., Chem. Commun.*, 1975, 188.
130. N. Singer, B. F. Studd and A. G. Swallow, *Chem. Commun.*, 1970, 342.
131. C. E. F. Rickard and D. C. Woollard, *Acta Crystallogr., Sect. B*, 1980, **36,** 292.
132. W. L. Smith and K. N. Raymond, *J. Am. Chem. Soc.*, 1981, **103,** 3341.
133. H. Noel and M. Potel, *Acta Crystallogr., Sect. B*, 1982, **38,** 2444.
134. D. Brown, D. G. Holah, C. E. F. Rickard and P. T. Moseley, UKAEA Report AERE-R-6907, 1971.
135. K. Abu-Dari and K. N. Raymond, *Inorg. Chem.*, 1982, **21,** 1676.
136. R. Gradl and N. Edelstein, US Report LBL-4000, pp. 344–346, 1975 (*Chem. Abstr.*, 1976, **85,** 153 301).
137. V. G. Sevast'yanov and M. A. Kop'eva, *Russ. J. Inorg. Chem. (Engl. Transl.)*, 1977, **22,** 1517.
138. A. A. Pinkerton, A. E. Storey and J.-M. Zellweger, *J. Chem. Soc., Dalton Trans.*, 1981, 1475.
139. R. M. Dell, in ref. 2, vol. 5, p. 319.
140. A. Cousson, H. Abazli, M. Pages and M. Gasperin, *Acta Crystallogr., Sect. C*, 1983, **39,** 425.
141. K. Mucker, G. S. Smith, Q. Johnson and R. E. Elson, *Acta Crystallogr., Sect. B*, 1969, **25,** 2362.
142. A. Zalkin, J. D. Forrester and D. H. Templeton, *Inorg. Chem.*, 1964, **3,** 639.
143. R. P. Dodge, G. S. Smith, Q. Johnson and R. E. Elson, *Acta Crystallogr., Sect. B*, 1968, **24,** 304.
144. D. Brown, in ref. 2, vol. 5, p. 151.
145. G. Brunton, *Acta Crystallogr.*, 1964, **21,** 814.
146. G. Brunton, *Acta Crystallogr., Sect. B*, 1969, **25,** 1919.
147. H. Abazli, A. Cousson, A. Tabuteau, M. Pagès and M. Gasperin, *Acta Crystallogr., Sect. B*, 1980, **36,** 2765.
148. F. H. Kruse, *J. Inorg. Nucl. Chem.*, 1971, **33,** 1625.
149. A. Cousson, A. Tabuteau, M. Pagès and M. Gasperin, *Acta Crystallogr., Sect. B*, 1979, **35,** 1198.
150. G. Brunton, *Acta Crystallogr., Sect. B*, 1969, **25,** 2163.
151. C. Keller and M. Salzer, *J. Inorg. Nucl. Chem.*, 1967, **29,** 2925.
152. H. Abazli, A. Cousson, J. Jové, M. Pagès and M. Gasparin, *J. Less-Common Met.*, 1984, **96,** 23.
153. W. H. Zachariasen, *J. Am. Chem. Soc.*, 1948, **70,** 2147.
154. A. Rosenzweig and D. T. Cromer, *Acta Crystallogr., Sect. B*, 1970, **26,** 38.
155. J. H. Burns, R. D. Ellison and H. A. Levy, *Acta Crystallogr., Sect. B*, 1968, **24,** 230.
156. M. Magette and J. Fuger, *Inorg. Nucl. Chem. Lett.*, 1977, **13,** 529.
157. D. Brown, B. Whittaker and P. E. Lidster, UKAEA Report AERE-R8035, 1975.
158. B. W. Berringer, J. B. Gruber, T. M. Loehr and G. P. O'Leary, *J. Chem. Phys.*, 1971, **55,** 4608.
159. R. H. Banks, N. M. Edelstein, R. R. Rietz, D. H. Templeton and A. Zalkin, *J. Am. Chem. Soc.*, 1978, **100,** 1957.
160. R. H. Banks, N. M. Edelstein, B. Spencer, D. H. Templeton and A. Zalkin, *J. Am. Chem. Soc.*, 1980, **102,** 620.
161. E. R. Bernstein, W. C. Hamilton, T. A. Keiderling, S. J. La Placa, S. J. Lippard and J. J. Mayerle, *Inorg. Chem.*, 1972, **11,** 3009.
162. R. Shinomoto, E. Gamp, N. M. Edelstein, D. H. Templeton and A. Zalkin, *Inorg. Chem.*, 1983, **22,** 2351.
163. S. Afshar and J. I. Bullock, *Inorg. Chim. Acta*, 1980, **38,** 145.
164. R. J. Hill and C. E. F. Rickard, *J. Inorg. Nucl. Chem.*, 1978, **40,** 2029.
165. R. J. Hill, C. E. F. Rickard and H. E. White, *J. Inorg. Nucl. Chem.*, 1981, **43,** 721.
166. W. Andruchow, Jr. and D. G. Karracker, *Inorg. Chem.*, 1973, **12,** 2194.
167. G. Paolucci, U. Castellato, G. Marangoni and R. Graziani, *Congr. Naz. Chim. Inorg. [Atti], 13th.*, 1980, 174 (*Chem. Abstr.*, 1981, **95,** 53 953).
168. F. Lux, O. F. Beck, H. Krauss, D. Brown and T. C. Tso, *Z. Naturforsch., Teil B*, 1980, **35,** 564.
169. T. Kobayashi, *Bull. Inst. Chem. Res., Kyoto Univ.*, 1978, **56,** 204.
170. A. Gieren and W. Hoppe, *Chem. Commun.*, 1971, 413.
171. R. Dubois, J. C. Carver, M. Tsutsui and F. Lux, *J. Coord. Chem.*, 1976, **6,** 123.
172. G. C. de Villardi, P. Charpin, R. M. Costes, G. Folcher, P. Plurien, P. Rigny and C. de Rango, *J. Chem. Soc., Chem. Commun.*, 1978, 90.
173. G. Bombieri, G. de Paoli and A. Immirzi, *J. Inorg. Nucl. Chem.*, 1978, **40,** 1889.
174. J. Selbin, N. Ahmad and M. J. Pribble, *J. Inorg. Nucl. Chem.*, 1970, **32,** 3249.
175. J. Selbin, C. J. Ballhausen and D. G. Durrett, *Inorg. Chem.*, 1972, **11,** 510.
176. M. Singh, *J. Less-Common Met.*, 1979, **65,** 279.
177. J. Selbin, D. G. Durrett, H. J. Sherrill, G. R. Newkome and M. Collins, *J. Inorg. Nucl. Chem.*, 1973, **35,** 3467.
178. J. A. Berry, J. H. Holloway and D. Brown, *Inorg. Nucl. Chem. Lett.*, 1981, **17,** 5.
179. (a) Yu. F. Volkov, G. I. Visyashcheva, I. I. Kapshukov, G. A. Simakin and G. N. Yakovlev, *Sov. Radiochem. (Engl. Transl.)*, 1976, **18,** 88; (b) S. V. Tomalin, Yu. F. Volkov, G. I. Visyashcheva and I. I. Kapshukov, *Radiokhimiya*, 1984, **26,** 734; *Soviet Radiochem. (Engl. Transl.)*, 1984, **26,** 696.
180. Yu. F. Volkov, G. I. Visyashcheva, S. V. Tomilin, I. I. Kapshukov and A. G. Rykov, *Sov. Radiochem. (Engl. Transl.)*, 1981, **23,** 195.
181. A. K. Solanki and A. M. Bhandari, *Radiochem. Radioanal. Lett.*, 1980, **43,** 279.

182. D. C. Bradley, *Adv. Inorg. Chem. Radiochem.*, 1972, **15,** 259.
183. A. J. Zozulin, D. C. Moody and R. R. Ryan, *Inorg. Chem.*, 1982, **21,** 3083.
184. A. V. Anan'ev, T. N. Bukhtiyarova and N. N. Krot, *Sov. Radiochem.* (*Engl. Transl.*), 1982, **24,** 88.
185. V. A. Vodovatov, L. G. Mashirov and D. N. Suglobov, *Sov. Radiochem.* (*Engl. Transl.*), 1975, **17,** 787.
186. A. A. Lychev, L. G. Mashirov, Yu. I. Smolin and D. N. Suglobov, *Sov. Radiochem.* (*Engl. Transl.*), 1980, **22,** 27.
187. J. H. Burns and C. Musikas, *Inorg. Chem.*, 1977, **16,** 1619.
188. Yu. F. Volkov, G. I. Visyashcheva, S. V. Tomilin, I. I. Kapshukov and A. G. Rykov, *Sov. Radiochem.* (*Engl. Transl.*), 1981, **23,** 200.
189. F. Nectoux, H. Abazli, J. Jové, A. Cousson, M. Pagès, M. Gasperin and G. Choppin, *J. Less-Common Met.*, 1984, **97,** 1.
190. G. Bombieri, D. Brown and C. Mealli, *J. Chem. Soc., Dalton Trans.*, 1976, 2025.
191. S. Dubey, A. M. Bhandari, S. N. Misra and R. N. Kapoor, *Indian J. Chem.*, 1970, **8,** 97.
192. W. H. Zachariasen, *Acta Crystallogr.*, 1949, **2,** 296, 390.
193. R. R. Ryan, R. A. Penneman, L. B. Asprey and R. T. Paine, *Acta Crystallogr., Sect. B*, 1976, **32,** 3311.
194. R. P. Dodge, G. S. Smith, Q. Johnson and R. E. Elson, *Acta Crystallogr.*, 1967, **22,** 85.
195. G. S. Smith, Q. Johnson and R. E. Elson, *Acta Crystallogr.*, 1967, **22,** 300.
196. D. Brown, T. Petcher and A. J. Smith, *Nature* (*London*), 1968, **217,** 738.
197. J.-C. Levet, M. Potel and J.-Y. le Marouille, *Acta Crystallogr., Sect. B*, 1977, **33,** 2542.
198. D. K. Sanyal, D. W. A. Sharp and J. M. Winfield, *J. Fluorine Chem.*, 1981, **19,** 55.
199. M. P. Eastman, P. G. Eller and G. W. Halstead, *J. Inorg. Nucl. Chem.*, 1981, **43,** 2839.
200. J. H. Burns, H. A. Levy and O. L. Keller, Jr., *Acta Crystallogr., Sect. B*, 1968, **24,** 1675.
201. D. Brown, S. F. A. Kettle and A. J. Smith, *J. Chem. Soc.* (*A*), 1967, 1429.
202. D. Brown, J. F. Easey and C. E. F. Rickard, *J. Chem. Soc.* (*A*), 1969, 1161.
203. I. G. Suglobova, V. L. Fedorov and D. E. Chirkst, *Russ. J. Inorg. Chem.* (*Engl. Transl.*), 1982, **27,** 106.
204. J. F. de Wet, M. R. Caira and B. J. Gellatly, *Acta Crystallogr., Sect. B*, 1978, **34,** 1121.
205. J. C. Taylor and A. B. Waugh, *Polyhedron*, 1983, **2,** 211.
206. V. A. Vodovatov, I. N. Ladygin, A. A. Lychev, L. G. Mashirov and D. N. Suglobov, *Sov. Radiochem.* (*Engl. Transl.*), 1975, **17,** 771.
207. R. F. Melkaya, Yu. F. Volkov, E. I. Sokolov, I. I. Kapshukov and A. G. Rykov, *Dokl. Chem.* (*Engl. Transl.*), 1982, **262/7,** 42.
208. R. F. Melkaya, Yu. F. Volkov, V. I. Spiriyakov and I. I. Kapshukov, *Sov. Radiochem.* (*Engl. Transl.*), 1982, **24,** 576.
209. S. P. McGlynn, J. K. Smith and W. Nealy, *J. Chem. Phys.*, 1961, **35,** 105.
210. J. I. Bullock, *J. Inorg. Nucl. Chem.*, 1967, **29,** 2257.
211. J. M. Haigh and D. A. Thornton, *J. Inorg. Nucl. Chem.*, 1971, **33,** 1787.
212. D. A. Johnson, J. C. Taylor and A. B. Waugh, *J. Inorg. Nucl. Chem.*, 1979, **41,** 827.
213. S. Bonatti, C. Cioni, M. Franzini and M. Troysi, *Atti Accad. Naz. Lincei, Nat. Rend. Cl. Sci. Fis. Mat.*, 1968, **44,** 427.
214. R. Cernătescu and M. Poni, *Ann. Sci. Univ. Jassy, Sect. I*, 1937, **23,** 196.
215. E. L. Ivanova, V. V. Skopenko and Kh. Keller, *Ukr. Khim. Zh.* (*Russ. Ed.*), 1981, **47,** 1024 (*Chem. Abstr.*, 1982, **96,** 45 232).
216. V. P. Markov and V. V. Tsapkin, *Russ. J. Inorg. Chem.* (*Engl. Transl.*), 1962, **7,** 250.
217. R. A. Bailey and T. W. Michelsen, *J. Inorg. Nucl. Chem.*, 1972, **34,** 2935.
218. V. G. Sles, A. I. Skoblo and D. N. Suglobov, *Sov. Radiochem.* (*Engl. Transl.*), 1974, **16,** 504.
219. N. W. Alcock, M. W. Roberts and D. Brown, *Acta Crystallogr., Sect. B*, 1982, **38,** 2870.
220. M. Mittal, S. R. Malhotra, K. Lal and S. P. Gupta, *Chim. Acta. Turc.*, 1980, **8,** 279 (*Chem. Abstr.*, 1981, **95,** 34 590).
221. R. Graziani, U. Casellato, P. A. Vigato, S. Tamburini and M. Vidali, *J. Chem. Soc., Dalton Trans.*, 1983, 697.
222. S. B. Ivanov, R. L. Davidovich, Yu. N. Mikhailov and R. N. Shchelokov, *Koord. Khim.*, 1982, **8,** 211.
223. J. Howatson, D. M. Grev and B. Morosin, *J. Inorg. Nucl. Chem.*, 1975, **37,** 1933.
224. N. C. Jayadevan and D. M. Chackraburtty, *Acta Crystallogr., Sect. B*, 1972, **28,** 3178.
225. E. Frasson, G. Bombieri and C. Panattoni, *Coord. Chem. Rev.*, 1966, **1,** 145.
226. A. Perrin, *J. Inorg. Nucl. Chem.*, 1977, **39,** 1169.
227. A. Perrin and J. Y. Le Marouille, *Acta Crystallogr., Sect. B*, 1977, **33,** 2477.
228. M. V. Susic and D. M. Minic, *Solid State Ionics*, 1982, **6,** 327.
229. B. Morosin, *Acta Crystallogr., Sect. B*, 1978, **34,** 3732.
230. Yu. Ya. Kharitonov and A. I. Moskvin, *Sov. Radiochem.* (*Engl. Transl.*), 1973, **15,** 177.
231. V. I. Spitsyn, L. M. Kovba, V. V. Tabachenko, N. V. Tabachenko and Yu. N. Mikhailov, *Bull. Acad. Sci. USSR, Div. Chem. Sci.*, 1982, **31,** 711.
232. H. R. Hoekstra and E. Gebert, *J. Inorg. Nucl. Chem.*, 1977, **39,** 2219.
233. A. D. Westland and M. T. H. Tarafder, *Inorg. Chem.*, 1981, **20,** 3992.
234. R. N. Shchelokov and N. A. Golubkova, *Sov. J. Coord. Chem.* (*Engl. Transl.*), 1982, **8,** 88.
235. D. A. Clemente, G. Bandoli, M. Vidali, P. A. Vigato, R. Portanova and L. Magon, *J. Cryst. Mol. Struct.*, 1973, **3,** 221.
236. P. I. Mackinnon and J. C. Taylor, *Polyhedron*, 1983, **2,** 217.
237. R. Graziani, M. Vidali, U. Casellato and P. A. Vigato, *Acta Crystallogr., Sect. B*, 1976, **32,** 1681.
238. G. Bandoli, D. A. Clemente, U. Croatto, M. Vidali and P. A. Vigato, *Inorg. Nucl. Chem. Lett.*, 1972, **8,** 961.
239. G. Bandoli, D. A. Clemente, U. Croatto, M. Vidali and P. A. Vigato, *Chem. Commun.*, 1971, 1330.
240. J. G. Reynolds, A. Zalkin and D. H. Templeton, *Inorg. Chem.*, 1977, **16,** 3357.
241. W. L. Smith and K. N. Raymond, *J. Inorg. Nucl. Chem.*, 1979, **41,** 1431.
242. G. M. Kramer, M. B. Dines, R. B. Hall, A. Kaldor, A. J. Jacobson and J. C. Scanlon, *Inorg. Chem.*, 1980, **19,** 1340.

243. N. W. Alcock, N. Herron, T. J. Kemp and C. W. Shoppee, *J. Chem. Soc., Chem. Commun.*, 1975, 785.
244. K. A. Avduevskaya, N. B. Ragulina, I. A. Rozanov, Yu. N. Mikhailov, A. S. Kanishcheva and T. G. Grevtseva, *Russ. J. Inorg. Chem.* (*Engl. Transl.*), 1981, **26**, 546.
245. S. L. Gupta, R. C. Saxena and A. N. Pandey, *Indian J. Pure Appl. Phys.*, 1977, **15**, 654.
246. R. Graziani, G. Bombieri, E. Forsellini, S. Degetto and G. Marangoni, *J. Chem. Soc., Dalton Trans.*, 1973, 451.
247. Yu. F. Volkov and I. I. Kapshukov, *Sov. Radiochem.* (*Engl. Transl.*), 1976, **18**, 257.
248. N. W. Alcock, M. M. Roberts and D. Brown, *J. Chem. Soc., Dalton Trans.*, 1982, 33.
249. V. M. Vdovenko, V. A. Vodovatov, L. G. Mashirov and D. N. Suglobov, *Proc. Moscow Symp. Chem. Transuranium Elem., 1972,* (*1976*) Pergamon, Oxford, p. 123.
250. K. A. Avduevskaya, N. B. Ragulina and I. A. Rozanov, *Inorg. Mater.* (*Engl. Transl.*), 1978, **14**, 1618.
251. S. A. Linde, Yu. E. Gorbunova and A. V. Lavrov, *Russ. J. Inorg. Chem.* (*Engl. Transl.*), 1980, **25**, 1105.
252. S. A. Linde, Yu. E. Gorbunova, A. V. Lavrov and V. G. Kuznetsov, *Dokl. Phys. Chem.* (*Engl. Transl.*), 1976, **229/231**, 991.
253. S. A. Linde, Yu. E. Gorbunova, A. V. Lavrov and V. G. Kuznetsov, *Dokl. Phys. Chem.* (*Engl. Transl.*), 1977, **232/237**, 678.
254. A. V. Lavrov, A. B. Pobedina, I. A. Rozanov and I. V. Tananaev, *Russ. J. Inorg. Chem.* (*Engl. Transl.*), 1981, **26**, 1001.
255. S. A. Linde, Yu. E. Gorbunova, A. V. Lavrov and A. B. Pobedina, *Inorg. Mater.* (*Engl. Transl.*), 1981, **17**, 776.
256. H. Barten and E. H. P. Cordfunke, *Thermochim. Acta,* 1980, **40**, 367.
257. V. N. Serezhkin, V. L. Balashov and L. B. Serezhkina, *Russ. J. Inorg. Chem.* (*Engl. Transl.*), 1980, **25**, 1052.
258. M. Hoshi and K. Ueno, *J. Nucl. Sci. Technol.*, 1978, **15**, 141.
259. R. N. Shchelokov, G. T. Bolotova and N. A. Golubkova, *Sov. J. Coord. Chem.* (*Engl. Transl.*), 1977, **3**, 810.
260. J. Martin Gil and F. J. Martin Gil, *Rev. Roum. Chim.*, 1978, **23**, 1397.
261. J. G. Ribas Bernat, J. Martin Gil and F. J. Martin Gil, *Quim. Anal.*, 1976, **30**, 95.
262. K. Ueno and M. Hoshi, *J. Inorg. Nucl. Chem.*, 1971, **33**, 2631.
263. L. Niinisto, J. Toivonen and J. Valkonen, *Acta Chem. Scand., Ser. A*, 1978, **32**, 647; 1979, **33**, 621.
264. Yu. N. Mikhailov, L. A. Kokh, V. G. Kuznetsov, T. G. Grevtseva, S. K. Sokol and G. V. Ellert, *Sov. J. Coord. Chem.* (*Engl. Transl.*), 1977, **3**, 388.
265. (a) V. N. Serezhkin, M. A. Soldatkina and V. A. Efremov, *J. Struct. Chem.* (*Engl. Transl.*), 1981, **22**, 454; (b) V. N. Serezhkin and M. A. Soldatkina, *Koord. Khim.*, 1985, **11**, 103.
266. V. N. Serezhkin and L. B. Serezhkina, *Russ. J. Inorg. Chem.* (*Engl. Transl.*), 1978, **23**, 414.
267. V. N. Serezhkin, M. A. Soldatkina and V. A. Efremov, *J. Struct. Chem.*, (*Engl. Transl.*), 1981, **22**, 451.
268. V. N. Serezhkin, L. B. Serezhkina and N. A. Rasshchepkina, *Sov. Radiochem.* (*Engl. Transl.*), 1978, **20**, 181.
269. G. Meunier and J. Galy, *Acta Crystallogr., Sect. B*, 1973, **29**, 1251.
270. J. Galy and G. Meunier, *Acta Crystallogr., Sect. B*, 1971, **27**, 608.
271. F. Brandstätter, *Z. Kristallogr.*, 1981, **155**, 193.
272. B. F. Mentzen, J.-P. Puaux and H. Loiseleur, *Acta Crystallogr., Sect. B*, 1977, **33**, 1848.
273. A. V. Anan'ev, T. N. Bukhtiyarova and N. N. Krot, *Sov. Radiochem.* (*Engl. Transl.*), 1983, **25**, 331.
274. S. D. LeRoux, A. van Tets and H. W. W. Adrian, *Acta Crystallogr., Sect. B*, 1979, **35**, 3056.
275. B. F. Mentzen, *Acta Crystallogr., Sect. B*, 1977, **33**, 2546.
276. B. F. Mentzen, J.-P. Puaux and H. Sautereau, *Acta Crystallogr., Sect. B*, 1978, **34**, 1846.
277. B. F. Mentzen, J.-P. Puaux and H. Sautereau, *Acta Crystallogr., Sect. B*, 1978, **34**, 2707.
278. (a) J. Howatson, D. M. Grev and B. Morosin, *J. Inorg. Nucl. Chem.*, 1975, **37**, 1933; (b) G. Micera, L. Erre Strinna, F. Cariati, D. A. Clemente, A. Marzotto and M. Biagini Cingi, *Inorg. Chim. Acta*, 1985, **109**, 135.
279. W. H. Zachariasen and H. A. Plettinger, *Acta Crystallogr.*, 1959, **12**, 526.
280. V. G. Albano, D. Braga, C. Concilio and N. Roveri, *Cryst. Struct. Commun.*, 1978, **7**, 133.
281. N. W. Alcock, T. J. Kemp and P. De Meester, *Acta Crystallogr., Sect. B*, 1982, **38**, 105.
282. A. Immirzi, G. Bombieri, S. Degetto and G. Marangoni, *Acta Crystallogr., Sect. B*, 1975, **31**, 1023.
283. G. Marangoni, S. Degetto, R. Graziani, G. Bombieri and E. Forsellini, *J. Inorg. Nucl. Chem.*, 1974, **36**, 1787.
284. G. Bombieri, S. Degetto, E. Forsellini, G. Marangoni and A. Immirzi, *Cryst. Struct. Commun.*, 1977, **6**, 115.
285. G. Bombieri, E. Forsellini, G. Tomat, L. Magon and R. Graziani, *Acta Crystallogr., Sect. B*, 1974, **30**, 2659.
286. G. A. Battiston, G. Sbrignadello, G. Bandoli, D. A. Clemente and G. Tomat, *J. Chem. Soc., Dalton Trans.*, 1979, 1965.
287. G. Bombieri, U. Croatto, R. Graziani, E. Forsellini and L. Magon, *Acta Crystallogr., Sect. B*, 1974, **30**, 407.
288. G. Bombieri, R. Graziani and E. Forsellini, *Inorg. Nucl. Chem. Lett.*, 1973, **9**, 551.
289. C. L. Christ, J. R. Clark and H. T. Evans, *Science*, 1955, **121**, 472.
290. R. Graziani, G. Bombieri and E. Forsellini, *J. Chem. Soc., Dalton Trans.*, 1972, 2059.
291. A. Anderson, C. Chieh, D. E. Irish and J. P. K. Tong, *Can. J. Chem.*, 1980, **58**, 1651.
292. N. C. Jayadevan and D. M. Chackraburtty, *Acta Crystallogr., Sect. B*, 1972, **28**, 3178.
293. N. W. Alcock, *J. Chem. Soc., Dalton Trans.*, 1973, 1614.
294. N. W. Alcock, *J. Chem. Soc., Dalton Trans.*, 1973, 1610.
295. N. W. Alcock, *J. Chem. Soc., Dalton Trans.*, 1973, 1616.
296. J.-P. Legros and Y. Jeannin, *Acta Crystallogr., Sect. B*, 1976, **32**, 2497.
297. R. M. Rojas, A. Del Pra, G. Bombieri and F. Benetollo, *J. Inorg. Nucl. Chem.*, 1979, **41**, 541.
298. G. Bombieri, F. Benetollo, E. Forsellini and A. Del Pra, *J. Inorg. Nucl. Chem.*, 1980, **42**, 1423.
299. G. Bombieri, F. Benetollo, A. Del Pra and R. Rojas, *J. Inorg. Nucl. Chem.*, 1979, **41**, 201.
300. F. Benetollo, G. Bombieri, J. A. Herrero and R. M. Rojas, *J. Inorg. Nucl. Chem.*, 1979, **41**, 195.
301. G. Bombieri, F. Benetollo, R. M. Rojas, M. L. de Paz and A. Del Pra, *Inorg. Chim. Acta*, 1982, **61**, 149.
302. B. F. Mentzen and H. Sautereau, *Acta Crystallogr., Sect. B*, 1980, **36**, 2051.
303. L. R. Nassimbeni, A. L. Rodgers and J. M. Haigh, *Inorg. Chim. Acta*, 1976, **20**, 149.
304. G. Bandoli and D. A. Clemente, *J. Inorg. Nucl. Chem.*, 1981, **43**, 2843.
305. R. N. Shchelokov, I. M. Orlova and A. V. Sergeev, *Dokl. Chem.* (*Engl. Transl.*), 1980, **250/5**, 573.

306. Yu. N. Mikhailov, I. M. Orlova, G. V. Podnebesnova, V. G. Kuznetsov and R. N. Shchelokov, *Sov. J. Coord. Chem.* (*Engl. Transl.*), 1976, **2**, 1298.
307. V. E. Mistryukov, Yu. N. Mikhailov, I. A. Yuranov, V. V. Kolesnik and K. M. Dunaeva, *Sov. J. Coord. Chem.* (*Engl. Transl.*), 1983, **9**, 163.
308. Yu. N. Mikhailov, S. B. Ivanov, I. M. Orlova, G. V. Podnebesnova, V. G. Kuznetsov and R. N. Shchelokov, *Sov. J. Coord. Chem.* (*Engl. Transl.*), 1976, **2**, 1212.
309. N. K. Dalley, M. H. Mueller and S. H. Simonsen, *Inorg. Chem.*, 1972, **11**, 1840.
310. Yu. N. Mikhailov, V. G. Kuznetsov and E. S. Kovaleva, *J. Struct. Chem.* (*Engl. Transl.*), 1968, **9**, 620.
311. V. E. Mistryukov, A. S. Kanishcheva, T. G. Grevtseva and Yu. N. Mikhailov, *Sov. J. Coord. Chem.* (*Engl. Transl.*), 1982, **8**, 860.
312. M. A. Soldatkina, V. N. Serezhkin and V. K. Trunov, *J. Struct. Chem.* (*Engl. Transl.*), 1981, **22**, 915.
313. V. N. Serezhkin, M. A. Soldatkina and V. K. Trunov, *Sov. Radiochem.* (*Engl. Transl.*), 1981, **23**, 551.
314. V. N. Serezhkin, M. A. Soldatkina and V. K. Trunov, *Koord. Khim.*, 1981, **7**, 1880.
315. E. Forsellini, G. Bombieri, R. Graziani and B. Zarli, *Inorg. Nucl. Chem. Lett.*, 1972, **8**, 461.
316. R. Julien, N. Rodier and P. Khodadad, *Acta Crystallogr., Sect. B*, 1977, **33**, 2411.
317. G. Bombieri, E. Forsellini, J. P. Day and W. I. Azeez, *J. Chem. Soc., Dalton Trans.*, 1978, 677.
318. N. W. Alcock, M. M. Roberts and D. Brown, *J. Chem. Soc., Dalton Trans.*, 1982, 25.
319. J. E. Fleming and H. Lynton, *Chem. Ind.* (*London*), 1959, 1409; 1960, 1415.
320. L. R. Nassimbeni and A. L. Rodgers, *Cryst. Struct. Commun.*, 1976, **5**, 301.
321. M. R. Caira, J. F. de Wet, J. G. H. du Preez, B. Busch and H. E. Rohwer, *Inorg. Chim. Acta*, 1983, **77**, L73.
322. C. Panattoni, R. Graziani, G. Bandoli, B. Zarli and G. Bombieri, *Inorg. Chem.*, 1969, **8**, 320.
323. (a) G. Bombieri, U. Croatto, E. Forsellini, B. Zarli and R. Graziani, *J. Chem. Soc., Dalton Trans.*, 1972, 560; (b) A. E. Storey, F. Zonnevijlle, A. A. Pinkerton and D. Schwarzenbach, *Inorg. Chim. Acta*, 1983, **75**, 103.
324. T. H. Lu, T. J. Lee, T. Y. Lee and C. Wong, *Inorg. Nucl. Chem. Lett.*, 1977, **13**, 363.
325. J. C. Taylor and A. B. Waugh, *J. Chem. Soc., Dalton Trans.*, 1977, 1630.
326. J. C. Taylor and A. B. Waugh, *J. Chem. Soc., Dalton Trans.*, 1977, 1636.
327. R. Graziani, B. Zarli, A. Cassol, G. Bombieri, E. Forsellini and E. Tondello, *Inorg. Chem.*, 1970, **9**, 2116.
328. B. Zarli, R. Graziani, E. Forsellini, U. Croatto and G. Bombieri, *Chem. Commun.*, 1971, 1501.
329. C. Panattoni, R. Graziani, U. Croatto, B. Zarli and G. Bombieri, *Inorg. Chim. Acta*, 1968, **2**, 43.
330. J. C. Dewan, A. J. Edwards, D. R. Slim, J. E. Guerchais and R. Kergoat, *J. Chem. Soc., Dalton Trans.*, 1975, 2171.
331. J. MacB. Harrowfield, D. L. Kepert, J. M. Patrick, A. H. White and S. F. Lincoln, *J. Chem. Soc., Dalton Trans.*, 1983, 393.
332. J. L. Katz and W. S. Arrington, Report RP1-2325-1, 1964 (*Chem. Abstr.*, 1965, **62**, 12 542).
333. D. L. Perry, D. H. Templeton and A. Zalkin, *Inorg. Chem.*, 1978, **17**, 3699.
334. D. L. Perry, D. H. Templeton and A. Zalkin, *Inorg. Chem.*, 1979, **18**, 879.
335. D. L. Perry, A. Zalkin, H. Ruben and D. H. Templeton, *Inorg. Chem.*, 1982, **21**, 237.
336. K. Bowman and Z. Dori, *Chem. Commun.*, 1968, 636.
337. A. S. R. Murty and M. B. Adi, *Proc. Nucl. Chem. Radiochem. Symp., 1980 (1981)*, p. 357 (*Chem. Abstr.*, 1982, **96**, 134 758).
338. K. Rediess and W. Sawodny, *Z. Naturforsch., Teil B*, 1982, **37**, 524.
339. R. T. Paine, R. R. Ryan and L. B. Asprey, *Inorg. Chem.*, 1975, **14**, 1113.
340. R. D. Peacock and N. Edelstein, *J. Inorg. Nucl. Chem.*, 1976, **38**, 771.
341. R. C. Burns and T. A. O'Donnell, *Inorg. Nucl. Chem. Lett.*, 1977, **13**, 657.
342. J. C. Taylor and P. W. Wilson, *J. Chem. Soc., Chem. Commun.*, 1974, 232.
343. R. Bougon, J. Fawcett, J. H. Holloway and D. R. Russell, *J. Chem. Soc., Dalton Trans.*, 1979, 1881.
344. W. W. Wilson and K. O. Christe, *Inorg. Chem.*, 1982, **21**, 2091.
345. R. Bougon, P. Charpin, J. P. Desmoulin and J. G. Malm, *Inorg. Chem.*, 1976, **15**, 2532.
346. (a) J. G. Malm, H. Selig and S. Siegel, *Inorg. Chem.*, 1966, **5**, 130; (b) B. Druzina, S. Milicev and J. Slivnik, *J. Chem. Soc., Chem. Commun.*, 1984, 363.
347. M. Iwasaki, N. Ishikawa, K. Ohwada and. T. Fujino, *Inorg. Chim. Acta*, 1981, **54**, L193.
348. K. W. Bagnall, J. G. H. du Preez, B. J. Gellatly and J. H. Holloway, *J. Chem. Soc., Dalton Trans.*, 1975, 1963.
349. (a) W. H. Zachariasen, *Acta Crystallogr.*, 1954, **7**, 783; (b) A. A. Lychev, L. G. Mashirov, Yu. I. Smolin and Yu. F. Shepelev, *Radiokhimiya*, 1984, **26**, 255; *Sov. Radiochem.* (*Engl. Transl.*), 1984, **26**, 239.
350. S. B. Ivanov, Yu. N. Mikhailov, V. G. Kuznetsov and R. L. Davidovich, *Koord. Khim.*, 1980, **6**, 1746.
351. Nguyen Quy Dao, S. Chouro and J. Heckly, *J. Inorg. Nucl. Chem.*, 1981, **43**, 1835.
352. S. B. Ivanov, Yu. N. Mikhailov, V. G. Kuznetsov and R. L. Davidovich, *J. Struct. Chem.* (*Engl. Transl.*), 1981, **22**, 302.
353. L. Di Sipio, A. Pasquetto, G. Pelizzi, G. Ingletto and A. Montenero, *Cryst. Struct. Commun.*, 1981, **10**, 1153.
354. F. J. Martin Gil, I. Escalante Roldan and J. G. Ribas Bernat, *An. Quim.*, 1976, **72**, 786.
355. J. P. Day and L. M. Venanzi, *J. Chem. Soc.* (*A*), 1966, 1363.
356. L. Di Sipio, E. Tondello, G. Pelizzi, G. Ingletto and A. Montenero, *Cryst. Struct. Commun.*, 1977, **6**, 723.
357. G. Bandoli, D. A. Clemente and M. Biagini Cingi, *J. Inorg. Nucl. Chem.*, 1975, **37**, 1709.
358. D. A. Clemente, G. Bandoli, F. Benetollo, M. Vidali, P. A. Vigato and U. Casellato, *J. Inorg. Nucl. Chem.*, 1974, **36**, 1999.
359. M. N. Akhtar, E. D. McKenzie, R. E. Paine and A. J. Smith, *Inorg. Nucl. Chem. Lett.*, 1969, **5**, 673.
360. (a) G. Bandoli and D. A. Clemente, *J. Chem. Soc., Dalton Trans.*, 1975, 612; (b) N. W. Alcock, D. J. Flanders, T. J. Kemp and M. A. Shand, *J. Chem. Soc., Dalton Trans.*, 1985, 517.
361. A. Marzotto, L. Garbin and F. Braga, *J. Inorg. Biochem.*, 1979, **10**, 257.
362. C. Keller and S. H. Eberle, *Radiochim. Acta*, 1967, **8**, 65.
363. D. H. Hall, A. D. Rae and T. N. Waters, *Acta Crystallogr.*, 1967, **22**, 258.
364. L. Baracco, G. Bombieri, S. Degetto, E. Forsellini, G. Marangoni, G. Paolucci and R. Graziani, *J. Chem. Soc., Dalton Trans.*, 1975, 2161.

365. K. M. Kanth and J. Sahu, *J. Indian Chem. Soc.*, 1978, **55**, 869.
366. C. A. Toso and A. Z. Nettle, *Afinidad*, 1981, **38**, 149.
367. V. W. Day, T. J. Marks and W. A. Wachter, *J. Am. Chem. Soc.*, 1975, **97**, 4519.
368. T. J. Marks and D. R. Stojakovic, *J. Chem. Soc., Chem. Commun.*, 1975, 28.
369. E. A. Cuellar and T. J. Marks, *Inorg. Chem.*, 1981, **20**, 3766.
370. P. G. Eller and R. A. Penneman, *Inorg. Chem.*, 1976, **15**, 2439.
371. D. L. Tomaja, *Inorg. Chim. Acta*, 1977, **21**, L31.
372. D. C. Moody and R. R. Ryan, *Cryst. Struct. Commun.*, 1979, **8**, 933.
373. F. A. Zakharova, M. M. Orlova and A. D. Gel'man, *Sov. Radiochem.* (*Engl. Transl.*), 1972, **14**, 121.
374. A. Y. Tsivadze and N. N. Krot, *Sov. Radiochem.* (*Engl. Transl.*), 1972, **14**, 646.
375. S. V. Tomilin, Yu. F. Volkov, I. I. Kapshukov and A. G. Rykov, *Sov. Radiochem.* (*Engl. Transl.*), 1981, **23**, 570, 695.
376. S. V. Tomilin, Yu. F. Volkov, G. I. Visyashcheva and I. I. Kapshukov, *Sov. Radiochem.* (*Engl. Transl.*), 1983, **25**, 56.
377. S. V. Tomilin, Yu. F. Volkov, I. Kapshukov and A. G. Rykov, *Sov. Radiochem.* (*Engl. Transl.*), 1981, **23**, (a) 570, (b) 574, (c) 695.
378. M. P. Mefod'eva, N. N. Krot, A. D. Gel'man and T. V. Afanas'eva, *Sov. Radiochem.* (*Engl. Transl.*), 1976, **18**, 85.
379. V. I. Dzyubenko, S. A. Karavaev and V. F. Peretrukhin, *Dokl. Phys. Chem.* (*Engl. Transl.*), 1981, **256/8**, 176.
380. A. A. Chaikhorskii, S. S. Zelentsov and E. V. Leikina, *Sov. Radiochem.* (*Engl. Transl.*), 1972, **14**, 636.
381. J. H. Burns, W. H. Baldwin and J. R. Stokeley, *Inorg. Chem.*, 1973, **12**, 466.
382. R. D. Baybarz, *J. Inorg. Nucl. Chem.*, 1973, **35**, 483.
383. R. D. Baybarz, L. B. Asprey, C. E. Strouse and E. Fukushima, *J. Inorg. Nucl. Chem.*, 1972, **34**, 3427.
384. J. R. Peterson, R. L. Fellows, J. P. Young and R. G. Haire, *Radiochem. Radioanal. Lett.*, 1977, **31**, 277.
385. J. R. Peterson and R. D. Baybarz, *Inorg. Nucl. Chem. Lett.*, 1972, **8**, 423.
386. J. F. Wild, E. K. Hulet, R. W. Lougheed and W. N. Hayes, Report UCRL-77709, 1975 (*Chem. Abstr.*, 1976, **85**, 40 215).
387. J. R. Peterson, D. D. Ensor, R. L. Fellows, R. G. Haire and J. P. Young, Report ORO-4447-075, CONF. 780823-3, 1978 (*Chem. Abstr.*, 1979, **90**, 214 394).
388. N. B. Mikheev, A. N. Kamenskaya, I. A. Rumer, N. A. Konovalova and S. A. Kulyukhin, *Sov. Radiochem.* (*Engl. Transl.*), 1983, **25**, 143.
389. C. Y. Marcus and D. Cohen, *Inorg. Chem.*, 1966, **5**, 1740.
390. J. H. Burns and J. R. Peterson, *Inorg. Chem.*, 1971, **10**, 147.
391. J. H. Burns and W. H. Baldwin, *Inorg. Chem.*, 1977, **16**, 289.
392. W. O. Milligan, M. L. Beasley, M. H. Lloyd and R. G. Haire, *Acta Crystallogr., Sect. B*, 1968, **24**, 979.
393. C. Keller and H. Schreck, *J. Inorg. Nucl. Chem.*, 1969, **31**, 1121.
394. M. D. Danford, J. H. Burns, C. E. Higgins, J. R. Stokeley and W. H. Baldwin, *Inorg. Chem.*, 1970, **9**, 1953.
395. E. V. Fedoseev, L. A. Ivanova, S. S. Travnikov, A. V. Davydov and B. F. Myasoedov, *Sov. Radiochem.* (*Engl. Transl.*), 1983, **25**, 343.
396. V. Ya. Mishin, E. M. Rubtsov and V. K. Isupov, *Sov. Radiochem.* (*Engl. Transl.*), 1980, **22**, 564.
397. R. Amano, A. Sato and S. Suzuki, *Bull. Chem. Soc. Jpn.*, 1982, **55**, 2594.
398. J. H. Burns and M. D. Danford, *Inorg. Chem.*, 1969, **8**, 1780.
399. W. H. Hale, Jr. and W. C. Mosley, *J. Inorg. Nucl. Chem.*, 1973, **35**, 165.
400. R. D. Baybarz, J. A. Fahey and R. G. Haire, *J. Inorg. Nucl. Chem.*, 1974, **36**, 2023.
401. (a) E. A. Erin, V. V. Kopytov, V. Ya. Vasil'ev and V. M. Vityutnev, *Sov. Radiochem.* (*Engl. Transl.*), 1977, **19**, 380; (b) V. G. Zubarev and N. N. Krot, *Radiokhimiya*, 1983, **25**, 639; *Sov. Radiochem.* (*Engl. Transl.*), 1983, **25**, 601; (c) V. G. Zubarev and N. N. Krot, *Radiokhimiya*, 1983, **25**, 631; *Sov. Radiochem.* (*Engl. Transl.*), 1983, **25**, 594.
402. S. Bouhlassa, Fr. Report IPNO-T-81-07, 1981 (*Chem. Abstr.*, 1983, **98**, 82 730).
403. T. Sekine and D. Dyrssen, *J. Inorg. Nucl. Chem.*, 1967, **29**, 1481.
404. A. V. Davydov, B. F. Myasoedov and S. S. Travnikov, *Dokl. Chem.* (*Engl. Transl.*), 1975, **220/5**, 672.
405. E. V. Fedoseev, S. Travnikov, A. V. Davydov and B. F. Myasoedov, *Radiochem. Radioanal. Lett.*, 1982, **51**, 271; *Sov. Radiochem.* (*Engl. Transl.*), 1983, **25**, 428.
406. D. D. Ensor, J. R. Peterson, R. G. Haire and J. P. Young, *J. Inorg. Nucl. Chem.*, 1981, **43**, 2425.
407. J. Jové and M. Pagès, *Inorg. Nucl. Chem. Lett.*, 1977, **13**, 329.
408. L. R. Morss and J. Fuger, *Inorg. Chem.*, 1969, **8**, 1433.
409. C. Keller, S. H. Eberle and K. Mosdzelewski, *Radiochim. Acta*, 1966, **5**, 185.
410. D. Feinauer and C. Keller, *Inorg. Nucl. Chem. Lett.*, 1969, **5**, 625.
411. W. Hagenberg, R. Gradl and F. Lux, *Ges. Dtsch. Chem. Hauptversammlung, Karlsruhe*, 1971, 87.
412. P. N. Moskalev and G. N. Shapkin, *Sov. Radiochem.* (*Engl. Transl.*), 1977, **19**, 294.
413. F. L. Moore, *Anal. Chem.*, 1966, **38**, 1872.
414. R. G. Haire and L. B. Asprey, *Inorg. Nucl. Chem. Lett.*, 1973, **9**, 869.
415. L. B. Asprey and R. G. Haire, *Inorg. Nucl. Chem. Lett.*, 1973, **9**, 1121.
416. H. O. Haug and R. D. Baybarz, *Inorg. Nucl. Chem. Lett.*, 1975, **11**, 847.
417. V. N. Kosyakov, E. A. Erin, V. M. Vityutnev, V. V. Kopytov and A. G. Rykov, *Sov. Radiochem.* (*Engl. Transl.*), 1982, **24**, 455.
418. V. G. Zubarev and N. N. Krot, *Sov. Radiochem.* (*Engl. Transl.*), 1982, **24**, 264.
419. C. Keller, L. Koch and K. H. Walter, *J. Inorg. Nucl. Chem.*, 1965, **27**, 1225.
420. G. A. Simakin, Yu. F. Volkov, G. I. Visyashcheva, I. I. Kapshukov, P. F. Baklanova and G. N. Yakovlev, *Sov. Radiochem.* (*Engl. Transl.*), 1974, **16**, 838.
421. D. Lawaldt, R. Marquart, G. D. Werner and F. Weigel, *J. Less-Common Met.*, 1982, **85**, 37.
422. F. Weigel and G. Hoffmann, *J. Less-Common Met.*, 1976, **44**, 133.

423. H. R. Hoekstra and E. Gebert, *Inorg. Nucl. Chem. Lett.*, 1978, **14,** 189.
424. A. M. Fedoseev and V. P. Perminov, *Sov. Radiochem.* (*Engl. Transl.*), 1983, **25,** 522.
425. S. H. Eberle and W. Robel, *Inorg. Nucl. Chem. Lett.*, 1970, **6,** 359.
426. T. K. Keenan, *Inorg. Nucl. Chem. Lett.*, 1968, **4,** 381.
427. D. Lawaldt, Ger. Report 1981, NP 3901798 (*Chem. Abstr.*, 1983, **99,** 114 873).
428. K. W. Bagnall, J. B. Laidler and M. A. A. Stewart, *J. Chem. Soc.* (*A*), 1968, 133.

36.1

Molybdenum: The Element and Aqueous Solution Chemistry

A. GEOFFREY SYKES
University of Newcastle upon Tyne, UK

36.1.1 THE ELEMENT

Molybdenum occurs chiefly as molybdenite, MoS_2, but also as molybdates $PbMoO_4$ and $MgMoO_4$. The largest known deposits are in Colorado (USA), but it is also found in Canada and Chile. The natural abundance in the earth's crust (~1.2 p.p.m.) is about the same as that of tungsten, but is much less than of chromium (122 p.p.m.). The name in fact originates from the Greek *molybdos* meaning lead.

The Swedish chemist Scheele (1778) produced the oxide of a new element from (black) MoS_2, thereby distinguishing the element from graphite with which it had been confused. The metal was isolated by Hjelm in Sweden three to four years later by heating the oxide with charcoal. In the procedure now used to isolate the metal, the MoS_2 component in ores is concentrated by flotation methods. The concentrate is then converted by roasting into MoO_3, which, after purification, is reduced with hydrogen. Reduction with carbon is avoided since carbides rather than the metal are obtained.

The chief use of molybdenum is in steels. The oxides and sulfides have some applications as catalysts. Molybdenum is the only element in the second and third transition series which appears to have a major role as a trace metal in enzymes. Several aspects of molybdenum chemistry have been widely studied in order to gain a better understanding of the biological relevance. Molybdenum is one of the few elements which currently has its own series of international conferences.[1]

Molybdenum and tungsten are similar chemically, although there are differences which it is difficult to explain. There is much less similarity in comparisons with chromium. In addition to the variety of oxidation states there is a wide range of stereochemistries, and the chemistry is amongst the most complex of the transition elements.

36.1.2 AQUEOUS SOLUTION CHEMISTRY

36.1.2.1 General Introduction

Molybdenum has an extensive aqueous solution chemistry for oxidation states II through VI. It is unique in having aqua or aqua/oxo ions for all five states in acidic solution (pH < 2). These are well defined in all but the Mo^{VI} case, the study of which is complicated by the existence of rapid equilibria involving protonated/deprotonated monomer/dimer (and higher) forms. The VI state is without question the most stable and in contrast to Cr^{VI} is only the mildest of oxidants. Compounds which have contributed to the development of the aqueous solution chemistry, including the aqua ions themselves, are considered under Section 36.1.2. It is only since 1971 that the aqua forms of oxidation state II–V ions have been identified, and preparative as well as structural features defined.[2] The chemistry of aqua ions generally has a somewhat elevated position, since they are often regarded as a point of reference or prototype for the behaviour of a particular oxidation state. Since they are difficult to crystallize, structures of derivative complexes are relevant. The simple aqua ions of oxidation states II–V are indicated in Table 1. The quite different structures of adjacent oxidation states are to be noted, which gives rise to an interesting and varied redox chemistry. All but $[Mo(H_2O)_6]^{3+}$ are diamagnetic and metal–metal bonding is a significant contributing feature.

Table 1 Summary of Aqua Ions of Molybdenum at pH < 2

Description	*Formula*	*Mo—Mo bonding*	*Colour*
Mo_2^{II}	$[Mo_2(H_2O)_8]^{4+}$	Quadruple	Red
Mo^{III}	$[Mo(H_2O)_6]^{3+}$	None	Pale yellow
Mo_2^{III}	$[Mo_2(OH)_2(H_2O)_8]^{4+}$	Triple	Green
Mo_3^{IV}	$[Mo_3O_4(H_2O)_9]^{4+}$	Single	Red
Mo_2^{V}	$[Mo_2O_4(H_2O)_6]^{2+}$	Single	Yellow
Mo_x^{VI}	Different forms $x = 1$ and 2	None	Colourless

The three most often used 'lead in' compounds for the chemistry described are sodium molybdate, $Na_2[MoO_4]\cdot 2H_2O$, molybdenum hexacarbonyl, $[Mo(CO)_6]$, and potassium hexachloromolybdate, $K_3[MoCl_6]$. Synthesis of II and III state complexes generally requires rigorous O_2-free techniques, using a range of methods from those involving Schlenk apparatus, to the use of N_2 or Ar gas, syringes, Teflon tubing and/or stainless steel needles, and rubber seals. In some cases solutions of IV and V state complexes must also be stored O_2 free. Perchlorate cannot be used with II and III aqua ions, and also appears to oxidize some trimeric Mo^{IV} ions on leaving overnight. Instead weakly coordinating, redox inactive and strongly acidic trifluoromethanesulfonic acid, CF_3SO_3H (abbreviated HTFMS or triflate), or *p*-toluenesulfonic acid, $C_6H_4(Me)SO_3H$ (abbreviated HPTS), finds wide usage. Methanesulfonic acid, $MeSO_3H$, has also been used.

36.1.2.2 Oxidation State II

The aqueous chemistry of the II state is dominated by dimeric complexes. The latter constitute the largest group of compounds of any element containing quadrupule bonds, a subject extensively covered in the Cotton and Walton text.[3] Soon after the recognition of the quadruple bond in $[Re_2Cl_8]^{2-}$ in 1964 the structure of $[Mo_2(O_2CMe)_4]$, Figure 1, was published (Mo—Mo distance 2.09 Å).[4] The $[Mo_2Cl_8]^{4-}$ complex, Figure 2, is obtained by treating $[Mo_2(O_2CMe)_4]$ in aqueous KCl solution with 12 M HCl at 0 °C when crystals of $K_4[Mo_2Cl_8]\cdot 2H_2O$ are obtained. Similarly the bromo analogue $(NH_4)_4[Mo_2Br_8]$ has been isolated. There are no bridging ligands and in both cases the Mo—Mo distance remains short (2.14 Å), consistent with retention of a quadruple metal–metal bond.[5] For comparison, typical M—M distances for quadruply bonded dimeric d^4 complexes are 2.21 Å (for $[W_2(O_2CCF_3)_4]$),[6] 2.19 Å (for $[Tc_2(O_2CCMe_3)_4Cl_2]$)[7] and 2.2 Å (for $K_2[Re_2Cl_8]\cdot 2H_2O$).[3,4b] The d^4–d^4 technetium complex is the only Tc_2^{III} example for which there is structural information, and the ease with which $[Tc_2Cl_8]^{3-}$ rather than $[Tc_2Cl_8]^{2-}$ is obtained is at present puzzling. The range of metal–metal distances in Mo_2^{II} complexes (2.04–2.18 Å) is much smaller than that observed for Cr_2^{II} (1.83–2.54 Å).[3]

Examples of monomeric Mo^{II} complexes are much less common. It has been demonstrated that the $[Mo(CN)_7]^{5-}$ ion has a pentagonal bipyramidal structure. Other examples are nitrile (*i.e.* isocyanide) complexes $[Mo(CNR)_7]^{2+}$, which has a capped trigonal prismatic structure, and the diarsine complex $[Mo(diars)_2X_2]$.

The acetate complex $[Mo_2(O_2CMe)_4]$ is obtained by heating $[Mo(CO)_6]$ with glacial acetic acid or a mixture of the acid and its anhydride.[8] Yields are low (15–20%), but can be improved (~80%) when diglyme is used as solvent.[9] The low yields are the result of the formation of trinuclear Mo^{IV} complexes of the type $[Mo_3X_2(O_2CMe)_6(H_2O)_3]^{4+}$, where the Mo_3X_2 unit is a trigonal bipyramid with capping groups X either O or CMe (alkylidene).

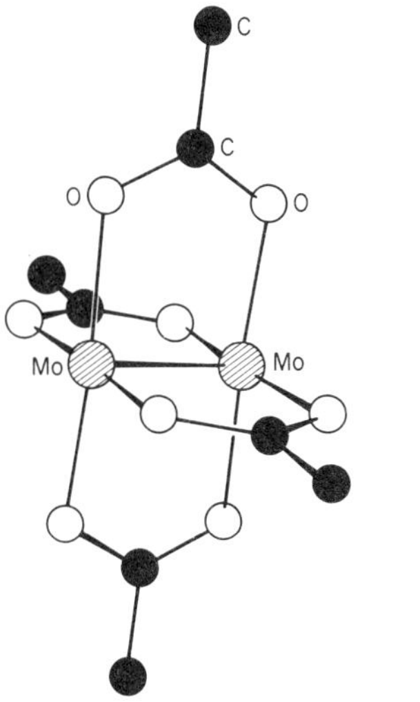

Figure 1 The structure of the Mo_2^{II} complex $[Mo_2(O_2CMe)_4]$[4a]

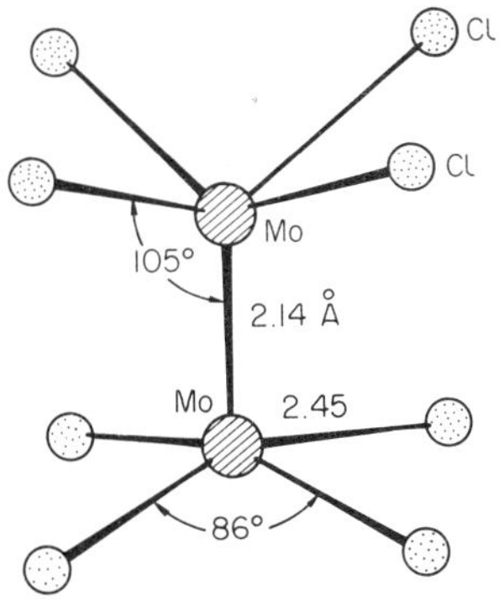

Figure 2 The structure of the Mo_2^{II} complex $[Mo_2Cl_8]^{4-}$ [5]

A listing of structural data of compounds containing Mo—Mo quadruple bonds with no bridging ligands (16 examples), with carboxylato bridges (23 examples), and a number of other compounds (25) up to 1981 has appeared.[2] A further listing of 67 structures from 1981 to 1984 has been reported.[10] The chemistry of this type of compound is therefore well established. As far as the bonding between two d^4 metals is concerned,[2] the eight electrons fill four bonding orbitals in a configuration $\sigma^2\pi^4\delta^2$. The order of energies of the molecular orbitals beginning with the most stable is $\sigma < \pi \ll \delta < \delta^* \ll \pi^* < \sigma^*$. Since there are eight bonding electrons and no non-bonding electrons the bond order is four. Features of the structures are the shortness of the M—M bond, and the tendency with no bridging ligands to get an eclipsed configuration. Whereas the σ and π bonds are cylindrically symmetrical the δ component is markedly angle sensitive. A rotation away from the eclipsed to the staggered conformation is expected to result in complete loss of the δ bond component. The δ bond is only weakly bonding, and consequently the δ^* orbital is only weakly antibonding. In the case of the $[Mo_2Cl_8]^{4-}$ complex there is mutual repulsion of the chloro ligands in one plane with those in another, with Mo—Mo—Cl angles 105° as indicated in Figure 2.

If more electrons are added the M—M bond order is reduced without loss of overall structure. Thus Rh^{II} dimers such as $[Rh_2(O_2CMe)_4(H_2O)_2]$ with an electron configuration $\sigma^2\pi^4\delta^2\delta^{*2}\pi^{*4}$ contain a single bond, and oxidation to $Rh^{II,III}$ gives a product of bond order 1.5. This has been shown to bring about a decrease in Rh—Rh bond length from 2.39 to 2.32 Å.[11] Similarly a large number of $Ru^{II,III}$ dimers $[Ru_2(O_2CR)_4]X$ are known which have an electron configuration $\sigma^2\pi^4\delta^2\delta^{*2}\pi^*$ (or $\sigma^2\pi^4\delta^2\pi^{*2}\delta^*$) and the bond order is 2.5.[12]

The mixed-metal dimer $[CrMo(O_2CMe)_4]$ has been prepared in ~30% yield by addition of $[Mo(CO)_6]$ in acetic acid, acetic anhydride and CH_2Cl_2, to a refluxing solution of $[Cr_2(O_2CMe)_4(H_2O)_2]$ in acetic acid and acetic anhydride.[13] The yellow product is volatile and gives the expected parent ion peak in the mass spectrum. It has a Raman active ν(Cr—Mo) mode at 394 cm^{-1} consistent with quadruple bonding. The crystal structure gives a metal–metal distance of 2.05 Å, which is shorter than in $[Mo_2(O_2CMe)_4]$ (2.09 Å) and $[Cr_2(O_2CMe)_4(H_2O)_2]$ (2.36 Å). A molybdenum–tungsten mixed-metal pivalato complex $[MoW(O_2CCMe_3)_4]$ has also been prepared.[14] The procedure involves refluxing a 1:3 mixture of $[Mo(CO)_6]$ and $[W(CO)_6]$ with pivalic acid in 1,2-dichlorobenzene. The purified product gives the $[MoW(O_2CCMe_3)_4]^+$ parent peak in the mass spectrum. A structure determination gives an Mo—W distance of 2.08 Å,[14] compared to 2.09 Å in $[Mo_2(O_2CCMe_3)_4]$. Attempts to prepare a mixed-metal octachloro complex from $[CrMo(O_2CMe)_4]$ were unsuccessful,[13] and the chemistry of these mixed-metal complexes is not at present extensive.

The carboxylate complexes $[Mo_2(O_2CR)_4]$ (R = Et, CMe_3 or Ph) are readily oxidized with iodine to give $[Mo_2(O_2CR)_4]I_3$ in 1,2-dichloroethane.[15] The complexes are paramagnetic, $\mu = 1.66$ BM for R = CMe_3, and give narrow ESR signals ($g = 1.93$) with hyperfine pattern consistent with two equivalent Mo(+2.5) nuclei. Oxidation of $[MoW(O_2CCMe_3)_4]$ with iodine to the 1+ ion was used in the separation from $[Mo_2(O_2CCMe_3)_4]$. The only example in which reduction of an Mo_2^{II} has been observed is in a pulse radiolysis study on $[Mo_2(O_2CCF_3)_4]$ in methanol.[16] The transient product believed to be the 1− ion with a peak at 780 nm undergoes rapid decay.

Studies on solutions containing 0.01 M $K_4[Mo_2Cl_8]$ in 0.10 M CF_3SO_3H have provided evidence for rapid loss of three Cl^- ions per $[Mo_2Cl_8]^{4-}$.[17] A number of compounds have been characterized in which limited replacement of halide ions of $[Mo_2X_8]^{4-}$ by H_2O has been achieved.[18] Preparative procedures include reacting the Mo_2^{III} compound $(NH_4)_3[Mo_2Cl_9]\cdot H_2O$ in hydrobromic acid with pyridinium bromide to give $(pyH)_3Mo_2Br_7\cdot 2H_2O$,[19] which has been shown to be the double salt $(pyH)_3[Mo_2Br_6(H_2O)_2]Br$. Other crystal structures of the 4-methylpyridinium salt $(pyH)_2[Mo_2X_6(H_2O)_2]$, and of $(picH)_2[Mo_2X_6(H_2O)_2]$ ($X^- = Br^-$ and I^-) have been reported.[20] All have the eclipsed configuration with one H_2O on each molybdenum. The Mo—Mo distance is ~0.02 Å shorter than in $[Mo_2X_8]^{4-}$ forms, possibly reflecting the decrease in anion charge.

The reaction of $K_4[Mo_2Cl_8]$ with acetic acid (up to 0.5 M) in solutions of HCl and HPTS of varying ratios, $[H^+] = 0.05$–1.95 M, has been studied.[21] The reaction proceeds in two kinetic stages which have been assigned to the entry of the first and third acetate ligands. Since there is the possible involvement of aqua intermediates, such studies are complicated. Further information on the manner of substitution on dimeric Mo^{II} will be considered below.

The kinetics of the reaction of $[Mo_2Cl_8]^{4-}$ and protons to produce the triply bridged Mo^{III} dimer $[Cl_3Mo(\mu\text{-}Cl)_2(\mu\text{-}H)MoCl_3]^{3-}$ have been studied at 25 °C in 6–12 M HCl.[22] The reaction is first order in $[Mo_2Cl_8]^{4-}$ and obeys a linear dependence with respect to the acidity function H_o. Decomposition of the hydrido-bridged product in HCl solutions (<3 M) to yield hydrogen and the green di-μ-hydroxo Mo^{III} dimer $Mo_2(OH)_2Cl_x^{(x-4)-}$ is observed, and the kinetics have again been determined.

The reverse of the $[Mo_2Cl_8]^{4-}$ to $[Mo_2Cl_8H]^{3-}$ conversion although not quite as facile has been accomplished.[23] Solutions of both the Mo_2^{III} complexes $[Mo_2Cl_8H]^{3-}$ and $[Mo_2Cl_9]^{3-}$ can be reduced with amalgamated zinc to yield red Mo_2^{II} solutions.

Electrochemical experiments on $[Mo_2Cl_8]^{4-}$ in 6 M HCl have also been reported.[24] In cyclic voltammetry studies a single oxidation peak is present at +0.5 V (*vs.* SCE), but except at high sweep rates (500 mV s^{-1}), the corresponding reduction peak is not observed. It is likely that $[Mo_2Cl_8H]^{3-}$ is formed in an irreversible step.

Replacement of chloro ligands on $K_4[Mo_2Cl_8]$ occurs on treating with H_2SO_4, and the structure $K_4[Mo_2(SO_4)_4]\cdot 2H_2O$ has been determined,[25] confirming an eclipsed structure with four bridging sulfates. An interesting feature is that the axial positions are occupied by O atoms of neighbouring sulfate groups. Air oxidation of a saturated solution of the red $K_4[Mo_2(SO_4)_4]\cdot 2H_2O$ in 2 M H_2SO_4 gives the blue Mo(+2.5) complex $K_3Mo_2(SO_4)_4\cdot 3.5H_2O$. The structure of the latter closely resembles that of $K_4[Mo_2(SO_4)_4]\cdot 2H_2O$ except that H_2O molecules are bound in each axial position.[26] The Mo—Mo distance is longer for the 3− ion (2.17 Å) as compared to the 4− ion (2.11 Å). No further oxidation is observed. In the corresponding reaction of $K_4[Mo_2Cl_8]$ with 2 M H_3PO_4 in air for 24 h in the presence of large cations, the purple crystalline μ-HPO_4 bridged dimeric Mo^{III} product is obtained,[27] which has triply bonded Mo's (2.23 Å). Analogous Mo_2^{II} and $Mo^{II,III}$ phosphato-bridged complexes do not appear to be formed.

Displacement of bridging acetate ligands of $[Mo_2(O_2CMe)_4]$ by treatment at 100 °C with methyl sulfonate (in diglycine) and trifluoromethanesulfonate gives the bridged dimers $[Mo_2(O_3SMe)_4]$ and $[Mo_2(O_3SCF_3)_4]$.[28] These complexes find use in the synthesis of other Mo_2^{II} complexes, since the ligands are good leaving groups.

On heating $K_4[Mo_2Cl_8]$ with ethylenediamine, and then adding hydrochloric acid to an aqueous solution of the crude product, orange crystals of $[Mo_2(en)_4]Cl_4$ are obtained.[17] These give an absorbance ($\delta \rightarrow \delta^*$) band at 475 nm. It is unlikely that the ethylenediamine ligand is bridging. Cyclic voltammetry measurements have shown that the complex exhibits an irreversible oxidation at +0.78 V (SCE).

The red dimeric Mo^{II} aqua ion is obtained under rigorous O_2-free conditions by adding Ba^{2+} to the sulfate complex $[Mo_2(SO_4)_4]^{4-}$.[17] Starting from $[Mo(CO)_6]$ the preparative route can be summarized as in Scheme 1. Following centrifugation to remove $BaSO_4$ the aqua Mo_2^{II} solution in 0.1 M CF_3SO_3H can be kept unchanged for up to 3 h (under N_2). As with $[MoCl_8]^{4-}$ and $[Mo_2(SO_4)_4]^{4-}$, solutions are red in colour and a comparison of UV–visible spectra, Figure 3, leaves little doubt that the aqua ion is also quadruply bonded. The absorbance peak at 510 nm (ε 370 $M^{-1}\,cm^{-1}$ per dimer) has been assigned to a $\delta \rightarrow \delta^*$ transition.[29] Although the aqua ion does not display a peak at ~370 nm, the absorbance is sufficiently intense in this region for the peak to have become concealed.[30] Although the aqua ion is generally formulated as $[Mo_2(H_2O)_8]^{4+}$, in solution weak additional coordination in the axial positions may be present in which case a more precise formulation is $[Mo_2(H_2O)_8(H_2O)_2]^{4+}$.

$$[Mo(CO)_6] \xrightarrow{MeCO_2H} [Mo_2(O_2CMe)_4] \xrightarrow{KCl} [Mo_2Cl_8]^{4+} \xrightarrow{K_2SO_4}$$

$$[Mo_2(SO_4)_4]^{4-} \xrightarrow{Ba^{2+}} [Mo_2(H_2O)_8]^{4+}$$

Scheme 1

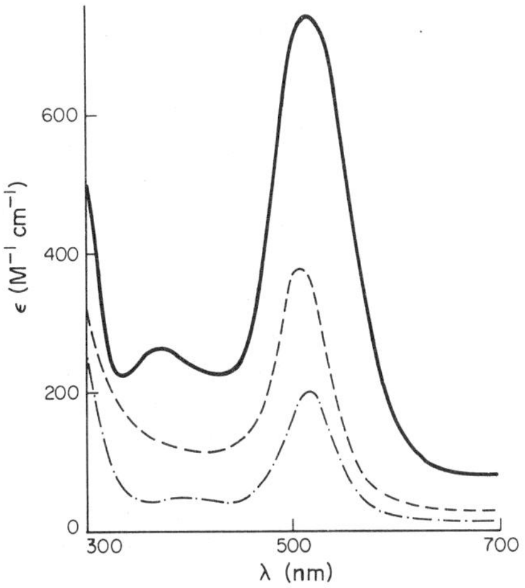

Figure 3 UV–visible spectra of $[Mo_2(H_2O)_8]^{4+}$ in 0.1 M CF_3SO_3H (– – –), $[Mo_2Cl_8]^{4-}$ in 2 M HCl (——) and $[Mo_2(SO_4)_4]^{4-}$ in 2 M H_2SO_4 (–·–)[30]

The kinetics of the 1:1 substitution of aqua Mo_2^{II} with NCS^- and $HC_2O_4^-$ have been studied in trifluoromethanesulfonic acid solutions, $I = 0.10$ M (CF_3SO_3Na), with the Mo_2^{II} reactant in greater than ten-fold excess (to avoid higher complex formation).[30] First-order equilibration rate constants, k_{eq}, determined by the stopped-flow method can be expressed as in equation (1). At 25 °C k_1 for the formation is 590 $M^{-1}\,s^{-1}$, and k_{-1} for the reverse reaction is 0.21 s^{-1}. With oxalate the rate law is equation (2), where K is the acid dissociation constant for $H_2C_2O_4$ to $HC_2O_4^-$, which is believed to be the reactant. In this case, at 25 °C, k_2 for formation is 43 $M^{-1}\,s^{-1}$ and k_{-2} is $4.7 \times 10^{-3}\,s^{-1}$.

$$k_{eq} = k_1[Mo_2^{4+}] + k_{-1} \tag{1}$$

$$k_{eq} = \frac{k_2K[Mo_2^{4+}]}{(K + [H^+])} + k_{-2} \tag{2}$$

The mechanism of substitution is believed to be similar to that of oxo ions such as VO^{2+} and TiO^{2+}.[31,32] Thus rapid (possibly approaching diffusion controlled) equilibration of NCS^- in the labile axial position (K) is followed by a subsequent slower step k leading to occupancy of an in-plane position on the complex (**1**). The observed rate constant is therefore the product of the two terms *kK*. Clearly a similar mechanism is possible for Mo_2^{II} (**2**). Hydrated $(NH_4)_4[Mo_2(NCS)_8]$ has N-bonded (monodentate) thiocyanate, and formation of the N-bonded product is likely in the kinetic study. In the case of $HC_2O_4^-$, bridging to give the μ-carboxylato product (**3**) can be presumed to occur in a rapid subsequent step.

(**1**)

(**2**)

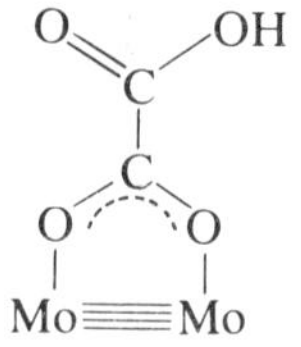

(**3**)

Values of rate constants and activation parameters for the VO^{2+}, TiO^{2+} and Mo_2^{4+} reactions with NCS^- are quite similar. The same mechanism has been proposed for the exchange of $CF_3CO_2^-$ with $[Mo_2(O_2CCF_3)_4]$ in acetonitrile from studies using ^{19}N NMR spectroscopy.[33]

Oxidation of aqua Mo_2^{II} in 1 M CF_3SO_3H to green aqua Mo_2^{III} with evolution of hydrogen is observed on irradiating at 254 nm.[34] More generally oxidation of Mo^{II} to Mo^{III} is difficult to bring about in a controlled manner. Studies on the oxidation of Mo_2^{II} in 1 M HCl have reported a new Mo^{III} dimer (λ_{max} 430 nm) possibly having a triple metal–metal bond and no bridging ligands.[35] The same product is not observed however on addition of Cl^- to the well-characterized green di-μ-hydroxo Mo^{III} aqua dimer.[36] The possible separate existence of dimeric Mo^{III} ions with and without bridging is of interest.

36.1.2.3 Oxidation State III

Hexachloromolybdate(III), prepared by electrolytic reduction of molybdate(VI) in 11 M HCl followed by addition of KCl,[38] provides a relatively easy route into Mo^{III} chemistry. The pink salt $K_3[MoCl_6]$ is stable indefinitely in dry air. In solution it aquates and is O_2 sensitive. The preparation of $[MoCl_5(H_2O)]^{2-}$ has also been described, and can be isolated as the ammonium salt, $(NH_4)_2[MoCl_5(H_2O)]$. Air-free aquation (two days) of $[MoCl_6]^{3-}$ followed by Dowex cation-exchange resin gives the pale yellow monomeric aqua ion.[17,39] In the earlier work, traces of O_2 resulted in contamination with the yellow Mo^V dimer, $Mo_2O_4^{2+}$, which because of its strong absorbance obscures the Mo^{III} spectrum.[40] The most recent spectrum in 2 M CF_3SO_3H gives peak positions λ/nm (ε/$M^{-1}\,cm^{-1}$) at 320 (19) and 386 (13.3) assigned as $^4A_{2g} \rightarrow {}^4T_{2g}$ and $^4A_{2g} \rightarrow a^4T_{1g}$ *d–d* ligand-field transitions.[41] The spectrum in CF_3SO_3H, which has only small background absorbance in the 250–300 nm range, contrasts with that of other inert electrolytes such as HPTS and HBF_4 which gives a peak at 310 nm (ε 23.2 $M^{-1}\,cm^{-1}$), Figure 4. A comparison of the spectrum with that of $[MoCl_6]^{3-}$ is consistent with the hexaaqua structure, Figure 4. This point has now been confirmed crystallographically following the isolation of the bright-yellow cesium molybdenum alum $Cs[Mo(H_2O)_6](SO_4)_2 \cdot 6H_2O$. The structure, belonging to the β class of alums,[42] gives an Mo—O distance of 2.09 Å. When exposed to air the crystals become brown within a few hours.

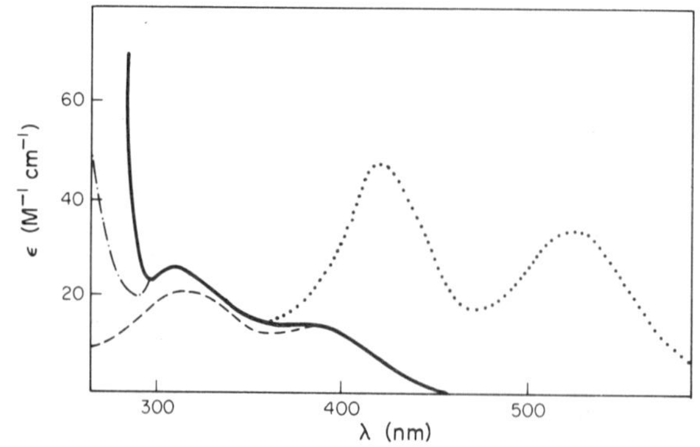

Figure 4 UV–visible spectra of $[Mo(H_2O)_6]^{3+}$ in 1.0 M CF_3SO_3H (– – –), *p*-$C_6H_5MeSO_3H$ (——) and HBF_4 (–·–·) compared with $[MoCl_6]^{3-}$ in 12 M HCl (· · · ·)[40,41]

An important advance in the above work was the isolation of a new compound, sodium hexakis(formato)molybdate(III), $Na_3[Mo(HCO_2)_6]$,[41] which can be aquated in strong acid to give concentrated solutions (~0.5 M) of $[Mo(H_2O)_6]^{3+}$ within a few minutes. To prepare the formato complex $(NH_4)_2[MoCl_5(H_2O)]$ (10 g) is dissolved in saturated sodium formate (~9 M) in 0.1 M formic acid. After one day the light green formato complex is filtered off. It gives in ~10 M saturated sodium formate a UV-vis spectrum with λ_{max}/nm (ε/M^{-1} cm^{-1}) at 385 (115) and an absorption minimum at 357 (97). The second ligand-field transition is obscured by this intense absorption.

From the energies of the two absorption bands for $[Mo(H_2O)_6]^{3+}$ in 2 M CF_3SO_3H, Figure 4, the values of the spectrochemical parameter separating the t_{2g} and e_g levels is 25 900 cm^{-1}, and the interelectronic repulsion parameter B is 476 cm^{-1}.[41] The shift of bands into the UV as compared to those for $[Cr(H_2O)_6]^{3+}$ at 408 and 587 nm is noted. An additional very weak band ($\varepsilon < 0.5$ M^{-1} cm^{-1}) has been observed for $[Mo(H_2O)_6]^{3+}$ between 560 and 600 nm, which may be the spin-forbidden $^4A_{2g} \rightarrow {}^2T_{2g}$ transition. At present $[Mo(H_2O)_6]^{3+}$ is the only monomeric and paramagnetic (3.69 BM) aqua ion of Mo. The acid dissociation constant of $[Mo(H_2O)_6]^{3+}$ has not yet been determined.

Conversion of $[MoCl_6]^{3-}$ (10 g) into $[Mo(NCS)_6]^{3-}$ can be achieved by dissolving in 50 ml of KNCS (7 M), and heating on a steam bath at 60 °C for 2 h.[43] A red to orange colour change is observed, and $K_3[Mo(NCS)_6]\cdot 4H_2O$ can be isolated. An X-ray crystal structure of the adduct $K_3[Mo(NCS)_6]\cdot MeCO_2\cdot H_2O$ has provided conclusive evidence for octahedral coordination, as well as for N-bonded thiocyanate.[44] Studies on the rate of exchange of NCS^- on $[Mo(NCS)_6]^{3-}$ have indicated an inert Mo^{III} centre. The potassium and pyridinium salts of the Mo^{IV} complex formulated as $[Mo(NCS)_6]^{2-}$ (and purple/red-brown in colour) have been obtained.[45] On further oxidation to the Mo^V state there are three possible products: $[MoO(NCS)_5]^{2-}$, $[Mo_2O_3(NCS)_8]^{4-}$ (μ-oxo) and $[Mo_2O_4(NCS)_6]^{4-}$ (di-μ-oxo). Oxidation of the Mo^{III} complex through Mo^{IV} to the Mo^V state therefore requires inclusion of at least one oxo ligand per Mo. In the oxidation of $[Mo(NCS)_6]^{3-}$ by $[IrCl_6]^{2-}$ formation of the Mo^{IV} ion is rapid.[46] The subsequent first-order stage monitored by stopped flow gives an inverse $[H^+]$ dependence which requires the presence of coordinated H_2O. It is suggested that the rate-determining step corresponds to loss of NCS^- from the hydroxo form of this complex, followed by (rapid) oxidation or disproportionation through to the Mo^V product. In studies relating to the one- and two-electron photooxidation of $[Mo(NCS)_6]^{3-}$ in which $[Mo(NCS)_6]^{2-}$ is first generated,[47] it has been demonstrated that the $[Mo(NCS)_6]^{2-}$ undergoes deproportionation in a mixed $MeCN/H_2O/CCl_4$ solvent (62:31:7 w/w) to give, in 0.2 M KNCS, $[Mo_2O_3(NCS)_6]^{4-}$ as the final Mo^V product.

The Mo^{III} cyano complex $K_4[Mo(CN)_7]\cdot 2H_2O$ has been isolated.[48] At room temperature it has a magnetic moment of 1.75 BM. The presence of only one unpaired electron has been attributed to d-orbital splitting in the low symmetry $[Mo(CN)_7]^{4-}$ or $[Mo(CN)_7(H_2O)]^{4-}$ ion. The complex is readily oxidized to $[Mo(CN)_8]^{4-}$.

The kinetics of the 1:1 substitution reactions of $[Mo(H_2O)_6]^{3+}$ with Cl^- and NCS^- have been investigated in HPTS, $I = 1.0$ M (LiPTS), and found to be independent of $[H^+]$ in the range 0.17–1.0 M.[49] At 25 °C formation rate constants are 4.6×10^{-3} M^{-1} s^{-1} for Cl^- ($\Delta H^{\neq} =$ 98.3 kJ mol^{-1}; $\Delta S^{\neq} = 40.2$ J K^{-1} mol^{-1}), and 0.28 M^{-1} s^{-1} for NCS^- ($\Delta H^{\neq} = 68.2$ J mol^{-1}; $\Delta S^{\neq} = -26.8$ J K^{-1} mol^{-1}) (1 cal = 4.2 J). The ratio of the two rate constants of 59 suggests an associative interchange (I_a) mechanism. This is also supported by the volume of activation ($\Delta V^{\neq}$) of -11.4 cm^3 mol^{-1} at 11.9 °C for the reaction of NCS^- with $[Mo(H_2O)_6]^{3+}$.[50] On replacing PTS^- by $TFMS^-$ the rate constant for the NCS^- reaction increases to 0.36 M^{-1} s^{-1}, and a similar small effect is observed for the Cl^- reaction. It is interesting that $[Mo(H_2O)_6]^{3+}$ is more labile than $[Cr(H_2O)_6]^{3+}$ by ~10^5 for both NCS^- and Cl^-, a trend which is in contrast to that observed with Co^{3+}, Rh^{3+} and Ir^{3+}. This is explained by the increased associative character of the reaction of $[Mo(H_2O)_6]^{3+}$, stemming from the low (d^3) electron population, and the size of 4d as opposed to 3d orbitals. While there is evidence for associative character in the reactions of Cr^{III} complexes, $\Delta V^{\neq} = -9.6$ cm^3 mol^{-1} for H_2O exchange on $[Cr(H_2O)_6]^{3+}$, this is less than observed for the d^1 ion $[Ti(H_2O)_6]^{3+}$ (-12.1 cm^3 mol^{-1}), and in some cases, *e.g.* with $[Cr(NH_3)_5H_2O]^{3+}$, Cr^{III} is best regarded as displaying borderline I_a/I_d properties.[5] Pathways involving the conjugate-base form, here $[Mo(H_2O)_5OH]^{2+}$, which often represent a more facile route for substitution involving an I_d mechanism, make little or no contribution alongside the dominant associative pathway involving $[Mo(H_2O)_6]^{3+}$. Alternatively, since Mo^{III} can adopt a higher coordination as in $[Mo(CN)_7]^{4-}$, an associative A-type mechanism is possible in the substitution of $[Mo(H_2O)_6]^{3+}$.

Other rate constants for 1:1 substitution into $[Mo(H_2O)_6]^{3+}$ include the value for $HC_2O_4^-$,[52] Table 2. Also the observation that the $[Co(C_2O_4)_3]^{3-}$ oxidation of $[Mo(H_2O)_6]^{3+}$ is independent of $[H^+]$, whereas that of $[IrCl_6]^{2-}$ displays an $[H^+]^{-1}$ dependence, suggests a substitution-controlled process.[53] The similarity of the rate constant to that for $HC_2O_4^-$ is noted. From kinetic studies on the O_2 oxidation of $[Mo(H_2O)_6]^{3+}$ two further rate constants, which may well be substitution-controlled processes, are obtained (see below). These are surprisingly large however, as are similar rate constants for the corresponding reactions of $[V(H_2O)_6]^{2+}$ (also d^3) with O_2 (2000 $M^{-1} s^{-1}$), and with $[V(H_2O)_5O_2]^{2+}$ (3700 $M^{-1} s^{-1}$). The latter are to be compared with values for NCS^- substitution (24 $M^{-1} s^{-1}$), and H_2O exchange (87 s^{-1}) on $[V(H_2O)_6]^{2+}$.[54]

Table 2 Rate Constants for 1:1 Substitution Reactions of $[Mo_2(H_2O)_6]^{3+}$ in HPTS at 25 °C, I = 1.0 M (LiPTS) Except as Indicated

Reactant	k ($M^{-1} s^{-1}$)	*Ref.*
Cl^-	0.0046	49
NCS^-	0.27	49
$HC_2O_4^-$	0.49	52
$[Co(C_2O_4)_3]^{3-}$	0.34[a]	53
O_2	180	57
MoO_2	42	57

[a] I = 2.0 M (LiPTS).

Kinetic studies have been reported with three strongly coloured one-equivalent oxidants $[IrCl_6]^{2-}$, $[Co(C_2O_4)_3]^{3-}$ and $[VO(H_2O)_5]^{2+}$,[53] which oxidize $[Mo(H_2O)_6]^{3+}$, *e.g.* $2Mo^{III} + 4Ir^{IV} \rightarrow Mo_2^V + 4Ir^{III}$, to give the Mo_2^V product $[Mo_2O_4(H_2O)_6]^{2+}$. Reactions were studied with $[Mo(H_2O)_6]^{3+}$ present in large excess. Rate laws are consistent with a rate-determining first stage as in equation (3), in which Mo^{IV} is formed. No monomeric aqua Mo^{IV} species has yet been identified, and no build-up of Mo^{IV} is apparent in this study. With $[IrCl_6]^{2-}$ (0.89 V; all reduction potentials are quoted against NHE) first-order rate constants from the stopped-flow decay of $[IrCl_6]^{2-}$ at 489 nm gave a dependence (4), where at 25 °C, I = 0.2 M (LiPTS), $k_a = 3.4 \times 10^4 M^{-1} s^{-1}$ and $k_b = 2.9 \times 10^4 s^{-1}$. On replacing p-toluenesulfonate (PTS^-) by trifluoromethanesulfonate ($TFMS^-$) k_{obs} shows little variation (~12% greater). The oxidant decay was also monitored in the $[Co(C_2O_4)_3]^{3-}$ (0.58 V) and $[VO(H_2O)_5]^{2+}$ (0.36 V) studies ($t_{1/2} > 1$ min). Here more complicated dependences are observed in the presence of PTS^-. At 25 °C, I = 2.0 M (LiPTS), the method of initial rates (R) gives dependence (5) for the $[Co(C_2O_4)_3]^{3-}$ reaction, with $k_c = 0.34 M^{-1} s^{-1}$. Similarly for the $[VO(H_2O)_5]^{2+}$ reaction (6) was obtained, with $k_f = 3.3 \times 10^{-3} M^{-1} s^{-1}$. On replacing PTS^- by $TFMS^-$ the oxidant dependent terms k_d ($1.87 \times 10^{-3} M^{-1} s^{-1}$) and k_e ($4.9 \times 10^{-5} s^{-1}$), which are common to both reactions, make no contribution. It seems likely therefore that k_d and k_e are spurious terms stemming from rate-determining processes involving PTS^-, which are not observed for the much faster $[IrCl_6]^{2-}$ reactions (or with $TFMS^-$ present). The rate constants for the $[IrCl_6]^{2-}$ reaction are much faster than those obtained for substitution into the $[Mo(H_2O)_6]^{3+}$ coordination sphere, and an outer-sphere mechanism is implied. With $[Co(C_2O_4)_3]^{3-}$ on the other hand, as already indicated, the oxidation is independent of $[H^+]$ and consistent with an inner-sphere substitution-controlled process, Table 3.

$$Mo^{III} + Ir^{IV} \longrightarrow Mo^{IV} + Ir^{III} \quad (3)$$

$$k_{obs}/2[Mo^{III}] = k_a + k_b[H^+]^{-1} \quad (4)$$

$$R = 2k_c[Mo^{III}][Co^{III}] + 4k_d[Mo^{III}]^2 + 2k_e[Mo^{III}] \quad (5)$$

$$R = 2k_f[Mo^{III}][V^{IV}] + 4k_d[Mo^{III}]^2 + 2k_e[Mo^{III}] \quad (6)$$

In a further study with $[Fe(H_2O)_6]^{3+}$ as oxidant, the concentration of $[Mo(H_2O)_6]^{3+}$ (initially 1.2×10^{-3} M) was monitored at 450 nm (millisec timescale) with the Fe^{3+} reactant (0.045 M) in large excess.[55] The rate constant (25 °C) of $1.3 \times 10^3 M^{-1} s^{-1}$, in I = 1.0 M (HPTS), is too fast for substitution into either coordination sphere to occur. The formation of Mo_2^V and its subsequent decay to Mo^{VI} can also be monitored (half-lives of order of seconds), Figure 5. What is particularly interesting is that the rate of formation of Mo_2^V does not match the decay

Table 3 Rate Constants (25 °C) for the First Stage in the Oxidation of $[Mo(H_2O)_6]^{3+}$

	k ($M^{-1}\,s^{-1}$)	*Conditions*	*Ref.*
$[IrCl_6]^{2-}$	3.4×10^4 [a]	$I = 0.2$ M (LiPTS)	53
$[Fe(H_2O)_6]^{3+}$	1.3×10^3	$I = 1.0$ M (HPTS)	55
O_2	180[b]	$I = 2.0$ M (HPTS)	57
$[Co(C_2O_4)_3]^{3-}$	0.34[c]	$I = 2.0$ M (LiPTS)	53
$[VO(H_2O)]^{2+}$	3.3×10^{-3}	$I = 2.0$ M (HPTS)	53

[a] Corresponds to k_a in a rate law dependence $k_a + k_b[H^+]^{-1}$.
[b] May be a substitution-controlled process.
[c] Assigned as a substitution-controlled process (no $[H^+]^{-1}$ dependence).

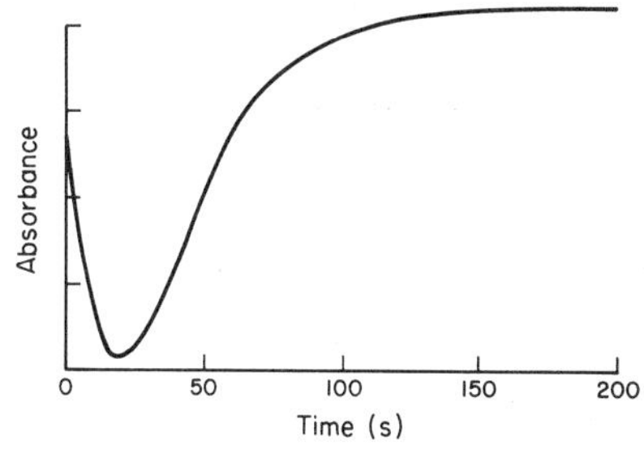

Figure 5 Oxidation of $[Mo(H_2O)_6]^{3+}$ (1.2×10^{-3} M) with $[Fe(H_2O)_6]^{3+}$ (0.045 M) monitored at λ 450 nm, 25 °C, $I = 1.0$ M (HPTS)[55]

of Mo^{III}, thus providing clearcut evidence for formation of an intermediate, Mo^{IV} and/or monomeric Mo^V. Further studies on this system would be of interest.

Perchlorate also oxidizes $[Mo(H_2O)_6]^{3+}$, but the features displayed are quite unusual and different to all previous reactions involving ClO_4^- and metal ions.[56] The formation of Mo_2^V with time is indicated in Figure 6, the reaction being characterized by induction, propagation and termination stages. Remarkably in such studies at 25 °C, $I = 1.0$ M (LiPTS), solutions of Mo^{3+}, $(2.2–8.6)\times10^{-3}$ M, give little or no detectable reaction with ClO_4^- (0.2–0.7 M) over a period of at least 0.5 h and as long as 3 h, after which complete reaction can be observed in as little as 1 h. Overall the reactions are faster at higher $[ClO_4^-]$, higher $[Mo(H_2O)_6^{3+}]$ and higher $[H^+]$. Details of the precise mechanism remain uncertain. Since perchlorate reactions are believed to involve O-atom transfer by either (7) or (8), substitution of ClO_4^- into the $[Mo(H_2O)_6]^{3+}$ coordination sphere is required. In an I_a substitution process a poor donor such as ClO_4^- is not expected to react very rapidly, and this no doubt contributes to the slowness of the reaction. The higher $[H^+]$ is known to assist O-atom transfer. The instability and transient nature of monomeric Mo^{IV} and/or Mo^V with a correspondingly high activation barrier for formation is also expected to contribute to the slowness of the reaction. Once under way the $ClO_4^- \rightarrow Cl^-$ eight-equivalent change, and presumably the ability of monomeric Mo^{IV} or Mo^V to reduce ClO_4^-, will account for the rapid propagation of the reaction. The reaction of the green aqua dimer of Mo_2^{III} with ClO_4^- exhibits similar behaviour.

$$ClO_4^- - e^- \longrightarrow ClO_3 + O^{2-} \quad (7)$$

$$ClO_4^- - 2e^- \longrightarrow ClO_3^- + O^{2-} \quad (8)$$

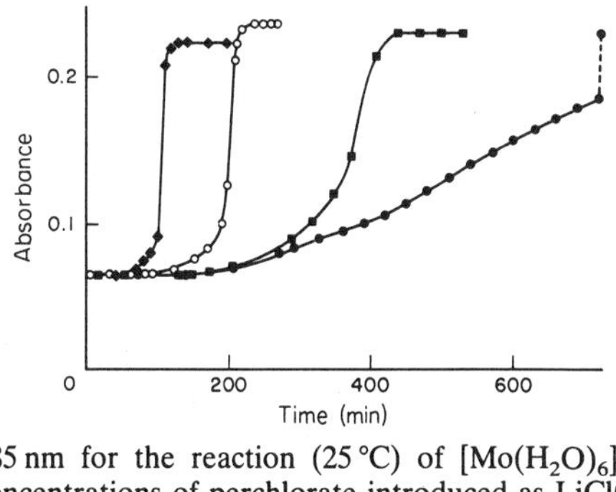

Figure 6 Absorbance changes at 385 nm for the reaction (25 °C) of $[Mo(H_2O)_6]^{3+}$ (4.3×10^{-3} M) with ClO_4^- at $[H^+] = 0.30$ M, $I = 1.0$ M (LiPTS). Concentrations of perchlorate introduced as $LiClO_4$ were 0.20 (●), 0.30 (■), 0.50 (○) and 0.70 (◆)[56]

The reaction of $[Mo(H_2O)_6]^{3+}$ with O_2 in *p*-toluenesulfonic acid solutions, $I = 2.0$ M (LiPTS), gives $[Mo_2O_4(H_2O)_6]^{2+}$ as product: $2Mo^{3+} + O_2 + 2H_2O \rightarrow Mo_2O_4^{2+} + 4H^+$.[57] Three stages are observed (Figure 7). Use of isotopically labelled O_2 followed by IR spectroscopy on the isolated $Mo_2O_4^{2+}$ product has demonstrated that ^{18}O is taken up in the μ-oxo positions. With Mo^{3+} in large excess an intense yellow intermediate MoO_2Mo^{6+} (or a related form) with a peak at 395 nm ($\varepsilon = 2300\ M^{-1}\ cm^{-1}$) is observed. At 25 °C stopped-flow studies on the decay of this intermediate give rate constants $k_{3(obs)}$ which can be expressed as in equation (9). Various constants were obtained as defined in equations (10)–(12), where values indicated are for 2.0 M HPTS.

$$k_{3(obs)} = \frac{K_1 K_2 k_3 [Mo^{3+}]^2}{1 + K_1[Mo^{3+}] + K_1 K_2 [Mo^{3+}]^2} \quad (9)$$

$$Mo^{3+} + O_2 \rightleftharpoons MoO_2^{3+} \quad (K_1\ 360\ M^{-1}) \quad (10)$$

$$MoO_2^{3+} + Mo^{3+} \rightleftharpoons MoO_2Mo^{6+} \quad (K_2\ 535\ M^{-1}) \quad (11)$$

$$MoO_2Mo^{6+} + 2H_2O \longrightarrow Mo_2O_4^{2+} + 4H^+ \quad (k_3\ 0.23\ s^{-1}) \quad (12)$$

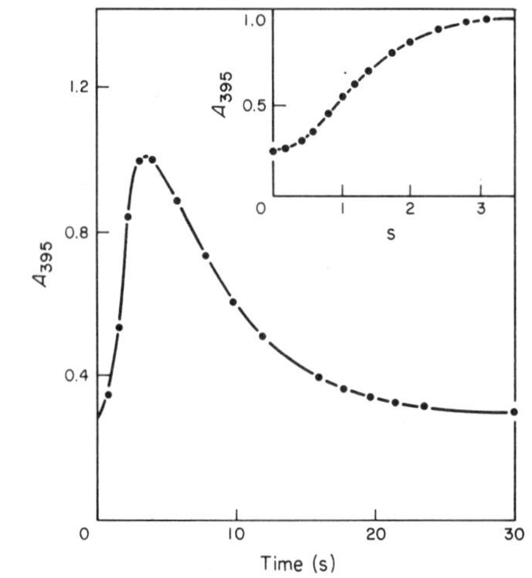

Figure 7 Identification of three stages in the reaction of $[Mo(H_2O)_6]^{3+}$ (10.6×10^{-3} M) with O_2 (5.3×10^{-4} M) monitored at 395 nm (1.67 cm pathlength), at 25 °C in 2.0 M HPTS[57]

Estimates of rate constants for reactions (10) and (11) were obtained from studies on the formation of MoO_2Mo^{6+}. All parameters were further refined by a fitting procedure using the program KINSIM for multistep reactions. The values of k_1 ($180\ M^{-1}\ s^{-1}$) and k_2 ($42\ M^{-1}\ s^{-1}$) are in excess of previous substitution rate constants for $[Mo(H_2O)_6]^{3+}$ as already mentioned. An alternative mechanism is that Mo^{IV} and O_2^- are generated in an outer-sphere pathway and that substitution then occurs. A similar process might also be relevant for $[V(H_2O)_6]^{2+}$ with O_2, where the substitution properties of $[V(H_2O)_6]^{3+}$ are already established and consistent with the results obtained. With O_2 in excess of Mo^{3+} (0.72×10^{-4} M) there was no evidence for MoO_2Mo^{6+}, and an alternative route for conversion of MoO_2^{3+} to $Mo_2O_4^{2+}$, which is first order in MoO_2^{3+} ($k_4 = 0.037\ s^{-1}$), was identified. In the refinement this step was included alongside equations (10)–(12). The reactions of MoO_2Mo^{6+} (k_3) and MoO_2^{3+} (k_4) exhibit $[H^+]^{-1}$ dependences consistent with an increase in hydrolysis of the metal on oxidation to a higher state.

A green Mo^{III} dimeric product is obtained on reduction of Mo^{VI} as $[MoO_4]^{2-}$, or Mo^V as $[Mo_2O_4(H_2O)_6]^{2+}$ in acidic solutions (not $HClO_4$), generally at $[H^+]$ in the range 0.5–2.0 M.[58] The reduction can be achieved using a Zn/Hg Jones reductor column, or by electrolytic reduction at a Hg-pool electrode at −0.5 V (*vs.* NHE). The product can be purified on a Dowex cation-exchange column. The charge per Mo has been determined as 2+ and the elution behaviour is consistent with a total charge of 4+.[58] By comparison of the UV–visible spectrum (Figure 8) with that of the Mo_2^{III}–edta complex,[36] the structure of which is known,[59] a di-μ-hydroxo structure is likely. This has been confirmed by EXAFS studies which indicate an Mo—Mo distance of 2.54 Å.[60] Two different Mo—O components in the fit correspond to bridging Mo—OH (2.06 Å), and terminal Mo—OH_2 (2.2 Å), bond distances.

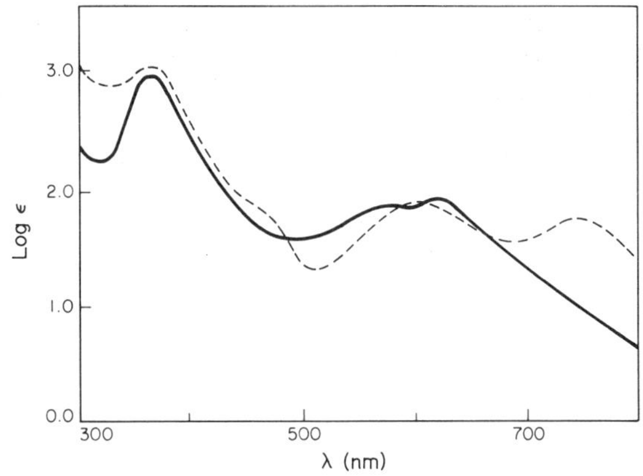

Figure 8 UV–visible spectra (ε values per dimer) for the Mo^{III} aqua dimer (——), $[H^+] = 1.0$ M (HPTS), and the Mo_2^{III}–edta complex $[Mo_2(OH)_2(O_2CMe)(edta)]^-$ (– – –) at pH 6[36]

From studies in aqueous HPTS there is some uncertainty concerning the applicability of Beer's law at $[Mo_2^{III}] < 4 \times 10^{-4}$ M, and comparative studies in $TFMS^-$ are called for. Kinetic studies on substitution reactions of Mo_2^{III} would also be of interest. Oxidation of Mo_2^{III} to Mo_2^{V} is observed with $[Co(C_2O_4)_3]^{3-}$ (stopped-flow range) and $[IrCl_6]^{2-}$ (fast). Again no build-up of other Mo states is evident even when there is a deficiency of oxidant. The faster rates for Mo_2^{III} as compared to the $[Mo(H_2O)_6]^{3+}$ implicate a more facile oxidation of the dimer, which is reasonable in view of the similar dimeric structures of the Mo_2^{III} and Mo_2^{V} ions. On oxidation of monomer $2Mo^{III} \rightarrow Mo_2^{V}$ the interplay of electron uptake ($4e^-$) and proton release ($8H^+$) is noted. In the corresponding $Mo_2^{III} \rightarrow Mo_2^{V}$ change $4e^-$ and $6H^+$ are involved. In the first stage a dimeric $Mo^{III,IV}$ transient is presumably formed. From preliminary studies at least two stages are observed in the reaction of Mo_2^{III} with O_2.

The preparation of the binuclear Mo_2^{III} complex $K[Mo_2(OH)_2(O_2CMe)(edta)]$ (edta = ethylenediaminetetraacetate) has been reported.[61] The procedure involves Zn—Hg reduction of the Mo_2^{V} complex $[Mo_2O_4(edta)]^{2-}$ in acetate buffer at pH 5. Incorporation of the bridging acetate was not anticipated. In the pH 3–4 range, with and without acetate, a water insoluble product $[Mo_2(OH)_2(H_2O)_2(edta)]$ with no charge is obtained.[61] The structure of the μ-acetato complex has been determined, Figure 9.[59] There are four bridging ligands in this complex with the edta coordinated to the binuclear unit in a 'basket' like conformation. The short Mo—Mo distance of 2.43 Å is possibly due to the presence of the μ-acetato ligand as well as the bridging edta. A bis(thiocyanato) complex $[Mo_2(OH)_2(NCS)_2(edta)]^{3-}$ has been isolated following treatment of the μ-acetato complex with thiocyanate. All three complexes are diamagnetic consistent with a triple metal–metal bond. A derivative with formate in place of acetate has been prepared. The μ-acetato complex is stable in air for only limited periods (a few hours), and in water it reacts rapidly with O_2 to give the Mo_2^{V} complex $[Mo_2O_4(edta)]^{2-}$.

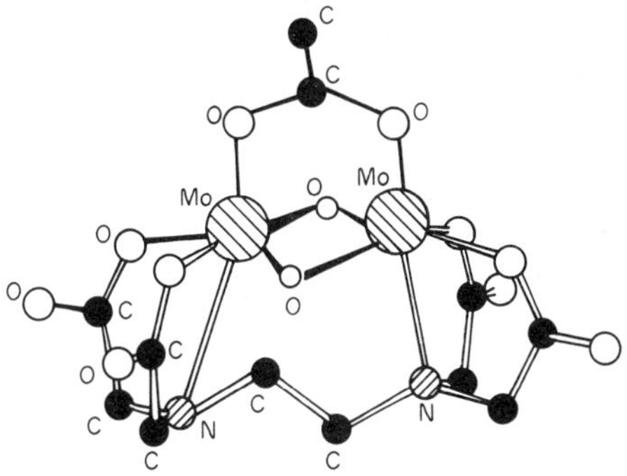

Figure 9 The structure of the Mo_2^{III} complex $[Mo_2(OH)_2(O_2CMe)(edta)]^-$ [59]

With the mild two-equivalent oxidizing agent azide as N_3^- (and with hydrazine) at pH > 6, the μ-acetato Mo_2^{III}–edta complex gives a maroon-red $Mo^{III,IV}$–edta complex.[62] The crystal structure has indicated a tetramer, $K_4[Mo_4O_4(OH)_2(edta)_2]\cdot 16H_2O$ (**4**), in which the four Mo atoms are in the same plane.[63] The number of oxidizing equivalents per Mo has been determined by titration, and indicates an average oxidation state of 3.5. The complex is diamagnetic and all bridging groups are symmetrically disposed (equal bond lengths) to adjacent Mo atoms. The short Mo—Mo distance of 2.41 Å bridged by the μ-oxo-μ-hydroxo ligands is consistent with metal–metal bonding. These two bridges can be distinguished from each other, the μ-oxo Mo—O bonds (1.93 Å) being shorter that the μ-hydroxo Mo—O bonds (2.08 Å). The single μ-oxo bridges have Mo—O bond lengths of 1.90 Å and an MoOMo angle of 164°. It is possible to comment on the mechanism of formation. The products of the reaction of azide are both the $Mo^{III,III,IV,IV}$ tetramer and Mo_2^V dimer. Quantitative formation of N_2 and NH_3 is observed, consistent with equations (13) and (14). A maximum yield (37%) of $Mo^{III,III,IV,IV}$ is observed for a 1:1 ratio of reactants. Competition between Mo_2^{III} and N_3^- for an intermediate is implied. Since azide is capable of a two-equivalent change we tentatively suggest that this intermediate is an Mo_2^{IV} dimer. Reaction Scheme 2 is proposed. No reaction is observed at pH 4.7 (the pK_a for HN_3 is 4.68) in the presence of free acetate, but reaction does occur slowly when no acetate is present. Displacement of acetate would therefore seem to favour reaction. As the pH is increased to 8 (even in the presence of acetate) the rate increases markedly. This corresponds to acid dissociation of one of the μ-hydroxo ligands determined spectrophotometrically as pK_a 7.73 at 0 °C (equation 15). Furthermore no reaction of the bis(thiocyanato) Mo_2^{III} complex is observed with N_3^- at pH 8. It is concluded that coordination of N_3^- precedes reaction.

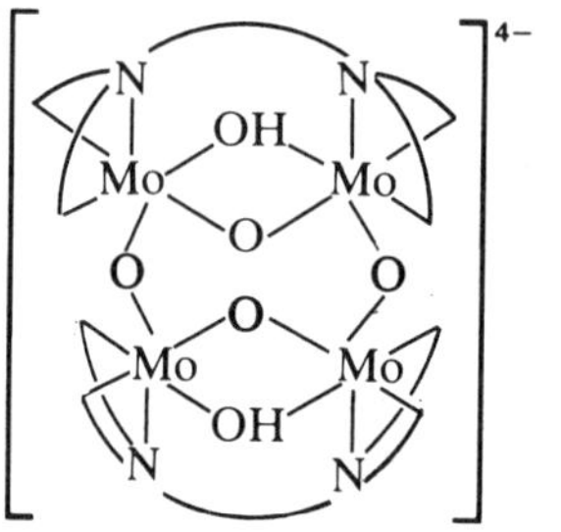

(**4**)

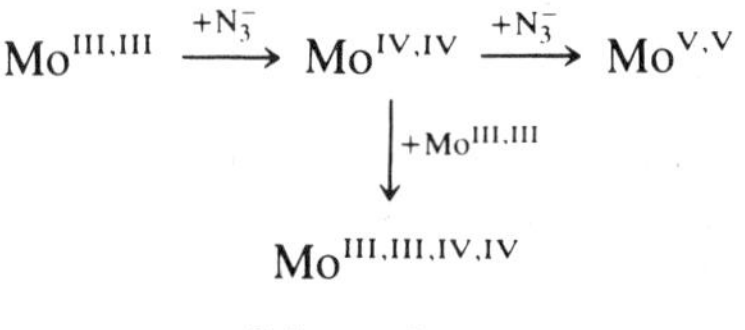

Scheme 2

$$2Mo_2^{III} + N_3^- \xrightarrow{H^+} Mo^{III,III,IV,IV} + N_2 + NH_3 \quad (13)$$

$$Mo_2^{III} + 2N_3^- \xrightarrow{H^+} Mo_2^V + 2N_2 + 2NH_3 \quad (14)$$

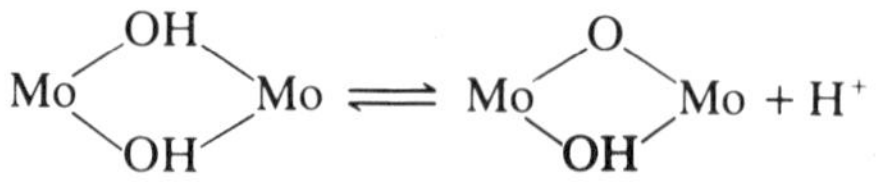

(15)

Controlled O_2 oxidation of the $Mo^{III,III,IV,IV}$ tetramer yields a product which titrates for Mo^{IV} and from an incomplete X-ray crystal study appears to be basically the same tetrameric structure, except that all the bridging ligands are now μ-oxo. This Mo^{IV} tetramer is also an intense red colour, Figure 10. The reduction potential for the couple involving the two tetramers is +0.07 V. Rapid reactions of both tetramers with O_2 are of interest in that there is no site for binding unless the coordination number increases. Mild oxidants such as $[Co(NH_3)_5Cl]^{2+}$ react rapidly ($t_{1/2} < 1$ min) with both tetramers to give the Mo_2^V dimer.

As already mentioned under Mo^{II} the reaction of $K_4[Mo_2Cl_8]$ with 2 M H_3PO_4 in air for 24 h yields a dimeric Mo^{III} product. The structures of $Cs_2[Mo_2(HPO_4)_4(H_2O)_2]$, which has axial H_2O molecules, and $(pyH)_3[Mo_2(HPO_4)_4Cl]$, in which there are infinite chains with shared axial

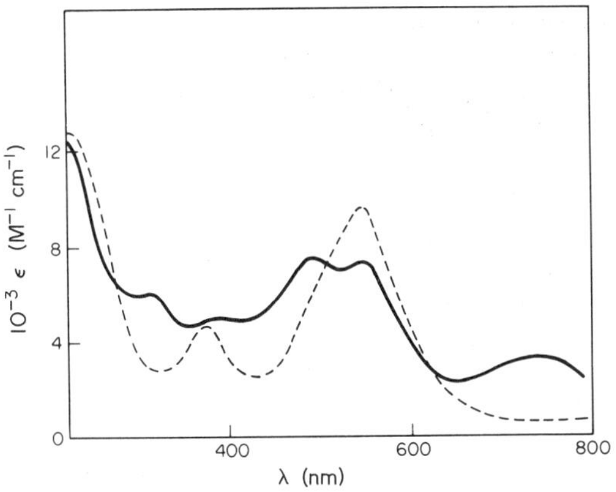

Figure 10 UV-visible spectra (ε values per tetramer) for the $Mo^{III,III,IV,IV}$ complex $[Mo_4O_4(OH)_2(edta)_2]^{4-}$ (——), and the related Mo_4^{IV} complex, $[Mo_4O_6(edta)_2]^{4-}$ (- - -) in H_2O[62]

chloride ligands, have been reported (Figure 11).[27] The Mo—Mo distance of 2.23 Å is consistent with a triple bond ($\sigma^2\pi^4$). Interestingly the positions of the protons can be identified as attached to outer O atoms of the μ-HPO_4 ligands, outer P—O distances being 1.48 Å as compared to 1.54 Å for P—OH.

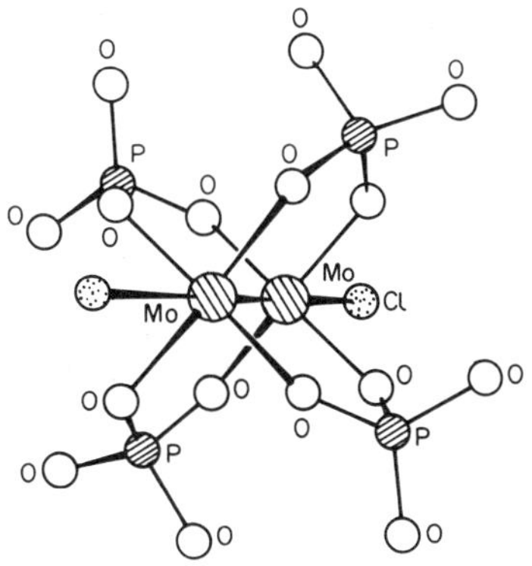

Figure 11 The structure of the Mo_2^{III} complex $[Mo_2(HPO_4)_4(Cl)]^{3-}$, where the chlorides are bridging[27]

Binuclear complexes of Mo^{III} have been prepared from paramagnetic mononuclear halogeno complexes of the type $LMoX_3$, where L = 1,4,7-triazacyclononane, sometimes referred to as 9-aneN_3.[64] Hydrolysis of $LMoCl_3$ in aqueous solution containing acetate or HCO_3^- yields $[LMo(OH)_2(O_2CMe)MoL]I_3 \cdot H_2O$ and $[LMo(OH)_2(CO_3)MoL]I_2 \cdot H_2O$. The acetato-bridged complex has been shown to have an Mo—Mo distance of 2.41 Å, which compares with 2.43 Å for $[Mo_2(OH)_2(O_2CMe)(edta)]$.[59] On treating the μ-carbonato complex with acids, products as indicated in Scheme 3 are obtained. A feature of the structure of the dichloro complex (Figure 12) is the *anti* (sometimes referred to as *trans*) dichloro arrangement. The $Mo_2(OH)_2$ ring is planar and the Mo—Mo distance 2.50 Å. On oxidation, *e.g.* with ClO_4^-, the corresponding red-purple Mo_2^V complex with *anti* terminal oxo ligands (and planar Mo_2O_2 ring) has been observed for the first time, and the structure verified.[65] Cleavage of a bridge in triply bridged complexes to give *anti* rather than *syn* structures has been observed previously with $[(NH_3)_3Co(OH)_3Co(NH_3)_3]^{3+}$.[66] A comparison of different metal–metal bond distances is made in Figure 13, indicating the influence of M—M bonding.[64] All di-μ-hydroxo Mo_2^{III} complexes are diamagnetic.

Kinetic studies on the reaction of $[L(H_2O)Mo(OH)_2Mo(H_2O)L]^{4+}$ with Cl^- (0.11 $M^{-1}\,s^{-1}$ at 25 °C) and ClO_4^- (a redox process) ($2.0 \times 10^{-3}\,M^{-1}\,s^{-1}$ at 21 °C) give rate constants independent of $[H^+]$ in the range 0.05–1.0 M. It is proposed that the ClO_4^- reaction is substitution controlled. This reaction proceeds $Mo_2^{III} \rightarrow$ *anti*-Mo_2^V (red-purple) $\rightarrow$ *syn*-Mo_2^V (yellow). No intermediates were identified. The reaction with NO_3^- has also been studied (again strictly air-free), when quantitative formation of the red-purple *anti*-$[L_2Mo_2O_4]^{2+}$ complex and nitrite is observed (equation 16).[67]

$$Mo_2^{III} + 2NO_3^- \longrightarrow Mo_2^V + 2NO_2^- \qquad (16)$$

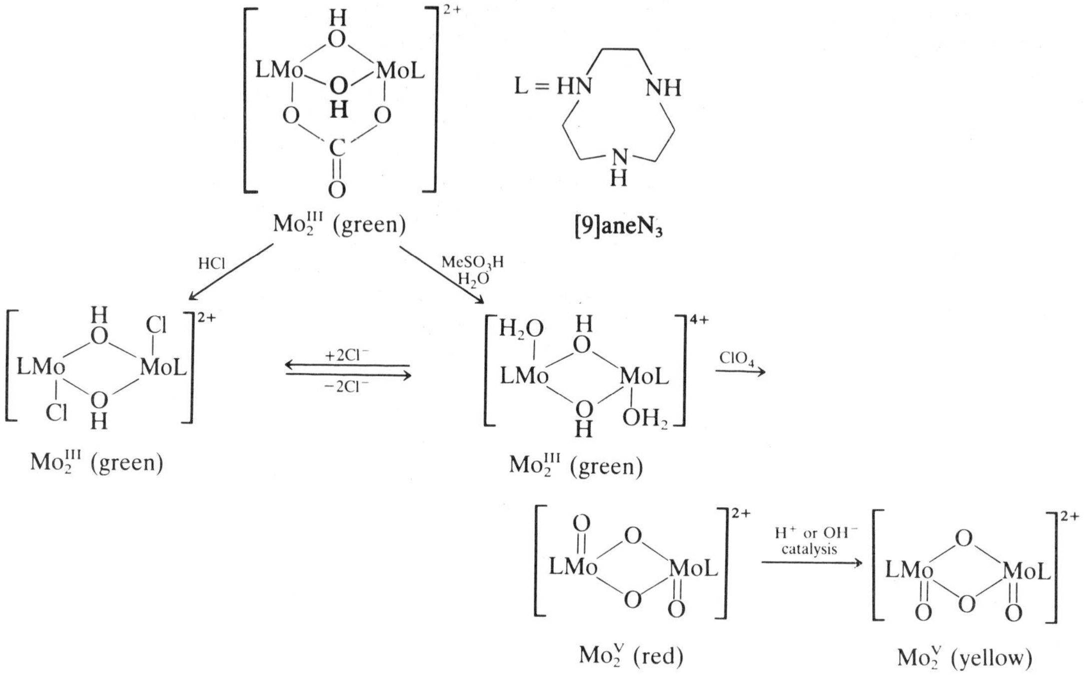

Scheme 3

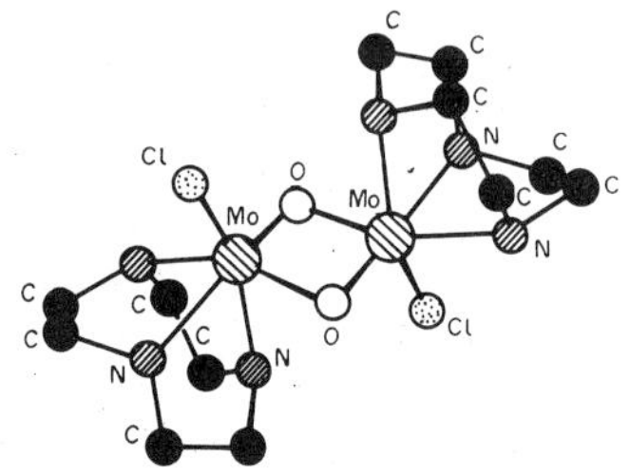

Figure 12 The structure of the Mo_2^{III} complex $[Mo_2L_2Cl_2(OH)_2]^{2+}$, where L is the [9]aneN_3 ligand 1,4,7-triazacyclononane[64a]

The formation of NO_2^- was demonstrated by carrying out the reaction in acid solution in the presence of sulfanilic acid and α-naphthylamine, neither of which reacts with the Mo complexes. The formation of the known red azo dye is diagnostic of NO_2^- formation. The stoichiometry was confirmed by determining the N_2 evolved (equation 17).

$$H_2NSO_3H + NO_2^- \xrightarrow{\text{fast}} HSO_4^- + N_2 + H_2O \qquad (17)$$

At 25°C the kinetics give a rate law (equation 18), with $k = 0.10\,M^{-1}\,s^{-1}$, and activation parameters $\Delta H^{\neq} = 69\,kJ\,mol^{-1}$ and $\Delta S^{\neq} = -33\,J\,mol^{-1}\,K^{-1}$, $I = 0.50\,M$ (CF_3SO_3Na). By using 90% ^{18}O-enriched NO_3^- it was demonstrated by IR that the terminal oxo ligands of the *anti*-$Mo_2O_4^{2+}$ product were labelled and that reaction is therefore by O atom transfer. The increase in k with $[H^+]$ is possibly related to the need to transfer an O atom as in ClO_4^- reactions.

$$-d[Mo_2^{III}]/dt = d[Mo_2^{V}]/dt = k[Mo_2^{III}][NO_3^-] \qquad (18)$$

A less extensive study of the reaction of $[Mo(H_2O)_6]^{3+}$ with NO_3^- has also been reported.[68] As in the above study rate constants are consistent with a substitution-controlled process.

By adjusting conditions, aqueous HCl solutions of $Na_2[MoO_4]$ can be reduced electrolytically to give $[Mo_2Cl_9]^{3-}$ rather than $[MoCl_6]^{3-}$.[69] An alternative procedure[70] yielding a purer product is by electrolytic oxidation of $[Mo_2Cl_8]^{4-}$. The W^{III} complex $[W_2Cl_9]^{3-}$ is known, but $[WCl_6]^{3-}$ is more difficult to prepare. The $[M_2X_9]^{3-}$ ($X^- = Cl^-, Br^-$) complexes have three bridging halide ions in a cofacial bioctahedral arrangement, which is also present in $[Cr_2Cl_9]^{3-}$.

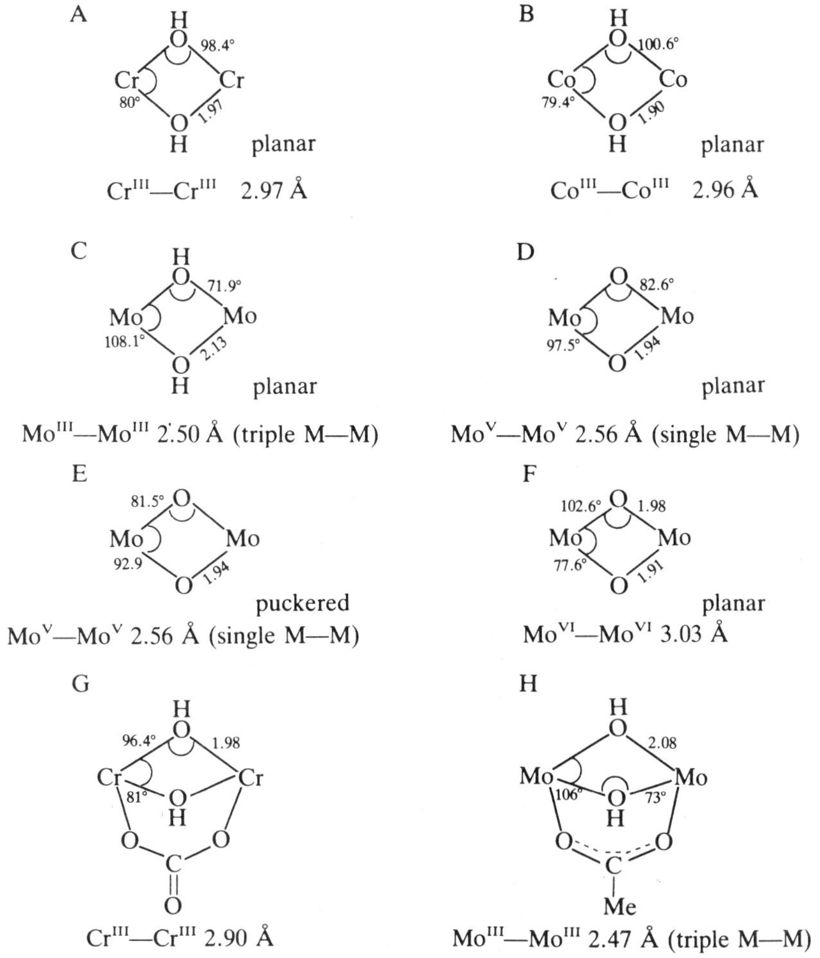

Figure 13 A comparison of M—M distances and bond angles in bridged binuclear complexes[64a]

However, the strength of interaction between two d^3 ions increases markedly in the series, Cr, Mo, W. There is no M—M bond in the case of Cr, and the Cr atoms actually repel each other. The magnetic and spectroscopic properties are essentially those of two unperturbed d^3 ions. In the case of $W_2Cl_9^{3-}$ the M—M interaction is very strong (2.45 Å),[71] with no unpaired electrons. The M—M distances in the case of $[Mo_2Cl_9]^{3-}$ (2.67 Å) and $[Mo_2Br_9]^{3-}$ (2.78 Å) indicate intermediate behaviour.[71] In the latter a temperature dependent magnetism has been observed suggesting that the Mo—Mo interaction is weak enough to allow some unpairing of electrons.

The reaction of $[Mo_2(O_2CMe)_4]$ and RbCl or CsCl in deoxygenated 12 M HCl at ~60 °C gives green-yellow products,[70] initially assigned formulae $Rb_3[Mo_2Cl_8]$ or $Cs_3[Mo_2Cl_8]$. Until they were demonstrated to be diamagnetic they were thought to be mixed-valent II,III complexes. Tritium labelling experiments have demonstrated the presence of an H atom, and the complexes are therefore reformulated as $[Mo_2X_8H]^{3-}$ ($X = Cl^-$ and Br^-) molybdenum(III) species. Consistent with Mo—H—Mo hydrido bridging, IR spectroscopy gives ν(Mo—H—Mo) at 1260 cm^{-1}.[72] The structure has been verified crystallographically for the pyridinium salt, Figure 14.[73]

36.1.2.4 Oxidation State IV

Souchay and colleagues were the first to report a red Mo^{IV} aqua ion.[74] Since that time preparative procedures have changed very little and generally involve heating Mo^{VI} as $Na_2[MoO_4]\cdot 2H_2O$, or Mo^V as the aqua dimer $Mo_2O_4^{2+}$, with $K_3[MoCl_6]$ in 2 M HPTS for 1–2 h at 80–90 °C.[2] The product can be stored in this form and purified by Dowex cation-exchange chromatography as required. Elution can be carried out with 2 M acid, but in perchloric acid

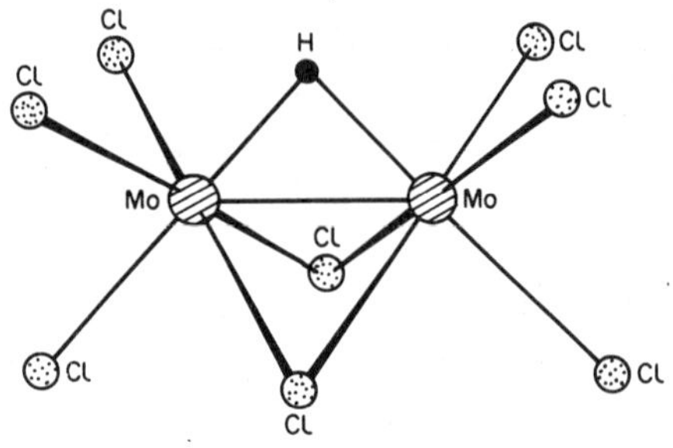

Figure 14 The structure of the Mo_2^{III} complex $[Mo_2Cl_6(Cl_2)(H)]^{3-}$ [73]

solutions accelerated conversion to yellow Mo_2^V occurs overnight. This process has not been studied and may have features similar to the reaction of $[Mo(H_2O)_6]^{3+}$ with ClO_4^-. X-Ray crystal structures of the oxalate, $Cs_2[Mo_3O_4(C_2O_4)_3(H_2O)_3]\cdot H_2O$[75] (Figure 15), the thiocyanate, $[Me_4N]_4[Mo_3O_4(NCS)_8(H_2O)]$,[76] and the *N*-methyliminatodiacetate complex, $Na_2[Mo_3O_4((O_2CCH_2)_2NMe)_3]\cdot 7H_2O$,[77] have indicated trinuclear structures. The Mo_3O_4 core has an incomplete cube structure with one corner unoccupied. In the case of the edta complex $Na_4[Mo_3O_4)_2(edta)_3]\cdot 4H_2O$, there are two trimeric Mo_3O_4 units per molecule, and the three edta ligands coordinate *via* two carboxylates and the imine group to an Mo of each $Mo_3O_4^{4+}$ cluster, giving three bridging edta ligands. This results in a prolate spheroid shape.[78] In much earlier work[79] the Mo^{IV} complex $K_2Mo_3O_4(C_2O_4)_3\cdot 5H_2O$ was reported, but full structural implications were not realized at the time.

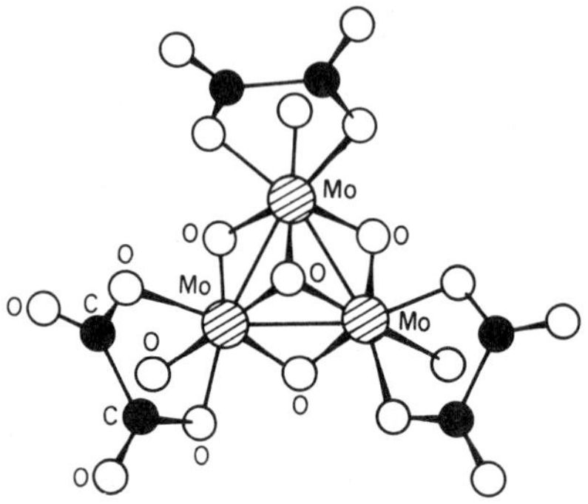

Figure 15 The structure of the Mo_3^{IV} complex $[Mo_3O_4(C_2O_4)_3(H_2O)_3]^{2-}$ [75]

The spectrum of the aqua $Mo_3O_4^{4+}$, peak at 505 nm ($\varepsilon = 63\ M^{-1}\ cm^{-1}$ per Mo), is shown in Figure 16. Substituents often have little effect, and equilibration with 4 M HCl followed by ion-exchange chromatography provides evidence for at least two chloro products, with visible spectra difficult to distinguish from that of the aqua ion. The colour originates therefore from the central $Mo_3O_4^{4+}$ core. With thiocyanate, on the other hand, changes in absorbance at 340 nm are observed, and substitution processes can be monitored. There is no evidence in the aqueous solution chemistry for the existence of stable monomer and dimer forms, which are transients only in redox processes. Thus $[Mo(H_2O)_6]^{3+}$ is oxidized to the Mo_2^V aqua dimer, and with a deficiency of oxidant unreacted $[Mo(H_2O)_6]^{3+}$ and Mo_2^V are the only identifiable Mo species present. Reports from a number of groups in the period 1973–1980 referring to the red aqua Mo^{IV} ion as dimeric or monomeric are now known to be incorrect.

The $Mo_3O_4^{4+}$ core has a coordination number of nine. Evidence for retention of the $Mo_3O_4^{4+}$ unit in the aqueous solution has come from ^{18}O studies, also directed at measuring the rate of oxygen-atom exchange.[80–82] The precipitating agent used was tetramethylammonium thiocyanate which gives $[Me_4N]_5[Mo_3O_4(NCS)_9]$ within 2 min at 0 °C. With ^{18}O enrichment of the solvent the amount of label transferred to the solid after minimal contact times was <5%. Furthermore it was possible to convert the aqua $Mo_3O_4^{4+}$ ion through the changes solid $[Mo_3O_4(NCS)_9]^{5-}$ → aqua $Mo_3O_4^{4+}$ → solid $[Mo_3O_4(C_2O_4)_3(H_2O)_3]^{2-}$ → aqua $Mo_3O_4^{4+}$ → solid $[Mo_3O_4(NCS)_9]^{5-}$ with <4% core O-atom exchange with solvent. Since the crystal structures of the oxalate and thiocyanate complexes have demonstrated that the $Mo_3O_4^{4+}$ core is present, all four core O atoms must exchange slowly in solution. Accordingly the aqua ion can be designated the $Mo_3O_4^{4+}$ structure, with nine water ligands attached (Figure 17). There

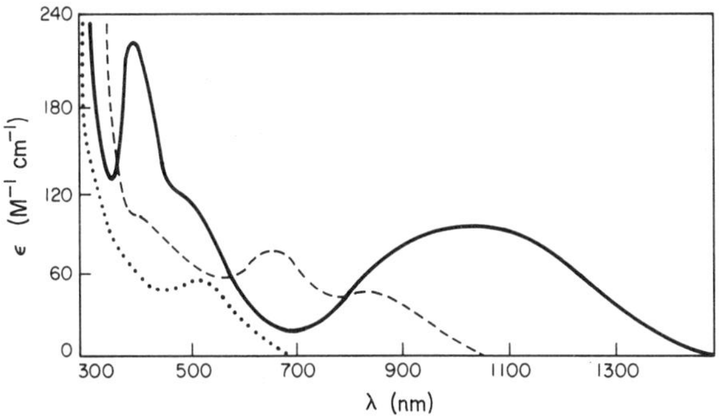

Figure 16 UV–visible–NIR spectra (ε values per Mo) for the Mo^{IV} aqua ion $[Mo_3O_4(H_2O)_9]^{4+}$ ($\cdot\cdot\cdot\cdot$), the Mo^{III} aqua trimer (– – –) and the mixed valence trimer $Mo^{III,III,IV}$ (——) in HPTS[84]

are four different types of O atom coordinated to each Mo. Present information suggests that the O atoms of type A exchange significantly faster than those of type B.[80–82] At 30 °C the half-life is about four days in 1.0 M $MeSO_3H$, and the number of O atoms exchanging at this rate was determined as 1.1. The rate was not significantly affected by small changes in $[H^+]$ (0.8–1.2 M) or $[Mo_3O_4^{4+}]$ (0.08–0.14 M). This result is difficult to explain because, in the process of exchange, a μ_3-oxo ligand (type A) should become a μ_2-oxo ligand (type B). At 25 °C the type B O atoms exchange very slowly and only a fraction of exchange occurs in 534 days in 1.1 M $MeSO_3H$. The half-life is estimated as 5 ± 2 years. To determine the exchange characteristics at type C positions the loss in enrichment of $[Mo_3O_4(H_2O^*)_6(H_2O^*)_3]^{4+}$ on conversion to $[Mo_3O_4(C_2O_4)_6(H_2O^*)_3]^{4+}$ was determined, where the latter can be precipitated with $[Pt(en)_2]^{2+}$. The crystal structure of the product was confirmed.[82] These experiments indicated three type C waters, with a half-life for exchange of ~1 h at 25 °C in 1.0 M $MeSO_3H$. Using $[Mo(CN)_8]^{4-}$ as a precipitating agent the half-life for type D waters was determined as 25 min at 0 °C in 1.2 M $MeSO_3H$. This is slower than might have been expected on the basis of NCS^- for H_2O substitution studies.

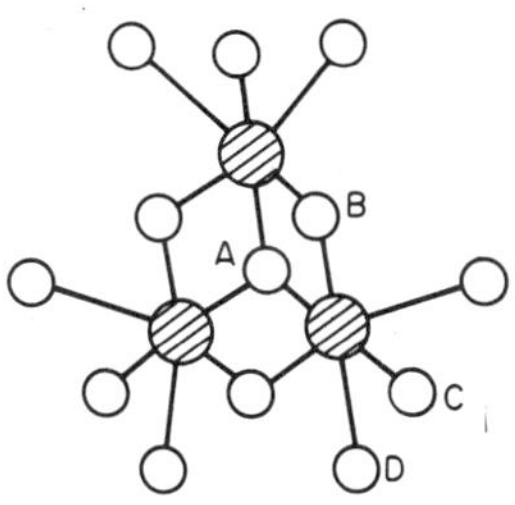

Figure 17 Structure of the Mo_3^{IV} aqua ion $[Mo_3O_4(H_2O)_9]^{4+}$ and designation of four different O atoms, A–D

Other relevant ^{18}O exchange studies have demonstrated that through the sequence of changes aqua $Mo_3O_4^{4+} \rightarrow Mo^{IV}$ (hydroxide precipitate) $\rightarrow$ aqua $Mo_3O_4^{4+}$, there is at least 96% core O-atom retention. Apparently the core is retained in the precipitate and no dramatic base-catalyzed core O-atom exchange is observed.

From crystal structure information Mo—O bond lengths for the different O atoms are 2.02 Å (A), 2.92 Å (B), 2.16 Å (C) and 2.26 Å (D). EXAFS data for aqua Mo^{IV} in 4 M $MeSO_3H$ are consistent with the accepted $Mo_3O_4^{4+}$ structure,[60] with an Mo—Mo distance of 2.49 Å compared to 2.51 Å from X-ray crystallography.[78]

The existence of equivalent and non-equivalent H_2O ligand sites on $Mo_3O_4^{4+}$ results in unusual kinetic situations when the complexing with NCS^- is studied.[83] For the conditions in which $[NCS^-]$ is in large (>10-fold) excess, a statistically corrected rate constant is obtained assuming three equivalent sites on the $Mo_3O_4^{4+}$ ion, and equilibration rate constants k_{eq} can be expressed as in equation (19). With $[Mo_3O_4(H_2O)_9]^{4+}$ in excess the reaction is constrained by the low $[NCS^-]$ to form only $Mo_3O_4(NCS)^{3+}$, and the rate law in equation (20) applies.

$$k_{eq} = (k_1/3)[NCS^-] + k_{-1} \quad (19)$$

$$k_{eq} = k_1[Mo_3O_4^{4+}] + k_{-1} \quad (20)$$

A comparison was made using the different anions, *p*-toluenesulfonate, trifluoromethanesulfonate and perchlorate ($I = 2.0$ M). At 25 °C forward and backward rate constants at $[H^+] = 2.0$ M are $k_1 = 0.78$, 1.02, 2.13 $M^{-1} s^{-1}$ and $10^3 k_{-1} = 1.57$, 1.94, 2.18 s^{-1}, for these three anions respectively. Rate constants k_1 and k_{-1} have $[H^+]$ dependences of the kind $a + b[H^+]^{-1}$, with b dominant for $[H^+] = 0.8$–2.0 M. In common with other conjugate-base pathways, b can be assigned an I_d mechanism. The large contribution from such a path for a low electron population metal (here d^2) is believed to be in some part due to the steric crowding around each Mo^{IV} resulting from Mo—Mo bonding. Thus with the inclusion of Mo—Mo bonding each Mo is eight-coordinate. Solutions containing higher complexes obtained by equilibrating $Mo_3O_4^{4+}$ (1.3×10^{-3} M) with NCS^- (5×10^{-3} M) give, on ion exchange, bands that in order of elution analyze (Mo and NCS^-) for $Mo_3O_4(NCS)_2^{2+}$, $Mo_3O_4(NCS)^{3+}$ and $Mo_3O_4^{4+}$. A fraction not held on the column at $[H^+] = 0.3$ M is believed to contain $Mo_3O_4(NCS)_3^+$ as well as free NCS^-. Two kinetic stages are observed in the aquation of $Mo_3O_4(NCS)_2^{2+}$ in 2 M HPTS, which are assigned to k_{-1} (here $1.7 \times 10^{-3} s^{-1}$), and k_{-2} ($0.24 \times 10^{-3} s^{-1}$).

The red aqua $Mo_3O_4^{4+}$ ion is reduced by amalgamated Zn under air-free conditions to a green Mo_3^{III} aqua ion.[84] The spectrum, Figure 16, is different from that of the green Mo_2^{III} aqua ion previously discussed. It is held more strongly on a Dowex column than $Mo_3O_4^{4+}$, which suggests that it may have a fully protonated core $Mo_3(OH)_4^{5+}$. This is an interesting point which requires confirmation. The oxalato complex $[Mo_3O_4(C_2O_4)_3(H_2O)_3]^{2-}$ can be similarly reduced.[85] Both nine-electron Mo_3^{III} complexes exhibit EPR spectra. On labelling the core with ^{18}O atoms the complete redox cycle can be accomplished with as high as 97% retention of the labelling.[82] In one experiment the Mo_3^{III} ion was allowed to exchange with solvent for two months at room temperature, and no (±3%) exchange was detected.[81]

Rapid (<1 min) equilibration of Mo_3^{III} with Mo_3^{IV} (2:1 amounts) gives quantitative formation of an orange pink $Mo^{III,III,IV}$ trimer (Figure 16) in 4.0 M HPTS (*p*-toluenesulfonic acid), 80% in 2.0 M HPTS, and 30% in 0.5 M HPTS.[84] At $[H^+] > 2.0$ M electrolytic reduction of Mo_3^{IV} at a potential of −0.25 V (*vs.* NHE) confirmed these results with spectrophotometric evidence for the formation of the same $Mo^{III,III,IV}$ ion. There is no evidence for the other mixed-valence state, $Mo^{III,IV,IV}$, which presumably rapidly disproportionates. Detailed electrochemical studies have suggested two forms of Mo_3^{III},[85] but have failed to identify the nature of these. From studies on the dependence of the Mo_3^{III}/Mo_3^{IV} equilibration on $[H^+]$ it is concluded that the $Mo^{III,III,IV}$ trimer is also protonated, and at least 4+ in charge.[84] A feature of the spectrum of the $Mo^{III,III,IV}$ form is the broad absorbance band at 1050 nm (ε 100 $M^{-1} cm^{-1}$ per Mo). Both the Mo_3^{IV} and $Mo^{III,III,IV}$ ions are diamagnetic.[83,85] Exposure of the Mo_3^{III} ion to O_2 results in oxidation to $Mo^{III,III,IV}$ and then Mo_3^{IV}.[84]

Whereas an oxidant such as $[Co(C_2O_4)_3]^{3-}$ readily converts $[Mo(H_2O)_6]^{3+}$ and aqua Mo_2^{III} through to the aqua Mo_2^V state, it is not able to oxidize Mo_3^{IV}.[86] The implication is that monomeric and dimeric Mo^{IV} containing forms generated as intermediates are readily oxidized, whereas Mo_3^{IV} is not. This is also borne out by electrochemical studies, in which it has been reported that it is difficult to electrooxidize Mo_3^{IV} to Mo_2^V. The trimeric form is oxidized by strong oxidants such as $[IrCl_6]^{2-}$ (0.89 V) and $[Fe(phen)_3]^{3+}$ (1.06 V), and conventional range kinetic studies have been reported.[86] Reactions proceed exclusively by an $[H^+]^{-1}$ dependent rate-law term, reflecting the greater degree of hydrolysis of the Mo^V product. In the $[IrCl_6]^{2-}$ case oxidation to Mo_2^V is faster than the subsequent $Mo_2^V \rightarrow Mo^{VI}$ oxidation. The converse holds in the case of $[Fe(phen)_3]^{3+}$. Because of these differences it is likely that the mechanisms are different and that the $[IrCl_6]^{2-}$ reaction is of the inner-sphere type. The nature of the phenanthroline ligands of $[Fe(phen)_3]^{3+}$ precludes an inner-sphere mechanism.

The synthesis of oxo–sulfido Mo_3^{IV} complexes in the series $Mo_3O_xS_{4-x}$ as aqua ions has been reported. Synthesis of the grey-coloured $Mo_3O_2S_2^{4+}$ μ_3-sulfido containing ion has been achieved by essentially the same route as for $Mo_3O_4^{4+}$, by reacting $[MoCl_6]^{3-}$ with the aqua di-μ-sulfido Mo_2^V complex $[Mo_2O_2S_2(H_2O)_6]^{2+}$ (**5**).[87] The reaction can be expressed as in equation (21), where the transient Mo^{IV} (or some related form) presumably also gives an Mo_3^{IV} product. The ~70% conversion of Mo_2^V to $Mo_3O_2S_2^{4+}$ after column chromatography is consistent with retention of the dimer unit and attachment of Mo^{III} to the dimer. The product has been characterized crystallographically as $(pyH)_5[Mo_3O_2S_2(NCS)_9]\cdot 2H_2O$.[87]

$$2Mo^{III} + Mo_2^V \longrightarrow Mo_3^{IV} + Mo^{IV} \qquad (21)$$

Using different approaches the dark green $Mo_3OS_3^{4+}$ ion has been obtained along with the

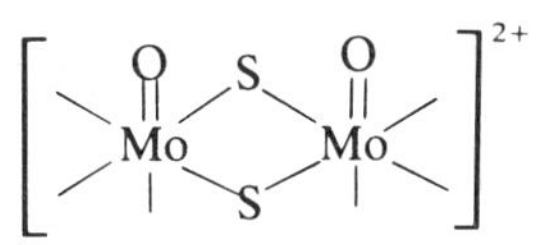

(5)

green cubic cluster complex $Mo_4S_4^{5+}$. Two procedures have been described involving either $NaBH_4$ or electrochemical reduction of the di-μ-sulfido Mo_2^V cysteinato complex, $[Mo_2O_2S_2(cys)_2]^{2-}$ (**6**), in HCl, under N_2 (or Ar), and followed by air oxidation.[88,89] The green cubic complex $Mo_4S_4^{5+}$ (**7**) yields $Mo_3S_4^{4+}$ (**8**) on heating solutions in 1 M HCl to ~90 °C for 3–4 h. Another preparative route yielding both (**7**) and (**8**) has been described in which $[Mo(CO)_6]$ is refluxed in acetic anhydride containing dry Na_2S under N_2. Water is added to hydrolyze the cooled reaction mixture.[90,91]

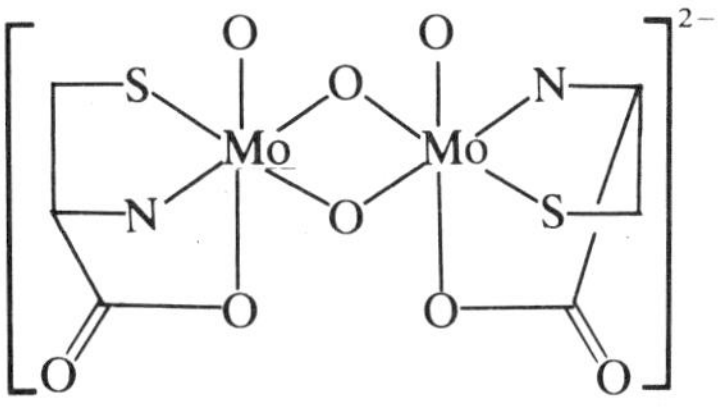

(**6**)

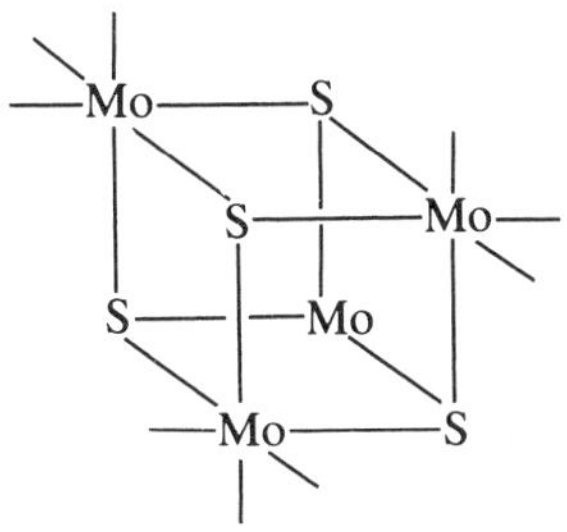

$[Mo_4S_4(H_2O)_{12}]^{5+}$
(**7**)

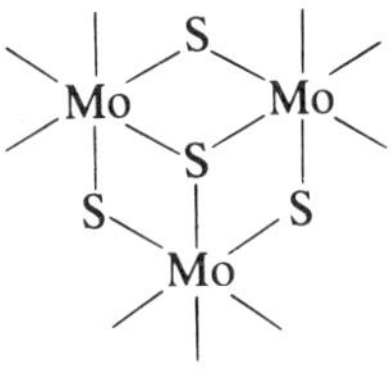

$[Mo_3S_4(H_2O)_9]^{4+}$
(**8**)

The existence of the $Mo_3S_4^{4+}$ core has been confirmed by X-ray structure analysis of the iminodiacetate and oxalate complexes, $Ca[Mo_3S_4(ida)_3]\cdot 11.5H_2O$ and $Cs_2[Mo_3S_4(C_2O_4)_3(H_2O)_3]\cdot 3H_2O$.[88,90] Selected mean bond distances from the first of these are Mo—Mo (2.7 Å), Mo—μ_3(S) (2.35 Å), and Mo—μ_2(S) (2.30 Å). Corresponding values for the $[Mo_3O_4(mida)_3]^{2-}$ complex are Mo—Mo (2.50 Å), Mo—μ_3(O) (2.04 Å) and Mo—μ_2(O) (1.92 Å).[77] The aqua ion of the $Mo_4S_4^{5+}$ cluster has formal oxidation states III, III, III and IV. X-Ray crystal structure determinations of both $Na_3[Mo_4S_4(edta)_2]$,[92] and the NCS^- complex $(NH_4)_6[Mo_4S_4(NCS)_{12}]\cdot 10H_2O$,[91] which appears to be a derivative of $Mo_4S_4^{6+}$, have been reported.

In the above preparations from di-μ-sulfido Mo_2^V, the cubic $Mo_4S_4^{5+}$ product is obtained in up to *ca.* 20% yield.[93] This suggests a mechanism involving dimerization of a reduced form of Mo_2^V at some stage to give the cubic Mo_4 product. Some subsequent (partial) decomposition to the trimer then occurs, although details of the reaction and the oxidation states involved have not been fully established.[93] If alternatively a mechanism involving addition of mononuclear Mo to the dimer occurred it is difficult to see how retention of all four bridging S^{2-} ligands could be achieved.

The μ-oxo-μ-sulfido, $[Mo_2O_3S(cys)_2]^{2-}$, and di-μ-oxo, $[Mo_2O_4(cys)_2]^{2-}$, complexes have also been used singly or in mixes in the above preparations, and other members of the $Mo_3O_xS_{4-x}^{4+}$ series have been identified.[93] Structures of $Mo_3OS_3^{4+}$ and $Mo_3O_3S^{4+}$, both with μ_3(S) ligands, have been determined.[94,95] Because, in the wider application of the preparative procedures using $[Mo_2O_4(cys)_2]^{2-}$, it has been found that there is a tendency for cysteine to remain attached to the Mo, there are advantages in using the corresponding Mo_2^V aqua ions, and mixtures of these.[93] Although $Mo_3O_4^{4+}$ can be prepared by this route, the formation of $Mo_4O_4^{5+}$ has to be confirmed. Complexes in the $Mo_3O_xS_{4-x}$ series with a μ_3-oxo ligand have not been prepared.

Studies on the 1:1 complexing of NCS^- for H_2O on aqua $Mo_3S_4^{4+}$ indicate an ~10^2 labilizing effect as compared to $Mo_3O_4^{4+}$.[96] A cyclic voltammogram of $[Mo_3S_4(ida)_3]^{2-}$ at a glassy carbon electrode shows two reversible one-electron waves indicating the existence of $Mo^{III,IV,IV}$ and $Mo^{III,III,IV}$ oxidation states.[88] This is the first time the $Mo^{III,IV,IV}$ state has been identified. Although evidence has been obtained in cyclic voltammetry for reduction of aqua $Mo_3S_4^{4+}$ the products have not as yet been prepared quantitatively. In 2 M HPTS the 4+ aqua ion at greater than 10^{-3} M is extremely stable. It reacts (*e.g.* 25 ml of 0.05 M) with metallic Fe (5 g) over a few hours to give the cubic mixed-metal complex $Mo_3FeS_4^{4+}$. From the crystal structure of $[Mo_3FeS_4(NH_3)_9(H_2O)]Cl_4$ (Figure 18), the Fe is seen to be four-coordinate whereas the three Mo atoms remain six-coordinate.[97]

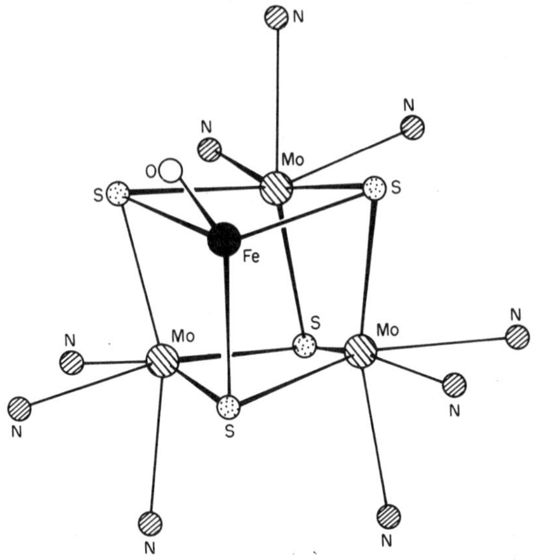

Figure 18 The structure of $[Mo_3FeS_4(NH_3)_9(H_2O)]Cl_4$[97]

The aqua ion of $Mo_4S_4^{5+}$ in acid (2 M HPTS) solutions is also stable,[89] but less so than with $Mo_3S_4^{4+}$, and at 50 °C (in air) gives $Mo_3S_4^{4+}$ with a half-life of approximately four days. The reaction is slower under N_2. It can be reduced to the orange 4+ ion, which is stable, whereas on oxidation the 6+ complex is unstable. UV–visible spectra of the 4+ and 5+ complexes are compared in Figure 19. With two edta ligands coordinated to $Mo_4S_4^{5+}$ the corresponding orange 4+ and red 6+ states are both stable, Figure 20.[98] Complexes having mixed oxo–sulfido cubic $Mo_4O_xS_{4-x}^{5+}$ clusters have not been isolated.

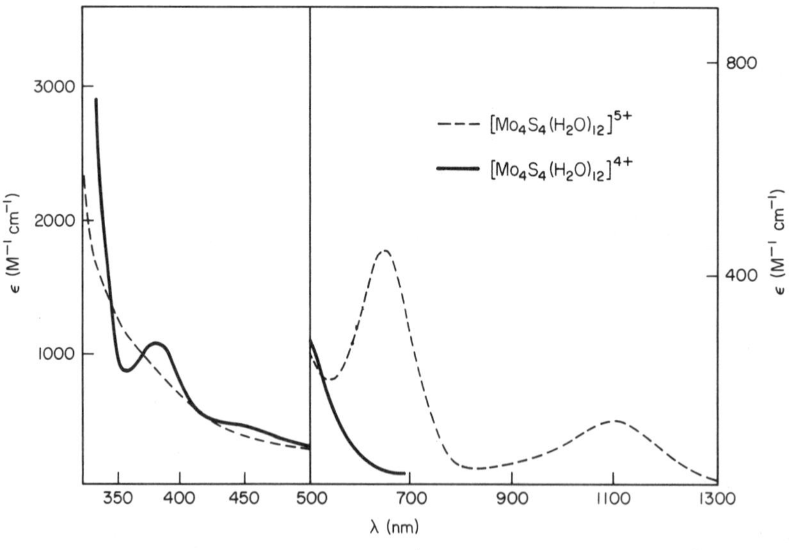

Figure 19 UV-visible-NIR spectra of $[Mo_4S_4(H_2O)_{12}]^{4+}$ (——) and $[Mo_4S_4(H_2O)_{12}]^{5+}$ (- - -) in 1 M HPTS

Cyano complexes $[Mo_3O_4(CN)_9]^{5-}$ and $[Mo_4S_4(CN)_{12}]^{8-}$, with $Mo_3S_4^{4+}$ and $Mo_4S_4^{4+}$ core structures, have been prepared by the reaction of amorphous MoS_3 with CN^- and characterized crystallographically.[99] It should also be mentioned that $[Mo_3S_4(\eta^5\text{-}C_5H_5)_3]^{+,0}$ and $[Mo_4S_4(\eta^5\text{-}C_5H_4Pr^i)_4]$ have also been obtained.[100–102]

In addition to the extensive trinuclear chemistry of Mo^{IV} there are a few examples of mononuclear complexes. These include $[Mo(CN)_8]^{4-}$ and *trans*-$[MoO(H_2O)(CN)_4]^{2-}$. The

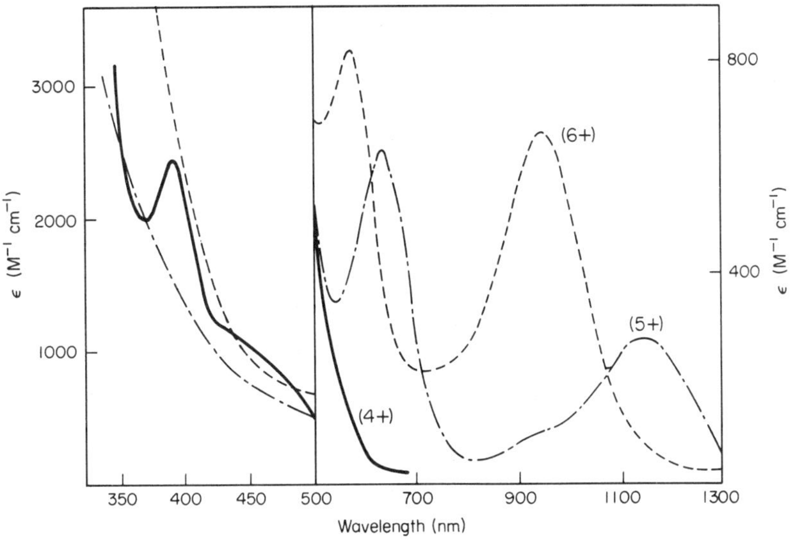

Figure 20 UV–visible–NIR spectra of $[Mo_4S_4(edta)_2]^{4+}$ complexes ($n = 4, 5, 6$) in H_2O[98]

variation of Mo—O bond lengths for the series of complexes $[Mo(O)_2(CN)_4]^{4-}$, $[MoO(OH)(CN)_4]^{3-}$ and $[MoO(OH_2)(CN)_4]^{2-}$ has been noted.[103] Increased protonation of one of the oxo groups results in an increase in Mo—O from 1.83 through 2.08 to 2.27 Å. Concomitantly the terminal Mo—O bond decreases from 1.83 through 1.70 to 1.67 Å. In the case of $[MoO(H_2O)(CN)_4]^{2-}$ blue salts with $[Ph_4As]^+$ and green salts with $[Ph_4P]^+$ provide an example of distortional isomerism.[104] The major structural difference is the Mo=O distance which is reported to be 1.72 Å in the blue and 1.60 Å in the green form.

A number of Mo^{IV} complexes of the type $[MoOCl_2(PR_3)_3]$ have also been isolated in both blue and green forms.[105] X-Ray structure determinations have shown that both forms possess the meridional stereochemical arrangement, with the oxo and the Cl^- ligand *trans* to each other.[106,107] The green form has a shorter Mo=O distance (1.68 Å) compared to the blue (1.80 Å) which is unusually long. No satisfactory explanation for the existence of distortion isomers in the crystalline state has been forthcoming.

Octahedrally coordinated mononuclear complexes with S- and P-donor atoms (**9**) and (**10**) have also been reported.[108,109]

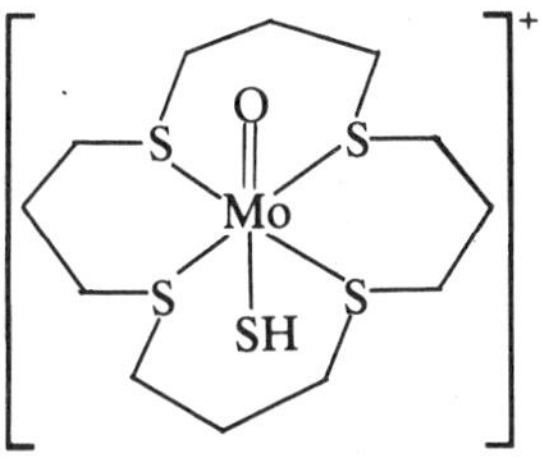

(9)

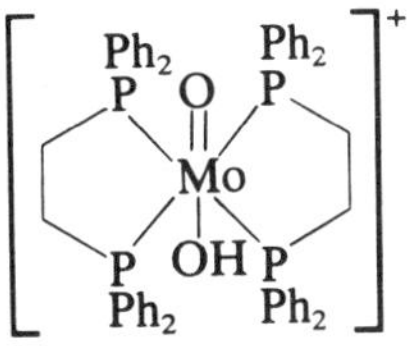

(10)

36.1.2.5 Oxidation State V

The Mo^V state is readily obtained as the binuclear orange-yellow $Mo_2O_4^{2+}$ ion by reduction of Mo^{VI} as $Na_2[MoO_4]\cdot 2H_2O$ in acidic solutions. Procedures used include reduction with hydrazine, shaking with mercury, or controlled potential reduction at a Hg-pool electrode. On addition of NCS^- three products $[MoO(NCS)_5]^{2-}$, $[Mo_2O_4(NCS)_6]^{4-}$ and $[Mo_2O_3(NCS)_8]^{4-}$ (*anti* form) are obtained, depending on the concentrations of NCS^- and Mo as well as the pH. These forms are representative of the three different Mo^V structures (**11**)–(**13**).

Probably the best known monomeric Mo^V form is the green chloro complex, which can be isolated as the solid $(NH_4)_2[MoOCl_5]$, and is retained in solution at >7 M HCl. The closely related $[MoOCl_4]^-$ and $[MoOCl_4H_2O]^-$ ions are also known, and chloro and iodo analogues

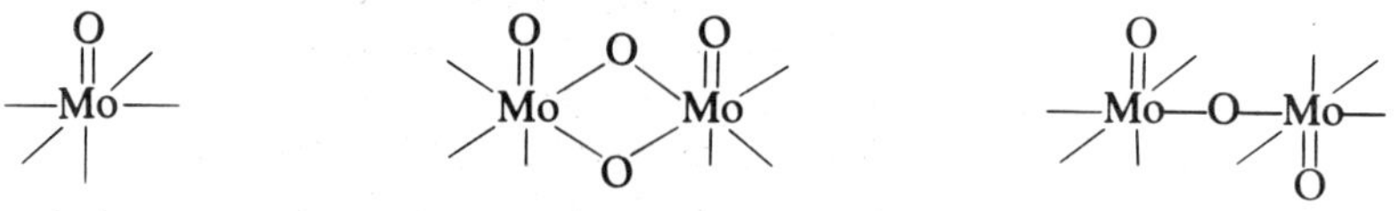

(11) (12) (13)

have been obtained.[110] Other preparative routes in addition to the reduction of $[MoO_4]^{2-}$ include the oxidation of $[Mo_2(O_2CMe)_4]$ in aqueous HX (X^- = halide), or by dissolving $MoCl_5$ in an appropriate acid. In aqueous solution, at lower HX concentrations, orange-yellow $Mo_2O_4^{2+}$ is obtained. Elution of $Mo_2O_4^{2+}$ from Dowex cation-exchange columns with 0.5–2.0 M $HClO_4$ gives the orange-yellow aqua ion $[Mo_2O_4(H_2O)_6]^{2+}$.

Monomeric oxo–Mo^V complexes have a characteristic EPR spectrum with a strong central band, and EPR provides the most direct and definitive means of distinguishing this form.[111] The $4d^1$ electron is believed to be located in an orbital in the plane perpendicular to the Mo=O bond. The bond *trans* to the terminal oxo ligand is relatively weak and five- as well as six-coordinate complexes are known. The structure of $K_2[MoOCl_5]$ gives four Mo—Cl bonds at 2.40 Å (mean) and one *trans* to the oxo at 2.95 Å.[112] The difference is similar to that for the VO^{2+} aqua ion (0.18 Å) in hydrated vanadyl sulfate. A bigger *trans* effect is expected in the case of the nitrido complex $[MoN(Cl)_5]^{3-}$.[113] The length of the Mo=O bond is shorter in five-coordinate complexes, and is 1.61 Å in $(AsPh_4)[MoOCl_4]$.[114] On addition of a sixth ligand as in $K_2[MoOCl_5]$ the Mo=O bond is ~1.67 Å. The presence of the terminal oxo results in doming as in $(AsPh_4)[MoOCl_4(H_2O)]$ (Figure 21),[115] and this is also a feature of Mo_2O_4 type structures with an O_tMoO_b angle between terminal and bridging oxo ligands of ~105°. The stabilization of $[MoO(SR)_4]^-$ type complexes using small peptides containing cysteine residues, and with bulky aromatic thiolate ligands has been reported.[116]

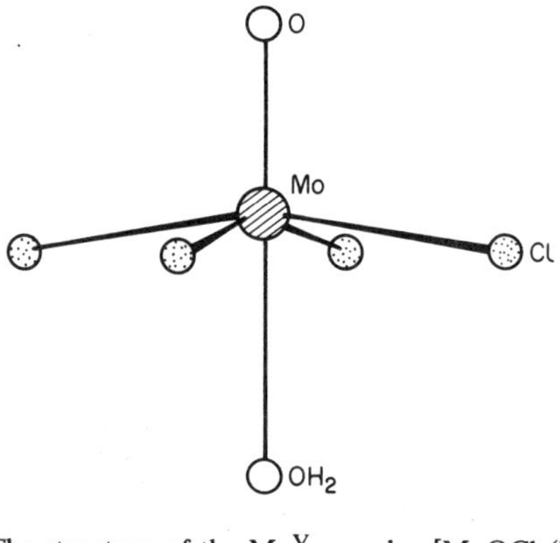

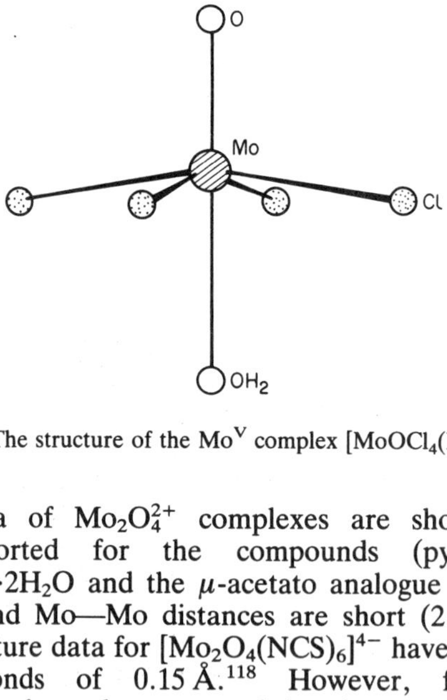

Figure 21 The structure of the Mo^V complex $[MoOCl_4(H_2O)]^-$ [115]

Typical UV–visible spectra of $Mo_2O_4^{2+}$ complexes are shown in Figure 22. Crystal structures have been reported for the compounds $(pyH)_2[Mo_2O_4Cl_4(H_2O)_2]\cdot H_2O$, $(pyH)_3[Mo_2O_4(HCO_2)(NCS)_4]\cdot 2H_2O$ and the μ-acetato analogue (Figures 23 and 24).[112] The complexes are diamagnetic and Mo—Mo distances are short (2.56–2.58 Å), consistent with metal–metal bonding.[117] Structure data for $[Mo_2O_4(NCS)_6]^{4-}$ have indicated a difference in the *cis* and *trans* Mo—N bonds of 0.15 Å.[118] However, for the oxalate complex $[Mo_2O_4(C_2O_4)_2(H_2O)_2]^{2-}$, it has been demonstrated that the waters are coordinated *cis* to each terminal oxo, with the oxalate *cis* and *trans* to the same oxo. No difference in the oxalate Mo—O distances is detected in this and the corresponding di-μ-sulfido complex $[Mo_2O_2S_2(C_2O_4)_2(H_2O)_2]^{2-}$ (**14**).[119] In a similar bipyridyl complex (**15**) the Mo—N bond length *trans* to the terminal oxo is 2.32 Å, while that in the *cis* position is 2.25 Å,[120] giving an intermediate elongation effect. The absence of a *trans* influence in the case of oxalate suggests it has a low *trans* susceptibility as compared with bipyridyl and other ligands. A more extensive list of compounds would be of interest to explain this effect. It has been noted that the ^{13}C NMR spectrum of $[Mo_2O_4(C_2O_4)_2(H_2O)_2]^{2-}$ in water at pH 4 gives only one sharp peak at 168 p.p.m. at 34 °C, which suggests that in solution there is some rapid exchange process taking place.[32,118]

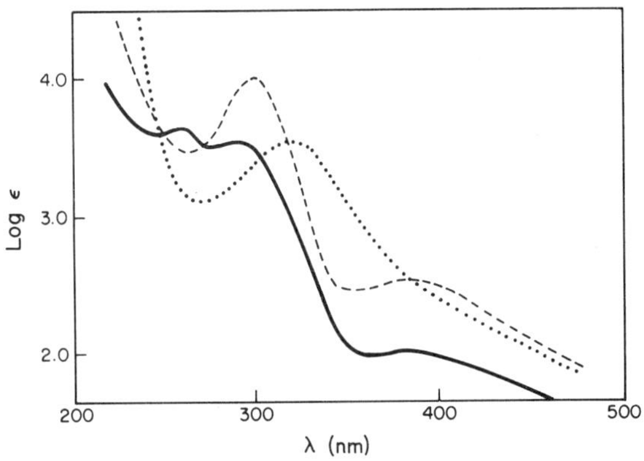

Figure 22 UV–visible spectra (ε values per dimer) of the Mo^{V} aqua dimer $[Mo_2O_4(H_2O)_6]^{2+}$ in 0.5 M $HClO_4$ (——), and of $[Mo_2O_4(edta)]^{2-}$ (– – –) at pH 5 (acetate buffer), and (ε per Mo) for the polymeric Mo^{V} ion (· · ·)[138b,136]

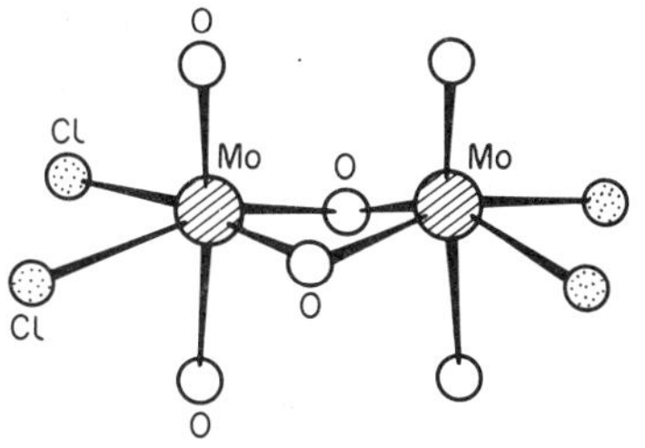

Figure 23 The structure of the Mo_2^{V} complex $[Mo_2O_4Cl_4(H_2O)_2]^{2-}$ [112]

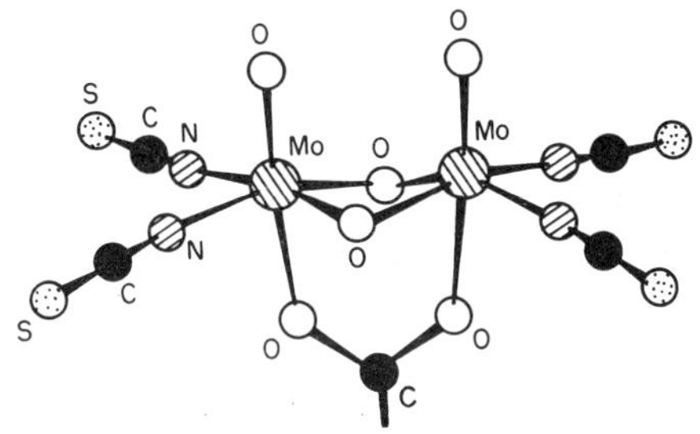

Figure 24 The structure of the Mo_2^{V} complex $[Mo_2O_4(HCO_2)(NCS)_4]^{3-}$ [112]

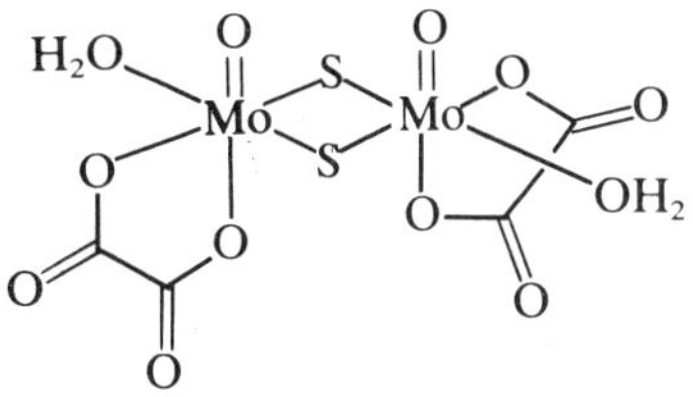

(14)

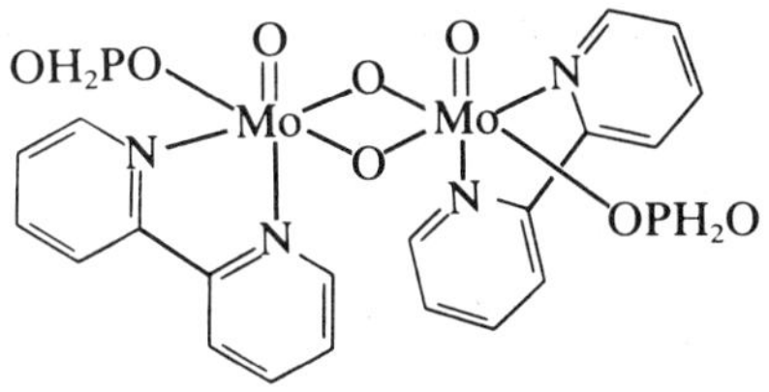

(15)

As in the case of the Mo_2^{III}–edta complex the edta ligand in $[Mo_2O_4(edta)]^{2-}$ is coordinated to the binuclear core in a 'basket' like conformation. Studies on complexes of $Mo_2O_4^{2+}$ having optically active amine-carboxylate ligands such as propylenediaminetetraacetate have been reported.[121]

Oxidation of the Mo_2^{III} complex $[L(H_2O)Mo(OH)_2Mo(H_2O)(L)]^{4+}$, where L is the cyclic triamine 1,4,7-triazacyclononane ligand, with (for example) ClO_4^- gives the red-purple coloured *anti*-$[Mo_2O_4L_2]$ complex (λ_{max} 525 nm), which undergoes irreversible acid-catalyzed isomerization to the yellow *syn*-$[Mo_2O_4L_2]$ complex (Figure 25).[64] If the 1,5,9-triazacyclododecane ligand L′ is used the *anti*-$[Mo_2O_4L'_2]$ form is likewise obtained, and in contrast to the behaviour observed with L, is stable for at least two days in 1 M $HClO_4$ at 25 °C.[65] Base-catalyzed isomerization is observed in both cases. Structures of the *anti* forms have been determined. For the L complex the short Mo—Mo distance (2.55 Å) is essentially the same as that for the yellow isomer (2.56 Å).[122] The diamagnetism in both cases is consistent with metal–metal bonding. The four-membered Mo_2O_2 ring in the *anti* form is planar whereas for the *syn* isomer it is puckered with an O_tMoO_b angle of 109°. Kinetic studies on the acid-catalyzed isomerization of the L complex (red-purple→yellow) give a dependence (equation 22), which is consistent with H^+-induced reaction (equations 23–25).

$$k_{obs} = \frac{k[H^+]}{[H^+] + K_a} \tag{22}$$

$$anti\text{-}[L_2Mo_2O_4]^{2+} + H^+ \underset{}{\overset{K_a^{-1}}{\rightleftharpoons}} anti\text{-}[L_2Mo_2O_4H]^{3+} \tag{23}$$

$$anti\text{-}[L_2Mo_2O_4H]^{3+} \xrightarrow{k} syn\text{-}[L_2Mo_2O_4H] \tag{24}$$

$$syn\text{-}[L_2Mo_2O_4H]^{3+} \xrightarrow{fast} syn\text{-}[L_2Mo_2O_4] + H^+ \tag{25}$$

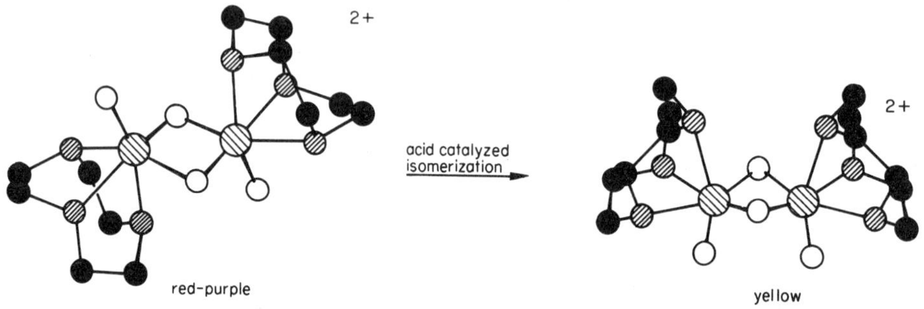

Figure 25 The structures and isomerization of *anti*- to *syn*-$[Mo_2O_4L_2]^{2+}$, L = 1,4,7-triazacyclononane[64a]

A mechanism involving protonation at a terminal oxo ligand followed by addition of a *trans* H_2O has been proposed.[64b] At 21 °C $K_a = 2.3$ M and $k = 2.4 \times 10^{-4}\ M^{-1}\ s^{-1}$. Theoretical calculations on *syn*- and *anti*-$[Mo_2S_4(S_2C_2H_4)_2]^{2-}$ indicate that the Mo—Mo bond is stronger in the *syn* isomer.[123] At present these are the only examples of the *anti* form.

The preparations of di-μ-sulfido and μ-oxo-μ-sulfido complexes of Mo^V with edta, cysteine and cysteine-ester ligands have been described.[124] Formulae with cysteine coordinated are shown in (**16**) and (**17**). To prepare (**16**) hydrogen sulfide is bubbled slowly through a solution of $Na_2MoO_4 \cdot 2H_2O$ in H_2O for ~2 h and after removal of excess H_2S and adjustment of the pH, the appropriate ligand is added. To prepare (**17**) H_2S is bubbled slowly through a solution of $MoCl_5$ in 3 M HCl for 2 h. Both types of complex are an orange-yellow colour. The ligands can be removed by addition of acid, *e.g.* 1–2 M HCl, followed by gel filtration chromatography to give the aqua ion analogues of $[Mo_2O_4(H_2O)_6]^{2+}$. Absorption bands in the UV–visible for $Mo_2O_2S_2^{2+}$ (277, 311, 340 and 463 nm) are at lower wavelengths than the corresponding bands for $Mo_2O_4^{2+}$ (280, 324, 365 and 493 nm). The edta and cysteine complexes undergo electrochemical reduction in a single four-electron step to Mo^{III} dimers in aqueous (buffered) solutions. Although the ease of reduction and electrochemical reversibility of the Mo_2^V/Mo_2^{III} couples increase with insertion of S^{2-} into the bridging positions, the Mo^{III} dimers become increasingly unstable for the S^{2-}-bridged complexes. With 1,1-dithiolate ligands such as *N,N*-diethyldithiocarbamate and diisopropyldithiophosphinate bridging and terminal oxo positions of the $Mo_2O_4^{2+}$ structure have been replaced to yield all the products from $Mo_2O_4(dtc)_2$ to $Mo_2S_4(dtc)_2$.[125] Using IR spectroscopy and ^{18}O for ^{16}O replacement, it has been demonstrated that substitution of O^{2-} by S^{2-} at the bridging positions is fairly facile. Much more extreme conditions using P_4S_{10} for example are required to effect terminal oxo substitution.

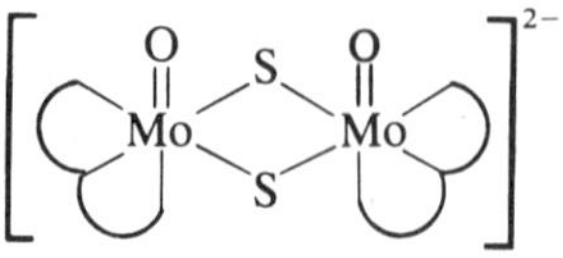

(**16**)

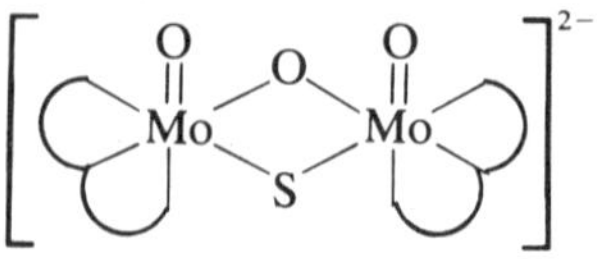

(**17**)

The second main category of dimeric Mo^{V} complexes having a single μ-oxo bridge can also give *syn* and *anti* (rotamer) forms. Structures of *anti*-$[Mo_2O_3(NCS)_4(C_2O_4)_2]^{4-}$ and *anti*-$[Mo_2O_3(NCS)_6(HCO_2)_2]^{4-}$ have been determined (Figures 26 and 27), and indicate a linear Mo—O—Mo bridge with Mo—O_b distances 1.85 and 1.86 Å, and Mo—O_t of 1.68 and 1.67 Å respectively.[112] Other examples of this type are with bidentate thiophosphate in $[Mo_2O_3\{(S_2POEt)_2\}_4]$ and $[Mo_2O_3(LL)_4]$, where LL represents *O*-thiopyridine.[126] Examples of the *syn* structure (**18**) are with bidentate sulfur-donor ligands in $[Mo_2O_3(S_2COEt)_4]$ and $[Mo_2O_3(S_2CNPr)_4]$. In the first of these (*O*-ethyldithiocarbonate) the Mo—Mo bond is 3.72 Å, and the Mo—O_t bond is 1.65 Å.[127] All known μ-oxo-bridged dimers are diamagnetic or exhibit only a small practically temperature independent paramagnetism.

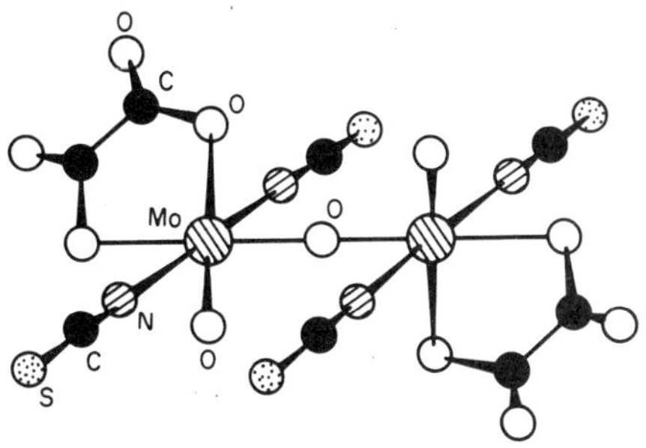

Figure 26 The structure of the Mo_2^{V} complex $[Mo_2O_3(NCS)_4(C_2O_4)_2]^{4-}$ [112]

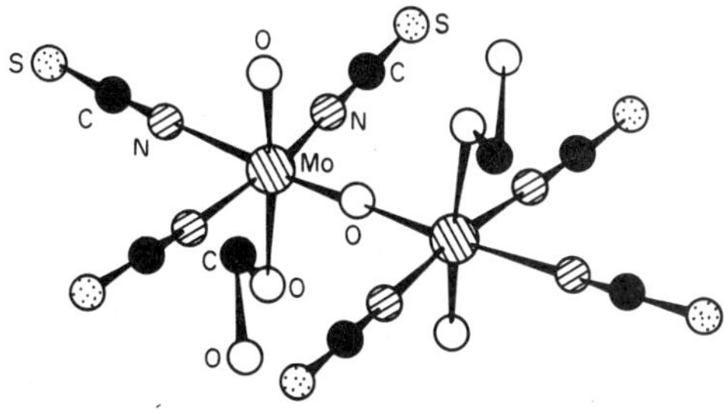

Figure 27 The structure of the Mo_2^{V} complex $[Mo_2O_3(NCS)_6(HCO_2)_2]^{4-}$ [112]

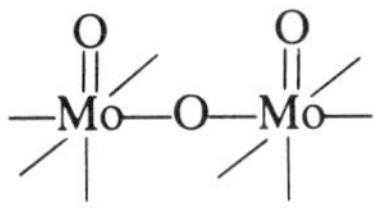

(**18**)

It is possible to use IR spectroscopy as a means of distinguishing between $Mo_2O_3^{4+}$ and $Mo_2O_4^{2+}$ complexes.[125] No interconversion of these two forms has been studied, and no aqua ion having the $Mo_2O_3^{4+}$ structure has yet been identified.

The staggered conformation of terminal oxo ligands observed for Mo^{VI} in $Mo_2O_5^{6+}$ is not observed for Mo^{V}.[128] However, in the tetraphenylporphyrin complex $[Mo_2O_3(TPP)_2]$ the linear Mo—O—Mo structure is retained, and the terminal oxo ligands lie *trans* to the μ-oxo group giving the opposed O_t—Mo—O_b—Mo—O_t structure.[129]

In a recent comprehensive study of the coordination of polypyrazolylborate, $HB(pz)_3^-$, to Mo^{V} a number of complexes have been prepared.[130] First by reacting $HB(pz)_3MoCl_3$ with O_2 the mononuclear complex $[HB(pz)_3MoOCl_2]$ is obtained (Figure 28). A pair of geometric isomers, $[Mo_2O_3Cl_2(H(Bpz)_3)_2]$ and $[Mo_2O_4(HB(pz)_3)_2]$, can be isolated from the reaction of $MoCl_5$ with $HB(pz)_3^-$ in dilute HCl. By further reaction of the latter with MeOH the zigzag tetramer $[Mo_4O_4(OMe)_2(HOMe)_2(HBpz_3)]_2$ is obtained (Figure 29). Crystal structures of all but the $Mo_2O_4^{2+}$ complex have been determined. The Mo_2O_3 geometric forms provide an example of distortional isomerism with terminal Mo═O distances of 1.67 and 1.78 Å respectively. Both are however yellow-brown in colour, unusual for $Mo_2O_3^{4+}$, which is generally purple with an intense band at ~525 nm. The latter was thought to be independent of the ligands present, a belief which it seems has now to be modified.

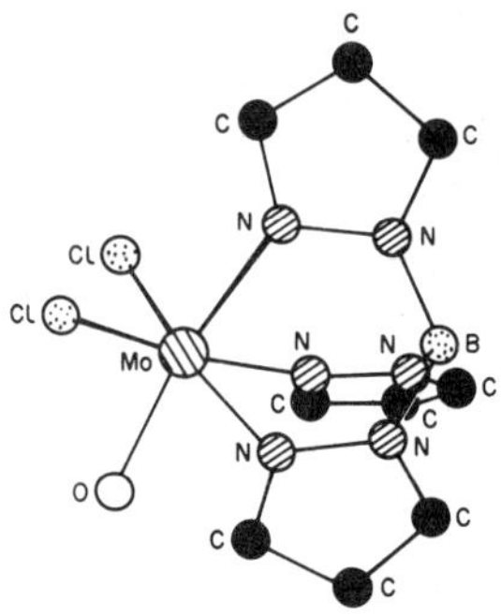

Figure 28 The structure of the hydrotripyrazolylborate, $HBpz_3^-$, Mo^V complex, $[MoOCl_2(HBpz_3)]$[130]

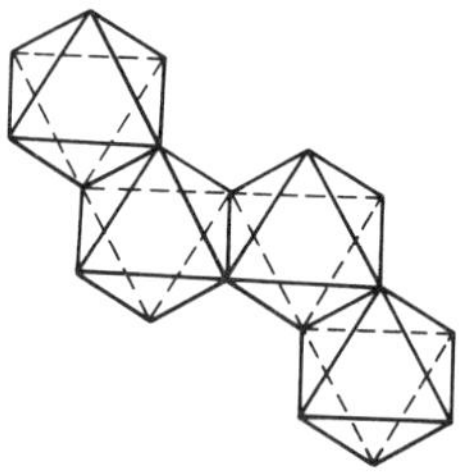

Figure 29 Representation of the tetrameric zigzag chain structure as found in $[Mo_4(HBpz_3)_2(O)_4(\mu\text{-}OMe)_2(HOMe)_2]$ [130]

Complexes reported as mixed-valence tetramers $Mo^{V,V,VI,VI}$, with formulae $[Mo_4O_6Cl_4(O\text{-}Pr^n)_6]$ and $[Mo_4O_8Cl_4(OEt)_4]^{2-}$,[131,132] have been reformulated as $[Mo_4O_6Cl_4(OPr^n)_4(HOPr)_2]$ and $[Mo_4O_8Cl_4(OEt)_2(HOEt)_2]^{2-}$,[130] indicating the importance in this chemistry of distinguishing between a coordinated alkoxide and coordinated alcohol. Both are accordingly tetrameric Mo^V, with close structure analogies to the above zigzag complex. The latter occurs also as a central unit in the $[Mo_6O_{10}(OPr^i)_{12}]$ complex.[133] A cyclic Mo^V octamer has been reported.[134] One example of a distorted cubane $(MoO)_4$ complex of Mo^V, $[Mo_4O_8(Me_2POS)_4]$, having four terminal oxo ligands has been reported (Figure 30).[135]

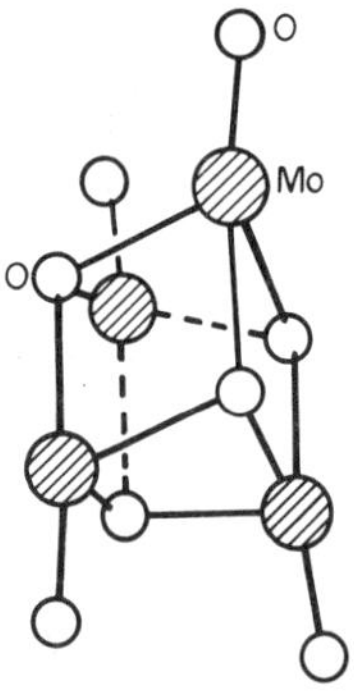

Figure 30 The structure of the Mo^V tetramer core in $[Mo_4O_8(Me_2POS)_4]$ showing the Mo_4O_4 cubic arangement with terminal oxo ligands attached to each Mo^V [135]

The structure of an amorphous orange-brown polymeric Mo^V form obtained from aqueous solutions has not been determined.[136] It has a distinctive peak at 318 nm (ε 3300 $M^{-1}\,cm^{-1}$ per Mo) and is stable in H_2O at pH ~6. On adjustment of $[H^+]$ to 0.17–0.50, $I = 0.50$ M ($H^+/LiClO_4$), aquation to $Mo_2O_4^{2+}$ proceeds by rapid protonation followed by a single rate-determining step. At 25 °C the protonation gives an acid dissociation constant $K_a = 0.13$ M, and $k = 5.5 \times 10^{-3}\,M^{-1}\,s^{-1}$ for the kinetic step.

The rate of ^{18}O exchange of O^{2-} in $[MoOCl_5]^{2-}$ has been studied under conditions of high Cl^- (~12 M), and found to be slow, with a half-life of 1–2 h.[81] One of the reaction paths is believed to involve formation of $[MoOCl_4(H_2O)]^-$ as an intermediate. In dichloromethane the substitution of Cl^-, Br^-, NO_2^- and NO_3^- for $OPPh_3$ (which is *trans* to the oxo group) in

$[MoOCl_3(OPPh_3)_2]$ is consistent with a dissociative (D) mechanism (equation 26). Rate constants obtained are $k_1 = 42.0$, 41.6, 51.8 and $40.3\,s^{-1}$ and $k_2/k_{-1} = 27$, 8, 47 and 31 respectively.[137] Thus in a non-coordinating solvent k_1 is seen to be relatively small.

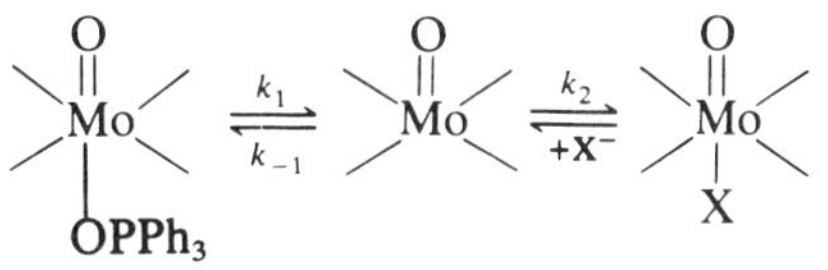

(26)

Measurement of H_2O solvent ^{18}O exchange into aqua $Mo_2O_4^{2+}$ has been possible, using a procedure in which complexing with edta was followed by precipitation of the $[Pt(en)_2]^{2+}$ salt.[81] The two types of oxo ligand were shown to have quite different exchange rates. Using vibrational spectroscopy it was demonstrated that the fast exchanging O atoms are those in the terminal positions. At 0 °C in 0.50 M H^+, these two O atoms exchange with a half-life of ~4 min. The other two O atoms have a half-life of ~70 h at 40 °C in 0.5 M H^+ and ~3 years at 0 °C in 1 M H^+. The rate law for the latter exchange has an $[H^+]^2$ dependence consistent with a mechanism involving Mo_2O_2 ring opening. An involvement of other ligand positions in the exchange of terminal oxo ligands is strongly implicated by the observation that $[Mo_2O_4(edta)]^{2-}$ gives <1% exchange of terminal ligands after 4.25 days. With $[Mo_2O_4(C_2O_4)_2(H_2O)_2]^{2-}$, four O atoms (the terminal oxo and H_2O ligands) exchange at indistinguishable rates followed by the $>10^3$ times slower exchange of bridging O atoms.[81]

Kinetic studies on the 1:1 complexing of NCS^- for H_2O on $[Mo_2O_4(H_2O)_6]^{2+}$,[138] $[Mo_2O_4(C_2O_4)(H_2O)_2]^{2-}$ [118] and di-μ-sulfido $[Mo_2O_2S_2(C_2O_4)_2(H_2O)_2]^{2-}$ [139] (the latter two at pH values of 3.0–4.5) have been reported. Formation rate constants ($10^{-4}\,k$) are respectively 2.9, 0.50 and 21 $M^{-1}\,s^{-1}$ at 25 °C, with I at either 2.0 (first study) or 1.0 M in perchlorate. With pyridine as the incoming ligand the rate constant of 0.50 decreases to 0.30 $M^{-1}\,s^{-1}$. The mechanistic interpretation has to take into account the observation that the oxalate of $[Mo_2O_4(C_2O_4)_2(H_2O)_2]^{2-}$ may be undergoing rapid rearrangement.[32] Therefore substitution is not necessarily confined to a basal plane site, and as in the case of $[VO(H_2O)_5]^{2+}$, rapid substitution at an axial position (equilibration process K) is believed to be followed by the slower exchange to a position in the plane of the complex. The not too dissimilar rate constants for $[Mo_2O_4(H_2O)_6]^{2+}$ and $[Mo_2O_4(C_2O_4)_2(H_2O)_2]^{2-}$ are noted. With the di-μ-sulfido complex a faster rate is observed and the μ-sulfido ligands produce a labilizing effect. The kinetics of the multistage equilibration reaction of $[Mo_2O_4(H_2O)_6]^{2+}$ with edta to give $[Mo_2O_4(edta)]^{2-}$ have been studied, $[H^+] = 0.5–2.0$ M, and are observed to be much slower ($t_{1/2} < 1$ min).

From kinetic studies the oxidation of aqua $Mo_2O_4^{2+}$ with $[IrCl_6]^{2-}$, $[Fe(phen)_3]^{3+}$ and $[(NH_3)_5Co(O_2)Co(NH_3)_5]^{5+}$ proceed by competing pathways.[140,141] The first is oxidant independent, and is the result of the dimer undergoing rate-determining change to a singly bridged form, or to monomeric Mo^V, which then reacts rapidly with the oxidant. Interestingly, this path has an $[H^+]^{-1}$ dependence suggesting involvement of a conjugate-base form of $Mo_2O_4^{2+}$, and gives $k = 3 \times 10^{-6}\,M\,s^{-1}$. This contrasts with the $[HCl]^2$ dependence (not at constant ionic strength) indicated for the ^{18}O exchange of bridging oxo ligands.[142] A second rate law term, first order in oxidant as well as $[Mo_2O_4^{2+}]$, and representing direct attack of the oxidant on the dimer, is observed. This has a dependence on $[H^+]^{-1}$ of the form $a + b[H^+]^{-1}$, where the contribution from b is presumably favoured because of the increase in extent of hydrolysis accompanying the $Mo^V \rightarrow Mo^{VI}$ change. The reactions with I_3^-, NO_3^- and $[PtCl_6]^{2-}$ as oxidants are very much slower.

In contrast to the above study the $[IrCl_6]^{2-}$ oxidation of $[Mo_2O_4(edta)]^{2-}$ gives evidence for an $Mo^{V,VI}$ intermediate.[143] In the presence of edta the $Mo^{V,VI}$ is presumably held together for a sufficiently long time for it to become kinetically significant. Thus on addition of the product $[IrCl_6]^{3-}$ the reverse reaction is seen to be effective, and there is a decrease in the rate of reaction, which can be quantified in terms of the mechanism shown in equations (27)–(29).

$$Mo_2^V + Ir^{IV} \rightleftharpoons Mo^{V,VI} + Ir^{III} \quad (27)$$

$$Mo^{V,VI} \longrightarrow Mo^V + Mo^{VI} \quad (28)$$

$$Mo^V + Ir^{IV} \longrightarrow Mo^{VI} + Ir^{III} \quad (29)$$

The reaction of $Mo_2O_4^{2+}$ with NO_3^- in tartrate buffer (which is capable of coordinating to the Mo_2^V) at pH 2.2–3.5 has also been studied.[144] The kinetics give a half-order dependence on

$[Mo_2O_4^{2+}]$ consistent with a mechanism shown in equations (30) and (31). From EPR measurements it has been confirmed that mononuclear Mo^V is present.

$$Mo_2^V \rightleftharpoons 2Mo^V \tag{30}$$

$$Mo^V + NO_3^- \longrightarrow \text{products} \tag{31}$$

Pulse radiolysis studies on $[Mo_2O_4(edta)]^{2-}$, $[Mo_2O_2S_2(edta)]^{2-}$, $[Mo_2O_4(cys)_2]^{2-}$ and $[Mo_2O_4(C_2O_4)_2(H_2O)_2]^{2-}$ at pH ~6, in which the Mo_2^V is reduced with e_{eq}^- and Zn^+ to the $Mo^{IV,V}$ mixed-valence form have been reported.[145] The latter absorbs in the visible range (ε 300–500 $M^{-1}\,cm^{-1}$ per dimer). In the absence of O_2 the decay is relatively slow ($t_{1/2}$ ~5–10 s), and does not result in Mo^{III} production. With O_2 present there is rapid (~$10^8\,M^{-1}\,s^{-1}$) oxidation back to Mo_2^V.

Inner-sphere paths are observed in the Cr^{2+} reduction of 10^{-4}–10^{-3} M solutions of Mo_2^V, $[Mo_2O_4(H_2O)_6]^{2+}$, and of Mo_3^{IV}, $[Mo_3O_4(H_2O)_9]^{4+}$, at 25 °C, $I = 2.0$ M (NaPTS).[146] In 1.9 M HPTS and with $[Cr^{2+}]$ in a greater than ten-fold excess of Mo_2^V, the reaction proceeds *via* a grey-green Cr-containing intermediate (~1 min), to give a product with the Mo_2^{III} spectrum (~24 h). The rate law for formation of the intermediate is of the form $k_1[Cr^{2+}]^2[Mo_2^V][H^+]$, with $k_1 = 9.1 \times 10^3\,M^{-3}\,s^{-1}$. Decay of the intermediate is independent of Cr^{2+} and can be expressed as k_2[intermediate]$[H^+]$. The rate constant $k_2 = 2.0 \times 10^{-5}\,M^{-1}\,s^{-1}$ is believed to correspond to a process involving loss of Cr^{III}. With a $Cr^{2+}:Mo_2^V$ ratio of reactants of 2:1 evidence for two intermediates is obtained (~1 min), one (green in colour) giving a Cr:Mo ratio of 1:1, and the other the same as that generated in the reaction of excess Cr^{2+} with Mo_2^V giving a Cr:Mo ratio of 2:1. Some 30% of the Mo remains as Mo_2^V. Over longer periods (~22 h) 60–70% overall conversion to Mo_3^{IV} is observed, and 24–30% of the Mo is present as Mo_2^V. In separate experiments, with Cr^{2+} in a greater than ten-fold excess, Mo_3^{IV} is reduced to $Mo^{III,III,IV}$ in a two-stage process, complete within 1 min, and then through to Mo_3^{III} (~40 min). Reduction of Mo_2^V to either Mo_2^{III} or Mo_3^{III} is observed therefore, depending on whether a single addition of excess Cr^{2+} or successive additions of Cr^{2+} are made.

36.1.2.6 Oxidation State VI

The solution chemistry of the Mo^{VI} state has been an area of intense research activity for several decades. It is dominated by the isopolymolybdate and heteropolymolybdate forms,[147] brief mention of which is included here. Such studies remain an active area of research with applications in the area of industrial catalysis.[1] More than 65 elements from all groups of the Periodic Table (except the rare gases) are implicated as heteroatoms in such structures.[147] In addition to the text by Pope,[147] a review of isopolyanions has appeared.[148] An overview on the hydrolysis of cations is also relevant.[149] At pH > 7 Mo^{VI} exists as the monomeric tetrahedral $[MoO_4]^{2-}$ ions. Polymers are generated by conversion of tetrahedral Mo^{VI} to octahedral forms in which there is edge, corner and (occasionally) face sharing of coordinated O^{2-} ions between adjacent Mo atoms. In the absence of heteropolyanions protonation to give $[HMoO_4]^-$, more precisely $[MoO_3(OH)]^-$ (pK_a 3.47 in 1 M HCl), is followed by a second protonation to give $[H_2MoO_4]$ (pK_a 3.7 in 1 M NaCl). The latter at least (possibly both) is believed to be octahedral and is sometimes referred to as $[Mo(OH)_6]$ although other formulations have been suggested.[150] The incidence of protonation at pH ~ 7 triggers polymerization. Polymeric forms play a dominant role in the chemistry of Mo^{VI} from pH 7 down to 2.[151] At pH 2 to 1 (depending on the concentration) break down of polymers to give dimeric and monomeric octahedral forms occurs, where the latter are often referred to as $[Mo(OH)_6]$. The first protonation constant for $[Mo(OH)_6]$ has been determined.[151b] Singly and doubly charged cationic species are present in such solutions,[151b,152] and formulae which have been suggested include *cis*-$[MoO_2(H_2O)_4]^{2+}$ and $[Mo(OH)_4(H_2O)_2]^{2+}$. The *cis*-dioxo structure is found in a number of mononuclear coordination complexes, including *cis*-$[MoO_2(Et_2dtc)_2]$ (Et_2dtc = *N,N*-diethyldithiocarbamate),[153] *cis*-$[MoO_2(tox)_2]$ (tox = thiooxine or 8-mercaptoquinolate),[154] Λ-*cis*-$[MoO_2\{(2R)\text{-cysOMe}\}_2]$ (cysOMe = 1-cysteine methyl ester)[155] and Λ-*cis*-$[Mo_2(S)\text{-penOMe})_2]$ (penOMe = penicillamine methyl ester).[156] Two *cis* oxos are also present in the oxalato μ-oxo complex $[(H_2O)(C_2O_4)O_2MoOMoO_2(C_2O_4)(H_2O)]^{2-}$ (Figure 31).[157] Facial trioxo coordination is on the other hand present in $[MoO_3dien]$ (dien = diethylenetriamine)[158] and the ethylenediaminetetraacetate complex $[O_3Mo(edta)MoO_3]^{4-}$ [159] (Figure 32). Features of both structures are the MoO_3 bond angles (~106°) which are approaching those of a regular tetrahedron, and the Mo—O(oxo) bond lengths of ~1.74 Å which indicate a bond order of two. The bond angles N—Mo—N (75°) for the dien complex, and N—Mo—O (73°) and O—Mo—O (75°) for the edta complex also suggest that the structures should be regarded as pseudotetrahedral. From Raman studies[160] it has been concluded that $[MoO_3dien]$ is largely dissociated in water (equation 32), and reforms rapidly and quantitatively as a solid on addition of alcohol at pH ~7. Thus although it is necessary to replace an oxo ligand on $[MoO_4]^{2-}$ the reaction occurs rapidly.

$$[MoO_3dien] + H_2O \rightleftharpoons [MoO_4]^{2-} + dienH_2^{2+} \quad (32)$$

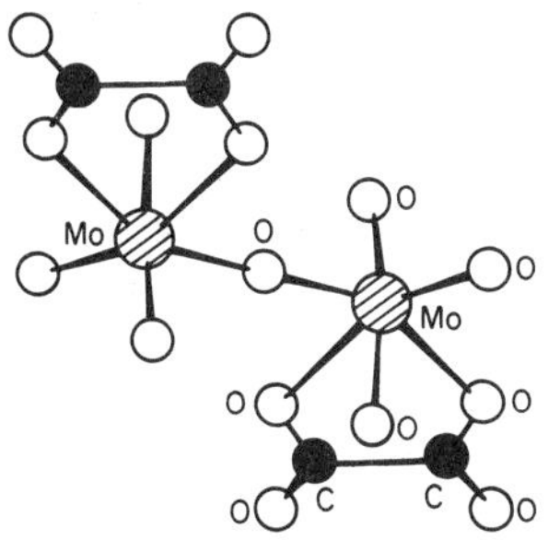

Figure 31 The structure of the Mo^{VI} complex $[Mo_2O_5(C_2O_4)_2(H_2O)_2]^{2-}$ [157]

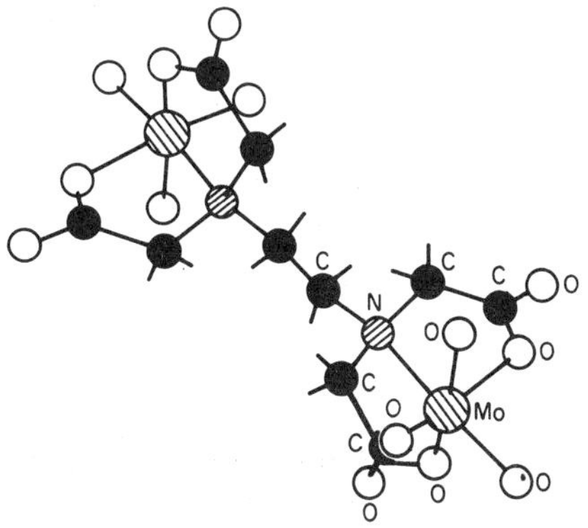

Figure 32 The structure of the Mo^{VI} complex $[O_3Mo(edta)MoO_3]^{4-}$ [159]

The polymeric forms obtained on decreasing the pH from 7 have been extensively investigated. For solutions with Mo^{VI} concentrations greater than 10^{-3} M, and pH in the range 3.0 to 5.5, the predominant form present is the heptamolybdate(VI) ion, $[Mo_7O_{24}]^{6-}$, sometimes referred to as paramolybdate (equation 33). Commercially available ammonium molybdate, $(NH_4)_6[Mo_7O_{24}]\cdot 4H_2O$, is obtained by crystallization of a solution of MoO_3 in aqueous NH_3. In spite of numerous attempts, no intermediate has been unambiguously characterized in the aqueous conversion of $[MoO_4]^{2-}$ to $[Mo_7O_{24}]^{6-}$. At pH < 4 the β-$[Mo_8O_{26}]^{4-}$ octamolybdate form is obtained (equation 34). Representations of $[Mo_7O_{24}]^{6-}$ and $[Mo_8O_{26}]^{4-}$ are shown in Figure 33. The existence of significant amounts of the $[Mo_8O_{26}]^{4-}$ ion has been controversial, but recent Raman and X-ray scattering[161] studies would seem to provide confirmation of its existence. Previously the interpretation of EMF data (pH 2–3) has been uncertain, because equally acceptable fits are given by a series of protonated $[Mo_7O_{24}]^{6-}$ forms, as by a mixture of protonated $[Mo_7O_{24}]^{6-}$ and $[Mo_8O_{26}]^{4-}$. From one set of stability constant determinations, Figure 34 has been obtained, indicating the distribution of different species. The ditetrahedral $[Mo_2O_7]^{2-}$ ion has recently been identified in $Mg[Mo_2O_7]$ and in the double salt $K_2[Mo_2O_7],KCl$.[162,163] These structures are to be contrasted with the crystalline product $(NH_4)_2[Mo_2O_7]$, obtained from hot aqueous ammonium molybdate solution after some hours, which consists of infinite chains of tetrahedral MoO_4 and octahedral MoO_6 units.[164]

$$7[MoO_4]^{2-} + 8H^+ \rightleftharpoons [Mo_7O_{24}]^{6-} + 4H_2O \quad (33)$$

$$8[MoCl]^{2-} + 12H^+ \rightleftharpoons [Mo_8O_{26}]^{4-} + 6H_2O \quad (34)$$

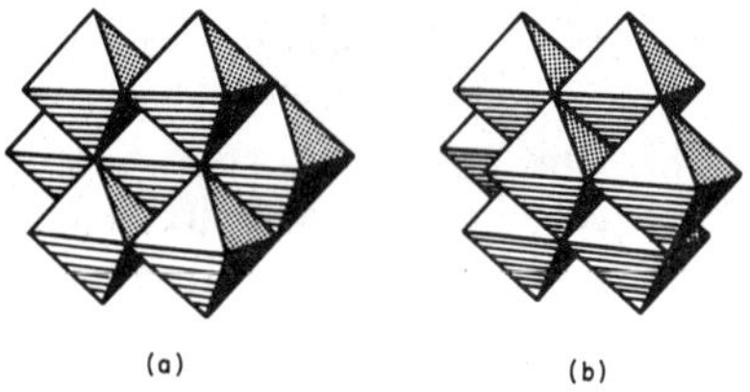

Figure 33 The structures of (a) heptamolybdate, $[Mo_2O_{24}]^{6-}$, and (b) octamolybdate, $[Mo_8O_{26}]^{4-}$ (one MoO_6 octahedron hidden)

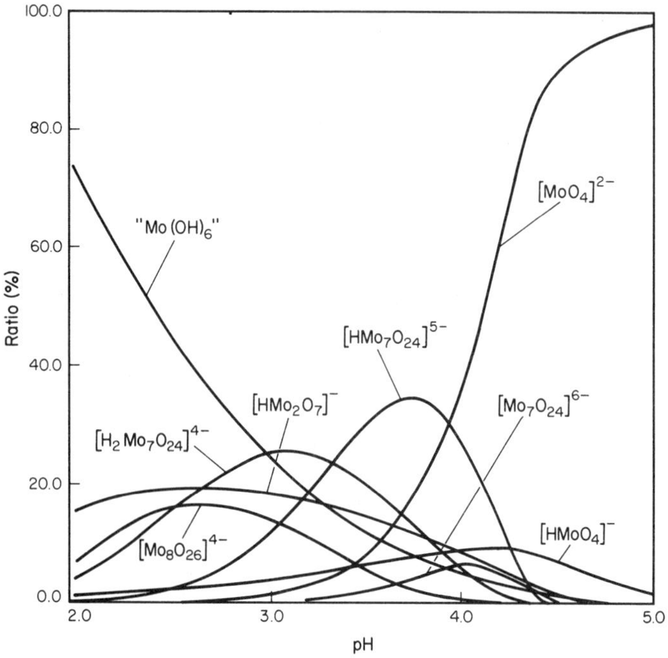

Figure 34 Distribution of Mo^{VI} forms with pH for a solution of 5×10^{-4} M molybdate at 25 °C, $I = 1.0$ M (NaCl), as calculated by A. Nagasawa and R. Iwata (unpublished work) from data in ref. 151

Solutions of Mo^{VI} acidified to $H^+/[MoO_4]^{2-} \sim 1.8$ contain one or more very large polymolybdate structures. Earlier measurements based on ultracentrifugation and EMF studies,[165] through to a recent structure report,[166] are consistent with a formula $[Mo_{36}O_{112}(H_2O)_{18}]^{8-}$. The reaction can be summarized as in equation (35). There are two seven-coordinate Mo^{VI} atoms in this structure. In all isopoly and heteropoly structures the metal ion does not lie at the centre of its polyhedron, but is displaced towards the exterior of the structure and towards a vortex or edge of its own polyhedron. Structures appear to be governed by electrostatic and radius-ratio principles as observed for extended ionic lattices. The Mo^{VI} tetrahedral radius is 0.55 Å (0.56 Å for W), the octahedral radius 0.73 Å (0.74 Å for W), and the radius of O^{2-} is 1.40 Å.[167]

$$36[MoO_4]^{2-} + 64H^+ \rightleftharpoons Mo_{36}O_{112}^{8-} + 32H_2O \qquad (35)$$

Important differences are observed in comparing the behaviour of Mo^{VI} and W^{VI}. Equilibria involving $[MoO_4]^{2-}$ and polymolybdates in aqueous solution are established rapidly and are complete in a matter of minutes, whereas those for W^{VI} can take several weeks. The polyanions of tungsten are made up of WO_6 octahedra, but in other respects the behaviour of W^{VI} solutions is quite different, and structurally the polyanions in one series do not have precise counterparts in the other. Known structures of isopolytungstates from aqueous solution include $[W_4O_{16}]^{8-}$, $[W_4O_{19}]^{2-}$, $[W_{10}O_{32}]^{4-}$, $[H_2W_{12}O_{42}]^{10-}$ and $[H_2W_{12}O_{40}]^{6-}$.[147]

A number of other isopolymolybdate anions including $[Mo_6O_{19}]^2$, α-$[Mo_8O_{26}]^{4-}$ and $[Mo_5O_{17}H]^{3-}$ can be stabilized in non-aqueous or mixed solvents.[168] The $[Mo_2O_7]^{2-}$ ion is also obtained from acetonitrile solution. Tetrahedral coordination is retained in the latter.[169] There are two distorted tetrahedra in α-$[Mo_8O_{26}]^{4-}$ above and below a crown of six octahedra formed by edge sharing. A tetrahedron spanning a ring of four edge- and face-shared octahedra is also believed to be present in $[Mo_5O_{17}H]^{3-}$.[170] The α and β structures can coexist in solution, and a cation dependent isomerization of α into β has been observed.[171]

Isopolymolybdates are generally colourless, and heteropoly forms can be coloured if another transition metal ion is present. Some other heteropolyanions can be coloured, for example that obtained by addition of ammonium molybdate to phosphoric acid to give ammonium 12-molybdophosphate, $[NH_4]_3[PMo_{12}O_{40}]$, which is yellow. The $[Mo_6O_{19}]^{2-}$ anion is unique for an isopolyanion in being yellow (λ_{max} 325 nm) and having a structure with a monocoordinated terminal oxo group rather than MoO_2 units. This ion is electrochemically reduced to the mixed valence $[Mo_6O_{19}]^{3-}$ and $[Mo_6O_{19}]^{4-}$ forms in DMF. Incorporation of the Mo^V state in blue-red and yellow mixed-valence species formulated as $[Mo_2^VMo_4^{VI}O_{18}]^{2-}$, $[Mo_3^VMo_3^{VI}O_{18}]^{2-}$ and $[Mo_2^VMo_4^{VI}O_{17}H]^{2-}$ respectively has also been reported. The reduction of addenda atoms in heteropolyanions results in the formation of heteropoly blues, and gives rise to a vast chemistry. The added electrons are delocalized according to varying timescales over certain atoms and/or regions of the structure.

Because the surfaces of polyanions have some similarities to those of metal oxides, it is thought they may have some relevance in the area of heterogeneous catalysis. As a result a whole new area concerned with the synthesis and study of organic and organometallic derivatives of polyanions is being investigated.[173]

A kinetic investigation of the ^{18}O exchange between water and molybdate $[MoO_4]^{2-}$ at pH > 11, $[OH^-] = 3 \times 10^{-3}$–0.15 M, $I = 1.00$ M ($NaClO_4$), has been carried out.[174] The rate law (equation 36) indicates paths involving reaction with H_2O and OH^- ($k_1 = 0.33\ s^{-1}$, $k_{OH} = 2.22\ M^{-1}\ s^{-1}$ at 25 °C). The corresponding k_1 values for $[CrO_4]^{2-}$ ($3.2 \times 10^{-7}\ s^{-1}$) and $[WO_4]^{2-}$ ($0.44\ s^{-1}$) provide one of the few kinetic comparisons for Cr, Mo and W. The difference in rate constants is attributable to the larger enthalpy of activation in the case of chromate, reflecting tighter Cr^{VI}—O bonds. It has to be borne in mind that there are other contributing paths for chromate, the full rate law having been determined as equation (37). The molybdate exchange cannot readily be studied by ^{18}O labelling at pH 11, because protonation is effective and much faster rates are observed.[174]

$$\text{Rate} = k_1[MoO_4^{2-}] + k_{OH}[MoO_4^{2-}][OH^-] \tag{36}$$

$$\text{Rate} = k_1[CrO_4^{2-}] + k_2[HCrO_4^-] + k_3[H^+][HCrO_4^-] + k_4[HCrO_4^-] + k_5[HCrO_4^-]^2 \tag{37}$$

Fast monomer–dimer equilibria involving Mo^{VI} in 0.2–0.3 M $HClO_4$, $I = 3.0$ M ($LiClO_4$), have been studied by the temperature-jump method.[176] A major pathway involves rapid dimerization of a monomer form, referred to as $[HMoO_3]^+$, which might alternatively be written $[MoO_2(OH)(H_2O)_3]^+$ or a related form. At 25 °C, $I = 3.0$ M $[LiClO_4]$, rate constants are $k_f = 1.71 \times 10^5\ M^{-1}\ s^{-1}$, and $k_b = 3.2 \times 10^3\ s^{-1}$. Information has been obtained by the stopped-flow method for the interconversion of tetrahedral chromate and dichromate ions.[177] At pH 2–4 hydrogen chromate, $[HCrO_4]^-$, is specified as the reactant. From this and other similar studies on the substitution reactions of $[HCrO_4]^-$ there is evidence for a dissociative process.[178] Further information concerning precise formulae of Mo^{VI} in acidic solutions is required to enable more extensive comparisons.

Temperature-jump studies of Mo^{VI} solutions at pH 5.50–6.75 have been reported for the monomer–heptamer and heptamer–octamer interconversions which are complete within a few milliseconds.[179] Molybdate as $[HMoO_4]^-$ (here assumed to be tetrahedral) reacts with bidentate 8-hydroxyquinoline[180] and catechol,[181] to give octahedral products in net 4→6 conversions.[180] Table 4 summarizes the kinetic data, where different degrees of protonation of the ligating groups are indicated. For the reaction with edta a path assigned to the reaction of $[H_2edta]^{2-}$ with $[HMoO_4]^-$ gives a rate constant of $2.3 \times 10^5\ M^{-1}\ s^{-1}$.[182] A reaction in which the tetrahedral geometry of $[MoO_4]^{2-}$ is retained is the rapid stopped-flow equilibration of molybdate with $[Co(NH_3)_5(H_2O)]^{3+}$ (pK_a 6.3) at pH 7–8 (equation 38).[183]

$$[Co(NH_3)_5(H_2O)]^{3+} + [MoO_4]^{2-} \rightleftharpoons [Co(NH_3)_5(OMoO_3)]^+ + H_2O \tag{38}$$

Substitution is at the Mo^{VI} centre with retention of the Co—O bond. The first-order $[H^+]$-dependent term is consistent with $[HMoO_4]^-$ as a reactant, and the $[H^+]$-independent term is assigned to the reaction of $[HMoO_4]^-$ with $[Co(NH_3)_5OH]^{2+}$. An interesting situation arises in the reaction of $[MoO_4]^{2-}$ with $[Cr(edta)(H_2O)]^-$ at pH 7.3–8.7.[184] As in the case of

Table 4 Second-order Rate Constants for Reactions of $[HMoO_4]^-$ at 25 °C

Ligand	*Ionic strength*	*Coordination change at Mo^{VI}*	k ($M^{-1} s^{-1}$)
H-oxine	0.20	$4 \rightarrow 6$	4.5×10^6
H-oxine-SO_3^-	0.20	$4 \rightarrow 6$	3.9×10^6
oxine$^-$	0.20	$4 \rightarrow 6$	1.5×10^8
oxine-SO_3^{2-}	0.20	$4 \rightarrow 6$	4.0×10^7
H-catechol$^-$	0.10	$4 \rightarrow 6$	1.9×10^8
H_2-edta	0.10	$4 \rightarrow 6$	2.3×10^5
$[Co(NH_3)_5H_2O]^{3+}$	1.0	$4 \rightarrow 4$	3.2×10^5
$[Co(NH_3)_5OH]^{2+}$	1.0	$4 \rightarrow 4$	6.6×10^4
$[Cr(edta)H_2O]^-$	1.0	$4 \rightarrow 4$	3.1×10^4
HS^-	0.50	$4 \rightarrow 4$	1.3×10^6

H-oxine = 8-hydroxyquinoline.
H-oxine-SO_3^- = 8-hydroxyquinoline-5-sulfonic acid.
H-catechol$^-$ = product of first acid dissociation of catechol.

$[Co(NH_3)_5(H_2O)]^{3+}$, there are two terms for complex formation in the rate law, one of which can be assigned to the reaction of $[Cr(edta)(H_2O)]^-$ with $[MoO_4]^{2-}$ (rate constant 21 $M^{-1} s^{-1}$). The second, which is first order in $[H^+]$, corresponds to replacement of the H_2O of $[Cr(edta)H_2O]^-$ by $[HMoO_4]^-$ (3.1×10^4 $M^{-1} s^{-1}$). Whereas the first is believed to correspond to replacement of the H_2O (labilized by the presence of the carboxylate) at the Cr^{III}, the latter as in other cases corresponds to substitution at the Mo^{VI} centre. There are proton ambiguities however because the first reaction could also be assigned to the reaction of $[Cr(edta)OH]^{2-}$ (pK_a 7.4) with $[HMoO_4]^-$. In all these examples the interpretation requires that there are pathways in which there is protonation of $[MoO_4]^{2-}$ prior to complexation. Protonation labilizes at least one oxo ligand and may itself induce a $4 \rightarrow 6$ change in coordination to give a more labile species. Consistent with this, rates are much more rapid than for $H_2{}^{18}O$ exchange with $[MoO_4]^{2-}$.

Tetrahedral sulfido or thiolate complexes are known for a number of d^0 transition metal ions including Mo^{VI}.[185] Preparative procedures involve the reaction of solutions of oxyanions with H_2S. In addition to $[MoS_4]^{2-}$ the mixed oxo/sulfido complexes $[MoOS_3]^{2-}$, $[MoO_2S_2]^{2-}$ and $[MoO_3S]^{2-}$ are well characterized. Although the latter is difficult to obtain in a pure crystalline state, X-ray crystal structure information on $(NH_4)_2[MoS_4]$ (Mo—S = 2.17 Å),[186] and $Cs_2[MoOS_3]$ (Mo—S = 2.18 Å; Mo—O = 1.79 Å)[187] have been reported, and the Mo—O distance is slightly longer than in Mo^{VI} complexes $K_2[MoO_4]$ (1.76 Å), $[MoO_3(dien)]$ (1.74 Å) and $[Mo_2O_5(C_2O_4)_2(H_2O)_2]^{2-}$ (Mo—O_t 1.69 Å). Bridged Mo^{VI} μ-sulfido complexes do not appear to be formed and there are no counterparts in the polymeric (and heteropoly) species found from $[MoO_4]^{2-}$. Kinetic studies on the interconversion of $[MoO_xS_{4-x}]^{2-}$ forms have been carried out.[188] The 1:1 equilibration of H_2S with $[MoO_4]^{2-}$ can be expressed as in equation (39).

$$[MoO_4]^{2-} + H^+ + HS^- \underset{k_{10}}{\overset{k_{01}}{\rightleftharpoons}} [MoO_3S]^{2-} + H_2O \qquad (39)$$

There are few quantitative studies with H_2S as a reactant, and some difficulties were experienced in choosing conditions appropriate to the study. At pH 9.2–10.2 (0.25 M NH_3/NH_4^+ buffer), $I = 0.50$ M (NaCl), plots of equilibration rate constants k_{eq} against $[MoO_4^{2-}]$ are in accord with equation (40).

$$k_{eq} = k_f[MoO_4^{2-}] + k_b \qquad (40)$$

The formation constant k_f is dependent on $[H^+]$ giving $k = 4.0 \times 10^9$ $M^{-2} s^{-1}$, and k_b for the reverse reaction is 6.5×10^{-3} s^{-1}. Interpretation of k in terms of a reaction between $[HMoO_4]^-$ and HS^- is preferred rather than the alternative of $[MoO_4]^{2-}$ with H_2S, and on this basis a second-order rate constant of 1.3×10^6 $M^{-1} s^{-1}$ is obtained, of similar magnitude to those listed in Table 4. Other formation rate constants for the different $[MoO_xS_{4-x}]^{2-}$ ions have to be left in the form of the experimental third-order rate constants since the relevant protonation constants are not known. For the rate law $k[MoO_xS_{4-x}^{2-}][HS^-][H^+]$, k values range from 4.0×10^9 $M^{-2} s^{-1}$ for the reaction of $[MoO_4]^{2-}$ to less than 1.6×10^6 $M^{-2} s^{-1}$ for $[MoOS_3]^{2-}$. Rate constants for aquation of $[MoO_3S]^{2-}$ (6.5×10^{-3} s^{-1}), $[MoOS_3]^{2-}$ ($\sim 5 \times 10^{-5}$ s^{-1}) and

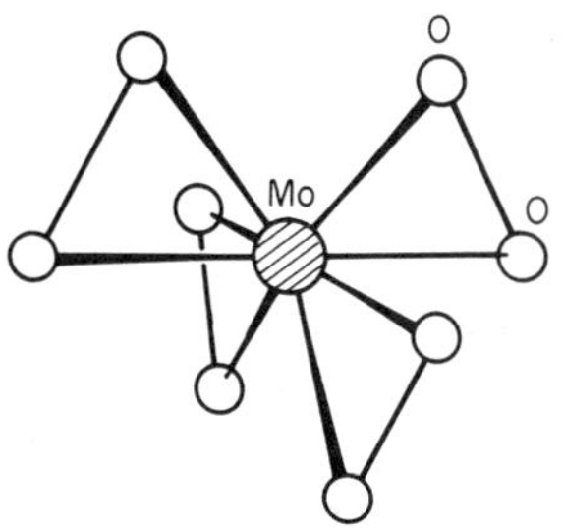

Figure 35 The structure of the Mo^{VI} peroxo complex, $[Mo(O_2)_4]^{2-}$ [190]

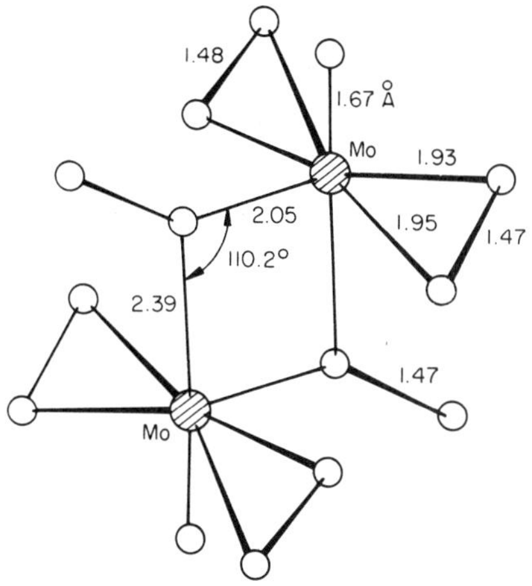

Figure 36 The structure of the Mo^{VI} peroxo/hydroperoxo complex $[(O_2)_2OMo(OOH)_2MoO(O_2)_2]^{2-}$ [196]

$[MoS_4]^{2-}$ ($1.6 \times 10^{-6}\,s^{-1}$) indicate a trend to smaller values as more sulfide ligands are introduced. The presence of sulfide does not appear to give a labilizing effect.

Different crystalline phases can be separated from aqueous solutions of potassium molybdate(VI) following reaction with varying amounts of hydrogen peroxide, pH 4–8.[189] Crystal structure determinations of $[Zn(NH_3)_4][Mo(O_2)_4]$,[190] $K_2[O\{MoO(O_2)_2(H_2O)\}_2]\cdot 2H_2O$,[190] $K_4[Mo_4O_{12}(O_2)_2]$[192] and $K_6[Mo_7O_{22}(O_2)_2]\cdot 8H_2O$[193] have for example been reported. In all cases the O_2^{2-} is bound sideways to the metal, with O—O distances varying from 1.38 to 1.55 Å, but generally around 1.48 Å, which is the expected value for peroxide. The peroxo ligands in the dark red $[Mo(O_2)_4]^{2-}$ compound (O—O distance 1.55 Å) are positioned tetrahedrally about the Mo^{VI} (Figure 35). A Cr^V analogue $[Cr(O_2)_4]^{3-}$, having a similar configuration (O—O distance 1.48 Å), is known. An interesting variation is the occurrence of hydroperoxide as a bridging ligand in $(pyH)_2[(O_2)_2OMo(OOH)_2MoO(O_2)_2]$,[194] where the bridge has a pendant —O(OH)— arrangement (Figure 36). Only one other example of this structure is known in the Co_2^{III} complex $[(en)_2Co(NH_2, O_2H)Co(en)_2]^{4+}$.[195] The complex $K_2[MoO(O_2)_2(C_2O_4)]$ is monomeric,[196] and to be regarded as five- or seven-coordinate depending whether the O_2^{2-} is assigned one or two coordination positions.

36.1.3 REFERENCES

1. See *e.g.* 'Proceedings of Climax International Conferences on the Chemistry and Uses of Molybdenum', 1973 (Reading), 1976 (Oxford), 1979 (Ann Arbor), 1982 (Colorado). The proceedings to the fifth of these meetings (Newcastle, 1985) have appeared in *Polyhedron,* 1986, **5,** 1–606.
2. (a) D. T. Richens and A. G. Sykes, *Commun. Inorg. Chem.*, 1981, **1,** 141; (b) *Inorg. Synth.*, 1985, **23,** 130.
3. F. A. Cotton and R. A. Walton, 'Multiple Bonds Between Metal Atoms', Wiley Interscience, New York, 1982.
4. (a) D. Lawton and R. Mason, *J. Am. Chem. Soc.*, 1965, **87,** 921; (b) F. A. Cotton and C. B. Harris, *Inorg. Chem.*, 1965, **4,** 330.
5. J. V. Bencic and F. A. Cotton, *Inorg. Chem.*, 1969, **8,** 7.
6. A. P. Sattelberger, K. W. McLaughlin and J. C. Huffman, *J. Am. Chem. Soc.*, 1981, **103,** 2880.
7. F. A. Cotton and L. D. Gage, *Nouv. J. Chim.*, 1977, **1,** 441.
8. T. A. Stephenson, E. Bannister and G. Wilkinson, *J. Chem. Soc.*, 1964, **2,** 538.

9. A. Bino, F. A. Cotton and Z. Dori, *J. Am. Chem. Soc.*, 1981, **103**, 243.
10. F. A. Cotton, *Polyhedron*, 1986, **5**, 3.
11. J. J. Ziolkowski, M. Moszner and T. Glowiak, *J. Chem. Soc., Chem. Commun.*, 1977, 760.
12. A. Bino, F. A. Cotton and T. R. Felthouse, *Inorg. Chem.*, 1979, **18**, 2599.
13. C. D. Garner, R. G. Senior and T. J. King, *J. Am. Chem. Soc.*, 1976, **98**, 647.
14. (a) V. Katovic, J. L. Templeton, R. J. Hoxmeier and R. E. McCarley, *J. Am. Chem. Soc.*, 1975, **97**, 5300; (b) V. Katovic and R. E. McCarley, *J. Am. Chem. Soc.*, 1978, **100**, 5586.
15. R. E. McCarley, J. L. Templeton, T. J. Colburn, V. Katovic and R. J. Hoxmeier, *Adv. Chem. Ser.*, 1976, **150**, 318.
16. J. H. Baxendale, C. D. Garner, R. G. Senior and P. Sharpe, *J. Am. Chem. Soc.*, 1976, **98**, 637.
17. A. R. Bowen and H. Taube, *Inorg. Chem.*, 1974, **13**, 2245.
18. J. V. Brencic and F. A. Cotton, *Inorg. Chem.*, 1970, **9**, 351.
19. J. V. Brencic, I. Leban and P. Segedin, *Z. Anorg. Allg. Chem.*, 1978, **444**, 211.
20. J. V. Brencic and P. Segedin, *Inorg. Chim. Acta*, 1978, **29**, L281.
21. R. J. Mureinik, *Inorg. Chim. Acta*, 1977, **23**, 103.
22. S. S. Miller and A. Haim, *J. Am. Chem. Soc.*, 1983, **105**, 5624.
23. A. Bino and D. Gibson, *J. Am. Chem. Soc.*, 1980, **102**, 4277.
24. F. A. Cotton and E. Pederson, *Inorg. Chem.*, 1975, **14**, 399.
25. F. A. Cotton, B. A. Frenz, E. Pedersen and T. R. Webb, *Inorg. Chem.*, 1975, **14**, 391.
26. F. A. Cotton, B. A. Frenz and T. R. Webb, *J. Am. Chem. Soc.*, 1973, **95**, 4431.
27. A. Bino and F. A. Cotton, *Inorg. Chem.*, 1979, **18**, 3562.
28. E. Hochberg and E. H. Abbott, *Inorg. Chem.*, 1978, **17**, 506.
29. W. C. Trogler and H. B. Gray, *Acc. Chem. Res.*, 1978, **11**, 232.
30. J. E. Finholt, P. Leupin and A. G. Sykes, *Inorg. Chem.*, 1983, **22**, 3315.
31. S. Ooi, M. Nishizawa, K. Matsumoto, H. Kuroya and K. Saito, *Bull. Chem. Soc. Jpn.*, 1979, **52**, 452.
32. K. Saito and Y. Sasaki, *Adv. Inorg. Bioinorg. Mech.*, 1982, **1**, 187.
33. K. Teramoto, Y. Sasaki, K. Migita, M. Iwaizumi and K. Saito, *Bull. Chem. Soc. Jpn.*, 1979, **52**, 446.
34. W. C. Trogler, D. K. Erwin, G. L. Geoffroy and H. B. Gray, *J. Am. Chem. Soc.*, 1978, **100**, 1160.
35. A. Bino, *Inorg. Chem.*, 1981, **20**, 623.
36. M. A. Harmer and A. G. Sykes, *Inorg. Chem.*, 1981, **20**, 3963.
37. M. G. B. Drew, P. C. Mitchell and C. F. Pygall, *J. Chem. Soc., Dalton Trans.*, 1979, 1213.
38. K. H. Lohmann and R. C. Young, *Inorg. Synth.*, 1953, **4**, 97.
39. A. R. Bowen and H. Taube, *J. Am. Chem. Soc.*, 1971, **93**, 3287.
40. Y. Sasaki and A. G. Sykes, *J. Chem. Soc., Chem. Commun.*, 1973, 767.
41. M. Brorson and C. E. Schaffer, *Acta Chem. Scand., Ser. A*, 1986, **40**, 358.
42. J. K. Beattie, S. P. Best, B. W. Skelton and A. H. White, *J. Chem. Soc., Dalton Trans.*, 1981, 2105.
43. J. Lewis, R. S. Nyholm and P. W. Smith, *J. Chem. Soc.*, 1961, 4590.
44. J. R. Knox and K. Eriks, *Inorg. Chem.*, 1968, **7**, 84.
45. C. J. Horn and T. M. Brain, *Inorg. Chem.*, 1972, **11**, 1970.
46. A. B. Soares, R. S. Taylor and A. G. Sykes, *Inorg. Chem.*, 1978, **17**, 496.
47. Q. Yao and A. W. Maverick, *J. Am. Chem. Soc.*, 1986, **108**, 5364.
48. G. R. Rossman, F.-D. Tsay and H. B. Gray, *Inorg. Chem.*, 1973, **12**, 824.
49. Y. Sasaki and A. G. Sykes, *J. Chem. Soc., Dalton Trans.*, 1975, 1048.
50. D. T. Richens, Y. Ducommun and A. E. Merbach, *J. Am. Chem. Soc.*, 1987, **109**, 603.
51. L. Monsted, T. Ramasami and A. G. Sykes, *Acta Chem. Scand., Ser. A*, 1985, **39**, 437.
52. H. M. Kelly, D. T. Richens and A. G. Sykes, *J. Chem. Soc., Dalton Trans.*, 1984, 1229.
53. D. T. Richens, M. A. Harmer and A. G. Sykes, *J. Chem. Soc., Dalton Trans.*, 1984, 2099.
54. J. D. Rush and B. H. Bielski, *Inorg. Chem.*, 1985, **24**, 4282.
55. H. Diebler and C. Millan, *Polyhedron*, 1986, **5**, 539.
56. E. F. Hills, C. Sharp and A. G. Sykes, *Inorg. Chem.*, 1986, **25**, 2566.
57. E. F. Hills, P. R. Norman, T. Ramasami, D. T. Richens and A. G. Sykes, *J. Chem. Soc., Dalton Trans.*, 1986, 157.
58. M. Ardon and A. Pernick, *Inorg. Chem.*, 1974, **13**, 2275.
59. G. G. Kneale, A. J. Geddes, Y. Sasaki, T. Shibahara and A. G. Sykes, *J. Chem. Soc., Chem. Commun.*, 1975, 356.
60. S. P. Cramer, P. K. Eiden, M. T. Paffett, J. R. Winkler, Z. Dori and H. B. Gray, *J. Am. Chem. Soc.*, 1983, **105**, 719.
61. T. Shibahara and A. G. Sykes, *J. Chem. Soc., Dalton Trans.*, 1978, 95.
62. T. Shibahara and A. G. Sykes, *J. Chem. Soc., Dalton Trans.*, 1978, 100.
63. T. Shibahara, B. Sheldrick and A. G. Sykes, *J. Chem. Soc., Chem. Commun.*, 1976, 523.
64. (a) K. Wieghardt, M. Hahn, W. Swiridoff and J. Weiss, *Inorg. Chem.*, 1984, **23**, 94; (b) M. Hahn and K. Wieghardt, *Inorg. Chem.*, 1985, **24**, 3151.
65. K. Wieghardt, M. Guffmann, P. Chaudhuri, W. Gebert, M. Minelli, C. G. Young and J. H. Enemark, *Inorg. Chem.*, 1985, **24**, 3151.
66. W. Jentsch, W. Schmidt, A. G. Sykes and K. Wieghardt, *Inorg. Chem.*, 1977, **16**, 1935.
67. K. Wieghardt, M. Woeste, P. S. Roy and P. Chaudhuri, *J. Am. Chem. Soc.*, 1985, **107**, 8276.
68. P. A. Ketchum, R. C. Taylor and D. C. Young, *Nature (London)*, 1976, **259**, 203.
69. J. Lewis, R. S. Nyholm and P. W. Smith, *J. Chem. Soc. (A)*, 1969, 57.
70. M. J. Bennett, J. V. Brencic and F. A. Cotton, *Inorg. Chem.*, 1969, **8**, 1060.
71. R. Saillant and R. A. D. Wentworth, *Inorg. Chem.*, 1969, **8**, 1226.
72. F. A. Cotton and B. J. Kalbacher, *Inorg. Chem.*, 1976, **15**, 522.
73. A. Bino and F. A. Cotton, *Angew. Chem., Int. Ed. Engl.*, 1979, **18**, 332.
74. P. Souchay, M. Cadiot and M. Duhameaux, *C.R. Hebd. Seances Acad. Sci.*, 1966, **262**, 1524.

75. A. Bino, F. A. Cotton and Z. Dori, *J. Am. Chem. Soc.*, 1978, **100**, 5252.
76. E. O. Schlemper, M. S. Hussein and R. K. Murmann, *Cryst. Struct. Commun.*, 1982, **11**, 89.
77. S. F. Gheller, T. W. Hambley, R. T. C. Brownlee, M. J. O'Connor and A. G. Wedd, *J. Am. Chem. Soc.*, 1983, **105**, 1527.
78. A. Bino, F. A. Cotton and Z. Dori, *J. Am. Chem. Soc.*, 1979, **101**, 3842.
79. H. M. Spittle and W. Wardlaw, *J. Chem. Soc.*, 1929, 792.
80. R. K. Murmann and M. E. Shelton, *J. Am. Chem. Soc.*, 1980, **102**, 3984.
81. G. D. Hinch, D. E. Wycoff and R. K. Murmann, *Polyhedron*, 1986, **5**, 487.
82. K. R. Rodger, R. K. Murmann, E. O. Schlemper and M. E. Shelton, *Inorg. Chem.*, 1985, **24**, 1313.
83. P. Kathirgamanathan, A. B. Soares, D. T. Richens and A. G. Sykes, *Inorg. Chem.*, 1985, **24**, 2950.
84. D. T. Richens and A. G. Sykes, *Inorg. Chem.*, 1982, **21**, 418.
85. M. T. Paffett and F. C. Anson, *Inorg. Chem.*, 1983, **22**, 1347.
86. M. A. Harmer, D. T. Richens, A. B. Soares, A. T. Thornton and A. G. Sykes, *Inorg. Chem.*, 1981, **20**, 4155.
87. T. Shibahara, T. Yamada, H. Kuroya, E. F. Hills, P. Kathirgamanathan and A. G. Sykes, *Inorg. Chim. Acta*, 1986, **113**, 221.
88. T. Shibahara and H. Kuroya, *Polyhedron*, 1986, **5**, 357.
89. P. Kathirgamanathan, M. Martinez and A. G. Sykes, *J. Chem. Soc., Chem. Commun.*, 1985, 953.
90. F. A. Cotton, Z. Dori, R. Llusar and W. Schwotzer, *J. Am. Chem. Soc.*, 1985, **107**, 6734.
91. F. A. Cotton, M. P. Diebold, Z. Dori, R. Llusar and W. Schwotzer, *J. Am. Chem. Soc.*, 1985, **107**, 6735.
92. T. Shibahara, H. Kuroya, K. Matsumoto and S. Ooi, *J. Am. Chem. Soc.*, 1984, **106**, 789.
93. M. Martinez, B.-L. Ooi and A. G. Sykes, *J. Am. Chem. Soc.*, 1987, in press.
94. T. Shibahara, H. Miyake, K. Kobayashi and H. Kuroya, *Chem. Lett.*, 1986, 139.
95. T. Shibahara, H. Hattori and H. Kuroya, *J. Am. Chem. Soc.*, 1984, **106**, 2710.
96. M. Martinez, B.-L. Ooi and A. G. Sykes, to be published.
97. T. Shibahara, H. Akashi and H. Kuroya, *J. Am. Chem. Soc.*, 1986, **108**, 1342.
98. T. Shibahara, H. Kuroya, K. Matsumoto and S. Ooi, *Inorg. Chim. Acta*, 1986, **116**, L25.
99. A. Müller, R. Jostes, W. Eltzner, C.-Si. Nie, E. Diemann, H. Bogge, M. Zimmerman, M. Dartmann, U. Reinsch-Vogell, S. Che, S. J. Cyvin and B. N. Cyvin, *Inorg. Chem.*, 1985, **24**, 2872.
100. P. J. Vergamini, H. Vahrenkamp and L. F. Dahl, *J. Am. Chem. Soc.*, 1971, **93**, 6327.
101. W. Beck, W. Danzer and G. Thiel, *Angew. Chem., Int. Ed. Engl.*, 1973, **85**, 625.
102. J. A. Bandy, C. E. Davies, J. C. Green, M. L. H. Green, K. Prout and D. P. S. Rogers, *J. Chem. Soc., Chem. Commun.*, 1983, 1395.
103. P. R. Robinson, E. O. Schlemper and R. K. Murmann, *Inorg. Chem.*, 1975, **14**, 2035.
104. K. Wieghardt, G. Backes-Dahmann, W. Holzback, W. J. Swiridoff and J. Weiss, *Z. Anorg. Allg. Chem.*, 1983, **499**, 44.
105. A. V. Butcher and J. Chatt, *J. Chem. Soc. (A)*, 1970, 2652.
106. L. Manojlovic-Muir, *J. Chem. Soc. (A)*, 1971, 2796.
107. L. Manojlovic-Muir and K. W. Muir, *J. Chem. Soc., Dalton Trans.*, 1972, 686.
108. R. E. Desimone and M. D. Glick, *Inorg. Chem.*, 1978, **17**, 3574.
109. M. R. Churchill and F. J. Rotella, *Inorg. Chem.*, 1978, **17**, 668.
110. A. Bino and F. A. Cotton, *Inorg. Chem.*, 1979, **18**, 2710.
111. M. I. Scullane, R. D. Taylor, M. Minelli, J. T. Spence, K. Yamanouchi, J. H. Enemark and N. D. Chasteen, *Inorg. Chem.*, 1979, **18**, 3213.
112. T. Glowiak, M. F. Rudolf, M. Sabat and B. Jezowska-Trzebiatowska, 'Proceedings of the Second International Conference on the Chemistry and Uses of Molybdenum', ed. P. C. H. Mitchell and A. Seaman, Climax Molybdenum Company, 1976, p. 17.
113. (a) W. Kolitsch and K. Dehnicke, *Z. Naturforsch., Teil B*, 1970, **25**, 1080; (b) K. Dehnicke and W. Kolitsch, *Z. Naturforsch., Teil B*, 1977, **32**, 1485.
114. C. D. Garner, L. H. Hill, F. E. Mabbs, D. L. McFadden and A. T. McPhail, *J. Chem. Soc., Dalton Trans.*, 1977, 853.
115. C. D. Garner, L. H. Hill, F. E. Mabbs, D. L. McFadden and A. T. McPhail, *J. Chem. Soc., Dalton Trans.*, 1977, 1202.
116. N. Ueyama, H. Zaima and A. Nakamura, *Chem. Lett.*, 1986, 1099.
117. B. Spivack and Z. Dori, *Coord. Chem. Rev.*, 1975, **17**, 99.
118. G. R. Cayley and A. G. Sykes, *Inorg. Chem.*, 1976, **15**, 2882.
119. W. S. McDonald, *Acta Crystallogr., Sect. B*, 1978, **34**, 2850.
120. B. H. Gatehouse and E. K. Nunn, *Acta Crystallogr., Sect. B*, 1976, **32**, 2627.
121. K. Z. Suzuki, Y. Sasaki, S. Ooi and K. Saito, *Bull. Chem. Soc. Jpn.*, 1980, **53**, 1288.
122. K. Wieghardt, M. Hann, W. Swiridoff and J. Weiss, *Angew. Chem., Int. Ed. Engl.*, 1983, **22**, 491.
123. T. Chandler, D. L. Lichtenberger and J. H. Enemark, *Inorg. Chem.*, 1981, **20**, 75.
124. V. R. Ott, D. S. Swieter and F. A. Schultz, *Inorg. Chem.*, 1977, **16**, 2538.
125. F. A. Schultz, V. R. Ott, D. S. Rolison, D. C. Bravard, J. W. McDonald and W. E. Newton, *Inorg. Chem.*, 1978, **17**, 1758.
126. F. A. Cotton, P. E. Fanwick and J. W. Finch, *Inorg. Chem.*, 1978, **17**, 3254.
127. A. V. Butcher and J. Chatt, *J. Chem. Soc. (A)*, 1971, 356.
128. A. J. Matheson and B. R. Penfold, *Acta Crystallogr., Sect. B*, 1979, **35**, 2707.
129. J. F. Johnson and W. R. Scheidt, *Inorg. Chem.*, 1978, **17**, 1280.
130. S. Lincoln and S. A. Koch, *Inorg. Chem.*, 1986, **25**, 1594.
131. J. A. Beaver and M. G. B. Drew, *J. Chem. Soc., Dalton Trans.*, 1973, 1376.
132. M. F. Belicchi, G. G. Fava and C. Pelizzi, *J. Chem. Soc., Dalton Trans.*, 1983, 65.
133. M. H. Chisholm, K. Folting, J. C. Huffman and C. C. Kirkpatrick, *Inorg. Chem.*, 1984, **23**, 1021.
134. D. J. Darensbourg, R. L. Gray and T. Delard, *Inorg. Chim. Acta*, 1985, **98**, L39.
135. R. Mattes and K. Muhlsiepen, *Z. Naturforsch., Teil B*, 1980, **35**, 265.

136. F. A. Armstrong and A. G. Sykes, *Polyhedron,* 1982, **1,** 109.
137. C. D. Garner, M. R. Hyde, F. E. Mabbs and V. I. Routledge, *J. Chem. Soc., Dalton Trans.*, 1975, 1175; 1977, 1198.
138. (a) Y. Sasaki, R. S. Taylor and A. G. Sykes, *J. Chem. Soc., Dalton Trans.*, 1975, 396; (b) Y. Sasaki and A. G. Sykes, *J. Chem. Soc., Dalton Trans.*, 1974, 1468.
139. F. A. Armstrong, T. Shibahara and A. G. Sykes, *Inorg. Chem.*, 1978, **17,** 189.
140. R. G. Cayley, R. S. Taylor, R. K. Wharton and A. G. Sykes, *Inorg. Chem.*, 1977, **16,** 1377.
141. Y. Sasaki, *Bull. Chem. Soc. Jpn.*, 1977, **50,** 1939.
142. R. K. Murmann, *Inorg. Chem.*, 1980, **19,** 1765.
143. R. K. Wharton, J. F. Ojo and A. G. Sykes, *J. Chem. Soc., Dalton Trans.*, 1975, 1526.
144. E. P. Guyman and J. T. Spence, *J. Phys. Chem.*, 1966, **70,** 1964.
145. J. D. Rush and B. H. Bielski, *Inorg. Chem.*, 1985, **24,** 3895.
146. E. F. Hills and A. G. Sykes, *J. Chem. Soc., Dalton Trans.*, in press.
147. M. T. Pope, 'Heteropoly and Isopoly Oxometalates', Springer-Verlag, Berlin, 1983, pp. 1–180.
148. K.-H. Tytko and O. Glemser, *Adv. Inorg. Chem. Radiochem.*, 1976, **19,** 239–316.
149. C. F. Baes and R. E. Mesmer, 'The Hydrolysis of Cations', Wiley-Interscience, New York, 1976.
150. R. R. Vold and R. L. Vold, *J. Magn. Reson.*, 1975, **19,** 365.
151. (a) J. J. Cruywagen, *Inorg. Chem.*, 1980, **19,** 552; (b) J. J. Cruywagen, J. B. B. Heyns and E. F. C. H. Rohwer, *J. Inorg. Nucl. Chem.*, 1976, **38,** 2033; 1978, **40,** 53.
152. L. Krumenacker, *Ann. Chim. (Paris*), 1972, **7,** 425; *Bull. Soc. Chim. Fr.*, 1974, **362,** 2820.
153. J. M. Berg and K. O. Hodgson, *Inorg. Chem.*, 1980, **19,** 2180.
154. K. Yamanouchi and J. H. Enemark, *Inorg. Chem.*, 1979, **18,** 1626.
155. T. Buchanan, M. Minelli, M. T. Ashby, T. J. King, J. H. Enemark and C. D. Garner, *Inorg. Chem.*, 1984, **23,** 495.
156. I. Buchanan, C. D. Garner and W. Clegg, *J. Chem. Soc., Dalton Trans.*, 1984, 1333.
157. F. A. Cotton, S. M. Morehouse and J. S. Wood, *Inorg. Chem.*, 1964, **3,** 1603.
158. F. A. Cotton and R. C. Elder, *Inorg. Chem.*, 1964, **3,** 397.
159. J. J. Park, M. D. Glick and J. L. Hoard, *J. Am. Chem. Soc.*, 1969, **91,** 301.
160. R. S. Taylor, P. Gans, P. F. Knowles and A. G. Sykes, *J. Chem. Soc., Dalton Trans.*, 1972, 24.
161. G. Johansson, L. Pettersson and N. Ingri, *Acta Chem. Scand., Ser. A,* 1979, **33,** 305.
162. K. Stadnicka, J. Haber and R. Kozlowski, *Acta Crystallogr., Sect. B,* 1977, **33,** 3859.
163. H. J. Bechner, N. J. Brockmeyer and U. Prigge, *J. Chem. Res. (S*), 1978, 117; (*M*), 1978, 1670.
164. A. W. Armour, M. G. B. Drew and P. C. H. Mitchell, *J. Chem. Soc., Dalton Trans.*, 1975, 1493.
165. (a) J. Aveston, E. W. Anacker and L. S. Johnson, *Inorg. Chem.*, 1964, **3,** 735; (b) Y. Sasaki and L. G. Sillen, *Ark. Kemi,* 1968, **29,** 253.
166. B. Krebs and I. Paulat-Boeschen, *Acta Crystallogr., Sect. B,* 1982, **38,** 1710.
167. Ref. 147, p. 18.
168. M. Filowitz, W. G. Klemperer and W. Shum, *J. Am. Chem. Soc.*, 1978, **100,** 2580.
169. V. W. Day, M. F. Fredrick, W. G. Klemperer and W. Shum, *J. Am. Chem. Soc.*, 1977, **99,** 6146.
170. B. Krebs and I. Paulat-Böschen, *Acta Crystallogr., Sect. B,* 1976, **32,** 1697.
171. W. G. Klemperer and W. J. Shum, *J. Am. Chem. Soc.*, 1976, **98,** 8291.
172. S. Ostrowetsky, *Bull. Soc. Chim. Fr.*, 1964, 1003.
173. Ref. 147, chapter 6, pp. 101–117.
174. H. von Felton, B. Wernli, H. Gamsjäger and P. Baertschi, *J. Chem. Soc., Dalton Trans.*, 1978, 496.
175. A. Okumura, M. Kitani, Y. Toyomi and N. Okazaki, *Bull. Chem. Soc. Jpn.*, 1980, **53,** 3143.
176. J. F. Ojo, R. S. Taylor and A. G. Sykes, *J. Chem. Soc., Dalton Trans.*, 1975, 500.
177. J. R. Pladziewicz and J. H. Espenson, *Inorg. Chem.*, 171, **10,** 634.
178. A. Haim, *Inorg. Chem.*, 1972, **11,** 3147.
179. D. S. Honig and K. Kustin, *Inorg. Chem.*, 1972, **11,** 65.
180. P. F. Knowles and H. Diebler, *Trans. Faraday Soc.*, 1968, **64,** 977.
181. K. Kustin and S.-T. Liu, *J. Am. Chem. Soc.*, 1973, **95,** 2487.
182. D. S. Honig and K. Kustin, *J. Am. Chem. Soc.*, 1973, **95,** 5525.
183. R. S. Taylor, *Inorg. Chem.*, 1977, **16,** 116.
184. Y. Sulfab, R. S. Taylor and A. G. Sykes, *Inorg. Chem.*, 1976, **15,** 2388.
185. M. A. Harmer and A. G. Sykes, in 'Molybdenum Chemistry of Biological Significance', ed. W. E. Newton and S. Otsuka, Plenum, New York, 1980, p. 401.
186. H. Shäfer , G. Shafer and A. Weiss, *Z. Naturforsch., Teil B,* 1964, **23,** 76.
187. B. Kreks, A. Müller and E. Kindler, *Z. Naturforsch., Teil B,* 1970, **25,** 222.
188. M. A. Harmer and A. G. Sykes, *Inorg. Chem.*, 1980, **19,** 2881.
189. R. Stomberg and L. Tysberg, *Acta Chem. Scand.*, 1969, **23,** 314.
190. R. Stomberg, *Acta Chem. Scand.*, 1969, **23,** 2755.
191. R. Stomberg, *Acta Chem. Scand.*, 1968, **22,** 1076.
192. R. Stomberg and L. Trysberg, *Acta Chem. Scand.*, 1970, **24,** 2678.
193. I. Larking and R. Stomberg, *Acta Chem. Scand.*, 1972, **26,** 3708.
194. A. Mitschler, J. M. Le Carpentier and R. Weiss, *Chem. Commun.*, 1968, 1260.
195. U. Thewalt and R. E. Marsh, *J. Am. Chem. Soc.*, 1967, **89,** 6364.
196. R. Stomberg, *Acta Chem. Scand.*, 1970, **24,** 2024.

36.2

Molybdenum(0), Molybdenum(I) and Molybdenum(II)

G. JEFFERY LEIGH and RAYMOND L. RICHARDS
AFRC Unit of Nitrogen Fixation, University of Sussex, Brighton, UK

36.2.1 INTRODUCTION

This survey of the lower oxidation states of molybdenum concentrates on those compounds which have a minimum of metal–carbon bonds. This is because there is an extensive literature of low-valent carbonyls, arene, alkene, alkyne and other organometallic complexes of molybdenum which has been thoroughly reviewed in 'Comprehensive Organometallic Chemistry'. We therefore exclude compounds of alkenes, alkynes and higher organic entities. Nevertheless, reference is made to significant compounds which contain molybdenum–carbon bonds where it is judged appropriate. Coverage of complexes containing molybdenum–molybdenum bonds is also kept to a minimum because these compounds are reviewed in Chapter 36.3.

In certain cases the nature of the ligands involved makes definition of the metal oxidation state difficult, *e.g.* NO, N_2R, dithiolenes, *etc.* For simplification, therefore, arbitrary assignment of ligand charge and therefore metal oxidation state has been made in the text, *e.g.* NO is defined as being NO^+ in its compounds and N_2R is arbitrarily regarded as N_2R^-. Hydride is regarded as H^-. Complexes of formal oxidation state below zero, *e.g.* the anion $[Mo(CO)_5]^{2-}$, are not reviewed here because of the high number of Mo—C bonds, as explained above.

36.2.2 MOLYBDENUM(0)

36.2.2.1 Carbon Ligands

36.2.2.1.1 Carbon monoxide

There is an extensive chemistry of Mo^0 formally based upon substitution of $[Mo(CO)_6]$ by various ligands, without oxidation of the metal. These include hydride; nitrogen donors such as bipy, phen, NR_3, RCN, NCS^-; oxygen donors such as OH^-, THF, RCO_2^-; sulfur donors such as R_2S, Me_2PhPS; selenium and tellurium donors such as R_2Se and R_2Te; and halogens. In general compounds of the type $[Mo(CO)_5L]$, $[Mo(CO)_4L_2]$, $[Mo(CO)_3L_3]$ and $[Mo(CO)_2L_4]$ result from direct substitution reactions using $[Mo(CO)_6]$, which are summarized in 'Comprehensive Organometallic Chemistry' (COMC).[1a]

A recent significant development has been the application of ^{95}Mo NMR to the study of these carbonyl compounds. Ranges of chemical shifts (p.p.m. relative to $[MoO_4]^{2-}$) which have been determined are: $[Mo(CO)_6]$ -1858; $[Mo(CO)_5L]$ (L = N, P, As or Sb donor) -1350 to -1870; $[Mo(CO)_4L_2]$ -1100 to -1600; $[Mo(CO)_3L_3]$ (here L may also be an arene or related ligand) -1100 to -1750; $[Mo(CO)_2L_4]$ -1400 to -1800.[2]

Alternative routes to carbonyl compounds are also known, *e.g.* by direct reduction of a higher oxidation state compound in presence of CO or by displacement of a more labile ligand by CO. Reaction (1) illustrates these routes.[1a] Mixed substituted complexes can also be obtained, *e.g.* $[Mo(CO)_2(PR_3)_2(CNMe)_2]$.[1a]

$$[MoCl_2(dppe)_2] + Mg \xrightarrow[CO]{THF} cis\text{-}[Mo(CO)_2(dppe)_2]$$

$$Mg \downarrow N_2 \qquad\qquad\qquad\qquad \uparrow \Delta$$

$$trans\text{-}[Mo(N_2)_2(dppe)_2] \xrightarrow{CO} trans\text{-}[Mo(CO)_2(dppe)_2] + 2N_2 \qquad (1)$$

A series of monocarbonyl complexes is known of which the five-coordinate complex $[Mo(CO)(dppe)_2]$ may be regarded as the parent. It has an almost square pyramidal structure and adds ligands L to give the series $[Mo(CO)(L)(dppe)_2]$ (L = N_2, DMF, C_2H_4, $4\text{-}ClC_6H_4CN$, PhCN, C_5H_5N, Me_2NCHO, CN^-, SCN^- or N_3^-).[3a] Further analogues of this series, $fac\text{-}[Mo(CO)\{PhP(CH_2CH_2PPh_2)_2\}(PMe_2Ph)_2]$ and $[Mo(CO)(PF_3)_4(dppe)]$, have also been prepared.[3] These compounds are notable for their low values of $v(CO)$ for mononuclear compounds, in the range 1690–1820 cm^{-1}.

36.2.2.1.2 Isocyanide ligands

Molybdenum carbonyl–isocyanide complexes may be obtained as in Section 36.2.2.1.1. In most respects isocyanides behave in a very similar way to carbon monoxide, with which they are isoelectronic. As for the carbonyls, the isocyanide complexes may be regarded as derivatives of $[Mo(CNR)_6]$ and have been reviewed comprehensively.[1a,b]

Isocyanides are rather more reactive than CO at low-valent molybdenum centres and may react with nucleophiles, *e.g.* $R'NH_2$, to give carbene complexes,[1a] *e.g.* equation (2), or with electrophiles to give carbyne complexes,[1,4] *e.g.* (3).

$$[Mo(CNR)_5(NO)]^+ + R'NH_2 = [Mo\{C(NHR)(NHR')\}(NO)(CNR)_4]^+ \qquad (2)$$

$$trans\text{-}[Mo(CNMe)_2(dppe)_2] \xrightarrow{R^+} trans\text{-}[Mo\{CN(R)Me\}(CNMe)(dppe)_2]^+ \qquad (3)$$

$$R^+ = H^+ \text{ or } Me^+$$

Simple oxidation/reduction has also been studied for a wide range of isocyanide complexes. Again those complexes containing mainly molybdenum–carbon bonds have been reviewed in

COMC and here we merely give as an example the series *trans*-$[Mo(CNR)_2(dppe)_2]$ (R = alkyl or aryl). They are not reducible under normal conditions, but they can be reversibly oxidized electrochemically[4,5] by one electron to give the series $[Mo(CNR)_2(dppe)_2]^+$ (see Section 36.2.3).[4,5] Complexes with aryl substituents are less easily oxidized than those with alkyl substituents, as might be expected and these compounds are notable for their low $\nu(NC)$ values, 1850–1920 cm^{-1}. The X-ray structure of *trans*-$[Mo(CNR)_2(dppe)_2]$ confirms the *trans* geometry of the series.[5]

36.2.2.2 Dinitrogen Complexes

These complexes are reviewed in a variety of articles[6,7] and are collected in Table 1.

In general they are prepared by the following methods.

(A) Reduction of a high-valent molybdenum species in presence of dinitrogen, *e.g.* reaction (4).

$$[MoCl_3(THF)_3] \xrightarrow[4PMe_2Ph]{Mg/THF/N_2} cis\text{-}[Mo(N_2)_2(PMe_2Ph)_4] \quad (4)$$

(B) Displacement of a ligand from a preformed complex by dinitrogen, free or coordinated, *e.g.* reactions (5) and (6).[6,8]

$$trans\text{-}[Mo(N_2)_2(PPr^n_2Ph)_4] + N_2 \longrightarrow mer\text{-}[Mo(N_2)_3(PPr^n_2Ph)_3] + PPr^n_2Ph \quad (5)$$

$$[MoH_4(Ph_2PCH_2CH_2PPh_2)_2 + N_2 \longrightarrow trans\text{-}[Mo(N_2)_2(Ph_2PCH_2CH_2PPh_2)_2] + 2H_2 \quad (6)$$

(C) Displacement of dinitrogen by a ligand, *e.g.*, reaction (7).[9]

$$trans\text{-}[Mo(N_2)_2(Ph_2PCH_2CH_2PPh_2)_2] + RCN \longrightarrow trans\text{-}[Mo(N_2)(RCN)(Ph_2PCH_2CH_2PPh_2)_2] \quad (7)$$

(D) Displacement of a coligand by a second coligand, *e.g.* reaction (8).[10]

$$trans\text{-}[Mo(N_2)_2(PPh_2Me)_4] + C_5H_5N \longrightarrow [Mo(N_2)_2(PPh_2Me)_3(C_5H_5N)] + PPh_2Me \quad (8)$$

The binding of dinitrogen to the metal centres in these mononuclear complexes is end-on to the molybdenum as shown, for example, by the X-ray structures of *trans*-$[Mo(N_2)_2(Ph_2PCH_2CH_2PPh_2)_2]$ (**1**)[11] and *mer*-$[Mo(N_2)_3(PPr^n_2Ph)_3]$ (**2**). (Data for other structurally characterized complexes are in Table 1).[8] Characteristic of this bonding mode for dinitrogen are slight elongation of the N≡N bond on binding (by about 0.01–0.02 Å from that of 1.0976 Å in free N_2) and a $\nu(N_2)$ value some 300–400 cm^{-1} less than the 2331 cm^{-1} value of free N_2 in its Raman spectrum.

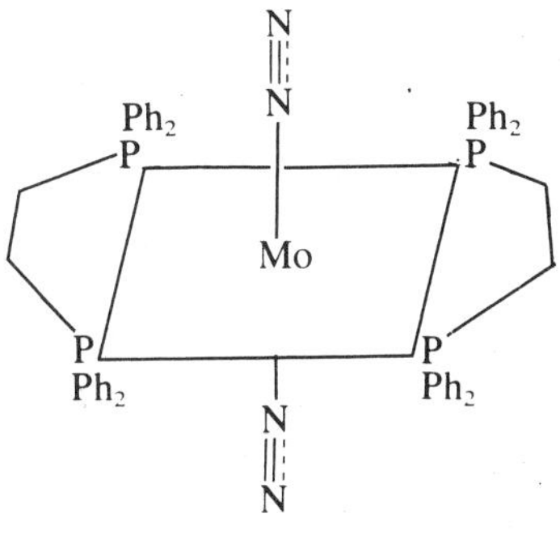

(**1**)

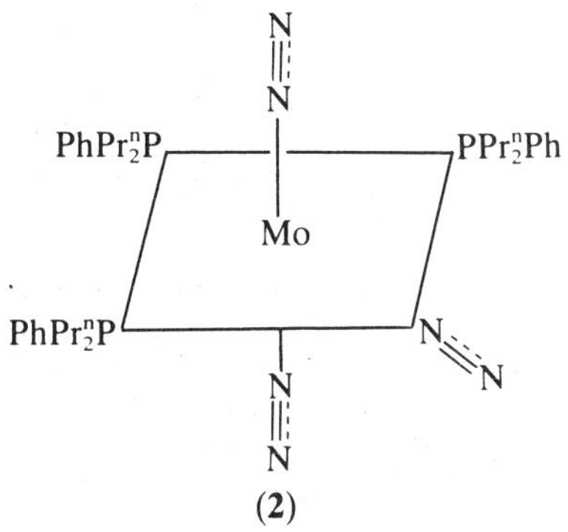

(**2**)

The majority of complexes contain tertiary phosphines or arsines as coligands but there is also a class which has π-arene coligands which is included in Table 1 for completion.

Despite the obvious relevance of dinitrogen complexes to the chemistry of the molybdenum–sulfur site of nitrogenase,[6] dinitrogen complexes of molybdenum with sulfur ligation have proved to be very elusive. They are confined to the compounds

Table 1 Dinitrogen Complexes of Molybdenum(0)

Complex	*Preparative method*[a]	$\nu(N_2)$ (cm^{-1})[b]	$\delta^{15}N$ (p.p.m.)[c] ($Mo—N_\alpha N_\beta$)	$\delta^{95}Mo$ (p.p.m.)[d]	$E_{1/2ox}$ (V)[e]	*Other data*	*Ref.*
$[Mo(N_2)(PMe_3)_5]$	A, B	1950[f]	−35.1, −41.2	−718	—	X-ray structure d(NN) 1.12(3) Å	1a, 2, 3
$[Mo(N_2)(triphos)L_2]$[g]	A	1962 (L_2 = 1,1-(η^5-$C_5H_4PPh_2)_2Fe$)	—	—	—	—	1b 1b
		1966 (L_2 = dmpm)	—	—	—	—	1b
		1978 (L_2 = dppm)	—	—	—	—	1b
		1955 (L_2 = dppe)	—	—	—	Two isomers	1b
		1960 (L_2 = 1,2-$(Me_2As)_2C_6H_4$)	—	—	—	Two isomers	1b
		1978 (L = PMe_2Ph)	—	—	—		1b
cis-$[Mo(N_2)_2(PMe_3)_4]$	A	2010[f] 1965[f]	−41.5, −34.3	−637	—	X-Ray structure d(NN) 1.15 Å 1.14 Å	3, 4
cis-$[Mo(N_2)_2(PMe_2Ph)_4]$	A	2025 1940	−39.2, −31.5	−455	—	—	5, 6
trans-$[Mo(N_2)_2(PPh_2Me)_4]$	A	1925	−37.6, −44.7	−464	—	X-Ray structure d(NN) 1.133(7) Å 1.154(7) Å	3, 5, 6, 7
trans-$[Mo(N_2)_2(PEt_2Ph)_4]$	A	1923[h]	—	—	—	—	8, 9
trans-$[Mo(N_2)_2(PPh_2Me)_2(dppe)_2]$	A, D	1950	—	—	—	—	9, 11
trans-$[Mo(N_2)_2(PPr^n_2Ph)_4]$	A	1920	—	—	—	Converts to *mer*-$[Mo(N_2)_3(PPr^n_2Ph)_3]$	10
$[Mo(N_2)_2(triphos)(PR_3)]$	A	1929 ($PR_3 = PMe_3$)	—	—	—	—	11
		1930 ($PR_3 = PMe_2Ph$)	—	—	—	—	11
		1950 ($PR_3 = PPh_2Me$)	—	—	—	—	11
		1955 ($PR_3 = PPh_3$)	—	—	—	—	11
		1955 ($PR_3 = P(C_6H_4$-OMe-4$)_3$)	—	—	—	—	11
mer-$[Mo(N_2)_3(PR_3)_3]$	A, B	2080, 1985, 1965 ($PR_3 = PPr^n_2Ph$)	−44.4, −25 −46.1, −40.3	−392	—	X-Ray structure d(NN) = 1.092(10), 1.124(9), 1.103(9) Å	10, 17
		1970, 1958, 1940 ($PR_3 = PEt_2Ph$)	—	—	—	Impure	10, 17
		2080, 1985, 1965 ($PR_3 = PBu^n_3$)	—	—	—	Impure	10, 17

trans,mer-$[Mo(N_2)_2(NC_5H_4R\text{-}4)(PPh_2Me)_3]$	D	1914 (R = H or Me)	—	—	—	—	18
trans-$[Mo(N_2)_2(PPh_2Me)_2(PPh_2CH_2CH_2SMe)]$	D	1942[i]	—	—	—	X-Ray structure *d*(NN) 104(3), 1.10(3) Å	19
$[Mo(N_2)_2(PMe_2Ph)_2(PhSCH_2CH_2SPh)]$	D	—	—	—	—	Decomposes above 20 °C	8
trans-$[Mo(N_2)(RCN)(diphos)_2]$ R = Me, Pr, $C_6H_4X\text{-}4$ where X = H, OMe, F, COMe; $(diphos)_2 = (dppe)_2$, $(depe)_2$, or (depe)(dppe)	C	Range 1920–1976	Range −32.2 to +3.1, −30.2 to −35.9	—	Range −0.16 to −0.58	—	20
trans-$[Mo(N_2)_2(dppe)_2]$	A	1976	−43.1, −42.8	−787	−0.16	X-Ray structure *d*(NN) 1.118(8) Å	6, 12
trans-$[Mo(N_2)_2(dppe)(depe)]$	A	1945	−41.2	−899	−0.28	—	6
trans-$[Mo(N_2)_2(depe)_2]$	A	1928	−42.0, −43.4	−1022	−0.43	—	6
trans-$[Mo(N_2)_2((-)\text{-}(2S,3S)bis(diphenylphosphino)butane)_2]$	A	1962	—	—	—	$[\alpha]^{25}_{389}$ + 671°	5
trans-$[Mo(N_2)_2((4\text{-}XC_6H_4P)_2CH_2CH_2P(C_6H_4\text{-}X\text{-}4)_2)_2]$ X = CF_3, Cl, Me, OMe	A	1990 (CF_3) 1979 (Cl) 1967 (Me) 1956 (OMe)	— — — —	−774 −785 −793 −798	+0.30 +0.05 −0.25 −0.30	— — — —	13, 14 13, 14 13, 14 13, 14
trans-$[Mo(N_2)_2(dmpte)_2]$	A	1960	—	−795	—	Good solubility	15
trans-$[Mo(N_2)_2(Ph_2P(CH_2)_nPPh_2)_2]$	A	1955 (*n* = 1) 1975 (*n* = 3)	—	—	—	Not pure (*n* = 1)	12, 15, 16
cis- and *trans*-$[Mo(N_2)_2(Ph_2PCH{=}CHPPh_2)_2]$	A	1970, 2002 (*cis*) 2000 (*trans*)	—	—	—	*cis* to *trans* isomerization	16
trans-$[Mo(N_2)_2(Ph_2PCH_2CH_2AsPh_2)_2]$	A	1964	—	—	—	—	15, 16
trans-$[Mo(N_2)_2(Ph_2AsCH_2CH_2AsPh_2)_2]$	A	1971	—	—	—	Impure	17
trans-$[Mo(N_2)(CO)(dppe)_2]$	C	2080, 2110	—	—	—	(CO) 1812, 1791 cm^{-1}	21
$[MoX(N_2)(dppe)_2]^-$	C	1890 (X = SCN) 1860 (X = N_3) 1900 (X = CN)	—	—	−0.96 −1.22 −1.0	(SCN) 2080 cm^{-1}	22
$[(Mo(\eta^6\text{-}C_6H_6)(PPh_3)_2)_2(N_2)]$	A, B	1910[j]	—	—	—	—	23

Table 1 *(continued)*

Complex	*Preparative method*[a]	$\nu(N_2)$ (cm^{-1})[b]	$\delta^{15}N$ (p.p.m.)[c] $(Mo—N_{\alpha}N_{\beta})$	$\delta^{95}Mo$ (p.p.m.)[d]	$E_{1/2ox}$ (V)[e]	*Other data*	*Ref.*
$[(Mo(\mu\text{-}\eta^1,\eta^6\text{-}C_6H_5PPh_2)(N_2)(PPh_3))_2]$	A	2005	—	—	—	X-Ray structure of Bu[t]NC analogue	24
$[(Mo(\eta^6\text{-}C_6H_3Me_3)(dmpe))_2(N_2)]$	A	1989	—	—	—	X-ray structure d(NN) 1.145(7) Å	25
$[(\eta^6\text{-}C_6H_5Me)(PPh_3)_2Mo(N_2)Fe(\eta^5\text{-}C_5H_5)(dmpe)]BF_4$	B	1930[j]	—	—	—	—	23
$[Mo(N_2AlMe_3)(N_2)(dppe)_2]$	B	1883	—	—	—	—	26

[a] See text. [b] Solid state spectra (Nujol or KBr) unless otherwise stated, only strong bands quoted. [c] Relative to CD_3NO_2. [d] Relative to $[MoO_4]^{2-}$. [e] Relative to standard calomel electrode. [f] Petroleum ether solution. [g] triphos = $PhP(CH_2CH_2PPh_2)_2$. [h] Benzene solution. [i] Toluene solution. [j] Raman spectrum.

1. (a) E. Carmona, J. M. Marin, M. L. Poveda, R. D. Rogers and J. L. Atwood, *J. Organomet. Chem.*, 1982, **238,** C63; (b) T. A. George and R. S. Tisdale, *Polyhedron*, 1986, **5,** 297; T. A. George and R. S. Tisdale, *J. Am. Chem. Soc.*, 1985, **107,** 5157.
2. M. Brookhart, K. Cox, F. G. N. Cloke, J. C. Green, M. L. H. Green, R. M. Hare, J. Bashakin, A. E. Derome and P. D. Grebenik, *J. Chem. Soc., Dalton Trans.*, 1985, 423.
3. E. Carmona, A. Galindo, M. Hughes, J. Mason and R. L. Richards, unpublished observations; M. Hughes, *Ph. D. Thesis,* Open University, 1985.
4. E. Carmona, J. M. Marin, M. L. Poveda, J. L. Atwood and R. D. Rogers, *J. Am. Chem. Soc.*, 1983, **105,** 3014.
5. T. A. George and C. D. Seibold, *J. Am. Chem. Soc.*, 1972, **94,** 6859; T. A. George and C. D. Seibold, *Inorg. Chem.*, 1973, **12,** 2544; G. E. Bossard, T. A. George, R. K. Lester, R. C. Tisdale and R. L. Turcotte, *Inorg. Chem.*, 1985, **24,** 1129.
6. S. Donovan-Mtunzi, J. Mason and R. L. Richards, *J. Chem. Soc., Dalton Trans.*, 1984, 469.
7. R. H. Morris, J. M. Ressner and J. F. Sawyer, *Acta Crystallogr.*, 1985, **41,** 1017.
8. M. Aresta and A. Sacco, *Gazz. Chim. Ital.*, 1972, **102,** 755.
9. J. Chatt, A. J. Pearman and R. L. Richards, *J. Chem. Soc., Dalton Trans.*, 1977, 2139.
10. S. N. Anderson, D. L. Hughes and R. L. Richards, *J. Chem. Soc., Chem. Commun.*, 1984, 958; S. N. Anderson, D. L. Hughes and R. L. Richards, *J. Chem. Soc., Dalton Trans.*, 1986, 1591.
11. T. A. George and R. A. Kover, *Inorg. Chem.*, 1981, **20,** 285; L. Dahlenburg and P. Pietsch, *Z. Naturforsch., Teil B,* 1986, **41,** 70.
12. Y. Uchida, T. Uchida, M. Hidai, and T. Kodama, *Acta Crystallogr., Sect. B,* 1975, **31,** 1197.
13. W. Hussain, G. J. Leigh, H. Mohd-Ali, C. J. Pickett and D. W. Rankin, *J. Chem. Soc., Dalton Trans.*, 1984, 1703.
14. S. Donovan-Mtunzi, M. Hughes, G. J. Leigh, J. Mason, H. Mohd-Ali and R. L. Richards, *J. Organomet. Chem.*, 1983, **246,** C1.
15. L. J. Archer and T. A. George, *J. Organomet. Chem.*, 1973, **54,** C25.
16. M. W. Anker, J. Chatt, G. J. Leigh and A. G. Wedd, *J. Chem. Soc., Dalton Trans.*, 1975, 2369.
17. T. A. George and L. D. Seibold, *Inorg. Nucl. Chem. Lett.*, 1972, **8,** 465.
18. R. H. Morris and J. M. Ressner, *J. Chem. Soc., Chem. Commun.*, 1983, 909.
19. R. H. Morris, J. M. Ressner, J. F. Sawyer and H. Shiralian, *J. Am. Chem. Soc.*, 1984, **106,** 3683.
20. S. Donovan-Mtunzi, J. Mason and R. L. Richards, *J. Chem. Soc., Dalton Trans.*, 1984, 2729; T. Tatsumi, M. Hidai and Y. Uchida, *Inorg. Chem.*, 1975, **14,** 2530; H. Dadkhah, J. R. Dilworth, K. Fairman, D. L. Hughes, Chi Tat Kan and R. L. Richards, *J. Chem. Soc., Dalton Trans.*, 1985, 1523.
21. T. Tatsumi, H. Tominaga, M. Hidai and Y. Uchida, *J. Organomet. Chem.*, 1976, **114,** C27.
22. J. Chatt, G. J. Leigh, H. Neukomm, C. J. Pickett and D. R. Stanley, *J. Chem. Soc., Dalton Trans.*, 1980, 121.
23. M. L. H. Green and W. E. Silverthorn, *J. Chem. Soc., Dalton Trans.*, 1973, 301.
24. R. Luck, R. H. Morris and J. F. Sawyer, *Organometallics,* 1984, **3,** 1009.
25. M. L. H. Green and W. E. Silverthorn, *J. Chem. Soc., Dalton Trans.*, 1974, 2164.
26. J. Chatt, R. H. Crabtree, E. A. Jeffrey and R. L. Richards, *J. Chem. Soc., Dalton Trans.*, 1973, 1167.

$[Mo(N_2)_2(PhSCH_2CH_2SPh)(PMe_2Ph)_2]$[12] (said to decompose at 20 °C) and $[Mo(N_2)_2(Ph_2PCH_2CH_2SMe)(PPh_2Me)_2]$.[13]

^{15}N NMR data can be used to characterize these compounds and to follow their reactions in solution. Characteristic resonances are shown in Table 1. Dinitrogen occupies a low position in a spectroscopic series of ligands based on ^{95}Mo NMR spectroscopy.[13,14]

Many dinitrogen complexes show a reversible one-electron oxidation in solution but are not reducible under normal conditions. The complexes $[Mo(N_2)(RCN)(Ph_2PCH_2CH_2PPh_2)_2]$ are exceptional in that they show two successive one-electron polarographic oxidation waves.[15] $E_{1/2}$ values ($0 \rightarrow 1$), shown in Table 1, can be used to deduce that dinitrogen is a relatively good electron-accepting ligand to molybdenum(0).[15] Dinitrogen complexes of molybdenum(I) are generally poorly stable (see Section 36.2.3).

Because of the electron-accepting properties of dinitrogen when bound to molybdenum in the phosphine complexes, it is susceptible to attack by acids (HX) to give ammonia *via* complexes of the type $[MoX(N_2H)(PR_3)_4]$ (X = anionic ligand, PR_3 = tertiary phosphine) and $[MoX(NNH_2)(PR_3)_4]X$. Hydrazine can also be formed, depending upon the solvent and acid used. Attack by organic reagents RX can also occur to give analogous compounds $[MoX(N_2R)(PR_3)_4]$ and $[MoX(NNR_2)(PR_3)_4]X$. The first step in this process can involve a radical mechanism for certain RX. The chemistry involved in these reactions has been reviewed in depth[16] and the diazenido complex products are described in Section 36.2.4.3.

Attack at terminal nitrogen by other electrophilic reagents can also be used to give binuclear complexes with bridging dinitrogen. Thus treatment of *trans*-$[Mo(N_2)_2(Ph_2PCH_2CH_2PPh_2)_2]$ with $AlMe_3 \cdot Et_2O$ gives *trans*-$[Mo(N_2AlMe_3)(N_2)(Ph_2PCH_2CH_2PPh_2)_2]$ and $[(\eta^6\text{-}C_6H_5Me)(PPh_3)_2Mo(N_2)Fe(\eta^5\text{-}C_5H_5)(Me_2PCH_2CH_2PMe_2)]BF_4$ is obtained by displacement of acetone from $[Fe(\eta^5\text{-}C_5H_5)(Me_2PCH_2CH_2PMe_2)(Me_2CO)]BF_4$ by the dinitrogen of $[Mo(\eta_6\text{-}C_6H_5Me)(N_2)(PPh_3)_2]$.[17]

Although displacement of one dinitrogen in *trans*-$[Mo(N_2)_2(dppe)_2]$ and its analogues can give rise to a new series of dinitrogen complexes *trans*-$[Mo(N_2)(RCN)(dppe)_2]$, more generally dinitrogen is completely displaced by such ligands as CO, H_2, RNC, CO_2 and alkenes to give new series of complexes, *e.g.* *trans*-$[Mo(RNC)_2(dppe)_2]$,[4] *trans*-$[Mo(C_2H_4)_2(PMe_3)_4]$,[7] $[Mo(CO_2)_2(PMe_3)_4]$[7] and *cis*-$[Mo(CO)_2(dppe)_2]$.[5] (See above and Section 36.2.2.6.)

36.2.2.3 Nitrogen Monoxide Complexes

With our definition of NO in its complexes as NO^+, a range of compounds is then formally of Mo^0. Often they have analogues containing the isoelectronic ligand N_2R, which we formally regard as N_2R^-. This consequently means that the N_2R analogues of NO complexes will appear in a section concerning the higher oxidation states and we cross refer to, or list, analogous complexes where this occurs. Nitrosyl complexes have been extensively reviewed.[18,19,20]

36.2.2.3.1 Complexes of the type $MoX_2(NO)_2$

These complexes may be regarded as the precursors of a wide range of derivatives, $[MoX_2(NO)_2L_2]$, *etc.* discussed in subsequent sections.

$[\{MoX_2(NO)_2\}_n]$ (X = Cl, Br or I) are generally prepared by the reaction of [NO]X with a carbonyl derivative, sometimes followed by metathesis (method A). Thus $[\{MoX_2(NO)_2\}_n]$ (X = Cl or Br) are obtained by treatment of $[Mo(CO)_6]$ with NOX, whereas $[\{MoI_2(NO)_2\}_n]$ is obtained from $[\{MoBr_2(NO)_2\}_n]$ and KI in acetone.[18,19,20] $[MoX_2(CO)_4]$ when treated with NO also gives these complexes (method B). Alternatively (C) the reduction of $MoCl_5$ or $MoOCl_3$ with NO in CH_2Cl_2, C_6H_5Cl, $CHCl_3$ or CCl_4 gives $[\{MoCl_2(NO)_2\}_n]$.[21] The structure of these compounds appears to be halide-bridged polymers, with a linked structure caused by a *cis* disposition of NO ligands [X = Cl, ν(NO), 1805 and 1609 cm^{-1}).[19,20]

Compounds of the type $[MoXCp(NO)_2]$ ($Cp = \eta^5 - C_5H_5$ or $\eta^5\text{-}C_5Me_5$) and $[MoX(Rpzb)(NO)_2]$ are also known ($Rpzb^-$ = anion tris(2-pyrazolyl)(R)borate)[18,22] and are monomeric in solution, the Cp derivatives having the 'three-legged piano stool' structure (see 5).[1a,19]

The ^{95}Mo chemical shifts of this class of compounds are in the range −800 to −985 p.p.m. (rel. $[MoO_4]^{2-}$).[22]

36.2.2.3.2 Complexes of the type $[MoX_2(NO)_2L_2]$ (L = monodentate or 1/2-bidentate neutral ligand)[15]

This is the largest class of nitrosyl complex and the members are shown in Table 2, together with notable physical properties. They are prepared by the following routes.

(A) By treatment of $[\{MoX_2(NO)_2\}_n]$ with neutral ligands. For example, treatment of $[\{MoX_2(NO)_2\}_n]$ with ligands L gives the series $[MoX_2(NO)_2L_2]$ (L = $MeC_6H_4NH_2$, $C_6H_{11}NH_2$, PPh_3, $AsPh_3$, py).[23]

Table 2 Dinitrosyl Complexes of Molybdenum

Compound	*Preparation*[a]	$\nu(NO)^b$ (cm^{-1})	*Ref.*
Uncharged complexes			
$[MoCl_2(NO)_2(PPh_3)_2]$[c]	A, B	1796, 1670	1, 2
$[MoBr_2(NO)_2(PPh_3)_2]$	A, B	1795, 1675	2
$[MoI_2(NO)_2(PPh_3)_2]$	A, B	1795, 1675	2
$[MoCl_2(NO)_2(PMe_2Ph)_2]$	A, B	1795, 1675	1, 3
$[MoCl_2(NO)_2(PEtPh_2)_2]$	B	1785, 1693	4
$[MoCl_2(NO)_2(PEt_3)_2]$	B	1775, 1660	4
$[MoCl_2(NO)_2(PBu_3)_2]$	B	1760, 1670	4
$[MoBr_2(NO)_2(PEtPh_2)_2]$	B	1795, 1677	4
$[MoCl_2(NO)_2(AsPh_3)_2]$	A, B	1795, 1680	2
$[MoCl_2(NO)_2(dppm)]$	B	1790, 1670	4, 5
$[MoCl_2(NO)_2(dppe)]$	A	1805, 1685	4
$[MoBr_2(NO)_2(dppm)]$	A	1770, 1650	5
$[MoCl_2(NO)_2(dpam)]$	A, B	1765, 1680	5
$[Mo(NCS)_2(NO)_2(dppe)]$	B	1787, 1675	4
$[Mo(NCS)_2(NO)_2(dppm)]$	B	1788, 1680	4
$[MoCl_2(NO)_2(OPPh_3)_2]$	B, C	1780, 1660	3, 6
$[MoBr_2(NO)_2(OPPh_3)_2]$	B, C	1780, 1660	3, 6
$[MoCl_2(NO)_2(Opy)_2]$	A, B	1785, 1675	3
$[MoBr_2(NO)_2(Me_2SO)_2]$[d]	A	1784, 1645	7
$[MoCl_2(NO)_2(EtOH)_2]$	A	1790, 1680	3
$[MoCl_2(NO)_2(H_2O)_2]$	C	1810, 1680	3
$[MoCl_2(NO)_2(PhNC)_2]$	A	1780, 1680	3
$[MoCl_2(NO)_2(py)_2]$	A, B	1790, 1680	3, 8
$[MoCl_2(NO)_2(bipy)_2]$	A, B	1785, 1675	3, 9
$[MoCl_2(NO)_2(phen)]$[f]	A, B	1780, 1670	3, 8
$[MoCl_2(NO)_2(en)]$	A, B	1790, 1680	10
$[Mo(NCS)_2(NO)_2(bipy)]$	See text	1785, 1672	11
$[Mo(NCS)_2(NO)_2(phen)]$	See text	1780, 1665	11
$[Mo(CN)_2(NO)_2(phen)]$	See text	1785, 1665	11
$[MoCl_2(NO)_2(MeCN)_2]$	A, B	1805, 1685	9
$[MoCl_2(NO)_2(CH_2{=}CHCN)_2]$	A	1800, 1680	9
$[MoCl_2(NO)_2(PhCN)_2]$	A	1800, 1690	9
$[MoCl_2(NO)_2(PhCH_2CN)_2]$	A, B	1795, 1695	3
$[MoCl_2(NO)_2(C_6H_{11}NH_2)_2]$	A, B	1785, 1665	3
$[MoCl_2(NO)_2(MeC_6H_4NH_2)_2]$	A, B	1775, 1665	3
$[MoBr_2(NO)_2(Me_2S)_2]$	A	1790, 1682	6
$[Mo(S_2CNMe_2)_2(NO)_2]$[g]	See text	1750, 1645	8
$[Mo(S_2CNEt_2)_2(NO)_2]$[h]	See text	1750, 1640	8
$[Mo(acac)_2(NO)_2]$[i]	See text	1703, 1645	8
$[Mo(oxine)_2(NO)_2]$	C	1700, 1640	8
$[Mo(mbt)_2(NO)_2]$	C	1740, 1620	8
$[Mo(mp)_2(NO)_2]$	C	1755, 1620	8
$[Mo(etg)_2(NO)_2]$	C	1715, 1620	8
$[Mo(ttp)(NO)_2]$[j]	C	1740, 1600	12
$[Mo(tacn)(NO)_2]$	A	1755, 1660	8
$[Mo(MeSC_6H_4S)_2(NO)_2]$	A	1765, 1665	8
Cationic complexes			
$[MoCl(NO)_2(MeCN)_3]^+$	D	1830, 1720	13
$[MoCl(NO)_2(CH_2{=}CHCN)_3]^+$	D	1830, 1720	13
$[MoCl(NO)_2(PhCN)_3]^+$	D	1830, 1720	13
$[MoCl(NO)_2(MeCN)(py)_2]^+$	D	1810, 1690	13, 14
$[MoCl(NO)_2(MeCN)(bipy)]^+$	D	1810, 1700	13
$[MoCl(NO)_2(diars)_2]^+$	A, B	1773, 1667	15
$[MoBr(NO)_2(tacn)]^+$	B	1750, 1650	16
$[MoH(NO)_2(tacn)]^+$	B	1770, 1640	16
$[Mo(acac)(NO)_2(MeCN)_2]$	See text	—	14

Table 2 *(continued)*

Compound	*Preparation*[a]	$\nu(NO)^b$ (cm^{-1})	*Ref.*
$[Mo(NO)_2(OPPh_3)_4]^{2+}$	B, D	1800, 1680	13
$[Mo(NO)_2(MeCN)_4]^{2+}$	D	1860, 1750	13
$[Mo(NO)_2(MeCN)_2(OPPh_3)_2]^{2+}$	B, D	1827, 1713	13
$[Mo(NO)_2(CH_2{=}CHCN)_4]^{2+}$	D	1850, 1730	13
$[Mo(NO)_2(PhCN)_4]^{2+}$	D	1850, 1750	13
$[Mo(NO)_2(MeNO_2)_4]^{2+}$	B, D	1853, 1743	13
$[Mo(NO)_2(MeCN)_2(py)_2]^{2+}$	D	1830, 1720	13
$[Mo(NO)_2(MeCN)_2(bipy)]^{2+}$	D	1830, 1730	13
$[Mo(NO)_2(diars)_2]^{2+}$	A, B	1816, 1733	15
$[Mo(NO)_2(bipy)_2]^{2+}$	B, D	1816, 1707	13
$[Mo(NO)_2(dppe)_2]^{2+}$	B, D	1800, 1676	13
Anionic complexes			
$[Mo(S_2C_6Cl_4)_2(NO)_2]^-$	See text	1773, 1658	17
$[Mo(S_2C_6H_3Me)_2(NO)_2]^-$	See text	1760, 1650	17
$[MoCl_4(NO)_2]^{2-}$ [k]	B	1720, 1600	8, 17
$[MoBr_4(NO)_2]^{2-}$	B	1720, 1610	17
$[MoCl_2Br_2(NO)_2]^{2-}$	B	1720, 1630	17
$[Mo(CN)_4(NO)_2]^{2-}$ [l]	B	1790, 1660	18
$[Mo(CNS)_4(NO)_2]^{2-}$ [m]	B	1786, 1658	11
$[Mo(OH)_4(NO)_2]^{2-}$	C	1740, 1610	3
$[Mo(S_2C_6Cl_4)_2(NO)_2]^{2-}$	See text	1704, 1595	17
$[Mo(MNT)_2(NO)_2]^{2-}$	See text	1742, 1633	17
$[Mo(NO)_2(ox)^{2-}$ [n]	See text	1785, 1630	8
$[Mo(NO)_2(C_6H_4S_2)_2]^{2-}$		1690, 1580	19

[a] See text. [b] Solid state spectra (Nujol or KBr). [c] δ^{95}Mo (DMF) −500 p.p.m. rel. $[MoO_4]^{2-}$; X-ray structure—see text. [d] X-ray structure; *d*(NO) 1.167, 1.153 Å; Me_2SO is O-bonded. [e] δ^{95}Mo (DMF) −263 p.p.m., X-ray structure—see text. [f] δ^{95}Mo (DMF) −265 p.p.m. [g] δ^{95}Mo (DMF) −438 p.p.m. [h] δ^{95}Mo (DMF) −430 p.p.m. [i] δ^{95}Mo (DMF) −298 p.p.m. [j] X-Ray structure—see text. [k] δ^{95}Mo (HCl) + 152 p.p.m. [l] δ^{95}Mo (DMF) −104 p.p.m. [m] X-Ray structure, *cis* NO's *d*(NO) 1.146 Å (av.). [n] δ^{95}Mo (DMF) +201 p.p.m.

ttp = *meso*-tetra-*p*-tolylporphyrinate; tacn = 1,4,7-triazacyclononane; oxine = 8-quinolinolate; mbt = 2-mercaptobenzothiazolate; mp = 2-pyridinethiolate; etg = ethylmercaptoacetate; dttd = 2,3,8,9-dibenzo-1,4,7,10-tetrathiadecane(2 −).

1. E. I. Stiefel, *Prog. Inorg. Chem.*, 1976, **22,** 1.
2. M. O. Visscher and K. G. Caulton, *J. Am. Chem. Soc.*, 1972, **94,** 5293; B. F. G. Johnson, *J. Chem. Soc. (A)*, 1967, 457.
3. F. A. Cotton and B. F. G. Johnson, *Inorg. Chem.*, 1964, **3,** 1609; U. Sartorelli, F. Zingales and F. Canziani, *Chim. Ind. (Milan)*, 1967, **49,** 751; K. H. Al-Obaidi and T. J. Al-Hassani, *Transition Met. Chem.*, 1978, **3,** 15; S. Bhattacharyya, N. N. Bandyopadhyay, S. Rakshit and P. Bandyopadhyay, *Z. Anorg. Allg. Chem.*, 1979, **449,** 181.
4. T. Nimry, M. A. Urbancic and R. A. Walton, *Inorg. Chem.*, 1979, **18,** 691.
5. J. A. Bowden, R. Colton and C. J. Commons, *Aust. J. Chem.*, 1972, **25,** 1393.
6. F. King and G. J. Leigh, *J. Chem. Soc., Dalton Trans.*, 1977, 429.
7. C. Schumacher, F. Weller and K. Dehnecke, *Z. Anorg. Allg. Chem.*, 1984, **508,** 79; 1985, **520,** 25.
8. M. Minelli, J. L. Hubbard and J. H. Enemark, *Inorg. Chem.*, 1984, **23,** 970; K. S. Nagaraja and M. R. Udupa, *Bull. Chem. Soc. Jpn.*, 1984, **57,** 1705; D. Sellman, L. Zapf, J. Keler and M. Moll, *J. Organomet. Chem.*, 1985, **289,** 71.
9. D. Ballivet-Tkatchenko, C. Bremard, F. Abraham and G. Nowogrocki, *J. Chem. Soc., Dalton Trans.*, 1983, 1137.
10. R. D. Feltham, W. Silverthorn and G. McPherson, *Inorg. Chem.*, 1969, **8,** 344.
11. R. Bhattachararya and G. P. Bhattachararjee, *J. Chem. Soc., Dalton Trans.*, 1983, 1593; with A. M. Saha, *Polyhedron*, 1985, 580; A. Müller, W. Eltzner, S. Sarkar, H. Bögge, P. J. Aymonino, M. Mohan, U. Sayer and P. Subramanian, *Z. Anorg. Allg. Chem.*, 1983, **503,** 22.
12. T. Diebold, M. Schappacher, B. Chevrier and R. Weiss, *J. Chem. Soc., Chem. Commun.*, 1979, 693.
13. D. Ballivet-Tkatchenko and C. Bremard, *J. Chem. Soc., Dalton Trans.*, 1983, 1143; P. Legzdins and J. C. Oxley, *Inorg. Chem.*, 1984, **23,** 1053.
14. B. F. G. Johnson, A. Khair, C. G. Savory and R. H. Walker, *J. Chem. Soc., Chem. Commun.*, 1974, 744.
15. N. G. Connelly and C. G. Gardner, *J. Chem. Soc., Dalton Trans.*, 1979, 609.
16. P. Chaudhuri, K. Wieghardt, Y. Tsai and C. Krüger, *Inorg. Chem.*, 1984, **23,** 427.
17. N. G. Connelly, J. Locke, J. A. McCleverty, D. A. Phipps and B. R. Ratcliff, *Inorg. Chem.*, 1970, **9,** 278.
18. R. Bhattachararya, G. P. Bhattacharajee and A. M. Saha, *Transition Met. Chem.*, 1983, **8,** 255.
19. D. Sellmann and L. Zapf, *Z. Naturforsch., Teil B*, 1985, **40,** 368; 380.

(B) By treatment of a molybdenum(II) compound such as $[MoX_2(CO)_3L_2]$ with NO. Typically $[MoX_2(CO)_3(PPh_3)_2]$ (X = Cl or Br) gives, on treatment with NO, $[MoX_2(NO)_2(PPh_3)_2]$.[24]

(C) By 'reductive nitrosylation', *i.e.* reaction of a high-valent molybdenum compound with NO in presence of a reductant and another ligand, *e.g.* $[MoCl_4(PPh_3)_2] + NO + PPh_3 + AlCl_2Et$ in hexane gives $[MoCl_2(NO)_2(OPPh_3)_2]$.

For completion, we list compounds of the class $[MoX(\eta^5\text{-}C_5H_5)(NO)_2L_2]$ in Table 2. Their analogues, $[MoCl\{Rpzb\}(NO)_2L_2]$ are also included.

As can be seen from the data of Table 2, these compounds have in general a *cis* arrangement of nitrosyl groups (two ν(NO) values) and the X-ray structure of $[MoCl_2(NO)_2(PPh_3)_2]$ confirms this[18,19] as shown in (**3**). The average Mo—N distance (1.19 Å) is consistent with the NO^- formalism and the Mo—N—O angle is significantly less than linear (162° average). Bending in nitrosyl complexes commonly occurs as a result of removal of the degeneracy of the π^*(NO) levels and has been discussed in detail elsewhere.[18] The *cis* NO structure has also been confirmed by X-rays for $[MoCl_2(NO)_2(bipy)]$,[18] and for $[MoBr_2(NO)_2(OSMe_2)_2]$.[18]

The reaction of $[\{MoCl_2(NO)_2\}_n]$ with diars and en gives $[MoCl_2(NO)_2(diars)]$ and $[MoCl_2(NO)_2(en)]$ but if these compounds are left in solution their colour changes from green to yellow and the formation of a hyponitrite-bridged complex, product (**4**), has been suggested.[25]

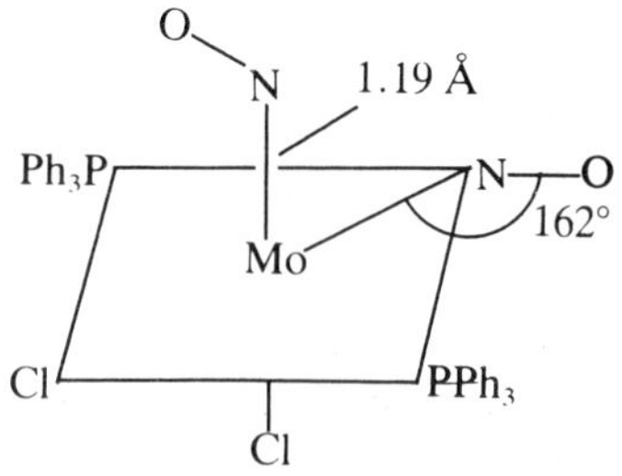

(**3**)

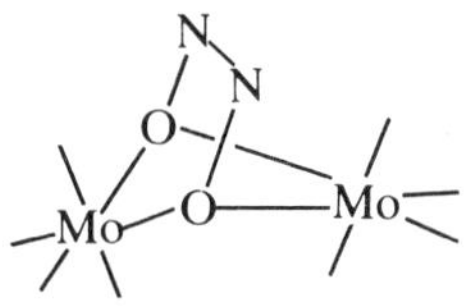

(**4**)

36.2.2.3.3 Other dinitrosyl complexes

Treatment of $[\{MoX_2(NO)_2\}_n]$ with sodium dialkyl dithiocarbamates (Nadtc) affords the dinitrosyl complexes *cis*-$[Mo(NO)_2(S_2CNR_2)_2]$ (R = Me or Et). These compounds may also be prepared by treatment of $[Mo(NO)_2(MeCN)_4][PF_6]_2$ with Nadtc and the complex $[Mo(NO)_2(acac)_2]$ is obtained using Na[acac].[26,27]

The cationic precursor complex $[Mo(NO)_2(MeCN)_4]^{2+}$ is itself prepared from $[Mo(CO)_3(MeCN)_3]$ and $[NO]PF_6$ in MeCN.[26] The exchange reactions of the MeCN in this dication have been studied by NMR, two MeCN ligands being relatively easily dissociated. It is assumed that these are *trans* to the *cis* NO groups. The cationic complexes $[Mo(acac)(NO)_2(MeCN)_2]^+$ and $[MoCl(NO)_2(MeCN)_3]^+$ have also been prepared, the former in which the MeCN groups are mutually *trans* does not dissociate MeCN, whereas the latter exchanges the one MeCN which is *trans* to NO.[28]

Two series of cationic complexes, $[Mo(NO)_2L_4]^{2+}$ (L = RCN, py, *etc.*) and $[MoCl(NO)_2L_3]^+$ have been prepared by halide abstraction from $[MoCl_2(NO)_2L_2]$ using silver salts in DME (method D). Some members of the series catalyze the polymerization of norbornadiene, whereas the uncharged parent compounds do not.[29] The series $[Mo(NO)_2L_4]^{2+}$ (L is a displaceable solvent such as THF) may also be obtained by treatment of $[Mo(CO)_6]$ with $NO[PF_6]$ followed by addition of further ligands such as bipy.[29] This latter reaction also gives $[\{Mo(NO)_2(PF_6)_2\}_n]$ which is assumed to be a polymer with weak PF_6 bridges.[29]

The reduction of $[Mo(ttp)Cl_2]$ (ttp = tetra-4-tolylporphinate) with zinc amalgam in benzene under NO followed by chromatography gave the dinitrosyl complex $[Mo(ttp)(NO)_2]\cdot C_6H_6$ which has the *cis* NO geometry established by X-rays. The NO groups are bent towards each other [NMoN = 78.4(5)° and OMoO = 60.0(3)°], a mutual NO attraction which is observed in other dinitrosyl compounds. In order to accommodate the *cis* geometry, the Mo is displaced 0.99 Å from the porphinato skeleton towards the nitrosyls.[30a]

A further class of dinitrosyl complex is that containing dithiolene and related ligands. Because of electronic delocalization, there are problems of assignment of oxidation state in these complexes,[27] *e.g.* in $[Mo(NO)_2(MNT)_2]^{2-}$ and $[Mo(NO)_2(S_2C_6Cl_4)_2]^{2-}$, but they are included in this section (see Table 2) for convenience. A further series of anionic complexes in this class are the green compounds $[MoX_4(NO)_2]^{2-}$ (X = Cl, Br or I), prepared from $[\{MoX_2(NO)_2\}_n]$ and *e.g.* $[NEt_4]X$, and $[MoClBr(NO)_2]^{2-}$, prepared from $[\{MoCl_2(NO)_2\}_n]$ and $[Ph_4P]Br$.[23] A binuclear, bromide-bridged dinitrosyl $[Mo_2Br_2(NO)_4]^{2-}$ has been structurally characterized by X-rays.[30b]

An important use of dinitrosyl complexes of molybdenum has been as catalyst precursors in the alkene disproportionation reaction. Thus $[MoCl_2(NO)_2L_2]$ (L = PPh_3 or py) together with $AlCl_2Et$ forms an active homogeneous catalyst for alkene disproportionation. The catalysis appears to require loss of NO from the metal and the field has been reviewed.[19,31]

36.2.2.3.4 Mononitrosyl complexes

These complexes are generally prepared by methods similar to those employed for the dinitrosyl compounds, treatment of a low-valent molybdenum compound with NO or a nitrosyl derivative being the principal method. Compounds and major physical properties are collected in Table 3.

Treatment with nitrosylating reagents of molybdenum–carbonyl precursors leads to a variety of carbonyl–nitrosyl complexes. Examples of these are included in Table 3, but they are not reviewed in depth for the reasons cited earlier. They have been reviewed in COMC (vol. 3, pp. 1112–1115), *e.g.* $[Mo(CO)_4(diars)]$ with $NO[PF_6]$ gives[32] $[Mo(diars)(CO)_3(NO)]^+$, and $[Mo(CO)_4(bipy)]$ gives $[MoX(CO)_2(NO)(bipy)]$, (X = Cl, Br or I)[32] (method A). This method has also given the series *trans*-$[MoX(NO)(dppe)_2]$ (X = F, Cl, Br, I, OH, SCN, CN, NCO, N_3, SPh or PPh_2), by treatment of *trans*-$[MoX(N_2H)(dppe)_2]$ with NO or *trans*-$[Mo(N_2)_2(dppe)_2]$ with $MeN(NO)C(O)NH_2$ (X = CNO) or Et_2NNO (X = OH).[33]

The *trans* structure of this latter series has been established by X-rays for X = OH, disorder preventing accurate determination of the NO distance.[33] The carbonyl analogues $[MoCl(NO)(CO)_2(diphos)_2]$ (diphos = dppe or dppm) have been obtained from $[Mo(CO)_3(MeCN)_3]$, diphos and NOCl.[32] Use of PPh_3 gives $[MoCl(NO)(CO)(MeCN)(PPh_3)_2]$. The compound $[MoCl(NO)(PPh_2H)_4]$ results from treatment of $[MoCl_5(NO)]^{2-}$ with PPh_2H and Mg.[31]

The monocationic complexes $[Mo(CO)(NO)(dppe)_2]^+$ (method A, $[Mo(CO)_4(dppe)]$ + $NO[PF_6]$) on treatment with NaS_2CNR_2 (R = Me or Et) give[34] $[Mo(CO)(NO)(S_2CNR_2)(dppe)]$; the analogues $[Mo(S_2CNR_2)(CO)(NO)(diars)]$ were similarly prepared[35] (Table 3). The anionic complex $[Mo(NO)(MNT)_2]^{2-}$ obtained[36] from $[\{MoCl_2(NO)_2\}_n]$ and Na_2MNT is also included in this section.

The compounds $[Mo(\eta^5\text{-}Cp)(NO)(CO)_2]$ (Cp = C_5H_5 or C_5Me_5) are also included in this section for completion (Table 3) (see COMC vol. 3 for details). Their analogues with pyrazolylborate ligands $[Mo(CO)_2L(NO)]$ [L = $HB(3,5\text{-}Me_2HC_3N_2)_3$, $HB(C_3H_3N_2)_3$ or $B(C_3H_3N_2)_4^-$] have been prepared by treatment of $[Mo(CO)_6]$ with the appropriate pyrazolylborate salt followed by NO or an NO source.[36] The complexes $[Mo(\eta^5\text{-}Cp)(NO)(CO)_2]$ have been shown to have the 'three-legged piano stool' structure (**5**) by X-ray spectroscopy[37] and presumably the pyrazolylborates have a similar structure. Treatment of the compound $[Mo(Bpz_4)(NO)(CO)_2]$ (pz = pyrazolyl) with RNC (R = C_6H_5MeCH, $PhCH_2$, Et) gives the adducts $[Mo(Bpz_4)(NO)(CO)(RNC)]$ which are enantiomeric and diastereoisomeric because of the asymmetry at molybdenum and in the isocyanide chain (Table 3).[35] Phosphine analogues $[Mo\{3,5\text{-}X_2pzb\}(NO)(CO)(PPh_2R')]$ (Table 3) are also diastereoisomeric.[35]

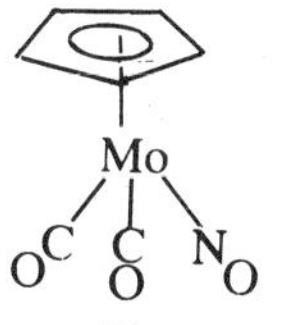

(**5**)

The complexes *trans*-$[MoX(NO)(dppe)_2]$ show a well-defined reversible one-electron oxidation in solution and the monocations can be isolated. They are described in Section 36.2.3.3. Treatment of *trans*-$[Mo(OH)(NO)(dppe)_2]$ with $Ag[BF_4]$ in MeCN gives $[Mo(NO)(NCMe)(dppe)_2][BF_4]$ as well as its dicationic, oxidized analogue.[32]

Table 3 Mononitrosyl Complexes of Molybdenum(0)

Complex	*Preparative method*	$\nu(NO)$ (cm^{-1})[a]	*Ref.*
trans-[$MoX(NO)(dppe)_2$][b] (X = F, Cl, Br, I, OH, SCN, CNO, CN, N_3, SPh, OPh)	A See text	1500–1600	1, 2
[$MoCl(NO)(PPh_2H)_4$]	See text	1542	3
[$MoCl(NO)(CO)_2(dppe)$]	A	1648	4
[$MoCl(NO)(CO)_2(dppm)$]	A	1632	4, 5
[$MoCl(NO)(CO)(PPh_3)(dppe)$]	A	1597	4
[$MoCl(NO)(CO)(PPh_3)_2(MeCN)$]	A	1617	5
[$MoCl(NO)(CO)(PPh_3)(py)_2$]	A	1595	5
[$MoX(NO)(CO)_2(diars)$] (X = Cl, Br, I)	A	1642–1649	6a
[$MoCl(NO)(CO)(dppe)_2$]	A	1601	4
[$MoCl(NO)(CO)(diars)_2$]	A	1605	6a
[$Mo(S_2CNMe_2)(NO)(CO)_2(diars)$] (R = Me or Et)	See text	1597	4
[$Mo(S_2CNMe_2)(NO)(CO)_2(diars)$]	See text	1589	6a
[$Mo(O_2PF_2)(NO)(CO)_2(diars)$]	See text	1658	6a
trans-[$MO(NO)(MeCN)(dppe)_2$]$^{+}$[c]	See text	1685	1, 2
mer-[$Mo(NO)(CO)_3(dppe)$]$^+$	A	1735	4
mer-[$Mo(NO)(CO)_3(diars)$]$^+$	A	1741	6a
[$Mo(NO)(CO)(dppe)$]$^+$	A	1645	4
[$Mo(NO)(CO)(diars)$]$^+$	A	1659	4
[$Mo(NO)(CO)_2(PMe_2Ph)(diars)$]$^+$	A	1671[d] 1703	6a
[$Mo(NO)(CO)_2\{P(OMe)_3\}(diars)$]$^+$	A	1708[d] 1659	6a
[$Mo(NO)(CO)_2(ttacn)$]$^+$	A	1665	6b
[$Mo(NO)(MNT)$]$^{2-}$	A	—	7
[$Mo(\eta^5\text{-}C_5H_5)(NO)(CO)_2$][e]	A	—	8, 9
[$Mo(\eta^5\text{-}C_5Me_5)(NO)(CO)_2$][f]	A	1642	8, 9
[$Mo(Me_2pzb)(NO)(CO)_2$][g]	A	1655	9
[$Mo(pzb)(NO)(CO)_2$][h]	A	1660	9
[$Mo\{B(H_3C_3N_2)_4\}(NO)(CO)_2$]	A	1665	10
[$Mo\{B(H_3C_3N_2)_4\}(NO)(CO)(RNC)$]	A	1620 (R = C_6H_5MeCH) 1620 (R = $C_6H_5CH_2$) 1615 (R = $MeCH_2$)	10 10 10
[$Mo(R\text{-}3,5\text{-}X_2pzb)(CO)(NO)L$][i] (R = pz, X = H; R = H, X = Me)		1600–1640	11
$K_4[Mo(CN)_5NO]$[j]		1455	12

[a] Solid state spectra (Nujol or Kbr). [b] X-Ray structure for X = OH confirms *trans* structure; δ ^{95}Mo −490 to −800 p.p.m. rel. [MoO_4]$^{2-}$ (THF). [c] δ ^{95}Mo −553 (CH_2Cl_2). [d] Two isomers. [e] X-Ray structure, three-legged piano stool, CO and NO disordered; δ ^{95}Mo −1584 (MeCN). [f] X-Ray structure as e; δ ^{95}Mo −1404 (MeCN). [g] δ ^{95}Mo −751 (MeCN). [i] L = $PPh_2NRCH(Me)Ph$ (R′ = H or Me). [j] X-Ray structure *d*(NO) 1.23 Å.
ttacn = *N,N′,N″*-trimethyl-1,4,7-triazacyclononane.

1. C. T. Kan, P. B. Hitchcock and R. L. Richards, *J. Chem. Soc., Dalton Trans.*, 1982, 79.
2. S. Donovan-Mtunzi, M. Hughes, G. J. Leigh, H. Mohd-Ali, J. Mason and R. L. Richards, *J. Organomet. Chem.*, 1983, **246,** C1.
3. F. King and G. J. Leigh, *J. Chem. Soc., Dalton Trans.*, 1977, 429.
4. W. R. Robinson and M. E. Swanson, *J. Organomet. Chem.*, 1972, **35,** 315.
5. N. G. Connelly, *J. Chem. Soc., Dalton Trans.*, 1973, 2183.
6. (a) N. G. Connelly and C. Gardner, *J. Chem. Soc., Dalton Trans.*, 1979, 609. (b) G. Backes-Dahmann and K. Wieghardt, *Inorg. Chem.*, 1985, **24,** 4044.
7. J. A. McCleverty, *Prog. Inorg. Chem.*, 1968, **10,** 145.
8. J. T. Malito, R. Shakir and J. L. Atwood, *J. Chem. Soc., Dalton Trans.*, 1980, 1253.
9. M. Minelli, J. L. Hubbard, K. A. Christensen and J. H. Enemark, *Inorg. Chem.*, 1983, **22,** 2652.
10. E. Frauendorfer and J. Puga, *J. Organomet. Chem.*, 1984, **265,** 259.
11. E. Frauendorfer and H. Brunner, *J. Organomet. Chem.*, 1982, **240,** 371.
12. A. Müller, W. Eltzner, S. Sarkar, H. Bögge, P. J. Aymonio, N. Mohan, U. Seyer and P. Subramanian, *Z. Anorg. Allg. Chem.*, 1983, **503,** 22 and references therein.

36.2.2.4 Nitrogen Donor Ligands

The complexes [$Mo(bipy)_3$] can be prepared by reduction of [$Mo(bipy)_3$]$^{3+}$ with Li_2bipy[37] or of [$MoCl_4(MeCN)_2$] with sodium in presence of bipy.[38] Formally, the oxidation state is zero, but there must be considerable distribution of electronic charge in orbitals which are ligand-based.

There are a considerable number of molybdenum carbonyl derivatives with nitrogen bases, *e.g.* of the types $[Mo(CO)_nL_{6-n}]$ (L = RCN, py, DMF, amines, various diamines, *etc.*, $n = 1$–4) which are not reviewed here but have received extensive coverage in COMC (vol. 3, p. 1087). We include $[Mo(PF_3)_2(bipy)_2]$ in this section.[38b] It is obtained by substitution of CO by PF_3 in $[Mo(CO)_2(bipy)_2]$.

36.2.2.5 Phosphorus Donor Ligands

These compounds, of the general formulation $[MoP_6]$ (P = PMe_3, PH_2Ph or $P(OMe)_3$, or P_2 = dmpe or dppm), may be synthesized by displacement of CO from $[Mo(CO)_6]$ under forcing conditions (for dmpe);[34] by reaction of molybdenum vapour with the ligand (for PMe_3 and dmpe);[40] by reduction of a molybdenum complex, in presence of the ligand, with $AlEt_3$ (for dppm) or Mg (for PMe_3, PH_2Ph or $P(OMe)_3$);[19] or by displacement of N_2 from *trans*-$[Mo(N_2)_2(dppe)_2]$. These compounds are red or brown, octahedral and react readily with dioxygen in solution. The X-ray structures of $[Mo(PMe_3)_6]$ and $[Mo(dmpe)_3]$ confirm the essentially octahedral structures (there is disorder in the case of $[Mo(PMe_3)_6]$) with d(Mo–P) 2.467(2) Å (for PMe_3) and 2.421(3) Å (P_2 = dmpe).[40] $[Mo(PMe_3)_6]$ is particularly reactive; thus PMe_3 is displaced by N_2, CO and C_2H_4 to give $[Mo(N_2)(PMe_3)_5]$,[21] $[Mo(PMe_3)_3(CO)_3]$ and $[Mo(C_2H_4)_2(PMe_3)_4]$ respectively; H_2 adds to give $[MoH_4(PMe_3)_4]$ in which a PMe_3 group will oxidatively add a C—H bond to the metal and a variety of organometallic and oxidized compounds may be obtained.[40] In contrast to this high reactivity, $[Mo(dppe)_3]$ is very inert to substitution and $[Mo\{P(OMe)_3\}_6]$ is only substituted when photolyzed.[40] This difference is reflected in the relatively long Mo—P bond in $[Mo(PMe_3)_6]$ (see above) compared to that of $[Mo(dmpe)_3]$, presumably because of greater steric pressure in the former complex. The complex $[Mo(triphos)(\eta^2\text{-dppFe})(\eta^1\text{-dppFe})]$ (dppFe = $Ph_2PC_5H_4FeC_5H_4PPh_2$) has also been prepared.[42]

36.2.2.6 Miscellaneous Compounds

36.2.2.6.1 Complexes with dithiolene ligands

Compounds of the type $[Mo\{S_2C_2(CF_3)_2\}_3]^{n-}$ ($n = 0$–2) present a problem of assignment of oxidation state because ligands of this type may be regarded as having two limiting structures (**6**) and (**7**) in their complexes. Thus the oxidation state of complexes if the limiting form is (**7**) would be low, *i.e.* (0), (I) and (II) in the above series. However, a range of spectroscopic and other investigations indicate that these compounds have oxidation states at the higher range [*e.g.* the above series could be viewed as (VI), (V) and (IV)] and therefore they are not included in this survey; they have been reviewed elsewhere.[19]

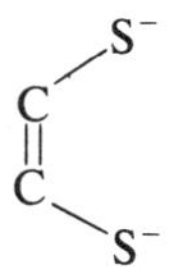

(**6**)

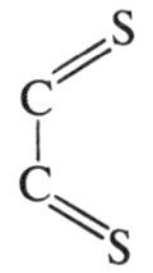

(**7**)

36.2.2.6.2 Complexes of CO_2, COS and SO_2

Complexes of these ligands are included in this section, again because of an ambiguity of oxidation state. This is exemplified by the X-ray structure of $[Mo(CO_2)_2(PMe_3)_3(CNPr^i)]$ (**8**) where the CO_2 ligands are C—O bonded to the metal. For the purposes of this review, CO_2 is regarded as a neutral ligand, and its bonding in the above complex resembles that of C_2H_4 in the analogue $[Mo(C_2H_4)_2(PMe_3)_4]$.[43] Compound (**8**) was obtained by treatment with RNC (R = Me, Pr^i or Bu^t) of $[Mo(CO_2)_2(PMe_3)_4]$, which in turn is obtained by displacement of N_2 from $[Mo(N_2)_2(PMe_3)_4]$ by CO_2.[43] The analogue $[Mo(CO_2)_2(dppe)_2]$ is similarly prepared from *trans*-$[Mo(N_2)_2(dppe)_2]$, but treatment of this complex with MeNC evolves CO_2 to give *trans*-$[Mo(CNMe)_2(dppe)_2]$.[44] Reaction of COS with $[Mo(N_2)_2(PMe_3)_4]$ gives the complex

$[Mo(S_2CO)(CO)_2(PMe_3)_2]$. The disproportionation of CO_2 and COS to give carbonato species commonly occurs at Mo^0 centres.[43]

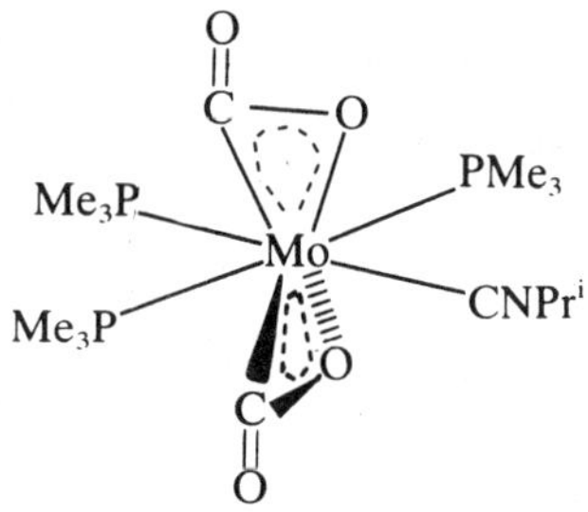

(8)

Several SO_2 complexes of molybdenum have been prepared which are assigned both the η^2-(S—O)— and η^1(S-bound)-bonding modes. Generally the metal coordination site has a mainly carbon environment and we give here only the examples *fac*-$[Mo(CO)_3(dmpe)(\eta^2\text{-}SO_2)]$; *mer*-$[Mo(CO)_3(dmpe)(\eta^1\text{-}SO_2)]$, *mer,trans*-$[Mo(CO)_3(PR_3)(\eta^1\text{-}SO_2)]$ and *trans*-$[Mo(CO)_2(dppe)(PR_3)(\eta^1\text{-}SO_2)]$ ($PR_3 = PMe_3$, PPh_2Me or PPr^i_3) and *fac*-$[Mo(CO)_3(MeCN)_2(\eta^2\text{-}SO_2)]$. They are generally prepared by displacement of a labile ligand, such as MeCN, from a precursor complex, *e.g.* $[Mo(CO)_3(MeCN)_3]$ by SO_2.[45]

36.2.3 MOLYBDENUM(I)

36.2.3.1 Carbon Ligands

36.2.3.1.1 Carbon monoxide

In this section we mention the paramagnetic monocationic complexes $[Mo(CO)_2(dppe)_2]^+$ and $[Mo(CO)_2(bipy)_2]^+$, both of which are produced by oxidation of the parent Mo^0 complexes. In the former case, the oxidants used include $[MeC_6H_4N_2][BF_4]$,[46] $NO[PF_6]$, I_2 and Ag^I [1a,47] and a variety of analogues $[Mo(CO)_2L_2]^+$ has been obtained (L_2 = dppm, dmpe, diars, *etc.*).[1a,47] Electrochemical oxidation has also been used, with other measurements, to show that a rapid *cis*→*trans* isomerization follows oxidation and an explanation for this phenomenon has been proposed on the basis of extended Hückel molecular orbital calculations, the stereochemical change being dependent on the number of valence electrons and the nature of the coligand π-donor or π-acceptor capacity. Similar studies have been made upon the compounds $[Mo(CO)_2L_2]^+$ (L = bipy or phen).[47]

36.2.3.1.2 Isocyanide ligands

The isocyanide analogues of the above carbonyls, *trans*-$[Mo(CNR)_2(dppe)_2]$ (R = alkyl or aryl) undergo a reversible one-electron oxidation in solution[4,5] and oxidation with I_2 or Ag^I salts allows isolation of the paramagnetic cation $[Mo(CNC_6H_4Me)_2(dppe)_2]^+$ ($\mu_{eff} = 2.01$ BM). With other isocyanides, substitution at the metal centre occurs on further oxidation.[48]

36.2.3.2 Dinitrogen Complexes

Oxidation of *trans*-$[Mo(N_2)_2(dppe)_2]$ with I_2 in methanol allows isolation of the unstable cationic complex *trans*-$[Mo(N_2)_2(dppe)_2]I_3$.[6] This cation has also been prepared electrochemically and the rate of its decomposition by loss of N_2 studied.[6] A range of dinitrogen complexes shows reversible one-electron oxidation in solution, but in general the monocations are too unstable to be isolated for molybdenum.[6]

Species of the type $[MoX(N_2)(dppe)_2]$ (X = Cl, Br or I) are thought to be intermediates in the alkylation reactions of coordinated dinitrogen in *trans*-$[Mo(N_2)_2(dppe)_2]$ and their isolation for X = Cl or Br has been claimed. The analogue $[MoCl(N_2)(PMe_3)_4]$ was also thought to exist but subsequent studies have indicated that the properties attributed to $[MoX(N_2)L_4]$ ($L = PMe_3$ or L_2 = dppe) can be reproduced by an equimolar mixture of $[MoX_2L_4]$ and $[Mo(N_2)_2L_4]$ and at best it appears that $[MoX(N_2)L_4]$ very rapidly disproportionates in solution.[49] However, the analogue $[Mo(SCN)(N_2)(dppe)_2]$ is stable at room temperature.[50]

36.2.3.3 Nitrogen Monoxide Complexes

These complexes are collected in Table 4.

The pale green paramagnetic complexes $[MoCl_2(NO)L_2]$ (Table 4) result from reaction[31] of alkyl phosphines with the red solution obtained from $MoCl_5$ and NO in CH_2Cl_2. Treatment of $[MoCl_2(NO)(PMePh_2)_2(PMePh_2O)]$ with $[NEt_4]Cl$ gives[31] the green paramagnetic salt $[NEt_4][MoCl_3(NO)(PMePh_2)_2]$. Other complexes in this series are generally obtained by oxidation of the formally Mo^0 nitrosyl (Section 36.2.2.3) or reduction of the formally Mo^{II} nitrosyl (Section 36.2.4.2).

Table 4 Nitrosyl Complexes of Molybdenum(I)

Complex	$\nu(NO)^a$ (cm^{-1})	*Other data*	*Ref.*
$[MoCl_2(NO)(PMePh_2)_2(PMePh_2O)]$	1585	μ_{eff} 1.91 BM	1
$[MoCl_2(NO)(PPr^nPh_2)_2(PPr^nPh_2O)]$	1600	μ_{eff} 1.93 BM	1
$[Mo(NCS)_2(NO)(bipy)]^b$	1678	μ_{eff} 2.1 BM $\langle g_{av}\rangle$ 1.99	2
$[Mo(NCS)_2(NO)(phen)]^b$	1665	μ_{eff} 1.8 BM $\langle g_{av}\rangle$ 1.99	2
$[Mo(NCS)_2(NO)(phen)]^c$	1600	μ_{eff} 1.9 BM $\langle g_{av}\rangle$ 1.97	2
$[MoBr_2(NO)(ttacn)]$	1565	μ_{eff} 1.83 BM	3
$[Mo(NCO)(S_2CNR_2)(OPPh_3]$ R = Me or Et	1640	—	4
$[Mo(ttp)(NO)(MeOH)]\cdot 2C_6H_6{}^d$	1540	μ_{eff} 1.71 BM $\langle g\rangle$ 1.968 $\langle A\ ^{95,97}Mo\rangle$ −99.66	5
$[Mo\{HB(Me_2C_3NH_2)_3\}I(NO)]$	—	$\langle g\rangle$1.998	6
$[MoX(NO)(dppe)_2]^+$ (X = Cl, Br, I, OH, H, NCO, NCS)	1645–1650	$\langle g_{av}\rangle$ (X = Cl) 2.02 μ_{eff} 1.9-2.02	7, 8
$[Mo(Me_2pzb)(NO)(MeCN)_2]^+$	1652	$\langle g\rangle$ 1.982	6
$[Mo(NO)(MeCN)(dppe)_2]^{2+}$	1685	μ_{eff} 2.2 BM $\langle g_{av}\rangle$ 2.02	7, 8
$[Mo(NO)(PhCN)(dppe)_2]^{2+}$	1700	μ_{eff} 2.1 BM	7
$[Mo(NO)(MeCN)_5]^{2+}$	1796	$\langle g_{av}\rangle$ 1.98	8
$[MoCl_3(NO)(PMePh_2)_2]^-$	1610	μ_{eff} 1.89	1
$[MoCl_4(NO)(OH_2)]^{2-}$	1624	d(NO) 1.158 Å	9
$[Mo(CN)_5(NO)]^{3-}$	1585	X-Ray structure disordered μ_{eff} 1.96 BM $\langle g_{av}\rangle$ 2.00	9
$[Mo(SCN)_4(NO)]^{2-}$	1676	—	9

[a] Solid state spectra (Nujol or KBr). [b] Square pyramidal structure. [c] Trigonal pyramidal structure. [d] X-ray structure—NO *trans* to MeOH, disorder precludes accurate d(NO).

ttacn = *N,N′,N″*-trimethyl-1,4,7-triazacyclononane; ttp = *meso*-tetra-*p*-tolylporphinate.

1. F. King and G. J. Leigh, *J. Chem. Soc., Dalton Trans.*, 1977, 429.
2. R. Bhattacharyya, G. P. Bhattacharjee, A. K. Mitra and A. B. Chatterjee, *J. Chem. Soc., Dalton Trans.*, 1984. 487.
3. G. Backes-Dahmann and K. Wieghardt, *Inorg. Chem.*, 1985, **24**, 4044.
4. K. H. Al-Obaidi and T. J. Al-Hassani, *Transition Met. Chem.*, 1978, **3**, 15.
5. T. Diebold, M. Schappacher, B. Chevrier and R. Weiss, *J. Chem. Soc., Chem. Commun.*, 1979, 693.
6. J. A. McCleverty, *Chem. Soc. Rev.*, 1983, **12**, 331.
7. P. B. Hitchcock, C. T. Kan and R. L. Richards, *J. Chem. Soc., Dalton Trans.*, 1982, 79.
8. S. Clamp, N. G. Connelly, G. E. Taylor and T. S. Louttit, *J. Chem. Soc., Dalton Trans.*, 1980, 2162.
9. A. Müller, W. Eltzner, S. Sarkar, H. Bögge, P. J. Aymonio, N. Mohan, U. Seyer and P. Subramanian, *Z. Anorg. Allg. Chem.*, 1983, **503**, 22 and references therein.

The compound $[Mo(NO)(NH_2O)(NCS)_2(phen)]$ on heating under reflux in acetophenone or on photolysis at 80 °C gives the five-coordinate, paramagnetic isomeric pair $[Mo(NCS)_2(NO)(phen)]$ (Table 4); a square pyramidal analogue $[Mo(NCS)_2(NO)(bipy)]$ was also obtained. In all compounds the NCS is N-bonded.[51]

The reaction of HNO_3 with $[Mo(Hacn)(CO)_3]$ (Hacn = *N,N′,N″*-trimethyl-1,4,7-triazacyclononane) gives the air- and water-stable complex cation $[Mo(CO)_2(NO)(Hacn)]^+$ which on bromination and reduction gives the paramagnetic complex $[MoBr_2(NO)(Hacn)]$

(Table 4). The complexes $[MoX_2(NO)(Hacn)]^+$ can be reduced electrochemically by one electron to give the formally Mo^I species in solution.[52] The complexes $[Mo(S_2CNR_2)_2(NO)(OPPh_3)]$ (R = Me or Et), obtained by treatment of $[Mo(S_2CNR_2)_2(CO)_2]$ with NO, are formulated solely on the basis of analysis and IR spectroscopy.[53] The compound $[Mo(ttp)(NO)(MeOH)]\cdot 2C_6H_6$ is obtained upon reduction of $[Mo(ttp)Cl_2]$ by zinc followed by chromatography on Al_2O_3 with CH_2Cl_2 containing traces of MeOH (ttp = *meso*-tetra-*p*-tolylporphinate). Its X-ray structure confirms octahedral coordination with NO *trans* to MeOH.[29]

Electrochemical reduction of $[Mo\{Me_2pzb\}NOI_2]$ (Me_2pzb = tris(3,5-dimethylpyrazolyl)-borate) gives the green, paramagnetic monoanion, which dissociates I^- to give the uncharged, green, paramagnetic $[Mo\{Me_2pzb\}NO]$.[54] A range of pyrazolylborate–Mo^{II} complexes can also be reduced electrochemically to give the Mo^I species in solution, *e.g.* the compounds $[Mo\{Me_2pzb\}XY(NO)]$ (X, Y = anions) and the bimetallic series {Mo}NHRNH{Mo} ({Mo} = $Mo(Me_2pzb)I(NO)$; R = $C_6H_4R'C_6H_4$; R' = CH_2, C_2H_4, SO_2, O) and $\{Mo\}NHC_6H_4Y$ (Y = NH_2 or OH).[54]

The magenta, paramagnetic complex $[Mo\{Me_2pzb\}(NO)I_2Li(OEt)_2]$ was also isolated by reduction of $[Mo\{Me_2pzb\}I_2(NO)]$ in ether with LiPh.[54]

An alternative procedure for the production of Mo^I nitrosyl species is to oxidize Mo^0 complexes. Thus oxidation[33] (electrochemical, Ag^I, Fe^{III} or Cu^{II}) of the series *trans*-$[MoX(NO)(dppe)_2]$ (X = F, Cl, Br, I, OH, SPh, CN, H, NCO, NCS or N_3) gives the violet or brown, monocationic series $[MoX(NO)(dppe)_2]^+$.

The green-brown, monocation $[Mo\{Me_2pzb\}(NO)(MeCN)_2]^+$ is obtained by oxidation of the uncharged parent compound with Ag^I in MeCN[54] and the dications $[Mo(NO)(RCN)(dppe)_2]^{2+}$ are similarly obtained from the monocationic precursors[33] (Table 4). Treatment of $[Mo(CO)_6]$ with an excess of $NO[PF_6]$ in MeCN gives[55] $[Mo(NO)(MeCN)_5]^{2+}$.

The complex anions $[Mo(CN)_5(NO)]^{3-}$ and $[MoCl_4(NO)(OH_2)]^{2-}$ have been prepared by oxidation of $[Mo(CN)_5(NO)]^{4-}$ with air (in conc. HCl in the latter case) and have been structurally characterized as the expected octahedral complexes, with NO *trans* to H_2O in the dianion.[56] A compound formulated as $[Mo(NCS)_4(NO)]^{2-}$ has also been prepared[56] by treatment of $[MoO_4]^{2-}$ with NH_2OH in presence of SCN^-.

36.2.3.4 Nitrogen Donor Ligands

The one member of this class is the cation $[Mo(bipy)_3]^+$ obtained by reduction of $MoCl_5$ with Mg in THF in presence of bipy, or by oxidation of $[Mo(bipy)_3]$.[38]

36.2.3.5 Phosphorus Donor Ligands

The green, paramagnetic cation $[Mo(dmpe)_3]^+$ is obtained by oxidation of $[Mo(dmpe)_3]$ with iodine or methyl iodide in toluene.[57]

36.2.4 MOLYBDENUM(II)

36.2.4.1 Carbon Ligands

36.2.4.1.1 Carbon monoxide

Here in general we exclude compounds with more than two Mo—C bonds as discussed earlier.

(i) Complexes of the type $[MoX_2(CO)_2L_2]$ and $[MoX(CO)_2L_4]^+$

These complexes (in which X = anionic, L = uncharged ligand) form the largest class of Mo^{II} carbonyls and examples are shown in Table 5 with pertinent physical properties. We have excluded complexes of the type $[\{MoX_2(CO)_4\}_2]$, $[MoX(CO)_4L_2]^+$ and related compounds for the reasons cited earlier and for the same reasons $[MoX_3(CO)_4]^-$, $[MoX(CO)_3L_2]^-$ and related anions are excluded. They are reviewed in COMC.[1a] Complexes of the type $[MoX_2(CO)_2L_2]$

where L is not a carbon ligand are prepared generally by two main routes: oxidation and substitution of a zero-valent carbonyl derivative with halogen (method A), *e.g.* reaction (9);[1a,19] and reduction of a high-valent molybdenum compound in presence of CO and another ligand if necessary (method B), *e.g.* reaction (10).[1a,19] Simple displacement of CO by another ligand may also be used (method C), *e.g.* reaction (11).

$$[Mo(CO)_4(diars)] + I_2 \longrightarrow [MoI_2(CO)_2(diars)] + 2CO \quad (9)$$

$$[MoCl_4(PPh_3)_2] + AlEtCl_2 + CO + PPh_3 \longrightarrow [MoCl_2(CO)_3(PPh_3)_2] \quad (10)$$

$$[\{MoBr_2(CO)_4\}_2] + 2dpam \longrightarrow [MoBr_2(CO)_2(dpam)_2] + 2CO \quad (11)$$

Table 5 Dicarbonyl Complexes of Molybdenum(II)

Compound	*Preparation*[a]	$\nu(CO)$ (cm^{-1})[b]	*Other data*	*Ref.*
$[MoX_2(CO)_2(PMe_3)_3]$ (X = Cl, Br, I, NCO, NCS)	See text	1920–1950 1820–1860	X = NCO, *mer*-PMe_3, CO caps one face of octahedron	1
$[MoCl_2(CO)_2(PMe_2Ph)_3]$	A	1934, 1842	Cap O/CTP	2
$[MoBr_2(CO)_2(PMe_2Ph)_3]$	A	1934, 1834	Cap O	2
$[MoBr_2(CO)_2(PPh_3)_2]$	A, B	1920, 1795	V, distorted octahedron	2
$[MoCl_2(CO)_2(py)_2]$	A	2010, 1920	—	2, 3
$[MoCl_2(CO)_2(btp)_2]$[c]	A	1950, 1880	—	2, 3
$[Mo(OBu^t)_2(CO)_2(py)_2]$	See text	1908, 1768	—	4
$[Mo(S_2CNR_2)_2(CO)_2]$[d] (R = Me, Et, Pr^i)	D	1925–1945 1855–1875	TP for R = Pr^i	5, 6
$[Mo(S_2CNR_2)_2(CO)_2L]$[e] (R = Me, Et, Pr^i)	D	1940–1850 R = Me, L = PPh_3	CTP for L = C_4H_8S	4, 5, 6
$[Mo\{S_2P(OEt)_2)_2\}(CO)_2PPh_3]$	D	1955, 1875	—	5
$[Mo(acac)Cl(CO)_2(PPh_3)_2]$	C, D	1933, 1832	—	7
$[Mo(O_2CH)_2(CO)_2(PEt_3)_2]$	C, D	1929, 1833	4:3 piano-stool, η^2- and η^1-O_2CH	7
$[MoCl_2(CO)_2(dpam)_2]$	C	1940, 1855	Cap O	2
$[MoBr_2(CO)_2(dpam)_2]$	A, B	1940, 1865	Cap O	2
$[MoCl_2(CO)_2(dppm)_2]$	A	1940, 1865	Cap O/CTP	2
$[MoI_2(CO)_2(triphos)]$[f]	A, B	1940, 1870	—	2
cis-$[MoI(CO)_2(dppm)_2]^+$	A	1940, 1865	Cap O	2
trans-$[MoI(CO)_2(dppm)_2]^+$	A	1890	—	2
$[MoCl(CO)_2(diars)_2]^+$	A	—	—	2
$[MoF(CO)_2(dppe)_2]^+$	A, B	1945, 1889	CTP	7
$[MoH(CO)_2(chel)_2]^+$ (chel = diphosphine, diarsine or mixed chelate ligand)	See text	*circa* 1880	Cap O	8
$[MoCl(CO)_2(bipy)_2]^+$	A, B	1965, 1875	—	2, 3
$[MoCl(CO)_2(phen)_2]^+$	A, B	1960, 1890	—	2, 3
$[Mo(CO)_2L(dppe)_2]^{2+}$ (L = MeCN or PhCN)	A, B	1965, 1905	—	9
cis-$[Mo(CO)_2(phen)_2(MeCN)]^{2+}$	A	1988, 1898	—	10
cis-$[Mo(CO)_2(bipy)(L)]^{2+}$ (L = MeCN, H_2O, Me_2CO)	A	1988 ± 4, 1908 ± 10	—	10

[a] See text. [b] Solid state spectra (Nujol, KBr), unless otherwise stated. [c] btp = *N*-*n*-butylthiopicolinimide. [d] $\delta(^{95}Mo)$ 390 p.p.m. (R = Et). [e] L = range of ligands, see text. [f] triphos = bis(*o*-diphenylphosphinophenyl)phenylphosphine.
Cap O = capped octahedron; CTP = capped trigonal prism, TP = trigonal prism.

1. E. Carmona, K. Doppert, J. M. Marin, L. Poveda, L. Sanchez and R. Sanchez-Delgado, *Inorg. Chem.*, 1984, **23**, 530.
2. M. G. B. Drew, *Prog. Inorg. Chem.*, 1977, **23**, 67 and references therein.
3. R. Colton, *Coord. Chem. Rev.*, 1971, **6**, 269 and references therein.
4. M. H. Chisholm, J. F. Huffmann and R. L. Kelly, *J. Am. Chem. Soc.*, 1979, **101**, 7615.
5. G. J. Chen, R. O. Yetton and J. W. McDonald, *Inorg. Chim. Acta*, 1977, **22**, 249.
6. C. G. Young and J. H. Enemark, *Aust. J. Chem.*, 1986, **39**, 997 and references therein.
7. D. C. Bromer, P. B. Winston, T. L. Tonker and J. L. Templetra, *Inorg. Chem.*, 1986, **25**, 2883.
8. S. Datta, B. Dezube, J. K. Kouba and S. S. Wreford, *J. Am. Chem. Soc.*, 1978, **100**, 4404.
9. M. R. Snow and F. L. Wimmer, *Aust. J. Chem.*, 1976, **29**, 2349.
10. J. A. Connor, E. J. James, N. El Murr and C. Overton, *J. Chem. Soc., Dalton Trans.*, 1984, 255.

The coordination number for these complexes may be six or seven and extensive X-ray structural studies have been carried out, particularly on the seven-coordinate complexes. The geometry and some other details of a representative selection of these compounds are shown in Table 5. In some cases, *e.g.* $[Mo(CO)_2Cl_2(dpam)_2]$, one potentially chelating ligand is monodentate to maintain seven-coordination.[58] Sulfur coordination may also be introduced at the metal centre by use of displacement metathesis by *e.g.* the dithiocarbamato ligand, see reaction (12) (method D).[59]

$$\tfrac{1}{2}[\{MoCl_2(CO)_4\}_2] + 2Na[Me_2NCS_2] \longrightarrow [Mo(CO)_3(Me_2NCS_2)_2] \rightleftharpoons [Mo(CO)_2(Me_2NCS_2)_2] + CO \qquad (12)$$

The reversible loss of CO from $[Mo(CO)_3(Me_2NCS_2)_2]$ is a property shown by other Mo^{II} carbonyls. For example, the complexes $[MoX_2(CO)_3(PR_3)_2]$ (X = Cl or Br) also lose one mole of CO reversibly to give blue, diamagnetic $[MoX_2(CO)_2(PR_3)_2]$, which (for X = Br) has been structurally characterized as a 'badly distorted octahedron'.[59,60] A common geometry for the seven-coordinate complexes is the capped octahedron, with a CO ligand in the capping position, although there are variations (Table 5), in particular $[MoCl(CO)_2(diars)_2]^+$ is a capped trigonal prism.[60]

A further series of dithiocarbamato complexes has been obtained from $[MoCl_2(CO)_2(PMe_3)_3]$ (obtained from $[MoCl_2(PMe_3)_4]$ and CO directly)[61] by treatment with $Na[S_2CNR_2]$. These are the dicarbonyls $[Mo(S_2CNR_2)_2(CO)_2(PMe_3)]$ and the monocarbonyls $[Mo(S_2CNR_2)_2(CO)(PMe_3)_2]$, the latter being obtained when a slight excess of PMe_3 is used in the reaction of $[MoCl_2(CO)_2(PMe_3)_3]$ with $Na[S_2CNR_2]$.[61] The analogues $[Mo(S_2CNEt_2)_2(CO)L_2]$ ($L = PMe_2Ph$ or $PMePh_2$; L_2 = dppe) have also been synthesized.[62] These monocarbonyl compounds are notable for their very low value of ν(CO) (1760–1735 cm^{-1}) implying a very electron-rich metal centre in these complexes.[62] They transform readily into the dicarbonyl complexes with CO.

The structures of the $[Mo(SCNR_2)_2(CO)_2(PR_3)]$ series appear to be analogous to that determined for $[W(S_2CNEt_2)_2(CO)_2(PPh_3)]$, namely a tetragonal base (4:3 'piano stool') geometry.[61] As is often observed for dithiocarbamate complexes, the alkyl groups of the S_2CNR_2 ligands show ^1H NMR equivalence at room temperature. The fluxional process required for this does not appear to involve phosphine dissociation.[61] The series $[Mo(S_2CNEt_2)_2(CO)_2L]$ derived from $[Mo(S_2CNEt_2)(CO)_2]$ has been investigated by ^{95}Mo NMR spectroscopy. The resonance region (310 to −430 p.p.m. relative to $[MoO_4]^{2-}$) has shown that all the resonances are shielded relative to $[Mo(S_2CNEt_2)_2(CO)_2]$ (390 p.p.m.) and the dependence of nuclear shielding upon L is $NH_2NHSO_2C_6H_4Me < Cl^- < OPPh_3 < \mu$-pyrazine < py < $NH_2NHCOPh < NH_2NMe_2 < \mu$-$NH_2NHMe < \mu$-$NH_2NH_2 < \mu$-$NH_2CH_2CH_2NH_2$ < NH_3 < N_3^- < F^- < $AsPh_3$ < PPh_3 < $SbPh_3$ < PPh_2Et < PPh_2Me < PMe_3 < $P(OPh)_3$ < $P(OEt)_3$ < $P(OMe)_3$ < CO.[63] The $OPPh_3$, halide and nitrogen donor ligand adducts participate in a dynamic equilibrium with $[Mo(S_2CNEt_2)_2(CO)_2]$ on the NMR timescale. The chelating oxygen donor series $[Mo(B–B)Y(CO)_2L_2]$ (B–B = acac, hexafluoroacetylacetonate, tropolonate or carboxylate, $L = PEt_3$ or PPh_3, Y = halide or carboxylate) has also been prepared from $[MoY_2(CO)_2L_2]$ by metathesis.[64] The structure of one member of the series $[Mo(O_2CH)_2(CO)_2(PEt_3)_2]$ shows the formate ligand to engage in monodentate (η^1) and bidentate (η^2) bonding modes in the solid state. These bonding modes exchange very rapidly in solution.[64] The series $[\{Mo(S_2CNEt_2)_2(CO)_2\}(\mu - L)]$ (L = pyrazine, NH_2NHMe, NH_2NH_2 or $NH_2CH_2CH_2NH_2$) noted above were obtained by treatment of $[Mo(S_2CNR_2)_2(CO)_3]$ with the appropriate difunctional ligand.

The hydride complexes $[MoH(CO)_2(chel)_2]^+$ (chel = variety of chelating diphosphines and arsines) are prepared by protonation of the parent Mo^0 dicarbonyls and appear to have a monocapped octahedral structure in solution, with the fluxional hydride moving around the possible capping positions. This process has been extensively studied by NMR.[65] A number of other cationic seven-coordinate analogues $[MoX(CO)_2(chel)_2]^+$ (X = halide or pseudohalide; chel = chelating ligand usually phosphine or arsine) are also known (Table 5).[1a]

The complexes $[Mo(CO)_2(py)_2(OR)_2]$ (R = Pr^i or Bu^t) have been prepared by treatment of $[Mo_2(OR)_8]$ with CO in presence of pyridine and their octahedral structure established by X-ray diffraction;[66] they have low ν(CO) values (1906, 1768 cm^{-1}). The (μ-OR) dimers $[Mo_2(OR)_8(CO)_2]$ also have low ν(CO) values.[66] Because of steric pressure in the potentially seven-coordinate complexes, a chelating ligand may also act as a bridge, to form dinuclear complexes, *e.g.* $[Mo_2(CO)_4(dppm)_3Br_4]$.[67]

The anions $[Mo(S_2CNR_2)X(CO)_2]^-$ (R = Me or Et; X = F or N_3) are obtained by treatment of $[Mo(S_2CNR_2)(CO)_2]$ with X^-; the crystal structure of $[NEt_4][Mo(S_2CNR_2)F(CO)_2]$ shows the geometry at molybdenum to be a fluoride-capped trigonal prism, very similar to that seen for $[MoF(CO)_2(dppe)_2][PF_6]$. Oxidation of $[Mo(CO)_2(chel)_2]$ (chel = bipy or phen) with Ag^+ in solution gives the dications $[Mo(CO)_2(chel)_2]^{2+}$ (Table 5).[68]

The zwitterion Et_3PCS_2 reacts with $[Mo(CO)_6]$ or $[Mo(C_7H_8)(CO)_3]$ to give $[\{Mo(CO)_2(PEt_3)(\mu\text{-}S_2CPEt_3)\}_2]$. X-Ray structural determination has shown that both zwitterions coordinate one metal through an η^3-S_2C linkage and the second metal through a single sulfur atom.[69]

The monocarbonyl complexes $[MoX(\eta^2\text{–}COR)(CO)(PMe_3)_3]$ (X = Cl, Br, I or NCO; R = Me, CH_2SiMe_3, CH_2CMe_3, CH_2CMe_2Ph) have been prepared by the action of the appropriate Grignard or lithium reagent with $[MoCl_2(CO)_2(PMe_3)_3]$ followed by metathesis. The ν(CO) values are low (in the region of 1800 cm^{-1}), and the η^2-aryl coordination has been confirmed by X-ray crystallography for $[MoCl(\eta^2\text{-}COCH_2SiMe_3)(CO)(PMe_3)_3]$. The analogue $[Mo(COMe)(S_2CNMe_2)(CO)(PMe_3)_2]$ attains an 18-electron configuration for molybdenum by a β-C—H interaction of the acetyl group (X-ray structure).[70]

36.2.4.1.2 *Isocyanide ligands*

The majority of complexes in this category have a high number of Mo—C bonds and they have been reviewed elsewhere.[1a,68] Typical examples are $[Mo(CNR)_6X]^+$ (X = Cl, Br, I, $SnCl_3$, or O_2CCF_3; R = aryl, Me, Bu^t or C_6H_{11}); $[Mo(CNR)_6(PR_3)]^{2+}$; $[Mo(CNR)_5X_2]$; $[Mo(CNR)_7]^{2+}$; $[Mo(CNR)_5(chel)]^{2+}$ (chel = dppm, dppe, bipy, *etc.*); $[MoCl(CNBu^t)_4(bipy)]^+$; $[Mo(CNC_6H_{11})_5(bipy)]^{2+}$; $[Mo(CNR)_3(bipy)_2]^{2+}$. They are generally prepared by displacement of CO, bipy, OAc, NO or other ligands from a suitable precursor by RNC and have capped octahedral or capped trigonal prismatic structures.[1a,70]

Protonation of the series $[Mo(CNR)_2(dppe)_2]$ (Section 36.2.2.1.2) gives the seven-coordinate hydrides $[MoH(CNR)_2(dppe)_2]^+$ which, like their carbonyl analogues (Section 36.2.4.1.1), are fluxional in solution with the hydride moving between face-capping sites in an octahedral structure.[71]

Oxidation of *trans*-$[Mo(CNR)_2(dppe)_2]$ with Ag^+ or halogen gives the cations $[Mo(CNR)_2(dppe)]^{2+}$ (R = Me or C_6H_4Me-4) and $[MoX(CNR)_2(dppe)_2]^+$ (X = Cl or Br).[72]

The alkyne complex $[MoI\{C(NHBu^t)\}_2(CNBu^t)_4]^+$ results from 'reductive coupling' upon reduction of $[MoI(CNBu^t)_6]^+$.[1a] The thiolate–isocyanide complex $[Mo(SBu^t)_2(CNBu^t)_4]$, prepared from $[Mo(SBu^t)_4]$ and $CNBu^t$, forms the adduct $[Mo(CNBu^t)_4(\mu\text{-}SBu^t)_2CuBr]$. All these compounds have been structurally characterized.[73]

Isocyanide acts as a reducing agent in the preparation of $[MoCl_2(CNMe)_2(N—N)]$ [N—N = $Me_2NCH_2CH_2NMe_2$, RN=CHCH=NR or R=NCMeCMe=NR (R = C_6H_4OMe-4)] from $[MoCl_3(THF)_3]$, MeCN and N–N.[74]

36.2.4.1.3 *Cyanide ligands*

Cyanide complexes of molybdenum(II) generally appear elsewhere under other classifications, where CN^- is an ancilliary ligand. The homoleptic anions $[Mo(CN)_6]^{4-}$ and $[Mo(CN)_7]^{5-}$ (pentagonal bipyramid) have been reviewed elsewhere.[1a]

36.2.4.2 Complexes of Group 15 and Group 16 Ligands

36.2.2.4.1 *Complexes of the type $[MoX_2L_4]$ (X = anionic ligand, L = monodentate uncharged ligand, L_2 = bidentate chelate ligand, L_3 = tridentate chelate ligand)*

The largest group of this class are those of general formula $[MoX_2P_4]$ (where X = H, Cl, Br, I, SR, O_2R and P = PR_3, P_2 = chelating diphosphine, P_3 = chelating triphosphine). The main route to their preparation (method A) is by reduction of a higher-valent molybdenum compound under argon or dinitrogen in presence of the ligand. Thus the complexes *trans*-$[MoCl_2(PR_3)_4]$ (PR_3 = PMe_3 or PMe_2Ph) are obtained by reduction of $[MoCl_3(PR_3)_3]$ in presence of PR_3 with zinc or sodium amalgam (PMe_3), or LiR (PMe_2Ph, R = Bu or Ph).[61,75] Both compounds have their *trans* octahedral structures confirmed by X-rays and have similar structural parameters [d(Mo—Cl) = 2.420, d(Mo—P) = 2.496 Å].[75,49] The compounds are paramagnetic, with μ_{eff} values in the region of 2.8 BM. The analogue $[MoI_2(PMe_3)_4]$ has also been prepared by interaction of molybdenum vapour with PMe_3 and MeI.[40]

The analogues *trans*-[$MoX_2(dppe)_2$] (X = Cl, Br or I) have been obtained by method (A) and their electrochemical reduction studied.[75] In the presence of ligands L (L = CO, N_2, RNC, H_2) the [$MoX_2(dppe)_2$] compounds are reduced to give [$MoL_2(dppe)_2$]; detailed mechanistic studies are reported.[76] The structure of *trans*-[$MoCl_2(dppe)_2$] has been confirmed as *trans* octahedral by X-rays and *cis*-[$MoCl_2(dppe)_2$] was also structurally characterized as a very distorted octahedron.[77] The preparation of the latter compound was by oxidation of *trans*-[$Mo(N_2)_2(dppe)_2$] using $SnCl_2Ph_2$ over a long period and the apparent distortion could be due to the presence of hydride ligands; HCl (from *e.g.* hydrolysis of $SnCl_2Ph_2$) is known[5] to produce [$MoH_2Cl_2(dppe)_2$] from *trans*-[$Mo(N_2)_2(dppe)_2$].[3] The compound [$MoCl_2(triphos)(PPh_3)$] has been obtained by route (A) and reacts with HCl to produce [$MoCl_4(Ph_2PCH_2CH_2PPhCH_2CH_2PHPh_2)$][$HCl_2$].[78]

The arsine analogues of this series, [$MoX_2(diars)_2$] (X = Cl, Br, I) have been obtained by the anaerobic reaction of acid molybdenum(III) halide solutions with diars in water and/or alcohol. They are paramagnetic (μ_{eff} 2.8–2.9 BM), with a possible d–d UV band at 11 000 cm^{-1} in [$MoCl_2(diars)_2$].[74]

A number of molybdenum(II) compounds have been obtained by oxidation of *trans*-[$Mo(N_2)_2(chel)_2$] (chel = chelating diphosphine). Thus treatment of *trans*-[$Mo(N_2)_2${(*S*,*S*)-chiraphos}$_2$] [(*S*,*S*)-chiraphos) = (−)-(2*S*,3*S*)-bis(diphenylphosphino)butane, $Ph_2PCH(Me)CH(Me)PPh_2$] with bromooctane gives *inter alia,* a small amount of [$MoBr_2${(*S*,*S*)-chiraphos}$_2$].[80] Treatment of *trans*-[$Mo(N_2)_2(diphos)_2$] (diphos = dppe or dmtpe) with RSH gives the diamagnetic compounds *trans*-[$Mo(SR)_2(diphos)_2$] [R = Me, Et, Pr^n, Bu^n or C_6H_4X-4 (X = Me, H, NH_2, OMe, F or Cl)], the *trans* octahedral structure being confirmed by X-rays for R = Bu^n.[81] If bulky thiols R′SH (R′ = Pr^i, Bu^i, 2,4,6-$Me_3C_6H_2$, 2,4,6-$Pr^i_3C_6H_2$ or 4-Br-$C_6H_2Pr^i_2$-2,6) are used in this reaction, the green diamagnetic hydride complexes [MoH(SR′)$(dppe)_2$] are obtained. Hydride complexes of this type appear to be intermediates in the formation of *trans*-[$Mo(SR)_2(dppe)_2$] described above. They have Mo—^{1}H chemical shifts in the range −4 to −5 p.p.m. (rel. $SiMe_4$).[82]

The diamagnetic dithiolate complexes *trans*-[$Mo(SR)_2(diphos)_2$] are reversibly oxidized to the Mo^{III} species and the $E^{ox}_{1/2}$ values for R = C_6H_4X-4 (X = Me, H, NH_2, OMe, F, Cl) show a linear dependence upon the Hammett σ function for X, as do the ^{31}P chemical shifts of these compounds.[81]

The hydride compound *trans*-[$MoH_2(dppe)_2$] was claimed from reaction of *trans*-[$Mo(N_2)_2(dppe)_2$] with H_2, but some doubt exists about its formulation because its low solubility prevents accurate ^{1}H NMR integration.[83] Preparation of a more soluble analogue with (2-$MeC_6H_4)_2PCH_2CH_2P(C_6H_4Me-2)_2$ allowed accurate integration and demonstrated the presence of four hydrides per Mo. By analogy, the above dihydride is probably [$MoH_4(dppe)_2$], although one could not exclude the existence of it and of [{*trans*-$MoH_2(dppe)_2$}$_2$(μ-dppe)] which also appeared to be formed by reaction of *trans*-[$Mo(N_2)_2(dppe)_2$] with H_2.[83]

A number of other hydride complexes have been prepared; the purple, diamagnetic hydride [$MoH(BH_4)(PMe_3)_4$] has been obtained by reaction of $NaBH_4$ with [$MoCl_3(PMe_3)_3$]. The essentially *trans* octahedral structure of this molecule has been determined by X-rays as shown in (**9**); the hydride, the molybdenum and the boron atom of the bidentate BH_4 group lie precisely on the intersection of two mirror planes in the molecule.[86]

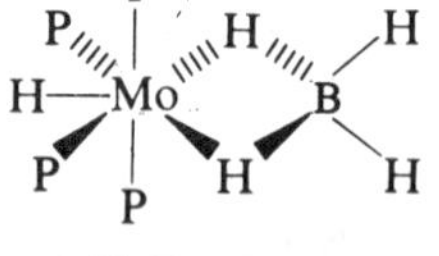

(**9**) P = PMe_3

In a related reaction, treatment of [$MoCl_2(PMe_3)_4$] with $LiAlH_4$ gives the bridged hydride aluminohydride complex [$(Me_3P)_4HMo(\mu$-H$)_2$AlH(μ-H$)_2$AlH(μ-H$)_2MoH(PMe_3)_4$], which has been shown by NMR to be non-rigid in solution. The central core of the structure is shown in (**10**).[85]

The carboxylato hydrides [$MoH(O_2CR)(PMe_3)_4$] (R = H, Et or CF_3) are obtained by treatment of [$MoH_2(PMe_3)_5$] (see Section 36.2.4.2.2) with CO_2 (R = H) or RCO_2H (R = Me or CF_3)[86] or by reaction of [$Mo(C_2H_4)_2(PMe_3)_4$] with CO_2 followed by H_2 (R = Et).[87] The X-ray structure of [$MoH(O_2CH)(PMe_3)_4$] is essentially pentagonal bipyramidal, with a chelating

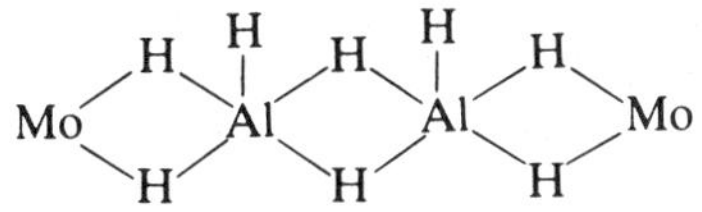

(10)

formate *trans* to hydride (**11**). The interaction of $[MoH_2(PMe_3)_5]$ with PhNCO gives a moderate yield of the analogous, red complex $[MoH\{PhNC(H)O\}(PMe_3)_4]$, formally by the insertion of the isocyanate into the Mo—H bond.[86]

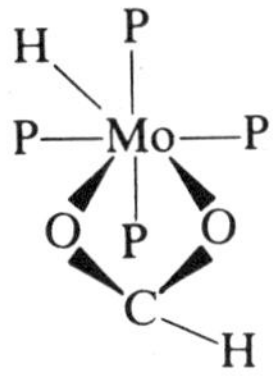

(**11**) P = PMe_3

The phosphite analogue, $[MoH(O_2CCF_3)\{P(OMe_3\}_4]$, is obtained by treatment of $[Mo\{P(OMe)_3\}_6]$ (see Section 36.2.2.5) with CF_3CO_2H; it has δ(MoH) at −4.6 p.p.m. (rel. $SiMe_4$). It has a similar solid-state structure to $[MoH(O_2CH)(PMe_3)_4]$, but is fluxional in solution with non-pairwise exchange of phosphorus-donor atoms being the dominant process.[88]

36.2.4.2.2 Complexes of the type $[MoH_2P_5]$ (P = tertiary phosphine or phosphite)

The reduction of $[MoCl_4(THF)_2]$ by magnesium in presence of PMe_3 in THF under dihydrogen gives a high yield of the yellow, air-sensitive hydride $[MoH_2(PMe_3)_5]$. Its crystal structure contains two crystallographically independent molecules which have essentially pentagonal bipyramidal goemetry, equal within the limits of experimental error. The Mo—P bond lengths vary in the structure. The axial Mo—P distances are very similar (2.424–2.428 Å) but one unique Mo—P(equatorial) bond in each molecule, which lies between two hydrogens, is some 0.6–0.7 Å shorter than the other two Mo—P(equatorial) bonds, probably because the long bond is almost *trans* to hydride.[89] The hydride resonance is a sextet at −5.23 p.p.m. (rel. $SiMe_4$) indicative of fluxionality at ambient temperature. $[MoH_2(PMe_3)_5]$ reacts with CO_2 and RCO_2H (R = Me or CF_3) to give the monohydrides $[MoH(CO_2CR)(PMe_3)_4]$ as described in Section 36.2.4.1.[86]

Photolysis of $[Mo\{P(OMe)_3\}_6]$ (Section 36.2.2.5) in presence of dihydrogen gives the off-white hydride $[MoH_2\{P(OMe)_3\}_5]$. The hydride resonance is indicative of fluxionality in solution [sextet at −5.8 p.p.m. (rel. $SiMe_4$)] down to −90 °C, and its structure is most likely a pentagonal bipyramid, analogous to that of $[MoH_2(PMe_3)_5]$ above. The analogous cation $[MoH\{P(OMe)_3\}_6]^+$, is obtained in low yield by treatment of $[Mo\{P(OMe)_3\}_6]$ with acids; the cation $[Mo\{P(OMe)_2\}\{P(OMe)_3\}_5]^+$ was also obtained by the same route. These products are not interconvertible and the isolation of the one or the other depends upon as yet undefined kinetic factors.[88]

36.2.4.2.3 Nitrosyl complexes

The largest group of molybdenum(II) complexes is constituted by the nitrosyls (oxidation states being assigned on the somewhat arbitrary assumption that nitrogen monoxide, as a ligand, is NO^+). Tables 6 and 7 give a selection of the compounds discussed.

The most important molybdenum nitrosyl, used in many studies as a starting material or as a catalyst, particularly for alkene metathesis, is $\{MoCl_3(NO)\}_n$.[89–91] It is produced[92] in a reasonably pure state by reaction of NOCl with $[Mo(CO)_4Cl_2]$, the chief contaminant probably being $\{MoCl_2(NO)_2\}_n$ or $K_4[Mo_2Cl_8]$.[93] The greenish-black hygroscopic, air-sensitive solid has ν(NO) at 1590 cm^{-1}. It reacts with acid species to eliminate chloride, for example with 1,2-$MeSC_6H_4SLi$ it generates $[Mo(NO)(SC_6H_4SMe)_3]$,[94] and forms adducts with neutral

Table 6 Nitrosyl Complexes of Molybdenum(II) (Excluding Pyrazolylborates)

Complex	$\nu(NO)$ (cm^{-1})
$\{Mo(NO)Cl_3\}_n$	1590 (Nujol);[1] 1739, 1851 (KBr)[2]
$[Mo(NO)(SC_6H_4SMe)_3]$	1655 (KBr)[3]
$[Mo(NO)Cl_3(OPPh_3)_2]$	1710 (Nujol),[1] 1705 (CH_2Cl_2),[2] 1710 (Nujol)[4]
$[Mo(NO)Cl_3(bipy)]$	1705 (Nujol),[1] 1706 (CH_2Cl_2)[2]
$[Mo(NO)Cl_3(MeCN)_2]$	1740 (KBr),[2] 1740 (Nujol)[4]
$[Mo(NO)Cl_3(py)_2]$	1707 (CH_2Cl_2)[2]
$[Mo(NO)Cl_3(HCONMe_2)_2]$	1713 (CH_2Cl_2)[2], 1712 (CH_2Cl_2)[4]
$[Mo(NO)Cl_3(OSMe_2)_2]$	1700 (CH_2Cl_2)[2], 1692 (CH_2Cl_2)[4]
$[Mo(NO)Cl_3\{OP(NMe_2)_3\}_2]$	1692 (CH_2Cl_2)[2]
$[Mo(NO)Cl_3(OPCl_3)_2]$	1725 (Nujol)[5]
$[Mo(NO)Cl_3(PPh_3)_2]$	1700 (CH_2Cl_2)[4]
$[Mo(NO)Cl_3(AsPh_3)_2]$	1720 (CH_2Cl_2)[4]
$[Mo(NO)Cl_3(OPMePh_2)_2]$	1695 (Nujol)[6]
$[Mo(NO)Cl_3(OPEt_2Ph)_2]$	1701 (Nujol)[6]
$[Mo(NO)Cl_3(PMe_2Ph)_2]$	1630 (Nujol)[6]
$[\{Mo(NO)Cl_3\}_2(dppe)_3]$	1660 (Nujol)[6]
$[Mo(NO)(CN)_3(triazacyclononane)]$	1660 (?)[7]
$[Mo(NO)(CN)_3(phen)]$	1700 (KBr)[8]
$[Mo(NO)(CN)_3(bipy)]$	1690 (KBr)[8]
$[Mo(NO)(N_3)_3(terpy)]$	1675 (KBr)[9]
$[Mo(NO)(SC_6H_2Pr^i_3)_3(NH_3)]$	?[10]
$[Mo(NO)(acac)_2\{ON{=}C(COMe)_2\}]$	1630 (CH_2Cl_2)[11]
$[Mo(NO)(SPh)_4]^-$	?[10,12]
$[Mo(NO)Cl(SBu^t)_3]^-$	?[12]
$[MO(NO)(SC_6H_2Me_3)_4]^-$	?[12]
$[Mo(NO)(SC_6H_4Me)_4]^-$	?[10]
$[Mo(NO)X_2(trimethyltriazacyclononane)]^+$	1565, 1545 (Br); 1695, 1640 (Cl); 1710 (I) (all KBr)[13]
$[Mo(NO)Cl(OEt)(trimethyltriazacyclononane)]^+$	1660, 1665 (KBr)[13]
$[Mo(NO)Cl(salen)]$	1660 (KBr)[2]
$[Mo(NO)(S_2CNR_2)_3]$	1640 (Me); 1635 (Et); 1630 (Pr^n); 1625 (Bu^n) (all $CHCl_3$)[14]
$[Mo(NO)(NCO)(S_2CNEt_2)_2X]^-$	1634 (NCO); 1630 (N_3); 1641 (NCS) (all KBr)[15]
$[Mo(NO)(NCO)(S_2CNEt_2)_2B]$	1648 (Me_2SO); 1651 (py); 1645 (NH_3); 1661 ($N_2H_4/2$) (all KBr)[15]
$[Mo(NO)(N_3)(S_2CNEt_2)_2(Me_2SO)]$	1648 (KBr)[15]
$[Mo(NO)(N_3)_2(S_2CNEt_2)_2]^-$	1626 (KBr)[15]
$[Mo(NO)(NCO)(S_2CNMe_2)_2(Me_2SO)]$	1633 (KBr)[15]
$[Mo(NO)(NCO)_2(S_2CNMe_2)_2]^-$	*ca.* 1635 (KBr)[15]
$[Mo(NO)(NCO)_2(S_2CNPr^i_2)_2]^-$	1633 (KBr)[15]
$[Mo(NO)Cl_4]^-$	1695,[16] 1715,[6] 1690 (all Nujol)[5]
$[Mo(NO)Br_4]^-$	1692 (Nujol)[16]
$[Mo(NO)Cl_5]^{2-}$	1676 (Nujol),[16] 1670 (Nujol),[6] 1670 (Nujol),[5] 1679 (KBr)[2]
$[Mo(NO)Br_5]^{2-}$	1680 (Nujol)[16]
$[Mo(NO)Cl_4L]^-$	*ca.* 1700 ($OPMe_2Ph$ or $OPEtPh_2$); *ca.* 1675 (PPh_3 or PMe_2Ph) (all Nujol)[6]
$[Mo(NO)Cl_4(O_2PCl_2)]^{2-}$	1688 (Nujol)[17]
$[Mo(NO)(CN)_5]^{2-}$	1645 (KBr),[18] 1650 (KBr)[8]
$[Mo(NO)(H_2NO)X_4]^{2-}$	1590 (Cl); 1620 (N_3) (all KBr)[18]
$[Mo(NO)(pyridine\text{-}2,6\text{-}dicarboxylate)(H_2NO)L]^{n-}$	1633 (H_2O), 1660 (CN), 1665 (N_3), 1680 (NCS) (all KBr)[19]
$[Mo(NO)(H_2NO)(H_2NOH)(terpy)]^{2+}$	1650 (KBr)[19]
$[Mo(NO)(H_2NO)_2(phen)]^+$	1625 (KBr)[19]
$[Mo(NO)(H_2NO)(phen)_2]^{2+}$	1690 (KBr)[19]
$[Mo(NO)(H_2NO)(pyridine\text{-}2\text{-}carboxylate)_2]$	1690 (KBr)[19]
$[Mo(NO)(H_2NO)(NCS)_4]^{2-}$	1658 (KBr)[20]
$[Mo(NO)(H_2NO)(NCS)_2(bipy)]$ (3 isomers)	1655; 1657; 1647 (KBr)[20]
$[Mo(NO)(H_2NO)(NCS)_2(phen)]$ (3 isomers)	1657; 1652; 1645 (KBr)[20]
$[Mo(NO)(H_2NO)(S_2CNEt_2)_2]$	1650 (KBr)[20]
$[Mo(NO)(H_2NO)_2(bipy)]^+$	1615 (?)[19]
$[Mo(NO)(HNO)(CN)(terpy)]$	1630 (?)[21]
$[Mo(NO)(Me_2C_3N_2H_2)_3(OC_6H_4O)]I_3$	1671 (KBr)[22]

1. R. Davis, B. F. G. Johnson and K. H. Al-Obaidi, *J. Chem. Soc., Dalton Trans.*, 1972, 508.
2. R. Taube and K. Seyferth, *Z. Anorg. Allg. Chem.*, 1977, **437,** 213.
3. D. Sellmann, L. Zapf, J. Keller and M. Moll, *J. Organomet. Chem.*, 1985, **289,** 71.
4. L. Bencze and J. Kohan, *Inorg. Chim. Acta*, 1982, **65,** L17; L. Bencze, J. Kohan, B. Mohai and L. Marko, *J. Organomet. Chem.*, 1974, **70,** 421.
5. K. Dehnicke, A. Lickett and F. Weller, *Z. Anorg. Allg. Chem.*, 1981, **474,** 83.

References to Table 6 (*continued*)

6. F. King and G. J. Leigh, *J. Chem. Soc., Dalton Trans.*, 1977, 429.
7. P. Chaudhuri, K. Wieghardt, Y. H. Tsay and H. Krüger, *Inorg. Chem.*, 1984, **23,** 427.
8. R. Bhattacharya, R. Bhattacharjee and A. M. Saha, *J. Inorg. Nucl. Chem.*, 1985, **4,** 583.
9. J. Beck and J. Strähle, *Z. Naturforsch, Teil B,* 1985, **40,** 891.
10. P. J. Blower, P. T. Bishop, J. R. Dilworth, T. C. Hsieh, J. Hutchinson, T. Nicholson and J. Zubieta, *Inorg. Chim. Acta,* 1985, **101,** 63.
11. S. Sarkar and P. Subramanian, *J. Inorg. Nucl. Chem.*, 1981, **43,** 202.
12. P. T. Bishop, J. R. Dilworth, J. Hutchinson and J. A. Zubieta, *Inorg. Chim. Acta,* 1984, **84,** L15.
13. G. Backes-Dahmann and K. Wieghardt, *Inorg. Chem.*, 1985, **24,** 4044.
14. B. F. G. Johnson, K. H. Al-Obaidi and J. A. McCleverty, *J. Chem. Soc. (A),* 1969, 1668.
15. J. A. Broomhead and J. R. Budge, *Aust. J. Chem.*, 1979, **32,** 1187.
16. S. Sarkar and A. Müller, *Angew. Chem., Int. Ed. Engl.*, 1977, **16,** 183.
17. A. Liebelt, F. Weller and K. Dehnicke, *Z. Anorg. Allg. Chem.*, 1980, **480,** 13.
18. K. Wieghardt, G. Backes-Dahmann, W. Swiridoff and J. Weiss, *Inorg. Chem.*, 1983, **22,** 1221.
19. K. Wieghardt, W. Holzbach and B. Prikner, *Angew. Chem., Int. Ed. Engl.*, 1979, **18,** 548; K. Wieghardt, W. Holzbach, B. Nuber and J. Weiss, *Chem. Ber.*, 1980, **113,** 629.
20. R. Bhattacharya and G. P. Bhattacharjee, *J. Chem. Soc., Dalton Trans.*, 1983, 1593.
21. K. Wieghardt, W. Holzbach and J. Weiss, *Z. Naturforsch., Teil B,* 1982, **37,** 680; K. Weighardt and W. Holzbach, *Angew. Chem., Int. Ed. Engl.*, 1979, **18,** 549.
22. H. Adams, N. A. Bailey, A. S. Drane and J. A. McCleverty, *Polyhedron,* 1983, **2,** 465.

Table 7 Molybdenum(II) Nitrosyls with Pyrazolylborates

Complex	$\nu(NO)$ (cm^{-1})
[Mo(NO)(Me_2pzb)X_2]	1688 (F) (KBr);[1] 1702 (Cl) (KBr);[1] 1702 (Br) (KBr);[1] 1700 (I) (KBr);[2] 1718, 1723 (Cl) (CH_2Cl_2)[3]
[Mo(NO)(pzb)X_2]	1700, 1710 (Cl); 1700, 1710 (Br); 1656, 1700 (I) (all KBr)[2]
[Mo(NO)(4-X-Me_2pzb)X_2]	1705 (Cl) (KBr);[4] 1722 (Br) ($CHCl_3$)[2]
[Mo(NO)(pzb)Br(OR)]	1677 (Me);[2] 1677 (Et)[2] (all $CHCl_3$)
[Mo(NO)(Me_2pzb)X(OEt)]	1666 (F) (KBr);[2] 1676 (Cl); 1673 (Br); 1678 (I) (all $CHCl_3$)[4]
[Mo(NO)(4-Cl-Me_2pzb)Cl(OR)]	1704 (Et); 1682 (Pr^i) (all $CHCl_3$)[4]
[Mo(NO)(Me_2pzb)X(OPr^i)]	1674 (Cl); 1674 (Br); 1673 (I) (all $CHCl_3$)[4]
[Mo(NO)(Me_2pzb)Br(OR)]	1676 (Bu^i); 1697 (C_3H_5) (all $CHCl_3$)[4]
[Mo(NO)(Me_2pzb)I(OR)]	1688 (CH_2CH_2OMe); 1675 (CH_2CH_2Cl); 1673 (CH_2CH_2OH); 1670 ($CH_2CH_2CH_2OH$); 1666 [$CH_2(CH_2)_2CH_2OH$]; 1672 [$CH_2(CH_2)_3CH_2OH$]; 1668 [$CH_2(CH_2)_4CH_2OH$] (all KBr);[5] (exact values not quoted[2] for R = CH_2OH or $CH_2CH_2CH_2Br$, *ca.* 1675); 1665 (H); 1675 (C_6H_{11}); 1675 ($CH_2C{\equiv}CCH_2OH$) (all KBr)[6]
[Mo(NO)(Me_2pzb)$(OR)_2$]	1640 (H); 1640 (Me); 1640 (Et); 1640 (Pr^i); 1640 (Bu^t) (all KBr)[6]
[Mo(NO)(Me_2pzb)(OR)(OR′)]	1640 (Et, Pr^i); 1640 (Pr^i, Bu^i) (all KBr)[6]
[Mo(NO)(Me_2pzb)I(OAr)]	1673 (Ph); 1672 (4-C_6H_4Me); 1686 (4-C_6H_4CN) (all KBr)[7]
[Mo(NO)(Me_2pzb)(OPh)(OR)]	1656 (Me); 1655 (Et); 1656 (Ph) (all KBr)[7]
[Mo(NO)(Me_2pzb)X(NHMe)]	1640 (F);[6] 1643 (Cl);[1] 1654 (Br);[6] 1640 (I);[8] 1625 (OMe);[6] 1628 (OEt);[6] 1637 (OPr^i);[6] 1650 (OPh)[7] (all KBr)
[Mo(NO)(Me_2pzb)X(NHEt)]	1649 (Cl);[1] 1655 (I);[8] 1633 (OMe);[6] 1630 (OEt);[6] 1631 (OPr^i)[6] (all KBr)
[Mo(NO)(Me_2pzb)X(NH_2)]	1630 (I);[8] 1635 (OMe);[6] 1630 (OEt);[6] 1637 (OPr^i)[6] (all KBr)
[Mo(NO)(Me_2pzb)X(NHPr)]	1630 (OEt, n); 1638 (OPr^i, i) (all KBr)[6]
[Mo(NO)(Me_2pzb)I(NR′R″)]	1646 (H, Bu^n); 1661 (H, Bu^i); 1658 (H, CH_2Ph); 1660 (H, C_6H_{11}); 1655 (H, C_3H_5); 1665 (CMe_2) (all KBr)[8]
[Mo(NO)(Me_2pzb)Br(NR′R″)]	1630 (H, H); 1615 (H, Pr^i); 1625 (H, CH_2Ph) (all KBr)[8]
[Mo(NO)(Me_2pzb)(OEt)(NR′R″)]	1625 (H, CH_2Ph); 1625 (H, C_6H_{11}) (all KBr)[6]
[Mo(NO)(Me_2pzb)(OPr^i)(NR′R″)]	1638 (H, Pr^i): 1633 (H, C_6H_{11}) (all KBr)[6]
[Mo(NO)(Me_2pzb)X(NMe_2)]	1672 (Cl)[2]; 1659 (Br)[2]; 1638 (I)[8] (all $CHCl_3$)
[Mo(NO)(Me_2pzb)Cl(NHR)]	Mean, R = Me, Et or Bu^n, 1646[2]; 1654 (Ph); 1653 (4-C_6H_4Me) (all KBr)[1]

Table 7 (*continued*)

Complex	*ν(NO)* (cm^{-1})
[Mo(NO)(Me_2pzb)I(NHAr)]	1659 (Ph); 1668 (4-C_6H_4Et); 1647 (4-C_6H_4OMe); 1654 (4-C_6H_4OEt); 1659 (4-C_6H_4F); 1661 (4-C_6H_4Cl); 1660 (4-C_6H_4Br); 1657 (4-C_6H_4I); 1646 (4-$C_6H_4CO_2$Me); 1666 (4-$C_6H_4NO_2$); 1666 (4-C_6H_4CN); 1654 ($C_{10}H_7$) (all KBr)[7]
[Mo(NO)(Me_2pzb)I(SR)]	1677 (Et); 1669 (Bu^n; 1685 ($C_{16}H_{33}$); 1672 (C_6H_{11}); 1672 (CH_2Ph); 1689 (Ph); 1692 (4-C_6H_4Me) (all KBr)[9]
[Mo(NO)(Me_2pzb)X(SBu^n)]	1653 (SBu^n);[9] 1625, 1636 (NHMe);[9] 1625 (NHEt);[9] 1634 ($NHPr^n$);[9] 1649 (OMe);[9] 1643 (OEt);[9] 1655 (Cl);[1] 1665 (Br)[1] (all KBr)
[Mo(NO)(Me_2pzb)X(SPh)]	1681 (SPh); 1624, 1635 (NHMe); 1630 (NHEt); 1652 ($NHPr^n$); 1637 ($NHBu^n$); 1658 (OMe); 1653 (OEt) (all KBr)[9]
[Mo(NO)(Me_2pzb)Cl(SR)]	1662 (Bu^n); 1671 (C_6H_{11}); 1671 (CH_2Ph); 1684 (Ph) (all KBr)[1]
[Mo(NO)(Me_2pzb)I{hydrazide(1 −)}]	1635 ($NHNH_2$); 1635 (NHNHMe); 1634 ($NHNMe_2$); 1630 (NHNHPh); 1639 ($NHNHC_6F_5$); 1640 ($NHNHC_6H_4$Me-4); 1636 (NHNMePh); 1654 ($NHNHCSNH_2$); 1652 ($NHNHCO_2$Me); 1623 (NHNHCOPh); 1658 (NHN = CMe_2) (all KBr)[10]
[Mo(NO)(Me_2pzb)Br{hydrazide(1 −)}]	1615 ($NHNH_2$); 1615 (NHNHMe); 1615 (NHNHPh); 1630 (NHN=CMe_2) (all KBr)[10]
[{Mo(NO)(Me_2pzb)I}$_2$O]	1663 (?)[11]
[Mo(NO)(Me_2pzb)I($NHC_6H_4XH_n$)]	1675 (N, 2); 1670 (O, 1); 1668 (S, 1) (all KBr)[12]
[Mo(NO)(Me_2pzb)I(SC_6H_3MeSH)]	1684 (KBr)[12]
[Mo(NO)(Me_2pzb)(NR′R″)$_2$]	1620 (Et, H); 1632 (Ph, H); 1628 (4-C_6H_4Me, H); 1682 [$CH_2(CH_2)_3CH_2$] (all KBr)[13]
[Mo(NO)(Me_2pzb)(NHPh)(NHC_6H_4Me)]	1638 (KBr)[13]
[Mo(NO)(pzb)Cl(N_2Ph)]	1710 (?)[14]

1. A. S. Drane and J. A. McCleverty, *Polyhedron,* 1983, **2,** 53.
2. J. A. McCleverty, *Chem. Soc. Rev.*, 1983, **12,** 331.
3. S. Trofimenko, *Inorg. Chem.*, 1969, **8,** 2675.
4. J. A. McCleverty, D. Seddon, N. A. Bailey and N. W. Walker, *J. Chem. Soc., Dalton Trans.*, 1976, 898.
5. G. Denti, J. A. McCleverty and A. Włodarczyk, *J. Chem. Soc., Dalton Trans.*, 1981, 2021.
6. J. A. McCleverty, E. A. Rae, I. Wołochowicz, N. A. Bailey and J. M. A. Smith, *J. Chem. Soc., Dalton Trans.*, 1982, 951.
7. J. A. McCleverty, G. Denti, S. J. Reynolds, A. S. Drane, N. El Murr, A. E. Rae, N. A. Badley, H. Adams and J. M. A. Smith, *J. Chem. Soc., Dalton Trans.*, 1983, 81.
8. J. A. McCleverty, E. A. Rae, I. Wołochowicz, N. A. Bailey and J. M. A. Smith, *J. Chem. Soc., Dalton Trans.*, 1982, 429.
9. J. A. McCleverty, A. S. Drane, N. A. Bailey and J. M. A. Smith, *J. Chem. Soc., Dalton Trans.*, 1983, 91.
10. J. A. McCleverty, A. E. Rae, I. Wołochowicz, N. A. Bailey and J. M. A. Smith, *J. Chem. Soc., Dalton Trans.*, 1983, 71.
11. H. Adams, N. A. Bailey, D. Gianfranco, J. A. McCleverty, J. M. A. Smith and A. Włodarcyzk, *J. Chem. Soc., Dalton Trans.*, 1983, 2287.
12. H. Adams, N. A. Bailey, A. S. Drane and J. A. McCleverty, *Polyhedron,* 1983, **2,** 465.
13. N. Al Obaidi, K. P. Brown, A. J. Edwards, S. A. Hollins, C. J. Jones, J. A. McCleverty and B. D. Neaves, *J. Chem. Soc., Chem. Commun.*, 1984, 690.
14. M. E. Deane and F. J. Lalor, *J. Organomet. Chem.*, 1973, **57,** C61.

donors, sometimes with oxidation. Thus Ph_3P and also $OPPh_3$ react to give [$MoCl_3$(NO)($OPPh_3$)$_2$],[92] and bipy reacts to generate [$MoCl_3$(NO)(bipy)].[92] The former complex is also attainable from [Mo(CO)$_4$(PPh_3)$_2$], or [Mo(CO)$_3$(PPh_3)$_3$] with NOCl.[89]

An alternative synthesis of {$MoCl_3$(NO)}$_n$ (described as brown-black) is from $MoCl_5$ and NO. The temperature and reaction time need to be controlled to avoid further reductive nitrosylation to the green Mo^0 species {$MoCl_2(NO)_2$}$_n$. Addition of pyridine to such a reaction mixture yields [$MoCl_3$(NO)py_2], and similarly $OPPh_3$ yields [$MoCl_3$(NO)($OPPh_3$)$_2$].[89] It is puzzling that the two sources for {$MoCl_3$(NO)}$_n$ quoted give different products with different IR spectra, yet the $OPPh_3$ complexes appear the same from either source, whereas the bipy complexes are different.

It is also pertinent that $MoCl_5$ and NO react in benzene rather than in dichloromethane to give a red solid. This is not {$MoCl_3$(NO)}$_n$, and may even be [$MoCl_5$(NO)], and has ν(NO) at 2000 cm^{-1}.[31] However, it is a good source of nitrosyl complexes such as [$MoCl_5$(NO)]$^{2-}$.

The complex $[Mo(NO)(SC_6H_4SMe)_3]$[94] is probably seven-coordinate, although its NMR spectrum in CD_2Cl_2 is more complex at 0 °C than one might expect. It reacts with nucleophiles, such as N_2H_4, PMe_3 and OH^-.[94]

The adducts of the general formula $[MoCl_3(NO)L_2]$ (see Table 6) are all probably obtainable by reaction of $\{MoCl_3(NO)\}_n$ with the neutral ligand, L.[89] However, reductive nitrosylation can be used directly. Thus, reaction of NOCl with $[Mo_2(O_2CMe)_4]$ followed by treatment with $OPPh_3$ gives $[MoCl_3(NO)(OPPh_3)_2]$ directly.[93] Alternative preparations have been reported.[95] Molybdenum(II) chloride and NOCl react in $POCl_3$ to yield $[MoCl_3(NO)(POCl_3)_2]$, which loses one $POCl_3$ in solution in CH_2Cl_2 to generate $[\{MoCl_3(NO)(POCl_3)\}_2]$.[96] The dinuclear species has chloro bridges, with chlorides in the axial positions and the $POCl_3$ molecules, NO ligands and remaining chlorides bound to give similar octahedral coordination about the molybdenum atoms, but all *trans* with respect to the pair of metal atoms. The MoNO system is approximately linear (178°), with Mo—N = 1.81 and N—O = 1.08 Å.[96] The mononuclear complex reacts with chloride to generate $[MoCl_4(NO)]^-$ and $[MoCl_5(NO)]^{2-}$.

All the complexes $[MoCl_3(NO)L_2]$ are presumably six-coordinate and octahedral. Clearly, this is not a limiting coordination number, since seven-coordination is allowed by electron-counting rules for molybdenum(II), and is observed among nitrosyls (*cf.* $[Mo(NO)(SC_6H_4SMe)_3]$). Specific examples based on this chloro system include $[MoCl_3(NO)(PMe_2Ph)_3]$ and $[\{MoCl_3(NO)\}_2(Ph_2PCH_2CH_2PPh_2)_3]$, prepared by variation of the routes discussed above.[31a]

Complete replacement of chloride to give complexes based upon $\{Mo(NO)X_3\}_n$ (X = CN,[97,98] N_3,[99] SR[100] or S_2CNR_2) has also been observed. For example, the compound $[Mo(CO)_3L]$ (L = 1,4,7-triazacyclononane) reacts with nitric acid to yield $[Mo(CO)_2(NO)L]^+$ which contains Mo^0, but inclusion of KCN in the reaction mixture causes formation of $[Mo(CN)_3(NO)L]$.[97] Probably all the nitrogens of the cyclononane are coordinated, making this complex seven-coordinate. Reaction of molybdate $[MoO_4]^{2-}$ at pH 8 in the presence of KCN and L (L = bipy or phen) with hydroxylamine can give rise to $[Mo(CN)_3(NO)L]\cdot H_2O$, though the precise products are a function of reaction conditions.[99] In an unusual reaction $[Mo(CO)_3(terpy)]$ and $MeSiN_3$ in nitromethane produce yellow $[Mo(NO)(N_3)_3(terpy)]$. This has a pentagonal bipyramidal structure with NO and a bent azide in axial positions. Mo—N is 1.777(4) and N—O 1.196(5) Å, and ∠MoNO is practically linear [176.2(4)°].[97]

The reaction of SR^- with $\{MoCl_3(NO)\}_n$ does not normally stop with $[Mo(NO)(SR)_3]$. Exceptions to this are $[Mo(NO)(SC_6H_4SMe)_3]$ cited above[94] and $[Mo(NO)(NH)(SC_6H_2Pr^i_3)_3]$, which has a thiolate ligand with exceptional steric requirements.[100] This is produced by reaction of $Pr^i_3C_6H_2S^-$ with 'Mo(NO)(NH₂O)'.[101] This is a genuinely five-coordinate, trigonal bipyramid with a thiolate in the trigonal plane and NO *trans* to NH_3. Again ∠MoNO is close to 180° [176.1(3)°] with Mo—N = 1.78(3) Å and N—O not reported. In contrast $Me_3C_6H_2S^-$ gives rise to $[Mo(NO)(SC_6H_2Me_3)_4]^-$,[100] Bu^tS^- produces $[MoCl(NO)(SBu^t)_3]^-$, and PhS^- yields $[Mo(NO)(SPh)_4]^-$.[102] The last complex has a trigonal bipyramidal structure (note that $[MoCl_4(NO)]^-$ is essentially square pyramidal), a linear NO occupying an axial position, Mo—N = 1.766(6), N—O = 1.188(8) Å and ∠MoNO = 178.5(5)°.[102] The steric influences at work are obvious, though not the precise way in which they operate.

We now discuss some compounds which may be formally considered to be based on $\{MoCl_3(NO)\}_n$, though they include a collection of miscellaneous species not easy to categorize. A complex described as $[Mo(NO)(acac)_2(hia)]$ [acac = 1,3-pentadionate, hia = hydroximinoacetylacetone (*sic*)] with $\nu(NO)$ at 1635 cm^{-1} (CH_2Cl_2) has been mentioned.[103] The complex $[LMo(CO)_3]$ (L = *N,N′,N″*-trimethyltriazacyclononane) reacts with nitric acid to give $[LMo(CO)_2(NO)]^+$, and this with HBr gives rise to $[LMoBr_2(NO)]^+$, isolated as the hexafluorophosphate. The dichloride and diiodide were obtained similarly. The former reacts with ethanol to produce $[LMoCl(NO)(OEt)]^+$.[52,98] The electrochemistry of the species has been reported, and the structures and NMR spectra rationalized. Several of these complexes are present in two isomeric forms. A comparison is made between L and tris(pyrazolyl)borate as a ligand.[52] Finally, $\{MoCl_3(NO)\}_n$ reacts with the Schiff base H_2salen to give [Mo(NO)Cl(salen)], which has not been characterized in detail.[89,104] The reaction is probably general for Schiff bases. A range of complexes can be synthesized from $[Mo(CO)_4$(Schiff base)] and NO in benzene. They analyze for $[Mo(NO)_2(NO_2)_2$(Schiff base)] and may contain Mo^{II}. The values of $\nu(NO)$, two bands in the ranges 1770–1785 and 1650–1670 cm^{-1}, would be consistent with this.

The seven-coordinate complexes $[Mo(NO)(S_2CNR_2)_3]$ (R = Me, Et, Pr^n or Bu^n) were originally prepared from $\{MoCl_3(NO)\}_n$ and the appropriate $Na(S_2CNR_2)$,[105–107] although irradiation of $[Mo(NO)_2(S_2CNR_2)_2]$ also leads to $[Mo(S_2CNR_2)_3]$.[107] They are volatile, and have NMR spectra consistent with the structures established by X-ray analysis.[108–110] The compound with $R = Bu^n$ is pentagonal bipyramidal, with NO in the axial position, ∠MoNO = 173.2(07)°, Mo—N = 1.731(8) and N—O = 1.154(9) Å. A detailed discussion of the structure has been provided,[109] and the ^{95}Mo NMR spectra have been reported for the complexes with R = Me, and R = Et.[111] The structure of $[Mo(NO)(C_2O_4)_3]$ has been reported. It is similar.[112]

Compounds formally related to these are $[Mo(NO)(NCO)(S_2CNEt_2)_2X]^-$ (X = Cl, N_3, NCS or NCO), $[Mo(NO)(NCO)(S_2CNEt_2)_2B]$ (B = NH_3, Me_2SO or py) and $[\{Mo(NO)(NCO)(S_2CNEt_2)_2\}_2N_2H_4]$.[113,114] These are also all reckoned to be seven-coordinate species, with one of two possible structures admitted as likely on the basis of NMR and IR data. The general relationships are indicated in Scheme 1. These compounds are of interest in the context of nitrogen fixation, and the observation that the azido ligand in $[Mo(NO)(N_3)(S_2CNEt_2)_2(Me_2SO)]$ reacts with acid to give ammonia (*via* a nitrido complex) and N_2 is particularly interesting in view of the reduction of azide by nitrogenase. A reasonably comprehensive list of compounds is given in Table 6.

The dithiocarbamato complex $[Mo(NO)(S_2CNR_2)_3]$ has a homologue based upon NS rather than NO. The preparation (for R = Me) is not related to that of the nitrosyl, but is by reaction of propylene sulfide or S on $[MoN(S_2CNMe_2)_3]$.[115,116] However, its structure is very similar, pentagonal bipyramidal with NS in an axial position, and ∠MoNS = 172.0(7)°, Mo—N = 1.738(11) and N—S = 1.592(11) Å. These complexes are part of a series of approximately isostructural compounds $[MoX(S_2CNR_2)_3]$ (X = N, O, NS or NO).[117] Thionitrosyl compounds with $R_2 = Et_2$ or $(CH_2)_4$ have also been prepared. A band assignable to ν(NS) could not be identified in the IR spectra, though it is characteristically found in other complexes at *ca.* 1160 cm^{-1}.[116]

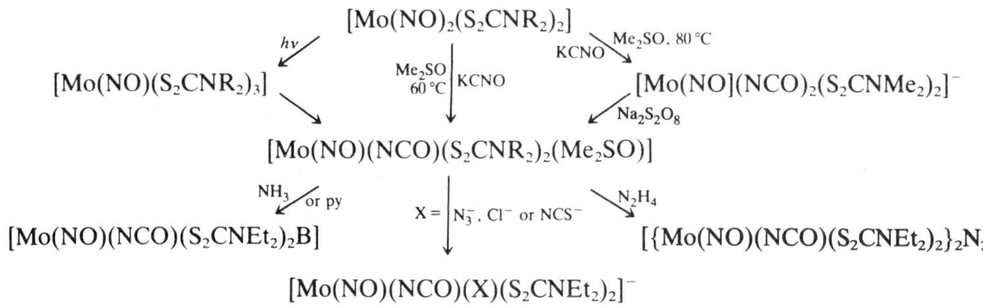

Scheme 1

The ^{95}Mo NMR spectra of $[Mo(NS)(S_2CNR_2)_3]$ (R = Me or Et) have been reported.[111]

The complex $\{MoCl_3(NO)\}_n$ is capable, formally, of adding chloride ions to form $[MoCl_4(NO)]^-$ and $[MoCl_5(NO)]^{2-}$. The six-coordinate species seems to crystallize as the cesium salt, and the five-coordinate as the tetraphenylphosphonium salt, from similar reaction mixtures which use $K_4[Mo(CN)_5(NO)]\cdot 2H_2O$ as the starting material.[118] There is an alternative route to the tetraphenylarsonium salts starting from $[MoCl_3(NO)(POCl_3)_2]$.[96] (The bromo analogues have also been reported.[118]) An alternative preparation from $MoCl_5 + NO$ seems rather more easy to use.[31a] The complex $[Et_4N][MoCl_4(NO)]$ reacts with tertiary phosphine and phosphine oxides (L) to give adducts $[Et_4N][MoCl_4(NO)L]$ (see Table 6). The other adduct of $[MoCl_4(NO)]^-$ which has been described is $[MoCl_4(NO)(O_2PCl_2)]^{2-}$, obtained by reaction of $[AsPh_4](O_2PCl_2)$ with $[MoCl_4(NO)]^-$. This mildly air-sensitive complex has a roughly octahedral structure, with the O_2PCl_2 binding through one oxygen only, ∠MoNO being 176.8(10)°, Mo—N = 1.701(13) and N—O = 1.182(11) Å.[119] A hint that $[Mo(CN)_4(NO)]^-$ can be synthesized[118] has not been amplified.

The complex $[MoCl_5(NO)]^{2-}$ has had its structure determined as the tetraphenylarsonium salt. It has an octahedral structure, with ∠MoNO = 173.9(25)°, Mo—N = 1.751(32) and N—O = 112.0(38) Å.[96] It is apparently a rather inert complex, said not to be of much value for synthetic work.[31a] It can be prepared from $K_4[Mo(CN)_5(NO)]\cdot 2H_2O$,[118] $MoCl_5$ and NO,[31a] $\{MoCl_3(NO)\}_n$[89] or $[NEt_4][Mo_2Cl_9]$ and nitrosyl chloride.[93] Finally, the ion $[Mo(NO)(CN)_5]^{2-}$ has been prepared by the action of cyanide on the complex material obtained from the reaction of sodium molybdate with hydroxylammonium chloride.[99,101] Some interesting nitrosyls, formally of molybdenum(II), have been obtained using this complex. These contain protonated nitrosyl as well as NO. Thus HCl reacts to give $[MoCl_4(NO)(H_2NO)]^{2-}$, and sodium azide produces $[Mo(NO)(H_2NO)(N_3)_4]^{2-}$.[101] The latter, as the tetraphenylphosphonium salt, has been shown to contain seven-coordinate molybdenum, with pentagonal bipyramidal coordination. The hydroxylamido(1 −) ligand is side-on, and the axial NO has ∠MoNO = 172.4(5)°, Mo—N = 1.761(8) and N—O = 1.21(1) Å.[101] The structure is very similar to that of $[Mo(NO)(H_2NO)(NCS)_4]^{2-}$,[120] and can be compared with $[Mo(NO)(H_2NO)_2(bipy)]^+$, which has two side-bonded hydroxylamido groups, $[Mo(NO)(H_2NO)(phen)_2]^{2+}$, which has one,[121] and $[Mo(NO)(H_2NO)(H_2O)(terpy)]^{2+}$, which also has one.[122] All these structures contain seven-coordinate pentagonal bipyramidal molybdenum, with linear, axial NO. The nitrosyl complexes containing hydroxylamine derivatives include $[Mo(NO)(H_2NO)(dipic)(L)]^{n-}$ (dipic = pyridine-2,6-dicarboxylate, $n = 0$ or 1, L = H_2O, CN, N_3 or SCN),[121] $[Mo(NO)(NH_2OH)(H_2NO)(terpy)]$,[121] $[Mo(NO)(H_2NO)(phen)_2]^{2+}$, $[Mo(NO)(H_2NO)_2(terpy)]^+$ [121] and $[Mo(NO)(H_2NO)(pic_2]$ (pic = pyridine-2-carboxylate).[121]

The molybdate/hydroxylamine reaction which gives rise to the species which is the source of these materials is clearly complex, and its constitution depends upon the pH of its preparation. For example, reaction in water at pH 4–4.5 can give rise to $[Mo(NO)(H_2NO)(NCS)_4]^{2-}$, $[Mo(NO)(H_2NO)(NCS)_2L]$ (L = bipy or phen) and $[Mo(NO)(H_2NO)(S_2CNEt_2)_2]$.[123] At higher pH the products are similar, but formally Mo^I or Mo^0. Possibly hydroxylamine derivatives could also be used, to give, for example, $[Mo(NO)(Et_2CNO)(NCS)_4]^{2-}$ [120] and $[Mo(NO)(Me_2CNO)(NCS)_4]^{2-}$ [124] which have actually been synthesized from the H_2NO precursor and a ketone. The structure of the latter has been shown to contain side-on hydroxylaminate.[112,120] They again fit the usual pattern, with axial NO. However, MeHNOH and molybdate do not give rise to a homologue, but to $[MoO_2(MeHNO)_2]$,[121] so that the reactivity pattern seems to depend upon the particular hydroxylamine used.

The H_2NO^- ligand can be deprotonated to give HNO^{2-} complexes. Thus, $[Mo(NO)(H_2NO)(terpy)(H_2O)]^{2+}$ and CN^- yield a complex [Mo(NO)(HNO)(CN)(terpy)].[122] The deprotonation is reversible and the proposed structure has been confirmed by X-ray analysis. In this case, too, the basic seven-coordinate pentagonal bipyramidal structure is observed, with NHO side-on, and NO axial, ∠MoNO = 178.0(7)°, Mo—N = 1.802(7) and N—O = 1.209(9) Å.[122]

Tris(pyrazolyl)borate has a formal resemblance to cyclopentadienide, in that it can behave as a six-electron donor and occupy three sites in a coordination sphere. Consequently, one might expect a similarity between complexes containing the Mo(pzb)(NO) group and those containing the Mo(Cp)(NO) group [pzb = tris(pyrazolyl)borate(1 −), cp = cyclopentadienide(1 −)]. This is indeed the case, and an extensive chemistry has developed for complexes of the types $[Mo(NO)(pzb)X_2]$ and [Mo(NO)(pzb)XY] (X and Y = anionic ligands). Much of this chemistry developed from an attempt to synthesize $[\{Mo(NO)(pzb)X_2\}_2]$, analogous to $[\{Mo(NO)(cp)X_2\}_2]$.[54] In this work, the differences between tris(pyrazolyl)borate and tris(3,5-dimethylpyrazolyl)borate (Me_2pzb) also became evident.

The compounds $[Mo(NO)(Me_2pzb)X_2]$ (listed in Table 7) are obtained by the general reaction of $[Mo(CO)_2(NO)(Me_2pzb)]$ with X_2. This is not always a straightforward reaction, and is sometimes accompanied by halogenation of the tris(pyrazolyl)borate. The Me_2pzb derivatives tend to be monomeric, whereas those based on pzb itself are dimeric.[125,126]

The dihalides react with alcohols leading to the formation of an enormous range of alkoxides and even aryloxides $[Mo(NO)(Me_2pzb)X(OR)]$ as detailed in Table 7. The crystal structure of $[Mo(NO)(Me_2pzb)Cl(OPr^i)]$ has been determined,[125] and the molybdenum shows normal octahedral six-coordination, albeit with a very short Mo–O bond.

Reaction of $[Mo(NO)(Me_2pzb)I(OR)]$ with silver acetate in the presence of the alcohol ROH leads to the bis(alkoxides) $[Mo(NO)(Me_2pzb)(OR)_2]$. The mixed alkoxides were similarly prepared.[126] The structures of several dialkoxides (Et, Pr^i; Pr^i, Pr^i; Et, Et) have been investigated in detail in view of the rather bulky character of the ligands at molybdenum.

The reaction of the dihalides with alcohols is paralleled by their reactions with other organic species containing acid hydrogens, such as phenols, amines and thiols. The resulting compounds are all detailed in Table 7. Representative structures include $[Mo(NO)(Me_2pzb)I(NHC_6H_4OMe)]$ and $[Mo(NO)(Me_2pzb)I(NHC_6H_4Me)]$[127] which are octahedral and very similar, with linear ∠MoNO systems [Mo—N = 1.754(11) and 1.78(2), N—O = 1.211(14) and 1.16(3) Å, respectively].[127] The thiolato complex $[Mo(NO)(Me_2pzb)I(SC_6H_{11})]$ is also distorted octahedral with a rather short Mo—S (2.31 Å); ∠MoNO = 176(3)°, Mo—N = 1.74(4) and N—O = 1.18(5) Å.[128]

The compounds with $NHNH_2$ are unusual and rare examples of hydrazide(1 −) as a ligand, and have end-on, rather than side-on, hydrazide coordination. The structures are otherwise as expected with ∠MoNO = 174(3) and 174.8(9)°, Mo—N = 1.70(3) and 1.80(2), and N—O = 1.24(4) and 1.17(3) Å for X, Y = I, NHNMePh and X, Y = I, $NHNMe_2$ in $[Mo(NO)(Me_2pzb)XY]$, respectively.[129] So far, the rich hydrazine/hydrazide chemistry of the cyclopentadienide homologues[130] has not been reproduced.

The NO stretching frequencies quoted in Table 7 fall into classes as shown $[Mo(NO)(Me_2pzb)XY]$: X, Y = I, SR (mean 1679 ± 5) > SR, SR (1675) > I, NHR (1667) > SR, OR (1656) > OR, OR (1646) > SR, NHR (1635) > OR, NHR (1632) (all cm^{-1}), generally consistent with the donor capacities of X, Y.[128] For a discussion of the reactions of the cyclopentadienyl analogues of some of these pyrazolyborate complexes, see the references cited.[54,131]

Whereas these reactions of dihalides with proton-active compounds seem reasonably simple, $[Mo(NO)(Me_2pzb)I_2]$ reacts with Me_2CO, MeEtCO or diacetone alcohol in complex fashion, affording some $[Mo(NO)(Me_2pzb)I(OEt)]$ but also some $[\{Mo(NO)(Me_2pzb)I\}_2O]$,[132,133] presumably *via* a reduced species such as the paramagnetic $Mo(NO)(Me_2pzb)I$.[133] The possible reaction course has been discussed, and it has been established that the dinuclear species has a slightly bent MoOMo system [171.0(15)°], with linear ∠MoNO (176(3)°, mean), Mo—N = 1.74(3) and N—O = 1.21(4) Å (mean values). The two molybdenum coordination shells are eclipsed with respect to each other.[133]

Less usual reactions of $[Mo(NO)(Me_2pzb)I_2]$ are with 1,2-$C_6H_4(XH)(YH)$ where X and Y bind replaceable (acid) protons. In general, only one group loses its proton, giving $[Mo(NO)(Me_2pzb)(XC_6H_4YH)I]$ for X = N, Y = N; X = N, Y = O; X = N, Y = S and X = S, Y = S. Why this should be isn't clear. Even more, for X = O, Y = O, the product is $[Mo(NO)(3,5\text{-}Me_2C_5N_2H_3)_3(OC_6H_4O)]I_3$ in which the catechol is still monodentate though dianionic, and from which a proton has been lost. The mechanism is again unclear, though the structure is unequivocal.[134] The molybdenum is still six-coordinate, with ∠MoNO = 173.3(12)°, Mo—N = 1.771(13) and N—O = 1.188(18) Å.[134]

The Mo^{II} diiodide reacts with heterocyclic compounds to give $[Mo(NO)(Me_2pzb)Z_2]^+$ (Z = pyridine, *etc.*), and with sodium amides to give diamido species, as detailed in Table 7. The product containing pyrollidides has the usual octahedral structure with ∠MoNO approaching linear, and N—O = 1.163 Å.[135]

Although 1,2-$C_6H_4(XH)(YH)$ (see above) do not form chelate or bridging (dinuclear) species, dinuclear species are formed if the groups XH and YH are far enough apart to enable the two centres to react with different metal ions quite independently. Thus 4-$H_2NC_6H_4CH_2C_6H_4NH_2$-4 forms $[Mo(NO)I(Me_2pzb)(NHC_6H_4CH_2C_6H_4NH_2)]$ and thence $[\{Mo(NO)(Me_2pzb)I\}_2NHC_6H_4CH_2C_6H_4NH]$. The species so formed, together with their reduction potentials, are reported in Table 8.[136] The two metal ions may be the same or different. Apart from those species listed in Table 8, $[Mo(NO)(Me_2pzb)I\text{-}\{NHC_6H_4Pd(PPh_3)_2I\}]$ and $[Mo(NO)(Me_2pzb)I\{NH(CH_2)_3PPh_2\}_2HgI_2]$ have also been mentioned.[136]

The electrochemical properties of these species have been studied, and it is apparent that there is no electrochemical influence of one metal atom upon the other if there are CH_2 groups in the bridge, However, if the bridge contains S, SO_2 or O, which allow transmission of electronic effects, two overlapping reduction waves are observed, suggesting some interaction. Finally, where the bridge is C_6H_4, the interaction is very strong. There is a linear correlation between $E_{1/2}$ and the Hammett σ *para* constants for compounds containing appropriate *para*-substituted groups.[136]

This study is part of a wider discussion of the electrochemistry of these Mo^{II} species.[137] In particular, $[Mo(NO)(Me_2pzb)I_2]$ is reduced reversibly in a range of solvents, S, to generate $[Mo(NO)(Me_2pzb)I(S)]$, in a one-electron process. The product is itself redox active. The $E_{1/2}$ value is +0.22 V (THF) whereas the corresponding dichloride is reversibly reduced at −0.09 V, under similar conditions. The reduction product is substantially more stable than $[Mo(NO)(Me_2pzb)I_2]^-$, and shows little sign of the production of free chloride. The only product isolated from the reduction of the diiodide is $[Mo(NO)(Me_2pzb)I_2\{Li(OEt_2)_2\}]$.[137]

36.2.4.2.4 Diazenido complexes

The principal interest of diazenido complexes of molybdenum(II) is that they are formed by the alkylation or protonation of coordinated dinitrogen. They thus represent a unique group of

Table 8 Reduction Properties of Some Bimetallic and Monometallic Mo^{II} Species

Complex	$E_{1/2}$ (V)[a]
Mo$NHC_6H_4NH_2$	−0.98, −1.52 (Ep/red)
MoNHC_6H_4NH*Mo*	−0.48
MoNHC_6H_4NH*W*	−0.74, −1.66 (Ep/red)
MoNHC_6H_4OH	−1.01 (Ep/red)
MoNHC_6H_4O*Mo*	−0.15, −1.16
MoNHC_6H_4O*W*	−0.28, −1.60 (Ep/red)
Mo$NHC_6H_4XC_6H_4NH$*M′*	
X = CH_2, M′ = H	−0.85
X = CH_2, M = *Mo*	−0.84 (two-electron process)
X = $(CH_2)_2$, M′ = H	−0.85
X = $(CH_2)_2$, M′ = *Mo*	−0.84
X = SO_2, M′ = H	−0.54
X = SO_2, M′ = *Mo*	−0.49, −0.60
X = O, M′ = H	−0.86
X = O, M′ = *Mo*	−0.88 (two-electron process)
X = O, M′ = *W*	−0.81, −1.37

[a] For one-electron process unless otherwise stated, *Mo* = Mo(NO)(Me_2pzb)I, *W* = W(NO)(Me_2pzb)I. See G. Denti, C. J. Jones, J. A. McCleverty, B. D. Neaves and S. J. Reynolds, *J. Chem. Soc., Chem. Commun.*, 1983, 474; and T. N. Briggs, C. J. Jones, J. A. McCleverty, B. D. Neaves, N. El Murr and H. M. Colquhoun, *J. Chem. Soc., Dalton Trans.*, 1985, 1249 for a complete discussion.

compounds, since well-defined species from the reactions of coordinated dinitrogen which contain dinitrogen hydride residues are uncommon. The general formula of the majority is $[Mo(N_2R)X(diphosphine)_2]$ (R = H or alkyl, X = halide, diphosphine usually dppe). They are listed in Table 9.

The base members of the series $[Mo(N_2H)X(dppe)_2]$ (X = F or Br) were obtained initially by treatment of $[Mo(N_2H_2)X(dppe)_2]^+$ (themselves obtained from reaction of $[Mo(N_2)_2(dppe)_2]$ with HX) with exactly one mole of NEt_3.[138] If an excess of base is used, reaction under N_2 regenerates the parent $[Mo(N_2)_2(dppe)_2]$. They react with acid, primarily to regenerate the hydrazide(2 −) complexes, and kinetic studies suggest that they are also formed as intermediates during the conversions of coordinated dinitrogen to hydrazide(2 −).[139]

$$Mo{-}N_2 \rightleftharpoons Mo{-}N_2H \rightleftharpoons Mo{-}N_2H_2 \quad (13)$$

Their proton base strengths are less than that of the dinitrogen complex, but greater than that of the hydrazido(2 −) complexes.[139] Their structures are likely to be similar to those of the alkyldiazenido complexes discussed below, that is *trans* octahedral with a linear MoNN system and NNH ≈ 120°, but definite evidence is lacking.

The alkyldiazenido complexes $[MoX(N_2R)(dppe)_2]$ are generally synthesized by reaction of alkyl halides with $[Mo(N_2)_2(dppe)_2]$. The reaction mechanism has been elucidated in some detail,[140] and is believed to involve a step in common with substitution reactions of the bis(dinitrogen) species, namely rate-controlling reversible thermal loss of N_2 (equation 14). The full reaction sequence is set out in equation (15).

$$[Mo(N_2)_2(dppe)_2] \rightleftharpoons [Mo(N_2)(dppe)_2] + N_2 \quad (14)$$

$$[Mo(N_2)_2(dppe)_2] \rightleftharpoons [Mo(N_2)(dppe)_2] \xrightarrow{RX} [Mo(N_2)(XR)(dppe)_2] \longrightarrow [Mo(N_2)X(dppe)_2] + R\cdot \quad (15)$$

The radical R· may then attack the solvent, as it does with THF, affording RH and a solvent radical. If R· is unreactive, it may eventually dimerize, yielding R_2 (*e.g.* for R = $PhCH_2$) and the principal complex product is $[MoX_2(dppe)_2)_2]$. If R· is unstable, then it can decay (*e.g.* $BrCH_2CH_2\cdot$ yields C_2H_4 and $[MoBr_2(dppe)_2]$). Finally, and in the most significant case, R· (or the solvent radical produced by it) can attack the remaining coordinated N_2 to produce $[Mo(N_2R)X(dppe)_2]$. The compounds obtained by this route are all listed in Table 9. In general, reactivity of alkyl halides falls in the sequence RI > RBr > RCl, but certain activated

Table 9 Diazenido Complexes of Molybdenum(II)

Complex	$\nu(N{=}N)$ (cm^{-1})
$[Mo(N_2H)X(dppe)_2]$	Not assigned (F); 1940 (Br) (Nujol)[1]
$[Mo(N_2Me)X(dppe)_2]$	1540 (Br);[2] 1535 (Br);[3] 1542 (I)[4] (all KBr); 1550 (I) (?),[5] not assigned in ref. 6
$[Mo(N_2Et)Br(dppe)_2]$	1560–1540;[3] 1555–1505[7] (all KBr)
$[Mo(N_2Et)Br\{(CF_3C_6H_4)_2PCH_2CH_2P(C_6H_4CF_3)_2\}_2]$	1560–1515 (KBr)[7]
$[Mo(N_2Et)Br\{(ClC_6H_4)_2PCH_2CH_2P(C_6H_4Cl)_2\}_2]$	1560–1515 (KBr)[7]
$[Mo(N_2Et)Br\{(MeC_6H_4)_2PCH_2CH_2P(C_6H_4Me)_2\}_2]$	1540–1500 (KBr)[7]
$[Mo(N_2Et)Br\{MeOC_6H_4)_2PCH_2CH_2P(C_6H_4OMe)_2\}_2]$	1535–1500 (KBr)[7]
$[Mo(N_2Bu)X(dppe)_2]$	1530, 1516 (Br) (CsI);[3] 1529, 1511 (I) (KBr);[4] 1535 (CN) (Nujol);[8] 1529 (CNS) (KBr);[3] 1525 (SCN) (Nujol);[8] 1523 (N_3) (KBr);[3] not assigned (OMe)[9]
$[Mo\{N_2CH_2(CH_2)_2CH_2OC(O)Me\}Br(dppe)_2]$	1540 (KBr)[10]
$[Mo(N_2CHOCH_2CH_2CH_2)Br(dppe)_2]$	1510 (KBr)[11]
$[Mo\{N_2CH_2(CH_2)_2CH_2OH\}Br(dppe)_2]$	1527 (KBr)[10]
$[Mo(N_2CH_2CH_2CMe_3)Br(dppe)_2]$	1525 (KBr)[3]
$[Mo(NC_6H_{11})X(dppe)_2]$[a]	1550 (I) (KBr);[4] 1550 (I) (?);[5] not assigned (I);[6] 1475 (OH) (CsI)[4]
$[Mo\{N_2CH_2CHCH_2(CH_2)_2CH_2\}Br(dppe)_2]$	Between 1516 and 1535, not assigned[3]
$[Mo(N_2C_8H_{17})I(dppe)_2]$	1535 (CsI);[4] not assigned in ref. 12
$[Mo(N_2C_8H_{17})Br\{2S,3S)(dppb)\}_2]$[b]	Not assigned[10]
$[Mo(N_2COR)Cl(dppe)_2]$	Not assigned (Me);[2] 1330 (Et) (Nujol);[2] 1356 (Ph) (KBr)[13]
$[Mo(N_2CONHPh)Br(dppe)_2]$	Not assigned[14]
$[Mo(N_2CO_2Et)Cl(dppe)_2]$	1632 (KBr)[15]
$[Mo(N_2COCF_3)(O_2CCF_3)(dppe)_2]$	1345 (?)[16]
$[Mo(N_2CH_2CO_2Et)X(dppe)_2]$	*ca.* 1508 (KBr) (for X = Cl, Br or I)[17]
$[Mo(N_2C_6H_4F)(bipy)(CO)_3]BF_4$	1622, 1570 (CH_2Cl_2)[18]
$[Mo(N_2Ar)(bipy)(CO)_2(PPh_3)]^+$	1632, 1561 (4-C_6H_4Me); 1628, 1565 (4-C_6H_4F); 1621, 1564 (3-C_6H_4F); 1623, 1560 (Ph) (all CH_2Cl_2)[18]
$[Mo(N_2Ar)(phen)(CO)_2(PPh_3)]^+$	1623, 1562 (4-C_6H_4Me); 1625, 1567 (4-C_6H_4F); 1628, 1656 (*sic*) (3-C_6H_4F); 1633, 1561 (Ph) (all CH_2Cl_2)[18]
$[Mo(N_2Ar)(bipy)(CO)_2X]$	1545 (4-C_6H_4Me, Cl); 1546 (4-C_6H_4Me, Br); 1550 (4-C_6H_4F, Cl); 1543 (4-C_6H_4F, Br); 1553 (4-C_6H_4F, I) (all CH_2Cl_2)[18]
$[Mo(N_2C_6H_4Me)(phen)(CO)_2X]$	1546 (Cl); 1547 (Br) (all CH_2Cl_2)[18]
$[Mo(N_2Ar)(pzb)(CO)_2]$	1559 (Ph); 1568 (4-C_6H_4F); 1538 (3-C_6H_4F); 1530 (4-$C_6H_4NO_2$); 1562 (2-C_6H_4Me); 1579 (2,5-$C_6H_3Me_2$) (solid state Raman);[19] IR values for $\nu(N_2)$ are *ca.* 1620[20]
$[Mo(N_2Ar)(pzpzb)(CO)_2]$[c]	1569 (Ph); 1582 (4-C_6H_4F); 1554 (3-C_6H_4F); 1528 (4-$C_6H_4NO_2$); (solid state Raman);[19] IR values for $\nu(N_2)$ *ca.* 1620[20]
$[Mo(N_2C_6H_4Me)(LL)(CO)(PPh_3)_2]^+$	1534 (bipy); 1535 (phen) (all CH_2Cl_2)[18]
$[Mo(N_2Ar)(pzb)(CO)(PPh_3)]$	1530, 1485 (4-C_6H_4Me); 1532, 1486 (4-C_6H_4F) (all CH_2Cl_2)[21]
$[Mo(N_2Ar)(pzb)(NO)Cl]$	probably Ar = Ph only, 1642 (?)[22]

[a] C_6H_{11} = cyclohexyl. [b] dppb = bis(diphenylphosphino)butane. [c] pzpzb = tetra(pyrazolyl)borate.

1. J. Chatt, A. J. Pearman and R. L. Richards, *J. Chem. Soc., Dalton Trans.*, 1976, 1520.
2. J. Chatt, A. A. Diamantis, G. A. Heath, N. E. Hooper and G. J. Leigh, *J. Chem. Soc., Dalton Trans.*, 1977, 688; A. A. Diamantis, J. Chatt, G. J. Leigh and G. A. Heath, *J. Organomet. Chem.*, 1975, **84,** C11.
3. G. E. Bossard, D. C. Busby, M. Chang, T. A. George and S. D. A. Iske, *J. Am. Chem. Soc.*, 1980, **102,** 1001.
4. D. C. Busby, T. A. George, S. D. A. Iske and S. D. Wagner, *Inorg. Chem.*, 1981, **20,** 22.
5. P. Brant and R. D. Feltham, *J. Organomet. Chem.*, 1976, **120,** C53.
6. V. W. Day, T. A. George and S. D. A. Iske, *J. Am. Chem. Soc.*, 1975, **97,** 4127.
7. W. Hussain, G. J. Leigh, H. Mohd.-Ali and C. J. Pickett, *J. Chem. Soc., Dalton Trans.*, 1986, 1473.
8. J. Chatt, G. J. Leigh, H. Neukomm, C. J. Pickett and D. R. Stanley, *J. Chem. Soc., Dalton Trans.*, 1980, 121.
9. D. C. Busby, C. D. Fendrick and T. A. George, *Proc. Third Int. Conf. Chem. and Uses Molybdenum*, 1979, 290.
10. G. E. Bossard, T. A. George, R. K. Lester, R. C. Tisdale and R. L. Turcotte, *Inorg. Chem.*, 1985, **24,** 1129.
11. P. C. Bevan, J. Chatt, A. A. Diamantis, R. A. Head, G. A. Heath and G. J. Leigh, *J. Chem. Soc., Dalton Trans.*, 1977, 1711; A. A. Diamantis, J. Chatt, G. A. Heath and G. J. Leigh, *J. Chem. Soc., Chem. Commun.*, 1975, 27; P. C. Bevan, J. Chatt, R. A. Head, P. B. Hitchcock and G. J. Leigh, *J. Chem. Soc., Chem. Commun.*, 1976, 509.
12. V. Day, T. A. George and S. D. A. Iske, *J. Organomet. Chem.*, 1976, **112,** C55.
13. M. Sato, T. Kodama, M. Hidai and Y. Uchida, *J. Organomet. Chem.*, 1978, **152,** 239; T. Tatsumi, M. Hidai and Y. Uchida, *Inorg. Chem.*, 1975, **14,** 2530.
14. K. Iwanami, Y. Mizobe, T. Takahachi, T. Kodama, Y. Uchida and M. Hidai, *Bull. Chem. Soc. Jpn.*, 1981, **54,** 1773.
15. G. Butler, J. Chatt, W. Hussain, G. J. Leigh and D. L. Hughes, *Inorg. Chim. Acta*, 1978, **30,** L287.
16. H. M. Colquhoun, *Transition Met. Chem.*, 1981, **6,** 57.
17. D. C. Busby and T. A. George, *J. Organomet. Chem.*, 1976, **118,** C16.
18. D. Condon, M. E. Deane, F. J. Lalor, N. G. Connelly and A. C. Lewis, *J. Chem. Soc., Dalton Trans.*, 1977, 925.
19. D. Sutton, *Can. J. Chem.*, 1974, **52,** 2634.
20. S. Trofimenko, *Inorg. Chem.*, 1969, **8,** 2675.
21. D. Condon, G. Ferguson, F. J. Lalor, M. Parvez and T. Spalding, *Inorg. Chem.*, 1982, **21,** 188.
22. M. E. Deane and F. J. Lalor, *J. Organomet. Chem.*, 1973, **57,** C61.

chlorides form diazenido complexes although most simply give $[MoCl_2(dppe)_2]$.[140] This mechanism has received support from studies involving chiral phosphines.[80,141] The rarer acylation reactions are believed to follow the same pattern.[140] An alternative reaction route for electron-rich dinitrogen complexes, such as $[Mo(N_2)(SCN)(dppe)_2]^-$ giving $[Mo(N_2Bu^n)(SCN)(dppe)_2]$, has been described,[50] as has a rather restricted reaction sequence for $[Mo(N_2)_2(dppe)_2]$ and $(CF_3CO)_2O$ yielding $[Mo(N_2COCF_3)(O_2CCF_3)(dppe)_2]$. This is believed to involve non-radical nucleophilic reaction of the coordinated N_2.[143]

The general reactions of the compounds are much as expected. They react with strong reducing agents to give ammonia and amines, with complete destruction of the complexes.[143–147] The halides are labile and can be substituted.[144] The compounds react with acids to form alkylhydrazido complexes,[148] and with alkyl halides to form dialkylhydrazido(2 –) complexes, in reactions which are typical S_N2 reactions,[149] though electrophilic attack (*viz.* by Me^+) is also a possibility.[150]

The structures of some of these complexes have been determined. They all have a *trans* configuration, with octahedral coordination around the molybdenum, with a linear MoNN system, and ∠NNC *ca.* 120°. For $[MoI(N_2C_6H_{11})(dppe)_2]$, ∠MoNN = 176(1)°, ∠NNC = 142(2)°, Mo—N = 1.95(1) and N—N = 0.91(1) Å.[151] This last value is almost certainly an artefact of the poor quality of the crystals used, and a value of *ca.* 1.15 Å might be expected. In $[MoCl(N_2CO_2Et)(dppe)_2]$, ∠MoNN = 178.9(5)°, ∠NNC = 117.0(6)°, Mo—N = 1.732(5) and N—N = 1.382(10) Å.[152] The acyldiazenido complex $[MoCl(N_2COPh)(dppe)_2]$ has ∠MoNN = 172.1(6)°, ∠NNC = 116.7(7)°, Mo—N = 1.813(7) and N—N = 1.255(10) Å.[9,153]

XPS data on $[MoI(N_2Me)(dppe)_2]$ and $[MoI(N_2C_6H_{11})(dppe)_2]$ have been reported, and in the latter case suggest the presence of two types of nitrogen, as expected, but for the former the two nitrogens give rise to only one signal.[154,155] ^{15}N NMR spectroscopy has been carried out on $[MoBr(N_2Et)(dppe)_2]$.[156]

The other diazenido complexes of Mo^{II} are generally carbonyl species and almost all are phenyldiazenido complexes. None is obtained *via* coordinated dinitrogen.

The complex $[Mo(CO)_3(N_2C_6H_4F)(bipy)]BF_4$ was prepared by reaction of $(p\text{-}FC_6H_4N_2)(BF_4)$ with $[Mo(CO)_4(bipy)]$.[157] This is the only compound of the class characterized, although the preparative route is suggested to be quite general for aryldiazonium salts. The complex shows two IR bands assigned to $\nu(CO)$, which were used as a basis for ascribing a *fac* conformation to the carbonyls.[157] The presence of two IR bands associated with $\nu(N{=}N)$ was explained by mixing of N=N and phenyl-ring vibrations, which finds support from ^{15}N-labelling studies.[157]

Dicarbonyl complexes $[Mo(CO)_2(N_2Ar)(L\text{–}L)(PPh_3)]^+$ (L–L = phen or bipy, Ar = aryl) can be prepared by reaction of the tricarbonyls with PPh_3, but the preferred route appears to involve the reaction of $[Mo(CO)_3(L\text{–}L)(PPh_3)]$ with the appropriate diazonium salt at −78 °C. The compounds so prepared are listed in Table 9. No discussion of structure is provided. These dicarbonyls react with triphenylphosphineiminium salts to lose phosphine and generate another series of dicarbonyls. $[Mo(CO)_2(N_2Ar)X(L\text{–}L)]$ (X = Cl, Br or I). Again, no structural discussion is provided. Finally $[Mo(CO)_2(N_2Ar)(L\text{–}L)(PPh_3)]^+$ react with PPh_3 in acetone at reflux to yield $[Mo(CO)(N_2AR)(L\text{–}L)(PPh_3)_2]^+$. Preparations and characterizations of all these compounds are fully described.[157] The structures (not fully discussed) are presumably the same as those of their nitrosyl analogues (classified under Mo^0).

The last considerable group of diazenido complexes contain tris(pyrazolyl)borate as a ligand. The complexes $[Mo(CO)_2(N_2Ar)(pzb)]$ and $[Mo(CO)_2(N_2Ar)(pzpzb)]$ (pzpzb = $B(pz)_4$, where pz = pyrazolyl) were first obtained by the reaction of the appropriate tricarbonyl(pyrazolylborato)molybdenum(1 –) with a diazonium salt.[158] The values of $\nu(CO)$ follow the trends expected from the electronic properties of the pyrazolylborate and arenediazenide groups. Where appropriate, the ^{19}F NMR spectra were discussed, but no specific values of $\nu(N{=}N)$ were reported. These compounds are remarkably stable, and $[Mo(CO)_2(N_2Ph)(pzb)]$ is unaffected by hot 70% sulfuric acid.[158] However, halogens produce very stable $\{Mo(N_2Ph)(pzb)X\}_n$ (X = Cl, Br or I).[159] Later Raman studies, involving labelling with ^2H and/or ^{15}N enabled assignments of $\nu(N{=}N)$ to be made with more precision. The band assigned to $\nu(N{=}N)$, or predominantly to (N=N), occurs in the region of 1550 cm^{-1} and is strongly Raman active.[160]

The crystal structure of $[Mo(CO)_2(N_2Ph)(pzb)]$ shows the expected octahedral coordination around molybdenum, with a 'singly bent' ∠NNPh and Mo—N = 1.83 Å.[161]

The pyrazolylborate dicarbonyls react with PPh_3 in refluxing xylene to generate monocarbonyls $[Mo(CO)(N_2Ar)(pzb)(PPh_3)]$, and with $(4\text{-}MeC_6H_4S)_2$ to produce $[Mo(N_2Ar)(pzb)(SC_6HMe)_2]$.[162] They also undergo oxidative additions with NOCl in CH_2Cl_2, to produce $[Mo(Cl)(N_2Ph)(pzb)(NO)]$, for example.[163] The chloride seems inert to Grignard reagents and to halide abstractions.

36.2.4.3 Miscellaneous

The five compounds in this group are unrelated, and even open to question to some degree. Thus, reaction between $[MoO_4]^{2-}$ and H_2S in the presence of cyanide yields $[Mo_2S_2(CN)_8]^{6-}$ which has a double sulfide bridge[164] and hence is formally Mo^{III}. However, if there is significant S—S interaction, this could be formulated as Mo^{II}. In the presence of O_2, this plus $[Mo(CN)_7]^{5-}$ yields $[Mo_2S(CN)_{12}]^{6-}$, which has a single Mo—S—Mo bridging system, the first of its kind reported, with short Mo—S [2.172(5) Å] and a linear MoSMo[169.5(3)°].[164]

The reaction of $[MoCl_4(bipy)]$ with Me_3SiN_3 results in a mixture of $[MoN(N_3)(bipy)]$ (isolated and characterized by X-ray structural analysis) and an insoluble material which analyzes for $MoCl_2(N_2)(bipy)$, and possesses no IR bands assignable to $\nu(Mo\equiv N)$, $\nu(N_3)$ or $\delta(N_3)$ [nor, apparently, to $\nu(N_2)$]. It has $\mu_{eff} = 2.2$ BM. If confirmed, it would be the first stable Mo^{II} dinitrogen complex.[165]

Carbon dioxide reacts with $[Mo(N_2)_2(PMe_2Ph)_4]$ to produce not a carbon dioxide complex, but a carbonato–carbonyl, whose formation involves reductive disproportionation of CO_2 (equation 16). The structure of the Mo^{II} product $[\{Mo(CO)(PMe_2Ph)_3\}_2(\mu\text{-}CO_3)_2]$ has been determined by X-ray structure analysis.[166]

$$4CO_2 + 4e^- \longrightarrow 2CO_3^{2-} + 2CO \qquad (16)$$

Nitrobenzene reacts with $[Mo(CO)_2(S_2CNEt_2)_2]$ to yield $[Mo(ONPh)O(S_2CNEt_2)_2]$ which is reduced by PPh_3 giving $[Mo(ONPh)(S_2CNEt_2)_2]$ and $OPPh_3$. This is regarded as a η^2-nitrosodurene complex on the basis of NMR evidence. If PhNO is regarded as a neutral ligand, this material contains Mo^{II}. It does not isomerize to $[MoO(NPh)(S_2CNEt_2)_2]$, a known compound, either thermally or photolytically.[167]

The complex $[Mo(CO)_2(S_2CNEt_2)_2]$ also reacts with acetylenes, losing one CO, to be replaced by acetylene, and with diazenes, yielding $[Mo(RN_2R')(S_2CNEt_2)_2]$ and $[Mo(RN_2R')_2(S_2CNEt_2)_2]$ $(R = R' = CO_2Et)$. Both complexes hydrolyze to give $EtO_2CNHNHCO_2Et$.[169] There is, of course, an extensive organometallic chemistry, some of it of Mo^{II}, based on the $Mo(S_2CNEt_2)_2$ moiety. Only the little discussed above is appropriate to this chapter.

36.2.5 REFERENCES

1. (a) G. Wilkinson, F. G. A. Stone and E. W. Abel (ed.), 'Comprehensive Organometallic Chemistry', Pergamon, Oxford, 1982, vol. 6, p. 1079; (b) D. A. Bohling, K. R. Mrun, S. Erjer, T. Gennelt, M. J. Warner and R. A. Walton, *Inorg. Chim. Acta,* 1985, **97,** L51 and references therein.
2. A. F. Masters, G. E. Bossard, T. A. George, R. T. C. Brownlee, M. J. O'Connor and A. G. Wedd, *Inorg. Chem.*, 1983, **22,** 908.
3. (a) M. Sato, T. Tatsumi, T. Kodama, M. Hidai and Y. Uchida, *J. Am. Chem. Soc.*, 1978, **100,** 4447; (b) T. A. George and R. C. Tisdale, *Polyhedron,* 1986, **5,** 297 and references therein.
4. A. J. L. Pombeiro and R. L. Richards, *J. Chem. Soc., Dalton Trans.*, 1980, 492; A. J. L. Pombeiro and R. L. Richards, *Transition Met. Chem.*, 1980, **5,** 55.
5. J. Chatt, C. M. Elson, A. J. L. Pombeiro, R. L. Richards and G. H. D. Royston, *J. Chem. Soc., Dalton Trans.*, 1978, 165.
6. J. Chatt, J. R. Dilworth and R. L. Richards, *Chem. Rev.*, 1978, **78,** 589; J. Chatt and R. L. Richards, *J. Organomet. Chem.*, 1982, **239,** 65 and references therein.
7. E. Carmona, A. Galindo, J. L. Atwood, L. G. Cavada and R. D. Rogers, *J. Organomet. Chem.*, 1984, **277,** 403.
8. S. N. Anderson, D. L. Hughes and R. L. Richards, *J. Chem. Soc., Dalton Trans.*, 1986, 1591.
9. T. Tatsumi, M. Hidai and Y. Uchida, *Inorg. Chem.*, 1975, **14,** 2530.
10. R. H. Morris and J. M. Ressner, *J. Chem. Soc., Chem. Commun.*, 1983, 909.
11. Y. Uchida, T. Uchida, M. Hidai and T. Kodama, *Acta Crystallogr., Sect. B,* 1975, **31,** 1197.
12. M. Aresta and A. Sacco, *Gazz. Chim. Ital.*, 1972, **102,** 755.
13. R. H. Morris, J. M. Ressner, J. F. Sawyer and M. Shiralian, *J. Am. Chem. Soc.*, 1984, **106,** 3683.
14. S. Donovan-Mtunzi, M. Hughes, G. J. Leigh, J. Mason, H. Mohd.-Ali and R. L. Richards, *J. Organomet. Chem.*, 1983, **246,** C1.
15. G. J. Leigh and C. J. Pickett, *J. Chem. Soc., Dalton Trans.*, 1977, 1797.
16. J. Chatt, C. T. Kan, G. J. Leigh, C. J. Pickett and D. R. Stanley, *J. Chem. Soc., Dalton Trans.*, 1980, 2032.
17. M. L. H. Green and W. Silverthorn, *J. Chem. Soc., Dalton Trans.*, 1973, 301.

18. J. H. Enemark and R. D. Feltham, *Coord. Chem. Rev.*, 1974, **13**, 339; D. Ballivet-Tkatchenko, C. Bremard, F. Abraham and G. Nowogrocki, *J. Chem. Soc., Dalton Trans.*, 1983, 1137; C. Schumacher, F. Weller and K. Dehnicke, *Z. Anorg. Allg. Chem.*, 1984, **508**, 79.
19. E. I. Stiefel, *Prog. Inorg. Chem.*, 1976, **22**, 1 and references therein.
20. N. G. Connelly, *Inorg. Chim. Acta, Rev.*, 1972, **6**, 48.
21. W. B. Hughes and E. A. Zuech, *Inorg. Chem.*, 1973, **12**, 471.
22. M. Minelli, J. L. Hubbard and J. H. Enemark, *Inorg. Chem.*, 1984, **23**, 970.
23. B. F. G. Johnson, *J. Chem. Soc., Dalton Trans.*, 1974, 475.
24. M. W. Anker, R. Colton and I. B. Tomkins, *Aust. J. Chem.*, 1968, **21**, 1149.
25. R. D. Feltham, W. Silverthorn and G. McPherson, *Inorg. Chem.*, 1969, **8**, 344.
26. M. Green and S. H. Taylor, *J. Chem. Soc., Dalton Trans.*, 1972, 2629.
27. N. G. Connelly, J. Locke, J. A. McCleverty, D. A. Phipps and B. Ratcliff, *Inorg. Chem.*, 1970, **9**, 278.
28. B. F. G. Johnson, A. Khair, C. G. Savory and R. H. Walter, *J. Chem. Soc., Chem. Commun.*, 1974, 744.
29. D. Ballivet-Tkatchenko and C. Bremard, *J. Chem. Soc., Dalton Trans.*, 1983, 1143.
30. (a) T. Diebold, M. Schappacher, B. Chevrier and R. Weiss, *J. Chem. Soc., Chem. Commun.*, 1979, 693; P. Legdzins and J. C. Oxley, *Inorg. Chem.*, 1984, **23**, 1053. (b) E. Herdtmeck, C. Schumacher and K. Dehnicke, *Z. Anorg. Allg. Chem.*, 1985, **526**, 93.
31. F. King and G. J. Leigh, *J. Chem. Soc., Dalton Trans.*, 1977, 429; L. Bencze and J. Engelhardt, *J. Mol. Catal.*, 1982, **15**, 123.
32. N. G. Connelly and C. Gardner, *J. Chem. Soc., Dalton Trans.*, 1979, 609, and references therein.
33. C. T. Kan, P. B. Hitchcock and R. L. Richards, *J. Chem. Soc., Dalton Trans.*, 1982, 79.
34. W. R. Robinson and M. E. Swanson, *J. Organomet. Chem.*, 1972, **35**, 315.
35. J. A. McCleverty, *Prog. Inorg. Chem.*, 1968, **10**, 145; E. Frauendorfer and J. Puga, *J. Organomet. Chem.*, 1984, **265**, 257.
36. M. Minelli, J. L. Hubbard, K. A. Christiensen and J. H. Enemark, *Inorg. Chem.*, 1983, **22**, 2652 and references therein.
37. J. T. Maleto, R. Stakir and J. L. Atwood, *J. Chem. Soc., Dalton Trans.*, 1980, 1253.
38. (a) S. Herzog and I. Schneider, *Z. Chem.*, 1962, **2**, 24; J. Qirk and G. Wilkinson, *Polyhedron*, 1982, **1**, 209; (b) J. A. Connor and C. Overton, *J. Chem. Soc., Dalton Trans.*, 1982, 2397.
39. J. Chatt and H. R. Watson, *J. Chem. Soc.*, 1962, 1343.
40. M. Brookhart, J. Bashkin, K. Cox, F. G. N. Cloke, A. E. Derome, J. C. Green, M. L. H. Green and P. M. Hare, *J. Chem. Soc., Dalton Trans.*, 1985, 423.
41. E. Minzoni, C. Pelizzi and G. Predieri, *Transition Met. Chem.*, 1982, **7**, 188.
42. T. A. George and R. C. Tisdale, *J. Am. Chem. Soc.*, 1985, **107**, 5157.
43. R. Alvarez, E. Carmona, E. Gutierrez-Puebla, J. M. Marin, A. Monge and M. L. Poveda, *J. Chem. Soc., Chem. Commun.*, 1984, 1326.
44. J. Chatt, W. Hussain and G. J. Leigh, *Transition Met. Chem.*, 1983, **8**, 383.
45. W. A. Schenk and F. E. Baumann, *J. Organomet. Chem.*, 1983, **256**, 261.
46. T. A. George and M. L. Timmer, *Inorg. Nucl. Chem. Lett.*, 1976, **12**, 773.
47. J. A. Connor, E. J. James, N. El Murr and C. Overton, *J. Chem. Soc., Dalton Trans.*, 1984, 255 and references therein.
48. A. J. L. Pombeiro and R. L. Richards, *J. Organomet. Chem.*, 1979, **179**, 459.
49. J. L. Atwood, E. Carmona, J. M. Maria, M. L. Poveda and R. D. Rogers, *Polyhedron*, 1983, **2**, 185.
50. J. Chatt, G. J. Leigh, H. Neukomm, C. J. Pickett and D. R. Stanley, *J. Chem. Soc., Dalton Trans.*, 1980, 121.
51. R. Bhattacharya, G. P. Bhattacharjee, A. K. Mitra and A. B. Chatterjee, *J. Chem. Soc., Dalton Trans.*, 1984, 487.
52. G. Backes-Dahmann and K. Wieghardt, *Inorg. Chem.*, 1985, **24**, 4044.
53. K. H. Al-Obaidi and T. J. Al-Hassani, *Transition Met. Chem.*, 1978, **3**, 15.
54. J. A. McCleverty, *Chem. Soc. Rev.*, 1983, **12**, 331 and references therein.
55. S. Clamp, N. G. Connelly, G. E. Taylor and T. S. Louttit, *J. Chem. Soc., Dalton Trans.*, 1980, 2162.
56. A. Müller, W. Eltzner, S. Sarkar, H. Bögge, P. J. Aymonio, N. Mohan, M. Seyer and P. Subramanian, *Z. Anorg. Allg. Chem.*, 1983, **503**, 22.
57. F. G. N. Cloke, P. J. Fyne, V. C. Gibson, M. L. H. Green, M. J. Ledoux, R. N. Perritz, A. Dix, A. Gourdon and K. Prout, *J. Organomet. Chem.*, 1984, **277**, 61.
58. M. G. B. Drew, A. P. Walters and I. B. Tomkins, *J. Chem. Soc., Dalton Trans.*, 1977, 974.
59. M. G. B. Drew, I. B. Tomkins and R. Colton, *Aust. J. Chem.*, 1970, **23**, 2517.
60. M. G. B. Drew and J. D. Wilkins, *J. Chem. Soc., Dalton Trans.*, 1973, 2664; M. G. B. Drew, *Prog. Inorg. Chem.*, 1977, **23**, 67.
61. E. Carmona, K. Doppert, J. M. Marin, M. L. Poveda, L. Sanchez and R. Sanchez-Delgado, *Inorg. Chem.*, 1984, **23**, 530.
62. B. A. Crichton, J. R. Dilworth, C. J. Pickett and J. Chatt, *J. Chem. Soc., Dalton Trans.*, 1981, 892.
63. C. G. Young and J. H. Enemark, *Aust. J. Chem.*, 1986, **39**, 997 and references therein.
64. D. C. Bromer, P. B. Winston, T. L. Tonker and J. L. Templeton, *Inorg. Chem.*, 1986, **25**, 2883.
65. S. Datta, B. Dezube, J. K. Konda and S. S. Wreford, *J. Am. Chem. Soc.*, 1978, **100**, 4404.
66. M. H. Chisholm, J. C. Hoffman and R. L. Kelly, *J. Am. Chem. Soc.*, 1979, **101**, 7615.
67. R. Colton, *Coord. Chem. Rev.*, 1971, **6**, 269 and references therein.
68. J. A. Connor, E. J. James, N. El Murr and C. Overton, *J. Chem. Soc., Dalton Trans.*, 1984, 255.
69. C. Bianchini, C. A. Ghilardi, A. Meli, S. Midollini and A. Orlandini, *J. Organomet. Chem.*, 1981, **219**, C13.
70. E. Carmona, L. Sanchez, J. M. Marin, M. L. Poveda, J. L. Atwood, R. D. Priester and R. D. Rogers, *J. Am. Chem. Soc.*, 1984, **106**, 3214.
71. J. Chatt, A. J. L. Pombeiro and R. L. Richards, *J. Chem. Soc., Dalton Trans.*, 1979, 1585.
72. A. J. L. Pombeiro and R. L. Richards, *J. Organomet. Chem.*, 1979, **179**, 459.
73. N. C. Payne, N. Okura and S. Otsuka, *J. Am. Chem. Soc.*, 1983, **105**, 245.

74. A. J. L. Pombeiro and R. L. Richards, *Transition Met. Chem.*, 1981, **6**, 255.
75. S. N. Anderson, D. L. Hughes and R. L. Richards, *Transition Met. Chem.*, 1985, **10**, 29.
76. T. Al-Salih and C. J. Pickett, *J. Chem. Soc., Dalton Trans.*, 1985, 1255.
77. G. Pelizzi and G. Prediere, *Gazz. Chim. Ital.*, 1982, **112**, 381.
78. T. A. George and D. B. Howell, *Inorg. Chem.*, 1984, **23**, 1503.
79. J. Lewis, R. S. Nyholm and P. W. Smith, *J. Chem. Soc.*, 1962, 2592.
80. G. E. Bossard and T. A. George, *Inorg. Chim. Acta*, 1981, **54**, L241.
81. D. C. Povey, R. L. Richards and C. Shortman, *Polyhedron*, 1986, **5**, 369.
82. R. L. Richards and C. Shortman, *J. Organomet. Chem.*, 1985, **286**, C3.
83. M. Hidai, K. Tominari and Y. Uchida, *J. Am. Chem. Soc.*, 1972, **94**, 110; L. J. Archer and T. A. George, *Inorg. Chem.*, 1979, **18**, 2079.
84. J. L. Atwood, E. Carmona-Guzman, W. E. Hunter and G. Wilkinson, *J. Chem. Soc., Dalton Trans.*, 1980, 467.
85. A. R. Barron, J. E. Salt and G. Wilkinson, *J. Chem. Soc., Dalton Trans.*, 1986, 1329.
86. M. B. Hursthouse, D. Lyons, M. Thornton-Pett and G. Wilkinson, *J. Chem. Soc., Dalton Trans.*, 1984, 695.
87. R. Alvarez, E. Carmona, D. J. Cole-Hamilton, A. Galindo, E. Gutierrez-Puebla, A. Monge, M. L. Poveda and C. Ruiz, *J. Am. Chem. Soc.*, 1985, **107**, 5529.
88. H. W. Choi and E. L. Muetterties, *J. Am. Chem. Soc.*, 1982, **104**, 153.
89. R. Taube and K. Seyferth, *Z. Anorg. Allg. Chem.*, 1977, **437**, 213.
90. K. Seyferth and R. Taube, *J. Mol. Catal.*, 1985, **28**, 53.
91. D. Schnurpfeil, G. Lauterback, K. Seyferth and R. Taube, *J. Prakt. Chem.*, 1984, **321**, 1025.
92. R. Davis, B. F. G. Johnson and K. H. Al-Obaidi, *J. Chem. Soc., Dalton Trans.*, 1972, 508.
93. K. E. Voes, J. D. Hudman and J. Kleinberg, *Inorg. Chim. Acta*, 1976, **20**, 79.
94. D. Sellmann, L. Zapf, J. Keller and M. Moll, *J. Organomet. Chem.*, 1985, **289**, 71.
95. L. Bencze and J. Kohan, *Inorg. Chim. Acta*, 1982, **65**, L17; L. Bencze, J. Kohan, B. Mohai and L. Marko, *J. Organomet. Chem.*, 1974, **70**, 421.
96. K. Dehnicke, A. Lickett and F. Weller, *Z. Anorg. Allg. Chem.*, 1981, **474**, 83.
97. P. Chaudhuri, K. Wieghardt, Y. H. Tsai and H. Krüger, *Inorg. Chem.*, 1984, **23**, 427.
98. R. Bhattacharya, R. Bhattacharjee and A. M. Saha, *Polyhedron*, 1985, **4**, 583.
99. J. Beck and J. Strähle, *Z. Naturforsch., Teil B*, 1985, **48**, 891.
100. P. J. Blower, P. T. Bishop, J. R. Dilworth, T. C. Hsieh, J. Hutchinson, T. Nicholson and J. Zubieta, *Inorg. Chem.*, 1985, **101**, 63.
101. K. Wieghardt, G. Backes-Dahmann, W. Swiridoff and J. Weiss, *Inorg. Chem.*, 1983, **22**, 1221.
102. P. T. Bishop, J. R. Dilworth, J. Hutchinson and J. A. Zubieta, *Inorg. Chim. Acta*, 1984, **84**, L15.
103. S. Sarkar and P. Subramanian, *J. Inorg. Nucl. Chem.*, 1981, **43**, 202.
104. S. C. Tripathi, S. C. Srivastava, A. K. Shrimal and O. P. Singh, *Transition Met. Chem.*, 1984, **9**, 478.
105. B. F. G. Johnson and K. H. Al-Obaidi, *Chem. Commun.*, 1968, 876.
106. B. F. G. Johnson, K. H. Al-Obaidi and J. A. McCleverty, *J. Chem. Soc. (A)*, 1969, 1688.
107. B. F. G. Johnson, A. K. Khair, G. G. Savory, R. H. Walter, K. H. Al-Obaidi and T. Y. Al-Hussain, *Transition Met. Chem.*, 1978, **3**, 81.
108. T. F. Brennan and I. Bernal, *Chem. Commun.*, 1970, 130.
109. T. F. Brennan and I. Bernal, *Inorg. Chim. Acta*, 1973, **7**, 283.
110. R. Davis, M. N. S. Hill, C. E. Holloway, B. F. G. Johnson and K. H. Al-Obaidi, *J. Chem. Soc. (A)*, 1971, 994.
111. F. Minelli, C. G. Young and J. H. Enemark, *Inorg. Chem.*, 1985, **24**, 1111.
112. A. Müller, S. Sarkar, N. Mohan and R. G. Bhattacharya, *Inorg. Chim. Acta*, 1980, **45**, L245.
113. J. A. Broomhead and J. R. Budge, *Aust. J. Chem.*, 1979, **32**, 1187.
114. J. A. Broomhead, J. R. Budge, W. D. Grumley, T. R. Norman and M. Sterns, *Aust. J. Chem.*, 1976, **29**, 2175.
115. J. Chatt and J. R. Dilworth, *Proc. Int. Conf. Coord. Chem., 16th, 1979*, 4.9; J. Chatt and J. R. Dilworth, *J. Chem. Soc., Chem. Commun.*, 1974, 517.
116. M. W. Bishop, J. Chatt and J. R. Dilworth, *J. Chem. Soc., Dalton Trans.*, 1979, 1.
117. M. B. Hursthouse and M. Motevalli, *J. Chem. Soc., Dalton Trans.*, 1979, 1362.
118. S. Sarkar and A. Müller, *Angew. Chem., Int. Ed. Engl.*, 1977, **16**, 183.
119. A. Liebelt, F. Weller and K. Dehnicke, *Z. Anorg. Allg. Chem.*, 1980, **480**, 13.
120. A. Müller, W. Eltzner, S. Sarkar, H. Bregge, P. J. Aymonino, N. Mohan, A. Seyer and P. Subramanian, *Z. Anorg. Allg. Chem.*, 1983, **503**, 22.
121. K. Wieghardt, W. Holzback, J. Weiss, B. Nuber and B. Prikner, *Angew. Chem., Int. Ed. Engl.*, 1979, **18**, 548; K. Wieghardt, W. Holzback, B. Nuber and J. Weiss, *Chem. Ber.*, 1980, **113**, 629.
122. K. Wieghardt, W. Holzback and J. Weiss, *Z. Naturforsch., Teil B*, 1982, **37**, 680; K. Wieghardt and W. Holzback, *Angew. Chem., Int. Ed. Engl.*, 1979, **18**, 549.
123. R. Bhattacharya and G. P. Bhattacharjee, *J. Chem. Soc., Dalton Trans.*, 1983, 1593.
124. A. Müller and N. Mohan, *Z. Anorg. Allg. Chem.*, 1981, **480**, 157.
125. J. A. McCleverty, D. Seddon, N. A. Bailey and N. W. Walker, *J. Chem. Soc., Dalton Trans.*, 1976, 898.
126. M. E. Deane and F. J. Lalor, *J. Organomet. Chem.*, 1974, **67**, C19.
127. J. A. McCleverty, G. Denti, S. J. Reynolds, A. S. Drane, N. El Murr, A. E. Rae, N. A. Bailey, H. Adams and J. M. A. Smith, *J. Chem. Soc., Dalton Trans.*, 1983, 81.
128. J. A. McCleverty, A. S. Drane, N. A. Bailey and J. M. A. Smith, *J. Chem. Soc., Dalton Trans.*, 1983, 91.
129. J. A. McCleverty, A. E. Rae, I. Wołochowicz, N. A. Bailey and J. M. A. Smith, *J. Chem. Soc., Dalton Trans.*, 1983, 71.
130. P. D. Frisch, M. M. Hunt, W. G. Kita, J. A. McCleverty, A. E. Rae, D. Seddon, D. Swann and J. Williams, *J. Chem. Soc., Dalton Trans.*, 1979, 1819.
131. See, for example, T. A. James and J. A. McCleverty, *J. Chem. Soc. (A)*, 1971, 1068; J. A. McCleverty and J. A. Williams, *Transition Met. Chem.*, 1976, **1**, 288.
132. H. Adams, N. A. Bailey, G. Denti, J. A. McCleverty, J. M. A. Smith and A. Włodarczyk, *J. Chem. Soc., Chem. Commun.*, 1981, 348.

133. H. Adams, N. A. Bailey, D. Gianfranco, J. A. McCleverty, J. M. A. Smith and A. Włodarczyk, *J. Chem. Soc., Dalton Trans.*, 1983, 2287.
134. H. Adams, N. A. Bailey, A. S. Drane and J. A. McCleverty, *Polyhedron,* 1983, **2,** 465.
135. K. H. Al-Obaidi, K. P. Brown, A. J. Edwards, S. A. Hollins, C. J. Jones, J. A. McCleverty and B. D. Neaves, *J. Chem. Soc., Chem. Commun.*, 1984, 690.
136. G. Denti, C. J. Jones, J. A. McCleverty, B. D. Neaves and S. J. Reynolds, *J. Chem. Soc., Chem. Commun.*, 1983, 474.
137. T. N. Briggs, C. J. Jones, J. A. McCleverty, B. D. Neaves, N. El-Murr and H. N. Colquhoun, *J. Chem. Soc., Dalton Trans.*, 1985, 1248.
138. J. Chatt, A. J. Pearman and R. L. Richards, *J. Chem. Soc., Dalton Trans.*, 1976, 1520.
139. R. A. Henderson, *J. Chem. Soc., Dalton Trans.*, 1982, 917; R. A. Henderson, *J. Chem. Soc., Dalton Trans.*, 1984, 2259; R. A. Henderson and J. Lane, *J. Chem. Soc., Dalton Trans.*, 1986, 2155.
140. J. Chatt, R. A. Head, G. J. Leigh and C. J. Pickett, *J. Chem. Soc., Dalton Trans.*, 1978, 1638.
141. G. E. Bossard, T. A. George, R. K. Lester, R. C. Tisdale and R. L. Turcotte, *Inorg. Chem.*, 1985, **24,** 1129.
142. H. M. Colquhoun, *Transition Met. Chem.*, 1981, **6,** 57.
143. P. C. Bevan, J. Chatt, G. J. Leigh and E. G. Leelamani, *J. Organomet. Chem.*, 1977, **139,** C59.
144. G. E. Bossard, D. C. Busby, M. Chang, T. A. George and S. D. A. Iske, *J. Am. Chem. Soc.*, 1980, **102,** 1001.
145. D. C. Busby and T. A. George, *Inorg. Chim. Acta,* 1978, **24,** L273.
146. V. Day, T. A. George and S. D. A. Iske, *J. Organomet. Chem.*, 1976, **112,** C55.
147. D. C. Busby, C. D. Fendrick and T. A. George, *Proc. Third Int. Conf. Chem. and Uses Molybdenum,* 1979, 290.
148. J. Chatt, A. A. Diamantis, G. A. Heath, N. E. Hooper and G. J. Leigh, *J. Chem. Soc., Dalton Trans.*, 1977, 688; J. Chatt, W. Hussain, G. J. Leigh, H. Neukomm, C. J. Pickett and D. A. Rankin, *J. Chem. Soc., Chem. Commun.*, 1980, 1024.
149. W. Hussain, G. J. Leigh, H. Mohd-Ali and C. J. Pickett, *J. Chem. Soc., Dalton Trans.*, 1986, 1473.
150. G. E. Bossard and T. A. George, *Inorg. Chim. Acta,* 1981, **54,** L239.
151. V. W. Day, T. A. George and S. D. A. Iske, *J. Am. Chem. Soc.*, 1975, **97,** 4127.
152. G. Butler, J. Chatt, W. Hussain, G. J. Leigh and D. L. Hughes, *Inorg. Chim. Acta,* 1978, **30,** L287.
153. M. Sato, T. Kodama, M. Hidai and Y. Uchida, *J. Organomet. Chem.*, 1978, **152,** 239.
154. P. Brant and R. D. Feltham, *J. Less-Common Met.*, 1977, **54,** 81.
155. P. Brant and R. D. Feltham, *J. Organomet. Chem.*, 1976, **120,** C53.
156. J. R. Dilworth, C. T. Kan, R. L. Richards, J. Mason and I. A. Stenhouse, *J. Organomet. Chem.*, 1980, **201,** C24.
157. D. Condon, M. E. Deane, F. J. Lalor, N. G. Connelly and A. C. Lewis, *J. Chem. Soc., Dalton Trans.*, 1977, 925.
158. S. Trofimenko, *Inorg. Chem.*, 1969, **8,** 2675.
159. M. E. Deane and F. J. Lalor, *J. Organomet. Chem.*, 1974, **67,** C19.
160. D. Sutton, *Can. J. Chem.*, 1974, **52,** 2634.
161. G. Avitabile, D. Ganis and M. Nemiroff, *Acta Crystallogr., Sect. B,* 1971, **27,** 725.
162. D. Condon, G. Ferguson, F. J. Lalor, M. Parvez and T. Spalding, *Inorg. Chem.*, 1982, **21,** 188.
163. M. E. Deane and F. J. Lalor, *J. Organomet. Chem.*, 1973, **57,** C61.
164. M. G. B. Drew, P. C. H. Mitchell and C. F. Pygall, *Angew. Chem., Int. Ed. Engl.*, 1976, **15,** 784.
165. E. Schweda and J. Strähle, *Z. Naturforsch., Teil B,* 1980, **35,** 1146.
166. J. Chatt, M. Kubota, G. J. Leigh, F. C. March, R. Mason and D. J. Yarrow, *J. Chem. Soc., Chem. Commun.*, 1974, 1033.
167. E. A. Maatta and R. A. D. Wentworth, *Inorg. Chem.*, 1980, **19,** 2597.
168. J. W. Macdonald, W. E. Newton, C. T. C. Creedy and J. L. Corbin, *J. Organomet. Chem.*, 1975, **92,** C25.

36.3

Molybdenum Complexes Containing One or More Metal–Metal Bonds

C. DAVID GARNER
University of Manchester, UK

36.3.1 INTRODUCTION

Molybdenum displays a remarkable propensity to form metal–metal bonds, in all of its oxidation states—even Mo^{VI} if one allows the close (*ca.* 2.7 Å) approach of metal atoms in complexes of $[MoS_4]^{2-}$ (see Section 36.6.1) to be considered as a direct metal–metal interaction. As with other second row transition metals, this tendency to form metal–metal bonds can be rationalized in terms of the effective covalent overlap that can be achieved using the 4*d* orbitals, especially for the metal in its lower oxidation states. As a clear indication of this, the heat of formation[1] of atomic Mo(g) is 658 kJ mol^{-1}, the bond dissociation energy of Mo_2(g) is[2] 404 ± 20 kJ mol^{-1} and the Mo—Mo separation 1.93 Å.[3] The electronic structure of this molecule has been discussed in some detail and potential energy curves calculated.[4] Simple MO considerations lead to the conclusion that the maximum Mo—Mo bond order will occur in Mo_2, which could be considered to possess the configuration $\sigma_g(s)^2\sigma_g(d)^2\pi_u(d)^4\delta_g(d)^4$, corresponding to a sextuple bond; however, correlation effects mean that such a simple description is inappropriate. These effects are also important in molybdenum compounds which involve a close approach of two metal atoms and, hence, there has been vigorous discussion concerning the nature, strength and the interpretation of spectroscopic properties of complexes involving multiple Mo—Mo bonds.

The formal bond order for Mo—Mo interactions in compounds ranges from 4 to <1, with fractional bond orders being well established. The distance between a pair of Mo atoms, in compounds where a definitive and direct interaction between two Mo atoms seems plausible, ranges from 2.037(3) Å in $[Mo(pyNC(O)Me)_4]$[5] to 3.267(6) Å in the guaiazulene (G) complex $[GMo_2(CO)_6]$.[6] Although the former is considered to involve a quadruple bond and the latter a single bond (perhaps lengthened by steric effects), the length of an Mo—Mo bond gives a general, rather than a precise, indication of the bond order. Discrete systems involving Mo—Mo bonds range from Mo_2 dimers, through Mo_3 and Mo_4 to Mo_6 clusters and molybdenum also displays considerable versatility in forming metal–metal bonds with other *d* transition metal atoms.

The preparation and properties of compounds containing metal–metal bonds, in which molybdenum features prominently or exclusively, have been the subject of several recent and authoritative reviews.[7–10] The following sections will, therefore, concentrate on the salient aspects of the coordination chemistry of these systems and the reader is referred to these reviews for further information.

36.3.2 DIMETALLIC COMPLEXES

36.3.2.1 Complexes Containing a Quadruple Metal–Metal Bond

36.3.2.1.1 Synthesis and structure

Aspects of this topic are included in Section 36.1.2. Complexes based on the Mo_2^{4+} centre presently constitute the largest group of compounds known to contain a quadruple bond (see Table 1). The ligation sites of this centre are illustrated by the structure of $[Mo_2(O_2CMe)_2(MeCN)_6][BF_4]_2$[11] (Figure 1). Each metal atom binds one axial and four equatorial ligands. There is a strong preference for an eclipsed arrangement of the two sets of equatorial ligands, even when there are no bridging ligands, *e.g.* $[Mo_2Cl_8]^{4-}$.[12] The length of the Mo—N$_{eq}$ and Mo—N$_{ax}$ bonds (2.148(4) and 2.759(19) Å, respectively) of $[Mo_2(O_2CMe)_2(MeCN)_6]^{2+}$ provide clear evidence of the static *trans* effect of an Mo—Mo quadruple bond and many Mo_2^{4+} complexes do not involve axial ligands.

$Mo_2(O_2CR)_4$ compounds are the most important systems containing an Mo—Mo quadruple bond because, traditionally, they have been the starting point for the synthesis of other Mo_2^{4+} derivatives. The popular preparative procedure is that developed by Wilkinson *et al.*;[13] $[Mo(CO)_6]$ is heated with the carboxylic acid and (if available) the anhydride, either without an additional solvent or dissolved in diglyme. Generally, the derivatives of mono-[13–15] and di-carboxylic[16] acids can be obtained by carboxylate exchange from $[Mo_2(O_2CMe)_4]$, excepting $[Mo_2(O_2CH)_4]$ which is obtained by reacting $[Mo(\eta^6\text{-PhMe})(\eta^3\text{-}C_3H_5)Cl]_2$ with anhydrous NaO_2CH in EtOH, followed by treatment with HCO_2H,[17] or by reacting $K_4[Mo_2Cl_8]\cdot 2H_2O$ with 90% formic acid and recrystallization of the resultant $K[Mo_2(O_2CH)_4Cl]$ from 90% formic acid.[18] Substitution of $K_4[Mo_2Cl_8]$ by glycine[19] or by DL-amino acids in aqueous solution has been used to prepare Mo_2^{4+} dimers in which four chiral ligands are arranged around the dimetallic centre in the cyclic order of DDLL.[20] The reaction of an aqueous solution of Mo_2^{4+} with D-mandelic acid, to form $[Mo_4(\text{D-mandelate})_4]$,[21] provides a further example of the synthetic routes to $[Mo_2(O_2CR)_4]$ compounds.

Many $[Mo_2(O_2CR)_4]$ compounds have been characterized by X-ray crystallography.[18–28] Although in $[Mo_2(O_2Cbiph)_4]$ (HO_2Cbiph = 2-phenylbenzoic acid) the bulky ligands prevent axial ligation, the general observation is that $[Mo_2(O_2CR)_4]$ compounds crystallize with the molecules linked in infinite chains formed by axial coordination of each molecule by the oxygen atoms of its neighbours. In the case of $[Mo_2(O_2CH)_4]$, three different intermolecular arrangements have been identified.[18] Alternatively, axial ligation may be provided by functional groups on the ligands, as established for $[Mo_2(O_2CC_6H_4OH)_4]$ (Figure 2),[28] or by a wide variety of ligands including halides, H_2O, DMSO, THF, diglyme, phosphine oxides, py, bipy, phosphines or arsines.[18–43] In all of these systems, the basic stereochemistry of the Mo_2^{4+} centre remains essentially as shown in Figure 1 and the length of the Mo—Mo quadruple bond (*ca.* 2.1 Å; Table 1) is insensitive to the nature of the carboxylate. There is, at most, a slight (*ca.* 0.01 Å) lengthening of this bond upon axial coordination and the absence of any significant effect is clearly demonstrated by comparison of the dimensions of $[Mo_2(O_2CCF_3)_4]$ in the gas phase[44] and solid state,[23] for which the Mo—Mo separations are 2.105(9) and 2.090(4) Å, respectively. In the case of $[Mo_2(O_2CMe)_4]$, the corresponding values are 2.079(3)[45] and 2.093(1) Å.[22]

Several reactions have been reported which involve the partial displacement of the carboxylate ligands of an $[Mo_2(O_2CR)_4]$ compound and a wide range of products have been characterized. Simple reactions include halide substitution; thus, $[Mo_2(O_2CMe)_4]$ reacts with dilute HCl to form $[Mo_2(O_2CCMe)_2Cl_4]^{2-}$ [46] and $[Mo_2(O_2CCF_3)_4]$ reacts with $[NBu_4]Br$ to form $[Mo_2(O_2CCF_3)_2Br_4]^{2-}$ [30] and both of these anions have a *trans* arrangement of the carboxylate groups. In contrast, $[Mo_2(O_2CMe)_2(acac)_2]$[46] and $[Mo_2(O_2CH(NH_3)R)_2(NCS)_4]\cdot nH_2O$ (R = H,

Table 1 A Representative Selection of Compounds Considered to Contain an Mo—Mo Quadruple Bond[7,9]

Compound	*Mo—Mo bond length* (Å)	*Ref.*
$[Mo_2(O_2CH)_4]$	2.091(2)	24
$[Mo_2(O_2CMe)_4]$	2.093(1)	22
$[Mo_2(O_2CCF_3)_4]$	2.090(4)	23
$[Mo_2(O_2CCH_2NH_3)_4]Cl_4\cdot 3H_2O$	2.112(1)	19
$K[Mo_2(O_2CH)_4Cl]$	2.105(4)	18
$[Mo_2(O_2CH)_4]\cdot 2H_2O$	2.100(1)	34
$[Mo_2(O_2CC_6H_4OH)_4]$	2.092(1), 2.094(1)	28
$[Mo_2(O_2CCF_3)_4(py)_2]$	2.129(2)	39
$[Mo_2(O_2CCF_3)_4(PBu_3)_2]$	2.105(1)	42
$[NBu_4]_2[Mo_2(O_2CCF_3)_4Br_2]$	2.134(2)	30
$[AsPh_4]_2[Mo_2(O_2CMe)_2Cl_4]\cdot 2MeOH$	2.086(2)	46
$[Mo_2(O_2CMe)_2\{(pz)_2BEt_2\}_2]$	2.129(1)	48
$[Mo_2(O_2CMe)_2(CH_2SiMe_2)_2(PMe_3)_2]$	2.098(1)	52
$[Mo_2(\mu\text{-}O_2CCF_3)_2(O_2CCF_3)_2(PPhEt_2)_2]$	2.100(1)	41
$[Mo_2(O_2CMe)\{(PhN)_2CMe\}_3]$	2.082(1)	67
$[Mo_2\{(PhN)_2CPh\}_4]$	2.090(1)	68
$[Mo_2(mhp)_4]$[a]	2.067(1)	69
$[Mo_2(map)_4]\cdot 2THF$[b]	2.070(1)	70
$[Mo_2(dmp)_4]$[c]	2.064(1)	71
$[Mo_2(ambt)_4]\cdot THF$[d]	2.103(1)	72
$[Mo_2(MeNC(PPh_2)_2S)_4]$	2.083(1)	73
$[Mo_2(\mu\text{-}PBu^t_2)_2(PBu^t)_2]$	2.209(1)	74
$[Mo_2(S_2CPh)_4]\cdot 2THF$	2.139(2)	75
$[Mo_2(\mu\text{-}S_2PEt_2)_2(SPEt_2)_2]$	2.137(1)	76
$K_4[Mo_2(SO_4)_4]\cdot 2H_2O$	2.110(3)	77
$[pyH]_2[Mo_2(HAsO_4)_4]\cdot 2H_2O$	2.265(1)	78
$[Mo_2(C_3H_5)_4]$	2.183(2)	79
$[Mo_2(C_8H_8)_3]$	2.302(2)	80
$[Mo_2\{(CH_2)_2PMe_2\}_4]$	2.082(2)	81
$K_4[Mo_2Cl_8]\cdot 2H_2O$	2.139(4)	12
$[NH_4]_4[Mo_2Br_8]$	2.135(2)	82
$[C_4H_{10}NO]_2[Mo_2Cl_6(H_2O)_2]$	2.118(1)	83
$[Mo_2Cl_4(dppe)_2]$[e]	2.183(3)	84
$[Mo_2Cl_4(pic)_4]\cdot CHCl_3$[f]	2.153(6)	85
$[Mo_2Cl_4(Et_2S)_4]$	2.144(1)	86
$[Mo_2(OPr^i)_4(Pr^iOH)_4]$	2.110(3)	87
$[Mo_2(OPr^i)_4(py)_4]$	2.195(1)	87
$[NH_4]_4[Mo_2(NCS)_8]\cdot 4H_2O$	2.162(1)	88
$Li_4[Mo_2Me_8]\cdot 4THF$	2.148	89
$[Mo_2Me_4(PMe_3)_4]$	2.153(1)	90
$[Mo_2H_4(PMe_3)_6]$	2.194(3)	91

[a] Hmhp = 2-hydroxy-6-methylpyridine. [b] Hmap = 2-amino-6-methylpyridine. [c] dmp = 2,6-dimethoxyphenyl. [d] Hambt = 2-amino-4-methylbenzothiazole. [e] dppe = 1,2-bis(diphenylphosphino)ethane. [f] pic = 4-methylpyridine.

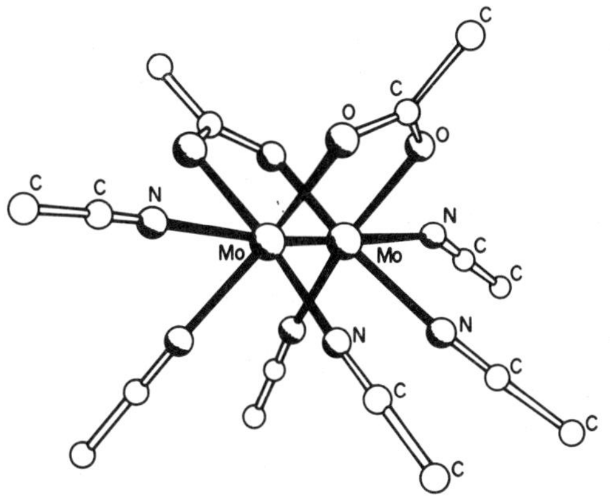

Figure 1 Structure of the $[Mo_2(O_2CMe)_2(MeCN)_6]^{2+}$ cation[11]

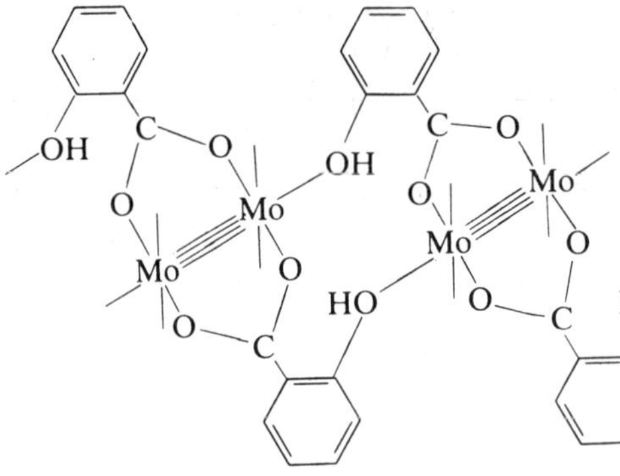

Figure 2 Molecular arrangement in $[Mo_2(O_2CC_6H_4OH)_4]\cdot C_6H_4Cl_2$-1,2[36]

$n = 1$; R = CH(Me)Et, $n = 4.5$)[47] involve a *cis* arrangement of the bridging carboxylato groups. Apart from the steric constraints imposed by the chelating ligands in the above and other[11,48–56] di-μ-carboxylato complexes, the factors which determine a *cis* or *trans* arrangement of the carboxylato ligands remain obscure. Further subtleties in the detailed coordination geometry of Mo_2^{4+} centres are revealed in the products of the reaction between $[Mo_2(O_2CR)_4]$ (R = Me or CMe_3) and lithium silylamides $[LiN(SiMe_3)_2$, $LiN(SiMe_2H)_2$ or $LiN(SiMe_3)(Me)]$ in the presence of tertiary phosphines (PMe_3, PMe_2Ph or PEt_3). Complexes of the type $[Mo_2(O_2CMe)_2(NR'_2)_2(PR''_3)_2]$ are formed and their stereochemistry has been probed by NMR and IR spectroscopy. The stereochemistry of these complexes is independent of the carboxylate or phosphine but dependent upon the nature of the silylamide. For R′ = $SiMe_2H$ the phosphines and silylamide groups are oriented *trans,* relative to the metal–metal bond. In contrast, for R′ = $SiMe_3$ or R'_2 = $(SiMe_3)(Me)$, the phosphine and silylamide groups are *trans* to each other on the same metal atom.[57]

Loss of carboxylate ligands from $[Mo_2(O_2CR)_4]$ compounds can be assisted by protic or Lewis acids. Thus, $[Mo_2(O_2CCF_3)_4]$ and bipy react in the presence of $[Et_3O][BF_4]$ to produce $[Mo_2(O_2CCF_3)_2(bipy)_2][BF_4]_2\cdot Et_2O$.[58] Also, replacement of two of the acetato groups of $[Mo_2(O_2CMe)_4]$ is readily achieved by reaction with strong acids;[11,59,60] thus, with $[Me_3O][BF_4]$ (*ca.* 1:4 mol ratio) in MeCN, $[Mo_2(O_2CMe)_2(MeCN)_6][BF_4]_2$ (Figure 1) is produced.[11] $[Mo_2(O_2CMe)_4]$ reacts with the zwitterion S_2CPEt_3 in the presence of $H[BF_4]$ to yield $[Mo_2(S_2CPEt_3)(O_2CMe)_3(OPEt_3)][BF_4]$.[61] As an example of an alternative means of removing carboxylate groups from an Mo_2^{4+} centre, treatment of $[Mo_2(O_2CCF_3)_4]$ in toluene with Me_3SiCl and EtCN at low temperature affords $[Mo_2(O_2CCF_3)_2(EtCN)_2Cl_2]\cdot[Mo_2(O_2CCF_3)_3(EtCN)Cl]$.[55]

The above examples of the partial substitution of an $[Mo_2(O_2CR)_4]$ molecule serve to demonstrate the considerable scope and variation that exists for the substituted derivatives of Mo_2^{4+}. This versatility is greatly increased by two other major factors; firstly, the large number of bidentate ligands capable[7] of bridging the metal atoms of the Mo_2^{4+} cation (Figure 3) and, secondly, the facile substitution reactions of $[Mo_2X_8]^{4-}$ (X = Cl, Br, NCS),[7] $[Mo_2(O_3SR)_4]$ (R = Me, CF_3),[62] $[Mo_2(MeCN)_8]^{4+}$ [63] or $Mo_{2(aq)}^{4+}$ [64] (see Section 36.1.2).

Mo_2^{4+} complexes can also be prepared by the reduction of molybdenum complexes in a higher oxidation state. Reduction of Mo^{VI} in HCl to form $[MoCl_6]^{3-}$, condensation of this anion to produce $[Mo_2Cl_9]^{3-}$, and quantitative reduction of the latter by a Jones reductor provides a convenient, if little exploited, route to Mo_2^{4+} complexes.[65] Furthermore, in principle, an Mo—Mo quadruple bond can be formed by reductive elimination from a compound containing an Mo—Mo triple bond. This potential has been realized for $[Mo_2R_2(NMe_2)_4]$ (R = an alkyl containing a β-hydrogen atom) upon reaction with CO_2 or 1,3-diaryltriazines, according to equations (1) and (2), respectively.[66]

$$[Mo_2R_2(NMe_2)_4] + CO_2 \longrightarrow [Mo_2(O_2CNMe_2)_4] + \text{alkene} + \text{alkane} \quad (1)$$

$$[Mo_2R_2(NMe_2)_4] + 4ArNNNHAr \longrightarrow [Mo_2(ArNNNAr)_4] + 4HNMe_2 + \text{alkene} + \text{alkane} \quad (2)$$

Table 1 illustrates the range of compounds known to contain an Mo—Mo quadruple bond and more comprehensive compilations have been provided.[7,9] This wide range of compounds manifests a variation in the length of the Mo—Mo quadruple bond (from 2.037(3)[5] to 2.302(2) Å[80]) and provides a specific reinforcement of the point that there is not a direct correlation between the length and (formal) bond order of a metal–metal bond.

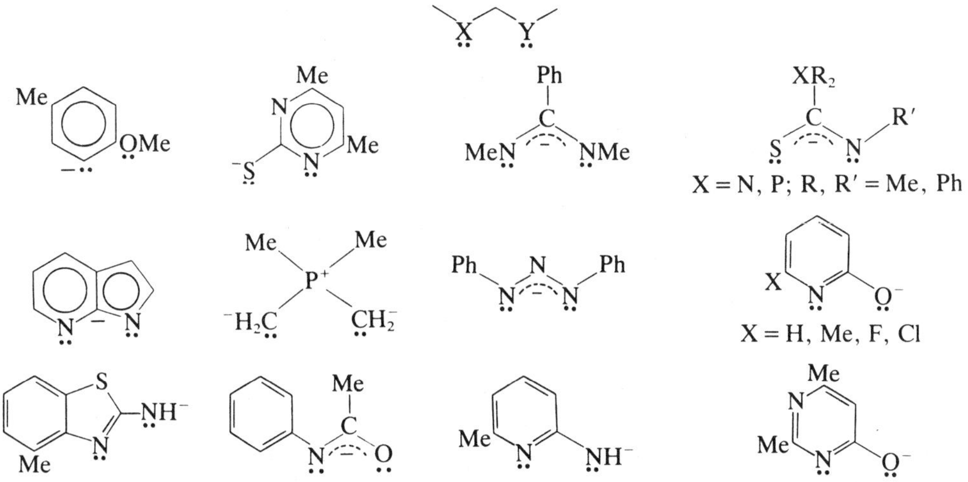

Figure 3 Examples of ligands capable of bridging the metal atoms of a Mo_2^{4+} centre[7]

As represented by the material covered so far in this review, much of the early investigations of Mo_2^{4+} systems involved μ-carboxylato ligands. However, there are a large number of other ligands capable of bridging a dimetal centre (Figure 3) and this is encouraged if: (i) the ligand is bidentate; (ii) the preferred, or only, conformation is such that the donor orbitals of X and Y are parallel; (iii) the distance between X and Y is in the range 2.0–2.5 Å.[7] These criteria are met by a large number of oxyanions, of which sulfate,[77] arsenate,[78] methylsulfonate, trifluoromethylsulfonate,[62] carbamate[66] and, of course, carboxylates have been established as binding to Mo_2^{4+}. Relatives of the carboxylates, with S or RN in place of one or both oxygen atoms, also form Mo_2^{4+} complexes[7,75,76] but perhaps the most important innovation in this area was the introduction of aromatic ligands capable of bridging across a dimetal centre *via* the carbon or nitrogen of the aromatic ring and another donor atom (usually oxygen or nitrogen) of a substituent in the 2-position. The most important sub-group of this latter class of ligands are the anions of the 2-hydroxypyridines, 2-HOC_6H_4-6-X (X = H, Hhp; X = Me, mhp; X = Cl, chp; X = F, fhp). It was with mhp that the first complete series of stable Group VI M_2^{4+} complexes was obtained.[69] Also, ligands of this type have permitted the properties of heteronuclear CrMo and MoW quadruple bonds to be determined.[92,93] These ligands, and their relatives (Figure 3), have a profound effect on the extent of axial ligation because of the location of the X groups (see Figure 4). This figure also shows the normal pattern of these unsymmetrical bridging ligands, *i.e.* the arrangement of the eight donor atoms of the square antiprismatic $\{Mo_2X_4Y_4\}$ central skeleton has D_{2d} symmetry. However, systems have been recognized in which a less symmetrical pattern of ligation occurs and one such example is $[Mo_2(fhp)_4]\cdot$THF, in which the four fhp ligands are all oriented in the same direction. Thus, one metal atom is coordinated entirely by nitrogen atoms and the other entirely by oxygen atoms; a THF molecule is axially coordinated to the latter, sterically unhindered, molybdenum atom.[94]

Reactions between Mo_2^{4+} centres and thiocarboxylates ($RCOS^-$),[95,96] dithiocarboxylates (RCS_2^-),[75,95,97] xanthates ($ROCS_2^-$),[96–98] thioxanthates ($RSCS_2^-$),[97] dithiophosphinates ($R_2PS_2^-$)[76,96,97] and dithiocarbamates ($R_2CNS_2^-$)[99,100] generally lead to products in which the S donor ligands are bridged across the dimetal centre. However, unusual behaviour has been observed and a notable example of this is the isolation of the green dimers '$Mo_2(S_2CNR_2)_2$' (R = Et, Pr^i), which are constituted as $[Mo_2S_2(\mu\text{-}S_2CNR_2)_2(SCNR_2)_2]$ with the bidentate thiocarboxamide ($SCNR_2$) ligands being formed by sulfur abstraction from dithiocarbamate groups.[99] Stoichiometric reactions of $K_4[Mo_2Cl_8]$ with $M(S_2PR_2)$ (M = K, Na; R = Et, Ph) in aqueous MeOH yield the corresponding green dimer $[Mo_2(S_2PR_2)_4]$. The structure of the complex with R = Et has been determined and shown to contain two bridging and two chelating ligands. Recrystallization of the compound from THF leads to a novel isomerization and a crystal structure has shown that $[Mo_2(S_2PEt_2)_4]\cdot$2THF possesses the usual quadruply bridged D_{4h} structure with a single axial THF ligand.[76] One of the products of reaction of $[Mo_2(O_2CMe)_4]$ with $[(pz)_2BEt_2]^-$, $[Mo_2\{(pz)_2BEt_2\}_4]$, is unusual in containing four chelating ligands coordinated to the Mo_2^{4+} core.[101] Similarly, unbridged Mo_2^{4+} centres in which each metal is coordinated to a porphyrin have been characterized.[102] Reaction of $Li(PBu^t_2)$ with $[Mo_2(O_2CMe)_4]$ (1:4) in

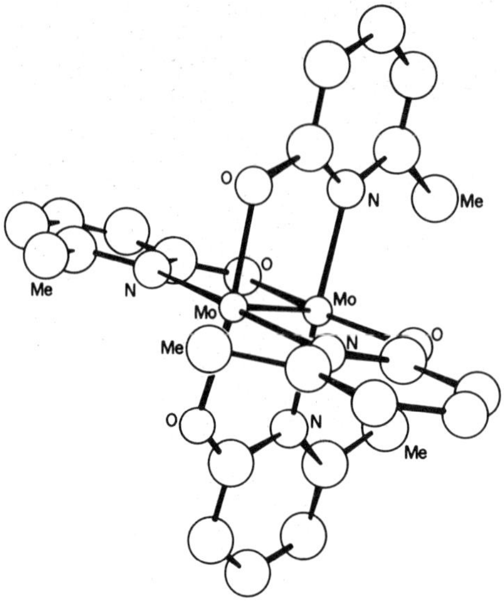

Figure 4 Structure of the $[Mo_2(mhp)_4]$ (Hmhp = 2-hydroxy-6-methylpyridine) molecule[69]

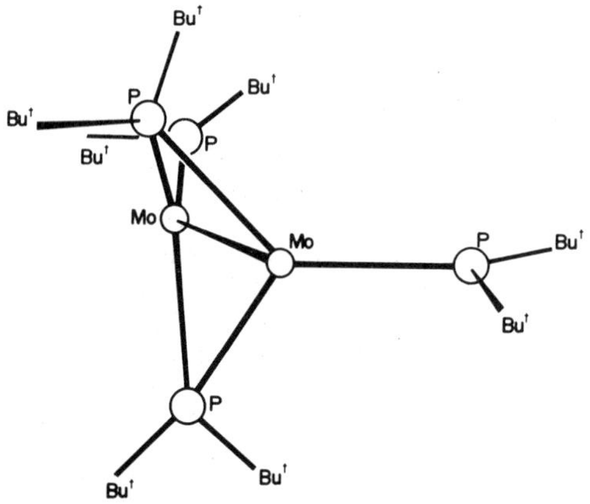

Figure 5 Structure of $[Mo_2(PBu^t_2)_2]$[74]

Et_2O at $-78\,°C$ gives $[Mo_2(PBu^t_2)_4]$. The molecule (Figure 5) possesses a central Mo_2P_2 butterfly and a terminal phosphide on each molybdenum completes a planar MoP_3 arrangement about each metal centre.[74]

$[Mo_2(C_3H_5)_4]$ ($C_3H_5 = \eta^3$-allyl) exists in solution as a mixture of two isomers. Each isomer has two inequivalent allyl groups which are mutually *trans* and bridge the Mo—Mo quadruple bond; the other allyl groups are each bonded to one metal. In the major (90%) isomer, these latter two allyls are *cis*, and in the minor isomer they are *trans*. At >80 °C, ^{1}H NMR studies showed that interconversion takes place *via* exchange of the *anti* protons of the non-bridging allyl group of the major isomer with the *anti* protons of the non-bridging allyl group of the minor isomer.[103] The crystal structure of $[Mo_2(C_3H_5)_4]$[79] is in agreement with the structural predictions made for the major isomer. This compound, and more especially $[Mo_2(C_8H_8)_3]$ (C_8H_8 = cyclooctatetraene),[80] involve longer Mo—Mo separations than normal (2.183(2) and 2.302(2) Å, respectively) but the reasons for this have not been established.

The conversion of $[Mo_2(O_2CMe)_4]$ to $K_4[Mo_2Cl_8]\cdot 2H_2O$[12] stimulated the synthesis of related anions,[7] including the centrosymmetric species $[Mo_2Br_8]^{4-}$ [82] and $[Mo_2X_6(H_2O)_2]^{2-}$ (X = Cl,[83] Br, I[104]). Two routes have been developed for the synthesis of $[Mo_2(NCS)_8]^{4-}$; the reaction of $[NH_4][NCS]$ with $[Mo_2(O_2CMe)_4]$[105] or with $K_4[Mo_2Cl_8]$.[88] The latter produces a mixture of the tetra- and hexa-hydrates of $[NH_4]_4[Mo_2(NCS)_8]$, both of which possess an anion with the expected eclipsed D_{4h} symmetry and approximately linear MoNCS groups.

Partially substituted derivatives of $[MoX_8]^{4-}$ (X = Cl, Br, NCS) abound[7] and, of these, the most interesting group are the neutral $[Mo_2X_4(PP)_2]$ (X = Cl, Br; PP = bidentate phosphine).[106]

These systems adopt two types of structure; one, designated α, has two chelating diphosphines and the other, designated β, has two bridging phosphines (Figure 6).[107] When the bidentate ligand possesses one carbon atom linking the phosphorus atoms, *e.g.* $Ph_2PCH_2PPh_2$, only the β isomer is formed with an eclipsed conformation about the Mo—Mo bond. With other, less rigid, bis(phosphines) both α and β isomers are obtained[108] and, for $R_2P(CH_2)_2PR_2$ systems, the former spontaneously isomerizes to the latter with first-order kinetics.[109] The α isomers have an essentially eclipsed conformation, but the β isomers display a variety of angles of internal rotation. The strength of the δ component of a quadruple metal–metal bond depends inversely on the angle of rotation (θ) away from the eclipsed conformation according to $\cos 2\theta$, *i.e.* the maximum overlap is achieved for $\theta = 0$ and zero overlap for $\theta = 45°$. A correlation between the Mo—Mo bond length and $\cos 2\theta$ has been defined and complete loss of the δ bond is associated with an increase of 0.097 Å in the bond length.[110]

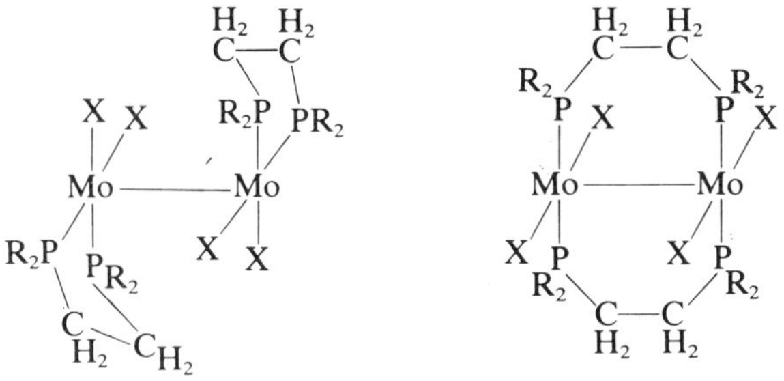

Figure 6 Arrangement of ligands in isomers of $[Mo_2X_4(PP)_2]$ (X = Cl, Br; (PP) = bidentate phosphine)

The alkoxides $[Mo_2(OR)_4L_4]$ ($R = Pr^i$, Bu^tCH_2; L = py, PMe_3, Pr^iOH, $HNMe_2$) have been prepared and subjected to structural scrutiny. The systems for L = py, PMe_3 involve a long (*ca.* 2.2 Å) Mo—Mo quadruple bond but the introduction of hydrogen bonds to the RO ligands ($L = Pr^iOH$, $HNMe_2$) causes a shortening of the Mo—Mo distance and a lengthening of the Mo—OR distance. These variations are considered to reflect the influence of the RO π donors on the Mo_2^{4+} centre. Thus, filled oxygen π orbitals interact with vacant Mo—Mo π^* and δ^* MOs, thereby weakening the bond and hydrogen bonding to the RO ligands reduces their π donating ability.[87]

Alkyl and aryl derivatives of Mo_2^{4+} centres are well known. The neatest example, $[Mo_2Me_8]^{4-}$, has been isolated as $Li_4[Mo_2Me_8]\cdot 4THF$ following the reaction of LiMe in Et_2O with $[Mo_2(O_2CMe)_4]$[89] or $[MoCl_3(THF)_3]$.[111] This anion adopts the usual eclipsed Mo_2L_8 structure and is a very reactive species and reaction with $[PMe_4]Cl$ provides a useful route to $[Mo_2\{(CH_2)_2PMe_2\}_4]$.[81,112] Several compounds involving alkyl groups and other ligands bonded to an Mo_2^{4+} centre are known, including: $[Mo_2R_4(PMe_3)_4]$ (R = Me,[90] CH_2CMe_3[113]), $[Mo_2(O_2CMe)_2(CH_2SiMe_3)_2(PMe_3)_2]$ and $[Mo_2(CH_2SiMe_3)_2\{(CH_2)_2SiMe_2\}(PMe_3)_3]$.[52,114] Also, $[Mo_2(O_2CMe)R_3(PMe_3)_3]$ (R = Ph, 4-FC_6H_4) have been prepared by the reaction of $[Mo_2(O_2CMe)_4]$ and MgR_2 in Et_2O containing an excess of PMe_3.[115]

The hydrido-bridged dimer, $[(Me_3P)_3HMo(\mu\text{-}H)_2MoH(PMe_3)_3]$, has been synthesized by reacting $[Mo_2(O_2CMe)_4]$ with Na/Hg in THF containing an excess of PMe_3 under H_2 (3 atm). This compound is pyrophoric and reacts rapidly with a variety of small molecules.[91]

The chemistry of the heteronuclear quadruple bonded species $CrMo^{4+}$ and MoW^{4+} is severely restricted, as compared to that of Mo_2^{4+}. $[CrMo(O_2CMe)_4]$ may be obtained by the slow addition of $[Mo(CO)_6]$ in $MeCO_2H/(MeCO)_2O/CH_2Cl_2$ to $[Cr_2(O_2CMe)_4(H_2O)_2]$ in refluxing $MeCO_2H/(MeCO)_2O$.[116] Synthesis of $[MoW(O_2CCMe_3)_4]$ can be achieved by refluxing $[Mo(CO)_6]$ and $[W(CO)_6]$ with pivalic acid in 1,2-dichlorobenzene. Recrystallization, followed by selective oxidation of the heteronuclear dimer by I_2 produces the insoluble $[MoW(O_2CCMe_3)_4]I$ and reduction of this precipitate by stirring with Zn in MeCN yields pure $[MoW(O_2CMe_3)_4]$.[117] These heteronuclear derivatives adopt the expected structure but the Cr—Mo and Mo—W distances (2.050(1) Å and 2.080(1) Å, respectively) are each shorter than the Mo—Mo quadruple bond in the corresponding $[Mo_2(O_2CR)_4]$ compound. $[MoW(mhp)_4]$ (Hmhp = 2-hydroxy-6-methylpyridine) has been prepared by refluxing a mixture of $[Mo(CO)_6]$ and $[W(CO)_6]$ with Hmhp in diglyme/heptane and small amounts of impure $[CrMo(mhp)_4]$ can be obtained in an analogous manner.[92] The Mo—W distance in $[MoW(mhp)_4]$ (2.091(1) Å) is intermediate between the corresponding Mo—Mo and W—W separations. The first examples of an unbridged heteronuclear quadruple metal–metal bond, $[MoWCl_4L_4]$ ($L_4 = (PMePh_2)_4$,

$(PMe_3)_2(PMePh_2)_2)$, were prepared by reacting $[Mo(\eta^6\text{-}PhPMePh)(PMePh_2)_3]$ in benzene with $[WCl_4(PPh_3)_2]$ suspended in benzene under N_2. This produces the former derivative which may be converted to the latter by reaction with PMe_3.[118]

36.3.2.1.2 Properties of Mo—Mo quadruple bonds

The high metal–metal bond order and short intermetallic distances in Mo_2^{4+} complexes, and the general observation that a wide range of ligand substitution reactions proceed without affecting this entity, imply the existence of a strong Mo—Mo bond. Various estimates of the bond strength have been advanced, including 480–550 kJ mol^{-1} for $[Mo_2X_8]^{4-}$ (X = Cl, Br) on the basis of a Birge–Sponer extrapolation of the frequencies observed in the ν(Mo—Mo) vibrational progressions.[119] However, thermochemical measurements for $[Mo_2(O_2CMe)_4]$ and $[Mo_2(O_2CMe)_2(acac)_2]$, together with assumptions about the transferability of bond enthalpy contributions, have led to the suggestion that D(Mo—Mo) = 334 kJ mol^{-1} in these molecules.[120] This value seems reasonable, especially with reference to D(Mo—Mo) for Mo_2(g) of 404 ± 20 kJ mol^{-1}.[2] Variable temperature ^{1}H NMR studies of *meso*-substituted porphyrin derivatives of Mo_2^{4+} have shown that the activation energy for rotation about the Mo—Mo bond is 42 ± 2 kJ mol^{-1}.[102]

The electronic structure of the Mo—Mo quadruple bond has been the subject of several investigations, especially by relating the theoretical interpretation[121–123] to the results of PES studies on $[Mo_2(O_2CR)_4]$,[17,124] $[Mo_2(mhp)_4]$,[92,93,125] $[Mo_2Cl_4(PMe_3)_4]$[126] and related molecules. There is general agreement about the essential nature of the ground state of the quadruple bond, $\sigma^2\pi^4\delta^2$, and that the first ionization occurs from the δ orbital and the second ionization, some 1.3 to 2.2 eV to higher energy, involves the π electrons. The position of the σ ionization has been controversial[7,92,93,124–126] but, especially due to the clarification provided by the PE spectrum of $[W_2(O_2CCF_3)_4]$,[127] there is now agreement[123,128] that the σ and π ionizations of Mo_2^{4+} compounds occur within the same[125] band envelope of the PE spectrum. The gas-phase core-electron ionization energies of $[Mo_2L_4]$ (L = O_2CH, O_2CMe, mhp) have been reported[129] and the carbon 1*s* spectrum shows an intense satellite peak close to the main 1*s* line which is assigned to a $\delta \rightarrow C2p_\pi$ transition of the core hole ion.

The ν(Mo—Mo) stretching frequency of Mo_2^{4+} complexes has been an important aspect of their spectroscopic characterization.[7] This vibration occurs in the range 336–425 cm^{-1}, with the unbridged species generally exhibiting a lower frequency than tetrabridged complexes (*e.g.* $K_4[Mo_2Cl_8]$[130] = 345 cm^{-1} and $[Mo_2(O_2CMe)_4]$[38] = 406 cm^{-1}); given the symmetry of these compounds, this mode is invariably IR inactive and Raman active. In the case of $[Mo_2(O_2CMe)_4]$, isotopic labelling has shown that this mode primarily involves the isolated ν(Mo—Mo) vibration.[131] Resonance Raman spectroscopy has been used to good effect for a selection of these compounds; in the case of $[Mo_2Cl_8]^{4-}$, 514.5 nm excitation produces a beautiful vibrational progression of ν_1(Mo—Mo), up to and including $11\nu_1$.[130] Such spectra allow an estimate of the dissociation limit of the Mo—Mo bond[119] and provide a useful confirmation of the assignment of the lowest energy absorption in the electronic absorption spectrum.

The lowest energy electronic absorption arises due to the $\delta \rightarrow \delta^*$ transition.[7] Polarized single crystal data, recorded for $K_4[Mo_2(SO_4)_4]\cdot 2H_2O$, are consistent with the $\delta \rightarrow \delta^*$ ($^1A_{1g} \rightarrow {}^1A_{2u}$) assignment; thus, the lowest energy transition (centred at *ca.* 525 nm) is *z* polarized and has an intensity which is temperature independent.[132] However, this particular $\delta \rightarrow \delta^*$ transition is unusual in showing no vibrational structure. The corresponding feature of the electronic spectrum of $K_4[Mo_2Cl_8]\cdot 2H_2O$ (also centred at *ca.* 525 cm^{-1}) displays vibronic coupling with ν(Mo—Mo) at low temperature and, as expected, the ν(Mo—Mo) vibration in the first electronic excited state is lower (by some 30 cm^{-1}) than in the ground state.[133] The assignment of the electronic absorption spectra of $[Mo_2(O_2CR)_4]$ compounds has been controversial. However, definitive studies[18,134] have established that the first observed electronic transition, with highly resolved vibrational structure, is due to a weak but electric dipole allowed $^1A_{1g} \rightarrow {}^1A_{2u}$ promotion. The relationship between the energy of the $\delta \rightarrow \delta^*$ promotion and the torsional twist about the Mo—Mo bond has been discussed for β-$[Mo_2X_4(PP)_2]$ (X = Cl, Br; PP = bidentate phosphine) complexes;[135] the CD spectra of these systems have been reported[136] and the principles established have led to the prediction[137] of a bridged structure for $[Mo_2(en)_4]^{4+}$. The emission spectra of $[Mo_2Cl_4(PMe_3)_4]$, β-$[Mo_2Cl_4(Me_2PCH_2CH_2PMe_2)_2]$, and several other Mo_2^{4+} complexes have been reported and the energy of the triplet $\delta\delta^*$ state determined.[138]

^1H and, especially for phosphine complexes, ^{31}P NMR spectroscopy have proved a useful means of characterizing Mo_2^{4+} complexes and assigning their stereochemistry. The diamagnetic anisotropy induced by Mo—Mo quadruple bonds has been discussed with special reference to the ^{31}P NMR spectra of $[Mo_2Cl_4(PR_3)_4]$ complexes.[139]

Several Mo_2^{4+} complexes have been investigated by ^{95}Mo NMR spectroscopy.[140] The resonances are broad ($\nu_{\frac{1}{2}} = 320–1440$ Hz) and the chemical shifts (2900 to 4200 p.p.m. relative to $[MoO_4]^{2-}$ at 0 p.p.m.) correspond to the lowest observed shielding of the ^{95}Mo nucleus; this is consistent with the extent of ^{95}Mo shielding decreasing steadily with an increase in Mo—Mo bond order.

36.3.2.1.3 Reactions of compounds containing an Mo—Mo quadruple bond

Several aspects of this topic are covered in Section 36.1.2 and this should be consulted for further information.

Ligand substitution reactions of Mo_2^{4+} centres have been investigated in some detail (Section 36.1.2) and evidence for a *trans* effect is seen in the phosphine exchange of $[Mo_2(O_2CBu^t)_2R_2(PMe_2Et)_2]$ (R = *e.g.* CH_2CMe_3, CH_2SiMe_3, $OSiMe_3$) complexes.[141] $[Mo_2Cl_8]^{4-}$ has been shown to act as a template for the self-condensation of 2-aminobenzaldehyde[142] and $[Mo_2O_2CCF_3)_4]$ observed to catalyze the anti-Markovnikov addition of CF_3CO_2H to propylene in benzene at 50 °C with a selectivity of >90%.[143]

Redox reactions of Mo_2^{4+} complexes are of interest, not least because the simple processes can involve changes in the population of the Mo—Mo orbitals and, therefore, a change in bond order.[7,144] $[Mo_2(O_2CR)_4]$ complexes are capable of undergoing one-electron reduction[145] or oxidation[146] and this latter aspect has been employed to good effect in the preparation of $[Mo_2(O_2CR)_4]I_3$ (R = Et, Bu^t, Ph), by I_2 oxidation. These paramagnetic complexes (μ = 1.66 BM for R = Bu^t) all display a sharp ESR signal and possess Mo—Mo bonds of order 3.5 with a $\sigma^2\pi^4\delta^1$ ground state. This description of the electronic structure of the Mo—Mo interaction is also appropriate to $[Mo_2(OC_6H_3\text{-}3,5\text{-}Me_2)_2(PMe_3)_4]^+$, prepared by one-electron electrochemical oxidation of its neutral parent; the ESR spectrum of this cation shows that the unpaired electron is delocalized and coupled to all four phosphorus nuclei.[147] $[Mo_2(SO_4)_4]^{3-}$ also involves a $\sigma^2\pi^4\delta^1$ ground state; this anion may be prepared by aerial oxidation of $K_4[Mo_2(SO_4)_4]$ in 2 M H_2SO_4,[148] UV irradiation of $K_4[Mo_2(SO_4)_4]$ in 5 M H_2SO_4,[149] or by H_2O_2 oxidation of $K_4[Mo_2Cl_8]$ in 2 M H_2SO_4 containing KCl.[150] Electrochemical studies of $K_4[Mo_2Cl_8]$[151] and $[Mo_2X_4L_4]$ (X = Cl, Br, NCS; L = PMe_3, PEt_3, PPr^n_3, $\frac{1}{2}$dppe, $\frac{1}{2}$dppm) have identified a one-electron oxidation process for all complexes, but reduction appears to be confined to $[Mo_2Cl_4(PMe_3)_4]$[147] and the phosphine–isothiocyanate complexes.[152]

$[Mo_4Cl_4L_4]$ (L = PEt_3, PPr^n_3, $\frac{1}{2}$dppe) undergo oxidation in refluxing CH_2Cl_2/CCl_4 to produce a phosphonium salt of $[Mo_2Cl_9]^{3-}$ [153] and, for monodentate phosphines, photochemical oxidation also appears to produce a compound with an Mo—Mo triple bond.[154] The most definitive studies of the conversion of an Mo—Mo quadruple bond to a triple bond are reactions of $[Mo_2(O_2CMe)_4]$ with HX (X = Cl, Br). This may be considered to occur in two stages; the initial process produces $[Mo_2X_8]^{4-}$ with retention of the quadruple bond but, at 60 °C, oxidative addition[155] of HX across the Mo—Mo bond occurs. The final product, $[Mo_2X_8H]^{3-}$, has a confacial (μ-H, X,X) bioctahedral structure[156] analogous to that of $[Mo_2Cl_9]^{3-}$ (Section 36.3.2.2). The corresponding iodo complex has been prepared by reacting $[Mo_2(O_2CMe)_4]$ and $[NEt_4]I$ in aqueous HI.[157] $[Mo_2Cl_4(RSCH_2CH_2SR)_2]$ (R = Et, Bu^n) reacts slowly in CH_2Cl_2 to form the corresponding $[Mo_2Cl_6(RSCH_2CH_2SR)_2]$ compound with an Mo—Mo triple bond;[158] this type of bond is also present in $[Mo_2Et(OPr^i)_5]$, formed when $[Mo_2(OPr^i)_4(Pr^iOH)_4]$ reacts with $HNMe_2$ in the presence of ethylene[87] and in $[Mo_2(HYO_4)_4]^{2-}$ (Y = P,[159] As[160]) obtained by O_2 oxidation of $[Mo_2Cl_8]^{4-}$ in H_3PO_4 or $[Mo_2(O_2CMe)_4]$ in H_3AsO_4, respectively. A reduction in metal–metal bond order can occur upon dimerization of compounds containing Mo—Mo quadruple bonds to form tetramers in which alternating triple and single Mo—Mo bonds occur (Section 36.3.2.3.2).

Reactions in which an Mo—Mo quadruple bond has been converted into a single bond include the following oxidations: $[Mo_2(S_2COEt)_4]$ by X_2 (X = Br, I) (forming $[Mo_2X_2(S_2COEt)_4]$);[161] $[Mo_2Cl_8]^{4-}$ by EtSSEt in the presence of $EtSCH_2CH_2SEt$ (dto) (forming $[Mo_2Cl_4(\mu\text{-}SEt)_2(dto)_2]$), and β-$[Mo_2Cl_4(dmpe)_2]$ (dmpe = bis(dimethylphosphino)ethane) by

EtSSEt (forming $[Mo_2Cl_4(\mu\text{-}SEt)_2(dmpe)_2]$).[162] Cleavage of Mo—Mo quadruple bonds has provided a useful synthetic route to several mononuclear compounds[7] and, of special significance, are the reactions of alkyl and aryl isocyanides with Mo_2^{4+} compounds.[163]

36.3.2.2 Complexes Containing a Triple Metal–Metal Bond

36.3.2.2.1 Synthesis and structure of $[Mo_2L_6]$ (L = CH_2SiMe_3, NMe_2, OR) molecules

These compounds are the parents of the most important class of complexes containing an Mo—Mo triple bond; not only do they exhibit a rich coordination chemistry, but also a versatile reactivity.[7,8,10,164] Many of the compounds of this class have been characterized by X-ray crystallography and a selection of these is presented in Table 2, together with their Mo—Mo separation. The chemistry of these systems has been developed in parallel to that of their tungsten analogues and the latter has helped illuminate the former.

Table 2 Examples of Complexes Containing an Mo—Mo Triple Bond

Complex	*Mo—Mo bond length* (Å)	*Ref.*
$[Mo_2(CH_2SiMe_3)_6]$	2.167	165
$[Mo_2(NMe_2)_6]$	2.214(2)	169
$[Mo_2(O_2CH_2CMe_3)_6]$	2.222(2)	171
$[Mo_2(SC_6H_2Me_3)_6]$	2.228(1)	199
$[Mo_2(MeN(CH_2)_2NMe)_3]$	2.190(1)	183
1,2-$[Mo_2(OC_6H_3\text{-}2,6\text{-}Me_2)_2(CH_2SiMe_3)_4]$	2.218(2)	174
1,2-$[Mo_2(OBu^t)_2(CH_2SiMe_3)_6]$	2.209(2)	175
1,2-$[Mo_2Cl_2(NMe_2)_4]$	2.201(2)	176
1,2-$[Mo_2Me_2(NMe_2)_4]$	2.201(1)	177
1,2-$[Mo_2(OSiMe_3)_6(NHMe_2)_2]$	2.242(1)	188
$[Mo_2(OBu^t)_4(O_2COBu^t)_2]$	2.241(1)	189
$[Mo_2(OBu^t)_4(O_2CPh)_2]$	2.236(2)	190
$[Mo_2(OPr^i)_4(N(Ph)C(O)OPr^i)_2]$	2.221(5)	191
$[Mo_2(OCH_2CMe_3)_4(acac)_2]$	2.237(1)	193
$[\{Mo_2F_2(OBu^t)_4\}_2]$	2.263(2)	195
1,2-$[Mo_2(SBu^t)_2(NMe_2)_4]$	2.217(1)	198
1,2-$[Mo_2(OPr^i)_2(SC_6H_2Me_3)_4]$	2.230(1)	200
1,2-$[Mo_2\{Sn(SnMe_3)\}_2(NMe_2)_4]$	2.201(2)	202

The first example of Mo_2L_6 compounds, $[Mo_2(CH_2SiMe_3)_6]$, and its neopentyl and benzyl analogues, were obtained by reacting $MoCl_5$ with Grignard reagents. $MoCl_5$ reacts with $LiCH_2SiMe_3$, or Me_3SiCH_2MgCl, to produce $[Mo_2(CH_2SiMe_3)_6]$, the molecules of which have a staggered, ethane-like, D_{3d}, Mo_2C_6 skeleton and a ^1H NMR spectrum which indicates that all the Me_3SiCH_2 groups are equivalent.[165] There is general agreement that the Mo—Mo triple bond in these Mo_2L_6 derivatives has the $\sigma^2\pi^4$ configuration.[7,164,166] Therefore, the metal–metal interaction has cylindrical symmetry and it is considered that the ligand–ligand repulsions are responsible for the adoption of a staggered conformation (see Figure 7).

$[Mo_2(NMe_2)_6]$ is conveniently prepared by adding $MoCl_3$ to a solution of $LiNMe_2$ in THF at 0 °C with subsequent warming; purification by sublimation provides the compound as a yellow crystalline solid in good yield.[167–169] $[Mo_2(NEt_2)_6]$ and $[Mo_2(NMeEt)_6]$ may be prepared in an analogous manner using the corresponding lithium amide. In the solid state[169] $[Mo_2(NMe_2)_6]$ molecules (Figure 7) adopt the staggered conformation with Mo—Mo = 2.214(2) Å; the NMe_2 groups are essentially planar, these planes contain the Mo—Mo axis, and Mo—N π-bonding interactions are important.[170] The orientations of the planar NC_2 moieties and the restricted rotation about the Mo—N bonds leads to the presence of two types of Me groups, classed as proximal or distal according to their locations with respect to the Mo—Mo bond. Variable temperature ^1H NMR studies have shown that the two types give rise to distinct signals at −71 °C, but at higher temperatures interconversion occurs and the signals coalesce at −30 °C. The large (*ca.* 2 p.p.m.) chemical shift separation between the resonances of the proximal and distal methyl protons arises due to the significant diamagnetic anisotropy of the Mo—Mo triple bond.[169]

Several $[Mo_2(OR)_6]$ compounds have been isolated and characterized and, in order to obtain

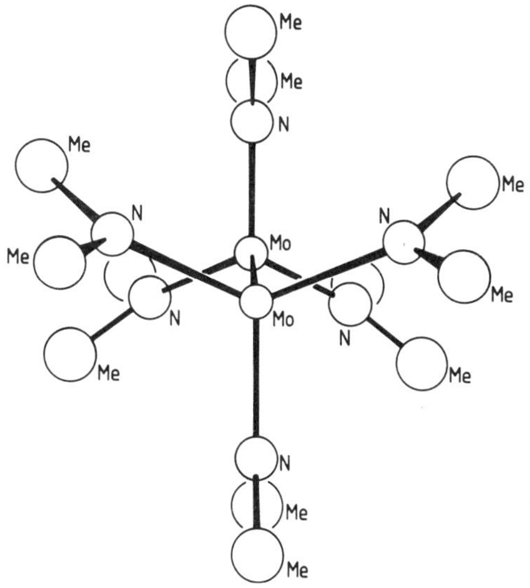

Figure 7 Structure of $[Mo_2(NMe_2)_6]$ viewed nearly along the Mo—Mo axis[169]

discrete molecular species, R is required to be a bulky group (*e.g.* R = Pr^i, Bu^t, CH(Me)Ph, CH_2CMe_3, Si_3Me_3, $SiEt_3$).[7,10] The molecules have been obtained by reaction of $[Mo_2(NMe_2)_6]$ with the alcohol or trialkylsilanol[168,171,172] and a crystal structure determination of $[Mo_2(OCH_2CMe_3)_6]$ has identified a staggered conformation and an Mo—Mo bond of length 2.222(2) Å.[171]

36.3.2.2.2 Substitution and adduct formation of $[Mo_2L_6]$ molecules

$[Mo_2(CH_2SiMe_3)_6]$ is very reactive and provides an excellent entry into chemistry of compounds with an Mo—Mo triple bond (Scheme 1). Compounds of the general formula $[Mo_2X_2R_4]$ (R = CH_2SiMe_3; X = Br, Me, OPr^i, OBu^t, 2,6-dimethylphenoxide, NMe_2)[173–75] have been reported; the pattern of substitution may be 1,1-$[X_2RMo{\equiv}MoR_3]$ or 1,2-$[XR_2Mo{\equiv}MoR_2X]$, depending upon the nature of X and the preparative route. These 1,1- and 1,2-$[Mo_2X_2R_4]$ compounds do not isomerize, but alkyl group transfer may occur during substitution reactions, *e.g.* 1,2-$[Mo_2Br_2R_4]$ and $LiNMe_2$ (1:2) yield 1,1-$[Mo_2(NMe_2)_2R_4]$. Variable temperature ^1H NMR spectra for 1,1- and 1,2-$[Mo_2X_2R_4]$ (X = NMe_2, OBu^t) are consistent with rotation about the Mo≡Mo bond; this occurs rapidly for the alkoxide (rotational barrier <33 kJ mol^{-1}) and significantly slower for the dialkylamide. These results are compatible with the rotational barrier arising due to steric and not electronic effects.[175]

$[Mo_2(NMe_2)_6]$ and $[Mo_2(OR)_6]$ molecules are versatile reagents (Scheme 1). The conversion of $[Mo_2(NMe_2)_6]$ to 1,2-$[Mo_2Cl_2(NMe_2)_4]$[176] and thence to 1,2-$[Mo_2R_2(NMe_2)_4]$[177] opens up a considerable field of chemistry. In each case, the ^1H NMR spectrum of 1,2-$[Mo_2R_2(NMe_2)_4]$ (R = Me, Et, Pr^i, Bu^n, Bu^t, CH_2CMe_3, CH_2SiMe_3) in toluene at −60 °C indicates the presence of a mixture of two rotamers, *anti* and *gauche*; the latter is always predominant and the preference is greater for bulky R groups. 1,2-$[Mo_2R_2(NMe_2)_4]$ compounds are thermally stable, implying that β-hydrogen elimination is unfavorable and the stabilizing influence has been attributed to $Me_2N \rightarrow Mo$ π bonding. This view is supported by studies accomplished on 1,2-$[Mo_2Ar_2(NMe_2)_4]$ (Ar = CH_2Ph, 2- or 4-tolyl); in these compounds the $MoNC_2$ units are planar with the NC_2 blades aligned along the Mo—Mo axis, and solution ^1H NMR studies have identified relatively high (*ca.* 59 kJ mol^{-1}) energy barriers for rotation about the Mo—N bonds.[178] 1,2-$[Mo_2R_2(NMe_2)_4]$ (R = alkyl) compounds react with CO_2 by selective insertion into the Mo—N bonds. Compounds containing alkyl groups lacking the β-hydrogen produce 1,2-$[Mo_2R_2(O_2CNMe_2)_4]$, in which the Mo—Mo triple bond is retained, but the other compounds react according to equation (1) to produce the quadruply bonded species $[Mo_2(O_2CNMe_2)_4]$.[66]

Addition of alcohols (R′OH) to 1,2-$[Mo_2R_2(NMe_2)_4]$ compounds yields either 1,2-$[Mo_2R_2(OR')_4]$ or $[Mo_2R(OR')_5]$ (Scheme 1). The latter reaction involves the elimination of the alkane RH and production of the quadruply bonded dimer which then undergoes oxidative

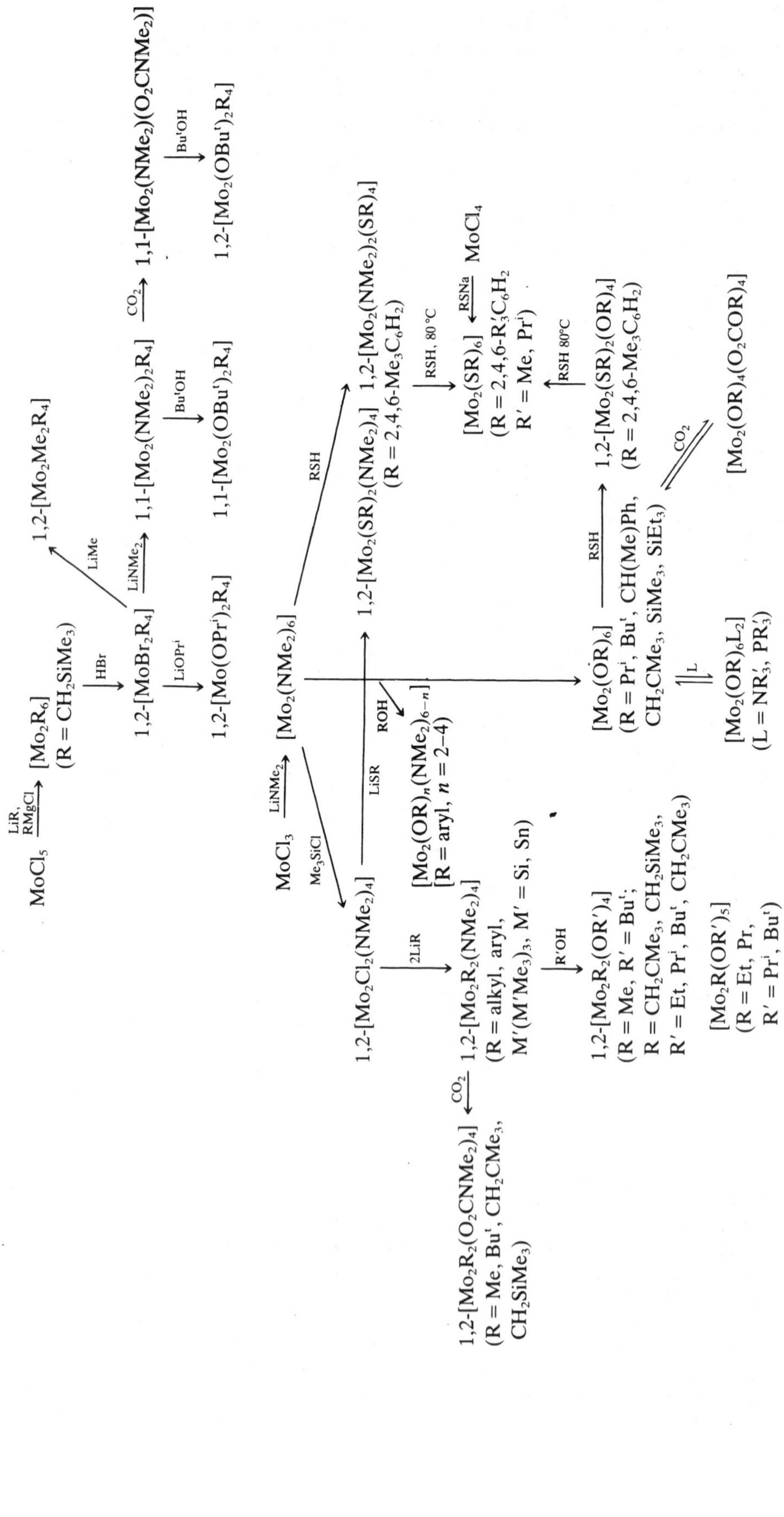

Scheme 1 Selected syntheses of compounds containing an Mo—Mo triple bond

addition with R′OH, to form (Mo≡Mo)(H)(OR′), and metal hydride alkene insertion. 1,2-$[Mo_2Me_2(OBu^t)_4]$ is coordinatively unsaturated and reacts with Lewis bases to form adducts, *e.g.* $[Mo_2Me_2(OBu^t)_4(py)_2]$.[179]

The reaction of $[Mo_2(NMe_2)_6]$ and 1,2-$[Mo_2Me_2(NMe_2)_4]$ with the precursors of several bridging ligands, *viz.* 6-methyl-2-hydroxypyridine (Hmhp), lithium 2,6-dimethylpyridine (Lidmp) and ditolyltriazine (Hdtz), have been described and the compounds isolated include: $[Mo_2(mhp)_2(NMe_2)_4]$, which contains two mutually *cis* bridging mhp ligands,[180] $[Mo_2(dmp)_2(NMe_2)_4]$, which exists as the non-bridged *gauche* isomer,[181] $[Mo_2(NMe_2)_4(dtz)_2]$ and $[Mo_2Me_2(NMe_2)_2(dtz)_2]$, which involve chelating and bridging dtz ligands, respectively.[182] 1,2-$[Mo_2Cl_2(NMe_2)_4]$ reacts with $Li(MeNH(CH_2)_2NMe)$ to produce $[Mo_2(MeNH(CH_2)_2NMe)_3]$, in which the three bridging *N,N′*-dimethylethylenediamido ligands give rise to an Mo_2N_6 skeleton with a near *eclipsed* conformation.[183] The Mo—Mo bond length (2.190(1) Å) is within the range (Table 2) found for the normal, staggered $[Mo_2L_6]$ compounds.

Phenols (ArOH = 2-*t*-butyl-6-methylphenol, 2,6-diphenylphenol, 2,6-dimethylphenol) react with $[Mo_2(NMe_2)_6]$ to produce $[Mo_2(OAr)_n(NMe_2)_{6-n}]$ ($n = 2$–4) depending on the nature of Ar; the molecular structure and dynamic behaviour of these molecules have been determined.[184] In contrast, the addition of 4-methylphenol (Ar′OH) to $[Mo_2(NMe_2)_6]$ produces $[Me_2NH_2][Mo_2(OAr')_7(NHMe_2)_2]$, which possesses a biconfacial octahedral structure with three μ-OAr′ groups and overall C_2 symmetry. The Mo—Mo separation (2.601(2) Å) is similar to that in $[Mo_2Cl_9]^{3-}$ (see Section 36.3.2.2.3) and, like this anion, the complex exhibits a slight paramagnetism at room temperature.[147,185] In a more conventional reaction, $[Mo_2(OPr^i)_2(OAr)_4]$ (ArOH = 2,6-dimethylphenol) has been prepared by the addition of an excess of ArOH to $[Mo_2(OPr^i)_6]$ and the structural characterization of this molecule has allowed a direct comparison of the bonding of alkoxide and aryloxide ligands to an Mo_2^{6+} centre.[186]

Tertiary phosphines react with 1,2-$[Mo_2Cl_2(NMe_2)_4]$ to give a variety of products depending on the phosphine. Monodentate phosphines (L = PMe_3, PMe_2Ph) induce ligand redistribution reactions, producing $[Mo_2(NMe_2)_6]$, $[Mo_2Cl_3(NMe_2)_3L_2]$ and $[Mo_2Cl_4(NMe_2)_2L_2]$; these last compounds undergo β-elimination from an NMe_2 ligand in the presence of added phosphine to give the quadruply bonded $[Mo_2Cl_4L_4]$ compound. Bidentate phosphines (LL = $Me_2P(CH_2)_nPMe_2$, $n = 1, 2$) produce stable 1:1 adducts, $[Mo_2Cl_2(NMe_2)_4(LL)]$, the ^1H and ^{31}P NMR spectra of which are consistent with a bridging phosphine and two MoN_2ClP planes with an eclipsed conformation.[187]

$[Mo_2(OR)_6]$ (R = CH_2CMe_3, $SiMe_3$) react rapidly with N and P donor ligands to give the adduct 1,2-$[Mo(OR)_6L_2]$.[171] The conformation about the Mo—Mo axis of 1,2-$[Mo(OSiMe_3)_6(NHMe_2)_2]$ is intermediate between staggered and eclipsed, but closer to the former.[188] This adduct formation is an equilibrium which has been monitored closely by NMR spectroscopy and the mono adducts, $[Mo_2(OR)_6L]$, shown to be unstable with respect to disproportionation to $[Mo_2(OR)_6]$ and $[Mo_2(OR)_6L_2]$.[10]

$[Mo_2(OR)_6]$ compounds react readily and reversibly, both in the solid state and solution, with CO_2 to produce $[Mo_2(OR)_4(O_2COR)_2]$ (Scheme 1). This insertion is considered to proceed by the direct attack of CO_2 on the Mo—OR bond. In $[Mo_2(OBu^t)_4(O_2COBu^t)_2]$, the bridging carbonato groups are mutually *cis* and the rotational conformation about the Mo—Mo bond is essentially eclipsed.[189] The related compound $[Mo_2(OBu^t)_4(O_2CPh)_2]$ has been prepared by the reaction of $[Mo_2(OBu^t)_6]$ with benzoyl peroxide and shown to have the same conformation.[190] Organic isocyanates will also insert into the Mo—O bond of $[Mo_2(OR)_6]$ compounds; thus, $[Mo_2(OPr^i)_6]$ reacts with PhNCO to produce $[Mo_2(OPr^i)_4(N(Ph)C(O)OPr^i)_2]$, which contains two *cis* bridging bidentate ligands.[191]

$[Mo_2(OR)_6]$ compounds react with bidentate ligands (HLL = 1,3-diaryltriazine, 2-hydroxypyridine) to produce $[Mo_2(OR)_4(LL)_2]$, in which the bidentate ligand spans the Mo—Mo triple bond.[192] However, reactions between $[Mo_2(OR)_6]$ and a β-diketonate produce $[Mo_2(OR)_4(R'COCHCOR'')_2]$ in which each β-diketonate is chelated to one molybdenum atom; in solution there is rapid rotation about the unbridged Mo—Mo triple bond.[193] $[Mo_2(OR)_4(acac)_2]$ (R = CH_2CMe_3) reacts with isocyanides to produce edge-shared $(\mu\text{-OR})_2$ octahedral complexes $[Mo_2(OR)_4(acac)_2(R'NC)_2]$. This is a novel conversion, thus: (i) an L_4MoMoL_4 compound is converted to an $L_4Mo(\mu\text{-L})_2L_4$ derivative; (ii) the multiple bond is *not* cleaved by isocyanide ligands; (iii) the unbridged $\sigma^2\pi^4$ triple bond is converted to an edge-bridged d^3–d^3 dimer with a formal $\sigma^2\pi^2\delta^2$ configuration. This change in electronic structure is accompanied by a significant increase in the Mo—Mo distance, from 2.237(1) to 2.508(2) Å (R′ = Bu^t).[194]

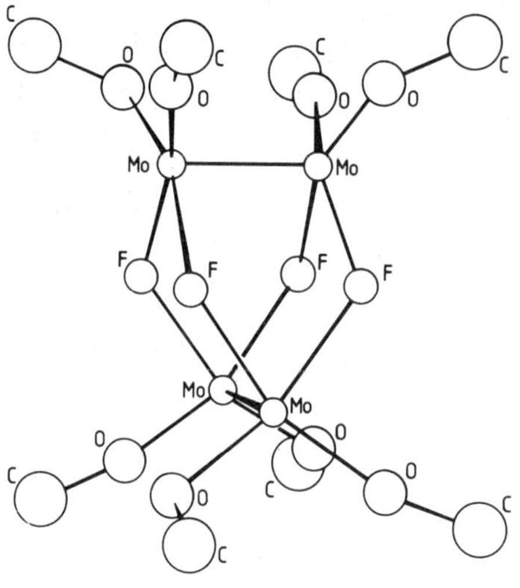

Figure 8 Structure of $[Mo_4F_4(OBu^t)_4]$[195]

$[Mo_2(OBu^t)_6]$ reacts with PF_3 to produce $[Mo_4F_4(OBu^t)_8]$ (Figure 8). This molecule consists of two Mo_2^{6+} centres linked by four bridging fluorides[195] which may be cleaved by PMe_3 to yield $[Mo_2F_2(OBu^t)_4(PMe_3)_2]$.[196]

Thiolato derivatives of Mo_2^{6+} were initially elusive,[197] but the use of 2,4,6-alkyl-substituted aryl thiolates has permitted the isolation of several such compounds. 1,2-$[Mo_2Cl_2(NMe_2)_4]$ reacts with LiSR (R = Me, Bu^t) to produce 1,2-$[Mo(SR)_2(NMe_2)_4]$[198] and this compound reacts with RSH (R = 2,4,6-$Me_3C_6H_2$) to produce $[Mo_2(SR)_6]$.[199] $[Mo_2L_6]$ (L = NMe_2, OPr^i, OBu^t, CH_2CMe_3) react with this thiol at room temperature to form the corresponding $[Mo_2L_2(SR)_4]$ derivatives, but at 80 °C complete substitution is achieved.[200] An alternative route to $[Mo_2(SR)_6]$ derivatives has been developed which involves the reaction of $MoCl_4$ with RSNa (R = 2,4,6-$Me_3C_6H_2$, 2,4,6-$Pr^i_3C_6H_2$) in 1,2-dimethoxyethane.[201]

Metathesis between 1,2-$[Mo_2Cl_2(NMe_2)_4]$ and $[(THF)_3Li][M'(M'Me_3)_3]$ (M' = Si, Sn) yields 1,2-$[Mo_2\{M'(M'Me_3)_3\}_2(NMe_2)_4]$, compounds which involve a branched chain of 10 ($M'_4Mo{\equiv}MoM'_4$) metal atoms. The compounds possess high barriers to rotation about the Mo—N bonds (67–80 kJ mol^{-1}) as a consequence of the severe steric crowding about the Mo_2^{6+} centre.[202]

36.3.2.2.3 Properties of Mo—Mo triple bonds

Thermochemical measurements have been made in an attempt to establish $D(Mo{\equiv}Mo)$. The interpretation of the data is dependent upon assumptions of the relevant metal–ligand bond energies and, in view of these uncertainties, ranges of values have been quoted; 592 ± 196 kJ mol^{-1} for $[Mo_2(NMe_2)_6]$ and 310–395 kJ mol^{-1} for $[Mo_2(OPr^i)_6]$.[203] MO calculations have indicated that the strength of the Mo—Mo triple bond is affected significantly by the nature of the attached ligands and π donors, such as NH_2, stabilize this bond.[164]

The $\sigma^2\pi^4$ description[7,164,166] of the electronic structure of the Mo—Mo triple bond in $[Mo_2L_6]$ molecules and their relatives has been reinforced by PES studies[170,204] for molecules with L = CH_2SiMe_3, NMe_2 and OR (R = Pr^i, Bu^t, CH_2CMe_3). The observed spectra have been interpreted successfully by calculations on simpler analogues. The electronic absorption spectra of these molecules have been described and the lowest energy absorption assigned to promotion of an $Mo{\equiv}Mo$ π electron into an orbital of e symmetry, which is an admixture of the $Mo{\equiv}Mo$ δ^* and π^* orbitals.[10]

Unlike their Mo_2^{4+} counterparts, the ν(Mo—Mo) stretching mode of Mo_2^{6+} complexes is not a distinctive feature of their Raman spectra. Rather, as shown by a study of $[Mo_2(NMe_2)_6]$ and $[Mo_2\{N(CD_3)_2\}_6]$,[169] this mode is strongly coupled with other internal vibrations.

Considerable attention has been paid to the dynamic behaviour of Mo_2^{6+} complexes in solution and ^1H NMR spectroscopy has been an essential probe. These properties have been reviewed[7,10] and emphasis given to the point that the barrier to rotation about the triple bond is determined only by steric factors associated with ligand–ligand interactions.[205] ^{95}Mo NMR spectroscopic studies of $[Mo_2L_6]$ (L = NMe_2, OPr^i, OBu^t, OCH_2CMe_3, CH_2SiMe_3) have been accomplished. The resonances all occur in a very deshielded portion of the known chemical shift range; for $[Mo_2(CH_2SiMe_3)_6]$ resonance occurs at 3624 p.p.m. and for the other molecules between 2430 and 2645 p.p.m.[206]

36.3.2.2.4 *Reactions of Mo—Mo triple bonds*

Compounds containing an Mo—Mo triple bond have been shown to exhibit novel reactivity[7,8,10,164] with a versatility which goes well beyond the Lewis base association, ligand substitution and CO_2 insertion reactions described in Section 36.3.2.2.2. These compounds undergo reductive elimination and oxidative addition reactions which result in a change in the order of the Mo—Mo bond; the conversion of $\{Mo{\equiv}Mo\}^{6+} \rightarrow \{Mo{=}Mo\}^{8+}$ has been mentioned in Section 36.3.2.1.1 and further information concerning the products of the transformations $\{Mo{\equiv}Mo\}^{6+} \rightarrow \{Mo{=}Mo\}^{8+}$ or $\{Mo{-}Mo\}^{10+}$ is provided in Sections 36.3.2.3 and 36.3.2.4, respectively. These reactions, and the interesting β hydrogen effects observed in the reactions of alcohols with 1,2-$[Mo_2R_2(NMe_2)_4]$ compounds,[179] have no precedence in the chemistry of mono-metal centres.

Oxidative additions to compounds containing an Mo—Mo triple bond can (equation 3) give products containing an Mo—Mo double bond.[175,190,207] Also, $[Mo_2(OCH_2Bu^t)_6]$ reacts with $PhCHBr_2$ in hexane/pyridine to form $[Mo_2(OCH_2Bu^t)_6Br_2(py)]$ which contains an Mo—Mo double bond.[208] The direct addition of halogens ($X_2 = Cl_2, Br_2, I_2$) to $[Mo_2(OPr^i)_6]$ gives $[Mo_2(OPr^i)_6X_4]$ which contains a pair of octahedrally coordinated molybdenum atoms linked by two μ-OPr^i groups and a single Mo—Mo bond[207] and $[Mo_2(OR)_6]$ compounds react with tetrachloroquinone to produce $[Mo_2(OR)_6(O_2C_6Cl_4)_2]$ which contains an Mo_2^V centre.[10]

$$[Mo_2(OPr^i)_6] + X_2 \longrightarrow [Mo_2(OPr^i)_6X_2] \quad (X = Pr^iO, PhCO_2) \tag{3}$$

Compounds containing an Mo—Mo triple bond have been shown to produce a variety of interesting products in reactions with molecules containing a C—O, C—N or C—C triple bond.[10,194] These may be considered to be oxidative addition reactions, in which electron density is removed from the Mo—Mo triple bond to form metal–ligand bonds. A solution of $[Mo_2(OBu^t)_6]$ in a hydrocarbon solvent reacts with CO; in the first instance, reversible addition to $[Mo_2(OBu^t)_6(CO)]$ occurs and, eventually, $[Mo(CO)_6]$ and $[Mo(OBu^t)_4]$ are formed.[209] Carbonylation of $[Mo_2(OPr^i)_6]$ proceeds in a similar manner and, in the presence of pyridine, $[Mo_2(OPr^i)_6(py)_2(CO)]$ is formed.[210] $[Mo_2(OR)_6]$ compounds form simple adducts with dimethyl- and diethyl-cyanamide (R'_2NCN), the process being reversible for R = Bu and R′ = Me and Et, in which an electronic redistribution occurs forming $\{R'_2NCN\}^{2-}$ and an Mo_2^{8+} centre.[10,211] Aryl-substituted diazomethanes react with $[Mo_2(OR)_6]$ compounds and are reduced to $\{Ar_2CN_2\}^{2-}$ moieties in compounds such as $[Mo_2(OPr^i)_6(N_2CPh_2)_2(py)]$.[212] Alkynes react with $[Mo_2(OR)_6]$, or their adducts $[Mo_2(OR)_6L_2]$, in hydrocarbon solvents at 25 °C to give a wide variety of products. Careful control of the conditions is required to isolate compounds such as $[Mo_2(OPr^i)_6(py)_2(\mu\text{-}C_2H_2)]$ since the alkyne adducts are labile towards C—C coupling. This coupling can lead to alkyne polymerization, but the interesting derivative $[Mo_2(OCH_2Bu^t)_6(\mu\text{-}C_4H_4)(py)]$ has been isolated and structurally characterized.[213]

Mo—Mo triple bonds may be cleaved by reactions with molecules containing multiple bonds and several interesting products have been isolated and characterized. $[Mo_2(OPr^i)_6]$ reacts with NO (1:2) to form 1,2-$[Mo_2(OPr^i)_6(NO)_2]$, in which the Mo—Mo distance of 3.335(2) Å is spanned by two μ-alkoxy groups. Therefore, the formation of the two Mo—NO triple bonds results in the loss of the Mo—Mo triple bond.[214] Reactions of 1,2-$[Mo_2(OPr^i)_6(NO)_2]$[215] and the formation of the mixed-metal dimer $[CrMo(OPr^i)_6(NO)_2]$[216] have been described. Hydrocarbon solutions of $[Mo_2(OBu^t)_6]$ react with aryl azides and molecular oxygen at room temperature according to equations (4) and (5), respectively.[217] $[Mo_2(OBu^t)_4(NC_7H_8)_4]$ has a structure which resembles that of $[Mo_2(OPr^i)_6(NO)_2]$, with an Mo—Mo separation of 3.247(1) Å bridged by two NC_7H_8 ligands. Although (equation 5) O_2 will cleave Mo—Mo triple bonds, interesting oligomers have been isolated in other reactions. Thus, $[Mo_2(OR)_6]$ (R = Pr^i, CH_2CMe_3) react with O_2 to form $[Mo_3O(OR)_{10}]$ initially, subsequently $[Mo_6O_{10}(OPr^i)_{12}]$, and finally $[MoO_2(OPr^i)_2]$.[218] $[Mo_2(OBu^t)_6]$ reacts with Ph_2CN_2 with cleavage of the Mo—Mo triple bond and formation of $[Mo(OBu^t)_4(N_2CPh_2)]$[212] and, in a similar manner, $[Mo_2(OPr^i)_6]$ reacts with bipy to produce $[Mo(OPr^i)_2(bipy)_2]$.[219]

$$[Mo_2(OBu^t)_6] + 4ArN_3 \longrightarrow [Mo_2(OBu^t)_4(NAr)_4] + 4N_2 + 2Bu^tOH \quad (4)$$

$$[Mo_2(OBu^t)_6] + 2O_2 \longrightarrow 2[MoO_2(OBu^t)_2] + 2Bu^tOH \quad (5)$$

36.3.2.2.5 Dicyclopentadienyltetracarbonyldimolybdenum

$[Cp_2Mo_2(CO)_4]$ is an important member of a group of cyclopentadienyl compounds and, like its chromium and tungsten analogues, this species is considered to involve a metal–metal triple bond.[7,220] These and related compounds manifest a rich and novel chemistry,[221] but this falls outside the field of this review.

36.3.2.2.6 Halide and related complexes[7]

$[Mo_2X_9]^{3-}$ (X = Cl, Br) anions are readily prepared by a variety of methods, including: electrolytic reduction of a solution of MoO_3 in aqueous HX,[222] thermally induced reactions of molybdenum(III) halides with alkali metal halides,[223] conproportion reactions of carbonyl halide anions of molybdenum with molybdenum halides,[224] and (Section 36.3.2.1.3) the oxidation of complexes containing an Mo—Mo quadruple bond in the presence of X^-. These $[Mo_2X_9]^{3-}$ anions consist of two MoX_6 octahedra sharing a face and, in their Cs^+ salt, the Mo—Mo separation is 2.655(11) and 2.816(9) Å for X = Cl and Br, respectively.[225] Although the potential for formation of an Mo—Mo triple bond exists in these anions, this is not realized; thus, as compared to their tungsten counterparts, the molybdenum atoms are shifted but little from the centre of the halide octahedron towards each other and the anions display effective magnetic moments at room temperature of 0.6 ($Cs_3[Mo_2Cl_9]$) and 0.8 BM ($Cs_3[Mo_2Br_9]$).[226] The substitution of μ-H for μ-X in these anions has a significant effect upon the Mo—Mo interaction. The $[Mo_2X_8H]^{3-}$ anions are diamagnetic and involve Mo—Mo separations which are some 0.3 (X = Cl) or 0.4 Å (X = Br)[227,228] shorter and, in contrast to $[Mo_2X_9]^{3-}$ (X = Cl, Br), the Mo—Mo distance in $[Mo_2X_8H]^{3-}$ (X = Cl, Br, I) anions is little affected by a change in X.[229] Initially, these anions were incorrectly formulated as $[Mo_2X_8]^{3-}$ species, but the presence of the μ-hydrido ligand was demonstrated by isotopic labelling, vibrational spectroscopy, and its location determined by X-ray crystallography and neutron diffraction; the Mo—H bonds in $[NMe_4]_3[Mo_2Cl_8H]$ are 1.73(1) and 1.823(7) Å.[230] The heteronuclear anion $[MoWCl_8H]^{3-}$, prepared by the reaction of $[MoW(O_2CMe)_4]$ with HCl, has the $[Mo_2Cl_8H]^{3-}$ structure and involves an Mo—W distance of 2.445(3) Å.[231]

$[NBu_4][Mo_2Br_6]$ has been obtained from the reaction of $[Mo_2(O_2CMe)_4]$ with HBr in MeOH in the presence of $[NBu_4]^+$ and this anion considered to possess a strong Mo—Mo interaction.[232] The synthesis, redox properties and reactivity towards phosphines of $[NBu_4]_2[Mo_2Br_6]$ have been investigated.[233]

The formation of $[Mo_2(HYO_4)_4]^{2-}$ (Y = P, As) anions has been described in Section 36.3.2.1.3; these anions contain an $\{Mo_2\}^{6+}$ centre spanned by four (μ-O_2YO_2H) ligands in an eclipsed conformation and involve Mo—Mo separations of 2.228(5)[159] and 2.265(1) Å,[160] respectively, consistent with the existence of a triple bond (Table 2).

36.3.2.3 Complexes Containing a Double Metal–Metal Bond

MoO_2 adopts a distorted rutile structure in which the metal atoms are drawn together in pairs some 2.511 Å apart,[234] a distance (*vide supra*) which is consistent with the presence of an Mo—Mo double bond.

$[Mo(NMe_2)_4]$ reacts readily with alcohols (ROH) to form the corresponding $[Mo(OR)_4]_n$ compound and, for R = $CHMe_2$ and Pr^i, it has been established that the diamagnetic, $[Mo_2(OR)_8]$, dimer is formed.[235] $[(Pr^iO)_3Mo(\mu\text{-}OPr^i)_2Mo(OPr^i)_3]$ is centrosymmetric, with an Mo—Mo separation of 2.523(1) Å spanned by two bridging isopropoxide ligands which form Mo—O bonds of length 1.958(3) and 2.111(3) Å.[236] Thermochemical measurements for $[Mo_2(OPr^i)_6]$ and $[Mo_2(OPr^i)_8]$ have led to a value in the range 837–962 kJ mol^{-1} being suggested for the Mo=Mo bond strength in the latter compound.[203]

Reactions of $[Mo_2L_6]$ compounds containing an Mo—Mo triple bond can lead to the formation of compounds with an Mo—Mo double bond (Section 36.3.2.2.4).[7,8,10,164,174,175,190,197,198,207–209,213] Thus, as an alternative to the above procedure, $[Mo_2(OPr^i)_8]$ may be prepared by the oxidative addition of Pr^iOOPr^i to $[Mo_2(OPr^i)_6]$,[207] and $[Mo_2(OR)_6]$ ($R = Pr^i, Bu^t$) reacts with $PhCO_2—O_2CPh$ to form $[Mo_2(OR)_6(O_2CPh)_2]$.[190] $[Mo_2(OBu^t)_6(CO)]$ separates as dark purple crystals when $[Mo_2(OBu^t)_6]$ is reacted with CO and the reaction mixture is cooled to −15 °C. Molecules of this compound possess a structure involving an Mo—Mo separation of 2.489(1) Å spanned by one μ-CO and two μ-OBu^t groups.[209] Carbonylation of $[Mo_2(OPr^i)_6]$ proceeds to give $[Mo(CO)_6]$, but in the presence of pyridine the adduct $[(Pr^iO)_2(py)Mo(\mu\text{-}OPr^i)_2(\mu\text{-}CO)Mo(py)(OPr^i)_2]$ can be isolated; this molecule involves an Mo—Mo separation of 2.486(2) Å consistent with the presence of a double bond.[210] This bond order is also considered to occur in the related compound $[Mo_2(OPr^i)_6(py)_2(\mu\text{-}C_2H_2)]$, in which the Mo—Mo separation is 2.554(1) Å.[213] $[Mo_2(OR)_6]$ compounds form 1:1 adducts with dialkylcyanamides, R'_2NCN, and a crystal structure determination for $R = Bu^t$ and $R' = Me$ has shown that the Me_2NCN ligand spans the Mo═Mo bond of distance 2.449(1) Å.[10,211] $[Mo_2(OCH_2Bu^t)_6(py)_2]$ and $PhCHBr_2$ react to form $[Mo_2(OCH_2Bu^t)_6Br_2(py)]$ which has a distorted confacial (μ-Br, $(\mu\text{-}OR)_2$) bioctahedral structure with an Mo═Mo bond of length 2.534(1) Å.[208] $[Mo_2(NMe_2)_6]$ and $[Mo_2(SBu^t)_2(NMe_2)_4]$ react with Bu^tSH to form $[(HNMe_2)(Bu^tS)_2Mo(\mu\text{-}S)_2Mo(SBu^t)_2(HNMe_2)]$ which involves an Mo—Mo separation of 2.730(1) Å.[197,198] $[Mo_2(OAr)_2(CH_2SiMe_3)_4]$ (OAr = 2,6-dimethylphenoxide) undergoes smooth loss of one equivalent of Me_4Si in the presence of pyridine to form $[Mo_2(\mu\text{-}H)(\mu\text{-}CSiMe_3)(CH_2SiMe_3)_2(OAr)_2(py)_2]$ which is considered to involve an $\{Mo_2\}^{8+}$ core and an Mo═Mo bond of length 2.380(2) Å.[174]

The variation in the length of Mo═Mo bonds is a demonstration of the general observation that there is no strict correlation between the bond order and the length of Mo—Mo bonds;[7] the molybdenum atoms in these compounds will have different electronic structures leading to different types of double bond (*e.g.* $\sigma^2\pi^2$ [174] or $\pi^2\delta^2$ [236]), a difference augmented by the particular stereochemical demands of the bridging ligands. With this perspective, it may be possible to view the formally Mo^{IV} dimers $[Mo_2S_2(\mu\text{-}S_2CNR_2)_2(SCNR_2)_2]$[99] as involving an Mo—Mo double bond, even though the metal–metal separation of 2.705 Å is longer than usual.

36.3.2.4 Complexes Containing a Single Metal–Metal Bond

The vast majority of these compounds are to be found in Section 36.4.3.5 which describes the chemistry of systems containing an $\{Mo_2\}^{10+}$ core. Several other complexes containing an Mo—Mo single bond have been prepared by the reaction of dimeric complexes possessing a quadruple (Section 36.3.2.1.3)[7,161,162] or triple (Section 36.3.2.2.4)[7,8,10,164] Mo—Mo bond. Also $[Mo_2(OPr^i)_6]$ reacts with X_2 (X = Cl, Br, I) to produce $[Mo_2(OPr^i)_6X_4]$; an X-ray crystallographic study of the chloro and bromo derivatives has shown that both compounds have central $Mo_2O_6X_4$ moieties with virtual D_{2h} symmetry and Mo—Mo separations of 2.731(1) and 2.739(1) Å, respectively.[207] $[Mo_2(OPr^i)_6(N_2CPh_2)_2(py)]$, formed by addition of diphenyldiazomethane to $[Mo_2(OPr^i)_6]$ in the presence of pyridine, involves an Mo—Mo single bond of length 2.662(1) Å spanned by three (μ-OPr^i) ligands.[212] The Mo—Mo separation in $[Mo_2(OCH_2Bu^t)_6(py)(\mu\text{-}C_4H_4)]$ of 2.69(1) Å approaches that of an Mo—Mo single bond, but there are alternatives to this description of the Mo—Mo interaction.[213] A similar situation pertains in $[(Me_2NCS_2)_2Mo(\mu\text{-}S)(\mu\text{-}EtC{\equiv}CEt)Mo(S_2CNMe_2)(SCNMe_2)]$, in which the Mo—Mo separation is 2.647(1) Å.[237] $[Cp_2Mo_2(CO)_6]$[238] and $[GMo_2(CO)_6]$[6] (G = guaiazulene) involve Mo—Mo bonds of length 3.235(1) and 3.267(6) Å, respectively, and are generally considered to involve an Mo—Mo single bond.

36.3.3 METAL CLUSTERS

36.3.3.1 Trimeric Clusters

Despite the proposal[239] in 1929 that molybdenum(IV) complexes such as $Mo_3O_4(C_2O_4)_3^{2+}$ contains an $\{Mo_3O_4\}^{4+}$ core, it was some 50 years later that the significance of trimeric molybdenum clusters was clearly appreciated.[9] Three types of Mo_3 cluster have been

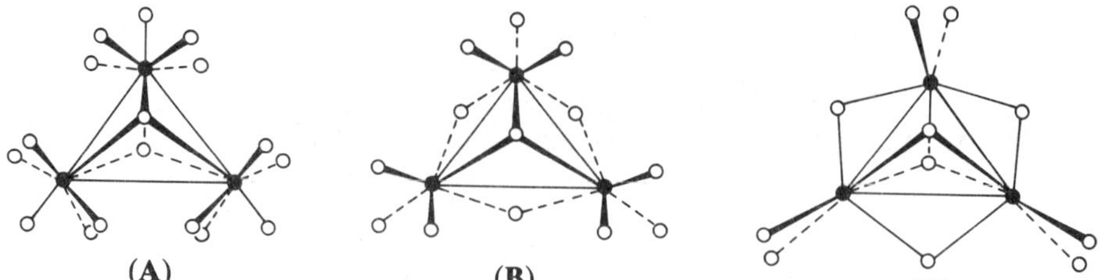

Figure 9 Structural arrangements recognized for Mo_3 clusters (● = Mo, ○ = ligand donor atom)[9]

recognized (Figure 9), involving differing patterns of ligation to an equilateral triangular array of molybdenum atoms. The common types are **A** and **B** and a fundamental understanding of the metal–metal bonding in these species has been achieved.[240]

The bicapped Mo_3X_{17} clusters (**A**) involve two orbitals per metal being used for metal–metal bonding and, therefore, a maximum of six electrons can be accommodated in the Mo—Mo bonding orbitals, thereby forming a metal–metal single bond over each edge of the triangle. Such a centre may be designated as $\{Mo_3\}^{12+}$. Several compounds exist with an oxidized version of this centre, *viz.* $\{Mo_3\}^{13+}$ or $\{Mo_3\}^{14+}$, which have Mo—Mo bond orders of $\frac{5}{6}$ or $\frac{2}{3}$, respectively, and the relative length of the Mo—Mo bonds (typically between 2.75 and 2.80 Å) are consistent with this MO picture. An important reaction in the context of these clusters is the formation of the $[Mo_2O_n(CMe)_{2-n}(O_2CMe)_6(H_2O)_3]^{m+}$ ($n = 0, 2, m = 2; n = 1, m = 1$) ions in the reaction of $MeCO_2H/(MeCO)_2O$ mixtures with $[Mo(CO)_6]$.[241–243] Although the initial report[241] was incorrect in some details, the successful isolation and characterization of these species was important as it stimulated further investigations which led to the synthesis of several $[Mo_3(\mu_3\text{-}X)_2(O_2CR)_6L_3]$ derivatives, in which X is an oxygen atom or an alkylidene group.[242–249] The presence of one μ_3-alkylidene ligand maintains the electron count of six electrons in the Mo—Mo bonding orbitals, as in $[Mo_3O(CMe)(O_2CMe)_6(H_2O_3]^+$, but the presence of two such ligands stabilizes the $\{Mo_3\}^{14+}$ core, as in $[Mo_3(CMe)_2(O_2CMe)_6(H_2O)_3]^{2+}$.

The monocapped Mo_3X_{13} clusters (**B**) have an electronic structure which can accommodate six Mo—Mo bonding electrons,[240] but one or two more electrons can be accommodated in an orbital that has little if any M—M antibonding character. Therefore, in addition to the $\{Mo_3\}^{12+}$ clusters, other type **B** systems based on $\{Mo_3\}^{11+}$ and $\{Mo_3\}^{10+}$ cores exist. Type **B** clusters are known for a wide variety of ligands and those containing $\{Mo_3O_4\}^{4+}$ and the corresponding partial or completely substituted sulfur cores have special importance; details of the synthesis, substitution and redox chemistry of many of these systems are described in Section 36.1.4 with particular attention being given to the trimeric species involved in the aqueous chemistry of Mo^{IV}.[250] The ^{95}Mo NMR chemical shift of the Mo^{IV} aqua ion in acidic media and those of the Mo^{IV} complexes containing oxalate,[239,251] 1,2-diaminoethanetetra-acetate[252] and *N*-methyliminatodiacetate, whose solid state structures are based on an $\{Mo_3O_4\}^{4+}$ core, fall in the relatively narrow range 990–1162 p.p.m. and this observation serves as a useful spectroscopic probe for the presence of this core in solutions of molybdenum complexes.[253] Thiocyanate ligands bind readily to the $\{Mo_3O_4\}^{4+}$ core and kinetic studies have indicated that there are three equivalent sites on this cation to which these ligands have access.[254] The synthesis and characterization of complexes involving $\{Mo_3O_nS_{4-n}\}^{4+}$ ($n = 3\text{–}0$) cores have been described[255–257] and, as the number of sulfur atoms in the core increases, so does the Mo—Mo separation, from *ca.* 2.5 Å in $\{Mo_3O_4\}^{4+}$ centres to *ca.* 2.75 Å in $\{Mo_3S_4\}^{4+}$ centres. $[Mo_3S_4(CN)_9]^{5-}$ has been structurally characterized[258,259] and electrochemical studies have identified the quasi-reversible one-electron oxidation of this anion. $[Mo_3(\mu_3\text{-}S)(S_2)_6]^{2-}$ (Figure 10, Section 36.6.1) is also based on an $\{Mo_3\}^{12+}$ core, but involves a (μ-S_2) ligand spanning each edge of the triangle and possesses one terminal disulfide ligand on each molybdenum.[260] $\{Mo_3(\mu_3\text{-}S)(\mu\text{-}S_2)_3\}^{4+}$ cores have been shown to bind chloride[261,262] and dialkyldioxo[263] and dialkyldithiophosphinate ligands;[264] in respect to the last ligation, the reaction of $[Mo(CO)_6]$ with $R_2P(S)—S_2—P(S)R_2$ (R = Et, Pr) affords $[Mo_3S_7(R_2PS_2)_3][R_2PS_2]$ and reaction of this cation with PPh_3 leads to sulfur abstraction and formation of the $[Mo_3S_4(R_2PS_2)_4]$ compound.[265]

As indicated above, type **B** clusters can accommodate one or two additional electrons, beyond the six needed to form the three Mo—Mo bonds of the triangular array. This redox versatility is clearly manifest in mixed-metal oxide phases of molybdenum. Examples include

$M_2Mo_3O_8$ (M = Mg, Mn, Fe, Co, Ni, Zn, Cd),[266] $LiZn_2M_3O_8$ and $Zn_3Mo_3O_8$[267] which, respectively, involve Mo_3O_{13} units with six ($\{Mo_3\}^{12+}$), seven ($\{Mo_3\}^{11+}$) and eight ($\{Mo_3\}^{10+}$) electrons in the metal-based orbitals; $La_3Mo_4SiO_{14}$[268] appears to involve six electrons in these orbitals. $[Mo_3(\mu_3\text{-}O)(\mu\text{-}Cl)_3(\mu\text{-}O_2CMe)_3(H_2O)_3]^{2+}$,[269] $[Mo_3(\mu_3\text{-}O)(\mu_2\text{-}Cl)_3(\mu\text{-}O_2CMe)_2Cl_5]^{2-}$,[261,270] $[Mo_3(\mu_3\text{-}O)(\mu_2\text{-}X)_3(\mu\text{-}O_2CR)_3Cl_3]^-$, (X = Cl, R = Me;[261] X = Br, R = H[271]) and $[Mo_3(\mu_3\text{-}CMe)_3(\mu_2\text{-}Br)_3(\mu\text{-}O_2CMe)_3(H_2O)_3]^+$ [272] are all based on an $\{Mo_3\}^{10+}$ core.

Type **C** (Mo_3X_{11}) clusters occur for $[Mo_3(\mu_3\text{-}O)(\mu_3\text{-}OR)(\mu\text{-}OR)_3(OR)_6]$ (R = Pr^i, CH_2CMe_3).[218,273,274] These species may be considered to involve an $\{Mo_3\}^{12+}$ centre and involve Mo—Mo separations of *ca.* 2.53 Å; their electronic structure has been described and related to their electronic spectra.[275] These species are obtained in the simple conproportionation reactions between the $[Mo_2(OR)_6]$ and $[MoO(OR)_4]$ derivatives and this route has permitted the synthesis of the *triangulo*-Mo_2W cluster.[274] The related systems $[Mo_3(\mu_3\text{-}S)_2(\mu\text{-}Cl)_3Cl_6]^{3-}$ and $[Mo_3(\mu_3\text{-}S)_2(\mu\text{-}Cl)_3\{S_2P(OEt)_2\}_3]$ involve an $\{Mo_3\}^{10+}$ centre and Mo—Mo bonds of *ca.* 2.6 Å[276] and a compound containing an $\{Mo_3(\mu_3\text{-}S)(\mu\text{-}S)_3\}$ core appended by diethoxythiophosphinate ligands has been shown to involve a triangular arrangement with two Mo—Mo separations of 2.808(1) and 2.839(1) Å and one longer distance of 3.337(1) Å.[276] The complex $[Mo_3S_8(N_2Me_2)_2]^{2-}$ has been synthesized by the reaction of $[Mo_2S_4(N_2Me_2)_2PPh_3]$ with LiSPh in MeCN. X-Ray structure analysis of the $[PPh_4]^+$ salt has revealed the anion to be a linear molybdenum trimer with two distinct coordination sites. The central metal atom enjoys pseudooctahedral geometry, through coordination of the N atoms of two end-on, linearly bound hydrazido(2−) groups and to four bridging sulfido groups and crystallographic symmetry which gives a rigorously planar Mo_2S_4 unit, with *trans* hydrazido-groups. The terminal molybdenum atoms are tetrahedrally coordinated to two bridging and two terminal sulfido groups. The unusual Mo—N and N—N bond lengths of 2.13(1) Å and 1.16(2) Å, respectively, suggest that the —$NNMe_2$ ligands are best described as isodiazene groups.[277]

Infinite chains of Mo_3 triangles occur in a variety of solid phases of molybdenum chalcogenides, including MMo_3S_3 (M = K, Rb, Cs) and $Tl_2Mo_2Se_6$; the electronic structure of the latter and related systems has been investigated.[278]

36.3.3.2 Tetrameric Clusters

Complexes involving four molybdenum atoms are known for several types of systems including pairs of $\{Mo_2O_4\}^{2+}$, $\{Mo_2(NR)_2S_2\}^{2+}$ and related systems (Section 36.4.3.5.4) and $[Mo_4S_4(NO)_4(CN)_8]^{8-}$ (Section 36.6.4). This section will describe moieties in which a framework of metal–metal bonds links all the molybdenum atoms.

$[Mo_4I_{11}]^{2-}$ has been synthesized by the attack of HI on $[Mo_2(O_2CMe)_4]$[279] and by the I_2 oxidation of $[Mo(CO)_4I_3]^-$;[280] the structure of this entity involves a distorted Mo_4 tetrahedron and may be viewed as containing an $\{Mo_4I_7\}^{2+}$ fragment of an $[Mo_4I_8]^{4+}$ cluster (Section 36.3.3.3) with an iodide bound as a terminal ligand to each molybdenum. $[Mo_4I_{11}]^{2-}$ reacts with diphos to produce $[Mo_4I_6(diphos)_2]$.[279] $[Mo_2Cl_4(PPh_3)_2(MeOH)_2]$, a typical quadruply bonded Mo^{II} dimer (Section 36.3.2.1), undergoes an interesting condensation reaction upon dissolution in benzene; thus, it loses MeOH to form $\{MoCl_2(PPh_3)\}_n$ which reacts with PR'_3 (R′ = Et, Bu^n) to form the corresponding, diamagnetic, tetramer $[Mo_4Cl_8(PR'_3)_4]$ (Figure 10). The Mo—Mo distances for the unbridged and bridged edges of the Mo_4 rectangle are 2.211(3) and 2.901(2) Å respectively, consistent with their formulation as Mo—Mo single and triple bonds respectively. Thus, these systems are molybdenum analogues of the cyclobutadiynes.[281] Alternative procedures may be used to prepare these and analogous tetramers; treatment of $K_4[Mo_2Cl_8]$ with the stoichiometric amount of PR'_3 in refluxing MeOH, or reaction of $[Mo_2(O_2CMe)_4]$ with $AlCl_3$ and PR'_3 (2:4:4) in THF, produces these systems and other derivatives corresponding to the general formula $[Mo_4X_8L_4]$ (X = Cl, Br, I; L = neutral donor ligand, *e.g.* PR'_3, PPh_3 MeOH, THF, RCN) have been isolated.[282] Also, $[Mo_4Cl_8\{P(OMe)_3\}_4]$ is formed when $[Mo_2Cl_4\{P(OMe)_3\}_4]$ is dissolved in CH_2Cl_2, or when $K_4[Mo_2Cl_8]$ reacts with $P(OMe)_3$ in MeOH.[283] The relationship of these tetrameric structures to β-$MoCl_2$ has been considered and the reaction of $[Mo_4Cl_8(PR'_3)_4]$ with $[Mo(CO)_4Cl_2]$ or $[Mo(CO)_6]$ in refluxing chlorobenzene shown to lead to phosphine extraction and the formation of the condensation products $[Mo_4Cl_8(PR_3)_2]_x$, for which the formulation $x = 2$ is favoured.[284] $[Mo_4X_3(OPr^i)_9]^-$ (X = Cl, Br), which have a structure similar to that of $[Mo_4F_4(OBu^t)_4]$ (Figure 8), react with MeCOX to form $[MoX_4(OPr^i)_8]$; an alternative preparative procedure is the reaction of $[Mo_2(OPr^i)_6]$ and MeCOX (1:2). The chloride derivative has a structure based on a square of molybdenum atoms, some 2.37 Å apart, with eight μ-OPr^i groups and four terminal chloride ligands. This system may be viewed as a molybdenum analogue of cyclobutadiene. However, the bromide involves a butterfly arrangement of molybdenum atoms with five Mo—Mo distances of *ca.* 2.5 Å and one of 3.1 Å; the bromide ligands are all terminal and two of the OPr^i groups are terminal, two μ- and three μ_3-bridging.[285]

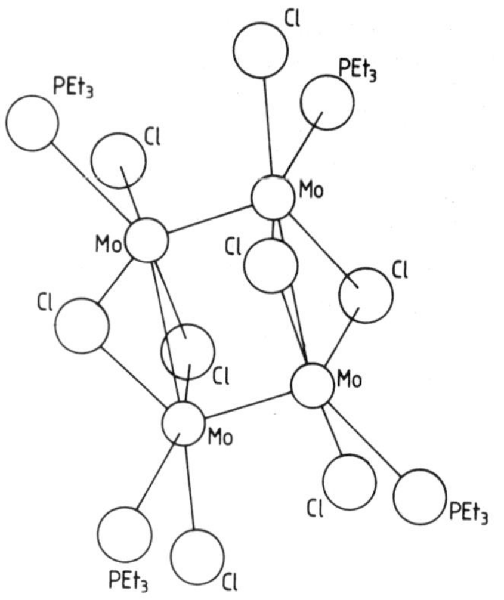

Figure 10 Structure of $[Mo_4Cl_8(PEt_3)_4]$[281]

$Ba_{1.13}Mo_8O_{16}$ contains two types of Mo_4O_{16} subunits which differ most significantly in their Mo—Mo distances and this difference can be correlated with the number of electrons involved in cluster bonding.[286,287] $[MoN(OBu^t)_3]$ reacts with $[Mo_2(OPr^i)_6]$ in hexane/Pr^{i}OH to form $[Mo_4(\mu_3\text{-}N)_2(\mu\text{-}OPr^i)_2(OPr^i)_{10}]$, the structure of which is shown in Figure 11 and involves contiguous Mo—Mo separations of 2.918(1), 2.552(1) and 2.918(1) Å.[288] The aquated $\{Mo_4O_4\}^{5+}$ ion has been claimed[256] and considered to possess a cubane-like structure, as identified in its sulfur analogue (*vide supra*).

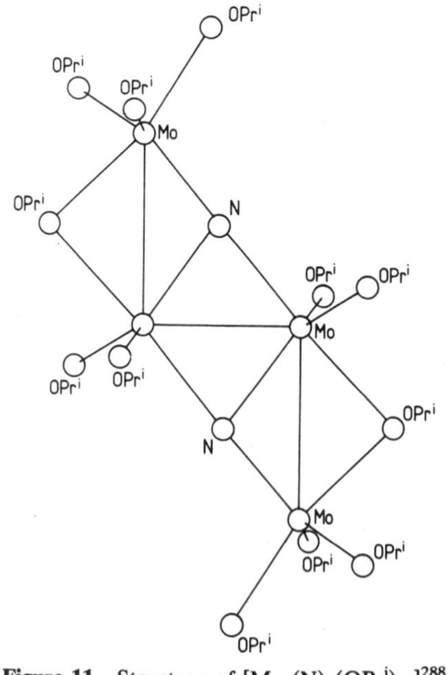

Figure 11 Structure of $[Mo_4(N)_2(OPr^i)_{12}]$[288]

Mo_4 clusters have been identified in many solid phases of molybdenum chalcogenides including $PbMo_6S_8$,[278] MoSX (X = Cl, Br, I)[289] and $GaMo_4(YY')$ (Y, Y′ = S, Se, Te);[290] the physical properties of these systems are of particular significance. Tetrahedral arrays of molybdenum atoms occur in $\{Mo_4S_4\}^{n+}$ cubane-like clusters which have been identified with a variety of ligands appended to the cluster. The number of electrons involved in the metal–metal bonding framework of these systems is variable and the most extensive, homologous, series is $[(Pr^iC_5H_4)_4Mo_4S_4]^{n+}$ ($n = 0, 1, 2$). These complexes are related by reversible, one-electron, redox reactions and oxidation causes only a slight decrease in the volume of the Mo_4 tetrahedra (Mo—Mo = 2.9–2.8 Å). The electronic structure of these systems has been discussed with reference to the MO scheme originally presented for $[Cp_4Fe_4S_4]$ and the PES data recorded for the neutral compound. The 12 electrons available for metal–metal bonding in $[(Pr^iC_5H_4)_4Mo_4S_4]$ are assigned to a_1, e and t_2 Mo—Mo bonding orbitals of the T_d cluster, with the a_1 and e orbitals being strongly bonding, but the t_2 orbital is less stable as manifest by its relatively low ionization potential.[291] $[Mo_4S_4(CN)_{12}]^{8-}$ also possesses a regular Mo_4 tetrahedron with 12 electrons in the Mo—Mo bonding orbitals and involves Mo—Mo separations of 2.855(1) Å.[259,292] An $\{Mo_4S_4\}^{5+}$ cluster has been prepared as the aquated ion and the complexes $[Mo_4S_4(NCS)_{12}]^{7-}$ [256] and $[Mo_4S_4(edta)_2]^{3-}$ [293] have been isolated; the latter complex approximates to S_4 symmetry with Mo—Mo distances which average 2.808 Å, possesses one unpaired electron, and is capable of undergoing a quasi-reversible one-electron oxidation or reduction. $\{Mo_4S_4\}^{6+}$ has been prepared as the aquated species and crystallized as its $[Mo_4S_4(NCS)_{12}]^{6-}$ salt; the central cluster possesses 10 electrons for metal–metal bonding and the observed change in structure from T_d to C_{3v} symmetry is consistent with a Jahn–Teller type distortion removing the degeneracy of the t_2 orbital to give a filled e and an empty a_2 orbital. The Mo_4 unit is a triangular pyramid with slant edges of length 2.791(1) Å and basal edges of 2.869(1) Å.[294] $[Mo_4S_4(S_2CNEt_2)_6]$ also involves an $\{Mo_4S_4\}^{6+}$ cubane-like cluster in which two of the dithiocarbamate ligands each form a bridge between the two molybdenum atoms on the opposite sides of the cube, the remaining four each coordinate to one molybdenum atom in a bidentate mode; the Mo—Mo distances range from 2.732(5) Å for two Mo atoms bridged *via* the same dithiocarbamate, to an average value of 2.861(6) Å for the other four Mo—Mo distances.[295] The cubane-like $\{Mo_4S_4\}^{6+}$ aquated ion is converted by aerial oxidation to the triangular $\{Mo_3S_4\}^{4+}$ aquated species (Section 36.3.3.1).[294]

36.3.3.3 Hexameric Clusters and Related Extended Arrays

The halides of molybdenum(II) were first described in 1848[296] and the structural characterization of the cluster species accomplished a century later for $[Mo_6Cl_8(OH)_4]\cdot 14H_2O$ and $[NH_4]_2[Mo_6Cl_{14}]\cdot H_2O$.[297] In idealized cubic symmetry, the $[Mo_6Cl_{14}]^{2-}$ ion consists of an octahedron of metal atoms (Mo—Mo is *ca.* 2.6 Å), ligated by eight μ_3 and six terminal chlorine atoms. This $\{Mo_6X_8\}^{4+}$ (X = Cl, Br, I) moiety has been demonstrated as a constituent of a large number of compounds,[298] including the dihalides Mo_6X_{12} which are isomorphous and possess a central Mo_6X_8 unit ligated by four μ and two terminal halides.[299] The reduction of higher molybdenum halides by aluminum in the appropriate $NaAlX_4$ (X = Cl, Br) melt provides a convenient and safe synthesis of the corresponding molybdenum(II) chloride or bromide in good yield.[300] The coordination chemistry of these $\{Mo_6X_8\}^{4+}$ species is well developed and includes the ligation of other halides, carboxylates, thiocyanates, alkoxide ions and a wide variety of neutral donor molecules at the terminal positions.[298,301] The bonding in these clusters has been defined and spectroscopic studies have shown that the $[Mo_6X_{14}]^{2-}$ (X = Cl, Br) ions are luminescent; these clusters undergo a one-electron oxidation and reduction and[303] electrogenerated $[Mo_6Cl_{14}]^{n-}$ ($n = 1$ or 3) ions react with electroactive donors or acceptors, respectively, to produce the luminescent state of $[Mo_6Cl_{14}]^{2-}$.[304]

The ions $[Mo_4I_{11}]^{2-}$ [279,280] and $[Mo_5Cl_{13}]^{2-}$ [305] may be considered as fragments of the $\{Mo_6X_8\}^{4+}$ cluster; the former involves a distorted tetrahedron, and the latter a square-based pyramid, of metal atoms. The mixed-metal halide cluster cations $\{Ta_5MoCl_{12}\}^{3+}$ and $\{Ta_4Mo_2Cl_{12}\}^{4+}$ have been obtained by reduction of $TaCl_5$–$MoCl_5$ mixtures with aluminum in fused $NaAlCl_4$–$AlCl_3$ at 325 °C and these cations are considered to be molybdenum-substituted derivatives of $\{Ta_6Cl_{12}\}^{n+}$ clusters.[306]

Substitution of sulfide and selenide into $\{Mo_6X_8\}^{4+}$ (X = Cl, Br, I) centres has been achieved in a host of solid phases and cores such as $\{Mo_6Cl_7S\}^{3+}$, $\{Mo_6Br_8S_2\}$, $\{Mo_6I_8Y_2\}$ (Y = S, Se), $\{Mo_6Br_6S_3\}$, $\{Mo_6Cl_2S_6\}$[290,307] and $MMo_6(S_{1-x}Se_x)_8$ (M = La, Yb, Sm, Ca, Sr, Ba, Eu, Sn, Pb)[308] have been characterized. Chevrel phases, *i.e.* ternary molybdenum chalcogenides, $M_xMo_6X_8$ (typically M = Pb, Sn or a rare earth element; X = S or Se) are of particular interest because of their physical properties—high superconducting critical temperatures, high critical fields and coexistence of magnetic order and superconductivity. Considerable flexibility in composition exists for these phases since, in addition to the variations indicated above, the ratio of Mo:X can be increased from the ideal 6:8 value. The fundamental structural unit in the Chevrel phases is an Mo_6 cluster and these units are coupled together to give linear arrays of the general composition $Mo_{3n}X_{3n+2}$ ($n = 2, 3, 4 \ldots \infty$; X = S, Se, Te).[309] A combined MO and crystal orbital analysis[310] has been presented for these systems and for $NaMo_4O_6$, which is also a metallic, infinite-chain polymer, derived from the condensation of Mo_6 octahedra.[311] Oxygen-donor ligands will also support discrete Mo_6 clusters; thus, Mo_6Cl_{12} when refluxed with NaOMe in MeOH forms $Na_2[\{Mo_6Cl_8\}(OMe)_6]$ and eventually $Na_2[\{Mo_6(OMe)_8\}\{OMe\}_6]$.[312]

36.3.4 REFERENCES

1. D. D. Wagman, W. H. Evans, V. B. Parker, I. Halow, S. M. Bailey and R. H. Schumm, *Nat. Bur. Stand. Tech. Note* 270-6, 1973. Washington, DC; E. R. Plante and A. B. Sessoms, *J. Res. Nat. Bur. Stand., Sect. A*, 1973, **77,** 237.
2. S. K. Gupta, R. M. Atkins and K. A. Gingerich, *Inorg. Chem.*, 1978, **17,** 3211.
3. Yu. M. Efremov, A. N. Samoilova, V. B. Kozhukhovsky and L. V. Gurvich, *J. Mol. Spectrosc.*, 1978, **73,** 430.
4. B. E. Bursten, F. A. Cotton and M. B. Hall, *J. Am. Chem. Soc.*, 1980, **102,** 6348; P. M. Atha and I. H. Hillier, *Mol. Phys.*, 1982, **45,** 285.
5. F. A. Cotton, W. H. Ilsley and W. Kaim, *Inorg. Chem.*, 1979, **18,** 2717.
6. M. R. Churchill and P. H. Bird, *Inorg. Chem.*, 1968, **7,** 1545.
7. F. A. Cotton and R. A. Walton, 'Multiple Bonds Between Metal Atoms', Wiley-Interscience, New York, 1982.
8. M. H. Chisholm (ed.), 'Reactivity of Metal–Metal Bonds', ACS Symposium Series, no. 155, American Chemical Society, Washington, DC, 1981.
9. F. A. Cotton, *Polyhedron,* 1986, **5,** 3.
10. M. H. Chisholm, *Polyhedron,* 1983, **2,** 681; M. H. Chisholm, D. M. Hoffman and J. C. Huffman, *Chem. Soc. Rev.*, 1985, **14,** 69.
11. G. Pimblett, C. D. Garner and W. Clegg, *J. Chem. Soc., Dalton Trans.*, 1986, 1257.
12. J. V. Brencic and F. A. Cotton, *Inorg. Chem.*, 1969, **8,** 7.
13. T. A. Stephenson, E. Bannister and G. Wilkinson, *J. Chem. Soc.*, 1964, 2538.
14. F. A. Cotton and J. G. Norman, Jr., *J. Coord. Chem.*, 1971, **1,** 161.
15. E. Hochberg, P. Walks and E. H. Abbott, *Inorg. Chem.*, 1974, **13,** 1824.
16. R. J. Mureinik, *J. Inorg. Nucl. Chem.*, 1976, **38,** 1275.
17. A. W. Coleman, J. C. Green, A. J. Hayes, E. A. Seddon, D. R. Lloyd and Y. Niwa, *J. Chem. Soc., Dalton Trans.*, 1979, 1057.
18. G. A. Robbins and D. S. Martin, *Inorg. Chem.*, 1984, **23,** 2086.
19. A. Bino, F. A. Cotton and P. E. Fanwick, *Inorg. Chem.*, 1979, **18,** 1719.
20. F. Apfelbaun-Tibika and A. Bino, *Inorg. Chem.*, 1984, **23,** 2902.
21. F. A. Cotton, L. R. Falvello and C. A. Murillo, *Inorg. Chem.*, 1983, **22,** 382.
22. D. Lawton and R. Mason, *J. Am. Chem. Soc.*, 1965, **87,** 921; F. A. Cotton, Z. C. Mester and T. R. Webb, *Acta Crystallogr., Sect. B,* 1974, **30,** 2768.
23. F. A. Cotton and J. G. Norman, Jr., *J. Coord. Chem.*, 1971, **1,** 161.
24. F. A. Cotton, J. G. Norman, Jr., B. R. Stults and T. R. Webb, *J. Coord. Chem.*, 1976, **5,** 217.
25. F. A. Cotton, M. Extine and D. M. Gage, *Inorg. Chem.*, 1978, **17,** 172.
26. A. Bino, F. A. Cotton and P. E. Fanwick, *Inorg. Chem.*, 1980, **19,** 1215.
27. F. A. Cotton and J. L. Thompson, *Inorg. Chem.*, 1981, **20,** 3887.
28. F. A. Cotton and G. N. Mott, *Inorg. Chim. Acta,* 1983, **70,** 159.
29. C. D. Garner and R. G. Senior, *J. Chem. Soc., Dalton Trans.*, 1975, 1171.
30. F. A. Cotton and P. E. Fanwick, *Inorg. Chem.*, 1983, **22,** 1327.
31. K. Jansen, K. Dehnicke and D. Fenske, *Z. Naturforsch., Teil B,* 1985, **40,** 13.
32. L. Golic, I. Leban and P. Segedin, *Croat. Chem. Acta,* 1984, **57,** 565.
33. A. Bino and F. A. Cotton, *J. Am. Chem. Soc.*, 1980, **102,** 3014.
34. P. A. Koz'min, M. D. Surazhskaya and T. B. Larina, *Russ. J. Inorg. Chem.* (*Engl. Transl.*), 1980, **25,** 702.
35. D. M. Collins, F. A. Cotton and C. A. Murillo, *Inorg. Chem.*, 1976, **15,** 2950.
36. G. Holste, *Z. Anorg. Allg. Chem.*, 1978, **438,** 125.
37. G. S. Girolami, V. V. Mainz and R. A. Andersen, *Inorg. Chem.*, 1980, **19,** 805.
38. A. P. Ketteringham and C. Oldham, *J. Chem. Soc., Dalton Trans.*, 1973, 1067.
39. F. A. Cotton and J. G. Norman, Jr., *J. Am. Chem. Soc.*, 1972, **94,** 5967.
40. J. San Filippo, Jr. and H. J. Snaidoch, *Inorg. Chem.*, 1976, **15,** 2209.
41. F. A. Cotton and D. G. Lay, *Inorg. Chem.*, 1981, **20,** 935.
42. D. J. Santure and A. P. Sattelberger, *Inorg. Chem.*, 1985, **24,** 3477.
43. J. Ribas, G. Jugie and R. Poilblanc, *Transition Met. Chem.* (*Weinheim, Ger.*), 1983, **8,** 93.
44. C. D. Garner, I. H. Hillier, I. B. Walton and B. Beagley, *J. Chem. Soc., Dalton Trans.*, 1979, 1279.
45. M. H. Kelley and M. Fink, *J. Chem. Phys.*, 1982, **76,** 1407.
46. C. D. Garner, S. Parkes, I. B. Walton and W. Clegg, *Inorg. Chim. Acta,* 1978, **31,** L451; W. Clegg, C. D. Garner, S. Parkes and I. B. Walton, *Inorg. Chem.*, 1979, **18,** 2250.
47. A. Bino and F. A. Cotton, *Inorg. Chem.*, 1979, **18,** 1381.
48. D. M. Collins, F. A. Cotton and C. A. Murillo, *Inorg. Chem.*, 1976, **15,** 1861.
49. F. A. Cotton, W. H. Ilsley and W. Kaim, *Inorg. Chim. Acta,* 1979, **37,** 267.

50. L. Hocks, P. Durbut and P. Teyssie, *J. Mol. Catal.*, 1980, **7**, 75; J. Lamotte, D. Dideberg, L. Dupont and P. Durbut, *Cryst. Struct. Commun.*, 1981, **10**, 59.
51. J. D. Arenivar, V. V. Mainz, H. Ruben, R. A. Andersen and A. Zalkin, *Inorg. Chem.*, 1982, **21**, 2649.
52. M. B. Hursthouse and K. M. A. Malik, *Acta Crystallogr., Sect. B*, 1979, **35**, 2709.
53. F. A. Cotton, W. H. Ilsley and W. Kaim, *Inorg. Chem.*, 1980, **19**, 3586.
54. R. A. Jones and G. Wilkinson, *J. Chem. Soc., Dalton Trans.*, 1979, 472.
55. F. A. Cotton and G. N. Mott, *Organometallics*, 1982, **1**, 302.
56. P. A. Agaskar and F. A. Cotton, *Inorg. Chim. Acta*, 1984, **83**, 33.
57. V. V. Mainz and R. A. Anderson, *Inorg. Chem.*, 1980, **19**, 2165.
58. C. D. Garner and R. G. Senior, *J. Chem. Soc., Dalton Trans.*, 1976, 1041.
59. J. Telser and R. S. Drago, *Inorg. Chem.*, 1984, **23**, 1798.
60. F. A. Cotton, A. H. Reid, Jr. and W. Schwotzer, *Inorg. Chem.*, 1985, **24**, 3965.
61. D. M. Baird, P. E. Fanwick and T. Barwick, *Inorg. Chem.*, 1985, **24**, 3753.
62. E. Hochberg and E. H. Abbott, *Inorg. Chem.*, 1978, **17**, 506.
63. J. M. Mayer and E. H. Abbott, *Inorg. Chem.*, 1983, **22**, 2774.
64. A. R. Bowen and H. Taube, *Inorg. Chem.*, 1974, **13**, 2245.
65. A. Bino and D. Gibson, *J. Am. Chem. Soc.*, 1980, **102**, 4277.
66. M. J. Chetcuti, M. H. Chisholm, K. Folting, D. A. Haitko and J. C. Huffman, *J. Am. Chem. Soc.*, 1982, **104**, 2138.
67. F. A. Cotton, W. H. Ilsley and W. Kaim, *Inorg. Chem.*, 1981, **20**, 930.
68. F. A. Cotton, T. Inglis, M. Kilner and T. R. Webb, *Inorg. Chem.*, 1975, **14**, 2023.
69. F. A. Cotton, P. E. Fanwick, R. H. Niswander and J. C. Sekutowski, *J. Am. Chem. Soc.*, 1978, **100**, 4725; W. Clegg, C. D. Garner, L. Akhter and M. H. Al-Samman, *Inorg. Chem.*, 1983, **22**, 2466.
70. F. A. Cotton, R. H. Niswander and J. C. Sekutowski, *Inorg. Chem.*, 1978, **17**, 3541.
71. F. A. Cotton, S. A. Koch and M. Millar, *Inorg. Chem.*, 1978, **17**, 2087.
72. F. A. Cotton and W. H. Ilsley, *Inorg. Chem.*, 1981, **20**, 572.
73. H. P. M. M. Ambrosius, F. A. Cotton, L. R. Falvello, H. T. J. M. Hintzen, T. J. Melton, W. Schwotzer, M. Thomas and J. G. M. van der Linden, *Inorg. Chem.*, 1984, **23**, 1611.
74. R. A. Jones, J. G. Lasch, N. C. Norman, B. R. Whittlesey and T. C. Wright, *J. Am. Chem. Soc.*, 1983, **105**, 6184.
75. F. A. Cotton, P. E. Fanwick, R. H. Niswander and J. C. Sekutowski, *Acta Chem. Scand., Ser. A*, 1978, **32**, 663.
76. J. H. Burk, G. E. Whitewell, II, J. T. Lemley and J. M. Burlitch, *Inorg. Chem.*, 1983, **22**, 1036.
77. F. A. Cotton, B. A. Frenz, E. Pedersen and T. R. Webb, *Inorg. Chem.*, 1975, **14**, 391.
78. J. Ribas, R. Poilblanc, C. Sourisseau, X. Solans, J. L. Brianso and C. Miravitlles, *Transition Met. Chem. (Weinheim, Ger.)*, 1983, **8**, 244.
79. F. A. Cotton and J. R. Pipal, *J. Am. Chem. Soc.*, 1971, **93**, 5441.
80. F. A. Cotton, S. A. Koch, A. J. Schultz and J. M. Williams, *Inorg. Chem.*, 1978, **17**, 2093.
81. F. A. Cotton, B. E. Hanson, W. H. Ilsley and G. W. Rice, *Inorg. Chem.*, 1979, **18**, 2713.
82. J. V. Brencic, I. Leban and P. Segedin, *Z. Anorg. Allg. Chem.*, 1976, **427**, 85.
83. J. V. Brencic, L. Golic and P. Segedin, *Inorg. Chim. Acta*, 1982, **57**, 247.
84. P. A. Agaskar and F. A. Cotton, *Inorg. Chem.*, 1984, **23**, 3383.
85. J. V. Brencic, L. Golic, I. Leban and P. Segedin, *Monatsh. Chem.*, 1979, **110**, 1221.
86. F. A. Cotton and P. E. Fanwick, *Acta Crystallogr., Sect. B*, 1980, **36**, 457.
87. M. H. Chisholm, K. Folting, J. C. Huffman and R. J. Tatz, *J. Am. Chem. Soc.*, 1984, **106**, 1153.
88. A. Bino, F. A. Cotton and P. E. Fanwick, *Inorg. Chem.*, 1979, **18**, 3558.
89. F. A. Cotton, J. M. Troup, T. R. Webb, D. H. Williamson and G. Wilkinson, *J. Am. Chem. Soc.*, 1974, **96**, 3824.
90. G. S. Girolami, V. V. Mainz, R. A. Anderson, S. H. Volmer and V. W. Day, *J. Am. Chem. Soc.*, 1981, **103**, 3953.
91. R. A. Jones, K. W. Chiu, G. Wilkinson, A. M. R. Galas and M. B. Hursthouse, *J. Chem. Soc., Chem. Commun.*, 1980, 408.
92. F. A. Cotton and B. E. Hanson, *Inorg. Chem.*, 1978, **17**, 3237.
93. B. E. Bursten, F. A. Cotton, A. H. Cowley, B. E. Hanson, M. Lattman and G. G. Stanley, *J. Am. Chem. Soc.*, 1979, **101**, 6244.
94. F. A. Cotton, L. R. Falvello, S. Han and W. Wang, *Inorg. Chem.*, 1983, **22**, 4106.
95. G. Holste, *Z. Anorg. Allg. Chem.*, 1976, **425**, 57.
96. D. E. Steele and T. A. Stephenson, *Inorg. Nucl. Chem. Lett.*, 1973, **9**, 777.
97. P. Vella and J. Zubieta, *J. Inorg. Nucl. Chem.*, 1978, **40**, 477.
98. L. Ricard, P. Karagiannidis and R. Weiss, *Inorg. Chem.*, 1973, **12**, 2179; J. A. Goodfellow and T. A. Stephenson, *Inorg. Chim. Acta*, 1980, **44**, L45.
99. L. Ricard, J. Estienne and R. Weiss, *J. Chem. Soc., Chem. Commun.*, 1972, 906; *Inorg. Chem.*, 1973, **12**, 2182.
100. R. D. Bereman, D. M. Baird and C. G. Moreland, *Polyhedron*, 1983, **2**, 59.
101. F. A. Cotton, B. W. S. Kolthammer and G. N. Mott, *Inorg. Chem.*, 1981, **20**, 3890.
102. J. P. Collman, C. E. Barnes and K. L. Woo, *Proc. Natl. Acad. Sci. USA*, 1983, **80**, 7684; J. P. Collman and K. L. Woo, *Proc. Natl. Acad. Sci. USA*, 1984, **81**, 2592.
103. R. Benn, A. Rufiniska and G. Schoth, *J. Organomet. Chem.*, 1981, **217**, 91.
104. J. V. Brencic, I. Leban and P. Segedin, *Z. Anorg. Allg. Chem.*, 1978, **444**, 211; J. V. Brencic and L. Golic, *Inorg. Chim. Acta*, 1978, **29**, L281.
105. E. Hochberg and E. H. Abbott, *Inorg. Chem.*, 1978, **17**, 506.
106. S. A. Best, T. J. Smith and R. A. Walton, *Inorg. Chem.*, 1978, **17**, 99.
107. P. A. Agaskar, F. A. Cotton, D. R. Derringer, G. L. Powell, D. R. Root and T. J. Smith, *Inorg. Chem.*, 1985, **24**, 2786 and references therein.

108. N. F. Cole, D. R. Derringer, E. A. Fiore, D. J. Knoechel, R. K. Schmitt and T. J. Smith, *Inorg. Chem.*, 1985, **24,** 1978.
109. J. F. Fraser, A. McVitie and R. D. Peacock, *J. Chem. Res. (S)*, 1984, 420.
110. F. L. Campbell, III, F. A. Cotton and G. L. Powell, *Inorg. Chem.*, 1985, **24,** 4384.
111. B. Heyn and C. Haroske, *Z. Chem.*, 1972, **12,** 338.
112. E. Kurras, H. Mennenga, G. Öhme, U. Rosenthal and G. Engelhardt, *J. Organomet. Chem.*, 1975, **84,** C13.
113. M. H. Chisholm and I. P. Rothwell, *J. Am. Chem. Soc.*, 1980, **102,** 5950.
114. R. A. Anderson, R. A. Jones and G. Wilkinson, *J. Chem. Soc., Dalton Trans.*, 1978, 446; R. A. Anderson, R. A. Jones, G. Wilkinson, M. B. Hursthouse and K. M. A. Malik, *J. Chem. Soc., Chem. Commun.*, 1977, 283.
115. R. A. Jones and G. Wilkinson, *J. Chem. Soc., Dalton Trans.*, 1979, 472.
116. C. D. Garner, R. G. Senior and T. J. King, *J. Am. Chem. Soc.*, 1976, **98,** 647.
117. V. Katovic, J. L. Templeton, R. J. Hoxmeier and R. E. McCarley, *J. Am. Chem. Soc.*, 1975, **97,** 5300; V. Katovic and R. E. McCarley, *J. Am. Chem. Soc.*, 1978, **100,** 5586.
118. R. L. Luck and R. H. Morris, *J. Am. Chem. Soc.*, 1984, **106,** 7978.
119. W. C. Trogler, C. D. Courman, H. B. Gray and F. A. Cotton, *J. Am. Chem. Soc.*, 1977, **99,** 2993; R. J. H. Clark and N. R. D'Urso, *J. Am. Chem. Soc.*, 1978, **100,** 3088.
120. K. J. Cavell, C. D. Garner, G. Pilcher and S. Parkes, *J. Chem. Soc., Dalton Trans.*, 1979, 1714.
121. J. G. Norman, Jr. and H. J. Kolari, *J. Chem. Soc., Chem. Commun.*, 1975, 649; J. G. Norman, H. J. Kolari, H. B. Gray and W. C. Trogler, *Inorg. Chem.*, 1977, **16,** 987; M. Benard, *J. Am. Chem. Soc.*, 1978, **100,** 2354.
122. I. H. Hillier, C. D. Garner, G. R. Mitcheson and M. F. Guest, *J. Chem. Soc., Chem. Commun.*, 1978, 204; P. M. Atha, I. H. Hillier and M. F. Guest, *Mol. Phys.*, 1982, **46,** 437.
123. T. Ziegler, *J. Am. Chem. Soc.*, 1985, **107,** 4453.
124. J. C. Green and A. J. Hayes, *Chem. Phys. Lett.*, 1975, **31,** 306; F. A. Cotton, J. G. Norman, Jr., B. R. Stults and T. R. Webb, *J. Coord. Chem.*, 1976, **5,** 217; D. L. Lichtenberger and C. H. Blevins, II, *J. Am. Chem. Soc.*, 1984, **106,** 1636.
125. C. D. Garner, I. H. Hillier, A. A. MacDowell, I. B. Walton and M. F. Guest, *J. Chem. Soc., Faraday Trans. 2,* 1979, **75,** 485; C. D. Garner, I. H. Hillier, M. J. Knight, A. A. MacDowell, I. B. Walton and M. F. Guest, *J. Chem. Soc., Faraday Trans. 2,* 1980, **76,** 885.
126. F. A. Cotton, J. L. Hubbard, D. L. Lichtenberger and I. Shim, *J. Am. Chem. Soc.*, 1982, **104,** 679.
127. G. M. Bancroft, E. Pellach, A. P. Sattelberger and K. W. McLaughlin, *J. Chem. Soc., Chem. Commun.*, 1982, 752.
128. E. M. Kober and D. L. Lichtenberger, *J. Am. Chem. Soc.*, 1985, **107,** 7199.
129. P. M. Atha, P. C. Ford, C. D. Garner, A. A. MacDowell, I. H. Hillier, M. F. Guest and V. R. Saunders, *Chem. Phys. Lett.*, 1981, **84,** 172; P. M. Atha, J. C. Campbell, C. D. Garner, I. H. Hillier and A. A. MacDowell, *J. Chem. Soc., Dalton Trans.*, 1983, 1085.
130. R. J. H. Clark and M. L. Franks, *J. Am. Chem. Soc.*, 1975, **97,** 2691.
131. B. Hutchinson, J. Morgan, C. B. Cooper, Y. Mathey and D. F. Shriver, *Inorg. Chem.*, 1979, **18,** 2048.
132. F. A. Cotton, D. S. Martin, P. E. Fanwick, T. J. Peters and T. R. Webb, *J. Am. Chem. Soc.*, 1976, **98,** 4681.
133. P. E. Fanwick, D. S. Martin, F. A. Cotton and T. R. Webb, *Inorg. Chem.*, 1977, **16,** 2013.
134. D. S. Martin, R. A. Newman and P. E. Fanwick, *Inorg. Chem.*, 1979, **18,** 2511.
135. F. L. Campbell, III, F. A. Cotton and G. L. Powell, *Inorg. Chem.*, 1985, **24,** 177.
136. P. A. Agaskar, F. A. Cotton, I. F. Fraser and R. D. Peacock, *J. Am. Chem. Soc.*, 1984, **106,** 1851; S. Christie, I. F. Fraser, A. McVitie and R. D. Peacock, *Polyhedron,* 1986, **5,** 35.
137. R. D. Peacock and I. F. Fraser, *Inorg. Chem.*, 1985, **24,** 988.
138. M. D. Hopkins and H. B. Gray, *J. Am. Chem. Soc.*, 1984, **106,** 2468; M. D. Hopkins, T. C. Zietlow, V. M. Miskowski and H. B. Gray, *J. Am. Chem. Soc.*, 1985, **107,** 510; I. F. Fraser and R. D. Peacock, *Chem. Phys. Lett.*, 1983, **98,** 620; P. E. Fanwick, *Inorg. Chem.*, 1985, **24,** 258.
139. J. San Filippo, Jr., *Inorg. Chem.*, 1972, **11,** 3140.
140. S. F. Gheller, T. W. Hambley, R. T. C. Brownlee, M. J. O'Connor, M. R. Snow and A. G. Wedd, *J. Am. Chem. Soc.*, 1983, **105,** 1527; R. A. Grieves and J. Mason, *Polyhedron,* 1986, **5,** 415.
141. G. S. Girolani, V. V. Mainz and R. A. Anderson, *J. Am. Chem. Soc.*, 1982, **104,** 2041.
142. A. Sahajpal and P. Thornton, *Polyhedron,* 1984, **3,** 257.
143. I. I. Moiseev, A. E. Gekhman and I. V. Kalechits, *Kinet. Katal.*, 1980, **21,** 284.
144. R. A. Walton, Ref. 8, p. 207.
145. J. H. Baxendale, C. D. Garner, R. G. Senior and P. Sharpe, *J. Am. Chem. Soc.*, 1976, **98,** 637.
146. R. E. McCarley, J. L. Templeton, T. J. Colburn, V. Katovic and R. J. Hoxmeier, *Adv. Chem. Ser.*, 1976, **150,** 318.
147. T. W. Coffindaffer, G. P. Niccolai, D. Powell, I. P. Rothwell and J. C. Huffman, *J. Am. Chem. Soc.*, 1985, **107,** 3572.
148. A. Pernick and M. Ardon, *J. Am. Chem. Soc.*, 1975, **97,** 1255.
149. D. K. Erwin, G. L. Geoffroy, H. B. Gray, G. S. Hammond, E. I. Solomon, W. C. Trogler and A. A. Zagars, *J. Am. Chem. Soc.*, 1977, **99,** 3620.
150. A. Bino and F. A. Cotton, *Inorg. Chem.*, 1979, **18,** 1159.
151. F. A. Cotton and E. Pedersen, *Inorg. Chem.*, 1975, **14,** 399.
152. T. Zietlow, D. D. Klendworth, T. Nimry, D. J. Salmon and R. A. Walton, *Inorg. Chem.*, 1981, **20,** 947.
153. H. D. Glickman, A. D. Hamer, T. J. Smith and R. A. Walton, *Inorg. Chem.*, 1976, **15,** 2205; S. A. Best, T. J. Smith and R. A. Walton, *Inorg. Chem.*, 1978, **17,** 99.
154. W. C. Trogler and H. B. Gray, *Nouv. J. Chim.*, 1977, **1,** 475.
155. F. A. Cotton and B. J. Kalbacher, *Inorg. Chem.*, 1976, **15,** 522; S. S. Miller and A. Haim, *J. Am. Chem. Soc.*, 1983, **105,** 5624.
156. A. Bino and F. A. Cotton, *Angew. Chem., Int. Ed. Engl.*, 1979, **18,** 332; *idem., J. Am. Chem. Soc.*, 1979, **101,** 4150.
157. A. Bino and S. Luski, *Inorg. Chim. Acta*, 1984, **86,** L35.

158. F. A. Cotton, P. E. Fanwick and J. W. Fitch, III, *Inorg. Chem.*, 1978, **17,** 3254.
159. A. Bino and F. A. Cotton, *Angew. Chem., Int. Ed. Engl.*, 1979, **18,** 462.
160. J. Ribas, R. Poilblanc, C. Sourisseau, X. Solans, J. L. Brianso and C. Miravitller, *Transition Met. Chem. (Weinheim, Ger.)*, 1983, **8,** 244.
161. F. A. Cotton, M. W. Extine and R. H. Niswander, *Inorg. Chem.*, 1978, **17,** 692.
162. F. A. Cotton and G. L. Powell, *J. Am. Chem. Soc.*, 1984, **106,** 3371.
163. W. S. Harwood, J.-S. Qi and R. A. Walton, *Polyhedron,* 1986, **5,** 15 and references therein.
164. M. H. Chisholm and I. P. Rothwell, *Prog. Inorg. Chem.*, 1982, **29,** 1; M. H. Chisholm, *Polyhedron,* 1986, **5,** 25.
165. G. Yaguspky, W. Mowat, A. Shortland and G. Wilkinson, *Chem. Commun.*, 1970, 1369; F. Huq, W. Mowat, A. Shortland, A. C. Skapski and G. Wilkinson, *Chem. Commun.*, 1971, 1079; W. Mowat, A. Shortland, G. Yagupsky, N. J. Hill, M. Yagupsky and G. Wilkinson, *J. Chem. Soc., Dalton Trans.*, 1972, 533.
166. T. A. Albright and R. Hoffmann, *J. Am. Chem. Soc.*, 1978, **100,** 7736; A. Dedieu, T. A. Albright and R. Hoffmann, *J. Am. Chem. Soc.*, 1979, **101,** 3141; M. B. Hall, *J. Am. Chem. Soc.*, 1980, **102,** 2104; R. A. Kok and M. B. Hall, *Inorg. Chem.*, 1983, **22,** 728; T. Ziegler, *J. Am. Chem. Soc.*, 1983, **105,** 7543; K. W. Dobbs, M. M. Francl and W. J. Hehre, *Inorg. Chem.*, 1984, **23,** 24.
167. M. H. Chisholm and W. W. Reichert, *J. Am. Chem. Soc.*, 1974, **96,** 1249.
168. M. H. Chisholm, D. A. Haitko and C. A. Murillo, *Inorg. Synth.*, 1982, **21,** 51.
169. M. H. Chisholm, F. A. Cotton, B. A. Frenz, W. W. Reichert, L. W. Shive and B. R. Stults, *J. Am. Chem. Soc.*, 1976, **98,** 4469.
170. B. E. Bursten, F. A. Cotton, J. C. Green, E. A. Seddon and G. G. Stanley, *J. Am. Chem. Soc.*, 1980, **102,** 4579.
171. M. H. Chisholm, F. A. Cotton, C. A. Murillo and W. W. Reichert, *Inorg. Chem.*, 1977, **16,** 1801.
172. M. H. Chisholm, W. W. Reichert, F. A. Cotton and C. A. Murillo, *J. Am. Chem. Soc.*, 1977, **99,** 1652.
173. M. H. Chisholm and I. P. Rothwell, *J. Chem. Soc., Chem. Commun.*, 1980, 985; *J. Am. Chem. Soc.*, 1980, **102,** 5950.
174. T. W. Coffindaffer, I. P. Rothwell and J. C. Huffman, *J. Chem. Soc., Chem. Commun.*, 1983, 1249.
175. M. H. Chisholm, K. Folting, J. C. Huffman and I. P. Rothwell, *Organometallics,* 1982, **1,** 251.
176. M. Akiyama, M. H. Chisholm, F. A. Cotton, M. W. Extine and C. A. Murillo, *Inorg. Chem.*, 1977, **16,** 2407.
177. M. H. Chisholm, F. A. Cotton, M. W. Extine and C. A. Murillo, *Inorg. Chem.*, 1978, **17,** 2338; M. H. Chisholm and D. A. Haitko, *J. Am. Chem. Soc.*, 1979, **101,** 6784.
178. M. J. Chetcuti, M. H. Chisholm, K. Folting, D. A. Haitko, J. C. Huffman and J. Janos, *J. Am. Chem. Soc.*, 1983, **105,** 1163.
179. M. H. Chisholm, J. C. Huffman and R. J. Tatz, *J. Am. Chem. Soc.*, 1983, **105,** 2075.
180. M. H. Chisholm, K. Folting, J. C. Huffman and I. P. Rothwell, *Inorg. Chem.*, 1981, **20,** 1854.
181. M. H. Chisholm, K. Folting, J. C. Huffman and I. P. Rothwell, *Inorg. Chem.*, 1981, **20,** 1496.
182. M. H. Chisholm, D. A. Haitko, J. C. Huffman and K. Folting, *Inorg. Chem.*, 1981, **20,** 171; *Inorg. Chem.*, 1981, **20,** 2211.
183. T. P. Blatchford, M. H. Chisholm, K. Folting and J. C. Huffman, *Inorg. Chem.*, 1980, **19,** 3175.
184. T. W. Coffindaffer, I. P. Rothwell and J. C. Huffman, *Inorg. Chem.*, 1984, **23,** 1433; 1985, **24,** 1643; J. W. Coffindaffer, W. M. Westler and I. P. Rothwell, *Inorg. Chem.*, 1985, **24,** 4565.
185. T. W. Coffindaffer, I. P. Rothwell and J. C. Huffman, *Inorg. Chem.*, 1983, **22,** 3178.
186. T. W. Coffindaffer, I. P. Rothwell and J. C. Huffman, *Inorg. Chem.*, 1983, **22,** 2906.
187. K. J. Ahmed, M. H. Chisholm, K. Folting and J. C. Huffman, *Inorg. Chem.*, 1985, **24,** 4039.
188. M. H. Chisholm, F. A. Cotton, M. W. Extine and W. W. Reichert, *J. Am. Chem. Soc.*, 1978, **100,** 153.
189. M. H. Chisholm, F. A. Cotton, M. W. Extine and W. W. Reichert, *J. Am. Chem. Soc.*, 1978, **100,** 1727.
190. M. H. Chisholm, J. C. Huffman and C. C. Kirkpatrick, *Inorg. Chem.*, 1983, **22,** 1704.
191. M. H. Chisholm, F. A. Cotton, K. Folting, J. C. Huffman, A. L. Ratermann and E. S. Shamshoum, *Inorg. Chem.*, 1984, **23,** 4423.
192. M. H. Chisholm, K. Folting, J. C. Huffman and I. P. Rothwell, *Inorg. Chem.*, 1981, **20,** 2215.
193. M. H. Chisholm, K. Folting, J. C. Huffman and A. L. Ratermann, *Inorg. Chem.*, 1984, **23,** 613.
194. M. H. Chisholm, J. F. Corning, K. Folting, J. C. Huffman, A. L. Ratermann, I. P. Rothwell and W. E. Streib, *Inorg. Chem.*, 1984, **23,** 1037.
195. M. H. Chisholm, J. C. Huffman and R. L. Kelly, *J. Am. Chem. Soc.*, 1979, **101,** 7100.
196. M. H. Chisholm, D. L. Clark and J. C. Huffman, *Polyhedron,* 1985, **4,** 1203.
197. M. H. Chisholm, J. F. Corning and J. C. Huffman, *Inorg. Chem.*, 1982, **21,** 286.
198. M. H. Chisholm, J. F. Corning and J. C. Huffman, *Inorg. Chem.*, 1983, **22,** 38.
199. M. H. Chisholm, J. F. Corning and J. C. Huffman, *J. Am. Chem. Soc.*, 1983, **105,** 5924.
200. M. H. Chisholm, J. F. Corning and J. C. Huffman, *Inorg. Chem.*, 1984, **23,** 754; M. H. Chisholm, J. F. Corning, K. Folting and J. C. Huffman, *Polyhedron,* 1985, **4,** 383.
201. P. J. Blower, J. R. Dilworth and J. Zubieta, *Inorg. Chem.*, 1985, **24,** 2866.
202. M. J. Chetcuti, M. H. Chisholm, H. T. Chiu and J. C. Huffman, *J. Am. Chem. Soc.*, 1983, **105,** 1060; M. H. Chisholm, H. T. Chiu, K. Folting and J. C. Huffman, *Inorg. Chem.*, 1984, **23,** 4097.
203. J. A. Connor, G. Pilcher, H. A. Skinner, M. H. Chisholm and F. A. Cotton, *J. Am. Chem. Soc.*, 1978, **100,** 7738; K. J. Cavell, J. A. Connor, G. Pilcher, M. A. V. Riveiro da Silva, H. A. Skinner, Y. Virmani and M. T. Zafarani-Moatter, *J. Chem. Soc., Faraday Trans. 1,* 1981, **77,** 1585.
204. F. A. Cotton, G. G. Stanley, B. J. Kalbacher, J. C. Green, E. Seddon and M. H. Chisholm, *Proc. Natl. Acad. Sci. USA,* 1977, **74,** 3109; E. M. Kober and D. L. Lichtenberger, *J. Am. Chem. Soc.*, 1985, **107,** 7199.
205. M. H. Chisholm, *Faraday Symp. Chem. Soc.*, 1980, **14,** 194.
206. C. G. Young, M. Minelli, J. H. Enemark, G. Miessler, N. Janietz, H. Kauermann and J. Wachter, *Polyhedron,* 1985, **5,** 407; M. Minelli, J. H. Enemark, R. T. C. Brownlee, M. J. O'Connor and A. G. Wedd, *Coord. Chem. Rev.*, 1985, **68,** 169.
207. M. H. Chisholm, C. C. Kirkpatrick and J. C. Huffman, *Inorg. Chem.*, 1981, **20,** 871.
208. M. H. Chisholm, J. C. Huffman, A. L. Ratermann and C. A. Smith, *Inorg. Chem.*, 1984, **23,** 1596.
209. M. H. Chisholm, F. A. Cotton, M. W. Extine and R. L. Kelly, *J. Am. Chem. Soc.*, 1979, **101,** 7645.

210. M. H. Chisholm, J. C. Huffman, J. Leonelli and I. P. Rothwell, *J. Am. Chem. Soc.*, 1982, **104,** 7030; F. A. Cotton and W. Schwotzer, *Inorg. Chem.*, 1983, **22,** 387.
211. M. H. Chisholm, J. C. Huffman and N. S. Marchant, *J. Am. Chem. Soc.*, 1983, **105,** 6162.
212. M. H. Chisholm, K. Folting, J. C. Huffman and A. L. Ratermann, *Inorg. Chem.*, 1984, **23,** 2303.
213. M. H. Chisholm, K. Folting, J. C. Huffman and I. P. Rothwell, *J. Am. Chem. Soc.*, 1982, **104,** 4389.
214. M. H. Chisholm, F. A. Cotton, M. W. Extine and R. L. Kelly, *J. Am. Chem. Soc.*, 1978, **100,** 3354.
215. M. H. Chisholm, J. C. Huffman and R. L. Kelly, *Inorg. Chem.*, 1980, **19,** 2762.
216. D. C. Bradley, C. W. Newing, M. H. Chisholm, R. L. Kelly, D. A. Haitko, D. Little, F. A. Cotton and P. E. Fanwick, *Inorg. Chem.*, 1980, **19,** 3010.
217. M. H. Chisholm, K. Folting, J. C. Huffman, C. C. Kirkpatrick and A. L. Ratermann, *J. Am. Chem. Soc.*, 1981, **103,** 1305.
218. M. H. Chisholm, K. Folting, J. C. Huffman and C. C. Kirkpatrick, *J. Chem. Soc., Chem. Commun.*, 1982, 188; M. H. Chisholm, K. Folting, J. C. Huffman and C. C. Kirkpatrick, *Inorg. Chem.*, 1984, **23,** 1021.
219. M. H. Chisholm, J. C. Huffman, I. P. Rothwell, P. G. Bradley, N. Kress and W. H. Woodruff, *J. Am. Chem. Soc.*, 1981, **103,** 4945.
220. R. B. King and M. B. Bisnette, *J. Organomet. Chem.*, 1967, **8,** 287; R. J. Klingler, W. M. Butler and M. D. Curtis, *J. Am. Chem. Soc.*, 1978, **100,** 5034.
221. M. D. Curtis, L. Messerle, N. A. Fotinos and R. F. Gerlach, in 'Reactivity of Metal–Metal Bonds', ed. M. H. Chisholm, ACS Symposium Series, no. 155, American Chemical Society, Washington, DC, 1981, p. 221; R. F. Gerlach, D. N. Duffy and M. D. Curtis, in 'Proceedings of the Fourth International Conference on the Chemistry and Uses of Molybdenum', ed. H. F. Barry and P. C. H. Mitchell, Climax Molybdenum Company, Ann Arbor, Michigan, 1982, p. 39; E. P. Kyba, J. D. Mather, K. L. Hassett, J. S. McKennis and R. E. Davis, *J. Am. Chem. Soc.*, 1984, **106,** 5371; I. Bernal, H. Brunner, W. Meier, H. Pfisterer, J. Wachter and M. L. Ziegler, *Angew. Chem., Int. Ed. Engl.*, 1984, **23,** 438; O. J. Scherer, H. Sitzmann and G. Wolmershäuser, *Angew. Chem., Int. Ed. Engl.*, 1984, **23,** 968; K. Blechsmitt, H. Pfisterer, T. Zahn and M. Ziegler, *Angew. Chem., Int. Ed. Engl.*, 1985, **24,** 66.
222. J. Lewis, R. S. Nyholm and P. W. Smith, *J. Chem. Soc. (A)*, 1969, 57; I. C. Grey and P. W. Smith, *Aust. J. Chem.*, 1969, **22,** 121.
223. R. Saillant and R. A. D. Wentworth, *Inorg. Chem.*, 1969, **8,** 1226.
224. W. H. Delphin and R. A. D. Wentworth, *J. Am. Chem. Soc.*, 1973, **95,** 7920; *Inorg. Chem.*, 1974, **13,** 2037; W. H. Delphin, R. A. D. Wentworth and M. S. Matson, *Inorg. Chem.*, 1974, **13,** 2552.
225. R. Saillant, P. B. Jackson, W. E. Streib, K. Folting and R. A. D. Wentworth, *Inorg. Chem.*, 1971, **10,** 1453.
226. R. Saillant and R. A. D. Wentworth, *Inorg. Chem.*, 1969, **8,** 1226; A. P. Ginsberg, *J. Am. Chem. Soc.*, 1980, **102,** 111; W. C. Trogler, *Inorg. Chem.*, 1980, **19,** 697.
227. F. A. Cotton and B. J. Kalbacher, *Inorg. Chem.*, 1976, **15,** 522.
228. M. J. Bennett, J. V. Brencic and F. A. Cotton, *Inorg. Chem.*, 1969, **8,** 1060; F. A. Cotton, B. A. Frenz and Z. C. Mester, *Acta Crystallogr., Sect. B*, 1973, **29,** 1515; A. Bino and F. A. Cotton, *Angew. Chem., Int. Ed. Engl.*, 1979, **18,** 332; *J. Am. Chem. Soc.*, 1979, **101,** 4150; A. Bino, *Inorg. Chim. Acta*, 1985, **101,** L7.
229. A. Bino and S. Luski, *Inorg. Chim. Acta*, 1984, **86,** L35.
230. F. A. Cotton, P. C. W. Leung, W. J. Roth, A. J. Schultz and J. M. Williams, *J. Am. Chem. Soc.*, 1984, **106,** 117.
231. V. Katovic and R. E. McCarley, *Inorg. Chem.*, 1978, **17,** 1268; *J. Am. Chem. Soc.*, 1978, **100,** 5586.
232. H. D. Glicksman and R. A. Walton, *Inorg. Chem.*, 1978, **17,** 3197.
233. J. Latorre, J. Soto, P. Salagre and J. E. Sueiras, *Transition Met. Chem. (Weinheim, Ger.)*, 1984, **9,** 447; P. Salagre, J. E. Sueiras, X. Solans and G. Germain, *J. Chem. Soc., Dalton Trans.*, 1985, 2263.
234. B. G. Brandt and A. C. Skapski, *Acta Chem. Scand.*, 1967, **21,** 661.
235. M. H. Chisholm, W. W. Reichert and P. Thornton, *J. Am. Chem. Soc.*, 1978, **100,** 2744.
236. M. H. Chisholm, F. A. Cotton, M. W. Extine and W. W. Reichert, *Inorg. Chem.*, 1978, **17,** 2944.
237. R. S. Herrick, S. J. Nieter-Burgmayer and J. L. Templeton, *J. Am. Chem. Soc.*, 1983, **105,** 2599.
238. R. D. Adams, D. M. Collins and F. A. Cotton, *Inorg. Chem.*, 1974, **13,** 1086.
239. H. M. Spittle and W. Wardlaw, *J. Chem. Soc.*, 1929, 792.
240. B. E. Bursten, F. A. Cotton, M. B. Hall and R. C. Najjar, *Inorg. Chem.*, 1982, **21,** 302.
241. A. Bino, M. Ardon, I. Maor, M. Kaftory and Z. Dori, *J. Am. Chem. Soc.*, 1976, **98,** 7093.
242. A. Bino, F. A. Cotton and Z. Dori, *J. Am. Chem. Soc.*, 1981, **103,** 243; A. Bino, F. A. Cotton, Z. Dori and B. W. S. Kolthammer, *J. Am. Chem. Soc.*, 1981, **103,** 5779.
243. M. Ardon, A. Bino, F. A. Cotton, Z. Dori, M. Kaftory, B. W. S. Kolthammer, M. Kapon and G. Reisner, *Inorg. Chem.*, 1981, **20,** 4083.
244. A. Bino and D. Gibson, *Inorg. Chim. Acta*, 1982, **65,** L37; *idem. J. Am. Chem. Soc.*, 1981, **103,** 6741; 1982, **104,** 4383; *Inorg. Chem.*, 1984, **23,** 109.
245. F. A. Cotton, Z. Dori, D. O. Marler and W. Schwotzer, *Inorg. Chem.*, 1983, **22,** 3104; 1984, **23,** 4033; 1984, **23,** 4738; R. E. Marsh, F. A. Cotton and W. Schwotzer, *Inorg. Chem.*, 1985, **24,** 3486.
246. A. Birnbaum, F. A. Cotton, Z. Dori, D. O. Marler, G. M. Reisner, W. Schwotzer and M. Shaia, *Inorg. Chem.*, 1983, **22,** 2723.
247. A. Birnbaum, F. A. Cotton, Z. Dori and M. Kapon, *Inorg. Chem.*, 1984, **23,** 1617.
248. A. Bino, F. A. Cotton, Z. Dori and B. W. S. Kolthammer, *J. Am. Chem. Soc.*, 1981, **103,** 5779.
249. E. O. Schlemper, M. S. Hussain and R. K. Murmann, *Cryst. Struct. Commun.*, 1982, **11,** 89.
250. R. K. Murmann and M. E. Shelton, *J. Am. Chem. Soc.*, 1980, **102,** 3984; K. R. Rodgers, R. K. Murmann, E. O. Schlemper and M. E. Shelton, *Inorg. Chem.*, 1985, **24,** 1313; G. D. Hinch, D. E. Wycoff and R. K. Murmann, *Polyhedron*, 1986, **5,** 487; S. P. Cramer, P. K. Eidem, M. T. Paffett, J. R. Winkler, Z. Dori and H. B. Gray, *J. Am. Chem. Soc.*, 1983, **105,** 799.
251. A. Bino, F. A. Cotton and Z. Dori, *J. Am. Chem. Soc.*, 1978, **100,** 5252; E. Benory, A. Bino, D. Gibson, F. A. Cotton and Z. Dori, *Inorg. Chim. Acta*, 1985, **99,** 137.
252. A. Bino, F. A. Cotton and Z. Dori, *J. Am. Chem. Soc.*, 1979, **101,** 3842.

253. S. F. Gheller, T. W. Hambley, R. T. C. Brownlee, M. J. O'Connor, M. R. Snow and A. G. Wedd, *J. Am. Chem. Soc.*, 1983, **105,** 1527.
254. P. Kathirgamanathan, A. B. Soares, D. T. Richens and A. G. Sykes, *Inorg. Chem.*, 1985, **24,** 2950.
255. T. Shibahara, H. Hattori and H. Kuroya, *J. Am. Chem. Soc.*, 1984, **106,** 2710; P. Kathirgamanathan, M. Martinez and A. G. Sykes, *J. Chem. Soc., Chem. Commun.*, 1985, 1437.
256. P. Kathirgamanathan, M. Martinez and A. G. Sykes, *J. Chem. Soc., Chem. Commun.*, 1985, 953; *Polyhedron,* 1986, **5,** 505.
257. T. R. Halbert, K. McGauley, W.-H. Pan, R. Czernuszewicz and E. I. Stiefel, *J. Am. Chem. Soc.*, 1984, **106,** 1849; F. A. Cotton, Z. Dori, R. Llusar and W. Schwotzer, *J. Am. Chem. Soc.*, 1985, **107,** 6734; F. A. Cotton, R. Llusar, D. O. Marler, W. Schwotzer and Z. Dori, *Inorg. Chim. Acta,* 1985, **102,** L25; T. Shibahara and H. Kuroya, *Polyhedron,* 1986, **5,** 357; X. Lin, Y. Lin, J. Huang and J. Huang, *Youji Huaxue,* 1985, 326 (*Chem. Abstr.* 1986, **103,** 152 620k).
258. N. C. Howlader, G. P. Haight, Jr., T. W. Hambley, G. A. Lawrance, K. M. Rahmoeller and M. R. Snow, *Aust. J. Chem.*, 1983, **36,** 377.
259. A. Müller, R. Jostes, W. Eltzner, C.-S. Nie, E. Diemann, H. Bögge, M. Zimmermann, M. Dartmann, U. Reinsch-Vogell, S. Che, S. J. Cyvin and B. N. Cyvin, *Inorg. Chem.*, 1985, **24,** 2872.
260. A. Müller, S. Sarkar, R. G. Bhattacharyya, S. Pohl and M. Dartmann, *Angew. Chem., Int. Ed. Engl.*, 1978, **17,** 535; A. Müller, S. Pohl, M. Dartmann, J. P. Cohen, J. M. Bennett and R. M. Kirchner, *Z. Naturforsch, Teil B,* **34,** 434; A. Müller, A. Bhattacharyya and B. Pfefferhorn, *Chem. Ber.*, 1979, **112,** 778.
261. H. Jinling, S. Maoyu, J. Huang, H. Zhuang, L. Shaofong and L. Jiaxi, *Jiegou Huaxue,* 1982, **1,** 1 (*Chem. Abstr.*, 1983, **99,** 186 292u).
262. H. Jinling, S. Maoyu and L. Jiaxi, *Sci. Sin. Ser. B* (*Engl. Ed.*) 1983, **26,** 1121 (*Chem. Abstr.*, 1984, **100,** 183677*b*).
263. S. Maoyu, H. Jinling, L. Jiaxi and C. S. Lu, *Acta Crystallogr., Sect. C,* 1984, **40,** 759.
264. H. Keck, W. Kuchen, J. Mathow, B. Meyer, D. Mootz and H. Wunderlich, *Angew. Chem., Int. Ed. Engl.*, 1981, **20,** 975.
265. H. Keck, W. Kuchen, J. Mathow and H. Wunderlich, *Angew. Chem., Int. Ed. Engl.*, 1982, **21,** 929.
266. W. H. McCarroll, L. Katz and R. Ward, *J. Am. Chem. Soc.*, 1957, **79,** 5410; G. B. Ansell and L. Katz, *Acta Crystallogr.*, 1966; **21,** 482.
267. C. C. Torardi and R. E. McCarley, *J. Am. Chem. Soc.*, 1979, **101,** 3963; C. C. Torardi, Ph.D. Dissertation, Iowa State University, Ames, Iowa, 1981.
268. P. W. Betteridge, A. K. Cheetham, J. A. K. Howard, G. Jakubicki and W. H. McCarroll, *Inorg. Chem.*, 1984, **23,** 737.
269. A. Bino, F. A. Cotton and Z. Dori, *Inorg. Chim. Acta,* 1979, **33,** L133.
270. S. Maoyu, H. Jinling, L. Jiaxi and C. S. Lu, *Acta Crystallogr., Sect. C,* 1984, **40,** 761.
271. D. Wu, J. Huang and J. Huang, *Acta Crystallogr., Sect. C,* 1985, **41,** 888.
272. A. Birnbaum, F. A. Cotton, Z. Dori, M. Kapon, D. Marler, G. M. Reisner and W. Schwotzer, *J. Am. Chem. Soc.*, 1985, **107,** 2405.
273. M. H. Chisholm, K. Folting, J. C. Huffman and C. C. Kirkpatrick, *J. Am. Chem. Soc.*, 1981, **103,** 5967; 1984, **23,** 1021.
274. M. H. Chisholm, K. Folting, J. C. Huffman and E. M. Kober, *Inorg. Chem.*, 1985, **24,** 241.
275. M. H. Chisholm, F. A. Cotton, A. Fang and E. M. Kober, *Inorg. Chem.*, 1984, **23,** 749.
276. H. Jinling, S. Maoyu, L. Shixiong and L. Jiaxi, *Sci. Sin. Ser. B.* (*Engl. Ed.*), 1982, **25,** 1269 (*Chem. Abstr.*, 1983, **98,** 99 186y); L. Shaofang, L. Yuhui, H. Jinling, L. Jiaxi and H. Jianquan, *Jiegou Huaxue,* 1984, **3,** 147 (*Chem. Abstr.*, 1986, **103,** 62 905e); L. Shaofang, H. Jianquan, H. Jinling and L. Jiaxi, *Jiegou Huaxue,* 1984, **3,** 151 (*Chem. Abstr.*, 1986, **103,** 62 906d).
277. J. R. Dilworth, J. Zubieta and J. R. Hyde, *J. Am. Chem. Soc.*, 1982, **104,** 365.
278. J. Huster, G. Schippers and W. Bronger, *J. Less-Common Met.*, 1983, **91,** 333; A. Le Beuze, M. A. Makyoun, R. Lissillour and M. Chernette, *Bull. Soc. Chim. Fr.*, 1981, **9–10,** 397.
279. H. D. Glicksman and R. A. Walton, *Inorg. Chem.*, 1978, **17,** 3197.
280. S. Stensvad, B. J. Helland, M. W. Babich, R. A. Jacobson and R. E. McCarley, *J. Am. Chem. Soc.*, 1978, **100,** 6257.
281. R. N. McGinnis, T. R. Ryan and R. E. McCarley, *J. Am. Chem. Soc.*, 1978, **100,** 7900.
282. T. R. Ryan and R. E. McCarley, *Inorg. Chem.*, 1982, **21,** 2072.
283. F. A. Cotton and G. L. Powell, *Inorg. Chem.*, 1983, **22,** 871.
284. W. W. Beers and R. E. McCarley, *Inorg. Chem.*, 1985, **24,** 472; W. W. Beers and R. E. McCarley, *Inorg. Chem.*, 1985, **24,** 468.
285. M. H. Chisholm, R. J. Errington, K. Folting and J. C. Huffman, *J. Am. Chem. Soc.*, 1982, **104,** 2025.
286. C. C. Torardi and R. E. McCarley, *J. Solid State Chem.*, 1981, **37,** 393; R. E. McCarley, M. H. Luly, T. R. Ryan and C. C. Torardi, in 'Reactivity of Metal–Metal Bonds', ed. M. H. Chisholm, ACS Symposium Series, no. 155, American Chemical Society, Washington, DC, 1981, p. 41.
287. F. A. Cotton and A. Fang, *J. Am. Chem. Soc.*, 1982, **104,** 113.
288. M. H. Chisholm, K. Folting, J. C. Huffman, J. Leonelli, N. S. Marchant, C. A. Smith and L. C. E. Taylor, *J. Am. Chem. Soc.*, 1985, **107,** 3722.
289. H. Ben Yaich, J. C. Jegaden, M. Potel, R. Chevrel, M. Sergent, A. Berton, J. Chaussy, A. K. Rastogi and R. Tournier, *J. Solid State Chem.*, 1984, **51,** 212.
290. C. Perrin, M. Sergent, J. C. Pilet, F. Le Traon and A. Le Traon, *Bull. Soc. Sci. Bretagne,* 1982, **54,** 69.
291. J. A. Bandy, C. E. Davies, J. C. Green, M. L. H. Green, K. Prout and D. P. S. Rodgers, *J. Chem. Soc., Chem. Commun.*, 1983, 1395.
292. A. Müller, W. Eltzner, H. Bögge and R. Jostes, *Angew. Chem., Int. Ed. Engl.*, 1982, **21,** 795.
293. T. Shibahara, H. Kuroya, K. Matsumoto and S. Ooi, *J. Am. Chem. Soc.*, 1984, **106,** 789.
294. F. A. Cotton, M. P. Diebold, Z. Dori, R. Llusar and W. Schwotzer, *J. Am. Chem. Soc.*, 1985, **107,** 6735.
295. T. C. Mak, K. S. Jasim and C. Chieh, *Inorg. Chem.*, 1985, **24,** 1587.
296. L. Svanberg and H. Struve, *Philos. Mag.*, 1848, **33,** 524; W. Blomstrand, *J. Prakt. Chem.*, 1859, **77,** 88.

297. C. Brosset, *Ark. Kemi, Mineral. Geol.*, 1945, **20A,** no. 7; 1947, **25A,** no. 19; P. A. Vaughan, *Proc. Natl. Acad. Sci. USA,* 1950, **36,** 461.
298. J. H. Canterford and R. Colton, 'Halides of the Second and Third Row Transition Metals', Wiley, New York, 1968, pp. 99–106; D. L. Kepert, 'The Early Transition Metals', Academic, London, 1972, pp. 351–356 and references therein.
299. H. Schäfer, H.-G. Von Schnering, J. Tillack, F. Kuhnen, H. Wöhrle and H. Baumann, *Z. Anorg. Allg. Chem.*, 1967, **353,** 281.
300. W. C. Dorman and R. E. McCarley, *Inorg. Chem.*, 1974, **13,** 491.
301. G. Holste and H. Schäfer, *Z. Anorg. Allg. Chem.*, 1972, **391,** 263; P. C. Healy, D. L. Kepert, D. Taylor and A. H. White, *J. Chem. Soc., Dalton Trans.*, 1973, 646.
302. F. A. Cotton and T. E. Haas, *Inorg. Chem.*, 1964, **3,** 10.
303. H. Schäfer, H. Plautz and H. Bauman, *Z. Anorg. Allg. Chem.*, 1973, **401,** 63; D. L. Kepert, R. E. Marshall and D. Taylor, *J. Chem. Soc., Dalton Trans.*, 1974, 506.
304. A. W. Maverick and H. B. Gray, *J. Am. Chem. Soc.*, 1981, **103,** 1298; A. W. Maverick, J. S. Najdzionek, D. MacKenzie, D. G. Nocera and H. B. Gray, *J. Am. Chem. Soc.*, 1983, **105,** 1878; D. G. Nocera and H. B. Gray, *J. Am. Chem. Soc.*, 1984, **106,** 824; R. D. Mussell and D. G. Nocera, *Polyhedron,* 1986, **5,** 47.
305. K. Jödden, H. G. von Schnering and H. Schäfer, *Angew. Chem., Int. Ed. Engl.*, 1975, **14,** 570.
306. J. L. Meyer and R. E. McCarley, *Inorg. Chem.*, 1978, **17,** 1867.
307. J. B. Michel, *Energy Res. Abstr.*, 1980, **5,** 16250 (*Chem. Abstr.*, 1980, **93,** 196 803); J. B. Michel and R. E. McCarley, *Inorg. Chem.*, 1982, **21,** 1864; C. Perrin and M. Sergent, *J. Chem. Res.* (*S*), 1983, 38; C. Perrin, M. Potel and M. Sergent, *Acta Crystallogr., Sect. C,* 1983, **39,** 415.
308. J.-M. Tarascon, D. C. Johnson and M. J. Sienko, *Inorg. Chem.*, 1983, **22,** 3769; D. C. Johnson, J.-M. Tarason and M. J. Sienko, *Inorg. Chem.*, 1983, **22,** 3773; 1985, **24,** 2598; 1985, **24,** 2808.
309. R. Chevrel, M. Hirrien and M. Sergent, *Polyhedron,* 1986, **5,** 87 and references therein.
310. T. Hughbanks and R. Hoffman, *J. Am. Chem. Soc.*, 1983, **105,** 1150; 1983, **105,** 3528.
311. C. C. Torardi and R. E. McCarley, *J. Am. Chem. Soc.*, 1979, **101,** 1963.
312. P. Nannelli and B. P. Block, *Inorg. Chem.*, 1968, **7,** 2423; M. H. Chisholm, J. A. Heppert and J. C. Huffman, *Polyhedron,* 1984, **3,** 475.

36.4

Molybdenum(III), (IV) and (V)

C. DAVID GARNER and JOHN M. CHARNOCK
University of Manchester, UK

36.4.1 MOLYBDENUM(III)

36.4.1.1 Introduction

Molybdenum(III) is a relatively rare oxidation state and is easily oxidized to more stable states by air and other mild oxidants. However, recent work has shown the existence of many Mo^{III}

compounds in both aqueous and nonaqueous systems. Also, Mo^{III} may be involved in catalytic and biological reactions of molybdenum, such as dinitrogen reduction, but there is no firm evidence for this.

This section omits the aqueous chemistry of Mo^{III} (see Section 36.1.2), Mo—Mo triple bonds, including $[Mo_2X_9]^{3-}$ (see Chapter 36.3), $[CpMo(SR)_2]$ and related compounds, and Mo—S clusters (see Section 36.6.5).

Molybdenum trihalides may be prepared either by reduction of higher halides or by direct combination of the elements.[1] They all adopt extended lattice structures: MoF_3 has a rhombohedral structure similar to VF_3 and CrF_3;[2] $MoCl_3$ exists in two polymorphic forms in which Mo atoms occupy octahedral holes in cubic (α) or hexagonal (β) close-packed arrays of chlorines;[3] $MoBr_3$ and MoI_3 have structures containing polymeric strings made up of face-sharing MoX_6 octahedra.[4] The magnetic moments of the trihalides given in Table 1[5] show an increase from Cl to I, indicative of antiferromagnetic coupling between adjacent Mo atoms which decreases with the increase in the metal–metal distance.

Table 1 Magnetic Moments of Molybdenum(III) Complexes

Compound	μ_{eff} (BM)	*T* (K)	*Ref.*
$MoCl_3$	0.7	a	1
$MoBr_3$	1.2	a	1
MoI_3	1.4	a	1
$Mo(SPh)_3$	0.77	293	6
$Mo(SC_5H_{11})_3$	1.17	293	6
$K_3[MoCl_6]$	3.79	300	37
$K_3[Mo(NCS)_6]\cdot 4H_2O$	3.79	300	37
$[Mo(en)_3](NCS)_3$	3.65	300	34
$MoI_2(phen)_2]\cdot I$	3.69	293	15
$[Mo(acac)_3]$	3.82	302	23
$[Mo(S_2PF_2)_3]$	3.53	b	28
$K_4[Mo(CN)_7]\cdot 2H_2O$	1.73	298	35

[a] Room temperature. [b] Temperature not specified.

Compounds of the stoichiometry $Mo(SR)_3$ can be prepared by oxidation of Mo^0 complexes using thiols or disulfides.[6] Their spectroscopic properties and low magnetic moments, ranging from 0.77 to 1.17 BM (Table 1), indicate a polymeric structure with considerable metal–metal interaction.

36.4.1.2 Monomeric Complexes

36.4.1.2.1 Halide complexes

$K_3[MoF_6]$ can be prepared by reacting $K_3[MoCl_6]$ or MoO_3 with KF or KHF_2; the anion contains Mo octahedrally coordinated with Mo—F bond lengths of 2.00(2) Å.[7] $M^I_3[MoCl_6]$ and $M^I_3[MoBr_6]$ ($M^I = Li^+, K^+, NH_4^+$) have been obtained by electrolytic reduction of MoO_3 in HCl or HBr solution respectively, followed by addition of the appropriate alkali metal halide. Other products isolated from these reactions include salts of the $[MoCl_5(H_2O)]^{2-}$ and $[MoBr_5(H_2O)]^{2-}$ anions.[8] Crystal structure determinations of salts containing these ions all show that the Mo^{III} possesses simple octahedral coordination with typical average bond lengths of, for example, 2.452(3) Å for Mo—Cl in $[Ph_2HP(CH_2)_2PHPh_2]_3[MoCl_6]_2\cdot 12H_2O$,[9] 2.591(1) Å for Mo—Br in $[LH]_2[H_5O_2][MoBr_6]$ (L = 2-amino-4,6-dihydroxypyrimidine),[10] and 2.155(3) Å or 2.185(3) Å for Mo—O in the hydrolysis products $[NH_4]_2[MoCl_5(H_2O)]$ and $[NH_4]_2[MoBr_5(H_2O)]$, respectively.[11] Other monosubstituted halide anions, such as the nitrile complexes $[MoCl_5(NCR)]^{2-}$,[12] can be prepared by ligand substitution from $[MoX_6]^{3-}$ species. $[NEt_3H][MoCl_4(THF)_2]$, obtained from the reduction of $[MoCl_4(THF)_2]$ in the presence of NEt_3, involves octahedral stereochemistry around the Mo, with the THF molecules adopting a *trans* geometry. This complex can be used as a starting material in the synthesis of other disubstituted anions.[13] A *trans* geometry has also been established for the anion of $[SMe_3][MoBr_4(SMe_2)_2]$, formed in the reaction of $MoBr_4$ with Me_2S.[14]

A large number of neutral complexes of the form $[MoX_3L_3]$ (X = halide; L = neutral ligand) are known. These are easily prepared by direct combination of MoX_3 and L, or by substitution reactions of $[MoX_6]^{3-}$. They can also be formed by a disproportionation reaction of the corresponding Mo^{II} complex $[Mo(CO)_4X_2]$, with L acting as the solvent,[15] or by reduction of the appropriate Mo^{IV} complex $[MoX_4L_2]$. This last reaction has been used to prepare, in high yield, $[MoCl_3(THF)_3]$, which is a useful starting material for the synthesis of other Mo^{III} compounds.[16] Generally, these complexes adopt a *meridional* configuration, even for the tridentate macrocyclic polythiaether 1,4,7,10,13,16-hexathiacyclooctadecane, which forms polymeric chains.[17] However, the *facial* isomer is formed with the tridentate ligand 1,4,7-trimethyl-1,4,7-triazacyclononane.[18] Crystal structure determinations have confirmed the *meridional* geometry of $[MoCl_3(py)_3]$ and $[MoX_3(pic)_3]\cdot 0.5pic$ (X = Cl, Br; pic = 4-methylpyridine).[19]

Mo^{III} halide complexes formed with chelating diazadienes or diamines (N—N) include $[MoCl_3(CNMe)(N—N)]$ species; with aromatic chelates such as 1,10-phenanthroline or 2,2′-bipyridine, salts containing the cation *cis*-$[MoCl_2(N—N)_2]^+$ have been isolated.[20] Compounds of stoichiometry $[MoX_3L_4]$ are known for X = Cl, Br; L = RCN and X = Br; L_2 = 2,6-lutidine,[15] but it is not clear whether these involve seven-coordinate Mo^{III}.

The reaction of potassium hydrotris(3,5-dimethyl-1-pyrazolyl)borate, $K[HB(3,5\text{-}Me_2pz)_3]$, with $[MoCl_3(THF)_3]$ gives the yellow crystalline compound $[K(THF)_3][HB(3,5\text{-}Me_2pz)_3MoCl_3]$. Its magnetic susceptibility (3.69 BM), electronic spectrum and ESR spectrum are consistent with a d^3, $S=\frac{3}{2}$ ground state. An X-ray diffraction study of $[NEt_4][HB(3,5\text{-}Me_2pz)_3MoCl_3]\cdot MeCN$ has revealed a monomeric anion with near octahedral *facial* coordination about the metal. This compound undergoes two one-electron oxidations: Mo^{III}/Mo^{IV} at +0.49 V and Mo^{IV}/Mo^{V} at +1.55 V (*vs.* SCE). As anticipated from the value of the former couple, $[HB(3,5\text{-}Me_2pz)_3MoCl_3]$ is readily reduced to the Mo^{III} anion by mild reductants such as sulfide and alcohols. The stabilization of the Mo^{III} by the polypyrazolylborate has been attributed to the interaction of the Mo d_π orbitals with the ligand π^* orbitals. This view receives support from the preparation of $[(HBpz_3)MoCl_2(pyrazole)]$ by reacting $MoCl_5$ with potassium hydrotris(1-pyrazolyl)borate, $K[HBpz_3]$, in aqueous HCl without the addition of a reducing agent. The structure of this Mo^{III} derivative has been determined and shown to be similar to that of $[HB(3,5\text{-}Me_2pz)_3MoCl_3]$, with replacement of a chloride by a pyrazole group.[21] $[NBu_4][MoBr_4(dppe)]$ has been obtained as a minor product of the reaction of $[NBu_4]_2[Mo_2Br_6]$ with dppe in Me_2CO and its crystal structure has been determined; the Mo atom displays a distorted octahedral coordination geometry.[22]

36.4.1.2.2 Complexes with oxygen and sulfur donor ligands

$[Mo(acac)_3]$ is conveniently prepared from $[MoCl_6]^{3-}$ and Na(acac) or by oxidation of $[Mo(CO)_6]$ with Hacac;[23] the molecule has an octahedral geometry of approximately D_3 symmetry, with Mo—O bonds of average length 2.04(1) Å.[24] Several other tris(β-diketonates) of Mo^{III} are known and, with an unsymmetrical ligand, *facial* and *meridional* isomers have been isolated.[25] Oxidative decarbonylation of $[Mo(CO)_6]$ has also been used to prepare the O,S-bound analogue tris(monothiadibenzoylmethanato)molybdenum(III).[26] A complex of the stoichiometry [Mo(HL)L] is formed in the reaction between the tridentate O,S,N ligand salicylidenethiosemicarbazone (H_2L), and $[Mo(NCS)_6]^{3-}$.[27]

Dithiophosphate[16,28] and dithiocarbamate[29] ligands, $(S—S)^-$, form pseudooctahedral complexes $[Mo(S—S)_3]$ with Mo^{III}; in $[Mo\{S_2P(OMe)_2\}_3]$, the Mo—S average distance is 2.507(2) Å.[16] The Mo^{IV} complex $[MoCl(S_2CNEt_2)_3]$ can be reduced by SO_2 to give $[Mo(S_2CNEt_2)_3(\eta^2\text{-}SO_2)]$, which has an eight-coordinate structure, with the SO_2 bonded through the sulfur and one oxygen atom.[30] Mixed thiolate diphosphine complexes *trans*-$[Mo(SR)_2(diphos)_2]^+$ can be oxidized by $FeCl_3$ or $CuCl_2$ to give $[Mo(SR)_2(diphos)_2]^+$, which has been isolated as the $[FeCl_4]^-$ or $[BPh_4]^-$ salt.[31]

36.4.1.2.3 Complexes with nitrogen donor ligands

A variety of salts containing the $[Mo(NCS)_6]^{3-}$ ion are known, in which the ligand is bonded through nitrogen with almost linear Mo—N—C arrangement. In $K_3[Mo(NCS)_6]\cdot H_2O\cdot MeCO_2H$ the average Mo—N distance is 2.09 Å.[32] A selenocyanate complex, $[NEt_4]_3[Mo(NCSe)_6]$, has also been isolated.[33] Substitution reactions between $[Mo(NCS)_6]^{3-}$ and bidentate nitrogen ligands, such as 1,2-diaminoethane, give various products containing one, two or three chelates per molybdenum.[34]

36.4.1.2.4 Cyanide complexes

The cyanide ligand, when present in excess, replaces chloride in $[MoCl_6]^{3-}$ to give the seven-coordinate complex $[Mo(CN)_7]^{4-}$. In aqueous solution a pentagonal bipyramidal geometry is adopted, but in the solid state spectroscopic studies indicate a lower symmetry for $K_4[Mo(CN)_7]\cdot 2H_2O$.[35] The crystal structure of $NaK_3[Mo(CN)_7]\cdot 2H_2O$ shows a relatively undistorted pentagonal bipyramidal coordination sphere, with average Mo—C distances of 2.160(3) Å.[36]

36.4.1.2.5 Magnetic and electronic properties

All the monomeric molybdenum(III) complexes are paramagnetic, and in Table 1[5] the magnetic moments of some typical examples are given. For six-coordinate compounds the values lie between 3.53 and 3.86 BM, with most in the region 3.7 to 3.8 BM as predicted for octahedral d^3 complexes.[37] The value of 1.73 BM in $K_4[Mo(CN)_7]\cdot 2H_2O$ is consistent with a spin-doublet ground state d^3 system.[35]

ESR spectroscopy has been used to examine molybdenum(III) complexes with Cl, O, S and N donor ligands.[38] At temperatures between 5 and 80 K, these compounds show broad axial or rhombic signals centred at $g = 2$ and 4. For $[Mo(acac)_3]$ the estimated A values are 10 ± 1 (x, y) and 12.2 ± 0.5 (z) mK. Complexes with sulfur atoms which are part of an extended π system have typical ESR spectra, with all g values near 2.

Electronic absorption spectra of Mo^{III} species can be interpreted using a ligand field approach for a d^3 ion with octahedral symmetry.[39] Three spin-allowed transitions should be seen, corresponding to excitation from the ground state $^4A_{2g}$ to $^4T_{2g}$, $^4T_{1g}$ (F) and $^4T_{1g}$ (P) states. Bands due to spin-forbidden transitions to ^2E_g, $^2T_{1g}$ and $^2T_{2g}$ states can also be detected because of spin–orbit coupling. Table 2 lists the absorption bands observed in some typical complexes.[5] For compounds of the form $[MoX_3L_3]$, the lower symmetry leads to a loss of degeneracy within the T states and this has been observed in some systems, and used to demonstrate the formation of *meridional* rather than *facial* isomers.[15]

Table 2 Electronic Absorption Spectra of Molybdenum(III) Complexes (transitions in cm^{-1}, molar extinction coefficients in parentheses)

Complex	$^4A_{2g}\rightarrow{}^2E_g, {}^2T_{1g}$	$^4A_{2g}\rightarrow{}^2T_{2g}$	$^4A_{2g}\rightarrow{}^4T_{2g}$	$^4A_{2g}\rightarrow{}^4T_{1g}$ (F)	*Ref.*
$[MoCl_6]^{3-}$	9500 (1.4)	14 800 (2.0)	19 400 (23)	24 200 (38)	39
$[MoCl_5(H_2O)]^{2-}$	8500 (4.0)	15 300 (3.1)	19 700 (29)	25 500 (51)	39
$[MoCl_3(py)_3]$	8800 (7.0)	14 500 (7.0)			39
$[Mo(S_2PF_2)_3]$			20 020 (205)	24 310 (99)	28

Phosphorescence has only been observed in $[MoX_3L_3]$ complexes with L = urea or thiourea. The emission lies between 9000 and 10 000 cm^{-1}, and has been assigned to the $^2E_g\rightarrow{}^4A_{2g}$ transition.[40]

36.4.1.3 Dimeric and Polymeric Complexes

The best characterized Mo^{III} dimers are those with a Mo—Mo triple bond alone between the metal atoms or bridged by halides, as in $[Mo_2X_9]^{3-}$, complexes which are discussed in Section 36.3.2.2.6. Substitution of three halides by neutral ligands gives complexes of the type $[Mo_2X_6L_3]$, which probably retain the halide bridges. One example is $[Mo_2Cl_6(THF)_3]$, which is formed from the monomeric $[MoCl_3(THF)_3]$ in solution.[41]

The reactions between $K_3[MoCl_6]$ and various thiols yield dimeric or polymeric complexes of uncertain stoichiometry, thought to be $[Mo_2Cl_4(SR)_2(H_2O)_3]$ or $[Mo_4Cl_7(SR)_5(H_2O)_7]$, with magnetic moments which indicate considerable metal–metal interaction.[42] Crystal structures have been determined for the chloride-bridged compounds $[PPh_4][(Me_2S)Cl_2Mo(\mu\text{-}Cl)_3MoCl_2(SMe_2)]$ and $[(Me_2S)Cl_2Mo(\mu\text{-}SMe_2)(\mu\text{-}Cl)_2MoCl_2(SMe_2)]$,[43] and of the bromide-bridged compound $[LBrMo(\mu\text{-}Br)(\mu\text{-}OH)MoBrL]Br_2{\cdot}4H_2O$ (L = 1,4,7-triazacyclononane).[44] In the chloride complexes, the structure around the Mo atoms is confacial bioctahedral, with three bridging atoms. For the bromide, each molybdenum is octahedrally coordinated and the two centres share a common edge. The short Mo—Mo distances in these compounds indicate strong metal–metal bonding.

Oxo-bridged dimers such as $[Mo_2O_3(H_2O)_5(DMF)]$ and $[Mo_2O_3(DMSO)_3]$ can be obtained by heating $[Mo(CO)_6]$ in wet DMF or DMSO.[45] Other dimeric species have been characterized with a variety of O and N donor ligands, including $[Mo_2O_2Cl_2(H_2O)_xL_y]$ (L = bipy, 4,4′-bipy, pyrrole, pyridazine, py),[46] $[Mo_2O(C_2O_4)L_2(H_2O)_4]$ (HL = phenylalanine),[47] and $[Mo_2O_2(HL)_4]^{2-}$ (H_2L = cysteine).[48] Low magnetic moments of these and analogous complexes, of between 0.4 and 0.6 BM, provide clear evidence for spin pairing between the metal centres.

Molybdenum(III) complexes of 1,4,7-triazacyclononane (L) dimerized by hydroxide bridges can be prepared by hydrolysis of the corresponding $[LMoCl_3]$ complexes.[49,50] Crystal structure determinations on $[LClMo(\mu\text{-}OH)_2MoClL]I_2$ and $[LMo(\mu\text{-}OH)_2(\mu\text{-}O_2CMe)MoL]I_3{\cdot}H_2O$ have revealed a distorted octahedral geometry around the Mo atoms in both cases, with short Mo—Mo distances thought to indicate a formal bond order of three.[49] In the complex $K[Mo_2(\mu\text{-}OH)_2(\mu\text{-}O_2CMe)(edta)]$, the hexadentate ethylenediaminetetraacetate also bridges the Mo—Mo bond *via* the two N atoms (see **9**, Section 36.1.2).[51] Edta is thought to bridge in a similar manner in the oxo-bridged complexes $MH[Mo_2(\mu\text{-}O)_2(edta)]{\cdot}nH_2O$ (M = Li, Na, K, NH_4).[52] A mixed oxidation state tetrameric molybdenum(III,IV) complex $Na_4[Mo_4O_4(OH)_2(edta)]{\cdot}12H_2O$ has been structurally characterized, and shows essentially two dimeric species connected *via* oxo bridges (see **4**, Section 36.1.2). With the optically active ligand (*R*)-propylenediaminetetraacetate [(*R*)-pdta], strong peaks are observed in the CD spectrum of $[Mo_2(\mu\text{-}OH)_2(\mu\text{-}O_2CMe)\{(R)\text{-}pdta\}]^-$, indicating that the four coordinated carboxylate oxygens arrange asymmetrically.[53]

Dithiocarbamate compounds of Mo^{III} can be prepared by oxidative decarbonylation of $[Mo(CO)_6]$, giving complexes of the form $[Mo_2(R_2dtc)_6]$.[54] Spectroscopic evidence is consistent with the presence of two bridging R_2dtc ligands. The crystal structure of $[(Et_2NCS_2)_2Mo(\mu\text{-}CSC(S)S)(\mu\text{-}S_3C_2NEt_2)Mo(S_2CNEt_2)]$, a product of the reaction between CS_2 and $[Mo(CO)_2(S_2CNEt_2)_2]$, has been determined; the bridging ligands appear to be formed by the insertion of CS into CS_2 and $Et_2NCS_2^-$.[55] Mixed oxidation state molybdenum(III,VI) complexes $[PPh_4]_2[S_2Mo(\mu\text{-}S)_2MoBr_3(SMe_2)]$ and $[PPh_4][S_2Mo(\mu\text{-}S)_2MoBr_2(NO)_2]$, in which the Mo atoms are linked by sulfido bridges, have been prepared by the reaction between $[PPh_4]_2[MoS_4]$ and $MoBr_4$ or $[MoBr_2(NO)_2]$.[56] A disulfide bridge and a sulfur dioxide bridge are found in the $[Mo_2(\mu_4\text{-}S_2)(\mu_2\text{-}SO_2)(CN)_8]^{4-}$ anion, the structure of which has been determined as its $[PPh_4]^+$ and K^+ salts; in the former the Mo—Mo separation is 2.730(4) Å, whereas the latter involves two different anions with Mo—Mo distances of 2.691(1) and 2.974(1) Å, respectively.[57] The salts $K_6[Mo_2(CN)_{12}S]{\cdot}4H_2O$[58,59] and $K_{6.68}[MoO_4]_{0.34}[Mo_2(CN)_{12}S]{\cdot}5.32H_2O$[58] have been isolated. The crystal structure of the latter has shown that the anion involves the two metal atoms, each in a pentagonal bipyramidal environment, bridged by an axial sulfur atom (Mo—S—Mo *ca.* 170°). The diamagnetic nature of the compounds and their electronic spectra are consistent with $S(p_\pi)\rightarrow Mo(d_\pi)$ bonding over the Mo—S bonds of length *ca.* 2.17 Å.

36.4.2 MOLYBDENUM(IV)

36.4.2.1 Introduction

The coordination chemistry of Mo^{IV} is diverse and involves an extensive range of monomeric and trimeric systems; the latter are described in Chapter 36.3 together with the few dimeric systems which involve a Mo—Mo double bond. Although there are many examples of monomeric oxo complexes, unlike the higher oxidation states, Mo^{IV} forms a comparable range of non-oxo complexes and exhibits a greater versatility in coordination number, with examples of four to eight, inclusive, being well established.

The most important binary systems which (formally) involve Mo^{IV} are the chalcogenides MoS_2 and $MoSe_2$; some comments on these systems are included in Section 36.6.3. The synthesis and properties of molybdenum(IV) halides have been thoroughly investigated.[1,60]

Fluorination of $[Mo(CO)_6]$ at −75 °C forms the olive green material Mo_2F_9, which decomposes on heating to the volatile MoF_5 and the light green, involatile MoF_4; reduction of MoF_5 with elemental silicon (4:1) provides an alternative route to MoF_4. The Raman spectrum of MoF_4 suggests that the material possesses a structure similar to that of NbF_4.[61] $MoCl_4$ exists in three forms. α-$MoCl_4$, conveniently prepared[62] by the prolonged refluxing of $MoCl_5$ and C_2Cl_4 in CCl_4, is isomorphous with $NbCl_4$ and comprises infinite chains of $MoCl_6$ octahedra sharing opposite edges over alternating long and short Mo—Mo distances, with magnetic exchange occurring down these chains leading to a room temperature effective magnetic moment of 0.85 BM. β-$MoCl_4$ is obtained by reduction of $MoCl_5$ by molybdenum metal under carefully controlled conditions;[3] the structure consists of cyclic $(MoCl_4)_6$ aggregates in which each molybdenum is octahedrally coordinated by two terminal and four bridging chlorines (Figure 1).[63] The closest approach between the Mo atoms (3.67 Å) does not lead to any significant spin-pairing (μ_{eff} *ca.* 2.4 BM). A third form of $MoCl_4$ has been obtained by treatment of MoO_2 in a stream of CCl_4 vapour.[64] $MoBr_4$ may be synthesized by the reaction between $MoBr_3$ and warm (60 °C) liquid bromine.[65] The tetrahalides are reactive, for example to air and water, and may be used as starting materials for the synthesis of Mo^{IV} complexes.

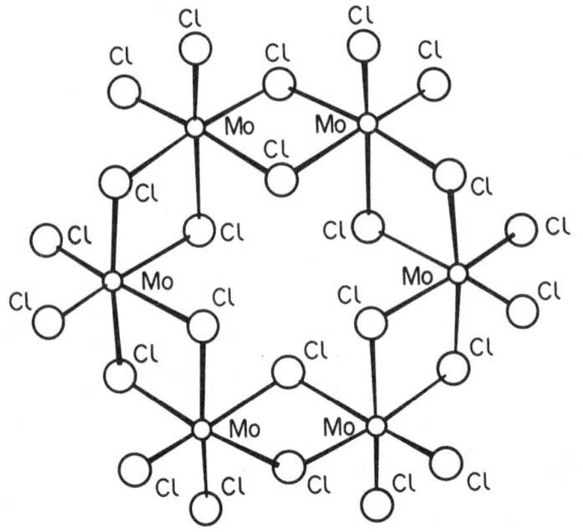

Figure 1 The structure of the $(MoCl_4)_6$ molecule in β-$MoCl_4$[63]

$MoOCl_2$ has been prepared by various chemical transport reactions[1] and $MoSCl_2$ and $MoSeCl_2$ have been described.[66] MoO_2 may be prepared by reduction of MoO_3 by H_2, the compound adopts a distorted rutile structure with Mo---Mo separations of 2.5106(5) and 3.1118(5) Å,[67] the former leading to spin pairing and the absence of a permanent magnetic moment on the metal centres.[68]

Mononuclear Mo^{IV} oxo complexes have special relevance to the molybdenum centres in the molybdenum-containing enzymes. $[MoO(S_2CNR_2)_2]$ complexes undergo addition reactions with a wide variety of unsaturated molecules.[69–79] These were once held to be of relevance to the activation of the substrates of nitrogenase prior to reduction; however, the nature of the molybdenum centre in these $[MoO(S_2CNR_2)_2]$ molecules has not been demonstrated to have any significant similarity to the catalytic site within the Fe—Mo—S cluster of the nitrogenases (see Section 36.6.2) and, therefore, detailed comparisons are unwarranted. $[MoO(S_2CNR_2)_2]$ and a selection of other Mo^{IV} complexes are special in respect of participating in reversible oxygen atom transfer, $MoO^{2+} + XO \rightarrow MoO_2^{2+} + X$, which have particular relevance to the catalyses effected at the molybdenum centres of oxomolybdoenzymes such as xanthine oxidase, sulfite oxidase or nitrate reductase.[80] The forward reaction was first observed in 1972 for $[MoO(S_2CNR_2)_2]$ complexes[81] and X = Ph_3P and this reaction has been developed and used for the preparation of MoO^{2+} complexes[82–84] and the mechanism defined.[85,86] A complicating feature of this reaction is the dimerization (equation 1) which is prevented in the oxomolybdoenzymes by structural constraints within the macromolecular assembly. Macrocyclic ligands have been employed in chemical studies to prevent such dimerization and, thereby, have allowed the development of more realistic chemical analogues of the oxomolybdoenzymes. These ligands include the anions of the molecules (Figure 2)[84,87–91] and these studies represent elegant developments in the coordination chemistry of molybdenum.

$$[MoO_2(S_2CNR_2)_2] + [MoO(S_2CNR_2)_2] \rightleftharpoons [Mo_2O_3(S_2CNR_2)_4] \quad (1)$$

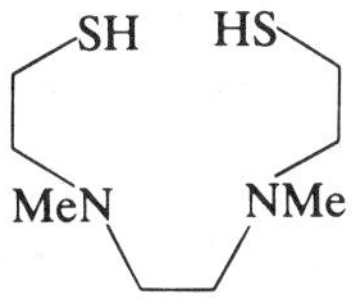

N,N'-Dimethyl-*N,N'*-bis 2'-mercaptoethyl-1,2-diaminoethane

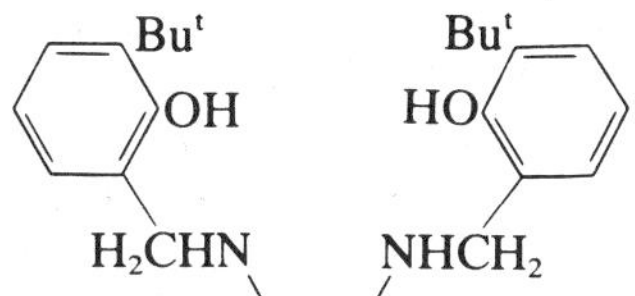

N,N'-Bis-(3',1'',1''-dimethylethyl-2'-hydroxyphenylmethyl)-1,2-diaminoethane

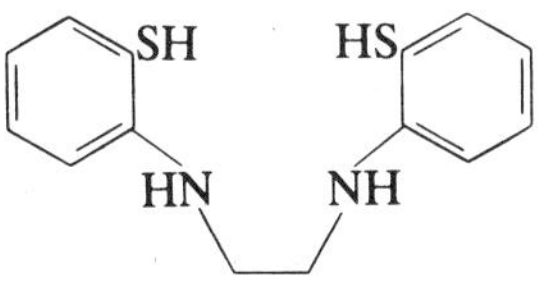

N,N'-Bis-2'-mercaptophenyl-1,2-diaminoethane

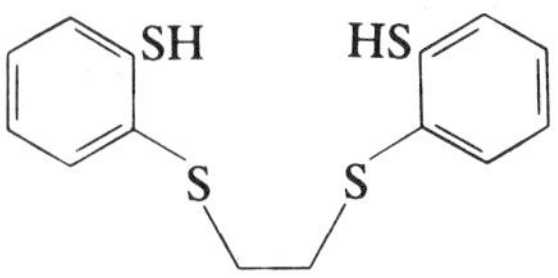

S,S'-Bis-2'-mercaptophenyl-1,2-dimercaptoethane

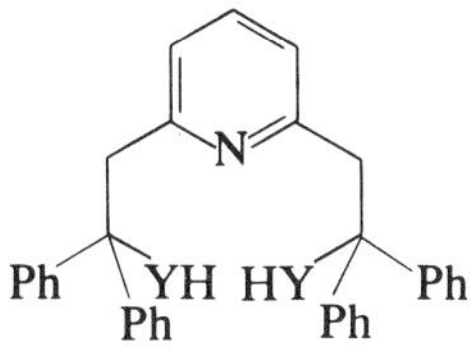

Y = O 2,6-Bis(2,2-diphenyl-2-hydroxoethanyl)pyridine

Y = S 2,6-Bis(2,2-diphenyl-2-mercaptoethanyl)pyridine

Figure 2 Precursors of macrocyclic ligands which allow redox reactions for monomeric molybdenum centres

36.4.2.2 Monomeric Complexes with Oxo Groups

36.4.2.2.1 General comments

Oxo complexes dominate the chemistry of molybdenum in its higher oxidation states,[5,80,92,93] and play a major role in the catalytic activity of molybdenum in biological systems (see Section 36.6.7). Mo^{IV} complexes have been characterized containing the monooxo $\{MoO\}^{2+}$ and the *trans*-dioxomolybdenum(IV) $\{MoO_2\}$ cores, but no compound has been isolated analogous to the *cis*-dioxomolybdenum(VI) species described in Section 36.5. Crystal structure determinations have been accomplished for a number of Mo^{IV} complexes, and some typical examples are listed in Table 3.[80]

Table 3 Molybdenum–Oxygen Bond Lengths of some Oxomolybdenum(IV) Complexes[80]

Compound	*Mo—O$_t$* (Å)	*Ref.*
$[MoO(S_2CNPr^n_2)_2]$	1.665	100
trans-$[MoO(hd)(SH)][CF_3SO_3]$[a]	1.667	103
cis,mer-$[MoOCl_2(PEt_2Ph)_3]$[b]	1.803	107
cis,mer-$[MoOCl_2(PMe_2Ph)_3]$[c]	1.676	108
$[MoOCl(dppe)_2][ZnCl_3(OCMe_2)]$	1.69	111
$NaK_3[MoO_2(CN)_4]$	1.834	120
$[Cr(en)_3][MoO(OH)(CN)_4]$	1.698	121
$[Pt(en)_2][MoO(OH_2)(CN)_4]$	1.668	121
$[MoOCl(CNMe)_4](I_3)$	1.636	126

[a] hd = 1,5,9,13-tetrathiacyclohexadecane. [b] Green isomer. [c] Blue isomer.

The nature of the bonding of oxo groups to molybdenum has been discussed in several reviews.[80,92,93] There are three components to molybdenum–oxygen terminal bonds (Mo—O$_t$), comprising a σ interaction along the bond axis (z axis) and two mutually perpendicular π components, each formed by the overlap of an oxgyen p_π orbital (p_x or p_y) with the appropriate molybdenum d_π orbital (d_{xz} or d_{yz}). Normally, each component is considered to involve donation of an electron pair from O^{2-} to a vacant orbital of the metal, and the formal bond order varies according to the degree of π interaction. Correlations have been observed between the bond order of a Mo—O_t bond and its length and force constant, with a high bond order corresponding to shorter and stronger bonds.[94-96] Generally, the Mo—O_t bond lengths in monooxomolybdenum(IV) compounds are ~1.67 Å (Table 3), consistent with a triple bond: for the *trans*-dioxo species a longer bond length of ~1.83 Å (Table 3) is found, indicative of a bond order between one and two.

In monooxomolybdenum(IV) centres the Mo—O_t bonding dominates the ligand field at the molybdenum, so that the two valence electrons of the metal are spin-paired in an orbital which is in the plane (xy) perpendicular to the Mo—O_t bond, and the complex has the $(d_{xy})^2$ configuration. The dioxo anion *trans*-$[MoO_2(CN)_4]^{4-}$ also possesses this configuration;[97] the two d electrons are in an orbital which is nonbonding with respect to the oxo groups and the σ orbitals of the cyano groups and, hence, at a lower energy than the other d orbitals. This situation may be further stabilized by back-bonding from $(d_{xy})^2$ into the π^* orbitals of the CN^- ligands. Similarly, ligand field stabilization may explain the formation of *trans*-$[MoO_2(CO)_4]$ in the photooxidation of $[Mo(CO)_6]$.[98]

The coordination sphere of molybdenum in the oxomolybdenum enzymes is believed to contain oxo and sulfur ligands,[80] and, therefore, interest has focussed on oxo complexes of Mo^{IV} with S-donor ligands as possible models for the reduced state of these systems. A number of oxomolybdenum(IV) complexes contain a five-coordinate $MoOS_4$ coordination sphere, including the dithiocarbamate species $[MoO(S_2CNR_2)_2]$. These can be prepared by reduction of the Mo^V dimers $[Mo_2O_3(S_2CNR_2)_4]$ using zinc dust or thiophenol, or by reduction of aqueous solutions of $Na_2[MoO_4]$ and $Na(S_2CNR_2)$ with dithionite.[99] An alternative route is by the reaction of PPh_3 with the corresponding Mo^{VI} complex, $[MoO_2(S_2CNR_2)_2]$,[81] in which one atom of oxygen is transferred from the molybdenum to the phosphorus in a reaction which may be analogous to the role played by the molybdenum in xanthine oxidase and other related enzymes.[80] The reverse reaction, involving abstraction of an oxygen atom from a substrate molecule by the monooxo Mo^{IV} complex to give the *cis*-dioxo Mo^{VI} complex, has been observed with dioxygen or pyridine *N*-oxide;[73] the claim that this also occurs with Ph_3PO appears to be erroneous. A kinetic study of the forward and reverse oxygen atom transfer reactions with $[MoO(S_2CNEt_2)_2]$ has shown that the dominant factor in determining reaction rate is the breaking of the bond to oxygen in the substrate.[85] The act of oxygen atom transfer has been suggested to involve a transition state comprising one molecule of each reactant, and the donation of the lone pair of electrons on phosphorus into the Mo═O π^* orbital, thus weakening the Mo—O bond and forming the P—O bond to produce the molybdenum(IV) complex $[MoO(Et_2dtc)_2(Ph_3PO)]$. Isomerization followed by the loss of Ph_3PO would produce the reaction products.[86] The Mo^{IV} complexes can also react with one of the oxo groups of $[MoO_2(S_2CNR_2)_2]$ to give the oxo-bridged Mo^V dimer $[Mo_2O_3(S_2CNR_2)_4]$ (equation 1).

The crystal structure of $[MoO(S_2CNPr^n_2)_2]$ has shown that the molecules possess a square pyramidal geometry, with the molybdenum displaced out of the plane of the four sulfur atoms in the direction of the axial oxo ligand;[100] the Mo—O_t distance is 1.695 Å. Complexes of this type are coordinatively unsaturated and react with ligands, such as pyridine, phosphines, alkynes and alkenes, to form 1:1 adducts. Spectroscopic evidence has indicated that the unsaturated molecules add to produce a seven-coordinate complex,[74] and this has been confirmed by a crystal structure determination of the tetracyanoethylene (TCNE) adduct of $[MoO(S_2CNPr^n_2)_2]$,[72] which shows distorted pentagonal bipyramidal coordination around the molybdenum. The equatorial plane contains the two ethene carbons of TCNE and three of the sulfur atoms; the fourth sulfur occupies an axial position *trans* to the oxo ligand, and the Mo—O_t bond length is 1.682(4) Å. Although the oxidation state of the molybdenum is difficult to define in these compounds, IR spectra of some ethyne adducts suggests that they are complexes of Mo^{VI}, formed by oxidative addition,[76] the d^2 electrons of the molybdenum being donated into the π^* orbital of the multiple bond. Kinetic studies of these addition reactions indicate that they proceed *via* a monocapped trigonal prismatic intermediate, followed by intramolecular rearrangement.[77,78]

$[MoO(S_2CNR_2)_2]$ forms 1:1 adducts with molecules containing nitrogen–nitrogen double bonds in which the —N═N— moiety is thought to be sideways bonded to the molybdenum *via* both nitrogen atoms. In contrast, an end-on nitrogen–molybdenum σ bond is formed in the reaction between molecules of the type $[MoO_2(S_2CNR_2)_2]$ and R′NNH, to produce the adducts $[(R'NN)Mo(S_2CNR_2)_3]$.[71] These complexes may be of interest as models for the intermediates formed during reduction of N_2 by the nitrogenases;[5] the adduct $[MoO(S_2CNEt_2)_2(R'NNR')]$

($R' = CO_2Et$) can be hydrolyzed in moist $CHCl_3$ to produce R′NHNHR′ and $[MoO_2(S_2CNEt_2)]$, showing that the N═N bond is activated by adduct formation.[69] The crystal structure of the adduct $[MoO(PhCONNCOPh)(S_2CNMe_2)_2]\cdot C_2H_4Cl$ has been determined.[79] The stereochemistry about the molybdenum is that of a distorted pentagonal bipyramid; the sulfur atoms of the dithiocarbamate ligands and a nitrogen atom of the benzoyldiazene ligand occupy the equatorial positions and the axial sites are occupied by the terminal oxo group and an oxygen atom of the diazene, which thus forms a five-membered chelate ring.

Studies of Mo^{IV}/Mo^{VI} redox reactions involving $[MoO(SCNR_2)_2]$ complexes are complicated by the formation of the Mo^V dimer, $[Mo_2O_3(S_2CNR_2)_4]$ (equation 1). In biological systems this type of dimerization is prevented by structutal constraints exercised by the enzyme, and attempts have been made to emulate this by preparing model compounds with chelates, some of which introduce steric constraints at the metal. Complexes with polydentate ligands of the form [MoOLL′] (L = salicylaldehyde-2-hydroxyanil, salicylaldehyde-2-mercaptoanil; L′ = bipy, phen) and [MoOL] (L = *N,N′*-bis(2-mercapto-2-methylpropyl)-1,2-diaminoethane) have been investigated.[84] They undergo electrochemical one-electron oxidation or reduction reactions, but can not be easily oxidized to Mo^{VI} species. $[MoO(PPh_2Et)(dttd)]$ (H_2dttd = *S,S′*-bis-2′-mercaptophenyl-1,2-dimercaptoethane, Figure 2), prepared by reduction of the Mo^{VI} complex $[MoO_2(dttd)]$ with PPh_2Et, loses the phosphine in DMF/$[Et_4N]Cl$ solution to form $[MoOCl(dttd)]^-$. This anion undergoes a reversible one-electron oxidation to produce [MoOCl(dttd)] but is not oxidized further to a Mo^{VI} species.[90]

The ligand 2,6-bis(2,2-diphenyl-2-mercaptoethanyl)pyridine (Figure 2) ($LN(SH)_2$) has been used to prepare the Mo^{IV} complex $[MoO(LNS_2)(DMF)]$ and the complementary dioxo Mo^{VI} complex $[MoO_2(LNS_2)]$; in both cases, the ligand acts as a tridentate N,S,S donor. The Mo^{IV} complex can be oxidized electrochemically to a Mo^V species but not to Mo^{VI}. However, $[MoO(LNS_2)(DMF)]$ reacts with DMSO to abstract an oxygen atom, forming $[MoO_2(LNS_2)]$: the reverse reaction can be accomplished using Ph_3P, generating Ph_3PO and $[MoO(LNS_2)(DMF)]$.[91] These reactions produce a catalytic system for the oxidation of Ph_3P by DMSO, and provide an interesting model for the reactions catalyzed by the oxomolybdoenzymes.

36.4.2.2.2 Complexes of oxygen and sulfur donor ligands

Reduction of $[NH_4]_2[MoOCl_5]$ with hydrazine, followed by extraction of the product with glacial acetic or malonic acid, gives the Mo^{IV} complexes $[MoO(O_2CMe)_2(H_2O)]$ or $[MoO(O_2CCH_2CO_2)(H_2O)_3]$, respectively.[101] Spectroscopic evidence has indicated that the molybdenum in these species is coordinated to six oxygen atoms. Other oxomolybdenum(IV) complexes with oxygen donor ligands include $[MoO(CN)_4(OH)]^{3-}$ and $[MoO(CN)_4(OH_2)]^{2-}$ (see Section 36.4.2.2.4), $[MoO(dppe)_4(OH)]^+$ and the aqua ions $MoO^{2+}(aq)$ and $MoO(OH)^+(aq)$ (see Section 36.1.4). $[MoO(dppe)_2(OH)]^+$ contains a linear O═Mo—OH group, with the two bidentate dppe ligands completing the octahedral coordination sphere.[102]

An analogous sulfur-bound complex has been isolated, using the macrocyclic tetrathioether 1,5,9,13-tetrathiocyclohexadecane (hd) to form $[MoO(hd)(SH)]^+$.[103] This cation contains a linear O═Mo—SH group, with a Mo—S distance of 2.486(1) Å and an Mo—O_t distance of 1.667(3) Å. The thioether acts as a tetradentate ligand to form a square planar arrangement around the molybdenum. Difluorodithiophosphate forms the five-coordinate complex, $[MoO(S_2PF_2)_2]$, from which the $(S_2PF_2)^-$ ligands are easily displaced by pyridine to give $[MoO(py)_4](S_2PF_2)_2$.[28] The bidentate S,O ligands, $(SCH_2CO_2)^{2-}$, $(SCH(Me_3)CO_2)^{2-}$ and $(SCH(CH_2CO_2H)CO_2)^{2-}$ (S—O), also form five-coordinate complexes of composition $[MoO(S—O)(H_2O)_2]$, which have been isolated as solids. A cysteine complex of MoO^{2+} has been identified in solution, but not characterized as a solid.[104] With the S,P-donor ligand $(SCH_2CH_2PPh_2)^-$, the coordination geometry around the molybdenum in $[MoO(SCH_2CH_2PPh_2)_2]$ has been shown to be intermediate between trigonal bipyramidal and square pyramidal.[105]

36.4.2.2.3 Complexes of phosphine ligands

In ethanol solution, tertiary phosphines react with $[MoCl_4(NCEt)_2]$ to yield *mer*-$[MoOCl_2(PR_3)_3]$, which can undergo exchange reactions to give *mer*-$[MoOX_2(PR_3)_3]$ (X = Br,

I, NCS, NCO).[106] Some of these complexes are green, with $\nu(Mo{=}O)$ stretching frequencies below 946 cm^{-1}, and others blue-green, with $\nu(Mo{=}O)$ above 946 cm^{-1}. In the case of $[MoOCl_2(PMe_2Ph)_3]$, both a blue isomer and a less stable green isomer have been isolated. Crystal structure determinations of green *cis,mer*-$[MoOCl_2(PEt_2Ph)_3]$[107] and blue *cis,mer*-$[MoOCl_2(PMe_2Ph)_3]$[108] have shown that both complexes have an octahedral geometry with notable differences in particular bond lengths and angles. Therefore, it is considered that a balance exists between metal–ligand attraction and ligand–ligand repulsion which, in this case, is very sensitive to the nature of the ligands and results in two possible distortional isomers of the same structural unit. Alternative routes to $[MoOCl_2(PR_3)_3]$ complexes have been developed and these include the reaction of the phosphine with $Na_2[MoO_4]$ in EtOH in the presence of HCl and Zn–Hg reduction of $[MoOCl_3(PR_3)_2]$ in THF.[109]

The reaction of an excess of a diphosphine (PP), such as dppe, with $[MoOCl_2(PR_3)_3]$ yields $[MoOCl(PP)_2]Cl$; the cation has been isolated as $[BF_4]^-$, $[BPh_4]^-$, $[NCS]^-$ and $[NCSe]^-$ salts.[109,110] When a stoichiometric (1:1) amount of (PP) is present, the intermediate, mixed phosphine complex $[MoOCl_2(PP)(PR_3)]$ can be prepared.[109] A crystal structure determination of $[MoOCl(dppe)_2][ZnCl_3(OCMe_2)]$ has shown that the cation contains octahedral Mo^{IV}.[111]

MoO^{2+} complexes with other bidentate ligands have also been prepared, and these include $[MoOCl_2L(PMePh_2)]$ (L = bipy, phen) and $[MoOL'(PMePh_2)]$ (H_2L' = 8-hydroxy- or 8-mercaptoquinoline). All, except the mercaptoquinoline system, undergo an essentially reversible one-electron oxidation.[112]

36.4.2.2.4 Complexes of nitrogen donor ligands

Complexes of Mo^{IV} with tetraphenylporphyrin (TPP) or tetratolylporphyrin (TTP) ligands are generally obtained by reduction of the corresponding Mo^V species. Zn–Hg reduction of $[MoOCl(TPP)]\cdot HCl$ yields $[MoO(TPP)]$, the structure of which has been determined.[113] The Mo^{IV} atom is in a square pyramidal geometry, slightly displaced from the plane formed by the four N ligands, in the direction of the apical oxo group. The Mo—O and Mo—N bond lengths are 1.656(6) and 2.110(6) Å, respectively.

ESR and electronic spectral data suggest that, in the reduction of $[Mo^VOBr(TPP)]$ by O_2^- to give $[Mo^{IV}O(TPP)]$, an intermediate dioxygen complex $[Mo^{IV}O(TPP)(O_2)]^-$ is formed, which is stable in solution at −72 °C.[114] The photochemical reduction of $[Mo^VOX(TPP)]$ (X = NCS, F, Cl, Br) to $[Mo^{IV}OX(TPP)]^-$ at 77 K has also been investigated. On warming the solution, X^- is released to give the $[Mo^{IV}O(TPP)]$ complex.[115]

The reaction of $[MoO(P)]$ (P = TPP, TTP) with HCl in benzene yields the corresponding dichloro complex, $[MoCl_2(P)]$, and these are the only Mo^{IV} complexes with porphyrin ligands which do not contain an oxo group.[113]

$[MoO(Pc)]$ (Pc = phthalocyanate dianion) can be prepared from the reaction between phthalodinitrile and $[Mo(CO)_6]$[116] or $[NH_4]_2[Mo_7O_{24}]\cdot 4H_2O$.[117] A crystal structure determination has shown that this complex possesses the same structure as $[MoO(TPP)]$, with a Mo—O bond length of 1.668(4) Å.

The reaction between MoX_3 (X = Cl, Br), aniline and 2,6-diacetylpyridine in butanol yields $[MoO\{py(anil)_2\}X_2]$, where $py(anil)_2$ is a tridentate Schiff base ligand formed *in situ*.[118] These complexes are isomorphous with $[VO\{py(anil)_2\}X_2]$ and, therefore, are considered to have octahedral coordination of the molybdenum.

36.4.2.2.5 Cyanide and isocyanide complexes

The photolysis of aqueous solutions of $K_4[Mo(CN)_8]$ gives compounds containing the *trans*-$[MoO_2(CN)_4]^{4-}$ ion.[119] These are easily protonated to given *trans*-$[MoO(OH)(CN)_4]^{3-}$ and *trans*-$[MoO(OH_2)(CN)_4]^{2-}$, and crystal structure determinations have been accomplished for compounds of each of these anions. In $NaK_3[MoO_2(CN)_4]$, the Mo—O_t bond length is 1.834(9) Å,[120] which is significantly longer than found in MoO^{2+} complexes (Table 3). In proceeding from $[Cr(en)_3][MoO(OH)(CN)_4]$ to $[Pt(en)_2][MoO(OH_2)(CN)_4]$, the Mo—$O_t$ bond length decreases from 1.698(7) to 1.668(5) Å, whilst the Mo—O distance for the protonated oxygen increases from 2.077(7) to 2.271(4) Å.[121] In the *trans*-$[MoO(OH_2)(CN)_4]^{2-}$ anion, a significant degree of distortional isomerism can occur, and compounds of this anion have been isolated from the reduction of aqueous solutions of $Na_2[MoO_4]$ in the presence of CN^-, in which the Mo—O_t distance varies between 1.60(2) Å, for the $[AsPh_4]^+$ salt, and 1.72(2) Å, for the $[PPh_4]^+$ salt.[122]

The only ligands, apart from cyanide, which are known to stabilize a dioxomolybdenum(IV) group are carbonyls, in the complex *trans*-$[MoO_2(CO)_4]$, which is formed in the photooxidation of $[Mo(CO)_6]$.[98] The strong π-acceptor capacity of both CN^- and CO removes electron density from the filled d_{xy} orbitals of the molybdenum, thereby increasing the (formal) positive charge on the metal and increasing the σ and π donation from the two terminal oxo groups. Protonation of an oxo group reduces this donation, resulting in the relative bond lengths Mo—O < Mo—OH < Mo—OH_2 with the concomitant decrease in the length of the bond to the *trans*-oxo ligand.

Monooxo complexes containing the $[MoO(CN)_5]^{3-}$ anion have also been isolated from reactions involving the reduction of $[MoO_4]^{2-}$ [122] or the photolysis of $[Mo(CN)_8]^{4-}$.[123] The crystal structure of $[PPh_4]_3[MoO(CN)_5]\cdot 7H_2O$ shows octahedral coordination around the Mo, with the Mo—O_t bond length being 1.705(4) Å.[122]

The reaction between $[MoO_2(CN)_4]^{4-}$ and 1,10-phenanthroline yields $[MoO(CN)_3(phen)]^-$. A crystal structure determination of $Na[MoO(CN)_3(phen)]\cdot 2phen\cdot MeOH\cdot H_2O$ has shown that one nitrogen of the phen ligand is *trans* to the oxo group, with the other nitrogen and the three CN^- ligands completing the octahedral coordination sphere of the Mo.[124] The Mo—N distance *trans* to Mo—O_t is 2.363(7) Å, as compared to a distance of 2.173(8) Å for the bond *trans* to a CN^- ligand, showing the large structural *trans* influence of the oxo group.

Mo^{IV} isocyanide complexes of the form $[MoOX(CNR)_4]Y$ (R = alkyl; X = Cl, Br; Y = Cl, Br, BrI_2, I_3, PF_6) have been prepared by several synthetic routes.[125] These complexes contain a *trans* O—Mo—X arrangement, and in $[MoOCl(CNMe)_4](I_3)$ the Mo—O_t distance is 1.64(4),[126] a length typical of the multiple terminal oxomolybdenum(IV) bond (Table 3).

36.4.2.2.6 Spectroscopic studies

Oxomolybdenum(IV) complexes have been studied by a wide variety of spectroscopic techniques, including NMR, vibrational, electronic and electrochemical measurements. Each of these provides information about different aspects of the compounds, and can be used in conjunction to illustrate various aspects of the bonding.

A survey of 38 complexes of MoO^{2+} has been made using ^{95}Mo NMR spectroscopy and some typical chemical shifts and line widths are given in Table 4.[127] The compounds display a large chemical shift range and highly variable line widths: the most deshielded resonance occurs for $[MoOCl_2(bipy)(PPh_2Me)]$, with δ 3160 p.p.m. (referenced to $[MoO_4^{2-}]$ of 0 p.p.m.), and the most shielded resonance is that of $[MoOCl(CNMe)_4]^+$, which is 1035 p.p.m. ^{17}O NMR spectra have been recorded for $[MoO(S_2CNEt_2)_2]$ and $[MoO(S_2PEt_2)_2]$, and compared with spectra of oxomolybdenum(V) and (VI) compounds. The ^{17}O chemical shifts vary with the strength and multiplicity of the Mo—O_t bond, but changing the oxidation state of molybdenum produces a bigger change, so comparisons between Mo—O bonds using ^{17}O NMR data can only be made within a particular oxidation state.[128]

Table 4 ^{95}Mo NMR data[127]

Compound	*Chemical shift* (p.p.m.)	*Line width* (Hz)
$[MoO(S_2CNEt_2)_2]$	2400	2700
$[MoOCl_2(PEt_2Ph)_3]$[a]	2200	1200
$[MoOCl_2(bipy)(PPh_2Me)]$	3160	800
$[MoOCl(dppe)_2][BPh_4]$	1260	1780
$K_4[MoO_2(CN)_4]\cdot 6H_2O$	1220	240
$K_3[MoO(OH)(CN)_4]$	1444	240
$[MoOCl(CNMe)_4][BPh_4]$	1035	330

[a] Green isomer.

Anomalies in the ^1H and ^{31}P NMR spectra of complexes *mer*-$[MoOX_2(PMe_2Ph)_3]$ (X = Cl, Br) have been interpreted as being due to second order effects caused by the small difference in chemical shifts between the two *trans* phosphines and the unique phosphine *trans* to Cl.[129]

The vibrational spectra of oxomolybdenum(IV) complexes all show a band due to Mo—O_t

stretching at *ca.* 900–1000 cm^{-1}. ^{18}O isotopic substitution has been used to make definitive assignment of this vibration for a series of complexes,[130] and some values are given in Table 5.[5] The frequency of the vibration is a sensitive indicator of the strength and formal order of the bond, and can distinguish between distortional isomers such as the green and blue forms of $[MoOCl_2(PMe_2Ph)_3]$.[106]

Table 5 Molybdenum–Oxygen Stretching Frequencies[5]

Compound	$\nu(Mo—O_t)$ (cm^{-1})	*Ref.*
$[MoO(S_2NCPr^n_2)_2]$	975	100
$[MoOCl_2(PEt_2Ph)_3]^a$	940	107
$[MoOCl_2(PMe_2Ph)_3]^b$	954	108
$[MoOCl(dppe)][ZnCl_3(OCMe_2)]$	948	111
$K_4[MoO_2(CN)_4]\cdot 6H_2O$	850^c	121
$[Cr(en)_3][MoO(OH)(CN)_4]$	900	121
$[PPh_4]Cs[MoO(OH_2)(CN)_4]\cdot 2H_2O$	980	122
$[MoOCl(NCMe)_4](I_3)$	946	125

[a] Green isomer. [b] Blue isomer. [c] Asymmetric stretch.

Electronic absorption spectra have been recorded for a large number of oxomolybdenum(IV) species but, in addition to *d–d* transitions, there is the possibility of two-electron excitations, charge transfer bands and splitting of degenerate energy levels due to low symmetry. These factors make it difficult to interpret or even to compare spectra.[5]

Electrochemistry has been used to investigate the redox properties of Mo^{IV} oxo complexes, particularly those used to model some aspects of molybdenum enzyme chemistry, such as complexes with a $\{MoOS_4\}$ core.[131] $[MoO(S_2NCR_2)_2]$ complexes undergo a reversible one-electron oxidation to the Mo^V species but cannot be further oxidized to Mo^{VI}. Complexes of the type [MoOLL′] (L = salicylaldehyde-2-hydroxyanil, salicylaldehyde-2-mercaptoanil; L′ = bipy, phen) can also be reversibly oxidized to the Mo^V species, and reduced to the Mo^{III} complex.[84] Other complexes which undergo reversible one-electron oxidation include $[MoOCl_2(bipy)(PMePh_2)]$, $[MoOCl_2(phen)(PMePh_2)]$ and $[MoOL(PMePh_2)]$ (H_2L = 8-hydroxyquinoline).[112]

36.4.2.3 Complexes Involving Multiple Molybdenum–Nitrogen Bonds

The chemistry of molybdenum with nitrogen ligands has been extensively studied, with particular attention being paid to those species which may serve as models of the nitrogen fixation process catalyzed by the nitrogenases. Complexes of Mo^{IV} which contain multiple Mo—N bonds are known with diazenide $(NNR)^-$, hydrazide $(NNRR')^{2-}$ or $(NHNRR')^-$, nitride $(N)^{3-}$, imide $(NR)^{2-}$ and azide $(N_3)^-$ ligands, and interest has focussed on reactions involving changes in the oxidation state of the molybdenum and the related reaction of the coordinated ligand.

Arenediazenide ligands, such as $(NNPh)^-$, are electronically analogous to NO and, therefore, can act as one- or three-electron donors. Thus, in mononuclear complexes they can coordinate to form linear Mo—N═N groups, in which the ligand can be regarded as $(NNAr)^+$ with the ligand acting as a three-electron donor, or to form bent Mo—N═N groups, with $(NNAr)^-$ acting as a one-electron donor.[132] The degree of multiple bonding in the Mo—N bond is reflected in the Mo—N—N interbond angle and the closer the angle is to 180° the higher is the bond order.

Diazenide complexes can be obtained by reacting a diazonium salt with a metal complex containing suitable leaving groups, such as carbonyls.[132] $[Mo(SC_6H_2Pr^i_3\text{-}2,4,6)_3(CO)_2]^-$ reacts with $[NNPh][BF_4]$ in MeCN to give $[Mo(NNPh)(SC_6H_2Pr^i_3\text{-}2,4,6)_3(NCMe)]$, in which the NNPh and MeCN ligands occupy the axial sites of a trigonal bipyramid with the sterically hindered thiolates equatorial.[133] The Mo—NNPh bond length is 1.78(1) Å, with a bond angle of 171(1)°, suggesting a considerable amount of Mo—N multiple bonding. This was the first reported example of a molybdenum–thiolate–diazenide complex and its existence supports the suggestion that bulky *ortho*-substituted arenethiolate ligands stabilize the binding of π-acceptor molecules to the molybdenum and, thereby, may provide interesting models for metal–sulfur sites that will interact with dinitrogen. Diazenide complexes can also be formed by reduction of hydrazide compounds. The Mo^{VI} complex $[Mo(NNHPh)_2L_2]\cdot H_2NNHPh$ (H_2L = 2,3-butanedithiol) reacts with PhSH in MeOH, in the presence of NEt_3, to give $[NHEt_3][Mo_2(NNPh)_4(SPh)_5]$.[134] In this dimer each Mo^{IV} atom is octahedrally coordinated to two terminal diazenides, one terminal thiolate and three bridging thiolates. The Mo—Mo distance is 3.527(1) Å, which is one of the largest separations identified in any simple molybdenum dimer.

For molybdenum dithiocarbamates, several compounds are known with both diazenide and hydrazide ligands attached. In $[Mo(NNCO_2Me)(NHNHCO_2Me)(S_2CNMe_2)_2]$ the overall geometry corresponds to that of a distorted pentagonal bipyramid, with the $NNCO_2Me$ group in one axial site and one sulfur atom from a S_2CNMe_2 group in the other. Two of the equatorial sites are occupied by the nitrogen atoms of the side-on hydrazide ligand, and the remaining three by sulfur atoms of the dithiocarbamates. The Mo—N distances of the hydrazide are 2.13(1) and 2.11(1) Å, whilst the Mo—N distance of the diazenide is 1.71(1) Å, this last value indicating considerable multiple bonding.[135] The reaction of alkyldithiocarbazates, $NH_2NHC(S)SR'$, with $[MoO_2(S_2CNR_2)_2]$ (R, R′ = Me, Et) yields mixed diazenidohydrazido complexes of composition $[Mo\{NNC(S)SR'\}\{NH_2NC(S)SR'\}\{S_2CNR_2\}_2]$. A structure determination of the complex R = Me, R′ = Et has identified pentagonal bipyramidal coordination at the molybdenum with the axial positions occupied by the diazenide ligand and one sulfur atom of a dithiocarbamate group. The hydrazide ligand is bound as a N,S chelate, *via* the NH_2 nitrogen and the unsubstituted sulfur atom, giving a five-membered ring. The other three equatorial positions are occupied by dithiocarbamate sulfur atoms. The Mo—N distance to the diazenide ligand is 1.77(1) Å, and to the hydrazide ligand is 2.250(8) Å.[136] Treatment of this complex with HCl in MeOH leads to the formation of the dimer $[(Me_2NCS_2)(O)Mo(\mu\text{-}NNC(S)SEt)_2Mo(S_2CNMe_2)]$, in which the bridging ligands are coordinated to both molybdenum atoms, *via* the terminal nitrogen, and are also bonded *via* the unsubstituted sulfur atoms to the non-oxo bound molybdenum.[137] The N—N distances in the bridging ligands are longer than found for bridging diazenido(1−) ligands in other compounds[138] and, hence, can be regarded as hydrazido(3−) groups with the formal oxidation state of each molybdenum being V.

The hydrazido(2−) complexes $[MoBr(NNRR')(dppe)_2]^+$ (R, R′ = H, alkyl) can be prepared in a two-stage process by the reaction of RBr with $[Mo(N_2)_2(dppe)_2]$ to give $[MoBr(NNR)(dppe)_2]$, which reacts with R′Br to give $[MoBr(NNRR')(dppe)_2]Br$.[139] Two-electron reduction of these complexes, followed by reaction with HBr, yields the imido–molybdenum(IV) species $[MoBr(NH)dppe)_2]Br$, which can be converted by treatment with base to the nitrido complex $[MoBr(N)(dppe)_2]$.[140] This series of reactions converts a terminally bonded dinitrogen ligand into a nitrido group, thereby breaking the N—N bond and forming an amine; the sequence provides a model of a possible mechanism for the fixation of dinitrogen. A crystal structure determination of $[MoBr(NH)(dppe)_2]Br\cdot MeOH$ has shown that the cation possesses octahedral geometry, with the bromide *trans* to the imido ligand. The Mo—N distance is 1.73(2) Å, typical of a multiple Mo—N bond.[141] The reaction of $Me_3Si(N_3)$ with $[Mo(N_2)_2(dppe)_2]$ yields the nitrido azido complex $[Mo(N)(N_3)(dppe)_2]$. The crystal structure of this compound shows a *trans* N—Mo—N_3 arrangement, with an unusually long Mo–nitrido distance of 1.79(2) Å and a Mo–azide distance of 2.20(2) Å. The Mo—N—N azide angle is 167.1(11)°, suggesting some π interaction in the metal–nitrogen bond.[141]

Complexes of molybdenum(IV) with 4-tolylimido ligands, $(Ntol)^{2-}$, have been prepared by reduction of Mo^{VI} and Mo^{V} species. Thus, $[MoO(Ntol)(S_2CNEt_2)_2]$ reacts with PPh_3 to form $[Mo(Ntol)(S_2CNEt)_2]$, which is easily oxidized by O_2 or DMSO to reform the starting material.[142,143] These oxygen transfer reactions are analogous to those of the *cis*-$[MoO_2(S_2CNR_2)_2]$/$[MoO(S_2CNR_2)_2]$ systems (see Section 36.4.2.2.1) and form a catalytic cycle for the oxidation of tertiary phosphines by O_2 or DMSO. Reduction of $[Mo(Ntol)Cl_3(PMe_3)_2]$ by Na—Hg in the presence of PMe_3 yields $[Mo(Ntol)Cl_2(PMe_3)_3]$.[143] The geometry of this complex is pseudooctahedral, with the phosphines in a *meridional* configuration and the chlorine atoms in mutually *cis* positions. The Mo—N distance is 1.739(2) Å, with an almost linear Mo—N—C arrangement.

^{15}N NMR spectroscopy has been used to characterize hydrazido, imido, and nitrido complexes of Mo^{IV}, and to monitor the reaction of $[MoI_2(^{15}N^{15}NH_2)(PMe_2Ph)_2]$ with HI in THF to give $^{15}NH_3$.[144] In the imido complexes, coupling between the ^1H and ^{15}N in the $(NH)^{2-}$ ligand was observed, confirming that the proton remains on the nitrogen, a fact which was not clear from other techniques.

36.4.2.4 Other Monomeric Complexes

36.4.2.4.1 Halide complexes

The simplest monomeric molybdenum(IV) halide complexes are the $[MoX_6]^{2-}$ anions. $Na_2[MoF_6]$ was first prepared by reacting MoF_6 with NaI in liquid SO_2, and was obtained as a water sensitive powder.[145] A crystal structure determination has been carried out on $Li_2[MoF_6]$, showing octahedral coordination with Mo—F distances between 1.927(2) and 1.945(2) Å.[146] Potassium salts of $[MoCl_6]^{2-}$ can be prepared by fusing $MoCl_6$ with KCl at 200 °C, or by reacting $[MoCl_5]$ with ICl and KCl at 150 °C.[147] $MoCl_5$ or $[NEt_4][MoCl_6]$ dissolved in either of the room temperature ionic liquids $AlCl_3$/*N*-*n*-butylpyridinium chloride or $AlCl_3$/1-methyl-3-ethylimidazolium chloride gives $[MoCl_6]^{2-}$ in solution as a stable entity.[148] No full crystallographic determination has been reported for an $[MoCl_6]^{2-}$ salt, but interpretation of the powder diffraction pattern of $K_2[MoCl_6]$ has given an Mo—Cl distance of 2.31(5) Å.[147] Hexabromomolybdates, $M_2[MoBr_6]$ (M = Rb, Cs), have been prepared by reacting $MoBr_3$, MBr and IBr in a sealed tube at 200 °C.[147]

Molybdenum(IV) tetrahalides react easily with a variety of ligands to form complexes of stoichiometry $[MoX_4L_2]$ (X = F, Cl, Br; L_2 = two neutral monodentate ligands or a neutral bidentate ligand). Examples of these complexes are known with N donor,[149–151] P donor,[106,150,152,153] As donor[150] and O donor ligands.[150] Complexes of this type can also be prepared by reduction of $MoCl_5$ with an excess of the ligand,[151] reaction between PR_3 and $[Mo_2(NCMe)_3Cl_6]$,[152] or by oxidation of $[Mo(CO)_4(PR_3)_2]$ with X_2 (R = Me, Et, Ph; X = Cl, Br).[153] An alternative and convenient synthesis is *via* $[MoCl_4(OEt_2)_2]$, which can be prepared from $MoCl_5$ and Et_2O in the presence of alkenes which act as Cl acceptors;[164] the Et_2O ligands of this complex are easily exchanged to give $[MoCl_4L_2]$. Vibrational studies have shown that the $[MoX_4(PR_3)_2]$ complexes adopt a *trans* octahedral configuration but, as expected in view of the span of the bidentate ligand, $[MoCl_4(dppe)]$ is obtained as the *cis* isomer.[106] Spectroscopic studies have indicated that $[MoCl_4L_2]$ (L = pyrazine, quinoxaline, 4,4′-bipyridyl or L_2 = 4,4′-bipyridyl) complexes contain a *cis*-$[MoCl_4N_2]$ chromophore.[151] In contrast, for $[NR_4]_2[MoCl_4(SnX_3)_2]$, prepared from $[MoCl_4]$ and $[NR_4][SnX_3]$ (R = H, Me, Et; X = Cl, Br), vibrational spectroscopic evidence favours a *trans* structure.[155]

Oxidation of $[Mo(CO)_3(PMe_2Ph)_3]$ with X_2 (X = Cl, Br) gives the seven-coordinate complexes $[MoX_4(PMe_2Ph)_3]$.[153] The structure of these is monocapped octahedral, the capped face consisting of three phosphorus atoms, while the cap and remaining three vertices are occupied by the halogen. A five-coordinate complex $[MoF_4(NEt_2Ph)]$ has been isolated, but not structurally characterized.[156]

The reaction of $[NR_4][LMo(CO)_3]$ with $SOCl_2$ yields $[LMoCl_3]$ (L = $\{HB(pyrazolyl)_2\}^-$, $\{HB(3,5\text{-}Me_2pyrazolyl)_3\}^-$), in which the tris(pyrazolyl)borate anion acts as a tridentate N donor ligand. An analogous compound is obtained from the oxidation of $[Mo(nane)(CO)_3]$ (nane = 1,4,7-trimethyl-1,4,7-triazacyclononane) by Br_2, which yields the cationic complex $[MoBr_3(nane)]^+$. This has been isolated as the $[PF_6]^-$ salt and a crystal structure determination accomplished; the geometry around the molybdenum is octahedral, with a *facial* configuration of the ligands and an average Mo—Br distance of 2.427(4) Å.[18]

Anionic ligands can replace chloride in $MoCl_4$, giving complexes of the form $[MoCl_2L_2]$, where L is a uninegative bidentate ligand. Thus, heating $MoCl_4$ or $MoCl_5$ with a β-diketone, HL, yields the corresponding monomeric $[MoCl_2L_2]$ complex.[157] Analogous complexes with the anions of 8-hydroxyquinoline and *N*-substituted salicylidenimines can be prepared by the direct substitution of chloride in $[MoCl_4(NCMe)_2]$, or of $(acac)^-$ in $[MoCl_2(acac)_2]$.[158,159] Although two bands are observed in the Mo—Cl stretching region of the IR spectra of these complexes, implying a *cis* geometry,[157–159] the crystal structure of $[MoCl_2$(*N*-methylsalicylaldimate)$_2]$ shows this complex to be the all *trans* isomer, with Mo—Cl, Mo—O and Mo—N bond lengths of 2.388(2), 1.953(6), and 2.137(8) Å, respectively.[160]

Complexes with tetradentate ligands L (L = $(salen)^{2-}$, $(acacen)^{2-}$), of the form $[MoCl_2L]$, have been prepared by reaction of $[MoCl_4(NCMe)_2]$ or $[MoCl_3(py)_3]$ with H_2L.[159] The reaction of HCl with [MoO(TPP)] (TPP = dianion of 5,10,15,20-tetra-4-tolylporphyrin) forms $[MoCl_2(TPP)]$; the structure of the latter involves a planar porphyrin ring with the *trans* chlorides at distances from the molybdenum of 2.347(4) and 2.276(4) Å, although they appear to be chemically equivalent.[113]

Seven-coordinate compounds $[MoClL_3]$ (L = anion of picolinic acid or a derivative of 8-quinolinol) have been prepared by prolonged reaction between $[MoO_2Cl_2]$ and HL; these complexes have been characterized by spectroscopic and magnetic measurements but no structural conclusions were possible.[161]

36.4.2.4.2 Dithiocarbamate and dithiocarboxylate ligands

Eight-coordinate dithiocarbamate complexes, $[Mo(S_2CNRR')_4]$, have been prepared by several different methods including: substitution reactions between $MoCl_4$ and $(S_2CNRR')^-$ salts;[162] oxidative decarbonylation of $[Mo(CO)_6]$ with $K(S_2CNRR')$,[163] and of $[Mo(CO)_6]$[164] or $[Mo(\eta^6\text{-}C_7H_8)(CO)_3]$[29] with $(RR'NCS_2)_2$. The crystal structure of $[Mo(S_2CNEt_2)_4]$ has shown that the molecule possesses a distorted square antiprismatic (or triangular dodecahedral) geometry around the central molybdenum atom.[165] The reaction between $MoCl_5$ and potassium pyrrole-*N*-carbodithioate in aqueous solution yields $[Mo\{S_2C(NC_4H_4)\}_4]$, which also has dodecahedral coordination about the metal, with the two different sulfur sites at average bond lengths of 2.541(6) and 2.507(6) Å.[166]

Dithiocarbamate ligands are also known to form seven-coordinate mixed-ligand complexes with Mo^{IV}. The complex $[Mo(SH)(S_2CNEt_2)_3]\cdot THF$ has been isolated as a product of the reaction between $[Mo(CO)_2(S_2CNEt_2)_2]$ and $[Fe_4(Cp)_4S_6]$.[167] The coordination around the molybdenum is pentagonal bipyramidal, with the axial sites occupied by the $(SH)^-$ ligand and one sulfur of a dithiocarbamate ligand. Complexes $[MoX(S_2CNR_2)_2L_2]Y$ and $[MoX_2(S_2CNR_2)_2L]$ ($R_2 = Me_2$, Et_2, $(CH_2)_5$; X = Cl, Br; Y = Cl, PF_6, BF_4, BPh_4; L = $\frac{1}{2}$(dppe), PPh_2Me, PMe_2Ph, PEt_2Ph) have been prepared by the reduction of the corresponding $[MoOX_2(S_2CNR_2)_2]$ complex with the phosphine. The crystal structures of $[MoCl(S_2CNEt_2)_2(PMePh_2)_2][BF_4]$ and $[MoCl(S_2CNEt_2)_2(dppe)][PF_6]$ both show pentagonal bipyramidal coordination around the molybdenum but, in the former, both axial positions are occupied by phosphorus atoms and, in the latter, the chlorine atom and one phosphorus of the dppe are in the axial positions.[168]

The reaction between $[MoCl_4(NCPr)_2]$ and dithiocarboxylic acids is a general route to the preparation of eight-coordinate $[Mo(S_2CR)_4]$ complexes.[169] The crystal structure of $[Mo(S_2CPh)_4]$ reveals these compounds to be isostructural with the dithiocarbamates, with a dodecahedral coordination around the molybdenum and average Mo—S distances of 2.475(1) and 2.543(1) Å to the two different sulfur sites.[170] Cyclic voltammetry has shown that in $[Mo(acda)_4]$ (Hacda = 2-aminocyclopent-1-ene-1-dithiocarboxylic acid) the Mo^{IV} can be reversibly oxidized and reversibly reduced in one-electron processes.[171]

The cyclopentadienedithiocarboxylate dianion, $(C_5H_4CS_2)^{2-}$, reacts with $[MoCl_4]$ to give the six-coordinate complex $[Mo(S_2CC_5H_4)_3]^{2-}$, which has been isolated as its $[NEt_4]^+$ salt. This has been characterized by spectroscopic measurements but the structure has not been determined.[172]

36.4.2.4.3 Eight-coordinate complexes with other bidentate ligands

In addition to dithiocarbamate and dithiocarboxylate compounds, Mo^{IV} forms eight-coordinate complexes with several other chelating ligands. For example, $[Mo(CO)_6]$ reacts with $\{PhCOCHC(Ph)S\}_2$ to form $[Mo\{OC(Ph)CHC(Ph)S\}_4]$, in which the monothio-β-diketonate ligands are bound to the metal *via* oxygen and sulfur atoms to form six-membered rings.[173] Compounds have also been prepared and studied with ligands containing both a heterocyclic nitrogen and anionic oxygen donor, such as 8-quinolinolate or picolate derivatives.[174,175] These are prepared by stepwise substitution of chlorides in $MoCl_4$ and, by isolating intermediate partially substituted species, it is possible to prepare mixed-ligand $[MoL_nL'_{4-n}]$ complexes, where L and L′ are anionic bidentate N,O-donor ligands.[175]

36.4.2.4.4 Alkoxy and thiolate ligands

Alkoxy complexes of stoichiometry $[Mo(OR)_4]$ can be prepared from the reaction of $[Mo(NMe_2)_4]$ with alcohols. With bulky groups ($R = Bu^t$, CH_2Bu^t) the complexes are monomeric, but for $R = Pr^i$, Et or Me polymeric species are formed.[176] The reaction of $[Mo(NMe_2)_4]$ with 1-adamantol (Hado) yields $[Mo(ado)_4(NHMe_2)]$, The structure of this compound is trigonal bipyramidal, with the amine ligand and one alkoxy group occupying the axial sites.[177] Phenoxy complexes can also be prepared by reacting $[Mo(NMe_2)_4]$ with phenols: the stoichiometry of the products obtained with a series of alkylated phenols was found to be dependent on the size of the substituent. $[Mo(NMe_2)_4]$ reacts with 4-methylphenol (Hmp) to yield $[Mo(mp)_4(NHMe_2)_2]$ but 2,6-diisopropylphenol (Hdpp) only substitutes three amido ligands to give $[Mo(dpp)_3(NMe_2)(NHMe_2)_2]$.[178] Trialkylsilanols also react with $[Mo(NMe_2)_4]$ to give the six-coordinate species $[Mo(OSiR_3)_4(NHMe_2)_2]$, (R = Me, Et). The crystal structure of $[Mo(OSiMe_3)_4(NHMe_2)_2]$ shows a *trans* octahedral geometry, with an average Mo—O bond length of 1.951(4) Å.[176]

$[Mo(SBu^t)_4]$ was the first monomeric four-coordinate thiolate complex of Mo^{IV} to be reported. The compound was prepared by treating $MoCl_4$ with $Li(SBu^t)$ and characterized by X-ray crystallography; the arrangement of the four sulfur atoms around the molybdenum has approximately D_{2d} symmetry, with the Mo—S distance 2.235(3) Å.[179] This complex is very labile and readily undergoes ligand exchange reactions with Bu^tNC, CO and PMe_2Ph to yield *cis*-$[Mo(SBu^t)_2(CNBu^t)_4]$, $[Mo_2(\mu\text{-}SBu^t)_2(CO)_8]$, and $[Mo_2(\mu\text{-}S)_2(SBu^t)_4(PMe_2Ph)_2]$, respectively.[180] The reaction between $[Mo(SBu^t)_4]$ and the peptide ligands Ac-Cys-OH, Z-Ala-Cys-OMe and Z-Cys-Ala-Ala-Cys-OMe (Z = carbobenzoxy) (in which the free amino groups at the N terminus is blocked to enforce coordination *via* the cysteine sulfur atoms) give $[Mo(Ac\text{-}Cys\text{-}OH)_4]$, $[Mo(SBu^t)_2(Z\text{-}Ala\text{-}Cys\text{-}OMe)_2]$ and $[Mo(Z\text{-}Cys\text{-}Ala\text{-}Ala\text{-}Cys\text{-}OMe)_2]$, respectively. Reaction of $[Mo(SBu^t)_4]$ with $[Fe_4S_4(SPr^i)_4]^{2-}$ has been followed by CD spectroscopy, and the formation of a Mo—Fe cluster observed.[181]

The bulky thiolate ligand 2,4,6-triisopropylbenzenethiolate (tipt) forms the stable complex $[Mo(tipt)_4]$, which gives monoadducts with alkynes, nitriles, CO and other small ligands and, unlike $[Mo(SBu^t)_4]$, the molybdenum does not appear to be reduced in these reactions.[182]

The reaction of $MoCl_5$ with $(HSCH_2CH_2)_2S_2$ (H_2mes) yields $[Mo(mes)_2]$, in which the $(mes)^{2-}$ dianions act as tridentate S donor ligands. The molecule possesses trigonal prismatic geometry with average Mo—S bond distances of 2.427(8) and 2.361(9) Å for the thioether and mercapto sulfurs, respectively. The complex undergoes a reversible one-electron oxidation and a reversible one-electron reduction;[183] an attempt to oxidize the complex with $Ag[PF_6]$ yielded the adduct $[\{Mo(mes)_2\}_2Ag][PF_6]\cdot\frac{1}{2}DMF$. In this latter compound the molybdenum atoms retain a distorted trigonal prismatic coordination, and one of the mercapto sulfur atoms of each $(mes)^{2-}$ ligand is also coordinated to Ag, forming a severely distorted tetrahedron.[184] The reaction between $[MoCl_4(PPh_3)_2]$ and $(HSCH_2CH_2)_2PPh$ gives $[Mo\{(SCH_2CH_2)_2P\}_2]$. The structure of this compound is analogous to that of $[Mo(mes)_2]$, with the ligands acting as tridentate P,S,S donors in a *trans* octahedral configuration; the average Mo—S distance is 2.348(9) Å.[155]

Addition of CS_2 to $[PPh_4]_2[(S_4)_2MoS]$ leads to insertion of the CS_2 into the Mo—S bonds and formation of perthiocarbonate ligands, $(CS_4)^{2-}$. The crystal structure of $[PPh_4]_2[(CS_4)_2MoS]$ shows that the anion possesses a square pyramidal geometry, with an apical sulfido ligand at a Mo—S distance 2.126(3) Å and two nearly planar CS_4^{2-} ligands in a *trans* arrangement with average Mo—S distances of 2.327(3) and 2.383(3) Å for the persulfido and sulfide S atoms, respectively.[186]

Molybdenum(IV) complexes with dithiolenes, of general formula $[Mo(S_2C_2R_2)_3]^{2-}$, adopt trigonal prismatic coordination. These compounds are discussed in Section 36.6.6.

36.4.2.4.5 Dialkylamide and other nitrogen donor ligands

The tetrakis(dialkylamido) complexes $[Mo(NR_2)_4]$ (R = Me, Et) have been synthesized by the reaction of $MoCl_5$ with $Li(NR_2)$ and spectroscopic evidence suggests a monomeric distorted tetrahedral geometry for the MoN_4 core. These compounds are highly reactive and the methyl derivative is readily converted into $[Mo(S_2CNMe_2)_4]$ by insertion of CS_2 into the Mo—N bonds,[187] and reactions with alcohols, phenols and silanols yield alkoxy, phenoxy or silanoxy complexes[176–178] (see Section 36.4.2.4.4).

The reaction of $[MoCl_4(NCMe)_2]$ with K(SCN) in MeCN solution yields $[Mo(NCS)_6]^{2-}$ and this anion has been isolated as its $[NBu_4]^+$ and $[AsPh_4]^+$ salts.[188] Vibrational spectroscopy has indicated N-bonded rather than S-bonded thiocyanate ligands. The adduct $[Mo(NCS)_4(bipy)]$ has also been obtained.[189]

36.4.2.4.6 Cyanide and isocyanide complexes

The octacyanomolybdate(IV) anion $[Mo(CN)_8]^{4-}$ has long been known and characterized and many aspects of its chemistry have been reviewed.[190,191] This complex is readily obtained from the addition of cyanide ions to aqueous solutions of Mo^{III} or Mo^{V} compounds. A simple preparation is by the reaction between molybdate(VI), $K[BH_4]$ and CN^- in the presence of acetic acid, followed by precipitation with ethanol.[192] The crystal structure of several complexes containing the $[Mo(CN)_8]^{4-}$ anion have been reported. In $K_4[Mo(CN)_8]\cdot 2H_2O$,[193] $[C_6H_6NO_2]_4[Mo(CN)_8]$[194] and $Rb_4[Mo(CN)_8]\cdot 3H_2O$[195] dodecahedral coordination has been found, but all the Mo—C bond lengths are equal, at 2.163(7) Å.[193] However, the molybdenum atoms in $[NH_4]_4[Mo(CN)_8]\cdot 0.5H_2O$ exist in two different coordination geometries, dodecahedral and square antiprismatic.[195] The structure of $H_4[Mo(CN)_8]\cdot 4HCl\cdot 12H_2O$ has not been completely determined, but the compound is isomorphous with the tungsten analogue, which is known to have a square antiprismatic geometry about the metal.[196] $[NHEt_3]_2[H_3O]_2[Mo(CN)_8]$ contains anions with a coordination polyhedron which is half-way between a dodecahedron and a bicapped trigonal prism.[197] Energy differences between the various topologies of $[Mo(CN)_8]^{4-}$ are small (see ref. 8, pp. 11–22), and rapid interconversion between them may be expected in solution, with crystal packing forces determining the more stable arrangement in the solid state.

Vibrational spectroscopy has been used to investigate the structure of aqueous solutions of $[Mo(CN)_8]^{4-}$. Raman spectra of $K_4[Mo(CN)_8]\cdot 2H_2O$ in the region 500–200 cm^{-1} are identical for solid and solution species and the number of totally symmetric modes is consistent with dodecahedral rather than square antiprismatic geometry.[198] ^{13}C NMR spectroscopy detects only one sharp carbon resonance even at temperatures of −160 °C, which may be due to the formation of square prismatic species, dodecahedral species with negligible differences in chemical shift between the two carbon sites, or rapid interconversion between the two.[199] Polarized UV absorption spectra of $[Mo(CN)_8]^{4-}$ in the temperature range 4.2–300 K have been recorded and show dodecahedral geometry, with appreciably less distortion from idealized geometry at low temperatures than at room temperature.[200]

Salts of stoichiometry $MFe[Mo(CN)_8]$ ($M = Cs, Rb, K, NH_4$) can be prepared by treating $Li_4[Mo(CN)_8]$ with $FeCl_3$ and MCl, and the corresponding $M_2Fe[Mo(CN)_6]$ complexes are obtained using $FeCl_2$. There is weak interaction between the $[Mo(CN)_8]^{4-}$ and the Fe^{III} or Fe^{II} atoms, and IR spectroscopy has indicated the presence of Mo—CN—Fe bridges.[201,202] Evidence for the formation of $Co_2[Mo(CN)_8]$, on mixing $CoCl_2$ and $K_4[Mo(CN)_8]$ solutions, has been obtained from radiometric titrations[203] and, in a series of complexes $M_2[Mo(CN)_8]\cdot xH_2O$ (M = Mn–Zn), magnetic studies have demonstrated weak electronic interactions between the Mo^{IV} and M^{II} centres.[204]

Photolysis of aqueous solutions of $[Mo(CN)_8]^{4-}$ produces $[MoO_2(CN)_4]^{4-}$, $[MoO(OH)(CN)_4]^{3-}$ or $[MoO(OH_2)(CN)_4]^{2-}$ species, depending upon the pH.[119] In liquid NH_3, photolysis of $[NH_4]_4[Mo(CN)_8]$ yields the six-coordinate $[Mo(CN)_4(NH_3)_2]$ complex.[205] The eight-coordinate complexes $K_2[Mo(CN)_6L]$ (L = bipy, phen) can be prepared *via* substitution reactions of $[Mo(CN)_8]^{4-}$.[206]

A number of isocyanide complexes, $[Mo(CN)_4(CNR)_4]$ (R = Me, Pr, Bu^t, allyl, $NCPh_2$), have been made by the action of RI on $Ag_4[Mo(CN)_8]$. The crystal structure of $[Mo(CN)_4(CNMe)_4]$ shows a dodecahedral geometry, with the CN^- ligands at 2.177(8) Å and the MeNC ligands at 2.148(8) Å from the molybdenum.[207] Both ^{13}C NMR and vibrational spectroscopy indicate that only one isomer is present in solution.

Treatment of an aqueous solution of $K_5[Mo(CN)_7]$ with acetic, hydrochloric or ascorbic acid, or with hydrogen sulfide gives the hydride complex $K_4[MoH(CN)_7]\cdot 2H_2O$.[208] This has been characterized by ^1H and ^{13}C NMR and by vibrational spectroscopy.

36.4.2.4.7 Magnetic and electronic properties

The magnetic moments of some non-oxo molybdenum(IV) compounds are listed in Table 6.[5] The six-coordinate complexes are all paramagnetic, with moments ranging between 1.9 and 2.8 BM. The temperature dependence of the magnetic susceptibility of these complexes is consistent with a configuration derived from a $^3T_{1g}$ ground state split by spin-orbit coupling and/or a ligand field of low symmetry. Electronic absorption spectra have been recorded for many six-coordinate species but assignments have generally been tentative.[159] Seven-coordinate

Table 6 Magnetic Moments of some Non-oxo Molybdenum(IV) Complexes[5]

Compound	μ_{eff} (BM)	*T* (K)	*Ref.*
$K_2[MoCl_6]$	2.28	300	147
$[MoCl_4(NCMe)_2]$	2.48	a	149
$[MoCl_4(PEt_2Ph)_2]$	2.61	300	106
$[MoCl_2(acac)_2]$	2.63	297	159
$[MoCl_2(salen)]$	2.61	292	159
$[MoCl(pic)_3]$	2.70	294	161
$[Mo(S_2CNEt_2)_4]$	0.68	a	164
$[Mo\{SC(Ph)CHC(Ph)O\}_4]$	Diamagnetic		173
$[Mo(SBu^t)_4]$	Diamagnetic		179
$[Mo(NMe_2)_4]$	Diamagnetic		187
$K_2[Mo(NCS)_6]$	3.02	a	188
$[Mo(CN)_4(CNMe)_4]$	Diamagnetic		207

[a] Temperature not reported.

species, such as $[MoCl(pic)_3]$, are also paramagnetic with magnetic moments slightly below the spin-only value for a d^2 system because of spin-orbit coupling.[161]

Most of the eight-coordinate complexes are diamagnetic; although some compounds have been reported to possess a magnetic moment, this may be due to impurities.[5] The dodecahedral geometry results in a *d* orbital splitting pattern that leaves the d_{xy} level lowest and this is thought to contain the two electrons in $[Mo(CN)_8]^{4-}$ and $[Mo(L\text{–}L)_4]$ complexes.[187] This interpretation is in agreement with X-ray PES studies and CNDO/INDO MO calculations.[209]

The four-coordinate complexes $[Mo(SBu^t)_4]$ and $[Mo(NR_2)_4]$ are also diamagnetic. The UV–PE spectrum of $[Mo(SBu^t)_4]$ exhibits a low ionization potential at 6.8 eV which has been assigned to the ionization of electrons with predominant molybdenum $4d_{z^2}$ character, on the basis of discrete variational-X_α MO calculations on the model compounds $[Mo(SH)_4]$ and $[Mo(SMe)_4]$.[210] For $[Mo(NR_2)_4]$ (R = Me, Et), the UV–PE spectrum contains a low energy ionization at 5.3 eV which has been attributed to ionization from the molybdenum $4d_{x^2-y^2}$ orbital. This assignment was based on Fenske–Hall calculations on $[Mo(NMe_2)_4]$.[211]

36.4.2.5 Cyclopentadienyl Complexes

Complexes of the general formulae $[Cp_2MoLL']^{2+}$, $[Cp_2MoLX]^+$ and $[Cp_2MoXY]$ (L, L′ = neutral ligand; X, Y = uninegative ligand) are known for a wide variety of ligands (ref. 212 and refs. therein). The structure of $[Cp_2MoCl_2]$ is a typical example of these complexes. The cyclopentadiene rings are staggered with respect to each other and lie on either side of the $MoCl_2$ plane; the angle between normals of the two Cp planes is *ca.* 135° and the overall geometry about the metal is that of a distorted tetrahedron.[213] Other compounds are known in which a second metal atom is coordinated to the Cp_2Mo centre *via* bridges formed by the other ligands, such as the hydrides in $[Cp_2MoH_2]$[214] or the thiols in $[Cp_2Mo(SR)_2]$[215] complexes, but there appears to be little direct metal–metal interaction in these systems. Generally, in the course of chemical reactions the two Cp ligands remain coordinated and unchanged while the other two sites are reactive.

36.4.2.6 Dimeric and Polymeric Complexes

Molybdenum(IV) forms a variety of dimeric and polymeric complexes with oxido, hydroxido and carboxylato bridges in aqueous solution, which are discussed in Section 36.1.4. Oxide bridges have also been found in non-aqueous systems. The reaction between $[MoCl_2(TTP)]$ (TPP = dianion of tetra-*p*-tolylporphyrin) and *N*-phenylhydroxylamine yields $[\{MoCl(TPP)\}_2O]$, in which each Mo is octahedrally coodinated and is displaced 0.08 Å from the N_4 plane towards the bridging oxygen. The complex is paramagnetic (μ = 2.82 BM per Mo) with no magnetic interaction between the metal centres.[216] A similar structure has been found in ethoxooxidobis(1,5,9,13-tetrathiacyclohexadecane)-μ-oxidodimolybdenum(IV) trifluoromethylsulfonate hydrate: the cation contains a linear OMoOMo═O unit, with each macrocycle forming a planar coordination around a molybdenum atom and staggered 43° relative to each other.[217]

The oxo-bridged mixed oxidation state Mo^{IV}, Mo^{V} compound $[\{Mo(S_2CNEt_2)_3\}_2O][BF_4]$ has been prepared by the reaction of $[MoO(S_2CNEt_2)_3][BF_4]$ and PPh_3. The cation contains two pentagonal bipyramidal units linked by an almost linear Mo—O—Mo bridge, with the oxygen occupying an axial site in each coordination sphere. The complex is paramagnetic (μ = 2.17 BM per Mo) and is the first example of a dimolybdenum oxo-bridged species with no terminal oxo or sulfido ligands.[218] The dimeric complexes $[Mo_2O_3(S_2CNR_2)_2]$ (R = alkyl) have been identified spectroscopically and isolated as intermediates in the oxidation of $[Mo_2(S_2CNR_2)_6]$ to Mo^{V} and Mo^{VI} compounds. They have low magnetic moments (1.29 BM per Mo), suggesting the presence of a significant metal–metal interaction.[54]

Molybdenum sulfides react with CN^- in aqueous solution to form a wide variety of oligomeric cyanothiomolybdate anions, which contain Mo_2S, Mo_2S_2, Mo_3S_4 and Mo_4S_4 cores[59] (see Section 36.1.4). These types of complexes are formed under conditions thought to occur in prebiotic times on this planet and, thus, they could be important in the evolution of molybdenum enzymes. Sulfur bridges are also important in other non-oxido dimolybdenum(IV) species. In $[\{(HS)Mo(hd)\}_2][CF_3SO_3]_2$ (hd = 1,5,9,13-hetrathiacyclohexadecane) there is an MoSMoS ring, in which each sulfur atom belongs to an intact macrocycle coordinated in the endo form.[219] An Mo_2S_2 ring is also found in the sulfido-bridged complex $[(Pr_2NCS_2)(Pr_2NCS)Mo(\mu\text{-}S)_2Mo(SCNPr_2)(S_2CNPr_2)]$, which also contains bidentate dithiocarbamates and thiocarboxamido ligands bound to the molybdenums in a side-on manner, and in $[Mo_2(\mu\text{-}S)_2(SBu^t)_4(HNMe_2)_2]$. These compounds involve Mo—Mo distances of 2.707(2) and 2.730(1) Å, respectively, consistent with the formation of a metal–metal bond.[220] A triply-bridging sulfur atom is found in $[Mo_3S_4(Cp)_3]^+$, in which there is a triangular cluster of Mo atoms triply-bridged by one sulfur atom above the centre of the triangle, with three doubly-bridging sulfur atoms below.[221]

$[NEt_4]_2[Cl_3Mo(\mu\text{-}S_2)(\mu\text{-}Cl)_2MoCl_3]$ contains a dimolybdenum(IV) centre bridged by one disulfide and two chloride ligands; the Mo—Mo distance is 2.763(2) Å and the low magnetic moment (1.40 BM per Mo) indicates formation of a metal–metal bond, leaving one unpaired electron per Mo.[222] Disulfide bridges are also found in $[AsPh_4]_2[Mo_2Br_6(S_2)_2(SMe_2)_2]$, which has been obtained from $[MoBr_4(SMe_2)_2]$ and $[AsPh_4]_2(S_7)$.[56]

Chlorides act as bridging ligands in $[\{MoCl_4(PhC_2Ph)\}_2]$ and $[PPh_4]_2[Mo_2Cl_{10}]$. The $[Mo_2Cl_{10}]^{2-}$ anion contains two bridging Cl groups, with a Mo—Mo distance of 3.80 Å and, therefore, there is no direct metal–metal interaction.[223] The bromide-bridged anion $[(NO)_2Br_2Mo(\mu\text{-}Br)_2MoBr_2(NO)_2]^{2-}$ also contains a long Mo—Mo distance (3.978 Å).[56]

36.4.3 MOLYBDENUM(V)

36.4.3.1 Introduction

Molybdenum(V) chemistry is dominated by oxo complexes: many of these exist as dimers, but monomeric species can be isolated from strongly acidic solutions or under nonaqueous conditions. Non-oxo compounds are also known, both as monomers and as dimers or polymers with halide or sulfur bridges. ESR spectroscopy has been used extensively to investigate the properties of monomeric Mo^{V} systems, and has shown the participation of this oxidation state in the reactions of the oxomolybdoenzymes (see Section 36.6.7).

Of the binary halide compounds of Mo^{V}, only fluoride and chloride species have been isolated and characterized. $[\{MoF_5\}_4]$ can be prepared by the reduction of MoF_6 with Si powder in the presence of dry HF; it is also a product of the reactions between $[MoF_6]$ and $[Mo(CO)_6]$ or Mo, and of the direct combination of Mo with F_2. The crystal structure shows formation of a tetramer, in which the four molybdenum atoms form a square plane bridged by four fluorines. Each molybdenum is octahedrally coordinated to four terminal fluorines with an average Mo—F bond length of 1.78(8) Å, and to two bridging fluorines at a distance of 2.06(4) Å.[224] Monomeric $[MoF_5]$ can be prepared by condensation of $[MoF_5]$ vapour in an argon matrix at liquid hydrogen temperature, or by UV photolysis of $[MoF_6]$ in an argon matrix. Vibrational spectra of $[MoF_5]$ formed by the condensation route are consistent with a D_{3h} trigonal bipyramidal structure, but the IR spectrum of the photolysis product has been interpreted assuming a C_{4v} square pyramidal geometry.[225] In the liquid and vapour states $[MoF_5]$ exists as a monomer: the electronic spectrum of liquid $[MoF_5]$ suggests a trigonal bipyramidal shape but ^{19}F NMR data are consistent with a square pyramidal structure.[226]

In the solid state $MoCl_5$ forms dimers of edge-sharing $MoCl_6$ octahedra, with the two bridging chlorides involving significantly longer Mo—Cl bonds (2.53(1) Å) than the terminal chlorides (2.24(1) Å).[227] The long Mo—Mo distance of 3.84(2) Å and magnetic moment of 1.64 BM per Mo are consistent with the absence of significant metal–metal interaction. In the gas phase monomeric [$MoCl_5$] is formed, which has been shown by electron diffraction and Raman studies to have a trigonal bipyramidal structure.[228]

The reaction of MoF_3 with Cl_2 at 120 °C yields MoF_3Cl_2. Magnetic measurements indicate antiferromagnetic exchange between Mo^V centres in a polymeric structure and mass spectroscopic data show the presence of [$MoFCl_4$], [$MoCl_5$], [$Mo_2F_{10-n}Cl_n$] ($n = 0$–6) and [$Mo_3F_{15-m}Cl_m$] ($m = 0$–7) in the vapour above solid MoF_3Cl_2.[229]

$MoOCl_3$(s) exists in two structural forms; the monoclinic form contains a terminal oxo group and the tetragonal form involves bridging Mo—O—Mo units. The former is obtained by the reduction of $MoOCl_4$ with Al at 130 °C, the thermal decomposition of $MoOCl_4$, or by reacting SO_2(liq) or SbO_3 with $MoCl_5$.[230–232] The structure is based on an infinite zigzag chain of molybdenum atoms, each possessing two pairs of μ_2-Cl atoms and *cis* terminal oxygen and chlorine atoms in a distorted octahedral environment; the length of the Mo—Cl bond (2.81 Å) *trans* to the oxo group is significantly longer than those *cis* to this group.[232] The room temperature magnetic moment of this material is *ca.* 1.65 BM. The latter form of $MoOCl_3$ is obtained by reduction of $MoOCl_4$ with HI in SO_2(liq) at −23 °C; the compound is isomorphous with $NbOCl_3$;[230] $MoOBr_3$ also adopts this structure.[233] $MoCl_5$ reacts with Sb_2Y_3 (Y = S, Se) at an elevated temperature, or with Sb_2S_3 in CS_2 solution at room temperature, to produce $MoYCl_3$.[66]

36.4.3.2 Monomeric Complexes with an Oxo Group

36.4.3.2.1 General comments

Molybdenum(V) forms many mononuclear complexes containing one oxo group but no analogues have been reported of the *cis*-dioxomolybdenum(VI) or *trans*-dioxomolybdenum(IV) species. The O_t group possesses a multiple bond (see Section 36.4.2.2.1) and dominates the ligand field at the molybdenum, so that the $4d^1$ electron is located in an orbital which is in the plane (xy) perpendicular to the oxo group (z).[80,92,93] Most monomeric Mo^V oxo complexes adopt a square pyramidal or octahedral geometry, and their electronic configuration is $(d_{xy})^1$.

36.4.3.2.2 Halide complexes

A large number of compounds are known containing the anions $[MoOX_5]^{2-}$ (X = F, Cl, Br). These complexes are stable in acidic solution, but magnetic measurements show that as the pH is increased towards neutrality the magnetic moment decreases, indicative of dimer formation, which allows an antiferromagnetic metal–metal interaction.[234] Various other equilibria exist in solution (see Section 36.1.5) and these produce, *inter alia,* the monomeric species $[MoOX_4]^-$ and $[MoOX_4(H_2O)]^-$ (X = Cl, Br), which have been characterized by ESR spectroscopy.[235]

Crystal structures have been obtained for several salts containing the six-coordinate $[MoOX_5]^{2-}$ and the five-coordinate $[MoOX_4]^-$ anions. In $K_2[MoOF_5]\cdot H_2O$ the molybdenum is octahedrally coordinated; the Mo—O_t distance is 1.66(2) Å. The length of the four *cis* Mo—F bonds (1.88(3) Å) is significantly shorter than that of the *trans* Mo—F (2.02(1) Å)[236] and the molybdenum is displaced out of the equatorial plane in the direction of the oxo ligand so that it is located close to the centre of the donor atom array; these are common features of all oxomolybdenum complexes. In $K_2[MoOCl_5]$, the Mo—O_t distance is 1.67 Å and the bond lengths to the *cis* and *trans* chlorides are 2.40 and 2.63 Å, respectively,[237] and $[AsPh_4][MoOCl_4(H_2O)]$ involves the following bond lengths: Mo—O = 1.672(15), Mo—OH_2 = 2.393(15) and Mo—Cl = 2.359(3) Å.[238] A shorter Mo—O_t bond is found in $[AsPh_4][MoOCl_4]$, which has a square pyramidal geometry with the oxo group axial; the Mo—O_t distance is 1.610(10) Å and the Mo—Cl distances are 2.333(3) Å.[239] Crystal structure determinations for related complexes include: $[MoOX_4(H_2O)]^-$ (X = Cl, Br, I),[240] $[MoOBr_4]^-$ [241] and $[MoOBr_5]^{2-}$.[242]

Complexes containing $[MoOX_4]^-$ anions (X = Cl, Br) are coordinatively unsaturated and react readily with neural or anionic ligands to give octahedral species. The mixed halide complex $[NEt_4]_2[MoOCl_4Br]$ can be obtained by treating $[NEt_4][MoOCl_4]$ with $[NEt_4]Br$; spectroscopic data indicate that the *trans* isomer is formed.[243]

Complexes of the form $[LH_2][MoOX_5]$ (L = bidentate N donor ligand; X = Cl, Br) can be isolated from the reaction between MoO_3 in hot HX and HI, followed by addition of L. These compounds lose HCl when dissolved in dry solvents, yielding $[MoOX_3L]$, which form as the *mer* or *fac* isomers depending on the conditions. With the ligand 2,2′-biquinolyl (biquin) an intermediate complex, $[biquinH][MoOCl_4]$, has also been isolated. Monodentate ligands L′ form the compounds $[L'H]_2[MoOCl_5]$, which can lose HCl to yield $[MoOCl_3L'_2]$.[244] For the ligand 8-hydroxyquinoline and its derivatives, IR spectra indicate that there is hydrogen bonding between $[LH]^+$ and $[MoOCl_5]^{2-}$ in $[LH]_2[MoOCl_5]$.[245]

Molybdenum halides and oxyhalides, including $[Mo_2Cl_{10}]$, $[Mo_2O_2Cl_6]$ and $[NH_4]_2[MoOCl_5]$, are useful starting materials for the preparation of monomeric Mo^V oxo compounds. A large number of these compounds are of the form $[MoOCl_3L_n]$ ($n = 1, 2$) and commonly involve O or S donor ligands.[246,247] $[Mo_2O_2Cl_6]$ reacts with THF to give $[MoOCl_3(THF)_2]$,[248,249] from which the loosely bound THF molecules are easily displaced by a wide variety of ligands, L, yielding $[MoOCl_3L_2]$. Many neutral ligands react directly with $[Mo_2O_2Cl_6]$ to give $[MoOCl_3L_2]$ or $[MoOCl_3L]$ species; $[Mo_2Cl_{10}]$ can also form oxo complexes by abstracting oxygen from the ligand or solvent. Equilibria exist in solution between the five- and six-coordinate products; for the oxygen-donor ligands Ph_3PO, $(Me_2N)_3PO$ or THF, the only product formed is the corresponding $[MoOCl_3L_2]$ compound, but for sulfur-donor ligands Ph_3PS, $(MeN)_2CS$ or Me_2S, the five-coordinate complex $[MoOCl_3L]$ predominates.[235] $[MoOCl_3L_2]$ (L = Ph_3PO; $P(NMe_2)_3O$) possess an octahedral coordination around the molybdenum, with the chloride ligands adopting a *mer* configuration and the phosphine oxides mutually *cis*. In the Ph_3PO adduct the Mo—O_t bond length is 1.662(13) Å, and the Mo—O distances to the phosphine oxide ligands *trans* to the oxo group and *trans* to chloride are 2.136(11) and 2.065(10) Å, respectively.[250] This compound reacts with halide ions and the rate-determining step is the loss of the Ph_3PO ligand *trans* to the oxo group; subsequent to halide substitution, exchange of the substituted halide with the other Ph_3PO molecule occurs.[251] $[MoOCl_3(SPPh_3)]$ adopts a geometry intermediate between square pyramidal, with an axial oxo group, and trigonal bipyramidal, with sulfur and chloride ligands axial. The Mo—O_t bond length is 1.647(3) Å.[252]

Molybdenum(V) forms a wide range of complexes with bidentate and polydentate ligands. The reaction of $[pyH]_2[MoOX_5]$ (X = Cl, Br) with acetylacetone yields $[pyH][MoOX_3(acac)]$ which, on heating, is converted to $[MoOX_2(acac)(py)]$; $[Mo_2Cl_{10}]$ reacts with Hacac and other β-diketones, HL, to give the disubstituted complex $[MoOClL_2]$.[253] In all of these compounds, spectroscopic evidence suggests that the diketonate acts as a bidentate ligand to form an octahedral mononuclear complex. Dithiocarbamates, R_2dtc, form mononuclear $[MoOX(R_2dtc)_2]$ (X = F, Cl, Br, I) products in reactions involving the reduction of MoO_3 by $N_2H_4{\cdot}HX$.[254] An analogous complex, [MoOCl(dttd)] (H_2dtdd = S,S′-bis(2′-mercaptophenyl)-1,2-mercaptoethane, Figure 2; Section 36.4.2.2.1), is formed in the reaction between $[NH_4]_2[MoOCl_5]$ and H_2dttd. This complex undergoes a reversible one-electron reduction to $[MoOCl(dttd)]^-$ but is not easily oxidized to an Mo^{VI} species.[90]

Interest in the nature of the molybdenum coordination sphere in the oxo–molybdenum enzymes has led to investigation of complexes formed by Mo^V with Schiff bases and related ligands. $[MoOCl_3(THF)_2]$ has been shown to react with a variety of Schiff bases (H_2L), including $salenH_2$ and its substituted derivatives, in the presence of base, with Li_2L or $L(SiMe_3)_2$ to give the corresponding [MoOLCl] complex, the configuration of which was assigned from ESR and IR data.[255] The anionic complex $[MoOCl_2L]^-$ (H_2L = *N*-2-hydroxysalicylidenimine) has been prepared from $[pyH]_2[MoOCl_5]$ and H_2L, and the crystal structure of its $[AsPh_4]^+$ salt has been determined. The molybdenum atom in the anion is in a distorted octahedral environment in which the planar tridentate Schiff base occupies *meridional* positions, with the nitrogen atom *trans* to the terminal oxo group; the Mo—O_t bond length is 1.673(3) Å.[256] ESR spectroscopy has been used to investigate equilibria in a solution of [MoOCl(salpn)] ($salpnH_2$ = *N,N′*-bis(salicylaldehydo)-1,3-diaminopropane); the results showed that considerable dimer formation occurred, with the dominant monomeric species being [MoOCl(salpn)].[257]

The reactions of $[MoOCl_3(THF)_2]$ with substituted quinolines yield a number of different types of complexes containing one or more quinoline ligands. The crystal structure of $[H_2ox][MoOCl_3(ox)]$ (Hox = 8-hydroxyquinoline) has been determined; the hydroxyquinolate acts as a bidentate O,N ligand, with the oxygen bound *trans* to the oxo group in an octahedral geometry.[258] $[Mo_2O_2Cl_6]$ reacts with 8-mercaptoquinoline, H(tox), to give $[MoOCl_2(tox)Q]$ (Q = solvent) in which tox is bidentate, $[MoOCl(tox)_2Q]$ in which one tox is bidentate and the other monodentate, and $[MoO(tox)_3]$ in which two tox ligands are bidentate and the third is monodentate. $[MoOCl(tox)_2]$ has been prepared by the reaction of $[MoCl_3(DMF)_3]$ and 8,8′-diquinolyl disulfide and its crystal structure determined. The molybdenum atom is in a distorted octahedral environment, with two bidentate tox ligands; the oxo and chloride ligands occupy *cis* positions, and the sulfur atoms of the tox ligands are *trans* to each other; the $Mo—O_t$ bond length is 1.716(4) Å.[259]

The reduction of nitrate and nitrite by a number of monomeric oxomolybdenum(V) complexes, including $[MoOCl_3(OPPh_3)_2]$, $[MoOCl_3L]$ (L = bipy, phen), and $[MoOClL'_2]$ (L′ = ox, tox), has been studied.[80,260–262] These reactions have relevance to the reduction of nitrate and the competitive inhibition by nitrite which occurs at the molybdenum centre of the nitrate reductases. The molybdenum centres of the MoO^{3+} complexes reduce nitrate in a one-electron step with concomitant transfer of an oxygen atom, producing NO_2 and the corresponding *cis*-MoO_2^{2+} complex. The initial substitution of nitrate occurs at the labile position *trans* to the oxo group, but the redox reaction proceeds only after ligand rearrangement so that the nitrate is coordinated by an oxygen atom *cis* to the oxo ligand.[260] Nitrate is also reduced to NO_2 by tripeptide (Gly-Gly-Met or Gly-Gly-His) molybdenum(V) complexes formed in the reaction of $[NH_4]_2[MoOCl_5]$ with the tripeptides immobilized *via* the terminal glycine to a polystyrene matrix. Analysis shows a molybdenum to tripeptide ratio of 1:1, but it is unclear how these potentially tridentate ligands bind to the molybdenum and which other ligands are present in the coordination sphere.[263] Molybdenum(V) chloro complexes of a cysteinyl-containing peptide have been prepared by reaction of $[MoOCl_3(THF)_2]$ with the peptide alone and in the presence of Et_3N. Although the compound produced under the former conditions is capable of reducing nitrate, this is not the case for the latter compound, perhaps because the chelation of the Mo^V prevents nitrate from ligating at a site *cis* to the oxo group.[264]

The properties of [MoOCl(TPP)] (H_2TPP = *meso*-tetraphenylporphyrin) and related complexes are covered in Section 36.4.3.2.4.

36.4.3.2.3 Complexes with sulfur and selenium donor ligands

Complexes of the type $[MoO(SR)_4]^-$ are important, both in their own right as interesting chemical species and as systems which provide information of relevance to the molybdenum centres of the oxomolybdoenzymes.[249,265–8] $[MoO(SAr)_4]^-$ (Ar = Ph, 4-MeC_6H_4, C_6Cl_5) and $[MoO(SePh)_4]^-$ have been prepared by the reaction of the corresponding thiol, or selenol, with $[MoOCl_3(THF)_2]$, $[pyH]_2[MoOCl_5]$ or $[pyH][MoOBr_4]$ in MeCN containing Et_3N and isolated as their quaternary ammonium salt. Their alkanethiolato analogues, $[MoO(SR)_4]^-$ (R = Et, CH_2Ph) have been prepared in solution at −60 °C. $[MoO(SPh)_4]^-$ has a square pyramidal structure with $Mo—O_t$ = 1.669(9) Å and Mo—S = 2.403(5) Å. The spectroscopic and magnetic properties of these species have been investigated in considerable detail[265] and chemically reversible one-electron reduction has been observed; oxidation leads to the formation of the dimer $[Mo_2O_2(XR)_6Q]$ (X = S, Se; Q = solvent) and RXXR.[266] An alternative synthesis to $[MoO(SPh)_4]^-$, which has also been used to prepare $[MoO\{S(CH_2)_nS\}_2]^-$ (n = 2, 3), involves the reaction of the Mo^{VI} butanediolato complex, $[MoO_2\{MeCH(O)CH(OH)Me\}_2]\cdot 2[MeCH(OH)CH(OH)Me]$ with the appropriate thiolato anion in MeOH. The crystal structure of $[PPh_4][MoO\{S(CH_2)_3S\}_2]$ has been determined and the anion (Figure 3) has a square pyramidal geometry with Mo—O = 1.667(8) Å and Mo—S = 2.389(4) Å. Although $[MoO(SPh)_4]^-$ reacts with R_2NNH_2 ($R_2 = Me_2$ or $(CH_2)_5$) in refluxing MeOH to produce $[MoO(NNR_2)(SPh)_3]^-$, it is inert to HN_3; however, $[MoO\{S(CH_2)_3S\}_2]^-$ reacts with HN_3 in MeOH to produce $[Mo_2O_2(\mu\text{-}N_3)\{S(CH_2)_3S\}_3]^-$.[267] $[MoOCl_4]^-$ in MeCN containing Et_3N reacts with the sterically hindered thiol, 2,4,6-$Pr^i_3C_6H_2S^-$ (Htipt) to produce $[MoO(tipt)_4]^-$.[268]

Oxomolybdenum(V) centres can form five- and seven-coordinate complexes with bidentate carbodithioate ligands. The reaction between $Na_2(C_5H_4CS_2)\cdot THF$ and $[Mo_2O_2Cl_6]$ in the presence of $[NEt_4]Br$ gives $[NEt_4][MoO(S_2CC_5H_4)_2]$. ESR spectra of this complex in solution and frozen solution indicate a d^1 monomer with square pyramidal C_{4v} symmetry.[172] Reductions of $[MoO_2L_2]$ with HL′ (HL′ = 2-aminocyclopent-1-ene-1-carbodithioic acid and its *N*-alkylated derivatives) yield the mixed ligand compounds $[MoOL_2L']$. The magnetic moment values (*ca.* 1.60 BM) and ESR spectra are consistent with the $(d_{xy})^1$ ground state.[269] $[MoO(NCS)(R_2dtc)_2]$

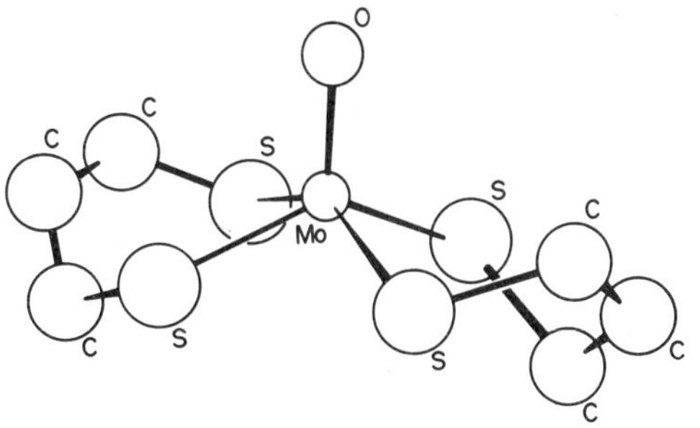

Figure 3 Structure of $[MoO(SCH_2CH_2CH_2S)_2]^-$ in its $[PPh_4]^+$ salt[266]

(R_2dtc = substituted dithiocarbamate) have been prepared and IR spectra show the thiocyanates to be N-bonded. ESR, electronic spectra and magnetic moment values (1.69 BM) confirm the monomeric nature of the complex. Cyclic voltammetry has been used to demonstrate reversible one-electron reduction to the corresponding Mo^{IV} species.[270]

Interesting redox behaviour is exhibited by the complexes $[MoOL]^-$ where L is a tetradentate *S,S′,N,N′* aromatic aminothiol group (*N,N′*-bis-2′-mercaptophenyl-1,2-diaminoethane (Figure 2) and its 1,2-diamino-1,2-dimethylethane analogue). In these complexes the amino groups are deprotonated, giving the ligand dithiolene characteristics. The compounds undergo both one-electron oxidation and reduction at a platinum electrode in DMF and show no tendency to form dimeric species due to the geometrical restraint of the ligand L.[271] The electrochemistry of a series of MoO^{3+} complexes, including $MoOClL_2$ (L = 8-mercaptoquinoline) and MoOClL (L = *N,N′*-dimethyl-*N,N′*-bis-2′-mercaptoethyl-1,2-diaminoethane (Figure 2) and *N,N′*-bis-2-mercapto-2-methylpropyl-1,2-diaminoethane), has been reported; the complexes undergo a facile reduction to an Mo^{IV} species but oxidation was not observed.[87]

36.4.3.2.4 Complexes with nitrogen donor ligands

Solutions containing $[MoOCl_5]^{2-}$ react with thiocyanic or selenocyanic acid to produce $[MoO(NCS)_5]^{2-}$ and $[MoO(NCSe)_5]^{2-}$, respectively.[272] Magnetic, vibrational and electronic spectra indicate that in both anions the ligands are N-bonded, forming monomeric complexes with close similarities to the corresponding halide $[MoOX_5]^{2-}$ species.

In HCl solution, in the presence of $SnCl_2$, Mo^V forms complexes $[MoO(Hd)_2]^+$ with the dioximes (H_2d) of cyclohexane-1,2-dione and cyclopentane-1,2-dione.[273] Several compounds of tetradentate Schiff bases[255,274] with oxomolybdenum(V) have been prepared and the crystal structure of *trans*-[MoO(salen)(MeOH)]Br has been determined.[274] The Mo—O_t distance is 1.666(10) Å, and there is close interaction in the crystal between each cation and an attendant bromide anion.

Tetraphenylporphyrin, H_2TPP, reacts with $[Mo(CO)_6]$ in refluxing decalin to give [MoO(TPP)(OH)]. Derivatives of the type [MoO(TPP)X] are known with a wide variety of uninegative ligands X, including Cl, O_2H, F, Br, BF_4, NCO, N_3 and NCS.[275] Vibrational, electronic and ESR spectroscopy confirm the presence of one unpaired electron and a central Mo—O_t group with a *trans* octahedral geometry of C_{4v} symmetry. The cyclic voltammograms of [MoO(TPP)X] (X = MeO, AcO, Cl) show a reversible one-electron reduction to a Mo^{IV} species, whereas that for the related corrolate complex, [MoO(mec)] (mec = trianion of 2,3,17,18-tetramethyl-7,8,12,13-tetraethylcorrole) shows both one-electron reduction and oxidation processes. In addition to one-electron reduction, [MoO(TPP)(OMe)] exhibits other redox couples, involving dissociation of $(OMe)^-$ and formation of Mo^{III} and Mo^{II} complexes.[276]

[MoO(TPP)Br] can be chemically reduced by superoxide to give [MoO(TPP)], *via* Mo^V and Mo^{IV} superoxide complexes, which are stable at low temperatures.[114] In the presence of superoxide, the substitution reaction of [MoO(TPP)Br] with MeOH to give [MoO(TPP)(OMe)] proceeds *via* an Mo^{IV} species rather than *via* $[MoO(TPP)(MeOH)]^+$, which is the pathway in the absence of superoxide.[277] Photochemical reduction of [MoO(TPP)(OMe)] yields [MoO(TPP)], which reacts with O_2 in MeOH to re-form [MoO(TPP)(OMe)] and produce H_2O_2, thus giving a catalytic route for the photoassisted reduction of molecular oxygen to hydrogen peroxide.[278]

36.4.3.2.5 Spectroscopic studies

A comprehensive tabulation of such data is available in ref. 5 and selected details will be presented here. Monomeric oxomolybdenum(V) complexes are paramagnetic, with magnetic moments in the range 1.65–1.73 BM (Table 7) due to the single unpaired electron. This makes Mo^V complexes accessible to investigation by ESR spectroscopy, and many studies have been reported, both with well-defined coordination compounds and with biological systems.[131,238,250,252,255,265,280–284] A monomeric Mo^V complex in solution which possesses sufficient thermal motion to average the molecular anisotropy on the ESR timescale has a characteristic ESR spectrum. This consists of a strong central line, due to resonance of molecules with even molybdenum isotopes ($I = 0$), flanked by six hyperfine lines due to interaction between the unpaired electron and the ^{95}Mo and ^{97}Mo nuclei ($I = \frac{5}{2}$). When molecular tumbling does not occur, for example in solids and frozen solutions, or is slow on the ESR timescale, the spectrum becomes more complex, displaying major features due to the principal g values. Monomeric Mo^V centres with axial symmetry should manifest two principal resonances, $g_{\parallel}$ (g_{zz}) and $g_{\perp}$ (g_{xx}, g_{yy}), whereas in systems of lower symmetry three principal resonances should be observed, each associated with six hyperfine lines.[80] A low-symmetry ligand field allows extensive d orbital mixing, which generally results in the principal molecular g values being displaced from the metal–ligand axes. ESR can provide a fingerprint of a Mo^V centre in a coordination complex or an enzyme, and Table 8 lists the ESR parameters recorded for some typical complexes. ESR spectra of Mo^V complexes can be grouped into two types according to the degree of g anisotropy.[283] One group is typical of octahedral and the other of square pyramidal complexes and, using this criterion, the Mo^V centres of the oxomolybdoenzymes would appear to be square pyramidal; this correlation is based on circumstantial evidence and, so far, no definitive structural prediction has been derived from ESR data alone.

Table 7 Magnetic Moments of some Oxomolybdenum(V) Complexes[5]

Compound	μ_{eff} (BM)	T (K)	*Ref.*
$[bipyH_2][MoOCl_5]$	1.73	303	243
$[NEt_4]_2[MoO(NCS)_5]$	1.67	294	272
$[MoOCl_3(THF)_2]$	1.73	a	248
$[MoO(R_2dtc)Cl]^b$	1.65	293	254
$[MoO(TPP)OH]^c$	1.70	79	275
$[NPr_4][MoOCl_4]$	1.71	297	279
$[NEt_4][MoO(SPh)_4]$	1.72	293	265

[a] Temperature not reported. [b] R_2dt = 4-morpholinodithiocarboxylate. [c] TPP = dianion of *meso*-tetraphenylporphyrin.

Table 8 ESR Data for some Oxomolybdenum(V) Complexes

Compound	*Medium*	*g values*[a]	*A values*[b]	*Ref.*
$[NH_4]_2[MoOCl_5]$	DMF	1.938, 1.938, 1.970	34.0, 34.0, 74.5	280
$[AsPh_4][MoOCl_4(H_2O)]^c$		1.935, 1.935, 1.970	45.5, 45.5, 72.8	238
$[AsPh_4][MoOCl_4]^c$		1.950, 1.950, 1.967	47.9, 47.9, 72.8	239
trans-$[MoOCl(acac)_2]$	DMF	1.950, 1.940, 1.927	33.4, 36.2, 77.6	280
cis-$[MoOCl(ox)_2]^d$	DMF	1.939, 1.953, 1.970	46.5, 9.1, 74.2	280
$[MoOCl(S_2CNEt_2)_2]^c$		1.945, 1.958, 1.984	24.1, 27.5, 61.4	281
$[MoO(S_2CNEt_2)_3]^c$		1.970, 1.978, 1.985	56.9, 33.6, 27.1	281
$[MoOCl_2(pz)]^e$	C_7H_8	1.934, 1.941, 1.971	38.1, 18.1, 71.4	282

[a] For axial systems in the sequence, g_{xx}, g_{yy}, g_{zz}. [b] $\times 10^{-4}$ cm^{-1}, for axial systems, the sequence is A_{xx}, A_{yy}, A_{zz}. [c] Single crystal. [d] Hox = 8-hydroxyquinoline. [e] pz = hydrotris(3,5-dimethylpyrazolyl)borate.

The use of ESR spectroscopy to probe Mo^V centres in biological systems has stimulated attempts to reproduce the ESR spectra of molybdoenzymes using well-defined chemical systems. The similarity between ESR parameters of species formed by reacting $Na_2[MoO_4]$ with S-donor ligands and those of xanthine oxidase imply that in the enzyme (cysteinyl)S ligation of molybdenum may occur.[284] However, no well-characterized Mo^V complex has an ESR spectrum which reproduces all the aspects of one observed for the Mo^V centre of an oxomolybdenum enzyme. A particular problem is the absence of superhyperfine coupling constants for well-defined Mo^V systems; a notable exception is the observation of ^{17}O

superhyperfine coupling in ^{17}O-enriched $[^{98}MoO(SPh)_4]^-$, a complex which exhibits exceptionally narrow ESR linewidths.[265]

According to molecular orbital calculations, for a ligand field of C_{4v} symmetry which is dominated by a strong molybdenum–oxygen interaction along the z axis, the d orbitals transform according to a_1 (d_{z^2}), b_1 $(d_{x^2-y^2})$, e (d_{xz}, d_{yz}) and b_2 (d_{xy}) representations, in order of decreasing energy.[285] The ground state configuration of a Mo^V complex is $^2B_2$ $(d_{xy})^1$, and three d–d transitions are predicted: $^2B_2 \rightarrow {}^2E$, $^2B_2 \rightarrow {}^2B_1$ and $^2B_2 \rightarrow {}^2A_1$. Several interpretations of the electronic spectra of MoO^{3+} complexes have been proposed.[238,285–288] The lowest energy band, which appears in the near IR region of the spectrum between 600 and 850 nm and (at low temperature) often manifests vibronic coupling with $\nu(Mo{-}O_t)$ of *ca.* 900 cm^{-1}, is due to the $^2B_2 \rightarrow {}^2E$ promotion. The other two d–d bands are often obscured by intense ligand to metal charge transfer bands,[265,288] but the $^2B_2 \rightarrow {}^2B_1$ transition is generally responsible for the second absorption band which is usually observed between 400 and 450 nm. In the case of $[AsPh_4][MoOCl_4]$, the $^2B_2 \rightarrow {}^2A_1$ promotion has been observed as a shoulder at *ca.* 380 nm on a intense Cl→Mo charge transfer band.[288]

The vibrational spectra of oxomolybdenum(V) complexes display an intense band in the range 940 to 1020 cm^{-1} due to $\nu(Mo{-}O_t)$;[5,80] the higher frequencies, corresponding to the shorter molybdenum–oxygen bonds, are found in square pyramidal complexes, in which there is no ligand *trans* to the oxo group.

36.4.3.3 Complexes Involving Multiple Molybdenum–Nitrogen Bonds

Monomeric non-oxo molybdenum(V) complexes have been isolated with nitride (N^{3-}), imide (NR^{2-}) and amide (NRR'^{-}) ligands, which may act as models for intermediates in the degradation of dinitrogen in nitrogenase. However, no molybdenum(V) compounds are yet known which correspond to the diazenide or hydrazide analogues of the molybdenum(IV) complexes described in Section 36.4.2.3.

$[PPh_3Me]_2[MoNCl_4]$ can be prepared by reduction of $[PPh_3Me][MoNCl_4]$ or $MoNCl_3$ with $[PPh_3Me]I$. A crystal structure determination has shown the presence of a square pyramidal coordination around the molybdenum in the anion, with an axial nitride ligand at a bond length of 1.634(6) Å.[289] $[MoNCl_2(PPh_3)_2]$ may be obtained by treating $[MoCl_4(THF)_2]$ or $[MoCl_3(THF)_3]$ with Me_3SiN_3, followed by addition of PPh_3.[290] The ESR spectrum of this latter compound at 20 °C shows superhyperfine coupling to two phosphorus nuclei which are equivalent on the ESR timescale. The analogous complexes $[MoNCl_2(bipy)]$[290] and $[MoNBr_2(bipy)]$[291] have also been prepared and characterized. The IR spectrum of all of these nitrido complexes shows a strong absorption in the range 950–1050 cm^{-1}, which is assigned to the stretching of a molybdenum–nitrogen triple bond.

Imido ligands bind to molybdenum in a manner equivalent to the binding of oxo groups, and several compounds have been synthesized which are analogues of known oxo species. 4-Tolylimido ($Ntol^{2-}$) complexes of molybdenum(V) can be prepared from the reduction of molybdenum(VI) compounds; for example, reaction of $[Mo(Ntol)Cl_4(THF)]$ with tertiary phosphines, L, yields $[Mo(Ntol)Cl_3L_2]$. These complexes display magnetic moments close to the spin-only value for one unpaired electron and show strong ESR signals at room temperature. The structure of $[Mo(Ntol)Cl_3(PEtPh_2)_2]$ displays a *meridional* arrangement of chlorine atoms and *trans* phosphines in octahedral geometry; the molybdenum–nitrogen bond length is 1.725(6) Å.[292] The reaction of $[Mo(CO)_4Cl_2]$ with $[NH_4][S_2P(OEt)_2]$ and RN_3 (R = Ph, 4-tolyl) gives $[Mo(NR)\{S_2P(OEt)_2\}_3]$ and $[Mo(NR)Cl\{S_2P(OEt)_2\}_2]$. The ESR spectra of these compounds show that the unpaired electron is coupled to ^{14}N, ^{31}P and $^{35,37}Cl$ nuclei and an unambiguous determination of the superhyperfine coupling constants was achieved using ^{15}N labelling.[293] ESR and IR spectroscopy have been used to monitor the reactions of $[Mo_2Cl_{10}]$ and $[MoOCl_4(DMF)]^-$ with $PhNH_2$. The reactions proceed to give $[Mo(NPh)(NHPh)_5]^{2-}$ *via* $[Mo(NPh)Cl_{4-n}(NHPh)_n(DMF)]$ (n = 0–3) intermediates.[294]

36.4.3.4 Other Monomeric Complexes

36.4.3.4.1 Halide complexes

The hexahalide ions $[MoF_6]^-$ and $[MoCl_6]^-$ form well-characterized compounds which are unstable towards hydrolysis. MoF_6 readily accepts electrons to form $[MoF_6]^-$ ions and will oxidize Ag metal or I_2 in MeCN at ambient temperatures, yielding the molybdenum(V) salts $[Ag(NCMe)_4][MoF_6]_2$ or $[I(NCMe)_2][MoF_6]$, respectively.[295] Simple salts of $[MoCl_6]^-$ are readily prepared, for example reaction of $[Mo_2Cl_{10}]$ with $[NEt_4]Cl$ gives $[NEt_4][MoCl_6]$.[296] $[MoCl_6]^-$ is also obtained in a KCl—$MoCl_5$ melt.[297]

The reactions of various ligands with $[Mo_2Cl_{10}]$ in dry solvents provide a general route for the preparation of non-oxo chloromolybdenum(V) complexes. $[Mo_2Cl_{10}]$ reacts with ICN, to produce $[MoCl_5(NCI)]$ in which the added molecule is N-bonded,[298] and with $Y(CN)_2$ (Y = S, Se) in CS_2 to give $[MoCl_5\{Y(CN)_2\}_2]$, in which the new ligand is coordinated through its Y atom.[299] The reactions of $[Mo_2Cl_{10}]$ or $[MoOCl_4(DMF)]^-$ with thiols, RSH, in py have been monitored by ESR and IR spectroscopy; $[MoSCl_{4-n}(SR)_n(py)]^-$ (n = 0–3) and $[MoS(SR)_5]^{2-}$ were suggested as possible products, in a reaction exactly analogous to that using $PhNH_2$ instead of RSH (see Section 36.4.3.3).[294] The dimer $[Mo_2S_2Cl_6]$ dissolves in MeCN to form $[MoSCl_3(NCMe)_2]$, which reacts with pyridine and HCl to give $[pyH][MoSCl_4(py)]$. Direct reaction of $[Mo_2S_2Cl_6]$ with pyridine gives the seven-coordinate compound $[MoSCl_3(py)_3]$.[300]

The chloro alkoxide complexes $[MoCl_4(OR)_2]^-$ (R = Me, Et) have been prepared by low-temperature reaction of $[Mo_2Cl_{10}]$ with ROH, and have been isolated as their $[NMe_4]^+$, $[pyH]^+$ and $[quinH]^+$ salts.[279] The vibrational, electronic and ESR spectra of these materials are consistent with a *trans* octahedral structure for these anions. $[MoCl_4(NPPrPh_2)(OPPrPh_2)]$ has been obtained by oxidation of $[MoCl_4(PPrPh_2)_2]$ with tolyl-4-sulfonyl azide.[301] An eight-coordinate complex $[MoCl_4(diars)_2]I_3$ has been prepared, and the crystal structure determined; the ligands adopt a dodecahedral geometry around the molybdenum, with arsenic atoms on the *A* sites, and chlorine atoms on the *B* sites.[302]

36.4.3.4.2 Complexes of sulfur donor ligands

Compounds containing the $[Mo(S_2CNEt_2)_4]^+$ cation can be prepared by the reaction of $[Mo_2O_2Cl_6]$ with $[Et_2NH][S_2CNEt_2]$,[303] or by oxidation of the corresponding Mo^{IV} or Mo^{III} complexes.[304,305] Crystal structures have been determined for salts with various anions, all of which display the same geometry for the cation. Crystals of $[Mo(S_2CNEt_2)_4]I_3$ are monoclinic, with dodecahedral coordination of sulfur atoms around the molybdenum; the mean Mo—S distances to the two sites are 2.545(2) and 2.500(2) Å.[305] The electrochemistry of several eight-coordinate complexes, including the dithiocarbamate, $[Mo(S_2CNEt_2)_4]^+$, dithiolenes, $[Mo(S_2C_2R_2)_4]^{3-}$, and the thioxanthate, $[Mo(S_2CSBu^t)_4]^+$, has been investigated.[306] All the systems exhibit a reversible one-electron reduction and, with the exception of the thioxanthate complex, a reversible one-electron oxidation.

Reaction of $[Mo_2O_4(S_2CNR_2)_2]$ (R = Et, Pr^i) with thiol-containing ligands 2-$NHR'C_6H_4SH$ or 1,2-$(HS)_2C_6H_4$ yields the monomeric complexes $[Mo(S_2CNR_2)(NRC_6H_4S)_2]$ or $[Mo(S_2CNR_2)(SC_6H_4S)_2]$, respectively.[307] A determination of the crystal structure of $[Mo(S_2CNEt_2)(NHC_6H_4S)_2]$ has shown a coordination geometry at the molybdenum midway between that of a trigonal prism and an octahedron, in which the two nitrogen atoms are mutually *trans*.[308]

$[NEt_4][MoO(S\text{-}4\text{-}tolyl)_4]$ reacts with *N*-(2-hydroxybenzyl)-2-mercaptoaniline (H_3hbma) to yield $[NEt_4][Mo(hbma)_2]$, the anions of which possess a highly distorted, all-*trans*, octahedral geometry. Cyclic voltammetry has been used to demonstrate reversible one-electron oxidation and reduction processes for this complex.[309]

36.4.3.4.3 Cyanide complexes

The octacyanide ion $[Mo(CN)_8]^{3-}$ can be prepared by oxidation of $[Mo(CN)_8]^{4-}$. Like the Mo^{IV} anion (Section 36.4.2.4.6) $[Mo(CN)_8]^{3-}$ has been structurally characterized in more than one geometry. In $[NBu_4]_3[Mo(CN)_8]$ the anion possesses a near-regular, D_{2d} triangular-dodecahedral structure, with an average Mo—C distance of 2.12(2) Å.[310] However, the eight cyano groups in $Cs_3[Mo(CN)_8]\cdot 2H_2O$ are arranged in a 4,4′-bicapped trigonal prism around the molybdenum, with an average Mo—C distance of 2.17(2) Å.[311] Furthermore, $K_3[Mo(CN)_8]\cdot H_2O$ is isomorphous with $K_3[W(CN)_8]\cdot H_2O$, which has square antiprismatic geometry, although a full structure was not determined for the molybdenum analogue as it decomposed under X-ray irradiation.[312] The ESR spectrum of $K_3[Mo(CN)_8]$ in glycerine solution indicates that all the cyanides are equivalent, consistent with a square antiprismatic structure, or with stereochemical non-rigidity on the ESR timescale.[313]

36.4.3.4.4 Spectroscopic studies

The magnetic moments of monomeric non-oxo molybdenum(V) complexes typically lie between 1.55 and 1.73 BM,[5] corresponding to the spin-only value for one unpaired electron. The ESR spectra of these complexes usually show one strong line with six hyperfine lines due to $^{95,97}Mo$ coupling; a typical example is $[Mo(S_2CNEt_2)_4]I_3$ with isotropic g and A values of 1.9794(1) and $37.1(1) \times 10^{-4}\ cm^{-1}$, respectively.[304] $[Mo(S_2CNEt_2)(NHC_6H_4S)_2]$ displays superhyperfine splitting for two equivalent nitrogen and two equivalent hydrogen atoms, with A values deduced from spectral simulation of $2.4 \times 10^{-4}\ cm^{-1}$ and $7.4 \times 10^{-4}\ cm^{-1}$, respectively.[307] These couplings are the same order of magnitude as those observed for superhyperfine splittings in the spectra of the molybdenum oxidases, demonstrating the possibility that these may be caused by coordinated N—H groups.

$[NEt_4][Mo(hbma)_2]$ (H_3hbme = *N*-(2-hydroxybenzyl)-2-mercaptoaniline) exhibits an axial ESR spectrum, with $g_\perp > g_\parallel$ and $A_\parallel > A_\perp$. Superhyperfine coupling to the two nitrogens was not detected over a temperature range from room temperature to 77 K.[309] The single crystal Q-band ESR spectrum of the D_{2d} anion of $[NBu_4]_3[Mo(CN)_8]$ has been determined at room temperature. The sense of the anisotropy in g values ($g_\parallel > g_\perp$) contradicts predictions based on simple crystal field theory, and possible explanations have been proposed.[314]

The electronic absorption spectrum of the $[MoCl_6]^-$ ion in solidified KCl—$MoCl_5$ melt shows an absorption band at $24\,100\ cm^{-1}$, assigned to the $(t_{2g})^1 \rightarrow (e_g)^1$ transition of an octahedral complex.[297] In $[NEt_4][MoCl_6]$ this band is split by symmetry lowering (as suggested from ESR data obtained for this salt)[315] and/or spin–orbit coupling.[296] Other, more intense, bands at higher energies in $[MoCl_6]^-$ complexes are due to ligand-to-metal charge transfer transitions. Low energy absorption bands in $[MoCl_4(OR)_2]$ (R = Me, Et) complexes can be interpreted assuming a *trans* octahedral structure.[279] The UV–vis spectrum has been reported for $[Mo(S_2CNEt)_4]I_3$, but no assignments have been proposed.[305]

36.4.3.5 Dimeric and Polymeric Complexes

36.4.3.5.1 General comments

Dimeric species play a major role in the chemistry of molybdenum(V), and of particular importance are compounds containing one or more oxygen bridges. Examples are also known of complexes bridged by sulfur, selenium or nitrogen, and halide bridges are found in $[\{MoF_5\}_4]$[225] and $[\{MoCl_5\}_2]$.[230] Many dimeric compounds are diamagnetic due to interaction between the two molybdenum(V) atoms, either directly and/or *via* the bridging atoms, which lead to spin-pairing of the two unpaired electrons. However, in some $(Mo^V)_2$ systems spin-pairing does not occur or is incomplete and ESR signals are observed characteristic of monomeric species, with one electron associated with each molybdenum.

36.4.3.5.2 Dimeric complexes with one bridging ligand

Almost all of these compounds are based upon an $\{Mo_2O_3\}^{4+}$ core and a representative selection of these compounds is given in Table 9. In these compounds, each molybdenum is coordinated to a terminal oxo ligand with a linear, or near-linear, Mo—O—Mo bridge. The terminal oxo groups are usually *cis* to the oxo bridge, directed either *cis* (*syn*) or *trans* (*anti*) to each other across the bridge, and no complexes are known in which the two terminal oxo ligands are *cis* to the Mo—O—Mo bridge in a skew arrangement, forming a dihedral angle of 90°. Bonding considerations for these $\{Mo_2O_3\}^{4+}$ complexes show that the molybdenum d_{xy} orbitals, which are nonbonding in mononuclear oxo complexes, are in proper alignment with the p_x orbital on the oxygen bridging atom to form three-centre delocalized molecular orbitals, producing a bonding, a nonbonding and an antibonding level.[5] This interaction is maximized at dihedral angles between the terminal oxo ligands of 0° (*syn*) and 180° (*anti*), but cannot occur at 90° (skew). Two electrons from the oxygen and one from each molybdenum fill the bonding and nonbonding orbitals, to give diamagnetic compounds. In the skew arrangement each molybdenum d_{xy} orbital would interact with a different oxygen p orbital, giving, for each Mo—μ-O π bond, one bonding and one antibonding level occupied by one molybdenum and two oxygen electrons; therefore, the electronic structure of the centre would involve the two degenerate antibonding orbitals each containing one unpaired electron, to give a paramagnetic complex with an $S = 1$ ground state. Although no crystal structures are known with the skew geometry, it is possible that in solution rotation about the Mo—O—Mo bonds may occur, giving rise to paramagnetic species. Bending of the Mo—O—Mo bridge would also reduce the Mo—O—Mo π interaction and lead to paramagnetism.

Table 9 A Representative Selection of Dimeric Molybdenum(V) Complexes Containing a $\{Mo_2O_3\}^{4+}$ Core

Complex	*Ref.*
$[Mo_2O_3Cl_4(DME)_2]$	316
$[Mo_2O_3(acac)_4]$	318
$[Mo_2O_3(acac)_2(facac)_2]$[a]	317
$[Mo_2O_3(H_2PO_2)_4]$	319
$[Mo_2O_3(O_2CNR_2)_4]$ (R = *e.g.* cyclohexyl, Me)	344
$[Mo_2O_3(S_2CNR_2)_4]$ (R = *e.g.* Me, Et)	100, 320, 321
$[Mo_2O_3(S_2CSPr)_4]$	323
$[Mo_2O_3(S_2COR)_4]$ (R = *e.g.* Me, Et, Pr^i)	324, 325
$[Mo_2O_3(S_2CR)_4]$ (R = *e.g.* hexyl, pentyl, Ph)	326
$[Mo_2O_3\{S_2P(OR)_2\}_4]$ (R = *e.g.* Et, Ph)	327, 328
$[Mo_2O_3$(R-L-cysteinate)$_4]$ (R = Me, Et)	329
$[Mo_2O_3\{S(CH_2CH_2S)_2\}_2]$, $[Mo_2O_3\{O(CH_2CH_2S)_2\}_2]$	330
$[Mo_2O_3\{MeN(CH_2CH_2S)_2\}_2]$	331
$[Mo_2O_3\{S(CH_2CH_2)NMe(C_2H_3R)NMe(CH_2CH_2)S\}_2]$	332
$[Mo_2O_3F_4(bipy)_2]$, $[Mo_2O_3F_4(phen)_2]$	333
$[Mo_2O_3Cl_4(glycine)_4]$, $[Mo_2O_3Cl_8]^{4-}$	334
$[Mo_2O_3(NCS)_8]^{4-}$	335
$[Mo_2O_3(NCS)_4(C_2O_4)_2]^{4-}$	336
$[Mo_2O_3(NCS)_4(bipy)_2]$, $[Mo_2O_3(NCS)_4(phen)_2]$	337
$[Mo_2O_3$(salicylaldiminate)$_4]$	338
$[Mo_2O_3Cl_2(ONN)_2]$[b]	339
$[Mo_2O_3$(8-hydroxyquinolinate)$_4]$	340–343
$[Mo_2O_3(NCS)_2$(8-hydroxyquinolinate)$]$	337

[a] Hfacac = 1,1,1,5,5,5-hexafluoropentane-2,4-dione.
[b] H(ONN) = $RC_6H_4C(O)NHNC(Me)C(Me)NOH$, 2-$H_2NNHC(O)C_6H_4NC(Me)C(Me)NOH$.

Rarely, a linear O═Mo—O—Mo═O arrangement is adopted, in which the molybdenum d_{xy} orbitals are orthogonal to the oxygen p_π orbitals and, therefore, the system is paramagnetic.

A typical example of a complex with a normal $\{Mo_2O_3\}^{4+}$ core in the *anti* arrangment is $[Mo_2O_3Cl_4(DME)_2]$, prepared by refluxing $[Mo_2Cl_{10}]$ with DME. Each molybdenum is octahedrally coordinated to a bridging oxo group, a terminal oxo group, two chloride ligands and the two oxygen atoms of a bidentate dimethoxyethane ligand; one oxygen of the DME is *trans* to the oxo bridge.[316] The crystal structure of $[Mo_2O_3(acac)_2(facac)_2]$ (Hfacac = 1,1,1,5,5,5-hexafluoropentane-2,4-dione) displays a similar geometry, with the terminal oxo groups *cis* to a linear Mo—O—Mo bridge in an *anti* conformation, and the diketonates completing the octahedral coordination spheres of the molybdenum.[317] Magnetic susceptibility measurements of $[Mo_2O_3(acac)_4]$ between 77 and 292 K show this material to be paramagnetic, with a temperature dependent magnetic moment suggesting some direct or indirect interaction between the molybdenums. The paramagnetism may be due to the presence of a bent Mo—O—Mo bridge, which would be consistent with the IR spectrum of this compound.[318]

There are a large number of well-characterized complexes with 1,1-dithiolate ligands which contain an $\{Mo_2O_3\}^{4+}$ core; these include compounds of dithiocarbamate $(S_2CNR_2)^-$, trithiocarbonate $(S_2CSR)^-$, dithiocarboxylate $(S_2CR)^-$ and dithiophosphate $(S_2P(OR)_2)^-$ ligands[320–328] (see Table 9). All of these complexes are diamagnetic, or have very low magnetic moments, and crystal structure determinations have shown examples of *syn* and *anti* conformations of the $\{Mo_2O_3\}^{4+}$ core; the oxo groups in $[Mo_2O_3\{S_2P(OEt)_2\}_4]$ adopt the *anti* configuration,[328] but in both $[Mo_2O_3(S_2COEt)_4]$[325] and $[Mo_2O_3(S_2CNEt_2)_4]$[100] the *syn* configuration is preferred. The length of the Mo—O_t bonds is significantly shorter than that of Mo—O_b bonds (*e.g.* in $[Mo_2O_3(S_2COEt)_4]$ the distances are 1.65(2) and 1.86(2) Å, respectively) and the longest Mo—S bonds are those *trans* to the terminal oxo ligands. The xanthate complexes, $[Mo_2O_3(S_2COR)_4]$ (R = alkyl), react with dialkylamines, alcohols or hydrogen sulfide to lose xanthate and produce $[Mo_2O_2S_2(S_2COR)_2]$, which contains two sulfido bridges.[324] The dithiocarbamates $[Mo_2O_3(S_2CNR_2)_4]$ disproportionate in solution to generate an equilibrium with Mo^{IV} and Mo^{VI} species (equation 1, Section 36.4.2.2.1), and can easily be prepared by reacting $[MoO_2(S_2CNR_2)_2]$ with $[MoO(S_2CNEt_2)_2]$.[70] The mixed Mo^V—Mo^{VI} dimer $[Mo_2O_3(S_2CNEt_2)_4]^-$ has been prepared and structurally characterized in a material also containing $[Mo_2O_3(S_2CNEt_2)_4]$ and (probably) $[H_5O_2]^+$.[322] The anion is paramagnetic, and ESR spectroscopy has shown the unpaired electron to be localized on one molybdenum, rather than over the whole molecule.

Complexes with an $\{Mo_2O_3\}^{4+}$ core which contain five-coordinate molybdenum atoms have been isolated with tridentate sulfur donor ligands. These complexes have the form $[Mo_2O_3L_2]$ ($L = S(C_2H_4S)_2^{2-}$, $O(C_2H_4S)_2^{2-}$, $MeN(C_2H_4S)_2^{2-}$), and crystallize with the terminal oxo groups in the *anti* configuration.[330,331] The coordination geometry around each molybdenum is that of a distorted trigonal bipyramid, with the axial positions occupied by the bridging oxygen and the central sulfur, oxygen[330] or nitrogen[331] atom of the tridentate ligand. With the tetradentate ligands $\{SCH_2CH_2N(Me)CH_2CHRN(Me)CH_2CH_2S\}^{2-}$, L, (R = H, Me) the molecules $[Mo_2O_3L_2]$ possess an *anti* configuration and the molybdenum atoms are octahedrally coordinated.[332]

Structure determinations of the isothiocyanate complexes $K_4[Mo_2O_3(NCS)_8]\cdot 4H_2O$[335] and $[pyH]_4[Mo_2O_3(NCS)_4(C_2O_4)_2]$[336] have shown that they adopt the *anti* configuration, with N-bound isothiocyanates. A report has suggested that $[Mo_2O_3(oxine)_4]$ (Hoxine = 8-hydroxyquinoline) can be isolated in four isomeric forms, one of which is strongly paramagnetic ($\mu = 1.83$ BM).[340] However, a later ^1H NMR study of this compound in DMSO showed no evidence of such isomer formation, and indicated that all the N atoms of the oxine ligands are coordinated *trans* to an oxo group.[341] The electrochemistry of $[Mo_2O_3(oxine)_4]$ in DMSO has shown that this molecule can be reduced in two reversible one-electron steps, but oxidation to an Mo^{VI}—Mo^V dimer causes decomposition.[342] An electrochemical study comparing Mo^V dimers with one, two or three bridging ligands has shown that the monooxo-bridged species are reduced by successive one-electron processes, rather than by a two-electron process which is observed for dioxo-bridged complexes.[345]

Assignments have been made of both terminal and bridging ν(Mo—O) stretching modes in the IR spectra of complexes containing the $\{Mo_2O_3\}^{4+}$ core.[130,346] In the Mo—O_t region (900–1000 cm^{-1}) one band should be observed for the *anti* and two for the *syn* conformer; however, only one band has been found in all cases, which may be due in the *syn* complexes to overlap of the two bands or to a low intensity of the antisymmetric mode. A distinctive feature of the electronic absorption spectra of $\{Mo_2O_3\}^{4+}$ complexes is an intense band at *ca.* 19 000 cm^{-1}, which is not appreciably ligand dependent and appears to be characteristic of a linear Mo—O—Mo bridge, as it is not present in the monomers or dimers with two bridging ligands.[5] The ^{17}O NMR chemical shifts of the oxo ligands of molybdenum(V) and (IV) complexes have been used to compare relative Mo—O π-bond strengths.[128]

A complex closely related to those possessing an $\{Mo_2O_3\}^{4+}$ core is $[Mo_2O(Ntol)_2(S_2CNEt_2)_4]$ (tol = 4-tolyl), prepared by reducing $[MoO(Ntol)(S_2CNEt_2)_2]$.[142] IR and ^{17}O NMR studies have indicated an oxo-bridged structure for the dimer; there is an intense absorption at 18 760 cm^{-1} in the electronic spectrum, which does not obey the Lambert–Beer law due to a disproportionation reaction which sets up an equilibrium with Mo^{IV} and Mo^{VI} species.

The only complexes which are known to contain an $\{Mo_2O_3\}^{4+}$ core in which the two terminal oxo groups are *trans* to the Mo—O—Mo bridge are $[Mo_2O_3(TPP)_2]$ and $[Mo_2O_3(OEP)_2]$ with tetradentate porphyrin ligands (TPP = tetraphenylporphyrin, OEP = octaethylporphyrin).[347,348] In $[Mo_2O_3(TPP)_2]$ the Mo—O_t and Mo—O_b bond lengths of 1.707(3) and 1.936(3) Å, respectively, are slightly longer than the corresponding distances observed in the other $\{Mo_2O_3\}^{4+}$ systems,[100,325,328] presumably because of the mutual *trans* effect of the oxo ligands.

A complex containing a mixed Mo^V—Mo^{VI} oxidation state $\{Mo_2O_5\}^+$ core can be obtained by reduction of $[Mo_2O_5(nane)_2][Br_3]_2$ (nane = *N,N',N''*-trimethyl-1,4,7-triazacyclononane). In the Mo^{VI} dimer there is a linear Mo—O—Mo bridge, and each molybdenum is also coordinated to two oxo ligands *cis* to the oxo bridge in an *anti* conformation, but the structure of the Mo^V—Mo^{VI} species is not known. An Mo^V—Mo^V dimer of this system can be produced electrochemically, but not by chemical reduction.[349] The reduction of $[MoO(S_2CNEt_2)_3][BF_4]$ results in the formation of $[Mo_2O(S_2CNEt_2)_6][BF_4]$, an Mo^V—Mo^{IV} dimer which contains a linear Mo—O—Mo bridge. Each molybdenum is seven-coordinate, and the equal Mo—O bond lengths and equivalent environments suggest that the unpaired electron is likely to be delocalized.[218]

Two types of complex are known with only one bridging ligand which is not an oxo group. In $[Mo_2O_2X_4(C_2O_4)]$ (X = F, Cl, Br) the oxalate is bidentate to each molybdenum; the coordination geometry around the molybdenum is square pyramidal, with axial oxo ligands, and the room temperature magnetic moments of 1.67 BM per molybdenum indicate that the interaction between the two molybdenum atoms is weak.[350] The reaction of $[MoCl_3(THF)_3]$ with 4-$C_6H_4(NO_2)(N_3)$ and $[Et_2NH_2][S_2P(OEt)_2]$ yields $[Mo_2(NC_6H_4N)\{S_2P(OEt)_2\}_6]$, which is bridged by the diimide, $(4\text{-}NC_6H_4N)^{4-}$, ligand. Each molybdenum atom is in a distorted octahedral environment, comprising an imido nitrogen and five sulfur atoms from one monodentate and two bidentate dithiophosphate ligands.[351]

36.4.3.5.3 Dimeric complexes with two bridging ligands

The majority of these complexes contain an $\{Mo_2Y_nY'_{4-n}\}^{2+}$ (Y = O; Y′ = S, Se; n = 4–0) core, and such systems constitute an important class of molybdenum compounds;[5,80,93] Table 10 provides a reasonably comprehensive summary of the known complexes and shows the extensive range possible, given the versatility in the nature of both the core and the appended ligands. These cores consist of two molybdenum atoms, each coordinated to one terminal and two bridging O, S or Se atoms and involve a relatively close approach (2.5–2.9 Å) of the two metals. Related dimeric systems with one (or two) terminal imido (RN^{2-}) ligands in place of the Y_t atom(s) are known. Also, Mo^V dimers, in which two Mo = Y_t groups are linked by two bridging ligands of another type, notably Cl^-, $[S_2]^{2-}$, $[SO_4]^{2-}$, have been characterized and are included in this section, as are complexes containing an $\{Mo_2Y_4\}^{2+}$ (Y = O, S) core, the bridging of which is supplemented by an additional *polydentate* ligand. Compounds involving oligomeric aggregates formed by the association of $\{Mo_2Y_4\}^{2+}$ (Y = O, S) cores are described in Section 36.4.3.5.4.

The first structural characterization of a compound containing a $\{Mo_2O_4\}^{2+}$ core was achieved[352] for $Ba[Mo_2O_4(C_2O_4)_2(H_2O)_2]\cdot 5H_2O$, the anion of which (Figure 4) consists of two distorted octahedra sharing an O—O edge. The anion possesses a C_2 axis, with the two terminal ligands *cis* to the oxo bridges to give the *syn* conformation, and each molybdenum centre has structural features identified earlier for MoO^{3+} and $\{Mo_2O_3\}^{4+}$ moieties. The shortest Mo—O bond is that involving the terminal oxo group (1.70(3) Å) and the bonds to the bridging oxygen atoms (1.88(3) and 1.93(3) Å) are shorter than those to the oxalato and aqua oxygens *cis* to the $Mo{=}O_t$ groups (2.14(4) and 2.22(4) Å, respectively), but there is no structural *trans* effect (Mo—O(oxalate) *trans* to $Mo{=}O_t$ = 2.11(3) Å). Each molybdenum lies out of the equatorial plane defined by the four donor atoms *cis* to $Mo{=}O_t$ and this results in a rough equality in the O---O distances at *ca.* 2.78 Å, close to the van der Waals separation for these atoms.[353] The atoms of the Mo_2O_2 bridge are not coplanar and there is a dihedral angle of 152° between the two MoO_2 planes. The terminal oxo groups of the $\{Mo_2O_4\}^{2+}$ core point away from each other; if they were not so aligned, their separation would be the same as the Mo—Mo separation (2.541 Å) and much closer than the van der Waals separation. This short Mo—Mo approach, and the essentially diamagnetic nature of the compound imply a direct Mo—Mo bonding interaction, and the bent bridge permits a closer approach of the Mo atoms without unduly enlarging the O_b—Mo—O_b, and/or contracting the Mo—O_b—Mo, angles. The qualitative description of the Mo—Mo bonding follows from the electronic structure of MoO^{3+} centres. Each molybdenum centre can be considered to possess a $(d_{xy})^1$ configuration and the geometry of the bridge allows a direct overlap of the two d_{xy} orbitals with the resultant bonding level accommodating the two electrons.[5,354]

$[Mo_2O_4(dane)_2][ClO_4]_2$ (dane = 1,5,9-triazacyclododecane) has been obtained both as a purple and a yellow material. The purple form contains[50] an $\{Mo_2O_4\}^{2+}$ core with the *anti* conformation in which the four-membered Mo_2O_2 ring is planar with an Mo—Mo separation of 2.586(1) Å. This separation, and the diamagnetic nature of the compound, are consistent with the formation of an Mo—Mo bond, as in a *syn* structure, by overlap of the molybdenum d_{xy} orbitals. The cation of the yellow form of this compound is considered to possess an $\{Mo_2O_4\}^{2+}$ core with the *syn* conformation by analogy with $[Mo_2O_4(nane)_2]^{2+}$ (nane = 1,4,7-triazacyclononane), which has been shown[355] to exist in both *syn* (*cis*) and *anti* (*trans*) conformers, with the immediate environment at molybdenum almost identical in the two forms and the Mo—Mo separation being 2.555(1) Å and 2.561(1) Å, respectively. *trans*-$[Mo_2O_4(nane)_4]^{2+}$ is

Table 10 Dimeric Molybdenum(V) Complexes Containing an $\{Mo_2Y_nY'_{4-n}\}^{2+}$ (Y = O; Y′ = S, Se; n = 4–0) Core

Complex	*Bridging atoms*	*Ref.*
$[Mo_2O_4Cl_4(H_2O)_2]^{2-}$, $[Mo_2O_4Br_4(H_2O)_2]^{2-}$, $[Mo_2O_4Cl_2(OH)_2(H_2O)_2]^{2-}$, $[Mo_2O_4Cl_2(OH)_4]^{4-}$	O,O	336, 354
$[Mo_2O_2S_2Cl_4(H_2O)_2]^{2-}$	S,S	384
$[Mo_2O_4Cl_4]^{2-}$	O,O	372
$[Mo_2O_4F_2(C_2O_4)_2]^{4-}$, $[Mo_2O_4(C_2O_4)_2(H_2O)_2]^{2-}$	O,O	352, 402
$[Mo_2O_2S_2(C_2O_4)_2(H_2O)_2]^{2-}$	S,S	384
$[Mo_2O_4(acac)_2(H_2O)_2]$	O,O	318
$[Mo_2O_4(catecholate)_2(H_2O)_2]^{2-}$	O,O	403
$[Mo_2O_4(O_2CNR_2)_2]$ (R = alkyl)	O,O	344
$[Mo_2O_3S(O_2CNR_2)_2]$ (R = alkyl)	O,S	344
$[Mo_2O_2S_2(O_2CNR_2)_2]$ (R = alkyl)	S,S	344
$[Mo_2O_4(OSCPh)_2]$, $[Mo_2O_4(OSCPh)_2(py)_2]$	O,O	383
$[Mo_2O_3S(OSCPh)_2]$	O,S	383
$[Mo_2O_4(S_2CNR_2)_2]$ (R = alkyl)	O,O	360–363
$[Mo_2O_3S(S_2CNR_2)_2]$ (R = alkyl)	O,S	362–365
$[Mo_2O_2S_2(S_2CNR_2)_2]$ (R = H, alkyl)	S,S	361–363, 365, 366
$[Mo_2OS_3(S_2CNR_2)_2]$ (R = alkyl)	S,S	363, 365, 367
$[Mo_2S_4(S_2CNR_2)_2]$ (R = alkyl)	S,S	362, 363, 365, 368–371
$[Mo_2Se_4(S_2CNR_2)_2]$ (R = alkyl)	Se,Se	404
$[Mo_2O_3S\{S_2P(OEt)_2\}_2]\cdot$pyrimidine	O,S	405
$[Mo_2S_4\{S_2P(OEt)_2\}_2]$	S,S	373
$[Mo_2O_2S_2(S_2COR)_2]$ (R = alkyl)	S,S	324
$[Mo_2O_2S_2(SCH_2CH_2S)_2]^{2-}$	S,S	406
$[Mo_2S_4(SCH_2CH_2S)_2]^{2-}$	S,S	357, 382, 406
$[Mo_2S_4(SPh)_4]^{2-}$	S,S	399
$[Mo_2O_4(SCH_2CH_2OH)_2]$	O,O	329
$[Mo_2O_2S_2(S_2C_2(CN)_2)_2]^{2-}$	S,S	358
$[Mo_2O_4(dmmpd)_2]$[a]	O,O	331
$[Mo_2O_4$(L-cysteinate)$_2]^{2-}$	O,O	329, 389
$[Mo_2O_2S_2$(L-cysteinate)$_2]^{2-}$	S,S	329, 390
$[Mo_2O_4$(Et-L-cysteinate)$_2]$	O,O	329, 391
$[Mo_2O_2S_2$(R-L-cysteinate)$_2]$ (R = Me, Et)	S,S	392
$[Mo_2O_4$(L-histidinate)$_2]$	O,O	393
$[Mo_2O_2S_2$(L-histidinate)$_2]$	S,S	394
$[Mo_2O_4(glycinate)_2(H_2O)_2]$	O,O	401
$[Mo_2O_2S_2(py)_4]^{2+}$	S,S	407
$[Mo_2O_4(mpr)_2(py)_2]$[b]	O,O	408
$[Mo_2O_4F_2(bipy)_2]$	O,O	333
$[Mo_2O_4Cl_2(bipy)_2]$	O,O	385
$[Mo_2O_4(O_2PH_2)_2(bipy)_2]$	O,O	386
$[Mo_2O_4F_2(phen)_2]$	O,O	333
$[Mo_2O_4Cl_2(phen)_2]$	O,O	329
$[Mo_2O_4(nane)_2]^{2+}$[c]	O,O	355
$[Mo_2O_4(dane)_2]^{2+}$[d]	O,O	50
$[Mo_2O_4(oxine)_2]$[e]	O,O	343
$[Mo_2O_3S(oxine)_2]$[e]	O,S	343
$[Mo_2O_2S_2(oxine)_2]$[e]	S,S	343
$[Mo_2O_4(NCS)_6]^{4-}$	O,O	387
$[Mo_2O_3S(NCS)_6]^{4-}$	O,S	409
$[Mo_2O_2S_2(NCS)_6]^{4-}$	S,S	409
$[Mo_2O_2S_2(S_2)_2]^{2-}$	S,S	375
$[Mo_2OS_3(S_2)_2]^{2-}$	S,S	376
$[Mo_2O_2S_2(S_2)(S_3O_2)]^{2-}$	S,S	374
$[Mo_2O_2S_2(S_2)(S_4)]^{2-}$	S,S	377
$[Mo_2S_4(S_2)_2]^{2-}$	S,S	378
$[Mo_2S_4(S_2)(S_4)]^{2-}$	S,S	379, 380
$[Mo_2S_4(S_4)_2]^{2-}$	S,S	380, 381
$[Mo_2O_2S_2(Cp)_2]$	S,S	410
$[Mo_2S_4(C_5H_{5-n}Me_n)_2]$ (n = 0–5)	S,S	359, 411

[a] Hdmmpd = *N,N′*-dimethyl-*N,N′*-bis(2-mercaptoethyl)propylenediamine.
[b] Hmpr = 2-mercaptopyrimidine. [c] nane = 1,4,7-triazacyclononane.
[d] dane = 1,5,9-triazacyclododecane. [e] Hoxine = 8-hydroxyquinoline and substituted derivatives.

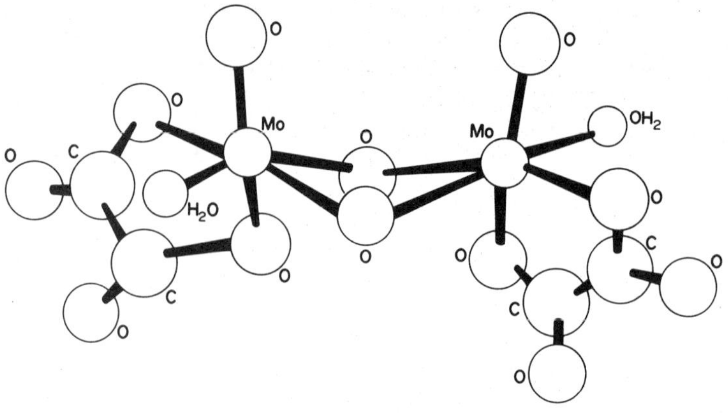

Figure 4 Structure of the anion of $Ba[Mo_2O_4(C_2O_4)_2(H_2O)_2]\cdot 5H_2O$[352]

irreversibly converted into *cis*-$[Mo_2O_4(nane)_4]^{2+}$ *via* acid–base catalysis and reasons for the relative thermodynamic stabilities have been attributed to the increased, local, O_t---O_b nonbonding distances in the latter as compared to the former.[356] Other factors may be important in determining the most stable conformation of $\{Mo_2O_4\}^{2+}$ and related complexes and the effects of Mo—Mo bonding[357] and steric interactions among coordinated ligands[358,359] have been considered.

The series of dithiocarbamato complexes $[Mo_2O_nS_{4-n}(S_2CNR_2)_2]$ ($n = 4$–0; R = alkyl)[360–371] demonstrate two important properties of these systems. Firstly, it is the μ-oxo groups which are substituted initially and, secondly, this substitution leads to an increase in the Mo—Mo separation, from 2.580(1) Å in $[Mo_2O_2(\mu\text{-}O)_2(S_2CNEt_2)_2]$,[360] through 2.673(3) Å in $[Mo_2O_2(\mu\text{-}O)(\mu\text{-}S)(S_2CNPr_2)_2]$,[364] to 2.820(1), 2.826(3) and 2.814(1) Å in $[Mo_2O_2(\mu\text{-}S)_2(S_2CNH_2)_2]$,[366] $[Mo_2OS(\mu\text{-}S)_2(S_2CNEt_2)_2]$[367] and $[Mo_2S_2(\mu\text{-}S)_2(S_2CNEt_2)_2]$,[369] respectively.

Molybdenum(V) dimers with $\{Mo_2Y_nY'_{4-n}\}^{2+}$ (Y = O, S; $n = 4$–0) may involve both molybdenum atoms with five-coordination or six-coordination (not including the Mo—Mo interaction). Examples of five-coordinate complexes include $[AsPh_4]_2[Mo_2O_4Cl_4]$,[372] $[Mo_2O_nS_{4-n}(S_2CNR_2)_2]$ ($n = 4$–0, R = H, alkyl),[360–371] $[Mo_2S_4\{S_2P(OEt)_2\}_2]$,[373] $[Mo_2O_2S_2(S_2)(S_3O_2)]^{2-}$,[374] and $[Mo_2O_nS_{4-n}(S_2)_{2-x}(S_4)_x]^{2-}$ ($n = 2$, $x = 0$, 1; $n = 1$, $x = 1$; $n = 0$, $x = 0$, 1, 2).[375–381] In these compounds the coordination around each molybdenum atom is square pyramidal, with the Mo—X_t (X = O, S) groups axial, *cis* to the two bridging ligands and in the *syn* conformation. The Mo_2X_2 bridging unit is not planar, but involves a dihedral angle between the two MoX_2 planes of 150–160°. The 1,2-dithiolate complex $[NEt_4]_2[Mo_2S_4(SC_2H_4S)_2]$ has been isolated and structurally characterized in both *syn* and *anti* conformations.[382] The *syn* form has a geometry similar to that of the corresponding dithiocarbamate complexes, with an Mo—Mo distance of 2.863(2) Å; in the *anti* conformation the Mo_2S_2 bridge is planar, and the Mo—Mo distance is 2.878(2) Å.

In the five-coordinate species with an $\{Mo_2Y_4\}^{2+}$ (Y = O, S) core there is a vacant coordination site *trans* to each terminal oxo or sulfido ligand. These can be filled to give species in which each molybdenum is six-coordinate; for example, recrystallizing $[Mo_2O_4(OSCPh)_2]$ from pyridine yields $[Mo_2O_4(OSCPh)_2(py)_2]$.[383] The crystal structures of $[Mo_2O_2X_2Cl_4(H_2O)_2]^{2-}$ (X = O, S) show octahedrally coordinated molybdenum atoms, with two bridging oxides or sulfides and *syn* terminal oxo ligands; the water molecules are *trans* to the oxo groups.[336,384] Other complexes with similar geometries include $[Mo_2O_4Cl_2(bipy)_2]$,[385] $[Mo_2O_4(O_2PH_2)_2(bipy)_2]$[386] and $[Mo_2O_4(NCS)_6]^{4-}$.[387] Also, $[Mo_2O_3S\{S_2P(OPr^i)_2\}_2]$ reacts with various ligands to form 1:1 adducts, in which the addended molecule is located in the 'molecular crevice' formed by the phosphorodithioato ligands[388] and forms a third bridging group.

Complexes of molybdenum(V) with biologically relevant ligands, such as L-cysteine, have been investigated as possible models for the molybdenum sites in enzymes. In $[Mo_2O_4(\text{L-cysteinate})_2]^{2-}$, the $\{Mo_2O_4\}^{2+}$ core possesses the *syn* conformation and each $(\text{cysteinate})^{2-}$ acts as a tridentate O,S,N donor ligand giving octahedral coordination at each molybdenum with a carboxylato oxygen atom *trans* to the terminal oxo group.[389] A similar geometry is found in $[Mo_2O_2S_2(\text{L-cysteinate})_2]^{2-}$, which contains two bridging sulfide atoms.[390] In the corresponding derivatives of the methyl and ethyl cysteinate esters, $(\text{Me-Cys})^-$ and $(\text{Et-Cys})^-$, a carboxylate oxygen does not bind to molybdenum and $[Mo_2O_4(\text{Et-Cys})_2]$ and $[Mo_2O_2S_2(\text{Me-Cys})_2]$ involve five-coordinate metal atoms;[391,392] in both cases the *syn* conformation is adopted and the geometry around each molybdenum is perhaps best described as distorted trigonal bipyramidal, with one bridging oxygen and the nitrogen of the ester in the axial positions. $[Mo_2O_4(\text{L-histidinate})_2]$ and $[Mo_2O_2S_2(\text{L-histidinate})_2]$ have structures closely resembling those of the corresponding $(\text{L-cysteinate})^{2-}$ complexes, with the amino acids acting as tridentate O,N,N donors and with the carboxylate oxygen *trans* to the terminal oxo ligands.[393,394]

IR and Raman studies of molybdenum(V) dimers with $\{Mo_2O_4\}^{2+}$, $\{Mo_2O_3S\}^{2+}$ and $\{Mo_2O_2S_2\}^{2+}$ cores have established that in complexes with *syn* geometry two bands are observed in the region 900–1000 cm^{-1} which can be assigned to molybdenum–terminal oxygen stretching.[130,395] Vibrations associated with the bridging framework give rise to weaker bands at lower frequencies, which are often obscured by vibrations of other ligands.

^{95}Mo NMR spectra have been measured for a number of oxo and sulfido-bridged spin-paired dimers; the majority of complexes examined to date involve the *syn*-$\{Mo_2O_2Y_2\}^{2+}$ (Y = O, S) core and possess a chemical shift in the range 320–982 p.p.m.[396,397] The chemical shifts are sensitive to the electronic environment of the molybdenum; for example the yellow *syn*-$[Mo_2O_4(\text{nane})_2]$ and the purple *anti*-$[Mo_2O_4(\text{nane})_2]$ (nane = 1,4,7-triazacyclononane) are readily distinguished by ^{95}Mo NMR spectroscopy, having chemical shifts of 586 and 342 p.p.m., respectively. Furthermore, their interconversion and its catalysis by acid are readily monitored by this technique.[50]

CD spectra have been recorded for complexes containing $\{MoO_2XY\}^{2+}$ cores (X, Y = O, S) liganded by optically active chelates, such as (*S*)-cysteinate and (*S*)-histidinate. These spectra manifest two distinctive peaks with opposite signs at *ca.* 26 000 and 33 000 cm^{-1}, which can be related to the nature of the asymmetric distortion around the Mo—Mo axis.[398] Unlike the $\{Mo_2O_3\}^{4+}$ dimers, those containing an $\{Mo_2O_4\}^{2+}$ core do not exhibit any characteristic absorption in their electronic spectra, although the compounds are generally yellow/orange in colour.

The electrochemistry of selected systems containing an $\{Mo_2Y_4\}^{2+}$ (Y = O, S) core has been extensively investigated. For the series of dithiocarbamato complexes $[Mo_2O_{4-n}S_n(S_2CNR_2)_2]$ (n = 0–4; R = alkyl) the μ-dioxo species are reduced by a two-electron process, whereas all other species are reduced by two successive one-electron processes. The substitution of oxygen by sulfur enhances the ease of reduction of the binuclear centre.[346,361,362,366,371] $[Mo_2O_2S_2(S_2CNH_2)_2]$ is reduced irreversibly in a two-electron process, with dissociation of the dithiocarbamates.[366] Both *syn* and *anti* isomers of $[NEt_4]_2[Mo_2S_4(SC_2H_4S)_2]$ undergo a reversible one-electron reduction in DMSO; they also exhibit two irreversible oxidation processes, each of which appear to involve two electrons per dimer.[382] Cyclic voltammetric studies of $[NEt_4]_2[Mo_2S_4(SC_6H_5)_4]$, $[NEt_4]_2[Mo_2S_4(2\text{-}SC_6H_4NH)_2]$ and $[Mo_2S_4\{SC(Me)_2CH_2NHMe\}_2]$ have shown that reversible one-electron reductions occur with complexes of the bidentate ligands, but that the thiophenolate complex is irreversibly reduced and decomposes.[399] The cysteinato complexes $[Mo_2O_{4-n}S_n(\text{L-cysteinate})_2]^{2-}$ (n = 0–2) undergo electrochemical reduction in a single four-electron step to Mo^{III} dimers in aqueous buffers. The ease of reduction and electrochemical reversibility of the Mo_2^V/Mo_2^{III} couple increase with insertion of sulfur into the bridge system. However, the corresponding ethyl cysteinate complexes $[Mo_2O_{4-n}S_n(\text{Et-Cys})_2]$ (n = 0–2) are reduced by two successive one-electron transfers to Mo_2^{IV} species.[400] $[Mo_2O_4(\text{glycinate})_2(H_2O)_2]$ is reduced in a two-electron transfer to give monomeric Mo^{IV} species, which can be electrochemically oxidized to M^V monomers that dimerize to re-form the initial complex. Proton-assisted oxidation of an aqueous solution of $[Mo_2O_4(\text{glycinate})_2(H_2O)_2]$ yields the Mo^V/Mo^{VI} dimer $[Mo_2O_4(\text{glycinate})_2(OH)(H_2O)_2]$, which can be electrochemically reduced in a one-electron step to a Mo_2^V dimer, which is further reduced in a two-electron step to monomeric Mo^{IV} species.[401]

The references in Table 10 should be consulted for other studies on $\{Mo_2Y_4\}^{2+}$ (Y = O, S, Se) systems. Dimeric complexes, analogous to those containing an $\{Mo_2Y_4\}^{2+}$ (Y = O, S) core, but with one or two terminal NR^{2-} ligands have been reported. $(Me_3CN)_2S$ reacts with $[\{CpMo(CO)_3\}_2]$ to form $[\{CpMo(NCMe_3)_2\}_2S_2]$, in which the central Mo_2S_2 moiety is planar with an Mo—Mo separation of 2.920(1) Å.[412] $[Mo_2S_2(\text{Ntol})_4\{S_2P(OEt)_2\}_4]$ reacts with $(S_2CNBu_2)^-$ to produce $[Mo_2(\text{Ntol})_2S_2(S_2CNBu_2)_2]$, which has a structure very similar to that of $[Mo_2O_2S_2(S_2CNBu_2)_2]$; the imido ligands are *cis* to the two sulfide bridges, in the *syn* conformation, and each molybdenum is five-coordinate with an approximately square pyramidal geometry. The Mo—Mo distance is 2.807(2) Å.[413] The acid-assisted hydrolysis of $[Mo_4S_4(Ntol)_4\{S_2P(OEt)_2\}_4]$ to give $[Mo_2O_2S_2\{S_2P(OEt)_2\}_2]$, proceeds *via* the binuclear intermediate $[Mo_2O(Ntol)S_2\{S_2P(OEt)_2\}_2]$, which contains one terminal oxo group and one terminal 4-tolylimido group, on different molybdenum atoms, and adopts a *syn* conformation, with an Mo—Mo distance of 2.812(1) Å.[414]

A crystal structure determination of $[pyH]_4[Mo_2O_3(SO_4)(NCS)_6]$ has shown the two molybdenum atoms to be bridged by one oxo and the sulfato ligand. The terminal oxygens are *syn* and each molybdenum has a distorted octahedral geometry.[415] Persulfide bridges have been identified in several dimeric Mo^V complexes. In $[NH_4]_2[Mo_2(S_2)_6]$ each molybdenum is coordinated by four S_2^{2-} ligands in a distorted dodecahedron, with two bridging and two terminal groups. The complex is diamagnetic, and has an Mo—Mo distance of 2.827(2) Å.[416] A similar geometry is found in $[PPh_3Me]_2[Mo_2(S_2)_2Cl_8]$, in which each molybdenum is coordinated to two bridging persulfides and four terminal chlorides; the Mo—Mo distance is 2.857(1) Å.[417]

$[Mo_2O_2Cl_6(OPCl_3)_2]$ contains two chloride bridges; each molybdenum is also coordinated to two terminal chlorides, a terminal oxo ligand and a phosphate oxide in an octahedral geometry. The oxo groups are *cis* to the chloride bridges, in an *anti* conformation, and are *trans* to the $OPCl_3$ ligands.[418] $[Mo_2Cl_2L_8]Cl_8$ ($L = H_2N(CH_2)_nNH_2$, $n = 2$–6, 8) can be prepared by reacting $[Mo_2Cl_{10}]$ with the appropriate diamine under an inert atmosphere. These complexes are diamagnetic, and IR data suggest that they possess a dimeric structure with two bridging chlorides and monodentate diamines. In $[Mo_2Cl_2L_4]Cl_8$ ($L = H_2N(CH_2)_nNH_2$, $n = 9$, 10) the diamines appear to be bidentate.[419]

Asymmetric dinuclear molybdenum(V) complexes with two nitrogen bridges are formed on reaction of $[MoO_2(S_2CNEt_2)_2]$ with the hydrazines $PhC(Y)NHNH_2$ (Y = O, S) in methanol. A structure determination of $[(Et_2NCS_2)Mo\{OC(Ph)NN\}_2MoO(S_2CNEt_2)]$ has shown that one molybdenum is six-coordinate, bound in a trigonal prismatic geometry to two sulfur atoms, two oxygen atoms, and two bridging nitrogen atoms of the $\{OC(Ph)NN\}^{3-}$; the other molybdenum is five-coordinate, bound to the two bridging nitrogens, two sulfurs and an oxo ligand which occupies the axial position of a square pyramid.[420] These dimers undergo a reversible one-electron reduction. $[Mo_2Cl_{10}]$ reacts with thioureas to give $[Mo_2Cl_4\{RNHC(S)NHR'\}_4]Cl_6$ (R = pyridyl, 5-nitropyridyl, 4-methylpyridyl, 6-methylpyridyl; R′ = Ph, 2-MeC_6H_4, 4-MeC_6H_4). Two thioureas are believed to be bridging, coordinated to one molybdenum *via* the sulfur atom and to the other *via* the pyridyl nitrogen atom; the other two thioureas are bidentate N,S donor ligands, and two chlorides on each molybdenum complete an octahedral coordination sphere.[421]

Complexes based on an $\{Mo_2Y_4\}^{2+}$ (Y = O, S) core may involve an additional ligand bridging the two Mo^V atoms. (Section 36.4.3.5.5). However, many of these complexes share characteristics with the dimeric complexes having only two bridging ligands, especially if the two sites *trans* to the terminal ligands are occupied by different atoms of a polydentate ligand. $[Mo_2O_3S\{S_2P(OPr^i)_2\}_2]$ reacts with pyridine or pyridazine (pydz) to give a 1:1 adduct. The structure of these two adducts are similar, in that the basic core of the parent compound is maintained, but there are major differences in the manner of the bonding of the py and pydz molecules; the py molecule is weakly bonded to both molybdenums through nitrogen with Mo—N distances of 2.97(2) and 2.93(1) Å, whereas each nitrogen atom of the pydz molecule coordinates to a different molybdenum atom with Mo—N distances of 2.59(2) and 2.58(1) Å.[388] Bidentate heterocyclic ligands bipy, phen, 5-NO_2phen and 2,9-Me_2phen (L_2) react with $[Mo_2O_4(ab)_2(py)_2]$ (Hab = picolinic acid, 8-hydroxyquinoline) to give the corresponding $[Mo_2O_4(L_2)(ab)_2]$ derivative, in which the bidentate ligand bridges the $\{Mo_2O_4\}^{2+}$ core. The products are diamagnetic, indicating retention of the metal–metal interaction.[422]

Crystal structure determinations of $[pyH]_3[Mo_2O_4(O_2CR)(NCS)_4]$ (R = H, Me) have revealed N-bonded thiocyanates, with the molybdenum atoms bridged by two oxo groups and a bidentate carboxylate, coordinated to each molybdenum through a different oxygen atom, *trans* to the terminal oxo groups. The Mo—Mo distances are 2.566(1) and 2.560(1) Å for the formate and acetate complexes, respectively.[336,423] A similar structure has been found for the $[Mo_2O_3(OEt)(O_2CC_6H_4OH)Cl_4]^{2-}$ anion, in which the bridging ligands are one oxo group, the ethoxide group and the salicylate; the Mo—Mo bond length is 2.646 Å.[424]

The potentially six-coordinate ligands $(edta)^{4-}$ and $(pdta)^{4-}$ (H_4edta = *N,N,N′,N′*-tetrakis(2-ethanoic acid)-1,2-diaminoethane, H_4pdta = *N,N,N′,N′*-tetrakis(2-ethanoic acid)-1,2-diaminopropane) add to dimeric molybdenum(V) centres to form complexes such as $[Mo_2O_2Y_2(edta)]^{2-}$ and $[MoO_2Y_2(pdta)]^{2-}$ (Y = O, S).[425–429] The structure of

$Cs_2[Mo_2O_2S_2(edta)]$ shows a *syn* $\{MoO_2S_2\}^{2+}$ core with sulfide bridges; the nitrogen atoms of the $(edta)^{4-}$ are coordinated *trans* to the terminal oxo groups, and the octahedral coordination spheres of the molybdenums are completed by oxygen atoms of the carboxylate groups.[426] The structures of $Na_2[Mo_2O_{4-n}S_n(R\text{-pdta})]$ ($n = 0$–2) are exactly analogous to this.[427] The low magnetic moments (*ca.* 0.5 BM) of $[Mo_2O_4(edta)]^{2-}$ and $[Mo_2O_2S_2(edta)]^{2-}$ indicate a strong interaction between the molybdenum atoms.[428] The kinetics of formation of $[Mo_2O_4(edta)]^{2-}$ in aqueous solution have been investigated, and a detailed mechanism for dissociation proposed.[429] Cyclic voltammetry has been used to study the electrochemistry of $[Mo_2O_2X_2(edta)]^{2-}$ (X = O, S) in aqueous buffers; the complexes undergo reduction in a single four-electron step to molybdenum(III) dimeric species, but no oxidation process was observed.[400]

The reaction of $[PPh_4]_2[MoS_4]$ with the mineral realgar, As_4S_4, yields $[PPh_4]_2[Mo_2O_2S_2(As_4S_{12})]$. The structure of the anion consists of a sulfido-bridged *syn*-$\{Mo_2O_2S_2\}^{2+}$ core and a tetradentate S-bonded $(As_4S_{12})^{4-}$ chelate. Each molybdenum is in a square pyramidal geometry with the terminal oxo groups in the axial positions. The $(As_4S_{12})^{4-}$ ligand contains a 1,5-(As_2S_8) eight-membered ring, which is broken in the reaction between $[PPh_4]_2[Mo_2O_2S_2(As_4S_{12})]$ and $[PPh_4]_2[MoO_2S_2]$, yielding $[PPh_4]_4[(Mo_2O_2S_2)_2(As_2S_5)_2]$. In this complex two $\{Mo_2O_2S_2\}^{2+}$ units are linked by two $(S_2AsSAsS_2)^{4-}$ ligands to give a symmetrical 16-membered macrocycle.[430]

36.4.3.5.4 Oligomeric complexes based on an $\{Mo_2O_4\}^{2+}$ and related cores

Oligomers may be formed by ligands linking two or more $\{Mo_2O_4\}^{2+}$ units together; alternatively ligands may join other molybdenum centres to an $\{Mo_2O_4\}^{2+}$ core.

The anions of $Cs_6[\{Mo_2O_3S(C_2O_4)_2\}_2(C_2O_4)]\cdot H_2O$ and $K_6[\{Mo_2O_3S(C_2O_4)_2\}_2(C_2O_4)]\cdot 10H_2O$ consist of two $\{Mo_2O_3S(C_2O_4)_2\}^{2-}$ dimers bridged by a quadridentate oxalate ligand, in which each carboxylate group bridges one dimeric unit. The Mo—O bond lengths to the quadridentate oxalates (av. 2.303(4) Å) are longer than those to the bidentate, terminal oxalates (av. 2.106(4) Å), due to the *trans* influence of the terminal oxo ligands.[431] A similar structure is adopted by the anion of $K_6[\{Mo_2O_4(O_2CCH_2CO_2)_2\}_2(O_2CCH_2CO_2)]\cdot 4H_2O$ in which one malonate dianion forms a bridge between two dimolybdenum species.[432]

Methoxide bridges between $\{Mo_2O_4\}^{2+}$ dimeric units are found in $[\{Mo_2O_4(PMe_3)\}_4(OMe)_8]$, in which four dimers are joined in an eight-membered Mo ring by four pairs of bridging $(OMe)^-$ ligands. In each dimer one molybdenum is square-pyramidally bound to a terminal oxo group, two bridging oxo groups, and two methoxides; the other molybdenum has an additional phosphine ligand, and is in an octahedral coordination geometry (Figure 5).[433] $[\{Mo_2O_4(HBpz_3)(MeOH)\}_2(OMe)_2]$ (pz = pyrazolyl) contains a zigzag arrangement of four molybdenum atoms composed of two dimeric units, joined by two methoxide bridges. Each dimer consists of one molybdenum octahedrally bonded to a tridentate tris(pyrazolyl)borate ligand, a terminal oxygen, and two bridging oxygens; the other molybdenum is also octahedrally coordinated to two bridging oxygens, one terminal methanol *trans* to a terminal oxo group, and to the two bridging methoxides.[434] In $[Mo_4O_6Cl_4(OPr)_6]$, propoxide bridges are present in two $\{Mo_2O_3(OPr)\}^{3+}$ cores (with Mo—Mo separations of 2.669(2) Å); these two cores are linked by two propoxide bridges and by two bridging oxygens of the $\{Mo_2O_3(OPr)\}^{3+}$ units (Figure 6). This complex formally possesses two Mo^V and two Mo^{VI} centres.[435] $[Mo_4O_8Cl_4(OEt)_4]^{2-}$ adopts a similar structure, but involves a shorter Mo—Mo separation (2.587(3) Å).[436] $[Mo_2(OPr^i)_6]$ in pyridine is oxidized by O_2 to yield, as a minor product, $[Mo_4O_8(OPr^i)_4(py)_4]$, which has been characterized by X-ray crystallography. The complex contains four terminal, two μ_3- and two μ-oxo groups, two terminal and two μ-bridging OPr^i ligands, plus two localized Mo—Mo single bonds.[437] $[Mo_4O_8(Me_2POS)_4]$ also involves the pairing of two $\{Mo_2O_4\}^{2+}$ cores and partial substitution of the Me_2POS ligands by Me_2PO_2 has been achieved.[438]

$[Mo_2O_4(CO_3)(OH)_4\{Mo(CO)(PMe_3)_3\}_2]$ involves a central $\{Mo_2O_4\}^{2+}$ unit in which each molybdenum is also bound to an $\{Mo^{II}(CO)(PMe_3)_2\}^{2+}$ group *via* two bridging hydroxides, with the carbonate ligand bridging all four metal atoms. The two outer Mo^{II} atoms are seven-coordinate and the Mo^V atoms are six-coordinate, with an Mo—Mo distance of 2.5522(9) Å.[439] Another tetranuclear arrangement occurs in $[Mo_2O_4(O_2CMe)_6\{MoOCl\}_2]$, with each of two $\{MoOCl\}^{2+}$ units coordinated to one molybdenum of the $\{Mo_2O_4\}^{2+}$ core through two acetates, to the other through one acetate, and directly bonded to one of the oxygen bridges (Figure 7). The Mo—Mo distance in the central core is 2.609(3) Å.[440] $Mo_6O_{10}(OPr^i)_{12}$, an intermediate in the reaction between $[Mo_2(OPr^i)_6]$ and O_2 consists of a serpentine chain of

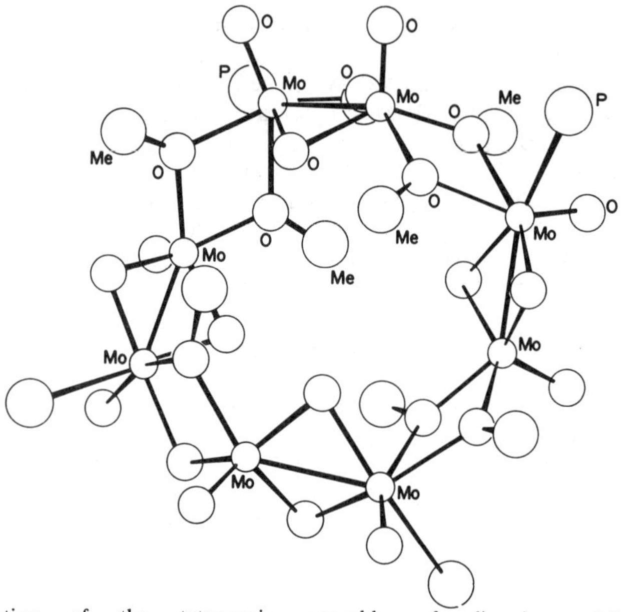

Figure 5 Representation of the tetrameric assembly of dimeric molybdenum(V) centres in $[\{Mo_2O_4(PMe_3)\}_4(OMe)_8]$[433]

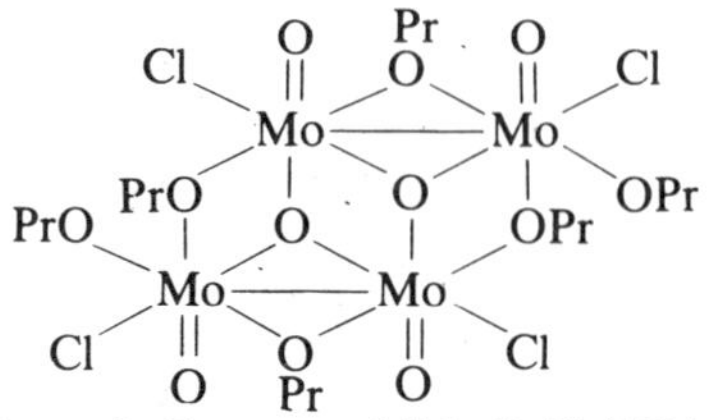

Figure 6 Structure of $[Mo_4O_6Cl_4(OPr)_6]$[435]

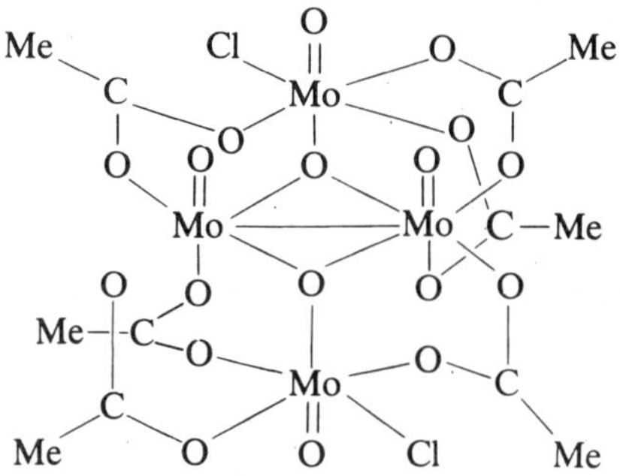

Figure 7 Representation of the structure of $[Mo_2O_4(O_2CMe)_6\{MoOCl\}_2]$[440]

molybdenum atoms, supported by μ-oxo and μ-OPr^i ligands, in which two Mo^V—Mo^V interactions (2.585(1) Å) occur.[441]

The tetranuclear complexes $[\{Mo_2Y(NR)S_2[S_2P(OEt)_2]_2\}_2]$ (Y = O, NR; R = Ph, 4-tolyl) contain two $\{MoY(NR)S_2\}^{2+}$ units linked by weak coordination of the bridging sulfides of one unit to the molybdenums of the other unit, in positions *trans* to the terminal oxo or imido ligands.[413,414] Addition of CF_3CO_2H to solutions of $[\{Mo_2(Ntol)_2S_2[S_2P(OEt)_2]_2\}_2]$ leads to formation of $[Mo_2(Ntol)_2S(SH)\{S_2P(OEt)_2\}_2(O_2CCF_3)]$, in which one of the bridging sulfides is protonated, and the trifluoroacetate bridges the dimolybdenum unit in the positions *trans* to the imido ligands. In solution two conformers have been detected by ^1H NMR spectroscopy, thought to be due to different orientations of the S—H bond.[438] Treatment of $[Mo_2(Ntol)_2S(SH)\{S_2P(OEt)_2\}_2(O_2CCF_3)]$ with triethylamine, followed by addition of MeBr, yields the methylated complex $[Mo_2(Ntol)_2S(SMe)\}S_2P(OEt)_2\}_2(O_2CCF_3)]$, which has an identical structure to the $(SH)^-$-bridged compound.[442]

36.4.3.5.5 Dimeric complexes with three bridging ligands

One of the products of the reaction between thiophenol and $[Mo_2O_4(S_2CNEt_2)_2]$ is $[Mo_2O_3(SPh)_2(S_2CNEt_2)_2]$. A crystal structure determination has shown that this dimer is bridged by two thiols and an oxo group, and that each molybdenum has distorted octahedral coordination (Figure 8). One of the bridging thiols is *trans* to the two *syn* terminal oxo groups, and the average Mo—S bond lengths to this thiol is *ca.* 0.2 Å longer than that to the *cis* ligand. The Mo—Mo distance is 2.678(9) Å. Prolonged exposure of $[Mo_2O_3(SPh)_2(S_2CNEt_2)_2]$ to $CHCl_3$ gives $[Mo_2O_2Cl(SPh)_2(S_2CNEt_2)_2]^+$, in which the three bridging ligands are a chloride and two thiophenolate anions. The chloride is *trans* to the terminal oxo groups, the two thiophenolate ligands are equivalent and the Mo—Mo distance is 2.822(2) Å.[443] The reaction of $[NBu_4][MoO(SPh)_4]$ with HCl yields $[NBu_4][Mo_2O_2Cl_5(SPh)_2]$, which has been characterized by X-ray crystallography. The structure of the anion is analogous to that of $[Mo_2O_2Cl(SPh)_2(S_2CNEt_2)_2]^+$, with one chloride and two thiolate bridges, the chloride being *trans* to each terminal oxo ligand. However, the Mo—Mo distance of $[Mo_2O_2Cl_5(SPh)_2]^-$ is 2.915(1), which is significantly longer than that of the related cation. The reduction of $[MoO_2(S_2CNMe_2)_2]$ by thiophenol yields $[Mo_2O_3(SPh)_2(S_2CNMe_2)_2]$, which is isostructural with its diethyldithiocarbamate analogue, and involves an Mo—Mo distance of 2.649(1) Å.[444]

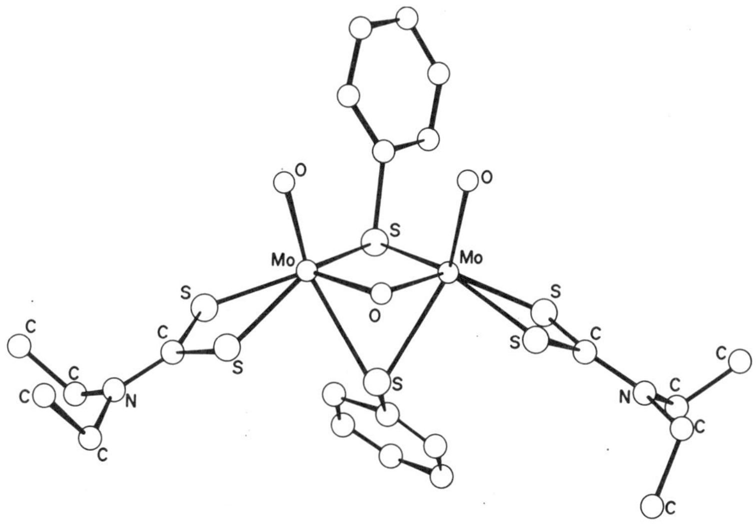

Figure 8 Structure of $[Mo_2O_3(SPh)_2(S_2CNEt_2)_2]$[443]

The reaction between $[MoOCl_3(THF)_2]$ and alkanethiols in the presence of base yields $[Mo_2O_2(SR)_7]^-$ anions, which have been isolated as their tetraalkylammonium salts. Physical data indicate that these binuclear monoanions contain three bridging thiolates. The arylthiolato anions $[MoO(SAr)_4]^-$ (Ar = Ph, 4-tolyl) react in the presence of an alkoxy or amido ligand, Z, to form $[Mo_2O_2(SAr)_6Z]$, in which the Z group and two thiolates act as bridging ligands.[445,446] A rich electrochemistry of $[Mo_2O_2(YR)_6Z]^-$ (Y = S, Se) has been demonstrated and stepwise reduction of these Mo_2^V centres to $Mo^{IV,V}$ and Mo_2^{IV} species has been observed.[266] The reaction between $[MoOCl_3(THF)_2]$ and Na(S-4-tolyl) in methanol, followed by addition of $[NEt_4]Br$, yields $[NEt_4][Mo_2O_2(S\text{-}4\text{-}tolyl)_6(OMe)]$, which has been structurally characterized. The anions involve an Mo—Mo separation of 2.919(5) Å, with two *syn* terminal oxo groups, four terminal and two bridging thiolate ligands, and a bridging methoxide ligand *trans* to each of the terminal oxygens.[444] Interconversion of $[MoO(SR)_4]^-$ (R = aryl, alkyl) and $[Mo_2O_2(SR)_6Z]^-$ (Z = OMe, OEt, SCH_2Ph) anions has been shown to occur *via* redox processes involving both the molybdenum and the thiolate redox centres.[266,448]

Several examples are known of triply bridged dimers in which one ligand provides two of the bridging atoms. The reaction of $[Mo_2O_4(py)_2(oxine)_2]$ (Hoxine = 8-hydroxyquinoline) with 2-hydroxyethanethiol gives $[Mo_2O_3(SCH_2CH_2O)(oxine)_2]$, the X-ray structure of which shows an Mo_2^V unit bridged by one oxo group and by the sulfur and oxygen atoms of the $(SCH_2CH_2O)^{2-}$ ligand. The two *syn* terminal oxo groups are *trans* to the oxygen atom of the mercaptoethanolate, with the nitrogen atoms of the oxines *trans* to the bridging sulfur.[449] A similar structure is found in $[Mo_2O_2S(SCH_2CH_2O)(S_2CNEt_2)_2]$, in which the three bridging atoms are the sulfide ligand and the oxygen and sulfur atoms of the $(SCH_2CH_2O)^-$ ligand; the bridging oxygen is *trans* to each of the terminal oxo groups.[450]

Reactions of 2-hydroxyethanethiol and various bases with $[MoOCl_4(H_2O)]^-$ or

$[MoOCl_3(THF)_2]$ yield many products, including the triply bridged dimers $[Mo_2O_2(SCH_2CH_2O)_2X_2Y]^-$ (X = Y = Cl; $X_2 = SCH_2CH_2O$, Y = Cl, SCH_2CH_2OH). These complexes involve bridging atoms which are the oxygen and sulfur of one $(SCH_2CH_2O)^{2-}$ ligand and the oxygen atom of another, the sulfur atom of which is coordinated to one of the molybdenum atoms in a terminal position. The terminal oxo groups are *trans* to the oxygen atom of the doubly bridging ligand. In each case the Mo—Mo distance is 2.734(6) Å.[445,446] The reaction of $[Mo_2O_2(SCH_2CH_2O)_2Cl_3]^-$ with excess 2-hydroxyethanethiol and base, B, yields $[BH]_2[Mo_2O_3(SCH_2CH_2O)_3]$ (B = piperidine, morpholine, pyrrolidine). The bridging atoms of the anion are one oxo group and the oxygen and sulfur of a $(SCH_2CH_2O)^{2-}$ ligand; each molybdenum atom is also coordinated to a terminal oxo group and a bidentate $(SCH_2CH_2O)^{2-}$ ligand. The Mo—Mo separation is 2.676(1) Å.[446]

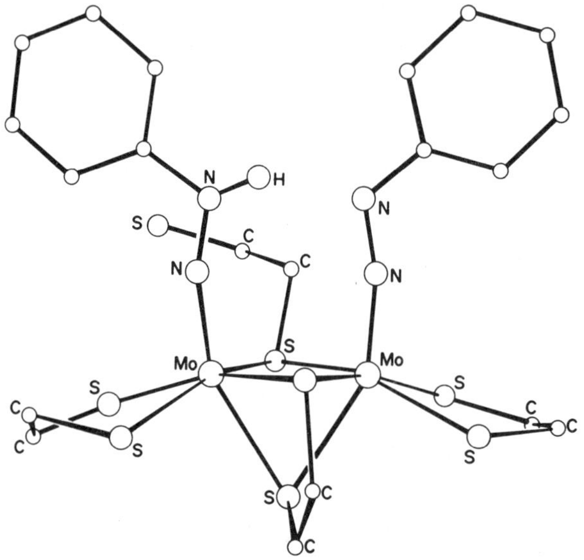

Figure 9 Structure of $[Mo_2(NNHPh)(NNPh)(SCH_2CH_2S)_3(SCH_2CH_2SH)]^{2-}$ [451]

Treating $[Mo(NNHPh)_2(butane\text{-}2,3\text{-}dithiolate)_2]$ with dimercaptoethane and trimethylamine in methanol gives $[NEt_3H]_2[Mo_2(NNPh)(NNHPh)(SCH_2CH_2S)_3(SCH_2CH_2SH)]$, which contains terminal and bridging bidentate thiolate ligands and a bridging monodentate thiolate ligand (Figure 9). The Mo—Mo distance is 2.837(2) Å, which is longer than in triply bridged molybdenum(V) dimers involving one or more bridging oxygens.[451] The anion $[MoO(SCH_2CH_2CH_2S)_2]^-$ reacts with HN_3 in methanol to give $[Mo_2O_2(N_3)(SCH_2CH_2CH_2S)_3]^-$. The crystal structure of this dimeric anion shows that the three bridging atoms are provided by a nitrogen atom of the azide, and by sulfur atoms of two different dithiolates, which are both chelated to the same molybdenum; the third dithiolate ligand is chelated to the other molybdenum. Two terminal oxo groups *trans* to the bridging azide complete octahedral coordination spheres for each molybdenum, and the Mo—Mo distance is 2.893(3) Å.[452]

36.4.4 REFERENCES

1. M. L. Larson, *Inorg. Synth.*, 1970, **12**, 165 *et seq.*
2. D. E. LaValle, R. M. Steele, M. K. Wilkinson and H. L. Yakel, Jr., *J. Am. Chem. Soc.*, 1960, **82**, 2433.
3. H. Schafer, H. G. von Schnering, J. Tillack, F. Kuhnen, H. Wohrle and H. Baumann, *Z. Anorg. Allg. Chem.*, 1967, **353**, 281.
4. D. Babel and W. Rudorff, *Naturwissenschaften*, 1964, **51**, 85.
5. E. I. Stiefel, *Prog. Inorg. Chem.*, 1977, **22**, 1.
6. D. A. Brown, W. K. Glass and C. O'Daly, *J. Chem. Soc., Dalton Trans.*, 1973, 1311.
7. L. M. Toth, G. D. Brunton and G. P. Smith, *Inorg. Chem.*, 1969, **8**, 2694.
8. D. L. Kepert, 'The Early Transition Metals', Academic, London, 1972, p. 339.
9. A. Herbowski and T. Lis, *Polyhedron*, 1985, **4**, 127.
10. P. E. Kazin, V. K. Bel'skii, A. I. Zhirov, N. A. Subbotina, M. G. Felin, V. V. Zelentsov and V. I. Spitsyn, *Zh. Neorg. Khim.*, 1985, **30**, 1426.
11. R. G. Cavell and J. W. Quail, *Inorg. Chem.*, 1983, **22**, 2597.

12. P. W. Smith and A. G. Wedd, *J. Chem. Soc. (A)*, 1968, 1377.
13. A. Hills, G. J. Leigh, J. Hutchinson and J. A. Zubieta, *J. Chem. Soc., Dalton Trans.*, 1985, 1069.
14. C. Schumacher, R. E. Schmidt and K. Dehnicke, *Z. Anorg. Allg. Chem.*, 1985, **520,** 25.
15. A. D. Westland and N. Muriithi, *Inorg. Chem.*, 1972, **11,** 2971.
16. J. R. Dilworth and J. A. Zubieta, *J. Chem. Soc., Dalton Trans.*, 1983, 397.
17. D. Sevdic and L. Fekete, *Polyhedron,* 1985, **4,** 1371.
18. G. Backes-Dahmann, W. Herrmann, K. Wieghardt and J. Weiss, *Inorg. Chem.*, 1985, **24,** 485.
19. J. V. Brencic, *Z. Anorg. Allg. Chem.*, 1974, **403,** 218; J. V. Brencic and I. Leban, *Z. Anorg. Allg. Chem.*, 1980, **465,** 173; J. V. Brencic, I. Leban and M. Slokar, *Acta Crystallogr., Sect. B,* 1980, **36,** 698.
20. A. J. L. Pombeiro and R. L. Richards, *Transition Met. Chem. (Weinheim, Ger.)*, 1981, **6,** 253; T. S. Morita, Y. Sasaki and K. Saito, *Bull. Chem. Soc. Jpn.*, 1981, **54,** 2678.
21. M. Miller, S. Lincoln and S. A. Koch, *J. Am. Chem. Soc.*, 1982, **104,** 288.
22. P. Salagre, J.-E. Sueiras, X. Solans and G. Germain, *J. Chem. Soc., Dalton Trans.*, 1985, 2263.
23. A. K. Gregson and M. Anker, *Aust. J. Chem.*, 1979, **32,** 503.
24. C. L. Raston and A. H. White, *Aust. J. Chem.*, 1979, **32,** 507.
25. G. R. Brubaker and J. A. Romberger, *Inorg. Chem.*, 1980, **19,** 1087.
26. P. Thomas, J. Schrekenbach and H. Hennig, *Z. Chem.*, 1984, **24,** 413.
27. S. P. Ghosh and K. M. Prasad, *J. Inorg. Nucl. Chem.*, 1978, **40,** 1963.
28. R. G. Cavell and A. R. Sanger, *Inorg. Chem.*, 1972, **11,** 2011.
29. D. A. Brown, W. K. Glass and K. S. Jasim, *Inorg. Chim. Acta,* 1980, **45,** 97.
30. J. A. Broomhead, N. S. Gill, B. C. Hammer and M. Sterns, *J. Chem. Soc., Chem. Commun.*, 1982, 1234.
31. D. C. Povey, R. L. Richards and C. Shortman, *Polyhedron,* 1986, **5,** 369.
32. J. R. Knox and K. Eriks, *Inorg. Chem.*, 1968, **7,** 84.
33. F. Pruchnik, S. Wajda and E. Kwaskowska-Chec, *Rocz. Chem.*, 1971, **45,** 537 (*Chem. Abstr.*, 1971, **75,** 69 006).
34. S. P. Ghosh and K. M. Prasad, in 'Proceedings of the First International Conference on Chemistry and Uses of Molybdenum', ed. P. C. H. Mitchell, Climax Molybdenum Co., Ann Arbor, MI, 1974, p. 119.
35. G. R. Rossman, F.-D. Tsay and H. B. Gray, *Inorg. Chem.*, 1973, **12,** 824.
36. M. B. Hursthouse, K. M. A. Malik, A. M. Soares, J. F. Gibson and W. P. Griffith, *Inorg. Chim. Acta,* 1980, **45,** L81.
37. B. M. Figgis, J. Lewis and F. E. Mabbs, *J. Chem. Soc.*, 1961, 3138.
38. B. A. Averill and W. H. Orme-Johnson, *Inorg. Chem.*, 1980, **19,** 1702.
39. C. Furlani and O. Piovesana, *Mol. Phys.*, 1965, **9,** 341.
40. H. L. Schlafer, H. Gausmann and H. Witzke, *J. Mol. Spectrosc.*, 1966, **21,** 125.
41. M. W. Anker, J. Chatt, G. J. Leigh and A. G. Wedd, *J. Chem. Soc., Dalton Trans.*, 1975, 2639.
42. L. F. Lindoy, S. E. Livingstone and T. M. Lockyer, *Aust. J. Chem.*, 1965, **18,** 1549.
43. P. M. Boorman, K. J. Moynihan and R. T. Oakley, *J. Chem. Soc., Chem. Commun.*, 1982, 899.
44. P. Chaudhuri, K. Wieghardt, I. Jibril and G. Huttner, *Z. Naturforsch., Teil B,* 1984, **39,** 1172.
45. V. I. Spitsyn, N. A. Subbotina, M. G. Felin, S. I. Pakhamov and A. I. Zhirov, *Dokl. Akad. Nauk SSSR,* 1979, **245,** 1130.
46. P. C. H. Mitchell and R. D. Scarle, *J. Chem. Soc., Dalton Trans.*, 1972, 1809.
47. S. Wajda and B. Kurzak, *Pol. J. Chem.*, 1980, **54,** 2131 (*Chem. Abstr.*, 1981, **95,** 68 689).
48. M. Lamache-Duhameaux, *J. Inorg. Nucl. Chem.*, 1981, **43,** 208.
49. K. Wieghardt, M. Hahn, W. Swiridoff and J. Weiss, *Inorg. Chem.*, 1984, **23,** 94.
50. K. Wieghardt, M. Guttmann, P. Chaudhuri, W. Gebert, M. Minelli, C. G. Young and J. H. Enemark, *Inorg. Chem.*, 1985, **24,** 3151.
51. G. G. Kneale, A. J. Geddes, Y. Sasaki, T. Shibahara and A. G. Sykes, *J. Chem. Soc., Chem. Commun.*, 1975, 356.
52. J. Koublek and J. Podlaha, *J. Inorg. Nucl. Chem.*, 1971, **33,** 2981.
53. Y. Sasaki, T. Tani and T. S. Morita, in 'Proceedings of the Fourth International Conference on Chemistry and Uses of Molybdenum', ed. H. F. Barry and P. C. H. Mitchell, Climax Molybdenum Co., Ann Arbor, MI, 1982, p. 89.
54. D. A. Brown, W. K. Glass and K. S. Jasim, in 'Proceedings of the Fourth International Conference on Chemistry and Uses of Molybdenum', ed. H. F. Barry and P. C. H. Mitchell, Climax Molybdenum Co., Ann Arbor, MI, 1982, p. 156.
55. S. J. N. Burgmayer and J. L. Templeton, *Inorg. Chem.*, 1985, **24,** 3939.
56. E. Herdtweck, C. Schumacher and K. Dehnicke, *Z. Anorg. Allg. Chem.*, 1985, **526,** 93.
57. C. Potvin, J.-M. Brégeault and J.-M. Manoli, *J. Chem. Soc., Chem. Commun.*, 1980, 664; J.-M. Manoli, J. M. Brégeault, C. Potvin and G. Chottard, *Inorg. Chim. Acta,* 1984, **88,** 75.
58. M. G. B. Drew, P. C. H. Mitchell and C. F. Pygall, *J. Chem. Soc., Dalton Trans.*, 1979, 1213.
59. A. Müller, R. Jostes, W. Eltzner, Chong-Shi Nie, E. Diemann, H. Bögge, M. Zimmermann, M. Dartmann, U. Reinsch-Vogell, Shun Che, S. J. Cyvin and B. N. Cyvin, *Inorg. Chem.*, 1985, **24,** 2872.
60. G. W. A. Fowles, *Prog. Inorg. Chem.*, 1964, **6,** 1.
61. R. D. Peacock, *J. Chem. Soc.*, 1957, 4684; R. T. Paine and L. B. Asprey, *Inorg. Chem.*, 1974, **13,** 1529; J. B. Bates, *Inorg. Nucl. Chem. Lett.*, 1971, **7,** 957.
62. D. L. Kepert and R. Mandyczewsky, *Inorg. Chem.*, 1968, **7,** 2091.
63. U. L. Müller, *Angew. Chem., Int. Ed. Engl.*, 1981, **20,** 692.
64. A. D. Westland and V. Uzelac, *Inorg. Chim. Acta,* 1977, **23,** L37.
65. P. J. H. Carnell, R. E. McCarley and R. D. Hogue, *Inorg. Synth.*, 1967, **10,** 49.
66. D. A. Rice, *Coord. Chem. Rev.*, 1978, **25,** 199.
67. B. G. Brandt and A. C. Skapski, *Acta Chem. Scand.*, 1967, **21,** 661.
68. J. Ghose, N. N. Greenwood, G. C. Hallam and D. A. Read, *J. Solid State Chem.*, 1976, **19,** 365.
69. P. W. Schneider, D. C. Bravard, J. W. McDonald and W. E. Newton, *J. Am. Chem. Soc.*, 1972, **94,** 8640.
70. W. E. Newton, J. L. Corbin, D. C. Bravard, J. E. Searles and J. W. McDonald, *Inorg. Chem.*, 1974, **13,** 1100.

71 M. W. Bishop, J. Chatt and J. R. Dilworth, *J. Organomet. Chem.*, 1974, **73,** C59.
72. L. Ricard and R. Weiss, *Inorg. Nucl. Chem. Lett.*, 1974, **10,** 217.
73. P. C. H. Mitchell and R. D. Scarle, *J. Chem. Soc., Dalton Trans.*, 1975, 2552.
74. E. A. Maatta and R. A. D. Wentworth, *Inorg. Chem.*, 1979, **18,** 524.
75. E. A. Maatta, R. A. D. Wentworth, W. E. Newton, J. W. McDonald and G. D. Watt, *J. Am. Chem. Soc.*, 1978, **100,** 1320.
76. W. E. Newton, J. W. McDonald, J. L. Corbin, L. Ricard and R. Weiss, *Inorg. Chem.*, 1980, **19,** 1997.
77. K. Tanaka, S. Miyaka and T. Tanaka, *Chem. Lett.*, 1981, 189.
78. T. Kashiwagi, K. Tanaka and T. Tanaka, *Chem. Lett.*, 1982, 851.
79. M. Tatsumisago, G. Matsubayashi, T. Tanaka, S. Nishigaki and K. Nakatsu, *J. Chem. Soc., Dalton Trans.*, 1982, 121.
80. C. D. Garner and S. Bristow, in 'Molybenum Enzymes', ed. T. G. Spiro, Wiley, New York, 1985, p. 343 *et seq.*
81. R. Barral, C. Bocard, I. Seree de Roch and L. Sajus, *Tetrahedron Lett.*, 1972, 1693.
82. G. J.-J. Chen, J. W. McDonald and W. E. Newton, *Inorg. Chem.*, 1976, **15,** 2612.
83. J. Hyde, K. Venkatasubramanian and J. Zubieta, *Inorg. Chem.*, 1978, **17,** 414.
84. I. W. Boyd and J. T. Spence, *Inorg. Chem.*, 1982, **21,** 1602.
85. M. S. Reynolds, J. M. Berg and R. H. Holm, *Inorg. Chem.*, 1984, **23,** 3057 and references therein.
86. R. Durant, C. D. Garner, M. R. Hyde and F. E. Mabbs, *J. Chem. Soc., Dalton Trans.*, 1977, 955.
87. R. D. Taylor, J. P. Street, M. Minelli and J. T. Spence, *Inorg. Chem.*, 1978, **17,** 3207.
88. C. Pickett, S. Kumar, P. A. Vella and J. Zubieta, *Inorg. Chem.*, 1982, **21,** 908.
89. A. Bruce, J. L. Corbin, P. L. Dahlstrom, J. R. Hyde, M. Minelli, E. I. Stiefel, J. T. Spence and J. Zubieta, *Inorg. Chem.*, 1982, **21,** 917.
90. B. H. Kaul, J. H. Enemark, S. L. Merbs and J. T. Spence, *J. Am. Chem. Soc.*, 1985, **107,** 2885.
91. J. M. Berg and R. H. Holm, *J. Am. Chem. Soc.*, 1984, **106,** 3035; 1985, **107,** 917; 1985, **107,** 925.
92. F. A. Cotton, in 'Proceedings of the First International Conference on Chemistry and Uses of Molybdenum', ed. P. C. H. Mitchell, Climax Molybdenum Co., Ann Arbor, MI, 1974, p. 6.
93. P. C. H. Mitchell, *Q. Rev., Chem. Soc.*, 1966, **20,** 103; B. Spivack and Z. Dori, *Coord. Chem. Rev.*, 1975, **17,** 99.
94. F. A. Cotton and R. M. Wing, *Inorg. Chem.*, 1965, **4,** 867.
95. W. E. Newton and J. W. McDonald, in 'Proceedings of the Second International Conference on Chemistry and Uses of Molybdenum', ed. P. C. H. Mitchell and A. Seaman, Climax Molybdenum Co., Ann Arbor, MI, 1976, p. 25.
96. K. F. Miller and R. A. D. Wentworth, *Inorg. Chem.*, 1979, **18,** 984.
97. W. P. Griffith, *Coord. Chem. Rev.*, 1970, **5,** 459.
98. J. A. Crayston, M. J. Almond, A. J. Downs, M. Poliakoff and J. J. Turner, *Inorg. Chem.*, 1984, **23,** 3051.
99. R. N. Jowitt and P. C. H. Mitchell, *J. Chem. Soc. (A)*, 1969, 2632.
100. L. Ricard, J. Estienne, P. Karagiannidis, P. Toledano, A. Mitschler and R. Weiss, *J. Coord. Chem.*, 1974, **3,** 277.
101. A. K. Banerjee and N. Banerjee, *J. Coord. Chem.*, 1981, **11,** 35.
102. M. R. Churchill and F. J. Rotella, *Inorg. Chem.*, 1978, **17,** 668.
103. R. E. DeSimone and M. D. Glick, *Inorg. Chem.*, 1978, **17,** 3574.
104. M. Lamache-Duhameaux, in 'Proceedings of the Second International Conference on Chemistry and Uses of Molybdenum', ed. P. C. H. Mitchell and A. Seaman, Climax Molybdenum Co., Ann Arbor, MI, 1976, p. 42.
105. J. Chatt, J. R. Dilworth, J. A. Schmutz and J. A. Zubieta, *J. Chem. Soc., Dalton Trans.*, 1979, 1595.
106. A. V. Butcher and J. Chatt, *J. Chem. Soc. (A)*, 1970, 2652.
107. L. Manojlovic-Muir and K. W. Muir, *J. Chem. Soc., Dalton Trans.*, 1972, 686.
108. L. Manojlovic-Muir, *J. Chem. Soc. (A)*, 1971, 2796.
109. L. K. Atkinson, A. H. Mawby and D. C. Smith, *Chem. Commun.*, 1970, 1399.
110. A. V. Butcher and J. Chatt, *J. Chem. Soc. (A)*, 1971, 2356.
111. B. T. Kilbourn, V. C. Adam and U. A. Gregory, *Chem. Commun.*, 1970, 1400.
112. C. A. Rice and J. T. Spence, *Inorg. Chem.*, 1980, **19,** 2845.
113. T. Diebold, B. Chevrier and R. Wiess, *Inorg. Chem.*, 1979, **18,** 1193.
114. T. Imamura, K. Hasegawa and M. Fujimoto, *Chem. Lett.*, 1983, **5,** 705.
115. T. Imamura, M. Takahashi, T. Tanaka, T. Jin, M. Fujimoto, S. Sawamura and M. Katayama, *Inorg. Chem.*, 1984, **23,** 3752.
116. V. Borschel and J. Strahle, *Z. Naturforsch., Teil B*, 1984, **39,** 1664.
117. S. J. Edmonson and P. C. H. Mitchell, *Polyhedron*, 1986, **5,** 315.
118. S. Midollini and M. Bacci, *J. Chem. Soc. (A)*, 1970, 2964.
119. R. L. Marks and S. E. Popielski, *Inorg. Nucl. Chem. Lett.*, 1974, **10,** 885.
120. V. W. Day and J. L. Hoard, *J. Am. Chem. Soc.*, 1968, **90,** 3374.
121. P. R. Robinson, E. O. Schlemper and R. K. Murmann, *Inorg. Chem.*, 1975, **14,** 2035.
122. K. Wieghardt, G. Backes-Dahmann, W. Holzbach, W. Swiridoff and J. Weiss, *Z. Anorg. Allg. Chem.*, 1983, **499,** 44.
123. M. Dudek and A. Samotus, *Transition Met. Chem. (Weinheim, Ger.)*, 1985, **10,** 271.
124. S. S. Basson, J. G. Liepoldt and I. M. Potgieter, *Inorg. Chim. Acta*, 1984, **90,** 57.
125. M. Novotny and S. J. Lippard, *Inorg. Chem.*, 1974, **13,** 828.
126. C. T. Lam, D. L. Lewis and S. J. Lippard, *Inorg. Chem.*, 1976, **15,** 989.
127. C. G. Young and J. H. Enemark, *Inorg. Chem.*, 1985, **24,** 4416.
128. K. F. Miller and R. A. D. Wentworth, *Inorg. Chem.*, 1980, **19,** 1818.
129. R. M. Lynden-Bell, G. G. Mather and A. Pidcock, *J. Chem. Soc., Dalton Trans.*, 1973, 715.
130. G. J.-J. Chen, J. W. McDonald, D. C. Bravard and W. E. Newton, *Inorg. Chem.*, 1985, **24,** 2327.
131. J. T. Spence, in 'Proceedings of the Third International Conference on the Chemistry and Uses of Molybdenum', ed. P. C. H. Mitchell and H. F. Barry, Climax Molybdenum Co., Ann Arbor, MI, 1979, p. 237.
132. D. Sutton, *Chem. Soc. Rev.*, 1975, **4,** 443.

133. P. J. Blower, J. R. Dilworth, J. Hutchinson, T. Nicholson and J. A. Zubieta, *J. Chem. Soc., Dalton Trans.*, 1985, 2639.
134. T. C. Hsieh and J. Zubieta, *Inorg. Chim. Acta*, 1985, **99,** L47.
135. J. R. Dilworth and J. A. Zubieta, *Transition Met. Chem.* (*Weinheim, Ger.*), 1984, **9,** 39.
136. R. Mattes and H. Scholand, *Angew. Chem., Int. Ed. Engl.*, 1983, **22,** 245.
137. R. Mattes and H. Scholand, *Angew. Chem., Int. Ed. Engl.*, 1984, **23,** 382.
138. C. F. Barrientos-Penna, F. W. B. Einstein, T. Jones and D. Sutton, *Inorg. Chem.*, 1983, **22,** 2614.
139. J. Chatt, R. A. Head, G. J. Leigh and C. J. Pickett, *J. Chem. Soc., Dalton Trans.*, 1978, 1638.
140. W. Hussain, G. J. Leigh and C. J. Pickett, *J. Chem. Soc., Chem. Commun.*, 1982, 747.
141. J. R. Dilworth, P. L. Dahlstrom, J. R. Hyde and J. Zubieta, *Inorg. Chim. Acta*, 1983, **71,** 21.
142. D. D. Devore and E. A. Maatta, *Inorg. Chem.*, 1985, **24,** 2846.
143. C. Y. Chou, D. D. Devore, S. C. Huckett, E. A. Maatta, J. C. Huffman and F. Takusagawa, *Polyhedron*, 1986, **5,** 301.
144. S. Donovan-Mtunzi, R. L. Richards and J. Mason, *J. Chem. Soc., Dalton Trans.*, 1984, 1329.
145. A. J. Edwards and R. D. Peacock, *Chem. Ind.* (*London*), 1960, 1441.
146. G. Brunton, *Mater. Res. Bull.*, 1971, **6,** 555.
147. A. J. Edwards, A. J. Peacock and A. Said, *J. Chem. Soc.*, 1962, 4643.
148. T. B. Scheffler, C. L. Hussey, K. R. Seddon, C. M. Kear and P. D. Armitage, *Inorg. Chem.*, 1983, **22,** 2099.
149. E. A. Allen, B. J. Brisdon and G. W. A. Fowles, *J. Chem. Soc.*, 1964, 4531.
150. E. A. Allen, K. Feeman and G. W. A. Fowles, *J. Chem. Soc.*, 1965, 1636.
151. W. M. Carmichael and D. A. Edwards, *J. Inorg. Nucl. Chem.*, 1972, **34,** 1181.
152. C. Miniscloux, G. Martino and L. Sajus, *Bull. Soc. Chim. Fr.*, 1973, 2179.
153. J. R. Moss and B. L. Shaw, *J. Chem. Soc.* (*A*), 1970, 595.
154. L. Castellani and M. C. Gallazzi, *Transition Met. Chem.* (*Weinheim, Ger.*), 1985, **10,** 194.
155. I. W. Boyd, G. P. Haight and N. C. Howlader, *Inorg. Chem.*, 1981, **20,** 3115.
156. E. L. Muetterties, *J. Am. Chem. Soc.*, 1960, **82,** 1082.
157. G. Doyle, *Inorg. Chem.*, 1971, **10,** 2348.
158. C.-T. Kan, *J. Chem. Soc., Dalton Trans.*, 1982, 2309.
159. A. van den Bergen, K. S. Murray and B. O. West, *Aust. J. Chem.*, 1972, **25,** 705.
160. J. E. Davies and B. M. Gatehouse, *J. Chem. Soc., Dalton Trans.*, 1974, 184.
161. C. J. Weber and R. D. Archer, *Inorg. Chem.*, 1983, **22,** 4149.
162. (a) T. M. Brown and J. N. Smith, *J. Chem. Soc., Dalton Trans.*, 1972, 1614; (b) T. M. Brown and J. N. Smith, *J. Chem. Soc., Dalton Trans.*, 1972, 1796.
163. R. Lozano, E. Alarcon, A. L. Doadrio, M. C. Ramirez and A. Doadrio, *An. Quim. Ser. B*, 1983, **79,** 41 (*Chem. Abstr.*, 1984, **100,** 44 335).
164. Z. B. Varadi and A. Nieuwpoort, *Inorg. Nucl. Chem. Lett.*, 1974, **10,** 801.
165. J. G. M. van der Aalsvoort and P. T. Beurskens, *Cryst. Struct. Commun.*, 1974, **3,** 653 (*Chem. Abstr.*, 1975, **82,** 50 060).
166. R. D. Bereman, D. M. Baird, C. T. Vance, J. Hutchinson and J. Zubieta, *Inorg. Chem.*, 1983, **22,** 2316.
167. N. Dupre, H. M. J. Hendriks and J. Jordanov, *J. Chem. Soc., Dalton Trans.*, 1984, 1463.
168. J. R. Dilworth, B. D. Neaves, C. J. Pickett, J. Chatt and J. A. Zubieta, *Inorg. Chem.*, 1983, **22,** 3524.
169. T. Roberie, A. E. Hoberman and J. Selbin, *J. Coord. Chem.*, 1979, **9,** 79.
170. M. Bonamico, G. Dessy, U. Fares and L. Scaramuzza, *J. Chem. Soc., Dalton Trans.*, 1975, 2079.
171. M. Chaudhury, *Inorg. Chem.*, 1984, **23,** 4434.
172. B. J. Kalbacher and R. D. Bereman, *Inorg. Chem.*, 1975, **14,** 1417.
173. E. Uhlemann and U. Eckelmann, *Z. Chem.*, 1972, **12,** 290.
174. R. D. Archer, C. J. Weber, and C. J. Donahue, in 'Proceedings of the Third International Conference on Chemistry and Uses of Molybdenum', ed. P. C. H. Mitchell and H. F. Barry, Climax Molybdenum Co., Ann Arbor, MI, 1979, p. 64.
175. R. D. Archer and C. J. Weber, *Inorg. Chem.*, 1984, **23,** 4158.
176. M. H. Chisholm, W. W. Reichert and P. Thornton, *J. Am. Chem. Soc.*, 1978, **100,** 2744.
177. M. Bochmann, G. Wilkinson, G. B. Young, M. B. Hursthouse and K. M. A. Malik, *J. Chem. Soc., Dalton Trans.*, 1980, 901.
178. S. Beshouri, T. W. Coffindaffer, S. Pirzad and I. P. Rothwell, *Inorg. Chim. Acta*, 1985, **103,** 111.
179. S. Otsuka, M. Kamata, K. Hirotsu and T. Higuchi, *J. Am. Chem. Soc.*, 1981, **103,** 3011.
180. M. Kamata, T. Yoshida, S. Otsuka, K. Hirotsu and T. Higuchi, *J. Am. Chem. Soc.*, 1981, **103,** 3572.
181. N. Ueyama, M. Nakata, A. Nakamura, M. Kamata and S. Otsuka, *Inorg. Chim. Acta*, 1983, **80,** 207.
182. E. Roland, E. C. Walborsky, J. C. Dewan and R. R. Schrock, *J. Am. Chem. Soc.*, 1985, **107,** 5795.
183. J. Hyde, L. Magin and J. Zubieta, *J. Chem. Soc., Chem. Commun.*, 1980, 204.
184. J. Hyde, J. Zubieta and N. Seeman, *Inorg. Chim. Acta*, 1981, **54,** L137.
185. P. J. Blower, J. R. Dilworth, G. J. Leigh, B. D. Neaves, F. B. Normanton, J. Hutchinson and J. A. Zubieta, *J. Chem. Soc., Dalton Trans.*, 1985, 2647.
186. D. Coucouvanis and M. Dragonjac, *J. Am. Chem. Soc.*, 1982, **104,** 6820.
187. D. C. Bradley and M. H. Chisholm, *J. Chem. Soc.* (*A*), 1971, 2741.
188. C. J. Horn and T. M. Brown, *Inorg. Chem.*, 1972, **11,** 1970.
189. C. J. Horn and T. M. Brown, *Inorg. Nucl. Chem. Lett.*, 1972, **8,** 377.
190. A. G. Sharpe, 'The Chemistry of Cyano Complexes of the Transition Metals', Academic, London, 1976, pp. 53–64.
191. S. J. Lippard, *Prog. Inorg. Chem.*, 1976, **21,** 91.
192. J. G. Leipoldt, L. D. C. Bok and P. J. Cilliers, *Z. Anorg. Allg. Chem.*, 1974, **407,** 350.
193. J. L. Hoard, T. A. Hamor and M. D. Glick, *J. Am. Chem. Soc.*, 1968, **90,** 3177.
194. S. S. Basson, J. G. Leipoldt and A. J. van Wyk, *Acta Crystallogr., Sect. B*, 1980, **36,** 2025.
195. D. I. Semenishin, T. Glowiak and M. G. Mys'kiv, *Koord. Khim.*, 1985, **11,** 122.

196. L. D. C. Bok, J. G. Leipoldt and S. S. Basson, *Z. Anorg. Allg. Chem.*, 1972, **392,** 303.
197. J. G. Leipoldt, S. S. Basson and L. D. C. Bok, *Inorg. Chim. Acta,* 1980, **44,** L99.
198. T. V. Long and G. A. Vernon, *J. Am. Chem. Soc.*, 1971, **93,** 1919.
199. E. L. Muetterties, *Inorg. Chem.*, 1973, **12,** 1963.
200. P. Morys and G. Glieman, *Z. Naturforsch., Teil B,* 1976, **31,** 1224.
201. D. I. Zubritskaya, D. I. Semenishin, Yu. A. Lyubechenko and K. N. Mikhalevich, *Zh. Neorg. Khim.*, 1973, **18,** 2289.
202. D. I. Zubritskaya, K. N. Mikhalevich, V. M. Litvinchuk and Yu. A. Lyubechenko, *Zh. Neorg. Khim.*, 1974, **19,** 706.
203. W. U. Malik, B. P. Singh and J. P. Jain, *J. Radioanal. Chem.*, 1972, **12,** 553.
204. G. F. McKnight and G. P. Haight, Jr., *Inorg. Chem.*, 1973, **12,** 3007.
205. R. D. Archer and D. A. Drum, *J. Inorg. Nucl. Chem.*, 1974, **36,** 1979.
206. H. K. Saha, B. C. Saha and M. Bagchi, *J. Indian Chem. Soc.*, 1974, **51,** 1055.
207. M. Novotny, D. F. Lewis and S. J. Lippard, *J. Am. Chem. Soc.*, 1972, **94,** 6961.
208. M. J. Mockford and W. P. Griffith, *J. Chem. Soc., Dalton Trans.*, 1985, 717.
209. P. G. Perkins and F. A. Schultz, *Inorg. Chem.*, 1983, **22,** 1133.
210. M. Takahashi, I. Watanabe, S. Ikeda, M. Kamata and S. Otsuka, *Bull. Chem. Soc. Jpn.*, 1982, **55,** 3757.
211. M. H. Chisholm, A. H. Cowley and M. Lattman, *J. Am. Chem. Soc.*, 1980, **102,** 46.
212. M. L. H. Green, *Pure Appl. Chem.*, 1972, **30,** 373.
213. K. Prout, T. S. Cameron, R. A. Forder, S. R. Critchley, B. Denton and G. V. Rees, *Acta Crystallogr., Sect. B,* 1974, **30,** 2290.
214. A. V. Aripovskii, B. M. Bulichev and V. B. Polyakova, *Zh. Neorg. Khim.*, 1981, **26,** 2722.
215. A. R. Dias and M. L. H. Green, *J. Chem. Soc. (A)*, 1971, 2808.
216. J. Colin, B. Chevrier, A. de Cian and R. Weiss, *Angew. Chem., Int. Ed. Engl.*, 1983, **22,** 247.
217. R. E. DeSimone, J. Cragel, Jr., W. H. Ilsley and M. D. Glick, *J. Coord. Chem.*, 1979, **9,** 167.
218. J. A. Broomhead, M. Sterns and C. G. Young, *Inorg. Chem.*, 1984, **23,** 729.
219. J. Cragel, V. B. Pett and R. DeSimone, *Inorg. Chem.*, 1978, **17,** 2885.
220. L. Ricard, J. Estienne and R. Weiss, *J. Chem. Soc., Chem. Commun.*, 1972, 906; *Inorg. Chem.*, 1973, **12,** 2182; M. H. Chisholm, J. F. Corning and J. C. Huffman, *Inorg. Chem.*, 1982, **21,** 286.
221. P. J. Vergamini, H. Vahrenkamp and L. F. Dahl, *J. Am. Chem. Soc.*, 1971, **93,** 6327.
222. U. Muller, P. Klingelhofer, C. Friebel and J. Pebler, *Angew. Chem., Int. Ed. Engl.*, 1985, **24,** 689.
223. E. Hey, F. Weller and K. Dehnicke, *Z. Anorg. Allg. Chem.*, 1984, **508,** 86.
224. R. T. Paine and L. B. Asprey, *Inorg. Synth.*, 1979, **19,** 137; A. J. Edwards, R. D. Peacock and R. W. H. Small, *J. Chem. Soc.*, 1962, 4486.
225. N. Acquista and S. Abramowitz, *J. Chem. Phys.*, 1973, **58,** 5484; O. V. Blinova and Yu. B. Predtechenskii, *Opt. Spektrosk.*, 1979, **47,** 1120 (*Chem. Abstr.*, 1980, **92,** 207 003).
226. R. D. Peacock and T. P. Sleigh, *J. Fluorine Chem.*, 1971, **1,** 243; A. M. Panich, V. K. Goncharuk, S. Gabuda and N. K. Moroz, *Zh. Strukt. Khim.*, 1979, **20,** 60.
227. D. E. Sands and A. Zalkin, *Acta Crystallogr.*, 1959, **12,** 723.
228. R. V. G. Ewens and M. W. Lister, *Trans. Faraday Soc.*, 1938, **34,** 1358; I. R. Beattie and G. A. Ozin, *J. Chem. Soc. (A)*, 1969, 1691.
229. S. A. Ryabov, A. S. Alikhanyan, V. D. Butskii, O. G. Ellert and V. S. Pervov, *Zh. Neorg. Khim.*, 1985, **30,** 1674.
230. M. Mercer, *J. Chem. Soc. (A)*, 1969, 2019.
231. R. Colton and I. B. Tomkins, *Aust. J. Chem.*, 1965, **18,** 447; D. A. Edwards, *J. Inorg. Nucl. Chem.*, 1963, **25,** 1198; P. C. Crouch, G. W. A. Fowles, I. B. Tomkins and R. A. Walton, *J. Chem. Soc. (A)*, 1969, 2412.
232. G. Ferguson, M. Mercer and D. W. A. Sharp, *J. Chem. Soc. (A)*, 1969, 2415.
233. G. W. A. Fowles and J. L. Frost, *J. Chem. Soc. (A)*, 1967, 671; P. C. Crouch, G. W. A. Fowles, J. L. Frost, P. R. Marshall and R. A. Walton, *J. Chem. Soc. (A)*, 1968, 1061.
234. M. C. Chakravorty and A. K. Bera, *Transition Met. Chem. (Weinheim, Ger.)*, 1983, **8,** 83.
235. P. M. Boorman, C. D. Garner and F. E. Mabbs, *J. Chem. Soc., Dalton Trans.*, 1975, 1299.
236. D. Grandjean and R. Weiss, *Bull. Soc. Chim. Fr.*, 1967, 3054.
237. V. V. Tkachev, O. N. Krasochka and L. O. Atovmyan, *Zh. Strukt. Khim.*, 1976, **17,** 940.
238. C. D. Garner, L. H. Hill, F. E. Mabbs, D. L. McFadden and A. T. McPhail, *J. Chem. Soc., Dalton Trans.*, 1977, 1202.
239. C. D. Garner, L. H. Hill, F. E. Mabbs, D. L. McFadden and A. T. McPhail, *J. Chem. Soc., Dalton Trans.*, 1977, 853.
240. A. Bino and F. A. Cotton, *J. Am. Chem. Soc.*, 1979, **101,** 4150; *Inorg. Chem.*, 1979, **18,** 2710.
241. C. Schumacher, F. Weller and K. Dehnicke, *Z. Anorg. Allg. Chem.*, 1982, **495,** 135.
242. L. O. Atovmyan, O. A. D'yanchenko and E. L. Lobkovskii, *Zh. Strukt. Khim.*, 1970, **11,** 469; J. E. Fergusson, A. M. Greenaway and B. R. Penfold, *Inorg. Chim. Acta*, 1983, **71,** 29.
243. J. P. Brunette and M. J. F. Leroy, *J. Inorg. Nucl. Chem.*, 1974, **36,** 289.
244. H. K. Saha and M. C. Halder, *J. Inorg. Nucl. Chem.*, 1971, **33,** 3719; H. K. Saha and K. Sikdar, *J. Indian Chem. Soc.*, 1981, **58,** 323.
245. R. Lozano, A. L. Doadrio, E. Alarcon, J. Roman and A. Doadrio Lopez, *An. Quim., Ser. B,* 1983, **79,** 190 (*Chem. Abstr.* 1985, **102,** 124 469).
246. R. A. Walton, *Prog. Inorg. Chem.*, 1972, **16,** 1.
247. P. C. H. Mitchell, *J. Inorg. Nucl. Chem.*, 1963, **25,** 963; S. M. Horner and S. Y. Tyree, *Inorg. Chem.*, 1962, **1,** 122; M. L. Larson and F. W. Moore, *Inorg. Chem.*, 1966, **5,** 801; J. Lewis and R. Whyman, *J. Chem. Soc.*, 1965, 6027; D. A. Edwards, *J. Inorg. Nucl. Chem.*, 1965, **27,** 303.
248. K. Feenan and G. W. A. Fowles, *Inorg. Chem.*, 1965, **4,** 310.
249. I. W. Boyd, I. G. Dance, K. S. Murray and A. G. Wedd, *Aust. J. Chem.*, 1978, **31,** 279.
250. C. D. Garner, P. Lambert, F. E. Mabbs and T. J. King, *J. Chem. Soc., Dalton Trans.*, 1977, 1191; C. D. Garner, N. C. Howlader, F. E. Mabbs, A. T. McPhail and K. D. Onan, *J. Chem. Soc., Dalton Trans.*, 1978, 1848.

251. C. D. Garner, M. R. Hyde, F. E. Mabbs and V. I. Routledge, *J. Chem. Soc., Dalton Trans.*, 1975, 1175.
252. C. D. Garner, N. C. Howlader, F. E. Mabbs, P. M. Boorman and T. J. King, *J. Chem. Soc., Dalton Trans.*, 1978, 1350.
253. H. K. Saha, M. Saha and T. K. R. Chaudhuri, *J. Indian Chem. Soc.*, 1983, **60,** 422; I. N. Marov, M. K. Pupkova and V. K. Belyaeva, *Zh. Neorg. Khim.*, 1983, **28,** 382.
254. S. Vasanthi, K. S. Nagaraja and M. J. Udupa, *Transition Met. Chem. (Weinheim, Ger.)*, 1984, **9,** 382.
255. J. R. Dilworth, C. A. McAuliffe and B. J. Sayle, *J. Chem. Soc., Dalton Trans.*, 1977, 849.
256. K. Yamanouchi, S. Yamada and J. H. Enemark, *Inorg. Chim. Acta,* 1984, **85,** 129.
257. J. R. Bradbury, G. R. Hanson, A. M. Bond and A. G. Wedd, *Inorg. Chem.*, 1984, **23,** 844.
258. C. A. McAuliffe, A. Hosseiny and F. P. McCullough, *Inorg. Chim. Acta,* 1977, **33,** 5; K. Yamanouchi, J. T. Huneke and J. H. Enemark, *Acta Crystallogr., Sect. B,* 1979, **35,** 2326.
259. M. M. Yunusov, G. M. Larin and P. M. Solozhenkin, *Dokl. Akad. Nauk Tadzh. SSR,* 1982, **25,** 671 (*Chem. Abstr.*, 1984, **99,** 47 000); K. Yamanouchi and J. H. Enemark, *Inorg. Chem.*, 1979, **18,** 1626.
260. C. D. Garner, M. R. Hyde, F. E. Mabbs and V. I. Routledge, *Nature (London)*, 1974, **252,** 579.
261. C. D. Garner, M. R. Hyde, F. E. Mabbs and V. I. Routledge, *J. Chem. Soc., Dalton Trans.*, 1975, 1180; M. R. Hyde and C. D. Garner, *J. Chem. Soc., Dalton Trans.*, 1975, 1186.
262. R. D. Taylor, P. G. Todd, N. D. Chasteen and J. T. Spence, *Inorg. Chem.*, 1979, **18,** 44.
263. J. Topich, *Inorg. Chim. Acta,* 1980, **46,** L97.
264. C. D. Garner, F. E. Mabbs and D. T. Richens, *J. Chem. Soc., Chem. Commun.*, 1979, 415.
265. J. R. Bradbury, M. F. MacKay and A. G. Wedd, *Aust. J. Chem.*, 1978, **31,** 2423; G. R. Hanson, A. A. Brunette, A. C. McDonell, K. S. Murray and A. G. Wedd, *J. Am. Chem. Soc.*, 1981, **103,** 1953.
266. J. R. Bradbury, A. F. Masters, A. C. McDonell, A. A. Brunette, A. M. Bond and A. G. Wedd, *J. Am. Chem. Soc.*, 1981, **103,** 1959.
267. R. J. Burt, J. R. Dilworth, G. J. Leigh and J. A. Zubieta, *J. Chem. Soc., Dalton Trans.*, 1982, 2295; P. T. Bishop, J. R. Dilworth, J. Hutchinson and J. A. Zubieta, *J. Chem. Soc., Chem. Commun.*, 1982, 1052.
268. P. T. Bishop, P. J. Blower, J. R. Dilworth and J. A. Zubieta, *Polyhedron,* 1986, **5,** 363.
269. M. Chaudhury, *Polyhedron,* 1986, **5,** 387.
270. K. S. Nagaraja and M. R. Udupa, *Transition Met. Chem. (Weinheim, Ger.)*, 1984, **9,** 290; K. S. Nagaraja and M. R. Udupa, *Polyhedron,* 1984, **3,** 907.
271. J. T. Spence, M. Minelli and P. Kroneck, *J. Am. Chem. Soc.*, 1980, **102,** 4538.
272. H. Sabat, M. F. Rudolf and B. Jezowska-Trzebiatowska, *Inorg. Chim. Acta,* 1973, **7,** 365; A. M. Golub, V. A. Grechnikhina, V. V. Trachevskii and N. V. Ul'ko, *Zh. Neorg. Khim.*, 1973, **18,** 2119.
273. I. M. Marov, E. S. Gur'eva and V. M. Peshkova, *Zh. Neorg. Khim.*, 1970, **15,** 3039.
274. S. F. Gheller, J. R. Bradbury, M. F. Mackay and A. G. Wedd, *Inorg. Chem.*, 1981, **20,** 3899.
275. R. A. Walton, P. C. Crouch and B. J. Brisdon, *Spectrochim. Acta, Part A,* 1968, 24, 601; E. B. Fleischer and T. S. Srivastava, *Inorg. Chim. Acta,* 1971, **5,** 151; T. Imamura, T. Numatatsu, M. Terui and M. Fujimoto, *Bull. Chem. Soc. Jpn.*, 1981, **54,** 170.
276. Y. Matsuda, S. Yamada and M. Yukito, *Inorg. Chem.*, 1981, **20,** 2239; K. M. Kadish, T. Malinski and H. Ledon, *Inorg. Chem.*, 1982, **21,** 2982.
277. T. Imamura, K. Hasegawa, T. Tanaka, W. Nakajima and M. Fujimoto, *Bull. Chem. Soc. Jpn.*, 1984, **57,** 194.
278. H. J. Ledon and M. Bonnet, *J. Am. Chem. Soc.*, 1981, **103,** 6209.
279. P. D. Rillema and C. H. Brubaker, Jr., *Inorg. Chem.*, 1969, **8,** 1645.
280. M. I. Scullane, R. D. Taylor, M. Minelli, J. T. Spence, K. Yamanouchi, J. H. Enemark and N. D. Chasteen, *Inorg. Chem.*, 1979, **18,** 3213.
281. B. Gahan, N. C. Howlader and F. E. Mabbs, *J. Chem. Soc., Dalton Trans.*, 1981, 142.
282. D. Collison, F. E. Mabbs, J. H. Enemark and W. E. Cleland, Jr., *Polyhedron,* 1986, **5,** 423.
283. T. Kikuchi, Y. Sigiura and H. Tanaka, *Inorg. Chim. Acta,* 1982, **66,** L5.
284. L. S. Meriwether, W. S. Marzluff and W. G. Hodgson, *Nature (London)*, 1966, **212,** 465.
285. H. B. Gray and C. R. Hare, *Inorg. Chem.*, 1962, **1,** 363; C. R. Hare, I. Bernal and H. B. Gray, *Inorg. Chem.*, 1962, **1,** 831.
286. J. Weber and C. D. Garner, *Inorg. Chem.*, 1980, **19,** 2206.
287. E. A. Allen, B. J. Brisdon, D. A. Edwards, G. W. A. Fowles and R. G. Williams, *J. Chem. Soc.*, 1963, 4649; O. V. Ziebartl and J. Selbin, *J. Inorg. Nucl. Chem.*, 1970, **32,** 849.
288. D. Collison, Ph.D. Thesis, University of Manchester, 1979.
289. J. Schmitte, C. Friebel, F. Weller and K. Dehnicke, *Z. Anorg. Allg. Chem.*, 1982, **495,** 148.
290. J. Chatt and J. R. Dilworth, *J. Chem. Soc., Chem. Commun.*, 1974, 517.
291. K. Dehnicke, H. Prinz, W. Hafitz and R. Kujanek, *Liebigs Ann. Chem.*, 1981, 20.
292. C. Y. Chou, J. C. Huffman and E. A. Maatta, *Inorg. Chem.*, 1986, **25,** 822.
293. A. W. Edelblut and R. A. D. Wentworth, *Inorg. Chem.*, 1980, **19,** 1110.
294. A. T. Pilipenko, V. V. Trachevskii and N. V. Rusetskaya, *Dokl. Akad. Nauk SSSR,* 1979, **244,** 112.
295. A. Prescott, D. W. A. Sharp and J. M. Winfield, *J. Chem. Soc., Chem. Commun.*, 1973, 667; G. M. Anderson, I. F. Fraser and J. M. Winfield, *J. Fluorine Chem.*, 1983, **23,** 403.
296. B. J. Brisdon, D. A. Edwards, D. H. Machin, K. S. Murray and R. A. Walton, *J. Chem. Soc. (A)*, 1967, 1825.
297. S. M. Horner and S. Y. Tyree, *Inorg. Chem.*, 1963, **2,** 568.
298. V. Fernandez, I. Yerga and U. Müller, *Z. Anorg. Allg. Chem.*, 1981, **477,** 205.
299. M. S. Delgado, F. Aguacil and V. Fernandez, *Z. Anorg. Allg. Chem.*, 1982, **493,** 187.
300. D. Britnell, G. W. A. Fowles and D. A. Rice, *J. Chem. Soc., Dalton Trans.*, 1975, 28.
301. D. Scott and A. G. Wedd, *J. Chem. Soc., Chem. Commun.*, 1974, 527.
302. M. G. B. Drew, G. M. Egginton and J. D. Wilkins, *Acta Crystallogr., Sect. B,* 1974, **30,** 1895.
303. C. A. McAuliffe and B. J. Sayle, *Inorg. Chim. Acta,* 1975, **12,** L7.
304. A. Nieuwpoort, *J. Less-Common Met.*, 1974, **36,** 271.
305. K. S. Jasim, G. B. Umbach, C. Chieh and T. C. W. Mak, *J. Crystallogr. Spectrosc. Res.*, 1985, **15,** 271 (*Chem. Abstr.*, 1985, **103,** 47 184).

306. D. A. Smith, J. W. McDonald, H. O. Finklea, V. R. Ott and F. A. Schultz, *Inorg. Chem.*, 1982, **21**, 3825.
307. N. Pariyadath, W. E. Newton and E. I. Stiefel, *J. Am. Chem. Soc.*, 1976, **98**, 5388.
308. K. Yamanouchi and J. H. Enemark, *Inorg. Chem.*, 1978, **17**, 1981.
309. O. A. Rajan, J. T. Spence, C. Leman, M. Minelli, M. Sato, J. H. Enemark, P. M. H. Kroneck and K. Sulger, *Inorg. Chem.*, 1983, **22**, 3065.
310. B. J. Corden, J. A. Cunningham and R. Eisenberg, *Inorg. Chem.*, 1970, **9**, 356.
311. S. S. Basson, J. G. Leipoldt, L. D. C. Bok, J. S. van Vollenhoven and P. J. Cilliers, *Acta Crystallogr., Sect. B,* 1980, **36**, 1765.
312. L. D. C. Bok, J. G. Leipoldt and S. S. Basson, *Acta Crystallogr., Sect. B,* 1970, **26**, 684.
313. B. R. McGarvey, *Inorg. Chem.*, 1966, **5**, 476.
314. C. D. Garner and F. E. Mabbs, *J. Inorg. Nucl. Chem.*, 1979, **41**, 1125.
315. R. D. Dowsing and J. F. Gibson, *J. Chem. Soc. (A)*, 1967, 655.
316. B. Kamenar, M. Penavic, B. Korpar-Colig and B. Markovic, *Inorg. Chim. Acta,* 1982, **65**, L245.
317. B. Kamenar, B. Korpar-Colig and M. Penavic, *Cryst. Struct. Commun.*, 1982, **11**, 1583.
318. J. Sobczak and J. J. Ziolkowski, *Transition Met. Chem. (Weinheim, Ger.)*, 1983, **8**, 333.
319. R. Kergoat, J. E. Guerchais and J. M. Luzunou, *Rev. Chim. Miner.*, 1971, **8**, 621.
320. F. W. Moore and M. L. Larson, *Inorg. Chem.*, 1967, **6**, 998.
321. A. T. Casey, D. J. Mackay, R. L. Martin and A. H. White, *Aust. J. Chem.*, 1972, **25**, 477.
322. C. D. Garner, N. C. Howlader, F. E. Mabbs, A. T. McPhail and K. D. Onan, *J. Chem. Soc., Dalton Trans.*, 1979, 962.
323. J. A. Zubieta and G. B. Maniloff, *Inorg. Nucl. Chem. Lett.*, 1976, **12**, 121.
324. W. E. Newton, J. L. Corbin and J. W. McDonald, *J. Chem. Soc., Dalton Trans.*, 1974, 1044.
325. A. B. Blake, F. A. Cotton and J. S. Wood, *J. Am. Chem. Soc.*, 1964, **86**, 3024.
326. A. L. Doadrio Villarejo and M. T. Calabuig Martinez, *An. R. Acad. Farm.*, 1985, **51**, 113 (*Chem. Abstr.*, 1985, **103**, 204 756).
327. G. J.-J. Chen, J. W. McDonald and W. E. Newton, *Inorg. Nucl. Chem. Lett.*, 1976, **12**, 697.
328. J. R. Knox and C. K. Prout, *Acta Crystallogr., Sect. B,* 1969, **25**, 2281.
329. A. Kay and P. C. H. Mitchell, *J. Chem. Soc., Dalton Trans.*, 1970, 2421; 1973, 1388.
330. J. R. Hyde and J. Zubieta, *Cryst. Struct. Commun.*, 1982, **11**, 929.
331. Y.-Y. P. Tsao, C. J. Fritchie, Jr. and H. A. Levy, *J. Am. Chem. Soc.*, 1978, **100**, 4090.
332. P. L. Dahlstrom, J. R. Hyde, P. A. Vella and J. Zubieta, *Inorg. Chem.*, 1982, **21**, 927.
333. M. C. Chakravorti and D. Bandyopadhyay, *J. Indian Chem. Soc.*, 1984, **61**, 836.
334. V. I. Spitsyn, M. G. Felin, N. A. Subbotina, M. I. Aizenberg and S. V. Mozgin, *Zh. Neorg. Khim.*, 1984, **29**, 706.
335. A. Bino, S. Cohen and L. Tsimering, *Inorg. Chim. Acta,* 1983, **77**, L79.
336. T. Glowiak, M. F. Rudolf, M. Sabat and B. Jezowska-Trzebiatowska, in 'Proceedings of the Second International Conference on Chemistry and Uses of Molybdenum', ed. P. C. H. Mitchell and A. Seaman, Climax Molybdenum Co., Ann Arbor, MI, 1976, p. 17.
337. O. S. Oh and J. D. Rhee, *Taehan Hawhak Chi,* 1982, **26**, 81 (*Chem. Abstr.*, 1982, **97**, 16 122).
338. K. Yamanouchi and S. Yamada, *Inorg. Chim. Acta,* 1974, **9**, 83.
339. S. N. Poddar, G. C. Samanta, G. Mukherjee and S. Ghosh, *J. Indian Chem. Soc.*, 1985, **62**, 7.
340. W. Andruchow, Jr. and R. D. Archer, *J. Inorg. Nucl. Chem.*, 1972, **34**, 3185.
341. L. W. Amos and D. T. Sawyer, *Inorg. Chem.*, 1974, **13**, 78.
342. A. F. Isbell, Jr. and D. T. Sawyer, *Inorg. Chem.*, 1971, **10**, 2449.
343. R. Lozano, J. Roman, E. Alarcon, A. L. Doadrio and A. Doadrio, *Rev. Chim. Miner.*, 1983, **20**, 173.
344. R. Lozano, J. Roman, M. D. Armada and A. Doadrio, *Polyhedron,* 1985, **4**, 1563.
345. J. K. Howie and D. T. Sawyer, *Inorg. Chem.*, 1976, **15**, 1892.
346. F. A. Cotton, D. L. Hunter, L. Ricard and R. Weiss, *J. Coord. Chem.*, 1974, **3**, 259.
347. J. W. Buchler and K. Rohbock, *Inorg. Nucl. Chem. Lett.*, 1972, **8**, 1073.
348. J. F. Johnson and W. R. Scheidt, *J. Am. Chem. Soc.*, 1977, **99**, 294.
349. K. Wieghardt, G. Backes-Dahmann, W. Herrmann and J. Weiss, *Angew. Chem., Int. Ed. Engl.*, 1984, **23**, 899.
350. R. Kergoat and J. E. Guerchais, *Z. Anorg. Allg. Chem.*, 1975, **416**, 174.
351. K. L. Wall, K. Folting, J. C. Huffman and R. A. D. Wentworth, *Inorg. Chim. Acta,* 1984, **86**, L25.
352. F. A. Cotton and S. M. Morehouse, *Inorg. Chem.*, 1965, **4**, 1377.
353. F. A. Cotton, S. M. Morehouse and J. S. Wood, *Inorg. Chem.*, 1965, **4**, 922; J. Donnohue, *Inorg. Chem.*, 1965, **4**, 921.
354. B. Jezowska-Trzebiatowska, M. F. Rudolf, L. Natkaniec and H. Sabat, *Inorg. Chem.*, 1974, **13**, 617.
355. K. Wieghardt, M. Hahn, W. Swiridoff and J. Weiss, *Angew. Chem., Int. Ed. Engl.*, 1983, **22**, 491.
356. M. Hahn and K. Wieghardt, *Inorg. Chem.*, 1984, **23**, 3977.
357. T. Chandler, D. L. Lichtenberger and J. H. Enemark, *Inorg. Chem.*, 1981, **20**, 75.
358. J. I. Gelder and J. H. Enemark, *Inorg. Chem.*, 1976, **15**, 1839.
359. J. M. Newsam and T. R. Halbert, *Inorg. Chem.*, 1985, **24**, 491.
360. L. Ricard, C. Martin, R. Wiest and R. Weiss, *Inorg. Chem.*, 1975, **14**, 2300.
361. L. J. de Hayes, H. C. Faulkner, W. H. Doub, Jr. and D. T. Sawyer, *Inorg. Chem.*, 1975, **14**, 2110.
362. W. E. Newton, G. J.-J. Chen and J. W. McDonald, *J. Am. Chem. Soc.*, 1976, **98**, 5387.
363. F. A. Schultz, V. R. Ott, D. R. Rolison, D. C. Bravard, J. W. McDonald and W. E. Newton, *Inorg. Chem.*, 1978, **17**, 1758.
364. J. Dirand-Colin, L. Ricard and R. Weiss, *Inorg. Chim. Acta,* 1976, **18**, L21.
365. K. Musha, Y. Ohashi, S. Yamazaki, S. Toda and S. Tanaka, *Nippon Kagaku Kaishi,* 1983, 653 (*Chem. Abstr.*, 1983, **99**, 15 475); K. Musha, S. Yamazaki and S. Toda, *Nippon Kagaku Kaishi,* 1984, 702 (*Chem. Abstr.*, 1984, **101**, 47 630).
366. R. Winograd, B. Spivack and Z. Dori, *Cryst. Struct. Commun.*, 1976, **5**, 373; N. C. Howlader, G. P. Haight, Jr., T. W. Hambley, M. R. Snow and G. A. Lawrance, *Inorg. Chem.*, 1984, **23**, 1811.

367. W. E. Newton, J. W. McDonald, K. Yamanouchi and J. H. Enemark, *Inorg. Chem.*, 1979, **18**, 1621.
368. B. Spivak, Z. Dori and E. I. Stiefel, *Inorg. Nucl. Chem. Lett.*, 1975, **11**, 501.
369. J. Huneke and J. H. Enemark, *Inorg. Chem.*, 1978, **17**, 3698.
370. A. Müller, R. G. Bhattacharyya, N. Mohan and B. Pfefferkorn, *Z. Anorg. Allg. Chem.*, 1979, **454**, 118.
371. K. S. Nagaraja and M. R. Udupa, *Polyhedron*, 1985, **4**, 649.
372. K. J. Moynihan, P. M. Boorman, J. M. Ball, V. O. Patel and K. A. Kerr, *Acta Crystallogr., Sect. B*, 1982, **38**, 2258.
373. S. Lu, M. Shang, W. Cheng, M. He and J. Huang, *Fenzi Kexueyu Huaxue Yanjiu*, 1984, **4**, 435 (*Chem. Abstr.*, 1985, **102**, 16 516).
374. A. Müller, U. Reinsch-Vogell, E. Krickemeyer and H. Bögge, *Angew. Chem., Int. Ed. Engl.*, 1982, **21**, 796.
375. W. Clegg, N. Mohan, A. Müller, A. Neumann, W. Rittner and G. M. Sheldrick, *Inorg. Chem.*, 1980, **19**, 2066; W. Clegg, G. M. Sheldrick, C. D. Garner and G. Christou, *Acta Crystallogr., Sect. B*, 1980, **36**, 2784.
376. X. Xin, N. L. Morris, G. B. Jameson and M. T. Pope, *Inorg. Chem.*, 1985, **24**, 3482.
377. L. Huang, B. Wang and X. Wu, *Huaxue Tongbao*, 1984, 15 (*Chem. Abstr.*, 1985, **102**, 124 459).
378. W.-H. Pan, M. A. Harmer, T. R. Halbert and E. I. Stiefel, *J. Am. Chem. Soc.*, 1984, **106**, 459.
379. W. Clegg, G. Christou, C. D. Garner and G. M. Sheldrick, *Inorg. Chem.*, 1981, **20**, 1562.
380. M. Draganjac, E. O. Simhon, L. T. Chan, M. Kanatzidis, N. C. Baenziger and D. Coucouvanis, *Inorg. Chem.*, 1982, **21**, 3321.
381. S. A. Cohen and E. I. Stiefel, *Inorg. Chem.*, 1985, **24**, 4657.
382. G. Bunzey, J. H. Enemark, J. K. Howie and D. T. Sawyer, *J. Am. Chem. Soc.*, 1977, **99**, 4168; G. Bunzey and J. H. Enemark, *Inorg. Chem.*, 1978, **17**, 682.
383. M. Rakowski DuBois, *Inorg. Chem.*, 1978, **17**, 2405.
384. K. Mennemann and R. Mattes, *J. Chem. Res. (S)*, 1979, 100.
385. J. Beck, W. Hiller, E. Schweda and J. Strähle, *Z. Naturforsch., Teil B*, 1984, 39, 1110.
386. B. M. Gatehouse, E. K. Nunn, J. E. Guerchais and R. Kergoat, *Inorg. Nucl. Chem. Lett.*, 1976, **12**, 23.
387. B. Kamenar and M. Penavic, *Z. Kristallogr.*, 1979, **150**, 327.
388. M. G. B. Drew, P. J. Baricelli, P. C. H. Mitchell and A. R. Read, *J. Chem. Soc., Dalton Trans.*, 1983, 649.
389. J. R. Knox and C. K. Prout, *Acta Crystallogr., Sect. B*, 1969, **25**, 1857.
390. D. H. Brown and A. D. Jeffreys, *J. Chem. Soc., Dalton Trans.*, 1973, 732.
391. M. G. B. Drew and A. Kay, *J. Chem. Soc. (A)*, 1971, 1846.
392. M. G. B. Drew and A. Kay, *J. Chem. Soc. (A)*, 1971, 1851.
393. L. T. J. Delbaere and C. K. Prout, *Chem. Commun.*, 1971, 162.
394. B. Spivack, A. P. Gaughan and Z. Dori, *J. Am. Chem. Soc.*, 1971, **93**, 5265.
395. N. Ueyama, M. Nakata, T. Araki, A. Nakamura, S. Yamashita and Y. Yamashita, *Inorg. Chem.*, 1981, **20**, 1934.
396. S. F. Gheller, T. W. Hambley, R. T. C. Brownlee, M. J. O'Connor, M. R. Snow and A. G. Wedd, *J. Am. Chem. Soc.*, 1983, **105**, 1527.
397. M. Minelli, J. H. Enemark, R. T. C. Brownlee, M. J. O'Connor and A. G. Wedd, *Coord. Chem. Rev.*, 1985, **68**, 169.
398. K. Suzuki, Y. Sasaki, S. Ooi and K. Saito, *Bull. Chem. Soc. Jpn.*, 1980, **53**, 1288.
399. K. F. Miller, A. E. Bruce, J. L. Corbin, S. Wherland and E. I. Stiefel, *J. Am. Chem. Soc.*, 1980, **102**, 5102.
400. V. R. Ott, D. S. Sweiter and F. A. Schultz, *Inorg. Chem.*, 1977, **16**, 2538.
401. M. Chaudhury, *J. Chem. Soc., Dalton Trans.*, 1983, 857.
402. M. C. Chakravorti and D. Bandyapadhyay, *Synth. React. Inorg. Met.-Org. Chem.*, 1984, **14**, 443.
403. L. M. Charney and F. A. Schultz, *Inorg. Chem.*, 1980, **19**, 1527.
404. K. S. Nagaraja and M. R. Udupa, *Transition Met. Chem. (Weinheim, Ger.)*, 1983, **8**, 191.
405. P. J. Baricelli, M. G. B. Drew and P. C. H. Mitchell, *Acta Crystallogr., Sect. C*, 1983, **39**, 843.
406. T. C. Hsieh, K. Gebreyes and J. Zubieta, *Transition Met. Chem. (Weinheim, Ger.)*, 1985, **10**, 81.
407. X. Jin, B. Bai and Y. Tang, *Jiegou Huaxue*, 1984, **3**, 139 (*Chem. Abstr.*, 1985, **103**, 63 853).
408. F. A. Cotton and W. H. Ilsley, *Inorg. Chim. Acta*, 1982, **59**, 213.
409. T. Shibahara, H. Kuroya, K. Matsumoto and S. Ooi, *Bull. Chem. Soc. Jpn.*, 1983, **56**, 2945.
410. D. L. Stevenson and L. F. Dahl, *J. Am. Chem. Soc.*, 1967, **89**, 3721.
411. J.-C. Daran, K. Prout, G. J. S. Adam, M. L. H. Green and J. Sala-Pala, *J. Organomet. Chem.*, 1977, **131**, C40; M. Rakowski DuBois, D. L. DuBois, M. C. VanDerveer and R. C. Haltiwanger, *Inorg. Chem.*, 1981, **20**, 3064.
412. L. F. Dahl, P. D. Frisch and G. R. Gust, in 'Proceedings of the First International Conference on the Chemistry and Uses of Molybdenum', ed. P. C. H. Mitchell, Climax Molybdenum Co., Ann Arbor, MI, 1974, p. 134.
413. A. E. Edelblut and R. A. D. Wentworth, *Inorg. Chem.*, 1980, **19**, 1117; A. W. Edelblut, K. Folting, J. C. Huffman and R. A. D. Wentworth, *J. Am. Chem. Soc.*, 1981, **103**, 1927; K. L. Wall, K. Folting, J. C. Huffman and R. A. D. Wentworth, *Inorg. Chem.*, 1983, **22**, 2366.
414. M. E. Noble, K. Folting, J. C. Huffman and R. A. D. Wentworth, *Inorg. Chem.*, 1983, **22**, 3671.
415. R. Yang and X. Yu, *Jiegou Huaxue*, 1984, **3**, 101 (*Chem. Abstr.*, 1985, **103**, 79 842).
416. A. Müller, W.-O. Nolte and B. Krebs, *Angew. Chem., Int. Ed. Engl.*, 1978, **17**, 279.
417. D. Fenske, B. Czeska, C. Schumacher, R. E. Schmidt and K. Dehnicke, *Z. Anorg. Allg. Chem.*, 1985, **520**, 7.
418. J. Strähle, G. Beyendorff, A. Liebelt and K. Dehnicke, *Z. Anorg. Allg. Chem.*, 1981, **474**, 171.
419. M. A. Banares, A. Angoso and J. J. Criado, *Synth. React. Inorg. Met.-Org. Chem.*, 1983, **13**, 145; 1983, **13**, 983.
420. M. W. Bishop, J. Chatt, J. R. Dilworth, G. Kaufman, S. Kim and J. Zubieta, *J. Chem. Soc., Chem. Commun.*, 1977, 70.
421. K. L. Madhok, *Indian J. Chem., Sect. A*, 1985, **24**, 122.
422. R. L. Dutta and B. Chatterjee, *J. Indian Chem. Soc.*, 1970, **47**, 657.
423. T. Glowiak, M. Sabat, H. Sabat and M. F. Rudolf, *J. Chem. Soc., Chem. Commun.*, 1975, 712.
424. B. Kamenar, M. Penavic and B. Markovic, *Croat. Chem. Acta*, 1984, **57**, 637.
425. J. Kloubek and J. Podlaha, *Inorg. Nucl. Chem. Lett.*, 1971, **7**, 67.
426. B. Spivack and Z. Dori, *J. Chem. Soc., Dalton Trans.*, 1973, 1173.
427. A. Kojima, S. Ooi, Y. Sasaki, K. Z. Suzuki, K. Saito and H. Kuroya, *Bull. Chem. Soc. Jpn.*, 1981, **54**, 2457.

428. H. Kobayashi, I. Tujikawa, T. Shibahara and N. Uryu, *Bull. Chem. Soc. Jpn.*, 1983, **56**, 108.
429. Y. Sasaki and A. G. Sykes, *J. Chem. Soc., Dalton Trans.*, 1974, 1468.
430. G. A. Zank, T. B. Rauchfuss and S. R. Wilson, *J. Am. Chem. Soc.*, 1984, **106**, 7621.
431. T. Shibahara, S. Ooi and H. Kuroya, *Bull. Chem. Soc. Jpn.*, 1982, **55**, 3742.
432. T. Shibahara and H. Kuroya, *Inorg. Chim. Acta*, 1981, **54**, L75.
433. D. J. Darensbourg, R. L. Grey and T. Delord, *Inorg. Chim. Acta*, 1985, **98**, L39.
434. S. A. Koch and S. Lincoln, *Inorg. Chem.*, 1982, **21**, 2904.
435. J. A. Beaver and M. G. B. Drew, *J. Chem. Soc., Dalton Trans.*, 1973, 1376.
436. M. F. Belicchi, G. G. Fava and C. Pelizzi, *J. Chem. Soc., Dalton Trans.*, 1983, 65.
437. M. H. Chisholm, J. C. Huffman, C. C. Kirkpatrick, J. Leonelli and K. Folting, *J. Am. Chem. Soc.*, 1981, **103**, 6093.
438. R. Maltes and K. Muhlsiepen, *Z. Naturforsch., Teil B*, 1980, **35**, 265.
439. E. Carmona, F. Gonzalez, M. L. Poveda, J. M. Marin, J. L. Atwood and R. D. Rogers, *J. Am. Chem. Soc.*, 1983, **105**, 3365.
440. B. Kamenar, B. Korpar-Colig and M. Penavic, *J. Chem. Soc., Dalton Trans.*, 1981, 311.
441. M. H. Chisholm, K. Folting, J. C. Huffman and C. C. Kirkpatrick, *J. Chem. Soc., Chem. Commun.*, 1982, 189.
442. M. E. Noble, J. C. Huffman and R. A. D. Wentworth, *Inorg. Chem.*, 1983, **22**, 1756; M. E. Noble, K. Folting, J. C. Huffman and R. A. D. Wentworth, *Inorg. Chem.*, 1984, **23**, 631.
443. K. Yamanouchi, J. H. Enemark, J. W. McDonald and W. E. Newton, *J. Am. Chem. Soc.*, 1977, **99**, 3529.
444. J. R. Dilworth, B. D. Neaves, P. Dahlstrom, J. Hyde and J. A. Zubieta, *Transition Met. Chem. (Weinheim, Ger.)*, 1982, **7**, 257.
445. I. W. Boyd, I. G. Dance, A. E. Landers and A. G. Wedd, *Inorg. Chem.*, 1979, **18**, 1875.
446. I. G. Dance and A. E. Landers, *Inorg. Chem.*, 1979, **18**, 3487.
447. I. Buchanan, W. Clegg, C. D. Garner and G. M. Sheldrick, *Inorg. Chem.*, 1983, **22**, 3657.
448. J. R. Bradbury, A. G. Wedd and A. M. Bond, *J. Chem. Soc., Chem. Commun.*, 1979, 1022.
449. J. I. Gelder, J. H. Enemark, G. Wolterman, A. Boston and G. P. Haight, *J. Am. Chem. Soc.*, 1975, **97**, 1616.
450. J. T. Huneke, K. Yamanouchi and J. H. Enemark, *Inorg. Chem.*, 1978, **17**, 3695.
451. T.-C. Hsieh, K. Gebreyes and J. Zubieta, *J. Chem. Soc., Chem. Commun.*, 1984, 1172.
452. P. T. Bishop, J. R. Dilworth, J. Hutchinson and J. A. Zubieta, *J. Chem. Soc., Chem. Commun.*, 1982, 1052.

36.5
Molybdenum(VI)

EDWARD I. STIEFEL
Exxon Research and Engineering Company, Annandale, NJ, USA

36.5.1 INTRODUCTION

The chemistry of hexavalent molybdenum is dominated by the oxo ligand and its analogs. This highest of Mo oxidation states requires for its stabilization the presence of ligands that are not only good σ donors but also good π donors. Thereby, sufficient charge density can be placed on the Mo to avoid violating the Pauling Electroneutrality Principle.[1] This requires ligands with filled p orbitals that are not otherwise engaged in bonding with other atoms. Formally, this engenders a situation in which a multiple bond is formed between the Mo and its donor ligand. About 10 years ago[2] the oxo ligand was virtually alone in stabilizing large numbers of formally Mo^{VI} complexes. This situation has changed dramatically in recent years. It is now clearly recognized that the sulfido ligand, S^{2-}, can, in a significant number of cases, isomorphously substitute for the oxo ligand, O^{2-}. Moreover, in certain cases selenido (Se^{2-}), peroxido (O_2^{2-}), persulfido (S_2^{2-}), imido (NR^{2-}), nitrido (N^{3-}), alkylcarbido (RC^{3-}, equivalent to alkylidyne), hydrazido(2−) (R_2NN^{2-}) and hydroxylamido (R_2NO^-) ions can form complexes with strict analogs in oxo chemistry. The oxo chemistry stands as a guidepost from which one can extrapolate to make predictions of the structural chemistry of its analog ligands.

Mononuclear examples are dominant in the simple coordination chemistry of molybdenum(VI). This is due in part to the absence of d electrons in Mo^{VI} complexes, which precludes the formation of Mo^{VI}—Mo^{VI} metal–metal bonds. However, note must be taken of the very large class of homo- and hetero-polymolybdates which contain polynuclear Mo. These are discussed elsewhere in this treatise. A substantial number of dinuclear complexes of Mo^{VI} are also known and these are discussed in this chapter.

In presenting the chemistry of Mo^{VI} we first discuss complexes in which oxo groups provide the sole π-donating ligands. We then discuss complexes in which sulfido, selenido, imido, nitrido, alkylcarbido, peroxido, disulfido, hydrazido and hydroxylamido can substitute in a direct way for oxo. Within the mononuclear class we make distinctions between complexes with four, three, two and one oxo ligands. Within the dinuclear class we distinguish between singly, doubly and triply bridged species. Finally, we discuss examples of non-oxo Mo^{VI} complexes. Aspects of Mo^{VI} chemistry have been considered in recent reviews.[2–5]

36.5.2 MOLYBDATE AND RELATED SPECIES

The tetrahedral molybdate ion is the key antecedent for much Mo^{VI} chemistry. This ion serves as the starting point for many of the preparative schemes in the chemistry of Mo^{VI} and, directly or indirectly, often for low oxidation state chemistry as well. The molybdate ion is found naturally in alkaline solutions such as sea water (pH = 8.0–8.5). It is the form in which Mo is often added to culture media used for the growth of microorganisms. In solutions of moderate concentrations the mononuclear ion is only present above pH = 7. Below pH = 7, the equilibria in equations (1) and (2) have considerable effect and the heptamolybdate ion predominates at moderate concentrations in aqueous solution. Detailed treatment of equilibria in aqueous Mo^{VI} solutions has been presented.[6] Although the chemistry of polyoxoanions of Mo is discussed elsewhere in this volume, it is worth pointing out that the six-coordinated octahedral MoO_2^{2+} and MoO^{4+} units can be viewed as the mononuclear building blocks of these polynuclear anions.[7] For example, $PMo_{12}O_{40}^{3-}$ has a central phosphate surrounded by an $Mo_{12}O_{36}$ unit in which each Mo^{VI} has a single terminal oxo in its coordination sphere. Similarly, $Mo_4O_{12}CH_2O_2OH^{3-}$ can be pictured as an Mo_4O_{12} ring containing dioxo Mo^{VI} with each Mo additionally bridged to all others by the OH^- and $CH_2O_2^{2-}$ anions.[7]

$$7MoO_4^{2-} + 8H^+ \rightleftharpoons Mo_7O_{24}^{6-} + 4H_2O \quad (1)$$

$$Mo_7O_{24}^{6-} + MoO_4^{2-} + 4H^+ \rightleftharpoons Mo_8O_{26}^{4-} + 2H_2O \quad (2)$$

At low concentrations hexavalent molybdenum in dilute acidic solutions is likely mononuclear.[8] At higher acidity the ion $MoO_2(H_2O)_4^{2+}$ is probably present.[8] In special situations the dimolybdate ion, $Mo_2O_7^{2-}$, is formed whose structure is analogous to that of the better known dichromate ion, $Cr_2O_7^{2-}$. (Dinuclear Mo^{VI} ions are discussed separately below.)

36.5.2.1 Structure of MoO_4^{2-}-containing Compounds

Molybdate, MoO_4^{2-}, is isolated in the form of salts of monovalent, divalent and trivalent cations. The salts of the simple monovalent cations are usually water soluble while salts with larger cations, *e.g.* *N*-propylammonium, *N*-ethylpyridinium and tetra-*n*-butyl ammonium may also have solubility in non-aqueous solvents.[9] The salts of di- and tri-valent cations are generally insoluble and form three-dimensional structures in the solid state. As discussed below, although many of these maintain the MoO_4^{2-} structural unit, some salts which stoichiometrically contain MoO_4^{2-} have octahedral six-coordinate Mo^{VI}.

In K_2MoO_4,[10] discrete MoO_4^{2-} ions have an Mo—O distance of 1.76 Å and the O—Mo—O bond angles of a regular tetrahedron. This distance is among the longest for *terminal oxo* molybdenum linkages, henceforth designated Mo—O_t. The $(NH_4)_2MoO_4$ salt is isomorphous to K_2MoO_4.[10]

The MoO_4^{2-} ion can act as a ligand to many metal ions. In some cases, as with higher valent metal ions containing labile coordination spheres, all or most of the original ligands around the second metal can be replaced by the O-donor MoO_4^{2-} leading to three-dimensional solid state structures. Recently studied examples, among many,[11,12] include $Fe_2(MoO_4)_3$,[13,14] $NiMoO_4$,[15] $MgMoO_4{\cdot}H_2O$,[16] $KDy(MoO_4)_2$,[17] $Tb_2(MoO_4)_2$,[18] Me_2SnMoO_4,[19] $MnMoO_4{\cdot}H_2O$[20] and $CdTh(MoO_4)_3$.[21] In the last six compounds the MoO_4^{2-} ion maintains an approximately tetrahedral four-coordination and performs a bridging role. However, in other cases these polymeric structures have six coordinate Mo^{VI}, *e.g.* $NiMoO_4$.[15]

In cases in which the full coordination sphere of the second metal is not accessible to substitution, MoO_4^{2-} can serve as a mono- or bi-dentate ligand. Examples of monodentate MoO_4^{2-} include $Co(NH_3)_5MoO_4^{+}$ [22] and $Cr(edta)MoO_4^{3-}$.[19] An example of bidentate MoO_4^{2-} is $Co(NH_3)_4MoO_4^{+}$. A rare example of bridging MoO_4^{2-} in a non-polymeric structure involves the D_{2d} symmetry $K(MoO_4)K$ unit which is isolated in a matrix of frozen N_2.[23] X-Ray structural studies are notably lacking for compounds containing MoO_4^{2-} as a ligand. One exception is the homopolymolybdate ion $Mo_8O_{26}(MoO_4)_2^{8-}$ which contains two monodentate MoO_4^{2-} units.[24] The tetranuclear complex anion $Mo_4O_8(OMe)_2(NNPh)_4^{2-}$ also contains bridging bidentate MoO_4^{2-} ligands.[25] The structure, shown in Figure 1, is roughly centrosymmetric in its inner coordination sphere. In this complex two $Mo(NNPh)_2^{2+}$ units are bridged by two *trans*-methoxy groups and by two MoO_4^{2-} groups *trans* to each other.[26] The non-molybdate Mo—Mo distance of 3.46 Å indicates the absence of metal–metal bonding. The terminal $Mo—O_t$ distance of 1.72 compares with 1.80 Å for the bridging $Mo—O_b$. The near-tetrahedral angles show that two MoO_4^{2-}-like ions are present.[25] This brief discussion of MoO_4^{2-} as a ligand contrasts with the extensive structural chemistry of MoS_4^{2-} as a ligand discussed elsewhere in this volume.

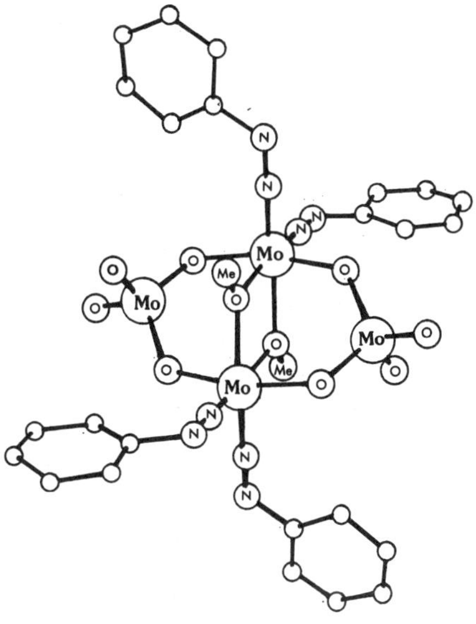

Figure 1 The structure of $Mo_4O_8(OMe)_2(NNPh)_4^{2-}$ in its $HNEt_4^{2-}$ salt (unlabelled atoms are carbon)[25]

36.5.2.2 Thio and Seleno Derivatives of MoO_4^{2-}

There are a significant number of strict analogs of MoO_4^{2-} that contain oxo-analog ligands in place of O while maintaining a tetrahedral or pseudotetrahedral structure about Mo^{VI}. For example, the treatment of basic solutions of molybdate with sulfide leads successively to thio substituted anions according to Scheme 1.

$$MoO_4^{2-} \longrightarrow MoO_3S^{2-} \longrightarrow MoO_2S_2^{2-} \longrightarrow MoO_3S^{2-} \longrightarrow MoS_4^{2-}$$

Scheme 1

Although the chemistry of these conversions had been known for many years, recent thermodynamic and kinetic work[27] has provided quantitative information. Similarly, treatment with Se^{2-} leads to $MoSe_4^{2-}$ [28–30] while treatment with H_2O_2 gives the $Mo(O_2)_2^{2-}$ ion discussed below. The compound MoO_2Cl_2 is also well known and, although containing six-coordinate Mo in the solid state, has a four-coordinate approximate tetrahedral structure in solution or in the gas phase.

Tetrathiomolybdate has been prepared with a number of organic cations including NEt_4^{+},[31] $NC_4H_9^{+}$ (pyrrolidinium),[32,34] $N_2C_4H_{11}^{+}$ (piperazinium), $NC_5H_{12}^{+}$ (piperidinium)[34,32] and $N_4C_6H_{13}$ (monoprotonated hexamethylene imine).[33,32] These salts are useful for structural and preparative studies. The structure of the MoS_4^{2-} ion was determined in its tetraethylammonium salt.[35] The average Mo—S distance of 2.177(6) and S—Mo—S angle of 107.5(9)° are found in a near perfect tetrahedral arrangement of sulfur atoms about the molybdenum.

The ion $MoOS_3^{2-}$ in Cs_2MoOS_3 [36] has Mo—O = 1.78 and Mo—S = 2.18 Å, values that are indistinguishable from the Mo—O and Mo—S distances found in MoO_4^{2-} and MoS_4^{2-}, respectively. Further, the angles in $MoOS_3^{2-}$ are all within 1° of the 109.5° tetrahedral angle. This identity indicates that Mo—O and Mo—S bonds are similar in their electronic and steric effects on the Mo^{VI} center (barring, of course, fortuitous cancellation of steric and electronic factors).

Reaction of $Mo_2O_7^{2-}$ with $(Me_3Si)_2S$ leads to the series $MoO_{3-x}S_x(OSiMe_3)^-$ ($x = 0$–3) with the $x = 3$ member of the series cleanly isolated. The anions are tetrahedral, derived from MoO_4^{2-}. $MoO_3(OSiMe_3)^-$ has Mo—O = 1.881(7), O—Si = 1.631(8) and Si—O—Mo = 146.6(4)° and Mo—S averaging 2.154.[37] Clearly $OSi(Me)_3^-$ can substitute for oxo in an isostructural series.[37]

36.5.2.3 Spectra and Bonding in the MoO_4^{2-} and Related Ions

In the UV region MoO_4^{2-} displays two band systems, one in the range 42000–46000 cm^{-1} and a second, more intense system, in the range 46000–53000 cm^{-1}. In single crystal studies (*e.g.* MoO_4^{2-} in Cs_2SO_4) both bands show vibrational structure.[38] The progression observed with $\nu_1 = 780\ cm^{-1}$ corresponds to $\nu(Mo—O_t)$ for the excited state compared to $\nu_1 = 897\ cm^{-1}$ for the ground state.

The bonding in the $4d^0$ tetrahedral molybdate and related systems has been treated by molecular orbital approaches.[39,40,41] An approximate energy level diagram in T_d symmetry is shown in Figure 2. The oxo (or thio or seleno) ligands are involved in both σ and π bonding with Mo. The difference between the $3t_2^*$ and $2e^*$ levels represents the Δ_t value for the system. Since there are no d electrons, this splitting is not discernible from d–d spectral transitions. It can, however, be estimated from the difference in energy of charge-transfer bands or X-ray absorptions, provided they can be properly assigned.

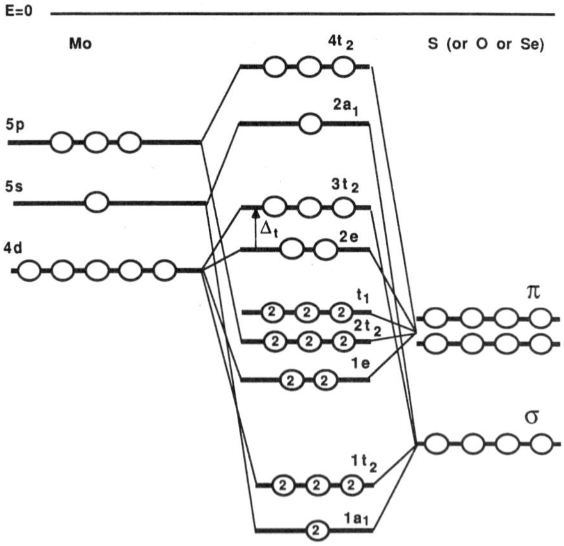

Figure 2 Qualitative energy level diagram for T_d ions such as MoO_4^{2-} and MoS_4^{2-}

The assignment of charge-transfer spectra in tetrahedral oxoanions and their tetrathio derivatives has been a controversial area. Electronic spectra have been studied in some detail.[42,43] The assignments of 'charge-transfer' bands for MoS_4^{2-} are likely to be $t_1 \rightarrow 2e^*(\nu_1)$, $2t_2 \rightarrow 2e^*(\nu_2)$, $t_1 \rightarrow 3t_2^*(\nu_3)$ and $2t_2 \rightarrow 4t_2^*(\nu_3)$.[41] The difference between ν_2 and ν_1 is the Δ_t value. Assignments differ somewhat for other ions. The assignments are supported by detailed electronic spectral studies,[44] valence photoelectron spectroscopy[45] and MCD data.[46]

Interestingly, for MoO_4^{2-} the theoretical studies[47] show that little O→Mo charge transfer occurs in going from the ground to excited states. It is nevertheless clear that the configuration changes from $4d^0$ to $4d^1$ due to excitation of an electron from mostly ligand-based into predominantly metal-based levels. However, electronic rearrangement in the remaining levels makes the net *charge* transfer negligible. It is therefore more appropriate to call these electron-transfer transitions as a guide to the configurational nature of the transition, since they do not necessarily reflect large net redistribution of charge.

The IR and Raman spectra of the tetrahedral molybdates, tetrathiomolybdates and tetraselenomolybdates have also received study.[28,48–53] For the T_d, tetrahedral dianion MoO_4^{2-}, the two stretching modes (A_1 and F_2) are Raman active and one (F_2) is IR active as well. Bending modes are of E and F_2 symmetry, and while both are Raman active, only F_2 is IR active. Bands at 897 and 837 cm^{-1} have been assigned as stretches, whereas those at 325 and 317 cm^{-1} are assigned to bends. Detailed vibrational analyses of MoO_4^{2-} have included $^{92}Mo/^{100}Mo$ isotopic substitution effects[17,53] and polarized IR measurements.[17] Studies on MoO_3S^{2-} and $MoO_2S_2^{2-}$ reveal the ν(Mo—O) bands expected for C_{3v} and C_{2v} structures, respectively.[42]

When MoO_4^{2-} coordinates to a metal ion, as in $[Co(NH_3)_5MoO_4]Cl$ or $[Co(NH_3)_4MoO_4]NO_3$, the IR and Raman spectra of the MoO_4^{2-} grouping are perturbed.[22] An instructive example considers the coordination isomers $[Co(NH_3)_5Cl]MoO_4$ and $[Co(NH_3)_5MoO_4]Cl$.[22]

36.5.2.4 Reactions of MoO_4^{2-}

In alkaline solutions the exchange reaction of equation (3) occurs in hydroxide-independent and hydroxide-dependent paths.[54] Rate constants, activation parameters and isotope effects have been reported.[54] Although proton transfer is implicated in the rate-determining step of the OH^--independent path and an association involving MoO_4^{2-} and OH^- is likely for the OH^--dependent pathway, no firm mechanistic conclusions can be drawn.

$$MoO_4^{2-} + {}^{18}OH_2 \longrightarrow Mo^{18}OO_3^{2-} + {}^{16}OH_2 \qquad (3)$$

The kinetics of reactions of MoO_4^{2-} leading to the binuclear complexes $Co(NH_3)_5OMoO_3^+$ and $Cr(edta)OMoO_3^{3-}$ have been studied around pH = 7.[55] Interestingly, the reaction mechanisms for the formation of these two species differ. The Co complex is formed by equation (4) where Mo—O bond breakage occurs. The Cr complex is formed as in equation (5) where the uncoordinated carboxylate of the $edta^{4-}$ ligand apparently plays a key role in labilizing the H_2O ligand on the otherwise normally inert Cr^{III} linkage.

$$Co(NH_3)_5OH_2^{3+} + MoO_4^{*2-} \longrightarrow Co(NH_3)_5OMoO_3^+ + H_2O^* \qquad (4)$$

$$Cr(edta)H_2O^{*-} + MoO_4^{2-} \longrightarrow Cr(edta)MoO_4^{3-} + H_2O^* \qquad (5)$$

These species are of interest as they represent examples of Mo bridged to a first transition row element. This is a suspected occurrence in nitrogenase where Mo bridged to Fe (through O or S) is implicated.[5]

36.5.3 COMPLEXES CONTAINING THE MoO_3 CORE

The MoO_3 structural unit is found in a few monomeric oxo Mo^{VI} complexes. The complex MoO_3(dien) has a six-coordinate octahedral structure[56,57] with the *cis*-trioxo arrangement, an Mo—O distance averaging 1.74 Å, O—Mo—O angles averaging 106° and Mo—N averaging 2.32 Å. MoO_3(dien) shows little stability in solution.[57,58] A similar six-coordinate monomeric structure with $Mo—O_t = 1.74$ Å is found in the $MoO_3(nta)^{3-}$ ion in $K_3[MoO_3(nta)]\cdot H_2O$.[59] Here, as shown in Figure 3, one of the nta^{3-} carboxylate arms is uncoordinated. Interestingly, the NMR spectrum shows that the coordinated and free carboxylate arms of nta^{3-} exchange rapidly.[60]

A related nominally dinuclear complex is the long studied edta complex found in $Na_4(MoO_3)_2edta\cdot 8H_2O$.[61] Here the two MoO_3 groups are connected only by a —NCH_2CH_2N— bridge. A six-coordinate structure with Mo—O = 1.74 Å makes the MoO_3 coordination similar to that in the nta^{3-} complex. Several amino acid complexes appear by spectroscopic criteria to have MoO_3 core structures.[62]

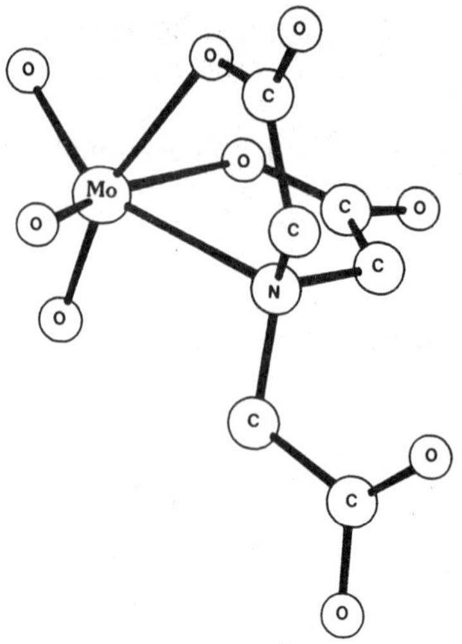

Figure 3 The structure of $MoO_3(nta)^{3-}$ in $K_3[MoO_3(nta)]\cdot H_2O$[59]

The MoO_3 unit is characterized by vibrational bands at ~890 and 840 cm^{-1}.[57,63,64] ^{17}O and ^{95}Mo NMR spectra may also be used to identify the MoO_3 core. In fact, ^{17}O NMR is sensitive to the nature of the group *trans* to oxo[66] as is also the case in MoO_2^{2+} complexes.[65]

In addition to the above structurally characterized complexes there are other cases where the presence of the MoO_3 unit seems possible on stoichiometric grounds. Reaction of K_2MoO_4 with $O{=}C(CF_2)_3C{=}O$ (with a bridging O between the two carbonyl carbons) (perfluoroglutaric anhydride) leads to $Na_2\{MoO_3[O_2C(CF_2)_3CO_2]\}$ which is likely to be polynuclear.[67] A band at 945 cm^{-1} due to Mo—O_t and bands at lower energy possibly attributable to Mo—O_b make the presence of the MoO_3 unit in this complex unlikely. Likewise, the stoichiometry $MoO_3(Hars)^{2-}$,[68] where Hars is alizarin red-S (9,10-dihydro-3,4-dihydroxy-9,10-dioxo-2-anthracene sulfonic acid), does not necessarily implicate a *cis*-MoO_3 core.

Interestingly the Mo enzyme formate dehydrogenase shows an EXAFS pattern that has been interpreted in terms of approximately three Mo—O = 1.74 Å distances indicating the likely presence of an MoO_3 unit bound to non-sulfur ligands.[69]

Of late there are also reports of molybdate-like monoanions, namely $ClMoO_3^-$, $R_3SiOMoO_3^-$ and $MeMoO_3^-$, which contain an MoO_3 grouping and a single monodentate monoanonic ligand. $ClMoO_3^-$ is prepared[70] as the Ph_4As^+ or Ph_4P^+ salts by reacting MoO_3 with $[Ph_4As]Cl$ or $[Ph_4P]Cl$, respectively. The Mo—O stretching vibrations occur at 930 and 894 cm^{-1}.[70] $R_3SiOMoO_3^-$ (with R = Ph or *t*-C_4H_9) is prepared[71] by the reaction of equation (6). The crystal structure of the NBu_4^{n+} salt of $Ph_3SiOMoO_3^-$ shows the expected tetrahedral (C_{3v}) structure and ν(Mo—O) stretching frequencies are identified at 906 and 884 cm^{-1}.[71] The complex anion $MeMoO_3^-$ prepared from $MoO_2MeBr(bipy)$ by base hydrolysis has only been characterized in solution.[72] It shows a single ^1H NMR peak at 0.611 p.p.m, downfield from TMS.[72] All of the $XMoO_3^-$ ions are colorless.

$$2R_3SiOH + Mo_2O_7^{2-} \longrightarrow 2R_3SiOMoO_3^- + 2H_2O \qquad (6)$$

36.5.4 COMPLEXES WITH THE MoO_2^{2+} CORE STRUCTURE

Complexes containing the MoO_2^{2+} core are by far the most common in Mo^{VI} chemistry. These complexes are usually six-coordinate although some four-coordinate compounds are known. The latter are tetrahedral or pseudotetrahedral and related to MoO_4^{2-}. Examples include $MoO_2S_2^{2-}$, $MoO_2Se_2^{2-}$, MoO_2Br_2 and $MoO_2(R_2NO)_2$. An example of a five-coordinate complex now exists.[73]

The vast majority of MoO_2^{2+} core complexes are six-coordinate and mostly of distorted octahedral structure. Some non-octahedral complexes are known. In the octahedral structure the two Mo—O_t bonds are invariably *cis* to each other. The strong σ- and π-donor nature of the oxo ligands makes it favorable for them to avoid competing for the same *p* and *d* orbitals. If the oxo groups were *trans* they would be forced to share two *d* orbitals and one *p* orbital. By residing on adjacent coordination sites they are forced to share only a single *d* orbital.

Repulsion between the short Mo—O_t bonds and between these and other bonds determines remaining details of geometry. Several articles have addressed structural trends in oxomolybdenum(VI) complexes.[2,74–78]

An MoO_2^{2+} unit is likely to be present in the oxidized form of sulfite oxidase and in the desulfo form of xanthine dehydrogenase.[20] The reactivity of the MoO_2^{2+} unit has been thoroughly studied, especially with regard to oxygen atom transfer reactions.

36.5.4.1 Structures Containing cis MoO_2^{2+} Cores

The MoO_2^{2+} unit has by far received the most attention in the structural chemistry of Mo^{VI}. There have been over 50 X-ray crystallographic structure determinations of six-coordinate complexes containing this structural unit. All but five of these determinations have revealed a near-octahedral structure with *cis* dioxo groups. The structural results in Table 1 clearly show the narrow range within which both the Mo—O_t distances and the O_t—Mo—O_t angles lie. The five exceptions shown at the bottom of the table are complexes in which a sterically demanding substituted amine thiolate ligand or a porphyrin ligand forbids the molybdenum from having the otherwise highly favored *cis*-octahedral structure. In addition to the mononuclear structures listed in Table 1, there are several dinuclear complexes discussed below and many polymeric structures (see *e.g.* ref. 79) which also have MoO_2^{2+} cores with dimensions similar to those found in the mononuclear complexes.

Table 1 Structural Studies on MoO_2^{2+} Complexes[a]

Complex	*Distance Mo—O* (Å)	*Angle O—Mo—O* (°)	*Ref.*
$MoO_2(PhCOCHCOPh)_2$	1.695(8), 1.697(8)	104.8(4)	92
$MoO_2(acac)_2$	1.71(2), 1.62(2)	111.1(9)	196
	1.72(2), 1.66(2)	104.3(8)	
	1.66, 1.70	105	197
$MoO_2(oxine)_2$	1.71	104	198
$MoO_2(Et_2dtc)_2$	1.703(2)	105.8(1)	80
$MoO_2(Pr^n_2dtc)_2$	1.695(5)	105.7(1)	196
	1.696(5)		
$K_2[MoO_2(O_2C_6H_4)_2]\cdot 2H_2O$	1.77	102	199
$MoO_2[H(OCH_2CH_2)_3N]$	1.75(4), 1.80(4)		200
	1.77(4), 1.83(4)		
$MoO_2(bipy)Br_2$	1.643(17), 1.826(18)	103.3	201
$K_2[MoO_2F_4]\cdot H_2O$	1.68(2), 1.73(1)	95(1)	202
$MoO_2Cl_2[OPPh_3]_2$	1.695(1), 1.673(1)	103.17(7)	74
$MoO_2Br_2[OPPh_3]_2$	1.69(1), 1.73(1)	103.2(5)	74
$MoO_2Cl_2(DMF)_2$	1.68(1)	102.2(7)	203
$(Cs)_2[MoO_2(Hmal)_2]$	1.708(9)	104.5(6)	204
$MoO_2(OCH_2CH_2OCH_2CH_2O)$	1.63(4), 1.73(2)	101(3)	205
$MoO_2[C_6H_3N\text{-}2,6(CH_2CH_2S)_2](C_4H_8SO)$	1.723(3), 1.694(3)	106.0(2)	176
$MoO_2[C_6H_3N\text{-}2,6(CH_2CPh_2O)_2][Me_2SO]$	1.702(3), 1.708(3)	105.4(2)	79
$MoO_2(OC_6H_4CHNMe)_2$	1.688(7)	107.7(4)	206
MoO_2(uramil-*N,N*-diacetato)$\cdot H_2O$	1.689(5)	106.6	207
$MoO_2[OC_6H_4CHNCH_2CHMeNCHC_6H_4O]$	1.714(4)	103.3	208
	1.701(4)		
$MoO_2Cl_2(phen)$	1.695(3)	106.8(2)	209
$MoO_2(NCS)_2(HMPA)_2$	1.68(1), 1.68(1)	101.4(5)	210
	1.67(1), 1.69(1)	100.9(5)	
$MoO_2Cl_2(HMPA)_2$	1.686(8)	102.46(46)	211
	1.686(8)		
$MoO_2Cl_2(MeOCH_2CH_2OMe)_2$	1.667(11)	105.0(5)	36
	1.673(10)		
$MoO_2(HOCH_2CH_2O)_2$	1.723(10), 1.729(10)	107.6(5)	213
	1.704(10), 1.737(11)	104.6(5)	
$MoO_2Cl_2[(Pr^iO)_2P(O)CH_2C(O)NEt_2]$	1.687(2)	102.8(1)	214
	1.677(2)		
$MoO_2Cl_2(H_2O)_2\cdot 2[C_5H_6N^+Cl^-]$	1.701(8)	103.0(5)	35
$MoO_2[OC_6H_4NCHMeCH_2NC_6H_4O]$	1.709(3)	103.2(1)	215
	1.710(3)		
$MoO_2[OC_6H_4CHNMe]_2$	1.682(8)	107.7(4)	216
$MoO_2[SC(C_6H_4\text{-}p\text{-}Me)NMeO]_2$	1.706(5)	103.7(2)	22
	1.713(5)		

Table 1 *(continued)*

Complex	*Distance Mo—O* (Å)	*Angle O—Mo—O* (°)	*Ref.*
$MoO_2[ONMeC(N\text{-}Bu^t)O]_2$	1.726		212
$MoO_2[OCMeN(C_6H_4\text{-}p\text{-}OEt)O]$	1.703(2)	103.94(11)	154
$MoO_2[OCMeN(Ph)O]_2$	1.704(3)	105.2(2)	154
	1.692(3)		
MoO_2Cl_2(9,10-phenanthroquinone)	1.671(3)	104.8(2)	129
	1.671(3)		
$MoO_2[OC_6H_3(Bu^t)CHNCH_2CH_2NCHC_6H_3(Bu^t)O]$	1.706(10), 1.718(9)	109.2(6)	190
	1.703(9), 1.708(10)	109.2(5)	
$MoO_2[(Pr^i)C_6H_3NCH_2CH_2NC_6H_3(Pr^i)O]$	1.706(10), 1.718(9)	109.2	94
$MoO_2(OC_6H_4C_7H_5N_2)_2$ [a]	1.70	103.5(2)	109
MoO_2(L-CysOMe)$_2$	1.714(4)	108.1(3)	217
MoO_2[(*S*)-PenOMe]$_2$	1.711(3), 1.720(4)	107.0(2)	218
	1.714(4), 1.703(3)	107.6(2)	
$MoO_2[SCH_2CH_2NMeCH_2CH_2NMeCH_2CH_2S]$	1.693(9), 1.684(7)	109.1(5)	219
	1.677(9), 1.750(10)	107.8(7)	219
$MoO_2[SCH_2CH_2NMeCH_2CH_2CH_2NMeCH_2CH_2S]$	1.704(6)	107.8(3)	219
$MoO_2[SC_6H_4NHCH_2CH_2NHC_6H_4S]$	1.708(3), 1.715(3)	109.9(1)	219
$MoO_2[SCH_2CH_2NHCH_2CH_2SCH_2CH_2S]$	1.721(3), 1.718(3)	106.0(1)	219
$MoO_2[SC_6H_4SCH_2CH_2SC_6H_4S]$	1.69(2), 1.70(2)	111.0(1)	81
$MoO_2[(SCH_2CH_2)_2NCH_2CH_2CH_2NMe_2]$	1.705(5), 1.699(2)	101.6(2)	220
			221
$MoO_2[(SCH_2CH_2)_2NCH_2CH_2SMe]$	1.694(5), 1.694(5)	101.8(2)	220
	1.709(6), 1.703(6)	100.8(2)	221
	1.688(5), 1.705(2)	100.1(2)	
$MoO_2[SCH_2Me_2CH_2NHCH_2CH_2NHCH_2CMe_2S]$	1.714(2), 1.710(2)	109.8(1)	85
$MoO_2[SCH_2CMe_2NH_2]_2$	1.705(3), 1.705(3)	106.7(2)	85
$MoO_2[MeNHCH_2CMe_2S]_2$	1.711(5), 1.723(5)	122.2(3)	84
			94
$MoO_2[MeNHCMe_2S]_2$	1.731(1), 1.727	120.8(1)	84
$MoO_2[Me_2NCH_2CMe_2S]_2$	1.720(3)	122.0(2)	84
$MoO_2[C_6H_3N\text{-}2,6\text{-}(CH_2CPh_2S)]$	1.69(6), 1.696(6)	110.5(2)	79
MoO_2(ttp)	1.709(9)	95.1(4)	91
	1.744(9)		

[a] The ligand is 2-*o*-hydroxyphenylbenzimidazole.

Representative structures are shown in Figures 4–9. In addition to the characteristic Mo—O distance and O—Mo—O angle, there are several other features that the octahedral complexes in Table 1 have in common. The bond *trans* to the Mo—O linkage experiences a structural *trans* influence and is significantly longer than corresponding bonds to the same donor type that are *cis* to the oxo. For example in the complex $MoO_2(Et_2dtc)_2$[80] shown in Figure 4, which has two-fold symmetry, the Mo—S distance *trans* to Mo—O is 2.639 Å, whereas that *cis* to oxo is 2.450 Å. This effect is clearly attributable to the competition for the σ and π orbitals along the shared axis between the oxo group and the sulfur atom *trans* to it.

The complex MoO_2L with $L = {}^-SC_6H_4SCH_2CH_2SC_6H_4S^-$ has the typical distorted octahedral structure[81] with the two thioether donors of the tetradentate ligand lying *trans* to $Mo—O_t$ at Mo—S = 2.69 and the two thiolate donors *trans* to each other with Mo—S = 2.41 Å. The longest known thioether Mo—S distance is 2.79 Å found[23] in the complex of the tripodal ligand $N(CH_2CH_2S)_2(CH_2CH_2SMe)^{2-}$ shown in Figure 5. The dimensions of this complex closely parallel those found by EXAFS for sulfite oxidase.[82]

In complexes where there is a choice, the general rule is that the weaker π-bonding donor atoms are found *trans* to $Mo—O_t$ since they do not compete for the empty *p* and *d* orbitals along the same axis.[74,80,83,84,85] For example in $MoO_2Cl_2[OPPh_3]_2$ the π-donating chloride ligands are found *trans* to each other and *cis* to the $Mo—O_t$ bonds. Likewise, in $MoO_2[(SCH_2CH_2)_2NCH_2CH_2NMe_2]$ the thiolate sulfur atoms are found *trans* to each other and *cis* to the terminal oxo groups.[41] After this preference has been exercised the trends in bond angles can be rationalized qualitatively in terms of interligand repulsions in the Mo coordination sphere.[80,83,84,85]

There are several complexes now known to possess a *cis* dioxo structure and to be non-octahedral in nature. In a number of these the coordination number about Mo is lower.

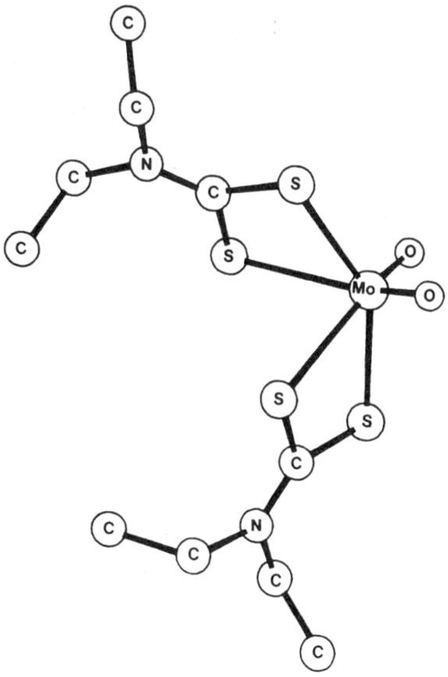

Figure 4 The structure of $MoO_2(Et_2dtc)_2$[80]

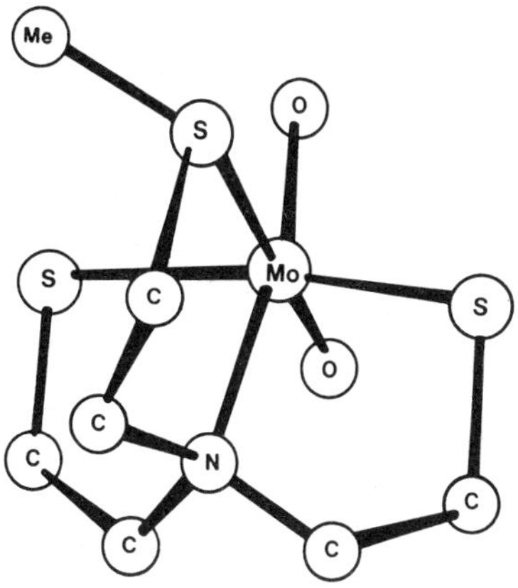

Figure 5 The structure of $MoO_2[N(CH_2CH_2S)_2CH_2CH_2SMe]$[23]

For example, four-coordinate $MoO_2S_2^{2-}$ and MoO_2Cl_2 have tetrahedral structures related to MoO_4^{2-}.

In some cases there may be *formal* six-coordination as in the complexes of deprotonated hydroxylamine ligands.[86–90] These complexes have, in addition to two oxo ligands, two deprotonated hydroxylamine ligands (R_2NO^-) in the Mo coordination sphere. The hydroxylamine(−H) ligands can be considered as bidentate N,O donors, in which case the complexes are formally six-coordinate. Alternatively, the complexes may be considered as four-coordinate with the hydroxylamine(−H) ligands occupying two of the tetrahedral sites about the Mo atom. If the center of the N—O bond is taken as representing the donor site for the ligand, a geometry very near to tetrahedral is indeed obtained.

The monodentate diatomic ligand formulation seems a structurally more consistent way of thinking about the deprotonated hydroxylamine ligands. As illustrated in Figure 6, if the structure of $MoO_2[Et_2NO]_2$ is considered to be six-coordinate then it is distinctly non-octahedral. The O—Mo—O angle of 113.7° is significantly larger than the angles found in the octahedral complexes in Table 1. The N and O atoms of the ligands and the Mo atom lie in a plane with the two oxo ligands roughly equidistant above and below that plane. The structure superficially resembles the skew trapezoidal bipyramid structure that is described more fully below. However, in the present case the trapezoid defined by the N_2O_2 ligand set does not contain the Mo atom within its boundary. The structure is therefore distinctly different from the skew trapezoidal structure discussed below and is really best represented as being tetrahedral with hydroxylamine(−H) ligands occupying two tetrahedral binding sites. Additional complexes containing hydroxylamine(−H) ligands are discussed separately below.

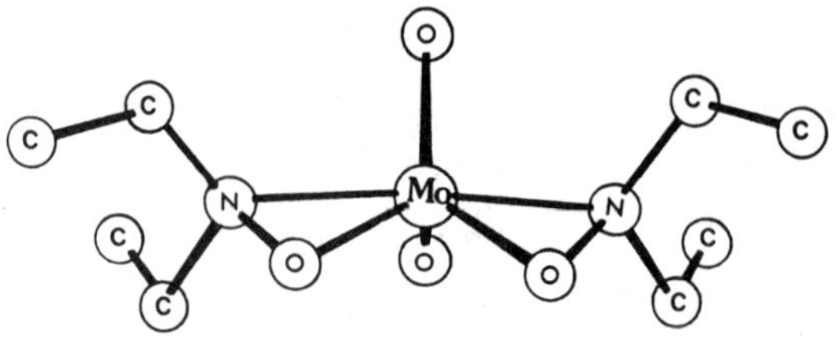

Figure 6 The structure of $MoO_2[Et_2NO]_2$[90]

There is one example[73] of a five-coordinated Mo^{VI} complex wherein the atypical coordination number is enforced by sterically bulky ligands. This structure[73] shown in Figure 7 is describable as a trigonal bipyramid in which two S-donor atoms of the tridentate ligand occupy the apical sites. The oxo groups and pyridine nitrogen of the tridentate ligand define the trigonal plane. Interestingly, the oxygen analog[79,73] of this sulfur donor ligand forms a conventional near-octahedral complex with MeOH or DMSO as the sixth ligand donor. Apparently, the difference in steric bulk and/or donor ability of two sulfur donors compared to two oxygen donors in the tridentate ligand is sufficient to cause the observed change in coordination tendency.

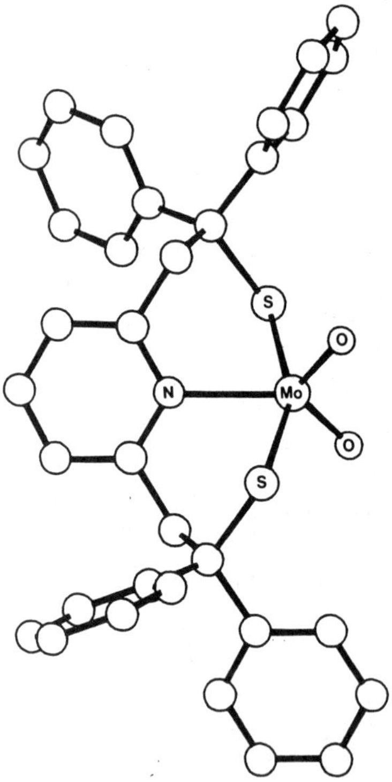

Figure 7 The structure of $MoO_2[NC_5H_3(CPh_2S)_2]$ (unlabelled atoms are carbon)[73]

There are two types of non-octahedral unequivocally six-coordinate formally Mo^{VI} complexes. The first type involves a complex of the ligand tetra(*p*-tolyl)porphyrin.[91] Here the structure shown in Figure 8 reveals the molybdenum atom to possess a distorted trigonal prismatic coordination in which the two oxo groups are found on the same side of a highly deformed porphyrin 'plane'. The O—Mo—O angle of 95.1° is the smallest known for any high resolution crystallographic determination of a *cis*-MoO_2^{2+} core. The simple alternative of having the oxo groups on opposite sides of the porphyrin ring is apparently not viable as this geometry is only known and expected for the $4d^2$ configuration of Mo^{IV}. The Mo atom in MoO_2(ttp) is fully 1.095 Å above the mean plane of the saddle-shaped porphyrin ligand. That the structure is stable, despite the fact that the small O_t—Mo—O_t angle must entail a significant oxo–oxo repulsion, is a testimony to the proclivity of Mo^{VI} to adopt the *cis* dioxo configuration.

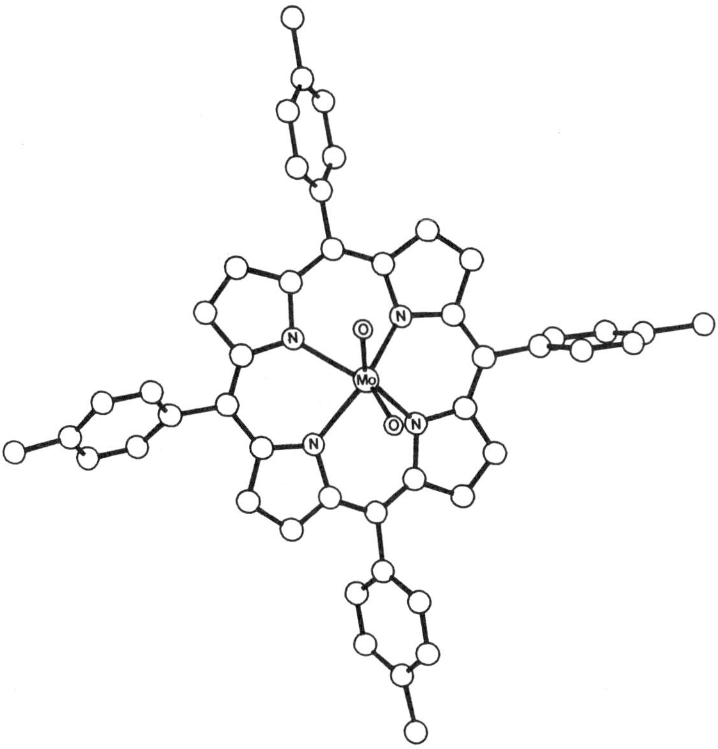

Figure 8 The structure of MoO_2(ttp) (unlabelled atoms are carbon)[91]

The second type of six-coordinate complex in which a distinctly non-octahedral structure is adopted involves *N*-substituted cysteamine ligands. When N is not substituted, a complex such as $MoO_2[NH_2CMe_2CH_2S]_2$ forms and adopts the expected octahedral structure[85] shown in Figure 9. In contrast, as illustrated in Figure 10, a skew trapezoidal bipyramidal structure[92] is found to be present[84,93,94] for $MoO_2[Me_2NCH_2CMe_2S]_2$. The two N- and two S-donor atoms form an approximate plane in which the Mo atom also lies. The oxo donors lie equidistant above and below this plane. The non-octahedral nature of the complex is seen by looking for the atom that is expected to be *trans* to oxo in the octahedral structure. Indeed, the largest angle about the oxo ligand is 122° and this is the O—Mo—O angle. This is the largest O_t—Mo—O_t angle which has been found in oxo Mo^{VI} complexes and it raises the interesting question as to whether these complexes are best formulated as containing Mo^{VI}. Indeed, it has been suggested that the observed short S--S contact of ~2.74 Å may represent the partial formation of a sulfur–sulfur (*i.e.* disulfide) bond. If a full disulfide bond had been formed, the complex would be formulated as one of Mo^{IV} with oxidized ligands. Such a complex would be

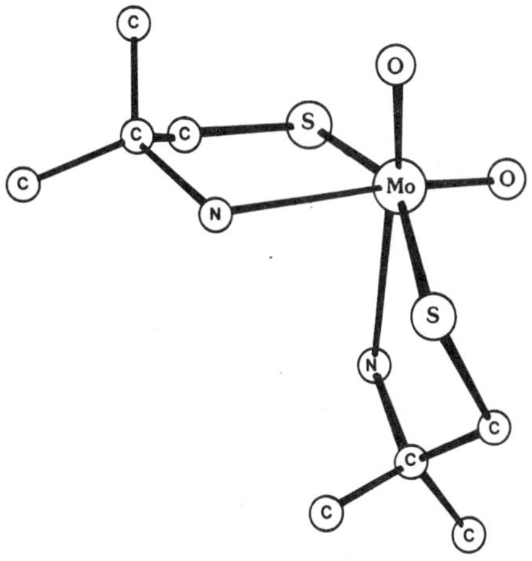

Figure 9 The structure of $MoO_2[NH_2CMe_2CH_2S]_2$[85]

expected to have a *trans* dioxo structure with O_t—Mo—O_t = 180°. However, the observed 'bond' length may indicate partial reduction of Mo. This would be consistent with the observed angle being increased from the roughly 105° expected to the larger value actually observed.

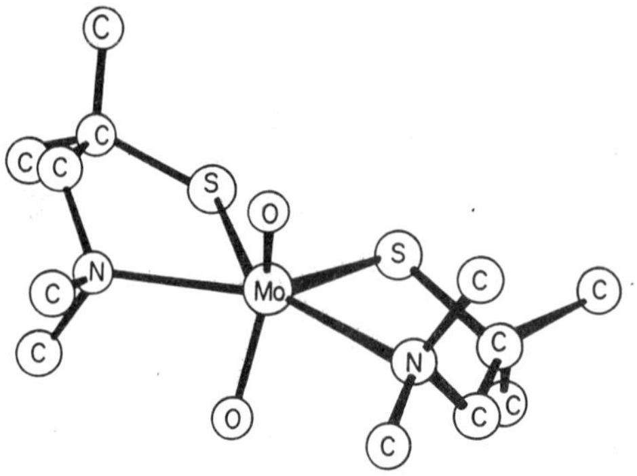

Figure 10 The structure of $MoO_2[Me_2NCH_2CMe_2S]_2$[84]

Clearly there is a universal tendency for dioxo Mo^{VI} complexes to adopt a *cis* configuration and there is a strong disposition toward octahedral structures. The distortions within these structures are clearly attributable to interligand repulsions. However, recent work reveals that appropriately rigid or sterically demanding ligands can enforce lower coordination numbers or non-octahedral coordination geometry on the Mo sphere.[73] In molybdoenzymes there is strong evidence for the presence of an MoO_2^{2+} grouping.[5,82] While octahedral coordination must be considered as a candidate structure for these enzymes, non-octahedral structures and lower coordination numbers cannot be eliminated. Indeed, proteins are capable of acting as extraordinarily sterically demanding ligands.

36.5.4.2 Spectra and Characterization of Complexes with MoO_2^{2+} Cores

The MoO_2^{2+} group manifests itself in the infrared and/or Raman spectrum in the form of an intense two-band pattern corresponding to the symmetric and asymmetric Mo—O stretching vibrations. Values for representative complexes are shown in Table 2 which also displays the lowest visible absorption bands for those complexes where these are available. The infrared (or Raman) absorptions are by far the most characteristic feature by which to identify the presence of the MoO_2^{2+} unit.

Several studies[95–98] of IR and Raman spectroscopy have concentrated on systematics of the MoO_2^{2+} core vibrations. Substitution of ^{18}O for ^{16}O is useful in confirming assignments and calculating forces constants.[96,98] The frequency shifts between complexes are dependent upon the donor ability of the remaining ligands, the geometry of the complex and the participation of the Mo—O_t linkage in H-bonding.[98] Often the IR spectrum along with elemental analysis data is sufficient to identify the MoO_2^{2+} core. Proton and carbon NMR data (see, for example, refs. 99–102) are often useful in a complementary way in establishing the disposition of the remaining non-oxo ligands. NMR can also be useful to establish the liability or fluxionality of the coordinated ligands.[100,102]

^{17}O NMR spectroscopy of oxomolybdenum complexes can give valuable structural information.[103–105] Since the sensitivity of the quadrupolar $I = \frac{5}{2}$ ^{17}O nucleus is inherently low, enrichment with ^{17}O is required to obtain signals in acceptable time periods. Apparently, ^{17}O can be incorporated by exchange into already formed MoO_2^{2+} core complexes.[65] The rates of exchange, although not quantitatively measured, appear to be qualitatively different for Mo—O_t groups in somewhat different ligand environments, *i.e.* with different *trans* groups.[65] In certain skew-trapezoidal bipyramidal structures[106] ^{17}O NMR was used to support the maintenance in solution of the structure found by single crystal X-ray diffraction in the solid.[84] The Mo—O_t resonance occurs at 850–1000 p.p.m. downfield from H_2O making it readily distinguishable from bridging (~400) and ligand oxygen atoms.[65,107]

^{95}Mo NMR has been applied to MoO_2^{2+} core complexes. The chemical shifts range over 900 p.p.m. and display potentially useful ligand dependence.[108–116] A comprehensive review of ^{95}Mo NMR summarizes key trends.[117]

Detailed electronic spectral studies for MoO_2Cl_2 and MoO_2Br_2 have been performed (ref.

Table 2 Vibrational and Electronic Spectra of MoO_2^{2+} Core Complexes

Complex	*IR or Raman (R)* $\nu(Mo—O_t)$ (cm^{-1})	λ (nm)	E (cm^{-1})	ε (M^{-1} cm^{-1})	*Ref.*
$MoO_2Cl[OPPh_3]_2$	947, 905				74
$MoO_2Br_2[OPPh_3]_2$	944, 903				74
$MoO_2Cl_2(bipy)$	935, 906				222
$MoO_2Cl_2(DMF)_2$	939, 905				223
$(C_5H_5)MoO_2Cl$	920, 887				224
$(NEt_4)MoO_2Cl_2(acac)$	931, 897				225
$MoO_2(acac)_2$	935, 905	320	31 200	3200	225
	925 (R), 898 (R)	270	37 000	22 000	
$MoO_2(oxine)_2$	926, 899	370	27 000		225
		250	39 100		226
		242	41 400		
$MoO_2(N\text{-}MeSal)_2$	922, 904				137
$MoO_2(salen)$	920, 885				138
$MoO_2(Cys\text{-}OMe)_2$	912, 884	351	28 500	5300	227
$Na_2[MoO_2(Cys)_2]\cdot DMF$	922, 892	306	32 700	12 000	227
$MoO_2(Me_2dtc)_2$	909, 875	377	26 500	3770	228
		290	34 500	6300	226
		252	39 700	15 000	
$MoO_2(Et_2dtc)_2$	905, 877	380	26 320	3700	228
$MoO_2(Bu^n_2dtc)_2$	909, 877	395	25 320	1020	228
$MoO_2(S_2PPh_2)_2$	925, 890	375	26 700	sh	229
$MoO_2(S_2PPr^i_2)_2$	925, 890	371	26 950	3270	229
$MoO_2(SCH_2CH_2NH_2)_2$	891, 866	354	28 250	4000	106
		272	36 760	4200	227
		246	40 650	4400	
		230	43 480	5100	
$MoO_2(SCMe_2CH_2NH_2)_2$	907, 872	355	28 170	5600	106
		260	38 460	7200	
		228	43 860	8000	
$MoO_2(SCMe_2CH_2NMe_2)_2$	893, 879	327	30 500	3200	106
		278	35 970	3980	
$MoO_2(SC_6H_4SCH_2CH_2SC_6H_4S)$	910, 885				124
$MoO_2[(SCH_2CH_2)_2NCH_2CH_2NMe_2]$	921, 893	391	25 600	2930	65
		325	30 800	sh	
		287	34 800	4430	
		236	42 400	9530	
$MoO_2[(SCH_2CH_2)_2NCH_2CH_2SMe]$	921, 891	366	27 300		65
		350	28 600		
		326	30 600		
		286	35 000		
		248	40 000		
$MoO_2(Et_2dtc)_2$	910, 877				96
	910 (R), 880 (R)				417
$MoO_2(CysOMe)_2$	912, 884	351	28 500	6130	227
$MoO_2(OC_6H_4NHCH_2CH_2NHC_6H_4O)$	930, 910	322	31 100	4570	189
$MoO_2[OC_6H_4CHNHC(O)C_5H_4N]H_2O$	922, 898				97
$MoO_2[OC_6H_4CHNHC(O)C_5H_4N]DMF$	932, 908				97
$MoO_2[OC_6H_4CHNHC(O)C_5H_4N]DMSO$	910, 894				97
$MoO_2[ON(Et)COMe]_2$	935, 900	274	36 500	19 000	153
$MoO_2Cl(9\text{-}ane)$	930, 900				192
$(NEt_4)_2[MoO_2(NCS)_4]$	925, 890				157
$Na_2[MoO_2Cl_4]$	964, 925				164
					230
$Na_2[MoO_2F_4]\cdot H_2O$	948, 912				231
	951 (R), 920 (R)				
$MoO_2Cl_4^{2-}$ in HCl	960, 922	310	32 300	5000	230
		226	33 200	7000	
MoO_2F_2 (gas)	1007, 987				157
	1009, 987				157
MoO_2Cl_2 (gas)	994, 972	306	32 680		157
		259	38 610		118
		225	44 440		
MoO_2Br_2 (gas)	991, 969	296	33 780		157
		252	39 680		118
		235	42 500		
MoO_2I_2 (gas)	972, 950				157

118 and refs. therein) and interpreted by SCF-Xα-SW calculations.[119] However, the electronic spectra of the six-coordinate species, which are the major group discussed in this section, have not received much attention.

EXAFS studies[120] have shown that the MoO_2^{2+} core is readily identifiable in a variety of known compounds. These have been used[120] to calibrate the EXAFS technique which was subsequently used to identify this core in sulfite oxidase and in desulfo xanthine oxidase.[82]

36.5.4.3 Preparation and Scope of MoO_2^{2+} Core Complexes

The method of choice for the preparation of many dioxo Mo^{VI} complexes involves the use of $MoO_2(acac)_2$ which is easily prepared from MoO_4^{2-} and acacH by adjusting the pH.[121,122] Thermodynamic data for the formation of $MoO_2(acac)_2$ have been presented.[123]

The reaction of equation (7)[124] is typical and proceeds readily in THF. Alcohols, especially MeOH, are also much used in related syntheses with base (often a tertiary amine) sometimes used to help ionize the incoming ligand. In the absence of base, the free acacH (b.p. = 139 °C) formed can be readily distilled under partial vacuum from the reaction solution along with the solvent.

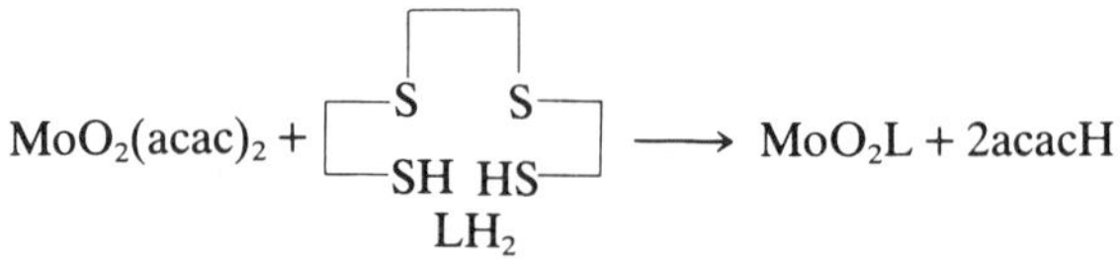

$$MoO_2(acac)_2 + \begin{array}{c} \overbrace{\text{S} \qquad \text{S}} \\ \underbrace{\text{SH} \quad \text{HS}} \\ LH_2 \end{array} \longrightarrow MoO_2L + 2acacH \qquad (7)$$

Alternative preparative approaches, while not systematically directed toward MoO_2^{2+} complexes can nevertheless be useful. For example, oxidation of $Mo(CO)_4(bipy)$ with Br_2 in $C_2H_5OH—CH_2Cl_2$ yields $MoO_2Br_2(bipy)$.[26] The corresponding dichloride can be prepared by ligand exchange in boiling acetone.[125] The complexes $MoO_2X(9\text{-ane})$ (X = Cl, Br; 9-ane = 1,4,7-trimethyl-1,4,7-triazanonane) are also prepared by oxidation of the low valent complex $Mo(CO)_3(9\text{-ane})$ with the appropriate halogen and subsequent hydrolysis.[192]

$MoO_2(R_2dtc)_2$ complexes can be prepared from the Mo^{IV} complexes, $MoO(R_2dtc)_2$. As the latter are often formed from the former, the route is not synthetically useful. However, since the oxygen atom donor can be DMSO (forming dimethyl sulfide)[126,127] the route is reminiscent of the enzymic reduction of biotin sulfoxide.[5] Other oxidants capable of effecting the conversion include N-oxides such as pyridine N-oxide or N-hexylmorpholine N-oxide[127] and oxygen[128,129] but not N_2O[130] as originally claimed.[127]

Certain classes of ligand deserve special mention as their MoO_2^{2+} complexes have received particular attention. The 1,1-dithiolate ligands, especially N,N-disubstituted dithiocarbamates, have been very thoroughly studied.[131-133] Evidence for the existence of the dithiophosphate complex $MoO_2[S_2P(OEt)_2]_2$ in solution has been obtained.[95] This and a number of other dithiolate complexes may be unstable toward internal redox in which the S ligand serves as the reductant to produce a lower valent Mo—S complex.

A related ligand is the ferrocenecarbodithioate ligand which behaves similarly to dithiocarbamates in forming an isolable series of Mo^{IV}, Mo^V and Mo^{VI} complexes.[134,135] The $MoO_2(S_2CC_5H_4FeC_5H_5)_2$ complex shows $\nu(Mo=O)$ at 920 and 889 cm^{-1} and gives a quasireversible reduction wave with $E_{1/2}$ of -0.89 V *vs.* SCE attributable to Mo reduction and a quasireversible oxidation wave at $+0.95$ V *vs.* SCE attributable to ligand (ferrocene) oxidation.[134,135]

The interesting ligand (1)[136] has two potential chelating modes upon ionization. Depending on its tautomeric state, either N,S or S,S coordination allows it to act as a bidentate chelating ligand. Apparently, it acts as a 2-aminocyclopent-1-ene-1-carbodithioate, *i.e.* as a 1,1-dithiolate, to form MoO_2L_2 complexes.[136]

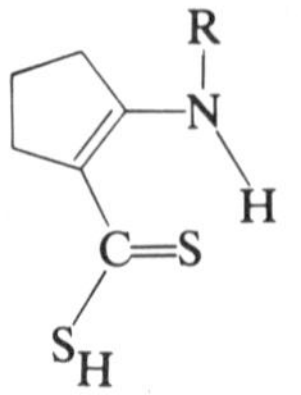

(1)

Complexes containing Schiff base chelating ligands have been extensively studied.[137–144] Here the ligands include bidentate, tridentate and tetradentate representatives where O or S donors supplement the imine nitrogen. Examples are listed in Tables 2 and 3. Sample ligands include (**2**), (**3**) and (**4**).

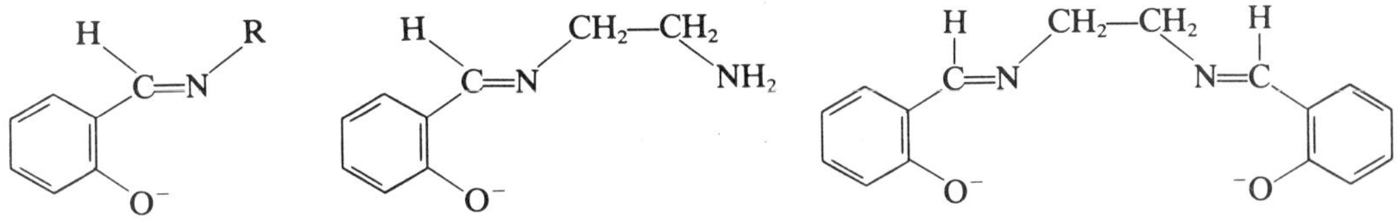

(**2**) L (**3**) L′ (**4**) Salen

The bidentate ligands (L) form MoO_2L_2 complexes while the tetradentate ligands form complexes such as MoO_2(Salen). The tridentate ligands (L′) form complexes $MoO_2L'(Y)$ where Y can be a variety of O-, N- or S-donor ligands, often the solvent from which the complex is isolated. Extensive studies have probed the effect of substitution on the aromatic ring, substitution for the hydrogen of the Schiff base carbon, and the length of the group pendent (or bridging) on N, the Schiff base nitrogen.[66,145–147] Mixed Fe, Mo complexes with Schiff base ligands have been reported[148] but not structurally characterized.

The complex anion, $MoO_2(cat)_2^{2-}$ (where cat = catecholate = benzene-1,2-diolate) has been extensively studied as the starting point for a series of complexes in the Mo^{III}, Mo^{IV} and Mo^{VI} oxidation states.[149,151,150] $MoO_2(cat)_2^{2-}$ in slightly acidic aqueous solutions undergoes sequential electrochemical one-electron and two-electron reductions leading to penta- and tri-valent Mo.[149] In slightly alkaline solution four oxidation states are electrochemically accessible.[151] Upon reduction the number of oxo groups per Mo decreases leading to reactivity and lability which may be related to the function of oxomolybdenum centers in enzymes.[150] Complexes containing 3,5-di-*tert*-butylcatecholate behave in a somewhat different fashion[152] involving an isolated dinuclear Mo^{VI} species.

The biologically active compounds glycine,[62] hydroxamate,[153] phenacetin[154] and acetanilid[154] each form MoO_2L_2 complexes. While these complexes help define the ways in which these organic molecules may serve as ligands, they have, at present, no obvious biological relevance. Several heterocyclic *N*-oxides also form MoO_2^{2+} complexes.[155]

Neutral complexes are known of the form MoO_2X_2 where X = F,[156] Cl,[118] Br,[118] I,[157] $OSiBu^t_3$,[158] R_2NO[86–89] and mesityl.[159] These complexes contain the MoO_2^{2+} core and two monodentate ligands. The halo complexes[2,118,119,156] are mononuclear in the gas phase and di- or poly-nuclear as solids or in solution. The $—OSiBu^t_3$[158] and $—C_6H_2Me_3$[159] complexes are apparently mononuclear. The siloxy compound[158] is made by the direct reaction of tris(*t*-butylsilanol) with MoO_3, presumably according to equation (8).

$$2Bu^t_3SiOH + MoO_3 \longrightarrow MoO_2[OSiBu^t_3]_2 + H_2O \tag{8}$$

The mesityl complex[159] is prepared by the reaction of the Grignard reagent and $MoO_2Cl_2(THF)_2$ in THF presumably according to Scheme 2.

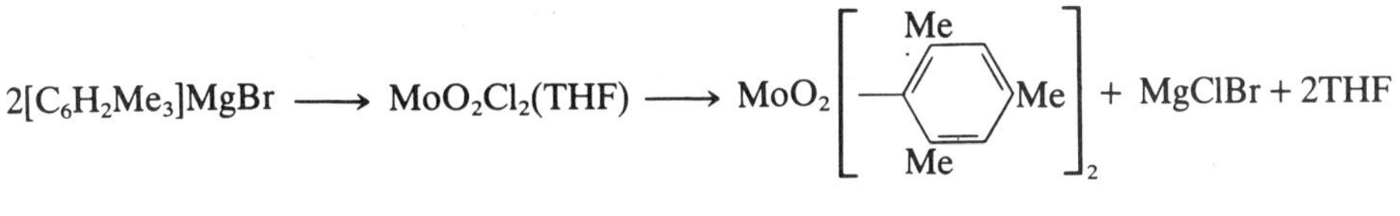

Scheme 2

These complexes likely have approximate tetrahedral structures and resemble MoO_4^{2-} rather than the largely octahedral MoO_2^{2+} complexes discussed in this section.

The series of complexes MoO_2X_2 (X = F, Cl, Br) can be used to prepare $MoO_2X_2L_2$ complexes where L_2 is py_2, bipy, $(DMSO)_2$ or other ligands. Examples of these complexes, which are formally adducts of MoO_2X_2, are given in Table 1. The structurally characterized complexes have the π-donor halides *trans* to each other and *cis* to the $Mo—O_t$ bonds.

36.5.4.4 Reactivity of Complexes with the MoO_2^{2+} Core

The dioxo core is susceptible to redox reactions which lead to reduction of the Mo center in conjunction with oxo removal,[131,160,161,162,163,164] Schematically, the core reaction in equation (9) gives the Mo^{VI} complex. The reactant and product of this reaction, however, can combine with comproportionation as in equation (10) to give the dinuclear Mo^{V} core.

$$\overset{VI}{MoO_2^{2+}} \underset{-[O]}{\overset{[O]}{\rightleftharpoons}} \overset{IV}{MoO^{2+}} \tag{9}$$

$$\overset{IV}{MoO^{2+}} + \overset{VI}{MoO_2^{2+}} \longrightarrow \overset{V}{Mo_2O_3^{4+}} \tag{10}$$

A general treatment has been published of oxygen atom transfer kinetics for complexes wherein the dimerization/comproportionation reaction is a complicating feature.[165] This treatment critically summarizes all pertinent earlier information. The assumption that dimerization is very fast with respect to oxygen atom transfer is born out by the agreement of the observed and calculated kinetic parameters.

The oxo removal reaction can be effected by a variety of trialkyl and triaryl group V reactants. For example, the reactivity of $MoO_2(Et_2dtc)_2$ with regard to oxo removal is reported[166] to qualitatively follow the order $Ph_3P \gg Et_3N \approx Ph_3As > Ph_3Sb > Ph_3Bi > Ph_3N$. For phosphines, the rates of reaction follow the expected order of nucleophilicity[165] $PEt_3 > PEt_2Ph > PEtPh_2 > PPh_3$.

As far as the ligand is concerned, toward oxidation of either PPh_3 or PhNHNHPh, the order of reactivity is[163] S_2CNEt_2 > thiooxinate > Cys-OMe. The oxidizing ability of MoO_2^{2+} complexes is enhanced by the presence of S-donor ligands.[163]

The reaction of $MoO_2(R_2dtc)_2$ (R = Et, Pr^n and Bu^i) with PPh_3 and other phosphines has been thoroughly studied.[167] The reaction is first order in both complex and phosphine which, along with the large negative entropy of activation, fits a simple bimolecular mechanism.[167] Similar studies have been performed on $MoO_2(Cys\text{-}OEt)_2$ with similar results. The general equation (11), (with the previously mentioned comproportionation as a complication) is clearly well established.[167,128] It has potential analytical applications for the determination of PPh_3 by monitoring the intensely colored $Mo_2O_3(Et_2dtc)_4$ formed in the comproportionation of equation (12).

$$Mo^{VI}O_2L + PR_3 \longrightarrow OPR_3 + Mo^{IV}OL_2 \tag{11}$$

$$MoO_2[Et_2dtc]_2 + MoO[Et_2dtc]_2 \longrightarrow Mo_2O_3[Et_2dtc]_2 \tag{12}$$

Kinetics of the reaction of certain MoO_2L complexes (L = S-containing salicylaldehyde Schiff base) with $PEtPh_2$ also show the reactions to be first order in complex and in phosphine,[145–147,168] The rate constants depend upon the substituent on the aromatic (salicylaldehyde derived) ring and correlate with reduction potentials of the complexes.[145]

The kinetics of the equilibrium reaction (13) was studied by concentration-jump relaxation techniques[168,169] for L = S_2CNEt_2, $SSeCNEt_2$, Se_2CNEt_2, $S_2P(OEt)_2$ and S_2PPh_2. The diethyl diselenocarbamate complex is found to be the most highly dissociated and the rate constant for its dissociation is the highest. The diphenyldithiophosphinate complex is the most stable and slowest to dissociate.[169,168]

$$Mo_2^{V}O_3L_4 \rightleftharpoons Mo^{IV}OL_2 + Mo^{VI}O_2L_2 \tag{13}$$

The porphyrin complex $MoO_2(tpp)$ also reacts with phosphines to form phosphine oxides[170] or with secondary alcohols to produce ketones.[171] The latter reaction appears to occur by one-electron (radical) steps showing the versatile reactivity of the Mo–oxo–porphyrin grouping.[171] A one-electron oxidation of $MoO_2(tpp)$ forms the radical cation which readily undergoes oxygen atom transfer to yield the Mo^{V} complex $MoO(tpp)^+$.[172] The reaction has some features in common with those of cytochrome P450.[172]

Reaction of $MoO_2(MeHNO)_2$ with heterocumulenes such as alkyl or aryl isocyanate yields[26] by equation (14) an adduct with a five-membered chelating ring. Related 'heterocumulenes' such as SCN⁻ and CS₂ yield similar structures (see below).

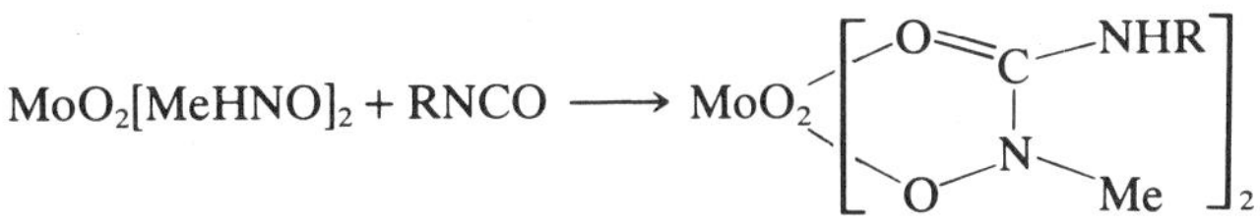

$$MoO_2[MeHNO]_2 + RNCO \longrightarrow MoO_2 \begin{bmatrix} O=C-NHR \\ | \qquad | \\ O-N-Me \end{bmatrix}_2 \qquad (14)$$

Reaction of $MoO_2(Et_2dtc)_2$ with $C_6H_{11}NCO$ forms the dinuclear $Mo_2O_3(Et_2dtc)_4$ complex. However, $MoO_2(Et_2dtc)_2$ or $MoO_2(acac)_2$ yields with ArNSO the oxo arylimido complexes $MoO(NAr)(Et_2dtc)_2$ and $MoO(NR)(acac)_2$, respectively.[130] The former compound has also been prepared by the reaction of $MoO(Et_2dtc)_2$ with ArN_3.[173]

The reactions[163] of equations (15) and (16) illustrate the substitution of a hydrazide(2−) ligand for an oxo ligand. Similar reactivity is seen for the complex $MoO_2(SCH_2CH_2NMeCH_2CH_2NMeCH_2CH_2S) = MoO_2L$ which yields $MoO(N_2R_2)L$ and $Mo(N_2R_2)_2L$.[174] Interestingly, the related complex, $MoO_2(SCH_2CH_2NMeCH_2CH_2CH_2-NMeCH_2CH_2S) = MoO_2L'$ reacts only with benzoylhydrazines to give $MoO(Ph-CONNH)_2L'$, a possible seven-coordinate MoO^{4+} core complex.[174]

$$MoO_2(R_2dtc)_2 + NH_2NHR \longrightarrow MoO(N_2HR)(R_2dtc)_2 \qquad (15)$$

$$MoO(N_2HR)(R_2dtc) + NH_2NHR \longrightarrow Mo(N_2HR)_2(R_2dtc)_2 \qquad (16)$$

Thermodynamic considerations indicate that the oxygen atom transfer reactions between $MoO_2(R_2dtc)_2$ and enzyme substrates should be largely irreversible.[175] For example, the hypothetical reaction of equation (17) has $\Delta H = -28.5 \pm 5.5 \, \text{kcal mol}^{-1} \, (= -119 \pm 23.0 \, \text{kJ mol}^{-1})$. The reaction should have a small ΔS and hence should have a negative ΔG and proceed spontaneously.[175] There are, however, complicating factors.

$$MoO_2(Et_2dtc)_2 + MeCHO \longrightarrow MoO(Et_2dtc)_2 + MeCO_2H \qquad (17)$$

The formal transfer of an oxygen atom is one way of describing the function of the Mo site in molybdoenzymes.[5] The formation of dinuclear reduction products is a complication that causes difficulty in trying to model the mononuclear site.[5,165] This difficulty can be overcome by the use of sterically demanding ligands that prevent the formation of the dinuclear complex.[73,79,125,176,177] For example, the cycle shown in Scheme 3 can be effected without dimerization. Further, in this case DMSO and the enzyme substrate biotin sulfoxide, can serve as the oxo donor to form the Mo^{VI} dioxo complex during the catalytic cycle.[79,177] The Mo^{VI} complex involved is discussed structurally above (Figure 7).

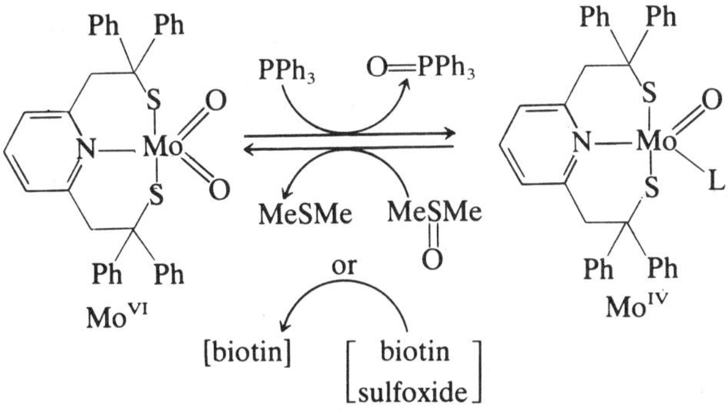

Scheme 3

$MoO_2(Cys-OEt)_2$ and other dioxo–Mo^{VI} complexes have also been studied as reactants toward enzyme substrates.[178,179,181,185] The original report of aldehyde oxidation to a carboxylic acid mimicking the action of the enzyme aldehyde oxidase[178] has not been substantiated,[179,181] Ligand, solvent and photochemical reactions are occurring[179] rather than the simple oxo transfer reaction postulated for the enzyme. Aldehydes react with $MoO_2(Cys-OEt)_2$ to form thiazolidines.[179] Reaction of aldehydes with $MoO_2(acac)_2$ or $MoO_2(Et_2dtc)_2$ require photochemical initiation and chlorinated solvents (CH_2Cl_2, $CHCl_3$) and are clearly not related to enzymic activity. Photolysis of $MoO_2(Et_2dtc)_2$[180] or MoO_2(dibenzoylmethanato)$_2$[182] leads to yet uncharacterized but interesting reduced Mo-containing products. $MoO_2(Et_2dtc)_2$ is reported to catalyze the reduction of NO_2^- to N_2O by HCO_2H in DMF.[183] This reaction involves the reduction of $MoO_2(Et_2dtc)_2$ by HCO_2H, giving $MoO(Et_2dtc)_2$,[183] a reaction that may be similar to the one catalyzed by formate dehydrogenase.

MoO_2(Cys-OMe)$_2$, $MoO_2(S_2CNEt_2)$ and other complexes are reported[99] to catalyze the air oxidation of benzoin to benzil. The complex MoO_2(Cys-OR)$_2$ (R = Me, Et, Pr^i and CH_2Ph) catalyzes the O_2 oxidation[128,184] of PPh_3 in DMF by a mechanistically involved process in which excess H_2O prevents the catalytic redox cycle.[184,185]

The electrochemistry of MoO_2^{2+} core complexes has been extensively studied.[65,73,106,136,126,151,153,149,174,186–193] Virtually all compounds studied show irreversible electrochemistry. However, one complex, $MoO_2(SCH_2CH_2NMeCH_2CH_2NMeCH_2CH_2S)$, displays a reversible, one-electron reduction wave.[174,106] The one-electron reduction product[194] has been characterized by EPR spectroscopy as $Mo^VO_2H(SCH_2CH_2NMeCH_2CH_2NMeCH_2CH_2S)^-$. This Mo^V complex displays proton super-hyperfine splitting reminiscent of that found in xanthine oxidase.[194]

The formation of MoO_2^{2+} complexes and their substitution, ligand exchange and rearrangement (fluxionality) reactions have been reviewed.[195]

36.5.5 COMPLEXES WITH MoO^{4+} CORES

The MoO^{4+} core structure is found in Mo^{VI} chemistry in a relatively small number of authenticated structures and in a few suggested examples shown in Table 3. The simplest set of complexes which potentially contains the MoO^{4+} core is $MoOX_4$(X = F, Cl, Br). While a square pyramidal structure seems probable for the chloro and bromo complexes,[232] the fluoro complex is polymeric in the solid state and in solution as well.[233–235]

Table 3 Complexes Containing MoO^{4+} Cores

Complex	*n(MO—O$_t$)* (cm^{-1})	*Mo—O$_t$* (Å)	*Ref.*
$MoOF_4^a$	1030	1.65(11)^a	232
$MoOCl_4$	1015		232
$MoOBr_4$	998		232
$MoOS_2(Pr^ndtc)_2$	917		240
$MoOF_2(Et_2dtc)_2$	945	1.701(4)	246
$MoOCl_2(Et_2dtc)_2$	947	1.701(4)	246
$MoOBr_2(Et_2dtc)_2$	960		246
$MoO(C_5H_{10}NO)_2(O_2C_6H_4)$	920	1.68(6)	247
$MoOCl_2$(acda)$_2$	955		136
MoO(abt)(acda)$_2$	940		136
MoO(MeNHO)$_2$[MeN(O)C(S)S]	928	1.689(6)	244
MoO(PhCONNH)[$SCH_2CH_2NMeCH_2CH_2CH_2NMeCH_2S$]	890		174
[$MoO(Et_2dtc)_3$]BF_4	935		239, 238
MoO(MeNHO)$_2$[HNC(S)NMeO]	926	1.677(8)	87
MoO(MeNHO)[HNC(S)NMeO][$H_2NC(S)N(Me)O$]		1.691(3)	87
$Cs_2[MoO(S_2)_2(SC(O)C(O)O)]$		1.660(8)	248

^{a} Polymeric solid.

The reaction of $MoO_2(R_2dtc)_2$ (R = Me, Et, Pr^n) with HF, HCl or HBr[246] eliminates H_2O and gives $MoOF_2(R_2dtc)_2$, $MoOCl_2(R_2dtc)_2$ and $MoOBr_2(R_2dtc)_2$, respectively. The structure of $MoOCl_2(Pr_2^ndtc)_2$, shown in Figure 11, reveals a seven-coordinate complex. An approximate pentagonal bipyramid is formed with an apical oxo and a *trans* chloro ligand. The four sulfur donors of the R_2dtc^- ligands and the second chloride ligand form the equatorial plane. An alternative route to $MoOCl_2(R_2dtc)_2$ involves oxidative addition of Cl_2 to $MoO(R_2dtc)_2$.[236] The ligand $C_5H_6NH_2CS_2^-$ = acda$^-$ [237] displays similar chemistry.[136]

The cation $MoO(R_2dtc)_3^+$ has been isolated in a number of salts[238,239] with BF_4^-, PF_6^-, ClO_4^-, $Mo_2O_4F_6^{2-}$ and $Mo_6O_{19}^{2-}$ as counterions. The structure of the $Mo_2O_4F_6^{2-}$ salt shows that the cation has a seven-coordinate pentagonal bipyramidal structure[238] with terminal oxo in an apical position. Proton NMR indicates this structure to be present in solution[239] although variable temperature work indicates significant fluxionality. The complex cation can be

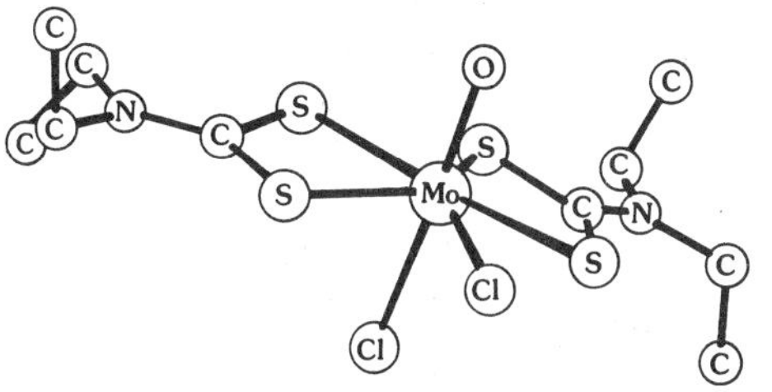

Figure 11 The structure of $MoOCl_2(Pr^n_2dtc)_2$[240]

prepared from $MoO_2(R_2dtc)_2$ and a hydrohalic acid. These reactions, although obviously complex, can give >50% yield of product with the remainder of the Mo presumably ending up as $Mo_6O_{19}^{2-}$. The characteristic Mo—O stretching band appears at ~935 cm^{-1} in the IR spectrum for $MoO(Et_2dtc)_3^+$.

The blue complexes $MoOS_2(R_2dtc)_2$ have been prepared by several routes.[240–242] The structure shown in Figure 12 again reveals a pentagonal bipyramid with S_2^{2-} in the equatorial plane, one R_2dtc^- ligand in the equatorial plane and the second R_2dtc^- ligand spanning axial and equatorial positions.

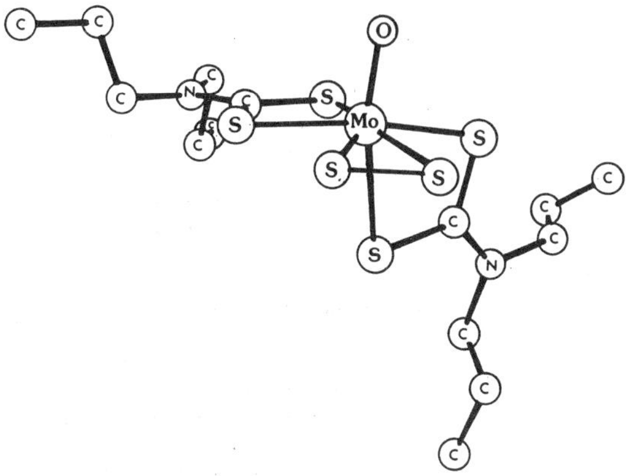

Figure 12 The structure of $MoOS_2(Pr^n_2dtc)_2$[240]

Among the reactions which lead to the $MoOS_2(R_2dtc)_2$ complex are those of equations (18) to (21).

$$MoO(R_2dtc)_2 + S_8 \longrightarrow MoOS_2(R_2dtc)_2 \qquad (18)^{50}$$

$$MoO_2(R_2dtc)_2 + H_2S \longrightarrow MoOS_2(R_2dtc)_2 \qquad (19)^{240,241}$$

$$MoO_2S_2^{2-} + R_2N\overset{\overset{\displaystyle S}{\|}}{C}—S—S—\overset{\overset{\displaystyle S}{\|}}{C}NR_2 \longrightarrow MoOS_2(R_2dtc)_2 \qquad (20)^{242}$$

$$Mo_2O_3(R_2dtc)_4 + Na_2S_4 \longrightarrow MoOS_2(R_2dtc)_3 \qquad (21)^{241}$$

The Mo^{IV} complex $MoO(R_2dtc)_2$ is a potential source of new MoO^{4+} complexes. Thus, reaction with S_8 gives $MoOS_2(R_2dtc)_2$[240] while reaction with Cl_2 gives $MoOCl_2(R_2dtc)_2$.[236] Reaction with $MeO_2CC{\equiv}CCO_2Me$ yields the acetylene adduct $MoO(R_2dtc)_2$-(MeO_2CCCCO_2Me) which, upon protonation with CF_3CO_2H, gives the complex $MoO(R_2dtc)_2$-($MeCO_2C{=}CHCO_2Me$)(O_2CCF_3) with coordinated trifluoroacetate[243] according to equation (22).

$$MoO(R_2dtc)_2(MeO_2CCCCO_2Me)_2 + CF_3CO_2H \longrightarrow MoO(R_2dtc)_2(MeO_2CC{=}CHCO_2Me)(O_2CCF_3) \qquad (22)$$

Complexes containing a single oxo and hydrazido(2−) ligand are discussed separately below.

The reaction of $MoO_2(MeNHO)_2$ in the presence of the hydroxylamine ligand and CS_2 proceeds according to Scheme 4.[244] The resultant formally seven-coordinate complex has a distorted pentagonal bipyramidal structure with the lone oxo group in the apical position. The five equatorial positions are occupied by two formally bidentate O,N hydroxylamido ligands of the starting material and the sulfur atom of the *N*-methyl-*N*-oxodithiocarbamate ligand. The oxygen of this ligand binds *trans* to Mo—O_t in the apical position with the O_t—Mo—O angle of 162.1°. As discussed below other heterocumulenes such as RCN, SCN^- and OCS are also reactive toward hydroxylamine to form related ligands and complexes.[244]

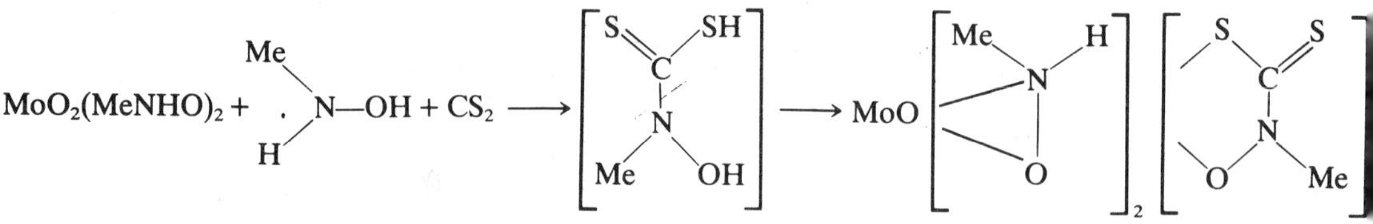

$MoO_2(MeNHO)_2$ +
Me
N—OH + CS_2
H
S
SH
C
N
Me
OH
MoO
Me
H
N
O
2
S
S
C
N
O
Me

Scheme 4

^{95}Mo NMR spectra of a representative sampling of MoO^{4+} complexes containing R_2dtc^- ligands reveal a relatively narrow distribution (55 to 183 p.p.m.) of chemical shifts. The complexes $MoO(R_2dtc)_3^+$ fall in the range 55–80 p.p.m. A combination of IR, ^{95}Mo and ^{17}O NMR spectra should prove powerful in characterizing MoO^{4+}-containing complexes in the future.

There is one reported example in Mo^{VI} chemistry of a MoS^{4+} and $MoSe^{4+}$ core. These are presumably found in $MoSF_4$[245] and $MoSeF_4$.[245] The band at 564 cm^{-1} in the former compound is assigned to the Mo—S stretch.[245]

36.5.6 NITRIDO COMPLEXES

The nitrido ligand N^{3-} is isoelectronic with the oxo ligand O^{2-}. Known nitrido complexes contain a single nitrido ligand and are often structurally related to analogous oxo complexes. However, unlike the situation for oxo complexes, there are as yet no examples of dinitrido or trinitrido complexes. The majority of known nitrido complexes are mononuclear and can be five-, six- or seven-coordinate. Such complexes possess square pyramidal, distorted octahedral or pentagonal bipyramidal structures, respectively. Structural results are given in Table 4 where the Mo—N distance can be seen to be extremely short, even shorter than Mo—O distances. This is due to stronger multiple bonding between Mo and N compared to O. For similar reasons ν(Mo—N) is usually above 1000 cm^{-1}, whereas ν(Mo—O_t) is usually found below 1000 cm^{-1}. The ability of the N^{3-} ligand to σ- and π-bond and hence donate very strongly to Mo may be the reason for the absence of multinitrido or oxo-nitrido complexes of Mo^{VI}.

Table 4 Nitrido Mo^{VI} Complexes

Complex	*Mo—N* (Å)	ν(*Mo—N*) (cm^{-1})	*Ref.*
$MoNCl_3(OPPh_3)_2$		1030	253
$MoN(N_3)_2Cl(terpy)$	1.662(7)		260
$MoN(N_3)_3(bipy)$	1.642(2)	948	261
$MoNCl_3(bipy)$		1010	252, 253
$MoNCl_3(OPPh_3)_2$		1020	253
$MoN(N_3)_3(py)$	1.643(3)		249, 251
$[AsPh_4][MoN(N_3)_4]$	1.630(6)	1040	250
$[AsPh_4][MoNF_4]$	[1.83(4)]	[969]	263
$[AsPh_4][MoNCl_4]$	1.66(4)	1054	264
$[PPh_4][MoNCl_4]$	1.637(4)		265
$[PPh_4][MoNBr_4]$	1.63(2)	1060	266
$[AsPh_4]_2[MoNCl_3(PO_2Cl_2)_2]$		1058	255
$MoN(Me_2dtc)_3$	1.641(9)	1019	267
$[MoN(tpp)]Br_3$		1014	254
$[MoNCl_3]_4$	1.638(11)		268
$[MoNCl_3{\cdot}OPCl_3]_4$	1.659(5) [2.150(5)]	1042	269
$[MoNBr_3]_4$		1025	257
$\{[MoN(Et_2dtc)_3]_2Mo(Et_2dtc)_3\}(PF_6)_3$	1.65(1) 1.66(1) 2.12(2) 2.14(2)		259

Structurally, the Mo nitride bond exerts a significant *trans* influence. In the extreme this leads to empty sites *trans* to Mo—N$_t$ such as in $MoNCl_4^-$ or $MoN(N_3)_3(py)$.[249] In the octahedral complex, $MoN(N_3)_3(bipy)$ the nitrogen of the bipy ligand *trans* to Mo—N$_t$ is 2.42 Å while the N of the same ligand *cis* to Mo—N$_t$ is 2.24 Å.[251] In the seven-coordinate pentagonal bipyramidal complex $MoN(Me_2dtc)_3$ the S *trans* to Mo—O$_t$ has Mo—*S* = 2.85 Å while S in the same ligand *cis* to Mo—N (in the equatorial plane) has Mo—S = 2.51 Å. Clearly, Mo—N$_t$ has a structural *trans* effect comparable to or greater than that of oxo in analogous complexes.

The most common preparative route for Mo^{VI} nitrido complexes involves the reaction of azides such as ClN_3 or Me_3SiN_3 with lower valent Mo complexes.[252] The resultant azido complexes are often unstable and eliminate N_2 to form the nitrido complex as shown in equations (23) and (24) or equation (25).

$$MoCl_5 + ClN_3 \longrightarrow MoCl_5N_3 + \tfrac{1}{2}Cl_2 \quad (23)$$

$$MoCl_5N_3 \longrightarrow MoNCl_3 + N_2 + Cl_2 \quad (24)$$

$$MoCl_4(MeCN)_2 + Me_3SiN_3 + bipy \longrightarrow MoNCl_3(bipy) + N_2 + 2MeCN + Me_3SiCl \quad (25)$$

Once the nitrido complexes are formed they are reactive toward ligand substitution as in equations (26) and (27). Many complexes have been prepared exploiting this substitutional reactivity,[251–257]

$$MoNCl_4^- + bipy \longrightarrow MoNCl_3(bipy) + Cl^- \quad (26)$$

$$MoNBr_3 + H_2tpp \longrightarrow [MoN(tpp)^+]Br_3^- + H_2 \quad (27)$$

The nitrido nitrogen is reactive nucleophilically leading in some cases to di- or tetra-nuclear complexes in which nitrido binds to a second molybdenum as a weak ligand. For example, the structure of $[MoNCl_3(OPCl_3)]_4$ is illustrated in Figure 13. The complex $MoN(R_2dtc)_3$ reacts with S_8 or propylene sulfide to give the thionitrosyl complex $MoNS(R_2dtc)_3$.[258]

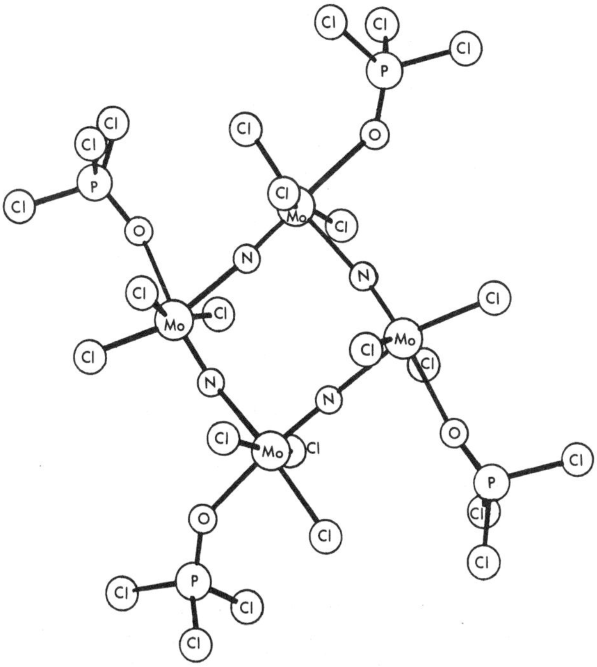

Figure 13 The structure of $[MoNCl_3(OPCl_3)]_4$[269]

Reaction of $MoN(Et_2dtc)_3$ with NH_2OSO_3H (in an attempt to make a hydrazido complex) yields an unusual trinuclear species.[259] The diamagnetic tricationic brown complex $[MoN(Et_2dtc)_3]_2Mo(Et_2dtc)_3$ formed[259] has been shown by X-ray studies to have two bridging nitrido ligands such that the central Mo is eight-coordinate (dodecahedral) while the terminal Mo atoms have pentagonal bipyramidal seven-coordinate structures. The Mo—N distances to terminal Mo atoms are comparable to those in $MoN(Me_2dtc)_3$ at 1.65(1) and 1.66(1) but the distances to the central Mo are significantly longer at 2.12 and 2.14 Å. The bridging is reminiscent of that found in $[MoNCl_3]_4$ and $[MoNCl_3(OPCl_3)]_4$.

36.5.7 IMIDO AND ARYL- AND ALKYL-IMIDO COMPLEXES

Imido (NH^{2-}) and arylimido (NR^{2-}) complexes have been prepared for Mo^{VI} as well as for lower oxidation states of Mo. The alternative formulation for the ligand is that of a nitrene wherein the ligand is neutral. This, of course, makes the formal oxidation state of the metal two units lower for each ligand compared to the corresponding arylimido formulation. We prefer the arylimido formulation as the resultant complexes all have analogs in the chemistry of oxo Mo^{VI} species and do not have analogs in lower oxidation state Mo chemistry.

The lone structurally characterized imido complex is $MoO(NH)Cl_2(OPEtPh_2)_2$.[270] Here the oxo and imido ligands are *cis* to each other with Mo—O = 1.66 and Mo—N = 1.70 Å, respectively. The N—H hydrogen has been located and the Mo—N—H angle is 157°. The orange-yellow complex was prepared by the reaction of $MoOCl_3$ with Me_3SiN_3 followed by the addition of the tertiary phosphine oxide.[270]

The arylimido complexes are prepared by the reaction of an arylazide with an appropriate Mo starting material. For example, the reaction of $Mo(CO)_2(R_2dtc)_2$ with phenyl azide[271] yields the complex $Mo(NPh)_2(R_2dtc)_2$ whose structure is shown in Figure 14. The structure is clearly analogous to that of $MoO_2(R_2dtc)_2$ with *cis* phenylimido ligands in place of the *cis* oxo ligands. The two phenylimido ligands differ significantly in their Mo—N—C angles and Mo—N bond lengths. These correspond to 139.4° and 1.789 Å, respectively, for one ligand and 169.4° and 1.754 Å, respectively, for the other. Interestingly, the phenylimido ligand does not appear to exert a significant *trans* influence in this or other complexes. The reaction of $MoO(R_2dtc)_2$ with phenyl azide gives the mixed oxo–phenylimido complex $MoO(NPh)(R_2dtc)_2$.[272] These complexes contain two oxo-equivalent ligands (i.e. one oxo and one arylimido) and approximate octahedral coordination geometries. In contrast to these complexes is the *p*-tolylimido complex $Mo(N\text{-}p\text{-}C_6H_4Me)Cl_4(THF)$[273] which has a single oxo-analog ligand and an octahedral six-coordination.

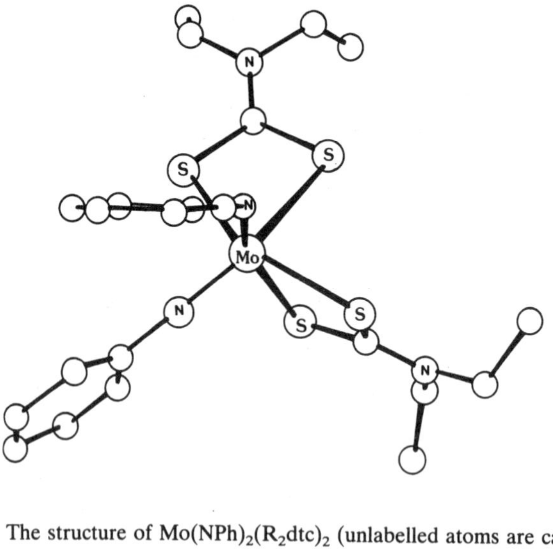

Figure 14 The structure of $Mo(NPh)_2(R_2dtc)_2$ (unlabelled atoms are carbon)[271]

Other aryl imido complexes seem to resemble more closely the seven-coordinate pentagonal bipyramidal complexes with a single oxo-equivalent ligand (in this case an arylimido or alkylimido ligand). The complex cation $Mo(NMe)(R_2dtc)_3^+$[274] is closely related both to the nitrido complex $MoN(R_2dtc)_3$ and to the oxo complex $MoO(R_2dtc)_3^+$. As anticipated, the Mo—N bond distance is, at 1.73 Å, significantly longer in the arylimido compared to either the corresponding nitrido or oxo complex. The complex $Mo(NPh)Cl_2(R_2dtc)_2$[275,272] has the pentagonal bipyramidal structure found for its oxo analog $MoOCl_2(R_2dtc)_2$.[246]

36.5.8 HYDRAZIDO(2−) COMPLEXES

Hydrazine is one of the suggested intermediates on the pathway of the reduction of dinitrogen.[276] Hydrazine can serve as a conventional ligand using the lone pairs on each N atom as donors to a transition metal. However, hydrazine can also be deprotonated by one or two steps to produce monoanionic and dianionic ligands, respectively. The double deprotonation of N_2H_4 leads to the hydrazido(2−) ion which, in its disubstituted manifestation $R_2N_2^{2-}$, is a ligand to a significant number of Mo^{VI} complexes. A majority of these have exact analogs in the chemistry of simple oxo complexes and so hydrazido(2−) joins the ranks of ligands that are oxo-like in character. In a significant number of cases a hydrazido(2−) complex can be prepared directly from the corresponding oxo complex by simply refluxing the oxo complex and the corresponding hydrazine in an appropriate solvent.

For example, reaction of $MoO_2(SCH_2CH_2NMeCH_2CH_2NMeCH_2CH_2S)$ with NH_2NPh_2 yields the hydrazido(2−) complex $MoO(N_2Ph_2)(SCH_2CH_2NMeCH_2CH_2NMeCH_2CH_2S)$ whose structure[277,278] is shown in Figure 15. Note the placement of the hydrazido(2−) ligand *cis* to oxo, the relatively short Mo—N distance of 1.778(3), and the O—Mo—N angle of 105.9(1)°. These features are reminiscent of the starting *cis* dioxo complex. The near linear Mo—N—N angle of 172.9(2)° and the N—N distance of 1.309 reveals some multiple binding between the N atoms of the hydrazide(2−) ligand.[277] The Mo—N distance at 2.464(3) *trans* to oxo and 2.359(3) *trans* to hydrazido(2−) reveal that the oxo group has a stronger *trans* bond lengthening influence than does the hydrazido(2−) ligand.[277] In most cases the Mo—N—N angle is near linear. A compilation of key distances in hydrazido(2−) complexes is given in Table 5.

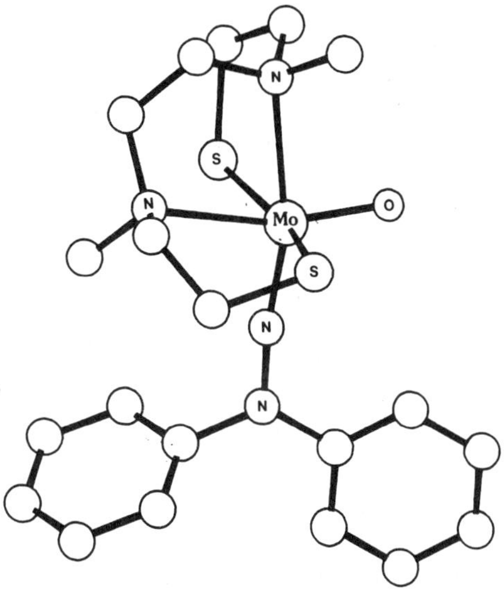

Figure 15 The structure of $MoO(N_2Ph_2)(SCH_2CH_2NMeCH_2CH_2NMeCH_2CH_2S)$ (unlabelled atoms are carbon)[277]

Table 5 Hydrazido(2−) Complexes of Mo^{VI}

Complex	*Mo—N* (Å)	*N—N* (Å)	*Mo—O* (Å)	*Ref.*
$\{Mo(NNPhEt)[(CH_2)_5dtc]_3\}BPh_4$	1.715(16)	1.37(2)	—	280
$Mo(NNMe_2)(SCH_2CH_2PPhCH_2CH_2S)_2$	1.775(6)	1.37	—	281
$[Mo(NNMePh)(NHNMePh)(Me_2dtc)_2]BPh_4$	1.752(10)	1.285(14)	—	279
$Mo(NNPh_2)_2(Me_2dtc)_2$	1.790(8)	1.31(1)	—	282
$Mo(NNMePh)_2(Me_2dtc)_2$	1.790(9)	1.30(1)	—	282
$[MoCl(NNMe_2)_2(PPh_3)_2]Cl$	1.763(2)	1.291(7)	—	282
	1.752(5)	1.276(8)	—	
$[Mo(NNMe_2)_2(bipy)_2](BPh_4)$	1.80(1)	1.27(3)	—	282
	1.79(1)	1.28(2)	—	
$MoO(NNMe_2)(Me_2dtc)_2$	1.85(2)	1.24(2)	—	283
$MoO(NNMe_2)(oxine)_2$	1.800(9)	1.28(1)	1.671(9)	284
$MoO(NNPh_2)(SCH_2CH_2NMeCH_2CH_2NMeCH_2CH_2S)_2$	1.778(3)	1.309(4)	1.696(2)	277
$[PPh_4][MoO(NNMe_2)(SPh)_3]\cdot Et_2O$	1.821(9)	1.292(14)	1.705(8)	285

Reaction of the Mo^V complex $MoO(SPh)_4^-$ with H_2NNMe_2 leads to $MoO(N_2Me_2)(SPh)_3$.[285] This complex anion has a distorted square pyramidal structure in the $[PPh_4][MoO(N_2Me_2)(SPh)_3]\cdot Et_2O$ salt with the oxo ligand in the apical position.[285] The complex has an unusually small Mo—N—N angle of 152.5(10)° in contrast to the near linear linkage Mo—N—N = 169°–180° found in all other complexes in Table 5.

The oxo hydrazido complex $MoO(N_2Me_2)(oxine)_2$ reacts with HX (X = Cl, Br, SPh, ½cat, ½tdt) to give $MoO(N_2Me_2)X_2(oxine)_2$.[284] This reactivity is analogous to that of the dioxo complexes reacting with HX to give the corresponding seven-coordinate oxo dihalogeno complexes.[246]

The bis[hydrazido(2−)] complex $Mo(NNMePh)_2(Me_2dtc)_2$ reacts with one equivalent of HCl to produce the cationic complex $Mo(NNMePh)(NHNMePh)(Me_2dtc)_2^+$ which contains one near-linear hydrazido(2−) and one hydrazido(1−) ligand bound to Mo through both N atoms.[279] The side-on bound hydrazido(1−) ligand has Mo—N = 2.069(8) and 2.185(9) and N—N = 1.388(12) Å. Other side-on bound hydrazido complexes are known.[262]

36.5.9 PEROXO COMPLEXES

Early work on peroxo compounds of molybdenum has been reviewed.[286] The stoichiometrically simplest Mo peroxo complex is the red-brown $Mo(O_2)_4^{2-}$ ion formed by the reaction of MoO_4^{2-} with H_2O_2. Although the complex is not stable in solution and decomposes slowly with the evolution of O_2, the anion can be crystallized as the $Zn(NH_3)_4^{2+}$ salt whose structure has been determined. In a sense, $Mo(O_2)_4^{2-}$ is an intermediate in the thermodynamically favored decomposition of hydrogen peroxide to water and oxygen.

As discussed above, the peroxido ligand, O_2^{2-}, is capable of substituting for the oxo ligand, O^{2-}, such that there is no gross change in geometry of the resultant complex. There is no more striking example of this than $Mo(O_2)_4^{2-}$. The structure of the anion in $[Zn(NH_3)_4][Mo(O_2)_4]$ is shown in Figure 16. It consists of a D_{2d} dodecahedron of eight oxygen atoms about Mo.[287] However, if the centroid of each O—O bond is taken, the four points describe a geometric figure close to the tetrahedron, which is, of course, the geometry of MoO_4^{2-}.

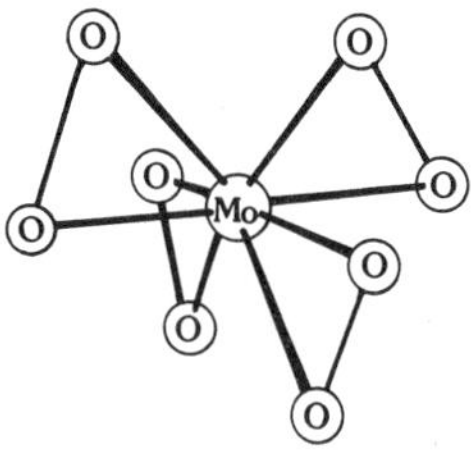

Figure 16 The structure of $Mo(O_2)_4^{2-}$ in $[Zn(NH_3)_4][Mo(O_2)_4]$[287]

A substantial number of Mo^{VI} peroxo complexes has been structurally characterized. These complexes all have nearly symmetrically side-on-bound peroxide. Among them are mononuclear, dinuclear, tetranuclear and heptanuclear compounds. The tetranuclear and heptanuclear complexes are related to the isopolymolybdates. Structural studies have identified $K_6[Mo_7O_{22}(O_2)_2]\cdot 8H_2O$ (O—O = 1.38 Å),[288] $K_5[HMo_7O_{22}(O_2)_2(H_2O)_6]$ (O—O = 1.38 Å)[288,289] and $K_4[Mo_4O_{12}(O_2)_2]$(O—O = 1.48 Å).[289]

The designation of coordination number in peroxo complexes is equivocal since a side-on-bound peroxide can be considered to occupy either one or two coordination sites. For the purpose of the following discussion we initially consider that O_2^{2-} occupies two coordination sites when bound side-on.

36.5.9.1 Mononuclear Peroxo Complexes

Mononuclear peroxo compounds of known crystal structure are listed in Table 6. Except for $Mo(O_2)_4^{2-}$ and $Mo(O_2)_2(ttp)$, all structurally characterized mononuclear peroxo complexes have one Mo—O_t in addition to one or two peroxo ligands. The Mo—O_t bond lengths fall in the range 1.63 to 1.73 Å. The Mo—O (peroxo) distances span from 1.83 to 1.96 Å. The peroxo ligand is always found *cis* to the oxo group and two peroxo ligands are always found *cis* to each other. In mixed oxo-peroxo complexes, the plane determined by the Mo—O—O ring is nearly perpendicular to the Mo—O_t direction. The overall coordination geometry is often describable as a pentagonal bipyramid with the oxo ligand in an apical position and peroxo groups in the equatorial plane.

Table 6 Structural Details for Mononuclear Peroxo Complexes

Complex	*O—O* (Å)	*Mo—O(peroxo)* (Å)	*Mo—O(oxo)* (Å)	*Ref.*
$[Zn(NH_3)_4][Mo(O_2)_4]$	1.55(5)	2.00(2) 1.93(3)		287
$(NH_4)_3[MoO(O_2)F_4]F$	1.36(3)			301
$K_2[MoO(O_2)F_4]\cdot H_2O$	1.44(3)	1.93(2) 1.94(2)	1.64(3)	332
$[MoO(O_2)_2(H_2O)_2]\cdot C_{12}H_{24}O_6\cdot H_2O$	1.445(6)	1.900(5) 1.920(5)	1.647(5)	290
$K_2[MoO(O_2)_2C_2O_4]$	1.47(2) 1.44(2)	1.95(2) 1.93(2)	1.68(1)	333, 292
$MoO(O_2)_2(Gly)H_2O$	1.464(1) 1.481(1)	1.947(1) 1.962(1) 1.943(1) 1.932(1)	1.680(1)	238
$MoO(O_2)_2(Pro)H_2O$	1.46–1.48	1.92–1.96	1.673(1)	238
$MoO(O_2)_2(O_7C_6H_6{}^a)H_2O$	1.482(12)	1.926(9) 1.975(8)	1.655(8)	293
$MoO(O_2)_2[OP(NMe_2)_3]H_2O$	1.478(3) 1.4748	1.929(5) 1.952(5) 1.952(5) 1.935	1.662(5)	308
$MoO(O_2)_2[OP(NMe_2)_3]py$	1.44(2)	1.94(1) 2.04(1)	1.66(1)	308
$MoO(O_2)_2(phen)$	1.459(6)	1.948(4) 1.912(5) 1.908(4) 1.953(4)	1.682(4)	344
$MoO(O_2)_2[(S)\text{-}MeCH(OH)CONMe_2]^b$	1.459(6)	1.935(5)	1.671(5)	335
$MoO(O_2)[PhN(O)C(O)Ph]_2$	1.212(5)	1.830(3)	1.733(2)	300
$MoO(O_2)Cl(C_5H_4NCO_2)[OP(NMe_2)_3]$	1.414(5)	1.916(3) 1.927(3)	1.663(3)	336
$H[Mo(O)(O_2)_2(C_5N_4NCO_2)]\cdot 2C_5H_4NCO_2H\cdot H_2O$	1.447(8)	1.907(5) 1.912(5)	1.670(5)	337
$Mo(O_2)_2[ttp]$	1.399(6)	1.958(4)		294

[a] The ligand is citrate bound through a single carboxylate oxygen and alcohol. [b] The chiral ligand is (*S*)-*N*,*N*-dimethyllactamide.

The simplest diperoxo structure is that of $MoO(O_2)_2(H_2O)_2$ which cocrystallizes with 18-crown-6 and H_2O.[290] In this pentagonal bipyramidal structure shown in Figure 17 the H_2O ligands occupy equatorial and axial sites at Mo—O_w = 2.325 and 2.084 Å, respectively. Interestingly, this product was formed from the reaction of MoO_3, THF and oxygen. The source of the reducing equivalents necessary to form the peroxo complex was not specified but is likely to be the THF molecule.[291] MoO_3 appears to serve as a free radical initiator causing THF and O_2 to form α-hydroperoxytetrahydrofuran, which leads in acid to H_2O_2.[291]

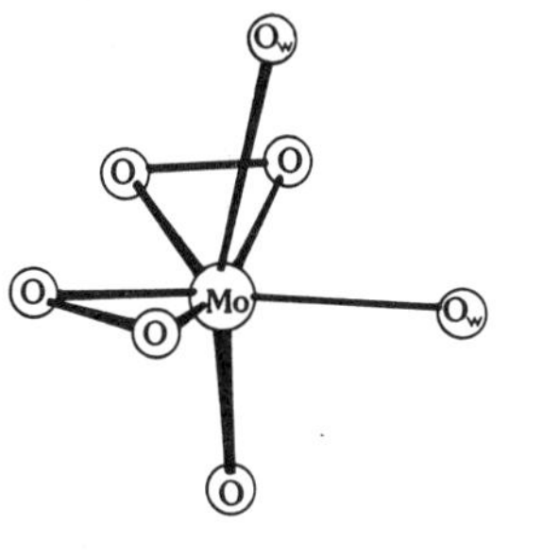

Figure 17 The structure of $MoO(O_2)_2(H_2O)_2$ (O_w is an oxygen atom from a water ligand)[290]

The structure of $MoO(O_2)_2(phen)$ is formally related to that of $MoO(O_2)_2(H_2O)_2$ by substitution of two water ligands by phen. Again the oxo ligand occupies one of the apical positions of the pentagonal bipyramid and the peroxo ligands occupy four of the five positions in the equatorial plane. The phen ligand spans the axial–equatorial edge. The structure of the yellow $MoO(O_2)_2(C_2O_4)^{2-}$ complex was recently redetermined[292] and is shown in Figure 18. It is representative in having a distorted pentagonal bipyramid around Mo with an oxo ligand in the apical position and the peroxo ligands and one O donor of the bidentate oxalate bound in the equatorial plane. The Mo—O stretch is identified at 970 cm^{-1} and ~865 cm^{-1} approximates the O—O stretching vibrational frequency.

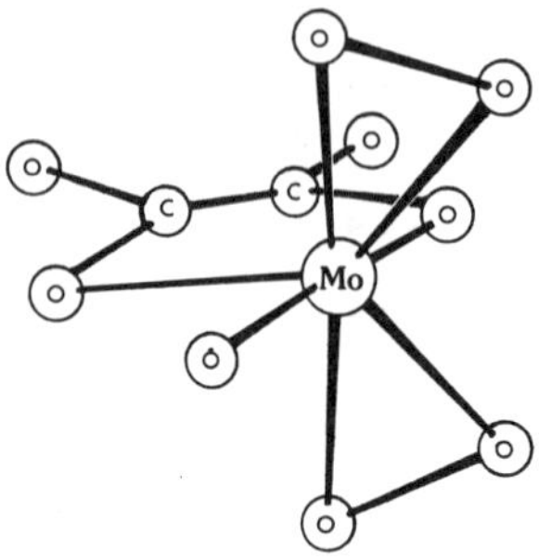

Figure 18 The structure of $MoO(O_2)_2(C_2O_4)^{2-}$ in its K^+ salt[292]

Interestingly, $MoO(O_2)_2(C_2O_4)^{2-}$ can be prepared from an aqueous solution of MoO_3 and H_2O_2 in which the only carbon source is malonate, $^-O_2CCH_2CO_2^-$ or malate, $^-O_2CCHOHCH_2CO_2^-$, but not succinate $^-O_2CCH_2CH_2CO_2^-$. The formation of the degraded two-carbon oxalate complex is proposed[292] to require a five- or six-membered ring chelated intermediate for the oxidation of the organic ligand to oxalate to occur. However, the isolation and structure determination of the apparently stable $MoO(O_2)_2(citrate)^{2-}$ complex[293] which has a five-membered chelate ring must be noted.

The structures of $MoO(O_2)_2(Pro)(H_2O)$[238] and $MoO(O_2)_2(Gly)(H_2O)$[238] also show the pentagonal bipyramidal geometry familiar to this class. The glycine is bound as a monodentate ligand through its carboxylate group in an equatorial position. The water ligand is found *trans* to the terminal oxo group. Although this is the first known amino acid peroxy molybdenum complex, its biological relevance is uncertain as Mo peroxy complexes are not known to play a biological role. The $Mo—O_t$ stretch is found at 970–978 while the O—O stretch appears in the 870–880 cm^{-1} region in the IR spectrum for these complexes.

Despite the utility of the bidentate formulation of O_2^{2-} for descriptive purposes, structural systematics of peroxo complexes are nicely understood by considering the O_2^{2-} ligand as an isomorphous replacement for an oxo ligand. As discussed previously, if the four oxo groups in MoO_4^{2-} are replaced by properly oriented peroxo ligands, the $Mo(O_2)_4^{2-}$ structure quite nearly obtains. In the monooxo, monoperoxo complexes the *cis* arrangement of the oxo and peroxo ligands is reminiscent of the *cis* dioxo structure. Moreover, the orientation of the O—O bond allows it to be a π donor analogously to an oxo ligand. The analogy weakens somewhat for the oxo–diperoxo complexes. Although the oxo and peroxo groups are mutually *cis* as in MoO_3 complexes, there are only two additional coordination sites filled, making the structures analogous to five-coordinate trigonal bipyramidal rather than the more common six-coordinate octahedral MoO_3 core compounds.

The unusual complex $Mo(O_2)_2(ttp)$ (ttp = tetra-*p*-tolylporphyrin) has been synthesized and structurally characterized.[294] As shown in Figure 19, the peroxo ligands are bound side-on on opposite sides of the planar heme group. The two O—O vectors eclipse a perpendicular pair of *trans* N—Mo—N linkages. The O—O distance of 1.40 Å is marginally shorter than that found in the other peroxo complexes. This is perhaps not surprising insofar as the porphyrin may participate significantly in delocalizing the charge placed on Mo by donation from O_2^{2-} π levels. This delocalization allows further $O_2^{2-} \rightarrow Mo$ donation and concomitant slight lengthening of the O—O bonds. The diperoxo porphyrin complex is photochemically active,[295] ejecting an O_2 ligand upon irradiation with a tungsten lamp to produce the *cis* dioxo complex $MoO_2(ttp)$. Interestingly, $Mo(O_2)_2(ttp)$ is thermally stable and reacts neither with cyclohexene or triphenylphosphine.[295] This complex is, however, electrochemically reactive displaying one quasireversible one-electron oxidation and two quasireversible one-electron reduction waves.[296]

The oxidation may involve extensive porphyrin contribution as a dramatic optical spectroscopic change is observed, reminiscent of porphyrin cation radical formation. Further, the EPR spectrum of the oxidized complex, presumably $Mo(O_2)_2(ttp)^+$, generated *in situ,* is devoid of observable $^{95,97}Mo$ hyperfine splitting. In contrast, the first one-electron reduction leads to an EPR active species with $g = 1.980$ and $^{95,97}Mo$ splitting characteristic of Mo^V.[296] Interestingly, neither redox process involves the peroxo ligands, a point to be pondered in attempting to understand the lack of reactivity of the ligands in this complex compared to other peroxo Mo^{VI} complexes. The complex $MoO(O_2)ttp$ reveals once again the oxo/peroxo interchangeability in the Mo coordination sphere.[171]

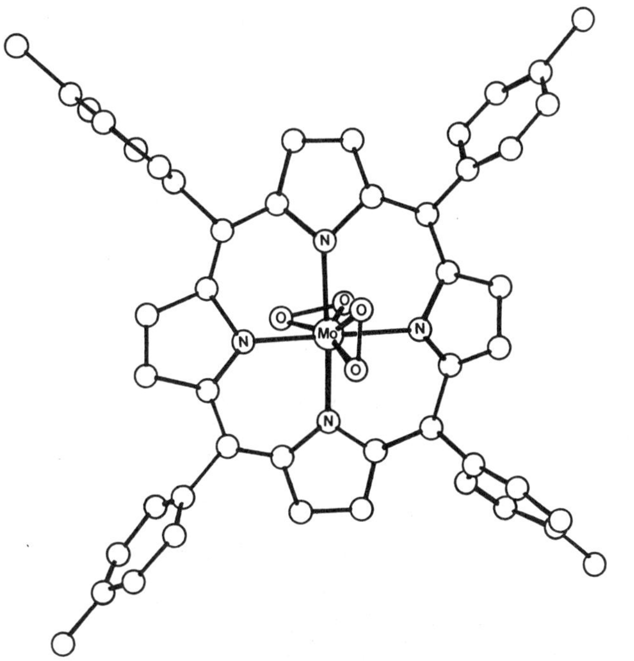

Figure 19 The structure of $Mo(O_2)_2(ttp)$ (unlabelled atoms are carbon)[294]

Structurally yet uncharacterized peroxo complexes include $MoO(O_2)(dtbc)_2$[297] ($dtbc^- = 3,5$-di-*tert*-butylcatecholato) prepared by reaction of $Mo_2O_2(dtbc)_2$ with $[NMe_4^+][O_2^-]$ and isolated as the $(Me_4N)_2[MoO(O_2)(dtbc)_2]\cdot 3DMSO$ salt. The reaction presumably involves disproportionation of O_2^- according to equation (28). The first step may involve O_2^- acting as a nucleophile followed by electron transfer from free O_2^- to bound O_2. $Mo_2O_2(dtbc)_2$ is known to be reactive to donor organic solvents.[297] The ν(Mo—O) at $\sim 950\ cm^{-1}$ is clearly distinguished from a band at $\sim 870\ cm^{-1}$ assigned to the ν(O—O) characteristic of side-on bound O_2^{2-}.

$$Mo_2O_2(dtbc)_2 + 4O_2^- \longrightarrow 2MoO(O_2)(dtbc)_2^{2-} + 2O_2 \tag{28}$$

Many complexes written as $MoO_5\cdot L_2$ or MoO_5L' (L' = bidentate ligand) undoubtedly contain the $MoO(O_2)_2$ unit. For example, L may be Me_3NO, Bu^n_3PO, Pr^n_3AsO, Ph_3AsO, C_5H_5NO, *etc.*[298] The analytical data and the observation of ν(Mo—O) at 950–970 and ν(O—O) at $\sim 850\ cm^{-1}$ leave little doubt about the formulation of these complexes.[298] The complexes stoichiometrically epoxidize alkenes or catalyze alkene epoxidation by *t*-butylhydroperoxide.[298]

$MoO(O_2)(oxinate)_2$ can be formed[299] from $MoO_2(oxinate)_2$ by treatment with H_2O_2. The *cis* dioxo and *cis* oxo peroxo complexes are electrochemically distinct with the latter converting to the former irreversibly upon electrochemical reduction.[299]

When O_2^{n-} is bound to a transition metal its formal oxidation state and that of the metal are ambiguous. In the compounds discussed above the Mo^{VI}–peroxide formulation is used. However, in the absence of structural data an Mo^{IV}–dioxygen or Mo^V–superoxide formulation is possible. The hypothetical reaction of Mo^{IV} and O_2 can be considered an oxidative addition, wherein the extent of charge transfer determines the proper formulation. In the complexes discussed here the O—O distance lies in the range 1.44 to 1.55 Å. Comparison of these distances with 1.44 Å in H_2O_2 and 1.21 Å in O_2 indicates that the Mo^{VI} peroxide formulation is quite appropriate for these compounds. Only $MoO(O_2)[PhN(O)C(O)Ph]_2$ with O—O distance 1.212(5) Å seems to be an exception.[300] The intermediate 1.36(3) Å distance reported[301] for $(NH_4)_3[MoO(O_2)F_4]$ may be inaccurate.

The electronic structure of $Mo(O_2)_4^{2-}$ has been studied using SCF-MS-Xα molecular orbital calculations.[302] The Mo^{VI}–peroxide formulation is appropriate as a configurational starting point for discussion of the bonding. However, there is significant lengthening (weakening) of the O—O bond in the complex compared to that in free O_2^{2-}, a feature that may be responsible for the reactivity of peroxo–molybdenum complexes in epoxidation reactions. Moreover, the significant electrophilic character of the bound O_2^{2-} (in the form of low-lying virtual σ-antibonding orbitals) accords well with the proclivity of the peroxo–Mo^{VI} complex to attack and epoxidize electron-rich alkenes.

36.5.9.2 Dinuclear Peroxo Complexes

Dinuclear complexes have been structurally characterized with a single bridging oxo or fluoride ligand or with a set of two bridging hydroperoxo ligands. The dinuclear ion $(H_2O)(O_2)_2OMoOMoO(O_2)_2(H_2O)^{2-}$ in its pyH^+ [303,304] and K^+ salts[305] has a non-linear monooxo bridge (136 and 148°, respectively). The bridging oxo is found to be *cis* to the Mo—O_t group on each Mo. The peroxo ligands on each Mo are *cis* to the Mo—O_t bonds as in the mononuclear compounds. The structure as illustrated in Figure 20 has two-fold symmetry and contains two pentagonal bipyramids bridged through an oxygen atom shared by their equatorial planes. It is clearly related to the $MoO(O_2)_2(H_2O)_2$ structure shown in Figure 17.

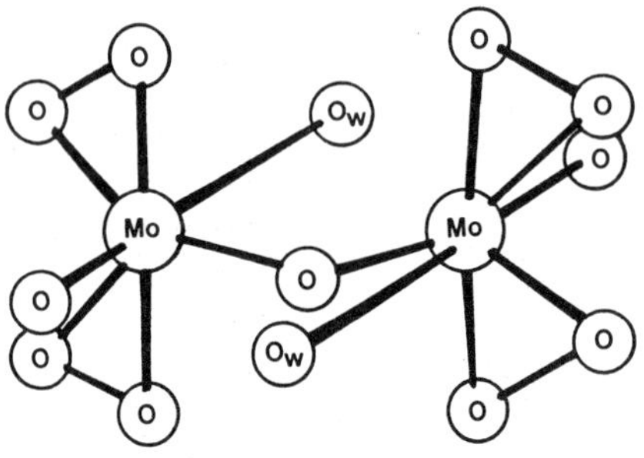

Figure 20 The structure of $(H_2O)(O_2)_2OMoOMoO(O_2)_2(H_2O)^{2-}$ in its K^+ salt (O_w is an oxygen atom from a water ligand)[305]

The unusual complex μ-F-$[MoO(O_2)(pydca)]_2^-$ (see **5**) is formed as the NEt_4^+ salt upon treatment of $Mo(O_2)(pydca)(OH_2)$ with F^- [418] in the presence of NEt_4^+. The complex, shown in Figure 21, is a centrosymmetric dinuclear ion with each Mo atom possessing pentagonal bipyramidal seven-coordination.[307] The linear $O_tMo^{VI}FMo^{VI}O_t$ unit is to date unique to this complex. The Mo—O_t distance is 1.659(3) Å and the Mo—F *trans* to it has the very long distance of 2.135(1) Å. The O—O distance of 1.43 Å is similar to that found in mononuclear peroxo compounds of Mo^{VI}.

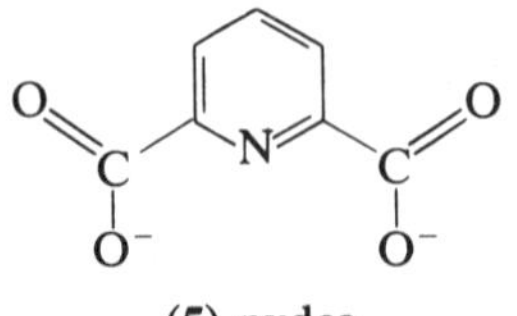

(5) pydca

The structure of $(pyH)_2[(O_2)_2MoO(O_2H)_2MoO(O_2)_2]$[304,308] shown in Figure 22 has two bridging hydroperoxo ligands. Each Mo coordination sphere is a heptacoordinate pentagonal bipyramid. Sharing of an axial–equatorial edge gives a centrosymmetric structure. The peroxo ligands occupy the remaining equatorial sites while the remaining axial position on each Mo contains a terminal oxo ligand. The Mo—O_b distance is 2.05 Å. The bridging and non-bridging peroxo groups both have distances near 1.47 Å consistent with their peroxide formulation.

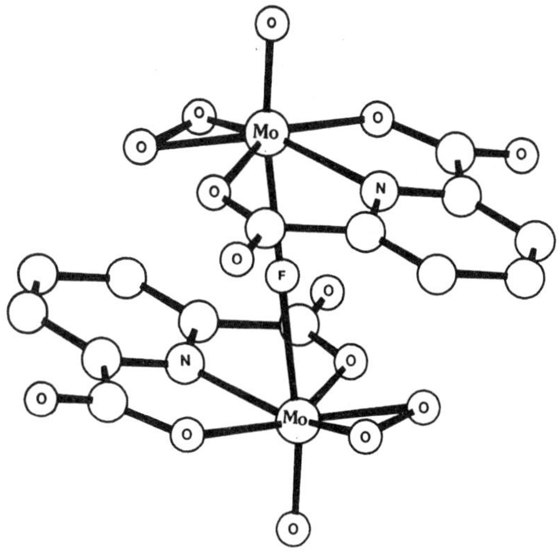

Figure 21 The structure of μ-F[$MoO(O_2)(pydca)]_2^-$ in its NEt_4^+ salt (unlabelled atoms are carbon)[418]

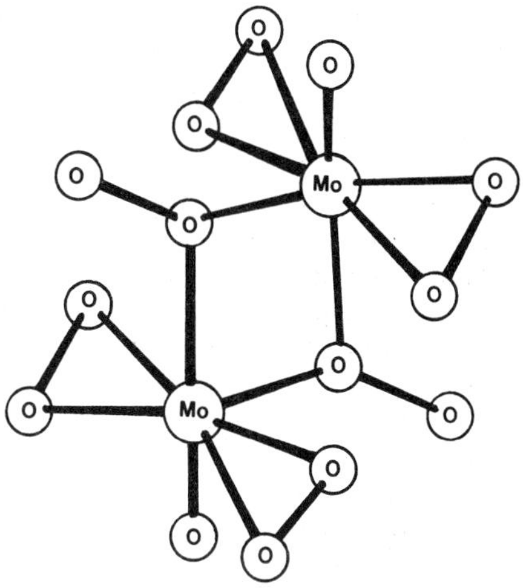

Figure 22 The structure of $(O_2)_2MoO(O_2H)_2MoO(O_2)^{2-}$ in its pyH^+ salt (terminal H atoms on bridging hydroperoxide are not shown)[304,308]

36.5.9.3 Spectral Indicators

Spectroscopically, in the IR region the O—O stretch can be identified between 845 and 895 cm^{-1}. ^{17}O NMR has proven useful insofar as the oxo and peroxo ligands are clearly distinguishable.[309] For example, in $MoO(O_2)(CN)_4^{2-}$ the peroxo signal (although broad) appears at 487 p.p.m., while the oxo signal appears at 705 p.p.m. downfield from H_2O. The latter value is well within the range found for oxo-molybdenum species.[105,310] The peroxo ^{17}O chemical shift in $MoO(O_2)_2(HMPA)$ at ~450 p.p.m. with oxo at 850 p.p.m.[311] again illustrates that oxo and peroxo are readily distinguished. Further, in this latter example oxo and peroxo are shown not to exchange their ^{17}O labels. Thus, ^{17}O NMR could prove a powerful tool in elucidation of reaction pathways in oxo–peroxo complexes. Although oxo–peroxo exchange is indeed slow on the NMR time scale, it can occur at a significant rate.[312] Indeed, Mo^{VI} compounds are reported to catalyze the exchange of ^{18}O between H_2O and H_2O_2.[312]

36.5.9.4 Reactivity Toward Organic Molecules

Peroxomolybdenum complexes have been used stoichiometrically to effect a variety of organic conversions and catalytically, usually with *t*-butylhydroperoxide or H_2O_2, to effect similar reactions.[313] The Mo^{VI} complexes have been anchored on polymers[137] and used under phase-transfer conditions.[314,315] Selective and asymmetric reactions have been reported.[272,316] We cannot even partially review this area which is part of a general thrust in reaction chemistry of early transition metal facilitated or catalyzed peroxidation/epoxidation/oxidation chemistry.[313,316,317] Reactions catalyzed or carried out include alkenes to epoxides,[156,215,318–322] sulfides to sulfoxides,[323] secondary alcohols to ketones,[316,324] aldehydes to carboxylic acids,[316] azo benzenes to azoxy benzenes,[325] N-heterocyclics to *N*-oxides,[326] amides to hydroxamates[154] and more complex reactions.[55,327–330] Theoretical studies offer potential mechanistic insights into reactivity of the Mo—O_2 group.[331]

There is no evidence that dioxygen or peroxides interact with Mo in enzymes in physiologically relevant processes. The evidence is strong that in Mo oxidases the dioxygen reagent interacts with the Fe—S, heme or flavin parts of the protein and not with the Mo site.[2,154] The reduced Mo states of enzymes, however, are susceptible to oxidation by O_2 *in vitro*.

36.5.10 HYDROXYLAMIDO(1−) AND HYDROXYLAMIDO(2−) COMPLEXES

Deprotonated substituted hydroxylamines have provided potent ligands to Mo^{VI}. The *N,N*-dialkyl hydroxylamine, R_2NOH, in its singly deprotonated form, R_2NO^-, *N,N*-dialkylhydroxamido(1−), serves as an N—O bound ligand with some similarity to peroxide. *N*-Alkylhydroxylamine, RNHOH, in its doubly deprotonated form, [RNO^{2-}, *N*-alkylhydroxamido(2−)] forms complexes quite analogous to those formed by peroxo or disulfido ligands. For example, the complex $MoO(R_2dtc)_2(ONR)$[338] is analogous to $MoO(R_2dtc)_2S_2$. The complex Mo(O)(pydca)(ONR)(HMPA)[338,339] whose structure is shown in Figure 23 has an exact analog in peroxo chemistry, $Mo(O)O_2$(pydca)(HMPA). The hydroxylamido complexes can be thought of as metallooxaziridines.[338,339] A systematic study of the one-electron reduction of $MoO(Et_2dtc)_2(ONR)$ complexes as a function of R and in comparison to $MoO_2(Et_2dtc)_2$ leads to the conclusion that the oxaziridine group is a better donor than the oxo group.[340]

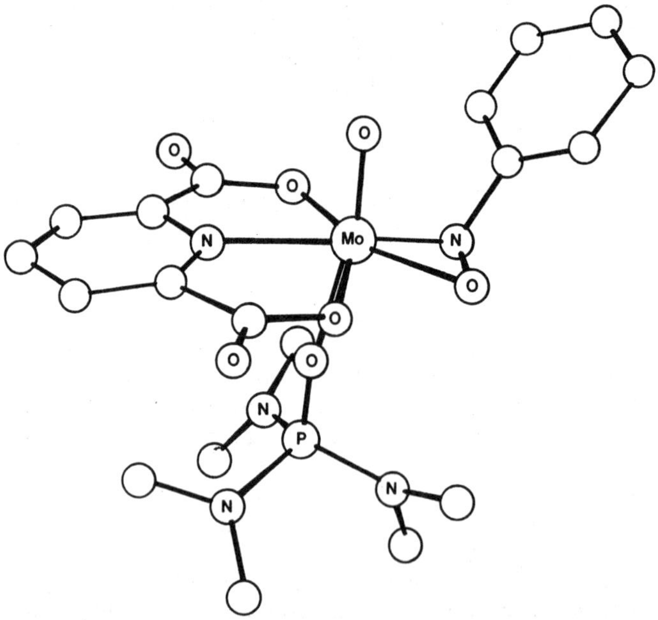

Figure 23 The structure of MoO(pydca)(ONR)(HMPA) (unlabelled atoms are carbon)[338]

The dioxo complexes containing two O,N-coordinated *N,N*-dialkylhydroxylamido(1−) ligands have been extensively studied.[86,87,89,90,191,244,341] They are prepared by the reactions of MoO_4^{2-} and *N,N*-dialkylhydroxylamine, usually in aqueous solution.

$MoO_2(C_5H_{10}NO)_2$ reacts in toluene with H_2S to give $MoOS(C_5H_{10}NO)_2$ and $MoS_2(C_5H_{10}NO)_2$.[342] With H_2Se only $MoOSe(C_5H_{10}NO)_2$ is formed. The structure of the dioxo, disulfido and oxo–sulfido complexes are known.[86,90,342,343] As discussed above, they can be viewed as distorted tetrahedral structures with each 'bidentate' hydroxylamido$(1-)$ ligand occupying a single coordination site. The hydroxylamido$(1-)$ complexes can therefore be considered as derived from MoO_4^{2-}. Alternatively, the structure shown in Figure 6 is describable in terms of non-octahedral six-coordination in which the N and O of the hydroxylamido(-1) ligand lie in a plane with O_t atoms equidisposed above and below the plane. It is related to the skew-trapezoidal six-coordination found in $MoO_2(SCMe_2CH_2NMe_2)_2$.[94,84] NMR studies of $MoO_2(R_2NO)_2$ and $MoS_2(R_2NO)_2$ reveal configurational lability attributed to rapid dissociation and reformation of the Mo—N bond.[341]

Spectroscopic characterization of the *cis*-$MoOS^{2+}$ unit in $MoOS(R_2NO)_2$ was used to discuss the occurrence of an $MoOS^{2+}$ center in oxidized xanthine oxidase.[343,191] Comparison of the voltammetric behavior of $MoXY(C_5H_{10}NO)_2$ (X = Y = S, O; X = S, Y = O) leads to the finding that the presence of the S atoms lowers the redox potential of the complex.[191] The great difference observed between the $MoOS(C_5H_{10}NO)_2$ and $MoO_2(C_5H_{10}NO)_2$ complexes (voltammetry differing in potential by 0.6 V)[191] contrasts with the tiny difference found between the proposed (inactive) MoO_2^{2+} and (active) $MoOS^{2+}$ forms of xanthine oxidase. This unusual behavior in the pseudotetrahedral complexes may be due to their significantly different electronic structure compared to that of the MoO_2^{2+} and $MoOS^{2+}$ cores found in xanthine oxidase.

The reaction of $MoO_2(R_2NO)_2$ with catechol derivatives yields complexes containing one oxo, two dialkylhydroxylamido$(1-)$ and one catecholate ligand. The structure of $MoO(C_5H_5NO)_2(C_6H_4O_2)$[247] is reminiscent of that of the analogous peroxo complex. Here the hydroxylamido (piperidinolato) ligand binds side-on to the Mo in the equatorial plane of a seven-coordinate Mo structure. The catecholate ligand bridges axial and equatorial positions with Mo—O_t at 1.69 Å typical of monooxo MoO^{4+} core complexes. The N—O distance of 1.38–1.39 Å is shorter than that in $MoO_2(C_5H_5NO)_2$ (1.43–1.47 Å), perhaps indicating a greater π donation by the 1-piperidinolato$(1-)$ group in this complex where it competes with only one oxo ligand. One remarkable feature of this structure is the virtual identity of Mo—O distances *cis* and *trans* to oxo in the same cathecholate ligand Mo—$O(cis)$ = 2.016(5) and Mo—$O(trans)$ = 2.012(5) Å. This complex also had its structure determined by 2D NMR techniques (^{1}H–^{13}C chemical shift correlations and ^{1}H–^{1}H scalar coupling correlations).[247]

Reaction of *N*-methylhydroxylamine and various unsaturated triatomic reactants yield chelating ligands that form yellow MoO_2L_2 complexes.[212] $MoO_2(MeHNO)_2$ reacts with the appropriate unsaturated molecule to yield complexes with five-membered chelate rings. For example, in equation (29) the resulting complex has the *cis* dioxo distorted octahedral structure.

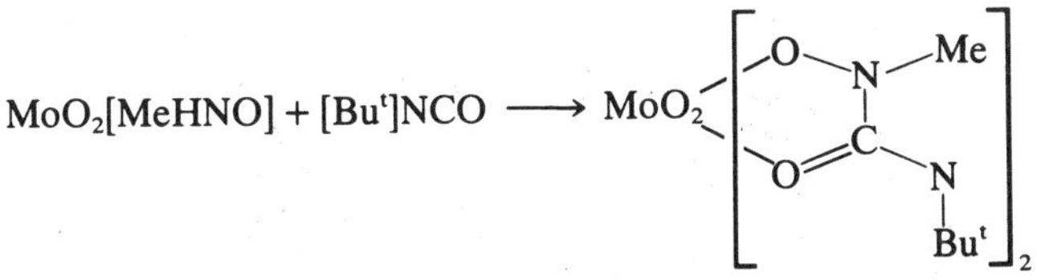

$$MoO_2[MeHNO] + [Bu^t]NCO \longrightarrow MoO_2\left[\begin{matrix} O-N-Me \\ | \quad\quad | \\ O-C-N \\ \quad\quad\quad | \\ \quad\quad\quad Bu^t \end{matrix}\right]_2 \tag{29}$$

Scheme 5 shows the relationship of the starting hydroxylamido ligand to the final ligand.[341]

Complexes that contain remaining hydroxylamido as well as these bidentate chelates are also formed (Table 3). For example, $MoO(MeNHO)_2[HNC(S)N(Me)O]$ forms by reaction of MoO_4^{2-}, MeNHOH and SCN^- at 60 °C. Its structure, illustrating the two ligand types, is shown in Figure 24.[87]

36.4.11 COMPOUNDS WITH Mo—C BONDS

Molybdenum(VI) compounds containing Mo—C bonds are a recent addition to the literature. These compounds include both oxo and non-oxo species. The latter have an alkylidyne (alternatively formulated as an alkylcarbido) group which is a bonding equivalent to the oxo group.

Several unexpectedly stable series of Mo—C bonded compounds[344,72,345,346] have been reported. The compounds are of the form $MoO_2RBr(bipy)$ (purple), $MoO_2R_2(bipy)$ (yellow) and MoO_2R^- (in solution, colorless). R can be Me, CH_2Me_3 or CH_2Ph, which, by lacking β-hydrogen atoms, are unable to undergo β elimination. The compounds are prepared by reacting $MoO_2Br_2(bipy)$ with a Grignard reagent according to Scheme 6.

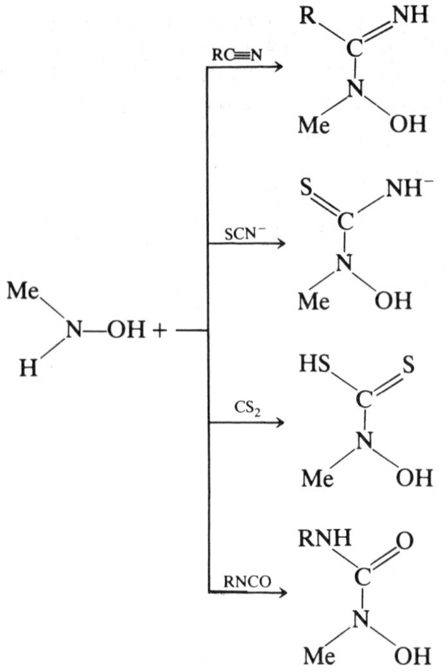

Scheme 5

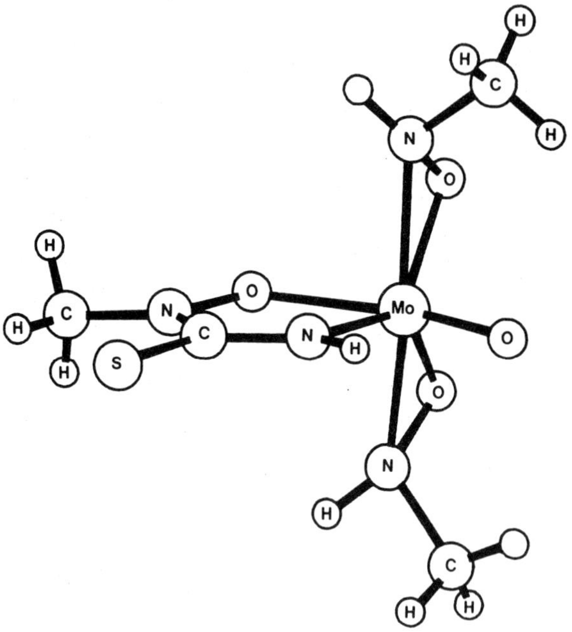

Figure 24 The structure of $MoO(MeNHO)_2[HNC(S)NMeO]$[87]

$$MoO_2Br_2(bipy) \longrightarrow RMoO_2Br(bipy) \longrightarrow MoO_2R_2(bipy) \xrightarrow{OH^-} RMoO_3^-$$

Scheme 6

The last complex, which has yet to be isolated as a salt, decomposes, albeit slowly ($t_{1/2}$ is 26.5 h at pH 11) in basic aqueous solutions at room temperature according to equation (30).

$$Me—MoO_3^- + OH^- \longrightarrow MoO_4^{2-} + CH_4 \quad (30)$$

The dimethyl complex, $MoO_2(Me)_2(bipy)$, shows the expected *cis* dioxo structure[345] with the methyl groups *trans* to each other and bent back from the oxo groups (C—Mo—C = 149°). The Mo—C distance is 2.19 Å in the R = Me complex and 2.20 Å in the R = CH_2Me_3 complex.

MoO_2Me_2(bipy) shows $\nu(Mo—O_t)$ at 934 and 905 cm^{-1} in the IR and has UV bands at 303 nm (24 900); 293 nm (17 000); and 245 nm (19 600). The near UV band may be responsible for the complexes' photosensitivity.[346] In the NMR the Mo—Me protons resonate at 0.58 p.p.m. downfield from TMS.[345] The neopentyl analog $MoO_2(CH_2CMe_3)_2$(bipy),[346] whose structure is shown in Figure 25, has high thermal stability (to 182 °C in air) attributed to the absence of β-hydrogen atoms. Structurally, it is very similar to the Me derivative with deviations attributed to repulsions due to the bulky neopentyl groups. Thermolysis of the complex in the presence of abstractable hydrogen atoms from silicone oil leads to formation of neopentane, presumably from neopentyl radicals. Reaction with alkaline $Na^+BH_4^-$, Zn in HCl or 1-thioglycerol yields neopentane, quantitatively, through reductive cleavage of the Mo—C bond. In H_2SO_4 a mixture of dineopentyl and neopentane is formed.

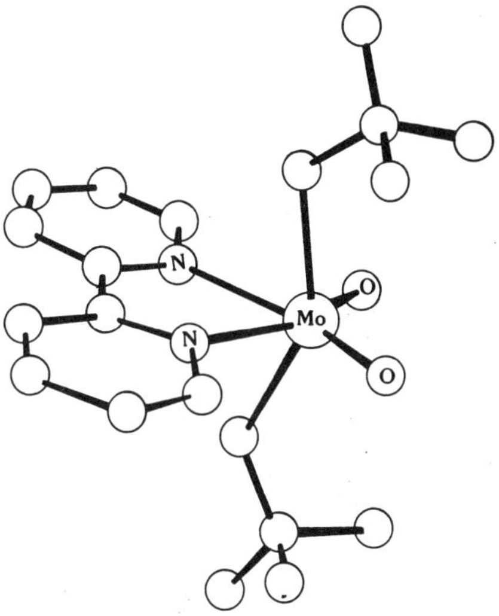

Figure 25 The structure of $MoO_2(CH_2CMe_3)_2$(bipy) (unlabelled atoms are carbon)[346]

Use has been made of Grignard reagents and organolithium reagents to prepare other alkyl Mo^{VI} complexes. For example, reaction of $MoO_2Cl_2{\cdot}2THF$ with (mesityl)MgBr (mesityl = 2,4,6-trimethylphenyl) gives the four-coordinate complex $MoO_2(mesityl)_2$.[159] This complex reacts with the ylide $Bu^n_3PCH_2$ to give Bu^n_3PO and (mesityl)MoO_2[C(mesityl)PBu^n_3] wherein a P-substituted carbene is bound to Mo.[347]

Reaction of $LiCH_2CMe_3$ with $MoCl_5$ gives the remarkable alkylidyne complex $(Me_3CCH_2)_3Mo(CCMe_3)$.[348] This complex can also be prepared from MoO_2Cl_2 and $(Me_3CCH_2)MgCl$ in ether.[353] Reaction of $Mo(CCMe_3)(CH_2CMe_3)_3$ with HX in DME (1,2-dimethoxyethane) gives $Mo(CCMe_3)X_3DME$ (X = Cl, Br).[353] These complexes are all formally Mo^{VI} if the alkylidyne ligand is considered as RC^{3-}.

The Mo^{III} complex $Mo_2(OCMe_3)_6$ reacts with phenylacetylene in pentane to give the alkylidyne complex $Mo(CC_6H_5)(OCMe_3)_3$ in 60% yield.[354] In the presence of quinuclidine (quin) both $Mo(CC_6H_5)(OCMe_3)_3$(quin) and $Mo(CH)(OCMe_3)_3$(quin) have been identified. These reactions are remarkable as equation (31) formally involves the breaking of metal–metal and carbon–carbon triple bonds and the formation of a metal–organic triple bond.[354]

$$(Me_3CO)_3Mo{\equiv}Mo(OCMe_3) + C_6H_5CCH \longrightarrow (Me_3CO)_3MoCC_6H_5 \qquad (31)$$

The complex $Mo(CCMe_3)(OR)_3$ is formally Mo^{VI} having a substituted molybdate structure. Here the alkylidyne ligand, formally R_3CC^{3-}, obviously displays the powerful σ- and π-donor ability required to stabilize Mo^{VI}. The complexes $MoOCl(CH_2CMe_3)_3$ and $MoO(CH_2CMe_3)_4$ have been isolated[355] by reaction of $MoOCl_4$ with bis(neopentyl)magnesium dioxane along with the non-oxo $Mo(CCMe_3)(CH_2CMe_3)_3$ complex prepared previously.[130]

The organic reactivity of the alkylidyne complexes has been extensively studied especially with regard to alkene and alkyne metathesis reactions.[353,356]

36.5.12 DINUCLEAR COMPLEXES OF Mo^{VI}

In addition to the multitude of mononuclear and heteropolynuclear complexes that contain Mo^{VI} there are a significant number of dinuclear complexes. These include complexes in which the Mo atoms only coincidentally find themselves in the same ion such as $(MoO_3)_2edta^{4-}$ [61] (where the two Mo atoms are more than 4 Å apart) to ions such as $Mo_2O_5(C_2O_4)_2(H_2O)_2^{2-}$ (where the two Mo atoms are bridged by a single oxygen atom). In no case are Mo—Mo bonds present in the $4d^0$, Mo^{VI}, dinuclear complexes.

The dinuclear complexes are diverse in structure. They can have a single, double or triple bridge. The single bridge is almost always a lone oxo group. The double bridge contains either a single oxo plus a non-oxo ligand or, more commonly, two non-oxo ligands. Besides this variety in the nature of the bridge, the remaining three, four or five ligands on Mo can also be a diverse mixture of oxo and non-oxo groups. As many as three oxo groups in $Mo_2O_7^{2-}$ [197] to one oxo group as in $Mo_2O_2(3,5\text{-dtbcat})_4$[349] may be present in non-bridging positions. There are even examples of dinuclear, formally Mo^{VI} complexes, containing no oxo groups such as $Mo_2(O_2C_6Cl_4)_6$[357] (see below under non-oxo complexes).

The simplest dinuclear ion is the dimolybdate ion. Discrete $Mo_2O_7^{2-}$ ions exist in $K_2Mo_2O_7\cdot KBr$,[350] in molten $K_2Mo_2O_7$ above 520 °C[233] and in $MgMo_2O_7$,[351,358] the latter having a dichromate-like structure with an Mo—O—Mo bridging angle of 160.7(3)°. However, the $Mo_2O_7^{2-}$ stoichiometry does not guarantee the presence of discrete $Mo_2O_7^{2-}$ ions as, for example, crystalline $K_2Mo_2O_7$ and $(NH_4)_2Mo_2O_7$ consist of infinite chains of edge-shared octahedra.[359]

The discrete $Mo_2O_7^{2-}$ anion has been structurally characterized as its NBu_4^+ [361,352] and $[(C_5H_4Me)Mo(CO)_2(\mu\text{-dppm})Pt(dppm)]^+$ [360] salts. In the latter salt, $Mo_2O_7^{2-}$ shows Mo—O_t distances from 1.68–1.79 and Mo—O_b distances of 1.82 and 1.95 Å with the Mo—O—Mo angle equal to 160°. In the $(NBu_4)_2Mo_2O_7$ complex, the Mo—O—Mo angle is 153.6°. This complex is soluble and stable in organic solvents and serves as a starting material in various preparations. Its stability in non-aqueous media compared to higher polynuclear species is attributed to counterion effects. ^{17}O NMR studies show peaks at 741 and 248 p.p.m, down field from H_2O in the expected 6:1 ratio assigned to the terminal and bridging O atoms, respectively.[361,352]

Other dinuclear complexes having a single oxo group as the sole bridge contain $Mo_2O_3^{6+}$ or $Mo_2O_5^{2+}$ cores. Of these, the latter is by far the most common. Complexes of known structure containing the $Mo_2O_5^{2+}$ core with a single bridging oxo ligand are listed in Table 7. The Mo—O—Mo angle varies from 136 to 180°. For example, the structure of $Mo_2O_5L_2$ (where L = hydrotris(3,5-dimethyl-1-pyrazolylborate) contains an $Mo_2O_5^{2+}$ core with a singly bridging O atom and an Mo—O—Mo angle of 167.1°. The MoO_2^{2+} unit of each Mo atom in the dimer is, as shown in Table 7, quite similar to the same unit found in mononuclear complexes displayed in Table 1.

Table 7 Structural Data on Oxo-bridged $Mo_2O_5^{2+}$ Complexes

Complex	*Distances* *Mo—O$_t$* (Å)	*Mo—O$_b$* (Å)	*Angles* *Mo—O—Mo* (°)	*Ref.*
$K_2[Mo_2O_5(O_2)_4(H_2O)_2]\cdot 2H_2O$	1.664(12), 1.662(10)	1.933(10)	136.1(4)	305
$Mo_2O_5[Me_2NCH_2CH_2NHCH_2CMe_2S]_2$	1.715(6), 1.715(6)	1.929(5)	143.8(2)	107
	1.714(5), 1.709(6)	1.917(5)		
$Mo_2O_5(Me_2NCH_2CH_2NHCH_2CH_2S)_2$	1.685(8), 1.737(7)	1.908(3)	147.0(5)	382
$Mo_2O_5(phen)_2(NCS)_2$	1.685(5), 1.688(4)	1.865(4)	162.7(2)	211
	1.691(6), 1.694(5)	1.885(4)		
$Mo_2O_5[HB(Me_2pz)_3]_2$	1.701(2), 1.696(2)	1.889(1)	167.1(2)	383
$Mo_2O_5(DMF)_4Cl_2$	1.68(3)	1.90(3)	171	199
$K_2[Mo_2O_5(C_2O_4)_2(H_2O)_2]$	1.680(19), 1.700(20)	1.876(2)	180	384
$Na_2[Mo_2O_5(Hnta)_2]\cdot 8H_2O$	1.680(6), 1.710(6)	1.880(1)	180	385
$(pyH)_5[Mo_2O_5(Hnta)_2]\cdot H_2O$	1.695(3), 1.695(3)	1.870(3)	167.0(2)	386
	1.700(3), 1.683(3)	1.881(3)		
$Mo_2O_5(DMSO)_4Cl_2$	1.691(5), 1.686(5)	1.864(1)	180	387
$Mo_2O_5(C_9H_{21}N_3)_2$[a]	1.694(7)	1.898(1)	180	388
	1.696(7)			

[a] The ligand is N′,N″,N‴-trimethyl-1,4,7-triazacyclononane.

In the salt $K_2[Mo_2O_5(C_2O_4)_2(H_2O)_2]$, the complex anion is crystallographically centrosymmetric.[384] As shown in Figure 26, the two terminal oxo ligands on each Mo are mutually *cis* and *cis* to the bridging oxo. One oxalate and one water ligand complete the six-coordination sphere of each Mo atom. The $Mo—O_t$ distance of 1.69 Å is similar to those found in mononuclear *cis* dioxo complexes. The $Mo—O_b$ distance is 1.88 Å. The long $Mo—OH_2$ distance (*trans* to $Mo—O_t$) of 2.33 Å illustrates the large structural *trans* effect of the O_t ligand. The two Mo—O distances to the oxalate ligand of 2.09 Å (*trans* to O_b) and 2.19 Å (*trans* to O_t) indicate that O_t has a larger *trans* bond-weakening effect than O_b. The largest O—Mo—O bond angles correspond to the shortest Mo—O bond distances in agreement with the idea that interligand or interbond repulsions determine the details of such coordination spheres. A centrosymmetric structure such as that in $Mo_2O_5(C_2O_4)_2(H_2O)_2^{2-}$ with an $Mo_2O_5^{2+}$ core is found in a number of complexes. In other cases, a structure closer to C_2 symmetry is found with the dioxo moieties on the two Mo atoms adopting a quasieclipsed conformation. In the complex $Mo_2O_5(phen)_2(NCS)_2$, the phenanthroline ligands are eclipsed,[211] i.e. stacked in almost parallel planes as shown in Figure 27.

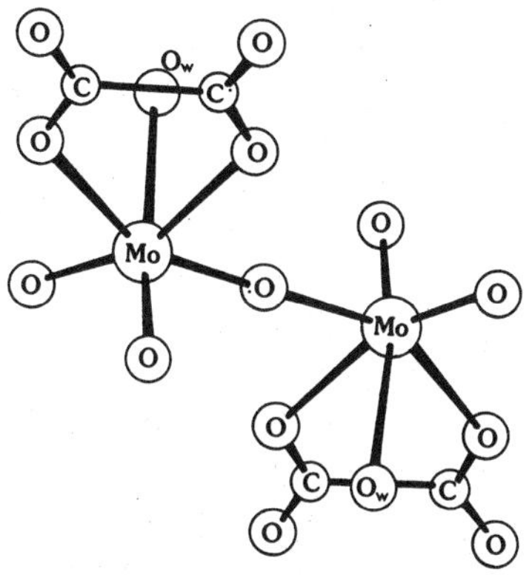

Figure 26 The structure of $Mo_2O_5(C_2O_4)_2(H_2O)_2^{2-}$ in its K^+ salt (O_w is an oxygen atom from a water ligand)[384]

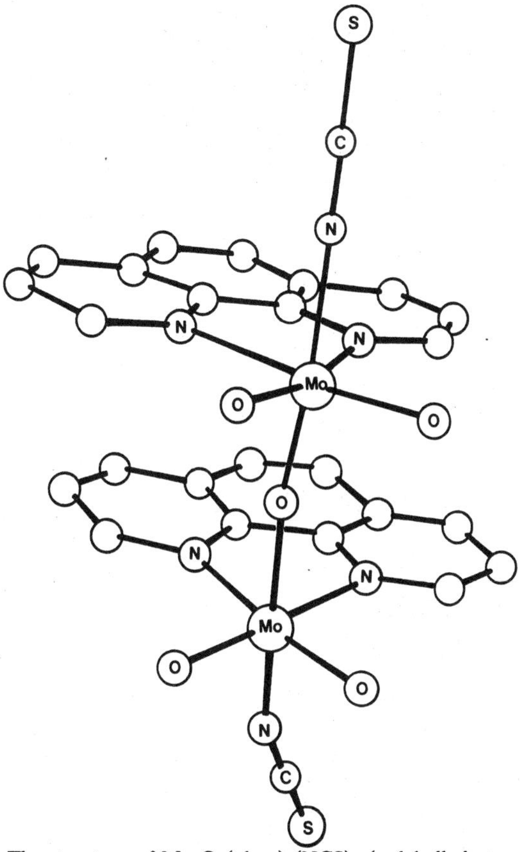

Figure 27 The structure of $Mo_2O_5(phen)_2(NCS)_2$ (unlabelled atoms are carbon)[211]

Complexes containing a triple bridge and an $Mo_2O_5^{2+}$ core are listed in Table 8. The prototypical complex $(NH_4)_2[Mo_2O_5(O_2C_6H_4)_2]\cdot 2H_2O$, shown in Figure 28,[362] has a triple bridge consisting of non-linear oxo and two oxygen atoms of different catecholate ligands. The structure can be considered as two shared distorted octahedra with the three bridging oxygen atoms at average distances of 1.92, 2.16 and 2.37 Å from each Mo. The catechol oxygen atoms do not bridge symmetrically as each catechol is chelated to a single Mo and bridges by one O atom to the second Mo. The two six-coordination spheres are completed by two terminal oxo groups and one non-bridging catechol oxygen on each Mo. The Mo—Mo distance is 3.13 Å as expected for this $4d^0$ system where no Mo—Mo bonding is possible. A similar triple bridge has been found in the complex $Mo_2O_5(PhenSQ)_2$ (where PhenSQ is the 9,10-phenanthrenequinone radical anion)[363] and in complexes containing mannitol,[364] dithiothreitol,[365] 2-mercatophenolato[366] and 4-xanthopterin.[274] The latter complex[274] shown in Figure 29 is the first example of Mo bound to a pterin ligand (as proposed for the molybdenum cofactor Moco). Interestingly, the pterin binds in a mode reminiscent of 8-hydroxyquinolinate ligands. However, the relevance of this binding mode to Moco is uncertain in view of the dinuclear nature of the complex, the proposed reduced nature of the pterin in Moco and the suggested sulfur binding by the Moco pterin.[5,274]

Table 8 Doubly and Triply Bridged Complexes Containing $Mo_2O_5^{2+}$

Complexes	*Mo—O$_t$* (Å)	*Mo—O$_b$—Mo* (°)	*Ref.*
$(NH_4)[Mo_2O_5(C_6H_{11}O_6)]\cdot H_2O$[a]	1.68(1), 1.72(1) 1.69(1), 1.71(1)	107.98(7)	364
$Na[Mo_2O_5(C_6H_{11}O_6)]\cdot 2H_2O$[a]	1.727(3), 1.693(7) 1.716(2), 1.703(2)	107.98(7)	389
$Mo_2O_5(phenSQ)_2$	1.691(5), 1.678(5)	112.7(2)	390
$[NBu^n_4]_2[Mo_2O_5(dbcat)_2]$	1.700(1), 1.707(10) 1.701(9), 1.674(10)	109.4(4)	363
$(NH_4)_2Mo_2O_5(cat)_2\cdot 2H_2O$	1.68(2), 1.69(2) 1.73(2), 1.71(2)	119.3	362 391
$Mo_2O_5(n\text{-pobb})_2$[b]	1.698(3), 1.691(3) 1.683(3), 1.699(3)		368
$[PPh_4]_2[Mo_2O_5(SC_6H_4O)_2]$	1.712(2) 1.699(2)		367
$(NEt_4)_2[Mo_2O_5(dtt)]$	1.702(12), 1.716(12) 1.674(9), 1.657(12)	85.3(3)	365
$[Na(DMSO)_2]_2[Mo_2O_5(xanthop)_2]$	1.691(7), 1.707(5)	146.2(3)	274

[a] The organic ligand is the ionized sugar, mannitolate. [b] *n*-pobb = 1-*n*-2α-hydroxybenzimidazole (deprotonated).

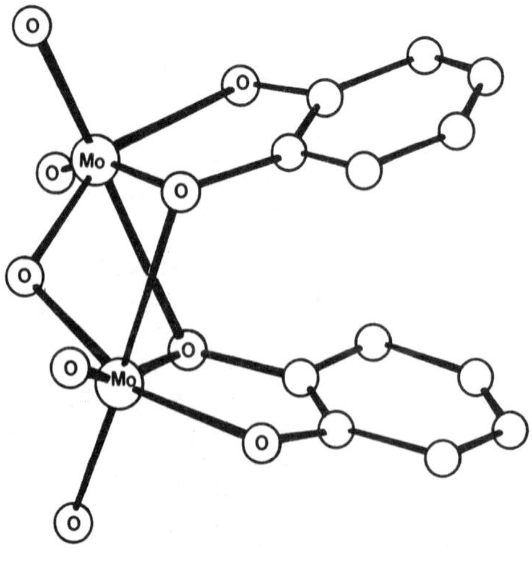

Figure 28 The structure of $Mo_2O_5(O_2C_6H_4)_2^{2-}$ in its NH_4^+ salt (unlabelled atoms are carbon)[362]

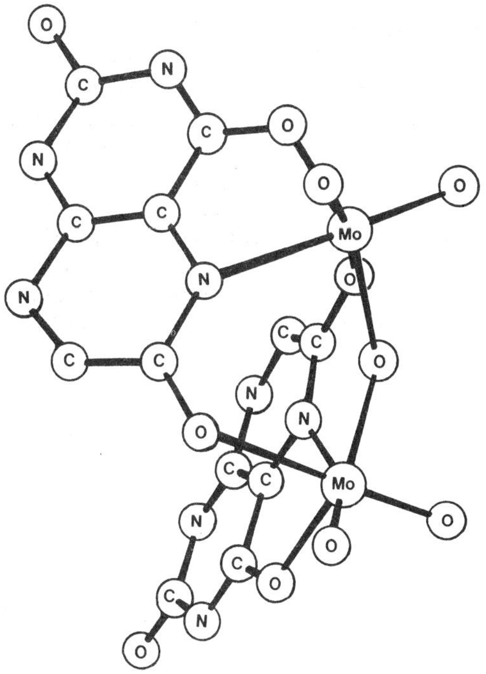

Figure 29 The structure of $Mo_2O_5(xanthopterinate)_2^{2-}$ as found in the salt $[Na(DMSO)_2]_2[Mo_2O_5(xanthopterinate)_2]$[274]

In certain $Mo_2O_5^{2+}$ complexes, the bridging ligand would appear to have a choice between an O and S bridge. In the case of the ligand $C_6H_4OS^{2-}$ (monothiocatecholate) the oxygen atoms of the ligand are the (non-oxo) bridging atoms.[367] Similarly, in $Mo_2O_5(dtt)^{2-}$ [365] again O rather than S bridging is found. In both cases this allows the S ligand to occupy a position *trans* to the bridging oxo rather than *trans* to the terminal oxo. This is the more favorable position for the strong π-donor S atom. This preference for not being *trans* to oxo may be the dominant factor in the adoption of the observed structure.

One interesting complex has a deprotonated 1-*n*-2α-hydroxybenzylbenzimidazole ligand whose O atom forms an additional bridge to an oxo.[368] The complex is unusual in having a double (rather than more common single or triple) bridge. Further, one of the Mo^{VI} atoms is five-coordinate while the other has the more conventional six-coordinate near-octahedral structure.[368]

A related tetranuclear complex having $Mo_2O_5^{2+}$-like units is formed between Mo^{VI} and malate, and has the formula $(NH_4)_2[(MoO_2)_4O_3(C_4H_3O_5)_2]\cdot H_2O$.[234]

Yellow complexes with the $Mo_2O_3^{4+}$ core are formed with substituted glycol ligands such as pinacol (pinH).[369] For example, the structure of $Mo_2O_3(pin)_2(Hpin)$ has a 162.2(9)° Mo—O—Mo bridging angle with intramolecular H bonding presumably helping to stabilize the complex.

There are a few examples in which only non-oxo bridges occur. In these cases alcoholate or phenolate type oxygen atoms form a double bridge. Some of the complexes have MoO_2^{2+} cores bridged doubly by alcoholate ligands. For example, as shown in Figure 30, $\{MoO_2[OCMe_2CMe_2OH]\}_2(-\mu\text{-}OMe)_2$ has a double methoxide bridge,[370] while $[Mo_2O_4(\mu\text{-}OCH_2CMe_2CH_2O)(H_2O)_2]$ has bridging oxygen from a chelating 2,2-dimethylpropanediolato ligand.[371] A related compound is $Mo_2I_2O_{16}^{6-}$, which can be thought of as containing MoO_2^{2+} groups bridged by IO_6^{5-} octahedra as represented in Figure 31.[372]

We have not dealt explicitly with preparative procedures for dinuclear complexes. These preparations have, in most cases, been fortuitous and no general protocol exists. Most preparations leading to dinuclear complexes did not seek them, whereas some efforts actually sought to avoid them!

One class of complexes, that of substituted catecholate ligands, has yielded a variety of mono- and di-nuclear Mo^{VI} complexes. $Mo(CO)_6$ reacts with the oxidized form of the 3,5-di-*tert*-butylcatechol (*i.e.* 3,5-di-*tert*-butyl-1,2-benzoquinone) to yield $Mo(dbcat)_3$ which reacts with O_2 to give the dinuclear complex $[MoO(dbcat)_2]_2$.[373] In contrast, reaction of $Mo(CO)_6$ with 3,4,5,6-tetrachloro-1,2-benzoquinone yields the dinuclear non-oxo complex $Mo_2(O_2C_6Cl)_6$.[357] The complex $Mo_2O_5(phenSQ)$ has already been discussed.[363] This complex

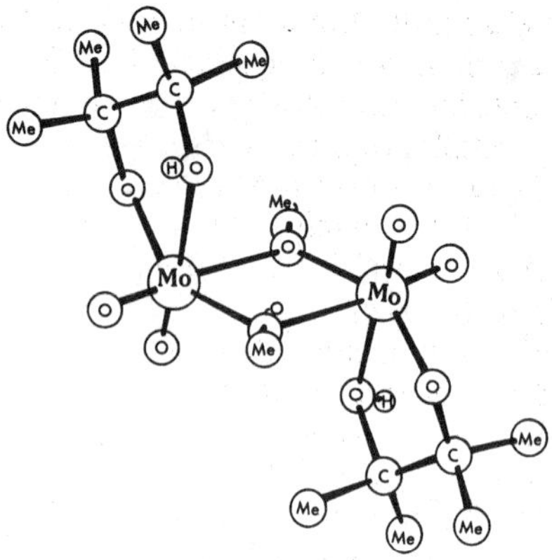

Figure 30 The structure of $\{MoO_2[OCMe_2CMe_2OH]\}(\mu\text{-}OMe)_2$[370]

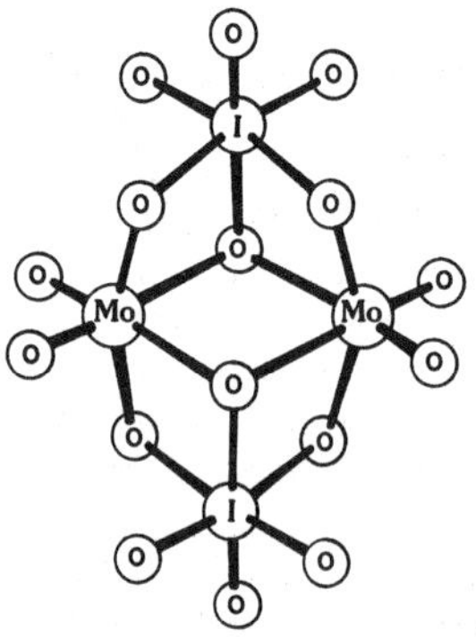

Figure 31 The structure of $Mo_2I_2O_{16}^{6-}$ as found in its K^+ salt[372]

differs by two electrons, *i.e.* is formally oxidized (neglecting the ligand difference) with regard to $Mo_2O_5(cat)^{2-}$ or $Mo_2O_5(dbcat)_2^{2-}$, each prepared from MoO_4^{2-} and the respective catechol. $Mo_2O_5(phenSQ)_2$ reacts by ligand substitution with 3,5-dbcatH_2 to give $Mo(dbcat)_3$. At least in this class of catecholate complexes, there seem to be some reactions that systematically interconnect the mononuclear and various dinuclear complexes.

Infrared and/or Raman spectroscopy are useful in identifying the core structures involved in dinuclear complexes. The Mo—O_t vibrations stand out as they do for mononuclear complexes and the Mo—O_b—Mo stretch can often be detected between 650 and 770 cm^{-1}. For example, in $Mo_2O_5[SCMe_2CCH_2NHCH_2CH_2NMe_2]_2$, ν(Mo—O_t) is found at 860 and 900 cm^{-1} while ν(Mo—O_b—Mo) occurs at 665 cm^{-1}. ^{17}O NMR spectroscopy on enriched samples is particularly informative. For example, in $Mo_2O_5[SCMe_2CCH_2NHCH_2CH_2NMe_2]_2$,[374] the ^{17}O resonance for Mo—O_t ligands occur at 873 and 857 while the bridging Mo—O_b—Mo resonance occurs at 366 p.p.m. downfield from H_2O. ^{95}Mo NMR spectra of dinuclear complexes are also observable[375] and, as more data become available, could also become a structurally useful technique.

In a number of studies spectroscopic techniques have led to the postulation of a dinuclear Mo^{VI} structure for complexes wherein crystallographic information is not yet available. For example, a fluoro-bridged structure has been assigned to $Mo_2O_2F_9^-$ by IR[377] and ^{19}F NMR[53] techniques. IR spectroscopy has been used to assign an $Mo_2O_5^{2+}$ structure with a single oxo bridge to $Cp_2Mo_2O_5$[367] (ν(Mo—O_t) = 920, 898 ν(Mo—O—Mo) = 770 cm^{-1}). Certain oxalate, citrate[378–380] and α-amino acid[381] complexes studied in solution (or isolated) are also indicated to be dinuclear.

36.5.13 COMPLEXES LACKING OXO-TYPE LIGANDS

Although the oxo group and its analogs dominate the chemistry of Mo^{VI}, there are several examples of compounds that lack the oxo ligand or any of its analogs. Perhaps the best known of these is molybdenum hexafluoride. MoF_6, prepared by reaction of Mo metal and F_2, has a regular octahedral structure. The volatile liquid (m.p. = 17.5, b.p. = 35 °C) is reactive toward addition, substitution and reduction. Lewis bases including F^-,[392] Cl^- [393] and Me_2O[392] form adducts with MoF_6[392] whose structures have not yet been determined. For example, K_2MoF_8 has been isolated[393] but not crystallographically characterized. Fluoride in MoF_6 can be substituted by F_5TeO^- or $CF_3CH_2O^-$ to form the series of complexes $MoF_n(OTeF_5)_{6-n}$[395] and $MoF_n(OCH_2CF_3)_{6-n}$ ($n = 1–6$),[394] respectively. As an oxidant, MoF_6 has a potential of +0.93 V *vs.* Cu/CuF_2(1.0 M NaF/HF) in HF[396] and +1.60 V *vs.* Ag^+/Ag^0 in MeCN[397] for the MoF_6/MoF_6^- couple as determined by cyclic voltammetry. MoF_6 is thus a strong oxidant and reacts with metals,[398] I_2,[397] NO[399] and many other molecules. For example, the explosively reactive amber $MoSF_4$ and $MoSeF_4$ molecules have been reported to be prepared from MoF_6 according to equation (32).[245] The vibrational spectrum of MoF_6 has been extensively studied[400] and exploited in laser isotope separation.[401] Although $MoCl_6$ has been briefly reported[402] its existence has not been confirmed and seems unlikely. $Mo(OMe)_6$ has been reported as a low-melting solid (m.p. = 67.6 °C).[403] The unusual complex $MoCl_4$(NSNSN)[404] can formally be considered to contain an Mo^{VI} core, which is consistent with a qualitative bonding scheme for the dithiatriazene ligand.[404]

$$3MoF_6 + Sb_2X_3 \longrightarrow 3MoXF_4 + 2SbF_3 \ (X = S, Se) \qquad (32)$$

The remaining non-oxo complexes contain O-, N- or S-donor polydentate ligands. The tris(dithiolene) complexes are the oldest known members of this class. For example $Mo(tdt)_3$,[53,419] $Mo(bdt)_3$[405,419] and $Mo(S_2C_2H_2)_3$[406] are six-coordinate trigonal prismatic complexes whose structural, spectroscopic and redox properties have been thoroughly investigated and reviewed,[2,407–409] The amidothiolate ligand $C_6H_4SNH^{2-}$ = abt^{2-} forms an analogous nearly trigonal prismatic complex $Mo(abt)_3$ with the *cis* arrangement of the ligands illustrated in Figure 32.[121,410] $MoO_2(acac)_2$ reacts with thioaroylhydrazines, e.g. $NH_2NHC(S)Ph$, to give $Mo(HNNHCSPh)_3$ which has a somewhat distorted trigonal prismatic coordination[411] related to that of $Mo(abt)_3$.[410,412] A related complex containing three chemically distinct N,S chelates has been reported.[291] Interestingly, these S,S and N,S donor ligands react with oxo molybdenum starting materials to give non-oxo products. Significantly, the N donors are always found in a deprotonated state such that they can serve as π as well as σ donors to the Mo^{VI} ion.[411,412] The amidothiolate complexes are more difficult to reduce than are the corresponding dithiolene complexes. Complexes of catecholate ligands such as $Mo_2(O_2C_6Cl_4)_6$[357] and $Mo(phenQ)_3$[413] have dinuclear and distorted trigonal prismatic coordination, respectively. These complexes are prepared from $Mo(CO)_6$ and the oxidized (*o*-quinone) form of the catecholate ligand.[413–415]

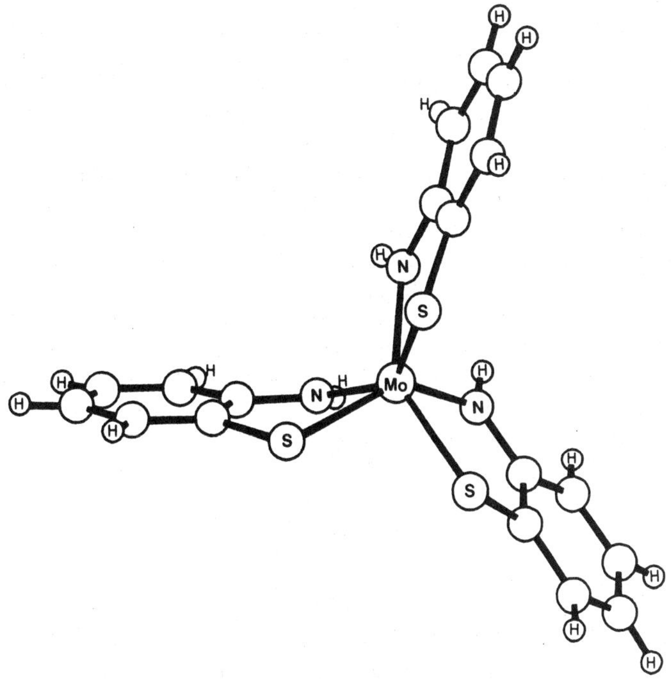

Figure 32 The structure of $Mo(abt)_3 = Mo(NHSC_6H_4)_3$ (unlabelled atoms are carbon)[410]

In addition to these complexes of bidentate ligands, several related tri- and tetra-dentate ligands are capable of removing all oxo groups from an Mo^{VI} coordination sphere. The tridentate ligand $hbma^{3-}$ forms $Mo(hbma)_2$ (**6**), a six-coordinate complex with a structure halfway between meridianal octahedral and trigonal prismatic geometries.[189] The related potentially tetradentate ligand H_2bpmp^{2-} (**7**) forms the yellow $MoO_2(H_2bpmp)$ complex[420] on reaction of the ligand with $MoO_2(acac)_2$. However, the brown complex $Mo(Hbpmp)_2{\cdot}MeOH$ formed by a less reliable procedure[420] has an octahedral structure with a meridianal arrangement of two of the ligands acting in a tridentate fashion. Each ligand has one uncoordinated phenolic OH group.[420] Complexes of $en(abt)_2^-$ of the form $Mo_2[en(abt)_2]_3$ [see (**8**)] have also been reported.[412]

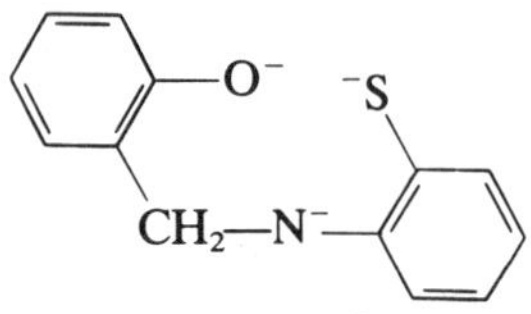

(**6**) $hbma^{3-}$

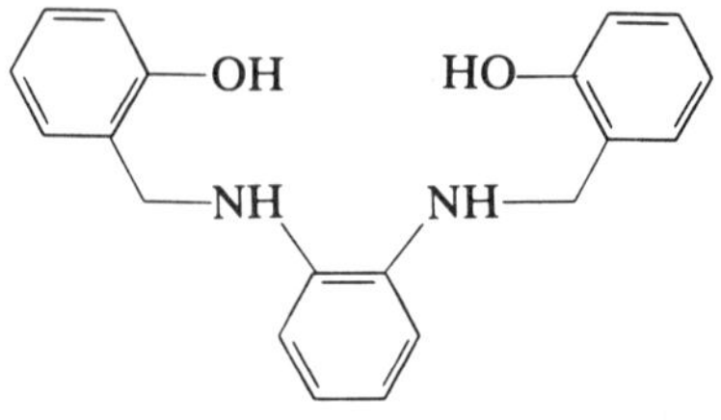

(**7**) H_4 bpmp

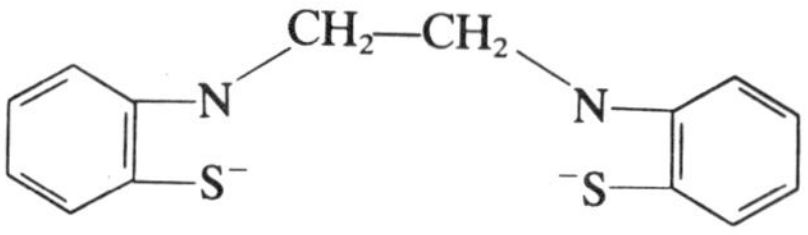

(**8**) $en(abt)_2$

The non-oxo Mo^{VI} complexes have in common the presence of good σ- and π- donor ligands. These serve the same purpose as the oxo group, namely the neutralization of high formal charge on Mo by strong donation.

Acknowledgement

I thank Professors Sharon J. N. Burgmayer, Karl Dehnicke, John H. Enemark, C. David Garner, Jean-Marie Manoli, Philip C. H. Mitchell, Achim Müller, Jon A. McCleverty, John W. McDonald, William E. Newton, Narayanan Kutty Pariyadath, Franklin A. Schultz, Dieter Sellman, Jack T. Spence, A. Geoffrey Sykes, Joseph L. Templeton, Richard A. Walton, Anthony G. Wedd, Karl Wieghardt and Jon Zubieta for providing me with results prior to publication. I thank Mary Reilly and Pat Deuel for outstanding word processing and Jeannette Stiefel for excellent proofreading of the manuscript.

36.5.14 REFERENCES

1. L. Pauling, 'The Nature of Chemical Bond', Cornell University Press, Ithaca, New York, 1960.
2. E. I. Stiefel, *Prog. Inorg. Chem.*, 1977, **22,** 1.
3. A. G. Wedd, in 'Sulfur: Its Significance in Chemistry, Geology, Biology, Cosmology and Technology', *Stud. Inorg. Chem.*, 1984, **5,** 81.
4. C. D. Garner and S. Bristow, in 'Molybdenum Enzymes', ed. T. G. Spiro, Wiley & Sons, New York, 1985, p. 343.
5. S. J. N. Burgmayer and E. I. Stiefel, *J. Chem. Educ.*, 1985, **62,** 943.
6. K. H. Tytko, G. Baethe and J. J. Cruywagen, *Inorg. Chem.*, 1985, **24,** 3132.
7. V. W. Day and W. G. Klemperer, *Science,* 1985, **228,** 533.
8. J. Burclova, J. Prasilova and P. Benes, *J. Inorg. Nucl. Chem.*, 1973, **35,** 909.
9. N. Pariyadath, personal communication.
10. B. M. Gatehouse and P. Leverett, *J. Chem. Soc. (A)*, 1969, 849.
11. G. A. Tsigdinos and F. W. Moore, 'Molybdenum Chemicals: Chemical Data Series', January 1981, Climax Molybdenum Company.
12. R. P. Rastogi, B. L. Dubey and I. Das, *J. Sci. Ind. Res.*, 1978, **37,** 122.
13. M. H. Rapposch, J. B. Anderson and E. Kostiner, *Inorg. Chem.*, 1980, **19,** 3531.
14. V. Massaretti, G. Flor and A. Marini, *J. Appl. Crystallogr.*, 1981, **14,** 64.
15. K. Sieber, R. Kershaw, K. Dwight and A. Wold, *Inorg. Chem.*, 1983, **22,** 2667.
16. P. O. Bars, J. Y. LeMarouille and D. Grandjean, *Acta Crystallogr., Sect. B,* 1977, **33,** 1155.
17. J. Hanuza and V. V. Fornitsev, *J. Mol. Struct.*, 1980, **66,** 1.
18. S. C. Abrahams, C. Svensson and J. L. Bernstein, *J. Chem. Phys.*, 1980, **72,** 4278.
19. Y. Sulfab, R. S. Taylor and A. G. Sykes, *Inorg. Chem.*, 1976, **15,** 2388.
20. A. Clearfield, A. Moini and P. R. Rudolf, *Inorg. Chem.*, 1985, **24,** 4606.
21. S. Launay and A. Rimski, *Acta Crystallogr., Sect. B,* 1980, **36,** 910.
22. R. Coomber and W. P. Griffith, *J. Chem. Soc. (A)*, 1968, 1128.
23. I. R. Beattie, J. S. Ogden and D. D. Price, *J. Chem. Soc., Dalton Trans.*, 1982, 506.
24. T. Yamase, *J. Chem. Soc., Dalton Trans.*, 1985, 2585.

25. T.-C. Hsieh, *Inorg. Chem.*, 1985, **24,** 1288.
26. Y. Hayashi, S. Shioi, M. Togani and T. Sakan, *Chem. Lett.*, 1973, 651.
27. M. A. Harmer and A. G. Sykes, *Inorg. Chem.*, 1980, **19,** 2881.
28. A. Müller, E. Diemann, R. Jostes and H. Bögge, *Angew. Chem., Int. Ed. Engl.*, 1983, **20,** 934.
29. A. Müller, B. Krebs and E. Diemann, *Angew. Chem., Int. Ed. Engl.*, 1967, **6,** 257.
30. A. Müller and E. Diemann, *Chem. Ber.*, 1969, **102,** 945.
31. J. W. McDonald, G. D. Friesen, L. D. Rosenhein and W. E. Newton, *Inorg. Chim. Acta,* 1983, **72,** 205.
32. D. Dembicka, S. Weglowski and A. Bartecki, *Rocz. Chim. Ann. Soc. Chim. Polonorum,* 1975, **40,** 1475.
33. M. R. Udupa, *Curr. Sci.*, 1975, **44,** 304.
34. M. R. Udupa and G. Aravamudan, *Curr. Sci.*, 1973, **42,** 676.
35. M. Kanatzidis and D. Concouvanis, *Acta Crystallogr., Sect. C,* 1983, **39,** 835.
36. B. Kamenar, M. Penavić, B. Korpar-Colig and B. Marković, *Inorg. Chim. Acta,* 1982, **65,** L245.
37. Y. Do, E. D. Simhon and R. H. Holm, *Inorg. Chem.*, 1985, **24,** 1831.
38. H. M. Nagarathna, L. Bencivenni and K. A. Gingerich, *J. Chem. Phys.*, 1984, **81,** 591.
39. T. Ziegler, A. Rauk and E. J. Baerends, *Chem. Phys.*, 1976, **16,** 209.
40. G. L. Gutsev and A. I. Baldyrev, *Chem. Phys. Lett.*, 1984, **108,** 255.
41. J. Bernholc and E. I. Stiefel, *Inorg. Chem.*, 1985, **24,** 1323.
42. A. Müller, E. Diemann and C. K. Jørgensen, *Struct. Bonding (Berlin),* 1973, **14,** 23.
43. R. J. H. Clark, T. J. Dines and M. J. Wolf, *J. Chem. Soc., Faraday Trans.*, 1982, **278,** 679.
44. R. Borromei, G. Ingletto and L. DiSipio, *Chem. Phys. Lett.*, 1984, **92,** 288.
45. A. Calabrese and R. G. Hayes, *Chem. Phys. Lett.*, 1976, **43,** 263.
46. R. H. Petit, B. Briat, A. Müller and E. Diemann, *Mol. Phys.*, 1974, **27,** 1373.
47. T. Ziegler, A. Rauk and E. J. Baerends, *Chem. Phys.*, 1976, **16,** 209.
48. N. Weinstock, H. Schulze and A. Müller, *J. Chem. Phys.*, 1973, **59,** 9.
49. G. J.-J. Chen, J. W. McDonald and W. E. Newton, *Inorg. Chim. Acta,* 1976, **19,** L67.
50. H. Siebert, *Z. Anorg. Allg. Chem.*, 1954, **275,** 225.
51. R. H. Busey and O. L. Keller, Jr., *J. Chem. Phys.*, 1964, **44,** 215.
52. E. Königer-Ahlborn and A. Müller, *Spectrochim. Acta, Part A,* 1977, **33,** 273.
53. H. J. Becher, F. Friedrich and H. Willner, *Z. Anorg. Allg. Chem.*, 1976, **426,** 15.
54. H. von Felton, B. Wernli, H. Gamsjäger and P. Baertschi, *J. Chem. Soc., Dalton Trans.*, 1978, 496.
55. K. Jitsukawa, K. Kaneda and S. Teranishi, *J. Org. Chem.*, 1984, **49,** 199.
56. F. A. Cotton and R. C. Elder, *Inorg. Chem.*, 1964, **3,** 397.
57. W. F. Marzluff, *Inorg. Chem.*, 1964, **3,** 395.
58. R. S. Taylor, P. Gans, P. F. Knowles and A. G. Sykes, *J. Chem. Soc., Dalton Trans.*, 1972, 24.
59. R. J. Butcher and B. R. Penfold, *J. Cryst. Mol. Struct.*, 1976, **6,** 13.
60. R. J. Kula, *Anal. Chem.*, 1966, **38,** 1581.
61. J. J. Park, M. D. Click and J. L. Hoard, *J. Am. Chem. Soc.*, 1959, **91,** 301.
62. R. J. Butcher, H. K. Powell, C. J. Wilkins and S. H. Yong, *J. Chem. Soc., Dalton Trans.*, 1976, 356.
63. W. P. Griffith and T. W. Wilkins, *J. Chem. Soc. (A)*, 1968, **400,** 653.
64. W. P. Griffith, *J. Chem. Soc. (A)*, 1969, 211.
65. J. L. Corbin, K. F. Miller, N. Pariyadath, S. Wherland, A. E. Bruce and E. I. Stiefel, *Inorg. Chim. Acta,* 1984, **90,** 419.
66. M. A. Freeman, F. A. Schultz and C. N. Reilly, *Inorg. Chem.*, 1982, **21,** 567.
67. B. M. Johnson, V. Thurston, B. Deatherage and G. L. Gard, *J. Fluorine Chem.*, 1984, **24,** 443.
68. T. Kato and T. Murayama, *Bull. Chem. Soc. Jpn.*, 1983, **56,** 2129.
69. S. P. Cramer, C. L. Lin, L. E. Mortenson, J. T. Spence, S.-M. Lin, I. Yamamoto and L. G. Ljungdahl, *J. Inorg. Biochem.*, 1985, **23,** 119.
70. E. Königer-Ahlborn and A. Müller, *Angew. Chem., Int. Ed. Engl.*, 1975, **14,** 574.
71. W. G. Klemperer, V. V. Mainz, R.-C. Wang and W. Shum, *Inorg. Chem.*, 1985, **24,** 1970.
72. G. N. Schrauzer, L. A. Hughes and N. Strampach, *Z. Naturforsch., Teil B,* 1982, **37,** 380.
73. J. M. Berg and R. H. Holm, *J. Am. Chem. Soc.*, 1985, **107,** 917.
74. R. J. Butcher, B. R. Penfold and E. Sinn, *J. Chem. Soc., Dalton Trans.*, 1979, 668.
75. B. Spivack and Z. Dori, *Coord. Chem. Rev.*, 1975, **17,** 99.
76. F. A. Schroeder, *Acta Crystallogr. Sect. B,* 1975, **31,** 2294.
77. J. C. J. Bart and Y. Ragaini, in 'Proceedings of the 3rd International Conference on the Chemistry and Uses of Molybdenum', ed. P. C. H. Mitchell and H. Barry, Climax Molybdenum Co., Ann Arbor, MI, 1979, p. 19.
78. J. C. J. Bart and V. Ragaini, *Inorg. Chim. Acta,* 1979, **36,** 261.
79. J. M. Berg and R. H. Holm, *J. Am. Chem. Soc.*, 1984, **106,** 3035.
80. J. M. Berg and K. O. Hodgson, *Inorg. Chem.*, 1980, **19,** 2180.
81. B. B. Kaul, P. Subramanian, J. H. Enemark and J. T. Spence, *J. Am. Chem. Soc.*, 1985, **107,** 2885.
82. S. P. Cramer, 'Advances in Inorganic and Bioinorganic Mechanisms', ed. A. G. Sykes, Academic, New York, 1983, vol. 2, p. 259.
83. K. Yamanouchi and J. H. Enemark, in 'Proceedings of the 3rd International Conference on the Chemistry and Uses of Molybdenum', ed. P. C. H. Mitchell and H. Barry, Climax Molybdenum Co., Ann Arbor, MI, 1979, p. 24.
84. J. M. Berg, D. J. Spira, K. O. Hodgson, A. E. Bruce, K. F. Miller, J. L. Corbin and E. I. Stiefel, *Inorg. Chem.*, 1984, **23,** 3412.
85. J. M. Berg, D. J. Spira, K. Wo, B. McCord, R. Lye, M. S. Co, J. Belmont, C. Barnes, K. Kosydar, S. Raybuck, K. O. Hodgson, A. E. Bruce, J. L. Corbin and E. I. Stiefel, *Inorg. Chim. Acta,* 1984, **90,** 35.
86. K. Wieghardt, W. Holzbach, J. Weiss, B. Nuber and B. Prinkner, *Angew. Chem., Int. Ed. Engl.*, 1979, **18,** 548.
87. K. Wieghardt, E. Hofer, W. Holzbach, B. Nuber and J. Weiss, *Inorg. Chem.*, 1980, **19,** 2927.
88. K. Wieghardt, W. Holzbach, E. Hofer and J. Weiss, *Inorg. Chem.*, 1981, **20,** 343.
89. K. Wieghardt, W. Holzbach, E. Hofer and J. Weiss, *Chem. Ber.*, 1981, **114,** 2700.

90. L. Saussine, H. Mimoun, A. Mitschler and J. Fisher, *Nouv. J. Chim.*, 1980, **4**, 235.
91. B. F. Mentzen, M. C. Bonnet and H. J. Ledon, *J. Inorg. Chem.*, 1980, **19**, 2061.
92. D. L. Kepert, *Prog. Inorg. Chem.*, 1977, **23**, 1.
93. E. I. Stiefel, K. F. Miller, A. E. Bruce, N. Pariyadath, J. Heinecke, J. L. Corbin, J. M. Berg and K. O. Hodgson, in 'Molybdenum Chemistry of Biological Significance', ed. W. E. Newton and S. Otsuka, Plenum, New York, 1980, 279.
94. E. I. Stiefel, K. F. Miller, A. E. Bruce, J. L. Corbin, J. M. Berg and K. O. Hodgson, *J. Am. Chem. Soc.*, 1980, **102**, 3624.
95. G. J.-J. Chen, J. W. McDonald and W. E. Newton, *Inorg. Nucl. Chem. Lett.*, 1976, **12**, 697.
96. W. E. Newton and J. W. McDonald, *J. Less-Common Met.*, 1977, **54**, 51.
97. C. P. Prabhakaran and B. G. Nair, *Transition Met. Chem.*, 1983, **8**, 368.
98. L. Willis, T. Loehr, A. E. Bruce, K. Miller and E. I. Stiefel, *Inorg. Chem.*, 1986, **25**, 4289.
99. N. Ueyama, K. Kamabuchi and A. Nakamura, *J. Chem. Soc., Dalton Trans.*, 1985, 635.
100. J. K. Howie, P. Bosserman and D. T. Sawyer, *Inorg. Chem.*, 1980, **19**, 2293.
101. B. M. Craven, K. Ramey and W. B. Wise, *Inorg. Chem.*, 1971, **10**, 2626.
102. T. J. Pinnavaia and W. R. Clements, *Inorg. Nucl. Chem. Lett.*, 1971, **7**, 1127.
103. K. F. Miller and R. A. D. Wentworth, *Inorg. Chem.*, 1980, **19**, 1818.
104. K. F. Miller and R. A. D. Wentworth, *Inorg. Chem.*, 1979, **18**, 984.
105. W. G. Klemperer, *Angew. Chem., Int. Ed. Engl.*, 1978, **17**, 246.
106. J. L. Corbin, K. F. Miller, N. Pariyadath, J. Heinecke, A. E. Bruce, S. Wherland and E. I. Stiefel, *Inorg. Chem.*, 1984, **23**, 3403.
107. C. P. Marabella, J. H. Enemark, K. F. Miller, A. E. Bruce, K. Pariyadath, J. L. Corbin and E. I. Stiefel, *Inorg. Chem.*, 1983, **22**, 3456.
108. K. A. Christensen, P. E. Miller, M. Minelli, T. W. Rickway and J. H. Enemark, *Inorg. Chim. Acta*, 1981, **56**, L279.
109. B. Piggott, S. D. Thorpe and R. N. Sheppard, *Inorg. Chim. Acta*, 1985, **103**, L3.
110. M. V. Capparelli, B. Piggott, S. D. Thorpe, S. F. Wong and R. N. Sheppard, *Inorg. Chim. Acta*, 1985, **106**, 19.
111. M. Minelli, K. Yamanouchi, J. H. Enemark, P. Subramanian, B. B. Kauland and J. T. Spence, *Inorg. Chem.*, 1984, **23**, 2554.
112. I. Buchanan, M. Minelli, M. T. Ashby, T. J. King, J. H. Enemark and C. D. Garner, *Inorg. Chem.*, 1984, **23**, 495.
113. E. C. Alyea and J. Topich, *Inorg. Chim. Acta*, 1982, **65**, L95.
114. S. F. Geller, T. W. Hambley, P. R. Traill, R. T. C. Brownlee, M. J. O'Connor, M. R. Snow and A. G. Wedd, *Aust. J. Chem.*, 1982, **35**, 2183.
115. S. F. Geller, P. A. Gazzana, A. F. Masters, R. T. C. Brownlee, M. J. C. O'Connor, A. G. Wedd, J. R. Rodgers and M. R. Snow, *Inorg. Chim. Acta*, 1981, **54**, L131.
116. M. Minelli, J. H. Enemark, K. Wieghardt and M. Hahn, *Inorg. Chem.*, 1983, **22**, 3952.
117. M. Minelli, J. H. Enemark, R. T. C. Brownlee, M. J. O'Connor and A. G. Wedd, *Coord. Chem. Rev.*, 1985, **68**, 169.
118. W. A. Levason, R. Narayanaswamy, J. S. Ogden, A. J. Rest and J. W. Turff, *J. Chem. Soc., Dalton Trans.*, 1982, 2009.
119. I. A. Topol, A. S. Chesnyi, V. M. Kovba and N. F. Stepanov, *Theoret. Chim. Acta*, 1982, **61**, 369.
120. C. Y. Chou, J. C. Huffman and E. A. Maatta, *J. Chem. Soc., Chem. Commun.*, 1984, 1184.
121. H. Gehrke, Jr. and J. Veal, *Inorg. Chim. Acta*, 1969, **3:4**, 623.
122. M. M. Jones, *J. Am. Chem. Soc.*, 1959, **81**, 3188.
123. G. Pilcher, K. J. Cavell, C. D. Garner and S. Parkes, *J. Chem. Soc., Dalton Trans.*, 1978, 1311.
124. D. Sellman and L. Zapf, *Z. Naturforsch., Teil B*, 1985, **40**, 368.
125. R. H. Holm and J. M. Berg, *Pure Appl. Chem.*, 1984, **56**, 1645.
126. L. J. DeHayes, H. C. Faulkner, W. H. Doub and D. T. Sawyer, *Inorg. Chem.*, 1975, **14**, 2110.
127. P. C. H. Mitchell and R. Scarle, *J. Chem. Soc., Dalton Trans.*, 1975, 2552.
128. G. Speier, *Inorg. Chim. Acta*, 1979, **32**, 139.
129. J. Deli and G. Speier, *Transition Met. Chem.*, 1981, **6**, 227.
130. S. Cenini and M. Pizzotti, *Inorg. Chim. Acta*, 1980, **42**, 65.
131. W. E. Newton, J. L. Corbin, D. C. Bravard, J. E. Searles and J. W. McDonald, *Inorg. Chem.*, 1974, **13**, 1100.
132. W. E. Newton and S. Otsuka (eds.), 'Molybdenum Chemistry of Biological Significance', Plenum, New York, 1980.
133. C. D. Garner and S. Bristow, in 'Molybdenum Enzymes', ed. T. G. Spiro, J. Wiley & Sons, New York, 1985, p. 343.
134. M. Nakamoto, K. Shimizu, K. Tanaka and T. Tanaka, *Inorg. Chim. Acta*, 1981, **53**, L51.
135. M. Nakamoto, K. Tanaka and T. Tanaka, *Bull. Chem. Soc. Jpn.* 1985, **58**, 1816.
136. M. Chaudhury, *J. Chem. Soc., Dalton Trans.*, 1984, 115.
137. K. Yamanouchi and S. Yamada, *Inorg. Chim. Acta*, 1974, **9**, 161.
138. Y. Yamanouchi and S. Yamada, *Inorg. Chim. Acta*, 1974, **9**, 83.
139. J. Topich and J. T. Lyons, III, *Polyhedron*, 1984, **3**, 55.
140. W.-K. Goh and M. C. Lim, *Aust. J. Chem.*, 1984, **37**, 2235.
141. K. Dey, R. K. Maiti and J. K. Bhar, *Transition Met. Chem.*, 1981, **6**, 346.
142. I. W. Boyd and J. T. Spence, *Inorg. Chem.*, 1982, **21**, 1602.
143. O. A. Rajan and A. Chakravorty, *Inorg. Chim. Acta*, 1979, **37**, L503.
144. P. K. Nath and K. C. Dash, *Transition Met. Chem.*, 1985, **10**, 262.
145. J. Topich and J. T. Lyon, III, *Polyhedron*, 1984, **3**, 61.
146. J. Topich and J. T. Lyon, III, *Inorg. Chim. Acta*, 1983, **80**, L41.
147. J. Topich and J. T. Lyon, III, *Inorg. Chem.*, 1984, **23**, 3202.
148. P. C. H. Mitchell and D. A. Parker, *J. Chem. Soc., Dalton Trans.*, 1976, 1821.

149. L. M. Charney and F. A. Schultz, *Inorg. Chem.*, 1980, **19,** 1527.
150. H. O. Finklea, S. K. Lahr and F. A. Schultz, *Adv. Chem. Ser.*, 1982, **201,** 709.
151. L. M. Charney, H. O. Finklea and F. A. Schultz, *Inorg. Chem.*, 1982, **21,** 549.
152. J. P. Wilshire, L. Leon, P. Bosserman and D. T. Sawyer, *J. Am. Chem. Soc.*, 1979, **101,** 3379.
153. P. Gosh and D. A. Chakravorty, *Inorg. Chem.*, 1983, **22,** 1322.
154. G. A. Brewer and E. Sinn, *Inorg. Chem.*, 1981, **20,** 1823.
155. R. G. Bhattacharyya and D. C. Bera, *J. Indian. Chem. Soc.*, 1975, **52,** 1212.
156. M. J. Atherton and J. H. Holloway, *J. Chem. Soc., Chem. Commun.*, 1978, 255.
157. B. G. Ward and F. E. Stafford, *Inorg. Chem.*, 1968, **7,** 2569.
158. M. Weidenbruch, C. Pierrard and H. Pesel, *Z. Naturforsch., Teil B*, 1978, **33,** 1468.
159. B. Heyn and R. Hoffmann, *Z. Chem.*, 1976, **16,** 195.
160. R. Barral, C. Bocard, I. Seree de Roch and L. Sajus, *Kinet. Catal.* (*Engl. Transl.*), 1973, **14,** 130.
161. R. Barral, C. Bocard, I. Seree de Roch and L. Sajus, *Tetrahedron Lett.*, 1972, 1693.
162. R. Durant, C. D. Garner, M. R. Hyde, F. E. Mabbs. J. R. Parsons and D. Richens, *J. Less-Common Met.*, 1977, **54,** 459.
163. A. Nakamura, M. Nakamura, K. Sugihashi and S. Otsuka, *Inorg. Chem.*, 1979, **18,** 394.
164. H. M. Neumann and N. C. Cook, *J. Am. Chem. Soc.*, 1957, **79,** 3026.
165. M. S. Reynolds, J. M. Berg and R. H. Holm, *Inorg. Chem.*, 1984, **23,** 3057.
166. X. Lu and J. Sun, *Synth. React. Inorg. Metal-Org. Chem.*, 1982, **12,** 427.
167. R. Durant, C. D. Garner, M. R. Hyde and F. E. Mabbs, *J. Chem. Soc., Dalton Trans.*, 1977, 955.
168. T. Tanaka, K. Tanaka, T. Matsuda and K. Hashi, in ref. 133, p. 361.
169. T. Matsuda, K. Tanaka and T. Tanaka, *Inorg. Chem.*, 1979, **18,** 454.
170. H. Ledon and M. Bonnet, *J. Mol. Catal.*, 1980, **7,** 309.
171. H. Ledon, F. Varescon, T. Malinski and K. M. Kadish, *Inorg. Chem.*, 1984, **23,** 261.
172. T. Malenski, H. Ledon and K. M. Kadish, *J. Chem. Soc., Chem. Commun.*, 1983, 1077.
173. E. A. Maatta and R. A. D. Wentworth, *Inorg. Chem.*, 1979, **18,** 2409.
174. C. Prickett, S. Kumar, P. A. Vella and J. Zubieta, *Inorg. Chem.*, 1982, **21,** 908.
175. G. D. Watt, J. W. McDonald and W. E. Newton, *J. Less-Common. Met.*, 1977, **54,** 415.
176. J. M. Berg and R. H. Holm, *Inorg. Chem.*, 1983, **22,** 1768.
177. J. M. Berg and R. H. Holm, *J. Am. Chem. Soc.*, 1985, **107,** 925.
178. J. T. Spence and P. Knoneck, *J. Less-Common Met.*, 1974, **36,** 465.
179. K. F. Miller and R. A. D. Wentworth, *Inorg. Chem.*, 1977, **16,** 3387.
180. M. W. Peterson and R. M. Richman, *Inorg. Chem.*, 1982, **21,** 2609.
181. C. D. Garner, R. Durant and F. E. Mabbs, *Inorg. Chim. Acta*, 1977, **24,** L29.
182. M. W. Peterson, R. B. van Atta and R. M. Richman, *Inorg. Chem.*, 1983, **22,** 1982.
183. K. Tanaka, M. Honjo and T. Tanaka, *Inorg. Chem.*, 1985, **24,** 2662.
184. N. Ueyama, M. Yano, H. Miyashita, A. Nakamura, M. Kamachi and S. I. Nozakura, *J. Chem. Soc., Dalton Trans.*, 1984, 1447.
185. N. Ueyma, E. Kamada and A. Nakamura, *Chem. Lett.*, 1982, 947.
186. A. F. Isbell and D. T. Sawyer, *Inorg. Chem.*, 1971, **10,** 2449.
187. O. A. Rajan and A. Chakravorty, *Inorg. Chem.*, 1981, **20,** 660.
188. J. Topich, *Inorg. Chem.*, 1981, **20,** 3704.
189. O. A. Rajan, J. T. Spence, C. Leman, M. Minelli, M. Sato, J. H. Enemark, P. M. H. Kroneck and S. Sulgar, *Inorg. Chem.*, 1983, **22,** 3065
190. P. Subramanian, J. T. Spence, R. Ortega and J. H. Enemark, *Inorg. Chem.*, 1984, **23,** 2564.
191. S. Bristow, C. D. Garner and C. J. Pickett, *J. Chem. Soc., Dalton Trans.*, 1984, 1617.
192. G. Backes-Dahmann, W. Herrmann, K. Wieghardt and J. Weiss, *Inorg. Chem.*, 1985, **24,** 485.
193. B. B. Kaul, J. H. Enemark, S. L. Merbs and J. T. Spence, *J. Am. Chem. Soc.*, 1985, **107,** 2885.
194. F. Farchione, G. R. Hanson, C. D. Rodriguez, T. D. Bailey, R. N. Bagchi, A. M. Bond, J. R. Pilbrow and A. G. Wedd, *J. Am. Chem. Soc.*, 1986, **108,** 831.
195. K. Saito and Y. Sasaki, *Adv. Inorg. Bioinorg. Mech.*, 1982, **1,** 179.
196. B. Kamenar, M. Penavic and C. K. Prout, *Cryst. Struct. Commun.*, 1973, **2,** 41.
197. O. N. Krasochka, Y. A. Sokolova and L. O. Atovmyan, *J. Struct. Chem.* (*Engl. Transl.*), 1975, **16,** 648. (*Zh. Strukt. Khim*, 1975, **16,** 696.)
198. L. O. Atovmyan and Y. A. Sokolova, *Chem. Commun.*, 1969, 6491.
199. L. O. Atovmyan, Y. A. Sokolova and V. V. Thachey, *Dokl. Akad. Nauk SSSR*, 1970, **195,** 1355.
200. L. O. Atovmyan and O. N. Krasochka, *Chem. Commun.*, 1970, 1670.
201. R. H. Fenn, *J. Chem. Soc.* (A), 1969, 1964.
202. R. G. W. Gingerich and R. J. Angelici, *J. Am. Chem. Soc.*, 1979, **101,** 5604.
203. L. R. Florian and E. R. Corey, *Inorg. Chem.*, 1968, **7,** 722.
204. C. B. Knobler, A. J. Wilson, R. N. Hider, I. W. Jensen, B. R. Penfold, W. T. Robinson and C. J. Wilkins, *J. Chem. Soc., Dalton Trans.*, 1983, 1299.
205. A. J. Wilson, B. R. Penfold and C. J. Wilkins, *Acta Crystallogr., Sect. C*, 1983, **39,** 329.
206. R. S. Taylor, *Inorg. Chem.*, 1977, **16,** 116.
207. R. J. Butcher and B. R. Penfold, *J. Cryst. Mol. Struct.*, 1976, **6,** 1.
208. A. Chiesi Villa, L. Goghi, A. Gaetani Manfredotti and C. Guastini, *Cryst. Struct. Commun.*, 1974, **3,** 551.
209. B. Viossat and N. Rodier, *Acta Crystallogr.*, 1979, 2715.
210. B. Viossat, P. Khodadad and N. Rodier, *Acta Crystallogr., Sect. B*, 1977, **33,** 3793.
211. B. Viossat, P. Khodadad and N. Rodier, *Acta Crystallogr., Sect. B*, 1981, **37,** 56.
212. M. Hahn, K. Wieghardt, W. Swiridoff and J. Weiss, *Inorg. Chim. Acta*, 1984, **89,** L31.
213. F. A. Schroder, J. Scherle and R. G. Hazell, *Acta Crystallogr., Sect. B*, 1975, **31,** 5311.
214. S. M. Bowen, E. N. Duesler, D. J. McCabe and R. T. Paine, *Inorg. Chem.*, 1985, **24,** 1191.
215. M. Gullotti, A. Pasini, G. M. Zanderighi, G. Ciani and A. Sironi, *J. Chem. Soc., Dalton Trans.*, 1981, 902.

216. K. Tsukuma, T. Kawaguchi and T. Watanabe, *Acta Crystallogr, Sect. B,* 1975, **31,** 2165.
217. I. Buchanan, M. Minelli, M. T. Ashby, T. J. King, J. H. Enemark and C. D. Garner, *Inorg. Chem.,* 1984, **23,** 495.
218. I. Buchanan, C. D. Garner and W. Clegg, *J. Chem. Soc., Dalton Trans.,* 1984, 1333.
219. A. Bruce, J. L. Corbin, P. L. Dahlstrom, J. R. Hyde, M. Minelli, E. I. Stiefel, J. T. Spence and J. Zubieta, *Inorg. Chem.,* 1982, **21,** 917.
220. J. M. Berg, K. O. Hodgson, S. P. Cramer, J. L. Corbin, A. Elsberry and N. Pariyadath, *J. Am. Chem. Soc.,* 1979, **101,** 2774.
221. J. M. Berg, K. O. Hodgson, A. E. Bruce, J. L. Corbin, N. Pariyadath and E. I. Stiefel, *Inorg. Chim. Acta,* 1984, **90,** 25.
222. C. G. Hull and M. H. B. Stiddard, *J. Chem. Soc. (A),* 1966, 1633.
223. M. L. Larson and F. W. Moore, *Inorg. Chem.,* 1966, **5,** 801.
224. R. Kergoat and J. E. Guerchais, *Rev. Chim. Miner.,* 1973, **10,** 585.
225. F. W. Moore and R. E. Rice, *Inorg. Chem.,* 1968, **7,** 2511.
226. A. Müller, V. V. K. Rao and E. Diemann, *Chem. Ber.,* 1971, **104,** 461.
227. A. Kay and P. C. H. Mitchell, *J. Chem. Soc. (A),* 1970, 2421.
228. F. W. Moore and M. L. Larson, *Inorg. Chem.,* 1967, **6,** 998.
229. G. J.-J. Chen, J. W. McDonald and W. E. Newton, *Inorg. Chem.,* 1976, **15,** 2612.
230. W. P. Griffith and T. D. Wickins, *J. Chem. Soc. (A),* 1967, 675.
231. W. P. Griffith and T. D. Wickins, *J. Chem. Soc. (A),* 1968, 400.
232. T. V. Irons and F. E. Stafford, *J. Am. Chem. Soc.,* 1966, **88,** 4819.
233. H. J. Becher, *J. Chem. Res. (S),* 1980, 92; *(M),* 1980, 1053.
234. J.-E. Berg and P.-E. Werner, *Z. Kristallogr.,* 1977, **145,** 3101.
235. A. J. Edwards and B. R. Stevenson, *J. Chem. Soc. (A),* 1968, 2503.
236. W. E. Newton, D. C. Bravard and J. W. McDonald, *Inorg. Nucl. Chem. Lett.,* 1974, **10,** 217.
237. M. Chaudhury, *Inorg. Chem.,* 1985, **24,** 3011.
238. J. Dirand, L. Ricard and R. Weiss, *Transition Met. Chem.,* 1975, **1,** 2.
239. C. G. Young, J. A. Broomhead and C. J. Boreham, *J. Chem. Soc., Dalton Trans.,* 1983, 2135.
240. J. Dirand, L. Richard and R. Weiss, *Inorg. Nucl. Chem. Lett.,* 1975, **11,** 661.
241. K. Leonard, K. Plute, R. C. Haltiwanger and M. Rakowski Dubois, *Inorg. Chem.,* 1979, **18,** 3246.
242. W.-H. Pan, T. R. Halbert, L. L. Hutchings and E. I. Stiefel, *J. Chem. Soc., Chem. Commun.,* 1985, 927.
243. G. J.-J. Chen, J. W. McDonald and W. E. Newton, *Organometallics,* 1985, **4,** 422.
244. W. Holzbach, K. Wieghardt and J. Weiss, *Z. Naturforsch., Teil B,* 1981, **36,** 289.
245. J. H. Holloway and D. Puddick, *Inorg. Nucl. Chem. Lett.,* 1979, **15,** 85.
246. J. Dirand, L. Richard and R. Weiss, *J. Chem. Soc., Dalton Trans.,* 1976, 278.
247. S. Bristow, J. H. Enemark, C. D. Garner, M. Minelli, G. A. Morris and R. B. Ortega, *Inorg. Chem.,* 1985, **24,** 4070.
248. K. Mennemann and R. Mattes, *J. Chem. Res.,* 1979, **5,** 102.
249. E. Schweda and J. Strähle, *Z. Naturforsch., Teil B,* 1981, **36,** 662.
250. K. Dehnicke, J. Schmitte and D. Fenske, *Z. Naturforsch., Teil B,* 1980, **35,** 1070.
251. E. Schweda and J. Strähle, *Z. Naturforsch., Teil B,* 1980, **35,** 1146.
252. K. Dehnicke and J. Strähle, *Angew. Chem., Int. Ed. Engl.,* 1981, **20,** 413.
253. K. Seyferth and R. Taube, *J. Organomet. Chem.,* 1982, **229,** C19.
254. K. Dehnicke, H. Prinz, W. Kafitz and R. Kujanek, *Liebigs Ann. Chem.,* 1981, 20.
255. M. El-Esawi, F. Weller and K. Dehnicke, *Z. Anorg. Allg. Chem.,* 1982, **486,** 147.
256. M. Minelli, C. G. Young and J. H. Enemark, *Inorg. Chem.,* 1985, **24,** 1111.
257. K. Dehnicke and N. Krüger, *Z. Naturforsch., Teil B,* 1978, **33,** 1242.
258. J. Chatt and J. R. Dilworth, *J. Chem. Soc., Chem. Commun.,* 1974, 517.
259. M. W. Bishop, J. Chatt, J. R. Dilworth, M. B. Hursthouse and M. Motevalli, *J. Chem. Soc., Chem. Commun.,* 1976, 780.
260. J. Beck, E. Schweda and J. Strähle, *Z. Naturforsch., Teil B,* 1985, **40,** 1073.
261. J. Sala-Pala and J. E. Guerchais, *J. Chem. Res. (S),* 1978, 227; *(M),* 3074.
262. J. R. Dilworth and J. A. Zubieta, *Transition Met. Chem.,* 1984, **9,** 39.
263. D. Fenske, W. Liebelt and K. Dehnicke, *Z. Naturforsch.,* 1980, **467,** 83.
264. B. Knopp, K. P. Lörcher and J. Strähle, *Z. Naturforsch., Teil B,* 1980, **32,** 1361.
265. U. Müller, E. Schweda and J. Strähle, *Z. Naturforsch., Teil B,* 1983, **38,** 1299.
266. K. Dehnicke, N. Krüger, R. Kujanek and F. Weller, *Z. Kristallogr.,* 1980, **153,** 181.
267. M. B. Hursthouse and M. Motevalli, *J. Chem. Soc., Dalton Trans.,* 1979, 1362.
268. J. Strähle, *Z. Anorg. Allg. Chem.,* 1970, **375,** 238.
269. J. Strähle, U. Weiher and K. Dehnicke, *Z. Naturforsch., Teil B,* **33,** 1978, 1347.
270. J. Chatt, R. Choukroun, J. R. Dilworth, J. Hyde, P. Vella and J. Zubieta, *Transition Met. Chem.,* 1979, **4,** 59.
271. B. L. Haymore, E. A. Maatta and R. A. D. Wentworth, *J. Am. Chem. Soc.,* 1979, **101,** 2063.
272. E. A. Maatta and R. A. D. Wentworth, *Inorg. Chem.,* 1979, **18,** 2409.
273. C. Y. Chou, J. C. Huffman and E. A. Maatta, *J. Chem. Soc., Chem. Commun.,* 1984, 1184.
274. S. J. N. Burgmayer and E. I. Stiefel, *J. Am. Chem. Soc.,* 1986, **108,** 8310.
275. E. A. Maatta, B. L. Haymore and R. A. D. Wentworth, *Inorg. Chem.,* 1980, **19,** 1055.
276. B. K. Burgess, in 'Molybdenum Enzymes', ed. T. G. Spiro, Wiley, New York, 1985, p. 161.
277. P. L. Dahlstrom, J. R. Dilworth, P. Shulman and J. Zubieta, *Inorg. Chem.,* 1982, **21,** 933.
278. R. E. Marsh and A. Toy, *Inorg. Chem.,* 1983, **22,** 1691.
279. J. Chatt, J. R. Dilworth, P. Dahlstrom and J. Zubieta, *J. Chem. Soc., Chem. Commun.,* 1980, 786.
280. F. C. March, R. Mason and K. M. Thomas, *J. Organomet. Chem.,* 1975, **96,** C43.
281. J. R. Dilworth, J. Hutchinson, L. Throop and J. Zubieta, *Inorg. Chem. Acta,* 1983, **79,** 208.

282. J. Chatt, B. A. L. Crichton, J. R. Dilworth, P. Dahlstrom, R. Gutkoska and J. Zubieta, *Inorg. Chem.*, 1982, **21,** 2383.
283. M. W. Bishop, J. Chatt, J. R. Dilworth, M. B. Hursthouse and M. Motevalle, *J. Chem. Soc., Dalton Trans.*, 1978, 1600.
284. J. Chatt, B. A. L. Chrichton, J. R. Dilworth, P. Dahlstrom and J. A. Zubieta, *J. Chem. Soc. Dalton Trans.*, 1982, 1041.
285. R. J. Burt, J. R. Dilworth, G. J. Leigh and J. A. Zubieta, *J. Chem. Soc., Dalton Trans.*, 1982, 2295.
286. G. A. Tsigdinos, 'Molybdenum Chemicals—Chemical Data Series', Bulltein Cd618, Climax Molybdenum Company, June 1973.
287. R. Stomberg, *Acta Chem Scand.*, 1969, **23,** 2755.
288. I. Larking and R. Stomberg, *Acta Chem. Scand.*, 1972, **24,** 3708.
289. R. Stomberg, L. Trysberg and I. Larking, *Acta Chem. Scand.*, 1970, **24,** 2678.
290. C. B. Shoemaker, D. P. Shoemaker, L. V. McAfee and C. W. DeKock, *Acta Crystallogr., Sect. C,* 1985, **41,** 347.
291. C. W. DeKock and L. V. McAfee, *Inorg. Chem.*, 1985, **24,** 4293.
292. C. Djordjevic, K. J. Covert and E. Sinn, *Inorg. Chim. Acta,* 1985, **101,** L37.
293. J. Flanagan, W. P. Griffith, A. C. Skapski and R. W. Wiggins, *Inorg. Chim. Acta,* 1985, **96,** L23.
294. B. Chevrier, T. Diebold and R. Weiss, *Inorg. Chim. Acta,* 1976, **19,** L57.
295. H. Ledon, M. Bonnet and J.-Y. Lallemand, *J. Chem. Soc., Chem. Commun,* 1979, 702.
296. K. M. Kadish, D. Chang, T. Malinski and H. Ledon, *Inorg. Chem.*, 1983, **22,** 3492.
297. M. C. Lim and D. T. Sawyer, *Inorg. Chem.*, 1982, **21,** 2839.
298. A. D. Westland, F. Hague and J. M. Bouchard, *Inorg. Chem.*, 1985, **19,** 2255.
299. E. Reisenhofer and G. Costa, *Gazz. Chim. Ital.*, 1984, **114,** 69.
300. H. Tomioka, K. Takai, K. Oshima, H. Nozak and K. Toriumi, *Tetrahedron Lett.*, 1980, **21,** 4843.
301. I. Larking and R. Stomberg, *Acta Chem. Scand.*, 1970, **24,** 2043.
302. M. Roch, J. Weber and A. F. Williams, *Inorg. Chem.*, 1984, **23,** 4571.
303. J. M. Le Carpentier, A. Mitschler and R. Weiss, *Acta Crystallogr., Sect. B.*, 1972, **28,** 1288.
304. A. Mitschler, J. M. Le Carpentier and R. Weiss, *Chem. Commun.*, 1968, 1260.
305. R. Stomberg, *Acta Chem. Scand.*, 1968, **22,** 1076.
306. R. Kergoat, J. E. Guerchais, A. J. Edwards and D. R. Slim, *J. Fluorine Chem.*, 1975, **6,** 67.
307. A. J. Edwards, D. R. Slim, J. E. Guerchais and R. Kergoat, *J. Chem. Soc., Dalton Trans.*, 1980, 289.
308. J. M. Le Carpentier, R. Schlupp and R. Weiss, *Acta Crystallogr., Sect. B,* **28,** 1972, 1278.
309. M. Pistel, C. Brevard, H. Arzoumanian and J. G. Riess, *J. Am. Chem. Soc.*, 1983, **105,** 4922.
310. K. F. Miller and R. D. Wentworth, *Inorg. Chem.*, 1979, **18,** 984.
311. R. Curci, G. Fusco, O. Sciacovelli and L. Troisi, *J. Mol. Catal.*, 1985, **32,** 251.
312. O. Bortolini, F. DiFuria and G. Modena, *J. Am. Chem. Soc.*, 1981, **103,** 3924.
313. H. Mimoun, *Angew. Chem., Int. Ed. Engl.*, 1982, **21,** 734.
314. C. Venturello, E. Aneri and M. Ricci, *J. Org. Chem.*, 1983, **48,** 3831.
315. F. DiFuria, R. Fornasier and U. Tonellato, *J. Mol. Catal.*, 1983, **19,** 81.
316. B. M. Trost and Y. Masuyama, *Isr. J. Chem.*, 1984, **24,** 134.
317. K. B. Sharpless, *Chemtech,* 692 (1985).
318. K. B. Sharpless, J. M. Townsend and D. R. Williams, *J. Am. Chem. Soc.*, 1972, **94,** 295.
319. K. F. Purcell, *J. Organomet. Chem.*, 1983, **252,** 181.
320. K. F. Purcell, *Organometallics,* 1985, **4,** 509.
321. O. Bortolini, Y. Conte, F. DiFuria and G. Modena, *J. Mol. Catal.*, 1983, **19,** 331.
322. H. J. Ledon, P. Durbut and F. Varescon, *J. Am. Chem. Soc.*, 1981, **103,** 3601.
323. A. Arcoria, F. P. Ballistreri, G. A. Tomaselli, F. DiFuria and G. Modena, *J. Mol. Catal.*, 1984, **24,** 189.
324. Y. Masuyama, M. Takahashi and Y. Kurusu, *Tetrahedron Lett.*, 1984, **25,** 4417.
325. N. A. Johnson and E. S. Gould, *J. Org. Chem.*, 1974, **39,** 407.
326. S. Brandänge, L. Lindbloom and D. Samuelson, *Acta Chem. Scand., Ser. B,* 1977, **31,** 907.
327. A. Frimer, *J. Chem. Soc., Chem. Commun.*, 1977, 205.
328. T. Kawasaki, C. S. Chien and M. Sakamoto, *Chem. Lett.*, 1983, 855.
329. M. M. Midland and S. B. Preston, *J. Org. Chem.*, 1980, **45,** 4514.
330. M. Baccouche, J. Ernst, J. H. Fuhrhop, R. Schlozer and H. Arzoumanian, *J. Chem. Soc., Chem. Commun.*, 1977, **621.**
331. J. Weber, M. Roch and A. F. Williams, *Chem. Phys. Lett.*, 1986, **123,** 246.
332. D. Grandjean and R. Weiss, *Bull. Soc. Chim. Fr.*, 1967, **34,** 3044.
333. R. Stomberg, *Acta Chem. Scand.*, 1970, **24,** 2024.
334. E. O. Schlemper, G. N. Schrauzer and L. A. Hughes, *Polyhedron,* 1984, **3,** 377.
335. W. Winter, C. Mark and V. Schurig, *Inorg. Chem.*, 1980, **19,** 2045.
336. P. Chaumette, H. Mimoun, L. Saussine, J. Fischer and A. Mitschler, *J. Organomet. Chem.*, 1983, **250,** 291.
337. S. E. Jacobson, R. Tang and F. Mares, *Inorg. Chem.*, 1978, **17,** 3055.
338. L. S. Liebeskind, K. B. Sharpless, R. D. Wilson and J. A. Ibers, *J. Am. Chem. Soc.*, 1978, **100,** 7061.
339. D. A. Muccigrosso, S. E. Jacobson, P. A. Apgar and F. Mares, *J. Am. Chem. Soc.*, 1978, **100,** 7063.
340. P. Ghosh, P. Bandyopadhyay and A. Chakravorty, *J. Chem. Soc., Dalton Trans.*, 1983, 401.
341. E. Hofer, W. Holzbach and K. Wieghardt, *Angew. Chem., Int. Ed. Engl.*, 1981, **20,** 282.
342. K. Wieghardt, M. Hahn, J. Weiss and W. Swiridoff, *Z. Anorg. Allg. Chem.*, 1982, **492,** 164.
343. S. Bristow, D. Collison, C. D. Garner and W. Clegg, *J. Chem. Soc., Dalton Trans.*, 1983, 2495.
344. G. N. Schrauzer, E. L. Moorehead, J. H. Grate and L. Hughes, *J. Am. Chem. Soc.*, 1978, **100,** 4760.
345. G. N. Schrauzer, L. A. Hughes, N. Strampach, P. R. Robinson, and E. O. Schlemper, *Organometallics,* 1982, **1,** 44.
346. G. N. Schrauzer, L. A. Hughes, N. Strampach, F. Ross, D. Ross and E. O. Schlemper, *Organometallics,* 1983, **2,** 481.

347. H. Arzoumanian, A. Baldy, R. Lai, J. Metzger, M. L. N. Peh and M. Pierrot, *J. Chem. Soc., Chem. Commun.*, 1985, 1151.
348. D. N. Clark and R. R. Schrock, *J. Am. Chem. Soc.*, 1978, **100,** 6774.
349. R. B. Buchanan and C. G. Pierpont, *Inorg. Chem.*, 1979, **18,** 1616.
350. H. J. Becher, H. J. Brockmeyer and U. Prigge, *J. Chem. Res. (S)*, 1978, 117; *(M)*, 1978, 1670.
351. K. Stadnicka, J. Haber and R. Kolowski, *Acta Crystallogr., Sect. B,* 1977, **33,** 3859.
352. W. G. Klemperer, W. Shum, V. W. Day and M. F. Fredrich, *J. Am. Chem. Soc.*, 1977, **99,** 6146.
353. L. G. McCullough, R. R. Schrock, J. C. Dewan and J. C. Murdzek, *J. Am. Chem. Soc.*, 1985, **107,** 5987.
354. H. Strutz and R. R. Schrock, *Organometallics,* 1984, **3,** 1600.
355. J. R. M. Kress, M. J. M. Russell, M. C. Wesolek and J. A. Osborn, *J. Chem. Soc., Chem. Commun.*, 1980, 431.
356. H. Strutz, J. C. Dewar and R. R. Schrock, *J. Am. Chem. Soc.*, 1985, **107,** 5999.
357. C. G. Pierpont, H. H. Downs and T. G. Rukavina, *J. Am. Chem. Soc.*, 1974, **96,** 5573.
358. K. Stadnicka, J. Haber and R. Kozlowski, *Acta Crystallogr., Sect. B,* 1977, **33,** 3859.
359. A. W. Armour, M. G. B. Dress and P. C. H. Mitchell, *J. Chem. Soc., Dalton Trans.*, 1975, 1493.
360. P. Braunstein, C. DeMeric de Bellefon, M. Lanfranchi and A. Tiripichio, *Organometallics,* 1984, **3,** 1772.
361. V. W. Day, M. F. Fredrich, W. G. Klemperer and W. Shum, *J. Am. Chem. Soc.*, 1977, **99,** 6146.
362. L. O. Atovmyan, V. V. Tkachev and T. G. Shishova, *Dokl. Akad. Nauk SSSR,* 1972, **205,** 609.
363. C. G. Pierpont and R. M. Buchanan, *Inorg. Chem.*, 1982, **21,** 652.
364. J. E. Godfrey and J. M. Waters, *Cryst. Struct. Commun.*, 1975, **4,** 5.
365. S. J. N. Burgmayer and E. I. Stiefel, submitted.
366. C. D. Garner, J. R. Nicholson and W. Clegg, *Angew. Chem.*, 1984, **96,** 961.
367. C. D. Garner, J. R. Nicholson and W. Clegg, *Angew. Chem., Int. Ed. Engl.*, 1984, **12,** 972.
368. B. Piggott, R. N. Sheppard and D. J. Williams, *Inorg. Chim. Acta,* 1984, **86,** L65.
369. C. Knobler, B. R. Penfold, W. T. Robinson, C. J. Wilkins and S. H. Yong, *J. Chem. Soc., Dalton Trans.*, 1980, 248.
370. C. Knobler, B. R. Penfold, W. T. Robinson and C. J. Wilkins, *Acta Crystallogr., Sect. B,* 1981, **37,** 942.
371. C. K. Chew and B. R. Penfold, *J. Cryst. Mol. Struct.*, 1975, **5,** 413.
372. R. Mattes, C. Matz and E. Sicking, *Z. Anorg. Allg. Chem.*, 1977, **435,** 207.
373. M. E. Cass and C. G. Pierpont, *Inorg. Chem.*, 1986, **25,** 123.
374. H. Mimoun, I. Seree de Roch and L. Sajus, *Bull. Soc. Chim. Fr.*, 1969, 1481.
375. B. Piggott, S. F. Wong and R. N. Sheppard, *Inorg. Chem.*, 1985, **107,** 97.
376. M. Cousins and M. L. H. Green, *J. Chem. Soc.*, 1964, 1567.
377. B. J. Brisdon and D. A. Edwards, *Inorg. Nucl. Chem. Lett.*, 1974, **10,** 301.
378. A. Beltran, F. Caturla, A. Cervilla and J. Beltran, *J. Inorg. Nucl. Chem.*, 1981, **43,** 3277.
379. H. J. Becher, G. Müller and N. Amsoneit, *Z. Anorg. Allg. Chem.*, 1976, **420,** 203.
380. H. J. Becher, N. Amsoneit, U. Prigge and G. Gatz, *Z. Anorg. Allg. Chem.*, 1977, **430,** 255.
381. M. Castillo and P. Palma, *Synth. React. Inorg. Metal-Org. Chem.*, 1984, **14,** 1173.
382. K. Gebreyes, S. N. Shaikh and J. Zubieta, *Acta Crystallogr.*, 1985, 871.
383. K. M. Barnhart and J. H. Enemark, *Acta Crystallogr., Sect. C,* 1984, **40,** 1362.
384. F. A. Cotton, S. M. Morehouse and J. S. Wood, *Inorg. Chem.*, 1964, **3,** 1603.
385. C. B. Knobler, W. T. Robinson, C. J. Wilkins and A. J. Wilson, *Acta Crystallogr., Sect. C,* 1983, **39,** 443.
386. K. Matsumoto, Y. Marutani and S.-I. Ooi, *Bull. Chem. Soc. Jpn.*, 1984, **57,** 2671.
387. T. Shibahara, H. Kuroya, S. Ooi and Y. Mori, *Inorg. Chim. Acta,* 1983, **76,** L315.
388. K. Wieghardt, G. Backes-Dahmann, W. Herrmann and J. Weiss, *Angew. Chem., Int. Ed. Engl.*, 1984, **23,** 899.
389. B. Hedman, *Acta Crystallogr., Sect. B,* 1977, **33,** 3077.
390. C. G. Pierpont and R. M. Buchanan, *J. Am. Chem. Soc.*, 1975, **97,** 6450.
391. V. V. Tkachev and L. O. Atovmyan, *Koord. Khim.*, 1976, **2,** 110.
392. S. Brownstein, B. H. Christian, G. Latrenouille and A. Steigel, *Can. J. Chem.*, 1976, **54,** 2543.
393. G. B. Hargreaves and R. D. Peacock, *J. Chem. Soc.*, 1958, 4390.
394. L. B. Handy, *J. Fluorine Chem.*, 1976, **7,** 641.
395. K. Schröder and F. Sladky, *Z. Anorg. Allg. Chem.*, 1981, **477,** 95.
396. A. M. Bond, I. Irvine and T. A. O'Donnell, *Inorg. Chem.*, 1977, **16,** 841.
397. G. M. Anderson and J. M. Winfield, *J. Chem. Soc., Dalton Trans.*, 1986, 337.
398. A. Prescott, D. W. A. Sharp and J. M. Winfield, *J. Chem. Soc., Dalton Trans.*, 1975, 936.
399. J. R. Geichman, E. A. Smith, S. S. Trond and P. R. Ogle, *Inorg. Chem.*, 1962, **7,** 661.
400. E. R. Bernstein and G. R. Meredith, *Chem. Phys.*, 1977, **24,** 289, 301, 311.
401. S. M. Freund and J. L. Lyman, *Chem. Phys. Lett.*, 1978, **55,** 435.
402. M. Mercer, *J. Chem. Soc., Chem. Commun.*, 1967, 1191.
403. E. Jacob, *Angew. Chem., Int. Ed. Engl.*, 1982, **21,** 143.
404. J. Anhaus, P. G. Jones, M. Noltemeyer, W. Pinkert, H. W. Roesky and G. M. Sheldrick, *Inorg. Chim. Acta,* 1985, **97,** L7.
405. M. Cowie and M. J. Bennett, *Inorg. Chem.*, 1976, **15,** 1584.
406. A. E. Smith, G. N. Schrauzer, V. P. Mayweg and W. Heinrich, *J. Am. Chem. Soc.*, 1965, **87,** 5798.
407. J. A. McCleverty, *Prog. Inorg. Chem.*, 1968, **10,** 145.
408. R. Eisenberg, *Prog. Inorg. Chem.*, 1970, **12,** 295.
409. R. H. Prince, *Adv. Inorg. Chem. Radiochem.*, 1979, **22,** 349.
410. K. Yamanouchi and J. H. Enemark, *Inorg. Chem.*, 1978, **17,** 2911.
411. J. R. Dilworth, J. Hyde, P. Lyford, P. Vella, K. Venkatasubramanian and J. A. Zubieta, *Inorg. Chem.*, 1979, **18,** 268.
412. J. K. Gardner, N. Pariyadath, J. L. Corbin and E. I. Stiefel, *Inorg. Chem.*, 1978, **17,** 897.
413. C. G. Pierpont and R. M. Buchanan, *J. Am. Chem. Soc.*, 1975, **97,** 4912.
414. C. G. Pierpont, H. H. Downs and T. G. Rukavina, *J. Am. Chem. Soc.*, 1974, 5573
415. C. G. Pierpont and H. H. Downs, *J. Am. Chem. Soc.*, 1975, **97,** 2123.
416. S. F. Gheller, T. W. Hambley, M. R. Snow, K. S. Murray and A. G. Wedd, *Aust. J. Chem.*, 1984, **37,** 911.

36.6

Molybdenum: Special Topics

C. DAVID GARNER
University of Manchester, UK

36.6.1 MOLYBDENUM–SULFUR SPECIES AS LIGANDS

36.6.1.1 Introduction

The thiomolybdate(VI) anions $[MoO_nS_{4-n}]^{2-}$ ($n = 0$–3) have been prepared by a number of methods[1–3] and, with the exception of $[MoO_3S]^{2-}$, act as versatile ligands to a wide variety of metal ions[4] (see Table 1). In each of the compounds listed in this table, the $[MoO_nS_{4-n}]^{2-}$ ion acts as a bidentate ligand to the metal and the resultant structural versatility has been clearly demonstrated for copper(I) (Table 2). In addition, thiomolybdate(VI) anions are of interest since they have special relevance to aspects of bioinorganic chemistry. Thus, $[MoS_4]^{2-}$ has been widely used in the synthesis of Fe—Mo—S clusters, examples of which serve as spectroscopic analogues for the iron–molybdenum cofactor of the nitrogenases (see Section 36.6.2), and Cu—Mo—S aggregates, which are relevant to the copper deficiency induced by molybdenum in ruminant animals.

36.6.1.2 Synthesis and Properties of $[MoO_nS_{4-n}]^{2-}$ ($n = 0$–2) Complexes

A variety of divalent *d* transition metal cations (M = Fe,[5] Co,[21–23] Ni,[23,24,48] Pd and Pt,[25,26] or Zn[23,24,45]) react with thiomolybdate(VI) anions to form the corresponding bis{thiomolybdate(VI)} complex, which has usually been isolated as its quaternary ammonium or tetraphenylphosphonium salt. Copper(II) reacts with $[NH_4]_2[MoS_4]$ to form a polymeric copper(I) derivative $[NH_4]_n[CuMoS_4]_n$, the anion of which is composed of chains of edge-sharing CuS_4 and MoS_4 tetrahedra.[27] Such polymerization of Cu^I and $Mo^{VI}S_4$ is inhibited by a variety of ligands which bind to the copper. Discrete aggregates with up to four Cu^I atoms bound to a thiomolybdate(VI) anion have been characterized (Table 2). $[(PhSCu)MoS_4]^{2-}$ (type I) and $[(PhSCu)_2MoS_4]^{2-}$ (type II) have been prepared[28] by the action of KSPh on the corresponding cyanide complexes and X-ray crystallography has shown that coordination to the copper causes relatively little perturbation of the MoS_4 unit. Reaction of CuCl and $[MoS_4]^{2-}$ (3:1) in MeCN produces $[(ClCu)_3MoS_4]^{2-}$ (type III), which has a T-shaped structure.[36] $[MoS_4]^{2-}$ reacts with an excess (>4 moles) of CuX (X = Cl, Br) in acetone to form $[(XCu)_4MoS_4]^{2-}$ (type IV). The bromo complex is a one-dimensional polymer, in which four edges of the MoS_4 tetrahedron are bridged by copper atoms; two *trans* copper atoms have trigonal planar stereochemistry with a terminal bromine atom, while the other two coppers are

Table 1 $[MoO_nS_{4-n}]^{2-}$ ($n = 0–2$) Complexes of the d Transition Elements

Complex	*Ref.*	*Complex*	*Ref.*
$[Fe(MoS_4)_2]^{3-}$	5	$[NH_4][CuMoS_4]$	27
$[(PhS)_2Fe(MoS_4)]^{2-}$	6, 7, 8	$[(PhSCu)MoS_4]^{2-}$	28
$[Cl_2FeMoS_4]^{2-}$	7, 9, 10	$[(NCCu)MoS_4]^{2-}$	29, 30, 31
$[(S_5)FeMoS_4]^{2-}$	8, 11, 12	$[(NCCu)MoOS_3]^{2-}$	29
$[Cl_2FeMo(O)S_4]^{2-}$	13	$[(PhSCu)_2MoS_4]^{2-}$	28
$[(PhO)_2FeMoS_4]^{2-}$	14	$[(NCCu)_2MoS_4]^{2-}$	29, 31
$[(NO)_2Fe(MoS_4)_2]^{2-}$	15	$[(Ph_3P)_3Cu_2MoS_4]$	32, 33
$[(Cl_2Fe)_2MoS_4]^{2-}$	16, 17	$[(Ph_3P)_3CuMoOS_3]$	34
$[Fe(MoO_2S_2)_2]^{2-}$	18	$[Cu_2MoS_4(py)_4]$	35
$[(4\text{-}MeC_6H_4S)_2FeS_2FeMoS_4]^{3-}$	19	$[(ClCu)_3MoS_4]^{2-}$	36
$[(bipy)_2RuMoS_4]$	20	$[(ClCu)_3MoOS_3]^{2-}$	34, 37
$[\{(bipy)_2Ru\}_2MoS_4]^{2+}$	20	$[\{Cu_3MoS_3Cl\}(PPh_3)_3S]$	32, 38, 39
$[Co(MoS_4)_2]^{3-}$	21, 22	$[\{Cu_3MoS_3Cl\}(PPh_3)_3O]$	32, 39
$[Co(MoOS_3)_2]^{2-}$	22	$[(BrCu)_4MoS_4]^{2-}$	40
$[Co(MoO_2S_2)]^{2-}$	23	$[(NCAg)MoS_4]^{2-}$	34
$[Ni(MoS_4)_2]^{2-}$	24	$[(Ph_3P)_3Ag_2MoS_4]$	39
$[Ni(MoOS_3)_2]^{2-}$	23	$[(Ph_3P_2Ag)_2MoS_4]$	41
$[Ni(MoO_2S_2)_2]^{2-}$	23	$[\{Ag_4Mo_2S_6\}(PPh_3)_4S_2]$	42, 43
$[Ni(S_2CNEt_2)(MoS_4)]^{-}$	25	$[(Et_3PAu)_2MoS_4]$	44
$[Pd(S_2CNEt_2)(MoS_4)]^{-}$	25	$[(Ph_3PAu)_2MoS_4]$	41
$[Pd(MoS_4)_2]^{2-}$	26	$[(Ph_3P)_3Au_2MoOS_3]$	41
$[Pd(MoOS_3)_2]^{2-}$	25	$[Zn(MoS_4)_2]^{2-}$	24, 45
$[Pt(MoS_4)_2]^{2-}$	26	$[Zn(MoOS_3)_2]^{2-}$	23
$[Pt(MoOS_3)_2]^{2-}$	25		

each tetrahedrally coordinated and are bonded to two bromines which are attached to the copper of an adjacent Cu_4MoS_4 unit.[40]

Compounds of the type $[\{Cu_3MoS_3Cl\}(PPh_3)_3Y]$ (type V) (Y = O, S), which contain a distorted Cu—S—Mo—Cl cubane-like cluster, have been prepared from the reaction between $[MoYS_3]^{2-}$, $CuCl_2$ and PPh_3 in CH_2Cl_2.[32,38,39] $[(ClCu)_3MoOS_3]^{2-}$ (type VIII) involves three copper atoms each bound across an S—S edge of the tetrahedral anion; the MoS_3Cu_3 core corresponds to a cubane-like framework with one position vacant.[34,37] Extraction of this complex with a solution of CuCN and PPh_3 in CH_2Cl_2 gives $[(Ph_3P)_3Cu_2MoOS_3]$ (type VII),[34] which can be considered to have a cubane-like framework with two vacancies. The analogous complex of $[MoS_4]^{2-}$, $[\{(Ph_3P)_2Cu\}(Ph_3PCu)MoS_4]$ (type IX),[32,33] and its silver analogue[41] possess a structure similar to $[(PhSCu)_2MoS_4]^{2-}$ (type II) but with one coinage metal in a trigonal planar and one in a tetrahedral environment, rather than two with a trigonal planar coordination geometry. $[\{(Ph_3P)_2Ag\}(Ph_3PAg)MoS_4]$ was obtained by reacting $[MoS_4]^{2-}$ with PPh_3 and $AgNO_3$ in CH_2Cl_2;[43] by adjusting the stoichiometry of the reaction $[\{Mo_2S_8Ag_4\}(PPh_3)_4]$ may be formed.[42,43] The central unit of this latter compound is a novel cage, arising from the fusion of two MoS_3Ag_2 rings. Gold(I) thiomolybdate(VI) derivatives have been prepared; $[(R_3PAu)_2MoYS_3]$ (Y = S, R = Et;[44] Y = O, S, R = Ph[41]) have been obtained by reacting $[MoYS_3]^{2-}$ and R_3PAuCl (1:2). $[(R_3PAu)_2MoOS_3]$ in CH_2Cl_2 reacts with $Cs_2[MoOS_3]$ to form $[\{(Ph_3P)_2Au\}(Ph_3PAu)MoOS_3]$,[41] which has a structure related to type VII of Table 1.

$[(Bipy)_2RuMoS_4]$ has been prepared by reacting $[MoS_4]^{2-}$ with $[(bipy)_2RuCl_2]$ (1:1) and this compound will bind a further ruthenium centre to form $[\{(bipy)_2Ru\}_2MoS_4]^{2+}$. Both complexes have been investigated as catalysts for C_2H_2 reduction and, in the presence of MeOH or other proton source, the electrochemically reduced species react to give C_2H_4 and C_2H_6.[20]

Thiomolybdate(VI) complexes of d^8 metal centres $[M'(MoYS_3)_2]^{2-}$ (M = Ni, Pd, Pt) have been prepared by the reaction of MCl_2 with $[MoYS_3]^{2-}$ (Y = O, S) (1:2) in H_2O/MeCN solution.[22–24,26,48] The 'mixed ligand' complexes $[(Et_2NCS_2)M'(MoS_4)]^{-}$ (M' = Ni, Pd) have been prepared by the reaction of $[MoS_4]^{2-}$ and $[M'(S_2CNEt_2)_2]$ (1:1) in acetone.[25] These systems involve a square planar geometry at the d^8 metal. $[Co(MoS_4)_2]^{3-}$, formally a Co^I complex, has been prepared by reacting $CoCl_2$, $[Et_3NH][SPh]$ and $[MoS_4]^{2-}$ in MeCN[21] and the magnetic properties of the complex are consistent with a tetrahedral environment about the cobalt.

Heating $[NH_4]_2[MoS_4]$ in DMF, alone or in the presence of benzenethiol, leads to the formation of $[Mo_3S_9]^{2-}$.[46] $[PPh_4]_2[MoO(MoS_4)_2]$ has been prepared by heating $[NH_4]_2[MoS_4]$ in

Table 2 Structural Diversity Observed for Thiomolybdate(VI) Complexes of Copper(I)

Complex	*Solid state structure*	*Type*	*Ref.*
$[(PhSCu)MoS_4]^{2-}$ $[(NCCu)MoS_4]^{2-}$	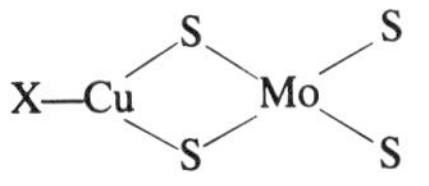	I	28 29
$[(PhSCu)_2MoS_4]^{2-}$ $[(BrCu)_2MoS_4]^{2-}$ $[(ClCu)_2MoS_4]^{2-}$ $[(NCCu)_2MoS_4]^{2-}$	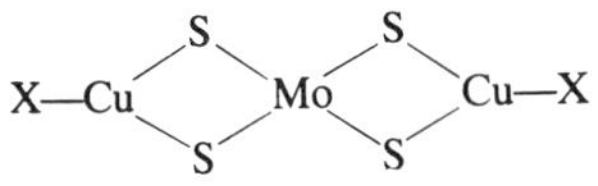	II	28
$[(BrCu)_3MoS_4]^{2-}$ $[(ClCu)_3MoS_4]^{2-}$	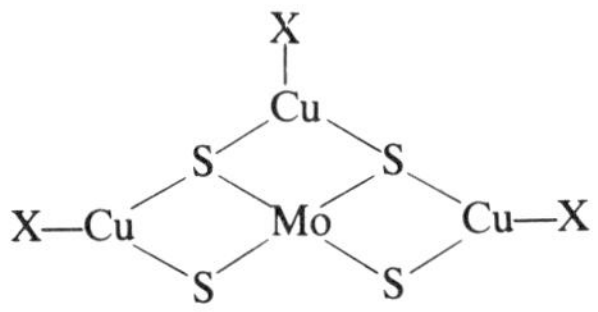	III	36
$[(BrCu)_4MoS_4]^{2-}$	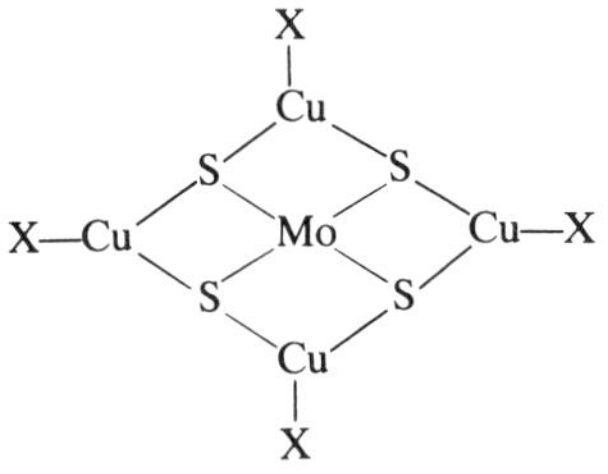	IV	40
$[\{Cu_3MoS_3Cl\}(PPh_3)_3S]$ $[\{Cu_3MoS_3Cl\}(PPh_3)_3O]$	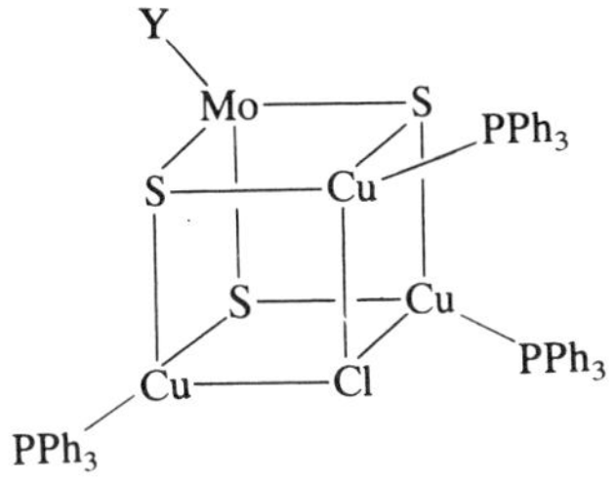	V	38, 39 32, 39
$[(NCCu)MoOS_3]^{2-}$	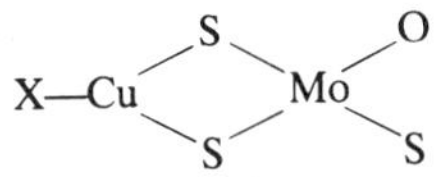	VI	29
$[\{(CuPPh_3)[Cu(PPh_3)_2]\}MoOS_3]$	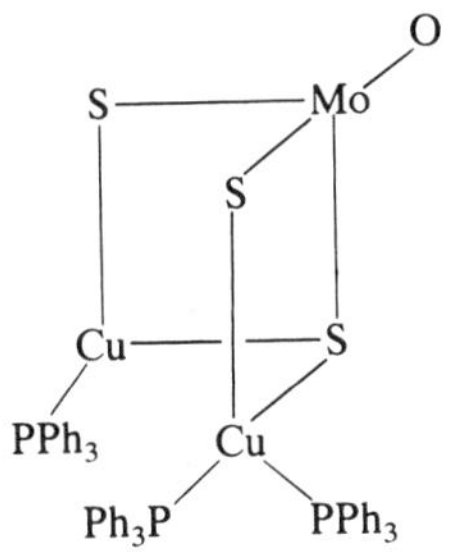	VII	32, 33, 34
$[(ClCu)_3MoOS_3]^{2-}$	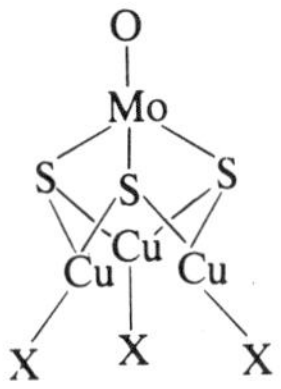	VIII	34, 37
$[(Ph_3P)_3Cu_2MoS_4]$	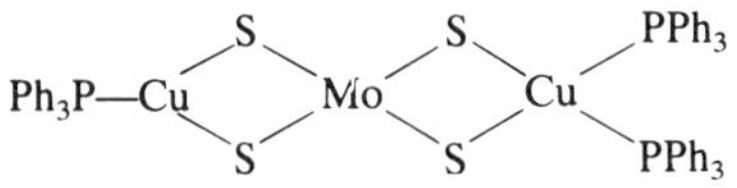	IX	32, 33

MeCN–EtOH in the presence of $Na[BH_4]$.[22] These anions involve a central $Mo^{IV}Y$ ($Y = O, S$) moiety coordinated by two bidentate $Mo^{VI}S_4$ groups and the central Mo^{IV} atom has a square pyramidal geometry. In the case of $[Mo_3S_9]^{2-}$ the central molybdenum is located 0.62 Å above the plane of the four thiomolybdate(VI) sulfurs; the shortest Mo—S bond (2.086(4) Å) is that of the Mo^{IV}—S group; within the $Mo^{VI}S_4$ groups, the terminal Mo—S bonds ($Mo—S_t$ = 2.231(3)–2.251(3) Å) are significantly shorter than the bridging bonds ($Mo—S_b$ = 2.369(3)–2.429(3) Å; $S_b—Mo—S_b$ (104.1(1) and 103.0(1)°) are reduced from the value in $[MoS_4]^{2-}$; and the Mo···Mo separations are *ca.* 2.95 Å.[46] Several of these features, *viz.* the lengthening of the $Mo—S_b$ bonds, the reduction in $S_b—Mo—S_b$ below the tetrahedral value, and the close approach of the metal atoms, are general to the complexes formed by $[MoS_4]^{2-}$ and related anions. For the 3*d* transition metals (M = Fe, Co, Ni and Cu) the Mo···M separations are in the range 2.63–2.80 Å. Thus, the Mo···M separations in the thiomolybdate(VI) complexes are sufficiently close to be consistent with the formation of an Mo—M bond; this could arise by donation of a pair of electrons from a filled orbital on M into an empty orbital on Mo.

$[MoO_nS_{4-n}]^{2-}$ ($n = 0$–2) and their complexes have been extensively investigated by IR and Raman spectroscopy, especially in respect of their ν(Mo—O) and ν(Mo—S) stretching vibrations, which occur in the ranges 800–1000 and 400–500 cm^{-1}, respectively. The Mo—S stretching modes of $[MoS_4]^{2-}$ have been assigned;[4] the totally symmetric (a_1) stretching mode occurs at 458 cm^{-1} and the triply degenerate, asymmetric (t_2) stretching mode at 472 cm^{-1}. The symmetry of the MoS_4 group is lowered upon complex formation and, in general, the number of IR and Raman active vibrations observed is consistent with the symmetry of the aggregate.[24–26,50] Resonance Raman spectra have been recorded for thiomolybdate(VI) complexes,[50,51] with excitation using laser lines of energy within the contour of the lowest energy sulfur-to-metal charge-transfer transition. The totally symmetric ν(Mo—S) stretches are easily identified as giving rise to the most intense bands and, especially for $[MoS_4]^{2-}$ complexes of Cu^I, the pattern (*i.e.* the number of bands, their frequencies and relative intensities) observed in the ν(Mo—S) region is distinctive for the number of Cu^I atoms coordinated to $[MoS_4]^{2-}$.[50]

The UV/visible absorption spectra of $[MoO_nS_{4-n}]^{2-}$ ($n = 0$–2) complexes are dominated by the sulfur-to-metal charge transfer transitions of the anion and these bands often serve to characterize the particular system, even though these absorptions may be correlated with those of the parent anion.[4] The reduction in symmetry due to complex formation often leads to a broadening or splitting of the bands and, in the majority of the *d* transition metal complexes, the lowest energy band is red-shifted by an amount which is dependent both upon the nature of the transition metal and the number of such metals bound per $Mo^{VI}O_nS_{4-n}$ group.

^{95}Mo NMR spectra have been reported for Cu^I, Ag^I and Au^I complexes of the $[MoO_nS_{4-n}]^{2-}$ ($n = 0$–2) ions.[28,29,32,41,49] The resonance position is characteristic of the complex and, as is seen clearly for the Cu^I systems, the chemical shift (from $[MoO_4]^{2-}$) decreases regularly (by *ca.* 500 p.p.m.) for each CuX unit bound to the thiomolybdate(VI) core.

Cyclic voltammetric investigation[4,26,52,53] of $[M(MoO_nS_{4-n})_2]^{2-}$ (M = Ni, $n = 0$–2; M = Pd, Pt, $n = 0$) complexes has shown that each can be reduced reversibly in a one-electron process and a further one-electron reduction is possible for some of the complexes. The first reduction is primarily centred on the d^8 metal but the second reduction is not so easily described, since the nature of it shows a sensitivity to both the nature of M and the thiomolybdate(VI) anion.[26] The electrochemical behaviour of $[Co(MoS_4)_2]^{3-}$ has also been investigated and two reversible processes, assigned to the 2−/3− and 3−/4− couples, have been observed.[21]

36.6.1.3 Other Mo—S Species as Ligands

The molecular fragment *cis*-$Mo(SBu^t)_2$ is capable of acting as a bidentate ligand; reaction of $[Mo(SBu^t)_4]$ or *cis*-$[Mo(SBu^t)_2(CNBu^t)_4]$ with $[CuBr(CNBu^t)_3]$ in acetone under dinitrogen produces $[(Bu^tNC)_4Mo(\mu\text{-}SBu^t)_2CuBr]$. Two isomers, differing in the relative arrangements of the *t*-butyl substituents on the thiolato bridges, have been crystallized and structurally characterized and their interconversion in solution demonstrated by ^1H NMR spectroscopy.[54] Similarly, $[Mo(SBu^t)_4]$ reacts with $[Fe_2(CO)_9]$ in toluene to give $[MoFe_2(SBu^t)_4(CO)_8]$ as the main product. Two carbonyls have been transferred to the molybdenum in the reaction and the resultant coordination about this metal corresponds to that of a distorted trigonal prism. The ^1H NMR spectrum shows two Bu^tS signals at room temperature which coalesce upon warming.[55] *cis*-$[Mo(SBu^t)_2(CNR)_4]$ (R = cyclohexyl, *t*-butyl) reacts with FeX_2 (X = Cl, Br) to give $[(RNC)_4Mo(\mu\text{-}SBu^t)_2FeX_2]$, and the related system $[Cp_2Mo(\mu\text{-}SBu^t)_2FeCl_2]$ involves an Mo · · · Fe separation of 3.660(3) Å.[56]

cis-$[MoS_2(pipNO)_2]$ (pipNO = piperidine-*N*-oxide) reacts with CuCl to form the chloride-bridged dimer $[\{(pipNO)_2MoS_2CuCl\}_2]$ (Figure 1). Coordination to the copper atom lengthens the Mo—S from 2.145(2) Å, in the parent Mo^{VI} complex, to 2.179(4) Å; concomitantly the S—Mo—S interbond angle is reduced from 115.0(1) to 110.3(1)°. The Cu · · · Mo separations are 2.639(1) Å.[57] All of these structural features are typical of the coordination of $[MoS_4]^{2-}$ molybdenum-sulfur centres as bidentate ligands to Cu^I and other metals.

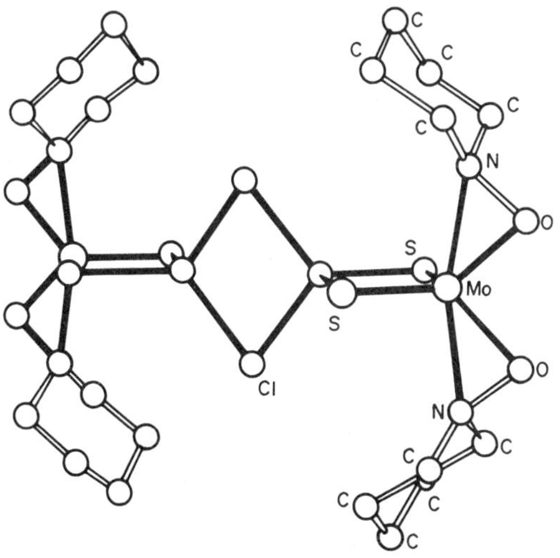

Figure 1 Structure of $[\{\text{piperidine } N\text{-oxide}(1-)\}_2MoS_2CuCl\}_2]$[57]

36.6.2 IRON–MOLYBDENUM–SULFUR CLUSTERS

Interest in the chemistry of Fe—Mo—S clusters has been stimulated by the discovery that the iron–molybdenum cofactor (FeMoco) of the nitrogenase enzymes is an aggregate of these elements, organized in a manner yet unidentified. The nitrogenases catalyze the reduction of dinitrogen to ammonia in nitrogen-fixing organisms. The enzyme complex consists of two types of protein, the Fe protein and the FeMo protein, and activity requires both proteins, ATP, a divalent cation (preferably Mg^{2+}) and a reductant. The Fe protein, which involves an Fe_4S_4-cubane-like cluster, acts as an electron transfer agent to the more complex FeMo protein which contains the catalytic site for substrate reduction.[58] There are two molybdenum atoms per FeMo protein molecule and each is involved with one FeMoco; these entities may be extracted from the protein,[59,60] have been shown to be common to molybdenum-containing nitrogenases of various organisms, and can activate extracts of the FeMo protein component of inactive mutant strains of different microorganisms.[61] In the organism *Klebsiella pneumoniae,* a cluster of 17 genes in seven transcriptional units has been associated with nitrogen fixation. One class of mutants, the *nif* V^- mutants, has altered substrate specificity and the enzyme from these mutants is relatively ineffective at dinitrogen reduction. This altered substrate specificity has been shown to be associated with FeMoco,[62] directly implicating this entity as part of the catalytically active site of the nitrogenases. The elemental composition of FeMoco has been estimated by a variety of analytical and spectroscopic procedures, leading to ranges of values for iron and sulfur per molybdenum of $MoFe_{6-8}S_{4-9}$.[59,60,63] There is no evidence for the presence of any organic entity associated with FeMoco, other than solvents[64] such as *N*-methylformamide (NMF) used in its abstraction.[59]

FeMoco, both as a constituent of the FeMo protein and an isolated entity, has been the subject of detailed spectroscopic examination.[65] ^{57}Fe Mössbauer and EPR studies of the cofactor have been interpreted in terms of an $S = \frac{3}{2}$ centre that contains one molybdenum and *ca.* six irons in a spin-coupled structure. The EPR signal serves as a valuable fingerprint of FeMoco; furthermore, release of FeMoco from the FeMo protein produces an EPR spectrum with broader features, but the same profile, thereby indicating that the core of this cluster is little changed by the extraction procedure. Treatment of FeMoco with *ca.* one equivalent of benzenethiol sharpens the EPR spectrum so that it more closely resembles that of the MoFe protein.[66,67] In the FeMo protein, FeMoco is capable of existing in at least three distinct levels of oxidation. The $S=\frac{3}{2}$ level is oxidized by the removal of one electron per molybdenum to an EPR-silent species and, during steady-state turnover, the cofactor is converted into an EPR-silent, super-reduced state.[60] The oxidized state can be reproduced by oxidation of FeMoco in NMF solution by methylene blue.[67]

The most definitive structural information concerning the molybdenum site in FeMoco and the FeMo protein has been provided by X-ray absorption near edge structure (XANES) and extended X-ray absorption fine structure (EXAFS) spectroscopy.[68–70] The profile of the molybdenum *K*-absorption-edge indicates the absence of terminal oxo groups and the position of this edge suggests an oxidation state of *ca.* +IV. Interpretation of the EXAFS has shown that, in the MoFe protein, the molybdenum possesses a first coordination sphere of three to four sulfur atoms 2.36 Å away, with a shell of two to three iron atoms at a distance of 2.68 Å; these structural features remain essentially unchanged for FeMoco in NMF, but there is evidence for an additional shell of two to three oxygen (or nitrogen) atoms at 2.10 Å. These latter atoms could originate from the solvent and IR spectroscopic studies of solutions of FeMoco in NMF have identified an intense band at 1600 cm^{-1}, which has been assigned to the C=O stretching of N-bonded, deprotonated solvent.[64]

The lack of precision in the structural definition of FeMoco has allowed much scope for the synthesis of possible chemical analogues and, thereby, stimulated the preparation and structural and spectroscopic characterization of Fe—Mo—S clusters.[71,72] The majority of these synthetic approaches employ $[MoS_4]^{2-}$ as a reagent and they can be classified in two main categories: (a) ligand exchange reactions between $Fe(L)_n$ complexes and $[MoS_4]^{2-}$ to give complexes containing tetrathiomolybdate(VI) anions acting as ligands[71] and (b) 'spontaneous self-assembly' reactions between $[MoS_4]^{2-}$ and $FeCl_3$, in the presence of a thiol, which yield Fe_3MoS_4 cubane-like clusters in dimeric species which, by subsequent manipulations, have been converted into monomeric species.[72]

Examples of complexes formed between iron and tetrathiomolybdate(VI) anions are given in Table 1,[5–19] and aspects of the vibrational and electronic spectra associated with their molybdenum centres are discussed in Section 36.6.1.2. Magnetic moments and ^{57}Fe Mössbauer spectra of $[S_2MoS_2FeS_2MoS_2]^{3-}$,[73] $[X_2FeS_2MoS_2]^{2-}$ (X = Ph, Cl, SPh, OPh; $X_2 = S_5$)[7,8,11,14] and $[Cl_2FeS_2MoS_2FeCl_2]^{2-}$ [16,17,74] have been recorded and interpreted in terms of Fe^{II} centres which are spin coupled to the $\{MoS_4\}^{2-}$ groups, with some transfer of electron density from the former to the latter. $[NEt_4]_2[(PhS)_2FeS_2FeS_2MoS_2]$ has a ^{57}Fe Mössbauer spectrum consistent with an Fe^{III} ($S=\frac{3}{2}$) centre antiferromagnetically coupled to a formally $\{Fe^{II}S_2Mo^{VI}S_2\}$ ($S=2$) unit to give an $S=\frac{1}{2}$ ground state.[19] The ^{57}Fe isomer shift in $[S_2MoS_2FeS_2MoS_2]^{3-}$ is smaller than that expected for an oxidation state of I or II for the iron, suggesting extensive Fe→Mo charge transfer; the EPR spectrum indicates an $S=\frac{3}{2}$ ground state, similar to that of FeMoco.[5,73] Electrochemical studies on $[X_2FeS_2MoS_2]^{2-}$ complexes (X = SPh,[7] OPh[14]) show an irreversible reduction at −1.34 V or −1.51 V *vs.* SCE for the thiophenolate and phenolate complexes, respectively. The $[(PhS)_2FeS_2FeS_2MoS_2]^{3-}$ ion displays an irreversible oxidation at 0.05 V and an irreversible reduction at −1.1 V *vs.* SCE.[19] The irreversible electrochemical reduction of $[Cl_2FeS_2MoS_2FeCl_2]^{2-}$ occurs at −0.56 V *vs.* SCE, with loss of a chloride to form $[MoS_4Fe_2Cl_3]^{2-}$, which may exist as a dimer in solution.[16] Mo *K*-edge EXAFS spectra have been recorded for the binuclear complexes $[X_2FeS_2MoS_2]^{2-}$ (X = Cl, SPh, OPh) and the trinuclear complexes $[(4\text{-}Me_3C_6H_4S)_2FeS_2FeS_2MoS_2]^{3-}$, $[S_2MoS_2FeS_2MoS_2]^{3-}$ and $[Cl_2FeS_2MoS_2Cl_2]^{2-}$.[75,76] The Mo—S and Mo—Fe distances determined agree with the X-ray crystallographic data to better than 0.5%. Comparisons of the Fourier transforms of the EXAFS spectra with that of the MoFe protein of nitrogenase show them to be significantly different, and suggest a first coordination sphere around molybdenum in the enzyme that is not simply tetrahedral.

Despite the undoubted chemical interest of the complexes involving $[MoS_4]^{2-}$ ligated to iron, the second category of Fe—Mo—S clusters, involving Fe_3MoS_4 cubane-like cores, appear to have more relevance to FeMoco. Reactions between $FeCl_3$, $[MoS_4]^{2-}$ and an excess of a thiolate anion in methanol solution yield several products based on cubane-like clusters containing one molybdenum and three iron atoms in a tetrahedral arrangement, with each face bridged by a μ_3-sulfide ligand.[72] By varying the stoichiometry, reaction times and cations present in solution, three different types of product have been isolated from this system: $[Fe_6Mo_2S_8(SR)_9]^{3-}$, $[Fe_6Mo_2S_8(SR)_6(OMe)_3]^{3-}$ and $[Fe_7Mo_2S_8(SR)_{12}]^{3-}$ (Figure 2). Each of these anions contains two $\{(RSFe)_3(\mu_3\text{-}S)_4Mo\}$ clusters with each molybdenum atom bonded to three atoms of a bridging system. The anions $[Fe_6Mo_2S_8(SR)_9]^{3-}$ ($R = Et, Ph, 4\text{-}ClC_6H_4, 4\text{-}MeC_6H_4$) and $[Fe_6Mo_2S_8(SPh)_6(OMe)_3]^{3-}$ contain three bridging $(SR)^-$ or $(OMe)^-$ ligands, which have been shown by ^1H NMR spectroscopy not to interchange with the terminal ligands in solution.[77,78] $[Me_3NCH_2Ph]_3[Fe_7Mo_2S_8(SEt)_{12}]$,[79] and the related (reduced) anion of $[NBu^n_4]_4[Fe_7Mo_2S_8(SCH_2Ph)_{12}]$,[78] involve a $(\mu\text{-}S)_3Fe(\mu\text{-}S)_3$ bridging arrangement containing a central Fe^{III}, or Fe^{II}, atom, respectively.

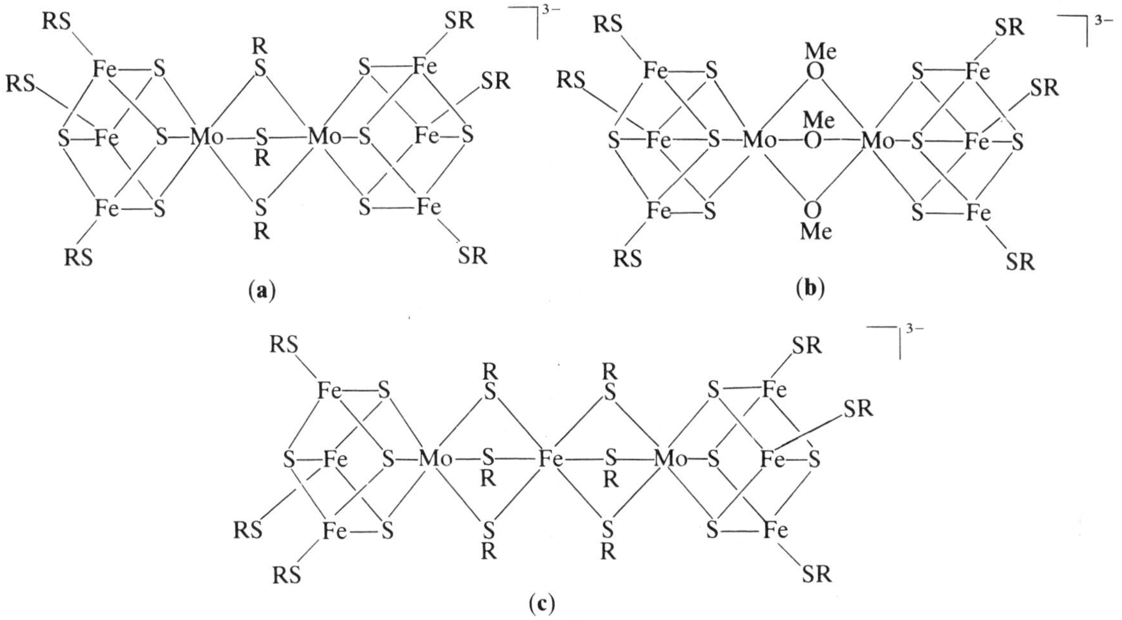

Figure 2 Structure of some dimeric assemblies of Fe_3MoS_4 cubane-like clusters;[72] (a) $[Fe_6Mo_2S_8(SR)_9]^{3-}$; (b) $[Fe_6Mo_2S_8(SR)_6(OMe)_3]^{3-}$; (c) $[Fe_7Mo_2S_8(SR)_{12}]^{3-}$

The bridging system in $[Fe_7Mo_2S_8(SEt)_{12}]^{3-}$ can be broken by reaction with catechol, $catH_2$, which yields $[Fe_4MoS_4(SEt)_3(cat)_3]^{3-}$ (Figure 3). The crystal structure of its $[NEt_4]^+$ salt shows this anion to consist of a single $\{(EtSFe)_3MoS_4\}$ cluster, with the molybdenum coordinated *via* three Mo—O—Fe bridges to an $Fe(cat)_3$ group acting as a tridentate ligand.[80] The reaction between $[NEt_4]_3[Fe_7Mo_2S_8(SEt)_{12}]$ and 3,6-dipropylcatechol, Pr_2catH_2, yields $[NEt_4]_4[Fe_6Mo_2S_8(SEt)_6(Pr_2cat)_2]$. This anion contains two $\{Fe_3MoS_4\}$ cubane-like clusters with each molybdenum chelated by a $(Pr_2cat)^{2-}$ ligand; two of the iron atoms in each cube are coordinated to one terminal ethanethiolate, and the third iron atom is bridged to the molybdenum atom of the other cube by a bridging ethanethiolate group. Similar products have been obtained with 3,6-diallylcatecholate, $(al_2cat)^{2-}$ (Figure 4) and tetrachlorocatecholate, $(Cl_4cat)^{2-}$, ligands.[81,82]

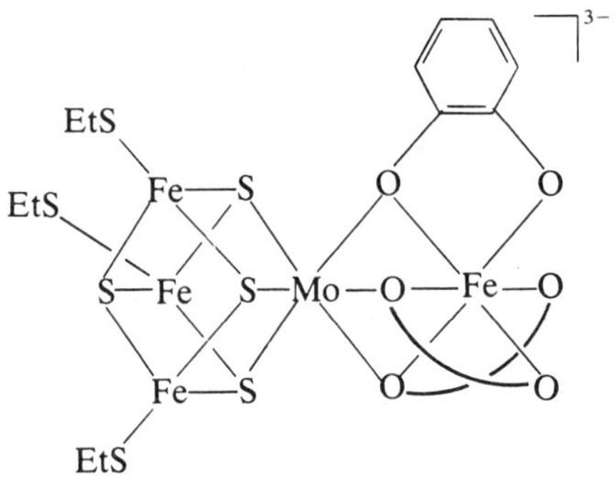

Figure 3 Structure of $[Fe_4MoS_4(SEt)_3(cat)_3]^{3-}$ [80]

Ligand exchange reactions between $[Fe_6Mo_2S_8(SEt)_9]^{3-}$ or $[Fe_7Mo_2S_8(SEt)_{12}]^{3-}$ anions and thiols, acetyl chloride, benzoyl chloride, benzoyl bromide or phenol have shown that the

COC3-TT

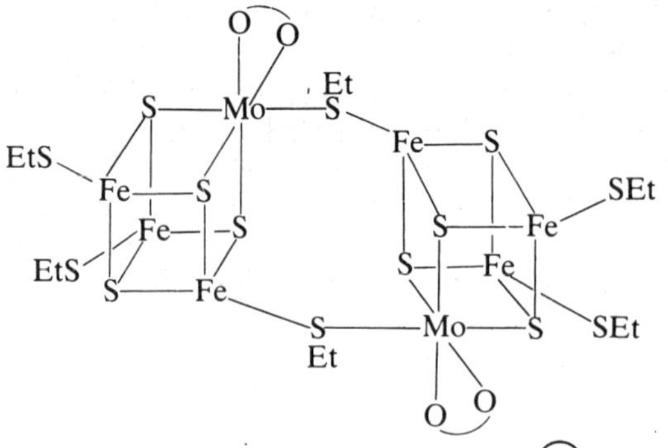

Figure 4 Structure of $[Fe_6Mo_2S_8(SEt)_6(3,6\text{-diallylcatecholate})_2]$ ($\widehat{O\ O}$ = 3,6-diallylcatecholate)[81]

terminal thiolate ligands are labile and easily replaced by other thiolate groups, halides, or phenolates; the bridging ligands are inert to substitution.[83–85] However, the reaction between arenethiols (RSH) and $[Fe_6Mo_2S_8(SEt)_6(R'_2cat)_2]^{4-}$ (see Figure 4) yields $[Fe_6Mo_2S_8(SR)_6(R'_2cat)_2]^{4-}$ (R = Ph, 4-MeC_6H_4, 4-ClC_6H_4; R′ = Pr, allyl) in which both bridging thiolates have been substituted as well as the terminal ligands. In coordinating solvents, including DMSO, DMF, THF and MeCN, $[Fe_6Mo_2S_8(SR)_6(R'_2cat)_2]^{4-}$ double cubane aggregates are cleaved to give the single Fe_3MoS_4 cubane-like cluster solvates $[Fe_3MoS_4(SR)_3(R'_2cat)(solvent)]^{2-}$ (see Figure 5). These systems possess specific reactivities at both the molybdenum and the iron sites.[81,82,86,87] Thus, the solvent molecule is readily replaced by a variety of ligands (RS^- (Figure 5), N_3^-, CN^-, RO^- and PR_3) and catecholate exchange can be effected at the molybdenum site. Normal thiolate exchange at the iron sites occurs and using acetyl chloride results in chloride for thiolate substitution. Reaction of the product $[Fe_3MoS_4Cl_3(cat)(solvent)]^{2-}$ with CN^- produces $[Fe_3MoS_4(CN)_4(cat)]^{3-}$, in which this substrate of nitrogenase is bound at each iron and molybdenum atom.

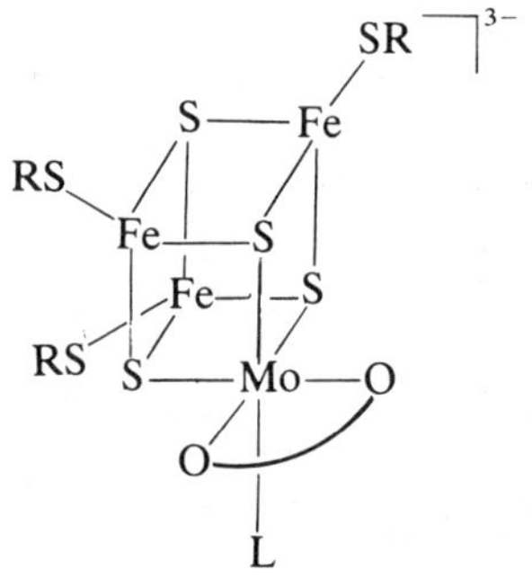

Figure 5 Structure of $[Fe_3MoS_4(SC_6H_4Cl\text{-}4)_4(3,6\text{-diallylcatecholate})]^{3-}$ (R = 4-ClC_6H_4; L = 4-ClC_6H_4S; ($\widehat{O\ O}$ = 3,6-diallylcatecholate)[86]

A novel double Fe_3MoS_4 cubane-like complex bridged by two persulfide groups has been prepared *via* an intermediate carbonyl cluster. Displacement of the chlorides from $[Cl_2FeS_2MoS_2]^{2-}$ by $[(CO)_6Fe_2S_2]^{2-}$ yields $[(CO)_6Fe_2S_2FeS_2MoS_2]^{2-}$; this reacts with (4-$ClC_6H_4S)_2$ to give $[Fe_6Mo_2S_8(S_2)_2(SC_6H_4Cl\text{-}4)_6]^{2-}$, in which the two persulfides form a bridge between the molybdenum atoms of two $\{Fe_3MoS_4\}$ units. This type of cluster can also be prepared in a self-assembly reaction from $[MoS_4]^{2-}$, $FeCl_3$, a thiolate and Na_2S_2 in methanol.[88]

The electronic structure of compounds containing the single and double Fe_3MoS_4 cubane-like clusters has been investigated by ^{57}Fe Mössbauer, EPR and NMR spectroscopy, and by magnetic moment measurements.[72,89–92] These results have shown that each $\{Fe_3MoS_4\}^{3+}$ framework is electronically delocalized and the coupling of electron spins produces an $S = \frac{3}{2}$ ground state. The EPR spectra of these single cubane systems show little dependence on the nature of the ligands bound to molybdenum and, of particular significance, closely resemble the EPR spectrum characteristic of FeMoco with the same magnetic ground state. Furthermore, none of these spectra manifest molybdenum hyperfine coupling. In the triply bridged double cubanes $[Fe_6Mo_2S_8(SR)_9]^{3-}$, the two $\{Fe_3MoS_4\}^{3+}$ sub-clusters are only weakly coupled, but in the doubly bridged anions $[Fe_6Mo_2S_8(SR)_6(R'_2cat)_2]^{4-}$ there is strong spin coupling between the cubanes to give an $S = 0$ ground state. The ^{57}Fe Mössbauer isomer shifts of compounds possessing one or two $\{Fe_3MoS_4\}^{3+}$ clusters compare closely with those for the $S = \frac{3}{2}$ state of FeMoco.

Electrochemical studies have shown that each $\{Fe_3MoS_4\}^{3+}$ cubane-like cluster exhibits a reversible one-electron reduction and oxidation process, giving rise, for example, in the dimeric species $[Fe_6Mo_2S_8(SR)_9]^{3-}$ to 1−/2−, 2−/3−, 3−/4− and 4−/5− couples.[72,85,86,89,91,93,94] ^{57}Fe Mössbauer isomer shifts indicate that the redox changes occur primarily at the iron centres.[91] Chemical reduction using sodium acenaphthylenide generates the reduced complexes in solution, and the doubly reduced complex $[Fe_6Mo_2S_8(SPh)_9]^{5-}$ has been crystallized as its $[NEt_4]^+$ salt. This anion retains the double cubane structure of its parent, but with an increased Mo—Mo separation.[94] The reduced clusters react with PhSH to produce dihydrogen, and can catalyze electrochemical reduction reactions, including the conversion of ethyne to ethene and reduction of dinitrogen to ammonia.[95–101]

Crystal structure determinations of the single and double cubane complexes have shown that the internal dimensions of the $\{Fe_3MoS_4\}$ clusters are independent of the nature of the other ligands present and, with Mo—Fe distances of 2.72 ± 0.01 Å and Mo—S bond lengths of 2.35 ± 0.01 Å, closely resemble the environment suggested for molybdenum in FeMoco on the basis of molybdenum *K*-edge EXAFS measurements. The molybdenum *K*-edge EXAFS of $[NBu_4]_3[Fe_6Mo_2S_8(SEt)_9]$, $[NEt_4]_3[Fe_6Mo_2S_9(SEt)_8]$ and $[NEt_3(CH_2Ph)]_3[Fe_6Mo_2S_8(SEt)_3(OPh)_6]$ agree well with crystallographic results, showing back-scattering from both a shell of sulfur atoms and a further shell of iron atoms. The molybdenum XANES of nitrogenases are similar to those of Fe_3MoS_4 cubane-like clusters possessing an MoS_3O_3 coordination.[70] However, unlike FeMoco, the iron *K*-edge EXAFS spectra of Fe_3MoS_4 clusters show no net back-scattering from the molybdenum and iron atoms due to destructive interference of the Fe–Fe and Fe–Mo waves.[102] Therefore, the Fe_3MoS_4 cubane-like clusters are deficient in this respect as models of FeMoco. A more fundamental objection is that their composition is not encompassed by the range of Fe:Mo:S values suggested for FeMoco.

Other approaches to the synthesis of Fe—Mo—S clusters have been made, especially using metal carbonyls as starting materials. The reaction between $[RMo(CO)_2]_2$ (R = Cp, MeCp) and $[FeS(CO)_3]_2$ yields two isomeric iron–molybdenum–sulfur clusters with different $\{Fe_2Mo_2S_2\}$ skeletons, but both products can be formulated as $[R_2Mo_2Fe_2(\mu_3\text{-}S)_2(\mu\text{-}CO)_2(CO)_6]$.[103,104] The major isomer (80%) has the expected butterfly arrangement, with the two molybdenum atoms in the hinge positions and a butterfly angle of 104°; each sulfide bridges an $\{Mo_2Fe\}$ face and two carbonyls bridge different Mo—Fe bonds; the Mo—Mo distance in the Cp product is 2.846(5) Å, and the average Mo—Fe distance is 2.817(56) Å.[103] The minor isomer (20%) is a centrosymmetric cluster (Figure 6) with a planar $\{Fe_2Mo_2\}$ skeleton, with the two (μ_3-S) groups bridging different $\{Mo_2Fe\}$ triangles above and below the plane; in the MeCp complex the Mo—Mo distance is 2.282(1) Å, indicating considerable metal–metal interaction; the average Mo—Fe distance is 2.791(11) Å.[104] $[R_2Mo_2S_2(SH)_2]$ reacts with $[Fe(CO)_5]$ in the presence of Me_3NO, or with $[Fe_2(CO)_9]$ to yield $[R_2Mo_2Fe_2(\mu_3\text{-}S)_2(\mu\text{-}S)_2(CO)_6]$, the structure of which (Figure 7) is related to that in Figure 6, by replacing the two μ-CO groups by μ-S^{2-} ligands.[105]

$[MoOCl_3(THF)_2]$ reacts with $[Fe_2(CO)_6S_2]^{2-}$ to give $[MoOFe_5S_6(CO)_{12}]^{2-}$, in which the central $\{MoO\}$ unit is bonded to two $\{Fe(CO)_3\}$ groups by two μ_3-sulfides, forming an $\{MoFe_2\}$ triangle, and to a tetrahedral $\{FeS_4\}$ group *via* two μ_2-sulfides; the other sulfides of the $\{FeS_4\}$ form μ_3-bridges to two $\{Fe(CO)_3\}$ groups.[106]

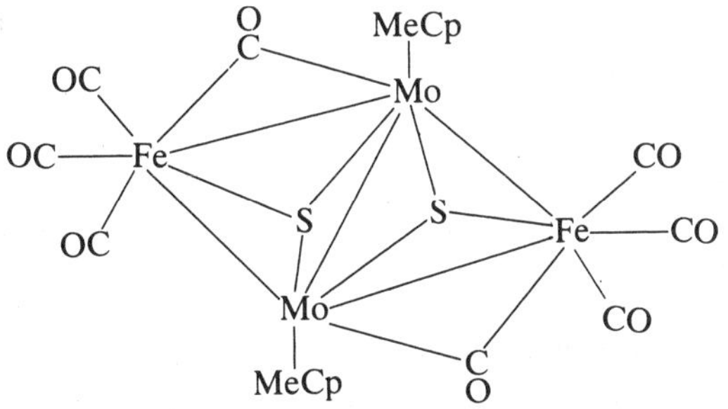

Figure 6 Structure of $[(MeCp)_2Mo_2Fe_2(\mu_3\text{-}S)_2(\mu\text{-}CO)_2(CO)_6]$[104]

The interaction of $[(Cp')_2Mo_2S_4]$ ($Cp' = C_5Me_5$) with $[Fe(NO)(CO)_3]^+$ yields $[(Cp')_2Mo_2Fe_2S_4(NO)_2]$ which, on the basis of spectroscopic evidence, is considered to have a

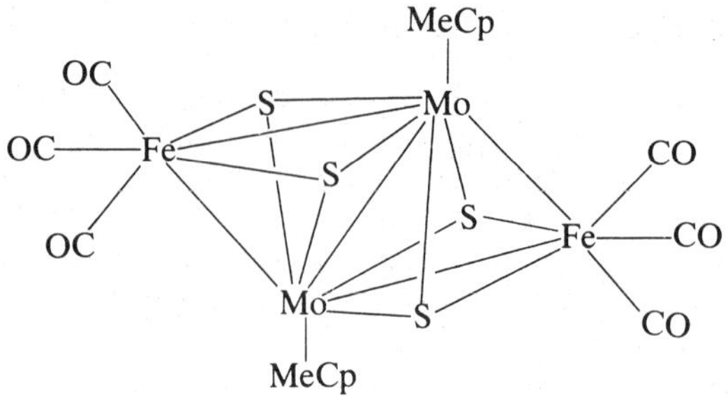

Figure 7 Structure of $[(MeCp)_2Mo_2Fe_2(\mu_3\text{-}S)_2(\mu\text{-}S)_2(CO)_6]$[105]

cubane-like $\{Mo_2Fe_2S_4\}$ core. This reaction demonstrates how attack of a metal–metal bond by an electrophilic metal species can be used to build up larger clusters.[107] $[(Cp')_2Mo_2S_4]$ reacts with $[Fe_2(CO)_9]$ in THF to form $[(Cp')_2MoS_4Fe(CO)_2]$, $[(Cp')_2Mo_2S_2(S_2CO)Fe(CO)_2]$ and $[(Cp')_2Mo_2S_4\{Fe(CO)_2\}_2]$, the structures of which have been confirmed by X-ray crystallography, the last compound having an $Mo_2Fe_2S_4$ cubane-like core. $[(MeC_5H_4)_2Mo_2Fe_2(CO)_6S_4]$ has also been reported.[108]

The addition of $\{Mo(CO)_3\}$ fragments to iron carbonyl anions has also been used to synthesize iron–molybdenum–sulfur compounds. Thus, the reaction between $[Mo(CO)_3(MeCN)_3]$ and $[Fe_6S_6X_6]^{3-}$ ($X = Cl, 4\text{-}OC_6H_4Me$) yields $[Fe_6S_6X_6\{Mo(CO)_3\}_2]^{n-}$ ($n = 3, 4$). A crystal structure determination accomplished for $[NEt_4]_4[Fe_6S_6Cl_6\{Mo(CO)_3\}_2]$ (Figure 8) has shown that the molybdenum atoms cap the top and bottom of the $\{Fe_6S_6Cl_6\}$ 'prismane' cluster. Derivatives containing such $\{Fe_6MoS_6\}$ and $\{Fe_7MoS_6\}$ units have been suggested as models for the Fe—Mo—S centre in nitrogenase but have not yet been synthesized.[109]

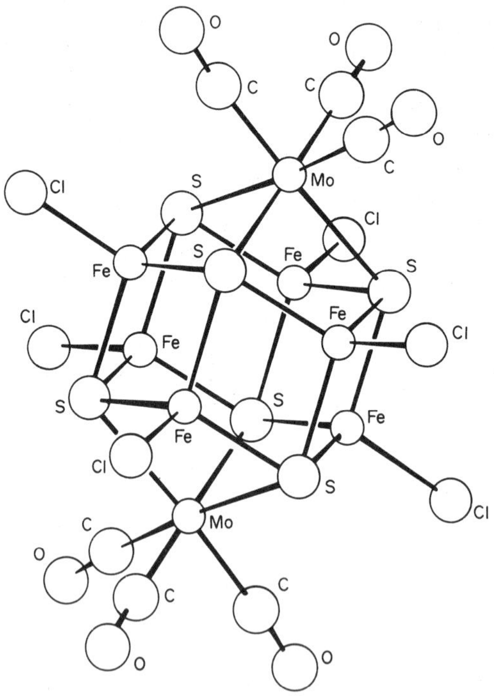

Figure 8 Structure of $[Fe_6S_6Cl_6\{Mo(CO)_3\}_2]^{4-}$ [108]

Several complexes have been prepared containing tetrahedral iron atoms coordinated to two molybdenums through two pairs of μ_2-sulfido ligands. $[NH_4]_2[MoS_4]$ reacts with $Na_2(SCH_2CH_2S)$ and $FeCl_3$; this mixture, on addition of $[NMe_4]Br$, yields $[NMe_4]_3[FeS_4\{MoS(SCH_2CH_2S)\}_2]$, the anions of which involve molybdenum atoms with a distorted square pyramidal geometry, and axial Mo—S ligands.[110] The catalytic activity of this and related complexes in the hydrogenation reaction of ethyne has been investigated.[111]

36.6.3 MOLYBDENUM CHALCOGENIDES

The principal ore of molybdenum is molybdenite, MoS_2, which occurs as fine crystals embedded in quartz. Molybdenum deposits exist in many parts of the world, but few are sufficiently rich to warrant the extensive processing required for commercial exploitation. The two most important molybdenum mines are at Climax and Henderson in Colorado, USA and the alternative source of molybdenum is as a by-product of copper mining. MoS_2 exists as an S—Mo—S · · · S—Mo—S layer lattice, in which each molybdenum atom is surrounded by a trigonal prism of six sulfur atoms. MoS_2 is of considerable commercial importance as a lubricant, catalyst precursor and support, precursor of intercalation superconductors, and as an electrode for photoelectrolysis. Therefore, the physical properties of MoS_2 and its derivatives have been extensively examined.[113] $MoSe_2$ and the low temperature form of $MoTe_2$ adopt the same structure as MoS_2, but in the high temperature form of $MoTe_2$ each molybdenum atom is located in an octahedral arrangement of tellurium atoms and is involved in two close Mo—Mo approaches.[114] The amorphous trichalcogenides, MoS_3 and $MoSe_3$, have interesting electrochemical and other physical properties;[115] molybdenum and selenium EXAFS studies have led to the proposal that an important structural unit of these systems is a binuclear metal site (Mo—Mo *ca.* 2.75 Å) with a triple bridge consisting of one Y^{2-} and one Y_2^{2-} (Y = S, Se) ion.[116]

Mo—Mo and Y—Y bonding and their interrelationships are a dominant feature of the chemistry of discrete molybdenum chalcogenide entities.[117–119] The clearest demonstration of this interrelationship is the *oxidation* of $[MoS_4]^{2-}$ by organic disulfides to form $[Mo_2S_8]^{2-}$, thereby achieving *reduction* of molybdenum (from Mo^{VI} to Mo^{V}), production of persulfide ions and the formation of an Mo—Mo bond.[120] Many other examples of oligomeric molybdenum systems bound to persulfide ligands exist;[118,119] two important examples are $[Mo_2(S_2)_6]^{2-}$ (Figure 9)[121] and $[Mo_3S(S_2)_6]^{2-}$ (Figure 10),[122] both as interesting systems in their own right and as synthetic precursors. Another important aspect of Mo—S chemistry is the demonstration that $[NEt_4]_2[MoS_4]$ reacts with either elemental sulfur or organic trisulfides (RSSSR) in MeCN to produce $[NEt_4]_2[MoS_9]$. This anion involves an $\{MoS\}^{2+}$ (Mo—S = 2.128(1) Å) core coordinated by two tetrasulfide ligands (Mo—S = 2.36(3) Å) to give a square pyramidal geometry at the metal. Heating an MeCN solution of $[MoS_9]^{2-}$ in air produces the $[MoO(S_4)_2]^{2-}$ analogue.[123] Oxomolybdenum centres ligated by groups containing one or more S—S bonds have been identified in many other systems, including: $[MoOS_2(S_2CNPr_2)_2]$,[124] $[Mo_2(S_2O)_2(S_2CNEt_2)_4]$,[125] $[Mo_2S_4(S_2)(S_3O)]^{2-}$ [126] and $[Mo_2O_nS_{4-n}(S_2)_{2-x}(S_4)_x]^{2-}$ ($n = 2$, $x = 0, 1$; $n = 1$, $x = 1$).[127] Anions of this last type, with $n = 0$ and $x = 0, 1, 2$ have been shown to participate in a complicated series of equilibria which arise in reactions of $[MoS_4]^{2-}$ with elemental sulfur, organic trisulfides or $[NH_4]_2S_x$, and thiols; these equilibria involve monomeric and dimeric molybdenum sulfide complexes and the equilibrium position depends upon several factors, including the nature of the solvent and the counterions.[128,129] One product of these reactions $[(S_4)Mo(S)(\mu\text{-}S)_2Mo(S)(S_4)]^{2-}$ (Figure 11) is an isomer of $[Mo_2(S_2)_6]^{2-}$ (Figure 9) with a remarkably different arrangement of sulfur atoms. Complexes involving S_3^{2-} and S_5^{2-} ions ligated to molybdenum have also been characterized.[118]

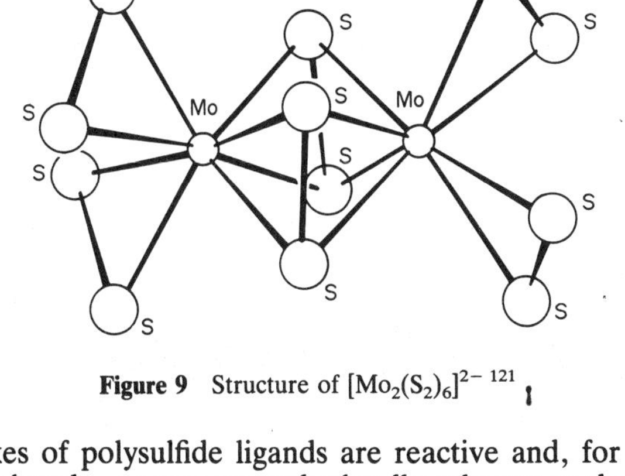

Figure 9 Structure of $[Mo_2(S_2)_6]^{2-}$ [121]

Molybdenum complexes of polysulfide ligands are reactive and, for example, readily make and break sulfur–carbon bonds, a property undoubtedly relevant to the involvement of Mo—S

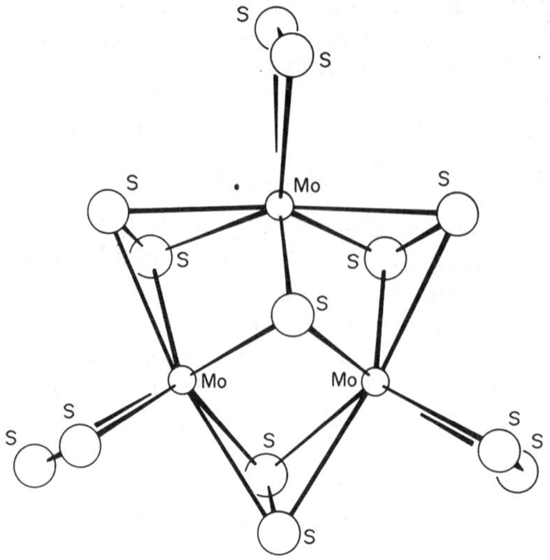

Figure 10 Structure of $[Mo_3S(S_2)_6]^{2-}$ [122]

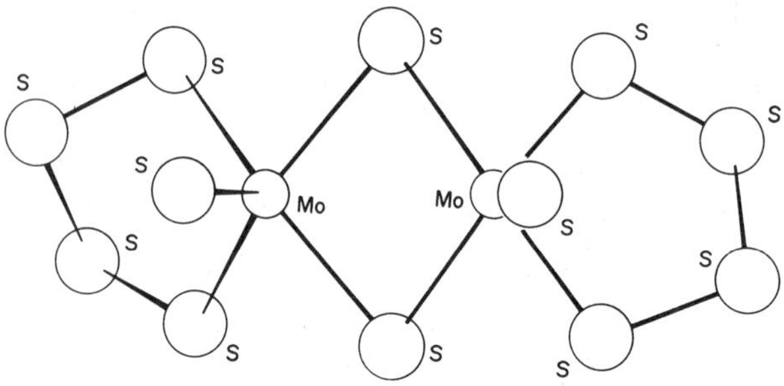

Figure 11 Structure of $[(S_4)Mo(S)(\mu\text{-}S)_2Mo(S)(S_4)]^{2-}$ [129]

centres in hydrodesulfurization catalysis. Addition of CS_2 to $[PPh_4]_2[MoS_9]$ and $[PPh_4]_2[(S_4)Mo(S)(\mu\text{-}S)_2Mo(S)(S_2)]$ produces $[PPh_4]_2[(CS_4)_2MoS]$ and $[PPh_4]_2[(CS_4)Mo(S)(\mu\text{-}S)_2Mo(S)(CS_4)]$, respectively, both products containing the perthiocarbonate ligand bound as an S,S chelate. The former anion contains an $\{MoS\}^{2+}$ core coordinated by two nearly planar $\{CS_4\}^{2-}$ ligands in a *trans* configuration; the latter anion consists of a *syn*-$\{Mo_2S_4\}^{2+}$ core with two $\{CS_4\}^{2-}$ ligands coordinated to the molybdenum atoms in a *cis* configuration.[130] 2-Butyne dimethanoate, $MeO_2CC{\equiv}CCO_2Me$, reacts with $[MoS_9]^{2-}$ to form, as one of the products, $[Mo\{S_2C_2(CO_2Me)_2\}_3]^{2-}$, a trigonal prismatic complex with three dithiolene ligands.[131] Similarly, $MeO_2CC{\equiv}CCO_2Me$ reacts with $[(S)_4Mo(S)(\mu\text{-}S)_2Mo(S)(S_4)]^{2-}$ or $[(CS_4)Mo(S)(\mu\text{-}S)_2Mo(S)(CS_4)]^{2-}$ to form $[Mo_2S_4\{S_2C_2(CO_2Me)_2\}_2]^{2-}$ and with $[MoO(S_4)_2]^{2-}$ to form $[MoO\{S_2C_2(CO_2Me)_2\}_2]^{2-}$. Reaction of $[(S_4)Mo(O)(\mu\text{-}S)_2Mo(O)(S_2)]^{2-}$ with $MeO_2C{\equiv}CCO_2Me$ results in the formation of the dithiolene complex $[Mo_2O_2S_2\{S_2C_2(CO_2Me)_2\}_2]^{2-}$, which is an isomer, in respect of the C—S framework, of the vinyl disulfide complex (Figure 12) obtained by reacting $[(S_2)Mo(O)(\mu\text{-}S)_2MoO(S_2)]^{2-}$ with

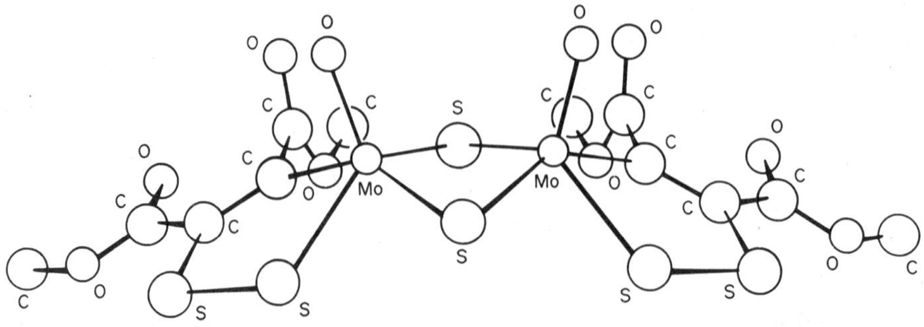

Figure 12 Structure of the bis(vinyl disulfide) complex $[Mo_2O_2S_2\{S_2C_2(CO_2Me)_2\}_2]^{2-}$ [133]

$MeO_2CC{\equiv}CCO_2Me$. Further important reactivities of Mo—S—C systems have been demonstrated by Rakowski DuBois *et al.*, for dimeric Mo—S complexes involving each metal ligated by a cyclopentadienyl ligand (Section 36.6.5).

36.6.4 NITROSYL AND CYANIDE MOLYBDENUM–SULFUR COMPLEXES

The structural chemistry of these systems is interesting and a considerable versatility exists, especially in respect of the nature of the Mo—S framework to which nitrosyl or cyanide ligands are attached.[118,134,135]

$[Mo_4(NO)_4(S_2)_6O]^{2-}$ can be obtained by the reaction of $[MoO_4]^{2-}$ with $NH_2OH{\cdot}HCl$ with H_2S in H_2O. This anion (Figure 13) consists of a central oxygen atom bound to four molybdenum atoms, arranged as a tetragonal bisphenoid over whose edges two handle-shaped and four roof-shaped S_2^{2-} ligands are located; each molybdenum is bound to a terminal nitrosyl ligand.[136] The related anion $[Mo_4(NO)_4S_{13}]^{4-}$ (Figure 14) involves a central sulfur bound to four $\{MoNO\}^{3+}$ moieties, five S_2^{2-} and two S^{2-} ligands (rather than the six S_2^{2-} ligands in Figure 13). $[Mo_4(NO)_4S_{13}]^{4-}$ is one product of the reaction of molybdate with ammonium polysulfide in aqueous solution in the presence of hydroxylamine.[134,137] $[Mo_2(NO)_2(S_2)_3(S_5)(OH)]^{3-}$ (Figure 15) has been obtained and involves an unusual $(\mu\text{-}S_5)^{2-}$ ligand.[138] Other studies of these and related anions have been reported.[139]

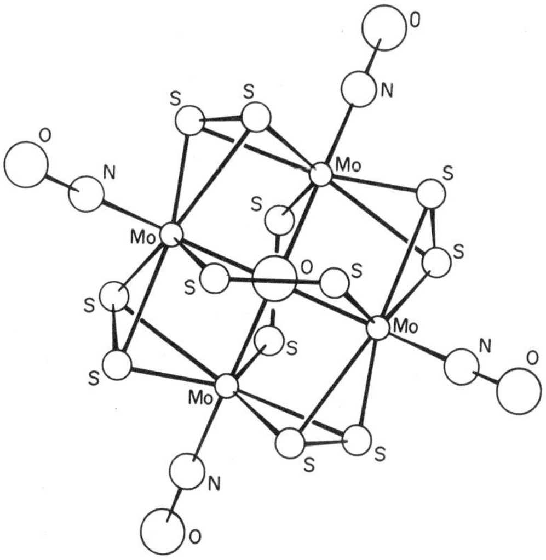

Figure 13 Structure of $[Mo_4(NO)_4(S_2)_6O]^{2-}$ [136]

Nitrosyl and cyanide ligands can coexist bound to a molybdenum–sulfur core, as in $[Mo_2S_2(NO)_2(CN)_6]^{6-}$ [135] and $[Mo_4S_4(NO)_4(CN)_8]^{8-}$.[140] The latter anion has been obtained by reacting $[Mo_4(NO)_4(S_2)_6O]^{2-}$ with KCN and it possesses a distorted cubane-like Mo_4S_4 framework, composed of two Mo_2S_2 quadrilaterals (Mo—Mo = 2.99(3) Å) linked with long (3.67(4) Å) Mo···Mo separations; each molybdenum is bound to one nitrosyl and two cyanide ligands external to the Mo_4S_4 cluster.[140] $[Mo_4S_4(CN)_{12}]^{8-}$ also contains an Mo_4S_4 cubane-like cluster, but the molybdenum and sulfur atoms are arranged as two, almost regular, interpenetrating tetrahedra with all the Mo—Mo separations essentially the same (2.85 Å); each molybdenum is octahedrally surrounded by three CN^- and three S^{2-} ligands.[141] $K_5[Mo_3S_4(CN)_9]{\cdot}3KCN{\cdot}4H_2O$ has been structurally characterized and shown to involve cyanide ligation of a normal $\{Mo_3S_4\}^{4+}$ core.[135] Cyanide ligation to $\{Mo_2S\}^{6+}$ (see Section 36.4.1.3), $\{Mo_2S_2\}^{2+}$ and $\{Mo_2S_2\}^{4+}$ cores has also been demonstrated; $[Mo_2S_2(CN)_8]^{4-}$ involves a significantly longer (2.758(7) *vs.* 2.644(5) Å) Mo—Mo bond than in $[Mo_2S_2(CN)_8]^{6-}$.[135,142]

The redox chemistries of di-, tri and tetra-nuclear Mo—S cyanide complexes have been discussed in relation to their electronic structures.[135] Passage of oxygen into an aqueous solution of $[Mo_2S_2(CN)_8]^{6-}$ leads to the formation of a dark violet mixed-crystal compound of the composition $K_{4+x}[Mo_2(SO_2)(S_2)(CN)_8]_x[Mo_2(SO_2)(S_2)(CN)_8]_{1-x}{\cdot}4H_2O$ ($x = 0.3$). In the crystal the two anions, whilst structurally similar, are located at crystallographically independent positions; each involves both molybdenum atoms surrounded by an approximately

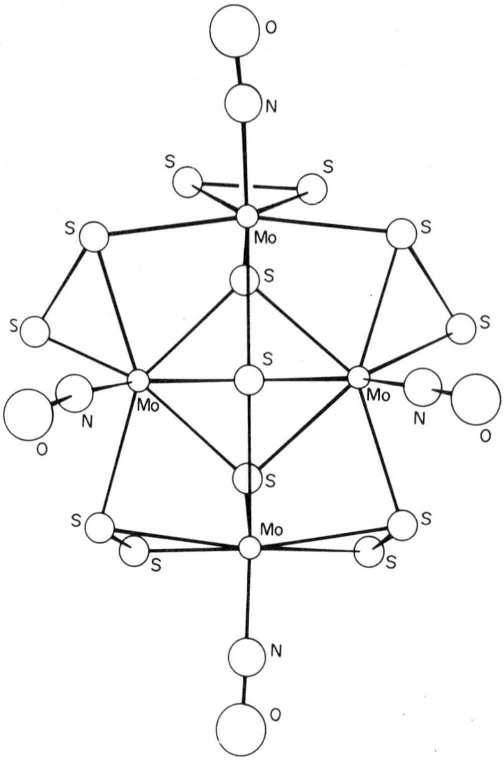

Figure 14 Structure of $[Mo_4(NO)_4S_{13}]^{2-}$ [137]

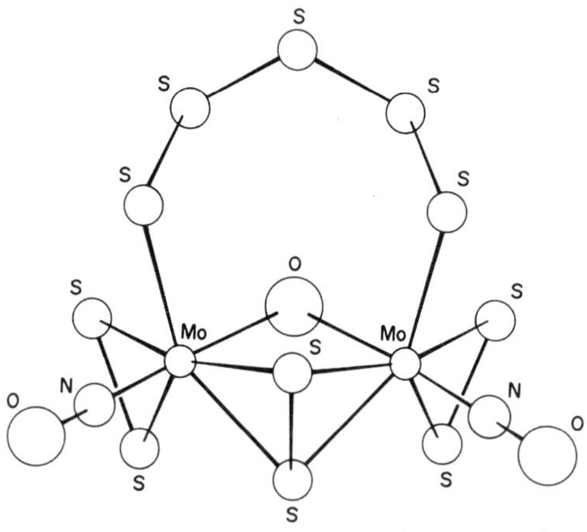

Figure 15 Structure of $[Mo_2(NO)_2(S_2)_3(S_5)(OH)]^{3-}$

pentagonal bipyramidal array of four terminal CN^- and three sulfur atoms of the bridging SO_2^{2-} and S_2^{2-} ligands. However, these anions involve the significantly different Mo—Mo separations of 2.790 and 2.684 Å, respectively, and the results of spectroscopic studies have led to the suggestion that the former is a mixed $Mo^{III}Mo^{IV}$ dimer whilst the latter involves two Mo^{IV} atoms.[143]

36.6.5 CYCLOPENTADIENYL MOLYBDENUM-SULFUR SYSTEMS

Cyclopentadienyl molybdenum–sulfur compounds are useful synthons for the preparation of Mo—M—S (*e.g.* M = Fe (see Section 36.6.2), Co, Ni) clusters[103–108,144] in reactions with metal carbonyls. However, the principal interest in molecules of this class has arisen because of the reactivity of the Mo—S system, primarily in respect of the making and breaking of S—H and/or S—C bonds, leading to novel chemistry and an illumination of processes relevant to hydrodesulfurization catalysis.

$[Cp_2Mo_2(CO)_6]$ reacts with elemental sulfur to form insoluble products,[145] but the introduction of methyl substituents on the cyclopentadienyl ring permits the isolation of the soluble products $[(Me_5C_5)_2Mo_2S_4]$, $[(MeC_5H_4)_2Mo_2S_4]$ and $[(Me_5C_5)_2Mo_2(S_2)_5]$. The first two products possess an *anti*-$\{Mo_2S_4\}^{2+}$ core (Section 36.4.3.5.3) and the last one involves two η^2-S_2 ligands bound to each metal with each bonded to a single sulfur atom of a μ_2-S_2 ligand.[146] Molecules of the composition $[R'_2Mo_2S_4]$ ($R' = Me_nC_5H_{5-n}$, $n = 0, 1, 5$) may exist in one of three isomeric forms (Figure 16) and photochemical isomerization between them has been demonstrated.[146–149] Dimers related to form I with oxo or imido ligands have been characterized,[150] the electronic structure of these molecules has been investigated,[151] and their intramolecular steric effects have been modelled.[152]

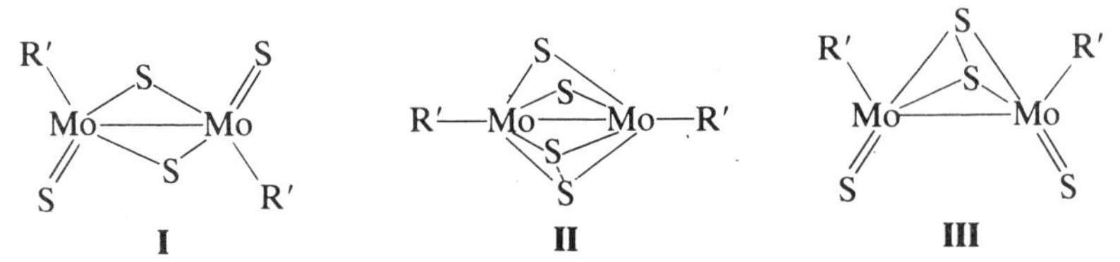

Figure 16 Isomers of molecules with the composition $[R'_2Mo_2S_4]$ ($R' = Me_nC_5H_{5-n}$, $n = 0, 1, 5$)[146–149]

The molecules $[R'_2Mo_2S_4]$ and their relatives manifest interesting reactivity towards H_2, alkenes, alkynes and other compounds with unsaturated linkages.[146,153–156] $[R'_2MoS_4]$ compounds react with H_2 under mild conditions to form a quadruply bridged structure with two hydrosulfido–ligands, $[R'_2Mo_2(\mu\text{-}S)_2(\mu\text{-}SH)_2]$. These systems catalyze H—D exchange between H_2 and D_2, as well as between H_2 and D_2O, and catalyze the reduction of elemental sulfur and SO_2.[153,154] Both alkene and arenethiols (RSH) react with $[R'_2Mo_2(\mu\text{-}S)_2(\mu\text{-}SH)_2]$ to release H_2S and form the corresponding $[R'_2Mo_2(\mu\text{-}S)_2(\mu\text{-}SR)_2]$ derivative; the structure of the $n = 1$, R = Me derivative has been determined and the molecules shown to be centrosymmetric with an Mo—Mo separation of 2.582(1) Å. $[R'_2Mo_2(\mu\text{-}S)_2(\mu\text{-}SH)_2]$ reacts with various unsaturated molecules, for example ethylene or benzyl isocyanide, releasing H_2 and forming dimers with two bridging ethane-1,2-dithiolate or dithiocarbonimidate groups, respectively.[153] The kinetics of the reactions of benzyl isocyanide with $[(MeCp)_2Mo_2(\mu\text{-}S)_2(\mu\text{-}SH)_2]$ and $[(MeCp)_2Mo_2(\mu\text{-}S)_2(\mu\text{-}SCH_2CH_2S)]$, to form the same dithiocarbonimidate complex, $[(MeCp)_2Mo_2(\mu\text{-}S_2CNCH_2Ph)_2]$, have been determined and the results suggested that the former reaction proceeds by an associative mechanism whilst the latter proceeds by a dissociative pathway.[155] Deprotonation of $[R'_2Mo_2(\mu\text{-}S)_2(\mu\text{-}SH)_2]$ with NaOMe (1:2) in the presence of excess CH_2Br_2 results in the formation of the dimer $[R'_2Mo_2S_2(S_2CH_2)]$ with a bridging methanedithiolate ligand;[156] this compound will catalyze the homogeneous hydrogenolysis of CS_2.[157]

Addition of C_2H_4 or C_2H_2 to $[R'_2Mo_2S_4]$ yields a quadruply bridged dimer with, respectively, two μ-ethane-1,2-dithiolate or μ-ethylene-1,2-dithiolate ligands.[146] Complexes related to the former, *viz.* $[(Cp)_2Mo(\mu\text{-}SC_nH_{2n}S)_2]$ ($n = 2$ and 3), have been prepared by the reaction of ethylene sulfide and propylene sulfide, respectively, with $[CpMoH(CO)_3]$ or $[(Cp)_2Mo_2(CO)_6]$ and their redox properties have been studied. These neutral dimeric complexes undergo a unique reaction with alkenes and alkynes, in which the hydrocarbon portion of the bridging dithiolate ligand is exchanged.[158] A series of these complexes, in which the molybdenum atoms are bridged by two inequivalent dithiolate ligands, have been synthesized and the relative tendencies of the ethanedithiolate ligands to eliminate ethylene investigated. Alkenes and cumulenes react with these complexes, displacing ethylene to form derivatives with dithiolene bridges.[156]

$[Cp_2Mo_2(CO)_6]$ reacts with $[Zn(S_3CPh)_2]$ to form $[CpMo(CO)_2(S_2CPh)]$ and $[Cp_2Mo_2(\mu\text{-}S)_2(S_2CPh)_2]$, and with 3,4-dimercaptotoluene to yield a purple dimer which, on irradiation, decomposes to produce the green compound $[(Cp)_2Mo_2(\mu\text{-}S_2C_6H_3Me_2)_2]$, which has a structure in which the core is similar to that of $[(Cp)_2Mo_2(\mu\text{-}SCHCHS)_2]$.[159] $[Cp_2Mo_2(\mu\text{-}SMe)_4]$ may be obtained by reacting $[Cp_2Mo_2(CO)_6]$ with MeSSMe and oxidation by $Ag[PF_6]$ yields the monocation; the neutral and cationic species possess the same geometry and oxidation leads to no essential change in dimensions.[160]

A wide variety of other cyclopentadienyl molybdenum–sulfur compounds are known including: monomeric $[Cp_2Mo(SR)_2]$,[161,162] $[Cp_2MoS_2]$ and $[Cp_2MoS_4]$,[163] dimeric

$[(BuCp)_2Mo_2(\mu\text{-}S_2)Cl_4]$ and $[CpMoCl_2(\mu\text{-}X)(\mu\text{-}SMe)_2ClMoCp]$ (X = OH, SMe),[164] trimeric $[Cp_3Mo_3(\mu\text{-}S)_3(\mu_3\text{-}S)]^{n+}$ ($n = 0$,[147] 1[165]) and $[Cp_3Mo_3(\mu_3\text{-}S)(CO)_6]^+$,[166] and the tetrameric $[(Pr^iCp)_4Mo_4(\mu_3\text{-}S)_4]^{n+}$ ($n = 0, 1, 2$)[167] and related systems (see Section 36.3.3.2).[168]

36.6.6 DITHIOLENE AND RELATED COMPLEXES

Transition metal complexes of unsaturated 1,2-dithiolates (metal dithiolenes) have attracted much attention because of their interesting structural and redox properties.[169] Molybdenum dithiolene complexes have featured prominently[170] in these studies and have special significance following the suggestion[171,172] that the molybdenum-containing cofactor of the oxomolybdoenzymes (Section 36.6.7) incorporates a molybdenum complex of an unsymmetrically substituted alkene-1,2-dithiolate.

The best characterized complexes are the tris(dithiolenes) which are known for several ligands, including: $\{S_2C_2(CN)_2\}^{2-}$ (mnt),[173–175] $\{S_2C_2R_2\}^{2-}$ (R = H, Me, CF_3, CO_2Me, Ph, 4-MeC_6H_4),[131,176–183] $\{S_2C_6H_4\}^{2-}$, $\{S_2C_6H_3\text{-4-Me}\}^{2-}$ (tdt), $\{S_2C_6H_2\text{-4,5-Me}_2\}^{2-}$, $\{S_2C_6F_4\}^{2-}$ [188] and quinoxaline-2,3-dithiolate.[189] These complexes are intensely coloured, due to S→Mo charge-transfer transitions and they have been synthesized by a variety of procedures, this versatility being required by the instability of some of the free ene-1,2-dithiols. *cis*-1,2-Dicyanoethane-1,2-dithiolate (maleonitriledithiolate, mnt), however, is a stable species and $MoCl_5$ reacts with Na_2mnt in THF to produce $[Mo(mnt)_3]^{2-}$.[173,174] Also, H_2tdt reacts with molybdate in acidic solution,[184] or with $MoCl_5$ in CCl_4,[182] to form $[Mo(tdt)_3]$; H_2tdt reacts with $[MoOCl_2(MePPh_2)_3]$ in MeCN in the presence of Et_3N to form $[Mo(tdt)_3]^{2-}$.[185] Bis(trifluoromethyl)dithieten, $CF_3C(S)(S)CCF_3$ is a convenient reagent and reacts with $[Mo(CO)_6]$ to form $[Mo\{S_2C_2(CF_3)_2\}_3]$;[176] reduction of this compound by hydrazine in ethanol affords the corresponding dianion.[177] Procedures for the *in situ* synthesis of dithiolene ligands have extended the range of metal–dithiolene chemistry. Thus, cyclic thiophosphate esters, prepared from benzoins or acyloins and P_4S_{10}, have yielded ene-1,2-dithiolate complexes when treated with transition metal salts under acidic conditions.[178] Also, hydrolysis of 2-(*N,N*-dialkylamino)-1,3-dithiolium salts affords a general method for the synthesis of ene-1,2-dithiolates and allows different substituents to be bound to the alkenic carbons.[190]

Molybdenum tris(dithiolene) complexes undergo electrochemically reversible, one-electron, redox processes. $[Mo(mnt)_3]^{n-}$ displays redox changes corresponding to the 0/1−, 1−/2−, 2−/3− and 3−/4− couples;[174] $[Mo(S_2C_2R_2)_3]^{n-}$ (R = H, Me, Ph, CF_3)[177,182,191] and $[Mo(tdt)_3]^{n-}$ [182] manifest 0/1−, 1−/2− and 2−/3− couples. These observations have encouraged chemical oxidations or reductions of these species, and stimulated a lively debate as to whether the processes are primarily metal- or ligand-based. The ligands are 'non-innocent' and this renders the precise definition of oxidation state difficult in these systems, but it is often assumed that each coordinated dithiolene contributes +II to the count of the metal's oxidation state. Discussions of the electronic structure of these systems have generally incorporated the information available from EPR studies of the paramagnetic species and X-ray crystallographic data.[131,175,181,183,189]

Metal trisdithiolenes were the first non-octahedral, six-coordinate transition metal complexes to be structurally characterized. Much of the interest shown in these complexes has concentrated on attempts to explain the observed trigonal prismatic geometry[192] and it is generally considered that the unusual structures of these complexes are due to electronic, rather than steric, effects. Two MO schemes have been proposed[179,182] which differ slightly in the ordering of the valence orbitals and that due to Gray *et al.*[182] gives better agreement with spectroscopic data. In this scheme, the stability of the trigonal prismatic geometry is attributed to two π-bonding interactions. Firstly, the π_h interaction which results from overlap of the metal d_{z^2} orbital with the sulfur sp^2 lone pair orbitals at 120° to the M—S and S—C bonds and, secondly, the π_v bonding which arises from overlap of the metal d_{xy} and $d_{x^2-y^2}$ orbitals with the butadienoid π orbitals of the ligand S_2C_2 groups. The optimal electron configuration for trigonal prismatic geometry is that of the neutral trisdithiolene complexes of the Group VI transition metals, *i.e.* the bonding π_h and π_v orbitals are filled and their antibonding counterparts unoccupied. The LUMO in these complexes is an antibonding π_h orbital; therefore, the occupation of this orbital in monoanionic and dianionic trisdithiolene complexes of the Group VI transition metals should destabilize the trigonal prismatic structure and allow distortion towards octahedral geometry. In support of this argument, the structure of $[Mo(mnt)_3]^{2-}$ is severely distorted from trigonal prismatic geometry.[175]

$[MoH(C_6H_4S_2\text{-}1,2)_3]^{3-}$ has been synthesized[193] and the selenium analogue of a trisdithiolene complex $[Mo\{Se_2C_2(CF_3)_2\}_6]$ shown to involve the six selenium atoms arranged in a trigonal prismatic manner about the molybdenum.[194]

A wide variety of other molybdenum dithiolene complexes have been prepared and characterized. Mononuclear complexes involving other ligands in the primary coordination sphere include: $[NBu_4][Mo(mnt)_2(S_2CNEt_2)]$, which possesses an essentially trigonal prismatic MoS_6 arrangement,[195] $[CpMo(mnt)_2]^-$, $[Cp_2Mo(mnt)]$ and $[Cp_2Mo(tdt)]$.[161,196] Several dimeric molybdenum complexes with dithiolene ligands are known and, of these, $[Mo_2(tdt)_5]$[197] is the only system based on dithiolene ligands alone. One of the products from the reaction of $[NH_4]_2[Mo_2O_7]$ with benzoin and P_4S_{10} is the binuclear dithiolene complex $[Mo_2(S_2)(S_2C_2Ph_2)_4]$. An X-ray analysis of the structure has revealed that each molybdenum atom is coordinated by seven sulfur atoms, in an approximately monocapped trigonal prismatic arrangement; the two metal atoms are bridged by four sulfur atoms of a shared face of the trigonal prisms, two of the bridging sulfur atoms are supplied by a bridging S_2 ligand and the other two are supplied by two bridging dithiolene ligands.[198] The structure of $[Mo_2O_2S_2\{S_2C_2(CN)_2\}_2]^{2-}$ has been determined; each molybdenum atom possesses distorted square pyramidal coordination geometry with the sulfur atoms forming the basal plane and an oxygen atom in the axial position.[199] Other dimeric molybdenum complexes with dithiolene ligands have been considered earlier in Sections 36.6.3 and 36.6.5.

36.6.7 OXOMOLYBDOENZYMES

The existence of a biological role for molybdenum was first recognized in 1930.[200] Since that time, an increasing volume of evidence has accumulated to show that molybdenum is essential for a significant number of biological processes. For example, molybdenum is required both for nitrogen fixation (Section 36.6.2) and nitrate reduction, and its participation in some animal hydroxylase enzymes and several fungal and bacterial enzymes is now firmly established.[201] The molybdoenzymes are all reasonably large proteins, the functions of which involve the molybdenum centres undergoing redox reactions in concert with other redox active constituents. An important aspect of the composition of the molybdoenzymes has been the demonstration of the existence of a low molecular weight prosthetic group (or cofactor) containing molybdenum necessary for catalytic activity and which may be dissociated reversibly from the main body of the protein. The original concept was of a cofactor common to all the molybdoenzymes.[202] Subsequently, studies demonstrated[59,203] that there are at least two such cofactors; nitrogenases yield FeMoco (Section 36.6.2) which is distinct from the molybdenum-containing cofactor, or Moco, of enzymes such as xanthine oxidase and dehydrogenase, sulfite oxidase and nitrate reductase.

Molybdoenzymes other than the nitrogenases are usually termed *oxo*molybdoenzymes. This prefix relates to the nature of the catalysis effected, *i.e.* the net effect of the conversion (xanthine to uric acid, sulfite to sulfate, nitrate to nitrite, or aldehyde to carboxylate) corresponds to the transfer of one oxygen atom to or from the substrate. Furthermore, molybdenum K-edge EXAFS studies have established that this metal is coordinated to one or more terminal oxo groups in each enzyme studied by this technique.[204]

EPR spectroscopy has been extensively applied to the oxomolybdoenzymes.[205] These studies have established that all of these enzymes have similar molybdenum centres, at which the substrate undergoes the catalytic conversion. The redox states of molybdenum, involved in the normal catalytic cycle of these enzymes, are Mo^{VI}, Mo^{V} and Mo^{IV} and the EPR characteristics of the Mo^{V} centres, and hence the Mo^{VI} and Mo^{IV} states, are mononuclear in molybdenum and coordinated to sulfur.[206] EPR spectroscopic studies have established that, at least in the EPR-detectable form of each of the oxomolybdoenzymes at low pH, there is a minimum of one proton magnetically coupled to the unpaired electron of the Mo^{V} centre which is chemically exchangeable with the protons of the external aqueous medium. Furthermore, for xanthine oxidase, EPR spectroscopy has provided direct evidence[207] in support of the hypothesis[208] that the mechanisms of all the molybdoenzymes involve coupled proton–electron transfer at the molybdenum centre. EPR spectra using ^{17}O-enriched H_2O have established that oxygen atoms are bound to the Mo^{V} centres[209] and studies with ^{33}S suggest that the difference between active and desulfo xanthine oxidase involves the replacement of an $Mo{=}S_t$ group in the former by the $Mo{=}O_t$ group in the latter.[210] The results of molybdenum K-edge EXAFS studies[204] are complementary to those of EPR spectroscopy, not only since it is easier to probe the EPR silent states by the former technique, but also because the results provide additional information, in respect of the nature, number and distance of the ligand donor atoms attached to molybdenum. For example, in oxidized sulfite oxidase molybdenum is bound to two oxygen atoms at 1.68 Å and two to three sulfur atoms at 2.41 Å; reduction to the Mo^{IV} state results in the loss of one of the oxo groups. The molybdenum centre of *Chlorella vulgaris* nitrate reductase resembles that in sulfite oxidase, but some important differences occur for molybdenum in *Escherichia coli* nitrate reductase.[211] Also, molybdenum *K*-edge EXAFS studies have strengthened the view that an $Mo{=}S_t$ group is present in active, but not desulfo, xanthine oxidase and dehydrogenase[212] and identified a close approach (*ca.* 3.0 Å) of Mo and As atoms in arsenite-inhibited xanthine oxidase.[213]

A major development in the definition of the constitution of Moco was the recognition[214] that an oxidized, inactive, form of this cofactor has the fluorescence properties characteristic of a pterin. Further studies indicated that the fluorescent material was derived from a novel, sulfur-containing reduced pterin substituted at the C-6 position.[171] Subsequently, a metabolic relationship was established between Moco and urothione, since persons genetically deficient in Moco are unique in having no detectable urothione in their urine. This metabolic relationship, and the chemical reactions of the cofactor, formed the basis from which Johnson and Rajagopalan proposed a structure (Figure 17) for Moco.[172]

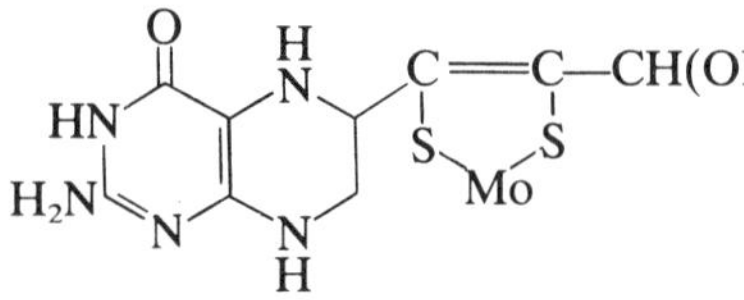

Figure 17 Proposed structure for the molybdenum cofactor of the oxomolybdoenzymes[172]

The biological importance of molybdenum, and the challenge to reproduce the spectroscopic properties and the reactions of the molybdenum centres of the oxomolybdoenzymes, have been considerable stimuli for the development of the coordination chemistry of this element (see Chapters 36.4 and 36.5).

36.6.8 HYDRIDE AND DIHYDROGEN COMPLEXES

The most extensive group of molybdenum hydride complexes are those which contain phosphine ligands, and mononuclear systems with one to six hydride ligands have been identified.

The reaction of $[Mo(acac)_3]$ and diphos in toluene with $AlEt_3$ under Ar affords $[Mo(C_2H_4)(diphos)_2]$ and $[MoH(acac)(diphos)_2]$. The former complex reacts with Hacac to produce the latter and this hydrido complex initiates the polymerization of acrylonitrile.[215] $[MoCl_3(THF)_3]$ reacts with PMe_3 in Et_2O or THF to form $[MoCl_3(PMe_3)_3]$ which reacts with $Na[BH_4]$ to yield $[MoH(BH_4)(PMe_3)_4]$, whose structure has been determined by X-ray crystallography; the molecular unit involves a distorted octahedral environment at the molybdenum, with the bidentate $[BH_4]^-$ group considered to occupy one vertex, and Mo—H = 1.63 Å.[216] The carbonyl hydrides of molybdenum include $[\mu\text{-}H\{Mo(CO)_4PMePh_2\}_2]^-$, which has been obtained by reacting $[\mu\text{-}H\{Mo(CO)_5\}_2]^-$ with a 10-fold excess of $PMePh_2$; its structure has been determined for the $[NEt_4]^+$ salt, the Mo···Mo separation in the three-centre, two-electron hydride bridge is 3.443(1) Å, significantly shorter than that of 3.7436(1) Å in $[\mu\text{-}H\{Mo_2(CO)_9PPh_3\}]^-$ and only 0.02 Å longer than in $[\mu\text{-}H\{Mo(CO)_5\}_2]^-$.[217] A combined structural and spectroscopic study has been performed to examine the stereochemical consequences resulting from the deprotonation of the bent Mo—H—Mo bond in $[(\mu\text{-}H)(\mu\text{-}PMe_2)\{CpMo(CO)_2\}_2]$; the most significant structural change is the reduction of the Mo—Mo separation from 3.262(2) Å in the hydride to 3.157(2) Å in $[(\mu\text{-}PMe_2)\{CpMo(CO)_2\}_2]^-$.[218]

Reduction of $[MoCl_4(THF)_2]$ by Mg under H_2 in the presence of PMe_3 gives $[MoH_2(PMe_3)_5]$ which possesses a distorted pentagonal bipyramidal stereochemistry with Mo—H bonds of length 1.67 Å. Both the ^1H and $^{31}P\{^1H\}$ NMR spectra of this compound are consistent with a fluxional structure; the former manifests a binomial sextet at $\delta = -5.23$ p.p.m.[219] $[MoH_4(PMe_2Ph)_4]$ reacts with $Ag[BF_4]$ (1:2) in MeCN to give

$[MoH_2(PMe_2Ph)_4(NCMe)_2][BF_4]_2$, the cation of which has been shown by spectroscopy and X-ray crystallography to possess a fluxional dodecahedral structure. The same salt is obtained by treating the tetrahydride with $H[BF_4]{\cdot}Et_2O$ in MeCN, but $[MoH_4(PMePh_2)_4]$ reacts with $Ag[BF_4]$ or $H[BF_4]{\cdot}Et_2O$ in MeCN to produce $[MoH_2(PMePh_2)_3(NCMe)_3]^{2+}$.[220] As a further variation on this theme, $[MoH_4(PMePh_2)_4]$ reacts with $H[BF_4]$ in THF to produce $[\{MoH_2(PMePh_2)_3\}_2(\mu\text{-}F)_3][BF_4]$, the structure of which has been determined by X-ray crystallography. Only one hydride per metal was revealed directly by the crystallographic analysis, but the presence of the second hydride was confirmed by IR and NMR studies; the ligands adopt a dodecahedral arrangement about each metal centre.[221] $[MoH_4(dppe)_2]$ reacts with Ph_3CCl in CH_2Cl_2 to form $[MoH_2Cl_2(dppe)_2]{\cdot}CH_2Cl_2$[222] and $[MoH_2(diphos)_2]$ has been prepared by reacting $[Mo(acac)_3]$ with $AlBu^i_3$ in benzene in the presence of diphos under an atmosphere of H_2,[223] and by reacting $[Mo(N_2)_2(diphos)_2]$ with $[FeH_4(PEtPh_2)_3]$.[224]

$[MoH_4L_4]$ (L = tertiary phosphine; L_2 = diphosphine) complexes are conveniently prepared by borohydride reduction of $[MoCl_4L_2]$ in the presence of the phosphine,[225] or by treatment of $[Mo(N_2)_2L_4]$ with H_2.[226] ^1H NMR spectroscopy has shown that $[MoH_4L_4]$ molecules are non-rigid; the nature of this non-rigidity has been discussed with reference to the molecular structure of $[MoH_4(PMePh_2)_4]$, which has been shown by X-ray analysis to contain a trigonal dodecahedral MoH_4P_4 framework, with the hydrogen atoms at the vertices of the elongated tetrahedron and the phosphorus ones at those of the flattened tetrahedron.[227] The dinuclear complex $[(PMe_3)_3HMo(\mu\text{-}H)_2MoH(PMe_3)_3]$ has been mentioned in Section 36.3.2.1.1.

$[MoH_6L_3]$ (L = PPr^i_3, $PPhPr^i_2$, PCy_3, PCy_2Ph; Cy = cyclohexyl) have been prepared by reacting $[MoCl_4(THF)_2]$, L and $Na[AlH_2(OCH_2CH_2OMe)_2]$ and characterized by ^1H and $^{31}P\{^1H\}$ NMR spectroscopy; the hydridic resonance occurs at *ca.* −4 p.p.m. These complexes have been claimed as the first examples of nine-coordinate mononuclear molybdenum, but attempts to prepare $[MoH_9]^{3-}$ were unsuccessful.[228]

An exciting recent development has been the isolation of transition metal complexes containing a coordinated dihydrogen molecule, bound as an η^2 ligand. Toluene solutions of the 16-electron compound $[Mo(CO)_3(PCy_3)_2]$ react readily and cleanly with H_2 (1 atm) to form *mer-trans*-$[Mo(CO)_3(PCy_3)_2(H_2)]$. The H_2 is extremely labile and, for the corresponding PPr^i_3 complex, the addition is reversible in solution. Neutron and X-ray diffraction performed for $[W(CO)_3(PPr^i_3)_2(H_2)]$ confirmed the η^2 bonding of the H_2 molecule and this is consistent with vibrational spectra obtained for the H_2, D_2 and HD forms.[229] These complexes are important, not only because of their relevance to the participation of metal centres in catalytic hydrogenation, but also since they pose an interesting theoretical challenge. In respect of this latter point, at present it is not clear why $[Mo(CO)_3(PR_3)_2(H_2)]$ complexes involve an η^2-H_2 molecule whereas, for example $[MoH_2(PR_3)_5]$ involve two hydride ligands.

Dimeric molybdenum halide complexes involving a μ-hydride ligand are described in Section 36.5.2.2.6. The novel dithiolene hydride complex $[MoH(C_6H_4S_2\text{-}1,2)_3]^{3-}$ has been characterized;[193] $K_4[MoH(CN)_7]{\cdot}2H_2O$ has been prepared by acidification of $K_5[Mo(CN)_7]$ and its reactivity investigated.[230] Hydrido-cyclopentadienyl,[231] -η^6-benzene[232] and -carbonyl[233] molybdenum complexes have a rich chemistry which is outside the scope of this review. $[Mo_2(OAr)_2(CH_2SiMe_3)_4]$ (OAr = 2,6-dimethylphenoxide) in the presence of pyridine undergoes the loss of Me_4Si to form $[Mo_2(\mu\text{-}H)(\mu\text{-}CSiMe_3)_2(OAr)_2(py)_2]$.[234]

36.6.9 REFERENCES

1. G. Krüss, *Liebigs Ann. Chem.*, 1884, **225,** 29.
2. J. W. McDonald, G. D. Friesen, L. D. Rosenheim and W. E. Newton, *Inorg. Chim. Acta,* 1983, **72,** 205.
3. A. Müller, H. Dornfeld, H. Schulze and R. C. Sharma, *Z. Anorg. Allg. Chem.*, 1980, **468,** 193.
4. A. Müller, E. Diemann, R. Jostes and H. Bögge, *Angew. Chem., Int. Ed. Engl.*, 1981, **20,** 934.
5. D. Coucouvanis, E. D. Simhon and N. C. Baenziger, *J. Am. Chem. Soc.*, 1980, **102,** 6644; J. W. McDonald, G. D. Friesen and W. E. Newton, *Inorg. Chim. Acta,* 1980, **49,** L79.
6. D. Coucouvanis, E. D. Simhon, D. Swenson and N. C. Baenziger, *J. Chem. Soc., Chem. Commun.*, 1979, 361.
7. R. H. Tieckelmann, H. C. Silvis, T. A. Kent, B. H. Huynh, J. V. Waszczak, B. K. Teo and B. A. Averill, *J. Am. Chem. Soc.*, 1980, **102,** 5550.
8. D. Coucouvanis, P. Stremple, E. D. Simhon, D. Swenson, N. C. Baenziger, M. Draganjac, L. T. Chan, A. Simopoulos, V. Papaefthymiou, A. Kostikas and V. Petrouleas, *Inorg. Chem.*, 1983, **22,** 293.
9. A. Müller, R. Jostes, H. G. Tolle, A. Trautwein and E. Bill, *Inorg. Chim. Acta,* 1980, **46,** L121.
10. J. A. Broomhead, J. R. Dilworth, J. Hutchinson and J. Zubieta, *Cryst. Struct. Commun.*, 1982, **11,** 1701.

11. D. Coucouvanis, N. C. Baenziger, E. D. Simhon, P. Stremple, D. Swenson, A. Kostikas, A. Simopoulos, V. Petrouleas and V. Papaefthymiou, *J. Am. Chem. Soc.*, 1980, **102,** 1730.
12. W. Clegg, C. D. Garner and P. A. Fletcher, *Acta Crystallogr., Sect. C,* 1984, **40,** 754.
13. A. Müller, S. Sarkar, H. Bögge, R. Jostes, A. Trautwein and U. Lauer, *Angew. Chem., Int. Ed. Engl.*, 1983, **22,** 561.
14. H. C. Silvis and B. A. Averill, *Inorg. Chim. Acta,* 1981, **54,** L57.
15. D. Coucouvanis, E. D. Simhon, P. Stremple and N. C. Baenziger, *Inorg. Chim. Acta,* 1981, **53,** L135.
16. D. Coucouvanis, E. D. Simhon, P. Stremple, M. Ryan, D. Swenson, N. C. Baenziger, A. Simopoulos, V. Papaefthymiou, A. Kostikas and V. Petrouleas, *Inorg. Chem.*, 1984, **23,** 741.
17. D. Coucouvanis, N. C. Baenziger, E. D. Simhon, P. Stremple, D. Swenson, A. Simopoulos, A. Kostikas, V. Petrouleas and V. Papaefthymiou, *J. Am. Chem. Soc.*, 1980, **102,** 1732.
18. B. Zhuang, J. W. McDonald and W. E. Newton, *Inorg. Chim. Acta,* 1983, **77,** L221.
19. B. K. Teo, M. R. Antonio, R. H. Tieckelmann, H. C. Silvis and B. A. Averill, *J. Am. Chem. Soc.*, 1982, **104,** 6126; B. A. Averill and R. H. Tieckelmann, *Inorg. Chim. Acta,* 1980, **46,** L35.
20. K. Tanaka, M. Morimoto and T. Tanaka, *Inorg. Chim. Acta,* 1981, **56,** L61.
21. W. H. Pan, D. C. Johnston, S. T. McKenna, R. R. Chianelli, T. R. Halbert, L. L. Hutchings and E. I. Stiefel, *Inorg. Chim. Acta,* 1985, **97,** L17.
22. A. Müller, W. Hellmann, C. Römer, M. Römer, H. Bögge, R. Jostes and U. Schimanski, *Inorg. Chim. Acta,* 1984, **83,** L75.
23. E. Königer-Ahlborn and A. Müller, *Angew. Chem., Int. Ed. Engl.*, 1974, **13,** 672; E. Königer-Ahlborn, A. D. Cormier, J. D. Brown, K. Nakamoto and A. Müller, *Inorg. Chem.*, 1974, **14,** 2009.
24. A. Müller, E. Königer-Ahlborn and H. H. Heinson, *Z. Anorg. Allg. Chem.*, 1971, **386,** 102.
25. K. P. Callahan and E. J. Cichon, *Inorg. Chem.*, 1981, **20,** 1941.
26. K. P. Callahan and P. A. Piliero, *Inorg. Chem.*, 1980, **19,** 2619.
27. W. P. Binnie, M. J. Redman and W. J. Mallio, *Inorg. Chem.*, 1970, **9,** 1449.
28. S. R. Acott, C. D. Garner, J. R. Nicholson and W. Clegg, *J. Chem. Soc., Dalton Trans.*, 1983, 713.
29. S. F. Gheller, T. W. Hambley, J. R. Rodgers, R. T. C. Brownlee, M. J. O'Connor, M. R. Snow and A. G. Wedd, *Inorg. Chem.*, 1984, **23,** 2519.
30. A. Müller, A. M. Dömrose, W. Jaegermann, E. Krickemeyer and S. Sarkar, *Angew. Chem., Int. Ed. Engl.*, 1981, **20,** 1061.
31. A. Müller, M. Dartmann, C. Römer, W. Clegg and G. M. Sheldrick, *Angew. Chem., Int. Ed. Engl.*, 1981, **20,** 1060.
32. A. Müller, H. Bögge, H. G. Tolle, R. Jostes, U. Schimanski and M. Dartmann, *Angew. Chem., Int. Ed. Engl.*, 1980, **19,** 654.
33. A. Müller, H. Bögge and U. Schimanski, *Inorg. Chim. Acta,* 1980, **45,** L249.
34. A. Müller, U. Schimanski and J. Schimanski, *Inorg. Chim. Acta,* 1983, **76,** L245.
35. X. Jin, K. Tang, L. Tong and Y. Tang, *Ziran Zazhi,* 1984, **7,** 314 (*Chem. Abstr.*, 1985, **102,** 55 011j).
36. W. Clegg, C. D. Garner and J. R. Nicholson, *Acta. Crystallogr., Sect. C,* 1983, **39,** 552; C. Potvin, J. M. Manoli, M. Salis and F. Secheresse, *Inorg. Chim. Acta,* 1984, **83,** L19.
37. W. Clegg, C. D. Garner, J. R. Nicholson and P. R. Raithby, *Acta. Crystallogr., Sect. C,* 1983, **39,** 1007.
38. A. Müller, H. Bögge and U. Schimanski, *J. Chem. Soc., Chem. Commun.*, 1980, 91.
39. A. Müller, H. Bögge and U. Schimanski, *Inorg. Chim. Acta,* 1983, **69,** 5.
40. J. R. Nicholson, A. C. Flood, C. D. Garner and W. Clegg, *J. Chem. Soc., Chem. Commun.*, 1983, 1179.
41. J. M. Charnock, S. Bristow, J. R. Nicholson, C. D. Garner and W. Clegg, *J. Chem. Soc., Dalton Trans.*, 1987, 303.
42. E. Königer-Ahlborn and A. Müller, *Angew. Chem., Int. Ed. Engl.*, 1976, **15,** 680.
43. A. Müller, H. Bögge, E. Königer-Ahlborn and W. Hellman, *Inorg. Chem.*, 1979, **18,** 2301.
44. E. M. Kinsch and D. W. Stephan, *Inorg. Chim. Acta,* 1985, **96,** L87.
45. A. Müller, H. H. Heinsen, K. Nakamoto, A. D. Cormier and N. Weinstock, *Spectrochim. Acta, Part A,* 1974, **30,** 1661.
46. W. H. Pan, M. E. Leonowiecz and E. I. Stiefel, *Inorg. Chem.*, 1983, **22,** 672.
47. Y. Do, E. D. Simhon and R. H. Holm, *Inorg. Chem.*, 1985, **24,** 1831.
48. I. Søtofte, *Acta Chem. Scand., Ser. A,* 1976, **30,** 157.
49. M. Minelli, J. H. Enemark, J. R. Nicholson and C. D. Garner, *Inorg. Chem.*, 1984, **23,** 4384.
50. R. J. H. Clark, S. Joss, M. Zvagulis, C. D. Garner and J. R. Nicholson, *J. Chem. Soc., Dalton Trans.*, 1986, 1595.
51. A. Müller, S. Sarkar, A. M. Dömrose and R. Filgueira, *Z. Naturforsch., Teil B,* 1980, **35,** 1592.
52. A. Müller, R. Jostes, V. Flemming and R. Potthast, *Inorg. Chim. Acta,* 1980, **44,** L33.
53. K. P. Callahan and P. A. Piliero, *J. Chem. Soc., Chem. Commun.*, 1979, 13.
54. N. C. Payne, N. Okura and S. Otsuka, *J. Am. Chem. Soc.*, 1983, **105,** 245.
55. S.-W. Lu, N. Okura, T. Yoshida, S. Otsuka, K. Hirotsu and T. Higuchi, *J. Am. Chem. Soc.*, 1983, **105,** 7470.
56. L. D. Rosenhein, J. W. McDonald and W. E. Newton, *Inorg. Chim. Acta,* 1984, **87,** L33; J. R. Knox and C. K. Prout, *Acta Crystallogr., Sect. C,* 1969, **25,** 2482.
57. S. Bristow, C. D. Garner and W. Clegg, *Inorg. Chim. Acta,* 1983, **76,** L261.
58. P. J. Stephens, in 'Molybdenum Enzymes', ed. T. G. Spiro, Wiley, New York, 1985, p. 117 *et seq.* and references therein.
59. V. K. Shah and W. J. Brill, *Proc. Natl. Acad. Sci. USA,* 1977, **74,** 3249.
60. E. I. Stiefel and S. P. Cramer, in 'Molybdenum Enzymes', ed. T. G. Spiro, Wiley, New York, p. 89 *et seq.* and references therein.
61. V. K. Shah, *Methods Enzymol.*, 1980, **69,** 792 and references therein.
62. T. R. Hawkes, P. A. McLean and B. E. Smith, *Biochem. J.*, 1984, **217,** 317.
63. B. E. Smith, in 'Molybdenum Chemistry of Biological Significance', ed. W. E. Newton and S. Otsuka, Plenum, New York, 1980, p. 179 *et seq.*; M. J. Nelson, M. A. Levy and W. H. Orme-Johnson, *Proc. Natl. Acad. Sci. USA,* 1987, **80,** 147.

64. M. A. Walters, S. K. Chapman and W. H. Orme-Johnson, *Polyhedron,* 1986, **5,** 561.
65. B. K. Burgess and W. E. Newton, in 'Nitrogen Fixation: The Chemical–Biochemical–Genetics Interface', ed. A. Müller and W. E. Newton, Plenum, New York, 1983, p. 83 *et seq.*; L. E. Mortenson and R. N. F. Thorneley, *Annu. Rev. Biochem.*, 1979, **48,** 387; R. R. Eady and B. E. Smith, in 'A Treatise on Dinitrogen Fixation', eds. R. W. F. Hardy, F. Bottomley and R. C. Burns, Wiley, New York, 1979, Section II, chap. 2; M. J. Nelson, P. A. Lindhal and W. H. Orme-Johnson, *Adv. Inorg. Biochem.*, 1982, **4,** 1; B. A. Averill, *Struct. Bonding (Berlin)*, 1983, **53,** 59.
66. J. Rawlings, V. K. Shah, J. R. Chisnell, W. J. Brill, R. Zimmermann, E. Münck and W. H. Orme-Johnson, *J. Biol. Chem.*, 1978, **253,** 1001; B. Huynh, M. T. Henzl, J. A. Christner, R. Zimmermann, W. H. Orme-Johnson and E. Münck, *Biochim. Biophys. Acta,* 1980, **623,** 124.
67. B. K. Burgess, E. I. Stiefel and W. E. Newton, *J. Biol. Chem.*, 1980, **255,** 353.
68. S. P. Cramer, K. O. Hodgson, W. O. Gillum and L. E. Mortenson, *J. Am. Chem. Soc.*, 1978, **100,** 3398; S. P. Cramer, W. O. Gillum, K. O. Hodgson, L. E. Mortenson, E. I. Stiefel, J. R. Chisnell, W. J. Brill and V. K. Shah, *J. Am. Chem. Soc.*, 1978, **100,** 3814; S. P. Cramer, A. M. Flank, M. Weininger and L. E. Mortenson, *J. Am. Chem. Soc.*, 1986, **108,** 1049.
69. M. K. Eidsness, A. M. Flank, B. E. Smith, A. C. Flood, C. D. Garner and S. P. Cramer, *J. Am. Chem. Soc.*, 1986, **108,** 2746.
70. S. D. Conradson, B. K. Burgess, W. E. Newton, K. O. Hodgson, J. W. McDonald, J. F. Rubinson, S. F. Gheller, L. E. Mortenson, M. W. W. Adams, P. K. Mascharak, W. A. Armstrong and R. H. Holm, *J. Am. Chem. Soc.*, 1985, **107,** 7935.
71. D. Coucouvanis, *Acc. Chem. Res.*, 1981, **14,** 201.
72. R. H. Holm, *Chem. Soc. Rev.*, 1981, **10,** 455; R. H. Holm and E. D. Simhon, in 'Molybdenum Enzymes', ed. T. G. Spiro, Wiley, New York, p. 1 *et seq.* and references therein.
73. G. D. Friesen, J. W. McDonald, W. E. Newton, W. B. Euler and B. M. Hoffman, *Inorg. Chem.*, 1983, **22,** 2202.
74. A. Simopoulos, V. Papaefthymiou, A. Kostikas, V. Petrouleas, D. Coucouvanis, E. D. Simhon and P. Stremple, *Chem. Phys. Lett.*, 1981, **81,** 261.
75. B.-K. Teo, M. R. Antonio and B. A. Averill, *J. Am. Chem. Soc.*, 1983, **105,** 3751.
76. B.-K. Teo, M. R. Antonio, D. Coucouvanis, E. D. Simhon and P. P. Stremple, *J. Am. Chem. Soc.*, 1983, **105,** 5767.
77. G. Christou and C. D. Garner, *J. Chem. Soc., Dalton Trans.*, 1980, 2354; S. R. Acott, G. Christou, C. D. Garner, T. J. King, F. E. Mabbs and R. M. Miller, *Inorg. Chim. Acta,* 1979, **35,** L337.
78. T. E. Wolff, J. M. Berg, K. O. Hodgson, R. B. Frankel and R. H. Holm, *J. Am. Chem. Soc.*, 1979, **101,** 4140.
79. T. E. Wolff, P. P. Power, R. B. Frankel and R. H. Holm, *J. Am. Chem. Soc.*, 1980, **102,** 4694; T. E. Wolff, J. M. Berg, P. P. Power, K. O. Hodgson and R. H. Holm, *Inorg. Chem.*, 1980, **19,** 430.
80. T. E. Wolff, J. M. Berg and R. H. Holm, *Inorg. Chem.*, 1981, **20,** 174.
81. W. H. Armstrong and R. H. Holm, *J. Am. Chem. Soc.*, 1981, **103,** 6246; W. H. Armstrong, P. K. Mascharak and R. H. Holm, *J. Am. Chem. Soc.*, 1982, **104,** 4373.
82. R. E. Palermo, R. Singh, J. K. Bashkin and R. H. Holm, *J. Am. Chem. Soc.*, 1984, **106,** 2600.
83. G. Christou, C. D. Garner, F. E. Mabbs and M. G. B. Drew, *J. Chem. Soc., Chem. Commun.*, 1979, 91; G. Christou and C. D. Garner, *J. Chem. Soc., Chem. Commun.*, 1980, 613.
84. W. E. Cleland, Jr. and B. A. Averill, *Inorg. Chim. Acta,* 1985, **107,** 187.
85. R. E. Palermo, P. P. Power and R. H. Holm, *Inorg. Chem.*, 1982, **21,** 173.
86. P. K. Mascharak, W. H. Armstrong, Y. Mizobe and R. H. Holm, *J. Am. Chem. Soc.*, 1983, **105,** 475.
87. R. E. Palermo and R. H. Holm, *J. Am. Chem. Soc.*, 1983, **105,** 4310.
88. J. A. Kovacs, J. K. Bashkin and R. H. Holm, *J. Am. Chem. Soc.*, 1985, **107,** 1784.
89. G. Christou, C. D. Garner, R. M. Miller, C. E. Johnson and J. D. Rush, *J. Chem. Soc., Dalton Trans.*, 1980, 2363.
90. G. Christou, D. Collison, C. D. Garner, S. R. Acott, F. E. Mabbs and V. Petrouleas, *J. Chem. Soc., Dalton Trans.*, 1982, 1575.
91. T. E. Wolff, P. P. Power, R. B. Frankel and R. H. Holm, *J. Am. Chem. Soc.*, 1980, **102,** 4694.
92. P. K. Mascharak, G. C. Papaefthymiou, W. H. Armstrong, S. Foner, R. B. Frankel and R. H. Holm, *Inorg. Chem.*, 1983, **22,** 2851.
93. C. D. Garner and R. M. Miller, *J. Chem. Soc., Dalton Trans.*, 1981, 1664.
94. G. Christou, P. K. Mascharak, W. H. Armstrong, G. C. Papaefthymiou, R. B. Frankel and R. H. Holm, *J. Am. Chem. Soc.*, 1982, **104,** 2820.
95. G. Christou, R. V. Hageman and R. H. Holm, *J. Am. Chem. Soc.*, 1980, **102,** 7601.
96. K. Tanaka, M. Tanaka and T. Tanaka, *Chem. Lett.*, 1981, 895.
97. K. Tanaka, Y. Hozumi and T. Tanaka, *Chem. Lett.*, 1982, 1203.
98. Y. Hozumi, Y. Imasaka, K. Tanaka and T. Tanaka, *Chem. Lett.*, 1983, 897.
99. Y. Imasaka, K. Tanaka and T. Tanaka, *Chem. Lett.*, 1983, 1477.
100. T. Yamamura, G. Christou and R. H. Holm, *Inorg. Chem.*, 1983, **22,** 939.
101. Y. Mizobe, P. K. Mascharak, R. E. Palermo and R. H. Holm, *Inorg. Chim. Acta,* 1983, **80,** L65.
102. M. R. Antonio, B.-K. Teo, W. E. Cleland and B. A. Averill, *J. Am. Chem. Soc.*, 1983, **105,** 3477.
103. P. Braunstein, J. M. Jud, A. Tiripicchio, M. Tiripicchio-Camellini and E. Sappa, *Angew. Chem., Int. Ed. Engl.*, 1982, **21,** 307.
104. P. D. Williams, M. D. Curtis, D. N. Duffy and W. M. Butler, *Organometallics,* 1983, **2,** 165.
105. B. Cowens, J. Noordik and M. Rakowski DuBois, *Organometallics,* 1983, **2,** 931; M. D. Curtis and P. D. Williams, *Inorg. Chem.*, 1983, **19,** 2661.
106. K. S. Bose, P. E. Lamberty, J. E. Kovacs, E. Sinn and B. A. Averill, *Polyhedron,* 1986, **5,** 393.
107. H. Brunner, H. Kauermann and J. Wachter, *Angew. Chem., Int. Ed. Engl.*, 1983, **22,** 549.
108. H. Brunner, N. Janietz, J. Wachter, T. Zahn and M. L. Ziegler, *Angew. Chem., Int. Ed. Engl.*, 1985, **24,** 133.

109. M. G. Kanatzidis and D. Coucouvanis, *J. Am. Chem. Soc.*, 1986, **108**, 337; D. Coucouvanis and M. G. Kanatzidis, *J. Am. Chem. Soc.*, 1985, **107**, 5005.
110. P. L. Dahlstrom, S. Kumar and J. Zubieta, *J. Chem. Soc., Chem. Commun.*, 1981, 411.
111. Z. Zhang, F. Wang and Y. Fan, *Kexue Tongbao* (*Foreign Lang. Ed.*), 1984, **29**, 1486 (*Chem. Abstr.*, 1985, **102**, 196 828); J. Xu, X. Liu, S. Niu, S. Li, Y. Fan, F. Wang and P. Lu, *Kexue Tongbao* (*Foreign Lang. Ed.*), 1984, **29**, 344 (*Chem. Abstr.*, 1985, **101**, 142 757).
112. R. G. Dickinson and L. Pauling, *J. Am. Chem. Soc.*, 1923, **45**, 1466.
113. E. R. Braithwaite, *Chem. Ind.* (*London*), 1973, 405; P. C. H. Mitchell, 'The Chemistry of Some Hydrodesulphurisation Catalysts Containing Molybdenum', Climax Molybdenum Co. Ltd., London, 1967; J. P. Farr, 'Molybdenum Disulphide in Lubrication', Climax Molybdenum Co. Ltd., London, 1974; J. Haber, 'The Role of Molybdenum in Catalysis', Climax Molybdenum Co. Ltd., London, 1981; L. F. Scheemeyer and M. J. Sienko, *Inorg. Chem.*, 1980, **19**, 789; D. Canfield and B. A. Parkinson, *J. Am. Chem. Soc.*, 1981, **103**, 1279; G. S. Calabrese and M. S. Wrighton, *J. Am. Chem. Soc.*, 1981, **103**, 6273; D. R. Huntley, T. G. Parham, R. P. Merrill and M. J. Sienko, *Inorg. Chem.*, 1983, **22**, 4144; H. Wise, *Polyhedron*, 1986, **5**, 145 and references therein.
114. A. F. Wells, 'Structural Inorganic Chemistry', 4th edn., Clarendon, Oxford, 1975, p. 613.
115. A. J. Jacobson, R. R. Chianelli, S. M. Rich and M. S. Whittingham, *Mater. Res. Bull.*, 1979, **14**, 1437.
116. S. P. Cramer, K. S. Liang, A. J. Jacobson, C. H. Chang and R. R. Chianelli, *Inorg. Chem.*, 1984, **23**, 1215.
117. M. A. Harmer, T. R. Halbert, W.-H. Pan, C. L. Coyle, S. A. Cohen and E. I. Stiefel, *Polyhedron*, 1986, **5**, 341.
118. A. Müller, *Polyhedron*, 1986, **5**, 323.
119. A. Müller, W. Jaegerman and J. H. Enemark, *Coord. Chem. Rev.*, 1982, **46**, 245.
120. W.-H. Pan, M. A. Harmer, T. R. Halbert and E. I. Stiefel, *J. Am. Chem. Soc.*, 1984, **106**, 459.
121. A. Müller, W. O. Nolte and B. Krebs, *Angew. Chem., Int. Ed. Engl.*, 1978, **17**, 279; *Inorg. Chem.*, 1980, **19**, 2835.
122. A. Müller, S. Sarkar, R. G. Bhattacharyya, S. Pohl and M. Dartmann, *Angew. Chem., Int. Ed. Engl.*, 1978, **17**, 535; A. Müller, S. Pohl, M. Dartmann, J. P. Cohen, J. M. Bennett and R. M. Kirchner, *Z. Naturforsch., Teil B*, 1979, **34**, 434; A. Müller, R. G. Bhattacharyya and B. Pfefferkorn, *Chem. Ber.*, 1979, **112**, 778.
123. E. D. Simhon, N. C. Baenziger, M. Kamatzidis, M. Draganjac and D. Coucouvanis, *J. Am. Chem. Soc.*, 1981, **103**, 1218.
124. J. Dirand, L. Ricard and R. Weiss, *Inorg. Nucl. Chem. Lett.*, 1975, **11**, 661.
125. J. Dirand-Colin, M. Schappacher, L. Ricard and R. Weiss, in 'Proceedings of the Second International Conference on the Chemistry and Uses of Molybdenum', ed. P. C. H. Mitchell and A. Seaman, Climax Molybdenum Co., Ann Arbor, Michigan, 1976, p. 46.
126. A. Müller, U. Reinsch-Vogell, E. Krickemeyer and H. Bögge, *Angew. Chem., Int. Ed., Engl.*, 1982, **21**, 796.
127. W. Clegg, N. Mohan, A. Müller, A. Neumann, W. Rittner and G. M. Sheldrick, *Inorg. Chem.*, 1980, **19**, 2066; W. Clegg, G. M. Sheldrick, C. D. Garner and G. Christou, *Acta Crystallogr., Sect. B*, 1980, **36**, 2784; X. Xin, N. L. Morris, G. B. Jameson and M. T. Pope, *Inorg. Chem.*, 1985, **24**, 3482.
128. W. Clegg, G. Christou, C. D. Garner and G. M. Sheldrick, *Inorg. Chem.*, 1981, **20**, 1562.
129. M. Draganjac, E. D. Simhon, L. T. Chan, M. Kanatzidis, N. C. Baenziger and D. Coucouvanis, *Inorg. Chem.*, 1982, **21**, 3321.
130. D. Coucouvanis and M. Draganjac, *J. Am. Chem. Soc.*, 1982, **104**, 6820.
131. M. Draganjac and D. Coucouvanis, *J. Am. Chem. Soc.*, 1983, **105**, 139.
132. D. Coucouvanis, A. Hadjikyriacou, M. Draganjac, M. G. Kanatzidis and O. Ileperuma, *Polyhedron*, 1986, **5**, 349.
133. T. R. Halbert, W.-H. Pan and E. I. Stiefel, *J. Am. Chem. Soc.*, 1983; **105**, 5476.
134. A. Müller, R. G. Bhattacharyya, W. Eltzner, N. Mohan, A. Neumann and S. Sarkar, in 'Proceedings of the Third International Conference on the Chemistry and Uses of Molybdenum', ed. H. F. Barry and P. C. H. Mitchell, Climax Molybdenum Co., Ann Arbor, Michigan, 1979, p. 59.
135. A. Müller, R. Jostes, W. Eltzner, C.-S. Nie, E. Diemann, H. Bögge, M. Zimmermann, M. Dartmann, U. Reinsch-Vogell, S. Che, S. J. Cyvin and B. N. Cyvin, *Inorg. Chem.*, 1985, **24**, 2872 and references therein.
136. A. Müller, W. Eltzner, H. Bögge and S. Sarkar, *Angew. Chem., Int. Ed. Engl.*, 1982, **21**, 535.
137. A. Müller, W. Eltzner and N. Mohan, *Angew. Chem., Int. Ed. Engl.*, 1979, **18**, 168.
138. A. Müller, W. Eltzner, H. Bögge and E. Krickenmeyer, *Angew. Chem., Int. Ed. Engl.*, 1983, **22**, 884.
139. X. Wu and B. Wang, *Jiegou Huaxue*, 1983, **2**, 29 (*Chem. Abstr.*, 1983, **99**, 205 048n); S. Lu, X. Wu, J. Huang and J. Lu, *Jiegou Huaxue*, 1984, **3**, 217 (*Chem. Abstr.*, 1985, **103**, 204 755e).
140. A. Müller, W. Eltzner, W. Clegg and G. M. Sheldrick, *Angew. Chem., Int. Ed. Engl.*, 1982, **21**, 536.
141. A. Müller, W. Eltzner, H. Bögge and R. Jostes, *Angew. Chem., Int. Ed. Engl.*, 1982, **21**, 795.
142. M. G. B. Drew, P. C. H. Mitchell and C. F. Pygall, *Angew. Chem., Int. Ed. Engl.*, 1976, **15**, 784.
143. A. Müller, W. Eltzner, R. Jostes, H. Bögge, E. Diemann, J. Schimanski and H. Lueken, *Angew. Chem., Int. Ed. Engl.*, 1984, **23**, 389.
144. H. Brunner and J. Wachter, *J. Organomet. Chem.*, 1982, **240**, C41; H. Brunner, H. Kauermann, U. Klement, J. Wachter, T. Zahn and M. F. Ziegler, *Angew. Chem., Int. Ed. Engl.*, 1985, **24**, 132.
145. R. A. Schunn, C. J. Fritchie, Jr. and C. T. Prewitt, *Inorg. Chem.*, 1966, **5**, 892.
146. M. Rakowski DuBois, D. L. DuBois, M. C. VanDerveer and R. C. Haltiwanger, *Inorg. Chem.*, 1981, **20**, 3064.
147. W. Beck, W. Danzer and G. Thiel, *Angew. Chem., Int. Ed. Engl.*, 1973, **12**, 582.
148. H. Brunner, W. Meier, J. Wachter, E. Guggolz, T. Zahn and M. L. Ziegler, *Organometallics*, 1982, **1**, 1107.
149. A. E. Bruce and D. R. Tyler, *Inorg. Chem.*, 1984, **23**, 3433.
150. L. F. Dahl and D. L. Stevenson, *J. Am. Chem. Soc.*, 1967, **89**, 3721; L. F. Dahl, P. D. Frisch and G. R. Gust, in 'Proceedings of the First International Conference on the Chemistry and Uses of Molybdenum', ed. P. C. H. Mitchell, Climax Molybdenum Co., Ann Arbor, Michigan, 1974, p. 134; J.-C. Daran, K. Prout, G. J. S. Adam, M. L. H. Green and J. Sala-Pala, *J. Organomet. Chem.*, 1977, **131**, C40.
151. T. Chandler, D. L. Lichtenberger and J. H. Enemark, *Inorg. Chem.*, 1981, **20**, 75; L. Szterenberg and B. Jezowska-Trzebiatowska, *Inorg. Chim. Acta*, 1982, **59**, 141.
152. J. M. Newsam and T. R. Halbert, *Inorg. Chem.*, 1985, **24**, 491.

153. M. Rakowski DuBois, M. C. VanDerveer, D. L. DuBois, R. C. Haltiwanger and W. K. Miller, *J. Am. Chem. Soc.*, 1980, **102,** 7456.
154. G. J. Kubas and R. R. Ryan, *J. Am. Chem. Soc.*, 1985, **107,** 6138.
155. D. L. DuBois, W. K. Miller and M. Rakowski DuBois, *J. Am. Chem. Soc.*, 1981, **103,** 3429.
156. M. McKenna, L. L. Wright, D. J. Miller, L. Tanner, R. C. Haltiwanger and M. Rakowski DuBois, *J. Am. Chem. Soc.*, 1983, **105,** 5329.
157. M. Rakowski DuBois, *J. Am. Chem. Soc.*, 1983, **105,** 3710.
158. M. Rakowski DuBois, C. R. Haltiwanger, D. J. Miller and G. Glatzmaier, *J. Am. Chem. Soc.*, 1979, **101,** 5245.
159. W. K. Miller, R. C. Haltiwanger, M. C. VanDerveer and M. Rakowski DuBois, *Inorg. Chem.*, 1983, **22,** 2973.
160. N. G. Connelly and L. F. Dahl, *J. Am. Chem. Soc.*, 1970, **92,** 7470.
161. M. L. H. Green and W. E. Lindsell, *J. Chem. Soc. (A)*, 1967, 1455.
162. M. A. A. F. de C. T. Carrondo, P. M. Matias and G. A. Jeffrey, *Acta Crystallogr., Sect. C*, 1984, **40,** 932.
163. H. Köpf, S. K. S. Hazari and M. Leitner, *Z. Naturforsch., Teil B*, 1978, **33,** 1398.
164. B. Meunier and K. Prout, *Acta Crystallogr., Sect. B*, 1979, **35,** 172; C. Couldwell, B. Meunier and K. Prout, *Acta Crystallogr., Sect. B*, 1979, **35,** 603.
165. P. J. Vergamini, H. Vahrenkamp and L. F. Dahl, *J. Am. Chem. Soc.*, 1971, **93,** 6327.
166. M. D. Curtis and W. M. Butler, *J. Chem. Soc., Chem. Commun.*, 1980, 998.
167. J. A. Bandy, C. E. Davies, J. C. Green, M. L. H. Green, K. Prout and D. P. S. Rodgers, *J. Chem. Soc., Chem. Commun.*, 1983, 1395.
168. H. Vahrenkamp, *Angew. Chem., Int. Ed. Engl.*, 1975, **14,** 322; B. Spivak and Z. Dori, *Coord. Chem. Rev.*, 1975, **17,** 99.
169. J. A. McCleverty, *Prog. Inorg. Chem.*, 1969, **10,** 49; G. N. Schrauzer, *Acc. Chem. Res.*, 1969, **2,** 72; R. Eisenberg, *Prog. Inorg. Chem.*, 1970, **12,** 295; R. P. Burns and C. A. McAuliffe, *Adv. Inorg. Chem. Radiochem.*, 1979, **22,** 203.
170. E. I. Stiefel, *Prog. Inorg. Chem.*, 1977, **22,** 1.
171. K. V. Rajagopalan, J. L. Johnson and B. E. Hainline, *Fed. Proc. Fed. Am. Soc. Exp. Biol.*, 1982, **41,** 2608; J. L. Johnson and K. V. Rajagopalan, *Proc. Natl. Acad. Sci. USA*, 1982, **79,** 6856; K. V. Rajagopalan, S. Kramer and S. Gardlik, *Polyhedron,* 1986, **5,** 573.
172. J. L. Johnson, B. E. Hainline, K. V. Rajagopalan and B. H. Arison, *J. Biol. Chem.*, 1984, **159,** 5414.
173. M. Gerloch, S. F. A. Kettle, J. Locke and J. A. McCleverty, *Chem. Commun.*, 1966, 29; J. A. McClverty, J. Locke, E. J. Wharton and M. Gerloch, *J. Chem. Soc. (A)*, 1968, 816.
174. E. I. Stiefel, L. E. Bennett, Z. Dori, T. H. Crawford, C. Simo and H. B. Gray, *Inorg. Chem.*, 1970, **9,** 281.
175. G. F. Brown and E. I. Stiefel, *Chem. Commun.*, 1970, 728; *Inorg. Chem.*, 1973, **12,** 2140.
176. R. B. King, *Inorg. Chem.*, 1963, **2,** 641.
177. A. Davison, N. Edelstein, R. H. Holm and A. H. Maki, *J. Am. Chem. Soc.*, 1964, **86,** 2799.
178. G. N. Schrauzer, V. P. Mayweg, H. W. Finck, U. Mueller-Westerhof and W. Heinrich, *Angew. Chem., Int. Ed. Engl.*, 1964, **3,** 381; G. N. Schrauzer, V. P. Mayweg and W. Heinrich, *Inorg. Chem.*, 1965, **4,** 1615.
179. G. N. Schrauzer and V. P. Mayweg, *J. Am. Chem. Soc.*, 1966, **88,** 3235.
180. E. Hoyer and W. Schroth, *Chem. Ind. (London)*, 1965, 652.
181. A. E. Smith, G. N. Schrauzer, V. P. Mayweg and W. Heinrich, *J. Am. Chem. Soc.*, 1965, **87,** 5798.
182. E. I. Stiefel, R. Eisenberg, R. C. Rosenberg and H. B. Gray, *J. Am. Chem. Soc.*, 1966, **88,** 2956.
183. M. J. Bennett, M. Cowie, J. L. Martin and J. Takats, *J. Am. Chem. Soc.*, 1973, **95,** 7504; M. Cowie and M. J. Bennett, *Inorg. Chem.*, 1976, **15,** 1584.
184. T. W. Gilbert, Jr. and E. B. Sandell, *J. Am. Chem. Soc.*, 1960, **82,** 1087.
185. C. A. Rice and J. T. Spence, *Inorg. Chem.*, 1980, **19,** 2845.
186. E. I. Stiefel and H. B. Gray, *J. Am. Chem. Soc.*, 1965, **87,** 4012.
187. J. Finster, N. Meusel, P. Müller, W. Dietzsch, A. Meisel and E. Hoyer, *Z. Chem.*, 1973, **13,** 146.
188. A. Callahan, A. J. Layton and R. S. Nyholm, *Chem. Commun.*, 1969, 399.
189. S. Boyde, C. D. Garner, J. H. Enemark and R. B. Ortega, *Polyhedron,* 1986, **5,** 377.
190. D. J. Rowe, C. D. Garner and J. A. Joule, *J. Chem. Soc., Perkin Trans. 1,* 1985, 1907.
191. D. C. Olson, V. P. Mayweg and G. N. Schrauzer, *J. Am. Chem. Soc.*, 1966, **88,** 4876.
192. E. I. Stiefel and G. F. Brown, *Inorg. Chem.*, 1972, **11,** 434.
193. D. Sellmann and L. Zapf, *Z. Naturforsch., Teil B*, 1985, **40,** 380.
194. C. G. Pierpont and R. Eisenberg, *J. Chem. Soc. (A)*, 1971, 2285.
195. W. P. Bosman and A. Nieuwpoort, *Inorg. Chem.*, 1976, **15,** 775.
196. M. R. Churchill and J. Cooke, *J. Chem. Soc. (A)*, 1970, 2048; J. R. Knox and C. K. Prout, *Acta Crystallogr., Sect. B*, 1969, **25,** 2013.
197. M. Koyama, K. Emoto, M. Kawashima and T. Fujinaga, *Chem. Anal. (Warsaw)*, 1972, **17,** 679 (*Chem. Abstr.*, 1973, **78,** 37 529r).
198. D. C. Bravard, W. E. Newton, J. Huneke, K. Yamanouchi and J. H. Enemark, *Inorg. Chem.*, 1982, **21,** 3795.
199. J. I. Gelder and J. H. Enemark, *Inorg. Chem.*, 1976, **15,** 1839.
200. H. Bortels, *Arch. Mikrobiol.*, 1930, **1,** 33.
201. R. C. Bray in 'Enzymes', 3rd edn., ed. P. D. Boyer, Academic, New York, 1975, vol. 12, p. 299; 'Molybdenum and Molybdenum-Containing Enzymes', ed. M. P. Coughlan, Pergamon, Oxford, 1980; 'Molybdenum Enzymes', ed. T. G. Spiro, Wiley, New York, 1985.
202. J. A. Pateman, D. J. Cove, B. M. Rever and D. B. Roberts, *Nature (London)*, 1964, **201,** 58; A. Nason, K. Y. Lee, S.-S. Pan, P. A. Ketchum, A. Lamberti and J. DeVries, *Proc. Natl. Acad. Sci. USA*, 1971, **68,** 3242.
203. P. T. Pienkos, V. K. Shah and W. J. Brill, *Proc. Natl. Acad. Sci. USA*, 1977, **74,** 5468.
204. S. P. Cramer, 'Advances in Inorganic and Bioinorganic Mechanisms', ed. A. G. Sykes, Academic, London, 1983, vol. 2, p. 259.
205. R. C. Bray, *Adv. Enzymol. Relat. Areas Mol. Biol.*, 1980, **51,** 107; R. C. Bray, *Polyhedron,* 1986, **5,** 591.
206. L. S. Meriwether, W. F. Marzluff and W. G. Hodgson, *Nature (London)*, 1966, **212,** 465.
207. S. Gutteridge, S. J. Tanner and R. C. Bray, *Biochem. J.*, 1978, **175,** 869.

208. E. I. Stiefel, *Proc. Natl. Acad. Sci. USA,* 1973, **70,** 988.
209. R. C. Bray and S. Gutteridge, *Biochemistry,* 1982, **21,** 5992.
210. S. Gutteridge, S. J. Tanner and R. C. Bray, *Biochem. J.*, 1978, **175,** 887; J. P. G. Malthouse and R. C. Bray, *Biochem. J.*, 1980, **191,** 265; J. P. G. Malthouse, G. N. George, D. J. Lowe and R. C. Bray, *Biochem. J.*, 1981, **199,** 629.
211. S. P. Cramer, L. P. Solomonson, M. W. W. Adams and L. E. Mortenson, *J. Am. Chem. Soc.*, 1984, **104,** 1467.
212. S. P. Cramer, R. Wahl and K. V. Rajagopalan, *J. Am. Chem. Soc.*, 1981, **103,** 7721; J. Bordas, R. C. Bray, C. D. Garner, S. Gutteridge and S. S. Hasnain, *Biochem. J.*, 1980, **191,** 499.
213. S. P. Cramer and R. Hille, *J. Am. Chem. Soc.*, 1985, **107,** 8164.
214. J. L. Johnson, B. E. Hainline and K. V. Rajagopalan, *J. Biol. Chem.*, 1980, **255,** 1783.
215. T. Ito, T. Kokuba, Y. Yamamato and S. I. Ikeda, *J. Chem. Soc., Chem. Commun.*, 1974, 136; *J. Chem. Soc., Dalton Trans.*, 1974, 1783.
216. J. L. Atwood, W. E. Hunter, E. Carmona-Guzman and G. Wilkinson, *J. Chem. Soc., Dalton Trans.*, 1980, 467.
217. M. Y. Darensbourg, J. L. Atwood, W. E. Hunter and R. R. Burch, Jr., *J. Am. Chem. Soc.*, 1980, **102,** 3290.
218. J. L. Peterson and R. P. Stewart, Jr., *Inorg. Chem.*, 1980, **19,** 186.
219. M. B. Hursthouse, D. Lyons, M. Thornton-Pett and G. Wilkinson, *J. Chem. Soc., Chem. Commun.*, 1983, 476.
220. L. F. Rhodes, J. D. Zubowski, K. Folting, J. C. Huffman and K. G. Caulton, *Inorg. Chem.*, 1982, **21,** 4185.
221. R. H. Crabtree, G. G. Hlatky and E. M. Holt, *J. Am. Chem. Soc.*, 1983, **105,** 7302.
222. E. B. Lobkovskii, V. D. Makhaev, A. P. Borisov, K. N. Semenenko, Yu. A. Antipin and Yu. T. Struchkov, *Koord. Khim.*, 1985, **11,** 983.
223. A. Frigo, G. Puosi and A. Turco, *Gazzetta,* 1971, **101,** 637.
224. B. Bell, J. Chatt and G. J. Leigh, *J. Chem. Soc., Dalton Trans.*, 1972, 2492.
225. F. Pennella, *Chem. Commun.*, 1971, 158; *Inorg. Synth.*, 1974, **15,** 42; R. H. Crabtree and G. G. Hlatky, *Inorg. Chem.*, 1982, **21,** 1273.
226. L. J. Archer and T. A. George, *J. Organomet. Chem.*, 1973, **54,** C25.
227. P. Meakin, L. J. Guggenberger, W. G. Peet, E. L. Muetterties and J. P. Jesson, *J. Am. Chem. Soc.*, 1973, **95,** 1467.
228. R. H. Crabtree and G. G. Hlatky, *Inorg. Chem.*, 1984, **23,** 2388.
229. G. J. Kubas, *J. Chem. Soc., Chem. Commun.*, 1980, 61; G. J. Kubas, R. R. Ryan, B. I. Swanson, P. J. Vergamini and H. J. Wasserman, *J. Am. Chem. Soc.*, 1984, **106,** 451; G. J. Kubas and R. R. Ryan, *Polyhedron,* 1986, **5,** 473.
230. M. J. Mockford, A. M. Soares and W. P. Griffith, *Transition Met. Chem. (Weinheim, Ger.)*, 1984, **9,** 40; M. J. Mockford and W. P. Griffith, *J. Chem. Soc., Dalton Trans.*, 1985, 717.
231. E. O. Fischer and Y. Hristridu, *Z. Naturforsch.*, 1960, **15б,** 135; M. L. H. Green, J. A. McCleverty, L. Pratt and G. Wilkinson, *J. Chem. Soc.*, 1961, 4854; N. J. Cooper, M. L. H. Green, C. Couldwell and K. Prout, *J. Chem. Soc., Chem. Commun.*, 1977, 145; J. C. Smart and C. J. Curtis, *Inorg. Chem.*, 1978, **17,** 3250.
232. M. L. H. Green, L. C. Mitchard and W. E. Silverthorn, *J. Chem. Soc., Dalton Trans.*, 1974, 1361.
233. H. D. Kaesz and R. B. Saillant, *Chem. Rev.*, 1972, **72,** 231; R. Bau, R. G. Teller, S. W. Kirtley and T. F. Koetzle, *Acc. Chem. Res.*, 1979, **112,** 176.
234. T. W. Coffindaffer, I. P. Rothwell and J. C. Huffman, *J. Chem. Soc., Chem. Commun.*, 1983, 1249.

Subject Index

Formula Index